Y0-BEE-988

American Men & Women of Science

1992-93 • 18th Edition

The 18th edition of *AMERICAN MEN & WOMEN OF SCIENCE* was prepared by the R.R. Bowker Database Publishing Group.

Stephen L. Torpie, Managing Editor
Judy Redel, Managing Editor, Research
Richard D. Lanam, Senior Editor
Tanya Hurst, Research Manager
Karen Hallard, Beth Tanis, Associate Editors

Peter Simon, Vice President, Database Publishing Group
Dean Hollister, Director, Database Planning
Edgar Adcock, Jr., Editorial Director, Directories

American Men & Women of Science

1992-93 • 18th Edition

A Biographical Directory of Today's Leaders in Physical, Biological and Related Sciences.

Volume 6 • Q-S

R. R. BOWKER
New Providence, New Jersey

Published by R.R. Bowker, a division of Reed Publishing, (USA) Inc.

International Standard Book Number

Set:	0-8352-3074-0
Volume I:	0-8352-3075-9
Volume II:	0-8352-3076-7
Volume III:	0-8352-3077-5
Volume IV:	0-8352-3078-3
Volume V:	0-8352-3079-1
Volume VI:	0-8352-3080-5
Volume VII:	0-8352-3081-3
Volume VIII:	0-8352-3082-1

International Standard Serial Number: 0192-8570
Library of Congress Catalog Card Number: 6-7326
Printed and bound in the United States of America.

8 Volume Set

ISBN 0-8352-3074-0

Contents

Advisory Committee

Dr. Robert F. Barnes
Executive Vice President
American Society of Agronomy

Dr. John Kistler Crum
Executive Director
American Chemical Society

Dr. Charles Henderson Dickens
Section Head, Survey & Analysis Section
Division of Science Resource Studies
National Science Foundation

Mr. Alan Edward Fechter
Executive Director
Office of Scientific & Engineering Personnel
National Academy of Science

Dr. Oscar Nicolas Garcia
Prof Electrical Engineering
Electrical Engineering & Computer Science Department
George Washington University

Dr. Charles George Groat
Executive Director
American Geological Institute

Dr. Richard E. Hallgren
Executive Director
American Meteorological Society

Dr. Michael J. Jackson
Executive Director
Federation of American Societies for Experimental Biology

Dr. William Howard Jaco
Executive Director
American Mathematical Society

Dr. Shirley Mahaley Malcom
Head, Directorate for Education and Human Resources Programs
American Association for the Advancement of Science

Mr. Daniel Melnick
Sr Advisor Research Methodologies
Sciences Resources Directorate
National Science Foundation

Ms. Beverly Fearn Porter
Division Manager
Education & Employment Statistics Division
American Institute of Physics

Dr. Terrence R. Russell
Manager
Office of Professional Services
American Chemical Society

Dr. Irwin Walter Sandberg
Holder, Cockrell Family Regent Chair
Department of Electrical & Computer Engineering
University of Texas

Dr. William Eldon Splinter
Interim Vice Chancellor for Research,
Dean, Graduate Studies
University of Nebraska

Ms. Betty M. Vetter
Executive Director, Science Manpower Comission
Commission on Professionals in Science & Technology

Dr. Dael Lee Wolfe
Professor Emeritus
Graduate School of Public Affairs
University of Washington

Preface

American Men and Women Of Science remains without peer as a chronicle of North American scientific endeavor and achievement. The present work is the eighteenth edition since it was first compiled as *American Men of Science* by J. Mckeen Cattell in 1906. In its eighty-six year history *American Men & Women of Science* has profiled the careers of over 300,000 scientists and engineers. Since the first edition, the number of American scientists and the fields they pursue have grown immensely. This edition alone lists full biographies for 122,817 engineers and scientists, 7021 of which are listed for the first time. Although the book has grown, our stated purpose is the same as when Dr. Cattell first undertook the task of producing a biographical directory of active American scientists. It was his intention to record educational, personal and career data which would make "a contribution to the organization of science in America" and "make men [and women] of science acquainted with one another and with one another's work." It is our hope that this edition will fulfill these goals.

The biographies of engineers and scientists constitute seven of the eight volumes and provide birthdates, birthplaces, field of specialty, education, honorary degrees, professional and concurrent experience, awards, memberships, research information and adresses for each entrant when applicable. The eighth volume, the discipline index, organizes biographees by field of activity. This index, adapted from the National Science Foundation's Taxonomy of Degree and Employment Specialties, classifies entrants by 171 subject specialties listed in the table of contents of Volume 8. For the first time, the index classifies scientists and engineers by state within each subject specialty, allowing the user to more easily locate a scientist in a given area. Also new to this edition is the inclusion of statistical information and recipients of theNobel Prizes, the Craaford Prize, the Charles Stark Draper Prize, and the National Medals of Science and Technology received since the last edition.

While the scientific fields covered by *American Men and Women Of Science* are comprehensive, no attempt has been made to include all American scientists. Entrants are meant to be limited to those who have made significant contributions in their field. The names of new entrants were submitted for consideration at the editors' request by current entrants and by leaders of academic, government and private research programs and associations. Those included met the following criteria:

1. Distinguished achievement, by reason of experience, training or accomplishment, including contributions to the literature, coupled with continuing activity in scientific work;

   or

2. Research activity of high quality in science as evidenced by publication in reputable scientific journals; or for those whose work cannot be published due to governmental or industrial security, research activity of high quality in science as evidenced by the judgement of the individual's peers;

   or

3. Attainment of a position of substantial responsibility requiring scientific training and experience.

This edition profiles living scientists in the physical and biological fields, as well as public health scientists, engineers, mathematicians, statisticians, and computer scientists. The information is collected by means of direct communication whenever possible. All entrants receive forms for corroboration and updating. New entrants receive questionaires and verification proofs before publication. The information submitted by entrants is included as completely as possible within the boundaries of editorial and space restrictions. If an entrant does not return the form and his or her current location can be verified in secondary sources, the full entry is repeated. References to the previous edition are given for those who do not return forms and cannot be located, but who are presumed to be still active in science or engineering. Entrants known to be deceased are noted as such and a reference to the previous edition is given. Scientists and engineers who are not citizens of the United States or Canada are included if a significant portion of their work was performed in North America.

The information in AMWS is also available on CD-ROM as part of *SciTech Reference Plus*. In adition to the convenience of searching scientists and engineers, *SciTech Reference Plus* also includes *The Directory of American Research & Technology*, *Corporate Technology Directory*, sci-tech and medical books and serials from *Books in Print* and *Bowker International Series*. *American Men and Women Of Science* is available for online searching through the subscription services of DIALOG Information Services, Inc. (3460 Hillview Ave, Palo Alto, CA 94304) and ORBIT Search Service (800 Westpark Dr, McLean, VA 22102). Both CD-Rom and the on-line subscription services allow all elements of an entry, including field of interest, experience, and location, to be accessed by key word. Tapes and mailing lists are also available through the Cahners Direct Mail (John Panza, List Manager, Bowker Files 245 W 17th St, New York, NY, 10011, Tel: 800-537-7930).

A project as large as publishing *American Men and Women Of Science* involves the efforts of a great many people. The editors take this opportunity to thank the eighteenth edition advisory committee for their guidance, encouragement and support. Appreciation is also expressed to the many scientific societies who provided their membership lists for the purpose of locating former entrants whose addresses had changed, and to the tens of thousands of scientists across the country who took time to provide us with biographical information. We also wish to thank Bruce Glaunert, Bonnie Walton, Val Lowman, Debbie Wilson, Mervaine Ricks and all those whose care and devotion to accurate research and editing assured successful production of this edition.

Comments, suggestions and nominations for the nineteenth edition are encouraged and should be directed to The Editors, *American Men and Women Of Science*, R.R. Bowker, 121 Chanlon Road, New Providence, New Jersey, 07974.

Edgar H. Adcock, Jr.
Editorial Director

Major Honors & Awards

Nobel Prizes

Nobel Foundation

The Nobel Prizes were established in 1900 (and first awarded in 1901) to recognize those people who "have conferred the greatest benefit on mankind."

1990 Recipients

Chemistry:

Elias James Corey

Awarded for his work in retrosynthetic analysis, the synthesizing of complex substances patterned after the molecular structures of natural compounds.

Physics:

Jerome Isaac Friedman

Henry Way Kendall

Richard Edward Taylor

Awarded for their breakthroughs in the understanding of matter.

Physiology or Medicine:

Joseph E. Murray

Edward Donnall Thomas

Awarded to Murray for his kidney transplantation achievements and to Thomas for bone marrow transplantation advances.

1991 Recipients

Chemistry:

Richard R. Ernst

Awarded for refinements in nuclear magnetic resonance spectroscopy.

Physics:

Pierre-Gilles de Gennes*

Awarded for his research on liquid crystals.

Physiology or Medicine:

Erwin Neher

Bert Sakmann*

Awarded for their discoveries in basic cell function and particularly for the development of the patch clamp technique.

Crafoord Prize

Royal Swedish Academy of Sciences
(Kungl. Vetenskapsakademien)

The Crafoord Prize was introduced in 1982 to award scientists in disciplines not covered by the Nobel Prize, namely mathematics, astronomy, geosciences and biosciences.

1990 Recipients

Paul Ralph Ehrlich

Edward Osborne Wilson

Awarded for their fundamental contributions to population biology and the conservation of biological diversity.

1991 Recipient

Allan Rex Sandage

Awarded for his fundamental contributions to extragalactic astronomy, including observational cosmology.

Charles Stark Draper Prize

National Academy of Engineering

The Draper Prize was introduced in 1989 to recognize engineering achievement. It is awarded biennially.

1991 Recipients

Hans Joachim Von Ohain

Frank Whittle

Awarded for their invention and development of the jet aircraft engine.

National Medal of Science

National Science Foundation

The National Medals of Science have been awarded by the President of the United States since 1962 to leading scientists in all fields.

1990 Recipients:

Baruj Benacerraf
Elkan Rogers Blout
Herbert Wayne Boyer
George Francis Carrier
Allan MacLeod Cormack
Mildred S. Dresselhaus
Karl August Folkers
Nick Holonyak Jr.
Leonid Hurwicz
Stephen Cole Kleene
Daniel Edward Koshland Jr.
Edward B. Lewis
John McCarthy
Edwin Mattison McMillan**
David G. Nathan
Robert Vivian Pound
Roger Randall Dougan Revelle**
John D. Roberts
Patrick Suppes
Edward Donnall Thomas

1991 Recipients

Mary Ellen Avery
Ronald Breslow
Alberto Pedro Calderon
Gertrude Belle Elion
George Harry Heilmeier
Dudley Robert Herschbach
George Evelyn Hutchinson**
Elvin Abraham Kabat
Robert Kates
Luna Bergere Leopold
Salvador Edward Luria**
Paul A. Marks
George Armitage Miller
Arthur Leonard Schawlow
Glenn Theodore Seaborg
Folke Skoog
H. Guyford Stever
Edward Carroll Stone Jr
Steven Weinberg
Paul Charles Zamecnik

National Medal of Technology

U.S. Department of Commerce,
Technology Administration

The National Medals of Technology, first awarded in 1985, are bestowed by the President of the United States to recognize individuals and companies for their development or commercialization of technology or for their contributions to the establishment of a technologically-trained workforce.

1990 Recipients

John Vincent Atanasoff
Marvin Camras
The du Pont Company
Donald Nelson Frey
Frederick W. Garry
Wilson Greatbatch
Jack St. Clair Kilby
John S. Mayo
Gordon Earle Moore
David B. Pall
Chauncey Starr

1991 Recipients

Stephen D. Bechtel Jr
C. Gordon Bell
Geoffrey Boothroyd
John Cocke
Peter Dewhurst
Carl Djerassi
James Duderstadt
Antonio L. Elias
Robert W. Galvin
David S. Hollingsworth
Grace Murray Hopper
F. Kenneth Iverson
Frederick M. Jones**
Robert Roland Lovell
Joseph A. Numero**
Charles Eli Reed
John Paul Stapp
David Walker Thompson

*These scientists' biographies do not appear in *American Men & Women of Science* because their work has been conducted exclusively outside the US and Canada.

**Deceased [Note that Frederick Jones died in 1961 and Joseph Numero in May 1991. Neither was ever listed in *American Men and Women of Science*.]

Statistics

Statistical distribution of entrants in *American Men & Women of Science* is illustrated on the following five pages. The regional scheme for geographical analysis is diagrammed in the map below. A table enumerating the geographic distribution can be found on page xvi, following the charts. The statistics are compiled by tallying all occurrences of a major index subject. Each scientist may choose to be indexed under as many as four categories; thus, the total number of subject references is greater than the number of entrants in *AMWS*.

All Disciplines

	Number	Percent
Northeast	58,325	34.99
Southeast	39,769	23.86
North Central	19,846	11.91
South Central	12,156	7.29
Mountain	11,029	6.62
Pacific	25,550	15.33
TOTAL	**166,675**	**100.00**

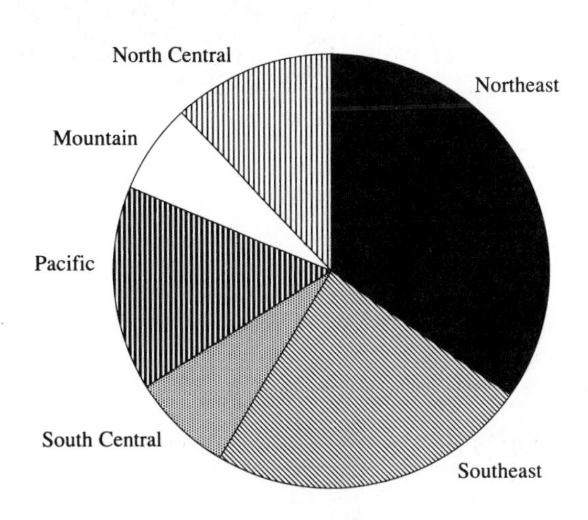

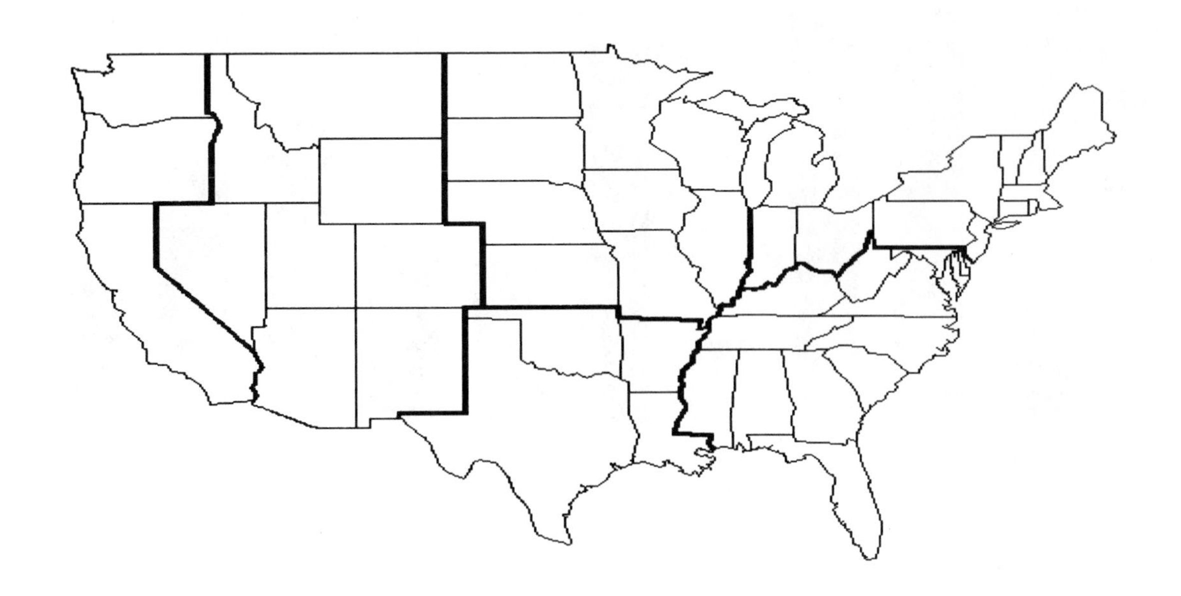

Age Distribution of American Men & Women of Science

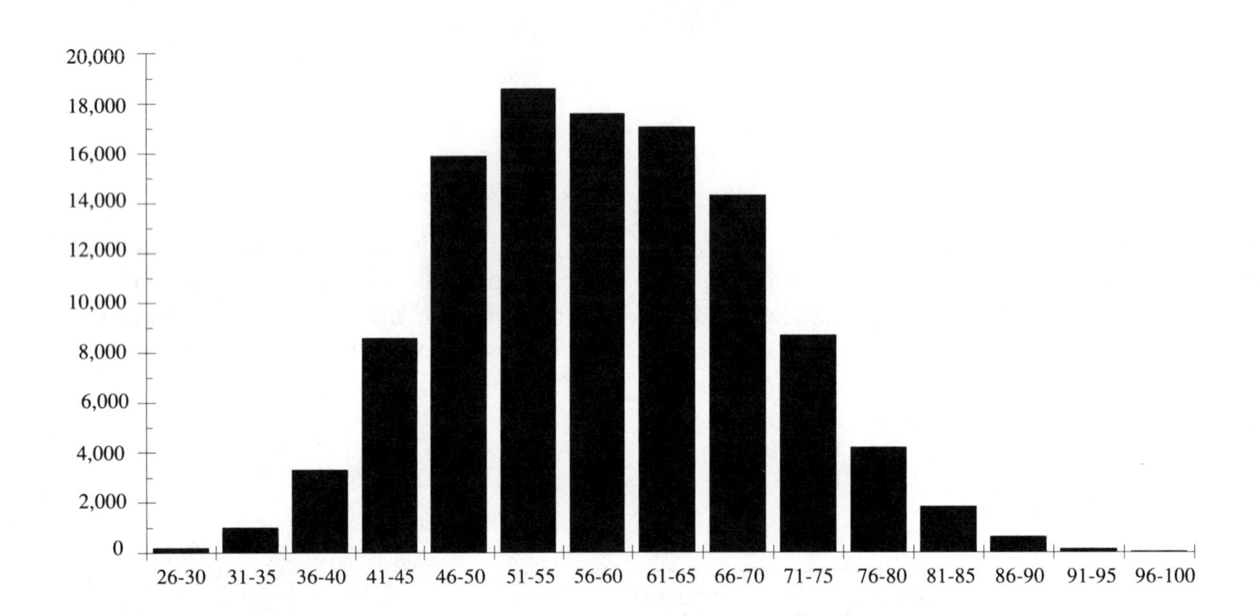

Number of Scientists in Each Discipline of Study

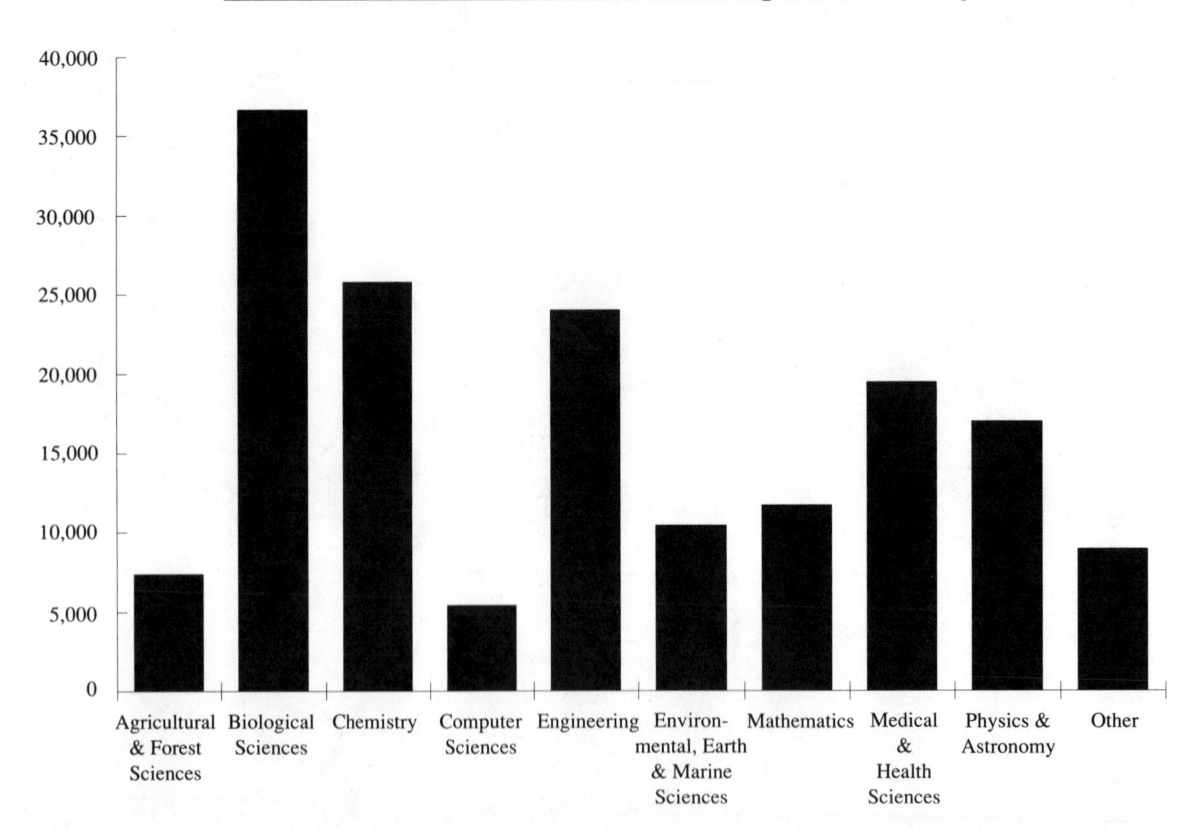

Agricultural & Forest Sciences

	Number	Percent
Northeast	1,574	21.39
Southeast	1,991	27.05
North Central	1,170	15.90
South Central	609	8.27
Mountain	719	9.77
Pacific	1,297	17.62
TOTAL	**7,360**	**100.00**

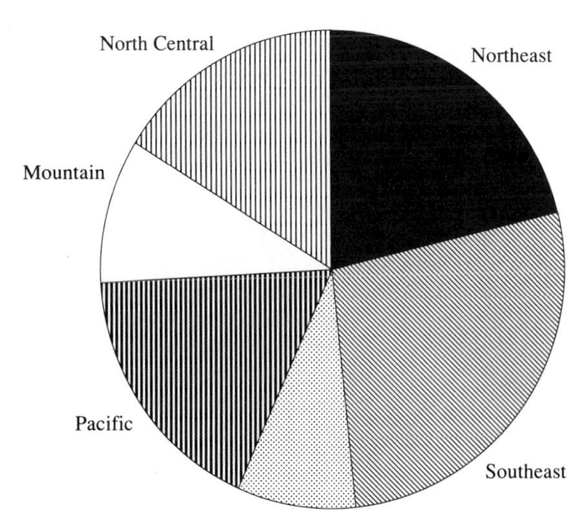

Biological Sciences

	Number	Percent
Northeast	12,162	33.23
Southeast	9,054	24.74
North Central	5,095	13.92
South Central	2,806	7.67
Mountain	2,038	5.57
Pacific	5,449	14.89
TOTAL	**36,604**	**100.00**

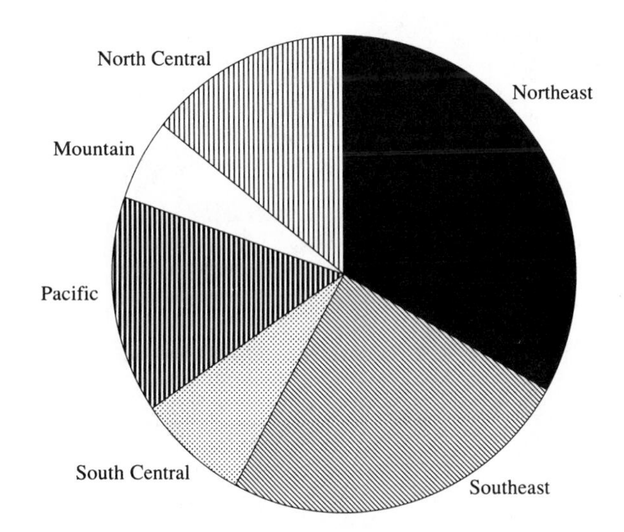

Chemistry

	Number	Percent
Northeast	10,343	40.15
Southeast	6,124	23.77
North Central	3,022	11.73
South Central	1,738	6.75
Mountain	1,300	5.05
Pacific	3,233	12.55
TOTAL	**25,760**	**100.00**

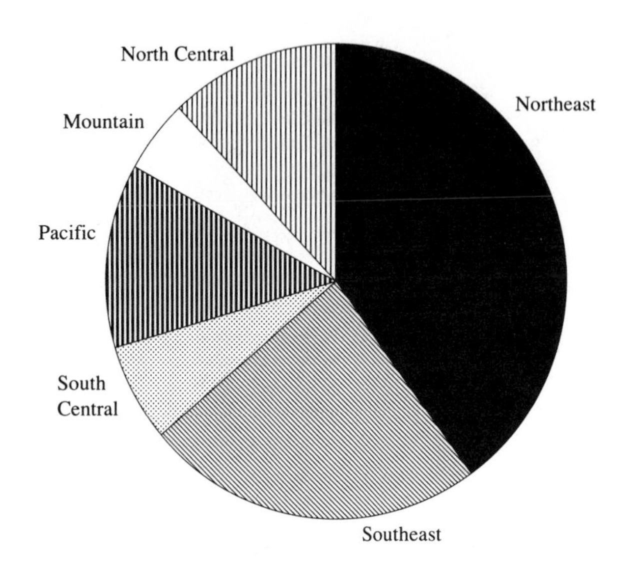

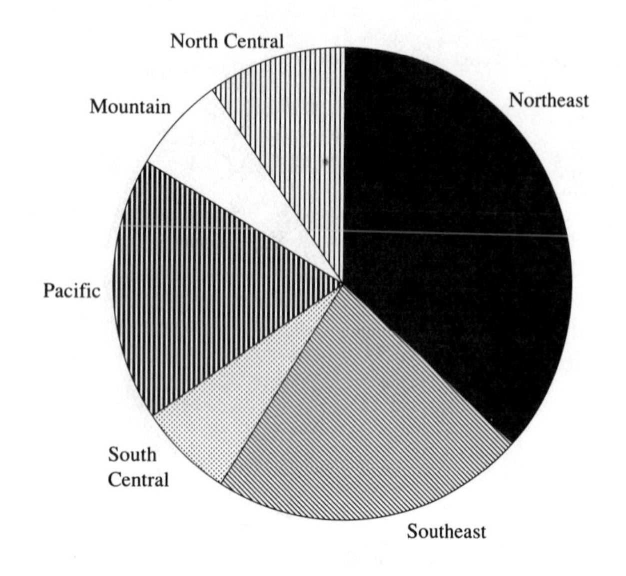

Computer Sciences

	Number	Percent
Northeast	1,987	36.76
Southeast	1,200	22.20
North Central	511	9.45
South Central	360	6.66
Mountain	372	6.88
Pacific	976	18.05
TOTAL	**5,406**	**100.00**

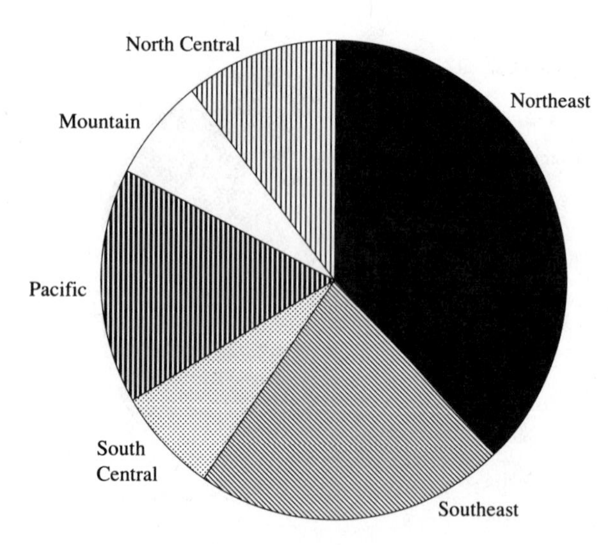

Engineering

	Number	Percent
Northeast	9,122	38.01
Southeast	5,202	21.68
North Central	2,510	10.46
South Central	1,710	7.13
Mountain	1,646	6.86
Pacific	3,807	15.86
TOTAL	**23,997**	**100.00**

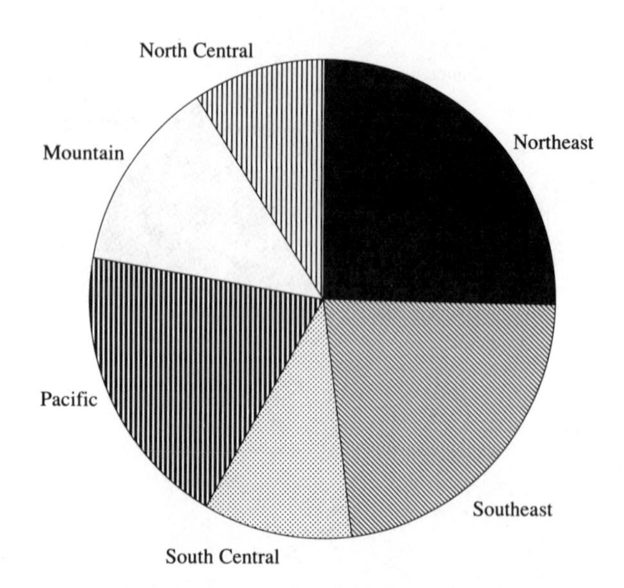

Environmental, Earth & Marine Sciences

	Number	Percent
Northeast	2,657	25.48
Southeast	2,361	22.64
North Central	953	9.14
South Central	1,075	10.31
Mountain	1,359	13.03
Pacific	2,022	19.39
TOTAL	**10,427**	**100.00**

Mathematics

	Number	Percent
Northeast	4,211	35.92
Southeast	2,609	22.26
North Central	1,511	12.89
South Central	884	7.54
Mountain	718	6.13
Pacific	1,789	15.26
TOTAL	**11,722**	**100.00**

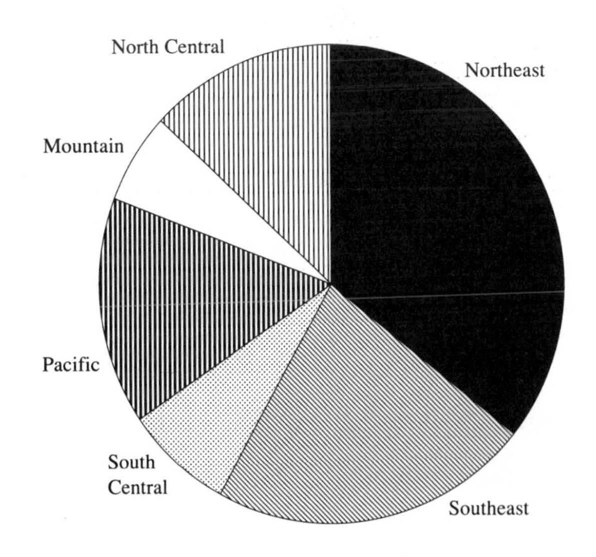

Medical & Health Sciences

	Number	Percent
Northeast	7,115	36.53
Southeast	5,004	25.69
North Central	2,577	13.23
South Central	1,516	7.78
Mountain	755	3.88
Pacific	2,509	12.88
TOTAL	**19,476**	**100.00**

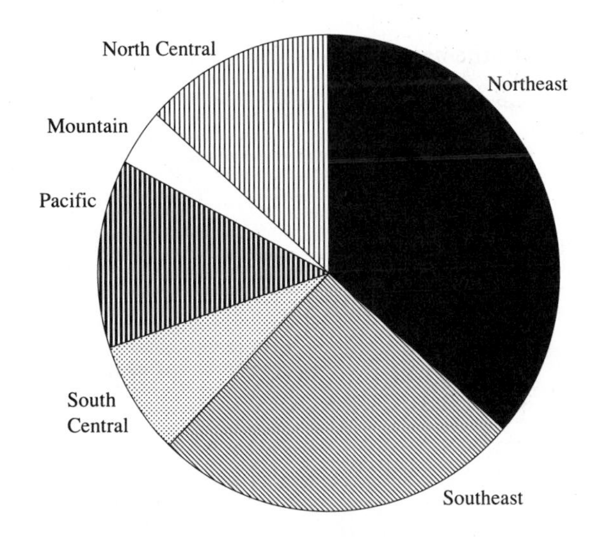

Physics & Astronomy

	Number	Percent
Northeast	5,961	35.12
Southeast	3,670	21.62
North Central	1,579	9.30
South Central	918	5.41
Mountain	1,607	9.47
Pacific	3,238	19.08
TOTAL	**16,973**	**100.00**

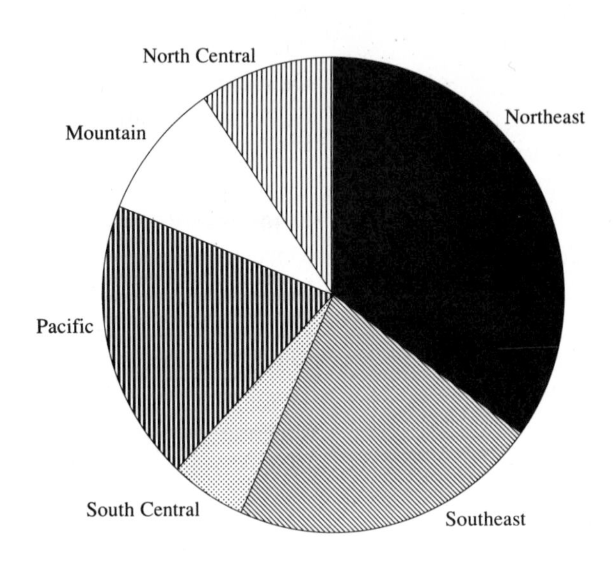

Geographic Distribution of Scientists by Discipline

	Northeast	Southeast	North Central	South Central	Mountain	Pacific	TOTAL
Agricultural & Forest Sciences	1,574	1,991	1,170	609	719	1,297	**7,360**
Biological Sciences	12,162	9,054	5,095	2,806	2,038	5,449	**36,604**
Chemistry	10,343	6,124	3,022	1,738	1,300	3,233	**25,760**
Computer Sciences	1,987	1,200	511	360	372	976	**5,406**
Engineering	9,122	5,202	2,510	1,710	1,646	3,807	**23,997**
Environmental, Earth & Marine Sciences	2,657	2,361	953	1,075	1,359	2,022	**10,427**
Mathematics	4,211	2,609	1,511	884	718	1,789	**11,722**
Medical & Health Sciences	7,115	5,004	2,577	1,516	755	2,509	**19,476**
Physics & Astronomy	5,961	3,670	1,579	918	1,607	3,238	**16,973**
Other Professional Fields	3,193	2,554	918	540	515	1,230	**8,950**
TOTAL	**58,325**	**39,769**	**19,846**	**12,156**	**11,029**	**25,550**	**166,675**

Geographic Definitions

Northeast
Connecticut
Indiana
Maine
Massachusetts
Michigan
New Hampshire
New Jersey
New York
Ohio
Pennsylvania
Rhode Island
Vermont

Southeast
Alabama
Delaware
District of Columbia
Florida
Georgia
Kentucky
Maryland
Mississippi
North Carolina
South Carolina
Tennessee
Virginia
West Virginia

North Central
Illinois
Iowa
Kansas
Minnesota
Missouri
Nebraska
North Dakota
South Dakota
Wisconsin

South Central
Arkansas
Louisiana
Texas
Oklahoma

Mountain
Arizona
Colorado
Idaho
Montana
Nevada
New Mexico
Utah
Wyoming

Pacific
Alaska
California
Hawaii
Oregon
Washington

Sample Entry

American Men & Women of Science (AMWS) is an extremely useful reference tool. The book is most often used in one of two ways: to find more information about a particular scientist or to locate a scientist in a specific field.

To locate information about an individual, the biographical section is most helpful. It encompasses the first seven volumes and lists scientists and engineers alphabetically by last name. The fictitious biographical listing shown below illustrates every type of information an entry may include.

The Discipline Index, volume 8, can be used to easily find a scientist in a specific subject specialty. This index is first classified by area of study, and within each specialty entrants are divided further by state of residence.

CARLETON, PHYLLIS B(ARBARA), b Glenham, SDak, April 1, 30. m 53, 69; c 2. ORGANIC CHEMISTRY. *Educ:* Univ Notre Dame, BSc, 52, MSc, 54, Vanderbilt Univ, PhD(chem), 57. *Hon Degrees:* DSc, Howard Univ, 79. *Prof Exp:* Res chemist, Acme Chem Corp, 54-59, sr res chemist, 59-60; from asst to assoc prof chem 60-63, prof chem, Kansas State Univ, 63-72; prof chem, Yale Univ, 73-89; CONSULT, CARLETON & ASSOCS, 89-. *Concurrent Pos:* Adj prof, Kansas State Univ 58-60; vis lect, Oxford Univ, 77, consult, Union Carbide, 74-80. *Honors & Awards:* Gold Medal, Am Chem Society, 81; *Mem:* AAAS, fel Am Chem Soc, Sigma Chi. *Res:* Organic synthesis, chemistry of natural products, water treatment and analysis. *Mailing Address:* Carleton & Assocs 21 E 34th St Boston MA 02108

Abbreviations

AAAS—American Association for the Advancement of Science
abnorm—abnormal
abstr—abstract
acad—academic, academy
acct—Account, accountant, accounting
acoust—acoustic(s), acoustical
ACTH—adrenocorticotrophic hormone
actg—acting
activ—activities, activity
addn—addition(s), additional
Add—Address
adj—adjunct, adjutant
adjust—adjustment
Adm—Admiral
admin—administration, administrative
adminr—administrator(s)
admis—admission(s)
adv—adviser(s), advisory
advan—advance(d), advancement
advert—advertisement, advertising
AEC—Atomic Energy Commission
aerodyn—aerodynamic
aeronaut—aeronautic(s), aeronautical
aerophys—aerophsical, aerophysics
aesthet—aesthetic
AFB—Air Force Base
affil—affiliate(s), affiliation
agr—agricultural, agriculture
agron—agronomic, agronomical, agronomy
agrost—agrostologic, agrostological, agrostology
agt—agent
AID—Agency for International Development
Ala—Alabama
allergol—allergological, allergology
alt—alternate
Alta—Alberta
Am—America, American
AMA—American Medical Association
anal—analysis, analytic, analytical
analog—analogue
anat—anatomic, anatomical, anatomy
anesthesiol—anesthesiology
angiol—angiology
Ann—Annal(s)
ann—annual
anthrop—anthropological, anthropology
anthropom—anthropometric, anthropometrical, anthropometry
antiq—antiquary, antiquities, antiquity
antiqn—antiquarian
apicult—apicultural, apiculture
APO—Army Post Office
app—appoint, appointed
appl—applied
appln—application
approx—approximate(ly)
Apr—April
apt—apartment(s)
aquacult—aquaculture
arbit—arbitration
arch—archives
archaeol—archaeological, archaeology
archit—architectural, architecture
Arg—Argentina, Argentine
Ariz—Arizona
Ark—Arkansas
artil—artillery
asn—association
assoc(s)—associate(s), associated
asst(s)—assistant(s), assistantship(s)
assyriol—Assyriology
astrodyn—astrodynamics
astron—astronomical, astronomy
astronaut—astonautical, astronautics
astronr—astronomer
astrophys—astrophysical, astrophysics
attend—attendant, attending
atty—attorney
audiol—audiology
Aug—August
auth—author
AV—audiovisual
Ave—Avenue
avicult—avicultural, aviculture

b—born
bact—bacterial, bacteriologic, bacteriological, bacteriology
BC—British Colombia
bd—board
behav—behavior(al)
Belg—Belgian, Belgium
Bibl—biblical
bibliog—bibliographic, bibliographical, bibliography
bibliogr—bibliographer
biochem—biochemical, biochemistry
biog—biographical, biography
biol—biological, biology
biomed—biomedical, biomedicine
biomet—biometric(s), biometrical, biometry
biophys—biophysical, biophysics
bk(s)—book(s)
bldg-building
Blvd—Boulevard
Bor—Borough
bot—botanical, botany
br—branch(es)
Brig—Brigadier
Brit—Britain, British
Bro(s)—Brother(s)
byrol—byrology
bull—Bulletin
bur—bureau
bus—business
BWI—British West Indies

c—children
Calif—California
Can—Canada, Canadian
cand—candidate
Capt—Captain
cardiol-cardiology
cardiovasc—cardiovascular
cartog—cartographic, cartographical, cartography
cartogr—cartographer
Cath—Catholic
CEngr—Corp of Engineers
cent—central
Cent Am—Central American
cert—certificate(s), certification, certified
chap—chapter
chem—chemical(s), chemistry
chemother—chemotherapy
chg—change
chmn—chairman
citricult—citriculture
class—classical
climat—climatological, climatology
clin(s)—clinic(s), clinical
cmndg—commanding
Co—County
co—Companies, Company
co-auth—coauthor
co-dir—co-director
co-ed—co-editor
co-educ—coeducation, coeducational
col(s)—college(s), collegiate, colonel
collab—collaboration, collaborative
collabr—collaborator
Colo—Colorado
com—commerce, commercial
Comdr—Commander

commun—communicable, communication(s)
comn(s)—commission(s), commissioned
comndg—commanding
comnr—commissioner
comp—comparitive
compos—composition
comput—computation, computer(s), computing
comt(s)—committee(s)
conchol—conchology
conf—conference
cong—congress, congressional
Conn—Connecticut
conserv—conservation, conservatory
consol—consolidated, consolidation
const—constitution, constitutional
construct—construction, constructive
consult(s)—consult, consultant(s), consultantship(s), consultation, consulting
contemp—contemporary
contrib—contribute, contributing, contribution(s)
contribr—contributor
conv—convention
coop—cooperating, cooperation, cooperative
coord—coordinate(d), coordinating, coordination
coordr—coordinator
corp—corporate, corporation(s)
corresp—correspondence, correspondent, corresponding
coun—council, counsel, counseling
counr—councilor, counselor
criminol—criminological, criminology
cryog—cryogenic(s)
crystallog—crystallographic, crystallographical, crystallography
crystallogr—crystallographer
Ct—Court
Ctr—Center
cult—cultural, culture
cur—curator
curric—curriculum
cybernet—cybernetic(s)
cytol—cytological, cytology
Czech—Czechoslovakia

DC—District of Columbia
Dec—December
Del—Delaware
deleg—delegate, delegation
delinq—delinquency, delinquent
dem—democrat(s), democratic
demog—demographic, demography
demogr—demographer
demonstr—demontrator
dendrol—dendrologic, dendrological, dendrology
dent—dental, dentistry
dep—deputy
dept—department
dermat—dermatologic, dermatological, dermatology
develop—developed, developing, development, developmental
diag—diagnosis, diagnostic
dialectol-dialectological, dialectology
dict—dictionaries, dictionary
Dig—Digest
dipl—diploma, diplomate
dir(s)—director(s), directories, directory
dis—disease(s), disorders
Diss Abst—Dissertation Abstracts
dist—district
distrib—distributed, distribution, distributive
distribr—distributor(s)
div—division, divisional, divorced
DNA—deoxyribonucleic acid
doc—document(s), documentary, documentation
Dom—Dominion
Dr—Drive
E—east
ecol—ecological, ecology
econ(s)—economic(s), economical, economy
economet—econometric(s)
ECT—electroconvulsive or electroshock therapy
ed—edition(s), editor(s), editorial
ed bd—editorial board
educ—education, educational
educr—educator(s)
EEG—electroencephalogram, electroencephalographic, electroencephalography
Egyptol—Egyptology
EKG—electrocardiogram
elec—elecvtric, electrical, electricity
electrochem-electrochemical, electrochemistry
electroph—electrophysical, electrophysics
elem—elementary
embryol—embryologic, embryological, embryology
emer—emeriti, emeritus
employ—employment
encour—encouragement
encycl—encyclopedia
endocrinol—endocrinologic, endocrinology
eng—engineering
Eng—England, English
engr(s)—engineer(s)
enol—enology
Ens—Ensign
entom—entomological, entomology
environ-environment(s), environmental
enzym—enzymology
epidemiol—epideiologic, epidemiological, epidemiology
equip—equipment
ERDA—Energy Research & Development Administration
ESEA—Elementary & Secondary Education Act
espec—especially
estab—established, establishment(s)
ethnog—ethnographic, ethnographical, ethnography
ethnogr—ethnographer
ethnol—ethnologic, ethnological, ethnology
Europ—European
eval—evaluation
Evangel—evangelical
eve—evening
exam—examination(s), examining
examr—examiner
except—exceptional
exec(s)—executive(s)
exeg—exegeses, exegesis, exegetic, exegetical
exhib(s)—exhibition(s), exhibit(s)
exp—experiment, experimental
exped(s)—expedition(s)
explor—exploration(s), exploratory
expos—exposition
exten—extension

fac—faculty
facil—facilities, facility
Feb—February
fed—federal
fedn—federation
fel(s)—fellow(s), fellowship(s)
fermentol—fermentology
fertil—fertility, fertilization
Fla—Florida
floricult—floricultural, floriculture
found—foundation
FPO—Fleet Post Office
Fr—French
Ft—Fort

Ga—Georgia
gastroenterol—gastroenterological, gastroenterology
gen—general
geneal—genealogical, genealogy
geod—geodesy, geodetic
geog—geographic, geographical, geography
geogr—geographer
geol—geologic, geological, geology
geom—geometric, geometrical, geometry
geomorphol—geomorphologic, geomorphology
geophys—geophysical, geophysics
Ger—German, Germanic, Germany
geriat—geriatric
geront—gerontological, gerontology
GES—Gesellschaft
glaciol—glaciology
gov—governing, governor(s)
govt—government, governmental
grad—graduate(d)
Gt Brit—Great Britain
guid—guidance
gym—gymnasium
gynec—gynecologic, gynecological, gynecology

handbk(s)—handbook(s)
helminth—helminthology
hemat—hematologic, hematological, hematology
herpet—herpetologic, herpetological, herpetology
HEW—Department of Health, Education & Welfare
Hisp—Hispanic, Hispania
hist—historic, historical, history
histol—histological, histology
HM—Her Majesty
hochsch—hochschule
homeop—homeopathic, homeopathy
hon(s)—honor(s), honorable, honorary
hort—horticultural, horticulture
hosp(s)—hospital(s), hospitalization
hq—headquarters

ABBREVIATIONS

HumRRO—Human Resources Research Office
husb—husbandry
Hwy—Highway
hydraul—hydraulic(s)
hydrodyn—hydrodynamic(s)
hydrol—hydrologic, hydrological, hydrologics
hyg—hygiene, hygienic(s)
hypn—hypnosis

ichthyol—ichthyological, ichthyology
Ill—Illinois
illum—illuminating, illumination
illus—illustrate, illustrated, illustration
illusr—illustrator
immunol—immunologic, immunological, immunology
Imp—Imperial
improv—improvement
Inc—Incorporated
in-chg—in charge
incl—include(s), including
Ind—Indiana
indust(s)—industrial, industries, industry
Inf—infantry
info—information
inorg—inorganic
ins—insurance
inst(s)—institute(s), institution(s)
instnl—institutional(ized)
instr(s)—instruct, instruction, instructor(s)
instrnl—instructional
int—international
intel—intellligence
introd—introduction
invert—invertebrate
invest(s)—investigation(s)
investr—investigator
irrig—irrigation
Ital—Italian

J—Journal
Jan—January
Jct—Junction
jour—journal, journalism
jr—junior
jurisp—jurisprudence
juv—juvenile

Kans—Kansas
Ky—Kentucky

La—Louisiana
lab(s)—laboratories, laboratory
lang—language(s)
laryngol—larygological, laryngology
lect—lecture(s)
lectr—lecturer(s)
legis—legislation, legislative, legislature
lett—letter(s)
lib—liberal
libr—libraries, library
librn—librarian
lic—license(d)
limnol—limnological, limnology
ling—linguistic(s), linguistical
lit—literary, literature
lithol—lithologic, lithological, lithology

Lt—Lieutenant
Ltd—Limited

m—married
mach—machine(s), machinery
mag—magazine(s)
maj—major
malacol—malacology
mammal—mammalogy
Man—Manitoba
Mar—March
Mariol—Mariology
Mass—Massechusetts
mat—material(s)
mat med—materia medica
math—mathematic(s), mathematical
Md—Maryland
mech—mechanic(s), mechanical
med—medical, medicinal, medicine
Mediter—Mediterranean
Mem—Memorial
mem—member(s), membership(s)
ment—mental(ly)
metab—metabolic, metabolism
metall—metallurgic, metallurgical, metallurgy
metallog—metallographic, metallography
metallogr—metallographer
metaphys—metaphysical, metaphysics
meteorol—meteorological, meteorology
metrol—metrological, metrology
metrop—metropolitan
Mex—Mexican, Mexico
mfg—manufacturing
mfr—manufacturer
mgr—manager
mgt—management
Mich—Michigan
microbiol—microbiological, microbiology
micros—microscopic, microscopical, microscopy
mid—middle
mil—military
mineral—mineralogical, mineralogy
Minn—Minnesota
Miss—Mississippi
mkt—market, marketing
Mo—Missouri
mod—modern
monogr—monograph
Mont—Montana
morphol—morphological, morphology
Mt—Mount
mult—multiple
munic—municipal, municipalities
mus—museum(s)
musicol—musicological, musicology
mycol—mycologic, mycology

N—north
NASA—National Aeronautics & Space Administration
nat—national, naturalized
NATO—North Atlantic Treaty Organization
navig—navigation(al)
NB—New Brunswick
NC—North Carolina
NDak—North Dakota
NDEA—National Defense Education Act
Nebr—Nebraska

nematol—nematological, nematology
nerv—nervous
Neth—Netherlands
neurol—neurological, neurology
neuropath—neuropathological, neuropathology
neuropsychiat—neuropsychiatric, neuropsychiatry
neurosurg—neurosurgical, neurosurgery
Nev—Nevada
New Eng—New England
New York—New York City
Nfld—Newfoundland
NH—New Hampshire
NIH—National Institute of Health
NIMH—National Institute of Mental Health
NJ—New Jersey
NMex—New Mexico
No—Number
nonres—nonresident
norm—normal
Norweg—Norwegian
Nov—November
NS—Nova Scotia
NSF—National Science Foundation
NSW—New South Wales
numis—numismatic(s)
nutrit—nutrition, nutritional
NY—New York State
NZ—New Zealand

observ—observatories, observatory
obstet—obstetric(s), obstetrical
occas—occasional(ly)
occup—occupation, occupational
oceanog—oceanographic, oceanographical, oceanography
oceanogr—oceanographer
Oct—October
odontol—odontology
OEEC—Organization for European Economic Cooperation
off—office, official
Okla—Oklahoma
olericult—olericulture
oncol—oncologic, oncology
Ont—Ontario
oper(s)—operation(s), operational, operative
ophthal—ophthalmologic, ophthalmological, ophthalmology
optom—optometric, optometrical, optometry
ord—ordnance
Ore—Oregon
org—organic
orgn—organization(s), organizational
orient—oriental
ornith—ornithological, ornithology
orthod—orthodontia, orthodontic(s)
orthop—orthopedic(s)
osteop—osteopathic, osteopathy
otol—otological, otology
otolaryngol—otolaryngological, otolaryngology
otorhinol—otorhinologic, otorhinology

Pa—Pennsylvania
Pac—Pacific
paleobot—paleobotanical, paleontology
paleont—paleontology

Pan-Am—Pan-American
parisitol—parasitology
partic—participant, participating
path—pathologic, pathological, pathology
pedag—pedagogic(s), pedagogical, pedagogy
pediat—pediatric(s)
PEI—Prince Edward Islands
penol—penological, penology
periodont—periodontal, periodontic(s)
petrog—petrographic, petrographical, petrography
petrogr—petrographer
petrol—petroleum, petrologic, petrological, petrology
pharm—pharmacy
pharmaceut—pharmaceutic(s), pharmaceutical(s)
pharmacog—pharmacognosy
pharamacol—pharmacologic, pharmacological, pharmacology
phenomenol—phenomenologic(al), phenomenology
philol—philological, philology
philos—philosophic, philosophical, philosophy
photog—photographic, photography
photogeog—photogeographic, photogeography
photogr—photographer(s)
photogram—photogrammetric, photogrammetry
photom—photometric, photometrical, photometry
phycol—phycology
phys—physical
physiog—physiographic, physiographical, physiography
physiol—physiological, phsysiology
Pkwy—Parkway
Pl—Place
polit—political, politics
polytech—polytechnic(s)
pomol—pomological, pomology
pontif—pontifical
pop—population
Port—Portugal, Portuguese
Pos:—Position
postgrad—postgraduate
PQ—Province of Quebec
PR—Puerto Rico
pract—practice
practr—practitioner
prehist—prehistoric, prehistory
prep—preparation, preparative, preparatory
pres—president
Presby—Presbyterian
preserv—preservation
prev—prevention, preventive
prin—principal
prob(s)—problem(s)
proc—proceedings
proctol—proctologic, proctological, proctology
prod—product(s), production, productive
prof—professional, professor, professorial
Prof Exp—Professional Experience
prog(s)—program(s), programmed, programming
proj—project(s), projection(al), projective
prom—promotion
protozool—protozoology
Prov—Province, Provincial
psychiat—psychiatric, psychiatry
psychoanal—psychoanalysis, psychoanalytic, psychoanalytical
psychol—psychological, psychology
psychomet—psychometric(s)
psychopath—psychopathologic, psycho pathology
psychophys—psychophysical, psychophysics
psychophysiol—psychophysiological, psychophysiology
psychosom—psychosomtic(s)
psychother—psychoterapeutic(s), psychotherapy
Pt—Point
pub—public
publ—publication(s), publish(ed), publisher, publishing
pvt—private

Qm—Quartermaster
Qm Gen—Quartermaster General
qual—qualitative, quality
quant—quantitative
quart—quarterly
Que—Quebec

radiol—radiological, radiology
RAF—Royal Air Force
RAFVR—Royal Air Force Volunteer Reserve
RAMC—Royal Army Medical Corps
RAMCR—Royal Army Medical Corps Reserve
RAOC—Royal Army Ornance Corps
RASC—Royal Army Service Corps
RASCR—Royal Army Service Corps Reserve
RCAF—Royal Canadian Air Force
RCAFR—Royal Canadian Air Force Reserve
RCAFVR—Royal Canadian Air Force Volunteer Reserve
RCAMC—Royal Canadian Army Medical Corps
RCAMCR—Royal Canadian Army Medical Corps Reserve
RCASC—Royal Canadian Army Service Corps
RCASCR—Royal Canadian Army Service Corps Reserve
RCEME—Royal Canadian Electrical & Mechanical Engineers
RCN—Royal Canadian Navy
RCNR—Royal Canadian Naval Reserve
RCNVR—Royal Canadian Naval Volunteer Reserve
Rd—Road
RD—Rural Delivery
rec—record(s), recording
redevelop—redevelopment
ref—reference(s)
refrig—refrigeration
regist—register(ed), registration
registr—registrar
regt—regiment(al)
rehab—rehabilitation
rel(s)—relation(s), relative
relig—religion, religious
REME—Royal Electrical & Mechanical Engineers
rep—represent, representative
Repub—Republic
req—requirements
res—research, reserve
rev—review, revised, revision
RFD—Rural Free Delivery
rhet-rhetoric, rhetorical
RI—Rhode Island
Rm—Room
RM—Royal Marines
RN—Royal Navy
RNA—ribonucleic acid
RNR—Royal Naval Reserve
RNVR—Royal Naval Volunteer Reserve
roentgenol—roentgenologic, roentgenological, roentgenology
RR—Railroad, Rural Route
Rte—Route
Russ—Russian
rwy—railway

S—south
SAfrica—South Africa
SAm—South America, South American
sanit—sanitary, sanitation
Sask—Saskatchewan
SC—South Carolina
Scand—Scandinavia(n)
sch(s)—school(s)
scholar—scholarship
sci—science(s), scientific
SDak—South Dakota
SEATO—Southeast Asia Treaty Organization
sec—secondary
sect—section
secy—secretary
seismog—seismograph, seismographic, seismography
seismogr—seismographer
seismol—seismological, seismology
sem—seminar, seminary
Sen—Senator, Senatorial
Sept—September
ser—serial, series
serol—serologic, serological, serology
serv—service(s), serving
silvicult—silvicultural, silviculture
soc(s)—societies, society
soc sci—social science
sociol—sociologic, sociological, sociology
Span—Spanish
spec—special
specif—specification(s)
spectrog—spectrograph, spectrographic, spectrography
spectrogr—spectrographer
spectrophotom—spectrophotometer, spectrophotometric, spectrophotometry
spectros—spectroscopic, spectroscopy
speleol—speleological, speleology
Sq—Square
sr—senior
St—Saint, Street(s)
sta(s)—station(s)
stand—standard(s), standardization
statist—statistical, statistics
Ste—Sainte
steril—sterility

stomatol—stomatology
stratig—stratigraphic, stratigraphy
stratigr—stratigrapher
struct—structural, structure(s)
stud—student(ship)
subcomt—subcommittee
subj—subject
subsid—subsidiary
substa—substation
super—superior
suppl—supplement(s), supplemental, supplementary
supt—superintendent
supv—supervising, supervision
supvr—supervisor
supvry—supervisory
surg—surgery, surgical
surv—survey, surveying
survr—surveyor
Swed—Swedish
Switz—Switzerland
symp—symposia, symposium(s)
syphil—syphilology
syst(s)—system(s), systematic(s), systematical

taxon—taxonomic, taxonomy
tech—technical, technique(s)
technol—technologic(al), technology
tel—telegraph(y), telephone
temp—temporary
Tenn—Tennessee
Terr—Terrace
Tex—Texas
textbk(s)—textbook(s)
text ed—text edition
theol—theological, theology
theoret—theoretic(al)
ther—therapy
therapeut—therapeutic(s)
thermodyn—thermodynamic(s)
topog—topographic, topographical, topography
topogr—topographer
toxicol—toxicologic, toxicological, toxicology
trans—transactions
transl—translated, translation(s)
translr—translator(s)
transp—transport, transportation
treas—treasurer, treasury
treat—treatment
trop—tropical
tuberc—tuberculosis
TV—television
Twp—Township

UAR—United Arab Republic
UK—United Kingdom
UN—United Nations
undergrad—undergraduate
unemploy—unemployment
UNESCO—United Nations Educational Scientific & Cultural Organization
UNICEF—United Nations International Childrens Fund
univ(s)—universities, university
UNRRA—United Nations Relief & Rehabilitation Administration
UNRWA—United Nations Relief & Works Agency
urol—urologic, urological, urology
US—United States
USAAF—US Army Air Force
USAAFR—US Army Air Force Reserve
USAF—US Air Force
USAFR—US Air Force Reserve
USAID—US Agency for International Development
USAR—US Army Reserve
USCG—US Coast Guard
USCGR—US Coast Guard Reserve
USDA—US Department of Agriculture
USMC—US Marine Corps
USMCR—US Marine Corps Reserve
USN—US Navy
USNAF—US Naval Air Force
USNAFR—US Naval Air Force Reserve
USNR—US Naval Reserve
USPHS—US Public Health Service
USPHSR—US Public Health Service Reserve
USSR—Union of Soviet Socialist Republics

Va—Virginia
var—various
veg—vegetable(s), vegetation
vent—ventilating, ventilation
vert—vertebrate
Vet—Veteran(s)
vet—veterinarian, veterinary
VI—Virgin Islands
vinicult—viniculture
virol—virological, virology
vis—visiting
voc—vocational
vocab—vocabulary
vol(s)—voluntary, volunteer(s), volume(s)
vpres—vice president
vs—versus
Vt—Vermont

W—west
Wash—Washington
WHO—World Health Organization
WI—West Indies
wid—widow, widowed, widower
Wis—Wisconsin
WVa—West Virginia
Wyo—Wyoming

Yearbk(s)—Yearbook(s)
YMCA—Young Men's Christian Association
YMHA—Young Men's Hebrew Association
Yr(s)—Year(s)
YT—Yukon Territory
YWCA—Young Women's Christian Association
YWHA—Young Women's Hebrew Association

zool—zoological, zoology

American Men & Women of Science

Q

QADRI, SYED M HUSSAIN, b Hyderabad, India, Feb 2, 42; m 68; c 3. MEDICAL MICROBIOLOGY. *Educ:* Univ Karachi, BS, 60, MS, 62; Univ Tex, Austin, PhD(microbiol, biochem), 68; Am Bd Med Microbiol, dipl. *Prof Exp:* Lectr microbiol, Univ Karachi, 62-64, asst prof microbiol, 68-69; microbiol specialist, Harris County Hosp Dist, 69-75; asst prof path, Univ Tex Med Sch, Houston, 75-78; assoc prof, 78-84, PROF PATH, UNIV OKLA MED SCH, OKLAHOMA CITY, 84-; DIR MICROBIOL, OKLA MEM HOSP & CLIN, 78- *Concurrent Pos:* R A Welch Found fel, Baylor Col Med, 69-74, adj asst prof path, microbiol & immunol, 75-; clin microbiologist, Hermann Hosp, Houston, 75-78. *Mem:* AAAS; Am Soc Microbiol; Am Soc Med Technol; Pakistan Asn Advan Sci; fel Am Acad Microbiol. *Res:* Biosynthesis of tripyrrole bacterial pigment, prodigiosin, produced by Serratia marcescens; rapid methods in identification of microorganisms. *Mailing Add:* Dept Path Univ Okla Health Sci Ctr PO Box 26901 Oklahoma City OK 73190

QASBA, PRADMANN K, b India, Feb 19, 38; US citizen; m; c 2. BIOCHEMISTRY. *Educ:* Birla Col Pilani, India, BPharm, 58; Munich Univ, WGer, PhD, 65. *Prof Exp:* Max-Planck fel, Max-Planck Inst Biochem, Munich, WGer, 65-68; asst prof, Dept Cell Biol & Pharmacol, Sch Med, Md Univ, Baltimore, 70-72; cancer expert, Lab Pathophysiol, Div Cancer Biol & Diag, Nat Cancer Inst, NIH, 79-82, sr res chemist, 82-84 & Lab Tumor Immunol & Biol, 85-86, SR RES CHEMIST, LAB MATH BIOL, DIV CANCER BIOL & DIAG, NAT CANCER INST, NIH, 86- *Concurrent Pos:* Vis assoc, Lab Biol & Viruses, Nat Inst Allergy & Infectious Dis, NIH, 68-70; vis scientist, Lab Biochem, Div Cancer Biol & Diag, Nat Cancer Inst, NIH, 72-74 & Lab Pathophysiol, 74-79; dir, Inst Biotechnol, New Delhi, India, 85; consult recombinant DNA technol, Nat Chem Lab, Poona, India, 86 & 87. *Mem:* Am Cancer Soc; Am Soc Biol Chemists; AAAS. *Res:* Cancer biology and diagnosis. *Mailing Add:* Lab Math Biol Nat Cancer Inst NIH Bldg 469 Rm 151 Frederick MD 21702-1201

QASIM, SYED REAZUL, b Allahabad, India, Dec 1, 38; m 67; c 2. SANITARY & ENVIRONMENTAL ENGINEERING. *Educ:* Aligarh Muslim Univ, India, BScEng, 57; WVa Univ, MSCE, 62, PhD(sanit eng), 65. *Prof Exp:* Apprentice eng, Irrig Dept, Govt Utter Pradesh, India, 57-58; sr lectr agr eng, Agr Inst, Allahabad, India, 58-59, asst dist eng, Munic Corp, 59-60; res asst, WVa Univ, 62-65; civil engr, Alden E Stilson & Assocs, Consult Engrs, Ohio, 66-68; sr civil engr, Battelle Mem Inst, 68-70; assoc prof civil eng, Polytech Inst New York, 71-73; assoc prof civil eng, 73-78, PROF CIVIL ENG, UNIV TEX, ARLINGTON, 78- *Concurrent Pos:* Fulbright fel, India, 86. *Mem:* Water Pollution Control Fedn; Am Soc Civil Engrs; Sigma Xi. *Res:* Wastewater treatment and control; water resources management; solid waste management; industrial and hazardous waste management; water supply; author of numerous publications on wastewater treatment; author of book on wastewater treatment. *Mailing Add:* Dept Civil Eng Univ Tex PO Box 19308 Arlington TX 76019

QAZI, QUTUBUDDIN H, b Pavas, India, June 15, 31; m 62; c 3. PEDIATRICS, GENETICS. *Educ:* Grant Med Col, Bombay, MB, BS, 56; Univ Toronto, MA, 65, PhD(genetics), 70. *Prof Exp:* House physician med & psychiat, J J Group Hosps, Bombay, 57-58; intern, Coney Island Hosp, Brooklyn, 59-60; from jr resident to chief resident pediat, King's County Hosp, Brooklyn, 60-63; fel genetics, Res Inst, Hosp Sick Children, Toronto, 63-68; asst prof, 69-73, assoc prof, 73-84, PROF PEDIAT, STATE UNIV NY DOWNSTATE MED CTR, 84- *Concurrent Pos:* Attend pediat, King's County Hosp, Brooklyn, 69- & State Univ Hosp, 69-; consult pediat, Methodist Hosp, Brooklyn, 74- *Mem:* Soc Pediat Res; fel Am Acad Pediat; Am Soc Human Genetics; Sigma Xi; Am Pediat Soc. *Res:* Microcephaly and mental retardation; congenital adrenal hyperplasia; lead poisoning; dermatoglyphics; cytogentic disorders; congenital malformations. *Mailing Add:* Dept Pediat Downstate Med Ctr State Univ NY 450 Clarkson Ave Brooklyn NY 11203

QIAN, RENYUAN, b Changshu, Jiangsu, China, Sept 19, 17; m 51, 61; c 2. POLYMER CHARACTERIZATION, CONDUCTING POLYMERS. *Educ:* Zhejiang Univ, BS, 39. *Prof Exp:* Assoc prof phys chem, Zhejiang Univ, 49-51; prof, Inst Phys Chem, 51-53, prof polymer phys chem, Inst Org Chem, 53-56, PROF POLYMER SCI, INST CHEM, ACAD SCINICA, BEIJING, CHINA, 56- *Concurrent Pos:* Exec pres, Chinese Chem Soc, 84-85, chmn, Polymer Div, 86-93; dep dir, Inst Chem, Acad Sinica, 77-81, dir, 81-85; assoc mem, Macromolecular Div, Int Union Pure & Appl Chem, 85-93. *Honors & Awards:* Sci Award, Nat Sci Cong, 78. *Mem:* Chinese Chem Soc (pres, 82-86); fel Am Inst Chemists; Sigma Xi. *Res:* Polymer solution properties; polymer characterization; inter-relation between processing, structure and properties of polymers; excimer formation; organic and polymeric conductors and photo-conductors. *Mailing Add:* Inst Chem Acad Sinica Beijing 100080 China

QUACKENBUSH, CARR LANE W, b Greensboro, NC, May 19, 46; m 69; c 2. CERAMIC TECHNOLOGY, GLASS TECHNOLOGY. *Educ:* Alfred Univ, BS, 68, MS, 69, PhD(ceramics), 73. *Prof Exp:* Prod engr refractories, Norton Co, Worcester, 69-70; asst prof glass & ceramics, Univ Erlangen, WGer, 73-74; mem tech staff glass & ceramics, 74-79, PRIN INVESTR INJECTION MOLDING CERAMICS, GTE LABS, 79- *Mem:* Am Ceramic Soc; Sigma Xi. *Res:* High performance ceramics, including fabrication and characterization of non-silicate refractory glasses; silicon nitride consolidation technology, hot pressing, sintering, thermomechanical properties, oxidation mechanisms and kinetics; silicon nitride shape making technology development; injection molding of turbine engine components, isostatic processing green machinery, process upscale. *Mailing Add:* 425 Great Elm Way Acton MA 01720

QUACKENBUSH, FORREST WARD, b Melrose, Wis, Aug 18, 07; m 37; c 2. BIOCHEMISTRY. *Educ:* Univ Wis, BS, 32, PhD(biochem), 37. *Prof Exp:* Asst biochem, Univ Wis, 34 & 36-37; Rockefeller Found fel, Kaiser-Wilhelm Inst & Univ Heidelberg, 38 & Rijk's Univ, Utrecht, 39; res fel, Univ Wis, 39-42; prof, 43-74, head dept, 43-65, EMER PROF BIOCHEM, PURDUE UNIV, WEST LAFAYETTE, 74- *Mem:* Am Chem Soc; Am Oil Chemist's Soc; Am Soc Biochem & Molecular Biol; Asn Off Anal Chem. *Res:* Biochemistry and nutrition of lipids; chemistry of carotenoids; biosynthesis of lipids. *Mailing Add:* Dept Biochem Purdue Univ West Lafayette IN 47907

QUACKENBUSH, ROBERT LEE, b Feb 27, 43; m; c 2. MICROBIAL GENETICS, EXTRACHROMOSOMAL ELEMENTS. *Educ:* Ind Univ, PhD(microbiol), 77. *Prof Exp:* CHIEF, BACT & MYCOL BR, DMID, NAT INST ALLERGY & INFECTIOUS DIS, NIH, 88- *Mem:* AAAS; Am Soc Microbiol. *Res:* Bacterial structure and function. *Mailing Add:* Bact/Mycol Br Westwood Bldg Rm 738 NIAID/NIH Bethesda MD 20892

QUADAGNO, DAVID MICHAEL, neuroendocrinology, for more information see previous edition

QUADE, CHARLES RICHARD, b Glasgow, Mont, June 18, 36; m 58; c 5. MOLECULAR PHYSICS. *Educ:* Univ Okla, BS, 58, MS, 60, PhD(physics), 62. *Prof Exp:* Asst prof physics, Univ Del, 62-65; from asst prof to assoc prof, 65-70, PROF PHYSICS, TEX TECH UNIV, 70- *Concurrent Pos:* Res grants, NSF, Res Corp & Univ Del Res Found, 63-65, Welch Found, Advan Res Proj Agency. *Mem:* Fel Am Phys Soc; Sigma Xi. *Res:* Vibration rotation interactions; microwave spectroscopy; magnetic susceptibilities; crystal field theory. *Mailing Add:* Dept Physics Tex Tech Univ Lubbock TX 79409

QUADE, DANA EDWARD ANTHONY, b Cardston, Alta, Jan 11, 35; US citizen; m 62; c 3. STATISTICS. *Educ:* Univ Calif, Los Angeles, BA, 55; Univ NC, PhD, 60. *Prof Exp:* Res assoc statist, Univ NC, 60; statistician, Communicable Dis Ctr, USPHS, Ga, 60-61 & Nat Inst Neurol Dis & Blindness, 61-62; from asst prof to assoc prof, 62-70, PROF BIOSTATIST, UNIV NC, CHAPEL HILL, 70- *Mem:* Fel Am Statist Asn; Inst Math Statist; Biomet Soc; Inter Statist Inst. *Res:* Statistical theory, especially nonparametric. *Mailing Add:* UNC - SPH - Biostatistics CB No 7400 Rosenau Hall Chapel Hill NC 27599-7400

QUADER, ATHER ABDUL, b Hyderabad, India, Oct 10, 41; m 68; c 3. MECHANICAL ENGINEERING. *Educ:* Osmania Univ, India, BE, 62; Univ Wis, PhD(mech eng), 69. *Prof Exp:* Res asst mech eng, Univ Wis, 63-68; from assoc res engr to staff res engr, 68-83, SR STAFF RES ENGR, FUELS & LUBRICANTS DEPT, GEN MOTORS RES LABS, 83- *Honors & Awards:* Horning Mem Award, Soc Automotive Engrs, 74, Arch T Colwell Award, 76 & 82, Oral Presentation Award, 76, 78, & 89 & Teetor Ind Lectr, 85. *Mem:* Soc Automotive Engrs; Combustion Inst; Sigma Xi. *Res:* Combustion; internal combustion engines; spectroscopic studies of engine combustion and pollutant formation; lean misfire limit in engines; stratified charge engines; cold start; alternate fuels; reformulated gasoline. *Mailing Add:* Fuels & Lubricants Dept Gen Motors Res Labs Tech Ctr Warren MI 48090-9055

QUADIR, TARIQ, b Karachi, Pakistan, Aug 26, 53; US citizen. TOUGHENED ZIRCONIA MATERIAL, SILICON NITRIDE MATERIAL. *Educ:* NED Univ Eng & Technol, BE, 78; Ohio State Univ, MS, PhD(ceramic eng), 84. *Prof Exp:* Sr ceramic engr, AC Spark Plug, Gen Motors, 84-87; SR CERAMIC ENGR, WR GRACE & CO, 87- *Mem:* Sigma Xi; Am Ceramic Soc; Am Soc Metals; Metall Soc. *Res:* Structural ceramics; developing toughened zirconia, zirconia toughened alumina composites, and silicon nitride for wear and high temperature applications; processing, forming techniques sintering, hiping of the above materials. *Mailing Add:* 10917 Brennan Ct Columbia MD 21044

QUADRI, SYED KALEEMULLAH, b Bidar, India. NEUROENDOCRINOLOGY. *Educ:* Osmania Univ, India, DVM, 60; Kans State Univ, MS, 66; Mich State Univ, MS, 70, PhD(neuroendocrinol), 73. *Prof Exp:* Vet asst surgeon, State Govt India, 60-64; res Ore Regional Primate Res Ctr, assoc, asst scientist neuroendocrinol, 76-77; asst prof physiol, Univ Ore Health Sci Ctr, 75-77; ASSOC PROF PHYSIOL, KANS STATE UNIV, 77- *Mem:* Sigma Xi; Soc Study Reproduction. *Res:* Primate and rodent neuroendocrinology, with special emphasis on prolactin, aging and breast cancer. *Mailing Add:* Dept Anat & Physiol VMS 228 Kans State Univ Manhattan KS 66506

QUADT, R(AYMOND) A(DOLPH), b Perth Amboy, NJ, Apr 16, 16; m 40; c 1. PHYSICAL METALLURGY. *Educ:* Rutgers Univ, BS, 39; Stevens Inst Technol, MS, 47. *Prof Exp:* Teacher high sch, NJ, 39-42; res metallurgist, Am Smelting & Refining Co, 42-48, mgr aluminum div, 48-50; dir res, Hunter Douglas Aluminum Co, 50-54, vpres res & develop, 54- 57, vpres res & develop, Bridgeport Brass Co, 57-60; pres, Reactive Metals, Inc, 60-65; vpres, Pascoe Steel Corp, 65-73; pres, Phoenix Cement Co, 73-80, consult, 80-84; RETIRED. *Honors & Awards:* Meritorious Pub Serv Citation, US Navy, 55. *Mem:* Am Soc Metals. *Res:* Physical and process metallurgy of aluminum, titanium, zirconium, columbium and tantalum. *Mailing Add:* 15 E San Miguel Phoenix AZ 85012

QUAGLIANO, JAMES VINCENT, b New York, NY, Nov 9, 15; m 61; c 2. INORGANIC CHEMISTRY. *Educ:* Polytech Inst Brooklyn, BS, 38, MS, 40; Univ Ill, PhD(inorg chem), 46. *Prof Exp:* Instr chem, Villanova Col, 40-43 & Univ Ill, 44-45; asst prof, Univ Md, 46-48; from asst prof to assoc prof, Univ Notre Dame, 48-58; from assoc prof to prof chem, Fla State Univ, 58-75; HUDSON PROF CHEM, AUBURN UNIV, 75- *Mem:* AAAS; Am Chem Soc; Am Inst Chemists; fel Royal Soc Chem; NY Acad Sci. *Res:* Inorganic complex compounds; reduction potentials of some inorganic coordination compounds; infrared absorption of inorganic coordination complexes. *Mailing Add:* Dept Chem Va Commonwealth Univ Box 2006 Richmond VA 23284

QUAIFE, MARY LOUISE, b Madison, Wis; m 69. BIOCHEMISTRY. *Educ:* Univ Mich, AB, 38, MS, 39; Univ Ill, PhD(anal chem), 42; Am Bd Clin Chem, dipl, 52. *Prof Exp:* Clin chemist, Univ Hosp, Vanderbilt, 39-40; asst chem, Univ Ill, 41-42, asst physiol, 42-43; sr res chemist, Distillation Prod Industs Div, Eastman Kodak Co, 43-54; res assoc nutrit, Sch Pub Health, Harvard Univ, 56-57, assoc, 57-59; res asst environ health, Sch Pub Health, Univ Mich, 60-61, res assoc obstet & gynec, Med Sch, 61-62; biochemist, Div Toxicol Eval, US Food & Drug Admin, 62-70; biochemist, toxicol br, hazard eval div, US Environ Protection Agency, 70-82. *Concurrent Pos:* Consult chem & toxicol, 82- *Mem:* Am Chem Soc; Am Soc Biol Chemists; Am Inst Nutrit; fel Am Inst Chemists; Soc Toxicol. *Res:* Toxicological evaluation of pesticides and food additives; analysis and biochemistry of vitamin E; modern analytical techniques. *Mailing Add:* 1506 33rd St NW Washington DC 20007

QUAIL, JOHN WILSON, b Brooklyn, NY, Mar 19, 36; m 59; c 4. CRYSTALLOGRAPHY. *Educ:* Univ BC, BSc, 59, MSc, 61; McMaster Univ, PhD(inorg chem), 63. *Prof Exp:* Fel inorg chem, McMaster Univ, 63-64; from asst prof to assoc prof, 64-83, PROF CHEM, UNIV SASK, 83- *Concurrent Pos:* Sabbatical leave, Univ Chem Lab, Cambridge Univ, 72-73, Dept Chem, Univ Alta, Edmonton, 81-82 & Dept Crystallog, Birkbeck Col, Univ London, 88-89. *Mem:* Chem Inst Can; Am Crystallog Asn. *Res:* Protein crystallography of small molecules; crystallography of drug molecules; protein crystallography. *Mailing Add:* Dept Chem Univ Sask Saskatoon SK S7N 0W0 Can

QUAIL, PETER HUGH, b Cooma, New SWales, Australia, Feb 4, 44; c 1. PLANT MOLECULAR BIOLOGY, MOLECULAR PHOTOBIOLOGY. *Educ:* Univ Sydney, BSc, 64, PhD(plant physiol), 68. *Prof Exp:* Res assoc, plant res lab, Mich State Univ, 68-71, Biol Inst, Univ Freiburg, 71-73; res fel, Res Sch Biol Sci, Australian Nat Univ, 73-77; sr fel, dept biol, Carnegie Inst Stanford, 77-79; assoc prof bot, Univ Wis-Madison, 79-84, prof bot & genetics, 84-87; PROF MOLECULAR PLANT BIOL, UNIV CALIF BERKELEY, 87-; RES DIR, PLANT GENE EXPRESSION CTR, ALBANY, CA, 87- *Mem:* Int Soc Plant Molecular Biol; Am Soc Plant Physiologists; Am Soc Photobiologists. *Res:* Molecular mechanism by which the plant regulatory photoreceptor, phytochrome, controls gene expression in response to light; structural studies on the phytochrome molecule and characterization of regulatory sequences in phytochrome-controlled genes. *Mailing Add:* Plant Gene Expression Ctr 800 Buchanan St Albany CA 94710

QUAILE, JAMES PATRICK, b Philadelphia, Pa, Jan 16, 43; m 68; c 2. MECHANICAL ENGINEERING, ENERGY SCIENCES. *Educ:* NJ Inst Technol, BS, 68; Lehigh Univ, MS, 69, PhD (mech eng) 72; State Univ NY, MBA, 85. *Prof Exp:* Pro engr transp, Transp Tech Ctr, 72-73, syst engr energy, 74-76, mgr, 76-77, liaison scientist consumer prod, 77-82, SYSTS ENGR INFO SYSTS LAB, CORP RES & DEVELOP, GEN ELEC CO, 82- *Concurrent Pos:* Adj prof, Union Col State Univ of NY. *Honors & Awards:* Fel, NSF. *Mem:* Am Soc Mech Engrs; fel NSF. *Res:* Energy; thermodynamics; heat transfer; fluid dynamics; interdisciplinary systems approach to the conception and development of novel computer systems for decision support and productivity improvement in engineering, manufacturing, marketing and financial areas; use of leading edge technologies such as artificial intelligence/expert systems combined with relational database concepts to implement these systems. *Mailing Add:* Corp Res & Develop PO Box Eight Schenectady NY 12301

QUALLS, CLIFFORD RAY, b Duncan, Okla, Oct 31, 36; m 59; c 2. MATHEMATICAL STATISTICS. *Educ:* Calif State Col Long Beach, BA, 61; Univ Calif, Riverside, MA, 64, PhD(math), 67. *Prof Exp:* Mathematician, US Naval Ord Lab, Calif, 56-61; reliability engr, Autonetics Div, NAm Aviation, Inc, 61-64; instr math, Calif State Col Fullerton, 64-67; asst prof, 67-74, assoc prof, 74-80, PROF MATH, UNIV NMEX, 80- *Mem:* Math Asn Am; Inst Math Statist; Inst Elec & Electronics Engrs. *Res:* Crossing problems for stationary stochastic processes; life testing; prediction theory for stochastic processes. *Mailing Add:* Dept Math & Statist Univ NMex Albuquerque NM 87131

QUALSET, CALVIN ODELL, b Newman Grove, Nebr, Apr 24, 37; m 57; c 3. AGRONOMY, CONSERVATION BIOLOGY. *Educ:* Univ Nebr, BS, 58; Univ Calif, Davis, MS, 60, PhD(genetics), 64. *Prof Exp:* Lab technician agron, Univ Calif, Davis, 60-64; asst prof, Univ Tenn, Knoxville, 64-67; from asst prof to assoc prof, 67-74, chmn, dept agron & range sci, 75-81, assoc dean, Col Agr & Environ Sci, 81-86, PROF AGRON, UNIV CALIF, DAVIS, 74-; DIR, CALIF GENETIC RESOURCES CONSERV PROG, 85- *Concurrent Pos:* Consult, ed bd Plant Breeding, Ger, 85-; Fulbright sr scholar, Australia, 76 & Yugoslavia, 84; mem, Nat Plant Genetic Resources Bd, 82-87, Res Adv Comt, US AID, 89-, Nat Res Coun Comt on Alternative Farming Practices, 86-88, Nat Res Coun Comt Global Mgt of Genetics Resources Workshop, chair on Rev Nat Plant Germplasm Syst, 87-89, Nat Res Coun Comt Sci Eval of Introd of Genetically Modified Microrganisms and Plants into the Environment; ed-in-chief, Crop Sci Soc Am, 80-83; chair, Agr Sect AAAS, 92. *Mem:* Fel AAAS; Am Genetics Asn; Genetics Soc Am; Genetics Soc Can; fel Am Soc Agron; fel Crop Sci Soc Am (pres, 89). *Res:* Analysis of quantitative genetic variation in plants; breeding for improvement in agronomic and quality characteristics in barley, wheat, oats, triticale and rye; genetic resources conservation in plants. *Mailing Add:* Dept Agron & Range Sci Univ Calif Davis CA 95616

QUAM, DAVID LAWRENCE, b Minneapolis, Minn, June 5, 42. AERODYNAMICS, CONTROL SYSTEMS. *Educ:* Univ Minn, BAE, 65; Univ Wash, MSAA, 70, PhD(aeronaut & astronaut), 75. *Prof Exp:* Assoc engr, Lockheed-Calif Co, 65; teaching assoc, Univ Wash, 68-75; assoc prof aerospace eng, Univ Dayton, 75-82; ASEOSPACE ENG CONSULT & PRES, QUAMAERO RES INC, 83- *Concurrent Pos:* Sr aerodynamicist, Boeing Commercial Airplane Co, 68-69; control systs analyst, Marine Systs Div, Honeywell, 72-75; fac researcher, Fac Res Prog, USAF-Am Soc Eng Educ, 77; chmn, Flight Simulation Course, Univ Dayton, 79-80; math modeling consult, Technol, Inc, 79-85; vis prof, Frank J Seiler Res Lab, USAF Acad, 80-81. *Mem:* Am Inst Aeronaut & Astronaut; Am Helicopter Soc; Am Soc Eng Educ. *Res:* Perception studies and myoelectric feedback for flight simulation; missile aerodynamics and control; digital control of spacecraft; experimental and theoretical work in unsteady aerodynamics; aircraft and ship simulator design. *Mailing Add:* 37 Seminary Ave Dayton OH 45403

QUAMME, GARY ARTHUR, b Moose Jaw, Sask, Oct 26, 44; m; c 3. MEDICINE. *Educ:* Univ Sask, BSc, 65, DVM, 69; Univ Ottawa, MSc, 70; McGill Univ, PhD(physiol), 74. *Prof Exp:* Res investr, Toxicol Res Div, Health Protection Br, Health & Welfare Can, 69-70; instr, Dept Physiol, Western Col Vet Med, Sask, 70-71; Med Res Coun fel, Centre d'Etudes Nucleaires de Saclay, Gif- sur-Yvette, France, 74-75; asst prof, Dept Physiol, McGill Univ, Montreal, 75-76; from asst prof to assoc prof, 76-86, PROF, DEPT MED, UNIV BC, 86- *Concurrent Pos:* Med Res Coun scholar, Dept Physiol, McGill Univ, 75-76 & Dept Med, Univ BC, 76-80; mem, Prog Comt, Am Soc Nephrology, 82 & 89, Grants Comt, Med Res Coun Can, 83 & Sci Coun, Kidney Found Can, 87-90; chmn, Symp Magnesium, Am Col Nutrit, 85 & Scanning Electron Micros Conf, Ont, 87; assoc ed, Can J Physiol & Pharmacol, 85-90; Med Res Coun vis scientist & prof, Inst Physiol, Univ Zurich, 87-88; Schering travel award, Can Soc Clin Invest, 87; vis scientist, Roche Res Found, 87-88; Izaak Walton Killam fac res fel, 88; ed, Magnesium Res, 88-; clin investr, Can Asn Gastroenterol, 89-; distinguished lectr, Fac Med, Univ BC, 89. *Mem:* Can Soc Nephrology; Can Physiol Soc; Am Fedn Clin Res; Am Soc Nephrology; Fedn Am Socs Exp Biol; Am Soc Clin Invest; Can Soc Clin Invest; Int Soc Nephrology; Am Physiol Soc; Int Union Physiol Sci. *Mailing Add:* Dept Med Univ BC 2211 Westbrook Mall Vancouver BC V6T 2B5 Can

QUAMME, HARVEY ALLEN, b Browniee, Sask, Apr 23, 40; m 63; c 2. HORTICULTURE. *Educ:* Univ Sask, BSA, 62, MSc, 64; Univ Minn, PhD(hort), 71. *Prof Exp:* From res officer freit res to res scientist, 63-74, res scientist I, 74-77, res scientist II, 78-82, MEM STAFF, RES STA, CAN DEPT AGR, 74-, RES SCIENTIST III, 88- *Honors & Awards:* Gourley Award Res Pomology, Am Soc Hort Sci, 73 & 83; Carroll R Miller Award, Nat Peach Coun, 79; Westdale Award, Can Soc Hort Sci, 87, Hoechst Award, 89. *Mem:* Am Soc Hort Sci; Can Soc Hort Sci; Int Soc Hort Sci; Agr Inst Can. *Res:* Evaluation and development of new roots tock varieties of tree fruits and grapes with emphasis on dwarfing precocity, cold hardiness and disease resistance; investigation of mechanisms of winter survival in fruit crops which involve deep supercooling. *Mailing Add:* Can Dept Agr Res Sta Summerland BC V0H 1Z0 Can

QUAN, STUART F, b San Francisco, Calif, May 16, 49; m 71; c 2. CRITICAL CARE MEDICINE, SLEEP DISORDERS. *Educ:* Univ Calif, Berkeley, AB, 70, San Francisco, MD, 74. *Prof Exp:* Med intern internal med, Univ Wis-Madison, 74-75, med resident, 75-77; fel critical care, Univ Calif, San Francisco, 77-78, fel emergency med, 78-79; fel pulmonary, Univ Ariz, 79-80, instr med, 80-81, asst prof, 81-86, ASSOC PROF MED, UNIV ARIZ, TUCSON, 86-, ASSOC PROF ANESTHESIOL, 87- *Concurrent Pos:* Mem, Anesthesiol & Respiratory Therapy Devices Panel, US Food & Drug Admin, 86-89, chmn, 88-89, consult, 89-; mem, Am Bd Sleep Med, 90- *Mem:* Am Thoracic Soc; Soc Critical Care Med; Am Col Chest Physicians; Clin Sleep Soc; Am Fedn Clin Res; Nat Asn Med Dirs Respiratory Care. *Res:* Mechanisms responsible for bronchial hyperreactivity after viral infections; epidemiology of sleep disorders; mechanical ventilation in acute respiratory failure. *Mailing Add:* Univ Ariz Col Med AHSC 1501 N Campbell Ave Tucson AZ 85724

QUAN, WILLIAM, b Aug 1, 48. ENVIRONMENTAL SCIENCE. *Educ:* Univ Calif, Berkeley, BS, 70; Ohio State Univ, MS, 72. *Prof Exp:* Chief formulating chemist, Lee Pharmaceut, S El Monte, Calif, 73-75; mem tech staff, Rocketdyne Div, Rockwell Int, Canoga Park, Calif, 75-77; pub health chemist, Air & Indust Hyg Lab, 77-80, coordr, State's Indust Waste Recycling Prog, 80-83, chief, Environ Assessment Unit, Toxic Substances Control Div, Calif Dept Health Serv, 83-88; PROJ MGR, SOLID WASTE MGT PROG, 88- *Res:* Waste reduction and environmental chemistry; polymer and analytical chemistry; environmental fate of contaminants; plastic mineral composite technology. *Mailing Add:* Off Chief Admin Officer Rm 271 City Hall San Francisco CA 94102

QUAN, XINA SHU-WEN, b Gloucester, NJ, Dec 23, 57; m 80; c 1. POLYMER CHEMISTRY, CHEMICAL ENGINEERING. *Educ:* Mass Inst Technol, SBChE, 80, SMChE, 80; Princeton Univ, PhD(chem eng), 86. *Prof Exp:* From assoc mem to mem, 80-91, DISTINGUISHED MEM TECH STAFF, AT&T BELL LABS, 91- *Mem:* Am Phys Soc; Am Chem Soc. *Res:* Multiphase polymer characterization; block copolymers; polymer blends; mechanical properties; small-angle scattering. *Mailing Add:* AT&T Bell Labs Rm 7F-212 600 Mountain Ave PO Box 636 Murray Hill NJ 07974-2070

QUANDT, EARL RAYMOND, JR, b Washington, DC, Feb 5, 34; m 56; c 4. CHEMICAL ENGINEERING. *Educ:* Univ Cincinnati, ChE, 56; Univ Pittsburgh, PhD(chem eng), 61. *Prof Exp:* Sr engr, Bettis Atomic Power Lab, Westinghouse Elec Corp, 56-63; res coordr appl mech, USN Marine Eng Lab, 63-67, HEAD POWER SYSTS DIV, USN SHIP RES & DEVELOP CTR, 67- *Mem:* Am Inst Chem Engrs; Sigma Xi. *Res:* Two phase flow; heat transfer; systems dynamics; reliability; marine propulsion. *Mailing Add:* 203 Winchester Rd SW 1605 Riverside Dr Annapolis MD 21401

QUANE, DENIS JOSEPH, inorganic chemistry; deceased, see previous edition for last biography

QUANSTROM, WALTER ROY, b Gary, Ind, Nov 20, 42; m 63; c 2. ANIMAL BEHAVIOR. *Educ:* Bethany Nazarene Col, BS, 64; Univ Okla, PhD(zool), 68. *Prof Exp:* NASA trainee, Univ Okla, 65-67; asst prof biol, Olivet Nazarene Col, 68-70; assoc prof & chmn dept, Northwest Nazarene Col, 70-74, chmn div natural sci & math, 72-74; staff ecologist, 74-77, dir ecol, 77-79, dir environ & energy conservation, 79-81, MGR, INDUST HYGIENE, TOXICOL & SAFETY, STANDARD OIL CO, IND, 81-; VPRES, ENVIRON AFFAIRS & SAFETY, AMOCO CORP. *Concurrent Pos:* Mem, Exec Develop Prog, Northwestern Univ, 80. *Mem:* Am Soc Mammalogists. *Res:* Ethoecology of Richardson's ground squirrel, Spermophilus richardsonii. *Mailing Add:* Amoco Corp MC 4905A 200 E Randolph Dr Chicago IL 60601

QUARLES, CARROLL ADAIR, JR, b Abilene, Tex, Nov 24, 38; m 71; c 2. ATOMIC PHYSICS. *Educ:* Tex Christian Univ, BA, 60; Princeton Univ, MA, 62, PhD(physics), 64. *Prof Exp:* Res fel physics, Brookhaven Nat Lab, 64-65, asst physicist, 65-67; from asst prof to assoc prof, 67-74, assoc dean, 74-78, chmn dept, 78-84, PROF PHYSICS, TEX CHRISTIAN UNIV, 74-, HOLDER W A MONCRIEF CHAIR PHYSICS, 87- *Mem:* AAAS; Am Phys Soc; Am Asn Physics Teachers. *Res:* Electron and atomic collisions; atomic field bremsstrahlung; inner shell ionization. *Mailing Add:* Dept Physics Tex Christian Univ Ft Worth TX 76129

QUARLES, GILFORD GODFREY, b Charlottesville, Va, Dec 24, 09; m 34; c 3. PHYSICS. *Educ:* Univ Va, BS, 30, MS, 33, PhD(physics), 34. *Prof Exp:* Instr physics, Univ Va, 31-33; actg prof, Mercer Univ, 34-35 & Univ Ala, 35-41; assoc prof, Furman Univ, 41-43; res assoc, Harvard Univ, 44-45; from assoc prof to prof eng res, Pa State Univ, 45-56; tech & sci consult to commanding gen, Ballistic Missile Agency, US Dept Army, 56-58, chief scientist, Ord Missile Command, Redstone Arsenal, Ala, 58-59, chief sci adv, CEngrs, 59-60; dir long-range mil planning, Bendix Corp, 60-61; chief sci adv, Off Chief Engrs, US Army CEngrs, 61-75; RETIRED. *Concurrent Pos:* Asst dir, Ord Res Lab, Pa State Univ, 47-52, dir, 52-56. *Mem:* AAAS. *Res:* Photographic latent image; electrooptical Kerr effect; underwater ordnance; sonic and ultrasonic vibrations; ballistic missiles. *Mailing Add:* 1244 Westerly Pkwy No 23 State College PA 16801-4168

QUARLES, JOHN MONROE, b Chattanooga, Tenn, May 24, 42; m 71; c 1. MEDICAL MICROBIOLOGY, VIROLOGY. *Educ:* Fla State Univ, BS, 63, MS, 65; Mich State Univ, PhD(microbiol), 73; Am Bd Microbiol, cert pub health & med lab microbiol, 75. *Prof Exp:* Res microbiologist, Ctr Dis Control, USPHS, 65-66; head, Serol-Virol Lab, Naval Med Sch, Nat Naval Med Ctr, US Navy, 66-69; fel, Univ Tenn & Oak Ridge Nat Lab, 73-74 & Nat Cancer Inst & Oak Ridge Nat Lab, 74-76; COORDR GRAD STUDIES, COL MED, TEX A&M UNIV, 89-, PROF MICROBIOL, 91- *Mem:* Am Soc Microbiol; Am Soc Virol; Soc Anal Cytol. *Res:* Transformation of mammalian cells by chemical carcinogens; rapid techniques for diagnosis of microorganisms; flow cytometry; dialysis culture of microorganisms and mammalian cells; animal virology; antiviral agents and vaccines; influenza. *Mailing Add:* Dept Med Microbiol Col Med Tex A&M Univ Col Sta TX 77843-1114

QUARLES, RICHARD HUDSON, b Baltimore, Md, Sept 23, 39; m 64; c 3. BIOCHEMISTRY, NEUROCHEMISTRY. *Educ:* Swarthmore Col, AB, 61; Harvard Univ, PhD(biochem), 66. *Prof Exp:* From staff fel to sr staff fel biochem, 68-73, res chemist, 73-77, HEAD, SECT MYELIN & BRAIN DEVELOP, NAT INST NEUROL & COMMUNICATIVE DIS & STROKE, 77- *Concurrent Pos:* NSF fel, Inst Animal Physiol, Cambridge, Eng, 66-67, Nat Inst Neurol Dis & Stroke fel, 67-68. *Mem:* AAAS; Am Chem Soc; Am Soc Neurochem; Int Soc Neurochem. *Res:* Metabolism of phospholipids, glycolipids, and glycoproteins; roles of lipids and proteins in membrane structure and function; developing brain; myelination. *Mailing Add:* NIH Park Bldg Rm 425 Bethesda MD 20898

QUARLES, RICHARD WINGFIELD, b Richmond, Va, Mar 21, 11; m 36; c 5. POLYMER CHEMISTRY. *Educ:* Univ Va, BSE, 31, PhD(phys chem), 35. *Prof Exp:* Asst phys chem, Univ Va, 35-37; res chemist, Union Carbide & Carbon Corp, WVa, 37-39; indust fel, Mellon Inst, 39-50, sr fel, 51-55; asst div head, Bakelite Co, 55-56, asst dir develop, Union Carbide Corp, 56, assoc dir develop, Plastics Div, 56-60, mgr patents & licenses, 60-64, dir polymer res & develop, 64-69, mgr patents, 69-76; CONSULT, 76- *Mem:* Am Chem Soc; Am Inst Chem Engrs; Sigma Xi; AAAS. *Res:* Vinyl resins with special emphasis on their use in surface coatings and adhesives; synthetic resins; adsorption of gases and surface phenomena. *Mailing Add:* 60 Marion Rd W Princeton NJ 08540

QUARONI, ANDREA, b Milano, Italy, Aug 9, 46; m 75; c 1. DEVELOPMENTAL BIOLOGY. *Educ:* Univ Pavia, Italy, PhD(biochem), 70. *Prof Exp:* Asst volontario biochem, Univ Pavia, Italy, 70-71; asst I biochem, Swiss Fed Inst Technol, Zurich, 72-75; res fel biochem, Harvard Med Sch & Mass Gen Hosp, 75-78; instr biochem, Harvard Med Sch, 78-80, asst prof, 80-81; DEPT PHYSIOL, CORNELL UNIV, 81- *Concurrent Pos:* Asst biochem, Mass Gen Hosp, 78-80. *Mem:* Soc Complex Carbohydrates; Swiss Biochem Soc; Swiss Biochem Soc; Tissue Culture Asn. *Res:* Structure and function of the intestinal epithelium; identification, purification and biosynthesis of surface membrane proteins and glycoproteins and cultured intestinal epithelial cells; structure and composition of the intestinal basement membrane. *Mailing Add:* Dept Physiol Cornell Univ Ithaca NY 14853

QUARRY, MARY ANN, b Philadelphia, Pa; m 89. ANALYTICAL CHEMISTRY. *Educ:* Villanova Univ, BS, 75, PhD(anal chem), 84. *Prof Exp:* Anal chemist, Arco Chem, 79-80; applications specialist, DuPont Instrument Systs, 81-84; res chemist, E I du Pont de Nemours & Co, 84-89, group leader, 89-91; RES MGR, DUPONT MERCK PHARMACEUT, 91- *Mem:* Sigma Xi; Am Chem Soc. *Res:* Pharmaceutical analysis; separation optimization by gradient elution HPLC; expert systems. *Mailing Add:* Dupont Merck Pharmaceut PO Box 80353 Wilmington DE 19880-0353

QUARTARARO, IGNATIUS NICHOLAS, b Brooklyn, NY, July 26, 26; m 52; c 4. DENTISTRY. *Educ:* NY Univ, DDS, 52; Am Bd Endodont, dipl, 59. *Prof Exp:* Asst endodontia, NY Univ, 52-54, from instr to asst prof, 54-68, assoc prof, 68-73, TRUSTEE, NY UNIV, 90- *Concurrent Pos:* Consult, Dept Surg, New York Infirmary Hosp, 61-66; dir endodont, Cath Med Ctr, Brooklyn & Queens, 72-85; pres, DSSNY, 89. *Mem:* Fel Am Asn Endodont; fel Am Col Dent; Am Acad Oral Med; Sigma Xi; fel Int Col Dentists. *Res:* Temporo-mandibular joint; fluoroscopy of sealing properties of endodontic cements; pathology of the periapical lesion. *Mailing Add:* 520 Franklin Ave Garden City NY 11530

QUARTERMAN, ELSIE, b Valdosta, Ga, Nov 28, 10. BOTANY, ECOLOGY. *Educ:* Ga State Col, AB, 32; Duke Univ, MA, 41, PhD(bot), 49. *Prof Exp:* Pub sch teacher, Ga, 32-43; from instr to assoc prof biol, 43-66, chmn div bact, bot & zool, 60-61, prof biol, 66-76, PROF EMER, VANDERBILT UNIV, 76- *Concurrent Pos:* Mem bd dirs, Tenn Bot Gardens, 71-72; US Fish & Wildlife Serv Tenn Coneflower Recovery Team, 80-83. *Honors & Awards:* Oakleaf Award, Nature Conservancy, 81; Sal Feinstone Environ Award, 82. *Mem:* Asn Southeastern Biologists; Sigma Xi. *Res:* Distribution of the Compositae in south Georgia; composition and structure of plant communities in middle Tennessee; ecology of bryophytes; climax forests of the coastal plain of southeastern United States; autecology of middle Tennessee endemics; conservation of ecosystems. *Mailing Add:* 1313 Belmont Park Ct Nashville TN 37215

QUASS, LA VERNE CARL, b Beloit, Wis, Jan 17, 37; m 59; c 3. INORGANIC CHEMISTRY. *Educ:* Luther Col, BA, 59; Univ Wis-Madison, MS, 64, PhD(chem), 69. *Prof Exp:* Instr chem, Univ Wis-Fox Valley Ctr, 63-66; asst prof chem, Univ Wis, Parkside, 69-76; assoc prof, 76-80, PROF CHEM, GRAND VIEW COL, 80-, CHMN, DIV NATURAL SCI, 86- *Mem:* AAAS; Am Chem Soc; The Chem Soc. *Res:* Organometallic and organosilicon chemistry. *Mailing Add:* Dept Chem Grand View Col 1200 Grandview Ave Des Moines IA 50316

QUAST, JAY CHARLES, b San Francisco, Calif, Sept 17, 23; m 49. ICHTHYOLOGY. *Educ:* Univ Calif, BA, 48, MA, 50 & 51, PhD(ichthyol), 60. *Prof Exp:* High sch teacher, Calif, 51-53; asst, Univ Calif, Los Angeles, 53-57; res biologist, Scripps Inst, Univ Calif, 58-61; supvry fishery biologist, Nat Marine Fisheries Serv, 61-74, proj scientist, Environ Res Labs, 74-76, nat sci ed, 77-80, RES BIOLOGIST, NAT MARINE FISHERIES SERV, NAT OCEANIC & ATMOSPHERIC ADMIN, 81- *Mem:* Am Fisheries Soc; Am Soc Ichthol & Herpet. *Res:* Fish ecology, taxonomy, variation, osteology and population dynamics; research management. *Mailing Add:* 1565 Jamestown St SE Salem OR 97302

QUASTEL, D M J, b Cardiff, UK, June 7, 36; Can citizen; m 59; c 2. PHYSIOLOGY. *Educ:* McGill Univ, BS, 55, MD, CM, 59, PhD(physiol), 63. *Prof Exp:* Med Res Coun Can fel, 61-63; Muscular Dystrophy Asn Can fel, 63-65; asst prof physiol, Dalhousie Univ, 65-69; assoc prof, 69-77, PROF PHARMACOL, UNIV BC, 77- *Mem:* Can Physiol Soc; Soc Neurosci. *Res:* Mechanisms of synaptic transmission. *Mailing Add:* Dept Pharmacol Univ BC 2194 Health Sci Mall Vancouver BC V6T 1W5 Can

QUASTEL, MICHAEL REUBEN, b Cardiff, Wales, June 30, 33; Can citizen; m 62; c 4. IMMUNOLOGY, RADIOBIOLOGY & NUCLEAR MEDICINE. *Educ:* McGill Univ, BSc, 53, MD & CM, 57; Univ Ottawa, PhD(immunology), 71. *Prof Exp:* Intern, Charity Hosp La, New Orleans, 57-58; jr res physician, Radiobiol & Nuclear Med Labs, Univ Calif, Los Angeles, 59-61; res assoc, Pasadena Found Med Res, 61-62; asst resident oncol & radiother, Hadassah Hosp, Jerusalem, 62-64; head biol sect, Radiation Protection Div, Dept Nat Health & Welfare, 65-71, asst chief human cytogenetics div & head environ mutagenesis sect, 71-74; head, Lab Clin Immunol, Isotope Dept, Soroka Med Ctr, 74-84, ASSOC PROF, SCH MED, BEN GURION UNIV, BEERSHEBA, ISRAEL, 74-, HEAD, INST NUCLEAR MED, SOROKA MED CTR, 81- *Concurrent Pos:* Instr, Dept Radiol, Univ Southern Calif, 61-62; res fel physiol, Med Sch, Hebrew Univ, 64-65; sr lectr, Sch Med, Univ Ottawa, 71-74; consult, Ottawa Civic Hosp, Can, 71-74; res & clin fel, Div Nuclear Med, Mass Gen Hosp & Harvard Med Sch, 80-81; Eleanor Roosevelt fel, Int Union Against Cancer, Univ BC, Vancouver, Can, 87-88. *Mem:* Can Soc Immunol; fel Royal Soc Health; Europ Asn Nuclear Med; NY Acad Sci; Israel Soc Immunol; Soc Nuclear Med. *Res:* Medical radiobiology and nuclear medicine; cell biology and tissue culture; human cytogenetics and environmental mutagenesis; clinical and cellular immunology. *Mailing Add:* Soroka Med Ctr PO Box 151 Beersheba 84101 Israel

QUATE, CALVIN F(ORREST), b Baker, Nev, Dec 7, 23; m 46; c 4. APPLIED PHYSICS, ELECTRICAL ENGINEERING. *Educ:* Univ Utah, BS, 44; Stanford Univ, PhD(elec eng), 50. *Prof Exp:* Tech staff mem, Bell Tel Labs, NJ, 49-58; dir & vpres res, Sandia Corp, NMex, 59-61; PROF APPL PHYSICS & ELEC ENG, STANFORD UNIV, 61-, LELAND T EDWARDS PROF ENG, 86- *Concurrent Pos:* Guggenheim fel & Fulbright scholar, Fac Sci, Montpellier, France, 68-69; chmn, Dept Appl Physics, Stanford Univ, 69-72 & 78-81, assoc dean, Sch Humanities & Sci, 72-74 & 82-83; sr res fel, Xerox Palo Alto Res Ctr, 84- *Honors & Awards:* Morris N Liebmann Award, Inst Elec & Electronics Engrs, 81, Rank Prize for Opto-Electronics, 82, Achievement Award, 86 & Medal of Hon, 88. *Mem:* Nat Acad Sci; Nat Acad Eng; Am Phys Soc; fel Inst Elec & Electronics Engrs; fel Acoust Soc Am; fel AAAS; fel Am Acad Arts & Sci; hon fel Royal Micros Soc. *Res:* Microwave electronics and solid state devices; physical acoustics; imaging microscopy and the storage of digital information; holder or co-holder of 42 patents; author or co-author of over 160 publications. *Mailing Add:* Dept Appl Physics Stanford Univ Stanford CA 94305

QUATRANO, RALPH STEPHEN, b Elmira, NY, Aug 3, 41; m 59; c 3. BOTANY. *Educ:* Colgate Univ, AB, 62; Ohio Univ, MS, 64; Yale Univ, PhD(biol), 68. *Prof Exp:* NSF undergrad res partic, Colgate Univ, 61-62; AEC res asst bot, plant physiol & biol, Ohio Univ, 62-64; teaching asst, Yale Univ, 64-65, NIH trainee develop biol, 65-67; from asst prof to assoc prof, 68-78, PROF BOT, ORE STATE UNIV, 78- & DIR, CTR GENE RES & BIOTECHNOL, 84- *Concurrent Pos:* Vis prof, Marine Biol Lab, Woods Hole, Stanford Univ & Fridge Harbor Lab, Univ Wash. *Mem:* Soc Develop Biol; Bot Soc Am; Am Soc Plant Physiologists; Int Plant Molecular Biol Soc; Am Soc Cell Biol. *Res:* Studies in physiology, cell and molecular biology of plant cell differentiation and embryogenesis in algae and angiosperms. *Mailing Add:* Dept Bot Ore State Univ Corvallis OR 97331

QUATTROCHI, DALE ANTHONY, b Cleveland, Ohio, Dec 3, 50; m 77. REMOTE SENSING & LANDSCAPE ECOLOGY. *Educ:* Ohio Univ, BS, 73; Univ Tenn, MS, 78; Univ Utah, PhD, 90. *Prof Exp:* Res assoc, Miss Remote Sensing Ctr, Miss State Univ, 77-80; GEOGRAPHER, NASA, SCI & TECHNOL LAB, JOHN C STENNIS SPACE CTR, 80- *Concurrent Pos:* Adj asst prof, dept geog, Univ New Orleans, 88-, Univ Southern Miss, 90- *Mem:* Asn Am Geographers; Am Soc Photogram & Remote Sensing; Sigma Xi; Int Asn Landscape Ecol. *Res:* Analysis of landscape environmental interrelationships; remote sensing data analysis and modeling of urban biophysical interrelationships; author of numerous articles and papers. *Mailing Add:* John C Stennis Space Ctr NASA Sci & Technol Lab Stennis Space Ctr MS 39529-6000

QUATTROPANI, STEVEN L, anatomy, reproductive biology; deceased, see previous edition for last biography

QUAY, JOHN FERGUSON, b Galion, Ohio, Feb 29, 32; m 57; c 3. BIOPHYSICS. *Educ:* Ohio State Univ, BS, 57, MS, 58; Ind Univ, PhD(physiol), 68. *Prof Exp:* Phys chemist, sr phys chemist, 68-75, res scientist, 75-84, SR RES SCIENTIST, ELI LILLY & CO RES, 84- *Mem:* Biophys Soc; Int Soc Study Xenobiotics; Am Physiol Soc; Sigma Xi; Am Soc Microbiol; Am Asn Pharmaceut Scientists. *Res:* Ion transport in the intestinal epithelium; intestinal absorption, permeability and pharmacokinetics of drugs. *Mailing Add:* Lilly Res Labs MC909 Lilly Corp Ctr Indianapolis IN 46285

QUAY, PAUL DOUGLAS, b New York, NY, Oct 10, 49. GEOCHEMISTRY, LIMNOLOGY. *Educ:* City Univ New York, BA, 71; Columbia Univ, PhD(geol), 77. *Prof Exp:* Res assoc geol, Quaternary Res Ctr, 77-80, ASST PROF, DEPT GEOL SCI & OCEANOG, UNIV WASH, 80- *Res:* Determining ocean mixing rates by using radiocarbon distribution; modeling global carbon dioxide and radiocarbon distributions in nature; studying the geochemistry of lakes, rivers and marine systems using naturally occurring radioisotopes; chemical oceanography. *Mailing Add:* Dept Geol Univ Wash Seattle WA 98195

QUAY, PAUL MICHAEL, b Chicago, Ill, Aug 24, 24. CAUSATION, ISOMORPHISMS IN PHYSICS. *Educ:* Loyola Univ, Ill, AB, 50; West Baden Col. lic phil, 52, lic theol, 62; Mass Inst Technol, BS, 55, PhD(statist thermodyn), 58. *Prof Exp:* Res assoc statist mech, Case Inst Technol, 62-63; vis prof physics, Loyola Univ, Ill, 65-67; from asst prof to assoc prof physics, St Louis Univ, 67-81; RES PROF PHILOS, LOYOLA UNIV, CHICAGO, 83- *Concurrent Pos:* Physicist, Nat Bur Standards, Colo, 66 & 67; assoc prof spirituality, Dept Theol Studies, St Louis Univ, 75-81; vis assoc prof physics, philos & theol, Loyola Univ, Chicago, 81-83. *Honors & Awards:* Thomas Linacre Award, Nat Fed Catholic Physicians' Guild. *Mem:* Am Phys Soc; Philos Sci Asn; Inst Theol Encounter with Sci & Technol; Am Cath Philos Asn; Jesuit Philos Asn; Soc Christian Cult. *Res:* Steady state nonequilibrium thermodynamics; nature of scientific explanation; parts and wholes and final causes in physics; philosophy of nature; ethics of brain death and of cessation of treatment; theory of spiritual development. *Mailing Add:* Dept Philos Loyola Univ Chicago IL 60626-5385

QUAY, THOMAS LAVELLE, b Mt Holly, NJ, Aug 23, 14; m 39; c 1. ANIMAL ECOLOGY. *Educ:* Univ Ark, BS, 38; NC State Col, MS, 40, PhD(zool), 48. *Prof Exp:* Asst, Univ Ark, 35-38; instr zool & entom, 46-48, from asst prof to assoc prof zool, 48-57, prof zool, NC State Univ, 57-80; ENVIRON CONSULT, 80- *Mem:* AAAS; Ecol Soc Am; Am Inst Biol Sci; Am Ornithologists Union; Wildlife Soc. *Res:* Ecological succession of birds; habitat associations and niche relationships of animals; animal behavior. *Mailing Add:* 2720 Vanderbuilt Ave Raleigh NC 27607

QUAY, WILBUR BROOKS, b Cleveland, Ohio, Mar 7, 27; m 53, 76; c 1. NEUROENDOCRINOLOGY, HISTOPHYSIOLOGY. *Educ:* Harvard Univ, AB, 50; Univ Mich, Ann Arbor, MS & PhD(zool), 52. *Prof Exp:* Instr anat, Med Sch, Univ Mich, 52-56; from asst prof to prof zool, Univ Calif, Berkeley, 56-73, Miller prof, 64-65 & 71-72; prof, Waisman Ctr Ment Retardation & Human Develop, Dept Zool & Endocrinol Reprod Physiol Prog. Univ Wis-Madison, 73-77; prof anat, Dept Anat & grad div biochem, Dept Human Biol Chem & Genetics, Univ Tex Med Br, 77-83; Sr scientist, Advan Develop Div, Healthdyne, Inc, 83-84; vis scholar & res assoc, Dept Physiol & Anat, Univ Calif, Berkeley, 83-90; INDEPENDENT RES, WRITING & CONSULT, 90- *Concurrent Pos:* Fel, Neth Orgn Advan Pure Res, Cent Inst Brain Res, Amsterdam, 66; res grants, NSF, NIH, Univ Wisc Sea Grant Prog, & Fisheries Res Bd Can; mem ment retardation res & training comt, Nat Inst Child Health & Human Develop, 70-73; co-founder & assoc ed, J Pineal Res, 83-; independant researcher, writer & consult, Bio-Res Lab, Napa, Calif, 83-90. *Honors & Awards:* Bissendorf lectr, 87. *Mem:* Endocrine Soc; Am Physiol Soc; Am Chem Soc; Am Asn Anat; Int Brain Res Orgn; Soc Neurosci. *Res:* Neuroendocrinology; chronobiology; regulatory, rhythmic and adaptive mechanisms in nervous, glandular, and reproductive systems; vertebrate pineal complex; biogenic amines; interrelations of pineal, retina and central nervous system; analytical methods and evaluations of avian male reproduction. *Mailing Add:* Bio Res Lab RR 1 Box 327 New Bloomfield MO 65063-9719

QUAZI, AZIZUL H(AQUE), b Rahimpur, Bangladesh, Jan 1, 35; m 67; c 2. ELECTRICAL ENGINEERING. *Educ:* Univ Dacca, BSEE, 56; Munich Tech Univ, DrEng, 63. *Prof Exp:* Asst engr, Govt Bangladesh & Utah Int Inc, USA, Chittagong H T, Bangladesh, 56-58; res assoc eng, Munich Tech Univ, 64-65; res electronic engr, Info Processing Div, US Navy Underwater Sound Lab, Conn, New London Lab, 65-69, team leader, 69-85, br head, 85-88, SR RES ENGR & CONSULT, INFO PROCESSING BR, NEW LONDON LAB, US NAVAL SYST CTR, 88- *Concurrent Pos:* Lectr, Dept Elec Eng, Univ RI, 66-68, Univ Conn, Hartford Grad Ctr, Univ New Haven; Ger Acad exchange scholar, 58-63. *Mem:* AAAS; Inst Elec & Electronics Engrs; fel Acoust Soc Am. *Res:* Research and development in the field of statistical theory of communication, especially in detection, localization and classification of underwater targets; semiconductor devices; high frequency; underwater acoustics; advanced systems technology, digital signal processing, underwater communications, sonar systems developments, and advanced concept developments. *Mailing Add:* Naval Underwater Syst Ctr New London Lab Ft Trumbull New London CT 06320

QUE, LAWRENCE, JR, b Manila, Philippines; US citizen. BIOINORGANIC CHEMISTRY, METALLOBIOCHEMISTRY. *Educ:* Ateneo Manila Univ, BSCH, 69. *Prof Exp:* Asst prof chem, Cornell Univ, 77-83; assoc prof, 83-87, PROF CHEM, UNIV MINN, 87- *Res:* Bioinorganic chemistry; metallobiochemistry; oxygen activation; nonheme iron enzymes; biomimetic chemistry. *Mailing Add:* Chem Dept Univ Minn 207 Pleasant St SE Minneapolis MN 55455

QUEBBEMANN, ALOYSIUS JOHN, b Chicago, Ill, Jan 19, 33; m 67; c 5. TOXICOLOGY. *Educ:* Univ Alaska, BS, 60; NMex Highlands Univ, MS, 64; State Univ NY, Buffalo, PhD(pharmacol), 68. *Prof Exp:* USPHS res fel pharmacol, Sch Med, State Univ NY, Buffalo, 68-69; Merck Found fac develop award, 69; asst prof, 69-75, ASSOC PROF PHARMACOL, SCH MED, UNIV MINN, MINNEAPOLIS, 75- *Mem:* Am Soc Nephrol. *Res:* Pharmacology of renal transport mechanisms. *Mailing Add:* Dept Pharmacol-3-260 Millard Hall Univ Minn Minneapolis MN 55455

QUEBEDEAUX, BRUNO, JR, plant physiology, crop science, for more information see previous edition

QUEDNAU, FRANZ WOLFGANG, b Dresden, Ger, Apr 27, 30; m 69; c 2. ENTOMOLOGY, TAXONOMIC ENTOMOLOGY. *Educ:* Free Univ, Berlin, BSc, 51, PhD(zool), 53. *Prof Exp:* Res asst entom, Biologische Bundesanstalt, Berlin, 53-60; prof officer, Plant Protection Res Inst, Pretoria, SAfrica, 60-63; RES SCIENTIST II ENTOM, CAN FORESTRY SERV, DEPT AGR, 64- *Concurrent Pos:* Fulbright fel travel grant & Ger Res Coun fel, Dept Biol Control, Univ Calif, Riverside, 58-59. *Mem:* Entom Soc Can. *Res:* Biological control of insect pests; ecology of hymenopterous parasites; toxonomy of trichogramma; taxonomy of aphids. *Mailing Add:* Laurentian Forest Res Ctr PO Box 3800 Ste Foy PQ G1V 4C7 Can

QUEEN, DANIEL, b Boston, Mass, Feb 15, 34; m 57; c 1. ELECTRONICS ENGINEERING, ACOUSTICS. *Prof Exp:* Engr, Magnecord Inc, Ill, 55-57; proj engr, 3M-Revere-Wollensak, 57-62; dir eng, Perma-Power Div, Chamberlain Mfg Co, 63-70; PRES, DANIEL QUEEN ASSOCS, 70- *Concurrent Pos:* Rep, Am Nat Standards Comt, 73-; standards mgr, Audio Eng Soc, 81-, chmn tech coun, 86- *Mem:* Acoust Soc Am; Inst Elec & Electronics Engrs; fel Audio Engr Soc; AAAS; Soc Motion Picture & TV

Engr. *Res:* Development of sound reinforcement technology for small meeting and teaching rooms; human factors research involving acoustical conditions and noise. *Mailing Add:* Daniel Queen Associates 239 W 23rd St New York NY 10011

QUEENAN, JOHN T, b Aurora, Ill, May 5, 33; m 57; c 2. OBSTETRICS & GYNECOLOGY. *Educ:* Univ Notre Dame, BS, 54; Cornell Univ, MD, 58; Am Bd Obstet & Gynec, dipl, 66. *Prof Exp:* From clin instr to clin assoc prof obstet & gynec, Med Col, Cornell Univ, 62-72; prof obstet & gynec & chmn dept, Univ Louisville, 72-80; PROF OBSTET & GYNEC & CHMN DEPT, SCH MED, GEORGETOWN UNIV, 80- *Concurrent Pos:* Dir Rh-clin & Lab, New York Hosp, 62-72; from asst attend to sr attend, Greenwich Hosp, Conn, 63-71; obstet & gynec-in-chief, Georgetown Univ Hosp, 80-; chief obstet & gynec, Norton-Children's Hosp, Louisville, 73-80; ed, Contemp Obstet-Gynec. *Mem:* Fel Am Col Obstetricians & Gynecologists; fel Am Col Surgeons; affil Royal Soc Med; fel Am Fertil Soc; NY Acad Sci. *Res:* Treatment of erythroblastosis fetalis by aminiocenteses and in intrauterine transfusions; care of the immunized obstetrical patient and treatment to prevent future immunization; perinatal medicine; management of the high risk pregnancy; perinatal ultrasound; study of intrauterine growth retardation. *Mailing Add:* Dept Obstet & Gynec Georgetown Univ Sch Med 3800 Reservoir Rd NW Washington DC 20007

QUEENER, SHERRY FREAM, b Muskogee, Okla, July 8, 43; m 67; c 2. BIOCHEMISTRY. *Educ:* Okla Baptist Univ, BS, 65; Univ Ill, Urbana, MS, 68, PhD(biochem), 70. *Prof Exp:* From instr to assoc prof, 71-84, PROF PHARMACOL, SCH MED, IND UNIV, INDIANAPOLIS, 84- *Concurrent Pos:* Woodrow Wilson scholar, 65; Am Cancer Soc fel pharmacol, Sch Med, Ind Univ, Indianapolis, 70-71. *Mem:* AAAS; Am Chem Soc; Am Soc Clin Res; NY Acad Sci; Am Soc Biol Chemists. *Res:* Regulation of metabolic pathways; enzyme regulation at the genetic level and as mediated by effector molecules; antibiotic therapy of Pneumocystis carinii. *Mailing Add:* Dept Pharmacol Ind Univ Sch Med, 635 Barnhill Dr Indianapolis IN 46202

QUEENER, STEPHEN WYATT, b Indianapolis, Ind, Jan 31, 43; m 67; c 2. MICROBIOLOGY. *Educ:* Wabash Col, BA, 65; Univ Ill, MS, 68, PhD(biochem), 70. *Prof Exp:* Fel, Univ Ill, 70; sr scientist, 70-74, res scientist, 74-80, SR RES SCIENTIST, ELI LILLY & CO, 81-, GROUP LEADER, CELL RES DEVELOP, 89- *Concurrent Pos:* Guest lectr, Czechoslovak Acad Sci & Europ Fedn Microbiol Sci, 81; mem, Int Sci Coun, World Congress Microbiol, 94. *Honors & Awards:* Serv Award Int Br Am Soc Microbiol. *Mem:* Am Soc Microbiol; Am Chem Soc; Soc Indust Microbiol (secy, 89-); Sigma Xi; Mycol Soc. *Res:* Development and application of techniques for efficient mutation and selection, strain breeding, and recombinant DNA manipulations. *Mailing Add:* Eli Lilly & Co Dept Ky 412 Bldg 314/5 Indianapolis IN 46285

QUE HEE, SHANE STEPHEN, b Sydney, Australia, Oct 11, 46. ANALYTICAL & ENVIRONMENTAL CHEMISTRY, ENVIRONMENTAL SCIENCES. *Educ:* Univ Queensland, BSc, 68, MSc, 71; Univ Sask, PhD(org chem), 76. *Prof Exp:* Res & teaching fel, McMaster Univ, 76-78; from asst prof to asoc prof environ health, Univ Cincinnati, 78-89; ASSOC PROF ENVIRON HEALTH SCI, UNIV CALIF, LOS ANGELES, 89- *Concurrent Pos:* Mem TOXNET, Nat Library Med, 85-89. *Honors & Awards:* Cert Outstanding Achievement, US Environ Protection Agency, 81. *Mem:* Weed Sci Soc Am; AAAS; Am Chem Soc; Indust Hyg Asn Am; Am Col Toxicol; Air Pollution Control Asn; NY Acad Sci; Am Conf Govt Indust Hyg; Am Publ Health Asn. *Res:* Microanalytical techniques; chemiluminescence; bioluminescence; industrial hygiene; pesticide chemistry; lipid chemistry; photochemistry; analytical chemistry; ecology; biological monitoring; clinical chemistry. *Mailing Add:* Univ of Calif, Dept Environ Health Sci, Sch Publ Health 10833 Le Conte Ave Los Angeles CA 90024-1772

QUENEAU, PAUL E(TIENNE), b Philadelphia, Pa, Mar 20, 11; m 39; c 2. EXTRACTIVE METALLURGY, MINERAL ENGINEERING. *Educ:* Columbia Univ, BA, 31, BSc, 32, EM, 33; Delft Univ Technol, DSc, 71. *Prof Exp:* Metall engr, Int Nickel Co, WVa, 34-37, res engr, Ont, 38-40, dir res, 41 & 46-48, metall engr, Exec Dept, NY, 49-57, vpres, 58-69, tech asst to pres, 60-66, asst to chmn & consult engr, 67-69; vis scientist, Delft Univ Technol, 70-71; prof eng, 71-87, EMER PROF ENG & RES, THAYER SCH ENG, DARTMOUTH COL, 87- *Concurrent Pos:* Geographer, Perry River Arctic Exped, 49; chmn adv comt arctic res, US Navy, 57; chmn int symposium extractive metall copper, nickel & cobalt, Am Inst Mining, Metall & Petrol Engrs, 60, dir, 67-70; mem, Columbia Eng Coun, 65-70; vis comt, Mass Inst Technol, 67-70, dir Eng Found, 66-76, chmn, 73-75; consult engr, 72-; vis prof, Inst Technol, Univ Minn, 74-75; vis prof, Univ Utah, 87-91. *Honors & Awards:* Egleston Medal, Columbia, 65; Douglas Gold Medal, Am Inst Mech Engrs, 68; Extractive Metall Lect Award, Am Inst Mining, Metall & Petrol Engrs, 77; Gold Medal, Brit Inst Mining & Metall, 80; Henry Krumb Lectr, Am Inst Mining & Eng, 84. *Mem:* Nat Acad Eng; fel Metall Soc (pres, 69); Nat Soc Prof Engrs; Can Inst Mining & Metall; Am Inst Mining, Metall & Petrol Engrs (vpres, 67-70); Australasian Inst Mining & Metall; Brit Inst Mining & Metall. *Res:* Utilization of mineral resources; patentee processes and apparatus employed in the pyrometallurgy, hydrometallurgy and vapometallurgy of nickel, copper, cobalt, lead and iron; extractive metallurgy oxygen technology; INCO oxygen flash smelting; oxygen top-blow rotary converter; lateritic ore matte smelting, nickel high pressure carbonyl and iron ore recovery processes; Lurgi QSL lead-making and Dravo oxygen sprinkle smelting copper processes. *Mailing Add:* Thayer Sch Eng Dartmouth Col Hanover NH 03755

QUENTIN, GEORGE HEINZ, b Rome, NY, Jan 25, 34; m 60; c 5. CHEMICAL ENGINEERING. *Educ:* Rensselaer Polytech Univ, BChE, 55; Iowa State Univ, MS, 62, PhD(chem eng), 65. *Prof Exp:* Chem engr, Ind Ord Works, E I du Pont de Nemours & Co, Inc, 55-57, process engr, 57-58; technologist, US Indust Chem Div, Nat Distillers & Chem Corp, Ill, 58-60; sr engr cent eng dept, Monsanto Co, Mo, 65-66, eng specialist, 66-69; asst prof chem eng, Univ NMex, 69-73; assoc prof control eng, Univ Tex, 73-77; PROJ MGR, GENERATION & STORAGE DIV, ELEC POWER RES INST, 77- *Concurrent Pos:* Consult process control, 70-77. *Mem:* Am Inst Chem Engrs; Am Chem Soc; Instrument Soc Am; Soc Comput Simulation. *Res:* Coal gasification research and development for electric power generation; experimental pilot plant studies of process dynamics; control analysis of advanced power plant technology by computer simulation; gas turbine advanced diagnostic instrumentation development and testing; expert systems for gas turbine operation and maintenance; knowledge-based interactive video training systems. *Mailing Add:* EPRI 3412 Hillview Ave Palo Alto CA 94303

QUERFELD, CHARLES WILLIAM, b Bloomington, Ill, Mar 29, 33; m 62; c 1. ATMOSPHERIC OPTICS. *Educ:* Harvard Univ, BA, 55; Clarkson Col Technol, MS, 69, PhD(physics), 70. *Prof Exp:* Field engr, Darin & Armstrong, 55-56; physicist, US Army Atmospheric Sci Lab, 56-66; res asst physics, Clarkson Col Technol, 66-69; scientist solar physics, High Altitude Observ, Nat Ctr Atmospheric Res, 70-81; scientist, E-Systs, 88-90; SCIENTIST, BALL AEROSPACE SYSTS, 82-88 & 90- *Concurrent Pos:* Consult, Atmospheric Sci Lab, White Sands, NMex, 72-73. *Honors & Awards:* Victor K LaMer Award, Am Chem Soc, 70. *Mem:* AAAS; Am Phys Soc; Sigma Xi. *Res:* Atmospheric optics. *Mailing Add:* 7183 Dry Creek Ct Longmont CO 80503-8581

QUERINJEAN, PIERRE JOSEPH, b Spa, Belg, Jan 13, 42; m 65; c 2. BIOTECHNOLOGY, MULTIMEDIA REMOTE TEACHING. *Educ:* Univ Louvain, PhD(biochem), 66; Inst Pasteur, Paris, cert immunol, 69. *Prof Exp:* IRSIA fel biochem, Univ Louvain, 63-66, FNRS res fel immunol, 67-75, res fel cancerology, 75-89; GEN MGR BIOTECHNOL INFO, NPMA BIOCLUB, 89- *Concurrent Pos:* Europ Molecular Biol Orgn fel immunol, Med Res Coun Lab, Cambridge Univ, 71-72; Int Union Against Cancer Fel immunol, Mt Sinai Sch Med, NY & Southern Med Sch, Dallas, 73-74; conceptor, Remote Teaching in Biol, 91- *Mem:* Belg Soc Biochem; Belg Soc Immunol; Am Asn Immunol; Brit Biochem Soc. *Res:* Systemic approach of biological problems with the help of computer science tools; cancer's reversibility including diagnostic and therapeutic approaches. *Mailing Add:* Ave Des Fauvettes 15 Ceroux-Mousty B-1341 Belgium

QUERRY, MARVIN RICHARD, b Butler, Mo, Nov 7, 35; m 57; c 3. OPTICAL PHYSICS. *Educ:* Univ Kansas City, BS(math) & BS(physics), 61; Kans State Univ, MS, 64, PhD(physics), 68. *Prof Exp:* Asst prof physics, Kans State Univ, 66-67, res assoc, 67-68; from asst prof to assoc prof, 68-75, prof physics, 75-87, CURATORS' PROF PHYSICS, 87-, CHMN DEPT PHYSICS, UNIV MO, KANSAS CITY, 87- *Concurrent Pos:* Res grants & contracts, Univ Mo, US Dept Interior, USAF, US Army, Dept Com & NASA, 69-; pres, Sci Metrics Inc, 75-; Univ Mo, Kansas City Trustees Fac Fel, 85-86. *Mem:* AAAS; fel Optical Soc Am; Am Phys Soc; Am Asn Physics Teachers. *Res:* Measurement of the optical properties and optical constants of liquids, solids and biological materials. *Mailing Add:* Dept Physics Univ Mo Kansas City MO 64110

QUERTERMUS, CARL JOHN, JR, b Chicago, Ill, June 28, 43; m 84; c 2. ANIMAL BEHAVIOR, AQUATIC ECOLOGY. *Educ:* Ill State Univ, BS, 65, MS, 67; Mich State Univ, PhD(zool), 72. *Prof Exp:* Instr biol, Ill High Sch, 67-69; from asst prof to assoc prof, 72-82, PROF BIOL, WGA COL, 83- *Mem:* Am Fisheries Soc. *Res:* Habitat selection and movement patterns of largemouth bass in reservoirs; distribution of fishes in Carroll County, Georgia; prior experience and fish habitat selection. *Mailing Add:* Dept Biol WGa Col Carrollton GA 30118

QUESADA, ANTONIO F, b San Jose, Costa Rica, Feb 25, 25; nat US; m 54. APPLIED MATHEMATICS. *Educ:* Mass Inst Technol, SB, 47; Harvard Univ, MS, 58, PhD, 64. *Prof Exp:* Instr physics, Univ Costa Rica, 42-44; sr res mathematician, Baird Atomic, Inc, 49-68 & Dynarand, Inc, 68-69; MATHEMATICIAN, AIR FORCE GEOPHYS LAB, 70- *Concurrent Pos:* Consult, Govt Costa Rica, 54-55. *Mem:* Am Math Soc; Soc Indust & Appl Math; Math Asn Am. *Res:* Differential equations; noise theory; magnetodydrodynamics. *Mailing Add:* Dept Math Sci Univ Akron Akron OH 44325

QUESENBERRY, CHARLES P, b Dugspur, Va, Apr 13, 31; m 53; c 4. STATISTICS, QUALITY CONTROL. *Educ:* Va Polytech Inst, 57, MS, 58, PhD(statist), 60. *Prof Exp:* From asst prof to assoc prof math, Mont State Univ, 60-66; assoc prof, 66-69, PROF MATH, NC STATE UNIV, 69- *Concurrent Pos:* NASA grant, 64-66; NSF grant, 77-79; indust consult, Gen Motors, Union Carbide & Interalia. *Mem:* Biomet Soc; fel Am Statist Asn; Inst Math Statist; Int Statist Inst; Am Soc Qual Control. *Res:* Statistical inference; nonparametric discrimination; goodness-of-fit, model discrimination and validity; quality control, particularly statistical process control. *Mailing Add:* Dept Statist NC State Univ Raleigh NC 27695-8203

QUESENBERRY, KENNETH HAYS, b Springfield, Tenn, Feb 28, 47; m 69; c 2. AGRONOMY, PLANT BREEDING. *Educ:* Western Ky Univ, BS, 69; Univ Ky, PhD(crop sci), 75. *Prof Exp:* D F Jones fel cytogenetics, Univ Ky, 72-75; from asst prof to assoc prof agron, 75-86, PROF AGRON, UNIV FLA, 86- *Concurrent Pos:* Vis prof agron dept, Univ Wis-Madison, 86- *Mem:* Am Soc Agron; Crop Sci Soc Am. *Res:* Breeding and cytogenetics of tropical forage legumes; selection of varieties producing greater yield, digestibility and animal performance; investigation of nitrogen fixing capacity of tropical legumes; germplasm evaluation for response to root-knot nematodes, soil flooding tolerance, disease resistance, seed production potential and drought stress tolerance; development of protocals for tissue culture and genetic transformation of forage legumes. *Mailing Add:* Dept Agron 2183 McCarty Hall Univ Fla Gainesville FL 32611

QUESNEL, DAVID JOHN, b Plattsburg, NY, Apr 5, 50; m 73; c 1. MATERIALS SCIENCE, MECHANICAL METALLURGY. *Educ:* State Univ NY, Stony Brook, BE, 72; Northwestern Univ, MS, 74, PhD(mat sci), 77. *Prof Exp:* From asst prof to assoc prof, 77-89, PROF MECH ENG & MAT SCI, RIVER CAMPUS, UNIV ROCHESTER, 89- *Concurrent Pos:* NSF res initiation grant, 78-; Alexander Von Humboldt fel, 85-86. *Honors & Awards:* Ralph Teetor Educ Award, Soc Advan Educ, 89. *Mem:* Sigma Xi; Metall Soc; Am Soc Mech Engrs; Am Soc Metals; Mat Res Soc. *Res:* Mechanical metallurgy with emphasis on fracture and fatigue of metals; cyclic response of metals to repetitive strain; author of technical papers in general area of mechanical behavior of materials. *Mailing Add:* Dept Mech Eng Univ Rochester Rochester NY 14627

QUESTAD, DAVID LEE, b Muskegon, Mich, Aug 22, 52. POLYMER SCIENCE. *Educ:* Pa State Univ, BS, 74; Rutgers Univ, MS, 78, PhD(mech & mat sci), 81. *Prof Exp:* Anal eng, Pratt & Whitney Aircraft, 74-75; asst, Rutgers Univ, 76-78, teaching asst statics & dynamics, 78-79; ASST PROF ENG SCI & MECH, PA STATE UNIV, 81- *Mem:* Am Phys Soc; Soc Plastics Engrs; Sigma Xi; NY Acad Sci. *Res:* Effects of hydrostatic pressure on physical and mechanical properties of polymers, specifically pressure, volume, temperature and dielectric measurements; large scale deformation and thermal aging of polymers. *Mailing Add:* 1317 Hillside Dr Vestal NY 13850

QUEVEDO, WALTER COLE, JR, b Brooklyn, NY, Jan 7, 30; m 55. ANIMAL GENETICS. *Educ:* St Francis Col, BS, 51; Marquette Univ, MS, 53; Brown Univ, PhD(biol), 56. *Prof Exp:* Asst, Marquette Univ, 51-53; teaching assoc, Brown Univ, 53-55; resident res assoc, Argonne Nat Lab, 56-58; sr cancer res scientist, Roswell Park Mem Inst, 58-61; from asst prof to assoc prof, 61-70, PROF BIOL, BROWN UNIV, 70- *Concurrent Pos:* Vis lectr dermat, Harvard Med Sch, 63- *Mem:* AAAS; Am Soc Photobiol; Soc Develop Biol; Genetics Soc Am; Soc Exp Biol & Med; Soc Investigative Dermat. *Res:* Mammalian genetics; radiation biology; physiological genetics of coat and skin coloration in mice; regulation of melanin formation in normal and neoplastic pigmented tissues. *Mailing Add:* Div Biol & Med Brown Univ Providence RI 02912

QUIBELL, CHARLES FOX, b Fresno, Calif, Jan 29, 36; m 70; c 2. PLANT ANATOMY, PLANT SYSTEMATICS. *Educ:* Pomona Col, BA, 58; Univ Calif, Berkeley, PhD(bot), 72. *Prof Exp:* Asst prof, 70-73, assoc prof, 73-81, PROF BIOL, SONOMA STATE UNIV, 81- *Mem:* AAAS; Bot Soc Am; Am Soc Plant Taxonomists; Am Inst Biol Sci; Soc Econ Bot; Sigma Xi. *Res:* Systematic anatomy of woody saxifrages; comparative wood anatomy of dicotyledonous plants. *Mailing Add:* 4682 Hidden Oaks Rd Santa Rosa CA 95404

QUICK, JAMES S, b Devils Lake, NDak, Oct 20, 40; m 68; c 3. PLANT BREEDING, GENETICS. *Educ:* NDak State Univ, BA, 62; Purdue Univ, MS, 65, PhD(plant breeding, genetics), 66. *Prof Exp:* Asst geneticist, Rockefeller Found, 66-69; assoc prof durum wheat breeding, NDak State Univ, 69-76, prof, 76-81; PROF WHEAT BREEDING, COLO STATE UNIV, 81- *Mem:* Crop Sci Soc Am. *Res:* Development of new varieties of wheat, improved breeding methods and genetic, pathological, physiological, entomological and agronomic research. *Mailing Add:* Dept Agron Colo State Univ Ft Collins CO 80523

QUICK, WILLIAM ANDREW, b Senlac, Sask, July 18, 25; m 53; c 3. BOTANY. *Educ:* Univ Sask, BA, 46, BEd, 51, MA, 60, PhD(plant physiol), 63. *Prof Exp:* Asst prof plant physiol, Univ Guelph, 63-67; assoc prof, 67-82, PROF BIOL, UNIV REGINA, 82 - *Mem:* AAAS; Am Soc Plant Physiologists; Can Soc Plant Physiologists; Weed Sci Soc Am. *Res:* Dormancy mechanisms in seeds; wild oat (Avena fatua L) seed dormancy and environmental constraints; physiology of herbicides. *Mailing Add:* Dept Biol Univ Regina Regina SK S4S 0A2 Can

QUIE, PAUL GERHARDT, b Dennison, Minn, Feb 3, 25; m 51; c 4. PEDIATRICS. *Educ:* St Olaf Col, BA, 50; Yale Univ, MD, 53; Am Bd Pediat, dipl. *Prof Exp:* Intern, Minneapolis Gen Hosp, 53-54; resident, Dept Pediat, Univ Minn, Minneapolis, 54-57, from instr to assoc prof pediat, 58-68, consult physician, Inst Child Develop Nursery Sch, 60-90, PROF PEDIAT & LAB MED, UNIV MINN, MINNEAPOLIS, 68-, PROF MICROBIOL & AM LEGION MEM HEART RES PROF, 74- *Concurrent Pos:* Res fels med, Hosp, Univ Minn, Minneapolis, 57-58; USPHS res fel, 60-61 & career develop award, 62-; John & Mary R Markle scholar med sci, 61-66; attend physician, Minneapolis Gen Hosp, 59-; guest investr, Rockefeller Inst, 62-64, Radcliffe Infirmary, Oxford, Eng, 71-72, Univ Cologne, Ger, 86 & 88 & Univ Bergen, Norway, 91; assoc mem comn streptococcal & staphylococcal dis, Armed Forces Epidemiol Bd, 65-; mem, Comt Control Infectious Dis, 70-76, Adv Comt, Inst Allergy & Infectious Dis, 72-74, Adv Coun, 75-79 & Bd Sci Counselors, 82-86; John Simon Guggenheim fel, 71; Alexander von Humboldt US sr scientist award, 85. *Honors & Awards:* Numerous Named Lectr, 73-90; Maxwell Finland Award, Infectious Dis Soc Am, 83. *Mem:* Inst Med-Nat Acad Sci; Am Fedn Clin Res; Am Soc Microbiol; NY Acad Sci; Infectious Dis Soc Am (pres, 85); Am Pediat Soc (pres, 88); AAAS; Asn Am Physicians. *Res:* Infectious diseases; author of numerous scientific publications. *Mailing Add:* Dept Pediat Mayo Mem Bldg Box 483 Univ Minn Hosp 420 Delaware St SE Minneapolis MN 55455

QUIGG, CHRIS, b Bainbridge, Md, Dec 15, 44; m 67; c 2. HIGH ENERGY PHYSICS, THEORETICAL PHYSICS. *Educ:* Yale Univ, BS, 66; Univ Calif, Berkeley, PhD(physics), 70. *Prof Exp:* Res assoc physics, State Univ NY, Stony Brook, 70-71, from asst prof to assoc prof, 71-74; head theoret physics dept, 77-87, dep dir SSC Central Design Group, 87-89, PHYSICIST, FERMI NAT ACCELERATOR LAB, 74- *Concurrent Pos:* Sloan Found fel, 74-78; vis scholar, Enrico Fermi Inst, Univ Chicago, 74-78; mem prog adv comt, Stanford Linear Accelerator Ctr, 75-77; mem high energy adv comt, Brookhaven Nat Lab, 78-80 & Lawrence Berkley Lab, 78-81; prof lectr, Univ Chicago, 78-82, prof, 82-; div assoc ed, Particles & Fields & Phys Review Lett, 81-83, assoc ed, Reviews Mod Physics, 81-; vis prof, Ecole Normale Superieure, Paris, 81-82; mem, Bd Overseers Superconducting Super Collider, 85-87; vis scientist, Lawrence Berkeley Lab, 89-90; scholar-in-residence, Bellagio Ctr, 90. *Mem:* Am Phys Soc; AAAS. *Res:* Phenomenology of elementary particles. *Mailing Add:* Theoretical Physics MS106 Fermilab PO Box 500 Batavia IL 60510-0500

QUIGG, RICHARD J, b Bethlehem, Pa, Nov 12, 30; div; c 3. PHYSICAL METALLURGY. *Educ:* Va Polytech Inst, BS, 52; Lehigh Univ, MS, 54; Case Inst Technol, PhD(phys metall), 59; Cleveland State Univ, JD, 66. *Prof Exp:* Metallurgist, E I du Pont de Nemours & Co, Inc, 52-53; res asst, Lehigh Univ, 53-54; res metallurgist, Rem-Cru Titanium, Inc, 54-56; res asst, Case Inst Technol, 56-59; res metallurgist, TRW Inc, 59-63, res supvr, 63-64, wrought metall mgr, Metals Div, 64-65, res sect mgr, 65-67, mgr mat & processes, 67-68, mgr res & develop, Metals Div, 68-70; exec vpres, Jetshapes, Inc, 70-73, pres, 73-78; sr staff engr, Pratt & Whitney Aircraft, 78-80; vpres mkt, 80-90, VPRES RES INT CANNON-MUSKEGON CORP, 90- *Mem:* Am Inst Mining, Metall & Petrol Engrs; Am Soc Metals; Am Soc Testing & Mat. *Res:* Titanium alloy development; hydrogen embrittlement; superalloy development; phase changes in nickel-base superalloys; law; casting and solidification. *Mailing Add:* Cannon-Muskegon Corp PO Box 506 Muskegon MI 49443

QUIGLEY, FRANK DOUGLAS, b Maysville, Ky, June 5, 28. MATHEMATICS. *Educ:* Harvard Univ, AB, 49; Univ Chicago, PhD(math), 53. *Prof Exp:* From instr to asst prof, Yale Univ, 53-59; assoc prof, 59-64, PROF MATH, TULANE UNIV, 64- *Mem:* Am Math Soc; Math Asn Am. *Res:* Algebraic geometry; commutative Banach algebra; several complex variables. *Mailing Add:* Dept Math Tulane Univ 406 B Gibson Hall 6328 St Charles Ave New Orleans LA 70118

QUIGLEY, GARY JOSEPH, b Syracuse, NY, Aug 1, 42; div. MOLECULAR BIOPHYSICS. *Educ:* State Univ NY Col Environ Sci & Forestry, BS, 64, PhD(chem), 69. *Prof Exp:* Fel & res assoc molecular biophys, 69-78, prin res scientist, Mass Inst Technol, 78-88; PROF CHEM, HUNTER COL, CUNY, 88- *Mem:* Am Crystallog Asn; Am Inst Physics; Am Chem Soc. *Res:* Determination of crystal structures of nucleic acids including transfer-RNA, Z-DNA and drug DNA complexes; ion, water and drug interactions with nucleic acids; molecular mechanics to understand nucleic acid structure and structure-function relationships; development of X-ray area detector for X-ray crystallographic data collection for macromolecules; DNA sequence analysis; RNA secondary structure. *Mailing Add:* Dept Chem Hunter Col CUNY 695 Park Ave New York NY 10021-5024

QUIGLEY, GERARD PAUL, b Boston, Mass, Jan 3, 42; m 66; c 2. LASER CHEMISTRY, LASER PHYSICS. *Educ:* Northeastern Univ, BSEE, 64, MS, 66; Polytech Inst Brooklyn, MS, 70; Cornell Univ, PhD(appl physics), 74. *Prof Exp:* asst group leader laser photochemistry, 75-86, PROJ LEADER, LASER PROPOGATION, LOS ALAMOS SCI LAB, 86- *Concurrent Pos:* US Steel fel, Cornell Univ, 74. *Mem:* Am Phys Soc; AAAS; Am Chem Soc. *Res:* Laser photochemistry and isotope separation; physics and kinetics of laser systems; optoacoustic spectroscopy of laser excited systems. *Mailing Add:* 600 Bajadaway Los Alamos NM 87544

QUIGLEY, HERBERT JOSEPH, JR, b Philadelphia, Pa, Mar 6, 37; m 64; c 1. PATHOLOGY, CHEMISTRY. *Educ:* Franklin & Marshall Col, BS, 58; Univ Pa, MD, 62; Am Bd Path, dipl, 68. *Prof Exp:* Resident path, Presby Hosp, New York, 62-66; chief path serv, DePoo Hosp, Key West, Fla, 66-68; from asst prof to assoc prof, 68-72, PROF PATH, CREIGHTON UNIV, 72-; CHIEF PATH SERV, VET ADMIN HOSP, 68- *Concurrent Pos:* NIH acad path career develop trainee, Col Physicians & Surgeons, Columbia Univ, 62-66; chief path serv, Monroe County Gen Hosp & US Naval Hosp, Key West, 66-68; porpoise pathologist & res consult, Off Naval Res Cetacean Lab, Key West, 66-68. *Honors & Awards:* Borden Prize for Med Res, Borden Corp, 62. *Mem:* Fel Col Am Path; fel Am Inst Chem; Am Chem Soc; fel Am Soc Clin Path. *Res:* Blood coagulation; disseminated intravascular coagulation; fibrinolysis; instrumental analytical chemistry. *Mailing Add:* Path Serv Vet Admin Hosp 4101 Woolworth Ave Omaha NE 68105

QUIGLEY, JAMES P, b New York, NY, Mar 18, 42; m 67; c 2. TUMOR CELL BIOLOGY, ENZYMOLOGY. *Educ:* Manhattan Col, New York, BS, 65; Johns Hopkins Univ, PhD(physiol chem), 69. *Prof Exp:* Fel chem biol, Rockefeller Univ, 70-73, asst prof, 73-74; asst prof, 74-77, ASSOC PROF MICROBIOL & IMMUNOL, DOWNSTATE MED CTR, STATE UNIV NY, 78- *Concurrent Pos:* Fel, Leukemia Soc Am, 70-72; scholar, Sinsheimer Found Scholar, 77; vis prof, Sch Path, Oxford Univ, 80-81. *Mem:* Am Asn Cancer Res; Am Soc Cell Biol; Harvey Soc; NY Acad Sci. *Res:* Biochemical examination of normal and malignant cells; role of tumor viruses in malignant transformation; mechanism of tumor cell invasion and metastasis. *Mailing Add:* Path Dept Health Sci Ctr State Univ NY Stonybrook NY 11794

QUIGLEY, ROBERT JAMES, b Cord, Ark, Feb 18, 40; div. VARIABLE STARS. *Educ:* Calif Inst Technol, BS, 61, MS, 62; Univ Calif, Riverside, MA, 64, PhD(physics), 68. *Prof Exp:* Res asst solid state physics, Univ Calif, Riverside, 65-68; asst prof physics, Ill Inst Technol, 68-70; from asst prof to assoc prof physics, 70-83, PROF PHYSICS & ASTRON, WESTERN WASH UNIV, 83- *Concurrent Pos:* Vis lectr, Inst Physics, Univ Frankfurt, 69-70; vis scholar astron, Univ Tex & McDonald Observ, 76-77; vis scientist, Sacramento Peak Observ, 80-81; vis scientist, Steward Observ, Univ Ariz, 84-85. *Mem:* Am Astron Soc; Astron Soc Pac. *Res:* Photoelectric photometry of cataclysmic variable stars. *Mailing Add:* Dept Physics & Astron Western Wash Univ Bellingham WA 98225-9064

QUIGLEY, ROBERT MURVIN, b Toronto, Ont, Jan 22, 34; m 57; c 3. SOIL MECHANICS, ENGINEERING GEOLOGY. *Educ:* Univ Toronto, BASc, 55, MASc, 56; Mass Inst Technol, PhD(soil mech), 61. *Prof Exp:* From jr to sr engr, Geocon Ltd, Ont & Que, 56-63; from asst prof to assoc prof, 63-70, PROF SOIL MECH, UNIV WESTERN ONT, 70-; DIR, GEOTECH RES

CTR, 85- *Concurrent Pos:* Consult, Conch Methane Serv, Ltd, Eng, 61-63 & Golder & Assocs Ltd, Can, 63-66; vis sr res fel, Univ Southampton, 71-72; assoc ed, Can Geotech J, 74-80; pres, R M Quigley, Inc, 74-; mem, Natural Sci & Eng Res Coun, Earth Sci Grant Selection Comn, 76-79, chmn, 78-79; ed, Can Geotech J, 80-84; vis scientist, Geotech Res Inst, McGill Univ, 81-82; vis prof, Univ Sydney, Australia, 89. *Mem:* Can Geotech Soc; Geol Asn Can; Eng Inst Can; Clay Minerals Soc; Am Soc Testing Mat; Am Soc Civil Engrs. *Res:* Application of mineralogy, geochemistry to problems in soil and rock mechanics; clay leachate compatibility re-waste disposal; coastal erosion and instability; pollutant migration through clay barriers. *Mailing Add:* Fac Eng Univ Western Ont London ON N6A 5B9 Can

QUILL, LAURENCE LARKIN, inorganic chemistry; deceased, see previous edition for last biography

QUILLEN, DANIEL G, b June 27, 40. MATHEMATICS. *Educ:* Harvard Univ, PhD(math), 69. *Prof Exp:* Norbert Wiener prof math, Mass Inst Technol, 73-88; PROF MATH, MATHS INST, OXFORD UNIV, 88- *Mem:* Nat Acad Sci; Am Math Soc. *Mailing Add:* Math Inst Oxford Univ 24 St Giles Oxford England

QUILLEN, EDMOND W, JR, b Feb 9, 53; m; c 3. NEUROHUMORAL REGULATIONS, REFLEX CONTROL. *Educ:* Univ Miss, PhD(physiol), 81. *Prof Exp:* ASST PROF OBSTET & GYNEC, MCGILL UNIV, 86- *Mem:* Am Physiol Soc. *Res:* Arterial pressure regulation; fluid and electrolyte balance; pregnancy; sheep; surgical techniques; computer science. *Mailing Add:* Dept Obstet Royal Victoria Hosp McGill Univ 687 Pine Ave W Montreal PQ H3A 1A1 Can

QUILLIGAN, JAMES JOSEPH, JR, b Philadelphia, Pa, Oct 18, 12; m 41; c 4. VIROLOGY. *Educ:* Ohio State Univ, BA, 36; Univ Cincinnati, MD, 40. *Prof Exp:* Instr pediat & res assoc epidemiol, Sch Pub Health, Univ Mich, 46-49, asst prof pediat, Univ & asst prof epidemiol, Sch Pub Health, 50; assoc prof pediat, Univ Tex Southwestern Med Sch Dallas, 51-54; assoc prof, 54-59, RES PROF PEDIAT, SCH MED, LOMA LINDA UNIV, 59-, DIR VIRUS LAB, 54- *Concurrent Pos:* Dir Labs, Children's Med Ctr, Dallas, 51-54; res career investr, Nat Inst Allergy & Infectious Dis, 63- *Mem:* Am Acad Microbiol; Soc Pediat Res; Am Acad Pediat; Am Asn Immunol; Infectious Dis Soc Am. *Res:* Influenza; herpes; hepatitis viruses; tumor viruses. *Mailing Add:* 2921 Via San Jacinto San Clemente CA 92672

QUILLIN, CHARLES ROBERT, b Crawfordsville, Ind, Jan 14, 38. CYTOLOGY. *Educ:* Wabash Col, AB, 60; Brown Univ, ScM, 63, PhD(bot), 66. *Prof Exp:* Sr asst, Wabash Col, 60; asst bot, Brown Univ, 60-62, asst biol, 62-65; from instr to asst prof, Colby Col, 65-70, assoc dean students, 67-70; fel, Off Inst Res, Mich State Univ, 70-71; dean students, Marshall Univ, 72-73, asst to vpres student affairs, 73-75; asst prof, 75-78, DEAN STUDENT DEVELOP, POINT PARK COL, 75-, ASSOC PROF, 78-, EXEC OFF, 86- *Concurrent Pos:* NSF vis scientist's prog lectr, 66- *Mem:* Sigma Xi. *Res:* Study of histone as related to deoxyribose nucleic acid, acid cycle in a cell. *Mailing Add:* Off Dean Student Develop Point Park Col Pittsburgh PA 15222

QUIMBY, FRED WILLIAM, b Providence, RI, Sept 19, 45; m 65; c 2. LYMPHOCYTE DIFFERENTIATION ANTIGENS, AUTOIMMUNITY. *Educ:* Univ Pa, VMD, 70, PhD(path), 74. *Prof Exp:* Fel hemat, New Eng Med Ctr, 74-75; from instr to asst prof path, Tufts Med Sch, 75-79; ASSOC PROF PATH, CORNELL UNIV, 79- *Concurrent Pos:* Vet, Springfield Animal Hosp, 70-74; consult, St Elizabeth Hosp, 74-79, Concord Field Sta, Harvard, 75-79 & Sidney Farber Cancer Ctr, Harvard, 78-79; dir, Lab Animal Med, Tufts New Eng Med Ctr, 75-79, Ctr Res Animal Resources, Cornell Univ, 79-; mem, Nat Acad Sci Comt on the Guide, 84-85, & Comt on Immunodeficient Rodents, 85-; ed, Lab Animal Sci, 91- *Honors & Awards:* Trum Award, New Eng Br, Am Asn Lab Animal Sci, 79. *Mem:* NY Acad Sci; Soc Vet Immunol; Am Vet Med Asn; Am Asn Lab Animal Sci (pres NE Br, 78-79). *Res:* Differentiation antigens on canine lymphocytes and immmunologic abnormalities in autoimmune disease; the etipathogenesis of bacterial toxic shock syndromes. *Mailing Add:* New York State Col Vet Med 221 VRT Res Tower Ithaca NY 14853-6401

QUIMBY, FREEMAN HENRY, b Battle Creek, Mich, June 11, 15; m 48; c 3. PHYSIOLOGY. *Educ:* Andrews Univ, BA, 38; Northwestern Univ, MS, 41; Univ Md, PhD(zool), 47. *Prof Exp:* Prof biol & head dept, Columbia Union Col, 41-47; chief investr, US Navy proj, Univ Md, 47-48; head physiol br, Off Naval Res, DC, 48-56, chief scientist, Calif, 56-59; chief res anal, Army Res Off, 59-60; asst dir life sci grants & contracts, Off Life Sci, NASA, 60-62, chief exobiol prog, Off Space Sci & Applns, 62-66; specialist, life sci, Sci Policy Res Div, Cong Res Serv, Lib Cong, 66-76; CONSULT, TRACOR JITCO, 76- *Mem:* AAAS; Am Physiol Soc. *Res:* Nutrition; aviation physiology; microbiology; endocrinology; science and public policy. *Mailing Add:* 3926 Rickover Rd Silver Spring MD 20902

QUIMPO, RAFAEL GONZALES, b Aklan, Philippines, Mar 23, 39; m 63; c 4. CIVIL ENGINEERING, HYDROLOGY. *Educ:* Feati Univ, Philippines, BS, 59; Seato Grad Sch Eng, Bangkok, ME, 62; Colo State Univ, PhD(civil eng), 66. *Prof Exp:* Civil engr, Am-Asia Eng Assocs, 62-63; res asst, Colo State Univ, 63-66; from asst prof to assoc prof civil eng, 66-75, PROF CIVIL ENG, UNIV PITTSBURGH, 75- *Concurrent Pos:* Res grants, Off Water Resources, US Dept Interior, Univ Pittsburgh, 67-70 & NSF, 70-72, 73-76, 79-80 & 84-86; consult, Mobay Chem Co, 72; vis prof, Fed Univ Rio de Janiero, Brazil, 72-73; vis scientist, Philippine Nat Sci Develop Bd, 75-76; NSF int travel grant, 76, 78, 80, 83 & 85; consult, US Army CEngrs, 79-80, J T Boyd Co, 85-86, USX Corp, 86-; NSF res grant, 86-88 & 90-91. *Mem:* AAAS; Am Geophys Union; Am Soc Civil Engrs; Am Soc Eng Educ; Int Asn Hydraul Res; Am Water Works Asn. *Res:* Water resources development; applied statistics; stochastic processes; stochastic hydrology; non-conventional energy sources; reliability of water distribution systems; remote sensing of the environment. *Mailing Add:* 940 Benedum Hall Univ Pittsburgh Pittsburgh PA 15261

QUIN, LOUIS DUBOSE, b Charleston, SC, Mar 5, 28; c 3. ORGANIC CHEMISTRY. *Educ:* The Citadel, BS, 47; Univ NC, MA, 49, PhD(org chem), 52. *Prof Exp:* Res chemist, Am Cyanamid Co, 49-50; res proj leader, Westvaco Chem Div, Food Mach & Chem Corp, 52-54 & 56; from res assoc to prof chem, 56-81, chmn dept, 70-76, James B Duke prof, Duke Univ, 81-86; PROF CHEM & HEAD DEPT, UNIV MASS, 86- *Concurrent Pos:* Ford Found fel, Woods Hole Oceanog Inst, 63-64. *Mem:* Fel AAAS; Am Chem Soc; Royal Soc Chem. *Res:* Organophosphorus and heterocyclic compounds; synthesis, stereochemistry, and spectral properties of cyclic phosphorus compounds; spectra-structure correlations of organophosphorus compounds. *Mailing Add:* Dept Chem Univ Mass Amherst MA 01003

QUINAN, JAMES ROGER, b Watervliet, NY, June 27, 21; m 50; c 1. INDUSTRIAL CHEMISTRY. *Educ:* State Univ NY Albany, AB, 42, AM, 48; Rensselaer Polytech Inst, PhD(infrared & Raman spectroscopy), 54. *Prof Exp:* Chemist & foreman, Adirondack Foundries & Steel, 42-45; res assoc biochem, Sterling-Winthrop Res Inst, 48-52; sr chemist, Behr-Manning Div, Norton Co, 54-57, group leader abrasive grain & electrostatics, 57-70, sr res prof, Coated Abrasive Div, 70-75; tech dir, mach div, Albany Int Corp, 77-79, proj engr, eng systs div, 79-83; RETIRED. *Concurrent Pos:* Consult, 75-77 & 83- *Mem:* Am Chem Soc; NY Acad Sci. *Res:* Chemical and physical properties of abrasives as related to coated abrasive products; electrostatics as applied to coated abrasives; metallurgical and high temperature materials research. *Mailing Add:* 57 Upper Loudon Rd Loudonville NY 12211

QUINE, WILLARD V, b Akron, Ohio, June 25, 08; m 30, 48; c 4. LOGIC, PHILOSOPHY OF SCIENCE. *Educ:* Oberlin Col, AB, 30; Harvard Univ, AM, 31, PhD(philos), 32. *Prof Exp:* EMER PROF, HARVARD COL, HARVARD UNIV, 78- *Concurrent Pos:* Vis prof, Oxford, Paris, Tokyo, Sao Paulo, Uppsala, Gerona. *Honors & Awards:* N M Butler Gold Medal, Columbia Univ, 70. *Mem:* Nat Acad Sci; Am Philos Soc; Am Acad Arts & Sci; Brit Acad; Norweg Acad; Institut de France. *Mailing Add:* Emerson Hall Harvard Col Harvard Univ Cambridge MA 02138

QUINLAN, DANIEL A, b Glen Ridge, NJ, Oct 8, 58; m 84; c 1. NOISE CONTROL, PERCEPTION OF NOISE. *Educ:* Univ NH, BS, 80; Pa State Univ, MS, 85. *Prof Exp:* Acoust engr, Genesis Physics Corp, 81-83; acoust consult, Bolt, Beranek & Newman, Inc, 85-86; MEM TECH STAFF, AT&T BELL LABS, 86- *Mem:* Inst Noise Control Engrs; Acoust Soc Am; Inst Elec & Electronics Engrs. *Res:* Physics of air-moving devices; measurement of acoustic and structural intensity; perception of noise and active control of sound. *Mailing Add:* AT&T Bell Labs Rm 2D-340 600 Mountain Ave Murray Hill NJ 07974-2070

QUINLAN, DENNIS CHARLES, b Detroit, Mich, Jan 29, 43; m 69; c 1. CELL BIOLOGY, BIOCHEMISTRY. *Educ:* Wayne State Univ, BS, 65, MS, 66; Univ Rochester, PhD(microbiol), 73. *Prof Exp:* Sr res assoc cell biol, Worcester Found Exp Biol, 73-76; asst prof, 76-80, ASSOC PROF BIOL, WVA UNIV, 80- *Concurrent Pos:* Prin investr, Am Cancer Soc grant, 77-79 & NIH grant, WVa Univ, 78-81. *Mem:* Am Soc Biol Chemists; Am Soc Zoologists; Develop Biol Soc; Int Soc Differentiation; AAAS; Sigma Xi. *Res:* Cell cycle regulation in cultured, mammalian cells; membrane dynamics of normal and tumor cells. *Mailing Add:* Dept Biol WVa Univ PO Box 6057 Morgantown WV 26506

QUINLAN, JOHN EDWARD, b Milwaukee, Wis, Aug 6, 30; m 57; c 3. PHYSICAL CHEMISTRY. *Educ:* Marquette Univ, BS, 52; Univ Ark, MS, 55; Univ Wis, PhD(chem), 59. *Prof Exp:* From instr to assoc prof, 58-69, dean admis & financial aid, 69-81, ASSOC PROF CHEM, & DIR ADMIN COMPUT POMONA COL, 69- *Concurrent Pos:* NSF fac fel, 64-65. *Mem:* AAAS. *Res:* Chemical kinetics and mechanisms of gas phase reactions. *Mailing Add:* Dir Admin Comput Pomona Col Claremont CA 91711

QUINLAN, KENNETH PAUL, b Somerville, Mass, Jan 13, 28; m 64; c 4. INORGANIC CHEMISTRY. *Educ:* Boston Univ, BA, 51; Tufts Univ, MS, 52; Univ Notre Dame, PhD(inorg chem), 59; Northeastern Univ, MS, 81. *Prof Exp:* Res chemist, US Army CEngrs, 52-53, Am Cyanamid Co, 53-54 & Nat Lead Co, Inc, 54-56; PHYS CHEMIST, AIR FORCE CAMBRIDGE RES LABS, 60- *Mem:* AAAS; Am Chem Soc; Electrochem Soc. *Res:* Photosynthesis; solar energy; solid state chemistry. *Mailing Add:* 70 Grasmere Newton MA 02158

QUINLIVAN, WILLIAM LESLIE G, b Waunfawr, Wales, Dec 20, 21; US citizen; m 50; c 3. OBSTETRICS & GYNECOLOGY. *Educ:* Univ London, MB, BS, 46, MD, 65; FRCS, 65; FRCOG, 66. *Prof Exp:* Asst prof obstet & gynec, Univ Pittsburgh, 62-65; assoc prof, 65-74, PROF OBSTET & GYNEC, UNIV CALIF, IRVINE, 74- *Concurrent Pos:* NIH res fel obstet & gynec, Cancer Res Inst, Univ Calif, San Francisco, 60-62; Health, Res & Serv Found grants, 62-65; NIH res grants, 63-72. *Mem:* Am Physiol Soc; fel Am Col Obstet & Gynec; Am Fertil Soc. *Res:* Methods for the pre-selection of sex in humans; immunological cause of infertility in the human female. *Mailing Add:* 660 W Via De Suenos Green Valley AZ 85614-1726

QUINN, B(AYARD) E(LMER), mechanical & design engineering; deceased, see previous edition for last biography

QUINN, BARRY GEORGE, b Rochelle Park, NJ, Dec 2, 34; div; c 2. AQUATIC BIOLOGY. *Educ:* Univ Utah, BS, 57, MA, 58; Univ Colo, PhD(biol), 62. *Prof Exp:* From asst prof to assoc prof, 62-72, PROF BIOL, WESTMINSTER COL, UTAH, 72-, HEAD DEPT, 63- *Concurrent Pos:* Scripps Inst Oceanography, 58-59; NIH fel, Marine Lab, Univ Miami, 63-66; chmn, Div Natural Sci & Math, Westminster Col, 66-71 & 74-77; mem, Eval Panel, NSF Undergrad Instr Sci Equip Prog, 65 & Steering Comt, Utah Conf Higher Educ, 72-74; partic, NSF Summer Inst Comp Anat, Univ Wash, 68. *Mem:* AAAS; Ecol Soc Am; Am Soc Limnol & Oceanog; Sigma Xi. *Res:* Comparative limnology of mountain lakes and streams; limnology of Great Salt Lake; ecology of coral reefs; behavior of marine invertebrates. *Mailing Add:* Westminster Col Salt Lake City UT 84105

QUINN, C JACK, b Westbaden, Ind, June 4, 29; m 53; c 4. ENERGY CONSERVATION & ALTERNATE ENERGY SOURCES, RECOVERY OF WASTE ENERGY. *Educ:* Ind Inst Technol, BSME, 56; Ball State Univ, MS, 61. *Prof Exp:* From instr to assoc prof mech eng, Ind Inst Technol, 56-69; from assoc prof to prof mech eng technol, 69-84, PROF & CHMN MFG TECHNOL, PURDUE UNIV FT WAYNE, 84- *Concurrent Pos:* Sr engr, Int Harvester, 63-66; consult, Franklin Elec Co, 68-70, Defense Civil Preparedness Agency, 71-81, Phelps Dodge, Inc, 82 & Kemtune, Inc, 83-; nat deleg, Am Soc Mech Engrs, 74 & 75; team chmn accreditation, Am Soc Heating, Refrigeration & Air-Conditioning Engrs, 82-89; prog evaluator for accreditation, Soc Mfg Engrs. *Mem:* Am Soc Mech Engrs; Am Soc Eng Educ; Am Soc Heating, Refrigeration & Air-Conditioning Engrs; Nat Soc Prof Engrs; Soc Mfg Engrs. *Res:* Energy conservation; waste heat recovery and utilization of alternate energy resources; environmental pollution. *Mailing Add:* 4726 N Webster Rd New Haven IN 46774

QUINN, COSMAS EDWARD, history of biology; deceased, see previous edition for last biography

QUINN, DAVID LEE, b Steubenville, Ohio, Nov 28, 38; m 59; c 3. NEUROENDOCRINOLOGY, REPRODUCTIVE PHYSIOLOGY. *Educ:* Washington & Jefferson Col, BA, 60; Purdue Univ, MS, 62, PhD(brain & ovulation), 64. *Prof Exp:* NIH fel neuroendocrinol, Sch Med, Duke Univ, 64-66; assoc prof, 66-77, PROF BIOL, MUSKINGUM COL, 77- *Mem:* AAAS; Sigma Xi; Endocrine Soc. *Res:* Comparative analysis of brain mechanisms controlling ovulation and prolactin secretion. *Mailing Add:* 180 Foxcreek Rd New Concord OH 43762

QUINN, DENNIS WAYNE, b West Grove, Pa, Apr 20, 47; m 67; c 2. APPLIED MATHEMATICS. *Educ:* Univ Del, BA, 69, MS, 71, PhD(math), 73. *Prof Exp:* Mathematician/programmer, E I du Pont de Nemours, 69-70; from res assoc to appl mathematician, Aerospace Res Lab, 74-75; appl mathematician res, Flight Dynamics Lab, 75-80, from asst prof to assoc prof, 80-86, PROF MATH & STATIST, AIR FORCE INST TECHNOL, WRIGHT-PATTERSON AFB, 87- *Concurrent Pos:* Nat Res Coun resident res assoc, Nat Acad Sci, 73-74. *Mem:* Soc Indust Appl Math; Am Math Soc. *Res:* Analysis of the behavior of solutions of singular, elliptic, partial differential equations; numerical solution of partial differential equations, particularly those arising in pharmacokinetics and beam physics; parameter identification of physiological parameters in pharmacokinetics. *Mailing Add:* Dept Math Air Force Inst Technol Wright-Patterson AFB OH 45433

QUINN, EDWIN JOHN, b Geneva, Ill, July 20, 27; m 64; c 3. ORGANIC CHEMISTRY, POLYMER CHEMISTRY. *Educ:* St Procopius Col, BSc, 51; Univ Ill, Urbana, MSc, 55; State Univ NY Col Forestry, Syracuse, PhD(org polymer chem), 62. *Prof Exp:* Res asst, US Govt Synthetic Rubber Prog, Univ Ill, Urbana, 53-55; res chemist, Blockson Chem Co div, Olin Mathieson Chem Corp, 55-57 & Naugatuck Chem Div, US Rubber Co, 60-64; res chemist, 64-81, SR RES SCIENTIST, ARMSTRONG WORLD INDUSTS, 81- *Mem:* Am Chem Soc. *Res:* Synthetic polymer chemistry; surfactants; carbohydrates and synthetic polyols; photochemistry; inorganic polymers and polyphosphazenes; polymer flammability and smoke evolution; polyvinyl chloride. *Mailing Add:* 1730 Santa Barbara Dr Lancaster PA 17601

QUINN, FRANK RUSSELL, b Washington, DC; c 1. MEDICINAL CHEMISTRY, INFORMATION SCIENCE. *Educ:* Cath Univ Am, AB, 50; Adelphi Univ, MS, 61; Am Univ, PhD(chem), 75. *Prof Exp:* Res textile chemist, Harris Res Labs, Gillette Corp, 55-58; res asst polymer chem, Adelphi Univ, 58-60; instr chem, Nassau Col, 60-63; chemist, US Food & Drug Admin, 63-66; asst ed, Chem Abstr Serv, Am Chem Soc, 67; RES CHEMIST MED CHEM, NAT CANCER INST, 68- *Concurrent Pos:* Adj res prof, Am Univ, 83-88. *Mem:* Am Chem Soc; Am Soc Info Sci; AAAS; Found Advan Educ Sci. *Res:* Linear free energy relationships; quantitative structure activity relationships in the design of anticancer agents; computer science; information theory and the design of scientific information systems; molecular modeling. *Mailing Add:* Pharmaceut Resources Br Nat Cancer Inst Exec Plaza N Rm 818 EPN Bethesda MD 20892

QUINN, FRANK S, b Havana, Cuba, June 3, 46; m; c 2. TOPOLOGY OF MANIFOLDS. *Educ:* Princeton Univ, PhD(math), 70. *Prof Exp:* Prof math, Princeton Univ, 71-73, Yale Univ, 73-76 & Rutgers Univ, 76-77; prof, 77-85, DISTINGUISHED PROF MATH, VA POLYTECH INST, 85- *Concurrent Pos:* Ed, Bull Am Math Soc. *Mem:* Am Math Soc. *Res:* Topology of manifolds. *Mailing Add:* Va Polytech Inst Blacksburg VA 24061

QUINN, GALEN WARREN, b Tama, SDak, Jan 28, 22; wid; c 8. ORTHODONTICS. *Educ:* Creighton Univ, DDS, 52; Univ Tenn, MS, 55; Am Bd Orthod, dipl, 69. *Prof Exp:* Elem sch teacher, 40-41; dep supply officer, Vet Admin, 45-47; pvt pract, 52-54; asst prof pedodontics, head dept & dir, postgrad & grad prog, Univ Tenn, 55-58, from assoc prof to prof orthod, 58-92, div chief, 64-84; RETIRED. *Concurrent Pos:* Dean, Sch Dent, Creighton Univ, 61-62; ed, Cleft Palate Bull, 59-62, NC Dent J, 78-80; consult, vet Admin Hosp, Durham, NC, State Bd Health, Voc Rehab, Site visits, Nat Inst Dent Res & US Army, Ft Benning, Ga & Ft Bragg, NC; mem, Craniofacial Biol Group, Int Asn Dent Res, 74. *Honors & Awards:* Pierre Fauchard Award. *Mem:* Am Soc Dent for Children; Am Dent Asn; assoc mem Am Asn Orthod; Am Cleft Palate Asn; fel AAAS; Int Asn Dent Res; Col Am Bd Orthod; fel Am Col Dentists; fel Int Col Dentists; Am Bd Orthod. *Res:* Etiology and treatment of congenital and acquired craniofacialorodental anomolies, i.e. cleft lip/palate, burns, tumors, trauma, arthritis, scoliosis, polio, caries, eruption, TMJ problems, etc; effects of soft tissue on growth, shape, position and posture of bone-especially face, jaws and causes of malocclusion; etiology and results of upper airway obstruction to cause mouth breathing treatment of upper airway; effect of mouthbreathing on muscle balance as a cause of orodentofacial deformities; non-surgical and surgical treatments of effects of mouthing; mandible and maxillahypo and hyperplasia-sinus blockage from nasal obstruction; developed a dental articulator and a rotational cineradiographic unit; author of numerous publications. *Mailing Add:* 806 E Forest Hills Blvd Durham NC 27707

QUINN, GEORGE DAVID, b Boston, Mass, Nov 28, 50; m 73; c 2. FAILURE ANALYSIS, HEAT TRANSFER. *Educ:* Northeastern Univ, BS, 73. *Prof Exp:* Ceramic engr, US Army Mat & Mech Res Ctr, 73-85; ceramic engr, US Army Mat Technol Lab, 85-87; Ger Aerospace Lab DFVLR, 87-88; CONSULT, 88- *Concurrent Pos:* Consult, Mat Res Coun, Army Res Proj Agency, 72-73 & Tech Coop Prog, 84-85. *Mem:* Am Ceramic Soc; US Naval Inst; Am Soc Testing & Mat. *Res:* Mechanical testing of ceramic materials; failure analysis of ceramic materials. *Mailing Add:* 30 Hawthorne St Watertown MA 02172

QUINN, GERTRUDE PATRICIA, b Bronxville, NY, Feb 6, 21. BIOCHEMICAL PHARMACOLOGY. *Educ:* Fordham Univ, BS, 43; Univ Rochester, MS, 48; George Washington Univ, PhD(pharmacol), 56. *Prof Exp:* Asst, Col Dent, NY Univ, 43-45, 48-49; res pharmacologist, Goldwater Mem Hosp, 49-52 & Lab Chem Pharmacol, Nat Heart Inst, 53-54; instr pharmacol, George Washington Univ, 54-55; res pharmacologist, Lab Chem Pharmacol, Nat Heart Inst, 55-59 & Res Labs, Ciba-Geigy Corp, 59-85; RETIRED. *Mem:* Am Soc Pharmacol & Exp Therapeut. *Res:* Physiological distribution and metabolism of drugs and their mechanism of action. *Mailing Add:* 31 Mullens Lane Bernardsville NJ 07924

QUINN, HELEN RHODA, b Melbourne, Australia, May 19, 43; US citizen; m 66; c 2. ELEMENTARY PARTICLE PHYSICS. *Educ:* Stanford Univ, BS, 63, MS, 64, PhD(physics), 67. *Prof Exp:* Res assoc physics, Stanford Linear Accelerator Ctr, 67-68; guest scientist, Ger Electron Synchrotron, Hamburg, Ger, 68-70; res fel, Harvard Univ, 71-72, from asst prof to assoc prof physics, 72-77; STAFF MEM, STANFORD LINEAR ACCELERATOR CTR, 79- *Concurrent Pos:* Alfred P Sloan fel, 74-77; vis assoc prof physics, Stanford Univ, 76-78; vis scientist, Stanford Linear Accelerator Ctr, 77-78, res assoc, 78-79. *Mem:* Fel Am Phys Soc. *Res:* Particle and theoretical physics; gauge field theories and their applications. *Mailing Add:* Stanford Linear Accelerator Ctr-Bin 81 PO Box 4349 Stanford CA 94305

QUINN, JAMES ALLEN, b Gary, Ind, Mar 29, 54; m 76; c 2. MATHEMATICAL MODELLING, STRUCTURE-ACTIVITY RELATIONS. *Educ:* Ohio Univ, BS, 75, MS, 77; Ohio State Univ, PhD(plant path), 80. *Prof Exp:* Res scientist agr chem, Rohm & Haas Co, 80-86; CONSULT, 86- *Mem:* Am Phytopath Soc; Mycol Soc Am; Can Phytopath Soc; AAAS. *Res:* Modelling the interactions between agricultural chemicals and pests, using statistics and computers; modelling of fungal morphology mathematically. *Mailing Add:* Dept Biol Rutgers Univ New Brunswick NJ 08903

QUINN, JAMES AMOS, b Chickasha, Okla, Aug 12, 39. PLANT ECOLOGY, POPULATION ECOLOGY. *Educ:* Panhandle State Col, BS, 61; Colo State Univ, MS, 63, PhD(bot sci), 66. *Prof Exp:* From asst prof to assoc prof, 66-77, assoc chair personnel, dept biol sci, 81-82, dir grad prog bot & mem exec coun, Grad Sch, 83-86, PROF BOT, RUTGERS UNIV, 77- *Concurrent Pos:* Travel grants, Rockefeller Found, Int Grassland Cong, 70, Bot Soc Am, Int Bot Cong, 75 & 87, Am Forage & Grassland Coun, Int Grassland Cong, 85 & 89; Rutgers Univ Res Coun fac fel, Australia, 72-73; vchmn, ecol sect, Bot Soc Am, 77, chmn, 78; vis scientist, Div Land Resources Mgt, CSIRO, Australia & Univ New Eng, Australia, 80-81; consult, Ont Coun Grad Studies, 82; assoc ed, Bull Torrey Bot Club, 83-85; sr ecologist, Bd Prof Cert, Ecol Soc Am, 90. *Mem:* Torrey Bot Club (pres, 82-83); Ecol Soc Am; Bot Soc Am; Am Inst Biol Sci; Soc Range Mgt; Am Forage & Grassland Coun. *Res:* Grassland ecology; population ecology; genetic differentiation within plant species; species interactions; evolutionary biology; reproductive biology of amphicarpic species; life histories and sex ratios in populations of dioecious species. *Mailing Add:* Dept Biol Sci Rutgers Univ Piscataway NJ 08855-1059

QUINN, JAMES GERARD, b Providence, RI, Oct 28, 38; m 65; c 3. MARINE ORGANIC CHEMISTRY. *Educ:* Providence Col, BS, 60; Univ RI, MS, 64; Univ Conn, PhD(biochem), 67. *Prof Exp:* USPHS training prog fel steroid biochem, Worcester Found Exp Biol, 67-68; from asst prof to assoc prof, 68-78, PROF CHEM OCEANOG, UNIV RI, 78- *Mem:* AAAS; Am Chem Soc; Am Soc Limnol & Oceanog; Int Asn Geochem & Cosmochem; Geochem Soc; Sigma Xi. *Res:* Marine organic chemistry; organic geochemistry of seawater and sediments; metal-organic and mineral-organic interactions; the biogeochemistry of organic pollutants in the marine environment. *Mailing Add:* Grad Sch Oceanog Univ RI Kingston RI 02881

QUINN, JARUS WILLIAM, b West Grove, Pa, Aug 25, 30; m 53; c 5. OPTICS. *Educ:* St Joseph's Col, Pa, BS, 52; Cath Univ Am, PhD(physics), 64. *Prof Exp:* Res assoc physics, Johns Hopkins Univ, 54-55; staff scientist, Res Inst Advan Study, Inc, 56-57; res assoc physics, Cath Univ Am, 58-60, from instr to asst prof, 61-69; EXEC DIR, OPTICAL SOC AM, 69- *Concurrent Pos:* Mem, Gov Bd, Am Inst Physics, 73- *Mem:* Am Phys Soc; fel Optical Soc Am; Coun Eng & Sci Soc Executives. *Res:* Science administration; optics. *Mailing Add:* Optical Soc Am 1816 Jefferson Pl NW Washington DC 20036

QUINN, JOHN A(LBERT), b Springfield, Ill, Sept 3, 32; m 57; c 3. CHEMICAL ENGINEERING. *Educ:* Univ Ill, Urbana, BS, 54; Princeton Univ, PhD(chem eng), 59. *Prof Exp:* Mem fac, Univ Ill, Urbana, 58-70, prof chem eng, 66-70; dept chmn, 80-85, PROF CHEM ENG, UNIV PA, 71-, ROBERT D BENT PROF, 78- *Concurrent Pos:* NSF sr fel, 65; vis prof, Imp Col, Univ London, 65-66 & 86; mem, Eng Res Bd, Comn Eng & Tech Systs & Comt Surv Chem Eng, Comn Phys Sci, Math & Resources, Nat Res Coun, 84-; Sherman Fairchild distinguished scholar, Calif Inst Technol, 85; bd chem sci & technol, NRC, 86-; sci adv bd, Sepracor, Inc, 84-, sci adv comt, Whitaker Found, 87- *Honors & Awards:* Colburn Award, Am Inst Chem Engrs, 66; Mason Lectr, Stanford Univ, 81; D L Katz Lectr, Univ Mich, 85; Reilly Lectr, Univ Notre Dame, 87. *Mem:* Nat Acad Eng; AAAS; Am Inst Chem Engrs; Am Chem Soc. *Res:* Interfacial phenomena; biotechnology; transport in biological systems; membrane structure and function. *Mailing Add:* Dept Chem Eng 311A Towne Bldg Univ Pa Philadelphia PA 19104-6393

QUINN, JOHN JOSEPH, b New York, NY, Sept 25, 33; m 58; c 4. THEORETICAL PHYSICS. *Educ:* St John's Univ, NY, BS, 54; Univ Md, PhD(physics), 58. *Prof Exp:* Res assoc, Univ Md, 58-59; mem tech staff, RCA Labs, 59-64; vis prof, Purdue Univ, 64-65; prof physics, Brown Univ, 65-91, Ford Found Prof, 85-91; PROF PHYSICS & ENG & CHANCELLOR, UNIV TENN, 89- *Concurrent Pos:* Vis lectr, Univ Pa, 61-62; vis prof, State Univ NY Stony Brook, 68-69. *Mem:* Am Phys Soc. *Res:* Solid state theory. *Mailing Add:* 527 Andy Holt Tower Univ Tenn Knoxville TN 37996

QUINN, LEBRIS SMITH, b Norwalk, Conn, Apr 13, 54; m 85. CELL DIFFERENTIATION, MUSCLE GROWTH & DEVELOPMENT. *Educ:* Swarthmore Col, BA, 76; Univ Wash, PhD(cell biol & anat), 82. *Prof Exp:* Grad res myogenesis, Med Sch, Univ Wash, 78-82, postdoctoral res myogenesis, 82-83, cellular aging, 84-85, res assoc biochem, 85-86, RES ASST PROF MYOGENESIS & ANAT, MED SCH, UNIV WASH, 86- *Concurrent Pos:* Postdoctoral res myogenesis, Univ Pa, 82-83; prin investr grants, indust & USDA, 86- *Res:* Factors whih control proliferation of vertebrate skeletal muscle precursor cells; cell lineage of myoblasts in development. *Mailing Add:* Biol Struct Dept Med Sch Univ Wash SM-20 Seattle WA 98195

QUINN, LOYD YOST, b Cutler, Ind, June 16, 17; m 45; c 4. BACTERIOLOGY. *Educ:* Purdue Univ, BS, 41, MS, 47, PhD(bact), 49. *Prof Exp:* From asst prof to assoc prof, 49-63, actg head dept, 57-59, PROF BACT, IOWA STATE UNIV, 63- *Mem:* Tissue Cult Asn; Am Soc Microbiol. *Res:* Antibody production by tissue cells grown in continuous culture; computerized feedback control of tissue cell culture conditions; aging in tissue cell cultures; effects of heavy metals on in vivo and tissue cell culture modes of immune response; genetic control of immune responses; computer graphics of protein structures. *Mailing Add:* Dept Microbiol 250 Sci I Iowa State Univ Ames IA 50011

QUINN, MICHAEL H, b S Fork, Pa, Feb 28, 43. POLYMER SYNTHESIS, COATING. *Educ:* Duquesne Univ, BS, 65; WVa Univ, MS, 67; Akron Univ, PhD(polymer sci), 73. *Prof Exp:* Sr chemist, Monsanto, 73-77; tech mgr, St Regis Paper Co, 77-80; tech dir, Frye Copysysts, Allied Signal, 80-84 & Coated Film Co, 84-86; GROUP MGR, WESLEY-JESSEN SCHERING PLOUGH, 86- *Mem:* Am Chem Soc. *Res:* Contact lenses; synthesis of new polymers and modification of existing polymers. *Mailing Add:* 3253 Thornhill Dr Valparaiso IN 46383-9081

QUINN, RICHARD PAUL, b Modesto, Calif, Oct 22, 42; m 64; c 3. IMMUNOCHEMISTRY. *Educ:* Univ San Francisco, BS, 64; Ore State Univ, PhD(biochem), 68. *Prof Exp:* USPHS fel, Biol Div, Oak Ridge Nat Lab, 68-70; res biochemist, 70-75, SR RES BIOCHEMIST, WELLCOME RES LABS, BURROUGHS WELLCOME CO, 75- *Mem:* Am Asn Immunologists. *Res:* Immunosuppressive and immunomodulating agents; immunochemical approaches to drug action; immunoassay development; metabolism of serum proteins; use of monoclonal antibodies for drug development. *Mailing Add:* Exp Ther Div Wellcome Res Labs 3030 Cornwallis Rd Research Triangle Park NC 27709

QUINN, ROBERT GEORGE, b Beaver Falls, Pa, June 14, 36; m 61; c 5. PLASMA PHYSICS, SPACE SCIENCES. *Educ:* Drexel Inst, BSEE, 59; Cath Univ, MS, 60, PhD(physics), 62. *Prof Exp:* Res assoc, Princeton Univ, 62-63; from asst prof to assoc prof, Dept Space Sci & Appl Physics, Cath Univ, 63-66; assoc prof, Ionosphere Res Lab, 66-71, dean acad instr commonwealth campuses, 71-74, PROF ENG, PA STATE UNIV, 72- *Concurrent Pos:* Res assoc, Goddard Space Flight Ctr, NASA, 65-66. *Mem:* Am Geophys Union; Am Phys Soc; Am Soc Testing & Mat; Int Elec & Electronics Engrs; Am Soc Eng Educ. *Res:* Ionospheric physics; dielectrics. *Mailing Add:* Dept Elec & Comp Engr Drexel Univ 32nd & Chestnut Sts Philadelphia PA 19104

QUINN, ROBERT M(ICHAEL), b Bedford, Ind, July 7, 41; m 61; c 5. ELECTRICAL ENGINEERING, ELECTROPHYSICS. *Educ:* Rensselaer Polytech Inst, BEE, 63, MEE, 65, PhD(elec eng), 68. *Prof Exp:* Staff engr, 68-76, ADV ENGR, IBM CORP, 76- *Res:* Instabilities in magnetoplasmas; integrated circuit device design and process technology. *Mailing Add:* Dept B-51 IBM Corp 1000 River St Bldg 967-1 Essex Junction VT 05452

QUINN, ROD KING, b Sherman, Tex, Oct 2, 38; m 60; c 2. PHYSICAL INORGANIC CHEMISTRY. *Educ:* Southern Methodist Univ, BS, 61, MS, 63; Univ Tex, PhD(inorg chem), 67. *Prof Exp:* Instr chem, Southern Methodist Univ, 62-63; lab instr anal chem, Univ Tex, 63-64; mem tech staff, 67-78, div supvr, Sandia Nat Lab, Sandia Corp, 78-85; assoc div leader, 86-88, CTR DIR, LOS ALAMOS NAT LAB, 88- *Mem:* Am Chem Soc; Electrochem Soc; Sigma Xi; fel Am Inst Chemists; Mat Res Soc; Am Ceramic Soc. *Res:* Solid state chemistry of inorganic materials such as ceramics, superconductors and semiconductors; spectroscopic techniques in solution, molten salt and high temperature chemistry; electrochemistry; semiconducting materials; thermal batteries; thermochemistry; studies of the discharge and aging mechanisms in lithium organic and inorganic electrochemical power sources; electron microscopic and spectroscopic techniques applied to electrochemical systems and solid state chemistry. *Mailing Add:* PO Box 2698 Santa Fe NM 87504-2698

QUINN, THOMAS PATRICK, b Freeland, Pa, Mar 20, 30; m 62; c 2. IONOSPHERIC PHYSICS. *Educ:* Pa State Univ, BS, 57, MS, 58, PhD(physics of ionosphere), 64. *Prof Exp:* Instr elec eng, Ionosphere Res Lab, Pa State Univ, 58-64; consult commun, Off Naval Res, 64-77, spec asst systs, Off Asst Secy Navy, 77-79; DEP ASST SECY, DEPT DEFENSE, 79- *Concurrent Pos:* Asst prof, Pa State Univ, 64-66; mem, Int Sci Radio Union, 64-; US mem, NATO Air Defense Electronics Environ Comt, 80-81, Commun & Info Systs Comt, 80-81, 84-, Sci Comt Nat Reps, 80-81, NATO Sr Nat Reps, 83-; chmn, NATO Panel Air Space Mgt & Control, 83-88, Sci Comt of Nat Rep; bd, Supreme Hq, Allied Powers Europe Tech Ctr, 81-86. *Honors & Awards:* Arthur S Flemming Award, Wash Jr Chamber Com, 67; Presidential Distinguished Exec Award, 84; Presidential Meritorious Exec Award, 89. *Mem:* Sigma Xi; Am Geophys Union; sr mem Inst Elec & Electronics Engrs; Armed Forces Commun & Electronics Asn. *Res:* Electromagnetic wave propagation; communications theory; radar systems; modulation and detection techniques. *Mailing Add:* 5399 Temple Hill Rd Temple Hills MD 20748

QUINN, WILLIAM HEWES, b Syracuse, NY, Sept 28, 18. PHYSICAL OCEANOGRAPHY, METEOROLOGY. *Educ:* Colgate Univ, AB, 40; Univ Mo, AM, 42; NY Univ, Cert(meteorol), 43; Univ Calif, Los Angeles, MS, 50; Ore State Univ, PhD(oceanog), 68. *Prof Exp:* Res assoc oceanog, 67-75, assoc prof & sr researcher, 76-80, EMER ASSOC PROF OCEANOG, ORE STATE UNIV, 81- *Concurrent Pos:* Asst dir, Nat Weather Serv Oceanog, 72-73. *Mem:* Am Meteorol Soc; Sigma Xi; Am Geophys Union; Nat Weather Asn. *Res:* Large scale air-sea interaction; climate change and its causes; long range ocean and weather forecasting; extension of records on El Niño occurrences and/or other Southern Oscillation; related climatic changes back to AD 622. *Mailing Add:* Col Oceanog Ore State Univ Corvallis OR 97331-5503

QUINNAN, GERALD VINCENT, JR, b Boston, Mass, Sept 7, 47. BIOLOGY. *Educ:* Col Holy Cross, Worcester, AB, 69; St Louis Univ Med Sch, MD, 73; Am Bd Internal Med, dipl. *Prof Exp:* Resident & fell, Boston Univ Med Ctr, 73-77; med officer, Bur Biologics, 77-80, dir, Herpes Virus Br, 80-81, actg dept dir, Div Virol, 81-88, DEPUTY DIR, CTR BIOLOGICS EVAL & RES, FOOD & DRUG ADMIN, 88- *Mem:* Infectious Dis Soc Am; Am Asn Immunol; Am Soc Clin Invest; Am Fedn Clin Res. *Res:* Immunology and pathogenesis of viral diseases with emphasis on cell mediated immunity; herpes viruses; retroviruses; vaccine safety. *Mailing Add:* Ctr Biologics Eval & Res 8800 Rockville Pike Bethesda MD 20892

QUINNEY, PAUL REED, b Haverhill, Mass, May 11, 24; m 47; c 4. ANALYTICAL CHEMISTRY. *Educ:* Univ NH, BS & MS, 49; Iowa State Col, PhD(chem), 54. *Prof Exp:* Fel chem, Mellon Inst, 54; sr chemist, Brown Co, 54-56 & Koppers Co, 56-58; from asst prof to prof, Butler Univ, 58-74, John Hume Reade prof chem, 74-88, head dept, 72-88; RETIRED. *Mem:* AAAS; Am Chem Soc; Am Inst Chemists. *Res:* Instrumental analysis. *Mailing Add:* Dept Chem Butler Univ Indianapolis IN 46208

QUINONES, FERDINAND ANTONIO, b Hormigueros, PR, May 30, 22; m 51; c 7. PLANT BREEDING. *Educ:* Univ PR, BSA, 46; Univ Minn, MS, 48, PhD(plant genetics), 54. *Prof Exp:* Instr hort, Univ PR, 48-50; res asst plant genetics, Univ Minn, 50-54; asst prof plant genetics, NMex State Univ, 64-63, assoc prof, 63-; RETIRED. *Mem:* Am Soc Agron. *Res:* Plant pathology; agronomy; horticulture; testing and breeding Indian ricegrass, western wheatgrass, tall wheatgrass and black gramagrass. *Mailing Add:* 1721 Calle De Saenos Las Cruces NM 88003

QUINONES, MARK A, b New York, NY, Jan 13, 31; m 52; c 2. PREVENTIVE MEDICINE, PUBLIC HEALTH. *Educ:* Southeastern La Univ, BA, 53; La State Univ, MA, 55, PhD(med sociol), 71; Wayne State Univ, MHA, 56; Columbia Univ, MPH, 73. *Prof Exp:* Health educr, Tuberculosis League of Pittsburgh, 56-57; consult, NJ Tuberculosis & Health Asn, 57-62; exec dir, Passaic County Heart Asn, NJ, 62-66; managing dir, Northwest Area Tuberculosis & Respirator Dis Asn, NJ, 66-69; coordr, Develop Dept, 69, adminr, Div Drug Abuse, 69-71, asst prof, 71-75, dir, Div Drug Abuse, 74-80, assoc prof, 76-80, PROF PREV MED & DIR, DIV GERIAT & SOCIAL MED, COL MED & DENT NJ, 80- *Concurrent Pos:* Vis prof, Sch Educ, Fairleigh Dickinson Univ, 71-74; health consult, PR Cong of NJ, 72- *Mem:* Am Pub Health Asn; Asn Teachers Prev Med; Soc Pub Health Educrs; Am Sociol Soc. *Res:* Areas concerned with the social aspects of health and medicine, particularly migrant health, criminal offenders, tuberculosis, asthma and allied health. *Mailing Add:* Dept Psychol/Prev Med/Community Health Univ Med & Dent NJ 185 S Orange Ave Newark NJ 07103

QUINSEY, VERNON LEWIS, b Flin Flon, Man, Oct 10, 44; Can Citizen; m 64; c 2. BEHAVIOR MODIFICATION, FORENSIC PSYCHOLOGY. *Educ:* Univ NDak, BSc, 66; Univ Mass, Amherst, MSc, 69, PhD (psychol), 70. *Prof Exp:* Fel, Dalhousie Univ, Halifax, NS, 70-71; psychologist, Mental Health Ctr, Penetanguishene, Ont, 71-75, dir res, 76-84 & 86-88; PROF PSYCHOL, QUEENS UNIV, KINGSTON, ONT, 88- *Concurrent Pos:* Vis scientist, Philippe Pinel Inst, Montreal, 84-86; adj assoc prof psychol dept, Concordia Univ, Montreal, 84-86; assoc prof psychiat dept, Univ Toronto, 86-88; chair, Nat Inst Mental Health Criminal & Violent Behav Res Comt, 86-88; chair, Ont Mental Health Found Res Comt, 80-82, Ont Mental Health Found Community & Social Serv Res Rev Comt, 87-; consult, Am Psychiat Asn Comt, 85-87. *Mem:* Fel Can Psychol Asn; Am Psychol Asn; Asn Advan Behav Ther; Int Acad Sex Res; Int Acad Law & Mental Health. *Res:* Antisocial behavior, applied decision making modification and program evaluation; psychophysiological assessment; sex offenders; forensic psychology. *Mailing Add:* Psychol Dept Queens Univ Kingston ON K7L 3N6 Can

QUINTANA, RONALD PRESTON, b New Orleans, La, Feb 23, 36; m 57; c 2. MEDICINAL CHEMISTRY. *Educ:* Loyola Univ, New Orleans, BS, 56; Univ Wis, MS, 58, PhD(pharmaceut chem), 61. *Prof Exp:* From instr to prof med chem, Col Pharm, Ctr Health Sci, Univ Tenn, 60-71, vchmn dept, 65-83, distinguished serv prof med chem, 71-83, prof periodont, Col Dent, 72-74; prin scientist, 83-84, ASST DIR, CONTACT LENS CARE RES, ALCON LABS INC, 85- *Mem:* Acad Pharm Sci; Am Chem Soc; Am Pharmaceut Asn; The Chem Soc. *Res:* Synthesis of, and surface-chemical studies on, compounds with biological significance. *Mailing Add:* Alcon Labs Inc 6201 S Freeway Ft Worth TX 76134-2001

QUINTANILHA, ALEXANDRE TIEDTKE, b Maputo, Mozambique, Aug 9, 45; Portuguese citizen. EXERCISE PHYSIOLOGY, OXYGEN TOXICITY. *Educ:* Witwatersrand Univ, BSc Hons, 68, PhD(solid state physics), 72. *Hon Degrees:* DSc, Ministry Educ, Portugal, 76. *Prof Exp:* Lectr physics, Witwatersrand Univ, 72-74; staff scientist, 74-84, asst head div, 84-87, DIR CTR ATMOSPHERIC & BIOSPHERIC EFFFECTS, LAWRENCE BERKELEY LAB, UNIV CALIF, 87- *Concurrent Pos:* Gulbenkian Found fel, 71; consult, Biomed Inst, Porto, Portugal, 74-78 & Children's Hosp, Oakland, Calif, 82-; Nat Acad Sci-Nat Res Coun fel, 78; adj prof physiol, Univ Calif, Berkeley, 82-; ed, John Wiley & Sons, 83-, CRC Press, 85- & Plenum Press, 86; vis prof biophys, Biomed Inst, Porto, Portugal, 78. *Mem:* Am Soc Biochem & Molecular Biol; Biophys Soc; AAAS; Am Soc Photobiol; Int Soc Magnetic Resonance; NY Acad Sci; Int Cell Res Orgn. *Res:* Oxidative and free radical mechanism of damage to biological systems; physical exercise; drug metabolism; photosensitization. *Mailing Add:* Lawrence Berkeley Lab 90-3026 Univ Calif Berkeley CA 94720

QUINTIERE, JAMES G, b Passaic, NJ, May 5, 40; c 2. FIRE CONTROL. *Educ:* Newark Col Eng, BS, 62; NY Univ, MS, 66, PhD(mech eng), 70. *Prof Exp:* Mech engr, nuclear rocket res, NASA Lewis Lab, 62-63; instr mech eng, NY Univ, 67-69; res scientist heat transfer, Am Standard Res & Develop Lab, 69-71; mech engr, Nat Inst Standards & Technol, 71-80, prog analyst, 85-86, group head, fire res, 80-85, chief, Fire Sci & Eng Div, 86-90; PROF, FPE, UNIV MD, 90- *Honors & Awards:* Bronze Medal, Nat Bur Standards, 76; Silver Medal, US Dept Com, 82; Emmans lectr in Fire, 86. *Mem:* Combustion Inst; Am Soc Mech Engrs. *Res:* Uncontrollable fire; enclosure fires, ignition and flame spread of materials and flammability test methods; natural convection heat transfer. *Mailing Add:* Dept FPE Eng Bldg 0147A Univ Md College Park MD 20742

QUINTO, ERIC TODD, b Indianapolis, Ind, May 10, 51. RADON TRANSFORMS, COMPUTED TOMOGRAPHY. *Educ:* Ind Univ, AB, 73; Mass Inst Technol, PhD(math), 78. *Prof Exp:* Lectr, 77-78, asst prof, 78-84, ASSOC PROF MATH, TUFTS UNIV, 84- *Concurrent Pos:* Vis scholar, Dept Math, Mass Inst Technol, 78-79; res grants, NSF, NIH; Humboldt res fel, 85, Vis Assoc Prof, Universitat Munster, Ber Ilan Univ. *Mem:* Am Math Soc; Math Asn Am; Asn Women Math. *Res:* Generalized radon transforms, a field of mathematics applicable to partial differential equations and computed tomography; when and how organs (functions) can be recovered from given tomographic data (integrals over surfaces). *Mailing Add:* Dept Math Tufts Univ Medford MA 02155

QUINTON, ARTHUR ROBERT, b Lowestoft, Eng, July 1, 24; m 46; c 3. EXPERIMENTAL NUCLEAR PHYSICS. *Educ:* Univ London, BSc, 44; Univ Western Ont, MSc, 51; Yale Univ, PhD, 54. *Prof Exp:* Res physicist, Mullard Radio Valve Co, 46-48; from instr to asst prof physics, Yale Univ, 54-63; assoc prof, Univ Fla, 63-66; PROF PHYSICS, UNIV MASS, AMHERST, 66- *Concurrent Pos:* Vis fel, Australian Nat Univ, 61-62 & vis prof, 72-73; mem Publ Comt, Am Asn Physics Teachers, 71-74. *Mem:* Am Phys Soc; Am Asn Physics Teachers. *Res:* Nuclear structure; low energy nuclear physics; fission; heavy ions; x-ray excitation by ion bombardment. *Mailing Add:* Dept Phys Astron Univ Mass Amherst MA 01003

QUINTON, DEE ARLINGTON, b Cardston, Alta, May 17, 39; m 66; c 6. RANGE NUTRITION, SECONDARY PRODUCTIVITY. *Educ:* Weber State Col, BS, 69; Colo State Univ, PhD(range sci), 72. *Prof Exp:* Res assoc range sci, Colo State Univ, 72-73; asst prof wildlife mgt, Tex Tech Univ, 73-75; RES SCIENTIST RANGE ECOL, AGR CAN, 76- *Mem:* Soc Range Mgt. *Res:* Secondary productivity of range lands; range improvements; range trend studies; beef production. *Mailing Add:* Res Sta Agr Can 3015 Ord Rd Kamloops BC V2B 8A9 Can

QUINTON-COX, ROBERT, histology, cytology; deceased, see previous edition for last biography

QUIOCHO, FLORANTE A, b Philippines, Oct 26, 37; m 59; c 3. BIOCHEMISTRY. *Educ:* Cent Philippines Univ, BS, 59; Howard Univ, MS, 61; Yale Univ, PhD(biochem), 66. *Prof Exp:* Mem res staff molecular biophys, Yale Univ, 64-66; res fel chem, Harvard Univ, 66-72; from asst prof to assoc prof, 72-81, PROF BIOCHEM, RICE UNIV, 81- *Concurrent Pos:* Mem, Cellular & Mollecular Basis Dis Rev Comt, Nat Inst Gen MedSci, NIH, 78-82; USPHS training fel biochem, Yale Univ, 62-64, res fel, European Molecular Biol Orgn, 80, fel, John Simon Guggenheim Mem Found, 80-81; vis scientist, Lab Molecular Biophysics, Oxford Univ, 80. *Honors & Awards:* Asian Chemist Award, Am Chem Soc, 61. *Mem:* Am Soc Biol Chemists; Sigma Xi. *Res:* Physical chemistry of biological macromolecules, especially proteins; x-ray crystallographic studies of proteins; mechanisms of enzyme action; chemical behavior of enzymes in the solid state. *Mailing Add:* Dept Biochem Rice Univ PO Box 1892 Houston TX 77251

QUIRK, JOHN THOMAS, b Dubuque, Iowa, Jan 21, 33; m 58; c 4. TREE PHYSIOLOGY, FOREST PRODUCTS. *Educ:* Iowa State Univ, BS, 56; Syracuse Univ, MS, 60; Univ Wis, PhD(forestry & forest prod), 67. *Prof Exp:* Res forester, Cent State Forest Exp Sta, USDA, 56-58; RES TECHNOLOGIST, FOREST PROD LAB, USDA, 60- *Mem:* AAAS; Soc Wood Sci & Technol; Soc Am Foresters; Electron Micros Soc Am; Am Inst Biol Sci. *Res:* Anatomy and morphology; structure-function; physiology-structure; structure-strength. *Mailing Add:* 117 N Franklin Ave Madison WI 53705

QUIRK, RODERIC P, b Detroit, Mich, Mar 26, 41; m 62; c 3. ORGANIC CHEMISTRY, SYNTHETIC INORGANIC & ORGANOMETALLIC CHEMISTRY. *Educ:* Rensselaer Polytech Inst, BS, 63; Univ Ill, MS, 65, PhD(chem), 67. *Prof Exp:* Anal chemist, Ethyl Corp, 63; res chemist, Minn Mining & Mfg Co, 64; res assoc, Univ Pittsburgh, 67-69; from asst prof to assoc prof chem, Univ Ark, 69-78; sr res scientist, Mich Molecular Inst, 79-83; PROF POLYMER SCI, INST POLYMER SCI, UNIV AKRON, 83- *Concurrent Pos:* Res chemist, Phillips Petrol Co, 74; vis prof, Inst Polymer Sci, Univ Akron, 76-77 & Tokyo Inst Technol, 90; adj assoc prof polymer chem, Case Western Reserve Univ, 79-83 & Cent Mich Univ, 80-83; NAm ed, Polymer Int. *Mem:* Am Chem Soc; Sigma Xi. *Res:* synthesis of functionalized polymers; solvation of alkyllithium compounds; synthesis of block copolymers. *Mailing Add:* Dept Polymer Sci Univ Akron Akron OH 44325

QUIRKE, TERENCE THOMAS, JR, b Minneapolis, Minn, Aug 18, 29; m 58; c 1. EXPLORATION GEOLOGY, COMPUTER APPLICATIONS IN EXPLORATION. *Educ:* Univ Ill, BS, 51; Univ Minn, MS, 53, PhD, 58. *Prof Exp:* Asst geol, Univ Minn, 52-53, asst & instr, 55-58; asst prof geol, Univ NDak, 58-60; geologist, Int Nickel Co Can, Ltd, 60-65, res geologist, 65-69, asst mgr, Western Region, Field Explor Dept, 69-71, regional mgr, 71-73, regional geologist, 73-75; dist geologist, Eastern US Region, Am Copper & Nickel Co, Inc, 75-79, supvr sr staff geologist, 79-90; RETIRED. *Mem:* Soc Econ Geologists; Geol Soc Am; Geol Asn Can; Can Inst Mining & Metall; Am Inst Mining & Metall. *Res:* Iron, copper and nickel in Canada; uranium, base and precious metal exploration; computer applications in exploration. *Mailing Add:* 2310 Juniper Ct Golden CO 80401-9107

QUIROS, CARLOS F, b Lima, Peru, Mar 17, 46; m 70; c 2. PLANT GENETICS, PLANT BREEDING. *Educ:* Agrarian Univ, Peru, BSc, 68; Univ NH, MS, 72; Univ Calif, Davis, PhD(genetics), 75. *Prof Exp:* Fel tomato genetics, Univ Calif, Davis, 75-76; res assoc breeding & genetics, Nat Inst Agr Res, Mex, 76-77; fel genetics, Univ Sherbrooke, 77-78; res assoc genetics & breeding, Univ Alta, 78-81; res scientist genetics & breeding, Int Plant Res Inst, 81-83; ASSOC PROF GENETICS & BREEDING, DEPT VEG CROPS, UNIV CALIF, DAVIS, 83- *Mem:* Sigma Xi; Am Soc Hort Sci; Econ Bot. *Res:* Evolution, genetics and breeding of crop plants, specifically solanaceas and cool season vegetables; germplasm collection and preservation. *Mailing Add:* Dept Veg Crops Univ Calif Davis CA 95616

QUIROZ, RODERICK S, b Ajo, Ariz, Nov 6, 23. METEOROLOGY. *Educ:* Univ Calif, Los Angeles, BA, 50; Univ Md, College Park, MS, 70. *Prof Exp:* Res meteorologist, US Air Weather Serv, 59-66; res meteorologist, Nat Meteorol Ctr, Nat Weather Serv, Nat Oceanic & Atmospheric Admin, 66-85; CONSULT, 85- *Concurrent Pos:* Mem, US Comt Exten to Standard Atmosphere, 62-85; comt chmn, Atmospheric Problems Aerospace Vehicles, 68-72; lectr, Von Karman Inst Fluid Dynamics, Brussels, 70; chmn, Am Meteorol Soc Comt Upper Atmosphere, 73-78. *Mem:* Fel Am Meteorol Soc; Am Geophys Union. *Res:* Structure and circulation of upper atmosphere, emphasizing interaction with the troposhere on long-wave and climatic time-scales; analysis of atmospheric measurements with rockets and satellites. *Mailing Add:* 4520 Yuma St NW Washington DC 20016

QUISENBERRY, DAN RAY, b Lake Co, Ind, Jan 3, 38; m 58; c 1. HEALTH PHYSICS. *Educ:* Univ Ky, AB, 61; Col William & Mary, MS, 66; World Open Univ, PhD(physics), 79. *Prof Exp:* Teacher sci & math, Ft Knox Dependent Schs, 61-66; asst prof sci, Brevard Jr Col, 66-67; asst prof, 68-80, chmn dept, 81-89, ASSOC PROF PHYSICS, MERCER UNIV, 80- *Mem:* Am Asn Physics Teachers; Am Phys Soc; Nat Geog Soc; Planetary Soc. *Res:* Environmental pollution, chiefly environmental effects of tritium; environmental monitoring of nuclear energy facilities; radon levels in buildings. *Mailing Add:* Dept Physics Mercer Univ Macon GA 31207

QUISENBERRY, KARL SPANGLER, JR, b Washington, DC, Apr 4, 26; m 49; c 3. PHYSICS. *Educ:* Univ Nebr, BS, 49; Univ Minn, MA, 52, PhD(physics), 55. *Prof Exp:* Asst physics, Univ Minn, 49-55, res assoc, 55-57; asst prof, Univ Pittsburgh, 57-60, assoc prof, 60-62; physicist, Knolls Atomic Power Lab, Gen Elec Co, 61-65, mgr advan exp physics, 65-74, spec critical facilities opers & safety, 71-74, mgr exp physics, 74-76, mgr advan develop activity, 76-77; dir nuclear, Schlumberger-Doll Res Ctr, 77-81, dir Houston Eng, Schlumberger Well Serv, 81-85; wireline coordr, EMR Photoelect, 85-86; CONSULT, SCI CONSULT, 86- *Mem:* Am Phys Soc. *Res:* Nuclear physics and spectroscopy; neutron and nuclear reactor physics; oil well logging; nuclear fuel. *Mailing Add:* 9105 Hilldale St Houston TX 77055-7425

QUISENBERRY, RICHARD KEITH, b Springfield, Ill, July 27, 34; m 57; c 4. SYNTHETIC POLYMERS, FINISHES. *Educ:* Millikin Univ, BA, 56; Univ Utah, PhD(chem), 61. *Prof Exp:* Res chemist, Dacron Res Lab, 60-62, sr res chemist, 62-63, supvr tech plant, 63-64, supvr res, 64-66, sr supvr, Nylon Technol Sect, Del, 66-68, res mgr, Benger Lab, 68-72, tech supt, Va, 72-74, prod supt, 74-75, prin consult, Corp Plans Dept, 75-77, mgr bus planning, Spunbonded Prod, 77-78, prod mgr indust fibers, 78-79, dir, Feedstock Res Div, 79-80, dir pioneering res-fibers, 80-81, dir res & develop, Fabrics & Finishes Dept, 81-84, VPRES RES, CENT RES DEPT, E I DU PONT DE NEMOURS & CO, INC, DEL, 84. *Mem:* Am Chem Soc; Sigma Xi; Indust Res Inst. *Res:* Synthesis and evaluation of polymeric materials for fibers, plastics, packaging, electronics and finishes; advanced materials; biotechnology. *Mailing Add:* 404 Way Rd Greenville DE 19807

QUISENBERRY, VIRGIL L, b Patesville, Ky, Sept 25, 46; m 67; c 1. SOIL PHYSICS. *Educ:* Univ Ky, BSA, 69, MS, 71, PhD(soil sci), 74. *Prof Exp:* From asst prof to assoc prof, 74-83, PROF SOIL PHYSICS, CLEMSON UNIV, 83- *Mem:* Am Soc Agron; Soil Sci Soc Am. *Res:* Flow of water and solutes in field soils with particular emphasis on macropore flow. *Mailing Add:* Dept Agron & Soils Clemson Univ Main Campus Clemson SC 29634

QUISENBERRY, WALTER BROWN, b Purman, Mo, June 24, 12; m 40; c 3. PREVENTIVE MEDICINE, CANCER. *Educ:* Loma Linda Univ, MD, 41; Westmont Col, BA, 42; Johns Hopkins Univ, MPH, 45; Am Bd Prev Med, dipl, 50. *Prof Exp:* Med intern, Henry Ford Hosp, 40-41; prof chem & sch health physician, Westmont Col, 41-42; venereal dis control officer, Southside Health Dist, Va, 42-44; asst med, Johns Hopkins Hosp, 44-45; dir div venereal dis, Nebr State Dept Health, 45-46; asst prof prev med & pub health, Sch Med, Loma Linda Univ, 46-54, actg head dept, 46-47; chief venereal dis & cancer control, Territorial Dept Health, Hawaii, 47-51, dir div prev med, 51-54; exec dir, Hawaii Cancer Soc, 54-58; dir div prev med, Hawaii State Dept Health, 58-63, dep dir health, 63-66, dir health, 66-74, assoc physician, 75-82, EMER PHYSICIAN, STRAUB CLIN & HOSP, HONOLULU, 82- *Concurrent Pos:* Rockefeller Found fel, Johns Hopkins Univ, 44-45; pvt pract, Long Beach, Calif, 41-42; staff physician, Coleman Med Group, Alhambra, Calif, 46-47; jr attend physician, Los Angeles County Hosp, 46-48; mem teaching staff pub health, Univ Hawaii, 47-75; attend staff, Queens, Kuakini, Kapiolani & Kauikeolani Children's Hosp, Honolulu, 49-; deleg, Nat Cancer

Conf, Mich, 56, Int Cancer Cong, Eng, 58, Int Conf Cancer Probs, Japan, 60; assoc clin prof, Sch Med, Loma Linda Univ, 60-75; lectr, Int Cancer Cong, Moscow, USSR, 62; Univ Sydney lectr oncol, Sydney, Adelaide, Melbourne & Brisbane, Australia & Auckland, NZ, 63; US rep int conf nasopharyngeal cancer, Int Union Against Cancer, Singapore, 64. *Mem:* Fel Am Pub Health Asn; AMA; fel Am Col Prev Med; fel Am Col Physicians; Royal Soc Health; Am Geriat Soc. *Res:* Cytologic diagnosis of cancer by the smear technique; ethnic differences in the incidence of cancer; methadone treatment of heroin addiction; epidemiology of cancer of the stomach, breast, liver and lung; epidemiology of venereal diseases; treatment of venereal diseases with antibiotics; twinning; sociocultural factors in cancer. *Mailing Add:* 2128 Kamehameha Ave Honolulu HI 96822

QUISMORIO, FRANCISCO P, JR, b Philippines, Jan 21, 41; m; c 2. RHEUMATOLOGY, CLINICAL IMMUNOLOGY. *Educ:* Univ Philippines, BS, 60, MD, 64; Am Bd Internal Med, cert internal med, 75, diagnostic lab immunol, 87; Am Bd Pediat, cert 87, Am Bd Internal med, rhematology, cert, 76. *Prof Exp:* Fel rheumatology, Univ Pa Hosp, 66-68; fel clin immunol, 68-70, from asst prof to assoc prof, 72-83, PROF, MED, UNIV SOUTHERN CALIF, 83-, PROF, PATH, 86- *Concurrent Pos:* Mem, med & sci comt, Arthritis Found, 79-86; assoc ed, Lupus Erythematosus; dir clin Rheumatology Lab, Univ Southern Calif Med Ctr, 80-; comn, Med Bd Calif, 81- *Mem:* Am Asn Immunologists; fel Am Col Physicians; Clin Immunol Soc; Am Fedn Clin Res; NY Acad Med; Am Col Rheumatol. *Res:* Immunopathology of systemic connective tissue diseases; significance of circulating auto antibodies; clin features and treatment of systemic lupus erythematosus and other rheumatic diseases. *Mailing Add:* Univ Southern Calif Sch Med HMR-715 2025 Zonal Ave Los Angeles CA 90033

QUISSELL, DAVID OLIN, b Pipestone, Minn, Oct 25, 44; m 72; c 1. REGULATORY BIOLOGY, METABOLIC REGULATION. *Educ:* Augustana Col, BA, 66; Univ Wis-Madison, PhD(biochem), 72. *Prof Exp:* Fel oncol & path, Univ Wis-Madison, 71-73; asst prof biochem, Univ Mo, Columbia, 73-77; asst prof pharmacol, 77-80, ASST PROF BIOCHEM, SCH MED, UNIV COLO, 80- *Concurrent Pos:* Res scholar award, Nat Cystic Fibrosis Found, 79. *Mem:* Am Chem Soc; Am Soc Cell Biol; Soc Complex Carbohydrates; Int Asn Dent Res; Sigma Xi. *Res:* Stimulus-secretion coupling mechanism in exocrine tissue; role of cyclic adenosine monophosphate, cyclic guanosine monophosphate, and calcium in the regulation of secretion; pathogenesis of cystic fibrosis. *Mailing Add:* 5183 S Jamaica Way Englewood CO 80111

QUIST, ARVIN SIGVARD, b Blair, Nebr, Nov 15, 33; m 57; c 3. PHYSICAL CHEMISTRY. *Educ:* Dana Col, Nebr, BS, 54; Univ Nebr, MS, 57, PhD(chem), 59, Univ Tenn, JD, 75. *Prof Exp:* Asst, Univ Nebr, 54-56; res assoc & instr, Univ Pittsburgh, 59-60; instr chem, Univ Nebr, 60-61; mem res staff, Oak Ridge Nat Lab, 61-76; staff mem, Off Waste Isolation, 76-78, staff mem, Gas Centrifuge Proj, Nuclear Div, Union Carbide Corp, 78-85; CLASSIFICATION OFFICER, MARTIN MARIETTA ENERGY SYST, INC, 85- *Mem:* Nat Classification Mgt Soc. *Res:* Physical chemistry of aqueous electrolyte solutions and molten salts; electrical conductances and Raman spectroscopy of aqueous solutions and molten salts at high temperatures and pressures; environmental and safety impacts of nuclear facilities. *Mailing Add:* 104 Neville Lane Oak Ridge TN 37830

QUIST, RAYMOND WILLARD, b Minneapolis, Minn, Nov 26, 34; m 57; c 2. SPEECH PATHOLOGY. *Educ:* Hamline Univ, BA, 52; Univ Minn, Minneapolis, MA, 66, PhD(speech path), 71. *Prof Exp:* Assoc prof speech path, Calif State Col (Pa), 68-71; assoc prof speech path, Madison Col, 71-74; assoc prof, 74-77, PROF SPEECH PATH, IND STATE UNIV, TERRE HAUTE, 77- *Mem:* Am Speech & Hearing Asn; Coun Except Children; Am Asn Univ Professors. *Res:* Behavior modification; stuttering; voice. *Mailing Add:* Dept Commun Dis Ind State Univ 217 N Sixth St Terre Haute IN 47809

QUIST, WILLIAM EDWARD, b Seattle, Wash, May 13, 35; m 66; c 2. ALUMINUM ALLOY FATIGUE & FRACTURE. *Educ:* Univ Wash, BS, 57, MS, 63, PhD(metall eng), 74. *Prof Exp:* Engr corrosion, Farwest Corrosion Control, 57; metallurgist, Pacific Car & Foundry Co, 58-59; PRIN ENGR, BOEING COM AIRPLANE CO, 59- *Concurrent Pos:* NSF res assoc grant, 67-68; lectr, Univ Wash, 72-73; co-chmn, Pac Northwest Metals & Minerals Conf, Am Soc Metals & Am Inst Mech Engrs, 77; co-prin investr, Study Develop Low Density Alloys, USAF & Boeing, 81-; gen chmn, Pac Northwest Mat Conf, Am Soc Metals, Am Welding Soc, Soc Advan Mat & Process Eng & Soc Nondestructive Testing, 83; chmn, Develop Al-Li Alloys, West IC-85, Am Soc Metals & Am Soc Mech Engrs, 85. *Mem:* Fel Am Soc Metals. *Res:* Phase transformation, fracture, fatigue, and corrosion studies in aluminum, titanium, and nickel base systems; aluminum alloy development, strengthening mechanisms and engineering property microstructural relationships. *Mailing Add:* 18215 NE 27th St Redmond WA 98052

QUISTAD, GARY BENNET, b Riverside, Calif, Jan 17, 47; m 82; c 3. AGRICULTURAL BIOCHEMISTRY, ORGANIC CHEMISTRY. *Educ:* Univ Calif, Riverside, BS, 69; Univ Calif, Los Angeles, PhD(chem), 72. *Prof Exp:* Sr chemist, Zoecon Corp, 73-80, group leader metab, 80-81, sect head, 82-85, sr sect head, 85-87; PRIN SCIENTIST, SANDOZ CROP PROTECTION CORP, 88- *Mem:* Am Chem Soc; Sigma Xi; AAAS. *Res:* Pesticide metabolism and environment degradation photochemistry; spider venoms, arthropod toxins; terpenoid biosynthesis. *Mailing Add:* Sandoz Crop Protection Corp 975 California Ave Palo Alto CA 94303-0859

QUITTNER, HOWARD, b Brooklyn, NY, Feb 1, 22; m 66; c 5. PATHOLOGY. *Educ:* Tulane Univ La, BS, 42, MD, 44; Am Bd Path, dipl, 51. *Prof Exp:* Adj pathologist, Beth Israel Hosp, 51-52; dir labs, Washington Hosp, Pa, 52-64; prof clin path, Sch Med & dir clin labs, Univ Ark, Little Rock, 64-76; prof path, Sch Med, Tulane Univ, 76-78, dir clin labs, Med Ctr, 76-78; PROF PATH, SCH MED, MARSHALL UNIV, 78-; DIR LAB SERV, VET ADMIN MED CTR, HUNTINGTON, 80- *Concurrent Pos:* Levy Found res fel, Beth Israel Hosp, New York, 49-52; adj prof biochem, Tulane Univ, 76-78. *Mem:* Am Soc Path; AMA; Asn Clin Sci (pres, 67). *Res:* Clinical pathology diagnostic techniques. *Mailing Add:* Lab Serv Vetr Admin Med Ctr 1540 Spring Valley Dr Huntington WV 25704

QUOCK, RAYMOND MARK, b San Francisco, Calif, June 9, 48; m 75; c 3. NEUROPHARMACOLOGY. *Educ:* Univ San Francisco, BS, 70; Univ Wash, PhD(pharmacol), 74. *Prof Exp:* Lab asst pharmacol, Sch Med, Univ Calif, San Francisco, 63-70; instr, Sch Med, Univ Wash, 74-75; asst prof physiol & pharmacol, Sch Pharm, Univ Pac, 75-79; from asst prof to prof pharmacol, Sch Dent, Marquette Univ, 79-89; ASSOC PROF, COL MED, UNIV ILL, ROCKFORD, 89- *Concurrent Pos:* Exam consult, Calif State Bd Pharm, 76-79; adj asst prof pharmacol & toxicol, Med Col Wis, 80-89; res assoc toxicol, C J Zablocki Vet Admin Med Ctr, 81-89, consult dent, 82-89. *Honors & Awards:* Dr Elwood Molseed Award, 70; Dr Leo Pinsky Award, 83, 86; Raymond B Allen Award, 90. *Mem:* Soc Neurosci; Western Pharmacol Soc; Am Soc Pharmacol & Exp Therapeut; Int Brain Res Orgn; Bioelectromagnetic Soc; Am Fedn Aging Res. *Res:* Mechanisms of pain and anxiety control; nitrous oxide; bioeffects of microwaves. *Mailing Add:* Dept Biomed Sci Col Med Univ Ill 1601 Parkview Ave Rockford IL 61107-1897

QUON, CHECK YUEN, b Canton, China, Nov 15, 49; US citizen; m 74; c 3. DRUG METABOLISM & PHARMACOKINETICS, BIOANALYTICAL CHEMISTRY. *Educ:* Univ Calif, Los Angeles, BS, 72, PhD(pharmacol), 77. *Prof Exp:* Res assoc, McArdle Lab Cancer Res, 77-79; res investr, Arnar-Stone, 79-81; sr res investr, Am Crit Care, 81-83, group leader, 83-85, sect head, 85-88; res mgr, Du Pont Pharmaceut, 88-90, DIR, DRUG METAB & PHARMAKINETICS, DU PONT MERCK PHARMACEUT CO, 91- *Mem:* Am Soc Pharmacol & Exp Therapeut; Int Soc Study Xenobiotics; Am Asn Pharmaceut Scientists; Pharmaceut Mfrs Asn. *Res:* Clinical and preclinical drug metabolism and pharmacokinetics; clinical pharmacology; bioanalytical chemistry; n-oxidation. *Mailing Add:* Du Pont Merck Pharmaceut Co Stine-Haskell Res Ctr PO Box 30 Elkton Rd Newark DE 19714

QUON, DAVID SHI HAUNG, b Canton, China, Dec 26, 31; Can citizen; m 60; c 2. MINERALOGY. *Educ:* Sun Yat Sen Univ, BSc, 49; Ohio State Univ, MSc, 59; Univ Mich, PhD(mineral), 65. *Prof Exp:* Geologist, Geotech Develop Co, 54-58; res asst mineral, Univ Mich, 60-64; mem sci staff, Res & Develop Labs, Northern Elec Co, 64-67 & 68-75; asst prof mineral, Lakehead Univ, 67-68; mem sci staff, Bell Northern Res, 68-75; SR RES SCIENTIST, CANMET ENERGY, MINES & RESOURCES, CAN, 75- *Concurrent Pos:* Nat Res Coun Can res grant, 68-69. *Mem:* Mineral Soc Am; Soc Econ Geologists; Am Ceramic Soc; Mineral Soc Can. *Res:* Mineralogy and geochemistry of carbonatites; crystallochemistry; mineralog synthetic ceramic materials; growing single crystals; mineral processing, extraction of alumina from non-bauxitic minerals and building materials research; solid electrolyte for energy storage; advanced ceramic materials. *Mailing Add:* Ten Sherk Cresent Kanata Ottawa ON K2K 2L4 Can

QURAISHI, MOHAMMED SAYEED, b Jodhpur, Rajasthan, India, June 23, 24; m 53; c 3. PESTICIDE TOXICOLOGY. *Educ:* St John's Col, Agra Univ, BSc, 42; Aligarh Muslim Univ, MSc, 44; Univ Mass, PhD(entom), 48. *Prof Exp:* Sr mem, UN WHO Team to Pakistan, 49-51; entomologist, Malaria Inst Pakistan, 51-55; sr res officer, Pakistan Coun Sci & Indust Res, 55-60; sr sci officer, Pakistan AEC, 60-64; assoc prof entom, Univ Man, 64-66; assoc prof entom, NDak State Univ, 66-70, prof, 70-74; chief sci biol, NY State Sci Serv, 74-75; entomologist-toxicologist chief, Pest Control & Consult Sect, 76-84, HEALTH SCIENTIST ADMINR, EXEC SECY, MIDRC, NAT INST ALLERGY & INFECTIOUS DIS, NIH, 84- *Concurrent Pos:* Assoc secy, Sci Comn Pakistan, 59-60; sr scientist, Cent Treaty Orgn Inst Nuclear Sci, Tehran, Iran, 60-64; prog mgr, Interdept Contract, Proj Themis, Dept of Defense, 68-74; mem, Publ Comt, Soc Environ Toxicol & Chem, 80-83. *Mem:* Entom Soc Am; Am Chem Soc; Soc Environ Toxicol & Chem; Sigma Xi. *Res:* Chemicals showing transient interference with vital phases of insect development. *Mailing Add:* NIH Westwood Bldg Rm 706 Bethesda MD 20892

QURESHI, A H, b Dagi, Pakistan, Oct 28, 32; m 61; c 3. ELECTRICAL ENGINEERING. *Educ:* Univ Peshawar, BEng, 55; Aachen Tech Univ, PhD(elec eng), 61. *Prof Exp:* Guest res collabr, Nuclear Res-Centre, Juelich, Ger, 58-61; asst prof elec eng, Univ Waterloo, 61-64; head dept, Col Eng, Riyadh, Saudi Arabia, 64-65; sessional lectr, Univ Calgary, 65-66; from asst prof to assoc prof, 66-70, PROF ELEC ENG & HEAD DEPT, UNIV WINDSOR, 70- *Res:* Magnetic and solid state materials; high voltage technology. *Mailing Add:* Dept Elec Eng Cleveland State Univ Cleveland OH 44115

QURESHI, NILOFER, b Karachi, Pakistan, July 13, 47; US citizen; m 79; c 1. LIPOPOLYSACCHARIDES, MYCOBACTERIAL LIPIDS. *Educ:* St Joseph's Col, BS, 67; Karachi Univ Pakistan, MS, 69; Univ Wis-Madison, PhD(physiol chem), 75. *Prof Exp:* Res assoc biochem, Univ Wis-Madison & Vet Admin Hosp, 76-77, proj assoc, 77-81; RES BIOCHEMIST, VET ADMIN HOSP, MADISON, 81- *Mem:* Am Soc Biol Chemists; Am Soc Microbiol. *Res:* Purification and structure of lipopolysaccharides and lipid A from Salmonella and E coli; lipid A, structure and function; structure and biosynthesis of mycolic acids in Mycobacteria; long-chain fatty acids in Mycobacteria structure and biosynthesis; purification and mechanism of action of B-hydroxy B-methylglutaryl-coenzyme A reductase (yeast). *Mailing Add:* Vet Admin Hosp D-2215 2500 Overlook Terr Madison WI 53705

QUTUB, MUSA Y, b Jerusalem, Palestine, June 2, 40; US citizen; m 70; c 4. WATER RESOURCES, GEOLOGY. *Educ:* Simpson Col, BA, 64; Colo State Univ, MS, 66; Iowa State Univ, PhD(geol & higher educ), 69. *Prof Exp:* Instr earth sci, Iowa State Univ, 66-69; asst prof, 69-72, assoc prof earth sci, 72-80, PROF HYDROGEOL, NORTHEASTERN ILL UNIV, 80- *Concurrent Pos:* NSF & NDEA fels, Northeastern Ill Univ, 69-72; NSF grants, 70-; consult, NSF, 72-, mem aerospace educ comt, 72; Off Environ Educ grant, 73-74; chmn, Six Nat Symposia Environ & Water Resources; Ministry Planning, Saudi Arabia, 77-78; leader, USA Environ Sci Deleg, People's Repub China, 84. *Mem:* AAAS; Int Asn Advan Earth & Environ Sci (pres); Nat Asn Geol Teachers (pres); Nat Sci Teacher Asn. *Res:* Geological education; science learning; ground water and application to city planning in northern Illinois. *Mailing Add:* Dept Geog & Environ Studies Northeastern Ill Univ Chicago IL 60625

QUYNN, RICHARD GRAYSON, b Newport News, Va, Jan 23, 28; m 53; c 2. POLYMER PHYSICS. *Educ:* Col William & Mary, BS, 47; Inst Textile Technol, MS, 49; Princeton Univ, AM, 52, PhD(physics, phys chem), 57. *Prof Exp:* Res physicist, Summit Res Labs, Celanese Corp Am, 53-54, sr res physicist, 57-63, res assoc physics, 63-65, sect head mat sci, Celanese Res Co, Celanese Corp, 65-70; mgr mat sci, Res Ctr, Burlington Industs, 70-72; asst dir mat sci, FRL Div, Albany Int Corp, 72-79; mem tech staff, Jet Propulsion Lab, Calif Inst Technol, 79; sr ed, High Technol Mag, 81; sr specialist engr, Boeing Mil Airplane Co, 82-85; RES PHYSICIST, RES, DEVELOP & ENG CTR, US ARMY, NATICK, 86- *Mem:* Am Phys Soc; Fiber Soc. *Res:* Physical structure of fibers and films; high polymer physics; spectroscopy; electro-optics. *Mailing Add:* 424 Lincoln St Duxbury MA 02332

R

RAAB, ALLEN ROBERT, structural mechanics, finite element analysis, for more information see previous edition

RAAB, FREDERICK HERBERT, b Ft Crook, Nebr, Feb 4, 46; m 70; c 1. RADIO FREQUENCY POWER AMPLIFIERS, COMMUNICATIONS. *Educ:* Iowa State Univ, Ames, BS, 68, MS, 70, PhD(elec eng), 72. *Prof Exp:* Engr, Collins Radio Co, Rockwell, 66-69; technologist, NASA Marshall Space Flight Ctr, 70; mem tech staff, Cincinnati Electronics Corp, 72-75; sr systs engr, Pozhemus Navig Sci, Inc, 75-80; PRES-OWNER, GREEN MOUNTAIN RADIO RES, 80- *Concurrent Pos:* Prog chmn, Radio Frequency Expo E Conf, Radio Frequency Design Mag, 90. *Mem:* Sigma Xi; sr mem Inst Elec & Electronics Engrs. *Res:* Radio frequency power amplifiers, especially high-efficiency; transmitters; communications, especially through-the-earth; signal processing; author and co-author of over 50 technical publications; reducing new theory to working prototype. *Mailing Add:* 240 Stanford Rd Burlington VT 05401

RAAB, HARRY FREDERICK, JR, b Johnstown, Pa, May 9, 26; m 51; c 3. REACTOR DESIGN, EXPERIMENTAL REACTOR PHYSICS. *Educ:* Mass Inst Technol, SB & SM, 51. *Prof Exp:* Control systs engr, Bettis Atomic Power Lab, Westinghouse Elec Corp, 51-55, mgr surface ship physics, 55-62, light water breeder physics, 62-72; CHIEF PHYSICIST, NAVY NUCLEAR PROPULSION DIRECTORATE, US DEPT ENERGY, 72- *Mem:* Fel Am Nuclear Soc; AAAS; Sigma Xi. *Res:* Developed light water breeder reactor; directed reactor physics development for the Navy. *Mailing Add:* 8202 Ector Ct Annandale VA 22003

RAAB, JACOB LEE, b Elkhart, Ind, Nov 29, 38; m 65; c 3. PHYSIOLOGY. *Educ:* Univ Chicago, BS, 60, MS, 65; Duke Univ, PhD(zool), 71. *Prof Exp:* Instr biol, Franklin & Marshall Col, 66-69; asst prof physiol, Rutgers Univ, Newark, 71-; sci dir, West Mountain Sci, Inc; RES SYST ANALYST, CIBA-GIEGY, 82- *Mem:* Am Physiol Soc; AAAS. *Res:* Relationship of the energetics of exercise to variables of temperature, humidity, terrain and the time of day. *Mailing Add:* 19 Laurel Ave Summit NJ 07901

RAAB, JOSEPH A, b Oshkosh, Wis, Dec 20, 34; m 55; c 6. MATHEMATICS. *Educ:* Wis State Univ, Oshkosh, BS, 57; Univ Ill, MS, 60; Univ Wis, PhD(math), 67. *Prof Exp:* Teacher pub schs, Wis, 57-59; assoc prof math, Wis State Univ, Oshkosh, 60-69; chmn dept math, 72-75, asst vpres acad affairs, 78-79, PROF MATH, METROP STATE COL, 69- *Concurrent Pos:* Teacher, Univ N Colo, 68; rep, Colo State Col & Univ Consortium, 75-77; coordr acad progs, Consortium State Cols in Colo, 77-78. *Mem:* Math Asn Am. *Res:* Fibonacci sequences; Pascal's triangle higher order continued fractions and associated algorithms; number theory; abstract algebra; analysis. *Mailing Add:* Dept Math Scis Box 38 Metrop State Col 1006 11th St Denver CO 80204

RAAB, WALLACE ALBERT, b Onawa, Iowa, Nov 4, 21; m 47; c 2. APPLIED MATHEMATICS. *Educ:* Morningside Col, BS, 49; Univ SDak, MA, 49; Iowa State Univ, PhD(math), 58. *Prof Exp:* Mathematician, US Naval Ord Test Sta, Calif, 51-52; sr dynamics engr, Gen Dynamics Corp, 55-58; prof math, Calif State Polytech Col, 57-64; PROF MATH, UNIV SDAK, 64- & COORD STATIST RES BUR, 80- *Mem:* Math Asn Am; Am Statist Asn; Soc Indust & Appl Math; Sigma Xi. *Res:* Fluid mechanics; differential equations; statistics. *Mailing Add:* Dept Math Univ SDak Vermillion SD 57069

RAABE, HERBERT P(AUL), b Halle, Ger, Aug 15, 09; nat US; m 56; c 4. ELECTRICAL ENGINEERING. *Educ:* Berlin Tech Univ, dipl, 36, Dr Ing, 39. *Prof Exp:* From instr to asst prof elec commun technol, Berlin Tech Univ, 36-45; mgr & tech consult, Asn Microfilm, Ger, 45-47; tech consult, Wright Air Develop Ctr, Ohio, 47-56; sr tech specialist, Litton Industs, Inc, 56-66 & Int Bus Mach Corp, NY, 66-68; sr engr, Fed Systs Div, IBM Corp, Gaithersburg, 68-74; CONSULT, 74- *Concurrent Pos:* Res engr, Heinrich Hertz Inst, 37-45; sci consult, Bur Commun Tech, 46-47. *Mem:* Sr mem Inst Elec & Electronics Engrs; Am Inst Aeronaut & Astronaut; hon mem Ger Soc Rocket Technol & Travel. *Res:* Electrical communication technique and radar; information theory; microwave theory and technique; antennas; propagation; electrical countermeasures; infrared technique; military reconnaissance; systems analysis. *Mailing Add:* 10121 Lloyd Rd Potomac MD 20854

RAABE, ROBERT DONALD, b Waukesha, Wis, May 8, 24; m 55; c 2. PLANT PATHOLOGY. *Educ:* Univ Wis, BS, 48, PhD(plant path), 51. *Prof Exp:* Collabr, USDA, Wis & Tex Agr Exp Stas, 51-52; from instr & jr plant pathologist to assoc prof plant path & assoc plant pathologist, Univ Calif, Berkeley, 52-63; plant pathologist, Univ Hawaii, 63-64; assoc prof & assoc plant pathologist, 64-68, PROF PLANT PATH & PLANT PATHOLOGIST, UNIV CALIF, BERKELEY, 68- *Mem:* Am Phytpath Soc; Mycol Soc Am. *Res:* Diseases of ornamentals; Armillaria root rot. *Mailing Add:* Dept Bioresource Sci Univ Calif Berkeley CA 94720

RAAEN, VERNON F, b Plentywood, Mont, Nov 8, 18; m 49; c 1. ORGANIC CHEMISTRY. *Educ:* Concordia Col, BA, 41; Univ Minn, MS, 50, PhD(chem) 58. *Prof Exp:* Chemist & opers supt, Columbia Powder Co, Olin Mathieson Chem Corp, 42-44; chemist, Union Carbide Nuclear Co Div, Oak Ridge Nat Lab, 50-79; RETIRED. *Mem:* AAAS; Am Chem Soc; Sigma Xi. *Res:* Organic reactions studies with the help of radiocarbon and tritium as tracers. *Mailing Add:* 111 Scenic Dr Oak Ridge TN 37830

RAAM, SHANTHI, b Madras, India, Nov 26, 41; US citizen. ONCOLOGY. *Educ:* Univ Madras, India, BS, 60, MS, 62; Univ Ga, PhD(immunol & microbiol), 73. *Prof Exp:* RES ASSOC, CANCER RES CTR, TUFTS UNIV, 73-, ASSOC PROF RES, SCH MED, TUFTS UNIV, 85-; DIR, ONCOL LAB, LEMUEL SHATTUCK HOSP, 77- *Concurrent Pos:* Consult radioimmunoassay, Leary Labs, Boston, Mass, 73-74; consult steroid receptors, New Eng Nuclear, Boston, Mass; researcher estrogen receptor breast cancer, Tufts Med Cancer Unit, Lemuel Shattuck Hosp, Jamaica Plain, Mass, 75-; prin investr, Am Cancer Soc NY res grants, 79-80 & 80-81; invited speaker consensus comt for steroid receptors in breast cancer, Nat Cancer Inst, 79; prin investr, Nat Cancer Inst, 83-, chmn ad hoc rev comt, 89, 90. *Mem:* Am Asn Cancer Res; AAAS; Am Asn Clin Oncologists; Am Asn Immunologists; Endocrine Soc. *Res:* Significance of steroid hormone receptors in cancer; search for tumor markers which may prove to be of prognostic and/or diagnostic value in cancer; study of functionally defective hormone receptors in hormone-therapy resistant breast cancers; immunoendocrinology. *Mailing Add:* Oncol Lab Tufts Univ Lemuel Shattuck Hosp 170 Morton St Boston MA 02130

RAAMOT, TONIS, b Tartu, Estonia, Jan 6, 32; US citizen; m 58; c 1. CIVIL ENGINEERING. *Educ:* Columbia Univ, BA, 53, BS, 54, MS, 56; Univ Ill, PhD(civil eng), 62. *Prof Exp:* From asst prof to prof civil eng & asst chmn dept, Newark Col Eng, 62-67; chief civil eng, RCP Div, Raymond Int Inc, NY, 67-69; PARTNER, RAAMOT ASSOCS, CONSULT ENGRS, 69-; PROF CIVIL ENG, NJ INST TECHNOL, 70- *Concurrent Pos:* Soils consult, Raymond Int Inc, 64; NSF res initiation grant, 64-66. *Honors & Awards:* Arthur Wellington Award, Am Soc Civil Engrs, 65. *Mem:* Am Soc Civil Engrs; Nat Soc Prof Engrs; Am Soc Eng Educ; Am Concrete Inst; Am Soc Testing & Mat. *Res:* Soil mechanics; foundation engineering; behavior of deep foundations; pile driving analysis. *Mailing Add:* Raamot Assoc Two Penn Plaza New York NY 10121

RAAPHORST, G PETER, b Holland; Can citizen; m 74; c 2. MEDICAL PHYSICS, RADIATION & HYPERTHERMIA BIOLOGY. *Educ:* Univ Waterloo, BSc, 72, MSc, 74, PhD(physics), 76. *Prof Exp:* Med Res Coun Can res fel, Colo State Univ, 76-78; res officer radiation biol, 77-85, HEAD RADIATION BIOL, WHITESHELL DIV, ATOMIC ENERGY CAN, 85-; CHIEF, MED PHYSICS, OTTAWA REGIONAL CANCER CTR, 85- *Concurrent Pos:* Prof, Dept Radiol, Univ Ottawa, 85-; adj prof physics, Carleton Univ, Ottawa, 86-; mem, allied sci staff, dept radiol, Civic Hosp, Ottawa, 86- *Mem:* Radiation Res Soc; Can Fedn Biol Sci; NAm Hyperthermia Group; Can Asn Radiation Oncol. *Res:* The effects of radiation chemotherapeutic agents and hyperthermia on cellular systems in vitro; flow cytometry; electron microscopy; high-performance liquid chromatography of normal and malignant cells given radiation drug hyperthermia or combined treatments; optimizing cancer treatment and determing mechanisms of cellular injury. *Mailing Add:* 1054 Deauvill Crescent Orleans ON K1C 5M1 Can

RAASCH, GILBERT O, b Milwaukee, Wis, May 27, 03; m 25; c 2. GEOLOGY, PALEONTOLOGY. *Educ:* Univ Wis, BA, 29, PhD(geol), 46. *Prof Exp:* Cur geol mus, Univ Wis, 29-36; surface geologist, Magnolia Petrol Corp, 36-37; geologist, Darby Petrol Co, 37-38, petrol explor, 38-40; supvr mus proj, Milwaukee Pub Mus, 40-42; from assoc geologist to geologist, Ill State Geol Surv, 46-53; paleontologist, Can Stratig Surv, 53-56; consult paleont, Shell Oil Co Can, 56-68; CONSULT GEOL, 68- *Mem:* Fel Geol Soc Am; Soc Econ Paleontologists & Mineralogists; Can Soc Petrol Geologists; Paleont Asn London. *Res:* Paleozoic paleontology and stratigraphy; paleozoic biostratigraphy of the Canadian Arctic, western Canada and Alaska. *Mailing Add:* Heritage Gardens Suite 5 2009 90th Ave SW Calgary AB T2V 0X4 Can

RAASCH, LOU REINHART, b Republican City, Nebr, Apr 27, 44; m 68; c 2. ANALYTICAL CHEMISTRY. *Educ:* Univ Nebr, Lincoln, BS, 65, PhD(chem), 71. *Prof Exp:* Asst prof chem, MacMurray Col, 69-75; ASST PROF CHEM, ETENN STATE UNIV, 75- *Mem:* Am Chem Soc. *Res:* Electrochemistry of coordination compounds; voltammetry involving charge transfers folowed by chemical reactions; non-aqueous solvents for electrochemical investigations. *Mailing Add:* Rte 1 Box 27 Johnson City TN 37601

RAASCH, MAYNARD STANLEY, b Castlewood, SDak, Feb 27, 15. ORGANIC CHEMISTRY. *Educ:* SDak Sch Mines, BS, 37; Ohio State Univ, MS, 38, PhD(org chem), 41. *Prof Exp:* Lab instr chem, SDak Sch Mines, 35-37; asst gen & org chem, Ohio State Univ, 37-39; res chemist, Exp Sta, E I du Pont de Nemours Co, 41-80; RETIRED. *Concurrent Pos:* Res assoc, Stetson Univ, 84; consult, 85-86; res scientist, Univ Ala, 87. *Mem:* Am Chem Soc. *Res:* Organic fluorine compounds; organic sulfur compounds; synthetic biologically active chemicals. *Mailing Add:* 2300 Inglewood Dr Wilmington DE 19803

RAB, PAUL ALEXIS, b Dayton, Ohio, Mar 2, 44; m 67; c 1. ZOOLOGY. *Educ:* Ohio State Univ, BSc, 66, MSc, 70, PhD(zool), 72. *Prof Exp:* instr, 72-80, PROF BIOL, SINCLAIR COMMUNITY COL, 80- & DEPT CHMN LIFE SCI, 80- *Res:* Social behavior; genetics of behavior. *Mailing Add:* Dept Biol Sinclair Community Col 444 W Third St Dayton OH 45402

RABA, CARL FRANZ, JR, b San Antonio, Tex, Dec 24, 37; c 5. GEOTECHNICAL ENGINEERING, CONSTRUCTION MATERIALS SCIENCE. *Educ:* Tex A&M Univ, BS, 61, ME, 62, PhD(civil eng), 68. *Prof Exp:* Res asst geotech, Tex Transp Inst Tex A&M Univ, 61-62; pres, Raba &

Assoc Consult Engrs, Inc, 67-78; pres, 79-80, CHMN, RABA-KISTNER CONSULT, INC, 80- *Concurrent Pos:* Mem comt, Am Soc Test & Mat, 72-; pres, Tex Coun Eng Labs, 74-75; chmn, Nat Comt Geotech Eng, Exam, Inst Cert Eng Technicians, 75-78. *Mem:* Sigma Xi; Am Soc Civil Engrs; Am Geophys Union; Am Soc Lubrication Engrs. *Res:* Geotechnical considerations of drilled pier and pile foundation systems, and stabilization aspects of fly ash in pavements and embankments. *Mailing Add:* 12821 W Golden Lane San Antonio TX 78249

RABALAIS, FRANCIS CLEO, b Bunkie, La, Aug 16, 37; m 59; c 2. PARASITOLOGY. *Educ:* Univ Southwestern La, BS, 61; La State Univ, MS, 63, PhD(zool), 67. *Prof Exp:* Instr zool, La State Univ, 66-67 & parasitol, Sch Med, Tulane Univ, 67-68; ASSOC PROF BIOL, BOWLING GREEN STATE UNIV, 68-; ASSOC PROF, HEALTH & COMMUNITY SERV, 73- *Concurrent Pos:* Adj assoc prof microbiol, Med Col Ohio; fel, Sch Trop Med, Tulane Univ, 67. *Mem:* Am Soc Trop Med & Hyg; Am Soc Parasitol. *Res:* Biology of trematodes of lower vertebrates; biology of filarial nematodes; host-parasite relationships of filarial nematodes. *Mailing Add:* Dept Biol Sci Bowling Green State Univ Bowling Green OH 43402

RABALAIS, JOHN WAYNE, b Bunkie, La, Sept 7, 44; m 66. PHYSICAL CHEMISTRY. *Educ:* Univ Southwestern La, BS, 66; La State Univ, PhD(phys chem), 70. *Prof Exp:* NATO fel electron spectros, Univ Uppsala, 70-71; asst prof phys chem, Univ Pittsburgh, 71-75; ASSOC PROF PHYS CHEM, UNIV HOUSTON, 75- *Mem:* Am Chem Soc; AAAS. *Res:* Ultraviolet and x-ray photoelectron spectroscopy and its applications to surfaces, catalysis, chemisorption, and adsorption; visible and ultraviolet absorption and emission spectroscopy; secondary ion mass spectrometry; applied quantum chemistry. *Mailing Add:* Dept Chem Univ Houston 4800 Calhoun Rd Houston TX 77204-5641

RABAN, MORTON, b St Louis, Mo, Oct 18, 40; m 64. ORGANIC CHEMISTRY, NUCLEAR MAGNETIC RESONANCE. *Educ:* Harvard Univ, AB, 62; Princeton Univ, MS, 66, PhD(org chem), 67. *Prof Exp:* Chemist, Res Inst Med & Chem, 62-63; instr chem, Princeton Univ, 66-67; from asst prof to assoc prof, 67-74, fac res fel, 68, PROF CHEM WAYNE STATE UNIV, 74- *Concurrent Pos:* Petrol Res Fund grant, 67-69 & 72-74; Res Corp grant-in-aid, 68-70; NIH res grant, 69-72; NSF res grant, 70-74; Sloan fel, 72-76. *Mem:* AAAS; Am Chem Soc; The Chem Soc. *Res:* Stereochemistry, including optical rotary dispersion-circular dispersion spectroscopy; asymmetric synthesis and determination of absolute configuration; organic chemistry and stereochemistry of sulfur and nitrogen compounds; dynamic nuclear magnetic resonance spectroscopy. *Mailing Add:* Dept Chem Wayne State Univ 335 Chem Bldg Detroit MI 48202

RABB, GEORGE BERNARD, b Charleston, SC, Jan 2, 30; m 53. ZOOLOGY, RESOURCE ADMINISTRATION. *Educ:* Col Charleston, BS, 51; Univ Mich, MA, 52, PhD(zool), 57. *Prof Exp:* Ed asst, Charleston Mus, 49; cur & coordr res, 56-64, assoc dir res & educ, 64-69, dept dir, 69-75, DIR, CHICAGO ZOOL PARK, 76- *Concurrent Pos:* Mem, Comt Evolutionary Biol, 69-; herpet ed, Copeia, Am Soc Ichthyol & Herpet, 64-68; res assoc, Field Mus Natural Hist, 65-, Univ Chicago, 60-67; rep, Am Asn Zool Parks & Aquariums, Species Surv Comn, Int Union, Conservation Nature, 78-; adv, Encyclopedia Britannica, Standard Educ Encyclopedia, 63-, Coronet Films, 64-; chmn, Int Union Conservation Nature's Species Survival Comn, 89- *Mem:* Fel AAAS; Am Soc Ichthyol & Herpet (pres, 78); Animal Behav Soc; Am Soc Mammal; Am Soc Naturalists; Sigma Xi. *Res:* Vertebrate behavior; evolution; ecology of reptiles; amphibians. *Mailing Add:* Chicago Zool Park Brookfield IL 60513

RABB, ROBERT LAMAR, b Lenoir, NC, Aug 6, 19; m 46; c 2. ENTOMOLOGY, POPULATION BIOLOGY. *Educ:* NC State Univ, BS, 47, MS, 50, PhD(entom), 53. *Prof Exp:* Asst entom, NC State Univ, 47-50, asst to exten entomologist, 51-52, from asst prof to assoc prof, 53-63, prof entom, 63-83; RETIRED. *Honors & Awards:* Ciba-Geigy Recognition Award, Entom Soc Am, 73, Founders Mem Award, 77, W N Reynolds Dist Prof, 81. *Mem:* Entom Soc Am; Ecol Soc Am; Am Inst Biol Sci; Int Orgn Biol Control; Sigma Xi. *Res:* Insect ecology and management of agricultural insect pests. *Mailing Add:* 1821 Pictou Rd Raleigh NC 27606

RABBIT, GUY, b Cairo, Egypt, Jan 30, 43; m 68; c 3. COMPUTER ARCHITECTURE. *Educ:* Queens Univ, Eng, BS, 67, MS, 68, PhD(elec eng), 71. *Prof Exp:* Design supv, Siemens, Ger, 68; asst lectr, 68-72, Adj prof, Queens Univ, 72-74; prog dir, IBM, 74-84; vpres, Austin Oper, CAE Syst Div, 84-86; Head, Elec & Electronics Eng, Gen Motors, 86-88; VPRES, RES & DEVELOP, MODCOMP/AGE, 88- *Concurrent Pos:* Chmn, Int Elec & Electronics Engrs conf, Circuits & Comput, 80, Comput Design, 83; ed-in-chief, Inst Elec & Electronics Engrs Circuits & Devices Mag, 84-86. *Honors & Awards:* Centennial Medal, Inst Elec & Electronics Engrs, 83. *Mem:* Am Automation Asn, (pres, 84-86); Inst Elec & Electronics Engrs. *Res:* Computer architecture, real time computing, software tools, operating systems, compilers and computer languages, VLSI automotive electronics. *Mailing Add:* Modcomp-AEG 1650 McNab Rd W PO Box 6099 Ft Lauderdale FL 33310

RABE, ALLEN E, b New Holstein, Wis, Nov 12, 31; m 56; c 5. CHEMICAL ENGINEERING. *Educ:* Univ Wis, BS, 54, MS, 55, PhD(chem eng), 58. *Prof Exp:* Develop engr, Linde Co, Union Carbide Corp, 58-60; chem engr, Elec Boat Div, Gen Dynamics Corp, 60-62; SR RES ENGR, E I DU PONT DE NEMOURS & CO, INC, 62 - *Mem:* Am Chem Soc; Am Inst Chem Engrs; Sigma Xi. *Res:* Development of research apparatus for reaction rate constants; measurement of reaction rate constants; analysis and interpretation of kinetic data. *Mailing Add:* Eng Dept E I du Pont de Nemours & Co Inc 104 Glennside Ave Wilmington DE 19803

RABE, AUSMA, b Daugavpils, Latvia, Jan 26, 26; Can citizen. PSYCHOBIOLOGY. *Educ:* Queen's Univ (Ont), BA, 53, MA, 54; Univ Mich, PhD(physiol psychol), 60. *Prof Exp:* Res scientist, Bur Res Neurol & Psychiat, NJ Neuropsychiat Inst, 60-72; vis scientist, 73-76, RES SCIENTIST VII, NY STATE INST BASIC RES DEVELOP DISABILITIES, 77-, CHMN DEPT PSYCHOBIOL, 85- *Concurrent Pos:* Vis prof, Grad Fac, New Sch Social Res, 63-81. *Mem:* AAAS; Am Psychol Asn; Can Psychol Asn; Soc Neurosci; Int Soc Develop Psychobiol; Teratology Soc. *Res:* Neural mechanisms of behavior; neuroteratology; behavioral teratology. *Mailing Add:* Neuroteratology NY State Inst Basic Res & Develop Disabilities 1050 Forest Hill Rd Staten Island NY 10314

RABE, EDWARD FREDERICK, b Watsontown, Pa, Nov 7, 18; m 43; c 4. PEDIATRIC NEUROLOGY. *Educ:* Bucknell Univ, BS, 39; Yale Univ, MD, 43. *Prof Exp:* Instr pediat, Sch Med, Yale Univ, 47-49; instr, Sch Med, Univ Kans, 49-50, asst prof, 50-51; chief dept, Geisinger Mem Hosp, Danville, Pa, 51-58; from asst prof to prof pediat, Sch Med, Tufts Univ, 61-86; consult & head sect pediat neurol, King Faisal Specialist Hosp & Res Ctr, Riyadh, Saudi Arabia, 86-89; RETIRED. *Concurrent Pos:* Fel neurol, Mass Gen Hosp, 58-61; dir seizure clin, Pa State Dept Health, 57-58; head sect pediat neurol, New Eng Med Ctr Hosps; pediat neurologist, Boston Floating Hosp; asst pediatrician, Mass Gen Hosp; consult, Paul A Dever State Sch, Mass & New Eng Rehab Inst, Woburn; mem, Task Force II-Minimal Brain Dysfunction, Nat Progs Learning Disabilities in Children, Comt Med & Health Related Servs, 66-69; mem neurol sci res training comt A, Nat Inst Neurol Dis & Stroke, 69-73; mem bd trustees, Easter Seal Res Found, 72-78, chmn, 74-78. *Mem:* Am Acad Pediat; Soc Pediat Res; Am Pediat Soc; Am Acad Neurol. *Res:* Cerebrospinal fluid dynamics; neurological aspects of minimal brain dysfunction; subdural fluid dynamics and treatment of cerebral dysrhythmia; pharmacokinetics of anticonvulsant drugs. *Mailing Add:* PO Box 73 North Whitefield ME 04353-0073

RABEL, FREDRIC M, b Mansfield, Ohio, May 29, 38. ANALYTICAL CHEMISTRY. *Educ:* Ohio Univ, BS, 60; Univ Wis-Madison, MS, 62; Univ Pa, PhD(chem), 67. *Prof Exp:* Sr chemist, J T Baker Chem Co, 67-71; sr scientist, H Reeve Angel & Co, Inc, 71-74, tech serv mgr, Whatman, Inc, 74-87; pres, ChromHELP, Inc, 87-90; PROD MGR, E M SEPARATIONS, 90- *Concurrent Pos:* Lectr, Sadtler Res Labs, 70-75, Ctr Prof Advan, 78- & Am Chem Soc, 81-84. *Mem:* AAAS; Am Chem Soc; Am Soc Testing & Mat. *Res:* Thin layer, column and high performance liquid chromatography, especially applications and materials development; extraction techniques; organometallic syntheses and bonding; development of analytical systems. *Mailing Add:* 406 Boyd St Boonton NJ 07005

RABEN, IRWIN A(BRAM), b New Orleans, La, Oct 26, 22; m 45; c 4. CHEMICAL ENGINEERING. *Educ:* Tulane Univ, BE, 42; La State Univ, MS, 47. *Prof Exp:* Chem engr, New Orleans Water Purification Plant, La, 42; sr process engr, Cities Serv Refining Corp, 47-55; sr process engr & supvr, Wyatt C Hedrick Corp, 55-58; mgr process develop & design, Southwest Res Inst, 58-60, mgr chem eng res, 60-64; supvr process eng, Bechtel Corp, 64-69, mgr air pollution control eng, 69-73; vpres, Western Opers, Combustion Equip Assocs, Inc, 73-79; pres, Iar Technol, Inc, 79-85; CONSULT, 86- *Mem:* Am Inst Chem Engrs; Air Pollution Control Asn; Am Chem Soc; Sigma Xi. *Res:* Development and design of chemical processes and equipment; heat transfer; economics; development and design of commercial size flue-gas desulphurization systems for utility boilers; process design of air quality control systems. *Mailing Add:* 130 Sandringham South Moraga CA 94556

RABENSTEIN, ALBERT LOUIS, b East Liverpool, Ohio, May, 20, 31. GENERAL COMPUTER SCIENCES. *Educ:* Washington & Jefferson Col, AB, 52; Univ WVa, MS, 53; Mass Inst Technol, PhD(math), 58. *Prof Exp:* Asst prof math, Allegheny Col, 59-61 & Pa State Univ, 61-64; asst prof, Macalester Col, 64-68, assoc prof, 68-72; assoc prof, 72-75, prof math, Washington & Jefferson Col, 75-90; CONSULT, 90- *Mem:* Math Asn Am; Asn Comput Mach. *Res:* Ordinary differential equations. *Mailing Add:* 632 E Beau St Washington PA 15301

RABENSTEIN, DALLAS LEROY, b Portland, Ore, June 13, 42; m 64; c 2. ANALYTICAL CHEMISTRY. *Educ:* Univ Wash, BS, 64; Univ Wis, PhD(anal chem), 68. *Prof Exp:* Res asst nuclear magnetic resonance, Univ Wis, 65-66, lectr, 67-68; res chemist, Chevron Res Co, 68-69; from asst prof to prof chem, Univ Alta, 69-85; PROF CHEM, UNIV CALIF, RIVERSIDE, 85-, CHMN DEPT CHEM, 89- *Concurrent Pos:* Res grants, NIH. *Honors & Awards:* Fisher Sci Award, Can, 84. *Mem:* AAAS; Am Chem Soc; fel Chem Inst Can. *Res:* Nuclear magnetic resonance spectroscopy; solution chemistry of metal-complexes; application of nuclear magnetic resonance to biochemistry; analytical biochemistry. *Mailing Add:* Dept Chem Univ Calif Riverside CA 92521

RABER, DOUGLAS JOHN, b New York, NY, Nov 13, 42; m 67. CHEMISTRY. *Educ:* Dartmouth Col, AB, 64; Univ Mich, PhD(org chem), 68. *Prof Exp:* NIH fel, Princeton Univ, 68-70; asst prof, 70-75, assoc prof, 75-80, PROF CHEM, UNIV SFLA, 80- *Mem:* AAAS; Am Chem Soc; Royal Soc Chem. *Res:* Synthetic organic chemistry, particularly the development of new methods and techniques; synthetic applications of physical organic chemistry. *Mailing Add:* Dept Chem Univ SFla 4202 Fowler Ave Tampa FL 33620

RABER, MARTIN NEWMAN, b New York, NY, Mar 29, 47; m 78; c 3. MEDICINE. *Educ:* Wash Univ, St Louis, AB, 68; Cath Univ Louvain, MD, 75; FRCPC, 79. *Prof Exp:* Intern resident internal med, Sch Med, Dalhousie Univ, 75-78; fel med oncol, M D Anderson Hosp, 78-80; from asst prof to assoc prof med, Sch Med, Univ Tex, Houston, 80-85; ASSOC PROF MED & CHIEF, SECT GEN ONCOL, M D ANDERSON HOSP & TUMOR INST, HOUSTON, 85- *Mem:* Fel Am Col Physicians; Am Soc Clin Oncol; NY Acad Sci; Soc Anal Cytol. *Res:* Clinical evaluation of new drugs and other therapeutic modalities in patients with cancer; application of flow cyotmetric analyses to the study of solid tumors. *Mailing Add:* Div Med M D Anderson Cancer Ctr PO Box 92 Houston TX 77030

RABIDEAU, PETER W, b Johnstown, Pa, Mar 4, 40; m 62, 86; c 6. ORGANIC CHEMISTRY. *Educ:* Loyola Univ, BS, 64; Case Inst Technol, MS, 67; Case Western Reserve Univ, PhD(org chem), 68. *Prof Exp:* Res assoc org chem, Ben May Lab Cancer Res, Univ Chicago, 68-69, instr, 69-70; from asst prof to prof chem, Ind Univ-Purdue Univ, Indianapolis, 70-89, chmn dept, 85-89; PROF CHEM, LA STATE UNIV, BATON ROUGE, 89-, DEAN, COL BASIC SCI, 89- *Mem:* Am Chem Soc. *Res:* Stereochemistry of cyclic hydrocarbons by nuclear magnetic resonance; metal ammonia reduction of aromatic compounds. *Mailing Add:* Off Dean Col Basic Sci La State Univ Baton Rouge LA 70803-1802

RABIDEAU, SHERMAN WEBBER, b Cloquet, Minn, May 9, 20; m 43; c 3. PHYSICAL CHEMISTRY. *Educ:* Univ Minn, BChem, 41; Univ Iowa, MS, 47, PhD(phys chem), 49. *Prof Exp:* Chemist, Firestone Tire & Rubber Co, 41-42; res chemist electrochem, US Naval Res Lab, 42-46; instr chem, Univ Iowa, 47-49; mem staff chem res, Los Alamos Nat Lab, Univ Calif, 49-82; CONSULT, 82- *Honors & Awards:* Clark Medal, Am Chem Soc. *Mem:* Fel AAAS; fel Am Inst Chemists; Am Chem Soc; Sigma Xi. *Res:* Laser induced chemistry; gas phase reaction kinetics; isotope separations. *Mailing Add:* 5913 Cubero Dr NE Albuquerque NM 87109-3113

RABIE, RONALD LEE, b Yakima, Wash, May 2, 46; m 70; c 2. SHOCK WAVE PHYSICS. *Educ:* Cent Wash Univ, BS, 72; Wash State Univ, PhD(shock wave physics), 77. *Prof Exp:* MEM STAFF SHOCK WAVE PHYSICS, LOS ALAMOS NAT LAB, 77- *Concurrent Pos:* Consult, Los Alamos Tech Assoc, 79- *Res:* Reactive flow in heterogeneous materials and vapor phase explosions both theoretical and experimental. *Mailing Add:* Los Alamos Nat Labs PO Box 1663 P952 Los Alamos NM 87545

RABIGER, DOROTHY JUNE, b Philadelphia, Pa, May 30, 35. ORGANIC CHEMISTRY. *Educ:* Ursinus Col, BS, 57; Univ Pa, MS, 60, PhD(org chem), 62. *Prof Exp:* Res chemist, Nat Renderers Asn-USDA, 57-58 & Rohm & Haas Co, 62-64; fel org chem & cancer chemother, Ravdin Inst, Hosp Univ Pa, 64-67, res assoc, 67-69; res chemist, Borden Chem Co, 69; instr pharmaceut chem, Sch Pharm, Temple Univ, 69-70, asst prof, 70-74; INDEP RES CHEM 1ST, 74- *Concurrent Pos:* Res grant-in-aid, Temple Univ, 70-71. *Mem:* AAAS; Am Chem Soc; Sigma Xi. *Res:* Cancer chemotherapy; effects of substituents and structural modifications on the properties of organic molecules; synthesis of novel organic compounds for biological applications; chemical topology. *Mailing Add:* 517 Boyer Rd Cheltenham PA 19012

RABII, JAMSHID, b Tehran, Iran, July 12, 46; m 68; c 3. ENDOCRINOLOGY, NEUROENDOCRINOLOGY. *Educ:* Univ Calif, Berkeley, BA, 70; Univ Calif, San Francisco, PhD(endocrinol), 75. *Prof Exp:* Res anatomist, Dept Anat & Brain Res, Univ Calif, Los Angeles, 75-76, fel, Mental Health Training Prog, 76-77; asst prof physiol, 77-83, ASSOC PROF BIOL SCI, RUTGERS UNIV, 83- *Concurrent Pos:* Prin investr grants, NIH, 78-83 & Rutgers Univ, 80- *Mem:* Endocrine Soc; Soc Neurosci; Int Soc Neuroendocrinol; Soc Study Reproduction. *Res:* Neuroendocrinology: longterm influences of opiates on various neuroendocrine phenomena; hypothalamic control of anterior pituitary hormone secretion; involvement of biogenic amines in the regulation of anterior pituitary hormone secretion in mammalian and avian species. *Mailing Add:* Dept Biol Sci Nelson Labs Rutgers State Univ Piscataway NJ 08855

RABII, SOHRAB, b Ahwaz, Iran, Dec 30, 37; div; c 2. THEORETICAL SOLID STATE PHYSICS. *Educ:* Univ Southern Calif, BS, 61; Mass Inst Technol, MS, 62, PhD(solid state physics), 66. *Hon Degrees:* MA, Univ Pa, 75. *Prof Exp:* Res fel solid state physics, Mass Inst Technol, 66-67; sr res physicist, Monsanto Co, Mo, 67-69; from asst prof to assoc prof elec eng, 69-78, chmn dept, 77-82, PROF ELEC ENG & SCI, MOORE SCH ELEC ENG, UNIV PA, 78 - *Concurrent Pos:* Fel, Max Planck Soc Advan Sci, Repub Ger, 75. *Mem:* Inst Elec & Electronics Engrs; Am Phys Soc; Col Art Asn Am; Am Soc Anesthetists. *Res:* Theoretical calculation of energy band structure and electronic properties of crystalline solids and disordered alloys; electronic structure of molecules and localized state in solids; relativistic effects in atoms molecules and solids. *Mailing Add:* Dept Elec Eng Univ Pa Philadelphia PA 19104

RABIN, BRUCE S, IMMUNOLOGY, PATHOLOGY. *Educ:* State Univ NY, MD & PhD, 69. *Prof Exp:* PROF PATH, UNIV PITTSBURGH PRESBY HOSP, 72- *Mailing Add:* Dept Path Univ Pittsburgh Presby Hosp 5725 Children's Tower Pittsburgh PA 15213

RABIN, ELIJAH ZEPHANIA, b Ottawa, Ont, Jan 30, 37; m 65; c 2. INTERNAL MEDICINE, BIOCHEMISTRY. *Educ:* Queen's Univ, Ont, MD, 61; McGill Univ, PhD(biochem), 71. *Prof Exp:* Assoc med officer, Prudential Assurance Eng, 68-74; asst prof med, Montreal Gen Hosp, McGill Univ, 70-74; ASSOC PROF MED, UNIV OTTAWA, 74- *Concurrent Pos:* Med Res Coun Can scholar, 70-74; chief med dir, Montreal Life Ins Co, 73-74. *Mem:* Can Soc Clin Invest; Am Soc Nephrology; Int Soc Nephrology. *Res:* Ribonuclease activity in renal failure; role of ribonuclease in uremic toxicology; biochemical structure of human urinary ribonuclease. *Mailing Add:* Div Nephrology Univ Ottawa 1053 Carling Ave Ottawa ON K1Y 4E9 Can

RABIN, ERWIN R, b St Louis, Mo, Oct 22, 30; m 54; c 3. PATHOLOGY, VIROLOGY. *Educ:* Wash Univ, AB, 52, MD, 56. *Prof Exp:* Intern path, Sch Med, Yale Univ, 56-57, resident, 57-59; asst pathologist, Sinai Hosp Baltimore, Inc, 61-62; asst prof path, Baylor Col Med, 62-66, assoc prof, 66-68; assoc prof path, Sch Med, Wash Univ, 68-74; mem staff, Lattimore-Fink Labs, Inc, 74-76; DIR PATH, HURON RD HOSP, OHIO, 76- *Concurrent Pos:* Life Ins Med Res Fund fel, 57-58 & USPHS trainee, 58-59; from asst to assoc pathologist, Jewish Hosp St Louis, 68-70, actg dir, 71-74. *Mem:* Col Am Pathologists; Am Asn Pathologists; Int Acad Path; Sigma Xi. *Res:* Ultrastructural changes in in-vivo viral infections as related to viral-host interaction and pathogenesis of the disease. *Mailing Add:* Huron Rd Hosp Two Bratenahl Pl No 110 Cleveland OH 44108-1172

RABIN, HARVEY, EXPERIMENTAL BIOLOGY. *Educ:* Temple Univ, PhD, 58. *Prof Exp:* ASSOC DIR VIRAL DIS RES, E I DU PONT DE NEMOURS & CO, INC, 89- *Mailing Add:* E I du Pont de Nemours & Co Inc Exp Sta E328 Wilmington DE 19880-0328

RABIN, HERBERT, b Milwaukee, Wis, Nov 14, 28; m 62; c 2. QUANTUM OPTICS, SOLID STATE PHYSICS. *Educ:* Univ Wis, BS, 50; Univ Ill, MS, 51; Univ Md, PhD(physics), 59. *Prof Exp:* Physicist, Elec Div, Naval Res Lab, 52-54, Solid State Physics Div, 54-62, head, Radiation Effects Sect, Dielec Br, 62-65, mat sci staff, 65-66, head, Radiation Physics Sect, Optical Mat Br, Optical Physics Div, 66-67, actg head, Appl Optics Br, 67-68, head, Quantum Optics Br, 68-70, assoc dir space & commun sci & technol, Naval Res Lab, 70-79; DEP ASST SECY NAVY, RES, APPL & SPACE TECHNOL, DEPT NAVY, WASHINGTON, DC, 79- *Concurrent Pos:* Vis scientist, Univ Stuttgart, 60-61; consult, Dept Physics, Univ Sao Paulo, 64 & 70; prof lectr, Dept Physics, George Washington Univ, 55-73; mem, Space Panel, Naval Studies Bd, Nat Acad Sci, 78-, Adv Panel NASA, 72-75. *Honors & Awards:* E O Hulburt Award, 70. *Mem:* AAAS; fel Am Phys Soc; Sigma Xi; Fedn Am Scientists; Optical Soc Am; corresp mem Brazilian Acad Sci. *Res:* Characterization of defect structure in insulating crystals; elucidation of nonlinear optical phenomena; space research and system developments. *Mailing Add:* 7109 Radnor Rd Bethesda MD 20817

RABIN, JEFFREY MARK, b Los Angeles, Calif, Aug 29, 55. STRING THEORY, GEOMETRY OF SUPERMANIFOLDS. *Educ:* Univ Calif, Los Angeles, BS, 76; Stanford Univ, MS, 78, PhD(physics), 81. *Prof Exp:* Res staff physicist & lectr physics, Yale Univ, 81-83; Enrico Fermi fel, Enrico Fermi Inst, Univ Chicago, 83-87; ASST PROF, DEPT MATH, UNIV CALIF-SAN DIEGO, 87- *Mem:* Am Math Soc. *Res:* Topology, geometry and integration theory on supermanifolds; string theory; two-dimensional quantum gravity. *Mailing Add:* Dept Math Univ Calif-San Diego La Jolla CA 92093-0112

RABIN, MICHAEL O, b Breslau, Ger, Sept 1, 31. THEORY OF ALGORITHMS. *Educ:* Hebrew Univ, Israel, MSc, 53; Princeton Univ, PhD(math), 56. *Prof Exp:* H B Fine instr, Princeton Univ, 56-58, mem, Inst Advan Study, 58; assoc prof, 58-65, PROF, HEBREW UNIV, JERUSALEM, 65-, ALBERT EINSTEIN CHAIR, 80-; THOMAS J WATSON SR PROF COMPUTER SCI, HARVARD UNIV, 83- *Concurrent Pos:* Consult, IBM, 57- & Bell Tel Labs, 60; vis assoc prof math, Univ Calif, Berkeley, 61-62, Mass Inst Technol, 62-63; chmn, Inst Math, Hebrew Univ, 64-66 & Computer Sci Dept, 70-71; lectr computer sci, Paris Univ, 65; vis prof math, Yale Univ, 67, math & computer sci, NY Univ, 70-71, appl math, Mass Inst Technol, 72-78, computer sci, Wash State Univ, 79 & Harvard Univ, 80-81; Gordon McKay prof computer sci, Harvard Univ, 81-83; Fairchild scholar, Calif Inst Technol, 87. *Honors & Awards:* C Weizmann Prize Exact Sci, 60; Rothschild Prize Math, 74; Turing Award Computer Sci, Asn Comput Mach, 76; Harvey Prize Sci & Technol, 80. *Mem:* Foreign assoc Nat Acad Sci; foreign hon mem Am Acad Arts & Sci; Israel Acad Sci & Humanities; foreign mem Am Philos Soc. *Res:* Theory of algorithms; randomized algorithms; complexity of compuatations; computer security. *Mailing Add:* Div Appl Sci Harvard Univ 29 Oxford St Cambridge MA 02138

RABIN, MONROE STEPHEN ZANE, b Brooklyn, NY, Dec 19, 39; m 65; c 2. ELEMENTARY PARTICLE PHYSICS, MEDICAL PHYSICS. *Educ:* Columbia Col, AB, 61; Rutgers Univ, MS, 64, PhD(physics), 67. *Prof Exp:* Physicist, Univ Calif, Lawrence Berkeley Lab, 67-72; assoc prof, 72-81, PROF PHYSICS, UNIV MASS, 81- *Concurrent Pos:* Vis physicist, Stanford Linear Accelerator Ctr, 79-80; vis scholar physics, Harvard Univ, 86-87; first soriano res scholar radiol physics, Dept Radiation Med, Mass Gen Hosp, 86-87. *Mem:* Am Phys Soc; Fedn Am Scientists; Sigma Xi. *Res:* Elementary particle physics: lifetimes of charmed particles, electromagnetic interactions, polarization and form factors of hyperons; hadronic production of heavy flavors; use of protons and other ionizing particles in cancer therapy. *Mailing Add:* Dept Physics Univ Mass Amherst MA 01003

RABIN, ROBERT, b Philadelphia, Pa, Aug 22, 28; m 53; c 3. MICROBIOLOGY, RESEARCH ADMINISTRATION. *Educ:* Philadelphia Col Pharm, BS, 50; Pa State Univ, MS, 52, PhD(bact, biochem), 55. *Prof Exp:* Asst, Pa State Univ, 52-55; asst scientist, Commun Dis Ctr, USPHS, 55-56; sr scientist & group leader, Sci Info Dept, Smith Kline & French Labs, 56-61; asst dir res, Albert Einstein Med Ctr, 61-62, assoc mem res labs, 62-69; grants assoc, Grants Assocs Prog, NIH, Md, 69-70; staff assoc, Off Interdisciplinary Res, NSF, 70-71, prog mgr, Div Environ Systs & Resources, 71-74, dep dir, Div Advan Environ Res & Technol, 74-75; dep assoc dir res & develop progs, Div Biomed & Environ Res, US Energy Res & Develop Admin, 75-76; exec asst, Directorate Biol Behav & Social Sci, NSF, 76, dep asst dir, 76-77, sr sci assoc, Off Dir, 77-79, dep asst dir, Directorate Biol. Behav & Social Sci, 79-85; ASST DIR LIFE SCI, OFF SCI & TECH POLICY, EXEC OFF PRES, 85- *Concurrent Pos:* USPHS fel, 62-65. *Mem:* AAAS. *Res:* Fatty acid and glyoxylate metabolism; amino acid biosynthesis; control mechanisms; environmental sciences; research and development administration. *Mailing Add:* US NSF Washington DC 20550

RABINER, LAWRENCE RICHARD, b Brooklyn, NY, Sept 28, 43; m 68; c 3. ELECTRICAL ENGINEERING, COMMUNICATIONS. *Educ:* Mass Inst Technol, BS & MS, 64, PhD(elec eng), 67. *Prof Exp:* MEM TECH STAFF, BELL TEL LABS, 67- *Honors & Awards:* Biennial Award, Acoust Soc Am, 74; Piori Award & Soc Award, Inst Elec & Electronics Engrs, 80. *Mem:* Nat Acad Sci; Nat Acad Eng; fel Inst Elec & Electronics Engrs; Acoust Soc Am. *Res:* Speech communications including recognition, synthesis, perception and analysis; digital filtering and computer applications. *Mailing Add:* Rm 2D-538 AT&T Bell Lab Murray Hill NJ 07974

RABINO, ISAAC, b Haifa, Israel, Dec 2, 38. BIOLOGICAL AND HEALTH SCIENCES. *Educ:* Hebrew Univ, Jerusalem, Israel, BSc, 62; Col Agr & Life Sci, Cornell Univ, Ithaca, NY, MS, 65; State Univ NY, Stony Brook, DPhil(biol sci), 76. *Prof Exp:* Asst prof biol, physiol, microbiol & environ sci, St Peter's Col, Jersey City, NJ, 77-81; asst prof embryol & develop biol, State

Univ NY, Stony Brook, 81-82; sci assoc environ affairs, Scientists & Engrs for Secure Energy, New York, 82-83; asst prof biol, Dept Natural Sci, Baruch Col, City Univ New York, 83-85; ASSOC PROF BIOL & HEALTH SCI, EMPIRE STATE COL, STATE UNIV NY, 85- *Concurrent Pos:* NSF Summer Res Grant, Columbia Univ, New York, 78-81; res grant, Richard Lounsbery Found, 87-90 & 87- *Mem:* NY Acad Sci; AAAS; Am Inst Biol Sci; Asn Politics & Life Sci; Soc Social Studies of Sci. *Res:* Impact of political advocacy on biotechnology research in the United States; AIDS and society; author of numerous scientific journal articles; drugs and society. *Mailing Add:* 92 Pinehurst Ave 5-D New York NY 10033

RABINOVICH, ELIEZER M, b Moscow, USSR, Apr 4, 37, US citizen; m 67; c 2. GLASS SCIENCE, GLASS-CERAMICS. *Educ:* Moscow Mendeleev Inst Chem Technol, MSc, 59, PhD(ceramic sci), 64. *Prof Exp:* Res engr glass for electronics, Res Inst Vacuum Electronics, Moscow, USSR, 59-64, sr res fel, 64-68, group supvr, 68-73; adj prof glass & ceramics, Technion-Israel Inst Technol, Haifa, Israel, 76-80; sr & prin investr, glass & ceramic sci, Israel Ceramic & Silicate Inst, Technion City, Haifa, Israel, 74-80; MEM TECH STAFF & PRIN INVESTR, GLASS & CERAMIC SCI, AT&T BELL LABS, MURRAY HILL, NJ, 81- *Concurrent Pos:* Vis prof, Paris Univ, Orsay, France, 89; mem ed adv comt, Am Ceramic Soc, 89-92. *Mem:* Am Ceramic Soc. *Res:* Broad experience in research and development of a variety of glass and ceramic materials, study of their properties, designing new materials and technological processes; extensive research in sol-gel preparation of materials, in glass ceramics, glasses for electronics; author of 90 papers and recipient of 8 patents. *Mailing Add:* AT&T Bell Labs Rm 7A-313 600 Mountain Ave Murray Hill NJ 07974

RABINOVICH, SERGIO ROSPIGLIOSI, b Lima, Peru, Apr 4, 28; m 53; c 4. INTERNAL MEDICINE, VIROLOGY. *Educ:* Univ Lima, BM & MD, 54. *Prof Exp:* Intern, Grasslands Hosp, Valhalla, NY, 54, asst med resident, 55- 56, chief med resident, 56-57; resident gastroenterol, Henry Ford Hosp, Detroit, 57-58; pvt pract, Peru, 58-59; prof med & head dept, Univ San Agustin, Peru, 60-61; physician-in-chg, Hosp Arzobispo Loayza, Lima, Peru, 63; assoc, Col Med, Univ Iowa, 63, from asst prof to assoc prof med, 65-73; chmn dept, 74-88, PROF MED & CHMN DIV INFECTIOUS DIS, SCH MED, SOUTHERN ILL UNIV, 73- *Concurrent Pos:* Consult, Arequipa Gen Hosp, Peru, 60-61; res fel med, Col Med, Univ Iowa, 63-65; res fel, Sch Med, Univ Kans, 64-65. *Mem:* Am Soc Microbiol; AMA; Sigma Xi; fel Am Col Physicians; fel Infectious Dis Soc Am. *Res:* Infectious disease, mycology, virology and antibiotics. *Mailing Add:* Sch Med Southern Ill Univ PO Box 3926 Springfield IL 62708

RABINOVITCH, B(ENTON) S(EYMOUR), b Montreal, Que, Feb 19, 19. PHYSICAL CHEMISTRY. *Educ:* McGill Univ, BSc, 39, PhD(phys chem), 42. *Prof Exp:* Res chemist, Chem Warfare Labs, 42; Royal Soc Can fel, Harvard Univ, 46-47, Milton fel, 47-48; from asst prof to prof, 48-85, EMER PROF CHEM, UNIV WASH, 85- *Concurrent Pos:* Guggenheim fel, 61; vis scientist, Nat Res Coun Can, 62; mem, various comts, Nat Acad Sci-Nat Res Coun, 65-; vis Sloan prof, Harvard Univ, 66; distinguished vis prof, Univ Ariz, 68 & Israel Inst Technol-Technion, 78; vis fel, Trinity Col, Oxford Univ, 71; ed, Ann Rev Phys Chem, 75-85; Frontiers Chem lectr, Wayne State Univ, 81, Frontiers Phys Chem lectr, Cambridge Univ, 83; Chmn, Puget Sound Sect, Am Chem Soc, 58 & Phys Chem Div, 67. *Honors & Awards:* Du Pont Lect, Univ Rochester, 64; Reilly Lectr, Univ Notre Dame, 68; Debye Lectr, Cornell Univ, 78; King Lectr, Kansas State Univ, 80; Res Prize, Sigma Xi, 81; Debye Award, Am Chem Soc, 84; Polanyi Medal, Royal Soc Chem London, 84. *Mem:* Fel Am Phys Soc; Am Chem Soc; Royal Soc Chem; fel Am Acad Arts & Sci; fel Royal Soc London. *Res:* Chemical kinetics; unimolecular reactions; chemical activation; non-equilibrium systems; energy transfer and relaxation. *Mailing Add:* Dept Chem BG-10 Univ Wash Seattle WA 98195

RABINOVITCH, MARLENE, b Montreal, Can, July 14, 46. PEDIATRICS. *Educ:* McGill Univ, BS, 67, MD, 71. *Prof Exp:* Instr pediat, Children's Hosp Med Ctr, Harvard Med Sch, Boston, 77-78, asst prof, 79-82; asst prof pediat & path, 82-88, GRAD FAC DEPT PATH, UNIV TORONTO, 82-, PROF PEDIAT & PATH, 88- *Concurrent Pos:* Asst, Dept Cardiol, Children's Hosp Med Ctr, Boston, 77-78, assoc, 78-82, sr assoc pediat & path, Hosp Sick Children, Toronto, 82-, actg dir cardiovasc res, Res Inst, 86-88, dir, 88-; reviewer, numerous journals; mem, Study Sect Young Investr, NIH, 81-, Pathobiochem, 90-93, Parent Comt Appln 90-91. *Mem:* Am Col Cardiol; Soc Pediat Res; Am Asn Path; Am Thoracic Soc; Am Soc Clin Invest. *Res:* Pulmonary circulation in congenital heart disease; pulmonary hypertension; vascular cellular and molecular biology; numerous publications. *Mailing Add:* Dept Pediat & Path Univ Toronto Hosp Sick Children 555 University Ave Toronto ON M5G 1X8 Can

RABINOVITCH, MICHEL PINKUS, b Sao Paulo, Brazil, Mar 22, 26; m 67; c 2. CELL BIOLOGY, EXPERIMENTAL MEDICINE. *Educ:* Univ Sao Paulo, MD, 49, Livre Docente, 53. *Prof Exp:* Res assoc cellular immunol, Rockefeller Univ, 64-65, asst prof, 65-69; assoc prof, 69-73, PROF CELL BIOL, SCH MED, NY UNIV, 73- *Concurrent Pos:* Rockefeller Found fel, Univ Chicago, Marine Biol Lab & Univ Calif, Berkeley, 53-54. *Mem:* Am Asn Immunol; Am Soc Cell Biol; Harvey Soc. *Res:* Nucleic acids content of tissues; RNA synthesis in amoeba; control of serum ribonuclease by the kidneys; phagocytic recognition by macrophages, tissue culture cells, insect hemocytes, Acanthamoeba; cell adhesion and spreading. *Mailing Add:* Dept Cell Biol NY Univ Sch Med 550 First Ave New York NY 10016

RABINOVITCH, PETER S, MITOTIC CELL CYCLE, CELL ACTIVATION. *Educ:* Univ Wash, MD, 79, PhD (genetics), 80. *Prof Exp:* ASST PROF PATH, UNIV WASH, 80- *Mailing Add:* Dept Path Univ Wash SM Seattle WA 98195

RABINOVITZ, MARCO, b Braila, Rumania, Dec 12, 23; US citizen; m 57; c 2. BIOCHEMISTRY. *Educ:* Univ Pa, BS, 44; Univ Minn, PhD(biochem), 50. *Prof Exp:* Am Cancer Soc fel, Comt on Growth, Nat Res Coun, Univ Calif, 50-52, res biochemist, Dept Physiol Chem, 52-58; BIOCHEMIST, NAT CANCER INST, 58- *Mem:* AAAS; Am Chem Soc; Am Soc Biol Chem; Soc Exp Biol & Med; Am Asn Cancer Res; Am Soc Cell Biol. *Res:* Protein biosynthesis; antimetabolites; biosynthetic control mechanisms; biochemical basis for cancer chemotherapy. *Mailing Add:* Nat Cancer Inst NIH Bldg 37 Rm 5C-25 Bethesda MD 20892

RABINOW, JACOB, b Karkov, Russia, Jan 8, 10; nat US; m 43; c 2. ENGINEERING. *Educ:* City Col New York, BS, 33, EE, 34. *Hon Degrees:* LHD, Towson State Univ, 83. *Prof Exp:* Radio technician, Sheffield Radio Co, NY, 34; radio engr, Sterling Radio Co, 34-35 & Halson Radio Mfg Co, 35-37; jr elec engr, Fed Power Comn, 37; eng draftsman, Gibbs & Hill Eng Co, 37-38; mech engr, Nat Bur Standards, 38-53; consult, Diamond Ord Fuze Labs, 53-54; pres, Rabinow Electronics Inc, Control Data Corp, 54-68, vpres, 68-72; chief res engr, Nat Eng Lab, Nat Bur Standards, 72-75, CONSULT, NAT INST STANDARDS & TECHNOL, 75- *Honors & Awards:* Pres Cert Merit, 48; Edward Longstreth Medal, Franklin Inst, 69; Harry Diamond Award, Inst Elec & Electronics Engrs, 77. *Mem:* Nat Acad Eng; Sigma Xi; fel Inst Elec & Electronics Engrs; fel AAAS. *Res:* Design of electronic equipment; design and development of ordnance devices such as guided missiles and fuzes; patents on special cameras, watch regulators, headlight dimmers; inventor of magnetic fluid clutch; optical character recognition machines; post office machinery. *Mailing Add:* Nat Inst Standards & Technol Bldg 411 Gaithersburg MD 20899

RABINOWICZ, ERNEST, b Berlin, Ger, Apr 22, 26; US citizen; m 53; c 3. MECHANICAL ENGINEERING. *Educ:* Cambridge Univ, BA, 47, PhD(phys chem), 50. *Prof Exp:* Mem staff, Div Indust Coop, 50-54, from asst prof to assoc prof, 54-67, PROF MECH ENG, MASS INST TECHNOL, 67- *Concurrent Pos:* Consult, IBM Corp, 61-75 & Asn Am Railroads, 75-83; vis prof, Israel Inst Technol, 69. *Honors & Awards:* Hodson Award, Am Soc Lubrication Engrs, 57; Ragnar Holm Award, Inst Elec & Electronics Engrs-Holm Conference on Elect Contacts, 83; Mayo D Hersey Award, Am Soc Mech Engr, 85; Nat Award, Soc Tribologists & Lubrication Engrs, 88. *Mem:* Am Phys Soc; fel Soc Tribologists & Lubrication Engrs; fel Am Soc Mech Engrs. *Res:* Tribology; surface properties of solids; experimentation and measurement techniques; accelerated testing. *Mailing Add:* Rm 35-010 Mass Inst Technol Cambridge MA 02139

RABINOWITCH, VICTOR, b London, Eng, 1934; US citizen; m 58; c 3. SCIENCE ADMINISTRATION. *Educ:* Univ Ill, BS, 65: Univ Wis, MS, 61, PhD(zool & int rels), 65. *Prof Exp:* Staff dir, Org Develop Bd, Nat Acad Sci, 65-68, Bd Sci & Technol Int Develop, Nat Res Coun, 70-82; dir, Ctr Study Sci & Soc, assoc prof sci & pub admin, State Univ, NY 69-70; exec dir, Off Int Affairs, Nat Res Coun, 81-; Dir, Comt Int Security & Arms Control, Nat Acad Sci, 85-; MEM STAFF, JOHN D & CATHERINE T MACARTHUR FOUND. *Concurrent Pos:* Ed, sci & technol, World Develop, 72-; dir, Comt Int Security & Arms Control, Nat Acad Sci, 85- *Mem:* Fel AAAS; Asn Advan Agr Sci Africa; Fedn Am Scientists; Int Ctr Insect Physiol & Ecol; Sigma Xi. *Res:* Numerous publications. *Mailing Add:* John D & Catherine T MacArthur Found 140 S Dearborn St Chicago IL 60603

RABINOWITZ, ISRAEL NATHAN, b New York, NY, Jan 24, 35; m 59; c 2. AGRICULTURAL & FOOD CHEMISTRY, CHEMICAL ENGINEERING. *Educ:* City Col New York, BS, 56; Univ Wash, MS, 62; Rutgers Univ, PhD(biochem), 65. *Prof Exp:* Damon Runyon fel, King's Col, London, 65-66; USPHS fel, 67-69, res assoc biophys, 69-70, physiol, 71-72, RES ASSOC PHYSIOL, SCH MED, STANFORD UNIV, 72-; CONSULT, VET ADMIN HOSP, PALO ALTO, 74- *Concurrent Pos:* Lectr, Col Notre Dame, Calif, 70-; vpres res & develop, M & T Labs, 85; pres ITD Corp, 87; lectur, biochem eng, Univ Calif, Santa Barbara, 81. *Mem:* AAAS; Am Crystallog Asn; Sigma Xi; Am Chem Soc; Inst Food Technol. *Res:* Biological structure and function; metal complexes; membrane biophysics; agricultural biotechnology. *Mailing Add:* 2534 Foothill Rd Santa Barbara CA 93105

RABINOWITZ, JACK GRANT, b New York, NY, July 9, 27; m 51, 72; c 5. RADIOLOGY. *Educ:* Univ Calif, BA, 49; Univ Berne, MD, 55. *Prof Exp:* Instr radiol, State Univ NY Downstate Med Ctr, 60-61; asst radiologist, Mt Sinai Hosp, 62-65; assoc prof radiol, Mt Sinai Sch Med, 65-67; radiologist-in-chief, Brooklyn-Cumberland Med Ctr, 67-70; prof radiol, State Univ NY Downstate Med Ctr, 70-74; prof diag radiol & chmn dept, Univ Tenn, Memphis, 74-78; PROF RADIOL & CHMN DEPT, MT SINAI SCH MED, 78- *Concurrent Pos:* Dir dept radiol, Kings County Hosp Ctr, 70-74; consult, US Vet Admin, Bronx. *Mem:* Fel Am Col Radiol; Radiol Soc NAm; AMA; Asn Univ Radiol. *Mailing Add:* Dept Radiol Mt Sinai Sch Med New York NY 10029

RABINOWITZ, JAMES ROBERT, b New York, NY, Apr 7, 42. MOLECULAR BIOPHYSICS. *Educ:* Alfred Univ, BA, 62; State Univ NY, Buffalo, PhD(physics), 72. *Prof Exp:* Res assoc, Ctr Theoret Biol, State Univ NY, Buffalo, 70-72, fel, 72-73; assoc res scientist, Inst Environ Med, New York Univ Med Ctr, 73-77; res scientist, NY Inst Technol, 77-80; RES PHYSICIST, HEALTH EFFECTS RES LAB, US ENVIRON PROTECTION AGENCY, 80- *Concurrent Pos:* NATO fel, Uppsala Univ, 69; guest scientist, Northeast Radiol Health Lab, Bur Pub Health, Food & Drug Admin, Dept Health, Educ & Welfare, 73. *Mem:* AAAS; Radiation Res Soc; Bioelectromagnetics Soc; Sigma Xi; Int Soc Quantum Biol. *Res:* The use of molecular modeling methods to predict chemical toxicity; structure activity in environmental health. *Mailing Add:* Genetic Toxicol Div MD-68 Health Effects Res Lab US Environ Protection Agency Triangle Park NC 27711

RABINOWITZ, JESSE CHARLES, b New York, NY, Apr 28, 25. BIOCHEMISTRY, ONE-CARBON METABOLISM. *Educ:* Polytech Inst Brooklyn, BS, 45; Univ Wis, MS, 47, PhD(biochem), 49. *Prof Exp:* Nat Heart Inst trainee, Univ Wis, 50-51; USPHS fel, Univ Calif, Berkeley, 51-53; chemist, Nat Inst Arthritis & Metab Dis, Md, 53-57; assoc prof, 57-62, chmn dept, 78-83, PROF BIOCHEM, UNIV CALIF, BERKELEY, 62- *Mem:* Nat Acad Sci; Am Soc Biol Chemists; Am Soc Microbiol. *Res:* Enzymology; purine fermentation; folic acid coenzymes; iron-sulfur proteins; protein biosynthesis. *Mailing Add:* Barker Hall Univ Calif Berkeley CA 94720

RABINOWITZ, JOSEPH LOSHAK, b Odessa, Ukraine, Nov 4, 23; US citizen; m 46; c 3. BIOCHEMISTRY. *Educ:* Philadelphia Col Pharm, BS, 44; Univ Pa, MSc, 48, PhD(org chem), 50. *Hon Degrees:* DSc, Univ Bordeaux, France, 79. *Prof Exp:* From res assoc to assoc prof, 53-70, PROF BIOCHEM, SCH DENT MED, UNIV PA, 70-; CHIEF RADIOISOTOPE RES, VET ADMIN HOSP, PHILADELPHIA, 58- *Concurrent Pos:* Fulbright fel, Carlsberg Lab, Copenhagen, Denmark, 58; fel, Eng, 68; prin scientist, Radioisotope Serv, Vet Admin Hosp, Philadelphia, 53-58; ed, Topics Med Chem. *Mem:* Am Chem Soc; Am Soc Biol Chem; Soc Nuclear Med; Soc Exp Biol & Med; Am Nuclear Soc. *Res:* Biochemistry of lipids, thyroid function, obesity and alchoholism; isotope methodology; transdermal delivery of medicinals; development and testing of hypocholesterolemic compounds. *Mailing Add:* 127 Juniper Rd Havertown PA 19083-5409

RABINOWITZ, LAWRENCE, b San Francisco, Calif, Apr 9, 33; m 59; c 3. PHYSIOLOGY. *Educ:* Univ Calif, Berkeley, AB, 56; Univ Calif, San Francisco, PhD(renal physiol), 61. *Prof Exp:* NIH res fel, 61-63; asst prof physiol, Univ NC, Chapel Hill, 63-68; from asst prof to assoc prof, 68-76, PROF HUMAN PHYSIOL, SCH MED, UNIV CALIF, DAVIS, 76- *Mem:* AAAS; Am Physiol Soc; Am Soc Nephrology. *Res:* Renal physiology, particularly the excretion of urea and related organic nonelectrolytes. *Mailing Add:* Dept Human Physiol Univ Calif Sch Med Davis CA 95616

RABINOWITZ, MARIO, b Mexico City, Mex, Oct 24, 36; US citizen. LOW TEMPERATURE PHYSICS, ELECTRICAL ENGINEERING. *Educ:* Univ Wash, BS, 59, MS, 60; Wash State Univ, PhD(physics), 64. *Prof Exp:* Electronics engr, Collins Radio Co, 57; res engr, Nuclear Physics Dept, Boeing Co, 58-61; res asst physics, Wash State Univ, 61-63; sr physicist, Plasma Prog, Westinghouse Res Labs, 63-66; mgr gas discharges & vacuum pump physics, Varian Assocs, 66-67; res physicist, Stanford Linear Accelerator Ctr, 67-74; mgr superconductivity & cryogenics, 74-80, SR SCIENTIST, ELEC POWER RES INST, 80- *Concurrent Pos:* George F Baker scholar, 56-57; assoc prof, San Jose State Univ, 73-76; adj prof, Boston Univ & Case Western Reserve Univ, 75-77, Ga Inst Technol, 88-, Univ Houston, 90- & Va Commonwealth Univ, 90- *Mem:* Am Phys Soc; Am Vacuum Soc; NY Acad Sci; Am Asn Physics Teachers; Am Nuclear Soc; Sigma Xi. *Res:* Radiation effects; electron and photodesorption of gases; electron emission; gas and metal-vapor arcs and plasmas; electrical discharges in vacuum; electrical explosion of metals; ultrahigh vacuum; superconducting generation, transmission; electric power; cryoresistive transmission; superconducting trapped magnetic fields; nuclear electromagnetic pulse; amorphous metals; cluster-impact fusion; physical electronics; classical tunneling. *Mailing Add:* 715 Lakemead Way Redwood City CA 94062

RABINOWITZ, PAUL H, b Newark, NJ, Nov 15, 39. NON-LINEAR FUNCTIONAL ANALYSIS. *Educ:* NY Univ, BA, 61, PhD(math), 66. *Prof Exp:* PROF MATH, UNIV WIS, 71- *Concurrent Pos:* Guggenheim fel, 78-79. *Mem:* Am Math Soc; Soc Indust & Appl Math. *Res:* Non-linear ordinary and partial differential equations; non-linear functional analysis. *Mailing Add:* Univ Wis 480 Lincoln Dr Madison WI 53706

RABINOWITZ, PHILIP, b Philadelphia, Pa, Aug 14, 26; m 51; c 5. NUMERICAL INTEGRATION. *Educ:* Univ Pa, AB, 46, AM, 48, PhD(math), 51. *Prof Exp:* Mathematician, Nat Bur Standards, 51-55 & 59-60; sr scientist, Weizmann Inst, 55-59, 60-64 & 65-67; vis assoc prof appl math, Brown Univ, 64-65; assoc prof, 67-86, PROF APPL MATH, WEIZMANN INST SCI, 86- *Concurrent Pos:* Vis prof, Hebrew Univ, Jerusalem, 67-68, Latrobe Univ, Australia, 77, Univ Witwatersrand, SAfrica, 78, 90 & Univ New South Wales, Australia, 80, 86, 89; prof, Bar-Illan Univ, Israel, 70-72; assoc ed, Math Comput, 70-87. *Honors & Awards:* Info Processing Asn Israel Prize, 68. *Mem:* Fel Japanese Soc Prom Sci. *Res:* Application of electronic digital computers to numerical analysis; numerical integration, theory and practice. *Mailing Add:* Dept Appl Math Weizmann Inst Sci Rehovot 76100 Israel

RABINOWITZ, RONALD, b Pittsburgh, Pa, Feb 24, 43; m 67; c 3. PEDIATRIC SURGERY, UROLOGY. *Educ:* Univ Pittsburgh, BS, 64, MD, 68. *Prof Exp:* Intern surg, hosps of Univ Health Ctr Pittsburgh, 68-69, resident, 69-70, resident urol surg, 72-75; resident & clin fel pediat urol surg, Hosp Sick Children, Toronto, 75-76; from asst prof to assoc prof, 76-87, PROF UROL SURG & PEDIAT, SCH MED, UNIV ROCHESTER, 87-; CHIEF UROL, ROCHESTER GEN HOSP, 76- *Concurrent Pos:* Attend pediat urologist, Rochester Gen Hosp, 76-; attend pediat urologist, Birth Defects Ctr, Univ Rochester & Strong Mem Hosps, 76-; proj surgeon, Artificial Urinary Sphincter Proj, NASA, 78- *Mem:* Am Urol Asn; Am Acad Pediat; Am Col Surgeons; Soc Pediat Urol; Soc Univ Urologists; Can Urol Asn; Am Soc Laser Surg & Med. *Res:* Development of an artificial urinary sphincter; laser welding in urinary tract reconstruction; experimental testis torsion. *Mailing Add:* Dept Urol Box 656 601 Elmwood Ave Rochester NY 14642

RABINS, MICHAEL J, b New York, NY, Feb 24, 32; m 56; c 3. MECHANICAL ENGINEERING, CONTROL SYSTEMS. *Educ:* Mass Inst Technol, BS, 53; Carnegie Inst Technol, MS, 54; Univ Wis, PhD(mech eng), 59. *Prof Exp:* Design engr, M W Kellogg Co, 53-54; hydraul engr, Repub Aviation Corp, 55; design engr, Atlantic Design Co, 55-56; asst prof mech eng, Univ Wis, 59-60; from asst prof to assoc prof, NY Univ, 60-70; prof syst eng & prog dir, Polytech Inst Brooklyn, 70-75; vis prof, Polytech Inst Grenoble, France, 75; dir, Off Univ Res, Off Secy Transp, Washington, DC, 75-77; chmn dept mech eng, 77-85, assoc dean eng res & grad progs, Wayne State Univ, 85-87; head mech eng dept, 87-89, Halliburton Prof, 87-88, TEX ENG EXP STA RES PROF, TEX A&M UNIV, 89- *Concurrent Pos:* NSF sci fac fel, Univ Calif, Berkeley, 67-68. *Mem:* Fel Am Soc Mech Engrs; sr mem Inst Elec & Electronics Engrs. *Res:* Nonlinear automatic controls; system engineering and design. *Mailing Add:* Mech Eng Dept Tex A&M Univ College Sta TX 77843-3123

RABITZ, HERSCHEL ALBERT, b Los Angeles, Calif, Apr 10, 44; m 70. PHYSICAL CHEMISTRY. *Educ:* Univ Calif, Berkeley, BS, 66; Harvard Univ, PhD(chem physics), 70. *Prof Exp:* from asst to assoc prof, 71-79, PROF CHEM, PRINCETON UNIV, 80- *Concurrent Pos:* Dreyfus Found teacher/scholar, 74; Sloan Found fel, 75. *Mem:* AAAS; Sigma Xi. *Res:* Theoretical chemistry; molecular collisions; time-dependent processes; chemical kinetics. *Mailing Add:* Frick Chem Lab Princeton Univ Princeton NJ 08540

RABJOHN, NORMAN, b Rochester, NY, May 1, 15; m 43; c 1. ORGANIC CHEMISTRY. *Educ:* Univ Rochester, BS, 37; Univ Ill, MS, 39, PhD(org chem), 42. *Prof Exp:* Chemist, Eastman Kodak Co, NY, 37-38; instr org chem, Univ Ill, 42-44; chemist, Goodyear Tire & Rubber Co, 44-48; from assoc prof to prof, 48-83, chmn dept chem, 58-61 & 66-69, EMER PROF CHEM, UNIV MO, COLUMBIA, 83- *Mem:* Fel AAAS; Am Chem Soc. *Res:* Synthesis; pharmaceuticals; hydrocarbons. *Mailing Add:* Dept Chem Univ Mo Columbia MO 65211

RABKIN, MITCHELL T, b Boston, Mass, Nov 27, 30; m 56; c 2. MEDICINE. *Educ:* Harvard Col, AB, 51; Harvard Med Sch, MD, 55; Am Bd Internal Med, 63. *Hon Degrees:* DSc, Brandeis Univ & Mass Col Pharm & Allied Health Sci, 83. *Prof Exp:* From intern to chief resident internal med, Mass Gen Hosp, 55-62; from instr to assoc prof, 62-82, PROF MED, HARVARD MED SCH, 83-; PRES & PHYSICIAN, BETH ISRAEL HOSP, BOSTON, 66- *Concurrent Pos:* Clin assoc, Nat Inst Arthritis & Metab Dis, NIH, 57-59; clin fel med, endocrine Unit, Mass Gen Hosp, 60-62, asst in med, 63-71, actg chief, Endocrine Clin, 64-66; dir, Mass Hosp Asn, 70-73, Med Found, 70-76 & Blue Cross Mass, Inc, 70-80; mem, Conf Boston Teaching Hosp, 74-, chmn, 75-76 & 80-81; secy, Coun Teaching Hosp, Asn Am Med Col, 79-80, chmn, 82; mem, Metrop Boston Hosp Coun, 76-84; mem, Study Group Grad Med Educ Joshiah Macy, Jr Found, 77-80; vis lectr health serv, Harvard Univ Sch Pub Health, 76-; med div chmn, United Way Mass Bay, 82; vpres, Commonwealth Health Care Corp, Inc, 81-82 & pres, 82-83; Task Force Organ Transplantation, Commonwealth Mass, 83-84; mem, med adv bd, Hadassah Med Orgn, 81-; mem adv comt, Clin Nurse Scholars Prog & Prog Prepaid Managed Health, Robert Wood Johnson Found, 82- & Off Technol Assessment, US Cong, 79-83; co rep Harvard Med Ctr, Asn Acad Health Ctr, 85-; affil, Bd Health Sci Policy, Inst Med, 86-; mem, health adv comt, US Gen Acctg Off, 87-; vis comt, sch mgt, Suffolk Univ, Boston, 87-; bd dirs, Partnership Inc, co-chmn, 85-88, Dead River Group, Portland, 84-86 & 88- & UST Corp, Boston, 88-; mem, Coun Res & Develop, Am Hosp Asn, 86- *Mem:* Inst Med-Nat Acad Sci; fel AAAS; fel Am Col Physician Execs; Soc Med Adminr; Am Clin & Climatol Asn; Am Fedn Clin Res; Am Med Asn; Boylston Med Soc. *Res:* Author of over 25 publications in medicine. *Mailing Add:* Beth Israel Hosp 330 Brookline Ave Boston MA 02215

RABL, ARI, b Ger, Feb 21, 42. ENERGY CONVERSION. *Educ:* Beloit Col, BSc, 63; Univ Calif, Berkeley, MA, 66, PhD(physics), 69. *Prof Exp:* Res assoc physics, Int Ctr Theoret Physics, Trieste, Italy, 69; res assoc, Weizmann Inst, Israel, 70-71; res assoc, Ohio State Univ, 72-73; asst physicist, Argonne Nat Lab, 74-77, engr, 77-78; prin scientist, Solar Energy Res Inst, Golden, Colo, 78-80; LECTR & RES SCIENTIST PRINCETON UNIV, 82- *Concurrent Pos:* Sr res assoc, Univ Chicago, 76-81; vis sr scientist, Princeton Univ, 80- *Mem:* Am Soc Mech Engrs; Am Phys Soc; Solar Energy Soc; Am Soc Heating, Refridgerating & Air-Conditioning Engrs. *Res:* Solar energy conversion; environmental problems; high energy physics. *Mailing Add:* Centre d' Energetique Ecole des Mines 60 Bd St Michel Paris 75272 CEDEX06 France

RABL, VERONIKA ARIANA, b Michalovce, Czech, Dec 16, 45. LOAD MANAGEMENT, ENERGY UTILIZATION. *Educ:* Weizmann Inst Sci, Israel, MSc, 71; Ohio State Univ, PhD(physics), 74. *Prof Exp:* Jr scientist physics, Weizmann Inst Sci, Israel, 71; res assoc, Syracuse Univ, 74-75; res assoc, Argonne Nat Lab, 75-77, asst scientist energy res, 77-81; subprog mgr, 81-88, prog mgr, Demand-Side Planning, 89-90, SR PROG MGR, DEMAND-SIDE MGT, ELEC POWER RES INST, 91- *Mem:* Am Soc Heating, Refrig & Air Conditioning Engrs; Inst Elec & Electronics Engrs. *Res:* Demand-side management; utility planning; optimization of energy supply systems; demand forecasting and analysis; energy storage systems; load management. *Mailing Add:* Electric Power Res Inst 3412 Hillview Ave Palo Alto CA 94304

RABOLD, GARY PAUL, b Providence, RI, July 10, 39; m 66; c 3. INDUSTRIAL CHEMISTRY. *Educ:* Harvard Univ, AB, 60; Northeastern Univ, PhD(org chem), 65. *Prof Exp:* Fel biophys, Univ Hawaii, 65-67; INDUST CHEMIST PROD DEVELOP, DOW CHEM CO, 67- *Mem:* Am Chem Soc. *Res:* New product development for organic chemicals; solvents and hydraulic fluids. *Mailing Add:* New Ventures Commercialization Dow Chem Co 2020 Dow Ctr Midland MI 48674

RABOLT, JOHN FRANCIS, b New York, NY, May 14, 49; m 90; c 1. POLYMER PHYSICS, CHEMICAL PHYSICS. *Educ:* State Univ NY Col, Oneonta, BS, 70; Southern Ill Univ, Carbondale, PhD(physics), 74. *Prof Exp:* Postdoctoral fel physics, Univ Mich, 74-75; Nat Res Coun, Nat Acad Sci res assoc, Nat Bur Standards, 76-77; RES STAFF MEM POLYMERS, IBM RES LAB, 78- *Honors & Awards:* Coblentz Award, 85; Williams-Wright Award, 90. *Mem:* Fel Am Phys Soc; Soc Appl Spectros; Am Chem Soc. *Res:* Use of Fourier transform (FT) infrared and FT and Conventional Raman spectroscopy to investigate crystal and molecular structure of long chain molecules and polymers; integrated optical techniques in conjunction with Raman Spectroscopy to investigate submicron polymer films and polymer surfaces; FTIR studies of self assembled and Langmoir-Blodgett films on metals & dielectrics; FT-Raman spectroscopy co-developer. *Mailing Add:* IBM Almaden Res Lab K93/801 650 Harry Rd San Jose CA 95120

RABOVSKY, JEAN, b Baltimore, Md. BIOCHEMISTRY. *Educ:* Univ Md, BS, 59; Brandeis Univ, PhD(biochem), 64. *Prof Exp:* Chemist, Nat Inst Occupational Safety & Health-Alosh, Morgantown, WV, 78-89; TOXICOLOGIST, CALIF, DEPT HEALTH SERV, SACRAMENTO,

CALIF, 89- *Mem:* Am Chem Soc; AAAS; NY Acad Sci; Sigma Xi. *Res:* Detoxication/bioactivation mechanisms; pulmonary cytochrome P450; role of cell types and cellular membranes; effect of particulates and organic compounds; relationship between detoxication mechanisms; environmental/ occupational health issues related to chemical exposure. *Mailing Add:* Calif State Dept Health Servs 714 P St, Rm 499 Sacramento CA 95814-2888

RABOY, SOL, b Ambridge, Pa, Feb 11, 20; m 48; c 7. NUCLEAR PHYSICS. *Educ:* Brooklyn Col, BA, 41; Carnegie Inst Technol, DSc(physics), 50. *Prof Exp:* Instr physics, Carnegie Inst Technol, 49-50, res assoc, 50-51; assoc physicist, Argonne Nat Lab, 51-65; chmn dept, 66-77, PROF PHYSICS, STATE UNIV NY BINGHAMTON, 65- *Mem:* Fel Am Phys Soc. *Res:* Mobility of electrons in insulators; angular distributions of nuclear radiation; spectroscopy of gamma rays; variation of electron mass with velocity; measurements of magnetic moments of excited states of nuclei; measurement of quadrupole moments of nuclei; nuclear structure; muonic x-rays; optics, quantum electronics and photonics. *Mailing Add:* Dept Physics State Univ NY Binghamton NY 13901

RABSON, ALAN S, b New York, NY, July 1, 26; m 50; c 1. PATHOLOGY. *Educ:* Univ Rochester, BA, 48; Long Island Col Med, MD, 50. *Prof Exp:* PATHOLOGIST, NAT CANCER INST, 55- *Mem:* Am Asn Path. *Res:* Oncogenic viruses and viral tumors. *Mailing Add:* Nat Cancer Inst 9000 Rockville Pike Bethesda MD 20892

RABSON, GUSTAVE, b New York, NY, Sept 28, 20; m 58; c 3. MATHEMATICS. *Educ:* Cornell Univ, AB, 41; Univ Mich, MA, 48, PhD(math), 52. *Prof Exp:* Engr, Tank-Automotive Ctr, Univ Mich, 42-44, mathematician, Ballistics Res Lab, 44-45; instr math, Purdue Univ, 49-53; from asst prof to assoc prof, Antioch Col, 53-57; sr mathematician, Am Optical Co, 57-59; res mathematician, Inst Sci & Technol, Univ Mich, 59-66; prof math, Tech Inst Aeronaut, Brazil, 66-67; EMER PROF, CLARKSON UNIV, 67- *Concurrent Pos:* NSF grant, 56; vis scientist, Mass Inst Technol, 78. *Mem:* Am Math Soc; Math Asn Am. *Res:* Topological groups; mathematical statistics; applied mathematics. *Mailing Add:* Clarkson Univ Potsdam NY 13676

RABSON, ROBERT, b Brooklyn, NY, Mar 4, 26; m 50; c 3. PLANT PHYSIOLOGY. *Educ:* Cornell Univ, BS, 51, PhD(plant physiol), 56. *Prof Exp:* Biologist, Oak Ridge Nat Lab, 56-58; from asst prof to assoc prof biol, Univ Houston, 58-64; biochemist, Div Biol & Med, US AEC, 63-67, asst chief biol br, 67-73, first officer, Food & Agr Orgn-Int Atomic Energy Agency Joint Div, Plant Breeding & Genetics Sect, Int Atomic Energy Agency, Vienna, Austria, 73-76; mem staff div biomed & environ res, Energy Res & Develop Admin, 76-78, DIR, DIV ENERGY BIOSCIENCES, OFF BASIC ENERGY SCI, US DEPT ENERGY, 78- *Honors & Awards:* Adoph Gude Award, Am Soc Plant Physiol, 86. *Mem:* Am Soc Plant Physiol; Crop Sci Soc; fel AAAS; Am Soc Photobiol; Am Soc Microbiol. *Res:* Genetics and biochemistry of protein synthesis in developing seeds. *Mailing Add:* Off Basic Energy Sci ER-17 US Dept Energy Washington DC 20545

RABSON, THOMAS A(VELYN), b Huston, Tex, July 31, 32; m 57; c 3. ELECTRICAL ENGINEERING, NUCLEAR PHYSICS. *Educ:* Rice Univ, BA, 54, BS, 55, MA, 57, PhD(nuclear physics), 59. *Prof Exp:* From asst prof to assoc prof, 59-70, chmn dept, 79-84, PROF ELEC ENG, RICE UNIV, 70- *Concurrent Pos:* NSF sci fac fel, 65-66. *Mem:* Am Phys Soc; Inst Elec & Electronics Engrs; Optical Soc Am; Soc Photo-Optical Instrumentation Engrs. *Res:* Semiconductor physics; lasers; ferroelectrics; nonvolatile memory arrays. *Mailing Add:* Dept Elec & Comp Eng Rice Univ PO Box 1892 Houston TX 77251

RABUCK, DAVID GLENN, b Geneva, Ill; m 64; c 1. HISTOLOGY-HISTOCHEMISTRY, BIOLOGY OF REPRODUCTION. *Educ:* NCent Col, BA, 63; Loyola Univ, MS, 66, PhD(anat), 76. *Prof Exp:* Instr anat, Pritzker Sch Med, Univ Chicago, 76; asst prof anat, Col Dent, Marquette Univ, 77-78 & Chicago Col Osteop Med, 78-82; vis asst prof embryol & histol, Univ Calif, Riverside, 82-83; sci tutor, 85-87; ADJ FAC ANAT, PHYSIOL & BIOL, OAKTON COMMUNITY COL, 87- *Concurrent Pos:* Teaching assoc anat, Med Sch, Rush Univ, 76; assoc anat, Med Col, Univ Ill, Chicago, 77, instr oral anat, Dent Sch, 90-91; anat & physiol tutor, Oakton Community Col, 89. *Mem:* AAAS; Am Soc Zoologists; assoc mem Soc Study Reproduction; Nat Educ Asn; Nat Space Soc. *Res:* Origin, morphogenesis, structure and endocrine mechanisms of gonadol tissues, especially the changes in the right gonad of overiectomized birds. *Mailing Add:* 3352 Crain St Skokie IL 60076-2408

RABUNG, JOHN RUSSELL, b Elyria, Ohio, July 22, 43; m 67; c 3. NUMBER THEORY. *Educ:* Univ Akron, BA, 65; Wash State Univ, MA, 67, PhD(number theory), 69. *Prof Exp:* Res mathematician number theory, Math Res Ctr, US Naval Res Lab, 69-70; instr & opers res analyst statist & oper res, US Army Logistics Mgt Ctr, 70-72; asst prof math, Randolph-Macon Col, 72-74; from asst prof to assoc prof math sci, Va Commonwealth Univ, 74-84; PROF COMPUT SCI, RANDOLPH MACON COL, ASHLAND, VA, 84- *Mem:* Am Math Soc; Math Asn Am. *Res:* Combinatorial problems in number theory, specifically, some aspects of Van der Waerden's theorem on arithmetic progressions. *Mailing Add:* Dept Computer Sci Randolph Macon Col Ashland VA 23005

RABUSSAY, DIETMAR PAUL, b Wolfsberg, Austria, Aug 9, 41; m 66; c 2. GENE REGULATION. *Educ:* Tech Univ Graz, Austria, MSc, 67; Univ Munich, PhD(biochem), 71. *Prof Exp:* Wissensch asst molecular biol, Max Planck Inst Biochem, 71-72; fel, Univ Calif, San Diego, 72-75, asst res biologist, 75-79; vis scientist, Max Planck Inst Biochem, 78-79; asst prof microbiol, Fla State Univ, 79-81; sect head, 81-82, res dir, 82-83, VPRES, BETHESDA RES LABS, LIFE TECHNOLOGIES INC, 83- *Concurrent Pos:* Adj prof, Univ Md, 81-86. *Mem:* AAAS; Am Soc Microbiol; Europ Molecular Biol Orgn; Am Soc Biol Chemists. *Res:* In vitro protein synthesis; mechanism and regulation of transcription; development of bacterial viruses; DNA enzymology; genetic engineering. *Mailing Add:* Molecular Biol Res & Develop Bethesda Res Lab Life Technol Inc 8717 Grovemont Circle Gaithersburg MD 20877

RABY, BRUCE ALAN, b Seattle, Wash, Aug 22, 30; m 54; c 4. TECHNICAL MANAGEMENT. *Educ:* Univ Wash, BS, 52; Univ Calif, Berkeley, MS, 54; Iowa State Univ, PhD(anal chem), 63. *Prof Exp:* Proj engr, Wright Air Develop Ctr, Wright-Patterson AFB, Ohio, 53-56; res asst anal chem, Inst Atomic Res, Ames Nat Lab, Iowa State Univ, 56-63; scientist, Rocketdyne Div, NAm Aviation, Inc, Calif, 63-64; chemist, Lawrence Radiation Lab, Univ Calif, Berkeley, 64-70; mem staff, Uthe Technol Int, 70-75, admin mgr res & applications, 75-80; supvr, Anal & Vacuum, 83-85, group leader, 86-87, sr engr, 87-88; CONSULT CHEM & TECHNOL, 80-83 & 88- *Mem:* Soc Appl Spectros; Am Chem Soc; Am Soc Mass Spectros; Sigma Xi; Am Vacuum Soc; Electrochem Soc. *Res:* Fluorescence; halogen fluorides as reagents; rocket fuels; magnesium alloys; nuclear reactors; production and separation of trans-lead isotopes; mass spectrometry; analytical chemistry; chemical vapor deposition; instrument design and fabrication; analytical software. *Mailing Add:* 1547 Arata Ct San Jose CA 95125

RABY, STUART, b Bronx, NY, May 18, 47; m 73; c 2. SUPERSYMMETRY, PHYSICS BEYOND THE STANDARD MODEL. *Educ:* Univ Rochester, BS, 69; Tel Aviv Univ, MS, 73, PhD(physics), 76. *Prof Exp:* Res assoc, Cornell Univ, 76-78; actg asst prof physics, Stanford Univ, 78-80, res assoc, Stanford Linear Accelerator Ctr, 80-81; staff mem, Los Alamos Nat Lab, 81-85, group leader, 85-89; PROF PHYSICS, OHIO STATE UNIV, 89- *Concurrent Pos:* Vis scientist, Univ Mich, 82-83. *Mem:* Fel Am Phys Soc. *Res:* High energy physics and particle; astrophysics; construct particle physics models to explain experimental data. *Mailing Add:* Dept Physics 4024 Smith Lab Ohio State Univ 174 W 18th Ave Columbus OH 43210

RACCAH, PAUL M(ORDECAI), b Tunis, Tunisia, June 24, 33; US citizen; m 54; c 4. MATERIAL SCIENCE, QUANTUM ELECTRONICS. *Educ:* Univ Paris, Exam physics, 56; Univ Lyons, Eng, 59; Univ Rennes, DrIng(phys chem), 62. *Prof Exp:* Group leader optimization, Compagnie d' Automatismes et d'Electronique, Plan Calcul, 62-64; staff mem, Lincoln Lab, Mass Inst Technol, 64-71; prof physics, Belfer Grad Sch, Yeshiva Univ, 71-76; PROF PHYSICS & HEAD DEPT, UNIV ILL, CHICAGO CIRCLE, 76- *Concurrent Pos:* Dir, Maybaum Inst Mat Sci & Quantum Electronics, 71-76. *Mem:* Fel Am Phys Soc. *Res:* Critical phenomena, Raman & Brillouin scattering; neutron diffraction; defects in crystalline solids; modulation spectroscopy; characterization of semiconducting alloys; laser mechanisms and tunability. *Mailing Add:* Dept Physics Univ Ill Chicago IL 60680

RACE, GEORGE JUSTICE, b Everman, Tex, Mar 2, 26; m 46; c 4. PATHOLOGY. *Educ:* Univ Tex, MD, 47; Univ NC, MSPH, 53; Baylor Univ, PhD(anat-microbiol), 69. *Prof Exp:* Instr & path, Duke Univ, 51-53 & instr, Harvard Med Sch, 53-54; from asst prof to prof, Cancer Ctr, 55-73, dir, 73-76; pathologist-in-chief & dir labs, Baylor Univ Med Ctr, 59-86, prof path & microbiol, Grad & Dent Sch, 62-68, prof & chmn dept path, 69-73; PROF ANAT, BAYLOR UNIV GRAD SCH, 71-, ADJ PROF BIOL, 81-; DEAN CONTINUING EDUC HEALTH SCI, BAYLOR UNIV MED CTR, 73-; PROF PATH & ASSOC DEAN CONTINUING EDUC, UNIV TEX SOUTHWESTERN MED SCH, 73- *Concurrent Pos:* Asst path, Peter Brent Brigham Hosp, Boston, 53-54; pathologist, St Anthony's Hosp, Fla, 54-55; consult path, Vet Admin Hosp, Dallas, 55-75; from asst to assoc pathologist, Children's Med Ctr, Dallas, Terrell's labs, & Parkland Mem Hosp, 55-59; lectr law, 71-75, Southern Methodist Univ, adj prof anthro & biol, 74-; vis pathologist, Guy's Hosp Med & Dent Sch, London, 72; chmn, bd dir, Baylor Res Found, 86-89. *Honors & Awards:* Caldwell Honor Award, Tex Soc Path, 73. *Mem:* Soc Med Educ Dirs Continuing Med Educ (secy-treas, 80-81, pres, 82-83); Sigma Xi; fel Col Am Pathologists; fel Am Soc Clin Pathologists; fel AAAS; fel Acad Clin Lab Physicians & Scientists. *Res:* Adrenal cortex functional zonation and hypertension; immunopathology of Trichinella spiralis and other parasites; anthropology. *Mailing Add:* 3429 Beverly Dr Dallas TX 75205

RACE, STUART RICE, b Glen Ridge, NJ, Sept 20, 26; m 57; c 4. ENTOMOLOGY. *Educ:* Gettysburg Col, AB, 51; Rutgers Univ, MS, 55, PhD(entom), 57. *Prof Exp:* Asst prof, NMex State Univ, 57-65; assoc prof, 65-70, EXTEN SPECIALIST ENTOM, RUTGERS UNIV, NEW BRUNSWICK, 70- *Mem:* Entom Soc Am. *Res:* Fruits, forage and field crops; livestock, poultry, and stored grain. *Mailing Add:* Cook Col Box 231 Rutgers Univ New Brunswick NJ 08903

RACETTE, GEORGE WILLIAM, b Schenectady, NY, June 2, 29; m 56; c 4. AEROSPACE MATERIALS, SEMICONDUCTORS. *Educ:* Siena Col, NY, BS, 51; Univ Rochester, MS, 54. *Prof Exp:* From jr to sr engr semiconductor devices, Philco Corp Res Div, 53-57, proj scientist infrared & photodevices, 57-64; proj scientist lasers, Philco-Ford Res Div, Ford Sci Lab, 64-66, eng specialist automotive electronics, Philco-Ford Res Div, 66-70; mgr, Whitemarsh Township, 70-74; PHYSICIST SEMICONDUCTOR DETECTORS, PHOTOVOLTAICS & AEROSPACE MAT, VALLEY FORGE SPACE CTR, GEN ELEC CO, 74- *Mem:* Am Phys Soc; Am Inst Aeronaut & Astronaut. *Res:* Semiconductor materials; infrared; photo detectors; lasers; high intensity light effects; vidicons; solar cells; vacuum deposition; aerospace materials; space environment. *Mailing Add:* Gen Elec Co Box 8555 Philadelphia PA 19101

RACEY, THOMAS JAMES, b Woodstock, Ont, Apr 27, 52. QUASI-ELASTIC LIGHT SCATTERING. *Educ:* Univ Waterloo, BSc, 75; Univ Guelph, MSc, 76, PhD(biophys), 82; Queen's Univ, BEd, 77. *Prof Exp:* Teacher physics & math, Frontenac County Bd Ed, 77-79; software specialist, Andyne Computing Ltd, 82-83; asst prof, 83-87, ASSOC PROF PHYSICS, ROYAL MIL COL CAN, 87- *Mem:* Can Asn Physicists; Am Asn Physics Teachers. *Res:* Optical signal processing; particle sizing from scattering information; analysis of Fourier transform information and time correlation techniques. *Mailing Add:* Royal Mil Col Can Kingston ON K7L 2W3 Can

RACH, RANDOLPH CARL, b Niles, Mich, Dec 9, 51; m 77; c 1. MICROWAVE TUBE ENGINEERING. *Educ:* Andrews Univ, BS(physics) & BS(math), 79. *Prof Exp:* Intel officer, Cent Intel Agency, 80-81; Mem tech staff, Electron Dynamics Div, Hughes Aircraft Co, 81-83; design & develop engr, Microwave & Power Tube Div, Raytheon Co, 83-84; sr engr, Teknadyn Div, Dynalectron Corp, 84-86; SR ENGR, MICROWAVE LABS INC, RALEIGH, NC, 86- *Concurrent Pos:* Consult, Ctr Appl Math, 81- *Mem:* Sigma Xi. *Res:* Microwave tube engineering; stochastic differential equations. *Mailing Add:* Dyna Lectron Corp PO Box I Alamogordo NM 88311

RACHELE, HENRY, b Helper, Utah, Aug 8, 29; m 59; c 5. ELECTRICAL ENGINEERING, MATHEMATICS. *Educ:* Utah State Univ, BS, 51; NMex State Univ, PhD(elec eng), 77. *Prof Exp:* Chief, Lower Atmospheric Res Tech Area, 64-69, actg chief, Atmospheric Sci Off, 69, chief, Lower Atmospheric Res Tech Area, 70-72, chief, Meteorol Satellite Tech Area, 72-73, chief, Meteorol Systs Tech Area, 73-74, dep/tech dir, 74-82, RES MATHEMATICIAN, 82-, US ARMY ATMOSPHERICS SCI LAB. *Concurrent Pos:* Lectr elec eng, NMex State Univ. *Honors & Awards:* Army Res & Develop Award, Dept Army, 68. *Mem:* Sigma Xi. *Res:* Meteorology; physics; unguided rocket ballistics; sound propagation through atmosphere; fog and cloud physics. *Mailing Add:* US Army Atmospheric Sci Lab White Sands Missile Range NM 88002

RACHFORD, HENRY HERBERT, JR, b El Dorado, Ark, June 14, 25; m 57; c 2. MATHEMATICS, ENGINEERING. *Educ:* Rice Inst, BS, 45, AM, 47; Mass Inst Technol, ScD, 50. *Prof Exp:* Res engr, Humble Oil & Ref Co, 49-56, asst div petrol engr, 56-57, res supvr, 57-64; PROF MATH & COMPUT SCI, RICE UNIV, 64- *Mem:* Am Math Soc; Soc Petrol Eng; Am Inst Mining, Metall & Petrol Eng; Am Inst Chem Eng. *Res:* Numerical techniques, especially for partial differential equations; use of digital computers; solution of engineering problems with mathematical methods. *Mailing Add:* 6150 Chevy Chase Dr Houston TX 77057

RACHFORD, THOMAS MILTON, b Bellevue, Ky, Mar 14, 42; m 64; c 2. CIVIL ENGINEERING. *Educ:* Univ Ky, BS, 64, MS, 66; Stanford Univ, PhD(civil eng), 72. *Prof Exp:* Asst prof civil eng, Pa State Univ, 69-73; PRIN ENGR, GANNETT FLEMING, CORDDRY & CARPENTER, 74- *Mem:* Water Pollution Control Fedn. *Res:* Civil engineering; hydrology and water resources; sanitary and environmental engineering. *Mailing Add:* 466 Woodcrest Ave Mechanicsburg PA 17050

RACHINSKY, MICHAEL RICHARD, b Stamford, Conn, Jan 2, 31; m 59; c 2. SCIENCE EDUCATION, BIOCHEMISTRY. *Educ:* Fordham Univ, BS, 52; Purdue Univ, PhD(chem), 60. *Prof Exp:* Res chemist, Res Ctr, Hercules Inc, 59-62; sci & technol consult, Becton, Dickinson & Co, 62-63; independent consult, 63-64 & 73-74; consult & mkt analyst, Hoffman-La Roche Inc, 65-66; instr chem & res asst, Rosemont Col, 67; assoc prof chem, West Chester State Col, 68-71 & Nyack Sr High Sch, 71; curric coordr, Westinghouse Learning Corp, 71-72; assoc prof chem & head natural sci & math div, St Thomas Aquinas Col, 75-77; CONSULT, 77- *Mem:* Am Chem Soc; Sigma Xi. *Res:* Chemistry and chemical marketing; improving curricula in undergraduate science education. *Mailing Add:* 511 Shippan Ave Stamford CT 06902

RACHLIN, JOSEPH WOLFE, b New York, NY, Jan 23, 36; m 60; c 2. AQUATIC BIOLOGY. *Educ:* City Col New York, BS, 57; NY Univ, MS, 62, PhD(aquatic biol), 67. *Prof Exp:* Lab technician chemother, Sloan-Kettering Inst Cancer Res, 57-58; biol sci asst environ med, US Army Med Res Lab, Ky, 58-60; biol sci trainee endocrinol, Sch Med, NY Univ, 60-64, fel radiol health, 64-65, trainee environ health, 65-66; asst prof, from instr to assoc prof, 67-77, PROF BIOL, LEHMAN COL, 77-; ADJ PROF ENVIRON MED, NY UNIV MED COL, 72- *Concurrent Pos:* Dir, City Univ New York Inst Marine & Atmospheric Sci, 76-77. *Mem:* AAAS; Am Soc Ichthyol & Herpet; Am Soc Limnol & Oceanog; Am Fisheries Soc; Am Soc Zoologists; Sigma Xi. *Res:* Effects of chemical pollutants on the fresh water and marine environments, particularly the effects of metals; fish cytogenetics. *Mailing Add:* Dept Biol Sci Lehman Col Bedford Park Blvd W Bronx NY 10468

RACHMELER, MARTIN, b Brooklyn, NY, Nov 21, 28; m 56; c 3. MICROBIOLOGY, GENETICS. *Educ:* Ind Univ, AB, 50; Western Reserve Univ, PhD(microbiol), 60. *Prof Exp:* Asst geneticist, Univ Calif, Berkeley, 61-62; asst prof, 62-67, ASSOC PROF MICROBIOL, MED SCH, NORTHWESTERN UNIV, CHICAGO, 67-, DIR, RES SERV ADMIN, EVANSTON, 77- *Concurrent Pos:* USPHS fel, Univ Calif, Berkeley, 59-61; vis fac, Baylor Col Med, 71-72. *Mem:* AAAS; Am Soc Microbiol; Sigma Xi. *Res:* Biochemistry of human genetic diseases; role of tumor viruses in cell transformation; regulation of cell growth. *Mailing Add:* 10695 Loire Ave San Diego CA 92131-1532

RACINE, MICHEL LOUIS, b Casselman, Ont, Jan 19, 45; m; c 1. ALGEBRA. *Educ:* Univ Ottawa, BSc, 60; Yale Univ, MPhil, 69, PhD(math), 71. *Prof Exp:* Res assoc math, Carleton Univ, 71-72; Nat Res Coun fel, Univ Wis-Madison, 72-73, MacDuffee fel, 73-74; from asst prof to assoc prof, 74-86, PROF MATH, UNIV OTTAWA, 86- *Concurrent Pos:* Alexander von Humboldt fel, Univ Munster, 80-81, SSHN fel, Univ Paris, 87-88. *Mem:* Am Math Soc; Math Asn Am; Can Math Soc. *Res:* Structure of Jordan algebras and related questions. *Mailing Add:* Dept Math Univ Ottawa Ottawa ON K1N 6N5 Can

RACINE, RENE, b Quebec City, Que, Oct 16, 39; m 63; c 2. ASTRONOMY. *Educ:* Laval Univ, BA, 58, BSc, 63; Univ Toronto, MA, 65, PhD(astron), 67. *Prof Exp:* Carnegie fel, Hale Observs, Calif, 67-69; from asst prof to assoc prof astron, Univ Toronto, 69-76; PROF, UNIV MONTREAL, 76- *Concurrent Pos:* Dir, Can-Fran-Haw Tel Corp, 80-84 & Observ Astron, Mont Megantic, 76-80 & 84-90. *Mem:* AAAS; Am Astron Soc; Can Astron Soc (pres 74-76); Royal Astron Soc Can. *Res:* Galactic structure; galaxies; open and globular clusters; optical instrumentation/telescopes. *Mailing Add:* Dept Physics Univ Montreal Montreal PQ H3C 3J7 Can

RACISZEWSKI, ZBIGNIEW, b Uchanie, Poland, Jan 20, 22. POLYMER CHEMISTRY. *Educ:* Univ Sask, MSc, 52; Univ Notre Dame, PhD(chem), 56. *Prof Exp:* Instr food technol, Agr Col Warsaw, Poland, 47-49; sr res chemist, Explor Org Chem Group, Pittsburgh Plate Glass Co, 55-60; res chemist, Consumer Prod Div, Union Carbide Corp, 60-63, Chem Div, WVa, 63-69; assoc ed, 69-74, sr assoc ed, 74-80, sr ed, Macromolecular Sect, Chem Abstr Serv, 80-85; RETIRED. *Res:* Mechanism of organic reactions; polymer chemistry. *Mailing Add:* 6619 Brock St Dublin OH 43017

RACK, EDWARD PAUL, b Reading, Pa, Dec 13, 31; m 65; c 2. CHEMISTRY, RADIOCHEMISTRY. *Educ:* Pa State Univ, BS, 54; Univ Mich, MS, 58, PhD(chem), 61. *Prof Exp:* Res fel chem, Inst Atomic Res, 61-62; from asst prof to assoc prof, 62-71, PROF CHEM, UNIV NEBR, LINCOLN, 71- *Concurrent Pos:* AEC teaching grant, 65, res contract, 66-; consult, Brookhaven Nat Lab, 71- & Vet Admin Med Ctr, Omaha, 72- *Mem:* Am Chem Soc; Sigma Xi. *Res:* Chemical effect of nuclear transformations, particularly involving halogens with solid, liquid and gaseous organic systems; neutron activation analysis of biological samples; development of analytical procedures for dertermination of parts-per-billion molecules in biological samples. *Mailing Add:* Dept Chem Univ Nebr Lincoln NE 68588-0304

RACK, HENRY JOHANN, b New York, NY, Nov 1, 42; c 2. MATERIAL SCIENCE, METALLURGY. *Educ:* Mass Inst Technol, SB, 64, SM, 65, ScD(metall), 68. *Prof Exp:* Scientist, Lockheed Ga Co, 68-72; mem tech staff, Sandia Labs, 72-81; prof, NMex Inst Mining & Technol, 75-81; mgr, metall dept, mat div, Exxon Enterprises, 81-82; mgr, metall silage opers, Arco Metals, 82-85, mgr, advan mat compos, 85; PROF MECH ENG & METALL, CLEMSON UNIV, 85- *Mem:* Am Soc Metals; Am Inst Mining, Metall & Petrol Engrs; Am Soc Testing & Mat. *Res:* Metal matrix composites; structural materials; fracture; structural reliability; nuclear waste management and transportation; solar materials and applications. *Mailing Add:* Dept Mech Eng Clemson Univ Clemson SC 29634-0921

RACKE, KENNETH DAVID, b Evergreen Park, Ill, July 6, 59; m 81; c 2. ENVIRONMENTAL CHEMISTRY. *Educ:* Trinity Christian Col, BA, 81; Univ Wis, MS, 84; Iowa State Univ, PhD(entom), 87. *Prof Exp:* Res assoc, Iowa State Univ, 85-87; asst soil scientist, Conn Agr Exp Sta, 87-8; SR RES CHEMIST, DOW CHEM CO, 88- *Honors & Awards:* Am Chem Soc Award, 87. *Mem:* Am Chem Soc; Am Sci Affil; Soil Sci Soc Am. *Res:* Fate and degradation of pesticides and other organic chemicals in soil, water, and waste materials; interaction between microorganisms and environmental pollutants. *Mailing Add:* Environ Chem Lab Dow Elanco Midland MI 48641-1706

RACKER, EFRAIM, oncology; deceased, see previous edition for last biography

RACKIS, JOSEPH JOHN, b Somersville, Conn, July 29, 22; m 54; c 2. BIOCHEMISTRY, FOOD SCIENCE. *Educ:* Univ Conn, BS, 50; Univ Iowa, PhD(biochem), 55. *Prof Exp:* Chemist, Northern Regional Res Ctr, Agr Res Serv, USDA, 55-60, prin chemist, 61-84; CONSULT, 84- *Honors & Awards:* Bond Award, Am Oil Chemists Soc, 65. *Mem:* Am Chem Soc; Am Soc Biol Chemists; Inst Food Technol; Am Asn Cereal Chem; Am Soybean Asn. *Res:* Plant biochemistry; physical organic chemistry of soybean proteins; chromatography; amino acids; plant analysis; lipids; nutritional, toxicological and physiological evaluation; food and feed uses of soybean products. *Mailing Add:* 3411 N Elmcroft Terrace Peoria IL 61604

RACKOFF, JEROME S, b Brooklyn, NY, Nov 14, 46; m 71; c 3. VERTEBRATE PALEONTOLOGY. *Educ:* Brooklyn Col, BS, 68; Yale Univ, MPhil, 73, PhD(geobiol), 76. *Prof Exp:* Teacher earth sci, Brooklyn Friends Sch, NY, 69-70; teacher biol, Friends Sem, NY, 70-72; asst prof biol, 75-78, found & govt rels officer, 78-82, assoc dir develop, 82-88, ASSOC VPRES, UNIV RELATIONS ADMIN, BUCKNELL UNIV, 88- *Mem:* Sigma Xi. *Res:* Functional morphology and evolution of Paleozoic fishes, particularly Crossopterygii and lower tetrapods; the origin of tetrapod limbs and terrestrial locomotion. *Mailing Add:* 229 N Second St Lewisburg PA 17837

RACKOW, ERIC C, MEDICINE, CRITICAL CARE MEDICINE. *Educ:* Franklin & Marshall Col, BA, 67; State Univ NY, MD, 71; Am Bd Internal Med, dipl, 75, dipl cardiovasc, 77, dipl critical care, 87. *Prof Exp:* Resident internal med, State Univ NY, Downstate Med Ctr & Kings County Hosp Ctr, 71-72, chief resident, 72-73, fel cardiol, 73-75; PROF MED & VCHMN, DEPT MED, NY MED COL, 89-; CHMN, DEPT MED, ST VINCENT'S HOSP & MED CTR, NY, 89- *Concurrent Pos:* Exec vpres, Inst Critical Care Med, Rancho Mirage, Calif, 90- *Mem:* Fel Am Col Physicians; fel Am Col Cardiol; fel Am Col Chest Physicians; fel Am Col Critical Care Med. *Res:* Critical care medicine; circulatory shock; sepsis and septic shock; fluid resuscitation; pulmonary edema; cardiopulmonary resuscitation; author of 300 technical publications. *Mailing Add:* St Vincent's Hosp & Med Ctr 153 W 11th St New York NY 10011

RACKOW, HERBERT, b New York, NY, June 17, 17; m 42. ANESTHESIOLOGY. *Educ:* Pa State Univ, BS, 39; Howard Univ, MD, 46. *Prof Exp:* Fel biochem, NY Univ, 48-50; from instr to assoc prof, Columbia Univ, 52-70, prof anesthesiol, Col Physicians & Surgeons, 70-77; RETIRED. *Res:* Respiratory physiology. *Mailing Add:* 147-01 Third Ave Whitestone NY 11357

RACLE, FRED ARNOLD, b Columbus, Ohio, Dec 16, 32; m 62. NATURAL SCIENCE. *Educ:* Ohio State Univ, BSc, 60, MSc, 62, PhD(bot), 65. *Prof Exp:* Assoc prof, 65-73, PROF NATURAL SCI, MICH STATE UNIV, 73- *Mem:* Sigma Xi. *Res:* Curriculum devlopment in general education, science. *Mailing Add:* Dept Natural Sci Mich State Univ East Lansing MI 48823

RACOTTA, RADU GHEORGHE, b Bucarest, Romania, Jan 5, 30; Mex citizen. CONTROL OF FOOD INTAKE, METABOLIC REGULATION. *Educ:* Univ Bucarest, BSc, 63; Inst Politech Nacional, Mex, DSc, 75. *Prof Exp:* Res agron, Inst Cult Corn, Romania, 59-62; res & diag physiol, Inst Physiol, Bucarest, Romania, 62-66; res physiol, Inst Biol, Bucarest, Romania, 66-69; TEACHING & RES PHYSIOL, NAT SCH BIOL SCI, IPN, MEXICO CITY, 70- *Mem:* Am Physiol Soc; Soc Study Ingestive Behav. *Res:* Food and water intake physiological control, specifically through hepatic receptors; catecholamines and energy metabolism; nervous control of the testes. *Mailing Add:* Dept Physiol ENCB Prol Carpio Y Plan de Ayala Mexico City 11340 Mexico

RACUNAS, BERNARD J, b June 12, 43; US citizen. CHEMICAL ENGINEERING. *Educ:* Univ Pittsburgh, BS, 65, MS, 67. *Prof Exp:* Engr, 66-74, sr engr, 74-76, staff engr, 76-78, SECT HEAD, ALCOA LABS, 78- *Mem:* Am Inst Chem Engrs; Am Inst Metall Engrs; Sigma Xi. *Res:* Aluminum smelting; carbon technology; molten salt technology. *Mailing Add:* 1075 Woodberry Rd New Kensington PA 15068

RACUSEN, DAVID, b Chicago, Ill, Feb 26, 25; m 69; c 3. PLANT BIOCHEMISTRY. *Educ:* Hobart Col, BS, 49; Iowa State Univ, PhD(plant physiol), 53. *Prof Exp:* Res fel, Calif Inst Technol, 53-54; plant biochemist, Shell Develop Co, 54-58; PROF BIOCHEM, UNIV VT, 58- *Mem:* AAAS. *Res:* Protein metabolism of leaves; plant glycoproteins. *Mailing Add:* Dept Agr Biochem Univ Vt Burlington VT 05405

RACUSEN, RICHARD HARRY, b Geneva, NY, July 26, 48; m 70. PLANT PHYSIOLOGY. *Educ:* Univ Vt, BS, 70, MS, 72, PhD(cell biol), 75. *Prof Exp:* Res fel cell biol, Unit Vt, 70-75; res fel plant physiol, Yale Univ, 75-78; ASST PROF BOT, UNIV MD, 78- *Mem:* AAAS; Sigma Xi; Am Soc Plant Physiol. *Res:* Ion transport, morphogenesis and plant bioelectric phenomena. *Mailing Add:* Dept Bot Univ Md College Park MD 20742

RAD, FRANZ N, b Zabol, Iran, Sept 25, 43; m 64; c 3. STRUCTURAL ENGINEERING. *Educ:* Univ Tex, Austin, BS, 68, MS, 69, PhD(civil eng), 73. *Prof Exp:* From asst prof, to assoc prof, 71-79, PROF CIVIL ENG & DEPT HEAD, PORTLAND STATE UNIV, 79- *Concurrent Pos:* Consult, industrialization, 72-; Western Elec Fund Award, Am Soc Eng Educ, 79; pres, Struct Eng Asn, Ore, 85-86. *Mem:* Fel Am Soc Civil Engrs; fel Am Concrete Inst; Am Soc Eng Educ; Nat Soc Prof Engrs; Post Tension Inst. *Res:* Limit states behavior of reinforced concrete members and structures. *Mailing Add:* Dept Civil Eng Portland State Univ Portland OR 97207-0751

RADABAUGH, DENNIS CHARLES, b Detroit, Mich, Sept 27, 42; m 67; c 2. BEHAVIOR OF PREDATORS & PREY, GALAPAGOS ECOLOGY. *Educ:* Albion Col, BA, 64; Ohio State Univ, MSc, 67, PhD(animal behav), 70. *Prof Exp:* Vis asst prof, 70-72, from asst prof to assoc prof, 70-81, PROF ZOOL, OHIO WESLEYAN UNIV, 82- *Mem:* AAAS; Animal Behav Soc; Am Soc Ichthyologists & Herpetologists; Am Soc Arachnologists. *Res:* Predator-prey behavioral interactions; effects of parasites on intermediate host behavior; behavior of fish; behavior of spiders. *Mailing Add:* Dept Zool Ohio Wesleyan Univ Delaware OH 43015

RADABAUGH, ROBERT EUGENE, geology; deceased, see previous edition for last biography

RADANOVICS, CHARLES, b Budapest, Hungary, Aug 9, 32; US citizen; m 60; c 3. FOOD SCIENCE. *Educ:* Univ Budapest, BS, 56; Univ Calif, Davis, MS, 63; Mich State Univ, PhD(food sci), 69. *Prof Exp:* Mgr qual control, Model Dairy, Melbourne, Australia, 57-60; sr chemist, Tarax Ale Co, Melbourne, 60-62; proj leader food res, Carnation Co, Calif, 63-65; sect mgr food res, Quaker Oats Co, 69-74; DIR RES & DEVELOP, JOHN SEXTON & CO, SUBSID S E RYKOFF CO, 74- *Concurrent Pos:* Adv panel, Univ Minn Food Sci Dept. *Mem:* Inst Food Technol; Am Chem Soc; Res & Develop Assoc (pres, 79). *Res:* Food service; new product development; regulatory activities; formulation of fabricated foods; process innovation; nutrition of foods; teaching food science. *Mailing Add:* John Sexton & Co 1800 Churchman Ave Indianapolis IN 46203

RADBILL, JOHN R(USSELL), b Upland, Pa, Apr 22, 32. NUMERICAL ANALYSIS, FLUID MECHANICS. *Educ:* Mass Inst Technol, BS & MS, 55, MechE, 56, ScD(mech eng), 58. *Prof Exp:* Asst mech eng, Mass Inst Technol, 54-58; develop engr ionic propulsion, Aerojet-Gen Corp, Gen Tire & Rubber Co, 58-61; tech specialist, Space & Info Systs Div, N Am Aviation, Inc, 61-66, sr tech specialist, Ocean Systs Opers, 66, mem tech staff, Autonetics Div, N Am Rockwell Corp, 66-70; MEM TECH STAFF APPL MATH, JET PROPULSION LAB, CALIF INST TECHNOL, 70- *Concurrent Pos:* Instr, Citrus Jr Col, 59-60; consult, Technol Assocs of Southern Calif, Inc, 72- *Mem:* Assoc Am Soc Mech Engrs; Am Inst Aeronaut & Astronaut; Asn Comput Mach; Sigma Xi. *Res:* Fluid mechanics; numerical analysis; numerical solution of nonlinear partial differential equations; tornado lifted missiles; blood flow in diseased arteries. *Mailing Add:* 10413 Haines Canyon Tujunga CA 91042

RADBRUCH-HALL, DOROTHY HILL, geology, for more information see previous edition

RADCLIFFE, ALEC, b Cleethorpes, Eng, Aug 28, 17; nat US; m 46. PHYSICS, SYSTEMS ANALYSIS. *Educ:* Univ London, BSc, 39. *Prof Exp:* Temp exp asst & exp officer, Mine Design Dept, Brit Navy, 40-46; from sci officer to prin sci officer, Nat Gas Turbine Estab, 46-54; physicist & asst dept supvr, Appl Physics Lab, Johns Hopkins Univ, 55-87; RETIRED. *Mem:* Fel Brit Inst Physics. *Res:* Magnetism; acoustics; combustion; propulsion; fuel injection; unsteady gas dynamics; operations research. *Mailing Add:* 1710 Highland Dr Silver Spring MD 20910

RADCLIFFE, EDWARD B, b Rapid City, Man, Oct 25, 36; US citizen; m 64; c 2. ENTOMOLOGY. *Educ:* Univ Man, BSA, 59; Univ Wis, MS, 61, PhD(entom), 63. *Prof Exp:* Res fel, 63-64, res assoc, 64-65, asst prof, 65-70, assoc prof, 70-76, PROF ENTOM, UNIV MINN, ST PAUL, 76- *Concurrent Pos:* Vis prof, Beijing Agr Univ & Jilin Agr Univ, 82, Univ PR, Mayaguez, 84-85; adj prof, IAV-Hassan II, Rabat, 87- *Mem:* AAAS; Entom Soc Am; Entom Soc Can; Potato Asn Am. *Res:* Resistance of plants to insect attack; integrated pest management. *Mailing Add:* Dept Entom Univ Minn St Paul MN 55108

RADCLIFFE, S VICTOR, b Eng, July 28, 27. MATERIALS ENGINEERING. *Educ:* Univ Liverpool, BEng, 48, PhD(metall), 56. *Prof Exp:* Res metallurgist, Lancashire Steel Corp, 48-53; res assoc metall, Mass Inst Technol, 56-62; res mgr mat, Manlabs, Inc, 62-63; from assoc prof to prof phys metall, Case Western Reserve Univ, 63-75, head dept metall & mat sci, 69-74; sr policy analyst, Sci & Technol Policy Off, Staff to Sci Adv to Pres Ford, 74-75, sr fel, Resources For Future, 76-79; VPRES CORP DEVELOP, NAT FORGE CO, 79- *Concurrent Pos:* Dir study mat sci & eng, Nat Acad Sci, 70-73; consult, 56-, UN, 77-; prin investr, Lunar Sci Prog NASA, 69-74. *Mem:* Am Inst Mining, Metall & Petrol Engrs; Soc Metals; fel Inst Metall; Am Phys Soc; Am Chem Soc; Nat Asn Bus Econ; Nat Econ Club; Cosmos Club. *Res:* Requirements for materials in relation to growth of national economies; interpretation of intercountry differences in terms of economic structure, intensity and technology; application to forecasting and public policy; international industrial competitiveness. *Mailing Add:* 2101 Connecticut Ave NW Washington DC 20008

RADD, F(REDERICK) J(OHN), b Greenfield, Mass, July 28, 21; m; c 3. METALLURGY. *Educ:* Univ Mo, BS, 43; Mass Inst Technol, ScD(metall), 49. *Prof Exp:* Asst metall, Res Lab, Gen Elec Co, 43; staff mem div indust coop, Mass Inst Technol, 43-45, res metallurgist, Atomic Energy Comn, 49-51; supvr, Boeing Airplane Co, 51-52; staff scientist, 52-77, SR RES ASSOC, CONTINENTAL OIL CO, 78- *Mem:* Am Soc Metals; Nat Asn Corrosion Engrs; Am Chem Soc; Am Inst Mining, Metall & Petrol Engrs. *Res:* Cryogenic and petroleum metallurgy; age hardening of metals; solidification theory; fatigue and corrosion fatigue behavior; high temperature reactions; corrosion processes; hydrogen-metal behaviors; electrochemistry; cemented carbides; coal seam geochemistry. *Mailing Add:* PO Box 2428 Ponca City OK 74602-2428

RADDING, CHARLES MEYER, b Springfield, Mass, June 18, 30; m 54; c 3. BIOCHEMISTRY, GENETICS. *Educ:* Harvard Univ, AB, 52, MD, 56. *Prof Exp:* Intern, Harvard Med Serv, Boston City Hosp, Mass, 56-57; res assoc metab, Nat Heart Inst, 57-59; Am Heart Asn advan res fel biochem, Sch Med, Stanford Univ, 59-62; asst prof human genetics, Univ Mich, 62-65, assoc prof, 65-67; assoc prof med, molecular biophys & biochem, 67-72, prof, 72-79, PROF HUMAN GENETICS, YALE UNIV, 79- *Concurrent Pos:* Vis Miller prof, Univ Calif, Berkeley, 77. *Mem:* Am Soc Biol Chemists. *Res:* Genetic recombination; molecular virology. *Mailing Add:* Dept Human Genetics Yale Univ Sch of Med 333 Cedar St PO Box 3333 New Haven CT 06510

RADEBAUGH, RAY, b South Bend, Ind, Nov 4, 39; div; c 4. CRYOGENICS, REFRIGERATION. *Educ:* Univ Mich, BSE, 62; Purdue Univ, MS, 65, PhD(physics), 66. *Prof Exp:* Res asst physics, Purdue Univ, 62-66; assoc, 66-68, PHYSICIST CRYOG, NAT BUR STANDARDS, 68- *Concurrent Pos:* Vis prof, Univ Tokyo, 72-73. *Honors & Awards:* Nat Bur Stand Superior Performance Award, 68, 84-87. *Mem:* Am Phys Soc; Sigma Xi. *Res:* Heat transfer, refrigeration, and thermometry at cryogenic temperatures. *Mailing Add:* MS 77330 Nat Bur of Standards Boulder CO 80303

RADEKA, VELJKO, b Zagreb, Yugoslavia, Nov 21, 30; m 58; c 2. INSTRUMENTATION SCIENCE. *Educ:* Univ Zagreb, Dipl Ing, 55, Dr Eng Sci(electronics), 61. *Prof Exp:* Scientist instrumentation, Ruder Boskovic Inst, Zagreb, 55-66; res assoc, 62-64, assoc scientist, 66-69, scientist, 69-73, SR SCIENTIST INSTRUMENTATION, BROOKHAVEN NAT LAB, 73-, DIV HEAD, 72- *Honors & Awards:* Merit Award, Inst Elec & Electronic Engrs, 83, Centennial Medal, 89. *Mem:* Fel Inst Elec & Electronics Engrs; Am Phys Soc. *Res:* Scientific instrumentation; nuclear detector signal processing. *Mailing Add:* Instrumentation Div Brookhaven Nat Lab Upton NY 11973

RADEL, STANLEY ROBERT, b New York, NY, July 6, 32; m 54; c 2. SCIENCE EDUCATION. *Educ:* NY Univ, AB, 53, MS, 56, PhD(phys chem), 63. *Hon Degrees:* FGS. *Prof Exp:* Tutor chem, Queens Col, NY, 57-59, lectr, 59-64; from instr to assoc prof, 64-91, PROF CHEM, CITY COL NY, 91- *Mem:* Am Chem Soc; Am Phys Soc; AAAS. *Res:* Intramolecular forces; molecular dynamics; quantum mechanics and spectroscopy. *Mailing Add:* Dept Chem City Col NY 138th St & Convent Ave New York NY 10031

RADEMACHER, LEO EDWARD, b Brush, Colo, Sept 3, 26; m 53; c 3. ORGANIC CHEMISTRY. *Educ:* Univ Colo, BA, 50, PhD(org chem), 56. *Prof Exp:* Jr chemist, Julius Hyman Co, Colo, 50-52; jr chemist, Shell Chem Co, 52-53; sr chemist, Plastics Div, 55-65, res specialist, Monsanto Polymers & Petrochem Co, 66-77, SR TECHNOL SPECIALIST, MONSANTO PLASTICS & RESINS CO, 77- *Mem:* Am Chem Soc; Soc Plastics Engrs. *Res:* Thermosetting surface coating resins; polymerization and applications of vinyl chloride; graft polymerization; polymer processing; composite systems. *Mailing Add:* 111 Meadowbrook Rd Springfield MA 01128-1332

RADER, CHARLES ALLEN, b Washington, DC, Sept 30, 32; m 56; c 3. SURFACE CHEMISTRY. *Educ:* Univ Md, BS, 55. *Prof Exp:* Chemist, Nat Bur Standards, 55-58; from chemist to sr chemist, Harris Res Labs, Inc, 58-64, res supvr, 64-67; group leader, 67-72, mgr biochem sci dept, 72-76, mgr, Phys Sci Dept, 77-80, DIR, GILLETTE RES INST, 80- *Mem:* Am Chem Soc; Am Asn Textile Chemists & Colorists; Am Inst Chemists; Fiber Soc; Soc Cosmetic Chemists. *Res:* Surface chemistry; detergents and surfactants; actinic degradation of polymers; cosmetic and personal products; aerosols; chemical and physical properties of skin and hair; textiles; chemical warfare. *Mailing Add:* Gillette Research Inst 401 Professional Dr Gaithersburg MD 20879-3400

RADER, CHARLES PHILLIP, b Greeneville, Tenn, Apr 9, 35; m 58; c 2. THERMOPLASTIC ELASTOMERS, RUBBER CHEMISTRY. *Educ:* Univ Tenn, BS, 57, MS, 60, PhD(chem), 61. *Prof Exp:* Instr chem, Univ Tenn, 59; sr res chemist, Monsanto Co, 61; org chemist, US Army Chem Ctr, 61-63; sr res chemist, Monsanto Co, 63-69, com develop proj mgr, 69-70, res group leader, 70-82, sr tech serv rep, 83-87, mkt tech serv prin, 88-90; MKT TECH SERV PRIN, ADVAN ELASTOMER SYSTS, LP, 91- *Concurrent Pos:* Asst prof, Univ Md, 62-63. *Honors & Awards:* Chmn, Rubber Div, Am Chem Soc, 86. *Mem:* Am Chem Soc; NY Acad Sci; Soc Plastic Engrs. *Res:* Applications of physical methods to organic chemistry; structure elucidation; conformational analysis; natural products; catalytic hydrogenation; rubber technology and tire technology; polymer chemistry; thermoplastic rubbers. *Mailing Add:* Advan Elastomer Systs LP 260 Springside Dr Akron OH 44334-0584

RADER, DENNIS, engineering mechanics, geophysics, for more information see previous edition

RADER, LOUIS T(ELEMACUS), b Frank, Alta, Aug 24, 11; nat US; m 38; c 2. ELECTRICAL ENGINEERING. *Educ:* Univ BC, BSc, 33; Calif Inst Technol, MS, 35, PhD(elec eng), 38. *Prof Exp:* Test engr, Gen Elec Co, NY, 37-38, adv eng prog, Gen Eng Dept, 38-39, sect head, Control Eng Dept, 43-45, div engr, Control Lab Div, 47-49, asst to mgr, 49-50, asst to mgr, Eng Div, 50-51, mgr eng, Control Div, 51-53, gen mgr, Specialty Control Dept, 53-59; dir elec eng dept & consult, Armour Res Found, Ill Inst Technol, 45-47; vpres, US Commercial Group & mem bd dirs, Int Tel & Tel Corp, 59-62; pres, Univac Div, Sperry Rand Corp, 62-64; vpres & gen mgr, Info Systs Div, Gen Elec Co, Charlottesville, 64-68; gen mgr, Indust Process Control Div, 68-69; prof elec eng & bus admin, 69-82, EMER PROF BUS ADMIN, UNIV VA, 82- *Concurrent Pos:* Bd visitors & gov, St Johns Col, Md, 61-70; trustee, Robert A Taft Inst Govt, NY, 63-80. *Mem:* Nat Acad Eng; Am Soc Eng Educ; fel Inst Elec & Electronic Engrs; Sigma Xi. *Res:* Principles of magnetic design; arc interruption. *Mailing Add:* 1200 Boxwood Circle Waynesboro VA 22980

RADER, RONALD ALAN, b Newark, NJ, May 28, 51. ANTIVIRAL DRUG & VACCINE DEVELOPMENT. *Educ:* Univ Md, BS, 73, MLS, 79. *Prof Exp:* User support coordr, Computer Sci Corp, 83-84; info scientist, Biospherics, Inc, 84; proj mgr, Expand Assoc, 84-85; ed & proj leader, Omec Int Inc, 85-90; PRES, BIOTECHNOL INFO INST, 90- *Concurrent Pos:* Mgr info serv, Porton Int Inc, 85-90; chem ed, Tech Resources Inc, 85; ed, Antiviral Agents Bull, 88- *Mem:* Am Chem Soc; Int Soc Antiviral Res; Drug Info Asn; Soc Indust Microbiol; Am Soc Microbiol; Am Soc Info Sci. *Res:* Biotechnology and pharmaceutical information resources design and development; acquired immunodeficiency syndrome and antiviral drug and vaccine development information resources; market and technology assessments. *Mailing Add:* Biotechnol Info Inst 1700 Rockville Pike Suite 400 Rockville MD 20852

RADER, WILLIAM AUSTIN, b Detroit, Mich, Aug 27, 16; m 42; c 4. VETERINARY TOXICOLOGY. *Educ:* Mich State Univ, DVM, 41. *Prof Exp:* Pvt practr vet med, 42-46 & 52-64; pub health off, Mich Dept Health, 46-47; dir res & develop, Vita-Vet Labs, Ind, 64-65; vet toxicologist, Petitions Rev Br, Bur Sci, Food & Drug Admin, US Dept Health, Educ & Welfare, 65-67, vet med off, Div Vet Med Rev, Bur Vet Med, 67-68, chief investr, New Animal Drug Br, New Animal Drugs Div, 68-73; chief toxicologist, Residue Planning & Eval Staff, Animal & Plant Health Inspection Serv, USDA, 73-75; RETIRED. *Concurrent Pos:* Pvt consult animal & plant health, 75-88. *Mem:* Am Vet Med Asn; Am Pub Health Asn; fel Am Col Vet Toxicologists; Am Soc Vet Physiologists & Pharmacologists; Soc Toxicologists. *Res:* Toxicological significance of pesticides, herbicides, fungicides, industrial environmental contaminants and chemicals and of oral and injectible drugs under conditions of use. *Mailing Add:* 4638 Bay Shore Rd Sarasota FL 33580

RADER, WILLIAM ERNEST, b Ellensburg, Wash, Aug 21, 16; m 38; c 2. PLANT PATHOLOGY. *Educ:* State Col Wash, BS, 39; Utah State Col, MS, 42; Cornell Univ, PhD(plant path), 46. *Prof Exp:* Asst, Utah State Col, 39-42 & Cornell Univ, 44-46; microbiologist, Biol Sci Res Ctr, Agr Lab, Shell Develop Co, 46-81; RETIRED. *Mem:* Am Phytopath Soc; Soc Nematol. *Res:* Pesticides and agricultural chemicals; biochemistry of fungicidal action; physiology and biochemistry of plant disease; chemical control of nematodes; biotreatment of industrial wastes. *Mailing Add:* 4901 Dale Rd Modesto CA 95356

RADERMACHER, REINHARD, b Heidelberg, Ger, Dec 21, 52; div; c 3. REFRIGERATION, HEAT TRANSFER. *Educ:* Tech Univ Munich, Ger, BS, 75, MS, 77, PhD(physics), 81. *Prof Exp:* NATO scholar res, Nat Inst Standards & Technol, 81-83; vis prof, 83-84, asst prof, 84-89, ASSOC PROF MECH ENG, UNIV MD, 89- *Concurrent Pos:* Prin investr, Nat Inst Standards & Technol, 83-, Sundstrand Corp, 84-87, Dept Energy, 86-89, NSF, 86-, GRI, 87, 88 & Environ Protection Agency, 89- *Mem:* Am Soc Heating Refrig & Air Conditioning Eng; Int Inst Refrig. *Res:* Energy conversion cycles, working fluid mixtures, environmentally safe fluids, household refrigerators; thermodynamics, heat transfer. *Mailing Add:* Dept Mech Eng Univ Md College Park MD 20742-3035

RADEWALD, JOHN DALE, b Niles, Mich, Feb 15, 29; m 53; c 2. PLANT PATHOLOGY. *Educ:* Ariz State Univ, BS, 52; Okla State Univ, MS, 56; Univ Calif, PhD(plant path), 61. *Prof Exp:* Res asst plant path, Okla State Univ, 55-56; res asst plant nematol, Univ Calif, 56-60; assoc plant pathologist, Pineapple Res Inst, Hawaii, 60-63; EXTEN NEMATOLOGIST, COOP EXTEN, UNIV CALIF, RIVERSIDE, 63-, LECTR NEMATOL, 74- *Mem:* Am Soc Nematologists; Europ Soc Nematologists; Am Phytopath Soc; Orgn Tropical Am Nematologists; Am Inst Biol Sci. *Res:* Plant nematology, primarily host-parasite relations; biology and physiology of plant parasitic nematodes. *Mailing Add:* Dept Nematology Univ Calif Riverside CA 92521

RADFORD, ALAN, b Chelmsford, Eng, Mar 31, 40; m 65; c 3. GENETICS. *Educ:* Univ Leeds, BSc, 62; McMaster Univ, MSc, 63, PhD(genetics), 66. *Prof Exp:* Res assoc biol, Stanford Univ, 66-70; asst prof zool, Univ Calif, Los Angeles, 70-71; Sci Res Coun fel, Univ Leeds, 71; lectr bot, Birkbeck Col, Univ London, 72-74; lectr, 74-81, SR LECTR GENETICS, UNIV LEEDS, 81- *Concurrent Pos:* Ed officer, British Genetical Soc, 84; vis prof, Univ Brasilia, 87; chmn, U K Interest group, Educ in Biotechnol. *Mem:* Brit Genetical Soc; Brit Inst Biol. *Res:* Regulation in pyrimidine metabolism; extracellular enzymes; fungal genetics; fungal biotechnology; gene fine structure and function; genetics education. *Mailing Add:* Dept Genetics Univ Leeds Leeds England

RADFORD, ALBERT ERNEST, b Augusta, Ga, Jan 25, 18; m 41; c 3. BOTANY. *Educ:* Furman Univ, BS, 39; Univ NC, PhD(bot), 48. *Prof Exp:* From instr to assoc prof, 47-59, PROF BOT, UNIV NC, CHAPEL HILL, 59- *Res:* Taxonomy of vascular plants; vascular flora of southeastern North America. *Mailing Add:* Dept Biol Univ NC CB No 3280 Coker Hall Chapel Hill NC 27599-3280

RADFORD, DAVID CLARKE, b Wellington, NZ, Mar 14, 54; m; c 2. HIGH-SPIN STATES, GAMMA-RAY SPECTROSCOPY. *Educ:* Univ Auckland, BSc, 75, PhD(physics), 79. *Prof Exp:* Res staff physicist, Wright Nuclear Structure Lab, Yale Univ, 78-81; vis res scientist, Nuclear Res Ctr, Strasbourg, France, 81-83; res assoc, Argonne Nat Lab, 83-84; asst res officer, 85, ASSOC RES OFFICER, CHALK RIVER NUCLEAR LAB, ATOMIC ENERGY CAN LTD, 86- *Res:* Nuclear structure reseach; high-spin states of nuclei using the techniques of Gamma-Ray spectroscopy. *Mailing Add:* Nuclear Physics Br Sta 49 Atomic Energy Can Ltd Chalk River ON K0J 1J0 Can

RADFORD, DAVID EUGENE, b Plattsburg, NY, June 4, 43; m 66; c 3. MATHEMATICS. *Educ:* Univ NC, Chapel Hill, BS, 65, MA, 68, PhD(math), 70. *Prof Exp:* Asst prof math, Lawrence Univ, 70-76; ASSOC PROF MATH, UNIV ILL CHICAGO CIRCLE, 76- *Concurrent Pos:* NSF grant, 74-75; vis lectr, Rutgers Univ, 75-76, vis assoc prof, 79-80. *Mem:* Am Math Soc. *Res:* Algebra; Hopf algebras, algebraic groups and co-algebras. *Mailing Add:* Dept Math M/C 249 Univ Ill Chicago IL 60680

RADFORD, EDWARD PARISH, b Springfield, Mass, Feb 21, 22; m 45, 68; c 5. ENVIRONMENTAL MEDICINE, EPIDEMIOLOGY. *Educ:* Harvard Univ, MD, 46. *Prof Exp:* Instr physiol, Med Sch, Harvard Univ, 50-52, assoc sch pub health, 52-55; physiologist, Haskell Lab, E I du Pont de Nemours & Co, 55-59; assoc prof physiol, Sch Pub Health, Harvard Univ, 59-65; prof environ med & dir Kettering Lab, Col Med, Univ Cincinnati, 65-68; prof environ med, Sch Hyg & Pub Health, Johns Hopkins Univ, 68-77; PROF EPIDEMIOL, UNIV PITTSBURGH GRAD SCH PUB HEALTH, 77- *Concurrent Pos:* Teaching fel physiol, Med Sch, Harvard Univ, 49-50; vis prof, Oxford Univ, 75-76; mem comt carbon monoxide, Nat Acad Sci, 76-77; consult med, Westvaco Corp, 77-; mem comt biol effects ionizing radiation, Nat Acad Sci, 70-72, chmn, 77-80; mem air pollution adv bd, State Md, 71-77; fac scholar, Macy Found, 75-76. *Mem:* Am Physiol Soc; Am Pub Health Asn; Radiation Res Soc; Soc Occup & Environ Health. *Res:* Radiation biology; toxicology; occupational medicine; pulmonary physiology; environmental epidemiology. *Mailing Add:* Old Sch House Church Rd Spelsbury Oxford 0X7 3JR England

RADFORD, HERSCHEL DONALD, b Butler, Mo, June 21, 11; m 39; c 2. ORGANIC CHEMISTRY. *Educ:* Park Col, AB, 33; Univ Mo, AM, 44, PhD(chem), 49. *Prof Exp:* Chemist, Res & Develop Dept, Pan-Am Refining Corp, Standard Oil Co, Ind, 41-44, group leader, Chem Res Sect, 44-51, head, Process Develop Sect, 51-57, dir, Process Div, Am Oil Co, 57-60, tech dir, Process Develop Labs, 60-62, asst dir res & develop dept, 62-73, dir process & eng develop, Amoco Oil Co, Standard Oil Co, Ind, 73-76; CONSULT, 76- *Mem:* Am Chem Soc; Am Inst Chem Engrs. *Res:* Alkylation of aromatic and aliphatic hydrocarbons; dealkylation of aromatic hydrocarbons, catalytic hydrogenation of substituted carbonyl and carbinol compounds; organic reactions catalyzed by anhydrous hydrofluoric acid; catalytic reforming; hydrocracking; desulfurization; isomerization; coking; catalytic cracking; synthetic fuels. *Mailing Add:* 1375 Dartmouth Rd Flossmoor IL 60422-0208

RADFORD, KENNETH CHARLES, b Manchester, Eng, July 1, 41; m 65; c 3. METALLURGY. *Educ:* Univ London, BSc, 63; Imp Col, dipl & ARSM, 63, PhD(metall), 67. *Prof Exp:* Sr scientist, 68-78, fel engr, 78-84, ADV ENG MGR CERAMICS, WESTINGHOUSE ELEC CORP, 84- *Mem:* Inst Metall; Am Ceramic Soc. *Res:* Dielectric properties; ceramics; nuclear fuel; electrical properties of ceramics; physical properties of ceramic powders; ceramic fabrication; optical ceramics diamond; ceramic matrix composites. *Mailing Add:* Sci & Technol Ctr Westinghouse Elec Corp Pittsburgh PA 15235

RADFORD, LOREN E, b Randolph, Nebr, Oct 4, 28; m 50; c 1. SOLID STATE PHYSICS. *Educ:* Univ Wash, BS, 50; Univ Va, MS, 60, PhD(physics), 62. *Prof Exp:* From asst prof to assoc prof physics, US Mil Acad, 61-74; prof physics & head div sci & math, WVa Northern Community Col, 74-; AT DEPT MATH & PHYSICS, BAPTIST COL. *Mem:* Am Asn Physics Teachers; Am Phys Soc; Sigma Xi. *Res:* Investigation of crystal defects using technique of electron paramagnetic resonance. *Mailing Add:* Dept Math & Physics Baptist Col PO Box 10087 Charleston SC 29411

RADFORD, TERENCE, b Sheffield, Eng, Apr 1, 39; m 72. ORGANIC CHEMISTRY. *Educ:* Liverpool Col Technol, ARIC, 62; Sheffield Univ, PhD(org chem), 65. *Prof Exp:* Asst lectr org chem, Sheffield Col Technol, 65-66; res fel, Ohio State Univ, 66-67 & Wayne State Univ, 67-68; res scientist, 68-80, SR RES SCIENTIST MASS SPECTROS, COCA-COLA CO, 80- *Mem:* Royal Soc Chem; Am Chem Soc. *Res:* The application of instrumental techniques, especially combined gas chromatography/mass spectrometry to the identification of natural products. *Mailing Add:* 5446 Redfield Rd Dunwoody GA 30338

RADFORTH, NORMAN WILLIAM, b Lancashire, Eng, Sept 22, 12; nat Can; m 39; c 2. PALEOBOTANY. *Educ:* Univ Toronto, BA, 36, MA, 37; Univ Glasgow, Scotland, PhD(paleobot), 39. *Prof Exp:* Lectr bot, Univ Toronto, 42-46; prof, McMaster Univ, 46-68, head dept, 46-53, chmn dept biol, 60-66, chmn org & assoc terrain res unit, 61-68, coordr acad develop, 65-66; prof biol & Muskeg Studies, Univ NB, 68-77, head dept biol, 68-70, dir, Muskeg Res Inst, 68-73; CONSULT, RADFORTH & ASSOCS, 77- *Concurrent Pos:* Dir, Royal Bot Gardens, Can, 46-53; mem, Assoc Comt Geotech Res, Nat Res Coun Can, 48-; prog chmn & secy, Int Cong Bot, 59; vpres, Int Orgn Paleont, 59. *Honors & Awards:* Silver Medal, Royal Soc Arts, 58. *Mem:* Royal Soc Can; fel Royal Soc Arts. *Res:* Experimental morphology and embryology of higher plants applying in vitro methods; micropaleobotany and northern peatland interpretation. *Mailing Add:* Radforth & Assocs Muskeg Lab Limbert Rd RR 3 Parry Sound ON P2A 2W9 Can

RADHAKRISHNAMURTHY, BHANDARU, b Andhra Pradesh, India, July 1, 28; m 53, 83; c 4. BIOCHEMISTRY. *Educ:* Osmania Univ, India, BS, 51, MS, 53, PhD(chem), 58. *Prof Exp:* Res chemist, Sirsilk, Ltd, India, 53-54; lectr chem, Osmania Univ, 55-61; from res assoc to assoc prof, 61-74, PROF MED & BIOCHEM, LA STATE UNIV MED CTR, NEW ORLEANS, 74- *Concurrent Pos:* Fulbright fel med & biochem, Sch Med, La State Univ, New Orleans, 61-62; mem coun atherosclerosis, Am Heart Asn, 71- *Mem:* NY Acad Sci; Am Chem Soc; Soc Exp Biol & Med; Am Soc Biol Chem; AAAS. *Res:* Biochemistry of connective tissue; proteoglycans and glycoproteins. *Mailing Add:* Dept Med La State Univ Med Ctr New Orleans LA 70112

RADHAKRISHNAN, CHITTUR VENKITASUBHAN, b Mannuthy, India, June 6, 37; m 65; c 3. VETERINARY MICROBIOLOGY. *Educ:* Univ Kerala, India, BVSc, 59; Univ Fla, PhD(vet parasitol), 71. *Prof Exp:* Lectr vet med, Col Vet Med, Univ Kerala, 59-64; sci officer parasitol, Hindustan Antibiotics Res Ctr, 64-68; res asst, Dept Vet Sci, Univ Fla, 68-71, teaching asst parasitol, 71-72; VET, BUR LABS, FLA DEPT AGR, 72- *Concurrent Pos:* Assoc prof pathobiol, Col Vet Med, Pahlavi Univ, Shiraz, Iran, 73-75. *Mem:* Am Oil Chemists' Soc; Am Asn Avian Pathologists; Am Asn Vet Lab Diagnosticians; Am Soc Parasitologists. *Res:* Symbiotic and competitive nature of microorganisms; pathobiology of sporozoa infection; avian respiratory viruses. *Mailing Add:* Bur Labs Box 1031 Fla Dept Agr Dade City FL 33525

RADHAKRISHNAN, EGYARAMAN, chemical engineering, environmental engineering, for more information see previous edition

RADICE, GARY PAUL, b Bay Village, OH, April 9, 52. MUSCLE DEVELOPMENT, MORPHOGENESIS. *Educ:* Yale Univ, PhD(biol), 79. *Prof Exp:* Res assoc, Ind Univ, 81-90; RES ASSOC, UNIV RICHMOND, 90- *Mailing Add:* Dept Biol Univ Richmond Richmond VA 23173

RADIMER, KENNETH JOHN, b Clifton, NJ, Mar 31, 20. INORGANIC CHEMISTRY. *Educ:* Mass Inst Technol, SB, 42, PhD(inorg & anal chem), 47. *Prof Exp:* Res chemist, Nat Res Corp, Mass, 42; asst physics & chem, Mass Inst Technol, 43-44; res chemist chg anal lab, Kellex Corp, NJ, 44-45; res asst, SAM Labs, Carbide & Carbon Chem Corp, 45-46; instr anal chem, Lehigh Univ, 47-48; asst prof inorg & anal chem, Ind Univ, 48-50; res chemist, Gen Chem Div, Allied Chem & Dye Corp, 50-51; sect leader, Vitro Corp Am, 51-54; res chemist, M W Kellog Co Div, Pullman, Inc, 54-57 & 59-62 & Minn Mining & Mfg Co, 57-58; chief chemist, CBS Labs, 58-59; mgr metals applns, FMC Corp, Princeton, 62-69, res assoc, Indust Chem Div, 69-82; RETIRED. *Mem:* Electrochem Soc. *Res:* Microrefractometry; fluorine; fluorocarbon analysis; freon synthesis; molten salt electrolysis; reactions of metals with chemicals; chemistry of hydrogen peroxide; electrolytic persulfate processes; production of soda ash, phosphorus and phosphates; corrosion. *Mailing Add:* 12 Martin Pl Little Falls NJ 07424

RADIN, CHARLES LEWIS, b New York, NY, Jan 15, 45; m 69; c 1. MATHEMATICAL PHYSICS. *Educ:* City Col New York, BS, 65; Univ Rochester, PhD(physics), 70. *Prof Exp:* Fel math physics, Univ Nijmegen, 70-71; res assoc, Princeton Univ, 71-73 & Rockefeller Univ, 73-74; instr math, Univ Pa, 74-76; from asst prof to assoc prof, 76-90, PROF MATH, UNIV TEX, AUSTIN, 90- *Mem:* Am Math Soc; Am Phys Soc; Int Asn Math Physics. *Res:* Qualitative dynamics of quantum systems, especially many-body systems; automorphisms of operator algebras; classical ground states; neural networks. *Mailing Add:* Dept Math Univ of Tex Austin TX 78712

RADIN, ERIC LEON, b New York, NY, Sept 14, 34; c 3. ORTHOPEDICS, BIOMEDICAL ENGINEERING. *Educ:* Amherst Col, BA, 56; Harvard Univ, MD, 60. *Prof Exp:* Teaching fel orthop surg, Med Sch, Harvard Univ, 65-66, instr, 68-70, from asst prof to assoc prof, 70-79; prof & chmn, dept orthop surg, Med Sch, WVa Univ, 79-89; DIR BONE & JOINT CTR, HENRY FORD HOSP, DETROIT, 89-; CLIN PROF ORTHOP SURG, UNIV MICH, 89- *Concurrent Pos:* Co-chief, arthritis clins, US Air Force Hosp, Andrews AFB, 66-68; instr orthop surg, Med Sch, George Washington Univ, 67-68; consult, Liberty Mutual Rehab Ctr, 69-79 & Ctr Law & Health Sci, Boston Univ, 71-79; lectr mech eng, Mass Inst Technol, 69-79; assoc, active staff, orthop surg, Beth Israel Hosp, 70-74 & Children's Hosp Med Ctr, 70-79; mem, active orthop staff, Mt Auburn Hosp, 74-79 & dir, orthop res, 78-79; assoc, Mus Comp Zool, Harvard Univ, 75-79; mem, Surg & Rehab Devices Panel, US Food & Drug Admin, 83-87, chmn, 85-87; mem, Orthop Study Sect, NIH, 87-90, bd trustees, Clin Orthop on Related Res, 87-90; mem, orthop residency rev comt, 89- *Mem:* Orthop Res Soc; Am Soc Mat & Testing; Sigma Xi; Int Soc Surg, Orthop & Trauma; Soc Biomat; Biomed Eng Soc; Am Bd Orthop Surg; Am Acad Orthop Surg; Am Orthop Asn; Asn Bone Joint Surg. *Res:* Joint degeneration; joint lubrication; biomechanics of the degenerative process. *Mailing Add:* Bone & Joint Ctr Henry Ford Hosp Detroit MI 48202

RADIN, JOHN WILLIAM, b New York, NY, Jan 8, 44; m 65; c 2. PLANT PHYSIOLOGY. *Educ:* Univ Calif, Davis, BS, 65, PhD(plant physiol), 70. *Prof Exp:* Assoc agron, Univ Calif, Davis, 70-71; PLANT PHYSIOLOGIST, WESTERN COTTON RES LAB, AGR RES SERV, USDA, 71- *Mem:* Am Soc Plant Physiologists; Crop Sci Soc Am. *Res:* General plant physiology; root physiology; nitrogen metabolism, especially nitrate reduction; hormonal control of growth and development in plants. *Mailing Add:* Western Cotton Res Lab USDA-Agr Res Serv 4135 E Broadway Phoenix AZ 85040

RADIN, NATHAN, b Brooklyn, NY, Jan 22, 19; m 46; c 2. CLINICAL CHEMISTRY. *Educ:* Univ Calif, BA, 41; Columbia Univ, MA, 47; Purdue Univ, PhD(anal chem), 51. *Prof Exp:* Proj engr, Eng Res & Develop Labs, US Army, 51-52; chemist, Lederle Labs, Am Cyanamid Co, 53-54; biochemist, Mt Sinai Hosp, New York, 55-56; instr chem, Rochester Inst Technol, 56-59; chief biochemist, Rochester Gen Hosp, 58-66; clin chemist, Harrisburg Hosp Inst Path & Res, Penn, 66-67; res chemist, Ctr Dis Control, 67-86; CLIN CHEM CONSULT, 86- *Mem:* Am Chem Soc; Am Asn Clin Chem; Sigma Xi. *Res:* Quality control, standards; reference materials; training. *Mailing Add:* 28215 Plantation Dr NE Atlanta GA 30324

RADIN, NORMAN SAMUEL, b New York, NY, July 20, 20; m 47; c 2. BIOCHEMISTRY. *Educ:* Columbia Univ, BA, 41, PhD(biochem), 49. *Prof Exp:* Asst res chemist, Off Sci Res & Develop, Pa, 42-45; fel, Univ Calif, 49-50; res scientist, Biochem Inst, Univ Tex, 50-52; res assoc, Med Sch, Northwestern Univ, 52-55, from asst prof to assoc prof, 55-60; prof biol chem in psychiat, 73-84, PROF NEUROCHEM IN PSYCHIAT, 84-, RES BIOCHEMIST, MENT HEALTH RES INST, UNIV MICH, ANN ARBOR, 60- *Concurrent Pos:* Prin scientist, Radioisotope Unit, Vet Admin Hosp, Hines, Ill, 52-54 & Res Hosp, Chicago, 54-57; ed, Anal Biochem; Sen Jacob Javits Res Investr, 84. *Mem:* Am Soc Biol Chemists; Am Soc Neurochem; Int Soc Neurochem. *Res:* Brain lipids; lipid methodology; glycolipid metabolism; sphingolipid enzyme inhibitors; Gaucher's disease. *Mailing Add:* 3544 Terhune Rd Ann Arbor MI 48104-5320

RADIN, SHELDEN HENRY, b Hartford, Conn, Dec 24, 36; m 60; c 3. PLASMA PHYSICS. *Educ:* Worcester Polytech Inst, BS, 58; Yale Univ, MS, 59, PhD(physics), 63. *Prof Exp:* From asst prof to assoc prof, 63-74, assoc chair, 84-86, PROF PHYSICS, LEHIGH UNIV, 74-,. *Concurrent Pos:* Mem, Exam Comt, Physics Achievement Test, Col Entrance Exam Bd, 70-78, chmn, 72-78; vis prof, Univ Rochester, 86-87; reviewer, ACT, 82-, GRE, 84, MCAT, 89-; vis comt, NJ Inst Technol, 83, Univ Cinncinnati, 90- *Mem:* Am Phys Soc; Am Asn Univ Profs; Am Asn Physics Teachers. *Res:* Statistical mechanics of plasmas; kinetic theory of nonequilibrium situations; numerical simulations of plasmas. *Mailing Add:* Dept Physics Bldg 16 Lehigh Univ Bethlehem PA 18015

RADKE, FREDERICK HERBERT, biochemistry; deceased, see previous edition for last biography

RADKE, LAWRENCE FREDERICK, b Seattle, Wash, Mar 19, 42. ATMOSPHERIC SCIENCE, CLOUD PHYSICS. *Educ:* Univ Wash, BSc, 64, MSc, 66, PhD(atmospheric sci), 68. *Prof Exp:* Res assoc, 68-70, res asst prof, 70-72, res assoc prof, 72-80, RES PROF ATMOSPHERIC SCI, UNIV WASH, 80- *Mem:* Am Meteorol Soc; AAAS; Sigma Xi. *Res:* Aircraft measurments and instrumental development; meteorology; cloud physics; atmospheric chemistry; air pollution. *Mailing Add:* 1215 E Boston Seattle WA 98102

RADKE, RODNEY OWEN, b Ripon, Wis, Feb 5, 42; m 63; c 3. WEED SCIENCE. *Educ:* Univ Wis-Madison, BS, 63, MS, 65, PhD(soil biochem), 67. *Prof Exp:* Plant physiologist, US Army Biol Res Labs, 67-69; sr res biologist, Agr Div, Monsanto Co, 69-74, res specialist, Monsanto Agr Prods Co, 74-75, sr res group leader, 75-78, mgr, 78-81, mgr res, 81-86; MGR RES, MONSANTO AGR CO, 86- *Mem:* Am Soc Agron; Weed Sci Soc Am. *Res:* Discovery and development of crop protection chemicals for control of weeds and plant diseases; evaluation of bioengineered crops; operation of research farms. *Mailing Add:* Monsanto Agr Co 800 N Lindbergh Blvd St Louis MO 63167

RADKE, WILLIAM JOHN, b Mankato, Minn, June 8, 47; m 70, 84; c 2. COMPARATIVE & HUMAN ANATOMY. *Educ:* Mankato State Univ, BS, 70, MA, 72; Univ Ariz, PhD(zool), 75. *Prof Exp:* PROF BIOL, CENT STATE UNIV, OKLA, 75- *Concurrent Pos:* Post doctorate, Wolfson Inst, Univ Hull, 82-83, Univ Ariz, 90-91. *Mem:* Sigma Xi. *Res:* Hypothalamic-hypophyseal-thyroid axis of the bird; avian air sacs and the relationship to spermatogenesis; avian emetics for stomach contents recovery. *Mailing Add:* Dept Biol Cent State Univ Edmond OK 73034-0177

RADKOWSKY, ALVIN, b Elizabeth, NJ, June 30, 15; m 50; c 1. REACTOR PHYSICS. *Educ:* City Col New York, BSE, 35; George Washington Univ, AM, 42; Cath Univ Am, PhD(physics), 47. *Prof Exp:* Chief scientist, US Naval Reactor Hq, Washington, DC, 50-72; PROF NUCLEAR ENG, TEL AVIV UNIV & BEN GURION UNIV, 72- *Honors & Awards:* Abromowitz-Zeitlin Award, 86. *Mem:* Nat Acad Eng; fel Am Phys Soc; fel Am Nuclear Soc; Sigma Xi. *Res:* Reactor physics; nuclear reactor design and concepts. *Mailing Add:* Dept Interdisciplinary Studies Tel-Aviv Univ Ramat Aviv Tel-Aviv 69978 Israel

RADLOFF, HAROLD DAVID, b Mellen, Wis, Aug 25, 37; m 60; c 2. DAIRY SCIENCE, BIOCHEMISTRY. *Educ:* Univ Wis, BS, 59, MS, 61, PhD(dairy sci, biochem), 64. *Prof Exp:* Res fel dairy sci, Univ Wis, 64-66; asst prof dairy husb, Univ Wyo, 66-70, assoc prof, 70-77, prof, 77-81, prof animal sci, 81-88, exten animal scientist, 81-88; MGR MKT, CONTINENTAL GRAIN, 88- *Mem:* Am Dairy Sci Asn; Am Soc Animal Sci. *Res:* Lipid metabolism in ruminants; milk fat synthesis; general dairy cow nutrition. *Mailing Add:* 522 Midlane Dr Crystal Lake IL 60012-3334

RADLOFF, ROGER JAMES, b Mason City, Iowa, Oct 16, 40; m 68; c 2. VIROLOGY. *Educ:* Iowa State Univ, BS, 62; Calif Inst Technol, PhD(biophys & chem), 68. *Prof Exp:* Fel, Biophys Lab, Univ Wis, 68-72; asst prof microbiol, Sch Med, Univ NMex, 72-81, ASSOC PROF, 82- *Concurrent Pos:* NIH fel, 68-70; NSF & NIH res grant, 74- *Mem:* Am Soc Microbiol; Am Soc Virol. *Res:* Structure and synthesis of encephalomyocarditis virus and DNA repair in neurospora. *Mailing Add:* Dept Microbiol Univ NMex Albuquerque NM 87131

RADLOW, JAMES, b New York, NY. APPLIED MATHEMATICS. *Educ:* NY Univ, PhD(math), 57. *Prof Exp:* Assoc prof math, Adelphi Univ, 59-62 & Purdue Univ, 62-65; prof appl math, 65-89, RES PROF, UNIV NH, 89- *Res:* Diffraction theory; singular integral equations; Hilbert space; partial differential equations; magnetohydrodynamics. *Mailing Add:* 47 Maple St Somersworth NH 03878

RADNER, ROY, b Chicago, Ill, June 29, 27; m; c 4. TECHNICAL STAFF ADMINISTRATION. *Educ:* Univ Chicago, PhB, 45, BS, 50, MS, 51, PhD(math statist), 56. *Prof Exp:* Res asst, Cowles Comn Res Econ, Univ Chicago, 51, res assoc, 51-54, asst prof, 54-55; asst prof, Dept Econ, Yale Univ, 55-57; from assoc prof to prof econ & statist, Univ Calif, Berkeley, 57-79, chmn, Dept Econ, 66-69; mem tech staff, 79-85, DISTINGUISHED MEM TECH STAFF, AT&T BELL LABS, 85-; RES PROF ECON, NY UNIV, 83- *Concurrent Pos:* Fel, Ctr Advan Study Behav Sci, 55-56; consult, Boeing Airplane Co, 56-57, Kaiser Found Psychol Res, 57-58, Rand Corp, 59-70, Soc Econ & Appl Math, 61-62, Syst Develop Corp, 62-65 & Mathematica, 56-66; consult, Maritime Cargo Transp Conf, Nat Acad Sci, 58-60, Comt Utilization of Sci & Eng Manpower, 62-63; assoc ed, Mgt Sci, 59-70, Econometrica, 61-68, J Econ Theory, 68- & Am Econ Rev, 79-82; Guggenheim Found fel, 61-62 & 65-66; mem, Econ Adv Panel, NSF, 63-65; mem, Tech Adv Comt, Carnegie Comn Future Higher Educ, 67-73, consult, 67-73; mem, Comt Status Teaching Asst, Am Asn Univ Profs, 68-70; overseas fel, Churchill Col, Cambridge, 69-70 & 90; mem math sci bd, Social Sci Res Coun, 70-74; mem, Nat Bur Econ Res Comt Econometrics & Math Econ, 71-; mem, Adv Comt Econ Educ, Nat Acad Educ, 72-73; mem, Adv Bd Off Math Sci, Nat Res Coun-Nat Acad Sci, 72-76, Comn Human Resources, 76-79, Comt Fundamental Res Relevant to Educ, 76-77, Assembly Behav & Social Sci, Nat Res Coun, 79-82, Comt Risk & Decision Making, 80-81, Comt Basic Res in Behav & Social Sci, Working Group Markets & Orgns, 85, Comt Contrib of Behav & Social Sci to Prev Nuclear War, 85-90; Taussig prof econ, Harvard Univ, 77-78; vis prof, Kennedy Sch Govt, Harvard Univ, 78-79. *Mem:* Nat Acad Sci; fel Am Acad Arts & Sci; fel AAAS; distinguished fel Am Econ Asn; fel Econometric Soc (vpres, 71-72, pres, 72-73); Fedn Am Scientists; Inst Math Statist. *Res:* Theories of decentralization, incentives, transfer pricing and budgeting; theories of sequential games with uncertainty; deregulation of long-distance telecommunications; strategic and economic aspects of national security; markets for data network products. *Mailing Add:* AT&T Bell Labs Rm 2C-176 600 Mountain Ave PO Box 636 Murray Hill NJ 07974-0636

RADNITZ, ALAN, b Miami Beach, Fla, Dec 16, 44; m 68; c 2. MATHEMATICS. *Educ:* Univ Calif, Los Angeles, AB, 66, PhD(math), 70. *Prof Exp:* asst prof, 70-75, assoc prof, 75-82, PROF MATH, CALIF STATE POLYTECH UNIV, POMONA, 82- *Mem:* Am Math Soc; Math Asn Am. *Res:* Differential equations in banach spaces; partial differential equations; functional analysis. *Mailing Add:* Dept Math Calif State Polytech Univ 3801 W Temple Ave Pomona CA 91768

RADO, GEORGE TIBOR, b Budapest, Hungary, July 22, 17; nat US; m 64; c 2. SOLID STATE PHYSICS. *Educ:* Mass Inst Technol, SB, 39, SM, 41, PhD(physics), 43. *Prof Exp:* Res assoc, Div Indust Coop, Mass Inst Technol, 42-43 & Radiation Lab, 44-45; physicist, Naval Res Lab, 45-55, head, Magnetism Br, 55-82, chief scientist magnetism, 82-83; RES PROF, JOHNS HOPKINS UNIV, 83- *Concurrent Pos:* Adj prof Univ Md, 62-83; mem comn magnetism, Int Union Pure & Appl Physics, 66-75, secy, 69-72, chmn, 72-75. *Honors & Awards:* Pure Sci Award, Naval Res Lab-Sci Res Soc Am; 57; E O Hulburt Sci Award, 65. *Mem:* Fel Am Phys Soc; Sigma Xi. *Res:* Light scattering; microwave propagation; saturation magnetization; magnetic spectra; domain theory; ferrites; Faraday effect; ferromagnetic resonance; magnetocrystalline anisotropy; magnetoelectric effects; ultra-thin magnetic films; magnetic surface anisotropy. *Mailing Add:* 818 Carrie Ct McLean VA 22101

RADOMSKI, JACK LONDON, b Milwaukee, Wis, Dec 10, 20; m 47, 70; c 4. CHEMICAL CARCINOGENESIS. *Educ:* Univ Wis, BS, 42; George Washington Univ, PhD, 50. *Prof Exp:* Chemist, Gen Aniline & Film Corp, 42-44; pharmacologist, Food & Drug Admin, Fed Security Agency, 44-53; from asst prof to assoc prof pharmacol, Sch Med, Univ Miami, 53-59, prof, 59-82; pres, Covington Chemtox, 82-87; CONSULT, 87- *Concurrent Pos:* Consult toxicol, WHO, 78-82. *Honors & Awards:* Acad Toxicol, Award 82. *Mem:* Am Soc Pharmacol & Exp Therapeut; Soc Toxicol; Am Asn Cancer Res; NY Acad Sci; Acad Toxicol Sci. *Res:* Toxicology and metabolism of drugs, chemicals and insecticides; environmental and occupational toxicology and carcinogenesis. *Mailing Add:* 6432 Driftwood Dr Hudson FL 34667

RADONOVICH, LEWIS JOSEPH, b Curtisville, Pa, July 2, 44; m 66; c 2. STRUCTURAL CHEMISTRY. *Educ:* Thiel Col, BA, 66; Wayne State Univ, PhD(phys chem), 70. *Prof Exp:* Res assoc inorg chem, Cornell Univ, 70-73; from asst prof to assoc prof, 73-84, PROF CHEM, UNIV NDAK, 84- *Concurrent Pos:* Res supvr inorg anal, Univ NDak Energy Res Ctr, 84-86; chmn, Dept Chem, 88- *Mem:* Am Chem Soc; Am Crystallog Asn. *Res:* Organometallic chemistry, x-ray crystallography and the structural chemistry of compounds of biological interest. *Mailing Add:* Dept Chem Univ NDak PO Box 7185 Grand Forks ND 58202

RADOSEVICH, LEE GEORGE, b Milwaukee, Wis, Nov 5, 38. ENERGY CONVERSION, SOLID STATE PHYSICS. *Educ:* Marquette Univ, BS, 60, MS, 62; Northwestern Univ, Ill, PhD(physics), 68. *Prof Exp:* Physicist, Allis-Chalmers Mfg Co, 62; res assoc physics, Univ Ill, Urbana, 67-69; MEM STAFF PHYSICS, SANDIA LABS, 69- *Mem:* Am Phys Soc. *Res:* Solar thermal power conversion. *Mailing Add:* Div 8166 Sandia Labs PO Box 969 Livermore CA 94550

RADOSKI, HENRY ROBERT, b Jersey City, NJ, Aug 18, 36; m 59; c 3. SPACE PHYSICS, PLASMA PHYSICS. *Educ:* Col Holy Cross, BS, 58; Mass Inst Technol, PhD(physics), 63. *Prof Exp:* Res asst plasma physics, Res Lab Electronics, Mass Inst Technol, 59-63; res assoc prof geophys, Weston Observ, Boston Col, 63-68; res physicist, Space Physics Lab, USAF Geophysics Lab, 68-76, PROG MGR, USAF OFF SCI RES, 76- *Mem:* Am Geophys Union; Am Phys Soc; Am Astron Soc; Fedn Am Scientists; Sigma Xi. *Res:* Magnetospheric physics; solar physics; astronomy; solid earth physics. *Mailing Add:* 2603 Turnbridge Lane Alexandria VA 22308

RADOVSKY, FRANK JAY, b Fall River, Mass, Jan 5, 29; div; c 2. ACAROLOGY, MEDICAL ENTOMOLOGY. *Educ:* Univ Colo, AB, 51; Univ Calif, Berkeley, MS, 59, PhD(parasitol), 64. *Prof Exp:* Actg asst prof entom, Univ Calif, Berkeley, 62-63; asst res parasitologist, Hooper Found, Med Ctr, Univ Calif, San Francisco, 63-69, lectr parasitol, Dept Int Health, 69; acarologist, Bishop Mus, 70-85, chmn dept entom, 72-85, asst to dir res, 73-76, actg dir, 76-77, asst dir, 77-85, L A Bishop distinguished chair zool, 84-86; vis prof entom, Ore State Univ, Corvallis, 87; DIR RES & COLLECTIONS, NC STATE MUS NATURAL SCI, 87- *Concurrent Pos:* Ed, J Med Entom, 70-85 & 87-88; secy, Int Cong Acarology, 71-78; mem bd mgt, Wau Ecol Inst, Papua, New Guinea, 72-85; mem, Hawaii Animal Species Adv Comn, 72-80, Sci Comt Entom, Pac Sci Asn, 82-87 & Hawaiian Natural Area Reserves Syst Comn, 85; managing ed, Pac Insects, 78-85; assoc ed, Annual Review Entom, 78-90, ed, 90-, Brimlegana, 88- *Mem:* AAAS; Entom Soc Am; Acarol Soc Am; Sigma Xi; Am Soc Trop Med & Hyg; Soc Vector Ecologists. *Res:* Biology and systematics of acarine parasites; adaptation and evolution of relationships between arthropod parasites and their hosts; arthropod vectors of disease agents. *Mailing Add:* NC State Mus Natural Sci PO Box 27647 Raleigh NC 27611-7647

RADSPINNER, JOHN ASA, b Vincennes, Ind, May 14, 17; m 42; c 2. PHYSICAL CHEMISTRY. *Educ:* Univ Richmond, BS, 37; Va Polytech Inst, MS, 38; Carnegie Inst Technol, DSc(phys chem), 42. *Prof Exp:* Res chemist, Pan-Am Refining Corp, Div Standard Oil Co Ind, 42-44, supvr operating dept, 44-54, asst dir indust rels, Am Oil Co Div, 54-56, personnel dir, Yorktown Refinery, 56-57; assoc prof, 57-62, actg dean, 69-70, chmn dept, 70-72, PROF CHEM, LYCOMING COL, 62- *Mem:* Am Chem Soc; AAAS. *Mailing Add:* 432 Oakland Ave Williamsport PA 17701

RADTKE, DOUGLAS DEAN, b New London, Wis, Nov 6, 38; m 62; c 2. PHYSICAL INORGANIC CHEMISTRY. *Educ:* Wis State Univ, Stevens Point, BS, 61; Univ Wis, PhD(phys chem), 66. *Prof Exp:* Asst prof, 66-69, assoc prof, 69-77, PROF CHEM, UNIV WIS, STEVENS POINT, 77- *Mem:* Am Chem Soc. *Res:* Molecular orbital calculations for transition metal complexes; preparation and structure of simple divalent rare earth compounds. *Mailing Add:* Dept Chem Univ Wis Stevens Point WI 54481

RADTKE, SCHRADE FRED, inorganic chemistry, metallurgy; deceased, see previous edition for last biography

RADWAN, MOHAMED AHMED, b Dakahlia, Egypt, Apr 16, 26; nat US; m 57. PLANT PHYSIOLOGY. *Educ:* Cairo Univ, Egypt, BSc, 46; MS, 50; Univ Calif, PhD(plant physiol), 56. *Prof Exp:* Asst lectr chem, Cairo Univ, 47-52, lectr plant physiol, 56-57; sr lab technician, Univ Calif, 57-58; instr chem, Sacramento City Col, 58-60; plant physiologist, 60-68, PRIN PLANT PHYSIOLOGIST, US FOREST SERV, 68- *Mem:* Am Chem Soc; Am Soc Plant Physiol; Bot Soc Am; Am Soc Agron; Soc Am Foresters. *Res:* Nutrition; fertilization; forest tree physiology. *Mailing Add:* 9130 Blomberg SW Olympia WA 98502

RADWANSKA, EWA, b Wilno, Poland, Oct 24, 38; c 1. OBSTETRICS & GYNECOLOGY, ENDOCRINOLOGY. *Educ:* Med Acad, Warsaw, MD, 62, Dr Med Sci, 69; Univ London, MPhil, 75. *Prof Exp:* Resident & instr obstet, gynec & endocrinol, Med Acad, Warsaw, 64-70; fel endocrinol, Univ Col Hosp, London, 70-75; registr, Hillingdon Hosp, London, 75-76; asst prof obstet, gynec & endocrinol, Univ NC, Chapel Hill, 77-79; mem staff, Univ Ark, Little Rock, 79-81; MEM STAFF, DEPT OBSTET & GYNEC, RUSH MED COL, CHICAGO, 81- *Concurrent Pos:* Med officer, City Clin, Warsaw, 64-70, Family Planning Asn, London, 70-76 & Marie Stopes Mem Birth Control Ctr, London, 70-76. *Mem:* Polish Endocrine Soc; Am Med Soc; Royal Col Obstetricians & Gynecologists; Am Fertil Soc; Soc Study Reprod. *Res:* Reproductive endocrinology, particularly induction of ovulation, ovarian failure, luteal deficiency, spontaneous abortions, infertility, progesterone assay, tubal sterilization. *Mailing Add:* Dept Obstet & Gynec Rush Med Col 600 S Paulina St Chicago IL 60612

RADWIN, HOWARD MARTIN, b New York, NY, Mar 13, 31; m 58; c 3. UROLOGY. *Educ:* Princeton Univ, AB, 52; Columbia Univ, MD, 56. *Prof Exp:* Asst prof urol, Tulane Univ, 64-68; PROF UROL & CHMN DEPT, UNIV TEX HEALTH SCI CTR SAN ANTONIO, 68- *Concurrent Pos:* Nat Cancer Inst fel, Tulane Univ, 61-62; consult, Brooke Army Hosp & Air Force Wilford Hall Hosp, 69- *Mem:* Fel Am Col Surg; Am Urol Asn; Soc Univ Urologists. *Res:* Prostate physiology; pyelonephritis; urologic cancer; application of renal physiology to urologic disease. *Mailing Add:* Dept Surg Univ Tex Health Sci Ctr 7703 Floyd Curl Dr San Antonio TX 78284-7845

RADZIALOWSKI, FREDERICK M, b Detroit, Mich, Mar 25, 39; m 60; c 2. PHARMACOLOGY, BIOCHEMISTRY. *Educ:* Wayne State Univ, BS, 60, MS, 64; Purdue Univ, PhD(pharmacol), 68. *Prof Exp:* SR RES INVESTR METAB, G D SEARLE & CO, 68- *Mem:* AAAS; Am Pharmaceut Asn. *Res:* Drug and lipid metabolism; obesity; circadian rhythms. *Mailing Add:* Biol Res Box 5110 C D Searle Co Chicago IL 60680

RADZIEMSKI, LEON JOSEPH, b Worcester, Mass, June 18, 37; c 2. ATOMIC SPECTROSCOPY. *Educ:* Col Holy Cross, BS, 58; Purdue Univ, MS, 61, PhD(physics), 64. *Prof Exp:* Lectr physics, USAF Inst Technol, 65-67; staff mem, Los Alamos Nat Lab, Univ Calif, 67-83; head physics dept, NMex State Univ, 83-88, assoc dean, Col Arts & Sci & dir, Arts & Sci Res Ctr, 88-90; DEAN, DIV SCI, COL ARTS & SCI, WASH STATE UNIV, 90- *Concurrent Pos:* Lectr, Wright State Univ, 66-67; vis scientist, Lab Aime Cotton, France, 74-75; mem comt line spectra of the elements, Nat Res Coun, 75-81; vis mem fac, Univ Fla, 78-79; vis scientist, Sandia Nat Lab, Livermore, 84-88. *Mem:* Fel Optical Soc Am; Am Phys Soc; Soc Applied Spectroscopy; sr mem Laser Inst Am (pres, 91). *Res:* Atomic spectroscopy; spectrochemical applications of lasers; remote, point, and in situ detection of toxic substances; laser-induced breakdown spectroscopy; laser guiding of electron beams; optical spectroscopy (conventional & laser); editor of 2 books, holds 2 patents and 65 publications. *Mailing Add:* Div Sci 208 Morrill Hall Wash State Univ Col Arts & Sci Pullman WA 99164-3520

RADZIKOWSKI, M ST ANTHONY, b Jermyn, Pa, Mar 10, 19. INORGANIC CHEMISTRY. *Educ:* Marywood Col, AB, 39; Univ Notre Dame, MS, 57, PhD(chem), 61. *Prof Exp:* Teacher parochial sch, Pa, 45-55; dean women, 55-58, chmn chem dept, 61-86, ASSOC PROF CHEM, MARYWOOD COL, 61- *Mem:* Am Chem Soc; Coblentz Soc. *Res:* Complexes of methyl esters of proline and sarcosine; infrared spectra of coordination compounds of amines with metal halides; spectroscopic studies of reactions with organic donor compounds. *Mailing Add:* Dept Chem Marywood Col 2300 Adams Ave Scranton PA 18509

RAE, PETER MURDOCH MACPHAIL, b Glasgow, Scotland, Jan 7, 44; US citizen; m 71; c 1. CELL BIOLOGY. *Educ:* Univ Calif, Davis, AB, 65, MA, 66; Univ Chicago, PhD(biol), 70. *Prof Exp:* Lectr biol, Harvard Univ, 70; Max-Planck Inst Biol, Tubingen, WGer, 71-72, vis scientist, 73; from asst prof to assoc prof biol, Yale Univ, 73-83; sr res scientist, 83-85, prin staff scientist, Molecular Diagnostics Inc, 85-90, PRIN STAFF SCIENTIST, MILES INC, 90- *Mem:* AAAS: Am Soc Cell Biol; Am Soc Cell Biol; Am Diabetes Asn. *Res:* Molecular biology of metabolic and autoimmune diseases. *Mailing Add:* Miles Res Ctr 400 Morgan Lane West Haven CT 06516

RAE, STEPHEN, b New York, NY, May, 44; m 65. ENVIRONMENTAL PHYSICS. *Educ:* Stevens Inst Technol, BS, 65; Univ Vt, MS, 69, PhD(physics), 73. *Prof Exp:* Eng physicist noise abatement, US Naval Marine Eng Lab, 65-66; anal physicist nuclear reactors, Knolls Atomic Power Lab, 66-67; teaching asst, 69-73, lectr physics, Univ Vt, 75-76; ASST PROF PHYSICS, WELLS COL, 76- *Concurrent Pos:* Pres sci & tech consult serv, N&R Assoc Inc, 75- *Mem:* Am Asn Physics Teachers; Fedn Am Scientists. *Res:* Application and teaching of physics related to environmental problems, currently in the area of energy; theoretical description of atomic collision processes. *Mailing Add:* Dept Physics Wells Col Aurora NY 13026

RAE, WILLIAM H, JR, b Tacoma, Wash, Nov 16, 27; m 53. AERODYNAMICS. *Educ:* Univ Wash, Seattle, BS, 53, MS, 59. *Prof Exp:* Supvr, Aeronaut Labs, 59-66, asst prof aeronaut & astronaut, 65-67, ASSOC PROF AERONAUT & ASTRONAUT, UNIV WASH, 67-, ASSOC DIR, AERONAUT LABS, 66- *Concurrent Pos:* Chmn, Subsonic Aerodynamic Testing Asn, 85-87. *Mem:* Am Inst Aeronaut & Astronaut; Subsonic Aerodynamic Testing Asn (chmn, 85-87). *Mailing Add:* Dept Aeronaut Univ Wash Seattle WA 98195

RAE, WILLIAM J, b Buffalo, NY, Sept 3, 29; m 57; c 4. AERONAUTICAL ENGINEERING, HEAT TRANSFER. *Educ:* Canisius Col, BA, 50; Cornell Univ, PhD(aeronaut eng), 60. *Prof Exp:* Computer, Cornell Aeronaut Lab, Inc, 50-53, jr mathematician, 53-54, jr aerodynamicist, 54-55, res asst, Cornell Univ, 55-59, res aerodynamicist, Cornell Aeronaut Lab, Inc, 59-63, prin aerodynamicist, 63-64; prin res engr, Advan Technol Ctr, Calspan Corp, 64-83; vis prof, 83-84, res prof, 84-85, PROF, DEPT MECH & AEROSPACE ENG, STATE UNIV NY, BUFFALO, 85- *Concurrent Pos:* Lectr, Medaille Col, 61-67, trustee, 69-76. *Mem:* Am Inst Aeronaut & Astronaut; Sigma Xi; Am Soc Mech Engrs; Nat Soc Prof Engrs. *Res:* Wing-body interference; viscous acoustics; boundary-layer flow; impact-generated shock-wave propagation in solids; low-density flow, environmental fluid mechanics; turbomachinery; computational fluid dynamics. *Mailing Add:* 215 Lamarck Dr Snyder NY 14226

RAE-GRANT, QUENTIN A, b Aberdeen, Scotland, Apr 5, 29; Can citizen; m 55; c 2. PSYCHIATRY. *Educ:* Aberdeen Univ, MB, ChB, 51; Univ London, dipl psychiat med, 58; FRCP(C), 73. *Prof Exp:* Intern med & surg, Aberdeen Univ, 52-53, resident psychiat, 53-54; resident, Maudsley Hosp, London, Eng, 55-58; dir child psychiat, Jewish Hosp, St Louis, Mo, 58-60; instr pediat & psychiat, Univ & psychiatrist, Univ Hosp, Johns Hopkins Univ, 60-61; dir ment health div, St Louis County Health Dept, 62-64; chief social psychiat sect, Community Res & Serv Br, NIMH, 65-66, dir & chief ment health study ctr, Md, 66-68; vchmn, dept psychiat, Univ Toronto, 71-82, prof & head div child psychiat, 68-87, prof & chmn dept behav sci, 84-87; psychiatrist-in-chief, Hosp Sick Children, 68-87; PROF CHILD PSYCHIAT, UNIV WESTERN ONT, 87- *Concurrent Pos:* Consult, Sinai Hosp, Baltimore & Rosewood State Hosp, Md, 60-61 & St Louis State Hosp, 62-64; asst prof, Wash Univ, 62-64; lectr, Sch Nursing, St Louis Univ, 62-64; asst prof, Johns Hopkins Univ, 65-68; consult, Comn Emotional & Learning Dis in Children, Clarke Inst Psychiat & St Michael's Hosp, Toronto, 68-70, assoc prof pediat, 73-81, prof, 81-87; chief examr, Royal Col Physicians & Surgeons, Can, 81-86. *Mem:* Am Psychiat Asn; Am Orthopsychiat Asn; Can Psychiat Asn (pres elect, 81-82, pres, 82-83); Royal Col Psychiat; Can Acad Child Psychiat (pres, 84-86). *Res:* Child and adolescent psychiatry; consultation service; new patterns of psychiatric service; evaluation of health care delivery system effectiveness; prevention of sexual abuse. *Mailing Add:* Dept Psychiat Univ Western Ont London ON N6A 5C1 Can

RAEL, EPPIE DAVID, b Cochiti, NMex, Jan 17, 43; m 71; c 2. IMMUNOLOGY, MICROBIOLOGY. *Educ:* Univ Albuquerque, BS, 65; NMex Highlands Univ, MS, 70; Univ Ariz, PhD(microbiol), 75. *Prof Exp:* Res asst clin immunol, Med Sch, Univ NMex, 69-71; asst microbiol, Univ Ariz, 71-75; from asst prof to assoc prof, 75-90, PROF IMMUNOL, UNIV TEX, EL PASO, 90- *Concurrent Pos:* Prin investr, Minority Biomed Support, Nat Cancer Inst, 75-77 & Minority Biomed Support, HEW, 77-; prog dir, Minority Biomed Support, 84-90; counr, Coun Undergrad Res. *Mem:* Am Soc Microbiol; AAAS; Sigma Xi; Int Asn Toxinologists; Am Asn Immunologists. *Res:* Monoclonal antibodies; complement; rattlesnake venom. *Mailing Add:* Dept Biol Sci Univ Tex El Paso TX 79968

RAEMER, HAROLD R, b Chicago, Ill, Apr 26, 24; m 47; c 3. PHYSICS, ELECTRICAL ENGINEERING. *Educ:* Northwestern Univ, BS, 48, MS, 49, PhD(physics), 59. *Prof Exp:* Asst math, Ind Univ, 49-50; asst physics, Northwestern Univ, 50-52; physicist, Res Labs, Bendix Aviation Co, Mich, 52-54, sr physicist, 54-55; sr engr, Cook Res Labs, Ill, 55-57, staff engr, 57-60; asst prof elec eng, Ill Inst Technol, 60; sr eng specialist, Appl Res Labs, Sylvania Electronic Systs, Gen Tel & Electronics Corp, 60-63; assoc prof, 63-66, chmn dept, 67-77, PROF ELEC ENG, NORTHEASTERN UNIV, 66- *Concurrent Pos:* Vis lectr, Harvard Univ, 62, hon res assoc, 72-73; consult, Sylvania Appl Res Labs, 63-71 & US Naval Res Lab, 69-; vis scientist, Mass Inst Technol, 84-85. *Mem:* AAAS; Am Phys Soc; sr mem Inst Elec & Electronics Engrs. *Res:* Electromagnetic radio wave propagation theory; statistical communication theory; plasma physics, particularly wave propagation in plasma. *Mailing Add:* Dept Elec Eng Northeastern Univ 360 Huntington Ave Boston MA 02115

RAESE, JOHN THOMAS, b West Chester, Pa, Apr 3, 30; m 53; c 4. AGRONOMY, PLANT PHYSIOLOGY. *Educ:* WVa Univ, BS, 52, MS, 59; Univ Md, PhD(agron), 63. *Prof Exp:* Instr agron, WVa Univ, 58-59; teacher high sch, Md, 62-63; res plant physiologist, Field Lab Tung Invests, La, 63-68, Tung Trees Lab, Fla, 68-71 & Pome Fruit Lab, Wash, 71-73, PLANT PHYSIOLOGIST, AGR RES SERV, USDA, 73- *Mem:* Soc Cryobiol; Am Soc Hort Sci; Sigma Xi. *Res:* Chemical analyses of soils and plant tissues; pasture and forage management; physiological and nutritional studies of the tung tree; plant nutrition, growth regulators, and cold hardiness of pome trees. *Mailing Add:* 1104 N Western Wenatchee WA 98801

RAESIDE, JAMES INGLIS, b Saskatoon, Sask, May 21, 26; m 54; c 4. PHYSIOLOGY. *Educ:* Glasgow Univ, BSc, 47; Univ Mo, MS, 50, PhD, 54. *Prof Exp:* Sr lectr animal physiol, NZ, 54-57; res fel, McGill Univ, 57-58; PROF PHYSIOL, ONT VET COL, UNIV GUELPH, 58- *Concurrent Pos:* Vis scientist, Karolinska Inst, Sweden, 64-65, Weizmann Inst, Israel, 82, INSERM U307, Lyon, France, 87. *Mem:* Am Soc Animal Sci; Can Biochem Soc; Brit Soc Study Fertil; Soc Study Reprod; AAAS; Am Soc Androl; Sigma Xi. *Res:* Comparative physiology of reproduction; endocrinology; steroid metabolism; hormone assay; animal production and behavior. *Mailing Add:* Dept of Biomed Sci Ont Vet Col Univ Guelph Guelph ON N1G 2W1 Can

RAETHER, MANFRED, b Stettin, Ger, Jan 22, 27; m 56; c 2. PLASMA PHYSICS. *Educ:* Univ Bonn, Dr rer nat(physics), 58. *Prof Exp:* Asst physics, Univ Bonn, 57-58; res engr, US Army Ballistic Missile Agency, Ala, 58-59; res asst prof, Coord Sci Lab, 59-61, assoc prof, Dept Physics & Coord Sci Lab, 61-67, PROF PHYSICS, UNIV ILL, URBANA, 67-, ASSOC HEAD DEPT, 80- *Mem:* Fel Am Phys Soc. *Res:* Plasma instabilities; plasma turbulence. *Mailing Add:* Dept Physics Univ Ill 215 Loomis Lab Urbana IL 61801

RAETZ, CHRISTIAN RUDOLF HUBERT, b Berlin, Ger, Nov 17, 46; US citizen; m 71; c 2. BIOCHEMISTRY, MEDICINE. *Educ:* Yale Univ, BS, 67; Harvard Univ, PhD(biochem) & MD, 73. *Prof Exp:* House officer, Peter Bent Brigham Hosp, Boston, Mass, 73-74; res assoc, NIH, 74-76; from asst prof to assoc prof, 76-82, prof biochem & dir, ctr membrane biosynthesis res, Univ Wis-Madison, 82-87; EXEC DIR, BIOCHEM, MERCK SHARP & DOHME RES LABS, 87- *Concurrent Pos:* Life Ins Found med scientist fel, 69-72; sr comn officer student training & extern prog fel, NIH, 72-73, res career develop award, 78-83, mem, physiol chem study sect, 78-81 & biochem study sect, 85; chmn, Membrane Lipid Metab, Gordon Conf, 83; H I Romnes fac fel, Univ Wis, 84. *Honors & Awards:* Harry & Evelyn Steenbock Award, Univ Wis, 76. *Mem:* Fel Japanese Soc Prom Sci; Am Soc Biol Chemists. *Res:* Synthesis and function of biological membranes; metabolism of phospholipids; genetics of bacteria and animal cells grown in tissue culture. *Mailing Add:* Merck Sharp & Dohme Res Labs PO Box 2000 Rahway NJ 07065

RAE-VENTER, BARBARA, b Auckland, NZ, July 17, 48; US citizen; m 68, 81, 90; c 1. BIOCHEMISTRY, MOLECULAR BIOLOGY. *Educ:* Univ Calif, San Diego, BA, 72, PhD(biol), 76; Univ Tex, Austin, JD, 85. *Prof Exp:* Fel, Roswell Park Mem Inst, 76-77, cancer res scientist II, 77-79; asst prof, Univ Tex Med Br, Galveston, 79-83; law clerk, Davis & Davis, Austin, Tex, 84-85; assoc patent atty, Richard, Harris, Medlock & Andrews, Dallas, 85-86; assoc patent atty, Leydig, Voit & Mayer, Palo Alto, Calif, 86-89; SPEC COUN, COOLEY GODWARD CASTRO HUDDLESON & TATUM, CALIF, 90- *Concurrent Pos:* Vis asst prof, Stanford Univ, 88-90. *Mem:* Endocrine Soc; Am Bar Asn. *Res:* The role of steroid and peptide hormones in tumorigenesis; control of peptide hormone receptor synthesis; elucidation of the subcellular events consequent to peptide hormone receptor binding. *Mailing Add:* Cooley Godward Castro Huddleson & Tatum S Palo Alto Square 4th Floor Palo Alto CA 94306

RAFAJKO, ROBERT RICHARD, b Chicago, Ill, Sept 3, 31; div; c 6. VIROLOGY. *Educ:* Coe Col, BA, 53; Univ Iowa, MS, 58, PhD(bact), 60. *Prof Exp:* Res assoc biol, Merck Sharp & Dohme, Inc, Pa, 60-61; virologist, Microbiol Assocs, Inc, Md, 61-66; vpres & gen mgr biol res, Med Res Consults Div, NAm Mogul Prod, Inc, 66-69, dir res & develop, Diag Div, 69-71, dir res & develop, NAm Biologicals, Inc, 71-74; PRES, BIOFLUIDS INC, 75- *Concurrent Pos:* Pres, Tysan Serum, Inc, 74- *Mem:* AAAS; NY Acad Sci; Tissue Cult Asn; Am Soc Microbiol. *Res:* Interferon induction and assay systems; development of killed equine virus vaccine; adenovirus strain differences and relatedness to oncogenicty; adeno-associated viruses; adenovirus induced transformation of mammalian cells; cell growth factors. *Mailing Add:* Biofluids Inc 1146 Taft St Rockville MD 20850

RAFANELLI, KENNETH R, b New York, NY, Nov 11, 37; m 61; c 1. THEORETICAL PHYSICS. *Educ:* Stevens Inst Technol, ME, 58, MS, 60, PhD(physics), 64. *Prof Exp:* From asst prof to assoc prof, 64-73, PROF PHYSICS, QUEENS COL, NY, 73- *Concurrent Pos:* Consult, TRW Systs, Calif, 65-68. *Mem:* Sigma Xi. *Res:* Theoretical research on the elementary particles. *Mailing Add:* Dept Physics Queens Col City Univ NY Flushing NY 11367

RAFELSKI, JOHANN, b Krakow, Poland, May 19, 50; Ger citizen; m 73; c 2. NUCLEAR & VACUUM STRUCTURE, NUCLEAR FUSION. *Educ:* Univ Frankfurt, dipl, 71, DPhilNat, 73. *Prof Exp:* Sci assoc, Univ Frankfurt, 71-73, assoc prof theoret physics, 79-83; sci assoc, Univ Pa, 73-74; staff, Argonne Nat Lab, 74-78; fel, Cern-Genera, 77-79; chair theoret physics, Univ Cape Town, 83-87; PROF THEORET PHYSICS, UNIV ARIZ, 87- *Mem:* Am Phys Soc; Europ Phys Soc. *Res:* Subnuclear theoretical physics; vacuum structure; nuclear collisions; muon catalyzed fusion; neural nets; low energy nuclear reactions. *Mailing Add:* Dept Physics Univ Ariz Tucson AZ 85721

RAFELSON, MAX EMANUEL, JR, b Detroit, Mich, June 17, 21; m 47, 73; c 2. BIOCHEMISTRY. *Educ:* Univ Mich, BS, 43; Univ Southern Calif, PhD(biochem), 51. *Prof Exp:* From asst prof to assoc prof, 53-61, PROF BIOL CHEM, UNIV ILL COL MED, 61-; vpres, Rush-Presby-St Luke's Med Ctr, 72-78; prof biochem, 70-88, EMER PROF & CHMN BIOCHEM, RUSH MED COL, 88- *Concurrent Pos:* USPHS res fel, Wenner Grens Inst, Stockholm, Sweden, 51-53; chmn dept biochem, Presby-St Luke's Hosp, 61-70; assoc dean biol & behav sci & serv, Rush Med Col, 70-72, vpres, mgt info sci, 72-78; vis prof, Univ Paris, 61, 77-78. *Mem:* AAAS; Am Chem Soc; Am Soc Biol Chem; Brit Biochem Soc; Nat Acad Clin Biochem. *Res:* Protein chemistry; enzymology; blood; platelets; endothelial cells; prostaglandins. *Mailing Add:* Dept Biochem Rush Med Sch Chicago IL 60612

RAFF, ALLAN MAURICE, b Chicago, Ill, May 21, 23; m 47; c 2. PHARMACY. *Educ:* Univ Ill, Chicago, BS, 49, MS, 53; Temple Univ, PhD(phys pharm), 64. *Prof Exp:* Mgr pharmaceut develop, Smith Kline & French Labs, 53-69 & med diag opers, Xerox, 69-70; dir res & develop, Barnes-Hind Pharmaceut Co, 71-72; vpres pharmaceut, Rachelle Labs, 72-83; vpres pharmaceut, Dey Labs, 83-90; MGR, ELSO PHARMACEUT CONSULT, INC, 90- *Mem:* Am Pharmaceut Asn; Am Chem Soc; Am Inst Chem Engr. *Res:* Effects of physical parameters on solid dosage forms. *Mailing Add:* 1500 Lincoln Lane Newport Beach CA 92660-4938

RAFF, HERSHEL, b Paterson, NJ, May 23, 53; m 76; c 1. NEUROENDOCRINOLOGY, SYSTEMS PHYSIOLOGY. *Educ:* Union Col, BA, 75; Johns Hopkins Univ, PhD(physiol), 81. *Prof Exp:* Teaching fel endocrinol, Univ Calif, San Francisco, 80-83; ASST PROF MED & PHYSIOL, MED COL WIS, 83-; DIR ENDOCRINE RES, ST LUKE'S HOSP, MILWAUKEE, 83- *Mem:* Am Physiol Soc; Endocrine Soc; Sigma Xi. *Res:* Neuroendocrine control; neuroendocrine responses to cardiopulmonary stimuli; control of vasopressin, ACTH and adrenal cortex. *Mailing Add:* 8500 N Seneca Rd Milwaukee WI 53217-2328

RAFF, HOWARD V, b Chicago, Ill, Aug 8, 50; m. INFECTIOUS DISEASE THERAPY. *Educ:* Univ Ill, Urbana, BS, 72; Wash State Univ, MS, 73, PhD, 77. *Prof Exp:* Postdoctoral fel clin immunol, Univ Calif, San Francisco, 77-78, res assoc, Howard Hughes Med Inst, 78-80; asst prof, 80-90, ASSOC PROF, DEPT MICROBIOL & IMMUNOL, UNIV WASH, SEATTLE, 91-; ASSOC DIR BIOL RES, BRISTOL-MYERS SQUIBB, PHARMACEUT RES INST, 91- *Concurrent Pos:* Pres bd, Food-Wise Nutrit Serv, Inc, 82-; Am Asn Immunologists travel award, 3rd Int Cong Immunol, Kyoto, Japan, 83; sr res investr, Bristol-Myers Squibb, Pharmaceut Res Inst, 84-90. *Mem:* Am Soc Microbiol; Am Asn Immunologists; Am Asn Lab Animal Sci. *Res:* Author of 38 technical publications. *Mailing Add:* Bristol-Myers Squibb Pharmaceut Res Inst 3005 First Ave Seattle WA 98121

RAFF, LIONEL M, b Mich, Nov 4, 34; m 55; c 2. CHEMICAL PHYSICS, PHYSICAL CHEMISTRY. *Educ:* Univ Okla, BS, 56, MS, 57; Univ Ill, PhD(phys chem), 62. *Prof Exp:* Chemist, Dow Chem Co, 57; NSF fel, Columbia Univ, 63-64; from asst prof to prof phys chem, 64-78, REGENTS PROF CHEM, OKLA STATE UNIV, 78- *Concurrent Pos:* Assoc ed, J Chem Physics, 80-83; Phys Chem Exam Comt, Am Chem Soc, 86-88. *Mem:* Am Chem Soc; Am Phys Soc. *Res:* Classical, quasiclassical, and quantum mechanical scattering calculations of inelastic and reactive gas-phase processes; classical and statistical mechanical studies of gas-surface interactions and of chemical processes under matrix-isolated conditions. *Mailing Add:* Dept Phys Chem Okla State Univ Stillwater OK 74078

RAFF, MARTIN JAY, b Brooklyn, NY, Mar 20, 37; m; c 6. INFECTIOUS DISEASES. *Educ:* Brandeis Univ, BA, 58; Univ Vt, MS, 60; Univ Tex Med Br Galveston, MD, 65; Univ Louisville, JD, 88. *Prof Exp:* From asst prof to prof med, 71-80, asst prof, 77-86, ASSOC PROF MICROBIOL & IMMUNOL, 86-, CHIEF, DIV INFECTIOUS DIS, 71- *Concurrent Pos:* NIH fel infectious dis, Col Med, Cornell Univ & New York Hosp, 66-67; consult, Vet Admin, Norton Childrens, Highlands Baptist, Audubon, Methodist, Suburan, Baptist East, St Mary, Elizabeth, Floyd City & Clark City Hops, Louisville, Ky, 71- & Ireland Army Hosp, Ft Knox, Ky, 71-; staff physician, Jewish Hosp, Louisville, Ky & Univ Louisville Hosp, 71-; chief, clin internal med, Univ Louisville, 87- *Mem:* Fel Am Col Physicians; fel Infectious Dis Soc Am; Am Soc Microbiol; AAAS; fel Am Col Chest Physicians; fel Am Col Legal Medicine. *Res:* Effects of steroids on the products of bacterial metabolism and on infection in animals; metabolic factors altering humoral and cellular host resistance mechanisms; antibiotic pharmacokinetics and therapeutic efficiency; infections of bones and joints; leukocyte function. *Mailing Add:* Sect Infectious Dis Sch Med Univ Louisville Louisville KY 40292

RAFF, MORTON SPENCER, b Chicago, Ill, Jan 12, 23; m 47; c 2. STATISTICS. *Educ:* Swarthmore Col, BA, 43; Yale Univ, cert, 48; American Univ, MA, 55. *Prof Exp:* Physicist, US Naval Res Lab, 43-44; physicist & mathematician, US Naval Ord Lab, 44-47; res asst hwy traffic, Yale Univ, 48-49; traffic res engr, Eno Found Hwy Traffic Control, 49-50; mathematician, US Bur Pub Roads, 50-55; math statistician, US Bur Lab Statist, 55-67 & 72-78 & Nat Heart & Lung Inst, 67-72; RETIRED. *Concurrent Pos:* Lectr, Johns Hopkins Univ, 56-59, USDA Grad Sch, 61-70 & Georgetown Univ, 67. *Res:* Medical and labor statistics; seasonal adjustment; probability theory applied to traffic behavior; approximations to the binomial distribution. *Mailing Add:* 3803 Montrose Driveway Chevy Chase MD 20815-4701

RAFF, RUDOLF ALBERT, b Shawinigan, Que, Nov 10, 41; US citizen; m 65; c 2. DEVELOPMENTAL BIOLOGY, EVOLUTION. *Educ:* Pa State Univ, BS, 63; Duke Univ, PhD(biochem), 67. *Prof Exp:* Officer, Armed Forces Radiobiol Res Inst, US Navy, 67-69; fel develop biol, Mass Inst Technol, 69-71; assoc prof biol, 71-80, prof, 80-, AT DEPT ZOOL, IND UNIV. *Concurrent Pos:* Instr-in-chief embryol, Marine Biol Lab, Woods Hole, Mass, 80-82. *Mem:* Am Soc Cell Biol; Soc Develop Biol; Am Soc Zoologists; Soc Study Evolution. *Res:* Molecular biology of early development; developmental genetics; role of developmental processes in evolution. *Mailing Add:* Dept Biol Ind Univ Jordan Hall Rm 142 Bloomington IN 47405

RAFF, SAMUEL J, b New York, NY, Nov 4, 20; m 85; c 6. OCEAN ACOUSTICS, NAVAL WARFARE. *Educ:* City Univ NY, BME, 43; Univ Md, MS, 50, PhD(physics), 57. *Prof Exp:* Head staff, USN Undersea Warfare Res & Develop Coun, 62-64; pres, Raff Assocs, 64-74; prog mgr, US NSF, 74-78; pres, Bethesda Corp, 78-86; prof elec eng & computer sci, George Washington Univ, 86-90; ED-IN-CHIEF, INT J COMPUTERS & OPERS RES, 73-; CONSULT, JOHNS HOPKINS APPL PHYSICS LAB, 85- *Mem:* Opers Res Soc Am; sr mem Inst Elec & Electronics Engrs; Soc Automotive Engrs. *Mailing Add:* 8312 Snug Hill Lane Potomac MD 20854

RAFFA, ROBERT B, ANALGESIA, BIOLOGY. *Educ:* Univ Del, BA & BChE, 71; Drexel Univ, MS, 79; Temple Univ, PhD(pharmacol), 82; Thomas Jefferson Univ, MS, 86. *Prof Exp:* Res asst prof, Jefferson Med Col, 85-86; res scientist, McNeil Pharmaceut, 86-87; res scientist, Janssen Res Found, 87-88, sr scientist, 88-90; sr scientist, 90-91, PRIN SCIENTIST, R W JOHNSON PHARMACEUT RES INST, 91- *Concurrent Pos:* Adj assoc prof, Dept Pharmacol, Med Sch, Temple Univ; adj asst prof, Dept Pharmacol, Jefferson Med Col; co-ed, Pharmacol Lett; Nat Res Serv awards, 83 & 84. *Mem:* Soc Neurosci; Am Soc Pharmacol & Exp Therapeut. *Res:* Author of more than 80 technical publications. *Mailing Add:* R W Johnson Pharmaceut Res Inst Welsh & McKean Rds Spring House PA 19477-0776

RAFFAUF, ROBERT FRANCIS, b Buffalo, NY, Jan 8, 16; m 50; c 2. ORGANIC CHEMISTRY. *Educ:* City Col New York, BS, 36; Columbia Univ, MA, 37; Univ Minn, PhD(org chem), 44. *Prof Exp:* Anal chemist, Tex Co, NY, 38-40; asst org chem, Univ Minn, 42-44; sr res chemist, Eaton Labs, Inc, NY, 44-47; asst, Univ Zurich, 47; res asst, Univ Basel, 47-48, fel, 49; res chemist, Nat Drug Co, 50-51; res assoc, Smith Kline & French Labs, 51-53; sr chemist, 53, lit scientist, 54-69; prof, 69-84, EMER PROF PHARMACOG, COL PHARM, NORTHEASTERN UNIV, 84- *Mem:* Fel AAAS; Am Chem Soc; NY Acad Sci; Swiss Chem Soc. *Res:* Natural products. *Mailing Add:* Col Pharm Northeastern Univ Boston MA 02115

RAFFEL, JACK I, b New York, NY, Apr 1, 30; m 59; c 2. INTEGRATED CIRCUITS. *Educ:* Columbia Univ, AB, 51, BS, 52; Mass Inst Technol, MS, 54. *Prof Exp:* Res asst digital comput lab, 52-54, staff mem, Lincoln Lab, 54-62, GROUP LEADER, LINCOLN LAB, MASS INST TECHNOL, 62- *Mem:* Inst Elec & Electronics Engrs. *Res:* Digital computer research and development; design and fabrication of very large-scale integrated circuits. *Mailing Add:* Lincoln Labs Mass Inst Technol Lexington MA 02173

RAFFEL, SIDNEY, b Baltimore, Md, Aug 24, 11; m 38; c 5. MEDICAL BACTERIOLOGY, IMMUNOLOGY. *Educ:* Johns Hopkins Univ, AB, 30, ScD(immunol), 33; Stanford Univ, MD, 43. *Prof Exp:* Asst immunol, Sch Hyg & Pub Health, Johns Hopkins Univ, 33 -35; from asst to assoc prof, 35-48, consult physician, Student Health Serv, 42-44, prof bact & exp path, Sch Med, 48-76, chmn dept med microbiol, 53-76, actg dean, 64-65, EMER CHMN DEPT MED MICROBIOL, SCH MED, STANFORD UNIV, 76-, EMER PROF, DEPT DERMAT, 77- 53- *Concurrent Pos:* Guggenheim fel, 49-50; lab dir, Palo Alto Hosp, 45-47; consult, Vet Admin, 47-; chmn study sect allergy & immunol, NIH, 56-59, mem training grant comt, 60- *Mem:* Am Soc Microbiol; Am Soc Exp Path; Am Thoracic Soc; Am Asn Immunol; Sigma Xi. *Res:* Immunology of tuberculosis; hypersensitivity; cellular immunity. *Mailing Add:* Dept Med Microbiol Stanford Univ Sch Med Stanford CA 94305

RAFFELSON, HAROLD, b Sheboygan, Wis, Oct 29, 20; m 48; c 1. PHARMACEUTICAL CHEMISTRY. *Educ:* Univ Wis, BS, 47; Univ Mich, PhD(pharmaceut chem), 51. *Prof Exp:* Org res chemist, Frederick Stearns & Co, 47; res chemist, Monsanto Co, 51-59, group leader, 59-61, res specialist, 61-73, sr res specialist, 73-85; RETIRED. *Mem:* Am Chem Soc. *Res:* Synthetic antispasmodics; steroid synthesis; organo-phosphorus compounds; process development; catalysis. *Mailing Add:* Seven Planters Dr Olivette MO 63132-3441

RAFFENETTI, RICHARD CHARLES, b Springfield, Mass, Oct 15, 42; m; c 2. SOFTWARE SYSTEMS. *Educ:* Tufts Univ, BS, 64; Iowa State Univ, PhD(phys chem), 71. *Prof Exp:* Res assoc quantum chem, Battelle Mem Inst, 71-73 & Johns Hopkins Univ, 73-74; vis scientist quantum chem, Inst Comput Applns Sci & Eng, NASA Langley Res Ctr, 74-76; asst chemist quantum chem, Chem Div, 76-79, COMPUT SCIENTIST, COMPUT SERVS, ARGONNE NAT LAB, 79- *Concurrent Pos:* Consult, VAX comput syst. *Honors & Awards:* Pacesetter Award. *Mem:* Sigma Xi. *Res:* Virtual Address Extension and Virtual Memory Operating Systems computing service project management and systems management and Virtual Address Extension cluster management; design and maintenance of networking applications software; system configuration and analysis; VAX/VMS system management. *Mailing Add:* Bldg 221 Argonne Nat Lab Argonne IL 60439

RAFFENSPERGER, EDGAR M, b Gettysburg, Pa, June 13, 26; m 53; c 3. ENTOMOLOGY. *Educ:* Pa State Univ, BS, 51, MS, 52; Univ Wis, PhD(entom), 55. *Prof Exp:* From asst prof to assoc prof entom, Va Polytech Inst, 55-61; assoc prof, 61-77, PROF ENTOM, CORNELL UNIV, 77- *Concurrent Pos:* Vis scientist, Norweg Agr Res Coun, Vollebekk, Norway, 68-69; lectr, Univ Oslo, 69; consult, stored prod insect control & arthropod pests in indust sites; instr trop insects, Insect Pop Mgt Res Unit, USDA, Kenya, Latin Am. *Mem:* AAAS; Entom Soc Am. *Res:* Taxonomy of Diptera and Mallophaga; insect transmission of fowl diseases; pesticides in agriculture; teaching in general entomology and cultural entomology; insect morphology and control; control of stored products insects; insect integrated pest management; management of the clusterfly (pollenia rudis). *Mailing Add:* Dept Entom Cornell Univ Ithaca NY 14850

RAFFENSPERGER, EDWARD COWELL, b Dickinson, Pa, July 9, 14; m 49. GASTROENTEROLOGY. *Educ:* Dickinson Col, BS, 36; Univ Pa, MD, 40. *Prof Exp:* Instr gastroenterol, Grad Sch Med, 48-58, assoc, Sch Med, 53-62, from asst prof to assoc prof, 62-71, PROF MED, SCH MED, UNIV PA, 71- *Concurrent Pos:* Res fel gastroenterol, Grad Hosp, Univ Pa, 46-48; consult, Vet Admin Hosp, Philadelphia & Polyclin Hosp, Harrisburg, 62-, Children's Hosp, Philadelphia, 64- & Lankenau Hosp, 69- *Mem:* Am Gastroenterol Asn; Am Fedn Clin Res. *Res:* Amino acids in nutrition; inflammatory bowel diseases. *Mailing Add:* 290 St James Pl Philadelphia PA 19106

RAFFERTY, FRANK THOMAS, b Greenville, Miss, Jan 28, 25; c 7. CHILD PSYCHIATRY. *Educ:* St Mary's Col, Minn, BS, 48; St Louis Univ, MD, 48; Univ Colo, MS, 53. *Prof Exp:* Resident psychiat, Colo Psychopath Hosp, Univ Colo Med Ctr, Denver, 49-50, chief resident, Ment Hyg Clin, 52-53; assoc prof psychiat ment health serv, Div Child Psychiat, Col Med, Univ Utah, 55-58; med dir, Ment Health Serv, Inc, 57-61; dir, Div Child Psychiat, Psychiat Inst, Univ Md, 61-71; PROF PSYCHIAT & DIR, INST JUV RES, ABRAHAM LINCOLN SCH MED, UNIV ILL, 71- *Concurrent Pos:* Fel child psychiat, Univ Colo Med Ctr, Denver, 51-53; mem comn child & adolescent psychiat, Am Psychiat Asn, 69-; Am Psychiat Asn Rep, Nat Consortium Ment Health serv to Children, 71-; mem, Nat Consortium Children Serv, 72. *Mem:* Fel Am Psychiat Asn; fel Am Acad Child Psychiat; Am Col Psychiatrists. *Mailing Add:* Oaks Family Ctr 1408 W Stassney Lane Austin TX 78745

RAFFERTY, KEEN ALEXANDER, JR, b Robinson, Ill, Mar 6, 26; m 53; c 2. EMBRYOLOGY, CELL CULTURE. *Educ:* Univ NMex, BS, 50; Univ Ill, MS, 51, PhD(zool), 55. *Prof Exp:* Asst zool, Univ Ill, 50-54; instr microbiol, Yale Univ, 57-58; from asst prof to assoc prof anat, Sch Med, Johns Hopkins Univ, 58-70; head dept, 70-77, PROF, DEPT ANAT, UNIV ILL MED CTR, 70- *Concurrent Pos:* NIH fel microbiol, Yale Univ, 55-57; Fogarty sr fel, 77-78. *Mem:* Am Asn Anat; Am Soc Cell Biol; Soc Develop Biol; Tissue Cult Asn. *Res:* Cellular differentiation and aging of cultured cells. *Mailing Add:* 1430 Golden Bell Ct Downers Grove IL 60515

RAFFERTY, NANCY S, b New York, NY, June 11, 30; m 53; c 2. CELL BIOLOGY, ANATOMY. *Educ:* Queens Col, NY, BS, 52; Univ Ill, MS, 53, PhD(zool), 58. *Prof Exp:* From instr to asst prof anat, Johns Hopkins Univ, 63-70; from asst prof to assoc prof, 72-76, PROF ANAT, MED & DENT SCH, NORTHWESTERN UNIV, 76- *Concurrent Pos:* NIH res fel, Johns Hopkins Univ, 58-60, fel anat, Sch Med, 60-63; NIH res grants, Johns Hopkins Univ, 65-71 & Northwestern Univ, 70-; mem, VISA study sect, 78-82; vis prof, Guy's Hosp, London, UK, 77-78 & 88-89. *Mem:* AAAS; Asn Res Vision & Ophthal; Am Asn Anat; Am Soc Cell Biol; Int Soc Eye Res. *Res:* Experimental cataract; wound healing; cell population kinetics; electron microscopy of lens; cellular dynamics of the proliferative response in injured frog, mouse and squirrel lens epithelium; etiology of senile cataract; lens aging; mechanism of lens accommodation; cytoskeleton of lens cells. *Mailing Add:* Dept Cell Molecular & Structural Biol Northwestern Univ Med & Dent Sch Chicago IL 60611

RAFLA, SAMEER, b Cairo, Egypt, Sept 3, 30; m 65; c 5. RADIATION MEDICINE, ONCOLOGY. *Educ:* Univ Cairo, BS, 47, MB & BCh, 53; London Univ, PhD(radiation med), 70. *Prof Exp:* DIR RADIATION THERAPY, METHODIST HOSP, 69-; DIR RADIOTHERAPY, LUTHERAN MED CTR, 77-; DIR RADIOTHERAPY, MAIMONIDES MED CTR, 79- *Concurrent Pos:* Radiotherapist, Manitoba Cancer Found, 67-69; prin investr, Cancer & Leukemia Group B, State Univ, NY, Downstate, Oncol Prog, Brooklyn Community Hosp, Nat Cancer Inst, Continuing Ed Radiotherapists, Am Cancer Soc Grant; clin prof radiation oncol, Downstate Med Ctr, State Univ NY, 81- *Mem:* Am Radium Soc; Royal Col Radiol; Am Soc Therapeut Radiologists; Soc Surg Oncol; Radiol Soc NAm. *Res:* The effect of radiation on the malignant and normal cell as well as its possible effect on the immune response; the treatment of certain cancers especially head and neck, kidney and lymphoma. *Mailing Add:* Methodist Hosp 506 Sixth St Brooklyn NY 11215

RAFOLS, JOSE ANTONIO, b Guantanamo, Cuba, July 7, 43; US citizen. ANATOMY. *Educ:* St Procopius Col, BS, 65; Univ Kans, PhD(anat), 69. *Prof Exp:* From instr to asst prof, 70-73, ASSOC PROF ANAT, SCH MED, WAYNE STATE UNIV, 73- *Concurrent Pos:* NIH trainee, Cajal Inst, Madrid, Spain, 71. *Mem:* AAAS; Pan-Am Asn Anat. *Res:* Golgi and electron microscopic analysis of the mammalian visual system and basal ganglia. *Mailing Add:* Dept Anat Sch Med Wayne State Univ 540 E Canfield Detroit MI 48201

RAFTER, GALE WILLIAM, b Seattle, Wash, Nov 3, 25; m 57; c 3. BIOCHEMISTRY. *Educ:* Univ Wash, BS, 48, PhD(biochem), 53. *Prof Exp:* Asst prof biochem, Sch Hyg & Pub Health, Johns Hopkins Univ, 55-59, asst prof microbiol, Sch Med, 59-65; assoc prof, 65-71, PROF BIOCHEM, SCH MED, WVA UNIV, 71- *Concurrent Pos:* Fel, McCollum-Pratt Inst, Johns Hopkins Univ, 53-55. *Mem:* Am Soc Biol Chem; Hist Sci Soc. *Res:* Mode of action of antirheumatic drugs; role of protein sulfhydryl groups in action of enzymes. *Mailing Add:* Dept Biochem WVa Univ Med Ctr Morgantown WV 26506

RAFTOPOULOS, DEMETRIOS D, b Argostolion, Greece, May 30, 26; US citizen; m 59; c 1. ENGINEERING MECHANICS. *Educ:* PMC Cols, BSCE, 59; Univ Del, MCE, 63; Pa State Univ, PhD(eng mech), 66. *Prof Exp:* Sr engr, Del State Hwy Dept, 59-61; instr eng, PMC Cols, 61-64; res asst eng mech, Pa State Univ, 64-67; assoc prof mech eng, 67-73, PROF MECH ENG, UNIV TOLEDO, 73- *Concurrent Pos:* Jr investr, Ballistic Res Lab, Aberdeen Proving Ground, Md, 64-67, proj dir, 67-68; prin investr, Naval Res Labs, Washington, DC, 67-69 & Atomic Energy Comn, Md, 68-71; reviewer, Appl Mech Res, 69-; vis prof mech, Nat Tech Univ Athens, 73-74, Univ Mich, Ann Arbor, 86-87. *Honors & Awards:* Sigma Xi Res Award, 83. *Mem:* Am Soc Eng Educ; Am Soc Civil Engrs; Am Soc Mech Engrs; Am Acad Mech; Sigma Xi. *Res:* Elasto-plastic stress waves; analysis of foundation interaction with nuclear power plants during earthquake loading; structure interaction with underwater shock waves; fracture mechanics; bio-mechanics. *Mailing Add:* 2801 W Bancroft St Univ Toledo Toledo OH 43616

RAFUSE, ROBERT P(ENDLETON), b Newton, Mass, Dec 7, 32; div; c 2. ELECTRICAL ENGINEERING. *Educ:* Tufts Col, BSEE, 54; Mass Inst Technol, SM, 57, ScD(elec eng), 60. *Prof Exp:* Teaching asst elec eng, Mass Inst Technol, 54-57, instr, Res Lab Electronics, 57-60, from asst prof to assoc prof, 60-70; pres, Rafuse Assocs, 70-75; mem staff, 75-78, MEM SR STAFF, LINCOLN LAB, MASS INST TECHNOL, 78- *Concurrent Pos:* Adv, NASA, NIH & Dept Defense, 54-66; Consult to many govt & indust orgns; mem, Nat Defense Exec Reserve, 69-, vchmn gov's comn emergency commun, Commonwealth Mass, 70-, mem gov's energy emergency comn, 71; chmn, Nat Acad Sci-Nat Res Coun eval panel for Nat Bur Standards-EMD, 72-78. *Mem:* AAAS; Soc Am Mil Engrs; Inst Elec & Electronics Engrs; Am Defense Preparedness Asn. *Res:* Sensor systems; microwave solid state circuits; space communications; management of energy resources. *Mailing Add:* Lincoln Lab Rm M180 Mass Inst Technol 244 Wood St Boston MA 02173

RAGAINI, RICHARD CHARLES, b Danbury, Conn, Feb 7, 42; div; c 2. ENVIRONMENTAL PROTECTION. *Educ:* Clark Univ, BA, 63; Mass Inst Technol, PhD(nuclear chem), 67. *Prof Exp:* Fel radiochem, Los Alamos Sci Lab, 67-69; res assoc chem, Brookhaven Nat Lab, 69-70; asst prof chem, Wash State Univ, 70-71; chemist, 71-75, sect leader, Radiochem Div, 75-77, dept div leader, 77-81, actg div leader, Environ Sci Div, 81-82, assoc div leader, Mech Eng Dept, 82-86, dept head, Environ Protection Dept, 86-90, ASSOC DEPT HEAD ENVIRON PROTECTION RES & DEVELOP, LAWRENCE LIVERMORE NAT LAB, 90- *Concurrent Pos:* Course dir, Innovative Cleanup Contaminated Soils & Groundwater, Ettore Majorana Ctr Sci Cult, Erice, Sicily, 90. *Mem:* Am Physics Soc; Am Chem Soc; Sigma Xi; Am Nuclear Soc. *Res:* Effects of trace elements and organics in the environment; methods for trace element and organic analysis; trace elements and organics from energy production. *Mailing Add:* L-192 Lawrence Livermore Lab Livermore CA 94550

RAGAN, CHARLES ELLIS, III, b Charleston, SC, Oct 19, 44; div; c 2. NUCLEAR PHYSICS. *Educ:* The Citadel, BS, 66; Duke Univ, PhD(nuclear physics), 71. *Prof Exp:* Res assoc nuclear physics, NC State Univ, 70; physicist, USAF Weapons Lab, 70-72; STAFF MEM NUCLEAR PHYSICS, LOS ALAMOS NAT LAB, 72- *Mem:* Am Phys Soc; Sigma Xi. *Res:* Precise equation-of-state measurements at pressures of 10-100 mega-bar using shock waves produced by underground nuclear explosions; nuclear physics experiments using neutrons produced by reactors, Van de Graaff linear accelerators, computer graphics and animation of 3-D data; nuclear explosions. *Mailing Add:* Group P-3 MS-D449 Los Alamos Sci Lab Los Alamos NM 87545

RAGAN, DONAL MACKENZIE, b Los Angeles, Calif, Oct 4, 29; m 52; c 2. STRUCTURAL GEOLOGY, ENGINEERING GEOLOGY. *Educ:* Occidental Col, BA, 51; Univ Southern Calif, MS, 54; Univ Wash, PhD(geol), 61; Univ London, DIC, 69. *Prof Exp:* From instr to assoc prof geol, Univ Alaska, 60-67; assoc prof, 67-70, PROF GEOL, ARIZ STATE UNIV, 70- *Concurrent Pos:* NSF res grants, 64-66; fac fel, Imp Col, London, 66-67. *Mem:* AAAS; Int Soc Rock Mech; Am Geophys Union; fel Geol Soc Am; fel Geol Soc London; Sigma Xi. *Res:* Structural geology; engineering geology. *Mailing Add:* Dept Geol Ariz State Univ Tempe AZ 85287-1404

RAGAN, HARVEY ALBERT, b Boise, Idaho, July 11, 29; m 51; c 3. TOXICOLOGY & HEMATOLOGY, RADIOBIOLOGY. *Educ:* Wash State Univ, BS, 56, DVM, 59; Am Bd Toxicol, dipl. *Prof Exp:* Pvt pract, 59-62; scientist radiobiol, Hanford Labs, Gen Elec Co, 62-65; res scientist radiobiol, Pac Northwest Labs, Battelle Mem Inst, 65-66, sr res scientist radiotoxicol, 66-67; NIH spec fel exp hemat, Col Med, Univ Utah, 67-69; sr res scientist hemat, 69-72, staff scientist hemat, 72-77, MGR EXP PATH SECT, PAC NORTHWEST LABS, BATTELLE MEM INST, 77- *Concurrent Pos:* Mem, Nat Coun Radiation Protection, 78-; adj prof, Joint Ctr Grad Studies, 81- *Mem:* Am Soc Vet Clin Path (pres 81-82); Int Soc Animal Biochem; Am Soc Hemat. *Res:* Effects of chemical and physical insults on the hematopoietic system; carcinogenesis; blood cell kinetics; immunology; iron metabolism; clinical pathology; inhalation toxicology. *Mailing Add:* Battelle Northwest Labs Richland WA 99352

RAGAN, ROBERT MALCOLM, b San Antonio, Tex, Dec 19, 32; m 55; c 3. HYDROLOGY. *Educ:* Va Mil Inst, BS, 55; Mass Inst Technol, MS, 59; Cornell Univ, PhD(civil eng), 65. *Prof Exp:* Designer, Whitman Requardt & Assocs, Md, 56-57; res asst sanit eng, Mass Inst Technol, 57-59; from asst prof to assoc prof, Univ Vermont, 59-67; assoc prof, 67-68, head dept, 69-76, PROF CIVIL ENG, UNIV MD, COL PARK, 69- *Mem:* Am Soc Civil Engrs; Am Geophys Union. *Res:* Watershed hydrology. *Mailing Add:* Dept Civil Eng Univ Md Rm 1179 College Park MD 20742

RAGENT, BORIS, b Cleveland, Ohio, Mar 2, 24; m 49; c 3. PHYSICS. *Educ:* Marquette Univ, BEE, 44; Univ Calif, Berkeley, PhD(physics), 54. *Prof Exp:* Engr electronics, Victoreen Instrument Co, Cleveland, 46-48; engr & res scientist electronics & physics, Radiation Lab, Univ Calif, Berkeley, 48-53, res scientist physics, Livermore, 53-56; res scientist, Broadview Res Corp, Burlingame, 56-59; staff scientist, Vidya Div, Itek Corp, Palo Alto, 59-66; chief electronic instrument develop br, 66-80, sr staff scientist, Space Sci Div, Ames Res Ctr, NASA, 80-87; SR RES ASSOC, SAN JOSE STATE UNIV FOUND, 87- *Concurrent Pos:* Lectr, Stanford Univ, 62 & 79. *Mem:* Am Phys Soc; AAAS. *Res:* Nuclear physics; instrumentation; plasma physics; planetary atmospherics. *Mailing Add:* 675 Edna Way San Mateo CA 94402

RAGEP, F JAMIL, b WVa, June 19, 50; m 73; c 2. ISLAMIC SCIENCE, HISTORY OF ASTRONOMY. *Educ:* Univ Mich, BA, 72, MA, 73; Harvard Univ, PhD(hist sci), 82. *Prof Exp:* Instr math, Long Island Univ, 82; instr hist sci, Stonehill Col, 89; asst prof hist sci, Brown Univ, 89-90; ASST PROF HIST SCI, UNIV OKLA, 90- *Concurrent Pos:* Lectr hist sci, Harvard Univ, 83-84 & 87-88. *Mem:* Hist Sci Soc. *Res:* History of ancient, medieval and early modern science, particularly in the way ancient astronomy was transformed in Islam and the repercussions of this for subsequent astronomy. *Mailing Add:* Dept Hist Sci Rm 622 Univ Okla 601 Elm Norman OK 73019-0315

RAGHAVACHARI, KRISHNAN, b Madras, India, Apr 3, 53; m; c 2. COMPUTATIONAL CHEMISTRY. *Educ:* Madras Univ, India, BSc, 73; Indian Inst Technol, MSc, 75; Carnegie-Mellon Univ, PhD(chem), 80. *Prof Exp:* DISTINGUISHED MEM TECH STAFF, MAT SCI, BELL LABS, MURRAY HILL, NJ, 81- *Mem:* Am Chem Soc; Am Physics Soc. *Res:* Development and application of new molecular orbital methods in quantum chemistry; theoretical study of atomic and molecular clusters. *Mailing Add:* Bell Labs 1A 362 600 Mountain Ave Murray Hill NJ 07974

RAGHAVAN, PRAMILA, b Bangalore, India; m 67. SOLAR NEUTRINOS, NUCLEAR PROPERTIES & HYPERFINE INTERACTIONS. *Educ:* Univ Mysore, India, BSc, 54, MSc, 56; Saha Inst, Univ Calcutta, India, assoc dipl, 58; Mass Inst Technol, PhD(physics), 67. *Prof Exp:* Lectr, Univ Mysore, India, 54-55 & 56-57; res assoc, Tata Inst Fundamental Res, India, 58-61; commonwealth scholar, Nuclear Physics Lab, Oxford Univ, 61-62; res asst, Mass Inst Technol, 62-67; guest prof, Univ Munchen, WGer, 67-69; asst, Technol Univ Munich, WGer, 70-72; fel, Rutgers Univ, 72-80, res assoc physics, 80-85, res prof, 85-87; RESIDENT VISITOR, BELL LABS, 72- *Mem:* Am Phys Soc. *Res:* Nuclear structure; nulear moments using radioactivity and nuclear reactions; interaction of nuclei with its environment; hyperfine interactions; applications to solid state physics, atomic physics and material science; solar neutrino detection. *Mailing Add:* AT&T Bell Labs Rm 1E-436 600 Mountain Ave Murray Hill NJ 07974-2070

RAGHAVAN, RAJAGOPAL, b Tiruchirappalli, India, July 26, 43. PETROLEUM ENGINEERING. *Educ:* Birla Inst Technol, Mesra, India, BSc, 66; Univ Birmingham, dipl, 67; Stanford Univ, PhD(petrol & mech eng), 70. *Prof Exp:* Res assoc petrol eng, 70-71, asst prof, Stanford Univ, 71-72; sr res engr, Amoco Prod Co, 72-75; from assoc prof to prof, Univ Tulsa, 75-80, prof petrol eng, 80-82, Mcman prof petrol eng, 82-89; prof petrol eng, Tex A&M, 89-90; SR STAFF ASSOC, RESRVIOR ENG, PHILLIPS PETROL CO, 90- *Concurrent Pos:* Tech ed, Soc Petrol Engrs, 78-79, 80-82. *Honors & Awards:* Distinguished Fac Award, Soc Petrol Engrs, 81. *Mem:* Soc Petrol Engrs; assoc fel, Brit Inst Petrol; NY Acad Sci; Sigma Xi; Am Geophys Union. *Res:* Unsteady state fluid flow and heat transfer in porous media, including well test analysis, stability of liquid interfaces, compaction and subsidence, geothermal energy and application of computers. *Mailing Add:* Phillips Petrol Co 245 GB Bartlesville OK 74004

RAGHAVAN, RAMASWAMY SRINIVASA, b Tanjore, India, Mar 31, 37; m 67. SOLID STATE PHYSICS, NUCLEAR ASTROPHYSICS. *Educ:* Univ Madras, India, MA, 57, MSc, 58; Purdue Univ, PhD(physics), 65. *Prof Exp:* Res asst, Tata Inst Fundamental Res, 59-62 & Purdue Univ, 62-65; fel, Bartol Res Found, 65-66; vis prof, Univ Bonn, Ger, 66-67; res assoc, Tech Univ, Munich, 67-72; mem staff, 72-89, DISTINGUISHED MEM STAFF PHYSICS, AT&T BELL LABS, 89- *Concurrent Pos:* Assoc grad fac, Rutgers Univ, 74- *Mem:* Am Phys Soc. *Res:* Neutrino physics; detection of solar neutrinos; nuclear electronics and detector hardware; nuclear structure; application of nuclear accelerators to geochronology and cosmochronology; nuclear interactions with matter; solid state physics. *Mailing Add:* AT&T Bell Labs 600 Mountain Ave 1E 432 Murray Hill NJ 07974

RAGHAVAN, SRINIVASA, b Madras, India, July 1, 40; nat US; m 77; c 2. BIOCHEMISTRY. *Educ:* Univ Madras, India, BSc, 60, MSc, 63; Indian Inst Sci, PhD(biochem), 70. *Prof Exp:* Res fel, Mass Gen Hosp, Boston, 70-73; res assoc, 73-74, sr res fel, 74-77, sr res assoc, 77-78, asst biochemist, 78-82, ASSOC BIOCHEMIST, E K SHRIVER CTR, MASS, 82- *Concurrent Pos:* Asst biochemist, neurol res, Mass Gen Hosp, 79-87, assoc biochemist, 88- *Mem:* Am Soc Neurochem. *Res:* Inherited neurological diseases of glycorphingolipid metabolism resulting from genetic deficiency of specific lysosomal hydrolases; animal models to understand the function of glycolipids in cell development differentiation myclination and demyelination in the nervous system. *Mailing Add:* E K Shriver Ctr Mental Retardation Inc 200 Trapelo Rd Waltham MA 02254

RAGHAVAN, THIRUKKANNAMANGAI E S, b Madras, India, Aug 5, 40; m 67; c 3. MATHEMATICS. *Educ:* Loyola Col, Madras, India, BSc, 60; Presidency Col, Madras, India, MSc, 62; Indian Statist Inst, Calcutta, PhD(statist, math), 66. *Prof Exp:* Lectr math, Univ Essex, 66-69; from asst prof, to assoc prof, 69-79, PROF MATH, UNIV ILL, CHICAGO CIRCLE, 79- *Res:* Positive operatons; non cooperative games; stochastic and differential games; matrix theory; statistical decision theory; mathematical economics; applied statistics. *Mailing Add:* Dept Math Univ Ill Box 4348 Chicago IL 60680

RAGHAVAN, VALAYAMGHAT, b Edavanakad, Cochin, India, Mar 19, 31; m 62; c 1. PLANT MORPHOGENESIS. *Educ:* Univ Madras, BS, 50; Benares Hindu Univ, MS, 52; Princeton Univ, PhD(biol), 61. *Prof Exp:* Res assoc biol, Harvard Univ, 61-63; reader bot, Univ Malaya, 63-70; guest investr biol, Rockefeller Univ, 66-67; vis prof, Dartmouth Col, 69-70; from asst prof to assoc prof, 70-77, PROF BOT, OHIO STATE UNIV, 77 - *Mem:* Bot Soc Am; Am Soc Plant Physiologists. *Res:* Developmental physiology of lower plants; photomorphogenesis and biochemical cytology of spore germination; experimental plant embryogenesis. *Mailing Add:* Dept Plant Biol Ohio State Univ Columbus OH 43210

RAGHEB, HUSSEIN S, b Cairo, Egypt, Jan 30, 24; m 56; c 1. MICROBIOLOGY, BIOCHEMISTRY. *Educ:* Cairo Univ, BS, 44, MS, 50; Mich State Univ, PhD(fermentation), 56. *Prof Exp:* Res asst microbiol, Mich State Univ, 53-56, res assoc, 57; fel food tech, Iowa State Univ, 56; asst prof biochem & microbiol, Ferris State Col, 57-61; ASSOC PROF BIOCHEM, PURDUE UNIV, LAFAYETTE, 61- *Mem:* Am Soc Microbiol; Asn Official Anal Chemists. *Res:* Microbial chemistry; mode of action, methods of assay and characterization of antibiotics. *Mailing Add:* Dept Biochem Purdue Univ W Lafayette IN 47907

RAGHU, SIVARAMAN, synthetic organic chemistry, for more information see previous edition

RAGHUVEER, MYSORE R, b Bangalore, India, June 17, 57; m 87; c 1. SIGNAL & IMAGE PROCESSING, COMMUNICATIONS. *Educ:* Mysore Univ, India, BE, 79; Indian Inst Sci, ME, 81; Univ Conn, PhD(elec eng), 84. *Prof Exp:* Mem tech staff, Advanced Micro Devices Inc, 85-87; ASSOC PROF SIGNAL & IMAGE PROCESSING, ROCHESTER INST TECHNOL, 87- *Concurrent Pos:* Prin investr, NSF grant, 89-91; assoc ed, Inst Elec & Electronics Engrs Trans on Signal Processing, 91-93; consult, RIT Res Corp & Orincon Corp. *Mem:* Inst Elec & Electronics Engrs; Inst Elec & Electronics Engrs Signal Processing Soc. *Res:* Digital image coding; digital image restoration and reconstruction; spectral analysis, especially with higher-order statistics; biomedical applications of signal processing. *Mailing Add:* Dept Elec Eng Rochester Inst Technol Rochester NY 14623

RAGHUVIR, NUGGEHALLI NARAYANA, b Bangalore, India, July 12, 30; m 57; c 1. ENTOMOLOGY. *Educ:* Univ Poona, India, BSc, 50; Karnatak Univ, MSc, 55; Utah State Univ, PhD(entom), 62. *Prof Exp:* Malaria supvr, Pub Health Dept, Poona, 50-51; res asst entom, Cent Food Tech Res Inst, Mysore, 56-58; instr zool, Duke Univ, 62-63; from instr to asst prof, 63-74, ASSOC PROF BIOL, UNIV BRIDGEPORT, 74- *Concurrent Pos:* Acad Year Exten res award, 64-65; consult entom, 78- *Mem:* Entom Soc Am; Sigma Xi. *Res:* Basic and applied aspects of insect physiology; general entomology, agricultural entomology and animal physiology. *Mailing Add:* 43 Bunker Hill Dr Trumbull CT 06611-1458

RAGINS, HERZL, b Tel Aviv, Israel, July 27, 29; US citizen; m 59; c 3. SURGERY. *Educ:* Univ Ill, BS, 47, MS & MD, 51; Univ Chicago, PhD(surg, gastric physiol), 56. *Prof Exp:* Instr surg, Univ Chicago, 59-60; instr, 60-62, from asst prof to assoc prof, 62-75, CLIN PROF SURG, ALBERT EINSTEIN COL MED, 75-; ATTEND SURG, BRONX MUNIC HOSP CTR, 68- *Concurrent Pos:* Am Cancer Soc fel, 57-58. *Mem:* Am Col Surg; Am Physiol Soc; Soc Surg Alimentary Tract; Am Gastroenterol Asn; Sigma Xi. *Res:* Gastric physiology; histochemistry of gastric mucosa; histamine metabolism; mast cells and parietal cell turn over in gastric mucosa; radiation effects on gastric mucosa; effect of intrajejunal amino acids on pancreatic secretion. *Mailing Add:* Albert Einstein Col Med Bronx NY 10461

RAGINS, NAOMI, b Chicago, Ill; m 55. ADULT & CHILD PSYCHIATRY, PSYCHOANALYSIS. *Educ:* Univ Chicago, PhB, 46, BS, 47, MD, 51; Am Bd Psychiat & Neurol, dipl, 59, cert child psychiat, 61; Am Psychoanal Assoc, cert adult & child psychoanal, 71. *Prof Exp:* Asst prof psychiat, 57-63, clin asst prof, 63-71, CLIN ASSOC PROF CHILD PSYCHIAT, SCH MED, UNIV PITTSBURGH, 71- *Concurrent Pos:* Fac psychoanal, Pittsburgh Psychoanal Inst, 67-, supv child analyst, 71-, training & supv analyst, 77-; teaching consult, Children's Hosp, Pittsburgh, 72-; consult, Child Develop Prog, Head Start, Pittsburgh Child Guid Ctr, 73-76. *Mem:* Am Psychoanal Asn; Am Psychiat Asn; Am Acad Child Psychiat; Asn Child Psychoanal; Am Orthopsychiat Asn. *Res:* Ego development in infancy and childhood. *Mailing Add:* Cathedral Mansions Suite 118-119 4716 Ellsworth Ave Pittsburgh PA 15213

RAGLAND, JOHN LEONARD, b Beaver Dam, Ky, Oct 30, 31; m 56; c 3. SOIL CHEMISTRY, PLANT NUTRITION. *Educ:* Univ Ky, BS, 55, MS, 56; NC State Univ, PhD(soil sci), 59. *Prof Exp:* Asst prof soil technol, Pa State Univ, 59-61; from asst prof to assoc prof, 61-66, chmn dept agron, 66-69, PROF AGRON, UNIV KY, 66-, ASSOC DEAN EXTEN & ASSOC DIR COOP EXTEN SERV, 69- *Concurrent Pos:* Chmn state comt rural community develop, USDA, 70- *Honors & Awards:* Thomas Poe Cooper Award Distinguished Agr Res, Univ Ky, 67. *Mem:* Soil Sci Soc Am; Am Soc Agron; Int Soc Soil Sci. *Res:* Interaction of plant nutrient availability with the microclimatic; cation exchange equilibria in soils. *Mailing Add:* Dept Agr-Agron Univ Ky Agr Sci Ctr N Bldg Off N122 Lexington KY 40506-0091

RAGLAND, PAUL C, b Lubbock, Tex, June 28, 36; m 58; c 2. GEOCHEMISTRY, PETROLOGY. *Educ:* Tex Tech Col, BS, 58; Rice Univ, MA, 61, PhD(geol), 62. *Prof Exp:* From asst prof to prof geol, Univ NC, Chapel Hill, 62-78; chmn natural sci area, 80-82, PROF GEOL & CHMN DEPT, FLA STATE UNIV, 78- *Concurrent Pos:* Consult, Sinclair Res, 65, US Naval Ord Labs, 68-69, Va Div Mineral Resources, 70-71, Dames & Moore, 73-77, E I du Pont de Nemours, 75-77, Ebasco, Inc, 75- & NUS Corp, 78; Adv Res Projs Agency Mat Res Ctr grant, 66-74; vis prof, Duke Univ, 68 & 75, Mineral Mus, Oslo, Norway, 69-70 & Univ Ky, 75; assoc dean res admin, Univ NC, 71, assoc chmn dept geol, 75; Dept Energy grant, 76-78 & NSF grant, 79-81. *Mem:* Fel Geol Soc Am; Geochem Soc. *Res:* Application of analytical chemical data to petrogenesis of igneous and metamorphic rocks; geochemical prospecting; trace elements in chemical weathering and diagenesis. *Mailing Add:* Dept Geol Fla State Univ Tallahassee FL 32306

RAGLAND, WILLIAM LAUMAN, III, b Richmond, Va, Aug 24, 34; m 61; c 3. IMMUNOPHARMACOLOGY. *Educ:* Col William & Mary, BS, 56; Univ Ga, DVM, 60; Wash State Univ, PhD(vet path & biochem), 66. *Prof Exp:* Res asst path, Tulane Univ, 60-61, instr, 61-62; Nat Cancer Inst spec res fel, 66-68, asst prof path & vet sci, McArdle Lab, Univ Wis, 68-70; assoc prof, 70-76, PROF, DEPTS AVIAN MED, PATH & MED MICROBIOL, COL VET MED, UNIV GA, 76- *Concurrent Pos:* Pres, Ragland Res Inc, Athens, Ga, 80-86; adj prof, Dept Path & Lab Med, Emory Univ, 84-; vis prof, Vet Fac, Univ Zagreb, Yugoslavia, 89. *Mem:* AAAS; Soc Toxicol; Int Acad Path; Am Asn Pathologists; Am Asn Cancer Res; Am Asn Immunologists. *Res:* Avian thymic hormones; immunoregulation and immunomodulation of chickens. *Mailing Add:* Poultry Dis Res Ctr 953 College Station Rd Athens GA 30605

RAGLE, JOHN LINN, b Colorado Springs, Colo, Feb 4, 33; m 69. PHYSICAL CHEMISTRY. *Educ:* Univ Calif, BS, 54; Wash State Univ, PhD(chem), 57. *Prof Exp:* Asst prof chem, Univ Mass, 57-60; fel, Cornell Univ, 60-62; mem res staff, Northrop Space Labs, 62-64; assoc prof, 64-69, PROF CHEM, UNIV MASS, AMHERST, 70- *Concurrent Pos:* Vis assoc prof chem, Univ BC, Vancouver, 69-70; preistrager, Alexander von Humboldt stiftung award, 75. *Mem:* Am Phys Soc. *Res:* Chemistry and physics of molecular structure. *Mailing Add:* Dept Chem Lgrt 102 Univ Mass Amherst MA 01003

RAGLE, RICHARD HARRISON, b Boston, Mass, June 11, 23; m 49; c 3. GEOLOGY. *Educ:* Middlebury Col, BA, 52; Dartmouth Col, MA, 58. *Prof Exp:* Geologist, Cold Regions Res & Eng Lab, US Army Corps Engrs, Greenland & Antarctic, 54-60; res scientist, Arctic Inst NAm, 60-64, staff scientist, 64-74; sr geologist, Dames & Moore, 74-77; mem staff, Naval Arctic Res Lab, 77-79; sr hydrogeologist, Northern Tech Serv, 80-81; CONSULT, 79- *Concurrent Pos:* Field sci leader, Icefield Ranges Res Proj, St Elias Mt, Yukon Terr, Can, 63-70, dir, 70-74. *Mem:* Fel Arctic Inst NAm; sr fel Geol Soc Am; Glaciol Soc; Am Inst Prof Geologists; Coastal Soc. *Res:* Glaciology, glacio-meteorology and climatology; ice and snow stratigraphy and metamorphism; sea ice mechanics and engineering. *Mailing Add:* 2419 Telequana Dr Anchorage AK 99517-1026

RAGONE, DAVID VINCENT, b New York, NY, May 16, 36; m 54; c 2. TEACHING, CONSULTING. *Educ:* Mass Inst Technol, SB, 51, SM, 52, ScD, 53. *Prof Exp:* From asst prof to prof, Dept Chem & Metall Eng, Univ Mich, Ann Arbor, 53-61; chmn, Metall Dept, John J Hopkins Lab Pure & Appl Sci Gen Atomic Div, Gen Dynamics, La Jolla, Calif, 62-67, asst dir, 65; Alcoa prof metall, Carnegie Mellon Univ, Pittsburgh, Pa, 67-69, assoc dean & prof eng, Sch Urban & Pub Affairs, 69-70; dean, Thayer Sch Eng, Dartmouth Col, 70-72; dean eng, Univ Mich, Ann Arbor, 72-80; pres, Case Western Reserve Univ & prof metall & mat sci, Cleveland, Ohio, 80-87; SR LECTR, DEPT MAT SCI & ENG, MASS INST TECHNOL, CAMBRIDGE, 87- *Concurrent Pos:* Mem bd dirs, var corp, 68-; mem bd trustees, var corp, 70-; Mem, adv comt Advan Automotive Power Systs, Coun Environ Qual, 70-76, chmn, 71-76, ad hoc panel Unconventional Engines, 69-70, US Dept Com Tech Adv Bd, 67-75, Panel on Automotive Air Pollution, 67-68, panel on housing technol, chmn, 68-69, panel automotive fuels, chmn, 70-71; mem, Nat Sci Bd, 84-; mem, White House Sci Coun Study Health of Univ, Exec Off President, 84-86; gen partner, Ampersand Specialty Mat Ventures, Wellesley, 88- *Mem:* Fel Am Soc Metals; Am Inst Mining & Metall Eng; Am Chem Soc; AAAS; Nat Soc Prof Engrs. *Res:* Metallurgical and chemical engineering. *Mailing Add:* Dept Mat Sci & Eng Mass Inst Technol Rm 8-301 Cambridge MA 02139

RAGOTZKIE, ROBERT AUSTIN, b Albany, NY, Sept 13, 24; m 49; c 3. METEOROLOGY, OCEANOGRAPHY. *Educ:* Rutgers Univ, BS, 48, MS, 50; Univ Wis-Madison, PhD(zool & meterol), 53. *Prof Exp:* Proj assoc meteorol, Univ Wis-Madison, 53; coord marine biol lab & asst prof biol, Univ Ga, 54-57, dir marine inst & assoc prof biol, 57-59; from asst prof to assoc prof, Univ Wis-Madison, 59-65, chmn dept meteorol, 64-67, dir, Marine Studies Ctr, 67-69, dir sea grant prog, 68-80, prof meteorol, 65-89, prof environ sci, 71-89, dir, Sea Grant Inst, 80-90; RETIRED. *Mem:* Fel AAAS; Am Soc Limnol & Oceanog; Am Meteorol Soc; Am Geophys Union; Int Asn Gt Lakes Res. *Res:* Physical limnology of Great Lakes, thermal structure and currents; Great Lakes as systems. *Mailing Add:* Dept Meteorol Univ Wis-Madison 1225 W Dayton St Madison WI 53706

RAGOZIN, DAVID LAWRENCE, b Brooklyn, NY, Apr 20, 41; m 70. MATHEMATICAL ANALYSIS, NUMERICAL ANALYSIS. *Educ:* Reed Col, BA, 62; Harvard Univ, AM, 63, PhD(math), 67. *Prof Exp:* Instr math, Mass Inst Technol, 67-69; asst prof, 69-75, ASSOC PROF MATH, UNIV WASH, 75- *Concurrent Pos:* NSF grant, Mass Inst Technol, 68-69, res assoc & NSF grant, 70-71; NSF grant, Univ Wash, 71-77. *Mem:* Am Math Soc; Math Asn Am. *Res:* Harmonic analysis on Lie groups and homogeneous spaces; applications of differential geometry; numerical analysis. *Mailing Add:* Dept Math Univ Wash Seattle WA 98195

RAGSDALE, DAVID WILLARD, b Boise, Idaho, Nov 8, 52; m 73; c 1. INTEGRATED PEST MANAGEMENT. *Educ:* Pt Loma Col, BA, 74; La State Univ, MS, 77, PhD(entom), 80. *Prof Exp:* Res assoc, La State Univ, 79-81; ASST PROF ENTOM, UNIV MINN, 81- *Mem:* Entom Soc Am; Sigma Xi. *Res:* Insects as vectors of plant disease agents; integrated pest management of field crops; use of serology in determining predator-prey relationships; production of monoclonal antibodies. *Mailing Add:* Dept Entom 219 Hodson Hall Univ Minn 1980 Folwell Ave St Paul MN 55108

RAGSDALE, HARVEY LARIMORE, b Atlanta, Ga, Mar 6, 40. ECOLOGY, BOTANY. *Educ:* Emory Univ, AB, 62; Univ Tenn, MS, 64, PhD(bot), 68. *Prof Exp:* Asst prof, 68-72, ASSOC PROF BIOL, EMORY UNIV, 72- *Concurrent Pos:* Consult, Allied Gen Nuclear Serv, 70-78 & Environ Div, Tex Instruments, Inc, 78; co-prin investr grants, US Energy Res & Develop Admin, 70-76 & NSF, 76-78; prin investr grants, US Dept Energy, 76-79. *Mem:* Am Inst Biol Sci; AAAS; Ecol Soc Am; Sigma Xi. *Res:* Ecological chemical element cycling; ecosystem modeling and simulation; radiation effects and cycling; deciduous forest community studies; solar energy from woody biomass fuel species. *Mailing Add:* 863 Barton Woods Rd Atlanta GA 30307

RAGSDALE, NANCY NEALY, b Griffin, Ga, Feb 5, 38; m 59; c 2. PESTICIDE CHEMISTRY, CELL PHYSIOLOGY. *Educ:* Cent Conn State Col, BS, 62; Univ Md, MS, 66, PhD(bot), 74. *Prof Exp:* Pesticide assessment specialist, Sci & Educ Admin-Chem Res, USDA, 78-80, pesticide coordr, Coop State Res Serv, 80-89; res assoc fungal physiol, Dept Bot, 74-78, ENVIRON COORDR, UNIV MD, 89- *Mem:* AAAS; Soc Toxicol; Am Chem Soc; Sigma Xi. *Res:* Mode of action of pesticides and environmental toxicology. *Mailing Add:* 13903 Overton Lane Silver Spring MD 20904

RAGSDALE, RONALD O, b Boise, Idaho, Dec 10, 32; m 56; c 3. INORGANIC CHEMISTRY. *Educ:* Brigham Young Univ, BS, 57; Univ Ill, MS, 59, PhD(chem), 60. *Prof Exp:* Res chemist, Gen Chem Div, Allied Chem Corp, 60-63; from asst prof to assoc prof, 63-72, PROF CHEM, UNIV UTAH, 72- *Mem:* Am Chem Soc; Royal Soc Chem. *Res:* Metal ion complexes; Lewis acid-base interactions; nuclear magnetic resonance. *Mailing Add:* Chem Dept Univ Utah 124 Eyring Bldg Salt Lake City UT 84112

RAGSDELL, KENNETH MARTIN, b Jacksonville, Ill, Sept 3, 42; m 62; c 3. MECHANICAL ENGINEERING. *Educ:* Univ Mo, Rolla, BS, 66, MS, 67; Univ Tex, Austin, PhD(mech eng), 72. *Prof Exp:* Instr eng, Okla State Univ, 67-68; mech engr, IBM Corp, 68-70; instr eng, Univ Tex, Austin, 70-72; asst prof mech eng, Purdue Univ, West Lafayette, 72-76, assoc prof, 76-; AT DEPT AEROSPACE & MECH ENG, UNIV ARIZ. *Concurrent Pos:* Consult,; pres, CAD Serv, Inc. *Res:* Computational aspects of design; optimization theory dynamics; computer aided design; engineering computation; optimization theory; design of dynamic mechanical systems. *Mailing Add:* 2314 W Las Lomitas Rd Tucson AZ 85704

RAHA, CHITTA RANJAN, b Faridpur, EBengal, Apr 1, 26; m 54; c 3. ORGANIC CHEMISTRY, BIOCHEMISTRY. *Educ:* Univ Calcutta, BSc, 45, MSc, 47, DPhil(chem), 54. *Hon Degrees:* FRIC, London. *Prof Exp:* Pool officer, Govt India, 61-65; asst prof oncol, Chicago Med Sch, 65-68; assoc prof biochem, Univ Nebr Med Ctr, Omaha, 68-88; RETIRED. *Concurrent Pos:* Int Agency Res Cancer travel fel, Wenner-Gren Inst, Univ Stockholm, 70. *Mem:* AAAS; Am Asn Cancer Res; Am Chem Soc; The Chem Soc. *Res:* Organic chemistry as applied to cancer research. *Mailing Add:* 7432 Spring St Omaha NE 68124

RAHAL, LEO JAMES, b Detroit, Mich, July 22, 39; m 71; c 2. NUCLEAR ENGINEERING, PLASMA PHYSICS. *Educ:* Univ Detroit, BS, 62, MS, 64; Univ NMex, PhD(physics), 78. *Prof Exp:* Physicist, LTV Aerospace, 68-73, Kirtland Weapons Lab, 73-76, Los Alamos Nat Lab, 76-77, Los Alamos tech assoc, 77-81; PHYSICIST, DIKEWOOD CORP, 81- *Res:* Plasma physics microinstability analysis: in the area of high density plasmas; nuclear waste management including waste disposal and air dispersion; nuclear reactor safety-hydrogen buildup in reactors and consequences. *Mailing Add:* 9417 Regal Ridge NE Albuquerque NM 87111

RAHE, JAMES EDWARD, b Muncie, Ind, Mar 12, 39; m 66; c 3. PLANT PATHOLOGY. *Educ:* Purdue Univ, BS, 61, PhD(biochem), 69. *Prof Exp:* Asst prof, 69-77, ASSOC PROF BIOL SCI, SIMON FRASER UNIV, 77- *Mem:* Am Phytopath Soc; Phytochem Soc NAm; Sigma Xi. *Res:* Biochemistry and physiology of host-parasite interaction; biological and integrated control of plant disease. *Mailing Add:* Dept Biol Sci Simon Fraser Univ Burnaby BC V5A 1S6 Can

RAHE, MAURICE HAMPTON, b Tucumcari, NMex, Jan 17, 44; m 73; c 2. MATHEMATICS. *Educ:* Pomona Col, BA, 65; Stanford Univ, MS, 70, PhD(math), 76. *Prof Exp:* Lectr & fel, Univ Toronto, 76-78; asst prof, 78-85, ASSOC PROF MATH, TEX A&M UNIV, 85- *Concurrent Pos:* Vis asst prof, Rice Univ, 81. *Mem:* Am Math Soc; Inst Elec & Electronics Engrs. *Res:* Ergodic theory; information theory; probability. *Mailing Add:* 3807 Westerman St Houston TX 77005

RAHE, RICHARD HENRY, b Seattle, Wash, May 28, 36; m 60; c 2. PSYCHIATRY. *Educ:* Univ Wash, MD, 61. *Prof Exp:* Intern med, Bellevue Hosp, New York, 61-62; from resident to chief resident psychiat, Univ Wash, 62-65; res psychiatrist, US Navy Neuropsychiat Res Univ, 65-68; head stress med div, US Naval Health Res Ctr, 70-76, comndg officer, 76-80; comndg officer, US Naval Hosp, Guam, 81-84; prof, US Univ Health Sci, Bethesda, 84-86; PROF PSYCHIAT & DIR NEV STRESS CTR, UNIV NEV, SCH MED, 86- *Concurrent Pos:* NIH spec fel, Karolinska Inst, Sweden, 68-69; adj prof psychiat, Neuropsychiat Inst; adj prof psychiat, Univ Calif, Los Angeles, 75- & Karolinska Inst, Sweden, 68-69; adj assoc prof psychiat, Univ Calif, San Diego & Univ Calif, Los Angeles, 70-74. *Honors & Awards:* McDonnell Prize, Univ Wash, 61. *Mem:* Fel Am Psychiat Asn; Am Psychosom Soc;

World Psychiat Asn; Asn Mil Surgeons US. *Res:* Life changes and illness onset; psychosocial aspects of physical illnesses; computer applications of stress and coping measures. *Mailing Add:* Dept Psychiat & Behav Sci Univ Nev Sch Med Manville Bldg Rm 9 Reno NV 89557-0046

RAHEEL, MASTURA, b Lahore, Pakistan, Mar 1, 38; m 59; c 2. TEXTILE SCIENCE, ORGANIC CHEMISTRY. *Educ:* Punjab Univ, Pakistan, BSc, 57, MSc, 59; Okla State Univ, MS, 62; Univ Minn, St Paul, PhD(textile sci), 71. *Prof Exp:* Asst prof & head textiles & clothing, Col Home Econ, Lahore, 60-77; lectr, Univ Minn, St Paul, 77-78; asst prof, 78-84, ASSOC PROF TEXTILE SCI & DIV CHMN, UNIV ILL, URBANA, 84- *Mem:* Am Asn Textile Chemists & Colorists; Am Home Econ Asn; Am Col Prof Textiles & Clothing; Sigma Xi. *Res:* Physical metrology of consumer textile products; chemical finishing of textiles; barrier properties of textiles toward toxic chemicals. *Mailing Add:* Div Textiles, Apparel & Interior Design Univ Ill 239 Bevier Hall Urbana IL 61801

RAHIMTOOLA, SHAHBUDIN HOOSEINALLY, b Bombay, India, Oct 17, 31; US citizen; m 67; c 3. CARDIOLOGY, INTERNAL MEDICINE. *Educ:* Univ Karachi, MB & BS, 56; MRCPE, 63; FRCP, 72. *Prof Exp:* Sr house officer, Barrowmore Chest Hosp, Chester, Eng, 56-57 & Whittington Hosp, London, 58-59; house physician, Cardiac Unit, London Chest Hosp, 59-60; Locum med registr, Whittington Hosp, London, 60; registr, Cardiac Unit, Wessex Reg Hosp, Southampton Eng, 60-63; co-dir, Cardiac Lab, Mayo Clin, Rochester, Minn, 65-66; sr registr cardiopulmonary dis, Dept Med, Queen Elizabeth Hosp, Birmingham, Eng, 66-67; res asst & hon sr registr, Dept Med, Royal Postgrad Med Sch & Hammersmith Hosp, London, 67-68; assoc prof med, Abraham Lincoln Sch Med, Col Med, Univ Ill, 69-72; prof med, Health Sci Ctr, Univ Ore, 72-80, dir res, 73-78; PROF MED & CHIEF SECT CARDIOL, UNIV SOUTHERN CALIF, 80-, GEORGE C GRIFFITH PROF CARDIOL, 84- *Concurrent Pos:* Co-dir, Dept Adult Cardiol, Cook County Hosp, Chicago, 69-70, dir, 70-72; consult cardiol, Madigan Gen Army Hosp, Ft Lewis, Wash, 72-; NIH grant, 72-77, 78-; Ore Heart Asn & Med Res Found Ore grants, 73-77; rep for Ore, Coun Clin Cardiol, Am Heart Asn, 75-77, mem exec comt, 77-, mem long range planning comt & mem nominating comt, 78-; consult, Nat Coop Study Valvular Heart Dis, 76-77; mem planning comt, Vet Admin, Washington, DC, 76-77, mem exec comt, 77-; mem circulatory systs devices panel, Food & Drug Admin, HEW, 76-80, chmn dept, 77-80; vis scientist, Cardiovasc Res Inst & vis prof med, Sch Med, Univ Calif, San Francisco, 78-79; grants, Coun Clin Cardiol, Am Heart Asn & Coun Circulation; mem adv panel cardiovascular drugs, US Pharmacocpia, Nat Forumlary, 81-; ed, Newsletter, Coun Clin Cardiol, An Heart Asn, 79-, Clin Cardiol, Am Med Asn, 80-83, Mod Concepts Cardiovasc Dis, Am Heart Asn, 85-88. *Mem:* Fel Am Col Cardiol; fel Am Col Chest Physicians; fel Am Col Physicians; Asn Univ Cardiologists. *Res:* Left ventricular performance in various disease states; coronary artery disease; valvular heart disease; cardiac electrophysiology. *Mailing Add:* Dept Med Univ Southern Calif 2025 Zonal Ave Los Angeles CA 90033

RAHLMANN, DONALD FREDERICK, b San Francisco, Calif, July 21, 23; m 49; c 3. GRAVITATIONAL PHYSIOLOGY. *Educ:* Univ Calif, Davis, BS, 55, MS, 56, PhD(animal physiol), 62. *Prof Exp:* Jr specialist, Dept Animal Husbandry, Univ Calif, Davis, 49-50, asst specialist reproductive physiology, 50-62; jr res physiologist, dept physiol & anat, Univ Calif, Berkeley, 62-63, asst res physiologist space physiol, 63-81, assoc res physiologist, Environ Physiol Lab, 81-85; RETIRED. *Concurrent Pos:* Consult, Lawrence Hall Sci, Univ Calif, Berkeley, 66-67; Int Govt Personnel Act, Ames Res Ctr, NASA, Moffett Field, Calif, 75-76; participant, Joint US-USSR Cosmos 1129 Flight Exp, 78-79. *Honors & Awards:* Cosmos Achievement Award, NASA, 81. *Mem:* Am Physiol Soc; Sigma Xi; AAAS; Am Asn Lab Animal Sci; Int Primatological Soc. *Res:* Animal experimentation in metabolism related to environmental changes; physiology of non-human primates; development of equipment and instrumentation for measurement of physiological response. *Mailing Add:* 1144 Nogales St Lafayette CA 94549

RAHM, DAVID CHARLES, b Ironwood, Mich, Dec 1, 27; m 51; c 2. PHYSICS. *Educ:* Univ Chicago, SB, 49; Univ Mich, MS, 51, PhD(physics), 56. *Prof Exp:* From asst physicist to assoc physicist, 55-62, physicist, 62-82, SR PHYSICIST, BROOKHAVEN NAT LAB, 82- *Concurrent Pos:* Physicist, Nuclear Res Ctr, Saclay, France, 60-61; vis scientist, Europ Orgn Nuclear Res, Geneva, 68-69, 75-80 & 85-88. *Mem:* Fel Am Phys Soc. *Res:* Particle physics; particle detectors; particle beams; superconducting magnets; accelerators. *Mailing Add:* Physics Dept Bldg 510 Brookhaven Nat Lab Upton NY 11973

RAHMAN, MD AZIZUR, b Santahar, Bangladesh, Jan 9, 41; Can citizen; m 63; c 3. PERMANENT MAGNETS, DIGITAL PROTECTION. *Educ:* Bangladesh Univ Eng, BSc, 62; Univ Toronto, MASc, 65; Carleton Univ, PhD(elec eng), 68. *Prof Exp:* Res scientist, Can Gen Elec Co, 68-69; from asst prof to assoc prof elec eng, Bangladesh Univ Eng, 69-74; Nuffield acad visitor, Imperial Col Sci & Technol, London, 74-75; sr engr & consult, Man Hydro, 75-76; assoc prof, 76-80, PROF ELEC ENG, MEM UNIV NFLD, 80- *Concurrent Pos:* Lectr elec eng, Bangladesh Univ Eng, 62-64; mem tech educ comn, E Pakistan, 69-70; res fel, Tech Univ, Eindhoven, Neth, 73-74; lectr, Univ Man, 75-76; consult, Nfld & Labrador Hydro, 77-78, 80-81 & 88-, Gen Elec Co, Schnectady, NY, 78-79; res prof, Univ Toronto, 84-85. *Honors & Awards:* G E Centennial Award; Inst Elec & Electronic Engrs Award. *Mem:* Fel Inst Engrs Bangladesh; Can Elec Asn; fel Inst Elec & Electronics Engrs; fel Inst Elec Engrs. *Res:* Design analysis of hysteresis, permanent magnet and superconducting motors; applications of supermagnets and superconductors in power apparatus and devices; delta modulated pulse width inverters and converters; digital protection of power transformers and reactors; published 225 papers. *Mailing Add:* Fac Eng & Appl Sci Mem Univ Nfld St John's NF A1B 3X5 Can

RAHMAN, MIZANUR, b Dhaka, Bangladesh, Sept 16, 32; Can citizen; m 61; c 2. BASIC HYPERGEOMETRIC SERIES, ORTHOGONAL POLYNOMIALS. *Educ:* Dhaka Univ, BSc, 53, MSc, 54; Cambridge Univ, BA, 58; Univ NB, PhD(math), 65. *Prof Exp:* Lectr math, Dhaka Univ, Bangladesh, 58-62; lectr, Univ NB, 62-65; from asst prof to assoc prof, 65-78, PROF MATH, CARLETON UNIV, OTTAWA, 78- *Mem:* Am Math Soc; Am Phys Soc; Can Math Soc; Can Appl Math Soc; Soc Inst Advan Mat. *Res:* Basic hypergeometric series; statistical mechanics and stochastic processes; author of approximately 70 publications in refreed journals in North America and United Kingdom; co-author of the book "Basic Hypergeometric Series", Cambridge University Press, 1990. *Mailing Add:* Dept Math & Statists Carleton Univ Ottawa ON K1S 5B6 Can

RAHMAN, TALAT SHAHNAZ, b Calcutta, India, Feb 5, 48; Pakistan citizen; m; c 1. SOLID STATE PHYSICS. *Educ:* Univ Karachi, BSc, 68, MSc, 69; Univ Rochester, PhD(physics), 77. *Prof Exp:* Res asst physics, Univ Rochester, 71-76, teaching asst, 73 & 76-77; res physicist, Univ Calif, Irvine, 77-82; AT DEPT PHYSICS, KANS STATE UNIV, 83- *Concurrent Pos:* Alexander von Humboldt Fel, 87-88. *Mem:* Am Phys Soc; Asn Women Sci. *Res:* Surface physics and optical properties of solids. *Mailing Add:* Dept Physics Kans State Univ Manhattan KS 66506

RAHMAN, YUEH ERH, b Canton, China, June 10, 30; m 56; c 1. MEDICINE, HEALTH SCIENCES. *Educ:* Univ Louvain, MD, 56. *Prof Exp:* Med officer, Belgian Leprosy Ctr, India, 57-58; asst res officer, Indian Cancer Res Ctr, 58-59; res assoc biochem cytol, Univ Louvain, 59-60; res assoc, Biol Div, Argonne Nat Lab, 60-63, asst biologist, 63-72, biologist, 72-81, sr biologist, 81-85; PROF, COL PHARM, DEPT PHARMACEUT, UNIV MINN, MINNEAPOLIS, 85-, DIR GRAD STUDIES, 88- *Concurrent Pos:* Vis scientist, Dept Biochem, Univ Utrecht, 68-69; adj assoc prof, Dept Biol Sci, Northern Ill Univ, 71-; mem rev group, Exp Therapeut Study Sect, NIH, 79-83; pres, Chicago Chap, Asn Women Sci, 79-80. *Honors & Awards:* Indust Res 100 Award, 76. *Mem:* Am Soc Cell Biol; Radiation Res Soc; NY Acad Sci; AAAS; Asn Women Sci; Am Asn Pharmaceut Scientists. *Res:* Cellular biochemistry; cell membranes; radiation and lysosomes; chemotherapy by use of liposome encapsulation of drugs, such as chelating agents, anti-tumor drugs and immunosuppressants. *Mailing Add:* Col Pharm Health Sci Unit F Univ Minn 308 Harvard St SE Minneapolis MN 55455

RAHMAT-SAMII, YAHYA, b Tehran, Iran, Aug 20, 48; US citizen. SATELLITE COMMUNICATION ANTENNAS, ELECTROMAGNETIC SCATTERING. *Educ:* Tehran Univ, BS, 70; Univ Ill, Urbana-Champaign, MS, 72, PhD(elec eng), 75. *Prof Exp:* Vis asst prof elec eng, Univ Ill, Urbana-Champaign, 75-78; sr res scientist, Jet Propulsion Lab, 78-88; PROF ELEC ENG, UNIV CALIF, LOS ANGELES, 88- *Concurrent Pos:* Dir, Electromagnetic Soc, 84- & Antennas Measurement Tech Asn, 90-; vis prof elec eng, Tech Univ Denmark, 86. *Mem:* Fel Inst Elec & Electronics Engrs; fel Int Union Radio Sci; fel Inst Advan Eng; Inst Elec & Electronics Engrs Antennas & Propagation Soc; Inst Elec & Electronics Engrs Microwave Theory & Tech Soc; Electromagnetics Soc; Antennas Measurement Tech Asn. *Res:* Novel space and ground antenna concepts; satellite communications; advanced antenna measurement and diagnostic techniques; radar cross section; asymptotic and high frequency diffraction methods; numerical modeling in electromagnetic scattering; author of over 180 technical publications. *Mailing Add:* Elec Eng Dept Eng IV Univ Calif 405 Hilgard Ave Los Angeles CA 90024-1594

RAHN, ARMIN, soldering technology, for more information see previous edition

RAHN, HERMANN, physiology; deceased, see previous edition for last biography

RAHN, JOAN ELMA, b Cleveland, Ohio, Feb 5, 29. PLANT MORPHOLOGY. *Educ:* Western Reserve Univ, BS, 50; Columbia Univ, AM, 52, PhD, 56. *Prof Exp:* From asst prof to assoc prof biol, Thiel Col, 56-59; instr bot, Ohio State Univ, 59-60; instr biol, Int Sch Am, 60-61; asst prof, Lake Forest Col, 61-67; SCI WRITING, 67- *Mem:* AAAS; Bot Soc Am; Am Inst Biol Sci; Sigma Xi. *Mailing Add:* 1656 Hickory St Highland Park IL 60035

RAHN, KENNETH A, b Hackensack, NJ, Aug 10, 40; wid; c 2. ATMOSPHERIC CHEMISTRY. *Educ:* Mass Inst Technol, BS, 62; Univ Mich, PhD(meteorol), 71. *Prof Exp:* Sci/math teacher, Classical High Sch, Barrington College, 63-68; res assoc atmospheric chem, Inst Nuclear Sci, Univ Ghent, 71-73; res assoc, Grad Sch Oceanog, Univ RI, 73-75; vis scientist, Max Planck Inst Chem, 75-76; res assoc atmospheric chem, grad sch oceanog, Univ RI, 76-; PROF CTR ATMOSPHERIC CHEM STUDIES, UNIV RI, 80- *Mem:* Am Chem Soc; AAAS; Am Meteorol Soc; Gesellschaft Fur Aerosolforschung. *Res:* Aerosols; arctic air chemistry; long-range transport. *Mailing Add:* Grad Sch Oceanog Univ RI Narragansett RI 02882-1197

RAHN, PERRY H, b Allentown, Pa, Oct 27, 36; m 62; c 4. HYDROLOGY, GEOMORPHOLOGY. *Educ:* Lafayette Col, BS & BA, 59; Pa State Univ, PhD(geol), 65. *Prof Exp:* Civil engr, Calif Dept Water Resources, 59-61; asst prof geol, Univ Conn, 65-68; from asst prof to assoc prof, 68-78, PROF GEOL, SDAK SCH MINES & TECHNOL, 78- *Concurrent Pos:* Hydrogeologist, Argonne Nat Lab, 77 & Bucknell Univ, 88. *Honors & Awards:* Claire P Holdredge Award, Asn Eng Geol, 87; E B Burwell Award, Geol Soc Am, 90. *Mem:* Fel Geol Soc Am; Int Asn Hydrogeol; Nat Water Well Asn; Am Quaternary Asn; Sigma Xi; Nat Soc Prof Engr. *Res:* Engineering geology; hydrology of glacial and limestone terranes; uranium tailing pond contamination; engineering geology. *Mailing Add:* Dept Geol & Geol Eng SDak Sch Mines & Technol 500 E St Joseph Rapid City SD 57701

RAHWAN, RALF GEORGE, b Egypt; Feb 28, 41; US citizen; c 1. TOXICOLOGY. *Educ:* Cairo Univ, BS, 61; Butler Univ, MS, 70; Purdue Univ, PhD(pharmacol), 72. *Prof Exp:* Retail pharmacist, 61-64; head, Sci Doc & Training Dept, Hoechst Orient Pharmaceut Co, 64-67; assoc pharmacologist, Human Health Res & Develop Labs, Dow Chem Co, 67-70; from asst prof to assoc prof, 72-80, PROF PHARMACOL, COL PHARM, OHIO STATE UNIV, 80- *Mem:* Am Soc Pharmacol & Exp Therapeut; Soc Toxicol; Sigma Xi; Am Soc Clin Pharmacol Therapeut. *Res:* Endocrine pharmacology, toxicology, calcium antagonists. *Mailing Add:* Div Pharmacol Ohio State Univ Col Pharm Columbus OH 43210-1291

RAI, AMARENDRA KUMAR, b Varanasi, UP, India, Oct 20, 52; m 71. ION BEAM PROCESSING OF MATERIALS, SCANNING TRANSMISSION ELECTRON MICROSCOPY. *Educ:* Gorakhpur Univ, BSc, 70; Banaras Hindu Univ, MSc, 72, PhD(physics), 77. *Prof Exp:* Jr res fel physics, Coun Sci & Indust Res, India, 72-75, sr res fel, 76-77, fel, 78-79; fel mats eng, NC State Univ, Raleigh, 79-81; scientist, 81-83, SR SCIENTIST, MATS RES DIV, UNIVERSAL ENERGY SYSTS, 83- *Concurrent Pos:* Vis scientist, Solid State Div, Oak Ridge Nat Lab, 81; prin investr, Dept Defense, 85-86 & 86-87. *Mem:* Am Phys Soc; Am Ceramic Soc; Mats Res Soc. *Res:* Surface modification of various materials employing ion beam processing; microstructural characterization of materials using scanning transmission electron microscopy; author of over 60 research papers in various scientific journals. *Mailing Add:* Universal Energy Systs Inc 4401 Dayton Xenia Rd Dayton OH 45423

RAI, CHARANJIT, b Barabanki, India, July 19, 29. CHEMICAL ENGINEERING, CHEMISTRY. *Educ:* Univ Agra, BSc, 48; Lucknow Univ, MSc, 50; Indian Inst Sci, Bangalore, dipl, 52, PhD(chem), 59; Univ Ill, MS, 56; Ill Inst Technol, PhD(chem eng), 60. *Prof Exp:* Res asst chem eng, Univ Ill, 54-56; instr, Ill Inst Technol, 56-58; sr res scientist, Pure Oil Co, 59-65; sr res supvr, Richardson Co, Ill, 65-66; dept head prod & process res, Cities Serv Oil Co, 66-76; prof, US Energy Res & Develop Agency, Morgantown Energy Res, 77-80; MEM FAC, DEPT CHEM & NATURAL GAS ENG, TEX A&I UNIV, 80- *Concurrent Pos:* Prof chem eng & chmn dept, Indian Inst Technol, Kanpur, 63-64. *Mem:* AAAS; Am Inst Chem Engrs; Am Chem Soc; Indian Inst Chem Engrs. *Res:* Mass transfer; thermodynamics; oxidation of hydrocarbons; petrochemicals; fuels and lubricants. *Mailing Add:* Dept Chem & Natural Gas Eng Tex A&I Univ Kingsville TX 78363-8203

RAI, DHANPAT, b June 12, 43; US citizen; m 72; c 2. SOIL CHEMISTRY, SOIL MINERALOGY. *Educ:* Panjab Agr Univ, BSc, 63, MSc, 65; Ore State Univ, PhD(soil sci), 70. *Prof Exp:* Res assoc soil sci, Ore State Univ, 70-71; fel, Colo State Univ, 72-73; res assoc & asst prof, NMex State Univ, 74-75; sr res scientist, 75-81, staff scientist, 81-90, SR STAFF SCIENTIST, BATTELLE, PAC NORTHWEST LABS, 90- *Mem:* Am Soc Agron; Soil Sci Soc Am; Int Soc Soil Sci; Res Soc NAm; Am Chem Soc. *Res:* Soil chemistry; environmental chemistry of actinides; geochemistry. *Mailing Add:* Battelle Pac Northwest Lab PO Box 999 Richland WA 99352

RAI, IQBAL SINGH, b Majali Kalan, Punjab, India, Jan 29, 36; m 62; c 2. STRUCTURAL ENGINEERING, ENGINEERING MECHANICS. *Educ:* Punjab Univ, India, BA, 56, BSc, 60; Univ Roorkee, ME, 66; Ohio State Univ, PhD(civil eng), 75. *Prof Exp:* From lectr to assoc prof struct eng, Guru Nanak Eng Col Ludhiana, India, 60-71; res assoc civil eng, Ohio State Univ, 71-77; SR RES ENGR & SECT HEAD, GOODYEAR TIRE & RUBBER CO, 77- *Concurrent Pos:* Consult, Nankana Sahib Educ Trust, Ludhiana, India, 63-71 & Columbus Aircraft Div, Rockwell Int, 76-77. *Mem:* Am Soc Civil Engrs; Soc Exp Stress Anal. *Res:* Finite element analysis; mechanics of structures; structural analysis and design. *Mailing Add:* 2507 Cardigan Dr Akron OH 44333-2948

RAI, KANTI R, b Jodhpur, India, May 10, 32; m 68; c 2. HEMATOLOGY, ONCOLOGY. *Educ:* Med Col, Univ Rajasthan, MB & BS, 55; Am Bd Pediat, dipl, 61. *Prof Exp:* Head exp med, Inst Nuclear Med, Delhi, India, 62-66; assoc scientist, Brookhaven Nat Lab, 66-70; assoc prof med, 72-80, PROF MED, SCH MED, STATE UNIV NY STONY BROOK, 80-; CHIEF, DIV HEMAT-ONCOL, LONG ISLAND JEWISH-HILLSIDE MED CTR, 81- *Concurrent Pos:* Leukemia res scholar, Nat Leukemia Asn, 66-67 & 75-77; attending physician hemat-oncol, Long Island Jewish-Hillside Med Ctr, 70-80. *Mem:* Am Soc Hemat; Am Soc Clin Oncol; Am Asn Cancer Res; Soc Nuclear Med; Soc Exp Biol & Med. *Res:* Natural history and biology of leukemias; cell kinetics in leukemias; new therapeutic approaches in the malignancies of blood. *Mailing Add:* 269-11 26th Ave New Hyde Park NY 11042

RAI, KARAMJIT SINGH, b Moranwali, Punjab, India, Mar 24, 31; m 56; c 5. CYTOGENETICS , GENETIC CONTROL. *Educ:* Punjab Univ, India, BSc, 53, MSc, 55; Univ Chicago, PhD(bot), 60. *Prof Exp:* Lectr bot, Khalsa Col, Amritsar, 55-56; head dept bot, Deshbandhu Col, Delhi, 56-58; Charles Hutchinson fel & Coulter res fel, Univ Chicago, 58-60; assoc, Chicago Natural Hist Mus, Ill, 60; res assoc, Radiation Lab & dept biol, 60-62, sr staff mem radiation lab, 62-77, from asst prof to assoc prof, 62-66, dir mosquito biol training prog, 69-75, PROF BIOL, UNIV NOTRE DAME, 70- *Concurrent Pos:* Consult, Ill Inst Technol, 64-68, Int Atomic Energy Agency, Vienna, 66 & 69 & WHO, Geneva, 66-75; adv, Govt Ceylon, 66 & Govt Brazil, 69; mem Int Atomic Energy Agency panels, 68 & 70; mem US-Japan panels on parasitic dis, 70,72,73; mem, planning & rev group, World Health Orgn, Res Unit Genetic Control Mosquitoes, New Delhi, 71-75; vis prof, Univ Pernambuco, Brazil, 69 & Guru Nanak Dev Univ, India, 73-74; co-prin investr, Mosquito Biol Unit, Mombasa, Nairobi, Kenya, 71-76; mem, Rockfeller Found, Conf Genetics Dis Vectors, Bellagio, Italy, 81; chmn, Conf Genetics & Molec Biol, Entomal Soc Am, 84. *Mem:* AAAS; Genetics Soc Am; Entom Soc Am; Am Mosquito Control Asn; Am Inst Biol Sci. *Res:* Cytogenetics and molecular genetics of Aedes mosquitoes, molecular organization and evolution of mosquito genomes; chromosomal rearrangements and insect population control; genetics of speciation in Aedes; mutagenesis; vector competence; genetic control of cell division. *Mailing Add:* Dept Biol Sci Univ Notre Dame Notre Dame IN 46556-0369

RAIBLE, ROBERT H(ENRY), b Cincinnati, Ohio, Aug 27, 35; m 58; c 1. ELECTRICAL ENGINEERING. *Educ:* Univ Cincinnati, EE, 58; Purdue Univ, PhD(elec eng), 64. *Prof Exp:* Proj engr, Cincinnati Milling Mach, 57-58; instr, Purdue Univ, 59-64; from asst prof to assoc prof, 64-76; PROF ELEC ENG, UNIV CINCINNATI, 77- *Concurrent Pos:* Consult, Spati Industs; Metcut Assoc. *Mem:* Inst Elec & Electronics Engrs. *Res:* Analysis and design of automatic control systems; theory and application of adaptive and learning systems. *Mailing Add:* Elec Engr Dept - LOC No 30 Univ Cincinnati Cincinnati OH 45221

RAICH, HENRY, b Philadelphia, Pa, June 11, 19; m 43; c 2. PHYSICAL CHEMISTRY. *Educ:* Rensselaer Polytech Inst, BS, 40 & 44, PhD, 49. *Prof Exp:* Chemist, Am Smelting & Refining Co, 40-43 & Los Alamos Sci Lab, 44-46; fel, Mellon Inst, 49-52; chemist, Nuodex Prod Co, Inc, 52-54; res assoc lubricants, Mobil Oil Corp, 54-83; RETIRED. *Concurrent Pos:* Consult, 83- *Mem:* Am Chem Soc; Am Soc Lubrication Engrs; Am Soc Testing & Mat; Nat Lubricating Grease Inst. *Res:* Lubricants; colloids; rheology; metal soaps. *Mailing Add:* 1620 Aster Dr Cherry Hill NJ 08003

RAICH, JOHN CARL, b Badgastein, Austria, May 9, 37; US citizen; m 63; c 2. SOLID STATE PHYSICS. *Educ:* Iowa State Univ, BS, 59, PhD(physics), 63. *Prof Exp:* Res assoc physics, Iowa State Univ, 64 & Purdue Univ, 64-66; from asst prof to prof physics, 66-78, chmn dept, 72-78, assoc dean, 78-85, DEAN, COL NAT SCI, COLO STATE UNIV, 85- *Concurrent Pos:* Consult, Los Alamos Sci Lab, 71-; NATO sr fel sci, NSF, 75, Humboldt fel, 77. *Mem:* Am Phys Soc. *Res:* Molecular crystals. *Mailing Add:* Colo State Univ Ft Collins CO 80523

RAICH, WILLIAM JUDD, polymer chemistry; deceased, see previous edition for last biography

RAICHEL, DANIEL R(ICHTER), b Paterson, NJ, Aug 22, 35; m 67; c 2. ROBOTICS, THEORETICAL PHYSICS. *Educ:* Rensselaer Polytech Inst, BME, 57; Mass Inst Technol, SM, 58; Columbia Univ, MechEngr, 62; NY Univ, EngScD, 70. *Prof Exp:* Asst proj engr, Curtiss-Wright Corp, 61-62; instr wind tunnel lab, Case Inst Technol, 62-63; eng consult, Polytech Design Corp, 63-64; asst res scientist aeronaut & astronaut, NY Univ, 64-65; instr mech eng, Newark Col Eng, 65-67; independent consult, 67-68; consult advan develop eng, Electro-Nucleonics, Inc, 69-71; pres & chief scientist med & lab instrumentation, Dathar Corp, 71-75; prin, Ingenieurs Int, 75-79; prof mech eng, Pratt Inst, 83-91; PRIN, RAICHEL TECHNOL GROUP & RAMAR CONSULT, 79-; PROF MECH ENG, COOPER UNION, 90- *Concurrent Pos:* Adj prof mech eng, NJ Inst Technol, 78-82. *Mem:* Am Soc Mech Engrs; Am Phys Soc; Acoust Soc Am; Audio Eng Soc; Sigma Xi; Am Soc Eng Educr. *Res:* Acoustics; fluid mechanics; materials science; molecular physics. *Mailing Add:* 532 Spencer Dr Wyckoff NJ 07481

RAICHLE, MARCUS EDWARD, b Hoquaim, Wash, Mar 15, 37; m 64; c 4. NEUROLOGY. *Educ:* Univ Wash, BS, 60, MD, 64. *Prof Exp:* Instr neurol, New York Hosp-Cornell Med Ctr, 68-69; consult neurol, Sch Aerospace Med, US Air Force, 69-71; asst prof, 71-75, assoc prof, 75-78, PROF NEUROL & RADIATION SCI, SCH MED, WASHINGTON UNIV, 78- *Concurrent Pos:* NIH teacher-investr award, Nat Inst Neurol & Commun Dis & Stroke, 71-; mem, Neurol A Study Sect, NIH, 75 & Cardiovasc D Res Study Comt, Am Heart Asn, 75- *Mem:* Am Neurol Asn; Am Acad Neurol; Am Physiol Soc; Soc Neuroscience; AMA. *Res:* In vivo measurement of brain hemodynamics, metabolism and exchange processes using trace kinetic techniques and positron-emitting, cyclotron-produced radioisotopes. *Mailing Add:* Dept Neurol & Radiol Campus Box 8131 Washington Univ Med St Louis MO 63110

RAIDER, STANLEY IRWIN, b New York, NY, July 21, 34; m 60; c 2. MATERIALS SCIENCE. *Educ:* Brooklyn Polytech Inst, BChE, 57; State Univ NY Syracuse, MS, 62; State Univ NY Stony Brook, PhD(chem), 67. *Prof Exp:* Chem engr, US Naval Powder Plant, 56 & Hooker Chem Co, 58-59; chemist, East Fishkill Facility, IBM Components Div, Hopewell Junction, NY, 67-75, CHEMIST, T J WATSON RES CTR, IBM CORP, 75- *Mem:* Electrochem Soc; Mat Res Soc. *Res:* Surface and interface chemistry; failure mechanisms in thin dielectric films; spectroscopy; superconducting materials; superconducting devices. *Mailing Add:* IBM T J Watson Res Ctr PO Box 218 Yorktown Heights NY 10598

RAIFORD, MORGAN B, b Franklin, Va, Oct 28, 12; m 49; c 3. OPHTHALMOLOGY. *Educ:* Guilford Col, BS, 33; Va Commonwealth Univ, MD, 37; Univ Pa, MSc, 49, DSc(med), 54; Am Bd Ophthal, dipl, 65. *Hon Degrees:* DHL, Pa Col Optom Med, 78. *Prof Exp:* FOUNDER & MED DIR, ATLANTA EYE CLIN, 63- *Concurrent Pos:* Mem bd, Atlanta Hosp, 67- & bd mem, Am Acad Prev Med, 73- *Honors & Awards:* Physicians Recognition Award, AMA, 73-79, 81-84. *Mem:* AMA; Pan Am Asn Ophthal; Am Acad Ophthal & Otolaryngol; fel Am Col Nuclear Med. *Mailing Add:* Atlanta Eye Clin 615 Peachtree St NE Atlanta GA 30308

RAIJMAN, LUISA J, b Cordoba, Argentina, Nov 2, 34; div. MITOCHONDRIAL METABOLISM, BIOGENESIS. *Educ:* Nat Univ Cordoba, MD, 57. *Prof Exp:* ASSOC PROF, DEPT BIOCHEM, SCH MED, UNIV SOUTHERN CALIF, 81- *Concurrent Pos:* Prin investr, 75, vis assoc prof, Univ NC, Chapel Hill, 85-; comt mem, Nat Sci Found, 80-85. *Mem:* Am Soc Biol Chemists; AAAS; Am Asn Univ Profs; Sigma Xi; Biochem Soc UK. *Res:* Structural and fuctional organization of enzymes in kinetic properties of enzymes in situ studies in fermeahilized cells and intochondria mammalian mitochondrial biogenesis. *Mailing Add:* Dept Biochem Univ Southern Calif Sch Med 2011 Zonal Ave Los Angeles CA 90033

RAIKHEL, NATASHA V, PLANT CELL & MOLECULAR BIOLOGY. *Educ:* Inst Cytol, Russia, PhD(biol), 75. *Prof Exp:* ASST RES SCIENTIST, UNIV GA, 68- *Res:* Tissue specificity of gene expression. *Mailing Add:* Dept Energy Plant Res Lab Mich State Univ East Lansing MI 48824-1312

RAIKOW, RADMILA BORUVKA, b Prague, Czech, Mar 20, 39; US citizen; m 66; c 2. AUTOIMMUNITY, CANCER BIOLOGY. *Educ:* NY Univ, BA, 60; Brooklyn Col, MA, 65; Univ Calif, Berkeley, PhD(genetics), 70. *Prof Exp:* Res assoc genetics, Univ Hawaii, 70-71; fel biochem, Univ Pittsburgh, 71-72; fel cancer biol, 75-78, RES SCIENTIST, CANCER RES, ALLEGHENY-SINGER RES INST, PITTSBURGH, 78- *Concurrent Pos:* Instr biol, Univ Pittsburgh, 83- *Mem:* AAAS; Sigma Xi; Am Asn Cancer Res; Clin Immunol Soc; Soc Exp Biol & Med; NY Acad Sci. *Res:* Autoimmunity in ophthalmopathy and immune functions in cancer etiology. *Mailing Add:* Allegheny-Singer Res Inst 320 E North Ave Pittsburgh PA 15212

RAIKOW, ROBERT JAY, b Detroit, Mich, May 28, 39; m 66; c 2. ANATOMY, ORNITHOLOGY. *Educ:* Wayne State Univ, BS, 61, MS, 64; Univ Calif, Berkeley, PhD(zool), 69. *Prof Exp:* Actg asst prof zool, Univ Calif, Berkeley, 69-70; NIH fel, Univ Hawaii, 70-71; from asst prof to assoc prof, 71-86, PROF BIOL SCI, UNIV PITTSBURGH, 86- *Concurrent Pos:* Rev ed, Wilson Bulletin, 74-84; prin investr NSF res grants, 74 -; res assoc, Carnegie Mus Natural Hist, 75 - *Mem:* Fel AAAS; Soc Syst Zool; Am Soc Zool; fel Am Ornith Union; Cooper Orinth Soc; Wilson Orinth Soc. *Res:* Avian anatomy and systematics; vertebrate functional anatomy; phylogenetic studies of birds based on cladistic analysis of morphology, primarily appendicular myology; systematic methodology. *Mailing Add:* Dept Biol Sci Univ Pittsburgh Pittsburgh PA 15260

RAIMI, RALPH ALEXIS, b Detroit, Mich, July 25, 24; m 47; c 2. MATHEMATICAL ANALYSIS. *Educ:* Univ Mich, BS, 47, MS, 48, PhD(math), 54. *Prof Exp:* Instr math, Univ Rochester, 52-55; Lloyd fel, Univ Mich, 55-56; from asst prof to assoc prof, 56-66, assoc dean grad studies, Col Arts & Sci, 67-75, chmn dept sociol, 83-86 PROF MATH, UNIV ROCHESTER, 66-,. *Concurrent Pos:* Fac ed, Coun Lib Learning, Asn Am Col. *Mem:* Math Asn Am. *Res:* Functional analysis; topological linear spaces; invariant measures and means. *Mailing Add:* Dept Math Univ Rochester Rochester NY 14627

RAIMONDI, ALBERT ANTHONY, b Plymouth, Mass, Mar 29, 25; m 89. MECHANICAL ENGINEERING. *Educ:* Tufts Col, BS, 45; Univ Pittsburgh, MS, 63, PhD(mech eng), 68. *Prof Exp:* Fel res engr, Mech Dept, Res Labs, Westinghouse Elec Corp, 46-68, mgr lubrication mech, 68-78, mgr tribology & exp struct mech, Westinghouse Res & Develop Ctr, 78-90, CONSULT ENGR, WESTINGHOUSE SCI & TECHNOL CTR, 90- *Concurrent Pos:* Ed, Soc Lubrication Engrs, 71. *Honors & Awards:* Hunt Mem Award, Soc Lubrication Engrs, 59, Nat Award, 68. *Mem:* Am Soc Mech Engrs; fel Soc Lubrication Engrs. *Res:* Tribology; constitutive relations; experimental mechanics; photoelasticity; bearing and seal design and application. *Mailing Add:* 125 Eighth St Turtle Creek PA 15145-1805

RAIMONDI, ANTHONY JOHN, b Chicago, Ill, July 16, 28; m 54; c 3. NEUROSURGERY, NEUROANATOMY. *Educ:* Univ Ill, BA & BS, 50; Univ Rome, MD, 54. *Prof Exp:* Instr neurosurg, Univ Chicago, 61-62; instr, Northwestern Univ, 62-64; clin asst prof, Univ Chicago, 64-66, clin assoc prof, 66-67; assoc prof neurosurg , Univ Ill Col Med, 67-69; PROF NEUROSURG & CHMN DIV, SCH MED, NORTHWESTERN UNIV, CHICAGO, 69-, PROF ANAT, 74- *Concurrent Pos:* Attend neurosurg, Children's Mem Hosp, 62-63, chmn div neurosurg, 69-; prof, Cook County Grad Sch Med, 63-70, chmn dept, Cook County Hosp, 63-70; mem fac adv bd, Chicago Med Sch Quart, 64-66; chmn neurosurg, Vet Res Hosp, Chicago, 72-; attend neurosurgeon, Passavant Mem Hosp, 69- & Northwestern Mem Hosp, 74-; attend physician neurosurg, Surg Serv, Vet Admin Hosp; consult-lectr, Great Lakes Naval Hosp; chmn med adv comt, Am Spina Bifida Asn, Epilepsy Fund Am & Asn Brain Tumor Res. *Mem:* Am Col Surg; Am Asn Neurosurg; Am Asn Neuropath; Int Soc Pediat Neurosurg (secy); Am Asn Surg of Trauma. *Res:* Ultrastructural characteristics of normal edematous, neoplastic and toxic glia; cerebral angiography in the newborn and infant; pediatric neurosurgery; pediatric neuroradiology. *Mailing Add:* Dept Anat & Cell Biol Northwestern Univ Med Sch 303 E Chicago Ave Chicago IL 60611

RAIMONDI, PIETRO, b Acqui, Italy, Feb 18, 29; m 56; c 6. CHEMICAL ENGINEERING. *Educ:* Univ Notre Dame, BS, 52, MS, 53; Carnegie Inst Technol, PhD(chem eng), 57. *Prof Exp:* From proj chem engr to sr res engr, 57-70, SECT SUPVR, GULF RES & DEVELOP CO, 70- *Mem:* Am Inst Chem Engrs; Am Inst Mining, Metall & Petrol Engrs. *Res:* Single and multiphase flow and diffusion and mixing of fluids in porous media; oil reservoir mechanics; synthetic fuel by in-situ method. *Mailing Add:* 6721 Quincy Dr Verona PA 15147

RAINA, ASHOK K, b Srinagar, Kashmir, India, Feb 28, 42; US citizen; m 60; c 2. BEHAVIORAL PHYSIOLOGY, INSECT-PLANT INTERACTIONS. *Educ:* Jammu & Kashmir Univ, India, BSc, 61; Aligarh Muslim Univ, India, MSc, 67; NDak State Univ, PhD(entom), 74. *Prof Exp:* Res asst, Commonwealth Inst Biol Control, Bangalore, India, 62-65; sr res asst, Regional Pulse Improv Proj, USAID, New Delhi, India, 67-70; res asst, NDak State Univ, 70-74; asst prof biol & elec micros, Minot State Col, Minot, NDak, 74-75; assoc entom, Va State Univ, 76-77; res scientist, Int Ctr Insect Physiol & Ecol, 78-79, prog coordr, 80-81; sr res assoc, dept entom, Univ Md, 81-85; RES ENTOMOLOGIST, AGRI RES SERV, USDA, BELTSVILLE, MD, 86- *Concurrent Pos:* Adj prof, Dept Entomol, Univ Md, 87- *Mem:* Entom Soc Am; fel Entom Soc India; Sigma Xi; Int Soc Chem Ecol. *Res:* Behavioral physiology, hormones and pheromones of insects; insect-plant interactions; endocrinology; electron microscopy; discovery of new insect hormones. *Mailing Add:* Insect Neurobiol & Hormone Lab Bldg 225 USDA-Agr Res Serv Beltsville Agr Res Ctr-E Beltsville MD 20705

RAINAL, ATTILIO JOSEPH, b Marion Heights, Pa, Feb 14, 30; m 57; c 2. NOISE THEORY & RANDOM PROCESSES, ELECTRICAL INTERCONNECTIONS & THEIR PERFORMANCE LIMITS. *Educ:* Pa State Univ, BS, 56; Drexel Univ, MS, 59; Johns Hopkins Univ, PhD(eng), 63. *Prof Exp:* Engr, Martin Co, Baltimore, 56-59; staff mem res, Carlyle Barton Lab, Johns Hopkins Univ, 59-64; mem tech staff, 64-83, DISTINGUISHED MEM TECH STAFF, AT&T BELL LABS, 83- *Mem:* Sr mem Inst Elec & Electronic Engrs; Inst Elec & Electronic Engrs Info Theory Soc; Inst Elec & Electronic Engrs Components Hybrids & Mfg Technol Soc. *Res:* Noise theory; signal detection and estimation; radiometry; radar; FM; first passage times of random processes; crosstalk; voltage breakdown; current carrying capacity of printed wires; performance limits of electrical interconnections. *Mailing Add:* 28 Woodruff Rd Morristown NJ 07960

RAINBOLT, MARY LOUISE, b Cleveland, Okla, June 21, 25. BIOLOGY. *Educ:* Okla Baptist Univ, BS, 46; Okla State Univ, MS, 48; Univ Okla, PhD, 63. *Prof Exp:* Prof biol, Southwestern State Col, Okla, 48-65; prof, 64-81, HITCHCOCK PROF BIOL, ILL COL, 81-, HEAD DEPT, 64- *Mem:* Am Soc Zool. *Res:* Physiology; endocrinology. *Mailing Add:* c/o J N Rainbolt Box 295 Rt 1 Seminole OK 74868

RAINBOW, ANDREW JAMES, b Essex, Eng, Dec 18, 43; Can citizen; m 72; c 2. MEDICAL BIOPHYSICS, MOLECULAR BIOLOGY. *Educ:* Univ Manchester, BSc, 65; Univ London, MSc, 67; McMaster Univ, PhD(biol), 70. *Prof Exp:* Fel biol, McMaster Univ, 70-71; radiol hosp physicist & assoc scientist, Royal Victoria Hosp, Montreal, 71-72; from asst prof to assoc prof, 72-84, assoc mem, dept biol, 73-85, PROF RADIOL & DIR REGIONAL RADIOL SCI PROG, MCMASTER UNIV, 84-, ASSOC MEM PHYSICS, 83-, PROF BIOL, 85- *Concurrent Pos:* Lectr radiol, McGill Univ, 71-72; lectr radiography, Dawson Col, 71-72; radiol physicist, Hamilton & Dist Hosps, 72-80; teaching master radiation physics, radiobiol & protection, Mohawk Col, Ont, 72-87; dir, regional radiol sci prog, Chedoke-McMaster Hosps, 80-84; vis scholar, dept zool, Univ Cambridge, UK, 83; chmn, Div Med & Biol Physics, Can Asn Physicists, 85-86; vis prof, Flinders Univ SAustralia, 88-89. *Mem:* Fel Can Col Physicists Med; Radiation Res Soc; Am Soc Photobiol; Can Asn Physicists. *Res:* Role of DNA damage and DNA repair in human cancer; radiobiology of viruses; patient exposure and quality assurance measurements in diagnostic radiology. *Mailing Add:* Depts Radiol & Biol McMaster Univ/Life Sci Bldg Rm 434 Hamilton ON L8S 4K1 Can

RAINE, CEDRIC STUART, b Eastbourne, Eng, May 11, 40; m 63; c 1. NEUROPATHOLOGY. *Educ:* Univ Durham, BSc, 62; Univ Newcastle, PhD(med), 67, DSc(med), 75; FRCPath, 88. *Prof Exp:* Sci officer neuropath, Demyelinating Dis Unit, Med Res Coun, Eng, 64-68; from asst prof to assoc prof path, 69-78, PROF PATH, ALBERT EINSTEIN COL MED, 78-, PROF NEUROSCI, 79- *Concurrent Pos:* NIH interdisciplinary fel, Albert Einstein Col Med, 68-69; NIH career develop award, 72-77; Javits Award 85-92; mem Neuro C Study Sect NIH, Nat MS Soc study sect, 87-92. *Honors & Awards:* Weil Award, Am Asn Neuropath, 69 & 75, Moore Award, 76. *Mem:* AAAS; Assoc Am Asn Neuropath; Soc Neurosci; NY Acad Sci; Brit Soc Neuropath; Am Soc Neurochem ARNMD; Soc Neuroimmunol. *Res:* Demyelinating conditions; nervous system development; ultrastructure; viral infections of nervous tissue; multiple sclerosis; in vitro studies of organized nervous tissue; myelin pathology; neuroimmunology. *Mailing Add:* Dept Path Albert Einstein Col Med Bronx NY 10461

RAINER, JOHN DAVID, b Brooklyn, NY, July 13, 21; m 44; c 2. PSYCHIATRY, MEDICAL GENETICS. *Educ:* Columbia Univ, AB, 41, MA, 44, MD, 51. *Hon Degrees:* LittD, Gallaudet Col, 68. *Prof Exp:* Assoc res scientist, 56-65, actg chief psychiat res, 65-68, CHIEF PSYCHIAT RES, NY STATE PSYCHIAT INST, 68- *Concurrent Pos:* From asst to assoc prof, Columbia Univ, 59-72, prof clin psychiat, 72- *Mem:* Asn Res Nerv Ment Dis; Am Soc Human Genetics; fel Am Psychiat Asn; Am Psychoanal Asn; Am Psychopath Asn. *Res:* Application of human genetics to psychiatry on molecular, cellular, chemical, psychological and social levels; psychiatric treatment of the deaf. *Mailing Add:* NY State Psychiat Inst 722 W 168th St New York NY 10032

RAINER, NORMAN BARRY, b New York, NY, May 14, 29. APPLIED CHEMISTRY. *Educ:* Univ Chicago, MS, 50; Univ Del, PhD(phys org chem), 56. *Prof Exp:* Res chemist polymers, Textile Fibers Div, E I du Pont de Nemours & Co, Inc, 56-61; res mgr polymers, Fibers Div, Allied Chem Corp, 61-68; SR SCIENTIST CATALYSIS & NATURAL PRODS, RES & DEVELOP, PHILIP MORRIS CORP, 68- *Mem:* Am Chem Soc. *Res:* Catalysis; fast organic reactions; inorganic chemistry; pyrolysis of cellulose. *Mailing Add:* 2008 Fon-Du-Lac Rd Richmond VA 23229

RAINES, ARTHUR, NEUROPHARMACOLOGY, CARDIOVASCULAR PHARMACOLOGY. *Educ:* Cornell Univ, PhD(pharmacol), 65. *Prof Exp:* PROF PHARMACOL & ACTG DEPT CHMN, SCH MED & DENT, GEORGETOWN UNIV, 69- *Mailing Add:* Pharmacol Dept Georgetown Univ 3900 Reservoir Rd NW Washington DC 20007

RAINES, GARY L, b Pocatello, Idaho, Jan 21, 46. ECONOMIC GEOLOGY. *Educ:* Univ Calif, Los Angeles, BA, 69; Colo Sch Mines, MA & PhD(geol), 71. *Prof Exp:* DEPUTY CHIEF, MINERAL RESOURCES, US GEOL SURV, 83-, GEOLOGIST, 88- *Mem:* Geol Soc Am. *Mailing Add:* US Geol Surv Reno Field Off Mackay Sch Mines Univ Nevada Reno NV 89557

RAINES, JEREMY KEITH, b Washington, DC, Nov 25, 47. ELECTROMAGNETIC ENGINEERING. *Educ:* Mass Inst Technol, BS, 69, PhD(electromagnetics), 74; Harvard Univ, MS, 70. *Prof Exp:* Instr elec eng, Mass Inst Technol, 69-73; CONSULT ANTENNAS, 73- *Concurrent Pos:* Elec engr commun, Naval Electronics Systs Command, 70; elec engr opers res, Naval Ship Res & Develop Ctr, Carderock, 71; elec engr electronics, Naval Electronics Lab Ctr, San Diego, 72; lectr antennas, George Washington Univ, 75- *Mem:* Asn Fed Commun Consult Engrs; Am Phys Soc; Fedn Am Scientists; Inst Elec & Electronics Engrs; Soc Am Mil Engrs. *Res:* Mathematical modeling of antennas and antenna arrays; analysis and design of communication and data transmission networks; radio wave propagation; electromagnetic theory; biological hazards of electromagnetic fields. *Mailing Add:* 13420 Cleveland Dr Rockville MD 20850-3603

RAINES, RONALD T, b Montclair, NJ, Aug 13, 58. PROTEIN DESIGN & ENGINEERING. *Educ:* Mass Inst Technol, BS(chem) & BS(biol), 80; Harvard Univ, MA, 82, PhD(chem), 86. *Prof Exp:* Postdoctoral fel, Univ Calif, San Francisco, 86-89; ASST PROF, DEPT BIOCHEM, UNIV WIS, MADISON, 89- *Concurrent Pos:* Searle scholar, Chicago Community Trust, 90; NSF presidential young investr, 90. *Mem:* Sigma Xi; Am Soc Biochem & Molecular Biol; AAAS; Am Chem Soc. *Res:* Protein design and engineering; protein folding; molecular recognition; heterologous gene expression. *Mailing Add:* Dept Biochem Univ Wis Madison WI 53706-1569

RAINES, THADDEUS JOSEPH, b Jersey City, NJ, Apr 17, 18; m 54; c 2. PHYSICAL CHEMISTRY. *Educ:* St Peter's Col, BS, 39; Columbia Univ, MA, 45, PhD(chem), 49; NY Univ, MBA, 63. *Prof Exp:* Asst prof, Boston Col, 49-52; group leader, Reaction Motors Div, Thiokol Chem Corp, 52-53, head propellant eval & analysis dept, 54-59, supvr mkt anal, 59-61; sr mkt analyst & staff consult res & develop planning, Air Reduction Co, 61-66; mgr prog develop, Am Gas Asn, Inc, 66-70; chmn dept, 70-73, PROF CHEM & BUS ADMIN, JERSEY CITY STATE COL, 70- *Mem:* Am Chem Soc; Sigma Xi; Chem Mkt Res Asn. *Res:* Physical chemistry of high polymer solutions; light scattering; physicochemical research on rocket propellants; research and development planning; commercial and technical intelligence; marketing research; chemicals from coal; business aspects of chemistry. *Mailing Add:* 610 Valley Rd Brielle NJ 08730-1229

RAINEY, DONALD PAUL, b Indianapolis, Ind, June 29, 40; m 79; c 3. AGRICULTURE. *Educ:* Butler Univ, BS, 62; Univ Wis-Madison, MS, 65, PhD(biochem), 67. *Prof Exp:* RES ASSOC, ELI LILLY & CO, 67- *Mem:* Am Chem Soc; Sigma Xi. *Res:* Environmental fate of agricultural chemicals in soil, water, plants and animals; pathways by which agricultural chemicals are degraded in the environment. *Mailing Add:* Rte 2 Box 283 PO Box 708 Greenfield IN 46140

RAINEY, JOHN MARION, JR, b Atlanta, Ga, June 20, 42; m 69; c 2. ANXIETY DISORDERS, FORENSIC PSYCHIATRY. *Educ:* Vanderbilt Univ, BA, 63, MD, 69, PhD(biochem), 72. *Prof Exp:* Intern path, 70-71, from resident to chief resident psychiat, Vanderbilt Univ Hosp, 71-74; asst prof psychiat, & dir res, Lafayette Clin, 74-87, ASSOC PROF PSYCHIAT, WAYNE STATE UNIV, 87-,; DIR MED EDUC & RES, DEPT PSYCHIAT, HARPER HOSP, DETROIT MED CTR, 87- *Concurrent Pos:* Consult, Vet Admin, Sci Adv Comn, 80-, Allen Park Vet Admin Hosp, 84-, NIMH, 84-, Mich Bd Pharm, 85- *Honors & Awards:* William C Menninger Award, Cent Neuropsychiat Asn, 74. *Mem:* Soc Biol Psychiat; Sigma Xi; Am Psychiat Asn; AMA. *Res:* Biological psychiatry and biochemistry; neurophysiology and psychobiology of sudden death in psychiatric disorders; cardiovascular effects of psychotropic drugs; physiology and biochemistry of anxiety disorders. *Mailing Add:* 766 Balfour Rd Grosse Pointe Park MI 48230

RAINEY, MARY LOUISE, b Flagler, Colo, Jan 20, 43; m 67; c 2. ANALYTICAL CHEMISTRY. *Educ:* Knox Col, BA, 64; Univ Md, PhD(analytical chem), 74. *Prof Exp:* Teacher math, Cardozo High Sch, Washington, DC, 67-69; master teacher, Urban Teacher Corps, Washington, DC, 68-69; teacher math, Walt Whitman High Sch, Bethesda, Md, 69-71; res spec, Analysis Labs, Dow Chem, 74-79, group leader, designed latexes & resins res, 79-84, mgr Fed Drug Admin Compliance, 84- 89, MGR HEALTH, ENVIRON & REGULATORY AFFAIRS, DOW CHEM, 89- *Mem:* Sigma Xi; Asn Women Sci; Am Chem Soc. *Res:* High performance liquid chromatography, gas chromatography. *Mailing Add:* Analysis Lab Dow Chem Midland MI 48640

RAINEY, ROBERT HAMRIC, b Charleston, Miss, Mar 23, 18; m 47; c 3. RADIOCHEMISTRY. *Educ:* Memphis State Univ, BS, 42. *Prof Exp:* Chemist, Oak Ridge Gaseous Diffusion Plant, 45-51; group leader process develop, Oak Ridge Nat Lab, 51-78, consult, 85-86; Oak Ridge Assoc Univs, 78-84; RETIRED. *Concurrent Pos:* Consult, through pvt co & Ger res lab. *Mem:* Am Nuclear Soc; Sigma Xi. *Res:* Nuclear reactor fuel recovery process development, separation and isolation of thorium, uranium, plutonium, protactinium, and americium by solvent extraction and ion exchange; environmental impact studies of nuclear reactors and nuclear fuel reprocessing facilities; fuel cycle economics. *Mailing Add:* 3635 Raliluna Ave, No B-6 Knoxville TN 37919

RAINIS, ALBERT EDWARD, b Chicago, Ill, May 15, 41; m 63; c 2. RADIATION PHYSICS. *Educ:* DePaul Univ, BS, 63, MS, 65; Univ Notre Dame, PhD(nuclear physics), 71, Central Mich Univ, MBA, 81. *Prof Exp:* Asst prof physics, Tri-State Col, 70-72 & WVa Univ, 72-75; PHYSICIST, BALLISTICS RES LAB, 75- *Concurrent Pos:* Geothermal consult, WVa Univ, 75-79. *Mem:* Am Phys Soc; Sigma Xi. *Res:* Nuclear radiation transport; geothermal phenomena; shielding calculations. *Mailing Add:* 1210 S Eads St Apt 502 Arlington VA 22202-2837

RAINIS, ANDREW, b Riga, Latvia, June 6, 40; Australian citizen; m 68; c 2. SURFACE CHEMISTRY, COAL PREPARATION. *Educ:* Univ New South Wales, Australia, BSc, 65, PhD(phys chem), 69. *Prof Exp:* Res assoc phys chem, Columbia Univ, 69-70 & Col Environ Sci & Forestry, State Univ NY, Syracuse, 70-75; proj mgr, Otisca Indust Ltd, 75-80; SR RES CHEMIST, CHEVRON RES CO, 80- *Mem:* Am Chem Soc; AAAS. *Res:* Mineral beneficiation; fossil fuel recovery and upgrading. *Mailing Add:* Chevron Res Co PO Box 1627 Richmond CA 94802-1792

RAINS, DONALD W, b Fairfield, Iowa, Dec 16, 37; m 59; c 3. PLANT NUTRITION, SOIL SCIENCE. *Educ:* Univ Calif, Davis, BS, 61, MS, 63, PhD(soil sci), 66. *Prof Exp:* NSF fel, 65-66; asst soil scientist, 66-70, assoc prof agron & range sci, 74-77, AGRONOMIST, UNIV CALIF, DAVIS, 70-, PROF AGRON, 77- *Concurrent Pos:* Consult, 65; lectr soil sci, 68-74; dir, Plant Growth Lab, 79-81; Chmn, Agron & Range Sci, 81-87. *Mem:* AAAS; Am Soc Plant Physiol; Am Soc Agron; Crop Sci Soc Am. *Res:* Ion transport and translocation in plants; plant nutrition and salinity; mineral cycling; heavy metal nutrition in soil-plant ecosystems; plant cell culture. *Mailing Add:* Dept Agron & Range Sci Univ Calif 207 Hunt Hall Davis CA 95616

RAINS, ROGER KERANEN, b Ann Arbor, Mich, May 11, 40; m 67; c 2. CHEMICAL PROCESS RESEARCH, RESEARCH MANAGEMENT. *Educ:* Univ Mich, BSE, 63, MSE, 64, MS, 65, PhD(chem eng), 68. *Prof Exp:* Sr res engr, Technol Dept, Monsanto Indust Chem Co, 68-72, res group leader, 72-78, sr res group leader, 78-82, res sec mgr, 83-85, RES MGR, MONSANTO CHEM CO, 86- *Mem:* Am Chem Soc; Am Inst Chem Engrs. *Res:* Process research and development, primarily separation processes and chemical reaction engineering; process research and development; rubber chemicals. *Mailing Add:* 3453 Timberwood Trail Richfield OH 44286

RAINS, THEODORE CONRAD, b Pleasureville, Ky, Jan 10, 25; m 47; c 3. ANALYTICAL CHEMISTRY. *Educ:* Eastern Ky Univ, BS, 50. *Prof Exp:* Teacher pub sch, Ky, 50-51; chemist, Ky Synthetic Rubber Co, 51-52 & Union Carbide Nuclear Co, 52-65; RES CHEMIST, NAT BUR STANDARDS, 65- *Concurrent Pos:* Vis prof, Univ Md, College Park, 75; column ed, J Appl Spectros, 75- *Honors & Awards:* Cert Recognition, E2 & E3, Am Soc Testing & Mat, 68. *Mem:* Am Chem Soc; Soc Appl Spectros (pres, 82). *Res:* Solvent extraction with applications for analytical chemistry; atomic absorption; emission and fluorescence spectrometry. *Mailing Add:* High Purity Standards PO Box 30188 Charleston SC 29417

RAINVILLE, DAVID PAUL, b Dover, NH, June, 21, 52. ORGANOTELLURIUM CHEMISTRY. *Educ:* Univ NH, BA, 74; Tex A&M Univ, PhD(chem), 79. *Prof Exp:* Res assoc, Tex A&M Univ, 79 & Univ NH, 80; asst prof chem, Austin Col, 81-82; asst prof, 82-86, ASSOC PROF, INORG CHEM, UNIV WIS, RIVER FALLS, 86- *Mem:* Am Chem Soc. *Res:* Synthesis of selenium and tellurium containing heterocycles for use in charge-transfer systems; electrolysis plating of metals via organometallics. *Mailing Add:* Dept Chem Univ Wis River Falls WI 54022

RAINWATER, DAVID LUTHER, b Phoenix, Ariz, Feb 13, 47; m 70; c 2. LIPOPROTEINS. *Educ:* Chapman Col, Orange, Calif, BA, 69; Univ Southern Calif, Los Angeles, PhD(cellular & molecular biol), 79. *Prof Exp:* Postdoctoral res assoc, Inst Biol Chem, Wash State Univ, 79-84; postdoctoral scientist, 84-87, ASST SCIENTIST, SOUTHWEST FOUND BIOMED RES, 87- *Concurrent Pos:* Mem, Coun Arteriosclerosis, Am Heart Asn. *Mem:* Fel Am Heart Asn; Am Soc Biochem & Molecular Biol; AAAS. *Res:* Genetic and dietary effects on primate lipoprotein phenotypes; lipoprotein chemistry and metabolism; animal models of atherosclerosis. *Mailing Add:* Dept Genetics Southwest Found Biomed Res PO Box 28147 San Antonio TX 78228

RAINWATER, JAMES CARLTON, b New York, NY, Jan 9, 46; m 74. STATISTICAL MECHANICS, KINETIC THEORY. *Educ:* Univ Colo, BA, 67, PhD(physics), 74. *Prof Exp:* Lectr physics, Univ Colo, Denver, 74; res fel chem, Univ BC, 75-76; res fel, 76-78, PHYSICIST, NAT INST STANDARDS & TECHNOL, BOULDER, COLO, 78- *Concurrent Pos:* Vis physicist, Nat Bur Standards, Washington, DC, 79; assoc adj prof physics, Univ Colo, 85- *Mem:* Am Asn Physics Teachers. *Res:* Classical and quantum statistical mechanics; kinetic theory; phase transitions and critical phenomena in mixtures; non-Newtonian liquids. *Mailing Add:* Nat Inst Standards & Technol Boulder CO 80303-3328

RAIRDEN, JOHN RUEL, III, b Denver, Colo, Apr 9, 30; m 50; c 2. METALLURGICAL ENGINEERING. *Educ:* Colo Sch Mines, MetE, 51; Rensselaer Polytech Inst, MMetE, 58. *Prof Exp:* Trainee engr, Gen Elec Co, 51-53, specialist, Res Lab, 53-57, metall engr, Res & Develop Ctr, 57-90; RETIRED. *Res:* Anodizing and surface treatment of metals; vacuum deposition and sputtering of thin films; oxidation and corrosion resistant coatings for high temperature alloys and plasma spray processing. *Mailing Add:* 2675 County Rd 1 Montrose CO 81401

RAISBECK, BARBARA, b Arlington, Mass, Feb 7, 28; m 48; c 5. DEVELOPMENTAL BIOLOGY, INSECT PHYSIOLOGY. *Educ:* Boston Univ, BS, 51; Brandeis Univ, PhD(biol), 69. *Prof Exp:* Res assoc biol, Tufts Univ, 69-71; Nat Res Coun vis scientist, Pioneering Res Labs, US Army Natick Labs, 71-73; instr, 70-71, asst prof, 73-74, res assoc biol, Northeastern Univ, 75-76; mem fac, Middlesex Community Col, 78-79; SCIENCE WRITER, 79- *Concurrent Pos:* Consult, Arthur D Little, 74-76. *Res:* Insect development; tissue culture; insect behavior; human anatomy and physiology. *Mailing Add:* 40 Deering St Portland ME 04104-2212

RAISBECK, GORDON, b New York, NY, May 4, 25; m 48; c 5. MANAGEMENT OF TECHNOLOGICAL INNOVATION. *Educ:* Stanford Univ, BA, 44; Mass Inst Technol, PhD(math), 49. *Prof Exp:* Asst, Stanford Univ, 43-44; instr math, Mass Inst Technol, 46-47 & 48-49; mem tech staff, Bell Tel Labs, Inc, 49-54, dir transmission line res, 54-61; sr staff mem systs eng, Arthur D Little, Inc, 61-64, dir systs eng, 65-72, dir phys sensor systs res, 72-75, vpres, 73-86; RETIRED. *Concurrent Pos:* Mem, Inst Defense Anal, 59-60. *Mem:* Fel Inst Elec & Electronics Engrs; Math Asn Am; fel Acoust Soc Am; Opers Res Soc Am. *Res:* Information theory; communication technology and system analysis; transmission lines; underwater acoustics; research and development planning; air traffic control. *Mailing Add:* 40 Deering St Portland ME 04101-2212

RAISEN, ELLIOTT, b New York, NY, Apr 24, 28; m 50; c 3. INORGANIC CHEMISTRY, PHYSICAL CHEMISTRY. *Educ:* City Col New York, BS, 50; Univ Cincinnati, MS, 52, PhD(inorg chem), 60. *Prof Exp:* Asst, Univ Cincinnati, 50-52; res chemist, Bell Aircraft Corp, NY, 54-56; res chemist, Ill Inst Technol Res Inst, 56-62, sr scientist, 62-72, mgr phys chem sect, 72-76; dir chem res div, Toth Aluminum Corp, New Orleans, 76-77; PRES, E&S ENTERPRISES, INC, 77-; PRES, CARDIO-RESPIRATORY HOME CARE INC, 79- *Mem:* Am Chem Soc; Am Inst Aeronaut & Astronaut; Am Ordnance Asn; Sigma Xi; NY Acad Sci. *Res:* Inorganic complexes; phosphate and high temperature chemistry; high temperature reactions; visible and infrared radiation from chemical reactions; water treatment; oxygen production. *Mailing Add:* 4721 Taft Park Metairie LA 70002

RAISZ, LAWRENCE GIDEON, b New York, NY, Nov 13, 25; m 48; c 5. INTERNAL MEDICINE, ENDOCRINOLOGY. *Educ:* Harvard Med Sch, MD, 47. *Prof Exp:* Instr physiol, Col Med, NY Univ-Bellevue Med Ctr, 48-50; resident med, Vet Admin Hosp, Boston, 52-54; instr, Sch Med, Boston Univ, 54-56; asst prof, Col Med, State Univ NY Upstate Med Ctr, 56-61; assoc prof pharmacol & med, Sch Med & Dent, Univ Rochester, 61-66, assoc prof med, 66-68, prof pharmacol & toxicol, 66-74; PROF MED & HEAD DIV ENDOCRINOL & METAB, SCH MED, UNIV CONN HEALTH CTR, FARMINGTON, 74- *Concurrent Pos:* USPHS spec fel, Strangeways Res Lab, Cambridge, Eng, 60-61, Nat Inst Dent Res, NIH, 71-72; asst chief radioisotopes, Vet Admin Hosp, Syracuse Univ, 56-57, clin investr, 57-61; physician, Strong Mem Hosp, 68-74. *Honors & Awards:* E B Astwood Lectr Award; Andre Lichtwitz Prize, 80; William F Neuman Award, 86. *Mem:* AMA; Am Fedn Clin Res; Am Soc Pharmacol & Exp Therapeut; Am Soc Clin Invest; Asn Am Physicians; Am Soc Bone & Mineral Res. *Res:* Parathyroid and calcium metabolism; clinical pharmacology; endocrinology and metabolism; laboratory studies on the hormonal and local regulation of bone formation and resorption and clinical studies on the pathogenesis; prevention and treatment of osteoporosis. *Mailing Add:* Div Endocrinol & Metab Sch Med Univ Conn Health Ctr Farmington CT 06032-9984

RAITT, RALPH JAMES, JR, b Santa Ana, Calif, Feb 9, 29; m 53; c 2. VERTEBRATE ZOOLOGY. *Educ:* Stanford Univ, AB, 50; Univ Calif, Berkeley, PhD(zool), 59. *Prof Exp:* Technician, Mus Vert Zool, Univ Calif, Berkeley, 53-55, teaching asst zool, 56-58; from instr to assoc prof, 58-68, PROF BIOL, NMEX STATE UNIV, 68- *Concurrent Pos:* Guggenheim Mem fel, 67; ed, The Condor, 69-71. *Mem:* Soc Study Evolution; Soc Syst Zool; Animal Behav Soc; Cooper Ornith Soc; Am Ornith Union. *Res:* Ecology, behavior, evolution and systematics of birds, especially those of southwestern United States and Latin America. *Mailing Add:* Dept Biol NMex State Univ Las Cruces NM 88003

RAIZADA, MOHAN K, b Fatehpur, India, Oct 21, 48; m 79. CELLULAR ENDOCRINOLOGY, CELLULAR BIOLOGY. *Educ:* Univ Lucknow, India, BS, 64, MSc, 66; Univ Kanpur, PhD(biol sci), 72. *Prof Exp:* Fel biochem, Med Col Wis, 73-74; assoc cell biol, Lady Davis Inst, Montreal, 74-76; res assoc, 76-78, from asst prof to assoc prof physiol & biochem, 79-86, PROF PHYSIOL, UNIV FLA, 86- *Honors & Awards:* Young Scientist Medal, Institut Nat de Systematique Appliquee Can, 74; Outstanding Scientist Award, Sigma Xi, 88. *Mem:* Endocrine Soc; Am Soc Cell Biol; Am Physiol Soc; AAAS. *Res:* Regulation of insulin receptors in cells cultured from nondiabetic and diabetic animals and humans; role of the central nervous system angiotensin-effector system in the development and maintenance of hypertension; molecular physiology and endocrinology of CNS mediated hypertension. *Mailing Add:* Dept Physiol Box J274 Col Med Univ Fla J Hillis Miller Health Ctr Gainsville FL 32610

RAIZEN, CAROL EILEEN, b Oklahoma City, Okla, May 10, 38; c 1. MICROBIAL GENETICS. *Educ:* Univ Okla, BS, 60, MS, 63; Univ Wis, PhD(bact), 67. *Prof Exp:* Fel microbiol, Sch Med, Univ Colo, 66-67; asst res prof, Univ Pittsburgh, 67-69; from asst prof to assoc prof biol, Duquesne Univ, 72-77; exec secy, Microbial Chem Study Sect, Div Res Grants, NIH, 77-79, Microbiol Physiol Study Sect, 79-80, Microbial Genetics Study Sect, 78-81, exec secy, Spec Study Sections, 81-87, health sci admin, Nat Ctr Nursing Res, 87-89, HEALTH SCI ADMIN, NAT CANCER INST, NIH, 89- *Concurrent Pos:* Asst prof adult nursing educ, Ohio Univ, 70-74; instr, Pa State Univ Continuing Educ, 74-77. *Mem:* AAAS; Am Soc Microbiol; Am Inst Biol Sci. *Res:* Microbial genetics, particularly bacterial pili-structure, function and biosynthesis; pili bacteriophages; cyanophyte genetics and cyanophages. *Mailing Add:* 2233 Chestertown Dr Vienna VA 22182-0537

RAIZEN, SENTA AMON, b Vienna, Austria, Oct 28, 24; US citizen; m 48; c 3. PHYSICAL CHEMISTRY. *Educ:* Guilford Col, BS, 44; Bryn Mawr Col, MA, 45. *Prof Exp:* Res chemist, Sun Oil Co, Pa, 45-48; staff asst chem, Nat Acad Sci-Nat Res Coun, 60-62; prof asst sci educ & admin, NSF, 62-65, asst prog dir, 65-68, assoc prog dir, 68-69, spec tech asst, 69-71; sr prog planner, Nat Inst Educ, 71-72; sr researcher, Domestic Prog Ctr, Rand Corp, 72-74; assoc dir, Nat Inst Educ, 74-78; independent consult, 78-80; study dir, Nat Acad Sci, 80-88; DIR, NAT CTR IMPROVING SCI EDUC, 88- *Concurrent Pos:* Abstractor, Chem Abstr, 46-60; consult, US Off Educ, NSF, pvt res firms. *Mem:* Fel AAAS; Am Chem Soc; Am Educ Res Asn. *Res:* Critical data compilations; strategies for educational and other domestic sector research, dissemination and utilization of research and development; federal education policy and support of research; evaluation of research and development programs; mathematics and science education; educational assessment and testing. *Mailing Add:* Network Inc 1920 L St NW Suite 202 Washington DC 20036

RAIZMAN, PAULA, b Vilnius, Lithuania, Jan 30, 11; US citizen; m 42; c 2. ORGANIC CHEMISTRY. *Educ:* Vilnius State Univ, MS, 35; Univ Paris, PhD(org chem), 54. *Prof Exp:* Asst chemist, Col of France, 36-39 & Med Fac, Paris, 38-42; res assoc org chem, French Nat Sci Res Ctr, 46-54; res chemist, Columbia Univ, 54-57; asst ed, Chem Abstr, 57-60; lit chemist, Cent Res Dept, Monsanto Co, 60-87; RETIRED. *Mem:* AAAS; NY Acad Sci; Am Chem Soc. *Res:* Inorganic analysis of radioactive minerals; hypoglycemic activity of insulin; synthesis of various organic compounds; medicinals, heterocycles. *Mailing Add:* 22 Crabapple Ct St Louis MO 63162

RAJ, BALDEV, b DI Kahn, Pakistan, Jan 8, 35; m 62; c 3. BOTANY. *Educ:* Panjab Univ, BSc, 57, MSc, 59; Univ Delhi, PhD(bot), 65. *Prof Exp:* Asst prof bot, Univ Delhi, 64-68; fel biol, Univ SC, 68-69; from asst prof to assoc prof, 69-76, PROF BIOL, JACKSON STATE UNIV, 76- *Res:* Embryology of vascular plants; isolation of plant protoplasts. *Mailing Add:* Dept Biol Jackson State Univ 1400 Lynch St Jackson MS 39217

RAJ, HARKISAN D, b Sehwan, Pakistan, Jan 1, 26; US citizen; m 56; c 2. MICROBIAL PHYSIOLOGY & SYSTEMATICS. *Educ:* Univ Bombay, BS hon, 47; Univ Puné, India, MS, 52, PhD(biochem & microbiol), 55. *Prof Exp:* Bacteriologist Pub Health Serv, India, 48-56; res fel biochem & nutrit, Tex A&M Univ, 56-57; instr microbiol, Ore State Univ, 57-58; asst prof, Univ Wash, 59-62; PROF MICROBIOL, CALIF STATE UNIV, LONG BEACH, 62- *Concurrent Pos:* Res grant, NIH, NSF & pvt corp, 60-; speciality expert, Food & Agr Orgn, UN, Rome, Italy, 67-70; mem adv bd, Advan Med Sci, Inc, Lawndale, Calif, 72-77; consult, dent serv, Vet Admin Hosp, Long Beach, Calif, 85-88. *Mem:* Electron Micros Soc Am; Am Soc Microbiol; Sigma Xi. *Res:* Author of 50 research and review articles concerning bacterial metabolism, physiology and ultrastructure and systematics; author of new bacteria Ancylobacter aquaticus, Microcyclus flavus (renamed Spirosoma linguale Raj), Cyclobacterium marinus Raj, WH-A and B and Runella slithyformis RPI. *Mailing Add:* Dept Microbiol Calif State Univ Long Beach CA 90840

RAJ, PRADEEP, b Meerut, India, Dec 15, 49; m 80; c 2. COMPUTATIONAL AERODYNAMICS. *Educ:* Indian Inst Sci, BE, 71, ME, 73; Ga Inst Technol, PhD(aerospace eng), 76. *Prof Exp:* Asst prof, Iowa State Univ, 76-78; asst prof mech & aerospace eng, Univ Mo-Rolla, 78-79; RES & DEVELOP ENGR, LOCKHEED AERONAUT SYSTS CO, BURBANK, 79- *Mem:* Am Inst Aeronaut & Astronaut; Am Helicopter Soc. *Res:* Applied computational aerodynamics; simulation of inviscid and viscous flow problems. *Mailing Add:* Dept 75-50 Bldg 80 Lockheed Aeronaut Systs Co Burbank CA 91520-7550

RAJ, RISHI S, b Moga, Punjab, India; US citizen; m 70; c 2. WAKES, TURBULENCE MODELING. *Educ:* Punjab Univ, BS Hons, 64; People's Friendship Univ, MS Hons, 69; Pa State Univ, PhD(aerospace), 74. *Prof Exp:* PROF MECH ENG, CITY UNIV NEW YORK, 75- *Concurrent Pos:* Consult, Curtis-Wright Corp, 76-84; fel, NASA-Langley Res Ctr, 89-90; distinguished fel, US Navy, 91. *Mem:* Am Soc Mech Engrs; Am Inst Aeronaut & Astronaut. *Res:* All aspects of steam and gas turbines; power plants; thermodynamics; deposition; erosion. *Mailing Add:* 86 Wortendyke Ave Emerson NJ 07630

RAJA, RAJENDRAN, b Guruvayur, India, July 14, 48; m 76; c 1. HIGH ENERGY PHYSICS. *Educ:* Univ Cambridge, Eng, BA, 70, PhD, 75. *Prof Exp:* Res assoc, 74-78, staff physicist, 78-83, Scientist I, 83-88, SCIENTIST II, FERMILAB, 88- *Concurrent Pos:* Fel Trinity Col, Cambridge, 73-79. *Mem:* Am Inst Phys Soc. *Res:* Collider physics; UA 1 and D0 experiments. *Mailing Add:* Dept Physics Fermi Lab PO Box 500 Batavia IL 60510

RAJAGOPAL, ATTIPAT KRISHNASWAMY, b Mysore City, India, June 3, 37; m 64; c 2. THEORETICAL PHYSICS. *Educ:* Lingaraj Col, India, BSc, 57; Indian Inst Sci, Bangalore, MSc, 60; Harvard Univ, PhD(appl physics), 65. *Prof Exp:* Res fel, Harvard Univ, 64-65; asst prof physics, Univ Calif, Riverside, 65-68; fel & reader theoret physics, Tata Inst Fundamental Res, India, 68-70; assoc prof, 70-72, prof physics, La State Univ, Baton Rouge, 72-84; RES PHYSICIST, NAVAL RES LAB, 85- *Concurrent Pos:* Prof, Ctr Theoret Studies, Indian Inst Sci, Bangalore, India, 74-75; consult, Naval Res Lab, Washington, DC, 80-81 & Oak Ridge Nat Lab, 81-85. *Honors & Awards:* Distinguished Res Master Award, La State Univ Coun Res, 84. *Mem:* Fel Am Phys Soc. *Res:* Quantum mechanics of two, three and many particle systems; solid state physics; mathematical physics. *Mailing Add:* Code 6860-1 Naval Res Lab Washington DC 20375

RAJAGOPAL, K R, b New Delhi, India, Nov 24, 50; m 74; c 2. MECHANICAL ENGINEERING. *Educ:* Indian Inst Technol, Madras, BTech, 73; Ill Inst Technol, MS, 74; Univ Minn, PhD(mech), 78. *Prof Exp:* Fel lectr mech, Univ Mich, Ann Arbor, 78-80; asst prof mech eng, Cath Univ Am, 80-82; from asst prof to prof mech eng, 82-85, PROF MECH ENG & MATH & STATIST, UNIV PITTSBURGH, 86- *Concurrent Pos:* Vis prof, Indian Inst Technol, Madras, 84; mem Constitutive Equations Comt, Appl Math Div, Am Soc Mech Engrs, 85 & Fluid Mechs Comt, 85; dir, Inst Appl & Computer Mech, Univ Pittsburgh, 85. *Mem:* Am Soc Mech Engr; Soc Natural Philos; Soc Rheology; Am Acad Mech. *Res:* Non-linear mechanics and applied non-linear analysis. *Mailing Add:* Dept Mech Eng Univ Pittsburgh Pittsburgh PA 15261

RAJAGOPAL, P K, b India, June 18, 36; c 1. FISH BIOLOGY. *Educ:* Annamalai Univ, Madras, BS, 57, MS, 58; Utah State Univ, PhD(fish biol), 75. *Prof Exp:* Sr res scholar, Zool Res Lab, Univ Madras, 58-62; res fel zool, Univ Col Rhodesia & Nyasaland, 62-65; sr res officer, Ghana Acad Sci, 66-71, instr gen biol for lab technicians, Inst Aquatic Biol, 66-70; res technologist, US-IBP Desert Biomed Prog, Utah Coop Fishery Unit, Utah State Univ, 71-72, res asst, Utah Coop Fishery Unit, 72-75, instr fish biol, Utah State Univ, 74-75; BIOLOGIST, STATE FISHERY EXP STA, 75- *Concurrent Pos:* Res assoc, Biol Dept, Univ Nev, Las Vegas, 69. *Mem:* Am Fisheries Soc; Am Inst Fishery Res Biologists; Am Inst Biol Sci; Freshwater Biol Asn UK. *Res:* General and aquatic biology; ecology and ecological physiology; fishery science; effects of pollution on aquatic animals; respiratory metabolism. *Mailing Add:* 1540 N 1600 E Logan UT 84321

RAJAGOPALAN, K V, b Mysore, India, Apr 11, 30; m 58; c 3. BIOCHEMISTRY. *Educ:* Presidency Col, Madras, BSc, 51; Univ Madras, MSc, 54, PhD(biochem), 57. *Prof Exp:* Asst res officer, Indian Coun Med Res, Madras, 58-59; fel biochem, Med Sch, 59-66, from asst prof to assoc prof, 67-77, PROF BIOCHEM, MED SCH, DUKE UNIV, 77- *Res:* Enzymology; metalloenzymes. *Mailing Add:* Dept Biochem Box 3711 Med Ctr Duke Univ Durham NC 27710

RAJAGOPALAN, PARTHASARATHI, b Mannargudi, India, Mar 13, 30; m 51; c 2. ORGANIC CHEMISTRY, MEDICINAL CHEMISTRY. *Educ:* Univ Madras, BS, 49; Univ Delhi, MS, 51; NY Univ, PhD(org chem), 60. *Prof Exp:* USPHS fel, Sch Med, NY Univ, 59-60; inst fel, Rockefeller Inst, 60-61; sr res scientist med chem, Ciba Res Ctr, Ciba India Ltd, Bombay, 62-67; sr res scientist, Endo Labs Inc, 67-73, res assoc, 73-80; MEM STAFF, EXP STA, E I DU PONT DE NEMOURS & CO, INC, 80- *Concurrent Pos:* Adj assoc prof, Queens Col, City Univ New York, 72- *Mem:* Am Chem Soc. *Res:* Heterocyclic chemistry; new, 1,3-dipolar cycloaddition reactions. *Mailing Add:* 4655 Norwood Dr Wilmington DE 19803-4811

RAJAGOPALAN, RAJ, TRANSPORT PHENOMENA. *Educ:* Indian Inst Technol, Madras, BS, 69; Syracuse Univ, MS, 71, PhD(chem eng), 75. *Prof Exp:* Asst prof chem eng, Syracuse Univ, NY, 75-76; from asst prof to prof, Rensselaer Polytech Inst, Troy, NY, 76-86; PROF CHEM ENG, UNIV HOUSTON, TEX, 86- *Concurrent Pos:* Prin investr, numerous res progs, 76-; consult, Gen Elec, Conn, 80, NSF, 83-85 & 87, T S Assocs, Md, 85-; adv, Comt Frontiers Chem Eng, Nat Res Coun, 85-86; prog dir, NSF, Washington, DC, 86- *Mem:* Am Chem Soc; Am Inst Chem Engrs; Sigma Xi; Am Phys Soc; Soc Indust & Appl Math; NY Acad Sci. *Res:* Statistical physics of supramolecular fluids; transport phenomena; condensed matter physics of colloids; separation and membrane phenomena. *Mailing Add:* Dept Chem Eng Univ Houston Houston TX 77204-4792

RAJAN, PERIASAMY KARIVARATHA, b Tamil Nadu, India, Sept 20, 42; nat US; m 71; c 1. DIGITAL SIGNAL PROCESSING, CIRCUITS AND SYSTEMS. *Educ:* Univ Madras, India, BS, 66; Indian Inst Technol, Madras, India, MTech, 69, PhD(elec eng), 75. *Prof Exp:* Lectr elec eng, Regional Eng Col, Trichy, India, 69-70; assoc lectr elec eng, Indian Inst Technol, Madras, India, 70-75, postdoctoral fel elec eng, 77-78; asst prof elec eng, State Univ NY Buffalo, 78-80; assoc prof elec eng, NDak State Univ, Fargo, 80-83; assoc prof elec eng, 83-85, PROF ELEC ENG, TENN TECHNOL UNIV, COOKEVILLE, 85- *Concurrent Pos:* Part time fac, Concordia Univ Montreal, 77-78, vis res asst prof, 79; grad prog dir, Dept Elec Eng, Tenn Technol Univ, 87- *Mem:* Int Elec & Electronic Engrs; Am Soc Eng Educ; Int Assoc Sci & Technol Develop; Sigma Xi. *Res:* Applications of symmetry for the design and application of two and three dimensional digital filters; spectral analysis and multidimensional fast fourier transform algorithm development; general digital signal processing algorithms and applications. *Mailing Add:* Dept Elec Eng Tenn Technol Univ Cookeville TN 38505

RAJAN, THIRUCHANDURAI VISWANATHAN, b Tanjore, India, Oct 1, 45; m 77; c 3. SOMATIC CELL GENETICS, IMMUNOGENETICS. *Educ:* All India Inst Med Sci, MB & BS, 69; Albert Einstein Col Med, PhD(cell biol), 74. *Prof Exp:* Asst prof path, 75-80, asst prof genetics, 78-80, ASSOC PROF PATH & GENETICS, ALBERT EINSTEIN COL MED, 80- *Concurrent Pos:* Attending pathologist, Bronx Municipal Hosp Ctr, 75- *Res:* Expression and function of transplantation antigens on surface of mouse leukemic cells by isolation mutants in their expression and evaluating the physiological consequences of such mutations. *Mailing Add:* Dept Path Albert Einstein Col Med 3301 Bainbridge Ave Bronx NY 10467

RAJANNA, BETTAIYA, b Bangalore, India; m 67; c 2. TOXICOLOGY, ENVIRONMENT POLLUTION. *Educ:* Mysore Univ, India, BS, 59 & 63; Miss State Univ, MS, 70, PhD(physiol), 72. *Hon Degrees:* DLett, Selma Univ, 89. *Prof Exp:* Res asst hort, Hort Dept, Mysore State, India, 63-64, asst dir, 64-67; res asst plant sci, Agron Dept, Miss State Univ, 67-72, fel ecol, Dept Zool, 73-75; assoc prof biol, 75-77, prof & chmn, Div Natural & Appl Sci, 77-87, ACAD DEAN & PROF, SELMA UNIV, 87- *Concurrent Pos:* Panelist, Sci Educ, NSF, 77-81 & NIH, 86-90; prin investr, MBS, Res Prog, NIH, 79-, consult, 80; consult, NIH, 80- *Honors & Awards:* White House Award Excellence in Sci & Technol. *Mem:* Am Soc Agron; Am Soc Crop Sci; Am Soc Plant Physiologists; Indian Soc Seed Technologists; Sigma Xi. *Res:* Interaction of heavy metals, cadmium, lead and mercury, with catecholamines uptake by rat brain and heart; biochemical changes due to aging in plant seeds; pollution ecology: effects of air pollutants on plants; seed physiology; 46 published papers and 81 presentations. *Mailing Add:* Acad Dean Selma Univ 1501 Lapsley St Selma AL 36701

RAJARAMAN, SRINIVASAN, b India, July 10, 43; US citizen; m 72; c 2. IMMUNOPATHOLOGY, NEPHROPATHOLOGY. *Educ:* Univ Madras, India, MB & BS, 65. *Prof Exp:* Med officer, captain, Indian Armed Forces, 65-69; res fel gen surg, Stanley Med Col, Madras, India, 69-71, asst surgeon, Govt Stanley Hosp, 71-74; resident gen surg, NJ Sch Med, Newark, 74-75 & resident path, Sch Med, WVa Univ, Morgantown, 75-79; spec fel immunopath, Cleveland Clin, Ohio, 79-80; asst prof, 80-86, ASSOC PROF PATH, UNIV TEX MED BR, GALVESTON, 86- *Mem:* Am Asn Pathologists; Int Acad Path; Am Soc Clin Pathologists; Am Soc Nephrology; AAAS; NY Acad Sci. *Res:* Molecular pathobiology of cell injury and regeneration; tumor biology; interrelationship between immune complexes and complement. *Mailing Add:* Renal Immunopath Keiller 201 Univ Tex Med Br Galveston TX 77550

RAJARATNAM, N(ALLAMUTHU), b Mukuperi, India, Dec 18, 34; m 61; c 2. HYDRAULIC ENGINEERING, FLUID MECHANICS. *Educ:* Univ Madras, BE, 57, MSc, 58; Indian Inst Sci, Bangalore, PhD(hydraul), 61. *Prof Exp:* Jr res engr, Irrig Res Sta, Madras, India, 58-59; sr sci officer hydraul, Indian Inst Sci, 60-63; Nat Res Coun Can fel, 63-65, session lectr, 65-66, from asst prof to assoc prof, 66-71, PROF HYDRAUL, UNIV ALTA, 71- *Concurrent Pos:* Nat Res Coun Can res grants, 65- *Mem:* Am Soc Civil Engrs; Eng Inst Can; Int Asn Hydraul Res. *Res:* Open channel flow; hydraulics of energy dissipations; turbulent boundary layers, jets and wakes; non-Newtonian flow; rivers; thermal and oil pollution problems; fishways. *Mailing Add:* Dept Civil Eng Univ Alta Fac Eng Edmonton AB T6G 2M7 Can

RAJCHMAN, JAN A(LEKSANDER), electronics; deceased, see previous edition for last biography

RAJENDRAN, VAZHAIKKURICHI M, b Vazhaikkurichi, India, Dec 2, 52; m 84; c 2. HORMONE REGULATION OF ION TRANSPORT, MOLECULAR IDENTIFICATION OF TRANSPORTER. *Educ:* Univ Madras, BS, 74, MS, 77, MPhil, 88, PhD(biochem), 81. *Prof Exp:* Postdoctoral fel, Med Col Wis, 83-85; assoc res scientist, 85-90, RES SCIENTIST, YALE UNIV SCH MED, 90- *Mem:* Am Physiol Soc. *Res:* Novel transport system localized in baselateral membranes; transport physiology; enzyme kinetics. *Mailing Add:* Dept Internal Med 89 LMP Yale Univ 333 Cedar St New Haven CT 06510

RAJESHWAR, KRISHNAN, b Trivandrum, India, Apr 15, 49; m 77; c 2. ELECTROCHEMISTRY. *Educ:* Univ Col, India, BSc, 69; Indian Inst Technol, Sc, 71, PhD(chem), 74. *Prof Exp:* Res chemist, Prods Formulation Group Foseco Int, India, 74-75; res fel chem, St Francis, Xavier Univ, Can, 75-76; res fel, Colo State Univ, 76-78, vis asst prof, 78-79 sr res assoc, 79-83; from asst prof to assoc prof, 83-89, PROF CHEM, TEX, ARLINGTON, 89- *Concurrent Pos:* Consult, Univ Wyo Res Corp & Forensic Labs, Tex. *Mem:* Am Chem Soc; Electrochem Soc. *Res:* Charge storage and transport mechanisms in a variety of materials; Semiconductors and polymers. *Mailing Add:* Dept Chem Univ Tex Arlington TX 76019-0065

RAJHATHY, TIBOR, b Pozsony, Hungary, Mar 27, 20; m 58; c 2. PLANT GENETICS & CYTOLOGY. *Educ:* Univ Tech & Econ Sci, Hungary, DSc(agr), 47. *Hon Degrees:* Dhc, Univ Hort & Food Sci, Hungary, 90. *Prof Exp:* Res assoc, Genetics Inst, Hungary, 39-43; asst prof, Univ Agr Sci Hungary, 47-49; head genetics dept, Hungarian Acad Sci, 49-56; cytogeneticist, Genetics & Plant Breeding Res Inst, 56-67, chief cytogenetics sect, Res Br, 67-76, dir, Ottawa Res Sta & Cent Exp Farm, Can Dept Agr, 76-84; RETIRED. *Concurrent Pos:* Consult, Minister Agr, 52-56. *Mem:* Genetics Soc Can; Hungarian Soc Agr Sci; fel Royal Soc Can. *Res:* Cytogenetics and radiation genetics of grasses and cereals; autopolyploids and allopolyploids; evolution of species. *Mailing Add:* 42 Farlane Blvd Nepean ON K2E 5H5 Can

RAJLICH, VACLAV THOMAS, b Praha, Czech, May 3, 39; m 68; c 4. SOFTWARE DEVELOPMENT METHODOLOGIES, TOOLS FOR SOFTWARE MAINTENANCE. *Educ:* Czech Tech Univ, MS, 62; Case Western Reserve Univ, PhD(math), 71. *Prof Exp:* Grad fel cybernet, Res Inst Math Mach, Prague, 66-68, res scientist software eng, 71-74, mgr, Dept Algorithms, 74-79; teaching asst math, Case Western Reserve Univ, 69-71; vis assoc prof computer sci, Calif State Univ, Fullerton, 80-81; assoc prof computer sci & eng, Univ Mich, 82-85; chair, 85-90, PROF COMPUTER SCI, WAYNE STATE UNIV, 85- *Concurrent Pos:* Mem fac, Dept Cybernet, Charles Univ, Prague, 74-79; prin investr grant, Int Bus Mach, 83-85, Chrysler Challenge Fund, 86-90; consult, Software Eng Develop Corp, 84; grantee, Inst Mfg Res, Wayne State Univ, 85-; vis scientist, Carnegie Mellon Univ, 87, Harvard Univ, 88; prog comt chair, Inst Elec & Electronics Engrs Conf on Software Maintenance, 91. *Mem:* Inst Elec & Electronics Engrs Computer Soc; Asn Comput Mach. *Res:* Software development methodologies and tools, particularly for object oriented systems; tools for maintenance of large software systems; parallel and graph grammars. *Mailing Add:* Dept Computer Sci Wayne State Univ Detroit MI 48202

RAJNAK, KATHERYN EDMONDS, b Kalamazoo, Mich, Apr 30, 37; m 61. ATOMIC PHYSICS. *Educ:* Kalamazoo Col, BA, 59; Univ Calif, Berkeley, PhD(chem), 63. *Prof Exp:* Fel chem, Lawrence Radiation Lab, 62-65; asst prof physics, Kalamazoo Col, 67-70; physicist, 74-75, CONSULT, LAWRENCE LIVERMORE LAB, 75- *Concurrent Pos:* Consult, Argonne Nat Lab, 66-; adj lectr physics, Kalamazoo Col, 76-85, adj assoc prof phys, 85-; vis prof, Univ Paris, IV, 79 & 80 & Univ Paris, Orsay, 79 & 81. *Mem:* Am Phys Soc; Am Asn Physics Teachers. *Res:* Theory and analysis of lanthanide and actinide spectra. *Mailing Add:* Dept Physics Kalamazoo Col Kalamazoo MI 49007

RAJNAK, STANLEY L, b Richmond, Calif, Apr 23, 36; m 61. MATHEMATICS. *Educ:* Univ Calif, Berkeley, AB, 60, PhD(math), 66. *Prof Exp:* From asst prof to assoc prof, 65-77, PROF MATH, KALAMAZOO COL, 77- *Mem:* Am Math Soc; Math Asn Am. *Res:* Analysis; linear topological spaces; distribution theory. *Mailing Add:* Dept Math Kalamazoo Col 1200 Acad St Kalamazoo MI 49007

RAJSUMAN, ROCHIT, b Meerut City, India, Mar 13, 64. COMPUTER ENGINEERING. *Educ:* KN Inst Technol, India, BTech, 84; Univ Okla, MS, 85; Colo State Univ, PhD(elec eng), 88. *Prof Exp:* Grad teaching asst elec eng, Colo State Univ, 86-88; ASST PROF COMPUTER ENG & SCI, CASE WESTERN RESERVE UNIV, 88-, ASST PROF ELEC ENG, 91- *Mem:* Inst Elec & Electronic Engrs Computer Soc Circuits & Syst Soc; Asn Computer Mach. *Res:* Digital hardware testing; design for testability; vlsi design; design automation; fault tolerant design; fault modeling; computer architecture; computer networks. *Mailing Add:* 521 Crawford Hall Case Western Reserve Univ Cleveland OH 44106

RAJU, IVATURY SANYASI, b Kakinada,India, Aug 9, 44; m 71; c 1. ENGINEERING, AERONAUTICS. *Educ:* Andhra Univ, India, BE, 65; Indian Inst Sci, ME, 67, PhD(aeronaut eng), 73; MEA, George Wash Univ, 82. *Prof Exp:* Sr eng, Vikram Sarabhai Space Ctr, Trivandrum, 71-75; res assoc, Nat Res Coun, 75-77, res assoc, 77-79, asst res prof, 79-82, assoc res prof, Joint Inst Flight Sciences, George Wash Univ, Langley Res Ctr, NASA, 82-83; sr scientist, Vigyan Res Assoc, Inc, 83-84; SR SCIENTIST & VPRES, ANAL SERVS & MATS, INC, 84- *Concurrent Pos:* Leverhulme overseas fel, Univ Liverpool, Eng, 73-74. *Mem:* Am Inst Aeronaut & Astronaut; Am Soc Testing & Mat. *Res:* Static; dynamic and stability analysis of aerospace structures; fracture mechanics; finite element methods; laminated composite structures; boundary element methods. *Mailing Add:* 107 Research Dr Hampton VA 23666

RAJU, MUDUNDI RAMAKRISHNA, b Bhimavaram, India, July 15, 31. RADIATION BIOPHYSICS, RADIOTHERAPY. *Educ:* Univ Madras, BSc, 52, MA, 54; Andhra Univ, India, MSc, 55,. *Hon Degrees:* DSC, Andhra Univ, 60. *Prof Exp:* Lectr nuclear instruments, Andhra Univ, India, 57-61; fel biophys, Mass Gen Hosp, Mass Inst Technol & Harvard Univ, 61-63; Donner fel, Donner Lab, Univ Calif, Berkeley, 63-64, biophysicist, 64-65; from asst prof to assoc prof biophys, Univ Tex, Dallas, 65-71; staff mem biophys, 71-79,

FEL, LOS ALAMOS NAT LAB, 80- *Concurrent Pos:* Vis scientist, Hammersmith Hosp, London, 67-68; guest scientist, Lawrence Radiation Lab, Univ Calif, 65- *Mem:* Radiation Res Soc. *Res:* Physics and radiobiology of new radiations, pi mesons, heavy charged particles and neutrons, and their potential applications in radiation therapy. *Mailing Add:* 1479 Big Rock Loop Los Alamos NM 87544

RAJU, NAMBOORI BHASKARA, b Pothumarru, India, Jan 1, 43; m 64; c 3. FUNGAL GENETICS, FUNGAL CYTOLOGY. *Educ:* Banaras Hindu Univ, BSc, 65, MSc, 67; Univ Guelph, PhD(genetics), 72. *Prof Exp:* Scientist, Coun Sci & Indust Res, India, 73-74; res assoc, 74-83, SR RES ASSOC, CYTOGENETICS, STANFORD UNIV, 84- *Mem:* Mycological Soc Am. *Res:* Cytology of fungi, especially Neurospora: meiotic and mitotic processes, including ascus development, behavior of chromosomes, nucleolus and spindle pole bodies; cytogentic behavior of Neurospora mutants that affect ascus and ascospore differentiation and chromosome rearrangements that involve the nucleolus organizer region. *Mailing Add:* Dept Biol Sci Stanford Univ Stanford CA 94305-5020

RAJU, PALANICHAMY PILLAI, b Theni, India, June 15, 37; US citizen; m 62; c 3. ENGINEERING MECHANICS, STRUCTURAL ENGINEERING. *Educ:* Madras Univ, BE, 60, MSc, 61; Univ Del, PhD(eng mech), 68. *Prof Exp:* Design engr, Larson & Toubro, 61-64; res fel aerospace, Univ Del, 64-68; lead engr, Westinghouse Nuclear Energy Systs, 68-76; consult engr, Teledyne Eng Serv, Teledyne Inc, 76-84; sr consult, Cygna Energy Servs, 84-87; RELIABILITY MGR, DIGITAL EQUIP CORP, 87- *Concurrent Pos:* Mem pressure vessel res comts & nuclear code comts, Am Soc Mech Engrs. *Mem:* Am Soc Mech Engrs; Am Soc Civil Engrs; Sigma Xi; Inst Elec & Electronic Engrs. *Res:* Shell theory of composite materials known for their anisotropic properties, both mechanical and thermal; safety and reliability of nuclear power plant components and piping systems; new analysis techniques; vibration and 3D analysis; fracture mechanics evaluation; reliability of computer hardware; life extension studies. *Mailing Add:* 26 Harvard Dr Sudbury MA 01776

RAJU, SATYANARAYANA G V, b Undi, India, Jan 8, 34; m 55; c 2. ENGINEERING, ELECTRICAL ENGINEERING. *Educ:* Andhra Univ, BS, 55; Banaras Univ, MS, 57; Indian Inst Technol, MTech, 59; Polytech Inst Brooklyn, PhD(elec eng), 65. *Prof Exp:* Res asst electronics, Phys Res Labs, Polytech Inst Brooklyn, 59-61, fel, 61-65; asst prof elec eng, Clarkson Col Technol, 65-67; from asst prof to prof elec eng, Ohio Univ, 72-91, chmn dept, 73-91. *Mem:* Sr mem Inst Elec & Electronics Engrs; Sigma Xi. *Res:* Control systems; design of control systems; adaptive control; system identification; pattern recognition; stability theory; air traffic control. *Mailing Add:* Div Eng Univ Tex San Antonio TX 78285-0665

RAKA, EUGENE CD, b Detroit, Mich, Aug 24, 24; m 66. PHYSICS. *Educ:* Univ Mich, BS, 49, PhD(physics), 53. *Prof Exp:* Res asst, Eng Res Inst, Univ Mich, 50-53; assoc physicist, 53-63, PHYSICIST, BROOKHAVEN NAT LAB, 63- *Mem:* Am Phys Soc. *Res:* High energy accelerator design, instrumentation and operation. *Mailing Add:* Brookhaven Nat Lab Bodg 911C Upton NY 11973

RAKE, ADRIAN VAUGHAN, b New York, NY, Mar 27, 34; m 61; c 3. GENETICS, MOLECULAR BIOLOGY. *Educ:* Swarthmore Col, BA, 56; Univ Pa, PhD(microbiol), 64; Pa State Univ, BS, 75. *Prof Exp:* NIH res fel biochem, Univ BC, 64-66; fel, Carnegie Inst, 66-68; asst prof biophys, Pa State Univ, 68-75; ASSOC PROF BIOL, WRIGHT STATE UNIV, 76- *Concurrent Pos:* Red Cross health vol, nat & int disaster. *Mem:* AAAS; NY Acad Sci; Genetics Soc Am; Sigma Xi; Biophys Soc. *Res:* Genome size in trisomy; altered nuclear organization. *Mailing Add:* 3824 Fowler Rd Springfield OH 45502

RAKER, CHARLES W, b Daylesford, Pa, July 19, 20; m 55; c 2. VETERINARY SURGERY. *Educ:* Univ Pa, DVM, 42. *Prof Exp:* Gen pract, 45-50; asst prof vet med, 50-53, asst prof surg, 54-55, assoc prof surg & chmn dept vet surg, 56-57, prof surg, 57-67, chief sect, 62-76, LAWRENCE BAKER SHEPPARD PROF SURG, SCH VET MED, UNIV PA, 67-, PROF COMP SURG, DIV GRAD MED, 58- *Concurrent Pos:* Dir clins Bolton Farm, Sch Vet Med, Univ Pa, 50-52 & New Bolton Ctr, 52-53, in charge large animal clin, 54-63, assoc chief staff & head surg serv, 64-, chmn dept surg, Sch Vet Med, S7-58, interim chief sect, 58-62. *Mem:* Am Asn Equine Practrs; Am Vet Med Asn; Am Asn Vet Clinicians; fel Am Col Vet Surg (pres, 75-76). *Res:* Orthopedic surgery; histology and histopathology in normal and diseased flexor tendons; surgical diseases of the equine upper respiratory tract. *Mailing Add:* Dept Surg Univ Pa Sch Vet Med Philadelphia PA 19104

RAKES, ALLEN HUFF, b Floyd, Va, Aug 19, 33; m 58; c 2. DAIRY NUTRITION. *Educ:* Va Polytech Inst, BS, 56, MS, 57; Cornell Univ, PhD(animal nutrit), 60. *Prof Exp:* Asst prof dairy sci, WVa Univ, 60-63; from asst prof to assoc prof, 63-73, PROF ANIMAL SCI, NC STATE UNIV, 73- *Mem:* Am Dairy Sci Asn; Am Soc Animal Sci. *Res:* Energy utilization of ruminants; voluntary feed intake control mechanisms. *Mailing Add:* Dept Animal Sci NC State Univ PO Box 7621 Raleigh NC 27695-7621

RAKES, JERRY MAX, b Bentonville, Ark, Dec 7, 32; m 49; c 3. ANIMAL SCIENCE. *Educ:* Univ Ark, BS, 54, MS, 55; Iowa State Univ, PhD(physiol), 58. *Prof Exp:* Asst prof dairy physiol, Iowa State Univ, 56-58; assoc prof, 58-63, PROF DAIRY GENETICS, UNIV ARK, FAYETTEVILLE, 63 - *Mem:* Am Soc Animal Sci; Am Dairy Sci Asn. *Res:* Dairy physiology and biochemistry; dairy cattle genetics. *Mailing Add:* Dept Animal Sci Univ Ark Fayetteville AR 72701

RAKESTRAW, JAMES WILLIAM, b Reidsville, NC, July 20, 36; m 61; c 2. ELECTRONIC SYSTEMS. *Educ:* Va Polytech Inst, BS, 59; Johns Hopkins Univ, PhD(physics), 64. *Prof Exp:* Engr, Philco Corp, 55-58; mem tech staff, Bell Tel Labs, 59-60; res asst physics, Johns Hopkins Univ, 61-64; mem staff, Westinghouse Defense & Space Ctr, 64-65, group leader systs anal, 65-66; mem tech staff systs anal group, Lincoln Lab, Mass Inst Technol, 68-72; mem tech staff, Electronics Systs Div, MRI, 72-77; mgr aerospace syst group & systs anal group, Comput Sci Corp, 77-78, sr prin engr, 77-80; CONSULTING SYSTEMS SCIENTIST, PLANNING RES CORP, 80- *Mem:* Inst Elec & Electronics Engrs; Commun Soc; Am Phys Soc; Sigma Xi. *Res:* Systems analysis; communications systems; digital computer applications and simulation; space systems; radar systems. *Mailing Add:* 2666 Reign St Fox Mill Estates Reston VA 22071

RAKESTRAW, ROY MARTIN, b Redding, Calif, Jan 31, 42; m 62; c 2. MATHEMATICS, COMPUTER SCIENCES. *Educ:* Okla State Univ, BS, 65, MS, 66, PhD(math), 69. *Prof Exp:* asst prof math, Univ Mo-Rolla, 69-76; assoc prof math, Wheaton Col, 76-; AT DEPT MATH, PHILLIPS UNIV. *Mem:* Math Asn Am; Am Math Soc; Am Sci Affil. *Res:* Convex sets; functional analysis. *Mailing Add:* Oral Roberts Univ 7777 S Lewis St Tulsa OK 74171

RAKHIT, GOPA, b India; US citizen; m 77. BIOCHEMISTRY, PHARMACOLOGY. *Educ:* Univ Calcutta, BSc, 65, MSc, 67; Univ Utah, PhD(chem physics), 76. *Prof Exp:* Vis fel, Nat Heart, Lung & Blood Inst, 76-78; STAFF FEL, BUR DRUGS, FOOD & DRUG ADMIN, 78- *Mem:* Biophys Soc; Am Chem Soc; AAAS. *Res:* Spectroscopic studies of enzyme-substrate, protein-ligand, and drug-biomolecular interaction; structure-function relation in proteins, nucleic acids and membranes; effects of radiation and the role of free radicals in drug-toxicty. *Mailing Add:* 5333 Westbard Ave NIH Rm 218B Bethesda MD 20892

RAKHIT, SUMANAS, b Banaras, India, Oct 17, 30; m 59; c 1. ORGANIC CHEMISTRY. *Educ:* Banaras Hindu Univ, MPharm, 53, PhD(pharmaceut), 57. *Prof Exp:* Coun Sci & Indust Res India sr res fel org chem, Cent Drug Res Inst, Univ Lucknow, 57-59; res assoc, Laval Univ, 59-60, Nat Res Coun Can fel, 60-62; staff scientist, Worcester Found Exp Biol, 62-65; sr res chemist, Ayerst Res Labs, 65-83; assoc prof, Univ Louis Pasteur, Strasbourg, France, 83-84; DIR CHEM, BIOMEGA, INC, 84- *Concurrent Pos:* Mem grants selection comt chem sect, Nat Sci & Eng Res Coun, Can, 81- *Mem:* Am Chem Soc; Chem Inst Can. *Res:* Structural determination; synthesis of steroids and other natural products; phospholipids; biosynthesis of steroids; synthesis of pharmacologically active compounds; antibiotics; peptide chemistry. *Mailing Add:* 2100 Cunard Chomedey PQ H7S 2G5 Can

RAKIC, PASKO, b Ruma, Yugoslavia, May 15, 33; m 69. NEUROSCIENCES, DEVELOPMENTAL BIOLOGY. *Educ:* Univ Belgrade, MD, 59, ScD(neuroembryol), 69. *Hon Degrees:* MS, Yale Univ, 78. *Prof Exp:* Intern, Univ Hosp, Belgrade Univ Med Sch, 59-60, resident neurosurg, 61-62, asst prof path physiol, Belgrade Univ, 60-61; asst prof, Inst Biol Res, Belgrade, 67-69; res fel neurosurg, Harvard Med Sch, 62-66, from asst prof to assoc prof neuropath, 69-77; chmn, Sect Neuroanat, 77-90, PROF NEUROSCI, SCH MED, YALE UNIV, 77-, DORYS MCCONNELL DUBERG CHAIR, 78-, CHMN, SECT NEUROBIOL, 90- *Concurrent Pos:* Prin investr grants, Nat Inst Neurol & Commun Dis & Stroke, NIH, 70-, Nat Eye Inst, 72-; mem study sects, NIH, 72-; consult, NSF, Atomic Energy Control Bd, Can, Med Res Coun, Can, March of Dimes, J S Guggenheim Mem Found, Huntington Chorea Found; assoc, Neurosci Res Prog, 80-89; Jacob K Javits Neurosci Investr Award, 84; dir, Sen Jacob Javits Ctr Excellence Neurosci, Yale Univ, 85-90; numerous invited lectrs. *Honors & Awards:* Grass Found Award & Lectr, 85; Karl Spencer Lashley Award, Am Philos Soc, 86; Pattison Award in Neurosci, 86. *Mem:* Nat Acad Sci; AAAS; Am Asn Anatomists; Am Asn Neuropathologists; Int Brain Res Orgn; Int Soc Develop Neurosci; NY Acad Sci; Sigma Xi; Soc Neurosci. *Res:* Developmental neurobiology; cellular and molecular mechanisms of neuronal migration, axonal navigation and synaptogenesis; genetic and epigenetic regulation of neuronal interactions during development. *Mailing Add:* Dept Neuroanat Sch Med Yale Univ PO Box 3333 New Haven CT 06510-8001

RAKITA, LOUIS, b Montreal, Que, US citizen; m 45; c 1. CARDIOVASCULAR DISEASES, INTERNAL MEDICINE. *Educ:* Sir George Williams Univ, BA, 42; McGill Univ, MD, CM, 49; Am Bd Internal Med, dipl, 56; Royal Col Physicians & Surgeons, Can, cert, 56; FACPS(C), 72. *Prof Exp:* Intern, Montreal Gen Hosp, 49-50; resident med, Jewish Gen Hosp, Montreal, 50-51; fel, Alton Ochsner Med Found, 51-52; chief resident, Cleveland City Hosp, 52-53; dir cardiol, Cleveland Metrop Gen Hosp, 66-87; from instr to assoc prof, 54-71, PROF MED, CASE WESTERN RESERVE UNIV, 71- *Concurrent Pos:* Am Heart Asn fel, Inst Med Res, Cedars Lebanon Hosp, Los Angeles, 53-54; Am Heart Asn Fel, Cleveland City Hosp, 54-55, advan fel, 59-61; USPHS sr res fel, Cleveland City Hosp, 61-62; USPHS res career develop award, Cleveland Metrop Gen Hosp, 62-69; asst vis physician, Cleveland City Hosp, 54-57, vis physician, 57-; vis cardiologist, Sunny Acres Hosp, 73- *Honors & Awards:* Gold Heart Award, Am Heart Asn, Cert of Merit. *Mem:* AAAS; Am Fedn Clin Res; Am Heart Asn; fel Am Col Physicians; fel Am Col Cardiol. *Res:* Electrophysiology and biochemistry of cardiac hypertrophy. *Mailing Add:* Cleveland Metrop Gen Hosp Cleveland OH 44100

RAKITA, PHILIP ERWIN, b Cleveland, Ohio, Sept 4, 44; m; c 2. ORGANOMETALLIC CHEMISTRY. *Educ:* Case Inst, BSc, 66; Mass Inst Technol, PhD(chem), 70. *Prof Exp:* Asst prof inorg chem, Univ NC, Chapel Hill, 70-75; grants officer, Indust Environ Res Lab, US Environ Protection Agency, 75-76; sr Fulbright prof, Moscow State Univ, USSR, 76; prof inorg chem, Univ Minn, 76-77; sr res chemist, Ferro Corp, 77-78, int prod mgr, 79-80; tech mgr, 81-83, mkt mgr, M&T Chem Inc, 84-89; DEPT HEAD, INDUST CHEM, ATOCHEM, PARIS, 89- *Concurrent Pos:* NSF & NIH fels. *Mem:* Am Chem Soc; Soc Plastics Engrs. *Res:* Organometallic chemistry; polymer additives. *Mailing Add:* 14 Rue Angelique Verien Neuilly-Sur-Seine 9220 France

RAKOFF, HENRY, b Brooklyn, NY, Nov 13, 24; m 50, 83; c 3. ORGANIC CHEMISTRY. *Educ:* City Col New York, BS, 44; Purdue Univ, MS, 48, PhD(chem), 50. *Prof Exp:* Asst chem, Purdue Univ, 46-48; petrol chemist, Natural Resources Res Inst, Univ Wyo, 50-52; sr res chemist, Velsicol Corp, Ill, 52-53; from asst prof to assoc prof chem, Tex A&M Univ, 53-66; prof, Parsons Col, 66-73, chmn dept, 67-73; res assoc, Univ Mo, Columbia, 73-74; RES CHEMIST, NAT CTR AGR UTILIZATION RES, AGR RES SERV, USDA, 74- *Mem:* AAAS; Am Chem Soc; Am Oil Chemists Soc. *Res:* Synthesis of pharmacologically active compounds; chemistry of fatty acids and glycerides. *Mailing Add:* Nat Ctr Agr Utilization Res USDA 1815 N Univ St Peoria IL 61604

RAKOFF, VIVIAN MORRIS, b Capetown, SAfrica, Apr 28, 28; Can citizen; m 59; c 3. PSYCHIATRY. *Educ:* Univ Capetown, BA, 47, MA, 49; Univ London, MB, BS, 57; McGill Univ, DPsych, 63; FRCP(C), 64. *Prof Exp:* Psychologist, Tavistock Clin, 50-51; house officer surg, St Charles Hosp, 57; house officer med, Victoria Hosp, 58; registr, Groote Schuur Hosp, 58-61; resident psychiat, McGill Univ, 61-63; assoc dir res, Jewish Gen Hosp, 63-67, asst prof & dir res, 67-68; from assoc prof to prof psychiat, 68-74, dir postgrad educ, 68-71, PROF PSYCHIAT EDUC, UNIV TORONTO, 74- *Concurrent Pos:* Prof & chmn dept psychiat, Univ Toronto, 80; dir & psychiatrist in chief, Clarke Inst Psychiat. *Honors & Awards:* Schonfeld Award, Am Soc Adolescent Psychiat, 88. *Mem:* Am Psychiat Asn; Can Psychiat Asn; Am Col Psychiatrists. *Res:* Patterns of mutual perception within the family; neurophysiological substrates of addictive behavior; adolescence and the family. *Mailing Add:* 250 College St Toronto ON M5T 1R8 Can

RAKOSKY, JOSEPH, JR, b Harrisburg, Pa, Apr 17, 21; m 44; c 4. FOOD SCIENCE, SANITATION. *Educ:* Univ Md, BS, 49, MS, 50; Pa State Univ, PhD(bact), 53. *Prof Exp:* Dairy technician, Univ Md, 49-50; asst bact, Pa State Univ, 50-51; sr microbiologist & group leader, Baxter Labs, Inc, 53-55; res microbiologist & group leader, Glidden Co, 55-58; res microbiologist, Cent Soya Co Inc, Chicago, leader, 58-64, asst div prod mgr, 64-65, tech serv mgr, 65-70, dir tech mkt 70-76; sr consult, Bernard Wolnak & Assoc, 76-77; FOOD INDUSTRY CONSULT, J RAKOSKY SERV INC, 78- *Mem:* AAAS; Am Chem Soc; Am Soc Microbiol; Inst Food Technol; Am Pub Health Asn. *Res:* Soy products, manufacture and use; regulatory matters, government liaison, technical literature and manuals, technical sales training; nutrition; marketing feasibility studies and surveys; plant sanitation. *Mailing Add:* 5836 Crain St Morton Grove IL 60053

RAKOWSKI, ROBERT F, b Rahway, NJ, Oct 8, 41; m 64; c 3. MEMBRANE BIOPHYSICS, IONIC CHANNELS & PUMPS. *Educ:* Cornell Univ, BChE, 64, MEng, 66; Univ Rochester, PhD(physiol), 72. *Prof Exp:* asst prof physiol, Wash Univ, 75-84; assoc prof, 84-89, PROF PHYSIOL, UNIV HEALTH SCI, CHICAGO MED SCH, 89- *Concurrent Pos:* Vis scientist, Max-Planck Inst Biophys, Frankfurt, Ger; bd of trustees res award, Univ Health Sci, Chicago Med Sch. *Mem:* Am Soc Zoologists; Biophys Soc; Soc Gen Physiologists; AAAS; NY Acad Sci. *Res:* Mechanism of voltage-dependent ion conductance changes in excitable cells; voltage-clamp studies of sodium pump activity in cells. *Mailing Add:* Univ Health Sci Chicago Med Sch 3333 Greenbay Rd North Chicago IL 60064-3095

RAKOWSKY, FREDERICK WILLIAM, b Cleveland, Ohio, Aug 24, 28; m 53; c 2. PHYSICAL CHEMISTRY. *Educ:* Baldwin-Wallace Col, BS, 50; Ohio State Univ, MS, 51, PhD(chem), 54. *Prof Exp:* Asst, Ohio State Univ, 50-54; RES ASSOC AMOCO OIL CO, 54- *Mem:* Am Chem Soc; Am Soc Testing & Mat. *Res:* Corrosion in hydrocarbon and water systems; hydrocarbon oxidation; air pollution. *Mailing Add:* Nine E 14th St Naperville IL 60540

RAKSIS, JOSEPH W, b Wilkes-Barre, Pa, March 9, 42; m 82; c 3. DESIGN & SYNTHESIS OF POLYMERIC MATERIALS. *Educ:* Wilkes Col, BS, 63; Univ Calif, Irvine, PhD(chem), 67. *Prof Exp:* Res chemist, Dow Chem Co, USA, 67-71, res mgr, 72-77; dir, 77-82, VPRES, RES DIV, W R GRACE & CO, 82- *Mem:* Am Chem Soc; Indust Res Inst. *Res:* Design and synthesis of polymeric materials having specific functional properties. *Mailing Add:* Res Dir W R Grace Co Rte 32 Columbia MD 21044-3310

RAKTOE, B LEO, statistics, experimental design, for more information see previous edition

RAKUSAN, KAREL JOSEF, b Slany, Czech, Jan 28, 35; Can citizen; m 61; c 2. PHYSIOLOGY. *Educ:* Charles Univ, Prague, MD, 60; Czech Acad Sci, PhD(physiol), 64. *Prof Exp:* Asst prof pathophysiol, Charles Univ, Prague, 60-64, assoc prof, 67-68; assoc prof, 69-74, PROF PHYSIOL, UNIV OTTAWA, 74- *Concurrent Pos:* Mem, Int Study Group Res Cardiac Metab; NIH fel cardiol, Wayne State Univ, 65-66; Med Res Coun grant, Univ Ottawa, 69-; vis scientist, Univ Ottawa, 68-69. *Honors & Awards:* Czech Acad Sci Award, 66. *Mem:* Am Physiol Soc; Can Physiol Soc; Int Soc Heart Res; Int Soc Oxygen Transport Tissue (pres). *Res:* Microcirculatory aspects of the oxygen supply; experimental cardiomegaly; oxygen in the heart muscle; developmental physiology and angiogenesis. *Mailing Add:* Dept Physiol Univ Ottawa Ottawa ON K1H 8M5 Can

RALEIGH, CECIL BARING, b Little Rock, Ark, Aug 11, 34; m 76, 81; c 4. GEOPHYSICS. *Educ:* Pomona Col, BA, 56, MA, 59; Univ Calif, Los Angeles, PhD(geol), 63. *Prof Exp:* Res fel geophys, Inst Advan Studies, Australian Nat Univ, 63-65, fel, 65-66; res geophysicist, Nat Ctr Earthquake Res, US Geol Surv, 66-73, br chief earthquake tectonics; DIR, LAMONT-DOHERTY GEOL OBSERV & PROF, DEPT GEOL SCI, COLUMBIA UNIV, 81- *Honors & Awards:* Interdisciplinary Award, Intersoc Comt Rock Mech, 69 & 74; Meritorious Serv Award, Dept Interior, 79. *Mem:* Fel Geol Soc Am; fel Am Geophys Union. *Res:* Experimental deformation of rocks at high pressure and temperature; studies of earthquakes triggered by fluid injection; plastic deformation of rock forming minerals; rock mechanics. *Mailing Add:* Sch Ocean & Earth Sci & Technol Univ Hawaii 1000 Pope Rd Honolulu HI 96822

RALEIGH, DOUGLAS OVERHOLT, b New York, NY, Aug 19, 29; m 60, 76; c 1. ELECTROCHEMISTRY. *Educ:* Rensselaer Polytech Inst, BS, 51; Columbia Univ, MA, 55, PhD(chem), 60. *Prof Exp:* Res chemist, Sylvania Elec Prod Corp, NY, 51-52; sr chemist, Atomics Int Div, 58-62, mem tech staff, 62-84, INDEPENDENT RES & DEVELOP IR & D MGR, SCI CTR, ROCKWELL INT, 84- *Concurrent Pos:* Invited lectr, Gordon Conf Electrochem, 68, Solid State, 73, Electrochem, 74 & high temperature chem, 78; vis assoc prof, Univ Utah, 69; invited tutorial lectr, NATO Advan Study Insts, Belgirate, Italy, 72 & Ajaccio, Corsica, 75, mem sci comt, 75; mem steering & prog comts, NBS Workshop on Electrocatalysis, 75; US-Japan Joint Sem, Defects & Diffusion in Solids, Tokyo, 76. *Res:* Electrochemical processes in molten salts and solid ionic conductors. *Mailing Add:* Rockwell Int Sci Ctr Thousand Oaks CA 91360

RALEIGH, JAMES ARTHUR, b Vancouver, BC, Feb 21, 38; m 62; c 4. ORGANIC CHEMISTRY, RADIATION BIOCHEMISTRY. *Educ:* Univ BC, BSc, 60, MSc, 62; Mass Inst Technol, PhD(org chem), 67. *Prof Exp:* Fel, Univ Sussex, 66-68; res off med biophys, Whiteshell Nuclear Res Estab, Atomic Energy Can Ltd, 68-78; SR SCIENTIST, CROSS CANCER INST, 78- *Concurrent Pos:* Adj assoc prof radiol, Univ Alta, Can, 78-84, adj prof radiol, 84-; assoc ed, radiation res, 83-86; mem, NIH radiation study sect, 83-86; chem counr, Radiation Res Soc, 86-89. *Mem:* Am Chem Soc; Radiation Res Soc; Chem Inst Can; Royal Soc Chem; Soc Free Radical Res. *Res:* Radiation chemistry of biologically important compounds; biochemistry of nitroaromatic compounds. *Mailing Add:* Dept Radiation Oncol CB No 7512 Univ NC Sch Med Chapel Hill NC 27599-7512

RALEY, CHARLES FRANCIS, JR, b Baltimore, Md, May 8, 23; m 47; c 3. ORGANIC CHEMISTRY. *Educ:* Univ Notre Dame, BS, 43, MS, 47; Univ Del, PhD(org chem), 50. *Prof Exp:* Org chemist, Southwest Res San Antonio, Tex, 50-56; org chemist, Dow Chem USA, 57-64, sr res chemist, 63-73, sr res specialist, Dow Chem Co, 73-75, assoc scientist, 75-82; RETIRED. *Concurrent Pos:* Adj instr chem, Saginaw Valley State Col, 84- *Mem:* Am Chem Soc; Sigma Xi. *Res:* Ignition-resistant plastics; thermal halogenations; high temperature reactions; free radical chemistry; halomethylation; Friedel-Crafts bromination. *Mailing Add:* 830 N Saginaw Rd Midland MI 48640

RALEY, FRANK AUSTIN, industrial & mechanical engineering, for more information see previous edition

RALEY, JOHN HOWARD, b Salt Lake City, Utah, Sept 28, 16; m 41; c 2. PHYSICAL ORGANIC CHEMISTRY. *Educ:* Univ Utah, AB, 37, AM, 39; Univ Rochester, PhD(chem), 42. *Prof Exp:* Chemist, Shell Develop Co, Calif, 42-51, refinery technologist, Shell Oil Co, Tex, 51-52, res supvr, Shell Develop Co, Calif, 52-68, dir phys sci, 68-69, res supvr, 69-72; chemist, Lawrence Livermore Nat Lab, Univ Calif, 75-81; RETIRED. *Concurrent Pos:* Chmn, Gordon Res Conf Hydrocarbon Chem, 63; consult, 81-90. *Mem:* Am Chem Soc. *Res:* Exploratory research petroleum and petrochemical processes products; hydrocarbon chemistry; oxidation; chemistry of reactive intermediates; metal complexes; homogeneous, heterogeneous catalysis; oil shale; shale oil. *Mailing Add:* 1040 Homestead Ave Walnut Creek CA 94598

RALL, DAVID PLATT, b Aurora, Ill, Aug 3, 26; m 54; c 2. PHARMACOLOGY. *Educ:* NCent Col (Ill), BA, 46; Northwestern Univ, MD & PhD(pharmacol), 51. *Prof Exp:* Res assoc pharmacol, Northwestern Univ, 51-52; intern, 2nd Med Div, Bellevue Hosp, Cornell Univ, 52-53; pharmacologist, Lab Chem Pharmacol, Nat Cancer Inst, 53-55, clin pharmacol & exp ther serv, 55-58, head, 58-63, chief, Lab Chem Pharmacol, 63-71, assoc sci dir exp therapeut, 65-71, dir, Nat Inst Environ Health Sci, 71-90, dir, Nat Toxicol Prog, 78-90; RETIRED. *Concurrent Pos:* Adj prof, Univ NC, Chapel Hill, 72-90; asst surgeon gen, USPHS, 90- *Honors & Awards:* Arnold J Lehman Award, Soc Toxicol, 83. *Mem:* Soc Toxicol; Am Soc Pharmacol & Exp Therapeut; Inst Med-Nat Acad Sci; Am Asn Cancer Res; Soc Occup & Environ Health. *Res:* Bacterial pyrogens; toxicology; drug distribution; cancer chemotherapy; environmental health. *Mailing Add:* 5302 Reno Rd NW Washington DC 20015

RALL, JACK ALAN, b Detroit, Mich, Apr 12, 44; m 67; c 2. MEDICAL PHYSIOLOGY. *Educ:* Olivet Col, Mich, BA, 66; Univ Iowa, PhD(physiol), 72. *Prof Exp:* Fel, Univ Calif, Los Angeles, 72-74; from asst prof to assoc prof, 74-85, PROF PHYSIOL, OHIO STATE UNIV, 85- *Concurrent Pos:* Fel, Muscular Dystrophy Asns Am, 72. *Mem:* Am Physiol Soc; Biophys Soc; Soc Gen Physiol. *Res:* Elucidation of the mechanism of muscle contraction with emphasis on the energetics of the contractile process. *Mailing Add:* Dept Physiol 333 W Tenth Ave Ohio State Univ Columbus OH 43210

RALL, JOSEPH EDWARD, b Naperville, Ill, Feb 3, 20; m 44, 78; c 2. ENDOCRINOLOGY, THYROIDOLOGY. *Educ:* NCent Col, BA, 40; Northwestern Univ, MS, 44, MD, 45; Univ Minn, PhD(med), 52. *Hon Degrees:* DSc, NCent Col, 66; Dr, Free Univ Brussels, 75; MD, Univ Naples, 85. *Prof Exp:* Asst pharmacol, Northwestern Univ, 41-44; asst prof med, Med Col, Cornell Univ, 50-55; chief clin endocrinol br, 55-62, dir, intramural res, Nat Inst Arthritis, Diabetes, Digestive & Kidney Dis, 62-83, actg dep dir sci, 81-82, DEP DIR, INTRAMURAL RES, NIH, 83- *Concurrent Pos:* Asst mem, Sloan-Kettering Inst, 50-51, assoc mem, 51-55; from asst attend physician to assoc attend physician, Med Serv, Mem Hosp, New York, 50-55; consult, Brookhaven Nat Lab, 50-; mem, Nat Res Coun, 60-; chmn, Coun Scientists, Human Frontier Sci Prog, 89. *Honors & Awards:* Van Meter Prize, Am Thyroid Asn, 50, Distinguished Service Award, 67; Fleming Award, 59; Robert H Williams Distinguished Leadership Award, 83; Super Serv Award, Dept Health, Educ & Welfare, 65, Distinguished Serv Award, 86. *Mem:* Nat Acad Sci; Endocrine Soc; Asn Am Physicians; Fr Soc Biol; Am Thyroid Asn; Am Acad Arts & Sci; Sigma Xi; Am Soc Clin Invest; AAAS; Am Physiol Soc; Radiation Effects Res Found. *Res:* Endocrinology. *Mailing Add:* NIH Bldg 1 Rm 126 Bethesda MD 20892

RALL, LLOYD L(OUIS), b Galesville, Wis, Dec 7, 16; m 52; c 4. ENGINEERING. *Educ:* Univ Wis, BSCE, 40. *Prof Exp:* Dept Engr forward area, US Army Strategic Air Force, Corps Engrs, US Army, 44-45, chief construct div, Far East Air Forces, Tokyo, 45-47, engr mem, Mil Surv Mission to Turkey, 47, Off Joint Chief Staff, Pentagon, 47-49, exec officer, Res & Develop Off, Chief Engrs, 49-51, asst dist engr, Seattle Dist, Wash, 52-54, dep engr, Commun Zone, France, 54-56, commanding officer, 540th combat engr group, 56-57, prof mil sci & tactics, Mo Sch Mines & Metall, 57-60, dep dir topog, Off Chief Engrs, 60-64, dir, Geod Intel & Mapping Res & Develop Agency, Ft Belvoir, Va, 64-66, dep asst dir, Defense Intel Agency Mapping, Charting & Geod, 66-69, asst dir, Defense Intel Agency Mapping & Charting, 69-72; DIR WASHINGTON OFF, OPTICAL SYSTS DIV, ITEK CORP, 77- *Concurrent Pos:* Mem, Nat Tech Adv Comt, Antarctica Mapping, 60-64. *Mem:* Am Soc Photogram; Am Cong Surv & Mapping; Am Inst Aerospace & Astronaut; Nat Space Club. *Res:* Mapping, charting and geodesy; geographic intelligence data. *Mailing Add:* 301 Cloverway Alexandria VA 22314

RALL, LOUIS BAKER, b Kansas City, Mo, Aug 1, 30; m 52; c 2. NUMERICAL ANALYSIS. *Educ:* Col Puget Sound, BS, 49; Ore State Col, MS, 54, PhD(math), 56. *Prof Exp:* Asst, Ore State Col, 53-56; mathematician, Shell Develop Co, 56-57; assoc prof math, Lamar State Col Technol, 57-60; from assoc prof to prof, Va Polytech Inst & State Univ, 60-62; from asst dir to assoc dir, 65-73, res mem, Math Res Ctr, 62-86, PROF MATH, UNIV WIS-MADISON, CTR MATH SCI, 86- *Concurrent Pos:* Vis prof, Innsbruck Univ, 70, Oxford Univ, 72-73, Univ Copenhagen & Tech Univ Denmark, 80 & Univ Karlsruhe, 85. *Mem:* Soc Indust & Appl Math; Am Math Soc; Math Asn Am; Inst Math & Its Appln; Int Asn Math & Comput in Simulation. *Res:* Functional and numerical analysis; integral equations; machine computing; interval analysis; theory of interval analysis and its applications to efficient, self-validating numerical solution of scientific and technological problems. *Mailing Add:* Math Res Ctr Math Res Ctr Van Vleck Hall Madison WI 53706

RALL, RAYMOND WALLACE, b Hanover, Ill, Mar 23, 26; m 49; c 4. EXPLORATION GEOLOGY. *Educ:* Univ Ill, BS, 50, MS, 51. *Prof Exp:* Res asst, Ill State Geol Surv, 50-51; geologist & stratigrapher, Pure Oil Co, 51-59; sr geologist, Tenneco Oil Co, 59-67, sr geologist, Tenneco Oil & Minerals Ltd, 67-74, geol specialist, Tenneco Oil Co, 74-78, sr geol specialist, 78-86; INDEPENDENT GEOLOGIST, 86- *Mem:* Soc Econ Paleont & Mineral; Am Asn Petrol Geologists; Can Soc Petrol Geologists. *Res:* Paleozoic stratigraphy of Texas, upper midwest United States, Williston Basin, northern Canada, east coast United States and Canada, Great Basin, Alaska and western interior basins of the United States. *Mailing Add:* 450 Shadycroft Dr Littleton CO 80120

RALL, STANLEY CARLTON, JR, b Seattle, Wash, May 18, 43; m 69. PROTEIN CHEMISTRY. *Educ:* Whitman Col, AB, 65; Univ Calif, Berkeley, PhD(biochem), 70. *Prof Exp:* Res asst biochem, Univ Calif, Berkeley, 65-70, fel, 71-74; fel biochem, Los Alamos Nat Lab, 75-78; SR SCIENTIST, GLADSTONE FOUND LABS, UNIV CALIF, SAN FRANCISCO, 79- *Concurrent Pos:* Fel, Am Cancer Soc, 71-73, Dernham Jr fel, 73. *Mem:* Fel Am Heart Asn; fel Am Soc Biochem & Molecular Biol; Protein Soc. *Res:* Structure and function of apolipoproteins. *Mailing Add:* Gladstone Found Labs for Cardiovasc Dis Univ Calif PO Box 40608 San Francisco CA 94140-0608

RALL, THEODORE WILLIAM, b Chicago, Ill, Apr 7, 28; m 49. PHARMACOLOGY. *Educ:* Univ Chicago, SB, 48, PhD(biochem), 52. *Prof Exp:* From res assoc to prof, Case Western Reserve Univ, 54-73, dir dept pharmacol, Sch Med, 73-75; PROF PHARMACOL, SCH MED, UNIV VA, 75- *Mem:* Am Soc Pharmacol & Exp Therapeut; Am Soc Biol Chemists; Soc Neurosci. *Res:* Hormonal regulatory mechanisms; neuropharmacology. *Mailing Add:* Dept Pharmacol Box 395 Univ VA Sch Med Charlottesville VA 22908

RALL, WALDO, b Los Angeles, Calif, Mar 20, 24; m; c 2. PHYSICS, RESEARCH ADMINISTRATION. *Educ:* Wash Univ, BA, 44; Ind Univ, MS, 48, PhD, 50. *Prof Exp:* Asst cyclotron lab, Wash Univ, 43-44; asst metall lab, Univ Chicago, 44; jr scientist, Clinton Lab, Tenn, 44-45; jr scientist, Los Alamos, NMex, 45-46; nuclear studies, 46-47; asst physics. Ind Univ, 47-49; from instr to asst prof, Yale Univ, 49-56; from asst div chief to div chief, Res Lab, US Steel Corp, 56-75, mgr anal & planning, 75-80, dir contract res, 80-85; RETIRED. *Concurrent Pos:* Consult, 85- *Mem:* AAAS; fel Am Phys Soc; Am Soc Metals; Sigma Xi. *Res:* Nuclear, instrumental, vacuum and metal physics; research planning and budgeting; marketing of research; research administration. *Mailing Add:* 100 Oxford Dr A-702 Monroeville PA 15146

RALL, WILFRID, b Los Angeles, Calif, Aug 29, 22; m 46, 83; c 2. BIOPHYSICS, NEUROPHYSIOLOGY,. *Educ:* Yale Univ, BS, 43; Univ Chicago, MS, 48; Univ NZ, PhD(physiol), 53. *Prof Exp:* Jr physicist, Manhattan Proj, Chicago, 43-46; lectr biophys, Med Sch, Otago, NZ, 49-51, sr lectr physiol, 51-56; head biophys div, Naval Med Res Inst, Nat Naval Med Ctr, 56-57; res biophysicist, 57-67, SR RES PHYSICIST, MATH RES BR, NAT INST DIABETES, DIGESTIVE & KIDNEY DIS, NIH, 67- *Concurrent Pos:* Rockefeller Found fel, Univ Col London. 54, Rockefeller Inst, 54-55; mem neurocommun & biophys panel, Int Brain Res Orgn, 60-, rep, Cent Coun, 68-73; mem Nat Res Coun nat comt, 72-76; comt brain sci, Nat Res Coun, 68-73; Sr Sci Performance Award, NIH, 83. *Mem:* Soc Neurosci; Biophys Soc; Physiol Soc UK; Am Physiol Soc. *Res:* Theoretical and experimental neurophysiology; dendritic branching; synaptic structure, function and integration; intracellular and extracellular potentials; computation with mathematical models; neurosciences. *Mailing Add:* Math Res Br NIDDK Room 4B-54 Bldg 31 NIH Bethesda MD 20892

RALL, WILLIAM FREDERICK, b Bayshore, NY, May 31, 51; m 74; c 2. CRYOBIOLOGY, EMBRYOLOGY. *Educ:* State Univ NY, BA, 73; Univ Tenn, Oak Ridge Grad Sch Biomed Sci, PhD(biomed sci), 79. *Prof Exp:* Vis scientist, div cell path, Clin Res Ctr, Med Res Coun, UK, 79-80; sr scientific off, Animal Res Sta, Inst Animal Physiol, Agr Res Coun, UK, 80-83; sr res fel, Cryobiol Lab, Am Red Cross Blood Res Lab, 83-84; sr scientist, Res & Develop Div, Rio Vista Int, Inc, 84-88; ASSOC SCIENTIST, AM TYPE CULTURE COLLECTION, ROCKVILLE, MD, 88-; CRYOBIOLOGIST, NAT ZOOL PARK, SMITHSONIAN INST, 89- *Concurrent Pos:* Vis scientist, health & technol div, Mass Inst Technol, 78-79; mem, Darwin Col, Cambridge, UK, 80-83; guest researcher, Embryo Cryopreserv Prog, Vet Resources Br, NIH, 84-86; adj assoc scientist, Southwest Found Biomed Res, San Antonio, Tex, 86- *Mem:* AAAS; Soc Cryobiol; Biophys Soc; Soc Low Temp Biol; Int Embryo Transfer Soc; Soc Study Fertility. *Res:* Cryobiology of mammalian embryos, spermatozoa and cell lines; low-temperature light microscopy; development and physiology of preimplantation mammalian embryos; germ plasm banking. *Mailing Add:* Dept Animal Health Nat Zool Park Washington DC 20008-2598

RALLEY, THOMAS G, b Chicago, Ill, July 10, 39; m 61; c 2. MATHEMATICS. *Educ:* Ill Inst Technol, BS, 61; Univ Ill, MS, 63, PhD(math), 66. *Prof Exp:* Asst prof, 67-73, ASSOC PROF MATH, OHIO STATE UNIV, 73- *Mem:* Am Math Soc. *Res:* Representations of finite groups and associative algebras. *Mailing Add:* Dept Math Rm 100 Ohio State Univ 231 W 18th Ave Columbus OH 43210

RALLS, JACK WARNER, b Los Angeles, Calif, Feb 1, 20; m 46; c 4. CHEMISTRY. *Educ:* Univ Calif, Los Angeles, BA, 43, MA, 44; Northwestern Univ, PhD(chem), 49. *Prof Exp:* Proj assoc, Univ Wis, 49-51; res chemist, G D Searle & Co, Ill, 51-55; res chemist, Calif Res Corp, 55-58; res chemist, Nat Canners Asn, 58-67, res mgr, Western Res Lab, 67-73, mgr res serv, 73-77. *Concurrent Pos:* Lectr exten div, Univ Calif, Berkeley, 58-71; collabr, Western Utilization Res & Develop Div, USDA, 58-64, res coordr, 64-67, res mgr, 67-73; sr res scientist, Eng Exp Sta, Ga Inst Technol, 77-78; gen mgr, Temp Tech Assocs, Kensington, 78-80; dir, Org Labs, Univ SC, 80-81. *Mem:* Am Chem Soc; Inst Food Technol. *Res:* Food processing; organic chemistry of thermally processed foods; flavor chemistry of fruits and vegetables; canned food technology; environmental chemistry. *Mailing Add:* Six Highgate Rd Kensington CA 94707-1141

RALLS, KATHERINE SMITH, b Oakland, Calif, Mar 21, 39; div; c 3. MAMMALOGY, CONSERVATION BIOLOGY. *Educ:* Stanford Univ, AB, 60; Radcliffe Col, MA, 62; Harvard Univ, PhD(biol), 65. *Prof Exp:* Fel animal behav, Univ Calif, Berkeley, 65-67; guest investr, Rockefeller Univ, 68-70; asst prof, Sarah Lawrence Col, 70-73; fel Radcliffe Inst, 73-74; fel, 73-75, RES ZOOLOGIST, SMITHSONIAN INST, 76-; ADJ PROF ENVIRON STUDIES, UNIV CALIF, SANTA CRUZ, 91- *Concurrent Pos:* Adj asst prof animal behav, Rockefeller Univ, 70-76; Am Asn Univ Women, 75-76; mem psychobiol adv comt, NSF, 81-83; sci adv comt, Marine Mammal Comn, 79-82; consult, Int Union Conserv Nature, Natural Res, Survival Serv Comn, Captive Breeding Specialist Group, 79-, Otter specialist group, 89-; Species Survival Plan Subcomt, Am Asn Zool Parks & Aquaria, 81-88; bd dir, Am Soc Mammalogists, 83-, second vpres, 90-91; mem gov bd, Soc Conserv Biol, 85-90; mem, Sea Otter Recovery Team, US Fish & Wildlife Serv, 89-, Condor Recovery Team, 90- *Mem:* Am Soc Mammal; Animal Behav Soc; Asn Women Sci; Soc Marine Mammal; Soc Conserv Biol; fel AAAS; Wildlife Soc. *Res:* Mammalian social behavior; genetics of small populations; marine mammals. *Mailing Add:* DZR Nat Zoo Smithsonian Inst Washington DC 20008

RALLS, KENNETH M(ICHAEL), b Salt Lake City, Utah, Feb 14, 38; m 78; c 4. MATERIALS SCIENCE, PHYSICAL METALLURGY. *Educ:* Stanford Univ, BS, 60; Mass Inst Technol, SM, 62, ScD(phys metall), 64. *Prof Exp:* Res assoc metall, Mass Inst Technol, 64-65; fel, Inorg Mat Res Div, Lawrence Radiation Lab, Calif, 65-67; from asst prof to assoc prof, 67-76, PROF MECH ENG, UNIV TEX, AUSTIN, 76- *Mem:* Am Inst Mining, Metall & Petrol Engrs (Metall Soc); Am Soc Metals; Am Phys Soc; fel Am Inst Chemists; Metal Soc Gr Brit; Sigma Xi. *Res:* Physical metallurgy of high magnetic field superconductive materials; fabrication and preparation of multifilamentary superconducting composites. *Mailing Add:* Dept Mech Eng ETC 5-160 Univ Tex Austin TX 78712

RALPH, C(LEMENT) JOHN, b Oakland, Calif, Sept, 3, 40; m 73; c 2. AVIAN ECOLOGY. *Educ:* Univ Calif, Berkeley, AB, 63; Calif State Univ, San Jose, MS, 69; Johns Hopkins Univ, ScD(pathobiol), 74. *Prof Exp:* Dir, Point Reyes Bird Observ, 66-69; asst prof ecol/behav, Dickinson Col, 73-76; RES ECOLOGIST, US FOREST SERV, 76- *Concurrent Pos:* Comnr, Animal Species Adv Comn, Hawaii, 80-82; ed, Elepaio. *Honors & Awards:* Tucker Award, Cooper Ornith Soc, 67; Wilson Award, Wilson Ornith Soc, 73; Roberts Award, Am Ornith Soc. *Mem:* Am Ornithologists Union; Ecol Soc Am; Animal Behav Soc; Wilson Ornith Soc; Sigma Xi; Cooper Ornith Soc (pres, 85-87). *Res:* Life history and ecological relationships of Hawaiian and Pacific Northwest forest birds, especially rare and endangered species. *Mailing Add:* 7000 Lamphers Rd Arcata CA 95521

RALPH, CHARLES LELAND, b Flint, Mich, Aug 16, 29; m 80; c 2. PHYSIOLOGY. *Educ:* Southeast Mo State Col, BS, 52; Northwestern Univ, MS, 53, PhD(biol), 55. *Prof Exp:* Spec prof personnel corps, US Army Chem Ctr, Md, 55-57; physiologist, USDA, 57-59; from asst prof to prof biol, Univ Pittsburgh, 59-74, chmn dept, 72-74; PROF ZOOL & ENTOM & CHMN DEPT, COLO STATE UNIV, 74- *Mem:* Fel AAAS; Am Soc Zool; Am Physiol Soc; Am Asn Anat; Am Inst Biol Sci. *Res:* Comparative physiology; neuroendocrinology; physiology of the pineal body; vertebrate color change. *Mailing Add:* Dept Zool & Entom Colo State Univ Ft Collins CO 80523

RALPH, PETER, b Brandon, Vt, Oct 7, 36. HEMATOLOGY, CANCER. *Educ:* Mass Inst Technol, PhD(biol), 68. *Prof Exp:* SR SCIENTIST & DIR CELL BIOL, CETUS CORP, EMERYVILLE, CALIF, 84- *Concurrent Pos:* Assoc mem, Sloan Kettering Inst Cancer Res, NY, 75-84. *Mem:* AAAS; Am Asn Immunologists; Am Soc Hemat. *Res:* Immunosuppression; hematopoiesis; cancer research; macrophages. *Mailing Add:* Cetus Corp 1400 53rd St Emeryville CA 94608

RALSTON, ANTHONY, b New York, NY, Dec 24, 30; m 58; c 4. MATHEMATICS EDUCATION. *Educ:* Mass Inst Technol, SB, 52, PhD(math), 56. *Prof Exp:* Mem tech staff, Bell Tel Labs, Inc, 56-59; lectr math, Univ Leeds, 59-60; mgr tech comput, Am Cyanamid Co, 60-61; from assoc prof to prof math, Stevens Inst Technol, 60-65, dir comput ctr, 60-65, dir comput serv, 65-70, chmn dept, 67-80,; PROF COMPUT SCI, STATE UNIV NY BUFFALO, 65- *Concurrent Pos:* Vis prof, Univ London, 71-72, 78-79 & 85-86; ed, Abacus, 83-88; mem, bd govs, Math Asn Am, 84-87 & Bd Math Sci Educ, Nat Res Coun, 85-89; consult, AT&T Bell Labs. *Honors & Awards:* Distinguished Service Award, Asn Comput Mach, 82. *Mem:* Asn Comput Mach (vpres, 70-72, pres, 72-74); Soc Indust & Appl Math; Am Fedn Info Processing Socs (pres, 75-76); Math Asn Am (Bd Govs, 84-87); fel AAAS. *Res:* Discrete mathematics; education in computer science and mathematics. *Mailing Add:* Dept Comput Sci 226 Bell Hall State Univ NY Buffalo Amherst NY 14260

RALSTON, DOUGLAS EDMUND, b Cherokee, Iowa, July 9, 32; m 53; c 2. BIOCHEMISTRY. *Educ:* Wayne State Col. BS. 55, MS, 57; SDak State Univ, MA, 59; Univ Minn, Minneapolis, PhD(biochem), 69. *Prof Exp:* Asst prof chem, Wayne State Col, 59-60; ASSOC PROF BIOCHEM, MANKATO STATE UNIV, 62- *Concurrent Pos:* Chmn, Nat Educ Asn Higher Educ Coun, 78-80. *Mem:* Am Chem Soc. *Res:* Membrane transport. *Mailing Add:* Dept Chem & Geol Mankato State Univ Box 40 S Roadland Ellis Ave Mankato MN 56001

RALSTON, ELIZABETH WALL, b Urbana, Ill, June 26, 45; m 69. COMPUTER ALGEBRA. *Educ:* Stanford Univ, BS, 66; Yale Univ, PhD(math), 70. *Prof Exp:* Instr math, Fordham Univ, 70-71; asst prof, Calif State Col, Dominguez Hills, 71-73; adj asst prof, Univ Calif, Los Angeles, 73-75; asst prof math, Fordham Univ, 75-77; mem technol staff, Aerospace Corp, 77-83; comput scientist, Transaction Technol, Inc, 83-84; MEM TECH STAFF, INFERENCE CORP, 84- *Mem:* Am Math Soc. *Res:* Computer algebra. *Mailing Add:* 550 N Continental Blvd El Segundo CA 90245

RALSTON, HENRY JAMES, b San Francisco, Calif, Feb 10, 06; m 34; c 3. PHYSIOLOGY. *Educ:* Univ Calif, AB, 29, PhD(zool), 34. *Prof Exp:* Lectr zool, Univ Calif, 34-35; instr physiol, San Francisco Jr Col, 35-39; instr, Col Dent, Univ Calif, 39-44, res assoc, 42-44; asst prof, Sch Med, Univ Tex, 44-45; from asst prof to assoc prof, 45-53, PROF PHYSIOL, UNIV PAC, 53-; RES PHYSIOLOGIST, SCH MED, UNIV CALIF, SAN FRANCISCO, 55- *Mem:* AAAS; Soc Exp Biol & Med; Am Physiol Soc. *Res:* Effects of x-rays on protozoa; dynamics of circulation; physiology of human muscle; energy expenditure in locomotion. *Mailing Add:* 184 Edgewood Ave San Francisco CA 94117

RALSTON, HENRY JAMES, III, b Berkeley, Calif, Mar 12, 35; m 60; c 2. NEUROANATOMY, ELECTRON MICROSCOPY. *Educ:* Univ Calif, Berkeley, AB, 56; Univ Calif, San Francisco, MD, 59. *Prof Exp:* Intern med, Mt Sinai Hosp, New York, 59-60; resident, Univ Calif, San Francisco, 60-61; asst prof anat, Sch Med, Stanford Univ, 65-69; assoc prof anat, Univ Wis-Madison, 69-73; PROF ANAT & CHMN DEPT, UNIV CALIF, SAN FRANCISCO, 73- *Concurrent Pos:* Nat Inst Neurol Dis & Blindness spec fel neuroanat, Univ Col, Univ London, 63-65; prin investr, Nat Inst Neurol & Communicative Dis & Stroke res grants, 65-; mem neurol A study sect, NIH, 77-81; mem, Nat Bd Med Examrs, 82-, chmn, Anat Test Comt, 85-; chmn, Acad Senate, Univ Calif, San Francisco, 86-88. *Honors & Awards:* Borden Award, 59. *Mem:* AAAS; Anat Soc Gt Brit & Ireland; Am Asn Anat (pres, 87-88); Soc Neurosci; Int Asn Study Pain. *Res:* Fine structural organization of mammalian nervous system; mechanisms of somatic sensation. *Mailing Add:* Dept Anat Univ Calif 513 Parnassus Ave San Francisco CA 94143

RALSTON, JAMES VICKROY, JR, b Elyria, Ohio, June 26, 43; m 69. MATHEMATICS. *Educ:* Harvard Univ, BA, 64; Stanford Univ, PhD(math), 69. *Prof Exp:* Vis mem, Courant Inst Math Sci, 68-70; asst prof math, NY Univ, 70-71; from asst prof to assoc prof, 71-77, PROF MATH, UNIV CALIF, LOS ANGELES, 77- *Concurrent Pos:* Fel, Alfred Sloan Found, 74-76. *Mem:* Am Math Soc. *Res:* Hyperbolic partial differential equations; scattering theory. *Mailing Add:* Dept Math Univ Calif Los Angeles CA 90024

RALSTON, JOHN PETER, b Reno, Nev, Aug 8, 51. HIGH ENERGY THEORY. *Educ:* Univ Nev, Reno, BS, 73; Univ Ore, PhD(physics), 80. *Prof Exp:* Engr instrument design, Hamilton Co, Reno, 74; res assoc high energy theory, McGill Univ, 80-82, fac lectr, 82; res assoc high energy theory, Argonne Nat Lab, 82-84; asst prof, 84-88, ASSOC PROF HIGH ENERGY THEORY, PHYSICS DEPT, UNIV KANS, 88- *Mem:* Sigma Xi. *Res:* Theoretical high energy physics; applications of quantum field theory to particle structure and interactions. *Mailing Add:* Dept Physics & Astron McGill Univ 3600 University St Montreal PQ H3A 2T8 Can

RALSTON, MARGARETE A, b Denver, Colo, July 15, 54. ELECTROMAGNETIC FIELD THEORY, TELECOMMUNICATIONS. *Educ:* Univ Denver, BA, 74; Univ Colo, MS, 77. *Prof Exp:* Proj engr, Vir James PC, 74-77; occup engr, Western Elec, 77-78; antenna design engr, Jampro Antennas, 78-79; engr, Broadcast Prod Div, Harris Corp, 79-80; sr engr, RCA Corp, 80-81 & Sci Atlanta, 82-84; staff engr, Martin Marietta Corp, 85-87; ASSOC PROF ELEC ENG, METROP STATE COL, DENVER, 87- *Concurrent Pos:* Rep, 2 degree spacing comt, space sta working group, Fed Commun Comn, 86; adj fac, Univ Denver, 89-; fac assoc, Inst Telecommun Sci, Nat Telecommun & Info Admin, 88-89. *Mem:* Sr mem Inst Elec & Electronic Engrs; Nat Soc Prof Engrs; Am Soc Eng Educ. *Res:* Communications, both analog and digital; satellites, antenna design and telecommunications; data communication networks; awarded one United States patent. *Mailing Add:* 3345 S Kendall St Denver CO 80227

RALSTON, ROBERT D, b Petersburg, NDak, July 7, 24; m 48; c 4. PLANT ECOLOGY. *Educ:* Mayville State Col, BS, 50; Univ Utah, MS, 60; Univ Sask, PhD(plant ecol), 68. *Prof Exp:* Teacher high sch, NDak, Nev & Minn, 50-59; PROF & CHMN BIOL, MAYVILLE STATE COL, 60- *Mem:* Ecol Soc Am. *Res:* Ecological research on native grasslands of northern Great Plains; phytosociological data and environmental factors correlated; part affects and fire recovery of woodlands in Little Missouri Badlands. *Mailing Add:* Mayville State Col Mayville ND 58257

RAM, BUDH, b Delhi, India, Jan 12, 35; m 64; c 2. PHYSICS. *Educ:* Univ Delhi, BS, 55, MS, 57; Univ Colo, PhD(physics), 63. *Prof Exp:* Lectr physics, Univ Delhi, 57-58, lectr, Panjab Univ, India, 58-59; teaching asst, Univ Colo, 59-62; res fel, Battersea Col Technol, Univ London, 63-64; res assoc, Univ NC, 64-66; from asst prof to assoc prof, 66-77, PROF PHYSICS, NMEX STATE UNIV, 77- *Mem:* Am Phys Soc. *Res:* Elementary particles; theoretical physics. *Mailing Add:* Dept Physics NMex State Univ Las Cruces NM 88003

RAM, C VENKATA S, b Machilipatnam, India, Oct 24, 48; US citizen; m 79; c 2. HYPERTENSION, CARDIOVASCULAR DISEASE. *Educ:* Govt Med Col-Marathwada Univ, BS, 66; Osmania Med Col, MD, 71; Am Bd Internal Med, cert. *Prof Exp:* Intern, Mercer Hosp, Trenton, NJ, 73; med intern, Brown Univ & RI Hosp, 74, resident internal med & teaching fel, Div Biol Sci, 75-76; fac assoc, 77-78, asst prof, 78-83, ASSOC PROF INTERNAL MED, UNIV TEX HEALTH SCI CTR, SOUTHWESTERN MED SCH, DALLAS, 84- *Concurrent Pos:* Instr, Univ Pa Sch Med, 76-77; attend physician, Parkland Mem Hosp, Dallas, Tex, 78-, dir, Hypertension Clin, 80-; attend physician, St Paul's Hosp, Dalllas, Tex, 79-, dir, Hypertension Clin, 79- *Mem:* Am Heart Asn; Am Col Chest Physicians; Am Col Physicians; Am Col Cardiol; Am Col Clin Pharmacol; AMA; Am Fedn Clin Res; Int Soc Hypertension; Am Soc Hypertension. *Res:* Clinical research mechanism and management of hypertension; development and application of new cardiovascular drugs. *Mailing Add:* Univ Tex Southwestern Med Ctr 5323 Harry Hines Blvd Dallas TX 75235-9030

RAM, J SRI, b India, Apr 5, 28; m 50; c 2. BIOCHEMISTRY, IMMUNOLOGY. *Educ:* Andhra Univ, India, BSc, 48; Univ Bombay, PhD(biochem), 52. *Prof Exp:* Res assoc, Columbia Univ, 54-55; res assoc immunochem, Univ Pittsburgh, 55-59, asst prof biochem, 59-61; asst prof biol chem, Univ Mich, 61-65; res biochemist, Nat Inst Arthritis, Metab & Digestive Dis, 65-74; exec secy, Pathobiol Chem Study Sect, Div Res Grants, Grants, NIH, 74-76; actg chief, 76-78, chief, Airways Dis Br, 78-89, SPEC ASST TO DIR, DIV LUNG DIS, NAT HEART, LUNG & BLOOD INST, NIH, 89- *Concurrent Pos:* Lady Tata scholar biochem, Indian Inst Sci, Bangalore, 52-53 & Fordham Univ, 53-54; Fulbright vis prof, India, 71. *Mem:* Am Asn Path; Am Soc Biochem & Molecular Biol; Am Asn Immunol. *Res:* Immunochemistry; mechanism of enzyme action; antigen-antibody interactions; protein modification; antigenicity of hormones and drugs; immunology and biochemistry of disease; aging; science policy; pulmonary disease research administration. *Mailing Add:* Nat Heart Lung & Blood Inst NIH Westwood Bldg Rm 6A-11 Bethesda MD 20892

RAM, JEFFREY L, b Newark, NJ, Sept 25, 45; m 77; c 3. INVERTEBRATE NEUROPHYSIOLOGY, ENDOCRINOLOGY. *Educ:* Univ Pa, BA, 67; Calif Inst Technol, PhD(biochem), 74. *Prof Exp:* Fel neurosci, Univ Calif, Santa Cruz, 73-77; asst prof, 77-82, ASSOC PROF NEUROPHYSIOL, WAYNE STATE UNIV, 82- *Concurrent Pos:* Stipendiary fel, Marine Biol Lab, Woods Hole, Mass, 75; vis prof, Technion, Haifa, Israel, 84-85. *Mem:* AAAS; Am Physiol Soc; Am Soc Neurosci. *Res:* Comparative aspects of gastropod egg hormones; modulatory effects of serotonin; activation of neurons by peptides; biophysics membranes; computer modeling. *Mailing Add:* Dept Physiol Sch Med Wayne State Univ Detroit MI 48201

RAM, MADHIRA DASARADHI, surgery, experimental pathology; deceased, see previous edition for last biography

RAM, MICHAEL, b Alexandria, Egypt, Dec 18, 36; m 59; c 3. ATMOSPHERIC PHYSICS, THEORETICAL PHYSICS. *Educ:* Israel Inst Technol, BSc, 60, MSc, 62; Columbia Univ, PhD(physics), 65. *Prof Exp:* Res assoc physics, Johns Hopkins Univ, 65-67; asst prof, 67-72, ASSOC PROF PHYSICS, STATE UNIV NY BUFFALO, 72- *Concurrent Pos:* Chmn dept, State Univ NY, 74-77. Dept, 74- *Mem:* Am Geophys Union. *Res:* Theoretical physics; atmospheric physics. *Mailing Add:* Dept Physics State Univ NY Buffalo NY 14260

RAM, MICHAEL JAY, b Newark, NJ, Dec 18, 40; m 64; c 3. MEDICAL DEVICES, PATENT LAW. *Educ:* Lafayette Col, BS, 62; Newark Col Eng, MS, 63, DSc(chem eng), 66; Seton Hall Univ, JD, 72. *Prof Exp:* Sr res engr, Celanese Res Co, Summit, 67-73; patent atty, Brooks, Haidt, Haffner, 73-74; dir tech liason, C R Bard Inc, 73-84; GEN COUN, PHARMACIA OPHTHALMICS INC, 85- *Mem:* Am Chem Soc; Am Inst Chem Engrs; Am Bar Asn. *Res:* Synthetic fibers; plastics; medical products; patent law; medical devices and the patent protection of products. *Mailing Add:* Sheldon & Mak 201 S Lake Ave 8th Floor Pasadena CA 91101

RAM, NEIL MARSHALL, b New York, NY, Feb 6, 52; m 74; c 2. HAZARDOUS WASTE ASSESSMENT, CHEMICAL FATE & TRANSPORT. *Educ:* Rutgers Univ, BS, 73, MS, 75; Harvard Univ, MS, 77 & PhD(environ eng), 79. *Prof Exp:* Post doctorate environ eng, Technion Inst Sci, 79-80; asst prof environ eng, Univ Mass, 80-84; lab mgr, Alliance Technol Corp, 84-87; sr prod mgr, Stone & Webster Eng Corp, 87-89 & ICF Kaiser Engrs, 89-91; MGR, REMEDIATION TECHNOL GROUP, GROUNDWATER TECHNOL INC, 91- *Concurrent Pos:* Co-ed, Assoc Environ Eng Profs, 82-84, standards methods comt, Standard Methods for Water & Wastewater, 84-88, proj advi comt, Am Water Works Asn Res Found, 87-88; mgr, Envirologic Inc, 91- *Mem:* Hazardous Mat & Control Res Inst. *Res:* Assessment, management and remediation of toxic or hazardous substances; occurence, treatment and significance of organic compounds in drinking water; determination of fate and transport for hazardous substances in environmental matrices; waste treatment technology development. *Mailing Add:* 41 Hemlock St Needham MA 02192-3501

RAMACHANDRAN, CHITTOOR KRISHNA, b Edavanakkad, Kerala, India, Nov 12, 45; US citizen; m 79; c 2. LIPID BIOCHEMISTRY. *Educ:* Univ Kerala, India, BSc, 66; Univ Baroda, India, MSc, 70; V P Chest Inst, Univ Delhi, India, PhD(biochem), 74. *Prof Exp:* Jr lectr chem, St Alberts Col, Ernakulam, 66-68; teaching fel neurochem, Brain Behav Res Ctr, Sonoma, Univ Calif, San Francisco, 74-76; res assoc cell biol, Univ Kans Med Ctr, 76-79; RES BIOCHEM CELL BIOL, VET ADMIN MED CTR, KANSAS CITY, MD, 79- *Concurrent Pos:* Adj asst prof microbiol, Univ Kans Med Ctr, Kansas City, Kans, 80- *Mem:* Am Soc Biol Chem; Tissue Cult Asn; Soc Exp Biol & Med. *Res:* Mode of action of glucocorhioids on plasma membrane components; lipid metabolism enzymes in cultured cells; cholesterol synthesis and dolichol-mediated glycosylation of proteins. *Mailing Add:* Dept Internal Med Rm 4C 216 Univ Utah Med Sch 50 N Medical Dr Salt Lake City UT 84132

RAMACHANDRAN, JANAKIRAMAN, b Bombay, India, June 12, 35; m 67; c 1. ENDOCRINOLOGY, BIOCHEMISTRY. *Educ:* Univ Madras, MA, 56; DePaul Univ, MS, 59; Univ Calif, Berkeley, PhD(biochem), 62. *Prof Exp:* Jr res biochemist, Hormone Res Lab, Univ Calif, Berkeley, 62-63, asst res biochemist, 63-68; lectr, Sch Med, Univ Calif, San Francisco, 64-68, from asst prof to prof biochem, Med Ctr, 68-83; sr scientist, Genentech, 83-88; VPRES RES, NEUREX CORP, 88- *Concurrent Pos:* Weizmann Mem fel biophys, Weizmann Inst, 65-66; adj prof biochem, Univ Calif, San Francisco, 83-88, adj prof physiol, 88-; dir, Astra Res Ctr India, Bangalore, 86- *Mem:* AAAS; NY Acad Sci; Endocrine Soc; Tissue Cult Asn; Am Soc Biol Chemists. *Res:* Study of the mode of action of polypeptide hormones. *Mailing Add:* Neurex Corp 3760 Haven Ave Menlo Park CA 94025

RAMACHANDRAN, MUTHUKRISHNAN, b Sivaganga, India, Feb 12, 43; m 74; c 2. SICKLE CELL DISEASE, RED CELL METABOLISM. *Educ:* Univ Madras, India, BS, 63, MS, 69; Univ Kerala, India, PhD(red cell metab), 81. *Prof Exp:* Demonstr & lectr biochem, Med Col, Kotcayam-8, Kerala, India, 70-74; lectr & biochemist, Jawaharlal PG Med Inst, Pondicherry-6, India, 74-80; sr lectr biochem, Col Med Sci, Univ Calabar, Nigeria, 80-85; sr res fel, 85-89, TECH SUPVR & ASST RES SCIENTIST, DEPT CELL & MOLECULAR BIOL, MED COL GA, AUGUSTA, 89- *Concurrent Pos:* Co-investr red cell membrane res, NIH, 85- *Mem:* Am Asn Clin Chemists; Am Soc Biochem & Molecular Biol; NY Acad Sci. *Res:* Abnormal hemoglobins; red cell membrane proteinkinase-c and its role in the pathogenesis of sickle cell disease; red cell metabolism; medical biochemistry; protein chemistry; DNA-protein interactions; post translational modifications. *Mailing Add:* Dept Cell & Molecular Biol Med Col Ga AC400 Laney-Walker Blvd Augusta GA 30912

RAMACHANDRAN, N, b Devicolam, India. BIOCHEMISTRY OF PROTEINS, STRUCTURE-ACTIVITY RELATIONSHIPS OF MACROMOLECULES. *Educ:* Univ Kerala, BSc, 61, MSc, 63; Univ Mo, Columbia, PhD(biochem), 73. *Prof Exp:* Res asst food technol, Cent Inst Fisheries Technol, Kerala, India, 63-65; sci officer biochem, Bhabha Atomic Res Ctr, Bombay, India, 65-67; scientist food technol, Hindustan Lever Res Ctr, Subsid Unilever Co, Bombay, 67-69; RES ASSOC PHARMACEUT, DUPONT-MERCK CO, 73- *Concurrent Pos:* Postdoctoral biochem, Univ Del, Newark, 74-79. *Res:* Cardiovascular research; structure-activity relationship of many proteins involved in the blood coagulation system; isolation & purification of pure proteins from their sources and from recombinant sources. *Mailing Add:* Dupont Merck Pharmaceuticals 500 S Ridgeway Ave Glenolden PA 19036

RAMACHANDRAN, PALLASSANA N, b Palghat, India; US citizen; m 66; c 1. PHYSICAL CHEMISTRY, SURFACE CHEMISTRY. *Educ:* Univ Bombay, BSc, 56; Temple Univ, MA, 62, PhD(phys chem), 65. *Prof Exp:* Fel chem, Textile Res Inst, NJ, 65-67; sr res chemist, 67-80, res assoc, 80-85, sr res assoc, 85-88, ASSOC RES FEL, COLGATE-PALMOLIVE RES CTR, 88- *Mem:* Am Chem Soc; Am Oil Chemists Soc; Fiber Soc. *Res:* Development and processing of household products. *Mailing Add:* Colgate-Palmolive Res Ctr 909 River Rd Piscataway NJ 08854

RAMACHANDRAN, SUBRAMANIA, b Madras, India, Jan 8, 38; m 68; c 1. ORGANIC CHEMISTRY, BIOCHEMISTRY. *Educ:* Annamalai Univ, Madras, 57, Hons, 59, MSc, 60; Ohio State Univ, MS, 64, PhD(biochem), 68. *Prof Exp:* Asst chem, Ohio State Univ, 61-68, fel physiol chem, 68-69; from asst mgr to mgr, Biochem Dept, 69-72, vpres res & develop, 72-74, MGR RES & DEVELOP, APPL SCI LABS, INC, 74- *Mem:* Am Oil Chem Soc. *Res:* Synthesis of lipids, including steroids; metabolism of lipids; chromatographic separation of organic compounds; analytical methods in clinical chemistry and pharmacology. *Mailing Add:* 26 Crickelwood Circle State College PA 16803

RAMACHANDRAN, VANGIPURAM S, b Bangalore, Mysore, India, Dec 30, 29; m 57; c 2. CONCRETE TECHNOLOGY, CLAY MINERALOGY. *Educ:* Mysore Univ, India, BSc, 49, DSc(cement chem), 81; Banaras Hindu Univ, MSc, 51; Calcutta Univ, PhD(catalysis), 56. *Hon Degrees:* DSc, Internation Univ Found, US, 87. *Prof Exp:* Sr res officer clay mineral, Cent Bldg Res Inst, Roorkee, India, 56-62, 65-68; res fel bldg sci, Nat Res Coun Can, 62-65, res officer bldg sci, 68-79, head, Bldg Mat Sect, 79-89, DISTINGUISHED RES, NAT RES COUN CAN, 89- *Concurrent Pos:* Mem bd, Ceramic Soc Abstracts, 63-; contrib ed, Cements Res Progress, Am Ceramic Soc, 74-83; chmn, Cements Div, Am Ceramic Soc, 81 & mem, Int Union Testing & Res Labs for Mat & Struct, France, 83-85; ed, J Mat & Struct, France, 81-; consult ser ed, Noyes Publ, NJ, 85-; res adv, Am Biog Inst, 86-; chief ed, J Mat Civil Eng, USA, 89- *Mem:* Fel Am Ceramic Soc; fel Royal Soc Chem, UK; fel Inst Ceramics, UK; Int Union Testing & Res Labs for Mat Struct, France; Can Standards Asn; Am Soc Testing & Mat. *Res:* Clay mineralogy; gypsum and cement chemistry; author of six books and 18 chapters in books. *Mailing Add:* 1079 Elmlea Dr Ottawa ON K1J 6W3 Can

RAMACHANDRAN, VENKATANARAYANA D, b Mysore City, India, May 3, 34; m 60; c 1. ELECTRICAL ENGINEERING. *Educ:* Cent Col, Bangalore, BSc, 53; Indian Inst Sci, Bangalore, BE, 56, ME, 58, PhD(elec eng), 65. *Prof Exp:* Sr res asst elec eng, Indian Inst Sci, Bangalore, 58-59, lectr, 59-65; asst prof, NS Tech Col, 66-69; assoc prof, 69-71, PROF ELEC ENG, CONCORDIA UNIV, 71- *Concurrent Pos:* Assoc ed, Can Elec Eng J, 81-83, ed, 83-85. *Honors & Awards:* Western Elec Fund Award, Am Soc Eng Educ, 83. *Mem:* Fel Inst Elec & Electronic Engrs; fel Inst Elec Engrs, UK; fel Inst Elec & Telecommun Engrs, India; fel Inst Eng, India; fel Eng Inst Can. *Res:* Circuit theory; active, lumped and multivariable networks. *Mailing Add:* Dept Elec Eng Concordia Univ Montreal PQ H3G 1M8 Can

RAMACHANDRAN, VILAYANUR SUBRAMANIAN, b Madras, India, Aug 10, 51; m 87. VISUAL PERCEPTION, NEUROPSYCHOLOGY. *Educ:* Stanley Med Col, MD, 74; Trinity Col, Eng, PhD(neurosci), 78. *Prof Exp:* Sr Rouse-Ball, Trinity Col, Cambridge Univ, Eng, 77-78; res fel, Calif Inst Technol, Pasadena, 79-81; PROF PSYCHOL, UNIV CALIF, SAN DIEGO, 83- *Concurrent Pos:* Vis assoc biol, Calif Inst Technol, 83- *Mem:* Asn Res Vision & Opthalmol. *Res:* Neuropsychology; visual perception; author of over 50 research papers. *Mailing Add:* Psychol Dept La Jolla CA 92093-0109

RAMADHYANI, SATISH, b Bangalore, India, Aug 1, 49; m 79. MECHANICAL ENGINEERING, HEAT TRANSFER. *Educ:* Indian Inst Tech, Madras, BTech, 71; Univ Minn, MS, 77, PhD(mech eng), 79. *Prof Exp:* Engr, Motor Industs Co, India, subsid Robert Bosch, 71-75; asst prof mech eng, Tufts Univ, Medford, Mass, 79-83; ASSOC PROF MECH ENG, PURDUE UNIV, WEST LAFAYETTE, IND, 83- *Concurrent Pos:* Lectr, Mass Inst Technol, 82; prin investr, United Eng Found, 82-83, IBM, 84-86, Whirlpool Corp, 85-87, NSF, 87-; consult, Whirlpool Corp, 84-85. *Honors & Awards:* President's Gold Medal, Indian Inst Technol, 71. *Mem:* Am Soc Mech Engrs. *Res:* Development of novel numerical techniques for prediction of heat transfer; experimental and numerical studies of solid-liquid phase change, compact heat exchangers and heat transfer augmentation in electronic packages; mathematical modeling of human thermal comfort. *Mailing Add:* 828 N Chauncey West Lafayette IN 47906

RAMAGE, COLIN STOKES, b Napier, NZ, Mar 3, 21; nat US; c 2. METEOROLOGY. *Educ:* Victoria Univ, NZ, BSc, 40, DSc, 61. *Prof Exp:* Meteorologist, Meteorol Serv. NZ, 41, sci officer, 46-53; dep dir, Royal Observ, Hong Kong, 54, actg dir, 55-56, assoc meteorologist, 56; assoc prof, Univ Hawaii, 57, prof meteorol, 58-88, chmn dept, 71-87; RETIRED. *Concurrent Pos:* Meteorol & oceanog, 60-62, geosci, 64-69, assoc dir, Hawaii Inst Geophys, 64-71; Commonwealth Fund fel, 53-54; consult, USAF, 56-61, US Navy, 69-71; sci dir, Indian Ocean Exped, 62-73; consult. *Mem:* fel Am Meteorol Soc; Am Geophys Union. *Res:* Meteorology of the tropics, south and southeast Asia; monsoons. *Mailing Add:* 1420 Acadia St Durham NC 27701-1302

RAMAGOPAL, MUDUMBI VIJAYA, b Tanuku, Andhra Pradesh, India, Jan 1, 52; m 79; c 2. CARDIOVASCULAR, AUTONOMIC. *Educ:* Andhra Univ, India, BS, 72; Univ Madras, India, MS, 75; Postgrad Inst, Chandigarh, India, PhD(cardiovasc pharmacol), 82. *Prof Exp:* Sci Officer pharmacol, Homeopathic Pharmacopoeia Labs, Govt India, 83-84, head pharmacol, 83-84; res asst, E Carolina Univ, 84-88; SCIENTIST PHARMACOL, GLAXO RES LABS, 88- *Concurrent Pos:* Prin investr, Indian Coun Med Res, Govt India, 76-83; mem, Drug Regulatory Affairs Comt, Govt India, 83-84. *Mem:* Am Soc Pharmacol & Exp Therapeut; Indian Pharmacol Soc; Asn Physiologists & Pharmacol India; Asn Scientists Indian Origin Am. *Res:* Cardiovascular pharmacology; specialization in isolated vascular smooth muscle biology; calcium homeostasis; pharmacological receptor modeling of agonist and antagonists; biochemical analysis of receptors; ligands; measuring of second messengers. *Mailing Add:* Div Pharmacol Glaxo Res Labs Five Moore Dr Research Triangle Park NC 27709

RAMAGOPAL, SUBBANAIDU, b Madras, India, July 5, 41; US citizen; div; c 2. PLANT MOLECULAR BIOLOGY, CELL BIOLOGY. *Educ:* Univ Madras, India, BS, 64; Utah State Univ, MS, 68; Univ Calif, Davis, PhD(plant biochem & physiol), 72. *Prof Exp:* Res fel molecular biol, Harvard Univ, 72-74; res assoc molecular biol & cell biol, Inst Cancer Res, Philadelphia, 74-77; sr res fel molecular biol, Roche Inst Molecular Biol, Nutley, NJ, 77-79; res scientist molecular biol & immunol, Sch Med, NY Univ & NJ Med Sch, 79-81; sr scientist molecular biol, cell biol & genetic eng, Armos Corp & Univ Calif, Berkeley, 81-83; PLANT PHYSIOLOGIST & LEAD SCIENTIST, PLANT MOLECULAR & CELL BIOL, AGR RES SERV, USDA, 83- *Concurrent Pos:* Adj grad prof, Univ Hawaii, 86-89, Univ Idaho, 90- *Honors & Awards:* Gold Medal, Food & Agr Orgn, 63. *Mem:* AAAS; Int Soc Plant Molecular Biol. *Res:* Plant molecular and celluar biology; plant genetic engineering and crop improvement; molecular bases of growth and development; cell differentiation; gene expression in response to abiotic stresses. *Mailing Add:* Agr Res Serv USDA PO Box 307 Aberdeen ID 83210

RAMAKER, DAVID ELLIS, b Sheboygan, Wis, Aug 11, 43; m 66; c 4. SURFACE CHEMISTRY, SOLID STATE PHYSICS. *Educ:* Univ Wis-Milwaukee, BS, 65; Univ Iowa, MS, 68, PhD(phys chem), 71. *Prof Exp:* Res physics, Sandia Labs, 70-72; res assoc & assoc instr, Univ Utah, 72-74; vis asst prof, Calvin Col, 74-75; from asst prof to assoc prof, 75-83, PROF PHYS CHEM, GEORGE WASHINGTON UNIV, 83-, CHAIR, DEPT CHEM, 88- *Concurrent Pos:* Consult res chem, Naval Res Lab, 76-; consult, Nat Bureau Standards, 82- *Honors & Awards:* Hildebrand Award, Am Chem Soc, 89. *Mem:* Am Chem Soc; Am Vacuum Soc; Combustion Inst. *Res:* Theoretical studies of surfaces and chemisorption; auger spectroscopy; electron and photon stimulated desorption. *Mailing Add:* Dept Chem George Washington Univ Washington DC 20052

RAMAKRISHAN, S, b Apr 11, 49. IMMUNOTOXINS, CANCER THERAPY. *Educ:* All India Inst Med Sci, New Delhi, PhD(biochem), 80. *Prof Exp:* Res assoc biochem, Univ Kans, 81-84; scientist, Protein Chem Dept, Cetus Corp, Emeryville, Calif, 85-86; ASST MED RES PROF, DUKE UNIV MED CTR, DURHAM, NC, 87- *Mem:* Am Soc Biochem & Molecular Biol; Am Asn Immunologists; NY Acad Sci. *Mailing Add:* Dept Pharmacol 3-249 Milard Hall 435 Delaware St SE Minneapolis MN 55455

RAMAKRISHNA, KILAPARTI, b Rajahmundry, Andhra Pradesh, Oct 13, 55; m 83. BRIDGING THE GAP BETWEEN SCIENCE & PUBLIC AFFAIRS. *Educ:* Andra Univ, SIndia, BSc, 73, BL, 76; Jawaharlal Nehru Univ, New Delhi, India, MPhil, 78, PhD(int environ law), 85. *Prof Exp:* Assoc ed, Indian J Int Law, 80-86; asst prof int law, Indian Soc Int Law, New Delhi, India, 80-85; vis scholar, Harvard Law Sch, 85-87; marine policy fel, Woods Hole Oceanog Inst, 86-89; SR ASSOC, INT ENVIRON LAW, WOODS HOLE RES CTR, 87-, DIR, PROF SCI PUB AFFAIRS, 91- *Concurrent Pos:* Vis prof int law, Boston Univ, Mass, 87-88 & 91. *Mem:* Sigma Xi. *Mailing Add:* Woods Hole Res Ctr PO Box 296 Woods Hole MA 02543

RAMAKRISHNAN, RAGHU, b Pudukkottai, India, Dec 2, 61; m 90; c 1. DATABASE SYSTEMS & THEORY, LOGIC PROGRAMMING & RULE-BASED SYSTEMS. *Educ:* Indian Inst Technol, Madras, BTech, 83; Univ Tex, Austin, PhD(computer sci), 87. *Prof Exp:* ASST PROF COMPUTER SCI, UNIV WIS-MADISON, 87- *Concurrent Pos:* Vis fac mem, Int Bus Mach Almaden Res Ctr & T J Watson Res Ctr, 88; prin investr, NSF & Int Bus Mach, 88-90; fac fel sci & technol, David & Lucille Packard Found, 89; NSF presidential young investr, 90. *Mem:* Asn Comput Mach; Inst Elec & Electronic Engrs; Asn Logic Prog. *Res:* Theory and implementation techniques to efficiently support declarative languages; extended query languages for relational databases and logic programming languages. *Mailing Add:* Computer Sci Dept Univ Wis 1210 W Dayton Madison WI 53706

RAMAKRISHNAN, TERIZHANDUR S, b Madras, India. FLUID DYNAMICS, TRANSPORT PHENOMENA. *Educ:* Indian Inst Technol, New Delhi, BTech, 80; Ill Inst Technol, Chicago, PhD(chem eng), 85. *Prof Exp:* RES SCIENTIST, SCHLUMBERGER-DOLL RES, 85- *Honors & Awards:* Silver Medal, Indian Inst Technol, 80; P C Ray Award, Indian Inst Chem Engrs, 80. *Mem:* Am Inst Chem Engrs; Soc Petrol Engrs; Sigma Xi; Am Chem Soc; Indian Inst Chem Engrs. *Res:* Creeping flow in porous media; chemically enhanced oil recovery; immiscible displacement in porous media; petrophysical properties of rocks; pressure transient analysis of petroleum formations; author of various publications. *Mailing Add:* Schlumberger-Doll Res Old Quarry Rd Ridgefield CT 06877

RAMAKRISHNAN, VENKATASWAMY, b Coimbatore, India, Feb 27, 29; m 62; c 2. STRUCTURAL ENGINEERING. *Educ:* Govt Col Technol, Coimbatore, India, BE, 52; PSG Col Technol, Coimbatore, dipl soc sci, 53; Univ London, PhD(civil eng), 60, Imp Col, dipl hydraul power, 56 & concrete technol, 57. *Prof Exp:* Jr engr, Madras Pub Works Dept, India, 52; asst lectr civil eng, PSG Col Technol, Coimbatore, 52-53, lectr, 53-60, asst prof, 60-61, prof & head dept, 61-69; PROF CIVIL ENG & DIR CONCRETE TECHNOL RES, SDAK SCH MINES & TECHNOL, 69- *Concurrent Pos:* Visitor, Bldg Res Inst, Prague Tech Univ, 67, Asian Inst Technol, Bangkok, SDak Sch Mines & Technol, Univs Colo, Ill, Chicago Circle & Mo-Columbia, 69, Norwegian Inst Tech, Swedish Cement & Concrete Res Inst, Univ West Indies & Inst Technol, Stockholm; coordr, Advan Summer Schs Struct Eng for Eng Col Teachers, India, 68 & 69; partic, Sem Recent Trends in Struct Design, 61, Ind Cong Appl of Math in Eng, Weimar, Ger, 67 & Int Conf Struct, Solid Mech & Eng Design in Civil Eng Mar, Southampton, Eng, 69; organizing secy & ed proc, Int Conf Shear, Torsion & Bond in Reinforced & Prestressed Concrete, 69; mem comt mech properties of concrete, Hwy Res Bd, Nat Acad Sci-Nat Res Coun; archit & struct eng consult; founding dir & guide prof, World Open Univ, 74-, vpres, 79- *Mem:* Am Concrete Inst; Am Soc Civil Engrs; Nat Soc Prof Engrs; Am Soc Eng Educ; Sigma Xi. *Res:* Concrete technology, particularly ultimate behavior and strength of reinforced concrete; materials technology; structural engineering and mechanics. *Mailing Add:* SDak Sch Mines & Technol 1809 Sheridan Lake Rd Rapid City SD 57701

RAMAKUMAR, RAMACHANDRA GUPTA, b Coimbatore, India, Oct 17, 36; m 63; c 2. POWER ENGINEERING, ENERGY. *Educ:* Univ Madras, India, BE, 56; Indian Inst Technol, Kharagpur, India, MTech, 57; Cornell Univ, PhD(elec eng), 62. *Prof Exp:* From asst lectr to lectr elec eng, Coimbatore Inst Technol, India, 57-62, asst prof, 62-67; vis assoc prof, 67-70, assoc prof, 70-76, PROF ELEC ENG, OKLA STATE UNIV, 76-, DIR, ENG ENERGY LAB, 87- *Concurrent Pos:* Consult, Jet Propulsion Lab, Calif, 78-79, Nat Sci Found, Washington, DC, 80, Florida Solar Energy Ctr, 81 , Kuwait Univ, 82, Mariah Inc, 84 & Dowell Schlumberger, 87-88; mem, Expert Group Energy Storage Develop Countries, UN Environ Prog, 83. *Mem:* Inst Elec & Electronics Engrs; Int Solar Energy Soc; Am Soc Eng Educr; Sigma Xi; Global Energy Soc. *Res:* Alternate energy sources development and application in developing countries for rural development; energy storage; energy conversion and power engineering; solar and wind energy systems. *Mailing Add:* 216 Eng S Okla State Univ Stillwater OK 74078-0321

RAMALEY, JAMES FRANCIS, b Columbus, Ohio, Oct 10, 41; m 67; c 2. MATHEMATICS. *Educ:* Ohio State Univ, BSc, 62; Univ Calif, Berkeley, MA, 64; Univ NMex, PhD(math), 67. *Prof Exp:* Reader math, Univ Calif, 63-64; res asst, Univ NMex, 64-65; lectr, Carnegie Inst Technol, 65-66; asst prof math, Bowling Green State Univ, 66-70; asst prof, Univ Pittsburgh, 70-73; systs analyst, On-Line Systs, Inc, 73-74; mgr info systs, 74-76, budget dir circulation, 76-82, VPRES, CIRCULATION SERV, ZIFF-DAVIS PUBL CO, 82- *Concurrent Pos:* Vis mem math res inst, Swiss Fed Inst Technol, 69; adj prof math, Univ Pittsburgh, 73-76. *Mem:* Am Math Soc; Math Asn Am; Opers Res Soc Am; Asn Comput Mach. *Res:* Category theory; logic; systems software; applications software. *Mailing Add:* 164 Revolutionary Rd Scarborough NY 10510

RAMALEY, JUDITH AITKEN, b Vincennes, Ind, Jan 11, 41; m 66; c 2. ENDOCRINOLOGY, REPRODUCTIVE BIOLOGY. *Educ:* Swarthmore Col, BA, 63; Univ Calif, Los Angeles, PhD(anat), 66. *Prof Exp:* Asst prof anat & physiol, Ind Univ, Bloomington, 69-72; from asst prof to prof physiol & biophys, Univ Nebr Med Ctr, Omaha, 72-82, asst vpres acad affairs, 81-82; vpres acad affairs, 82-87, State Univ NY, Albany, 82-87, actg pres, 84-85, exec vpres academic affairs, 85-87; exec vchancellor, Univ Kans Lawrence, 87-90; PRES, PORTLAND STATE UNIV, ORE, 90- *Concurrent Pos:* NIH fel, Ctr Neurol Sci, Ind Univ, Bloomington, 67-68, NIH fel chem, 68; mem, NSF, Regulatory Biol Panel, 78-81; mem, Biochem Endocrinol Study Sect, NIH, 81-84; chair, Acad Affairs Coun, Nat Asn State Univ & Land Grant Col, & Comn Women Higher Educ, Am Coun Educ, 87-88. *Mem:* Am Physiol Soc; Am Asn Anat; Soc Neurosci; Endocrine Soc; Soc Study Reproduction. *Res:* Physiology of puberty; control of male and female fertility. *Mailing Add:* Portland State Univ PO Box 751 Portland OR 97207

RAMALEY, LOUIS, b El Paso, Tex, Oct 7, 37; m 64; c 2. MASS SPECTROMETRY, ELECTROCHEMISTRY. *Educ:* Univ Colo, BA, 59; Princeton Univ, MA. 61, PhD(electrochem), 64. *Prof Exp:* Assoc, Univ Ill, 63-64; asst prof chem, Univ Ariz, 64-70; ASSOC PROF CHEM, DALHOUSIE UNIV, 70- *Mem:* AAAS; Am Chem Soc; Am Soc Mass Spectrometry; Chem Inst Can. *Res:* Chemical instrumentation; electrochemistry; electroanalytical and surface chemistry; mass spectrometry. *Mailing Add:* Dept Chem Dalhousie Univ Halifax NS B3H 3J5 Can

RAMALEY, ROBERT FOLK, b Colorado Springs, Colo, Dec 15, 35; div; c 2. BIOCHEMISTRY, MICROBIOLOGY. *Educ:* Ohio State Univ, BS, 59, MS, 62; Univ Minn, PhD, 64. *Prof Exp:* Asst prof microbiol, Ind Univ, Bloomington, 66-72; assoc prof, 72-78, PROF BIOCHEM, UNIV NEBR MED CTR, OMAHA, 78-, PROF PATH & MICROBIOL, 83- *Concurrent Pos:* USPHS fel, 64-66; lectr, Am Soc Microbiol Found; res grants, NIH & NSF. *Mem:* Am Soc Microbiol; Am Soc Biol Chem. *Res:* Physiology of sporulation; control of intermediate metabolism and enzyme intermediates; thermophilic microorganism and medical microbiology. *Mailing Add:* Dept Biochem Univ Nebr Med Ctr Omaha NE 68105

RAMALINGAM, MYSORE LOGANATHAN, b Mysore, Karnataka, India, Dec 12, 54; m 81; c 2. ENERGY CONVERSION SCIENCES, HEAT TRANSFER & MATERIAL CHARACTERISTICS. *Educ:* Bangalore Univ, India, BE, 75; Indian Inst Sci, India, ME, 77; Ariz State Univ, PhD(mech eng), 86. *Prof Exp:* Asst engr, Jyoti Pumps, Inc, India, 77; engr SC/SD, Indian Space Res Orgn, India, 77-82; grad asst thermodynamics, Ariz State Univ, 82-83, res asst, 83-86; res scientist, 86-89, SR SCIENTIST, UNIVERSAL ENERGY SYSTS, INC, 89- *Concurrent Pos:* Heat pipe expert, Gujarat Rotating Mach Corp, India, 77, Gas Turbine Res Est, India, 81-82; thermal consult, Vikran Sarabhai Space Ctr, India, 79-80, K-tron Int, Inc, Phoenix, 83-84; res assoc, Ariz State Univ, 85-86; thermionics expert, Wright Patterson AFB, 87-; mem, ASME/AESD thermionics, thermoelectrics Comt, 90- *Mem:* Assoc mem Inst Engrs, India; Sr mem Am Inst Aeronaut & Astronaut; Am Soc Metals. *Res:* Thermionic, thermophysical and material properties act high temperature of refractory metals and alloy slated for applications in thermionic fuel; performance characteristics of low and high temperature heat pipes with emphasis on boiling heat transfer at the evaporator. *Mailing Add:* Universal Energy Systs 4401 Dayton-Xenia Rd Dayton OH 45432-1894

RAMAMOORTHY, CHITTOOR V, b Henzada, Burma, May 5, 26; US citizen; m 57; c 3. COMPUTER SCIENCES, ELECTRICAL ENGINEERING. *Educ:* Univ Madras, BS, 46 & 49; Univ Calif, Berkeley, MS, 51, MechEng, 53; Harvard Univ, AM & PhD(appl math & comput theory), 64. *Prof Exp:* Res engr, Honeywell Inc, 56-57, sr engr, Electronic Data Processing Div, 58-60, staff engr, 61-65, sr staff scientist, 65-67; prof elec eng, Univ Tex, Austin, 67-72, prof comput sci, 68-72; PROF ELEC ENG & COMPUT SCI, UNIV CALIF, BERKELEY, 72- *Concurrent Pos:* Res fel appl math, Harvard Univ, 66-67. *Mem:* Asn Comput Mach; fel Inst Elec & Electronics Engrs. *Res:* Computer theory, design, use and applications information sciences. *Mailing Add:* Dept Elec Eng & Comput Sci Univ Calif Berkeley CA 94720

RAMAMOORTHY, PANAPAKKAM A, b India, Dec 20, 49; m 72; c 2. DIGITAL SIGNAL PROCESSING, NEURAL NETWORKS. *Educ:* Univ Madras, India, BS, 71; Indian Inst Technol, MS, 74; Univ Calgary, PhD(elec eng), 77. *Prof Exp:* Fel, Elec Eng Dept, Univ Calgary, 77-79; asst prof elec & comput eng, New Eng Col, Mass, 79-81; asst prof, elec & comput eng dept, Wayne State Univ, 81-82; asst prof, 82-85, ASSOC PROF, UNIV CINCINNATI, 85- *Concurrent Pos:* Consult, M B Electronics, 79-82; fac fel, NASA Lewis Res Ctr, Cleveland, 84-85, Rome Air Develop Ctr, Rome, NY, 87, Wright Patterson Svionics Lab, Dayton, Ohio, 88; prog dir, Circuits & Signal Processing, Nat Sci Found, 89-90. *Honors & Awards:* Elec & Comput Eng Res Award, Dept Elec & Comput Eng, Univ Cincinnati, 88. *Mem:* Inst Elec & Electron Engrs; Soc Photo-Instrumentation Engrs. *Res:* Digital signal and image processing algorithms, architectures and applications; optical signal processing and computing; neural networks; research administration; technical management; science policy. *Mailing Add:* Dept Elec & Comput Eng MS 30 Univ Cincinnati Cincinnati OH 45221-0030

RAMAMURTHY, AMURTHUR C, Indian citizen. ELECTROCHEMISTRY & ELECTRODEPOSITION, IMPEDANCE SPECTROSCOPY. *Educ:* Indian Inst Technol, Bombay, BS, 72, MS, 74; Indian Inst Sci, Bangalore, PhD(chem), 78. *Prof Exp:* Res assoc chem, Ohio State Univ, 80-81; chemist, Occidental Chem Corp, 82-84; sr res chemist, OMI Int Inc, Mich, 84-86; RES ASSOC ELECTROCHEM, COATINGS & INKS DIV, BASF CORP, 86- *Concurrent Pos:* Vis lectr, Caninus Col, Buffalo, NY, 83, Oakland Univ, Rochester, Mich, 84-86. *Mem:* Nat Asn Corrosion Engrs; Am Soc Testing Mat. *Res:* Fundamentals of corrosion beneath organic coatings and development of new techniques; fundamental aspects of depositon of polymer films; study of viscoelastic properties of coatings. *Mailing Add:* 26701 Telegraph Rd PO Box 5009 Southfield MI 48086

RAMAMURTI, KRISHNAMURTI, b Erthangal, Madras, India, Nov 11, 19; nat US; m 45; c 4. CEMENT CHEMISTRY, ASPHALT DETERIORATION. *Educ:* Madras Univ, BSc, 37; Annamalai Univ, MA, 40; Columbia Univ, MS, 55, DEd, 56; Manchester Univ, Eng, MSc Tech, 60; Kansas State Univ, PhD(foods & nutrit), 75. *Prof Exp:* From asst prof to prof chem, Birla Col, India, 48-54; teaching asst, Columbia Univ, 54-55, res assoc, 55-56; lectr org chem, Manchester Col Sci & Technol, UK, 57-58, sr res fel biochem, 58-61; sci officer food res, Cent Food Technol Res Inst, Mysore, India, 61-63; prof chem & chmn, Fac Sci, Univ Libya, Tripoli, 63-70; lectr nutrit Brooklyn Col & Kings County Med Ctr, City Univ New York, 70-71; res asst foods & nutrit, Kans State Univ, 72- 75; res chemist chem sect, Planning & Develop Dept, 76-80, head chem res unit, Bur Mat & Res, Kans Dept Transp, 80-86; RETIRED. *Concurrent Pos:* Seagrams Int fel, 50; ed, Proc Rajasthan Acad Sci, 52-54; hon reader biochem, Manchester Univ, UK, 58-60; prin investr, Kans Dept Transp, 77-80. *Mem:* Rajasthan Acad Sci India (secy, 52-54); fel Royal Inst Chem; fel Royal Soc Chem; fel Am Inst Chemists. *Res:* Protein rich foods; nutritional studies on fish flour; effect of processing ground beef on its fatty acid pattern; bacterial cellulose as a grown-in-place sealant in glycerol impregnated concrete bridge decks; microbial deterioration of asphalt; biosynthesis of cellulose. *Mailing Add:* 11103 W 121st Terr Overland Park KS 66213-1942

RAMAN, ARAVAMUDHAN, b Madras, India, Oct 13, 37; m 65; c 3. CORROSION, PHYSICAL METALLURGY. *Educ:* St Joseph's Col, India, MA, 58; Indian Inst Sci, Bangalore, BEng, 60; Tech Univ Stuttgart, Dr rer Nat(phys metall), 64. *Prof Exp:* Assoc lectr metall, Indian Inst Technol, Bombay, 61; res assoc phys metall, Univ Ill, Urbana, 64-65; fel mat sci, Univ Tex, Austin, 65-66; PROF MAT SCI, LA STATE UNIV, BATON ROUGE, 66- *Concurrent Pos:* NASA res grant, 66-68; Sea res grant, 78-81; transportation res grant, 83-86. *Mem:* Am Soc Metals; Sigma Xi; Nat Asn Corrosion Engrs. *Res:* X-ray metallography; x-ray crystallography; crystal and alloy chemistry of metallic phases; low temperature physical properties of alloys; structural imperfections and stacking faults in alloys; corrosion science and engineering; metallic coatings. *Mailing Add:* 6919 N Rothmer Dr La State Univ Baton Rouge LA 70808

RAMAN, SUBRAMANIAN, b North Parur, India, Apr 2, 38; US citizen; m 67; c 3. NUCLEAR PHYSICS. *Educ:* Univ Madras, BE, 59; Rensselaer Polytech Inst, MEE, 61; Pa State Univ, University Park, PhD(physics), 66. *Prof Exp:* res staff mem, 66-80, SR RES STAFF MEM, OAK RIDGE NAT LAB, 80- *Mem:* fel Am Phys Soc. *Res:* Nuclear spectroscopy and reactions; data compilations; heavy ion applications. *Mailing Add:* Physics Div Bldg 6010 MS 6354 Oak Ridge Nat Lab PO Box 2008 Oak Ridge TN 37831-6354

RAMAN, VARADARAJA VENKATA, b Calcutta, India; m 62; c 2. THEORETICAL PHYSICS, HISTORY OF SCIENCE. *Educ:* St Xavier's Col, India, BS, 52; Univ Calcutta, MS, 54; Univ Paris, PhD(theoret physics), 58. *Prof Exp:* Res assoc physics, Saha Inst Nuclear Physics, India, 59-60; assoc prof, Univ PR, Mayaguez, 60-63; chmn dept, Inst Telecommun, Columbia, 63-64; UNESCO expert appl math, Nat Polytech Sch, Univ Algiers, 64-66; assoc prof, 66-77, PROF PHYSICS, ROCHESTER INST TECHNOL, 77- *Mem:* Am Asn Physics Teachers; Hist Sci Soc. *Res:* Historical aspects of physics. *Mailing Add:* Dept Physics Rochester Inst Technol One Lomb Mem Dr Rochester NY 14623

RAMANAN, V R V, b Madras, India, July 5, 52; m 83; c 1. FERROMAGNETISM, AMORPHOUS MATERIALS. *Educ:* Univ Delhi, BSc, 71, MSc, 73, Carnegie-Mellon Univ, MS, 75, PhD(physics), 79. *Prof Exp:* RES ASSOC, METALS & CERAMICS LAB, ALLIED-SIGNAL INC, 79- *Mem:* Inst Elec & Electronics Engrs; Am Phys Soc; Mat Res Soc; Am Soc Metals; Metall Soc; Am Inst Mining, Metall & Petrol Engrs. *Res:* Ferromagnetic behavior, thermal and magnetic stabilities; structure-property relationships in metallic glasses; design and optimization of new magnetic materials for specific applications; stability of intermetallic phases in Al-based alloys. *Mailing Add:* Allied-Signal Inc PO Box 1021R Morristown NJ 07962-1021

RAMANATHAN, GANAPATHIAGRAHARAM V, b Madras, India; m 74; c 1. APPLIED MATHEMATICS, STATISTICAL MECHANICS. *Educ:* Madras Univ, BE, 57; Princeton Univ, PhD(aerospace), 66. *Prof Exp:* Asst lectr mech eng, Govt Col Technol, Coimbatore, 57-58; sci officer nuclear eng, Atomic Energy Estab, Bombay, 59-60; assoc res scientist math, Courant Inst Math Sci, NY Univ, 65-66; Nat Acad Sci res assoc plasma physics, Goddard Space Flight Ctr, 66-68; assoc res scientist math, Courant Inst Math Sci, NY Univ, 68-69, asst prof, 69-70; assoc prof, 70-83, PROF MATH, UNIV ILL, CHICAGO, 83- *Res:* Singular and secular perturbation theories. *Mailing Add:* Dept Math Univ Ill Box 4348 Chicago IL 60680

RAMANATHAN, M, US citizen. ENVIRONMENTAL ENGINEERING. *Educ:* Madras Univ, BE, 58, MSc, 59; Case Inst Technol, MS, 63; Okla State Univ, PhD(environ eng), 66; Am Acad Environ Engrs, dipl. *Prof Exp:* Mem, Coun Sci & Indust Res, India, 59-60; mem staff, John G Reutter Assoc, 69-71, Weston, 71-79, Environ Quality Syst, Inc, Rockville, Md, 79-80; ENGR, WESTON, 80- *Mem:* Sigma Xi; Am Soc Civil Engrs; mem, Water Pollution Control Fedn. *Res:* Wastewater treatment process development and design; conceptual process design; water quality; field surveys, including sampling, analysis and pilot-plant evaluations; and design installation; hazardous waste treatment; remedial investigations. *Mailing Add:* Weston Way West Chester PA 19380

RAMANATHAN, VEERABHADRAN, b Madras, India, Nov 24, 44; m 73; c 3. ATMOSPHERIC SCIENCE. *Educ:* Annamalai Univ, India, BE, Hons, 65; Indian Inst Sci, Bangalore, India, MSc, 70; State Univ NY, Stony Brook, PhD(atmospheric sci), 74. *Prof Exp:* Nat Acad Sci-Nat Res Coun fel atmospheric sci, NASA Langley Res Ctr, 74-75, vis scientist, 75-76; vis scientist climate, 76-77, sr scientist, 82-90, STAFF SCIENTIST, NAT CTR ATMOSPHERIC RES, 77-, LEADER CLOUD CLIMATE INTERACTIONS GROUP, 81- *Concurrent Pos:* Mem panel, comt impacts stratospheric change, AMPS, Nat Acad Sci, 78- & comt solar-terrestrial res, Geophys Res Bd, 78-; assoc ed, J Atmospheric Sci, 79-82; mem sci team, Earth Radiation Budget Satellite Exp, NASA, 79-84; mem, comt earth sci, Nat Res Coun, 81-84; mem, climate res comt, Nat Acad Sci, 83-, Panel Int Satellite Cloud Climate Proj, 84-; fac affil, Colo State Univ, Fort Collins, 85-; prof, Univ Chicago, 86. *Mem:* fel Am Meteorol Soc; AAAS; Am Geophys Union. *Res:* Climate, especially theory, modeling and cloud feedback mechanisms; stratospheric research, especially radiative-dynamic interactions, troposphere-stratosphere interactions, ozone-climate effects; atmospheric radiation, especially greenhouse effects of atmospheric trace gases such as ozone, carbon-dioxide, and chlorofluoromethanes. *Mailing Add:* Scripps Inst Oceanog Calif Space Inst Univ Calif San Diego Mierenderg Hall Rm 325 8605 La Jolla Shores Dr La Jolla CA 92093

RAMANI, RAJA VENKAT, b Madras, India, Aug 4, 38; US citizen; m 72; c 2. MINING, COMPUTER SCIENCE. *Educ:* Ranchi Univ, India, BS, 62; Indian Sch Mines, Dhanbad, AISM, 62; Pa State Univ, University Park, MS, 68, PhD(mining), 70. *Prof Exp:* Safety officer, vent officer & prod mgr, Bengal Coal Co, Andrew Yule, India, 62-66; from asst prof to assoc prof, 70-78, CHMN, MINERAL ENG MGT, PA STATE UNIV, UNIVERSITY PARK, 74-, PROF MINING ENG, 78-, HEAD DEPT, 87- *Concurrent Pos:* Proj dir develop mine vent similator, US Bur Mines, 73-77, proj dir appln total systs simulator to surface coal mining, 75-78; proj dir, Premining Planning Manual Eastern Surface Coal Mining, Environ Protection Agency, 75-78; chmn, Comt Underground Mine Disaster Survival & Rescue, Nat Acad Sci, 79-81; vis prof, Mo Sch Mines, Rolla, 80, Tech Univ Berlin, 88, Univ Rome, 88, Univ Queensland, 88; proj dir, Integration Surface Mining & Lane Use Planning, US Off Surface Mining, 79-82; dir, Ctr Excellence Longwall Mining, SOHIO, 83-89, Generic Technol Ctr Respirable Dust, 83-; co-dir, Nat Mined Land Res Ctr, 89- *Honors & Awards:* Distinguished Mem Award, Soc Mining Engrs, USA, 89; APCOM Distinguished Achievement Award, Int Coun Appln of Computers to Mineral Indust, 89; Fulbright Lectr Award Soviet Union, Ctr Int Exchange Scholars, 89-90; Environ Conserv Award, Am Inst Mining, Metall & Petrol Engrs, 90; Howard N Eavenson Award, Soc Mining, Metall & Explor, Inc, 91. *Mem:* Am Inst Mining, Metall & Petrol Engrs; Inst Mgt Sci; Am Soc Eng Educ; Mine Ventilation Soc SAfrica. *Res:* Surface mining and underground mining methods; ventilation; health and safety; computer-oriented planning and control; management; resource management; technical management. *Mailing Add:* Pa State Univ 104 Mineral Sci Bldg University Park PA 16802

RAMANUJAM, V M SADAGOPA, b July 2, 46; m 74; c 2. ORGANIC CHEMISTRY, ENVIRONMENTAL CHEMISTRY. *Educ:* Univ Madras, India, BSc, 66, MSc, 68, PhD(org chem), 73. *Prof Exp:* Instr chem, Vivekananda Col, Madras, India, 68-72; develop chemist, Res Div, Greaves Foseco, Ltd, Calcutta, India, 73; Robert A Welch Found fel, 74-78, asst prof, 79-84, ASSOC PROF, DEPT PREV MED & COMMUNITY HEALTH, ENVIRON HEALTH LAB, UNIV TEX MED BR, GALVESTON, 84- *Concurrent Pos:* Consult, Nat Acad Sci, 79-; sci adv, US Environ Protection Agency, 80-81. *Honors & Awards:* Merit Award, Govt India, 66-68. *Mem:* Sigma Xi; Am Chem Soc. *Res:* Physico-chemical characterization of toxins from Gymnodinium breve Davis; structure-activity relationship studies on drugs; carcinogens and mutagens; development of analytical methods for drugs, toxins and environmental pollutants; mutagenicity studies on atomatic hydrocarbons and amines; oxidation reaction mechanisms and syntheses. *Mailing Add:* Dept Prev Med-Comn Univ Tex Med Sch 700 Strand St Rte J10 Galveston TX 77550

RAMANUJAN, MELAPALAYAM SRINIVASAN, b Coimbatore, India, July 16, 31; m 65. MATHEMATICS. *Educ:* Annamalai Univ, Madras, BS, 51, MA, 52, MSc, 53, DSc(math), 58. *Prof Exp:* Res assoc math, Ramanujan Inst Math, 57-58; lectr, Aligarh Muslim Univ, India, 58-59; from instr to assoc prof, 59-72, PROF MATH, UNIV MICH, ANN ARBOR, 72- *Concurrent Pos:* Reader, Ramanujan Inst Math, 61-63; Humboldt fel, Univ Frankfurt, 69-70, 83 & 87. *Honors & Awards:* Narasinga Rao Gold Medal, Indian Math Soc, 53. *Mem:* Am Math Soc; Math Asn Am; Indian Math Soc (secy, 62-63). *Res:* Summability; moment problems; topological vector spaces; duality theory; abstract sequence spaces. *Mailing Add:* 2435 Praire Ann Arbor MI 48105

RAMAPRASAD, K R (RAM), b Bangalore, India, Dec 8, 38; c 1. ELECTROCHEMISTRY. *Educ:* Univ Mysore, Bangalore, India, BSc(hons), 58; NY Univ, MS, 71, PhD(phys chem), 72. *Prof Exp:* Res assoc, dept chem, Princeton Univ, 74-77, mem res staff, dept chem eng, 77-79; SR SCIENTIST, CHRONAR CORP, PRINCETON, NJ, 79- *Mem:* Am Chem Soc; Sigma Xi. *Res:* Photochemistry; spectroscopy. *Mailing Add:* Four Cresthill Rd Lawrenceville NJ 08648

RAMASASTRY, SAI SUDARSHAN, b Channapatna, India, May 3, 45; US citizen; m 82. MICROSURGERY, HAND SURGERY. *Educ:* Nat Col, Bangalore Univ, India, AB, 61; Bangalore Med Col, MBBS, 67; FRCSEd. *Prof Exp:* Resident surg, Univ Conn Hosp, 75-79, chief resident, 79-80; chief resident plastic surg, 80-82, res fel, 82-83, clin instr surg, 83-84, ASST PROF SURG, SCH MED, UNIV PITTSBURGH, 84- *Concurrent Pos:* Chmn, Hyperbaric Oxygen Prog Task Force, Presby-Univ Hosp, 85-87, Hyperbaric Oxygen Comt, 87- & mem, Trauma Comt, 87-; assoc ed, J Clin Anat, 87-; mem, Falk Clin Mgt Comt, 87-; mem, Surg Res Lab Comt, Univ Pittsburgh, 87- & Anat Lab Comt, 88- *Mem:* Am Soc Plastic & Reconstructive Surgeons; Am Asn Clin Anatomists; Undersea & Hyperbaric Med Soc; AMA. *Res:* Etiology and prevention of necrosis of flaps; skin, muscle and myocutaneous, to enhance their survival and facilitate reconstruction of major tissue defects; anatomic studies to develop new flap applications. *Mailing Add:* Div Plastic Surg Sch Med Univ Pittsburgh 1117 Scaife Hall Pittsburgh PA 15261

RAMASWAMI, DEVABHAKTUNI, b Pedapudi, India, Apr 4, 33; m; c 1. CHEMICAL ENGINEERING. *Educ:* Andhra, India, BSc, 53, MSc, 54, DSc, 58; Univ Wis, PhD(chem eng), 61. *Prof Exp:* Res scholar chem eng, Andhra, India, 54-56; Indian Inst Technol, Kharagpur, 56-57; asst prof, Banaras Hindu Univ, 57-58; res asst, Univ Wis, 58-61; res engr, Int Bus Mach

Corp, 61-62; res assoc, 62, CHEM ENGR, ARGONNE NAT LAB, 62- *Honors & Awards:* Am Chem Soc Award, 60. *Mem:* Fel Am Inst Chem Engrs. *Res:* Nuclear reactor core; development; engineering; author or coauthor of over 93 publications; industrial chemical reactions and petroleum refining. *Mailing Add:* PO Box 3029 Westmont IL 60559-8029

RAMASWAMI, VAIDYANATHAN, b Kerala, India, Feb 24, 50; m 77; c 2. OPERATIONS RESEARCH, STATISTICS. *Educ:* Univ Madras, BSc, 69, MSc, 71; Purdue Univ, MS, 76, PhD(opers res), 78. *Prof Exp:* Lectr statist, Loyola Col, Madras, India, 71-74; ASST PROF MATH, DREXEL UNIV, 78- *Concurrent Pos:* Statist consult, Madras, Ctr Soc Med & Community Health, Jawaharlal Nehru Univ, New Delhi, India, 72-74. *Mem:* Opers Res Soc Am. *Res:* Stochastic processes; computational probability; queueing theory; mathematical programming; discrete optimization. *Mailing Add:* 575 Hillcrest Dr Neshanic Station NJ 08853

RAMASWAMY, H N, b Honnavally, India, Oct 30, 37; m 66; c 2. INORGANIC CHEMISTRY, ANALYTICAL CHEMISTRY. *Educ:* Univ Mysore, BSc, 58; Karnatak Univ, India, MSc, 61; Tulane Univ, PhD(inorg chem), 67. *Prof Exp:* Teacher, Govt High Sch, India, 58-59; lectr chem, APS Col, Bangalore, 61-63; lectr, Tulane Univ, 63-67; res assoc, Southern Regional Res Lab, USDA, La, 67-69; sr chemist, Thiokol Chem Corp, Ga, 69-70; head anal labs, AZS Chem Corp, 70-79, supt process eng & process develop, 79-86; DEVELOP ASSOC, NAT STARCH & CHEM CO, 86- *Concurrent Pos:* NSF fel, 67-69. *Mem:* Am Chem Soc; Sigma Xi. *Res:* Spectroscopy; infrared chemical analysis; pyrolysis and gas-liquid chromatography; textile chemicals and polymers; liquid chromatography, amines, alkyd resins, hydrogenation, distillation product and process development, thermal analysis, glycidyl ethers; process improvement and development; scale-up. *Mailing Add:* Nat Starch & Chem Co Cedar Springs Rd Salisbury NC 28144

RAMASWAMY, KIZHANATHAM V, b Jalarpet, India, July 17, 35; m 65; c 1. INDUSTRIAL ENGINEERING. *Educ:* Univ Madras, BE, 57; Tex Tech Univ, MS, 59, PhD(indust eng), 71. *Prof Exp:* Trainee nuclear eng, Bhaba Atomic Res Ctr, Bombay, India, 58-59, design engr, 59-61, fabrication engr, 61-64, plant engr, Radiochem Plant, 64-66, asst plant supt, 66-67; asst prof prod mgt, 71-80, ASST PROF GEN BUS, TEX SOUTHERN UNIV, 80- *Mem:* Am Inst Indust Engrs; Soc Mfg Engrs; Indian Inst Eng. *Res:* Manufacturing science; operations research; engineering analysis and design. *Mailing Add:* Dept Bus Tex Southern Univ 3100 Cleburne Ave Houston TX 77004

RAMASWAMY, VENKATACHALAM, b Madras, India, Apr 28, 55; m 86. ATMOSPHERIC RADIATION, CLIMATE PROCESSES. *Educ:* Univ Delhi, India, BSc, 75, MSc, 77; State Univ Ny, Albany, PhD(atmospheric sci), 82. *Prof Exp:* Res assoc, State Univ NY, 83, fel, Advan Study Prog, 83-85; res staff mem, 85-89, RES SCIENTIST, GEOPHYS FLUID DYNAMICS LAB, PRINCETON UNIV, 89- *Concurrent Pos:* Mem, World Meteorol Orgn Proj Intercomparison Radiation Codes Climate Models, 83-, Int Aerosol Climatol Proj, 87-, Ozone Trends Panel, World Meteorol Soc & NASA, 88-, Earth Observ Systs Mission, 88- & Alternative Fluorocarbon Environ Acceptability Study, 89. *Mem:* Am Geophys Union; Optical Soc Am. *Res:* Transfer of radiation in scattering-absorbing atmospheres; radiative and climatic effects of aerosols, clouds and gases; interaction of radiation with other mechanisms such as microphysical, chemical and dynamical processes; cloud-climate interactions; general circulation modeling of the earth's atmosphere and atmospheric transport of species; investigation of past, present and future climatic perturbations, naturally and anthropogenically induced. *Mailing Add:* Atmospheric & Oceanic Sci Prog Princeton Univ PO Box 308 Princeton NJ 08542

RAMATY, REUVEN, b Timisoara, Rumania, Feb 25, 37; m 61; c 2. ASTROPHYSICS. *Educ:* Tel-Aviv Univ, BSc, 61; Univ Calif, Los Angeles, PhD(space sci), 66. *Prof Exp:* Asst res geophysicist, Inst Geophys & Planetary Physics, Univ Calif, Los Angeles, 66-67; Nat Res Coun resident res assoc astrophys, 67-69, astrophysicist, 69-80, HEAD, THEORY OFF, LAB HIGH ENERGY ASTROPHYS, GODDARD SPACE FLIGHT CTR, NASA, 80- *Concurrent Pos:* Vis scientist, Stanford Univ, 72; vis prof physics, Wash Univ, 78; Fairchild Scientist, Calif Inst Technol, 79. *Honors & Awards:* Lindsay Award, Goddard Space Flight Ctr, NASA, 80, Exceptional Sci Achievement Medal, 81; Sr US Scientist Award, Alexander von Humboldt Found, Fed Repub Ger, 75. *Mem:* Am Astron Soc; Int Astron Union; fel Am Phys Soc. *Res:* High energy astrophysics; solar physics; gamma-ray astronomy. *Mailing Add:* Goddard Space Flight Ctr NASA Code 665 Greenbelt MD 20771

RAMAYYA, AKUNURI V, b Bezwada, India, Aug 15, 38; m 65; c 1. EXPERIMENTAL NUCLEAR PHYSICS. *Educ:* Andhra Univ. India, BSc, 57, MSc, 58; Ind Univ, PhD(physics), 64. *Prof Exp:* Asst physics, Ind Univ, Bloomington, 60-64; res assoc, 64-70, asst prof, 70-75, assoc prof, 75-80, PROF NUCLEAR PHYSICS, VANDERBILT UNIV, 80- *Concurrent Pos:* Alexander von Humboldt fel, 81-82. *Mem:* Sigma Xi; Am Phys Soc. *Res:* Heavy ion nuclear physics. *Mailing Add:* Box 1807 Stat B Vanderbilt Univ Nashville TN 37235

RAMAZZOTTO, LOUIS JOHN, b New York, NY, Dec 18, 40; m 66; c 2. PHYSIOLOGY. *Educ:* Fairleigh Dickinson Univ, BS, 62; Fordham Univ, MS, 64, PhD(physiol), 66. *Prof Exp:* Lab instr biol, Fairleigh Dickinson Univ, 62-63; lectr, St Peters Col, NJ, 63-64; lectr physiol, Hunter Col, 64-66; asst prof, Marymount Col, NY, 66-67; from asst prof to prof physiol, Sch Dent & Grad Sch, Fairleigh Dickinson Univ, 74-88, chmn dept, 67-88; DIR, RES SERV, LONG ISLAND JEWISH MED CTR, 88- *Concurrent Pos:* Coun Accrediation, Am Asn Accrediation Lab Animal Care. *Mem:* AAAS; NY Acad Sci; Am Phys Soc; Fed Am Soc Exp Biol; Int Asn Dent Res; Am Asn Lab Animal Sci. *Res:* Effects of nitrous oxide and other inhalation anesthetics on blood and reproductive system. *Mailing Add:* Long Island Jewish Med Ctr Res Serv Rm 133 New Hyde Park NY 11042

RAMBAUT, PAUL CHRISTOPHER, b Southampton, Eng, May 23, 40; US citizen; m 76. FOREIGN POLICY, SCIENCE ADMINISTRATION. *Educ:* McGill Univ, BSc, 62, MSc, 64; Mass Inst Technol, ScD, 66; Harvard Univ, MPH, 68. *Prof Exp:* Instr nutrit, Mass Inst Technol, 66; biochemist, Miami Valley Labs, Procter & Gamble Co, 66-67; biochemist, Johnson Space Ctr, NASA, 68-75; from asst to assoc dir, Bur Foods, Food & Drug Admin, 75-76; chief, Med Res Br, Johnson Space Ctr, 76-79, mgr Biomed Res, Hq, NASA, 79-85, sr physiologist, Off of Space Sta, 85-86; dep dir, Div Extramural Activities, Nat Cancer Inst, NIH, 86-89; DEP ASST SECY GEN, SCI & ENVIRON, NATO, BRUSSELS, BELGIUM, 89- *Honors & Awards:* Underwood Prescott Award, Mass Inst of Technol, 74; Eric Liljencrantz Award, Aerospace Med Asn, 82; Blue Pencil Award, Nat Med Asn, 82. *Mem:* Fel Aerospace Med Asn; Am Foreign Serv Asn; Am Physiol Soc; Am Soc Clin Nutrit; Int Acad Astronaut; Int Acad Aviation & Space Med. *Res:* Space medicine; nutrition and musculoskeletal function. *Mailing Add:* US Mission NATO Box 88 APO New York NY 09667

RAMBERG, CHARLES F, JR, BIOMEDICAL ENGINEERING. *Educ:* Rutgers Univ, BS; Univ Pa, VMD. *Prof Exp:* PROF NUTRIT, DEPT CLIN STUDIES, SCH VET MED, UNIV PA, 82- *Concurrent Pos:* Chief Sect Nutrit, Sch Vet Med, Univ Pa, dir, Ctr Animal Health & Productivity, lectr, course organizer; chair, Computerization Task Force, New Bolton Ctr. *Res:* Mathematical modeling; kinetic analysis; computer applications in veterinary medicine, agriculture and biology; mineral and trace element nutrition; disorders of metabolism and homeostasis; numerous publications. *Mailing Add:* Sch Vet Med Univ Pa 382 West Street Rd Kennett Square PA 19348

RAMBERG, STEVEN ERIC, b Boston, Mass, Jan 4, 48; m 67; c 2. MECHANICAL ENGINEERING, FLUID MECHANICS. *Educ:* Univ Lowell, BS, 70, MS, 72; Cath Univ Am, PhD(mech eng), 78. *Prof Exp:* Res engr fluid mech, Naval Res Lab, 72-88; RES MGR OCEAN ENG, OFF NAVAL RES, 88- *Honors & Awards:* Moisseif Award, Am Soc Civil Engrs, 79. *Mem:* Am Soc Mech Engrs; Sigma Xi. *Res:* Flow-induced vibrations; bluff body wakes; ocean wave forces; cable dynamics; wind-wave growth; stratified flows. *Mailing Add:* 7611 Range Rd Alexandria VA 22306

RAMBOSEK, G(EORGE) M(ORRIS), chemical engineering, for more information see previous edition

RAMDAS, ANANT KRISHNA, b Poona, India, May 19, 30; m 56. SOLID STATE PHYSICS, OPTICS. *Educ:* Univ Poona, BSc, 50, MSc, 53, PhD(physics), 56. *Prof Exp:* Res assoc physics, 56-60, from asst prof to assoc prof, 60-67, PROF PHYSICS, PURDUE UNIV, LAFAYETTE, 67- *Concurrent Pos:* Alexander von Humboldt US sr scientist, 77-78. *Mem:* Fel Am Phys Soc; fel Indian Acad Sci. *Res:* Spectroscopy; application of spectroscopic techniques to solid state physics; electronic and vibrational spectra of solids studied by absorption and emission spectra in the visible and the infrared and by laser Raman spectroscopy. *Mailing Add:* Dept Physics Purdue Univ Lafayette IN 47907

RAMER, LUTHER GRIMM, b Pawpaw, Ill, May 24, 08; m 35; c 1. ACOUSTICS. *Educ:* Univ Ill, BS, 30, MS, 34. *Prof Exp:* Engr, Bell Tel Labs, 30-32; res engr, Univ Ill, 34-36; res physicist, Riverbank Acoust Labs, Armour Res Found, 36-47, lab supvr acoust, 47-54; res engr, Mech Div, Gen Mills, Inc, 54-60; mgr acoust lab, Wood Conversion Co, 60-62; sr res engr, Trane Co, 62-74; ACOUSTICAL CONSULT, 74- *Mem:* Acoust Soc Am. *Res:* Developmental research in architectural acoustics and acoustical materials; acoustics related to sounds of air conditioning equipment. *Mailing Add:* 373 Lunar Dr Ft Myers FL 33908

RAMETTE, RICHARD WALES, b Stafford Springs, Conn, Oct 9, 27; m 49; c 5. COULOMETRY, SOLUBILITY. *Educ:* Wesleyan Univ, BA, 50; Univ Minn. PhD(chem), 54. *Prof Exp:* From asst prof to assoc prof chem, 54-65, chmn dept, 60-72, dir off sci activ, 69-72, PROF CHEM, CARLETON COL, 65- *Concurrent Pos:* Vis scholar, St Olaf Col, 62-63; resident res assoc, Argonne Nat Lab, 66-67; sci adv, US Food & Drug Admin, 69-80 & Oak Ridge Nat Lab, 83-84; vis prof, Univ Fla, 75-76; chmn, Am Chem Soc, Div of Chem Educ, 77. *Honors & Awards:* Col Chem Teachers Award, Mfr Chemists Asn, 66. *Mem:* Am Chem Soc. *Res:* Aqueous equilibria; solution thermodynamics. *Mailing Add:* Dept Chem Carleton Col Northfield MN 55057

RAMEY, CHESTER EUGENE, b Santa Maria, Calif, Jan 15, 43; m 64; c 3. ORGANIC CHEMISTRY. *Educ:* Univ Calif, Berkeley, BS, 64; Univ Ore, PhD(org chem), 68. *Prof Exp:* Sr res chemist, Plastics & Additives Div, Ciba-Geigy Corp, 68-76; group leader synthesis, Ferro Corp, 76-82, res mgr, Bedford Chem Div, 82-84, tech dir, Bedford Chem Div, 84-88; TECH MGR, NEW VENTURES GROUP, LUBRIZOL CORP, 88- *Mem:* Am Chem Soc; Soc Plastics Engrs. *Res:* Polymer additives; antioxidants; ultraviolet stabilizers; heat stabilizers; stabilization and degradation of polymers. *Mailing Add:* Lubrizol Corp 29400 Lakeland Blvd Wickliffe OH 44092

RAMEY, DANIEL BRUCE, b Shelby, Mich, Dec 11, 49; m 75; c 1. CLUSTER ANALYSIS. *Educ:* Mich State Univ, BA, 71, MS, 73; Yale Univ, MPhil, 79, PhD(statist), 82. *Prof Exp:* Analyst, Gerber Prod Co, 74-77; res asst, Yale Univ, 79-81; SCIENTIST, LOCKHEED-EMSCO, 81- *Concurrent Pos:* Lectr, Southern Conn State Univ, 80 & Univ Houston, 81- *Mem:* Am Statist Asn; Inst Math Statist; Royal Statist Soc. *Res:* Cluster analysis techniques and computing algorithms; application of statistical techniques to remote sensor data. *Mailing Add:* Lockheed Corp 4500 Park-Grenada Blvd Calabasas CA 91399

RAMEY, ESTELLE R, b Detroit, Mich, Aug 23, 17; m 41; c 2. PHYSIOLOGY, ENDOCRINOLOGY. *Educ:* Columbia Univ, MA, 40; Univ Chicago, PhD(physiol), 50. *Hon Degrees:* Numerous from US univs. *Prof Exp:* Tutor chem, Queens Col, NY, 38-41; lectr, Univ Tenn, 42-47; instr physiol, Univ Chicago, 51-54, asst prof, 54-58; from asst prof to assoc prof, 56-66, prof biophys, 80-87, PROF PHYSIOL, SCH MED, GEORGETOWN

UNIV, 66-, EMER PROF BIOPHYS, 87- *Concurrent Pos:* USPHS fel, Univ Chicago, 50-51; Mem adv bd, Planned Parenthood, Dir of NIH & Health & Human Serv; mem bd dirs, Asn Women Sci & Admiral H G Rickover Found; pres, Asn Women Sci, 72-74; founder & pres, Asn Women Sci Educ Found; mem, Comt for Women Vet, US Vet Admin, President's Adv Comt Women, Exec Adv Panel to the Chief of Naval Opers & Gen Med Study Sect, NIH; vis prof & lectr at several universities. *Mem:* Am Physiol Soc; Am Chem Soc; Endocrine Soc; Am Diabetes Asn; Am Acad Neurol. *Res:* Endocrinology metabolism chiefly in the field of adrenal function; sex hormones and longevity. *Mailing Add:* Dept Physiol 247 Basic Georgetown Univ 37th & O Sts NW Washington DC 20057

RAMEY, H(ENRY) J(ACKSON), JR, b Pittsburgh, Pa, Nov 30, 25; m 48; c 4. CHEMICAL ENGINEERING. *Educ:* Purdue Univ, BS, 49, PhD(chem eng), 52. *Prof Exp:* Asst chem eng, Unit Opers Lab, Purdue Univ, 49; asst radiant heat transfer from gases, 51-52; sr res technologist, petrol prod res, Magnolia Petrol Co, Socony Mobil Oil Co, Inc, 52-55, proj engr, Gen Petrol Corp, 55-60, staff reservoir engr, Mobil Oil Co Div, 60-63; prof petrol eng, Tex A&M Univ, 63-66; Keleen & Carlton Beal prof petrol eng & chmn dept, 66-86, PROF PETROL & ENG, STANFORD UNIV, 86- *Concurrent Pos:* Consult, Chinese Petrol Corp, Taiwan, 62-63; Endowed prof, 81. *Honors & Awards:* Ferguson Medal, Am Inst Mining, Metall & Petrol Engrs, 59, Lucas Gold Medal, 83; Uren Award, Soc Petrol Engrs, 73, Carll Award, 75; Nat Acad Eng mem, 81. *Mem:* Nat Acad Eng; hon mem Am Inst Mining, Metall & Petrol Engrs; Am Inst Chem Engrs; Soc Petrol Engrs. *Res:* Heat transfer, thermodynamics; fluid flow; petroleum production. *Mailing Add:* Dept Petrol Eng Stanford Univ Stanford CA 94305

RAMEY, HARMON HOBSON, JR, b Russell, Ark, Dec 4, 30; m 54; c 2. FIBER SCIENCE. *Educ:* Univ Ark, BSA, 51, MS, 52; NC State Col, PhD(plant breeding, genetics), 59. *Prof Exp:* Asst, Univ Ark, 51-52; asst cotton geneticist, Delta Br Exp Sta, Miss State Univ, 55-57 & 59-61; asst, NC State Col, 57-59; geneticist & fiber scientist, Nat Cotton Coun Am, Tenn, 61-70; res geneticist, Agr Res Serv, USDA, 70-84; AGR MKT SPECIALIST, AMS, USDA, 84- *Mem:* Am Soc Qual Control; Fiber Soc; Sigma Xi. *Res:* fiber & textiles technology; relationship of fiber properties to processing performance and product quality; fiber property measurement. *Mailing Add:* USDA AMS CN Div 4841 Summer Ave Memphis TN 38122

RAMEY, MELVIN RICHARD, b Pittsburgh, Pa, Sept 13, 38; m 64; c 2. CIVIL ENGINEERING, BIOMECHANICS. *Educ:* Pa State Univ, BS, 60; Carnegie-Mellon Univ, MS, 65, PhD(civil eng), 67. *Prof Exp:* Bridge design engr, Pa State Dept Hwys, 60-63; res asst, Carnegie-Mellon Univ, 63-67; from asst prof to assoc prof, 67-73, PROF CIVIL ENG, UNIV CALIF, DAVIS, 73- *Concurrent Pos:* Consult, Calif State Div Hwys, 68-69, Murray & McCormick Consult Engrs, 69, Fireman's Fund Am Ins Co, 70 & var archit design firms, 71-; consult, various architectural design firms, 70- *Mem:* Am Soc Civil Engrs; Am Concrete Inst; Int Soc Biomech in Sports. *Res:* Structural design and analysis; materials behavior and testing; biomechanics with applications to human movement and sports; computer aided structural design; fiber reinforced concrete. *Mailing Add:* Dept Civil Eng Univ Calif Davis CA 95616

RAMEY, ROBERT LEE, b Middletown, Ohio, June 26, 22; m 46; c 2. ENGINEERING PHYSICS. *Educ:* Duke Univ, BSEE, 45; Univ Cincinnati, MS, 47; NC State Col, PhD(elec eng, physics), 54. *Prof Exp:* Asst elec eng, Univ Cincinnati, 46-48; instr, NC State Col, 49-54; res lab dir, Wright Mach Div, Sperry-Rand Corp, 54-56; assoc prof elec eng, 56-62, PROF ELEC ENG, UNIV VA, 62- *Concurrent Pos:* Ed, Encycl Sci & Technol, 59; NASA res grant, 62-70. *Mem:* Am Phys Soc; Inst Elec & Electronics Engrs. *Res:* Physical electronics, including vacuum, gaseous and solid state. *Mailing Add:* Dept Elec Eng Thornton Hall C257 Univ Va Charlottesville VA 22903

RAMEY, ROY RICHARD, b Kansas City, Mo, July 11, 47; m 70; c 2. TECHNICAL CERAMICS. *Educ:* Univ Mo, Rolla, BS, 70, MS, 72, PhD(ceramic eng), 74. *Prof Exp:* Res engr, Inland Steel Co, 74-79; SR RES MGR, A P GREEN REFRACTORIES CO, 79- *Concurrent Pos:* Mem Ceramic Educ Coun, 85- *Mem:* Am Ceramic Soc; Nat Inst Ceramic Engrs; Keramos; Am Soc Testing & Mat. *Res:* Sintering of fine grained oxides to near theoretical density; development of dense structural ceramics. *Mailing Add:* A P Green Refractories Co Green Blvd Mexico MO 65265

RAMEZAN, MASSOOD, US citizen; m 87. COMBUSTION, THERMAL FLUID SCIENCE. *Educ:* WVa Univ, BS, 77, MS, 79, PhD(mech eng), 84. *Prof Exp:* Consult engr, Hosp Utility Inc, 79-81; res fel engr, WVa Univ, 81-84, asst prof, 84-86; res engr, ORAU/METC, 86-88; sr engr, 88-90, PRIN ENGR, BURNS & ROE SERV CORP, 90- *Mem:* Fel Am Soc Mech Engrs; Soc Automotive Engrs. *Res:* Conducted research in areas of fluid, heat transfer, combustion and its engineering applications; numerical modeling and computer simulation. *Mailing Add:* 116 Copperwood Dr Bethel Park PA 15102

RAMFJORD, SIGURD, b Kolvereid, Norway, June 6, 11; nat US; m 56; c 1. DENTISTRY, PERIODONTOLOGY. *Educ:* Univ Mich, MS, 48, PhD, 51. *Hon Degrees:* DMD, Univ Geneva, 78, Dr Odontol, Gothenburg Univ, 80 & Oslo Univ, 81. *Prof Exp:* Pvt pract, Oslo, Norway, 34-46; prof dent, 58-80, EMER PROF, SCH DENT, UNIV MICH, ANN ARBOR, 80- *Concurrent Pos:* Consult, Vet Admin Hosp, Ann Arbor & WHO, India, emer nat consult, US Air Force. *Honors & Awards:* Basic Res Award, Int Asn Dent Res, 68; William J Gies Found Award, 71. *Mem:* Am Dent Asn; Am Acad Periodont; Am Acad Oral Path; NY Acad Sci; Int Asn Dent Res; Sigma Xi. *Res:* Periodontics; occlusion; electromyography; radioisotopes. *Mailing Add:* 393 Lake Park Lane Sch Dent Ann Arbor MI 48103

RAMIG, ROBERT E, b McGrew, Nebr, June 22, 22; m 43; c 3. SOIL CONSERVATION, SOIL FERTILITY. *Educ:* Univ Nebr, BSc, 43, PhD(soils), 60; Wash State Univ, MSc, 48. *Prof Exp:* Asst agronomist, Exp Sta, Univ Nebr, 48-51; coop agent, Exp Sta, Univ Nebr & USDA, 51-57; soil scientist, Agr Res Serv, 57-71, dir, Columbia Plateau Conserv Res Ctr, 71-81, RES SOIL SCIENTIST, AGR RES SERV, USDA, 81- *Mem:* Am Soc Agron; Soil Sci Soc Am; Soil Conserv Soc Am; AAAS. *Res:* Soil and water conservation using balanced fertility to give maximum production per unit of water. *Mailing Add:* 1208 NW John Ave Pendleton OR 97801-1261

RAMILINGAM, SUBBIAH, b Udumalpet, India, June 15, 35; m 67. MECHANICAL ENGINEERING, MATERIALS ENGINEERING. *Educ:* Indian Inst Technol, Khapagpur, India, BTech Hons, 56; Univ Ill, Urbana, MS, 61, PhD(mech eng), 67. *Prof Exp:* Instr, Univ Ill, Urbana, 61-67, asst prof, 67-68; from asst prof to prof mech eng, State Univ NY, Buffalo, 68-77; PROF MECH ENG, GA INST TECHNOL, 77- *Concurrent Pos:* Vis prof, Monash Univ, Australia, 75-76. *Mem:* Am Inst Mining, Metall & Petrol Engrs; Am Soc Metals; Soc Mfg Engrs; Japan Soc Precision Engrs; Am Soc Mech Engrs. *Res:* Machining theory; theory of tool wear; tribology; deformation processing; alloy design for processing; thin film science and technology; materials conservation through thin film technology; electron microscopy of metals; magnetron melting and plasma processing. *Mailing Add:* Dept Mech Eng Univ Minn 111 Church St SE Rm 215 Minneapolis MN 55455

RAMIREZ, ARTHUR P, b Amityville, NY, Aug 4, 56; m 86; c 1. SOLID STATE PHYSICS. *Educ:* Yale Univ, BS, 78, PhD(physics), 84. *Prof Exp:* Postdoctoral mem tech staff, 84-86, MEM TECH STAFF, AT&T BELL LABS, MURRAY HILL, NJ, 86- *Mem:* Am Phys Soc. *Res:* Condensed matter experimental physics, especially in the field of magnetism and superconductivity in novel materials; heavy fermion systems; low-dimensional magnetism; spin glass; high-Tc superconductivity. *Mailing Add:* AT&T Bell Labs Rm 1B-120 600 Mountain Ave Murray Hill NJ 07974-2070

RAMIREZ, DONALD EDWARD, b New Orleans, La, May 21, 43; div; c 3. ABSTRACT HARMONIC ANALYSIS, COMPUTATIONAL STATISTICS. *Educ:* Tulane Univ, BS, 63, PhD(math), 66. *Prof Exp:* Off Naval Res fel & res assoc, Univ Wash, 66-67; asst prof, 67-71, ASSOC PROF MATH, UNIV VA, 71- *Mem:* Am Status Asn. *Res:* Measure algebras factor analysis. *Mailing Add:* Dept Math Univ Va Charlottesville VA 22903

RAMIREZ, FAUSTO, b Zulueta, Cuba, June 15, 23; nat US; m 47; c 2. ORGANIC CHEMISTRY. *Educ:* Univ Mich, BS, 46, MS, 47, PhD(org chem), 49. *Prof Exp:* McConnell fel, Univ Va, 49-50; from instr to asst prof chem, Columbia Univ, 50-58; assoc prof, Ill Inst Technol, 58-59; prof, 59-85, EMER PROF CHEM, STATE UNIV NY, STONY BROOK, 85- *Concurrent Pos:* Lectr, Gordon Res Conf Org Reactions Processes, 59, Org Reactions, 62 & Heterocyclics, 66; Sloan fel, 61-63; lectr, symp organophosphorus compounds, Int Union Pure & Appl Chem, Ger, 64; colloquium phosphorus chem, Nat Ctr Sci Res, Toulouse, France, 65; NSF fel, 65-66; distinguished res fel, Res Found, State Univ NY, 67; plenary lectr, Int Colloquium Phosphorus, Paris, 69; frontier-in-chem lectr, Case Western Reserve Univ, 70; lectr, Conf Org Reaction Mech, Univ Calif, Santa Cruz, 70; plenary lectr, All Union Conf Organophosphorous Chem, Moscow, 72, Int Colloquium Phosphorus, Gdansk, Poland & Int Conf on Oligonucleatide Synthesis, Poznan, Poland, 74, Dymaczewo, Poland, 76, Int Conf on Phosphorus, Halle, East Ger, 79 & Int Conf on Phosphorus Chem, Durham, NC, 81; Alexander von Humboldt Found award, Munich Tech Univ, 73-74. *Honors & Awards:* Silver Medal, City of Paris, 69; A Cresy-Morrison Award, NY Acad Sci, 68. *Mem:* Am Chem Soc; fel NY Acad Sci. *Res:* Theoretical and practical aspects of the chemistry of phosphorus and sulfur compounds; organic synthesis; molecular biology. *Mailing Add:* 526 Fearington Post Staunton VA 24401-2160

RAMIREZ, FRANCISCO, HUMAN MOLECULAR GENETICS, GENE EVOLUTION. *Educ:* Univ Sci, Palarmo, Italy, PhD(genetics), 69. *Prof Exp:* Assoc prof molecular genetics, Rutgers Univ, 79-86; PROF HUMAN GENETICS, STATE UNIV NY HEALTH SCI CTR, 86- *Res:* Connective tissue disorders. *Mailing Add:* Dept Microbiol State Univ NY Health Sci Ctr 450 Clarkson Ave Brooklyn NY 11203

RAMIREZ, GUILLERMO, b Bogota, Colombia, Sept 19, 34; US citizen; m 57; c 2. ONCOLOGY. *Educ:* Nat Col St Bartholomew, BS, 51; Nat Univ Colombia, MD, 58. *Prof Exp:* ASSOC PROF HUMAN ONCOL, SCH MED, UNIV WIS-MADISON, 71- *Concurrent Pos:* Consult, Vet Admin Hosps, 68-; prin investr, Cent Oncol Group, 72- *Mem:* Am Asn Cancer Res; Am Soc Clin Oncol; Int Asn Study Lung Cancer; NY Acad Sci; Am Asn Study Neoplastic Dis. *Res:* Clinical-pharmacological studies; phase I, II and III drug studies. *Mailing Add:* Dept Oncol Univ Wis Sch Clin Ctr 600 Highland Ave Madison WI 53792

RAMIREZ, J ROBERTO, b Ponce, PR, Feb 17, 41; US citizen; m 72; c 1. BIO-ORGANIC CHEMISTRY. *Educ:* Univ Notre Dame, BSc, 63; Univ PR, MSc, 66; Univ Karlsruhe, Ger, Dr rer nat, 70. *Prof Exp:* Fel, Swiss Fed Inst Technol, 71-72; CHMN & ASSOC PROF CHEM, UNIV PR, RIO PIEDRAS CAMPUS, 75- *Mem:* Soc Chemists PR (secy, 73-74, pres-elect, 74-75, pres, 75-); Am Chem Soc. *Res:* Biosynthesis of acyclic carotenes; synthesis of carotenoids and model compounds. *Mailing Add:* 32 Belen St Alturas de San Patricio Rio Piedras PR 00920

RAMIREZ, W FRED, b New Orleans, La, Feb 19, 41; m 63; c 3. CHEMICAL ENGINEERING. *Educ:* Tulane Univ, BS, 62, MS, 64, PhD(chem eng), 65. *Prof Exp:* From asst prof to assoc prof, 65-75, chmn dept, 71-79, Croft res prof, 80, PROF CHEM ENG, UNIV COLO, BOULDER, 75- *Concurrent Pos:* Fulbright res fel, France, 76; fac fel, Univ Colo, 85; vis scientist, Mass Inst Technol, 85-; fel, Acad Sci Exchange, to Soviet Union, 87. *Honors & Awards:* Dow Award, Am Soc Eng Educ, 74; Levey Award, Tulane Univ, 74; Western Elec Award, Am Soc Eng Educ, 80; Col Eng Fac Res Award, Univ Colo, 86; Fel, Am Inst Chem Engrs, 90. *Mem:* Am Inst Chem Engrs; Am Soc Eng Educ; Soc Petrol Engrs. *Res:* Optimal control of chemical, biochemical and energy processes. *Mailing Add:* Dept Chem Eng Univ Colo Boulder CO 80309-0424

RAMIREZ-RONDA, CARLOS HECTOR, b Mayaquez, PR, Jan 24, 43; US citizen; m 63; c 2. INFECTIOUS DISEASES. *Educ:* Northwestern Univ, Chicago, BSM, 64, MD, 67. *Prof Exp:* Res fel infectious dis, Southwestern Med Sch, Univ Tex, 73-75; asst prof med, Sch Med, Univ Puerto Rico, 75-78; assoc chief staff res & develop, Vet Admin Med Ctr, San Juan, 75-90; assoc prof med, 78-84, DIR INFECTIOUS DIS, SCH MED, VET ADMIN HOSP, 78-, PROF MED, 84-, CHIEF DEPT MED, SAN JUAN DVA MED CTR, 90-; CHIEF, INFECTIOUS DIS RES LAB, VET ADMIN MED CTR, 76- *Concurrent Pos:* Vis prof, Autonomous Univ, Mex, 78, 79 & 80; mem bacteriol & mycol study sect, NIH, 81-85; consult infectious dis, San Juan City Hosp, 76-; investr, Am Heart Asn & PR Heart Asn, 77-; assoc ed, Puerto Rico Med Asn J, 78-; prog dir infectious dis, Univ Hosp, San Juan, 78-; vis prof, Univ Cent Colombia, Bogotá, 88; mem, Adv Comt Immunization Pract, 90-94. *Mem:* Infectious Dis Soc; Am Fedn Clin Res; Am Soc Microbiol; Am Col Physicians; Int Soc Infectious Dis; Asn Panam Infectol; Am Med Asn; NY Acad Sci; AAAS; Am Pub Health Asn. *Res:* Pathogenesis of bacterial diseases especially adherence and bacterial endocarditis; clinical microbiology; microbial susceptibility and resistance; antibiotic pharmacology; seroepidemiology; clinical studies on new antimicrobials. *Mailing Add:* Vet Admin Med Ctr One Vet Plaza 151 San Juan PR 00927-5800

RAMKE, THOMAS FRANKLIN, b Bancker, La, Jan 1, 17; m 41; c 4. FORESTRY. *Educ:* La State Univ, BS, 40. *Prof Exp:* Forester, La Dept Conserv, 40-41, asst dist forester, 41-42; forester, Tenn Valley Auth, 42-61, asst dist forester, 48-49, chief forestry field br, 61-66, forest mgt br, Norris, 66-67, tributary area rep, 67-74, dist mgr, Off Tributary Area Develop, 74-79; RETIRED. *Mem:* Soc Am Foresters. *Res:* Factors related to the application of forest and watershed management and skillful use of forest reserves; elements related to improving community structure for effective citizen participation; community planning, evaluation and development. *Mailing Add:* 4107 Fulton Rd Knoxville TN 37918-4314

RAMLER, EDWARD OTTO, b Washington, DC, Sept 25, 16; m 42; c 3. ORGANIC CHEMISTRY. *Educ:* Cath Univ Am, BS, 38; Pa State Col, MS, 40, PhD(org chem), 42. *Prof Exp:* Asst chem, Pa State Col, 39-41, instr, 41-42; res chemist, Plastics Dept, E I du Pont de Nemours & Co, Inc, 42-46, tech investr, Textile Fibers Dept, 46-53, supvr, Patent Div, 53-60, patent adminr, Int Dept, 60-70, mgr, Patents Trademarks & Contracts Sect, 70-80; RETIRED. *Mem:* Am Chem Soc. *Res:* Organic chemistry of fluorine; synthesis of vinyl type monomers and polymers; plastics technology; reactions catalyzed by hydrogen fluoride. *Mailing Add:* 513 Woodside Ave Woodside Hills Wilmington DE 19809-1636

RAMLER, W(ARREN) J(OSEPH), b Joliet, Ill, Jan 1, 21; c 3. INDUSTRIAL RADIATION SYSTEMS. *Educ:* Ill Inst Technol, BS, 43, MS, 51. *Prof Exp:* Student engr, Westinghouse Elec Corp, 43; instr elec eng, Carnegie Inst Technol, 43-44 & 46; asst elec eng, Tenn Eastman Corp, 44-46; assoc elec eng, Argonne Nat Lab, 46-49, asst group leader, Cyclotron Proj, 49-56, group leader, 56-59, sr scientist, 59, group leader low energy accelerators, 59-73; gen mgr, PPG Industs, 73-81; sr vpres, RPC Industs, Inc, 81-86; PRES, WJR CONSULTS, 86-; PRES, AETEK INT, INC, 88- *Concurrent Pos:* Consult, Argonne Nat Lab, 73- *Mem:* Inst Elec & Electronics Engrs; Am Phys Soc; NY Acad Sci; Am Mgt Asn. *Res:* Development and systems design and construction of radiation generating equipment for industrial use and laboratory research; Dc and cyclic accelerators, linacs, and ultraviolet processors. *Mailing Add:* 155 Canyon Diablo Rd Sedona AZ 86336

RAMM, ALEXANDER G, b Leningrad, USSR, Jan 13, 40; US citizen; m; c 2. GEOPHYSICS. *Educ:* Univ Leningrad, MS, 61; Univ Moscow, PhD(math physics), 64; Inst Math Acad Sci, Minsk, DrSci, 72. *Prof Exp:* Prof Math, Univ Mich, 79-81; PROF MATH, KANS STATE UNIV, 81- *Concurrent Pos:* vis scientist, Schlumberger Doll Res, 83; vis prof math, Univ Göteborg, 82; Univ Manchester, 84; Univ London, 85; Acad Sinica, Taipei, 86; Univ Bonn, Ger, 84; Univ Stuttgart, Ger, 83; Univ Heidelberg, Ger, 87; Indian Inst Sci, Bangalore, 87, Univ Uppsala, Royal Inst Technol, 87, Concordia Univ, 90; consult, Dikewood Corp, 84; Standard Oil Prod Co, 85; elected mem, Electromagnetic Acad, Mich Inst Technol, 90; Fulbright prof, 90. *Mem:* Am Math Soc. *Res:* Spectral and scattering theory in classical physics and quantum mechanics; inverse problems, theoretical numerical analysis and ill posed problems; random fields estimation and signal processing; wave propagation and nonlinear passive networks. *Mailing Add:* Dept Math Kansas State Univ Manhattan KS 66506

RAMM, DIETOLF, b Berlin, Ger, June 17, 42; US citizen; m 66. COMPUTER SCIENCE. *Educ:* Cornell Univ, BA, 64; Duke Univ, PhD(physics), 69. *Prof Exp:* Assoc community health sci, Med Ctr, Duke Univ, 69-70, asst prof community health sci, univ, 70-71, asst prof info sci in psychiat, 70-76 & comput sci, 71-76, ASSOC MED RES PROF PSYCHIAT, MED CTR, DUKE UNIV, 76-, DIR GERIAT COMPUT CTR, CTR STUDY AGING & HUMAN DEVELOP, 69- *Concurrent Pos:* Lectr comput sci, Duke Univ, 76- *Mem:* AAAS; Asn Comput Mach; Geront Soc; Am Phys Soc. *Res:* Medical applications for computing; micro-computers in the laboratory; human-machine interface problems; interactive computing; computers in psychiatry and the study of aging. *Mailing Add:* 3538 Hamstead Ct Durham NC 27707

RAMMER, IRWYN ALDEN, b Stockton, Calif, Aug 15, 28; m 56; c 3. PESTICIDE DEVELOPMENT. *Educ:* Univ Calif, BS, 51, MS, 52, PhD(entom), 60. *Prof Exp:* Res asst entom, Univ Calif, 56-59; res assoc, Agr Chem Group, FMC CORP, 59-86; AGR CONSULT, 86- *Mem:* Entom Soc Am. *Res:* Pesticides for control of agricultural pests. *Mailing Add:* 2682 Moraga Dr Pinole CA 94564

RAMMING, DAVID WILBUR, b Oklahoma City, Okla, Oct 31, 46; m 75; c 3. PLANT BREEDING. *Educ:* Okla State Univ, BA, 68, MA, 72; Rutgers Univ, PhD(hort), 76. *Prof Exp:* Teaching asst crop sci, Okla State Univ, 68-69, asst hort, 71-72; asst fruit breeding, Rutgers Univ, 72-75; res leader fruit breeding, Western Region, Sci & Educ Admin, 75-86, RES HORTICULTURIST, AGR RES, USDA, 86- *Mem:* Am Soc Hort Sci; Am Pomol Soc; Sigma Xi. *Res:* Fruit breeding, development of improved stone fruit and grape varieties, cytological analysis of Prunus chromosomes and pollen tube incompatibility; embryo culture of prunus and vitis. *Mailing Add:* 207 N Locan Ave Fresno CA 93727

RAM-MOHAN, L RAMDAS, b Poona, India, July 21, 44; US citizen. SOLID STATE PHYSICS. *Educ:* Univ Delhi, BSc, 64; Purdue Univ, MS, 67, PhD(physics), 71. *Prof Exp:* Instr physics, Purdue Univ, 75-78; from asst prof to assoc prof, 78-85, PROF, WORCESTER POLYTECH INST, 85- *Concurrent Pos:* Consult, MIT & Purdue Univ. *Mem:* Am Phys Soc. *Res:* Theory of electromagnetic properties of metals; optical properties of semiconductors; many-body theory; quantum field theory. *Mailing Add:* Dept Physics Worcester Polytech Inst Worcester MA 01609

RAMO, SIMON, b Salt Lake City, Utah, May 7, 13; m 37; c 2. ELECTRICAL ENGINEERING, PHYSICS. *Educ:* Univ Utah, BS, 33; Calif Inst Technol, PhD(elec eng, physics), 36. *Hon Degrees:* DEng, Case Western Reserve Univ, 60, Univ Mich, 66 & Polytech Inst New York, 71; DSc, Univ Utah, 61, Union Col, 63, Worcester Polytech Inst, 68, Univ Akron, 69 & Cleveland State Univ, 76; LLD, Carnegie-Mellon Univ, 70 & Univ Southern Calif, 72, Conzaga Univ, 83, Occidental Col, 84, Claremont Univ, 85. *Prof Exp:* Res engr, Gen Elec Co, NY, 36-46; vpres & dir opers, Hughes Aircraft Co, 46-53; co-founder, vpres, Ramo-Wooldridge Corp, 53-58, sci dir, US Intercontinental Ballistic Missile Prog, 54-58; exec vpres, TRW Inc, 58-61, vchmn bd, 61-78, chmn exec comt, 69-78, dir, 54-85, chmn bd, TRW-Fujitsu Co, 80-83; RETIRED. *Concurrent Pos:* Pres & dir, Bunker Ramo Corp, 64-66; dir, Union Bank, Times Mirror, Atlantic Richfield; mem, White House Energy Res & Develop Coun, 73-75, adv comt sci & foreign affairs, US Dept State, 73-75 & bd dirs, Los Angeles World Affairs Coun, 73-85; trustee, Calif Inst Technol; chmn, President's Comt Sci & Technol, 76-77, co-chmn, Transition Task Force Sci & Technol, 80-81; mem, Secy's Adv Coun, Dept Commerce, 76-77, consult to adminr, ERDA, 76-77; vis prof mgt sci, Calif Inst Technol, 78-; fac fel, John F Kennedy Sch Govt, Harvard Univ, 80-84; mem bd adv sci & technol, Repub China, 81-84; Regent's lectr, Univ Calif, Los Angeles, 81-82. *Honors & Awards:* Presidential Medal of Freedom; Nat Medal Sci, Franklin Inst, 78. *Mem:* Nat Acad Sci; Nat Acad Eng; fel Am Inst Aeronaut & Astronaut; fel Am Phys Soc; fel Inst Elec & Electronics Engrs; Am Philos Soc; Am Acad Arts & Sci; Int Acad Astronaut. *Res:* Electronics; microwaves. *Mailing Add:* 9200 Sunset Blvd Suite 401 Los Angeles CA 90069

RAMOHALLI, KUMAR NANJUNDA RAO, b Karnataka, India, Nov 12, 45; m 77. COMBUSTION, ACOUSTICS. *Educ:* Univ Col Eng, BE, 67; Indian Inst Sci, ME, 68; Mass Inst Technol, PhD(propulsion), 71. *Prof Exp:* Res fel propulsion, Guggenheim Jet Propulsion Ctr, Calif Inst Technol, 71-74, sr res fel, 74-75, sr sect mem tech staff advan technol, 75-79, group leader & group supvr thermal & chem, Jet Propulsion Lab, 79-, res engr, 81-; PROF, DEPT AEROSPACE & MECH ENG, UNIV ARIZ, 82-; PRIN INVESTR, SPACE ENG RES CTR, NASA, 88- *Concurrent Pos:* Vis scientist, Indian Inst Sci, Indian Space Res Orgn, 78; mgr, Sunfuels, Jet Propulsion Lab, Calif Inst Technol, 80-81; vis scientist, Beijing Inst Technol, 87; vis scientist, Northwestern Polytech Univ, Xiam, China, 87; adj prof, Penn State Univ, 89; vis assoc, Harvard Univ, 89-90. *Honors & Awards:* Exceptional Serv Medal, NASA, 84. *Mem:* Assoc fel Am Inst Aeronaut & Astronaut; Combustion Inst; Sigma Xi. *Res:* Combustion involving solids and gases; theory of hybrid combustion; composite solid propellant combustion including nitramines; novel perforated porous plate analogue for heterogeneous combustion; acoustic diagnostics of burners; graphite composts and their hazards alleviation through gasification; space junk cleanup; light weight, deployable solar collectors; nine certificates of recognition from NASA; production of propellants & other useful materials extraterrestrially for cost effectiveness of future space missions. *Mailing Add:* Aerospace & Mech Eng Univ Tucson AZ 85721

RAMON, SERAFIN, b Feb 3, 34; US citizen; m 58; c 3. CYTOGENETICS. *Educ:* Panhandle Agr & Mech Col, BS, 57; Univ N Mex MS, 62; Univ Kans, PhD(bot), 67. *Prof Exp:* From instr to asst prof biol, Panhandle Agr & Mech Col, 59-65; assoc prof biol, 67-71, head dept, 69-85, head dept sci, 73-75, PROF BIOL, PANHANDLE STATE UNIV, 71-, HEAD DEPT, 77- *Concurrent Pos:* NSF sci fel, 66-67. *Mem:* AAAS; Genetics Soc Am; Bot Soc Am; Am Genetic Asn. *Res:* Plant morphology and root anatomy; cytogenetic and biosystematics of selected Compositae. *Mailing Add:* Dept Biol Panhandle State Univ Goodwell OK 73939

RAMOND, PIERRE MICHEL, b Neuilly-Seine, France, Jan 31, 43; US citizen; m 67; c 3. ELEMENTARY PARTICLE PHYSICS. *Educ:* Newark Col Eng, BSEE, 65; Syracuse Univ, PhD(physics), 69. *Prof Exp:* Res assoc physics, Fermi Lab, 69-71; instr, Yale Univ, 71-73, asst prof, 73-76; Millikan fel, Calif Inst Technol, 75-80; PROF PHYSICS, UNIV FLA, 80- *Concurrent Pos:* Trustee, Aspen Ctr Physics, Univ Fla, 80-; div assoc ed, Phys Rev Letts, 84-; Guggenheim fel, 85. *Mem:* Am Phys Soc; AAAS. *Res:* Grand unified theories; unification of gravity with elementary particles. *Mailing Add:* Dept Physics Univ Fla Gainesville FL 32611

RAMON-MOLINER, ENRIQUE, b Murcia, Spain, July 11, 27; Can citizen; m 57; c 4. NEUROANATOMY. *Educ:* Inst Cajal, Madrid, MD, 56; McGill Univ, PhD, 59. *Prof Exp:* Asst res prof anat, Univ Md, 59-63; from asst prof to assoc prof physiol, Laval Univ, 63-68; from assoc prof to prof anat, Sch Med, Univ Sherbrooke, 74-; RETIRED. *Concurrent Pos:* Assoc, Med Res Coun Can, 63-81. *Mem:* Am Asn Anat; Can Asn Anat; Can Physiol Soc; Int Brain Res Orgn; Soc Neurosci. *Res:* Histology and cytology of the central nervous system; structure of the cerebral cortex; morphological varieties and classification of nerve cells; correlation between dendritic morphology and function of nerve cells; ultrastructure of the central nervous system; neurohistochemistry; corticothalamic connections. *Mailing Add:* Dept Cellular Biol Univ Sherbrooke 2500 L'Universite Blvd Sherbrooke PQ J1K 2R1 Can

RAMOS, HAROLD SMITH, b Atlanta, Ga, July 20, 28; m 54; c 3. MEDICINE. *Educ:* Johns Hopkins Univ, AB, 48; Med Col Ga, MD, 54. *Prof Exp:* From asst prof to assoc prof med, Sch Med, Emory Univ, 63-75; chief med & dir med educ, 63-85, MED DIR , CRAWFORD W LONG MEM HOSP, 85-; PROF MED, SCH MED, EMORY UNIV, 75-, ASST DEAN SCH MED, 72- *Concurrent Pos:* Fel hemat & renal dis, Walter Reed Army Inst Res, Washington, DC, 58-59. *Mem:* AMA; Am Col Physicians; NY Acad Sci. *Res:* Medical education; cardiology. *Mailing Add:* 550 Peachtree St Atlanta GA 30365

RAMOS, JUAN IGNACIO, b Bernardos, Spain, Jan 28, 53; m 89; c 1. MECHANICAL ENGINEERING. *Educ:* Madrid Polytech Univ, BAEng, 75; Princeton Univ, MA, 79, PhD(mech eng), 80, Madrid Polytech Univ, PhD(eng), 83. *Prof Exp:* Res engr, Aeronaut Constructs Ltd, Spain, 76-77; from instr to asst prof, 80-89, PROF MECH ENG, CARNEGIE MELLON UNIV, 89- *Concurrent Pos:* Consult, PPG Industs, 82-84; prin investr, NASA Lewis Res Ctr, 80-, NSF, 81-84, AFOSR 84-86; Guggenheim fel, 77-78; Van Ness Lothrop fel, 79-80; NASA Faculty fel, 82, 88; consult, Software Eng Inst, 86; vis prof, Univ Rome, 88, Univ Malaga, Spain, 90-91. *Honors & Awards:* Ralph R Teetor Award, Soc Automotive Eng, 81; Aeronaut Eng Medal, 77. *Mem:* Soc Indust & Appl Math. *Res:* Numerical modeling of internal combustion engines and gas turbines (combustion and fluid mechanics); numerical analysis finite elements; heat transfer and ignition, thermal sciences; applied mathematics, wave propagation; liquid curtains; chemical reactors. *Mailing Add:* Dept Mech Eng Carnegie-Mellon Univ Pittsburgh PA 15213-3890

RAMOS, LILLIAN, b Ponce, PR; US citizen. CHEMISTRY, SCIENCE EDUCATION. *Educ:* Cath Univ PR, BS, 54; Fordham Univ, MSEd, 60, PhD, 71. *Prof Exp:* Prof chem, biol, phys sci & math, 55-70, prof educ, 70-78, dir grad studies educ, 71-77, PROF SCH ADMIN, CATH UNIV PR, 70- *Mem:* Am Chem Soc; Col Chem PR. *Res:* School administration. *Mailing Add:* Cath Univ PR Ponce PR 00731

RAMP, FLOYD LESTER, b Newman, Ill, Mar 6, 23; m 48; c 4. ORGANIC CHEMISTRY. *Educ:* Univ Ill, BS, 44; Univ Minn, PhD, 50. *Prof Exp:* Du Pont fel, Mass Inst Technol, 51; res chemist, 51-62, sr res assoc, 62-69, RES FEL CHEM, RES CTR, B F GOODRICH CO, BRECKSVILLE, 69- *Mem:* Am Chem Soc. *Res:* Chemical reactions of high polymers; electrochemistry. *Mailing Add:* BF Goodrich Res & Develop Ctr 3948 Humphery Rd Richfield OH 44286

RAMP, WARREN KIBBY, b New York, NY, Aug 19, 39; m 63; c 2. BONE METABOLISM, ORAL BIOLOGY. *Educ:* State Univ NY, Oneonta, BS, 63; Colo State Univ, MS, 64; Univ Ky, PhD (physiol, biochem), 67. *Prof Exp:* From asst prof to assoc prof oral biol & pharmacol, Univ NC, Chapel Hill, 70-79; assoc prof, 79-82, chmn oral biol, 83-86, PROF ORAL BIOL, PAHRMACOL & TOXICOL, UNIV LOUISVILLE, 82- *Concurrent Pos:* Nat Inst Dent Res fel, Univ Rochester, 67-70; adj scientist, Emory Univ, 87-88; vis scientist, Univ NC, Wilmington, 89- *Mem:* Am Physiol Soc; Soc Exp Biol & Med; Am Soc Bone & Mineral Res; AAAS; Int Asn Dent Res; Sigma Xi. *Res:* Calcium metabolism; bone metabolism; effects of humoral and nutritional factors on calcium homeostasis and connective tissues. *Mailing Add:* Dept Oral Health Health Sci Ctr Univ Louisville Louisville KY 40292

RAMPACEK, CARL, b Omaha, Nebr, Aug 7, 13; m 39; c 2. METALLURGY, MINERAL RESOURCES. *Educ:* Creighton Univ, BS, 35, MS, 37. *Prof Exp:* Chemist, Phillips Petrol Co, Okla, 39-41; mineral technologist, US Bur Mines, Ala, 41-43, phys chemist, 43-45, metallurgist, Ariz, 45-51, chief, Process Develop & Res Br, Metall Div, 51-54, supvry metallurgist, Southwest Exp Sta, 54-60, res dir, Tuscaloosa Metall Res Ctr, 60-63, asst dir admin, 63-67, res dir, Col Park Metall Res Ctr, 67-69, asst dir metall, US Bur Mines, 69-75; dir, Mineral Resources Inst, Univ Ala, 76-83, assoc dir, 83-89; RETIRED. *Honors & Awards:* Henry Krumb Lectr Metall, Am Inst Mining & Metall Engrs, 77; Robert Earll McConnell Award, Am Inst Mining, Metall & Petrol Engrs, 78. *Mem:* Am Chem Soc; Am Inst Mining, Metall & Petrol Engrs; AAAS; Fedn Mat Soc (vpres-pres elect, 78). *Res:* Metallurgical research. *Mailing Add:* 4901 Emeral Bay Dr Northport AL 35476

RAMPINO, MICHAEL ROBERT, b Brooklyn, NY, Feb 8, 48. CLIMATE CHANGE, GEOPHYSICS. *Educ:* Hunter Col, BA, 68; Columbia Univ, PhD(geol), 78. *Prof Exp:* Instr geol, Hunter Col, 72-74 & Rutgers Univ, 76-78; Nat Acad Sci res assoc, 78-80, res assoc climatol, 80-85, RES CONSULT, GODDARD INST SPACE STUDIES, NASA, 85-; asst prof, 85-90, ASSOC PROF, EARTH SYSTS GROUP, DEPT APPL SCI, NY UNIV, 91- *Concurrent Pos:* Adj instr geol, Lehman Col, NY, 74-77; instr, Earth Sci Dept, Fairleigh-Dickinson Univ, 78-79; vis asst prof, Dartmouth Col, 80, 82; lectr geol, Columbia Univ, 79-83; adj instr, Sch Visual Arts, 80-86; adj asst prof, Barnard Col, 82-83; ed, Climate History, Periodicity & Predictability, 87; adj asst prof, Sch Continuing Educ, NY Univ, 83-; res consult, Ctr Study Global Habitability, Columbia Univ, 85-; chair, Geol Sci Sect, NY Acad Sci, 89-90; rep, Int Climate Comn, Int Geosphere-Biosphere Proj, 87; mem, Nat Oceanic & Atmospheric Admin Joint US-USSR Working Group VIII on Climate Change, 89- *Mem:* AAAS; Am Geophys Union; Geol Soc Am; Soc Sedimentary Geol; NY Acad Sci; Int Soc Study Origin Life; Sigma Xi. *Res:* Climatic change especially the effects of volcanic eruptions and extraterrestrial impactors on climate and the environment; causes of mass extinctions, comet showers, and episodic volcanism; periodicity in the geologic record. *Mailing Add:* Earth Systs Group Dept Appl Sci New York Univ New York NY 10003

RAMPONE, ALFRED JOSEPH, b Kelowna, BC, May 21, 25; nat US; m 57; c 4. PHYSIOLOGY. *Educ:* Univ BC, BA, 47, MA, 50; Northwestern Univ, PhD, 54. *Prof Exp:* Res assoc physiol, Sch Med, Northwestern Univ, 54-55; instr, St Louis Univ, 55; from instr to prof, 55-71, actg chmn, Dept Physiol, 79-81, PROF PHYSIOL, MED SCH, UNIV ORE, 71- *Mem:* Am Physiol Soc; Soc Exp Biol & Med. *Res:* Intestinal transport of lipids; energy metabolism. *Mailing Add:* Dept Physiol Sch Med Ore Health Sci Univ 3181 SW Sam Jackson Park Rd Portland OR 97201

RAMPP, DONALD L, b Meramac, Okla, Feb 10, 35; m; c 2. SPEECH PATHOLOGY. *Educ:* Northeastern State Col, BAEd, 57; Ohio State Univ, MA, 58; Univ Okla, PhD(speech path), 67. *Prof Exp:* Speech pathologist, Pub Sch, Okla, 57-58; asst prof speech path, Northeastern State Col, 58-62; chief speech path & audiol, Child Develop Ctr, Med Units, Univ Tenn, 66-69; assoc prof speech path & coordr med serv, Memphis State Univ, 69-74; PROF & HEAD DEPT AUDIOL & SPEECH PATH, LA STATE UNIV MED CTR, NEW ORLEANS, 74- & PROF DEPT OTOLARYNGOL & BIOCOMMUN, SCH MED, 74- *Concurrent Pos:* Speech pathologist, Med Ctr, Univ Okla, 66-68; supvr speech, lang & hearing, Collaborative Perinatal Res Proj, Med Units, Univ Tenn, Memphis, 69-, prof, 71-; consult, Vet Admin Hosp, Memphis, Tenn, 70- *Mem:* Am Cleft Palate Asn; Am Speech & Hearing Asn. *Res:* Auditory processing of verbal stimuli and its relationship to learning disabilities in children; voice quality characteristics of cleft palate persons. *Mailing Add:* Dept Commun Dis La State Univ 1900 Grazier St New Orleans LA 70112

RAMRAS, MARK BERNARD, b Brooklyn, NY, May 18, 41. MATHEMATICS. *Educ:* Cornell Univ, BA, 62; Brandeis Univ, MA, 64, PhD(math), 67. *Prof Exp:* Asst prof math, Harvard Univ, 67-70 & Boston Col, 70-74; assoc prof math, Univ Mass, Boston, 74-75; ASSOC PROF MATH, NORTHEASTERN UNIV, 75- *Mem:* Am Math Soc; Math Asn Am. *Res:* Ring theory; homological algebra. *Mailing Add:* Dept Math Northeastern Univ Boston MA 02115

RAMSAY, ARLAN (BRUCE), b Dodge City, Kans, July 1, 37; m 58; c 2. GROUPOIDS, REPRESENTATION THEORY. *Educ:* Univ Kans, BA, 58; Harvard Univ, AM, 59, PhD(math), 62. *Prof Exp:* Instr math, Mass Inst Technol, 62-64; vis asst prof, Brandeis Univ, 64-65; asst prof, Univ Rochester, 65-68; assoc prof, 68-72, PROF MATH, UNIV COLO, BOULDER, 72- *Mem:* Am Math Soc. *Res:* Locally compact groups; representation theory; groupoids in analysis; orthmodular lattices. *Mailing Add:* Dept Math Box 426 Univ Colo Boulder CO 80309-0426

RAMSAY, DAVID JOHN, b Hornchurch, Essex, Eng, Apr 20, 39; m 65; c 3. MEDICAL PHYSIOLOGY. *Educ:* Oxford Univ, BA, 60, MA & DPhil(renal physiol), 63, BM, BCh, 66. *Prof Exp:* Demonstr physiol, Oxford Univ, 63-66, med tutor, Corpus Christi Col & Univ lectr, 66-75; vis prof, 75, assoc prof, 75-78, PROF PHYSIOL, SCH MED, UNIV CALIF, SAN FRANCISCO, 78- *Mem:* Am Physiol Soc; Endocrine Soc; Soc Neurosci; Am Soc Nephrol; Brit Physiol Soc. *Res:* The control of fluid intake and output; the role of the renin-angiotensin system and the etiology of edema in congestive cardiac failure. *Mailing Add:* Dept Physiol Sch Med Univ Calif San Francisco CA 94143-0400

RAMSAY, DONALD ALLAN, b London, Eng, July 11, 22; Can citizen; m 46; c 4. MOLECULAR SPECTROSCOPY. *Educ:* Cambridge Univ, BA, 43, MA & PhD, 47, ScD, 76. *Hon Degrees:* Dsc, Univ Reims, 69; Fil Hed Dr, Univ Stockholm, 82. *Prof Exp:* Jr res officer, 47-49, asst res officer, 50-54, assoc res officer, 55-60, sr res officer, 61-67, prin res officer, Nat Res Coun Can, 68-87; RETIRED. *Concurrent Pos:* Guest lectr, Univ Ottawa, 55-67; vis prof, Univ Minn, 64, Univ Orsay, 66, 75, Univ Stockholm, 67, 71, 74, Univ Sao Paulo, 72, 78 & Univ Western Australia & Australian Nat Univ, 76; regents lectr, Univ Calif, Irvine, 70; adv prof, East China Normal Univ, Shanghai, 87. *Honors & Awards:* Queen Elizabeth II Silver Jubilee Medal, 77; Centenary Medal, Royal Soc Can, 82. *Mem:* Fel Am Phys Soc; fel Chem Inst Can; Royal Soc Can (vpres acad sci, 75-76, hon treas, 76-79); fel Royal Soc; Can Asn Physicists. *Res:* Molecular spectroscopy, especially the spectra of free radicals. *Mailing Add:* 1578 Drake Ave Ottawa ON K1G 0L8 Can

RAMSAY, JOHN BARADA, b Phoenix, Ariz, Dec 28, 29; m 53; c 4. DETONATION PHYSICS, PHYSICAL CHEMISTRY. *Educ:* Univ Tex, El Paso, BS, 50; Univ Wis, PhD(anal chem), 55. *Prof Exp:* Staff mem anal chem, Los Alamos Sci Lab, 54-57, staff mem detonation physics, 57-70; assoc prof anal chem, Univ Petrol & Minerals, Saudi Arabia, 70-73; MEM STAFF DETONATION PHYSICS, LOS ALAMOS NAT LAB, 73- *Concurrent Pos:* Lectr, Univ NMex, Los Alamos Campus, 80- *Mem:* AAAS; Sigma Xi. *Res:* Explosive initiation and related phenomena; saline deposits of arid areas. *Mailing Add:* Six Erie Lane Los Alamos NM 87544

RAMSAY, JOHN MARTIN, b Bethlehem, Pa, Apr 9, 30; m 54; c 2. ANIMAL BREEDING. *Educ:* Berea Col, BS, 52; Iowa State Univ, MS, 64, PhD(animal breeding), 66. *Prof Exp:* Instr agr, Warren Wilson Jr Col, 52-55; assoc dir rural life, John C Campbell Folk Sch, 66-67, dir, 67-73; asst prof recreation exten, 74-76, DIR RECREATION & ASST PROF ANIMAL SCI, DEPT ANIMAL SCI, BEREA COL, 76- *Mem:* Am Dairy Sci Asn. *Res:* Use of identical twins in dairy breeding research; genetic interpretation of heterogeneous variance of milk production; economic feasibility of crossing and upgrading a Jersey herd to Holsteins. *Mailing Add:* 112 Adams St Berea KY 40403

RAMSAY, MAYNARD JACK, b Buffalo, NY, Nov 22, 14; m 41; c 5. ENTOMOLOGY. *Educ:* Univ Buffalo, AB, 36, AM, 38; Cornell Univ, PhD, 42. *Prof Exp:* Asst zool, Univ Buffalo, 36-38; asst insect morphol & insect physiol, Cornell Univ, 39-40, biol, 40-42; hort inspector, Bur Plant Indust, State Dept Agr & Mkts, NY, 42-43; plant quarantine inspector, Plant Quarantine Div, Agr Res Serv, USDA, 43-50, port entomologist, Plant Importations Br, 50-56, training off, 56-66, head post-entry quarantine sect, 66-67, AGRICULTURIST, ANIMAL & PLANT HEALTH INSPECTION

SERV, PLANT PROTECTION & QUARANTINE PROGS, PLANT QUARANTINE DIV, AGR RES SERV, USDA, 67-, STAFF OFFICER NAT PROG PLANNING STAFF, 75- *Concurrent Pos:* Head publ coop econ insect rep weekly, Nat Econ Insect Surv, 73- *Honors & Awards:* Superior Serv Award, USDA, 56. *Mem:* Entom Soc Am. *Res:* Coleoptera of Allegany State Park, NY; Mexican bean beetle control; Dutch elm disease control; international plant quarantine; survey methods for economic insects; losses due to pests. *Mailing Add:* 3806 Viser Ct Bowie MD 20715

RAMSAY, OGDEN BERTRAND, b Baltimore, Md, Sept 24, 32; m 62; c 1. ORGANIC CHEMISTRY, HISTORY OF CHEMISTRY. *Educ:* Washington & Lee Univ, BS, 55; Univ Pa, PhD(org chem), 60. *Prof Exp:* Fel org chem, Ga Inst Technol, 59-61; asst prof chem, Univ of the Pacific, 61-63; res fel org chem, Northwestern Univ, 63-64, instr chem, 64-65; from asst prof to assoc prof, Eastern Mich Univ, 65-74, actg head dept, 80-82, head dept, 82-86, PROF CHEM, EASTERN MICH UNIV, 74- *Concurrent Pos:* NSF sci fac fel, Dept Chem, Univ Wis-Madison, 68-69. *Mem:* Am Chem Soc; Royal Soc Chem. *Res:* History of chemistry; chemical information retrieval. *Mailing Add:* Dept Chem Eastern Mich Univ Ypsilanti MI 48197

RAMSAY, WILLIAM CHARLES, b Jamaica, NY, Nov 6, 30; m 51, 66, 88; c 4. RESOURCE MANAGEMENT. *Educ:* Univ Colo, BA, 52; Univ Calif, Los Angeles, MA, 57, PhD(physics), 62. *Prof Exp:* NSF fel, Univ Calif, San Diego, 62-63, res assoc physics, 63-64; asst prof, Univ Calif, Santa Barbara, 64-67; sr staff scientist, Systs Assocs, Inc, 67-72; sr environ economist, Atomic Energy Comn, 72-75; tech adv, Nuclear Regulatory Comn, 75-76; sr fel, Resources for the Future, 76-83; sr fel, Ctr for Stategic & Int Studies, Georgetown Univ, 83-85; sr staff officer, Nat Acad Sci, 85-86; CONSULT & WRITER, 86- *Mem:* Am Phys Soc; Am Astron Soc; Int Asn Energy Economists. *Res:* Energy strategies; environmental management. *Mailing Add:* 4003 Oliver St Chevy Chase MD 20815

RAMSDALE, DAN JERRY, b El Paso, Tex, Dec 12, 42; m 69; c 2. UNDERWATER ACOUSTICS, SIGNAL PROCESSING. *Educ:* Univ Tex, El Paso, BS, 64; Kans State Univ, PhD(physics), 69. *Prof Exp:* Res dir, Gus Mfg, Inc, Globe Universal Sci, Inc, 69-74; res physicist, Acoust div, US Naval Res Lab, Naval Ocean Res & Develop Activ, 74-77, res physicist, 77-81, sr prin investr, Array Effects Br, 81-85, tech prog mgr arctic environ acoust & head arctic acoust br, Ocean Acoust Div, 85-88, ASST DIR, OCEAN ACOUST & TECHNOL DIRECTORATE, NAVAL OCEANOG & ATMOSPHERIC RES LAB, 88- *Concurrent Pos:* Adj prof, El Paso Community Col, 71-74; adj prof physics, Univ New Orleans, 81-; past chmn, Tech Comt Underwater. *Mem:* Acoust Soc Am; Sigma Xi; Am Geophys Union; Inst Elec & Electronic Engrs; Am Inst Physics; Oceanic Eng Soc. *Res:* Atmospheric acoustics, infrasonics, acoustic echo-sounding and the acoustic grenade sounding technique; electroacoustics and electrostatic transducer design; theoretical atomic physics, Auger and x-ray transition rates; seismic waves; underwater acoustics, especially low frequency propagation studies and fluctuations; the use of acoustic arrays as measurement tools; propagation and scattering of underwater acoustic energy in the Arctic Ocean. *Mailing Add:* Naval Oceanog & Atmospheric Res Lab Ocean Acoust & Technol Directorate Stennis Space Ctr MS 39529

RAMSDELL, DONALD CHARLES, b Yuba City, Calif, Dec 28, 38; m 60; c 2. PLANT PATHOLOGY. *Educ:* Univ Calif, Davis, BS, 60, MS, 70, PhD(plant path), 71. *Prof Exp:* Res asst plant path, Univ Calif, Davis, 68-71; asst prof, 72-75, assoc prof, 75-78, PROF PLANT PATH, DEPT BOT & PLANT PATH, MICH STATE UNIV, 78- *Concurrent Pos:* Mem, Int Coun Study Virus & Virus-Like Dis of Grapevine, 75- *Honors & Awards:* Lee M Hutchins Award, Am Phytopath Soc. *Mem:* Am Phytopath Soc; Brit Asn Appl Biol; Int Soc Hort Sci. *Res:* Etiology, epidemiology and control of viral and fungal diseases of small fruits crops. *Mailing Add:* Box 162 Biol Res Ctr Mich State Univ East Lansing MI 48824

RAMSDELL, ROBERT COLE, b Trenton, NJ, July 8, 20; m 46; c 1. GEOLOGY. *Educ:* Lehigh Univ, BA, 43; Rutgers Univ, MS, 48; Princeton Univ, MA, 50. *Prof Exp:* With State Bur Mineral Res, NJ, 48-50; from instr to asst prof geol, Williams Col, 50-61; asst prof, Rutgers Univ, 61-66; assoc prof, Geosci Div, Montclair State Col, 66-; RETIRED. *Mem:* AAAS; Geol Soc Am; Paleont Soc; Soc Econ Paleont & Mineral; Nat Asn Geol Teachers. *Res:* Paleontology and stratigraphy of Atlantic coastal plain; Silurian and Devonian Appalachian paleontology and stratigraphy. *Mailing Add:* 226 Winding Way Morrisville PA 19067

RAMSDEN, HUGH EDWIN, b Amesbury, Mass, May 30, 21; m 46; c 3. ORGANIC CHEMISTRY. *Educ:* Mass Inst Technol, SB, 43, PhD(org chem), 46. *Prof Exp:* Res chemist, E I du Pont de Nemours & Co, NJ, 46-48; res chemist, Metal & Thermit Corp, 49-52, res supvr & head dept org chem, 52-59; res chemist, Esso Res & Eng Co, 59-61, res assoc, 61-70, Esso Agr Prod Lab Div, 66-70; res assoc, R T Vanderbilt Co, 70-71; mem staff, Rhodia Inc, 71-79; scientist, 79-85, PRIN SCIENTIST, J T BAKER CHEM CO, 85- *Mem:* Am Chem Soc; AAAS; NY Acad Sci. *Res:* Sugars; organoalkali reagents; fluorine chemistry; condensation polymers; plasticizers; coordination compounds; rubber; reaction of rubbers with organometallic compounds; organic synthesis; vinyl grignards; gasoline additives; pesticides synthesis; terpenes, perfume and flavor, fine chemicals; solid phases for L C of bioengineered products, proteins, RNA, DNA, and monoclonals. *Mailing Add:* 2080 Wood Rd Scotch Plains NJ 07076-1299

RAMSEUR, GEORGE SHUFORD, b Burke Co, NC, July 19, 26; m 53; c 3. BOTANY. *Educ:* Elon Col, AB, 48; Univ NC, MEd, 53, PhD(bot), 59. *Prof Exp:* Teacher high sch, NC, 49-54; from instr to assoc prof, 58-73, PROF BOT, UNIV OF THE SOUTH, 73- *Mem:* AAAS; Bot Soc Am; Am Soc Plant Taxon. *Res:* Taxonomy of vascular plants; southern Appalachian flora. *Mailing Add:* Dept Biol Univ of the South Box 1218 Sewanee TN 37375

RAMSEY, ALAN T, b Madison, Wis, May 23, 38; m 60; c 2. PLASMA SPECTROSCOPY. *Educ:* Princeton Univ, AB, 60; Univ Wis, MS, 62, PhD(physics), 64. *Prof Exp:* Physicist, Lawrence Radiation Lab, Univ Calif, 64-67; asst prof physics, Brandeis Univ, 67-73; proj scientist, Am Sci & Eng, 74-76; res scientist, Mass Inst Technol, 76-78; RES SCIENTIST, PRINCETON PLASMA PHYSICS LAB, 79- *Res:* Optical pumping and atomic beam research on atomic and nuclear structure; x-ray astronomy; plasma diagnostics; spectroscopic instrumentation; medical instrumentation and research. *Mailing Add:* Princeton Plasma Physics Lab PO Box 451 Princeton NJ 08543

RAMSEY, ARTHUR ALBERT, b Schenectady, NY, Apr 11, 40; m 62; c 3. PESTICIDE CHEMISTRY. *Educ:* Albany Col Pharm, BS, 62; Univ Kans, PhD(med chem), 68. *Prof Exp:* Sr res chemist, Agr Chem Div, 68-77, mgr, compound acquisition, 77-88, SR METAB CHEMIST, AGR CHEM GROUP, FMC CORP, 88- *Mem:* Am Chem Soc. *Res:* Metabolism and environmental fate studies of agricultural chemicals; physiological chemistry of plants, fungi and insects. *Mailing Add:* Agr Chem Group FMC Corp Box 8 Princeton NJ 08543

RAMSEY, BRIAN GAINES, b Union, SC, Mar 17, 37; m 76; c 3. PHYSICAL ORGANIC CHEMISTRY, MOLECULAR SPECTROSCOPY. *Educ:* Univ SC, BSc, 56; Univ Wis, MSc, 58; Fla State Univ, PhD(chem), 62. *Prof Exp:* Fel, Pa State Univ, 62-64; asst prof, Univ Akron, 64-69; from assoc prof to prof chem, San Francisco State Univ, 69-80; PROF CHEM, ROLLINS COL, 80- *Concurrent Pos:* Sr Fulbright res fel, Ger, 72-73; Alexander von Humboldt Award, Germany, 72 & 73. *Mem:* Am Chem Soc. *Res:* Spectroscopic investigations of reactive intermediates in organic chemistry; electronic transitions in organometallics; chemistry of organoboranes. *Mailing Add:* Dept Chem Rollins Col Winter Park FL 32789-4499

RAMSEY, CLOVIS BOYD, b Sneedville, Tenn, Aug 1, 34; m 58; c 2. ANIMAL SCIENCE. *Educ:* Univ Tenn, BS, 56; Univ Ky, MS, 57, PhD(meats), 60. *Prof Exp:* Asst prof meat sci, Univ Tenn, 60-68; PROF MEAT SCI, TEX TECH UNIV, 68- *Mem:* Am Meat Sci Asn; Am Soc Animal Sci; Inst Food Technol; Sigma Xi. *Res:* Physical, chemical and organoleptic properties of beef, lamb and pork; live-animal carcass evaluation; meat processing methods; factors affecting meat quality and quantity. *Mailing Add:* Dept Animal Sci Tex Tech Univ Lubbock TX 79409-4169

RAMSEY, DERO SAUNDERS, b Starkville, Miss, June 17, 28; m 50; c 2. DAIRY SCIENCE. *Educ:* Miss State Univ, BS, 50, MS, 53; Univ Wis, PhD(dairy husb), 57. *Prof Exp:* From asst prof to prof, 56-90, EMER PROF DAIRY SCI, MISS STATE UNIV, 90- *Mem:* AAAS; Am Dairy Sci Asn; Am Soc Animal Sci. *Res:* Physiology and nutrition of dairy cattle; animal waste management. *Mailing Add:* Dept Dairy Sci Miss State Univ Drawer DD Mississippi State MS 39762

RAMSEY, ELIZABETH MAPELSDEN, b New York, NY, Feb 17, 06; m 34. PLACENTOLOGY, PATHOLOGY. *Educ:* Mills Col, BA, 28; Yale Univ, MD, 32; Med Col Pa, DSc, 65. *Prof Exp:* Asst path, Yale Univ, 33-34; guest investr, Carnegie Inst Dept Embryol, 34-49, res assoc, 49-51, staff mem placentology & path, 51-71; Mamie A Jessup vis prof obstet & gynec, Sch Med, Univ Va, 71-76; RES ASSOC, DEPT EMBRYOL, CARNEGIE INST, 75- *Concurrent Pos:* Assoc, George Washington Univ, 34-41, prof lectr, 41-55; asst info off, Off Med Info, Nat Res Coun, 42-45; prof lectr obstet & gynec, Med Sch, Georgetown Univ, 81- *Honors & Awards:* Mosse Mem Lectr, Rotunda, Dublin, Ireland, 70; Dipl Hon, Int Fedn Gynec Infantile & Juv, France, 72; Pres Distinguished Scientist Award, Soc Gynec Invest, 87. *Mem:* AAAS; Perinatal Res Soc; Soc Gynec Invest; Am Col Obstet & Gynec; Am Gynecol Soc; Sigma Xi. *Res:* Vasculature of pregnant endometrium; radioangiography of placenta in rhesus monkey; placental circulation; placentation; myometrial activity in pregnancy. *Mailing Add:* 3420 Que St NW Washington DC 20007

RAMSEY, FRED LAWRENCE, b Ames, Iowa, Mar 3, 39; m 66. MATHEMATICAL STATISTICS. *Educ:* Univ Ore, BA, 61; Iowa State Univ, MS, 63, PhD(statist), 64. *Prof Exp:* Asst prof statist, Iowa State Univ, 64; NIH fel, Johns Hopkins Univ, 65-66; asst prof, 66-72, ASSOC PROF STATIST, ORE STATE UNIV, 72- *Mem:* Inst Math Statist. *Res:* Time series analysis; non-parametric statistics. *Mailing Add:* Dept Statist Ore State Univ Corvallis OR 97331

RAMSEY, GWYNN W, b Drexel, NC, Nov 13, 31; m 52; c 3. PLANT SYSTEMATICS. *Educ:* Appalachian State Teachers Col, BS, 55, MA, 58; Univ Tenn, PhD(bot), 65. *Prof Exp:* Teacher high sch, NC, 55-58; instr biol & bot, Lees-McRae Col, 58-61; chmn dept, 68-71 & 83-85, PROF BIOL, LYNCHBURG COL, 65-, CUR HERBARIUM, 66- *Mem:* Bot Soc Am; Am Soc Plant Taxon. *Res:* Biosystematics of the genus Cimicifuga; Virginia flora. *Mailing Add:* Dept Bot Lynchburg Col Lynchburg VA 24503

RAMSEY, HAROLD ARCH, b Ft Scott, Kans, Sept 16, 27; m 51; c 3. ANIMAL NUTRITION. *Educ:* Kans State Univ, BS, 50; NC State Univ, MS, 53, PhD(animal nutrit), 55. *Prof Exp:* From asst prof to assoc prof animal sci, NC State Univ, 55-62; vis assoc prof dairy sci, Univ Ill, 62-63; assoc & prof animal sci, 63-65, prof animal sci & head dairy husb sect, 65-70, PROF ANIMAL SCI, NC STATE UNIV, 70- *Mem:* Am Dairy Sci Asn; Am Inst Nutrit. *Res:* Nutritional requirements of ruminants; nutritional value of soy protein for newborn calves. *Mailing Add:* Dept Animal Sci Box 7621 NC State Univ Raleigh NC 27695-7621

RAMSEY, JAMES MARVIN, b Wilmington, Ohio, May 21, 24; div; c 3. ENERGY METABOLISM. *Educ:* Wilmington Col, BS, 48; Miami Univ, MS, 51. *Prof Exp:* Instr biol sci, Cedarville Col, 48-52; res assoc skin allergy & toxicol, Col Med, Univ Cincinnati, 52-53; instr physiol, Miami Univ, 55-60; assoc prof, 64-80, PROF BIOL SCI, UNIV DAYTON, 88- *Concurrent Pos:* NSF instnl res grants, 66-68 & 71-; NIH res grant, 68-71. *Mem:* AAAS; Sigma Xi; Physiol Soc. *Res:* Carbon monoxide toxicology; red blood cell metabolism;

the response of erythrocytic 2, 3-diphosphoglycerate to hypoxic stress; the relation of non-specific stress to asthmatic bronchoconstriction; regulation of plasma glucose and lipoproteins; author of textbook on pathophysiology. *Mailing Add:* Metab Lab Dept Biol Univ Dayton 300 College Park Ave Dayton OH 45469

RAMSEY, JED JUNIOR, b Dighton, Kans, Oct 17, 25; m 48; c 4. ZOOLOGY. *Educ:* Kans State Univ, BS, 49; Kans State Teachers Col, MS, 62; Okla State Univ, PhD(zool), 66. *Prof Exp:* Teacher high schs, Kans, 50-63; assoc prof, 65-72, PROF BIOL, LAMAR UNIV, 72- *Mem:* Sigma Xi; Am Ornith Union; Wilson Ornith Soc; Cooper Ornith Soc. *Res:* Metabolic changes in Chimney Swifts at lowered environmental temperatures; ecology and behavior of ciconiiform birds; avifauna of the Beaumont unit of the Big Thicket. *Mailing Add:* 875 Belvedere Beaumont TX 77710

RAMSEY, JERRY DWAIN, b Tulia, Tex, Nov 6, 33; m 56; c 4. ERGONOMICS, HUMAN FACTORS ENGINEERING. *Educ:* Tex A&M Univ, BS, 55, MS, 60; Tex Tech Univ, PhD(indust eng), 67. *Prof Exp:* Indust eng trainee, Square D Co, Wis, 53, Mich, 54; engr, Great Western Drilling Co, Tex, 55; indust engr, Collins Radio Co, Tex, 57-58; instr indust eng, Tex A&M Univ, 58-60, asst prof, 60-61; tech staff mem, Sandia Corp, NMex, 61-65; from asst prof to assoc prof, 67-75, PROF INDUST ENG, TEX TECH UNIV, 75-, ASSOC V PRES, 77- *Concurrent Pos:* Mem, President's Comt Employment of Handicapped; consult, Occup Health & Safety Admin, 72-, Nat Inst Occup Health & Safety, 73-, Us Bureau Int Labor Affairs, 74-, US Consumer Prod Safety Comn, 76-; chmn, Nat Standards Adv Comt Heat Stress, 73-74; exec comt, Nat Safety Coun Pub Employees, 74-90; bd dirs, Tex Safety Asn, 76- & Southwest Lighthouse for the Blind, 82-; ed, Int J Indust Ergonomics, 85- *Honors & Awards:* Citation Outstanding Contrib Ergonomics, Inst Indust Eng, 85. *Mem:* Nat Soc Prof Engrs; Inst Indust Engrs; Human Factors Soc; Nat Safety Coun; Am Indust Hyg Asn; Am Soc Safety Engrs. *Res:* Ergonomics; human factors engineering; product safety; occupational safety and health; management systems and optimization techniques; effects of environmental stressors on human performance; safety behavior and psychology, ergonomics applications for the disabled. *Mailing Add:* Dept Indust Eng Tex Tech Univ Lubbock TX 79409

RAMSEY, JERRY WARREN, b Springfield, Ill, June 30, 32; m 52; c 3. RESEARCH ADMINISTRATION, SYNTHETIC FUELS. *Educ:* Ill Col, AB, 56; Agr & Mech Col, Tex, MS, 58. *Prof Exp:* Instr gen chem, Tex Western Col, 58-59; instr, Pa State Univ, 59-61, instr org chem, 61-62; chemist, Anthracite Res Ctr, US Bur Mines, 62-64, res chemist, 64-65, Laramie Petrol Res Ctr, Wyo, 65-67, chemist, Off Dir Petrol Res, 67-70, staff chemist & asst to chief, Div Shale Oil, 70-75, br chief oil shale conversion, Div Oil, Gas & Shale Technol, 75-80, asst dir, Div Oil Shale, 80-82, MGR OIL SHALE PROG, US DEPT ENERGY, 82- *Concurrent Pos:* Consult, El Paso Natural Gas Co, 58-59; chemist, Anthracite Res Ctr, US Bur Mines, 60-62; instr, Pottsville Hosp Sch Nursing, Pa, 61-62; mem task force, Prototype Oil Shale Leasing Prog, US Dept Interior, 72-73; chmn oil shale task group, Synthetic Fuels Commercialization Prog, Off Mgt & Budget, 75. *Mem:* Am Chem Soc. *Res:* All areas of oil shale and shale oil research management. *Mailing Add:* US Dept Energy, FE-33 1000 Independence SW Washington DC 20585

RAMSEY, JOHN CHARLES, b Yakima, Wash, June 19, 33. TOXICOLOGY. *Educ:* Univ Puget Sound, BS, 55; Ore State Univ, PhD(plant biochem), 64; Am Bd Toxicol, dipl. *Prof Exp:* Res chemist pesticide residues, Dept Agr-Chemicals, 64-71, res toxicologist indust & agr chem, 71-83, TOXICOLOGIST, ANAL & ENVIRON CHEM, DEPT TOXICOL, DOW CHEM CO, 83- *Mem:* Soc Toxicol. *Res:* Pharmacokinetics and toxicology of agricultural and industrial chemicals; mathematical modeling of chemicals in biological systems. *Mailing Add:* 616 Gerald Ct Apt B Midland MI 48640

RAMSEY, JOHN SCOTT, b Tsingtao, China, June 23, 39; US citizen. FISH BIOLOGY. *Educ:* Cornell Univ, BS, 60; Tulane Univ, PhD(ichthyol), 65. *Prof Exp:* Assoc investr, Inst Marine Biol & asst prof biol, Univ PR, 65-67; unit leader Ala Coop Fishery Res Unit, US Fish & Wildlife Serv & res assoc prof zool & fisheries, Auburn Univ, 67-86; ASST LEADER FISHERIES, IOWA FISH & WILDLIFE RES UNIT, FISH & WILDLIFE SERV, 86-, PROF ANIMAL ECOL, IOWA STATE UNIV, 86- *Honors & Awards:* Commendation Award, Sporting Fishing Inst, 74. *Mem:* Am Fisheries Soc; Am Soc Ichthyol & Herpet; Asn SE Biologists; Alpha Zeta; Sigma Xi; Iowa Acad Sci; Mississippi River Res Coun; SE Fishes Coun. *Res:* Systematics and ecology of fishes, especially those limited to large rivers of the world. *Mailing Add:* Iowa Coop Fish & Wildlife Res Unit 124 Science II Iowa State Univ Ames IA 50011-3221

RAMSEY, LAWRENCE WILLIAM, b Louisville, Ky, Mar 14, 45; m 70. STELLAR SPECTROSCOPY. *Educ:* Univ Mo, St Louis, BA, 68; Kans State Univ, MS, 72; Ind Univ, PhD(astron), 76. *Prof Exp:* Simulation systs design engr, 66-70, res asst solar physics, Kitt Peak Nat Observ, 72-73; assoc prof astron, 76-88, PROF ASTRON & ASTROPHYS, PA STATE UNIV, 88- *Mem:* Am Astron Soc; Int Astron Union; Astron Soc Pac. *Res:* Cool star atmospheres; solar-like phenomenon on other stars; stellar spectroscopy and astronomical instrumentation. *Mailing Add:* Pa State Univ 525 Davey Lab University Park PA 16802

RAMSEY, LLOYD HAMILTON, b Lexington, Ky, June 10, 21; wid; c 4. MEDICAL ADMINISTRATION, INTERNAL MEDICINE. *Educ:* Univ Ky, BS, 42; Wash Univ, MD, 50; Am Bd Internal Med, dipl, 59. *Prof Exp:* Intern med, Duke Univ, 50-51; asst resident, Peter Bent Brigham Hosp, Boston, 51-52, asst, 52-53; instr, 54-55, asst prof, 55-63, investr, Howard Hughes Med Inst, 55-65, assoc prof, 63-78, PROF MED, SCH MED, VANDERBILT UNIV, 78-, ASSOC DEAN, 75- *Concurrent Pos:* Fel, Harvard Med Sch, 52-53; res fel, Sch Med, Vanderbilt Univ, 53-54; chief resident physician, Univ Hosp, Vanderbilt Univ, 54-55; consult, Mid Tenn State Tuberc Hosp, Nashville, 55-65. *Mem:* AAAS; fel Am Col Physicians; Am Clin & Climat Asn; Sigma Xi. *Res:* Pulmonary physiology, especially gas diffusion and relationships of external respiration to blood flow. *Mailing Add:* Vanderbilt Med Sch D-3309 Nashville TN 37232

RAMSEY, MAYNARD, III, b Birmingham, Ala, Aug 28, 43; m 69; c 2. BIOMEDICAL ENGINEERING. *Educ:* Emory Univ, BA, 65; Duke Univ, MD, 69, PhD(biomed eng), 75. *Prof Exp:* Dir res, Appl Med Res Corp, 75-79; vpres res & develop, 79-82, VPRES SCI & TECHNOL, CRITIKON, 82- *Mem:* Inst Elec & Electronics Engrs; Asn Advan Med Instrumentation. *Res:* Principles and mechanism of indirect and direct measurement of blood pressure; body surface electrocardiography and its implementation for clinical use. *Mailing Add:* Critikon Inc PO Box 31800 Tampa FL 33631-3800

RAMSEY, NORMAN FOSTER, JR, b Washington, DC, Aug 27, 15; m 40, 85; c 4. PHYSICS. *Educ:* Columbia Univ, AB, 35, PhD(physics), 40; Cambridge Univ, MA, 41, DSc, 54. *Hon Degrees:* MA, Harvard Univ, 47; DSc, Case Western Reserve Univ, 68, Middlebury Col, 69, Oxford Univ, 73, Rockefeller Univ, 86 & Univ Chicago, 89; DCL, Oxford Univ, 89. *Prof Exp:* Assoc physics, Univ Ill, 40-42; assoc prof, Columbia Univ, 42-47; from assoc prof to prof, 47-66, dir nuclear lab, 48-50 & 52, HIGGINS PROF PHYSICS, HARVARD UNIV, 66- *Concurrent Pos:* Consult, Off Sci Res & Develop & Nat Defense Res Comt, 40-45; consult US Secy War, 42-45; res assoc, Radiation Lab, Mass Inst Technol, 40-42; group leader & assoc div head, Atomic Energy Proj Lab, Los Alamos Sci Lab, Univ Calif, 43-45; head physics dept, Brookhaven Nat Lab, 46-47; mem sci adv bd, US Dept Air Force, 49-56 & US Dept Defense, 54-58; trustee, Brookhaven Nat Lab, 52-56, Carnegie Endowment Int Peace, 63- & Rockefeller Univ, 76-; Guggenheim fel, 54-55; sci adv, NATO, 58-59; mem gen adv comt, AEC, 60-72; chmn high energy physics panel, Sci Adv Bd, Off of the President, 63; dir, Varian Assocs, 63-66, sr fel, Harvard Soc Fels, 69-; pres, Univs Res Asn, 66-81 & chmn physics sect, AAAS, 77-78; Eastman prof, Oxford Univ, 73-74; chmn bd govrs, Am Inst Physics, 80-86; vis prof, Colo, 86-87, Chicago, 88, Mich, 89- *Honors & Awards:* Nobel Prize, 89; Lawrence Award & Medal, 60; Davisson-Germer Prize, Am Phys Soc, 74; Medal Hon, Inst Elec & Electronics Engrs, 84; Rabi Prize, 85; Monie Ferst Prize, Sigma Xi, 85; Karl Compton Prize, Am Inst Physics, 85; Rumford Prize, 85; Nat Medal Sci, 85; Oersted Medal, Am Assoc Physics Teachers, 88. *Mem:* Nat Acad Sci; AAAS; fel Am Phys Soc (pres, 78-79); Am Philos Soc; French Acad Sci. *Res:* Nuclear moments; molecular beams; high energy particles; nuclear interactions in molecules; deuteron quadrupole moment; molecular structure atomic clocks; diamagnetism; thermodynamics; proton-proton scattering; high energy accelerators; atomic masers; electron scattering; neutrons. *Mailing Add:* Lyman Physics Lab Harvard Univ Cambridge MA 02138

RAMSEY, O C, JR, mathematics, for more information see previous edition

RAMSEY, PAUL ROGER, b Lake Charles, La, July 27, 45; m 67; c 2. POPULATION BIOLOGY, GENETICS. *Educ:* Tex Tech Univ, BS, 67, MS, 69; Univ Ga, PhD(zool & ecol), 74; La Tech Univ, BA, 89. *Prof Exp:* Asst prof biol, Presby Col, 73-75; assoc prof, 75-78, PROF ZOOL, LA TECH UNIV, 78- *Concurrent Pos:* Adj asst prof biol, Fla Inst Technol, 78-; vis prof, Livestock Res Inst, Univ Novi Sad, Yugoslavia, 89-90; Fulbright fel, Yugoslavia. *Mem:* Am Soc Mammalogists; Soc Study Evolution. *Res:* Ecological genetics and protein variation of marine fish and small mammals. *Mailing Add:* Dept Biol Sci La Tech Univ PO Box 3179 Tech Sta Ruston LA 71272

RAMSEY, PAUL W, b Wilkinsburg, Pa, Feb 17, 19; m 42; c 3. METALLURGY. *Educ:* Carnegie Inst Technol, BS, 40; Univ Wis, MS, 56. *Prof Exp:* Sr investr, NJ Zinc Co, Pa, 40-51; proj engr, A O Smith Corp, 51-56, supvr, 56-65, mgr welding, 65-81, mgr welding & metall res & develop, 81-; RETIRED. *Honors & Awards:* Nat Award, Am Welding Soc, 71; S W Miller Mem Award, Am Welding Soc, 80. *Mem:* Am Welding Soc (vpres, 72-75, pres, 75-76); Am Inst Mining, Metall & Petrol Eng; fel Am Soc Metals; Soc Automotive Engrs; Soc Metall Engrs. *Res:* Welding metallurgy, mechanical testing; welding controls; power sources; physical metallurgy. *Mailing Add:* 3016 E Newport Ct Milwaukee WI 53211

RAMSEY, RICHARD HAROLD, b San Francisco, Calif, Nov 21, 36; m 60; c 2. PHYTOPATHOLOGY, MYCOLOGY. *Educ:* Univ Calif, Davis, 58, PhD(phytopath), 66. *Prof Exp:* From asst prof to prof biol, Rocky Mountain Col, 75-85; vpres & dean, Northland Col, 85-90. *Res:* Genetics of Pleospora herbarum. *Mailing Add:* 429 A Bayside Rd Arcata CA 59102

RAMSEY, ROBERT BRUCE, b Moline, Ill, Jan 4, 44; m 67; c 2. COMPUTATIONAL CHEMISTRY, NEUROCHEMISTRY. *Educ:* Augustana Col, BA, 66; St Louis Univ, PhD(biochem), 71, MBA, 84. *Prof Exp:* From asst prof to assoc prof neurol, Sch Med, St Louis, 72-79; prod mgr, Sherwood Med, 79-85; mgr planning & develop, McDonnell Douglas Health Systs, Co, 85-87, mgr, mergers & acquisitions, Info Syst Group, 87-88, mgr planning & mkt res, 88-90; ASSOC CLIN PROF, ST LOUIS UNIV, 79-; MGR MKT PLANNING, TRIPOS ASSOCS, 90- *Concurrent Pos:* NIH fel, Inst Neurol, Univ London, 71-73. *Mem:* AAAS; Am Chem Soc; Brit Biochem Soc; Am Soc Biol Chemists. *Res:* Molecular modeling. *Mailing Add:* 1133 Ridgelynn Dr St Louis MO 63124-1219

RAMSHAW, JOHN DAVID, b Salt Lake City, Utah, Mar 20, 44; div; c 2. STATISTICAL MECHANICS, FLUID DYNAMICS. *Educ:* Col Idaho, BS, 65; Mass Inst Technol, PhD(chem physics), 70. *Prof Exp:* Res assoc & asst instr, Univ Utah, 71-72; staff scientist physics & eng, Appl Theory Inc, 72-73; assoc scientist, Aerojet Nuclear Co, 73-75; staff mem theoret div, Los Alamos Nat Lab, 75-86; SCI & ENG FEL, IDAHO NAT ENG LAB, 86- *Concurrent Pos:* Air Force Off Sci Res-Nat Res Coun res award chem physics, Univ Md, 70-71. *Mem:* Am Phys Soc; Soc Indust & Apply Math. *Res:* Equilibrium and nonequilibrium statistical mechanics; dielectrics; liquids; nonlinear stochastic processes; transport far from equilibrium; analytical and numerical fluid dynamics; turbulence; multicomponent flow; chemically reactive flow; two-phase flow. *Mailing Add:* Idaho Nat Eng Lab PO Box 1625 Idaho Falls ID 83415-2516

RAMSLEY, ALVIN OLSEN, b North Bergen, NJ, Feb 6, 20; m 49; c 2. PHYSICAL CHEMISTRY. *Educ:* Houghton Col, BS, 43; Columbia Univ, MA, 48. *Prof Exp:* Chemist, Gen Elec Co, 48-50; chemist, Gen Test Labs, 50-53, res chemist, Natick Labs, US Army QM, 53-84, chief Countersurveillance Sect, Natick Res & Develop Command, 74-84; RETIRED. *Mem:* Am Chem Soc; Inter-Soc Color Coun; Sigma Xi. *Res:* Physical chemistry of excited states as related to chemical structure, absorption spectra, and luminescence; photodegradation of dyes; materials research for visual and non-visual counter-surveillance measures; vision, color measurement and colorant formulation. *Mailing Add:* 15 Farm Rd Sherborn MA 01770

RAMSPOTT, LAWRENCE DEWEY, b Jacksonville, Fla, Dec 9, 34. PROJECT MANAGEMENT. *Educ:* Principia Col, BS, 56; Pa State Univ, PhD(geol), 62. *Prof Exp:* Asst prof geol, Univ Ga, 62-67; sr geologist, Lawrence Livermore Nat Lab, 67-70, group leader geol & geophys, 70-74, sect leader geol, 74-75, sr scientist, 75-76, proj leader nuclear waste mgt, 76-88, ASSOC PROG LEADER ENERGY PROG, LAWRENCE LIVERMORE NAT LAB, 88- *Concurrent Pos:* NSF grant, 63-66. *Mem:* AAAS; Geol Soc Am; Am Nuclear Soc. *Res:* Petrology, mineralogy and structural geology; applied geology and geophysics; relation between site geology and containment of radioactivity and seismic coupling from an underground nuclear explosion; underground radionuclide migration; high-level nuclear waste disposal. *Mailing Add:* Lawrence Livermore Nat Lab L209 PO Box 808 Livermore CA 94550

RAMSTAD, PAUL ELLERTSON, b Minneapolis, Minn, Jan 30, 18; m 41, 88; c 4. FOOD SCIENCE. *Educ:* Univ Minn, BS, 39, PhD(agr biochem), 42. *Prof Exp:* Coop agent, USDA, 39-42; res chemist, Gen Mills, Inc, Minn, 42-46, head cereal res sect, 47-48; assoc prof biochem, Sch Nutrit, Cornell Univ, 48-53; asst dir res, Oscar Mayer & Co, Wis, 53-55; tech dir dept qual control, Gen Mills, Inc, 55-65; vpres res & develop, Am Maize-Prod Co, 65-68, pres, Corn Processing Processing Div, 69-75, pres, Co, 76-78; CONSULT, 78- *Concurrent Pos:* Ed, Cereal Sci Today, 57-62; sci ed, Cereal Chem, 78-84. *Mem:* AAAS; Am Chem Soc; Am Asn Cereal Chemists (pres, 64-65); Inst Food Technol. *Res:* Research administration; cereal chemistry and technology; grain storage; starches; syrups; vegetable gums; packaged foods; nutrition. *Mailing Add:* PO Box 841 Ithaca NY 14851-0841

RAMULU, MAMIDALA, b Andhra Pradesh, India, Sept 19, 49; m 85; c 2. MECHANICS OF COMPOSITE MACHINING, MECHANICS OF ABRASIVE WATERJET. *Educ:* Osmania Univ, India, BE, 74; Indian Inst Technol, New Delhi, MTech, 77; Univ Wash, PhD(mech eng), 82. *Prof Exp:* Postdoctoral fel mech eng, Univ Wash, 82, res asst prof, 82-85, asst prof, 85-90, ASSOC PROF MECH ENG, UNIV WASH, 90- *Concurrent Pos:* Mem, Fracture Tech Activ Comt & Res Papers Rev Comt, Soc Exp Mech, 86-; lectr, Indian Inst Metals, Am Soc Metals, 86; chmn, Fatigue Tech Comt, Soc Exp Mech, 88-, Mat Processing Comt, Am Soc Mech Engrs, 89-; NSF presidential young investr award, 89; AT&T Found award, Am Soc Eng Educ, 89. *Honors & Awards:* Ralph R Teetor Award, Soc Automotive Engrs, 87. *Mem:* Am Soc Eng Educ; Am Soc Mech Engrs; Soc Exp Mech; Soc Mfg Engrs; Soc Automotive Engrs; Am Soc Metals. *Res:* Development of production methods, modeling, and optimization of production processes; traditional and nontraditional machining methods to process engineered materials; EDM and ultrasonic machining process; fracture mechanics; author of numerous technical publications. *Mailing Add:* Dept Mech Eng FU-10 Univ Wash Seattle WA 98195

RAMUS, JOSEPH S, b Detroit, Mich, May 7, 40; m 81; c 3. ALGAE, PHYSIOLOGY. *Educ:* Univ Calif, Berkeley, AB, 63, PhD(bot), 68. *Prof Exp:* From asst prof to assoc prof biol, Yale Univ, 68-78; assoc prof, 78-85, asst dir, 81-89 actg dir, 89-90, PROF BOT, DUKE UNIV, 85-, DIR MARINE LAB, 90- *Mem:* Am Soc Limnol & Oceanog; Phycol Soc Am; Am Asn Univ Prof; Am Geophys Union. *Res:* Algal ecological physiology; estuarine dynamics; carbon partitioning. *Mailing Add:* Marine Lab Duke Univ Beaufort NC 28516

RAMWELL, PETER WILLIAM, b US citizen. PHARMACOLOGY, PHYSIOLOGY. *Educ:* Univ Sheffield, BS, 51; Univ Leeds, PhD(anesthesiol), 58. *Prof Exp:* Mem sci staff, Med Res Coun, Univ Leeds, 55-58; sr lectr pharmacol, Univ Bradford, 58-60; mem sci staff, Med Res Coun, Univ Birmingham & Oxford Univ, 60-64; sr scientist, Worcester Found Exp Biol, Mass, 64-69; assoc prof physiol, Stanford Univ, 69-74; PROF PHYSIOL & BIOPHYS, SCH MED, GEORGETOWN UNIV, 74- *Concurrent Pos:* Indust consult, UK & US, 60-69; dir, NIMH Postdoctoral Training Prog, 67-69. *Mem:* Am Physiol Soc; Physiol Soc; Am Pharmacol Soc; Brit Pharmacol Soc. *Res:* Prostaglandins and cell biology. *Mailing Add:* Dept Physiol 225 Basic Sci Georgetown Univ Med Ctr Reservoir Rd NW Washington DC 20007

RANA, MOHAMMAD A, b Lahore, Pakistan, May 1, 49; US citizen; m 89; c 2. PLANT CYTO-HISTOCHEMISTRY, BOTANY-ECOLOGY. *Educ:* Punjab Univ, Pakistan, BSc, 69, MSc, 71; Univ London, Eng, PhD(cell biol), 82. *Hon Degrees:* MIB, Inst Biol London, 81. *Prof Exp:* Lectr biol, Shiraz Univ, Sharaz, Iran, 73-79; demonstr biol, Univ London, Eng, 79-81; sr lectr bot, Univ Port Harcourt, Nigeria, 82-86; chairperson, Sci Dept, Mother Cabrini Inst, 86-90; ASST PROF ECOL, ST JOSEPH'S COL, 90- *Concurrent Pos:* Environ consult, Univ Port Harcourt, Nigeria, 83-86. *Res:* Idenfification of plant secondary products and their interaction with plant and animal life; searching resistant varieties of halophytes, a possible solution for oil spill problem. *Mailing Add:* 870 W 181 St No 5 New York NY 10033

RANA, MOHAMMED WAHEEDUZ-ZAMAN, b Lahore, Pakistan, May 28, 34; US citizen; m 65; c 4. HUMAN ANATOMY. *Educ:* Olivet Col, BA, 64; Wayne State Univ MS, 66, PhD(anat), 68. *Prof Exp:* From instr to asst prof, 68-78, ASSOC PROF ANAT, SCH MED, ST LOUIS UNIV, 78- *Mem:* Am Asn Anat; Soc Exp Biol & Med. *Res:* Study of RPE-retinal complex. *Mailing Add:* Dept Anat St Louis Sch Med St Louis MO 63104

RANA, RAM S, b Delhi, India, Aug 7, 28; m 59; c 1. SOLID STATE PHYSICS, SPECTROSCOPY. *Educ:* Univ Delhi, MSc, 51; Johns Hopkins Univ, PhD(physics), 59. *Prof Exp:* Lectr physics, D J Col, Baraut, India, 51-53; lectr, Vaish Col, Bhiwani, 53-54; jr instr, Johns Hopkins Univ, 54-58, res asst, 58-59; asst prof, Valparaiso Univ, 59-60; sci officer, Atomic Energy Estab, Bombay, India, 60-61; from asst prof to assoc prof physics, 61-69, PROF PHYSICS, COL OF THE HOLY CROSS, 69- *Concurrent Pos:* Vis prof, Univ Paris & Nat Ctr Sci Res, Paris, France, 66-67; consult scientist, Argonne Nat Lab, 67-77; guest scientist, Coun Sci & Indust Res, New Delhi, India, 69-70; vis scholar, Howard Univ, Cambridge, 83-84. *Mem:* Am Asn Physics Teachers. *Res:* Solid state spectroscopy. *Mailing Add:* Dept Physics Col of the Holy Cross Worcester MA 01610

RANADE, MADHAV (ARUN) BHASKAR, b Indore City, India, Sept 27, 42; m 68; c 1. CHEMICAL ENGINEERING, PARTICLE TECHNOLOGY. *Educ:* Nagpur Univ, BTech, 64; Ill Inst Technol, MS, 68, PhD(chem eng), 74. *Prof Exp:* Res engr chem eng, Chicago Bridge & Iron Co, 68-70; res engr particle technol, IIT Res Inst, 72-77; sr engr, Res Triangle Inst, 77-78, sect head particle technol, 78-87, mgr, Environ Technol Dept, 80, dir, Ctr Separation Processes, 84-87; PRES & OWNER, PARTICLE TECH, INC, 87- *Concurrent Pos:* Adj assoc prof, dept environ eng, Univ NC, 80- *Mem:* Fine Particle Soc; Am Inst Chem Engrs; Am Chem Soc; Sigma Xi; Air Pollution Control Asn; AAAS; Int Asn Colloid & Interface Scientists. *Res:* Fine particles technology; aerosol science; air pollution measurement and control; particle size measurement; powder technology; colloid and surface science. *Mailing Add:* Particle Tech Inc Bldg 335 Paint Branch Dr College Park MD 20742

RANADE, MADHUKAR G, b Bombay, India, Sept 8, 53; m 79; c 2. PROCESS METALLURGY, TECHNOLOGY DEVELOPMENT & TRANSFER. *Educ:* Indian Inst Technol, Bombay, BTech, 75; Univ Calif, Berkeley, MS, 77. *Prof Exp:* Res asst, Univ Calif, 75-77; engr, Inland Steel Co, 77-79, res engr, 79-82, sr res engr, 82-85, sect mgr, 85-89, MGR, INLAND STEEL CO, 89- *Concurrent Pos:* Reviewer, Ironmaking & Steelmaking J, 80-82; leader US deleg, ISO Meeting, Ottawa, Can, 82; key reader, Transactions of Iron & Steel Soc, 83-85; prog comt, Iron & Steel Soc, Am Inst Mining, Metall & Petrol Engrs, 83-; invited lectr, Univ Minn, Duluth, 87; direct steelmaking prog adv, Am Iron & Steel Inst, 89- *Honors & Awards:* Silver Medal, Indian Inst Technol, Bombay, India, 75; J E Johnson Award, Iron & Steel Soc, 89. *Mem:* Am Soc Testing & Mat; Iron & Steel Inst Japan; Am Inst Mech Engrs Iron & Steel Soc. *Res:* Development of new processes and improvement of existing processes for production of raw materials, iron and steel; responsible for leading and managing process research and development, and technology transfer and implementation for a major steel manufacturer. *Mailing Add:* Inland Steel Res 9-000 East Chicago IN 46312

RANADE, VINAYAK VASUDEO, b Wani, India, Feb 5, 38; m 64; c 1. MEDICINAL CHEMISTRY. *Educ:* Univ Bombay, BSc, 58, MSc, 61, PhD(org chem), 65. *Prof Exp:* Res assoc med chem, Col Pharm, Univ Mich, Ann Arbor, 65-75, univ fel, 65-68; SR PHARMACOLOGIST, ABBOTT LABS, 75- *Honors & Awards:* GC Prize for Nuclear Med & Radio Pharmacol, Tel-Aviv Univ, Israel, 74. *Mem:* Am Chem Soc; Am Pharmaceut Asn; Acad Pharmaceut Sci; fel Am Inst Chemists; Sigma Xi. *Res:* Synthetic medicinal chemistry and synthesis of radio pharmaceuticals; drug metabolism, biotransformation. *Mailing Add:* 1219 Deer Trail Libertyville IL 60048-1394

RANADIVE, NARENDRANATH SANTURAM, b Bombay, India, Sept 9, 30; m 60; c 2. IMMUNOLOGY, EXPERIMENTAL PATHOLOGY. *Educ:* Univ Bombay, BSc, 52, MSc, 57; McGill Univ, PhD(biochem), 65. *Prof Exp:* Res asst biophys, Indian Cancer Res Ctr, Bombay, India, 58-59; analyst, Glaxo Labs, Bombay, 59-60; asst prof, 69-74, ASSOC PROF PATH, DEPT PATH UNIV TORONTO, 74- *Concurrent Pos:* Fel, Scripps Clin & Res Found, Calif, 66-69; Med Res Coun Can scholar, 69-72. *Mem:* Am Asn Pathologists; Am Asn Immunol; Am Soc Photobiol; Int Pigment Cell Soc; NY Acad Sci. *Res:* Cellular mechanism in anaphylaxis; immunologic tissue injury; chemical mediators in neutrophil lysosomes; mechanism of the release of lysosomal constituents; free radicals in phototoxic reactions; phototoxic cutaneous reactions. *Mailing Add:* Dept Path Med Sci Bldg Univ Toronto Toronto ON M5S 1A8 Can

RANALLI, ANTHONY WILLIAM, b Portchester, NY, Jan 13, 30; m 64; c 3. FOOD TECHNOLOGY. *Educ:* Syracuse Univ, BS, 52, MS, 53, PhD(microbiol), 59. *Prof Exp:* Res supvr, Continental Baking Co, 59-63; tech dir refined syrups & sugars div, CPC Int, Inc, 63-70; asst to pres, Pepperidge Farm, Inc, 70-73, dir qual control, 70-77; dir qual assurance, Quaker Oats Co, Chicago, 77-79; VPRES QUAL ASSURANCE, PEPPERIDGE FARMS INC, 79- *Mem:* Am Chem Soc; Inst Food Technol; fel Am Inst Chem. *Res:* Chemistry; product development; quality control. *Mailing Add:* 46 Undercliff Rd Trumbull CT 06611-2547

RANCK, JAMES BYRNE, JR, NEUROSCIENCE, SPATIAL BEHAVIOR. *Educ:* Columbia Univ, MD, 55. *Prof Exp:* PROF NEUROSCI, STATE UNIV NY, DOWNSTATE MED CTR, 75- *Res:* Hippocampus. *Mailing Add:* Physiol - Box 31 Downstate Med Ctr Clarkson Ave Brooklyn NY 11203

RANCK, JOHN PHILIP, b Needmore, Pa, Aug 20, 36; m 61; c 2. MOLECULAR STRUCTURE. *Educ:* Elizabethtown Col, BS, 58; Princeton Univ, MA, 60, PhD(chem), 62. *Prof Exp:* Instr chem, Upsala Col, 62-63; from asst prof to assoc prof, 63-69, PROF CHEM, ELIZABETHTOWN COL, 69- *Concurrent Pos:* Schering Found award for grad teaching asst, Princeton Univ, 62; NSF partic, Univ Calif, Berkeley, 64, NSF fac fel, H C Orsted Inst, Copenhagen Univ, 70-71. *Mem:* AAAS; Am Chem Soc; Am Asn Physics Teachers; fel Sigma Xi. *Res:* Molecular orbital theory; molecular spectroscopy; equilibria of transition metal ions with asymmetric ligands; computers in chemistry; computer interfacing; mathematical modeling and arms control. *Mailing Add:* Dept Chem Elizabethtown Col One Alpha Way Elizabethtown PA 17022-2298

RANCK, RALPH OLIVER, organic chemistry; deceased, see previous edition for last biography

RAND, ARTHUR GORHAM, JR, b Boston, Mass, Sept 29, 35; m 60; c 3. FOOD SCIENCE. *Educ:* Univ NH, BS, 58; Univ Wis, MS, 61, PhD(dairy & food indust, biochem), 64. *Prof Exp:* Res asst dairy & food indust, Univ Wis, 58-63; instr animal & dairy sci, Univ RI, 63-65, from asst prof to assoc prof animal sci & food & resource chem, 65-75, chmn dept, 81-90, PROF FOOD SCI & NUTRIT, UNIV RI, 75- *Concurrent Pos:* Vis prof, Univ New South Wales, 71 & Rutgers Univ, 90. *Mem:* Am Wine Soc; Inst Food Technologists; Am Soc Enol & Viticult; NY Acad Sci; Am Inst Nutrit. *Res:* Food enzyme technology; food bioprocessing; seafood quality; enology. *Mailing Add:* Food Sci & Nutrit Res Ctr 530 Liberty Lane Univ RI West Kingston RI 02892

RAND, AUSTIN STANLEY, b Seneca Falls, NY, Sept 19, 32; m 61; c 3. EVOLUTIONARY BIOLOGY. *Educ:* De Pauw Univ, BA, 55; Harvard Univ, PhD(biol), 61. *Prof Exp:* Asst mammal, Field Mus Natural Hist, Chicago, 57; res asst herpet, Mus Comp Zool, Cambridge Univ, 61-62; zoologist, Secy Agr, Sao Paulo, 62-64; BIOLOGIST HERPET, SMITHSONIAN TROP RES INST, 64-70, 71- *Concurrent Pos:* Adj assoc prof, Univ Pa, 70-75. *Mem:* Am Soc Naturalists; Soc Study Evolution; Am Soc Ichthyologists & Herpetologists; Soc Study Amphibians & Reptiles; Herpetologists League. *Res:* Studies of behavior and ecology of reptiles and amphibians, particularly social behavior and communication as adaptations to resource partitioning in lizards and frogs in tropical environments. *Mailing Add:* Smithsonian Trop Res Inst PO Box 2072 Balboa Panama

RAND, JAMES LELAND, b Ft Worth, Tex, Dec 23, 35; m 74; c 6. AEROSPACE ENGINEERING, MECHANICAL ENGINEERING. *Educ:* Univ Md, College Park, BS, 61, MS, 63, PhD(mech eng), 67. *Prof Exp:* Instr eng, Univ Md, College Park, 61-62; res engr, Naval Ord Lab, 62-68; from asst prof to prof struct mech, Tex A&M Univ, 68-78; mgr, Dynamic Anal Southwest Res Inst, 78-83; PRES, WINZEN INT INC, 83- *Concurrent Pos:* Dir, Balloon Eng Lab, 77-78; vis mem, Grad Fac, Tex A&M Univ, 78- *Mem:* Am Inst Aeronaut & Astronaut; Am Soc Mech Engrs; Am Acad Mech; Am Soc Eng Educ. *Res:* Stress wave propagation; internal ballistics; hypervelocity impact effects shock wave propagation; structural mechanics; dynamic plasticity; scientific balloon design-analysis. *Mailing Add:* Winzen Int Inc 12001 Network Blvd Suite 200 San Antonio TX 78249

RAND, LEON, b Boston, Mass, Oct 8, 30; m 59; c 3. ORGANIC CHEMISTRY. *Educ:* Northeastern Univ, BS, 53; Univ Tex, MA, 56, PhD(chem), 58. *Prof Exp:* Fel, Purdue Univ, 58-59; from asst prof to assoc prof chem, Univ Detroit, 59-68; prof chem, Youngstown State Univ, 68-81, dean grad studies & res, 73-81; prof chem & vchancellor acad affairs, Pembroke State Univ, 81-85; CHANCELLOR, IND UNIV SOUTHEAST, 85- *Mem:* Am Chem Soc; Am Inst Chemists. *Res:* Steric effects and use of potassium fluoride in organic chemistry; anodic oxidation reactions; urethane chemistry; carbonium ion processes; halide catalysis. *Mailing Add:* Ind Univ SE 4201 Grant Line Rd New Albany IN 47150

RAND, PATRICIA JUNE, b St Paul, Minn, June 6, 26. PLANT ECOLOGY. *Educ:* Univ Minn, Minneapolis, BS, 47, MS, 53; Duke Univ, PhD(bot), 65. *Prof Exp:* Teaching asst, res asst & instr bot, Univ Minn, Minneapolis, 47-53; instr biol, Hamline Univ, 53-58; teaching asst bot, Duke Univ, 58-60; from instr to asst prof, Univ Ark, Fayetteville, 62-66; asst prof, Univ Nebr, Lincoln, 66-73; SR SCI ADV ECOL, ATLANTIC RICHFIELD CO, 73- *Concurrent Pos:* Seed technologist, Northrup, King & Co, 51-52; collabr, US Nat Park Serv, 59-70; adj prof, Univ Nebr, Lincoln, 73-76. *Mem:* Fel AAAS; Ecol Soc Am; Bot Soc Am; Am Inst Biol Sci. *Res:* Reclamation of disturbed areas; land use management; physiological ecology of woody plants. *Mailing Add:* Health Safety & Environ Protection Atlantic Richfield Co 515 S Flower St Los Angeles CA 90071

RAND, PETER W, b Boston, Mass, Oct 26, 29; m 53; c 2. CARDIOVASCULAR PHYSIOLOGY, HEMATOLOGY. *Educ:* Harvard Univ, AB, 51, MD, 55. *Prof Exp:* Intern, Maine Med Ctr, 55-56, resident internal med, 56-57 & 59-60, dir Res Dept, 65-89, CHMN PRE-MED, MAINE MED CTR, 74-; ASST PROF MED, COL MED, UNIV VT, 81- *Concurrent Pos:* Fel cardiol, Maine Med Ctr, 60-61, USPHS fel, 61-63, NIH grant, 63; asst clin prof med, Col Med, Univ Vt, 72-80; adj prof, appl immunol, Univ SMaine, 88- *Mem:* Am Fedn Clin Res; Am Physiol Soc; Am Heart Asn; Soc Rheol; Microcirc Soc. *Res:* Blood rheology and viscosity; oxygen transport; hemodynamics. *Mailing Add:* Dept Med Univ Maine 22 Bramhall St Portland ME 04102

RAND, PHILLIP GORDON, b Meredith, NH, Nov 5, 34; m 55; c 3. BIOCHEMISTRY. *Educ:* John Brown Univ, BA, 56; Univ Wyo, MS, 58; Purdue Univ, PhD(biochem), 63. *Prof Exp:* Res biochemist, 63-71, sr res biochemist & sect head, 71-73, prod develop mgr, 73-77, prin res scientist, 77-83, SR DEVELOP SCIENTIST, MILES INC, 83- *Mem:* Am Chem Soc; Am Oil Chem Soc. *Res:* Nutritional effects of fats; cholesterol metabolism; clinical diagnosis of disease. *Mailing Add:* 1320 W Lexington Ave Elkhart IN 46514-2048

RAND, RICHARD PETER, b Can, Jan 31, 37; c 3. MEMBRANE BIOLOGY, ELECTROPHYSIOLOGY. *Educ:* Carleton Univ, BSc, 59; Univ Western Ont, MSc, 61, PhD(biophysics), 65. *Prof Exp:* Fel, Med Res Coun Can, Nat Ctr Sci Res, Paris, 64-66; from asst prof to assoc prof, 66-73, PROF BIOL SIC, BROCK UNIV, 73- *Concurrent Pos:* Chmn, Brock Univ, 73-; assoc mem, Dept Biol, McMaster Univ, & Dept Physics, Guelph Univ, 76-; mem, Grant Selection Comt, Natural Sci & Eng Res Coun Can, 78-80, grants & scholar comt, Nat Sci & Eng Res Coun Can, 86-89. *Mem:* Biophys Soc. *Res:* Measurements of the forces of interaction between model and biological cell membranes and mechanisms of membrane fusion; developmental aspects of the electrical properties of nerve cells in culture. *Mailing Add:* Biol Sci Brock Univ Merrittville Hwy St Catharines ON L2S 3A1 Can

RAND, ROBERT COLLOM, b Pittsburgh, Pa, Aug 24, 17; m 42; c 3. MATHEMATICS. *Educ:* Duke Univ, AB, 39, AM, 40; Univ Md, PhD(math), 43. *Prof Exp:* Asst, Univ Md, 40-43; stress analyst, Eng & Res Corp, 42-44; instr math, US Naval Acad, 46-48; sr mathematician, Appl Physics Lab, Johns Hopkins Univ, 48-82; RETIRED. *Mem:* Math Asn Am. *Res:* Shock waves in non-steady flow; supersonic aerodynamics and shock waves; rectilinear motion of a gas subsequent to an internal explosion; dynamics of guided missiles; radar countermeasures; reliability; transportation systems. *Mailing Add:* 7018 Pindell School Rd Fulton MD 20759

RAND, SALVATORE JOHN, b Brooklyn, NY, Dec 1, 33; m 56; c 5. ANALYTICAL CHEMISTRY, PHYSICAL CHEMISTRY. *Educ:* Fordham Univ, BS, 56; Rensselaer Polytech Inst, PhD(phys chem), 60. *Prof Exp:* Sr res chemist, Colgate-Palmolive Co, 60-62; res scientist, Res Labs, United Technols, 62-67; res chemist, 67-70, CONSULT, FUELS RES, RES & DEVELOP DEPT, TEXACO INC. *Concurrent Pos:* Mem adj fac, Grad Sch Chem, St Joseph Col, Conn, 64-67. *Mem:* AAAS; Am Chem Soc; Am Phys Soc. *Res:* Photochemistry; spectroscopy; kinetics of gas and liquid phase reactions; lasers; radiochemistry; lubrication chemistry; hydrocarbon analysis; gas and liquid chromatography; fuels analysis and distribution. *Mailing Add:* Res & Develop Dept Texaco Inc PO Box 509 Beacon NY 12508

RAND, STEPHEN COLBY, b Seattle, Wash, Nov 20, 49; m 75; c 2. QUANTUM ELECTRONICS. *Educ:* McMaster Univ, BSc, 72; Univ Toronto, MSc, 74, PhD(physics), 78. *Prof Exp:* Res assoc, Varian Labs, Stanford Univ, 80-82; mem tech staff, Hughes Res Lab, 82-87; ASSOC PROF, UNIV MICH, 87- *Mem:* Am Phys Soc; Optical Soc Am. *Res:* Upconversion and color center lasers; cooperative nonlinear dynamics in rare earth materials; four-wave mixing spectroscopy; growth and characterization of diamond-like materials. *Mailing Add:* Dept Physics Univ Mich Ann Arbor MI 48109-2122

RAND, WILLIAM MEDDEN, b Seneca Falls, NY, June 26, 38; m 67; c 1. BIOSTATISTICS. *Educ:* Ind Univ, BA, 59; Brandeis Univ, MA, 61; Univ Calif, Los Angeles, PhD(biostatist), 69. *Prof Exp:* Engr, Jet Propulsion Labs, 62-64; res assoc med, Univ Southern Calif, 65-68; from asst prof to assoc prof biostatist, Mass Inst Technol, 69-77, lectr, 77-88, dir, Infoods, 83-88; PROF BIOSTATIST, TUFTS UNIV SCH MED, 88- *Mem:* Biomet Soc; Am Statist Asn; Am Inst Nutrit. *Res:* Mathematical and statistical biomedicine. *Mailing Add:* Tufts Univ Sch Med 136 Harrison Ave Boston MA 02111

RANDA, JAMES P, b Chicago, Ill, Jan 26, 47; m 70; c 1. THEORETICAL PHYSICS. *Educ:* Ill Benedictine Col, BSc, 69; Univ Ill, Urbana, MSc, 70, PhD(physics), 74. *Prof Exp:* Vis asst prof, Dept Physics, Tex A&M Univ, 74-75; res fel, Dept Theoret Physics, Univ Manchester, 75-78; vis asst prof, dept physics, Univ Colo, Boulder, 78-80, asst prof, 80-83; PHYSICIST, ELECTROMAGNETIC FIELDS DIV, NAT INST STANDARDS & TECHNOL, BOULDER, 83- *Concurrent Pos:* Lectr, Dept Physics, Univ Colo, Boulder, 85-89. *Mem:* Am Phys Soc; Inst Elec & Electronic Engrs; Int Union Radio Sci. *Res:* Mechanisms for symmetry breaking and mass generation in gauge theories for elementary particles; characterization of complex electromagnetic environments; antenna metrology; electric-field probe development. *Mailing Add:* Nat Inst Standards & Technol 723 03 325 Broadway Boulder CO 80303-3328

RANDALL, BARBARA FEUCHT, b Buffalo, NY, Jan 7, 25; m 49; c 4. PHYSIOLOGY. *Educ:* State Univ NY, BS, 45; Univ Iowa, MS, 48, PhD, 51. *Prof Exp:* Asst physiol, Univ Iowa, 47-52; res assoc phys med, Ohio State Univ, 53-55; instr med, Med Ctr, Univ Mo, 55-62; lectr phys med, Med Sch, Northwestern Univ, 64-69; asst prof, 69-88, EMER PROF PHYSIOL, IND UNIV, BLOOMINGTON, 88- *Mem:* Am Phys Ther Asn. *Mailing Add:* 609 S Jordan Ave Bloomington IN 47401-5121

RANDALL, CHARLES ADDISON, JR, b Daytona Beach, Fla, Sept 12, 15; m 41; c 2. COSMIC RAY PHYSICS. *Educ:* Kalamazoo Col, AB, 36; Cornell Univ, MA, 39; Univ Mich, PhD(physics), 51. *Prof Exp:* Instr physics, Allen Acad, Tex, 39-40; instr, Wayland Jr Col & Acad, 40-42; res physicist, Fairbanks Morse & Co, 42-45; asst, Univ Mich, 46-50; from asst prof to assoc prof, 50-60, PROF PHYSICS, OHIO UNIV, 60- *Concurrent Pos:* Off Europ Econ Coop sr sci fel, NSF Europ Orgn Nuclear Res Lab, 60-61; res partic, Oak Ridge Nat Lab, 52; instr, Goodyear Atomic Corp, 54; consult, Los Alamos Sci Lab, 56-; consult nuclear emulsion inst, Univ Chicago, 56; on sabbatical leave, Sandia Corp, NMex & Atomic Energy Res Estab, Eng, 68-69; mem int comt of the forum on physics & soc, Am Phys Soc, 75- *Mem:* AAAS; Am Geophys Union; fel Am Phys Soc; Am Asn Physics Teachers. *Res:* Cosmic rays; fundamental particles. *Mailing Add:* 14 Palmer Lane Palm Coast FL 32037

RANDALL, CHARLES CHANDLER, b Cedar Rapids, Iowa, Mar 27, 13; m 41; c 2. MICROBIOLOGY. *Educ:* Univ Ky, BS, 36; Vanderbilt Univ, MD, 40; Am Bd Path, dipl. *Prof Exp:* NIH fel, Vanderbilt Univ, 48-49, instr path, Sch Med, 49-51, asst prof path & bact, 51-52, assoc prof bact, 52-55, prof microbiol & actg head dept, 55-57; prof & chmn dept, 57-78, EMER PROF MICROBIOL, SCH MED, UNIV MISS, 78- *Mem:* Am Asn Immunologists; Am Soc Microbiol; Am Asn Pathol; Am Soc Cell Biol. *Res:* Virology. *Mailing Add:* Dept Microbiol Univ Miss Med Ctr Jackson MS 39216

RANDALL, CHARLES HAMILTON, mathematics, physics; deceased, see previous edition for last biography

RANDALL, CHARLES MCWILLIAMS, b Visalia, Calif, May 3, 38; m 60; c 2. ATMOSPHERIC PHYSICS. *Educ:* Union Col, Nebr, BA, 60; Mich State Univ, MS, 62, PhD(physics), 64. *Prof Exp:* Asst prof physics, Mich State Univ, 64-65; mem tech staff, 65-84, DEPT HEAD, SPACE SCI LAB, AEROSPACE CORP, 84- *Mem:* Am Meteorol Soc; Optical Soc Am; Sigma Xi. *Res:* Infrared properties of the atmosphere; solar energy resource assessment; application of computers to satellite data analysis. *Mailing Add:* 4925 Calle de Arboles Torrance CA 90505

RANDALL, CLIFFORD W(ENDELL), b Somerset, Ky, May 1, 36; m 59; c 2. POLLUTION CONTROL. *Educ:* Univ Ky, BSCE, 59, MSCE, 63; Univ Tex, Austin, PhD(environ health eng), 66. *Prof Exp:* Asst prof civil eng, Univ Tex, Arlington, 65-68; from asst prof sanit eng to prof sanit eng, 68-81, CHARLES LUNSFORD PROF CIVIL ENG & CHMN, ENVIRON ENG & SCI PROGS, VA POLYTECH INST & STATE UNIV, 81. *Concurrent Pos:* Res specialist, Aerobiol Lab, Southwest Med Sch, Univ Tex, 66-68; res dir, San Antonio River Auth, 67; consult pollution control, United Piece Dye Works, Inc, 69-, Blue Ridge Winkler, 76-89, Hester Industs, 78-82, Holly Farms Inc, 79-82 & Celanese Inc, 81-; consult waste treatment & munitions, Hercules, Inc, & Radford Army Ammunitions Plant, Va, 70-74; consult munic, Wiley & Wilson, Inc, Lynchburg, 70-71 & Harwood Beebe, Inc, Spartanburg, SC; consult indust, E I du Pont de Nemours & Co, Inc, Martinsville, 70, 72 & 78, Waynesboro, Va, 77-; consult, Mead Corp, Lynchburg, 70-71, & Belding Corticelli Fiber Glass Fabrics Co, Bedford, Va, 71-73; res & training grant consult, Environ Protection Agency, 70-71; chmn watershed monitoring subcomt & dir, Occoquan Watershed Water Qual Monitoring Prog, State Water Control Bd Va, 71-; affiliated consult, George A Jeffreys & Co, Inc, Salem, Va, 72-78, Black & Veatch Inc, 86-, & Innovatech, Inc, 84-; consult indust waste treat, Am Cyanamid, 72-78, Va Bd Cert Water & Wastewater Works Operators, 78-86, Va-NC Chowan River Basin Tech Panel; scientist & tech adv comt, Chesapeake Bay Prog, 85-; chmn, US Nat Comt Int Asn Water Pollution Res & Control, 86-88, mem, gov; proj anal water supply, Southern Baptist Foreign Mission Bd, Kenya, 83; consult, WHO, New Delhi, India, 83-84. *Honors & Awards:* Bedell Award, Water Pollution Control Fedn, 83, Phillip F Morgan Cert, 81. *Mem:* Water Pollution Control Fedn; Am Soc Civil Engrs; Am Water Works Asn; Int Asn Water Pollution Res; Asn Environ Eng Professors (secy-treas, 79-80). *Res:* Mechanisms of sludge dewatering; munitions wastes treatment; reservoir eutrophication; sanitary microbiology and stormwater runoff pollution control, industrial waste treatment; biological nutrient removal wastewater treatment. *Mailing Add:* Dept Civil Eng Va Polytech Inst & State Univ Blacksburg VA 24061

RANDALL, DAVID CLARK, b St Louis, Mo, Apr 23, 45; m 68; c 3. AUTONOMIC NERVOUS SYSTEMS, BEHAVIORAL MEDICINE. *Educ:* Taylor Univ, AB, 67; Univ Wash, PhD(physiol & biophys), 71. *Prof Exp:* Asst prof behav biol, Sch Med, Johns Hopkins Univ, 72-75; from asst prof to assoc prof, 75-85, PROF PHYSIOL, SCH MED, UNIV KY, 85- *Mem:* Am Physiol Soc; Fedn Am Soc Exp Biol; Am Sci Affil; Soc Exp Biol & Med; Pavlovian Soc NAm. *Res:* Nervous control of the heart and coronary circulation during periods of behavioral and environmental stress in unanesthetized non-human primates and dogs; gravitational effects on circulation; behavioral medicine. *Mailing Add:* Dept Physiol & Biophys Univ Ky Sch Med Lexington KY 40536-0084

RANDALL, DAVID JOHN, b London, Eng, Sept 15, 38; Can citizen; m 62; c 5. ZOOLOGY, PHYSIOLOGY. *Educ:* Univ Southampton, BSc, 60, PhD(physiol), 63; FRSC, 81. *Prof Exp:* From asst prof to assoc prof, 63-73, PROF ZOOL, UNIV BC, 73-, ASSOC DEAN GRAD STUDIES, 90- *Concurrent Pos:* Vis lectr, Bristol Univ, 68-69; Guggenheim Found fel, 68-69; vis scientist, Marine Labs, Univ Tex, 70 & Zool Sta, Naples, 73; chmn animal biol comt, Nat Res Coun, Can, 74; NATO vis scientist, Acadia Univ, 75 & Marine Lab, Univ Tex, 77; chief scientist, Alpha Helix Amazon Exped, 76; mem adv bd, J Comp Physiol, 77- & J Exp Biol, 81-84; assoc ed, Marine Behavior Physiol. *Mem:* Can Soc Zoologists; Soc Exp Biologists; fel Royal Soc Can, 81. *Res:* Respiration and circulation in fish and amphibia with an emphasis on oxygen and carbon dioxide transfer and hydrogen ion regulation across the gills of fish. *Mailing Add:* Univ BC Dept Zool 6270 University Blvd Vancouver BC V6T 2A9 Can

RANDALL, EILEEN LOUISE, medical microbiology; deceased, see previous edition for last biography

RANDALL, ERIC A, b Silver Springs, NY, June 12, 46; m 71; c 2. BRYOLOGY, AGROFORESTRY. *Educ:* State Univ NY Oswego, BS, 68; Pa State Univ, PhD(bot), 73. *Prof Exp:* from asst prof to assoc prof biol, 73-85, PROF BIOL, STATE UNIV NY BUFFALO, 85-, CHMN, DEPT BIOL, 91- *Mem:* Am Bryol & Lichenological Soc; Am Soc Plant Taxon; Ecol Soc Am; Int Asn Plant Taxon; Sigma Xi. *Res:* Bryological and phytogeographic studies of Atlantic northern United States and Canada; microfiltration of maple (Acer saccharum) sap. *Mailing Add:* Biol Dept State Univ Col 1300 Elmwood Ave Buffalo NY 14222

RANDALL, FRANCIS JAMES, b Williston, NDak, Feb 17, 42. POLYMER CHEMISTRY. *Educ:* Dickinson State Col, NDak, BA, 63; Univ Sask, MS, 68; Univ NDak, PhD(chem), 70. *Prof Exp:* CHEMIST PROD DEVELOP, S C JOHNSON & SON, INC, 69-, SR RES ASSOC, S C JOHNSON WAX. *Mem:* Am Chem Soc; AAAS; Sigma Xi. *Res:* Polymer coatings. *Mailing Add:* SC Johnson Wax Inc 1525 Howel St MS115 Racine WI 53403-5011

RANDALL, GYLES WADE, b Rochester, Minn, Jan 3, 42; m 66; c 2. SOIL SCIENCE. *Educ:* Univ Minn, St Paul, BS, 63, MS, 68; Univ Wis-Madison, PhD(soils), 72. *Prof Exp:* Res asst soils, Univ Minn, 64-65, res fel, 65-69; res asst, Univ Wis, 69-72; from asst prof to assoc prof, 72-80, PROF SOILS, FAC SOIL SCI, SOUTHERN EXP STA, UNIV MINN, 80- *Mem:* Am Soc Agron; Soil Sci Soc Am; Soil Conserv Soc Am; Sigma Xi; Coun Agr Sci & Technol. *Res:* Tillage, soil fertility and plant nutrition with emphasis on nutrient accumulation and movement. *Mailing Add:* 123 Eighth St SE Waseca MN 56093

RANDALL, HENRY THOMAS, b New York, NY, Aug 29, 14; m 40; c 3. NUTRITION. *Educ:* Princeton Univ, AB, 37; Columbia Univ, MD, 41, ScD(surg), 50. *Hon Degrees:* MA, Brown Univ, 68. *Prof Exp:* From instr to asst prof surg, Col Physicians & Surgeons, Columbia Univ, 50-51; assoc prof, Med Col, Cornell Univ, 51-52, prof, Sloan-Kettering Div, 52-55, prof, Med Col, 55-67; prof med sci, 67-79, chmn, Sect Surg, Div Biol & Med Sci, 71-79, EMER PROF MED SCI, BROWN UNIV, 79- *Concurrent Pos:* Asst attend surgeon, Presby Hosp & asst vis surgeon, Delafield Hosp, 50-51; clin dir, Mem Hosp, 51-61, chmn dept surg, 51-66, attend surgeon, 51-67, med dir & vpres med affairs, 61-65; vis surgeon, James Ewing Hosp, 51-67; surgeon-in-chg, Div Surg Res, RI Hosp, 67-75, surgeon-in-chief, 70-79, consult surgeon, 79; Am Col Surgeons, 77; Am Cancer Soc, 85. *Honors & Awards:* Distinguished Serv Award, Am Col Surgeons, 77. *Mem:* Soc Univ Surg (pres, 59-60); Am Soc Clin Surg; Am Surg Asn; fel Am Col Surg; Int Surg Soc; Sigma Xi; Am Cancer Soc. *Res:* Metabolic response to surgery, especially electrolyte and water balance; renal and gastrointestinal tract physiology; surgical and nutritional problems of patients with cancer; nutrition of surgical patients. *Mailing Add:* RI Hosp 593 Eddy St Providence RI 02902

RANDALL, HOWARD M, b Rockville Ctr, NY, May 5, 36; m 62, 78; c 2. PHYSIOLOGY. *Educ:* Univ RI, BS, 58; Univ Rochester, PhD(physiol), 65. *Prof Exp:* Instr, 65-68, from asst prof to assoc prof, 68-83, PROF PHYSIOL, SCH MED, LA STATE UNIV MED CTR, NEW ORLEANS, 83- *Concurrent Pos:* La Heart Asn grant, 66-67; NIH grant, 68-; assoc dean student affairs & records, Sch Med, La State Univ. *Mem:* AAAS; Am Physiol Soc; Biophys Soc; Am Asn Med Cols; Sigma Xi. *Res:* Relationships between physiological functions and metabolism in the kidney. *Mailing Add:* 5711 Pratt Dr New Orleans LA 70122

RANDALL, J MALCOM, b East St Louis, Ill, Aug 9, 16; m 72. HEALTH CARE ADMINISTRATION. *Educ:* McKendree Col, Lebanon, Ill, AB, 39; St Louis Univ, MHA, 55. *Prof Exp:* Chief spec serv, Vet Admin Hosp, St Louis, 53-56; asst dir, Vet Admin Hosp, Spokane, 56-57, Vet Admin Res Hosp, Chicago, 57-58, Vet Admin Hosp, Indianapolis, 58-60 & Vet Admin Ctr, Milwaukee, 60-64; dir, Vet Admin Hosp, Miles City Mont, 64-66; assoc prof, 66-74, PROF HEALTH & HOSP ADMIN, COL HEALTH RELATED PROFESSIONS, UNIV FLA, 75-; DIR, VET ADMIN MED CTR, GAINESVILLE, 66-, MED DIST DIR, MED CENTERS & OUTPATIENTS CLINICS FLA, 83- *Concurrent Pos:* Mem. bd dirs, Am Health Planning Asn, Coun Teaching Hosps, Asn Am Med Cols, Coun Regents, Am Col Hosp Adminr & Nat Adv Comts, Vet Admin; partic, US-Yugoslavia Health Scientist Exchange Prog, Univ Clin Ctr, Ljubljana, Yugoslavia, 83, 84, 85, 86, 87, 88 & 89, US- Hungary Health Scientist Exchange Prog, 89 & 90. *Honors & Awards:* Presidential Rank Award, 83. *Mem:* Inst Med-Nat Acad Sci; fel Am Col Hosp Admin; Am Hosp Asn; Am Soc Pub Admin. *Res:* Numerous published articles in med journals. *Mailing Add:* Vet Admin Med Ctr Archer Rd Gainesville FL 32602

RANDALL, JAMES CARLTON, JR, b Florence, SC, May 26, 37; m 60; c 4. PHYSICAL CHEMISTRY. *Educ:* Univ SC, BS, 59; Emory Univ, MS, 61; Duke Univ, PhD(phys chem), 64. *Prof Exp:* Res chemist, Chemstrand Res Ctr, Inc, Monsanto Co, 64-68; res assoc, Phillips Petrol Co, 68-85; sr res assoc, Exxon, 85-88; mgr polymer sci, Machelen, Belg, 88-90; SR RES ASSOC, EXXON CORP RES, 91- *Mem:* Am Chem Soc. *Res:* Nuclear magnetic resonance of polymers; structure and properties of polymers; blend science. *Mailing Add:* 170 LC Exxon Corp Res Hwy 22E Clinton NJ 08801

RANDALL, JAMES EDWIN, b Bloomington, Ind, July 23, 24; m 49; c 4. BIOPHYSICS, PHYSIOLOGY. *Educ:* Purdue Univ, BSEE, 47; State Univ Iowa, 52; Ohio State Univ, PhD(biophys), 55. *Prof Exp:* Electronics eng, Collins Radio Co, Cedar Rapids, 47-49; from asst prof to assoc prof physiol, Univ Mo, 55-63; prof, Northwestern Univ, 63-68; prof physiol, 68-89, EMER PROF, IND UNIV, BLOOMINGTON, 89- *Concurrent Pos:* Consult, Nat Health & Lung Prog Proj Comt, 70-74. *Mem:* Am Physiol Soc. *Res:* Physiological variables as statistical signals; physiological simulations with microcomputers; physiological time series; laboratory digital computation. *Mailing Add:* 609 S Jordan Ave Bloomington IN 47401-5121

RANDALL, JANET ANN, b Twin Falls, Idaho, July 3, 43; m 85. COMMUNICATION, SOCIAL ORGANIZATION. *Educ:* Univ Idaho, BS, 65; Univ Wash, MEd, 69; Wash State Univ, PhD(zool), 77. *Prof Exp:* Postdoctoral fel biopsychol, Univ Tex, Austin, 77-79; from asst prof to asoc prof biol, Cent Mo State Univ, 79-87; ASSOC PROF BIOL, SAN FRANCISCO STATE UNIV, 87- *Concurrent Pos:* Vis assoc prof psychol, Cornell Univ, 84-85; prin investr grants, NSF, 86- *Mem:* Animal Behav Soc; Am Soc Zoologists; Am Soc Mammalogists; Int Soc Behav Ecol. *Res:* Comparison of communications and social organization of desert rodents-kangaroo rats. *Mailing Add:* Dept Biol San Francisco State Univ San Francisco CA 94132

RANDALL, JOHN D(EL), b Whittier, Calif, Nov 19, 32; m 53; c 4. NUCLEAR ENGINEERING. *Educ:* Univ Calif, BS, 55, MS, 56; Tex A&M Univ, PhD(nuclear eng), 65. *Prof Exp:* Physicist, Lawrence Radiation Lab, Calif, 55-56; nuclear engr, Nucleonics Div, Aerojet Gen Corp, Gen Tire & Rubber Co, 56-58; asst prof nuclear eng, Tex A&M Univ, 58-63, assoc head, Nuclear Sci Ctr, 63-65, prof nuclear eng & dir, Nuclear Sci Ctr, 65-83; sr exec consult, NUS Corp, 83-85; prin engr, Gen Physics Corp, 85-87; prog dir, NY Energy Res & Develop Authority, 87-90; EXEC DEP CHMN, NY STATE LOW-LEVEL RADIOACTIVE WASTE SITING COMN, 90- *Concurrent Pos:* Vis lectr, Am Inst Biol Sci, 65- *Mem:* Fel Am Nuclear Soc. *Res:* Neutron activation analysis; utilization of radioactive and non-radioactive tracers; industrial applications of nuclear energy; forensic neutron activation analysis. *Mailing Add:* Three Nicklaus Dr Gansevoort NY 12831

RANDALL, JOHN DOUGLAS, b Corning, NY, July 23, 42; m 69; c 3. APPLIED MATHEMATICS, SOFTWARE SYSTEMS. *Educ:* Cornell Univ, BMchE, 65, PhD(fluid dynamics), 72; Clarkson Col, MSME, 67. *Prof Exp:* Instr thermodynamics, Clarkson Col, 66-67; sr engr, Appl Physics Lab, Johns Hopkins Univ, 73-80; SECT LEADER, US NUCLEAR REGULATORY COMN, 80- *Mem:* AAAS; Am Soc Mech Engrs; Soc Indust & Appl Math; Am Geophys Union. *Res:* Management of development and coordination of research projects to provide information for regulation of radioactive waste disposal. *Mailing Add:* 6318 Dry Stone Gate Columbia MD 21045-2888

RANDALL, JOHN ERNEST, b Los Angeles, Calif, May 22, 24; m 51; c 2. MARINE BIOLOGY. *Educ:* Univ Calif, Los Angeles, BA, 50; Univ Hawaii, PhD(marine zool), 55. *Prof Exp:* Asst zool, Univ Calif, Los Angeles, 50; asst, Univ Hawaii, 50-53; Bishop Mus fel, Yale Univ, 55-56; res asst prof ichthyol, Marine Lab, Univ Miami, 57-61; prof biol & dir inst marine biol, Univ PR, 61-65; MARINE ZOOLOGIST & SR ICHTHYOLOGIST, BERNICE P BISHOP MUS, 65- *Concurrent Pos:* Dir, Oceanic Inst, Waimanalo, 65-66; marine biologist, Inst Marine Biol, Univ Hawaii, 67-69; mem subcomt conserv ecosysts, Int Biol Prog; Great Barrier Reef Comt. *Honors & Awards:* Stoye Award Ichthyol, Am Soc Ichthyol & Herpet; Distinguished Fel Am Soc Ichthyol & Herpet; Gibbs Award, Syst Ichthyol, 90. *Mem:* Am Soc Ichthyol & Herpet; Ichthyol Soc Japan; Australian Coral Reef Soc; Europ Ichthyol Union; Int Soc Reef Studies; Explorers Club. *Res:* Tropical marine ichthyology and biology. *Mailing Add:* Div Ichthyol Box 19000-A Bernice P Bishop Mus Honolulu HI 96817-0916

RANDALL, JOHN FRANK, b Walnut, NC, Aug 2, 18; m 50; c 1. BIOLOGY. *Educ:* Univ NC, AB, 41; Univ Mich, MS, 50; Univ SC, PhD(biol), 57. *Prof Exp:* Instr biol, Alpena Community Col, 52-55; from assoc prof to prof zool, 57-77, PROF BIOL, APPALACHIAN STATE UNIV, 77- *Mem:* AAAS; Ecol Soc Am; Sigma Xi. *Res:* Vertebrate ecology; ornithology; ichthyology. *Mailing Add:* Dept Biol Appalachian State Univ Boone NC 28607

RANDALL, JOSEPH LINDSAY, b Clanton, Ala, Dec 7, 32; m 54; c 4. LASERS, COMMUNICATIONS. *Educ:* Univ Ala, BS, 54, MS, 56, PhD(physics), 60. *Prof Exp:* Res physicist, Lockheed Aircraft, 60-61 & Redstone Arsenal, 61-62; res physicist, 62-67, BR CHIEF, TECHNOL DIV, ASTRIONICS LAB, MARSHALL SPACE FLIGHT CTR, NASA, 67-, DIR, INFO & ELECTRONICS SYST LAB. *Honors & Awards:* NASA Exceptional Sci Achievement Medal. *Mem:* Am Phys Soc; Optical Soc Am. *Res:* Laser communication systems for space application; atmospheric effects on optical communication. *Mailing Add:* 2212 Shadecrest Rd Huntsville AL 35801

RANDALL, LAWRENCE KESSLER, JR, b Rochester, NY, Oct 21, 38; m 63; c 1. LARGE TELESCOPE DESIGN. *Educ:* Univ Ariz, BS, 62, MS, 73. *Prof Exp:* Res assoc, Solar Div, Kitt Peak Nat Observ, 62-65, sr engr, 65-68, eng mgr 4-M telescopes, 68-71; sr scientist, Europ Orgn Nuclear Res, 72-73; dir eng, Kitt Peak Nat Observ, 73-77; prog mgr radio astron, Div Astron Sci, NSF, 78-80, sect head, Astron Ctr, 81-82; dep dir, Ctr Astrophysics & Space Sci, Univ Calif, San Diego, 82-91, proj mgr, Hubble Space Telescope Faint Object Spectrog, 84-91; PROG DIR, GEMINI PROJ EIGHT-METER TELESCOPE PROJ, NAT OPTICAL ASTRON OBSERV, 91- *Concurrent Pos:* Consult, La State Univ, 67-71, Europ Southern Observ, 70-73, Univ Hawaii, 75-79, NASA, 75-81, Univ Calif, 76-80 & Keck 10m telescope, 83-; sci leader, US Antarctic Res Prog, South Pole, 78-79; staff, Pres Comn Nat Agenda 80's, 80-81. *Mem:* Am Astron Soc; Am Soc Mech Engrs; Soc Photo-Optical Instrumentation Engrs; Am Soc Tool & Mfg Engrs. *Res:* Space craft instrument design. *Mailing Add:* Nat Optical Astron Observ 950 N Cherry Ave Tucson AZ 85701

RANDALL, LINDA LEA, b Montclair, NJ, Aug 7, 46; m 70. MEMBRANE BIOLOGY. *Educ:* Colo State Univ, BS, 68; Univ Wis, Madison, PhD(molecular biol), 71. *Prof Exp:* Fel, Inst Pasteur, France, 71-73; res assoc molecular biol, Univ Uppsala, 73-75, asst prof, 75-81; assoc prof, 81-83, PROF BIOCHEM, WASH STATE UNIV, 83- *Honors & Awards:* Eli Lilly Award, 84. *Mem:* Am Soc Microbiol; AAAS; Am Soc Biol Chemists. *Res:* Molecular mechanism of export of protein through biological membranes using proteins in envelope of E coli. *Mailing Add:* Dept Biochem & Biophysics Wash State Univ Pullman WA 99164-4660

RANDALL, PETER, b Philadelphia, Pa, Mar 29, 23; m 48; c 4. PLASTIC SURGERY. *Educ:* Princeton Univ, AB, 44; Johns Hopkins Univ, MD, 46; Am Bd Plastic Surg, dipl, 55. *Prof Exp:* Intern, Union Mem Hosp, Baltimore, Md, 46-47; resident, US Naval Hosp, Philadelphia, 47-48; asst instr surg, Sch Med & resident, Hosp, Univ Pa, 49-50; asst instr surg, Sch Med, Wash Univ, 50-53; instr, Sch Med, Univ Pa, 53-56, assoc, Hosp, 53-59, assoc, Sch Med, 56-59, from asst prof to assoc prof, 59-70, chief div, 80-89, PROF PLASTIC SURG, SCH MED & HOSP, UNIV PA, 70- *Concurrent Pos:* Resident, Barnes & St Louis Children's Hosp, 52-53; from asst surgeon to sr surgeon, Children's Hosp, Philadelphia, 53-, chief Div Plastic Surg, 62-81; attend plastic surgeon, Vet Admin Hosp, 54-; chief dept plastic surg, Lankenau Hosp, 71; bd mem, Am Bd Plastic Surg, 71. *Mem:* Am Soc Plastic & Reconstruct Surg (secy, 66-69, vpres, 75-76, pres, 77-78); Am Asn Plastic Surg; Am Col Surg; Am Cleft Palate Asn (pres, 66-67); Plastic Surg Res Coun (pres, 64-65). *Res:* Cleft lip and palate. *Mailing Add:* Plastic Surg G12 Univ Pa Ten Penn Tower 34th & Spruce St Philadelphia PA 19104

RANDALL, RAYMOND VICTOR, b Washington, DC, Aug 1, 20; m 46; c 3. MEDICINE, ENDOCRINOLOGY & METABOLISM. *Educ:* Harvard Univ, AB, 42, MD, 45; Univ Minn, MS, 51. *Prof Exp:* House officer med, Mass Gen Hosp, 45-46; instr, 54-59, from asst prof to assoc prof, 59-70, head sect internal med & endocrinol, 63-74, PROF MED, MAYO GRAD SCH MED, UNIV MINN, 70- *Concurrent Pos:* Clin & res fel, Mass Gen Hosp, 51-52; teaching fel pediat, Harvard Med Sch, 51, teaching fel med, 52; resident physician, House of the Good Samaritan, Boston, 51; asst to staff, Mayo Clin, 52; consult, St Mary's Hosp & Rochester Methodist Hosp, 53-; consult, Mayo Clin, 53-74, sr consult 74- *Mem:* Endocrine Soc; Am Fedn Clin Res; Int Endocrine Soc; Int Soc Neuroendocrinol; Int Soc Psychoneuroendocrinol. *Res:* Endocrine and metabolic diseases. *Mailing Add:* Mayo Clin Rochester MN 55905

RANDALL, WALTER CLARK, b Akeley, Pa, Dec 12, 16; m 43; c 4. PHYSIOLOGY. *Educ:* Taylor Univ, AB, 38; Purdue Univ, MS, 40, PhD(physiol), 42. *Prof Exp:* Asst biol, Purdue Univ, 38-42; from instr to assoc prof, Sch Med, St Louis Univ, 43-54; chmn dept, Loyola Univ, Chicago, Ill, 54-75; RES PROF NATURAL SCI, TAYLOR UNIV, UPLAND, IND, 87- *Concurrent Pos:* Fel physiol, Sch Med, Western Reserve Univ, 42-43; mem prog-proj comt, Nat Heart & Lung Inst, 63-67 & 68-72; chmn, Nat Bd Med Examr, 68; mem field mus, Smithsonian Inst; adj prof physiol, Univ Ky, 85-, Univ S Ala, 86-; adj prof physiol, stritch ShH Med, Loyola Univ, Chicago, 87- *Honors & Awards:* Wiggers Award, Am Physiol Soc, 79. *Mem:* Hon fel Am Col Cardiol; Am Inst Biol Sci; Am Physiol Soc (pres, 82-83); Soc Exp Biol & Med; Sigma Xi. *Res:* Sweating; temperature regulation; circulation and autonomic nervous system; nervous control of the heart. *Mailing Add:* Dept Biol Taylor Univ Upland IN 46989

RANDALL, WILLIAM CARL, b Hampton, Iowa, Jan 27, 41. PHYSICAL CHEMISTRY. *Educ:* Iowa State Univ, BS, 63; Univ Wis, PhD(phys chem), 67. *Prof Exp:* Res fel, 67-80, SR RES FEL, MERCK SHARP & DOHME RES LABS, WEST POINT, 80- *Mem:* AAAS; Am Chem Soc. *Res:* Kinetics of reactions in solution and the application of physical chemistry to medicinal chemistry. *Mailing Add:* 1735 Supplee Rd Lansdale PA 19446-5457

RANDAZZO, ANTHONY FRANK, b Staten Island, NY, Sept 20, 41; m 65; c 2. GEOLOGY, RESEARCH ADMINISTRATION. *Educ:* City Col New York, BS, 63; Univ NC, Chapel Hill, MS, 65, PhD(geol), 68. *Prof Exp:* From asst prof to assoc prof, 67-77, asst dir, 77-80, assoc dean sponsored res, 80-82, dir res admin, 82-88, PROF GEOL, UNIV FLA, 77-, CHMN, DEPT GEOL, 88- *Concurrent Pos:* Vis prof, Brigham Young Univ, 72. *Mem:* Fel Geol Soc Am; Soc Econ Paleont & Mineral; Sigma Xi; Am Asn Petrol Geologists; Nat Coun Univ Res Adminr. *Res:* petrography and geohydrology of limestones of Florida and the Caribbean; reduction of paperwork in contract and grant administration. *Mailing Add:* Dept Geol Univ Fla Gainesville FL 32611

RANDELL, RICHARD, b Fairfield, Iowa, Aug 23, 46. TOPOLOGY, SINGULARITY THEORY. *Educ:* Univ Wis, PhD(math), 73. *Prof Exp:* From asst prof to assoc prof math, 81-87, PROF MATH, UNIV WIS, 87- *Mailing Add:* Dept Math Univ Iowa Iowa City IA 52242

RANDELS, JAMES BENNETT, b Detroit, Mich, June 13, 31; m 56; c 2. COMPUTER SCIENCE. *Educ:* Univ Calif, Los Angeles, BA, 53; Ohio State Univ, MA, 58, PhD(math), 65. *Prof Exp:* Res engr, Univ Calif, Los Angeles, 54; instr math, Univ Dayton, 56-57; res asst, 57-58, res assoc, 61-63, chief systs programmer, 63-65, asst prof math, 65-66, asst prof comput sci, 66-70, chmn comput coord comt, 70-73, asst dir learning resources comput ctr, 70-72, asst dir univ systs, 72-75, assoc prof comput & info sci, 70-80, assoc dir comput systs programming, 75-80, SR PROGRAMMER ANALYST, COMPUT CTR & SR COMPUT SPECIALIST, UNIV SYSTS, OHIO STATE UNIV, 82- *Concurrent Pos:* Mem tech staff comput & data reduction, Space Tech Labs, Inc, 58-60, head spec proj group, 60-61; assoc prof, Denison Univ, 80-81. *Mem:* Asn Comput Mach; Math Asn Am; Sigma Xi. *Res:* Digital computer programming; computer operating systems; simulation of systems. *Mailing Add:* 999 Greenridge Rd Columbus OH 43235-3417

RANDERATH, KURT, b Dusseldorf, Ger, Aug 2, 29; m 62. BIOCHEMISTRY. *Educ:* Univ Heidelberg, DrMed, 55, dipl, 58. *Prof Exp:* Asst org chem, Darmstadt Tech, 59-62; from res assoc to asst prof, Harvard Med Sch, 64-71; from assoc prof to prof pharmacol, 71-89, PROF & HEAD, DIV TOXICOL, BAYLOR COL MED, 89- *Concurrent Pos:* Res fel biol chem, Harvard Med Sch, 63-64; Nat Cancer Inst res career develop award, 69; Am Cancer Soc fac res award, 72. *Mem:* AAAS; Am Soc Biol Chem; Am Asn Cancer Res; Am Chem Soc. *Res:* Analysis of nucleic acids and derivatives; drug effects on nucleic acids; separation methods, particularly thin-layer chromatography; chemical carcinogenesis; DNA damage and repair. *Mailing Add:* Dept Pharmacol Baylor Col Med Houston TX 77030

RANDERSON, DARRYL, b Houston, Tex, July 8, 37; m 61; c 2. METEOROLOGY. *Educ:* Tex A&M Univ, BS, 60, MS, 62, PhD(meteorol), 68. *Prof Exp:* Instr meteorol, Tex A&M Univ, 62-65; res scientist air pollution, Tex A&M Res Found, 67-68, res assoc, 68-69; res meteorologist, 69-81, supvr meteorologist, 81-84, DEP METEOROLOGIST-IN-CHG, NAT WEATHER SERV NUCLEAR SUPPORT OFF, 85- *Concurrent Pos:* Res asst, Tex A&M Univ, 61-65; traineeship, NIH, 65-67; vis prof meteorol, Univ Nev, Las Vegas, 70- *Honors & Awards:* NASA Group Achievement Award, Johnson Space Ctr, Houston, Tex, 74; Nat Oceanic & Atmospheric Admin Spec Achievement Award, Nat Weather Serv, Las Vegas, Nev, 75. *Mem:* Fel Am Meteorol Soc. *Res:* Numerical modeling; thunderstorms; weather forecasting; satellite meteorology; air pollution meteorology; radar meteorology. *Mailing Add:* Nat Weather Serv PO Box 94227 Las Vegas NV 89193-4227

RANDERSON, SHERMAN, genetics; deceased, see previous edition for last biography

RANDHAWA, JAGIR SINGH, b Vahila, India, Nov 1, 22; US citizen; m 54; c 2. PHYSICS. *Educ:* Univ Punjab, WPakistan, BS, 45, MS, 46; Univ Colo, Boulder, MS, 59; NMex State Univ, PhD(physics), 64. *Prof Exp:* Lectr physics, Educ Dept, Univ Punjab, India, 50-57; teaching asst, Univ Colo, Boulder, 57-59; res assoc, NMex State Univ, 59-64; res physicist, Atmospheric Sci Lab, White Sands Missile Range, 64-91; RETIRED. *Concurrent Pos:* Spec Act Award, US Army, 67-72. *Mem:* Am Phys Soc; Am Geophys Union; Am Meteorol Soc; Am Inst Aeronaut & Astronaut; Sigma Xi. *Res:* Physics of upper atmosphere; aeronomy; meteorology; photochemistry of ozone. *Mailing Add:* 5830 S Bethel Ave Del Rey CA 93616

RANDIC, MILAN, b Belgrade, Yugoslavia, Oct 1, 30; US citizen; m 60; c 1. CHEMICAL GRAPH THEORY, CHEMICAL STRUCTURE DOCUMENTATION. *Educ:* Univ Zagreb, Yugoslavia, BA, 54; Univ Cambridge, Eng, PhD(spectros), 58. *Prof Exp:* Assoc theoret chem, Inst Rugjer Boskovic, 60-65; from assoc prof to prof chem, Univ Zagreb, 65-70; vis prof physics, Univ Utah, 71-72; fac guest, dept chem, Harvard Univ, 72-73; vis prof chem, Tufts Univ, 73-74; from assoc prof to prof, 80-87, DISTINGUISHED PROF MATH, DRAKE UNIV, 88- *Concurrent Pos:* Mem adv bd, Croatica Chem Act & J Math Chem. *Honors & Awards:* Boris

Kidric Found Award, Slovene Nat Assembly, 87. *Mem:* Int Soc Math Chem (pres); Math Asn Am; Am Chem Soc; Croatian Chem Soc. *Res:* Mathematical modeling of chemical structure with an emphasis on combinatorial and topological aspects of a structure; structure-property and structure-activity studies; drug design. *Mailing Add:* Dept Math Drake Univ 25th St Des Moines IA 50311

RANDIC, MIRJANA, b Ogulin, Yugoslavia, Oct 12, 34; m 60; c 1. NEUROPHYSIOLOGY. *Educ:* Univ Zagreb, MD, 59, PhD(pathophysiol), 62. *Prof Exp:* From asst to assoc prof neurophysiol, Rudjer Boskovic Inst, Yugoslavia, 59-70; asst prof, McGill Univ, 64-65; assoc prof neuropharmacol, Dept Biochem & Pharmacol, Sch Med, Tufts Univ, 72-75; assoc prof, 75-77, PROF PHARMACOL, DEPT VET ANAT PHARMACOL & PHYSIOL, IOWA STATE UNIV, 77- *Mem:* Int Asn Study Pain; Brit Physiol Soc; Brit Pharmacol Soc; Int Brain Res Orgn; Soc Neurosci. *Res:* Chemical synaptic transmission; the physiological role of peptides, especially substance P, endorphins and somatostatin in nocioceptive pathways. *Mailing Add:* Dept Vet Anat Pharmacol & Physiol Iowa State Univ Ames IA 50011

RANDINITIS, EDWARD J, b Scranton, Pa, Apr 7, 40; m 72; c 2. BIOPHARMACEUTICS. *Educ:* Wayne State Univ, Detroit, BS, 62, MS, 64, PhD(pharm), 69. *Prof Exp:* Asst pharm, Wayne State Univ, 62-69; res pharmacist, 69-72, res scientist, 72-76, RES ASSOC PHARM, PARKE-DAVIS & CO, WARNER-LAMBERT, 76- *Mem:* Am Pharmaceut Asn; Acad Pharmaceut Sci; Sigma Xi; Am Asn Pharmaceut Scientists. *Res:* Assessment of pharmaceutical formulations regarding bioavailability; assay of biological fluids for drugs and metabolites; development of assay for such. *Mailing Add:* 11943 Beacon Hill Plymouth MI 48170

RANDLE, ROBERT JAMES, b Detroit, Mich, July 19, 23; m 50; c 2. PHYSIOLOGICAL OPTICS. *Educ:* Stanford Univ, AB, 51; Calif State Univ, San Jose, MA, 56; Univ Calif, Berkeley, MS, 73. *Prof Exp:* Res psychologist, Aero-Med Lab, Wright Air Develop Centre, US Air Force, 56-59; eng psychologist, Human Factors Off, Pac Missile Range, Calif, 59-61; sr engr, Philco Western Develop Lab, Calif, 61-62; res scientist, Biotech Div, Ames Res Ctr, NASA, 62-83; CONSULT, 84- *Res:* Application of experimental techniques of psychology to the definition and solution of problems in vision and optics in the operations of aerospace vehicles and systems. *Mailing Add:* 1543 Mallard Way Sunnyvale CA 94087

RANDLES, CHESTER, b Painesville, Ohio, Sept 14, 18; m 41; c 3. MICROBIOLOGY. *Educ:* Kent State Univ, BS, 42; Ohio State Univ, PhD(bact), 47. *Prof Exp:* Asst bact, Ohio State Univ, 42-44; asst prof, Rutgers Univ, 47-49; from asst prof to assoc prof, microbiol, Ohio State Univ, 49-71, prof, 71-; RETIRED. *Mem:* AAAS; Am Soc Microbiol. *Res:* Organic and inorganic aspects of bacterial physiology, including respiratory mechanisms, nutrition and polysaccharide production; relation of physiology to virulence and to viral duplication; sulfur bacteria. *Mailing Add:* 180 W South St Worthington OH 43085

RANDLES, RONALD HERMAN, b Canton, Ohio, Sept 4, 42; m 68; c 2. STATISTICS. *Educ:* Col Wooster, BA, 64; Fla State Univ, MS, 66, PhD(statist), 69. *Prof Exp:* From asst prof to assoc prof, Univ Iowa, 69-78, prof statist, 78-81; PROF STATIST, UNIV FLA, 81-, DEPT CHAIR, 89- *Concurrent Pos:* Assoc ed, Am Statistician, 74-76, J Am Statist Asn, 79-85, Comn Statist, 86- & J Nonpar Statist, 90- *Mem:* Fel Am Statist Asn; Inst Math Statist. *Res:* Nonparametrics and large sample distribution theory. *Mailing Add:* Dept Statist Univ Fla Gainesville FL 32611

RANDLETT, HERBERT ELDRIDGE, JR, b Centralia, Wash, July 17, 17; m 46; c 2. COMPUTER SYSTEMS, CHEMICAL ENGINEERING. *Educ:* Univ Calif, BS, 39. *Prof Exp:* Mgr exp lab, Shell Oil Co, 54-64, systs mgr mfg, 64-74, staff systs rep, Rep Opers, 47-79; RETIRED. *Res:* Computer systems applications. *Mailing Add:* 1324 Chardonnay Houston TX 77077

RANDOL, BURTON, b New York, NY, Sept 16, 37; m 64. MATHEMATICS. *Educ:* Rice Univ, BA, 59; Princeton Univ, PhD(math), 62. *Prof Exp:* Instr math, Princeton Univ, 62-63; lectr, Yale Univ, 63-64, asst prof, 64-69; assoc prof, 69-74, PROF MATH, GRAD CTR, CITY UNIV NY, 75- *Mem:* Am Math Soc; Soc Math France. *Res:* Analysis. *Mailing Add:* Dept Math City Univ NY Grad Ctr New York NY 10036

RANDOLPH, ALAN DEAN, b Muskogee, Okla, Mar 25, 34; m 57; c 3. CHEMICAL ENGINEERING. *Educ:* Univ Colo, BSChE, 56; Iowa State Univ, MSChE, 59, PhD(crystallization), 62. *Prof Exp:* Asst technologist, Shell Chem Corp, 56-58; res proj engr, Am Potash & Chem Corp, Calif, 62-65, head crystallization sect, Res Dept, 65; assoc prof chem eng, Univ Fla, 65-68; assoc prof, 68-70, PROF CHEM ENG, UNIV ARIZ, 70- *Concurrent Pos:* NSF res grant crystallization, 66-; consult, Dow Chem Co, 68-70 & Kerr-McGee Corp, 69-; mem, Consult Comt Nuclear Waste Immobilization, Atlantic Richfield Hanford Co, 71-79; consult, Los Alamos Sci Lab, 78-81, US Borax, 78-82 & E I DuPont de Nemours & Co Inc, 78-83; vis prof, Univ Col, London, 81. *Mem:* Am Inst Chem Engrs; Am Chem Soc. *Res:* Mathematical simulation, description and control of particulate systems, especially crystallization processes, theoretical and experimental study of nucleation-growth rate kinetics and residence-time distributions of particulate systems. *Mailing Add:* 2131 Rainbow Vista Dr Tucson AZ 85712-2910

RANDOLPH, JAMES COLLIER, b Knox City, Tex, Mar 26, 44; m 68, 85. ECOLOGY. *Educ:* Univ Tex, Austin, BA, 66, MA, 68; Carleton Univ, PhD(biol), 71. *Prof Exp:* Res scientist, Ecol Sci Div, Oak Ridge Nat Lab, 72-74; from asst prof to assoc prof, 74-82, assoc dean, 86-90, PROF BIOL, SCH PUB & ENVIRON AFFAIRS, IND UNIV, BLOOMINGTON, 82-; DIR, MIDWESTERN CTR GLOBAL ENVIRON CHANGE, 90- *Mem:* AAAS; Am Inst Biol Sci; Ecol Soc Am; Am Soc Mammal; Am Soc Pub Admin; Am Soc Photo Rem Sensing. *Res:* Physiological ecology and applied ecology. *Mailing Add:* Sch Pub & Environ Affairs Ind Univ Bloomington IN 47405

RANDOLPH, JAMES E, b Los Angeles, Calif, Jan 19, 40; m 86; c 3. COMPUTER AIDED DESIGN. *Educ:* Calif State Univ, Los Angeles, BS, 64; Univ Southern Calif, MS, 67. *Prof Exp:* Systs engr, Viking Mars Studies, 70-72, leader, Mission Planning Team, 72-75, mission engr & sci integration team mgr, Voyager, 75-77, mgr, Starprobe Mission Study, 76-91; MGR, ORBITER & ROVER CONTRACT STUDIES, MARS & ROVER SAMPLE RETURN PROJ, JET PROPULSION LAB, CALIF INST TECHNOL, 87- *Concurrent Pos:* Mgr planetary observ studies, Jet Propulsion Lab, Calif Inst Technol, 83-85, mgr Mars Rover Studies, 86-87. *Mem:* Am Inst Aeronaut & Astronaut; Am Astron Soc; Am Geophys Union. *Res:* Application of advanced mission and systems engineering techniques to the design, implementation, and management of interplanetary missions to optimize the return of scientific data. *Mailing Add:* Jet Propulsion Lab Calif Inst Technol MS 301-170U 4800 Oak Grove Dr Pasadena CA 91109

RANDOLPH, JOHN FITZ, mathematics; deceased, see previous edition for last biography

RANDOLPH, JUDSON GRAVES, b Macon, Ga, July 19, 27; m 52; c 5. SURGERY. *Educ:* Vanderbilt Univ, BA, 50, MD, 53. *Prof Exp:* Teaching fel, Harvard Med Sch, 60-61; asst surgeon, Children's Hosp Med Ctr, Boston, 61-63; assoc prof, 64-68, PROF SURG, SCH MED, GEORGE WASHINGTON UNIV, 68-, PROF CHILD HEALTH & DEVELOP, 71-; SURGEON-IN-CHIEF, CHILDREN'S HOSP, WASHINGTON, DC, 64- *Concurrent Pos:* Instr, Harvard Med Sch, 62-63; consult, Nat Naval Med Ctr, Bethesda, 64- & Walter Reed Army Med Ctr, DC, 65-; mem staff, NIH, 65- *Mem:* Am Col Surg; Am Acad Pediat; Am Asn Thoracic Surg; Am Pediat Surg Asn; Soc Univ Surg. *Res:* Burns in children; surgical metabolism in infants; jejunoileal bypass in adolescents; esophageal surgery in infants and children. *Mailing Add:* Childrens Hosp Nat Med Ctr Dept Surgery 111 Michigan Ave NW Washington DC 20010

RANDOLPH, KENNETH NORRIS, b Oxford, Miss, Dec 27, 38; m 61; c 3. AQUACULTURE, ZOOLOGY. *Educ:* Delta State Univ, BS, 60; Memphis State Univ, MS, 69; Univ Okla, PhD(zool), 75. *Prof Exp:* Crop reporter, Agr Stability & Conserv Serv, Miss Dept Agr, 60-61; lab technician, US Army Recruiting Serv, US Army, 61-63; inspector, Seafood Inspection & Cert Unit, US Dept Com, 64-67; ASST PROF, DEPT FISHERIES, AUBURN UNIV, 76- *Concurrent Pos:* Consult, US Agency Int Develop Mission-Jamaica, Kingston, 77-79; vis lectr, Jamaica Sch Agr, Kingston, 77-; proj mgr inland fisheries, Ministry Agr, Jamaica, 78- *Mem:* Am Fisheries Soc; World Maricult Soc; Catfish Farmers Am. *Res:* Applied research in the field of aquaculture, especially in the areas of fish behavior, fish nutrition and fish physiology. *Mailing Add:* Rte 1 Box 178 Potts Camp MS 38659

RANDOLPH, LYNWOOD PARKER, b Richmond, Va, May 21, 38; m 60; c 4. INFORMATION RESOURCE MANAGEMENT, TECHNICAL MANAGEMENT. *Educ:* Va State Univ, BS, 59; Howard Univ, MS, 64, PhD(physics), 72. *Prof Exp:* Physicist, Harry Diamond Labs, 64-68, res physicist, 68-75; prog mgr, 75-80, MGR, NASA, 80- *Concurrent Pos:* Lectr, Univ DC, 72-80; adj prof, Univ DC & Howard Univ, 80- *Mem:* Am Inst Aeronaut & Astronaut; Am Phys Soc; AAAS; Inst Elec & Electronics Engrs. *Res:* Solar cells and other devices which convert sunlight into electrical energy; lasers to be used in the future to transmit power or propel spacecraft into space; radiation-induced effects in optoelectronic materials. *Mailing Add:* 3000 Fairhill Ct Suitland MD 20746

RANDOLPH, MALCOLM LOGAN, b West Palm Beach, Fla, Oct 11, 20; m 49, 83; c 5. BIOPHYSICS, RADIOBIOLOGY. *Educ:* Univ Va, BA, 43, MS, 46, PhD(physics), 47. *Prof Exp:* Instr, Univ Va, 41-44, res assoc, Manhattan Dist & Off Sci Res & Develop, 42-46; Nat Acad Sci predoctoral fel, Univ Va, 46-47; from inst to asst prof physics, Tulane Univ, 47-53; biophysicist, Biol Div, Oak Ridge Nat Lab, 52-77 & Health & Safety Res Div, 77-85, consult, 85; RETIRED. *Mem:* AAAS; Am Phys Soc; Biophys Soc; Health Physics Soc; Radiation Res Soc. *Res:* Technology assessments; radiological physics; radiobiology; electron spin resonance; ultra centrifugation; molecular biology. *Mailing Add:* 358 East Dr Oak Ridge TN 37830

RANDOLPH, PAUL HERBERT, b Jamestown, NY, Jan 14, 25; m 48; c 6. OPERATIONS RESEARCH. *Educ:* Univ Minn, BA, 48, MA, 49, PhD(statist), 55. *Prof Exp:* Instr math, Bethany Lutheran Col, 49-50; instr bus admin, Univ Minn, 50-53; asst prof indust eng, Ill Inst Technol, 54-57; assoc prof math & statist, Purdue Univ, 57-66; from assoc prof to prof math, NMex State Univ, 66-74; prof eng, Iowa State Univ, 74-76; opers res analyst, Dept Energy, 76-77; vpres energy econ, Chase Manhattan Bank, 77-79; sr assoc engr, Mobil Res & Develop Corp, 79-81; PROF, INFO SYSTS & QUANT SCI, COL BUS ADMIN, TEX TECH UNIV, 81- *Concurrent Pos:* Guest prof, Univ Heidelberg, 64-65; sci adv, Norsk Regnesentral, 65-66; consult, Braddock, Dunn & McDonald, 66-69; consult, White Sands Missile Range, 69-73; vis prof, Mid East Tech Univ, Turkey, 70-71; vis prof, Inst Math, Univ Oslo, 73-74; sr Fulbright prof, Mid East Tech Univ, Ankara, Turkey, 80-81; vis distinguished prof, Opers Res, US Army Logistics Mgt Col, Ft Lee, VA, 86-87. *Mem:* Opers Res Soc Am; Inst Mgt Sci; Am Inst Indust Engrs; Asn Comput Mach; Data Processing Mgt Asn; Sigma Xi. *Res:* Optimization techniques; expert systems; decision support systems. *Mailing Add:* Info Systs & Quant Sci Tex Tech Univ Lubbock TX 79409-2101

RANDOLPH, PHILIP L, b Casper, Wyo, Feb 25, 31; m 52; c 2. PHYSICS, RESEARCH ADMINISTRATION. *Educ:* Univ Wash, BS, 52, PhD(physics), 58. *Prof Exp:* Physicist, Lawrence Radiation Lab, Univ Calif, Livermore, 58-61, dep tech dir, Proj Gnome, 61-62, tech dir salmon event, Proj Dribble, 62-63 & 64-66, assoc div leader, 66-68; mgr nuclear group, El Paso Natural Gas Co, 68-74, dir res, 74-77; assoc dir, 77-79, DIR UNCONVENTIONAL SUPPLY RES, INST GAS TECHNOL, 79- *Mem:* Am Phys Soc; Am Nuclear Soc; Soc Petrol Engr. *Res:* Nuclear explosive test execution; use of nuclear explosives to stimulate natural gas production and produce underground storage for natural gas; massive hydraulic fracturing of tight natural gas reservoirs; improving quantitative understanding of natural gas well completions and reservoir rock properties; producing natural gas from aquifers and coal seams; underground gas storage. *Mailing Add:* 1713 Crestwood Dr Texas City TX 77591

RANDRUP, JORGEN, b Aarhus, Denmark, Sept 23, 46. HEAVY-ION REACTIONS, NUCLEAR DYNAMICS. *Educ:* Univ Aarhus, Cand Scient, 70, Lic Scient, 72. *Prof Exp:* Res fel, Physics Dept, Univ Aarhus, Denmark, 72-75; postdoctoral fel, Nuclear Sci Div, Lawrence Berkeley Lab, 73-75; res fel, Niels Bohr Inst, Copenhagen, 75-76, Nordita, 76-79; div fel, 78-81, SR PHYSICIST, NUCLEAR SCI DIV, LAWRENCE BERKELEY LAB, UNIV CALIF, BERKELEY, 81- *Concurrent Pos:* Vis researcher, Nuclear Sci Div, Lawrence Berkeley Lab, Univ Calif, Berkeley, 72-73; vis prof, GSI, Darmstadt, Ger, 74-75 & Nordita, Copenhagen, 84-85; res assoc, Caltech, Pasadena, Calif, 77-78; sci dir, Nuclear Theory & Data Eval, Nuclear Sci Div, Lawrence Berkeley Lab, 87-91; lectr, Physics Dept, Univ Calif Berkeley, 88; assoc div ed, Physics Rev Lett, 89-91; vis scientist, GSI, Darmstadt, Ger, 91-92. *Honors & Awards:* Sr Scientist Award, Alexander von Humboldt Stiftung, 89. *Mem:* Am Phys Soc. *Res:* Theoretical nuclear physics especially heavy ion reactions; nuclear physics. *Mailing Add:* Nuclear Sci Div Lawrence Berkeley Lab 70A-3307 Berkeley CA 94720

RANDT, CLARK THORP, b Lakewood, Ohio, Nov 18, 17; m 44; c 3. NEUROLOGY. *Educ:* Colgate Univ, AB, 40; Western Reserve Univ, MD, 43. *Prof Exp:* Intern & asst resident internal med, Univ Hosps Cleveland, 43-45; demonstr med, Sch Med, Western Reserve Univ, 45; asst resident & resident neurol, Neurol Inst, Columbia-Presby Med Ctr, 47-50; from sr instr to assoc prof, Sch Med, Western Reserve Univ, 50-56; dir div neurol, Univ Hosps Cleveland, 56-59; dir off life sci, NASA, 60-61; PROF NEUROL & CHMN DEPT, SCH MED & DIR, BELLEVUE NEUROL SERV, NY UNIV, 62- *Concurrent Pos:* Sr consult, NY Vet Admin Hosp; mem comt bioastronaut, Armed Forces-Nat Res Coun, mem panel manned space flight, NASA; mem, NASA-US Dept Defense Aeronaut & Astronaut Coord Bd. *Mem:* Soc Exp Biol & Med; Am Neurol Asn; Aerospace Med Asn (vpres, 60); fel Am Acad Neurol. *Res:* Physiology of sensory systems; neurophysiological effects of anesthetic agents; memory; applications of computer technology; brain and behavior in early life; nutritional deprivation; mental retardation. *Mailing Add:* 59 Husted Lane Greenwich CT 06830

RANDY, HARRY ANTHONY, b Dumont, New Jersey, Oct 20, 45; m 69; c 2. CATTLE NUTRITION, GOAT NUTRITION. *Educ:* Rutgers Univ, BS, 67, MS, 69; Va Polytechnic Inst, PhD(physiol), 72. *Prof Exp:* Dir res, 81-83, exec vpres, 83-88, PRES, WILLIAM MINER AGR RES INST, 88- *Mem:* Am Dairy Sci Assoc; AAAS. *Res:* Horse management. *Mailing Add:* Miner Inst Box 90 Chazy NY 12921

RANEY, GEORGE NEAL, b Portland, Ore, Oct 14, 22; m 52; c 2. MATHEMATICS. *Educ:* Queens Col, NY, BS, 43; Columbia Univ, PhD(math), 53. *Prof Exp:* Instr math, Mass Inst Technol, 43-44; lectr, Columbia Univ, 46-50, instr, 50-53; instr, Brooklyn Col, 53-55; from asst prof to assoc prof, Pa State Univ, 55-61; vis assoc prof, Wesleyan Univ, 61-63; assoc prof, 63-66, prof, 66-85, EMER PROF MATH, UNIV CONN, 85- *Mem:* Am Math Soc. *Res:* Lattice theory; combinatorial analysis; automata theory; continued fractions. *Mailing Add:* Dept Math Univ Conn Storrs CT 06268

RANEY, HARLEY GENE, entomology; deceased, see previous edition for last biography

RANEY, J(OHN) P(HILIP), b Kendallville, Ind, Jan 20, 31. MECHANICAL ENGINEERING. *Educ:* Purdue Univ, BS, 54, MS, 57, PhD(mech eng), 59. *Prof Exp:* Instr mech eng, Purdue Univ, 57-59; assoc prof, Univ Va, 59-63; head, Anal Dynamics Sect, 63-78, head, Nastran Mgt Off, 70-73, head, Noise Technol Br, 78-81, HEAD, NOISE PREDICTION BR, LANGLEY RES CTR, NASA, 81- *Mem:* Am Soc Mech Engrs. *Res:* Mechanical vibrations and dynamics; aircraft noise. *Mailing Add:* Space Sta Freedom Off MS288 Langley Res Ctr NASA Hampton VA 23665

RANEY, RICHARD BEVERLY, b Raleigh, NC, July 21, 06; m 38; c 2. ORTHOPEDIC SURGERY. *Educ:* Univ NC, BA, 26; Harvard Univ, MD, 30. *Prof Exp:* Instr surg, Med Sch, Univ Rochester, 33-34; from instr to asst prof orthop surg, Sch Med, Duke Univ, 34-52; prof & chmn div, 52-67, clin in prof, 67-77, EMER PROF ORTHOP SURG, SCH MED, UNIV NC, CHAPEL HILL, 78- *Concurrent Pos:* Orthop surgeon, NC Mem Hosp, 52-67; writer & consult, 77- *Mem:* Am Orthop Asn; Orthop Res Soc; Am Med Asn; Am Col Surg; Am Acad Orthop Surg. *Res:* Orthopedic surgical education. *Mailing Add:* Univ NC Sch Med Chapel Hill NC 27515-2467

RANEY, RUSSELL KEITH, b Auburn, NY, Dec 26, 37; Can citizen. REMOTE SENSING. *Educ:* Harvard Univ, BS, 60; Purdue Univ, MS, 62; Univ Mich, Ann Arbor, PhD(comput, info & control eng), 68. *Prof Exp:* Asst engr, Zenith Radio Corp, Ill, 60; engr, Systs Div, Bendix Corp, Mich, 62-63; res engr, Willow Run Labs, Univ Mich, Ann Arbor, 62-74, res engr, Environ Res Inst Mich, 74-76; chief radar scientist, Radarsat, 84-89, RES SCIENTIST, DEPT ENERGY, MINES & RESOURCES, FED GOVT CAN, 76-, CHIEF RADAR SCIENTIST, CAN CENTRE FOR REMOTE SENSING. *Concurrent Pos:* Consult, NASA, 74- & Europ Space Agency, 77. *Honors & Awards:* Distinguished Achievement & Outstanding Serv Award, Geosci & RSM Sensing Soc, Inst Elec & Electronic Engrs. *Mem:* Can Remote Sensing Soc; fel Inst Elec & Electronic Engrs; Int Remote Sensing Soc. *Res:* Application of radar remote sensing systems to environmental surveillance, synthetic aperture radar research; dissemination and utilization of scientific knowledge; oceanic reflectivity; systems development; planetary radar. *Mailing Add:* Can Ctr Remote Sensing 2464 Sheffield Rd Ottawa ON K1A 0E4 Can

RANEY, WILLIAM PERIN, b Neenah, Wis, June 27, 27; m 53; c 4. PHYSICS, ACOUSTICS. *Educ:* Harvard Univ, AB, 50; Brown Univ, ScM, 53, PhD(physics), 55. *Hon Degrees:* DSc, Lawrence Univ, 77. *Prof Exp:* Res assoc physics, Brown Univ, 54-55; res fel acoustics, Harvard Univ, 55-56, asst prof appl physics, 56-60; assoc prof elec eng, Univ Minn, 60-62; exec secy, Comt Undersea Warfare, Nat Acad Sci-Nat Res Coun, 62-64; spec asst to asst secy navy for res & develop, 64-72; dep & chief scientist, Off Naval Res, 72-76; sr policy analyst, Off Sci & Tech Policy, NASA, 77-78, asst assoc admin, Space Science & Appln, 78-84, dir utilization & req, 84-87, SPEC ASST TO DIR, SPACE STA, NASA, 87- *Concurrent Pos:* Mem tech staff, Bell Tel Labs, 59-60. *Mem:* AAAS; Acoust Soc Am; Am Phys Soc; Sigma Xi. *Res:* Physical acoustics; sound propagation in inhomogeneous media; finite amplitude effects; propagation at hypersonic frequencies; linear systems analysis. *Mailing Add:* 5946 Wilton Rd Alexandria VA 22310-2150

RANFTL, ROBERT M(ATTHEW), b Milwaukee, Wis, May 31, 25; m 46. TECHNOLOGICAL PRODUCTIVITY & CREATIVITY. *Educ:* Univ Mich, BSEE, 46. *Prof Exp:* Prod engr, Russell Elec Co, 46-47; head, Eng Dept, Radio Inst Chicago, 47-50; sr proj engr, Webster Chicago Corp, 50-51; prod design engr, 51-53, head equip design group, 53-54, head electronic equip sect, 54-55, mgr, Prod Eng Dept, 55-58, mgr reliability & qual control, 58-59, mgr admin, 59-61, mgr, Prod Effectiveness Lab, 61-77; corp dir configuration mgt, data mgt & design rev, Hughes Aircraft Co, 64-74, corp dir eng & design mgt, 74-84, asst dir prod integrity, 77-84, corp dir managerial prod, 84-86; PRES, RANFTL ENTERPRISES INC, 81- *Concurrent Pos:* Teacher res & develop mgt & prod, Hughes Aircraft Co, 75-86; consult, res & develop mgt, creativity & prod, 78-; instr eng & mgt prog, Univ Calif, Los Angeles, 87- *Mem:* AAAS; Am Inst Aeronaut & Astronaut; sr mem Inst Elec & Electronics Eng; NY Acad Sci. *Res:* Means of improving creativity and productivity in technology-based organizations. *Mailing Add:* Ranftl Enterprises Inc PO Box 49892 Los Angeles CA 90049-0892

RANG, EDWARD ROY, b Milwaukee, Wis, Dec 23, 27; m 87; c 2. APPLIED MATHEMATICS, GENERAL COMPUTER SCIENCES. *Educ:* Univ Wis-Madison, BS, 49, MS, 50; Univ Minn, Minneapolis, PhD(math), 57. *Prof Exp:* Asst prof mech, Univ Minn, 57-60; res eng, Honeywell, 60-86; PROF MATH & COMPUTER, UNIV WIS-RIVER FALLS, 86- *Concurrent Pos:* Prof, US Naval Postgrad Sch, 68-69. *Mem:* Soc Indust & Appl Math; Inst Elec & Electronics Engrs; Math Asn Am; Sigma Xi; Asn Comput Mach. *Res:* Automatic control theory; applied mathematics. *Mailing Add:* 505 Tower Rd Hudson WI 54016

RANGANATHAN, BRAHMANPALLI NARASIMHAMURTHY, b Madras, India, Feb 14, 45; US citizen; m 72; c 1. METALLURGICAL ENGINEERING. *Educ:* Indian Inst Sci, BE, 68; Ga Tech, PhD(metall), 72. *Prof Exp:* Group engr, Martin Marietta Michoud Aerospace, 81-84, chief res progs, Advan Qual Technol, 84-90, SR SCIENTIST, MARTIN MARIETTA CORP LABS, 90- *Mem:* Am Soc Metals; Am Soc Nondestructive Testing. *Res:* Physical metallurgy; characterization of materials; composites. *Mailing Add:* 5704 Columbia Rd Columbia MD 21044

RANGANAYAKI, RAMBABU POTHIREDDY, b Secunderabad, India, Jan 15, 42; m 68; c 2. GEOPHYSICS. *Educ:* Osmania Univ, India, MSc, 64; Univ Hawaii, Honolulu, MS, 72; Mass Inst Technol, PhD(geophys), 78. *Prof Exp:* Sr sci asst geomagnetism, Nat Geophys Res Inst, India, 64-69, scientist, 69-70; res asst geophys, Mass Inst Technol, 72-78; res assoc geophys, Carnegie Inst Washington, 78-80; sr staff geophysicist, Mobil Res & Develop Corp, 80-84, res geophysicist, 84-85, sr res geophysicist, 85-89, ASSOC, MOBIL RES & DEVELOP CORP, 89- *Mem:* Am Geophys Union; Soc Explor Geophysicists; Sigma Xi. *Res:* Magnetotelluric depth sounding; geomagnetic depth sounding; data analysis and interpretation; model development; paleomagnetism; induced polarization; vertical seismic profiling; enhanced oil recovery monitoring; prestack migration. *Mailing Add:* Mobil Res & Develop Corp 13777 Midway Rd Dallas TX 75244

RANGE, R MICHAEL, b WGer, Aug 7, 44; m 69; c 3. CAUCHY-RIEMANN EQUATIONS, INTEGRAL REPRESENTATIONS. *Educ:* Univ Gottingen, dipl, 68; Univ Calif-Los Angeles, PhD(math), 71. *Prof Exp:* JW Gibbs instr math, Yale Univ, 71-73; from asst prof to assoc prof, 73-83, PROF MATH, STATE UNIV NY, ALBANY, 83- *Concurrent Pos:* Vis asst prof, Univ Wash, Seattle, 75-76; res prof, Univ Bonn, WGer, 80 & Max Planck Inst Math, 84 & 86. *Mem:* Am Math Soc; Ger Math Soc; Math Asn Am. *Res:* Multidimensional complex analysis; analytic methods and problems involving integral representations. *Mailing Add:* Dept Math State Univ NY Albany NY 12222

RANGER, KEITH BRIAN, b Salisbury, Eng, Aug 11, 35; m 59; c 2. APPLIED MATHEMATICS. *Educ:* Univ London, BSc, 56, PhD, 59. *Prof Exp:* Asst lectr math, Bedford Col, London, 58-61; lectr, 61-63, from asst prof to assoc prof, 63-70, PROF MATH, UNIV TORONTO, 70- *Mem:* Am Math Soc; Soc Indust & Appl Math; Can Math Cong. *Res:* Axially symmetric potentials; slow motion of a viscous fluid; magnetohydrodynamics. *Mailing Add:* Dept Math St George Campus Univ Toronto 100 St George St Toronto ON M5S 1A1 Can

RANGO, ALBERT, b Cleveland, Ohio, Nov 7, 42; m 66; c 1. WATERSHED MANAGEMENT, REMOTE SENSING. *Educ:* Pa State Univ, BS, 65, MS, 66; Colo State Univ, PhD(watershed mgt), 69. *Prof Exp:* Asst prof meteorol, Pa State Univ, University Park, 69-72; sr hydrologist, 72-80, head, Hydrological Sci Br, 80-83, CHIEF, HYDROL LAB, AGR RES SERV, NASA, 83- *Concurrent Pos:* Consult, Environ Serv Oper, E G & G, 68-69; vis instr, State Univ NY Col Buffalo, 70. *Mem:* Am Meteorol Soc; Am Geophys Union; Am Water Resources Asn. *Res:* Weather modification effects; fluvial geomorphology; snow hydrology; bioclimatology; watershed modelling; watershed physiography; floodplain mapping; soil moisture; meteorology; snowmelt-runoff modeling. *Mailing Add:* 127 Southwood Ave Silver Spring MD 20901

RANHAND, JON M, b New York, NY, Feb 11, 39; m 63; c 2. HEALTH SCIENTIST ADMINISTRATION. *Educ:* City Col New York, BS, 61; Johns Hopkins Univ, MS, 64; Univ Cincinnati, PhD(microbiol), 68. *Prof Exp:* Fel, 68-69, staff fel, Lab Microbiol, NIH, 69-73, SR SCIENTIST, USPHS, 73- *Mailing Add:* 17 Barrington Fare Rockville MD 20850

RANHOTRA, GURBACHAN SINGH, b Abbotabad, India, Aug 8, 35; m 60; c 2. NUTRITIONAL BIOCHEMISTRY. *Educ:* Agra Univ, BVSc, 58, MS, 60; Univ Minn, PhD(nutrit), 64. *Prof Exp:* Fel nutrit biochem, Univ Ill, Urbana, 64-65; assoc prof nutrit, Punjab Agr Univ, 65-68; assoc mem biochem, Univ Okla, 68-69; group leader, 69-80, DIR NUTRIT, AM INST BAKING, 81- *Concurrent Pos:* Adj prof foods & nutrit, Kans State Univ; assoc ed, Cereal Chem, 79-83, J Food Science, 84-87, ed tech bulletins, Am Inst Baking, 88- *Mem:* Am Inst Nutrit; Am Asn Cereal Chem; Inst Food Technologists. *Res:* Fiber lipids; minerals; cereal-based foods. *Mailing Add:* Nutrit Res Am Inst Baking 1213 Bakers Way Manhattan KS 66502

RANIERI, RICHARD LEO, b Chicago Heights, Ill, Oct 30, 43; m 71. ORGANIC CHEMISTRY. *Educ:* Univ Ill, BS, 65; Univ Toledo, PhD(chem), 73. *Prof Exp:* Fel, Dept Med Chem & Pharmacog, Purdue Univ, 73-76; res chem, Wash Res Ctr, W R Grace & Co, 76-80; mem staff, Sodyeco Div, Martin Marietta Co, 80-; AT SANDOZ CHEM CORP. *Mem:* Am Soc Pharmacog; Am Chem Soc. *Res:* Structure elucidation by spectroscopy; tetrahydrolsoquinoline synthesis; chromatography; organophosphorus chemistry; natural products chemistry. *Mailing Add:* 3238 Sunnybrook Dr Charlotte NC 28210-3229

RANK, GERALD HENRY, b Man, Sept 30, 40; m 68; c 2. GENETICS MOLECULAR BIOLOGY, FOOD SCIENCE & TECHNOLOGY. *Educ:* Univ Man, BScAg, 63, MSc, 64; Univ BC, PhD(genetics), 70. *Prof Exp:* PROF BIOL, UNIV SASK, 75- *Mem:* Can Soc Genetics. *Res:* Molecular evolution; breeding of the alfalfa leafcutter bee; gene transfer in industrial Saccaromyces yeasts; expression of antisense RNA. *Mailing Add:* Dept Biol Univ Sask Saskatoon SK S7N 0W0 Can

RANKEL, LILLIAN ANN, b New York, NY; m; c 3. INORGANIC CHEMISTRY. *Educ:* Molloy Col, BS, 66; Fordham Univ, MS, 68; Princeton Univ, PhD(inorg chem), 77. *Prof Exp:* Assoc mem technol staff chem, Bell Labs, Murray Hill, NJ, 68-73; res asst, Princeton Univ, 73-77; ASSOC CHEM CATALYSIS, MOBIL RES & DEVELOP CORP, 77-88. *Mem:* Am Chem Soc. *Res:* Upgrading and demetallation of heavy petroleum-based oils; characterization of processed oils, deasphalted oils and coked oils; study of petroporphyrins and metalloporphyrins contained in heavy crudes. *Mailing Add:* Mobil Res & Develop Corp PO Box 1025 Princeton NJ 08540

RANKEN, WILLIAM ALLISON, b Amityville, NY, Jan 18, 28; m 51; c 3. ENERGY CONVERSION. *Educ:* Yale Univ, BS, 49; Rice Univ, MS, 56, PhD(physics), 58. *Prof Exp:* Res asst weapons develop, Los Alamos Sci Lab, 50-52, staff mem, 52-54; res physicist, Union Carbide Nuclear Co, NY, 57-58; staff mem Advan Propulsion Group, 58-62, from asst group leader to group leader, 62-73, group leader, Advan Heat Transfer Group, 73-77, ALT GROUP LEADER, REACTOR & ADVAN HEAT TRANSFER TECHNOL GROUP, LOS ALAMOS SCI LAB, 77- *Mem:* AAAS; Am Phys Soc; Sigma Xi. *Res:* Heat pipe technology; thermionic conversion; radiation damage; low energy nuclear physics; nuclear reactor development. *Mailing Add:* 166 Navajo Rd Los Alamos NM 87544

RANKIN, ALEXANDER DONALD, b Troy, NY, July 5, 16; m 43; c 3. VETERINARY MEDICINE. *Educ:* Cornell Univ, DVM, 39, MS, 40. *Prof Exp:* Asst vet physiol, Cornell Univ, 39-42, 46-47, actg prof, 53-54; assoc prof physiol, Colo State Univ, 47-48, prof & head dept, 48-55; from asst clin res dir to assoc clin res dir, Squibb Inst Med Res Div, Olin Mathieson Chem Corp, 56-64, dir agr sci sect, 64-66, dir animal health res, 66-67; exec dir animal sci res, Merck Sharp & Dohme Res Labs, 67-73; exec dir res & develop, Lambert Kay Div, Carter-Wallace, Inc, 73-75; CONSULT, INDUST VET MED, 75- *Concurrent Pos:* Mem animal health comt, Nat Res Coun, 69-73, Inter-Soc Comt Drugs & Chem, 69-74, Indust Vet Asn del, Am Vet Med Asn, 69-75, bd dirs, Agr Res Inst, Nat Acad Sci-Nat Res Coun, 71-74, chmn feed adjuvants comt, 72; pvt vet pract, 75- *Mem:* Am Soc Vet Physiol & Pharmacol (pres, 62-63); Am Soc Animal Sci; Am Vet Med Asn; Am Asn Vet Nutritionists (pres, 75-76); Indust Vet Asn (pres, 60-61). *Res:* Ruminant physiology; role of antibiotic feed additives in meat production; endocrinological aspects of animal reproduction; application of pharmaceutical developments to the veterinary field; medical and health sciences. *Mailing Add:* 11731 Heathcliff Dr Santa Ana CA 92705

RANKIN, DAVID, physics, geophysics, for more information see previous edition

RANKIN, DOUGLAS WHITING, b Wilmington, Del, Sept 9, 31; m 56; c 2. TECTONICS, VOLCANOLOGY. *Educ:* Colgate Univ, BA, 53; Harvard Univ, MA, 55, PhD, 61. *Prof Exp:* Field asst, US Geol Surv, Colo, 54; asst prof geol, Vanderbilt Univ, 58-61; geologist & teacher, AID, Washington, DC, 61-62; geologist, US Geol Surv, 62-78, supvry geologist, 78-83; CONSULT - *Concurrent Pos:* Staff scientist, Lunar Sample Off, NASA, 72; coordr, Charleston Invests, SC, 76-78; chief, Br Eastern Regional Geol, 78- *Mem:* AAAS; fel Geol Soc Am; Mineral Soc Am; Am Geophys Union. *Res:* Paleovolcanology and tectonics of the Appalachian orogenic belt; geology of the Absaroka volcanic field in Wyoming; intraplate earthquakes. *Mailing Add:* US Geol Surv Stop 926 Nat Ctr Reston VA 22092

RANKIN, GARY O'NEAL, b Little Rock, Ark, Oct 6, 49. CHEMICAL-INDUCED TOXICITY, NEPHROTOXICITY. *Educ:* Univ Ark, Little Rock, BS, 72; Univ Miss, PhD(med chem), 76. *Prof Exp:* Teaching assoc, pharmacol, Med Col Ohio, 76-78; from asst prof to assoc prof, 78-86, PROF & CHMN DEPT PHARMACOL, MARSHALL UNIV SCH MED, 86-, ASSOC DEAN BIOMED GRAD EDUC & RES DEVELOP, 89- *Concurrent Pos:* Mem subcomt, Prof Utilization & Training, Am Soc Pharmacol & Exp Therapeut, 86-, chair, 89; ad hoc rev, Toxicol Study Sect, NIH, 87; jour reviewer. *Mem:* Am Soc Pharmacol & Exp Therapeut; Soc Toxicol; NY Acad Sci; Soc Exp Biol & Med; Int Soc Study Xenobiotics; Genetic Toxicol Asn; Sigma Xi; Am Chem Soc; AAAS; Asn Med Sch Pharmacol. *Res:* Increasing understanding of why chemicals are toxic, examining chemical structure and its relationship to toxicity with select kidney toxins. *Mailing Add:* Dept Pharmacol Marshall Univ Sch Med Huntington WV 25755-9310

RANKIN, JOANNA MARIE, b Denver, Colo. RADIO ASTRONOMY. *Educ:* Southern Methodist Univ, BS, 65; Tulane Univ, MS, 66; Univ Iowa, PhD(astrophys), 70. *Prof Exp:* Res assoc radio astronomy, Univ Iowa, 70-74; asst prof astron, Cornell Univ, 74-78, sr res assoc hist, 78-80; assoc prof, 80-88, PROF PHYSICS, UNIV VT, 88- *Concurrent Pos:* Vis scientist, Arecibo Observ, PR, 69-78; Am Philos Soc res grant, Radiophys Div, Commonwealth Sci Indust Res Orgn, Sydney, Australia, 72; actg head, Arecibo Observ Comput Dept, 75; grants, NSF & Res Corp, 73-; Indo-Am fel, Raman Res Inst, Bangalore, India, 90-91. *Mem:* Int Union Radio Sci; Int Astron Union; Am Astron Soc. *Res:* Observational properties of pulsars and the interstellar medium; history and philosophy of contemporary physical science; feminist perpsectives on contemporary science. *Mailing Add:* Dept Physics Cook Bldg A405 Univ Vt Burlington VT 05405

RANKIN, JOEL SENDER, b Brockton, Mass, Sept 13, 31; wid; c 2. OBSTETRICS & GYNECOLOGY. *Educ:* Yale Univ, BA, 53; Boston Univ, MD, 57; Am Bd Obstet & Gynec, dipl, 65. *Prof Exp:* Intern med, Beth Israel Hosp, Boston, 57-58; resident surg, Mass Mem Hosp, 58-59; resident obstet & gynec, Boston City Hosp, 59-62; officer in chg, Castle AFB Hosp, Calif, 62-64; asst clin prof obstet & gynec, Western Reserve Univ, 66-69; ASSOC PROF OBSTET & GYNEC, SCH MED, BOSTON UNIV, 69-; DIR OBSTET & GYNEC, FRAMINGHAM UNION HOSP, 75-; ASSOC PROF OBSTET & GYNEC, UNIV MASS MED SCH, 80- *Concurrent Pos:* USPHS fel endocrinol, Jefferson Med Col, 64-66; asst dir obstet & gynec, Mt Sinai Hosp Cleveland, Ohio; asst obstetrician & gynecologist, Cleveland Metrop Gen Hosp, dir endocrine-sterility clin, 66-69, asst vis obstetrician & gynecologist, 67-69; dir gynec-infertility clin, Boston City Hosp, Mass, 69-75, assoc vis surg, 69-; assoc vis gynecologist, Univ Hosp, Boston, 69- *Mem:* Am Col Obstet & Gynec; Am Fertil Soc. *Mailing Add:* 115 Lincoln St Framingham MA 01701

RANKIN, JOHN, internal medicine, physiology; deceased, see previous edition for last biography

RANKIN, JOHN CARTER, b Knoxville, Tenn, Dec 21, 19; m 43; c 3. CEREAL CHEMISTRY. *Educ:* Bradley Univ, BS, 42, MS, 55. *Prof Exp:* Chemist carbohydrate & starch chem, 46-58, res chemist, 58-67, prof leader cereal chem, Northern Regional Res Lab, Agr Res Serv, 67-79, CONSULT STARCH CHEM, USDA, 80- *Mem:* Am Chem Soc; Sigma Xi. *Res:* Reactions of carbohydrates; structure of starch and dextrans; starch and flour derivatives; utilization of cereal grains and fractions therefrom in films, paper, and textiles. *Mailing Add:* 1734 E Maple Ridge Dr Peoria IL 61614.

RANKIN, JOHN HORSLEY GREY, b Gosforth, Eng, May 1, 37; US citizen; m 64; c 2. PHYSIOLOGY, OBSTETRICS & GYNECOLOGY. *Educ:* Univ Melbourne, BSc, 59; Univ Tenn, Knoxville, MS, 63; Univ Ore, PhD(physiol), 68. *Prof Exp:* Teacher biol, Hobart High Sch, Tasmania, 64-65; from instr to asst prof physiol, Univ Colo Med Ctr, Denver, 68-71; asst prof, 71-74, assoc prof, 74-77, PROF PHYSIOL & OBSTET & GYNEC, DIV FETAL-PERINATAL MED, UNIV WIS-MADISON, 77- *Mem:* Perinatal Res Soc; Soc Gynec Invest; Am Physiol Soc; Int Soc Oxygen Transport to Tissue; Soc Exp Biol & Med. *Res:* Fetal physiology and physiology of the placenta. *Mailing Add:* Obstet-Gynec 7224E Madison Gen Hosp Univ Wis Madison WI 53706

RANKIN, JOSEPH EUGENE, b Washington, DC, Jan 13, 20; m 43; c 3. PSYCHIATRY. *Educ:* Cath Univ Am, BS, 42; George Washington Univ, MD, 46; Wash Psychoanal Inst, grad, 66. *Prof Exp:* Intern, US Naval Hosp, Oakland, Calif, 47; intern, US Naval Hosp, Bethesda, Md, 47, resident neurol, 48; resident psychiat, St Elizabeth's Hosp, Washington, DC, 50-52; mem staff, Child Guid Clin, Cath Univ Am, 52-56; PROF PSYCHIAT, SCH MED, GEORGE WASHINGTON UNIV, 56- *Concurrent Pos:* Mem psychother dept, St Elizabeth's Hosp, Washington, DC, 52-56, consult, 57-62; consult, DC Gen Hosp, 57-61 & Crownsville State Hosp, Md, 62-81; pvt pract psychiat & psychoanal, 72-; mem bd dirs, Psychiat Inst Found, Washington, DC. *Mem:* Fel Am Psychiat Asn; Am Psychoanal Asn. *Res:* Psychoanalysis; adolescence. *Mailing Add:* 11527 Shipsveiw Rd Annapolis MD 21401

RANKIN, SIDNEY, b Baltimore, Md, Dec 18, 31; m 60; c 1. CHEMICAL ENGINEERING. *Educ:* Johns Hopkins Univ, BE, 53; Univ Del, MChE, 55, PhD(chem eng), 61. *Prof Exp:* Sr process engr, Kordite Co, NY, 57-60; chem engr, Monsanto Chem Co, Mass, 60-62; process develop & design engr, Silicone Prod Dept, Gen Elec Co, NY, 62-70; process supt, Borden Chem Co, Mass, 71-74; dir res & eng, Dacor Inc, 74-75; sr staff engr, GAF Corp, 75-77; sr res engr, Celanese Res Co, Summit, NJ, 77-84; ASST RES PROF, RUTGERS UNIV, 84- *Mem:* Am Inst Chem Engrs; Soc Plastic Engrs. *Res:* Product and process development, especially aspects of commercial development and process design for chemicals, polymers, specialties and agricultural products; solid waste management and recycling technology. *Mailing Add:* 137 Tulip St Summit NJ 07901

RANNELS, DONALD EUGENE, JR, b Lancaster, Pa, Apr 5, 46; m 67; c 1. PHYSIOLOGY. *Educ:* Pa State Univ, BS, 68, PhD(physiol), 72. *Prof Exp:* Res assoc, 71-73, from instr to assoc prof, 73-83, PROF CELLULAR & MOLECULAR PHYSIOL, HERSHEY MED CTR, PA STATE UNIV, 83-, SR RES ASSOC ANESTHESIA, 78- *Concurrent Pos:* Vis prof, Duke Univ Col Med, 85-86; pres, assembly cell biol, Am Lung Asn, 88-89; ed bd, Biochem J, 85- & numerous physiol journals; assoc ed, Am J Physiol, Endocrinol & Metab, 88-; mem res training rev comt, Nat Heart Liver Blood Inst, 90- *Mem:* Am Soc Biol Chem; Am Thoracic Soc; Am Physiol Soc; Am Heart Asn; Am Lung Asn; Biochem Soc; Am Soc Cell Biol. *Res:* Regulation of lung growth and development role of type II pulmouary epithelial cells in these processes and in the response of the lung injury, role of extracellular matrix in cellular differentiation, metabolic consequences of membrane defarmation or stretch regulation of protein turnover; effects of oxygen deprivation on myocardial metabolism; control of protein metabolism in pulmonary aviolar macrophages; compartmentation of intracellular amino acids; lung growth and metabolism. *Mailing Add:* Dept Physiol Hershey Med Ctr Pa State Univ PO Box 850 Hershey PA 17033

RANNEY, BROOKS, b Daytona Beach, Fla, Jan 31, 15; m 30; c 3. OBSTETRICS & GYNECOLOGY. *Educ:* Oberlin Col, AB, 36; Northwestern Univ, BM, 40, MD, 41, MS, 48. *Prof Exp:* Lab instr physiol, Med Sch, Northwestern Univ, 47-48; clin asst prof, 48-51, clin prof, 51-76, PROF OBSTET & GYNEC, 76-, CHMN DEPT, SCH MED, UNIV SDAK, YANKTON CLIN, 51- *Mem:* Am Col Obstet & Gynec (pres, 81-82); Am Col Surg; Cent Asn Obstet & Gynec (pres, 74-75); Sigma Xi. *Res:* Obstetric analgesia and anesthesia; congenital incompetence of the cervix; paracervical block analgesia for primigravidas; diagnosis and treatment of enterocele; family planning and sex education; external cephalic version; prenatal care studies; clinical studies of endometriosis; advantages of local anesthesia for cesarean section; uterine hypertrophy ovarian function after hysterectomy; volume reduction of uterus during hysterectomy; decreasing numbers of patients for vaginal hysterectomy; sequessae of incomplete gynecological operations. *Mailing Add:* Yankton Med Clin-AC 1104 W Eighth PO Box 590 Yankton SD 57078

RANNEY, CARLETON DAVID, b Jackson, Minn, Jan 23, 28; m 49; c 2. PLANT PATHOLOGY. *Educ:* Agr & Mech Col, Tex, BS, 54, MS, 55, PhD(plant path), 59. *Prof Exp:* Asst plant path, Agr & Mech Col, Tex, 54-55; agent, Field Crops Res Br, 55-57, plant pathologist, Crops Res Div, Cotton & Cordage Fibers Res Br, Tex, 57-58, plant pathologist, Delta Exp Sta, Miss, 58-70, leader cotton path invest, Plant Indust Sta, Md, 70-72, asst area dir, Agr Res Serv, Ala-N Miss Area, 72-74, area dir, Delta States Area, 74-82, area dir, Mid South Area, Agr Res Serv, USDA, 82-87; ASST DIR & HEAD DELTA RES & EXTEN CTR, MISS AGR & FORESTRY EXP STA, 87- *Concurrent Pos:* Adj prof, Dept Agron, Miss State Univ, 74-; chmn, Cotton Dis Coun, 61-62. *Mem:* Cotton Dis Coun; Am Soc Agron; Crop Sci Soc Am; Sigma Xi; Coun Agr Sci & Technol. *Res:* Chemical control of plant diseases; physiology of disease resistance; nature of host parasite relationships. *Mailing Add:* Head Delta Res & Exten Ctr Delta Br Sta PO Box 197 Stoneville MS 38776

RANNEY, DAVID FRANCIS, b Chicago, Ill, Feb 14, 43. IMMUNOBIOLOGY, ONCOLOGY. *Educ:* Oberlin Col, BA, 65; Case Western Reserve Univ, MD, 69. *Prof Exp:* From intern to resident surg, Stanford Univ Hosp, 69-71; res assoc immunol, Nat Inst Dent Res, 71-73; res assoc immunol, Dept Surg, 73-75, asst prof microbiol-immunol & surg, Med Sch, Northwestern Univ, 75-78; ASST PROF PATH, MED SCH, UNIV TEX HEALTH SCI CTR, DALLAS, 78- *Concurrent Pos:* Consult, Natural Prod Sect, Drug Res & Develop Br, NIH, 75- *Res:* Regulation of the immune response by natural products from normal and malignant tissues and by synthetic drugs; regulation of lymphoid neoplasias and autoimmunity by natural products; surgery. *Mailing Add:* Dept Path Univ Tex Health Sci Ctr 5323 Harry Hines Blvd Dallas TX 75235

RANNEY, HELEN M, b Summer Hill, NY, Apr 12, 20. INTERNAL MEDICINE, HEMATOLOGY. *Educ:* Barnard Col, AB, 41; Columbia Univ, MD, 47. *Hon Degrees:* DSc, Univ Southern Calif, 79. *Prof Exp:* Asst prof clin med, Columbia Univ, 58-60; from assoc prof to prof med, Albert Einstein Col Med, Yeshiva Univ, 60-70; prof, State Univ NY Buffalo, 70-73; chmn dept med, Univ Calif, San Diego, 73-86, prof med, 73-90; distinguished physician, Vet Admin Med Ctr, San Diego, 86-91; RETIRED. *Concurrent Pos:* Mem bd dirs, Squibb Corp, 75-89; master, Am Col Physicians, 80. *Honors & Awards:* Joseph Mather Smith Prize, Columbia Univ, 55; Gold Medal, Col Physicians & surgeons Alumni Asn, Columbia Univ, 78; May H Soley Res Award, Western Soc Clin Invest, 87. *Mem:* Nat Acad Sci; Inst Med; Asn Am Physicians; fel AAAS; Am Acad Arts & Sci; Am Col Physicians; Am Soc Clin Invest; Am Soc Hemat; Am Physiol Soc. *Res:* Relationship of hemoglobin and red cell membranes in sickle cell disease and red cell survival. *Mailing Add:* 6229 La Jolla Mesa Dr San Diego CA 92037

RANNEY, J BUCKMINSTER, b Brattleboro, Vt, Dec 26, 19; m 80. COMMUNICATIONS SCIENCE, AUDIOLOGY. *Educ:* NY Univ, BA, 46, MA, 47; Ohio State Univ, PhD(speech), 57. *Prof Exp:* From asst prof to assoc prof speech, Ohio Northern Univ, 48-57; clin dir commun sci, Auburn Univ, 57-69; exec secy, Critical Design Rev Com, NIH, 69-75, chief, Sci Eval Br, 75-78, dep dir commun dis prog, Nat Inst Neurol & Commun Dis & Stroke Admin, 78-; RETIRED. *Concurrent Pos:* Nat Inst Neurol & Commun Disorders & Stroke Spec fel, Johns Hopkins Univ Hosp, 62-63; audiol consult, State Ala Maternal & Child Health, 64-69; chief audio & speech path, Vet Admin Hosp, San Juan, PR, 66-67; assoc ed, Deafness, Speech, Hearing Abstr, 68-82; actg dir, HEW/NIH/Nat Inst Neurol & Commun Disorders & Stroke/Extramural Activ Prog, 78. *Mem:* Fel Am Speech & Hearing Asn; AAAS; Alexander Graham Bell Asn. *Res:* Deafness; language pathology and speech pathology. *Mailing Add:* 6646 Hillandale Ave Chevy Chase MD 20815

RANNEY, RICHARD RAYMOND, b Atlanta, Ga, Jul 11, 39; c 4. PERIODONTOLOGY, DENTAL EDUCATION. *Educ:* Univ Iowa, DDS, 63; Univ Rochester, MS, 69; Eastman Dental Ctr, cert, 69. *Prof Exp:* Asst prof periodontics, Univ Ore Dental Sch, 69-72, assoc prof, Va Commonwealth Univ, 72-78, dir grad periodon, 72-76, chmn periodon, 74-77, asst dean res, sch dentistry, 77-86, prof periodontics, Va Commonwealth Univ, 78-86, prof & dean, 86-89, PROF PERIODONTICS, UNIV ALA, BIRMINGHAM SCH DENTISTRY, 86- *Concurrent Pos:* Intern, US Public Health Serv Hosp, San Francisco, 63-64, chief dental officer, US Pub Health Serv Outpatient Clin, 64-66, prin investr, Nat Inst Health grants res, 70-86, dir clin res ctr, Va Commonwealth Univ, 78-86, bd sci counselors, Nat Inst Dental Res, 81-85, consult, Coun Dental Therapeut, Am Dental Asn, 84-, grants & allocations comn, Am Fund Dental Health, 86-90, Nat Affairs Comn, bd dir, Am Asn Dental Res, 87-91. *Honors & Awards:* Balint Orban Prize, Am Acad Periodontol; Basic Res Award Periodontol Int Asn Dental Res. *Mem:* Am Acad Periodontol; Int Asn Dental Res; Am Asn Dental Res (pres, 90-91); Am Asn Dental Sch; Am Dental Asn; Am Soc Microbiol; Int Col Dentists; Am Col Dentists; AAAS. *Res:* Etiology and pathogenesis of periodontal diseases particularly in relating microbiology, immunology and other host defense mechanisms, and human genetics to clinical disease. *Mailing Add:* UAB Station Birmingham AL 35294

RANOV, THEODOR, b Campina, Romania, Oct 15, 10; nat US; m 41; c 3. ENGINEERING MECHANICS. *Educ:* Tech Univ, Berlin, Dipl Ing, 35, Dr Ing, 37. *Prof Exp:* Instr math & mech, Tech Col Wellington, NZ, 46-48; from asst prof to prof, 49-75, EMER PROF ENG, STATE UNIV NY, BUFFALO, 76- *Concurrent Pos:* Consult, Buffalo Forge Co. *Mem:* Fel Am Soc Mech Engrs; Soc Lubrication Engrs; Am Soc Eng Educ; Sigma Xi; Am Inst Aeronaut & Astronaut. *Res:* Fluid mechanics; lubrication; turbomachinery; diffusion flow. *Mailing Add:* 414 53rd St West Palm Beach FL 33407

RANSFORD, GEORGE HENRY, b Detroit, Mich, Oct 26, 41; m 66; c 2. BROMINE CHEMISTRY. *Educ:* Albion Col, BA, 63; Wayne State Univ, PhD(org chem), 70. *Prof Exp:* Res chemist, Ash Stevens, Inc, 64-65, sr res chemist, 70-76; res chemist, Ethyl Corp, 76-83, sr res chemist, 83-88, res & develop specialist, 88-89, SR RES & DEVELOP SPECIALIST, ETHYL CORP, 89- *Concurrent Pos:* Secy & treas, Ouachita Valley Sect, Am Chem Soc, 91- *Mem:* Am Chem Soc; Sigma Xi; Int Union of Pure & Appl Chem. *Res:* Organic synthesis; nucleoside, nucleotide, cardiac glycoside synthesis; halogen chemicals; process research; carbohydrate chemistry; phase transfer catalysis; heterocycles; condensed aromatics; technical service to manufacturing. *Mailing Add:* Ethyl Corp PO Box 729 Magnolia AR 71753

RANSIL, BERNARD J(EROME), b Pittsburgh, Pa, Nov 15, 29. RESEARCH METHODOLOGY. *Educ:* Duquesne Univ, BS, 51; Cath Univ Am, PhD(phys chem), 55; Univ Chicago, MD, 64. *Prof Exp:* Res assoc, Lab Molecular Struct & Spectra, Univ Chicago, 56-62; intern, Harbor Gen Hosp, Univ Calif, Los Angeles, 64-65; staff Boston City Hosp, 66-74; from assoc to prin assoc 71-82, ASSOC PROF MED, HARVARD MED SCH, 82-; DIR CORE LAB CLIN RES CTR & ASST PHYSICIAN, BETH ISRAEL HOSP, 74- *Concurrent Pos:* Nat Res Coun-Nat Acad Sci fel, Nat Bur Standards, 55-56; Guggenheim fel, 65-66; consult, exobiol proj, NASA & Nat Bur Standards, 56-62; consult, prophet proj, NIH, 71-88; consult, Howard Hughes Med Inst, 79-80; consult, Cooperative Cataract Res Group, 81-83; vis scientist, Rockefeller Univ, 85, Scripps Found, 86, Calif State Univ, 86 & Univ Pittsburg Med Sch, 87. *Mem:* Am Chem Soc; Am Inst Chem; AMA; Sigma Xi; NY Acad Sci; AAAS. *Res:* AB initio calculation of properties and structures of diatomic molecules; statistical comparison of computed and experimental molecular properties; charge density and chemical bonding; biomedical data analysis and inference; research methodology. *Mailing Add:* Dept Med Beth Israel Hosp Boston MA 02215

RANSLEBEN, GUIDO E(RNST), JR, b Comfort, Tex, Oct 19, 25; m 51; c 4. AERONAUTICAL ENGINEERING. *Educ:* Tex A&M Col, BS, 50. *Prof Exp:* Traffic analyst, US Air Force Security Serv, 50-51; draftsman, Douglas Aircraft Corp, 51; engr, Reynolds Andricks Consult Eng, 51-52; res engr, Southwest Res Inst, 52-56; aircraft eng supvr, Mfg Div, Howard Aero, Inc, 56-58; sr res engr, 58-85, STAFF ENGR, SOUTHWEST RES INST, 85- *Mem:* Am Inst Aeronaut & Astronaut. *Res:* Vibrations; dynamics; aeroelasticity; hydro-elasticity; wind tunnel and towing tank testing; instrumentation; flow component testing. *Mailing Add:* Southwest Res Inst PO Drawer 28510 San Antonio TX 78228-0510

RANSOHOFF, JOSEPH, b Cincinnati, Ohio, July 1, 15; m 37; c 2. NEUROSURGERY. *Educ:* Harvard Univ, BS, 38; Univ Chicago, MD, 41; Am Bd Neurol Surg, dipl, 51. *Prof Exp:* Instr surg, Med Sch, Univ Cincinnati, 43-44; asst neurol surg, Col Physicians & Surgeons, Columbia Univ, 49-50, instr neurol surg & assoc neurol surgeon, 50-52, asst prof clin neurol surg, 54-58, assoc prof, 59-61; PROF NEUROSURG & CHMN DEPT, MED CTR, NY UNIV, 61-; DIR NEUROSURG, BELLEVUE HOSP CTR, 61- *Concurrent Pos:* From asst attend neurol surgeon to attend neurol surgeon, Neurol Inst, Presby Hosp, New York, 49-61, chief clin neurosurg, Vanderbilt Clin, 53-55; consult, St John's Riverside & Yonkers Gen Hosps, 61-; consult, St Francis Hosp, Port Jervis; chief consult, Manhattan Vet Admin Hosp. *Mem:* Am Acad Neurol; Cong Rehab Med; AMA; Am Acad Neurol Surg; Soc Neurol Surg (pres). *Res:* Surgery of intracranial tumors, vascular malformations and aneurysms; epilepsy and hydrocephalus. *Mailing Add:* Dept Neurosurg Sch Med NY Univ 550 First Ave New York NY 10016

RANSOM, BRUCE DAVIS, b Binghamton, NY, Apr 15, 51; m 78. PHYSICAL CHEMISTRY. *Educ:* State Univ NY, Binghamton, BA, 73, PhD(chem), 78. *Prof Exp:* Assoc chem, Rensselaer Polytech Inst, 78-80; VPRES, SKAARLAND HOMES, 80- *Mem:* Am Chem Soc; Sigma Xi. *Res:* Excited electronic states of large polyatomic molecules; excited states of butadiene and pyridazine; naphthalene type heterocyles. *Mailing Add:* RR 2 Box 66 Canajoharie NY 13317-9346

RANSOM, C J, b Denison, Tex, June 26, 40. COMPUTER INTEGRATED MANUFACTURING, COMPUTER AIDED ENGINEERING. *Educ:* Univ Tex Austin, BSc, 62, MS, 65, PhD(plasma physics), 68. *Prof Exp:* Asst prof physics, Univ Tex, 67-68; chief, comput mfg, Gen Dynamics, 68-85; MGR, CAD/CAM ANAL, BELL HELICOPTER TEXTRON, 85- *Concurrent Pos:* Lectr, Nat Mgt Asn, 72-85. *Mem:* Am Phys Soc; Am Helicopter Soc. *Res:* Computer integrated manufacturing and mathematical analysis for aerospace. *Mailing Add:* Bell Helicopter Textron 0126 A87 PO Box 482 Ft Worth TX 76101

RANSOM, J(OHN) T(HOMPSON), b Philadelphia, Pa, Aug 4, 20; m 45; c 3. PHYSICAL METALLURGY. *Educ:* Lehigh Univ, BS, 42; Carnegie Inst Technol, DSc(metall), 50. *Prof Exp:* Instr metall, Carnegie Inst Technol, 46-48; res engr, 48-55, res supvr, 55-62, res mgr, 62-63, res sect mgr, 63-66, sr eng assoc, 66-81, Du Pont res fel, E I du Pont de Nemours & Co, Inc, 81; consult, 81-82; RETIRED. *Mem:* Am Soc Metals. *Res:* Materials engineering and fabrication. *Mailing Add:* PO Box 5 Yorklyn DE 19736

RANSOM, PRESTON LEE, b Peoria, Ill, Jan 2, 36; m 62; c 2. ELECTRICAL ENGINEERING. *Educ:* Univ Ill, Urbana, BS, 62, MS, 65, PhD(elec eng), 69. *Prof Exp:* Technician electronics, Res Div, Caterpillar Tractor Co, Peoria, 59-60; student technician, Antenna Lab, Univ Ill, 60-62; elec engr microwave antennas, Raytheon Co, Bedford, 62-63; res asst, Antenna Lab, 63-67, instr,

67-69, res assoc, 69-70, from asst prof to assoc prof, 70-87, PROF ELEC ENG, UNIV ILL, 87- *Concurrent Pos:* Paul V Galvin teaching fel, Univ Ill, 67-68; prin investr, NSF Grant, 71-72 & 73-74; hon res fel, Univ Col London, 76; mem adv comt, USSR & Eastern Europe, Nat Acad Sci, 76- *Mem:* Inst Elec & Electronics Engrs; Optical Soc Am; Am Soc Eng Educ; Sigma Xi. *Res:* Coherent optical processing; holography; diffraction theory; frequency independent antennas. *Mailing Add:* Elec-Comput Eng Univ Ill 1406 W Green St Urbana IL 61801

RANSOME, RONALD DEAN, b Pueblo, Colo, June 9, 54; m 86; c 1. INTERMEDIATE ENERGY NUCLEAR PHYSICS. *Educ:* Colo Sch Mines, BS, 76; Univ Tex, Austin, PhD(physics), 81. *Prof Exp:* Res assoc, Max-Planck Inst Physics, 81-85; asst prof, 85-91, ASSOC PROF PHYSICS, RUTGERS UNIV, 91- *Mem:* Am Phys Soc. *Res:* Nucleon-nucleon and antinucleon-nucleon interaction; pion absorption; delta resonance in nuclei. *Mailing Add:* Physics Dept Rutgers Univ PO Box 849 Piscataway NJ 08855-0849

RANT, WILLIAM HOWARD, b Dothan, Ala, May 24, 45. MATHEMATICS. *Educ:* Univ Ala, BS, 66, MA, 68, PhD(math), 70. *Prof Exp:* Asst prof, Jacksonville State Univ, 70-74; ASST PROF MATH, LINCOLN UNIV, MO, 74- *Mem:* Math Asn Am. *Res:* Ring theory with emphasis on perfect rings and theory of modules. *Mailing Add:* 201 Cityview Dr Jefferson City MO 65101

RANU, HARCHARAN SINGH, b India. ORTHOPEDIC & SPINE BIOMECHANICS, BIOTRIBOLOGY & BIOMATERIALS. *Educ:* Leicester Polytech, Eng, BSc, 63; Univ Surrey, Eng, MSc, 68; Middlesex Hosp Med Sch & Polytech Cent London, PhD(biomed eng), 76. *Prof Exp:* Demonstr mech eng, Leicester Polytech, Eng, 63-64; asst to chief engr, Fabricom, Belg, 65-66; biomed res scientist, Med Res Coun, London, 67-70 & Plastics Res Asn Gt Brit, 77; asst prof bioeng & mech eng, Wayne State Univ, 77-81; prof biomed eng, La Tech Univ, 82-85; ADJ PROF BIOMED, COL PHYSICIANS & SURGEONS, COLUMBIA UNIV, 88-; CHMN & PROF ORTHOP BIOMECH, NY COL OSTEOP MED, 89- *Concurrent Pos:* Biomed scientist, NATO, 82-; consult, Lincoln Gen Hosp, 82-85, La State Univ Med Ctr, 82-, St Luke's & Roosevelt Hosp Ctr, 88- & NY Scientists' Inst Pub Info, 89-; mem, grad fac, La Tech Univ, 82-85 & NY Inst Technol, 89-; vis biomed scientist, Dryburn Hosp, Eng, 86-87; vis prof, Indian Inst Technol, New Delhi, Postgrad Inst Med Educ, Chandigarh, India & King's Col Hosp Med Sch, Univ London, 89-, USSR Acad Sci, Moscow, 90 & Polytech Cent, London, 91-; guest ed, Inst Elec & Electronics Engrs Eng Med & Biol, 90-92; mem staff, Nassau County Med Ctr, 89- *Honors & Awards:* Clayton Award, Inst Mech Eng, London; President's Award, Biol Eng Soc, London. *Mem:* Fel Am Soc Mech Engrs; Am Soc Biomech; Orthop Res Soc; fel Inst Mech Eng London; Biol Eng Soc London; Biomed Eng Soc. *Res:* Orthopedic and spine biomechanics; biomaterials; biotribology; human gait analysis; laser applications; modeling in biomechanics; blood flow mechanics; skin biomechanics; rehabilitation biomedical sciences; author of over 130 publications; research administration. *Mailing Add:* Nelson A Rockefeller Acad Ctr NY Col Osteop Med Old Westbury NY 11568

RANU, RAJINDER S, b Jallan, India, July 26, 40; US citizen; m 70; c 2. MICROBIOLOGY. *Educ:* Punjab Univ, DVM, 61; Univ Pa, MS, 66, PhD(microbiol), 71. *Prof Exp:* Spec Fel Biochem, Univ Chicago, 71-73; res assoc biol, Mass Inst Technol, 74-78; assoc prof microbiol, 79-88, ASSOC PROF PLANT PATH, COLO STATE UNIV, 88- *Concurrent Pos:* Veterinarian; Am Career Soc young & talented investr award, Univ Chicago, 73. *Mem:* Am Soc Microbiol; Am Soc Genetics; NY Acad Sci; Am Soc Biochem & Molecular Biol. *Res:* Translational regulation of gene expression using rabbit reticulocyte lysates and caulimovirusu as model systems. *Mailing Add:* Plant Path Dept Colo State Univ Ft Collins CO 80523

RANZ, WILLIAM E(DWIN), b Blue Ash, Ohio, June 3, 22; m 52; c 4. CHEMICAL ENGINEERING. *Educ:* Univ Cincinnati, ChE, 47; Univ Wis, PhD(chem eng), 50. *Prof Exp:* Res assoc, Univ Ill, 50-51, asst prof chem eng, 51-53; from assoc prof to prof eng res, Pa State Univ, 53-58; PROF CHEM ENG, UNIV MINN, MINNEAPOLIS, 58- *Concurrent Pos:* NSF fel, Cambridge Univ, 52-53. *Mem:* Am Chem Soc; Am Inst Chem Engrs; fel, AAAS. *Res:* Aerosols; sprays; heat and mass transfer; fluid mechanics. *Mailing Add:* Dept Chem Eng & Mat Sci Univ Minn Minneapolis MN 55455-0132

RANZONI, FRANCIS VERNE, b Los Angeles, Calif, Nov 29, 16; c 2. MYCOLOGY. *Educ:* Univ Calif, PhD(mycol), 50. *Prof Exp:* Asst prof biol, Eastern Wash Col, 50-53; asst res botanist, Univ Calif, 53-55; from asst prof bot to assoc prof plant sci, 55-62, chmn, Dept Plant Sci, 60-63, chmn, Dept Biol, 65-68, prof, 62-82, EMER PROF BIOL, VASSAR COL, 82- *Concurrent Pos:* NSF sci fac fel, 64-65. *Mem:* Bot Soc Am; Mycol Soc Am; Ecol Soc Am; NY Acad Sci; Sigma Xi. *Res:* Fungi imperfecti; ascomycetes; plant physiology. *Mailing Add:* 6176 Henderson Rd Sanibel FL 33957

RAO, ANANDA G, b Quilon, India, Dec 27, 30; US citizen; m 62; c 3. BIOCHEMISTRY. *Educ:* Univ Kerala, BSc, 52, MSc, 54; Univ Tex Southwestern Med Sch Dallas, 62-66, PhD(biochem), 66. *Prof Exp:* Lectr chem, Sree Sankara Col, Kalady, India, 54-55; res scholar, Univ Kerala, 55-58; asst res off biochem, Indian Coun Med Res, 58-59; res asst, Wellcome Res Lab, Vellore, 59-62; Welch Found fel, Univ Tex Southwestern Med Sch, Dallas, 62-67; res assoc, Tex A&M Univ, 67-69, res scientist, 69-71; RES BIOCHEMIST, VET ADMIN MED CTR, MARTINEZ, 71- *Concurrent Pos:* NIH res grant, 70-73; Vet Admin res grant, 71-; assoc ed, 86-88, ed, Biochem Archives, 89- *Mem:* Res Soc Alcoholism; Am Inst Nutrit; Int Soc Biomed Res Alcoholism. *Res:* Lipid metabolism; hematology, erythropoiesis; role of drugs and diet fat on tissue lipid composition; iron deficiency; alcoholic fatty liver damage; dietary control of alcoholic liver damage; role of inadequate nutrition on effects attributed to chronic alcoholism. *Mailing Add:* Vet Admin Med Ctr Martinez CA 94553

RAO, B SESHAGIRI, b Masulipatam, India, Apr 18, 36; US citizen; m 63; c 3. SPECTROSCOPY, OPTICS. *Educ:* Andhra Univ, BSc, 53; Banaras Hindu Univ, MSc, 56; Pa State Univ, PhD(physics), 63. *Prof Exp:* Instr physics, Pa State Univ, 64; sr scientist, Warner & Swasey Co, 64-65; asst prof, Duquesne Univ, 65-66; from asst prof to assoc prof, 66-78, PROF PHYSICS, UNIV NDAK, 78-, CHMN DEPT, 87- *Mem:* Am Phys Soc; Optical Soc Am. *Res:* Dipole moment functions of simple molecules; optical properties of matter. *Mailing Add:* Physics Dept PO Box 8008 Univ NDak Grand Forks ND 58202

RAO, BALAKRISHNA RAGHAVENDRA, b Udupi, India, Sept 15, 36; m 65; c 3. INSECT PHYSIOLOGY. *Educ:* Banaras Hindu Univ, BScAg, 57; Karnatak Univ, India, MScAgr, 59; Ohio State Univ, PhD(entom), 64. *Prof Exp:* Lectr agr entom, Col Agr, Dharwar, India, 59-60; res assoc, Johns Hopkins Univ, 64-65 & Univ Conn, 65-67; from assoc prof entom to prof entom, 67-80, PROF BIOL, EAST STROUDSBURG UNIV, 80- *Concurrent Pos:* Res fel, Marine Biol Lab, Woods Hole, Mass, 64-67. *Mem:* Entom Soc Am. *Res:* Reproductive physiology of cockroaches. *Mailing Add:* Dept Biol Sci East Stroudsburg Univ East Stroudsburg PA 18301

RAO, CHALAMALASETTY VENKATESWARA, b Bantumelli, India, Dec 26, 41; m 71; c 2. REPRODUCTIVE ENDOCRINOLOGY. *Educ:* Sri Venkateswara Univ, BVSc, 64; Wash State Univ, MS, 66, PhD(animal sci), 69. *Prof Exp:* Res asst animal sci, Wash State Univ, 66-69; res fel biochem & assoc urol, Albert Einstein Col Med, 69-70; res fel reproductive endocrinol, Med Col, Cornell Univ, 70-72; from asst prof to assoc prof obstet & gynec, 72-79, fac assoc biochem, 72-76, PROF OBSTET & GYNECOL, SCH MED, UNIV LOUISVILLE, 79-, DIR ENDOCRINE LAB & RES, 87- *Mem:* Am Soc Biol Chemists & Molecular Biol; Endocrine Soc; Soc Study Reproduction; Am Fertil Soc; Am Physiol Soc; Soc Gynec Invest; Am Soc Cell Biol. *Res:* Mechanism of action of prostaglandins, gonadotropins, and growth factors in reproductive tissues; identification and characterization of receptors for these agents and molecular mechanisms involved beyond receptor binding. *Mailing Add:* Dept Obstet & Gynec Univ Louisville Sch Med Louisville KY 40292

RAO, DESIRAJU BHAVANARAYANA, b Visakhapatnam, India, Dec 8, 36; US citizen; m 89; c 2. ATMOSPHERIC DYNAMICS, OCEANOGRAPHY. *Educ:* Andhra Univ, India, BSc, 56, MSc, 59; Univ Chicago, MS, 62, PhD(geophys), 65. *Prof Exp:* Fel, Nat Ctr Atmospheric Res, Boulder, 65-67; res scientist oceanog, Marine Sci Br, Dept Energy, Mines & Resources, Ottawa, 67-68; asst prof meteorol, Dept Atmospheric Sci, Colo State Univ, 68-71; vis assoc prof oceanog, Dept Physics, Univ Wis-Milwaukee, 71-72, assoc prof, Dept Energetics, 72-74, prof, 74-75; head phys limnol & meteorol, Great Lakes Environ Res Lab, Nat Oceanic & Atmospheric Admin, Ann Arbor, 75-80; head oceans & ice br, Goddard Space Flight Ctr, Md, 80-84; CHIEF MARINE PRODS BR, NAT METEOROL CTR, NAT OCEANIC & ATMOSPHERIC ADMIN, 84- *Concurrent Pos:* Consult, Can Ctr Inland Waters, Burlington, Ont, 71 & Marine Environ Data Serv, Dept Environ, Ottawa, 74; adj prof limnol & meteorol, Univ Mich, 77-80; adj prof meteorol, Univ Md, Col Park, 81- *Mem:* Sigma Xi; fel Am Meteorol Soc; Am Soc Limnol & Oceanog; Oceanog Soc; Am Geophys Union. *Res:* Oscillations and circulations in lakes; numerical modeling of lake and atmospheric phenomena; waves on continental shelves. *Mailing Add:* Nat Meteorol Ctr Rm 206 Washington DC 20233

RAO, DEVULAPALLI V G L N, b Pithapuram, India, July 6, 33; m 62; c 3. SOLID STATE PHYSICS, LASERS. *Educ:* Andhra Univ, India, BSc(Hons), 53, MSc, 54, DSc, 58. *Prof Exp:* Lectr physics, Andhra Univ, India, 57-59 & 61-63; res assoc, Duke Univ, 59-61; sr scientist, Solid State Physics Lab, Govt of India, Delhi, 63-66; res physicist, Maser Optics Inc, 66-68; mgr, Spacerays Inc, 68; assoc prof, 68-75, chmn dept, 78-81, PROF PHYSICS, UNIV MASS, BOSTON, 75- *Concurrent Pos:* Res fel eng & appl physics, Harvard Univ, 67-69; dir, Laser Nucleonics Inc, 68-75; NSF res grant, Univ Mass, Boston, 70-74; grad prog dir, Univ Mass, 86-88; Battelle res contract, 90. *Mem:* Am Phys Soc; Optical Soc Am. *Res:* Nonlinear optics; liquid crystals; polymers. *Mailing Add:* Dept Physics Univ Mass Boston MA 02125

RAO, GANDIKOTA V, b Vizianagram, India, July 15, 34; m 65; c 2. METEOROLOGY. *Educ:* Andhra Univ, BS, 54, MSc, 55; Indian Inst Technol, Kharagpur, MTech, 58; Univ Chicago, MS, 61, PhD(meteorol), 65. *Prof Exp:* Res assoc meteorol, Univ Chicago, 65; Environ Sci Serv Admin fel & asst prof, Nat Hurricane Res Labs, Univ Miami, 65-68; Nat Res Coun Can fel, Can Meteorol Serv, 68-70 & Univ Waterloo, 70-71; NSF fel & asst prof, 71-74, assoc prof, 74-79, PROF METEOROL, ST LOUIS UNIV, 79-, DIR METEOROL & ASSOC CHMN, 80- *Mem:* Am Meteorol Soc; Sigma Xi; Am Geophys Union. *Res:* Mesometeorology; tropical meteorology; attacking meteorological problems with sound dynamical and numerical techniques; analyzing structure of convection over the Indian Ocean during a southwest monsoon season and the structure of typhoons using satellite data. *Mailing Add:* 3507 Laclede Ave St Louis MO 63103

RAO, GHANTA NAGESWARA, biochemistry, veterinary science, for more information see previous edition

RAO, GIRIMAJI J SATHYANARAYANA, b Bethamangala, India, Feb 13, 34; m 64; c 2. ENZYMOLOGY, GENETICS. *Educ:* Univ Mysore, BSc, 53; Indian Inst Sci, MSc, 59, PhD(biochem), 64. *Prof Exp:* Lectr chem, Univ Mysore, 53-55; asst prof pediat, 71-76, ASSOC PROF PEDIAT, MED SCH, NORTHWESTERN UNIV, CHICAGO, 76-; RES ASSOC BIOCHEM GENETICS, CHILDREN'S MEM HOSP, 69-; ASST PROF PATH, BROWN UNIV, 80- *Concurrent Pos:* W B Lawson fel, NY State Dept Health, Albany, 64-67; S M & O M Rosen fel, Albert Einstein Col Med, 67-69; Robert & Mary Wood innovative res fel in cystic fibrosis, 72; trainer in biochem, Prog in Human Biochem Genetics, Children's Mem Hosp, 70-; asst dir clin chem, RI Hosp, 80- *Mem:* AAAS; Soc Pediat Res; Am Chem Soc; Am Soc Human Genetics. *Res:* Enzymology of cystic fibrosis; control mechanisms in cultured cells; chemical modification of enzymes and proteins. *Mailing Add:* 40 Dartmouth Dr Cranston RI 02902

RAO, GOPAL SUBBA, b India, Aug 12, 38; US citizen; m 72; c 1. PERIODONTAL DISEASES, ORAL BIOLOGY. *Educ:* Madras Univ, India, BSc, 58; Howard Univ, MS, 65; Univ Mich, Ann Arbor, PhD(pharmaceut chem), 69. *Prof Exp:* Chemist forensic chem, Lab Chem & examiner to the Govt Mysore, Pub Health Inst, Bangalore, 58-61; instr biomed chem, Col Pharm, Howard Univ, 62-65; res asst, Dept Pharmaceut, Col Pharm, Univ Mich, Ann Arbor, 65-69; NIH fel, Lab Chem, Nat Heart, Lung & Blood Inst, NIH, 69-72, NIH fel, Lab Chem Pharmacol, 72-74; dir & chief res scientist, Div Biochem, 78-85, chief, lab pharmacol, res inst, Am Dent Asn Health Found, Chicago, 74-85; CLIN PROF, DEPT BIOCHEM, DENT SCH, LOYOLA UNIV, MAYWOOD, ILL, 85- *Concurrent Pos:* Prin investr grants & contracts, NIH & Am Fund Dent Health, 74-; Sigma Xi lectr, Med Ctr, Univ Miss, 82. *Mem:* Am Soc Pharmacol & Exp Therapeut; Soc Toxicol; Am Chem Soc; Am Pharmaceut Asn; Am & Int Asn Dent Res; Am Col Toxicol; AAAS; Am Soc Pharmacog; Sigma Xi; Int Soc Study Xenobiotics. *Res:* Biochemical etiology of periodontal and oral diseases; development of new diagnostic methods and novel drugs and procedures useful in the treatment of oral diseases; salivary nitrite and carcinogenic nitrosamine formation; occupational hazards in dental practice; oral health effects of smokeless tobacco usage. *Mailing Add:* Dept Biochem Loyola Univ Dent Sch Maywood IL 60153

RAO, GOPALAKRISHNA M, b Udupi, India, Mar 17, 44; US citizen; m; c 1. PHYSICAL CHEMISTRY, ELECTROCHEMISTRY. *Educ:* Mysore Univ, India, BSc, 64, MSc, 66; Mem Univ Nfld, St John's, PhD(phys chem & electrochem), 73. *Prof Exp:* Fel metall eng, Queen's Univ, Kingston, Ont, 77-78; res assoc, Ctr Mat Res, Standford Univ, 78-81; res specialist inorg & metals res, 81-86, res specialist inorg chem, Dow Chem Co, Freeport, Tex, 85-86; Welch vis scholar, Rice Univ, Dept Chem, Houston, Tex, 86-87; consult, Advan Clin Prod, Inc, League City, Tex, 87-88; guest scientist, Nat Res Coun, USAF Acad, Colorado Springs, Colo, 88-89; PROG OFFICER, NASA/GODDARD SPACE FLIGHT CTR, GREENBELT, MD, 89- *Concurrent Pos:* Welch Vis Fel, 86-87; Fel, Coun Sci Indust Res, India, 87; Sr Res Assoc, Nat Res Coun, 88-89. *Mem:* Electrochem Soc; Chem Inst Can; Am Asn Crystal Growth; Am Chem Soc; Am Soc Metals; Mat Res Soc; Metall Soc. *Res:* Electrochemistry of fused salts; electrosyntheses of metals and semi-conductors; electro deposition and dissolution of metals, semi-conductors and minerals; chlor-alkali technology; bioelectrochemistry and energies of electron transfer processes in living systems; charge transfer at metal-solution interface; diffusion and permeation of hydrogen isotopes through metals; aqueous and non-aqueous batteries; fuel cell; clinical chemistry; aerospace battery. *Mailing Add:* 15528 Norge Ct Bowie MD 20716-1378

RAO, GUNDU HIRISAVE RAMA, b Tumkur, India, Apr 17, 38; US citizen; m 65; c 2. BIOCHEMICAL PHARMACOLOGY. *Educ:* Univ Mysore, India, BS, 57; Univ Poona, India, BS Hons, 58, MS, 59; Kans State Univ, PhD(entom), 68. *Prof Exp:* Res fel, Commonwealth Inst Biol Control, 59-61 & Cent Food Technol Res Inst, 61-65; res asst, Kans State Univ, 65-68; res fel, Tex A&M Univ, 68-70; res fel, St Paul, 70-71, NIH fel, Minneapolis, 71-72, asst scientist, 72-73, scientist, 73-75, from asst prof to assoc prof, 75-88, PROF, DEPT LAB MED & PATH, UNIV MINN, MINNEAPOLIS, 88- *Concurrent Pos:* Coun Sci Ind Res fel, India, 60, sr fel, 62; mem Nat Thrombosis Coun, Am Heart Asn. *Mem:* Am Asn Pathologists; Int Soc Thrombosis & Haemostasis; Am Heart Asn; Am Soc Hemat; NY Acad Sci; Am Assoc Biol Chem & Molecular Biol; Biochem Soc London. *Res:* Elucidation of mechanisms involved in atherosclerosis, thrombosis, strokes and hemostasis; experimental pathology. *Mailing Add:* Dept Lab Med & Path Box 198 Univ Minn Hosp & Clins Univ Minn Med Sch 420 Delaware St SE Minneapolis MN 55455

RAO, JAGANMOHAN BOPPANA LAKSHMI, b Raghavapuram, India, Aug 6, 36; US citizen; m 60; c 3. ANTENNAS, RADAR SYSTEMS. *Educ:* Andhra Univ, India, BSc, 56; Madras Inst Technol, DMIT, 59; Univ Wash, MS, 63, PhD(elec eng), 66. *Prof Exp:* Asst res engr antennas, Univ Mich, 66-68; staff engr microwave antennas, Northrop Corp, 68-70; asst prof elec eng, Savannah State Col, 70-71; res assoc, NASA, 71-73; ELECTRONICS ENGR RADAR, NAVAL RES LAB, 74- *Concurrent Pos:* Asst prof elec eng, Howard Univ, 73. *Mem:* Sr mem, Inst Elec & Electronics Engrs. *Res:* Antennas; electromagnetic theory; radar systems and radar signal processing. *Mailing Add:* Radar Div Code 5317 Naval Res Lab Washington DC 20375-5000

RAO, JAMMALAMADAKA S, b Munipalle, India, Dec 7, nat US; m 72; c 2. DIRECTIONAL DATA ANALYSIS. *Educ:* Indian Statist Inst, Calcutta, BA, 64, MA, 65, PhD(statist), 69. *Prof Exp:* Res scholar statist, Indian Statist Inst, 65-69; vis asst prof probability & statist, Ind Univ, 69-70, asst prof, 70-76; from asst prof to assoc prof, 80-82, PROF PROBABILITY & STATIST, UNIV CALIF, SANTA BARBARA, 82- *Concurrent Pos:* Vis prof, Univ Wis-Madison, 75-76 & Univ Leeds, 76; dir, Statist Consult Ctr, Univ Calif, Santa Barbara, 80- *Mem:* Am Statist Asn; fel Inst Math Statist; Royal Statist Soc; Int Statist Inst; fel Inst Combinatorics & Applications. *Res:* Nonparametric statistical methods; inference based on sample spacings; large sample theory; efficiencies of test procedures; statistics of directional data. *Mailing Add:* Dept Statist & Appl Probability Univ Calif Santa Barbara CA 93106

RAO, JONNAGADDA NALINI KANTH, b Eluru, India, May 16, 37; m 65; c 2. STATISTICS. *Educ:* Andhra Univ, India, BA, 54; Univ Bombay, MA, 58; Iowa State Univ, PhD(statist), 61. *Prof Exp:* Asst prof statist, Iowa State Univ, 61-63; assoc prof, Grad Res Ctr Southwest, 64-65; from assoc prof to prof, Tex A&M Univ, 65-69; prof, Univ Man, 69-73; PROF STATIST, CARLETON UNIV, 73- *Concurrent Pos:* Consult, Statist Can, 74- *Mem:* Fel Inst Math Statist; elected mem Int Statist Inst; Can Statist Asn; Biomet Soc; fel Am Statist Asn. *Res:* Sample survey theory and practice; linear models and variance components; time series. *Mailing Add:* Dept Math & Statist Carleton Univ Ottawa ON K1S 5B6 Can

RAO, K V N, b Visakhapatnam, India, June 22, 33; US citizen; m 67. PHYSICS, ELECTRICAL ENGINEERING. *Educ:* Andhra Univ, India, BSc, 52; Indian Inst Sci, dipl, 55; Univ Ill, Urbana, PhD(elec eng), 62. *Prof Exp:* From res asst to res assoc elec eng, Univ Ill, Urbana, 55-63; res physicist, Air Force Cambridge Res Labs, 63-76; RES PHYSICIST, ROME AIR DEVELOP CTR/ELECTROMAGNETIC SCI DIV, HANSCOM AFB, 76- *Concurrent Pos:* Adj prof, Northeastern Univ, 65-69; adv, Rensselaer Polytech Inst, 70- *Mem:* Inst Elec & Electronics Eng; Am Inst Physics. *Res:* Reentry plasma physics; gaseous electronics; microwave interaction with gyrotropic media; propagation in ionosphere; basic atomic and molecular physics; laser interaction with solid dielectrics radar systems, EM transmission; radar systems; environmental effects; EM scattering; land, sea terrain features. *Mailing Add:* One Old Bellerica Rd Hanscom AFB Bedford MA 01730

RAO, KALIPATNAPU NARASIMHA, b Naraspur, India, Mar 7, 37; US citizen; m 65; c 3. EXPERIMENTAL NUTRITION. *Educ:* Bombay Univ, BS, 58; Nagpur Univ, MS, 60; Indian Agr Res Inst, PhD(biochem), 65. *Prof Exp:* Res officer biochem, Nat Inst Commun Dis, India, 64-71; instr, 71-76, ASST PROF PATH, UNIV PITTSBURGH, 76-; ASSOC MEM CANCER RES, PITTSBURGH CANCER INST, 86- *Concurrent Pos:* NIH res grant, 80; vis prof, Univ Cagliari, Sardinia, Italy, 83; Nat Dairy Coun res grant, 85. *Mem:* Am Pancreatic Asn; Am Inst Nutrit; Am Asn Pathologists; Soc Toxicol Pathologists. *Res:* Biochemical pathology of gastrointestinal tract; chemical carcinogenesis; experimental nutrition; regulation of cholesterol metabolism. *Mailing Add:* Dept Path Scaife Hall 744A Univ Pittsburgh 4200 Fifth Ave Pittsburgh PA 15261

RAO, KANDARPA NARAHARI, b Kovvur, India, Sept 5, 21; m 52; c 1. PHYSICS, ASTROPHYSICS. *Educ:* Andhra Univ, India, BSc, 41, MSc, 42; Univ Chicago, PhD, 49. *Prof Exp:* With Govt Meteorol Serv, India, 42-46; with sci off, Nat Phys Lab, 50-52; res assoc physics, Duke Univ, 52-53; res assoc & asst prof, Univ Tenn, 53-54; res assoc, 54-60, lectr, 59-60, assoc prof, 60-63, PROF PHYSICS, OHIO STATE UNIV, 63- *Concurrent Pos:* Consult, Nat Oceanic & Atmospheric Admin, 73-; ed J Molecular Spectros. *Mem:* Am Astron Soc; fel Optical Soc Am; fel Am Phys Soc; Int Astron Union; Coblentz Soc. *Res:* Structures of molecules, especially their electronic spectra in the ultraviolet and high resolution absorption and emission spectra in the infrared. *Mailing Add:* Dept Physics Ohio State Univ 174 W 18th Ave Columbus OH 43210

RAO, KROTHAPALLI RANGA, b Amartaluru, India, Sept 24, 41; m 65; c 2. CRUSTACEAN PHYSIOLOGY, INVERTEBRATE ENDOCRINOLOGY. *Educ:* Andhra Univ, India, BS, 58, MS, 61, PhD(zool), 67. *Prof Exp:* Demonstr biol, Andhra Christian Col, India, 58-59, demonstr zool, Andhra Univ, 61-62, res fel, 62-65; res assoc biol, Tulane Univ, 66-72; from asst prof to assoc prof, 72-78, PROF BIOL, UNIV W FLA, 78-, DISTINGUISHED UNIV RES PROF, 86- *Concurrent Pos:* Assoc ed, J exper zool, 86-88; mem NSF rev panel for Presidential Young Invest Award prog cellular biol; mem EPA Environ Biol rev panel, 88. *Mem:* Am Soc Zool; fel AAAS; Crustacean Soc; Int Pigment Cell Soc; Sigma Xi. *Res:* Biochemistry and functions of invertebrate neuropeptides; control of color changes and molting in crustaceans; pollution physiology of marine animals; isolation and characterization of the crustacean neurohormones; comparative endocrinology. *Mailing Add:* Fac Biol Univ WFla Pensacola FL 32514-5751

RAO, M S(AMBASIVA), b Chiluvur, India, Oct 19, 42; m 68; c 2. SURGICAL PATHOLOGY. *Educ:* Osmania Univ, MB & BS, 65; Andhra Univ, MD, 72. *Prof Exp:* From asst prof to assoc prof, 77-87, PROF PATH, NORTHWESTERN UNIV MED SCH, 87- *Concurrent Pos:* Prin investr, NIH, 83-90. *Mem:* Am Asn Cancer Res; Int Acad Path; Am Asn Path. *Res:* Peroxisome proliferators-induced hepatocarcinogenesis and transdifferentiation of pancreatic cells into hepatocytes. *Mailing Add:* Dept Path Northwestern Univ Med Sch Chicago IL 60611

RAO, MALEMPATI MADHUSUDANA, b June 6, 29; Indian citizen; m; c 2. MATHEMATICAL STATISTICS. *Educ:* Andhra Univ, India, BA, 49; Univ Madras, MA, 52, MSc, 55; Univ Minn, PhD, 59. *Prof Exp:* Lectr math, Univ Col Andhra Univ, India, 52-53; res mathematician, Carnegie-Mellon Univ, 59-60, from asst prof to prof, 60-72; PROF MATH, UNIV CALIF, RIVERSIDE, 72- *Concurrent Pos:* NSF grants, 60-62, 63-65 & 66-79; Air Force grant award, 68-69; vis mem, Inst Advan Study, 70-72 & 84-85; res awards, Off Naval Res, 80- *Mem:* Am Math Soc; fel Inst Math Statist; AAAS; Int Statist Inst. *Res:* Probability; function spaces; related areas in analysis. *Mailing Add:* Dept Math Univ Calif Riverside CA 92521

RAO, MAMIDANNA S, b Kaikalur, India, June 21, 31; US citizen; m 76; c 2. BIOSTATISTICS. *Educ:* Univ Madras, BSc, 51; Univ Punjab, MA, 60; Univ Pittsburgh Sch Pub Health, MSHyg, 68, ScD(biostatist), 70. *Prof Exp:* Statistician biostatist, Venereal Dis Training Ctr, 55-64 & Indian Coun Med Res, 64-67; biostatistician, St Francis Gen Hosp, Community Ment Health Ctr, 69-70; asst prof biostatist, Univ Tex Med Br, 70-71; asst dir biostatist, Montefiore Hosp & Med Ctr, 71-72; statistician biostatist, Pan Am Health Orgn, WHO, 72-76; assoc prof, 76-88, PROF BIOSTATIST, COL MED, HOWARD UNIV, 88- *Concurrent Pos:* Biostatistician & liaison officer, USAID & Govt Sudan proj schistosomiasis control Sudan, Africa, 82-84. *Mem:* Sigma Xi; Am Pub Health Asn; fel Royal Soc Trop Med & Hyg; Am Statist Asn; Biomet Soc. *Res:* Teaching of biostatistics; consultation and research; design of experiments; sample surveys; computer applications. *Mailing Add:* Howard Univ Col Med 520 W St NW Rm 2400 Washington DC 20059

RAO, MENTREDDI ANANDHA, b Dornakal, India, July 4, 37; US citizen; m 70; c 1. FOOD ENGINEERING. *Educ:* Osmania Univ, India, BChE, 58; Univ Cincinnati, MS, 63; Ohio State Univ, PhD(chem eng), 69. *Prof Exp:* Res assoc chem eng, USAF, Dayton, Ohio, 63-65; proj engr, Am Standards, Inc, 69-71; prof & head food eng, Univ Campinas, Brazil, 71-73; from asst prof to assoc prof, 73-86, PROF FOOD SCI, CORNELL UNIV, 86- *Concurrent*

Pos: Fulbright res scholar, Brazil, 80-81, Portugal, 88-89. *Mem:* Inst Food Technologists; Am Soc Agr Engrs; Am Inst Chem Engrs; Soc Rheology; Instrument Soc Am. *Res:* Energy use and conservation for foods; rheology of fluid foods; heat transfer in food processing. *Mailing Add:* Dept Food Sci Cornell Univ Geneva NY 14456

RAO, MRINALINI CHATTA, b Bangalore, India; m; c 2. MEMBRANE TRANSPORT, HORMONE ACTION. *Educ:* Univ Delhi, BSC, 69, MSC, 71; Univ Mich, MS, 74, PhD(cell & molecular biol), 77. *Prof Exp:* Fel, Univ Chicago, 77-80, asst prof & res assoc, 80-83; asst prof, 84-89, ASSOC PROF, UNIV ILL, CHICAGO, 89- *Concurrent Pos:* Teaching asst fel, Univ Mich, 74. *Mem:* Am Soc Cell Biologists; Am Physiol Soc; Am Gastroenterol Asn. *Res:* Molecular mechanisms involved in agonist (hormone) and mediator dependent regulation of cell function; regulation of ion transport in epithelial cells; role of protein phosphorylation; signal transduction. *Mailing Add:* Dept Physiol & Biophys Univ Ill Chicago PO Box 6998 M-C901 Chicago IL 60680

RAO, NUTAKKI GOURI SANKARA, b Tenali, India, Dec 2, 33; m 60; c 3. TOXICOLOGY. *Educ:* Univ Saugar, BSc, 55, MSc, 57, BPharm, 58; St Louis Col Pharm, MS, 62; NDak State Univ, PhD(pharmaceut chem), 66; dipl, Am Bd Forensic Toxicol. *Prof Exp:* Chemist, Ciba Pharmaceut Ltd, 58-59; assoc prof 70-76, CHMN DEPT TOXICOL, NDAK STAT UNIV, 75, PROF TOXICOL, 76. *Concurrent Pos:* State toxicologist, NDak, 73- *Mem:* Am Acad Clin Toxicologists; Am Acad Forensic Sci; Int Asn Forensic Toxicologists; Sigma Xi; Soc Forensic Toxicologists; Am Soc Crime Lab Dirs. *Res:* Detection and quantitation of drugs and metabolites from biological tissues; analytical and clinical toxicology. *Mailing Add:* Pharm Dept NDak State Univ Fargo ND 58102

RAO, P KRISHNA, b Kapileswarapuram, India, Mar 26, 30; m 54; c 2. METEOROLOGY, OCEANOGRAPHY. *Educ:* Andhra Univ, India, BS, 50, MS, 52; Fla State Univ, MS, 57; NY Univ, PhD(meteorol, oceanog), 68. *Prof Exp:* Asst res scientist, NY Univ, 56-60; meteorologist, Can Meteorol Serv, 60-61; DIR, OFF RES & APPL NAT ENVIRON SATELLITE SERV, NAT OCEANIC & ATMOSPHERIC ADMIN, 61- *Concurrent Pos:* Dept Com sci & technol fel & mem, Rann Prog, NSF, 71-72; expert meteorol satellites, World Meteorol Orgn, Geneva, 74-76; Fullbright fel, 85-86. *Mem:* Fel Am Meteorol Soc; fel Royal Meteorol Soc; fel NY Acad Sci. *Res:* Satellite meteorology and oceanography. *Mailing Add:* 15824 Buena Vista Dr Rockville MD 20855

RAO, PALAKURTHI SURESH CHANDRA, b Warangal, India, Feb 15, 47; m; c 1. SOIL PHYSICS, GROUNDWATER QUALITY. *Educ:* Andhra Pradesh Agr Univ, Hyderabad, India, BSc, 67; Colo State Univ, MS, 69; Univ Hawaii, PhD(soil sci), 74. *Prof Exp:* Asst res scientist, 77-79, from asst prof to assoc prof, 79-85, PROF SOIL SCI, UNIV FLA, 85- *Concurrent Pos:* assoc ed, J Contaminant Quality, 80-83, Water Res, 90-92; ed, J Contaminant Hydrol, 85-; mem water sci technol bd, Nat Res Coun. *Mem:* Fel Soil Sci Soc Am; fel Am Soc Agron; Int Soil Sci Soc; Am Geophys Union; Am Chem Soc; Soc Environ Toxicol Chem. *Res:* Environmental quality; computer modeling of soil-water- plant systems; groundwater contamination. *Mailing Add:* 2171 McCarty Hall Univ Fla Gainesville FL 32611-0151

RAO, PAPINENI SEETHAPATHI, b Vetapalam, India, Apr 19, 37; m 67; c 3. REPRODUCTIVE PHYSIOLOGY. *Educ:* Andhra Vet Col, India, BVSc, 59; Univ Mo, Columbia, MS, 61, PhD(reproductive physiol), 65. *Prof Exp:* Vet asst surgeon, Dept Vet Med, Andhra, India, 59-60; res asst, Univ Mo, Columbia, 62-65, res assoc, 65-66; from instr to assoc prof gynec, obstet & physiol, Sch Med, St Louis Univ, 66-79; assoc prof gynec, obstet & physiol, 79-85, PROF GYNEC & OBSTET, UNIV S FLA, 85- *Honors & Awards:* Kiepe Scholar, Univ MO, 61-62. *Mem:* AAAS; NY Acad Sci; Soc Study Reproduction; Am Physiol Soc; Shock Soc; Am Asn Univ Prof; Int Endotoxin Soc. *Res:* Clinical veterinary medicine; conception and contraception; endotoxic shock; toxemia of pregnancy; cardiovascular physiology. *Mailing Add:* Dept Obstet/Gynec Box 18 Univ SFla Col Med Tampa FL 33612

RAO, PEJAVER VISHWAMBER, b Udipi, India, June 11, 35; m 62; c 2. NONPARAMETRICS, BIOSTATISTICS. *Educ:* Univ Madras, BA, 54; Univ Bombay, MA, 56; Univ Ga, PhD(statist), 63. *Prof Exp:* Lectr statist, Col Sci, Univ Nagpur, 56-60; asst statistician, Univ Ga, 62-64, asst prof math, 63-64; from asst prof to assoc prof, 64-72, PROF STATIST, UNIV FLA, 72- *Concurrent Pos:* Vis fel, Australian Nat Univ, 85. *Mem:* Inst Math Statist; Biomet Soc; Am Statist Asn; Int Statist Inst. *Res:* Nonparametric estimation; censored data analysis. *Mailing Add:* Dept Statist Univ Fla 480 Little Hall Gainesville FL 32611

RAO, PEMMARAJU NARASIMHA, b Rajahmundry, India, Dec 20, 28; m 53; c 3. ORGANIC CHEMISTRY. *Educ:* Andhra Univ, India, BSc, 48, MSc, 50; Univ Calcutta, PhD(chem), 54. *Prof Exp:* Fulbright travel grant & fel steroid chem, Sch Med, Univ Rochester, 54-55; res assoc org chem, Indian Inst Sci, Bangalore, 55-56; jr sci officer, Nat Chem Lab, Poona, 56-58; res assoc steroid chem, 58-61, chief org chem sect, 61-65, chmn dept org chem, 65-67, asst found scientist, 67-73, found scientist org chem, 73-77, CHMN, DEPT ORG CHEM, SOUTHWEST FOUND BIOMED RES, 78- *Concurrent Pos:* Res prof, St Mary's Univ, Tex, 60-; adj prof, Incarnate Word Col, San Antonio, TX, 84- *Mem:* Am Chem Soc; Royal Soc Chem; Endocrine Soc; Am Inst Chemists; Am Asn Clin Chem. *Res:* Organic synthesis; natural products chemistry, particularly steroids and diterpenes; synthesis of polycyclic hydrocarbons; steroid radioimmunoassays; synthesis of unnatural aminoacids; steroid metabolism. *Mailing Add:* Southwest Found Biomed Res PO Box 28147 San Antonio TX 78284

RAO, PEMMARAJU VENUGOPALA, b Tirupatipuram, India, Sept 1, 32; m 58; c 2. NUCLEAR PHYSICS, ATOMIC PHYSICS. *Educ:* Andhra Univ, India, BSc, 53, MSc, 54; Univ Ore, PhD(physics), 64. *Prof Exp:* Demonstr physics, Andhra Univ, India, 55-57, lectr, 58-59; res assoc, Univ Ore, 64-66; res assoc, Ga Inst Technol, 66-67; asst prof, 67-71, ASSOC PROF PHYSICS, EMORY UNIV, 71- *Concurrent Pos:* Ed, Vijnana Patrika, 70-76. *Mem:* AAAS; Sigma Xi; Am Phys Soc; Am Asn Physics Teachers. *Res:* Nuclear spectroscopy; fast neutron reactions, atomic fluorescence yields and inner shell ionization. *Mailing Add:* Dept Physics Emory Univ Atlanta GA 30322

RAO, PEMMASANI DHARMA, b Burripalem, Andhra Pradesh, Apr 15, 33; c 3. MINERAL ENGINEERING. *Educ:* Andhra Univ, India, BSc, 52, MSc, 54; Pa State Univ, MS, 59, PhD(mineral prep), 61. *Prof Exp:* Jr sci asst petrol, Ore Dressing Div, Nat Metall Lab, Jamshedpur, India, 55-57; asst mineral prep, Pa State Univ, 58-61; tech adv coal & mineral processing, McNally-Bird Eng Co Ltd, India, 62-66; from asst prof to assoc prof coal technol, 66-76, PROF COAL TECHNOL, UNIV ALASKA, FAIRBANKS, 76-, ASSOC DIR, MINERAL INDUST RES LAB, 85- *Mem:* Am Inst Mining, Metall & Petrol Engrs. *Res:* Coal petrology; coal characterization and utilization; ore microscopy; mineral processing. *Mailing Add:* Mineral Indust Res Lab Univ Alaska 212 E O'Neill Resources Bldg Fairbanks AK 99775-1180

RAO, PODURI S R S, b Kakinada, India, Dec 13, 34; m 70; c 2. MATHEMATICS, STATISTICS. *Educ:* Andhra Univ, India, BA, 55; Karnatak Univ, India, MA, 57; Harvard Univ, PhD(statist), 65. *Prof Exp:* Demonstr & lectr statist, Univ Bombay, 57-60; asst prof, Univ Rochester, 64-66; sr math statistician, Info Res Assocs Inc, 66-67; assoc prof statist, 67-73, PROF STATIST, UNIV ROCHESTER, 74- *Concurrent Pos:* Fulbright travel grant, 60; J N Tata endowment; elect mem, Int Statist Inst. *Mem:* Am Statist Asn; Inst Math Statist. *Res:* Sampling; linear models; mullivariate analysis; variance components. *Mailing Add:* Dept Statist Univ Rochester Rochester NY 14627

RAO, POTU NARASIMHA, b Muppalla, India, July 1, 30; m 57; c 3. CELL BIOLOGY, CYTOGENETICS. *Educ:* Andhra Univ, India, BSc, 52; Univ Ky, PhD(cytogenetics), 63. *Prof Exp:* Instr cellular physiol, Univ Ky, 64-66, asst prof, 66-68; asst prof cell biol, Sch Med, Univ Colo, 68-71; assoc prof cell biol, 71-77, CHIEF SECT CELLULAR PHARMACOL, UNIV TEX M D ANDERSON CANCER CTR, 76-, PROF CELL BIOL, 77- *Mem:* AAAS; Am Asn Cancer Res; Am Soc Cell Biol. *Res:* Biochemical processes related to the regulation of DNA synthesis and mitosis in mammalian cells in culture; monoclonal antibodies specific to Mitotic cells. *Mailing Add:* Dept Med Oncol M D Anderson Cancer Ctr Houston TX 77030

RAO, PRAKASH, b Bangalore, India, Oct 16, 41. MATERIALS SCIENCE, ENGINEERING. *Educ:* Univ Delhi, BS, 61; Univ Bombay, BS, 63; Indian Inst Sci, BEng, 65; Univ Calif, Berkeley, MS, 66, PhD(mat sci & eng), 69. *Prof Exp:* Res asst inorg mat res div, Lawrence Radiation Lab, Univ Calif, Berkeley, 65-69; mem tech staff, Ingersoll-Rand Res, Inc, 70-71; instr mat sci & eng & res assoc, Cornell Univ, 71-72; staff mat scientist, 72-77, actg mgr, Chem & Structural Anal Br, 77-78, MGR, TECH RELATIONS, ENERGY SCI & ENG, CORP RES & DEVELOP, GEN ELEC CO, 78- *Mem:* Am Inst Mining, Metall & Petrol Engrs; Am Soc Metals; Electron Micros Soc Am. *Res:* High voltage transmission electron microscopy and electron diffraction techniques for the study of irradiation effects in metals and alloys; metallurgy. *Mailing Add:* 1808 Meadowood Ct S Brooklyn Park MN 55444

RAO, R(AMACHANDRA) A, b Kanakapura, India; m 71; c 3. HYDROLOGY, WATER RESOURCES. *Educ:* Univ Mysore, BE, 60; Univ Minn, MSCE, 64; Univ Ill, PhD(civil eng), 68. *Prof Exp:* Lectr civil eng, Univ Mysore, 60-62; instr lang, Univ Minn, 62-63, res asst hydraul, 63-64; res asst hydrol, Univ Ill, 64-68; from asst prof to assoc prof, 68-79, PROF CIVIL ENG, PURDUE UNIV, WEST LAFAYETTE, 79- *Concurrent Pos:* Consult to several orgns. *Mem:* AAAS; Am Geophys Union; Am Soc Civil Engrs; Int Asn Hydraul Res; Int Asn Sci Hydrol. *Res:* Urban and stochastic hydrology. *Mailing Add:* Sch Civil Eng Purdue Univ West Lafayette IN 47907

RAO, R(ANGAIYA) A(SWATHANARAYANA), b Bangalore, India, Feb 27, 34; m 67. ELECTRONICS & MICROWAVES, SOLID STATE PHYSICS. *Educ:* Univ Mysore, BSc, 53; Indian Inst Sci, Bangalore, dipl elec commun eng, 57; Univ Calif, Berkeley, MS, 61, PhD(elec eng), 66. *Prof Exp:* Sir Dorabji Tata scholar, Indian Inst Sci, Bangalore, 56-57; jr sci officer microwave tubes. Cent Electronic Eng Res Inst, Pilani, India, 57-59; grad res engr, Electronics Res Lab, Univ Calif, Berkeley, 59-66; mem tech staff, Fairchild Semiconductor Res & Develop Lab, Calif, 66-68; asst prof solid state electronics, 68-74, assoc prof elec eng, 74-80, PROF ELEC ENG, CALIF STATE UNIV, SAN JOSE, 80- *Concurrent Pos:* Vis scholar, Stanford Univ, 77-78; consult, SKE Power Interface Devices, 79, Avantek Inc, Santa Clara, 84; USAF fel & res grant, Wright Patterson AFB, 79-80; fel, Lawrence Livermore Nat Labs, AWU-Dept Energy, 82, Japan Soc Prom Sci, Kyoto Univ, 88-89, fac res fel, Naval Ocean Systs Ctr, San Diego, 89, 90. *Mem:* Inst Elec & Electronics Engrs; Am Soc Eng Educ. *Res:* Electron beams and guns; microwave tubes; solid state devices; microwave semiconductor devices; device and circuit modeling and simulation; monolithic microwave integrated circuits. *Mailing Add:* Dept Elec Eng Calif State Univ Washington Sq San Jose CA 95192

RAO, RAMACHANDRA M R, b Bangalore City, India, Oct 30, 31. FOOD SCIENCE & TECHNOLOGY. *Educ:* Univ Mysore, BSc, 59; Univ Houston, BS, 62; La State Univ, Baton Rouge, MS, 63, PhD(food sci), 66. *Prof Exp:* Chemist, Savage Labs, Houston, 60-62; dir res, AME Enterprises, NJ, 66-67; asst prof food sci, 67-70, assoc prof, Food Preservation, 70-80, PROF FOOD SCI, LA STATE UNIV, BATON ROUGE, 80- *Concurrent Pos:* Allen Prod grant, La State Univ, Baton Rouge, 67-68; fel, Int Atomic Energy Agency, SVietnam & adv, Vienna, Austria, 70. *Honors & Awards:* US AEC Award, 66. *Mem:* Assoc Am Inst Chem Eng; Inst Food Technol; Can Inst Food Sci & Technol. *Res:* Food processing and preservation; water and air pollution; waste utilization and disposal; fermentation technology. *Mailing Add:* 921 Woodstone Dr Baton Rouge LA 70808

RAO, RAMGOPAL P, b India, Aug 15, 42; US citizen; m 69; c 3. OPHTHALMIC INSTRUMENTATION, SCIENTIFIC INSTRUMENTATION. *Educ:* Regional Eng Col, India, BSEE, 65; Okla State Univ, Stillwater, MSEE, 66; Northeastern Univ, MBA, 72. *Prof Exp:* Sr syst engr, Honeywell Inc, 66-70; proj engr, Digilab Inc, 70-72, dir res & develop, 72-75, dir customer serv, 75-77, vpres mfg, 77-79; div mgr, Ophthalmic Div, Biorad Labs Inc, 81-90; CHIEF EXEC OFFICER, TOMEY SCI INC, 91- *Mem:* Inst Elec & Electronics Engrs. *Mailing Add:* 31 Stewart Rd Needham MA 02192

RAO, SALEM S, b Bangalore, India, Apr 12, 34; Can citizen; m 64; c 2. MICROBIOLOGY, BACTERIOLOGY. *Educ:* Univ Mysore, PhD(zool), 64; Royal Soc London, FRSH, 65. *Prof Exp:* Res microbiologist, Rensselaer Polytech Inst, 61-64; res scientist, Ont Res Found, 64-65; scientist bacteriol, Ont Dept Health, 65-73; RES SCIENTIST, NAT WATER RES INST, CAN, 73- *Concurrent Pos:* Teacher environ microbiol/ecol; hon res assoc & adj prof, Brock Univ & Univ Toronto; vis prof, Univ Chile, Santiago, Chile; vis scientist, Nat Univ Laplata, Arg. *Mem:* Royal Soc Can; Sigma Xi; Am Soc Testing & Mat. *Res:* Limnology of aquatic ecosystems; cold water bacteriology; assimilation under low temperatures; bacterial nutrient relationships under different temperatures; acid rain-stress bacteriology (biodegradation and bioassimilation); acid lake recovery study; contaminat-biotic interactions and transport in flurial systems; ecotoxicology; microbiology toxicity. *Mailing Add:* Ecotoxicol & Microbiol Lab Rivers Res Br Nat Water Res Inst Burlington ON L7R 4A6 Can

RAO, SAMOHINEEVEESU TRIVIKRAMA, b India, July 2, 44; US citizen; m 74; c 2. AIR POLLUTION METEOROLOGY, ATMOSPHERIC MODELING. *Educ:* Andhra Univ, India, BSc, 62, MSc, 65; State Univ NY at Albany, PhD(atmospheric sci), 73. *Prof Exp:* Lectr physics, Govt Arts Col, Nizamabad, India, 65-66; sr sci asst meteorol, Inst Tropical Meteorol, Govt India, 66-69; res fel, State Univ NY at Albany, 69-72, res assoc atmospheric sci, 73-74; res scientist air pollution & meteorol & chief atmospheric modeling sect & math modeling sect, 74-84, DIR RES, NY STATE DEPT ENVIRON CONSERV, 84- *Concurrent Pos:* Prin investr, Funded Res Proj, US Environ Protection Agency & other fed agencies, 75-; Consult, State Univ NY at Albany, 75-76, M B Assoc, San Francisco, 77-78; adj prof, State Univ NY at Albany, 78-; panel mem, Nat Coop Highway Res Prog, Nat Acad Sci, 78-82; mem, Tech Coun Meteorol & Educ, Air & Waste Mgt Asn, 79-; cert consult meteorologist, Am Meteorol Soc, 80-; consult, Indian Inst Technol, New Delhi, India. *Mem:* Am Meteorol Soc; Air & Waste Mgt Asn. *Res:* Mathematical modeling of transport and diffusion of pollutants in the atmosphere; develop and validate air pollution dispersion models; statistical analysis of air pollution data. *Mailing Add:* 30 Huntwood Dr Clifton Park NY 12065

RAO, SHANKARANARAYANA RAMOHALLINANJUNDA, b Hiriyur, India, July 11, 23; m 58. CIVIL ENGINEERING, SOIL MECHANICS. *Educ:* Univ Mysore, BE, 46; Univ Roorkee, ME, 58; Univ Conn, MS, 61; Rutgers Univ, PhD(soil mech), 64. *Prof Exp:* Supvr irrig-works, Mysore Pub Works Dept, 46-50; tech asst, River Valley Projs, 50-57; asst engr, Mysore Eng Res Sta, 58-59; assoc prof, 64-65, chmn dept, 64-84, PROF CIVIL ENG, PRAIRIE VIEW A&M UNIV, 65-, ASSOC DEAN ENG, 85- *Mem:* Am Soc Civil Engrs; Am Soc Eng Educ; Am Soc Testing & Mat; Concrete Inst; Sigma Xi; Int Soc Soil Mech & Found Eng; Nat Soc Prof Engrs. *Res:* Field study of restriction of evaporation from open water surfaces; problem of canal lining; phenomena of frost action in highways; permeability of soils; structural strength of brick masonry prisms; correlation of data from field for remote sensing of wheat canopy; nature of fracture failure in composite materials. *Mailing Add:* Col Eng PB 936 Prairie View A&M Univ Prairie View TX 77446-2345

RAO, SREEDHAR P, b Jan 2, 43; US citizen. PEDIATRICS, HEMATOLOGY-ONCOLOGY. *Educ:* Kakatiya Med Col, Warangal, India, MBBS, 66; Am Bd Pediat, cert pediat hemat-oncol, 76. *Prof Exp:* Internship med-surg, Bergen Pines County Hosp, Paramus, NJ, 67-68; resident pediat, 68-71, fel pediat-hemat-oncol, 71-73, fel pediat-oncol, 73-74, instr pediat, 74-75, asst prof pediat, 75-81, ASSOC PROF CLIN PEDIAT, STATE UNIV NY DOWNSTATE MED CTR, 81- *Concurrent Pos:* Attend physician pediat, Kings County Hosp, 73-; consult pediat-hemat-oncol, Staten Island Hosp, NY, 83. *Mem:* Assoc fel Am Acad Pediat; Am Soc Hemat; Am Soc Pediat Hemat-Oncol; Am Soc Clin Oncol. *Res:* Sickle cell disease in children; infections; osteomyelitis versus bone infarction in sickle cell disease; leukemia in children; neutropenia in children; parvovirus big infection in children with sickle cell disease, leukemia, solid tumor or with AIDS. *Mailing Add:* Dept Pediat Box 49 450 Clarkson Ave Brooklyn NY 11203

RAO, SURYANARAYANA K, b Hyderabad, India, Feb 20, 39; c 2. TOXICOLOGY. *Educ:* Osmania Univ, India, DVM, 61; Magadh Univ, India, MS, 63, PhD(toxicol), 68. *Prof Exp:* Res fel toxicol, Magadh Univ, India, 64-67; asst res officer, Nutrit Res Labs, Hyderabad, 67-68; fel physiol, Case Western Reserve Univ, 68-69; res assoc pharmacol, Mich State Univ, 69-71; res investr, Dept Path-Toxicol, Searle Labs, 71-72, sr res investr toxicol, 72-77; res specialist, 77-80, res leader teratology & reproduction, Dept Toxicol, 80-89, MGR REGULATORY AFFAIRS, DOW CHEM CO, 89- *Mem:* Soc Toxicol; Teratology Soc. *Res:* Carbon tetrachloride hepatotoxicity in the rat; aflatoxin induced hepatotoxicity in the Rhesus monkey; physical examination procedures, teratology, reproduction and mutagenicity testing, monitoring cardiovascular parameters and data recording methodology employed in safety studies. *Mailing Add:* PO Box 1965 Midland MI 48641-1965

RAO, V UDAYA S, b Visakhapatnam, India, Aug 4, 38; m 66; c 1. CATALYSIS, SOLID STATE CHEMISTRY. *Educ:* Andhra Univ, BS, 58; Tata Inst Fundamental Res, PhD(physics), 67. *Prof Exp:* Res fel, Tata Inst Fundamental Res, India, 59-68; res assoc, 69-71, asst prof superconductivity of intermetallics, Univ Pittsburgh, 71-78; res chemist, 78-85, br chief, 85-88, PROJ MGR, US DEPT ENERGY, 89- *Mem:* Am Chem Soc; Catalysis Soc. *Res:* Indirect liquefaction of coal; conversion of syngas to olefins and gasoline; low temperature synthesis of methanol from syngas. *Mailing Add:* Liquid Fuels Div Pitsburgh Energy Technol Ctr DOE PO Box 10940 Pittsburgh PA 15236

RAO, VALLURU BHAVANARAYANA, b Tenali, India, June 27, 34; m 56; c 1. ALGORITHMS, QUEUEING THEORY. *Educ:* Andhra, India, BSc, 54, MSc, 56; Univ Ill, Urbana, MS, 65; Wash Univ, DSc(appl math & comput sci), 67. *Prof Exp:* Sr res investr, directorate econ & statist, Ministry Food & Agr, India, 57-62; from asst prof to assoc prof math, 67-83, chmn dept math & co-chmn dept comput sci, 77-79, ASSOC PROF COMPUT SCI, UNIV BRIDGEPORT, 84- *Concurrent Pos:* Res grant, Univ Bridgeport, 68-70; consult, Glendinning Co, 67-69, Marena Systs, 79, Honeywell Info Systs, 81 & Philips Med Systs, 84; Monsanto fel, Wash Univ, 65-67. *Mem:* Math Asn Am; Am Asn Univ Prof; Opers Res Soc Am; fel AAAS; Asn Comput Mach. *Res:* Algorithms, computation; integer, dynamic and stochastic programming. *Mailing Add:* Ten James St Fairfield CT 06430-6420

RAO, VASAN N, b India, Mar 19, 29; m 65; c 2. ELECTRON OPTICS, THIN FILMS. *Educ:* Univ Madras, India, BSc, 49; Univ Bombay, MSc, 52; Univ Tubingen, WGer, DSc, 64. *Prof Exp:* Tech supvr, Elec & Mech Engrs Corp, 53-58; asst prof eng, Pa State Univ, 66-67; mgr develop eng, 67-83, TECH MGR, NORTH AM PHILIPS, 83- *Mem:* Inst Elec & Electronics Engrs; Electron Microscope Soc WGer. *Res:* Imaging using electron and ion beams used in microlithography; thermal imaging; invention of the diode gun used in tv camera tubes. *Mailing Add:* 123 Central Pike Foster RI 02825

RAO, VELDANDA VENUGOPAL, pure mathematics; deceased, see previous edition for last biography

RAO, VIJAY MADAN, b Delhi, India, Sept 15, 50; m 74; c 2. DIAGNOSTIC RADIOLOGY. *Educ:* Hindu Col, Univ Delhi, BA, 68; All India Inst Med Sci, MD, 73. *Prof Exp:* Intern surg med, Albert Einstein Med Ctr, 73-75; resident diagnostic radiol, 75-91, from instr to assoc prof, 78-87, PROF RADIOL, MED COL, THOMAS JEFFERSON UNIV HOSP, 91- *Mem:* Radiol Soc N Am; Am Roentgen Ray Soc; Am Asn Univ Radiologists; Sigma Xi; AMA; Am Soc Head & Neck Radiologists. *Res:* Radiographic manifestations of sickle cell disease; diagnostic imaging of head and neck diseases. *Mailing Add:* Dept Radiol 1029 Main Bldg Thomas Jefferson Univ Hosp Philadelphia PA 19107

RAO, YALAMANCHILI A K, b Godavarru, India, Sept 15, 43; m 70; c 2. ALLERGY, IMMUNOLOGY. *Educ:* Guntur Med Col, India, MD, 69; Columbia Univ, MS, 71. *Prof Exp:* Resident pediat, 71-73, fel allergy & immunol, 73-75, resident internal med, 75-77, ASST DIR ALLERGY & IMMUNOL, LONG ISLAND COL HOSP, 77- *Concurrent Pos:* Asst clin prof, Downstate Med Ctr, Brooklyn. *Mem:* Fel Am Acad Allergy & Immunol; fel Am Col Allergists; assoc Am Col Physicians; fel Am Acad Pediat. *Res:* Pharmacology of theophylline; AIDS research; chronic asthma; environmental health. *Mailing Add:* 159 Clinton St A1-21 Brooklyn NY 11201

RAO, YALAMANCHILI KRISHNA, b India, May 09, 41; US citizen; m 69; c 3. METALLURGICAL ENGINEERING, PHYSICAL CHEMISTRY. *Educ:* Banaras Hindu Univ, BS, 62; Univ Pa, PhD(metall), 65. *Prof Exp:* Res engr, Inland Steel Co, 65-67; res metallurgist, Corning Glass Works, 67-68; from asst prof to assoc prof extractive metall, Columbia Univ, 68-76; assoc prof, 76-80, PROF, UNIV WASH, 80- *Honors & Awards:* Fel Inst of Mining & Metall, 84. *Mem:* Am Inst Mining, Metall & Petrol Engrs; Am Chem Soc; Iron & Steel Inst of Japan; Inst Min Metall; Indian Inst Metals. *Res:* Physical chemistry of metal extraction; reaction kinetics and mass transfer; thermodynamics of metallurgical systems; research publications in the fields of catalysis, carbon gasification, intrinsic rates and effectiveness factors, thermodynamics of complex systems, Gibbs Phase rule, diffusion and equilibria. *Mailing Add:* Dept Mat Sci & Eng Univ Wash Roberts Hall FB-10 Seattle WA 98195

RAO, YEDAVALLI SHYAMSUNDER, b Rajahmundry, India, Nov 15, 30; m 52; c 4. ORGANIC CHEMISTRY. *Educ:* Andhra Univ, India, BSc, 50; Osmania Univ, India, MSc, 54; Ill Inst Technol, PhD(org chem), 63. *Prof Exp:* USPHS fel, Ill Inst Technol, 62-63; res chemist, Richardson Co, Melrose Park, 63-65 & CPC Int Co, Argo, 68-70; assoc prof chem, 70-77, PROF CHEM, KENNEDY-KING COL, 77- *Concurrent Pos:* USPHS fel, Ill Inst Technol, 67-68, asst prof, Eve Div, 67- *Mem:* Am Chem Soc; Royal Soc Chem. *Res:* Heterocycles; fluorine chemistry; amino acids. *Mailing Add:* 931 S Euclid Oak Park IL 60304-2066

RAOUF, ABDUL, b Jullundur, India, Jan 15, 29; Can citizen; m 62. INDUSTRIAL ENGINEERING. *Educ:* WPakistan Univ Eng & Technol, EE, 50; Univ Toledo, MSIE, 66; Univ Windsor, PhD, 70. *Prof Exp:* Eng positions of various responsibility, Ministry of Defense, Govt Pakistan, 51-61; indust engr, Can Acme Screw & Gear Co, 62-64; teaching fel, Univ Toledo, 65-66; from asst prof to assoc prof human factors eng, Univ Windsor, 66-76, head dept, 76-81, prof indust eng, 76-84; PROF SYSTS ENG, UNIV PETROL & MINERALS, DHAHRAN, SAUDI ARABIA, 84- *Concurrent Pos:* Nat Res Coun Can grants, 66-88. *Mem:* Sr mem Am Inst Indust Engrs; sr mem Am Soc Qual Control; Am Soc Eng Educ; Eng Inst Can. *Res:* Human performance, including prediction, reliability and optimization in man machine systems. *Mailing Add:* Dept Systs Eng King Fahd Univ Petrol & Minerals Box 128 Dhahran 31261 Saudi Arabia

RAPACZ, JAN, b Lubien, Poland, Jan 21, 28; US citizen; m 69; c 2. IMMUNOGENETICS. *Educ:* Jagiellonian Univ, BS, 53, MS, 55, PhD(immunogenetics), 59. *Prof Exp:* Res assoc, Jagiellonian Univ, 58-61; fel, Univ Wis-Madison, 61-63; head, Polish Zootech Inst, Krakow, 63-65; vis prof immunogenetics, Univ Wis-Madison, 65-68; dir, Polish Zootech Inst, 68-70; vis prof, 70-71, assoc prof, 72-78, PROF IMMUNOGENETICS, UNIV WIS-MADISON, 78- *Mem:* Am Genetic Asn; Am Heart Asn. *Res:* Serum protein polymorphisms; immunogenetics of immunoglobulins; active and passive immunity; lipoprotein immunogenetics and atherogenesis; genetic susceptibility to pathogens and cell receptors. *Mailing Add:* Dept Genetics 666 Animal Sci Bldg Univ Wis Madison WI 53706

RAPAKA, RAO SAMBASIVA, b Rapaka, India, June 16, 43; US citizen; m 71; c 3. PHARMACEUTICAL CHEMISTRY. *Educ:* Andhra Univ, India, BS, 63, MS, 64; Univ Calif, San Francisco, MS, 68, PhD(pharmaceut chem), 70. *Prof Exp:* Res chemist, Univ Calif, San Francisco, 70, asst res biochemist, 71-75; res assoc radiation biol, Albert Einstein Med Ctr, Philadelphia, 75-76;

res assoc, Lab Molecular Biophys, Med Ctr, Univ Ala, Birmingham, 76-78; mem staff, Biopharmaceut Lab, Food & Drug Admin, 78-84; PROG ADMIN, NAT INST DRUG ABUSE, ROCKVILLE, MD, 84- *Mem:* Am Chem Soc; Am Asn Pharmaceut Scientists. *Res:* Synthesis of polymeric peptides as models for enzymes, and as models for collagen; studies on collagen chemistry, biochemistry and radiobiology; synthesis of peptides as models for elastin and mechanisms of arterial wall calcification; gas chromatographic, high-performance liquid chromatography, studies on thyroidal amino acids; studies on bioequivalency of drugs; drug analysis; opioid peptides; edit books; organize conferences; medical chemistry of drugs and abuse. *Mailing Add:* 15109 Gravenstein North Potomac MD 20878

RAPAPORT, ELLIOT, b Los Angeles, Calif, Nov 22, 24; m 43; c 3. CARDIOLOGY. *Educ:* Univ Calif, San Francisco, AB, 44, MD, 46. *Prof Exp:* Intern, San Francisco Gen Hosp, 46-47; asst resident med, Univ Hosp, Univ Calif, 47-50; instr, Albany Med Col, 55-56, asst prof, 56-57; dir cardiopulmonary lab, Mt Zion Hosp, San Francisco, Calif, 57-60; PROF MED & ASSOC DEAN CARDIOL, UNIV CALIF, SAN FRANCISCO, 60-; CHIEF CARDIOL, SAN FRANCISCO GEN HOSP, 60- *Concurrent Pos:* Res fel, Med Sch, Univ Calif, 50-51 & Harvard Med Sch, 53-55; dir res, Vet Admin Hosp, Albany, NY, 55-57; consult, USPHS Hosp, San Francisco, 58- & Letterman Army Hosp, 59-; coordr regional med progs, Area I, Calif, 67-73. *Mem:* Am Heart Asn (pres, 74-75); Am Fedn Clin Res; Am Soc Clin Invest; Am Physiol Soc; Asn Am Physicians. *Res:* Cardiovascular physiology, particularly blood flows and regional volumes; coronary artery disease; creatine kinase. *Mailing Add:* Dept Med Univ Calif San Francisco Sch Med San Francisco Gen Hosp 1001 Potrero Ave San Francisco CA 04110

RAPAPORT, FELIX THEODOSIUS, b Munich, Ger, Sept 27, 29; m 69; c 5. SURGERY, TRANSPLANTATION IMMUNOLOGY. *Educ:* NY Univ, AB, 51, MD, 54. *Prof Exp:* Asst med, Med Ctr, NY Univ, 58-60, from instr to prof surg, 60-77, dir, transplantation & Immunol Div, 65-77; PROF SURG, DEP CHMN DEPT, DIR TRANSPLANTATION DIV & ATTEND SURGEON SURG SERV, STATE UNIV NY, STONY BROOK, 77- *Concurrent Pos:* Res fel med & surg, Sch Med, NY Univ, 54-55, res fel path, 56; USPHS res career develop award, 61-62; City of New York Health Res Coun career scientist award, 63-72; intern, Mt Sinai Hosp, New York, 55-56; asst resident & chief resident, Med Ctr, NY Univ, 58-62; mem, adv comt collab res in transplantation & immunol, Nat Inst Allergy & Infectious Dis, 64-68, chmn, arthritis & metab dis prog proj adv comt, Div Res Grants, NIH, Md, 68-72; consult, Vet Admin Hosp, Manhattan, 64-77; assoc attend surg, NY Univ Hosp, 69-77; vis surgeon, NY Univ Surg Div, Bellevue Hosp, NY, 70-77; mem, sci adv bd, Nat Kidney Found, 72-; mem, Am Inst Biol Sci Adv Bd to Off Naval Res, 74-76; mem, merit rev bd immunol, Vet Admin, 74-78; mem, sci adv bd, Am Cancer Soc, 75-77; mem, Surg Study Sect B, Div of Res Grants, NIH, Md, 75-76 & Surg Study Sect A, 76-79; ed-in-chief, Transplantation Proc, 76-79; consult surgeon, Vet Admin Hosp, Northport, NY, 77-; attend surg & dir, Transplantation Serv, Univ Hosp, State Univ NY, Stony Brook, 77- *Honors & Awards:* Comdr, Order of Sci Merit of France, 68; Chevalier, Nat Order of Merit of France, 70; Gold Medal, City Paris, 79; Grand Croix, Palmes Acad France, 81. *Mem:* AAAS; Transplantation Soc (secy, 66-74, vpres, 74-76, pres elect, 76-78, pres, 78-); Am Burn Asn; Am Asn Immunol; Soc Univ Surg; Am Surg Asn; Am Asn Transplant Surg; Am Asn Transplant Physicians; Soc for Exp Biol & Med; Sigma Xi. *Res:* Transplantation biology and medicine; trauma; burns. *Mailing Add:* 18 Lawson Lane Great Neck NY 11794

RAPAPORT, IRVING, b New York, NY, May 21, 25; m 52; c 4. MINING GEOLOGY. *Educ:* Univ Minn, BS, 49; Grad Studies, Columbia Univ, 55-56. *Prof Exp:* Explor geologist, AEC, Colo, 49-50, proj chief explor, Utah & NMex, 50-52; mgr, Hanosh Mines Corp, 52; OWNER, FOUR CORNERS EXPLOR CO, 53- *Concurrent Pos:* Consult, Chilean Nitrate Corp, 52-53, J H Whitney & Co, Kerr McGee Corp, Santa Fe Railway Corp, Spencer Chem Corp & others, 53- *Honors & Awards:* Small Mining Co Award, Mining World, 59. *Mem:* Am Inst Mining, Metall & Petrol Engrs; fel Geol Soc Am. *Res:* Exploration and development of uranium ore bodies. *Mailing Add:* Four Corners Explor Co PO Box 116 Grants NM 87020

RAPAPORT, JACOBO, b Santiago, Chile, Nov 30, 30; m 58; c 2. EXPERIMENTAL NUCLEAR PHYSICS. *Educ:* Univ Chile, Engr, 56; Univ Fla, MSc, 57; Mass Inst Technol, PhD(physics), 63. *Prof Exp:* Instr physics, Univ Chile, 57-60, prof, 63-65, dir inst physics, 64; asst prof, Mass Inst Technol, 65-69; assoc prof, 69-73, prof, 73-81, DISTINGUISHED PROF PHYSICS, OHIO UNIV, 81- *Mem:* Fel Am Phys Soc. *Res:* Experimental information on nuclear spectroscopy obtained by means of low energy nuclear reactions; nuclear reaction studies induced with high energy neutrons; charge exchange (pion, muon); reactions at intermediate energies. *Mailing Add:* Dept Physics Ohio Univ Athens OH 45701-2979

RAPAPORT, SAMUEL I, b Los Angeles, Calif, Nov 19, 21; m 51; c 4. INTERNAL MEDICINE, HEMATOLOGY. *Educ:* Univ Southern Calif, MD, 45. *Prof Exp:* Instr physiol, Univ Southern Calif, 48-49, vis asst prof, 49-55; chief hemat sect, Vet Admin Hosp, Long Beach, Calif, 55-57; assoc prof med, Med Ctr, Univ Calif, Los Angeles, 57-58; from assoc prof to prof, Sch Med, Univ Southern Calif, 58-74; PROF MED, UNIV CALIF, SAN DIEGO, 74-, CO-HEAD HEMAT/ONCOL DIV, DEPT MED, 78-, PROF PATH & DIR, HEMAT LAB, MED CTR, 80- *Concurrent Pos:* Chief hemat sect, Vet Admin Hosp, Long Beach, 50-53, consult, 59-68; Fulbright res scholar, Univ Oslo, Norway, 53-54; head hemat sect, Med Serv, Los Angeles County Hosp-Univ Med Ctr, 66-74; mem, Hematol Training Grant Study Sect, Nat Inst Arthritis & Metab Dis, 69-72; mem, Bd Gov, Am Bd Internal Med, 73-80, chmn & secy-tres, Subspeciality Comt Hemat, 78-80; chief med serv, San Diego Vet Admin Hosp, 74-78; mem, Hemat Study Sect A, NIH, 84- *Honors & Awards:* Stratton Lectr, Am Soc Hemat, 84. *Mem:* Am Soc Hemat (pres, 77); Am Soc Clin Invest; fel Am Col Physicians; Am Asn Physicians; Int Soc Hemat. *Res:* Coagulation; hemostasis and thrombosis. *Mailing Add:* Univ Hosp H811K 225 Dickinson San Diego CA 92103

RAPAPORT, WILLIAM JOSEPH, b Brooklyn, NY, Sept 30, 46; div. COGNITIVE SCIENCE, ARTIFICIAL INTELLIGENCE. *Educ:* Univ Rochester, BA, 68; Ind Univ, AM, 74, PhD (philos), 76; State Univ NY, Buffalo, MS, 84. *Prof Exp:* Teacher math, Inwood Jr High Sch, 68-69 & Walden Sch, 69-71; assoc instr philos, Ind Univ, 71-72 & 75, asst ed, Dept Philos, 72-75; from asst prof to assoc prof philos, State Univ NY, Fredonia, 76-84; asst prof, 84-88, ASSOC PROF COMPUTER SCI, STATE UNIV NY, BUFFALO, 88- *Concurrent Pos:* Prin investr, NSF, NEH & State Univ NY Res Found, 86-89; Master's scholar award, Northeastern Asn Grad Schs, 87; Steelman vis scientist, Lenoir-Rhyne Col, 88. *Mem:* Am Asn Artificial Intel; Am Philos Asn; Asn Comput Ling; Asn Comput Mach; Cognitive Sci Soc; Asn Symbolic Logic. *Res:* Knowledge representation; natural-language understanding; philosophical foundations of cognitive science. *Mailing Add:* Dept Computer Sci State Univ NY Buffalo NY 14260

RAPER, CARLENE ALLEN, b Plattsburgh, NY, Jan 9, 25; m 49; c 2. MYCOLOGY, DEVELOPMENTAL GENETICS. *Educ:* Univ Chicago, BS, 46, MS, 48; Harvard Univ, PhD(biol), 77. *Prof Exp:* Res scientist radiation biol, Argonne Nat Labs, Chicago, 47-53; res scientist fungal genetics develop, 61-74, lectr & res assoc, Dept Biol, Harvard Univ, 74-78; lectr microbiol, Bridgewater State Col, Mass, 78; asst prof biol & genetics develop biol, dept biol sci, Wellesley Col, Mass, 78-83; res asst prof, Dept Microbiol, 83-87, RES ASSOC PROF, DEPT MICROBIOL & MOLECULAR GENETICS, UNIV VT, 87- *Concurrent Pos:* Coop investr, US-Israel Binat Sci Orgn, 74-77; res grants, Campbell Inst Agr Res, 74-78, Maria Moores Cabot Fount, Harvard Univ, 75-77, Neth Orgn Advan Pure Res, 77 & Res Corp, 77-81; res fel, Univ Groningen, Neth, 77, NSF, 83-87 & NIH, 87-90; assoc ed, Exp Mycol, 79- & Mycologia, 87-; co-vice-chmn, Gordon Conf Fungal Metab, 82, co-chmn, 84. *Honors & Awards:* NSF fac award, 82. *Mem:* Genetics Soc Am; Mycol Soc Am. *Res:* Genetic regulation of morphogenesis in the sexual cycle of higher fungi; comparative biology, sexuality and breeding of edible fungi, especially Agaricus species. *Mailing Add:* Dept Microbiol & Molecular Genetics Univ Vt Given Bldg Burlington VT 05405

RAPHAEL, LOUISE ARAKELIAN, b New York, NY, Oct 24, 37; c 2. MATHEMATICAL ANALYSIS, APPLIED MATHEMATICS. *Educ:* St Johns Univ, BS, 59; Cath Univ Am, MA, 62, PhD(math), 67. *Prof Exp:* Asst prof math, Howard Univ, 66-70; assoc prof math, Clark Col, 71-82; assoc prof, 82-86, PROF MATH, HOWARD UNIV, 86- *Concurrent Pos:* NSF res grant, 75-76 & 89-91, Army Res Off Grants, 81-88; vis assoc prof math, Mass Inst Technol, 77-78 & 89-90; actg admin officer, Conf Bd Math Sci, 85-86; NSF fel, 63, 64, NASA fel, 65-66; assoc prog dir, sci & math directorate, NSF, 86-87, prog dir, div math sci, 87-88. *Mem:* Am Math Soc; Math Asn Am; AAAS; Sigma Xi; Nat Coun Teachers Math; Soc Indust Appl Math. *Res:* Differential equations; approximation theory. *Mailing Add:* Seven Thompson St Annapolis MD 21401

RAPHAEL, ROBERT B, theoretical physics, for more information see previous edition

RAPHAEL, THOMAS, b Somerville, Mass, June 9, 22; m 48; c 4. APPLIED CHEMISTRY, PHOTOGRAPHIC CHEMISTRY. *Educ:* Harvard Univ, AB, 44; Univ Chicago, PhSc, 45. *Prof Exp:* Div res mgr, Dewey & Almy Chem Co, W R Grace & Co, 46-56; group leader, Arthur D Little, Inc, 56-66; asst mgr, Tech Control Ctr, Polaroid Corp, 66-69, mgr spec projs, 69-74, sr tech mgr, Appl Technol Div, 74-81, sr mgr mat develop, Tech Photo Div, 81-85; RETIRED. *Mem:* Electrochem Soc; Am Chem Soc; Soc Plastics Indust; Soc Plastics Eng; Tech Asn Pulp & Paper Indust; Am Nat Standards Inst; Int Standards Orgn. *Res:* Process and product development in photographic products, plastics, rubber, paper and textiles. *Mailing Add:* 90 Grove St Winchester MA 01890-3845

RAPIN, ISABELLE (MRS HAROLD OAKLANDER), b Lausanne, Switz, Dec 4, 27; US citizen; m 59; c 4. NEUROLOGY, PEDIATRIC NEUROLOGY. *Educ:* Univ Lausanne, Swiss Fed physician dipl, 52, Dr(med), 55. *Prof Exp:* Intern pediat, NY Univ-Bellevue Med Ctr, 53-54; asst resident neurol, Columbia-Presby Med Ctr, 54-57; instr neurol, 58-61, from asst to assoc prof neurol & pediat, 61-72, dir pediat neurol training prog, 60-68, 70-74, PROF NEUROL & PEDIAT, ALBERT EINSTEIN COL MED, 72- *Concurrent Pos:* Fel pediat neurol & asst neurologist, Columbia-Presby Med Ctr, 57-58; asst vis physician, Bronx Munic Hosp Ctr, 58-63, assoc vis physician, 63-69, vis neurologist, 69-; prin investr, Nat Inst Neurol Dis & Blindness res grant, 59-74 & 85-; prin investr, Children's Bur, USPHS grant, 68-74; adj vis physician, Montefiore Hosp, 64-; assoc vis pediatrician, Lincoln Hosp, 64-74; vis physician, Hosp, Albert Einstein Col Med, 66-, vis neurologist, 69-; mem nat adv coun, Nat Inst, Neurol Commun Disorders & Stroke, 84-88; prin investr, Nat Inst, Commun Disorders & Stroke, 85- *Honors & Awards:* Hower Award, Child Neurol Soc, 87; Frank R Ford Mem lectr, Int Child Neurol Asn, 89. *Mem:* AAAS; Int Neuropsychol Soc; Am Neurol Asn (vpres, 81-82); Child Neurol Soc; Int Child Neurol Asn (secy gen, 79-82, vpres, 82-86); fel Am Acad Neurol. *Res:* Deaf and nonverbal children; brain damage in children; degenerative diseases of the nervous system in children. *Mailing Add:* Rm 807 Kennedy Ctr Albert Einstein Col Med Bronx NY 10461

RAPISARDI, SALVATORE C, b New York, NY, July 8, 41; m 79. NEUROSCIENCE, ANATOMY. *Educ:* Duke Univ, BA, 63; Temple Univ, MA, 66; Univ Calif, Riverside, PhD(physiol psychol), 71. *Prof Exp:* Fel anat, Dept Neurol & Anat, Sch Med, Stanford Univ, 72-74; instr, Univ Calif, Berkeley, 77; fel, Med Sch, Univ Calif, San Francisco, 74-77; ASST PROF ANAT, COL MED, HOWARD UNIV, 77- *Concurrent Pos:* Prin investr, Nat Eye Inst, 79-82 & Biomed Interdisplinary Proj, NIH, 79- *Mem:* Am Soc Neurosci; AAAS; Am Asn Anatomists; Europ Soc Neurosci. *Res:* Synaptology of the dorsal lateral geniculate in cat and monkey by analysis of series of consecutive thin sections; reconstruction of neural processes from consecutive thin sections. *Mailing Add:* Dept Anat Howard Univ 520 W St NW Washington DC 20059

RAPKIN, MYRON COLMAN, b Rochester, NY, Nov 24, 38; m 66; c 2. CLINICAL CHEMISTRY, REAGENT STRIPS. *Educ:* Rochester Inst Technol, BS, 63. *Prof Exp:* Clin chemist, Genesee Hosp, 65-66; med lab technologist, Wilson Mem Hosp, 66-68; develop chemist emulsions & color photog, Gen Aniline & Film, 68-70; sr assoc res scientist, prod res & develop, Ames Co Div, Miles Lab, 70-82; MGR RES & DEVELOP, SMITHKLINE BECKMAN, 82-; RES INVESTR, BOEHRINGER-MANNHEIM, 82- *Concurrent Pos:* chmn, Indianapolis sect, CDIC. *Mem:* Fedn Socs Coatings Technol; Tech Asn Pulp & Paper Indust; Technol Transfer Soc. *Res:* Diagnostic systems; confirmatory tests; clinical and urinalysis control systems; performance and proficiency testing; solid phase or dry chemistry systems; patents; technology transfer. *Mailing Add:* 6231 Oakland Ave N Indianapolis IN 46220

RAPOPORT, ABRAHAM, b Toronto, Ont, Aug 18, 26; m 52; c 4. MEDICINE. *Educ:* Univ Toronto, MD, 49, MA, 52; FRCP(C), 54. *Prof Exp:* Dir dept biochem, Toronto Western Hosp, 57-71, dir metab renal unit, 59-73, physician in chief, 73-88; assoc prof med & path chem, 65-71, PROF MED, UNIV TORONTO, 71- *Mem:* Fel Am Col Physicians. *Res:* Renal and metabolic diseases; hypertension. *Mailing Add:* Toronto Western Hosp 399 Bathurst St Toronto ON M5T 2S8 Can

RAPOPORT, ANATOL, b Lozovaya, Russia, May 22, 11; US citizen; m 49; c 3. EXPERIMENTAL SOCIAL PSYCHOLOGY, THEORY OF GAMES. *Educ:* Univ Chicago, SB, 38, SM, 40 & PhD(math), 41. *Hon Degrees:* LHD, Univ Western Mich & LLD, Univ Toronto, 86. *Prof Exp:* Instr math, Ill Inst Technol, 46-47; res assoc math biol, Univ Chicago, 47-48, asst prof, 48-54; from assoc prof to prof, math biol, Univ Mich, 55-70; prof psychol & math, Univ Toronto, 70-80; dir, Inst Adv Studies, Vienna, 80-84; PROF PEACE STUDIES, UNIV TORONTO, 84- *Honors & Awards:* Lenz Int Peace Res Award, 75; Harold Losswell Award, 86. *Mem:* Am Math Soc; Soc Math Biol; Int Soc Gen Semantics (pres, 53-55); Soc Gen Systs Res (pres, 65-66); Can Peace Res & Educ Asn (pres, 72-75); Sci Peace (pres, 84-86). *Res:* Strategic aspects of conflict and cooperation using experimental games in studying behavior in situations including choice between individually rational and collectively rational strategies. *Mailing Add:* 38 Wychwood Park Toronto ON M6G 2V5 Can

RAPOPORT, HENRY, b Brooklyn, NY, Nov 16, 18; m 44; c 3. ORGANIC CHEMISTRY. *Educ:* Mass Inst Technol, SB, 40, PhD(org chem), 43. *Prof Exp:* Chemist, Heyden Chem Corp, NJ, 43-45; Nat Res Coun fel, NIH, 46; from instr to assoc prof, 46-57, PROF CHEM, UNIV CALIF, BERKELEY, 57- *Honors & Awards:* Res Achievement Award, Acad Pharmaceut Sci, 72. *Mem:* Am Chem Soc. *Res:* Alkaloids; heterocyclic compounds; natural products; pigments; biosynthesis. *Mailing Add:* Dept Chem Univ Calif 2120 Oxford St Berkeley CA 94720

RAPOPORT, JUDITH LIVANT, b New York, NY, July 12, 33; m 61; c 2. CHILD PSYCHIATRY, PSYCHIATRY. *Educ:* Swarthmore Col, BA, 55; Harvard Univ, MD, 59; Am Bd Psychiat & Neurol, dipl & cert child psychiat, 69. *Prof Exp:* Intern, Mt Sinai Hosp, 59-60; resident psychiat, Mass Ment Health Ctr, Boston, 60-61 & St Elizabeth's Hosp, Washington, DC, 61-62; NIMH health fel, Psychol Inst, Uppsala & Karolinska Hosp, Stockholm, Sweden, 62-64; NIMH fel child psychiat, Children's Hosp, Washington, DC, 64-66; spec fel, Lab Psychol, NIMH, 66-67; clin assoc prof pediat & psychiat, Med Sch, Georgetown Univ, 67-77; MEM STAFF, NIMH, 77- *Concurrent Pos:* Prin investr, USPHS grant biol factors in hyperactivity, 72-76. *Mem:* Fel Am Psychiat Asn; Am Col Neuropsychopharmacol; Psychiat Res Soc; fel Am Acad Child Psychiat. *Res:* Biological child psychiatry; pediatric psychopharmacology; learning disabilities. *Mailing Add:* NIMH Bldg 10 Rm 3N204 Bethesda MD 20014

RAPOPORT, LORENCE, b Springfield, Mass, Oct 8, 19; m 46; c 3. ORGANIC POLYMER CHEMISTRY. *Educ:* Harvard Univ, AB, 41; Duke Univ, PhD(org chem), 44. *Prof Exp:* Asst, Boston Woven Hose & Rubber Co, 41; lab instr, Duke Univ, 41-44; B F Goodrich fel, Ohio State Univ, 44-45, Upjohn fel, 45-46; res chemist, Am Cyanamid Co, 46-54, group leader, 54-63; dir, Res Sect, 64-67, dir, Res & Develop Sect, 67-70, dir new prod res & develop, 70-75, MGR PROD & PROCESS DEVELOP, OLIN CORP, 75- *Concurrent Pos:* Res assoc, Off Sci Res & Develop, Ohio State Univ, 45; vis scientist, Cath Univ Louvain, 60-61. *Mem:* Am Chem Soc. *Res:* Synthesis of detergents; carcinogenic hydrocarbons; acrylonitrile and derivatives; nitrogen compounds; s-triazines; cellophane; cellulose derivatives; polymers; coatings; plastic films. *Mailing Add:* 175 Windsor Rd Asheville NC 28804

RAPOPORT, STANLEY I, b New York, NY, Nov 24, 32; m 61; c 2. MEDICINE, PHYSIOLOGY. *Educ:* Princeton Univ, AB, 54; Harvard Med Sch, MD, 59. *Prof Exp:* Intern med, Bellevue Hosp, New York, 59-60; res scientist neurophysiol, NIMH, 60-62, res scientist physiol, 64-78; CHIEF LAB NEUROSCI, GERONT RES CTR, NAT INST AGING, BALTIMORE, MD, 78- *Concurrent Pos:* NSF fel biophys, Physiol Inst, Uppsala, Sweden, 62-64; prof lectr, Georgetown Univ Sch Med, 71- *Mem:* AAAS; Biophys Soc; Soc Neurosci; Soc Gen Physiol; Am Physiol Soc. *Res:* Physiology of blood brain barrier; aging of nervous system membrane phenomena; excitation-contraction coupling in muscle. *Mailing Add:* NIA NIH Bldg 10 Rm 6C103 Bethesda MD 20892

RAPP, DONALD, b Brooklyn, NY, Sept 27, 34; m 56; c 2. SPACECRAFT TECHNOLOGY, ASTROPHYSICS. *Educ:* Cooper Union, BS, 55; Princeton Univ, MS, 56; Univ Calif, Berkeley, PhD(phys chem), 60. *Prof Exp:* Staff scientist, Lockheed Palo Alto Res Labs, 59-65; assoc prof chem, Polytech Inst Brooklyn, 65-69; assoc prof chem, Univ Tex, Dallas, 69-73, prof solar energy, 73-79; SR RES SCIENTIST & DIV TECHNOLOGIST, JET PROPULSION LAB, CALIF INST TECHNOL, 79- *Concurrent Pos:* Consult, Solar Energy Proj, Am Technol Univ, 74-79. *Res:* Technology for advanced telescopes in space; IR astronomy; advanced composite structures; extra-solar planets. *Mailing Add:* 1445 Indiana Ave S Pasadena CA 91030

RAPP, DOROTHY GLAVES, b Sheffield, Eng, Auug 14, 43; m 78. EXPERIMENTAL CANCER METASTASIS. *Educ:* Univ London, BSc, 65; Univ Nottingham, PhD(tumor immunol), 69. *Prof Exp:* Res officer, Cancer Res Campaign, Univ Nottingham, 69-72; CANCER RES SCIENTIST, NY STATE DEPT HEALTH, ROSWELL PARK MEM INST, 72- *Mem:* AAAS; NY Acad Sci. *Res:* Characterization of host and tumor related factors which determine the fate of cancer cells during the post-intravasation phases of metastasis. *Mailing Add:* Dept Exp Path Roswell Park Mem Inst 666 Elm St Buffalo NY 14263

RAPP, FRED, b Fulda, Ger, Mar 13, 29; nat US; c 3. VIROLOGY. *Educ:* Brooklyn Col, BS, 51; Union Univ, NY, MS, 56; Univ Southern Calif, PhD(med microbiol), 58; Am Bd Med Microbiol, dipl. *Prof Exp:* Jr bacteriologist, Div Labs & Res, NY State Dept Health, 52-53, bacteriologist, 53-55; instr med microbiol, Sch Med, Univ Southern Calif, 56-59; virologist, Philip D Wilson Res Found, 59-62; asst prof microbiol & immunol, Med Col, Cornell Univ, 61-62; from assoc prof to prof virol & epidemiol, Baylor Col Med, 62-69; assoc provost, dean health affairs & dir, Specialized Cancer Res Ctr, 73-84, prof, 69-78, EVAN PUGH PROF MICROBIOL, COL MED, PA STATE UNIV, 78-, ASSOC DEAN ACAD AFFAIRS, RES & GRAD STUDIES, 87- *Concurrent Pos:* Consult supvry microbiologist, Hosp Spec Surg, NY, 59-62; res career prof virol, Am Cancer Soc, 66-69, prof, 77-; sr mem grad sch & mem res ref reagents comt, Nat Inst Allergy & Infectious Dis, NIH, 67-71, chmn, Atlantic Coast Tumor Virol Group, 71, mem,Virol Study Sect, 72-76 & Virol Task Force, 76-; mem, Nat Bladder Cancer Proj, 73-76; chmn, Gordon Res Conf Cancer, 75, Herpes Simplex Virus Vaccine, Nat Inst Allergy & Infections Dis, 81; mem, deleg viral oncol, US-USSR Joint Comt Health Coop & US-France & advbd, Cancer Info Dissemination & Anal Ctr, 76-; consult, Virus Cancaer Prog, Nat Cancer Inst, NIH, 71-75; mem viral oncol, US-USSR Joint Comt Health Coop; Am Cancer Soc career prof, 66-69 & 77-; mem, Herpesviruses Study Group, NIH, 81-84, Herpes Resource Ctr, Am Soc Health Asn, 83-, coun, Soc Exp Biol & Med, 83-, Coun Res & Clin Invest, Am Soc Virol, 84-; chmn, DNA Viruses Div, Am Soc Microbiol, 81-82, mem Med Microbiol & Immunol Comt Pub & Sci Affairs Bd, 86-88; mem sci adv comt, Wilmot fel, Univ Rochester Med Ctr, 81-; mem adv comt res etiology, diag, natural Hist, prev & ther multiple sclerosis, Nat Multiple Sclerosis Soc, 85-88; mem sci adv bd, Showa Univ Res Inst Biomed Fla, 85-88; Am Soc Microbiol rep, US Nat Comt, Int Union Microbiol Soc, 86-90; mem DNA adv comt, Working Group Transgenic Animals, NIH, 88. *Mem:* AAAS; Am Soc Microbiol; Am Asn Immunol; Am Asn Cancer Res; Soc Exp Biol; Soc Gen Microbiol; Am Asn Univ Prof; Sigma Xi. *Res:* Replication of viruses; transformation of mammalian cells by viruses; viral genetics and immunology; tumor viruses; herpes viruses; measles virus. *Mailing Add:* Dept Microbiol & Immunol Pa State Univ Col Med PO Box 850 Hershey PA 17033

RAPP, GEORGE ROBERT, JR, b Toledo, Ohio, Sept 19, 30; m 56; c 2. ARCHAEOLOGY. *Educ:* Univ Minn, BA, 52; Pa State Univ, PhD(geochem), 60. *Prof Exp:* Asst prof mineral, SDak Sch Mines & Technol, 57-60, assoc prof & curator mineral, 61-65; assoc prof mineral, Univ Minn, Minneapolis, 65-75, assoc chmn dept, 69-72; Fulbright sr scholar, Fulbright-Hays Prog, 72-73; dean, Col Lett & Sci, 75-84, PROF GEOL & ARCHAEOL, UNIV MINN, DULUTH, 76-, DEAN, COL SCI & ENG, 84- *Concurrent Pos:* NSF fel, 63-64; assoc dir, Minn Messenia Exped, 69-78; chmn, Coun Educ Geol Sci, 69-73; archaeom dir, Tel Michal Excavation, 77-80; nat lectr, Sigma Xi, 79-81. *Honors & Awards:* Archaeol Geol Award, Geol Soc Am; Nat Award, Am Fedn Mineral Soc. *Mem:* Fel Geol Soc Am; fel Mineral Soc Am; Nat Asn Geol Teachers (pres, 69-70); Archaeol Inst Am; Asn Field Archaeol (pres, 79-81); Soc Prof Archaeologists; Am Inst of Mining, Metall & Petrol Engrs; fel AAAS; Soc Archaeol Sci (pres, 83-84); Sigma Xi; Am Soc Eng Educ; Soc Am Archaeologists. *Res:* Archaeological geochemistry; archaeological geology; environmental geology and geochemistry. *Mailing Add:* Col Sci & Eng Univ Minn Duluth MN 55812-2496

RAPP, JOHN P, b New York, NY, Dec 22, 34; m 65. PATHOLOGY. *Educ:* Cornell Univ, DVM & MS, 59; Univ Pa, PhD(path), 64. *Prof Exp:* Res assoc physiol, Penrose Res Lab, Zool Soc Philadelphia, 62-76; assoc prof, 76-80, PROF, DEPT MED, MED COL OHIO, 80- *Concurrent Pos:* NIH fels, Univ Pa, 62-63 & Univ Utah, 64-65. *Res:* Biomedical research in experimental hypertension and function of renal juxtaglomerular apparatus; quantitative methods in steroid biochemistry; biochemical genetics of blood pressure regulation. *Mailing Add:* Dept Med-Path Med Col Ohio CS 10008 Toledo OH 43699

RAPP, PAUL ERNEST, b Chicago, Ill, Sept 2, 49; div. THEORETICAL NEUROPHYSIOLOGY. *Educ:* Univ Ill, Urbana-Champaign, BS(physiol) & BS(physics), 72; Cambridge Univ, Eng, PhD(math), 75. *Prof Exp:* Fel math, Caius Col, Cambridge Univ, 75-79; asst prof, 79-82, ASSOC PROF PHYSIOL, MED COL, PA, 82- *Concurrent Pos:* Vis fac, dept math, Rutgers Univ, 78, vis fac mem, dept math, Univ Western Australia, 85, 86, 88; Winston Churchill scholar, Cambridge Univ; mem bd dirs, Soc Math Biol. *Mem:* Am Math Soc; Soc Math Biol; Soc Psychoanal Psychother. *Res:* Mathematical investigations of biochemical and biophysical control systems; large dimension nonlinear differential equations; theoretical neurobiology. *Mailing Add:* Dept Physiol Med Col Pa 3300 Henry Ave Philadelphia PA 19129

RAPP, RICHARD HENRY, b Danbury, Conn, Aug 26, 37; m 65; c 2. GEODESY. *Educ:* Rensselaer Polytech Inst, BS, 59; Ohio State Univ, MSc, 61, PhD(geod sci), 64. *Prof Exp:* Res assoc, 61-65, asst supvr, 65-66, instr geod, 63-64, from asst prof to assoc prof, 64-71, PROF GEOD SCI, OHIO STATE UNIV, 71-, RES SUPVR, RES CTR, 66- *Concurrent Pos:* Assoc ed, J Geophys Res, 77-79; pres, Sect 5, Int Asn Geod, 79-83; chmn, Comt Geod, Nat Acad Sci, 84-87. *Honors & Awards:* W H Heiskanen Award, 65. *Mem:* Fel Am Geophys Union (pres, geod sect, 82-84); Am Cong Surv & Mapping. *Res:* Gravimetric, geometric and satellite geodesy. *Mailing Add:* Dept Geod 413 Cockins Ohio State Univ Columbus OH 43210-1247

RAPP, ROBERT, b Toronto, Ont, Oct 10, 29; m 66; c 2. PEDIATRIC DENTISTRY, HISTOLOGY. *Educ:* Univ Toronto, DDS, 53; Univ Mich, MS, 56; FRCD(C), 66. *Prof Exp:* Intern, Hosp for Sick Children, 53-54; res assoc, Univ Mich, 56-57; from asst to assoc hist, Univ Toronto, 59-62; asst prof pediat dent, Univ Mich, 62-65; assoc prof pediat dent, Univ Pittsburgh, 65-68, chmn dept, 65-73, dir, Dent Asst Utilization Prog, 67-71, PROF PEDIAT DENT, UNIV PITTSBURGH, 68- *Concurrent Pos:* Res grants, 62-67, 68-; WHO fel, Eng, Scand & USSR, 71-; consult, Pan Am Health Orgn, Brazil & Mex; mem, Am Asn Dent Schs. *Mem:* Am Acad Pedodont; Int Asn Dent Res; Am Asn Dent Res; fel Am Col Dentists; fel Int Col Dentists. *Res:* Histology and histochemistry of innervation of dental tissues; circulation in dental pulp; pathology; evaluation of audio-visual teaching procedures; nature of distribution of odontoblastic processes in dentine of primary teeth. *Mailing Add:* Dent Med Res Off 623 Salk Hall Univ Pittsburgh 4200 Fifth Ave Pittsburgh PA 15261

RAPP, ROBERT, b Reading, Pa, Apr 22, 21; m 45. RADIOLOGY. *Educ:* Ursinus Col, BS, 42; Temple Univ, MD, 46; Am Bd Radiol, dipl. *Prof Exp:* From instr to ossoc prof 53-80, EMER ASSOC PROF RADIOL, UNIV MICH, ANN ARBOR, 80- *Mem:* Am Col Radiol; Radiol Soc NAm. *Res:* Clinical radiology; teaching. *Mailing Add:* 1460 Cedar Bend Dr Ann Arbor MI 48105

RAPP, ROBERT ANTHONY, b Lafayette, Ind, Feb 21, 34; m 60; c 4. METALLURGY. *Educ:* Purdue Univ, BS, 56; Carnegie Inst Technol, MS, 59, PhD(metall eng), 60. *Prof Exp:* Fulbright fel, Max Planck Inst Phys Chem, 59-60; from asst prof to assoc prof, 63-69, PROF METALL ENG, OHIO STATE UNIV, 69- *Concurrent Pos:* Guggenheim fel, Univ Grenoble, 72-73. *Honors & Awards:* Stoughton Young Teacher Award, Am Soc Metals, 67, Howe Gold Medal, 73; Willis R Whitney Award, Nat Asn Corrosion Engr, 86. *Mem:* Fel Am Inst Mining, Metall & Petrol Engrs; fel Am Soc Metals; Electrochem Soc; Nat Asn Corrosion Engrs. *Res:* Oxidation of metals and alloys; thermodynamics; electrochemistry, point defects in compounds, hot corrosion of materials. *Mailing Add:* 1379 Southport Dr Columbus OH 43235

RAPP, ROBERT DIETRICH, b Reading, Pa, Dec 21, 30; m 53; c 2. ORGANIC CHEMISTRY. *Educ:* Tufts Univ, BS, 55; Lehigh Univ, PhD(org chem), 67. *Prof Exp:* Chemist, Glidden Co, 55-57 & Polymer Corp, 57; biochemist, Reading Hosp, 57-64; asst chem, Lehigh Univ, 64-65, res asst, 65-66; instr, Lafayette Col, 66-67; from asst prof to assoc prof, 67-79, PROF CHEM, ALBRIGHT COL, 79- *Concurrent Pos:* Sigma Xi res grant, 67. *Mem:* AAAS; Am Chem Soc; Am Soc Pharmacog; NY Acad Sci. *Res:* Synthesis of the macrolide antibiotics; peracid oxidation of enol ethers; synthesis of antimalarial compounds; structure and synthesis of pyrrolizidine alkaloids. *Mailing Add:* Dept Chem Albright Col Reading PA 19612-5234

RAPP, WALDEAN G, b Oakley, Kans, Mar 30, 36; m 57; c 3. BIOCHEMISTRY, FOOD SCI & TECHNOLOGY. *Educ:* Univ Ottawa, BS, 59; Univ Ark, MS, 61, PhD(chem), 63. *Prof Exp:* Res chemist, Chem Res Dept, Food Div, Anderson, Clayton & Co, 63-67; chief chemist, Hubinger Co, Iowa, 67-70; tech dir, Cargill, Inc, Corn Starch & Syrup, 70-82, MGR, BIOTECHNOL RES & ANALYTIC SERV, CARGILL, INC, 82- *Mem:* Am Chem Soc; Am Asn Cereal Chem; Am Soc Brewing Chemists; Am Oil Chem Soc; Inst Food Technol; Soc Soft Drink Technologists. *Res:* Hormonal effects on the acid soluble nucleotides of liver; flavor studies of various food and food products, including sensory evaluation of foods; basic food science; corn wet-milling; biotechnology and fermentation research. *Mailing Add:* Cargill Inc Res Bldg PO Box 5600 Minneapolis MN 55440

RAPP, WILLIAM RODGER, b Dover, NJ, May 18, 36; m 66; c 2. PATHOLOGY. *Educ:* Kans State Univ, BS, 64, DVM, 66, MS, 70. *Prof Exp:* Res assoc neuropath, Kans State Univ, 66-67, instr path, 67-72; dir path, Bio Dynamics Inc, East Millstone, NJ, 72-76; consult pathologist, 76-87; DIAG PATHOLOGIST, 87- *Mem:* Am Vet Med Asn; Am Asn Vet Lab Diagnosticians. *Res:* Comparative pathology; veterinary pathology; experimental toxicology and neoplasia epidemiology. *Mailing Add:* PO Box 279 Arden NC 28704

RAPPAPORT, ARON M, b Sereth, Austria, June 7, 04; m 50; c 2. PHYSIOLOGY, SURGERY. *Educ:* Ger Univ Prague, Czech, MD, 29; Univ Paris, dipl, 34; Univ Toronto, PhD(physiol), 52. *Prof Exp:* Resident surg, Friedrichshain Hosp, Berlin, Ger, 29-33; clin asst surg, Cochin Hosp, Paris, France, 33-34; head surg, Jewish Community Hosp, Botoshani, Rumania, 35-41; surgeon, Hosp Love-of-Men, Bucharest, 42-48; res assoc, Banting & Best Dept Med Res, 48, lectr, 52, from assoc prof to prof, 55-73, EMER PROF PHYSIOL, FAC MED, UNIV TORONTO, 73- *Concurrent Pos:* Clin asst, Toronto Gen Hosp, 52; sr res scientist, Dept Med, Sunnybrook Med Ctr, Univ Toronto, 78-83. *Honors & Awards:* Sixth Sarazin Lectr, 82; Hon Lectr, Japanese Soc Microcirculation, 85; Gold Medal Award, Can Liver Found, 87; Distinguished Achievement Award, Am Asn Study Liver Dis, 89. *Mem:* Microcirc Soc; Can Physiol Soc; Am Asn Study Liver Dis; Europ Microcirc Soc; Can Microcirc Soc. *Res:* Structure, function and experimental surgery and pathology of liver; hepatic microcirculation and radiology; pancreas and cardiovascular system. *Mailing Add:* 160 Lytton Toronto ON M4R 1L4 Can

RAPPAPORT, DAVID, b Kiev, Russia, June 13, 07; m 33; c 2. MATHEMATICS EDUCATION. *Educ:* Univ Chicago, BS, 28; Northwestern Univ, MA, 54, EdD(educ), 57. *Prof Exp:* Teacher, Pub Schs, Ill, 37-40; teacher, High Sch, 40-56; PROF MATH EDUC, NORTHEASTERN ILL UNIV, 57- *Concurrent Pos:* Math consult, Am Educ Publs, 71-72. *Res:* Problems of teaching elementary school mathematics. *Mailing Add:* 2747 Coyle Ave Chicago IL 60645

RAPPAPORT, HARRY P, b Los Angeles, Calif, Oct 9, 27. BIOLOGY. *Educ:* Calif Inst Technol, BS, 51; Yale Univ, PhD(physics), 56. *Prof Exp:* USPHS fel, Yale Univ, 56-58, asst prof biophys, 59-65; assoc prof biol, 65-71, PROF BIOL, TEMPLE UNIV, 71- *Mem:* Am Chem Soc; Am Soc Biol Chem. *Res:* Ligand-protein interactions; bacterial secretion of proteins. *Mailing Add:* Dept Biol Temple Univ Board & Montgomery Philadelphia PA 19122-2585

RAPPAPORT, IRVING, b New York, NY, Sept 4, 23; m 48; c 3. IMMUNOLOGY. *Educ:* Cornell Univ, AB, 48; Calif Inst Technol, PhD(immunol, biochem), 53. *Prof Exp:* Res botanist, Univ Calif, Los Angeles, 53-61; asst prof microbiol, Univ Chicago, 61-64; actg chmn dept, 66-67, PROF MICROBIOL, NEW YORK MED COL, 64- *Mem:* Fel AAAS; Am Asn Immunol; Reticuloendothelial Soc; Genetics Soc Am; Am Soc Microbiol; Sigma Xi. *Res:* Antigenic structure of viruses; antibody synthesis in vitro. *Mailing Add:* 200 Waverley Rd Scarsdale NY 10583

RAPPAPORT, LAWRENCE, b New York, NY, May 28, 28; m 53; c 3. PLANT PHYSIOLOGY, HORTICULTURE. *Educ:* Univ Idaho, BS, 50; Mich State Univ, MS, 51, PhD(hort), 56. *Prof Exp:* From appt at jr olericulturist to assoc olericulturist, 56- 68, actg dir, Plant Growth Lab, 74-76, dir, 78-79, chmn, dept veg crops, 78-84, PROF & OLERICULTURIST VEG CROPS, UNIV CALIF, DAVIS, 68- *Concurrent Pos:* Res fel, Calif Inst Technol, 58; NIH spec vis prof & fel org chem, Univ Bristol, 70-71; vis prof, dept biol, Univ Calif, San Diego, 78, Div Hort, Commonwealth Sci Indust Res Orgn, Adelaide, S Australia, 84 & dept chem, City Univ London, 85; vis scientist, Friedrich Miescher Inst, Basel, 85; chmn bot group, Univ Calif, Davis, 86-90. *Honors & Awards:* Fulbright & Guggenheim Fels, Hebrew Univ & Univ Tokyo, 63-64; Fel Am Soc Hort Sci. *Mem:* Am Soc Hort Sci; Am Soc Plant Physiol; Japanese Soc Plant Physiol; Soc Exp Biol; Tissue Cult Asn; AAAS. *Res:* Growth and development; plant growth regulators; gibberellins; plant cell culture; somatic cell biology; cell and plant selection for disease resistance; hormonal regulation; mode of action of plant hormones. *Mailing Add:* Dept Veg Crops Univ Calif Plant Growth Lab Davis CA 95616

RAPPAPORT, MAURICE, b New York, NY, Feb 9, 26; c 6. PSYCHIATRY, NEUROPSYCHOLOGY. *Educ:* Stanford Univ, MD, 62; Ohio State Univ, PhD(psychol), 54. *Prof Exp:* Chief res, Agnews State Hosp, 62-72; RESEARCHER PSYCHIAT, UNIV CALIF, 72- *Mem:* AAAS; Am Med Asn; Human Factors Soc; Am Psychiat Asn. *Res:* Neuropsychiatry; psychopharmacology. *Mailing Add:* 1120 McKendrie St San Jose CA 95126

RAPPAPORT, RAYMOND, JR, b North Bergen, NJ, May 21, 22; m 48; c 3. EMBRYOLOGY. *Educ:* Bethany Col, WVa, BS, 48; Univ Mich, MS, 48; Yale Univ, PhD, 52. *Prof Exp:* Lab instr gen biol, Yale Univ, 51-52; from asst prof to assoc prof biol, 52-61, PROF BIOL, UNION COL, NY, 61- *Concurrent Pos:* Trustee, Mt Desert Island Biol Lab, 54-70, dir 56-59, pres, 79-81. *Mem:* Fel AAAS; Am Soc Zool; Am Micros Soc; Am Soc Cell Biol; Soc Develop Biol; Sigma Xi. *Res:* Animal cell division; role of water in growth and differentiation. *Mailing Add:* Mt Desert Island Biol Lab Salisbury Cove ME 04672

RAPPAPORT, STEPHEN MORRIS, b San Antonio, Tex, Jan 6, 48; m 70. INDUSTRIAL HYGIENE. *Educ:* Univ Ill, BS, 69; Univ NC, Chapel Hill, MSPH, 73, PhD(indust hyg), 74; Am Bd Indust Hyg, cert, 75. *Prof Exp:* Anal chemist, Hazleton Labs, Inc, 69-71; staff mem indust hyg, Los Alamos Sci Lab, Univ Calif, 74-76; asst prof, 76-82, ASSOC PROF INDUST HYG, SCH PUB HEALTH, UNIV CALIF, BERKELEY, 82- *Mem:* Am Indust Hyg Asn; Am Acad Indust Hyg; Am Conf Gov Indust Hygienists; Am Chem Soc. *Mailing Add:* Dept Pub Health Sci Univ Calif 2120 Oxford St Berkeley CA 94720

RAPPAPORT, STEPHEN S, b New York, NY, Sept 26, 38; m 66; c 2. COMMUNICATIONS AND SYSTEMS ENGINEERING. *Educ:* Cooper Union, BEE, 60; Univ Southern Calif, MSEE, 62; NY Univ, PhD(elec eng), 65. *Prof Exp:* Mem tech staff, Hughes Aircraft Co, 60-62; instr elec eng, NY Univ, 64-65; mem tech staff, Bell Tel Labs, 65-68; from asst prof to assoc prof, 68-81, dir grad prog, 72-73, 78-79 & 84-85, PROF ENG, STATE UNIV NY, STONY BROOK, 81- *Concurrent Pos:* Indust consult, 69-; NSF grant commun tech, 76-85; chmn, Undergrad Prog Comt, Col Eng, State Univ NY, Stony Brook, 77-78, chmn, Data Commun Syst Comt, Inst Elec & Electronic Engrs, 84-86; assoc ed, Inst Elec & Electronics Engrs Trans Commun, 82-85; prin investr, Off Nav Res, 85-; mem bd gov, Inst Elec & Electronics Engrs Commun Soc, 87-90; chmn, Awards Comt, Long Island Sect Inst Elec & Electronics Engrs, 88-91. *Mem:* Sr mem & fel Inst Elec & Electronics Engrs. *Res:* Communications systems and theory; analytical modelling and simulation; network architecture, multiple access techniques, and communications traffic; mobile communication systems; data, voice, and computer communications; personal communication networks; spread spectrum techniques; 8c code division multi-access. *Mailing Add:* Dept Elec Eng State Univ NY Stony Brook NY 11794-2350

RAPPERPORT, EUGENE J, b St Louis, Mo, Mar 13, 30; m 50; c 3. METALLURGY, MECHANICAL ENGINEERING. *Educ:* Mass Inst Technol, BS, 52, ScD(metall), 55. *Prof Exp:* Group leader metall, Nuclear Metals, Inc, 51-6; sr scientist, Ledgemont Labs, Kennecott Copper Corp, 63-73; prof, New Col, 73-74; CONSULT, E RAPPERPORT & ASSOCS, 74- *Mem:* Am Inst Mining, Metall & Petrol Engrs; Am Welding Soc; Am Soc Metals; Am Soc Testing & Mat; AAAS. *Res:* Diffusion; x-ray physics; refractory metal technology; phase equilibrium and electron microprobe studies; deformation of metals; chemical thermodynamics and kinetics; computer applications; mass spectrometry; nuclear fusion engineering; magnetohydrodynamics engineering; magnet design and construction. *Mailing Add:* E Rapperport & Assoc 209 Old County Rd Lincoln MA 01773

RAPPORT, DAVID JOSEPH, b Omaha, Nebr, Feb 16, 39; m 65, 86; c 3. ECOLOGY, THEORETICAL BIOLOGY. *Educ:* Univ Mich, BBA, 60, MA, 66, PhD(econ), 67. *Prof Exp:* Res assoc econ, Univ Mich, 67-68; fisheries res bd fel, Univ Toronto, 68-69, fel ecol, 69-70; vis asst prof & Can Coun Killam sr res fel, Simon Fraser Univ, 70-74; environmentalist, Statist Can, 74-81; Titular prof zool, Univ Toronto, 77-81; Sci Adv, Statist Can, 82-89; RES COORD, INST RES ENVIRON & ECOL, UNIV OTTAWA, 89- *Concurrent Pos:* Vis scientist, Statistics Sweden, Stockholm. *Mem:* Inst Ecol; Soc Am Naturalists; fel Linwean Soc (London). *Res:* Data for development; common foundations for economics and society; behavior of ecosystems under stress; framework for environmental statistics; state of environment reporting; economic environment linkages. *Mailing Add:* 308 First Ave Ottawa ON K1S 2G8 Can

RAPPORT, MAURICE M, b New York, NY, Sept 23, 19; m 42; c 2. BIOCHEMISTRY. *Educ:* City Col New York, BS, 40; Calif Inst Technol, PhD(org chem), 46. *Prof Exp:* Technician, Rockefeller Inst, 40-41; asst, Off Sci Res & Develop, Calif Inst Technol, 42-45; mem res staff, Cleveland Clin Found, 46-48; res assoc, Col Physicians & Surgeons, Columbia Univ, 48-51; assoc res scientist immunol, Div Labs & Res, State Dept Health, NY, 51-58; prof biochem, Albert Einstein Col Med, 58-61, Am Cancer Soc prof, 62-67; PROF BIOCHEM, COL PHYSICIANS & SURGEONS, COLUMBIA UNIV, 67-; CHIEF DIV NEUROSCI, NY STATE PSYCHIAT INST, 68- *Concurrent Pos:* Fulbright scholar, Inst Superiore Sanita, Italy, 52; head immunol sect, Sloan-Kettering Inst, 54-58. *Mem:* Am Chem Soc; Am Soc Biol Chem; Soc Exp Biol & Med; Am Soc Neurochem; fel AAAS; fel NY Acad Sci; Biochem Soc; Int Soc Neurochem. *Res:* Serotonin; lipid haptens; brain proteins; plasmalogens; lipid-protein interactions; gangliosides; antibodies as interventive agents in studies of central nervous system function. *Mailing Add:* Dept Biochem 1300 Morris Park Ave Bronx NY 10461-1924

RAPUNDALO, STEPHEN T, b Sudbury, Ont, Aug 22, 58; m 82; c 2. CARDIOVASCULAR. *Educ:* Laurentian Univ, Sudbury, BSc Hons, 79; Med Col Va, PhD(physiol), 83. *Prof Exp:* Postdoctoral fel pharmacol, Col Med, Univ Cincinnati, 83-85, res assoc, 85-87; scientist, 87-90, SR SCIENTIST PHARMACOL, PARKE-DAVIS PHARMACEUT RES, 90- *Mem:* Am Physiol Soc; Biophys Soc; Am Soc Pharmacol & Exp Therapeut; Cardiac Muscle Soc; AAAS. *Res:* Cellular mechanisms related to cardiac pathophysiology; characterization of membrane receptor structure and function, signal transduction, cellular and molecular actions, and processing enzymes of vasoactive peptides; development and evaluation of discovery strategies for cardiovascular therapies. *Mailing Add:* Dept Pharmacol Parke-Davis Pharmaceut Res 2800 Plymouth Rd Ann Arbor MI 48105

RARD, JOSEPH ANTOINE, b St Louis, Mo, Sept 29, 45; m 71; c 2. PHYSICAL CHEMISTRY, CHEMICAL THERMODYNAMICS. *Educ:* Southern Ill Univ, Edwardsville, BA, 67; Iowa State Univ, PhD(phys chem), 73. *Prof Exp:* Fel, Ames Lab, Dept Energy, 73-76; vis res asst prof geol, Univ Ill, Urbana-Champaign, 77-78; CHEMIST THERMODYN & TRANSP PROPERTIES, LAWRENCE LIVERMORE NAT LAB, UNIV CALIF, 79- *Mem:* Sigma Xi; Am Chem Soc. *Res:* Experimental determination of the thermodynamic and transport properties of aqueous solutions, especially brine salts, rare earth electrolytes and transition metal electrolytes; critical reviews of chemical thermodynamics for rare earths and second transition series (fission products) elements. *Mailing Add:* Lawrence Livermore Nat Lab PO Box 808 Livermore CA 94550

RARIDON, RICHARD JAY, b Newton, Iowa, Oct 25, 31; m 56; c 2. PHYSICAL CHEMISTRY. *Educ:* Grinnell Col, BA, 53; Vanderbilt Univ, MA, 55, PhD(chem), 59. *Prof Exp:* Assoc prof physics, Memphis State Univ, 58-62; res chemist, 62-71, COMPUT SPECIALIST, OAK RIDGE NAT LAB, 72- *Concurrent Pos:* Res specialist, Coop Sci Educ Ctr, 71-72. *Mem:* Fel AAAS; Sigma Xi; Asn Acad Sci (secy-treas, 71-75, pres, 77). *Res:* Temperature dependence of chemical equilibria; physical properties of solutions; water desalination; plasma physics; environmental modeling. *Mailing Add:* 111 Columbia Dr Oak Ridge TN 37830-7745

RARITA, WILLIAM ROLAND, b Bordeaux, France, Mar 21, 07; nat US; m 49; c 1. ELEMENTARY PARTICLE PHYSICS. *Educ:* City Col NY, BS, 27, EE, 29; Columbia Univ, MA, 30, PhD(physics), 37. *Prof Exp:* From instr to prof physics, Brooklyn Col, 30-62; res physicist, Space Sci Lab, 62-63, VIS SCIENTIST, LAWRENCE BERKELEY LAB, UNIV CALIF, 63- *Concurrent Pos:* Sr scientist, Manhattan Proj, 44-46; consult, Res Inst Advan Study, 56-62. *Mem:* Fel Am Phys Soc. *Res:* Photoelectric effect; beta activity; deuteron and triton; high energy nucleon-nucleon interaction; Regge poles. *Mailing Add:* 752 Grizzly Peak Blvd Berkeley CA 94708

RAS, ZBIGNIEW WIESLAW, b Warsaw, Poland, June 17, 47; m 87; c 1. KNOWLEDGE REPRESENTATION, DISTRIBUTED SYSTEMS. *Educ:* Univ Warsaw, MS, 70, PhD(computer sci), 73. *Prof Exp:* Programmer, Polish Acad Sci, 73-74; asst prof math & computer sci, Univ Warsaw, 73-83 & Jagiellonian Univ, Poland, 76-78; res assoc, Columbia Univ, 75-76; assoc prof, Univ Tenn, Knoxville, 85-86; PROF COMPUTER SCI, UNIV NC, CHARLOTTE, 81- *Concurrent Pos:* Vis asst prof math, Univ Fla, Gainesville, 78-79; vis prof, Univ Bonn, Ger, 87 & Linkoping Univ, Sweden, 80. *Mem:* Asn Comput Mach; Am Asn Artificial Intel; Sigma Xi; Inst Elec & Electronics Engrs Computer Soc. *Res:* Knowledge representation and distributed information systems. *Mailing Add:* 9029-60 JM Keynes Dr Charlotte NC 28262

RASAIAH, JAYENDRAN C, b Colombo, Ceylon, Apr 1, 34. PHYSICAL CHEMISTRY. *Educ:* Univ Ceylon, BSc, 57; Univ Pittsburgh, PhD(chem), 65. *Prof Exp:* Off Saline Water fel, State Univ NY Stony Brook, 65-68, instr chem, 68-69; from asst prof to assoc prof, 69-78, PROF CHEM, UNIV MAINE, ORONO, 78- *Concurrent Pos:* Sr vis fel, Sci Res Coun, Oxford Univ & London Univ, 75-76; vis fel, Dept Appl Math, Australian Nat Univ, Canberra; vis prof, Dept Chem, Univ NSW, Australia, 80. *Mem:* Am Chem Soc; Am Phys Soc; Royal Soc Chem. *Res:* Statistical mechanics of electrolyte solutions; perturbation theories of polar and non-polar fluids; experimental determination of activity coefficients in heavy water; computer simulation studies of fluids and solutions; non linear effects in polar fluids. *Mailing Add:* Dept Chem Univ Maine Orono ME 04469

RASBAND, S NEIL, b Ogden, Utah, June 21, 39; m 63; c 3. PHYSICS. *Educ:* Univ Utah, BA, 64, PhD(physics), 69. *Prof Exp:* AEC fel, Princeton Univ, 69-70; res assoc physics, La State Univ, Baton Rouge, 70-71, vis asst prof, 71-72; asst prof, 72-74, ASSOC PROF PHYSICS, BRIGHAM YOUNG UNIV, 74- *Mem:* Am Phys Soc; Am Astron Soc. *Res:* Plasma physics; magnetic confinement fusion. *Mailing Add:* Dept Physics Brigham Young Univ Provo UT 84601

RASBERRY, STANLEY DEXTER, b Lubbock, Tex, July 23, 41; m 61; c 2. METROLOGY. *Educ:* Johns Hopkins Univ, BA, 63. *Prof Exp:* Physicist, 63-75, exec asst, 75-79, dept chief, 79-83, CHIEF, NAT BUR STANDARDS, 83- *Concurrent Pos:* Ed, Ref Mat Column, Am Lab, 80-; deleg, US Pharmacopoeial Conv, 80-; lectr, Montgomery Col, 82-85. *Mem:* Am Phys Soc; Am Chem Soc; Soc Appl Spectros; Am Soc Testing & Mat; Int Standardization Orgn. *Res:* Development of spectrometric methods for materials analysis including co-development of Rasberry-Heinrich correction for interelement effects in x-ray fluorescence analysis; development of standard reference materials as calibrants for chemical analysis or for physical measurement. *Mailing Add:* Nat Bur Standards B-354 Physics Gaithersburg MD 20899

RASCH, ELLEN M, b Chicago Heights, Ill, Jan 31, 27; m 50; c 1. CELL BIOLOGY, CYTOCHEMISTRY. *Educ:* Univ Chicago, PhB, 45, BS, 47, MS, 48, PhD(bot), 50. *Prof Exp:* Asst histologist, Am Meat Inst Found, Chicago, Ill, 50-51; USPHS fel zool, Univ Chicago, 51-53, from res asst to res assoc, 53-59; instr civil defense, Milwaukee Civil Defense Dept, 60-62; from res assoc prof biol, Marquette Univ, 62-78; PROF BIOPHYS, QUILLEN-DISHNER COL MED, E TENN STATE UNIV, 78-, CHMN, 86- *Concurrent Pos:* Nat Inst Gen Med Sci Res Career Develop Award, 67-72. *Mem:* Fel AAAS; Am Soc Zool; Royal Micros Soc; Am Soc Cell Biol; Histochem Soc (secy, 75-79, treas, 84-86). *Res:* Quantitative cytophotometry; nucleoprotein synthesis in dipteran polytene chromosomes; evolutionary biology of unisexual fish. *Mailing Add:* Dept Biophys Quillen Col Med ETenn State Univ Johnson City TN 37614-0002

RASCH, ROBERT, b Chicago, Ill, Nov 19, 26; m 50; c 1. PHYSIOLOGY. *Educ:* Univ Chicago, PhB, 46, PhD(physiol), 59; Northwestern Univ, MD, 51. *Prof Exp:* Instr, 59-61, from asst prof to assoc prof physiol, Med Col Wis, 61-77; prof & chmn, 77-87, EMER PROF PHYSIOL, QUILLEN-DISHNER COL MED, 87- *Mem:* AAAS; Am Soc Cell Biol; NY Acad Sci. *Res:* Chromosome structure and nucleocytoplasmic interaction; compensatory physiological mechanisms, chromosome structure and nucleocytoplasmic interaction; physiological simulations on microprocessors. *Mailing Add:* Dept Physiol Quillen-Dishner Med Col Box 19780A Johnson City TN 39614

RASCHE, JOHN FREDERICK, b Bonne Terre, Mo, Apr 14, 36; m 58; c 3. STATISTICAL PROCESS CONTROL. *Educ:* Univ Mo, Rolla, BS, 58. *Prof Exp:* Assoc develop chem engr, 58-61, develop engr, 61-65, sr develop engr, 65-75, group leader process res & develop, 75-79, process engr mgr, 79-87, ENG FEL, A E STALEY MFG CO, 87-, MGR, STATIST PROCESS ANALYSIS, 88- *Mem:* Am Inst Chem Engrs. *Res:* Catalysis of carbohydrates using soluble and immobilized enzymes; purification and separation processes applied to corn and soybean processing. *Mailing Add:* 1821 Burning Tree Dr Decatur IL 62521

RASCO, BARBARA A, b Pittsburgh, Pa, Apr 3, 57; m 84. FOOD REGULATIONS, FISHERIES TECHNOLOGY. *Educ:* Univ Pa, BSE, 79; Univ Mass, PhD(food sci & nutrit), 83. *Prof Exp:* Biochem engr, Cargill Inc, 82-83, res chemist, 83-84; asst prof, 84-89, ASSOC PROF FOOD SCI, INST FOOD SCI & TECHNOL, SCH FISHERIES, UNIV WASH, 89- *Mem:* Inst Food Technologists; Am Chem Soc. *Res:* Development of rapid analytical methods for fisheries technology and aquaculture; applied enzymology; food product/process development, food adulteration. *Mailing Add:* 4025 Dayton Ave N Seattle WA 98103

RASE, HOWARD F, b Buffalo, NY, Oct 18, 21; m 54; c 2. CHEMICAL ENGINEERING. *Educ:* Univ Tex, BS, 42; Univ Wis, MS, 50, PhD(chem eng), 52. *Prof Exp:* Chem engr res & develop, Dow Chem Co, 42-44; process engr, Eastern State Petrol Co, 44; process engr & proj engr, Foster Wheeler Corp, 44-49; from asst prof to prof, 52-74, chmn dept, 63-68, W A CUNNINGHAM PROF CHEM ENG, UNIV TEX, AUSTIN, 74- *Concurrent Pos:* Fulbright lectr, Tech Univ Denmark, 57; consult, catalysis & reactor design. *Mem:* Am Chem Soc; fel Am Inst Chem Engrs. *Res:* Applied kinetics and reactor design; homogeneous and heterogeneous catalysis; catalyst development; process design techniques; enzyme catalysis. *Mailing Add:* Dept Chem Eng Univ Tex Austin TX 78712

RASEMAN, CHAD J(OSEPH), b Detroit, Mich, Aug 29, 18; m 45; c 5. CHEMICAL & NUCLEAR ENGINEERING. *Educ:* Wayne State Univ, BS, 41; Univ Mich, MS, 44; Cornell Univ, PhD(chem eng), 51. *Hon Degrees:* EngD, Wayne State Univ, 58. *Prof Exp:* Chief chem engr, R P Scherer Corp, 41-44 & Boyle-Midway, Inc, 46-47; head, Oak Ridge Chem Eng Div, Kellex Corp, 47-48; fel, Brookhaven Nat Lab, 48-51, group leader, Nuclear Eng Dept, 51-58, asst div head, 58-65, tech asst to dept chmn, Dept Appl Sci, 65-70, group leader rotating fluidized bed reactor, Space Nuclear Propulsion Proj, 70-73; vpres, 67-88, PRES, SOLAR SUNSTILL, INC, 88- *Concurrent Pos:* Asst, Cornell Univ, 48-49; chief eval sect, Div Reactor Develop, US AEC, 60-62; dir, Nat Agr Plastics Asn, 72-76. *Mem:* AAAS; Am Chem Soc; fel Am Inst Chem; Am Nuclear Soc; Int Solar Energy Soc. *Res:* Design, evaluation and testing of the fluidized bed concept for space nuclear propulsion and unique chemical reactions; co-inventor of solar distillation units and special coatings that control condensation and light transmission. *Mailing Add:* Solar Sunstill Inc 644 W San Francisco Santa Fe NM 87501

RASENICK, MARK M, b Chicago, Ill, Sept 5, 49; m; c 3. NEUROBIOLOGY, NEUROTRANSMITTER RESPONSE. *Educ:* Wesleyan Univ, PhD(develop biol), 77. *Prof Exp:* Assoc res scientist, dept neurol, Sch Med, Yale Univ, 81-83; asst prof, 83-88, ASSOC PROF PHYSIOL & BIOPHYS, COL MED, UNIV ILL, 88- *Concurrent Pos:* Mental Health Clin Res Ctr; Ill State Psychiatric Inst, 83-86; postdoctoral, Yale Univ, 81; mem, Cellular Neurosci Panel, NSF. *Honors & Awards:* Res Scientist Develop Award, NIMH. *Mem:* NY Acad Sci; Soc Neurosci; AAAS; Am Soc Biol Chemists; Union Concerned Scientists. *Res:* Neurotransmitter receptors, cyclic nucleotides and coupling proteins in synaptic function; influence of the cytoskeleton on neurotransmitter responsiveness in synaptic membranes and cultured neural cells; cellular and molecular basis of mental and neurological dysfunction; molecular basis of antidepressant and other neuroleptic drug action. *Mailing Add:* Dept Physiol & Biophys Col Med Univ Ill PO Box 6998 Chicago IL 60680

RASERA, ROBERT LOUIS, b New York, NY, July 25, 39; m 61; c 2. SOLID STATE PHYSICS, NUCLEAR PHYSICS. *Educ:* Wheaton Col, BS, 60; Purdue Univ, PhD(physics), 65. *Prof Exp:* Res assoc, Purdue Univ, 65; guest prof, Inst Radiation & Nuclear Physics, Univ Bonn, 65-66; asst prof, Univ Pa, 66-71; assoc prof, 71-81, PROF PHYSICS, UNIV MD, BALTIMORE COUNTY, 81- *Mem:* Am Phys Soc; Am Asn Physics Teachers; Sigma Xi. *Res:* Perturbed angular correlations of nuclear radiations; hyperfine fields in solids. *Mailing Add:* Dept Physics Univ Md Baltimore County Baltimore MD 21228

RASEY, JANET SUE, b Fremont, Mich, June 13, 42. CANCER, RADIOBIOLOGY. *Educ:* Univ Mich, BS, 64; Ore State Univ, MS, 65; Univ Ore, PhD(biol), 70. *Prof Exp:* NIH fel, Cell & Radiation Biol Lab, Allegheny Gen Hosp, Pittsburgh, 70-72; assoc prof, 72-83, PROF TUMOR RADIATION & CELL BIOL, DEPT RADIATION ONCOL, MED SCH, UNIV WASH, 83- *Mem:* Cell Kinetics Soc; Radiation Res Soc; Am Soc Therapeut Radiol & Oncol. *Res:* Tumor radiation response; nuclear imaging of tumors. *Mailing Add:* Dept Radiation Oncol Univ Wash Med Sch 1959 NE Pacific St Seattle WA 98195

RASH, FRED HOWARD, b Elkin, NC, Jan 21, 41; m 67; c 2. PHOTOGRAPHIC CHEMICALS MANUFACTURING. *Educ:* Wake Forest Univ, BS, 62; Duke Univ, PhD(chem), 67. *Prof Exp:* Chemist, Tenn Eastman Co, 66-72, sr chemist, 73-77, group leader, 77-89, RES ASSOC, TENN EASTMAN CO, 81- *Mem:* Am Chem Soc; AAAS. *Res:* Product and process research and development in industrial chemicals. *Mailing Add:* Eastman Chem Div PO Box 1972 Bldg 150B Kingsport TN 37662

RASH, JAY JUSTEN, b Iowa Falls, Iowa, Dec 26, 41; m 61; c 3. METABOLISM. *Educ:* Northwest Mo State Univ, BS, 66; Univ Mo-Columbia, MS, 67, PhD(anal biochem), 71. *Prof Exp:* Res scientist metab, Cent Res Labs, 72-81, DIR REGULATORY AFFAIRS, PFIZER, INC, 81- *Mem:* Am Chem Soc; Sigma Xi. *Res:* Development of analytical methods in tissues and biological fluids for drugs, metabolites and important biological compounds; the determination of the metabolic pathways of xenobiotics in biological systems. *Mailing Add:* 120 Highland Dr Pleasant Hill MO 64080

RASH, JOHN EDWARD, b Dallas, Tex, Feb 16, 43; m 65. NEUROBIOLOGY. *Educ:* Univ Tex, BA, 65, MA, 67, PhD(zool), 69. *Prof Exp:* Teaching asst biol, Univ Tex, 66-67; fel surg & cancer res, Sch Med, Johns Hopkins Univ, 69-70; asst investr, Dept Embryol, Carnegie Inst Wash, 70-72; res assoc, Univ Colo, 72-74; asst prof, 74-77, assoc prof pharmacol & exp therapeut, Sch Med, Univ Md, 77-79; ASSOC PROF ANAT, COLO STATE UNIV, 79- *Concurrent Pos:* Muscular Dystrophy Asn grant-in-aid, 75; prin investr, NIH grants; instr, Marine Biol Lab, Woods Hole, Mass. *Mem:* AAAS; Am Soc Cell Biol; Soc Neurosci. *Res:* Myogenesis; membrane differentiation; electron microscopy; freeze-fracture of mammalian neuromuscular junctions. *Mailing Add:* Dept Anatomy Colo State Univ Ft Collins CO 80523

RASHEED, SURAIYA, b Hyderabad, India; US citizen; m; c 2. GENETICS, MOLECULAR BIOLOGY. *Educ:* Osmania Univ, India, BSc, 53, MSc, 55, PhD, 58; London Univ, PhD, 64. *Prof Exp:* Res assoc, Dept Cancer Res, Mt Vernon Hosp, Eng, 64-70; instr, path, 70-72, from asst prof to assoc prof 72-82, PROF PATH, UNIV SOUTHERN CALIF, SCH MED, 82-, DIR, VIRAL ONCOL & AIDS RES. *Concurrent Pos:* Consult, Int Ctr Med Res & Training, 85-, Abbott Lab, Diagnostic Div, Ill, Prince Agha Khan Univ, 86-, Alpha Therapeut, Los Angeles, 86-; mem, Int adv bd, Future Trends Chemotherapy, Italy, 87-, special adv comt, NCI-NIH AIDS-Antiviral Drug Screening Prog, 86-90, virology core comt, AIDS Clin Trials Group, NIAID, NIH, 86-91; mem, comt rev res grants & res career grant appl, Nat Cancer Inst, NIH, 80-, special reviewer, Nat Large Bowel Projs, 81, Cancer Biol-Immunology Contract rev comt, 86-90, special tech sci rev group, Small Bus Innovation Res Prog, 86-90; hon professorships, Peoples Repub China. *Mem:* Int Asn Comp Res Leukemia & Related Dis; Am Soc Microbiol; Am Soc Virologists; AAAS; Am Asn Univ Professors; Int AIDS Soc. *Res:* Molecular mechanisms of oncogenesis & the Acquired Immune Deficiency Syndrome; viral oncology; AIDS retrovirology; immunology. *Mailing Add:* Lab Viral Oncol & Aids Res Univ Southern Calif 1840 N Soto St Los Angeles CA 90032-3626

RASHKIN, JAY ARTHUR, b New York, NY, Sept 21, 33; div; c 2. PHYSICAL CHEMISTRY. *Educ:* NY Univ, BA, 55; Princeton Univ, MA, 57, PhD(phys chem), 61. *Prof Exp:* Res chemist, E I du Pont de Nemours & Co, 59-62; mem tech staff, Space Tech Labs, 62-64; res specialist, Monsanto Co, 64-66; sr res chemist, Cities Serv Oil Co, 66-76; res assoc, Halcon Catalyst Indust, 77-82 & Halcon Res, 82-86; ASSOC PROF CHEM, SULLIVAN COUNTY COMMUNITY COL, 86- *Mem:* Am Chem Soc; Am Phys Soc; AAAS. *Res:* Heterogeneous catalysis; differential thermal analysis and x-ray structure studies of catalysts; catalytic oxidation of hydrocarbons; catalytic processes in petroleum refining and petrochemicals; surface chemistry; electron spectroscopy for chemical analysis; surface analysis; chemisorption; temperature programmed desorption. *Mailing Add:* Sullivan County Comm Col Loch Sheldrake NY 12759

RASHKIND, WILLIAM JACOBSON, pediatrics; deceased, see previous edition for last biography

RASK, NORMAN, b Duanesburg, NY, June 26, 33; m 55; c 4. AGRICULTURAL ECONOMICS. *Educ:* Cornell Univ, BS, 55, MS, 60; Univ Wis, PhD(agr econ), 64. *Prof Exp:* Exten assoc agr econ, Cornell Univ, 59-60; asst prof, Univ Wis, 64-65; from asst prof to assoc prof, 65-73, PROF AGR ECON, OHIO STATE UNIV, 73- *Concurrent Pos:* Consult, World Bank, Food & Agr Orgn & US Agency Int Develop. *Res:* Energy policy; energy from biomass; economics of farm size, and world food population problems. *Mailing Add:* Agr Econ 103 Agr Admin Ohio State Univ Columbus OH 43210

RASKA, KAREL FRANTISEK, JR, b Prague, Czech, May 26, 39; US citizen; m 60; c 2. PATHOLOGY, MICROBIOLOGY. *Educ:* Charles Univ, MD, 62; Czech Acad Sci, PhD(biochem), 65. *Prof Exp:* Commonwealth Fund res fel pharmacol, Sch Med, Yale Univ, 65-66; res assoc, Inst Microbiol, Rutgers Univ, New Brunswick, 66-67; scientist, Inst Org Chem & Biochem, Czech Acad Sci, 67-68; from asst prof to assoc prof microbiol, Univ Med & Dent NJ, Robt Wood Johnson, 71-78, from assoc prof to prof path, 73-89, prof microbiol, 78-89; CHMN, DEPT LAB MED & PATH, UNIV MED & DENT NJ, NJ MED SCH, NEWARK, 89- *Concurrent Pos:* Pathologist, Robt Wood Johnson Univ Hosp 77-; dir, Univ Diag Lab, 82-; chief serv, Univ Hosp, Newark, 89- *Honors & Awards:* Prize, Czech Acad Sci, 65. *Mem:* Am Soc Virology; Am Soc Microbiol; Am Asn Pathologists; Am Asn Immunologists; Am Asn Cancer Res; Soc Clin Immunol; Col Am Path; Soc Exp Biol Med; Sigma Xi. *Res:* Molecular biology of animal viruses; cell transformation by oncogenic viruses; cancer chemotherapy; immunopathology. *Mailing Add:* Dept Lab Med Path C-579 MSB Univ Med & Dent NJ NJ Med Sch 1855 Orange Ave Newark NJ 07103-2714

RASKAS, HESCHEL JOSHUA, b St Louis, Mo, June 11, 41; m 62; c 4. BIOCHEMISTRY. *Educ:* Mass Inst Technol, BS, 62; Harvard Univ, PhD(biochem, molecular biol), 67. *Prof Exp:* Res fel, Inst Molecular Virol, 67-69, asst prof molecular virol & path, Sch Med, Inst Molecular Virol & Dept Path, St Louis Univ, 69-73; assoc prof, 73-77, dir, Ctr Basic Cancer Res, 77-80, prof path & microbiol, 77-85, VIS PROF PATH, SCH MED, WASH UNIV, 85- *Concurrent Pos:* Assoc ed, Virol, 75-78 & 82-85; pres, Raskas Foods, Inc, 82- *Mem:* AAAS; Am Soc Microbiol. *Res:* Regulation of oncogenic viral gene expression in eukaryotic cells. *Mailing Add:* 722 Brittany Lane St Louis MO 63130

RASKI, DEWEY JOHN, b Kenilworth, Utah, Dec 12, 17; m 43; c 3. NEMATOLOGY. *Educ:* Univ Calif, BS, 41, PhD(nematol), 48. *Prof Exp:* Instr & jr nematologist, 48-50, from lectr & asst nematologist to lectr & assoc nematologist, 50-60, PROF NEMATOL & NEMATOLOGIST, UNIV CALIF, DAVIS, 60- *Mem:* Am Phytopath Soc; Am Soc Nematol; Soc Europ Nematol. *Res:* Plant parasitic nematodes; virus-nematode vector relationships; biology, control and systematics. *Mailing Add:* Dept Nematol Univ Calif Davis CA 95616

RASKIN, BETTY LOU, b Baltimore, Md, Apr 9, 24. PLASTICS ENGINEERING, CONSUMER BEHAVIOR. *Educ:* Goucher Col, AB, 44; Johns Hopkins Univ, MA, 47, PhD, 68. *Prof Exp:* Lab asst Manhattan Proj, Johns Hopkins Univ, 44-45, res chemist, Inst Coop Res, 45-50, res staff asst, 51-54, res assoc & head plastics res & develop, Radiation Lab, 54-64; asst prof psychol, 67-69, ASSOC PROF PSYCHOL, TOWSON STATE COL, 69-; CONSULT FOAMED PLASTICS, 64- *Mem:* Fel AAAS; fel Am Inst Chemists; Am Psychol Asn. *Res:* Foamed plastics; reinforced plastics; foamed plastic particles and smoke; consumer psychology; teenage drinking and driving. *Mailing Add:* Dept Psychol Towson State Col Baltimore MD 21204

RASKIN, JOAN, b Baltimore, Md, Aug 11, 30. MEDICINE, DERMATOLOGY. *Educ:* Goucher Col, BA, 51; Univ Md, MD, 55; Am Bd Dermat, dipl, 61. *Prof Exp:* Intern, Hosp, Univ Md, 55-56, asst resident med, 56-57, resident dermat, 57-58; assoc med, Sch Med, 60-63, lectr, Sch Phys Ther, 61-66, asst prof, Sch Med, 63-66, ASSOC PROF MED, DIV DERMAT, SCH MED, UNIV MD, BALTIMORE CITY, 66- *Concurrent Pos:* Fel, Hosp, Univ Md, 58-59; fel, Hosps, Univ Minn, 59-60; NIH grants, 61-65; Bressler Fund grant, 62-66; mem courtesy staff, South Baltimore Gen Hosp, 61-71, sr staff, 72-, head, Div Dermat, 80-; mem courtesy staff, Sinai Hosp, 61-64, active staff, 64-86; mem staff, Mercy Hosp, 74-; dermatologist, Md Sch for Blind, 74-; active staff, Union Mem Hosp, 75-; consult, 86- *Mem:* Am Acad Dermat; Am Med Asn; Am Med Women's Asn. *Res:* Dermatologic areas of autoimmune diseases and virology. *Mailing Add:* 3506 Calvert St N Baltimore MD 21218

RASKIN, NEIL HUGH, b New York, NY, Jan 16, 35. NEUROLOGY, NEUROCHEMISTRY. *Educ:* Dartmouth Col, AB, 56; Harvard Med Sch, MD, 59. *Prof Exp:* Resident neurol, Columbia Univ Col Physicians & Surgeons, 61-64, res fel cerebral metab, 64-65; chief of serv neurol, US Naval Hosp, Philadelphia, 65-66; res assoc cerebral metab, NIH, 66-68; from asst prof to assoc prof, 68-79, PROF NEUROL & VCHMN DEPT, UNIV CALIF, SAN FRANCISCO, 79- *Concurrent Pos:* NIH res career develop awardee, 68-73; res grant, Nat Inst for Alcohol Abuse & Alcoholism, 68-73; mem, Res Group Migraine & Headache, World Fedn Neurol, 76-; dir neurol outpatient & consult serv, Sch Med, Univ Calif, San Francisco, 77- *Mem:* Am Neurol Asn; Am Soc Neurochem; Asn Res Nerv & Ment Dis; Int Asn Study Pain; Am Acad Neurol. *Res:* Biochemical effects of alcohol upon the nervous system; mechanisms of migraine; metabolic neurologic disorders. *Mailing Add:* Dept Neurol Univ Calif Sch Med San Francisco CA 94143

RASKOVA, JANA D, b Prague, Czech, Oct 18, 40; US citizen; m 60; c 2. PATHOLOGY, MEDICINE. *Educ:* Charles Univ, Prague, MD, 63. *Prof Exp:* Res fel pharmacol, Sch Med, Yale Univ, 65-66; vis investr immunol, Inst Microbiol, Rutgers Univ, 66-67; fel genetics, Inst Exp Biol & Genetics, Czech Acad Sci, Prague, 67-68; res assoc immunol, Inst Microbiol, Rutgers Univ, 69-70, asst prof path, 70-78, assoc prof path, 78-87, PROF PATH, UNIV MED & DENT NJ, ROBERT WOOD JOHNSON MED SCH, 87- *Mem:* AAAS; Am Asn Pathologists; Am Soc Cell Biol; Int Acad Path. *Res:* Immunopathology and pathology of chronic renal failure. *Mailing Add:* Dept Path Univ Med & Dent NJ Robert Wood Johnson Med Sch PO Box 101 Piscataway NJ 08854

RASLEAR, THOMAS G, b New York, NY, Nov 25, 47; m 71. BEHAVIORAL TOXICOLOGY, COMPARATIVE PSYCHOPHYSICS. *Educ:* City Col NY, BS, 69; Brown Univ, ScM, 72, PhD(psychol), 74. *Prof Exp:* Asst prof psychol, Boston Univ, 74-75 & Wilkes Col, Wilkes-Barre, 75-79; res psychologist, Dept Med Neurosci, 79-89, SR RES PSYCHOLOGIST, DEPT MICROWAVE RES, WALTER REED ARMY INST RES, 89- *Concurrent Pos:* Dep chief, Physiol & Behav Br, Dept Med Neurosci, Walter Reed Inst Res, 85-89. *Mem:* Am Psychol Soc; Am Psychol Asn; Acoust Soc Am; AAAS. *Res:* Use methods of behavioral psychology and psychophysics to measure the effects of potentially toxic substances, such as microwave radiation, on cognitive function, circadium rhythms, sensation and perception in animal subjects. *Mailing Add:* Dept Microwave Res Walter Reed Army Inst Res Washington DC 20307-5100

RASMUSEN, BENJAMIN ARTHUR, b Somonauk, Ill, Nov 29, 26; m 57; c 3. ANIMAL GENETICS. *Educ:* Univ Ill, BS, 49; Cornell Univ, MS, 51, DVM, 55; Univ Calif, Davis, PhD(genetics), 58. *Prof Exp:* Asst animal genetics, Dept Poultry Husb, Cornell Univ, 49-54; asst specialist, Univ Calif, Davis, 55-58; from assoc prof to prof, 58-83, EMER PROF ANIMAL GENETICS, UNIV ILL, URBANA, 83- *Concurrent Pos:* Sabbatical, Animal Breeding Res Orgn, Edinburgh, Scotland, 65-66 & Inst Animal Physiol, Cambridge, Eng, 72-73. *Mem:* Int Soc Animal Blood Group Res (pres, 76-84). *Res:* Blood groups in pigs and sheep. *Mailing Add:* 4517 E 23rd Leland IL 60531

RASMUSSEN, ARLETTE IRENE, b Thief River Falls, Minn. NUTRITION. *Educ:* Northwestern Univ, Ba, 56; Univ Wis, MS, 58, PhD(human nutrit & biochem), 61. *Prof Exp:* Res asst, Univ Wis, 56-61; asst prof, 61-65, ASSOC PROF NUTRIT, UNIV DEL, 65- *Concurrent Pos:* Sr fel, Dept Med, Hematol Div, Sch Med, Univ Wash, 70-71; co-chair, Northeast Res Prog, Steering Comt Nutrit & Food Safety, US Dept Agr & State Agr Exp Sta, 77-82; exec bd mem, Nat Nutrit Consortium 77-80; actg chair, Dept Food Sci & Human Nutrit, Univ Del, 78-79; nutritionist, Coop State Res Serv, US Dept Agr, 79-80. *Mem:* Am Inst Nutrit; Am Dietetic Asn; Soc Nutrit Educ; Sigma Xi; AAAS. *Res:* Protein evaluation; amino acid utilization; vitamin B-6; mineral nutrition; human nutrition and metabolism; nutrient interactions. *Mailing Add:* Col Human Resources Univ Del Newark DE 19716

RASMUSSEN, CHRIS ROYCE, b Trenton, Nebr, Feb 19, 31; m 52; c 4. MEDICINAL CHEMISTRY. *Educ:* Ft Hays Kans State Col, AB, 54; Univ Kans, PhD(org & pharmaceut chem), 62. *Prof Exp:* Res assoc, Ind Univ, 62-63; res scientist, 64-70, group leader, 70-76, RES FEL, MCNEIL LABS, INC, 76- *Mem:* Am Chem Soc. *Res:* Hypoglycemic agents; central nervous system drugs, specifically muscle relaxants, anticonvulsants, anti-anxiety agents, cardiovascular-anti-anginal, antihypertensive and antiarrhythmic drugs; gastro-intestinal-antisecretory agents; anti-irritable bowel. *Mailing Add:* R W Johnson Pharm Res Inst Spring House PA 19477-1100

RASMUSSEN, DAVID IRVIN, b Ogden, Utah, Dec 11, 34; m 54; c 4. EVOLUTIONARY BIOLOGY, GENETICS. *Educ:* Univ Utah, BS, 56, MS, 58; Univ Mich, PhD(zool), 62. *Prof Exp:* NIH fel genetics, Univ Calif, Berkeley, 62-63; from asst prof to assoc prof zool, 63-75, PROF ZOOL, ARIZ STATE UNIV, 75- *Concurrent Pos:* Prin investr fac res grant, Ariz State Univ, 63-64, 66-67 & 75-76 & NIH res grants, 64-72. *Mem:* Am Soc Mammal; Soc Study Evolution; Am Soc Nat; Sigma Xi. *Res:* Genetic polymorphisms in natural populations. *Mailing Add:* Dept Zool Ariz State Univ Tempe AZ 85287

RASMUSSEN, DAVID TAB, b Salt Lake City, Utah, June 17, 58. PRIMATOLOGY, VERTEBRATE PALEONTOLOGY. *Educ:* Colo Col, BA, 80; Duke Univ, PhD(anthrop), 86. *Prof Exp:* Res assoc paleont, Primate Ctr, Duke Univ, 86-87; vis asst prof anthrop, Rice Univ, 87-88; asst prof anthrop, Univ Calif, Los Angeles, 88-91; ASST PROF ANTHROP, WASH UNIV, ST LOUIS, 91- *Concurrent Pos:* Res assoc, Natural Hist Mus Los Angeles County, 88. *Mem:* Am Asn Phys Anthropologists; Am Soc Naturalists; Soc Avian Paleont & Evolution; Soc Vertebrate Paleont. *Res:* Primate evolution and paleontology; prosimian adaptations and life history studies; anthropoid origins; field paleontology in Africa and North America; comparative anatomy and biology of primates; mammalian and avian paleontology; hyracoid evolution. *Mailing Add:* Dept Anthrop Wash Univ St Louis MO 63130

RASMUSSEN, DON HENRY, b Wild Rose, Wis, Sept 20, 44; m 66; c 3. CHEMICAL METALLURGY, MATERIALS SCIENCE. *Educ:* Univ Wis-Madison, BA, 67, MS, 71, PhD(mat sci), 74. *Prof Exp:* Res assoc, Cryobiol Res Inst of Am Found, Madison, 67-75; res assoc metall, Dept Metals & Minerals Eng, Univ Wis-Madison, 75-78; asst prof, 78-80, PROF CHEM ENG, CLARKSON UNIV, 80- *Mem:* Am Chem Soc; Am Inst Mining, Metall & Petrol Engrs; Am Inst Chem Engrs. *Res:* Surface properties of metals; phase transformations; nucleation phenomena; chemical vapor deposition of metals. *Mailing Add:* Dept Chem Eng Clarkson Univ Potsdam NY 13676

RASMUSSEN, HARRY PAUL, b Tremonton, Utah, July 18, 39; m 58; c 4. PLANT PHYSIOLOGY, PLANT NUTRITION. *Educ:* Utah State Univ, BS, 61; Mich State Univ, MS, 62, PhD(hort), 65. *Prof Exp:* Asst prof plant physiol, Conn Agr Exp Sta, 65-66; from asst prof to assoc prof hort, Mich State Univ, 66-73, prof, 73-80; CHMN DEPT HORT & LANDSCAPE ARCHIT, WASH STATE UNIV, 80- *Concurrent Pos:* Dir, Agr Exp Sta, Utah State Univ. *Honors & Awards:* Alex Laurie Award, 75. *Mem:* Fel Am Soc Hort Sci. *Res:* Leaf and fruit abscission; botanical histochemistry; electron microprobe x-ray analysis and scanning electron microscopy of plant tissues and cells. *Mailing Add:* Agr Exp Sta Utah State Univ Logan UT 84322-4810

RASMUSSEN, HOWARD, b Harrisburg, Pa, Mar 1, 25; m 50; c 4. MEDICINE, CELL BIOLOGY. *Educ:* Gettysburg Col, AB, 48; Rockefeller Inst, PhD, 59; Gettysburg Col, DSc, 66. *Hon Degrees:* MA, Univ Pa, 71. *Prof Exp:* Asst prof physiol, Rockefeller Inst & assoc physician, Hosp, 59-61; assoc prof biochem, Univ Wis, 61-64, prof, 64-65; chmn dept biochem, Med Sch, Univ Pa, 65-71, prof pediat, biochem & biophys, 65-77; PROF MED & CELL BIOL, SCH MED, YALE UNIV, 77- *Concurrent Pos:* NSF sr fel, Cambridge Univ, 71-72; mem endocrine study sect, NIH, 63-65, mem cardiovasc study sect, 73-74; mem fel panel, NSF, 69; sr physician, Children's Hosp Philadelphia, 70-77; trustee, Gettysburg Col, 71-77; mem, Gen Med B Study Sect, 74-77; assoc endocrinologist, Children's Hosp, 75-; consult metab bone dis, Hosp Lariboisiere, Paris. *Honors & Awards:* Andre Lichwitz Prize, France, 71. *Mem:* AAAS; Endocrine Soc; Soc Gen Physiol; Am Soc Cell Biol; Am Soc Biol Chem. *Res:* Biochemistry and physiology of ion transport and of peptide and steroid hormone action; cellular basis of oxygen-2 toxicity. *Mailing Add:* Cell Biol-Phys Med Yale Univ 333 Cedar St New Haven CT 06510

RASMUSSEN, JEWELL J, b Palo Alto, Calif, Jan 30, 40. CERAMICS. *Educ:* Univ Utah, BS, 64; Mass Inst Technol, PhD(ceramics), 69. *Prof Exp:* Res asst ceramics, Mass Inst Technol, 65-68; sr res scientist, Pac Northwest Lab, Battelle Mem Inst, 68-76; MGR APPL RES, MONT ENERGY & MONT HOUSING & DEVELOP RES & DEVELOP INST, 76- *Concurrent Pos:* Tech consult, Manlabs, Inc, 65-67; adj assoc prof, Mont Col Mineral Sci & Technol, 77- *Mem:* Am Ceramic Soc. *Res:* Growth; properties of molten oxides; materials development and characterization; fuel cell power systems; materials development for industrial products. *Mailing Add:* 2357 St Mary's Dr Salt Lake City UT 84108

RASMUSSEN, JOHN OSCAR, JR, b St Petersburg, Fla, Aug 8, 26; m 50; c 4. NUCLEAR CHEMISTRY. *Educ:* Calif Inst Technol, BS, 48; Univ Calif, Berkeley, PhD(chem), 52. *Hon Degrees:* AM, Yale Univ, 69. *Prof Exp:* From instr to prof chem, Univ Calif, Berkeley, 52-69; prof chem, Yale Univ, 69-73; PROF CHEM, UNIV CALIF, BERKELEY, 73- *Concurrent Pos:* Vis prof, Nobel Inst Physics, Stockholm, Sweden, 53; NSF sr fel, Niels Bohr Inst, Copenhagen, Denmark, 61-62; vis prof, Fudan Univ, Shanghai, 79, 84, hon prof, 84; sr sci, Von Humboldt, Munich, 91. *Honors & Awards:* E O Lawrence Award, 67; Am Chem Soc Award for Nuclear Applns in Chem, 76. *Mem:* Am Chem Soc; fel Am Phys Soc; fel AAAS; Fedn Am Scientists. *Res:* Nuclear structure theory and experiment; heavy ion nuclear reactions. *Mailing Add:* MS 70A-3307 Lawrence Berkeley Lab Berkeley CA 94720

RASMUSSEN, KATHLEEN GOERTZ, b Bastrop, Tex, July 10, 58; m 79. ENDOCRINOLOGY-IMMUNE SYSTEM INTERACTIONS, VIRUS-INDUCED NEUROLOGICAL DAMAGE. *Educ:* Tex A&M Univ, BS, 80, MS, 83; Tex Tech Univ, PhD(med biochem), 88. *Prof Exp:* Postdoctoral molecular biol, Tex Tech Univ, 88-89, res instr, 88-90; POSTDOCTORAL IMMUNOL & INFECTIOUS DIS, UTAH STATE UNIV, 90- *Mem:* Soc Cell Biol; AAAS; Am Asn Parasitologists. *Res:* Infectious diseases; effects of the diseases on the immune systems of the hosts; molecular approaches for the control of these diseases; cryptosporidiosis, an opportunistic infection of acquired immune deficiency syndrome patients; autism. *Mailing Add:* 1512 Lynnwood Logan UT 84321

RASMUSSEN, KATHLEEN MAHER, b Dayton, Ohio, Mar 1, 48; m 70; c 1. REPRODUCTIVE PHYSIOLOGY. *Educ:* Brown Univ, AB, 70; Harvard Univ, ScM, 75, ScD(nutrit), 78. *Prof Exp:* Teacher sci, Cape Hatteras Sch, 71-72; anal chemist, Berkley Machine Works & Foundry Co, Inc, 72-73; res fel, Harvard Univ, 78; res fel, Cornell Univ, 78-80, res assoc, 81, instr, 81-83, asst prof, 83-88, ASSOC PROF NUTRIT, CORNELL UNIV, 88- *Mem:* Am Inst Nutrit; Am Soc Clin Nutrit; Int Soc Res Human & Milk & Lactation. *Res:* Nutrition and reproduction; maternal and child nutrition; pregnancy; lactation. *Mailing Add:* Div Nutrit Sci 111 Savage Hall Cornell Univ Ithaca NY 14853-6301

RASMUSSEN, LOIS E LITTLE, b Summit, NJ, Nov 11, 38; m 61; c 2. ANIMAL PHEROMONE, NEUROCHEMISTRY. *Educ:* Stanford Univ, BA, 60; Wash Univ, PhD(neurochem), 64. *Prof Exp:* NIH fel, Dept Psychiatry, Med Sch, Wash Univ, 61-64; NIH staff fel, Sect Chem Neuropath, Nat Inst Neurol Dis & Blindness, NIH, 64-66; proj biochemist, Dow Corning Corp, 66-68; adj scientist, Dept Zool, Wash State Univ, 70-77; SR SCIENTIST, ORE GRAD CTR, 77-, MEM FAC, 85- *Concurrent Pos:* Lectr, Saginaw Valley Col, 60-69; instr, invertebrate & vertebrate embryol & fresh water ecol, Saginaw Valley Col, 68-69; res assoc & consult, Dept Zool, Wash State Univ, 75-77; mem, Asian elephant spec group, Species Survival Comn, Int Union Conserv Nature & Natural Resources. *Mem:* Am Fisheries Soc; Am Soc Mammologists; AAAS; Am Soc Neurochem; Soc Protection Old Fishes; Asn Chemoreception Sci. *Res:* Elephant pheromone study; comparative study of temporal gland secretions in Asian and African elephants; chemical characterization of the sex pheromones in Asian elephants. *Mailing Add:* Ore Grad Ctr 19600 Von Neumann Dr Beaverton OR 97006

RASMUSSEN, LOWELL W, b Redmond, Utah, Mar 21, 10; m 38; c 2. AGRONOMY. *Educ:* Utah State Univ, BS, 40, MS, 41; Iowa State Univ, PhD(plant physiol), 47. *Prof Exp:* Field supvr, Exp Sta, Utah State Col, 37-41; asst county supvr, Farm Security Admin, 41; jr agronomist, Soil Conserv Serv, USDA, 41-42; county agr agent, Exten Serv, Univ Utah, 42-45; from asst prof & asst agronomist to prof & agronomist, 47-56, from asst dir res to assoc dir res, 56-75, EMER ASSOC DIR RES, COL AGR, WASH STATE UNIV, 75- *Concurrent Pos:* Ford Found adv, Ministry Agr, Saudi Arabia, 65-67. *Mem:* AAAS; Weed Sci Soc Am; Am Soc Plant Physiol. *Res:* Weed control; action of growth regulator herbicides; techniques in weed research; research administration. *Mailing Add:* SE 910 Glen Echo Rd Pullman WA 99163

RASMUSSEN, MAURICE L, b Coon Rapids, Iowa, June 2, 35; m 61; c 2. AERONAUTICS, ASTRONAUTICS. *Educ:* Ore State Univ, BS, 57, MS, 59; Stanford Univ, PhD(aeronaut, astronaut), 64. *Prof Exp:* Res engr, Ames Res Ctr, NASA, 58-59; res asst aeronaut & astronaut, Stanford Univ, 59-64, lectr & res assoc, gas dynamics, 64-65, actg asst prof, 64-67; assoc prof, 67-70, PROF GAS DYNAMICS, UNIV OKLA, 70- *Concurrent Pos:* Vis res scientist, Air Force Armaments Lab, 80-81; David Ross Boyd prof, 88- *Honors & Awards:* Halliburton Distinguished Lectr Award. *Mem:* Am Inst Aeronaut & Astronaut; Am Phys Soc; Am Soc Mech Engr. *Res:* Gas dynamics; rarefied plasma dynamics; nonlinear oscillations; hypersonic aerodynamics. *Mailing Add:* Sch Aerospace & Mech Eng 208A Felgar Hall Univ Okla Norman OK 73019

RASMUSSEN, NORMAN CARL, b Harrisburg, Pa, Nov 12, 27; m 54; c 2. NUCLEAR ENGINEERING. *Educ:* Gettysburg Col, BA, 50; Mass Inst Technol, PhD(physics), 56. *Hon Degrees:* Dr, Gettysburg Col, 79, Cath Univ, Leuven, Belg, 80. *Prof Exp:* From instr to assoc prof, 56-65, dept head, 75-81, PROF NUCLEAR ENG, MASS INST TECHNOL, 65-, MCAFEE PROF ENG, 83- *Concurrent Pos:* Consult, Am Nuclear Insurers, 58-, Nuclear Regulatory Comn, 72-82, Cabot, Corp, 73-, EG&G Idaho, Inc, 76-84 & Nuclear Utility Serv Corp, 77-84; mem, Defense Sci Bd, 74-77, Presidential Adv Group on Contributions of Technol to Econ Strength, 75 & bd trustees, Northeast Utilities, 77-; mem, Nat Sci Bd, 82-88. *Honors & Awards:* Theos & Thompson Award, Am Nuclear Soc, 81-; Enrico Fermi Award, US Dept Energy, 85. *Mem:* Nat Acad Sci; Nat Acad Eng; AAAS; fel Am Nuclear Soc; Health Physics Soc; fel Am Acad Arts & Sci; Soc Risk Anal. *Res:* Early research in activation analysis; low-level counting techniques and gamma-ray spectroscopy; nuclear safety and environmental impact of nuclear power; reliability analysis and risk assessment. *Mailing Add:* Mass Inst Technol 77 Massachusetts Ave Rm 24-205 Cambridge MA 02139

RASMUSSEN, PAUL G, b Chicago, Ill, Jan 27, 39; m 60; c 3. INORGANIC POLYMER CHEMISTRY. *Educ:* St Olaf Col, BA, 60; Mich State Univ, PhD, 64. *Prof Exp:* From asst prof to assoc prof chem, Univ Mich, Ann Arbor, 64-75, assoc dean res & facil, prof, 75-82. *Mem:* Am Chem Soc. *Res:* Electron delocalization in transition metal complexes; polymers based on cyanoimidazoles. *Mailing Add:* Dept Chem Univ Mich Ann Arbor MI 48109

RASMUSSEN, REINHOLD ALBERT, b Brockton, Mass, Nov 4, 36; m 61. PLANT PHYSIOLOGY, CHEMICAL ENGINEERING. *Educ:* Univ Mass, BS, 58, MEd, 60; Wash Univ, PhD(bot), 64. *Prof Exp:* Plant biochemist, Agr Div, Monsanto Co, 62-63; clin biochemist, Walter Reed Army Med Ctr, 64-67; res proj chemist & head plant sci sect, Biomed Res Lab, Dow-Corning Corp, 67-69; assoc plant physiologist, Air Pollution Res Sect, Col Eng, 69-74, prof, 74-75, sect head, Ar Resources Sect, Chem Eng Br, Wash State Univ, 75-77; PROF ENVIRON TECHNOL, ORE GRAD CTR, 77- *Concurrent Pos:* Consult, US Army Tropic Test Ctr, 64-67 & Nat Ctr Atmospheric Res, 67-; lectr, Saginaw Valley Col, 67-; consult, Vapor Phase Org Air Pollutants from Hydrocarbons, Nat Acad Sci-Nat Res Coun, 72, consult, Panel Ammonia; consult, Air Pollution Physics & Chem Adv Comt, Environ Protection Agency, 72-75; mem, Intersoc Comt D-5 Hydrocarbon Anal; affil prof, Inst Environ Studies, Univ Wash, 75- *Mem:* Am Chem Soc; Am Meteorol Soc. *Res:* Role of naturally occurring organic volatiles in the atmosphere, especially their chemical identification, photochemistry and biological interactions with the environment and man. *Mailing Add:* Dept Environ Sci Ore Grad Ctr 19600 NW Walker Rd Beaverton OR 97006

RASMUSSEN, ROBERT A, b St Peter, Minn, Aug 31, 33; m 61; c 4. PHYCOLOGY. *Educ:* Col St Thomas, BS, 56; Univ Minn, Minneapolis, MS, 62; Univ Canterbury, PhD(zool), 65; Humboldt State Univ, MATU, 85. *Prof Exp:* Res assoc, Friday Harbor Marine Labs, Univ Wash, 65-66; from asst prof to assoc prof bot, 66-77, PROF BOT, HUMBOLDT STATE UNIV, 77- *Concurrent Pos:* Mem Environ Consults, Pac Marine Eng, Inc, 72-74; vpres, Environ Res Consults, 73-78. *Mem:* Int Phycol Soc; Phycol Soc Am; Brit Phycol Soc; Asn Teachers Tech Writing. *Res:* Seaweed autoecology. *Mailing Add:* Dept Biol Humboldt State Univ Arcata CA 95521

RASMUSSEN, RUSSELL LEE, b Allen, Nebr; m 65; c 2. INORGANIC CHEMISTRY, BIO-ORGANIC CHEMISTRY. *Educ:* Univ Nebr, BS, 60, PhD(inorg biochem), 70. *Prof Exp:* Asst prof chem, Stout State Univ, 66-67; from asst prof to assoc prof, 69-74, head Dept Phys Sci, 72-74, PROF CHEM, WAYNE STATE COL, 74- *Concurrent Pos:* Vis prof, Univ Nebr, Lincoln, 81; vis prof, Utah State Univ, 85-86, 88. *Mem:* Am Chem Soc; Int Union Pure Appl Chem. *Res:* Model enzyme systems; coordination compounds; thiol esters of amino acids; pyridoxal containing enzymes; relationship of verbal ability and scientific achievement; the two cultures. *Mailing Add:* Div Math Sci Wayne State Col Wayne NE 68787-1486

RASMUSSEN, THEODORE BROWN, b Provo, Utah, Apr 28, 10; m 47; c 4. NEUROLOGY, NEUROSURGERY. *Educ:* Univ Minn, BS & MB, 34, MD, 35, MS, 39; FRCS(C). *Hon Degrees:* DrMed, Edinburgh Univ, 80, Umea Univ, Sweden, 88. *Prof Exp:* Lectr neurol & neurosurg, McGill Univ, 46-47; prof neurol surg, Univ Chicago, 47-54; neurosurgeon, Montreal Neurol Inst, 54-80; prof, 54-80, EMER PROF NEUROL & NEUROSURG, MCGILL UNIV, 80- *Concurrent Pos:* Assoc neurosurgeon, Montreal Neurol Inst, 46-47, dir, 60-72, sr neurosurg consult, 72-; mem, Int Brain Res Orgn. *Honors & Awards:* Ambassador Award, Epilepsy Int, 79; Penfield Award, Can League Against Epilepsy, 82; Lennox Award, Am Epilepsy Soc, 86; Distinguished Serv Award, Soc Neurol Surg, 89. *Mem:* AAAS; Neurosurg Soc Am; Am Asn Neurol Surg; Soc Neurol Surg; AMA; Am Neurol Asn. *Res:* Cerebral circulation; localization of cortical function; effects of radiation on cerebral tissue; focal epilepsy; surgical treatment of epilepsy. *Mailing Add:* 3801 University St Montreal PQ H3A 2B4 Can

RASMUSSEN, V PHILIP, JR, b Logan, Utah, Apr 3, 50; m 74; c 5. SOILS & SOIL SCIENCE. *Educ:* Utah State Univ, BS, 74, MS, 76; Kans State Univ, PhD(agron), 79. *Prof Exp:* Res assoc soil physics, Utah State Univ, 74-76, crop modeling, Kans State Univ, 76-78; asst prof, Ricks Col, 78-81; ASSOC PROF SOIL PHYSICS, UTAH STATE UNIV, 81- *Concurrent Pos:* Consult, Campbell Sci, Inc, 76-81, Omnidata Int, 80-; lectr, Prof Farmers Inst, Data Processing Mgt Asn & Farmers Home Admin, USDA. *Mem:* Sigma Xi; Am Soc Agron; Soil Sci Soc Am; Int Soc Soil Sci. *Res:* On farm microcomputers; plant growth modelling. *Mailing Add:* UMC 48 Utah State Univ Logan UT 84322-4820

RASMUSSEN, WILLIAM OTTO, b Burley, Idaho, Jan 29, 42; m 64; c 2. MODELING, COMPUTER SIMULATION. *Educ:* Univ Idaho, BS, 64, MS, 66; Univ Ariz, PhD(watershed mgt), 73. *Prof Exp:* Explor geophysicist, Heinrichs Geoexplor Co, 68-70; dir Lab Remote Sensing & Comput Mapping, Univ Ariz, 73-82; sr scientist, Bell Tech Oper, Tucson, 81-87; asst res prof comput mapping & remote sensing, 73-82, dir Western Computor Consortium, 82-87, ASSOC PROF BIOSYSTS ENG, UNIV ARIZ, 87- *Concurrent Pos:* Dir remote sensing & comput mapping, Univ Ariz, 73-, remote sensing exten specialist, 75- *Mem:* Am Water Resources Asn; AAAS; Sigma Xi; Am Asn Univ Prof. *Res:* Simulation of electromagnetic radiation-environment interaction; scene synthesis; simulation of natural resource systems; water resources quality and quantity; atmospheric optics; ecosystem modeling; simulation and prediction of strip mine resources, mass balance, and reclamation; development of geographic data analysis and display systems; visibility analysis in vegetated scenes; computer software design and development; electronic warfare simulation. *Mailing Add:* Dept Bio Syst Eng Univ Ariz 507 Shantz Tucson AZ 85721

RASMUSSON, DONALD C, b Ephraim, Utah, May 28, 31; m 50; c 5. PLANT GENETICS, PLANT BREEDING. *Educ:* Utah State Univ, BS, 53, MS, 56; Univ Calif, PhD(genetics), 58. *Prof Exp:* Res assoc, 58-72, PROF PLANT GENETICS, UNIV MINN, ST PAUL, 72- *Mem:* Fel Am Soc Agron; Crop Sci Soc Am. *Res:* Plant breeding and genetics. *Mailing Add:* Agron & Plant Gen Univ Minn 410 Borlaug St Paul MN 55108

RASMUSSON, DOUGLAS DEAN, b Denver, Colo, Aug 4, 46; Can citizen. NEUROPHYSIOLOGY, SOMATOSENSORY SYSTEM. *Educ:* Colo Col, BA, 68; Dalhousie Univ, MA, 70, PhD(physiol), 75. *Prof Exp:* Fel neurophysiol, Univ Toronto, 75-77; ASST PROF PHYSIOL & BIOPHYS, DALHOUSIE UNIV, 77- *Mem:* Can Physiol Soc; Soc Neurosci; Can Asn Neuroscientists. *Res:* Organization and function of somatosensory cortex; plasticity due to changes in peripheral nervous system. *Mailing Add:* Dept Physiol & Biophysics Dalhousie Univ Halifax NS B3H 4H6 Can

RASMUSSON, EUGENE MARTIN, b Lindsborg, Kans, Feb 27, 29; m 60; c 4. METEOROLOGY. *Educ:* Kans State Univ, BS, 50; St Louis Univ, MS, 63; Mass Inst Technol, PhD(meteorol), 66. *Prof Exp:* Design engr, Kans State Hwy Comn, 50-51; meteorologist, USAF, 51-55; engr, Pac Tel Co, 55-56; meteorologist & hydrologist, US Weather Bur, 56-60, meteorologist forecaster, 60-63; res meteorologist, Nat Oceanic & Atmospheric Admin, Rockville, Md, 63-86; SR RES SCIENTIST, UNIV MD, 86- *Concurrent Pos:* Mem steering comt, Int Field Year for Great Lakes, 67-71, US co-chmn lake meteorol panel, 69-79; mem, Trop Ocean & Global Atmosphere panel, Nat Acad Sci, 83-85, Sci Steering Group, World Climate Res Prog, 85-88, Nat Res Coun Comt on US Geol Surv Water Res, 88-, Nat Res Coun Expert Task Group, Strategic Hwy Res Prog, 90- *Honors & Awards:* Silver Medal, Dept Com; Jule Charney Award, Am Meteorol Soc. *Mem:* AAAS; Am Geophys Union; fel Am Meteorol Soc. *Res:* Atmospheric general circulation; large scale water balance; air-sea interactions; tropical meteorology; climate variability. *Mailing Add:* 10005 Autumnwood Way Potamac MD 20854

RASMUSSON, GARY HENRY, b Clark, SDak, Aug 2, 36; m 58; c 4. MEDICINAL CHEMISTRY. *Educ:* St Olaf Col, BA, 58; Mass Inst Technol, PhD(org chem), 62. *Prof Exp:* NSF res fel, Stanford Univ, 62-63; NIH res fel, 63-64; sr chemist, 64-72, res fel, 72-77, sr res fel, 77-82, SR INVESTR, MERCK SHARP & DOHME RES LABS, RAHWAY, 82- *Mem:* Am Chem Soc. *Res:* Synthetic organic and medicinal chemistry in the areas of heterocycles, steroids and pharmaceutically active compounds. *Mailing Add:* Merck Sharpe & Dohme Res Labs Rahway NJ 07065

RASNAKE, MONROE, b Buchanan County, Va, Feb 8, 42; m 68; c 3. SOIL CHEMISTRY, SOIL FERTILITY. *Educ:* Berea Col, BS, 65; Va Polytech Inst, MS, 67; Univ Ky, PhD(soil chem), 73. *Prof Exp:* High sch teacher math sci & biol, Grundy, Va, 67-70; grad asst res lab instr, Univ Ky, 70-73; asst prof, Va Polytech Inst & State Univ, 73-77; supvry mgt agronomist, Dept Army, Blackstone, Va, 77-78; EXTEN AGRON SPECIALIST, UNIV KY, 78- *Honors & Awards:* Merit Cert Award, Am Forage & Grasslands Coun, 88. *Mem:* Coun Agr Sci & Technol; Am Soc Agron; Sigma Xi; Am Forage & Grasslands Coun. *Res:* Fire-cured tobacco information; effect of soil fertility level on ozone injury of tobacco; effect of agricultural drainage on water quality; effect of tobacco nutrition on occurrence of insect pests; soil erosion as related to soybean production practices; no-till corn fertilization; effect of soil pH and lime on crop rotations. *Mailing Add:* Univ Ky PO Box 469 Princeton KY 42445

RASOR, NED S(HAURER), b Dayton, Ohio, Jan 2, 27; m 47; c 2. ENERGY CONVERSION, CARDIOVASCULAR DEVICES. *Educ:* Ohio State Univ, BS, 48; Univ Ill, MS, 51; Case Inst Technol, PhD(physics), 54. *Prof Exp:* Res engr, NAm Aviation, Inc, 48-50, sr res engr, 51-52, sr res specialist, Atomics Int Div, 54-56, proj engr, 56-59, group leader, 59-60, dir energy conversion dept, 60-62; dir res, Thermo Electron Corp, Mass, 62-63, vpres, 63-65; consult, 65-72; PRES, RASOR ASSOCS, INC, 71- *Mem:* Am Phys Soc; assoc fel Am Inst Aeronaut & Astronaut; sr mem Inst Elec & Electronics Engrs. *Res:* Surface physics; gaseous discharge; thermionic energy conversion; thermal properties; high temperature materials; nuclear space power; bioengineering. *Mailing Add:* 253 Humboldt Ct Sunnyvale CA 94089-1382

RASSIN, DAVID KEITH, b Liverpool, Eng, Dec 1, 42; m 65; c 2. NEUROCHEMISTRY, NUTRITION. *Educ:* Columbia Univ, AB, 65; City Univ New York, PhD(pharmacol), 74. *Prof Exp:* Res fel neurol, Columbia Univ, 66, res asst, 66-67; asst res scientist, NY State Inst Basic Res Ment Retardation, 67-70, res scientist, 70-74, sr res scientist human develop & nutrit, 74-77, res scientist IV, 77-79, res scientist V, 79-80; assoc prof, 80-85, PROF, UNIV TEX MED BR, GALVESTON, 85- *Concurrent Pos:* Adj asst prof Mt Sinai Sch Med, 76-80 & adj assoc prof, Col Staten Island, City Univ New York, 79-80. *Mem:* Int Soc Neurochem; Am Soc Neurochem; Am Soc Clin Nutrit; Soc Pediat Res; Am Soc Exp Pharmacol & Therapeut. *Res:* Neurochemistry of amino acids, especially sulfur containing amino acids, as they relate to inborn errors of metabolism and nutrition in children. *Mailing Add:* Dept Pediat Univ Tex Med Br Galveston TX 77550

RASSWEILER, MERRILL (PAUL), physics; deceased, see previous edition for last biography

RAST, HOWARD EUGENE, JR, b Mexia, Tex, June 8, 34; m 58; c 2. SOLID STATE SCIENCE, FIBER OPTICS. *Educ:* Univ Tex, BA, 56; Univ Ore, PhD(phys chem), 64, Univ Southern Calif, MS, 77. *Prof Exp:* Res chemist, Calif Ink Co, Div Tenneco Inc, 58-60; res asst chem, Univ Ore, 60-64; phys chemist, US Naval Weapons Ctr, 64-70; res physicist, 70-80, supvry scientist, 80-86, MGR, SOLID STATE ELECTRONICS DIV, NAVAL OCEAN SYSTS CTR, 86- *Mem:* Am Phys Soc; Optical Soc Am. *Res:* Spectra and optical properties of materials; lattice dynamics; military optical counter-measures; electro-optics and optical technology; fiber optics. *Mailing Add:* Code 55 Naval Ocean Systs Ctr San Diego CA 92152-5000

RAST, NICHOLAS, b Teheran, Iran, June 20, 27; c 5. STRUCTURAL GEOLOGY. *Educ:* Univ Col, London, BSc, 52; Univ Glasgow, PhD(geol), 56. *Prof Exp:* Lectr geol, Univ Wales, 55-59; lectr, 59-61, sr lectr, 61-65, reader, Univ Liverpool, 65-71; chmn dept, Univ NB, Can, 71-79; HUDNALL PROF GEOL, UNIV KY, LEXINGTON, 79-, CHMN DEPT, 81- *Concurrent Pos:* Prof geol, Univ Mex, 70; asst lectr, Univ Wales, 54-55, Univ Glasgow, 52-54. *Mem:* AAAS; Geol Soc Am; Geol Asn Can; Hist Earth Sci Soc; Nat Asn Geol Teachers; Sigma Xi. *Res:* Structural and tectonic geology of the Scottish Highlands, England and Wales and the Appalachian Chain of North America - both Canada and the USA; volcanic rocks of Wales. *Mailing Add:* Dept Geol Sci Univ Ky Bowman Hall Lexington KY 40546

RAST, WALTER, JR, b San Antonio, Tex, Jan 14, 44; m 71; c 1. AQUATIC CHEMISTRY, LIMNOLOGY. *Educ:* Univ Tex, Austin, BA, 70; Univ Tex, Dallas, MS(molecular biol), 74, MS(environ sci), 76, PhD(environ sci), 78. *Prof Exp:* Res asst, Genetics Res Found, Univ Tex, Austin, 69-70; hydrol field asst stream flow and quality monitoring, Water Resources Div, US Geol Surv, 70-71; res asst, Molecular Biol Inst, Univ Tex, Dallas, 71-74, res asst aquatic chem, Environ Sci Prog, 74-75, teaching asst, Environ Sci Prog, 75-77; limnologist, Great Lakes Water Qual, Great Lakes Regional Off, US/Can Int Joint Comn, Windsor, Ont, 77-79, environ advr, US Sect, Int Joint Comn, Washington, DC, 79-82; res hydrologist, Sacramento, Calif, 82-85, RES HYDROLOGIST, WATER RESOURCES DIV, US GEOL SURV, US DEPT INTERIOR, AUSTIN, TEX, 85- *Concurrent Pos:* Assoc, Enviroqual Consult & Labs, Inc, 74-77; adj asst prof, Biol Dept, Wayne State Univ, 79-82. *Mem:* AAAS; Am Soc Limnol & Oceanog; Am Water Works Asn; Soc Environ Toxicol Chem; Water Pollution Control Fedn. *Res:* Nutrient load-lake response relationships and trophic status indices in natural waters; environmental chemistry of toxic and hazardous substances; environmental modelling of effects of eutrophication and toxic and hazardous substances on water quality and biota; watershed land use activities--lake water quality relationships. *Mailing Add:* US Dept Interior US Geol Surv Water Resources Div 8011 Cameron Rd Austin TX 78753

RASTALL, PETER, b Washingborough, Eng, Nov 18, 31. THEORETICAL PHYSICS. *Educ:* Univ Manchester, BSc, 52, PhD(theoret physics), 55. *Prof Exp:* With aerodyn dept, Rolls Royce, Ltd, 55-57; lectr physics, 57-58, from instr to assoc prof, 58-71, PROF PHYSICS, UNIV BC, 71- *Res:* Field theory; quantum mechanics. *Mailing Add:* Dept Physics Univ BC Vancouver BC V6T 1W5 Can

RASTANI, KASRA, b Tehran, Iran, May 16, 59. INTEGRATED & FIBER OPTICS, MICROOPTICS. *Educ:* Univ Southern Calif, BSEE, 81, PhD(integrated optics), 88; Stanford Univ, MSEE, 82. *Prof Exp:* Postdoctoral fel, Univ Southern Calif, 88-89; MEM TECH STAFF, BELL COMMUN RES, 89- *Concurrent Pos:* Engr, Wave Theory, Stanford Univ, 84. *Mem:* Optical Soc Am; Am Inst Physics. *Res:* Integrated and fiber optics, microoptics and diffraction gratings and their use to interconnects. *Mailing Add:* Bellcore NVC 3X118 331 Newman Springs Rd Red Bank NJ 07701

RASTETTER, WILLIAM HARRY, organic chemistry, for more information see previous edition

RASTOGI, PRABHAT KUMAR, b Chapra, India, June 7, 44; m 69; c 2. MATERIAL SCIENCE, METALLURGY. *Educ:* Indian Inst Technol, Kanpur, India, BTech, 65; State Univ NY, Stony Brook, MS, 67; Calif Inst Technol, PhD(mat sci), 70. *Prof Exp:* Res engr, Inland Steel Co, Ind, 70-75, sr res engr, 75-79, staff res engr magnetism & phys metall, 79-; ASSOC PROF PHYSICS, CASE WESTERN RESERVE UNIV. *Mem:* Am Soc Metals. *Res:* Metal physics; magnetism; physical metallurgy; development of new and improved products for electromagnetic applications. *Mailing Add:* Dept Elec Eng & Appl Physics Case Western Reserve Univ Glennan Bldg Rm 715 Cleveland OH 44106

RASTOGI, SURESH CHANDRA, b Sambhal, India, July 7, 37; US citizen; m 66; c 4. MATHEMATICAL STATISTICS, BIOSTATISTICS. *Educ:* Univ Lucknow, BSc, 57, MSc, 60; Univ Iowa, PhD(statist), 65. *Prof Exp:* Lectr statist, Univ Lucknow, 60-62; statistician, Univ Md, Baltimore City, 66-67; asst prof statist, Univ Md, College Park, 67-72; mgr statist, Hq, US Postal Serv, 72-73; MATH STATISTICIAN, BUR BIOLOGICS, FOOD & DRUG ADMIN, DEPT HEW, 73- *Concurrent Pos:* Consult div biologics stand, NIH, 67-70. *Mem:* Am Statist Asn; Biomet Soc; Int Asn Survey Statist. *Res:* Multivariate statistical methods; regression and analysis of variance; sample survey techniques as used in applied fields; bioassay. *Mailing Add:* Ctr Biologics Eval & Res Off Dir Bldg 29 Rm 100 Bethesda MD 20892

RASWEILER, JOHN JACOB, IV, b Newport, RI, June 4, 43; m 64; c 1. REPRODUCTIVE PHYSIOLOGY, ANATOMY. *Educ:* Colgate Univ, BA, 65; Cornell Univ, PhD(physiol), 70. *Prof Exp:* Vis asst prof morphol, Div Health, Univ Valle, Colombia, 70-72; asst prof anat, Col Physicians & Surgeons, Columbia Univ, 72-78; asst prof anat, 78-84, asst prof, 84-86 Dir, Invitro Fertiliz Labs, 87-88, ASSOC PROF REPRODUCTIVE BIOL, MED COL, CORNELL UNIV, 86- *Concurrent Pos:* Pop Coun fel, Univ Valle, Colombia, 70-72; NIH grant, Int Ctr Med Res & Training, Tulane Univ, 71-72, vis scientist, 72-74. *Mem:* Soc Study Reproduction; Am Asn Anatomists; Am Soc Mammal. *Res:* Reproductive physiology; development of bats as laboratory models. *Mailing Add:* Dept Obstet & Gynec Med Col Cornell Univ New York NY 10021

RATAJCZAK, HELEN VOSSKUHLER, b Tucson, Ariz, April 9, 38; div; c 4. CELL MEDIATED IMMUNITY, IMMUNOTOXICOLOGY. *Educ:* Univ Ariz, BS, 59, MS, 70, PhD(molecular biol), 76. *Prof Exp:* Asst res, SW Clin & Res Inst, Univ Ariz, 67-72, NIH trainee, Health Sci Ctr, 73-74, Am Thoracic Soc fel, 74-76; asst res scientist, Col Med, Univ Iowa, 76-78; instr, Eye & Ear Hosp, Univ Pittsburg, 78-80, res assoc, Med Sch, 80-81; asst prof, Med Sch, Loyola Univ, 81-83; res immunol, 83-86, STAFF MEM IMMUNOL, IIT RES INST, 86- *Concurrent Pos:* Lectr, Occup Safety & Health Admin, 84- *Mem:* Sigma Xi; Am Thoracic Soc; Am Asn Immunologists; NY Acad Sci; AAAS; Soc Toxicol. *Res:* Definition of alteration of immune competence by the interactions of foreign substances with the host or a modification of self. *Mailing Add:* Inst Technol Res Ten W 35th St Chicago IL 60616

RATCHES, JAMES ARTHUR, b Hartford, Conn, May 12, 42; m 73; c 3. ELECTRO-OPTICS, SYSTEMS ANALYSIS. *Educ:* Trinity Col, Conn, BA, 64; Worcester Polytechnic Inst, MS, 66, PhD(physics), 69. *Prof Exp:* Res physicist, Ctr for Night Vision & Electro-Optics, 69-81, dir, Visionics Div, 81-87, dir, Spec Proj Off, 87-88, ASSOC DIR SCI & TECHNOL, CTR FOR NIGHT VISION & ELECTRO-OPTICS, 88- *Res:* Development of target acquisition models for thermal imaging devices; analysis and design evaluation of electro-optical sensors; atmospheric propagation effects on electro-optical devices. *Mailing Add:* Ctr Night Vision & Electro-Optics Ft Belvoir VA 22060-5677

RATCHFORD, JOSEPH THOMAS, b Kinstree, SC, Sept 30, 35; m 60; c 4. SOLID STATE PHYSICS. *Educ:* Davidson Col, BS, 57; Univ Va, MA, 59, PhD(physics), 61. *Prof Exp:* Staff mem, Sandia Corp, 59; asst prof physics, Washington & Lee Univ, 61-64; proj scientist, Solid State Sci Div, Air Force Off Sci Res, Arlington, 64-70; sci consult, Comt Sci & Technol, US House Rep, 70-77; ASSOC EXEC OFFICER, AAAS, WASHINGTON, DC, 77- *Concurrent Pos:* Res physicist, US Naval Ord Lab, 63-64; Am Polit Sci Asn cong fel, 68-69; res scholar, Int Inst Appl Systs Anal, Laxenburg, Austria, 76; chmn, Res Coord Panel, Gas Res Inst, Chicago, Ill, 76-79, mem, 76-; chmn, Adv Panels to US Congressional Off Technol Assessment, Solar Energy, Energy Conserv & Energy Biol Processes, 78-80; chmn, Adv Comt Int Progs, NSF, 84-87; bd mem, Int Develop Conf, 77- *Mem:* Fel AAAS; Am Phys Soc; Sigma Xi. *Res:* Science and government; materials sciences; energy technology and policy; international science cooperation and policy. *Mailing Add:* Assoc Dir Policy Int Affairs Off Sci Technol Exec Office of the Pres Washington DC 20506

RATCHFORD, ROBERT JAMES, b Firesteel, SDak, Nov 16, 24. PHYSICAL CHEMISTRY. *Educ:* Spring Hill Col, BS, 53; Cath Univ, PhD(chem), 58. *Prof Exp:* Res asst, Max Planck Inst Phys Chem & Karlsruhe Tech Univ, 62-63; asst prof chem, Loyola Univ, La, 64-80, dean arts & sci, 75-79; RETIRED. *Mem:* Am Chem Soc; Electrochem Soc. *Res:* Solution electrochemistry; electrochemistry of solid state electrolytes. *Mailing Add:* 6363 St Charles Ave New Orleans LA 70118-6195

RATCLIFF, BLAIR NORMAN, b Grinnell, Iowa, Sept 26, 44; m 76. ELEMENTARY PARTICLE PHYSICS. *Educ:* Grinnell Col, BA, 66; Stanford Univ, MS, 68, PhD(physics), 71. *Prof Exp:* Res assoc, Rutherford High Energy Lab, 71-74; res assoc, 75-79, STAFF PHYSICIST, STANFORD LINEAR ACCELERATOR CTR, 74- *Concurrent Pos:* Vis scientist, European Orgn Nuclear Res, 71-74. *Mem:* Am Phys Soc; AAAS. *Res:* Experimental high energy physics; meson spectroscopy; strange baryon spectroscopy; hadrom production mechanisms; lepton pair production. *Mailing Add:* Stanford Linear Accelerator Ctr Box 4349 Stanford CA 94305

RATCLIFF, KEITH FREDERICK, b Drexel Hill, Pa, Nov 15, 38; m; c 3. ASTROPHYSICS, THEORETICAL PHYSICS. *Educ:* Northwestern Univ, Evanston, BA, 60; Univ Pittsburgh, PhD(physics), 65. *Prof Exp:* Res assoc physics, Univ Rochester, 65-67 & Mass Inst Technol, 67-68, instr, 68-69; from asst prof to assoc prof, 69-90, PROF PHYSICS, STATE UNIV NY, ALBANY, 90- *Concurrent Pos:* Vis assoc prof, State Univ NY, Stony Brook, 72-73; scientist, Space Astron Lab, 75-89. *Mem:* AAAS; Am Phys Soc; Am Asn Physics Teachers. *Res:* Theoretical investigations of nuclear structure and nuclear reactions; neutron star evolution; cosmic dust problems; many-body theory; astronautics. *Mailing Add:* Dept Physics State Univ NY Albany NY 12222

RATCLIFF, MILTON, JR, b Memphis, Tenn, Apr 19, 44; m 73. ANALYTICAL CHEMISTRY, ORGANIC CHEMISTRY. *Educ:* Southwestern at Memphis, BS, 66; Case Western Reserve Univ, PhD(org chem), 70. *Prof Exp:* Fel geochem, Indiana Univ, 70-72; proj leader chem, Jet Propulsion Lab, Calif Inst Technol, 72-74; sr chemist geochem, US Geol Surv, 74-76; mgr anal res & residue methods anal chem, Zoecon Corp-Subsid Hooker Chem Co, 76-80; PROJ LEADER COMPUT SCI, NELSON ANAL

INC, 80- *Mem:* Am Chem Soc; Asn Comput Mach. *Res:* Gas chromatography; mass spectrometry; trace analysis; chemical applications of data processing; instrument interfacing; capillary column development. *Mailing Add:* PE Nelson Syst 10040 Bubb Rd Cupertino CA 95014

RATCLIFFE, CHARLES THOMAS, b Malad, Idaho, Nov 18, 38; m 64; c 2. INORGANIC CHEMISTRY. *Educ:* Univ Idaho, BS, 61, MS, 63, PhD(inorg chem), 67. *Prof Exp:* Sci Res Coun fel chem, Glasgow Univ, 67-68; res chemist, Corp Res Ctr, Allied Chem Corp, 68-72, sr res chemist, 72-76, res group leader, 76-80; with Exxon Res Eng Co, 80-; SUPVR, PROC RES, UNION OIL OF CALIF. *Concurrent Pos:* Sci Res Coun fel, Great Britain, 67-68. *Mem:* Sigma Xi; Am Chem Soc; fel NY Acad Sci. *Res:* Catalysis reactions with coal, sulfur; coal chemistry; heterogeneous sulfur containing catalysts; reduction reactions. *Mailing Add:* 271 Avenida Santa Barbara La Habre CA 90631

RATCLIFFE, NICHOLAS MORLEY, b Bryn Mawr, Pa, Feb 4, 38; m 60; c 1. GEOLOGY. *Educ:* Williams Col, BA, 60; Pa State Univ, PhD(geol), 65. *Prof Exp:* Assoc prof geol, City Col New York, 65-74, prof earth & planetary sci, 74-80; STAFF MEM, US GEOL SURV, 80- *Mem:* Geol Soc Am. *Res:* Structural geology and petrology; igneous and metamorphic petrology. *Mailing Add:* US Geol Surv 925 National Ctr Reston VA 22092

RATEAVER, BARGYLA, b Ft Dauphin, Madagascar, Aug 3, 16; US citizen; c 1. CONSERVATION, SOIL FERTILITY. *Educ:* Univ Calif, Berkeley, AB, 43, MSLS, 59; Univ Mich, Ann Arbor, MS, 50, PhD(bot), 51, Eminence Credential, Einstein Clause, Calif educ code, 69. *Prof Exp:* Specialist lima bean maturity standards, Univ Calif, Davis, 51-52, technician photo period determinations floral crops, hort dept, Los Angeles, 53-54; pvt plant propagator, 55-56; organizer, biol libr, Kaiser Labs, Univ Calif, Berkeley, 58-59; tech abstractor, G C Rocket Co, 59-60; organizer & dir, histol lab, Univ Alta, 59-64; librn, D Victor Co, 61 & Marin County Sch Syst, Calif, 67-68; regional rep sci & technol books, J S Stacey, 63; automation analyst, Space & Info Systs, NAm Aviation, 64; Instr, Manpower Develop Training Prog, 65-67; instr org gardening & farming, Jr cols, Univ Calif & Calif State Univs, 70-78; ed & publ Conserv Gardening & Farming Series, The Rateavers, 73-80; CONSULT, 80- *Concurrent Pos:* Longwood Gardens grant, 55; lit res, Libr Sch, Univ Calif, Berkeley, 58-59; lectr org gardening, 65-; assoc prof, Calif State Univ, Sacramento, 72-73, instr, Long Beach, 75; organizer, First Int Conf Org Method Farm & Garden, San Francisco, 73; mem, Coord Comt, Int Fedn Org Agr Movements & Working Groups Info Educ, 78-82. *Honors & Awards:* Am Acad Arts & Sci Award, 55. *Mem:* Int Fedn Org Agr Movements. *Res:* Development and introduction of first US course in organic gardening and farming; madagascar plants; tropical agriculture; world literature on tropical plants; biological agriculture and horticulture; science education in conservation; soil fertility of field crops, tree and vegetable crops; utilization of weeds; fertilizer comparisons; author of publications and books on organic method of gardening and farming. *Mailing Add:* Int Consult 9049 Covina St San Diego CA 92126

RATH, BHAKTA BHUSAN, b Banki, India, Oct 28, 34; m 63; c 1. METALLURGY. *Educ:* Utkal, India, BSc, 55; Mich Technol Univ, MS, 58; Ill Inst Technol, PhD(metall), 62. *Prof Exp:* Res assoc metall, Ill Inst Technol, 60-61; asst prof & res metallurgist, Wash State Univ, 61-65; res scientist, E C Bain Lab Fundamental Res, US Steel Corp, 65-71; mem res staff, McDonnell Douglas Res Labs, St Louis, 72-76; supt, Mat Sci Div, 77-82, HEAD, MAT SCI & COMP TECH DIRECTORATE & ASSOC DIR RES, US NAVAL RES LAB, 82- *Concurrent Pos:* Adj fac mem, Carnegie-Mellon Univ. *Mem:* AAAS; fel Am Soc Metals; Am Inst Mining, Metall & Petrol Engrs; Brit Inst Metals; Phys Soc Japan; Mat Res Soc India. *Res:* Recovery, recrystallization and grain growth in metals; teaching physics of metals; x-ray diffraction; theory of solids; crystallography; deformation and recrystallization textures; micro-calorimetry; solid state physics. *Mailing Add:* Mat Sci & Comp Tech Directorate US Naval Res Lab Code 6000 Washington DC 20375-5000

RATH, CHARLES E, b Philippines, Aug 14, 19; m 41; c 4. HEMATOLOGY. *Educ:* Col Wooster, AB, 40; Western Reserve Univ, MD, 43; Am Bd Internal Med, dipl, 51. *Prof Exp:* From instr to assoc prof med, 49-61, dir hemat & blood bank, 49-61, PROF MED, UNIV & DIR LABS, UNIV HOSP, GEORGETOWN UNIV, 61- *Concurrent Pos:* Consult, NIH Clin Ctr, 53- & Bethesda Naval Hosp, 54- *Mem:* Am Soc Hemat; AMA; Am Fedn Clin Res; fel Am Col Physicians; Int Soc Hemat. *Res:* Iron metabolism. *Mailing Add:* Georgetown Univ Hosp Washington DC 20007

RATH, NIGAM PRASAD, b Berhampur, India, Mar 24, 58; m 85; c 1. SINGLE CRYSTAL X-RAY CRYSTALLOGRAPHY, SYNTHETIC ORGANOMETALLIC CHEMISTRY. *Educ:* Berhampur Univ, India, BSc Hons, 77, MSc, 79; Okla State Univ, Stillwater, PhD(chem), 85. *Prof Exp:* Postdoctoral res assoc chem, Univ Notre Dame, 86-87, asst fac fel, 87-89; RES ASST PROF CHEM, UNIV MO, ST LOUIS, 89- *Mem:* Am Crystallog Asn; Am Phys Soc. *Res:* Single crystal x-ray diffraction studies; synthesis and characterization of organometallic compounds; use of crystallographic data bases and molecular modelling studies; structure-activity relationships. *Mailing Add:* Chem Dept Univ Mo 8001 Natural Bridge Rd St Louis MO 63121-4499

RATHBUN, EDWIN ROY, JR, b Kansas City, Mo, Apr 6, 22; m 49; c 5. PHYSICS, ELECTRONICS ENGINEERING. *Educ:* Iowa State Col, BS, 48, MS, 50. *Prof Exp:* Asst electronics engr physics instr, Argonne Nat Lab, 54-56; staff engr nuclear physics, Cook Elec Co, Skokie, Ill, 56-64; consult engr nuclear weapons effects, Gen Elec Co, Philadelphia, 65-68, specialist consult engr nucleonics electromagnetics, Syracuse, NY, 68-69; chief electromagnetic pulse branch, Naval Ord Lab, 69-74; proj engr nuclear vulnerability & hardening, Naval Surface Weapons Ctr, 74-84; RETIRED. *Mem:* Inst Elec & Electronics Engrs. *Res:* Nuclear electromagnetic pulse; nuclear vulnerability of naval systems; effects of nuclear weapons; nuclear hardening of military systems; nuclear radiation effects. *Mailing Add:* 212 Mowbray Rd Silver Springs MD 20984

RATHBUN, TED ALLAN, b Ellsworth, Kans, Apr 11, 42; m 64; c 1. PHYSICAL & FORENSIC ANTHROPOLOGY, PATHOLOGY. *Educ:* Univ Kans, BA, 64, MA, 66, PhD(anthrop), 71. *Prof Exp:* Instr eng, Ahwaz Agr Col, Peace Corps, 66-68; from instr to assoc prof, 70-84, chmn, 87-89, PROF ANTHROP, UNIV SC, 84- *Concurrent Pos:* Res assoc, Inst Archaeol, Univ SC, 71-; Comt Res & Prod Scholar grant, 71 & 78; consulting phys anthropologist, Off State Med Examr, Med Univ SC, 72-; Field Res Projs grant, Field Mus Natural Hist & Univ SC, 72-73; dipl, Am Bd Forensic Anthrop, 82- & mem bd Am Bd Anthrop; biomed grant, Afroamerican biohistory, Univ SC, 85; Dep State Archaeologist Forensics, NSF Grant, 87, Venture Fund, 90. *Mem:* Am Anthrop Asn; Am Asn Phys Anthrop; dipl Am Bd Forensic Anthropol; Am Acad Forensic Sci; Paleopath Asn. *Res:* Osteology; Bronze and Iron Ages in Southwest Asia; social structure and fertility; physical anthropology of groups in Southwest Asia; forensic anthropology; paleopathology; non-metric variation; colonial and 19th century South Carolina; Egypt-early dynastic; forensic video superimposition; bone chemistry; pathology. *Mailing Add:* Dept Anthrop Univ SC Columbia SC 29208

RATHBUN, WILLIAM B, b Wisconsin Dells, Wis, June 20, 32; m; c 2. LENS & CORNEA BIOCHEMISTRY. *Educ:* Univ Wis, BS, 54, MS, 55; Univ Minn, Minneapolis, PhD(biochem), 63. *Prof Exp:* Res assoc, 64-67, asst prof, 67-73, ASSOC PROF OPHTHAL BIOCHEM, UNIV MINN, MINNEAPOLIS, 73- *Concurrent Pos:* NIH fel ophthal biochem, Univ Minn, Minneapolis, 63-64, grants, 69-71, 72-86 & 87-92. *Mem:* Asn Res Vision & Ophthal; Sigma Xi; Int Soc Eye Res; Asn Eye Res; Int Soc Ocular Toxicol. *Res:* Lens enzymes; biochemistry of cataracts; biosynthesis and metabolism of glutathione; biochemistry of cornea. *Mailing Add:* Dept Ophthal Univ Minn Minneapolis MN 55455

RATHBURN, CARLISLE BAXTER, JR, b Fulton, NY, Apr 23, 24; m 51; c 2. PUBLIC HEALTH ENTOMOLOGY. *Educ:* Syracuse Univ, AB, 46; Univ Fla, BSA, 50, MAg, 51; Cornell Univ, PhD(entom), 64. *Prof Exp:* Res asst entom, Cornell Univ, 53-57; entomologist, Entom Res Ctr, Fla State Dept Health & Rehab Serv, 57-64, entomologist, W Fla Arthropod Res Lab, 64-86; biol adminr, John A Mulrennan Sr Res Lab, 86-90; RETIRED. *Concurrent Pos:* Prin investr, USPHS res grant, 60-66; ed, J Florida Mosquito Control Asn, 79- *Honors & Awards:* Maurice W Provost Mem Award, 90. *Mem:* Sigma Xi; Entom Soc Am; Am Mosquito Control Asn. *Res:* Biology and control of mosquitoes and other insects of medical importance. *Mailing Add:* 3340 Robinson Bayou Cir Panama City FL 32405

RATHCKE, BEVERLY JEAN, b Wadena, Minn, July 12, 45. ECOLOGY. *Educ:* Gustavus Adolphus Col, BA, 67; Imp Col, Univ London, MSc, 68; Univ Ill, Urbana-Champaign, PhD(ecol), 73. *Prof Exp:* Student ecol, Cornell Univ, 73-75; res assoc ecol, Brown Univ, 75-78; asst prof, 78-85, ASSOC PROF BIOL, UNIV MICH, 85- *Concurrent Pos:* Ed, J Ecol Soc Am, 75-82; NATO fel, 79. *Mem:* Ecol Soc Am; Brit Ecol Soc; Soc Study Evol; Soc Am Naturalists. *Res:* Animal-plant interactions and community ecology. *Mailing Add:* Dept Biol Univ Mich Ann Arbor MI 48109-1048

RATHER, JAMES B, JR, b Bryan, Tex, Feb 2, 11; m 35; c 3. CHEMICAL ENGINEERING. *Educ:* Lehigh Univ, ChE, 32. *Prof Exp:* Asst chief chemist, Magnolia Petrol Co, 35-37, supvr eng, Tech Serv Div, Socony Mobil Oil Co, 38-41, asst mgr div, 43-56, admin dir, 56-60, mgr toxicol & pollution, Res Dept, Mobil Oil Corp, 60-67, corp asst air & water conserv coord, 67-72; PRES, SCH HOUSE ENTERPRISES LTD, 72- *Concurrent Pos:* Mem, Ny State Action for Clean Air Comt. *Mem:* Air Pollution Control Asn; Am Nat Standards Insts; Am Soc Testing & Mat (vpres, 64, pres, 66); Am Chem Soc. *Res:* Petroleum technology; air and water pollution control; standardization; toxicology. *Mailing Add:* PO Box 600 Winter Harbor ME 04693

RATHER, LELLAND JOSEPH, pathology; deceased, see previous edition for last biography

RATHJEN, WARREN FRANCIS, fisheries; deceased, see previous edition for last biography

RATHKE, JEROME WILLIAM, b Humboldt, Iowa, July 10, 47; m 68; c 2. CATALYSIS. *Educ:* Iowa State Univ, BS, 69; Ind Univ, PhD(inorg chem), 73. *Prof Exp:* Fel chem, Cornell Univ, 73-75; CHEMIST, ARGONNE NAT LAB, 75- *Concurrent Pos:* Group leader, Argonne Nat Lab, 81- *Mem:* Am Chem Soc; NY Acad Sci; AAAS; Sigma Xi; Am Inst Chemists. *Res:* Catalysis; organometallic chemistry; coal conversion chemistry. *Mailing Add:* Chem Eng Div Argonne Nat Lab Argonne IL 60439

RATHKE, MICHAEL WILLIAM, b Humboldt, Iowa, Aug 13, 41; m 65; c 1. ORGANIC CHEMISTRY. *Educ:* Iowa State Univ, BS, 63; Purdue Univ, PhD(chem), 67. *Prof Exp:* NSF fel, Purdue Univ, 67-68; from asst prof to assoc prof, 68-78, PROF CHEM, MICH STATE UNIV, 78- *Mem:* Am Chem Soc. *Res:* Synthetic organic chemistry; organometallic chemistry, particularly boron organic chemistry. *Mailing Add:* 6125 Horizon Dr East Lansing MI 48823-2238

RATHMANN, CARL ERICH, b Chicago, Ill, June 27, 45; m 68; c 3. THERMODYNAMICS, CHAOTIC DYNAMICS. *Educ:* Northwestern Univ, BS, 68, MS, 70, PhD(mech eng), 75. *Prof Exp:* Engr, Res Div, Gen Am Transp Corp, 65-75; vis asst prof mech eng, Northwestern Univ, 76-77; staff mem, Mission Res Corp, 77-78; asst prof physics, Westmont Col, 79; PROF MECH ENG, DIR & ASSOC DEAN, ENG GRAD STUDIES, COL ENG, CALIF STATE POLYTECH UNIV, 79- *Concurrent Pos:* Instr mech eng, Univ Calif, Santa Barbara, 79; consult, Gen Dynamics Corp, 80-84; adj assoc prof, Harvey Mudd Col, 81-86. *Mem:* Sigma Xi; Am Soc Mech Engrs; Am Phys Soc; Am Soc Eng Educ. *Res:* Numerical simulation of plasma phenomena including development of methods for simulation long-time-scale wave-particle (resonance) interactions; investigation of spectral decomposition techniques appropriate to plasma parameters; investigation of chaotic transition to turbulence. *Mailing Add:* Col Eng Calif State Polytech Univ 3801 Temple Pomona CA 91768

RATHMANN, FRANZ HEINRICH, b Gotha, Fla, Apr 8, 04; m 49; c 2. CHEMISTRY, ASTRONOMY. *Educ:* Univ Minn, BA, 24, MA, 27; Univ Gottingen, PhD(org chem), 41. *Prof Exp:* Asst, Univ Minn, 25-27; asst prof chem, Millikin Univ, 29-31; sr sci worker, Inst Chem Physics, Leningrad, Russia, 31-35 & Inst Food-Pharm Res, Leningrad-Moscow, 35-37; res assoc phys chem, Univ Minn, 47-51; assoc prof chem, Univ Omaha, 51-55; from assoc prof to prof, 55-74, EMER PROF CHEM, NDAK STATE UNIV, 74- *Concurrent Pos:* Rockefeller Found fel, Univ Minn, 42; NSF res grants, 59-60; grad prof org chem, Univ Saigon, 62-63; adj prof astron, Moorhead State Univ, Minn, 74-76, NDak State Univ, 85-; mem coun, AAAS, 58-76, Am Chem Soc, 79-82. *Mem:* AAAS; Am Chem Soc; Asn Acad Sci (pres, 73-76). *Res:* Hydroximic and hydroxamic acids; mechanism and kinetics of tautomerization reactions; mass spectra and purines; vitamins B-1 and E; isoxazoles; solar eclipses; Halley's comet. *Mailing Add:* 4530 12th St Fargo ND 58102

RATHNAM, PREMILA, b India, Jan 7, 36; m. BIOCHEMISTRY, ENDOCRINOLOGY. *Educ:* Univ Madras, BSc, 55; Univ Wis, MS, 62; Seton Hall Univ, PhD(biochem), 66. *Prof Exp:* Res asst biochem & home econ, Univ Wis, 61-62; res asst chem, Seton Hall Univ, 62-66; fel endocrinol in med, 66-69, instr biochem in med, 69-71, asst prof, 71-78, ASSOC PROF BIOCHEM IN MED & ENDOCRINOL IN OBSTET & GYNEC, MED COL, CORNELL UNIV, 78- *Concurrent Pos:* Lalor Found fel, Med Col, Cornell Univ, 66-68, USPHS trainee, 66-69; Int Cong Biochem travel awards, 70, 73 & 88, Endocrine Soc travel award, 76, 80 & 88. *Mem:* NY Acad Sci; Am Chem Soc; Endocrine Soc; Am Soc Biol Chemists. *Res:* Isolation, characterization and structure function relationships of human anterior pituitary hormones and its receptors; one patent. *Mailing Add:* Div of Reproductive Endocrinol Dept Obstet & Gynec Cornell Univ Med Col 1300 York Ave New York NY 10021

RATLIFF, FLOYD, b La Junta, Colo, May 1, 19; m 42; c 1. NEUROPHYSIOLOGY, VISION. *Educ:* Colo Col, AB, 47; Brown Univ, MSc, 49, PhD(psychol), 50. *Hon Degrees:* DSc, Colo Col, 75. *Prof Exp:* Nat Res Coun fel, Johns Hopkins Univ, 50-51; instr psychol, Harvard Univ, 51-52, asst prof, 52-54; from assoc to assoc prof biophys, Rockefeller Univ, 54-66, prof, 66-89, pres, Harry Frank Guggenheim Found, 83-89; RETIRED. *Concurrent Pos:* William James fel, Am Psychol Soc, 89. *Honors & Awards:* Warren Medal, Soc Exp Psychol, 66; Tillyer Award, Optical Soc Am, 76; Pisart Vision Award, NY Asn Blind, 83; Distinguished Sci Contrib Award, Am Psychol Asn, 84. *Mem:* Nat Acad Sci; fel Am Acad Arts & Sci; Am Psychol Soc; Am Philos Soc. *Res:* Neurophysiology of vision. *Mailing Add:* 2215 Calle Cacique Sante Fe NM 87505-4944

RATLIFF, FRANCIS TENNEY, b Bogalusa, La, Oct 6, 19; m 44; c 1. PAPER TECHNOLOGY. *Educ:* La State Univ, BS, 40; Lawrence Col, MS, 42, PhD(paper chem), 48. *Prof Exp:* Res chemist, Standard Oil Develop Co, La, 42-45 & Johns-Manville Co, 48-51; paper technologist, Personal Prod Co, 51-54, tech dir, Paper Div, 54-57; asst res dir, Rhinelander Paper Co, 57-60, res dir, Rhinelander Div, St Regis Paper Co, 60-71, prod develop mgr, 71-73, asst tech dir, 73-74, res dir, 74-83, asst tech dir, 77-83; RETIRED. *Mem:* Tech Asn Pulp & Paper Indust; Am Soc Qual Control; Can Pulp & Paper Asn; Brit Paper & Bd Makers' Asn. *Res:* Physical properties of wood pulps; roofing papers; creped tissues; glassine, greaseproof, reprographic and packaging papers. *Mailing Add:* 534 Southeastern Ave Rhinelander WI 54501

RATLIFF, LARRY E, geology, geophysics, for more information see previous edition

RATLIFF, LOUIS JACKSON, JR, b Cedar Rapids, Iowa, Sept 1, 31. MATHEMATICS. *Educ:* Univ Iowa, BA, 53, MA, 58, PhD(math), 61. *Prof Exp:* Lectr math, Ind Univ, 61-63; lectr, 63-64, from asst prof to assoc prof, 64-69, PROF MATH, UNIV CALIF, RIVERSIDE, 69- *Concurrent Pos:* NSF grants, 64-68, 70-72 & res grant, 72-86. *Mem:* Am Math Soc. *Res:* Commutative algebra; local ring theory. *Mailing Add:* Dept Math Univ Calif Riverside CA 92521

RATLIFF, NORMAN B, JR, b Winchester, Ky, Aug 28, 38. PATHOLOGY. *Educ:* Duke Univ, MD, 62. *Prof Exp:* PATHOLOGIST & HEAD, AUTOPSY SERV, CLEVELAND CLIN, 79- *Mem:* Am Asn Pathol. *Mailing Add:* 2861 Pakton Rd Shaker Heights OH 44120

RATLIFF, PRISCILLA N, b Uniontown, Pa, Dec 26, 40; m 69; c 1. CHEMICAL INFORMATION SCIENCE. *Educ:* Maryville Col, BS, 62; Vanderbilt Univ, MS, 64. *Prof Exp:* Asst ed, Chem Abstr Serv, 64-67; info scientist, Battelle Columbus Labs, 67-73; tech writer, Warren-Teed Pharmaceut, Inc, 73-76; res chemist, 76-78, SUPVR LIBR & INFO SERV, ASHLAND CHEM CO, 78- *Mem:* Am Chem Soc; Am Soc Info Sci; Spec Libr Asn. *Res:* Computerized and manual information retrieval systems; patent searching. *Mailing Add:* Ashland Chem Co PO Box 2219 Columbus OH 43216

RATLIFF, ROBERT L, b Shawnee, Okla, Dec 1, 31; m 59; c 2. BIOCHEMISTRY. *Educ:* Univ Santa Clara, BS, 56; St Louis Univ, PhD(biochem), 60. *Prof Exp:* Am Cancer Soc fel enzymol, Univ Wis, 60-63; STAFF MEM BIOMED RES GROUP ENZYM, LOS ALAMOS NAT LAB, 63- *Mem:* Fedn Am Soc Exp Biol; AAAS; Am Soc Biol Chem; Sigma Xi. *Res:* Steroid--metabolism of bile acids; enzymes--purification of enzymes nucleoside diphosphokinases; nucleic acids; enzymatic synthesis of DNA and RNA. *Mailing Add:* 252 Lacueva Los Alamos NM 87544

RATNAYAKE, WALISUNDERA MUDIYANSELAGE NIMAL, b Kandy, Sri Lanka, Feb 22, 49; Can citizen; m 76; c 2. FATS & OILS CHEMISTRY, LIPID CHEMISTRY. *Educ:* Univ Sri Lanka, BSc, 72; Dalhousie Univ, MSc, 78, PhD(org chem), 80. *Prof Exp:* Asst lectr phys chem, Univ Sri Lanka, 72-73; res officer lipid chem, Ceylon Inst Sci & Indust Res, Sri Lanka, 73-82; res assoc Lipid Chem, Can Inst Fisheries Technol, Tech Univ NS, Halifax, 82-88; RES SCIENTIST, FOOD DIRECTORATE HEALTH & WELFARE CAN, OTTAWA, 89- *Concurrent Pos:* Vis lectr, lab technician course, Chem Inst Sri Lanka, 74-76; Can Commonwealth scholar, 76-80. *Mem:* Can Inst Chem; Chem Inst Sri Lanka; Chemists Soc. *Res:* Vegetable and marine oils chemistry; partial hydrogenation of fats and oils; heated and oxidized oils; directed interesterification of marine oils; large scale isolation of n3-polyunsaturated fatty acids in fish oils; analytical techniques in lipid chemistry; lipid nutrition. *Mailing Add:* Bur Nutrit Sci Food Directorate Health Protection Br Tunney's Pasture Ottawa ON K1A 0L2 Can

RATNER, ALBERT, b Brooklyn, NY, Sept 10, 37; m 67; c 2. ENDOCRINOLOGY, PHYSIOLOGY. *Educ:* Brooklyn Col, BS, 59; Mich State Univ, 62, PhD(physiol), 65. *Prof Exp:* PROF PHYSIOL, SCH MED, UNIV NMEX, 67- *Concurrent Pos:* NIH fel physiol, Univ Tex Southwestern Med Sch Dallas, 65-67; NSF grant, 68-75. *Mem:* Psychoneuroendocrine Soc; Am Physiol Soc; Biol Reprod Soc; Am Neurosci Soc; Endocrine Soc; Am Fertil Soc. *Res:* Neuroendocrine control of anterior pituitary and gonodal function. *Mailing Add:* Dept Physiol Univ NMex Sch Med Albuquerque NM 87131

RATNER, BUDDY DENNIS, b Brooklyn, NY, Jan 19, 47; m 68; c 1. MATERIALS SCIENCE ENGINEERING. *Educ:* Brooklyn Col, BS, 67; Polytech Inst Brooklyn, PhD(polymer chem), 72. *Prof Exp:* from res assoc to res asst prof, 72-84, assoc prof, 84-86, PROF CHEM ENG, UNIV WASH, 86- *Concurrent Pos:* NIH prin investr; asst ed, J Biomed Mat Res; consult; dir, Nat Electron Spectros Chem Anal & Surface Anal Ctr Biomed Probs. *Mem:* Am Chem Soc; AAAS; Soc Biomat; Adhesion Soc. *Res:* Interaction of biological systems with synthetic polymeric materials; materials for blood-contact and ophthalmologic applications; surface analysis of materials; plasma deposition. *Mailing Add:* Dept Chem Eng BF-10 Univ Wash Seattle WA 98195

RATNER, LAWRENCE THEODORE, b Philadelphia, Pa, Feb 16, 23; m 47. MATHEMATICS. *Educ:* Univ Calif, Los Angeles, AB, 44, MA, 45, PhD(math), 49. *Prof Exp:* Asst math, Univ Calif, Los Angeles, 44-45 & 47-49; asst prof, 49-54, ASSOC PROF MATH, VANDERBILT UNIV, 54- *Mem:* Am Math Soc; Math Asn Am; Sigma Xi. *Res:* Analysis in abstract spaces; topology; probability. *Mailing Add:* Vanderbilt Univ Box 43 Sta B Nashville TN 37235

RATNER, LAZARUS GERSHON, b Chicago, Ill, Sept 14, 23; m 53; c 3. HIGH ENERGY PHYSICS, ACCELERATOR PHYSICS. *Educ:* Univ Calif, Berkeley, AB, 48, MA, 50. *Prof Exp:* Asst physicist accelerator physics, Lawrence Berkeley Lab, Univ Calif, 50-60; physicist accelerator & high energy physics, Argonne Nat Lab, 60-81; PHYSICIST ACCELERATOR, BROOKHAVEN NAT LAB, 81- *Concurrent Pos:* Vis scientist, Ctr Europ Nuclear Res, Geneva, Switz, 71-72. *Mem:* AAAS; Sigma Xi. *Res:* Strong interaction physics; polarization phenomena in high energy scattering acceleration of polarized beams; design of high energy accelerators. *Mailing Add:* 911-B Brookhaven Nat Lab Upton NY 11973

RATNER, MARK A, b Cleveland, Ohio, Dec 8, 42; m 69. PHYSICAL CHEMISTRY. *Educ:* Harvard Univ, AB, 64; Northwestern Univ, Evanston, PhD(chem), 69. *Prof Exp:* Amanuensis chem, Aarhus Univ, Denmark, 69-70; asst, Munich Tech Univ, 70; from asst prof to assoc prof, NY Univ, 70-75; assoc prof, 75-80, PROF CHEM, NORTHWESTERN UNIV, 80- *Concurrent Pos:* A P Sloan fel; fel, Jerusalem Advan Study Inst. *Mem:* AAAS; fel Am Phys Soc; NY Acad Sci; The Chem Soc; Am Chem Soc. *Res:* Theoretical chemistry; nonadiabatic problems; hydrogen bonding; kinetics; spectra; green functions; electron transfer; reaction dynamics, non-linear optics. *Mailing Add:* Dept Chem Northwestern Univ Evanston IL 60208

RATNER, MICHAEL IRA, b New York, NY, June 30, 49; m 75; c 2. RADIO ASTRONOMY. *Educ:* Yale Col, BS, 71; Univ Colo, PhD(astro-geophysics), 76. *Prof Exp:* res assoc radio astron, Dept Earth & Planetary Sci, Mass Inst Technol, 76-82; RES ASSOC, RADIO & GEOASTRON DIV, HARVARD-SMITHSONIAN CTR ASTROPHYS, 83- *Mem:* Am Astron Soc. *Res:* Computerized analyses of interferometric observations of radio signals from spacecraft and natural radio sources; observational tests of gravitation theory; motions and natures of radio sources. *Mailing Add:* Ctr Astrophys Harvard Univ Cambridge MA 02138

RATNER, ROBERT (STEPHEN), b Newark, NJ, Apr 13, 41; m 64; c 1. TRANSPORTATION, SYSTEMS ENGINEERING. *Educ:* Mass Inst Technol, BS; Stanford Univ, MS, 65, PhD(elec eng), 68. *Prof Exp:* Res engr & sr res engr, Stanford Res Inst, SRI Int, 68-71, group mgr transp eng & control, 71-76, dir, Transp & Indust Systs Ctr, 76-81, vpres, Systs Consult Div, 82-86; PRES, RATNER ASSOC, INC. *Mem:* Opers Res Soc Am. *Res:* Air transportation and railroad operations and management consulting; air traffic control; transportation systems. *Mailing Add:* 5150 El Camino Real Suite E21 Los Altos CA 94022

RATNER, SARAH, b New York, NY, June 9, 03. BIOCHEMISTRY. *Educ:* Cornell Univ, BA, Columbia Univ, MA, 27, PhD(biochem), 37. *Hon Degrees:* DSc, Univ NC, 81, Northwestern Univ, Evanston, Ill, 82 & State Univ NY, Stony Brook, 84. *Prof Exp:* Asst biochem, Col Physicians & Surgeons, Columbia Univ, 30-31, 32-34, Macy res fel, 37-39, from instr to asst prof, 39-46; asst prof pharmacol, Col Med, NY Univ, 46-53, assoc prof, 53-54; assoc mem, Div Nutrit & Physiol, 54-57, MEM, DEPT BIOCHEM, PUB HEALTH RES INST CITY NEW YORK, INC, 57- *Concurrent Pos:* Ed, J Biol Chem, 59- & Anal Biochem, 74-; res prof, Col Med, NY Univ; Fogarty scholar-in-residence, NIH, 78-79. *Honors & Awards:* Schoenheimer lectr, 56; Neuberg Medal, 59; Garvan Medal, Am Chem Soc, 61; L & B Freedman Found Award, NY Acad Sci, 75. *Mem:* Nat Acad Sci; Am Acad Arts & Sci; Am Soc Biol Chem; fel Harvey Soc; fel NY Acad Sci; Am Chem Soc; Sigma Xi; fel AAAS. *Res:* Metabolism and chemistry of amino acids; application of isotopes to intermediary metabolism; enzymatic mechanisms of arginine biosynthesis and urea formation and other nitrogen transferring reactions; regulation and structure function relationships in enzymes of arginine biosynthesis from citrulline. *Mailing Add:* Dept Biochem NYC Pub Health Res Inst Inc 455 First Ave New York NY 10016

RATNEY, RONALD STEVEN, b Brooklyn, NY, June 1, 32; m 57; c 3. INDUSTRIAL HYGIENE. *Educ:* Calif Inst Technol, BS, 54; Yale Univ, PhD(chem), 59; Harvard Sch Pub Health, MS, 72. *Prof Exp:* Chemist, Trubek Labs, 58-60; from asst prof to assoc prof chem, Hood Col, 60-67; assoc prof chem, Bentley Col, 67-71; chemist, Mass Div Occup Hyg, 72-75; indust hygienist, 75-84, ASST REGIONAL ADMIN, OCCUP SAFETY & HEALTH ADMIN, 84- *Concurrent Pos:* Chmn, Subcomt Dusts & Inorg, Threshold Limit Value Comt. *Mem:* Am Acad Indust Hyg; Sigma Xi; Am Indust Hyg Soc; Am Conf Govt Indust Hygienists; NY Acad Sci. *Res:* Toxicology; industrial hygiene. *Mailing Add:* 167 Old Billerica Rd Bedford MA 01730

RATNOFF, OSCAR DAVIS, b New York, NY, Aug 23, 16; m 45; c 2. INTERNAL MEDICINE. *Educ:* Columbia Univ, AB, 36, MD, 39. *Hon Degrees:* LLD, Univ Aberdeen, Scotland, 80. *Prof Exp:* Intern, Med Serv, Johns Hopkins Hosp, 39-40; asst resident med, Montefiore Hosp, New York, 42; asst med, Col Physicians & Surgeons, Columbia Univ, 42-46; instr, Johns Hopkins Univ, 48-50; from asst prof to assoc prof, 50-61, asst vis physician 52-57, assoc vis physician, 57-67, PROF MED, SCH MED, CASE WESTERN RESERVE UNIV, 61-, VIS PHYSICIAN, UNIV HOSPS, 67- *Concurrent Pos:* Austin teaching fel physiol, Harvard Med Sch, 40-41; fel, Sch Med, Johns Hopkins Univ, 46-48; resident, Res Serv Chronic Dis, Goldwater Mem Hosp, 42-43; assoc, Mt Sinai Hosp, Cleveland, Ohio, 50-52; career investr, Am Heart Asn, 60-86. *Honors & Awards:* Thelin Award, Nat Hemophilia Found, 71; Dameshek Award, Am Soc Hemat, 72; John Phillips Award, Am Col Physicians, 74; John Smith Prize, Columbia Univ, 76; Grant Award, Int Soc Thrombosis, 81; Kovalenko Award, Nat Acad Sci, 85; Kober Medal, Asn Am Physicians, 88. *Mem:* Nat Acad Sci; Am Soc Hemat; master, Am Col Physicians; Cent Soc Clin Res (pres, 70-71); Asn Am Physicians; fel AAAS. *Res:* Hemostatic mechanisms. *Mailing Add:* 2916 Sedgewick Rd Shaker Heights OH 44120

RATTAN, KULDIP SINGH, b India, Apr 25, 48; m 76; c 2. ELECTRICAL ENGINEERING, COMPUTER ENGINEERING. *Educ:* Punjab Eng Col, India, BSc, 69; Univ Ky, MS, 72, PhD(elec eng), 75. *Prof Exp:* Fel bio-eng, Univ Ky, 76, res assoc, 77-78; asst prof, 79-82, ASSOC PROF ELEC ENG, WRIGHT STATE UNIV, 82- *Concurrent Pos:* Res assoc & SCEE fel, SCEEE Fac Res Prog, USAF, 80-84; prin investr, USAF Off Sci Res, 81-; vis prof, Carnegie-Mellon Univ, 87-88. *Mem:* Inst Elec & Electronics Engrs; Am Inst Aeronaut & Astronaut. *Res:* Design and analysis of digital control systems; develop a computer aided method for redesign of existing continuous control systems; robotics and computer control; computer aided design; model reduction. *Mailing Add:* Dept Elec Systems Eng Wright State Univ Dayton OH 45435

RATTAZZI, MARIO CRISTIANO, b Naples, Italy, Oct 1, 35; m 62; c 1. HUMAN GENETICS, BIOCHEMICAL GENETICS. *Educ:* Univ Naples, MD, 61. *Prof Exp:* Res asst prof human genetics, Univ Leiden, Neth, 62-69; from res asst prof to prof pediat, State Univ NY, Buffalo, 69-84; PROF PEDIAT, NORTH SHORE UNIV HOSP, CORNELL UNIV MED COL, 84- *Concurrent Pos:* Adv, WHO, 63; res career develop award, Nat Inst Gen Med Sci, 73; co-ed, Isozymes: Current Trends Biol & Med Res, 76-88; mem, exec comt, NY State Genetic Dis Prog, 82-87; mem, Ment Retardation Res Comt, Nat Inst Child Health & Human Develop, NIH, 84-88, chmn, 87-88; prog dir & prin investr, biomed res support grant, NIH, North Shore Univ Hosp, 85- *Mem:* Am Soc Human Genetics; AAAS; NY Acad Sci; Soc Pediat Res; Soc Inherited Metab Dis. *Res:* Human lysosomal storage diseases; biochemical genetic diagnostic and therapeutic aspects; animal models of human diseases. *Mailing Add:* Div Human Genetics Dept Pediat North Shore Univ Hosp Manhasset NY 11030

RATTÉ, CHARLES A, b Brattleboro, Vt, Mar 3, 27; m 55; c 2. GEOLOGY. *Educ:* Middlebury Col, BA, 53; Dartmouth Col, MA, 55; Univ Ariz, PhD(geol), 63. *Prof Exp:* Teaching asst geol, Dartmouth Col, 53-55 & Univ Ariz, 55-56; geologist, US Steel Corp, 56-61; instr geol, Univ Ariz, 61-63; prof geol & chmn dept, Windham Col, 63-76, chmn div sci, 64-66; VT STATE GEOLOGIST, 76- *Mem:* Am Geol Inst; Geol Soc Am; Asn Am State Geologists. *Res:* Groundwater investigations; igneous, metamorphic and glacial terrains; Vermont mineral resources assessment; Caribbean petrology; slope stability of Vermont. *Mailing Add:* Agency Natural Resources Ctr Bldg 103 S Main St Waterbury VT 05676

RATTE, JAMES C, b Hartford, Conn, Nov 21, 25. GEOLOGY. *Educ:* Michigan State Univ, BS, 50; Darthmouth Col, MA, 52. *Prof Exp:* GEOLOGIST, US GEOL SURV, 53- *Mem:* Fel Geol Soc Am; Soc Econ Geol; Colo Sci Soc. *Res:* Research in mid territory of caldera complexes. *Mailing Add:* US Geol Surv Fed Ctr Mail Stop 905 Box 25046 Denver CO 80225

RATTI, JOGINDAR SINGH, b Rajoya, India, Jan 1, 35; m 62; c 2. ANALYTICAL MATHEMATICS. *Educ:* Univ Bombay, BSc, 55, MSc, 58; Wayne State Univ, PhD(math), 66. *Prof Exp:* Lectr math, Khalsa Col, India, 58-60 & Nat Col, Bombay, India, 60-61; instr, Nev Southern Univ, 63-65 & Wayne State Univ, 65-66; asst prof, Oakland Univ, 66-67; assoc prof, 67-69, chmn dept, 69-77, PROF MATH, UNIV S FLA, 69- *Mem:* Am Math Soc; Math Asn Am. *Res:* Graph theory; summability; univalent functions; polynomials; graphs of semigroups and semirings. *Mailing Add:* Dept Math Univ SFla 4202 Fowler Ave Tampa FL 33620

RATTNER, BARNETT ALVIN, b Washington, DC, Oct 4, 50; m 78; c 1. REPRODUCTIVE PHYSIOLOGY. *Educ:* Univ Md, BS, 72, MS, 74, PhD(environ & reprod physiol), 77. *Prof Exp:* Teaching asst zool, Univ Md, 72-76, instr, 76-77; Nat Res Coun res assoc, Naval Med Res Inst, Nat Naval Med Ctr, Bethesda, Md, 77-78; res physiologist, Physiol Sect, 78-85, leader, Wildlife Toxicol Sect, 85-88, DEP CHIEF, ENVIRON CONTAMINANTS RES BR, PATUXENT, WILDLIFE RES CTR, 88- *Mem:* Am Physiol Soc; Am Soc Zoologists; Soc Study Reprod; Soc Exp Biol & Med; Soc Environ Toxicol & Chem. *Res:* Biochemical indicators of pollutant exposure in wildlife; hypobaric and hyperbaric physiology; toxicology; environmental and nutritional effects on endocrine and reproductive function. *Mailing Add:* Environ Contaminants Res Br US Fish & Wildlife Serv Laurel MD 20708

RATTNER, JEROME BERNARD, b Cincinnati, Ohio, Aug 12, 45; m 73; c 2. ANATOMY, MOLECULAR BIOLOGY. *Educ:* Miami Univ, BS, 67; Univ Tex, MS, 69; Washington Univ, PhD(biol), 73. *Prof Exp:* Fel cell biol, Univ Calif, Irvine, 73-75, res asst, 76-81; NATO fel biol, Nat Ctr Sci Res, France, 75-76; asst prof, 81-85, ASSOC PROF ANAT, DEPT MED BIOCHEM, UNIV CALGARY, 85- *Concurrent Pos:* Ed, Chromosoma. *Mem:* Am Soc Cell Biol; Genetics Soc Can. *Res:* Organization of chromatin and chromosomes in eukaryotic cells with special reference to the kinetochore. *Mailing Add:* Univ Calgary 3330 Hosp Dr NW Calgary AB T2N 4N1 Can

RATTO, PETER ANGELO, b San Francisco, Calif, Jan 31, 30; m 66; c 1. PHARMACEUTICAL CHEMISTRY. *Educ:* Univ Calif, BS, 51, MS, 56, PhD(pharmaceut chem), 58. *Prof Exp:* Asst prof pharm & pharmaceut chem, Loyola Univ, 58-63; sr res scientist, Bristol Labs, 63-71; res pharmacist, Norwich-Eaton Pharmaceut Co, 71-80; MEM FAC SCH PHARM, SOUTHWESTERN OKLA STATE UNIV, 80- *Mem:* Am Pharmaceut Asn; Am Soc Hosp Pharmacist; Sigma Xi; Am Asn Col Pharmacist. *Res:* Synthesis of organic medicinal agents; physical chemical evaluation and preparation of pharmaceutical dosage formulations. *Mailing Add:* 1533 Steiner Rd Weatherford OK 73096

RATTRAY, BASIL ANDREW, b Iron Hill, Que, Nov 16, 27; m 55. MATHEMATICS. *Educ:* McGill Univ, BSc, 48, MSc, 49; Princeton Univ, PhD(math), 54. *Prof Exp:* Lectr math, Univ NB, 52-54; from asst prof to assoc prof, 54-78, PROF MATH, MCGILL UNIV, 78- *Mem:* Am Math Soc; Can Math Cong. *Res:* Topology. *Mailing Add:* Dept Math McGill Univ 805 Sherbrooke St W Montreal PQ H3A 2K6 Can

RATTRAY, MAURICE, JR, b Seattle, Wash, Sept 16, 22; m 51; c 3. HYDRODYNAMICS, PHYSICAL OCEANOGRAPHY. *Educ:* Calif Inst Technol, BS, 44, MS, 47, PhD(physics), 51. *Prof Exp:* From asst prof to prof, 50-85, chmn dept, 68-78, EMER PROF OCEANOG, UNIV WASH, 85- *Concurrent Pos:* Rossby fel, Woods Hole Oceanog Inst, 66-67; mem, Adv Panel Earth Sci, NSF, 66-68; consult, US Naval Oceanog Off, 68-74; consult sci adv comt, US Coast Guard, 69-72; mem ocean sci comt, Nat Acad Sci, 71-74; mem environ pollutant movement & transformation adv comt, Environ Protection Agency, 76-78; mem, comt rev outer continental shelf, Environ Studies Prog, NAS/NRC, 87-, chmn phys oceanog panel. *Mem:* Am Soc Limnol & Oceanog (vpres, 63-64, pres-elect, 64-65, pres, 65-66); Am Geophys Union. *Res:* Dynamics of estuarine and oceanic current systems. *Mailing Add:* Dept Oceanog Univ Wash WB-10 Seattle WA 98195

RATTS, KENNETH WAYNE, b Martinsville, Ill, July 7, 32; m 59; c 3. ORGANIC CHEMISTRY. *Educ:* Univ Eastern Ill, BS, 54; Ohio State Univ, PhD, 59. *Prof Exp:* Asst chem, Ohio State Univ, 54-55, asst org chem, 55-57, asst instr, 57-59; res chemist, Monsanto Co, 59-62, sr res specialist, 63-65, sr group leader, 66, sci fel, 66-75, mgr res, 75-78, res dir process technol, 78-80, dir chem res, 80-83, dir prod res, 83-87, dir environ sci & support technol, 87-90; RETIRED. *Mem:* NY Acad Sci; Sigma Xi; Am Chem Soc. *Res:* 5-halobenzo (a) biphenylenes and transformations; agricultural chemicals; ylid and phosphorus chemistry; sulfur chemistry. *Mailing Add:* 15194 Strollways Dr Baldwin MO 63011

RATTY, FRANK JOHN, JR, b San Diego, Calif, June 26, 23; wid; c 1. GENETICS. *Educ:* San Diego State Col, BA, 48; Univ Utah, MS, 49, PhD(genetics), 52. *Prof Exp:* Instr biol, Univ Utah, 52-53; res geneticist poultry husb, Univ Calif, 53-54; from instr to assoc prof zool, 55-62, PROF BIOL, SAN DIEGO STATE UNIV, 62- *Mem:* AAAS; Genetics Soc Am. *Res:* Mutation; cytogenetics. *Mailing Add:* 899 Van Horn Rd El Cajon CA 92019

RATZ, H(ERBERT) C(HARLES), b Hamilton, Ont, July 23, 27; m 55; c 3. ELECTRICAL ENGINEERING. *Educ:* Univ Toronto, BASc, 50; Mass Inst Technol, SM, 52; Univ Sask, PhD(elec eng), 63. *Prof Exp:* Res engr, Ferranti-Electronics, Ltd, Ont, 52-55, proj engr, 56-57; design engr, Fischer & Porter (Can) Ltd, 57-59; asst prof elec eng, Univ Sask, 60-63; from assoc prof to prof, 63-86, assoc dean, 68-71 & 87-89, PROF ELEC ENG & DIR EXCHANGE PROGS, UNIV WATERLOO, 89- *Concurrent Pos:* Nat Res Coun Can res grants, 62-72; consult, Ferranti-Electronics, Ltd, 64-66, Can Westinghouse Co, 65 & Naval Res Estab, 67-71; vis prof, NS Tech Col, 71; sr indust fel, Nat Sci Eng Res Coun Can, Bell Northern Res, Can, 84-85. *Mem:* sr mem Inst Elec & Electronic Engrs. *Res:* Signal analysis and communications; computer communications networks; safety of computer control. *Mailing Add:* 288 Lourdes Crescent Waterloo ON N2L 1P5 Can

RATZ, JOHN LOUIS, b Aurora, Ill, May 18, 47; m 69; c 3. DERMATOLOGY. *Educ:* Aurora Col, BS, 70; Case Western Reserve, MD, 75. *Prof Exp:* Intern, Cleveland Clin Found, 75-76, resident dermat, 76-79, fel dermat surg & oncol & supvr, Psoriasis Ctr, 79-80; from asst prof to assoc prof dermat, Univ Cincinnati Med Ctr, 80-83; STAFF DERMATOLOGIST, DEPT DERMAT, CLEVELAND CLIN FOUND, 84- *Concurrent Pos:* Chief dermat sect, Vet Admin Med Ctr, 80-82; med staff, dept dermat, Jewish Hosp, 81-83; mem, Comt Mohs Surg Facil, Am Col Chemosurg, 86-88, Sci Prog Comt, 86-88, & comt Public Info, Am Soc Dermat Surg, 86-87; vis prof, Billings Clinic, 87, Northwestern Univ Med Sch, 87, Mayo Clinic, 87, & Ohio State Univ, 88. *Res:* Clinical evaluation of dermatologic conditions; dermatologic surgery; laser surgery. *Mailing Add:* Ctr for Dermat, Cosmetic & Laser Surg 4050 Healthway Dr Suite 220 Aurora IL 60504

RATZLAFF, KERMIT O, b Hillsboro, Kans, Dec 26, 21; m 44; c 5. PHYSIOLOGY, ZOOLOGY. *Educ:* Univ Calif, Los Angeles, AB, 49, MA, 51, PhD(zool), 62. *Prof Exp:* Res technician, White Mem Hosp, Los Angeles, Calif, 52-56; teaching asst zool, Univ Calif, Los Angeles, 56-58; res physiologist, Sch Med, Univ Calif, 59-62; asst prof physiol & zool, 62-74, ASSOC PROF BIOL SCI, SOUTHERN ILL UNIV, EDWARDSVILLE, 74- *Concurrent Pos:* Instr, Biola Col, 50-54; consult, Vet Admin Ctr, Los Angeles, 60-62 & Vet Admin Hosp, Long Beach, 62. *Mem:* AAAS; Am Sci Affil; Sigma Xi. *Res:* Mechanisms of functioning of sense organs, particularly regeneration of visual pigments; problems related to ocular metabolism. *Mailing Add:* Dept Biol Sci Southern Ill Univ Edwardsville IL 62025

RATZLAFF, MARC HENRY, b Bakersfield, Calif, Feb 21, 42; m 63; c 2. ANATOMY, VETERINARY MEDICINE. *Educ:* Univ Calif, Davis, AB, 64, MA, 66, PhD(anat), 69; Mich State Univ, DVM, 74. *Prof Exp:* Asst prof anat, Mich State Univ, 69-74; pvt vet pract, Eagle, Idaho, 74-75; ASSOC PROF ANAT, WASH STATE UNIV, 76- *Mem:* Am Asn Vet Anatomists; Am Vet Med Asn. *Res:* Equine locomotion; biomechanics. *Mailing Add:* Dept Vet Anat Pharm & Physiol Wash State Univ Pullman WA 99164

RATZLAFF, WILLIS, b Fairview, Okla, June 1, 26; m 49; c 4. LIMNOLOGY. *Educ:* Kans State Teachers Col, BS, 50, MS, 51; Univ Kans, PhD, 72. *Prof Exp:* Teacher high schs, Kans, 51-53; supv prin pub sch, Ohio, 53-56; teacher high sch, Kans, 56-58; supt Princeton schs, Kans, 59-61; assoc prof, 63-72, PROF LIFE SCI, MILLERSVILLE STATE COL, 72- *Concurrent Pos:* Consult, Sci Teachers Inst, Kans State Teachers Col, 59. *Mem:* AAAS; Am Soc Limnol & Oceanog; Ecol Soc Am; Am Fisheries Soc; Am Inst Biol Sci; Sigma Xi. *Res:* Ecology of small and ephemeral bodies of water; zooplankton population dynamics; ecological effects of electric power generation on the lower Susquehanna River. *Mailing Add:* Dept Biol Millersville State Col Millersville PA 17551

RAU, A RAVI PRAKASH, b Calcutta, India, Aug 9, 45; m 69, 85; c 2. ATOMIC PHYSICS. *Educ:* Univ Delhi, BSc, 64, MSc, 66; Univ Chicago, PhD(atomic physics), 70. *Prof Exp:* Res assoc physics, Univ Chicago, 70; assoc res scientist physics, NY Univ, 70-72; vis fel theoret physics, Tata Inst Fundamental Res, Bombay, India, 72-73; from asst prof to assoc prof, 74-81, PROF PHYSICS, LA STATE UNIV, BATON ROUGE, 81- *Concurrent Pos:* Conf fel, III Int Conf Atomic Physics, Boulder, Colo, 72; vis assoc prof, Yale Univ, 78-79; Alfred P Sloan Found fel; vis fel, Joint Inst Lab Astrophys, Univ Colo, 84 & Australian Nat Univ, 87-88. *Honors & Awards:* Hannan Rosenthal Mem Lectr, Yale & Columbia Univs, 85. *Mem:* Sigma Xi; Fel Am Phys Soc. *Res:* Two-electron phenomena, structure and properties of atoms in intense magnetic fields, such as those on pulsars; development and application of general variational principles in atomic physics and in other areas of physics. *Mailing Add:* Dept Physics La State Univ Baton Rouge LA 70803-4001

RAU, ALLEN H, b Pittsburgh, Pa, Apr 6, 58; m 89; c 1. SURFACTANT CHEMISTRY, TABLET PROCESSING. *Educ:* Rutgers Univ, BS, 79; Univ Cincinnati, MBA, 83. *Prof Exp:* Staff engr, Procter & Gamble, 79-81, group leader, 81-85; sr proj mgr, 85-87, dir prod develop, 87-90, SR RES SCIENTIST, ANDREW JERGENS CO, 90- *Mem:* Soc Cosmetic Chemists; Am Inst Chem Engrs; Am Chem Soc. *Res:* Development of new and improved consumer products; surfactant-based and emulsion products; tablet products. *Mailing Add:* 8681 Twilight Tear Lane Cincinnati OH 45249

RAU, BANTWAL RAMAKRISHNA, b Lucknow, India, Feb 13, 51; m 75; c 1. COMPUTER SCIENCE, COMPUTER ENGINEERING. *Educ:* Indian Inst Technol, Madras, BTech, 72; Stanford Univ, MS, 73, PhD(elec eng), 77. *Prof Exp:* Mem tech staff comput design, Palyn Assocs, Inc, 73-74; res asst elec eng, Stanford Univ, 75-77; asst prof elec eng, Univ Ill, Urbana-Champaign, 77-; MGR PERFORMANCE ACCELERATORS, ELXSI, SANTA CLARA, CALIF. *Concurrent Pos:* Assoc consult, Palyn Assocs, Inc, 78- *Mem:* Inst Elec & Electronics Engrs; Asn Comput Mach. *Res:* Computer architecture; computer performance evaluation; applied queueing theory. *Mailing Add:* Mgr Performance Accelerators 3410 Cent Expressway Santa Clara CA 95051

RAU, CHARLES ALFRED, JR, b Philadelphia, Pa, Jan 28, 42; m 65; c 3. FAILURE ANALYSIS, STRUCTURAL RELIABILITY. *Educ:* Lafayette Col, BS, 63; Stanford Univ, MS, 65, PhD(mat sci), 67. *Prof Exp:* Res asst metal fracture, Stanford Univ, 64-67; sr res assoc mech behav, Advan Mat Res & Develop Lab, Pratt & Whitney Aircraft, 67-70, supvr life prediction methods group, Mat Eng & Res Lab, United Technol Corp, 70-74; gen mgr, 74-76, vpres & prin engr, 76-80, EXEC VPRES, FAILURE ANALYSIS ASSOC, 80- *Concurrent Pos:* Lectr metall, Univ Conn, 67-68; lectr fracture mech, Univ Calif, Los Angeles, 78- *Mem:* Am Inst Mining, Metall & Petrol Engrs; Am Soc Metals; Soc Exp Mech; Am Soc Testing & Mat; fel Am Soc Metals; Am Soc Mech Engrs; Nat Asn Corrosion Engrs. *Res:* Fatigue and fracture of materials, failure analysis, metallurgy and corrosion, fracture mechanics analysis and testing, mechanical reliability and risk prediction. *Mailing Add:* Failure Analysis Assoc PO Box 3015 149 Commonwealth Dr Menlo Park CA 94025

RAU, ERIC, b Weissenfels, Ger, Sept 25, 28; US citizen; m 55; c 2. INDUSTRIAL CHEMISTRY, QUALITY ASSURANCE. *Educ:* NY Univ, BA, 51, PhD(phys chem), 55. *Prof Exp:* Chemist, US Naval Air Rocket Test Sta, 50-52; from engr to sr engr, Bettis Atomic Power Lab, Westinghouse Elec Corp, 55-60; res chemist, Cent Res FMC Corp, Princeton, NJ, 60-63, supvr, Inorg Chem Div, 63-65, mgr, Phys Chem Sect, Inorg Res & Develop Dept, 65-74, asst dir process develop, Indust Chem Div, Res & Develop Dept, 74-78; dir res & develop, 78-81, vpres, 81-84, DIR TECHNOL, CONVERSION SYSTS INC, 81-; ASST DIR, LAB & QUAL ASSURANCE, NJ DEPT ENVIRON PROTECTION, 87- *Concurrent Pos:* Dir, Chem Div, Spex Indust, 85; asst vpres, EA Labs, 86. *Mem:* Am Chem Soc; Asn Res Dirs; fel Am Inst Chemists. *Res:* Phosphorus; alkali salts; corrosion; high temperature reactions; coal; electrochemistry; waste management and ultimate disposal; pozzalanic reactions; alkaline earth salts; quality assurance. *Mailing Add:* 17 Pine Knoll Lawrenceville NJ 08648

RAU, GREGORY HUDSON, b Tacoma, Wash, Dec 14, 48. BIOGEOCHEMISTRY. *Educ:* Western Wash Univ, BA, 71; Univ Wash, MS, 74, PhD(ecol), 79. *Prof Exp:* Res asst, Western Wash State Col, 71; res asst, Univ Wash, 73-78; res assoc, Ore State Univ, 79; fel scholar, Univ Calif, Los Angeles, 79-81; RES ASSOC, AMES RES CTR, NASA, 81-; ASSOC RES SCIENTIST, UNIV CALIF, SANTA CRUZ, 85- *Concurrent Pos:* Consult, Coop Fisheries Unit, Univ Wash, 72, Global Geochem Corp, 80-81, Southern Calif Coastal Water Res Proj, 81, Aqua Resources Inc, 84, Us Geol Surv, 85, Westec Serv Inc, 85 & Intl Atomic Energy Agency, 86-88; Assoc Res Scientist, Tiburon Ctr Environ Studies, 83-85. *Honors & Awards:* Antarctic Serv Medal, NSF, 86. *Mem:* Am Soc Limnol & Oceanog; Am Geophys Union; AAAS; Geochem Soc; Sigma Xi. *Res:* Carbon and nitrogen flow and cycling in past and present environments; microbial, plant, animal and human biology, ecology, and nutrition; interpretation of preceding based on stable isotope natural abundances. *Mailing Add:* Ames Res Ctr NASA Mail Stop 239-4 Moffett Field CA 94035

RAU, JON LLEWELLYN, geology, for more information see previous edition

RAU, MANFRED ERNST, b May 30, 42; m 69; c 1. PARASITOLOGY, BEHAVIORAL ECOLOGY. *Educ:* Univ Western Ont, BSc, 65, PhD(parasitol), 70. *Prof Exp:* Fel, 69-70, res assoc, 70-75, asst prof, 75-79, ASSOC PROF, INST PARASITOL, MCGILL UNIV, 79- *Mem:* Can Soc Zool; Am Soc Parasitologists; Am Mosquito Control Asn. *Res:* Behavioral ecology of parasite transmission; biological control of mosquitoes. *Mailing Add:* Inst Parasitol MacDonald Col McGill Univ 21111 Lakeshore Rd Ste Anne de Bellevue PQ H9X 1C0 Can

RAU, R RONALD, b Tacoma, Wash, Sept 1, 20; m 44; c 2. PHYSICS, ACCELERATORS. *Educ:* Col Puget Sound, BS, 41, Calif Inst Technol, MS, 43, PhD(physics), 48. *Prof Exp:* Physicist, Calif Inst Technol, 43-45; from instr to asst prof physics, Princeton Univ, 47-56; from assoc physicist to sr physicist, 56-70, chmn dept physics, 66-70, assoc dir high energy physics, 70-81, SR PHYSICIST, BROOKHAVEN NAT LAB, 81- *Concurrent Pos:* Fulbright prof, France, 54-55; mem policy comt, Stanford Linear Accelerator Ctr, 67-73; mem prog comt, Los Alamos Meson Physics Fac, 70-74; chmn high energy adv comt, Brookhaven Nat Lab, 70-81; mem high energy physics adv panel, AEC, 70-74; adj prof, Univ Wyo, 70-84; mem bd trustees, Univ Puget Sound, 78-84; vis prof, Desy Lab, Hamburg, Ger, 84-85 & Mass Inst Technol, 84-85. *Honors & Awards:* Alexander von Humboldt Sr US Scientist Award, 88. *Mem:* Am Phys Soc; AAAS; NY Acad Sci. *Res:* Cloud chamber studies of cosmic rays at high altitudes and of heavy mesons; experimental high energy particle physics; bubble chambers; strong interactions; resonance production; particle accelerators; superconducting magnets. *Mailing Add:* Brookhaven Nat Lab Upton NY 11973

RAU, RICHARD RAYMOND, b Philadelphia, Pa, Apr 17, 28; m 54; c 4. PHYSICS, COMPONENT ENGINEERING. *Educ:* Muhlenberg Col, BS, 49; Yale Univ, MS, 50; Univ Pa, PhD(solid state physics), 55. *Prof Exp:* Sr engr semiconductor develop, Sperry Rand Corp, 56-59; dir eng, Nat Semiconductor Corp, 59-67; mgr eng digital integrated circuit, Transitron Electronic Corp, 67-70; mgr prod procurement electronic components, Raytheon Co, 71-72; mgr eng hybrid circuits, Control Prod Div, Bell & Howell Co, 72-78; SR STAFF ENGR COMPONENTS & CIRCUITS, PERKIN-ELMER CO, DANBURY, 78- *Concurrent Pos:* Asst instr, Univ Bridgeport, 59; lectr, Univ Conn, 65-67. *Mem:* Am Phys Soc; Inst Elec & Electronics Engrs; Res Soc Am; Int Soc Hybrid Mfrs. *Res:* Silicon devices. *Mailing Add:* 17 Longview Heights Rd Newtown CT 06470

RAU, WELDON WILLIS, b Tacoma, Wash, Jan 20, 21; m 44; c 1. MICROPALEONTOLOGY, BIOSTRATIGRAPHY. *Educ:* Col Puget Sound, BS, 43; Univ Iowa, MS, 46, PhD(paleont), 50. *Prof Exp:* Asst geol, Univ Iowa, 43-47; from instr to asst prof, Col Puget Sound, 47-50; geologist stratig micropaleont, Geol Div, US Geol Surv, 50-60, consult geol, 60-82; GEOLOGIST, DIV GEOL & EARTH RESOURCES, WASH STATE DEPT NATURAL RESOURCES, 60-; CONSULT, PAC, NORTHWEST & ALASKA BIOSTRATIGRAPHY, 82- *Mem:* Paleont Soc; Geol Soc Am; Soc Econ Paleont & Mineral; Am Asn Petrol Geol. *Res:* Tertiary Foraminifera of the Pacific Northwest and Southeast Alaska; stratigraphic micropaleontology of West Coast Tertiary rocks. *Mailing Add:* 3035 Edgewood Dr Olympia WA 98501

RAUB, HARRY LYMAN, III, b Lancaster, Pa, Oct 22, 19; m 47; c 2. PHYSICS. *Educ:* Franklin & Marshall Col, 41; Cornell Univ, PhD(exp physics), 47. *Prof Exp:* Asst, Cornell Univ, 41-47, res assoc, 47; from asst prof to assoc prof, 47-56, prof, 56-84, EMER PROF PHYSICS, MUHLENBERG COL, 84. *Mem:* Am Phys Soc; Am Asn Physics Teachers; Sigma Xi. *Res:* Elastic losses in high polymers; study of sound velocities with acoustic interferometer; x-ray diffraction. *Mailing Add:* 2872 Reading Rd Allentown PA 18103

RAUB, THOMAS JEFFREY, MEMBRANE PROTEIN, ENDOCYTOSIS. *Educ:* Univ Fla, PhD(cell biol), 82. *Prof Exp:* RES SCIENTIST, DRUG DELIVERY SYSTS RES, UPJOHN CO, 85- *Res:* Protein sorting. *Mailing Add:* Cell Biologist DDSR 7271-259-6 Upjohn Co Kalamazoo MI 49001

RAUB, WILLIAM F, b Alden Station, Pa, Nov 25, 39; m 64; c 3. PHYSIOLOGY, COMPUTER SCIENCE. *Educ:* Wilkes Col, AB, 61; Univ Pa, PhD(physiol), 65. *Prof Exp:* NSF predoctoral fel, Univ Pa, 61-64, Pa Plan fel, 64-66; health sci adminr, Div Res Facil & Resources, NIH, 66-69, chief, Biotechnol Resources Br, Div Res Resources, 69-75, assoc dir extramural & collab progs, Nat Eye Inst, 75-78, assoc dir extramural res, 78-83, dep dir extramural res, 83-86, DEP DIR, NIH, BETHESDA, MD, 86- *Res:* Automated information-handling in biology and medicine and respiratory physiology. *Mailing Add:* 11408 Rolling House Rockville MD 20852

RAUBER, LAUREN A, b St Louis, Mo, May 18, 46. THEORETICAL PHYSICS, SOLID STATE PHYSICS. *Educ:* Emory Univ, BS, 68; Univ Md, MS, 72, PhD(solid state physics), 76. *Prof Exp:* staff scientist, G&G Corp, 77-85; COMPUT PHYSICIST, APPL THEORET PHYSICS, LOS ALAMOS NAT LAB, 85- *Mem:* Am Phys Soc; Am Asn Physics Teachers. *Mailing Add:* Los Alamos Nat Lab M/S F669 Los Alamos NM 87545

RAUCH, DONALD J(OHN), b St Peters, Mo, Oct 20, 35; m 58; c 2. ELECTRICAL ENGINEERING. *Educ:* Washington Univ, St Louis, BS, 57; Mich State Univ, MS, 60, PhD(elec eng), 63. *Prof Exp:* Radar engr, Emerson Elec Mfg Co, 57-58; sr lectr elec eng, SDak State Col, 58-59; systs res assoc, Planning Res Corp, 63-67; sr engr guid & control systs, Litton Industs, 67-71; vpres & dir, Sysdyne, Inc, 71-76, pres, 76-78; PRES, EVOLVING TECHNOL CO, 79- *Concurrent Pos:* Sr lectr, Univ Southern Calif, 65-; chmn session on state space synthesis, Asilomar Conf Circuits & Systs, 70; mem tech prog comt, Joint Nat Conf Major Systs, 71; chmn, Los Angeles Prof Group Circuit Theory. *Mem:* Sr mem Inst Elec & Electronics Engrs. *Res:* Radar cross sections, extended diffraction scattering model for electromagnetic waves to include plasma or multilayered media; adaptive control processes; development of techniques for synthesizing physical systems from a time domain model. *Mailing Add:* Evolving Tech Co 3725 Talbot St Suite F San Diego CA 92106

RAUCH, EMIL BRUNO, b Friedland, Czech, mar 19, 19; nat US; m 60; c 2. ORGANIC CHEMISTRY. *Educ:* Univ Heidelberg, PhD, 53. *Prof Exp:* Res assoc org chem, Wash Univ, 53-55; res chemist, Indust Photo Div, GAF Corp, 55-64, res assoc, 64-67, mgr color res & develop, 67-74, dir res & develop, 74-77, sr scientist, 77-80; sr scientist, Anitec Image Corp, 80-88; RETIRED. *Mem:* Am Chem Soc; Soc Ger Chem; Soc Photog Sci & Engr. *Mailing Add:* Five Perkins Ave Binghamton NY 13901-1822

RAUCH, FRED D, b Rainier, Ore, Jan 17, 31; m 58; c 2. HORTICULTURE, PLANT PHYSIOLOGY. *Educ:* Ore State Univ, BS, 56, MS, 63; Iowa State Univ, PhD(hort, plant physiol), 67. *Prof Exp:* Exp farm technician, Ore State Univ, 58-60, asst hort, Mid-Columbia Exp Sta, Hood River, 60-63; instr, Iowa State Univ, 63-67; asst prof, Miss State Univ, 67-70; from asst specialist to assoc specialist, 70-80, SPECIALIST, DEPT HORT, UNIV HAWAII, 80- *Mem:* Int Plant Propagators Soc; Am Soc Hort Sci; Int Palm Soc. *Res:* Stock-scion relationships; plant nutrition; growth regulators; herbicides. *Mailing Add:* Dept Hort 3190 Maile Way Univ Hawaii Honolulu HI 96822

RAUCH, GARY CLARK, b Dayton, Ohio, Oct 14, 42; m 67; c 2. PHYSICAL METALLURGY. *Educ:* Mass Inst Technol, BS, 64, PhD(metall), 68; Univ Ill, Urbana, MS, 65. *Prof Exp:* Scientist, Fundamental Res Lab, US Steel Corp, Pa, 68-71; sr engr, Magnetics Dept, Westinghouse Res & Develop Ctr, 72-85; SR CONSULT ENGR, ADVAN RECORDING MEDIA DEVELOP, DIGITAL EQUIP CORP, COLORADO SPRINGS, COLO, 85- *Mem:* Am Soc Metals; Metall Soc of Am Inst Mining, Metall & Petrol Engrs; Inst Elec & Electronics Engrs; Mat Res Soc. *Res:* Phase transformations and precipitation in iron-base alloys; mechanical properties of martensite; physical metallurgy of ferromagnetic materials; core loss phenomena in transformer steels; magnetic domains; phase transformations and precipitation in iron-base alloys; processing-structure-properties relationships in magnetic recording materials; tribology of magnetic recording media. *Mailing Add:* Digital Equip Corp 4755 Forge Rd Colorado Springs CO 80907

RAUCH, HAROLD, b New York, NY, Oct 13, 25; m 52; c 3. ZOOLOGY. *Educ:* Queens Col, NY, BS, 44; Univ Ill, MS, 47; Brown Univ, PhD(biol), 50. *Prof Exp:* From instr to assoc prof, 50-60, chmn dept, 71-74, PROF ZOOL, UNIV MASS, AMHERST, 60- *Mem:* Genetics Soc Am; Am Soc Zool. *Res:* Mammalian physiological genetics; nervous system, pigmentation and copper metabolism in the mouse. *Mailing Add:* 110 Red Gate Lane Amherst MA 01002

RAUCH, HELENE COBEN, b Los Angeles, Calif; c 4. IMMUNOLOGY, NEUROIMMUNOLOGY. *Educ:* Univ Calif, Los Angeles, BA, 51, PhD(microbiol), 58. *Prof Exp:* Bacteriologist, Los Angeles County Dept Pub Health, 51-52; part-time asst, Univ Calif, Los Angeles, 53-57; res assoc, Dept Allergy & Immunol, Palo Alto Med Res Found, 57-60; res assoc, Div Infectious Dis, Sch Med, Stanford Univ, 60-61, prin investr, Nat Inst Neurol Dis & Stroke Grant & res assoc, Dept Med Microbiol, 64-75; ASSOC PROF IMMUNOL, WAYNE STATE UNIV, 75- *Concurrent Pos:* Nat Multiple Sclerosis Soc fel, Sch Med, Stanford Univ, 61-64. *Mem:* AAAS; Am Soc Microbiol; Sigma Xi; Am Asn Immunologists; Tissue Cult Asn. *Res:* Auto-allergic diseases, especially experimental allergic encephalomyelitis; immunopathology; cellular hypersensitivity; multiple sclerosis. *Mailing Add:* Dept Immunol & Microbiol Wayne State Univ 540 E Caufield Detroit MI 48201

RAUCH, HENRY WILLIAM, b Amsterdam, NY, Oct 23, 42; m 71; c 1. HYDROGEOLOGY, GEOCHEMISTRY. *Educ:* Alfred Univ, BA, 65; Pa State Univ, PhD(geochem), 72. *Prof Exp:* PROF GEOL, WVA UNIV, 70- *Concurrent Pos:* Hydrogeol consult; mem, WVa Reclamation Bd Rev, 78-84. *Mem:* Am Chem Soc; Am Water Resources Asn; Nat Water Well Asn; fel Nat Speleol Soc. *Res:* Ground water hydrology and aqueous geochemistry; karst hydrogeology; ground water pollution; ground water exploration; effects of coal mining on ground water; hydrogeologic methods for natural gas exploration; lineaments. *Mailing Add:* 423 White Hall WVa Univ Morgantown WV 26506

RAUCH, HERBERT EMIL, b St Louis, Mo, Oct 6, 35; m 61; c 4. SIGNAL PROCESSING, OPTIMAL CONTROL. *Educ:* Calif Inst Technol, BS, 57; Stanford Univ, MS, 58, PhD(elec eng), 62. *Prof Exp:* Mem tech staff, Hughes Aircraft Co, 58-62; STAFF SCIENTIST & SR MEM RES LAB, LOCKHEED PALO ALTO RES LAB, 62- *Concurrent Pos:* Assoc prof, San Jose State Univ, 68-70; ed, J Astronaut Sci, 80-86 & Inst Elec & Electronics Engrs Control Systs Magazine, 85-; chmn, Math Control Comt, Int Fedn Automatic Control, 84-87; mem publ comt, Am Inst Aeronaut & Astronaut, 80- *Honors & Awards:* Centennial Medal, Inst Elec & Electronics Engrs, 84; Space Shuttle Award, Am Inst Aeronaut & Astronaut, 84. *Mem:* Sr mem Inst Elec & Electronics Engrs; assoc fel Am Inst Aeronaut & Astronaut; fel Am Astronaut Soc (vpres, 80-84). *Res:* Optimal estimation and control; signal processing; astrodynamics; guidance; expert systems; neural networks. *Mailing Add:* 401 Dracena Lane Los Altos CA 94022

RAUCH, JEFFREY BARON, b New York, NY, Nov 29, 45; m; c 1. MATHEMATICAL PHYSICS. *Educ:* Harvard Univ, AB, 67; NY Univ, PhD(math), 71. *Prof Exp:* Instr math, NY Univ, 68-71; from asst prof to assoc prof, 71-81, PROF MATH, UNIV MICH, ANN ARBOR, 81-, CHMN, DEPT MATH, 90- *Concurrent Pos:* Vis mem, Inst Advan Study, 75-76, Inst Advan Sci Studies, France, 78-79; assoc prof, Higher Normal Sch, France, 85-86. *Mem:* Am Math Soc. *Res:* Hyperbolic partial differential equations. *Mailing Add:* Dept Math Angell Hall Univ Mich Ann Arbor MI 48104

RAUCH, LAWRENCE L(EE), b Los Angeles, Calif, May 1, 19; m 61; c 2. APPLIED MATHEMATICS. *Educ:* Univ Southern Calif, AB, 41; Princeton Univ, MA, 48, PhD(math), 49. *Prof Exp:* Asst, Uranium Separation Proj, Princeton Univ, 42, & anti-aircraft fire control, 42-43, instr math, 42-49, instr pre-radar sch, 43-44, res supvr radio telemetering systs for aircraft, 44-46, supvr air blast telemetering, Oper Crossroads, Bikini, 46, nonlinear differential equations proj, US Off Naval Res, 47-49; from asst prof to prof aeronaut eng, Univ Mich, 49-79, chmn nuclear eng prog, 51-52, instrumentation prog, 52-63 & mgt sci prog, 58-59, chmn comput, info & control eng, 71-76, assoc chmn dept elec & comput eng, 72-76, EMER PROF ENG, UNIV MICH, ANN ARBOR, 79- *Concurrent Pos:* Consult, 45-; mem res adv comt commun, instrumentation & data processing, NASA, 63-70; vis prof, Ecole Nationale Supérieure de l'Aéronatique et de l'Espace, Toulouse, France, 70, Univ Tokyo, 78 & Calif Inst Technol, 77-85. *Honors & Awards:* Spec Award, Inst Elec & Electronics Engrs, 57; Annual Award, Nat Telemetering Conf, 60; Eckman Award, Instrument Soc Am, 66. *Mem:* Fel AAAS; Am Math Soc; fel Inst Elec & Electronics Engrs; Am Inst Aeronaut & Astronaut. *Res:* Radio telemetry and communication theory; mathematical models and physical realizations of communication operations; communications engineering. *Mailing Add:* 759 N Citrus Ave Los Angeles CA 90038-3401

RAUCH, RICHARD TRAVIS, b New Orleans, La, Nov 18, 55. GRAVITATION THEORY & COSMOLOGY, SYSTEMS ENGINEERING-ANALYSIS & ADVANCED CONCEPT DEVELOPMENT. *Educ:* La State Univ, Baton Rouge, BS, 77; State Univ NY, Stony Brook, MA, 79, PhD(theoret physics), 82. *Prof Exp:* Exp res asst, Dept Physics & Astron, La State Univ, 74-77; teaching asst, Dept Physics, State Univ NY, Stony Brook, 77-78, res asst physics, Inst Theoret Physics, 78-82; physicist, R&D Assocs, 83-88; SR SCIENTIST, DEFENSE GROUP INC, 88- *Concurrent Pos:* Consult to R&D Assocs, Space Shuttle Prog, Rockwell Int, 87-88, independent consult, 88. *Honors & Awards:* Sustained Serv Award, Am Inst Aeronaut & Astronaut, 86. *Mem:* Am Phys Soc; Am Inst Aeronaut & Astronaut; Space Studies Inst. *Res:* Systems engineering-analysis and advanced concept development for a variety of space, strategic and tactical systems; development of extended theories of gravitation and an assessment of their astrophysical and cosmological consequences and their general experimental viability. *Mailing Add:* Defense Group Inc 606 Wilshire Blvd Suite 706 Santa Monica CA 90401

RAUCH, SOL, b Horodenko, Poland, Apr 22, 40; Can citizen; m 71; c 1. APPLY TECHNOLOGY TO DEFINE PRODUCT DEVELOPMENT OPPORTUNITIES, SIGNAL PROCESSING IN RADIO NAVIGATION SYSTEMS. *Educ:* McGill Univ, BS, 62, MS, 66. *Prof Exp:* Design engr, Avionics Div, Can Marconi Co, 62-65, proj engr, 66-71, prod mgr, 71-78, group mgr, 78-90, GEN MGR, AVIONICS DIV, CAN MARCONI CO, 90- *Concurrent Pos:* Lectr, Loyola Col, Montreal, 73-78; auxilary prof, McGill Univ, 75-78. *Mem:* Inst Elec & Electronics Engrs; Soc Info Displays; Asn Prof Engrs Ont. *Res:* Display systems for high ambient illumination environments; ruggedness, small size, intelligence and high brightness and contrast ratio. *Mailing Add:* 16 Parkmount Crescent Nepean ON K2H 5T4 Can

RAUCH, STEWART EMMART, JR, b Bethlehem, Pa, Aug 29, 21; m 47; c 4. ORGANIC CHEMISTRY. *Educ:* Moravian Col, BSc, 42; Lehigh Univ, MS, 47. *Prof Exp:* Instr meteorol, Univ Va, 42-43; from asst prof to assoc prof chem, Moravian Col, 47-62; res engr, Bethlehem Steel Corp, 62-77; RETIRED. *Concurrent Pos:* Consult, 77-88. *Mem:* AAAS. *Res:* Antimalarials; developments commercially marketed; photoelectric colorimeter; portable potentiometer; portable wheatstone bridge; self-contained portable gas chromatograph; electronic clinical thermometer; chemical instrumentation; internal ballistics. *Mailing Add:* 5521 Old Bethlehem Pike Bethlehem PA 18015

RAUCHER, STANLEY, b St Paul, Minn, Nov 4, 48. SYNTHETIC ORGANIC CHEMISTRY. *Educ:* Univ Minn, BA, 70, PhD(chem), 73. *Prof Exp:* Asst prof, 75-81, ASSOC PROF CHEM, UNIV WASH, 81- *Mem:* Am Chem Soc. *Res:* Synthesis of natural products. *Mailing Add:* Dept Chem Univ Wash Seattle WA 98195

RAUCHFUSS, THOMAS BIGLEY, b Baltimore, Md, Sept 11, 49; m 77; c 2. INORGANIC CHEMISTRY, CATALYSIS. *Educ:* Univ Puget Sound, BS, 71; Wash State Univ, PhD(inorg chem), 75. *Prof Exp:* RES FEL CHEM, AUSTRALIAN NAT UNIV, 75- *Mem:* Am Chem Soc; Royal Soc Chem. *Res:* Preparative inorganic; homogeneous catalysis; organometallic; ligand design and synthesis; organic synthesis with metal ions; phosphorus and sulfur chemistry. *Mailing Add:* Noyes Lab Box 21 Univ Ill 505 S Mathews St Urbana IL 61801

RAUCKHORST, WILLIAM H, b Covington, Ky, Sept 9, 40; m 64; c 3. LOW TEMPERATURE PHYSICS. *Educ:* Thomas More Col, BA, 62; Univ Cincinnati, PhD(physics), 67. *Prof Exp:* Asst prof physics & chmn dept, Bellarmine Col, Ky, 67-71, assoc prof physics & chmn div natural sci, 71-72; planner phys sci prog, Sangamon State Univ, 72-73, coordr, 73-75, assoc prof phys sci, 72-76; prog mgr, Fac Develop Prog, US Dept Energy, Washington, DC, 76-78; head, Col/Univ Progs Sect, Argonne Nat Lab, 78-85; ASSOC DEAN & PROF PHYSICS, MIAMI UNIV, 85-; PROG DIR, UNDERGRAD SCI, ENG & MATH EDUC, NSF, 90- *Mem:* Am Phys Soc; Am Asn Physics Teachers; Sigma Xi. *Res:* Specific heat studies; superconductivity; energy education. *Mailing Add:* Undergrad Sci Educ NSF 1800 G St NW Washington DC 20550

RAUDORF, THOMAS WALTER, b Berlin, Ger, May 6, 43; m 75; c 1. PHYSICS, NUCLEAR INSTRUMENTATION. *Educ:* Concordia Univ, BSc, 64; McGill Univ, MSc, 67, PhD(physics), 71. *Prof Exp:* Staff physicist, Simtec Industs Ltd, 71-73 & Electronic Assocs Can Ltd, 73-75; staff physicist, 75-85, MGR, DETECTOR RES & DEVELOP, EG&G ORTEC INC, 85- *Mem:* Can Asn Physicists; Am Phys Soc; sr mem Inst Elec & Electronics Engrs. *Res:* Rectifying and ohmic contacts on semiconductors; charge carrier transport in semiconductors; semiconductor surface physics; deep levels in semiconductors; semiconductor crystal growth. *Mailing Add:* EG&G Ortec Inc 100 Midland Rd Oak Ridge TN 37830

RAUH, ROBERT DAVID, JR, b Medford, Mass, Nov 15, 43; m 69; c 1. ELECTROCHEMISTRY, PHOTOCHEMISTRY. *Educ:* Bowdoin Col, AB, 65; Wesleyan Univ, MA, 68; Princeton Univ, PhD(chem), 72. *Prof Exp:* Fel chem, Brandeis Univ, 72-74; sr scientist, 74-76, prin scientist chem, 76-79, DIR RES, EIC LAB, INC, 79- *Mem:* Am Chem Soc; Electrochem Soc; Sigma Xi. *Res:* Chemical aspects of energy conversion; high energy density batteries; photochemical storage of solar energy; photovoltaic devices; chemical sensors. *Mailing Add:* 17 Converse Ave Newton MA 02158

RAUHUT, MICHAEL MCKAY, b New York, NY, Dec 4, 30; m 57; c 3. ORGANIC CHEMISTRY, PHOSPHORUS & PHOTO CHEMISTRY. *Educ:* Univ Nev, BS, 52; Univ NC, PhD(chem), 55. *Prof Exp:* Chemist, Cent Res Div, Am Cyanamid Co, Conn, 55-62, sr res chemist, 62, group leader, 63-70, res mgr, Chem Res Div, 70-81, mgr venture technol, Bound Brook, 81-82, mgr, plastics additives & resins res, Stamford Labs, 83-84, mgr regulatory affairs, 85-89, MGR NEW TECHNOL APPRAISAL, AM CYANAMID CO, 90- *Honors & Awards:* IR-100 Award, Indust Res, Inc, 70. *Mem:* Am Chem Soc. *Res:* Physical organic chemistry; mechanisms of aromatic nucleophilic substitution and organophosphorus reactions; mechanisms of chemiluminescent reactions. *Mailing Add:* 201 Range Rd Wilton CT 06897

RAUK, ARVI, b Estonia, Sept 30, 42; Can citizen; m 68; c 2. CHEMISTRY. *Educ:* Queen's Univ, Ont, BSc, 65, PhD(chem), 68. *Prof Exp:* Nat Res Coun Can fel, Princeton Univ, 68-70; asst prof chem, 70-77, assoc prof, 77-80, PROF CHEM, UNIV CALGARY, 80- *Mem:* Am Chem Soc; Am Phys Soc; Chem Inst Can. *Res:* Theoretical and experimental studies of optical activity of organic and organometallic systems. *Mailing Add:* Dept Chem Univ Calgary Calgary AB T2N 1N4 Can

RAULET, DAVID HENRI, b June 3, 54. T-CELL ACTIVATION & DEVELOPMENT. *Educ:* Mass Inst Technol, PhD(biol), 81. *Prof Exp:* ASSOC PROF BIOL, MASS INST TECHNOL, 83- *Concurrent Pos:* Cancer Res Inst investr award. *Mem:* Am Asn Immunologists. *Mailing Add:* Mass Inst Technol E17 128 77 Massachusetts Ave Cambridge MA 02139

RAULINS, NANCY REBECCA, b Dickson, Tenn, Oct 17, 16. ORGANIC CHEMISTRY. *Educ:* Tulane Univ, BA, 36, MS, 38; Univ Wyo, PhD(chem), 53. *Prof Exp:* Teacher high sch, La, 38-39, sci & math, All Saints Col, 39-41; teacher physics & chem, Spartanburg Jr Col, 41-42; chemist, Chickasaw Ord Works, 42-43; head dept chem, Lambuth Col, 43-46; from instr to prof, 48-79, EMER PROF CHEM, UNIV WYO, 80- *Mem:* Am Chem Soc. *Res:* Physical and chemical analytical methods in organic research; mechanism of reactions; hydrogen bonding in heterocyclic systems. *Mailing Add:* 466 N Ninth St Laramie WY 82070

RAULSTON, JAMES CHESTER, b Enid, Okla, Nov 24, 40. HORTICULTURE. *Educ:* Okla State Univ, BS, 62; Univ MD, MS, 66, PhD(hort), 69. *Prof Exp:* Instr plant sci, Inst Appl Agr, Univ Md, 65-66; asst prof ornamental hort, Agr Res & Educ Ctr, Univ Fla, 69-72; assoc prof ornamental hort, Tex A&M Univ, 72-75; PROF ORNAMENTAL HORT, NC STATE UNIV, 76- *Mem:* Am Soc Hort Sci; Int Soc Hort Sci. *Res:* Research in production and utilization of perennial landscape plants, including production, marketing and establishment of woody plants; production and use of native plants; physiology of landscape plants. *Mailing Add:* Dept Hort Sci NC State Univ Box 7609 Raleigh NC 27695-7609

RAUN, ARTHUR PHILLIP, b Upland, Nebr, Apr, 28, 34; m 58; c 2. ANIMAL NUTRITION. *Educ:* Univ Nebr, BS, 55; Iowa State Univ, MS, 56, PhD(animal nutrit), 58. *Prof Exp:* Asst prof chem & physiol, USAF Acad, 59-62; sr scientist, 62-68, res scientist, 69-70, head animal sci field res, 71-72, HEAD ANIMAL NUTRIT RES, LILLY RES LABS, ELI LILLY & CO, 72- *Mem:* Am Soc Animal Sci. *Res:* Ruminant nutrition, specifically action of anabolic compounds in ruminants; ruminant bloat and rumen microbial metabolism. *Mailing Add:* Oak Dr New Palestine IN 46163

RAUN, EARLE SPANGLER, b Sioux City, Iowa, Aug 28, 24; m 46; c 3. ENVIRONMENTAL SCIENCE GENERAL, AGRICULTURE GENERAL. *Educ:* Iowa State Col, BS, 46, MS, 50, PhD, 54. *Prof Exp:* Entomologist, Bur Entom & Plant Quarantine, USDA, Fla, 46-48; exten entomologist, Iowa State Univ, 48-60; res entomologist corn borer invests, Entom Res Div, Agr Res Serv, USDA, 61-66; prof entom, Univ Nebr, Lincoln, 66-74, assoc dir, Nebr Coop Exten Serv, 70-74, chmn dept entom, 66-70; entomologist, Pest Mgt Consults, Inc, 74-83, pres, 80-83; OWNER & PRES, PEST MGT CO, 83- *Concurrent Pos:* Consult, Off Technol Assessment, US Cong, 78-80; Pesticide Users Adv Comt, Environ Protection Agency, 81-; pres, Registry Prof Entom, 81; pres, Entom Soc Am, 83-84. *Honors & Awards:* Distinguished Entomologist, Am Registry Prof Entomologists, 86. *Mem:* Entom Soc Am; Nat Alliance Independent Crop Consults (pres, 78-80); Am Registry Prof Entomologists (pres, 80). *Res:* Integration of insect population management techniques for grower use in corn, alfalfa, soybeans, and grain sorghum. *Mailing Add:* 3036 Prairie Rd Lincoln NE 68506

RAUN, NED S, b Upland, Nebr, Feb 10, 25; m 46; c 6. BIOCHEMISTRY. *Educ:* Univ Nebr, BS, 48; Iowa State Univ, PhD(nutrit), 61. *Hon Degrees:* DSc, Univ Nebr, 84. *Prof Exp:* Animal scientist, Rockefeller Found, 61-64; assoc prof animal nutrit, Okla State Univ, 64-65; animal scientist, Rockefeller Found, 65-76, dir animal sci, Int Ctr Trop Agr, Cali, Colombia, 69-76; chief livestock div, AID, Dept State, 76-78; vpres projs, 78-84, actg pres, 84-85, REGIONAL REP WASH, WINROCK INST, 85- *Honors & Awards:* Int Animal Agr Award, Am Soc Animal Sci, 84. *Mem:* Fel Am Soc Animal Sci; Fedn Am Soc Exp Biol; Am Inst Nutrit; Latin Am Asn Animal Prod; fel AAAS. *Res:* Ruminant nutrition; pasture and forage utilization. *Mailing Add:* 6865 Williamsburg Pond Ct Falls Church VA 22043

RAUNIO, ELMER KAUNO, organic chemistry, for more information see previous edition

RAUP, DAVID MALCOLM, b Boston, Mass, Apr 24, 33; m 56, 87; c 1. INVERTEBRATE PALEONTOLOGY. *Educ:* Univ Chicago, SB, 53; Harvard Univ, AM, 55, PhD(geol), 57. *Prof Exp:* Instr invertebrate paleont, Calif Inst Technol, 56-57; from asst prof to assoc prof geol, Johns Hopkins Univ, 57-65; from assoc prof to prof geol, Univ Rochester, 66-78, chmn dept, 69-71; res assoc, 78-80, chmn dept, 82-85, PROF GEOPHYS SCI, UNIV CHICAGO, 80-, PROF CONCEPTUAL FOUND SCI, 81-, PROF EVOLUTIONARY BIOL, 82- *Concurrent Pos:* Grants, Am Asn Petrol Geol, 57, Am Philos Soc, 59, NSF, 61-64, 66-67, 75-81 & Am Chem Soc, 65-71, NASA, 83-; resident mem staff, Morgan State Col, 58; vis prof, Univ T06bingen, 65 & 72 & Univ Chicago, 77 & 78; mem adv panel earth sci, NSF, 70-73 & adv bd syst biol, 82-83; mem vis comt geol sci, Harvard Univ, 77-81 & Univ Colo, 83 & vis comt organismic & evolutionary biol, Harvard Univ, 82-; cur & chmn, dept geol, Field Mus Hatural Hist , 78-80, dean sci, 80-82; mem, geol sci bd, Nat Acad Sci-Nat Res Coun, 80-82, Space Sci Bd, 86-89, Comt Biodiversity, 87-90, Comn Phys Sci, Math & Resources, 87-90; Swell L Avery Distinguished Serv Prof, Univ Chicago, 84- *Honors & Awards:* Schuchert Award, Paleont Soc, 73. *Mem:* Nat Acad Sci; fel Geol Soc Am; Soc Study Evolution; Paleont Soc (pres, 76-77); fel AAAS; Am Soc Naturalists (vpres, 83); Paleont Asn; Sigma Xi; Soc Syst Zool. *Res:* Skeletal mineralogy and crystallography; theoretical morphology; paleoecology; computer applications; evolution; extinction. *Mailing Add:* Dept Geophys Sci Univ Chicago Chicago IL 60637i

RAUP, HUGH MILLER, b Springfield, Ohio, Feb 4, 01; m 25; c 2. BOTANY. *Educ:* Wittenberg Col, AB, 23; Univ Pittsburgh, AM, 25, PhD, 28. *Hon Degrees:* DSc, Wittenberg Univ, 68. *Prof Exp:* From instr to asst prof biol, Wittenburg Col, 23-32; asst, Arnold Arboretum, 32-34, res assoc, 34-38, asst prof plant ecol, 38-45, assoc prof plant geog, 45-49, prof bot, 49-60, prof forestry, 60-67, EMER PROF FORESTRY, HARVARD UNIV, 67-, DIR HARVARD FOREST, 46-; vis prof, Johns Hopkins Univ, 67-70. *Concurrent Pos:* Harvard Univ, 29-30; fel, Nat Res Coun, 30-32; mem bot surv expeds, Mackenzie River Basin, 26-30, 32, 35, 39, Alcan Hwy, 43, 44, Southwestern Yukon, 48 & Northeast Greenland, 56-58, 60, 64; vis prof, Johns Hopkins Univ, 67-70. *Mem:* AAAS; Ecol Soc Am; Asn Am Geog; Arctic Inst N Am. *Res:* Plant ecology and geography. *Mailing Add:* PO Box 325 Petersham MA 01366

RAUP, OMER BEAVER, b Washington, DC, July 14, 30; m 60; c 2. GEOLOGY, GEOCHEMISTRY. *Educ:* Am Univ, BS, 52; Univ Colo, PhD(geol), 62. *Prof Exp:* Geologist, 53-72, chief, Chem Resources Br, 72-76, RES GEOLOGIST, US GEOL SURVEY, 76- *Mem:* Am Asn Petrol Geologists; Geol Soc Am; Soc Econ Paleont & Mineral. *Res:* Geology of plateau uranium deposits; quadrangle and reconnaissance mapping of Washington and Oregon; clay mineralogy of redbeds; geology and mineralogy of marine evaporites. *Mailing Add:* 12295 Applewood Knolls Dr Lakewood CO 80215

RAUP, ROBERT BRUCE, JR, b New York, NY, Nov 4, 29; m 55; c 3. GEOLOGY. *Educ:* Columbia Univ, BA, 51; Univ Mich, MA, 52. *Prof Exp:* Geologist, Geol Div, 52-70, asst chief geologist environ geol, 70-74, GEOLOGIST, GEOL DIV, US GEOL SURV, 74- *Mem:* Geol Soc Am. *Res:* Geology of uranium in Precambrian rocks of central Arizona; geology and mineral deposits of southern Arizona; environmental geologic aspects of coal and oil shale, northwestern Colorado. *Mailing Add:* 6570 S Crestbrook Dr Morrison CO 80465

RAUSCH, DAVID JOHN, b Aurora, Ill, Oct 24, 40; m 62; c 6. ORGANIC CHEMISTRY, BIOCHEMISTRY. *Educ:* St Procopius Col, BS, 62; Iowa State Univ, PhD(org chem), 66. *Prof Exp:* Res assoc chem, Univ Wis, 65-66; from asst prof to assoc prof, 66-71, chmn dept, 77-82, PROF CHEM, ILL BENEDICTINE COL, 71- *Concurrent Pos:* Consult, Continental Can Co, 66-67, Argonne Nat Lab, 67-78, Res Corp, 70-72, Amoco Chem Co, 87-, McIntyre Chem, 89- *Mem:* AAAS; Am Chem Soc; Royal Soc Chem. *Res:* Organic reaction mechanisms, including beta-elimination reactions, free radical substitution reactions and pyrolysis reactions; photochemistry of aromatic compounds; nuclear magnetic resonance and spectral interpretation; stereochemistry and conformational analysis. *Mailing Add:* Dept Chem Ill Benedictine Col 5700 College Rd Lisle IL 60532

RAUSCH, DAVID LEON, b Marysville, Ohio, Apr 3, 37; m 61; c 3. AGRICULTURAL ENGINEERING, WATER QUALITY. *Educ:* Ohio State Univ, BAgrE, 60, MS, 61. *Prof Exp:* agr engr reservoirs, Agr Res, USDA, 65-87; FACIL DESIGN ENGR, OHIO STATE UNIV, 87- *Mem:* Am Soc Agr Engrs; Soil & Water Conserv Soc. *Res:* Control of reservoir sedimentation and eutrophication and how they are effected by the recently developed Automatic Bottom-Withdrawal Spillway; how water management effects crop yield and water quality of outflows. *Mailing Add:* Ohio State Univ-PREC PO Box 549 Piketon OH 45661-0549

RAUSCH, DOUGLAS ALFRED, b Ft Wayne, Ind, July 26, 28; m 53; c 3. ORGANIC CHEMISTRY. *Educ:* Univ Ind, BS, 50; Univ Colo, PhD(chem), 54. *Prof Exp:* Asst, Univ Colo, 50-53; res chemist, E I du Pont de Nemours & Co, 53-54; res mgr, 56-79, DIR PROD STEWARDSHIP, DOW CHEM CO, 79- *Mem:* AAAS; Am Chem Soc; Sigma Xi. *Res:* Fluorine and inorganic chemistry. *Mailing Add:* 4518 James Dr Midland MI 48640

RAUSCH, DOYLE W, b Dover, Ohio, May 3, 31; m 64; c 2. METAL FIBERS, WIRE REINFORCEMENT. *Educ:* Ohio State Univ, BMetE, 61, PhD(metall eng), 65. *Prof Exp:* Asst to plant mgr, US Ceramic Tile Co, 51-56, res asst, 58; tech asst process metall, Battelle Mem Inst, 59-60; res fel metall eng, Ohio State Univ Res Found, 60-64; asst prof, Ill Inst Technol, 65-68; assoc div chief, Mat Processing Div, Battelle Mem Inst, 68-73; asst dir res, 73-76, DIR RES, NAT-STANDARD CO, 76- *Mem:* Am Soc Metals; Am Inst Mining, Metall & Petrol Engrs; Am Soc Testing & Mat; Nat Asn Corrosion Engrs; Electrochem Soc. *Res:* Teflon processing; thermochemical and metallurgical process development; wire technology, metal fibers, wire reinforcement of polymeric materials; inorganic coating technology. *Mailing Add:* Nat-Standard Co Niles MI 49120

RAUSCH, GERALD, b New Hampton, Iowa, Mar 18; 38; m 60; c 1. ORGANIC CHEMISTRY. *Educ:* Univ Iowa, MS, 62, PhD(org chem), 63. *Prof Exp:* Res chemist, Chem Div, Union Carbide Corp, 63-65; from asst prof to assoc prof, 65-69, PROF ORG & ANALYTICAL CHEM, UNIV WIS-LA CROSSE, 69- *Mem:* Am Chem Soc. *Res:* Organic synthesis; N-acyl lactam chemistry; amines and amine derivatives; analytic separations, alcohols. *Mailing Add:* 2540 Sherwood Dr La Crosse WI 54601

RAUSCH, JAMES PETER, b Ravenna, Ohio, Aug 17, 38; m 65; c 3. PHYSIOLOGY, BIOMATERIALS RESEARCH. *Educ:* Kent State Univ, BA, 64, MA, 66, PhD(physiol), 71. *Prof Exp:* Instr physiol, Kent State Univ, 68-69; from asst prof to assoc prof, 69-82, PROF BIOL, ALFRED UNIV, 82-, CHMN, BIOL DIV, 81- *Mem:* Soc Biomat. *Res:* Ceramic-tissue interface response; physiology of marine and freshwater invertebrates; respiratory physiology. *Mailing Add:* Dept Biol Alfred Univ Alfred NY 14802

RAUSCH, MARVIN D, b Topeka, Kans, June 27, 30; m 83; c 1. ORGANOMETALLIC CHEMISTRY, ORGANIC CHEMISTRY. *Educ:* Univ Kans, BS, 52, PhD, 55. *Prof Exp:* NSF fel, Univ Munich, 57-59; sr res chemist, Monsanto Co, 59-63; assoc prof, 63-68, PROF CHEM, UNIV MASS, AMHERST, 68- *Concurrent Pos:* Alexander von Humboldt fel & vis prof, Univ Munich, 69-70 & 77, US Sr Scientist Award, Univ Bayreuth, 84-85, 90. *Mem:* Royal Soc Chem; Am Chem Soc. *Res:* Organic derivatives of transition metals; metallocene chemistry; organometallic catalysis; organometallic polymers; functionally substituted cyclopentadienyl-metal compounds. *Mailing Add:* Dept Chem Univ Mass Amherst MA 01003

RAUSCH, ROBERT LLOYD, b Marion, Ohio, July 20, 21; m 53; c 1. VETERINARY PARASITOLOGY. *Educ:* Ohio State Univ, BA, 42, DVM, 45; Mich State Col, MS, 46; Univ Wis, PhD(parasitol), 49. *Hon Degrees:* LLD, Univ Sask, 85; DSc, Univ Alaska, Fairbanks, 87. *Prof Exp:* Asst, Ohio State Univ, 43-45 & Mich State Col, 45-46; from asst to instr, Univ Wis, 46-49; parasitologist, Arctic Health Res Ctr, USPHS, Alaska, 49-51, chief, Infectious Dis Sect, 51-74; prof parasitol, Western Col Vet Med, Univ Sask, 75-78; assoc dir, Div Animal Med, 86-89, PROF PATHOBIOL & COMPARATIVE MED, UNIV WASH, 78- *Concurrent Pos:* Field work, Arctic & Subartic, 49- *Honors & Awards:* Arctic Sci Prize, 84. *Mem:* AAAS; Am Soc Parasitol; Am Micros Soc; Am Soc Mammal; Am Soc Trop Med Hyg; Am Vet Med Asn. *Res:* Host-parasite ecology and epizootiology of helminths and diseases in wildlife; diseases of wildlife in relation to public health; hydatid disease. *Mailing Add:* Dept Comp Med SB42 Sch Med Univ Wash Seattle WA 98195

RAUSCH, STEVEN K, b Aurora, Ill, Nov 17, 55. BIOPROCESSING RESEARCH. *Educ:* Purdue Univ, BA, 77; Univ Ill, PhD(biochem), 82. *Prof Exp:* SR RES SCIENTIST, BIOPROCESSING RES DEPT, PITMAN-MOORE INC, 82- *Mem:* Am Chem Soc; AAAS; Am Soc Biochem & Molecular Biol; Protein Soc; Sigma Xi. *Mailing Add:* Bioprocessing Res Dept Pitman-Moore Inc PO Box 207 Terre Haute IN 47808

RAUSCHER, FRANK JOSEPH, JR, b Hellertown, Pa, May 24, 31; m 55; c 5. MICROBIOLOGY. *Educ:* Moravian Col, BS, 53; Rutgers Univ, PhD(virol), 57. *Prof Exp:* Res asst virol, Rutgers Univ, 55-57, res assoc, 57-58, asst prof, 58-59; microbiologist, Lab Viral Oncol, Nat Cancer Inst, 59-64, head, Viral Oncol Sect, 64-66, chmn spec virus cancer prog, 64-70, chief, Viral Leukemia & Lymphoma Br, 66-67, assoc sci dir viral oncol, 67-70, sci dir etiol, 70-72, dir inst, 72-76; SR VPRES RES, AM CANCER SOC, 76- *Concurrent Pos:* Vis instr, Trinity Col, 59-; hon investr, Rutgers Univ, 62-67; mem fel adv bd, Nat Acad Sci, 70; sci coun, Int Agency Res Cancer, 72-74; mem expert adv panel on cancer, WHO, 72-77; mem bd dirs, Whittaker Corp, Los Angeles, 77- *Honors & Awards:* Selman A Waksman Hon Lectureship Award, Nat Acad Sci, 67; Arthur S Fleming Award, 68; Man of Sci Award, ARCS Found, 75; Charles R Drew Mem Cancer Award, Howard Univ, 77; David A Karnovsky Mem Award, Am Soc Clin Oncologists, 77; Papanicolaou Award, Papanicolaou Cancer Res Inst, 78; Selman A Waksman Award, Rutgers Univ, 78; Janeway Award, Am Radium Soc, 78. *Mem:* Am Asn Cancer Res; Am Asn Immunologists; Am Acad Microbiol; World Soc Leukemia & Related Dis; hon mem Am Soc Clin Oncol. *Res:* Comparative host responses to the necrotizing and to the tumorigenic viruses; known tumor-virus-host model systems to the search for etiologic agents in human neoplasms. *Mailing Add:* 112 Valley Forge Rd Weston CT 06883

RAUSCHER, GRANT K, b Sherrill, NY, Jan 5, 22; m 43; c 4. PHYSICAL CHEMISTRY. *Educ:* Colgate Univ, BA, 43; Rensselaer Polytech Inst, PhD(phys chem), 49. *Prof Exp:* Sr chemist, Air Reduction Co, 49-54; group leader chem res, Behr-Manning Div, Norton Co, NY, 55-58; group leader paper res, Int Paper Co, 58-66; tech dir, Eaton-Dikeman Div, 67-74, RES ASSOC, KNOWLTON BROS, INC, 74- *Honors & Awards:* Bolton-Emerson Award, 65. *Mem:* Am Chem Soc; Tech Asn Pulp & Paper Indust. *Mailing Add:* Knowlton Bros Inc Box 9552 Chattanooga TN 37412

RAUSCHER, TOMLINSON GENE, b Oneida, NY, May 27, 46; c 4. SOFTWARE ENGINEERING MANAGEMENT. *Educ:* Yale Univ, BS, 68; Univ NC, MS, 71; Univ MD, PhD(comput sci), 75; Univ Rochester, MBA, 84. *Prof Exp:* Software engr, Naval Res Lab, 72-75; mgr, software, Nat Cash Regist Corp, 75-76; sr comput archit, Amdahl Corp, 76-77; mgr, software, Gen Tel Elec, 77-78; MGR, SOFTWARE, XEROX CORP, 78- *Mem:* Inst Elec & Electronics Engrs; Asn Comput Mach. *Res:* Software engineering management; author of two books based on microcomputing. *Mailing Add:* Xerox Corp 800 Phillips Rd 105-47C Webster NY 14580

RAUSCHKOLB, ROY SIMPSON, b St Louis, Mo, Apr 18, 33; m 53; c 3. SOIL FERTILITY. *Educ:* Ariz State Univ, BA, 61; Univ Ariz, MS, 63, PhD(agr chem, soils), 68. *Prof Exp:* Cotton specialist, Agr Exten, Univ Ariz, 65-66, supt exp sta & adminr exp farms opers, Col Agr, 66-67, soil specialist, Agr Exten 67-69; soil specialist, Agr Exten & res assoc, Exp Sta, Univ Calif, Davis, 69-77, asst dir, 77-81, assoc dean, Col Agr & dir, coop exten resource sci & eng, 81-; AT DEPT SOILS, WATER & ENG, UNIV ARIZ, TUCSON. *Concurrent Pos:* Consult, United Nations Food & Agr Orgn, 71, 74; mem intergovt personnel act assignment staff, US Environ Protection Agency, R S Kerr Lab, 76-77. *Mem:* fel Soil Sci Soc Am; Am Soc Agron. *Res:* Soil-plant relationships; plant nutrition; soil and plant tissue testing; soil pollution; soil incorporation and recycling of plant residues and animal wastes; reactions and movement of plant nutrients in soils. *Mailing Add:* Dept Soils Water & Eng Univ Ariz Tucson AZ 85721

RAUSEN, AARON REUBEN, b Jersey City, NJ, June 30, 30; m 68; c 3. PEDIATRICS, HEMATOLOGY. *Educ:* Dartmouth Col, 47-50; State Univ NY Downstate Med Ctr, MD, 54. *Prof Exp:* USPHS fel pediat hemat, Children's Hosp, Boston & Harvard Med Sch, 59-61; from assoc prof to prof pediat, Mt Sinai Sch Med, 66-81; dir pediat, Beth Israel Med Ctr, 73-81; PROF PEDIAT & DIR PEDIAT ONCOL, NY UNIV MED CTR, 81- *Concurrent Pos:* Chief pediat, Greenpoint Hosp, 62-64 & City Hosp Ctr, Elmhurst, 64-73; from asst attend pediatrician to attend pediatrician, Mt Sinai Hosp, 63-81; mem acute leukemia study group B, NIH, 64-81; consult, USPHS Hosp, 71-81, Beekman-Downtown Hosp, 78- & Hackensack Hosp, 78-; vis physician, Rockefeller Univ Hosp, 78-; consult pediat hemat, Lenox Hill Hosp, 81-; responsible investr, Children Cancer Study Group. *Mem:* Am Asn Cancer Res; Am Pediat Soc; Am Soc Hemat; Am Acad Pediat; Am Soc Clin Oncol. *Res:* Disorders of blood in children and cancer in childhood; oncology. *Mailing Add:* NY Univ Med Ctr 530 First Ave New York NY 10016

RAUSER, WILFRIED ERNST, b Arlesheim, Switz, Oct 11, 36; Can citizen; m 61; c 2. PLANT PHYSIOLOGY, PLANT BIOCHEMISTRY. *Educ:* Univ Toronto, BSA, 59, MSA, 61; Univ Ill, Urbana, PhD(agron), 65. *Prof Exp:* Res officer agron, Exp Farm, Can Dept Agr, Sask, 61-62, res scientist plant physiol, Res Sta, Man, 65-66; asst prof plant physiol, 67-72, assoc prof, 72-82, PROF BOT, UNIV GUELPH, 82- *Concurrent Pos:* Fel, Div Biosci, Nat Res Coun Can, 66-67. *Mem:* Am Soc Plant Physiol; Can Soc Plant Physiol. *Res:* Physiological effects of excess metal ions on plants; mechanisms of metal ion toxicity and tolerance; plant metal binding proteins. *Mailing Add:* Dept Bot Univ Guelph Guelph ON N1G 2W1 Can

RAUSHEL, FRANK MICHAEL, b Hibbing, Minn, Dec 12, 49; m 75; c 3. ENZYMOLOGY, REACTION MECHANISMS. *Educ:* Univ Wis-Madison, PhD(biochem), 76. *Prof Exp:* Assoc prof, 86-89, PROF ORG CHEM, TEX A&M UNIV, 89- *Mem:* Sigma Xi; Am Chem Soc. *Res:* Analysis of enzyme reaction mechanisms by kinetic, magnetic resonance and genetic techniques. *Mailing Add:* Dept Chem Tex A&M Univ College Station TX 77843

RAUT, KAMALAKAR BALKRISHNA, b Bombay, India, Aug 10, 20; m 45; c 5. ORGANIC CHEMISTRY. *Educ:* Univ Bombay, BSc, 41, BA, 42, MSc, 46; Univ Okla, PhD(chem, pharmaceut chem), 59; Univ Ga, certs, 65, 66 & 67. *Prof Exp:* Res chemist, India Pharmaceut Labs, 46-55; asst, Univ Okla, 55-59; vis instr org chem, ETex State Univ, 59-60; sci officer, Cent Drug Res Inst, Govt India, Lucknow, 60-61; asst prof, Indian Inst Technol, Kanpur, 62-64; PROF CHEM, SAVANNAH STATE COL, 64- *Concurrent Pos:* Abstractor, Chem Abstr Serv, 59-; counr, Am Chem Soc, 71-73 & cong sci counr, 73-75, 80-88. *Honors & Awards:* Outstanding Chem award, Am Chem Soc, 87. *Mem:* Am Chem Soc; Indian Sci Cong Asn; Int Union Pure & Appl Chem. *Res:* Synthetic dyes; natural products; flavons; chalcones; chromones; synthetic medicinal products; reaction mechanisms; computer programming; chemical education; use of computers in chemical education; air pollution; water pollution. *Mailing Add:* Dept Chem Savannah State Col Savannah GA 31404

RAUTAHARJU, PENTTI M, b Tuusniemi, Finland, Dec 23, 32; m 57; c 4. PATHOPHYSIOLOGY, ELECTROCARDIOLOGY. *Educ:* Univ Helsinki, MD, 59; Univ Minn, PhD(biophys), 63. *Hon Degrees:* DrMed, Univ Kuopio, 78. *Prof Exp:* Res assoc electrophysiol, Int Inst Occup Health, 55-58; res fel electrocardiography, Univ Minn, 58-61, res assoc biophys, 61-62; asst prof physiol, 62-64, assoc prof, 64-66, PROF PHYSIOL & BIOPHYS, DALHOUSIE UNIV, 66- *Concurrent Pos:* Med Res Coun Can res scholar, 63-66; Med Res Coun Can res assoc, 66- *Mem:* Fel Am Col Cardiol; fel Coun Epidemiol. *Res:* Electrocardiology; computer electrocardiology; development of diagnostic criteria and new risk functions for identification of person athigh risk of future heat attacks and sudden death; development of electrocardiology software for clinical trials and health surveys; exercise electrocardiology; high resolution electrocardiology; ambulator electrocardiology; modeling of cardiac excitation and repolarization. *Mailing Add:* Heart Dis Res Ctr Four G Tupper Bldg Dalhousie Univ Halifax NS B3H 4H6 Can

RAUTENBERG, THEODORE HERMAN, b Cleveland, Ohio, May 14, 30; m 54. PHYSICS. *Educ:* Amherst Col, BA, 52. *Prof Exp:* Physicist, Light Prod Physics Br, Res & Develop Ctr, Gen Elec Co, 53-72, physicist, Plasma Physics Br, 72-74, physicist, Electronic Power Systs Br, 74-81, physicist, Lighting Systs Prog, 81-90; CONSULT, 90- *Mem:* Am Phys Soc; Optical Soc Am; Inst Elec & Electronics Engrs; Audio Eng Soc. *Res:* Fundamental studies of chemical and gas discharge light sources; experimental optical spectroscopy and photometry; electrooptical devices, application to light control systems; electroacoustics, design of sound reinforcement systems. *Mailing Add:* Res & Develop Ctr Gen Elec Co PO Box Eight Schenectady NY 12301

RAUTENSTRAUCH, CARL PETER, b New York, NY, Sept 19, 36; m 59; c 2. APPLIED MATHEMATICS. *Educ:* Univ Fla, BS, 58; Univ Ala, MA, 60; Auburn Univ, PhD(math), 67. *Prof Exp:* Instr math, Auburn Univ, 63-66; asst prof, Univ Tex, Arlington, 67-68; asst prof, 68-72, ASSOC PROF MATH, UNIV CENT FLA, 72- *Mem:* Am Math Soc; Soc Indust & Appl Math. *Res:* Special functions; complex analysis; differential equations. *Mailing Add:* 1490 Tuskawilla Rd Oviedo FL 32765

RAUTENSTRAUS, R(OLAND) C(URT), b Gothenburg, Nebr, Feb 27, 24; m 46; c 1. CIVIL ENGINEERING. *Educ:* Univ Colo, BS, 46, MS, 49. *Hon Degrees:* DL, Univ NMex, 76. *Prof Exp:* Instr civil eng, 47-50, from asst prof to assoc prof, 50-57, head, Dept Civil Eng, 59-64, assoc dean fac, 64-68, vpres educ & student rels, 68-70, vpres univ rels, 70-73, exec vpres, 73-74, PROF CIVIL & ENVIRON ENG, UNIV COLO, BOULDER, 57-, PRES, 74- *Concurrent Pos:* Consult, Travelers Ins Co, 58-59; mem educ panel, Esso Refining Humble Oil Co, 60; mem bd dirs, Northwest Eng Co, 69-; mem, Gov's Energy Task Force, 73-74; mem, Regional Adv Bd, Inst Int Educ, 75- *Honors & Awards:* Lincoln Gold Medal, Am Welding Soc; Robert L Stearns Award; Norlen Medal. *Mem:* Am Soc Photogram; Am Soc Eng Educ; Am Soc Civil Engrs. *Res:* Altimetry; photogrammetry; highway engineering. *Mailing Add:* Dept Civil Environ & Arch Eng Univ Colo Boulder CO 80309-0428

RAUTH, ANDREW MICHAEL, b Rochester, NY, Oct 8, 35; m 68; c 2. BIOPHYSICS. *Educ:* Brown Univ, BSc, 58; Yale Univ, PhD(biophys), 62. *Prof Exp:* Nat Cancer Inst grant, Ont Cancer Inst, Toronto, 62-65; asst prof biophys, 65-74, assoc prof, 74-79, PROF BIOPHYS, UNIV TORONTO, 79-; PHYSICIST, ONT CANCER INST, TORONTO, 65- *Mem:* Radiation Res Soc; Can Soc Cell Biol; Sigma Xi. *Res:* Radiation biology; mechanisms of drug action; somatic cell genetics; mutational processes in mammalian cells in vitro; radiobiology of solid tumors. *Mailing Add:* Exp Therapeut Ont Cancer Inst Toronto ON M4X 1K9 Can

RAVAL, DILIP N, b Bombay, India, June 3, 33; US citizen; m 61. PHYSICAL BIOCHEMISTRY. *Educ:* Univ Bombay, BS, 53, MS, 55; Univ Ore, PhD(chem), 62. *Prof Exp:* NIH fel, Univ Ore, 61-62; fel virus res, Univ Calif, Berkeley, 62-64; res scientist, Palo Alto Med Res Inst, 64-66; mgr res, Varian Assocs, 66-68; dir clin labs, Med Ctr, Univ Calif, San Francisco, 68-70; dir res, 70-72, gen mgr, Sci & Technol Div, 72-73, VPRES, RES & DEVELOP DIV, ALCON LABS, 73- *Mem:* Acad Clin Lab Physicians & Sci. *Res:* Enzyme kinetics protein structure; medical instrument development; pharmaceutical drug development. *Mailing Add:* 4405 Overton Terr Ft Worth TX 76109

RAVE, TERENCE WILLIAM, b Mendota, Ill, Aug 23, 38; m 65; c 3. SYNTHETIC POLYOLEFIN PULPS, PAPER CHEMISTRY. *Educ:* Bradley Univ, BS, 60; Univ Wis, PhD(org chem), 65. *Prof Exp:* Res chemist, Procter & Gamble Co, 65-67; from res chemist to sr res chemist, Hercules Inc, 67-76, res scientist, 76-80, res assoc, 80-84, mkt dir, 84-87, dir res, 88- 89, VPRES TECHNOL, HERCULES INC, 89- *Mem:* Am Chem Soc; Tech Asn Pulp & Paper Indust. *Res:* Organic nitrogen and phosphorous chemistry; paper chemistry; polymer synthesis and modification, synthetic polyolefin pulps; preparation, modification and applications of synthetic polyolefin pulps; paper wet and dry strength resins; polyolefin films; polyolefin fibers; reactive injection molding systems. *Mailing Add:* 2523 Blackwood Rd Wilmington DE 19810

RAVECHE, ELIZABETH MARIE, b Stuttgart, Ger, Nov 21, 50; US citizen; m 74; c 4. AUTOIMMUNITY, CELL CYCLE KINETICS. *Educ:* Seton Hill Col, BS, 72; George Washington Univ, PhD(genetics), 77. *Prof Exp:* Res scientist immunol, NIH, 77-85; ASSOC PROF IMMUNOL, ALBANY MED COL, 85- *Mem:* Am Asn Immunologists; Am Rheumatism Asn; Am Asn Pathologists; Tissue Cult Asn. *Res:* Murine models of autoimmunity; modes of inheritance of immunologic abnormalities and modulation by sex hormones; the study of mechanisms of B cell activation using flow cytometric techniques; abnormal lymphocyte differentiation and the development of aneuploidy in autoimmunity. *Mailing Add:* Dept Microbiol & Immunol Albany Med Col 47 New Scotland Ave Albany NY 12208

RAVECHE, HAROLD JOSEPH, b New York, NY, Mar 18, 43; m 74; c 4. CHEMICAL PHYSICS. *Educ:* Hofstra Univ, BA, 63; Univ Calif, San Diego, PhD(chem physics), 68. *Prof Exp:* Nat Res Coun-Nat Acad Sci assoc statist physics, Nat Bur Standards, 68-70, res chemist, 70-78, chief, thermophysics div, 78-85; dean sci, Rensselaer Polytech Inst, 85-88; PRES, STEVENS INST TECHNOL, 88- *Honors & Awards:* Electroendosmosis Award, Nat Bur Standards. *Mem:* Soc Indust & Appl Math; Am Phys Soc; AAAS. *Res:* Statistical mechanics of equilibrium and non-equilibrium phenomena; computer simulation. *Mailing Add:* Pres Stevens Inst Technol Castle Point on the Hudson Hoboken NJ 07030

RAVEED, DAN, b Baltimore, Md, Aug 12, 21; m 46; c 2. PLANT PHYSIOLOGY, BIOCHEMISTRY. *Educ:* Univ Calif, Berkeley, BS, 56; Univ Calif, Davis, PhD(plant physiol), 65. *Prof Exp:* Lab head soil & water relations, Water Authority, Israel, 54-55; lab technician, Univ Calif, Davis, 56-59, res chemist, 59-61, teaching asst plant physiol, 61-64, fel ion uptake, 67-68; assoc biochemist, Negative Inst, Beer Sheva, Israel, 64-67; staff scientist, Photobiol Br, C F Kettering Res Lab, 68-72, head, Electron Micros Lab, 72-77; asst prof path, Dept Electron Microscope Lab, Ind Sch Med, 77-80; pesticide chem, Ind State Bd Health, 80-86; qual assurance officer, Off Ind Dept Environ Mgt, 86-89; TECH DIR LAB PACK, HERITAGE ENVIRON SERV, 89- *Mem:* AAAS; Am Soc Cell Biologists; Japanese Soc Plant Physiol; Scand Soc Plant Physiol; Electron Micros Soc Am. *Res:* Biochemical ultrastructure of functional enzyme complexes from membranes of chloroplasts, bacterial chromatophores, retinas; immuno-electron microscopy; membrane response to stress, pollutants and inhibitors; physiology of photosynthesis; ion uptake by plants. *Mailing Add:* Heritage Environ Serv 5459 B Char Dr Indianapolis IN 46241

RAVEENDRAN, EKARATH, b India, Jan 1, 50; m 78; c 2. ANALYTICAL CHEMISTRY, POLLUTION MONITORING. *Educ:* Univ Kerala, India, BSc, 70. *Prof Exp:* Chemist, Therapeut Res Corp, India, 70 & Hoodlass Nerelae Paints, Bombay, India, 71-77; analyst, Bahrain Petrol Co, India, 77-83; SR CHEMIST, ENVIRON PROTECTION, BAHRAIN, INDIA, 83- *Mem:* Am Chem Soc; Royal Soc Can; Fel Asn Inst Chem; Mem Inst Water Pollution Control. *Res:* Environmental pollution monitoring of air, land and water; public health hazards; toxic waste survey and disposal. *Mailing Add:* Environ Protection PO Box 26909 Adliya Bahrain India

RAVEL, JOANNE MACOW, b Austin, Tex, July 28, 24; m 46; c 2. BIOCHEMISTRY,MOLECULAR BIOLOGY. *Educ:* Univ Tex, BS, 44, MA, 46, PhD(chem), 54. *Prof Exp:* Res scientist, Clayton Found Biochem Inst, 44-53, Hite fel, 54-56, res scientist, 56-70, asst dir, 70-85, from assoc prof to prof chem, Univ Tex, Austin, 72-85, RES SCIENTIST, CLAYTON FOUND BIOCHEM INST, UNIV TEX, AUSTIN, 85- *Mem:* Am Soc Biochem & Molecular Biol; Am Chem Soc; Am Soc Plant Physiologists; Am Soc Microbiol. *Res:* Eukaryotic protein synthesis and its regulation; biological control mechanisms. *Mailing Add:* Dept Chem Univ Tex Austin TX 78712

RAVELING, DENNIS GRAFF, b Devils Lake, NDak, Feb 28, 39; m 62. WILDLIFE BIOLOGY. *Educ:* Southern Ill Univ, BA, 60; Univ Minn, MA, 63; Southern Ill Univ, PhD(zool, physiol), 67. *Prof Exp:* Res biologist, Ont Dept Lands & Forests, 67; res scientist, Can Wildlife Serv, 67-70; from asst prof to assoc prof, 71-80, chmn dept, 80-83, PROF WILDLIFE BIOL, UNIV CALIF, DAVIS, 80- *Honors & Awards:* Spec Recognition Award for Leadership in Wildlife Teaching & Res, Wildlife Soc, 89; Publ Award, Wildlife Soc, 90. *Mem:* Fel AAAS; Wildlife Soc; Am Ornith Union; Wilson Ornith Soc; Am Soc Naturalists. *Res:* Taxonomy; zoogeography; physiology; behavior; population dynamics and management of birds. *Mailing Add:* Dept Wildlife & Fisheries Biol Univ Calif Davis CA 95616

RAVEN, CLARA, b Russia, May 9, 07. FORENSIC PATHOLOGY, BACTERIOLOGY. *Educ:* Univ Mich, BA, 27, MS, 28; Northwestern Univ, MD, 38. *Prof Exp:* Colonel, US Army, 43-65; emer dep chief med examr, Wayne County, 59-72; RETIRED. *Honors & Awards:* Elizabeth B Blackwell Award, Am Med Writers Asn, 82. *Mem:* Am Pub Health Asn; Am Acad Forensic Sci; AAAS; AMA; Am Med Writers Asn; Am Med Womens Asn; Am Soc Clin Pathologists; Int Asn Pathologists; Am Soc Pathologists & Bacteriologists; Am Soc Trop Med & Hyg Acteriol. *Res:* Dissociation of bacteria; methods of water analysis and coliform organisms; stress ulcers secondary to trauma; streptococci; leptospirosis; gonococcus; teratoma of mediastinum; epidemic hemorrhagic fever; parasitism in the Army; virus pneumonitis; sudden deaths in infants. *Mailing Add:* 1419 Nicolet Pl Detroit MI 48207

RAVEN, FRANCIS HARVEY, b Erie, Pa, July 29, 28; m 52; c 7. MECHANICAL ENGINEERING. *Educ:* Gannon Col, BS, 48; Pa State Univ, BS, 50, MS, 51; Cornell Univ, PhD(mech eng), 58. *Prof Exp:* Anal design engr, Hamilton Standard Div, United Aircraft Corp, 50-54; instr mech eng, Cornell Univ, 54-58; from asst prof to assoc prof, 58-66, PROF MECH ENG, UNIV NOTRE DAME, 66- *Mem:* Assoc Am Soc Mech Engrs; Am Soc Eng Educ. *Res:* Kinematics and automatic control systems. *Mailing Add:* Dept Aerospace & Mech Eng Univ Notre Dame Notre Dame IN 46556

RAVEN, PETER BERNARD, CARDIOVASCULAR RESPONSES. *Educ:* Univ Ore, Eugene, PhD(phys educ), 69. *Prof Exp:* Assoc prof, 77-86, PROF PHYSIOL, TEX COL OSTEOP MED, 86- *Res:* Exercise physiology; work physiology; cardiovascular responses to exercise. *Mailing Add:* Dept Physiol Tex Col Osteop Med Camp Bowie at Montgomery Ft Worth TX 76107

RAVEN, PETER HAMILTON, b Shanghai, China, June 13, 36; US citizen; m 58, 68; c 4. BOTANY, POPULATION BIOLOGY. *Educ:* Univ Calif, Berkeley, AB, 57; Univ Calif, Los Angeles, PhD(bot), 60. *Hon Degrees:* DSc, St Louis Univ, 82, Knox Col, 83, Southern Ill Univ, Edwardsville, 83, Miami Univ, 86, Univ. Gutenberg, 87 & Rutgers Univ, 88. *Prof Exp:* NSF fel, Brit Mus Nat Hist, 60-61; botanist, Rancho Santa Ana Bot Garden, 61-62; from asst prof to assoc prof biol sci, Stanford Univ, 62-71; ENGELMANN PROF BOT, WASH UNIV, 71-; DIR, MO BOT GARDEN, 71- *Concurrent Pos:* NSF grants, 61-; Guggenheim fel & sr res fel, Dept Sci Indust Res, NZ, 69-70; res assoc, Calif Acad Sci, 70-; mem comt res & explor, Nat Geog Soc, 82-; mem bd dirs, World Wildlife Fund-US, 83-; res assoc bot, Bernice P Bishop Mus, 85-88; John D & Catherine T MacArthur Found fel, 85-89; hon dir, Inst Bot Acad Sinica, Beijing, 88- *Honors & Awards:* Int Environ Leadership Medal, UN Environ Prog, 82; Intern Prize Biol, Govt Japan, 86. *Mem:* Nat Acad Sci(home secy, 87-); Am Acad Arts & Sci; Soc Study Evolution (vpres, 68, 72, pres, 78); Am Soc Naturalists (pres, 83); Am Soc Plant Taxon (pres, 72); fel AAAS; Am Inst Biol Sci (pres, 83-84); foreign mem Royal Danish Acad Sci & Lett; Orgn Trop Studies (treas, 81-84, vpres, 84-85, pres, 85-86); foreign mem Royal Swed Acad Sci; Am Philos Soc. *Res:* Taxonomy, especially Onagraceae; general botany; biogeography; taxonomic theory; biosystematics; cytogenetics; geography; ethnobotany; taxonomic theory; conservation biology; cytogenetics; pollination systems. *Mailing Add:* Mo Bot Garden PO Box 299 St Louis MO 63166

RAVENEL, DOUGLAS CONNER, b Alexandria, Va, Feb 17, 47; m 83; c 4. MATHEMATICS. *Educ:* Oberlin Col, BA, 69; Brandeis Univ, MA, 69, PhD(math), 72. *Prof Exp:* Instr math, Mass Inst Technol, 71-73; asst prof, Columbia Univ, 73-76; mem staff, Inst Advan Study, 74-75; from asst prof to prof math, Univ Wash, 76-88; PROF MATH, UNIV ROCHESTER, 88- *Concurrent Pos:* Alfred P Sloan Found res fel, 77-79. *Mem:* Am Math Soc. *Res:* Algebraic topology; complex cobordism theory; homotopy theory. *Mailing Add:* Dept Math Univ Rochester Rochester NY 14627

RAVENHALL, DAVID GEOFFREY, b Birmingham, Eng, Mar 4, 27; US citizen; m 52; c 2. THEORETICAL PHYSICS. *Educ:* Univ Birmingham, BSc, 47, PhD(electrodynamics), 50. *Prof Exp:* Dept Sci & Indust Res sr fel theoret physics, Univ Birmingham, 50-51; res physicist, Carnegie Inst Technol, 51-52;

mem, Inst Advan Study, 52-53; res assoc theoret physics, Stanford Univ, 53-57; from asst prof to assoc prof, 57-63, PROF THEORET PHYSICS, UNIV ILL, URBANA, 63- *Concurrent Pos:* NSF sr fel, 63-64. *Mem:* Am Phys Soc. *Res:* Theoretical nuclear physics at intermediate energies; particle physics, mesoscopic physics, dense matter. *Mailing Add:* Dept Physics 237B Loomis Lab Univ Ill 1110 W Green St Urbana IL 61801

RAVENHOLT, REIMERT THOROLF, b Milltown, Wis, Mar 9, 25; m 48, 81; c 4. EPIDEMIOLOGY, PUBLIC HEALTH. *Educ:* Univ Minn, BS, 48, MB, 51, MD, 52; Univ Calif, Berkeley, MPH, 56; Am Bd Prev Med, dipl & cert pub health, 60. *Prof Exp:* Intern, USPHS Hosp, San Francisco, 51-52; mem staff, Epidemic Intel Serv, Nat Commun Dis Ctr, USPHS, Ga, 52-54; dir epidemiol & commun dis div, Seattle-King County Dept Pub Health, 54-61; epidemiol consult, Europ Area, USPHS, Am Embassy, Paris, France, 61-63; assoc prof prev med, Sch Med, Univ Wash, 63-66; chief pop br, Health Serv, Off Tech Coop & Res, Develop Support Bur, 66-67, dir pop serv, Off War on Hunger, 67-69, dir, Off Pop, Tech Assistance Bur, 69-72, dir, Off Pop, Pop & Humanitarian Assistance, 72-77, dir, Off Pop, 77-79; dir, World Health Surv, Ctrs Dis Control, Rockville, Md, 80-82; asst dir res, Nat Inst Drug Abuse, Rockville, Md, 82-84; CHIEF, EPIDEMIOL BR, OFF EPIDEMIOL & BIOSTATIST, FOOD & DRUG ADMIN, ROCKVILLE, MD, 84- *Concurrent Pos:* Originator & mem prog steering comn, World Fertil Surv, 71- *Honors & Awards:* John J Sippy Mem Award, Am Pub Health Asn, 61; Distinguished Honor Award, AID, 72; Hugh Moore Mem Award, 74; Carl Schultz Award, Am Pub Health Asn, 79. *Mem:* AAAS; fel Am Pub Health Asn; fel Am Col Epidemiol. *Res:* Population; preventive medicine; infectious diseases; immunization; diseases of smoking and ionizing radiation; malignant cellular evolution; mortality and fertility patterns; contraceptive development; population dynamics. *Mailing Add:* 3156 E Laurelhurst Dr NE Seattle WA 98105

RAVENTOS, ANTOLIN, IV, b Wilmette, Ill, June 3, 25; m 76. RADIATION ONCOLOGY. *Educ:* Univ Chicago, SB, 45, MD, 47; Univ Pa, MSc, 55. *Prof Exp:* From asst instr to prof radiol, Sch Med, Univ Pa, 51-70; chmn dept, 70-80, PROF RADIOL, SCH MED, UNIV CALIF, DAVIS, 70- *Concurrent Pos:* Spec consult, Nat Cancer Inst, Surgeon Gen Army, 61-62 & Armed Forces Radiobiol Res Inst, 64-67; consociate mem, Nat Coun Radiation Protection & Measurements; pres, Am Registry Radiol Technol, 66. *Mem:* Am Med Writers Asn; Am Radium Soc (pres, 72); Radiol Soc NAm; Radiation Res Soc; Am Soc Cancer Educ. *Res:* Radiation therapy; radioactive isotopes in medicine; radiobiology. *Mailing Add:* Vet Admin Med Ctr Univ Calif Sch Med 150 Muir Rd Martinez CA 94553

RAVENTÓS-SUÁREZ, CARMEN ELVIRA, b Lima, Peru, May 27, 47; m 81; c 1. ANALYTICAL CYTOLOGY, CELL CYCLE ANALYSIS. *Educ:* San Marcos Univ, Lima, Peru, BASc, 70, biologist, 73, PhD(biol), 85; Univ Chile, Santiago, lic in sci, 76. *Prof Exp:* Tech assoc, Bact Inst Chile, 70-71; vol virol, Virol Unit, Sch Med, Univ Chile, 72-73, instr, 74-77; researcher exp hemat, Albert Einstein Col Med, NY, 78-79, assoc, 79-85; RES SCIENTIST ONCOL & IMMUNOL, MED RES DIV, AM CYANAMID CO, 85- *Concurrent Pos:* Vol neoplastic dis, Atran Labs, Mt Sinai, NY, 78; researcher, NY Blood Ctr, 78. *Mem:* Int Soc Anal Cytol; Am Asn Immunologists. *Res:* Characterization of early progenitor cells from the mouse bone marrow and their interaction with cytokines by fluorescence activated cell sorter methodology, response and activation by cytotoxic and immunomodulator drugs. *Mailing Add:* Med Res Div-Oncol & Immunol Am Cyanamid Co Pearl River NY 10965

RAVICZ, ARTHUR EUGENE, b New Rochelle, NY, Oct 28, 30; m 57; c 3. CHEMICAL ENGINEERING. *Educ:* Univ Colo, BS, 52; Univ Tex, MS, 55; Univ Mich, PhD(chem eng), 59. *Prof Exp:* Res engr, Calif Res Corp, 52 & 58-63, sr res engr, 63-67, sr eng assoc, 67-81, supv process engr, 81-88, SR STAFF PROCESS ENGR, CHEVRON CORP, 88- *Concurrent Pos:* Instr, Exten Div, Univ Calif, 63-64, lectr, 70. *Mem:* Am Inst Chem Engrs. *Res:* Distillation; applications of automatic computers in chemical engineering; petroleum and petrochemical process design. *Mailing Add:* Chevron Res & Technol Co 100 Chevron Way Richmond CA 94802-0627

RAVILLE, MILTON E(DWARD), b Malone, NY, July 12, 21; m 43; c 6. ENGINEERING. *Educ:* Norwich Univ, BS, 43; Kans State Univ, MS, 47; Univ Wis, PhD(mech), 55. *Prof Exp:* Instr appl mech, Kans State Univ, 47-50, from asst prof to assoc prof, 50-56, prof & head dept, 56-62; PROF & DIR SCH ENG SCI & MECH, GA INST TECHNOL, 62- *Concurrent Pos:* Res engr, Forest Prod Lab, Wis, 54-55 & Gen Dynamics, 63. *Mem:* Am Soc Eng Educ; Nat Soc Prof Engrs; Soc Eng Sci; Am Acad Mech. *Res:* Stress analysis and vibrations of solid bodies; analysis of sandwich structures. *Mailing Add:* Sch Eng Sci/Mech Ga State Tech Atlanta GA 30332

RAVIN, LOUIS JOSEPH, pharmacy, for more information see previous edition

RAVINDRA, NUGGEHALLI MUTHANNA, b Hyderabad, India, Oct 1, 55; m 84; c 2. MICROELECTRONICS & OPTOELECTRONICS. *Educ:* Bangalore Univ, BS Hons, 74, MS, 76; Roorkee Univ, PhD(physics), 82. *Prof Exp:* Res scientist, Ctr Nat Res Sci, 82-85 & Int Ctr Theoret Physics, 83-85; vis scientist, Microelectronics Ctr NC & NC State Univ, 85-86; res assoc prof mat sci, Vanderbilt Univ, 86-87; ASSOC PROF PHYSICS, NJ INST TECHNOL, 87- *Concurrent Pos:* Prin investr, Bell Commun Res, 88-, Sematech Ctr Excellence, 88-90 & NJ Comn Sci & Technol, 90- *Mem:* Inst Elec & Electronics Engrs; Soc Photo-Optical Instrumentation Engrs; Mat Res Soc; Sigma Xi; Electrochem Soc. *Res:* Material science and technology; applications in infra-red detectors, solar cells, mos devices and cmos device technology; silicon processing and technology. *Mailing Add:* Dept Physics NJ Inst Technol Newark NJ 07102

RAVINDRA, RAVI, b Patiala, India, Jan 14, 39; m 65; c 2. GEOPHYSICS, COSMOLOGY. *Educ:* Indian Inst Technol, Kharagpur, BSc, 59, MTech, 61; Univ Toronto, MSc, 62, PhD(physics), 65; Dalhousie Univ, MA, 68. *Prof Exp:* From asst prof to assoc prof physics & philos, Dalhousie Univ, 66-73; vis fel, Hist & Philosophy of Sci, Princeton Univ, 68-69; vis scholar, Dept Relig, Columbia Univ, 73-74; assoc prof, 75-79, PROF PHYSICS & RELIG DALHOUSIE UNIV, 79- *Concurrent Pos:* Res grants, Nat Res Coun Can, 66-72, Geol Surv Can, 66- & Dom Observ Can, 66-; Can Coun fel philos & Killam res fel, 68-69; vis fel philos sci, Princeton Univ, 68-69; Can Coun res grant, 72-77; cross-disciplinary fel, Soc Relig Higher Educ, Columbia Univ, 73-74; Can Coun fel, 73-74; Can Coun Leave fel, 77-78; Shastri Indo-Can Inst sr fel, 77-78; Soc Sci & Humanities Res Coun Can res grant, 77-82; dir, Threshhold Award, 78-80; vis mem Inst Advan Study, Princeton Univ, 77. *Mem:* Am Asn Physics Teachers; Am Soc Study Relig; Can Soc Study Relig. *Res:* Relativistic cosmology; yoga and consciousness; philosophy of religion, particularly spiritual traditions; philosophy and history of science; comparative study of cultures. *Mailing Add:* Depts Physics & Relig Dalhousie Univ Halifax NS B3H 3J5 Can

RAVINDRAN, NAIR NARAYANAN, b Vechoor, India, Nov 25, 34; m 66; c 1. ORGANIC CHEMISTRY. *Educ:* Univ Kerala, India, BS, 58, MS, 60; Purdue Univ, PhD(chem), 72. *Prof Exp:* Sci officer radiochem, Bhabha Atomic Res Ctr, Bombay, 58-67; res assoc org chem, Purdue Univ, 72-74; res chemist, 74-80, SR RES CHEMIST, EASTMAN KODAK CO, 80- *Mem:* Am Chem Soc. *Res:* Design and synthesis of photographically useful organic compounds to meet the needs of future photographic products of the company; design and building of color photographic film products. *Mailing Add:* 148 Montmor Ency Dr Rochester NY 14612

RAVIOLA D'ELIA, GIUSEPPINA E(NRICA), anatomy, animal physiology; deceased, see previous edition for last biography

RAVITCH, MARK MITCHELL, surgery; deceased, see previous edition for last biography

RAVITSKY, CHARLES, b New York, NY, May 25, 17; m 40; c 2. APPLIED PHYSICS, OPTICS. *Educ:* City Col NY, BS, 38, MS, 39. *Prof Exp:* Teacher high sch, NY, 38; from jr physicist to prin physicist, Nat Bur Stand, 41-53; prin physicist, Diamond Ord Fuze Labs, 53-54, chief systs res sect, 54-58, res supvr, 58-62; chief scientist & chief physics br, US Army Res & Develop Group, Europe, 62-67; sr prog mgr, 67-72, asst to dir tactical technol, Defense Advan Res Projs Agency, Off Secy Defense, 72-75; physicist, Cerberonics, Inc, 77-78; STAFF SCIENTIST, SWL INC, 78- *Concurrent Pos:* Assoc physics, George Washington Univ, 43-50, lectr, 50-54; asst chmn electronic scientist panel, Bd Civil Serv Exam, Nat Bur Standards, 50-55, chmn, 55-59, mem bd, 55-61; mem, US Deleg, NATO Panel IV Optics & Infrared, 71-75; consult, Battelle Columbus Labs, 75-88. *Honors & Awards:* US Naval Ord Develop Award, 45; Awards, Diamond Ord Fuze Labs, 55, US Army Sci Conf, 62 & Harry Diamond Labs, 65; Outstanding Performance, Defense Advan Res Proj Agency, 75. *Mem:* Am Phys Soc; Optical Soc Am; Soc Photo-Optical Instrumentation Engrs. *Res:* Administration of research and development programs; communications research and development; surveillance; military systems utilizing infrared or visible radiation; development of submersible vehicles; remote sensors; anti-submarine warfare; intrusion detection sensors; airborne reconnaissance systems. *Mailing Add:* 1505 Drexel St Takoma Park MD 20912-7032

RAVITZ, LEONARD J, JR, b Cuyahoga City, Ohio, Apr 17, 25. ELECTROMAGNETIC FIELD MEASUREMENTS, HYPNOSIS. *Educ:* Case Western Reserve Univ, BS, 44; Wayne State Univ, MD, 46; Yale Univ, MS, 50: Am Bd Psychiat & Neurol, cert, 52. *Prof Exp:* Intern, St Elizabeths Hosp, Washington, DC, 46-47; asst resident psychiat, Yale-New Haven Hosp, 47-49; res fel Neuro-Anat Sect, Yale Med Sch, 49-50; sr resident neuropsychiat, Duke Univ Hosp, Durham, NC, 50-51, assoc, Pvt Diag Clin, 51-53; asst dir, prof educ, Dorrey Vet Admin Hosp, North Chicago, Ill, 53-54; assoc, Dept Psychiat, Sch Med & Hosp, Univ Pa, 55-58; dir training & res, Eastern State Hosp, Williamsburg, Va, 58-60; psychiatrist & consult, Div Alcohol Studies & Rehab, Va Dept Mental Health & Ment Retardation, Norfolk Alcohol Serv, 61-81; consult, Nat Inst Rehab Therapy, Butler, NJ, 82-83; CLIN ASST PROF PSYCHIAT, STATE UNIV NY HEALTH SCI CTR & DOWNSTATE MENT HYG ASN, 83- *Concurrent Pos:* Pvt consult, Cleveland, Ohio, 60-69; mem staff Med Ctr Hosp, Norfolk Gen Div, Va, 61-; lectr, Int Conf Rhythmic Functions in the Living Systs, NY Acad Sci, 61; lectr sociol, Old Dominion Univ, Norfolk, Va, 61-62, consult Nutrit Res Proj & res prof psychol, 78-90; consult, Tidewater Epilepsy Found, Chesapeake, 62-68, spec med consult, Frederick Mil Acad, Portsmouth, 63-71, US Pub Health Hosp Alcohol Unit, Norfolk, VA, 80-81; guest lectr, Fourth Int Conf Hypn & Paychosom Med, Guttenburg Univ, Mainz, WGer, 70; featured lectr, Significance Field Measurements in Hypn, Health & Dis, 14th Ann Meeting, Am Soc Clin Hypn, Chicago, 71; asst ed, J Am Soc Psychosom, Dent & Med, Brooklyn, 80-83; psychiatrist, Greenpoint Multiserv Ctr, Brooklyn, 83-87, 17th St Clin, Manhattan, 87-; teacher, Nursing Sch Affil, Univ Wyo. *Mem:* Fel NY Acad Sci; AAAS; fel Am Phychiat Asn; Sigma Xi; Am Asn Clin Hypn; fel Royal Soc Health. *Res:* Medical disorders and aging; electrocyclic phenomena in humans which parallel those of other life forms, earth and atmosphere; electromagnetic field measurements of meditation and hypnosis; health, aging, emotional states, psychiatric and medical disorders; author and coauthor of numerous articles in journals, magazines and textbooks. *Mailing Add:* Dept Psychiat Med Sch State Univ Ny Health Sci Ctr 450 Clareborn Ave Box 1203 Brooklyn NY 11203

RAVIV, JOSEF, b Slonim, Poland, Mar 11, 34; US citizen; m 56; c 3. APPLIED MATHEMATICS, COMPUTER SCIENCE. *Educ:* Stanford Univ, BS, 55, MS, 60; Univ Calif, Berkeley, MA, 63, PhD(elec eng), 64. *Prof Exp:* Res staff mem, T J Watson Res Ctr, 64-72, mgr, IBM Israel Sci Ctr, 72-85, MGR, IBM ISRAEL SCI & TECHNOL, HAIFA, 85- *Concurrent Pos:* Lectr, Univ Conn, 65-; assoc prof, Technion, Israel, 71-86; pres, Info Processing Asn Israel. *Mem:* Fel Inst Elec & Electronics Engrs; Asn Comput Mach. *Res:* Decision making; pattern recognition; data compaction; recognition of continuous speech. *Mailing Add:* IBM Israel Sci & Technol Technion City Haifa 32000 Israel

RAW, CECIL JOHN GOUGH, b Ixopo, SAfrica, Oct 20, 29; US citizen; m 56; c 4. PHYSICAL CHEMISTRY. *Educ:* Univ Natal, BSc, 51, MSc, 52, PhD(phys chem), 56. *Prof Exp:* Lectr chem, Univ Natal, 54-57, sr lectr, 58-59; res assoc, Univ Minn, 59-60; from asst prof to assoc prof chem, 60-66, chmn dept, 83-88, PROF CHEM, ST LOUIS UNIV, 66- *Concurrent Pos:* African Explosives & Chem Industs res fel, 57-59; summer vis asst prof, Univ Minn, 60 & Univ Md, 62. *Mem:* Am Chem Soc; Am Asn Univ Professors; Sigma Xi. *Res:* Microcomputers in chemistry; chaotic and oscillating chemical systems. *Mailing Add:* Dept Chem St Louis Univ St Louis MO 63103

RAWAL, KANTI M, b Karachi, Pakistan, Sept 25, 40; m 72; c 2. CROP IMPROVEMENT, SCIENTIFIC INFORMATION SYSTEMS. *Educ:* Gujarat Univ, India, BSc, 61, MSc, 64; Univ Ill, PhD(genetics), 69. *Prof Exp:* Fel biochem & genetics, Univ Ill, 69-70; asst prof genetics, Univ Ibadan, Nigeria, 70-72; geneticist plant genetics, Int Inst Trop Agr, Nigeria, 72-75; chief scientist theory orgn, Lab Info Sci & Genetic Resources, Univ Colo, Boulder, 75-78, plant breeding, 78-80; PLANT BREEDER, DEL MONTE CORP, R J REYNOLDS INDUST, 80- *Concurrent Pos:* Mem, Int Wheat Descriptor Comt, Food & Agr Orgn, Rome, 75-80, Int Sorghum Germplasm Comt, 76-80 & Tech Adv Comt Wheat, Sorghum, Peas, Beans & Tomatoes, Sea Sci & Educ Admin, Agr Res, USDA, 78-80. *Mem:* Fel Linnaean Soc; Am Genetic Asn; Am Soc Agron; Crop Sci Soc Am; Soc Econ Bot. *Res:* Plant breeding of tomatoes, dry legumes, cucumbers; germplasm resources exploration, utilization and management; population biology of native plants; computerized information management for agriculture; statistics; biosystematics; crop evolution and tropical agricultural ecosystems. *Mailing Add:* 571 Mitchell Ave San Leandro CA 94577

RAWALAY, SURJAN SINGH, organic chemistry, biochemistry, for more information see previous edition

RAWAT, ARUN KUMAR, b Uttar Pradesh, India, Sept 19, 45; m 74. BIOCHEMISTRY. *Educ:* Univ Lucknow, BSc, 62, MSc, 64; Univ Copenhagen, DSc(biochem), 69. *Prof Exp:* NIH fel, 69-70; instr med, City Univ New York, 70; asst prof psychiat, State Univ NY Downstate Med Ctr, 70-72; assoc prof psychiat & biochem, Med Col Ohio, 73-78; PROF PHARMACOL, UNIV TOLEDO, 78- *Concurrent Pos:* Dir div neurochem, Dept Psychiat, State Univ NY Downstate Med Ctr, 70-72; dir, Alcohol Res Ctr, 73- *Mem:* Am Soc Biol Chemists; Am Soc Neurochem; Am Soc Clin Res. *Res:* Neurochemistry of alcoholism; effects of alcohol on fetus; mechanisms of addiction; effect of alcohol on protein synthesis in brain. *Mailing Add:* Dept Pharmacol Univ Toledo 4235 Monroe St Toledo OH 43613-5888

RAWAT, BANMALI SINGH, b Garhwal, UP, India, July 2, 47; US citizen; m 77; c 2. MICROWAVES, OPTICAL FIBER COMMUNICATIONS. *Educ:* Banaras Hindu Univ, India, BS, 68, MS, 70; Sri Venkasteswara Univ, India, PhD(elec eng), 76. *Prof Exp:* Instr engr, W Coast Paper Mills, India, 71-72; sr res fel, BITS-Pilani, SV Univ Tirupati, 72-75; sr scientist res, Defense Res & Develop, Govt India, 75-78; assoc prof elec eng, Univ Gorakhpur, India, 78-80, prof & head, 80-81; assoc prof elec eng, Univ NDak, 81-86, prof, 86-88; PROF & HEAD ELEC ENG, UNIV NEV, RENO, 88- *Concurrent Pos:* Prin investr, IBM, E F Johnson Co, II Morrow Projs, 82-90; consult, E F Johnson Co, Waseca, Minn, 82-88, UPS/II Morrow Inc, Salem, Ore, 89-90; chmn adv comt, ISRAMT, 89-92. *Mem:* Sr mem Inst Elec & Electronics Engrs; Am Soc Eng Educ; Electromagnetics Acad; Sigma Xi. *Res:* Microwave integrated circuits; mm-waves; dielectric waveguides; microstrip antennas; mobile communication filters; dielectric resonators; EM numerical techniques; optical fibers. *Mailing Add:* Elec Eng Dept Univ Nev Reno NV 89557

RAWITCH, ALLEN BARRY, b Chicago, Ill, Dec 29, 40; m 62; c 2. BIOCHEMISTRY, FORENSIC CHEMISTRY. *Educ:* Univ Calif, Los Angeles, BS, 63, PhD(biochem), 67. *Prof Exp:* Res chemist, Wadsworth Vet Admin Hosp, Los Angeles, 62-63; res assoc biochem, Univ Ill, Urbana, 67-69; from asst prof to assoc prof chem, Kent State Univ, 69-75; assoc prof, 75-82, PROF BIOCHEM, MED SCH, UNIV KANS, 82- *Concurrent Pos:* Fel, Univ Ill, Urbana, 67-69; scientist, Mid Am Cancer Ctr Prog; dir Med Biochem, Univ Kans Med Ctr, 76-87, vchmn, 81-; dir, Biotech Support Facil, 88-91. *Honors & Awards:* Res Career Develop Award, NIH, 72. *Mem:* AAAS; Am Chem Soc; Am Soc Biol Chemists; Sigma Xi; Am Thyroid Asn. *Res:* Physical and chemical properties of proteins; application of fluorescence spectroscopy to study macromolecules; structure of thyroid proteins; thyroid hormone biosynthesis; structure of pancreatic hormones; comparative endocrinology; drug analyses techniques; gas-phase micro sequencing of peptide and proteins; solid-phase peptide synthesis. *Mailing Add:* Dept Biochem & Molecular Biol Univ Kans Med Ctr Kansas City KS 66103

RAWITSCHER, GEORGE HEINRICH, b Freiburg, Ger, Feb 27, 28; US citizen; m 57, 82; c 2. THEORETICAL NUCLEAR PHYSICS. *Educ:* Univ Sao Paulo, Brazil, AB, 49; Stanford Univ, PhD(physics), 56. *Prof Exp:* Instr physics, Brazil Ctr Invest Physics, 50-52; asst, Stanford Univ, 53-56; instr, Univ Rochester, 56-58; instr, Yale Univ, 58-61, asst prof, 61-64; assoc prof, 66-72, PROF PHYSICS, UNIV CONN, 72- *Concurrent Pos:* Alexander V Humboldt Stiftung fel, Max Planck Inst Nuclear Physics, 64-66; Brazilian Army Res, 50-71. *Mem:* Am Phys Soc. *Res:* Elementary particle and nuclear physics; mu mesons; scattering and reaction theory in nuclear physics. *Mailing Add:* Dept Physics Univ Conn Storrs CT 06268

RAWLING, FRANK L(ESLIE), JR, b Lowell, Mass, Dec 2, 35; m 66; c 2. CHEMICAL ENGINEERING. *Educ:* Lowell Tech Inst, BS, 59; Univ Maine, MS, 61; Iowa State Univ, PhD(chem eng), 64. *Prof Exp:* Res asst chem eng, Iowa State Univ, 61-64; res engr, 64-68, sr res engr, Textile Fibers Dept, 68-80, proj engr, 81-83, CONSULT, ENG DEPT, E I DU PONT DE NEMOURS & CO, INC, 83- *Mem:* Am Inst Chem Engrs. *Res:* Polymer processing, especially with regard to man-made fibers; mass transfer; mixing technology, process design. *Mailing Add:* 32 Carriage Lane Covered Bridge Farms Newark DE 19711

RAWLINGS, CHARLES ADRIAN, b Paducah, Ky, Nov 11, 36. INSTRUMENTATION. *Educ:* Univ Ill, BS, 59; Southern Ill Univ, MS, 65, PhD(eng & physiol), 74. *Prof Exp:* Engr, Sperry Utah Eng Labs, div Sperry Rand Corp, 59-61; field eng training rep, Autonetics Div, NAm Aviation, 61-65, mem tech staff, clin eng, 66-69, sr logistics field engr, Space & Info Syst, 65; from lectr to asst prof, 64-77, ASSOC PROF, DEPT ELEC ENG, SOUTHERN ILL UNIV, CARBONDALE, 77-, DIR, BIOMED ENG, 81- *Concurrent Pos:* Mem, Bd Examrs Cert Biomed Equip Technicians, 73-75, chmn, 75-79, chmn Cert Comn, 79-81; qualified instr, Defense Civil Preparedness Agency, 70-; mem bd dir, Asn Advan Med Instrumentation Foundations, 85-, pres, Asn Advan Med Instrumentation, 81-83, chmn bd dir, 83-84, ed, Med Instrumentation, 87-; dir Sem Biomed Instrumentation, 72-; asst chmn, Dept Elec Eng, Southern Ill Univ, Carbondale, 87- *Mem:* Sr mem Instrument Soc Am; Inst Elec & Electronics Engrs; Asn Advan Med Instrumentation (pres, 81-83); Am Soc Hosp Eng. *Res:* Medical instrumentation, especially that related to cardiovascular system; rehabilitation engineering; sensory physiology; effects of electricity on humans. *Mailing Add:* Elect Eng Southern Ill Univ Carbondale IL 62901

RAWLINGS, CLARENCE ALVIN, b Olney, Ill, Apr 18, 43; m 67; c 2. VETERINARY SURGERY. *Educ:* Univ Ill, BS, 65, DVM, 67; Colo State Univ, MS, 69; Univ Wis, PhD(vet med), 74; Am Col Vet Surgeons, dipl. *Prof Exp:* Staff vet, Humane Soc Mo, St Louis, 67-68; Nat Defense Educ Act fel surg, Col Vet Med, Colo State Univ, 68-69; chief & asst chief exp surg, USAF Sch Aerospace Med, Brooks AFB, Tex, 69-72; fel phsyiol, Univ Wis, 72-74; ASSOC PROF SURG, UNIV GA, 72- *Mem:* Am Vet Med Asn; Acad Vet Cardiol; Am Soc Vet Anesthesiol; Am Soc Vet Physiologists & Pharmacologists; Am Heart Asn. *Res:* Understanding cardiopulmonary function in clinical conditions; heartworm disease, anesthesia, shock, cardiac tamponade, and congenital heart disease; prospective clinical studies involve soft tissue surgery. *Mailing Add:* Col Vet Med Univ Ga Athens GA 30602

RAWLINGS, GARY DON, b Houston, Tex, Feb 6, 48; m 69, 82; c 3. ENVIRONMENTAL SCIENCE, PHYSICS. *Educ:* Southwest Tex State Univ, BS, 70, MS, 71; Tex A&M Univ, PhD(environ sci, eng), 74. *Prof Exp:* Res asst & grant, Tex Exp Sta, Tex A&M Univ, 73-74; contract mgr & eng specialist environ res & develop, Dayton Lab, 74-83, int mkt mgr, nuclear sources, 83-84, BUS DEVELOP MGR, PHOSPHATE FIBER, MONSANTO, CORP, 84- *Mem:* Air Pollution Control Asn; Am Inst Chem Engrs; Sigma Xi; Soc Advan Mat & Process Eng; Soc Plastic Indust. *Res:* Solution to environmentally related problems; detection systems; analytical techniques; control technology alternatives; new business development. *Mailing Add:* Monsanto 800 N Lindbergh Blvd St Louis MO 63167

RAWLINGS, JOHN OREN, b Archer, Nebr, July 26, 32; m 52; c 3. BIOMETRICS. *Educ:* Univ Nebr, BS, 53, MS, 57; NC State Col, PhD, 60. *Prof Exp:* Geneticist, Agr Res Serv, USDA, 59-60; asst statist, 60-61, from asst prof to assoc prof, 61-68, PROF STATIST, DEPT STATIST, NC STATE UNIV, 68- *Concurrent Pos:* Assoc ed, Biometrics, Biomet Soc, 75; statist consult, Nat Crop Loss Assessment Network, 81-89; chmn, Environ Protection Agency acid precipation rev, Am Statist Asn, 83-84. *Mem:* Am Soc Agron; Crop Sci Soc Am; Biomet Soc; Am Statist Asn; AAAS. *Res:* Design and analysis of experiments. *Mailing Add:* Dept Statist NC State Univ Box 8203 Raleigh NC 27695-8203

RAWLINGS, SAMUEL CRAIG, b Wichita, Kans, Sept 7, 38; m; c 3. BEHAVIORAL SCIENCES, VISION RESEARCH. *Educ:* Calif State Univ, Fullerton, BS, 64; Univ Miami, MS, 68 & PhD(psychol), 70. *Prof Exp:* Asst prof, Univ Houston, 71-74, dir, glaucoma & training, Nat Eye Inst, NIH, 75-77, asst prof & dir res, ophthal, Univ Tex Health Sci Ctr, 77-80, exec secy, human develop & aging study, Sect Div Grants, NIH, 80-86, CHIEF, BEHAV & NEUROL SCI REV SECT, DIV RES GRANTS, NIH, 86- *Concurrent Pos:* Ed bd, Peer Rev Notes, Nat Inst Health, 86- *Mem:* Am Psychol Asn. *Res:* Supervise behavioral and neurosciences review section which is composed of 18 Nat Inst Health study sections. *Mailing Add:* Div Res Grants NIH 5333 Westbard Ave Westwood Bldg Rm 310 Bethesda MD 20892

RAWLINS, NOLAN OMRI, b McRae, Ga, Nov 30, 38; m 64; c 4. AGRICULTURAL ECONOMICS. *Educ:* Univ Ga, BSA, 61, MS, 63; Tex A&M Univ, PhD(agr econ), 68. *Prof Exp:* Asst prof econ & sociol, Middle Ga Col, 67-68; from asst prof to assoc prof, 68-81, PROF AGR, MID TENN STATE UNIV, 81- *Mem:* Nat Asn Col & Teachers Agr. *Mailing Add:* Dept Agr Mid Tenn State Univ Murfreesboro TN 37132

RAWLINS, STEPHEN LAST, b Lewiston, Utah, May 29, 32; m 52; c 7. SOILS, PHYSICS. *Educ:* Brigham Young Univ, BS, 54; Washington State Univ, MS, 56, PhD(soils), 61. *Prof Exp:* Soil physicist, Conn Agr Exp Sta, 60-64; res soil scientist, Soil & Water Conserv Res Div, USDA, 64-71, supvry soil scientist, US Salinity Lab, Sci & Educ Admin, 71-87, NAT PROG LEADER SOIL EROSION/GLOBAL CHANGE, USDA, 87- *Concurrent Pos:* Adj prof, Univ Calif, Riverside, 71-80. *Mem:* Am Soc Agron; Am Soc Plant Physiol. *Res:* Physics of water movement in soils and plants; instrument for measurement of energy status of water in soils and plants; modeling soil-plant-water system. *Mailing Add:* USDA-ARS NPS Rm 233 Bldg 005 BARC-West Beltsville MD 20705

RAWLINS, WILSON TERRY, b Edinburg, Tex, Nov 8, 49; m 75; c 2. PHYSICAL CHEMISTRY, CHEMICAL KINETICS. *Educ:* Univ Tex, Austin, BS & BA, 72; Univ Pittsburgh, PhD(chem), 77. *Prof Exp:* head aeronomy & surface sci group, Aerospace Sci Area, 86-87, PRIN SCIENTIST, PHYS SCI INC, 77-, MGR CHEM SCI, 87- *Mem:* Am Geophys Union; Combustion Inst; Soc Photo-Optical Instrumentation Engrs. *Res:* Gas phase kinetics, photochemistry and spectroscopy; chemistry of planetary atmospheres; combustion chemistry; gas-surface interactions. *Mailing Add:* Physical Sci Inc 20 New England Business Ctr Andover MA 01810

RAWLINSON, DAVID JOHN, b Manchester, Eng, May 14, 35. ORGANIC CHEMISTRY. *Educ:* Oxford Univ, BA, 57, PhD(chem), 63. *Prof Exp:* Res assoc, Univ Ore, 59-61; tech officer, Plant Protection Ltd, Eng, 61-63; chemist, Agr Chem Div, Shell Develop Co, US, 63-65; fel, Ill Inst Technol, 65-66; fel & lectr, Univ Wis-Milwaukee, 66-68; from asst prof to assoc prof, 68-77, PROF CHEM, WESTERN ILL UNIV, 77- *Res:* Free radical chemistry; chemistry of peroxides. *Mailing Add:* Dept Chem Western Ill Univ Macomb IL 61455

RAWLINSON, JOHN ALAN, b Liverpool, Eng, Nov 30, 40; Can citizen; m 67; c 3. MEDICAL PHYSICS. *Educ:* Univ London, BSc Hons, 63; Univ Toronto, MSc, 70. *Prof Exp:* Clin physicist, Ont Cancer Inst, 65-68, sr clin physicist, 70-77; head physicist, Inst Radiother, Brazil, 77-78; PHYSICIST-IN-CHG HIGH ENERGY SECT, ONT CANCER INST, 78- *Concurrent Pos:* Mem tech comt, 62, Can Nat Comt, Int Electrotech Comn, 76- *Mem:* Can Asn Physicists; Am Asn Physicists in Med; Brazilian Asn Med Physics. *Res:* Improvement in the radiation characteristics of equipment used in radiation therapy; improvements in treatment techniques in radiation therapy. *Mailing Add:* Eight Dunhill Toronto ON M1C 1Y4 Can

RAWLS, HENRY RALPH, b Chattahoochee, Fla, Nov 19, 35; m 78; c 1. BIOMEDICAL MATERIALS, CONTROLLED RELEASE. *Educ:* La State Univ, BS, 57; Fla State Univ, PhD(phys chem), 64. *Prof Exp:* Res chemist, Unilever Res Lab, Unilever NV, Holland, 64-67; res chemist, Gulf S Res Inst, 68-73, sr res chemist & mgr physics dept, 73-76; from asst prof to assoc prof biomat, Sch Dent, La State Univ, 77-85; staff res & develop scientist, Gillette Co, Boston, 85-87; PROF, UNIV TEX HEALTH SCI CTR, SAN ANTONIO, 87- *Concurrent Pos:* Vis scientist, Forsyth Dent Ctr, Boston, 78 & staff assoc, 85-87; Fogarty sr fel, Mat Tech Lab, Univ Groningen, Holland, 80; consult, Johnson & Johnson, 72-79, Gulf Res Inst, 76-86, Univ Mich Dent Sch, 79-84, Bausch & Lomb Co, 89- & Oral-B Labs, 90-; adj assoc prof biomed eng, Tulane Univ, 80-84; prin investr res grants, NIH, 74-, res career awardee, 77-79; res assoc, Polymer Res Inst, State Univ NY, Syracuse, 82- *Mem:* Am Asn Dent Res; Am Chem Soc; Int Asn Dent Res; Acad Dent Mat; Europ Orgn Caries Res. *Res:* Applications of physical, surface and polymer chemistry to biomedical problems; oral diseases; development of restorative materials and oral-care products for dentistry. *Mailing Add:* Div Biomat Univ Tex Health Sci Ctr 7703 Floyd Curl Dr San Antonio TX 78284-7890

RAWLS, JOHN MARVIN, JR, b Madison, Tenn, May 12, 46; m 69; c 2. GENE STRUCTURE & EXPRESSION. *Educ:* Univ S Ala, BS, 69; Univ NC, Chapel Hill, PhD(zool), 73. *Prof Exp:* Postdoctoral fel genetics, Univ Calif, Berkeley, 73-75; from asst prof to assoc prof, 75-90, PROF BIOL SCI, T H MORGAN SCH BIOL SCI, UNIV KY, 90- *Concurrent Pos:* Vis scientist, Eucaryotes Molecular Genetics Lab, Strasbourg, France, 81-82, Sch Biol Sci, Univ Sussex, UK, 88; dir, Sch Biol Sci, Univ Ky, 88- *Mem:* Genetics Soc Am. *Res:* Molecular genetics of coordinate gene expression in animal cells, using the pyrimidine biosynthesis genes in Drosophila melanogaster as a model experiment station. *Mailing Add:* 101 Morgan Bldg Univ Ky Lexington KY 40506

RAWLS, WILLIAM EDGAR, virology, epidemiology; deceased, see previous edition for last biography

RAWNSLEY, HOWARD MELODY, b Long Branch, NJ, Nov 20, 25; m 67; c 2. PATHOLOGY. *Educ:* Haverford Col, AB, 49; Univ Pa, MD, 52; Am Bd Path, cert anat path, 57, clin path, 58. *Prof Exp:* From assoc to assoc prof clin path, Sch Med, Univ Pa, 57-69, prof path, 69-75, from asst dir to dir, William Pepper Lab, 60-75, assoc dir clin res ctr, Univ Hosp, 62-70; vchmn dept, 75-79, PROF PATH, DARTMOUTH MED SCH, 75-, CHMN DEPT, 79- *Concurrent Pos:* Trustee, Am Bd Path. *Mem:* AAAS; AMA; Am Asn Clin Chem; Col Am Path; Am Soc Clin Path; Am Bd Path. *Res:* Clinical chemistry; liver disease; serum proteins; interpretation of laboratory information. *Mailing Add:* Dept Path Dartmouth Med Sch Hanover NH 03756

RAWSON, ERIC GORDON, b Saskatoon, Sask, Mar 4, 37; m 66; c 3. OPTICS. *Educ:* Univ Sask, BA, 59, MA, 60; Univ Toronto, PhD(physics), 66. *Prof Exp:* Mem tech staff optics, Bell Tel Labs, 66-73; mem res staff optics, 73-80, MGR I&TT AREA, OPTICS, XEROX PALO ALTO RES CTR, 80- *Mem:* Fel Optical Soc Am; Inst Elec & Electronics Engrs; Soc Photo Instrumentation Engrs. *Res:* Fiber-optical waveguides and systems; optics of display systems; light scattering; 3-dimensional displays. *Mailing Add:* Xerox Palo Alto Res Ctr 3333 Coyote Hill Rd Palo City CA 94304

RAWSON, JAMES RULON YOUNG, b Boston, Mass, July 28, 43; m 70; c 2. PLANT MOLECULAR BIOLOGY. *Educ:* Cornell Univ, BS, 65; Northwestern Univ, PhD(biol), 69. *Prof Exp:* NIH fel, Univ Chicago, 69-71, trainee biophys, 71-72; from asst prof to assoc prof bot & biochem, Univ Ga, 72-83, prof bot & genetics, 83-84; res assoc, 84-89, SR RES ASSOC, BP AM, 89- *Concurrent Pos:* NSF res grants, 73-75, 75-77, 77-79, 79-81, 80-82 & 82-85; Res Corp grant, 75-76; USDA grant, 80-82. *Mem:* Am Soc Biochem & Molecular Biol; Am Chem Soc. *Res:* Microbial genetics; biochemistry. *Mailing Add:* BP Am 4440 Warrensville Center Rd Cleveland OH 44128

RAWSON, RICHARD RAY, b Loma Linda, Calif, Dec 31, 28; m 52; c 6. GEOLOGY. *Educ:* Brigham Young Univ, BS, 56, MS, 57; Univ Wis, PhD(geol), 66. *Prof Exp:* Res geologist, Continental Oil Co, 57-63; asst prof geol, Emory Univ, 66-67; from asst prof to assoc prof geol, Northern Ariz Univ, 67-80; sr geologist, 80-82, district geologist, 82-84, div staff geologist, 84-88, REGIONAL STAFF GEOLOGIST, MARATHON OIL CO, 88- *Mem:* Am Asn Petrol Geol; Soc Econ Paleont & Mineral. *Res:* Depositional environments carbonate sediments; stratigraphy; sedimentation; basin analysis; petroleum exploration. *Mailing Add:* 16214 Peach Bough Lane Houston TX 77095

RAWSON, ROBERT ORRIN, b E Saint Louis, Ill, Apr 25, 17; c 3. PHYSIOLOGY. *Educ:* Univ Ill, Urbana, BS, 40; Loyola Univ, Ill, PhD(physiol), 60. *Prof Exp:* Res metallurgist, Am Zinc Co, Ill, 40-44; radio & TV broadcasting, St Louis & Chicago, 44-56; instr biol, Univ Ill, Chicago Circle, 56-58; instr physiol, 61-68, asst prof epidemiol, 68-75, SR RES ASSOC & LECTR, SCH MED, YALE UNIV, 75- *Concurrent Pos:* Fel physiol, Loyola Univ, Ill, 60-61; from asst fel physiol to fel, John B Pierce Found, Conn, 61-82, fel emer, 82. *Mem:* Am Physiol Soc. *Res:* Physiology of temperature regulation; nervous control of circulation; physiology and pharmacology of the autonomic nervous system. *Mailing Add:* John B Pierce Found Conn 273 Legend Hill Madison CT 06443

RAWSON, RULON WELLS, medicine; deceased, see previous edition for last biography

RAY, AJIT KUMAR, b Calcutta, India, Feb 1, 25; m 56; c 2. APPLIED MATHEMATICS, FLUID DYNAMICS. *Educ:* Univ Calcutta, BSc, 44, MSc, 47; Univ Göttingen, DSc(math, natural sci), 55. *Prof Exp:* Prof appl math, Asutosh Col, Univ Calcutta, 48-56; reader aeronaut sci, Indian Inst Sci, Bangalore Univ, 56-60; assoc res officer, Nat Aeronaut Estab Nat Res Coun Can, 61-64; math adv & res scientist, Dept Transport, Govt Can, 64-65; asst prof appl math, Clarkson Tech, 65-66; prof appl math & dir res, Univ Ottawa, 62-82; RETIRED. *Concurrent Pos:* Reviewer, Appl Mech Rev, 66-84; vis prof, Ctr Advan Study Appl Math, Univ Calcutta, 70 & 76; sect chmn boundary layer theory, Third Canadian Appl Mech Cong, Calgary, Can, 71; sessional chmn, Euromech Colloquium-27, Poland, 72; vis res prof, Indian Inst Sci, Bangalore & Indian Inst Technol, Kanpur, Jadavpore Univ, India, 76; consult, 80-; adv, Fundmental Res Inst, Can, 84; invited lectr, NATO-Adv Studies Inst, Liege, Belg 10th Int Cong Aeronaut Sci, Ottawa, Can, 76; Int Cong, Math, Warsaw, Poland, 83; Int Conf Immunol & differential equations, Toronto, Can, 86; invited lectr, Dept Appl Math, Indian Inst Sci Banglore, Calcutta Univ; adv World Comn Environ Develop, Ottawa, Can, 86; Alexander von Humbolt Scholar, Gottingen Univ, WGer, 52-54; sem lectr, Nat Inst Oceanog, India, 88 & Chinese Acad Sci, Beijing, 90. *Honors & Awards:* Spec Foreign Scholar, Calcutta Univ India, 54; Cert of Merit, Am Inst Aeronaut & Astronaut, 84; Fifth Wright Mem Lectr, Indian Aeronaut Soc, Bangalore Br, 88; Sir G I Taylor Mem Lectr, 35th Cong Indian Soc Theoret & Appl Mech, Madras, 90. *Mem:* Fel AAAS; Am Math Soc; assoc fel Am Inst Aeronaut & Astronaut; fel Inst Math & Appln UK; fel Royal Aeronaut Soc. *Res:* Non-linear partial differential equations in fluid dynamics and aerodynamics; numerical mathematics class, functional analysis and discrete mathematics. *Mailing Add:* Innes Park 2767 Innes Rd Suite 318-A Gloucester ON K1B 4L4 Can

RAY, ALDEN E(ARL), b Centralia, Ill, Feb 14, 31; m 52; c 4. PHYSICAL METALLURGY. *Educ:* Southern Ill Univ, BA, 53; Iowa State Univ, PhD(metall), 59. *Prof Exp:* Jr chemist, Ames Lab, Iowa State Univ, 53-56, res asst, 56-59; res metallurgist, Res & Eng Div, Monsanto Chem Co, 59-61; assoc prof, Univ Dayton, 61-71, dir grad prog mat sci, 71-76, supvr, Metals & Cramics Div, 74-87, SR METALLURGIST, RES INST, UNIV DAYTON, 61-, PROF, SCH ENG, 71-, DIR MAGNETICS LAB, 89- *Mem:* Fel Am Soc Metals; Sigma Xi; Am Soc Testing & Mat; sr mem Inst Elec & Electronics Engrs; Magnetics Soc. *Res:* Structure-property relationships of intermetallic phases, especially of rare earth transition metal alloys; magnetic properties; phase stability; phase diagrams. *Mailing Add:* Res Inst Univ Dayton Dayton OH 45469-0170

RAY, ALLEN COBBLE, b Jacksonville, Tex, Nov 17, 41; m 73. VETERINARY TOXICOLOGY, ANALYTICAL TOXICOLOGY. *Educ:* Univ Tex, Austin, BS, 64, PhD(chem), 71. *Prof Exp:* Teaching asst chem, Univ Tex, Austin, 64-67; res scientist, Clayton Found, Biochem Inst, Univ Tex, 67-71, res assoc, 72-73; vet toxicologist, 73-87, HEAD DRUG TESTING LAB, TEX VET MED DIAG LAB, TEX A&M UNIV, 87- *Concurrent Pos:* Vis mem, Dept Vet Physiol & Pharmacol, Tex A&M Univ, 78-; mem, safety, Am Asn Vet Diag Comm, 87-, anal toxicol, 88-; assoc ref, Asn Off Anal Chem, 87-89; chair, SW Asn Toxicol, 87- *Mem:* Am Chem Soc; Am Asn Vet Lab Diag; Soc Toxicol; Am Inst Chem; Asn Off Anal Chem. *Res:* Development of analytical and diagnostic methods in veterinary and human toxicology; chemistry modes of action and metabolism of naturally occurring and environmental toxins. *Mailing Add:* TV MDL PO Drawer 3040 College Station TX 77840

RAY, APURBA KANTI, b Calcutta, India, Sept 9, 43; US citizen; m 81; c 1. HIGH INTENSITY SWEETENERS. *Educ:* Jadavpur Univ, Calcutta, BSc, 61, MSc, 63, PhD(kinetics & solution chem), 69. *Prof Exp:* Instr chem, Pratt Inst, 69-70 & Brooklyn Col, City Univ NY 70-71-; interdisciplinary res fel neurol, Albert Einstein Col Med, Bronx, NY, 71-73, res assoc lung surfactant physico-chem, 73-79; res scientist gum & emulsions, 79-82, SR SCIENTIST, GUMS, EMULSIONS & SWEETENERS, COCA-COLA CO, ATLANTA, 82- *Mem:* Am Chem Soc; fel Am Inst Chemists; Inst Food Technologists; Int Asn Colloid & Interface Scientists. *Res:* Emulsion science and technology; sensory analyses of high intensity sweeteners relevant to beverage systems. *Mailing Add:* Corp Res & Develop Coca-Cola Co PO Drawer 1734 Atlanta GA 30301

RAY, ASIT KUMAR, b Calcutta, India, Jan 23, 54; m 85. MATHEMATICAL MODELING, HEAT & MASS TRANSFER. *Educ:* Indian Inst Technol, BTech, 75; Clarkson Col Technol, MS, 77, PhD(chem eng), 80. *Prof Exp:* Fel, Univ NMex, 80; from asst prof to assoc prof, 80-88, PROF CHEM ENG, UNIV KY, 88- *Honors & Awards:* Kenneth T Whitby Award, Am Asn Aerosol Res, 84; Presidential Young Investr Award, NSF, 84. *Mem:* Sigma Xi; Am Inst Chem Engrs; NY Acad Sci; Am Asn Aerosol Res; Fine Particle Soc; Sigma Xi. *Res:* Formation and growth of aerosol particles; experimental and theoretical studies on single particle systems to understand mass transfer phenomena in continuum and non-continuum regimes. *Mailing Add:* 500 Lake Tower Dr Unit 101 Lexington KY 40502

RAY, CHARLES, JR, b Baltimore, Md, Dec 6, 11; m 37; c 1. GENETICS. *Educ:* Lafayette Col, AB, 37; Univ Va, PhD(genetics), 41. *Prof Exp:* Geneticist, Plant Res Dept, Cent Fibre Corp, 41-52; from asst prof to prof, 52-80, EMER PROF BIOL, EMORY UNIV, 80- *Concurrent Pos:* Mem, Marine Biol Lab, Woods Hole Oceanog Inst. *Mem:* Genetics Soc Am. *Res:* Population genetics; computer simulation. *Mailing Add:* 532 Princeton Way Atlanta GA 30307

RAY, CHARLES DEAN, b Americus, Ga, Aug 1, 27; m 52; c 4. NEUROSURGERY, SURGERY OF THE SPINE. *Educ:* Emory Univ, AB, 50; Univ Miami, MS, 52; Med Col Ga, MD, 56; Am Col Surgeons, FACS, 69. *Hon Degrees:* FRSH, Royal Soc Health, London, 70. *Prof Exp:* Res fel bioeng, Mayo Found & Clin, 62-64; asst prof, neurosurg & lectr bioeng, Johns Hopkins Hosp & Univ, 64-68; vdir, med eng, Roche, Inc, Basel, Switz, 68-73; vpres, neurol devices, Medtronic, Inc, 73-80; assoc prof neurosurg, Univ Minn, 73-83; ASSOC DIR SPINE SURG, INST LOW BACK CARE, 80-; PRES & CHMN BD RES & DEVELOP, SPINE, CEDAR SURG, INC, 86- *Concurrent Pos:* Ed, Med Eng, 68-74, Med Progress Through Technol, 69-74; lectr, Univ Basel, Switz, 68-72; consult, WGer Armed Forces Med Soc, 69-71; Europ Reg dir, Inst Elec & Electronics Engrs, 69-72; indust rep, Food & Drug Admin Panel on Neurol Devices, 76-79; chief deleg, ISO Comt on Neurol Devices, 76-80; comt chair, World Fedn Neurosurg Socs, 80-91. *Honors & Awards:* Sci Award, Bausch & Lomb, Inc, 47; Golden Spine Award, Challenge of the Lumbar Spine Soc, 87. *Mem:* AMA; fel Am Col Surgeons; fel Royal Soc Health Eng; sr mem Inst Elec & Electronics Engrs; Sigma Xi; fel NAm Spine Soc (pres, 91-92). *Res:* Author of 235 publications; granted 32 patents; spinal surgical devices and methods; artificial human disc. *Mailing Add:* 19550 Cedarhurst Wayzata MN 55391

RAY, CLARENCE THORPE, b Hutto, Tex, July 17, 16; m 42; c 1. INTERNAL MEDICINE. *Educ:* Univ Tex, BA, 37, MD, 41; Am Bd Internal Med & Am Bd Cardiovasc Dis, dipl. *Prof Exp:* Intern, Scott & White Hosp, Temple, Tex, 41-42; resident, Parkland Hosp, Dallas, 42-44, dir outpatient dispensary, 43-44; from instr to assoc prof med, Sch Med, Tulane Univ, 45-58, dir heart sta, 47-58; prof med & chmn dept, Sch Med & physician & chief med serv, Hosp, Univ Mo-Columbia, 58-67; dir educ & res, Alton Ochsner Med Found, 67-75; chmn dept, 75-82, PROF MED, SCH MED, TULANE UNIV, 75- *Concurrent Pos:* Instr, Southwestern Med Found, Inc, 43-44; from asst vis physician to sr vis physician, Charity Hosp, 45-58; consult, Vet Admin, 53-58, USPHS, 55-58 & Vet Admin Hosps, New Orleans & Alexandria, La, 57-58; head sect cardiol, Dept Med, Ochsner Clin, New Orleans, La, 68-75, trustee & vpres, Alton Ochsner Med Found, 70-75. *Mem:* Am Soc Clin Invest; AMA; fel Am Col Physicians. *Res:* Cardiovascular disease. *Mailing Add:* 473 Woodvine Ave Metairie LA 70005

RAY, CLAYTON EDWARD, b New Castle, Ind, Feb 6, 33; m 53; c 4. VERTEBRATE PALEONTOLOGY. *Educ:* Harvard Univ, BA, 55, MA, 58, PhD(geol), 62. *Prof Exp:* Asst cur vert paleont, State Mus & asst prof biol, Univ Fla, 59-63; assoc cur later cenozoic mammals, 64-68, CUR QUATERNARY & MARINE MAMMALS, NAT MUS NATURAL HIST, SMITHSONIAN INST, 69- *Mem:* Am Soc Mammal; Soc Vert Paleont. *Res:* Systematics, morphology, distribution and evolution of Cenozoic mammals including living ones; emphasis on pinnipeds. *Mailing Add:* Nat Mus Natural Hist Smithsonian Inst Washington DC 20560

RAY, DALE C(ARNEY), b Highland Park, Mich, Aug 31, 33; m 53; c 4. ELECTRICAL ENGINEERING, SOLID STATE PHYSICS. *Educ:* Univ Mich, BSE, 56, MSE, 57, PhD(cryomagnetics), 62. *Prof Exp:* Res asst comput design & technol, Eng Res Inst, Univ Mich, 56-57, instr elec eng, Univ, 57-62, admin dir solid state devices lab, 60-61, asst prof elec eng, Univ, 62-66; assoc prof, 66-77, assoc dean, Div Grad Studies & Res, 69-77, PROF ELEC ENG, GA INST TECHNOL, 77- *Concurrent Pos:* Consult, Off Res Admin, Univ Mich, 57-, Power Equip Div, Lear Siegler, Inc, 62-63, Sensor Dynamics, Inc, 63-64 & Lockheed-Ga Co, 67-; Ford Found fel, 62-63; prin engr, Raytheon Corp, 63-64. *Mem:* AAAS; Am Phys Soc; Inst Elec & Electronics Engrs; Am Asn Physics Teachers; Am Soc Eng Educ. *Res:* Cryomagnetics; magnetic measurements and application; heat transfer; non-electromechanical energy conversion; microwave integrated circuit technology and studies in sociotechnology. *Mailing Add:* 1607 Barclay Pl Atlanta GA 30306

RAY, DAN S, b Memphis, Tenn, Dec 27, 37; c 2. MOLECULAR BIOLOGY, BIOPHYSICS. *Educ:* Memphis State Univ, BS, 59; Western Reserve Univ, MS, 61; Stanford Univ, PhD(biophysics), 65. *Prof Exp:* USPHS fel, biochem virol, Max Planck Inst Biochem, 65-66; from asst prof to assoc prof, 66-73, PROF MOLECULAR BIOL, UNIV CALIF, LOS ANGELES, 73- *Concurrent Pos:* USPHS res grants, 67-70 & 72-89, Am Cancer Soc grant, 78-80, WHO grant, 80-82; found scientist, Int Genetic Eng, Inc, 81. *Mem:* AAAS; Biophys Soc; Am Soc Microbiol. *Res:* Structure and replication of DNA; genetic control of DNA replication; molecular cloning of DNA; transposable genetic elements; DNA sequence analysis; replication mechanisms of small viruses; genetic and biochemical analysis of tryponasomes. *Mailing Add:* Dept Biol & Molecular Biol Inst Univ Calif 405 Helgard Ave Los Angeles CA 90024-1570

RAY, DAVID SCOTT, b New Haven, Conn, Oct 5, 30; m 56; c 4. MATHEMATICS. *Educ:* Washington & Jefferson Col, AB, 52; Univ Mich, MA, 56; Univ Tenn, PhD(math), 64. *Prof Exp:* Teaching fel math, Univ Mich, 56-58; instr, Univ Tenn, 58-64; assoc prof, 64-70, PROF MATH & CHMN DEPT, BUCKNELL UNIV, 70- *Mem:* Am Math Soc; Math Asn Am. *Res:* Topology, especially problems associated with the imbedding of Peano continua in Euclidean spaces. *Mailing Add:* Dept Math Bucknell Univ Lewisburg PA 17837

RAY, EARL ELMER, b Burnsville, NC, Aug 5, 29; m 55; c 1. MEAT SCIENCE. *Educ:* NC State Col, BS, 52, MS, 54; Ore State Col, PhD(genetics), 58. *Prof Exp:* Asst, NC State Col, 52-54; animal husbandman sheep breeding, Agr Res Serv, USDA, NMex, 59-61; asst prof, 61-72, assoc prof, 72-74, PROF ANIMAL SCI NMEX STATE UNIV, 74- *Res:* Improvement of beef tenderness by postmortem treatments; palatability characteristics of hot-boned, prerigor meat. *Mailing Add:* Dept Animal Range & Wildlife Sci NMex State Univ Las Cruces NM 88003

RAY, EVA K, b Zagreb, Yugoslavia, Mar 5, 33; US citizen; m 52; c 4. BIOTECHNOLOGY, GRAVITATIONAL BIOLOGY & LIFE SCIENCE. *Educ:* Cornell Univ, BA, 55; Bryn Mawr Col, MA, 65, PhD(biochem), 73. *Prof Exp:* Instr, Med Sch, Univ Pa, 73-76; asst prof bio chem, Med Col Pa, 76-82, dir prog for women, 80-83; TECHNOL CONSULT, STEG, RAY & ASSOCS, 83- *Concurrent Pos:* Fel, Med Sch, Univ Pa, 73-75, res ass, SCEIE Eye Inst, 74-76; prof & lectr biochem, St Joseph Univ, 75-76; vchmn, gravitational effects on living systs, Gordon Res Conf, 88, chmn, 90. *Mem:* Am Soc Microbiol; Sigma Xi; AAAS; Asn Women Sci; Am Inst Aeronaut & Astronaut; Aerospace Med Asn. *Res:* Lipid biochemistry - effects of lipoproteins on cholesterol metabolism; biogenesis of herpes simplex virus envelope (glycosylation); basic and applied research in space related living systems; cell biology. *Mailing Add:* 1222 Prospect Hill Rd Villanova PA 19085

RAY, FREDERICK KALB, b Zanesville, Ohio, Mar 23, 44; m 71; c 3. FOOD SCIENCE & TECHNOLOGY. *Educ:* Ohio State Univ, BScAg, 67, MS, 74; Purdue Univ, PhD(animal sci), 78. *Prof Exp:* Prod supvr meat packing, Dinner Bell Foods, Defiance, Ohio, 70-72; EXTEN ANIMAL FOODS SPECIALIST MEAT & DAIRY PROD, DEPT ANIMAL SCI, OKLA STATE UNIV, 78- *Mem:* Am Soc Animal Sci; Inst Food Technologists; Am Meat Sci Asn. *Res:* Emulsion technology; factors effecting emulsion stability; evaluation of meat emulsions using scanning electron microscopy to observe the structure of fat and protein. *Mailing Add:* Dept Animal Sci 103 Animal Sci Bldg Stillwater OK 74078-0425

RAY, G CARLETON, b New York, NY, Aug 15, 28; m 59; c 2. ZOOLOGY, ECOLOGY. *Educ:* Yale Univ, BS, 50; Univ Calif, MA, 53; Columbia Univ, PhD(zool), 60. *Prof Exp:* Asst to dir, NY Aquarium, 57-59, cur, 60-66; from asst prof to assoc prof pathobiol, Sch Hyg & Pub Health, Johns Hopkins Univ, 67-79; RES PROF, UNIV VA, 79- *Mem:* Soc Marine Mammal; Am Soc Limnol & Oceanog; Am Soc Naturalists; Ecol Soc Am; Soc Conserv Biol. *Res:* Role of large organisms, especially marine mammals, in marine ecosystems; ecology and thermoregulation and underwater studies in marine mammals; acoustics as related to behavior in marine mammals, principally in their natural environment; marine ecology, especially related to marine conservation. *Mailing Add:* Dept Environ Sci Clark Hall Univ Va Charlottesville VA 22903

RAY, HOWARD EUGENE, b Iola, Kans, Aug 15, 26; m 46; c 3. COMMUNICATION SCIENCE, SOIL FERTILITY. *Educ:* Kans State Univ, BS, 49, MS, 50; Univ Minn, PhD(soils), 56. *Prof Exp:* Instr soils, Kans State Univ, 49-50; instr high sch, Kans, 50-51; exten soils specialist, Univ Ariz, 51-54; asst soils, Univ Minn, 54-56; asst soils chemist, Everglades Exp Sta, Fla, 56-58; agronomist, Agr Ext Serv, Univ Ariz, 58-66; dep team leader, Ford Found IADP Prog, 67-73; prog leader, Acad Educ Develop, Basic Village Educ Proj, Guatemala, 73-76; RURAL DEVELOP ADV & DIR, ASIAN PROGS, ACAD EDUC DEVELOP, 76- *Mem:* Am Soc Agron; Soil Sci Soc Am; Soc Int Develop; Indian Soil Sci Soc. *Res:* Plant nutrition; nitrogen movement and transformations in soil; cotton production; intensive agricultural development; use of communication technology in development programs. *Mailing Add:* 2526 E Blanton Dr Tucson AZ 85716

RAY, JAMES ALTON, b Hicksville, Ohio, Sept 29, 32; m 60; c 2. VETERINARY PATHOLOGY, VETERINARY MICROBIOLOGY. *Educ:* Ohio State Univ, DVM, 57; Mich State Univ, MS, 61, PhD(path), 66. *Prof Exp:* Instr microbiol & pub health & leader microbiol diag lab, Mich State Univ, 57-59, co-dir tuberculosis res, 59-66; res assoc, Upjohn Co, 66-72; sect head, Ethicon Res Found, Somerville, NJ, 72, mgr, 72-73, dir, 74-87; DIR PRECLIN & COMPLIANCE, JANSSEN PHARMACEUT, 87- *Mem:* Am Vet Med Asn; Am Asn Lab Animal Med; AAAS; NY Acad Sci; Soc Biomat. *Res:* Industrial pathology; toxicology; experimental surgery; histology; histochemistry product development and research. *Mailing Add:* 40 Kingsbridge Rd Piscataway NJ 08855

RAY, JAMES DAVIS, JR, botany; deceased, see previous edition for last biography

RAY, JAMES P, b New York, NY, Jan 16, 44; m 66. CORAL REEF ECOLOGY, CRUSTACEAN SYSTEMATICS. *Educ:* Univ Miami, BS, 66; Tex A&M Univ, MS, 70, PhD(biol oceanog), 74. *Prof Exp:* Sr staff scientist marine biol, Corp Environ Affairs, 74-82, MGR ENVIRON SCI SUPPORT, SHELL OIL CO, 82-; SCI ADV COMT, MINERAL MGT SERV, 87- *Concurrent Pos:* Contrib author, Petrol in the Marine Environ, Nat Acad Sci, 81; panel mem, Fate & Effects of Drilling Muds in the Marine Environ, Nat Acad Sci, 81-83; mem Arctic Marine Sci Comt, Polar Res Bd, Nat Acad Sci, 84-85, Task Force on Particulate Dispersion in Oceans, 86-87. *Mem:* Int Soc Petrol Indust Biologists (pres, 81-82). *Res:* Petroleum industry research on fate and effects of drilling fluids and petroleum hydrocarbons in marine environment. *Mailing Add:* Environ Affairs Dept Shell Oil Co PO Box 4320 Houston TX 77210

RAY, JESSE PAUL, b Central Lake, Mich, Nov 8, 16; m 39, 82; c 1. EDUCATIONAL ADMINISTRATION, ANALYTICAL CHEMISTRY. *Educ:* Asbury Col, AB, 39; Univ Syracuse, PhD(anal chem), 47. *Prof Exp:* Teacher high sch, Ill, 39-42; chief chemist, Chem & Metall Lab, Remington Arms Co, Inc, 42-44; asst chem, Univ Syracuse, 44-45; res engr, Battelle Mem Inst, 46-47; from assoc prof to prof chem, Asbury Col, 47-82; RETIRED. *Concurrent Pos:* Consult chemist, 47-80; educ adminr, 82- *Mem:* Fel Am Chem Soc. *Res:* Physical chemistry of metals; determination of metal phase diagrams; development of analytical methods of steel analysis; instrumentation. *Mailing Add:* Asst to Pres Asbury Col Wilmore KY 40390

RAY, JOHN DELBERT, b Murphysboro, Ill, Aug 21, 30; m 56; c 3. MECHANICAL ENGINEERING. *Educ:* Univ Ill, BS, 56, MS, 57; Univ Okla, PhD(mech eng), 68. *Prof Exp:* Engr, McDonnell Aircraft Corp, 57-62; res dir, Res Inst, Univ Okla, 62-68; from asst prof to prof mech eng, Col Eng, Memphis State Univ, 68-89, chmn dept & dir grad studies & res, 77-89;

DEAN, HEFF COL, 89- *Mem:* Am Soc Mech Engrs. *Res:* Vibrations; structural dynamics; design; high temperature metal erosion and fatigue; composite material. *Mailing Add:* Heff Col Eng Memphis State Univ Memphis TN 38152

RAY, JOHN ROBERT, b Beckley, WVa, Jan 27, 39; m 65; c 2. PHYSICS. *Educ:* Rose-Hulman Inst Technol, BS, 61; Univ Ohio, PhD(physics), 64. *Prof Exp:* Asst prof physics, Auburn Univ, 64-65; res assoc, Coord Sci Lab, Univ Ill, 65-66; from asst prof to assoc prof, 66-77, PROF PHYSICS, CLEMSON UNIV, 77- *Concurrent Pos:* Res fel, Marshall Space Flight Ctr, Huntsville, Ala, 81-82; res partic, Argonne Nat Lab, Ill, 83-85. *Mem:* Am Phys Soc. *Res:* Relativity; field theory; theoretical physics; molecular dynamic computer simulation of solids. *Mailing Add:* Dept Physics & Astron Clemson Univ Clemson SC 29634-1911

RAY, JOHN ROBERT, b Alderson, WVa, Aug 27, 21; m 48. GEOGRAPHIC INFORMATION SYSTEMS. *Educ:* Ind Univ, AB, 54, MA, 55; Ohio State Univ, PhD(geog), 72. *Prof Exp:* From instr to asst prof geog, Univ Miami, 55-64; from instr to prof, 64-84, chmn dept, 74-84, EMER PROF GEOG, WRIGHT STATE UNIV, 84- *Concurrent Pos:* Wright State Univ Res Coun grant, 74-75; vis prof civil eng, Ohio State Univ, 80-81; mem bd dirs, Am Soc Photogram & Remote Sensing, 81-84; mem bd trustees, Soc Photog Scientists & Engrs, 85-86 & Int Geog Info Found. *Mem:* Am Soc Photogram & Remote Sensing; Asn Am Geogrs; Int Geog Info Found (pres, 87-91). *Res:* Analysis of environmental problems with remote sensing techniques; terrain analysis, using aerial photo interpretation; measuring human attitudes toward environmental insults; conservation of resources, geographic information systems applications; remote sensing of Earth Resources. *Mailing Add:* 710 Leisure Lane Waverly OH 45690-1513

RAY, OAKLEY S, b Altoona, Pa, Feb 6, 31; m 53; c 4. PSYCHOPHARMACOLOGY. *Educ:* Cornell Univ, BA, 52; Univ Pittsburgh, MEd, 54, PhD(psychol), 58. *Prof Exp:* Assoc prof psychol pharmacol, Univ Pittsburgh, 60-70; chief ment health unit patient care, Vet Admin Ctr, Nashville, 73-87; PROF PSYCHOL, VANDERBILT UNIV, 70-, ASSOC PROF PHARMACOL, MED CTR, 70-, PROF PSYCHIAT, 82- *Concurrent Pos:* Fel, Neuropharmacol Res Labs, NIMH, Pittsburgh, 58-60; chief psychol serv, Vet Admin Med Ctr, Nashville, 70-; chief, Psychol Res Lab, Vet Admin Med Ctr, Nashville, 70-; mem, Grad Neurobiol Res Training Progs, NIMH, 74-78. *Mem:* Fel Am Col Neuropsychopharmacol; fel Am Psychol Asn; Am Soc Pharmacol & Exp Therapeut. *Res:* Research centers around the genetic and developmental determinants of brain function in animals, and on the effects of CNS drugs on behavior during development and at maturity. *Mailing Add:* 2100 Hampton Ave Nashville TN 37215

RAY, PAUL DEAN, b Monmouth, Ill, Dec 7, 34; m 57; c 4. BIOCHEMISTRY. *Educ:* Monmouth Col, Ill, AB, 56; St Louis Univ, PhD(biochem), 62. *Prof Exp:* Am Cancer Soc fel, Enzyme Inst, Univ Wis-Madison, 62-65, univ fel, 65-67; from asst prof to assoc prof, 67-73, PROF BIOCHEM, SCH MED, UNIV NDAK, 73- *Concurrent Pos:* Estab investr, Am Heart Asn, 67-72, mem Great Plains regional res rev & adv comn, 74-78; mem study sect arthritis metab & digestive dis, NIH, 69-73. *Mem:* Sigma Xi; Am Chem Soc; Am Soc Biol Chem. *Res:* Carbohydrate metabolism, gluconeogenesis, and metabolic regulation; endocrinology. *Mailing Add:* Dept Biochem Univ NDak Sch Med Grand Forks ND 58202

RAY, PETER MARTIN, b San Jose, Calif, Dec 17, 31; m 54; c 2. PLANT PHYSIOLOGY. *Educ:* Univ Calif, AB, 51; Harvard Univ, PhD(biol), 55. *Prof Exp:* Jr fel, Soc Fels, Harvard Univ, 55-58; from asst prof to prof bot, Univ Mich, Ann Arbor, 58-65; prof biol, Univ Calif, Santa Cruz, 66-68; PROF BIOL SCI, STANFORD UNIV, 68- *Honors & Awards:* Charles Albert Schull Award, Am Soc Plant Physiol, 71. *Mem:* Am Soc Plant Physiol. *Res:* Physiology and biochemistry of plant growth and development; plant hormones. *Mailing Add:* Dept Biol Sci Stanford Univ Stanford CA 94305

RAY, PETER SAWIN, b Iowa City, Iowa, July 26, 44; m 70; c 1. METEOROLOGY. *Educ:* Iowa State Univ, BS, 66; Fla State Univ, MS, 70, PhD(meteorol), 73. *Prof Exp:* Res meteorologist, 73-80, CHIEF, METEOROL RES GROUP, NAT SEVERE STORMS LAB, NAT OCEANIC & ATMOSPHERIC ADMIN, 80- *Concurrent Pos:* Nat Res Coun fel, Nat Severe Storms Lab, 73-74; adj asst prof, Depts Meteorol & Elec Eng, Univ Okla, 74- *Mem:* Am Meteorol Soc; Am Geophys Union; Am Inst Physics. *Res:* Scattering physics of radiation and hydrometeors; using models and observations, the study of the morphology and dynamics of severe storms. *Mailing Add:* Dept Meteorol Fla State Univ Tallahassee FL 32306

RAY, PRASANTA K, b Calcutta, India, Sept 29, 41; m 68; c 2. IMMUNOTOXICOLOGY. *Educ:* Univ Calcutta, BS, 62, MS, 64, PhD(biochem), 68, DSc, 74. *Prof Exp:* Sr scientist res & teaching tumor immunol, Biomed Group, Bhabha Atomic Res Ctr, 73-76; dir admin & res cancer, Chittaranjan Nat Res Ctr, India, 76-77; res dir, Bengal Immunol Res Inst, Calcutta, 78-84; DIR, INDUST TOXICOL RES CTR, MAHATMA GANDHI MARG, LUCKNOW, INDIA, 84- *Concurrent Pos:* Prin investr, Med Col Pa & Hosp, 78-; chief, Comt Nutrit & Immunol, Int Fedn Biosocial Develop & Human Health, NY, 80-; prof, dept surg & microbiol & dir, Alma Dea Morani Lab Surg Immunobiol, Med Col, 84- *Honors & Awards:* Gold Medal, Soc Toxicol India, 85; Ranbaxy Nat Award Excellence in Med Res, 86. *Mem:* NY Acad Sci; Int Soc Detection & Prev Cancer; Am Asn Immunologists; Am Asn Apheresis; Indian Immunol Soc; Am Asn Cancer Res; Soc Biol Chem India; Mutagen Soc India; Indian Asn Cancer Res; fel Inst Biol London; fel Soc Toxicol India; fel Indian Col Allergy & Appl Immunol. *Res:* Investigations to remove blocking factors extracorporeally or ex vivo from both animal and human tumor-bearing hosts to study its therapeutic benefit; mechanism of immunosuppression in cancer; mechanism of prevention; studies of toxicology of chemicals and immunotoxicology of chemicals; immunotherapy of cancer. *Mailing Add:* Indust Toxicol Res Ctr Mahatma Gandhi Marg PO Box 80 Lucknow 226001 India

RAY, RICHARD SCHELL, b Antwerp, Ohio, May 21, 28; m 54; c 3. VETERINARY PHARMACOLOGY. *Educ:* Ohio State Univ, BA, 50, DVM, 55, MSc, 58, PhD(physiol, pharmacol), 63. *Prof Exp:* From instr to assoc prof, 55-73, prof vet clin sci & dir, pre & post race testing labs, 73-84, EMER PROF, OHIO STATE UNIV, 84- *Concurrent Pos:* Grants, NY Racing Asn & Jockey Club, 65-66, Harness Racing Inst, 66-67 & Thoroughbred Racing Fund, 68- *Mem:* Am Asn Equine Practitioners; Am Soc Vet Physiol & Pharmacol; World Asn Physiologists, Pharmacologists, Biochemists; Asn Drug Detection Labs; fel Am Col Vet Pharmacol & Therapeut. *Res:* Intermediate metabolism and diseases related to metabolism; detection of illegally used drugs; serum transaminase changes related to disease; research and development of drug detection methods in biological fluids; general physiology. *Mailing Add:* 2752 Folkstone Rd Columbus OH 43220

RAY, ROBERT ALLEN, b Scottsbluff, Nebr, Dec 19, 39; m 60; c 4. CLINICAL CHEMISTRY. *Educ:* Univ Nebr, BS, 61, MS, 63, PhD(biochem), 66. *Prof Exp:* Res chemist, 66-74, eng mgr, 74-78, PROG MGR, BECKMAN INSTRUMENTS, 81- *Mem:* AAAS; Am Chem Soc; Am Asn Clin Chemists. *Res:* Chemical and biomedical instrumentation for the assay of enzymes or their substrates. *Mailing Add:* 601 W 20th Hialeah FL 33012

RAY, ROBERT DURANT, b Cleveland, Ohio, Sept 21, 14; m 53; c 5. ORTHOPEDIC SURGERY. *Educ:* Univ Calif, BA, 36, MA, 38, PhD(anat), 48; Harvard Univ, MD, 43. *Hon Degrees:* MedDrSci, Royal Univ Umea, 72. *Prof Exp:* Asst anat, Med Sch, Univ Calif, 37-38; intern surg, Peter Bent Brigham Hosp, Boston, 43; asst orthop surg, Harvard Med Sch, 44-45; instr anat, Med Sch, Univ Calif, 47-48; asst prof orthop surg, Sch Med, Wash Univ, 48-51, assoc prof & head dept, 54-56; prof orthop surg & head dept, Col Med, Univ Ill, Chicago, 56-81; RETIRED. *Concurrent Pos:* Mem staff, Ravens Wood Hosp, 71- *Mem:* Am Orthop Res Soc (pres, 59); Am Asn Anat; Am Orthop Asn; Am Col Surg; Am Acad Orthop Surg. *Res:* Bone growth, maturation and metabolism; influence of intrinsic and extrinsic factors on these processes including endocrines and radiation; kinetics of bone-seeking radioactive isotopes. *Mailing Add:* 2200 Laguna Vista Dr Novato CA 94945

RAY, ROSE MARIE, b Hayward, Calif, Mar 30, 43; m 73; c 3. MATHEMATICAL STATISTICS. *Educ:* Univ Calif, Berkeley, BA, 65, PhD(statist), 72. *Prof Exp:* From actg instr to actg asst prof statist, Univ Calif, Berkeley, 71-72; asst prof math, Northwestern Univ, 72-74; asst prof statist, Univ Fla, 74-76; statist consult, 76; personnel res statistician, Pac Gas & Elec Co, San Francisco, 78-88; MANAGING SCIENTIST, FAILURE ANALYSIS ASSOC, PALO ALTO, 88- *Concurrent Pos:* Consult mkt res, Montgomery Ward Co, Chicago, 74; consult, Biostatist Unit, J Hillis Miller Health Ctr, Univ Fla, 74-76; sr statistician, Sci Comput Serv, Univ Calif, San Francisco, 76-78; sr biostatistician, Contraceptive Drug Study, Kaiser Found Hosp, Walnut Creek, Calif, 76-78; lectr, Dept Statist, Univ Calif, Berkeley, 77- *Honors & Awards:* Evelyn Fix Biostatistics Award, 70. *Mem:* Inst Math Statist; Am Statist Asn; AAAS; Sigma Xi; Biomet Soc. *Res:* Risk analysis development and application of special stochastic models for research in medicine, biology, marketing, manpower planning and cost analysis; research in the theory of C-alpha tests. *Mailing Add:* Failure Analysis Assoc 149 Commonwealth Dr PO Box 3015 Menlo Park CA 94025

RAY, SAMMY MEHEDY, b Mulberry, Kans, Feb 25, 19; m 43; c 4. MARINE BIOLOGY. *Educ:* La State Univ, BS, 42; Rice Inst, MA, 52, PhD(biol), 54. *Prof Exp:* Fishery res biologist, US Fish & Wildlife Serv, 54-59; asst prof oceanog & meteorol, 59-63, assoc prof oceanog, 63-69, PROF BIOL, OCEANOG, MARINE SCI & WILDLIFE FISHERIES SCI, TEX A&M UNIV, 69-, DIR, MARINE LAB, 63-, HEAD DEPT MARINE SCI, 74-, dir sch marine technol, Tex A&M Univ, 77- *Mem:* AAAS; Nat Shellfisheries Asn; Am Soc Limnol & Oceanog; Phycol Soc Am; Am Inst Fishery Res Biol. *Res:* Oyster biology; marine microbiology, phytoplankton and pollution. *Mailing Add:* 7213 Yucca Dr Galveston TX 77551

RAY, SIBA PRASAD, b Dinhata, India, Jan 4, 44; m 77; c 2. MATERIALS SCIENCE, METALLURGY. *Educ:* Univ Calcutta, BE, 64; Columbia Univ, MS, 70, DEngSc, 74. *Prof Exp:* Sci officer, Bhabha Atomic Res Ctr, India, 65-69; scientist, 77-78, sr scientist mat sci, 78-82, SCI ASSOC, ALCOA LABS, 82- *Concurrent Pos:* Res assoc, Pa State Univ, 74-77. *Mem:* Am Ceramic Soc; Sigma Xi; Am Inst Mining, Metall & Petrol Engrs. *Res:* High temperature materials; conducting ceramics; aluminum smelting process development; refractory electrodes; oxygen sensors; ceramic engineering; solid state chemistry; ceramic composites, squeeze casting; ceramic matrix composites. *Mailing Add:* Alcoa Tech Ctr Alcoa Center PA 15069

RAY, SYLVIAN RICHARD, b Pineville, La, Aug 26, 31; m 59; c 2. COMPUTER SCIENCE, ELECTRICAL ENGINEERING. *Educ:* Southwestern La Inst, BS, 51; Univ Ill, MS, 57, PhD(elec eng), 60. *Prof Exp:* Elec scientist, Naval Res Lab, 51-54; res assoc elec eng, 60-61, from asst prof to assoc prof, 61-76, PROF ELEC ENG, UNIV ILL, URBANA, 76- *Mem:* Sigma Xi; Asn Comput Mach. *Res:* Artificial intelligence; biomedical applications of computers; image processing. *Mailing Add:* Dept Comput Sci 2323 DCL Univ Ill Urbana IL 61801

RAY, VERNE A, b Portsmouth, NH, July 28, 29; m 52; c 2. MICROBIOLOGY, BIOCHEMISTRY. *Educ:* Univ NH, BS, 51, MS, 55; Univ Tex, PhD(bact), 59. *Prof Exp:* Sr res scientist, Chas Pfizer & Co, 59-67, proj leader, 67-72, mgr, 72-75, asst dir, Dept Drug Safety Eval, 75-90, SR TECH ADV, CENT RES, PFIZER INC, 90- *Mem:* Am Soc Microbiol. *Res:* Molecular biology of virus infectious process; factors controlling induced mutation frequency in microorganisms; elaboration products of microorganisms and methods of increasing yield. *Mailing Add:* 60 Beach Pond Rd Groton CT 06340

RAY, W(ILLIS) HARMON, b Washington, DC, Apr 4, 40; m 62; c 3. CHEMICAL ENGINEERING. *Educ:* Rice Univ, BA, 62, BSChE, 63; Univ Minn, PhD(chem eng), 66. *Prof Exp:* Asst prof chem eng, Univ Waterloo, 66-69, assoc prof, 69-70; from assoc prof to prof, State Univ NY Buffalo, 70-76; chmn dept, 81-83, Steenbock chair eng, 86, PROF CHEM ENG, UNIV WIS-MADISON, 76-, STEENBOCK PROF ENG, 86- *Concurrent Pos:* Indust consult, 67-; vis prof, Rijksuniversiteit Gent & Univ Leuven, 73-74, Tech Univ Stuttgart, WGer, 74, Dept Chem Eng, Univ Minn, 86 & Cornell Univ, 91; Guggenheim mem fel, Europe, 73-74; C-I-L distinguished vis lectr, Univ Alta, Can, 82 & McMaster Univ, Ont, 85. *Honors & Awards:* Eckman Award, Am Automatic Control Coun, 69; A K Doolittle Award, Am Chem Soc, 81; Prof Prog Award, Am Inst Chem Eng, 82; Distinguished Reilly Lect Award, Univ Notre Dame, Ind, 84; W N Lacey Lectr, Calif Inst Technol, Pasadena, 88; Educ Award, Am Automatic Control Coun, 89. *Mem:* Nat Acad Eng; Am Chem Soc; Am Soc Eng Educ; Soc Indust & Appl Math; Chem Inst Can; fel Am Inst Chem Engrs; Am Asn Artificial Intel; Inst Elec & Electronics Engrs; Sigma Xi. *Res:* Chemical reactor engineering including polymerization processes; process modeling, optimization, dynamics and control. *Mailing Add:* Dept Chem Eng Univ Wis Madison WI 53706

RAY, WILLIAM J, b Birmingham, Ala, Sept 3, 45; c 2. CLINICAL PSYCHOLOGY, PSYCHOPHYSIOLOGY. *Educ:* Eckerd Col, St Petersburg, Fla, BA, 67; Vanderbilt Univ, MA, 69, PhD(clin psychol), 71. *Prof Exp:* Fel med psychol, Langley Porter Neuropsychiat Inst, 71-72; PROF PSYCHOL, PA STATE UNIV, 72- *Concurrent Pos:* Vis prof, Univ Hawaii, 87, Univ Tübingen, WGer, 89; dir, Health Psychol Prog, 85- *Honors & Awards:* Nat Media Award, Am Psychol Found, 76 & 78. *Mem:* AAAS; Am Psychosomatic Soc; Am Psychol Asn. *Res:* Psychophysiological assessment-brain/behavior relationships; chaos and EEG; behavioral medicine; interdependence of mental, emotional and physiological/motor activities. *Mailing Add:* Dept Psychol Pa State Univ University Park PA 16802

RAY, WILLIAM JACKSON, JR, b Bradenton, Fla, Mar 19, 32; m 54; c 2. ORGANIC CHEMISTRY. *Educ:* Bethany-Nazarene Col, BS, 49; Purdue Univ, PhD(org chem), 57. *Prof Exp:* Res assoc biochem, Brookhaven Nat Lab, 57-59; asst prof, Rockefeller Inst, 59-61; from asst prof to assoc prof, 61-70, PROF BIOCHEM, PURDUE UNIV, LAFAYETTE, 70- *Mem:* Am Chem Soc; Am Soc Biol Chem. *Res:* Mechanism of enzyme action. *Mailing Add:* Dept Biol Sci Purdue Univ West Lafayette IN 47907

RAYBON, GREGORY, b Port Arthur, Tex, Dec 29, 61; m 87; c 1. ELECTRICAL ENGINEERING, OPTICS. *Educ:* Pa State Univ, BS, 84; Stevens Inst Technol, MS, 90. *Prof Exp:* Prod engr, Nat Semiconductor, 84-85; MEM TECH STAFF, AT&T BELL LABS, 85- *Mem:* Optical Soc Am. *Res:* Short optical pulse generation from semiconductor photonic devices, such as lasers, novel modulation schemes and optical thin films, which are directly applicable to future optical communication systems for long haul and local loop exchange. *Mailing Add:* 19 Monmouth Ave Leonardo NJ 07737

RAYBORN, GRAYSON HANKS, b Columbia, Miss, May 26, 39; m 65. ATOMIC PHYSICS. *Educ:* Rensselaer Polytech Inst, BS, 61; Univ Fla, PhD(physics), 69. *Prof Exp:* Asst engr, Sperry Rand Corp, 61-62; res asst physics, Univ Fla, 68-69; asst prof physics, Old Dominion Univ, 69-70; asst prof, 70-75, assoc prof, 75-79, PROF PHYSICS, UNIV SOUTHERN MISS, 79- *Concurrent Pos:* Consult, Langley Res Ctr, NASA, 71. *Mem:* Am Asn Physics Teachers; Am Phys Soc; Fedn Am Scientists; Am Chem Soc. *Res:* Photoionization and dissociative photoionization; deconvolution and inverse digital filtering; free electron magnetometer; development of novel apparatus for use by undergraduate physics majors. *Mailing Add:* 2012 Eddy St Hattiesburg MS 39401

RAYBURN, LOUIS ALFRED, b Columbus, Ga, Dec 10, 21; m 48; c 4. PHYSICS. *Educ:* Univ Ky, BS, 48, MS, 50, PhD(physics), 54. *Prof Exp:* Instr physics, Univ Ky, 48-50; assoc physicist, Oak Ridge Nat Lab, 50-52; res asst physics, Univ Ky, 52-54; asst physicist, Argonne Nat Lab, 54-59; prof physics & head dept physics & astron, Univ Ga, 59-64; head univ participation officer, Oak Ridge Assoc Univs, 64-70; PROF PHYSICS & CHMN DEPT, UNIV TEX, ARLINGTON, 70- *Mem:* AAAS; Am Phys Soc. *Res:* Low energy nuclear physics. *Mailing Add:* Dept Physics Univ Tex Arlington TX 76019

RAYBURN, MARLON CECIL, JR, b Clay, Ky, Sept 29, 31; m 62; c 3. TOPOLOGY. *Educ:* Evansville Col, BA, 52; Auburn Univ, MS, 56; Univ Ky, PhD(math), 69. *Prof Exp:* Instr math & physics, Earlham Col, 59-62; asst prof math, State Univ NY Col Geneseo, 65-68; from asst prof to assoc prof, 69-82, PROF MATH, UNIV MAN, 83- *Mem:* Can Math Cong; Am Math Soc; Math Asn Am. *Res:* General topology; compactifications and realcompactifications; uniformities; proximities; applications to analysis; sigma algebras. *Mailing Add:* Winnipeg MB R3T 2N2 Can

RAYBURN, WILLIAM REED, b St Louis, Mo, Apr 7, 40; m 67; c 2. BOTANY & MICROBIOLOGY. *Educ:* Wash Univ, BA, 63; Ind Univ, Bloomington, MA, 67, PhD(bot, microbiol), 71. *Prof Exp:* Instr, Wash State Univ, 67-69, asst prof bot & gen biol, 69-78, assoc prof bact, pub health & bot, 78-85, chair gen biol, 82-87, assoc dean, Grad Sch, 87-89, PROF MICROBIOL, BOT & GEN BIOL, WASH STATE UNIV, 85-, ASSOC VPROVOST RES, 89-, ASSOC DIR, WASH TECHNOL CTR, 89- *Concurrent Pos:* Interim dir, Nuclear Radiation Ctr, Wash State Univ, 89-90 & Plant Biotechnol Ctr, Wash Technol Ctr, 90- *Mem:* Bot Soc Am; Phycol Soc Am; Brit Phycol Soc; Int Phycol Soc; Am Soc Microbiol; Am Inst Biol Sci. *Res:* Sexuality of algae; ecology of soil algae; microbial extra-cellular polysaccharides. *Mailing Add:* Grad Sch Wash State Univ Pullman WA 99164-1030

RAYCHAUDHURI, ANILBARAN, b Tipperah, India; m 64; c 3. IMMUNOLOGY. *Educ:* Univ Calcutta, BSc, 51; Univ Rangoon, MS, 58; Univ Cincinnati, PhD(biol sci), 63. *Prof Exp:* Res chemist geront, Vet Admin Hosp, Baltimore, Md, 65-67; res assoc, Georgetown Univ, 67-70; sr scientist, Merrell-Nat Lab, Cincinnati, 70-78; SR SCIENTIST IMMUNOL & INFLAMMATION, CIBA-GEIGY CORP, SUMMIT, NJ, 78- *Concurrent Pos:* Vis fel, NIH, Bethesda, Md, 63-65; consult, Vet Admin Hosp, Baltimore, 64-65. *Mem:* NY Acad Sci; AAAS; Sigma Xi. *Res:* Immunology in general and cell-mediated immunity in particular; application of chemotherapy to rheumatoid arthritis involving a unique type of compound designated as disease modifying anti-rheumatic drugs. *Mailing Add:* Inflammation Osteoarthritis Res Ciba-Geigy Corp 556 Morris Ave Summit NJ 07928

RAY-CHAUDHURI, DILIP K, b Dacca, E Pakistan, Sept 4, 29; m 61; c 2. POLYMER CHEMISTRY, ORGANIC CHEMISTRY. *Educ:* Univ Dacca, BSc, 48, MSc, 49; Univ Calcutta, PhD(chem), 56. *Prof Exp:* Jr res asst chem jute cellulose, Tech Res Lab, India, 56-58; fel, Cellulose Res Inst, State Univ NY Col Forestry, Syracuse, 58-61; res assoc, Ont Res Found Can, 61-64; proj supvr org chem res, 64-67, sect leader, 67-72, mgr cent res, 72-77, dir corp res, 77-81, DIV VPRES & DIR CORP RES, NAT STARCH & CHEM CO, 81- *Mem:* AAAS; Am Chem Soc; NY Acad Sci; Sigma Xi. *Res:* Polyelectrolytes; adhesives; wet strength additives; polyurethanes and polyesters. *Mailing Add:* Nat Starch & Chem Co Ten Finderne Ave Bridgewater NJ 08807

RAY-CHAUDHURI, DWIJENDRA KUMAR, b Narayangang, Bangladesh, Nov 1, 33; m 62; c 3. DISCRETE MATHEMATICS, COMBINATORICS. *Educ:* Presidency Col, Calcutta, BSc, 53; Calcutta Univ, MSc, 55; Univ NC, Chapel Hill, PhD(math statist), 59. *Prof Exp:* Res assoc, Case Inst Technol, 59-60; asst prof math statist, Univ NC, Chapel Hill, 60-61; reader, Indian Statist Inst, 61-62; mem res staff math, T J Watson Res Ctr, IBM Corp, NY, 62-64; consult statist & math, Cornell Med Ctr & Sloan-Kettering Inst, 64-65; vis assoc prof math, Math Res Ctr, Univ Wis-Madison, 65-66; chmn dept, 80-83, PROF MATH, OHIO STATE UNIV, 66- *Concurrent Pos:* Invited speaker, Int Math Cong, 70; vis prof, Gottingen Univ, WGer, 72 & Erlanger Univ, 76; sr Sci Res Coun fel, Univ London, 84; assoc ed, J Comb Theory B, Combinatorica & J Statist Inference & Planning. *Honors & Awards:* Alexander von Humboldt Award, WGer, 84. *Mem:* Am Math Soc; Math Asn Am. *Res:* Combinatorial mathematics; finite geometry; graph and information theory; error-correcting codes; statistical design experiments. *Mailing Add:* Dept Math 231 W 18th Ave Ohio State Univ Columbus OH 43210

RAYCHAUDHURI, KAMAL KUMAR, b Dinapore, India, Nov 11, 47; m 74; c 2. TELECOMMUNICATIONS. *Educ:* Univ Calcutta, India, BSc, 67, MSc, 69; Univ Pa, PhD(physics), 77, MSE, 78. *Prof Exp:* Assoc, Saha Inst Nuclear Physics, Calcutta, India, 71; res fel physics, Univ Pa, 77-78; res assoc, Univ Mass, Amherst, 78-79, asst prof physics, 80-83; AT AT&T BELL LABS, 83- *Mem:* Inst Elec & Electronics Engrs. *Res:* Experimental high energy physics: hyperon studies and studies of new particles using high speed electronic detectors. *Mailing Add:* AT&T Bell Labs Crawfords Corner Rd Rm Ho 2G-520 Holmdel NJ 07733

RAYCHOWDHURY, PRATIP NATH, b Calcutta, India, Jan 1, 32. APPLIED MATHEMATICS, MATHEMATICAL PHYSICS. *Educ:* Univ Calcutta, BS, 51; Univ Col, BA, 55; Brigham Young Univ, MS, 58; George Washington Univ, PhD(math physics), 66. *Prof Exp:* Asst prof physics, Rutgers Univ, 64-65; asst prof math, Royal Mil Col Can, 65-67; assoc prof lectr eng, George Washington Univ, 67, sr staff scientist, 67-68; assoc prof physics, NY Inst Tech, 68-69; assoc prof, 69-74, PROF MATH SCI, VA COMMONWEALTH UNIV, 74- *Mem:* Am Math Soc; Soc Indust & Appl Math; Am Phys Soc; Int Asn Math Physics. *Res:* Excited states of many-body systems; shock waves and high velocity plasma; group theory and application in physics. *Mailing Add:* Dept Math Sci Va Commonwealth Univ Box 2014 Richmond VA 23284

RAYFIELD, GEORGE W, b San Francisco, Calif, Feb 17, 36; m 59; c 1. PHYSICS. *Educ:* Stanford Univ, BS, 58; Univ Calif, Berkeley, MS & PhD(physics), 64. *Prof Exp:* Res asst eng sci, Univ Calif, Berkeley, 60-61, physics, 61-64; asst prof, Univ Pa, 64-68; ASSOC PROF PHYSICS, UNIV ORE, 68- *Mem:* Am Phys Soc. *Res:* Solid state physics; liquid helium; ionic probes in liquid helium; microwave tubes; electron beams. *Mailing Add:* Dept Physics Univ Ore Eugene OR 97403

RAYFORD, PHILLIP LEON, b Roanoke, Va, July 25, 27; m 52. PHYSIOLOGY, ENDOCRINOLOGY. *Educ:* NC A&T State Univ, BS, 49; Univ Md, College Park, MS, 69, PhD(reproductive endocrinol), 73. *Hon Degrees:* LHD, NC A&T State Univ, 85. *Prof Exp:* Supvry biologist endocrinol, Nat Cancer Inst, 55-62, supvry biologist, NIHMR, 62-64, supvry biologist radioimmunoassay, Nat Cancer Inst, 64-70, supvry biology, NICHO, 70-73; from asst prof to assoc prof biochem, Div Human Cell Biol & Genetics, Univ Tex Med Br Galveston, 73-77, prof & dir, Biochem Lab, Dept Surg, 77-80; PROF & CHMN DEPT PHYSIOL, UNIV ARK SCH MED, LITTLE ROCK, 80- *Mem:* Endocrinol Soc; Am Physiol Soc; Am Gastroenterol Asn; Soc Exp Biol & Med; NY Acad Sci. *Res:* Metabolism and catabolism of gastrointestinal and pancreatic hormones in man and dogs; research and development of radioimmunoassay systems for measuring hormones of gastrointestinal, pancreatic and pituitary origin. *Mailing Add:* Dept Physiol & Biophys 505 4301 W Markham Little Rock AR 72205

RAYLE, DAVID LEE, b Pasadena, Calif, Oct 22, 42; m 67; c 2. PLANT PHYSIOLOGY. *Educ:* Univ Calif, Santa Barbara, BA, 64, PhD(biol), 67. *Prof Exp:* NSF fel & res assoc bot, Mich State Univ-Atomic Energy Comn Plant Res Lab, Mich State Univ, 67-68; res assoc, Univ Wash, 68-70; from asst prof to assoc prof, 70-75, chmn dept, 74-80, PROF BOT, SAN DIEGO STATE UNIV, 80- *Mem:* Am Soc Plant Physiol. *Res:* Mechanism of action of plant growth hormones; physical properties of plant cell walls; plant growth and development. *Mailing Add:* Dept Bot San Diego State Univ San Diego CA 92182

RAYLE, RICHARD EUGENE, b Freesoil, Mich, Apr 5, 39; m 63; c 1. GENETICS. *Educ:* Mich State Univ, BS, 62; Univ Ill, Urbana, PhD(genetics), 67. *Prof Exp:* NIH trainee, Univ Calif, Davis, 67-69; res assoc zool, Univ NC, Chapel Hill, 69-70, vis asst prof, 70; ASST PROF ZOOL, MIAMI UNIV, 70- *Mem:* AAAS; Genetics Soc Am. *Res:* Structural and functional organization of eukaryotic genetic systems. *Mailing Add:* Dept Zool Miami Univ Oxford OH 45056

RAYMAN, MOHAMAD KHALIL, b Guyana, SAm, Feb 23, 38; Can citizen; m 59; c 2. FOOD MICROBIOLOGY. *Educ:* McGill Univ, BSc, 66, PhD(microbiol), 70. *Prof Exp:* Med Res Coun Can fel, Univ Toronto, 70-73; RES SCIENTIST, HEALTH & WELFARE, CAN, 73- *Mem:* Can Soc Microbiol; Asn Off Anal Chem. *Res:* Methodology related to isolation and identification of food poisoning organisms; mechanism of succinate transport into membrane vesicles of Escherichia coli; mechanism of thermal injury in Salmonella; testing replacement for nitrite in food preservation, development of genetic probes for identification of food-borne microorganisms. *Mailing Add:* Health Protection Br-Microbiol Res Div Tunney's Pasture Sir FG Banting Bldg Ottawa ON K1A 0L2 Can

RAYMON, LOUIS, b New Brunswick, NJ, Oct 17, 39; m 62; c 4. MATHEMATICS. *Educ:* Yeshiva Col, BA, 60; Yeshiva Univ, MA, 61, PhD(math), 66. *Prof Exp:* from asst prof to assoc prof, 66-77, PROF MATH, TEMPLE UNIV, 77- *Mem:* Am Math Soc. *Res:* Classical problems in real and complex analysis, especially approximation theory. *Mailing Add:* Dept Math Temple Univ Philadelphia PA 19122

RAYMOND, ARTHUR E(MMONS), b Boston, Mass, Mar 24, 99; m 21; c 1. AEROSPACE ENGINEERING. *Educ:* Harvard Univ, BS, 20; Mass Inst Technol, MS, 21. *Hon Degrees:* DSc, Polytech Inst Brooklyn, 47. *Prof Exp:* Engr, Douglas Aircraft Co, 25-34, vpres eng, 34-60; consult, Rand Corp, 60-85; RETIRED. *Concurrent Pos:* Mem, Nat Adv Comt Aeronaut, 46-56, consult, NASA, 62-68; trustee, Aerospace Corp, 60-71 & Res Anal Corp, 65-71. *Mem:* Nat Acad Sci; Nat Acad Eng; hon fel Am Inst Aeronaut & Astronaut. *Res:* Aeronautics; astronautics. *Mailing Add:* 65 Oakmont Dr Los Angeles CA 90049

RAYMOND, CHARLES FOREST, b St Louis, Mo, Oct 31, 39; m 65. GEOPHYSICS. *Educ:* Univ Calif, Berkeley, BA, 61; Calif Inst Technol, PhD(geophys), 69. *Prof Exp:* from asst prof to assoc prof, 69-79, PROF GEOPHYS, UNIV WASH, 79- *Mem:* Am Geophys Union; Int Glaciol Soc. *Res:* Rheology of earth materials; flow and structure of glaciers. *Mailing Add:* Geophys Prog Univ Wash Seattle WA 98195

RAYMOND, DALE RODNEY, b Farmington, Maine, Mar 8, 49; m 74; c 3. CHEMICAL ENGINEERING, TECHNICAL MANAGEMENT. *Educ:* Univ Maine, BS, 71, MS, 73, PhD(chem eng), 75. *Prof Exp:* Instr chem eng, Univ Maine, Orono, 73-75; Res scientist, Papermaking Prog, 73-79, group leader papermaking, 79-83, sect head eng areas, 83-85, TECH DIR PROCESS, ENVIRON & TECH SERV, RES & DEVELOP, UNION CAMP, 85- *Mem:* Sigma Xi; Tech Asn Pulp & Paper Indust; Am Inst Chem Engrs. *Res:* Forming, pressing and drying on the paper machine and in the chemical recovery area to develop high solids firing; process engineering and environmental engineering. *Mailing Add:* 220 Deer Run Dr PO Box 570 Prattville AL 36067

RAYMOND, DAVID JAMES, b Hammond, Ind, Oct 24, 43. PHYSICS, METEOROLOGY. *Educ:* Rensselaer Polytech Inst, BS, 65; Stanford Univ, PhD(physics), 70. *Prof Exp:* Asst prof meteorol, Univ Hawaii, 70-73; res assoc, 73-75, asst prof, 75-79, ASSOC PROF PHYSICS, NMEX INST MINING & TECHNOL, 79- *Mem:* Am Phys Soc; Am Meteorol Soc. *Res:* Mesoscale meteorology; turbulence in geophysical flows. *Mailing Add:* Dept Physics NMex Inst Mining & Technol Socorro NM 87801

RAYMOND, GERALD PATRICK, b Bagdad, Iraq, June 25, 33; Can citizen; m 59; c 2. CIVIL ENGINEERING. *Educ:* Univ London, BSc, 56, PhD(soil mech), 65, DSc(eng), 73; Princeton Univ, MSE, 57. *Prof Exp:* Eng asst, Howard Humphries & Sons, 51-54 & Kennedy & Donkin, 54-56; engr, Procter & Redfern, 57-58; dep city engr, North Bay, Ont, 58-59; lectr civil eng, Univ Sydney, 59-61; from asst prof to assoc prof, 61-72, chmn grad studies, 75-77, PROF CIVIL ENG, QUEEN'S UNIV, ONT, 72- *Concurrent Pos:* Current mem comts soil & rock properties, mech of earth masses & layered systs & track struct syst design, Transp Res Bd, Nat Res Coun; current chmn, comt geotextiles, Can Geotech Soc; current mem, comt track maintenance, Asn Am Railways. *Honors & Awards:* Walmsley Mem Prize, 56. *Mem:* Am Soc Civil Engrs; Am Rwy Eng Asn; Can Geotech Soc. *Res:* Consolidation of clays and settlement of foundations on clays; bearing capacity of peat; stresses and deformations under dynamic and static load systems in railroad track structure and support. *Mailing Add:* Ellis Hall Queen's Univ Kingston ON K7L 3N6 Can

RAYMOND, HOWARD LAWRENCE, b Seattle, Wash, Aug 2, 29; m 70; c 2. FISHERIES. *Educ:* Univ Wash, BS, 53. *Prof Exp:* Fishery biologist res, Bur Com Fisheries, 54-57; design engr statist, Boeing Co, 57-60; supvry fishery biologist res, Nat Marine Fisheries Serv, 60- 61; RETIRED. *Concurrent Pos:* Consult, Tech Adv Comt, Columbia Basin Fisheries, 75- *Mem:* Am Fisheries Soc; Am Inst Fisheries Res Biologists. *Res:* Development of methodology for protecting migrating anadromous fish in dammed and impounded rivers. *Mailing Add:* 11105 317th Ave Carnation WA 98014

RAYMOND, JOHN CHARLES, b Edgerton, Wis, Nov 28, 48; m 75; c 1. ASTROPHYSICS. *Educ:* Univ Wis, Madison, BA, 70, PhD(physics), 76. *Prof Exp:* RES FEL ASTROPHYS, HARVARD COL OBSERV, 76- *Mem:* Am Astron Soc. *Res:* Ultraviolet astronomy; solar physics; interstellar medium. *Mailing Add:* Ctr Astrophys Harvard Univ Cambridge MA 02138

RAYMOND, KENNETH NORMAN, b Astoria, Ore, Jan 7, 42; m 65, 75, 77; c 4. INORGANIC CHEMISTRY, CRYSTALLOGRAPHY. *Educ:* Reed Col, BA, 64; Northwestern Univ, Evanston, PhD(chem), 68. *Prof Exp:* from asst prof to assoc prof, 68-78, PROF INORG CHEM, UNIV CALIF, BERKELEY, 78- *Concurrent Pos:* Vis prof, Stanford Univ, Australian Nat Univ, Univ Sydney, Univ Strasbourg, Univ Rennes, Queensland Univ; Miller prof, Univ Calif, 77-78; Guggenheim fel, 80-81. *Honors & Awards:* E O Lawrence Award, 84. *Mem:* AAAS; Am Chem Soc; Am Crystallog Asn; Sigma Xi. *Res:* Chemistry of transition metal coordination compounds. *Mailing Add:* Dept Chem Univ Calif Berkeley CA 94720

RAYMOND, LAWRENCE W, b Buffalo, NY, Feb 14, 35. MEDICINE. *Educ:* Manhattan Col, BCE, 56; Harvard Univ, SM, 57; Cornell Univ, MD, 64. *Prof Exp:* Res & engr, Exxon, 51-60; intern, Georgetown Med Ctr, 64-65; res physician, Naval Med Res Inst, 65-67 & 72-74; resident med, Bethesda Naval Hosp, 67-70, chief pulmonary med, 74-77; pulmonary fel, Univ Calif, San Francisco, 70-72; assoc prof med, Yale Univ, 77-79; asst med dir, 79-84, med dir res & eng, 84-90, ASSOC MED DIR, EXXON CO USA, 90- *Mem:* AMA; Am Thoracic Soc; fel Am Col Physicians; fel Am Col Occup Med; Am Physiol Soc; Am Asn Accredited Scientists; Sigma Xi. *Mailing Add:* Exxon Co USA PO Box 2180 Houston TX 77252

RAYMOND, LOREN ARTHUR, b Sebastopol, Calif, Nov 23, 43; m 65; c 1. STRUCTURAL GEOLOGY, PETROLOGY. *Educ:* San Jose State Col, BS, 67, MS, 69; Univ Calif, Davis, PhD(geol), 73. *Prof Exp:* Instr, Appalachian State Univ, 72-73, asst prof geol, 73-76 & 77-78; asst prof geol, S Ore State Col, 76-77; assoc prof, 78-82, PROF GEOL, APPALACHIAN STATE UNIV, 82- *Concurrent Pos:* Chmn geol dept, Appalachian State Univ, 82-83; pres, Geol Servs Int, 76- *Mem:* Geol Soc Am; Am Geophys Union; Mineral Soc Am; Asn Geoscientists Int Develop. *Res:* Understanding the deformational and metamorphic processes in subduction and convergent zones as revealed by the Franciscan Complex of California and Ashe Metamorphic Suite of North Carolina. *Mailing Add:* Dept Geol Appalachian State Univ Boone NC 28608

RAYMOND, LOUIS, b Natrona, Pa, Nov 18, 34; m 57; c 4. METALLURGY. *Educ:* Carnegie Inst Technol, BS, 56, MS, 58; Univ Calif, Berkeley, PhD(metall), 63. *Prof Exp:* Methods engr mat process, Pittsburgh Plate Glass Co, 55; res engr stainless steel, Allegheny Ludlum Steel Co, 56-58; mech metall, Inst Eng Res, Univ Calif, Berkeley, 58-63; sr res engr strength mech, Aeronutronic Div, Ford Motor Co, 63-65; MEM TECH STAFF, AEROSPACE CORP, 65-, HEAD METALL RES, 67-, STAFF SCIENTIST, 77-; prof, Calif State Univ, Long Beach, 79- *Concurrent Pos:* Mem fac, Calif State Univ, Long Beach, 63-; spec consult, UNESCO-UN Develop Prog Proj, Higher Mining Eng Sch, Oviedo, Spain; lectr exten course, Univ Calif, Los Angeles; consult failure anal, Dept Transp, USCG, 75-77; consult struct integrity offshore platforms, US Geol Surv, Dept Interior, 76; consult life prediction anal, Dept Transp, Fed Railroad Admin, 77-; mem fracture toughness testing comt, Nat Mat Adv Bd-Nat Acad Sci, 75-76, mem fracture toughness requirements in design comt, 77- *Honors & Awards:* Am Soc Testing & Mat Award, 63-64; Space Processing Invention Award, NASA, 78. *Mem:* Am Inst Mining, Metall & Petrol Engrs; fel Inst Advan Eng; Sigma Xi; Am Soc Testing & Mat; AAAS. *Res:* Mechanical metallurgy; thermal mechanical processing; strengthening mechanisms; fracture toughness; hydrogen embrittlement; corrosion-fatigue; space processing; failure analysis. *Mailing Add:* 915 Celtis Pl Eastbluff Newport Beach CA 92660

RAYMOND, MAURICE A, b New Bedford, Mass, Jan 8, 38; m 63; c 2. ORGANIC CHEMISTRY, POLYMER CHEMISTRY. *Educ:* Providence Col, BS, 58; Univ Fla, PhD(org chem), 62. *Prof Exp:* Sr res chemist, Olin Corp, New Haven, 62-64, group supvr, 64-67, sect mgr, 67-70, tech mgr, 70-77, mkt mgr rigid urethanes, 77-78, bus mgr chem specialites, 78-85; CORP DIR RES & DEVELOP, TREMCO, INC, 85- *Mem:* Am Chem Soc; Sigma Xi; Soc Cosmetics Chemists; Soc Plastics Eng. *Res:* Cyclopolymerization; fluoraromatics; nitrenes and carbenes; plasticizers; functional fluids; thermally stable elastomers; homogeneous catalysis; urethane foam machinery and chemical systems. *Mailing Add:* 6006 Parkland Dr Chagrin Falls OH 44022

RAYMOND, SAMUEL, b Chester, Pa, Feb 7, 20; m 51; c 2. MEDICINE, COMPUTERS. *Educ:* Swarthmore Col, BA, 41; Univ Pa, MA & PhD(chem), 45; Columbia Univ, MD, 57. *Prof Exp:* Asst instr, Univ Pa, 41-45; asst, Col Physicians & Surgeons, Columbia Univ, 47-48, instr, 49-52; from asst prof to assoc prof clin path, Univ Pa, 58-90; RETIRED. *Concurrent Pos:* Dir, Am Bd Clin Chem, 71-76; mem lab adv bd, Pa State Dept Health. *Mem:* AAAS; Am Chem Soc; AMA; Am Asn Clin Chem; Asn Comput Mach. *Res:* Electrophoresis; medical applications of computers. *Mailing Add:* 31 Bar Neck Rd Woods Hole MA 02543

RAYMONDA, JOHN WARREN, b Wickenburg, Ariz, May 2, 39; m 63; c 2. PHYSICAL CHEMISTRY. *Educ:* Cornell Univ, BA, 61; Univ Wash, Seattle, PhD(chem), 66. *Prof Exp:* Res assoc molecular beam spectros, Harvard Univ, 66-68; asst prof phys chem, Univ Ariz, 68-72; prin chemist, Aerodyn Res Dept, Calspan Corp, 72-76; RES CHEMIST, HIGH ENERGY LASER TECHNOL DEPT, BELL AEROSPACE TEXTRON, 76- *Mem:* Sigma Xi. *Res:* Electronic spectroscopy of sigma bonded systems; molecular beam spectroscopy of high temperature species; primary events in photochemical processes; chemical lasers, laser induced chemical reactions, high temperature thermodynamics, laser radar; chemical laser modeling and development, laser diagnostics using nonlinear optics. *Mailing Add:* Textron Bell Aerospace Dept 902 PO Box One Buffalo NY 14240

RAYMOND-SAVAGE, ANNE, b Scituate, RI, Apr 10, 39; div; c 4. MARINE BIOLOGY. *Educ:* Univ RI, BS, 59, MA, 69; Ore State Univ, PhD(sci educ), 71. *Prof Exp:* Teacher sci, Coventry Schs, RI, 59-61 & West Warwick Schs, 64-66; instr sci educ, Univ RI, 67-69; ASSOC PROF SCI EDUC & MARINE BIOL, OLD DOMINION UNIV, NORFOLK, VA, 71-, ASSOC VPRES

ACAD AFFAIRS, 85-; ASSOC VPRES, LIFELONG LEARNING & ACAD TV SERV, 91- *Concurrent Pos:* Consult, WHRO-TV, Norfolk, Va, 73- & Corp Pub Broadcasting, Dept Defense. *Mem:* Nat Sci Teachers Asn; Asn Educ Commun Technol. *Res:* Marine biology education models; coral reef ecology. *Mailing Add:* Hughes Hall 101 Old Dominion Univ Norfolk VA 23529

RAYMUND, MAHLON, b Columbus, Ohio, June 10, 32; m 56; c 3. FRACTURE, FINITE ELEMENTS. *Educ:* Univ Chicago, AB, 51, SB, 54, SM, 60, PhD(physics), 63. *Prof Exp:* Res assoc hyperfragments, Enrico Fermi Inst, Univ Chicago, 63-64, Kaonproton scattering, 65-68; sr scientist, Nutron Cross Sect Data & Reactor Comput, Nuclear Energy Systs, Westinghouse Elec Corp, 68-81, fel scientist methods, supercomput, 81-86; SR LECTR MATH & STATIST DEPT, UNIV PITTSBURGH, 86- *Concurrent Pos:* Dept Sci & Indust Res sr vis fel, Univ Col, Univ London, 64-65; res collabr, Nat Neutron Cross Sect Ctr, Brookhaven Nat Lab, 70-71. *Mem:* Am Phys Soc; Soc Indust & Appl Math. *Res:* Finite element methods for fracture and seismic analysis; computer applications; computational fluid dynamics. *Mailing Add:* Univ Pittsburgh 301 Mackerey Hall Pittsburgh PA 15260

RAYNAL, DUDLEY JONES, b Greenville, SC, Jan 1, 47; m 71; c 2. PLANT ECOLOGY. *Educ:* Clemson Univ, SC, BS, 69; Univ Ill, Urbana, PhD(bot), 74. *Prof Exp:* Vis lectr bot, Univ Ill, Urbana, 74; from asst prof to assoc prof, 74-84, PROF BOT, COL ENVIRON SCI & FORESTRY, STATE UNIV NY, SYRACUSE, 84- *Concurrent Pos:* Chmn, tech comt, Nat Atmospheric Deposition Prog, NADP, 85-86; exec chair fac, State Univ NY-Environ Sci & Forestry, Syracuse, 90-92. *Honors & Awards:* Outstanding Res Award, Sigma Xi. *Mem:* Ecol Soc Am; Bot Soc Am; Brit Ecol Soc; AAAS; Sigma Xi. *Res:* Plant population and community ecology; plant succession plant life history studies; role of man-induced disturbance on terrestrial ecosystems; atmospheric deposition effects on forests. *Mailing Add:* Fac Environ & Forest Biol Col Environ Sci & Forestry State Univ NY Syracuse NY 13210

RAYNE, JOHN A, b Sydney, Australia, Mar 22, 27; m 54; c 3. PHYSICS. *Educ:* Univ Sydney, BSc, 48, BE, 50; Univ Chicago, MS, 51, PhD, 54. *Prof Exp:* Sci officer, Commonwealth Sci & Indust Res Orgn, Australia, 54-56; res engr, Westinghouse Elec Co, Pa, 56-61, adv engr, 61-64; assoc prof, 64-65, PROF PHYSICS, CARNEGIE-MELLON UNIV, 65- *Mem:* Am Phys Soc. *Res:* Cryogenics; physics of metals; alloy theory. *Mailing Add:* Dept Physics Carnegie-Mellon Univ 5000 Forbes Ave Pittsburgh PA 15213

RAYNER, JOHN NORMAN, b Worstead, Eng, Mar 12, 36; m 57; c 3. DYNAMIC CLIMATOLOGY. *Educ:* Univ Birmingham, Eng, BA, 58; McGill Univ, MS, 61; Univ Canterbury, NZ, PhD(geog), 65. *Prof Exp:* Lectr geog, Univ Canterbury, NZ, 61-65; from asst prof to assoc prof geog, 66-71, assoc prof physics, 68, PROF GEOG, OHIO STATE UNIV, 68-, CHMN DEPT, 75-, DIR ATOMOSPHERIC SCI PROG, 85-; State Climatologist, Ohio, 77-86. *Concurrent Pos:* Res assoc climat, Inst Polar Studies, Ohio State Univ, 66. *Honors & Awards:* Honors Award, Asn Am Geographers, 90. *Mem:* Am Meteorol Soc; Royal Meteorol Soc; Asn Am Geographers; Nat Coun Geog Educ; Nat Weather Asn. *Res:* Computer modelling of atmospheric systems; quantitative analysis of form-shape and of N-dimensional patterns. *Mailing Add:* Dept Geog Ohio State Univ 109 N Oval Mall Columbus OH 43210-1361

RAYNER-CANHAM, GEOFFREY WILLIAM, b London, Eng, 1944; Can citizen. CHEMICAL EDUCATION, HISTORY OF CHEMISTRY. *Educ:* Univ London, BSc, 66, DIC, 69, PhD(inorg chem), 69. *Prof Exp:* Fel, Simon Fraser Univ, 69-71, York Univ, 71-72 & Simon Fraser Univ, 72-73; vis asst prof inorg chem, Univ Victoria, 73-74; vis asst prof, Bishop's Univ, 74-75; asst prof, 75-80, assoc prof, 80-88, PROF INORG CHEM, MEM UNIV, NFLD, 88- *Concurrent Pos:* Vis assoc prof, Colo Sch Mines, 81-82; res assoc, Univ Calif, Santa Cruz, 89; vis scholar, New Col, Univ SFla, 90. *Honors & Awards:* Polysar Award, Chem Inst Can, 80; Catalyst Award, 85. *Mem:* Fel Chem Inst Can; Royal Soc Chem. *Res:* Program development for preparatory college chemistry students; development of chemistry in late 18th and early 19th century; pioneer women in nuclear science. *Mailing Add:* Dept Chem Sir Wilfred Grenfell Col Mem Univ Nfld Corner Brook NF A2H 6P9 Can

RAYNES, BERTRAM C(HESTER), b Jersey City, NJ, Mar 12, 24; m 44. CHEMICAL ENGINEERING. *Educ:* Pa State Univ, BS, 44; Union Univ, NY, MS, 49. *Prof Exp:* Asst res lab, Gen Elec Co, 44-50; develop engr, Brush Beryllium Co, 50-51; head process eng, Horizons, Inc, 51-62; vpres appl res, Rand Develop Corp, 62-70; head environ eng, Trygve Hoff & Assocs Consult Engrs, 70-72; CONSULT CHEM ENG, 72- *Mem:* Am Chem Soc; Am Inst Chem Engrs; Water Pollution Control Fedn. *Res:* Process research and development; fused salt electrolysis of refractory metals; high temperature ceramics; water pollution control; nonbiologic waste water treatment; land use management; environmental controls. *Mailing Add:* PO Box LL Jackson WY 83001

RAYNOLDS, PETER WEBB, b East Orange, NJ, Feb 3, 51; m 81; c 2. COLLOID CHEMISTRY. *Educ:* Hope Col, Holland, Mich, BS, 72; Ohio State Univ, PhD(org chem), 77. *Prof Exp:* Res asst, Univ Gronigen, Neth, 73 & Univ Zürich, Switz, 73-74; res fel, Univ Minn, 77-79; res chemist, 79-82, sr res chemist, 82-90, ASSOC RES CHEMIST, TENN EASTMAN CO, 90- *Mem:* Am Chem Soc. *Res:* Emulsion polymerization chemistry; colloid chemistry. *Mailing Add:* PO Box 1972 Kingsport TN 37662

RAYNOLDS, STUART, b Chicago, Ill, Oct 29, 27; m 61; c 3. ORGANIC CHEMISTRY, POLYMER CHEMISTRY. *Educ:* Cornell Univ, AB, 50; Univ Pittsburgh, MS, 55, PhD(org chem), 59. *Prof Exp:* Asst assayer, US Bur Mint, DC, 50-51; jr fel, Mellon Inst, 51-55; chemist, Jackson Labs, 59-65, res supvr, 65-69, res assoc, 69-72, RES FEL, E I DU PONT DE NEMOURS & CO, INC, 72- *Mem:* AAAS; Am Chem Soc. *Res:* Tar base and textile chemistry; polymers; colloid chemistry. *Mailing Add:* 2415 Ramblewood Dr Wilmington DE 19810

RAYNOR, SUSANNE, b Philadelphia, Pa, May 18, 48; m 72. CHEMICAL DYNAMICS. *Educ:* Duke Univ, BS, 70; Georgetown Univ, PhD(chem), 76. *Prof Exp:* Res assoc chem, Univ Toronto, 76-78; res assoc, Harvard Univ, 78-82; res asst prof, 82-88, ASSOC PROF CHEM, RUTGERS UNIV, NEWARK, 88- *Concurrent Pos:* Lectr, Univ Toronto, New Col, 78. *Mem:* Am Phys Soc; Am Chem Soc. *Res:* Ab initio quantum mechanics of molecules and solids; theoretical study of the dynamics and kinetics of molecular energy transfer and reaction. *Mailing Add:* Dept Chem Rutgers Univ Newark NJ 07102

RAYPORT, MARK, b Kharkov, Russia, Sept 6, 22; US citizen; m 51; c 3. NEUROSURGERY. *Educ:* Earlham Col, BA, 43; McGill Univ, MD, CM, 48, PhD(neurophysiol), 58. *Prof Exp:* Neurosurg resident, Montreal Neurol Inst, 54-57; asst prof neurosurg, Albert Einstein Col Med, 58-61, assoc prof, 61-68, asst prof physiol, 58-68; asst chief surg, Neurol Inst, Mt Zion Hosp & Med Ctr, San Francisco, 68-69; PROF NEUROL SURG & CO-CHMN NEUROSCI, MED COL OHIO, 69- *Concurrent Pos:* Duggan fel neuropath, Montreal Neurol Inst, 53, res fel, 55-56 & 58; res fel, USPHS, 55-56 & 58; NIMH Interdisciplinary Prog spec sr fel, 58-61; career scientist award, Health Res Coun NY, 62-68; vis prof, Univ Paris, 67-68. *Mem:* AAAS; Asn Res Nerv & Ment Dis; Am Acad Neurol Surg; Am Electroencephalog Soc; Soc Neurosci. *Res:* Basic approaches to clinical problems; neurosurgical treatment of epilepsy and pain; neurophysiology of mammalian and human cortex; interdisciplinary studies of brain and behavior. *Mailing Add:* Neurol Surg Med Col Ohio Toledo OH 43699-0003

RAYSIDE, JOHN STUART, b Quebec, Que, Can, Aug 24, 42; m 67; c 2. INSTRUMENTATION. *Educ:* Carleton Univ, Ottawa, BS, 67; Univ Minn, MS, 69, PhD(physics), 73. *Prof Exp:* Res assoc optical spectroscopy, Dept Chem, Univ Tenn, 72-76; DEVELOP ENGR, UNION CARBIDE NUCLEAR DIV, 76- *Concurrent Pos:* Consult, NIH, 73-74. *Mem:* Optical Soc Am; Am Asn Physics Teachers. *Res:* Raman spectroscopy; optics; computer interfacing; instrumentation in general. *Mailing Add:* 129 Morningside Dr Oak Ridge TN 37830

RAYSON, BARBARA M, KIDNEYS, ION TRANSPORT. *Educ:* Univ Melbourne, Australia, PhD(physiol & biochem), 76. *Prof Exp:* ASST PROF PHYSIOL MED, MED COL, CORNELL UNIV, 83- *Mailing Add:* Dept Physiol & Biophys Med Col Cornell Univ 1300 York Ave New York NY 10021

RAYUDU, GARIMELLA V S, b Andhra Pradesh, India, Oct 1, 36; m 65; c 4. NUCLEAR CHEMISTRY, NUCLEAR MEDICINE. *Educ:* Andhra Univ, India, BSc, 56, MSc, 57; McGill Univ, PhD(nuclear chem), 61; Am Bd Radiol, cert, 77; Am Bd Nuclear Med, cert, 79. *Prof Exp:* Res asst health physics, Atomic Energy Estab, Bombay, India, 57-58; res assoc, Carnegie Inst Technol, 61-65; sr res assoc nuclear activation anal, Univ Toronto, 65-67; asst prof radiochem & nuclear chem, Loyola Univ, La, 67-68; from asst prof to assoc prof, 68-88, PROF NUCLEAR MED, MED SCH, RUSH UNIV, 88-; ASSOC PROF, RUSH UNIV, 76- *Concurrent Pos:* US AEC grant nuclear & cosmochem, Carnegie Inst Technol, 61-65; Food & Drug Directorate Can pub health grant, Univ Toronto, 65-67; sr scientist, Rush-Presby St Luke's Med Ctr, 68- *Mem:* AAAS; Am Chem Soc; Royal Soc Chem; Am Asn Physicists in Med; Soc Nuclear Med. *Res:* Organ imaging radiopharmaceuticals; trace elements in liver, lung, pancreas, muscle and kidney; radiochemistry; cosmochemistry; instrumental analytical chemistry. *Mailing Add:* Dept Diag Radiol NUC Med Rush Univ 600 S Paulina St Chicago IL 60612

RAZ, AVRAHAM, b Bucharest, Romania, Mar 3, 45; c 5. HUMAN METASTASIS. *Educ:* Ben-Gurion Univ, BSc, 70, MSc, 72; Weismann Inst Sci, Rehovot, Israel, PhD, 78. *Prof Exp:* Vis scientist, Frederick Cancer Res, Nat Cancer Inst, 78-80; res fel, Dept Cell Biol, Weismann Inst, 80-81, sr scientist, 81-86, assoc prof, 86-88; MEM & DIR CANCER METASTASIS, MICH CANCER FOUND, DETROIT, MICH, 87- *Concurrent Pos:* Adj prof, Dept Radiol Oncol, Wayne State Univ, 88; ad hoc mem, Path H Study Sect, NIH, Bethesda, Md, 88, consult, 89; dir, Tumor Biol Prog, M L Prentis Comprehensive Cancer Ctr, 90. *Honors & Awards:* Bondi Mem Award, Weismann Inst, 76, H Dudley Wright Res Award, 85. *Mem:* Am Asn Cancer Res; Am Asn Cell Biol; AAAS; Europ Asn Cancer Res; Int Metastasis Soc; Israel Biochem Soc. *Res:* Role of tumor ecell antigens and genes in the spread of cancer in the body. *Mailing Add:* 110 E Warren Ave Detroit MI 48201

RAZAK, CHARLES KENNETH, b Collyer, Kans, Sept 15, 18; m 40; c 2. FORENSIC ENGINEERING, ENGINEERING. *Educ:* Univ Kans, BS, 39, MS, 42. *Prof Exp:* From instr to asst prof aeronaut eng, Univ Kans, 29-43; 39-43; assoc prof & head dept, Wichita Univ, 43-48, prof, 48-64, dir dept eng, 48-51, actg dean col bus admin, 51-53, dean sch eng & dir eng res, 53-64; prof eng & dir indust exten serv, Kans State Univ, 64-70; ENG & MGT CONSULT, 70- *Concurrent Pos:* Mem bd dirs, KARD-TV, Wichita, Kans, Kans Invest Co & Midland Metal Craft, Inc; pres, Pane, Inc, Aerial Distributors, Inc, 64 & Managers, Inc, 66- *Mem:* Soc Automotive Engrs; Am Asn Automotive Med; Int Asn Automotive Med; Rotary Int. *Res:* Expert testimony; lowspeed aerodynamics; aircraft design; transportation safety. *Mailing Add:* 7717 Killarney Ct Wichita KS 67206

RAZDAN, MOHAN KISHEN, b Srinagar, India. COMBUSTION, FLUID MECHANICS. *Educ:* Regional Eng Col, India, BE, 71; Indian Inst Technol, Kanpur, MTech, 74; Pa State Univ, MS, 76, PhD(mech eng), 79. *Prof Exp:* Res asst, Pa State Univ, 77-79, asst prof mech eng, 79-80; STAFF ENGR, EXXON RES & ENG CO, 80- *Mem:* Combustion Inst; assoc mem Am Soc Mech Engrs; Am Inst Aeronaut & Astronaut; Sigma Xi. *Res:* Effects of fluid mechanics on combustion processes and heat transfer in turbulent reacting flows with an emphasis on reduction of pollutants; fluidized bed reactor modeling; design of solid-fluid systems. *Mailing Add:* 7005 Andre Dr Indianapolis IN 46278-1533

RAZDAN, RAJ KUMAR, b Simla, India, Dec 19, 29; m 56; c 2. ORGANIC CHEMISTRY, MEDICINAL CHEMISTRY. *Educ:* Univ Delhi, BSc, 48; Indian Inst Sci, Bangalore, Dipl, 51; Univ Glasgow, PhD(chem), 54. *Prof Exp:* Jr sci officer, Nat Chem Lab, India, 54-56; sci officer, Glaxo Labs, Ltd, Eng, 56-58, joint works mgr fine chem prod, India, 58-63; res assoc org chem, Univ Mich, 63-64; sr staff scientist, Arthur D Little, Inc, Mass, 64-70; vpres res, 70-80, pres, SISA Inst Res & SISA Inc, 81-83, dir toxicol, SISA Inc, 81-83, prin scientist, SISA Pharmaceut Labs Inc, 84-86; CHIEF EXEC OFFICER, ORGANIX INC, 86-; AFFIL PROF, DEPT PHARMACOL & TOXICOL, MED COL VA, RICHMOND, 86- *Concurrent Pos:* Consult, Nat Inst Drug Abuse, Nat Cancer Inst & Sarabhai Industs, India. *Mem:* Am Chem Soc; Royal Soc Chem; AAAS. *Res:* Cannabinoids; terpenes; steroids; central nervous system active drugs; molecular rearrangements; morphine chemistry; narcotic antagonists; lysergic acid chemistry. *Mailing Add:* 76 Lawrence Lane Belmont MA 02178

RAZGAITIS, RICHARD A, b Chicago, Ill, Jan 13, 44; m 67; c 5. MECHANICAL ENGINEERING. *Educ:* Univ Ill, BS, 65; Univ Fla, MS, 69; Southern Methodist Univ, PhD(mech eng), 74; Ohio State Univ, MBA, 90. *Prof Exp:* Mem Apollo launch team, Cape Kennedy Space Ctr, McDonnell-Douglas Corp, 65-69; asst prof, Univ Portland, 69-72; asst prof eng, Ohio State Univ, 74-81; MGR VP COM DEVELOP, BATTELLE, 81- *Concurrent Pos:* Dresser fel, Southern Methodist Univ, 72-74; Dupont asst prof, Ohio State Univ, 75-76. *Mem:* Am Soc Mech Engrs; Am Soc Eng Educ; Licensing Exec Soc. *Res:* Swirl flow heat transfer; aerosol mechanics; cyclonic separation techniques of particulates and steam; thermal and fluid sciences. *Mailing Add:* 2070 W Lane Ave Columbus OH 43221

RAZNIAK, STEPHEN L, b Detroit, Mich, May 23, 34; m 65; c 3. ORGANIC CHEMISTRY. *Educ:* Wayne State Univ, BS, 55; Wash State Univ, PhD(chem), 59. *Prof Exp:* NSF fel org chem, Brown Univ, 59-60, instr, 60-61; from asst prof to assoc prof, 61-64, head dept, 74-79, PROF ORG CHEM, ETEX STATE UNIV, 64- *Mem:* Am Chem Soc; Sigma Xi. *Res:* Organic sulfur chemistry; environmental chemistry. *Mailing Add:* Dept Chem ETex State Univ Commerce TX 75429

RAZOUK, RASHAD ELIAS, b Dumiat, Egypt, Aug 22, 11; US citizen; m 46; c 2. CHEMISTRY. *Educ:* Cairo Univ, BSc, 33, MSc, 36, PhD(chem), 39. *Prof Exp:* Asst prof chem, Cairo Univ, 39-47, assoc prof, 47-50; prof & chmn dept, Ain Shams Univ, Cairo, 50-66, vdean, 54-60; prof, Am Univ Cairo, 66-68; prof, 68-78, EMER PROF CHEM, CALIF STATE UNIV, LOS ANGELES, 79- *Concurrent Pos:* Actg dir, Div Colloid & Surface Chem, Nat Res Ctr, Cairo, 50-68. *Mem:* Am Inst Chem; Am Chem Soc; Royal Soc Chem. *Res:* Surface chemistry; adsorption on carbons and active solids; solid reactions; surface tension and contact angles; wetting and wettability. *Mailing Add:* 1140 Keats St Manhattan Beach CA 90266

RAZZELL, WILFRED EDWIN, microbiology, biochemistry, for more information see previous edition

RE, RICHARD N, b Palisade, NJ, Sept 4, 44; m 79; c 3. CELL BIOLOGY. *Educ:* Harvard Col, AB, 65; Harvard Med Sch, MD, 69; cert, Am Bd Internal Med, Am Bd Endocrinol & Metab. *Prof Exp:* Med intern & resident, Mass Gen Hosp, 69-71; clin & res fel endocrinol, 71-74; clin asst med, 74-76; asst prof med, Harvard Med Sch, 77-79; chmn clin invests comt, Alton Ochsner Med Found, 80-86, HEAD SECT HYPERTENSIVE DIS, OCHSNER CLIN, 81-, VPRES & DIR RES, ALTON OCHSNER MED FOUND, 85-, STAFF MEM, 79- *Concurrent Pos:* Res fel, Harvard Med Sch, 71-74; instr med, 75-76; chief, Hypertension Clinic, Mass Gen Hosp, 75-79, & asst in med, 76-79; assoc clin prof med, Tulane Univ Sch Med, & assoc prof med, La State Univ Sch Med, 80-; fel Coun High Blood Pressure, Am Heart Asn. *Mem:* Fel, Am Col Physicians; Am Heart Asn; Int Soc Hypertension; Am Fed Clin Res; AAAS; NY Acad Sci; Soc Exp Biol & Med. *Res:* Cellular biology of the Renin-angiotensin systems and the sequeiae of hypertension. *Mailing Add:* Alton Ochsner Med Found 1516 Jefferson Hwy New Orleans LA 70121

REA, DAVID KENERSON, b Pittsburgh, Pa, June 2, 42; m 67; c 2. GEOLOGICAL OCEANOGRAPHY. *Educ:* Princeton Univ, AB, 64; Univ Ariz, MS, 67; Ore State Univ, PhD(oceanog), 74. *Prof Exp:* Asst prof oceanog, Sch Oceanog, Ore State Univ, 74-75; from asst prof to prof, dept atmospheric & oceanic sci, 75-87, PROF, DEPT GEOL SCI, UNIV MICH, 87- *Concurrent Pos:* Assoc dir, Climate Dynamics Prog, NSF, Washington, DC, 86-87. *Mem:* Geol Soc Am; Am Geophys Union; AAAS; Sigma Xi. *Res:* Paleoclimatology and paleoceanography; marine and lacustrine sediments and sedimentation. *Mailing Add:* Dept Geol Sci Univ Mich Ann Arbor MI 48109-1063

REA, DAVID RICHARD, b Indianapolis, Ind, May 4, 40; m 64; c 4. CHEMICAL ENGINEERING. *Educ:* Purdue Univ, BS, 62; Princeton Univ, MA, 64, PhD(chem eng), 67. *Prof Exp:* TECH SUPT, PLASTIC PRODS & RESINS DEPT, E I DU PONT DE NEMOURS & CO, INC, 66- *Mem:* Am Inst Chem Engrs; Am Chem Soc. *Res:* Rheology as applied to plastics; process development of fluorocarbon chemistry products; low-density polyethylene; nylon intermediates work; engineering thermoplastics. *Mailing Add:* 4297 Sunningdale Dr Bloomfield Township MI 48013-7306

REA, DONALD GEORGE, b Portage La Prairie, Man, Sept 21, 29; nat US. PLANETARY SCIENCES. *Educ:* Univ Man, BSc, 50, MSc, 51; Mass Inst Technol, PhD(chem), 54. *Hon Degrees:* DSc, Univ Man, 80. *Prof Exp:* Nat Res Coun Can fel, Oxford Univ, 54-55; res chemist, Calif Res Corp, 55-61; assoc res chemist, Space Sci Lab, Univ Calif, Berkeley, 61-68; dep dir planetary progs, Off Space Sci & Appln, NASA Hq, 68-70; asst lab dir sci, Jet Propulsion Lab, Calif Inst Technol, 70-76, dept asst lab dir, 76-80, asst lab dir technol & space prog develop, 80-91; CONSULT SCIENTIST, MITRE CORP, VA, 91- *Concurrent Pos:* Res fel, John F Kennedy Sch Govt, Harvard Univ, 79-80. *Honors & Awards:* Exceptional Sci Achievement Award, NASA, 69 & Outstanding Leadership Medal, 85. *Mem:* Am Chem Soc; Am Phys Soc; Optical Soc Am; Am Astron Soc; Am Geophys Union. *Res:* Molecular spectroscopy of planetary atmospheres; remote sensing of planetary surfaces; space exploration advancement. *Mailing Add:* Mitre Corp 7525 Culshire Dr McLean VA 22102-3481

REA, KENNETH HAROLD, b Red Oak, Iowa, Aug 20, 46; m 67; c 4. PLANT SYNECOLOGY, COMPUTER SCIENCE. *Educ:* NMex State Univ, BS, 69, MS, 72; Utah State Univ, PhD(ecol), 76. *Prof Exp:* MEM STAFF ECOL, LOS ALAMOS NAT LAB, 76- *Mem:* AAAS; Soc Range Mgt. *Res:* Plant demography; impacts of geothermal energy development; hazardous waste management. *Mailing Add:* Royal Crest Trailer Park Los Alamos NM 87544

READ, ALBERT JAMES, b Albany, NY, June 8, 26; m 53; c 1. PHYSICS, SCIENCE EDUCATION. *Educ:* State Univ NY Col Educ, Albany, BA, 47, MA, 54. *Prof Exp:* Instr physics, Rensselaer Polytech Inst, 47-50; asst prof, Morrisville Agr & Tech Inst, 52-57; from assoc prof to prof physics, 57-85, DIR SCI DISCOVERY CTR, STATE UNIV NY COL ONEONTA, 85- *Mem:* Int Solar Energy Soc; Am Asn Physics Teachers; Nat Sci Teachers Asn; Hist Sci Soc; Soc Hist Technol. *Mailing Add:* Dept Physics State Univ NY Col Oneonta Oneonta NY 13820

READ, CHARLES H, b Amherst, NS, July 22, 18; m 42; c 5. PEDIATRICS, ENDOCRINOLOGY. *Educ:* Acadia Univ, BSc, 39; McGill Univ, MD & CM, 43. *Prof Exp:* Rutherford Caverhill fel, Fac Med, McGill Univ, 47-49; Commonwealth Fund fel, Harvard Med Sch & Mass Gen Hosp, Boston, 49-51; asst prof pediat, Fac Med, Univ Man, 51-52, assoc prof, 52-54; from asst prof to assoc prof, 54-59, PROF PEDIAT, COL MED, UNIV IOWA, 59- *Mem:* NY Acad Sci; Soc Pediat Res; Endocrine Soc; Am Diabetes Asn; Am Acad Pediat. *Mailing Add:* Dept Pediat 2504 JCP Univ Hosps Univ Iowa Iowa City IA 52242

READ, D E, b Calgary, Alta, July 17, 24; m 47; c 1. SCIENCE POLICY. *Educ:* Mt Allison Univ, BSc, 45; McGill Univ, PhD(org chem), 49. *Prof Exp:* Res chemist, Cent Res Lab, Can Industs Ltd, 49-56, res coordr, Chem Div, Que, 56-67; process specialist, Sandwell & Co Ltd, 67-72; sr policy adv, Ministry State Sci & Technol, 72-75, proj dir, 75-86; RETIRED. *Mem:* Can Pulp & Paper Asn; Chem Inst Can. *Res:* Forestry, pulp and paper, chlorohydrocarbons; environment. *Mailing Add:* 2205-900 Dynes Rd Ottawa ON K2C 3L6 Can

READ, DAVID HADLEY, b Seattle, Wash, May 20, 21; m 51, 63; c 5. PHYSICAL ORGANIC CHEMISTRY. *Educ:* Seattle Univ, BS, 42; Univ Ill, MS, 44; Univ Notre Dame, PhD(org chem), 49. *Prof Exp:* Asst prof chem, Univ Seattle, 48-51; res chemist, Am-Marietta Co, 51-54; assoc prof, 54-65, PROF CHEM, SEATTLE UNIV, 65- *Mem:* AAAS; Am Chem Soc; The Chem Soc. *Res:* Vinyl polymerization; electronic effects in rigid systems; clinical separations by high-performance liquid chromatography. *Mailing Add:* 911 11th E Seattle Univ 900 12th & E Columbia Seattle WA 98102-4573

READ, DAVID THOMAS, b Seattle, Wash, Sept 17, 47; m 72; c 4. FRACTURE MECHANICS, PHYSICAL METALLURGY. *Educ:* Univ Santa Clara, BS, 69; Univ Ill, MS, 71, PhD(physics), 75. *Prof Exp:* PHYSICIST MECH PROPERTIES, US NAT BUR STANDARDS, 75- *Concurrent Pos:* Nat Res Coun fel, Nat Bur Standards, 75-76. *Mem:* Am Phys Soc; Am Soc Testing & Mat. *Res:* Low-temperature mechanical properties of metals; fracture mechanics; measurements of the J contour integral; mechanical properties of thin films. *Mailing Add:* Mat Reliability Div 430 Nat Inst Standards & Technol 325 Broadway Boulder CO 80303-3328

READ, FLOYD M, b Ray City, Ga, Oct 4, 24; m 43; c 1. PHYSICS. *Educ:* Univ Fla, BSEd, 52, MEd, 56; NY Univ, PhD, 69. *Prof Exp:* Asst prof phys sci, 57-60, from asst prof to assoc prof physics, 60-75, PROF PHYSICS, E CAROLINA UNIV, 75-, AT DEPT SCI EDUC. *Mem:* Nat Sci Teachers Asn; Am Asn Physics Teachers; Nat Asn Res Sci Teaching; Sigma Xi. *Res:* Solar radiometry; teaching of physics. *Mailing Add:* 1804 Fairview Way Greenville NC 27858

READ, GEORGE WESLEY, b Los Angeles, Calif, June 24, 34; m 54; c 2. PHARMACOLOGY, AUTACOIDS. *Educ:* Stanford Univ, BA, 59, MS, 62; Univ Hawaii, PhD, 69. *Prof Exp:* Instr gen sci, Univ Hawaii, Hilo, 63-64; res asst, 64-68, from asst prof to assoc prof, 68-86, PROF PHARMACOL, UNIV HAWAII, MANOA, 86- *Concurrent Pos:* Vis scientist, Univ Tex, Dallas, 75 & 84, Univ Wash, 76, NIH, 78 & St Louis Univ, 83. *Mem:* Am Soc Pharmacol & Exp Therapeut; NAm Histamine Res Soc. *Res:* Pharmacology of histamine release, excitation-secretion coupling; computer-assisted instruction. *Mailing Add:* Dept Pharmacol Univ Hawaii Sch Med Honolulu HI 96822

READ, JOHN FREDERICK, b Reading, Eng, Apr 11, 40; m 63; c 3. PHYSICAL CHEMISTRY. *Educ:* Univ Nottingham, BSc, 61, PhD(phys chem), 64. *Prof Exp:* Fel chem, Northwestern Univ, 64-65; teaching fel, Hope Col, 65-66; from asst prof to prof, 66-74, from asst dean to assoc dean, 70-78, DEAN, COL ARTS & SCI, MT ALLISON UNIV, 78- *Mem:* Chem Inst Can. *Res:* Heterogeneous catalysis of simple gas phase reactions by rare-earth compounds. *Mailing Add:* Dean Col Arts & Sci Mt Allison Univ Sackville NB E0A 3C0 Can

READ, JOHN HAMILTON, b Joliette, Que, Feb 20, 24; m 48; c 3. PEDIATRICS, PREVENTIVE MEDICINE. *Educ:* McGill Univ, BSc, 48, MD, CM, 50; Univ Toronto, DPH, 52. *Prof Exp:* Med officer, Simcoe Co, Ont, 52-54; resident pediat, Univ Mich, 54-56, instr, 56-58; asst prof pediat & prev med, Univ BC, 58-62; prof prev med & head dept & asst prof pediat, Queen's Univ, Ont, 62-68, PROF & HEAD DIV COMMUN HEALTH SCI & PROF PEDIAT, FAC MED, UNIV CALGARY, 68- *Mem:* Am Pub Health Asn. *Res:* Preventive pediatrics. *Mailing Add:* 3712 Underhill Dr NW Calgary AB T2N 4N1 Can

READ, MARSHA H, b Salt Lake City, Utah. NUTRITION. *Educ:* Univ Nev, BS, 68, MS, 69; Utah State Univ, PhD(nutrit), 77. *Prof Exp:* From instr to assoc prof, 69-84, PROF NUTRIT, UNIV NEV, RENO, 84- *Mem:* Am Diet Asn; Am Inst Nutrit. *Res:* Micronutrient supplementation patterns and effects; nutrition and physical performance; dietary compliance. *Mailing Add:* Dept Nutrit 142 Univ Nev Reno NV 89557

READ, MERRILL STAFFORD, b Baltimore, Md, June 3, 28; div; c 2. BIOCHEMISTRY, NUTRITION. *Educ:* Northwestern Univ, BS, 49; Ohio State Univ, MS, 51, PhD(biochem), 56. *Prof Exp:* Asst biochem, Ohio State Univ, 49-52; chief, Irradiated Food Br, Med Res & Nutrit Lab, Fitzsimons Army Hosp, Denver, 54-59; tech coordr, Radiation Preservation of Food Prog, Off Surgeon Gen, US Army, 59; vis prof biochem & nutrit, Va Polytech Inst & State Univ, 59-60; dir nutrit res, Nat Dairy Coun, 60-65, exec asst to pres, 64-66; nutrit prog admnr, Nat Inst Child Health & Human Develop, 66, dir, Growth & Develop Br, 66-76, actg dep dir, Ctr Res Mothers & Children, 74-76; adv nutrit res, Div Family Health, Nutrit Unit, Pan Am Health Orgn, 76-79; chief, Clin Nutrit & Early Develop Br, Nat Inst Child Health & Human Develop, 80-85; PROF & CHMN, DEPT HUMAN NUTRIT & FOOD SYSTS, COL HUMAN ECOL, UNIV MD, 85- *Concurrent Pos:* Mem adv comt food irradiation, Am Inst Biol Sci-AEC, 60-62; mem comn Ill, Int Union Nutrit Sci, 67-76; vis scientist, Mass Inst Technol, 70; mem & chmn, Nat Adv Coun, NY State Col Human Ecol, Cornell Univ, 71-76; mem subcomt malnutrit, brain develop & behav, Nat Acad Sci-Nat Res Coun, 75-81; mem, Gov Coun, Am Pub Health Asn, 81-83; chmn, Comt Nutrit Educ & Training, Am Soc Clin Nutrit, 80-84, mem, 85-87. *Honors & Awards:* Dir Award Nutrit, NIH, 76. *Mem:* AAAS; Am Inst Nutrit; Am Soc Clin Nutrit; Am Pub Health Asn; Soc Nutrit Educ; Latin Am Soc Nutrit. *Res:* Child growth and development; mental development; maternal health; nutritional surveillance; nutrition education. *Mailing Add:* Dept Human Nutrit & Food Syst Col Human Educ Univ Md Marie Mount Hall Rm 3304 College Park MD 20742

READ, PAUL EUGENE, b Canandaigua, NY, July 13, 37; m 89; c 3. HORTICULTURE, PLANT TISSUE CULTURE. *Educ:* Cornell Univ, BS, 59, MS, 64; Univ Del, PhD(Biol sci), 67. *Prof Exp:* County 4-H Club agent, Fulton County Exten Serv Asn, NY, 59-62; teaching asst hort, Cornell Univ, 62-64; res assoc, Univ Del, 64-67; from asst prof to assoc prof, 67-78, prof hort, Univ Minn, St Paul, 78-87; PROF & HEAD DEPT HORT, UNIV NEBR, LINCOLN, 87- *Concurrent Pos:* Consult, Teaching & Res, Peoples Repub China, Zambia, Morocco, Australia, Eastern Europe. *Mem:* Fel Am Soc Hort Sci (vpres, 86-87); Int Asn Plant Tissue Cult; Am Hort Soc; Bot Soc Am; Tissue Cult Asn. *Res:* Plant tissue culture; plant propagation; chemical plant growth regulation; nutrition of horticultural plants; administration. *Mailing Add:* Dept Hort Univ Nebr Lincoln NE 68583-0724

READ, PHILIP LLOYD, b Flint, Mich, Jan 9, 32; m 56; c 3. PHYSICS, INSTRUMENTATION. *Educ:* Oberlin Col, AB, 53; Univ Mich, MS, 54, PhD(physics), 61. *Prof Exp:* Physicist, Res Lab, Gen Elec Co, NY, 60-67, mgr x-ray components eng, X-Ray Dept, Wis, 67-70, mgr cardio-surg systs sect, Med Systs Div, 70-75; vpres & gen mgr, Prod Systs Div, 75-81, chief oper officer, 81-83, sr vpres, Computervision Corp, Bedford, 81-88; CONSULT, 88- *Mem:* Am Phys Soc. *Res:* Ionic conduction in solids; circuit theory; thermionic emission; ultrahigh vacuum; electrical transport properties of insulator surfaces; biophysics; computer graphics; Computer-Aided Design/ Computer-Aided Manufacturing. *Mailing Add:* 80 Witherell Dr Sudbury MA 01776

READ, RAYMOND CHARLES, b London, Eng, Jan 26, 24; nat US; m 46; c 3. SURGERY. *Educ:* Cambridge Univ, MA, 44, MB, BCh, 47; Univ Minn, MB, 46, MD, 51, MS, 57, PhD, 58. *Prof Exp:* Intern & resident surg, Kings Col, Univ London, Harvard Univ & Univ Hosps, Univ Minn, 46-51; Harvey Cushing res fel, Harvard Med Sch, 51-53; Life Ins Med res fel, Med Sch, Univ Minn, Minneapolis, 56-58, asst prof surg, 58-61; assoc prof, Wayne State Univ, 61-66; PROF SURG, UNIV ARK, LITTLE ROCK, 66-; CHIEF SURG SERV & STAFF SURGEON, VET ADMIN HOSP, 66- *Concurrent Pos:* Staff surgeon, Vet Admin Hosp, Minneapolis, 58-61; mem sr staff, Detroit Gen Hosp, 61-66. *Mem:* AAAS; Soc Exp Biol & Med; AMA; Sigma Xi. *Res:* Fundamental and surgical cardiovascular physiology. *Mailing Add:* 4300 W Seventh St Univ Ark Little Rock AR 72205

READ, ROBERT E, b Stoke on Trent, Eng, Jan 30, 33; US citizen; m 55; c 2. ORGANIC CHEMISTRY. *Educ:* Haverford Col, BS, 55; Univ Del, MS, 57, PhD(org chem), 60. *Prof Exp:* Chemist, Chem Dyes & Pigments Dept, 60-80, tech prog mgr, Cent Res & Develop Dept, 81-85, tech prog mgr, Biomed Prod Dept, 85-90, MGR SUPPORT & DEVELOP, MED PROD DEPT, E I DU PONT DE NEMOURS & CO, INC, 90- *Mem:* Am Chem Soc; Am Asn Textile Chemists & Colorists; Sigma Xi. *Res:* Urethanes, organic carbodiimides and isocyanate related chemistry; polymer chemistry; development of chemical finishing agents for textiles; development of medical instrumentation. *Mailing Add:* 1537 Old Coach Rd Newark DE 19711

READ, ROBERT G, b Kingston, NY, Apr 2, 18; m 47; c 6. METEOROLOGY, OCEANOGRAPHY. *Educ:* US Naval Postgrad Sch, BS, 53, MS, 61. *Prof Exp:* Instr meteorol & oceanog, US Naval Postgrad Sch, 59-61; from asst prof to prof, 61-88, EMER PROF METEOROL, SAN JOSE STATE UNIV, 88- *Concurrent Pos:* Assoc dir, NSF Summer Inst Earth Sci, 64 & Partic in-serv inst oceanog, Moss Landing Marine Labs & instnl grant marine influences on potential evaporation in coastal Calif, 68-69; investr, Sea Trout Prog, Moss Landing Marine Labs, Calif, 70-71; estab environ measurement network on Barro Colorado Island, CZ, Smithsonian Inst Trop Res, 71; joint researcher with Mid Am Res Unit, NIH; vis scientist, Nat Ctr Atmospheric Res, Boulder, Colo, 66; prin investr, NIH, 73-79; meteorol consult, Fed Univ Rio de Janeiro & Fed Univ Rio Grande do Sol, 82; vis scholar, Smithsonian Trop Res Inst, Panama, 84; Fulbright scholar, Univ Panama, 85. *Mem:* AAAS; Am Meteorol Soc; Am Geophys Union. *Res:* Problems in evaporation in the tropics and in the marine coastal atmosphere; marine meteorology and the energy transport across the air-ocean interface; general synoptic meteorology; author numerous publications. *Mailing Add:* Dept Meteorol San Jose State Univ Washington Sq San Jose CA 95192

READ, ROBERT H, b Jacksonville, Ill, Feb 15, 28; m 50; c 3. METALLURGY, PHYSICS. *Educ:* Ill Col, AB, 52; Pa State Univ, MS, 53, PhD(metall), 55. *Prof Exp:* Res metallurgist, Armour Res Found, Ill Inst Technol, 56-57, supvr powder metals res, 57-58, supvr phys metall, 59-62; dir res, Atlas Steels Co, Rio Algom Mines Ltd, Ont, 62-64, mgr technol, 64-69, vpres res & metall, Co, 69-72, mgr corp planning, 72-73; sr vpres technol & sales, Teledyne Vasco, 73-76, exec vpres, 76-80, PRES, TELEDYNE PORTLAND FORGE, 80-, GROUP EXEC. *Mem:* Am Iron & Steel Inst; Am Soc Metals; Am Inst Mining, Metall & Petrol Engrs; Can Inst Mining & Metall. *Mailing Add:* Teledyne Portland Forge RR 1 Portland IN 47371

READ, ROBERT RICHARD, b Columbus, Ohio, Oct 5, 29; m 63; c 3. MATHEMATICAL STATISTICS. *Educ:* Ohio State Univ, BS, 51; Univ Calif, PhD(math statist), 57. *Prof Exp:* Asst res statistician, Univ Calif, 57-60; vis asst prof statist, Univ Chicago, 60-61; assoc prof, 61-71, PROF PROBABILITY & STATIST, NAVAL POSTGRAD SCH, 71- *Concurrent Pos:* Lectr, Univ Calif, 58-59; consult, Maritime Cargo Transportation Conf, Nat Acad Sci-Nat Res Coun, Lockheed Calif Co & ARRO Res Corp. *Mem:* Inst Math Statist. *Res:* Probability; statistics; operations research. *Mailing Add:* Dept Opers Res Naval Postgrad Sch Monterey CA 93940

READ, RONALD CEDRIC, b London, Eng, Dec 19, 24; m 49, 87; c 2. MATHEMATICS, COMPUTER SCIENCE. *Educ:* Cambridge Univ, BA, 48, MA, 53; Univ London, PhD(math), 58. *Prof Exp:* Lectr math, Univ WI, 50-60, sr lectr, 60-66, reader, 66-67, prof, 67-70; prof, 70-90, ADJ PROF MATH, UNIV WATERLOO, 90- *Concurrent Pos:* USAF Off Sci Res grant, Univ WI, 65-68; Nat Res Coun Can grant, Univ Waterloo, 71-; ed, J Asn Comput Mach, 71-75. *Mem:* Asn Comput Mach. *Res:* Enumerative graph theory; applications of computers to graph theoretical and combinatorial problems. *Mailing Add:* Dept Combinatorics & Optimization Univ Waterloo Waterloo ON N2L 3G1 Can

READ, THOMAS THORNTON, b Philadelphia, Pa, Jan 24, 43; m 67; c 1. DIFFERENTIAL EQUATIONS. *Educ:* Oberlin Col, BA, 63; Yale Univ, MA, 65, PhD(math), 69. *Prof Exp:* From asst prof to assoc prof, 67-76, PROF MATH, WESTERN WASH UNIV, 76- *Concurrent Pos:* Vis lectr, Chalmers Univ Technol, Gothenburg, Sweden, 73-74; vis prof, Univ Groningen, Neth, 78-80. *Mem:* Am Math Soc; London Math Soc. *Res:* Linear differential equations, especially spectral theory. *Mailing Add:* Dept Math Western Wash Univ Bellingham WA 98225

READ, VIRGINIA HALL, b Louisville, Miss, Oct 15, 37; m 60; c 3. BIOCHEMISTRY, ENDOCRINOLOGY. *Educ:* Univ Miss, BS, 59, PhD(biochem), 64. *Prof Exp:* Instr biochem, Sch Med, Univ Miss, 65-66, asst prof, 66-68; asst prof biochem, 70-74, assoc prof clin lab sci, 79-88, ASSOC PROF BIOCHEM, SCH MED, UNIV MISS, 74-, ASSOC PROF PATH, 88- *Concurrent Pos:* NIH spec fel endocrinol, Sch Med, Univ Ala, Birmingham, 68-70; Miss Heart Asn fel, 65-67. *Mem:* Sigma Xi; Am Chem Soc; Endocrine Soc; Am Asn Clin Chem. *Res:* Control of aldosterone secretion; mechanism of aldosterone action and its relationship to diseases; relationship of adrenal steroids and thyroid hormones to electrolyte and water metabolism; radioimmunoassay of hormones. *Mailing Add:* Dept Path Div Lab Med Univ Miss Med Sch Jackson MS 39216-4505

READ, WILLIAM F, b Chicago, Ill, Oct 18, 15. GEOLOGY. *Educ:* Harvard Univ, BA, 36; Univ Chicago, PhD(geol), 42. *Prof Exp:* Prof, 41-80, EMER PROF GEOL, LAWRENCE UNIV, 80- *Mem:* Geol Soc Am; Metall Soc; Am Geol Soc Teachers. *Mailing Add:* 1905 N Alexander St Appleton WI 54911

READ, WILLIAM GEORGE, b Stratton, Colo, Apr 13, 21; m 48. NUCLEAR PHYSICS. *Educ:* Kans State Teachers Col, Ft Hays, BS, 43, MS, 48; Univ Kans, PhD(physics), 56. *Prof Exp:* Instr physics, Kans State Teachers Col, Ft Hays, 47-48; from asst prof to prof, 49-59, head dept, 59-70, vpres acad affairs & dean of fac, 70-78, PROF, MURRAY STATE UNIV, 78- *Mem:* AAAS; Am Phys Soc; Nat Asn Physics Teachers; Sigma Xi. *Res:* Experimental nuclear physics and physical properties of soils; electronics. *Mailing Add:* Col Sci Murray State Univ Murray KY 42071

READ-CONNOLE, ELIZABETH LEE, b Atlanta, Ga, Feb 22, 52; m 84; c 2. VIROLOGY. *Educ:* Mary Baldwin Col, Staunton, Va, BA, 74. *Prof Exp:* Lab technician, Litton Bionetics, 74-77; chemist microbiol & infectious dis, Nat Inst Allergy & Infectious Dis, 78-79, biologist, Biol Response Modifier Prog, 80-82, BIOLOGIST, LAB TUMOR CELL BIOL, NAT CANCER INST, NIH, 82- *Res:* Isolation of human T-cell lymphotropic virus type 3/ lymphadenopa thy-associated virus which causes aquired immunodeficiency syndrome; development of continuously infected lymphoid lines for studying cell biology of human T-cell lymphotropic virus type 3/lymphadenopathy-associated virus infection to develop treatments and vaccine. *Mailing Add:* Lab Tumor Cell Biol NIH Bldg 37 Bethesda MD 20892

READDY, ARTHUR F, JR, b Jersey City, NJ, Mar 4, 28; m 60; c 3. PHYSICAL CHEMISTRY, MATERIALS SCIENCE ENGINEERING. *Educ:* St Peters Col, BS, 49; Stevens Inst Technol, MS, 57. *Prof Exp:* Res chemist, Onyx Oil & Chem Co, 49-51, Theobald Industs, 51-53 & Colgate-Palmolive Co, 53-56; head chem sect, Appl Sci Br, Naval Supply Res & Develop Facility, 56-66; scientist, Plastics Tech Eval Ctr, Armament Res & Develop Command, US Dept Defense, Dover, 66-88; CONSULT, MAT UTILIZATION, 88- *Concurrent Pos:* Vis prof, Jersey City State Col, 59-61. *Res:* Surface chemistry and physics; radiations effects and applications; composite materials; protection-decontamination of equipment; electrical, electronic and other properties of materials; nuclear, chemical, biological agents decontamination; materials specifications. *Mailing Add:* Nine Lenox Ave Cranford NJ 07016

READE, MAXWELL OSSIAN, b Philadelphia, Pa, Apr 11, 16; m 66; c 3. MATHEMATICS. *Educ:* Brooklyn Col, BS, 36; Harvard Univ, MA, 37; Rice Inst, PhD(math), 40. *Prof Exp:* Instr math, Ohio State Univ, 40-42 & Purdue Univ, 42-44 & 46; from asst prof to prof, 46-86, EMER PROF MATH, UNIV MICH, ANN ARBOR, 86- *Mem:* AAAS; assoc Am Math Soc; assoc Math Asn Am. *Res:* Theory of functions of one complex variable. *Mailing Add:* Dept Math 3220 Angell Hall Univ Mich Ann Arbor MI 48109-1003

READER, GEORGE GORDON, b Brooklyn, NY, Feb 8, 19; m 42; c 4. MEDICINE, PUBLIC HEALTH. *Educ:* Cornell Univ, BA, 40, MD, 43. *Hon Degrees:* DSc, Drew Univ, 88. *Prof Exp:* Intern med, NY Hosp, 44, res fel, 46-47, asst res physician, 47-49; from instr to prof med, 49-72, LIVINGSTON FARRAND PROF PUB HEALTH & CHMN DEPT, MED COL, CORNELL UNIV, 72- *Concurrent Pos:* Chmn human ecol study sect, NIH, 61-65; ed, Milbank Mem Fund Quart, Health & Soc, 72-76. *Mem:* Sr mem, Inst Med of Nat Acad Sci; fel Am Pub Health Asn; fel Am Col Physicians; fel Am Col Prev Med; Am Sociol Asn. *Res:* Medical education and medical care; medical sociology. *Mailing Add:* NY Hosp Cornell Med Ctr 1300 York Ave A-631 New York NY 10021

READER, JOSEPH, b Chicago, Ill, Dec 1, 34; m 56; c 3. PHYSICS. *Educ:* Purdue Univ, BS, 56, MS, 57; Univ Calif, Berkeley, PhD(physics), 62. *Prof Exp:* Res assoc physics, Argonne Nat Lab, 62-63; STAFF PHYSICIST, NAT INST STANDARDS & TECHNOL, 63- *Honors & Awards:* Gold Medal, Dept Com. *Mem:* Fel Am Phys Soc; fel Optical Soc Am. *Res:* Experimental atomic physics; optical spectroscopy; hyperfine structure; electronic structure of highly ionized atoms; wave length standards; ionization energies of atoms and ions. *Mailing Add:* Phys A153 Nat Inst Standards & Technol Gaithersburg MD 20899

READER, WAYNE TRUMAN, b Danville, Pa, May 3, 39; m 61; c 5. ACOUSTICAL MATERIALS. *Educ:* Univ Md BS, 63; Cath Univ Am, MS, 68, PhD(appl physics), 70. *Prof Exp:* Proj scientist, Structure Acoust Br, David Taylor Res Ctr, 63-73, br head, Target Acoustic Br, 73-77, sr res scientist, Ship Acoust Dept, 77-88; PRIN RES SCIENTIST, VECTOR RES CO, INC, 88- *Concurrent Pos:* Res assoc, Cath Univ Am, 70-71, instr, 71-72, lectr, 78-85. *Mem:* Sigma Xi; Acoust Soc Am. *Res:* Determining and characterizing the mechanisms by which submerged structures radiate and scatter sound; devising and developing techniques for reducing the radiated and scattered fields. *Mailing Add:* 3541 Clarksburg Rd Monrovia MD 21770

READEY, DENNIS W(ILLIAM), b Aurora, Ill, Aug 6, 37; m 58; c 2. CERAMICS, METALLURGY. *Educ:* Univ Notre Dame, BS, 59; Mass Inst Technol, ScD(ceramics), 62. *Prof Exp:* Group leader phys ceramics, Argonne Nat Lab, 64-67; mgr mat processing lab, Res Div, Raytheon Co, 67-74; ceramist mat sci prog, Div Res, US Energy Res & Develop Admin, 74-77; assoc prof, 77-79, PROF CERAMIC ENG, OHIO STATE UNIV, 79-, DEPT CHMN, 82- *Honors & Awards:* Ross Coffin Purdy Award, Am Ceramic Soc. *Mem:* Am Ceramic Soc; AAAS; Am Inst Mining & Metall Engrs; Mat Res Soc. *Res:* Ceramics processing; optical materials; electrical and electronic properties; ceramics corrosion; transport properties. *Mailing Add:* Dept Ceramic Eng Ohio State Univ 177 Watts Hall Columbus OH 43210

READHEAD, CAROL WINIFRED, b Johannesburg, SAfrica, Sept 9, 47; m 69; c 2. BIOLOGY. *Educ:* Univ Witwatersrand, BSc hons, 70; Cambridge Univ, PhD(biol), 77. *Prof Exp:* Res asst biochem, Cambridge Univ, 70-72; fel biol, 77-79, sr res fel, 80-83, STAFF SCIENTIST, CALIF INST TECHNOL, 83- *Concurrent Pos:* Asst prof, Univ Southern Calif, 79-80. *Mem:* Sigma Xi. *Res:* Recombinant DNA research on immunoglobulin genes; molecular biology of neurological mouse mutants, genetic engineering. *Mailing Add:* Dept Biol Calif Inst Technol 147-75 Pasadena CA 91107

READING, ANTHONY JOHN, b Sydney, Australia, Sept 10, 33; m 75; c 2. PSYCHIATRY. *Educ:* Univ Sydney, MB & BS, 57; Johns Hopkins Univ, MPH, 61, ScD, 64; Am Bd Psychiat & Neurol, cert, 71. *Prof Exp:* Jr resident med officer, Sydney Hosp, Australia, 57-58, sr resident med officer, 58-59; pvt pract, 59-60; instr pathobiol & ment hyg, Sch Med, Johns Hopkins Univ, 62 -65, asst prof pathobiol, 64-65, asst resident psychiat, 65-68, instr med, 68-69, from asst prof to assoc prof psychiat, 68-75, asst prof med, 69-75, dir, Psychiat Liaison Serv, Johns Hopkins Hosp, 74-75 & Comprehensive Alcoholism Prog, 72-75; PROF & CHAIRPERSON PSYCHIAT, UNIV SFLA, TAMPA, 75-, PSYCHIATRIST-IN-CHIEF, PSYCHIAT CTR, 85- *Concurrent Pos:* Physician-in-charge, Alcoholism Clin, John Hopkins Hosp, 69 -75 & Psychosomatic Clin, 74-75. *Mem:* AAAS; Am Psychosomatic Soc; AMA; Am Psychiat Asn. *Res:* Psychosocial aspects of illness; psychobiology. *Mailing Add:* Dept Psychiat 3515 E Fletcher Ave Tampa FL 33613

READING, JAMES CARDON, b Ogden, Utah, Dec 7, 37; m 64; c 4. BIOSTATISTICS, MATHEMATICAL STATISTICS. *Educ:* Stanford Univ, BS, 60, MS, 66, PhD(statist), 70. *Prof Exp:* Comput programmer & math analyst, Lockheed Missiles & Space Co, 62-67; instr, 70-72, asst prof biostatist & adj asst prof math, 72-76, ASSOC PROF BIOSTATIST & ADJ ASSOC PROF MATH, 76-, CHMN DIV BIOSTATIST, 77- *Mem:* Inst Math Statist; Am Statist Asn; Biomet Soc; Soc Clin Trials. *Res:* Methodological aspects of the analysis of clinical trials including design, conduct and statistical methods; classification problems. *Mailing Add:* Dept Fam & Comm Med Univ Utah 50 N Medical Dr Salt Lake City UT 84132

READING, JOHN FRANK, b West Bromwich, Eng, Oct 19, 39; m 63; c 4. THEORETICAL PHYSICS. *Educ:* Christ Church, Oxford Univ, BA, 60, MA, 63; Univ Birmingham, dipl math physics, 61, PhD(physics), 64. *Prof Exp:* Instr theoret physics, Mass Inst Technol, 64-66; sr res assoc, Univ Wash, 66-68; Harwell fel, Atomic Energy Res Estab, Harwell, Eng, 68-69; assoc prof, Northeastern Univ, 69-71; assoc prof, 71-81, PROF THEORET PHYSICS, TEX A&M UNIV, 81- *Concurrent Pos:* Consult, Oak Ridge Nat Lab; ed, Bienneal Conf, cross sect fusion & other applns. *Mem:* Fel Am Phys Soc. *Res:* Scattering theory in atomic, nuclear and solid state physics; numerical calculations of ion-atom collisions producing excitation, charge transfer and ionization. *Mailing Add:* Dept Physics Tex A&M Univ College Station TX 77843

READING, ROGERS W, b Toledo, Ohio, Apr 2, 34; m 61; c 1. PHYSIOLOGICAL OPTICS, OPTOMETRY. *Educ:* Ind Univ, BS, 56, MO, 57, PhD(physiol optics), 68. *Prof Exp:* From instr to assoc prof, 64-82, PROF OPTOM, IND UNIV, BLOOMINGTON, 82- *Concurrent Pos:* Asst investr, Extrahoropteral Stereopsis in Tracking Performance, US Army grant, 64-68; prin investr, Some Time Factors in Stereopsis, NASA grant, 68-70; vpres, Nat Bd Examr Optom, 81-83; prin investr fusional stimuli & corresp pts, Nat Eye Inst, NIH, 81-85; author, Binocular Vision: Found & Appl. *Mem:* Am Acad Optom; Sigma Xi. *Res:* Binocular vision; photometry; psychophysics. *Mailing Add:* 2403 Browncliff Bloomington IN 47401

READNOUR, JERRY MICHAEL, b Muncie, Ind, Oct 19, 40; m 62; c 2. PHYSICAL INORGANIC CHEMISTRY, CHEMICAL EDUCATION. *Educ:* Ball State Teachers Col, BS, 62; Purdue Univ, PhD(chem), 69. *Prof Exp:* From asst prof to assoc prof, 68-79, PROF CHEM, SOUTHEAST MO STATE UNIV, 79- *Concurrent Pos:* Mem, Gen Chem Test Comt, Am Chem Soc, 79-89. *Mem:* Am Chem Soc. *Res:* Thermodynamic properties of aqueous solutions; stability constants; learning theory. *Mailing Add:* Dept Chem Southeast Mo State Univ Cape Girardeau MO 63701-4799

READY, JOHN FETSCH, b St Paul, Minn, July 13, 32; m 53; c 6. LASERS & ELECTRO-OPTICS. *Educ:* Col St Thomas, BS, 54; Univ Minn, MS, 56. *Prof Exp:* From res scientist to sr res scientist, Honeywell Corp Res Ctr & Mat Sci Ctr, 58-66, sr prin res scientist, 66-78, staff scientist, 78-79, group leader, Honeywell Corp Technol Ctr, 79-82, sect chief, 82-86, SR RES FEL, HONEYWELL SYSTS & RES CTR, 86- *Mem:* Am Phys Soc; Laser Inst Am. *Res:* Nuclear radiation damage; infrared materials and components; lasers, particularly effects of laser radiation and laser applications; laser based material processing and sensors. *Mailing Add:* 4401 Gilford Dr Edina MN 55435

REAGAN, DARYL DAVID, b Longview, Wash, Apr 29, 25; m 59; c 2. PHYSICS. *Educ:* Stanford Univ, MS, 49, PhD(physics), 55. *Prof Exp:* Sr physicist plasma physics, Lawrence Livermore Lab, 55-63; res assoc, Oxford Univ, 58-59; PHYSICIST ACCELERATOR PHYSICS, STANFORD UNIV, 63- *Mem:* Am Phys Soc. *Res:* Photonuclear physics; shock hydrodynamics; plasma physics; accelerator physics; magnetic measurements; synchrotron radiation. *Mailing Add:* 967 Moreno Ave Palo Alto CA 94303

REAGAN, JAMES OLIVER, b Lampasas, Tex, Nov 13, 45; m 82; c 2. MEAT SCIENCES. *Educ:* Tex A&M Univ, BS, 68, MS, 70, PhD(animal sci), 74. *Prof Exp:* Instr animal sci, Tex A&M Univ, 72-73; asst prof, Ore State Univ, 73-74; from asst prof to assoc prof, 74-85, PROF ANIMAL & DAIRY SCI, UNIV GA, 85- *Concurrent Pos:* Mem, Coun Agr Sci & Technol. *Honors & Awards:* Creative Res Award, Outstanding Young Scientist, AGHON. *Mem:* Am Meat Sci Asn; Am Soc Animal Sci; Inst Food Technologists; Sigma Xi; Coun Agr Sci & Technol. *Res:* Quantitative and qualitative evaluation of meat animals, methods of extending the caselife of fresh meats; biochemical and physical attributes of pre-and post-rigor meat. *Mailing Add:* Dept Food Sci Univ Ga Athens GA 30602

REAGAN, JAMES W, b Oakmont, Pa, Aug 6, 18; m 44; c 5. PATHOLOGY. *Educ:* Univ Pittsburgh, BS, 42, MD, 43. *Prof Exp:* From demonstr to assoc prof, 46-58, prof path & reproductive biol, 58-86, EMER PROF PATH, CASE WESTERN RESERVE UNIV, 87- *Concurrent Pos:* Fel, Univ Hosps Cleveland, 46-49, assoc pathologist, 49-, pathologist in chg cytol lab, 49-70, pathologist in charge surg path, 57-83, dir path anat, 83-; assoc dir tutorials clin cytol, Univ Chicago, 74- & Int Acad Cytol, 74- *Honors & Awards:* Goldblatt Award, Int Acad Cytol, 64; Papanicolaou Award, Am Soc Cytol, 69; Ward Burdick Award, Am Soc Clin Path, 73. *Mem:* Am Soc Cytol (pres, 64); Am Soc Clin Path; fel Col Am Path; assoc fel Am Col Obstet & Gynec; Int Acad Cytol. *Mailing Add:* Case Western Reserve Univ 2085 Adelbert Rd Cleveland OH 44106

REAGAN, JOHN ALBERT, b Grandview, Mo, May 2, 41; m 66; c 2. ATMOSPHERIC RADIOMETRY. *Educ:* Univ Mo, Rolla, BS, 63, MS, 64; Univ Wis, PhD(elec eng), 67. *Prof Exp:* Elec engr, IBM Systs Develop Div, 64; from asst prof to assoc prof elec eng, Univ Ariz, 67-70, prof, 76; CONSULT, 76- *Concurrent Pos:* Vis scientist, NASA Langley Res Ctr, aerosol measurement res branch, 78-79; mem, NASA LITE P, 85-89, Nat Oceanic & Atmospheric Admin Profiler Adv Comt, 87-88; vis prof, dept meteorol, Pa State Univ, 88-89; NASA Lite Sci Steering Group, 89-91; guest ed, Opt Engr Lidar Special Issue, 91; mem adv comt, Geosci & Remote Sensing Soc, Inst Elec & Electronics Engrs, 80-93, secy-treas, 84-87, vpres, 88-89, pres, 90-91. *Honors & Awards:* Outstanding Serv Award, Geosci & Remote Sensing Soc, Inst Elec & Electronics Engrs, 88. *Mem:* Fel Inst Elec & Electronics Engrs; Am Meteorol Soc; Int Soc Optical Eng. *Res:* Atmospheric remote sensing of aerosols and trace gases by laser radar; solar radiometry and microwave radiometry including both experimental implentation and signal processing. *Mailing Add:* Dept Elec & Comput Eng ECE Bldg Univ Ariz Tucson AZ 85721

REAGAN, THOMAS EUGENE, b Tylertown, Miss, Jan 12, 47; m 68; c 3. ENTOMOLOGY. *Educ:* La State Univ, Baton Rouge, BS, 70, MS, 72; NC State Univ, PhD(entom, ecol, statist), 75. *Prof Exp:* asst prof & exten specialist entom, NC State Univ, 75-77; from asst prof to assoc prof, 77-85, PROF ENTOM, LA STATE UNIV, 85- *Mem:* Sigma Xi; Entom Soc Am. *Res:* Applied and basic research on ecology and pest management of sugar cane insects; interdisciplinary aspects of pest management; pest/beneficial arthropod interactions with cultural practices; management of plant virus disease transmission by insect vectors; pesticide impact assesment, insecticide resistance; predictive mathematical modeling. *Mailing Add:* 402 Life Sci Bldg Entom Dept La State Univ Baton Rouge LA 70803

REAGAN, WILLIAM JOSEPH, b Salem, Mass, Nov 16, 43; m 65; c 3. INORGANIC CHEMISTRY. *Educ:* Boston Col, BS, 65; Mich State Univ, PhD(inorg chem), 70. *Prof Exp:* Res asst, Mich State Univ, 66-69; res assoc, Univ Southern Calif, 70; sr res chemist, Mobil Res & Develop Corp, 70-78; group leader catalyst res, Englehard Minerals & Chem Corp, 78-88, res assoc, Englehard Corp, 81-88; RES CHEMIST, AMOCO OIL RES & DEVELOP, 88- *Mem:* Am Chem Soc. *Res:* Synthesis and characterization of molybdenum and tungsten compounds and fluorocarbon phosphine transition element compounds; transition metals in catalysis; solid state chemistry of heterogeneous catalysts; zeolite synthesis and catalysis. *Mailing Add:* Amoco Res Ctr Amoco Oil Co PO Box 400 Naperville IL 60566-0400

REAGOR, JOHN CHARLES, b Llano, Tex, Mar 25, 38; m 61; c 2. TOXICOLOGY, BIOCHEMISTRY. *Educ:* Tex A&M Univ, BS, 60, MS, 63, PhD(biochem, nutrit), 66. *Prof Exp:* Asst prof biochem, 65-71, VIS MEM, DEPT VET PHYSIOL & PHARMACOL, UNIV & HEAD DEPT TOXICOL, TEX VET MED DIAG LAB, TEX A&M UNIV, 69- *Mem:* Asn Off Anal Chemists; Am Asn Vet Lab Diagnosticians. *Res:* Development of analytical methods and techniques for forensic analyses of biological materials; study of poisonous plants including diagnostic methods. *Mailing Add:* PO Drawer 3040 College Station TX 77840

REAL, LESLIE ALLAN, b Philadelphia, Pa, June 22, 50. ECOLOGY, POPULATION BIOLOGY. *Educ:* Ind Univ, BA, 72; Univ Mich, MS, 75, PhD(zool), 77. *Prof Exp:* Fel biol, Univ Miami, 77-78; ASST PROF ZOOL & BIOMATH, NC STATE UNIV, 78-; PROF BIOL, UNIV NC, CHAPEL HILL. *Concurrent Pos:* Rosenstiel fel, Univ Miami, 77. *Honors & Awards:* Prof Develop Award, NC State Univ, 78. *Mem:* Soc Study Evolution; Ecol Soc Am; AAAS. *Res:* Theoretical population biology; pollination ecology; animal behavior; ecological genetics. *Mailing Add:* Biol Dept Univ NC Coker Hall CB 3280 Chapel Hill NC 27599-3280

REALS, WILLIAM JOSEPH, b Hot Springs, SDak, June 22, 20; m 44; c 5. PATHOLOGY. *Educ:* Creighton Univ, BS, 44, MD, 45, MS, 49. *Prof Exp:* Instr path, Sch Med, Creighton Univ, 49-50; PATHOLOGIST & DIR LABS, ST JOSEPH'S HOSP & PROF PATH & DEAN, SCH MED, UNIV KANS, 50- *Concurrent Pos:* Asst pathologist, Creighton Mem-St Joseph's Hosp, Omaha, Nebr, 49-50; consult, Wichita Vet Admin Hosp, 50-; lectr, Med Ctr, Univ Kans, 55-72; consult, Surgeon Gen, USAF, DC, 59-77 & Civil Air Surgeon, Fed Aviation Agency, 60-75. *Mem:* Am Soc Clin Path; Asn Mil Surg US; Aerospace Med Asn; Col Am Path (pres, 71-73). *Res:* Oncology; hormone chemistry, especially thyroid diseases; forensic pathology related to aircraft accidents. *Mailing Add:* 1010 N Kansas Wichita KS 67214

REAM, BERNARD CLAUDE, b Johnstown, Pa, Jan 30, 39; m 67; c 2. ORGANIC CHEMISTRY, CATALYSIS. *Educ:* Seton Hall Univ, BS, 61; Ohio State Univ, MS, 63, PhD(org chem), 65. *Prof Exp:* Res chemist, 66-76, proj scientist, 76-80, res scientist, 80-87, SR RES SCIENTIST, UNION CARBIDE CORP, 87- *Mem:* Am Chem Soc. *Res:* Metal-catalyzed synthesis of organic compounds; ethylene oxide/ethylene glycol process chemistry; syn-gas chemistry, catalysis; ethyleneamines and ethanolamines process chemistry. *Mailing Add:* 1807 Rolling Hills Rd Charleston WV 25314

REAM, LLOYD WALTER, JR, b Chester, Pa, Mar 20, 53; m 75. MICROBIAL GENETICS, PLANT MOLECULAR BIOLOGY. *Educ:* Vanderbilt Univ, BA, 75; Univ Calif, Berkeley, PhD(molecular biol), 81. *Prof Exp:* Sr res fel microbiol, dept microbiol, Univ Wash, 80-83; asst prof molecular biol, Dept Biol, Ind Univ, 83-88; ASSOC PROF, ORE STATE UNIV, 88- *Concurrent Pos:* Res asst molecular genetics, Inst Genetics, Univ Kölu, Ger, 75; grad student molecular biol, Univ Calif, Berkeley, 75-80; consult ed, Plant Molecular Biol, 84-; Inst fel, Inst Molecular & Cellular Biol, Ind Univ, 84- *Mem:* Genetics Soc Am; Am Soc Microbiol; AAAS; Int Soc Plant Molecular Biol; Am Cancer Soc. *Res:* The mechanism of T-DNA transmission during crown gall tumorigenesis; inducible virus resistance in plants; bateria-plant interactions; responses of plants to stress; genetic recombination. *Mailing Add:* Dept Agr Chem Ore State Univ Corvallis OR 97331

REAMES, DONALD VERNON, b West Palm Beach, Fla, Dec 30, 36. ASTROPHYSICS, PHYSICS. *Educ:* Univ Calif, Berkeley, AB, 58, PhD(physics), 64. *Prof Exp:* ASTROPHYSICIST COSMIC RAYS, NASA, GODDARD SPACE FLIGHT CTR, 64- *Mem:* Am Phys Soc; Inst Elec & Electronics Engrs Comput Soc; Am Geophys Union. *Res:* Cosmic rays, solar physics; interplanetary particles and fields; particle detectors; nuclear physics; microprocessor controlled experiments. *Mailing Add:* 11306 Sherrington Ct Upper Marlboro MD 20772

REAMS, MAX WARREN, b Virgil, Kans, Mar 10, 38; m 61; c 3. GEOMORPHOLOGY, SEDIMENTARY PETROLOGY. *Educ:* Univ Kans, BA & BS, 61, MS, 63; Wash Univ, PhD(geol), 68. *Prof Exp:* From asst prof to assoc prof, 67-77, CHMN DEPT GEOL SCI, OLIVET NAZARENE UNIV, 69-, PROF GEOL, 77-, CHMN DIV NATURAL SCI, 75- *Mem:* Geol Soc Am; Am Quaternary Asn; Clay Minerals Soc; Sigma Xi. *Res:* Geology and geochemistry of caves and cave sediments; weathering zone precipitates; mineralogy and petrology of speleothems; fractals in Karst. *Mailing Add:* Dept Geol Sci Olivet Nazarene Univ Kankakee IL 60901

REAMS, WILLIE MATHEWS, JR, b Richmond, Va, Feb 23, 30; m 56; c 3. DEVELOPMENTAL BIOLOGY. *Educ:* Univ Richmond, BS, 51; Johns Hopkins Univ, PhD(embryol), 56. *Prof Exp:* Jr instr biol, Johns Hopkins Univ, 52-55; instr anat, Med Col Va, 52-60; asst prof zool, La State Univ, 60-64; assoc prof, 64-70, PROF BIOL, UNIV RICHMOND, 70-, DIR LORAROBINS GALLERY DESIGN FROM NATURE, 87- *Concurrent Pos:* Res asst clin prof dermat, Med Col Va, 66-87. *Mem:* Am Asn Anatomists; Soc Invest Dermat; Sigma Xi; Am Asn Museums. *Res:* Experimental pigment cell differentiation in mice. *Mailing Add:* Dept Biol Univ Richmond Richmond VA 23173

REAP, JAMES JOHN, b Hazelton, Pa, Jan 22, 48; m 70; c 2. ORGANIC CHEMISTRY. *Educ:* Villanova Univ, BS, 69; Univ Pittsburgh, PhD(org chem), 75. *Prof Exp:* Teacher physics & chem, Gloucester Cath Sr High Sch, 69-70; RES CHEMIST BIOCHEM DEPT, E I DU PONT DE NEMOURS & CO, INC, 75- *Mem:* Am Chem Soc. *Res:* Synthesis of novel biologically active organic chemicals. *Mailing Add:* 2505 Bona Rd Wilmington DE 19810

REARDEN, CAROLE ANN, b Belleville, Ont, June 11, 46; US citizen. IMMUNOHEMATOLOGY, TRANSPLANTATION. *Educ:* McGill Univ, BSc, 69, MSc, 71, MDCM, 71. *Prof Exp:* Intern pediat, Children's Mem Hosp, Chicago, 71-72, fel biochem genetics, 72-73; res pediat, 74, res clin path, 75-76, fel immunohemat, 76-79, asst prof path, 79-86, ASSOC PROF PATH, UNIV CALIF, SAN DIEGO, 86-, DIR, HISTOCOMPATIBILITY & IMMUNOGENETICS LAB, 79-, HEAD, DIV LAB MED, 89- *Concurrent Pos:* Prin investr, NIH grant, 83-86; mem, comt transplantation, Am Asn Blood Banks, 82-86. *Mem:* Am Asn Path; Am Fedn Clin Res; Am Soc Hemat; Am Asn Blood Banks; Am Asn Histocompatability & Immunogenetics; Int Soc Blood Transfusion. *Res:* Immunogenetics of major red cell membrane proteins (glycophorins). *Mailing Add:* Dept Path 0612 Univ Calif San Diego 9500 Gilman Dr La Jolla CA 92093-0612

REARDON, ANNA JOYCE, b East St Louis, Ill, Jan 22, 10. PHYSICS. *Educ:* Col St Teresa, BA, 30; St Louis Univ, MS, 33, PhD(physics), 37. *Prof Exp:* High sch instr, Minn, 30-31 & Mo, 32-35; instr physics & Math, Ursuline Col, La, 36-37, Mt St Scholastica Col, 37-39, Col St Teresa, 39-40 & Loretto Heights Col, 40-41; from instr to prof & head dept, 41-75, EMER PROF PHYSICS, UNIV NC, GREENSBORO, 75- *Concurrent Pos:* Chmn, NC Comt High Sch Physics, 67-76, instr spec course in physics for x-ray technician students, Moses Cone Hosp. *Mem:* Am Phys Soc; Am Asn Physics Teachers; Sigma Xi; Asn Women Sci. *Res:* Theoretical photography. *Mailing Add:* 1105 Dover Rd Greensboro NC 27408-7313

REARDON, EDWARD JOSEPH, JR, b Southbridge, Mass, Apr 24, 43; m 78. RADIATION CURED COATINGS, PHOTORESISTS. *Educ:* Brown Univ, ScB, 65; Seton Hall Univ, MS, 67, PhD(org chem), 69. *Prof Exp:* NIH fel, Ind Univ, Bloomington, 69-70; assoc, Univ NC, Chapel Hill, 70-72; vis asst prof org chem, Bucknell Univ, 72-73; sr res assoc, Photohorizons Div, Horizons Res, 74-75; sr chemist, 76-77, group leader, 78-79, res mgr, 79-84, tech dir,84-86, VPRES TECH OPERS, DYNACHEM ELECTRONIC MAT GROUP, MORTON INT INC, 86- *Mem:* Am Chem Soc; Soc Photographic Scientists & Engrs. *Res:* Applied research and development of radiation sensitive coatings, primarily resists for fabrication of electronic circuits; synthesis of polymers and sensitizers; formulation, applications, manufacturing, and quality control methodologies. *Mailing Add:* Dynachem Electronic Mat Group Morton Int Inc 2631 Michelle Dr Tustin CA 92680

REARDON, FREDERICK H(ENRY), b Philadelphia, Pa, Oct 22, 32; m 56; c 3. COMBUSTION & PROPULSION, ENGINEERING EDUCATION. *Educ:* Univ Pa, BS, 54, MS, 56; Princeton Univ, PhD(aeronaut eng), 61. *Prof Exp:* Instr mech eng, Univ Pa, 54-56; res engr, Princeton Univ, 56-61; supvr, Liquid Rocket Opers, Aerojet-Gen Corp, Sacramento, 61-64, tech specialist, 64-65, sr res engr, 65-66; asst prof mech eng, 66-68, assoc prof eng, 68-73, chmn dept, 76-81, assoc dean, 81-86, PROF ENG, CALIF STATE UNIV, SACRAMENTO, 73- *Concurrent Pos:* Mem working group combustion, Joint Army-Navy-NASA-Air Force, 64-; consult, Liquid Rocket Oper, Aerojet-Gen Corp, 66-, USAF, 76-86 & Chem Systs Div, United Technol Corp, 80-85. *Mem:* Sigma Xi; Fel Am Soc Mech Engrs; assoc fel Am Inst Aeronaut & Astronaut; Combustion Inst; Am Soc Eng Educ. *Res:* Combustion and flow processes in engines, furnaces and waste recovery systems; automobile economy and emissions control; impact of scientific technology on society; development of educational simulations and games. *Mailing Add:* 5619 Haskell Ave Carmichael CA 95608

REARDON, JOHN JOSEPH, b Westerly, RI, July 27, 21; m 47; c 2. ECOLOGY. BIOLOGY. *Educ:* Univ Mich, BS, 48, MA, 49; Univ Ore, PhD(biol), 50. *Prof Exp:* Asst prof biol & head dept, Col Charleston, 49-52; asst prof, Los Angeles State Col, 54-58; assoc prof & chmn dept, San Fernando Valley State Col, 58-62; assoc prof, Eastern Ore Col, 62-64; asst prof zool, Univ Toronto, 64-65; chmn dept, 65-74, PROF BIOL, SOUTHEASTERN MASS UNIV, 65- *Concurrent Pos:* Team leader, Westinghouse Ocean Res Lab Starfish Reef Exped, Micronesia, 68. *Mem:* AAAS; Ecol Soc Am; Am Soc Mammal; Am Inst Biol Sci; Sigma Xi. *Res:* Biology of mammals; ecology of dune inhabiting organisms; radioisotope pollution of environment; ecology of reef environments; ecology and behavior of Birgus latro, the coconut crab. *Mailing Add:* Dept Biol Southeastern Mass Univ North Dartmouth MA 02747

REARDON, JOSEPH DANIEL, b Buffalo, NY, Aug 24, 44; m 68. PHYSICAL CHEMISTRY. *Educ:* Univ Rochester, BS, 66; Univ Conn, PhD(phys chem), 70. *Prof Exp:* Res & develop engr, GTE Sylvania Inc, 70-73; assoc dir res, Quantum, Inc, 73-77; SUPVR MAT ENG, METCO INC, 77- *Mem:* Am Soc Metals; Am Chem Soc. *Res:* Development of composite powders for plasma and vacuum plasma flame spray applications especially for sprayed abradable coatings; mass spectrometry and high vacuum technology; processing parameters for high performance plastics; development of synthetic membranes. *Mailing Add:* Metco Inc 1101 Prospect Ave Westbury NY 11590-2724

REARDON, JOSEPH EDWARD, b Albany, NY, Jan 25, 38; m 60; c 4. ORGANIC CHEMISTRY, POLYMER CHEMISTRY. *Educ:* Canisius Col, BS, 61; Univ Notre Dame, PhD(org chem), 67. *Prof Exp:* Chemist, Carborundum Co, 61-63; res chemist, Plastics Dept, 66-67, res chemist, Electrochem Dept, 67-72, res chemist, Plastics Dept, 72-80, RES ASSOC, F & FP DEPT, E I DU PONT DE NEMOURS & CO, INC, 85- *Mem:* Am Chem Soc. *Res:* Dispersion chemistry and polymer synthesis, primarily methacrylates. *Mailing Add:* Nine Lyells Ct Wilmington DE 19808

REARDON, JOSEPH PATRICK, b Pittston, Pa, Sept 26, 40; m 69; c 2. PHYSICAL CHEMISTRY, MATERIALS SCIENCE. *Educ:* Spring Hill Col, BS, 65; Am Univ, MS, 69, PhD(chem), 75. *Prof Exp:* Instr chem, Georgetown Prep Sch, 65-67; res chemist, Naval Res Lab, 71-79; mgr mat develop, Pure Carbon Co, 79-80; mgr mat process & develop, Tribon Bearing Co, 80-86; MGR MAT PROCESS & DEVELOP, DEXTER COMPOSITES, 86- *Concurrent Pos:* Chmn, N Ohio chap, Soc Aerospace Mat & Process Engrs, 89-90. *Mem:* Soc Aerospace Mat & Process Engrs. *Res:* Low surface energy polymers; adhesives; carbon fibers and graphite intercalation chemistry; electrets and piezoelectric polymers; electrically conductive organic polymers; high temperature composites; solid self-lubricating bearing materials; novel chemical modifications and processing techniques for polyimide resins, molding compounds and molded components, in order to reduce the cost of advanced composites to feasible industrial levels. *Mailing Add:* Dexter Composites 17960 Englewood Dr Cleveland OH 44130

REARICK, DAVID F, b Danville, Ill, Aug 5, 32; div. MATHEMATICS. *Educ:* Univ Fla, BS, 54; Adelphi Univ, MS, 56; Calif Inst Technol, PhD(math), 60. *Prof Exp:* Instr math, Univ BC, 60-61; asst prof, 61-66, ASSOC PROF MATH, UNIV COLO, BOULDER, 66- *Mem:* Am Math Soc; Math Asn Am. *Res:* Analytic number theory; arithmetic functions. *Mailing Add:* Dept Math Univ Colo Box 462 Boulder CO 80309

REASENBERG, ROBERT DAVID, b New York, NY, Apr 27, 42; m 65; c 1. PHYSICS, ASTRONOMY. *Educ:* Polytech Inst Brooklyn, BS, 63; Brown Univ, PhD(physics), 70. *Prof Exp:* Res assoc, Mass Inst Technol, 69-71, res staff mem, 71-79, prin res scientist, 81-82; PHYSICIST, SMITHSONIAN ASTROPHYS OBSERV, 83- *Concurrent Pos:* Consult, Lincoln Lab, Mass Inst Technol, 70-76, C S Draper Lab, Inc, 78-80 & Ames Res Ctr, NASA,81-83; assoc mem, Viking Radio Sci Team, Mariner-Venus-Mercury Radio Sci Team, Mariner-9 Celestial Mech Team & Pioneer Venus Science Steering Group; lectr, Int Sch Cosmology & Gravitation, 77, 79, 82 & 85; consult, NASA Ames Res Ctr, 81-83; chmn, Comt Gravitation & Relativity, Starprobe Mission, 80-81; mem, Planetary Systs Sci Working Group, 88-, Ad Hoc Comt Gravitation Physics & Astron, 89- & Interferometry Panel, Astron & Astrophys Surv, 89- *Honors & Awards:* Newcomb Cleveland Award, AAAS, 77. *Mem:* Am Phys Soc; AAAS; Am Astron Soc; Am Geophys Union; Sigma Xi; Int Astron Union; Int Soc Gen Relativity & Gravitation. *Res:* Gravity research, especially tests of theories of gravitation, determination of solar-system constants and planetary ephermerides, and determination of the structure, gravitational potential and rotational motion of planets; optical interferometry and space astrometry; author of over 100 publications. *Mailing Add:* 16 Garfield St Lexington MA 02173

REASER, DONALD FREDERICK, b Wichita Falls, Tex, Sept 30, 31; m 75. GEOLOGY, TECTONICS. *Educ:* Southern Methodist Univ, BS, 53, MS, 58; Univ Tex, Austin, PhD(geol), 74. *Prof Exp:* Geol asst, De Golyer & MacNaughton, Dallas, 56-57; instr introductory & struct geol, Arlington State Col, 61-64; petrol geologist, Humble Oil & Refining Co, 65-66; asst prof, WTex State Univ, 67-68; ASSOC PROF STRUCT GEOL & TECTONICS, UNIV TEX, ARLINGTON, 68- *Concurrent Pos:* Grant-in-aid res, Sigma Xi, 60; field trip consult, Shell Develop Co, 64-; res asst, Bur Economic Geol, Austin, 67-68; consult geologist, Cor Labs, Dallas, 74-79, Gearhart Indust, Ft Worth, 82-84; instr, Inst Energy Develop, Ft Worth, 78-79; Tex Elec Serv Co, 81-85, Ebasco Serv, Austin, 87, Southwest Labs, Dallas, 87, Earth Technol Corp, Long Beach, Calif, 89-90. *Mem:* Am Asn Petrol Geologists; Geol Soc Am; Soc Economic Paleontologists & Mineralogists; Sigma Xi. *Res:* Structure and regional tectonics of West Texas and Northern Mexico; stratigraphy and structure of Upper Cretaceous rocks in North-central Texas; petroleum possibilities of subsurface Mesozoic rocks in Northeast Texas; geologic studies of Texas site for superconducting super collider (SSC). *Mailing Add:* Dept Geol Univ Tex Arlington TX 76010

REASONER, JOHN W, b Winona, Mo, Feb 28, 40; m 61; c 3. PHOTOCHEMISTRY, FUEL SCIENCE. *Educ:* Southeast Mo State Col, BS, 61; Iowa State Univ, PhD(org chem), 65. *Prof Exp:* PROF CHEM, WESTERN KY UNIV, 65- *Mem:* Am Chem Soc; Royal Soc Chem; Sigma Xi; InterAm Photochem Soc. *Res:* Organic structure of coal and coal plasticity; technique of analytical pyrolysis as a tool for the study of plastic coals. *Mailing Add:* Dept Chem Western Ky Univ Bowling Green KY 42101

REASONS, KENT M, b Dyersburg, Tenn, June 24, 40. PLANT PHYSIOLOGY. *Educ:* Univ Tenn, MS, 64. *Prof Exp:* RES & FIELD DEVELOP, MKT & BUS MGT, DOMESTIC & OVERSEAS, AGR PROD DEPT, E I DU PONT DE NEMOURS & CO, INC, 64- *Mem:* Wheat Sci Soc Am; Agron Soc; Nat Agr Chem Asn; Entom Soc Am; Coun Agr Sci & Technol. *Mailing Add:* Agr Prod Dept E I du Pont de Nemours & Co Inc Barley Hill Plaza WM-1-266 Wilmington DE 19898

REASOR, MARK JAE, b Evansville, Ind, Nov 3, 45; m 67; c 2. PULMONARY TOXICOLOGY, IMMUNOTOXICOLOGY. *Educ:* Purdue Univ, BS, 67; Duke Univ, MA, 69; Johns Hopkins Univ, PhD(toxicol), 75; Am Bd Toxicol, dipl, 81. *Prof Exp:* Res fel pharmacol, Nat Inst Environ Health Sci, NIH, 75-76; from asst prof to assoc prof, 76-84, PROF PHARMACOL & TOXICOL, MED CTR, WVA UNIV, 84- *Concurrent Pos:* Prin investr, NIH res grant, 79-83; vis scholar, Univ Calif, San Diego, 83; consult, Ctr Environ Health & Human Toxicol, 84-89. *Mem:* Soc Toxicol; Am Soc Pharmacol & Exp Therapeut. *Res:* Toxicity of cationic amphiphilic drugs in humans and animals with an emphasis on impairment in pulmonary and immune functions. *Mailing Add:* Dept Pharmacol & Toxicol Health Sci Ctr WVa Univ Morgantown WV 26506

REAVEN, EVE P, b Kosice, Czech, Jan 18, 28. CELL SECRETION, ELECTRON MICROSCOPY. *Educ:* Univ Chicago, PhD(anat), 52. *Prof Exp:* AT VET ADMIN HOSP, PALO ALTO, 70- *Mem:* Am Soc Cell Biol; Am Soc Endocrinol; Europ Asn Study Diabetes. *Mailing Add:* Vet Admin Hosp 3801 Miranda Ave 182-B Palo Alto CA 94304

REAVES, GIBSON, b Chicago, Ill, Dec 26, 23; m 55; c 1. ASTRONOMY. *Educ:* Univ Calif, Los Angeles, BA, 47; Univ Calif, Berkeley, PhD(astron), 52. *Prof Exp:* Asst astron, Univ Calif, 47-49; from instr to assoc prof, 52-65, chmn dept, 69-74, PROF ASTRON, UNIV SOUTHERN CALIF, 65- *Concurrent Pos:* Assoc meritus, Lowell Observ, 85- *Mem:* Am Astron Soc; Royal Astron Soc; Int Astron Union. *Res:* Extragalactic problems; history of astronomy. *Mailing Add:* Dept Astron Univ Southern Calif Los Angeles CA 90089-1342

REAVES, HARRY LEE, b Clarksburg, WVa, Feb 25, 27; c 3. PHYSICS, MATHEMATICS. *Educ:* WVa Univ, AB & MS, 49; Va Polytech Inst, PhD, 63. *Prof Exp:* Asst prof physics, Clemson Col, 49-52; asst prof physics & math, Hampden-Sydney Col, 52-56; asst prof physics, Va Polytech Inst, 56-63; physicist, Humble Res Lab, Houston, 63-64; physicist, Houston Res Lab, Shell Oil Co, 64-71; ASST PROF PHYSICS & MATH, SAN JACINTO COL, 71- *Mem:* Am Asn Physics Teachers; Sigma Xi. *Res:* Nuclear magnetic resonance and solid state physics. *Mailing Add:* San Jacinto Col Pasadena TX 77505-2007

REAVEY-CANTWELL, NELSON HENRY, b Buffalo, NY, May 8, 26; m; c 3. MEDICINE. *Educ:* Canisius Col, BS, 44; Fordham Univ, MSc, 48, PhD(phys org chem), 52; Columbia Univ, MD, 59; Henry George Sch, dipl, 58. *Hon Degrees:* HCD, Univ Tokyo, 81; ScD, Univ Philippines, 81. *Prof Exp:* From asst to instr chem, Fordham Univ, 47-52; from instr to asst prof, Yale Univ, 52-55; intern, Bellevue Hosp & Mem Ctr Cancer & Allied Dis, 59-60; resident internal med, Vet Admin Hosp, Manhattan, NY, 60-61; asst dir med res, Merck Inst, 61-68; asst prof, 68-71, ASSOC PROF CLIN MED, THOMAS JEFFERSON UNIV, 71- *Concurrent Pos:* Fel, Harvard Univ, 49; attend physician, Curtis Clin, Thomas Jefferson Univ Hosp, 68-; dir dev res, William H Rorer, Inc, 68-69, vpres res, 69-78, vpres med & sci affairs, Rorer Int, 78- *Honors & Awards:* Merck Award, 66. *Mem:* Am Soc Clin Pharmacol & Therapeut; Am Soc Internal Med; fel Am Col Clin Pharmacol; Am Fedn Clin Res; fel Royal Soc Med. *Res:* Internal medicine; endocrinology and immunology; kinetics and mechanisms of metabolic reactions; steroid and electrolyte metabolism. *Mailing Add:* Spring Valley Rd Box 258 Furlong PA 18925-0258

REAY, JOHN R, b Pocatello, Idaho, Oct 27, 34; m 58; c 3. MATHEMATICS. *Educ:* Pac Lutheran Univ, BA, 56; Univ Idaho, MS, 58; Univ Wash, PhD(math), 63. *Prof Exp:* Assoc prof, 63-68, PROF MATH, WESTERN WASH UNIV, 68- *Mem:* Am Math Soc; Math Asn Am. *Res:* Convexity and geometry. *Mailing Add:* Dept Math Western Wash Univ Bellingham WA 98225

REAZIN, GEORGE HARVEY, JR, b Chicago, Ill, Feb 3, 28; m 50; c 3. PLANT PHYSIOLOGY, BIOCHEMISTRY. *Educ:* Northwestern Univ, BS, 49; Univ Mich, MS, 51, PhD(plant physiol), 55. *Prof Exp:* Res assoc, Brookhaven Nat Lab, 54-56; res scientist, Joseph E Seagram & Sons, Inc, 56-62, head biochem sect, 62-71, head chem sect, 71-82, MGR CHEM RES & SERV, JOSEPH E SEAGRAM & SONS, INC, WHITE PLAINS, NY, 82- *Honors & Awards:* Guymon Mem Lectr, Am Soc Enologists, 81. *Mem:* Bot Soc Am; Am Chem Soc; Am Soc Plant Physiologists. *Res:* Chemistry of flavors and biochemistry of their formation; chemistry laboratory management using computer laboratory information management systems (LIMS). *Mailing Add:* 25 Brantwood Lane Stamford CT 06903

REBA, RICHARD CHARNEY, b Milwaukee, Wis, July 1, 32; m 54, 83; c 2. MEDICINE, NUCLEAR MEDICINE. *Educ:* Univ Md, MD, 57; Am Bd Internal Med, dipl, 64; Am Bd Nuclear Med, dipl, 72. *Prof Exp:* Asst resident med, Univ Hosp, Baltimore, Md, 59-61; fel nuclear med, Johns Hopkins Univ, 61-62; sr investr, Walter Reed Army Inst Res, 62-65, actg chief dept isotope metab, 64, chief, 64-65, chief med serv, 85th Evacuation Hosp, Vietnam, 65-66; from asst prof to assoc prof radiol & radiol sci, Sch Hyg & Pub Health, Johns Hopkins Univ, 66-70, asst prof internal med, 67-70, assoc radiol sci, 70-78; chmn dept nuclear med, Wash Hosp Ctr, 70-76; PROF RADIOL, SCH MED, GEORGE WASHINGTON UNIV, 71-, PROF MED, SCH MED, 76-, DIR DIV NUCLEAR MED, MED CTR, 76- *Concurrent Pos:* Asst chief gen med, Sect Four, Walter Reed Army Gen Hosp, 64-65; chief clin nuclear med sect, Johns Hopkins Med Insts, 68-70; clin prof med, Sch Med, Georgetown Univ, 71-79. *Mem:* AAAS; Soc Nuclear Med; fel Am Col Nuclear Physicians (pres, 77-); fel Am Col Physicians; Am Fedn Clin Res. *Res:* Diagnostic and research applications of radioisotopes in clinical medicine. *Mailing Add:* George Washington Univ Med Ctr 901 23rd St NW Washington DC 20037

REBACH, STEVE, b New York, NY, Nov 15, 42; m 68; c 2. BIOLOGICAL RHYTHMS, MIGRATION. *Educ:* City Col New York, BS, 63; Univ RI, PhD(oceanog), 70. *Prof Exp:* Instr oceanog, Grad Sch Oceanog, Univ RI, 69-70; asst prof, St Mary's Col Md, 70-72; from asst prof to assoc prof biol, 72-85, PROF BIOL, UNIV MD, EASTERN SHORE, 85- *Concurrent Pos:* USDA & NSF res grants maricult, behav & ecol; chmn grad prog marine, estuarine & environ sci, Univ Md, 81-; prin investr, Crab Ecol & Maricult Proj, Am Inst Biol Sci. *Mem:* AAAS; Animal Behav Soc; Am Inst Biol Sci; Sigma Xi. *Res:* Biological rhythms, particularly in relation to marine organisms, navigation, orientation, and migration; mariculture of crustacea; optimal foraging in crustacea. *Mailing Add:* Dept Biol Univ Md Eastern Shore Princess Anne MD 21853

REBBI, CLAUDIO, b Trieste, Italy, Mar 1, 43; m 67; c 2. PARTICLE THEORY, QUANTUM FIELD THEORY. *Educ:* Univ Torino, Italy, Laurea, 65, PhD(nuclear physics), 67. *Prof Exp:* Prof, Univ Trieste, 70-72; res assoc, Europ Orgn Nuclear Res, 72-74; vis assoc prof, Mass Inst Technol, 74-77; scientist, Brookhaven Nat Lab, 77-87; PROF, BOSTON UNIV, 86- *Concurrent Pos:* Fel, Calif Inst Technol, 68-69; vis, Europ Orgn Nuclear Res, 80-81 & 84-85. *Mem:* Fel Am Phys Soc. *Res:* Elementary particle theory; quantum field theory; computational physics. *Mailing Add:* Physics Dept Boston MA 02215

REBEC, GEORGE VINCENT, b Harrisburg, Pa, Apr 6, 49. NEUROPHARMACOLOGY, NEUROBIOLOGY. *Educ:* Villanova Univ, AB, 71; Univ Colo, Boulder, MA, 74, PhD(biopsych), 75. *Prof Exp:* NIMH res fel, Univ Calif, San Diego, 75-77; from asst prof to assoc prof, 77-85, PROF & DIR PROG NEURAL SCI, IND UNIV, BLOOMINGTON, 85- *Concurrent Pos:* Prin investr, Nat Inst Drug Abuse grant, 79-, NSF grant, 85- *Honors & Awards:* Lilly Award, Lilly Endowment, 79. *Mem:* Soc Neurosci; Int Brain Res Orgn; AAAS; Am Psychol Soc; Int Basal Ganglia Soc. *Res:* Neurochemical and neurophysiological systems underlying the behavioral response to certain drugs of abuse and to the antipsychotic drugs; electrochemical and electrophysiological recordings obtained from specific regions of the central nervous system. *Mailing Add:* Dept Psychol Prog Neural Sci Ind Univ Bloomington IN 47405

REBEK, JULIUS, JR, b Beregszasz, Hungary, Apr 11, 44; US citizen; c 2. ORGANIC CHEMISTRY. *Educ:* Univ Kans, BA, 66; Mass Inst Technol, PhD(chem), 70. *Prof Exp:* Am Chem Soc-Petrol Res Fund fel, 70-73, Eli Lilly res fel, 72-74, asst prof chem, Univ Calif, Los Angeles, 70-76; assoc prof chem, Univ Pittsburgh, 76-80, prof, 80-89; PROF CHEM, MASS INST TECHNOL, 89- *Concurrent Pos:* A P Sloan fel, 76-78, Alexander von Humboldt fel, 81; Guggenheim fel, 86. *Mem:* Am Chem Soc; AAAS. *Res:* Enzyme models; organic reaction mechanisms; molecular recognition; self-replicating systems. *Mailing Add:* Dept Chem Mass Inst Technol Cambridge MA 02139

REBEL, WILLIAM J, b Troy, NY, Mar 15, 34. POLYMER CHEMISTRY. *Educ:* State Univ NY Albany, BS, 57, MS, 59; Univ Alta, PhD(org chem), 63. *Prof Exp:* Fel org chem, Univ Rochester, 63-64; chemist, Chem Div, Union Carbide Corp, 64-66; chemist, Polymer Technol Div, 66-74, TECH ASSOC, MFG TECHNOL DIV, EASTMAN KODAK CO, 74- *Mem:* Am Chem Soc; NY Acad Sci. *Res:* Decomposition of iodonium salts; characterization and synthesis of cellulose esters; synthesis of addition polymers; preparation of new polymers; dispersion of pigments and radiation curing of monomer/polymer systems. *Mailing Add:* 770 Van Voorsis Ave Rochester NY 14617-2170

REBENFELD, LUDWIG, b Czech, July 10, 28; nat US; m 56. ORGANIC CHEMISTRY. *Educ:* Lowell Tech Inst, BS, 51; Princeton Univ, MA, 53, PhD(org chem), 55. *Hon Degrees:* DSc, Philadelphia Col, 79. *Prof Exp:* Asst instr chem, Lowell Tech Inst, 49-51; sr chemist, 54-55, group leader, 55-60, assoc res dir, 60-65, vpres educ & res, 66-70, PRES & DIR, TEXTILE RES INST, 71- *Concurrent Pos:* Vis prof, Princeton Univ, 65-; chmn bd & life trustee, Philadelphia Col Textiles & Sci. *Honors & Awards:* Smith Medal, Am Soc Testing & Mat, 74; Inst Medal, Textile Inst Eng, 76; Distinguished Achievement Award, Fiber Soc, 68; Olney Medal, Am Asn Textile Chemists & Colorists, 87. *Mem:* Fiber Soc (secy-treas); Am Chem Soc; Am Asn Textile Chemists & Colorists; fel Brit Textile Inst. *Res:* Chemistry of cellulose and cellulose derivatives; chemical and physical properties of textile fibers; cotton fiber technology; keratin fiber deformation processes; structure and properties of synthetic polymers. *Mailing Add:* 49 Pardoe Rd Princeton NJ 08540

REBER, ELWOOD FRANK, b Reading, Pa, June 24, 19; m 42; c 3. NUTRITION, BIOCHEMISTRY. *Educ:* Berea Col, AB, 44; Cornell Univ, MNS, 48; Okla State Univ, MS, 50, PhD(chem), 51. *Prof Exp:* Food inspector, Kroger Grocery & Baking Co, 44-45; lab asst biochem, Okla State Univ, 48-49, asst microbiol assays, 49-51; res chemist, Swift & Co, 51-52; from asst prof to prof vet physiol & pharmacol, Univ Ill, 52-64; prof foods & nutrit & head dept, Univ Mass, Amherst, 64-68 & Purdue Univ, Lafayette, 68-74; prof nutrit & dean col nutrit, textiles & human develop, 74-78, prof nutrition & food sci, 78-86, EMER PROF NUTRIT & FOOD SCI, TEX WOMAN'S UNIV, 86- *Concurrent Pos:* Fulbright res fel, Col Agr & Vet Med, Copenhagen, Denmark, 61-62. *Mem:* Am Chem Soc; Inst Food Technologists; Am Inst Nutrit. *Res:* Human nutrition and foods; food science; wholesomeness of irradiated foods; nutrition and disease; metabolic disorders; use of cottonseed in foods for humans. *Mailing Add:* 1824 Concord Lane Denton TX 76205

REBER, JERRY D, b Lebanon, Pa, May 25, 39; m 63; c 2. NUCLEAR PHYSICS. *Educ:* Franklin & Marshall Col, AB, 61; Univ Ky, MS, 64, PhD(nuclear physics), 67. *Prof Exp:* NSF fel, Univ Ky, 65; res assoc & asst prof physics, Univ Va, 68-69; assoc prof, 69-79, PROF PHYSICS, STATE UNIV NY COL GENESEO, 80-, CHMN DEPT, 69- *Mem:* Am Phys Soc. *Res:* Neutron scattering studies of calcium, potassium, lanthanum and holmium; isobaric spin impurities of fluorine; ion induced x-ray fluorescence; neutron-proton scattering; ion beam lithography. *Mailing Add:* Dept Physics & Astron State Univ NY Col Geneseo NY 14454

REBER, RAYMOND ANDREW, b Apr 16, 42; m 63; c 3. ADSORPTION SEPARATION SYSTEMS DEVELOPMENT, TECHNOLOGY LICENSING. *Educ:* NY Univ, BS, 63, MS, 66. *Prof Exp:* Process develop engr, M W Kellogg, 66-70; supvr process develop, Union Carbide, 70-75, mgr, licensing petrol refinery technol, 75-77, molecular sieve process, 77-82, molecular sieve catalysis, 82-85 & molecular sieve adsorption technol, 85-88; dir, molecular sieve adsorption technol, 88-89, DIR, NEW VENTURES DEVELOP, UNIV OF PAC, 89- *Mem:* Am Inst Chem Engrs; Nat Soc Prof Engrs; Com Develop Asn. *Res:* New applications for molecular sieves including heterogeneous catalysis, molecular separations, ion exchangers and additives for property modification of films, polymers and paints. *Mailing Add:* Ten Bonnie Hollow Lane Montrose NY 10548

REBERS, PAUL ARMAND, b Minneapolis, Minn, Jan 24, 23; m 52; c 3. IMMUNOCHEMISTRY. *Educ:* Univ Minn, Minneapolis, BS, 43, MS, 46; Univ Minn, St Paul, PhD(agr biochem), 53. *Prof Exp:* Process engr, Rohm & Haas, 46-49; res chemist, Nat Animal Disease Lab, 61-88; mem grad fac, Iowa State Univ, 66-88, assoc prof biochem, 70-88; RETIRED. *Mem:* AAAS; Am Chem Soc; Am Soc Microbiol; Am Asn Immunol; Soc Exp Biol & Med; Sigma Xi. *Res:* Purification, isolation and structure of carbohydrate antigens; colorimetric analysis of sugars; immunology of fowl cholera. *Mailing Add:* 627 14th Street Pl Nevada IA 50201

REBHUHN, DEBORAH, b Heidenheim, Ger, Oct 23, 46; US citizen; m 80; c 1. APPLIED MATHEMATICS. *Educ:* Cornell Univ, AB, 68; Univ Ill, MS, 70, PhD(math), 74. *Prof Exp:* Teaching asst math, Univ Ill, 68-73; asst prof math, Vassar Col, NY, 73-79; mem tech staff, Bell Labs, 80-83, DIST MGR, BELL COMMUN RES, 84- *Mem:* Asn Women Math. *Res:* Qualitative properties of control systems, particularly the study of properties of control systems that are stable under sufficiently small perturbations; systems engineering and systems analysis to support development of large software systems. *Mailing Add:* Bell Commun Res 33 Knightsbridge Rd Piscataway NJ 08854

REBHUN, LIONEL ISRAEL, b Bronx, NY, Apr 19, 26; m 49; c 2. ZOOLOGY, CELL BIOLOGY. *Educ:* City Col New York, BS, 49; Univ Chicago, MS, 51, PhD(zool), 55. *Prof Exp:* From instr to asst prof anat, Col Med, Univ Ill, 55-58; from asst prof to assoc prof biol, Princeton Univ, 58-69; prof, 69-77, COMMONWEALTH PROF BIOL, UNIV VA, 77- *Concurrent Pos:* Lalor fel, 56; Guggenheim fel, 62. *Mem:* Am Soc Cell Biologists; Biophys Soc; Electron Micros Soc Am; Soc Develop Biol; Soc Gen Physiologists. *Res:* Control of cell division; cellular motility; cell ultrastructure; cryobiology. *Mailing Add:* Dept Biol Gilmer Hall 50 Univ Va Charlottesville VA 22903

REBICK, CHARLES, b Halifax, NS, Oct 21, 44; m 71; c 2. PHYSICAL CHEMISTRY. *Educ:* Univ Toronto, BSc, 67; Mass Inst Technol, PhD(phys chem), 71. *Prof Exp:* Fel phys chem, Hebrew Univ, Jerusalem, 72 & Theoret Chem Inst, Univ Wis-Madison, 73; res chemist, Exxon Res Eng Co, 73-82, proj leader, Exxon Res & Develop Labs, Baton Rouge, La, 80-82, lab dir, Catalysis Sci Lab, 84-86, sr staff assoc planning, 86-88, SR RES ASSOC, SENSORS & ANALYZERS, CORP RES, EXXON RES ENG CO, EXXON CORP, 88- *Honors & Awards:* Sigma Xi Award, 71. *Mem:* Am Chem Soc. *Res:* Kinetics and mechanisms of free radical reactions, especially thermal reactions of hydrocarbons; heterogeneous catalysis; sensors and analyzers. *Mailing Add:* 355 Whispering Hills Dr Annandale NJ 08801

REBMAN, KENNETH RALPH, b Mishawaka, Ind, Oct 4, 40; m 64. EDUCATIONAL ADMINISTRATION. *Educ:* Oberlin Col, AB, 62; Univ Mich, Ann Arbor, MA, 64, PhD(math), 69. *Prof Exp:* From asst prof to assoc prof, 69-77, dept chmn, 80-85, PROF MATH & COMPUT SCI, CALIF STATE UNIV, HAYWARD, 77-, DEAN, SCH SCI, 87- *Concurrent Pos:* Bd gov, Math Asn Am, 78-81. *Mem:* Math Asn Am; Sigma Xi. *Res:* Mathematical optimization; combinatorics. *Mailing Add:* Sch Sci Calif State Univ Hayward CA 94542

REBOUCHE, CHARLES JOSEPH, b New Orleans, La, Dec 27, 48. METABOLISM. *Educ:* Tulane Univ, BS, 70; Vanderbilt Univ, PhD(biochem), 74. *Prof Exp:* Post-doctoral res assoc biochem, Univ Tex, 74-75; res fel, Mayo Found, 75-80, assoc consult neurol, 80-84; asst prof, 84-88, ASSOC PROF PEDIAT, UNIV IOWA, 88- *Concurrent Pos:* Pfizer travelling fel, Clin Res Inst, Montreal, 83. *Mem:* Am Chem Soc; Am Inst Nutrit; Am Soc Biochem & Molecular Biol. *Res:* Carnitine biosynthesis, metabolism and function; amino acid metabolism; human nutrition. *Mailing Add:* Dept Pediat Univ Hosp Univ Iowa Iowa City IA 52242

REBOUL, THEO TODD, III, b New Orleans, La, Oct 3, 22; m 52; c 2. SOLID STATE PHYSICS. *Educ:* Tulane Univ, BS, 44, MS, 48; Univ Pa, PhD(physics), 53. *Prof Exp:* Instr, Gen & Intermediate Physics Labs, Tulane Univ, 46-48; instr gen physics, Univ Pa, 48-50 & Drexel Inst, 51-53; physicist, Photo Prod Dept, E I du Pont de Nemours & Co, 53-59; sr engr, Radio Corp Am, 59-63, staff engr, 63-69; staff tech adv, 69-74, dir, Corp Contrib Progs, 80-87, CHMN EDUC AID COMT, RCA CORP, 74- *Concurrent Pos:* Mem, Overseas Schs Adv Coun, US State Dept, 74-86. *Mem:* Am Phys Soc; Sigma Xi. *Res:* Photographic properties; plasma physics; electro-optics; imaging sensors; administration. *Mailing Add:* 132 Ramblewood Rd Moorestown NJ 08057-2628

REBSTOCK, THEODORE LYNN, b Elkhart, Ind, June 24, 25; m 57; c 2. PLANT BIOCHEMISTRY. *Educ:* NCent Col, BA, 49; Mich State Univ, MS, 51, PhD(chem), 56. *Prof Exp:* From asst to asst prof agr chem, Mich State Univ, 49-59; from assoc prof to prof chem, Westmar Col, 59-84, chmn dept, 63-84, dir nat sci div, 80-83; lab mgr, Harkers Inc, 85-90; RETIRED. *Mem:* AAAS; Am Chem Soc; Sigma Xi. *Res:* Mechanism of action and synthesis and isolation of plant growth regulators. *Mailing Add:* 1026 Sixth Ave SE LeMars IA 51031

REBUCK, ERNEST C(HARLES), b Klingerstown, Pa, Sept 24, 44; div; c 2. HYDROGEOLOGY. *Educ:* Pa State Univ, BSAE, 66, MS, 67; Univ Ariz, PhD(hydrol), 72. *Prof Exp:* Asst prof agr eng, Univ Md, 71-74; geologist, Md Dept Natural Resources, 74-78, div chief, 78-79, prog dir, 79-81; sr water resources engr, Greenhorne & O'Mara, 81-83; asst prof, Agr Eng, Univ Md, 83-85; bur chief, 85-86, PROG MGR, NMEX ENVIRON IMPROV DIV, 86- *Mem:* Am Soc Agr Engrs. *Res:* Groundwater hydraulics; groundwater modeling; surface water resources. *Mailing Add:* NMex Environ Improv Div 1190 St Francis Dr Santa Fe NM 87503

REBUCK, JOHN WALTER, b Minneapolis, Minn, Nov 24, 14; m 43. HEMATOLOGY. *Educ:* Creighton Univ, AB, 35; Univ Minn, MA, 40, MB, MD, 43, PhD(hemat), 47. *Prof Exp:* Asst hemat, Univ Minn, 38-42; intern, Henry Ford Hosp, 43; hematopathologist, Army Inst Path, 46; pathologist & chief Div Lab Hemat, Henry Ford Hosp, 47-76, sr hematopathologist, Labs, 76-84; RETIRED. *Concurrent Pos:* Ed, R E S, 65-74; prof, Wayne State Univ, 70-81, emer prof, 81-; prof, Univ Mich, 71-81. *Honors & Awards:* H P Smith Award, Am Soc Clin Path, 82. *Mem:* Fel Am Soc Hemat (secy, 58-61); Am Soc Clin Path; Reticuloendothelial Soc (pres, 58-60); Am Asn Pathologists & Bacteriologists; fel Int Soc Hemat. *Res:* Electron microscopy of blood cells; functions of leukocytes; cytology of inflammatory exudate; ultrastructure of sickle cells; hematopathology. *Mailing Add:* 22447 N Nottingham Birmingham MI 48010

REBUFFE-SCRIVE, MARIELLE FRANCOISE, b Paris, France. STEROID HORMONES, ADIPOSE TISSUE METABOLISM. *Educ:* Univ Paris VI, France, MSc, 69, PhD(biochem), 77; Univ Goteborg, Sweden, PhD(med sci), 86. *Prof Exp:* Asst biol, Hotel-Dieu Med Clin, Paris, 71-82; res fel, Dept Med, Salgren's Hosp, Goteborg, Sweden, 82-88; RES SCIENTIST, DEPT PSYCHOL, YALE UNIV, NEW HAVEN, 91- *Concurrent Pos:* Vis res scientist, Dept Psychol, Yale Univ, New Haven, 88-91. *Mem:* Am Inst Nutrit; Europ Soc Clin Invest. *Res:* Hormonal and behavioral determinants of regional fat distribution and metabolism and its associated diseases. *Mailing Add:* Dept Psychol Yale Univ PO Box 11A Yale Sta New Haven CT 06520

RECANT, LILLIAN, b New York, NY, Mar 7, 22; m 55. MEDICINE. *Educ:* Hunter Col, BA, 41; Columbia Univ, MD, 46. *Prof Exp:* Intern med, Col Physicians & Surgeons, Columbia Univ & Presby Hosp, 46-48; asst med & endocrinol, Peter Bent Brigham Hosp, 48-49; asst resident, Col Physicians & Surgeons, Columbia Univ & Presby Hosp, 49-50, Commonwealth fel biochem, 50-51; Commonwealth fel biochem, Sch Med, Washington Univ, 51-53, from asst prof to assoc prof med & prev med, 53-66; PROF MED, SCH MED, GEORGETOWN UNIV, 66-; CHIEF DIABETES RES LAB, VET ADMIN HOSP, 66- *Mem:* Am Soc Clin Invest; Endocrine Soc; Asn Am Physicians; Asn Teachers Prev Med; Soc Exp Biol & Med; Am Diabetes Asn. *Res:* Diabetes, metabolism and endocrinology; nephrosis; liver diseases. *Mailing Add:* 3502 Alton Pl Washington DC 20016

RECH, RICHARD HOWARD, b Irvington, NJ, Mar 20, 28; m 52; c 3. NEUROPHARMACOLOGY, PSYCHOPHARMACOLOGY. *Educ:* Rutgers Univ, BSc, 52; Univ Mich, MSc, 55, PhD(pharmacol), 60. *Prof Exp:* Pharmacist, Univ Hosp, Univ Mich, 55-56, asst pharmacol, Univ, 57-58; from instr to assoc prof, Dartmouth Med Sch, 61-71; PROF PHARMACOL, MICH STATE UNIV, 71-, PROF TOXICOL, 77- *Concurrent Pos:* USPHS fel, Univ Utah, 59-61; NIH Fogarty fel (sr res fel), Mario Negri, Milan, Italy, 78-79. *Mem:* AAAS; Am Soc Pharmacol & Exp Therapeut; fel Am Col Neuropsychopharmacol; Soc Neurosci. *Res:* Pharmacology. *Mailing Add:* Dept Pharmacol Mich State Univ B 440 Life Sci East Lansing MI 48824-1317

RECHARD, OTTIS WILLIAM, b Laramie, Wyo, Nov 13, 24; m 43; c 4. COMPUTER SCIENCE, MATHEMATICS. *Educ:* Univ Wyo, BA, 43; Univ Wis, MA, 46, PhD(math), 48. *Prof Exp:* Asst, Alumni Res Found, Univ Wis, 45-48, instr math, 48; from instr to asst prof, Ohio State Univ, 48-51; staff mem, Los Alamos Sci Lab, Univ Calif, 51-56; assoc prof math, Wash State Univ, 56-61, dir comput ctr, 56-68, prof comput sci & math, 61-76, chmn dept comput sci, 63-76, dir systs & comput, 68-70; dir comput serv, 76-79, PROF MATH & COMPUT SCI, UNIV DENVER, 76- *Concurrent Pos:* Consult, Los Alamos Sci Lab, Univ Calif, 56-59, 75-79, vis staff mem, 74-75; consult, Atomic Energy Div, Phillips Petrol Co, 59-70; consult, NSF, 63-64 & 65-, dir comput sci prog, 64-65. *Mem:* Fel AAAS; Am Math Soc; Soc Indust & Appl Math; Math Asn Am; Asn Comput Mach. *Res:* Electronic computers; computer architecture and operating systems; design and analysis of algorithms. *Mailing Add:* Univ Denver Denver CO 80208

RECHARD, PAUL A(LBERT), b Laramie, Wyo, June 4, 27; m 49; c 2. HYDROLOGIC ENGINEERING, CIVIL ENGINEERING. *Educ:* Univ Wyo, BS, 48, MS, 49, CE, 55. *Prof Exp:* Civil engr, US Bur Reclamation, 48-49, hydraul engr, 49-54, asst proj hydrologist, 53-54; dir water resources & interstate streams comnr, Wyo Natural Resource Bd, 54-58; prin hydraul engr, Upper Colo River Comn, 58-64; water resources res engr, Univ Wyo, 64-66, asst dir, Natural Resources Res Inst, 66-71, dir, Water Resources Res Inst, 66-81, prof civil eng, 64-82; PRES, WESTERN WATER CONSULTS, INC, 80- *Concurrent Pos:* Off Water Resources Res grants, 66-81; grants, Environ Protection Agency, 73-81. *Mem:* Int Comn Irrig & Drainage; Am Water Works Asn; Am Geophys Union; fel Am Soc Civil Engrs; Am Water Resources Asn; Nat Soc Prof Engrs; Nat Water Well Asn; Sigma Xi. *Res:* Hydrologic research dealing with precipitation-runoff relationships; consumptive use by agricultural and municipal and users; water planning criteria; hydrologic aspects of surface mining; snow hydrology; precipitation measurement. *Mailing Add:* Western Water Consults Inc 611 Skyline Rd Laramie WY 82070

RECHCIGL, MILOSLAV, JR, b Mlada Boleslav, Czech, July 30, 30; nat US; m 53; c 2. NUTRITIONAL BIOCHEMISTRY, BIOTECHNOLOGY. *Educ:* Cornell Univ, BS, 54, MNS, 55, PhD, 58. *Prof Exp:* Asst, Cornell Univ, 52-57, asst, Grad Sch Nutrit, 57-58, res assoc biochem & nutrit, 58; USPHS fel, Lab Biochem, Nat Cancer Inst, 58-60, chemist, Enzyme & Metab Sect, 60-61, res biochemist, Tumor-Host Rels Sect, 62-64, sr investr, 64-68, sr investr, Biosynthesis sect, 68, USPHS grants assoc, 68-69; spec asst nutrit & health, Off Dir, Regional Med Progs Serv, Health Serv & Ment Health Admin, 69-70; nutrit adv, 70, chief, Res & Inst Grants Div, 70-73, asst dir, Off Res & Inst Grants, 73-74, actg dir, 74-75, dir interregional res staff, 75-78, chief res & methodology div, 79-83, RES DIR, OFF SCI ADV, AID, US DEPT STATE, WASHINGTON, DC, 83- *Concurrent Pos:* Nat Acad Sci travel grant, 62; mem educ comn, Nat Cancer Inst Assembly of Scientists, 62-63, chmn commun, mem comn & mem coun, 63-65; deleg, White House Conf Food, Nutrit & Health, 69; consult, Off Secy, USDA, 69-70, US Dept Treas, 73-74, Off Technol Assessment, 77-79, Food & Drug Admin, 79-81 & Nat Acad Sci & Nat Res Coun, 85-; exec secy nutrit prog adv comt, Health Serv & Ment Health Admin, 69-70; exec secy, Res & Inst Grants Coun, Res Adv Comn, 71-83 & Rep, Consult Group Gen Res, US Dept State, 71-78; AID rep to USC/FAR Comt, 72-78; ed-in-chief, series in Nutrit & Food, 77- *Mem:* Am Chem Soc; Am Soc Biol Chem; Am Inst Nutrit; Soc Int Develop; hon mem Czech Soc Arts & Sci (pres, 74-78); Am Inst Biol Sci. *Res:* Amino acid nutrition; vitamin A and protein metabolism; tumor-host relationship; regulatory mechanisms of enzyme activity; enzyme degradation and turnover; catalase; bibliography, history and historiography of science; research management; science administration; international development; biotechnology. *Mailing Add:* 1703 Mark Lane Rockville MD 20852

RECHNITZ, GARRY ARTHUR, b Berlin, Ger, Jan 1, 36; US citizen; m 58. ANALYTICAL BIOCHEMISTRY. *Educ:* Univ Mich, BS, 58; Univ Ill, MS, 59, PhD(chem), 61. *Prof Exp:* Asst prof chem, Univ Pa, 61-66; from assoc prof to prof, State Univ NY, Buffalo, 66-77, assoc provost natural sci & math, 67-69; prof chem & biotechnol, Univ Del, 78-89; PROF CHEM, UNIV HAWAII, 89- *Concurrent Pos:* Sloan Found res fel, 66-68; vis prof path, Scripps Clin & Res Found, La Jolla, Calif, 87-88. *Honors & Awards:* Van Slyke Award, 78; Am Chem Soc Award, 83; Iddles Lectr, 88; Gardinier Lectr Award, 89. *Mem:* AAAS; Am Chem Soc. *Res:* Biosensors membrane electrodes; ion selectivity in liquid and crystal membranes; clinical instrumentation. *Mailing Add:* Dept Chem Univ Hawaii Honolulu HI 96822

RECHNITZER, ANDREAS BUCHWALD, b Escondido, Calif, Nov 30, 24; m 46; c 4. BUSINESS MANAGEMENT, ADVANCED TECHNOLOGIES CONSULTING. *Educ:* Mich State Univ, BS, 47; Univ Calif, Los Angeles, MA, 51, PhD(oceanog), 56. *Prof Exp:* Adj prof oceanog, USN Post Grad Sch, 87-88; ADJ PROF OCEANOG, SAN DIEGO STATE UNIV, 86-; RES ASSOC, CALIF ACAD SCI, 90- *Concurrent Pos:* Sci & technol adv, Oceanogr Navy Off, Wash, DC, 73-85; sr scientist, Sci Applications Int Corp, 85-; pres, Viking Oceanog, 85-; consult, Unique Mobility, Inc, Marine Develop Assocs & Int Maritime, Inc, 85- *Mem:* Marine Technol Soc; Am Geophys Union; Cedam Int (pres, 67-76). *Res:* Scientist-in-charge of the world's record descent by man to 35,800 feet into the Marianas Trench, January 23, 1960; specialist on the effects of explosions and hyperbaric pressure on marine life; marine ecology; animal behavior; consulting in the field of ocean engineering, including materials, power sources, navigation and controls, field operations and advanced systems. *Mailing Add:* 1345 Lomita Rd El Cajon CA 92020

RECHT, HOWARD LEONARD, b Pittsburgh, Pa, July 16, 27; m 52; c 4. WATER CHEMISTRY, ELECTROCHEMISTRY. *Educ:* Carnegie Inst Technol, BS, 48; Cornell Univ, PhD(phys chem), 54. *Prof Exp:* Res chemist, Res & Develop Div, Consol Coal Co, Pa, 54-59; sr chemist, Atomics Int Div, NAm Aviation, Inc, Calif, 59-60, res specialist, 60-61; head electrochem, Astropower, Inc, Douglas Aircraft Corp, 61-62; supvr electrochem, 62-70, proj engr water technol, Atomics Int Div, NAm Rockwell Corp, 70-77, MEM TECH STAFF, ENERGY SYSTS GROUP, ROCKWELL INT CORP, 77- *Mem:* AAAS; Am Chem Soc; Am Phys Soc; Electrochem Soc; Sigma Xi. *Res:* Electrochemical energy conversion, including high temperature fuel cells, radiation effects on batteries; sodium vapor-graphite interaction; corrosion, water and wastewater treatment; phosphate removal; nuclear waste disposal. *Mailing Add:* 11002 Garden Grove Ave Northridge CA 91326

RECHT, RODNEY F(RANK), impact dynamics, mechanics of materials, for more information see previous edition

RECHTIEN, RICHARD DOUGLAS, b St Louis, Mo, Sept 10, 33; m 53; c 2. GEOPHYSICS. *Educ:* Wash Univ, BS, 58, MA, 59, PhD(geophys), 64. *Prof Exp:* Engr, McDonnell Aircraft Corp, Mo, 59-62; fluid dynamicist, Marshall Space Flight Ctr, NASA, 62-66; asst prof, 66-70, ASSOC PROF GEOPHYS, UNIV MO, ROLLA, 70- *Mem:* Soc Explor Geophys; Acoust Soc Am. *Res:* Noise generation in turbulent flow; rocket acoustics; random noise theory; nonlineal wave propagation; geomagnetism; magnetohydrodynamics of the earth's core; interplanetary magnetic fields. *Mailing Add:* Dept Geophys Univ Mo Box 249 Rolla MO 65401

RECHTIN, EBERHARDT, b Orange, NJ, Jan 16, 26; m 51; c 5. INDUSTRIAL & MANUFACTURING ENGINEERING. *Educ:* Calif Inst Technol, BS, 46, PhD, 50. *Prof Exp:* Res engr, Jet Propulsion Lab, Calif Inst Technol, 49-52, group supvr secure commun, 52-54, sect chief commun res, 54-57, dir chief guid res, 57-59, dir chief telecommun, 59-60, dir, NASA Deep Space Instrumentation Prog, 60-63, asst dir tracking & data acquisition, 63-67; dir, Advan Res Projs Agency, US Dept Defense, 67-70, prin dep dir, defense res & eng, 70-72, asst secy defense, Telecommun, 72-73; chief engr, Hewlett-Packard, 73-77; pres, Aerospace Corp, 77-87; PROF ENG, UNIV SOUTHERN CALIF, 88- *Concurrent Pos:* Var Govt & Nat Res Coun Comts, 56- *Honors & Awards:* NASA Medal Sci, 65; Aerospace Commun Award, Am Inst Aeronaut & Astronaut, 69, von Karman Lectureship in Astronaut, 85; Alexander Bell Award, Inst Elec & Electronics Engrs, 77; Gold Medal Eng, Armed Forces Commun & Electronics Asn. *Mem:* Nat Acad Eng; fel Am Inst Aeronaut & Astronaut; fel Inst Elec & Electronics Engrs; Int Acad Astronaut; hon fel Inst Environ Sci. *Res:* Information and communication theory; space and secure communication; international cooperative space research; systems architecture; electronics engineering. *Mailing Add:* Univ Southern Calif University Park Los Angeles CA 90089

RECHTIN, MICHAEL DAVID, b Ft Smith, Ark, Feb 1, 44; m 66; c 3. MATERIALS SCIENCE, LAW. *Educ:* Univ Ill, Urbana, BS, 66; Mass Inst Technol, PhD(metall & mat sci), 70; Ill Inst Technol, JD, 82. *Prof Exp:* Res assoc metall & mat sci, Mass Inst Technol, 70-73; staff scientist solid state sci, Res Lab, Tex Instruments Inc, 73-75; assos scientist mat sci, Argonne Nat Lab, 75-; MANAGING PARTNER & PATENT ATTY, REINHART BOERNER & VAN DUEREN, 88- *Concurrent Pos:* IBM fel, Mass Inst Technol, 70-71; patent lawyer, STD oil, 81; mem staff, Standard Oil Ind, Naperville, Ill; bd dirs, High Tech Mat Res Corp, Chicago, Ill. *Mem:* Am Ceramic Soc; Am Phys Soc. *Res:* Electron microscopy of defect structures in silicon and ion bombarded glasses and insulators; structure and computer models of amorphous materials, infrared optical properties of glasses; electronic and catalytic properties of insulators and magnetic and structural phase transformations in insulators. *Mailing Add:* Reinhart Boerner Van Dueren 111E Wisconsin Ave Suite 1800 Milwaukee WI 53202

RECHTSCHAFFEN, ALLAN, b New York, NY, Dec 8, 27; m 80; c 3. SLEEP, DREAMS. *Educ:* City Col New York, BSS, 49, MA, 51; Northwestern Univ, PhD(psychol), 56. *Prof Exp:* Psychologist, Fergus Falls State Hosp, 51-53; lectr psychol, Northwestern Univ, 56-57; researcher, Vet Admin, 56-57; instr, 57-58, from asst prof to assoc prof, 58-68, PROF PSYCHOL, UNIV CHICAGO, 68- *Honors & Awards:* Kleitman Award, Asn Sleep Dis Clins, 85; Distinguished Scientist Award, Sleep Res Soc, 89. *Mem:* Sleep Res Soc (pres, 79-82). *Res:* Function and physiology of sleep; psychophysiology of dreaming. *Mailing Add:* Univ Chicago 5741 S Drexel Ave & 5801 Ellis Ave Chicago IL 60637

RECK, GENE PAUL, b Chicago, Ill, Sept 12, 37; c 2. PHYSICAL CHEMISTRY. *Educ:* Univ Ill, Urbana, BS, 59; Univ Minn, PhD(phys chem), 63. *Prof Exp:* Instr phys chem, Univ Minn, 63-64; res assoc, Brown Univ, 64-65; from asst prof to assoc prof, 65-75, PROF PHYS CHEM, WAYNE STATE UNIV, 75- *Concurrent Pos:* Eastman Kodak sci award, Univ Minn, 62. *Mem:* Am Phys Soc; Am Chem Soc; Sigma Xi. *Res:* Atomic and molecular scattering; laser driven chemistry; laser spectroscopy of chemically reacting systems. *Mailing Add:* Dept Chem Wayne State Univ Detroit MI 48202

RECK, RUTH ANNETTE, b Rolla, Mo; div; c 2. ATMOSPHERIC PHYSICS, ATMOSPHERIC CHEMISTRY. *Educ:* Mankato State Univ, BA, 54; Univ Minn, PhD(phys chem), 64. *Prof Exp:* Instr physics & chem, Wis State Univ, River Falls, 59-61; res assoc chem, Brown Univ, 64-65; assoc sr res physicist, 65-70, assoc sr res chemist, 71-74, sr res chemist, 74-76, sr scientist, 76-79, STAFF RES SCIENTIST, GEN MOTORS RES LABS, 79- *Concurrent Pos:* Consult, Environ Movement & Transformation Adv Comt, Environ Protection Agency, 75-76, mem comt, 76-81, mem, adv bd, 77-; mem, Climate Modeling Group VIII Exchange with USSR, 76; US rep CO2 Workshop, Int Inst Appl Systs Anal, Austria, 78, Workshop Extended Clouds, World Meterol Orgn, Oxford Univ, 78, Int Asn Meterol & Atmospheric Physics, 81 & ad hoc working group aerosols, Lille, France, 82; mem, comt chem & biol sensors, Nat Res Coun, 83 & sci adv bd, Int Joint Comn, 85-86, chmn, air pollution indicators task force, 83-84; mem, Radiol & Environ Rev Bd, Argonne Nat Lab, 83-86; mem, panels climatic effects trace gases & surface radiation budget climate appln, NASA, 85. *Honors & Awards:* Gold Award Eng. *Mem:* AAAS; Am Chem Soc; Am Phys Soc; Sigma Xi; Am Geophys Union. *Res:* Statistics of climate change, atmospheric radiative transfer and related thermal effects from trace gases and particles; statistical properties of polymeric and magnetic materials; transport-limited rate processes in random systems. *Mailing Add:* 7279 Westchester West Bloomfield MI 48322

RECKASE, MARK DANIEL, b Chicago, Ill, Aug 31, 44; m 68; c 2. ITEM RESPONSE THEORY, PSYCHOMETRICS. *Educ:* Univ Ill, Urbana, BA, 66; Syracuse Univ, MS, 71 & PhD(psychol), 72. *Prof Exp:* Res assoc statist, Adult Develop Study, Syracuse Univ, 70-71; 1st lieutenant, US Army, 71-72; from asst prof to assoc prof educ psychol, Univ Mo, Columbia, 72-81; dir resident prog, 81-84, ASST VPRES ASSESSMENT PROG, ACT, 84- *Concurrent Pos:* Consult, Multipurpose Arthritis Ctr, Univ Mo Sch Med, 80-81; mem, CAT & DAC adv comt, Dept Defense, 81-84, 89-, tech adv comt, Calif Assessment Prog, 84 & Nat Coun Measurement Educ. *Mem:* Fel Am Psychol Soc; Am Educ Res Asn; Psychometric Soc (secy, 86-); Brit Psychol Soc; Soc Multivariate Exp Psychol. *Res:* Modeling of the interaction of persons and test items; multidimensional models of the persons item interaction; computer applications to measurement of cognitive skills. *Mailing Add:* ACT PO Box 168 Iowa City IA 52243

RECKEL, RUDOLPH P, b New York, NY, Feb 14, 34; m 57; c 3. IMMUNOLOGY, BIOCHEMISTRY. *Educ:* Brooklyn Col, BS, 56; Georgetown Univ, MS, 62; Rutgers Univ, PhD(biochem), 68. *Prof Exp:* Biochemist, Pioneer Blood Antigen Lab, USDA, Md, 59-62; biochemist, Ortho Res Inst Med Sci, 63-75, dir div clin immunol, Ortho Res Found, 70-75, immunochemist & dir div immunol, 75-80; dir div clin immunol, Ortho Diag Systs Inc, 81-; AT DEPT IMMUNOL, IMMUNOMEDICS INC. *Concurrent Pos:* Mem, Nat Comt Clin Lab Standards, 68-; mem work party, Int Comt Standardization Hematol, Standardization Anti-Human Globulin Reagents, 74- *Mem:* Am Chem Soc; NY Acad Sci; fel Am Inst Chemists; Am Asn Blood Banks. *Res:* Immunopathology of immune complexes in connective tissue diseases and cancer, especially detection methods, analysis and diagnostic algorithms. *Mailing Add:* 605 Stanley Pl Landing NJ 07850

RECKHOW, KENNETH HOWLAND, b San Francisco, Calif, Feb 7, 48; m 75; c 2. WATER QUALITY MODELING, APPLIED STATISTICS. *Educ:* Cornell Univ, SB, 71; Harvard Univ, SM, 72, PhD(environ systs), 77. *Prof Exp:* Asst prof water resources, Dept Resource Develop, Mich State Univ, 77-80; ASSOC PROF WATER RESOURCE SYSTS, SCH ENVIRON, DEPT CIVIL ENG, INST STATIST & DECISION SCI, DUKE UNIV, 80- *Concurrent Pos:* Assoc ed, J Water Resources Res, 80-85; consult, US Army CEngr, US Environ Protection Agency, 79- & US Soil Conserv Serv; prin investr, Nat Oceanic & Atmospheric Admin, 79-83; assoc ed, Water Resources Bull, 84-89; prin investr, US Environ Protection Agency, 85-88, NSF, 90- *Mem:* Am Geophys Union; Am Water Resources Asn; Am Statist Asn; Soc Risk Anal; NAm Lake Mgt Soc (secy, 85). *Res:* Mathematical and statistical methods employed in water quality management, including techniques for mathematical model confirmation, risk analysis, uncertainty analysis and statistical descriptions of space-time variability. *Mailing Add:* Sch Environ Duke Univ Durham NC 27706

RECKHOW, WARREN ADDISON, b Brooklyn, NY, Mar 29, 21; m 45; c 4. ORGANIC CHEMISTRY, INFORMATION SCIENCE. *Educ:* Drew Univ, BA, 43; Univ Rochester, PhD(chem), 50. *Prof Exp:* Chemist, 46-47, res chemist, 50-55, RES ASSOC, RES LAB, EASTMAN KODAK CO, 55- *Concurrent Pos:* Lectr, Univ Rochester, 51-52. *Mem:* Am Chem Soc. *Res:* Fries rearrangement reactions; synthetic curariform compounds; color photographic chemistry; the use of microform in scientific information. *Mailing Add:* 54 Coronado Dr Rochester NY 14617

RECKNAGEL, RICHARD OTTO, b Springfield, Mo, Jan 11, 16; m 43; c 2. PHYSIOLOGY, TOXICOLOGY. *Educ:* Wayne State Univ, BS, 38; Univ Pa, PhD(zool), 49. *Prof Exp:* Nat Res Coun fel, McArdle Lab Cancer Res, Univ Wis-Madison, 49-51; PROF PHYSIOL, SCH MED, CASE WESTERN RESERVE UNIV, 74-, DIR DEPT, 77- *Mem:* Am Soc Biol Chem; Am Physiol Soc; Am Asn Study Liver Dis. *Res:* Toxic liver injury. *Mailing Add:* Dept Physiol Sch Med Case Western Reserve Univ 2040 Adelbert Rd Cleveland OH 44106

RECKTENWALD, GERALD WILLIAM, b Lexington, Ky, June 28, 29; m 55; c 2. PHYSICAL CHEMISTRY, RESEARCH ADMINISTRATION. *Educ:* Univ Ky, BS, 49, MS, 50; Ind Univ, PhD(phys chem), 55. *Prof Exp:* Asst chem, Ind Univ, 52-54; radio chemist, Dow Chem Co, Mich, 54-55; chemist, Major Appliance Lab, Gen Elec Co, 55-65; mgr res & develop rigid foams, Olin Mathieson Chem Corp, 65-69, plant mgr plastics div, Cellular Prod Dept, Olin Corp, 69-72; proj mgr, Air Prod & Chem, 72-74, mgr new ventures, 74-78, dir com develop, 78-81, mgr r&d facil, 81-87; RETIRED. *Concurrent Pos:* Lectr, Ind Univ, 57-58, Univ Louisville, 58 & Nazareth Col, Ky, 59-60. *Mem:* Am Chem Soc; Soc Plastics Indust; Soc Plastic Engrs. *Res:* Polyurethane foams and hermetic systems; surface active agents; wire enamels; radiochemistry; chemical instrumentation; barrier properties of plastics; medical instrumentation. *Mailing Add:* 314 N 28th St Allentown PA 18104

RECORD, M THOMAS, JR, b Exeter, NH, Dec 18, 42; c 2. BIOPHYSICAL CHEMISTRY. *Educ:* Yale Univ, BA, 64; Univ Calif, San Diego, PhD(chem), 67. *Prof Exp:* NSF fel biochem, Stanford Univ, 68-70; from asst prof to prof chem, 70-82, PROF CHEM & BIOCHEM, UNIV WIS-MADISON, 82- *Concurrent Pos:* NSF & NIH grants. *Mem:* AAAS; Am Chem Soc; Biophys Soc; Am Soc Biol Chemists. *Res:* Physical chemistry of nucleic acids and proteins; protein-nucleic acid interactions, physical chemical basis of control of gene expression. *Mailing Add:* Dept Chem Univ Wis 1101 University Ave Madison WI 53706-1322

RECORDS, RAYMOND EDWIN, b Ft Morgan, Colo, May 30, 30; div; c 1. HUMAN EYE. *Educ:* Univ Denver, BS, 56; St Louis Univ, MD, 61. *Prof Exp:* Instr ophthal surg, Sch Med, Univ Colo, 65-67, asst prof, 67-70; PROF OPHTHAL & CHMN DEPT, COL MED, UNIV NEBR, 70- *Concurrent Pos:* Consult & sect chief, Vet Admin Ctr, 70-; consult, Bishop Clarkson Mem Hosp, 73- *Mem:* Am Acad Ophthal; Asn Res Vision & Ophthal; Asn Univ Prof Ophthal. *Res:* Physiology of the human eye and visual system. *Mailing Add:* Dept Ophthal Univ Nebr Med Ctr Omaha NE 68105

RECSEI, ANDREW A, b Kula, Yugoslavia, July 22, 02; nat US; m 42; c 3. ORGANIC CHEMISTRY. *Educ:* Univ Vienna, Austria, BA, 22; Univ Brno, Czech, MA, 24, PhD, 26. *Prof Exp:* Res chemist, Dr Honsig Chem Lab, 26-30; plant mgr, Pharmador Pty Ltd, SAfrica, 31-39; pres, Recsei Labs, Calif, 40-42; res assoc, Univ Calif, Los Angeles, 42-43; res chemist, Calif Inst Technol, 43-44 & Printing Arts Res Lab, 44-46; PRES, RECSEI LABS, 46- *Concurrent Pos:* Instr, Univ Calif, Santa Barbara, 54-56, assoc, 56-65. *Mem:* Am Chem Soc. *Res:* Antihistamines; allergy; nutrition; antimitotic compounds and anti-tumor agents. *Mailing Add:* 633 Tabor Lane Santa Barbara CA 93108

RECSEI, PAUL ANDOR, b Santa Barbara, Calif, Feb 24, 45. BIOCHEMISTRY. *Educ:* Harvard Univ, BS, 67; Univ Calif, PhD(biochem), 72. *Prof Exp:* Fel biochem, Univ Calif, 73-75; VPRES PHARMACEUT, RECSEI LABS, 75-, PRES; RES SCIENTIST, DEPT PLASMID BIOL, PUB HEALTH RES INST. *Res:* Development and design of pharmaceuticals. *Mailing Add:* 330 S Kellogg Ave Goleta CA 93117

RECTOR, CHARLES WILLSON, b Sioux City, Iowa, Apr 29, 26; m 54; c 2. SOLID STATE PHYSICS. *Educ:* Univ Chicago, PhB, 46, SB, 49; Franklin & Marshall Col, MS, 59; Johns Hopkins Univ, PhD(physics), 66. *Prof Exp:* Design engr, Tube Div, Radio Corp Am, 54-59, mem tech staff, RCA Labs, 60-62; assoc prof, 66-75, PROF PHYSICS, US NAVAL ACAD, 75- *Concurrent Pos:* Instr, Elizabethtown Col, 57-59; res assoc physics, Johns Hopkins Univ, 66. *Mem:* Am Asn Physics Teachers; Am Phys Soc. *Res:* Photoconductivity; semiconductor smokes; rare earth ions in crystals; electron paramagnetic resonance; optical spectroscopy; laser damage effects. *Mailing Add:* Dept Physics US Naval Acad Annapolis MD 21402

RECTOR, FLOYD CLINTON, JR, b Slaton, Tex, Jan 28, 29; m 50; c 3. NEPHROLOGY. *Educ:* Tex Tech Univ, BS, 50; Univ Tex Southwestern Med Sch, MD, 54. *Prof Exp:* Intern, Parkland Mem Hosp, Dallas, 55, resident, 56; instr internal med, Univ Tex Southwestern Med Sch, 58-59, asst prof, 59-63, assoc prof nephrol, 63-66, prof & dir div, 66-73; SR SCIENTIST, CARDIOVASC RES INST, PROF MED & PHYSIOL & DIR DIV NEPHROLOGY, UNIV CALIF, SAN FRANCISCO, 77- *Concurrent Pos:* Mem & chmn, Cardiovasc Study Sect, NIH, 65-69, mem, Nephrol & Urol Fel & Training Grants Comt, 71-76, chmn, 72-73; chmn, Fel & Res Grants Comt, Nat Kidney Found, 68-71, chmn, Sci Adv Bd, 71-73. *Mem:* Am Soc Clin Invest; Am Asn Physicians; Am Physiol Soc; Am Soc Nephrol (secy-treas, 73-76, pres, 76-77); Biophys Soc. *Res:* Mechanisms of ion and water transport by renal tubules. *Mailing Add:* Dept Med & Philos Univ Calif San Francisco CA 94143-0532

REDALIEU, ELLIOT, b Bronx, NY, Dec 12, 39; m 62; c 3. DRUG METABOLISM, MEDICINAL CHEMISTRY. *Educ:* Fordham Univ, BS, 61; Univ Mich, MS, 63, PhD(pharmaceut chem), 66. *Prof Exp:* Teaching asst pharmaceut anal, Univ Mich, 61-62; fel, Stanford Res Inst, 66-68; res biochemist, Pharmaceut Div, Geigy Chem Corp, 68-71; sr scientist, 71-80, sr scientist II, 75-80, mgr, bioavailability & pharmacokinetics, 80-86, ASST DIR CLIN PHARMACOKINETICS & DISPOSITION, PHARMACEUT DIV, CIBA-GEIGY CORP, 87- *Mem:* AAAS; Am Chem Soc; NY Acad Sci; Am Asn Pharmaceut Scientists; Int Soc Study Xenobiotics. *Res:* Metabolism of drugs; organic synthesis; analgesics; cardiovascular compounds; radiotracer techniques; gas chromatography; liquid chromatography; bioavailability; pharmacokinetics. *Mailing Add:* Clin Pharmacokinetics & Disposition Subdivision Ciba-Geigy Saw Mill River Rd Ardsley NY 10502

REDBURN, DIANNA AMMONS, b Many, La, Oct 21, 43; m 65; c 1. NEUROBIOLOGY. *Educ:* Centenary Col, BS, 64; Univ Kans, PhD(neurobiol), 72. *Prof Exp:* Fel psychobiol, Univ Calif, Irvine, 72-74; from asst prof to PROF NEUROBIOL, MED SCH UNIV TEX, HOUSTON, 74-, ASST DEAN RES TRAINING, 88- *Mem:* Asn Res Vision Opthal; Soc Neurosci; Neurochem Soc; Soc Cell Biol; Am Soc Biol Chem; Asn Women in Sci; AAAS; Int Soc Develop Neurosci. *Res:* The nature of chemical transmission in neuronal tissue; mechanisms of neurotransmitter release, identification of functional neurotransmitters, modulation of chemical transmission by intrinsic and extrinsic factors; development of in vitro techniques for biochemical analysis of neurotransmitter systems; neurochemistry of the retina. *Mailing Add:* Dept Neurobiol & Anat Univ Tex Med Sch Houston TX 77025

REDDAN, JOHN R, b Trenton, NJ, Apr 18, 39; m 62; c 4. CELL PHYSIOLOGY. *Educ:* St Michaels Col, BA, 61; Univ Vt, PhD(zool), 66. *Prof Exp:* NIH fel cell biol, 65-67, from asst prof to assoc prof, 67-78, PROF BIOL, OAKLAND UNIV, 78- *Mem:* AAAS; Am Soc Cell Biol; Asn Res Vision & Ophthal; Am Soc Zool. *Res:* Metabolic and cytological, light and electron microscopic changes which precede and accompany the initiation of cell division in normal, injured and cultured mammalian lenses epitherial cells. *Mailing Add:* Dept Biol Sci Oakland Univ Rochester MI 48063

REDDAN, WILLIAM GERALD, b St Louis, Mo, Aug 29, 27; m 52; c 3. PHYSIOLOGY. *Educ:* Univ Mo-Columbia, BS, 51; Univ Wis-Madison, MS, 55, PhD(biodynamics), 65. *Prof Exp:* Teacher, Spring Green Pub Schs, 55-60; res asst physiol, 60-62, proj assoc pulmonary physiol, 62-64, instr, 64-65, asst prof environ physiol, 65-72, ASSOC PROF PREV MED, MED SCH, UNIV WIS-MADISON, 72-, ASSOC PROF ANAT, 77- *Mem:* Am Physiol Soc. *Res:* Gas exchange in the lung; pulmonary physiology applied to occupational and environmental lung disease; physical education. *Mailing Add:* Dept Prev Med Univ Wis 204-504N Walnut Madison WI 53706

REDDELL, DONALD LEE, b Tulia, Tex, Sept 28, 37; m 57; c 3. GROUND WATER HYDROLOGY & QUALITY. *Educ:* Tex Technol Col, BS, 60; Colo State Univ, MS, 67, PhD(agr eng), 69. *Prof Exp:* Jr engr, High Plains Underground Water Conserv Dist, Lubbock, Tex, 60-62, agr engr, 62-64, engr, 64-65; from asst prof to assoc prof, 69-77, PROF & HEAD AGR ENG, TEX A&M UNIV, 77- *Mem:* Am Soc Agr Engrs; Am Geophys Union; Soil Conserv Soc Am. *Res:* Groundwater hydrology, hydraulics and quality; heat transfer in groundwater aquifers; mathematical modeling of hydrological systems; irrigation and drainage; pollution problems in agriculture; contaminant transport in ground water; operation of farm irrigation systems. *Mailing Add:* Dept Agr Eng Tex A&M Univ College Station TX 77843-2117

REDDEN, JACK A, b Rossville, Ill, Sept 24, 26; m 51; c 2. GEOLOGY. *Educ:* Dartmouth Col, AB, 48; Harvard Univ, MA, 50, PhD, 55. *Prof Exp:* Assoc prof geol, Va Polytech Inst & State Univ, 58-64; assoc prof geol & dir, Wright State Univ, 64-68, assoc dean sci & eng, 68-69; dir eng & mining exp sta, 69-83, PROF GEOL, SDAK SCH MINES & TECHNOL, 69- *Concurrent Pos:* Geologist, US Geol Surv, 48- *Mem:* Geol Soc Am; Mineral Soc Am. *Res:* Petrology; structure and ore deposits; specialist on stratigraphy, structure and ore deposits of Precambrian rocks of the Black Hills including uraniferous conglomerates. *Mailing Add:* SDak Sch Mines & Technol Rapid City SD 57701

REDDEN, PATRICIA ANN, b New York, NY, Sept 10, 41. PHYSICAL CHEMISTRY, EDUCATION. *Educ:* Cabrini Col, BS, 62; Fordham Univ, PhD(phys chem), 68. *Prof Exp:* Asst prof analytical chem, 68-74, assoc prof chem, 74-80, PROF CHEM, ST PETER'S COL, NJ, 80-, CHMN DEPT, 77- *Mem:* Am Chem Soc; Am Asn Univ Professors; Sigma Xi; NY Acad Sci. *Res:* Food analysis; lab safety. *Mailing Add:* 94 Duncan Ave Jersey City NJ 07306

REDDICK, BRADFORD BEVERLY, b Portsmouth, Va, May 28, 54; m 76. PLANT VIROLOGY. *Educ:* Randolph Macon Col, BS, 76; Clemson Univ, MS, 78, PhD(plant path), 81. *Prof Exp:* Res assoc, dept plant path, Univ Wis-Madison, 81-83; ASST PROF PLANT VIROL, DEPT ENTOM & PLANT PHYSIOL, UNIV TENN, KNOXVILLE, 83- *Mem:* Am Phytopath Soc; AAAS. *Res:* Plant virus problems on basic and applied levels. *Mailing Add:* Dept Entom Univ Tenn Knoxville TN 37996

REDDICK-MITCHUM, RHODA ANNE, b Waynesboro, Ga, Nov 2, 37; m 70. MEDICAL MICROBIOLOGY. *Educ:* Ga Col, AB, 58; Emory Univ, cert med technol, 59; Med Col Ga, MS, 65, PhD(med microbiol), 68; Am Bd Med Microbiol, dipl, 73. *Prof Exp:* Med technologist, Emory Univ Hosp, 59-61 & Eugene Talmadge Hosp, Med Col Ga, 62-64; fel med microbiol, Ctr Dis Control, USPHS, Atlanta, Ga, 68-70; dir, Div Diag Microbiol, Bur Labs, 70-75, DIR, DIV LAB IMPROV, SC DEPT HEALTH & ENVIRON CONTROL, 75- *Mem:* Am Soc Microbiol; Am Pub Health Asn; fel Am Acad Microbiol; Am Veneral Dis Asn; NY Acad Sci. *Res:* Development of new diagnostic procedures in medical microbiology and evaluation of products available to diagnostic microbiology laboratories. *Mailing Add:* 2605 Pine Lake Dr W Cola Columbia SC 29169

REDDING, FOSTER KINYON, b Owatonna, Minn, July 22, 29; m 60; c 4. NEUROLOGY, NEUROPHYSIOLOGY. *Educ:* Univ Pa, MD, 54; McGill Univ, PhD(neurophysiol), 64. *Prof Exp:* Physician & surgeon, Palen Clin, Minneapolis, 55-56; asst prof neurol, Sch Med, Univ Ill, 64-67; neurologist & neurophysiologist, Henry Ford Hosp, Detroit, 67-73, chief neurol, 72-73; prof neurol, Sch Med, Wayne State Univ, 73-82. *Concurrent Pos:* USPHS res grant, Sch Med, Univ Ill, 66-69. *Mem:* Am Acad Neurol; Am EEG Soc; Am Epilepsy Soc. *Res:* Relationship of rhinencephalon with brain stem reticular formation. *Mailing Add:* 18430 Mack Ave Grosse Pointe MI 48236

REDDING, JOSEPH STAFFORD, b Macon, Ga, May 29, 21; m 49; c 5. ANESTHESIOLOGY. *Educ:* Univ NC, Chapel Hill, BA, 43; Univ Md, Baltimore City, MD, 48. *Prof Exp:* Resident anesthesiol, Univ NC, 56-58; from instr to assoc prof anesthesiol, Johns Hopkins Univ, 58-70; prof anesthesiol, Univ Nebr Med Ctr, Omaha, 70-74; PROF ANESTHESIOL & HEAD, DIV RESPIRATORY/CRIT CARE, MED UNIV SC, 74- *Concurrent Pos:* Nat Heart Inst res grant, Baltimore City Hosps, 60-69, from asst chief to chief anesthesiol, Baltimore City Hosps, 58-70; assoc prof, Univ Md, 63-67, prof, 67-70. *Mem:* Am Soc Anesthesiol; Int Anesthesia Res Soc; Soc Crit Care Med; fel Am Col Physicians; Royal Soc Med. *Res:* Cardiopulmonary resuscitation; physiology of sudden death; life support measures; critical care medicine. *Mailing Add:* 236 Hobsaw Dr Charleston SC 29464

REDDING, RICHARD WILLIAM, b Toledo, Ohio, Mar 18, 23; m 46; c 3. VETERINARY PHYSIOLOGY, PHARMACOLOGY. *Educ:* Ohio State Univ, DVM, 46, MSc, 50, PhD(vet physiol & pharmacol), 57. *Prof Exp:* Instr vet surg, Ohio State Univ, 48-50; asst prof, Univ Calif, 50-51 & Univ Ga, 51-53; from instr to assoc prof vet physiol & pharmacol, Ohio State Univ, 53-63, prof, Col Vet Med & Grad Sch, 63-68; prof small animal surg & med & physiol & pharmacol, Auburn Univ, 68-85; RETIRED. *Concurrent Pos:* Consult, Martin Co, 63 & USN Radiol Defense Labs, 66- *Mem:* AAAS; Am Vet Med Asn; Am Soc Vet Physiol & Pharmacol; Am Asn Vet Neurol (secy-treas, 72-). *Res:* Techniques, application and use of electroencephalograph in canine diagnosis; surgical control of behavior in the canine species. *Mailing Add:* 449 Camellia Dr Auburn AL 36830

REDDING, ROGERS WALKER, b Louisville, Ky, July 15, 42; m 66; c 2. PHYSICAL CHEMISTRY, PHYSICS. *Educ:* Ga Inst Technol, BS, 65; Vanderbilt Univ, PhD(chem), 69. *Prof Exp:* Nat Acad Sci fel, Nat Bur Standards, Washington, DC, 69-70; from asst prof to assoc prof, Univ NTex, 70-79, chmn dept, 80-87, dir, 87-89, PROF PHYSICS, TEX ACAD MATH & SCI, UNIV NTEX, 79- *Mem:* AAAS; Am Phys Soc; Am Asn Physics Teachers. *Res:* Theoretical molecular physics; molecular spectroscopy. *Mailing Add:* Col Arts & Sci Box 5187 Univ NTex Denton TX 76203-5187

REDDISH, PAUL SIGMAN, b Durham, NC, June 8, 10; m 36. BIOLOGY. *Educ:* Duke Univ, AB, 33, AM, 35. *Prof Exp:* Head sci dept, NC High Sch, 35-43; instr eng physics, NC State Col, 43-44, asst, Off Sci Res & Develop, 44; assoc cult dept, Carolina Biol Supply Co, 44-45; from asst prof to prof, Elon Col, 45-76, chmn dept, 73-75; RETIRED. *Concurrent Pos:* Consult microbiol, 76- *Mem:* Am Micros Soc. *Res:* Ecology of protozoa; physiology; bacteriology. *Mailing Add:* PO Box 275 Elon College NC 27244

REDDISH, ROBERT LEE, animal science; deceased, see previous edition for last biography

REDDOCH, ALLAN HARVEY, b Montreal, Que, Jan 19, 31; m 70. CHEMICAL PHYSICS, SEMICONDUCTOR MATERIALS. *Educ:* Queen's Univ, Ont, BSc, 53, MSc, 55; Univ Calif, Berkeley, PhD(phys chem), 60. *Prof Exp:* Fel chem, Nat Res Coun Can, 59-61, from asst res officer to assoc res officer, Div Chem, 61-74, sr res officer, 74-86, SR RES OFFICER, INST MICROSTRUCT SCI, NAT RES COUN CAN, 86- *Concurrent Pos:* Lectr, Univ Ottawa, 61-65; adj prof, Univ Waterloo, 86-90. *Mem:* Am Phys Soc; Chem Inst Can. *Res:* Electron spin resonance of organic radicals in solution; organic charge-transfer crystals; hydrogen-bonded ferroelectrics; semimagnetic semiconductors; spin dynamics phase transitions. *Mailing Add:* Inst Microstruct Sci Nat Res Coun Can Ottawa ON K1A 0R6 Can

REDDY, BANDARU SIVARAMA, b Nellore, India, Dec 30, 32; m 62; c 3. BIOCHEMISTRY, NUTRITION. *Educ:* Madras Univ, BVSc, 55; Univ NH, MS, 60; Mich State Univ, PhD(biochem), 63. *Prof Exp:* Vet surg & med, Andhra Pradesh Govt, India, 55-58; res assoc biochem, 63-65, res scientist, 65-68, assoc res prof, Lobund Lab, Univ Notre Dame, 68-72; HEAD DEPT NUTRIT BIOCHEM, AM HEALTH FOUND, 71-, ASSOC CHIEF, DIV NUTRIT, 79-; RES PROF MICROBIOL, NY MED COL, 76- *Concurrent Pos:* NIH res grants, 65. *Honors & Awards:* Tana Award, 85; Tokten Prog, United Nations, 87. *Mem:* Am Inst Nutrit; Am Asn Pathologists; Asn Gnotobiotics; Soc Exp Biol & Med; Am Asn Cancer Res; Soc Toxicol. *Res:* Effect of intestinal microflora on the nutritional biochemistry of host; role of bile acids and microflora on the etiology of colon cancer; chemical carcinogenesis; nutrition and cancer; nutritional toxicology; mechanism of carcinogenesis; primary and secondary prevention of colon cancer. *Mailing Add:* Am Health Found Mo One Dana Rd Valhalla NY 10595

REDDY, CHURKU MOHAN, b Kothapalli, India, Aug 3, 42; c 2. PEDIATRIC ENDOCRINOLOGY. *Educ:* Osmania Univ, India, MB, BS, 66; Am Bd Pediat, dipl, 75, cert endocrinol, 83. *Prof Exp:* Med officer, Primary Health Ctr, India, 68-70; asst instr pediat, State Univ NY, Downstate Med Ctr, 71-75; from asst prof to assoc prof, 75-82, PROF PEDIAT, MEHARRY MED COL, 82-, DIR DIV ENDOCRINOL & METAB, 75- *Concurrent Pos:* Fel pediat endocrinol & metab, State Univ NY, Downstate Med Ctr, Kings County Hosp Ctr, 73-75. *Mem:* Fel Am Acad Pediat; AMA. *Res:* Clinical research. *Mailing Add:* Dept Pediat Meharry Med Col Nashville TN 37208

REDDY, GADE SUBBARAMI, b Aluru, India, May 20, 35; m 56; c 4. PHYSICAL CHEMISTRY. *Educ:* Andhra Univ, India, BSc, 54; Benares Hindu Univ, MSc, 56; Emory Univ, PhD(phys chem), 60. *Prof Exp:* Govt India sr res scholar phys chem, Benares Hindu Univ, 56-57; res assoc spectros, Emory Univ, 60-62; res chemist, Eastern Lab, 62-66, RES CHEMIST, CENT RES DEPT, EXP STA, E I DU PONT DE NEMOURS & CO, INC, 66- *Mem:* Am Chem Soc. *Res:* Nuclear magnetic resonance spectroscopy; nuclear magnetic resonance in liquid crystals; kinetics and reaction mechanisms by nuclear magnetic resonance; nuclear magnetic resonance in biological systems. *Mailing Add:* Cent Res Dept Exp Sta E I du Pont de Nemours & Co Inc Wilmington DE 19898

REDDY, GUNDA, b Cheekodu, India; US citizen; m 76. ENTOMOLOGY, PHARMACOLOGY. *Educ:* Osmania Univ, India, BSc, 60, MSc, 62, PhD(zool, entom), 68; Am Bd Toxicol, dipl, 84. *Prof Exp:* Fel pesticide & residues, dept entom, Univ Ky, 69-70; res assoc metab ammonia & calcium, dept biol, Rice Univ, 70-71; vis scholar insect hormones, dept biol, Marquette Univ, 71-73; res assoc pesticide metab, dept biol, Univ Ill, 73-76; res assoc insect hormones, Marquette Univ, 76-78; res assoc PCB metab & path, dept path, Univ Wis-Madison, 78-83; RES TOXICOLOGIST, METAB MUNITION COMPOUNDS, HEALTH & ENVIRON EFFECTS, US ARMY, FT DETRICK, MD, 83- *Mem:* Entom Soc Am; Soc Toxicol; Am Chem Soc; Indian Sci Cong; Am Col Toxicol; Soc Environ Toxicol & Chem. *Res:* Action and metabolism of pesticides; insect endocrine interactions; environmental and health effect of pollutants and munition chemicals; biochemical toxicology. *Mailing Add:* Health Effects Res Div US Army Biomed Res & Develop Lab Ft Detrick Frederick MD 21702-5010

REDDY, JANARDAN K, b India, Oct 7, 38; m 62; c 2. PATHOLOGY. *Educ:* Osmania Univ, India, MB, BS, 61; All India Inst Med Sci, MD, 65. *Prof Exp:* From asst prof to assoc prof path & oncol, Univ Kans Med Ctr, Kansas City, 70-76; PROF PATH, NORTHWESTERN UNIV MED SCH, CHICAGO, 76- *Concurrent Pos:* United Nations Tokten Scholar to India; Assoc Ed, Cancer Res, J Toxicol, Env Health. *Honors & Awards:* Fel Yamagina-Yoshida Cancer; NIH Merit Award. *Mem:* AAAS; Am Asn Pathologists; Soc Exp Biol & Med; Histochem Soc; Am Soc Cell Biol; Biochemical Soc, UK. *Res:* Experimental chemcial carcinogenesis; effect of drugs and carcinogens on the structure and function of liver and pancreas; role of hypolipidemic drugs on the induction of peroxisome proliferation in liver cells. *Mailing Add:* Dept Path Sch Med Northwestern Univ 303 E Chicago Ave Chicago IL 60611

REDDY, JUNUTHULA N, b Warangal, India, Aug 12, 45; m 68; c 2. MECHANICAL ENGINEERING, APPLIED MECHANICS. *Educ:* Osmania Univ, India, BE, 68; Okla State Univ, MS, 70; Univ Ala, Huntsville, PhD(appl mech), 73. *Prof Exp:* Res assoc mech, Univ Ala, 70-73; fel, Univ Tex, Austin, 73-74; res assoc aeromech, Lockheed Missiles & Space Co, 74-75; from asst prof to assoc prof mech eng, Univ Okla, 75-80; prof, 80-85, CLIFTON C GARVIN PROF, VA POLYTECH INST, 85- *Concurrent Pos:* Consult, Battelle Columbus Labs, Ohio, Alcoa Tech Labs & Gen Dynamics; res grants, NSF, Off Naval Res, NASA, Army Res Off & Air Force Off Sci Res. *Honors & Awards:* Teetor Award Res & Teaching, Soc Automotive Engrs, 76; Univ Okla Res Award, 79; Huber Prize, Am Soc Civil Engrs, 84. *Mem:* Am Acad Mech; fel Am Soc Mech Engrs; Soc Eng Sci; Am Soc Civil Engrs. *Res:* Analysis of composite structural components; computational fluid mechanics; theory and application of the finite element method in solid and fluid mechanics; variational methods and composite materials. *Mailing Add:* Dept Eng Sci Va Polytech Inst 220 Norris Blacksburg VA 24061

REDDY, KALLURU JAYARAMI, b Pidugupalli, Andhra Pradesh, Apr 15, 53; m 84; c 1. GENE EXPRESSION, BIOLOGICAL NITROGEN FIXATION. *Educ:* Sri Venkateswara Univ, BSc, 72, MSc, 74; Univ Miami, PhD(marine biol), 84. *Prof Exp:* Jr res fel, Indian Inst Technol, 74-76; jr plant physiologist, Indian Agr Res Inst, New Delhi, 76-80; res asst, Univ Miami, 80-84; fel molecular biol, Univ Mo, Columbia, 85-89; res scientist, Purdue Univ, West Lafayette, 89-90; ASST PROF MICROBIOL, STATE UNIV NY, BINGHAMTON, 90- *Mem:* Am Soc Microbiol; Am Soc Plant Physiol. *Res:* Molecular genetics of nitrogen fixation in unicellular cyanobacteria. *Mailing Add:* Dept Biol Sci State Univ NY Binghamton NY 13902-6000

REDDY, KAPULURU CHANDRASEKHARA, b Nellore, India, Aug 20, 42; m 67; c 1. APPLIED MATHEMATICS, FLUID MECHANICS. *Educ:* V R Col, India, BA, 59; Sri Venkateswara Univ, India, MSc, 61; Indian Inst Technol, Kharagpur, MTech, 62, PhD(appl math), 65. *Prof Exp:* Assoc lectr math, Indian Inst Technol, Kharagpur, 64-65; instr aerospace eng, Univ Md, College Park, 65-66; from asst prof to assoc prof, 66-75, PROF MATH, SPACE INST, UNIV TENN, TULLAHOMA, 75-, DEAN, ACAD AFFAIRS, 90- *Concurrent Pos:* Staff engr, Lockheed Electronics Co, Houston, 69 & 70; consult, US Army Res Off, Durham, 73 & Lockheed Ga Co, Marietta, Ga, 76; res engr, AROC, Inc & CALSPAN, Arnold Eng Develop Ctr, 77- *Mem:* Am Inst Aeronaut & Astronaut; Sigma Xi; Soc Indust & Appl Math. *Res:* Computational fluid mechanics; transonic flow problems; numerical analysis; boundary layers. *Mailing Add:* Dean's Off Space Inst Univ Tenn Tullahoma TN 37388

REDDY, MOHAN MUTHIREVAL, b Chittor, India, Apr 25, 42; US citizen; m 72; c 2. IMMUNOLOGY, AIDS. *Educ:* Univ Chicago, PhD(biol), 69. *Prof Exp:* DIR, ALLERGY & CLIN IMMUNOL, ST LUKES ROOSEVELT HOSP CTR, 79- *Mem:* Am Asn Immunologist; Am Asn Clin Pathologists. *Res:* Rol of immunology in diseases. *Mailing Add:* Dept Allergy & Immunol St Lukes Roosevelt Hosp Ctr 428 W 59th St New York NY 10019

REDDY, NARENDER PABBATHI, b Karimnagar, India, May 5, 47; m 76; c 2. BIOMEDICAL ENGINEERING, MECHANICAL ENGINEERING. *Educ:* Osmania Univ, India, BE, 69; Univ Miss, MS, 71; Tex A&M Univ, PhD(bioeng), 74. *Prof Exp:* Res asst, Tex A&M Univ, 71-74, res assoc bioeng, 74-75; res assoc rehab eng, Baylor Col Med, 75-76; res physiologist, Univ Calif, San Francisco, 77-78; sr res scientist, Biomech Res Unit, Helen Hays Hosp, NY, 78-81; assoc prof, 81-89, PROF, DEPT BIOMED ENG, UNIV AKRON, 89- *Concurrent Pos:* Res fel, Cardiovasc Res Inst, Univ Calif, San Francisco, 77-78; adj assoc prof, Rensselaer Polytech Inst, 79-81; adj staff, Edwin Shaw Hosp, Akron, Ohio, 84-; chmn, tech sessions, numerous nat & int sci meetings & confs. *Mem:* Biomed Eng Soc; Am Soc Mech Engrs; Am Soc Eng Educ; Rehab Eng Soc NAm. *Res:* Biomechanical engineering; computer modeling; cardiopulmonary biomechanics; medical devices; rehabilitation engineering; orthopedic biomechanics. *Mailing Add:* Biomed Eng Dept Col Eng Univ Akron Akron OH 44325-0302

REDDY, PADALA VYKUNTHA, b Eletipadu, India, May 7, 46; m 75; c 2. PHOSPHOLIPID METABOLISM. *Educ:* Andhra Pradesh Agr Univ, BS, 68; G B Pant Univ Agr & Technol, 72; Indian Inst Sci, Bangalore, PhD(biochem), 77. *Prof Exp:* Res fel endocrinol, Dept Obstet & Gynec, Hormel Inst, Ann Arbor, Mich, 78-79, res fel lipid biochem, Austin, Minn, 79-81, res assoc, 82-86; asst prof lipid biochem, Dept Biochem, Univ Minn, Duluth, Minn, 86-88; SR SCIENTIST, CIBA-CORNING DIAGNOSTICS, OBERLIN, OHIO, 88- *Honors & Awards:* Mrs Hanuman-Tha Rao Medal, Coun Indian Inst Sci, Bangalore, 80. *Res:* Metabolism of complex lipids in the brain, heart, platelets and peripheral nerve; myelination and demyelination; nutritional stress, diabetes and heart disease; phospholipases, acyltransferases base exchange and transacylation reactions involving phospholipids; clinical diagnostics assay development. *Mailing Add:* Ciba Corning Diagnostics 132 Artino St Oberlin OH 44074-1293

REDDY, PARVATHAREDDY BALARAMI, b Nellore, India, Dec 1, 42; US citizen; m 67; c 2. CONTROL SYSTEMS, ELECTRICAL ENGINEERING. *Educ:* Sri Venkateswara Univ, BE hons, 65; Indian Inst Technol, MTech, 67; Rutgers Univ, PhD(elec eng controls), 73. *Prof Exp:* Res & teaching asst elec eng, Rutgers Univ, 69-73; sr systs analyst control & navig, Dynamics Res Corp, 73-78; mem tech staff navig & guid, Litton Guid & Control Systs, 78-80; res scientist, Teledyne Systs Co, 80-85; PRES, APEX TECHNOL INC, 85- *Mem:* Inst Elec & Electronics Engrs. *Res:* Guidance, navigation, statistical modeling; modeling and simulation with general modeling interest; strapdown systems using ring laser gyros. *Mailing Add:* Apex Technol Inc 347 Bell Canyon Rd Bell Canyon CA 91307-1116

REDDY, RAJ, b Katoor, India, June 13, 37; US citizen; m 66; c 2. COMPUTER SCIENCE. *Educ:* Univ Madras, BE, 58; Univ New South Wales, MTech, 61; Stanford Univ, PhD(comput sci), 66. *Prof Exp:* Appl sci rep, IBM Corp, Australia, 60-63; asst prof computer sci, Stanford Univ, 66-69; from assoc prof to prof, 69-84, UNIV PROF, CARNEGIE-MELLON UNIV, PA, 84-, DIR, ROBOTICS INST, 80- *Concurrent Pos:* Consult, Litton Indust, 68-69, Stanford Res Inst, 70-71, Palo Alto Res Ctr, Xerox Corp, 70-78, NSF, 72-76, ITT, 78-79, Jet Propulsion Labs, 78-79, Gen Motors, 78-80, Rand Corp, 78-80 & var other co, 78-; John Guggenheim fel, 75-76; chmn, bd trustees, Int Joint Coun Artificial Intel, 77-79; mem, gov bd, Cognitive Sci Soc, 79-84, adv bd, Trans Pattern Anal & Mach Intel, Inst Elec & Electronics Engrs, 79-80, bd dirs, Robot Inst Am, 82 & mfg studies bd, Nat Res Coun, 83-86; vchmn study group mach intel & robotics, NASA, 79-80; vpres & chief scientist, World Ctr Comput Sci & Human Resources, 82-; pres, Am Asn Artificial Intel, 87-89. *Mem:* Nat Acad Eng; fel Acoust Soc Am; fel Inst Elec & Electronics Engrs; Asn Comput Mach; Asn Comput Ling. *Res:* Speech and visual input to computers; graphics; man-machine communication; artificial intelligence; robotics; computer science. *Mailing Add:* Sch Computer Sci Carnegie-Mellon Univ Pittsburgh PA 15213

REDDY, RAMAKRISHNA PASHUVULA, b KalliaKurthy, India, June 30, 36; US citizen; m 61; c 2. ANIMAL HEALTH, IMMUNOLOGY. *Educ:* Osmania Univ, BVSc & AH, 59; Indian Vet Res Inst, MVSc, 68, Va Polytechnic Inst & State Univ, MS, 71, PhD(genetics), 76. *Prof Exp:* Dir gen, 77-80, dir res & develop, 80-87, VPRES & DIR RES & DEVELOP, PETERSON INDUSTS INC, 87- *Concurrent Pos:* Pres & exec comt, Poultry Breeders Am, 79-83; sect chairperson genetics, Poultry Sci Asn, 82, 91; exec comt & pres, Nat Breeders Roundtable, 86-89; assoc ed, Poultry Sci J, 87-92; chmn, Nat Broiler Coun Poultry Sci Asn Res Award, 89-91. *Mem:* Sigma Xi; Poultry Sci Asn; World Poultry Sci Asn; Am Registry Cert Animal Scientists. *Res:* Broiler breeding research involving the genetic improvement of growth, behavior, reproductive efficiency for commercial production of broiler meat; balancing the fitness and nonfitness characteristics for the most efficient and low cost production of broiler meat under diverse environmental conditions. *Mailing Add:* Peterson Industs Res & Develop PO Box 248 Decatur AR 72722

REDDY, REGINALD JAMES, b Staten Island, NY, July 16, 34. MOLECULAR SPECTROSCOPY. *Educ:* St Bonaventure Univ, BA, 57; Holy Name Col, STM, 61; Univ SC, PhD(physics), 77. *Prof Exp:* Instr physics & math, Archbishop Walsh High Sch, 61-65; teaching asst, Univ SC, 68-73; asst prof, 73-89, ASSOC PROF PHYSICS, SIENA COL, NY, 89- *Mem:* Am Asn Physics Teachers; Nat Sci Teachers Asn; Sigma Xi. *Res:* Flash photolysis studies of various polycyclic hydrocarbons including triplet-triplet transitions, quenching of the triplet states by various impurities, energy transfer between triplets. *Mailing Add:* Dept Physics Siena Col Loudonville NY 12211

REDDY, SATTI PADDI, b Sept 1, 32; Can citizen; m 55; c 3. PHYSICS, MOLECULAR SPECTROSCOPY. *Educ:* Andhra Univ, India, BSc, 54, MSc, 55. *Hon Degrees:* DSc, Andhra Univ, India, 59. *Prof Exp:* Lectr physics, Andhra Univ, India, 59-61; res assoc, Univ Toronto, 61-63; res fel, 63-64, from asst prof to assoc prof, 64-72, PROF PHYSICS, MEM UNIV NFLD, 72- *Concurrent Pos:* Sr res fel, Andhra Univ, India, 60-61; vis prof physics, Ohio State Univ, 77-78. *Mem:* Fel Brit Inst Physics; Can Asn Physicists; fel Am Phys Soc. *Res:* Molecular physics; infrared, laser and optical spectroscopy. *Mailing Add:* Dept Physics Mem Univ Nfld St John's NF A1B 3X7 Can

REDDY, SUDHAKAR M, b Gadwal, India, Jan 5, 38; m 63. ELECTRICAL ENGINEERING. *Educ:* Osmania Univ, India, BS, 58 & 62; Indian Inst Sci, MS, 63; Univ Iowa, PhD(elec eng), 68. *Prof Exp:* Asst prof elec eng, 68-77, PROF INFO ENG, UNIV IOWA, 77- *Concurrent Pos:* NSF grant, 69-71. *Mem:* Inst Elec & Electronics Engrs; Sigma Xi. *Res:* Coding theory; digital systems. *Mailing Add:* Dept Elec Eng Univ Iowa Iowa City IA 52240

REDDY, THOMAS BRADLEY, b Amesbury, Mass, Sept 11, 33; m 88; c 3. LITHIUM BATTERIES. *Educ:* Yale Univ, BS, 55; Univ Minn, PhD(phys chem), 60. *Prof Exp:* Res assoc chem, Univ Ill, 59-61; mem tech staff electrochem, Bell Tel Labs, 61-65; proj leader, Am Cyanamid Co, 65-74, prin res chemist, Stamford Res Labs, 74-79; dir-battery-technol, Stonehart Assocs, Inc, Madison, Conn, 79-80; dir technol, 80-88, dir mil mkt, 88-89, VPRES, POWER CONVERSION, INC, SADDLE BROOK, NJ, 89- *Concurrent Pos:* Rice fel, G E Educ & Charilabe Found, 58-59. *Mem:* Am Chem Soc; Electrochem Soc. *Res:* Lithium battery technology; environmental science; fused salts; electrochromic display devices. *Mailing Add:* 30 Elm Rock Rd Bronxville NY 10708

REDDY, VENKAT N, b Hyderabad, India, Nov 4, 22; nat US; m 55; c 2. BIOCHEMISTRY. *Educ:* Madras Univ, BSc, 45; Fordham Univ, MS, 49, PhD(biochem), 52. *Prof Exp:* Asst chem, Fordham Univ, 48-50, sr asst biochem, 50-52; res asst obstet & gynec, Columbia Univ, 52-54, res assoc, 54-56, res fel, Banting & Best Inst, Can, 56; asst prof ophthalmic biochem, Kresge Eye Inst, 57-61, assoc prof ophthal, 61-77, PROF BIOMED SCI & DIR EYE RES INST, OAKLAND UNIV, 75- *Concurrent Pos:* Mem visual sci study sect, NIH, 66-70, consult, Cataract Workshop, 73; asst dir inst biol sci, Oakland Univ, 68-75; Nat Acad Sci-Nat Res Coun Comt Vision, 71-; consult, Nat Adv Eye Coun Vision Res Prog Planning Comt, 74; bd sci counr, Nat Eye Inst, 78-82. *Honors & Awards:* Fight for Sight Citation, Nat Coun Combat Blindness, 68; Friedenwald Award, Asn Res Vision & Ophthal, 79. *Mem:* AAAS; Am Soc Biol Chemists; NY Acad Sci; Asn Res Vision & Ophthal; Brit Biochem Soc; Sigma Xi. *Res:* Transport mechanisms; aqueous humor dynamics; metabolism of ocular tissues; cataract; biochemistry of the lens; monoclonal antibodies. *Mailing Add:* Eye Res Inst Oakland Univ Rochester MI 48063

REDDY, WILLIAM L, b Albany, NY, Dec 15, 38; div; c 1. TOPOLOGY. *Educ:* Siena Col, BS, 60; Syracuse Univ, MA, 62, PhD(math), 64. *Prof Exp:* Asst prof math, Webster Col, 64-65 & State Univ NY Albany, 67-68; from asst prof to assoc prof, 68-75, chmn dept, 74 & 77-80, assoc dir, Grad Summer Sch, 76-79, PROF MATH, WESLEYAN UNIV, 75-, ACTG CHMN, COMPUT CTR, 80- *Concurrent Pos:* Mem, Inst Advan Study, 69. *Mem:* AAAS; Math Asn Am; Am Math Soc. *Res:* Branched coverings; topological dynamics. *Mailing Add:* Dept Math Wesleyan Univ Middletown CT 06457

REDEI, GYORGY PAL, b Vienna, Austria, June 14, 21; m 53; c 1. GENETICS, GENETIC ENGINEERING. *Educ:* Magyarovar Acad Agr, Hungary, dipl, 48; Univ Agr Sci, Hungary, dipl, 49; Hungarian Acad Sci, CSc, 55. *Prof Exp:* Res asst, Nat Inst Plant Breeding, Magyarovar, Hungary, 48, Inst Genetics, Hungarian Acad Sci, 49 & Agr Exp Sta, Kisvarda, 50; res adminr, Ministry Agr, Budapest, 51; res assoc, Inst Genetics, Hungarian Acad Sci, 52-56; from asst prof to assoc prof, 57-69, PROF GENETICS, UNIV MO, COLUMBIA, 69-, CHMN GENETICS PROG, 90- *Concurrent Pos:* Grants, NSF, 59, 61, 63, 65, 69 & 89, NIH, 63 & 64, AEC, 66, 67 & 68, NATO, 75 & 79 & Environ Protection Agency, 79, 80, 83, 85, USDA, 86; vis prof, Max-Planck Inst, Cologne, Ger, 86. *Honors & Awards:* Sr Res Award, Sigma Xi, 89. *Mem:* Genetics Soc Am; foreign mem, Hungarian Nat Acad Sci. *Res:* Physiological genetics of Arabidopsis; thiamine auxotrophy; regulation of gene activity by metabolites and antimetabolites; genetics of organelles; mutation; detection of carcinogens and mutagens; transformation; author of textbook, Genetics, 82. *Mailing Add:* 117 Curtis Hall Univ Mo Columbia MO 65211

REDEKER, ALLAN GRANT, b Lincoln, Nebr, Sept 10, 24; m 50, 79; c 3. MEDICINE. *Educ:* Northwestern Univ, BS, 49, MD, 52. *Prof Exp:* From instr to assoc prof, 58-69, PROF MED, UNIV SOUTHERN CALIF, 69- *Concurrent Pos:* Schweppe Found res fel, Sch Med, Univ Southern Calif, 54-56; Bank of Am-Giannini Found res fel, 56-57; Bank of Am-Giannini Found traveling res fel, Minn, London & Malmo Clins, Sweden, 57; Lederle med fac award, 59-62; USPHS career develop award, 62-69; mem attend staff, Los Angeles County Hosp, 55- & Rancho Los Amigos Hosp. *Mem:* Am Fedn Clin Res; Am Soc Clin Invest; Am Asn Study Liver Dis (pres, 71); Int Soc Study Liver; Asn Am Physicians; Am Gastroenterol Asn. *Res:* Hepatic physiology and diseases of the liver; bilirubin and pyrrole pigment metabolism. *Mailing Add:* Dept Med Univ Southern Calif Sch Med 2025 Zonal Ave Los Angeles CA 90033

REDENTE, EDWARD FRANCIS, b Derby, Conn, Feb 18, 51; m 73; c 3. LAND RESTORATION, PLANT ECOLOGY. *Educ:* Western Mich Univ, BA, 72; Colo State Univ, MS, 74, PhD(range ecol), 80. *Prof Exp:* Environ engr, Utah Int Inc, 74-76; res assoc, 76-79, instr, 79-80, asst prof mined land reclamation, 80-87, PROF RANGE SCI, RANGE SCI DEPT, COLO STATE UNIV, 88- *Concurrent Pos:* Consult, Thorne Ecol Inst, 76-77, Colo State Dept Natural Resources, 80-81; prin investr, Oil Shale Reclamation Proj, 80-84, mechanisms of secondary succession, US Dept Energy, 84- *Mem:* Soc Range Mgt; Ecol Soc Am; Sigma Xi; Am Soc Surface Mining & Reclamation. *Res:* Primary and secondary successional processes that occur on arid and semiarid land disturbed by energy development; determine effects of revegetation practices on rate and direction of plant succession and interaction between succession and soil microbial processes. *Mailing Add:* Dept Range Sci Colo State Univ Ft Collins CO 80523

REDER, FRIEDRICH H, b Garsten, Austria, Dec 9, 19; US citizen; m 52; c 1. PHYSICS. *Educ:* Graz Univ, MS, 47, PhD(physics), 49. *Prof Exp:* Asst physics, Graz Univ, 48-52; physicist, Frequency Control Div, US Army Electronics Labs, 53-62, sr scientist, Inst Explor Res, 62-71, chief antennas & geophys res area, Electronics Technol & Devices Lab, 71-73, chief commun res technol area, Commun/ADP Lab, 73-78; res physicist, Ctr Commun Systs, Commun Res & Develop Command, 78-80; CONSULT, OMEGA RADIO NAVIG, 80- *Concurrent Pos:* UNESCO fel plasma physics, Radiation Lab Electronics, Mass Inst Technol, 50-51, US Indust fel, 51; mem laser comt, US Dept Defense, 61-62. *Honors & Awards:* Res & Develop Ann Achievement Award, Army Materiel Command, 70. *Res:* Atomic frequency, time control and propagation of very low frequency electromagnetic waves. *Mailing Add:* 480 Marvin Dr Long Branch NJ 07740

REDETZKI, HELMUT M, b Memel, Lithuania, Sept 23, 21; nat US; m 57. PHARMACOLOGY. *Educ:* Univ Hamburg, MD, 48; Am Bd Med Toxicol, dipl, 75. *Prof Exp:* Intern, Med Sch, Univ Hamburg, 47-48, res assoc biochem, 48-51, res assoc virol, Poliomyelitis Res Inst, 51-52; resident internal med, St George Hosp, Hamburg, Ger, 52-56; instr pharmacol, Univ Tex Med Br Galveston, 59-60, asst prof, 60-61; assoc prof, Sch Med, La State Univ, New Orleans, 61-66, prof 66-68; PROF PHARMACOL & HEAD DEPT, SCH MED, LA STATE UNIV, SHREVEPORT, 68- *Concurrent Pos:* McLaughlin Found fel, Tissue Metab Res Lab, Univ Tex Med Br Galveston, 56-59; proj & med dir, La Regional Poison Control Ctr, 77- *Honors & Awards:* Dehnecke Medal & Award, Univ Hamburg, 55; H M Hub Cotton Faculty Excellence Award, 75. *Mem:* Fel Am Acad Clin Toxicol; Soc Exp Biol & Med; Am Heart Asn; fel Am Col Physicians; Am Soc Clin Pharmacol & Therapeut; Am Acad Clin Toxicol (pres, 84-86). *Res:* Alcohol metabolism; cancer chemotherapy; drug-enzyme interactions; clinical toxicology. *Mailing Add:* Dept Pharmacol & Therapeut Sch Med La State Univ PO Box 33932 Shreveport LA 71130

REDFEARN, PAUL LESLIE, JR, b Sanford, Fla, Oct 5, 26; m 49; c 2. BRYOLOGY. *Educ:* Fla Southern Col, BS, 48; Univ Tenn, MS, 49; Fla State Univ, PhD, 57. *Prof Exp:* Instr bot, Univ Fla, 50-51; PROF LIFE SCI, SOUTHWEST MO STATE UNIV, 57- *Concurrent Pos:* Res assoc, Mo Bot Garden, 74. *Honors & Awards:* Burlington Res Award, Southwest Mo State Univ, 87. *Mem:* Fel AAAS; Am Bryol & Lichenological Soc (pres, 71-73); Ecol Soc Am; Am Soc Plant Taxonomists. *Res:* Taxonomy and ecology of bryophytes; interior highlands of NA, China. *Mailing Add:* Dept Life Sci Southwest Mo State Univ Springfield MO 65802

REDFERN, ROBERT EARL, b Blackburn, Ark, July 29, 29; m 55; c 4. ENTOMOLOGY. *Educ:* Univ Ark, BS, 59, MS, 61. *Prof Exp:* Entomologist, Fruit & Veg Res Br, Ind, USDA, 60-62, entomologist-in-charge, Pesticide Chem Res Br, Tex, 63-68, Md, 68-72, chief biol eval chem lab, 72-74, res entomologist, biol eval chem lab, Agr Environ Qual Inst, 74-89; RETIRED. *Mem:* Entom Soc Am. *Res:* Insects affecting deciduous fruits; primary screening of juvenile and molting hormones. *Mailing Add:* Sci & Educ Admin Beltsville Agr Res Ctr-East Beltsville MD 20705

REDFIELD, ALFRED GUILLOU, b Boston, Mass, Mar 11, 29; m 60; c 3. PHYSICAL BIOCHEMISTRY. *Educ:* Harvard Univ, BA, 50; Univ Ill, MA, 52, PhD(physics), 53. *Prof Exp:* Univ fel appl sci, 54-55; physicist, Watson Lab, IBM Corp, 55-71; PROF PHYSICS & BIOCHEM, BRANDEIS UNIV, 72-, PROF, ROSENSTIEL BASIC MED SCI RES CTR, 77- *Concurrent Pos:* Assoc, Columbia Univ, 55-61, adj asst prof, 61-63, adj prof, 63-71; vis physicist, AEC, Saclay, France, 60-61; Miller vis prof, Univ Ill, Urbana, 61; mem corp, Woods Hole Oceanog Inst, 64-; vis physicist, Univ Calif, Berkeley, 70-71, NSF sr fel, 71-72; Little vis prof, Mass Inst Technol, 72; staff mem, Rosensteil Ctr Basic Med Res, Brandeis Univ, 72- *Honors & Awards:* Patterson lectr, Inst Cancer Res, 81. *Mem:* Nat Acad Sci; fel Am Phys Soc; Am Soc Biol Chemists; Am Chem Soc; Am Acad Sci. *Res:* Nuclear magnetic resonance and relaxation; protein catalysis and electronic structure; superconductivity. *Mailing Add:* Dept Biophys Brandeis Univ Waltham MA 02154

REDFIELD, DAVID, b New York, NY, Sept 20, 25; m 50; c 2. OPTOELECTRONIC MATERIALS, SOLAR ENERGY. *Educ:* Univ Calif, Los Angeles, AB, 48; Univ Md, MS, 53; Univ Pa, PhD(physics), 56. *Prof Exp:* Electronic scientist, Nat Bur Standards, 49-52; sr res physicist, Union Carbide Corp, 55-64; assoc prof elec eng, Columbia Univ, 64-67; res physicist, RCA Corp, 67-85; CONSULT PROF, STANFORD UNIV, 85- *Mem:* AAAS; fel Am Phys Soc; sr mem Inst Elec & Electronics Engrs; Fedn Am Scientists. *Res:* Optical and electronic properties of solids; effects of surfaces, defects; optical and transport properties of disordered semiconductors; solar energy. *Mailing Add:* Dept Mat Sci & Eng Stanford Univ Stanford CA 94305-2205

REDFIELD, JOHN A(LDEN), b Orange, NJ, Mar 18, 33; m 54; c 3. NUCLEAR ENGINEERING. *Educ:* Univ Cincinnati, BS, 55; Univ Pittsburgh, MS, 60, PhD(chem eng), 63. *Prof Exp:* Assoc engr, 57-59, from engr to sr engr, 59-66, fel engr, 66-69, adv engr, 69-71, mgr thermal-hydraulic develop, 71-73, mgr A4W reactor eng, 73-75, mgr LWBR technol, 75-77, mgr reactor technol, 77-78, mgr advan water breeder, 78, MGR LIGHT WATER BREEDER REACTOR, BETTIS ATOMIC POWER LAB, WESTINGHOUSE ELEC CORP, 78- *Mem:* Am Inst Chem Engrs; Am Nuclear Soc; Inst Elec & Electronics Engrs; Sigma Xi. *Res:* Heat transfer; fluid flow; systems analysis; reactor plant kinetics; dynamics of physical systems. *Mailing Add:* 2620 Quail Hill Dr Pittsburgh PA 15241

REDFIELD, ROSEMARY JEANNE, b Vancouver, BC, Aug 26, 48. EVOLUTION OF SEX, MOLECULAR EVOLUTION. *Educ:* Monash Univ, Australia, BSc, 77; McMaster Univ, Can, 80; Stanford Univ, PhD(biol sci), 86. *Prof Exp:* Postdoctoral evol, Harvard Univ, 87; postdoctoral molecular biol, Sch Med, Johns Hopkins Univ, 88-90; ASST PROF BIOCHEM, UNIV BC, CAN, 90- *Concurrent Pos:* Scholar, Can Inst Advan Res, 90- *Mem:* Genetics Soc Am; Am Soc Microbiol. *Res:* Regulation of competence in naturally transformable bacteria; mechanisms of genetic exchange and recombination in bacteria; chromosome structure and organization; evolution of sex; deep phylogeny. *Mailing Add:* Dept Biochem Univ BC Vancouver BC V6T 1Z3 Can

REDFIELD, WILLIAM DAVID, b Portland, Ore, July 26, 15; m 41, 67; c 2. MICROBIOLOGY, ORGANIC GEOCHEMISTRY. *Educ:* Univ Calif, Los Angeles, AB, 39, MA, 40; Scripps Inst Oceanog, PhD(microbiol), 45. *Prof Exp:* Asst bact, Univ Calif, Los Angeles, 39-40; bacteriologist, Hunt Foods, Calif, 41-43; res asst microbiol, Scripps Inst Oceanog, Univ Calif, San Diego, 43-45, res assoc, 45-47; res microbiologist, Chevron Oil Field Res Co, 47-59, sr res biochemist, 59-66, sr res assoc, 66-85; RETIRED. *Mem:* AAAS; Soc Gen Microbiol; Am Soc Microbiol; Soc Indust Microbiol; fel Am Acad Microbiol; Sigma Xi. *Res:* Geomicrobiology; petroleum microbiology; organic geochemistry. *Mailing Add:* 1403 Sunny Crest Dr Fullerton CA 92635

REDFORD, JOHN W B, b Victoria, BC, Aug 7, 28; m 54; c 6. REHABILITATION MEDICINE. *Educ:* Univ BC, BA, 49; Univ Toronto, MD, 53; Mayo Clin & Mayo Found, MS, 58; Am Bd Phys Med & Rehab, dipl, 62. *Prof Exp:* From instr to asst prof phys med & rehab, Sch Med, Univ Wash, 58-63; prof & chmn dept, Med Col Va, 63-67; prof rehab med & dir sch, Univ Alta, 67-72, chmn dept phys med & rehab, Univ Hosp, 67-74; prof rehab med & chmn dept, Univ Kans Med Ctr, Kansas City, 74-88; CONSULT, 88- *Mem:* Am Geriat Soc; AMA; Am Acad Phys Med & Rehab; Am Cong Rehab Med; Am Asn Electrodiag. *Mailing Add:* Dept Rehab Med A-34 Univ Kans Col Health 39th St & Rainbow Blvd Kansas City KS 66103

REDGATE, EDWARD STEWART, b Yonkers, NY, Mar 13, 25; m 54; c 3. NEUROPHYSIOLOGY, NEUROENDOCRINOLOGY. *Educ:* Bethany Col, BS, 49; Univ Minn, MS, 52, PhD(physiol), 54. *Prof Exp:* Asst, Univ Minn, 49-54; instr, Western Reserve Univ, 57-59, asst prof physiol, 59-62; ASSOC PROF PHYSIOL, SCH MED, UNIV PITTSBURGH, 62- *Concurrent Pos:* Fel neurophysiol, Univ Minn, 55-56; USPHS fel, Western Reserve Univ, 56-57. *Mem:* Am Physiol Soc; Endocrine Soc. *Res:* Brain stem regulations; cardiovascular, respiratory and autonomic nervous systems; viscera; norepinephrine; physiology of the hypothalamus; neural control of adrenocorticotropic hormone release; interaction of hypothalamic neuropeptides. *Mailing Add:* Dept Physiol Univ Pittsburgh Sch Med Pittsburgh PA 15261

REDHEAD, PAUL AVELING, b Brighton, Eng, May 25, 24; m 48; c 2. SURFACE PHYSICS, VACUUM PHYSICS. *Educ:* Cambridge Univ, BA, 44, MA, 48, PhD, 69. *Prof Exp:* Sci officer, Serv Electronics Res Lab, Brit Admiralty, 44-47; from res officer to prin res officer, Nat Res Coun, 47-71, dir prog planning & anal, 70-72, dir gen planning, 72-73, dir div physics, 73-86, group dir, Phys & Chem Sci Lab, 74-83, chmn, Comt Dirs, 81-86, secy, Sci & Technol Policy Comt, 86-89, EMER RESEARCHER, NAT RES COUN, 89- *Concurrent Pos:* Ed, J Vacuum Sci & Technol, 69-74; asst ed-in-chief, Can J Res, 74-87; dir, Ont Ctr Mat Res, 89-91. *Honors & Awards:* Welch Award, Am Vacuum Soc, 75; Jubilee Medal, 77; Achievement in Physics Medal, Can Asn Physicists, 89. *Mem:* Fel Am Phys Soc; Can Asn Physicists; fel Inst Elec & Electronics Engrs; hon mem Am Vacuum Soc (pres,

68); fel Royal Soc Can. *Res:* Electron physics; chemical adsorbtion and interaction of electrons with absorbed layers; vacuum physics; ion trapping; electron-plasmas. *Mailing Add:* Inst Microstruct Sci Nat Res Coun Ottawa ON K1A 0R6 Can

REDHEAD, SCOTT ALAN, b Regina, Sask, Can, Dec 26, 50; m 72; c 4. TAXONOMY, ECOLOGY. *Educ:* Univ BC, BSc, 72, MSc, 74; Univ Toronto, PhD(mycol), 79. *Prof Exp:* Biologist, 77-79, RES SCIENTIST, BIOSYSTS RES CTR, AGR CAN, 79- *Concurrent Pos:* Ed, Fungi Canadenses, Agr Can. *Mem:* Mycol Soc Am; Mycol Soc Japan; Int Asn Plant Taxonomy; Can Bot Asn; NAm Mycol Asn. *Res:* The biogeography, taxonomy, nomenclature, ecology (pathogenicity, mycorrhizal associations, symbioses and wood decay types) and the influences on human affairs (legality, toxicity, edibility, drug production) of the mushroom flora of Canada. *Mailing Add:* Biosysts Res Ctr Agr Can Ottawa ON K1A 0C6 Can

REDHEFFER, RAYMOND MOOS, b Chicago, Ill, Apr 17, 21; m 51; c 1. MATHEMATICS. *Educ:* Mass Inst Technol, SB, 43, SM, 46, PhD(math), 48. *Prof Exp:* Mem staff, Radiation Lab, Mass Inst Technol, 42-46, res assoc, Lab Electronics, 46-48; instr math, Harvard Univ, 48-50; from instr to assoc prof, 50-60, PROF MATH, UNIV CALIF, LOS ANGELES, 60- *Concurrent Pos:* Peince fel, Harvard Univ, 48-50; NSF sr fel, Univ Göttingen, 56; Fulbright fel, Univ Vienna, 57; Fulbright res fel, Univ Hamburg, 61-62, guest prof, 66; guest lectr, Tech Univ Berlin, 62; guest prof, Univ Karlsruhe, 71-72, 81, 85 & 88. *Honors & Awards:* Humboldt Found Sr US Scientist Award, 76 & 85. *Mem:* Math Asn Am; Am Math Soc. *Res:* Differential and integral inequalities. *Mailing Add:* 176 N Kenter Ave Los Angeles CA 90049

REDI, MARTHA HARPER, b Bryn Mawr, Pa; m 63; c 1. SOLID STATE PHYSICS, BIOPHYSICS. *Educ:* Mass Inst Technol, BS, 64; Rutgers Univ, MS, 66, PhD(physics), 69. *Prof Exp:* Fel physics, Rutgers Univ, 69-70; fel quantum chem, Princeton Univ, 76-77, vis res fel physics, 77-80, res staff mem, Dept Physics, 80, Geophys Fluid Dynamics Lab, 80-82, prof sci & eng staff mem, Plasma Physics Lab, 82-90, STAFF RES PHYSICIST, PLASMA PHYSICS LAB, PRINCETON UNIV, 90- *Concurrent Pos:* NIH vis res fel, 77-80. *Mem:* Am Phys Soc; AAAS; Am Geophys Union. *Res:* Transport modeling in plasma physics; oceanographic modeling; superconductivity; magnetic interactions; biophysics of electron transfer; hemaglobin action; oceanography. *Mailing Add:* Plasma Physics Lab Forrestal Campus Princeton Univ Box 451 Princeton NJ 08544

REDI, OLAV, b Tallinn, Estonia, May 29, 38; US citizen; m 63; c 1. EXPERIMENTAL ATOMIC PHYSICS. *Educ:* Rensselaer Polytech Inst, BS, 60; Mass Inst Technol, PhD(physics), 65. *Prof Exp:* Res assoc physics, Princeton Univ, 64-66, from instr to asst prof, 66-71; MEM FAC PHYSICS, NY UNIV, 71- *Mem:* Am Phys Soc. *Res:* Nuclear moments and isotope shifts; atomic level-crossing spectroscopy; optical and atomic beam hyperfine structure studies. *Mailing Add:* 124 Fisher Pl Princeton NJ 08540

REDICK, MARK LANKFORD, b Seattle, Wash, Nov 30, 54. ENDOCRINOLOGY, ELECTRON MICROSCOPY. *Educ:* Southwest Mo Univ, BA; Univ Kans, PhD(anat), 85. *Prof Exp:* RES ASSOC, VET ADMIN MED CTR, 86- *Mailing Add:* Dept Anat 2008 Whe Kans Univ Med Ctr Kansas City KS 66103

REDICK, THOMAS FERGUSON, b Youngstown, Ohio, Oct 20, 21; m 52. PHYSIOLOGY. *Educ:* Miami Univ, BA, 48; Univ Pittsburgh, MS, 52, PhD, 55. *Prof Exp:* Asst zool, Univ Pittsburgh, 50-52; jr fel, Mellon Inst, 52-54; from instr to asst prof physiol, Sch Med, Univ Pittsburgh, assoc prof, State Univ NY, 63-66; PROF PHYSIOL & BIOL, FROSTBURG STATE COL, 66- *Concurrent Pos:* Dept HEW fel neuropharmacol, Leech Farm Vet Hosp, Pittsburgh, Pa, 62-63; vis investr, Radiobiol Lab, US Bur Fisheries, NC, 70-72; mem staff marine sci, Univ La Laguna, Spain, 71-72, res prof, 72- res fel, NASA Langley Field, 77; unicate res, La Laguna Tenerife, 84. *Mem:* Soc Syst Zool. *Res:* Mollusks; cardiovascular research; neuropharmacology in tunicates. *Mailing Add:* Beall's Lane Frostburg MD 21532

REDIKER, ROBERT HARMON, b Brooklyn, NY, June 7, 24; m 80; c 2. ELECTRO-OPTIC DEVICES, SEMICONDUCTOR LASERS. *Educ:* Mass Inst Technol, BS, 47, PhD(physics), 50. *Prof Exp:* Asst physics, Mass Inst Technol, 48-50, res assoc, 50-51, mem staff, Lincoln Lab, 51-52; res assoc, Ind Univ, 52-53; mem staff, Lincoln Lab, 53-57, asst group leader, 57-59, group leader appl physics, 59-66, prof, dept elec eng, 66-76, assoc head, Optics Div, 70-72, head, 72-80, adj prof, dept elec eng, 76-82, SR STAFF, LINCOLN LAB, 80-, SR RES SCIENTIST, DEPT ELEC ENG, MASS INST TECHNOL, 82- *Concurrent Pos:* Mem Nat Acad Sci evaluation panel for Nat Bur Standards. *Honors & Awards:* David Sarnoff Award, Inst Elec & Electronics Engrs, 69. *Mem:* Nat Acad Eng; fel Optical Soc Am; fel Inst Elec & Electronics Engrs; fel Am Phys Soc. *Res:* Solid state devices; optics; guided-wave optics; semiconductor lasers and light emitters; semiconductor devices. *Mailing Add:* Lincoln Lab Mass Inst Technol PO Box 73 Lexington MA 02173-0073

REDIN, ROBERT DANIEL, b Rockford, Ill, Jan 18, 28. THERMAL & ELECTRICAL CONDUCTIVITY. *Educ:* Iowa State Univ, BS, 52, MS, 55, PhD(physics), 57. *Prof Exp:* Asst physics, Iowa State Univ, 54-57; physicist, Electronics Lab, US Dept Navy, 57-62; PROF PHYSICS, SDAK SCH MINES & TECHNOL, 62- *Mem:* Am Asn Physics Teachers; Am Phys Soc; Sigma Xi. *Res:* Thermal conductivity in solids; semiconductors; alloys; transport properties of solids. *Mailing Add:* SDak Sch Mines & Technol 501 E St Joseph St Rapid City SD 57701-3995

REDINBO, G ROBERT, b Lafayette, Ind, July 11, 39; m 61; c 3. ELECTRICAL ENGINEERING. *Educ:* Purdue Univ, BS, 62, MS, 66, PhD(elec eng), 70. *Prof Exp:* Instr elec eng, Purdue Univ, 65-70; asst prof, Univ Wis-Madison, 70-76; assoc prof elec & systs eng, Rensselaer Polytechnic Inst, 76-; PROF, DEPT ELEC ENG & COMPUTER SCI, UNIV CALIF, DAVIS, 84- *Concurrent Pos:* Trustee, Nat Electronics, Inc, 72-; NASA-Am Soc Eng Educ fel, Goddard Space Flight Ctr, 72; NSF fel, Univ Wis, 72-73; adv comt, Inst Elec & Electronics Engrs Signal Processing Soc, 76-80. *Mem:* Inst Elec & Electronics Engrs; Asn Comput Mach; Sigma Xi. *Res:* Algebraic coding theory; communication theory; generalized transforms; digital filtering in communications systems. *Mailing Add:* Dept Elec Eng & Computer Sci Univ Calif Davis CA 95616

REDINGER, RICHARD NORMAN, b E Windsor, Sandwich, Ont, Feb, 18, 38; m 65; c 4. GASTROENTEROLOGY. *Educ:* Univ Western Ont, BA, 60, MD, 62; FRCP(C), 68 & 75. *Prof Exp:* Res assoc med, Sch Med, Boston Univ, 68-71; mem med staff, Univ Hosp, 72-78, dir gastrointestinal lab, 73-78; chief gastrointestinal res, Univ Hosp, 78-81, assoc prof dept med, Boston Univ Med Ctr, 78-81, assoc vis physician & assoc mem, Evans Mem, 78-81; PROF MED, DEPT MED, UNIV LOUISVILLE, 81-, ASSOC PROF DEPT BIOCHEM, 81-, VCHMN DEPT MED, 84- *Concurrent Pos:* Consult gastrointestinal dis, Westminster Hosp, 71-; asst prof med, Univ Western Ont, 71, assoc prof, 75-78; NIH res grant, 78-81 & 84-; chief, digestive dis & nutrit, Univ Louisville, 81- *Mem:* Can Med Asn; Can Soc Clin Invest; Am Fedn Clin Res; Am Gastroenterol Asn; Am Asn Study Liver Dis; Am Soc Gastrointestinal Endoscopy; AMA. *Res:* Effects of phenobarbital and cholesterol lowering agents on biliary lipid composition and gallstone formation and/or dissolution; hepatobiliary disposal of cholesterol in baboons. *Mailing Add:* Dept Med ACB-3rd Floor Univ Louisville 550 S Jackson St Louisville KY 40292

REDINGTON, CHARLES BAHR, b Elyria, Ohio, Mar 20, 42; m 64; c 2. PLANT PHYSIOLOGY, APPLIED ENVIRONMENTAL ASSESSMENTS. *Educ:* Baldwin-Wallace Col, BS, 64; Rutgers Univ, MS, 66, PhD(plant physiol, path), 69. *Prof Exp:* Teaching & res fel biol, Rutgers Univ, 64-66, res fel plant path, physiol, 66-69; from asst prof to assoc prof, 69-83, PROF BIOL, SPRINGFIELD COL, 83-, DIR, ENVIRON HEALTH TECH PROG, 85- *Concurrent Pos:* Vis prof, Western New Eng Col, 84-; consult environ biol, Baystate Environ Consults, 85-; exped leader, Springfield Col & EAfrican Safaris, Ltd, 85- *Mem:* Sigma Xi. *Res:* Use of atomic absorption spectrophotometry in assessing nutrient deficiency in bud blasting; plant communication; bomb calorimetric evaluation of factors increasing crop yield. *Mailing Add:* Bemis Hall Off 107 Springfield Col Alden St Springfield MA 01109

REDINGTON, RICHARD LEE, b Minneapolis, Minn, May 16, 33; m 57; c 1. PHYSICAL CHEMISTRY. *Educ:* Univ Minn, BA, 55; Univ Wash, PhD(phys chem), 61. *Prof Exp:* Res fel, Mellon Inst, 61-64; asst prof chem, Utah State Univ, 64-67; from asst prof to assoc prof, 67-73, PROF CHEM, TEX TECH UNIV, 73- *Concurrent Pos:* Vis scientist, Mass Inst Technol, 83-90. *Mem:* Am Chem Soc; Optical Soc Am; Am Phys Soc. *Res:* Molecular spectroscopy; molecular structure. *Mailing Add:* Dept Chem Tex Tech Univ Lubbock TX 79412

REDINGTON, ROWLAND WELLS, b Otego, NY, Sept 26, 24; m 47; c 2. MEDICAL DIAGNOSTIC IMAGING. *Educ:* Stevens Inst Technol, ME, 45; Cornell Univ, PhD(exp physics), 51. *Prof Exp:* Teaching asst, Cornell Univ, 46-51; physicist, Gen Elec Res & Develop Ctr, 51-66, prog mgr, 66-76, br mgr, 76-88, lab mgr, 88-91; RETIRED. *Concurrent Pos:* Adj prof, Rensselaer Polytech Inst, 64-66; adj prof, Univ Calif, San Francisco Med Ctr, 79-80. *Honors & Awards:* Indust Applications Physics Prize, Am Inst Phys. *Mem:* Am Phys Soc; Nat Acad Eng; Sigma Xi. *Res:* Electronic imaging devices; photoconductivity; photoemission; stereographic display; computed tomography; medical diagnostic imaging systems, especially magnetic resonance imaging, in vino NMR spectroscopy; x-ray computed tomography and phased array ultrasound. *Mailing Add:* 1169 Mohawk Rd Schenectady NY 12309

REDISCH, WALTER, b Prague, Czech, Sept 26, 98; nat US; m 35; c 2. PHYSIOLOGY. *Educ:* Ger Univ, Prague, 22; Am Bd Internal Med, dipl, 42. *Prof Exp:* Res fel, 38-39, from asst to asst prof med, 40-52, assoc prof clin med, 53-78, RES PROF PHYSIOL, SCH MED, NY UNIV, 78- *Concurrent Pos:* Assoc attend physician, Hosp, NY Univ, 53-; vis physician & res assoc, Res Serv, Goldwater Mem Hosp; vis res prof, NY Med Col, 69-; assoc vis physician, Bellevue Hosp; chief circulation res unit, NY Med Col; consult, Vet Admin Hosp, NY. *Mem:* Geront Soc; Am Physiol Soc; Harvey Soc; AMA; Am Heart Asn. *Res:* Circulation. *Mailing Add:* Dept Physiol Med Sci Bldg NY Univ Med Col Ten East End Ave Apt 3E New York NY 10021

REDISH, EDWARD FREDERICK, b New York, NY, Apr 1, 42; m 67; c 2. THEORETICAL NUCLEAR PHYSICS,. *Educ:* Princeton Univ, AB, 63; Mass Inst Technol, PhD(physics), 68. *Prof Exp:* Fel, Ctr Theoret Physics, Univ Md, 68-70, from asst prof to assoc prof, 70-79, chmn, Dept Physics & Astron, 82-85, PROF PHYSICS, UNIV MD, COLLEGE PARK, 79- *Concurrent Pos:* Vis scientist, Saclay Nuclear Res Ctr, France, 73-74; Nat Acad Sci-Nat Res Coun sr resident res assoc, Goddard Space Flight Ctr, NASA, 77-78; vis prof, Ind Univ & Cyclotron Facil, 85-86; prin investr, Md Univ, Proj Physics & Educ Technol, 85-; mem, Cyclotron Facil Prog Adv Comt, Ind Univ, 85, chmn, 86, Nuclear Sci Adv Comt, 87-89, Steering Comt, Am Phys Soc Trop Group on Few Body Syst & Multiparticle Dynamics, 87-89, chmn, 89-90; mem, Bonner Prize Comt, Div Nuclear Physics, Am Phys Soc, 88-90, chmn, 89-90; co-chair, Conf Computers Physics Instr, Raleigh, NC, 81-88. *Honors & Awards:* Inst Medal, Cent Res Inst Physics, Budapest, Hungary, 79; Leo Schubert Award, Wash Acad Sci, 88. *Mem:* AAAS; Sigma Xi; fel Am Phys Soc; Am Asn Phys Teachers. *Res:* Many body theory of nuclear reactions; three body aspects of nuclear scattering problems; off-energy-shell effects in direct nuclear reactions; computers in physics education. *Mailing Add:* Dept Physics & Astron Univ Md College Park MD 20742-4111

REDISH, JANICE COPEN, b Newark, NJ, Aug 12, 41; m 67; c 2. LINGUISTICS, PLAIN ENGLISH. *Educ:* Bryn Mawr Col, AB, 63; Harvard Univ, PhD(ling), 69. *Prof Exp:* Res assoc, Ctr Appl Ling, 71-72, Educ Study Ctr, 72-73, Asn Renewal Educ, 75-77; res scientist, 77-79, dir, Doc Design

Ctr, 79-90, VPRES, AM INST RES, 84- *Concurrent Pos:* Fulbright Hayes scholar Slavic ling, Univ Amsterdam, 63-64; Vis lectr, George Washington Univ, 74. *Mem:* Ling Soc Am; Soc Tech Commun; Asn Comput Mach; Asn Teachers Tech Writing. *Res:* Writing in the workplace; readability and usability of technical documents; development of usable software interfaces and online software; testing products and manuals for usability. *Mailing Add:* Am Insts Res 3333 K St NW Washington DC 20007

REDISH, KENNETH ADAIR, b London, Eng, May 6, 26; m 50; c 7. COMPUTER SCIENCE. *Educ:* Univ London, BSc, 53. *Prof Exp:* Asst lectr math, Woolwich Polytech Inst Eng, 53-55; lectr, Battersea Col Adv Technol, 55-59; lectr comput, Univ Birmingham, 60-62, sr lectr, 62-67, dir comput serv, 64-67; ASSOC PROF APPL MATH, MCMASTER UNIV, 67- *Mem:* Asn Comput Mach; fel Brit Comput Soc. *Res:* Numerical analysis. *Mailing Add:* Dept Appl Math McMaster Univ 1280 Main St W Hamilton ON L8S 4L8 Can

REDLICH, FREDRICK CARL, b Vienna, Austria, June 2, 10; nat US; m 37, 55; c 2. PSYCHIATRY. *Educ:* Univ Vienna, MD, 35. *Prof Exp:* Intern, Allgem Krankenhaus, Vienna, 35-36; resident, Univ Psychiat Clin, Univ Vienna, 36-38; asst physician, State Hosp, Iowa, 38-40; resident, Neurol Unit, Boston City Hosp, 40-42; from instr to assoc prof psychiat, Sch Med, Yale Univ, 42-50, exec officer, 47-50, prof psychiat & chmn dept, 50-67, assoc provost med affairs & dean, 67-72, prof psychiat, Sch Med, 72-77, dir, Behav Sci Study Ctr, 73-77; assoc chief of staff educ, Vet Admin Med Ctr, Brentwood, 77-82; emer prof psychiat & neuropsychiat, Yale Univ, 77-; MEM STAFF, NEUROPSYCHIAT INST, UNIV CALIF, LOS ANGELES, 82- *Concurrent Pos:* Teaching fel, Harvard Med Sch, 41-42; dir, Conn Ment Health Ctr, 64-67; consult, NIMH & Off Surgeon Gen, US Army. *Mem:* Inst Med-Nat Acad Sci; Am Psychosom Soc; fel Am Psychiat Asn; Am Orthopsychiat Asn; AAAS. *Res:* Personality theory; social structure and mental disorder; scientific methodology in psychiatry. *Mailing Add:* Neuropsychiat Inst Univ Calif Los Angeles CA 90024

REDLICH, MARTIN GEORGE, b Vienna, Austria, Dec 1, 28; US citizen. PHYSICS. *Educ:* Univ Calif, Berkeley, AB, 48; Princeton Univ, PhD(physics), 54. *Prof Exp:* Res asst physics, Princeton Univ, 54; proj assoc, Univ Wis-Madison, 54-56; res assoc, Wash Univ, 56-57; res, Mass Inst Technol, 57-59; res, 61-62, MEM NUCLEAR THEORY GROUP, LAWRENCE BERKELEY LAB, UNIV CALIF, 62- *Concurrent Pos:* NSF fel, 57-58. *Mem:* Am Phys Soc. *Res:* Theory of the structure of the nucleus. *Mailing Add:* Bldg 70A Nuclear Sci Div Lawrence Berkeley Lab Univ Calif Berkeley CA 94720

REDLICH, ROBERT WALTER, b Lima, Peru, Sept 20, 28; US citizen; m 60; c 2. ELECTRONICS ENGINEERING, ELECTROMAGNETISM. *Educ:* Rensselaer Polytech Inst, BS, 50, PhD(physics), 60; Mass Inst Technol, MS, 51. *Prof Exp:* Develop engr gyroscopes, Gen Elec Co, 51-52; res & develop engr inertial navig, US Army Ballistic Missiles Agency, 54-56; from asst prof to assoc prof elec eng, Clarkson Tech Univ, 60-65; sr res fel, Univ Sydney, 65-68; from assoc prof to prof, Ohio Univ, 68-75; chief engr, Gorman-Redlich Mfg Co, 74-87; CONSULT, 79. *Concurrent Pos:* adj prof, Ohio Univ, 81- *Mem:* Inst Elec & Electronics Engrs; NY Acad Sci; Sigma Xi. *Res:* Electromagnetic theory, especially radiation from aerials and moving charges; electronics; linear electric motors and generators; linear motion transducers. *Mailing Add:* Nine Grand Park Blvd Athens OH 45701

REDLINGER, LEONARD MAURICE, b Keota, Iowa, Dec 4, 22; m 49; c 6. COMMODITY CHEMICAL PROTECTANTS, FUMIGATION. *Educ:* Iowa Wesleyan Col, BS, 46; Kans State Col, MS, 47. *Prof Exp:* Lab instr zool & cur insect collection, Iowa Wesleyan Col, 43-46; entomologist, Mosquito Control Proj, Bur Entom & Plant Quarantine, USDA, Alaska, 47, Stored-Prod Insects Sect, 48-54 & Mkt Res Div, Agr Mkt Serv, 54-56; entomologist, Pfeffer & Son Warehouse Co, Tex, 56-60; sta leader, Stored-Prod Insects Br, Mkt Qual Res Div, USDA, 60-65, leader, Peanut & Southern Corn Insects Invests, 65-73, res entomologist, Stored-Prod Insects Res & Develop Lab, Chem Control Res Unit, 73-86; RETIRED. *Concurrent Pos:* Secy-treas, Ga Entom Soc, 64-70, pres, 71-72, mem, bd dirs, 72-73. *Honors & Awards:* Golden Peanut Res & Educ Award, USDA, 87. *Mem:* Entom Soc Am; Am Peanut Res & Educ Soc; Sigma Xi. *Res:* Applied and developmental research to improve existing and to devise new chemical methods of controlling or preventing insect infestations of post harvest agricultural commodities in the marketing channels. *Mailing Add:* 3910 Doster Rd Monroe NC 28112-9684

REDMAN, CHARLES EDWIN, b Pawtucket, RI, Aug 1, 31; m 53; c 4. BIOLOGY, STATISTICS. *Educ:* Univ Mass, BS, 54, MS, 56; Univ Minn, PhD(statist), 60. *Prof Exp:* Sr biometrician, Eli Lilly & Co, 60-62, dept head statist, 62-64, asst head, 64-65, head statist & rec, 65-69, asst dir sci serv, 69-72, dir sci info, 72-80, dir med info systs, 80-83; PRES, HCA MED RES CO, 84- *Mem:* Am Statist Asn; Biomet Soc; Drug Info Asn; Pharmaceut Mfrs Asn; Sigma Xi. *Res:* Population genetics and associated statistical design problems; design, analysis and interpretation of screening and physiology studies; clinical research management, phases I, II, III, and IV. *Mailing Add:* 4403 Charleston Pl Nashville TN 37215

REDMAN, COLVIN MANUEL, b Dominican Repub, Jan 26, 35; US citizen; m 65; c 3. BIOCHEMISTRY, MOLECULAR BIOLOGY. *Educ:* McGill Univ, BS, 57; Univ Wis-Madison, PhD(physiol chem), 62. *Prof Exp:* Fel biochem, Univ Wis-Madison, 62-63 & Rockefeller Univ, 63-65; staff fel, Addiction Res Ctr, NIMH, Ky, 65-66, res biochemist, 67; assoc investr, 67-70, investr, 70-76, SR INVESTR, NY BLOOD CTR, 76- *Mem:* Am Soc Biol Chem; Am Soc Cell Biol; Harvey Soc; Am Chem Soc. *Res:* Membrane structure and function; kell blood group system; fibrinogen. *Mailing Add:* NY Blood Ctr 310 E 67th St New York NY 10021

REDMAN, DONALD ROGER, b Eaton, Ohio, Jan 21, 36; m 61; c 3. VETERINARY MEDICINE, VETERINARY SURGERY. *Educ:* Ohio State Univ, BS, 58, DVM, 62, MS, 66, PhD(prev med), 73. *Prof Exp:* Clin instr 62-73, from asst prof to assoc prof, 72-82, PROF PREV MED, OHIO AGR RES & DEVELOP CTR, OHIO STATE UNIV, 82- *Mem:* Am Vet Med Asn; Int Embryo Transfer Soc; Am Asn Swine Practitioners. *Res:* Developmental and immune response of porcine fetus exposed to viral agents; reproductive efficiency of domestic animals; embryo transfer techniques; enteric diseases of neonatal calves and cryptosporidiosis in animals. *Mailing Add:* 1733 Messner Rd Wooster OH 44691

REDMAN, KENNETH, pharmacognosy; deceased, see previous edition for last biography

REDMAN, ROBERT SHELTON, b Fargo, NDak, Aug 1, 35; m 58; c 1. DENTISTRY, PATHOLOGY. *Educ:* Univ Minn, Minneapolis, BS & DDS, 59, MSD, 63; Univ Wash, PhD (exp path), 69; Am Bd Oral Path, cert, 73. *Prof Exp:* Clin asst prof oral diag, Sch Dent, Univ Minn, Minneapolis, 63-64; instr oral biol, Sch Dent, Univ Wash, 68-69; assoc prof oral biol, Sch Dent, Univ Minn, Minneapolis, 69-75; res oral pathologist, Vet Admin Hosp, Denver, Colo, 75-78; STAFF ORAL PATHOLOGIST & CHIEF, ORAL PATH RES LAB, DEPT VET AFFAIRS MED CTR, WASHINGTON, DC, 78- *Concurrent Pos:* Am Cancer Soc Clin fels, 62-64; Nat Inst Dent Res trainee path, Univ Wash, 64-68, Nat Inst Dent Res career develop award oral biol, 68-69; Nat Inst Dent Res career develop award, Sch Dent, Univ Minn, Minneapolis, 71-75; consult, Children's Orthop Hosp, Seattle, Wash, 66-69; assoc prof oral biol, Sch Dent, Univ Colo Med Ctr, Denver, 75-78; Vet Admin Res Career Prog Clin Investr, Vet Admin Med Ctr, Denver Colo, 76-78; prog spec oral biol, Vet Admin Dept Med Surg, 83-87; prog chmn, Salivary Res Group, Int Asn for Dent Res, 82-86. *Mem:* Am Dent Asn; Am Acad Oral Path; Int Asn Dent Res; Am Inst Nutrit; Tissue Cult Asn. *Res:* Salivary gland growth and development; morphogenesis of salivary gland neoplasms; etiology of tongue diseases; etiology of dental caries; experimental carcinogenesis. *Mailing Add:* Dent Serv Dept Vet Affairs 50 Irving St NW Washington DC 20422

REDMAN, WILLIAM CHARLES, b Washington, DC, June 19, 23; m 48; c 3. REACTOR PHYSICS, MAGNETOHYDRODYNAMICS. *Educ:* Georgetown Univ, BS, 43; Yale Univ, MS, 47, PhD(physics), 49. *Prof Exp:* Jr physicist, Dept Terrestrial Magnetism, Carnegie Inst & Nat Bur Standards, 43; asst instr physics, Yale Univ, 43-44; jr physicist, Manhattan Dist, Argonne Nat Lab, Univ Chicago, 44-46; asst instr physics, Yale Univ, 47-48; assoc physicist, Argonne Nat Lab, 49-54; pres, Fournier Inst Technol, 54-55; assoc physicist, Argonne Nat Lab, 55-59, sr physicist, 59-63, assoc dir Appl Physics Div, 63-72, sr physicist, 72-74, dep dir Energy Conversion Progs, 74-85, consult, energy systs, 85-87; RETIRED. *Concurrent Pos:* Lectr, Fournier Inst Technol, 52-54. *Mem:* Fel Am Phys Soc; fel Am Nuclear Soc. *Res:* Physics of neutrons and nuclear reactors; instrumentation; environmental planning and impact, and energy conversion systems; atomic and molecular physics. *Mailing Add:* 608-S Monroe St Hinsdale IL 60521

REDMANN, ROBERT EMANUEL, b Jamestown, NDak, Nov 24, 41; m 63; c 4. PLANT ECOLOGY. *Educ:* Univ NDak, BA, 64; Univ Ill, Urbana, MS & PhD(bot), 68. *Prof Exp:* From asst prof to assoc prof, 68-78, PROF PLANT ECOL, UNIV SASK, 78- *Concurrent Pos:* Vis prof, N E Norm Univ, Changchun, People's Repub China. *Mem:* Sigma Xi; Ecol Soc Am; Soc Range Mgt. *Res:* Grassland ecology; plant carbon dioxide exchange; plant water relations; salt tolerance in plants; pollution ecology. *Mailing Add:* Dept Crop Sci Plant Ecol Univ Sask Saskatoon SK S7N 0W0 Can

REDMON, JOHN KING, b Lexington, Ky, Nov 4, 20; m 67; c 5. ELECTRICAL ENGINEERING. *Educ:* Newark Col Eng, BS, 42; Stevens Inst Technol, MS, 49; Worcester Polytech Inst, MS, 70. *Prof Exp:* Sales engr, Westinghouse Elec Corp, 45-51, consult & appln engr, 53-60; from asst prof to assoc prof elec eng, Newark Col Eng, 60-70; educ adminr, Pa Power & Light Co, 70-81; prof elec eng, Grad Sch, Lehigh Univ, 70-82; CONSULT EDUC, REDMON ASSOCS, 81- *Concurrent Pos:* Consult, Pub Serv Elec & Gas Co, 61-66; adj prof, Northampton County Area Community Col, 75-85. *Mem:* Inst Elec & Electronics Engrs; Am Soc Eng Educ. *Res:* Application of power equipment to industrial use; electrical distribution and lightning protection; power systems and circuit analysis. *Mailing Add:* 325 N Cool Spring St Fayetteville NC 28301-5160

REDMON, MICHAEL JAMES, b Long Beach, Calif, Oct 3, 41; m 75. PHYSICAL CHEMISTRY, ATOMIC PHYSICS. *Educ:* Fla State Univ, BS, 63; Rollins Col, MS, 68; Univ Fla, PhD(chem), 73. *Prof Exp:* Res engr physics, Orlando Div, Martin Co, 63-67; asst prof, Valdosta State Col, 67-73; res assoc chem, Univ Tex, Austin, 74-77; res scientist chem physics, Columbus Labs, Battelle Mem Inst, 77-80; AT CHEM DYNAMICS CORP, 80- *Concurrent Pos:* Res asst, Univ Fla, 68-74; Robert Welch Found fel, 75. *Mem:* Am Phys Soc; Am Chem Soc. *Res:* Atomic and molecular collisions; quantum chemistry; chemical physics. *Mailing Add:* Calvert Inst PO Box 1324 Prince Frederick MD 20678-1324

REDMOND, BILLY LEE, b Franklin, Tenn, May 25, 42. ENTOMOLOGY, ELECTRON MICROSCOPY. *Educ:* Univ Tenn, Martin, BS, 64; Univ Ill, Urbana, MS, 67; Cornell Univ, PhD(zool), 71. *Prof Exp:* Lab instr zool, Univ Tenn, Martin, 64-65; NIH fel entom & zool, Ohio State Univ, 71-72; asst prof gen biol & anat, Otterbein Col, 72-73; from asst prof to assoc prof, 73-87, PROF BIOL, STATE UNIV NY NEW PALTZ, 87-, DIR, ELECTRON MICROS LAB, 77- *Concurrent Pos:* Ed, NAm Registry Electron Micros Courses, 80- *Mem:* Electron Micros Soc Am; Entom Soc Am; Acarological Soc Am; Sigma Xi; Nat Sci Teachers Asn. *Res:* Developmental morphology and fine structure of tissues, especially those of the arthropods; electron microscopy. *Mailing Add:* Dept Biol State Univ NY New Paltz NY 12561

REDMOND, DONALD EUGENE, JR, b San Antonio, Tex, June 17, 39. BIOLOGICAL PSYCHIATRY, NEUROPHARMACOLOGY. *Educ:* Southern Methodist Univ, BA, 61; Baylor Col Med, MD, 68; Am Bd Psychiat & Neurol, cert, 77; Yale Univ, MA, 87. *Prof Exp:* Intern, Ben Taub Hosp, 68-69; resident res, Ill State Psychiat Inst, 69-72; clin assoc, NIMH, 72-74; from asst prof toassoc prof, 74-87, PROF, MED SCH, YALE UNIV, 87-, HEAD, NEUROBEHAV LAB, 85- *Concurrent Pos:* Falk fel, Am Psychiat Asn, 70-71; mem fac biol sci training grant, NIH, 74-; Guggenheim Found grant, 77-85; NIMH grant, 78- & Nat Inst Drug Abuse grant, 79-86; career develop award, Nat Inst Drug Abuse, 80-85; career res sci, NIMH, 86-91, prog dir, neural grafts prog proj, 86-89. *Honors & Awards:* Res Found Fund Prize, Am Psychiat Asn, 81. *Mem:* Am Psychiat Asn; Am Psychosomatic Med Soc; Soc Neurosci; Am Soc Primatologists; Am Cong Neuropsychopharmacol. *Res:* Biology of depression, anxiety, opioid drugs; two neurochemically defined brain systems, norepinephrine and dopamine; role of central norepinephrine in anxiety and drug-withdrawal syndromes; the dopamine system work aims to determine the causes of Parkinson's disease and possible treatment using transplanted brain dopamine cells. *Mailing Add:* Neurobehav Lab PO Box 3333 New Haven CT 06510

REDMOND, DONALD MICHAEL, b San Francisco, Calif, Feb 5, 48; m 77. MATHEMATICS, NUMBER THEORY. *Educ:* Univ Santa Clara, BS, 70; Univ Ill, MS, 73, PhD(math), 76. *Prof Exp:* Asst prof, 77-83, ASSOC PROF MATH, SOUTHERN ILL UNIV, 83- *Mem:* Am Math Soc; Math Asn Am; Sigma Xi; Nat Coun Teachers Math. *Res:* Analytic number theory, particularly the properties of Dirichlet series satisfy functional equation involving gamma factors. *Mailing Add:* Dept Math Neckers A 0259 Southern Ill Univ Carbondale IL 62901-4408

REDMOND, DOUGLAS ROLLEN, b Upper Musquodoboit, NS, Aug 30, 18; m 43; c 4. FORESTRY. *Educ:* Univ NB, BScF, 49; Yale Univ, MF, 50, PhD(forest path), 54. *Prof Exp:* Prov forest pathologist, NS, 50-51; officer chg forest path invests, Sci Serv, Can Dept Agr, 51-57; chief forest res div, Forestry Br, Dept Northern Affairs & Nat Resources, 57-60; dir forest res br, Dept Forestry, 60-65, sci adv, 65-69, dir forestry rels, Dept Fisheries & Forestry, 69-71; dir forestry rels, Dept Environ, 71-75, dir, Nat Forestry Insts, 75-79; consult forestry, 79-90; RETIRED. *Concurrent Pos:* Hon lectr, Univ NB, 51-57; mem permanent comt, Int Union Forest Res Orgn, 61-71, vpres, 72-76. *Honors & Awards:* Fernow Award, Am Forestry Asn, 75. *Mem:* Can Inst Forestry (pres, 78-79); hon mem Int Union Forest Res Orgn; Can Forestry Asn (pres, 80-82). *Res:* Forest ecology. *Mailing Add:* 643 Tillbury Ave Ottawa ON K2A 0Z9 Can

REDMOND, JAMES RONALD, b Ohio, July 14, 28; m 49; c 2. COMPARATIVE PHYSIOLOGY. *Educ:* Univ Cincinnati, BS, 49; Univ Calif, Los Angeles, PhD(zool), 54. *Prof Exp:* Asst prof biol, Univ Fla, 56-62; assoc prof, 62-67, actg chmn, 83-84, PROF ZOOL, IOWA STATE UNIV, 67- *Concurrent Pos:* Chief scientist, Alpha Helix Nautilus Expedition, 74. *Mem:* Fel AAAS; Am Soc Zoologists; Sigma Xi. *Res:* Blood pigments; circulatory and respiratory physiology of invertebrates. *Mailing Add:* Dept Zool Iowa State Univ Ames IA 50011

REDMOND, JOHN PETER, b Camden, NJ, Aug 1, 25; m 50; c 6. PLASTICS CHEMISTRY. *Educ:* Cath Univ Am, BS, 50, MS, 55; Pa State Univ, PhD(fuel tech), 59. *Prof Exp:* Phys chemist, Chem Div, US Naval Res Lab, 51-55; asst, Pa State Univ, 55-59; sr chemist, Appl Physics Lab, Johns Hopkins Univ, 59-65; sr scientist, Gen Tech Corp, 65-66; res dir & vpres, Commonwealth Sci Corp, 66-69; res assoc, Res Div, 69-73, mgr, plastics lab, 79, anal lab, 83, SR RES ASSOC, AMP, INC, 76-, MGR SPEC PRODS. *Mem:* Am Chem Soc; Soc Plastics Engrs. *Res:* Chemical kinetics; vapor deposition; plastics, injection and transfer molding; rheology studies; resin selection and compounding for electronic industry; electrical connector development; 15 patents. *Mailing Add:* MS 21-01 AMP Inc Box 3608 Harrisburg PA 17105

REDMOND, PETER JOHN, b Southampton, Eng, July 1, 29; US citizen; m 54; c 4. PHYSICS. *Educ:* Cooper Union, BEE, 51; Univ Birmingham, PhD(physics), 54. *Prof Exp:* Instr physics, Columbia Univ, 54-57; res physicist, Univ Calif, Berkeley, 57-59, assoc prof, Univ Calif, Santa Barbara, 59-63; mem tech staff physics, Defense Res Corp, 63-72, MEM TECH STAFF PHYSICS, GEN RES CORP, 72- *Concurrent Pos:* Fulbright scholar, 51-53; consult, Rand corp, 56-62 & Gen Elec Co, Tempo, Calif, 60-63. *Mem:* Am Phys Soc. *Res:* Atomic and molecular scattering; lasers; quantum field theory; nuclear physics; electromagnetic scattering theory. *Mailing Add:* 1809 Anapaca St Santa Barbara CA 93101

REDMOND, ROBERT F(RANCIS), b Indianapolis, Ind, July 15, 27; m 52; c 5. NUCLEAR ENGINEERING, ENERGY CONVERSION. *Educ:* Purdue Univ, BS, 50; Univ Tenn, MS, 55; Ohio State Univ, PhD(physics), 61. *Prof Exp:* Engr, Oak Ridge Nat Lab, 50-53; fel, Anal Physics Div, Battelle Mem Inst, 53-70; prof nuclear eng & chmn dept, Ohio State Univ, 70-77, ASSOC DEAN, COL ENG, OHIO STATE UNIV, 77-, DIR ENG EXP STA, 77-, ACTG DEAN, 90- *Concurrent Pos:* Mem bd trustees, Argonne Univs Asn, 72-80; mem, Ohio Power Siting Comn, 78-82; mem bd dir, Nat Regulatory Res Inst, 87-, Edison Welding Inst, 87-; pres & mem bd dirs, TRC, Inc, 90- *Mem:* Am Nuclear Soc; Am Soc Eng Educ; AAAS. *Res:* Nuclear and reactor physics and engineering; research engineering. *Mailing Add:* Col Eng 2070 Neil Ave Ohio State Univ Columbus OH 43210

REDMORE, DEREK, b Horncastle, Eng, Aug 8, 38; m 65; c 2. ORGANIC CHEMISTRY. *Educ:* Univ Nottingham, BSc, 59, PhD(org chem), 62. *Prof Exp:* Res assoc chem, Wash Univ, 62-65; res chemist, 65-67, sect leader, 67-76, sect mgr org chem, Tretolite Div, 76-90, DIR TECHNOL SUPPORT, PETROLITE CORP, 90- *Honors & Awards:* St Louis Award, Am Chem Soc, 82. *Mem:* Am Chem Soc; Royal Soc Chem; AAAS. *Res:* Organic, organophosphorus, alicyclic and heterocyclic chemistry. *Mailing Add:* 300 Park Rd St Louis MO 63119

REDMOUNT, MELVIN B(ERNARD), b Lakewood, Pa, Oct 11, 26; m 52; c 3. CHEMICAL ENGINEERING, ECONOMICS. *Educ:* Pa State Univ, BChE, 48; Polytech Inst Brooklyn, MChE, 52. *Prof Exp:* Chem engr, Develop Dept, Tidewater Oil Co, 51-53; group leader & res chem engr, Columbia Univ, 53-57; resident supvr develop, Speer Carbon Co, 57-64, resident mgr develop, 64-66, mgr new prod com develop, 66-70, mgr planning, Airco Speer Carbon-Graphite Div, 70-74, DIR PLANNING, AIRCO SPEER CARBON-GRAPHITE DIV, BOC GROUP INC, 74- *Mem:* Am Chem Soc; Am Inst Mining, Metall & Petrol Engrs. *Res:* Carbon; graphite; high temperature materials; continuous fermentation; minerals processing; economics analyses. *Mailing Add:* 310 Jackson Ave Ridgeway PA 15853-1916

REDNER, SIDNEY, b Hamilton, Can, Nov 10, 51; m 77; c 2. PHYSICS. *Educ:* Univ Calif, Berkeley, AB, 72; Mass Inst Technol, PhD(physics), 77. *Prof Exp:* Vis asst prof, 78-79, asst prof, 79-84, ASSOC PROF PHYSICS, BOSTON UNIV, 84- *Concurrent Pos:* Fel, Univ Toronto, 77-78; vis scientist, Schlumberger-Noll Res, 84-85. *Mem:* Am Phys Soc. *Res:* Chemical, kinetics phase transitions and critical phenomena; percolation theory; polymers; computer simulations. *Mailing Add:* Dept Physics Boston Univ Boston MA 02215

REDO, SAVERIO FRANK, b Brooklyn, NY, Dec 28, 20; m 48; c 2. SURGERY. *Educ:* Queens Col, NY, BS, 42; Cornell Univ, MD, 50; Am Bd Surg & Am Bd Thoracic Surg, dipl. *Prof Exp:* From intern to resident surg, 50-57, asst attend surgeon, 59-60, ASSOC ATTEND SURGEON & DIR DEPT PEDIAT SURG, NY HOSP, 60-; PROF SURG, MED COL, CORNELL UNIV, 73- *Concurrent Pos:* Ledyard fel, NY Hosp, 57-59; asst prof surg, Med Col, Cornell Univ, 59-61, clin assoc prof, 61- *Mem:* Soc Univ Surg; fel Am Col Surg; Am Acad Pediat; Am Fedn Clin Res; Am Surg Asn; Sigma Xi. *Res:* Etiology and methods of surgical management of peptic esophagitis and esophageal pathology; cardiovascular research, including development of an artificial heart and procedures for myocardial revascularization. *Mailing Add:* Dept Surg 525 E 68th St New York NY 10021

REDONDO, ANTONIO, b Guatemala City, Guatemala, Dec 10, 48; US & Span citizen; m 71; c 3. SURFACE SCIENCE. *Educ:* Utah State Univ, BSc, 71; Calif Inst Technol, MSc, 72, PhD(appl physics), 77. *Prof Exp:* Asst prof physics & surface sci, Univ Los Andes, Merida, Venezuela, 77-80; vis assoc, 80-81, res assoc chem, Calif Inst Technol, 81-83; STAFF MEM, LOS ALAMOS NAT LAB, 83- *Mem:* Am Phys Soc; Am Chem Soc. *Res:* Applications of quantum mechanics to the study of surfaces, interfaces and solids. *Mailing Add:* Los Alamos Nat Lab Group MEE-11 MS D429 Los Alamos NM 87545

REDSHAW, PEGGY ANN, b Beardstown, Ill, Sept 4, 48; m 85. MICROBIAL GENETICS. *Educ:* Quincy Col, BS, 70; Ill State Univ, PhD(biol sci), 74. *Prof Exp:* Fel microbiol, Sch Med, St Louis Univ, 74-77; asst prof biol, Wilson Col, 77-79; PROF BIOL, AUSTIN COL, 79- *Concurrent Pos:* Primary researcher, Cottrell Col Sci grant, Res Corp, 78-83; mem, Action/Adv Comt, Proj Kaleidoscope, 90- *Mem:* Am Soc Microbiol; AAAS; Coun Undergrad Res. *Res:* Genetics of streptomyces; isozyme analysis of ferns. *Mailing Add:* Dept Biol Austin Col PO Box 1177 Sherman TX 75091-1177

REDWINE, ROBERT PAGE, b Raleigh, NC, Dec 3, 47; m 86; c 1. NUCLEAR PHYSICS, PARTICLE PHYSICS. *Educ:* Cornell Univ, BA, 69; Northwestern Univ, PhD(physics), 73. *Prof Exp:* Res assoc physics, Los Alamos Sci Lab, 73-74 & Univ Bern, 74-75; res assoc, Los Alamos Sci Lab, 75-77, staff scientist, 77-79; from asst prof to assoc prof, 79-90, PROF PHYSICS, MASS INST TECHNOL, 90- *Mem:* Am Phys Soc; Am Asn Adv Sci. *Res:* Intermediate energy nuclear and particle physics. *Mailing Add:* Dept Physics Mass Inst Technol Cambridge MA 02139

REDWOOD, R(ICHARD) G(EORGE), b Dorset, Eng, May 1, 36; Can citizen; m 68; c 2. CIVIL ENGINEERING, STRUCTURAL ENGINEERING. *Educ:* Bristol Univ, BSc, 57, PhD(civil eng), 64; Univ Toronto, MASc, 60. *Prof Exp:* Asst engr, Shawinigan Eng Co, 57-58 & W S Atkins & Assoc, 58-59; lectr civil eng, Bristol Univ, 64-65; from asst prof to assoc prof, 65-70, chmn, 84-89, PROF CIVIL ENG, MCGILL UNIV, 74- *Mem:* Can Soc Civil Eng; fel Brit Inst Struct Eng; Brit Inst Civil Eng. *Res:* Structural engineering; behaviour and design of metal structures; numerical analysis. *Mailing Add:* Dept Civil Eng McGill Univ 817 Sherbrooke St W Montreal PQ H3A 2K6 Can

REE, ALEXIUS TAIKYUE, b Korea, Jan 26, 02; m 32; c 4. PHYSICAL CHEMISTRY. *Educ:* Kyoto Imp Univ, Japan, BS, 27, ScD(chem), 31. *Hon Degrees:* DSc, Seoul Nat Univ, 64, Sogang Univ, Seoul, 76, Korea Univ, Seoul, 79. *Prof Exp:* Instr chem, Kyoto Imp Univ, 31-36, from asst prof to prof, 36-45; prof chem & dean col lib arts & sci, Seoul Nat Univ, 45-48; AT KOREA ADVAN INST SCI & TECHNOL. *Concurrent Pos:* Distinguished prof, Korea Advan Inst Sci, Seoul, 73-; from res prof to emer prof chem, Univ Utah, 70- *Honors & Awards:* First Order Civil Merit, Govt Repub Korea; Award, Korean Nat Acad, 60. *Mem:* Am Chem Soc; Am Phys Soc; Korean Chem Soc (pres, 46-48); Korean Nat Acad. *Res:* Chemical kinetics; catalysis; rheology; quantum chemistry; theory of liquids. *Mailing Add:* Korea Advan Inst Sci & Technol PO Box 150 Chong Yang 131- 650 Seoul Republic of Korea

REE, BUREN RUSSEL, b Bentley, Alta, Feb 15, 43; m 66; c 1. ORGANIC CHEMISTRY. *Educ:* Univ Alta, BSc, 64; Univ Ill, MS, 66, PhD(chem), 69. *Prof Exp:* Sr res chemist, 69-73, RES SPECIALIST, 3M CO, 73- *Mem:* Am Chem Soc. *Res:* Molecular weight modification in vinyl polymers. *Mailing Add:* 1790 Neal Ave N Stillwater MN 55082-1704

REE, FRANCIS H, b Kyoto, Japan, Sept 6, 36; US citizen. THEORETICAL PHYSICS. *Educ:* Univ Utah, BA, 57, PhD(physics), 60. *Prof Exp:* RES PHYSICIST, LAWRENCE LIVERMORE LAB, UNIV CALIF, 60- *Mem:* Am Phys Soc; Biophys Soc. *Res:* Classical statistical mechanics, thermodynamics; theoretical studies on thermodynamic and statistical mechanical properties of gases, liquids and solids; high pressure physics; quantum chemistry. *Mailing Add:* Lawrence Livermore Lab Univ Calif L-321 PO Box 808 Livermore CA 94550

REE, WILLIAM O(SCAR), b South Milwaukee, Wis, Mar 13, 13; m 48; c 1. CIVIL ENGINEERING. *Educ:* Univ Wis, BS, 35. *Prof Exp:* Draftsman, Bucyrus Erie Co, Wis, 29-31; draftsman, Soil Conserv Serv, USDA, 35-37, jr engr, SC, 37-38, proj supvr, 38-41, proj supvr, Okla, 41-54, proj supvr, Agr Res Serv, 54-64, res invest leader, 64-75; hydraul eng consult, 75-78; RETIRED. *Mem:* Am Soc Agr Engrs; Am Soc Civil Engrs; Int Asn Hydraul Res. *Res:* Hydraulics; hydraulics of conservation structures, hydrology and sedimentology. *Mailing Add:* 2015 W Tenth Stillwater OK 74074

REEBER, ROBERT RICHARD, b Flushing, NY, Jan 22, 37; m 72; c 4. MATERIALS ENGINEERING, MINERAL PHYSICS. *Educ:* NY Univ, BE, 58, MS, 60; Ohio State Univ, PhD(indust mineral), 68. *Prof Exp:* Mat engr transistors, Radio Corp Am, Somerville, Mass, 59-60; mat engr ceramics, Aerospace Res Labs, 60-63, vis res assoc phase transformations, 63-69; asst prof mat sci, Mich State Univ, 69-70; asst prof metall eng, Arya-Mehr Indust Univ, 70-71; res assoc crystallog, T H Aachen, 72-73; sr res asst mineral, Cambridge Univ, 73-74; prog mgr geochem eng, Dept Energy, 76-81; MAT ENGR, ARMY RES OFF, 81- *Concurrent Pos:* Adj asst prof, Mich State Univ, 71-73; sr res Fulbright, Inst Crystallog, T H Aachen, Ger, 72; vis sr researcher, Electron Micros Inst, Fritz Haber Inst Max Planck Soc, Berlin, 75, 86; vis researcher, Inorg Mat Div, Nat Bur Standards, 78-80; joint adj assoc prof geol & physics, Univ NC-Chapel Hill, 83-89, adj prof geol, 90- *Mem:* Fel AAAS; Mat Res Soc; Mineral Soc Am; Am Ceramic Soc; Am Crystallog Asn; Am Physics Soc. *Res:* Crystal chemistry; thermalexpansion; phase transformations in solids; materials engineering; materials policy and planning; X-ray diffraction; tribology. *Mailing Add:* Army Res Off PO Box 12211 Research Triangle Park NC 27709-2211

REEBURGH, WILLIAM SCOTT, b Port Arthur, Tex, Feb 25, 40; m 63; c 3. CHEMICAL OCEANOGRAPHY. *Educ:* Univ Okla, BS, 61; Johns Hopkins Univ, MA, 64, PhD(oceanog), 67. *Prof Exp:* Res asst, Chesapeake Bay Inst, Johns Hopkins Univ, 61-68; from asst prof to assoc prof, 68-77, PROF MARINE SCI, INST MARINE SCI, UNIV ALASKA, 77- *Mem:* Am Chem Soc; Geochem Soc; Am Soc Limnol & Oceanog; Am Soc Microbiol; Am Geophys Union. *Res:* Physical and chemical properties of sea water; gases in natural waters; composition of interstitial waters, microbial ecology, marine and terrestrial methane biogeochemistry. *Mailing Add:* Inst Marine Sci Univ Alaska Fairbanks AK 99775-1080

REECE, JOE WILSON, b Elkin, NC, Mar 1, 35; m 55; c 2. MECHANICAL ENGINEERING, APPLIED MATHEMATICS. *Educ:* NC State Univ, BSNE, 57, MS, 61; Univ Fla, PhD(eng mech), 63. *Prof Exp:* Instr eng mech, NC State Univ, 58-61; from asst prof to assoc prof, Auburn Univ, 64-76, prof mech eng, 76-83; PRES, REECE ENG ASSOC, 70-; PROF ENG, SURRY COMM COL, 73- *Concurrent Pos:* Dep dir, Div Oper Reactors, US Nuclear Regulatory Comn, 76-78; consult combustion eng, US Army, E I du Pont de Nemours & Co, US Nuclear Regulatory Comn & Westinghouse. *Mem:* Soc Eng Educ; Am Soc Mech Engrs. *Res:* Inviscid unsteady flow; steam generator mechanics; reactor hydraulics. *Mailing Add:* Rte 1 Box 1-F Boonville NC 27011

REECK, GERALD RUSSELL, b Tacoma, Wash, Dec 28, 45; m 67; c 2. BIOCHEMISTRY. *Educ:* Seattle Pac Col, Wash, BA, 67; Univ Wash, PhD(biochem), 71. *Prof Exp:* Res assoc develop biochem, Lab Nutrit & Endocrinol, Nat Inst Arthritis, Metab & Digestive Dis, NIH, Bethesda, Md, 71-74; asst prof, 74-78, ASSOC PROF BIOCHEM, KANS STATE UNIV, MANHATTAN, KANS, 78- *Mem:* Am Chem Soc. *Res:* Structure and function of chromatin proteins, especially the nonhistone chromatin proteins; trypsin inhibitors. *Mailing Add:* Dept Biochem Willard Hall Kans State Univ Manhattan KS 66506

REED, A THOMAS, b Anderson, Ind, Apr 10, 46. INORGANIC CHEMISTRY. *Educ:* Ball State Univ, Ind, BS, 69. *Prof Exp:* Asst city chemist, Anderson Munic Water Works, Anderson, Ind, 65-69; NSF grant chem, Univ Nebr, Lincoln, 75-76; MEM STAFF, DEPT CHEM, MIAMI UNIV, OXFORD, 76- *Mem:* Am Chem Soc; Am Crystallog Asn; Sigma Xi. *Res:* Crystal and molecular structure of metal-dicarboxylate including the synthesis of compounds, growth of single crystals and x-ray determination. *Mailing Add:* 2633 McLain Ct Grove City OH 43123

REED, ADRIAN FARAGHER, medicine, anatomy; deceased, see previous edition for last biography

REED, ALLAN HUBERT, b Youngstown, Ohio, Jan 4, 41; m 66; c 2. ELECTROCHEMISTRY, PHYSICAL CHEMISTRY. *Educ:* Thiel Col, AB, 63; Case Western Reserve Univ, MS, 64, PhD(phys chem), 68. *Prof Exp:* Res asst phys chem, Case Western Reserve Univ, 66-68; res chemist, Columbus Labs, Battelle Mem Inst, 68-73; eng scientist res div, AMP, Inc, 73-81; mgr res, Electrochem, 81-90; MGR RES, OHIO DIV, TECHNIC INC, 90- *Mem:* Electrochem Soc; Am Chem Soc; Am Electroplaters Soc; Sigma Xi. *Res:* Electrode processes; batteries, electrodeposition of precious metals; electroforming; high-rate electrochemical processes; electrochemical synthesis; internal reflection spectroscopy. *Mailing Add:* 19 Centennial Dr Poland OH 44514-1707

REED, BRENT C, INSULIN RECEPTOR METABOLISM, PROTEIN TRAFFICKING. *Educ:* Univ Utah, PhD(biochem), 76. *Prof Exp:* Asst prof biochem, Health Sci Ctr, Univ Tex, 80-88; ASSOC PROF BIOCHEM, DEPT BIOCHEM & MOLECULAR BIOL, MED CTR, LA STATE UNIV, 88- *Mailing Add:* Dept Biochem & Molecular Biol La State Univ 1501 Kings Hwy Shreveport LA 71130

REED, BRUCE LORING, b Jackman, Maine, July 31, 34; m 60; c 4. GEOLOGY. *Educ:* Univ Maine, BA, 56; Wash State Univ, MS, 58; Harvard Univ, PhD(geol), 66. *Prof Exp:* GEOLOGIST, ALASKAN BR, US GEOL SURV, 62- *Mem:* Geol Soc Am; Asn Explor Geochem; Soc Econ Geol; Am Geophys Union. *Res:* Geology of Alaska; economic geology and the structural control of ore deposits; tin commodity specialist; exploration geochemistry; mechanics of igneous intrusion; geochronology, chemistry and generation of batholiths in the circum-Pacific region. *Mailing Add:* US Geol Surv 4200 Univ Dr Anchorage AK 99508-4667

REED, CHARLES ALLEN, b Portland, Ore, June 6, 12; m 51; c 3. AGRICULTURAL ORIGINS. *Educ:* Univ Ore, BS, 37; Univ Calif, PhD(zool), 43. *Prof Exp:* Instr zool, Univ Ore, 36-37; asst, Univ Calif, 37-42, asst anat, Med Sch, 43; instr biol, Reed Col, 43-46; from instr to asst prof zool, Univ Ariz, 46-49; from asst prof to assoc prof, Col Pharm, Univ Ill, 49-61; assoc prof biol, Yale Univ & cur mammal & herpet, Peabody Mus, 61-66; prof anthrop & biol sci, Univ Ill, Chicago, 66-67; actg head dept, 67-70, prof, 67-80, EMER PROF ANTHROP, UNIV ILL, CHICAGO, 80- *Concurrent Pos:* Mem, Univ Ore Archaeol Exped, Catlow Caves, 37, John Day Archaeol Exped, 46 & Univ Chicago Oriental Inst Prehist Proj, Iraq, 54-55, Iran, 60 & Turkey, 70; dir, Yale Prehist Exped, Nubia, 62-65; res assoc vert anat, Field Mus, 66- *Honors & Awards:* Archaeol Inst Am Sci Contrib Archaeol, 85. *Mem:* AAAS; Am Asn Phys Anthropologists; Am Anthrop Asn; Am Soc Zoologists; Soc Vert Paleont; Sigma Xi. *Res:* Origins of agriculture, human evolution; evolutionary anatomy; prehistory of Near East. *Mailing Add:* Dept Anthrop Univ Ill Chicago IL 60680

REED, CHARLES E(LI), b Findlay, Ohio, Aug 11, 13. CHEMISTRY, CHEMICAL ENGINEERING. *Educ:* Case Inst Technol, BS, 34; Mass Inst Technol, ScD(chem eng), 37. *Prof Exp:* Asst prof chem eng, Mass Inst Technol, 37-42; res assoc res lab, Gen Elec Co, 42-45, eng mgr, Chem Div, 45-52, gen mgr, Silicone Prod Dept, 52-59 & Metall Prod Dept, 59-62, vpres & gen mgr, Chem & Metall Div, 62-68, vpres & group exec, Components & Mat Group, 68-71, sr vpres corp technol, 71-79; RETIRED. *Honors & Awards:* Nat Medal of Technol, 91. *Mem:* Nat Acad Eng; AAAS; Am Chem Soc; fel Am Inst Chem Engrs; Am Inst Chemists. *Res:* Colloid chemistry; high polymers; distillation. *Mailing Add:* 3200 Park Ave Bridgeport CT 06604

REED, CHARLES E, b Boulder, Colo, Mar 13, 22. INTERNAL MEDICINE, ALLERGY. *Educ:* Columbia Univ, MD, 45. *Prof Exp:* Intern, Sch Med, Univ Colo, 45-46; resident med, Roosevelt Hosp, 48-51; clin instr med, Med Sch, Univ Ore, 51-58, clin asst prof, 58-61; from asst prof to prof med, Univ Wis-Madison, 61-78; PROF MED MAYO GRAD SCH MED, UNIV MINN, ROCHESTER, 78- *Concurrent Pos:* Pvt pract, Ore, 51-61. *Mem:* Am Acad Allergy (pres, 76); Am Col Physicians; Am Asn Immunol; Am Fedn Clin Res; Cent Soc Clin Res. *Res:* Bronchial asthma; nonasthmatic allergic diseases of the lung. *Mailing Add:* 200 First St SW Rochester MN 55905

REED, CHRISTOPHER ALAN, b Auckland, NZ, Feb 25, 47; c 2. BIOINORGANIC, COORDINATION & ORGANOMETALLIC CHEMISTRY. *Educ:* Auckland Univ, NZ, BSc, 68, MSc hon, 69, PhD(chem), 71. *Prof Exp:* Res assoc chem, Stanford Univ, 71-73; from asst prof to assoc prof, 73-81, PROF CHEM, UNIV SOUTHERN CALIF, LOS ANGELES, 81- *Honors & Awards:* A P Sloan fel, 76; Dreyfus Teacher-Scholar Award, 76. *Mem:* Am Chem Soc; NZ Inst Chem. *Res:* Bioinorganic chemistry of the transition metals particularly the structure and the reactivity of hemes, oxygen carriers and cytochromes; organo-transition metal chemistry and coordination chemistry; magnetic interactions. *Mailing Add:* Dept Chem Univ Southern Calif Los Angeles CA 90089-0744

REED, COKE S, b Austin, Tex, Mar 8, 40; m 61; c 1. MATHEMATICS. *Educ:* Univ Tex, BS, 62, MA, 65, PhD(math), 66. *Prof Exp:* Asst prof math, Ga Inst Tech, 66-67; asst prof, 67-75, assoc prof, 75-81, PROF MATH, AUBURN UNIV, 81- *Mem:* Am Math Soc. *Res:* Real variables; topological dynamics. *Mailing Add:* Supercomput Res Ctr 17100 Science Dr Bowie MD 20715

REED, DALE HARDY, b Houston, Tex, Aug 5, 30; m 54; c 4. EXPLORATION GEOPHYSICS, ELECTRICAL ENGINEERING. *Educ:* Rice Univ, BA, 52, BSEE, 53. *Prof Exp:* Res engr, Atlantic Richfield Co, 53-72, res dir, 64-72, res assoc, 72-82, mgr explor spec projs, 82-85; RETIRED. *Concurrent Pos:* Consult geophys & expert witness, UN. *Mem:* Soc Explor Geophysicists; Europ Asn Explor Geophysicists. *Res:* Borehole logging; reflection seismic; magnetics; geochemistry; planning and evaluation; marine seismic systems; exploration on basalt covered terrain; shear wave seismology. *Mailing Add:* 10415 Coleridge Dallas TX 75218

REED, DANIEL A, b Wichita, Kans, June 12, 57; m 87. PARALLEL PROCESSING, PERFORMANCE ANALYSIS. *Educ:* Univ Mo, Rolla, BS, 78; Purdue Univ, MS, 80, PhD(computer sci), 83. *Prof Exp:* Asst prof computer sci, Univ NC, Chapel Hill, 83-84; from asst prof to assoc prof computer sci, 84-91, SR SOFTWARE ENGR, CTR SUPERCOMPUT RES & DEVELOP, UNIV ILL, URBANA-CHAMPAIGN, 86-, PROF COMPUTER SCI, 91- *Concurrent Pos:* Prin investr, NSF, 84-; consult, ICASE, NASA Langley Res Ctr, 84-; IBM fac develop award, 84-85; NSF presidential young investr award, 87-92; mem, Sigmetrics Bd Dirs, Asn Comput Mach, 91-93. *Mem:* Asn Comput Mach; Inst Elec & Electronics Engrs Computer Soc; AAAS. *Res:* Interaction of architecture, software, and application algorithms across a range of high performance computing systems; parallel computer design. *Mailing Add:* 2413 Digital Computer Lab Univ Ill 1304 W Springfield Ave Urbana IL 61801

REED, DAVID WILLIAM, b Opelousas, La, Dec 7, 52; m 74; c 1. HORTICULTURE, FLORICULTURE. *Educ:* Univ Southwestern La, BS, 74; Cornell Univ, MS, 77, PhD(hort), 79. *Prof Exp:* ASST PROF HORT, TEX A&M UNIV, 78- *Mem:* Am Soc Hort Sci. *Res:* Penetration of foliar-applied compounds into the leafs of plants; structure, development and function of the plant cuticle of leaves; mineral nutrition of plants. *Mailing Add:* Dept Hort Sci Tex A&M Univ College Station TX 77843

REED, DONAL J, b Riverdale, Calif, Apr 24, 24; m 45; c 5. PHARMACOLOGY. *Educ:* Col Idaho, BA, 48; Univ Calif, PhD(physiol), 59. *Prof Exp:* Asst physiol, Univ Calif, 56-57 & 58-59; from instr to prof pharmacol, 62-88, actg chmn dept, 80-88, EMER PROF, COL MED, UNIV UTAH, 88- *Concurrent Pos:* NIH res fel pharmacol, Col Med, Univ Utah, 59-62. *Mem:* AAAS; Am Soc Pharmacol & Exp Therapeut; Am Physiol Soc. *Res:* Mechanism and kinetics of distribution of electrolytes and other substances among the blood, cerebrospinal fluid and the brain; neuropharmacology; pharmaco-kinetics; adrenal steroids. *Mailing Add:* Dept Pharmacol Col Med Univ Utah Salt Lake City UT 84132

REED, DONALD JAMES, b Montrose, Kans, Sept 26, 30; m 49; c 6. BIOCHEMISTRY, ENVIRONMENTAL HEALTH. *Educ:* Col Idaho, BS, 53; Ore State Univ, MS, 55, PhD(chem), 57. *Prof Exp:* Asst, Ore State Univ, 53-55; assoc biochemist cereal invests, Western Regional Res Lab, Agr Res Serv, USDA, Calif, 57-58; asst prof chem, Mont State Univ, 58-62; from asst prof to assoc prof, 62-72, PROF BIOCHEM, ORE STATE UNIV, 72-, DIR ENVIRON HEALTH SCI CTR, 81- *Concurrent Pos:* USPHS spec res fel, NIH, 69-70, mem toxicol study sect, 71-75, mem, Ad Hoc Rev Comt, Nat Cancer Inst, 75-87; Eleanor Roosevelt Am Cancer Soc Int Cancer fel, Karolinska Inst, Stockholm, 76-77; environ sci review panel health res, Environ Protection Agency, 81; environ health sci review comt, Nat Inst Environ Health Sci, 82-85; assoc ed, J Toxicol & Environ Health, 80-84, Toxicol & Appl Pharmacol, 81-84, Cancer Res, 87- & ed, Cell Biol & Toxicol; Burroughs Wellcome Toxicol Scholar Award Selection Comt, 84-87; vis prof, Burroughs Wellcome Travel Grant, MRC Toxicol Unit, Carshalton, Eng, 84; sabbatical scientist, Nat Cancer Inst, NIH, Bethesda, Md, 84-85; mem, Task Group Health Criteria, Int Prog Chem Safety, WHO, 86 & Biochem & Carcinogenesis Rev Comt, Am Cancer Soc, 84-; consult, Univ Calif, San Francisco, 85- *Mem:* Am Soc Biol Chem; Am Soc Pharmacol & Exp Therapeut; Soc Toxicol; Am Asn Cancer Res; Sigma Xi. *Res:* Biological oxidations; environmental toxicology; biochemical anticancer drugs; protective mechanisms of glutathione functions; vitamin E status. *Mailing Add:* Dept Biochem & Biophys Ore State Univ Corvallis OR 97331-6503

REED, DWAYNE MILTON, b Kinsley, Kans, Dec 10, 33; c 2. ENVIRONMENTAL HEALTH. *Educ:* Univ Calif, Berkeley, BA, 55, MPH, 62, PhD(epidemiol), 69; Univ Calif, San Francisco, MD, 60. *Prof Exp:* Asst chief, epidemiol br, Nat Inst Neurol Dis & Blindness, NIH, Guam, 62-64; chief, field unit, Arctic Health Res Ctr, Anchorage, 64-66; assoc res, epidemiol, Sch Pub Health, Univ Calif, 66-69; assoc prof, Sch Pub Health, Univ Tex, 69-71; Dep chief, Nat Inst Neurol Dis & Stroke, NIH, 71-74; epidemiologist, South Pacific Comm, 74-75; chief, epidemiol br, Nat Inst Child Health & Human develop, NIH, 75-78; med Epidemiologist, Calif State Dept Health, 78-79; DIR, HONOLULU HEART PROG, NAT HEART, BLOOD, LUNG INST, NIH, 80- *Concurrent Pos:* Consult, Hawaii Heart Asn, 83-, Japan Heart Found, 84, Int Soc Hypertension, 84, WHO, Philippines, 86, Brunei, 87; liaison officer, US Japan Agreement Cardiovascular Dis, 82- *Honors & Awards:* Commendation Medal, USPHS, 87. *Mem:* Am Pub Health; Soc Epidemiol Res; Am Epidemiol Soc; fel Am Heart Asn. *Res:* Public health and research in epidemiology. *Mailing Add:* Honolulu Heart Prog 347 N Kuakini St Honolulu HI 96817

REED, EDWARD BRANDT, b Longmont, Colo, Jan 11, 20; m 46; c 2. LIMNOLOGY, ECOLOGY. *Educ:* Colo State Univ, BS, 53, MS, 55; Univ Sask, PhD(biol), 59. *Prof Exp:* Fisheries biologist, Sask Dept Natural Resources, 56-59; from asst prof to prof zool, 59-74, mem affil fac zool, Colo State Univ, Pres Ecol Consults, Inc, 74-81; RETIRED. *Concurrent Pos:* Dir, NSF Summer Inst, 60-64; environ consult, Nat Park Serv, 67 & Western Solo Resource Study, 70-71; Environ Protection Agency res grant, 70-71. *Mem:* Fel AAAS; Int Asn Theoret & Appl Limnol; Am Soc Limnol & Oceanog. *Res:* Application of limnological principles to problems associated with water resource use; abatement of ecological problems associated with disturbed lands and resource development; systematics of freshwater cyclopoid copepeds. *Mailing Add:* 1901 Stover St Ft Collins CO 80525

REED, ELIZABETH WAGNER, b Baguio, Philippines, Aug 27, 12; m 40, 46; c 3. HUMAN GENETICS. *Educ:* Ohio State Univ, AB, 33, MA, 34, PhD(plant physiol), 36. *Prof Exp:* Prof biol, Atlantic Christian Col, 38-40; res assoc, Res Found, Ohio State Univ, 43-44; instr plant sci, Vassar Col, 44-45; asst prof bot, Ohio Wesleyan Univ, 45-46; res assoc, Dight Inst Human Genetics, 40-66; asst prof curriculum res staff, Math & Sci Teaching Ctr, 66-69; lectr, Human Genetics, Continuing Educ Women, Univ Minn, Minneapolis, 69-82, asst prof, Women's Studies Dept, 76-82; RETIRED. *Concurrent Pos:* Lectr biol & independent res scholar, Hist Women Sci, 82- *Mem:* Fel AAAS. *Res:* Effects of chemical dusts on transpiration; production and testing of penicillin; population studies on Drosophila; inheritance of mental retardation; school curricula; biology of women; early women in science. *Mailing Add:* 1588 Vincent St St Paul MN 55108

REED, ELLEN ELIZABETH, b Covina, Calif, Sept 16, 40. MATHEMATICS. *Educ:* Gonzaga Univ, BA, 62; Univ Colo, Boulder, MA, 64, PhD(math), 66. *Prof Exp:* From asst prof to assoc prof math, Univ Mass, Amherst, 66-77; mem staff, Dept Math, St Mary's Col, 77-80; CONSULT, 80- *Concurrent Pos:* Lectr, Smith Col, 72; mem staff, Dept Math, Notre Dame, 77 & Col Eng, 79-81. *Mem:* Am Math Soc. *Res:* Uniform spaces and generalizations; extensions and compactifications of topological spaces and convergence spaces. *Mailing Add:* 515 E Angela South Bend IN 46617

REED, EUGENE D, b Vienna, Austria, Oct 12, 19; nat US. ELECTRICAL ENGINEERING. *Educ:* Univ London, BSEE, 42; Columbia Univ, PhD(elec eng), 53. *Prof Exp:* Exec, Bell Labs, 47-75; RETIRED. *Mem:* Nat Acad Eng; Inst Elec & Electronics Engrs. *Mailing Add:* 3125 Middle Ranch Rd Pebble Beach CA 93953

REED, F(LOOD) EVERETT, b North Stonington, Conn, Aug 23, 14; m 39; c 4. MECHANICAL & MARINE ENGINEERING. *Educ:* Webb Inst Naval Archit, BS, 36; Mass Inst Technol, MS, 51. *Prof Exp:* From asst marine engr to sr marine engr, US Maritime Comn, 39-46; asst prof mech eng, Mass Inst Technol, 46-49; mech engr, Arthur D Little, Inc, 49-51, leader, Appl Mech Group, 51-55, sr mech engr & head appl mech, Tech Opers, Inc, 55-59; partner, Conesco Consults, 59-61, tech vpres, Conesco, Inc & Flow Corp, 62; PRES & TREAS, LITTLETON RES & ENG CORP, 62- *Concurrent Pos:* Chmn, SNAME M20 Mach Vibr Panel. *Mem:* Am Soc Lubrication Engrs; Am Soc Mech Engrs; Soc Exp Mech; Soc Naval Archit & Marine Engrs; Acoust Soc Am; Soc Eng Educ; fel AAAS. *Res:* Vibration; stress analysis; applied mechanics; hydrodynamics; ship structure vibration; mechanical vibration; applied mechanics. *Mailing Add:* 95 Russell St PO Box 128 Littleton MA 01460

REED, FRED DEWITT, JR, b Port Arthur, Tex, Apr 25, 37; m 58. MEDICINAL CHEMISTRY. *Educ:* Univ Tex, PhD(med chem), 70. *Prof Exp:* Tex pharmacol res fel med chem, Col Pharm, Univ Tex, 68-69; Phillip Morris res fel, Univ Va, 69-70; Ger sci asst, Org Chem Inst, Univ Heidelberg, 70-72; asst prof med chem, Col Pharm, Idaho State Univ, 73-77, assoc prof, 77-79; CONSULT BIOMED, 79- *Mem:* Am Chem Soc; Am Pharmaceut Asn. *Res:* Design and synthesis of potentially active medicinal agents; natural products. *Mailing Add:* Rte 4 Box 855 Flagstaff AZ 86001

REED, GEORGE FARRELL, b Oswego, NY, Oct 25, 22; m 47; c 4. OTOLARYNGOLOGY, MEDICAL EDUCATION. *Educ:* Colgate Univ, AB, 44; Syracuse Univ, MD, 46. *Prof Exp:* Asst clin prof, Mass Eye & Ear Infirmary, 52-65; prof otolaryngol & chmn dept, State Univ NY Upstate Med Ctr & chief otolaryngol, State Univ Hosp, 65-76, dean, Col Med & Exec vpres, 76-86, PROF OTOLARYNGOL, STATE UNIV NY UPSTATE MED CTR, 65- *Concurrent Pos:* Asst, Am Bd Otolaryngol, 52-56, vpres, 72, pres, 76 & secy-treas, 81-86; sr fac mem, Home Study Courses, Am Acad Ophthal & Otolaryngol, 53-60, consult, 60-, secy continuing educ, 70-80; consult, Mass Health Ctr, Boston, 53-65; consult to surgeon gen, Otolaryngol Training Grant Comt, USPHS, 56-60; consult, Norwood Hosp, Mass, 63-65; consult, Crouse Irving, Syracuse Psychiat, St Joseph's, Vet Admin Hosps & VanDuyn Home & Hosp, Syracuse, NY, 65-; chmn med bd, State Univ Hosp, 68-72; mem sci info prog activities comt, Nat Inst Neurol Dis & Stroke, 70-, chmn, 72-73. *Honors & Awards:* Award of Merit, Am Acad Ophthal & Otolaryngol, 61. *Mem:* AMA; Am Acad Ophthal & Otolaryngol; fel Am Laryngol, Rhinol & Otol Soc; fel Am Col Surg; fel Am Acad Facial Plastic & Reconstruct Surg. *Res:* Clinical and basic research, especially oncology related to the head and neck. *Mailing Add:* 750 E Adams St State Univ NY Health Sci Ctr Syracuse NY 13210

REED, GEORGE HENRY, b Muncie, Ind, Aug 29, 42. BIOPHYSICS, BIOCHEMISTRY. *Educ:* Purdue Univ, BS, 64; Univ Wis, PhD(chem), 68. *Prof Exp:* Lectr chem, Univ Wis, 67-68; asst prof biophys, UnivPa, 71-76, assoc prof biochem & biophys, 76--; PROF, INST FOR ENZYME RES, UNIV WIS. *Concurrent Pos:* NIH fel, Univ Pa, 69-71 & USPHS career develop award, 72-77. *Mem:* Am Chem Soc; Am Soc Biol Chemists; Sigma Xi. *Res:* Spectroscopic investigation of enzyme-substrate complexes; applications of electron paramagnetic resonance and nuclear magnetic resonance spectroscopy in biological chemistry; interactions of inorganic cations in biological processes. *Mailing Add:* 7733 Sundown Dr Verona WI 53593

REED, GEORGE W, JR, b Washington, DC, Sept 25, 20; m 45; c 4. RADIO CHEMISTRY, METEORITICS. *Educ:* Howard Univ, BS, 42, MS, 44; Univ Chicago, PhD(chem), 52. *Prof Exp:* Asst chemist, SAM Labs, Columbia Univ & Metall Labs, Univ Chicago, 44-47; assoc chemist, 52-68, SR SCIENTIST, ARGONNE NAT LAB, 68- *Concurrent Pos:* Sr res assoc, Univ Chicago, 74. *Mem:* Am Chem Soc; Am Geophys Union; Sigma Xi. Geochem Soc; Meteoritical Soc. *Res:* Lunar and meteoritic science; radiochemistry; geocosmochemistry. *Mailing Add:* 5227 University Ave Chicago IL 60615

REED, GERALD, b Berlin, Germany, Mar 14, 13; nat US; m 38; c 1. CHEMISTRY. *Educ:* Prague Univ, PhD(chem), 37. *Prof Exp:* Oderberger Chem Works fel inst tech, Prague Univ, 37; res assoc, Dent Col, Univ Chicago, 38-41; res chemist, Upjohn Co, 41-44; head res dept, Libby, McNeill & Libby, 44-47; mem tech sales, Rohm and Haas Co, 47-56; vpres corp develop, Amber Labs, 78-83; dir res, Red Star Yeast & Prod Co, 56-66, vpres res, 66-78, CONSULT, UNIVERSAL FOODS CORP, 83- *Mem:* Am Chem Soc; Am Soc Enol; Am Soc Bakery Eng; fel Am Asn Cereal Chemists; fel Inst Food Technol. *Res:* Polarography; spectroscopy; fat soluble vitamins; infant nutrition; enzyme technology; industrial microbiology. *Mailing Add:* 1016 Monmouth Ave Durham NC 27701

REED, HAZELL, b Heth, Ark. HORTICULTURE. *Educ:* AM&N Col, BS, 68; Pa State Univ, MS, 73; Univ Ark, PhD(plant sci), 83. *Prof Exp:* Instr ornamental hort, Ark AM&N Col, 68-71; instr hort, Univ Ark, Pine Bluff, 73-76, exten horticulturist specialist, Coop Exten, 76-79 & 83-85, instr horticulture, 79-83, exten horticulturist, Coop Exten Serv, 85, dean agr & home econ, 85-89, VCHANCELLOR, UNIV ARK, PINE BLUFF, 89- *Concurrent Pos:* Mem, exec comt, SE Consortium Int Develop, 87-, chmn bd, 89-90; comt mem, Int Comt Orgn & Policy, 87- & State Found & Agr Coun, 88- *Mem:* Sigma Xi; Am Soc Hort Sci; Southern Region Am Soc Hort Sci; Southern Asn Agr Scientists. *Mailing Add:* Univ Ark PO Box 82 Pine Bluff AR 71601

REED, HORACE BEECHER, b Etowah, Tenn, July 8, 23; m 64. MEDICAL ENTOMOLOGY, INSECT ECOLOGY. *Educ:* Univ Tenn, AB, 46, MS, 48, PhD(entom), 53; Univ Mich, MA, 52. *Prof Exp:* Instr zool, Univ Maine, 48-49; asst, Univ Tenn, 52-53, asst entom, 54, asst, bact, 57; head sci dept high sch, Ga, 59; prof biol & head dept, Shorter Col, Ga, 59-63; from asst prof to assoc prof biol, Mid Tenn State Univ, 64-84; RETIRED. *Res:* Ecology of medically important arthropods. *Mailing Add:* 215 City View Dr Murfreesboro TN 37130

REED, IRVING STOY, b Seattle, Wash, Nov 12, 23. ELECTRICAL ENGINEERING, COMPUTER SCIENCE. *Educ:* Calif Inst Technol, BS, 44, PhD(math), 49. *Prof Exp:* Lincoln Lab, Mass Inst Technol, 51-60; sr staff mem, RAND Corp, 60-63; PROF ELEC ENG & COMPUTER SCI, UNIV CALIF, LOS ANGELES, 63- *Concurrent Pos:* Consult, RAND, MITRE Corp; Dir, Adaptive Sensors Inc. *Mem:* Nat Acad Sci; fel Inst Elec & Electronics Engrs. *Res:* Mathematics; computer design; coding theory; stochastic processes; information theory. *Mailing Add:* Univ Southern Calif 510 Powell Hall Los Angeles CA 90089-0272

REED, JACK WILSON, b Corning, Iowa, Sept 24, 23; m 44; c 1. METEOROLOGY. *Educ:* Univ NMex, BS, 48. *Prof Exp:* Meteorologist, Sandia Nat Labs, 48-51, USAF Air Weather Serv, 51-53; meteorologist, Sandia Nat Labs, 53-89; CONSULT METEOROLOGIST, JWR, INC, 89- *Concurrent Pos:* Chmn working group, Atmospheric Blast Effects, Am Nat Standards Inst, 52- *Mem:* AAAS; Am Meteorol Soc; Am Geophys Union; Am Nuclear Soc; Acoust Soc Am; Air & Waste Mgt Asn. *Res:* Atmospheric propagation of explosion effluents; wind power climatology; meteorological statistics. *Mailing Add:* 1128 Monroe SE Albuquerque NM 87108

REED, JAMES ROBERT, JR, b Wayland, NY, Apr 29, 40; m 65; c 2. FISH BIOLOGY, AQUATIC ECOLOGY. *Educ:* Harvard Univ, AB, 62; Cornell Univ, MS, 64; Tulane Univ, PhD(biol), 66. *Prof Exp:* Fel environ sci, Oak Ridge Nat Lab, 66-68; asst prof, Va Commonwealth Univ, 68-73, assoc prof biol, 73-77; PRES, JAMES R REED & ASSOCS, INC, 77- *Mem:* Am Soc Ichthyologists & Herpetologists; Am Fisheries Soc; Ecol Soc Am; Am Coun Independent Labs; Am Soc Testing & Mat. *Res:* Ecology of fishes with respect to effects of pollution; aquaculture; fish behavior; environmental monitoring. *Mailing Add:* 486 Windmill Pt Hampton VA 23664

REED, JAMES STALFORD, b Jamestown, NY, June 7, 38; m 61; c 1. CERAMICS. *Educ:* Pa State Univ, BS, 60; Alfred Univ, PhD(ceramic sci), 65. *Prof Exp:* Res engr, Harbison-Walker Refractories Co, 60-62; from asst prof to assoc prof, 66-78, chairperson div, 87-90, PROF CERAMIC ENG, NY STATE COL CERAMICS, ALFRED UNIV, 78- *Mem:* Am Ceramic Soc; Nat Inst Ceramic Engrs; Soc Coating Technol. *Res:* Mechanics of fabrication processes; mechanical properties of ceramics; firing whiteware ceramics. *Mailing Add:* Dept Ceramic Eng Alfred Univ Alfred NY 14802

REED, JOEL, b Pittsburgh, Pa, Nov 6, 42. ADVANCED EDUCATION. *Educ:* Pa State Univ, BA, 65, Univ Pittsburgh, MED, 70, PhD, 72. *Prof Exp:* Dir univ commun educ prog, Col Art & Sci, 75-86, DIR ORIENTATION & RECRUTMENT & INSTR MATH, COL GEN STUDIES, UNIV PITTSBURG, 86- *Mem:* Am Educ Res Asn; Nat Coun Educ Opportunity Prog. *Res:* Familiarized minority children with the educational project. *Mailing Add:* Col Gen Studies 419 CL Univ Pittsburgh Pittsburgh PA 15200

REED, JOHN CALVIN, JR, b Erie, Pa, July 24, 30; m 60; c 2. STRUCTURAL GEOLOGY. *Educ:* Johns Hopkins Univ, PhD, 54. *Prof Exp:* Geologist, 54-55 & 57-74, chief, Off Environ Geol, 74-79, GEOLOGIST, US GEOL SURV, 79- *Concurrent Pos:* Chief, Eastern States Br, US Geol Surv, Md, 65-69; mem, Adv Comn Basic Res, Nat Res Coun-US Army, 73-; mem, Comt Stratigraphic Names, US Geol Surv, 74-86; ed, Map & Chart Series, Geol Soc Am; vpres 28th Int Geol Cong; G K Gilbert Fel, US Geol Surv, 84-86; mem adv comt, Geol Div, US Geol Surv, 91-93. *Mem:* Fel Geol Soc Am. *Res:* Geology of crystalline rocks of central and southern Appalachians; geology of central Alaska Range; Appalachian structure and tectonics; geochronology; Precambrian crystalline rocks of Wyoming, Colorado and New Mexico; Precambrian rocks of US; geologic map of North America. *Mailing Add:* US Geol Surv Fed Ctr MS 913 Box 25046 Denver CO 80225

REED, JOHN FRANCIS, physical chemistry, for more information see previous edition

REED, JOHN FREDERICK, b Rockport, Maine, Nov 18, 11; m 34, 69; c 3. BOTANY. *Educ:* Dartmouth Col, AB, 33; Duke Univ, MA, 35, PhD(bot), 36. *Prof Exp:* Asst bot, Duke Univ, 33-34; instr natural sci, Amarillo Col, 36-38; from instr to assoc prof biol, Baldwin-Wallace Col, 38-46, dean men, 42-46; from asst prof to prof bot, Univ Wyo, 46-56; prof, Univ NH, 56-62, dean grad sch, 56-60 & col lib arts, 58-60, vpres, 61, actg pres, 61-62; prof bot & pres, Ft Lewis Col, 62-69; sect head ecol & syst biol, Div Biol & Med Sci, NSF, 69-70; prof ecosysts anal & chmn dept, 70-75, dean acad affairs, 75-77, prof, 75-82, EMER PROF ENVIRON SCI, UNIV WIS, GREEN BAY, 82- *Concurrent Pos:* Technician Pedo Bot Mission to Ruanda-Urundi, Africa, Econ Coop Admin, 51-52; mem environ biol panel, NSF, 56-59; mem coun, Nat Inst Dent Res, 59-63; comn plans & objectives higher educ, Am Coun Educ, 63-66; comt radioactive waste mgt, Nat Res Coun, 70-74; chmn, US Nat Comt Int Biol Prog, 72-74; mem, US Nat Comt Man & the Biosphere, 73-75; mem, US Nat Comn, UNESCO, 73-78. *Mem:* Fel AAAS; Brit Ecol Soc; Am Soc Range Mgt; Ecol Soc Am (secy, 53-57, pres, 63); fel Explorers Club. *Res:* Plant ecology; forest ecology of the Rocky Mountains. *Mailing Add:* 201 W Park Ave Durango CO 81301

REED, JOHN J R, b Eunice, La, Oct 2, 32; m 65. ANALYTICAL CHEMISTRY. *Educ:* Southwestern La Univ, BS, 57; Tulane Univ, PhD(chem), 63. *Prof Exp:* Sr res chemist, Exxon Res & Eng Co, 61-82; CONSULT ANALYTICAL CHEMIST, PVT PRACT, 83- *Res:* Infrared spectroscopy; nuclear magnetic resonance; quality assurance; computer science. *Mailing Add:* PO Box 3762 Baytown TX 77522-3762

REED, JOSEPH, b Bradenton, Fla, Jan 18, 44; m 66. ORGANOMETALLIC CHEMISTRY. *Educ:* Lincoln Univ, AB, 66; Temple Univ, MA, 71; Brown Univ, PhD(chem), 74. *Prof Exp:* Chemist, E I du Pont de Nemours & Co, Inc, 66-70 & Rohm and Haas, 70; chemist, Bell Labs, 74-77; CHEMIST, EXXON RES & ENG CO, 77- *Concurrent Pos:* Adj prof chem, Lincoln Univ, 79- *Mem:* Am Chem Soc; Nat Orgn Prof Advan Black Chemist & Chem Engrs; NY Acad Sci. *Res:* Chemistry and homogeneous catalytic approach to hydrodesulfurization with organometallic complexes. *Mailing Add:* NSF 1800 G St NW Chem Div Washington DC 20550

REED, JOSEPH, b New York, NY, Dec 11, 20; m 51; c 4. CIRCUIT DESIGN, SERVO-MECHANISM. *Educ:* Cooper Union, BEE, 44; Polytech Inst Brooklyn, MEE, 51; Polytech Inst NY, DEE (control physics & math), 75. *Prof Exp:* Asst plant supt, RCA Commun Inc, 47-53; servo-lab head, AM Mach & Foundry Lab, 53-54; head, Radar Dept, Gen Precision Labs, 54-62; vpres res & dir eng, Litcom Div, Litton Syst Inc, 62-72; tech dir, ITT Corp, 72-86; TECH DIR, ITT DEFENSE INC, 86- *Concurrent Pos:* Lectr, Col City New York, 63-68. *Mem:* Fel Inst Elec & Electronics Engrs; fel AAAS. *Mailing Add:* ITT Defense Inc 1000 Wilson Blvd Arlington VA 22209

REED, JOSEPH RAYMOND, b Pittsburgh, Pa, Aug 15, 30; m 60; c 3. HYDRAULIC ENGINEERING. *Educ:* Pa State Col, BS, 52; Pa State Univ, MS, 55; Cornell Univ, PhD(civil eng), 71. *Prof Exp:* Asst fluid mech, dept civil eng, Pa State Col, 52-53; lieutenant-captain & liaison engr, US Army CEngr, Southwestern Div, Dallas, Tex, USAF, 56-59; res assist civil eng, Cornell Univ, 64-66; instr fluids & surv, dept civil eng, Pa State Univ, 53-55 & 59-60, asst prof civil eng, 60-64 & 67-72, assoc prof, 72-87, PROF CIVIL ENG, PA STATE UNIV, 87-, ACAD OFFICER, 89- *Concurrent Pos:* Asst engr, George H McGinness Assocs, 52-55; consult, Ketron Inc, Westvaco, McGraw-Hill Bk Co & others, 63-; publ referee, hydraul div, Am Soc Civil Engrs, 69-; prin investr projs, Fed Hwy Admin, Fed Aviation Admin, Pa Coal Res Bd & others, 71-; chmn, storm water authority, State Col Borough, Pa, 74-78; proposal reviewer, NSF, 80-81; NSF sci fac fel, Cornell Univ, 66-67. *Mem:* Am Soc Civil Engrs; Am Soc Eng Educ; Int Asn Hydraul Res; Sigma Xi. *Res:* Rainfall runoff from highways and runways as it affects the hydroplaning potential of the pavements; dredging mechanics of large river dredges, friction slope modeling in steady nonuniform channel flow. *Mailing Add:* 1394 Penfield Rd State College PA 16801

REED, JUTA KUTTIS, Can citizen. BIOCHEMISTRY. *Educ:* Queen's Univ, BA, 66; Univ Western Ont, MSc, 67; Univ Wis, PhD(biochem), 72. *Prof Exp:* Res fel, Calif Inst Technol, 73-75; asst prof, 75-81, ASSOC PROF CHEM, ERINDALE COL, UNIV TORONTO, 81- *Mem:* Am Chem Soc; Can Biochem Soc. *Res:* Pharmacology of purines in the cns and pns; membrane proteins and developmental control mechanisms in neurogenesis; fluorescence spectroscopy of biopolymers and molecular assemblies. *Mailing Add:* Dept Chem Erindale Col Univ Toronto Mississauga ON L5L 1C6 Can

REED, KENNETH PAUL, b Covington, Ky, Aug 30, 37; m 66; c 3. ANALYTICAL CHEMISTRY, INDUSTRIAL HYGIENE. *Educ:* Thomas More Col, AB, 57; Xavier Univ, MS, 59; La State Univ, PhD(chem), 68. *Prof Exp:* From instr to prof chem, Thomas More Col, 59-78, chmn dept, 68-72, dir freshman studies prog, 72-75, dir develop, 75-78; staff consult, Actus Environ Serv, Florence, 78-80, gen mgr, 80; pres, Chemalytics, 80-89; PRES, REED ENVIRON SERV, 89- *Concurrent Pos:* NSF Sci Fac Fel, 66-67. *Mem:* Am Chem Soc; Am Indust Hyg Asn; Am Conf Govt Indust Hygienists; Sigma Xi; Am Asn Higher Educ. *Res:* Analytical separations; fractional entrainment sublimation; metal chelates; organic synthesis of chelating agents; sampling methods for industrial toxicology; environmental chemistry. *Mailing Add:* Doctors Bldg 33 E Seventh St Rm 300 Covington KY 41011

REED, LESTER JAMES, b New Orleans, La, Jan 3, 25; m 48; c 4. ENZYMOLOGY. *Educ:* Tulane Univ, BS, 43; Univ Ill, PhD(chem), 46. *Hon Degrees:* DSc, Tulane Univ, 77. *Prof Exp:* Asst, Nat Defense Res Comt, Univ Ill, 44-46; res assoc biochem, Med Col, Cornell Univ, 46-48; from asst prof to assoc prof, 48-58, Univ Tex, Austin; assoc dir, 62-63, RES SCIENTIST, CLAYTON FOUND BIOCHEM INST, 49-, DIR, 63-; PROF CHEM, UNIV TEX, AUSTIN, 58-, ASHBEL SMITH PROF, 84- *Honors & Awards:* Lilly Award, Am Chem Soc, 58. *Mem:* Nat Acad Sci; AAAS; Am Chem Soc; Am Soc Biochem & Molecular Biol; Am Acad Arts & Sci. *Res:* Chemistry and function of lipoic acid; enzyme chemistry; structure, function and regulation of multienzyme complexes. *Mailing Add:* Dept Chem & Biochem Univ Tex Austin TX 78712-1096

REED, LESTER W, b Elgin, Iowa, Sept 4, 17; m 41; c 3. SOIL CHEMISTRY, MINERALOGY. *Educ:* Okla Agr & Mech Col, BS, 41, MS, 47; Univ Mo, PhD, 53. *Prof Exp:* From asst prof to assoc prof, 47-57, PROF AGRON, OKLA STATE UNIV, 57- *Mem:* Fel AAAS; Am Soc Agron; Soil Sci Soc Am; Crop Sci Soc Am; Clay Minerals Soc. *Res:* Clay mineralogy; phosphorous chemistry of soils; soil testing; water chemistry and pollution; amino acid composition of plants; laboratory instrumentation. *Mailing Add:* 807 Brooke Lane Stillwater OK 74075

REED, MARION GUY, b Osceola, Iowa, June 26, 31; m 51; c 2. SOIL CHEMISTRY. *Educ:* Iowa State Univ, BS, 57, PhD(soil chem), 63. *Prof Exp:* Res assoc soils, Iowa State Univ, 57-62; res chemist, Chevron Res Co, Calif, 62-69, sr res chemist, 69-73, SR RES ASSOC, CHEVRON OIL FIELD RES CO, 73- *Mem:* Am Soc Agron; Soil Sci Soc Am; Clay Minerals Soc (vpres, 85, pres, 86); Soc Petrol Eng; Am Chem Soc; Soc Core Analysts. *Res:* Physical chemistry of clays and clay-fluid interactions as related to petroleum, uranium and geothermal energy production; kinetics and mechanism of potassium release from micaceous minerals. *Mailing Add:* Chevron Oil Field Res Co PO Box 446 LaHabra CA 90631

REED, MARK ARTHUR, b Suffern, NY, Jan 4, 55. NANOELECTRONICS, ARTIFICIALLY STRUCTURED MATERIALS. *Educ:* Syracuse Univ, BS, 77, MS, 79, PhD(physics), 83. *Prof Exp:* Univ fel, physics dept, Syracuse Univ, 81-83; mem tech staff, 83-88, SR MEM TECH STAFF, ADVAN CONCEPTS BR, CENT RES LAB, TEX INSTRUMENTS INC, 88- *Concurrent Pos:* Jour referee, Appl Physics Lett, J Appl Physics, 85-, Electron Device Lett, NSF, 85-; adj prof physics, Tex A&M Univ, 86- *Mem:* Sr mem Inst Elec & Electronics Engrs; Am Phys Soc; Optical Soc Am; Sigma Xi. *Res:* Investigating ultra-submicron quantum size effect devices; physics of mesoscopic systems, tunneling, heterostructures, quantum wells and superlattices; novel quantum confined semiconductor structures. *Mailing Add:* Tex Inst CRL M/S 154 PO Box 225936 Dallas TX 75265

REED, MELVIN LEROY, b Kalamazoo, Mich, Oct 28, 29; m; c 2. INTERNAL MEDICINE, ONCOLOGY. *Educ:* Kalamazoo Col, AB, 51; Univ Mich, MD, 55. *Prof Exp:* Intern med, City Mem Hosp, Winston-Salem, NC, 55-56; resident, Henry Ford Hosp, Detroit, 58-60, assoc physician med oncol, 62-63; ASSOC PROF ONCOL, SCH MED, WAYNE STATE UNIV, 63- *Concurrent Pos:* Nat Cancer Inst fel clin cancer res, Henry Ford Hosp, Detroit, 60-62; assoc dir, Darling Mem Ctr, Mich Cancer Found, 63-75; dir, Southeastern Mich Regional Cancer Prog, 71-76. *Mem:* Am Fedn Clin Res; Am Soc Clin Oncol; fel Am Col Physicians; Am Asn Cancer Res. *Res:* Experimental drugs and methods in therapy of human malignant diseases. *Mailing Add:* Dept Internal Med 350 Harper H Sch Med Wayne State Univ 540 E Canfield Detroit MI 48201

REED, MICHAEL CHARLES, b Kalamazoo, Mich, May 7, 42; m 73; c 3. MATHEMATICAL PHYSICS. *Educ:* Yale Univ, BS, 63; Stanford Univ, MS, 66, PhD(math), 69. *Prof Exp:* From instr to asst prof math, Princeton Univ, 68-74; PROF MATH, DUKE UNIV, 74-, CHMN DEPT, 86- *Concurrent Pos:* Ed, Duke J Math, 74-80; co-dir, Ctr Math & Comput in Life Sci & Med, 86-; chmn math dept, Duke Univ, 82-85. *Mem:* Am Math Soc. *Res:* Problems in nonlinear harmonic analysis and partial differential equations, especially scattering theory and propagation of singularities; mathematical problems in physiology. *Mailing Add:* Dept Math Duke Univ Durham NC 27701

REED, NORMAN D, b Lyons, Kans, July 6, 35; m 62; c 4. IMMUNOLOGY. *Educ:* Kans State Univ, BS, 59, MS, 62; Montana State Univ, PhD(microbiol), 66. *Prof Exp:* Bacteriologist, Kans State Bd Health, 61-62; res virologist, Mont Vet Res Lab, 62-63; asst prof microbiol, Univ Nebr, Lincoln, 66-70; from asst prof to assoc prof, 70-76, PROF MICROBIOL, MONT STATE UNIV, 76-, HEAD DEPT, 77- *Concurrent Pos:* consult, Dorsey Labs, Nebr, 66-68; career develop award, USPHS, Nat Inst Allergy & Infectious Dis, Dept HEW, 72-77. *Honors & Awards:* Sigma Xi res award, Kans State Univ, 62. *Mem:* Fel AAAS; Am Soc Microbiol; fel Am Acad Microbiol; Am Asn Pathologists; Am Asn Immunol. *Res:* Immunological tolerance; functions of thymus gland; generation and regulation of immune responses; immunoparasitology. *Mailing Add:* Dept Microbiol Mont State Univ Bozeman MT 59717

REED, PETER WILLIAM, b White Plains, NY, July 1, 39. PHARMACOLOGY, GRADUATE STUDIES. *Educ:* Syracuse Univ, BA, 61; State Univ NY Upstate Med Ctr, PhD(pharmacol), 68. *Prof Exp:* Asst res prof, Inst Enzyme Res, Univ Wis-Madison, 70-73; from asst prof to assoc prof physiol, Med Ctr, Univ Mass, 73-75; ASSOC PROF PHARMACOL, MED SCH, 75-, ASSOC DEAN GRAD STUDIES & RES, VANDERBILT UNIV, 84- *Concurrent Pos:* NIH fel, Inst Enzyme Res, Univ Wis-Madison, 68-70; estab investr, Am Heart Asn, 72-77. *Mem:* Am Soc Biol Chemists; Am Soc Pharmacol & Exp Therapeut. *Res:* Leukocyte metabolism; mitochondria and antibiotics; calcium and magnesium in metabolism; toxic antibiotic effects on cells. *Mailing Add:* 411 Kirkland Hall Vanderbilt Univ Nashville TN 37240

REED, R(OBERT) M(ARION LAFOLLETTE), b Bend, Ore, Sept 13, 06; m 32; c 3. CHEMICAL ENGINEERING. *Educ:* Univ Wash, BS, 29, MS, 30, PhD(chem), 35. *Prof Exp:* Res chemist, Procter & Gamble Co, 30-37; chemist, Swann & Co, 37-38; chemist, Gas Processes Div, Girdler Corp, 38-40, tech dir, 40-43, chief process engr, 43-45, chief chem engr, 45-49, tech dir, 49-55, proj mgr, 55-57, tech adv, 57-60, tech dir, 60-69, CONSULT, GAS PROCESSES DIV, C&I/GIRDLER, INC, 70- *Mem:* Am Chem Soc; Am Inst Chem Engrs; Sigma Xi. *Res:* Essential oils and soap perfumes; detergents; gas purification; hydrogen and hydrogen synthetic sulfide manufacturing; heavy water production; ammonia manufacturing, storage and transportation; nitric acid and ammonium nitrate production; urea production. *Mailing Add:* 331 Scottsdale House 4800 N 68th St Scottsdale AZ 85251-1102

REED, RANDALL R, b Davis, WVa, Sept 1, 21; m 48; c 2. ANIMAL PHYSIOLOGY. *Educ:* WVa Univ, BSc, 49, MSc, 52; Ohio State Univ, PhD, 60. *Prof Exp:* County agr agent, WVa Univ, 49-50, res asst animal sci, 50-51; res assoc animal sci, 52-55, asst prof, 55-57, res asst animal physiol, 57-59, assoc prof 60-65, PROF ANIMAL SCI, RUTGERS UNIV, 65- *Concurrent Pos:* State exten meat specialist, 59-60; dist supvr, Coop Exten Serv, 65- *Mem:* Am Soc Animal Sci; Am Meat Sci Asn. *Res:* Animal reproductive physiology and parasitology; meat and carcass evaluation studies; beef industry education. *Mailing Add:* Seven St Dunstans Garth Baltimore MD 21212

REED, RAYMOND EDGAR, b Kankakee, Ill, May 11, 22; m 46; c 2. ANIMAL PATHOLOGY. *Educ:* Wash State Univ, BS, 50, DVM, 51; Am Col Vet Path, dipl. *Prof Exp:* From asst prof to assoc prof, 52-59, actg head dept, 64-65, head dept, 65-77, PROF VET SCI, UNIV ARIZ, 59- *Mem:* Am Vet Med Asn; Am Col Vet Path; Sigma Xi. *Res:* Pathology of animal diseases. *Mailing Add:* 3441 N Olsen Ave Tucson AZ 85719

REED, RICHARD JAY, b Gilmer, Tex, July 23, 28; m 53; c 3. MEDICINE, PATHOLOGY. *Educ:* Tulane Univ, MD, 52; Am Bd Path, dipl, 61. *Prof Exp:* Gen pract, Tex, 53-55; from instr to prof, Sch Med, Tulane Univ, 57-75, clin prof, 75-85, prof path, 85-90; STAFF MEM, REED LAB SKIN PATH, 90- *Concurrent Pos:* Fel, Tulane Univ, 57-60; fel surg path, Barnes Hosp, St Louis, Mo, 60-61; fel, Warren Found Path Lab, Tulsa, Okla, 70-71; consult, USPHS Hosp, New Orleans, 62-; consult, Ochsner Found Hosp, 65-; physician, Charity Hosp, New Orleans, 71-90; surg path, Touro Infirmary & partner, Dermatopath Lab, New Orleans, 75-85. *Mem:* Fel Col Am Path; AMA; Int Acad Path; Am Soc Clin Pathologists. *Res:* Dermatopathology; orthopedic and surgical pathology. *Mailing Add:* 1401 Foucher St New Orleans LA 70115

REED, RICHARD JOHN, b Braintree, Mass, June 18, 22; m 50; c 3. METEOROLOGY. *Educ:* Calif Inst Technol, BS, 45; Mass Inst Technol, ScD(meteorol), 49. *Prof Exp:* Mem staff, Mass Inst Technol, 49-54; from asst prof to assoc prof, 54-63, PROF METEOROL, UNIV WASH, 63- *Concurrent Pos:* Ed, J Appl Meteorol, 66-68; Carl Gustaf Rossby Res Metal, Am Meteorol Soc, 89. *Honors & Awards:* Meisinger Award, Am Meteorol Soc, 64, 2nd Half Century Award, 72; Charles Franklin Brooks Award, Am Meteorol Soc, 84. *Mem:* Nat Acad Sci; Am Meteorol Soc (pres, 72); Am Geophys Union. *Res:* Weather analysis and forecasting; stratospheric and tropical meteorology. *Mailing Add:* Dept Atmospheric Sci Univ Wash Seattle WA 98195

REED, RICHARD P, b Hammond, Ind, May 17, 34; m 80; c 3. PHYSICAL METALLURGY. *Educ:* Purdue Univ, BS, 56; Univ Colo, MS, 58; Colo Sch Mines, MS, 62; Univ Denver, PhD(metall), 66. *Prof Exp:* Supvry metallurgist, Nat Bureau Standards, 57-79, chief, Fracture & Deformation Div, 79-90; CONSULT, 90- *Concurrent Pos:* Ed, Advances Cryog Eng Mat & Cryogenic Mat Series. *Honors & Awards:* Silver Medal & Gold Medal, Nat Bur Standards. *Mem:* Am Phys Soc; Am Inst Mining, Metall & Petrol Engrs; Am Soc Testing & Mat; Am Welding Soc. *Res:* Deformation, fracture, phase transformations, materials at low temperatures. *Mailing Add:* 2625 Iliff Boulder CO 80303

REED, ROBERT MARSHALL, b Berea, Ohio, June 29, 41; m 66. PLANT ECOLOGY. *Educ:* Duke Univ, BA, 63; Wash State Univ, PhD(plant ecol), 69. *Prof Exp:* Asst prof biol, Univ Ottawa, 69-77; res assoc, 77-80, staff res mem, 80-83, GROUP LEADER, OAK RIDGE NAT LAB, 83- *Mem:* Ecol Soc Am; Am Inst Biol Scientists; Soil Conserv Soc Am; AAAS. *Res:* Synecology, with emphasis on forest communities and relationship of vegetation to soil; environmental assessment of energy and defense-related projects; environmental regulatory analysis. *Mailing Add:* 104 E Morningside Dr Oak Ridge TN 37831

REED, ROBERT WILLARD, b Fountain Hill, Pa, Jan 31, 41; m 63. LOW TEMPERATURE PHYSICS. *Educ:* Lafayette Col, BS, 63; Pa State Univ, PhD(physics), 68. *Prof Exp:* Instr, Pa State Univ, 68-70, asst prof physics, 70-83; res engr, 83-85, sr res eng, 85-88, SUPVR, UNITED TECHNOL RES CTR, 88- *Mem:* Acoust Soc Am; Am Phys Soc; AAAS. *Res:* Low temperature physics; ultrasonics and hypersonics; fermi surface studies; propagation of sound in liquid helium; superconductivity; nondestructive testing and evaluation; underwater sound propagation. *Mailing Add:* United Technol Res Ctr Mail Stop 86 East Hartford CT 01608

REED, ROBERTA GABLE, b Baltimore, Md, Sept 18, 45; m 67. BIOCHEMISTRY. *Educ:* Lebanon Valley Col, Pa, BS, 67; Wesleyan Univ, MA, 69, PhD(org chem), 71. *Prof Exp:* Res asst chem, Res Triangle Inst, NC, 71-72; instr chem, State Univ NY Col Oneonta, 72-73; RES BIOCHEMIST, MARY IMOGENE BASSETT HOSP, COOPERSTOWN, NY, 73- *Concurrent Pos:* Vis biologist, Ind Univ, Bloomington, 79-80; vis scientist, John Hopkins Univ Sch Med, Baltimore, 87. *Mem:* Am Soc Biol & Molecular Biology; Am Chem Soc; Am Asn Clin Chem; Protein Soc. *Res:* Dynamic and structural relationships in binding of ions and small organic molecules to serum transport proteins. *Mailing Add:* Mary Imogene Bassett Hosp Cooperstown NY 13326

REED, RONALD KEITH, b Mountain Top, Ark, May 6, 32; m 65; c 2. OCEANOGRAPHY. *Educ:* Ark Polytech Col, BS, 58; Ore State Univ, MS, 73. *Prof Exp:* Oceanogr, Coast & Geodetic Surv, 58-65, OCEANOGR, US DEPT COM, NAT OCEANIC & ATMOSPHERIC ADMIN, 65- *Mem:* Am Geophys Union; Am Meteorol Soc. *Res:* Circulation of offshore and inshore currents in the subarctic Pacific and Gulf of Alaska; relation of flow to important ecosystems. *Mailing Add:* Pac Marine Environ Lab 7600 Sand Point Way NE Seattle WA 98115

REED, RUSSELL, JR, b Glendale, Calif, Dec 25, 22; m 56; c 3. PROPELLANT CHEMISTRY, ORGANIC CHEMISTRY. *Educ:* Univ Calif, Los Angeles, BS, 44, PhD(org chem), 50. *Prof Exp:* Res chemist, US Naval Ord, China Lake Test Sta, 51-58; head org & polymer res sect, Hughes Tool Co, Culver City, Calif, 58-59; head, Dept Chem, Rocket Power Inc, Mesa, Ariz, 59-64; head propellant res sect, Thiokol Corp, Brigham City, Utah, 64-72; head appl res & processing div, 72-76, SR RES SCIENTIST, US NAVAL WEAPONS CTR, NWC, 76- *Honors & Awards:* William B McLean Award. *Mem:* Am Chem Soc; Am Inst Aeronaut & Astronaut; sr fel Naval Weapons Ctr. *Res:* Propellant chemistry combustion; pryotechnic and explosive chemistry; processing of energetic materials; polymer chemistry. *Mailing Add:* 2026 S Mono Ridgecrest CA 93555-4977

REED, RUTH ELIZABETH, b Ames, Iowa, Dec 14, 46; m 72. CHEMISTRY EDUCATION. *Educ:* Winthrop Col, BA, 68; Va Polytech Inst & State Univ, PhD(biochem), 74. *Prof Exp:* Fel biochem, Dept Physiol Chem, Sch Med, Johns Hopkins Univ, 74-76; from asst prof to assoc prof, 76-88, PROF CHEM, JUNIATA COL, 88- *Concurrent Pos:* Nat Heart, Lung & Blood Inst res fel, Johns Hopkins Univ, 75-76; Fulbright award, Univ Gottingen, 68-69; Fulbright inter-univ exchange, Polytech de Lille, France, 81; vis assoc prof chem, Univ NC, Chapel Hill, 85, 86. *Mem:* Am Chem Soc. *Res:* Regulation of carbon and nitrogen metabolism in photosynthetic tissue of higher plants. *Mailing Add:* Dept Chem Juniata Col 1700 Moore St Huntingdon PA 16652

REED, SHERMAN KENNEDY, b Chicago, Ill, Apr 11, 19; m 43; c 3. CHEMISTRY. *Educ:* Univ Ill, BS, 40, Cornell Univ, PhD(chem), 49. *Prof Exp:* Asst chem, Cornell Univ, 40-43, Nat Defense Res Comt, 43; res scientist, SAM Labs, Columbia Univ, 43-46; from instr to asst prof chem, Bucknell Univ, 46-50; group leader, Res & Develop Dept, Westvaco Chem Div, FMC Corp, 50-53, asst dir res, Niagara Chem Div, 53-57, dir res & develop, Chem & Plastics Div, 57-60, dir cent res dept, 60-72, vpres technol, Chem Group, 72-74, vpres, 74-84; RETIRED. *Concurrent Pos:* Consult, Smith Kline Corp, 47-50; dir, Franklin Inst, Birkett Mills, Avicon, Inc & Indust Res Inst; chmn, Franklin Res Ctr. *Mem:* Am Chem Soc. *Res:* Fluorine phosphorus and agricultural chemistry; plastics; computer applications to chemistry; toxicology. *Mailing Add:* 14 Sailfish Rd Vero Beach FL 32960-5279

REED, STUART ARTHUR, b Erie, Pa, Jan 29, 30; m 53; c 4. ZOOLOGY, PHYSIOLOGY. *Educ:* Kent State Univ, BS, 51, MS, 53; Mich State Univ, PhD(zool), 62. *Prof Exp:* Instr physiol, Mich State Univ, 57-58, zool, 59-60, biol, 61-63, asst prof, 63-69; assoc prof, 69-73, PROF ZOOL, UNIV HAWAII, 73- *Concurrent Pos:* NSF sci fac fel, 66-67. *Mem:* AAAS; Nat Asn Biol Teachers; Am Soc Zoologists; Sigma Xi; Am Inst Biol Sci. *Res:* Invertebrate zoology; curriculum developmental in marine biology. *Mailing Add:* Dept Zool Univ Hawaii Edmondson Hall Honolulu HI 96822

REED, TERRY EUGENE, b Mechanicsburg, Pa, Oct 7, 45; m 73; c 3. MEDICAL GENETICS, EPIDEMIOLOGY. *Educ:* Juniata Col, BS, 67; Ind Univ, PhD(med genetics), 71; Univ Pittsburgh. MPh, 82, Am Bd Med Genetics, dipl, 82. *Prof Exp:* Fel, 71-72, from instr to assoc prof, 72-91, PROF MED GENETICS, SCH MED, IND UNIV, 91- *Concurrent Pos:* Dir, Genetic Serv, Indianapolis Comprehensive Sickle Cell Ctr, 73-77; bd dir, Am Dermatoglyphics Asn, 76-79; consult, Genetic Dis Sect, Maternal & Child Health Div, Ind State Bd Health, 85-87, birth defects surveillance comt, Great Lakes Regional Serv Network, 86-; data base mgr, Vet Twin Study Coord Ctr, Nat Heart, Lung & Blood Inst, 85-91. *Mem:* Am Soc Human Genetics; Int Dermatoglyphics Asn; Am Dermatoglyphics Asn; Sigma Xi. *Res:* Studies of dermatoglyphics in man to better understand their inheritance and usefulness as a developmental marker in twins and individuals with congenital abnormalities; the role of environmental factors and genetic predisposition to inherited disease. *Mailing Add:* Dept Med & Molecular Genetics Sch Med Ind Univ Indianapolis IN 46202-5251

REED, THEODORE H(AROLD), b Washington, DC, July 25, 22; m 80; c 2. VETERINARY MEDICINE. *Educ:* Kans State Col, DVM, 45. *Prof Exp:* Instr vet path, Kans State Col, 45-46; asst state vet, Ore, 46-48; vet, Portland Zoo, 49-55; vet, 55-56, actg dir, 56-58, dir, Nat Zool Park, 58-; RETIRED. *Concurrent Pos:* Pvt pract, Idaho & Ore, 49-55; mem, Mayor's Zoo Comn, Portland Zoo, 51-55. *Honors & Awards:* Arthur S Flemming Award, 62. *Mem:* Fel Am Asn Zool Parks & Aquariums (pres, 63-64). *Res:* Zoo administration and zoo veterinary medicine. *Mailing Add:* 104 Cherry St Milford DE 19963

REED, THOMAS BINNINGTON, b Chicago, Ill, Jan 2, 26; m 47; c 4. SYNTHETIC FUEL TECHNOLOGY. *Educ:* Northwestern Univ, BS, 47; Univ Minn, PhD(phys chem), 52. *Prof Exp:* Chemist, Oil Explor & Develop Lab, Shell Oil Co, 47-48; res chemist, Linde Air Prod Co, Union Carbide & Carbon Corp, 52-59; mem staff, Solid State Div, Lincoln Lab, Mass Inst Technol, 59-77; sr staff, Biochem Conversion Br, 77-80, PRIN SCIENTIST, THERMOCHEM & ELECTROCHEM RES BR, NAT SOLAR ENERGY RES INST, 80-; RES PROF, DEPT CHEM ENG, COLO SCH MINES, 86- *Concurrent Pos:* Sci Res Coun sr fel inorg chem, Oxford Univ, 65-66; mem high temperature chem comt, NSF; grant, Methanol as a Synthetic Fuel, J B Hawley Found, 74-; mem renewable resources comt, NSF, 75- *Mem:* Am Chem Soc. *Res:* X-ray crystallography; thermal plasmas; solid state chemistry; high temperature processes; heat mirrors for solar insulation; synthetic fuel manufacture and use, biomass conversion processes; oil spill control. *Mailing Add:* 1810 Smith Rd Golden CO 80401

REED, THOMAS EDWARD, b Gadsen, Ala, Nov 12, 23; m 49; c 2. HUMAN GENETICS, BEHAVIORAL GENETICS. *Educ:* Univ Calif, AB, 48; Univ London, PhD(zool), 52. *Prof Exp:* Jr geneticist, Univ Mich, 52-56, asst prof, 56-57, res assoc, 57-60; assoc prof zool & pediat, 60-69, PROF ZOOL & ANTHROP, UNIV TORONTO, 69- *Concurrent Pos:* Vis assoc prof, Sch Pub Health, Univ Calif, Berkeley, 65-66. *Mem:* AAAS; Am Soc Human Genetics; Genetics Soc Am; Int Soc Biomed Res Alcoholism; Behav Genetics Asn; Can Soc Phys Anthrop. *Res:* Human behavioral genetics; genetics of responses to alcohol. *Mailing Add:* Dept Zool Univ Toronto Toronto ON M5S 1A1 Can

REED, THOMAS FREEMAN, b Webster Co, Iowa, Jan 9, 37; m 58; c 2. POLYMER CHEMISTRY, RUBBER CHEMISTRY. *Educ:* Univ Iowa, BS, 59; Univ Akron, MS, 65, PhD(polymer sci), 70. *Prof Exp:* Prod engr, B F Goodrich Co, 59-65, res chemist, 65-66; res chemist, Univ Akron, 66-68; mem tech staff polymer sci, Bell Tel Labs, 70-73; res scientist, 73-78, sect head tire cord adhesives, 78-80, MGR RUBBER RES, GEN CORP RES DIV, 80- *Mem:* Am Chem Soc; Sigma Xi; Adhesion Soc. *Res:* Light scattering of polymers in solution; diffusion of polymers in solution; physical properties of polymers; rubber dynamic properties; rubber to cord adhesives. *Mailing Add:* Duinweg 2 Wassenaar 2243 DH Netherlands

REED, WARREN DOUGLAS, b Boulder, Colo, June 4, 38. BIOCHEMISTRY, TOXICOLOGY. *Educ:* Univ Calif, Riverside, BA, 61, PhD(toxicol & biochem), 68. *Prof Exp:* NIH staff fel, Geront Res Ctr, NIH, Md, 68-71; fel physiol chem, Sch Med, Johns Hopkins Univ, 72-74; mem staff, Dept Pediat Res, 74-81, assoc prof pediat res, Sch Med, Univ Md, Baltimore, 81-8581-; vis prof, Univ Calif, San Francisco, 85-87; DEP RADIATION SAFETY OFF, VET ADMIN MED CTR, SAN FRANCISCO, 88- *Mem:* Am Chem Soc; Entom Soc Am. *Res:* Enzyme control and mechanisms. *Mailing Add:* Pediat Res Univ Md Sch Med 655 W Baltimore St Baltimore MD 21201

REED, WILLIAM, CELL BIOLOGY, CYTOSKELETON. *Educ:* Univ Calif, San Francisco, PhD(biophysics), 84. *Prof Exp:* Fel, Ctr Neural Biol & Behav Col Physicians & Surgeons, Columbia Univ, NY, 84-86; FEL BIOPHYSICS, UNIV NC, 86- *Res:* Cell surface receptors. *Mailing Add:* Div Rheumatol & Immunol Univ NC 932 Fac Lab Off Bldg 931-H Chapel Hill NC 27514

REED, WILLIAM ALFRED, b Rochester, NY, Jan 5, 36; m 58; c 2. EXPERIMENTAL SOLID STATE PHYSICS. *Educ:* Oberlin Col, BA, 57; Northwestern Univ, PhD(physics), 62. *Prof Exp:* Mem tech staff, Metal Physics Dept, Bell Labs, 61-71, mem staff, Condensed State Res Dept, 71-79, Glass Res Dept, 79-85; AT BELL TEL LABS. *Mem:* Am Phys Soc; Am Optom Soc. *Res:* Galvanomagnetic effects; Fermi surface studies; resistive behavior of superconductors; Compton scattering of x-rays. *Mailing Add:* Bell Tel Labs Rm 6D 209 Murray Hill NJ 07974

REED, WILLIAM DOYLE, b Eupora, Miss, Sept 25, 97; m 26. ENTOMOLOGY. *Educ:* Miss A&M Col, BS, 22. *Prof Exp:* Entomologist, fruit insects, USDA, Fresno, Calif, 25-30, tobacco pests, Richmond, Va, 30-42, insects & other pests, US Army, 42-65; RETIRED. *Mem:* AAAS; hon mem Entom Soc Am; Smithsonian Assocs. *Res:* Control measures for pests of fruits and tobacco; control of insects and other arthropods that are vectors of diseases affecting humans. *Mailing Add:* 1330 Massachusetts Ave NW Washington DC 20005

REED, WILLIAM EDWARD, b Columbia, La, July 15, 14; m 42; c 3. SOIL CHEMISTRY. *Educ:* Southern Univ, BS, 37; Iowa State Univ, MS, 41; Cornell Univ, PhD(soil sci), 46. *Prof Exp:* Asst agr engr, Soil Conserv Serv, USDA, 36-37; county agent, La State Univ, 37-40; instr soil sci & chem, Southern Univ, 42-47; agr res specialist, US Dept State, Liberia, 47-49; dean, Sch Agr, Agr & Tech Col NC, 48-57; chief field staff, Int Coop Admin contract, Int Develop Serv, Inc, Ghana, 57-59, Int Coop Admin rep, Lome, 61, opers officer, Ibadan, 61, asst dir, Lagos, 61-68, dep dir mission to Ethiopia, 68-72, officer-in-residence, USAID, 72-74, spec asst to chancellor int progs, 74-76, assoc dean res & spec projs, 78, dir, int progs & spec projs, NC A&T State Univ, 78-84; RETIRED. *Concurrent Pos:* Mem US deleg, Soviet Union, 55; mem US deleg, UN Conf Sci & Technol, 63. *Mem:* AAAS; Am Chem Soc; Am Soc Agron; Am Foreign Serv Asn; Soil Sci Soc Am. *Res:* Soil genesis, fertility and morphology; agronomy. *Mailing Add:* 2711 McConnell Rd Greensboro NC 27401

REED, WILLIAM J, b Hastings, UK, Jan 19, 46; Can citizen; m 74; c 2. STOCHASTIC MODELLING, OPTIMIZATION. *Educ:* Imp Col, Univ London, BSc, 68; McGill Univ, MSc, 70; Univ BC, PhD(appl math), 75. *Prof Exp:* Res assoc ecol, Univ BC, 77-79; PROF STATIST, UNIV VICTORIA, 79- *Concurrent Pos:* Lectr statist, Portmouth Polytech, 70-71. *Mem:* Resource Modeling Asn; Soc Indust & Appl Math; Biomet Soc; Inst Math Statist; Statist Soc Can. *Res:* Resource management modelling; management modeling and statistics in forestry and fisheries; resource economics. *Mailing Add:* Dept Math & Statist Univ Victoria PO Box 3045 Victoria BC V8W 3P4 Can

REED, WILLIAM ROBERT, b Graham, Tex, Oct 21, 22; m 46; c 4. PHYSICAL INORGANIC CHEMISTRY. *Educ:* Univ Okla, BS, 48, MS, 50; Mich State Univ, PhD, 54. *Prof Exp:* From asst ed to sr assoc ed, 54-69, sr ed, Chem Abstracts Serv, 69-87; RETIRED. *Mem:* Am Chem Soc. *Res:* Scientific literature; literature storage and retrieval. *Mailing Add:* 1612 Penworth Dr Columbus OH 43229-5213

REEDER, CHARLES EDGAR, b Fairfield, Iowa, July 20, 27; m 56; c 3. PHYSICAL CHEMISTRY, ORGANIC CHEMISTRY. *Educ:* Wheaton Col, BS, 51; Iowa State Col, PhD(chem), 55. *Prof Exp:* Draftsman, Louden Mach Co, 45-47; asst, Parsons Col, 48, Wheaton Col, 49-50 & Iowa State Col, 51-55; instr chem, Bates Col, 55-57; asst prof, 57-64, ASSOC PROF CHEM, BAYLOR UNIV, 64- *Honors & Awards:* Am Inst Chemists Award. *Mem:* Am Chem Soc; Sigma Xi. *Res:* Reaction mechanisms; catalysis; molecular structure; coordination compounds; chemistry teaching methods and effectiveness. *Mailing Add:* 717 Falcon Dr Waco TX 76712

REEDER, CLYDE, b Huntingdon, Pa, Mar 2, 24; m 46; c 3. CHEMICAL ENGINEERING. *Educ:* Juniata Col, BS, 48; Ohio State Univ, PhD(chem eng), 51. *Prof Exp:* Res engr, Org Chem Dept, Jackson Lab, E I du Pont de Nemours & Co, 51-60, sr res engr, Elastomer Chem Dept, Exp Sta, 60-63; mgr process develop, Agr Chem, Atlanta Res Ctr, Armour Agr Chem Co, 63-68, lab mgr, USS Agrichem, Inc, 68-69; technologist, Mobay Chem Co, 69-87; consult, 87-89; RETIRED. *Mem:* Am Chem Soc; Am Inst Chem Engrs. *Res:* Fluorinated elastomers; process development; kinetics; economic analyses of chemical processes; elastomer chemical and engineering technology; urethanes; agricultural chemicals. *Mailing Add:* Four Hartle Lane Pittsburgh PA 15228

REEDER, DON DAVID, b Dubuque, Iowa, Aug 18, 35; m 58; c 2. HIGH ENERGY PHYSICS. *Educ:* Univ Ill, BS, 58; Univ Wis, MS, 62, PhD(physics), 66. *Prof Exp:* From asst prof to assoc prof, 66-72, PROF PHYSICS, UNIV WIS-MADISON, 72- *Concurrent Pos:* Fel, NATO, 74. *Mem:* Fel Am Phys Soc; AAAS. *Res:* Experimental elementary particle physics. *Mailing Add:* Dept Physics Sterling Hall Univ Wis Madison WI 53706

REEDER, JOHN HAMILTON, b Baltimore, Md, Jan 25, 44; m 74. MATHEMATICS. *Educ:* Stevens Inst Technol, BS, 66; Northwestern Univ, MS, 67, PhD(math), 72. *Prof Exp:* Vis asst prof math, Ore State Univ, 71-72; vis scholar, Univ Victoria, 72-73; ASST PROF MATH, UNIV MO, COLUMBIA, 73- *Mem:* Am Math Soc. *Res:* Water waves; kinetic theory of fluids; Wiener-Hopf operators. *Mailing Add:* Dept Math Univ Mo 202 Math Sci Bldg Columbia MO 65211

REEDER, JOHN RAYMOND, b Grand Ledge, Mich, July 29, 14; m 41. BOTANY. *Educ:* Mich State Col, BS, 39; Northwestern Univ, MS, 40; Harvard Univ, MA, 46, PhD(syst bot), 47. *Prof Exp:* From instr to assoc prof, Cur Herbarium, Yale Univ, 47-68; prof & cur, Rocky Mountain Herbarium, Univ Wyo, 68-76; SCHOLAR, UNIV ARIZ, 76- *Concurrent Pos:* Vis prof, Univ Venezuela, 64, hon prof, 66-; ed, Brittonia, 67-71. *Mem:* Am Soc Naturalists; Int Asn Plant Taxon; Torrey Bot Club; Sigma Xi; fel Linnean Soc London; Am Soc Plant Taxon; Botanical Soc Am. *Res:* Taxonomy of vascular plants; taxonomy and phylogeny of Gramineae. *Mailing Add:* Herbarium Rm 113 Shantz Bldg Univ Ariz Tucson AZ 85721

REEDER, PAUL LORENZ, b Dayton, Ohio, Sept 28, 36; m 84; c 4. NUCLEAR CHEMISTRY, FISSION PRODUCTS. *Educ:* Col Wooster, Ohio, AB, 58; Univ Calif, Berkeley, PhD(nuclear chem), 63. *Prof Exp:* Fel nuclear chem, Brookhaven Nat Lab, Upton, NY, 63-65; asst prof chem, Washington Univ, St Louis, 65-70; SR RES SCIENTIST NUCLEAR CHEM, PAC NORTHWEST LABS, BATTELLE MEM INST,

RICHLAND, WASH, 70- *Concurrent Pos:* NATO fel, Orsay, France, 68; sabbatical, Lawrence Berkeley Lab, 90. *Mem:* Am Phys Soc; Am Chem Soc. *Res:* Fission yields; nuclear decay properties of short-lived isotopes; on-line mass spectrometry; instruments for radiation detection. *Mailing Add:* Pac Northwest Labs Battelle Mem Inst Battelle Blvd Richland WA 99352

REEDER, RAY R, b Cleveland, Ohio, July 18, 43; m 65; c 1. PHYSICAL CHEMISTRY, INORGANIC CHEMISTRY. *Educ:* Case Inst Technol, BS, 65; Brown Univ, PhD(chem), 70. *Prof Exp:* From instr to assoc prof chem, 69-76, interim dean fac, 76-77, dir, Ctr Community Educ, 76-78, dept chmn, 81-89, ASSOC PROF CHEM, ELIZABETHTOWN COL, 78- *Mem:* Am Chem Soc. *Res:* Inorganic systems through use of physical techniques; computer applications in undergraduate chemical education; computer simulation and graphics. *Mailing Add:* Dept Chem Elizabethtown Col Elizabethtown PA 17022

REEDER, RONALD HOWARD, b Denver, Colo, Sept 7, 39; m 54. MOLECULAR BIOLOGY. *Educ:* Columbia Union Col, Md, BS, 61; Mass Inst Technol, PhD(biol), 65. *Prof Exp:* Staff mem, dept embryol, Carnegie Inst Washington, 69-78; MEM & SECT CHIEF, HUTCHINSON CANCER RES CTR, 78- *Concurrent Pos:* Mem, NIH Cell Biol Study Sect, 75-77; ed, J Cell Biol, 77- & J Biol Chem, 79- *Mem:* Am Soc Cell Biol; Am Soc Biol Chemists. *Res:* Study of the primary structure of the ribosomal genes and the proteins which bind to them. *Mailing Add:* Hutchinson Cancer Res Ctr 1124 Columbia St Seattle WA 98104

REEDER, WILLIAM GLASE, b Los Angeles, Calif, Feb 4, 29; m 51; c 4. ANIMAL ECOLOGY, VERTEBRATE PALEONTOLOGY. *Educ:* Univ Calif, Los Angeles, BA, 50; Univ Mich, MS, 53, PhD(zool), 57. *Prof Exp:* Instr zool, Univ Calif, Los Angeles, 55-56; from instr to prof zool, Univ Wis-Madison, 56-78; PROF ZOOL, UNIV TEX, AUSTIN, 78-; DIR TEX MEM MUS, 78- *Mem:* AAAS; Am Soc Mammal; Am Soc Archnology. *Res:* Vertebrate ecology, especially of North American deserts, the Arctic and tropics, especially Galapagos Islands; problems of thermoregulation; paleoecology, especially of rodents; human ecology; biogeography and ecology of arachnids. *Mailing Add:* Tex Mem Mus 2401 San Jacinto Austin TX 78705

REED-HILL, ROBERT E(LLIS), b Detroit, Mich, Nov 19, 13; m 38; c 2. METALLURGY. *Educ:* Univ Mich, BS, 36, MS, 38; Yale Univ, DEng, 56. *Prof Exp:* From instr to assoc prof eng, USCG Acad, 39-60; from assoc prof to prof metall, Dept Mat Sci & Eng, Univ Fla, 60-84; RETIRED. *Mem:* Am Inst Mining Metall & Petrol Engrs; fel Am Soc Metals; Brit Inst Metals; Am Soc Testing & Mat; Sigma Xi. *Res:* Mechanical metallurgy; deformation twinning; dynamic strain aging; slow strain-rate embrittlement. *Mailing Add:* 11100 NW 11th Ave Gainesville FL 32606

REEDS, LLOYD GEORGE, b Lindsay, Ont, July 11, 17; m 49; c 3. AGRICULTURAL GEOGRAPHY. *Educ:* Univ Toronto, BA, 40, MA, 42, PhD(geog), 56. *Prof Exp:* Soil survr, Ont Agr Col, 42-43; lectr geog, Univ Toronto, 45-48; prof geog, 48-85, EMER PROF, MCMASTER UNIV, 85- *Honors & Awards:* Senvile Award, Can Asn Geog, 80. *Mem:* Asn Am Geog; Can Asn Geog (pres, 61-62); fel Royal Can Geog Soc. *Res:* Agricultural land-use problems in Southern Ontario. *Mailing Add:* Dept Geog McMaster Univ 1280 Main St W Hamilton ON L8S 4K1 Can

REEDY, JOHN JOSEPH, b Buffalo, NY, May 14, 27; m 52; c 4. BIOLOGY. *Educ:* Niagara Univ, BSNS, 48; Notre Dame Univ, MS, 50, PhD(zool), 52; Bridgewater State Col, MEd, 60. *Prof Exp:* Instr, Univ Detroit, 52-53; from asst prof to prof zool, Stonehill Col, 53-60; PROF ZOOL, NIAGARA UNIV, 60- *Mem:* AAAS; Soc Syst Zool; Am Genetics Soc; Soc Study Evolution; Sigma Xi. *Res:* Human genetics; radiobiology; Drosophila genetics dealing with Epistasis. *Mailing Add:* Dept Biol Niagara Univ Niagara Falls NY 14109

REEDY, MICHAEL K, b Seattle, Wash, June 19, 34; m 74; c 3. CELL BIOLOGY. *Educ:* Univ Wash, BA, 58, MD, 62. *Prof Exp:* Intern path, Univ Hosp, Seattle, Wash, 62-63; NIH res fel, 63-66; asst prof physiol, Med Ctr, Univ Calif, Los Angeles, 66-68, assoc prof, 68-69; assoc prof, 69-86, prof anat, 86-88, PROF CELL BIOL, SCH MED, DUKE UNIV, 88- *Concurrent Pos:* NIH res career develop award, 67-69 & 72-75; Alexander von Humboldt US Sr Scientist Award & Guggenheim fel, 78. *Honors & Awards:* Newcomb S Cleveland Prize, AAAS, 67. *Mem:* AAAS; Am Soc Cell Biol; Biophys Soc. *Res:* Ultrastructure of myofibrils; mechanisms of contractility; preparative methods for biological electron microscopy; mass measurements by microscopy. *Mailing Add:* Dept Cell Biol Duke Univ Sch Med Box 3011 Durham NC 27710

REEDY, ROBERT CHALLENGER, b Summit, NJ, Mar 5, 42; m 69; c 2. NUCLEAR CHEMISTRY, COSMOCHEMISTRY. *Educ:* Colgate Univ, BA, 64; Columbia Univ, PhD(chem physics), 69. *Prof Exp:* Res assoc nuclear chem, Columbia Univ, 64-69; res chemist, Univ Calif, San Diego, 69-72; STAFF MEM, LOS ALAMOS NAT LAB, 72- *Concurrent Pos:* Consult, NASA, 74-76; guest scientist, Max-Planck Inst, Mainz, WGer, 82-83. *Mem:* Am Phys Soc; Meteoritical Soc; Am Geophys Soc. *Res:* Nuclear interactions and nuclear reactions in the moon and other extraterrestrial matter; chemistry of planets from orbit by gamma-ray spectroscopy. *Mailing Add:* Group SST-8 Mail Stop D 438 Los Alamos Nat Lab Los Alamos NM 87545

REEKE, GEORGE NORMAN, JR, b Green Bay, Wis, Oct 16, 43; m 69. CRYSTALLOGRAPHY, NEUROBIOLOGY. *Educ:* Calif Inst Technol, BS, 64; Harvard Univ, MA, 66, PhD(chem), 69. *Prof Exp:* Res asst chem, Harvard Univ, 69-70; asst prof biochem, 70-76, ASSOC PROF DEVELOP & MOLECULAR BIOL, ROCKEFELLER UNIV, 76- *Concurrent Pos:* Res fel, Alfred P Sloan Found, 75-77. *Mem:* AAAS; Am Soc Biol Chemists; Am Crystallog Asn. *Res:* Synthetic neural modelling protein crystallography. *Mailing Add:* Rockefeller Univ Box 287 New York NY 10021

REEKER, LARRY HENRY, b Spokane, Wash, Feb 2, 43; m 64, 89; c 5. COMPUTER SCIENCE & COMPUTATIONAL LINGUISTICS, KNOWLEDGE & DATA MANAGEMENT. *Educ:* Yale Univ, BA, 64; Carnegie-Mellon Univ, PhD(comput sci), 74. *Prof Exp:* Asst prof comput & info sci & ling, Ohio State Univ, 68-73; asst prof comput sci, Univ Ore, 73-75; assoc prof, Univ Ariz, 75-78; reader & head comput sci, Univ Queensland, 78-82; prof & head computer sci, Tulane Univ, 82-85; vpres & sr prin scientist, BDM Corp, 85-89; RES STAFF, INST DEFENSE ANALYSIS, 89- *Concurrent Pos:* Consult, Tech/Ops Corp, 67-68; vis lectr comput sci, Univ Pittsburgh, 68; asst dir, Ling Inst, Ling Soc Am, 70; ed consult, Barnes & Noble Inc, 70-71; ed, SIGACT News, Asn Comput Mach, 70-78; vis res scientist, Chemical Abstracts Serv, 82; vis scientist, Naval Res Lab, 84-85; coun rep, Capital Region, Asn Comput Mach, 86-89. *Mem:* Asn Comput Mach; Asn Comput Ling; Ling Soc Am; Europ Asn Theoret Comput Sci; Australian Comput Soc; Am Asn Artificial Intel; Inst Elec & Electronics Engrs Computer Soc. *Res:* Computer modeling of language acquisition; formal language theory and related theory of computing; programming languages and environments; automatic information extraction from texts; knowledge acquisition and representation. *Mailing Add:* Inst Defense Analysis 1801 N Beauregard St Alexandria VA 22311-1772

REEL, JERRY ROYCE, b Washington, Ind, May 4, 38; m 65; c 1. ENDOCRINOLOGY. *Educ:* Ind State Univ, BA, 60; Univ Ill, MS, 63, PhD(physiol), 66. *Prof Exp:* USPHS fel, Oak Ridge Nat Lab, 66, Am Cancer Soc fel, 66-68; res scientist, 68, endocrinologist biochemist, 68-70, sect dir endocrinol, 70-75, sr res scientist, Parke Davis & Co, 75-78; sect mgr endocrinol, 78-80, dir life sci & toxicol, div, Res Triangle Inst, 80-85; dir endocrinol dept, 85-87, DIR, ENDOCRINE PHARMACOL & ASST DIR, PHARMACOL DEPT, STERLING-WINTHROP RES INST, 87- *Concurrent Pos:* Adj assoc prof, Wayne State Univ Sch Med, 74-78; diplomate, Am Bd Toxicol, 80- *Mem:* Am Chem Soc; Soc Biol Reproduction; Endocrine Soc; Am Physiol Soc; Tissue Cult Asn; Soc Toxicol. *Res:* Reproduction; teratology; cell culture; hormone and drug binding; hormone and drug action; hypothalamic releasing factors and their antagonists; reproductive toxicology. *Mailing Add:* 150 Lockwood W Cary NC 27511

REEMTSMA, KEITH, b Madera, Calif, Dec 5, 25; m; c 2. SURGERY. *Educ:* Idaho State Univ, BS, 48; Univ Pa, MD, 49; Am Bd Surg, dipl, 58; Am Bd Thoracic Surg, dipl, 60. *Hon Degrees:* DSc, Columbia Univ, 58. *Prof Exp:* From intern to resident, Presby Hosp, New York, 50-57; asst, Col Physicians & Surgeons, Columbia Univ, 57; asst prof, Sch Med, Tulane Univ, 57-62, assoc prof surg, 62-66; prof & head dept, Sch Med, Univ Utah, 66-71; PROF SURG & CHMN DEPT, COL PHYSICIANS & SURGEONS, COLUMBIA UNIV, 71-; DIR SURG SERV, PRESBY HOSP, 71- *Concurrent Pos:* Mem study sect A, Surg, NIH, 65- *Mem:* Soc Clin Surg; Am Surg Asn; Soc Univ Surgeons; Am Col Surgeons; Am Fedn Clin Res. *Res:* Transplantation; cardiovascular surgery. *Mailing Add:* Dept Surg Columbia-Presby Med Ctr New York NY 10032-3784

REENSTRA, ARTHUR LEONARD, b Clifton, NJ, Mar 5, 36; m 59; c 3. ELECTRICAL ENGINEERING, SOLID STATE ELECTRONICS. *Educ:* Bucknell Univ, BS, 59; Carnegie Inst Technol, MS, 60; Purdue Univ, PhD(elec eng), 67. *Prof Exp:* Res engr, Res Labs, US Rubber Co, 60-62; instr elec eng, Purdue Univ, 62-64; mem tech staff, Res & Develop Dept, Mat & Electronic Controls Group, Tex Instruments, Inc, 67-76; dir prod develop, Beede Elec Instrument Co, 76-77, dir eng, 77-78, vpres, 79-80; DIR ENG, PACKAGING & WEIGHT DIV, FRANKLIN ELEC CO, 80- *Mem:* Soc Automotive Engrs; Inst Elec & Electronics Engrs. *Res:* Positive temperature coefficient thermistor; thermistoic networks; control sensors and actuators. *Mailing Add:* Eng Mgr John Chatillon & Sons Inc Force Measurement Div 7609 Business Pk Dr Greensboro NC 27409

REENTS, WILLIAM DAVID, JR, b Portsmouth, Va, Jan 18, 54; m 77; c 2. MASS SPECTROMETRY, GAS PHASE CLUSTER CHEMISTRY. *Educ:* Monmouth Col, BS, 76; Purdue Univ, PhD(chem), 80. *Prof Exp:* Mem tech staff, 80-87, DISTINGUISHED MEM TECH STAFF, AT&T BELL LABS, 87- *Mem:* Am Soc Mass Spectrometry; Am Chem Soc. *Res:* Ion-molecule reactions of gas phase clusters using Fourier transform mass spectrometry; purity of hazardous reagents for semiconductor device manufacture; application of ion-molecule reaction chemistry to analysis of hazardous reagents. *Mailing Add:* AT&T Bell Labs Rm 1A 216 600 Mountain Ave Murray Hill NJ 07974

REES, ALLAN W, b Piqua, Ohio, Dec 23, 33; m 59; c 2. PHYSICAL BIOCHEMISTRY. *Educ:* Univ Cincinnati, ChE, 56, MS, 63, PhD(biochem), 67. *Prof Exp:* Res chemist, Cincinnati Milling Mach Co, 56-62; asst prof biochem, Univ Tex Health Sci Ctr, 69-77, ASST PROF, UNIV TEX DIV EARTH SCI, SAN ANTONIO, 77- *Concurrent Pos:* Nat Cancer Inst res fel, 67-68. *Mem:* AAAS; Fedn Am Soc Exp Biol; Am Chem Soc; Biophys Soc. *Res:* Structure and conformation of DNA and nucleoproteins; centrifugal and hydrodynamic techniques. *Mailing Add:* 14319 123rd Ave NE No D Kirkland WA 98034-1440

REES, ALUN HYWEL, b Pontypridd, Wales, Aug 30, 28; m 51; c 2. ORGANIC CHEMISTRY, PHARMACEUTICAL CHEMISTRY. *Educ:* Cambridge Univ, BA, 49, MA, 53, PhD(chem), 68; Univ London, PhD(org chem), 58; Oxford Univ, MA, 65. *Prof Exp:* Res chemist, Roche Prod Ltd, Eng, 49-53; lectr chem, Univ Ibadan, 53-62, sr lectr, 62-64; res fel, Oxford Univ, 64-66; assoc prof, 66-74, prof, 74-92, chmn, Dept Chem, 89-91, EMER PROF, TRENT UNIV, 92- *Concurrent Pos:* Fulbright scholar & res fel, Harvard Univ, 59 & 63; vis prof pharmaceut chem, Univ Ife, Nigeria, 74-75; vis prof org & appl chem, Univ Port, Harcourt, Nigeria, 81; vis res fel, Yale Univ, 85, Univ Wales, Cardiff, 74 & 89. *Mem:* Chem Inst Can; Am Chem Soc; Royal Soc Chem. *Res:* Heterocyclic, medicinal and natural product chemistry; azepines. *Mailing Add:* Dept Chem Trent Univ Peterborough ON K9J 7B8 Can

REES, CHARLES SPARKS, b Dallas, Tex, Nov 21, 40; m 62; c 2. MATHEMATICS. *Educ:* La State Univ, BS, 62; Univ Kans, MA, 63, PhD(math), 67. *Prof Exp:* Asst prof math, Univ Tenn, Knoxville, 67-70; from asst prof to assoc prof math, 70-77, PROF MATH, UNIV NEW ORLEANS, 77- *Mem:* Am Math Soc; Math Asn Am. *Res:* Integration and summability of Fourier series. *Mailing Add:* Dept Math Univ New Orleans New Orleans LA 70148

REES, EARL DOUGLAS, b Cleveland, Ohio, May 1, 28;; c 4. ONCOLOGY, METABOLISM. *Educ:* Harvard Univ, AB, 50; Yale Univ, MD, 54. *Prof Exp:* Intern, Hosps, Ohio State Univ, 55-56; mem, Biophys Sect, Armed Forces Inst Path, 56-57, chief, Biochem Sect, 57-58; instr, Ben May Lab, Univ Chicago, 58-60; asst prof med, 60-65, from assoc prof to prof med & pharmacol, 65-86, EMER PROF MED & PHARMACOL, MED CTR, UNIV KY, 86-, DIR, CLIN RES LAB, 66- *Concurrent Pos:* Nat Found fel, Yale Univ, 54-55; pvt pract, endocrinol. *Honors & Awards:* Res Found Award, Univ Ky, 71. *Mem:* AAAS; Am Asn Clin Chemists; Am Physiol Soc; Soc Exp Biol & Med; Am Fedn Clin Res; Sigma Xi. *Res:* Endocrine and cytogenetic aspects of carcinogenesis; protein and biophysical chemistry; clinical lipid metabolism and pharmacology; clinical endocrinology. *Mailing Add:* 1780 Nicholasville Rd Suite 203 Lexington KY 40503

REES, EBERHARD F M, b Trossingen, Ger, Apr 28, 08; US citizen; m 47. AEROSPACE TECHNOLOGY. *Educ:* Stuttgart Tech Univ, BS; Dresden Inst Technol, MS, 34. *Hon Degrees:* DSc, Rollins Col, 59, Univ Ala, Huntsville, 72. *Prof Exp:* Tech asst, Meier & Weichelt Foundry & Steel Mill, Leipzig, 34-40; plant mgr, Guided Missile Ctr, Peenemuende, Ger, 40-45; aerodyn develop engr, Ord Res & Develop, Ft Bliss, 45-50; dep chief, Guided Missile Develop Div, Huntsville, Ala, 50-56, dep dir, Army Ballistic Missile Agency, 56-60, dep dir, G C Marshall Space Flight Ctr, NASA, 60-70, dir, 70-73; RETIRED. *Honors & Awards:* Medal Outstanding Leadership, NASA, 66. *Mem:* Nat Acad Eng; fel Am Astronaut Soc; hon mem Hermann Oberth Soc; fel Am Inst Aeronaut & Astronaut. *Res:* Rocketry; space flight technology. *Mailing Add:* 69 Revere Way Huntsville AL 35801

REES, HORACE BENNER, JR, b Big Lake, Tex, July 25, 26; m 54; c 1. VIROLOGY. *Educ:* Univ Tex, BA, 49, MA, 50; George Washington Univ, PhD, 66. *Prof Exp:* Asst immunologist, Tex State Dept Health, 50-52; asst dir, Regional Lab, San Antonio City Health Dept, 52; dir, Abilene-Taylor County Health Unit Lab, 53; asst dir, Wene's Poultry Labs, Pleasantville, NJ, 54; microbiologist, Viral & Rickettsial Div, US Army Biol Ctr, Ft Detrick, Md, 54-66; chief, Ecol & Technol Br, Life Sci Lab Div, US Army Biol Labs, 66-83; RETIRED. *Concurrent Pos:* Author & consult, 83- *Mem:* Am Soc Microbiol; Tissue Cult Asn; NY Acad Sci; Sigma Xi. *Res:* Rickettsiae; medical bacteriology. *Mailing Add:* 3584 Millstream Lane Salt Lake City UT 84109-3254

REES, JOHN, b Bangor, Wales, Mar 25, 48; m 71; c 3. ECONOMIC DEVELOPMENT-SCIENCE BASED. *Educ:* Univ Wales, BA Hons, 69; Univ Cincinnati, MA, 71; London Sch Econ, PhD(geog), 77. *Prof Exp:* From asst prof to assoc prof polit econ, Univ Tex, Dallas, 75-83; assoc prof geog, Syracuse Univ, 83-87; PROF GEOG & HEAD DEPT, UNIV NC, GREENSBORO, 87- *Concurrent Pos:* Prin investr, NSF, 76-90; consult, Joint Econ Comt, US Cong, 78-84, Pres Carter's Comn, Nat Agenda US, 79-80 & Off Technol Assessment, US Cong, 83-85; Nat Res Coun fel, 82-83; vis scientist, Int Inst Appl Systs Anal, Austria, 82; mem, Tissot Econ Found, Switz, 87-; sr fel, Kenan Inst, Univ NC, Chapel Hill, 89- *Mem:* Asn Am Geographers; Regional Sci Asn; Sigma Xi. *Res:* Impact of industry and government policy on regional development; science and technology policy in US, Europe and Asia. *Mailing Add:* Dept Geog Univ NC Greensboro NC 27412

REES, JOHN DAVID, b Los Angeles, Calif, Mar 16, 32. BIOGEOGRAPHY, CULTURAL GEOGRAPHY. *Educ:* Univ Calif, Los Angeles, BA, 55, MA, 61, PhD(geog), 71. *Prof Exp:* Asst prof, 65-66 & 68-77, ASSOC PROF GEOG, CALIF STATE UNIV, LOS ANGELES, 77- *Mem:* AAAS; Asn Am Geographers; Soc Econ Botanists. *Res:* Peasant utilization of biotic resources in middle America; rural economy; neotropical biogeography and conservation; regional geography of middle America. *Mailing Add:* Dept Geol Calif State Univ 5151 State University Dr Los Angeles CA 90032

REES, JOHN ROBERT, b Peru, Ind, Feb 17, 30; m 56; c 2. HIGH ENERGY PHYSICS. *Educ:* Ind Univ, AB, 51, MS, 54, PhD(physics), 57. *Prof Exp:* Res fel physics, Harvard Univ, 56-65; staff mem, Stanford Linear Accelerator Ctr, 65-67; chief advan accelerators br, High Energy Physics Prog Div Res, US AEC, DC, 67-69; adj prof, 69-80, ASSOC DIR, STANFORD LINEAR ACCELERATOR CTR, 80- *Concurrent Pos:* Instr, Northeastern Univ, 59-60. *Mem:* Fel Am Phys Soc. *Res:* Particle accelerators. *Mailing Add:* Stanford Linear Accelerator Ctr PO Box 4349 Stanford CA 94305

REES, MANFRED HUGH, b Ger, June 29, 26; nat US; m 49; c 2. AERONOMY. *Educ:* WVa Univ, BSEE, 48, Univ Colo, MS, 56, PhD(physics), 58. *Prof Exp:* Engr, Nat Adv Comt Aeronaut, 48-49; proj engr, Sperry Gyroscope Co, 51-53; instr physics, Univ Colo, 53-58; asst prof geophys, Geophys Inst, Univ Alaska, 58-60; sr res fel appl math, Queen's Univ, Belfast, 60-61; res physicist, Univ Colo, 61-65; assoc prof physics & head dept, Univ Alaska, 65-66; mem staff, Lab Atmospheric & Space Physics, Univ Colo, Boulder, 66-75, lectr astro-geophys, 70-75; PROF GEOPHYS, UNIV ALASKA, 75- *Concurrent Pos:* Physicist, Nat Bur Standards, 56-58. *Mem:* Am Geophys Union. *Res:* Upper atmosphere physics; aurora and airglow; zodiacal light; spectroscopy; atomic and molecular collision processes; solar-terrestrial relations. *Mailing Add:* Geophys Inst Univ Alaska 701 Elvey Fairbanks AK 99775-0800

REES, MARTIN J, b York, Eng, June 23, 42. ASTROPHYSICS. *Educ:* Cambridge Univ, BA, 63, MA & PhD(astrophys), 67. *Hon Degrees:* DSc, Sussex, 90. *Prof Exp:* Prof astron, Sussex Univ, 72-73; PROF ASTRON & EXP PHILOS, CAMBRIDGE UNIV, 73-, DIR INST ASTRON, 77- *Concurrent Pos:* Vis assoc, Calif Tech Inst, 70; vis prof, Harvard Univ, 72, 86-88 & Inst Astrophys, Princeton Univ, 82; Regents fel, Smithsonian Inst, 84-88. *Honors & Awards:* Heinemann Prize, Am Inst Physics, 84; Golo Medal, Royal Astron Soc, 87; Bolzan Prize, 89; Robinson Prize, 90. *Mem:* Foreign mem Nat Acad Sci; foreign hon mem Am Acad Arts & Sci. *Res:* Astrophysics; cosmology; space research. *Mailing Add:* Inst Astron Madingley Rd Cambridge CB3 OHA England

REES, MINA S, b Cleveland, Ohio, Aug 2, 02; m 55. MATHEMATICS. *Educ:* Hunter Col, AB, 23; Columbia Univ, AM, 25; Univ Chicago, PhD(math), 31. *Hon Degrees:* Eighteen from US cols & univs. *Prof Exp:* From instr to prof math, Hunter Col, 26-61, dean fac, 53-61; dean grad studies, 61-68, provost, Grad Div, 68-69, pres grad sch, 69-72, EMER PROF MATH, CITY UNIV NEW YORK, 72-, EMER PRES GRAD SCH, 72- *Concurrent Pos:* Tech aide & exec asst to chief appl math panel, Nat Defense Res Comt, Off Sci Res & Develop, 43-46; head math br, Off Naval Res, 46-49, dir math sci div, 49-52, dep sci dir, 52-53; mem math div, Nat Res Coun, 53-56, mem exec comt, 54-56, comt surv math in US, 54-57; chmn adv comt math, Nat Bur Standards, 54-57, mem, 54-58; adv panel math, NSF, 55-58; Nat Sci Bd, 64-70; subcomt, Sci & Tech Manpower, Nat Manpower Adv Comt; mem, NY State Adv Coun Grad Educ, 62-72; chmn, Coun Grad Sch United States, 70; bd dir, NY Assoc Hosp Serv, 62-74, Health Serv Improvement Fund, 74-83; dir, Inst Math & Soc, 73- *Honors & Awards:* Pub Welfare Medal, Nat Acad Sci, 83. *Mem:* Fel AAAS; Am Math Soc; Math Asn Am (second vpres, 63-65); Soc Indust & Appl Math; fel NY Acad Sci; Sigma Xi. *Res:* Linear algebras; numerical analysis; history of computers. *Mailing Add:* 301 E 66th St New York NY 10021

REES, PAUL KLEIN, b Center Point, Tex, June 10, 02; m 35; c 2. MATHEMATICS. *Educ:* Southwestern Univ, Tex, AB, 23; Univ Tex, MA, 25; Rice Inst, PhD(math), 33. *Prof Exp:* Teacher high sch, Tex, 23-26; instr math, Tex Tech Col, 26-28; asst prof, Univ Miss, 28-30; instr math, Tex Tech Col, 34-35; asst prof, NMex Col, 35-39; from asst prof to assoc prof, Southern Methodist Univ, 39-43; prof, Southwestern La Inst, 43-46; from assoc prof to prof, 46-67, actg head dept, 47-48, EMER PROF MATH, LA STATE UNIV, BATON ROUGE, 67- *Mem:* Am Math Soc; Math Asn Am. *Res:* Fuchsian groups; automorphic functions; algebra; trigonometry; analytic geometry; mathematics of finance; calculus. *Mailing Add:* 345 Centenary Dr Baton Rouge LA 70808

REES, REES BYNON, b Bakersfield, Calif, Feb 2, 15; m 57; c 2. PHARMACOLOGY, THERAPEUTICS. *Educ:* Univ Calif, Berkeley, AB, 36; Univ Calif, San Francisco, MD, 40. *Prof Exp:* From instr to prof dermat, 43-86, EMER CLIN PROF DERMAT & RADIOL, UNIV CALIF, SAN FRANCISCO, 86- *Concurrent Pos:* Consult, St Luke's Hosp, 46-, Shriners Hosp, 47-80, Mt Zion Hosp, 54-80, US Army, 66-80, Am Bd Dermat, 64-74 & Santa Rosa Mem Hosp, 83-; vis prof, Tulane Univ & La State Univ, 63-83, Henry Ford Hosp & Wayne State Univ, 67-82 & Emory Univ & Miami Univ, 78-82; hon staff mem, St Luke's Hosp, 80- *Honors & Awards:* Finnerud Award, Dermat Found, 74. *Mem:* Am Dermat Asn (pres, 75); hon mem Am Acad Dermat (pres, 78); AMA; hon mem Danish Dermat Soc; hon mem Bulgarian Dermat Soc; hon mem Polish Dermat Soc. *Res:* Antifolics for psoriasis; antimalarials in dermatology; biology and treatment of warts. *Mailing Add:* 461 Seventh St W Santa rosa CA 95476

REES, RICHARD WILHELM A, chemistry, medicinal chemistry, for more information see previous edition

REES, ROBERTS M, b Akron, Ohio, Mar 15, 20; m 45; c 2. MEDICINE. *Educ:* Temple Univ, MD, 45. *Prof Exp:* Pvt pract, Akron, Ohio & Beverly Hills, Calif, 52-58; med dir, Pfizer Labs, 60-64, dir clin res, Chas Pfizer & Co, Inc, 61-64; dir clin res, Winthrop Prod Inc, 64-66; dir clin res, Winthrop Labs, Sterling Drug Inc, 66-68, vpres, 67-73, dir med res div, Sterling-Winthrop Res Inst, 68-73, corp med officer, Sterling Drug Inc, 70-73; pres, Clin Resources Inc, 78-84; vpres, Med Affairs, Zenith Lab, 84-88; VPRES, MED & REGULATORY AFFAIRS, CLIN RES INT, 89- *Concurrent Pos:* Spec consult to Exec Off of President, Spec Action Off Drug Abuse Prev, 71-72. *Mem:* AMA; Am Soc Clin Pharmacol & Therapeut; Drug Info Asn; AMA. *Mailing Add:* 557 Weathersfield Pittsboro NC 27312

REES, ROLF STEPHEN, b St Johns, Nfld, Jan 30, 60. COMBINATORICS & FINITE MATHEMATICS. *Educ:* Mem Univ Nfld, 80, B Med Sc, 82; Queens Univ, PhD(math), 86. *Prof Exp:* Fel dept combinatorics & optimization, Univ Waterloo, 86-87; asst prof math, Dept Math & Comput Sci, Mt Allison Univ, 87-90; ASSOC PROF MATH, DEPT MATH & STATIST, MEM UNIV NFLD, 90- *Mem:* Can Math Soc. *Res:* Construction and uses of pairwise balanced designs. *Mailing Add:* Dept Math & Statist Mem Univ Nfld St John's NF A1B 3Y1 Can

REES, THOMAS CHARLES, b Pottsville, Pa, June 6, 39; m 65; c 2. ORGANIC CHEMISTRY. *Educ:* Mt St Mary's Col, Md, BS, 61; Pa State Univ, PhD(org chem), 66. *Prof Exp:* Fel, Radiation Chem Div, Max Planck Inst Coal Res, 65-66; sr chemist, Sherwin-Williams Co, Chicago, 67-72, group leader pigment dispersions, 72-76, group leader plastics additives, 76-81, mgr, Polymer Additives Lab, 81-85; MGR POLYMER CHEM, PMC SPECIALTIES GROUP, CHICAGO, 85- *Mem:* Soc Plastics Engrs; Am Chem Soc. *Res:* Ultraviolet and heat stabilization of polymers; organometallic chemistry, including Grignard and lithium reagents and transition metal organometallics; photochemistry and color properties of pigments; flame retardants. *Mailing Add:* PMC Specialties Group 735 E 115th St Chicago IL 60628

REES, WILLIAM JAMES, b Kansas City, Mo, July 13, 22; m 50; c 5. MEDICINE, DERMATOLOGY. *Educ:* Rockhurst Col, AB, 42; St Louis Univ, MD, 46; Univ Minn, MPH, 50. *Prof Exp:* Pvt pract, Mo, 50-52; scientist admin med, Sci Liaison & Adv Group, DC, 52-57; Europ rep, Upjohn Co, 57-58; scientist admin med, Sci Liaison & Adv Group, Ger, 58-61; clin investr, Abbott Labs, Ill, 61-63; resident dermat, Med Ctr, Univ Calif, San

Francisco, 63-65; assoc exec dir life sci res & chmn biomed res, Stanford Res Inst, 65-69, dir life sci activities, Europ Off, Switz, 69-71, asst dir, DC Off, 71-72; dir clin res, Int Med Affairs, G D Searle & Co, 72-73; dir clin res, Chemex Pharmaceuts, Denver, 83-86; PRES & MANAGING DIR, RHYS INT ASSOCS, 88- *Concurrent Pos:* Consult, Div Occup Health, Calif Dept Pub Health, 63-68; clin instr dermat, Med Ctr, Univ Calif, San Francisco, 65-69, asst clin prof, 69-71; colonel, Med Corps, USAR, 72-87. *Mem:* Am Acad Clin Toxicol; assoc Am Acad Dermat; Am Pub Health Asn; Asn Mil Surgeons US; Royal Soc Health. *Res:* Dermatology, especially occupational and tropical aspects. *Mailing Add:* 550 Seamont Lane Edmonds WA 98020

REES, WILLIAM SMITH, JR, b Quanah, Tex, Nov 2, 59; m 86. SYNTHETIC MECHANISTIC & STRUCTURAL MAIN GROUP COMPOUNDS, ELECTRONIC MATERIALS BY CHEMICAL VAPOR DEPOSITION. *Educ:* Tex Tech Univ, BS, 80; Univ Calif, Los Angeles, PhD(inorg chem), 86. *Prof Exp:* From teaching asst to res asst chem, Tex Tech Univ, 78-81; res chemist, Cosden Oil & Chem Co, 81; teaching consult chem, Off Instrnl Develop, Univ Calif, Los Angeles, 82-83, instr chem, Exten, 82-83, teaching assoc chem, 82-83, vis instr chem, 83 & res assoc chem, 83-86; postdoctoral fel chem, Mass Inst Technol, 86-89; vis asst prof, 84-89, ASST PROF CHEM, FLA STATE UNIV, 89- *Concurrent Pos:* John von Neumann fel, NSF, 76; univ scholar, Tex Tech Univ, 77; mem, Fla High Technol & Indust Coun, Microelectronics & Mat Subcomt, 90-; co-organizer, Fla Advan Mat Chem Conf, 91- *Mem:* Am Chem Soc; Mat Res Soc; Am Ceramic Soc; Int Union Pure & Appl Chem; AAAS; Sigma Xi. *Res:* Synthetic, mechanistic and structural study of main group inorganic and organometallic compounds useful for the chemical vapor deposition of thin films of electronic materials-superconductors, conductors, semiconductors and insulators. *Mailing Add:* Dittmer Labs Chem Unit 1 MS B-164 Fla State Univ Mat Res & Technol Ctr Tallahassee FL 32306-3006

REES, WILLIAM WENDELL, b Albany, NY, Dec 21, 33; m 80; c 3. ORGANIC CHEMISTRY. *Educ:* Amherst Col, BA, 55; Mass Inst Technol, PhD(chem), 58. *Prof Exp:* Asst, Amherst Col, 53-55 & Mass Inst Technol, 55-57; res chemist, 58-60, photog scientist, 60-65, admin asst, Gen Mgt Staff, 65-67, lab head, Res Labs, 67-69, sr lab head, 69-72, asst div dir, 72-77, DIV DIR, RES LABS, EASTMAN KODAK CO, 77- *Mem:* Soc Photog Sci & Eng; Am Chem Soc; Sigma Xi. *Res:* Physical organic chemistry; kinetics and mechanisms of organic reactions; photographic process; research administration. *Mailing Add:* Eastman Kodak Co Res Lab 35 N Country Club Dr Rochester NY 14618

REESE, ANDY CLARE, b Wann, Okla, June 22, 42; m 65. IMMUNOLOGY. *Educ:* Univ Okla, BS, 64; Univ Mo, PhD(biochem), 71; Augusta Col, MBA, 90. *Prof Exp:* Assoc chemist, Skelly Oil Co, 64-66; res assoc med, Sch Med, Case Western Reserve Univ, 70-72, fel microbiol, 72-74; asst res prof path, Mt Sinai Sch Med, 74-75; asst prof, 75-81, ASSOC PROF IMMUNOL, MED COL GA, 81-, DIR, MD-PHD PROG, 91. *Mem:* Am Asn Immunologists; Reticuloendothelial Soc; Soc Exp Biol & Med; NY Acad Sci; Sigma Xi. *Res:* Macrophage function in control of immune responses; neuroendocrine-immune interactions; computer assisted instruction. *Mailing Add:* Dept Immunol & Microbiol Med Col Ga Augusta GA 30912

REESE, BRUCE ALAN, b Provo, Utah, Aug 3, 23; m 45; c 3. AEROSPACE ENGINEERING, PROPULSION. *Educ:* Univ NMex, BS, 44; Purdue Univ, MS, 48, PhD(mech eng), 53. *Prof Exp:* From asst prof to prof mech eng, Purdue Univ, 53-73, dir, Jet Propulsion Ctr, 65-73, head, Sch Aeronaut & Astronaut, 73-79; chief scientist, USAF Arnold Eng Develop Ctr, Tullahoma, Tenn, 79-83; dep comdr eng, 83-86, CONSULT AEROSPACE ENG, US ARMY STRATEGIC DEFENSE COMMAND, 86- *Concurrent Pos:* Dep dir, Nike Zeus Res & Develop, US Army, 61-62 & tech dir, Nike X Proj Off, US Army, 62-63; consult, Sci Adv Panel, US Army, 67-79; mem Sci Adv Bd, US Air Force, 69-76, Adv Group Foreign Tech Div, 70-74 & Div Adv Group Aeronaut Systs Div, 73-76; chmn, Tank & Automotive Command Sci Adv Group, 70-74; mem, Missile Command Adv Comt, 72-79 & Navel Res Adv Comt, 78-79; adv, Assembly Engrs, Nat Res Coun, 77; consult, Aerospace Govt Agencys & Industs, 53- *Mem:* Fel Am Inst Aeronaut & Astronaut; Sigma Xi. *Res:* Heat transfer; gas dynamics; hybrid fueled rockets; laser diffusers; slurry fuel combustion. *Mailing Add:* 5804 Macon Dr Huntsville AL 35802

REESE, CECIL EVERETT, b Benson, Utah, July 3, 21; m 46; c 5. POLYMER CHEMISTRY. *Educ:* Univ Utah, BS, 45, MS, 47, PhD(chem), 49. *Prof Exp:* Asst prof chem, Kans State Col, 49-50; res chemist, 50-59, res assoc, 59-67, RES FEL, E I DU PONT DE NEMOURS & CO, INC, 67- *Mem:* Am Chem Soc. *Res:* Specific heats of organic liquids; diffusion and membrane permeability; visco-elastic properties of hair and other fibers; mechanical behavior of polyacrylonitrile fibers in the presence of an external plasticizer. *Mailing Add:* 1211 Stockton Rd Kinston NC 28501

REESE, ELWYN THOMAS, b Scranton, Pa, Jan 16, 12; m 40; c 2. MYCOLOGY. *Educ:* Pa State Col, BS, 33, MS, 38, PhD(mycol), 46. *Prof Exp:* Teacher high sch, Pa, 34-39; mycologist, Knaust Bros Mushroom Co, NY, 42-43; bacteriologist, Phila Qm Corps Depot, 44-45; mycologist, J T Baker Chem Co, 46-48; mycologist lab, Qm Res & Develop Ctr, Mass, 48-72, MYCOLOGIST SATD LAB, US ARMY NATICK LABS, 72- *Concurrent Pos:* Secy Army fel, 64-65. *Mem:* Mycol Soc Am; Am Chem Soc. *Res:* Decomposition of cellulose by microorganisms; physiology of fungi; polysacchararases. *Mailing Add:* Seven Charles St Cochituate MA 01778-4711

REESE, ERNST S, b Madison, Wis, Jan 26, 31; m 61. ANIMAL BEHAVIOR, ECOLOGY. *Educ:* Princeton Univ, AB, 53; Univ Calif, Los Angeles, PhD(zool), 60. *Prof Exp:* Assoc prof, 60-71, PROF ZOOL, UNIV HAWAII, 71-, NSF RES GRANT, 63- *Concurrent Pos:* NSF fel, Univ Groningen, 61-62; mem adv sci comt, Charles Darwin Found for Galapagos Isles, 66- *Mem:* AAAS; Ecol Soc Am; Animal Behav Soc (pres-elect). *Res:* Ecology and behavior of marine animals, especially crustacea and fish; emphasis on comparative developmental aspects and ecological significance behavior. *Mailing Add:* Zool Dept-Edmondson Hall 2538 The Mall Univ Hawaii Manoa Honolulu HI 96822

REESE, FLOYD ERNEST, b Ransomville, NY, Nov 8, 17; m 43; c 3. BIOCHEMISTRY, ORGANIC CHEMISTRY. *Educ:* Greenville Col, BS, 41; Purdue Univ, MS, 44, PhD(biochem), 47. *Prof Exp:* Res chemist biochem, Upjohn Co, Mich, 47-48; from assoc prof to prof chem, Houghton Col, 48-56; from asst prof to assoc prof, Calif State Univ, Chico, 56-65, head dept, 63-67, prof chem, 65-81; RETIRED. *Mem:* Am Chem Soc. *Res:* Metabolism of amino acids and guanidino acids; gamma irradiation of proteins. *Mailing Add:* 5081 Wilderness Way Paradise CA 95969

REESE, LYMON C(LIFTON), b Murfreesboro, Ark, Apr 27, 17; m 48; c 3. CIVIL ENGINEERING. *Educ:* Univ Tex, BS, 49, MS, 50; Univ Calif, PhD, 55. *Prof Exp:* Rodman, Int Boundary Comn, 39-41; layout engr, E I du Pont de Nemours & Co, 41-42; field engr, Assoc Contractors & Engrs, 45; draftsman, Phillips Petrol Co, 46-48; res scientist, Univ Tex, 48-50; asst prof civil eng, Miss State Col, 50-51, 53-55; from asst prof to assoc prof, Univ Tex, Austin, 55-64, prof civil eng, 64-87, chmn dept, 65-72, T U Taylor prof & assoc dean res, 72-79, Rashid chair, 81-84, EMER NASSER I AL-RASHID CHAIR, UNIV TEX, AUSTIN, 87-; OWNER, LCRNA, INC & ENSOFT, INC, 87- *Concurrent Pos:* Consult, var co & govt agencies, 55- *Honors & Awards:* Middlebrooks Award, Am Soc Civil Engrs, 58; Karl Terzaghi Lectr, 76. *Mem:* Nat Acad Eng; Am Soc Eng Educ; Am Soc Civil Engrs. *Res:* Soil mechanics; interaction between soils and structures; analysis and design of foundations for offshore structures; behavior of laterally loaded piles. *Mailing Add:* LCRNA Inc PO Box 180348 Austin TX 78718

REESE, MILLARD GRIFFIN, JR, b Dinwiddie, Va, Aug 14, 31; m 58; c 1. ORGANIC CHEMISTRY. *Educ:* Randolph-Macon Col, BS, 52; Univ Va, MS, 55, PhD(org chem), 57. *Prof Exp:* Res chemist, Dacron Res Lab, E I du Pont de Nemours & Co, Inc, 57-85; RETIRED. *Mem:* Sigma Xi. *Res:* Unsaturated diketones, reactions; reaction mechanisms and derivatives; condensation polymerization; properties of high polymers; textile fibers. *Mailing Add:* 2705 Carey Rd Kinston NC 28501

REESE, ROBERT TRAFTON, b Norfolk, Va, Apr 11, 42; m 71. IMMUNOLOGY, BIOCHEMISTRY &BIOPHYSICS. *Educ:* Va Polytech Inst, BS, 64; Yale Univ, PhD(biophysics), 71. *Prof Exp:* Fel immunol, Johns Hopkins Univ, 70-74; res assoc, Rockefeller Univ, 74, asst prof immunol, 74-80; assoc mem fac, Dept Immunol, Res Inst Scripps Clin, 80-87; PRIN SCIENTIST, AGOURON INST, 87- *Concurrent Pos:* Vis prof, Univ Ibadon, Nigeria, 78; WHO adv, Immunobiol of Malaria, 78, 80, 89; UN adv, ICGEB, 89; Fulbright scholar, 90-91. *Mem:* Sigma Xi; Am Asn Immunologists; Am Asn Biochemists; Protein Soc; NY Acad Sci; Am Asn Microbiologists. *Res:* Regulation of the immune response; immunology of malaria. *Mailing Add:* Agouron Inst 505 Coast Blvd S La Jolla CA 92037-4696

REESE, RONALD MALCOLM, b Oakland, Calif, Oct 2, 38; m 60; c 2. SHIP PROTECTION, SHIP DESIGN. *Educ:* US Naval Acad, BS, 60; Mass Inst Technol, MS, 70, PhD(ocean eng), 72. *Prof Exp:* Chief engr, US Naval Destroyer, 62-64, mat officer, USN Destroyer Squadron, 65-67; ship design mgr patrol hydrofoil, Naval Ship Eng Ctr, Hyattsville, Md, 72-73, ship design mgr aircraft carriers, Newport News, Va, 75, supvr shipbuilding, conversion & repair, 75-79; force maintenance officer, Naval Surface Force, Atlantic Fleet, Norfolk, Va, 79-80; HEAD SYSTS ANALYSIS BR, NKF ENG, INC, 80-, HEAD, SYSTS ENG DIV, 80- *Mem:* Sigma Xi; Soc Naval Architects & Marine Engrs; Am Soc Naval Engrs. *Res:* Design of propulsion plants for deep submersibles; engineering design of complex warships and application of new technology to ship design. *Mailing Add:* 11424 Vale Rd Oakton VA 22124

REESE, THOMAS SARGENT, b Cleveland, Ohio, May 20, 35; m 75; c 2. CELL BIOLOGY, NEUROSCIENCE. *Educ:* Harvard Univ, BA, 57; Columbia Univ, MD, 62. *Prof Exp:* Res asst, Psycho-Acoust Lab, Harvard Univ, 57-58; intern, Boston City Hosp, 62-63; res fel anat, Sch Med, Harvard Univ, 65-66; res assoc, 63-65, res med officer, 66-70, CHIEF LAB NEUROBIOL, NAT INST NEUROL & COMMUN DIS & STROKE, NIH, 82- *Concurrent Pos:* Instr neurobiol, Marine Biol Labs, Woods Hole, Mass, 74- *Honors & Awards:* C Judson Herrick Award; Mathilde Solowey Award; Joseph Mather Smith Prize, Columbia Univ, 85. *Mem:* Nat Acad Sci; Am Soc Cell Biol; Soc Neurosci; Biophys Soc. *Res:* Membrane structure and function in neural cells, axonal transport synapses. *Mailing Add:* Bldg 36 Rm 2A-29 NIH Bethesda MD 20892

REESE, WELDON HAROLD, b Stonewall Co, Tex, May 5, 27; m 48; c 2. POLLUTION BIOLOGY, PHYCOLOGY. *Educ:* Tex Tech Col, BS, 52, MS, 61; Ore State Univ, PhD(pollution biol, phycol), 66. *Prof Exp:* Teacher sec schs, 52-61; NSF fel pollution res, 61-62; res asst, Ore State Univ, 62-63, fel pollution res & res asst, 63-64; assoc prof, 66-77, PROF BIOL, WAYLAND COL, 77-, HEAD DEPT BIOL & CHMN DIV SCI, 66- *Concurrent Pos:* Lectr, Jamaica, 67-68; Terra-Rite Corp res grant, 67-; res grant, Tex Agr Exp Sta, 68-; mem int comn inter-col educ, Southern Asn Baptist Col & Schs, 68- *Mem:* AAAS; Am Soc Limnol & Oceanog; Phycol Soc Am. *Res:* Effects of community imbalance in lentic and lotic environments caused by organic enrichment and other pollutants; periphyton community. *Mailing Add:* RR 2 No 15 Floydada TX 79235

REESE, WILLIAM, b Kansas City, Mo, Feb 24, 37; m 58; c 1. THERMAL PHYSICS. *Educ:* Reed Col, BA, 58; Univ Ill, MS, 60, PhD(physics), 62. *Prof Exp:* Res assoc physics, Univ Ill, 62-63; from asst prof to assoc prof, 63-73, PROF PHYSICS, US NAVAL POSTGRAD SCH, 73- *Mem:* Am Phys Soc; Sigma Xi. *Res:* Low temperature physics; thermal properties of polymers; second order phase transitions; military systems analysis. *Mailing Add:* 8806 Four Seasons Ct Alexandria VA 22309

REESE, WILLIAM DEAN, b Baltimore, Md, Sept 10, 28; m 50; c 4. BIOLOGY. *Educ:* Univ Md, BS, 53; Fla State Univ, MS, 55, PhD(bot), 57. *Prof Exp:* Asst bot, Fla State Univ, 53-55; from asst prof to assoc prof, 57-66, chmn dept, 74-77, PROF BIOL, UNIV SOUTHWESTERN LA, 66-, HEAD DEPT, 81- *Concurrent Pos:* Ed, The Bryologist, 70-74. *Mem:* AAAS; Soc Bot

Mex; Am Bryol & Lichenological Soc (vpres, 77-79, pres, 79-81); Int Asn Plant Taxon; Brit Bryol Soc. *Res:* Taxonomy, distribution and ecology of Musci. *Mailing Add:* Dept Biol Univ Southwestern La PO Box 42451 Lafayette LA 70504

REESE, WILLIAM GEORGE, b Lewiston, Utah, Apr 2, 17; m 42; c 3. PSYCHIATRY. *Educ:* Univ Idaho, BS & MS, 38; Washington Univ, MD, 42. *Hon Degrees:* DSc, Univ Ariz. *Prof Exp:* Intern internal med, Barnes Hosp, St Louis, 42-43; resident psychiat, Johns Hopkins Hosp, 46-48, instr psychiatrist, 48-51; prof 51-85, EMER PROF PSYCHIAT & CHMN DEPT, COL MED, UNIV ARK MED SCH, LITTLE ROCK, 87- *Concurrent Pos:* Commonwealth fel, Johns Hopkins Univ, 46-48; dir prof educ, Vet Admin Hosp, Perry Point, Md, 48-51, consult, North Little Rock. *Mem:* AMA; fel Am Psychiat Asn; Pavlovian Soc; fel Am Col Psychiat; Sigma Xi. *Res:* Basic psychophysiology. *Mailing Add:* Dept Psychiat Univ Ark Med Sci 4301 W Markham Little Rock AR 72205-7199

REESMAN, ARTHUR LEE, b Eldorado, Ill, Feb 20, 33; m 63; c 3. GEOLOGY. *Educ:* Eureka Col, BS, 55; Univ Mo-Columbia, AM, 61, PhD(geol), 66. *Prof Exp:* Asst geol, Univ Mo-Columbia, 59-62; asst prof, Univ NDak, 62-63 & Western Mich Univ, 66-67; asst prof, 68-70, chmn, 76-79, ASSOC PROF GEOL, VANDERBILT UNIV, 70- *Mem:* Clay Minerals Soc; Geochem Soc. *Res:* Chemical weathering of rocks and minerals; genesis of clay minerals; evolution of landscapes through weathering. *Mailing Add:* Dept Geol Vanderbilt Univ Nashville TN 37240

REESOR, JOHN ELGIN, b Saskatoon, Sask, June 2, 20; m 54; c 3. GEOLOGY. *Educ:* Univ BC, BASc, 49; Princeton Univ, PhD(geol), 52. *Prof Exp:* Geologist, Can Geol Surv, 52-85; RETIRED. *Mem:* Fel Geol Soc Am. *Mailing Add:* 1166 Bonnie Crescent Ottawa ON K2C 1Z5 Can

REETHOF, GERHARD, b Teplice, Czech, July 1, 22; US citizen; m 56; c 3. MECHANICAL ENGINEERING, ACOUSTICS. *Educ:* Mass Inst Technol, SB, 47, SM, 49, ScD(mech eng), 54. *Prof Exp:* Proj engr, Sperry Gyroscope Co, NY, 49-50; asst prof fluid power & mech eng, Dynamic Anal & Controls Lab, Mass Inst Technol, 50-55; chief res, Vickers Inc, Mich, 55-58; mgr acoust eng, Flight Propulsion Div, Gen Elec Co, Ohio, 58-67; Alcoa prof mech eng, Pa State Univ, 67-76, prof mech eng & dor, Noise Control Lab, 76-88; RES PROF, DREXEL UNIV, 88- *Concurrent Pos:* Fulbright prof, Finland Inst Technol, Helsinki, 53-54; consult, Electronics Div, Gen Dynamics Corp, 67-70, US Nuclear Regulatory Comn, 68-80 & Army Mat Command, 71-82; expert witness, du Pont Atomic Energy Div, 75-85. *Mem:* Fel Am Soc Mech Engrs; Acoust Soc Am; Inst Noise Control; Am Soc Eng Educ; Am Inst Aeronaut & Astronaut. *Res:* Acoustics and noise control; jet noise; fan noise; propagation and attenuation of complex acoustic waves in ducts; valve noise; high intensity acoustics; acoustic agglom of submicro wave particles; transportation noise; probabilistic methods in design and reliability. *Mailing Add:* 704 St Francis Dr Broomall PA 19008

REETZ, HAROLD FRANK, JR, b Watseka, Ill, Mar 10, 48; m 73; c 3. CROP PRODUCTION, COMPUTER SIMULATION. *Educ:* Univ Ill, BS, 70; Purdue Univ, MS, 72, PhD(agron, crop physiol), 76. *Prof Exp:* Exten res agronomist grain crop prod, Dept Agron, Purdue Univ, 74-82; REGIONAL DIR, POTASH & PHOSPHATE INST, 82- *Concurrent Pos:* Consult, Control Data Corp & Int Harvester, 78- *Mem:* Am Soc Agron; Crop Sci Soc Am; Coun Agr Scientists & Technologists; Soil Sci Soc Am. *Res:* Computer simulation of crop production; physiological reactions and production impact of environmental stresses; cultural practices in grain crop production; computer applications in agribusiness and in farming. *Mailing Add:* Potash & Phosphate Inst RR 2 Box 13 Monticello IL 61856

REEVE, AUBREY C, b Staines, Eng, June 23, 37; m 59; c 3. ENVIRONMENTAL CONTROL, CHEMICAL ENGINEERING. *Educ:* Univ Birmingham, BSc, 58, PhD(gasification of hydrocarbons), 61. *Prof Exp:* Res fel movement of particles, McGill Univ & Pulp & Paper Res Inst Can, 61-63; sr chem engr, Res Div, Carrier Corp, 63-73, mgr systs develop, 73-74, proj control coordr, 74-75; res engr, 75-78, SR RES ENGR, AMOCO CHEM CORP, 78- *Mem:* Assoc mem Am Inst Chem Engrs. *Res:* By-product aromatics during gasification of hydrocarbons; motion of particles through force fields; removal of odors; air pollution control; gas absorption in packed columns; process development. *Mailing Add:* 29 W 150 Old Farm Lane Warrenville IL 60555

REEVE, ERNEST BASIL, b Liverpool, Eng, May 5, 12; m 84; c 4. PHYSIOLOGY. *Educ:* Oxford Univ, BA, 35, BM, BCh, 38. *Prof Exp:* Clin asst & mem staff, Clin Res Unit, Guys Hosp, London, 40-52; mem permanent staff, Med Res Coun, London, 46-52; vis prof physiol, Columbia Univ, 53; from assoc prof to prof, 53-83, EMER PROF MED, SCH MED, UNIV COLO, DENVER, 83- *Mem:* Am Physiol Soc; Brit Physiol Soc. *Res:* Metabolism of plasma proteins, especially clotting proteins; regulation of body water, sodium, calcium and phosphate; systems analysis of physiological functions. *Mailing Add:* 182 Race Denver CO 80206

REEVE, JOHN, b Nelson, Eng, Oct 11, 36; Can citizen; c 2. ELECTRICAL ENGINEERING. *Educ:* Univ Manchester, BSc, 58, MSc, 59, PhD(elec eng), 66. *Hon Degrees:* DSc, Univ Manchester, 74. *Prof Exp:* Develop engr, Eng Elec Protection Co, Stafford, Eng, 58-61; lectr, elec eng, Univ Manchester Inst Sci Technol, 61-67; assoc prof, elec engr, 67-71, PROF, UNIV WATERLOO, CAN, 71- *Concurrent Pos:* Vis prof, Monash Univ, Australia, 74-75; vis prof, Univ Canterbury, NZ, 75; proj mgr, elec syst, Elec Power Res Inst, 80-81; pres, John Reeve Consult Ltd, 71-; mem, Hydro-Que Inst Res, 89-90. *Mem:* Fel Inst Elec & Electronics Engrs. *Res:* Transmission power by direct current, control, protection, system analysis and simulation; high power electronics. *Mailing Add:* Elec & Computer Eng Dept Univ Waterloo Waterloo ON N2L 3G1 Can

REEVE, MARIAN E, b Placerville, Calif, Aug 28, 20; m 41; c 4. PLANT ANATOMY. *Educ:* Univ Calif, Berkeley, AB, 40, PhD(bot), 49. *Prof Exp:* Prof biol, Merritt Col, 55-81; RETIRED. *Mem:* AAAS; Bot Soc Am; Sigma Xi. *Mailing Add:* 4325 Mountain View Ave Oakland CA 94605

REEVE, PETER, b London, Eng, Jan, 3, 34; m 62, 86; c 4. IMMUNOLOGY, MICROBIOLOGY. *Educ:* London Univ, BSc, 55, PhD(anat), 61. *Prof Exp:* Virologist, Med Res Coun, 58-67; asst prof microbiol, San Francisco Med Sch, 63-64; Wellcome lectr virol, Royal Postgrad Med Sch, 67-72; lectr bact, Univ Col Hosp Med Sch, London, 72-75; lab mgr, Sandoz Res Inst, Vienna, 75-79; sr scientist, Nat Inst Biol Standards, London, 79-81; sci mgr, 81-85, GROUP DIR, BIOTECHNOL PLANNING & DEVELOP, SMITH KLINE & FRENCH LABS, 85- *Concurrent Pos:* Mem intersci conf antimicrobial agents & chemother prog comt, Am Soc Microbiol; vpres, Prod Develop, Invitron Corp. *Mem:* Inst Biol; fel Royal Col Path; Soc Gen Microbiol. *Res:* Mechanisms of viral pathoglinicity; chlamydial infections; influenza, parainfluenza and polioviruses. *Mailing Add:* Two Bennett Rd Redwood City CA 94063

REEVE, RONALD C(ROPPER), b Hinckley, Utah, Mar 27, 20; m 40; c 3. AGRICULTURAL ENGINEERING. *Educ:* Utah State Univ, BS, 43; Iowa State Univ, MS, 49. *Prof Exp:* Engr, Boeing Aircraft Co, Wash, 43-45; irrig & drainage engr, Salinity Lab, USDA, 46-55, tech staff specialist, Salinity Lab & Western Soil & Water Mgt Br, Soil & Water Conserv Res Div, Agr Res Serv, 55-57, agr engr, Salinity Lab, 57-64, res invests leader, 64-74, staff scientist water mgt, Nat Prog Staff, Soil, Water & Air Sci, 73-77; Tech dir, Advan Drainage Systs, Inc, 77-82; RETIRED. *Concurrent Pos:* Consult, Abaca invests, div cotton & other fiber crops & dis, Bur Plant Indust, Soils & Agr Eng, Costa Rica, 51; drainage res proj consult, PR Agr Exp Sta & Univ PR, 55 & 56; consult, Col Agr Eng, Punjab Agr Univ, 66, Int Eng Co, Inc, Ankara, Turkey, 66 & Food & Agr Orgn, UN, Mexico City, 68; adj prof agr eng, Ohio State Univ, 70-74. *Mem:* Fel Am Soc Agr Engrs; Am Soc Civil Engrs; Int Comn Irrig & Drainage. *Res:* Water management, especially irrigation, drainage and salinity control. *Mailing Add:* 1222 Southport Dr Columbus OH 43235

REEVER, RICHARD EUGENE, COSMETIC & TOILETRIES, DRUGS. *Educ:* Millersville State Univ, BS, 70. *Prof Exp:* Assoc dir res & develop, Alberto Culver Co, 73-77; VPRES RES & DEVELOP, MINNETONKA INC, 77- *Mem:* Am Chem Soc; Soc Cosmetic Chemists. *Res:* Drugs: research and development; quality assurance. *Mailing Add:* Softsoap Enterprises PO Box G Chaska MN 55318

REEVES, ANDREW LOUIS, b Budapest, Hungary, Oct 13, 24; nat US; m 51; c 3. TOXICOLOGY. *Educ:* Univ Munich, dipl, 51-53; Wayne State Univ, PhD(physiol chem), 59. *Prof Exp:* Res assoc indust med & hyg, 55-63, from asst prof to assoc prof, 63-72, PROF OCCUP & ENVIRON HEALTH, SCH MED, WAYNE STATE UNIV, 72- *Concurrent Pos:* Sr sci fel, Univ Milan, 73; mem permanent comn, Int Asn Occup Health; vis prof, Univ Wurzburg, 80-81. *Mem:* AAAS; Am Chem Soc; Am Indust Hyg Asn; NY Acad Sci. *Res:* Biochemical aspects of industrial toxicology; mechanism of action of air pollutants; etiology of pulmonary carcinogenesis; toxicology of arsenic, asbestos, barium, beryllium. *Mailing Add:* 573 Pemberton Grosse Point MI 48230-1711

REEVES, BARRY L(UCAS), b St Louis, Mo, Jan 11, 35; m 54; c 3. MECHANICAL ENGINEERING, PHYSICS. *Educ:* Wash Univ, St Louis, BS, 56, MS, 58, DSc(eng, physics), 60. *Prof Exp:* Res assoc, McDonnell Aircraft Corp, Mo, 59-60; Air Force Off Sci Res fel, Aeronaut Dept, Calif Inst Technol, 60-64; sr staff scientist, 64-72, sr consult scientist, 72-78, prin scientist, Avco Systs Div, Avco Corp, 78-85; PRIN SCIENTIST, TEXTRON DEFENSE SYSTS, 85- *Concurrent Pos:* Consult, Space Gen Corp, 62-63 & Nat Eng & Sci Co, 63-64; fel, NSF, 59. *Mem:* Sigma Xi; Am Phys Soc; Am Inst Aeronaut & Astronaut. *Res:* Hypersonic wakes and boundary layers; laminar and turbulent separated flows; massive turbulent ablation; shock wave-boundary layer interactions; reentry physics and reentry vehicle plasma environments; gas-dynamic lasers; boundary layer transition. *Mailing Add:* Ten Hillcrest Pkwy Winchester MA 01890

REEVES, C C, JR, b Cincinnati, Ohio, May 19, 30; div; c 6. GEOLOGY. *Educ:* Univ Okla, BS, 55, MS, 57; Tex Tech Univ, PhD(geol), 70. *Prof Exp:* Geologist, Texaco, Inc, 56-57; FROM INSTR TO PROF GEOL, TEX TECH UNIV, 72- *Mem:* Am Inst Prof Geol; Geol Soc Am. *Res:* Hydrogeology of Ogallala of WTex; tertiary-quaternary of Tex-NMex; qualified expert witness; experienced well driller. *Mailing Add:* PO Box 4516 Lubbock TX 79409

REEVES, DALE LESLIE, b Norton, Kans, Mar 24, 36; m 57; c 3. PLANT BREEDING, AGRONOMY. *Educ:* Kans State Univ, BS, 58, MS, 63; Colo State Univ, PhD(plant genetics), 69. *Prof Exp:* Asst county agent, Kans State Univ, 62-63; instr field crops, Colo State Univ, 63-70; asst prof, 70-74, assoc prof, 74-80, PROF PLANT BREEDING, SDAK STATE UNIV, 80- *Concurrent Pos:* Mem, USAID proj, Botswana, Africa. *Mem:* Am Soc Agron; Crop Sci Soc Am. *Res:* Oat improvement; improving yield, straw strength and rust resistance in oats; increasing protein levels by breeding; rye variety development; semi dwarf rye. *Mailing Add:* Dept Plant Sci S Dak State Univ Brookings SD 57007

REEVES, EDMOND MORDEN, b London, Ont, Jan 14, 34; m 56; c 2. SPACEFLIGHT INSTRUMENTATION, SOLAR PHYSICS. *Educ:* Western Ontario Univ, BSc, 56, MSc, 57, PhD(physics), 59. *Prof Exp:* Nat Res Coun Can overseas fel, Imp Col, Univ London, 59-61; res physicist & lectr astron, Harvard Univ, 61-78, sr res assoc, Harvard Col Observ, 68-78, assoc, 78-80, sect head, Nat Ctr Atmospheric Res, High Altitude Observ, 78-82; chief, Astrophys Br, Spacelab Flight Div, 82-88, DEP DIR, FLIGHT SYSTS DIV, OFF SPACE SCI & APPLICATIONS, NASA, 88- *Concurrent Pos:* Mem solar physics panel, Astron Missions Bd, NASA, 68-71, consult, 68-77 & 79-81, mem space shuttle opers mgt working group, 72-76; physicist, Smithsonian Astrophys Observ, 73-78. *Honors & Awards:* Except Sci Achievement Medal, NASA, 74. *Mem:* Int Astron Union; Am Astron Soc; fel Optical Soc Am; Int Acad Astronautics. *Res:* Vacuum ultraviolet spectroscopy of the sun from rockets and satellites; related problems in laboratory astrophysics using laboratory plasmas from the far ultraviolet to the visible. *Mailing Add:* NASA Hq Code SM 600 Independence Ave SW Washington DC 20546

REEVES, FONTAINE BRENT, JR, b Eufaula, Ala, May 16, 39; m 64; c 2. MYCOLOGY. *Educ:* Tulane Univ, BS, 61, MS, 63; Univ Ill, PhD(bot), 66. *Prof Exp:* From asst prof to assoc prof bot, 66-79, PROF PLANT PATH, COLO STATE UNIV, 79- *Mem:* AAAS; Mycol Soc Am; Bot Soc Am; Sigma Xi. *Res:* Cytology and evolution of fungi; ultrastructure of fungi; mycorrhizae. *Mailing Add:* Dept Bot & Plant Path Colo State Univ Ft Collins CO 80523

REEVES, GEOFFREY D, b Boston, Mass, Dec 23, 61; m 91. SPACE PLASMA PHYSICS, MAGNETOSPHERIC PHYSICS. *Educ:* Univ Colo, BA, 83; Stanford Univ, PhD(appl physics), 88. *Prof Exp:* STAFF SCIENTIST, LOS ALAMOS NAT LAB, 89- *Mem:* Am Geophys Union; Am Inst Physics. *Res:* Space plasma physics, shuttle-based electron beam-plasma-wave interactions in the ionosphere; energetic particle data in the investigation of magnetospheric substorms. *Mailing Add:* Mail Stop D-438 Los Alamos Nat Lab Los Alamos NM 87545

REEVES, HENRY COURTLAND, b Camden, NJ, Nov 25, 33; m 52; c 2. BIOCHEMISTRY, BACTERIAL PHYSIOLOGY. *Educ:* Franklin & Marshall Col, BS, 55; Vanderbilt Univ, MA, 56, PhD(biochem), 59. *Prof Exp:* USPHS fels bact physiol, Walter Reed Army Inst Res, Washington, DC, 60-61 & biochem, Albert Einstein Med Ctr, 61-63, USPHS res career develop awardee, 62-69; vis investr, Max Planck Inst Cell Chem, 65-66; mem fac biochem, Albert Einstein Med Ctr, 66-69; chmn dept bot & microbiol, 72-77, PROF MICROBIOL, ARIZ STATE UNIV, 69-, VPRES RES, 85- *Concurrent Pos:* Vis investr Biochem Inst, Freiburg, Ger, 75-76; dir Div Physiol, Cell & Molecular Biol, NSF, 77-79. *Mem:* Sigma Xi; AAAS; Am Soc Biol Chem; Am Chem Soc; Am Soc Microbiol. *Res:* Enzymology; intermediary metabolism; bacterial genetics. *Mailing Add:* Dept Bot & Microbiol Ariz State Univ Tempe AZ 85287-0001

REEVES, HOMER EUGENE, b Atoka, Okla, Dec 4, 28; m 60; c 3. AGRONOMY. *Educ:* Okla State Univ, BS, 55, MS, 57; Kans State Univ, PhD, 71. *Prof Exp:* Asst, Okla State Univ, 55-57; from asst prof to assoc prof agron, 57-73, prof agron & biol Sci, 73-85, CHMN AGRIBUS, PANHANDLE STATE UNIV, 83-, CHMN & PROF AGRON, 85- *Concurrent Pos:* Asst, Kans State Univ, 65-67, vis asst prof, 70-72; mem, Coun Agr Sci & Technol. *Mem:* Am Soc Agron. *Res:* Crop ecology; crop and soil management; water management. *Mailing Add:* PO Box 338 Goodwell OK 73939-0338

REEVES, JAMES BLANCHETTE, b Beaumont, Tex, Jan 24, 24; m 47; c 2. MEDICAL MICROBIOLOGY. *Educ:* La State Univ, BS, 48, MS, 49; Univ Tex, PhD, 64. *Prof Exp:* Lab dir, State Health Dept, Tex, 50-55; from asst prof to assoc prof biol sci, Univ Tex, El Paso, 55-61, prof & head dept, 61-70, coordr health related progs, 70-75; asst dean, Tex Col Osteop Med, 75-77, prof microbiol, 77-84; RETIRED. *Concurrent Pos:* USPHS res grant, 58-59; Res Corp grant, 62-63. *Mem:* AAAS; Am Soc Microbiol. *Res:* Sanitary bacteriology; composting of sewage sludge; growth initiation and inhibition; public health bacteriology; proteases from pseudomonas aeruginosa burn strains. *Mailing Add:* 5433 Northcrest Ft Worth TX 76107

REEVES, JERRY JOHN, b Watsonville, Calif, Oct 5, 43; m 66; c 2. ENDOCRINOLOGY, REPRODUCTIVE PHYSIOLOGY. *Educ:* Ore State Univ, BS, 65, MS, 67; Univ Nebr, Lincoln, PhD(reprod physiol & animal sci), 69. *Prof Exp:* Res asst animal sci, Univ Nebr, Lincoln, 67-69; instr med & NIH fel, Med Sch, Tulane Univ, La & endocrine & polypeptide lab, Vet Admin Hosp, New Orleans, La, 69-70; from asst prof to assoc prof, 70-80, PROF ANIMAL SCI, WASH STATE UNIV, 80- *Concurrent Pos:* Consult, Int Atomic Energy Agency, 84, W R Grace & Co, 85-, Monoclonal Anti-bodies Inc, 84- & Norden Labs, 85-; feature teacher agr, Wash State Univ, 76. *Honors & Awards:* Young Sci Award, Am Soc Animal Sci, 77. *Mem:* Am Soc Animal Sci; Soc Study Reproduction; Endocrine Soc; Soc Exp Biol & Med; Domestic Animal Endocrinol. *Res:* Neuroendocrinology concerning hypothalamic control of reproduction, growth and lactation; immunization of animals against their own hormones to control reproduction. *Mailing Add:* Dept Animal Sci Clark Hall Wash State Univ Pullman WA 99163-6332

REEVES, JOHN PAUL, b Bryn Mawr, Pa, June 16, 42; m 67; c 3. BIOCHEMISTRY, PHYSIOLOGY. *Educ:* Juniata Col, BS, 64; Mass Inst Technol, PhD(biol), 69. *Prof Exp:* Asst prof biol, Allen Univ, 69-70; NIH fel, Rutgers Univ, 70-72; guest worker biochem, Roche Inst Molecular Biol, 72-73; from asst prof to assoc prof physiol, Univ Tex Health Sci Ctr, Dallas, 73-81; ASSOC RESEARCHER, ROCHE INST MOLECULAR BIOL, NJ, 81- *Concurrent Pos:* NIH grantee, 74-81 & NSF, 78-80. *Mem:* AAAS; Biophys Soc; Am Physiol Soc; Soc Gen Physiol; Physiol Soc Eng. *Res:* Permeability properties and transport activities of biological membranes; Na-Ca exchange in heart cell membranes. *Mailing Add:* Dept Biochem Roche Inst Molecular Biol Nutley NJ 07110

REEVES, JOHN T, b Hazard, Ky, Nov 17, 28; c 2. CARDIOLOGY. *Educ:* Mass Inst Technol, BS, 50; Univ Pa, MD, 54. *Prof Exp:* USPHS res fel cardiol, Univ Colo, 57-58, Colo Heart Asn res fel, 58-59, Am Heart Asn advan res fel, 59-61; from asst prof to prof med, Univ Ky, 61-72; PROF MED, UNIV COLO MED CTR, DENVER, 72- *Mem:* Am Physiol Soc; Soc Exp Biol Med; Am Col Chest Physicians; Am Thoracic Soc. *Mailing Add:* Dept Med Univ Colo Med Ctr 4200 E Ninth Ave Denver CO 80220

REEVES, LEONARD WALLACE, b Bristol, Eng, Feb 8, 30; m 54. PHYSICAL CHEMISTRY. *Educ:* Bristol Univ, BSc, 51, PhD(phys chem), 54. *Hon Degrees:* DSc, Bristol Univ, 65. *Prof Exp:* Res asst, Univ Calif, 54-56; fel, Nat Res Coun Can, 56-57; fel, Mellon Inst, 57-58; from asst prof to prof chem, Univ BC, 58-69; chmn dept, 69-71, PROF CHEM, UNIV WATERLOO, 69- *Concurrent Pos:* Vis prof, Univ Sao Paulo, 67-; Noranda lectr, 69. *Mem:* AAAS; Am Chem Soc; fel Chem Inst Can; Am Phys Soc; Int Soc Magnetic Resonance; fel Royal Soc Can; corresp mem Nat Acad Sci Brasil. *Res:* Metal ions in aqueous and ordered environments; lyotropic liquid crystals and membranes; chemical exchange and reaction mechanisms; pulsed and continuous wave nuclear magnetic resonance; molecular structure and intermolecular forces. *Mailing Add:* Rm 131 Dept Chem Univ Waterloo Waterloo ON N2L 3G1 Can

REEVES, PERRY CLAYTON, b Brady, Tex, Nov 9, 42; m 64; c 2. CHEMISTRY. *Educ:* Abilene Christian Col, BS, 65; Univ Tex, Austin, PhD(chem), 69. *Prof Exp:* From asst prof to prof chem, Southern Methodist Univ, 69-80; dean, Col Natural & Appl Sci, 81-87, PROF CHEM, ABILENE CHRISTIAN UNIV, 80- *Mem:* Am Chem Soc; Royal Soc Chem. *Res:* Organic synthesis via organometallic compounds. *Mailing Add:* 810 Green Valley Abilene TX 79601

REEVES, R C, b Columbus, Miss, Dec 27, 48; m 72; c 3. CARDIOVASCULAR MEDICINE. *Educ:* Univ Ala, BS, 70; Univ Ala Sch Med, MD, 74. *Prof Exp:* AT UNIV ALA, BIRMINGHAM. *Mailing Add:* Baptist Med Ctr 880 Montclair Rd Suite 170 Birmingham AL 35213

REEVES, RAYMOND, b St Louis, Mo, June 29, 43; div; c 2. NUCLEIC ACID BIOCHEMISTRY. *Educ:* Univ Calif, Berkeley, BA, 66, PhD, 71. *Prof Exp:* Fel, Oxford Univ, Eng, 71-72; fel, Med Res Coun Lab Molecular Biol, Cambridge, 72-73; asst prof, Univ BC, 73-78; PROF BIOCHEM, BIOPHYS, MOLECULAR GENETICS & CELL BIOL, WASH STATE UNIV, 79- *Mem:* Am Soc Biol Chemists; Soc Develop Biol; Am Soc Cell Biol; Int Soc Develop Biol. *Res:* Biochemistry of eukaryotic gene regulation; lymphokine regulation of lymphocyte growth and response of lymphokine genes to viral infections; chromatin structure and function. *Mailing Add:* Dept Biochem & Genetics Wash State Univ 431 Heald Hall Pullman WA 99164

REEVES, RICHARD EDWIN, b Lincoln, Nebr, Oct 28, 12. ORGANIC CHEMISTRY. *Educ:* Doane Col, BA, 33; Yale Univ, PhD(chem), 36. *Prof Exp:* Nat Tuberc Asn res fel, Yale Univ, 36-37, Wood mem fel, 40-41; fel chem, Rockefeller Inst, 37-38; res chemist, Boyce Thompson Inst Plant Res, 38-40; from assoc chemist to chemist, Southern Regional Res Lab, Bur Agr & Indust Chem, USDA, 41-53, Southern Utilization Res Br, Agr Res Serv, 53-54; res assoc, 54-57, assoc prof, 57-63, PROF BIOCHEM, SCH MED, LA STATE UNIV, NEW ORLEANS, 63-, PROF TROP MED & MED PARASITOL, 74- *Mem:* Am Chem Soc; Am Soc Trop Med & Hyg; Am Soc Biol Chem. *Res:* Bacterial lipids and carbohydrates; cellulose chemistry; fine structure of glycosides; nutrition and biochemistry of Entamoeba Histolytica. *Mailing Add:* 3336 Esplanade New Orleans LA 70119

REEVES, ROBERT BLAKE, b Philadelphia, Pa, July 26, 30; m 67; c 2. PHYSIOLOGY. *Educ:* Swarthmore Col, AB, 52; Harvard Univ, PhD(physiol), 59. *Prof Exp:* Res assoc physiol, Univ Pa, 59-60; res assoc biophys, Sch Med, State Univ NY, Buffalo, 60-61; asst prof physiol, Cornell Univ, 61-66; assoc prof, 66-76, PROF PHYSIOL, STATE UNIV NY, BUFFALO, 76- *Concurrent Pos:* Jr fel, Soc Fels, Harvard Univ, 57-60. *Mem:* Am Physiol Soc. *Res:* Acid-base balance; oxygen transport; red cell gas kinetics. *Mailing Add:* Dept Physiol State Univ NY Health Sci Ctr 3435 Main St Sherman Hall Rm 117 Buffalo NY 14214

REEVES, ROBERT DONALD, b Lubbock, Tex, Jan 14, 42; m 67; c 2. NUTRITION, BIOCHEMISTRY. *Educ:* Tex Tech Univ, BA, 64, MS, 65; Iowa State Univ, PhD(nutrit), 71. *Prof Exp:* Res assoc nutrit, Agr Exp Sta, Iowa State Univ, 65-71; res & teaching nephrology med, Vet Admin Hosp, Little Rock, Ark, 71-77; from instr to asst prof med & biochem, Med Sci, Univ Ark, Little Rock, 74-77; assoc prof, 77-86, PROF NUTRIT, DEPT FOODS & NUTRIT, KANS STATE UNIV, 86- *Concurrent Pos:* Scientist nutrit, Agr Exp Sta, Kans State Univ, 77- *Mem:* Am Inst Nutrit; Am Fedn Clin Res; Am Dietetic Asn; Sigma Xi; fel Am Col Nutrit; Am Soc Clin Nutrit. *Res:* Clinical nutrition and nutritional aspects of metabolic disease; nutritional factors influencing somatomedin activity; dietary fiber and lipid metabolism. *Mailing Add:* Dept Foods & Nutrit Kans State Univ Manhattan KS 66506

REEVES, ROBERT GRIER (LEFEVRE), b York, Pa, May 30, 20; m 42; c 2. ECONOMIC GEOLOGY, GEOPHYSICS. *Educ:* Univ Nev, BS, 49; Stanford Univ, MS, 50, PhD(econ geol, geochem), 65. *Prof Exp:* Geophysicist, US Geol Surv, 49-50, geologist, 50-69; sr Fulbright lectr, Univ Adelaide, 69; prof geol, Colo Sch Mines, 69-73; phys scientist, US Geol Surv, 73-78; prof geol, Univ Tex, Permian Basin, 78-85, dean, Col Sci & Eng, 79-84, fac geol, 78-83, prof geol, 78-85, dir, Div Eng, 84- 85; GEOL & MINING CONSULT, 85- *Concurrent Pos:* Co-leader, Int Field Inst Brazil, 66; vis geologist, Boston Col & Boston Univ, 67; mem, Coun Educ Geol Sci, Am Geol Inst, 68-74; ed-in-chief, Manual Remote Sensing, Am Soc Photogram, 70-75; leader, People to People Mining & Econ Geol Delegation to Brazil & Peru, 85. *Honors & Awards:* Fischer Res Award, 50. *Mem:* Fel Geol Soc Am; Am Inst Mining, Metall & Petrol Engrs; Soc Econ Geologists; Soc Independent Prof Earth Scientists. *Res:* Application of electromagnetic remote sensor and geophysical data to geology and mineral exploration; economic geology of iron ore, ferrous and non-ferrous metals, and precious metals; application of electromagnetic remote sensor and geophysical data to geology and mineral exploration; economic geology and geochemistry of iron and manganese. *Mailing Add:* 4025 Lakeside Dr Odessa TX 79762

REEVES, ROBERT LLOYD, dentistry, for more information see previous edition

REEVES, ROBERT R, b Albany, NY, June 16, 30; m 55; c 3. PHYSICAL CHEMISTRY, CHEMICAL ENGINEERING. *Educ:* Rensselaer Polytech Inst, BChe, 51, MChe, 52, PhD(chem eng), 54. *Prof Exp:* Res assoc chem, 54-71, assoc prof, 71-76, PROF CHEM, RENSSELAER POLYTECH INST, 76- *Concurrent Pos:* Prog mgr, US Dept Energy, 77, consult, 78- *Mem:* Sigma Xi; Am Chem Soc. *Res:* Photochemistry; atmospheric phenomena and reaction kinetics; laser studies. *Mailing Add:* Dept Chem Rensselaer Polytech Inst Troy NY 12181

REEVES, ROBERT WILLIAM, b Morristown, NJ, Aug 12, 39; m 64; c 2. TROPICAL METEOROLOGY. *Educ:* NY Univ, BS, 61, MS, 65; Univ Wash, PhD(meterol), 80. *Prof Exp:* Res meteorologist, Nat Hurricane Res Lab, 65-69; METEOROLOGIST, CTR ENVIRON ASSESSMENT SERV, 69- *Concurrent Pos:* Coordr aircraft opers, Barbados Oceanog & Meteorol Exp, 69; Rawinsonda data qual expert, Workshop Global Atmospheric Res

Prog, Atlantic Trop Exp, 77. *Mem:* Am Meteorol Soc. *Res:* Investigation of the dynamics of the disturbed and undisturbed tradewind atmosphere; implementation of existing coastal ocean circulation modeling to aid in marine environmental assessment. *Mailing Add:* 5400 Waneta Rd Bethesda MD 20816

REEVES, ROGER MARCEL, b Rochester, NY, Feb 7, 34; m 63; c 2. ENTOMOLOGY, ACAROLOGY. *Educ:* State Univ NY Col Forestry, Syracuse Univ, BS, 57, PhD(entom), 65; Cornell Univ, MS, 61. *Prof Exp:* Asst prof, 64-74, ASSOC PROF ENTOM & FORESTRY, UNIV NH, 74- *Mem:* Entom Soc Am; Acarological Soc Am; Am Entom Soc. *Res:* Taxonomy and ecology of Acari, particularly the Cryptostigmata; Carabidae, spiders. *Mailing Add:* Dept Entom Univ NH Durham NH 03824

REEVES, ROY FRANKLIN, b Warrensburg, Mo, July 8, 22; m 51; c 5. MATHEMATICS. *Educ:* Univ Colo, BS, 47; Iowa State Col, PhD(math), 51. *Prof Exp:* Instr math, Univ Colo, 47-48; instr elec eng, Iowa State Col, 48-50, instr math, 50-51; prof math, Ohio State Univ, 51-81, dir, Computer Ctr, 55-81; PROF MATH SCI, OTTERBEIN COL, 81- *Mem:* Am Mgt Asn; NY Acad Sci; AAAS; Am Math Soc; Math Asn Am. *Res:* Numerical analysis and computing; computer science. *Mailing Add:* 145 Elmwood Pl Westerville OH 43081-2001

REEVES, T JOSEPH, b Waco, Tex, Apr 22, 23;; c 3. CARDIOVASCULAR PHYSIOLOGY, MEDICINE. *Educ:* Baylor Univ, BS, 43, MD, 46. *Prof Exp:* Intern med, Parkland Hosp, Dallas, Tex, 46-47, resident, 49-51; pvt pract, Beaumont, Tex, 52-54; from asst prof to prof med, Sch Med, Univ Ala, Birmingham, 54-73, dir cardiovasc div, 58-66, assoc prof physiol & biophys, 58-73, dir cardiovasc res & training prog, 66-73, chmn dept med, 70-73; DIR CARDIOVASC LAB, ST ELIZABETH HOSP, 73-; CLIN PROF MED, UNIV TEX, HOUSTON, 81- *Concurrent Pos:* Fel heart dis, Univ Ala, 51-52; Nat Heart Inst spec fel, Queen Elizabeth Hosp, Birmingham, Eng, 57-58; chief med serv, Vet Admin Hosp, Birmingham, Ala, 54-55, consult, 56-73; mem exec comt, Coun Circulation, Am Heart Asn, 63-66; mem nat adv comt, Heart Dis Control Prog & chmn prog proj comt, Nat Heart Inst, 65-67; chmn adv comt cardiol, Nat Heart & Lung Inst, 74-77. *Mem:* Am Soc Clin Invest; Am Physiol Soc; Am Fedn Clin Res; Sigma Xi. *Res:* Myocardial contraction; physiology of muscular exercise; cardiac diagnosis. *Mailing Add:* 2929 Calder Suite 310 Beaumont TX 77702

REEVES, W PRESTON, b Handley, Tex, Sept 22, 35; m 60; c 1. ORGANIC CHEMISTRY. *Educ:* Tex Christian Univ, BS, 57, MA, 59; Univ Tex, Austin, PhD(chem), 66. *Prof Exp:* Instr chem, Arlington State Col, 61-62; From asst prof to assoc prof, 65-77, PROF CHEM, TEX LUTHERAN COL, 77- *Mem:* Am Chem Soc. *Res:* Thermal reactions of strained rings; allene chemistry; phase transfer catalysis; ultrasound. *Mailing Add:* Dept Chem Tex Lutheran Col Seguin TX 78155

REEVES, WILLIAM CARLISLE, b Riverside, Calif, Dec 2, 16; m 40; c 3. EPIDEMIOLOGY. *Educ:* Univ Calif, BS, 38, PhD(entom), 43, MPH, 48. *Prof Exp:* Lab asst, Entom Div, Univ Calif, 38-42, entomologist, Hooper Found, 41-42, asst epidemiol, Entom Div, 42-46, asst, Med Sch, 45-48, res asst, 46-49, lectr, Sch Pub Health, 47-54, assoc prof pub health, 49-54, from actg dean to dean, Sch Pub Health, 67-71, prof, 54-87, EMER PROF EPIDEMIOL, SCH PUB HEALTH, UNIV CALIF, BERKELEY, 87- *Concurrent Pos:* Consult, Commun Dis Ctr, USPHS & State Dept Health, Calif. *Mem:* AAAS; Am Soc Trop Med & Hyg; Entom Soc Am; Am Pub Health Asn; Am Mosquito Control Asn. *Res:* Epidemiology of the arthropod-borne virus encephalitides; mosquito biology and systematics; avian malaria epidemiology. *Mailing Add:* Sch Pub Health Univ Calif Berkeley CA 94720

REEVES, WILSON ALVIN, b Mittie, La, July 14, 19; m 42; c 4. TEXTILE CHEMISTRY. *Educ:* Southwestern La Inst, BS, 41; Tulane Univ, La, MS, 50;. *Hon Degrees:* DSc, Clemson Univ, 69. *Prof Exp:* Chemist, Southern Regional Res Lab, Agr Res Serv, USDA, 42-75; prof, La State Univ, 76-85; RETIRED. *Concurrent Pos:* Consult, 75-76. *Honors & Awards:* John Scott Award, 65; Olney Medal, 66; Honor Award, Am Inst Chem, 70, 76, 78. *Mem:* Am Chem Soc; Sigma Xi; Am Asn Textile Chem & Colorists; Fiber Soc. *Res:* Organic, polymer and cellulose chemistry; fire retardant textiles. *Mailing Add:* 715 Marguerite Rd Metairie LA 70003

REFFES, HOWARD ALLEN, b New York, NY, Sept 18, 28. ANALYTICAL CHEMISTRY. *Educ:* Queens Col, NY, BS, 50; Stevens Inst Technol, MS, 56. *Prof Exp:* Res analytical chemis, Colgate-Palmolive Co, 52-57; sr chemist & head anal develop, Wallace & Tiernan, Inc, 57-61; mgr qual control, Flavor Div, Int Flavors & Fragrances, Inc, NJ, 61-68, plant mgr, 68-73; indust mgr, North Am Dairy Industs, 73-81; vpres & gen mgr, Food Div, Goodhost Foods, Ltd, 81-84; DIR, TECH SERV, NESTLE ENTERPRISES LTD, 84- *Concurrent Pos:* Instr, NY Community Col, 56-68. *Mem:* AAAS; Am Chem Soc; Inst Food Technol; Am Soc Qual Control; fel Am Inst Chem. *Res:* Methods development for food additives and residues, pharmaceuticals, plasticizers and fatty acids; analytical research in condensed phosphates; polymeric anhydrides; analytical instrumentation; quality control of flavors; flavor manufacturing. *Mailing Add:* 110 Shaftesbury Ave Toronto ON M4T 1A5 Can

REFFNER, JOHN A, b Akron, Ohio, Jan 5, 35; m 57; c 2. CHEMICAL MICROSCOPY, INFRARED SPECTROSCOPY. *Educ:* Univ Akron, BS, 56; Ill Inst Technol, MS, 60; Univ Conn, PhD(polymer sci), 75. *Prof Exp:* Mat engr, B F Goodrich Co, 55-57; dir res, W C McCrone Assoc, 58-66; asst dir, Inst Mat Sci, Univ Conn, 66-77; prin res scientist, Am Cyanamid Res Labs, 77-86; CORP FEL, SPECTRA-TECH INC, 86- *Concurrent Pos:* Sci consult, Conn State Police, 74-; adj fac, John Jay Col, 85- *Mem:* Am Chem Soc; Electron Micros Soc Am; Northeastern Asn Forensic Sci; Am Soc Testing & Mat; Microbeam Soc Am; Am Acad Forensic Sci. *Res:* Chemical microscopy; infrared spectroscopy; ultramicro analysis; polymer science; forensic science; failure analysis. *Mailing Add:* Spectra-Tech Inc 652 Glenbrook Rd Stamford CT 06906

REFOJO, MIGUEL FERNANDEZ, b Santiago, Spain, July 6, 28; US citizen; m 59, 86; c 2. BIOMATERIALS, OPHTHALMOLOGY. *Educ:* Univ Santiago, Spain, Lic Sc, 53, DSc(org chem), 56. *Hon Degrees:* Dr, Univ Santiago, Spain, 88. *Prof Exp:* Fel, Yale Univ, 56-59; res chemist, Tech Dept, DuPont Can, Ont, 59-63; res assoc, Mass Eye & Ear Infirmary, Boston, 63-64; res assoc, 64-70, assoc, 70-71, sr scientist, 64-71, HEAD POLYMER CHEM UNIT, EYE RES INST RETINA FOUND, 71-; ASSOC PROF OPHTHAL, HARVARD MED SCH, 82- *Concurrent Pos:* Dir, Corneal Sci, Inc, Boston, 72-79; prin assoc ophthal & biochem, Harvard Med Sch, 75-82; vis prof, Col Optom, Univ Houston, 84; adj assoc prof, Sch Optom, Univ Mo, St Louis, 90- *Honors & Awards:* Hon Mem, Academia Medico-Quisurgica, Santiago, Spain, 88. *Mem:* AAAS; Am Chem Soc; Asn Res Vision & Ophthal; Sigma Xi; Soc Biomat; Int Soc Contact Lens Res (pres, 84-86). *Res:* Ophthalmology; synthetic polymers in medicine and surgery; hydrogels; contact lenses; drug delivery. *Mailing Add:* Eye Res Inst Retina Found 20 Staniford St Boston MA 02114

REFT, CHESTER STANLEY, b Pittsburgh, Pa, Sept 29, 44; m 68; c 3. MEDICAL PHYSICS, RADIATION DOSIMETRY. *Educ:* Carnegie-Melon Univ, BS, 66; Univ Pittsburgh, PhD(physics), 73. *Prof Exp:* Res physicist, US Army-Harry Diamond Lab, 72-75; res assoc, Old Dominion Univ, 75-77, asst prof physics, 77-78, res assoc, 78-79, asst prof, 79-80; res assoc, 80-88, ASST PROF PHYSICS, UNIV CHICAGO, 88- *Mem:* Am Phys Soc; Am Asn Physicists Med. *Res:* Improving the dosimetry of electron, photon and neutron teletheupy units for radyation therapy. *Mailing Add:* 246 Indiana St Park Forest IL 60466

REGAL, JEAN FRANCES, ALLERGY, LUNG PATHOLOGY. *Educ:* Univ Minn, PhD(pharmacol), 77. *Prof Exp:* ASSOC PROF PHARMACOL, UNIV MINN, 85- *Res:* Immediate hypersensitivity. *Mailing Add:* Dept Pharmacol Univ Minn 2400 Oakland Ave Duluth MN 55812

REGAL, PHILIP JOE, b Los Angeles, Calif, Dec 2, 39. BEHAVIORAL BIOLOGY, EVOLUTION. *Educ:* San Diego State Col, BA, 62; Univ Calif, Los Angeles, MA, 66, PhD(zool), 68. *Prof Exp:* Trainee, NIMH Brain Res Inst-Univ Calif, Los Angeles-Univ Calif, San Diego-Scripps Inst Oceanog, 68-70; ASSOC PROF ECOL & BEHAV BIOL & CUR HERPET, MUS NATURAL HIST, UNIV MINN, MINNEAPOLIS, 70- *Concurrent Pos:* Mem US directorate UNESCO Man in Biosphere Prog. *Mem:* AAAS; Animal Behav Soc; Am Soc Ichthyologists & Herpetologists; Soc Study Amphibians & Reptiles; Ecol Soc Am. *Res:* Behavioral and physiological adaptations; behavioral temperature regulation; evolutionary processes; evolutionary trends in vertebrates, particularly amphibians and reptiles, evolutionary ecology of plants. *Mailing Add:* Dept Ecol & Biol Univ Minn 318 Church St SE 109 2001 Bldg Minneapolis MN 55455-0302

REGAN, FRANCIS, b Vigo Co, Ind, Jan 10, 03. MATHEMATICS. *Educ:* Ind State Univ, AB, 22; LaSalle Exten Univ, LLB, 26; Ind Univ, AM, 30; Univ Mich, PhD(math), 32. *Prof Exp:* Teacher high sch, Ind, 23-25; prof com & asst prof math, Columbus Col, 25-29; asst prof math, Colo State Univ, 29-30; from instr to prof, 32-71, dir dept, 50-71, EMER PROF MATH, ST LOUIS UNIV, 71- *Honors & Awards:* C C MacDuffee Award, 70; Nancy McNeir Ring Award, 71. *Mem:* Am Math Soc; Math Asn Am; Sigma Xi. *Res:* Infinite series; foundations of probability; mathematical analysis and statistics. *Mailing Add:* 5230 N Ragan Pl West Terre Haute IN 47885

REGAN, GERALD THOMAS, b Omaha, Nebr, Apr 19, 31. ECOLOGY. *Educ:* St Louis Univ, AB, 55, PhL, 57, MS, 62; Univ Kans, PhD(zool), 72. *Prof Exp:* Instr biol, Creighton Univ, 68-71; from asst prof to assoc prof, 72-82, PROF BIOL, SPRING HILL COL, 82- *Concurrent Pos:* Chmn, sci div, 82-85, prog comt Marine Environ Sci Consortium, 85-86; Coord, Ala Marine Mammal Stranding Network, 90- *Mem:* Soc Marine Mammal; Am Cetacean Soc. *Res:* Evolution; ecological biogeography. *Mailing Add:* Dept Biol Spring Hill Col Mobile AL 36608-1791

REGAN, JAMES DALE, b Lancaster, Ohio, May 23, 31; m 55; c 2. HUMAN GENETICS, UV PHOTOBIOLOGY. *Educ:* Univ Ohio, BS, 53; Univ Miami, Fla, MS, 59; Univ Hawaii, PhD(genetics), 64. *Prof Exp:* Asst prof biol, Chaminade Col Honolulu, 62-64; Nat Cancer Inst fel, 64-66, BIOLOGIST, OAK RIDGE NAT LAB, 66-, SR INVESTR & DIR, MOLECULAR & GENETIC GRANTS. *Concurrent Pos:* Am Cancer Soc Hawaii Div res grant, 62-64; vis lectr, Univ Tenn, 66- *Mem:* AAAS; Am Asn Cancer Res; Am Soc Human Genetics. *Res:* Genetics of human cells in vitro; repair of DNA and repair-deficient diseases; DNA repair and carcinogenesis; UV-B and ozone depletion. *Mailing Add:* Biol Div Oak Ridge Nat Lab Oak Ridge TN 37830

REGAN, RAYMOND WESLEY, b New York, NY, Aug 30, 43; m 68; c 7. BIOLOGICAL WASTE TREATMENT SYSTEMS. *Educ:* Manhattan Col, BEChE, 65, ME, 66; Kansas Univ, Lawrence, PhD(environ health eng), 72. *Prof Exp:* Proj engr, Environ Eng Grad Prog, Manhattan Col, 66-70; asst prof, 72-78, ASSOC PROF CIVIL ENG, PA STATE UNIV, 78-, CO-DIR, OFF HAZARDOUS & TOXIC WASTE MGT, INST RES LAND & WATER RESOURCES, 81- *Concurrent Pos:* Assoc, Environ Sci Div, Oak Ridge Nat Lab, Tenn, 82 & Toxic & Hazardous Mat Agency, US Army, Aberdeen, Md, 85. *Mem:* Am Soc Civil Engrs; Water Pollution Control Fedn. *Res:* Advancing technologies for improved hazardous waste management for smaller industries, including electroplaters-metal finishers, plastics and synthetics, paint and allied products; foundries. *Mailing Add:* Pa State Univ 134 Land & Water Bldg University Park PA 16802

REGAN, THOMAS HARTIN, b Pittsburgh, Pa, June 7, 25; m 51; c 6. ORGANIC CHEMISTRY. *Educ:* Duquesne Univ, 49, MS, 51; Mass Inst Technol, PhD(org chem), 55. *Prof Exp:* Res chemist, Explosives Dept, E I du Pont de Nemours & Co, 55-64; sr res chemist, 64-66, RES ASSOC, EASTMAN KODAK CO, ROCHESTER, 66- *Mem:* Am Chem Soc. *Res:* Kinetics of ester aminolysis and nitrous acid oxidation; synthetic organic chemistry; nuclear magnetic resonance; application to structure determination of organic molecules. *Mailing Add:* Three Manor Hill Dr Fairport NY 14450-2519

REGAN, THOMAS M(ICHAEL), b New Orleans, La, Nov 28, 41; m 64; c 2. CHEMICAL ENGINEERING, BIOENGINEERING. *Educ:* Tulane Univ, BS, 63, PhD, 67. *Prof Exp:* From asst prof to assoc prof, 66-76, PROF CHEM ENG, UNIV MD, COLLEGE PARK, 76- *Mem:* Am Inst Chem Engrs; Am Chem Soc. *Res:* Optimization of artificial kidney systems; membrane test cell design; testing of blood-gas exchangers; ionic and membrane diffusion. *Mailing Add:* Dept Chem Eng Univ Md College Park MD 20742

REGAN, TIMOTHY JOSEPH, b Boston, Mass, July 24, 24; c 4. INTERNAL MEDICINE, CARDIOLOGY. *Educ:* Boston Col, AB, 48; Boston Univ, MD, 52; Am Bd Internal Med, dipl, 60. *Prof Exp:* Rotating intern, City Detroit Receiving Hosp, 56-57; instr med, Sch Med, Wayne State Univ, 57-59, asst prof, 59-60; from asst prof to assoc prof, 60-66, PROF MED, UNIV MED & DENT NJ, 66-, DIR DIV CARDIOVASC DIS, 65- *Concurrent Pos:* Jr assoc med, City Detroit Receiving Hosp, 57-60; assoc in med, Children's Hosp, Detroit, 59-60; assoc dir, T J White Cardiopulmonary Inst, B S Pollak Hosp, Jersey City, 60-65, dir, 65-71, attend physician, Hosp, 60-71; estab investr, Am Heart Asn, 61-66; attend physician, Vet Admin Hosp, East Orange, NJ & Univ Med & Dent NJ-Univ Hosp, Newark, 65-; chmn subcomt diag procedures in heart dis, NJ Regional Med Prog, 69-; mem cardiovasc study sect, Nat Adv Coun & Comts, NIH, 69-73. *Mem:* Am Fedn Clin Res; Am Physiol Soc; Am Heart Asn; Am Soc Clin Invest; Am Diabetes Asn; Asn Am Physicians. *Res:* Myocardial metabolism in disease. *Mailing Add:* Dept Med/Div Cardiovasc Dis Univ Med & Dent NJ Med Sch 185 S Orange Ave Newark NJ 07103-2757

REGELSON, WILLIAM, b New York, NY, July 12, 25; m 48; c 6. MEDICINE. *Educ:* Univ NC, AB, 48; State Univ NY, MD, 52. *Prof Exp:* Intern med, Maimonides Hosp, Brooklyn, NY, 52-53; from asst resident to sr resident, Mem Ctr Cancer & Appl Dis, New York, 53-55; spec fel cancer res, Roswell Park Mem Inst, 55-56; sr cancer res internist, 56-57, assoc cancer res internist, 57-59, assoc chief med, 59-67; asst res prof med, State Univ NY, Buffalo, 65-67; chief div med oncol, 67-76, prof med, Med Col Va, 67-; AT DEPT MED, VA COMMONWEALTH UNIV. *Concurrent Pos:* Consult, Monsanto Chem Co, 59, A H Robins Co, Inc, 69-71, Hercules, Inc, 71-74 & Merrell-Nat Labs, 71-76. *Mem:* Reticuloendothelial Soc; Am Heart Asn; Am Asn Cancer Res; Am Fedn Clin Res; Am Soc Clin Oncol. *Res:* Effect of polyelectrolytes on cell growth and differentiation and on enzyme and viral function; effects of various chemotherapeutic and immunotherapeutic agents on cancer in man and animal; production and prevention of tumor growth; aging. *Mailing Add:* Med Col Va Box 273 Richmond VA 23298

REGEN, DAVID MARVIN, b Nashville, Tenn, Mar 18, 34; m 58; c 3. PHYSIOLOGY, BIOCHEMISTRY. *Educ:* Davidson Col, BS, 56; Vanderbilt Univ, PhD(physiol), 62. *Prof Exp:* Instr physiol, Vanderbilt Univ, 62-63; guest investr, Max Planck Inst Cell Chem, 63-64; asst prof, 64-69, assoc prof, 69-76, PROF PHYSIOL, VANDERBILT UNIV, 76- *Concurrent Pos:* Howard Hughes Med Inst fel, 63-64, investr, 64-71; NIH res grant, 65- *Mem:* Am Physiol Soc. *Res:* Mechanism and kinetics of glucose transport; regulatory effects of insulin, diabetes, anoxia and work on glucose utilization in muscle; control of hepatic cholesterol synthesis; monocarboxylate transport; regulation of glucose transport in lymphocytes; brain glucose metabolism; brain ketone-body metabolism; cardiac dynamics. *Mailing Add:* Dept Molecular Physiol & Biophys Vanderbilt Univ Nashville TN 37232

REGENBRECHT, D(OUGLAS) E(DWARD), b Bryan, Tex, June 8, 24; m 58; c 3. SPACE THERMAL SYSTEMS, OPTICAL THERMAL STABILITY. *Educ:* Tex A&M Univ, BSME, 48; Purdue Univ, MSME, 51, PhD(mech eng), 62. *Prof Exp:* Instr mech eng, Purdue Univ, 50-57; asst prof, Univ Tulsa, 57-63, assoc prof & head dept, 63-67; sr mem tech staff, Ball Bros Res Corp, 67-77, staff scientist, 77-80; prin thermal engr, Ball Aerospace Systs Group, 80-89; CONSULT, 89. *Concurrent Pos:* Consult, Space & Info Systs Div, NAm Aviation, Inc, Tulsa, 62-66. *Mem:* Am Soc Mech Engrs; Am Soc Eng Educ; Am Inst Aeronaut & Astronaut; Soc Packaging & Handling Engrs; Sigma Xi. *Res:* Fluid mechanics; convective and radiative heat transfer; thermal energy conversion systems; mathematical modeling. *Mailing Add:* 870 Gilpin Dr Boulder CO 80303

REGENER, VICTOR H, b Berlin, Ger, Aug 25, 13; nat US; m 41; c 2. PHYSICS. *Educ:* Univ Stuttgart, DrEng, 38. *Prof Exp:* Res fel, Padova Univ, 38-40; res fel, Univ Chicago, 40-42, instr physics, 42-46; from assoc prof to prof physics, Univ NMex, 46-57, chmn dept, 46-57 & 62-79, res prof, 57-76; OWNER, VHR SYSTS, 79- *Concurrent Pos:* Assoc cur, Mus Sci & Indust, Univ Chicago, 45-46; hon prof, Univ Mayor de San Andres, La Paz, Bolivia. *Mem:* Fel Am Phys Soc; Am Astron Soc; Am Geophys Union; fel NY Acad Sci. *Res:* Atmospheric ozone; cosmic radiation; zodiacal light; balloon and satellite experiments; optical studies of pulsars; electronics; optics. *Mailing Add:* VHR Systs 7200 Jefferson NE Albuquerque NM 87109

REGENSTEIN, JOE MAC, b Brooklyn, NY, Sept 22, 43; m 66; c 2. FOOD SCIENCE. *Educ:* Cornell Univ, BA, 65, MS, 66; Brandeis Univ, PhD(biophys), 72. *Prof Exp:* Fel muscle, Children's Cancer Res Ctr, Boston, 73 & Brandeis Univ, 73-74; asst prof poultry sci, 74-80, from asst prof to assoc prof food sci, 75-87, PROF POULTRY SCI & FOOD SCI, CORNELL UNIV, 87-, DIR, CORNELL KOSHER FOOD PROG, 91- *Concurrent Pos:* Sabbatical leave, Torry Res Sta, Aberdeen, Scotland, 80-81; guest fel, New Zealand Inst Food Sci & Technol; counr, Inst Food Technologists. *Honors & Awards:* Earl P McFee Award, Atlantic Fisheries Technol Soc. *Mem:* Am Chem Soc; Poultry Sci Asn; Inst Food Technologists; AAAS; Am Meat Sci Asn. *Res:* Functional properties of muscle proteins, especially water retention properties and emulsification; frozen storage changes in gadoid fish; shelf-life extension of fresh poultry and fish; new product development from minced fish and poultry; kosher foods; waste management in fisheries. *Mailing Add:* Dept Food Sci 112 Rice Hall Cornell Univ Ithaca NY 14853-5601

REGER, BONNIE JANE, b Fargo, NDak, Sept 29, 40. PLANT PHYSIOLOGY. *Educ:* NDak State Univ, BS, 62; Univ Md, MS, 65; PhD(plant physiol), 68. *Prof Exp:* Fel biochem, Oak Ridge Nat Lab, 68-71; plant physiologist, Southern Weed Sci Lab, Agr Res Serv, Stoneville, Miss, 71-76; PLANT PHYSIOLOGIST, RUSSELL RES CTR, SCI & EDUC ADMIN, USDA, 76- *Concurrent Pos:* USPHS fel, 69-70; NATO travel grant, Spetsai, Greece, 69-; adv, Nat Res Coun-Agr Res Postdoctoral Res Associateship Prog, 73-76; consult, Substitute Chem Prog, Environ Protection Agency. *Mem:* Am Soc Plant Physiologists; Am Chem Soc; AAAS; Weed Sci Soc Am; Sigma Xi. *Res:* Photosynthesis especially regulation of RuBP carboxylase synthesis, cell separation of Cfour plants and enzymatic characterization, chloroplast biogenesis; intergeneric crosses especially methods for overcoming rejection of incompatible pollen. *Mailing Add:* Russell Res Ctr PO Box 5677 Athens GA 30613

REGER, DANIEL LEWIS, b Mineral Wells, Tex, Sept 16, 45; m 68; c 2. ORGANOMETALLIC CHEMISTRY. *Educ:* Dickinson Col, BS, 67; Mass Inst Technol, PhD(chem), 72. *Prof Exp:* From asst prof to assoc prof, 72-84, PROF CHEM, UNIV SC, 84- *Mem:* Am Chem Soc; Sigma Xi. *Res:* Transition metal organometallic synthesis; organic synthesis via organometallics; homogeneous catalysis. *Mailing Add:* Dept Chem Univ SC Columbia SC 29208

REGER, JAMES FREDERICK, b Norway, Iowa, Oct 27, 24; m 46; c 3. CYTOLOGY. *Educ:* Univ Iowa, PhD(zool), 54. *Prof Exp:* Asst, Univ Iowa, 50-54; asst prof zool, Ariz State Univ, 54-55; from res assoc to asst prof anat, Sch Med, Univ Colo, 55-65; assoc prof, 65-70, PROF ANAT, MED UNITS, UNIV TENN, MEMPHIS, 70- *Concurrent Pos:* USPHS career develop award, 59-65. *Mem:* AAAS; Am Soc Zoologists; Soc Protozool; Am Asn Anatomists; Am Soc Cell Biol. *Res:* Cytology of spinal ganglion cells; electron microscopy of euglena, myoneural junction, the synapse, kidney, muscle, oocytes; spermatozoa. *Mailing Add:* Dept Anat Univ Tenn Health Sci Ctr 800 Madison Ave Memphis TN 38103

REGER, RICHARD DAVID, b Chico, Calif, May 10, 39; m 68; c 2. GLACIAL GEOLOGY, QUATERNARY GEOLOGY. *Educ:* Univ Alaska, Fairbanks, BS, 63, MS, 64; Ariz State Univ, PhD(geol), 75. *Prof Exp:* Instr geol, Ariz State Univ, 72-73; sr geologist, R & M Consult, Inc, 73-75; GEOLOGIST III, ALASKA DIV GEOL & GEOPHYS SURV, 75- *Res:* Mapping surficial deposits, especially glacial deposits, throughout Alaska for a comprehensive environmental evaluation. *Mailing Add:* 1983 Kittiwake Dr Fairbanks AK 99709-4609

REGEZI, JOSEPH ALBERTS, b Grand Rapids, Mich, May 14, 43; m 64; c 2. ORAL PATHOLOGY. *Educ:* Univ Mich, DDS, 68, MS, 71. *Prof Exp:* Chmn dept path, David Grant Med Ctr, USAF, 71-73; asst prof oral path, 73-77, ASST PROF DENT, DEPT HOSP DENT, UNIV MICH HOSP, 73-, ASSOC PROF ORAL PATH, SCH DENT, UNIV MICH, ANN ARBOR, 77-, ASST PROF PATH, MED SCH, 76- *Mem:* Fel Am Acad Oral Path; Int Asn Dent Res. *Res:* Light and electron microscopic studies of head and neck neoplasms, especially salivary gland and odontogenic tumors, in regard to their classification, diagnosis, and histogenesis. *Mailing Add:* 2209 Woodside Rd Ann Arbor MI 48104

REGIER, HENRY ABRAHAM, b Brainerd, Alta, Mar 5, 30; m 56; c 3. ECOLOGY, FISHERIES. *Educ:* Queen's Univ, BA, 54; Cornell Univ, MS, 59, PhD(fishery biol), 62. *Prof Exp:* Teacher sci, Stamford Collegiate Inst, Ont, 55-57; res scientist, Ont Dept Lands & Forests, 61-63; res assoc biomet, Cornell Univ, 63-64, asst prof conserv & asst leader, NY Coop Fish Unit, 64-66; from asst prof to assoc prof, 66-73, PROF ZOOL, UNIV TORONTO, 73- *Concurrent Pos:* Chief resource eval br fisheries, Food & Agr Orgn, UN, Rome, 70-71; trustee, Inst Ecol, 73-75; res plan consult, Fisheries Res Bd Can, 73-76; comnr, Great Lakes Fishery Comn, 80- *Honors & Awards:* Centenary Medal, Royal Soc Can, 86. *Mem:* Int Union Theoret & Appl Limnol; Int Asn Gt Lakes Res; Am Fisheries Soc (pres, 78-79); Int Asn Ecol. *Res:* Ecology of aquatic ecosystems, particularly large-scale responses of fish communities to major cultural stresses; screening and assessing ecological models and methods for interdisciplinary application; sustainable redevelopment of the Great Lakes Basin. *Mailing Add:* Dept Zool Univ Toronto Toronto ON M5S 1A1 Can

REGIER, LLOYD WESLEY, b Hillsboro, Kans, Sept 19, 28; m 53, 83; c 4. FOOD SCIENCE, FISHERIES. *Educ:* Univ Calif, Berkeley, BS, 50; Univ Calif, Davis, PhD(agr chem), 61. *Prof Exp:* Food technologist, Procter & Gamble Co, Ohio, 56-59; chemist, Western Res Labs, Nat Canners Asn, Calif, 59-63; assoc prof environ chem, Univ NC, Chapel Hill, 63-68; group leader process & prod res, fisheries & marine sci, Fisheries Res Bd Can, 68-71, res scientist, Fish Utilization Prog, 71-75, prog mgr & res scientist, fish utilization, Halifax Lab, 75-78; chief develop div, Charleston Lab, SC, 78-87, res food technologist, Exp Process Field Facil, 87-90, SCIENTIFIC STAFF OFFICER, NAT SEAFOOD INSPECTION LAB, NAT MARINE FISHERIES SERV, US DEPT COM, PASCAGOULA, MISS, 87- *Mem:* Inst Food Technol; Can Inst Food Sci & Technol. *Res:* Fishery science; processing of and products from underutilized fish species; preservation by refrigeration; drying, chemicals or solvent extraction; by-product characterization; utilization of fish oils; seafood inspection systems. *Mailing Add:* Nat Marine Fisheries Serv PO Drawer 1207 Pascagoula MS 39567

REGISTER, RICHARD ALAN, b Cheverly, Md, Sept 6, 63; m 89. MULTIPHASE POLYMERS, POLYMER MORPHOLOGY. *Educ:* Mass Inst Technol, BS, 83, BS, 84, MS, 85; Univ Wis-Madison, PhD(chem eng), 89. *Prof Exp:* ASST PROF CHEM ENG, PRINCETON UNIV, 90- *Concurrent Pos:* Fac mem, Princeton Mat Inst, 90- *Mem:* Am Chem Soc; Am Phys Soc; Am Inst Chem Engrs; Mat Res Soc; Soc Plastics Engrs. *Res:* Morphology and properties of multiphase polymeric materials, particularly block copolymers, polymer blends, and ionomers; dynamics of block copolymers in bulk; relaxation behavior of multiphase polymers; semicrystalline block copolymers; applications of small-angle scattering. *Mailing Add:* Dept Chem Eng Princeton Univ Princeton NJ 08544

REGISTER, ULMA DOYLE, b West Monroe, La, Feb 4, 20; m 42; c 3. BIOCHEMISTRY, NUTRITION. *Educ:* Madison Col, BS, 42; Vanderbilt Univ, MS, 44; Univ Wis, PhD(biochem), 50. *Prof Exp:* Fel, Sch Med, Tulane Univ, 50-51; from instr to assoc prof biochem, 51-67, chmn grad prog nutrit, 69, chmn dept nutrit, dietetics, 72-82, prof nutrit & chmn dept, 67-84, PROF NUTRIT, LOMA LINDA UNIV, 84- *Concurrent Pos:* Commonwealth Fund fel, Karolinska Inst, Sweden, 63-64. *Mem:* Am Inst Nutrit; Am Dietetic Asn; Am Pub Health Asn; Am Soc Clin Nutrit; Sigma Xi. *Res:* Vitamin metabolism; protein nutrition; diet and alcohol. *Mailing Add:* Dept Nutrit Loma Linda Univ Loma Linda CA 92354

REGNA, PETER P, b Hoboken, NJ, May 26, 09; c 2. ORGANIC BIOCHEMISTRY. *Educ:* Polytech Inst NY, BS, 32, MS, 37, PhD(phys org chem), 42. *Prof Exp:* Res group leader, Pfizer, Inc, 45-50, tech asst to dir res, 50-54, coordr cancer prog, 54-57, spec projs officer, 57-61; dir res planning, Squibb Inst Med Res, NJ, 61-70; sr partner, 70-72, MANAGING PARTNER, HARRINGTON RES CO, 72- *Concurrent Pos:* Consult; pres, Indust Chem Soc. *Honors & Awards:* Perkin Medal, Am Sect, Soc Chem Indust, 86; Kohnstamm Award, Columbia Univ, 88; Honor Scroll Award, Am Inst Chemist, 72, 86. *Mem:* AAAS; Am Chem Soc; fel Am Inst Chemists; fel NY Acad Sci; fel Am Col Clin Pharmacol. *Res:* Kinetic reactions; carbohydrate chemistry; synthetic vitamins; structure antibiotics; chemotherapeutic agents; development of pharmaceuticals; antineoplastic substances; physical chemistry; physical organic. *Mailing Add:* 110 Roberts Rd Englewood Cliffs NJ 07632-2315

REGNER, JOHN LAVERNE, b Columbus, Ohio, Oct 24, 46; m 68; c 2. NUCLEAR PHYSICS, SYSTEMS ANALYSIS. *Educ:* Ohio State Univ, BS, 69, MS, 69, PhD(physics), 76. *Prof Exp:* Systs analyst missile systs, NAm Rockwell, 69-73; res assoc nuclear physics, Ohio State Univ, 73-76, assoc, Van de Graaff Lab, 76-77; systs analyst defense systs, Inst Defense Anal, 77-83; MISSION/SYST ANALYST, TELEDYNE BROWN ENG, 83- *Concurrent Pos:* Adj asst prof physics, Univ Ala-Huntsville, 84- *Mem:* Am Phys Soc. *Res:* Nuclear physics utilizing polarized particles; nuclear spectroscopy; communication systems for command and control; strategic force exchange analysis; ballistic missile defense analysis. *Mailing Add:* 132 Thomas Rd Madison AL 35758

REGNERY, DAVID COOK, b La Grange, Ill, June 26, 18; m 45; c 3. BIOLOGY. *Educ:* Stanford Univ, AB, 41; Calif Inst Technol, PhD(genetics), 47. *Prof Exp:* From instr to assoc prof, 47-53, PROF BIOL, STANFORD UNIV, 53- *Res:* Wildlife diseases, pox viruses, myxomatosis. *Mailing Add:* 488 Westridge Dr Menlo Park CA 94028

REGNIER, FREDERICK EUGENE, b Fairbury, Nebr, July 7, 38; m 60; c 1. BIOCHEMISTRY. *Educ:* Nebr State Teachers Col, Peru, BS, 60; Okla State Univ, PhD(chem), 65. *Prof Exp:* Fel, Okla State Univ, 65-66 & Univ Chicago, 66-68; from asst prof to assoc prof, 68-76, PROF BIOCHEM, PURDUE UNIV, LAFAYETTE, 76- *Mem:* Am Chem Soc; Am Soc Biol Chemists; Sigma Xi. *Res:* Pheromones; hormones; instrumental analysis. *Mailing Add:* Dept Biochem Purdue Univ Lafayette IN 47907

REGO, VERNON J, PERFORMANCE, SIMULATION. *Educ:* Mich State Univ, MS, 83, PhD(computer sci), 85. *Prof Exp:* Asst prof, 85-91, ASSOC PROF COMPUTER SCI, DEPT COMPUTER SCI, PURDUE UNIV, 91- *Concurrent Pos:* Res visitor computer sci, Univ Stuttgart, Ger, 88; fac visitor computer sci, Oak Ridge Nat Lab, Tenn, 90 & 91. *Mem:* Inst Elec & Electronics Engrs. *Res:* Performance evaluation; stochastic modelling; simulation; parallel simulation; computer networks; distributed systems; software engineering and reliabilty. *Mailing Add:* Dept Comput Sci Purdue Univ West Lafayette IN 47906

REGOLI, DOMENICO, b Lucca, Italy, May 16, 33; Can citizen; m 68; c 4. PEPTIDES. *Educ:* Liceo Classico Carducci Grosseto-Baccal, 53; Univ Siena, Italy, MD, 59; Univ Lausanne, Switz, Priv Docent(pharmacol), 67. *Prof Exp:* Asst prof med, Univ Siena, 59-60; fel pharmacol, Ciba Biol Labor, 60-63; fel, Royal Col Surgeons, 63-64; from asst prof to assoc prof, Univ Lausanne, 65-68; PROF & CHMN PHARMACOL, UNIV SHERBROOKE, 68-, CAREER INVESTR MED RES COUN CAN, 73- *Honors & Awards:* M Sarrazin Prize, Med Res, 87. *Mem:* Am Soc Pharmacol & Therapeut; Brit Pharmacol Soc; Pharmacol Soc Can; Int Soc Hypertension; NY Acad Sci. *Res:* Pharmacology of peptide hormones and antagonists; isolation and identification of naturally occurring peptides; chemical synthesis and purification of peptide analogues; development of anti-hypertensive, anti-inflammatory and analgesic drugs. *Mailing Add:* Dept Pharmacol Univ Sherbrooke Sherbrooke PQ J1H 5N4 Can

REGULSKI, THOMAS WALTER, b Detroit, Mich, Dec 13, 43; m 63; c 3. POLYMER CHEMISTRY. *Educ:* Wayne State Univ, BS, 67; Univ Nebr, PhD(org chem), 71. *Prof Exp:* Assoc, Univ Toronto, 71-73; SR RES CHEMIST, DOW CHEM CO, 73- *Mem:* Am Chem Soc. *Res:* Area of polymer synthesis for industrial and biomedical applications. *Mailing Add:* Dow Chem Co 2800 Mitchell Dr Walnut Creek CA 94598

REGUNATHAN, PERIALWAR, b Samugarengapuram, Tamil Nadu, India, Feb 23, 40; m 63; c 3. WATER PURIFICATION, WATER FILTRATION. *Educ:* Univ Madras, India, BE, 61, MSc, 63; Iowa State Univ, MS, 65, PhD(sanitary eng), 67. *Prof Exp:* Asst lectr civil eng, Col Eng, Univ Madras, India, 62-63; instr sanitary eng, Iowa State Univ, 67; engr-scientist, Eng-Sci, Inc, Oakland, 67-68; supvr res, 68-70, mgr res, 70-78, res & develop, 78-80, VPRES RES & DEVELOP, EVERPURE, INC, BEATRICE, CO, WESTMONT, ILL, 80- *Concurrent Pos:* Mem, sci adv comn, Water Qual Asn, 80-, chmn, Drinking Water Comt, 85-; indust rep, Environ Protection Agency, 77-80 & 84-86. *Honors & Awards:* Spes Hominum Award, Nat Sanitation Found, 85. *Mem:* Am Water Works Asn; Filtration Soc; Am Soc Testing & Mat; Water Qual Asn. *Res:* Development of processes and products employing advanced ideas in filtration adsorption, disinfection and destabilization areas of water treatment for use by final user in home or other establishments such as restaurants. *Mailing Add:* 1490 Jasper Dr Wheaton IL 60187

REH, THOMAS ANDREW, b Chicago, Ill, Feb 17, 55; m 81. DEVELOPMENTAL NEUROBIOLOGY. *Educ:* Univ Ill, Champaign, BS(biochem) & BS(physiol), 77; Univ Wis-Madison, PhD(neurosci), 81. *Prof Exp:* vis fel, Neurosci, Princeton Univ, 81-84; prof med physiol, Univ Calgary, 88; ASSOC PROF BIOL STRUCT, UNIV WASH, SEATTLE, 88- *Concurrent Pos:* Reviewer ad hoc grant, NSF, 82- *Honors & Awards:* Jerzy Rose Award, Univ Wis-Madison, 82; Recipiant AHFMR Scholar; Recipiant Alfred East Sloan Scholar, 87-88. *Mem:* Am Soc Cell Biol. *Res:* Development of the retina. *Mailing Add:* Dept Biol Struct SM-20 Univ Wash Seattle WA 98195

REHAK, MATTHEW JOSEPH, b Baltimore, Md, Apr 24, 29; m 53; c 5. CLINICAL CHEMISTRY, TOXICOLOGY. *Educ:* Loyola Col, AB, 50; Univ Md, MS, 55, PhD(biochem, pharmacol), 58. *Prof Exp:* Chemist, Off Chief Med Exam, Md, 50-51, asst, 53-54; asst pharmacol, Med Sch, Univ Md, 54-57; state toxicologist, Conn State Health Dept, 57-60; CLIN CHEMIST, ST AGNES HOSP, 60- *Concurrent Pos:* Consult clin chem, Dept of State, Washington, DC, 68-73. *Mem:* Nat Acad Clin Biochem; Clin Radioassay Soc; Am Chem Soc; Am Asn Clin Chem. *Res:* New analytical methods for drugs and metabolites; detection and determination of drug effects on enzyme systems. *Mailing Add:* St Agnes Hosp Caton & Wilkins Ave Baltimore MD 21229

REHAK, PAVEL, b Prague, Czech, Dec 5, 45; m 77. HIGH ENERGY PHYSICS, ELECTRODYNAMICS. *Educ:* Charles Univ, Prague, RNDr(nuclear physics), 69; State Univ Col Pisa, PhD(elem particle physics), 72. *Prof Exp:* Res physicist particle physics, Kernforschungscentrum, Karlsruhe, WGer, 72-73; res assoc, Yale Univ, 73-76; RES PHYSICIST PARTICLE PHYSICS, BROOKHAVEN NAT LAB, 76- *Res:* Experiments in elementary particle physics; detector and particle detection system development; electrodynamics in strong magnetic field. *Mailing Add:* Physics Dept Brookhaven Nat Lab Bldg 510C Upton NY 11973

REHBERG, CHESSIE ELMER, b Cairo, Ga, Aug 15, 11; m 42; c 3. ORGANIC CHEMISTRY. *Educ:* Ga State Col for Men, BS, 33; Emory Univ, MS, 37; Univ Tex, PhD(org chem), 41. *Prof Exp:* Teacher high sch, Ga, 34-36; asst & instr, Emory Univ, 36-38; instr math & physics, Emory Jr Col, Valdosta, 38; instr chem, Univ Tex. 38-41; org res chemist, Sharples Chems, Inc, Mich, 41-42; org res chemist, Eastern Regional Lab, Bur Agr & Indust Chem, USDA, 42-51, tech analyst & patents, Eastern Utilization Res Br, 51-56; group leader, Patent Dept, Dow Chem USA, 56-68, sr patent agent, 68-75; RETIRED. *Mem:* Am Chem Soc; Sci Res Soc Am. *Res:* Synthetic resins and plasticizers; esters of lactic and acrylic acids; phenacyl carbinamines; alkylene oxides; polyglycols; polyurethanes. *Mailing Add:* Rte Four Box 215 Hendersonville NC 28739-4395

REHDER, HARALD ALFRED, b Boston, Mass, June 5, 07; m 38; c 2. INVERTEBRATE ZOOLOGY. *Educ:* Bowdoin Col, AB, 29; Harvard Univ, AM, 33; George Washington Univ, PhD(zool), 34. *Prof Exp:* Sr sci aide, 32-34, from asst cur to assoc cur, 34-46, actg cur, 46, cur, 46-65, sr zoologist, Div Mollusks, 65-76, EMER ZOOLOGIST, DEPT INVERT ZOOL, NAT MUS NATURAL HIST, SMITHSONIAN INST, 76- *Concurrent Pos:* Co-ed, Indo-Pac Mollusca; mem field expeds to French Polynesia, Yucatan & Marshall Islands; adj prof biol sci, George Washington Univ, 70- *Mem:* AAAS; Paleont Soc; Soc Syst Zool; fel Am Malacol Union (pres, 40); Unitas Malacologica; Int Soc Reef Studies. *Res:* Systematic malacology; geographical distribution of mollusks; marine mollusks of Indo-Pacific, especially Polynesia. *Mailing Add:* 3900 Watson Pl No G Washington DC 20016

REHDER, KAI, b Hohenwestedt, WGer, Dec 17, 28; m 58; c 4. ANESTHESIOLOGY, PHYSIOLOGY. *Educ:* Univ Freiburg, MD, 53. *Prof Exp:* Intern pediat, Univ Hosp, Freiburg, 53-54, resident pharmacol, 56-57; resident oncol, Jeanes Hosp, Philadelphia, Pa, 54-55; resident internal med, Mayo Grad Sch Med, Univ Minn & Mayo Clin, 57-58, resident anesthesiol, 58-60, res asst physiol, 60-61; docent anesthesiol, Univ Würzburg, 62, head dept, Univ Hosp, 62-65; asst prof anesthesiol, Mayo Grad Sch Med, Univ Minn & consult, Mayo Clin & Mayo Found, 66-77, PROF PHYSIOL & ANESTHESIOL, MAYO CLIN, 77-, PROF MAYO GRAD SCH, 76- *Mem:* Am Soc Anesthesiologists; Am Physiol Soc. *Res:* Pulmonary physiology. *Mailing Add:* Dept Anesthesiol Stm Three-A Mayo Grad Sch Med Rochester MN 55905

REHFIELD, DAVID MICHAEL, b Mason City, Iowa, Aug 19, 42. BETA-RAY SPECTROSCOPY, HIGHER EDUCATION. *Educ:* Seattle Univ, BSc, 64; Univ Ariz, MSc, 67; McGill Univ, PhD(physics), 77. *Prof Exp:* Fel, II Physikalishes Inst Justus Liebig Univ, Ger, 77-79; res assoc, Foster Radiation Lab, McGill Univ, 79-81; instr, Vanier Col, Montreal, 81-82; asst prof, Swarthmore Col, 81-82; ASST PROF PHYSICS, LAFAYETTE COL, 82- *Concurrent Pos:* Vis scientist, Brookhaven Nat Lab, 80-; res collabr, Nat Res Coun Can, 81-83. *Mem:* Sigma Xi; Am Asn Physics Teachers; Am Phys Soc; Am Nuclear Soc. *Res:* Nuclear physics, with emphasis on beta-ray spectroscopy, involving solid-state detectors and the development of superconducting-solenoid beta-ray spectrometers, and data-analysis programs; study of short-lived fission products. *Mailing Add:* Dept Physics Lafayette Col Easton PA 18042

REHFIELD, LAWRENCE WILMER, b Miami, Fla, Feb 1, 38. AEROSPACE ENGINEERING. *Educ:* Ga Inst Technol, BAeroE, 61; Mass Inst Technol, MS, 62; Stanford Univ, PhD(aeronaut, astronaut), 65. *Prof Exp:* Res asst aeronaut & astronaut, Stanford Univ, 63-65; asst prof aeronaut & astronaut & fel eng, Mass Inst Technol, 65-67; from asst prof to assoc prof aerospace eng, 67-77, PROF AEROSPACE ENG, GA INST TECHNOL, 77- *Mem:* Am Inst Aeronaut & Astronaut; Soc Exp Stress Analysis; Am Inst Ultrasonics Med. *Res:* Structural mechanics as applied to vehicle technology; stability of shell structures. *Mailing Add:* Dept Aerospace Eng Ga Inst Technol 225 N Ave NW Atlanta GA 30332

REHFUSS, MARY, b Albany, NY, Sept 16, 27. ORGANIC CHEMISTRY. *Educ:* Col St Rose, BS, 49; St Louis Univ, PhD(org chem), 58. *Prof Exp:* Instr, 49-50, 53-54, from asst prof to assoc prof, 58-65, PROF CHEM, COL ST ROSE, 65- *Mem:* Sigma Xi; Am Chem Soc. *Res:* Kinetics of molecular rearrangements; applications of chemistry to conservation of art objects. *Mailing Add:* Dept Chem Col St Rose Albany NY 12203

REHKUGLER, GERALD E(DWIN), b Lyons, NY, Apr 11, 35. AGRICULTURAL & FOOD ENGINEERING. *Educ:* Cornell Univ, BS, 57, MS, 58; Iowa State Univ, PhD, 66. *Prof Exp:* From asst prof to assoc prof, 58-77, chmn, Dept Agr & Biol Eng, 84-90, PROF AGR & BIOL ENG, CORNELL UNIV, 77-, ASSOC DEAN, COL ENG, 90- *Concurrent Pos:* NSF sci fac fel, 64; vis prof, Mich State Univ, 74. *Mem:* Am Soc Agr Engrs; Am Soc Eng Educ. *Res:* Design and development of machinery for handling, harvesting and processing of food and agricultural products; dynamics of agricultural vehicles; computer control and simulation in agricultural and food processing machinery. *Mailing Add:* Col Eng Cornell Univ 223 Carpenter Hall Ithaca NY 14853-2201

REHM, ALLAN STANLEY, b Chicago, Ill, May 2, 36; m 59; c 2. POLITICO-MILITARY GAMING & SIMULATIONS, MODELING COMBAT. *Educ:* Univ Ill, BS, 58, MS, 59, PhD(math), 64. *Prof Exp:* Grad asst math, Univ Ill, 58-63; sr res engr, NAm Aviation, 63-85; asst prof math, Clarkson Col Technol, 65-67; tech staff, Ketron, Inc, 73-75; br chief, US Cent Intel Agency, 75-83; div mgr, SAIC, 83-85; tech staff, Ctr Naval Analysis, 67-73 & 85-90; LEAD SCIENTIST, MITRE CORP, 90- *Concurrent Pos:* Instr, Dept Admin & Bus, Col Continuing Educ, Johns Hopkins Univ, 80-91; Consult, Independent Consult, 83-90; distinguished vis analyst, US Army Concepts Anal Agency, 88-89. *Mem:* Am Math Soc; Math Asn Am; Soc Indust & Appl Math; Inst Mgt Sci; Opers Res Soc Am. *Res:* Analysis of communications systems; political-military gaming and simulations; soviet military applications of operations research; graphical techniques of information display; quantitative data from historical combat, battles, campaigns; economic warfare and history. *Mailing Add:* 13320 Tuckaway Dr Fairfax VA 22033

REHM, GEORGE W, b St Clairsville, Ohio, Oct 1, 41; m 64; c 3. SOIL FERTILITY, AGRONOMY. *Educ:* Ohio State Univ, BS, 63; Univ Minn, MS, 65, PhD(soil sci), 69. *Prof Exp:* Teaching asst soil sci, Univ Minn, 63-69; dist exten agronomist, Univ Nebr, 69-80, exten soils specialist, 80-83; EXTEN SCIENTIST SOIL FERTIL, UNIV MINN, 83- *Mem:* Am Soc Agron; Soil Sci Soc Am. *Res:* Soil fertility, particularly nitrogen-sulfur interactions; plant nutrition; soil-plant relationships; plant nutrition research with corn and forage crops. *Mailing Add:* Dept Soil Sci Univ Minn St Paul MN 55108

REHM, LYNN P, b Chicago, Ill, May 20, 41; m 64; c 2. CLINICAL PSYCHOLOGY, COGNITIVE-BEHAVIOR THERAPY. *Educ:* Univ Southern Calif, BA, 63; Univ Wis-Madison, MA, 66, PhD(clin psychol), 70. *Prof Exp:* From asst prof to assoc prof psychol, Univ Pittsburgh, 70-79, assoc prof psychiat, 77-79; PROF PSYCHOL, UNIV HOUSTON, 79- *Concurrent Pos:* Intern, Wood Vet Admin, Milwaukee, Wis, 67-68; actg instr psychol, Dept Psychiat, Univ Calif, Los Angeles, 68-69, asst prof, 69-70; ed, The Tex Psychologist, 85-88; consult, Tex Dept Corrections, 86-; chmn bd, Coun Univ Dirs Clin Psychol, 86-88. *Mem:* Fel Am Psychol Asn; Asn Advan Behav Ther; fel Am Psychopath Asn; Soc Res Psychopath; Behav Ther & Res Soc. *Res:* Psychotherapy and psychopathology theory and research. *Mailing Add:* Dept Psychol Univ Houston Houston TX 77204-5341

REHM, RONALD GEORGE, b Chicago, Ill, Nov 6, 38; m 59; c 5. APPLIED MATHEMATICS, FLUID DYNAMICS. *Educ:* Purdue Univ, BS, 60; Mass Inst Technol, PhD(appl math), 65. *Prof Exp:* Prin engr appl math & fluid dynamics, Cornell Aeronaut Lab, Buffalo, NY, 65-75; mathematician appl math & fluid dynamics, 75-87, FEL, NAT BUR STANDARDS, 87- *Honors & Awards:* Gold Medal, US Dept Com, 85. *Mem:* Am Phys Soc; Soc Indust Appl Math; AAAS; Combustion Inst. *Res:* Waves in stratified fluids; buoyancy-induced fluid flows; fluid flows induced by laser heating of materials; fire-induced fluid flows; combustion. *Mailing Add:* Nat Bur Standards Admin Bldg A 302 Gaithersburg MD 20899

REHM, THOMAS R(OGER), b Los Angeles, Calif, Nov 11, 29; m 57; c 1. CHEMICAL ENGINEERING. *Educ:* Univ Wash, BSChE, 52, PhD(chem eng), 60. *Prof Exp:* From asst prof to assoc prof chem eng, Univ Denver, 60-66; from asst prof to assoc prof, 66-76, PROF CHEM ENG, UNIV ARIZ, 76- *Concurrent Pos:* Res engr, Denver Res Inst, 60-66; consult, Walvoord, Inc, Colo, 61, Thermo-Tech, Inc, 62-65, Monsanto Co, 73 & Criterion Anal, 77-78; dir property develop, Rehm & Condon, Inc, 65-; abstr nuclear tech, Chem Abstr, 65-76. *Mem:* Am Inst Chem Engrs. *Res:* Boiling heat transfer; suspended solid-liquid mass transfer; turbulent fluid dynamics. *Mailing Add:* Dept Chem Eng Univ Ariz Tucson AZ 85721

REHN, LYNN EDUARD, b Detroit, Mich, Sept 12, 45; m 75; c 4. PHYSICS, MATERIALS SCIENCE. *Educ:* Albion Col, BA, 67; Univ Ill, Urbana, MS, 69, PhD(physics), 73. *Prof Exp:* Scientist solid state, Kernforschungsanlage, Julich, WGer, 73-76; asst physicist, 76-80, physicist mat sci, 80-82, GROUP LEADER, IRRADIATION & KINETIC EFFECTS, ARGONNE NAT LAB, 82- *Honors & Awards:* Sustained Outstanding Res, US Dept Energy, 84. *Mem:* Am Phys Soc; Sigma Xi; Mat Res Soc; Am Soc Metals. *Res:* Defects in metals; ultrasonics; Auger electron spectroscopy; internal friction; radiation damage; ion implantation. *Mailing Add:* Mat Sci Div Argonne Nat Lab 9700 S Cass Ave Argonne IL 60439

REHN, VICTOR LEONARD, b Ophiem, Ill, Aug 14, 27; m 54; c 3. EXPERIMENTAL SOLID STATE PHYSICS. *Educ:* Univ Calif, Berkeley, AB, 53; Univ Pittsburgh, PhD(physics), 62. *Prof Exp:* Physicist, Westinghouse Res Labs, 53-55; res physicist, Armour Res Found, Ill Inst Technol, 60-62; res assoc, Inst Study Metals, Univ Chicago, 62-65; res physicist, US Naval Ord Test Sta, 66-67, head electron struct solids, 67-80, HEAD SEMICONDUCTOR & SURFACE SCI, PHYSICS DIV, MICHELSON LAB, NAVAL WEAPONS CTR, 80-, SPOKESMAN, MICHELSON LABS SYNCHROTRON RADIATION PROJ, 72- *Concurrent Pos:* Vis sr res assoc, Stanford Synchrotron Radiation Lab; guest scientist, Lawrence Berkeley Labs, 77-79. *Honors & Awards:* L T E Thompson Award, 81. *Mem:* AAAS; Am Phys Soc; Am Vacuum Soc; Sigma Xi. *Res:* Electroreflectance studies of electronic energy-band structures of semiconductors, insulators and metals; ultraviolet and soft x-ray properties of semiconductors and insulators using synchrotron radiation; surface science of molecular and semiconducting materials. *Mailing Add:* 819 Mamie St Ridgecrest CA 93555

REHR, JOHN JACOB, b Carlisle, Pa, May 6, 45; m 66; c 2. SOLID STATE PHYSICS. *Educ:* Univ Mich, BSE, 67; Cornell Univ, PhD(theoret physics), 72. *Prof Exp:* Scholar physics, Univ Calif, San Diego, 73-75; from asst prof to assoc prof, 75-85, PROF PHYSICS, UNIV WASH, 85- *Concurrent Pos:* NATO fel, King's Col, London, Eng, 72-73; Humboldt fel, Max Planck Inst Solid State, Stuttgart, W Ger, 78; vis scientist, Cornell Univ, Ithaca, NY, 87-88. *Mem:* Am Phys Soc. *Res:* Condensed matter; critical phenomena. *Mailing Add:* Dept Physics FM-15 Univ Wash Seattle WA 98195

REHWALDT, CHARLES A, b Kewanee, Ill, Sept 7, 25; m 52; c 2. GENETICS, PLANT PHYSIOLOGY. *Educ:* Mankato State Univ, BA, 51; Univ Minn, MS, 53; State Univ NY Col Forestry, Syracuse Univ, PhD(genetics), 65. *Prof Exp:* Teacher high sch, Minn, 55-57; instr biol, Austin Community Col, 57-62; head dept, 68-78, prof, 65-86, EMER PROF BIOL, ST CLOUD STATE UNIV, 86- *Concurrent Pos:* Lectr, NSF Summer Insts, 67, 68. *Mem:* Genetics Soc Am; Am Soc Human Genetics; Sigma Xi. *Res:* Genetics and physiology of seed dormancy. *Mailing Add:* Dept Biol Sci St Cloud State Univ St Cloud MN 56301

REHWOLDT, ROBERT E, analytical chemistry; deceased, see previous edition for last biography

REIBEL, KURT, b Vienna, Austria, May 23, 26; nat US; m 54; c 3. PHYSICS. *Educ:* Temple Univ, BA, 54; Univ Pa, MS, 56, PhD(physics), 59. *Prof Exp:* Jr res assoc physics, Brookhaven Nat Lab, 57-59; res assoc, Univ Pa, 59-61; from asst prof to assoc prof, 61-70, PROF PHYSICS, OHIO STATE UNIV, 70- *Mem:* AAAS; Am Phys Soc; Fedn Am Sci. *Res:* High energy physics; particle detectors; instrumentation; polarized targets. *Mailing Add:* Dept Physics Ohio State Univ Columbus OH 43210-1106

REIBEL-SHINFELD, DIANE KAREN, CARDIAC HYPOTROPHY, MYOCARDIAL ISCHEMIA. *Educ:* Thomas Jefferson Univ, PhD(physiol), 78. *Prof Exp:* ASST PROF CARDIOVASC PHYSIOL, THOMAS JEFFERSON UNIV, 81- *Mailing Add:* Thomas Jefferson Univ 1025 Walnut St Philadelphia PA 19107

REICE, SETH ROBERT, b Brooklyn, NY, June 30, 47; m 71; c 1. STREAM ECOLOGY, DECOMPOSITION. *Educ:* Univ Rochester, BA, 69; Mich State Univ, PhD(zool), 73. *Prof Exp:* Asst prof, 73-79, ASSOC PROF BIOL, UNIV NC, CHAPEL HILL, 79- *Concurrent Pos:* Vis prof zool, Hebrew Univ, Jerusalem, 81. *Honors & Awards:* Hamilton Award & Charles Award, 69. *Mem:* Ecol Soc Am; Am Inst Biol Sci; Sigma Xi; Int Limnol Asn. *Res:* Regulation of benthic community structure and litter decomposition in woodland streams; roles of substrate type, disturbance and predation. *Mailing Add:* Dept Biol Coker Hall-0Ten-A Univ NC Chapel Hill NC 27514

REICH, BRIAN M, b Pretoria, SAfrica, May 16, 27; m 52; c 3. CIVIL ENGINEERING, HYDROLOGY. *Educ:* Univ Witwatersrand, BSc, 51; Iowa State Univ, MS, 59; Colo State Univ, PhD(civil eng), 62. *Prof Exp:* Conserv officer, Dept Conserv, SRhodesia, 51-53; engr, Dept Agr Tech Serv, SAfrica, 53-62; invests leader hydrol res, Natal Region, SAfrica, 62-64; asst prof civil eng, Colo State Univ, 64-66; from assoc prof to prof, Pa State Univ, 66-74; flood plain engr, City Tucson, 75-77; flood plain mgr, Pima Co, 78-80; flood plain engr, City Tucson, 80-84; MEM STAFF DEPT HYDROL & WATER RES, UNIV ARIZ. *Concurrent Pos:* Mem, US working group floods & their comput, Int Hydrol Decade; consult engr & hydrologist, 84-; mem, US deleg on estimation of extreme floods, People's Repub China. *Mem:* Fel Am Soc Civil Engrs; Am Soc Agr Engrs; Am Geophys Union; Am Water Resources Asn. *Res:* Rainstorms; floods from rural area; urban hydrology; planning of open space in suburban watersheds; personal programmable calculators. *Mailing Add:* 2635 E Cerrada Adelita Tucson AZ 85718

REICH, CHARLES, b Minneapolis, Minn, Aug 2, 42; m 63; c 3. RESEARCH ADMINISTRATION. *Educ:* Univ Minn, Minneapolis, BS, 64; Univ Wis-Madison, PhD(org chem), 68. *Prof Exp:* NIH fel, Mass Inst Technol, 68; res specialist, Org Chem Res Lab, 3M Co, 68-74, tech mgr, 74-78, tech dir, Bldg Servs & Cleaning Prods Div, 78-82, managing dir, 3M Switz, 82-87, exec dir res & develop, 87-89, VPRES DENT PROD DIV, 3M CO, 89- *Mem:* Am Chem Soc. *Res:* Organometallic synthesis; polymer chemistry; catalyst synthesis. *Mailing Add:* 1292 Sylvandale Rd Mendota Heights MN 55118

REICH, CHARLES WILLIAM, b Oklahoma City, Okla, Sept 12, 30; m 52; c 3. NUCLEAR PHYSICS. *Educ:* Univ Okla, BS, 52; Rice Inst, MA, 54, PhD(physics), 56. *Prof Exp:* Predoctoral fel, NSF, 54-55, group leader, Radioactivity & Decay Schemes Group, Atomic Energy Div, Phillips Petrol Co, 59-66; group leader, Idaho Nuclear Corp, 66-71; group leader, Radioactivity & Decay Schemes Group, Aerojet Nuclear Co, 71-74, sect chief, 74-76; sect chief, Nuclear Struct Sect, 76-83, prin scientist, 81-82, SCI & ENG FEL, EG&G IDAHO, INC, 82- *Concurrent Pos:* Prin investr, Dept Energy Res Progs, 60-, mem, Transplutonium Prog Comt, 78-; guest scientist, Niels Bohr Inst, Copenhagen, Denmark, 64-65; chmn, Decay-Data Subcomt, Cross Sect Eval Working Group, 73-; US coordr & rep, Int Atomic Energy Agency Coord Res Prog, Measurement Actinide Decay Data, 77-84; mem, Transplutonium Prog Comt, US Dept Energy, 78-, Task Force on Decay Heat Predictions, Nuclear Energy Agency, Nuclear Data Comt, 78-; adj prof, Utah State Univ, 68-82; physics curric coordr, Idaho Nat Eng Lab Educ Prog, Univ

Idaho, 79- *Mem:* AAAS; fel Am Phys Soc; Sigma Xi; NY Acad Sci; Am Chem Soc. *Res:* Experimental investigation and analysis of nuclear energy level structure; compilation and evaluation of nuclear data. *Mailing Add:* Idaho Nat Eng Lab PO Box 1625 Idaho Falls ID 83415-2219

REICH, CLAUDE VIRGIL, b Reading, Pa, May 18, 21; m 48; c 1. MEDICAL MICROBIOLOGY, LEPROLOGY. *Educ:* Pa State Univ, BS, 53, PhD(bact), 58; Univ Wis, MS, 54. *Prof Exp:* Instr vet sci, Pa State Univ, 54-58; instr bact, Col Med, Univ Ill, 58-59; assoc bacteriologist, Johns Hopkins Univ, 59-62, lab dir & microbiologist, Leonard Wood Mem Lab, Cebu, Philippines, 62-83, chief lab br, Philippines Div, 63-83, asst prof pathobiol, Sch Hyg, 59-72, chief opers, 73-83; RES PROF, DEPT MICROBIOL, GEORGE WASHINGTON UNIV, 81- *Concurrent Pos:* Mem leprosy expert panel, WHO, 68- *Mem:* Am Soc Microbiol; Conf Res Workers Animal Dis; Int Leprosy Asn. *Res:* Immunologic and nutritional aspects of bacteria related to infectious infertility; factors associated with non-cultivable states of mycobacteria; clinical chemistry and bacteriology of leprosy; animal transmission of mycobacterial diseases; leprosy immunology. *Mailing Add:* 1516 N 14th St Reading PA 19604

REICH, DANIEL, b New York, NY, June 25, 41; m 65; c 1. MATHEMATICS. *Educ:* Cornell Univ, AB, 62; Princeton Univ, MA, 64, PhD, 66. *Prof Exp:* Instr math, Johns Hopkins Univ, 66-68, asst prof, 68-70; ASST PROF MATH, TEMPLE UNIV, 70- *Mem:* Am Math Soc. *Res:* Number theory; algebraic geometry. *Mailing Add:* Dept Math Temple Univ Philadelphia PA 19122

REICH, DONALD ARTHUR, b Quincy, Ill, Dec 14, 29; m 51; c 1. ORGANIC CHEMISTRY. *Educ:* Millikin Univ, BA, 52; Univ Mo, PhD(chem), 56. *Prof Exp:* Supvr org chem, Chem Div, Pittsburgh Plate Glass Co, Ohio, 56-81, SR SUPVR ORG RES, PPG INDUSTS, INC, 67- *Mem:* Am Chem Soc; Am Soc Testing Mats. *Res:* Application research; process and product development. *Mailing Add:* PPG Indust Chem Div PO Box 1000 Lake Charles LA 70602

REICH, EDGAR, b Vienna, Austria, June 7, 27; nat US; m 49; c 2. MATHEMATICS. *Educ:* Polytech Inst Brooklyn, BEE, 47; Mass Inst Technol, MS, 49; Univ Calif, Los Angeles, PhD(math), 54. *Prof Exp:* Asst servomech lab, Mass Inst Technol, 47-49; mathematician, Rand Corp, 49-56; from asst prof to assoc prof, 56-61, PROF MATH, UNIV MINN, MINNEAPOLIS, 61- *Concurrent Pos:* NSF fel, 54-55; mem, Inst Advan Study, 54-55, Math Res Inst, Zurich, Switz, 71-72 & 78-79; consult, Rand Corp, 56-66; Guggenheim fel & Fulbright res grant, Math Inst, Aarhus Univ, 60-61; vis prof, Israel Inst Technol, 65-66, Swiss Fed Inst Technol, Zurich, 82-83. *Mem:* Am Math Soc; foreign mem Finnish Acad Sci & Letters. *Res:* Complex analysis. *Mailing Add:* Sch Math Inst Technol Univ Minn Minneapolis MN 55455

REICH, GEORGE ARTHUR, b Los Angeles, Calif, Jan 18, 33; m 57; c 7. MEDICINE, EPIDEMIOLOGY. *Educ:* Univ Fla, BS, 56; Univ Iowa, MD, 62; Univ NC, MPH, 69. *Prof Exp:* Intern, Hosp, USPHS, Norfolk, Va, 62-63, staff physician, Outpatient Clin, Miami, Fla, 63-64, chief med serv, 64-65, field epidemiologist, Commun Study on Pesticides, Bur State Serv, 65-66, asst chief commun studies, Pesticides Prog, Nat Commun Dis Ctr, 66-69; chief epidemiologist, Div Commun Studies, Food & Drug Admin, 69-70; dir health maintenance orgn serv, Polk County, Fla, 71-73, dir, PSRO Prog, 73-74, dir div qual & standards, 74, regional health adminr, region IV, 74-85, health officer, 85-88; med dir, Polk County Dept Human Serv, Bartow, Fla, 88-89, dir, Office Substance Abuse, Bartow, 89-90. *Mem:* AMA. *Res:* Delivery of health care. *Mailing Add:* Polk County Health Dept PO Box 1480 Winter Haven FL 33882-1480

REICH, HANS JURGEN, b Danzig, Ger, May 6, 43; Can citizen; m 69. ORGANIC CHEMISTRY, MAIN GROUP ORGANOMETALLIC CHEMISTRY. *Educ:* Univ Alta, BSc, 64; Univ Calif, Los Angeles, PhD(org chem), 68. *Prof Exp:* Nat Res Coun Can fel, Calif Inst Technol, 68-69 & Harvard Univ, 69-70; from asst prof to assoc prof, 70-79, PROF CHEM, UNIV WIS-MADISION, 79- *Concurrent Pos:* Res grants, Petrol Res Found, 70, 73, 76 & 81, 85, 90, Res Corp, 72, NSF, 74, 77, 81, 85 & 88 & NIH, 78 & 81; Sloan Found fel, 75; vis prof, Phillips-Univ, Marburg, WGer, 79, Universitd Louis Pasteur, Strasbourg, France, 87. *Mem:* Am Chem Soc; Royal Soc Chem; Sigma Xi. *Res:* Organic and organometalloid chemistry; synthetic applications, stereochemistry and mechanism in organosulfur-selenium-iodine and silicon chemistry; synthesis of theoretically interesting molecules; nuclear magnetic resonance spectroscopy; organolithium chemistry. *Mailing Add:* Dept Chem Univ Wis Madison WI 53706

REICH, HERBERT JOSEPH, b Staten Island, NY, Oct 25, 00; m 26; c 2. ELECTRONICS, ELECTRICAL ENGINEERING. *Educ:* Cornell Univ, BE, 24, PhD(physics), 28. *Prof Exp:* Instr mach design, Cornell Univ, 24-25, instr physics, 25-29; from asst prof to prof elec eng, Univ Ill, 29-46; prof elec eng, 46-64, prof eng & appl sci, 64-69, EMER PROF ENG & APPL SCI, YALE UNIV, 69- *Concurrent Pos:* Spec res assoc, Radio Res Lab, Harvard Univ, 44-46; mem adv group electron tubes, Off Asst Secy Defense, Res & Develop, 51-59; US deleg tech comt electron tubes, Int Electrotech Comn, 60-71, chmn subcomt microwave tubes, 65-72; prof physics & math, Deep Springs Col, Calif, 76-79. *Mem:* Fel Am Phys Soc; fel Inst Elec & Electronics Engrs. *Res:* Electron devices and electron-device circuits; microwave devices and microwave-device circuits electronics; stabilized cathode ray oscilloscope; author of three college textbooks and co-author and editor of four textbooks. *Mailing Add:* Eight Park St Groveland MA 01834

REICH, IEVA LAZDINS, b Riga, Latvia, June 30, 42; US citizen; m 69. ORGANIC CHEMISTRY. *Educ:* Univ Wash, BS, 64; Univ Calif, Los Angeles, PhD(org chem), 69. *Prof Exp:* NIH fel, Harvard Univ, 69-70; res assoc org chem, 70-80, lectr gen chem, 81-82, lectr org chem, 85-90, ASSOC SCIENTIST, UNIV WIS-MADISON, 82- *Concurrent Pos:* Consult, Miles Labs, Madison, 74-79. *Mem:* Am Chem Soc. *Res:* Synthesis of polychlorinated biphenyl arene oxides; synthetic methods involving organo-selenium, organo-sulfur and organo-tin compounds; mechanism of lithium-organometalloid exchange; functional group manipulation involving steroids. *Mailing Add:* Dept Chem Univ Wis Madison WI 53706

REICH, ISMAR M(EYER), b New York, NY, Aug 13, 24; m 56; c 3. CHEMICAL ENGINEERING, FOOD TECHNOLOGY. *Educ:* City Col New York, BChE, 45; Polytech Inst Brooklyn, MChE, 55. *Prof Exp:* Chem engr, Fleischmann Labs, Standard Brands, Inc, 45-48, head pilot plant dept, 48-53, head process develop div, 53-60, dir res, Coffee Instants, Inc, 60-65, dir mfg & res, 65-66, vpres mfg, 66-69; vpres, 69-80, gen mgr, Sol Cafe Div, 76-80, tech dir, 80-82, VPRES, RES & DEVELOP, CHOCK FULL O'NUTS CORP, NEW YORK, NY, 82- *Mem:* AAAS; Am Chem Soc; Am Inst Chem Engrs; NY Acad Sci; fel Am Inst Chem. *Res:* Food technology; extraction; dehydration; instrumentation; agglomeration. *Mailing Add:* 2136 Holland Way Merrick NY 11566

REICH, JAMES HARRY, b New York, NY, Oct 25, 50. PERSONALITY DISORDER VALIDATION, ANXIETY DISORDER RESEARCH. *Educ:* Univ Calif, Berkeley, BA, 73; Univ Colo Med Ctr, MD, 78; Yale Univ, MPh, 84. *Prof Exp:* Resident psychiat, Univ Calif Med Ctr, Davis, 78-82; clin instr, Yale Univ, 82-84; asst profpsychiat, Univ Iowa Med Ctr, 84-87; ASST PROF PSYCHIAT, HARVARD DEPT PSYCHIAT, MASS MENT HEALTH CTR, 87- *Concurrent Pos:* Bush & NIMH res fel, Yale Univ, 82-84; staff physician, dept psychiat, Va Brockton, W Roxburry, 87-90; med dir psychiat, Va Worchester Mental Hygiene Clin, 87-90. *Mem:* affl mem Royal Soc Med; Am Psycho Path Asn; Am Acad Clin Psychiat. *Mailing Add:* 138 Fuller No 4 Brookline MA 02146

REICH, LEO, b Brooklyn, NY, June 23, 24; m 54; c 1. POLYMER CHEMISTRY. *Educ:* Polytech Inst Brooklyn, MS, 49; Stevens Inst Technol, PhD(chem), 59. *Prof Exp:* Sr develop chemist, Nepera Chem Co, Inc, 49-57; instr chem eng, Stevens Inst Technol, 57-59; sr res chemist, Air Reduction Co, Inc, 59-61; polymer res chemist, Picatinny Arsenal, Dover, 61-74; ADJ PROF, STEVENS INST TECHNOL, 72- *Mem:* Am Chem Soc. *Res:* Reaction kinetics; electric discharge phenomena; photochemistry; thermal and thermooxidative degradation of polymers. *Mailing Add:* Three Wessman Dr West Orange NJ 07052

REICH, MARVIN FRED, b Brooklyn, NY, Dec 30, 47. MEDICINAL CHEMISTRY. *Educ:* Polytech Inst, Brooklyn, BS, 68; NY Univ, MS, 72; Univ Ill, PhD(org chem), 78. *Prof Exp:* Chemist, Lederle Labs Div, Am Cyanamid Co, 69-73; teaching asst org chem, Univ Ill, 73-75, res asst, 75-78; staff fel, NIH, 78-79; res chemist, Bound Brook Res Labs, 79-80, SR RES CHEMIST, LEDERLE LAB DIV, AM CYANAMID CO, 80- *Mem:* Am Chem Soc. *Res:* Medicinal and organic chemistry; design, synthesis and characterization of new pharmaceutical agents; anti-inflammatory, anti-allergy and anti-fungal agents; cardiovascular drugs. *Mailing Add:* Med Res Div Am Cyanamid Co Pearl River NY 10965

REICH, MELVIN, b New York, NY, July 17, 32; m 63; c 2. MICROBIOLOGY, BIOCHEMISTRY. *Educ:* City Col New York, BS, 53; Rutgers Univ, PhD(biochem, physiol), 60. *Prof Exp:* Asst biochem, Rutgers Univ, 55-60; asst res prof pharmacol, 60-64, asst prof microbiol, 64-68, assoc prof, 68-79, PROF MICROBIOL, SCH MED, GEORGE WASHINGTON UNIV, 79- *Mem:* AAAS; Am Soc Microbiol; Sigma Xi; Am Asn Univ Professors. *Res:* Bacterial physiology; mycobacteria; antimicrobials. *Mailing Add:* Dept Microbiol Sch Med George Washington Univ 2300 I St NW Washington DC 20037

REICH, MURRAY H, b Brooklyn, NY, May 29, 22; m 50; c 3. POLYMER CHEMISTRY. *Educ:* City Col New York, BS, 43; Univ Akron, MS, 54; Trenton State Col, MED, 74; Columbia Univ, DEd, 82. *Prof Exp:* Process engr elastomers, US Govt Labs, 47-52, group leader, 52-55, chief engr, 55-56; res chemist, FMC Corp, 56-62; res chemist, Princeton Chem Res, Inc, 62-64, lab mgr res & develop lab, 64-70, 74-76, tech dir, 76-77, consult, 77-81 & 85-89, DEVELOP CHEMIST, PRINCETON CHEM RES, INC, 70-; VPRES, BIOLAN CORP, 89- *Concurrent Pos:* Preretirement counr, gerontologist & dir, Premac Assocs; adj prof, gerontology; consult, div aging, Off Ombudsmen. *Mem:* Am Chem Soc. *Res:* Emulsion and solution polymerization of dienes; rubber and epoxy development; polyacetal and polyolefin research and degradable plastics; role of mentors in the careers of executives, both men and women; commercialization of degradable plastics, work has resulted in a commercial degradable agriculture mulch film, called Biolan film. *Mailing Add:* 184 Loomis Ct Princeton NJ 08540

REICH, NATHANIEL EDWIN, b New York, NY, May 19, 07; m 43; c 2. MEDICINE. *Educ:* NY Univ, BS, 27; Univ Chicago, MD, 32; Am Bd Internal Med, dipl, 42. *Prof Exp:* Instr phys diag, 38-42, assoc prof, 52-74, prof, 74-77, EMER PROF CLIN MED, COL MED, STATE UNIV NY DOWNSTATE MED CTR, 77- *Concurrent Pos:* Asst attend physician, NY Postgrad Med Sch, Columbia Univ, 38-40; attend physician, Kings County Hosp, 47- & State Univ NY Downstate Med Ctr; impartial specialist, NY State Dept Labor, 52-58; consult, Long Beach Mem Hosps, 54-; vis prof, Fac Med, San Marcos Univ, Peru, 68, Medico, Afghanistan, 70 & Indonesia, 72; consult, US Dept Health & Health Serv, US RR Retirement Bd, & NY State Disability Determ. *Honors & Awards:* Am Col Angiol Res Awards, 57 & 58. *Mem:* Fel Royal Soc Med; fel Am Col Physicians; fel Am Col Chest Physicians; fel Am Col Cardiol; fel Am Col Angiol. *Res:* Cardiology. *Mailing Add:* 135 Eastern Pkwy Brooklyn NY 11238

REICH, SIMEON, b Cracow, Poland, Aug 12, 48; US citizen; m 74; c 3. NONLINEAR FUNCTIONAL ANALYSIS, NONLINEAR EVOLUTION EQUATIONS. *Educ:* Israel Inst Technol, BSc, 70, DSc, 73. *Prof Exp:* Lectr math, Tel Aviv Univ, 73-75; Dickson instr, Univ Chicago, 75-77; from asst prof to assoc prof, 77-84, PROF MATH, UNIV SOUTHERN CALIF, 84- *Concurrent Pos:* Vis scientist, Argonne Nat Lab, 78; consult, Math Res Ctr, Madison, Wis, 78 & 80; vis assoc prof, Univ Calif,

Berkeley, 81; actg chmn, Math Dept, Univ Southern Calif, 83-84; prof, Technion-Israel Inst Tech, 87-88, 90. *Mem:* Am Math Soc; Math Asn Am; Soc Indust & Appl Math. *Res:* Nonlinear analysis: fixed point theory, asymptotic behavior of nonlinear semigroups, constructive solvability of nonlinear equations, properties of accretive and monotone operators in Banach spaces, integral equations, the Hilbert ball, nonlinear identification problems. *Mailing Add:* Dept Math Univ Southern Calif Los Angeles CA 90089-1113

REICH, VERNON HENRY, b Rushville, Ill Apr 30, 39; m 66. AGRONOMY. *Educ:* Univ Ill, BS, 61, MS, 65; Iowa State Univ, PhD(agron, plant breeding), 68. *Prof Exp:* Res assoc agron, Iowa State Univ, 65-68; asst prof agron, 68-76, ASSOC PROF PLANT & SOIL SCI, UNIV TENN KNOXVILLE, 76- *Mem:* Am Soc Agron; Crop Sci Soc Am; Genetics Soc Am; Am Genetic Asn. *Res:* Plant breeding and genetics. *Mailing Add:* Plant & Soil Sci Univ Tenn PO Box 1071 Knoxville TN 37901-1071

REICHARD, GRANT WESLEY, b Chicago, Ill, Apr 9, 38; m 61; c 2. RADIOMETRY, PEN DRIVES. *Educ:* Univ Ill, Urbana, Bs, 61, MS, 64; Univ Chicago, MBA, 69. *Prof Exp:* Eng trainee, Borg-Warner Corp, 63-66; proj engr, Bastian-Blessing Corp, 66-72; sr prof engr, Stewart-Warner Corp, 72-74; CHIEF ENGR, DICKSON CO, 74- *Concurrent Pos:* Mem, Am Nat Standards Comt B 40, 76-91. *Mem:* Am Soc Mech Engrs; Instrument Soc Am. *Res:* Emissive properties of materials; infrared polarimetry. *Mailing Add:* 1708 S Clifton Ave Park Ridge IL 60068

REICHARD, H(AROLD) F(ORREST), b Easton, Pa, Apr 15, 20; m 44; c 3. CHEMICAL ENGINEERING. *Educ:* Lafayette Col, BS, 41. *Prof Exp:* Develop chem engr, E I du Pont de Nemours & Co, 41-44 & 46-48; supvr chem eng res, prod contracts, US AEC, 48-51; proj & group leader & supvr lab & pilot res, Vitro Corp, 51-56; asst dir res & develop, Mining & Mat Dept, Mining & Metals Div, Union Carbide Corp, 56-68, prod mgr spec alloys, 68-71, gen mgr alloys prod, 71-81; mgr prod spec metals, Elkem Metals Co, 81-83; RETIRED. *Mem:* Fel Am Inst Chemists; Am Inst Mining, Metall & Petrol Engrs; Am Inst Chem Engrs. *Res:* Unit operations of distilling, drying, grinding, crystallizing and extracting; process development in dyestuffs, vitamins, fine chemicals, resin polymerization, organic chlorination, uranium chemistry and physical metallurgy; hydrometallurgy, ion exchange, adsorption, industrial minerals processing and applications; electrolytic process metallurgy. *Mailing Add:* Four Gen Howard Rd South Yarmouth MD 02664

REICHARD, RONNAL PAUL, b Troy, NY, Nov 25, 50; c 1. COASTAL OCEANOGRAPHY, SMALL CRAFT DESIGN. *Educ:* Univ NH, BS, 73, MS, 76, PhD(eng mech), 80. *Prof Exp:* Lectr civil eng, Univ NH, 78-79; res fel, Univ Wash, 79-81; asst prof, 81-85, chmn ocean eng, 86-88, ASSOC PROF OCEAN ENG & DIR STRUCT COMPOSITES LAB, FLA INST TECHNOL, 85- *Concurrent Pos:* Consult, Ctr Inst & Indust Develop, Univ NH, 74-79, Normandeau Assocs Inc, 75-79, Struct Composites Inc, 88- *Mem:* Soc Naval Architects & Marine Engrs; Am Soc Naval Engrs; Soc Advan Mat Processes & Eng; Am Soc Mat. *Res:* Composite materials; measurement, analysis, and design of composite material structures. *Mailing Add:* Dept Oceanog & Ocean Eng Fla Inst Technol Melbourne FL 32901

REICHARD, SHERWOOD MARSHALL, b Easton, Pa, June 24, 28; m 54; c 3. RADIOBIOLOGY, PHYSIOLOGY. *Educ:* Lafayette Col, BA, 48; NY Univ, MS, 50, PhD(endocrine physiol), 55. *Prof Exp:* Res collabr, Dept Biol, Brookhaven Nat Lab, 53-55, res assoc, 55; Muscular Dystrophy Asn Am Lilienthal Mem fel, McCollum-Pratt Inst, Johns Hopkins Univ, 57-58, advan res fel, Am Heart Asn, 58-60; asst prof physiol, Fla State Univ, 60-64; assoc prof, 64-69, dir, Div Radiobiol, 69-76, PROF RADIOL & PHYSIOL, MED COL GA, 69-, REGENTS PROF, 79- *Concurrent Pos:* Vis investr, Dept Radiobiol, Armed Forces Inst Path, 58-60; consult, Off Tech Utilization, NASA. *Honors & Awards:* Founders Day Award, NY Univ, 56; Zool Medal, Int Cong Zool, 63; Fred Conrad Koch Travel Award, Endocrine Soc, 65. *Mem:* Fel AAAS; Am Physiol Soc; Am Soc Zoologists; Radiation Res Soc; Endocrine Soc; Sigma Xi. *Res:* Physiology of reticuloendothelial system; endocrine inter-relations; protection against traumatic shock and x-irradiation; radiation effects; vitamin E and electron transport; hormones and enzymes and terminal respiration. *Mailing Add:* Med Col Ga Augusta GA 30912

REICHARDT, JOHN WILLIAM, b Imperial, Nebr, Nov 18, 40. PHYSICS, DEVICE ENGINEERING. *Educ:* Univ Denver, BS, 62; Univ Wichita, MS, 64; Univ Va, PhD(physics), 67. *Prof Exp:* Staff scientist res & develop, Sandia Labs, 67-73; staff scientist, 73-75, prod mgr neutron gererators, 75-84, DIR ENG, KAMAN INSTRUMENTATION CORP, 84-, VPRES, ENG & QUAL ASSURANCE, 87- *Res:* Neutron production and instrumentation for borehole geophysics; neutron activation analysis; small accelerator design; vacuum tube design; vacuum technology; materials and process technology; physics and chemistry of solid surface. *Mailing Add:* Kaman Instrumentation Corp 1500 Garden of the Gods Rd Colorado Springs CO 80907

REICHART, CHARLES VALERIAN, b Zanesville, Ohio, Feb 8, 10. ENTOMOLOGY. *Educ:* St Thomas Aquinas C, BA, 35; Ohio State Univ, MSc, 45, PhD(entom), 47. *Hon Degrees:* Providence Col, MA, 58. *Prof Exp:* Instr, Aquinas Col High Sch, Ohio, 39-43; instr biol, Providence Col, 43-44; instr zool, Ohio State Univ, 44-47; from assoc prof to prof biol, 47-78, head dept natural sci, 48-55, chmn dept biol, 55-71, EMER PROF BIOL, PROVIDENCE COL, 78- *Mem:* AAAS; Entom Soc Am; Entom Soc Can. *Res:* Notonectid taxonomy; genus anisops. *Mailing Add:* Dept Biol Providence Col Providence RI 02918-0001

REICHBERG, SAMUEL BRINGEISSEN, b Santiago, Chile, Aug 30, 46; m 72; c 2. CLINICAL LABORATORY MEDICINE, CELL BIOLOGY. *Educ:* Univ Chile, LicMed, 70, MD, 71; Yale Univ, MPhil, 73, PhD(biochem), 75. *Prof Exp:* Fel human genetics, Med Sch, Yale Univ, 72-74, med, 74-76, residency lab med, 76-78; ASST PROF PATH, NY MED COL, 78- *Concurrent Pos:* Dir, Clin Lab Dept, Brookdale Hosp, 83-; fac, Med Sch, NY Univ, Downstate, 82- *Mem:* AAAS; NY Acad Sci. *Res:* Regulation of cell plasma membrane transport activity; correlation between transport effects and chemical changes in membrane composition produced by the hormones insulin and glucocorticoids and by growth regulatory compounds. *Mailing Add:* 22 Lakeview Ave N Tarrytown NY 10591

REICHE, LUDWIG P(ERCY), b Germany, Dec 26, 19; m 51; c 3. SATELLITE TELECOMMUNICATIONS, ELECTRONICS INSTRUMENTATION. *Educ:* NY Univ, BEE, 48. *Prof Exp:* Proj engr, Victor Div, Radio Corp Am, Calif, 50-52; staff engr, Int Telemeter Corp, 52-53; sr proj engr, Hoffman Labs, 53-54; sr res engr, Radio Systs Lab, Stanford Res Inst, 54-60; mgr microwave commun, Melabs, 60-64; mem staff, 64-72, sr proj engr, 72-80, PROJ MGR, HUGHES AIRCRAFT CO, 80- *Concurrent Pos:* Coordr, Indonesian Pub Tel Enterprise, 75-77. *Mem:* Inst Elec & Electronics Engrs. *Mailing Add:* 843 Via Campobello Santa Barbara CA 93111

REICHEL, WILLIAM LOUIS, b Philadelphia, Pa, July 10, 27; m 53; c 2. ENVIRONMENTAL CHEMISTRY. *Educ:* Philadelphia Col Pharm, BSc, 52. *Prof Exp:* Org chemist, Philadelphia Naval Shipyard, 52-59; res chemist, 59-66, CHIEF CHEMIST, BUR SPORT FISHERIES & WILDLIFE, US FISH & WILDLIFE SERV, 66- *Mem:* Am Chem Soc. *Res:* Development of procedures for isolation, identification and quantitative measurement of pesticide residues in animal tissues and their environment. *Mailing Add:* 9145 Winding Way Ellicott City MD 21043

REICHELDERFER, CHARLES FRANKLIN, invertebrate pathology, entomology; deceased, see previous edition for last biography

REICHELDERFER, THOMAS ELMER, b Newark, NJ, Aug 31, 16; m 43; c 3. PEDIATRICS, PUBLIC HEALTH. *Educ:* Rutgers Univ, BS, 39; Johns Hopkins Univ, MD, 50, MPH, 56. *Prof Exp:* From asst to instr pediat, Johns Hopkins Univ, 51-54, instr pediat & pub health admin, 54-56; asst prof pediat, Univ Minn, 56-57; chief gen study unit, Lab Infectious Dis, Nat Inst Allergy & Infectious Dis, 57-58; chief med officer, DC Gen Hosp, 58-72; assoc prof, 73-85, EMER ASSOC PROF PEDIAT, JOHNS HOPKINS UNIV, 85- *Concurrent Pos:* Asst resident, Johns Hopkins Hosp, Baltimore, Md, 51-52, resident pediatrician, 52-54; consult, Nat Naval Med Ctr, 53-54; pediatrician, Johns Hopkins Hosp, Baltimore, Md, 54-56 & 73-; clin prof pediat, Howard Univ, 58-72; assoc clin prof, George Washington & Georgetown Univs, 58-72. *Mem:* AMA; fel Am Acad Pediat. *Res:* Infectious diseases; newborn infants. *Mailing Add:* 2029 Chesapeake Rd Annapolis MD 21401

REICHENBACH, GEORGE SHERIDAN, b Waterbury, Conn, May 25, 29; m 56; c 3. MECHANICAL ENGINEERING. *Educ:* Yale Univ, BME, 51; Mass Inst Technol, MS, 52, ScD, 56. *Prof Exp:* From asst prof to assoc prof mech eng, Mass Inst Technol, 56-66; asst dir res, Norton Co, 66-69, dir res & develop, 69-74, vpres & gen mgr org prod, Grinding Wheel Div, 74-79, vpres & gen mgr, Mat Div, 79-81, vpres bonded abrasives, 81-86; vpres, 86-88, SR VPRES, ADVENT INT, 88- *Honors & Awards:* Alfred Noble Prize, Joint Eng Socs, 60; Yale Eng Award, 61. *Mem:* Am Soc Mech Engrs. *Res:* Materials, lubrication and metal processing. *Mailing Add:* 123 West St Carlisle MA 01741

REICHENBACHER, PAUL H, b Aurora, Ill, Feb 4, 40; m 63; c 4. REINFORCED PLASTICS, ORGANIC CHEMISTRY. *Educ:* St Mary's Col, BA, 62; Pa State Univ, PhD(chem), 67. *Prof Exp:* Res chemist, Corp Res Ctr, 67-72, group leader, 72-77, DIR TECHNOL, NORPLEX/OAK INC, ALLIED SIGNAL CO, 77- *Concurrent Pos:* Adj prof, Ill Benedictine Col, 69-70 & Concordia Teachers Col, 70-71; dir, Alumni Bd, St Mary's Col, 78-84, pres, 80-82. *Mem:* Am Chem Soc; Inst Interconnecting & Packaging Electronic Circuits; Sigma Xi; Int Electronic Packaging Soc. *Res:* Reinforced thermoset laminates; electronic interconnects; composite materials; electrical properties of materials; polymerization reactions; specialty organic chemicals; physical organic chemistry. *Mailing Add:* Norplex/Oak Inc 1300 Norplex Dr PO Box 1448 LaCrosse WI 54602-1448

REICHENBECHER, VERNON EDGAR, JR, b Meyersdale, Pa, Mar 29, 48; m 76; c 2. MONOCLONAL ANTIBODY PRODUCTION. *Educ:* WVa Univ, BS, 70; Duke Univ, PhD(biochem, 76. *Prof Exp:* Fel med genetics, Baylor Col Med, 76-79, res assoc, 80-81; asst prof, 81-86, ASSOC PROF BIOCHEM, SCH MED, MARSHALL UNIV, 86- *Concurrent Pos:* Mem grad fac, WVa Univ, 82- *Mem:* Am Soc Cell Biol; Genetics Soc Am; Sigma Xi; AAAS. *Res:* Production of monoclonal antibodies; structure and function of mammalian ribosomes; toxic plant lectins; hypertension; somatic cell genetics. *Mailing Add:* Dept Biochem Sch Med Marshall Univ Huntington WV 25704

REICHERT, JOHN DOUGLAS, b Cameron, Tex, Nov 29, 38; m 61; c 2. THEORETICAL PHYSICS, OPTICAL PHYSICS. *Educ:* Univ Tex, Austin, BS & BA, 61; Calif Inst Technol, PhD(theoret physics), 65. *Prof Exp:* Fel theoret physics, Relativity Ctr, Univ Tex, Austin, 65-66; res assoc, Univ Southern Calif, 66-67, asst prof physics, 67-71; ASSOC PROF ELEC ENG, TEX TECH UNIV, 71- *Concurrent Pos:* Adj prof, Optical Sci Ctr, Univ Ariz, 71- *Mem:* Optical Soc Am; Am Phys Soc; Sigma Xi. *Res:* Laser beam propagation; optical resonators; diffraction and scattering theory; nonlinear interactions of light with matter; Fourier optics; quantum theory. *Mailing Add:* 1003 Warm Sands Trail SE Albuquerque NM 87123

REICHERT, JONATHAN F, b Cincinnati, Ohio, Aug 29, 31; m 53; c 3. SOLID STATE PHYSICS, MAGNETIC RESONANCE. *Educ:* Case Western Reserve Univ, BS, 53; Wash Univ, PhD(physics), 62. *Prof Exp:* Fel physics, Harvard Univ, 63-65; asst prof, Case Western Reserve Univ, 65-70; assoc prof, 70-90, PROF PHYSICS, STATE UNIV NY BUFFALO, 91. *Concurrent Pos:* Air Force res grant, 66-69; NY State Res Found grant, 72; NSF grant ions in liquid helium, 74-76 & 76-78; NY Res Found grant, 75-76; chmn fac senate, State Univ NY Buffalo, 76-78; vis assoc prof, Princeton

Univ, 78-79; dir, Nuclear War Prev Studies, 87-91. *Mem:* Am Phys Soc; Am Asn Physics Teachers; Union Concerned Scientists. *Res:* Mossbauer spectroscopy and liquid helium; nuclear and electronic magnetic resonance and solid state physics of defects; science education. *Mailing Add:* Physics Rm 128 Fronczak Hall State Univ NY North Campus Buffalo NY 14260

REICHERT, LEO E, JR, b New York, NY, Jan 9, 32; m 57; c 4. BIOCHEMISTRY, ENDOCRINOLOGY. *Educ:* Manhattan Col, BS, 55; Loyola Univ Chicago, MS, 57, PhD(biochem), 60. *Prof Exp:* From instr to assoc prof biochem, Emory Univ, 60-71, prof, 71-79; chmn, 79-89, PROF, ALBANY MED COL, 79- *Concurrent Pos:* Secy gonadotropin subcomt, Nat Pituitary Agency, 68-74, mem med adv bd, Agency, 71-74; mem reproductive biol study sect, NIH, 71-75; mem Adv Panel Cellular Physiol, NSF, 83-86; mem Expert Adv Panel Biol Standardization, WHO, 84-91. *Honors & Awards:* Ayerst Award, Endocrine Soc, 71. *Mem:* AAAS; Am Soc Biol Chemists; Endocrine Soc. *Res:* Biochemistry and physiology of gonadotropin hormones and their receptors; relationship of structure to function; authored 300 original articles, chapters, abstracts. *Mailing Add:* Dept Biochem Albany Med Col 43 New Scotland Ave Albany NY 12208

REICHES, NANCY A, b Cleveland, Ohio, Jan 26, 49. EPIDEMIOLOGY, BIOSTATISTICS. *Educ:* Univ Colo, Boulder, BA, 71, MA, 72; Ohio State Univ, PhD(prev med), 77. *Prof Exp:* Lectr commun theory, Univ Nev, Las Vegas, 72-73; res asst biostatist, Biomet, Ohio State Univ, 73-76, res assoc epidemiol, Comprehensive Cancer Ctr, 76-77, sr res assoc epidemiol & biostatist, 77-84; dir res, Riverside Methodist Hosp, 84-90; PVT CONSULT, 90- *Concurrent Pos:* Consult, Battelle Mem Inst, Columbus, Ohio, 77- *Mem:* Am Pub Health Asn; Soc Epidemiol Res; Am Soc Prev Oncol. *Res:* Cancer epidemiology; environmental epidemiology; biostatistical methods; biomedical computing. *Mailing Add:* 91 S Roosevelt Columbus OH 43209

REICHGOTT, MICHAEL JOEL, b Newark, NJ, July 26, 40; m 62; c 3. MEDICINE, CLINICAL PHARMACOLOGY. *Educ:* Gettysburg Col, AB, 61; Albert Einstein Col Med, MD, 65; Univ Calif, San Francisco, PhD(pharmacol), 73. *Prof Exp:* Trainee clin pharmacol, Med Ctr, Univ Calif, San Francisco, 69-72; assoc med, 72-73, asst prof med, 73-81, dir, Outpatient Clins, 75-78, dir, Dept Pract, 78-80, assoc chief staff ambulatory care, Med Ctr, 80-81, assoc prof med, Hosp, Univ Pa, 73-, chief, Sect Gen Med, Med Ctr, 81-84; med die, Bronx Municipal Hosp Ctr, 84-89; asst dean, 84-89, ASSOC DEAN, STUDENTS & GRAD MED EDUC, ALBERT EINSTEIN COL MED, 89- *Mem:* Fel Am Col Physicians; AAAS; Am Soc Pharmacol & Exp Therapeut; Am Fedn Clin Res; Soc Gen Internal Med. *Res:* Compliance, health services delivery; ambulatory care; medical education. *Mailing Add:* Dept Med Albert Einstein Col Med Bronx NY 10461

REICHLE, ALFRED DOUGLAS, b Port Arthur, Tex, Dec 19, 20; m 43; c 2. PETROLEUM REFINING, CATALYSIS. *Educ:* Rice Univ, BS, 42, MS, 43; Univ Wis, PhD(chem eng), 48. *Prof Exp:* Chemist petrol refining, Shell Develop Co, 43-57; staff asst process develop, Phillips Petrol Co, 57-59; engr, 59-60, sr res engr, 60-62, eng assoc, 62-63, sect head, 63-67, sr eng assoc, 67-71, ENG ADV PETROL REFINING, EXXON RES & DEVELOP CO, 71- *Mem:* Am Inst Chem Engrs. *Res:* Petroleum process development; catalytic refining processes; catalyst development. *Mailing Add:* 1025 Broadmoor Circle Baton Rouge LA 70815

REICHLE, DAVID EDWARD, b Cincinnati, Ohio, Oct 19, 38; m 61; c 3. ENVIRONMENTAL SCIENCES. *Educ:* Muskingum Col, BS, 60; Northwestern Univ, MS, 61, PhD(biol sci), 64. *Prof Exp:* AEC fel, Oak Ridge Nat Lab, 64-66; ecologist, 66-72, assoc dir, 72-86, DIR, ENVIRON SCI DIV, OAK RIDGE NAT LAB, 86- *Concurrent Pos:* Consult, ecosyst analysis panel, NSF; mem environ studies & toxicol & environ health bds, Nat Acad Sci, US SCOPE Comt & directorate, US MAB Prog; prof grad prog ecol, Univ Tenn; Danforth Assoc; mem nat bd govs, Nature Conservancy. *Mem:* Fel AAAS; Ecol Soc Am; Am Inst Biol Sci; Nature Conservancy; Int Asn Ecol. *Res:* Environmental geochemistry and health; terrestrial invertebrate ecology, bioenergetics, structure and function of arthropod communities; radioecology, movement of radioisotopes through food chains, effects of ionizing radiation upon natural arthropod communities; mineral cycling; ecosystem analysis. *Mailing Add:* Environ Sci Div Environ Life & Social Sci Directorate Oak Ridge TN 37831-6253

REICHLE, FREDERICK ADOLPH, b Neshaminy, Pa, Apr 20, 35. SURGERY. *Educ:* Temple Univ, BA, 57, MS, 61 & 66, MD, 61. *Prof Exp:* Intern med, Abington Mem Hosp, 61-62; resident, 62-66, from instr to assoc prof, 66-76, PROF SURG, HEALTH SCI CTR, TEMPLE UNIV, 76-CHIEF SECT PERIPHERAL VASCULAR SURG, 74-; CHMN & PROF SURG, PRESBY-UNIV PA MED CTR, 80- *Concurrent Pos:* Fel vascular surg, Health Sci Ctr, Temple Univ, 66-67; asst attend surgeon, Episcopal & St Christopher's Hosps, 66-; mem coun thrombosis, Am Heart Asn, 71-; mem, Nat Kidney Found, 72; consult, Vet Admin Hosp, Wilkes-Barre. *Mem:* Fel Am Col Surgeons; Am Surg Asn; Soc Surg Alimentary Tract; AAAS; Int Soc Thrombosis & Haemostasis. *Res:* Vascular surgery; liver metabolism; amino acid metabolism; cancer; thrombosis. *Mailing Add:* Presby-Penn Med Ctr 51 N 89th St Philadelphia PA 19104

REICHLE, WALTER THOMAS, b Cleveland, Ohio, Aug 2, 28; m 53; c 3. CHEMISTRY. *Educ:* NJ Inst Technol, BS, 53; Ohio State Univ, PhD(chem), 58. *Prof Exp:* Res chemist, Res Labs, Plastics Div, 58-68, sr res scientist chem & plastics, Res & Develop Lab, 69-84, CORP FEL, UNION CARBIDE CORP, 84- *Concurrent Pos:* Vis fel, Princeton Univ, 73-74. *Mem:* Am Chem Soc; Sci Res Soc Am; NY Acad Sci. *Res:* Organo-metallic and inorganic chemistry; catalytic agents; homogeneous and heterogeneous catalysis. *Mailing Add:* Tech Ctr Union Carbide Corp PO Box 670 Bound Brook NJ 08805

REICHLER, ROBERT JAY, b Bronx, NY, Nov 22, 37; c 2. CHILD PSYCHIATRY. *Educ:* Univ Chicago, BA & BS, 57; Albert Einstein Col Med, MD, 61; Am Bd Psychiat & Neurol, dipl, 71. *Prof Exp:* Instr psychiat, Strong Mem Hosp, Univ Rochester, 64-65; clin asst prof, Med Col Charleston, 68-69; asst prof, Sch Med, Univ NC, Chapel Hill, 69-72, assoc prof child psychiat & co-dir, Div Teach, 72-76; prof psychiat & head, Div Child Psychol, Sch Med, Univ Wash, 76-79; DIR RES, HARBORVIEW MED CTR, 78- *Concurrent Pos:* NIMH fel psychiat, Strong Mem Hosp, Univ Rochester, 62-65; NIMH fel child psychiat, NC Mem Hosp, Univ NC, 65-67; from asst surgeon to sr surgeon, USPHS, 67-69; mem child & family develop res rev comt, Off Child Develop, Dept HEW, 72-74; mem prof adv bd, Nat Soc Autistic Children, 72-, chmn, 77-; mem intervention comt, NC Coun Develop Disabilities, 72-75; consult, Div Neuropharm, Food & Drug Admin, 76-, Off Sci Eval, Bur Drugs, Food & Drug Admin & Div Neuropharmacol, & NIMH, 76-; dir, Dept Behav Sci, Children's Orthop Hosp & Med Ctr, 76-79; adj prof pediat, Sch Med, Univ Wash, 77-; actg head inpatient psychiat treat div, Children's Orthop Hosp & Med Ctr, 78; pres, Wash State Coun Child Psychiat, 88-89; dir, Child at Risk Proj, 84-89; co-dir, Ctr Anxiety & Depression, 87-90; mem, Fac Senate Exec Comt, 87-89. *Honors & Awards:* Gold Achievement Award, Am Psychiat Asn, 72. *Mem:* AAAS; Am Psychopathol Asn; Am Psychiat Asn; Am Acad Child Psychiat. *Res:* Psychotic and communication disordered children; parent and professional estimates of current and future abilities; psychophysiological parameters in childhood psychosis; nosology for child psychopathology; developmental psychopharmacology; anxiety and depression disorders; high risk children. *Mailing Add:* Dept Psychol & Behav Sci Univ Wash Seattle WA 98195

REICHLIN, MORRIS, b Toledo, Ohio, Feb 2, 34; m 58; c 2. MEDICINE. *Educ:* Washington Univ, AB, 55, MD, 59. *Prof Exp:* Fel biochem, Brandeis Univ, 61-63 & Univ Rome, 63-64; instr med, Col Med, Univ Vt, 64-65; from asst prof to assoc prof med & biochem, State Univ NY Buffalo, 65-71, prof, 71-81; MEM & HEAD, ARTHRITIS & IMMUNOL LAB, OKLA MED RES FOUND, PROF & CHIEF, IMMUNOL SECT, COL MED, OKLA HEALTH SCI CTR, 81- *Mem:* Am Soc Biol Chemists; Am Asn Immunologists. *Res:* Autoimmunity and chemistry of autoantigens. *Mailing Add:* Okla Med Res Found 825 NE 13th St Oklahoma City OK 73104

REICHLIN, SEYMOUR, b New York, NY, May 31, 24; m 51; c 3. INTERNAL MEDICINE, PHYSIOLOGY. *Educ:* Antioch Col, AB, 45; Washington Univ, MD, 48; Univ London, PhD, 54. *Prof Exp:* Commonwealth Fund fel, 52-54; instr med & psychiat, Sch Med, Washington Univ, 54-56, asst prof, 56-61; assoc prof med, Sch Med & Dent, Univ Rochester, 62-66, prof, 66-69; prof med & chmn dept med & pediat specialties, Sch Med, Univ Conn, 69-71, prof physiol & head dept, 71-72; PROF MED, SCH MED, TUFTS UNIV, 72- *Concurrent Pos:* Palmer Fund fel, 54-56; mem endocrinol study sect, USPHS, 66-70. *Honors & Awards:* Eli Lilly Award, Endocrine Soc, 72; Berthold Award, Ger Endocrine Soc, 83. *Mem:* Endocrine Soc (pres, 75-76); Am Physiol Soc; Am Psychosom Soc; Asn Res Nerv & Ment Dis (pres, 77); Asn Am Physicians. *Res:* Endocrinology; neuroendocrinology; thyroid and pituitary physiology. *Mailing Add:* New Eng Med Ctr Box 275 750 Wash St Boston MA 02111

REICHMAN, OMER JAMES, b Tampa, Fla, Jan 4, 47; m 68. ECOLOGY. *Educ:* Tex Tech Univ, BA, 68, MS, 70; Northern Ariz Univ, PhD(biol), 74. *Prof Exp:* Instr biol, Northern Ariz Univ, 72-74; res asst prof biol, Univ Utah, 74-75; RES ECOLOGIST BIOL & ASST DIR, MUS NORTHERN ARIZ, 75- *Concurrent Pos:* Fel, Univ Utah, 74-75. *Mem:* Ecol Soc Am; Am Soc Mammal; Soc Study Evolution; Am Soc Nat; Sigma Xi. *Res:* Plant and animal interactions; resource distribution particularly seeds; pocket gopher ecology; resource utilization by rodents. *Mailing Add:* Ecol Prog Rm 215 NSF 1800 G St NW Washington DC 20550

REICHMAN, SANDOR, b Nov 24, 41; US citizen; m 66; c 3. PHYSICAL CHEMISTRY. *Educ:* City Col NY, BS, 63; NY Univ, PhD(phys chem), 67. *Prof Exp:* Fel infrared spectros, Univ Minn, 66-68; asst prof, 68-72, assoc prof, 72-76, PROF PHYS CHEM, CALIF STATE UNIV, NORTHRIDGE, 76- *Concurrent Pos:* Vis prof, Hebrew Univ, Jerusalem, 80. *Mem:* Am Phys Soc. *Res:* Infrared spectroscopy; molecular dynamics; high resolution infrared spectroscopy; anharmonicity calculations. *Mailing Add:* Dept Chem Calif State Univ 1811 Nordhoff St Northridge CA 91330

REICHMANIS, ELSA, b Melbourne, Australia, Dec 9, 53; US citizen; m 79; c 4. LITHOGRAPHIC MATERIALS. *Educ:* Syracuse Univ, BS, 72, PhD(org chem), 75. *Prof Exp:* Intern org chem, Syracuse Univ, 75-76, Chaim Weizmann fel sci res, 76-78; mem tech staff org chem, 78-84, SUPVR, RADIATION SENSITIVE MAT & APPLN GROUP, AT&T BELL LABS, 84- *Mem:* Am Chem Soc; AAAS; Soc Photo-Optical Instrumentation Engrs. *Res:* Chemistry; properties and application of radiation sensitive materials; electronic materials; microlithography; photochemistry; synthesis; polymers for electronic applications. *Mailing Add:* 550 St Marks Ave Westfield NJ 07090

REICHMANN, MANFRED ELIEZER, b Trencin, Czech, Apr 16, 25; m 57; c 3. VIROLOGY. *Educ:* Hebrew Univ, Israel, MSc, 49, PhD(biochem), 51. *Prof Exp:* USPHS fel, Harvard Univ, 51-53; Nat Res Coun Can fel, 53-55; res officer, Plant Virus Inst, Res Br, Can Dept Agr, 55-64; prof bot, 64-71, PROF MICROBIOL, UNIV ILL, URBANA, 71- *Concurrent Pos:* Prof biochem, Univ BC, 62-64; assoc mem, Ctr Adv Studies, Univ Ill, 77-78; scholar, Am Cancer Soc, 77-78. *Mem:* Am Soc Biol Chemists; Am Soc Microbiol; AAAS. *Res:* Physiochemical studies of viruses; chemical composition of viral proteins and nucleic acids; defective interfering particles; mechanism of autointerference; Vesicular Stomatitis virus, Papilloma viruses; molecular biology; biochemistry. *Mailing Add:* Dept Microbiol Univ Ill 131 Burrill Hall Urbana IL 61801

REICHSMAN, FRANZ KARL, b Vienna, Austria, Sept 26, 13; nat US; m 45; c 5. PSYCHOSOMATIC MEDICINE. *Educ:* Univ Vienna, MD, 38; Am Bd Internal Med, dipl. *Prof Exp:* Res asst, Johns Hopkins Hosp, 39-40; intern med, Sinai Hosp, Baltimore, 40-41, asst path, 41-42, asst resident med, 42-43; asst prof, Univ Tex Southwestern Med Sch Dallas, 46-49, clin asst prof, 49-52; from instr to assoc prof med & psychiat, Sch Med, Univ Rochester, 52-64; prof med & psychiat, State Univ NY Downstate Med Ctr, 64-; RETIRED. *Concurrent Pos:* Dazian Found fel, Bowman Gray Sch Med, 43-44 & Univ Tex Southwestern Med Sch Dallas, 44-45; Commonwealth Fund fel, Sch Med, Univ Rochester, 52-54; chief chest serv, Vet Admin Hosp, McKinney, Tex, 47-50, asst chief med serv, 50-52; career investr, USPHS, 56-61; vis physician, Kings County Hosp, 64-; assoc physician & assoc psychiatrist, Strong Mem Hosp; vis prof, Oxford Univ, 71-72. *Mem:* Royal Soc Med; Am Psychiat Asn; Group Advan Psychiat; Am Psychosom Soc (pres elect, 80-81). *Res:* Dynamics of congestive heart failure; arterial hypertension; adaptation to hemodialysis; emotions and gastric function. *Mailing Add:* Box 275 Marlboro VT 05344

REID, ALLEN FRANCIS, b Deer River, Minn, July 31, 17; m 43; c 2. BIOPHYSICS. *Educ:* Univ Minn, BCh, 40; Columbia Univ, AM, 42, PhD(chem), 43; Southwest Sch Med, Univ Tex, MD, 59. *Prof Exp:* Asst chem, Columbia Univ, 40-42, res scientist in chg radioactivity labs, 42-46; indust consult, Sun Oil Co, Pa, 46-47; from assoc prof to prof biophys, Southwest Med Sch, Univ Tex, 47-60, chmn dept, 47-60; prof biol & chmn dept, Univ Dallas, 60-68; dir, Clin Biochem, Brooklyn-Cumberland Med Ctr, 68-74; prof biol & chmn dept, State Univ NY, Geneseo, 74-82; RETIRED. *Concurrent Pos:* Prof & chmn dept biophys & phys chem, Grad Res Inst, Baylor Univ, 47-50, biophysicist, Univ Hosp, 47-50; consult, Oak Ridge Inst Nuclear Studies, 50-52; biophysicist, Parkland Mem Hosp, Dallas, 50-58; consult, US Vet Admin, 50-67; clin prof path, State Univ NY Downstate Med Ctr, 68-74; dir path, Brooklyn-Cumberland Med Ctr, 70-74. *Mem:* Am Asn Cancer Res; fel Am Inst Chemists; Am Chem Soc; Am Phys Soc; Am Physiol Soc; AMA; Marine Technol Soc. *Res:* Energy recovery; tracer work in chemical and biological systems; radioactivity; methods of fractionation; reaction and biologic mechanisms; desalination. *Mailing Add:* 4736 Reservoir Rd Geneseo NY 14454

REID, ARCHIBALD, IV, b Janesville, Wis, Nov 23, 30; m 57; c 2. PLANT ECOLOGY. *Educ:* Univ Wis, Platteville, BS, 57; MS, 59, PhD(bot), 62. *Prof Exp:* Asst prof bot, Univ Wyo, 61-66; assoc prof landscape archit & regional planning, Univ Pa, 66-68; assoc prof, 68-72, PROF BIOL, STATE UNIV NY COL, GENESEO, 72- *Mem:* Am Mus Natural Hist; Ecol Soc Am; Wilderness Soc; Nature Conservancy. *Res:* Growth inhibitors produced by vascular plants; ecology as the base for landscape architecture and regional planning; structure of plant communities; environmental measurements. *Mailing Add:* Dept Biol State Univ NY Col Geneseo NY 14454

REID, BOBBY LEROY, b San Benito, Tex, Aug 28, 29; m 50; c 3. BIOCHEMISTRY, NUTRITION. *Educ:* Agr & Mech Col, Tex, BS, 50, MS, 52, PhD, 55. *Prof Exp:* Asst prof, Agr & Mech Col, Tex, 55-59; dir prod res, Feed Div, Pillsbury Co, 59-60; PROF POULTRY SCI, UNIV ARIZ, 60-, HEAD DEPT & POULTRY SCIENTIST, AGR EXP STA, 74-, MEM STAFF DEPT NUTRIT & ANIMAL SCI. *Concurrent Pos:* Biochemist, Nutrit Surv Haiti, Williams-Waterman Found, 58; spec consult, Nutrit Surv Turkish Armed Forces, Interdept Comt Nutrit for Nat Defense, 57. *Mem:* Soc Exp Biol & Med; Poultry Sci Asn; Animal Nutrit Res Coun. *Res:* Poultry nutrition research on unidentified factors, antibiotics, vitamin A, caloric-protein relationships, fat metabolism and B vitamins. *Mailing Add:* Dept Nutrit & Animal Sci Univ Ariz Tucson AZ 85721

REID, BRIAN ROBERT, b Gillingham, Eng, Nov 14, 38; m 61; c 2. BIOCHEMISTRY, STRUCTURAL CHEMISTRY. *Educ:* Cambridge Univ, BA, 60; Univ Calif, Berkeley, PhD(biochem), 65. *Prof Exp:* Jane Coffin Childs Mem fel biochem, Dartmouth Med Sch, 64-66; from asst prof to assoc prof, 66-75, prof biochem, Univ Calif, Riverside, 75-80; PROF CHEM, DEPT CHEM & BIOCHEM, UNIV WASH, 80- *Concurrent Pos:* Guggenheim Found fel, Med Res Coun Lab Molecular Biol, Cambridge Univ, 72-73; Fogarty sr int fel, Univ Oxford, 79-80. *Mem:* Am Soc Biol Chemists; Am Chem Soc. *Res:* Protein biosynthesis; nucleic acids; structure and function of transfer RNA and DNA. *Mailing Add:* Dept Chem & Biochem Univ Wash BG-Ten Seattle WA 98195

REID, CHARLES FEDER, b New York, NY, Feb 28, 33; m 66; c 2. VETERINARY RADIOLOGY. *Educ:* Cornell Univ, DVM, 56, MS, 60; Univ Pa, MA, 71. *Prof Exp:* Radiologist & head radioisotope lab, Animal Med Ctr, New York, 61-63; assoc prof radiol, 70-76, assoc prof radiol sci, 73-78, PROF RADIOL, SCH VET MED, UNIV PA, 76- ASSOC PROF RADIOL SCI, 78- *Concurrent Pos:* Res consult radiol, Hosp Spec Surg & Margaret M Caspary Clin, New York, NY, 63-75. *Mem:* Am Vet Med Asn; Radiol Soc NAm; Am Asn Equine Practrs; Am Vet Radiol Soc; Am Col Vet Radiol. *Res:* Thyroid function studies in the horse; use of high kilovoltage x-ray diagnostics procedures in large animals; radiation therapy in large animals. *Mailing Add:* 1307 S 49th St Philadelphia PA 19143

REID, CHARLES PHILLIP PATRICK, b Columbia, Mo, Jan 8, 40; m 61; c 2. FOREST ECOLOGY, PHYSIOLOGY. *Educ:* Univ Mo, BSF, 61; Duke Univ, MF, 66, PhD(forest ecol), 68. *Prof Exp:* Nat Acad Sci-Nat Res Coun assoc herbicide res, Plant Sci Lab, Ft Detrick, Dept Army, 67-69; from asst prof to prof, Colo State Univ, 69-85; PROF & CHMN DEPT FORESTRY, UNIV FLA, 86- *Concurrent Pos:* Fulbright lectureship, Univ Innsbruck, Austria, 85-86; mem, US-Australia Coop Sci Prog, S Australia, 76-77. *Mem:* AAAS; Soc Am Foresters; Ecol Soc Am. *Res:* Mycorrhizae of forest trees; plant-microbiol interactions; ecosystem nutrient cycling; plant-water relations; iron nutrition of plants. *Mailing Add:* Dept Forestry Univ Fla Newins-Ziegler Hall Gainesville FL 32611-0303

REID, CLARICE D, b Birmingham, Ala; m; c 4. SICKLE CELL DISEASE RESEARCH. *Educ:* Talledega Col, BS, 52; Univ Cincinnati, MD, 59; Am Bd Pediat, dipl, 64. *Prof Exp:* Pvt pract pediat, 62-68; dir pediat educ, Jewish Hosp, 68-69, chmn, Dept Pediat, 69-70; med consult, Pub Health Serv, Nat Ctr Family Planning, Health Serv & Ment Health Admin, 72-73; dep dir, Sickle Cell Dis Prog, Bur Commun Health Serv, Health Serv Admin, 73-76; actg dir, Div Blood Dis & Resources, 88-89, NAT COORDR, SICKLE CELL DIS PROG & CHIEF SICKLE CELL DIS BR, NAT HEART LUNG & BLOOD INST, NIH, 76- *Concurrent Pos:* Assoc attend pediat, Jewish Hosp, Cincinnati, 62-68, Children's Hosp Med Ctr, 62-70, Catherine Booth Hosp, 62-68; clin instr pediat, Univ Cincinnati, Col Med, 64-68; attend pediat, Bethesda Hosp, 63-70; pediatrician, Maternal & Child Health-Babies Milk Fund, Cincinnati, 63-70, Hamilton County Welfare Dept, 64-69; pediat consult, Ohio Dept Health, 66-70; asst prof pediat & asst chief, Family Care Prog, Univ Cincinnati, Sch Med, 68-70; clin asst prof pediat, Howard Univ Col Med, 79- *Mem:* Fel Am Soc Pediat; Am Acad Pediat; Nat Med Asn; AAAS; Am Soc Hemat; NY Acad Sci. *Res:* Sickle cell disease; pediatrics. *Mailing Add:* Nat Heart Lung & Blood Inst NIH 7500 Wisconsin Ave Rm 508 Bethesda MD 20892

REID, CYRIL, chemical physics, for more information see previous edition

REID, DAVID MAYNE, b Belfast, Northern Ireland, Dec 7, 40; m 67. PLANT PHYSIOLOGY. *Educ:* Queen's Univ, Belfast, BSc, 64, PhD(plant physiol), 67. *Prof Exp:* Asst lectr bot, Queen's Univ, Belfast, 67-68; from asst prof to assoc prof, 68-76, PROF BOT, UNIV CALGARY, 76- *Mem:* Brit Photobiol Soc; Can Soc Plant Physiol; Am Soc Plant Physiol; Sigma Xi; Soc Exp Biol. *Res:* Sites of synthesis of hormones in plants; interactions of hormones and phytochrome; root-shoot relations. *Mailing Add:* Biol Dept Univ Calgary 6824 Bow Crescent NW Calgary AB T2N 1N4 Can

REID, DONALD EUGENE, b Brookville, Pa, July 18, 30; m 52; c 3. POLYMER CHEMISTRY. *Educ:* Franklin & Marshall Col, BS, 52; Ohio State Univ, PhD(org chem), 57. *Prof Exp:* Res chemist, 57-73, sr res chemist, 73-86, RES SCIENTIST, HERCULES INC, WILMINGTON, 86- *Res:* Fatty and resin acids; coatings; polyolefins; rubber; polymer stabilization; polypropylene film; film metallization. *Mailing Add:* Packaging Films Div Hercules Inc Hercules Res Ctr Bldg 8136 Wilmington DE 19894

REID, DONALD HOUSE, b Phillipsburg, Pa, May 31, 35; m 64; c 2. AVIATION & SPACE PHYSIOLOGY. *Educ:* Cornell Univ, BS, 58; SDak State Univ, MS, 60; Univ Southern Calif, PhD(physiol), 68. *Prof Exp:* Res asst nutrit, SDak State Univ, 58-60; Aerospace physiologist, US Navy, 60-80; mgr, Sci Mgt, Gen Elec Co, 80-84; SR LIFE SCIENTIST, BOEING CORP, 84- *Honors & Awards:* Fred A Hitchcock Award Excellence Aerospace Physiol, 73. *Mem:* Fel Aerospace Med Asn; Soc Fed Med Agencies; Am Physiol Soc. *Res:* Stress physiology of humans particularly related to space flight and to the physiology of flying high performance aircraft, physiology of parachuting, and the development of aeromedically more acceptable aircrew life support equipment; physiology of parachuting; development of aeromedically more acceptable aircrew life support equipment for military and civilian aircraft. *Mailing Add:* 5611 134th Ave SE Bellevue WA 98006

REID, DONALD J, b Whitney, Tex, Jan 29, 38; m 61; c 2. AGRONOMY, FIELD CROPS. *Educ:* Calif State Polytech Inst, BS, 59; Cornell Univ, MS, 61, PhD(agron), 64. *Prof Exp:* Tech rep, Agr Chem Div, Shell Chem Co, 64-66; asst prof crop & soil sci, Mich State Univ, 66-75; assoc prof & exten agronomist-crops, SDak State Univ, 75-80, prof, 80-81; ASSOC PROF AGRON, ETEX STATE UNIV, 81- *Mem:* Am Soc Agron; Crop Sci Soc Am. *Res:* Information dissemination on small grains and row crops in the eastern half of South Dakota. *Mailing Add:* Dept Agr Sci ETex State Univ Commerce TX 75429

REID, EVANS BURTON, b Brock Twp, Ont, Mar 29, 13; nat US; m 42, 63; c 1. ORGANIC CHEMISTRY. *Educ:* McGill Univ, BSc, 37, PhD(org chem), 40. *Prof Exp:* Grad asst chem eng, McGill Univ, 37-38, demonstr org chem, 38-40; res chemist, Dominion Tar & Chem Co, Montreal, 40-41; instr chem, Middlebury Col, 41-43, asst prof, 43-46; asst prof chem, Johns Hopkins Univ, 46-54; actg dean fac, Colby Col, 67-68, dir, NSF Summer Inst sci, 58-60, 61-67, 68-73, Merrill prof & chmn dept chem, 54-78, EMER PROF, COLBY COL, 78- *Concurrent Pos:* Consult, Tainton Prod, Baltimore, 51-54; Smith-Mundt vis prof, Univ Baghdad, 60-61; consult, Comt Educ & Personnel, NSF, 63-65; corporator, Maine Med Care Develop, Inc, 68-76; chmn, Maine Sect Am Chem Soc, 56, 62, 71. *Res:* Chemistry of tetronic acids; Michael condensation; cyclobutane acids; plant growth hormones; chemistry of neurospora; dimeric ketenes and cyclobutanediones; sesquiterpenes; organic alicyclics; natural products. *Mailing Add:* 11 Highland Ave Waterville ME 04901

REID, F JOSEPH, b Lancaster, Ohio, Mar 19, 30; m 54; c 6. SOLID STATE PHYSICS. *Educ:* Ohio State Univ, BS, 52, MS, 57. *Prof Exp:* Prin physicist, Phys Chem Div, Battelle Mem Inst, Ohio, 56-59, proj leader, 59-62, assoc chief, 62-67; group mgr, 67-69, dept mgr, 69-72, res mgr, 72-78, DIR GTE LABS, INC, GEN TEL & ELECTRONICS CORP, 78- *Mem:* Electrochem Soc; Am Phys Soc; Am Ceramic Soc. *Res:* Materials science. *Mailing Add:* PO Box 660 Belmont MA 02178-0005

REID, GEORGE KELL, b Fitzgerald, Ga, Mar 23, 18; m 49; c 2. AQUATIC ECOLOGY. *Educ:* Presby Col, BS, 40; Univ Fla, MS, 49, PhD(zool), 52. *Prof Exp:* Instr biol sci, Univ Fla, 49-52; asst prof biol, Col William & Mary, 52-53; asst prof wildlife mgt, Tex Agr & Mech Univ, 53-56; asst prof zool, Rutgers Univ, 56-60; prof biol, 60-83, EMER PROF BIOL, ECKERD COL, 83- *Concurrent Pos:* Chmn, aquatic ecol sect, Ecol Soc Am, 64-66. *Mem:* Ecol Soc Am; Int Asn Theoret & Appl Limnol; Am Soc Limnol & Oceanog; Am Soc Zool; fel AAAS; Sigma Xi; Nat Audubon Soc. *Res:* Ichthyology; limnology; ecology of mangrove communities; intertidal zones, lakes and streams. *Mailing Add:* 6079 Town Colony Dr No 1022 Boca Raton FL 33433

REID, GEORGE W(ILLARD), b Indianapolis, Ind, Dec 18, 17; m 45; c 4. CIVIL & SANITARY ENGINEERING. *Educ:* Purdue Univ, BSCE, 42, CE, 50; Harvard Univ, SM, 43. *Prof Exp:* Instr civil eng, Purdue Univ, 42; asst prof, Univ Fla, 43-44; sanit engr, Ind State Health Dept, 44-45; indust hyg engr, Bell Aircraft Co, 45; sanit engr, USPHS, 45-46; assoc prof civil eng, Ga Inst Technol, 46-50; from assoc prof to prof, 50-72, REGENTS PROF CIVIL ENG, UNIV OKLA, 72-, DIR, 59- *Concurrent Pos:* Consult, USPHS, 50-, USAF, 56-, USN & US Army; mem, Interstate Water Comn, Okla, 58-; mem, Sen Select Comt Water Resources, 59-60; WHO traveling fel, 60; gov confs water res & long range planning, Okla, 65- *Honors & Awards:* Rudolf Herring Prize, Am Soc Civil Engrs, 74. *Mem:* Am Soc Civil Engrs; Sigma Xi; Am Pub Health Asn. *Res:* Water resources; radiation; radioisotope applications to sanitary engineering; environmental sanitation; biokinetics; systems research. *Mailing Add:* RBNSNS LNDG Elgen OK 73538

REID, H(ARRY) F(RANCIS), JR, b Coshocton, Ohio, Sept 15, 17; m 60; c 2. METALLURGY. *Educ:* Geneva Col, BS, 39; Ohio State Univ, MS, 48. *Prof Exp:* Chemist, Ceramic Color & Chem Mfg Co, Pa, 39-42; res engr, Battelle Mem Inst, 42-51; mgr tech serv div, McKay Co, 51-68, asst to vpres mkt, 68-73, asst vpres tech serv, 73-81, SR STAFF ENGR, AM WELDING SOC, 81- *Mem:* Am Welding Soc; Am Soc Metals. *Res:* Development of coating for ferrous and nonferrous arc welding electrodes. *Mailing Add:* Am Welding Soc 550 LeJeune Rd Miami FL 33126

REID, HAY BRUCE, JR, b Hyannis, Mass, Sept 15, 39; m 74; c 2. PLANT PHYSIOLOGY, BOTANY. *Educ:* Drew Univ, AB, 61; Univ Mass, Amherst, MA, 63; Univ Calif, Los Angeles, PhD(bot), 66. *Prof Exp:* Scholar, Univ Calif, Los Angeles, 66-67; asst prof bot, Rutgers Univ, New Brunswick, 67-73; ASSOC PROF BIOL, KEAN COL, NJ, 73- *Mem:* Am Asn Plant Physiologists. *Res:* Photoperiodism; physiology of flowering. *Mailing Add:* Dept Biol Kean Col Union NJ 07083

REID, IAN ANDREW, b Hobart, Australia, Aug 31, 43; m 71; c 2. PHYSIOLOGY. *Educ:* Univ Melbourne, BAgSc, 65; Monash Univ, PhD(renal physiol), 69. *Prof Exp:* Sr teaching fel physiol, Monash Univ, 69-70; fel physiol, 70-72, lectr & asst res physiologist, 72-73, adj asst prof, 73-77, ASSOC PROF PHYSIOL, UNIV CALIF, SAN FRANCISCO, 77- *Concurrent Pos:* NIH res career develop award, 75; consult, Hypertension Task Force, Nat Heart & Lung Inst, 76-77. *Mem:* Am Physiol Soc; Endocrine Soc; AAAS; Am Fedn Clin Res; Soc Exp Biol Med. *Res:* Regulation of fluid and electrolyte balance and blood pressure with emphasis on the role of the renin-angiotensin system. *Mailing Add:* Dept Physiol Univ Calif San Francisco CA 94143-0444

REID, JACK RICHARD, b Youngstown, Ohio, Oct 31, 47; m 69; c 2. ORGANIC CHEMISTRY, MEDICINAL CHEMISTRY. *Educ:* Lebanon Valley Col, Pa, BS, 69; Lehigh Univ, MS, 72, PhD(chem), 73. *Prof Exp:* Air pollution chemist analytical chem, US Army Environ Hyg Agency, 70-72; sr res chemist org chem, 75-87, MGR ORG CHEM, LORILLARD DIV, LOEWS THEATRE, INC, 87- *Concurrent Pos:* Res assoc med chem, Univ Kans, 73-75. *Mem:* Am Chem Soc; Sigma Xi. *Res:* Natural products, terpenes, alkaloids; heterocyclic chemistry; phosphorous chemistry; flavor chemistry. *Mailing Add:* Lorillard Res Ctr Lorillard Inc Box 21688 Greensboro NC 27420

REID, JAMES CUTLER, b Akron, Ohio, Apr 17, 18. ORGANIC CHEMISTRY. *Educ:* Univ Pa, BS, 39; Pa State Col, MS, 40; Univ Calif, PhD(org chem), 44. *Prof Exp:* Instr chem, Bowling Green State Univ, 40-42; asst, Univ Calif, 42-44, res assoc, 44-45, mem sci staff, Radiation Lab, 45-49; SR CHEMIST, PHYSIOL LAB, NAT CANCER INST, 49- *Mem:* AAAS; Am Chem Soc; Am Soc Biol Chem. *Res:* Metabolism of compounds labeled with carbon 14 with special reference to cancer; tracer and other metabolic studies with special reference to tumor-host relationships. *Mailing Add:* 6330 Annapolis Lane Dallas TX 75214

REID, JAMES DOLAN, b Augusta, Ga, June 24, 30; m 59; c 3. MATHEMATICS. *Educ:* Fordham Univ, BS, 52, MA, 53; Univ Wash, PhD(math), 60. *Hon Degrees:* MA, Wesleyan Univ, 72. *Prof Exp:* Instr math, Univ Wash, 59-60; asst prof, Syracuse Univ, 60-61 & Amherst Col, 62-63; from asst prof to assoc prof, Syracuse Univ, 63-69; assoc prof, 69-71, PROF MATH, WESLEYAN UNIV, 71- *Concurrent Pos:* Off Naval Res res fel, Yale Univ, 61-62. *Mem:* Soc Indust & Appl Math; Am Math Soc; Math Asn Am. *Res:* Algebra. *Mailing Add:* Dept Math Wesleyan Univ Middletown CT 06457

REID, JOHN DAVID, b Portland, Ore, Feb 19, 09; m 31; c 2. TEXTILE CHEMISTRY. *Educ:* State Col Wash, BS, 30; George Washington Univ, MA, 32; American Univ, PhD(chem), 37. *Prof Exp:* Jr chemist, Color & Farm Waste Div, Bur Chem & Soils, USDA, 30-37, asst chemist, Agr By-prods Lab, Iowa State Col, 37-40, assoc chemist, Southern Regional Res Lab, Bur Agr & Indust Chem, 40-42, chemist, 42-44, sr chemist, 44-45, in charge, Finishing Unit, Cotton Chem Processing Sect, Southern Utilization Res Br, 54-58, head chem finishing invests, Cotton Chem Lab, Southern Utilization Res & Develop Div, 58-75; CONSULT TEXTILE CHEM, 75- *Mem:* Am Chem Soc; Am Inst Chem. *Res:* Chemical modification of cotton fibers; creative problem solving; creative thinking and efficiency in research. *Mailing Add:* 4519 Banks St New Orleans LA 70119

REID, JOHN MITCHELL, b Minneapolis, Minn, June 8, 26; m; c 3. MEDICAL IMAGING, ELECTRICAL ENGINEERING. *Educ:* Univ Minn, BS, 50, MS, 57; Univ Penn, PhD(elec eng), 65. *Prof Exp:* Res fel surg, Univ Minn, 50-51, res fel elec eng, 51-53; chief res engr, St Barnabas Hosp, 54-57; res assoc biomed, Univ Penn, 57-65; res assoc prof physiol & biomed engr, Univ Wash, 66-71; asst dir physiol res, Providence Hosp, Seattle, 73-79 & Inst Appl Physiol Med, 71-81; CALHOUN PROF ELEC & BIOMED ENG, DREXEL UNIV, 82-; PROF RADIOL, THOMAS JEFFERSON UNIV, 83- *Concurrent Pos:* Reviewer, NIH, 64-; affil assoc prof elec eng, Univ Wash, 73-79; mem, Int Electrotech Comn, 79-, subcomt ultrasonics. *Honors & Awards:* Pioneer Award, Am Inst Ultrasound Med. *Mem:* Fel Acoust Soc Am; fel Am Inst Ultrasound Med; fel Inst Elec & Electronics Engrs; AAAS; Biomed Eng Soc. *Res:* Wave energy for imaging and diagnosis, particularly ultrasonic energy; development of transducers & circuits; extraction of parameters that characterize the medium being imaged. *Mailing Add:* Biomed Eng & Sci Inst Drexel Univ Philadelphia PA 19104

REID, JOHN REYNOLDS, JR, b Melrose, Mass, Jan 4, 33; m 56; c 4. GLACIOLOGY, GEOLOGY. *Educ:* Tufts Univ, BS, 55; Univ Mich, MS, 57, PhD(geol), 61. *Prof Exp:* Asst prof geol, Mt Union Col, 59-60; lectr, Univ Mich, 60-61; from asst prof to assoc prof, 61-71, assoc dean, Col Arts & Sci, 67-78, PROF GEOL, UNIV NDAK, 71- *Concurrent Pos:* Res geologist, Great Lakes Res Div, Inst Sci Technol, 60-61; fel, Quaternary Res Ctr, Univ Wash, 69-70; assoc dir, NDak Regional Environ Assessment Prog, 75-77, Interim Dir, 77-78; consult, US Army Corp Eng, 80-; vis prof, Univ Bergen, Norway, 86 & Edinburg Univ, Scotland, 87. *Mem:* AAAS; fel Geol Soc Am; Nat Asn Geol Teachers; Am Quaternary Asn; Sigma Xi. *Res:* Glacial geology; geomorphology of North Dakota and Alaska; glacial deposits; regimen of Alaska glaciers; structural glaciology of Antarctic firn folds; environmental geology of potential impact areas; shoreline erosion processes of Lakes Sakakawea and Oahe. *Mailing Add:* PO Box 8068 Univ Sta Grand Forks ND 58202-8068

REID, JOSEPH LEE, b Franklin, Tex, Feb 7, 23; m 53; c 2. OCEANOGRAPHY. *Educ:* Univ Tex, BA, 42; Univ Calif, MS, 50. *Prof Exp:* Res oceanographer & lectr, 50-74, dir marine Llife res group, 74- 87, PROF OCEANOG, SCRIPPS INST OCEANOG, UNIV CALIF, SAN DIEGO, 74- *Honors & Awards:* Nat Oceanog Data Ctr Award, Wash DC, 84; Spec Creativity Award, NSF, 88-90. *Mem:* Fel Am Geophys Union; Am Soc Limnol & Oceanog; fel AAAS. *Res:* Ocean circulation; distribution of temperature, salinity, oxygen, phosphate and marine life; wind-driven and thermohaline circulation; exchange between oceans; Antarctic Ocean circulation; California current; descriptive physical oceanography; world ocean. *Mailing Add:* Scripps Inst Oceanog Univ Calif La Jolla CA 92093-0230

REID, KARL NEVELLE, JR, b Fayetteville, Ark, Oct 10, 34; m 56; c 2. MECHANICAL ENGINEERING. *Educ:* Okla State Univ, BS, 56, MS, 58;. *Hon Degrees:* DSc, Mass Inst Technol, 64. *Prof Exp:* Asst mech eng, Mass Inst Technol, 58-60, instr, 60-64, lab coordr, 61-62; from asst prof to assoc prof mech & aero-eng, 64-70, dir ctr systs sci, 68-72, PROF MECH & AERO-ENG, OKLA STATE UNIV, 70-, HEAD SCH, 72- *Concurrent Pos:* Tech consult, Scovill Mfg, Westinghouse Elec Co; educ consult to NSF & USAID; chmn components comt, Am Automatic Control Coun, 72-; US mem tech comt components, Int Fedn Automatic Control; assoc ed, Trans/J Dynamic Systs, Measurements & Control, 71-73, ed, 74-76, sr tech ed, 76- *Mem:* Am Soc Mech Engrs; Am Soc Eng Educ; Fluid Power Soc; Nat Soc Prof Engrs. *Res:* Systems dynamics; analysis and design of fluid control systems; automatic control systems; fluidics; biomedical engineering. *Mailing Add:* Okla State Univ Stillwater OK 74078

REID, KENNETH BROOKS, b Jacksonville, Fla, Mar 2, 43; m 66; c 2. MATHEMATICS. *Educ:* Univ Calif, Berkeley, BA, 64; Univ Ill, Urbana, MA, 66, PhD(math), 68. *Prof Exp:* From asst prof to assoc prof, 68-80, PROF MATH, LA STATE UNIV, BATON ROUGE, 80-, CHMN, 87- *Concurrent Pos:* Vis assoc prof, Univ Waterloo, Can, 74; vis fel, Inst Advan Studies, Australian Nat Univ, 75; vis prof, Johns Hopkins Univ & Ga Inst Technol, 82; W K Kellogg Nat fel, 81-84; vis lectr, Math Asn Am, 75- *Mem:* Am Math Soc; Math Asn Am; Soc Indust & Appl Math; Australian Math Soc. *Res:* Combinatorial theory, particularly graph theory; enumeration of various discrete structures; existence problems concerning structure in graphs and directed graphs, particularly tournaments; extremal problems in graphs and directed graphs; combinatorics of permutation groups; special voting problems. *Mailing Add:* Founding Fac Calif State Univ 820 W Los Vallecitos Blvd C San Marcos CA 92069-1477

REID, LLOYD DUFF, b North Bay, Ont, Jan 6, 42; m 64; c 3. FLIGHT SIMULATION, FLIGHT MECHANICS. *Educ:* Univ Toronto, BASc, 64, MASc, 65, PhD(aerospace), 69. *Prof Exp:* From asst prof to assoc prof, 69-82, PROF & ASSOC DIR, INST AEROSPACE STUDIES, UNIV TORONTO, 82- *Concurrent Pos:* Engr oper res, deHavilland Aircraft Can, 69; consult, var co, 69-; first vpres, Aerospace Eng & Res Consults Ltd, 73-75, pres, 75-82; vis prof, Delft Univ Technol, 78-79. *Mem:* Am Inst Aeronaut & Astronaut; Can Aeronaut & Space Inst. *Res:* Flight simulator motion and visual systems; human pilot performance; aircraft flight through wind shear and turbulence; car driver-vehicle interaction; mathematical models of human operators. *Mailing Add:* Inst Aerospace Studies Univ Toronto 4925 Dufferin St Downsview ON M3H 5T6 Can

REID, LOIS JEAN, b Portsmouth, Va, Sept 23, 37. MATHEMATICS. *Educ:* Col William & Mary, BS, 59; Duke Univ, MA, 61, PhD(math), 67. *Prof Exp:* Instr math, Duke Univ, 60-61; from instr to asst prof, Mary Wash Col, Univ Va, 63-66; asst prof, Univ NC, Greensboro, 67-70; assoc prof, Longwood Col, 70-74; MATHEMATICIAN, NAVAL SURFACE WEAPONS CTR, DAHLGREN LAB, 74- *Mem:* Math Asn Am. *Res:* Algebra and number theory, particularly symmetric q-polynomials, their generalizations and applications. *Mailing Add:* Box 628 Dahlgren VA 22448

REID, LOLA CYNTHIA MCADAMS, b Charlotte, NC, May 25, 45; m 82. BIOCHEMISTRY. *Educ:* Univ NC, BA, 68, PhD(neuroendocriol), 74. *Prof Exp:* Technician physiol ecol, Univ NC, Chapel Hill, 68, teaching asst, 73; fel cancer biol, Univ Calif, San Diego, 74-77; ASST PROF DIFFERENTIATION, 77-, ASSOC PROF DEPT MOLECULAR PHARMACOL, MICROBIOL & IMMUNOL, ALBERT EINSTEIN COL. *Concurrent Pos:* Res fel, Bar Harbor, Maine, 75, Pasteur Inst, 76; prin investr, var res proj, 77-; consult, var co, 76- *Honors & Awards:* Sinsheimer Career Develop Award, 77. *Mem:* Am Asn Cancer Res; AAAS; Am Soc Zoologists. *Res:* Regulation of differentiation in normal and neoplastic cells especially by the synergistic interactions of hormones and exgracellular matrix. *Mailing Add:* Albert Einstein Col Med 1300 Morris Park Ave Chanin 601 Bronx NY 10461

REID, MICHAEL BARON, b Ft Worth, Tex, Aug 28, 52; m 84; c 2. RESPIRATORY MUSCLES, CHEST WALL MECHANICS. *Educ:* Univ Tex, Arlington, BS, 74, Dallas, PhD(physiol), 80. *Prof Exp:* Res fel, Health Sci Ctr, Univ Tex, Dallas, 80-81; res fel, Sch Pub Health, Harvard Univ, 81-83, res assoc, 83-87, asst prof pysiol, 87-89; ASST PROF MED, BAYLOR COL MED, 89- *Concurrent Pos:* Respiratory therapist, Harris Hosp, Ft Worth, 70-75, St Paul's Hosp, Dallas, 75-76, Presby Hosp, Dallas, 77-81 & Beth Israel Hosp, Boston, 81-83; adj fac, Simmons Col, Boston, 81-83; res physiologist, Vet Admin, 84-89; vis prof, Catholic Univ Leuven, Belgium; mem Cardiopulmonary Coun, Am Heart Asn. *Mem:* Am Physiol Soc; Am Thoracic Soc; Am Heart Asn; Europ Respiratory Soc. *Res:* Respiratory muscle physiology and pharmacology; chest wall mechanics; proprioceptive respiratory reflexes. *Mailing Add:* Pulmonary Med Rm 520B Baylor Col Med One Baylor Plaza Houston TX 77030

REID, PARLANE JOHN, b Long Beach, Calif, Apr 14, 37; m 59; c 2. BIOCHEMISTRY, MICROBIAL GENETICS. *Educ:* Univ Calif, Santa Barbara, BA, 60, MA, 61, PhD(biol), 66. *Prof Exp:* Res assoc genetics, Univ Calif, Santa Barbara, 60-61; instr biol, Calif State Polytech Col, 61-62; fel biochem, Sch Med, Stanford Univ, 66-68; ASST PROF BIOCHEM, SCH MED, UNIV CONN, 68- *Res:* Mechanisms of RNA and protein synthesis in microbes and bacteriophage. *Mailing Add:* 1524 Danforth Lane Osprey FL 34229

REID, PHILIP DEAN, b Ypsilanti, Mich, Mar 6, 37; m 84; c 2. PLANT PHYSIOLOGY. *Educ:* Eastern Mich Univ, BA, 62; Univ Mo, MS, 64; Univ Mass, PhD(bot), 70. *Prof Exp:* Res biol, Uniroyal Corp, 64-66; res fel, Univ Calif, Riverside, 70-71; from asst prof to assoc prof biol sci, Smith Col, 71-83. asst to pres campus planning, 79-88, chmn, dept biol sci, 86-89, PROF BIOL SCI, SMITH COL, 83- *Mem:* AAAS; Sigma Xi; Am Soc Plant Physiologists; NY Acad Sci; Plant Growth Regulator Soc Am. *Res:* Hormonal control of plant metabolism and development. *Mailing Add:* Dept Biol Sci Smith Col Northampton MA 01063

REID, PRESTON HARDING, b Akron, Colo, Nov 15, 23; m 48; c 1. SOIL CHEMISTRY, SOIL FERTILITY. *Educ:* Colo State Univ, BS, 49; NC State Univ, MS, 51, PhD(soils), 56. *Prof Exp:* Res asst prof soils, NC State Univ, 56-60, assoc prof, 60-66, prof soil sci, 66-69, dir soil test div, NC Dept Agr, 65-69; dir, Tidewater Res & Continuing Educ Ctr, Va Polytech Inst & State Univ, 69-86; RETIRED. *Concurrent Pos:* Soils adv to Peru, 61-63; ed, Peanut Science, J Am Peanut Res & Educ Asn, 74-75; exten specialist, soybeans, Va Polytech Inst & State Univ, 75- *Mem:* Am Soc Agron; Soil Sci Soc Am; Am Soybean Asn; Sigma Xi. *Res:* Soil Testing. *Mailing Add:* 3104 SE Balboa Dr Vancouver WA 98684

REID, RALPH R, b Topeka, Kans, Nov 19, 34; c 2. RADAR SYSTEMS, MICROWAVE RECEIVING. *Educ:* Washburn Univ, BS, 56. *Prof Exp:* Vpres eng, Goodyear Aerospace Corp, 63-87; SR VPRES ENG, LORAL DEFENSE SYSTS, 87- *Res:* Radar systems. *Mailing Add:* 3853 W Port Royale Lane Phoenix AZ 85023

REID, RICHARD J(AMES), b Burlington, Iowa, Oct 29, 32; m 52; c 5. ELECTRICAL ENGINEERING. *Educ:* Iowa State Univ, BS, 55, MS, 56; Mich State Univ, PhD, 59. *Prof Exp:* Instr elec eng, Iowa State Univ, 55-56; from instr to assoc prof, 56-70, PROF COMPUTER SCI, MICH STATE UNIV, 70- *Mem:* Inst Elec & Electronics Engrs; Asn Comput Mach. *Res:* Artificial intelligence. *Mailing Add:* Dept Computer Sci Mich State Univ East Lansing MI 48824

REID, ROBERT C(LARK), b Denver, Colo, June 11, 24; m 50; c 2. CHEMICAL ENGINEERING. *Educ:* Purdue Univ, BS, 50, MS, 51; Mass Inst Technol, ScD(chem eng), 54. *Prof Exp:* From asst prof to assoc prof, Mass Inst Technol, 54-64, prof chem eng, 64-81, Chevron prof, 81-87, EMER PROF, MASS INST TECHNOL, 87- *Concurrent Pos:* Am Soc Eng Educ lectr chem eng, 77; Olaf A Hougen prof, Univ Wis-Madison, 80-81. *Honors & Awards:* Warren K Lewis Award, Am Inst Chem Engrs, 76; Founders Award, Am Inst Chem Engrs, 86. *Mem:* Nat Acad Eng; Am Inst Chem Engrs. *Res:* Thermodynamics; cryogenics; liquefied natural gas and petroleum gas; safety; critical point extraction; migration of chemicals from polymer food wraps to food; superheated liquid explosions; botantical engineering. *Mailing Add:* Dept Chem Eng 66-544 Mass Inst Technol Cambridge MA 02139

REID, ROBERT LELON, b Detroit, Mich, May 20, 42; m 62; c 3. SOLAR ENERGY, HEAT TRANSFER. *Educ:* Univ Mich, BSE, 63; Southern Methodist Univ, MSE, 66, PhD(mech eng), 69. *Prof Exp:* Res engr, Atlantic Richfield Corp, 64-65; staff engr, Linde Div, Union Carbide Corp, 66-68; prof, Exxon Prod Res Corp, 72 & 73, NASA, 70 & 76; assoc prof mech eng, Univ Tenn, Knoxville, 69-75, Cleveland State Univ, 75-77; prof mech eng & asst dir, Energy, Environ & Resources Ctr, Univ Tenn, Knoxville, 77-82; chmn, Mech & Indust Eng Dept, Univ Tex, El Paso, 82-88; DEAN, SCH ENG, MARQUETTE UNIV, 88- *Concurrent Pos:* Prin investr contracts, Exxon Corp, 72-73, Union Carbide, 74-82, NASA, 77-78, NSF, 77-79, Tenn Valley Authority, 79-82, US Bur Reclamation, 83-; consult, Oak Ridge Nat Lab, 80-81; assoc ed, J Solar Energy Eng, Am Soc Mech Engrs, 81- *Honors & Awards:* Centennial Medallion, Am Soc Mech Engrs. *Mem:* Fel Am Soc Mech Engrs; Am Soc Heating, Refrig & Air Conditioning Engrs; Int Solar Energy Soc; Am Soc Engr Educ. *Res:* Solar energy; energy conservation; heat transfer; author or coauthor of over 60 publications. *Mailing Add:* Dean Eng Sch Marquette Univ 1217 W Wisconsin Ave Milwaukee WI 53233-2290

REID, ROBERT LESLIE, RUMINANT NUTRITION, FORAGE UTILIZATION. *Educ:* Aberdeen Univ, Scotland, PhD(agr biochem), 57. *Prof Exp:* PROF ANIMAL SCI, DIV ANIMAL & VET SCI, WVA UNIV, 69- *Res:* Mineral metabolism. *Mailing Add:* Dept Animal Sci Box 6108 WVa Univ Morgantown WV 26506

REID, ROBERT LESLIE, b Belleville, Ont, Jan 26, 51; m 76; c 3. REPRODUCTIVE ENDOCRINOLOGY & INFERTILITY. *Educ:* Queen's Univ, MD, 74. *Prof Exp:* Fel reproductive endocrinol, Dept Obstet & Gynec, Univ Calif, San Diego, 79-81; from asst prof to assoc prof, 81-90, PROF, DEPT OBSTET & GYNEC, QUEEN'S UNIV, KINGSTON, 90- *Concurrent Pos:* Chmn, Joint Can Fertil & Andrology Soc & Soc Obstetricians & Gynecologists, Can Ethics Comt New Reproductive Technologies, 87-91; pres, Can Fertil & Andrology Soc, 88-89; FIGO 1944 Organizing Comt mem, Fedn Int Gynecologie Obstetrie, 88-94. *Mem:* Soc Gynec Invest; Soc Obstetricians & Gynecologists Can; Can Fertil & Andrology Soc; Am Fertil Soc. *Res:* Menstrual cycle related mood and physical disorders; hormonal effects on mood in menopausal hormone replacement therapy; photodynamic ablation of endometrim/endometriosis. *Mailing Add:* Etherington Hall Queens Univ Kingston ON K7L 3N6 Can

REID, ROBERT OSBORNE, b Milford, Conn, Aug 24, 21; m 47; c 6. OCEANOGRAPHY. *Educ:* Univ Southern Calif, BE, 46; Univ Calif, MS, 48. *Hon Degrees:* DSc, Old Dominion Univ, 88. *Prof Exp:* Asst, Scripps Inst, Univ Calif, 46-47, oceanogr, 48-51; meteorologist, US Naval Electronics Lab, Calif, 47-48; from asst prof to assoc prof oceanog & meteorol, Tex A&M Univ, 51-59, prof oceanog & civil eng, 59-78, distinguished prof oceanog, 78-87, EMER DISTINGUISHED PROF OCEANOG, TEX A&M UNIV, 87- *Concurrent Pos:* Assoc ed, J Geophys Res, 61-73 & J Marine Res, 61-73 & 83-85; consult, US Army CEngrs, 65-78 & Hydraul Div, Waterways Exp Sta, Vicksburg, 75-; ed-in-chief, J Phys Oceanog, Am Meteorol Soc, 70-80; mem, Ad Hoc Panel Comput Resources & Facil Ocean Circulation Modeling, Nat Acad Sci, 79-80, Comt Coastal Flooding, 80-84, US Nat Comt Int Union Geod & Geophys, 80-84, Storm Surge Prog Rev Bd, Nat Oceanic & Atmospheric Admin, 81-83, subcomt, Nat Marine Bd, Nat Acad Sci, 86-88 & Coastal Eng Res Bd, US Army CEngr, 88- *Honors & Awards:* Spec Award, Am Meteorol Soc, 75. *Mem:* Nat Acad Eng; fel Am Meteorol Soc; fel Am Geophys Union; Int Asn Hydraul Res; Sigma Xi. *Res:* Physical oceanography, especially problems in ocean waves; storm tides and circulation; author of numerous technical publications. *Mailing Add:* Dept Oceanog Tex A&M Univ College Station TX 77843

REID, RODERICK VINCENT, JR, b Charlotte, NC, Oct 17, 32; m 59; c 3. THEORETICAL PHYSICS. *Educ:* Univ Denver, AB, 58, MS, 59; Cornell Univ, PhD(physics), 68. *Prof Exp:* Physicist, Mass Inst Technol, 67-69; asst prof, 69-74, ASSOC PROF PHYSICS, UNIV CALIF, DAVIS, 74- *Mem:* Am Inst Physics. *Res:* Nuclear theory. *Mailing Add:* Dept Physics Univ Calif Davis CA 95616

REID, ROLLAND RAMSAY, b Wilbur, Wash, Nov 12, 26; m 47, 75; c 3. GEOLOGY. *Educ:* Univ Wash, PhD(geol), 59. *Prof Exp:* Instr geol, Mont Sch Mines, 53-55; asst prof, 55-60, assoc prof geol & head dept geol & geog, 60-65, actg dean, 63-65, actg head dept geol & geog, 59-60, dean, Col Mines, 65-74, PROF GEOL, UNIV IDAHO, 65- *Concurrent Pos:* NSF fac fel, 58-59; dep asst secy, Energy & Minerals, US Dept Interior, 75-76. *Mem:* Geol Soc Am; Soc Econ Geologists. *Res:* Metamorphic and structural petrology; structural geology. *Mailing Add:* 621 East C St Moscow ID 83843

REID, RUSSELL MARTIN, b St Louis, Mo, July 30, 41; m 64; c 2. PHYSICAL ANTHROPOLOGY, BIOLOGICAL ANTHROPOLOGY. *Educ:* Univ Ill, BS, 63, PhD(anthrop), 71. *Prof Exp:* Asst prof anthrop, Univ Tex, Austin, 69-76; asst prof, 76-78, assoc prof anthrop & chmn, dept anthrop, Univ Houston, 78-83; PROF & CHAIR, DEPT ANTHROP, UNIV LOUISVILLE, 83- *Concurrent Pos:* NSF-Univ Sci Develop Prog grant, Univ Tex, Austin Field Res, Ceylon, 69-73, Haiti, 78. *Mem:* Am Anthrop Asn; Am Asn Phys Anthrop; Brit Soc Study Human Biol; Soc Study Social Biol. *Res:* Human population genetics, especially the role of social organization on genetic structure of populations, particularly consanguinity and inbreeding in human populations; dietary behavior and nutrition. *Mailing Add:* Dept Anthrop Univ Louisville Box 35260 Louisville KY 40292

REID, SIDNEY GEORGE, b Glamis, Scotland, Sept 21, 23; Can citizen; m 51; c 2. ORGANIC CHEMISTRY. *Educ:* Univ St Andrews, BSc, 44, PhD(chem), 49. *Prof Exp:* Fel, Univ Edinburgh, 48-51; res fel lignin chem, Ont Res Found, 51, asst dept dir, 57-63, dir dept appl chem, 63-85; RETIRED. *Mem:* Fel Chem Inst Can; Can Pulp & Paper Asn; Tech Asn Pulp & Paper Inst. *Res:* Wood, cellulose and lignin chemistry; paper additives; utilization of pulp and paper wastes and industrial wastes. *Mailing Add:* 291 Balsam Dr Oakville ON L6J 3X7 Can

REID, STANLEY LYLE, b Royal Oak, Mich, June 8, 30; m 54; c 2. ORGANIC CHEMISTRY. *Educ:* Univ Mich, PhD(chem), 57. *Prof Exp:* CHEMIST, MONSANTO CO, 57- *Mem:* Am Chem Soc. *Res:* Natural products; organic synthesis; oxidation. *Mailing Add:* 333 Grant St Jonesville MI 49250

REID, TED WARREN, b Cayuga, Ind, Sept 26, 39; m 61; c 2. BIOCHEMISTRY. *Educ:* Occidental Col, BS, 61; Univ Ariz, MS, 63; Univ Calif, Los Angeles, PhD(chem), 67. *Prof Exp:* Asst prof molecular biophys & biochem, Med Sch, Yale Univ, 70-75, assoc prof molecular biophys & biochem, ophthal & visual sci, 75-84; DIR RES, DEPT OPHTHAL, SCH MED, UNIV CALIF, 84- *Concurrent Pos:* Res to Prevent Blindness res prof, 74; vis fel, Imp Cancer Res Found, London, 80. *Mem:* Am Chem Soc; Asn Res Vision & Ophthal; Am Soc Cell Biol. *Res:* Biochemistry of the retina, cornea and cell growth factors. *Mailing Add:* Ophthal Dept Univ Calif-Davis 1603 Alhambra Blvd Sacramento CA 95816-7051

REID, THOMAS S, b Trenton, NJ, Dec 20, 11; m 42; c 2. ORGANIC CHEMISTRY. *Educ:* Rutgers Univ, BS, 36, MS, 38; Univ Minn, PhD(biochem), 42. *Prof Exp:* Res chemist, Eastern Regional Res Lab, USDA, Pa, 42-44; res chemist, 44-51, sect leader, Org Sect, 51-55, assoc dir, Cent Res Dept, 55-60, coordr, Div Res, 60-62, mgr biochem res, 62-64, dir biochem res, 64-71, dir biosci res lab, 71-77, CONSULT, MINN MINING & MFG CO, 77- *Mem:* AAAS; Am Chem Soc; Am Ceramic Soc. *Res:* Biochemistry, polymer and fluorine chemistry; medicinal and health sciences. *Mailing Add:* 735 E County Rd B2 St Paul MN 55117

REID, WILLARD MALCOLM, b Ft Morgan, Colo, Oct 9, 10; m 37; c 5. PARASITOLOGY, POULTRY SCIENCE. *Educ:* Monmouth Col, BS, 32; Kans State Univ, MS, 37, PhD, 41. *Hon Degrees:* DSc, Monmouth Col, 59. *Prof Exp:* Instr biol & head dept, Assiut Col, Egypt, 33-35; prof & head dept, Monmouth Col, 38-51; head poultry unit, US State Dept, Egypt, 52-55; from assoc prof to prof, 55-78, EMER PROF AVIAN PARASITOL, UNIV GA, 78- *Concurrent Pos:* Consult, Eli Lilly & Co, Greenfield, Ind, 68-79 & Mathtech Poultry Proj, Egypt, 78- *Mem:* Fel AAAS; Am Soc Parasitologists; Am Soc Zoologists; Am Asn Avian Pathologists; Poultry Sci Asn. *Res:* Parasitic diseases of poultry and their control using pharmaceutical drugs and biological control measures. *Mailing Add:* 240 Burnett St Athens GA 30605

REID, WILLIAM BRADLEY, b Indianapolis, Ind, Aug 2, 20; m 45; c 2. PHARMACEUTICALS. *Educ:* Butler Univ, BS, 42; Ind Univ, MA, 44, PhD(org chem), 46. *Prof Exp:* res chemist, Upjohn Co, 46-51, sect head, 52-83; RETIRED. *Concurrent Pos:* Mgr, Facilities Planning & Environ Regulatory Afffairs, 74- *Mem:* Am Chem Soc; Am Soc Testing & Mat. *Res:* Synthetic organic chemistry; bacteriology; administration. *Mailing Add:* 1821 Nichols Rd Kalamazoo MI 49007

REID, WILLIAM HARPER, b Bridgeport, Conn, Nov 16, 33. POPULATION ECOLOGY. *Educ:* Univ Mo, BS, 59; Univ Colo, PhD(biol), 74. *Prof Exp:* Design engr cryog, Beech Aircraft Corp, 59-61; develop engr liquid rockets, Aerojet-Gen Corp, 61-62; group engr spacecraft systs, Beech Aircraft Corp, 62-66; chief struct aircrew systs, Stanley Aviation Corp, 66-67; sr engr spacecraft thermodyn, Martin-Marietta Corp, 67-68; asst prof, 75-81, ASSOC PROF, DEPT BIOL SCI, UNIV TEX, EL PASO, 81- *Mem:* AAAS; Ecol Soc Am; Sigma Xi; Asn Trop Studies. *Res:* Ecology, adaptation and evolution of sympatric populations; field studies and computer simulation of single and multiple species sets; Chihuahuan desert ecology; dune field community structure and successional processes; pollination biology of xerophytes; biogeography. *Mailing Add:* 465 S Bluff 238 St George UT 84770

REID, WILLIAM HILL, b Oakland, Calif, Sept 10, 26; m 62; c 1. APPLIED MATHEMATICS. *Educ:* Cambridge Univ, PhD(math), 55; Brown Univ, AM ad eundem, 61. *Hon Degrees:* ScD, Cambridge Univ, 67. *Prof Exp:* NSF fel math, Yerkes Observ, Univ Chicago, 57, res assoc astron, 58; from asst prof to assoc prof appl math, Brown Univ, 58-63; from assoc prof to prof, 63-89, EMER PROF APPL MATH, UNIV CHICAGO, 89-; PROF MATH SCI, IND UNIV-PURDUE UNIV, INDIANAPOLIS, 89- *Concurrent Pos:* Lectr, Johns Hopkins Univ, 55-56; consult, Gen Motors Corp, 60-73; Fulbright res grant math, Australian Nat Univ, 64-65. *Mem:* Fel Am Phys Soc; Am Meteorol Soc; Am Math Soc; Soc Indust & Appl Math. *Res:* Fluid mechanics; hydrodynamic stability; asymptotic analysis. *Mailing Add:* Dept Math Sci Ind Univ-Purdue Univ Indianapolis IN 46205-2810

REID, WILLIAM JAMES, b Abbeville, SC, Nov 2, 27; m 64; c 3. SCIENCE EDUCATION. *Educ:* Erskine Col, AB, 49; Duke Univ, MA, 58; Clemson Univ, PhD(physics), 67. *Prof Exp:* Instr chem, Erskine Col, 49-51, field rep, 51-52, asst prof physics & math, 56-62, assoc prof physics, 66-68; pres fac senate, 71-72, 76-77 & 88-89, PROF PHYSICS & HEAD DEPT, JACKSONVILLE STATE UNIV, 68- *Mem:* Am Phys Soc; Sigma Xi; Am Asn Physics Teachers. *Res:* Superconductivity in thin films; pedagogy of physics; physics lecture demonstrations. *Mailing Add:* Dept Physics Jacksonville State Univ Jacksonville AL 36265

REID, WILLIAM JOHN, b Dublin, Ireland, Jan 29, 45; Can citizen; m 69; c 2. POLYMER CHEMISTRY, PHYSICAL CHEMISTRY. *Educ:* Univ Dublin, BS, 67; Univ BC, MS, 70, PhD(chem), 72. *Prof Exp:* Res assoc polymer chem, Ecoplastics, Univ Toronto, 72-73; res chemist, Res Lab, Uniroyal Ltd, 73-76, res chemist, Uniroyal Inc, 76-78; technol mgr, Plastics & Additives Div, Ciba-Geigy Corp, 78-; MEM STAFF CONLIAM RESOURCES PTY LTD, AUSTRALIA. *Mem:* Am Chem Soc. *Res:* Stabilization and degradation mechanisms in polymers; relationship of properties of polymers to molecular structure and morphology. *Mailing Add:* Plastic Ser PTD Ltd 13 Salisbury St Cutteslue 6011 Western Australia

REID, WILLIAM SHAW, b Slate Springs, Miss, Feb 24, 38; m 60; c 3. SOIL FERTILITY. *Educ:* Miss State Univ, BS, 59, MS, 61, PhD(agron, soils), 65. *Prof Exp:* Proj officer radiation biol, Air Force Weapons Lab, Kirtland AFB, NM, 63-66; from asst prof to assoc prof, 66-79, dept agron ext leader, 75-87, PROF SOIL SCI, CORNELL UNIV, 79- *Mem:* Am Soc Agron; Soil Sci Soc Am. *Res:* Interactions of fertilizers and climate on the uptake of plant nutrients and plant growth with respect to efficient agricultural production. *Mailing Add:* 333 Snyder Hill Rd Ithaca NY 14850

REID, WILLIAM T(HOMAS), b Racine, Wis, Feb 14, 07; m 30; c 3. ENERGY CONVERSION. *Educ:* Univ Wash, BS, 29. *Prof Exp:* Jr fuel engr, US Bur Mines, Pa, 29-34, from asst fuel engr to assoc fuel engr, 34-42, fuel engr, 42-43, supv engr, 43-46; asst supvr combustion res, Battelle Mem Inst, 46-47, supvr combustion res, 47-50, supvr graphic arts res, 50-53, asst tech dir, 53-65, sr fel mech eng, 65-72; RETIRED. *Concurrent Pos:* Energy conversion consult, 72-84. *Honors & Awards:* Melchett Medal, Brit Inst Fuel, 69; George Westinghouse Gold Medal, Am Soc Mech Engrs, 82; Percy Nicholls Award, Am Soc Mech Engrs, 68. *Mem:* Hon mem Am Soc Mech Engrs (vpres, 71-73); Brit Inst Energy. *Res:* Fuel-cell technology; energy sources for electric automobiles; corrosion and deposits in boiler furnaces; coal-ash technology. *Mailing Add:* 2470 Dorset Rd Columbus OH 43221

REIDENBERG, MARCUS MILTON, b Philadelphia, Pa, Jan 3, 34; m 57; c 3. PHARMACOLOGY, MEDICINE. *Educ:* Temple Univ, MD, 58. *Prof Exp:* From instr to asst prof pharmacol, Temple Univ, 62-72, assoc prof pharmacol & med, 72-75, chief sect clin pharmacol, 72-75; assoc prof med, 76-80, PROF PHARMACOL & HEAD DIV, MED COL, CORNELL UNIV, 76-, PROF MED, 80- *Concurrent Pos:* NIH fel, Temple Univ, 59-60, NIH res career develop award, 71-74. *Honors & Awards:* Rawls Palmer Award, Am Soc Clin Pharmacol & Therapeut, 81; Exp Therapeut Award, Am Soc Pharmacol, 83. *Mem:* Am Soc Pharmacol & Exp Therapeut; Am Col Physicians; Am Soc Clin Invest; Am Soc Clin Pharmacol & Therapeut; Am Fedn Clin Res; Asn Am Physicians. *Res:* Clinical pharmacology; drug metabolism; adverse drug reactions. *Mailing Add:* Cornell Univ Med Col 1300 York Ave New York NY 10021

REIDER, MALCOLM JOHN, b Reading, Pa, Feb 16, 14; m 37; c 4. ORGANIC CHEMISTRY, FORENSIC CHEMISTRY. *Educ:* Albright Col, BS, 36; Columbia Univ, PhD(org chem), 41. *Prof Exp:* Res chemist, Am Cyanamid Co, Conn, 40-41; chief chemist, George W Bollman & Co, 41-42, res dir, 42-52; pres, Pagoda Industs, Inc, 51-56; PRES, M J REIDER ASSOCS INC, 56- *Concurrent Pos:* Indust consult & consult chemist, 52-56; pres, Nat Stand Test Labs, Inc, 58- *Mem:* AAAS; Am Chem Soc; Am Acad Forensic Sci; Am Soc Qual Control; Am Asn Textile Chem & Colorists. *Res:* Synthetic tanning agents; deodorizing of tung oil; surface active agents; studies in the pyridine series; technology of long vegetable fibers; air, land and water pollution and control; legal chemistry, arson and fire investigations. *Mailing Add:* 1115 Thrush Rd Reading PA 19610

REIDER, PAUL JOSEPH, b New York, NY, June 7, 51; m 74. SYNTHETIC ORGANIC CHEMISTRY. *Educ:* Washington Square Col, AB, 72; Univ Vt, PhD(org chem), 78. *Prof Exp:* NIH fel, Colo State Univ, 78-80; SR RES CHEMIST, MERCK, SHARP & DOHME RES LABS, 80- *Mem:* Am Chem Soc. *Res:* Synthesis of complex organic molecules of pharmacological interest; development of viable processes for such synthesis. *Mailing Add:* 621 Kimball Ave Westfield NJ 07090

REIDER, RICHARD GARY, b Denver, Colo, Feb 7, 41. PHYSICAL GEOGRAPHY. *Educ:* Univ Northern Colo, BA, 63, MA, 65; Univ Nebr, PhD(geog), 71. *Prof Exp:* Instr geog, Ind Univ, Pa, 65-66; instr, Univ Nebr, Lincoln, 66-69; from asst prof to assoc prof, 69-83, PROF GEOG, UNIV WYO, 83- *Concurrent Pos:* NSF grant & Sigma Xi grant, 74-83; soils consult, Dept of Anthrop, Univ Wyo & Smithsonian Inst, 75-85; ed, Great Plains-Rocky Mountain Geog J, 76-80. *Mem:* Am Quaternary Asn; Sigma Xi; Geol Soc Am. *Res:* Geomorphology and soils geography; geoarchaeology. *Mailing Add:* Dept Geog Univ Wyo Laramie WY 82071-3371

REIDIES, ARNO H, b Tilsit, Ger, July 31, 25; US citizen; m 53; c 3. INDUSTRIAL ELECTROCHEMISTRY. *Educ:* Univ Freiburg, Dipl, 54. *Prof Exp:* Res chemist, 54-58, chief res chemist, 58-64, res mgr, 64-68, dir res, 68-83, div technologist & consult, 83-90, PRIN SCIENTIST, CARUS CHEM CO, 90- *Mem:* Am Chem Soc; Electrochem Soc; Ger Chem Soc; Am Asn Textile Chemists & Colorists. *Res:* Chemistry of manganese, specifically manganese oxides, manganates and permanganates, quinones and hydroquinones; environmental chemistry of manganese and other transition metals. *Mailing Add:* Carus Chem Co Eighth St LaSalle IL 61301

REIDINGER, RUSSELL FREDERICK, JR, b Reading, Pa, June 19, 45; m 73; c 4. FISH & WILDLIFE SCIENCES. *Educ:* Albright Col, BS, 67; Univ Ariz, PhD(zool), 72. *Prof Exp:* Asst prof biol, Augustana Col, 71-74; res physiologist, Philippines, 74-78, asst mem, Monell Chem Senses Ctr & wildlife biologist, 78-86, DIR, DENVER WILDLIFE RES CTR, US FISH & WILDLIFE SERV, PHILADELPHIA, 87- *Concurrent Pos:* Vis prof, Dept Zool, Univ Philippines, 75-78; consult, Bangladesh Agr Res Coun, USAID, 77, Ministry Agr Develop & Agrarian Reform, Nicaragua, 81. *Mem:* Am Soc Mammalogists. *Mailing Add:* Denver Wildlife Res Ctr Fed Ctr Bldg 16 PO Box 25266 Denver CO 80225-0266

REIDLINGER, ANTHONY A, b Islip, NY, Nov 30, 26; m 59. GENERAL CHEMISTRY. *Educ:* Hofstra Univ, BA, 49; NY Univ, MS, 51, PhD(org chem), 55. *Prof Exp:* Assoc eng scientist, Res Div, Col Eng, NY Univ, 55-57; assoc prof org chem, Grad Sch, St John's Univ, NY, 56-60; from assoc prof to prof, 60-89, chmn dept chem, 66-80, EMER PROF ORG CHEM, LONG ISLAND UNIV, 89- *Concurrent Pos:* Consult, Evans Chemetics, 59-60. *Mem:* Am Chem Soc; Electrochem Soc; Sigma Xi. *Res:* Polarography of organic compounds; nitro derivatives of naphthalene. *Mailing Add:* Dept Chem 24-71 Bellmore Ave Bellmore NY 11710

REIDY, JAMES JOSEPH, b Tulsa, Okla, June 12, 36; m 62; c 3. ANTIPROTON INTERACTIONS. *Educ:* Univ Notre Dame, BS, 58, PhD(nuclear physics), 63. *Prof Exp:* Instr physics, Univ Notre Dame, 62-63; res assoc, Univ Mich, 63-65, asst prof, 65-71; vis prof, Franklin & Marshall Col, 71-72; assoc prof, 72-82, PROF PHYSICS, UNIV MISS, 82-, ACTG CHMN, 87- *Concurrent Pos:* Vis prof, Tech Univ, Munich, Ger, 76-77; vis staff mem, Los Alamos Nat Lab, 72-; res assoc, Cern, Geneva, Switz, 85; mem, Working Group on State Initiatives in Appl Res, Nat Gov Assoc, 88- *Mem:* Am Asn Physicists Med; Am Phys Soc; Am Asn Physics Teachers; Sigma Xi. *Res:* Physics of negative muon and pion capture by atoms; interaction of negative pions and antiprotons at near rest with nuclei; chemical effects in negative pion and muon atomic capture; determination of nuclear decay schemes. *Mailing Add:* Dept Physics & Astron Univ Miss University MS 38677

REIERSON, JAMES (DUTTON), b Seward, Nebr, Oct 14, 41; m 75. NUCLEAR PHYSICS, OPERATIONS RESEARCH. *Educ:* Univ Nebr, Lincoln, BS, 63; Iowa State Univ, PhD(physics), 69; Marymont Univ, Va, MBA, 84. *Prof Exp:* Systs analyst, Analytical Serv, Inc, 69-73; lectr physics, Univ South Pac, 73-75; comput scientist, Comput Sci Corp, 76-78; LEAD ENGR, MITRE CORP, 78- *Concurrent Pos:* Mem, Inst Phys, George Mason Univ, 79. *Mem:* Am Phys Soc; Am Asn Artificial Intel. *Res:* Systems engineering; analysis and application of artificial intelligence for air traffic control, energy and military problems. *Mailing Add:* 3311 N George Mason Dr Arlington VA 22207

REIF, ARNOLD E, b Vienna, Austria, July 15, 24; US citizen; m 50, 79; c 3. IMMUNOLOGY. *Educ:* Cambridge Univ, BA, 45, MA, 49; Univ London, BSc, 46; Carnegie-Mellon Univ, MS, 49. *Hon Degrees:* DSc, Carnegie-Mellon Univ, 50. *Prof Exp:* From jr sci officer to sci officer, Dept Sci & Indust Res, Gt Brit, 44-47; res fel, Carnegie-Mellon Univ, 47-50; McArdle Mem Lab

Cancer Res fel oncol, Sch Med, Univ Wis, 50-53; res assoc, Lovelace Found, NMex, 53-57; asst prof surg & biochem, 57-69, assoc prof surg, Sch Med, Tufts Univ, 69-75; RES PROF PATH, SCH MED, BOSTON UNIV, 75- *Concurrent Pos:* Res pathologist, Mallory Inst Path, Boston City Hosp, 73- *Mem:* Am Asn Immunol; Am Asn Cancer Res; NY Acad Sci. *Res:* Cancer research; immunology; health education. *Mailing Add:* 39 College Rd Wellesley MA 02131

REIF, CHARLES BRADDOCK, b Washington, DC, July 31, 12; m 47. ZOOLOGY. *Educ:* Univ Minn, BA, 35, MA, 38, PhD(zool), 41. *Prof Exp:* Asst zool, Univ Minn, 35-41; cur educ, Mus Natural Hist, Minn, 41-42; from asst prof to prof, 42-85, EMER PROF ZOOL, WILKES COL, 85- *Concurrent Pos:* At US Forest Serv, 36 & 38. *Mem:* Am Soc Limnol & Oceanog; Phycol Soc Am; Micros Soc Am; Sigma Xi. *Res:* Limnology; biology. *Mailing Add:* Dept Biol Wilkes Univ Wilkes-Barre PA 18766

REIF, DONALD JOHN, b Oshkosh, Wis, Feb 6, 31; m 52; c 2. NUCLEAR CHEMISTRY. *Educ:* Univ Wis, BS, 53; Mass Inst Technol, PhD(org chem), 57. *Prof Exp:* Asst, Mass Inst Technol, 53-57; res chemist, 57-65, sr res chemist, Old Hickory, Tenn, 65-77, staff chemist, Res & Develop Labs, 77-83, RES STAFF CHEMIST, SAVANNAH RIVER LAB, E I DU PONT DE NEMOURS & CO, INC, AIKEN, SC, 83- *Res:* Spent nuclear fuel reprocessing. *Mailing Add:* 722 Ravenel Rd Augusta GA 30909

REIF, FREDERICK, b Vienna, Austria, Apr 24, 27; nat US; m 69. PHYSICS, SCIENCE EDUCATION. *Educ:* Columbia Univ, AB, 48; Harvard Univ, AM, 49, PhD(physics), 53. *Prof Exp:* From instr to asst prof physics, Univ Chicago, 53-60; assoc prof, Univ Calif, Berkeley, 60-64, Miller prof, 64-65, prof physics, 64-89, prof educ, 83-89; PROF PHYSICS & PSYCHOL, CARNEGIE-MELLON UNIV, 89- *Concurrent Pos:* Alfred P Sloan fel, Univ Chicago, 55-59. *Mem:* AAAS; Am Phys Soc; Am Educ Res Asn; Cognitive Sci Soc. *Res:* Nuclear magnetic resonance; solid state and low temperature physics; superconductivity; superfluidity of liquid helium; educational research and development; cognitive science. *Mailing Add:* CDEC Carnegie-Mellon Univ Pittsburgh PA 15213

REIF, JOHN H, b Madison, Wis, Aug 4, 51; m. COMBINATORIAL ALGORITHMS, PARALLEL ALGORITHMS. *Educ:* Harvard Univ, PhD(comput sci), 77. *Prof Exp:* Assoc prof comput sci, Harvard Univ, 79-86; PROF COMPUT SCI, DUKE UNIV, 86- *Mem:* Am Math Soc; Asn Comput Mach; Soc Indust & Appl Math. *Res:* Games, robotics, combinatorics, graphs and program optimization; algorithms for real time synchronization of parallel distributed computer systems, using probabilistic methods; parallel algorithms for sorting, searching and solution of linear systems; optical computing; robotic movement planning; data compression. *Mailing Add:* Comput Sci Dept Duke Univ Durham NC 27706

REIF, L RAFAEL, b Maracaibo, Venezuela, Aug 21, 50; div; c 1. SEMICONDUCTORS, INTEGRATED CIRCUITS. *Educ:* Univ Carabobo, Venezuela, Ingeniero Electrico, 73; Stanford Univ, MS, 75, PhD(elec eng), 79. *Prof Exp:* Asst prof elec eng, Univ Simon Bolivar, Venezuela, 73-74; vis asst prof elec eng, Stanford Univ, 78-79; from asst prof to assoc prof, 80-88, PROF ELEC ENG, MASS INST TECHNOL, 88- *Concurrent Pos:* Consult, Mass Inst Technol, Lincoln Lab, SPIRE Corp, Digital Infrared Industs, Aerodyne Res, Gen Motors Res Labs, M/A Com & Advantage Corp; Presidential Young Investr Award, NSF, 84-89. *Mem:* Electrochem Soc; Inst Elec & Electronics Engrs; Metall Soc-Am Inst Mining, Metall & Petrol Engrs; Mat Res Soc. *Res:* Fabrication technology of advanced integrated circuits, eg low temperature thin film epitaxial technology for Si, Si-Ge, GaAs, GaP, IC manufacturing equipment; polysilicon thin film transistor technology. *Mailing Add:* Mass Inst Technol Rm 39-321 Cambridge MA 02139

REIF, THOMAS HENRY, biomechanical engineering, fluid mechanics, for more information see previous edition

REIF, VAN DALE, b Burlington, Iowa, Feb 18, 47; m 81; c 1. PHARMACEUTICAL CHEMISTRY. *Educ:* Univ Iowa, BS, 70; Univ Mich, MS, 71, PhD(pharmaceut chem), 75. *Prof Exp:* Sr sci assoc pharmaceut anal, Drug Res & Testing Lab, US Pharmacopeia, 75-78; unit supvr, 78-85, MGR, ANALYTICAL RES & DEVELOP, WYETH LABS, 85- *Mem:* Am Chem Soc; Am Asn Pharmaceut Scientists. *Res:* Pharmaceutical analysis, drug metabolism, analytical chemistry, chromatography. *Mailing Add:* K 11 3 Schering Plough 2000 Gallopeng Hill Rd Kenilworth NJ 07033-1310

REIFENBERG, GERALD H, b Brooklyn, NY, Jan 18, 31; m 60; c 2. ORGANOMETALLIC CHEMISTRY. *Educ:* City Col New York, BS, 55; NY Univ, PhD(chem), 62. *Prof Exp:* Analytical chemist, Refining Unincorp, 55-57; chemist, NY Naval Shipyard, 57-58; teaching fel org chem, NY Univ, 58-59; chemist, Gen Chem Div, Allied Chem Corp, 61-62; res chemist, M&T Chem, Inc, NJ, 63-68, supvr specialty chem, 68-69; scientist, Am Cyanamid Co, 69-71; dir res, Magic Marker Corp, 71-73; res chemist, 73-78, TECH MGR PLASTIC ADDITIVES, PENNWALT CORP, 79- *Concurrent Pos:* Instr, Rutgers Univ, 64-65 & Newark Col Eng, 65-67. *Mem:* Am Chem Soc; Sigma Xi. *Res:* Organometallic chemistry, especially organotins; radioisotopes in tracer work; organofluorine, organophosphorus and flame retardant chemistry. *Mailing Add:* Pennwalt Corp 900 First Ave King of Prussia PA 19406

REIFENRATH, WILLIAM GERALD, b Crofton, Nebr, Mar 2, 47; m 79. DERMATOLOGY. *Educ:* Univ Nebr-Lincoln, BS, 69, MS, 72, PhD(med chem), 75. *Prof Exp:* RES CHEMIST, DIV CUTANEOUS HAZARDS, LETTERMAN ARMY INST RES, 76- *Mem:* Am Chem Soc. *Res:* Interaction of chemicals with skin; animal and in vitro models and analytical procedures for studying mechanisms of percutaneous penetration. *Mailing Add:* Six Christopher Ct Novato CA 94947-2831

REIFF, GLENN AUSTIN, b Newton, Kans, Nov 18, 23; m 47; c 2. ELECTRONICS ENGINEERING. *Educ:* US Naval Acad, BS, 45; US Naval Postgrad Sch, BS, 52, MS, 53. *Prof Exp:* Officer, USN, 42-61; prog mgr & engr, NASA, 62-70; engr, US Dept Transp, 71-78; PROF IN RESIDENCE ENG TECH, UNIV SOUTHERN COLO, 78- *Honors & Awards:* Except Serv Award, NASA, 71. *Mem:* Sr mem Inst Elec & Electronics Engrs. *Mailing Add:* 59 Villa Dr Pueblo CO 81001

REIFF, HARRY ELMER, b Allentown, Pa, Apr 19, 24; m 47; c 3. ORGANIC CHEMISTRY. *Educ:* Lehigh Univ, BSChE, 49, MS, 50; Univ Minn, PhD(org chem), 55. *Prof Exp:* Asst, Lehigh Univ, 49-50; jr chemist, Merck & Co, Inc, 50-52,; asst, Univ Minn, 52-54; sr chemist, Smith Kline & French Labs, 55-58, group leader, 58-60, asst sect head, 60-62, head org chem sect, 62-67, dir chem support & lab animal sci, Res & Develop Div, 67-72, dir phys sci mfg & planning, 72-78, dir tech planning, 78-81, dir tech assurance, 81-84; consult, 84-86; RETIRED. *Mem:* Fel AAAS; Am Chem Soc; fel Am Inst Chem; Sigma Xi. *Res:* Physical sciences research and development involving pharmaceuticals and drugs, particularly synthetic organic chemistry. *Mailing Add:* 1601 Clair Martin Place Ambler PA 19002

REIFF, PATRICIA HOFER, b Oklahoma City, Okla, Mar 14, 50; m 76. MAGNETOSPHERIC PHYSICS. *Educ:* Okla State Univ, BS, 71; Rice Univ, MS, 74, PhD(space physics, astron), 75. *Prof Exp:* Res assoc magnetospheric physics, Rice Univ, 75; Nat Acad Sci-Nat Res Coun resident res assoc magnetospheric physics, Marshall Space Flight Ctr, NASA, 75-76; asst prof magnetospheric physics, 76-81, ASSOC RES SCIENTIST, CTR SPACE PHYSICS, RICE UNIV, 81-, ASST CHMN, DEPT SPACE PHYSICS & ASTRON, 79- *Concurrent Pos:* US deleg, Int Union Geod & Geophys, 75 & 81; fac assoc, Nat Res Coun-Nat Acad Sci, 79, mem comt solar-terrestrial res, 79- *Mem:* Am Geophys Union; AAAS; Sigma Xi; Audubon Soc. *Res:* Study of solar wind, magnetosphere and ionosphere interactions, theoretically and using on atmosphere explorer satellites -C and -D; low-energy electron experiment data and Apollo 14 charged particle data; co-investigator on high altitude plasma instrument on dynamic explorer spacecraft; computer simulation of magnetospheric convertion. *Mailing Add:* Dept Space Physics & Astron 4214 Southwestern St Houston TX 77005

REIFF, WILLIAM MICHAEL, b Binghamton, NY, Mar 9, 42; m 67; c 2. INORGANIC CHEMISTRY. *Educ:* State Univ NY Binghamton, AB, 64; Syracuse Univ, PhD(chem), 68. *Prof Exp:* NSF fel & fac assoc chem & physics, Univ Tex, Austin, 68-70; asst prof, 70-77, PROF CHEM, NORTHEASTERN UNIV, 78- *Concurrent Pos:* Vis scientist, Francis Bitter Nat Magnetic Lab. *Mem:* Am Chem Soc. *Res:* Physical inorganic chemistry; magnetically perturbed Mossbauer spectroscopy; magnetochemistry, electronic structure of coordination compounds and solid state materials. *Mailing Add:* Dept Chem Northeastern Univ Boston MA 02115-5096

REIFFEL, LEONARD, b Chicago, Ill, Sept 30, 27; m 71; c 2. PHYSICS. *Educ:* Ill Inst Technol, BSc, 47, MSc, 48, PhD, 53. *Prof Exp:* Physicist, Perkin-Elmer Corp, 48; eng physicist, Univ Chicago, Inst Nuclear Studies, 48-49; dir physics res, Ill Inst Technol Res Inst, 49-63, vpres physics div, 63-65; chmn bd, Instruct Dynamics Inc, 65-81; CHMN BD, INTERAND CORP, 69- *Concurrent Pos:* Consult & dep dir Apollo Pro, NASA Headquarters, 65-70, tech dir, Manned Space Flight Exper Bd, 66-70; sci ed, WBBM-CBS Radio Chicago, 71-72, host Backyard Safari, WBBM-TV, 71-73; sci ed & feature broadcaster, WEEI-CBS Radio, Boston, 65-75; syndicated news columnist, Universal Sci News, 66-72, Los Angeles Times Syndicate, 72-78; sci consult, CBS Network, 67-71; consult, Korean Govt Estab Atomic Energy Res Prog, 58-60, US Army, 79. *Honors & Awards:* IR-100 Award, 70, 72, 73, 85. *Mem:* Fel Am Phys Soc; AAAS; Sigma Xi. *Res:* Author of book. *Mailing Add:* 602 Deming Pl Chicago IL 60614

REIFFEN, BARNEY, b Brooklyn, NY, Oct 5, 27; m 50; c 2. ELECTRICAL ENGINEERING. *Educ:* Cooper Union, BS, 49; Polytech Inst Brooklyn, MS, 53; Mass Inst Technol, PhD(elec eng), 60. *Prof Exp:* Engr, Harvey-Wells Electronics, Inc, 51-53 & Balco Res Labs, 53-55; staff mem, 55-61, group leader, 61-73, DIV HEAD, LINCOLN LAB, MASS INST TECHNOL, 73- *Concurrent Pos:* Prin assoc med, Beth Israel Hosp, Harvard Med Sch, 69-73. *Mem:* Sr mem Inst Elec & Electronics Engrs. *Res:* Satellite communications systems; information systems. *Mailing Add:* 26 Peacock Farm Rd Lexington MA 02173

REIFFENSTEIN, RHODERIC JOHN, b Montreal, Que, Nov 1, 38; m 61; c 4. PHARMACOLOGY. *Educ:* McGill Univ, BSc, 59; Univ Man, PhD(pharmacol), 65. *Prof Exp:* Teaching fel pharmacol, Univ Man, 59-64; from asst prof to assoc prof, 64-83, PROF PHARMACOL, UNIV ALTA, 83- *Concurrent Pos:* Med Res Coun Can vis scientist, Sch Pharm, Univ London, 71-72; vis prof, Anaesthesia Res, McGill Univ, 79-80. *Mem:* NY Acad Sci; European Soc Neurochem; Brit Pharmacol Soc; Pharmacol Soc Can; Soc Neurosci; Can Asn Neurosci. *Res:* Release of amino acids from hippocampus; Alzheimer's disease; cellular mechanism of focal cortical epilepsy; toxicity of sedative/hypnotic and ethanol combinations; alpha 2 adrenoreceptors in locus coeruleus; behavioral and single cell actions of amphetamine-hallucinogens; computer assisted instruction of pharmacology; hydrogen sulfide neurotoxicity. *Mailing Add:* Dept Pharmacol Univ Alta Edmonton AB T6G 2H7 Can

REIF-LEHRER, LIANE, b Vienna, Austria, Nov 14, 34; US citizen; m 60; c 2. BIOCHEMISTRY. *Educ:* Barnard Col, Columbia Univ, BA, 56; Univ Calif, Berkeley, PhD(phys org chem), 60. *Prof Exp:* Staff scientist, Res & Adv Develop Div, Avco Corp, Mass, 60-62; res fel biochem, Harvard Univ, 63-66, instr ophthal res, 66-71, from asst to assoc prof biochem ophthal, Harvard Med Sch, 71-85, dir Off Acad Careers, 81-82; PRES, ERIMON ASSOCS, 85- *Concurrent Pos:* NIH fel, 64-66, res grant, 68-85; staff scientist, Boston Biomed Res Inst, Harvard Univ, 72-85; sr scientist, Eye Res Inst, 75-85; mem grant rev bd, NIH, 76-78. *Mem:* Asn Res Vision & Ophthal; NY Acad Sci; AAAS; Am Soc Biol Chemists. *Res:* Effects of excitotoxic amino acids on neural retina; control mechanisms in animal cells; biochemical controls in the normal and diseased retina; effects of ingestion of excess monosodium glutamate on humans; grant writing, expository writing, time management writing and author of over 50 publications. *Mailing Add:* Erimon Assocs PO Box 645 Belmont MA 02178

REIFLER, CLIFFORD BRUCE, b Chicago, Ill, Dec 28, 31; m 54; c 3. PSYCHIATRY, MEDICAL ADMINISTRATION. *Educ:* Univ Chicago, AB, 51; Northwestern Univ, Evanston, BS, 53; Yale Univ, MD, 57; Univ NC, Chapel Hill, MPH, 67. *Prof Exp:* USPHS fel & resident psychiat, Strong Mem Hosp, Rochester, NY, 58-61; from instr to assoc prof psychiat, Sch Med, Univ NC, Chapel Hill, 63-70, from asst prof to assoc prof ment health, Sch Pub Health, 68-70; PROF HEALTH SERV, PSYCHIAT, PREV MED & COMMUNITY HEALTH & DIR UNIV HEALTH SERV, UNIV ROCHESTER, 70- *Concurrent Pos:* Assoc physician-in-chg psychiat, Student Health Serv, Univ NC, Chapel Hill, 63-67, sr psychiatrist, 67-70; consult psychiat, Rochester Inst Technol, 76-83; interim vpres student affairs, Univ Rochester, 80-81; mem vis comt, Harvard Univ Health Serv, 78-84; med dir, Strong Mem Hosp, 83-85; assoc dean clin affairs, Sch Med & Dent, Univ Rochester, 83-85, actg chmn dept health serv, 83-85. *Honors & Awards:* Edward Hitchcock Award, Am Col Health Asn, 81; Ruth Boynton Award, Am Col Health Asn, 88. *Mem:* Am Col Physician Execs; fel Am Pub Health Asn; fel Am Psychiat Asn; fel Am Col Health Asn (pres, 76-77); Am Acad Med Dirs; Int Union Sch & Univ Health & Med (vpres, 85-88). *Res:* Psychiatric epidemiology; health care delivery systems. *Mailing Add:* Univ Rochester Rochester NY 14627

REIFSCHNEIDER, WALTER, b Vienna, Austria, July 29, 26; nat US; m 59; c 1. ORGANIC CHEMISTRY. *Educ:* Univ Vienna, PhD(chem), 53. *Prof Exp:* Res assoc, Univ Ill, 53-57; res chemist, Dow Chem Co, 57-61, group leader, 61-70, assoc scientist, 70-81, res scientist, 81-90; FEL, DOWELANCO, 91- *Mem:* AAAS; Am Chem Soc; fel Sci Res Soc Am. *Res:* General organic synthesis; heterocycles; sulfur compounds; pesticide chemistry; natural products. *Mailing Add:* 3538 Bayberry Dr Walnut Creek CA 94598

REIFSNIDER, KENNETH LEONARD, b Baltimore, Md, Feb 19, 40; m 63; c 2. MATERIAL SCIENCE, MECHANICS. *Educ:* Western Md Col, BA, 63; Johns Hopkins Univ, BES, 64, MSE, 65, PhD(metall), 68. *Prof Exp:* From asst prof to assoc prof eng mech, 68-75, PROF ENG MECH, VA POLYTECH INST & STATE UNIV, 75-, CHMN MAT ENG SCI PROG, 71- *Concurrent Pos:* NATO mat sci consult, 69 & 78; sabbatical leave, Univ Calif, Livermore, 81. *Mem:* Am Inst Mining, Metall & Petrol Engrs; Am Soc Metals; Am Soc Testing & Mat. *Res:* Composite materials; nondestructive testing and evaluation; mechanics of inhomogeneous deformation including fatigue and fracture; continuum theories of material defects. *Mailing Add:* ESM Dept Va Polytec Inst State Univ Blacksburg VA 24061

REIFSNYDER, WILLIAM EDWARD, b Ridgway, Pa, Mar 29, 24; m 54; c 3. FOREST METEOROLOGY. *Educ:* NY Univ, BS, 44; Univ Calif, MF, 49; Yale Univ, PhD(forest meteorol), 54. *Prof Exp:* Meteorologist, Calif Forest & Range Exp Sta, 51-54; from asst prof to assoc prof, 55-65, prof forest meteorol, 65-90, pub health, 67-90, EMER PROF FOREST METEOROL, YALE UNIV, 90- *Concurrent Pos:* Mem & chmn adv comt climate, US Weather Bur, Nat Acad Sci-Nat Res Coun, 57-63, mem sci task force atmospheric sci, 61, panel educ, Comt Atmospheric Sci, 62-64, chmn adv comt biometeorol; vis scientist, Soc Am Foresters-NSF, 61-65 & 70, assoc ed, Canadian J Forest Res, 71-; vis prof, Univ Munich, 68; consult forest meteorol, World Meteorol Orgn, 73, chmn working group on the applications of meteorol to forestry, 75-80; vis scientist, Swed Univ Agr Sci, 81; sr res sci, Nat Oceanic & Atmospheric Admin, 84-85, vpres, Int Soc Biometrol, 84-88; ed-in-chief, Agr & Forest Metrol Int J, 80- *Honors & Awards:* Achievement Award, Am Meteorol Soc. *Mem:* Fel AAAS; Soc Am Foresters; Am Meteorol Soc; Int Soc Biometeorol; Sigma Xi. *Res:* Energy budgets; air pollution meteorology. *Mailing Add:* Lama Star Rte Box 3 Questa NM 87556

REIGEL, EARL WILLIAM, b Summit Hill, Pa, Aug 18, 35; m 55; c 6. COMPUTER ARCHITECTURE, SYSTEM INTEGRATION. *Educ:* Pa State Univ, BS, 59, MS, 61; Univ Pa, PhD, 69. *Prof Exp:* Res asst comput software & hardware, Pa State Univ, 59-61; sr engr, Burroughs Corp, 61-64; sr programmer comput software, Gen Elec, 64-67; mgr comput archit, Adv Develop, Burroughs Unisys Syst Eng, 67-80, dir comput res, Paoli Res Ctr, 80-86, dir comput eng, 86-88, DIR ADVAN TECHNOL, UNISYS SYST ENG, 88- *Concurrent Pos:* Adj prof, Univ Pa, 72-82 & Villanova Univ, 78-82; bd adv, Widener Univ, 81-87; eval bd, Adv Tech Ctr Southeast Pa, 85-88; nat lectr, Asn Comput Mach, 72; distinguished visitor, Inst Elec & Electronics Engrs, 72. *Mem:* Asn Comput Mach; fel Inst Elec & Electronics Engrs. *Res:* Computer architecture: parallelism-algorithms and software to identify parallelism and hardware to exploit it; microprogramming-hardware architectures and their use: emulation and application adaptation; artificial intelligence and application specific languages. *Mailing Add:* 1417 Highland Ave Downingtown PA 19335

REIHER, HAROLD FREDERICK, b Detroit, Mich, July 8, 27; m 53; c 3. ACOUSTICS. *Educ:* Univ Mich, BS, 50. *Prof Exp:* Lab asst, Eng Res Inst, Univ Mich, 49-51, res assoc, 51-56, assoc res engr, 56-57; res engr, 57-64, vpres, 64-89, PRES, GEIGER & HAMME, INC, 89- *Mem:* Inst Elec & Electronics Engrs; Am Soc Test & Mat. *Res:* Architectural acoustics; measurement and control of sound and vibration; evaluation of acoustical properties and performance of materials, structures and equipment; electronic instrumentation. *Mailing Add:* 1835 Knight Rd Ann Arbor MI 48103-9303

REILING, GILBERT HENRY, b St Paul, Minn, Sept 19, 28; m 51; c 8. ENGINEERING PHYSICS. *Educ:* Col St Thomas, BS, 51; Univ NDak, MS, 52; Univ Mo, PhD(physics), 57. *Prof Exp:* Instr physics, Univ Mo, 53; physicist, 57-61, MGR ENG, LIGHTING RES & TECH SERV, GEN ELEC CO, 61- *Concurrent Pos:* Lectr, Siena Col, 58- *Mem:* Am Phys Soc; Inst Elec & Electronics Eng; fel Illum Eng Soc. *Res:* Optical properties of solid state materials; fundamental processes in low pressure arcs. *Mailing Add:* Lighting Res & Tech Serv O 1301 Gen Elec Co Nela Park Cleveland OH 44112

REILLY, BERNARD EDWARD, b Meadville, Pa, June 9, 35; m 70. PLASMID BIOLOGY, VIROLOGY. *Educ:* Westminster Col, BS, 58; Case Western Reserv Univ, PhD(microbiol), 65. *Prof Exp:* Res fel microbiol, Univ Minn, Minneapolis, 62-65; res assoc microbiol, Scripps Clin & Res Found, 65-68; asst prof biol, NMex State Univ, 68-69; vis prof genetics, 69-71, asst prof, 72-75, ASSOC PROF MICROBIOL, SCH DENT, UNIV MINN, MINNEAPOLIS, 75- *Res:* Bacteriophage and microbial genetics, viral and post transcriptional function modification of proteins; viral assemlidy; viral assemlidy. *Mailing Add:* 632 Third Ave Minneapolis MN 55414

REILLY, CHARLES AUSTIN, b Summerside, PEI, May 18, 16; nat US; m 42. CHEMICAL PHYSICS. *Educ:* Dalhousie Univ, BSc, 39, MSc, 40; Harvard Univ, AM, 46, PhD(chem physics), 50. *Prof Exp:* Asst res physicist, Nat Res Coun Can, 40-43; asst prof chem, Dalhousie Univ, 48-51; physicist, Belleaire, Shell Develop Co, 51-80, sr staff res physicist, 80-85; RETIRED. *Mem:* Fel Am Phys Soc; Am Chem Soc; fel Am Inst Chemists. *Res:* Molecular beams; nuclear magnetic resonance; field emission spectroscopy. *Mailing Add:* 2622 Country Club Blvd Sugar Land TX 77478

REILLY, CHARLES BERNARD, b New York, NY, Dec 7, 29; m 55; c 6. ORGANIC CHEMISTRY, POLYMER CHEMISTRY. *Educ:* Queen's Col, NY, BS, 53; Univ Cincinnati, MS, 55, PhD(chem), 57. *Prof Exp:* Res & develop chemist, Gen Elec Co, 57-65, sr chemist, 65-66, mgr process develop, 66-69; group leader, 69-80, SR CHEMIST RUBBER COMPOUNDINGS, URETHANE APPLNS, GOODYEAR TIRE & RUBBER CO, 80- *Mem:* Am Chem Soc; AAAS. *Res:* Polymer chemistry specializing in thermoset resins as epoxies and urethanes; organic synthesis of polymer intermediates; physical properties of polymers. *Mailing Add:* 3256 Linden St Uniontown OH 44685

REILLY, CHARLES CONRAD, b Cornwall, NY, Dec 3, 40; m 61; c 5. HOST-PARASITE INTERACTIONS. *Educ:* Auston Peay State Univ, BS, 71; Univ Ill, MS, 74, PhD(plant path), 77. *Prof Exp:* Grad Res Asst plant path, dept plant path, Univ Ill, 71-77, res assoc, dept hort, 77-78; res assoc, USDA, Am Res Serv, Metab & Radiation Res Lab, 78-80; RES PLANT PATHOL, USDA, SOUTHEASTERN FRUIT & TREE NUT RES LAB, BYRON, GA, 80- *Concurrent Pos:* Adj fac, Fort Valley State Col, Ga, 80-; Adj prof, dept plant path, Univ Ga, Athens, 83- *Mem:* Sigma Xi; Am Phytopath Soc. *Res:* Production limiting problems of pecan trees; chemical or biological control of diseases and insects; physiological interactions of the host and parasites; diseases of pecan. *Mailing Add:* USDA-ARS Southeastern Fruit & Tree Nut Res Lab Byron GA 31008

REILLY, CHRISTOPHER ALOYSIUS, JR, b Tucson, Ariz, Aug 8, 42; m 68; c 3. VIROLOGY. *Educ:* Loyola Univ, Los Angeles, BS, 64; Univ Ariz, MS, 66, PhD(microbiol), 68. *Prof Exp:* Fel virol, Argonne Nat Lab, 68-69, asst microbiologist, 69-73, dep prog mgr, Synfuels Environ Res Prog, 80-85, assoc div dir & dir, Ctr Environ Res, 87-89, MICROBIOLOGIST, ARGONNE NAT LAB, 73- & DIV DIR, ENVIRON RES DIV, 89- *Concurrent Pos:* Adj assoc prof biol sci, Northern Ill Univ, 75- *Mem:* Am Soc Microbiol; Soc Toxicol; Sci Exp Biol & Med; AAAS; Sigma Xi. *Res:* Characterization of the environmental effects of chemical pollutants, with particular emphasis on carcinogenesis and chemical toxicology. *Mailing Add:* Environ Res Div Argonne Nat Lab 9700 S Cass Ave Argonne IL 60439

REILLY, CHRISTOPHER F, PHARMACOLOGY. *Prof Exp:* RES FEL, MERCK SHARP & DOHME RES LABS, 87- *Mailing Add:* Dept Pharmacol Merck Sharp & Dohme Res Labs WP42-300 West Point PA 19486

REILLY, EDWIN DAVID, JR, b Troy, NY, Apr 27, 32; m 54; c 6. COMPUTER SCIENCE, PHYSICS. *Educ:* Rensselaer Polytech Inst, BS, 54, MS, 58, PhD(physics), 69. *Prof Exp:* Mathematician, Knolls Atomic Power Lab, Gen Elec Co, 56-61, mgr digital anal & comp, 61-65; dir comput ctr, 65-70, chmn dept, 67-73, ASSOC PROF COMPUT SCI, STATE UNIV NY ALBANY, 67- *Concurrent Pos:* Supvr, Town of Niskayuna, NY, 70-79, 89-; pres, Cybernetic Info Systs, 80- *Mem:* AAAS; Asn Comput Mach; Am Phys Soc; Inst Elec & Electronics Engrs; Sigma Xi; Math Asn Am. *Res:* Application of computers to the humanities; computer organization; cryptography; scattering of electromagnetic waves from nonspherical targets. *Mailing Add:* Dept Comput Sci Rm Li67 State Univ NY Albany NY 12222

REILLY, EMMETT B, b Los Angeles, Calif, Aug 19, 20; m 55; c 6. PATHOLOGY, HEMATOLOGY. *Educ:* Loyola Univ, Calif, BS, 42; Univ Southern Calif, MD, 46; Am Bd Path, dipl, 52, cert clin path, 56, cert hemat, 63, cert radioisotopic path, 74. *Prof Exp:* Chief clin path, Vet Admin Hosp, Long Beach, 50-57; chief pathologist, Orange County Hosp, 57-65; chief pathologist, Daniel Freeman Hosp, 65-90; CLIN PROF, UNIV SOUTHERN CALIF, 65- *Concurrent Pos:* Consult, Vet Admin Hosp, Long Beach, Calif, 57- *Mem:* Fel Am Soc Clin Pathologists; NY Acad Sci; fel Col Am Pathologists. *Res:* Coagulation of blood; neoplasia. *Mailing Add:* 7000 Arizona Ave Los Angeles CA 90045

REILLY, EUGENE PATRICK, b New York, NY, Mar 12, 39; m 69; c 2. ORGANIC CHEMISTRY. *Educ:* St Peter's Col, NJ, BS, 63; Fordham Univ, PhD(org chem), 68. *Prof Exp:* Res chemist plastics, acrylics & acetals, E I du Pont de Nemours Co, Inc, Wilmington, 67-71; develop chemist polyesters, Plastics Dept, Gen Elec Co, Pittsfield, Mass, 71-73; res scientist terpen chem, Union Camp Corp, Princeton, 73-76; SR RES SCIENTIST RESOURCE RECOVERY, AM CAN CO, 76- *Mem:* Am Inst Chem Engrs; Tech Asn Pulp & Paper Indust. *Res:* Plastics especially acrylics, acetals, polyesters, thermosets; laboratory to plant scale; upgrading of natural products especially terpenes and talloils to useful products; resource recovery, utilization of natural organic waste materials from paper making processes. *Mailing Add:* 243 Glenn Ave Trenton NJ 08648

REILLY, FRANK DANIEL, b Fairborn, Ohio, Aug 20, 49; m 69; c 1. HUMAN ANATOMY, MEDICAL PHYSIOLOGY. *Educ:* Ohio State Univ, BS, 71; Univ Cincinnati, PhD(anat), 75. *Prof Exp:* Asst prof anat, Col Med, Univ Cincinnati, 75-78; assoc prof, 78-85, PROF ANAT, SCH MED, WVA UNIV, 85- *Mem:* Am Asn Anatomists; Microvascular Soc; Sigma Xi; AAAS. *Res:* Hematology and microvascular physiology; morphology and pharmacology. *Mailing Add:* Dept Anat WVa Univ Morgantown WV 26506

REILLY, GEORGE JOSEPH, physical organic chemistry, for more information see previous edition

REILLY, HILDA CHRISTINE, b New Brunswick, NJ, Feb 25, 20. MICROBIOLOGY. *Educ:* Rutgers Univ, BSc, 41, PhD(microbiol), 46. *Prof Exp:* Lab technician chem, E R Squibb & Sons, NJ, 41-42; chemist, White Labs, Inc, 42-43; asst, Rutgers Univ, 46-47; asst prof microbiol, Sloan-Kettering Div, Med Col, Cornell Univ, 52-56, assoc prof, 56-71; mem faculty, Dept Bact, 71-75, ASSOC PROF BIOL SCI, DOUGLASS COL, RUTGERS UNIV, 75- *Concurrent Pos:* Asst, Sloan-Kettering Inst, 47-49, assoc, 49-60, assoc mem, 60-71, head microbiol sect, 56-71. *Mem:* AAAS; Am Soc Microbiol; Harvey Soc; Am Asn Cancer Res; fel Am Acad Microbiol; fel NY Acad Sci. *Res:* Production, isolation and mode of action of antibiotics; microbial physiology; experimental cancer chemotherapy. *Mailing Add:* Biol Sci Douglass Col New Brunswick NJ 08903

REILLY, HUGH THOMAS, b New York, NY, Apr 19, 25. PHYSICAL CHEMISTRY, CHEMICAL ENGINEERING. *Educ:* St Peter's Col, NJ, BS, 50; Polytech Inst NY, MS, 53. *Prof Exp:* Res & teaching fel, Polytech Inst Brooklyn, 50-51; proj engr, Thiokol Chem Corp, Elkton, Md, 57-59, mgr space propulsion systs, 59-60, mgr new prod develop, 60-63; tech adv & proj engr, Land Warfare Lab, US Army Chem & Armaments Mat Command, 63-74, chief lethal group, Edgewood Arsenal, 74-76, team leader pipeline gas prog, 76-78, chief, Dept Energy Support Off & Supvry Chem Engr, 78-81, dir, Environ Technol Chem Res & Develop Ctr, 81-87; PRIN INVESTR, RES ENG CORP, 87- *Mem:* Fel Am Inst Aeronaut & Astronaut; Am Chem Soc; fel Am Inst Chemists. *Res:* Atmospheric sensing; chemical process; chemical detection; pollution; cloud physics, coal process. *Mailing Add:* 425 Delaware Ave Elkton MD 21921

REILLY, JAMES PATRICK, b Mt Vernon, NY, Aug 29, 50. PHYSICAL CHEMISTRY. *Educ:* Princeton Univ, AB, 72; Cambridge Univ, CPGS, 73; Univ Calif, Berkeley, PhD(chem), 77. *Prof Exp:* Guest researcher, Max Planck Inst, Garching, 77-79; ASST PROF CHEM, IND UNIV, 79- *Concurrent Pos:* Alfred P Sloan fel, Alfred P Sloan Found, 82. *Mem:* Sigma Xi. *Res:* Probing optical transitions in transient species and excited molecular states; investigation of the chemistry of excited molecules; the interaction of radiation and matter. *Mailing Add:* Dept Chem Ind Univ Bloomington IN 47405

REILLY, JAMES PATRICK, b Jersey City, NJ, Nov 17, 37; m 62; c 4. AERONAUTICAL & ASTRONNAUTICAL ENGINEERING. *Educ:* Univ Detroit, BAE, 61; Mass Inst Technol, MS(aeronaut & astronaut) & MS(mech eng), 63, ScD, 67. *Prof Exp:* Res staff mem, Div Sponsored Res, Mass Inst Technol, 67-68; vpres appl technol, Avco Everett Res Lab, 68-78; VPRES SCI & ENG, W J SCHAFER ASSOC INC, 78- *Concurrent Pos:* Lectr, Mass Inst Technol, 72-; mem tech adv bd, Int Gas Flow & Chem Laser Conf, 78-90. *Honors & Awards:* Consejo Superior Investigaciones Cientificas, Spain, 90. *Mem:* Assoc fel Am Inst Aeronaut & Astronaut; Sigma Xi; Soc Photo Optical Instrumentation Engrs. *Res:* Experimental & theoretical aspects of high power repetitively; pulsed electrically excited gas lasers; charged particle accelerator technology; free electron lasers; particle beam technology; high power microwave emitters; radiometric sensors; laser radars; gas discharges. *Mailing Add:* 25 Walnut St Lexington MA 02173

REILLY, JAMES PATRICK, b Barnesboro, Pa, Nov 30, 33; m 56; c 4. GEODESY, SURVEYING. *Educ:* Pa State Univ, BS, 56; Ohio State Univ, MS, 69, PhD(geodetic sci), 74. *Prof Exp:* Assoc prof surv, Iowa State Univ, 74-77; nat prod mgr, Wild Heerbrugg Inst, Inc, 77-83; pres, Dudley-Reilly Assocs, 83-86 & Geodetic Enterprises, Inc, 86-90; PROF & DEPT HEAD SURV, NMEX STATE UNIV, 90- *Mem:* Am Cong Surv & Mapping; Am Soc Photogram & Remote Sensing; Inst Navig. *Res:* Surveying; author of various publications. *Mailing Add:* Dept 3SUR NMex State Univ Las Cruces NM 88003

REILLY, JAMES WILLIAM, b Jersey City, NJ, Oct 18, 35; m 56; c 4. CHEMICAL ENGINEERING, PHYSICAL CHEMISTRY. *Educ:* Seton Hall Col, BS, 56; Stevens Inst Technol, MS, 59, ScD(chem eng), 65. *Prof Exp:* Develop engr, Turbomotor Div, Curtiss-Wright Corp, 56-57; sr res engr, Tex-US Chem Co, 57-60; proj engr, Plastics Div, Koppers Co, 60; teaching asst, Mass Inst Technol, 60-61; asst prof chem & chem eng, Newark Col Eng, 61-65; sr develop engr, Bloomfield Div, Eng Develop Ctr, Lummus Co, Combustion Eng, Inc, 65-74, prin engr, 74-81, mgr proj eng, 81-86; MGR RES & DEVELOP, ABB LUMMUS CREST INC, 87- *Mem:* Am Inst Chem Engrs; Sigma Xi. *Res:* Engineering process research and development for proprietary processes in catalyst, hydrogenation and process design/scale up. *Mailing Add:* 1205 Boulevard Westfield NJ 07090

REILLY, JOSEPH F, b Waucoma, Iowa, May 14, 15; m 48; c 5. PHARMACOLOGY, TOXICOLOGY. *Educ:* Univ Ill, BA, 37; Harvard Univ, MA, 39; Univ Chicago, PhD(pharmacol), 47. *Prof Exp:* Chemist, Chem Res Dept, Armour Labs, Ill, 39-43; res assoc, Anti-Malarial Prog & asst pharmacol, Univ Chicago, 43-47; pharmacologist, US Army Chem Ctr, Md, 47-48; res fel, Med Sch, Cornell Univ, 48-49, from instr to asst prof pharmacol, 49-54, asst prof pharmacol & psychiat, 54-62; chief pharmacodyn sect, Div Pharmacol, Bur Sci, 63-70, CHIEF DRUG BIOANALYSIS BR, DIV RES & TESTING, CTR DRUG EVAL & RES, US FOOD & DRUG ADMIN, DC, 70- *Concurrent Pos:* Chief pharmacologist, Payne Whitney Clin, NY Hosp-Cornell Med Ctr, 54-62; consult, Coun Drugs, AMA, 58 & 65; collabr, US Pharmacopoeia Standards, 70; actg dep dir, Div Drug Biol, Bur Drugs, US Food & Drug Admin, Wash, DC, 78-79. *Mem:* Am Soc Pharmacol & Exp Therapeut; Soc Toxicol; Soc Exp Biol & Med; Harvey Soc; Sigma Xi. *Res:* Anti-malarials; plasma enzymes; diethylstilbestrol-enzymes; acetylstrophanthidin-fluoroacetate-heart; experimental arrythmias; catecholamines-psychiatric patients; reserpine; glutathione reductase-carbon tetrachloride; shellfish toxin; desmethylimipramine toxicity; age, sex, ulcers; age-catecholamines; organochlorine and monosodium glutamate effects. *Mailing Add:* Div Res & Testing Ctr Drug Eval & Res HFD 470 US Food & Drug Admin 200 C St SW Washington DC 20204

REILLY, KEVIN DENIS, b Omaha, Nebr, Sept 12, 37; m 61; c 3. BIOMATHEMATICS. *Educ:* Creighton Univ, BS, 59; Univ Nebr, Lincoln, MS, 62; Univ Chicago, PhD(math biol), 66. *Prof Exp:* Res scientist, Univ Calif, Los Angeles, 66-70, lectr comp sci, 69-70; PROF COMPUT & INFO SCI, UNIV ALA, BIRMINGHAM, 70- *Concurrent Pos:* Sr lectr, Sch Bus, Univ Southern Calif, 69-70. *Mem:* Soc Math Biol; Inst Elec & Electronics Engrs; Asn Comput Mach; Soc Comput Simulation Int; Neural Networks Soc. *Res:* Digital modeling and simulation, discrete-event programming systems; artificial intelligence systems; software systems; programming systems; mathematical biology; biophysics. *Mailing Add:* Dept Comput & Info Sci Univ Ala Birmingham AL 35294

REILLY, MARGARET ANNE, b Port Chester, NY, 1937. NEUROPHARMACOLOGY, PSYCHOPHARMACOLOGY. *Educ:* Col New Rochelle, BA, 59; NY Med Col, MS, 78, PhD(pharmacol), 81. *Prof Exp:* Sr lab technician, Sloan-Kettering Inst, 59-64; lab & teaching biol asst, Hunter Col, City Univ NY, 65; RES SCIENTIST, NATHAN KLINE INST PSYCHIAT RES, 66- *Concurrent Pos:* Adj assoc prof pharmacol, Sch Nursing, Col New Rochelle, 78-; bk rev ed, Asn Women in Sci, 84-; adj asst prof pharmacol, Concordia Col, Bronxville, 88-; mem, Coun Res Scientists, NY State Off Ment Health. *Mem:* Histamine Res Soc NAm (secy-treas, 80-); Asn Women Sci; Am Soc Pharmacol & Exp Therapeut; NY Acad Sci; Women Neurosci. *Res:* Investigations of the interactions of various substances such as aspartame and acetyl-L-carnitine with binding characteristics at central nervous system neurotransmitter receptor systems. *Mailing Add:* Nathan Kline Res Inst Orangeburg NY 10962

REILLY, MARGUERITE, b Troy, NY, Mar 20, 19. PROTOZOOLOGY, MICROBIOLOGY. *Educ:* Col St Rose, BA, 53, MS, 60; St John's Univ, NY, PhD(zool), 63. *Prof Exp:* Teacher high schs, Albany & Syracuse Diocese, 51-60; chmn, Natural Sci Div, 77-85, PROF BIOL, COL ST ROSE, 63-, CHMN DEPT, 66- *Mem:* AAAS; Am Inst Biol Sci. *Res:* Nutrition of Protozoa and the effect of radiation on Protozoa grown under varying conditions. *Mailing Add:* Dept Biol Col St Rose 432 Western Ave Albany NY 12203

REILLY, MICHAEL HUNT, b Rochester, NY, Dec 23, 39; m 65; c 2. THEORETICAL PHYSICS. *Educ:* Univ Rochester, BS, 61, PhD(solid state physics), 67. *Prof Exp:* RES PHYSICIST MATH PHYSICS, NAVAL RES LAB, 66- *Concurrent Pos:* Nat Acad Sci-Nat Res Coun fel, Naval Res Lab, 66-68. *Mem:* Am Phys Soc; Am Geophys Union. *Res:* Space science; communications science; fluid dynamics. *Mailing Add:* Advan Space Technol Code 8340-R Naval Res Lab Washington DC 20375

REILLY, NORMAN RAYMUND, b Glasgow, Scotland, Jan 30, 40; m 66; c 3. SEMIGROUPS, ORDERED ALGEBRAIC SYSTEMS. *Educ:* Univ Glasgow, BSc Hons, 61, PhD(math), 65. *Prof Exp:* Asst lectr math, Univ Glasgow, 64-65; vis asst prof math, Newcomb Col, Tulane Univ, New Orleans, 65-66; asst prof, 66-69, assoc prof, 69-74, chmn, 76-78, actg assoc vpres, 78-79, PROF MATH, SIMON FRASER UNIV, 74- *Concurrent Pos:* Vis prof math, Monash Univ, Melbourne, Australia, 71. *Mem:* Can Math Soc; Am Math Soc; Soc Actuaries. *Res:* Algebra especially in the structure, representations and varieties of algebras. *Mailing Add:* Dept Math Simon Fraser Univ Burnaby BC V5A 1S6 Can

REILLY, PARK MCKNIGHT, b Welland, Ont, May 14, 20; m 45; c 4. STATISTICS, CHEMICAL ENGINEERING. *Educ:* Univ Toronto, BASc, 43; Univ London, PhD(statist) & dipl, Imp Col, 62. *Prof Exp:* Jr chem engr, Welland Chem Works, 41-45; lectr chem eng, Ajax Div, Univ Toronto, 45-47; prin chem engr, Polysar Ltd, 47-67; prof chem eng, 67-, EMER PROF, UNIV WATERLOO. *Concurrent Pos:* Adj prof, Univ Waterloo, 64-67; consult several chem industs, 67- *Mem:* Fel Can Soc Chem Eng; fel Royal Statist Soc. *Res:* Application of statistical methods to chemical engineering research and plant operation; design of experiments; model discrimination. *Mailing Add:* 48 Culpepper Dr Waterloo ON N2L 5L1 Can

REILLY, PETER JOHN, b Newark, NJ, Dec 26, 38; m 65, 76; c 2. CHEMICAL ENGINEERING. *Educ:* Princeton Univ, AB, 60; Univ Pa, PhD(chem eng), 64. *Prof Exp:* Res engr, Org Chem Dept, Jackson Lab, E I du Pont de Nemours & Co, Inc, 64-68; asst prof chem eng, Univ Nebr, Lincoln, 68-74; assoc prof, 74-79, PROF CHEM ENG, IOWA STATE UNIV, 79- *Concurrent Pos:* Invited prof, Swiss Fed Inst Tech, Lausanne, Switz, 83-84. *Mem:* Am Chem Soc; Am Inst Chem Engrs. *Res:* Biochemical engineering; enzyme kinetics; agricultural residue utilization; carbohydrate chromatography. *Mailing Add:* Dept Chem Eng Iowa State Univ Ames IA 50011

REILLY, RICHARD J, b La Crosse, Wis, Jan 15, 30; m 54; c 1. HEAT TRANSFER. *Educ:* Univ Minn, BS, 51. *Prof Exp:* Jr res engr, Rosemount Res Facility, Univ Minn, 51; res engr, Aeronaut Res Div, Gen Mills, Inc, 51-54; res engr, Northrop Aircraft Inc, 54-57; sr develop engr, Aeronaut Div, Honeywell Inc, Minn, 57-60, sr res scientist fluid mech, Mil Prod Res, 60-63, prin res scientist, 63-64, supvr res, 64-65, res sect head, Systs & Res Ctr, 65-67, fluid sci res mgr, 67-69; exec vpres, Cytec Corp, 69-74; PRES, GALILEO CO, 74-; PRES, CUYUNA CORP, 78- *Concurrent Pos:* Pvt consult, 59-64; consult lectr, Adv Group Aerospace Res & Develop, NATO, 66-70; dir, BMT, Inc, 76-79; adj instr, St Thomas Univ Bus Sch, 88- *Honors & Awards:* USAF Systs Command Award, 63. *Mem:* Assoc fel Am Inst Aeronaut & Astronaut. *Res:* Aerodynamics; compressible flow; boundary layer flows; supersonic inlets; fluidics; alternate energy sources; aircraft propulsion; flight testing techniques. *Mailing Add:* 1759 Venus Ave St Paul MN 55112

REILLY, THOMAS E, b New York, NY. GROUND-WATER HYDROLOGY. *Educ:* Villanova Univ, BCE, 74; Princeton Univ, MSE, 75; Polytech Univ, PhD(civil eng), 86. *Prof Exp:* HYDROLOGIST, US GEOL SURV, 75- *Mem:* Am Geophys Union; Asn Ground Water Scientists & Engrs. *Res:* Three-dimensional simulation; aquifer test analysis; solute transport; salt water - fresh water interaction in ground-water systems. *Mailing Add:* US Geol Surv 411 Nat Ctr Reston VA 22092

REILY, WILLIAM SINGER, b Chicago, Ill, June 13, 24; m 51; c 6. ORGANIC CHEMISTRY, POLYMER CHEMISTRY. *Educ:* Roosevelt Univ, BS, 48; De Paul Univ, MS, 50. *Prof Exp:* Develop chemist, Bauer & Black Lab, Kendall Co, 52-53; res chemist, Baxter Labs, Inc, 53-54; develop chemist, G D Searle & Co, 54-57; proj engr, Amphenol Electronics Co, 57-58; develop chemist, Du Kane Corp, 58-67; proj leader, US Gypsum, 67-87; RETIRED. *Mem:* Am Chem Soc. *Res:* Applied and industrial chemistry; process and product development; trouble shooting and environmental testing. *Mailing Add:* 884 Horne Terr Des Plaines IL 60016

REIM, ROBERT E, b Abrams, Wis. ELECTROCHEMISTRY. *Educ:* Univ Wis, OshKosh, BS, 71, Milwaukee, MS, 74. *Prof Exp:* SR RES CHEMIST, DOW CHEM CO, 74- *Mem:* Am Chem Soc; Soc Appl Spectros; Sigma Xi. *Res:* Development and application of analytical techniques including electroanalytical chemistry, spectroscopy and chromatography. *Mailing Add:* Dow Chem-Mich Div Bldg 1897 Midland MI 48674

REIMANN, BERNHARD ERWIN FERDINAND, b Berlin, Ger, May 30, 22; m 49; c 1. ELECTRON MICROSCOPY. *Educ:* Free Univ Berlin, Dr rer nat, 59. *Prof Exp:* Scientist asst, Bot Inst, Marburg, 58-60; asst res biologist, Scripps Inst, Univ Calif, San Diego, 61-67, supvr electron microscope facil, 64-67; chief electron micros, William Beaumont Army Med Ctr, US Army, 67-87; PVT CONSULT, 87- *Concurrent Pos:* NSF grant, 65-67 & 68-69; assoc prof, NMex State Univ, 67-; assoc, Grad Fac, Univ Tex, El Paso, 68-71; assoc clin prof path, Tex Tech Med Sch. *Mem:* Emer fel AAAS; emer mem Electron Micros Soc Am. *Res:* Cytology and ultrastructure of mineral deposition in biological systems; histopathology diagnostic at ultrastructure level. *Mailing Add:* PO Box 44 Capitan NM 88316

REIMANN, ERWIN M, b Parkston, SD, July 26, 42; m 61; c 2. BIOCHEMISTRY. *Educ:* Augustana Col, SDak, BA, 64; Univ Wis-Madison, PhD(biochem), 68. *Prof Exp:* Fel biochem, Univ Calif, Davis, 68-70; from asst prof to assoc prof, 70-82, PROF BIOCHEM, MED COL OHIO, TOLEDO, 82- *Concurrent Pos:* Vis asst prof physiol, Vanderbilt Univ, Nashville, 74; vis scientist, Univ Wash, Seattle, 82-83. *Mem:* Sigma Xi; Am Chem Soc; AAAS; Am Soc Biol Chem. *Res:* The role of cyclic adenosine monophosphate and protein kinases in cellular function, especially their roles in glycogen metabolism and in gastric secretion. *Mailing Add:* Dept Biochem Med Col Ohio Caller Sernce No 1009 Toledo OH 43699

REIMANN, HANS, b Vienna, Austria, Dec 28, 30; nat US; m 57; c 1. ORGANIC CHEMISTRY, REGULATORY AFFAIRS. *Educ:* Univ Calif, Los Angeles, BS, 51, PhD(chem), 57. *Prof Exp:* From chemist to sr chemist, 57-68, unit head, 68-71, asst to dir, Chem & Microbiol Develop, 71-73, coordr, Corp Prod Develop, 73-74, mgr prod planning & control, 74-77, assoc dir, Tech Regulatory Doc & Actives Control, 78-80, dir, Regulatory Affairs-Tech, 80-87, ASSOC DIR, REGULATORY AFFAIRS, SCHERING CORP, 87- *Mem:* Am Chem Soc; Am Soc Microbiol; Regulatory Affairs Prof Soc; Drug Info Asn. *Res:* Natural products; steroids; antibiotics. *Mailing Add:* Schering Corp 2000 Galloping Hill Rd Kenilworth NJ 07033

REIMANN, HOBART ANSTETH, medicine; deceased, see previous edition for last biography

REIMER, CHARLES BLAISDELL, immunology; deceased, see previous edition for last biography

REIMER, DENNIS D, b Corn, Okla, May 20, 40; m 61; c 2. MATHEMATICS, COMPUTER SCIENCE. *Educ:* Southwestern Okla State Univ, BSEd, 62; Okla State Univ, MS, 64; NTex State Univ, EdD(math), 69. *Prof Exp:* Instr, 63-65, PROF MATH, SOUTHWESTERN OKLA STATE UNIV, 67- *Mem:* Nat Coun Teachers Math; Math Asn Am. *Mailing Add:* Rt 4 Box 128 Weatherford OK 73096

REIMER, DIEDRICH, b Altona, Man, May 6, 25; US citizen; m 51; c 3. GENETICS, ANIMAL SCIENCE. *Educ:* Univ Man, BScA, 50; Univ Minn, MS, 55, PhD(genetics, animal sci), 59. *Prof Exp:* Vet agr instr, US Dept Vet Affairs, 50-53; from instr to assoc prof, Univ Minn, 55-64; assoc prof, 64-72, PROF ANIMAL SCI, UNIV HAWAII, 72-, ANIMAL SCIENTIST, HAWAII INST TROP AGR & HUMAN RESOURCES, 74- *Concurrent Pos:* Supt, Hawaii Agr Exp Sta, Hawaii Br, 78-80. *Mem:* AAAS; Am Genetic Asn; Am Soc Animal Sci. *Res:* Improvement of beef cattle, swine and sheep through breeding methods; swine nutrition and livestock management; improvement of beef cattle production under tropical range conditions through the application of breeding methods; development of a synthetic line of beef cattle selected for improved reproduction, growth rate, carcass quality and adaptability to tropical environments. *Mailing Add:* Beamont Ctr Univ Hawaii 461 W Lenikaule Hilo HI 96720

REIMERS, THOMAS JOHN, b West Point, Nebr. ENDOCRINOLOGY, REPRODUCTIVE PHYSIOLOGY. *Educ:* Univ Nebr, BS, 67; Univ Ill, MS, 69, PhD(animal sci), 74. *Prof Exp:* Res asst animal sci, Univ Ill, 67-69 & 71-74; fel physiol, Colo State Univ, 74-75, res assoc, 75-78; ASST PROF ENDOCRINOL, COL VET MED, CORNELL UNIV, 78- *Mem:* Soc Study Reproduction; Am Soc Animal Sci; AAAS; Sigma Xi. *Res:* Reproduction physiology in domestic animals; clinical endocrinology in large and small domestic animals. *Mailing Add:* 131 Diagnostic Lab NY State Col Vet Med Ithaca NY 14853

REIMOLD, ROBERT J, b Greenville, Pa, Nov 15, 41; m 63; c 3. ENVIRONMENTAL SCIENCE. *Educ:* Thiel Col, BA, 63; Univ Del, MA, 65, PhD(biol sci), 68. *Prof Exp:* Teaching asst biol, Thiel Col, 62-63; res asst salt marsh ecol, Marine Labs, Univ Del, 63-68; res assoc, Marine Inst, Univ Ga, 68-69, asst prof zool & marine inst, 69-74, ecologist, Marine Resources Ext Ctr, 75-77; dir, Coastal Resources Div, Ga Dept Natural Resources, 77-81; PRIN BIOLOGIST, METCALF & EDDY, INC, BOSTON, 82- *Concurrent Pos:* Univ Ga Marine Inst res fel, 68-69; vis prof, W I Lab, Fairleigh Dickinson Univ; sci consult, Encycl Britannica Corp, 74-; ecology comt, Water Pollution Control Fedn, marine water qual comt. *Mem:* AAAS; Ecol Soc Am; Brit Ecol Soc; Estuarine Res Fedn (past pres); Water Pollution Control Fedn; Sigma Xi. *Res:* Applied environmental engineering specializing in coastal, aquatic and wetland sciences supporting major infrastructure engineering projects. *Mailing Add:* Metcalf & Eddy Inc 30 Harvard Mill Sq Wakefield MA 01880

REIMSCHUSSEL, ERNEST F, b Poischwitz, Germany, July 21, 17; US citizen; m 40; c 5. HORTICULTURE. *Educ:* Brigham Young Univ, BA, 40, MS, 51. *Prof Exp:* Gardner, Brigham Young Univ, 41-42, asst land archit, 42-47, instr, 47-54, asst hort, 47-54, from instr to assoc prof 54-72, chmn dept, 58-66, assoc prof agron & hort, 72-82; RETIRED. *Mem:* Sigma Xi. *Res:* Ornamental woody plants; landscape architecture; trees. *Mailing Add:* 835 No 300 W Provo UT 84604

REIN, ALAN JAMES, b New York, NY, Nov 1, 48; m 80. PHYSICAL CHEMISTRY, SPECTROSCOPY. *Educ:* Rutgers Univ, BA, 70, MS, 73, PhD(phys chem), 74. *Prof Exp:* Sr res chemist phys & anal chem, Merck Sharp & Dohme Res Labs, 73-77, res fel phys chem, 77-79; adv scientist appl sci, 79-80, SR SCIENTIST & TECH MGR, IBM INSTRUMENTS, INC, 80- *Mem:* Sigma Xi; Soc Appl Spectros; Am Chem Soc. *Res:* Vibrational spectroscopy of inorganic, organometallic and biochemical species; laser raman spectroscopy; Fourier transform infrared spectroscopy. *Mailing Add:* 25 Lauree Dr Englewood Cliffs NJ 07632-1862

REIN, DIANE CARLA, ENZYMOLOGY, BIOCHEMISTRY. *Educ:* Univ Cincinnati, PhD(develop biol), 77. *Prof Exp:* SR RES ASSOC, UNIV CINCINNATI, 79- *Res:* Molecular genetics. *Mailing Add:* Dept Biol Sci Univ Cincinnati Cincinnati OH 45221

REIN, JAMES EARL, b Chicago, Ill, Mar 24, 23; m 48; c 2. ANALYTICAL CHEMISTRY. *Educ:* Univ Ill, BS, 44, PhD(chem), 49. *Prof Exp:* Res chemist, Los Alamos Sci Lab, 49-53; group leader anal & radiochem methods develop, Atomic Energy Div, Phillips Petrol Co, 53-66; sect leader analytical chem develop, Idaho Nuclear Corp, 66-69; MEM STAFF, LOS ALAMOS NAT LAB, UNIV CALIF, 69- *Mem:* Am Chem Soc. *Res:* Analytical methods development; radiochemistry; remote handling techniques for highly radioactive samples; analytical chemistry of the transuranic elements; statistics and quality control. *Mailing Add:* RR 1 Homer IL 61849

REIN, ROBERT, b Ada, Yugoslavia, June 1, 28; m 70. QUANTUM CHEMISTRY, BIOPHYSICS. *Educ:* Hebrew Univ, Israel, MSc, 55, PhD(phys chem), 60. *Prof Exp:* Res asst, Weizmann Inst, 56-60; sr res scientist, Quantum Chem Group, Univ Uppsala, 63-65; asst prof theoret biol, 65-66, assoc res prof theoret biol & biophys, 66-68, RES PROF BIOPHYS SCI, SCH PHARM, STATE UNIV NY BUFFALO, 68-; PRIN CANCER RES SCIENTIST, ROSWELL PARK MEM INST, STATE UNIV NY BUFFALO, 67-, RES PROF & CHMN, BIOMET DEPT, 80- *Mem:* AAAS; Am Chem Soc; Biophys Soc; Int Soc Quantum Biol. *Res:* Molecular orbital theory of organic and biomolecules; quantum theory of intermolecular interactions and their application to molecular recognition in biology; electronic and physicochemical aspects of biopolymers. *Mailing Add:* Roswell Park Mem Inst 666 Elm St Buffalo NY 14263

REIN, ROBERT G, JR, b Detroit, Mich, Jan 16, 40; m 67; c 2. CHEMICAL ENGINEERING, RHEOLOGY. *Educ:* Mass Inst Technol, SB, 61; Univ Okla, PhD(eng), 67. *Prof Exp:* Res assoc, Res Inst, Univ Okla, 67-75; sr engr, Univ Engrs, Norman, Okla, 75-76; sr res assoc, Quaternary Res Ctr, Univ Wash, 76-82; CONSULT, 82- *Concurrent Pos:* Adj asst prof, Univ Okla, 71-75. *Mem:* Am Chem Soc; Soc Rheology. *Res:* Rheological properties of fluids and solids; properties of frozen soil; behavior of freezing and thawing ground; high pressure phenomena; properties of lubricants at high pressures; fire research. *Mailing Add:* 3106 128th NE Bellevue WA 98005

REINBERG, ALAN R, b New York, NY, Oct 19, 31; m 54; c 4. SOLID STATE PHYSICS. *Educ:* Univ Chicago, BA, 52; Ill Inst Technol, BS, 57, MS, 59, PhD(physics), 61. *Prof Exp:* Res physicist, Res Inst, Ill Inst Technol, 60-63; sr physicist, Lear Siegler Inc, 63-64; mem tech staff physics, Tex Instruments Inc, 64-80; Perkin Elmer Corp, 80-90. *Mem:* Inst Elec & Electronics Engrs. *Res:* Properties of point defects by magnetic resonance; optical properties; plasma chemistry; x-ray lithography. *Mailing Add:* Two Charbeth Lane Westport CT 06880

REINBERGS, ERNESTS, b Latvia, Mar 1, 20; nat Can; m 44; c 2. PLANT BREEDING. *Educ:* Univ Toronto, MSA, 54; Univ Man, PhD(cytogenetics, plant breeding), 57. *Prof Exp:* Lectr field husb, 54-57, from asst prof to prof, 57-85, EMER PROF CROP SCI, ONT AGR COL, UNIV GUELPH, 86- *Honors & Awards:* Grindley Medal, Agr Inst Can, 77; Outstanding Res Award, Can Soc Agron, 87. *Mem:* Can Soc Agron; Genetics Soc Can; Agr Inst Can; Sigma Xi. *Res:* Barley, oat and triticale breeding; double haploids and breeding methods in barley; cytogenetics; disease resistance. *Mailing Add:* Dept Crop Sci Ont Agr Col Univ Guelph Guelph ON N1G 2W1 Can

REINBOLD, GEORGE W, b Williamsport, Pa, Apr 10, 19; m 42; c 3. BACTERIOLOGY, DAIRY INDUSTRY. *Educ:* Pa State Univ, BS, 42; Univ Ill, MS, 47, PhD(dairy mfg), 49. *Prof Exp:* Proj dir bact, Kraft Foods Co, Nat Dairy Prod Corp, 49-53, prod technician dairy indust, 53-58; prod mgr, Tolibia Cheese Mfg Corp, 58-59; exten specialist, Iowa State Univ, 59-60, from assoc prof to prof dairy bact, 60-74; VPRES RES & DEVELOP,

LEPRINO FOODS, 74- *Honors & Awards:* Pfizer Cheese Res Award, 70; Dairy Res Inc Award, 77. *Mem:* Am Dairy Sci Asn; Am Soc Microbiol; Int Asn Milk, Food & Environ Sanitarians. *Res:* Dairy microbiology, especially indicator organisms, sanitation and cheese microbiology and manufacture. *Mailing Add:* 4180 Dudley St Wheat Ridge CO 80033

REINBOLD, PAUL EARL, b Vincennes, Ind, Oct 21, 43; m 65. ANALYTICAL CHEMISTRY, INORGANIC CHEMISTRY. *Educ:* Olivet Nazarene Col, AB, 65; Purdue Univ, West Lafayette, Ind, 68; Tex A&M Univ, PhD (analytical chem), 70. *Prof Exp:* Analytical chemist, Armour Pharmaceut Co, Ill, 64-65; Robert A Welch-Tex A&M Res Coun fel, Tex A&M Univ, 69-70; from asst prof to assoc prof, Bethany Nazarene Col, 70-76, PROF CHEM & HEAD DEPT, SOUTHERN NAZARENE UNIV, 76- *Mem:* Am Chem Soc. *Res:* Optically active inorganic coordination complexes; kinetics and mechanisms of inorganic exchange reactions; kinetics of electrode deposition processes from metal ion complexes; microcomputers in chemical education. *Mailing Add:* Dept Chem Southern Nazarene Univ 6729 NW 39th Expressway Bethany OK 73008

REINCKE, URSULA, radiobiology, cell biology, for more information see previous edition

REINDERS, VICTOR A, b Mallard, Iowa, Dec 27, 06; m 42. CHEMISTRY. *Educ:* Carroll Col, Wis, BA, 29; Univ Wis, MA, 30, PhD(inorg chem), 35. *Prof Exp:* Instr, 30-33, from instr to prof, 35-72, EMER PROF CHEM, EXTEN DIV, UNIV WIS-MILWAUKEE, 72- *Res:* Chemistry of rhenium in its lower valence state. *Mailing Add:* 2047 Dixie Dr No 1 Waukesha WI 53186

REINECCIUS, GARY (AUBREY), b Webster, Wis, Jan 12, 44; m 64; c 2. FOOD SCIENCE. *Educ:* Univ Minn, BS, 64, MS, 67; Pa State Univ, PhD(food sci), 70. *Prof Exp:* Assoc prof, 70-80, PROF FOOD SCI, UNIV MINN, ST PAUL, 80- *Mem:* Am Chem Soc; Inst Food Technol; hon mem Soc Flavor Chemists. *Res:* Chemistry of food flavor, including biogenesis and chemical composition of flavor. *Mailing Add:* Dept Food Sci Univ Minn St Paul MN 55101

REINECKE, MANFRED GORDON, b Milwaukee, Wis, May 19, 35; m 57; c 3. ORGANIC CHEMISTRY. *Educ:* Univ Wis, BS, 56; Univ Calif, Berkeley, PhD(org chem), 60. *Prof Exp:* Asst org chem, Univ Calif, Berkeley, 56-57, instr, 59-60; asst prof, Univ Calif, Riverside, 60-64; from asst prof to assoc prof, 64-74, PROF ORG CHEM, TEX CHRISTIAN UNIV, 74- *Concurrent Pos:* NSF fac fel, Univ Tubingen, 71-72; Nat Acad Sci exchange scientist, Acad Wissenschaften, Ger Dem Repub, 79 & 90; vis prof, Univ BC, Vancouver Can, 87. *Honors & Awards:* W T Doherty Award, Am Chem Soc. *Mem:* AAAS; Am Chem Soc; Royal Soc Chem. *Res:* Chemistry of behavior; natural products; medicinal herbs. *Mailing Add:* Dept Chem Tex Christian Univ Ft Worth TX 76129

REINECKE, ROBERT DALE, b Ft Scott, Kans, Mar 26, 29; m 52; c 1. OPHTHALMOLOGY. *Educ:* Ill Col Optom, OD, 51; Univ Kans, AB, 55, MD, 59. *Prof Exp:* From asst instr to asst prof, Harvard Univ, 64-69, sci dir, Vision Info Ctr, 67-70; prof ophthal & chmn dept, Albany Med Col, 70-81; ophthal-in-chief, Wills Eye Hosp, 81-85, chmn ophthal, Thomas Jefferson Med Col, 81-85, PROF OPHTHAL, THOMAS JEFFERSON MED COL, 81-, DIR, FOERDERER EYE MOVEMENT CTR, WILLS EYE HOSP, 85- *Concurrent Pos:* Teaching fel ophthal, Harvard Med Sch, 63-64; grant, Albany Med Col, 72-; instr eye anat, Simmons Col, 62-68; chief instr Infirmary, 63-67, dir ocular motility clin, 67-69, asst instr, Infirmary, 63-69; asst instr, Mass Gen Hosp, 63-69; mem bd dirs, Conrad Berens Int Eye Film Libr, 70-; mem visual sci study sect, NIH, 71-75; chmn med adv comt, Comn Blind & Visually Handicapped, 71-76; mem comt vision, Nat Acad Sci-Nat Res Coun, 77-80; chmn panel ophthal devices, Food & Drug Admin, 74-78; trustee, Asn Res Vision & Ophthal, 86-91. *Honors & Awards:* Howe Award, 81; Sr Honor Award, Am Acad Ophthal, 86. *Mem:* AAAS; AMA; Asn Res Vision & Ophthal; Am Acad Ophthal (pres, 89); Am Asn Pediat Ophthal & Strobisms. *Res:* Eye movement; visual acuity; stereopsis; nystagmus; amblyopia. *Mailing Add:* Wills Eye Hosp Ninth & Walnut St Philadelphia PA 19107

REINECKE, THOMAS LEONARD, b Park Falls, Wis, Sept 14, 45. ELECTRONIC PROPERTIES OF SOLIDS, PHASE TRANSITIONS. *Educ:* Ripon Col, Wis, BA, 68; Oxford Univ, PhD(physics), 72. *Prof Exp:* Res assoc & lectr, Dept Physics, Brown Univ, Providence, RI, 72-74; res assoc, 74-76, res physicist, 76-80, head, Theory Sect Semiconductors Br, 80-90, HEAD, ELEC & OPTICAL PROPERTIES SECT, NAVAL RES LAB, 90- *Concurrent Pos:* Panel mem, NSF Res Initiation & Support Prog, 76-77; mem steering comt, Greater Wash Solid State Physics Colloquia, 78-, secy, 83-85; guest scientist, Max Planck Inst Solid State Res, Stuttgart, W Ger, 79; panel on artificially structured mat, Nat Acad Sci, 84-85; selection comt for Rhodes Scholars, Va, 84-; mem, Solid State Sci Comt, Forum Nat Res Coun Nat Acad Sci, 84-; mem, NSF Eval Comt, Ctr Sci & Technol, 88-89. *Mem:* Fel Am Phys Soc; Sigma Xi. *Res:* Solid state theory including interacting electronic systems, phase trasitions, surfaces and interfaces; various systems including semiconductors, and magnetic and ferroelectric materials. *Mailing Add:* Code 6877 Naval Res Lab Washington DC 20375-5000

REINECKE, WILLIAM GERALD, b Indianapolis, Ind, July 21, 35; m 59; c 2. AERODYNAMICS. *Educ:* Purdue Univ, BS, 57; Princeton Univ, MA & PhD(aeronaut eng), 61. *Prof Exp:* Sr staff scientist, Res & Develop Div, Avco Corp, 64-65, group leader exp aeodyne Ballistic Range Group, Res & Technol Labs, 65-66, sect chief, Exp Fluid Physics Sect, 66-71, sr consult, Technol Directorate, 71-77, mgr ballistics lab, Avco Systs Div, 77-78, dir Ballistics & Ordnance Technol, 78-90; PRIN RES SCIENTIST, PHYS SCI INC, 90- *Concurrent Pos:* Mem, Aeroballistic Range Asn. *Mem:* Assoc fel Am Inst Aeronaut & Astronaut; Am Defense Preparedness Asn. *Res:* Aerodynamics, especially high speed flows, high speed erosion and internal and terminal ballistics. *Mailing Add:* Phys Sci Inc 20 New Eng Bus Ctr Andover MA 01810

REINEKE, CHARLES EVERETT, organic chemistry; deceased, see previous edition for last biography

REINEMUND, JOHN ADAM, b Muscatine, Iowa, Jan 14, 19; m 43. GEOLOGY. *Educ:* Augustana Col, BA, 40; Univ Chicago, BA, 42, 51. *Hon Degrees:* DHumL, Augustana Col, 67. *Prof Exp:* Geologist, Strategic Mineral Invests, US Geol Serv, 42-44, Oceanog Res, Off Sci Res & Develop, 44-45, Mineral Fuel Invests, 46-49, Coal Surv, Econ Coop Admin, Korea, 49-50, asst chief, Eastern Invest Sect, Fuels Br, 51-53, regional supvr, Midcontinent Region, 53-56, geol adv, AID, Geol Surv, Pakistan, 56-64, chief, Br Foreign Geol, DC, 64-69, chief, Off Int Geol, 69-84, EXEC DIR, CIRCUM-PACIFIC COUN ENERGY & MINERAL RESOURCES, US GEOL SURV, 84- *Concurrent Pos:* Bd mem, Int Geol Corr Prog, 73-79; treas, Int Union Geol Sci, 79-89. *Honors & Awards:* Distinguished Serv Award, US Dept of Interior, 88. *Mem:* Geol Soc Am; Am Asn Petrol Geol; Am Geophys Union. *Res:* Structural geology; sedimentary petrology; geology of fuels; origin of mineral and fuel resources in relation to sedimentary and tectonic processes; principal research has dealt with structural and sedimentational controls for the origin and distribution of fossil fuels in tectonic basins and continental margin belts; recent activity has been mainly concerned with organizing programs for the Pacific and Atlantic Basins. *Mailing Add:* PO Box 890 Leesburg VA 22075

REINER, ALBEY M, b Brooklyn, NY, Aug 11, 41; m 65; c 2. MICROBIOLOGY, GENETICS. *Educ:* Princeton Univ, BS, 62; Oxford Univ, cert math statist, 63; Univ Wis-Madison, MS, 64; Harvard Univ, PhD(molecular biol), 69. *Prof Exp:* Rothschild Found fel, Hebrew Univ, Israel, 68-69; NIH fel, Univ Calif, Berkeley, 69-70; asst prof, 71-77, ASSOC PROF MICROBIOL, UNIV MASS, AMHERST, 77- *Res:* Microbial genetics. *Mailing Add:* Dept Microbiol Univ Mass Amherst MA 01003

REINER, CHARLES BRAILOVE, b Ellenville, NY, Dec 3, 20; m 51; c 3. PATHOLOGY, PEDIATRICS. *Educ:* Temple Univ, AB, 42, MD, 45, MSc, 53; Am Bd Pediat, dipl, 54; Am Bd Path, dipl, 66. *Prof Exp:* Fel histochem, Univ Chicago, 51-52; physician-in-chg, Pediat Outpatient Clin, Univ Hosp, Temple Univ, 52-55; asst prof path, Col Med, State Univ NY Downstate Med Ctr, 56-59; from asst prof to assoc prof path & pediat, Col Med, Ohio State Univ, 69-86; chief div anod path, Dept Lab Med, Children's Hosp, Columbus, 72-85, pathologist, 85-86; RETIRED. *Concurrent Pos:* Assoc pathologist, Children's Hosp, Columbus, 59-72; pathologist, Inst Perinatal Studies, 60-64; consult, ped path, 86-88. *Mem:* AMA. *Res:* Pediatric pathology; perinatal problems, especially hyaline membrane syndrome; heparin in human tissues; laboratory aspects of blood coagulation; sudden infant death syndrome. *Mailing Add:* 3555 Piatt Rd Delaware OH 43015-9622

REINER, IRMA MOSES, b Newburgh, NY, Mar 3, 22; wid; c 2. MATHEMATICS. *Educ:* Cornell Univ, AB, 42, AM, 44, PhD(algebra, geom, physics), 46. *Prof Exp:* Instr math, Temple Univ, 46-48, Univ Ill, Urbana, 48-49 & Danville Community Col, Ill, 49-50; instr 56-57, ASST PROF MATH, UNIV ILL, URBANA, 57- *Concurrent Pos:* Erastus Brooks fel, Cornell Univ, 42. *Mem:* Am Math Soc; Math Asn Am; Asn Women Math; Sigma Xi. *Res:* Theory of numbers. *Mailing Add:* Dept Math Univ Ill 1409 W Green St Urbana IL 61801

REINER, JOHN MAXIMILIAN, b Boston, Mass, Apr 19, 12; m 34, 63; c 1. BIOCHEMISTRY & BIOPHYSICS, PHYSIOLOGY. *Educ:* Univ Chicago, BSc, 38, MSc, 39; Univ Minn, PhD(physiol), 46. *Prof Exp:* Asst math biophys, Univ Chicago, 36-39; instr physics, City Col New York, 40-41; instr, Univ Minn, 43, res assoc physiol, 43-44, instr, 44-46; Nat Res Coun fel, Wash Univ, 46, Am Cancer Soc fel, 47; res assoc, Med Sch, Tufts Col, 48-49, asst prof, 49-50; asst prof biochem, Col Physicians & Surgeons, Columbia Univ & sr enzyme chemist, Inst Cancer Res, 50-54; res dir, Simon Baruch Res Labs, Saratoga Spa, 54-57; biochemist, Radioisotope Labs, Vet Admin Hosp, Albany, NY, 57-58; assoc prof microbiol, Sch Med, Emory Univ, 58-68; PROF BIOCHEM & RES PROF PATH, ALBANY MED COL, 68- *Concurrent Pos:* Assoc, Albany Med Co, 54-57. *Mem:* Soc Develop Biol; Soc Math Biol; Soc Indust & Appl Math; Am Math Soc; Sigma Xi. *Res:* Molecular biology of growth, differentiation and senescence; enzyme kinetics; transport processes; population dynamics; neural models of behavior; philosophy of science. *Mailing Add:* 111 Emerston St Denver CO 80218-3779

REINER, LEOPOLD, b Leipzig, Ger, Jan 22, 11; nat US; m 46. MEDICINE. *Educ:* Univ Vienna, MD, 36; Am Bd Path, dipl, 52. *Prof Exp:* Intern, Rothschild Hosp, Vienna, Austria, 36-38; from resident path to asst pathologist, W Jersey Hosp, Camden, NJ, 41-46; instr path, Harvard Med Sch, 50-53, clin assoc, 53-56; vis assoc prof, 56-72, prof, 72-79, EMER PROF PATH, ALBERT EINSTEIN COL MED, 79-; PATHOLOGIST & CONSULT, BRONX-LEBANON HOSP CTR, 82- *Concurrent Pos:* From resident path to actg pathologist, Beth Israel Hosp, Boston, 46-56; pathologist & dir labs, Bronx Lebanon Hosp Ctr, 56-82. *Mem:* Histochem Soc; Am Asn Path; NY Acad Med; Int Acad Path; Am Soc Clin Path. *Res:* Pathology, especially cardiovascular pathology; coronary arterial and mesenteric arterial circulation; cardiac hypertrophy. *Mailing Add:* 277 Old Colony Rd Hartsdale NY 10530-9980

REINERS, WILLIAM A, b Chicago, Ill, June 10, 37; m 62; c 2. ECOLOGY. *Educ:* Knox Col, BA, 59; Rutgers Univ, MS, 62, PhD(bot), 64. *Prof Exp:* From instr to asst prof bot, Univ Minn, 64-67; from asst prof to prof biol, Dartmouth Col, 76-83; PROF BOT, UNIV WYO, 83- *Honors & Awards:* Henry J Oosting lectr, Duke Univ. *Mem:* AAAS; Ecol Soc Am (treas, 81-); Am Inst Biol Sci. *Res:* Biogeochemistry of terrestrial ecosystems with special emphasis on succession and landscape relations; global ecology. *Mailing Add:* Dept Bot Univ Wyo Laramie WY 82071-3165

REINERT, JAMES A, b Enid, Okla, Jan 26, 44; m 63; c 6. TURFGRASS ENTOMOLOGY, ORNAMENTAL PLANT ENTOMOLOGY. *Educ:* Okla State Univ, BS, 66; Clemson Univ, MS, 68, PhD(entom), 70. *Prof Exp:* Res asst entom, Clemson Univ, 66-70; entomologist, State Bd Agr, Univ Md, College Park, 70; from asst prof to prof entom, Ft Lauderdale Res & Educ Ctr, Univ Fla, 70-84; PROF ENTOM & RESIDENT DIR RES, TEX A&M UNIV RES & EXTEN CTR, TEX AGR EXP STA, 84- *Concurrent Pos:*

UNIX Comput Sch. *Honors & Awards:* Porter Henninger Mem Hort Award; Fla Entomologist of the Year. *Mem:* Entom Soc Am; Int Turfgrass Soc; Hort Soc Am. *Res:* Urban agriculture with specialty in insects and mites in turfgrass and ornamental plants; host plant resistance and pest management of turfgrass insects and mites; biology, behavior and control by chemical or biological agents of insect and mite pests. *Mailing Add:* Tex Agr Exp Sta 17360 Coit Rd Dallas TX 75252-6599

REINERT, RICHARD ALLYN, b Elkhorn, Wis, June 3, 35; m 59; c 3. PLANT PATHOLOGY, HORTICULTURE. *Educ:* Univ Wis, BS, 58, PhD(plant path, hort), 62. *Prof Exp:* Asst prof plant path, Univ Ky, 62-67; res plant pathologist, R A Taft Sanit Eng Ctr, USDA-USPHS, Ohio, 67-69, Nat Environ Res Ctr, Plant Sci Res Div, USDA-Environ Protection Agency, 69-73; adj assoc prof, Univ NC, 69-73, assoc prof, Sci & Admin-Agr Res, 73-77, PROF PLANT PATH, SOUTHERN REGION, AGR RES SERV, USDA & NC STATE UNIV, 77- *Mem:* AAAS; Am Phytopath Soc; Am Soc Hort Sci. *Res:* Plant virology and tissue culture; physiology of growth and development; effects of air pollutants on cultivated plants and diseases of horticulture and ornamental crops. *Mailing Add:* Dept Plant Path Gardner Hall NC State Univ Raleigh NC 27650

REINES, DANIEL, GENETICS, ENZYMOLOGY. *Educ:* Albert Einstein Col Med, PhD(molecular biol), 85. *Prof Exp:* Fel biochem, Univ Calif, Berkeley, 85-90. *Mailing Add:* Dept Biochem Sch Med Emory Univ Atlanta GA 30322

REINES, FREDERICK, b Paterson, NJ, Mar 16, 18; m 40; c 2. ELEMENTARY PARTICLE PHYSICS, COSMIC RAY PHYSICS. *Educ:* Stevens Inst Technol, ME, 39, MS, 41; NY Univ, PhD(theoret physics), 44. *Hon Degrees:* DSc, Univ Witwatersrand, 66; Dr, Stevens Inst Technol, 84. *Prof Exp:* Mem staff & group leader, theoret div, Los Alamos Sci Lab, 44-49, dir exp, Operation Greenhouse Eniwetok, 51; prof physics & head dept, Case Western Inst Technol, 59-66; dean phys sci, 66-74, PROF PHYSICS, UNIV CALIF, IRVINE, 66- *Concurrent Pos:* Lectr, Exten Div, Univ Calif, 49; consult, Armed Forces spec weapons proj, 49-53 & Rand Corp, 50; centennial lectr, Univ Md, 56; fels, Guggenheim Found, 58-59 & Sloan Found, 59-63; mem, NASA Electrophys Adv Comt, 63-64 & Fulbright Physics Screening Comt, 64-66; consult, Inst Defense Anal, 65-69 & Los Alamos Sci Lab; trustee, Argonne Univ Assocs, 66. *Honors & Awards:* J Robert Oppenheimer Mem Prize, 81; Nat Medal Sci, 83; Michelson-Morley Award, 90. *Mem:* Nat Acad Sci; fel Am Phys Soc; fel Am Acad Arts & Sci; Am Soc Physics Teachers; fel AAAS; Nat Medal Sci. *Res:* Nuclear fission; physics of nuclear weapons and effects; scintillation detectors; free neutrino; cosmic rays. *Mailing Add:* Dept Physics Univ Calif Irvine CA 92717

REINESS, GARY, b Pittsburgh, Pa, Aug 20, 45; div; c 1. NEUROBIOLOGY. *Educ:* Johns Hopkins Univ, BA, 67; Columbia Univ, MPhil, 74, PhD(biol), 75. *Prof Exp:* Fel, Dept Neurobiol, Harvard Med Sch, 75-76; scholar, Dept Physiol, Univ Calif, San Francisco, 76-81; asst prof, 81-85, chair, 88-90, ASSOC PROF BIOL, POMONA COL, 85-, ASSOC DEAN, 90- *Mem:* Soc Neurosci; AAAS. *Res:* Development of neuromuscular junction, particularly regarding the regulation and properties of acetylcholine receptors and acetylcholinesterase; development of sympathetic nervous system and its regulation by nerve growth factor. *Mailing Add:* Dept Biol Pomona Col Claremont CA 91711

REINFURT, DONALD WILLIAM, b Wilkes-Barre, Pa, Aug 30, 38; m 65; c 2. APPLIED STATISTICS. *Educ:* State Univ NY Albany, BS, 60; State Univ NY Buffalo, MA, 63; NC State Univ, PhD(statist), 70. *Prof Exp:* Staff assoc, 68-80, ASSOC DIR, ANALYTICAL STUDIES, HWY SAFETY RES CTR, UNIV NC, 80-, ADJ ASSOC PROF, DEPT BIOSTATIST, 78- *Mem:* Sigma Xi; Asn Advan Automotive Med; Am Statist Asn; Am Pub Health Asn. *Res:* Application of statistical methods, particularly categorical data analysis, to traffic safety problems. *Mailing Add:* 403 Highview Dr Chapel Hill NC 27514

REINGOLD, EDWARD MARTIN, b Chicago, Ill, Nov 12, 45; m 68; c 4. ANALYSIS ALGORITHMS, DATA STRUCTURES. *Educ:* Ill Inst Technol, BS, 67; Cornell Univ, MS, 69, PhD(comput sci), 71. *Prof Exp:* From asst prof to assoc prof, 70-82, PROF COMPUTER SCI, UNIV ILL, URBANA-CHAMPAIGN, 82- *Mem:* Asn Computer Mach; Soc Indust & Appl Math; Am Math Soc; Math Asn Am. *Res:* Design and analysis of algorithms and data structures for non-numerical problems such as sorting, searching, graph and tree manipulation and exhaustive search. *Mailing Add:* Dept Computer Sci Univ Ill 1304 W Springfield Ave Urbana IL 61801-2987

REINGOLD, HAIM, b Lodz, Poland, Mar 16, 10; nat US; m 66; c 3. MATHEMATICS. *Educ:* Univ Cincinnati, AB, 33, AM, 34, PhD(math), 38. *Prof Exp:* Instr math, Univ Cincinnati, 35-36; prof & head dept, Our Lady Cincinnati Col, 38-42; supvr instr, Signal Corps Training Schs, 42-43, from asst prof to assoc prof, 43-56, PROF MATH, ILL INST TECHNOL, 56-, DIR EVE DIV, 46-, CHMN DEPT MATH, 54-; PROF MATH, MUNDELEIN COL, 84- *Concurrent Pos:* Actg chmn dept math, Ill Inst Technol, 51-54; prof math, Ind Univ Northwest & 75-82, Purdue Univ Calumet, 82-84. *Mem:* AAAS; Am Math Soc; Am Soc Eng Educ; Math Asn Am. *Res:* Invariants of a system of linear homogeneous differential equations of the second order; generalized determinants of Vandermonde; basic mathematics for engineers and scientists. *Mailing Add:* 1329 E 55th St Chicago IL 60615

REINGOLD, IRVING, b Newark, NJ, Nov 13, 21; m 48; c 2. ELECTRONICS ENGINEERING. *Educ:* Newark Col Eng, BS, 42. *Prof Exp:* Elec mfg engr, Westinghouse Elec Co, NJ, 43-45; proj engr, Air Force Watson Labs, NJ, 45-51; proj engr, Chief Switching Devices Sect, Microwave Tubes Br, Electronics Technol & Devices Lab, NJ, 51-60, dep br chief, 60-66, br chief, Pickup, Display & Storage Devices Br, 66-75, dir, Beam Plasma & Display Div, 75-81, dep dir, elec technol & devices lab, US Army elec res & develop command, 81-85; adj prof & consult, SE Ctr Elec Eng Educ, 85-89; ADJ PROF & CONSULT, GEO-CENTERS, INC, 89- *Concurrent Pos:* Mem adv group on electron devices, US Dept Defense, 81-85; chmn, NJ Coast Sect, Hons & Awards Comt, Inst Elec & Electronics Engrs, 85-90; mem, MIT Electromagnetics Acad, 90-, NJ Inst Technol Workforce 2000 Action Coun, 90- *Honors & Awards:* Tech Leadership Award, US Army Electronic Res & Develop Lab, 62; Soc Info Francis Rice Darne Mem Award, 78; Beatrice Winner Mem Award, Soc Info Display, 88; Commander's Award, Dept Army, 85, Medal Meritorious Civilian Serv; Region One Award, Elec Eng Mgt, Inst Elec & Electronics Engrs, 85, Prof Achievement Award, 90. *Mem:* Fel Inst Elec & Electronics Engrs; fel Soc Info Display; Sigma Xi. *Res:* Research and development in the fields of microwave tubes and devices, pulsers and display devices; microelectronics; integrated circuits; solid state microwave devices. *Mailing Add:* 409 Runyon Ave Deal Park Deal NJ 07723

REINGOLD, IVER DAVID, b Concord, NH, Aug 29, 49; m 74; c 2. THEORETICAL CHEMISTRY. *Educ:* Dartmouth Col, AB, 71; Univ Ore, PhD(chem),76. *Prof Exp:* Res assoc chem, Univ Alta, 77-78; asst prof, Haverford Col, 78-79 & Middlebury Col, 79-86; ASSOC PROF CHEM, JUNIATA COL, 88- *Concurrent Pos:* Vis asst prof, Univ Chicago, 83-84; vis assoc prof chem, Lewis & Clark Col, 86-88. *Mem:* Am Chem Soc. *Res:* Synthetic organic chemistry; theoretically interesting molecules; strained molecules. *Mailing Add:* Dept Chem Juniata Col Huntingdon PA 16652

REINHARD, EDWARD HUMPHREY, b St Louis, Mo, Dec 9, 13; m 40, 76; c 4. MEDICINE. *Educ:* Wash Univ, AB, 35, MD, 39. *Prof Exp:* From instr to prof med, Sch Med, Wash Univ, 43-80, emer prof, 81-; RETIRED. *Mem:* Am Soc Hemat; fel Am Col Physicians; Asn Am Physicians; Int Soc Hemat. *Res:* Hematology; therapy of malignant diseases; treatment of leukemia; anemia associated with malignant diseases. *Mailing Add:* 42 Frederick Lane St Louis MO 63122

REINHARD, JOHN FREDERICK, pharmacology; deceased, see previous edition for last biography

REINHARD, JOHN FREDERICK, JR, b Bronxville, NY, Sept 2, 51; m 76; c 2. NEUROCHEMISTRY. *Educ:* Mass Inst Technol, MS, 77, PhD(neural & endocrine regulation), 80. *Prof Exp:* Res fel pharmacol, Sch Med, Yale Univ, 80-82; SR BIOCHEMIST, WELLCOME RES LABS, 82- *Mem:* Int Soc Neurochem; Soc Neurosci; Am Soc Pharmacol & Exp Therapeut. *Res:* Compensatory neuronal mechanisms which occur in response to damage within the central nervous system; disorders of amino acid metabolism. *Mailing Add:* Wellcome Res Labs 3030 Cornwallis Rd Research Triangle Park NC 27709

REINHARD, KARL R, b Coplay, Pa, Jan 13, 16; m 45; c 3. VETERINARY MEDICINE. *Educ:* Muhlenberg Col, BS, 36; Pa State Univ, MS, 40; Cornell Univ, DVM, 49, PhD(microbiol), 50. *Prof Exp:* Asst animal path res, Pa State Univ, 39-41; prof bact, Univ Ky, 50-51; chief leptospirosis res, Rocky Mountain Lab, USPHS, 51-54, chief infectious dis prog, Arctic Health Res Ctr, Alaska, 54-60, exec secy, Gen Med Study Sect, Div Res Grants, NIH, 60-63, asst to chief, 63-66, chief eval staff, Bur Dis Prev & Environ Control, 66-67, chief prog eval, Div Indian Health, 67-68; prof microbiol & dean col vet med, Okla State Univ, 68-69; chief health status surveillance, Health Prog Systs Ctr, Indian Health Serv, USPHS, 69-79; LECTR, DEPT FAMILY & COMMUNITY MED, UNIV ARIZ, 70- *Concurrent Pos:* Consult, WHO, 73, 75, 76, 77, 80 & 81; NIH Res Fel, 49-50. *Honors & Awards:* Borden Award, 49. *Mem:* Arctic Inst NAm. *Res:* Microbiology; experimental pathogenesis of infectious diseases; semiotics of health documentation; taxonomy of health problems; ecology of disease; research administration; health programs planning and evaluation; lay reporting of disease; group health statistics from automated health records systems. *Mailing Add:* 4911 Hidden Valley Rd Tucson AZ 85715

REINHARDT, CHARLES FRANCIS, b Spring Grove, Ind, Nov 25, 33; m 56; c 4. OCCUPATIONAL MEDICINE, TOXICOLOGY. *Educ:* Wabash Col, BA, 55; Ind Univ, MD, 59; Ohio State Univ, MSc, 64; Am Bd Prev Med, dipl & cert occup med, 67; Am Bd Toxicol, dipl & cert gen toxicol, 80. *Prof Exp:* Plant physician, Chambers Works, 64-66, physiologist, 66-69, chief physiol sect, 69-70, res mgr environ sci, 70-71, asst dir, 71-74, assoc dir, 74-76, DIR, HASKELL LAB TOXICOL & INDUST MED, E I DU PONT DE NEMOURS & CO, INC, 76- *Mem:* AMA; Am Indust Hyg Asn; Am Occup Med Asn; Am Acad Occup Med; Soc Toxicol. *Mailing Add:* Haskell Lab E I du Pont de Nemours & Co Inc PO Box 50 Elkton Rd Newark DE 19714

REINHARDT, DONALD JOSEPH, b New York, NY, Dec 6, 38; m 68; c 3. MICROBIOLOGY, MYCOLOGY. *Educ:* Manhattan Col, BS, 60; Columbia Univ, MA, 62, PhD(microbiol), 66; Am Bd Microbiol, dipl. *Prof Exp:* Teaching asst mycol, Columbia Univ, 61-66; res fel med microbiol, Ctr Dis Control, USPHS, Ga, 66-69; ASSOC PROF MICROBIOL, GA STATE UNIV, 69- *Concurrent Pos:* Expert witness, indust consult, trouble shooting infection control indust settings. *Mem:* Soc Indust Microbiol; Med Mycol Soc Am; fel Am Soc Microbiol; fel Am Acad Microbiol. *Res:* Cell development and physiology of Amebae, fungi, bacteria and algae; clinical medical microbiology; chemical and medical microbiology; epidemiology and infection control in hospitals; thirty publications, two books, excellence in speaking/teaching. *Mailing Add:* Dept Biol Ga State Univ PO Box 4010 Atlanta GA 30302-4010

REINHARDT, HOWARD EARL, b Nezperce, Idaho, Mar 16, 27; m 56; c 3. MATHEMATICAL STATISTICS. *Educ:* Univ Idaho, BS, 49; State Col Wash, MA, 51; Univ Mich, PhD(math), 59. *Prof Exp:* Instr math, State Col Wash, 52-53; from asst prof to prof math, Univ Mont, 57-82, chmn dept, 66-73, dean arts & sci, 82-85 CONSULT, 85- *Mem:* Math Asn Am. *Res:* Statistical inference, particularly parametric and non-parametric hypothesis testing techniques. *Mailing Add:* Dept Math Univ Mont Missoula MT 59812

REINHARDT, JUERGEN, b Eutingen-Baden, WGer, Oct 27, 46; US citizen; m 68; c 2. GEOLOGY, SEDIMENTOLOGY. *Educ:* Brown Univ, AB, 68; Johns Hopkins Univ, PhD(geol), 73. *Prof Exp:* Geologist, Md Geol Surv, 73-75; GEOLOGIST, US GEOL SURV, 75- *Concurrent Pos:* Lectr sedimentology, Johns Hopkins Univ, 73-74. *Honors & Awards:* Coastal Barrier Task Force Unit Award, Dept Interior. *Mem:* Geol Soc Am; Int Soc Sedimentologists; Soc Sedimentary Geol; Am Asn Petrol Geologists. *Res:* Comparative sedimentology of sedimentary basins; carbonate-clastic rock transitions; Cenozoic tectonics; cretaceous stratigraphy; penecontemporaneous deformation structures. *Mailing Add:* US Geol Surv 926 Nat Ctr Reston VA 22092

REINHARDT, RICHARD ALAN, b Berkeley, Calif, Oct 18, 22. INORGANIC CHEMISTRY. *Educ:* Univ Calif, BS, 43, PhD(chem), 47. *Prof Exp:* Asst chem, Univ Calif, 43-44; jr scientist, Univ Chicago, 44-45; jr scientist, Los Alamos Sci Lab, 45-46; asst chem, Univ Calif, 46; instr chem, Cornell Univ, 47-51; res chemist, Wright Air Develop Ctr, Wright-Patterson AFB, 51-53; from asst prof to prof, 54-86, EMER PROF CHEM, NAVAL POSTGRAD SCH, 86- *Mem:* Am Chem Soc; Sigma Xi. *Res:* Kinetics of inorganic redox reactions; transition-metal complexes; thermodynamics of internal explosions. *Mailing Add:* 25045 Valley Pl Carmel CA 93923

REINHARDT, ROBERT MILTON, b New Orleans, La, 1927; m 51; c 2. ORGANIC CHEMISTRY. *Educ:* Tulane Univ, BS, 47. *Prof Exp:* Chemist, Chem Properties Sect, Cotton Fiber Div, Southern Regional Res Ctr, USDA, 47-54, chemist, Chem Finishing Invests, Cotton Chem Lab, 54-61, Wash-Wear Invests, Cotton Finishes Lab, 61-76, sr res chemist, Cotton Textile Chem Lab, 76-85, LEAD SCIENTIST, DYEING & FINISHING, TEXTILE FINISHING CHEM RES, SOUTHERN REGIONAL RES CTR, USDA, 85- *Mem:* Am Chem Soc; Sigma Xi; Am Asn Textile Chemists & Colorists. *Res:* Chemical modification and finishing of cotton and cellulose derivatives; chemistry of crosslinking agents for cellulose; properties of chemically modified cottons; free radical modification of cellulose; dyeing crosslinked cotton. *Mailing Add:* Southern Regional Res Ctr PO Box 19687 New Orleans LA 70179-0687

REINHARDT, WALTER ALBERT, applied physics, aeronautics; deceased, see previous edition for last biography

REINHARDT, WILLIAM NELSON, b Bartlesville, Okla, May 12, 39; div; c 2. MATHEMATICS. *Educ:* Col Wooster, BA, 61; Univ Calif, Berkeley, PhD(math), 67. *Prof Exp:* Asst prof, 67-73, ASSOC PROF MATH, UNIV COLO, BOULDER, 73- *Concurrent Pos:* NSF grants, 67-73; vis prof, Univ Amsterdam, 72-73; mem, Inst Advan Study, 73-74. *Mem:* Am Math Soc; Math Asn Am. *Res:* Set theory and foundations of mathematics; logic; model theory; philosophical logic. *Mailing Add:* Dept Math Campus Box 426 Univ Colo Boulder CO 80309-0426

REINHARDT, WILLIAM PARKER, b San Francisco, Calif, May 22, 42; m 79; c 2. CHEMICAL PHYSICS. *Educ:* Univ Calif, Berkeley, BS, 64; Harvard Univ, AM, 66, PhD(chem physics), 68. *Hon Degrees:* MA, Univ Pa, 85. *Prof Exp:* From instr to assoc prof chem, Harvard Univ, 67-74; prof chem, Univ Colo, Boulder, 74-, chmn dept, 77-80; prof, Dept Chem, Univ Pa, 84-91, chmn dept, 85-88, D M Crow prof, 86-91; PROF CHEM, UNIV WASH, 91- *Concurrent Pos:* Vis fel, Joint Inst Lab Astrophys, 72, fel, 74-; Sloan Found fel, 72, Dreyfus Found teacher scholar, 72-77; Guggenheim Mem fel, 78; fac fel, Coun Res & Creative Work, Univ Colo, 78; Nat Lectr, Sigma Xi, 80-82. *Mem:* Am Chem Soc; fel AAAS; fel Am Phys Soc. *Res:* Atomic and molecular structure; scattering processes; many-body theory as applied to chemical problems; classicial and semiclassical theories of highly excited electronic and vibrational states; atoms in intense fields; classical and quantum chaos; simulation of molecular fluids and surfaces. *Mailing Add:* Chem Dept Univ Wash Seattle WA 98195

REINHART, BRUCE LLOYD, topology; deceased, see previous edition for last biography

REINHART, GREGORY DUNCAN, b Chicago, Ill, Nov 1, 51; m 76; c 1. BIOCHEMISTRY. *Educ:* Univ Ill, BS, 73; Univ Wis, Madison, PhD(biochem), 79. *Prof Exp:* Res fel, Mayo Clin Found, 79-80, res assoc, 80-81, assoc consult biochem, 81-83; instr biochem, Mayo Med Sch, 81-83; ASST PROF, DEPT CHEM, UNIV OKLA, 83- *Mem:* Biophys Soc; NY Acad Sci; Sigma Xi; Am Soc Biol Chemists; Am Chem Soc. *Res:* Regulation of enzyme activity; biophysical properties of enzymes; regulation of carbohydrate metabolism. *Mailing Add:* Dept Chem Univ Okla Norman OK 73019

REINHART, JOHN BELVIN, b Merrill, Wis, Dec 22, 17; m 49; c 6. PSYCHIATRY, PEDIATRICS. *Educ:* Duke Univ, AB, 39; Wake Forest Col, MD, 43. *Prof Exp:* Instr pediat, Bowman Gray Sch Med, Wake Forest Col, 50-52; dir dept psychiat, Children's Hosp, 56-74; dir div behav sci, Children's Hosp, Pittsburgh, 74-83; from asst prof to prof, 56-83, EMER PROF PEDIAT & CHILD PSYCHIAT, SCH MED, UNIV PITTSBURGH, 83-; CONSULT CHILD PSYCHIAT, TREND MENT HEALTH SERV, 83- *Honors & Awards:* Simon Wile Award, Am Acad Child Adolescent Psychiat, 90. *Mem:* Am Psychiat Asn; Am Acad Pediat; Am Acad Child Psychiat. *Res:* Child abuse and neglect; consultation-liaison to psychiatry; failure to thrive; psychosomatic disease in children; brief pediatric-child psychiatry liaison. *Mailing Add:* 34 Hunters Lane Hendersonville NC 28739

REINHART, MICHAEL P, b Lancaster, Pa, Sept 9, 52. BIOCHEMISTRY. *Educ:* Norrisville State Univ, BS, 74; Bryn-Mawr Col, PhD(biol), 80. *Prof Exp:* Postdoctoral fel, Univ Pa, 80-83; asst prof biol, Philadelphia Col Pharmacol & Sci, 83-85; vis scientist, E I du Pont de Nemours & Co, Inc, 85-88; RES CHEMIST, DEPT BIOCHEM LIPIDS, AGR RES SERV, USDA, PHILADELPHIA, 88- *Concurrent Pos:* Nat res serv award, Nat Cancer Inst, NIH, 81, young investr award, 83. *Mem:* Am Asn Accredited Scientists; Am Soc Biochem & Molecular Biol. *Mailing Add:* Dept Biochem Lipids USDA-ARS-ERRC 600 E Mermaid Lane Philadelphia PA 19118

REINHART, RICHARD D, b Austin, Minn, Nov 12, 29; m 80; c 4. FOOD PRODUCT DEVELOPMENT, FOOD ENGINEERING. *Educ:* Univ Minn, BS & BChE, 53. *Prof Exp:* Jr develop engr, Union Carbide Nuclear Co, 53-54, assoc engr, 56-57, sect leader prog planning, 57-59; res engr, 59-63, staff asst prog rev, 63-65, group leader food res, 65-66, dept head spec develop, 66-70, dept head non-cereal breakfast prod develop, 70-80, dept head new ventures develop, 80-83, dept head, prod develop, 83-86, DEPT HEAD DESSERTS DEVELOP, GEN MILLS, INC, 86- *Mem:* Inst Food Technol. *Res:* Gaseous diffusion barrier theory and mechanics; cascade theory and economics; engineering economics; food research and process development; research planning and project evaluation; management; new product development. *Mailing Add:* Gen Mills Inc 9000 Plymouth Ave N Minneapolis MN 55427

REINHART, ROY HERBERT, b Cincinnati, Ohio, Sept 11, 19; m 41; c 3. GEOLOGY, VERTEBRATE PALEONTOLOGY. *Educ:* Miami Univ, AB, 41; Univ Chicago, MS, 49; Univ Calif, Berkeley, PhD(paleont), 52. *Prof Exp:* Asst geol, WTex State Univ, 50-51; from asst prof to assoc prof, 51-62, actg chmn dept, 64-65, PROF GEOL, MIAMI UNIV, 62- *Concurrent Pos:* Fel, Miami Univ, 59; res assoc, Univ Fla, 61-66. *Mem:* Soc Vert Paleont. *Res:* Fossil marine mammals, especially orders Sirenia and Desmostylia of world; correlation of Cenozoic stratigraphy of world. *Mailing Add:* 841 S Maple Ave Oxford OH 45056

REINHART, STANLEY E, JR, b Cincinnati, Ohio, Apr 25, 28; m 52; c 5. ELECTRICAL ENGINEERING. *Educ:* US Mil Acad, BS, 50; Ga Inst Technol, MS, 64, PhD(electromagnetic theory), 66. *Prof Exp:* US Army, 46-, from instr to assoc prof elec eng, 53-77, PROF ELEC ENG & ACTG HEAD DEPT, US MIL ACAD, 77- *Mem:* Inst Elec & Electronics Engrs; Am Soc Eng Educ. *Res:* Electromagnetic theory; near fields of antennas; numerical calculation of fields; undergraduate electrical engineering curricula. *Mailing Add:* 12 Wilson Rd West Point NY 10996

REINHEIMER, JOHN DAVID, b Springfield, Ohio, Dec 23, 20; m 44; c 5. ORGANIC CHEMISTRY. *Educ:* Kenyon Col, AB, 42; Johns Hopkins Univ, AM, 44, PhD(chem), 48. *Prof Exp:* From instr to assoc prof, 48-58, Col Wooster, prof chem, 58-85; RETIRED. *Concurrent Pos:* Von Humboldt fel, 63; vis prof, Univ NC, 58-59, Univ Munich, 62-63; vis scientist, UCSB, 68-69, 73-74, 79-80, 85. *Mem:* Am Chem Soc; Sigma Xi. *Res:* Qualitative and physical organic chemistry; kinetics of organic reactions; kinetics of the aromatic nucleophilic substitution reaction; nuclear magnetic resonance studies on simple molecules and biochemical systems; ring opening reactions. *Mailing Add:* 5750 Vial Real No 303 Carpinteria CA 93013

REINHEIMER, JULIAN, b Philadelphia, Pa, Oct 19, 25; m 56; c 1. PHYSICS. *Educ:* Pa State Col, BS, 49; Univ Minn, MS, 50; NY Univ, PhD(physics), 68. *Prof Exp:* Physicist, Minn Mining & Mfg Co, 50-52; proj engr, Fisher Sci Co, 52-53; res physicist, Inst Coop Res, Univ Pa, 53-58; from staff scientist to sect head, Repub Aviation Corp, 58-64; sect mgr, 64-68, assoc dept head, 68-72, assoc prog dir, 72-76, systs dir, 76-86, SR PROJ ENGR, AEROSPACE CORP, EL SEGUNDO, 86- *Concurrent Pos:* Ed staff, Siam J, 54-56; lectr, Univ Calif, Riverside, Exten, 67. *Mem:* Am Phys Soc; Optical Soc Am; Am Asn Physics Teachers. *Res:* Optics; mathematical physics; solid state physics, radiation effects. *Mailing Add:* 4112 Quinlin Dr Palos Verdes Peninsula CA 90274

REINHOLD, VERNON NYE, b Beverly, Mass, May 13, 31; m 53; c 5. ANALYTICAL BIOCHEMISTRY. *Educ:* Univ NH, BS, 59, MS, 61; Univ Vt, PhD(biochem), 65. *Prof Exp:* AEC fel protein chem, Brookhaven Nat Lab, NY, 65-67; Helen Hay Whitney fel, Mass Inst Technol & Harvard Med Sch, 67-71, jr res assoc chem, Mass Inst Technol, 71-76; LECTR DEPT BIOL CHEM, HARVARD MED SCH, 76- *Mem:* AAAS; Soc Complex Carbohydrates; Am Chem Soc. *Res:* Gas chromatography-mass spectrometry; computer assisted analysis of biochemical components; glycoprotein structure and carbohydrate sequence analysis via gas liquid chromatography-mass spectrometry; protein and organic chemistry. *Mailing Add:* Harvard Sch Pub Health 665 Huntington Ave Boston MA 02115

REINIG, JAMES WILLIAM, b Augusta, Ga, May 20, 54. MAGNETIC RESONANCE IMAGING, DIAGNOSTIC RADIOLOGY. *Educ:* Harvard Col, AB, 76; Med Univ SC, MD, 80; Am Bd Radiol, cert radiol, 84; Am Bd Nuclear Med, cert nuclear med, 84. *Prof Exp:* Fac diagnostic radiol, NIH, 84-86; CLIN DIR, ANNE ARUNDEL MAGNETIC RESONANCE IMAGING, 86- *Concurrent Pos:* Vis fac, Dept Diagnostic Radiol, NIH, 86- *Mem:* Soc Magnetic Resonance Med; Radiol Soc of NAm; Am Col Radiol; Am Roentgen Ray Soc; AMA. *Res:* Application of clinical magnetic resonance imaging of the body. *Mailing Add:* 235 Jennifer Rd Annapolis MD 21401

REINIG, WILLIAM CHARLES, b New York, NY, June 5, 24; m 49; c 2. HEALTH PHYSICS. *Educ:* Polytech Inst Brooklyn, BME, 45; Am Bd Health Physics, dipl. *Prof Exp:* Med engr, Hanford Works, Gen Elec Co, 46-48; assoc health physicist, Brookhaven Nat Lab, 48-51; area supvr health physics, Savannah River Plant, E I du Pont de Nemours & Co, Inc, 51-61, chief tech supvr, 61-65, sr res supvr environ effects, 65-76, res mgr environ anal & planning, 76-78, supt health protection dept, 78-88, gen supt tech dept, 88-89, DEP GEN MGR, SAVANNAH RIVER SITE, WESTINGHOUSE SAVANNAH RIVER CO, 89- *Concurrent Pos:* Chmn, Am Bd Health Physics, 74-76; chmn comt tritium measurement & comt nuclear decommissioning, Nat Coun Radiation Protection & Measurements. *Mem:* Health Physics Soc (secy, 64-66, pres, 80-81); Am Acad Health Physics (dir, 85-87). *Res:* Environmental radiation and radioactivity; radiological health. *Mailing Add:* Savannah River Site Westinghouse Savannah River Co Aiken SC 29808

REINING, PRISCILLA COPELAND, b Chicago, Ill, Mar 11, 23; m 44, 84; c 3. ANTHROPOLOGY, AGRARIAN SYSTEMS. *Educ:* Univ Chicago, AB, 45, AM, 49, PhD(anthrop), 67. *Prof Exp:* Sr res fel, E African Inst Social Res, 51-55; lectr, Univ Minn, 56-59 & Howard Univ, 60-65; res assoc, Cath Univ, 66-68, Smithsonian Inst, 66 & 68-70; consult, proj res, Int Bank Reconstuct & Develop, 72 & AID, 73; res assoc, 74, PROG DIR, ADMIN, AAAS, 75- *Concurrent Pos:* Lectr, Comt Space Res, Brazil, 74; mem, US deleg UN Conf Desertification, Nairobi, Kenya, 77; mem, Bd Sci & Technol Int Develop, Nat Acad Sci, 78-80 & adv comt Sahel, 79-81; mem, experts group desertification, UN Environ Prog, 79 & 81, IV Int Conf AIDS, Stockholm, 88. *Mem:* Fel AAAS; fel African Studies Asn; fel Am Anthrop Asn. *Res:* Africa land tenure and land use, population and kinship; desertification processes in Africa and other Third World countries; decentralized energy systems in villages; use of remote sensing methodology in the social sciences. *Mailing Add:* 3601 Rittenhouse St Washington DC 20005

REININGER, EDWARD JOSEPH, b Chicago, Ill, Dec 30, 29; c 2. PHYSIOLOGY. *Educ:* Univ Ill, BS, 50, MS, 52; Ohio State Univ, PhD(physiol), 57. *Prof Exp:* From asst to instr physiol, Ohio State Univ, 52-58; lectr, McGill Univ, 58-60, asst prof, 60-71; assoc prof, Sch Med, Ind Univ, Terre Haute, 71-74; prof physiol, Sch Med, Southern Ill Univ, Springfield, 74-76; prof & chmn dept physiol & pharm, 77-78, PROF PHYSIOL, SCH MED, UNIV CENT CARIBE, 78- *Concurrent Pos:* Grants, Que Heart Found, 59-65 & Med Res Coun Can, 66-70; summer res prog, Minority Hypertension, Univ Fla, 81-83. *Mem:* Am Physiol Soc; Natural Hist Soc PR (treas, 84-85 & vpres, 86). *Res:* Cardiovascular effects of cardiac pacing and feeding; sighing in man and spontaneous gasps and post-gasp apnea in dogs; mechanism that prevents atelectasis; measurement of cardiac output; minicomputer programming for teaching and research. *Mailing Add:* 330 SE Second St No 403G Hallandale FL 33009

REINISCH, RONALD FABIAN, chemistry, for more information see previous edition

REINKE, DAVID ALBERT, b Manitowoc, Wis, May 15, 33; m 56; c 2. PHARMACOLOGY, PHYSIOLOGY. *Educ:* Univ Wis, BS, 55; Univ Mich, MBA, 60, PhD(pharmacol), 64. *Prof Exp:* Res chemist, Dow Corning Corp, 55-58; ASSOC PROF PHARMACOL & TOXICOL, MICH STATE UNIV 64-, VICE CHAIRPERSON, DEPT PHARMACOL & TOXICOL, 87- *Concurrent Pos:* Mem, Medicinal Chem Div, Am Chem Soc. *Mem:* Am Soc Pharmacol & Exp Therapeut. *Mailing Add:* Dept Pharmacol Mich State Univ East Lansing MI 48824-1317

REINKE, LESTER ALLEN, b Davenport, Nebr, Sept 29, 46; m 68; c 2. ALCOHOL, FREE RADICALS. *Educ:* Univ Nebr-Lincoln, BS, 69, MS, 75; Univ Nebr Med Ctr-Omaha, PhD(biomed chem), 77. *Prof Exp:* Fel pharmacol, Univ NC, 77-80, res asst prof, 80; ASSOC PROF PHARMACOL, COL MED, UNIV OKLA, 80- *Concurrent Pos:* Provost res award, Univ Okla Health Sci Ctr, 84; hon lectr, Mid-Am State Univ Asn, 86-87. *Mem:* Am Soc Pharmacol & Exp Therapeut; Int Soc Biomed Res Alcoholism; Int Soc Study Xenobiotics. *Res:* Detection of free radical intermediates in biological samples, their biological effects; mechanisms of cellular protection against radicals; alcohol and free radicals. *Mailing Add:* Dept Pharmacol BMSB Univ Okla Health Sci Ctr Oklahoma City OK 73190

REINKE, WILLIAM ANDREW, b Cleveland, Ohio, Aug 10, 28; div; c 4. BIOSTATISTICS. *Educ:* Kenyon Col, BA, 49; Univ Pa, MBA, 50; Case Western Reserve Univ, PhD(statist), 61. *Prof Exp:* Staff asst to controller, Warner & Swasey Co, 50-55; syst analyst, US Steel Corp, 55-56; statistician, Union Carbide Corp, 56-59; instr statist, Case Western Reserve Univ, 59-61; sr res mathematician, Corning Glass Works, 61-63; asst prof biostatist, Univ Md, 63-64; asst dean, 74-76, assoc dean, 76-77, PROF INT HEALTH, SCH HYG, JOHNS HOPKINS UNIV, 70- *Concurrent Pos:* Assoc ed, Opers Res, 71-74; treas, Univ Assoc for Int Health Inc, 73-; mem comt tech consult, Nat Ctr Health Statist, 73; mem, Nursing Res & Educ Adv Comt, 74-78. *Mem:* Inst Mgt Sci; fel Am Pub Health Asn; Am Statist Asn. *Res:* Health planning methodology; health practice research in relation to health services delivery. *Mailing Add:* Dept Int Health Johns Hopkins Univ Sch Hyg 615 N Wolfe St Baltimore MD 21205

REINKING, LARRY NORMAN, STOMACH-RENAL INTERACTION. *Educ:* Univ Mont, PhD(zool), 78. *Prof Exp:* ASSOC PROF BIOL, MILLERSVILLE UNIV, 81- *Res:* Upper urinary tract. *Mailing Add:* Dept Biol Millersville Univ Millersville PA 17551

REINMUTH, OSCAR MCNAUGHTON, b Lincoln, Nebr, Oct 23, 27; m 51, 80; c 3. NEUROLOGY, INTERNAL MEDICINE. *Educ:* Univ Tex, AB, 48; Duke Univ, MD, 52. *Prof Exp:* Intern, Duke Hosp, 52-53; asst resident, New Haven Med Ctr, 53-55; asst resident to chief resident, Boston City Hosp, 55-57; from assoc prof to prof neurol, Sch Med, Univ Miami, 58-77; PROF & CHMN DEPT NEUROL, SCH MED, UNIV PITTSBURGH, 77- *Concurrent Pos:* NIH trainee, Med Sch, Yale Univ, 54-55; lectr neurol, Sargent Col, Boston Univ, 55-56; teaching fel, Harvard Med Sch, 56-57; NIH spec trainee, Nat Hosp, Queen's Square, London, 57-58; consult, Nat Inst Neurol Dis & Stroke, 68-; consult adv comt, Sect Head Injury & Stroke, NIH, 73-; mem coun stroke, Am Heart Asn, ed, Stroke, 87- *Mem:* Am Acad Neurol (vpres, 71-75); Am Neurol Asn (vpres, 78-79); fel Am Col Physicians; fel Am Heart Asn; Soc Neurosci; Sigma Xi. *Res:* Cerebral circulation and metabolism in humans and experimental animals; cerebral vascular disease; movement disorders. *Mailing Add:* Univ Pittsburgh Sch Med Dept Neurol 322 Scaife Hall Pittsburgh PA 15261

REINMUTH, WILLIAM HENRY, b Baltimore, Md, Sept 1, 32; m 64; c 1. ANALYTICAL CHEMISTRY. *Educ:* Univ Chicago, AB, 52, MS, 54; Mass Inst Technol, PhD(chem), 57. *Prof Exp:* From instr to assoc prof, 57-64, PROF CHEM, COLUMBIA UNIV, 64- *Concurrent Pos:* Sloan Found fel, 62-66; Guggenheim fel, 66-67. *Mem:* Am Chem Soc; Electrochem Soc. *Res:* Electroanalytical chemistry. *Mailing Add:* Riverside Dr No 4C New York NY 10025

REINSBOROUGH, VINCENT CONRAD, b Buctouche, NB, May 14, 35; m 77; c 3. PHYSICAL CHEMISTRY. *Educ:* Univ Toronto, BA, 58, MA, 59, STB, 64; Univ Tasmania, PhD(chem), 69. *Prof Exp:* Teacher high sch, Ont, 59-61; lectr chem, Univ St Michael's Col, 61-65; Nat Res Coun Can fel, Univ Toronto, 69-70; from asst prof to assoc prof, 70-83, PROF CHEM, MT ALLISON UNIV, 83-, DEPT HEAD, 87- *Concurrent Pos:* Vis prof Fritz-Haber Inst, Berlin, 84-85. *Mem:* Chem Inst Can; Royal Soc Chem; Electrochem Soc; Am Chem Soc. *Res:* Solubilization, kinetics and catalysis in micellar solutions; dynamics in solution of cyclodextrin inclusions. *Mailing Add:* Dept Chem Mt Allison Univ Sackville NB E0A 3C0 Can

REINSCHMIDT, KENNETH F(RANK), b Cincinnati, Ohio, Mar 26, 38; m 67. CIVIL ENGINEERING, ENERGY ECONOMICS. *Educ:* Mass Inst Technol, SB, 60, SM, 62, PhD(civil eng), 65. *Prof Exp:* From asst prof to assoc prof civil eng, Mass Inst Technol, 65-73, sr res assoc, 73-75; consult engr, 75-80, VPRES & SR CONSULT ENGR, STONE & WEBSTER ENG CORP, 80-, PRES, STONE & WEBSTER ADVAN SYSTS DEVELOP SERV, 88- *Mem:* Nat Acad Eng; Am Soc Civil Engrs; Opers Res Soc Am; Inst Mgt Sci; Sigma Xi; AAAS. *Res:* Historical development of building science; expert systems and artificial intelligence; computer modeling and operations research; mathematical programming and optimization; computer-aided design; neural networks; three-dimensional computer graphics; engineering, manufacturing, construction, production planning, scheduling, plant operations and control. *Mailing Add:* 20 Tahattawan Rd Littleton MA 01460

REINSEL, GREGORY CHARLES, b Wilkinsburg, Pa, Mar 10, 48; m 76; c 2. STATISTICAL TIME SERIES ANALYSIS, MULTIVARIATE STATISTICAL METHODS. *Educ:* Univ Pittsburgh, BS, 70, MA, 72, PhD(math & statist), 76. *Prof Exp:* From asst prof to assoc prof, 76-87, PROF STATIST, UNIV WIS- MADISON, 87- *Concurrent Pos:* Prin investr, NASA, Nat Oceanic & Atmospheric Admin & Chem Mfrs Asn, 81-; res assoc, Grad Sch Bus, Univ Chicago, 84. *Mem:* Am Statist Asn; Inst Math Statist; Royal Statist Soc; Am Geophys Union. *Res:* Development of useful statistical methods and related theory for the analysis of time series data, especially multivariate time series; trend analysis of atmospheric ozone and temperature time series data for global changes. *Mailing Add:* Dept Statist Univ Wis 1210 Dayton St Madison WI 53706

REINSTEIN, LAWRENCE ELLIOT, b New York, NY, Apr 18, 45; m 68; c 3. MEDICAL PHYSICS. *Educ:* Brooklyn Col, BS, 66; Yale Univ, MS, 68; Boston Univ, PhD(physics), 75. *Prof Exp:* Fel med physics, Mem Sloan-Kettering Cancer Inst, 75-76; physicist radiation oncol, RI Hosp, Providence, 76-; MEM STAFF DEPT RADIATION MED, BROWN UNIV. *Concurrent Pos:* Asst prof bio-med, Brown Univ, 76- *Mem:* Am Phys Soc; Am Asn Physicists Med. *Res:* Electron beam radiation as used in treatment of cancer; 3- dimensional treatment planning. *Mailing Add:* Dept Radiation Med SUNY Health Sci Ctr Stony Brook NY 11794

REINTJES, J FRANCIS, b Troy, NY, Feb 19, 12; m 42; c 3. ELECTRICAL ENGINEERING. *Educ:* Rensselaer Polytech Inst, BS, 33, ME, 34. *Prof Exp:* Elec engr, Gen Motors Corp, 36-37; asst prof elec eng, Manhattan Col, 37-43; vis asst prof elec eng, Mass Inst Technol, 43-45; elec engr, Gen Elec Co, 46-47; prof, 47-77, EMER PROF & SR LECTR ELEC ENG, MASS INST TECHNOL, 77- *Concurrent Pos:* Dir, Electronic Systs Lab, Mass Inst Technol, 53-73 & Co-Op Prog Elec Eng, 60-67. *Mem:* Fel Inst Elec & Electronics Engrs; Am Soc Elec Eng. *Res:* Pulse doppler radar; information storage and retrieval; digital encoding of images. *Mailing Add:* Mass Inst Technol Rm 35-418 77 Mass Ave Cambridge MA 02139

REINTJES, JOHN FRANCIS, JR, b Boston, Mass, Dec 7, 45; m 71; c 1. LASERS, QUANTUM OPTICS. *Educ:* Mass Inst Technol, BS, 66; Harvard Univ, PhD(appl physics), 72. *Prof Exp:* Postdoctoral fel nonlinear optics res, Int Bus Mach Corp Watson Res Ctr, 71-73; RES PHYSICIST, NAVAL RES LAB, 73- *Concurrent Pos:* Consult, Lawrence Livermore Lab, 71; lectr, Catholic Univ, 84- *Mem:* Am Phys Soc; fel Optical Soc Am; Sigma Xi; affil Inst Elec & Electronics Engrs. *Res:* Investigation of the use of nonlinear optics for extreme ultraviolet generation, higher order harmonic generation, Raman beam cleanup and optical phase conjugation; low light level image amplification. *Mailing Add:* Naval Res Lab Washington DC 20375

REINTJES, MARTEN, b Meeden, Netherlands, Mar 13, 32; US citizen; m 57; c 2. ORGANIC CHEMISTRY. *Educ:* Univ Calif, Riverside, BA, 59, PhD(chem), 66. *Prof Exp:* Analyst, Chemische Fabriek Flebo, Netherlands, 49-52; asst chemist, Orange County Sanit Dist, Calif, 55-57; res chemist, Sunkist Growers, Inc, 59-62 & Arapahoe Chem Div, Syntex Corp, Colo, 65-68; from res chemist to res group leader, 68-73, sect leader, 73-77, sect supvr, Olympic Res Div, 78-82, res assoc, 83, mgr tech admin, res ctr, 84-88 & PROD SAFETY ADMINR, 89- *Mem:* Soc Petrol Engrs; Am Forestry Asn; Am Chem Soc. *Res:* Organic synthesis and natural products; process development; organo-boron, boron-hydride and carborane chemistry; silvichemicals. *Mailing Add:* ITT Rayonier Inc Res Ctr Shelton WA 98584

REIS, ARTHUR HENRY, JR, b Chicago, Ill, Nov 6, 46; m 70; c 2. INORGANIC CHEMISTRY. *Educ:* Cornell Col, Iowa, BA, 68; Harvard Univ, MA, 69, PhD(chem), 72. *Prof Exp:* Teaching fel org chem, Harvard Univ, 72; space systs analyst satellite tracking, US Air Force, Ent AFB, Colo, 72-73; space oper officer satellite tracking, USAF, Thule AFB, Greenland, 73-74; appointee, Argonne Nat Lab, Argonne, Ill, 74-75, res assoc inorg chem, 75-76, asst chemist, 76-79; adminr chem dept, Brandeis Univ, 79-82, assoc prof chem, 80-83, dir, 82-86, assoc dean, sci resources & planning, 86-89, ASSOC PROVOST, BRANDEIS UNIV, 90- *Concurrent Pos:* Consult, Picker X-Ray Corp, 71-72; assoc proj dir, Undergrad Res Particip Proj, NSF-GTE Corp, 81; Nat Coun Univ Res Adminr, 81; dir Forefront topics in sci proj, Brandeis Univ, 83-; New Eng Coun, 84. *Mem:* Am Chem Soc; Am Crystallog Asn; Coun Chem Res. *Res:* Synthesis and structural characterization by x-ray and neutron diffraction of one-dimensional transition metal conductors, pulsed neutron diffraction and exafs studies of metal complexes in zeolites. *Mailing Add:* Off Provost Brandeis Univ Irving 104 Waltham MA 02254-9110

REIS, DONALD J, b New York, NY, Sept 9, 31. NEUROLOGY, NEUROBIOLOGY. *Educ:* Cornell Univ, AB, 53, MD, 56. *Prof Exp:* Res anatomist, Univ Calif, Los Angeles, 54-55; intern med, New York Hosp, 56-57; resident neurol, Boston City Hosp, 57-59; res assoc neurophysiol, NIMH, 60-62; from asst prof to prof neurol, 63-81, GEORGE C COTIAS DISTINGUISHED PROF NEUROL, CORNELL UNIV, 81- *Concurrent Pos:* Teaching fel, Harvard Med Sch, 57-59; United Cerebral Palsy Found fel brain res, Nat Hosp, London, Eng, 59-60; Nat Inst Neurol Dis & Blindness spec fel, Nobel Neurophysiol Inst, Karolinska Inst, Sweden, 62-63; Nat Inst Neurol Dis & Blindness career develop res award, 66-; USPHS career develop award; mem, Karolinska Inst, Sweden, 59-60; vis scientist, Chiba Univ Med Sch, Japan, 63; vis scientist, Lab Clin Sci, NIMH, 70. *Honors & Awards:* CIBA Medal, Am Heart Asn, 87. *Mem:* Am Acad Neurol; Am Physiol Soc; Am Soc Pharmacol & Exp Therapeut; Am Soc Clin Invest; Am Soc Neurochem; Sigma Xi; Am Asn Physicians. *Res:* Central neural autonomic regulation; molecular mechanisms in central nervous regulation of cardiovascular function; neural mechanisms of emotive behavior; central neurotransmitters; brain monoamines and behavior. *Mailing Add:* Cornell Univ Med Col 1300 York Ave New York NY 10021

REIS, IRVIN L, b Lincoln, Nebr, Oct 5, 26; m 51; c 2. MECHANICAL & INDUSTRIAL ENGINEERING. *Educ:* Univ Nebr, BSME, 49, MSME, 50; Univ Ill, PhD(indust eng), 57. *Prof Exp:* Job analyst, Univ Nebr, 49, supvr insts, 50-53; lectr indust eng, Univ Ill, 53-57; res engr, Lincoln Steel Corp, 57; assoc prof mech eng, Univ Nebr, 57-59; prof indust eng & head dept, Kans State Univ, 59-62; prof, Univ Ark, 62-64; vis prof mech eng, Univ Tex, 64-66; prof mech eng & head dept, Mont State Univ, 66-70; prof mech eng, Lamar Univ, 70-80, dept head, 77-80; mem fac, Indust Eng Dept, Mont State Univ, 80-84; RETIRED. *Concurrent Pos:* Indust training consult, 50-62; consult, City of Lincoln, Nebr, 58-59 & Bayer & McElrath, Mich, 64-; prof indust eng & head dept, Mont State Univ, 66-67; partner, Mgt Insts Unlimited, 75- *Mem:* Am Soc Eng Educ; Am Inst Indust Engrs; Am Soc Mech Engrs. *Res:* Probabilistic models; conveyer theory; economics. *Mailing Add:* Rte 1 Box 160 Hinsdale AR 72738

REIS, PAUL G(EORGE), b St Cloud, Minn, May 3, 25; m 53; c 6. CHEMICAL ENGINEERING. *Educ:* Northwestern Univ, BS, 45, MS, 49; Univ Wis, PhD(chem eng), 54. *Prof Exp:* Chem engr, process develop, E I du Pont de Nemours & Co, 45 & 47-48; asst, Univ Wis, 49-52; CHEM ENGR PROCESS DEVELOP, PIGMENTS DEPT, E I DU PONT DE NEMOURS & CO, INC, 54- *Mem:* Am Inst Chem Engrs; Am Chem Soc. *Res:* Chemical equilibrium and kinetics. *Mailing Add:* Four Toby Ct Sherwood Park 11 Wilmington DE 19808

REIS, WALTER JOSEPH, b Worzburg, Ger, Aug 5, 18; nat US; m 43; c 3. PSYCHIATRY, CLINICAL PSYCHOLOGY. *Educ:* Univ Gonzaga, BPh, 47; City Col New York, BS, 47; Western Reserve Univ, PhD(psychol), 51; Emory Univ, MD, 55; Am Bd Psychiat & Neurol, dipl, 62; Pittsburgh Psychanal Inst, cert, 67. *Prof Exp:* Intern, USPHS Hosp, Norfolk, Va, 55-56; fel, Western Psychiat Inst, Pa, 56-59; clin instr psychiat, Univ Pittsburgh, 59-63, clin asst prof, 63-69; PSYCHIATRIST & PSYCHOANALYST, PSYCHIAT ASSOCS, 69-; CLIN PROF PSYCHIAT, SCH MED, UNIV PITTSBURGH, 90- *Concurrent Pos:* Mem fac, Pittsburgh Psychoanal Inst, 67-, pvt practr psychiat, 59-; consult, Dixmont State Hosp, Sewickley, Pa, 68-73; training analyst, Pittsburgh Psychoanal Inst, Pa, 72-; clin assoc prof, Sch Med, Univ Pittsburgh, 74-90. *Mem:* Fel Am Psychiat Asn; Am Psychol Asn; Soc Personality Assessment. *Res:* Psychotherapy; psychoanalysis. *Mailing Add:* 230 N Craig St Pittsburgh PA 15213

REISA, JAMES JOSEPH, JR, b Oak Park, Ill, Dec 13, 41; div. ENVIRONMENTAL BIOLOGY, SCIENCE POLICY. *Educ:* Loyola Univ, Chicago, BS, 66; Northwestern Univ, Evanston, Ill, MS, 68, PhD(biol sci), 71. *Prof Exp:* US Environ Protection Agency res fel, Dept Biol Sci, Northwestern Univ, 71-72; staff biologist, Argonne Nat Lab, Argonne, Ill, 72-74; staff mem, Coun Environ Qual, Exec Off President, Washington, DC, 74-75, coordr environ monitoring prog, 75-77, sr staff mem, 77-78; dir environ rev div, Off Toxic Substances, US Environ Protection Agency, 78-79, assoc dep asst adminr for toxic substances, 79-81, dir, Off Explor Res, 81-82; vpres, IDEA Tech Assoc, Alexandria, Va, 82-86; assoc dir, 86-88, DIR, BD ENVIRON STUDIES & TOXICOL, NAT RES COUN, NAT ACAD SCI, 88- *Concurrent Pos:* Vis lectr biol, Mundelein Col, Chicago, 70-71, vis asst prof biol, 71-72; chmn Fed Interagency Task Force Air Qual Indicators, 75-77; chmn, President's Task Force Environ Data & Monitoring, 77-78; chmn, Toxics Res Comt, US Environ Protection Agency, 79-81; chmn, appl ecol sect, Ecol Soc Am, 80-82; mem, bd dir, Soc Environ Toxicol & Chem, 81-82. *Honors & Awards:* Bronze Medal, US Environ Protection Agency, 80. *Mem:* Ecol Soc Am; AAAS; Am Soc Limnol & Oceanog; Am Inst Biol Sci; Sigma Xi; Soc Environ Toxicol & Chem. *Res:* Assessment of environmental fate and effects of toxic chemicals; science policy and regulatory decision making; environmental data systems. *Mailing Add:* Bd Environ Studies & Toxicol MH354 Nat Acad Sci 2101 Constitution Ave NW Washington DC 20418

REISBERG, BORIS ELLIOTT, b New York, NY, Dec 12, 35. MEDICINE, INFECTIOUS DISEASES. *Educ:* Brown Univ, AB, 57; State Univ NY, MD, 61; Am Bd Internal Med, dipl, 68, cert infectious dis, 74. *Prof Exp:* Instr internal med, New Eng Ctr Hosp, Tufts Univ, 64-66; NIH fel, 66-67; assoc, 68-69, ASST PROF INTERNAL MED, NORTHWESTERN UNIV, CHICAGO, 69- *Mailing Add:* Northwestern Mem Hosp 251 E Chicago Ave Chicago IL 60611

REISBERG, JOSEPH, b New York, NY, May 10, 21; m 53. CHEMISTRY. *Educ:* City Col New York, BS, 43. *Prof Exp:* Res chemist, SAM Lab, Columbia Univ, 43-44, Los Alamos Sci Lab, Univ Calif, 44-47 & Colgate-Palmolive-Peet Corp, NJ, 47-48; res chemist, explor & prod res lab, 49-66, res assoc, 66-70, Shell Lab, Rijswijk, Holland, 70-71, sr res assoc, 71-74, res assoc, Shell Develop Co, Tex, 74-; RETIRED. *Mem:* AAAS; Am Chem Soc; Am Inst Mining, Metall & Petrol Engrs; NY Acad Sci. *Res:* Surface and colloid chemistry; unconventional methods for petroleum recovery. *Mailing Add:* 5508 Shadowcrest Houston TX 77096

REISBIG, RONALD LUTHER, b Kalamazoo, Mich, Jan 31, 38; m 58; c 4. MECHANICAL ENGINEERING, THERMODYNAMICS. *Educ:* Mich State Univ, BSME, 60, PhD(mech eng), 66; Univ Wash, MSME, 63. *Prof Exp:* Res engr, Boeing Co, 60-63; asst teaching, Mich State Univ, 63-64; asst prof eng, Western Mich Univ, 65-66; asst prof mech eng, Wayne State Univ, 66-69; assoc prof, Univ Mo-Rolla, 69-75; prof & dean, Victoria Campus, Univ Houston, 75-77; dean eng, Western New Eng Col, 77-84; dean, 84-85, vchancellor acad affairs, 85-89, PROF AEROSPACE ENG, COL ENG & AVIATION, EMBRY-RIDDLE AERONAUT UNIV, 89- *Concurrent Pos:* Consult, Nat Waterlift Co, Mich, 65-66; Westinghouse scholar, Mich State Univ, 59-60. *Mem:* Am Soc Mech Engrs; Am Soc Eng Educ; Instrument Soc Am; Sigma Xi. *Res:* Thermal science; interferometric holography; solar energy. *Mailing Add:* Col Eng-Aviation Embry-Riddle Aeronaut Univ Daytona Beach FL 32114

REISCH, BRUCE IRVING, b New York, NY, July 23, 55; m 84; c 1. PLANT BREEDING, GENETICS & TISSUE CULTURE. *Educ:* Cornell Univ, BS, 76; Univ Wis, Madison, MS, 78, PhD(plant breeding), 80. *Prof Exp:* asst prof, 80-86, ASSOC PROF VITICULT, NY STATE AGR EXP STA, CORNELL UNIV, 86- *Concurrent Pos:* Mem, Grape Comm Adv Comt, Nat Plant Germplasm Syst; invited lectr, NATO Advan Study Inst Plant Biotechnol, 87; vis prof, Univ Calif, Riverside, 90. *Mem:* Int Asn Plant Tissue Culture; Am Soc Enol; Crop Sci Soc Am; Am Soc Hort Sci; Sigma Xi. *Res:* Application of new technology to plant breeding, including gene transformation, tissue culture, grape genetics and grape breeding. *Mailing Add:* 57 High St Geneva NY 14456

REISCH, KENNETH WILLIAM, b Southington, Conn, Oct 7, 29; m 52; c 4. HORTICULTURE. *Educ:* Univ Conn, BSc, 52; Ohio State Univ, MSc, 53, PhD(hort), 56. *Prof Exp:* From instr to assoc prof, 53-66, PROF HORT, OHIO STATE UNIV, 66-, ASSOC DEAN COL AGR & HOME ECON, 72- *Concurrent Pos:* Chmn, Res Instr, Comt Org & Policy, Nat Asn State Univ, & Land Grant Col, 86. *Mem:* Am Soc Hort Sci; Sigma Xi. *Res:* Physiological and taxonomical studies with woody ornamental plants, especially growth, reproduction and nutrition. *Mailing Add:* 6528 Masefield St Columbus OH 43085

REISCHMAN, MICHAEL MACK, b Barnesville, Ohio, Sept 26, 42; m 84; c 3. EDUCATION ADMINISTRATION. *Educ:* NMex State Univ, BS, 67, MS, 69; Okla State Univ, PhD(mech engr), 73. *Prof Exp:* Res assoc, Nat Res Coun, Nat Acad Sci, 73-74; res engr, Naval Ocean Systs Ctr, 74-78, br head, 78-83; sci officer, 83-86, PROG DIR, OFF NAVAL RES, 86- *Mem:* Am Soc Mech Engrs; Am Phys Soc. *Res:* Experimental fluid dynamics; laser anemometry; turbulence; drag reduction; hydroacoustics; flow visualization; image processing. *Mailing Add:* Off Naval Res 800 N Quincy Code 1132F Arlington VA 22217

REISCHMAN, PLACIDUS GEORGE, b South Bend, Wash, Sept 15, 26. ZOOLOGY. *Educ:* St Martin's Col, BA, 50; Cath Univ Am, MSc, 57, PhD, 60. *Prof Exp:* Instr, 55-56, PROF BIOL, ST MARTIN'S COL, 59- *Mem:* AAAS. *Res:* Marine invertebrate zoology; parasitic Crustacea; Rhizocephala. *Mailing Add:* Dept Math & Sci St Martin's Col Lacey WA 98503

REISEL, ROBERT BENEDICT, b Chicago, Ill, Apr 27, 25; m 61; c 3. MATHEMATICS. *Educ:* DePaul Univ, BS, 49; Univ Chicago, MS, 51; Northwestern Univ, PhD(math), 54. *Prof Exp:* From instr to asst prof, 54-63, ASSOC PROF MATH, LOYOLA UNIV CHICAGO, 63- *Mem:* Am Math Soc; Math Asn Am. *Res:* Associative algebras. *Mailing Add:* 5052 N Nordica Ave Chicago IL 60656

REISEN, WILLIAM KENNETH, b Jersey City, NJ, Feb 11, 46; m 71; c 2. MEDICAL ENTOMOLOGY, VECTOR ECOLOGY. *Educ:* Univ Del, BS, 67; Clemson Univ, MS, 68; Univ Okla, PhD(zool), 74. *Prof Exp:* Teaching asst zool, dept Entomol & zool, Clemson Univ, 67-68; capt US Armed Forces & med entomologist, 1st Med Serv Wing, Clark Air Base, Philippines, 69-71; teaching asst zool, dept zool, Univ Okla, 71-74; asst prof med, Int Health Prog, Sch Med, Univ Md, 74-80; RES ENTOMOLOGIST & DIR, ARBOVIRUS FIELD ST, SCH PUB HEALTH, UNIV CALIF, BERKELEY, 80- *Concurrent Pos:* Investr ecol & control arboviruses, NAIAD, NIH, 84-; consult mosquito vector field studies, Bd Sci & Technol Int Develop & US Agency Int Develop, 83-, Nepal Malaria Eradication Orgn Vector Biol & Control Proj, 86-, S Calif Mosquito Control Orgn, 87-; co-ed, J Med Entomol, 88-; co-investr epidemiol & control arboviruses, Univ Calif Mosquito Res Funds. *Mem:* Royal Soc Trop Med Hyg; Am Soc Trop Med & Hyg; Entomol Soc Am; Soc Vector Ecologists; Am Mosquito Control Asn. *Res:* Ecological relationships among environmental factors, mosquito bionomics and the prevalance of mosquito-borne disease (malaria and encephalitis); range of specific topics, from mosquito demography to control. *Mailing Add:* Arbovirus Field Sta 4705 Allen Rd Bakersfield CA 93312

REISENAUER, HUBERT MICHAEL, b Portland, Ore, Mar 11, 20; m 45; c 2. SOIL SCIENCE. *Educ:* Univ Idaho, BS, 41; NC State Univ, PhD(agr), 49. *Prof Exp:* Prof soils, Wash State Univ, 49-62; PROF SOIL SCIENCE & SOIL SCIENTIST, UNIV CALIF, DAVIS, 62- *Mem:* AAAS; Am Soc Agron; Soil Sci Soc Am; Int Soil Sci Soc; Sigma Xi. *Res:* Plant nutrition; micronutrients; soil fertility and soil-plant interrelationships. *Mailing Add:* Land Air & Water Res Univ Calif Davis CA 95616

REISER, CASTLE O, b Berthoud, Colo, Dec 21, 12; m 35; c 2. CHEMICAL & NUCLEAR ENGINEERING. *Educ:* Colo Agr & Mech Col, BS, 34; Colo Sch Mines, PE, 38; Univ Wis, PhD(chem eng), 45. *Prof Exp:* Res engr, Pilot Plant, Standard Oil Develop Co, 38-41; instr & res assoc chem eng, Univ Wis, 41-45; asst prof, Univ Colo, 45-46; assoc prof, Okla Agr & Mech Col, 46-47; prof & head dept, Univ Idaho, 47-53; pilot plant supvr, Chem Res Ctr, Food Mach & Chem Corp, 53-55; sr res engr, Atomics Int Div, N Am Aviation, Inc, 55-56; prof chem eng & chmn dept, Univ Seattle, 56-58; prof eng & chmn dept, 58-78, EMER PROF, DEPT CHEM ENG, ARIZ STATE UNIV, 80- *Concurrent Pos:* Consult, Water Planning for Israel, 67-68. *Mem:* Am Chem

Soc; fel Am Inst Chem Engrs; Am Soc Eng Educ; Am Nuclear Soc. *Res:* Industrial wastes; nitrogen fixation; process design; evaporation control; pollution abatement; environmental control; nuclear fuel cycle. *Mailing Add:* Univ Ariz Tempe AZ 85287

REISER, H JOSEPH, b Bad Kissingen, WGermany, July 1, 46; m 70; c 2. RESEARCH ADMINISTRATION, MEDICAL SCIENCES. *Educ:* Ind Univ, MS, 74, PhD(physiol), 76. *Prof Exp:* EXEC DIR RES & DEVELOP, BERLEX LABS, INC, 85- *Concurrent Pos:* Adj assoc prof med & physiol, Likoff CV Inst, Hahnemann Univ, 82-; mem circulation coun, Am Heart Asn. *Mem:* Am Col Cardiol; Am Heart Asn; Int Soc Heart Res; Am Physiol Soc; Am Heart Asn. *Mailing Add:* Berlex Labs Inc 110 E Hanover Ave Cedar Knolls NJ 07927

REISER, MORTON FRANCIS, b Cincinnati, Ohio, Aug 22, 19; m 76; c 3. PSYCHIATRY. *Educ:* Univ Cincinnati, BS, 40, MD, 43. *Prof Exp:* Instr psychiat & internal med, Med Sch, Cincinnati Gen Hosp, 49-50, asst prof, 50-52; res psychiatrist, Neuropsychiat Div, Walter Reed Army Inst Res, 54-55; dir res psychiat, Albert Einstein Col Med, Yale Univ, 55-65, from assoc prof to prof, 55-69, prof psychiat & chem dept, 69-86, Albert E Kent prof, 86-90, EMER ALBERT E KENT PROF PSYCHIAT, SCH MED, YALE UNIV, 90- *Concurrent Pos:* Fel psychiat, Cincinnati Gen Hosp, 47-50; vis psychiatrist, Bronx Munic Hosp, 54-69; mem small grants comt, NIMH, 56-58, mem career investr comt, 59-63; consult, Walter Reed Army Inst Res, Washington, DC, 57-58; prof lectr, State Univ NY Downstate Med Ctr, 59-65; chief div psychiat, Montefiore Hosp & Med Ctr, 65-69; mem fac, Western New Eng Inst Psychoanal, 69-; mem, Jerusalem Ment Health Ctr, 72-; mem, Comn Present Condition & Future Acad Psychiat, Josiah Macy Jr Found, 77; mem exec coun, Acad Behav Med Res, 78. *Honors & Awards:* William Meninger Award, Am Col Physicians, NY City, 88. *Mem:* Am Psychosom Soc (secy-treas, 56-59, pres, 60); Am Soc Clin Invest; fel Am Psychiat Asn; fel Am Col Psychiatrists; Int Col Psychosom Med (pres, 75-77); Am Psychoanal Asn (pres, 82-84); Benjamin Rush Soc. *Res:* Psychoanalysis and psychophysiology; psychosomatic medicine. *Mailing Add:* Dept Psychiat Yale Univ Sch Med 25 Park St New Haven CT 06519

REISER, PETER JACOB, b Feb 6, 53. CARDIAC MUSCLE & DEVELOPING SKELETAL MUSCLE PHYSIOLOGY. *Educ:* Ohio State Univ, PhD(physiol), 81. *Prof Exp:* NIH fel Physiol, Univ Wis-Madison, 84-86, assoc researcher, 86-88; ASST PROF PHYSIOL, UNIV ILL, 88- *Concurrent Pos:* Am Heart Asn Postdoctoral Fel, Case Western Reserve Univ, 81-83. *Mem:* Am Physiol Soc; Biophys Soc. *Mailing Add:* Univ Ill Box 6998 M/C 901 Chicago IL 60680

REISER, RAYMOND, b Philadelphia, Pa, July 28, 06; m 39; c 2. BIOCHEMISTRY. *Educ:* Western Reserve Univ, BA, 29; Ohio State Univ, PhD(agr chem), 36. *Prof Exp:* Hanes fel med, Duke Univ, 36-40; from asst chemist to assoc chemist, Div Chem Exp Sta, Tex A&M Univ, 40-48, from assoc prof to prof, 48-65, distinguished prof, 65-76, DISTINGUISHED EMER PROF BIOCHEM & BIOPHYS, TEX A&M UNIV, 76- *Concurrent Pos:* NIH res career award, 62, mem, Nutrit Study Sect, 63-67. *Honors & Awards:* Glycerine Producer's Asn Award, 52; Southwest Regional Award, Am Chem Soc, 64; Can Award, Am Oil Chem S No, 63; Dr Norman E Borlaug Award, 73; Alton Bailey Medal, Am Oil Chemists' Soc, 76. *Mem:* AAAS; Am Chem Soc; Am Oil Chem Soc (vpres, 66, pres, 67); Am Soc Biol Chemists; fel Am Inst Nutrit. *Res:* Fat absorption; glyceride and essential fatty acid metabolism; lipid analysis; fats in nutrition. *Mailing Add:* Dept Biochem & Biophys Tex A&M Univ College Station TX 77843

REISER, SHELDON, b New York, NY, Oct 13, 30; m 55; c 2. BIOCHEMISTRY, ANIMAL NUTRITION. *Educ:* City Col New York, BS, 53; Univ Wis, MS, 57, PhD(biochem), 60. *Prof Exp:* Assoc prof biochem & med, Med Ctr, Ind Univ, Indianapolis, 60-73; LAB CHIEF, NUTRIT INST, USDA, 73-, RES BIOCHEMIST, 77- *Concurrent Pos:* Mem staff, Vet Admin Hosp, Indianapolis, 60-73. *Mem:* Am Inst Nutrit; Am Chem Soc; Biophys Soc; Am Soc Biol Chem. *Res:* The effects of the type and amount of dietary carbohydrate consumed by animals and humans and carbohydrate and lipid metabolism, intestinal absorption and digestion, and hormone responses. *Mailing Add:* Agr Res Serv E Bldg 307 Rm 315 Beltsville MD 20705

REISERT, PATRICIA, b New York, NY, July 2, 37; m 62; c 2. MICROBIOLOGY, BOTANY. *Educ:* Manhattanville Col, BA, 59; Brown Univ, MA, 61, PhD(bot), 65. *Prof Exp:* Instr biol, St Joseph's Col, Pa, 65-66, asst prof, 66-68; asst prof, Villanova Univ, 69-71 & Worcester Polytech Inst, 74-75; from asst prof to assoc prof natural sci, 75-86, PROF BIOL, ASSUMPTION COL, 86-, CHAIR, DIV NATURAL SCI, 88- *Concurrent Pos:* Fac res fel, St Joseph's Col, Pa, 67; affil asst prof life sci, Worcester Polytech Inst, 75-80; sabbatical leave, Med Sch Pharmacol, Univ Mass, 81-82, 88-89, affil assoc prof pharmacol, 84-; NSF fac develop fel, 81-82; NSF planning grant, 89-90. *Mem:* AAAS; Am Inst Biol Sci. *Res:* Cell surface phenomena in plants; biology of Epstein Barr Virus. *Mailing Add:* Assumption Col 500 Salisbury St Worester MA 01609

REISFELD, RALPH ALFRED, b Suttgart, Ger, Apr 23, 26; US citizen; m 56; c 2. IMMUNOCHEMISTRY, BIOCHEMISTRY. *Educ:* Rutgers Univ, BS, 52; Ohio State Univ, PhD, 57. *Prof Exp:* Biochemist, Endocrinol Br, Nat Cancer Inst, 57-59; sr chemist, Merck & Co, 59-63; biochemist, Immunol Lab, Nat Inst Allergy & Infectious Dis, Md, 63-70; mem dept exp path, 70-74, MEM DEPT MOLECULAR IMMUNOL, SCRIPPS CLIN & RES FOUND, 74- *Concurrent Pos:* Adj prof, Univ Calif, San Diego, 72. *Mem:* AAAS; Am Chem Soc; Soc Exp Biol & Med; Am Asn Immunol. *Res:* Isolation and biological characterization of transplantation antigens from inbred guinea pigs; isolation and biochemical characterization of human lencocyte-a, b, c and human leucocyte-DR antigens; expression and biosynthesis of cell surface antigens on human and murine lymphoid cells; biosynthesis and structure of human melanoma associated antigens. *Mailing Add:* Scripps Clin & Res Found 10666 N Torrey Pines La Jolla CA 92037

REISH, DONALD JAMES, b Corvallis, Ore, June 15, 24; m 52; c 3. MARINE ZOOLOGY. *Educ:* Univ Ore, BS, 46; Ore State Col, MA, 49; Univ Southern Calif, PhD(zool), 52. *Prof Exp:* Res asst, Hancock Found, Univ Southern Calif, 53, res assoc, 53-58; from asst prof to prof, 58-66, EMER PROF BIOL, CALIF STATE UNIV, LONG BEACH, 66- *Concurrent Pos:* Mem staff, Arctic Res Lab, 53 & Eniwetok Marine Biol Lab, 57-58; mem Pac expeds, Hancock Found, Univ Southern Calif, 49 & 53, res assoc, Univ, 73-88. *Mem:* AAAS; Soc Toxicol Chem; Water Pollution Control Fedn; Marine Biol Asn UK. *Res:* Systematics and biology of polychaetous annelids; marine ecology and pollution. *Mailing Add:* Dept Biol Calif State Univ 1250 Bellflower Blvd Long Beach CA 90840-3702

REISING, RICHARD F, b St Louis, Mo, Nov 18, 34; m 59; c 4. RADIOISOTOPIC METHODS. *Educ:* Princeton Univ, BA, 56; Wash Univ, PhD(chem), 63. *Prof Exp:* Mem staff, Argonne Nat Lab, 63-65; res scientist, McDonnell Aircraft Corp, Mo, 65-67; assoc sr res chemist, 67-76, SR STAFF RES SCIENTIST, GEN MOTORS RES LABS, 76- *Res:* Radioisotope applications: use of radioisotopes in solving unique scientific and engineering problems. *Mailing Add:* Gen Motors Res Lab Analysis Chem Dept 30500 Mound Rd Warren MI 48090-9055

REISINGER, JOSEPH G, b New York, NY, Dec 22, 29. PHYSICS. *Educ:* Hofstra Univ, BS, 55; NY Univ, MS, 59, PhD(physics), 64. *Prof Exp:* Instr physics, Queens Col, 56-57; health physicist, Mem Ctr Cancer, 57-58; instr physics, Queens Col, 59-60; res asst environ radiation, NY Univ, 60-64; asst prof, 64-80, PROF PHYSICS, HOFSTRA UNIV, 80- *Res:* Charge distribution in irradiated materials; environmental radiation. *Mailing Add:* Dept Physics Hofstra Univ 1000 Fulton Ave Hempstead NY 11550

REISKIN, ALLAN B, b New York, NY, Apr 24, 36; m 62; c 2. RADIOLOGICAL SCIENCES, ONCOLOGY. *Educ:* City Col New York, BA, 63; Univ Pa, DDS, 63; Oxford Univ, DPhil, 66. *Prof Exp:* Am Cancer Soc Brit-Am fel, 63-66; from asst biologist to assoc biologist, Argonne Nat Lab, 68-70; prof oral radiol, Univ Conn, 70-90; CONSULT, 90- *Concurrent Pos:* Res assoc, Zoller Dent Clin & asst prof path, Univ Chicago, 68-70; consult clin assoc prof, Loyola Univ Chicago, 70; consult, Am Dent Asn. *Mem:* Am Asn Cancer Res; Radiol Soc NAm; Am Acad Dent Radiol; Radiation Res Soc; Brit Inst Radiol. *Res:* Diagnostic imaging; carcinogenesis; radiation safety. *Mailing Add:* Sinai Imaging Assocs 580 Cottage Grove Rd Bloomfield CT 06002

REISKIND, JONATHAN, b Staten Island, NY, May 27, 40; m 66; c 2. EVOLUTIONARY BIOLOGY, ARACHNOLOGY. *Educ:* Amherst Col, AB, 62; Harvard Univ, MA, 65, PhD(biol), 68. *Prof Exp:* Asst prof, 67-72, ASSOC PROF ZOOL, UNIV FLA, 72-, ASSOC DIR, UNIV HONORS PROG, 88- *Concurrent Pos:* Res assoc, Fla State Collection Arthropods, 68- *Mem:* AAAS; Soc Study Evolution; Am Arachnol Soc (pres, 81-83); Asn Trop Biol; Soc Syst Zool. *Res:* Systematics and biology of spiders; mimicry in spiders; spider-plant associations; ethology of arachnids; tropical biology; ecology; biogeography; amber (fossils). *Mailing Add:* Dept Zool Univ Fla Gainesville FL 32611

REISLER, DONALD LAURENCE, b Brooklyn, NY, May 28, 41; m 64; c 1. INFORMATION SCIENCE, PHYSICS. *Educ:* Rutgers Univ, AB, 63; Yale Univ, MS, 65, PhD(physics), 67. *Prof Exp:* Staff analyst, Res Analysis Corp, 67-70 & Lambda Corp, Va, 70-73; PRES & CHMN BD, DBS CORP, 73- *Mem:* Inst Elec & Electronics Engrs; NY Acad Sci. *Res:* Design and installation of information systems; mathematical formulation and solution of organizational problems and decisions. *Mailing Add:* 360 Glyndon St NE Vienna VA 22180

REISLER, HANNA, b Tel-Aviv, Israel, July 12, 43; m 66; c 1. CHEMICAL KINETICS, PHOTOCHEMISTRY. *Educ:* Hebrew Univ, BSc, 64, MSc, 66; Weizmann Inst Sci, PhD(phys chem), 72. *Prof Exp:* Fel chem, Johns Hopkins Univ, 72-74; sr scientist, Soreg Nuclear Res Ctr, 74-77; res scientist elec eng, 77-79, res asst prof elec eng & chem, 79-83, res assoc prof chem, 78-87, ASSOC PROF CHEM, UNIV SOUTHERN CALIF, 87- *Mem:* Am Phys Soc; Am Chem Soc. *Res:* Kinetics and dynamics of elementary processes in the gas phase; laser kinetic spectroscopy of free radicals; multi photon ionization and dissociation; unimolecular reactions of molecules and ions. *Mailing Add:* SSC 619 University Park 0482 Univ Southern Calif Los Angeles CA 90089

REISMAN, ABRAHAM JOSEPH, b Springfield, Mass, Dec 28, 25; m 49; c 2. POLYMER CHEMISTRY. *Educ:* Univ Mass, BS, 59, MS, 73. *Prof Exp:* SR RES CHEMIST, MONSANTO CO, INDIAN ORCHARD, 59- *Mem:* Am Chem Soc. *Res:* Color technology. *Mailing Add:* 51 Emerson St Springfield MA 01118-1732

REISMAN, ARNOLD, b New York, NY, June 12, 27; m 48; c 4. PHYSICAL CHEMISTRY, INORGANIC CHEMISTRY. *Educ:* Brooklyn Col, MA, 53; Polytech Inst Brooklyn, PhD(chem), 58. *Prof Exp:* Control chemist, City New York, 49-51; res staff mem, US Govt, 51-53; MEM RES STAFF, T J WATSON RES CTR, INT BUS MACH CORP, 53- *Concurrent Pos:* Assoc ed, J Electronics Materials; Solid State Sci Panel, Nat Res Coun. *Mem:* Am Chem Soc; Electrochem Soc; fel Am Inst Chemists; Am Inst Mining, Metall & Petrol Engrs. *Res:* Materials science of semiconductors and solid-gas reaction phenomena; epitaxial growth via chemical transport reactions; high pressure reactions; plasma enhanced reactions; radiation induced insulator defects. *Mailing Add:* 816 Thatcher Way Raleigh NC 27609

REISMAN, ARNOLD, b Lodz, Poland, Aug 2, 34; nat US; m 54, 81; c 4. OPERATIONS RESEARCH, INDUSTRIAL ENGINEERING. *Educ:* Univ Calif, Los Angeles, BS, 55, MS, 57, PhD(eng), 63. *Prof Exp:* Asst mech engr, Los Angeles Dept Water & Power, 55-57; from asst prof to assoc prof eng, Calif State Col Los Angeles, 57-66; vis prof eng & bus admin, Univ Wis-Milwaukee, 66-68; PROF OPERS RES, CASE WESTERN RESERVE UNIV, 68- *Concurrent Pos:* Consult, 57-; NSF fac fel, 62-63; assoc res engr,

Western Mgt Sci Inst, Univ Calif, Los Angeles, 64-65; vpres, Univ Assocs, Inc, 69-74; mem, Coun AAAS & prog coordr, AAAS-Inst Mgt Sci, 73-75; vis prof, Hebrew Univ of Jerusalem, 74-75; mem, Japan-Am Inst Mgt Sci, Honolulu & mem, Inst Planning Comt & Bd Trustees, 75-; mem, Rev Bd, Lake Erie Regional Transp Authority, 75-76. *Honors & Awards:* Engr of Year Award, var eng socs, 73. *Mem:* Opers Res Soc Am; Inst Mgt Sci; fel AAAS; sr mem Am Inst Indust Engrs; NY Acad Sci. *Res:* Engineering economy; systems analysis applications to operations management problems in health care delivery, industry and educational institutions; basic research in manpower planning, decision analysis and countertrade analysis; authored 12 books in engineering economy, systems analysis, materials management, health care planning and one nonfiction-Wellcome Tomorrow. *Mailing Add:* Dept Opers Res Case Western Reserve Univ 2040 Adelbert Rd Cleveland OH 44106

REISMAN, ELIAS, b New York, NY, Mar 12, 26; m 48; c 3. EXPERIMENTAL PHYSICS. *Educ:* Cornell Univ, AB, 50, PhD(exp physics), 57. *Prof Exp:* Asst mass spectros, Cornell Univ, 50-55; sr physicist, Radiation Lab, Univ Calif, 57-60; sr scientist, Aeronutronic Div, Philco-Ford Corp, 60-; RETIRED. *Mem:* Am Phys Soc; Sigma Xi; Optical Soc Am. *Res:* Atmospherics; propagation and scattering; quantum electronics and laser applications. *Mailing Add:* 839 E Palmdale Ave Orange CA 92665

REISMAN, HAROLD BERNARD, b Brooklyn, NY, Oct 29, 35; m 60; c 2. BIOCHEMICAL ENGINEERING. *Educ:* Columbia Univ, BS, 56, PhD(chem eng), 65; Cornell Univ, MS, 59. *Prof Exp:* Chem engr, Merck & Co, 61-64, sr chem eng, Merck Sharp & Dohme Res Labs, Rahway, 64-67, sect mgr biochem eng, 67-73; dir, Bioeng Lab, Stauffer Chem Co, 73-75, plant mgr, 75-76, dir mfg, Food Ingredients Div, 76-89; VPRES OPERS, ORGANOGENESIS INC, 89- *Concurrent Pos:* Auth, Economic Analysis Fermentation Processes, 88. *Mem:* Am Inst Chem Engrs; Am Chem Soc; Am Technion Soc; Inst Food Technologists. *Res:* Biochemical engineering, especially fermentation, tissue engineering; natural product isolation and purification; design of fermentors and auxiliaries; pilot plant operations; process development; operations management. *Mailing Add:* 15 October Dr Weston CT 06883

REISMAN, HOWARD MAURICE, b Syracuse, NY, May 23, 37; m 64; c 3. ICHTHYOLOGY. *Educ:* Syracuse Univ, BA, 59, MA, 61; Univ Calif, Santa Barbara, PhD(biol), 67. *Prof Exp:* NIH fel, Cornell Univ, 67-69; asst prof biol, 69-73, assoc prof, 75-80, PROF BIOL & MARINE SCI, SOUTHAMPTON COL, LONG ISLAND UNIV, 80- *Concurrent Pos:* Vis res prof, Tiergarten Schönbrunn, Vienna, Austria, 87-89. *Mem:* AAAS; Am Soc Ichthyol & Herpet; Animal Behav Soc; Am Inst Biol Sci. *Res:* Ichthyology; general marine biology. *Mailing Add:* Div Natural Sci Southampton Col Southampton NY 11968

REISMAN, OTTO, b Vienna, Austria, July 29, 28; US citizen; m 58; c 2. NUCLEAR ENGINEERING, PHYSICS. *Educ:* City Col New York, BS, 58; NY Univ, MS, 60, PhD(nuclear eng), 73. *Prof Exp:* Jr engr physics, Weston Elec Instruments, 57-58; proj engr elec eng, Bendix Aviation, 58-61; instr physics, St Peter's Col, 61-62; ASST PROF PHYSICS, NJ INST TECHNOL, 62- *Mem:* Sigma Xi; Am Nuclear Soc. *Res:* Nuclear reactor heat transfer. *Mailing Add:* Dept Physics NJ Inst Technol 323 High St Newark NJ 07102

REISMAN, STANLEY S, b New York, NY, June 11, 41; c 2. ELECTRICAL ENGINEERING, BIOENGINEERING. *Educ:* Polytech Inst NY, BS, 62, PhD(bioeng), 74; Mass Inst Technol, MS, 63. *Prof Exp:* Mem tech staff, Bell Tel Labs Inc, 63-68; from instr to assoc prof, 68-85, PROF ELEC ENG, NJ INST TECHNOL, 85- *Concurrent Pos:* Lectr elec eng, City Col New York, 68. *Mem:* Inst Elec & Electronics Engrs; Am Soc Eng Educ. *Res:* Mathematical and computer simulation of physiologic systems; biomedical instrumentation. *Mailing Add:* 59 Eastbrook Terr Livingston NJ 07039

REISMANN, HERBERT, b Vienna, Austria, Jan 26, 26; US citizen; m 53; c 2. SOLID MECHANICS, AERONAUTICAL ENGINEERING. *Educ:* Ill Inst Technol, BS, 47, MS, 49; Univ Colo, PhD(eng mech), 62. *Prof Exp:* Instr mech, Ill Inst Technol, 47-50; proj struct engr, Gen Dynamics Corp, 51-53; prin systs engr, Repub Aviation Corp, 54-56; sect chief solid mech, Martin-Marietta Corp, 57-64; PROF ENG, STATE UNIV NY, BUFFALO, 64-, DIR AEROSPACE PROG, 80- *Concurrent Pos:* Grants shell dynamics, USAF Off Sci Res & Army Res Off, 65-; consult, Bell Aerosysts Co, 65- *Honors & Awards:* Outstanding Aerospace Achievement Award, Am Inst Aeronaut & Astronaut, 88. *Mem:* AAAS; assoc fel Am Inst Aeronaut & Astronaut; Am Soc Mech Engrs; Int Asn Bridge & Struct Engrs. *Res:* Elasticity theory, particularly the dynamics of plates and shells; aeroelasticity; elastokinetics; dynamics of elastic bodies; geophysics. *Mailing Add:* 71 Chaumont Dr Williamsville NY 14221

REISNER, GERALD SEYMOUR, b Brooklyn, NY, Apr 10, 26; m 49; c 4. MICROBIOLOGY. *Educ:* State Univ NY Col Educ Albany, AB, 49, MA, 51; Cornell Univ, MS, 55, PhD(plant physiol), 56. *Prof Exp:* Teacher high sch, NY, 49-52; plant physiologist, Plant, Soil & Nutrit Lab, Soil & Water Conserv Res Div, Agr Res Serv, USDA, 52-56; mem fac biol, Goddard Col, 56-58; asst prof, 58-65, assoc prof microbiol, 63-70, PROF BIOL, ALLEGHENY COL, 70- *Concurrent Pos:* Nat Acad Sci res assoc, USDA, 64-65; NIH fel, Ctr Biol Natural Systs, Wash Univ, 71-72; bacteriologist, Bd Health, City Meadville, Pa, 58-75; chmn, microbiol dept, Allegheny Col, 81-86. *Mem:* Sigma Xi; Am Soc Microbiol; NY Acad Sci; Am Asn Univ Professors. *Res:* The application of asymbiotic nitrogen fixing bacteria to the growth of crop plants; effects of cations on bacterial enzymes. *Mailing Add:* Dept Biol Allegheny Col Meadville PA 16335

REISNER, RONALD M, b Buffalo, NY, May 2, 29; m 72; c 2. DERMATOLOGY. *Educ:* Univ Calif, Los Angeles, BA, 52, MD, 56; Am Bd Dermat, dipl, 61. *Prof Exp:* From asst resident to resident dermat, UCLA, 57-59, res trainee, 60, chief div dermat, Harbor Gen Hosp, 62-73; from asst prof to assoc prof dermat, 62-73, coordr dermat, Complex Affil Insts, 73-77, PROF & CHIEF DIV, SCH MED, UNIV CALIF, LOS ANGELES, 73-; DIR, COMBINED UNIV CALIF-VET ADMIN WADSWORTH MED CTR DERMAT PROG, 77- *Concurrent Pos:* Consult, USAF Ballistics Missile Div Med Facil, 62-72, US Naval Regional Hosp, San Diego, Calif, 76-80; chief dermat serv, Vet Admin Wadsworth Med Ctr, Los Angeles, 77-; pres, Pac Dermat Asn, 85-86; mem bd dirs, Soc Invest Dermat, 74-79. *Mem:* Fel Am Acad Dermat; Soc Invest Dermat; Am Dermat Asn; Am Asn Prof Dermat (secy-treas, 74-76); Int Soc Trop Dermat; fel Pan-Am Med Asn; Am Dermat Asn. *Res:* Pathogenesis of acne. *Mailing Add:* Div Dermat Sch Med Univ Calif Los Angeles CA 90024

REISS, CAROL S, b Boston, Mass, Mar 14, 50; c 2. CELLULAR IMMUNOLOGY. *Educ:* City Univ NY, PhD(biomed sci), 78. *Prof Exp:* From instr to asst prof, 80-88, ASSOC PROF CELLULAR IMMUNOL, DANA-FARBER CANCER INST, 88- *Concurrent Pos:* Dir Animal Facil, Dana- Farber Cancer Inst, 81-89. *Mem:* AAAS; NY Acad Sci; Asn Women Sci; Am Soc Microbiologists; Am Soc Immunologists. *Res:* Regulation, specificity immune response to viruses. *Mailing Add:* Div Pediat Oncol Dana-Farber Cancer Inst Boston MA 02115

REISS, DIANA, b Philadelphia, Pa, Nov 1, 48. BIOACOUSTICS, ANIMAL COGNITION. *Educ:* Temple Univ, BS, 72, PhD(commun & speech), 81. *Prof Exp:* Admin asst, Animal Sonar Systs Symp, NATO, 78-79; lectr speech & commun, Temple Univ, 79-80; researcher biomed res, Stanford Res Inst Int, 81-82; fac mem animal commun, human commun & speech, San Francisco State Univ, 83-88; PRIN INVESTR & PROJ DIR, DOLPHIN COMT RES, MARINE WORLD FOUND, 81- *Concurrent Pos:* dir & bd mem, Marine World Found, 85-88; consult, Time-Life Books-Bioastron Series, 88- *Mem:* AAAS; Soc Marine Mammal; Am Asn Zool Parks & Aquariums. *Res:* Dolphin communication and cognition; ontogeny of dolphin behavior and communication in a captive environment. *Mailing Add:* 7656 Stonewood Ct Rosewood CA 95661

REISS, ERIC, medicine; deceased, see previous edition for last biography

REISS, ERROL, b New York, NY, Jan 16, 42; m 68; c 2. MICROBIAL IMMUNOCHEMISTRY. *Educ:* City Col New York, BSc, 63; Rutgers Univ, PhD(microbiol), 72. *Prof Exp:* Bacteriologist, Vet Admin Hosp, Washington, DC, 66-67; NIH fel, 68 & 72-74; res microbiologist, 74-80, HEAD IMMUNOCHEM LABS, DIV MYCOTIC DIS, CTR DIS CONTROL, ATLANTA, 80- *Concurrent Pos:* Instr, grad prog, NIH, 72-74, & dept biol, Atlanta Univ, 80-82; adj prof, dept lab pract, Sch Pub Health, Univ NC, Chapel Hill, 78-, dept biol, Ga State Univ, Atlanta, 83- & Med Sch, Emory Univ, 87-; guest lectr, Morehouse Med Sch, Atlanta, 81-83. *Mem:* Am Asn Immunol; Am Chem Soc; Am Acad Microbiol; Am Soc Microbiol; Int Soc Human Animal Mycol; Med Mycol Soc Am. *Res:* Molecular immunology and molecular biology of microbial infections; mycotic infections; microbial immunochemistry; monoclonal antibodies; genetic probes; enzyme immunoassays; animal models of infection and immunity cellular immunology; cell wall chemistry; immunity to respiratory infections. *Mailing Add:* 3642 Castaway Ct Chamblee GA 30341-4602

REISS, HOWARD, b New York, NY, Apr 5, 22; m 45; c 2. PHYSICAL CHEMISTRY. *Educ:* NY Univ, AB, 43; Columbia Univ, PhD(phys chem), 49. *Prof Exp:* Chemist, Tenn Eastman Corp, Tenn, 44-45; instr chem, Boston Univ, 49-51; chemist, Celanese Corp Am, 51-52 & Bell Tel Labs, Inc, 52-60; from assoc dir to dir res dept, Atomics Int Div, NAm Aviation, Inc, 60-62, pres & dir sci ctr & vpres NAm Aviation, Thousand Oaks, 62-68; PROF CHEM, UNIV CALIF, LOS ANGELES, 68- *Concurrent Pos:* Corp rep, Am Inst Physics, 63-66; mem, Physics Res Eval Group, Air Force Off Sci Res, 66-, Reactor Chem Eval Comt, Oak Ridge Nat Lab, 66-68, Mat Res Coun, Advan Res Proj Agency, 68- & Adv Comt Math & Phys Sci, NSF, 70-74; Guggenheim fel, 78-79; mem comn socio-tech systs, Nat Res Coun; ed, J Statist Physics, 68-78. *Honors & Awards:* Tolman Medal, Am Chem Soc, 73; Herbert Newby McCoy Award, 74; Colloid & Surface Chem Prize, Am Chem Soc, 80, J H Hildebrand Award, 91. *Mem:* Nat Acad Sci; fel Am Phys Soc; Sigma Xi; fel AAAS; Am Chem Soc. *Res:* Semiconductors; statistical mechanics; solid state chemistry; thermodynamics; nucleation theory; information theory; electrochemistry; polymers science. *Mailing Add:* Dept Chem & Biochem Univ Calif Los Angeles CA 90024

REISS, HOWARD R, b Brooklyn, NY, July 29, 29; m 83; c 2. INTENSE-FIELD PHENOMENA. *Educ:* Polytech Inst Brooklyn, BAE, 50, MAE, 51; Univ Md, PhD(physics), 58. *Prof Exp:* Asst, Polytech Inst Brooklyn, 51; physicist, David W Taylor Model Basin, Bur Ships, US Dept Navy, 51-55, Naval Ord Lab, 55-58, chief nuclear physics div, 58-69; mem fac, Dept Physics, 78-81, res prof, Ariz Res Lab, Univ Ariz, 81-86; PROF PHYSICS, AM UNIV, 69- *Concurrent Pos:* Lectr, Univ Md, 59-63; vis scientist, Univ Torino, Italy, 63-64; adj prof physics, Am Univ, 67-69; vis prof physics, Univ Ariz, 75-81; consult, Naval Res Lab, Wash DC, 74-75, Standard Oil Co, Ind, 80-82, Los Alamos Nat Lab, 85- 87, Naval Surface Weapons Ctr, White Oaks, Md, 86-90. *Mem:* Fel Am Phys Soc; AAAS; Optical Soc Am. *Res:* Development of theoretical methods for very intense electromagnetic fields, and applications to interaction of intense fields with atoms nuclei and solids. *Mailing Add:* Dept Physics Am Univ Washington DC 20016

REISS, KEITH WESTCOTT, b Washington DC, July 22, 45; m 66; c 1. MICROWAVE PHYSICS, ENGINEERING PHYSICS. *Educ:* Univ Va, BS, 66; Wake Forest Univ, MA, 68; Duke Univ, PhD(physics), 71. *Prof Exp:* Chmn natural sci & math div & dept head phys sci, Truett-McConnell Col, 71-73; PHYSICIST USN SUPPORT, VITRO LABS DIV, AUTOMATION INDUST, 73- *Mem:* Am Phys Soc; Sigma Xi. *Res:* Development and analysis of US Navy surface-to-surface missile systems. *Mailing Add:* 3522 Laurel Leaf Lane Fairfax VA 22031

REISS, OSCAR KULLY, b Bad-Duerkheim, Ger, May 6, 21; nat US; m 44; c 3. BIOCHEMISTRY. *Educ:* Univ Chicago, BS, 50, PhD(biochem), 54. *Prof Exp:* Instr physiol chem, Sch Med, Johns Hopkins Univ, 57-58, asst prof, 58-59; asst prof, 59-67, ASSOC PROF BIOCHEM, SCH MED, UNIV COLO, DENVER, 67- *Mem:* Am Soc Biol Chemists; Am Chem Soc; Brit Biochem Soc. *Res:* Intermediary and lipid metabolism of lung and other tissues; structure and function of membranes; effects of organothiophosphates on pulmonary enzyme systems. *Mailing Add:* Dept Biochem Sch Med Univ Colo05855135x Denver CO 80262

REISS, WILLIAM DEAN, b Breese, Ill, July 24, 37; m 61; c 2. AGRONOMY. *Educ:* Southern Ill Univ, BS, 60, MS, 61; Univ Ill, PhD(crop prod), 67. *Prof Exp:* Res asst crop prod, Southern Ill Univ, 60-61; asst, Univ Ill, 61-65; exten agronomist, 65-74, RES AGRONOMIST, PURDUE UNIV, WEST LAFAYETTE, 74- *Concurrent Pos:* Mem, Bd Dirs, Asn Off Seed Certifying Agencies, 67- *Mem:* Am Soc Agron. *Res:* Influence of density of stand, row width, planting date and plant nutrients on behavior of corn, sorghum and soybeans; factors affecting seed quality of soybeans and small grains; seed certification. *Mailing Add:* 30 Oriole Dr Lafayette IN 47905

REISSE, ROBERT ALAN, b Philadelphia, Pa, Apr 9, 46; m 73; c 2. PHYSICS. *Educ:* Wesleyan Univ, BA, 67; Univ Md, MS, 70, PhD(physics), 76. *Prof Exp:* Res assoc, Physics Dept, Univ Md, 76-77; mem tech staff physics, Sperry Res Ctr, Sperry Rand Corp, 77-81; res asst prof & mem staff, Ariz Res Labs & Santa Catalina Lab Exp Relativity By Astrometry, Univ Ariz, 81-82; at CGR Med Corp, Baltimore, Md, 82-85; mem staff, ITE Inc, Beltsville, Md, 85-88; mentor, Technologies Inc, Rockville, Md, 88-91; PRIN SCIENTIST, SCI INQUIRIES, INC, 91- *Mem:* Am Phys Soc; Optical Soc Am. *Res:* Optical information processing; quantum electronics; experimental general relativity; solar astrophysics. *Mailing Add:* 312 Patleigh Rd Catonsville MD 21228-5630

REISSIG, JOSE LUIS, genetics, molecular biology, for more information see previous edition

REISSIG, MAGDALENA, b Buenos Aires, Arg, Sept 1, 23. ELECTRON MICROSCOPY, VIROLOGY. *Educ:* Univ Buenos Aires, Arg, MD, 50. *Prof Exp:* Res asst cell ultrastruct, Inst Biol Sci, Uruguay, 50-53; res asst prev med, Sch Med, Yale Univ, 53-56, res assoc, 56-58; res assoc microbiol, Albert Einstein Med Ctr, 58-61; from asst prof to assoc prof pathobiol, 61-82, assoc prof, 83-88, EMER PROF IMMUNOL & INFECTIOUS DIS, JOHNS HOPKINS UNIV, 88- *Mem:* Electron Micros Soc Am; Am Asn Immunologists; Am Soc Cell Biol. *Res:* Cytology; cytopathology and pathogenesis of virus infections; ultrastructure of parasitic helminths. *Mailing Add:* Dept Immunol & Infectious Dis Sch Hyg & Pub Health Johns Hopkins Univ 615 N Wolfe St Baltimore MD 21205

REISSMANN, THOMAS LINCOLN, b Wilmington, Del, Feb 12, 20; m 51; c 2. ANALYTICAL CHEMISTRY. *Educ:* Pa State Univ, BS, 42, MS, 47, PhD(chem), 49. *Prof Exp:* Chemist, Atlas Powder Co, 42 & 45-46; asst chief chemist, Ky Ord Works, 42-45; asst chem, Pa State Col, 46-49; res chemist, Ethicon Inc, 49-58, dept mgr collagen res, 58-69, absorbable suture res & develop, 70-72, dept mgr chem, 73-80, asst to corp dir, Qual Assurance, 80-82; RETIRED. *Concurrent Pos:* Consult, 82- *Mem:* AAAS; Am Chem Soc; Am Soc Qual Control; NY Acad Sci. *Res:* Collagen chemistry; fibers and films; medical products; chemical quality of drugs; application of computerized systems to the quality assurance function. *Mailing Add:* 24 Hillcrest Rd Normandy Beach NJ 08836

REISSNER, ERIC, b Aachen, Ger, Jan 5, 13; US citizen; m 38; c 2. APPLIED MECHANICS, APPLIED MATHEMATICS. *Educ:* Tech Univ, Berlin, Dipl Ing, 35, Dr Ing (civil eng), 36; Mass Inst Technol, PhD(math), 38. *Hon Degrees:* Dr Ing(mech eng), Hannover Univ, 64. *Prof Exp:* From instr to prof math, Mass Inst Technol, 39-69; prof, 70-78, EMER PROF APPL MECH, UNIV CALIF, SAN DIEGO, 79- *Concurrent Pos:* Guggenheim fel, 62; NSF sr fel, 67. *Honors & Awards:* Theodore von Karman Medal, Am Soc Civil Eng, 64; Timoshenko Medal, Am Soc Mech Eng, 73, Medal, 88; Struct, Struct Dynamics & Mat Award, Am Inst Aeronaut & Astronaut, 84. *Mem:* Nat Acad Eng; Am Acad Arts & Sci; Am Soc Mech Eng; Am Inst Aeronaut & Astronaut; Am Math Soc; Am Acad Mech. *Res:* Theory of elasticity, especially development of variational methods, and behavior of beams, plates and shells. *Mailing Add:* Dept Appl Mech & Eng Sci 0411 Univ Calif San Diego La Jolla CA 92093-0411

REIST, ELMER JOSEPH, b Can, Aug 29, 30; nat US; m 54; c 2. ORGANIC CHEMISTRY. *Educ:* Univ Alta, BSc, 52; Univ Calif, PhD(org chem), 55. *Prof Exp:* Fel, Nat Res Coun Can, 55-56; asst dir, 56-80, ASSOC DIR BIOORG CHEM, STANFORD RES INST INT, 80- *Mem:* Am Chem Soc; AAAS; Am Soc Microbiol. *Res:* Carbohydrates; synthesis; nitrogen mustards; neuraminic acid; enzyme chemistry; nitrosamines; chemical carcinogenesis; nucleic acids; antiviral agents. *Mailing Add:* Life Sci Div SRI Int Menlo Park CA 94025-3493

REIST, PARKER CRAMER, b Williamsport, Pa, Mar 3, 33; m 55; c 2. AEROSOL SCIENCE. *Educ:* Pa State Univ, BS, 55; Mass Inst Technol, SM, 57; Harvard Univ, SMHyg, 63, ScD(radiol health), 66. *Prof Exp:* Lectr indust hyg, Harvard Univ, 65-66, from asst prof to assoc prof, 66-72; PROF AIR & INDUST HYG, SCH PUB HEALTH, UNIV NC, CHAPEL HILL, 72- *Concurrent Pos:* Mem, NC State Bd Refrigeration Examrs, 75-; Fulbright res scholarship, 84; vis prof, Univ Col, Galway, Ireland, 84-85. *Mem:* Am Bd Indust Hyg; Am Indust Hyg Asn. *Res:* Industrial hygiene; respiratory protection; air pollution control; aerosol technology and particle behavior; radiation protection; disposal of radioactive gaseous and particle wastes. *Mailing Add:* Sch Pub Hyg Univ NC Chapel Hill NC 27511

REISTAD, GORDON MACKENZE, b Philipsburg, Mont, Oct 21, 44; m 65; c 2. THERMODYNAMICS, HEATING & AIR CONDITIONING. *Educ:* Mont State Univ, BS, 66; Univ Wis-Madison, MS, 67, PhD(mech eng), 70. *Prof Exp:* Instr mech eng, Univ Wis-Madison, 69-70; from asst prof to assoc prof, 70-81, PROF MECH ENG, ORE STATE UNIV, 81-, HEAD DEPT, 86- *Concurrent Pos:* Consult, Battelle Northwest, Lawrence Livermore Labs, Elec Power Res Inst, Nat Bur Standards, Int Dist Heating Asn & others; chmn, Nat Prog Comt, Am Soc Heating & Air-Conditioning Engrs. *Mem:* Fel Am Soc Mech Engrs; Am Soc Heating, & Air-Conditioning Engrs; Am Soc Engr Educ. *Res:* Energy systems evaluation and design; second law of thermodynamics; geothermal energy systems analysis and design. *Mailing Add:* 4370 NW Queens Ave Corvallis OR 97330

REISTER, DAVID B(RYAN), b Los Angeles, Calif, Feb 22, 42; m 63; c 2. ENERGY MODELS, ROBOTICS. *Educ:* Univ Calif, Berkeley, BS, 64, MS, 66, PhD(eng sci), 69. *Prof Exp:* Lectr eng sci, State Univ NY Buffalo, 68-69, asst prof, 69-74; scientist, Inst Energy Anal, 74-85; SCIENTIST, OAK RIDGE NAT LAB, 85- *Mem:* AAAS; Inst Elec & Electronics Engrs; Int Asn Energy Economists; Sigma Xi. *Res:* robotics systems. *Mailing Add:* PO Box 2008 Oak Ridge TN 37831

REISWIG, HENRY MICHAEL, b St Paul, Minn, July 8, 36; m 63; c 3. INVERTEBRATE ZOOLOGY, BIOLOGY OF SPONGES. *Educ:* Univ Calif, Berkeley, BA, 58, MA, 66; Yale Univ, MSc, 68, PhD(biol), 71. *Prof Exp:* staff res, Dept Biol, Yale Univ, 71, asst res, Peabody Mus, 71-72; asst prof, Dept Biol, 72-77, ASSOC PROF BIOL, REDPATH MUS, MCGILL UNIV, 77- *Mem:* Am Soc Zoologists; Can Soc Zool; Am Soc Limnol & Oceanog; Asn Marine Labs Caribbean; Sigma Xi. *Res:* Anatomy and taxonomy of hexactinellida and spongillidae (porifera). *Mailing Add:* Redpath Mus McGill Univ 859 Sherbrooke St W Montreal PQ H3A 2K6 Can

REISWIG, ROBERT D(AVID), b Wichita, Kans, July 14, 29; m 51; c 2. METALLURGY. *Educ:* Univ Kans, BS, 51; Univ Wis, MS, 53, PhD, 56. *Prof Exp:* Res engr, Battelle Mem Inst, 51-52; STAFF MEM, LOS ALAMOS SCI LAB, 55- *Mem:* Am Inst Mining, Metall & Petrol Engrs; Am Soc Metals. *Res:* Titanium casting; pyrophoric alloys; transformations; phase equilibria; thermal conductivity; temperature measurement; microstructures; carbons and graphites; carbide-carbon composites; corrosion; shaped charge liners. *Mailing Add:* 90 Tecolote Los Alamos NM 87544

REIT, BARRY, b New York, NY, Oct 27, 42. ENDOCRINOLOGY, PHYSIOLOGY. *Educ:* Mich State Univ, BS, 64; Univ Tenn, Knoxville, MS, 68; Univ Ky, PhD(animal physiol), 71. *Prof Exp:* Vis scientist metab endocrinol, Div Biol Standards, Nat Inst Med Res, London, Eng, 71-72, mem sci staff, 72-74; res assoc orthop surg, Children's Hosp Med Ctr, 74-78; INSTR MED, HARVARD MED SCH, 78- *Concurrent Pos:* Asst biol, Mass Gen Hosp, 78- *Mem:* Soc Endocrinol; Brit Bone & Tooth Soc; Sigma Xi; Am Soc Bone & Mineral Res; AAAS. *Res:* Calcium physiology; endocrine influences on calcified tissue and mineral metabolism measured by bioassay and intravenous administration of parathyroid hormone; mechanism of action of parathyroid hormone; metabolic endocrinology. *Mailing Add:* 245-88 62nd Ave Douglaston NY 11363

REIT, ERNEST MARVIN I, b New York, NY, July 3, 32; m 56; c 4. PHARMACOLOGY. *Educ:* Cornell Univ, BS, 53, DVM, 57; Yale Univ, PhD(pharmacol), 64. *Prof Exp:* Asst prof, 66-69, ASSOC PROF PHARMACOL, COL MED, UNIV VT, 69- *Concurrent Pos:* Spec fel, Nat Inst Med Res, Mill Hill, Eng, 64-65; USPHS career develop award, 67-72. *Mem:* AAAS; Am Soc Pharmacol & Exp Therapeut; Brit Pharmacol Soc. *Res:* Neuropharmacology, especially chemical mediation and modulation of the transmission of nerve impulses. *Mailing Add:* Dept Pharmacol Univ Vt Col Med 85 S Prospect St Burlington VT 05405

REITAN, DANIEL KINSETH, b Duluth, Minn, Aug 13, 21; m 46; c 2. COMPUTER APPLICATIONS. *Educ:* NDak State Univ, BSEE, 46, Univ Wis, MSEE, 49, PhD(elec & comput engr), 52. *Prof Exp:* Res engr, Gen Eng & Res Lab, Gen Elec Co, Schenectady, 46-48; toll line engr, Gen Telephone Co, Madison, Wis, 49-50; prof, elec & comput eng, Univ Wis, Madison, 52-83; PRIVATE CONSULT, 83- *Concurrent Pos:* Chmn, Basic Sci Comt, Inst Elec & Electronics Engrs, 55; dir, Wis Utilities AC Network Calculator Lab, 58-68, Univ Wis Power Systs Simulation Lab, 68-83, wind energy res, Univ Wis Energy Ctr, 72-83; consult, Nat Bur Standards, 79-83. *Honors & Awards:* Centennial Medal & Cert, Inst Elec & Electronics Engrs. *Mem:* Sigma Xi; fel Inst Elec & Electronics Engrs; Am Soc Eng Educ. *Res:* Published fifty articles in various journals. *Mailing Add:* PO Box 213 Pelican Rapids MI 56572

REITAN, PAUL HARTMAN, b Kanawha, Iowa, Aug 18, 28; m 62; c 2. GEOCHEMISTRY, PETROLOGY. *Educ:* Univ Chicago, AB, 53; Univ Oslo, PhD(geol), 59. *Prof Exp:* Geologist, US Geol Surv, 53-56; state geologist, Geol Surv Norway, 56-60; asst prof mineral, Stanford Univ, 60-66; assoc prof, State Univ NY, Buffalo, 66-69, assoc provost, Fac Natural Sci & Math, 70-75, dir, Natural Sci & Math Res Inst, 70-80, actg provost, 75-76, provost, 76-78, dean, Fac Natural Sci & Math, 78-80, PROF GEOL, STATE UNIV NY, BUFFALO, 69- *Concurrent Pos:* Fulbright sr lectr, Indian Sch Mines, Dhanbad, India, 86. *Mem:* AAAS; fel Geol Soc Am; fel Mineral Soc Am; Int Asn Geochem & Cosmochem. *Res:* Metamorphic petrology and recrystallization; fractionation of elements between coexisting minerals; temperatures of metamorphism; cycling of elements in earth's crust; genesis of pegmatites and magma in crust. *Mailing Add:* Dept Geol State Univ NY Buffalo NY 14260

REITAN, PHILLIP JENNINGS, b Grove City, Minn, July 14, 29; m 53; c 5. ZOOLOGY. *Educ:* Concordia Col, Moorhead, Minn, BA, 52; Univ Wis, MS, 54, PhD(zool), 58. *Prof Exp:* Teacher pub sch, NDak, 49-50; from instr to assoc prof biol, Wagner Col, 57-62; assoc prof, 62-67, chmn dept, 62-72, PROF BIOL, LUTHER COL, IOWA, 67- *Concurrent Pos:* Co-prin investr, NIH res grant, 58-63; NSF sci fac fel, Harvard Univ, 65-66; vis scholar, Hopkins Marine Sta, 73, Pac Lutheran Univ, 80; hon fel zool, Univ Wis, 81;

volunteer, Page Mus, Los Angeles, 88. *Mem:* AAAS; Am Soc Zoologists; Am Asn Univ Profs; Sigma Xi. *Res:* Drosophila development, especially the relationship between inherited abnormalities and normal embryology; relationship between dehydration and radiation effects in Drosophila embryology; function of avian amnion in development; impact of Darwinism. *Mailing Add:* 606 Center St Decorah IA 52101

REITAN, RALPH MELDAHL, b Beresford, SDak, Aug 29, 22; div; c 5. NEUROPSYCHOLOGY. *Educ:* Univ Chicago, PhD(psychol), 50. *Prof Exp:* Lectr psychol, South Bend Exten Ctr, Ind Univ, Indianapolis, 48-51, from asst prof to assoc prof surg, Sch Med, 51-60, prof psychol & dir sect neuropsychol, 60-70; prof psychol & neurol surg, Univ Wash, 70-77; prof psychol, Univ Ariz, 77-86; PRES, REITAN NEUROPSYCHOL LABS INC. *Concurrent Pos:* Instr, Univ Chicago, 49-51; consult, Space Med Adv Group, NASA, 64-65, Off Assoc Dir, Nat Inst Neurol Dis & Blindness, 61-76 ,& Vet Admin Hosp, Tucson, 77-85. *Honors & Awards:* Gordon Barrows Mem Award, 65. *Mem:* Am Psychol Asn; Am Neurol Asn; Am Acad Neurol. *Res:* Brain localization of abilities, neuropsychology; brain-behavior relationships. *Mailing Add:* Neuropsychol Lab 2920 S Fourth Ave South Tucson AZ 85713

REITEMEIER, RICHARD JOSEPH, b Pueblo, Colo, Jan 2, 23; m 51; c 7. GASTROENTEROLOGY, ONCOLOGY. *Educ:* Univ Denver, AB, 43; Univ Colo, MD, 46; Univ Minn, MS, 54; cert, Am Bd Intern Med, 54. *Prof Exp:* From instr to assoc prof med, Mayo Med Sch, 57-67, chmn, Dept Internal Med, Mayo Clin, 67-74, emer prof med, 71-87, consult, Mayo Med Sch & Found, 54-87; MED DIR, PHOENIX ALLIANCE, INC, ST PAUL, MINN, 91- *Concurrent Pos:* Vis prof, Univ Iowa, 71, Scott White Clin, Temple Tex, 76; mem, Bd Develop, Mayo Found, 86; exec & sci dir, Ludwig Inst Cancer Res, 88- *Honors & Awards:* Irving S Cutter Gold Medal, 86; Alfred Stengel Award, Am Col Physicians, 90. *Mem:* Inst Med-Nat Acad Sci; Am Col Physicians (pres, 83-84); Am Gastroenterol Asn; AMA; Am Clin & Climotol Asn; Am Fedn Clin Res; Am Soc Clin Oncol. *Mailing Add:* 707 12th Ave SW Rochester MN 55902

REITER, ELMAR RUDOLF, b Wels, Austria, Feb 22, 28; m 54; c 3. CLIMATOLOGY. *Educ:* Univ Innsbruck, PhD(meteorol, geophys), 53. *Prof Exp:* Res asst meteorol, Univ Chicago, 52-53, res assoc & instr, 54-56; instr, NATO Officer's Sch, Germ 53-54; res assoc meteorol & geophys, Univ Innsbruck, 56-59, asst prof, 59-61; assoc prof, Colo State Univ, 61-65, head dept, 68-74, prof, Atmospheric Sci, 65-88, PROF CIVIL ENG, COLO STATE UNIV, 85-; PRES, WELS RES CORP, 88- *Concurrent Pos:* Res fel, Deutscher Wetterdienst, Repub Ger, 59; lectr, SEATO Grad Sch, Bangkok, 63; consult, Ger Lufthansa Airlines, 59-61, Meteorol Res Inc, Calif, 63-65, Univ Melbourne, 63, Litton Industs, Minn, 64, Boeing Aircraft Co, Seattle, 65, NASA-Marshall Space Flight Ctr, Huntsville & US Army Missile Commmand, 65- & Nat Comt Clear Turbulence, 66, Inst Defense Analysis, 73-75 & Nat Acad Sci, 73-75; pres meteorol sect, Am Geophys Union, 80-82; mem adv bd, Geophys Inst, Univ Alaska. *Honors & Awards:* Advan Sci Award, Govt Upper Austria, 62; Silver Anniversary Medal Distinguished Serv, Univ Innsbruck, 70; Robert M Losey Award, Am Inst Aeronaut & Astronaut, 67. *Mem:* Fel Am Meteorol Soc; fel Am Geophys Union; Am Inst Aeronaut & Astronaut; Royal Meteorol Soc; Meteorol Soc Japan; Math Soc Austria; Austrian Acad Sci. *Res:* Numerical and heuristic computer modeling; aviation meteorology, especially clear-air turbulence; computerized decision support systems. *Mailing Add:* Wels Res Corp 3300 28th St No 200 Boulder CO 80301

REITER, HAROLD BRAUN, b Jackson, Tenn, Oct 14, 42; m 66. MATHEMATICS. *Educ:* La State Univ, Baton Rouge, BS, 64; Clemson Univ, MS, 65, PhD(math), 69. *Prof Exp:* Asst prof math, Univ Hawaii, 69-72; ASSOC PROF MATH, UNIV NC, CHARLOTTE, 72- *Concurrent Pos:* Assoc chmn, dept comput sci, Univ Md, College Park. *Mem:* Am Math Soc; Math Asn Am. *Res:* Comparison of topologies; function algebras; convexity; game theory; recreational mathematics, including magic geograms, knight interchange problems and teacher education. *Mailing Add:* Dept Math Univ NC Charlotte NC 28223

REITER, MARSHALL ALLAN, b Pittsburgh, Pa, Sept 11, 42; m 64; c 3. GEOPHYSICS. *Educ:* Univ Pittsburgh, BS, 65; Va Polytech Inst, PhD(geophys), 70. *Prof Exp:* From asst prof to assoc prof geophys, NMex Inst Mining & Technol, 70-75; geophysicist, 75-79, SR GEOPHYSICIST, NMEX BUR MINES & MINERAL RESOURCES, 79- *Concurrent Pos:* Adj assoc prof geophys, NMex Inst Mining & Technol, 75-80, adj prof, 80- *Mem:* Am Geophys Union; fel Geol Soc Am; Soc Explor Geophysicists; Sigma Xi. *Res:* Geothermal studies; define the geographic variation of terrestrial heat flux in the Southwestern United States with borehole temperature measurements, locate geothermal areas, thermal conditions in the crust and upper mantle. *Mailing Add:* NMex Bur Mines Socorro NM 87801

REITER, RAYMOND, b Toronto, Ont, June 12, 39. COMPUTER SCIENCE. *Educ:* Univ Toronto, BA, MA, 63; Univ Mich, Ann Arbor, PhD(comput sci), 67. *Prof Exp:* Nat Res Coun Can fel, Univ London, 68-69; asst prof, 69-74, ASSOC PROF COMPUT SCI, UNIV BC, 74- *Mem:* Asn Comput Mach; Asn Symbolic Logic. *Res:* Artificial intelligence; mechanical theorem proving; theory of computation. *Mailing Add:* Dept Comput Sci Univ Toronto Toronto ON M5S 1A4 Can

REITER, RUSSEL JOSEPH, b St Cloud, Minn, Sept 22, 36; m 62, 90; c 2. NEUROENDOCRINOLOGY, CELL & MOLECULAR BIOLOGY. *Educ:* St John's Univ, Minn, BA, 59; Bowman Gray Sch Med, MS, 61, PhD(anat), 64. *Prof Exp:* Exp endocrinologist, Edgewood Arsenal, Md, 66; asst prof anat, Med Ctr, Univ Rochester, 66-69, assoc prof, 69-71; assoc prof, 71-73, PROF NEUROENDOCRINOL, UNIV TEX HEALTH SCI CTR, SAN ANTONIO, 73- *Concurrent Pos:* NIH career develop award, 69-74; ed, J Pineal Res, Neurosci Lett & Pineal Res Rev. *Mem:* Endocrine Soc; Am Physiol Soc; NY Acad Sci; Soc Neurosci; Int Soc Chronobiol; Sigma Xi. *Res:* Neuroendocrinology, especially the pineal gland, cell biology and reproductive physiology; brain chemistry and behavior; author of six books. *Mailing Add:* Dept Cellular & Struct Biol Univ Tex Health Sci Ctr 7703 Floyd Curl Dr San Antonio TX 78284

REITER, STANLEY, b Apr 26, 25. MATHEMATICS, ECONOMICS. *Educ:* Queens Col, AB, 47; Univ Chicago, MA, 50, PhD(econ), 55. *Prof Exp:* Instr econ, Stanford Univ, 50-53; from asst prof to prof econ & math, Purdue Univ, 54-67; PROF ECON & MATH, NORTHWESTERN UNIV, 67- *Concurrent Pos:* From res asst to res assoc, Cowles Comt, Res in Econ, Univ Chicago, 48-50; res assoc, Appl Math & Statist Lab, Stanford Univ, 53-54; prin investr, Off Naval Res Contract Math Econ, Purdue Univ, 58-67; vis fel, Churchill Col, Cambridge Univ, 61-62; chmn, Managerial Econ & Decision Sci, Northwestern Univ, 73-74, dir, Ctr Math Studies Econ & Mgt Sci, 74-; sr mem, Inst Math & Applications, Univ Minn, 83-84; Morrison prof econ & math, Col Arts & Sci, Dept Managerial Econ & Decision Sci, Grad Sch Mgt, Northwestern Univ, 78-; vis fac assoc, Calif Inst Technol, 84-85; mem, Adv Comt, Div Info, NSF & vis team eval, Dulce Univ proposal Sci & Technol Ctr, 88; chmn, vis team eval, Univ Ariz proposal Sci & Technol Ctr, 88; reviewer, Sci & Technol Ctr Proposals, NSF; chmn, Nat Res Coun; assoc ed, Complex Systs. *Mailing Add:* Leverone Hall Rm 3-014 Northwestern Univ 2001 Sheridan Rd Evanston IL 60208

REITER, WILLIAM FREDERICK, JR, b Egg Harbor City, NJ, July 20, 38; m 72; c 2. MECHANICAL ENGINEERING. *Educ:* Rutgers Univ, BS, 61; Auburn Univ, MS, 66; NC State Univ, PhD(mech eng), 73. *Prof Exp:* Instr mech eng, Auburn Univ, 66-69; asst prof, 73-77, ASSOC PROF MECH ENG, NC STATE UNIV, 78- *Mem:* Am Soc Mech Engrs; Soc Automotive Engrs. *Res:* Dynamics of rotating machinery; vibration and sound radiation from structures; sound and vibration signal processing. *Mailing Add:* Dept 9B7 IBM Corp PO Box 3025 Boca Raton FL 33432

REITH, MAARTEN E A, b Utrecht, Neth, Dec 29, 46; m 80; c 1. NEUROCHEMISTRY, NEUROPHARMACOLOGY. *Educ:* State Univ Utrecht, Neth, BS, 68, MS, 71, PhD(neurochem-pharmacol), 75. *Prof Exp:* Res fel neurochem, Ctr Neurochem, Fac Med, Strasbourg, France, 71; res scientist biochem, Rudolf Magnus Inst Pharmacol, State Univ Utrecht, Neth, 71-74, sr res scientist biochem, Int Molecular Biol, 74-78; SR RES SCIENTIST, CTR NEUROCHEM, N S KLINE INST, NEW YORK, 78- *Concurrent Pos:* Prin investr, NY State Health Res Coun, 81-82 & Nat Inst Drug Abuse, 83-90. *Mem:* Am Soc Biol Chemists; Am Soc Neurochem; Int Soc Neurochem. *Res:* Receptors for neurotransmitters and psychoactive drugs; sodium channels of excitable tissues. *Mailing Add:* Ctr Neurochem Wards Island New York NY 10035

REITMEYER, WILLIAM L, b Deer Lodge, Mont, Apr 15, 28; m 62; c 5. ASTRONOMY, COMPUTER SYSTEMS. *Educ:* St Louis Univ, BS, 52; Marquette Univ, MS, 60; Univ Ariz, PhD(astron, aerospace eng), 64. *Prof Exp:* From instr to asst prof eng, Marquette Univ, 54-65; assoc prof mech eng, NMex State Univ, 65-66, assoc prof earth sci, 66-69; staff assoc astron, NSF, 69-70; chmn dept, NMex State Univ, 70-74, prof astron, 70-89. *Concurrent Pos:* Grants, NSF, 63-65 & NASA, 65-66. *Mem:* Am Astron Soc. *Res:* Photoelectric photometry and associated astrophysical problems of interstellar medium and galaxies; educational computer programs. *Mailing Add:* PO Box 26 Port Hedland 6721 West Australia

REITNOUR, CLARENCE MELVIN, b Spring City, Pa, Oct 17, 33; m 69. ANIMAL NUTRITION. *Educ:* Pa State Univ, BS, 59; Univ Ky, MS, 62, PhD, 68. *Prof Exp:* Res asst, Univ Ky, 60-62; res asst, Univ Md, 62-64; exten specialist, Univ Ky, 64-66, res asst, 66-68; assoc prof, 68-80, PROF, UNIV DEL, 80-, EQUINE SPECIALIST, 68- *Mem:* Am Soc Animal Sci. *Res:* Equine nitrogen metabolism. *Mailing Add:* Dept Animal Sci & Agr Biochem Univ Del Newark DE 19711

REITSEMA, ROBERT HAROLD, b Grand Rapids, Mich, Jun 25, 20; c 4. GEOCHEMISTRY. *Educ:* Calvin Col, AB, 42; Univ Ill, PhD(org chem), 45. *Prof Exp:* Res chemist, Upjohn Co, 46-48; chief chemist, A M Todd Co, 48-57; sr res chemist, 57-60, sr planning assoc, 60-62, mgr, Chem Develop Div, Ohio, 62-69, RES ASSOC, DENVER RES CTR, MARATHON OIL CO, 69- *Concurrent Pos:* Adj prof, Findlay Col, 63-69, chmn natural sci div, 68-69. *Mem:* Am Chem Soc; European Asn Org Geochem; Sigma Xi; Geochem Soc. *Res:* Synthetic pharmaceuticals; essential oils; terpenes; petrochemical processes; plant design; organic geochemistry; origin and migration of natural gases and petroleum. *Mailing Add:* 2766 S Clayton Denver CO 80210

REITZ, ALLEN BERNARD, b Alameda, Calif, Apr 7, 56; m 78; c 2. STEREOSELECTIVE SYNTHESIS METHODOLOGY, ANTI-PSYCHOTIC RESEARCH. *Educ:* Univ Calif, Santa Barbara, BA, 77; Univ Calif, San Diego, MS, 79, PhD(chem), 82. *Prof Exp:* Postdoctoral fel, McNeil Pharmaceut, Johnson & Johnson, 82-83, res scientist, 83-84, sr scientist, 84-87, prin scientist, Janssen Res Found, 87-89, PRIN SCIENTIST, R W JOHNSON PHARMACEUT RES INST, 90- *Concurrent Pos:* Lectr, numerous univs & res conferences; reviewer, J Org Chem, Carbohydrate Res, Heteroatom Chem, Tetrahedron Lett. *Mem:* Am Chem Soc; Sigma Xi; AAAS. *Res:* Systematic development of structure-activity relationships, attempting to maximize a desired biological profile; discovery of new and useful stereoselective synthetic methodology in organic chemistry, including cycloadditions and novel methods for aminosugar synthesis; author of numerous publications; four US patents. *Mailing Add:* 109 Greenbriar Rd Lansdale PA 19446

REITZ, HERMAN J, b Belle Plaine, Kans, July 5, 16; m 45; c 2. HORTICULTURE. *Educ:* Kans State Col, BS, 39; Ohio State Univ, MS, 40, PhD(hort), 49. *Prof Exp:* From assoc horticulturist to horticulturist, 46-56, horticulturist chg, 57-65, DIR, FLA AGR RES & EDUC CTR, 65- *Mem:* Fel AAAS; Int Soc Citricult (pres, 73-77); Int Soc Hort Sci; fel Am Soc Hort Sci. *Res:* Mineral nutrition of citrus; citrus growing, harvesting and processing. *Mailing Add:* 290 Park Lane W Lake Alfred FL 33850

REITZ, JOHN RICHARD, b Lakewood, Ohio, Feb 7, 23; m 47; c 4. THEORETICAL PHYSICS. *Educ:* Case Inst Technol, BS, 43; Univ Chicago, MS, 47, PhD(physics), 49. *Prof Exp:* Res assoc acoust, Underwater Sound Lab, Harvard Univ, 43-45; mem staff theoret physics, Los Alamos Sci Lab, Univ Calif, 49-52; sci liaison officer physics, Off Naval Res, London, 52-54; from asst prof to prof, Case Inst Technol, 54-65; mgr physics dept, Sci Lab, Ford Motor Co, 65-87; RETIRED. *Mem:* Fel Am Phys Soc. *Res:* Cohesion of solids; imperfection in alkalihalide crystals; electronic structure and transport properties of solids; plasma physics; high speed ground transportation; electricity and magnetism. *Mailing Add:* 2260 Chaucer Ct Ann Arbor MI 48103

REITZ, RICHARD ELMER, b Buffalo, NY, Sept 18, 38; m 60; c 2. ENDOCRINOLOGY, CALCIUM METABOLISM. *Educ:* Heidelberg Col, Tiffin, Ohio, BS, 60; State Univ NY, Buffalo, MD, 64. *Prof Exp:* Intern med, Hartford Hosp, Conn, 64-65; resident, Yale Univ Hosp, 65-66 & Hartford Hosp, 66-67; res assoc endocrinol, Nat Heart Inst, NIH, Bethesda, Md, 67-68; res fel endocrinol, Harvard Med Sch, 67-69 & Mass Gen Hosp, 68-69; asst dir, Clin Invest Ctr, Naval Regional Med Ctr, Oakland, Calif, 69-71; asst prof, Univ Calif-San Francisco, 71-76, assoc clin prof, 76-86, CLIN PROF MED, UNIV CALIF, DAVIS, 86-; DIR ENDOCRINE METAB CTR, OAKLAND, CALIF, 73- *Concurrent Pos:* Chief endocrinol, Providence Hosp, Oakland, 73- *Mem:* Endocrine Soc; Am Soc Bone & Mineral Res; Am Fedn Clin Res; Am Fertil Soc; AAAS; Am Soc Int Med. *Res:* Cytoreceptor assay for 1,25 dihydroxy vitamin D; metabolic bone disease of renal failure. *Mailing Add:* 3100 Summit St Box 23020 Oakland CA 94623

REITZ, RICHARD HENRY, b Minneapolis, Minn, Sept 1, 40; m 63; c 1. BIOCHEMISTRY, TOXICOLOGY. *Educ:* DePauw Univ, BS, 62; Northwestern Univ, Evanston, PhD(biochem), 66, Am Bd Toxicol, dipl. *Prof Exp:* Res biochemist drug develop, Human Health Res & Develop, 66-74, sr res biochemist fermentation, Cent Res Lab, 74-78, RES SPECIALIST, TOXICOL RES LAB, DOW CHEM CO, 78- *Mem:* Am Chem Soc; Am Soc Microbiol; AAAS. *Res:* Chemical carcinogenesis and mutagenesis; microbiological production of chemicals; asthma and hypersensitivity; psychopharmacology. *Mailing Add:* 4105 Chelsea Midland MI 48640

REITZ, ROBERT ALAN, b Lakewood, Ohio, Sept 8, 26; m 48; c 4. PHYSICS. *Educ:* Case Inst Technol, BS, 49; Univ Ill, MS, 51, PhD(physics), 55. *Prof Exp:* Asst physics, Univ Ill, 49-51 & 52-54; from asst prof to assoc prof, 54-65, chmn dept, 57-71, PROF PHYSICS, CARLETON COL, 65- *Concurrent Pos:* NSF fac fel, Ger, 63-64; vis scholar, Stanford Univ, 71-72 & Univ Calif, San Diego, 79-80. *Mem:* Am Phys Soc; Am Asn Physics Teachers; Sigma Xi. *Res:* Solid state physics. *Mailing Add:* Dept Physics Carleton Col Northfield MN 55057

REITZ, ROBERT REX, b Oklahoma City, Okla, Dec 18, 43; m 67; c 2. ORGANIC CHEMISTRY. *Educ:* Austin Col, BA, 66; Kans State Univ, PhD(chem), 71. *Prof Exp:* Assoc org chem, Ohio State Univ, 72-73; RES CHEMIST, ELASTOMER CHEM DEPT, E I DU PONT DE NEMOURS & CO, INC, 73- *Mem:* Am Chem Soc. *Res:* Research and development of new synthetic rubber and rubber chemicals. *Mailing Add:* 727 Folly Hill Rd West Chester PA 19382-6909

REITZ, RONALD CHARLES, b Dallas, Tex, Feb 27, 39; m 65; c 2. BIOCHEMISTRY. *Educ:* Tex A&M Univ, BS, 61; Tulane Univ, PhD(biochem), 66. *Prof Exp:* USPHS fel, Univ Mich, Ann Arbor, 66-69; asst prof biochem, Univ NC, Chapel Hill, 69-75; assoc prof, 75-80, PROF BIOCHEM, UNIV NEV, RENO, 80- *Concurrent Pos:* Vis scientist, Unilever Res Lab, Frythe, Welwyn, Eng, 68, Du Pont de Nemours & Co, Del, 84; sci consult, NC Ctr Alcohol Studies, 74-76. *Mem:* Am Soc Biol Chemists; Am Soc Pharmacol & Exp Ther; AAAS; Am Oil Chemists Soc; Res Soc Alcoholism. *Res:* Lipid metabolism; effects of chronic alcoholism on membrane lipids and on membrane function; membrane lipids of tumor tissue; sex pheromone synthesis in houseflies; synthesis of cuticular hydrocarbons in insects. *Mailing Add:* Dept Biochem Univ Nev Reno NV 89577

REIVICH, MARTIN, b Philadelphia, Pa, Mar 2, 33; m 60; c 2. NEUROLOGY. *Educ:* Univ Pa, BS, 54, MD, 58. *Hon Degrees:* Dr, Semmelweis Med Univ, 90. *Prof Exp:* Intern med, King County Hosp, Seattle, Wash, 58-59; resident neurol, Hosp Univ Pa, 59-61; clin clerk, Nat Hosp, London, Eng, 61-62; instr, Univ Pa, 62-65, instr pharmacol, 63-64, asst prof neurol, 66-68, assoc prof, 68-72, dir, Cerebrovascular Res Lab, 66-73, PROF NEUROL, SCH MED, UNIV PA, 72-, DIR, CEREBROVASCULAR RES CTR, 73- *Concurrent Pos:* NIH res fel, 60-61, training grant, 63; Fulbright fel, 61-62; USPHS career res develop award, 66-75; res assoc physiol, NIMH, 64-65; mem coun cerebrovascular dis, Am Heart Asn, 67-; mem, Int Study Group Cerebral Circulation, 68-; mem neurol A study sect, NIH, 71-75; consult, Vet Admin Hosp, Philadelphia, 68- & US Naval Air Develop Ctr, 70-72; co-dir stroke ctr, Philadelphia Gen Hosp, 71-72; mem task force stroke, Nat Heart, Blood Vessel, Lung & Blood Prog; mem, NSP-A study sect, NIH, 86-90. *Honors & Awards:* Semmelweis Award, 77. *Mem:* Am Neurol Asn; Am Physiol Soc; Soc Neurosci; Am Acad Neurol; Asn Res Nerv & Ment Dis; Soc Nuclear Med; Fedn Neurol. *Res:* Cerebral circulation and metabolism; cerebrovascular disease. *Mailing Add:* Dept Neurol Univ Pa Hosp Philadelphia PA 19104-4283

REIZER, JONATHAN, b Haifa, Israel, Feb 16, 40; US citizen; m 68; c 2. REGULATION OF SUGAR TRANSPORT, METABOLIC PATHWAYS OF SUGAR METABOLISM. *Educ:* Hebrew Univ, Jerusalem, BSc, 64, MSc, 67, PhD(microbiol), 78. *Prof Exp:* Res assoc, Thomas Jefferson Univ, 79-80, Brown Univ, 80-81 & Univ Calif, San Diego, 82-85; sr staff fel, NIH, 85-88; RES BIOLOGIST, UNIV CALIF, SAN DIEGO, 88- *Mem:* Am Soc Microbiol; Am Soc Biochem & Molecular Biol. *Res:* Biochemistry and physiology of transmembrane sugar transport in bacteria; contributed articles to scientific journals. *Mailing Add:* Dept Biol Univ Calif San Diego C-016 La Jolla CA 92093

REJALI, ABBAS MOSTAFAVI, b Aug 19, 21; US citizen; m 54; c 1. RADIOLOGY, NUCLEAR MEDICINE. *Educ:* State Univ NY Downstate Med Ctr, MD, 51; Am Bd Radiol, dipl, 55. *Prof Exp:* Intern med, Grasslands Hosp, Valhalla, NY, 51-52; resident radiol, Univ Hosps, Cleveland, 52-55, teaching fel radiation ther & actg dir radiation ther, 56-57; assoc chief radiol, Roswell Park Mem Inst, 59-61; assoc prof radiol & nuclear med, 63-68, asst prof radiol, 68-76, ASSOC PROF RADIOL, CASE WESTERN RESERVE UNIV, 76-, DIR DEPT NUCLEAR MED, 74- *Concurrent Pos:* Atomic Energy Proj fel, Case Western Reserve Univ, 55-57, sect assoc radiation biol, 56-57; US rep, Int Atomic Energy Agency, 68 & 70; consult radiologist, Highland View Hosp, Metrop Gen Hosp & Huron Rd Hosp. *Mem:* Radiol Soc NAm; fel Am Col Nuclear Med; Soc Nuclear Med; fel Am Col Radiol. *Res:* Diagnostic use of radioisotope tracer; development of radioisotopic instrumentation. *Mailing Add:* 2853 Montgomery Rd Cleveland OH 44122

REJTO, PETER A, b Budapest, Hungary, Apr 28, 34; US citizen; m 60; c 1. MATHEMATICS. *Educ:* NY Univ, PhD(math), 59. *Prof Exp:* Asst prof, NY Univ, 60-64 & Math Res Ctr, Univ Wis, 64-65; asst prof, 65-77, PROF MATH, SCH MATH, UNIV MINN, MINNEAPOLIS, 77- *Mem:* Am Math Asn; Am Math Soc; Soc Indust & Appl Math. *Res:* Spectral theory of operators in Hilbert space. *Mailing Add:* Dept Math 127 Vincent Hall Univ Minn 206 Church St SE Minneapolis MN 55455-0487

REKASIUS, ZENONAS V, b Lithuania, Jan 1, 28; m 60; c 3. ELECTRICAL ENGINEERING. *Educ:* Wayne State Univ, BS, 54, MS, 56; Purdue Univ, PhD(elec eng), 60. *Prof Exp:* From asst prof to assoc prof elec eng, Purdue Univ, 60-64; assoc prof, 64-68, PROF ELEC ENG, TECHNOL INST, NORTHWESTERN UNIV, 68- *Mem:* Inst Elec & Electronics Engrs. *Res:* Control systems; stability; optimization. *Mailing Add:* Dept Elec Eng Northwestern Univ Evanston IL 60201

REKERS, ROBERT GEORGE, b Rochester, NY, Feb 1, 20; m 51; c 3. ANALYTICAL CHEMISTRY. *Educ:* Univ Rochester, BSc, 42; Univ Colo, PhD(chem), 51. *Prof Exp:* Asst foundry chemist, Gen Rwy Signal Co, 41; anal chemist & group leader, Eastman Kodak Co, 42-47; asst phys chem, Univ Colo, 47-51; spectroscopist, US Naval Ord Test Sta, Calif, 51-55; from asst prof to assoc prof chem, Text Tech Univ, 55-86 asst chmn dept, 69-75; RETIRED. *Mem:* AAAS; Am Chem Soc; Colbentz Soc; Sigma Xi. *Res:* Spectroscopy of flames; instrumental methods of analysis. *Mailing Add:* 1717 Norfolk Ave No 2479 Lubbock TX 79416

REKLAITIS, GINTARAS VICTOR, b Posen, Poland, Oct 20, 42; US citizen; m 66; c 2. COMPUTER AIDED DESIGN, PROCESS OPERATIONS. *Educ:* Ill Inst Technol, BS, 65; Stanford Univ, MS, 69, PhD(chem eng), 69. *Prof Exp:* NSF fel, Inst Opers Res, Zurich, Switz, 69-70; from asst prof to assoc prof, 70-80, PROF CHEM ENG, PURDUE UNIV, WEST LAFAYETTE, 80-, ASST DEAN ENG, 85-, HEAD CHEM ENG, 87- *Concurrent Pos:* Fulbright sr lectr, Lithuania, 80; dir, Comput & Systs Technol Div, Am Inst Chem Engrs, 80-83, chmn, Prog Bd, 86-89, div 2nd vchmn, 89, 1st vchmn, 90, chmn, 91; secy, Comput Aids Chem Eng Educ Corp, 82-84, vpres, 84-86, pres, 86-88. *Honors & Awards:* Comput Chem Eng Award, Am Inst Chem Engrs, 84. *Mem:* Am Chem Soc; Math Prog Soc; Opers Res Soc Am; Am Inst Chem Engrs; AAAS. *Res:* Optimization theory; process simulation; computer aided design; scheduling and design of batch processes; computer aided process operations. *Mailing Add:* Sch Chem Eng Purdue Univ West Lafayette IN 47907

REKOFF, M(ICHAEL) G(EORGE), JR, b Galveston, Tex, July 27, 29; m 51; c 3. ELECTRICAL ENGINEERING. *Educ:* Agr & Mech Col, Tex, BSEE, 51, MSEE, 55; Univ Wis, PhD(elec eng), 61. *Prof Exp:* Instr elec eng, Agr & Mech Col, Tex, 54-56 & Univ Wis, 56-59; from asst prof to assoc prof, Tex A&M Univ, 59-69; mem tech staff, TRW Systs Group, 69-70, staff engr, Telelyn Brown Eng, 70-76; dir elec res & develop, Onan Corp, 76-79; EMER PROF, UNIV TENN, CHATTANOOGA, 79- *Mem:* Inst Elec & Electronics Engrs; Instrument Soc Am; Soc Mfg Engrs. *Res:* Servo-mechanisms; computing; machines; systems engineering. *Mailing Add:* Dept Elec Eng Univ Al Birmingham AL 35294

RELMAN, ARNOLD SEYMOUR, b New York, NY, June 17, 23; m 53; c 3. MEDICINE. *Educ:* Cornell Univ, AB, 43; Columbia Univ, MD, 46; Am Bd Internal Med, dipl, 52, 74. *Hon Degrees:* MA, Univ Pa, 75; ScD, Med Col Wis, 80 & Albany Med Col, 83; DMSc, Brown Univ, 81; DHumL, Downstate Med Ctr, 83. *Prof Exp:* Intern, assist resident, assoc resident, med, New Haven Hosp, 46-49; asst med, Sch Med, Yale Univ, 47-49, Boston Univ, 49-50; from asst prof to prof med, Sch Med, Boston Univ, 50-67, Conrad Wesselhoeft Prof Med, 67-68; head renal & electrolyte sect, Evans Mem Dept Clin Res, Mass Mem Hosps, 50-67; dir V & VI med serv, Boston City Hosp, 67-68; Frank Wister Thomas Prof Med & chmn dept, Sch Med, Univ Pa, 68-77; dir med serv, Hosp Univ Pa, 68-77; PROF MED, HARVARD MED SCH, 77-; SR PHYSICIAN, BRIGHAM & WOMEN'S HOSP, BOSTON, 77- *Concurrent Pos:* Nat Res Coun fel med sci, Evans Mem Dept Clin Res, Mass Mem Hosps, 49-50; Res Career Award, NIH, 61-67; ed, J Clin Invest, Am Soc Clin Invest, 62-67, New England J Med, 77-88, ed-in-chief, 88-; res assoc biol chem, Harvard Med Sch, 65-66; consult, Coun Int Exchange Sch, Nat Acad Sci, 73-75, Inst Med Adv Comt Teaching Hosps, 74-75, Food & Drug Admin Adv Rev Panel Over-the-Counter-Antacid Drugs, HEW, Pub Health Serv, 72-73, div med, Bur Health Manpower, 76-77; vis prof, numerous Am & foreign univs, 70-85; mem adv panel, Study Cost Educ Health Prof, Nat Acad Sci, 72-73, mem comt, Biomed Res Vet Admin, 74-75, chmn, comt study Health-Related Effects Cannabis & Derivatives, 80-82; attend physician, Brigham & Women's Hosp, Boston; vis mem, Merton Col, Univ Oxford, 75-76. *Honors & Awards:* John Phillips Mem Award, Am Col Physicians, 85; William C Menninger Lectr, Am Psychiat Asn, 85; McGovern Award Lectr, Med Library Asn, 85; Charles V Chapin Orator, RI Med Soc, 85; Distinguished Serv Award, Am Col Cardiol, 87. *Mem:* Inst Med-Nat Acad Sci; fel Am Acad Arts & Sci; master Am Col Physicians; Am Fedn Clin Res (vpres, 59-60, pres, 60-61); Am Soc Clin Invest (pres, 68-69); fel AAAS; fel Royal Col Physicians; Am Soc Nephrology; Asn Am Physicians. *Res:* Kidney physiology and disease; acid-base and electrolyte physiology; internal medicine and medical education. *Mailing Add:* Ten Shattuck St Boston MA 02115

RELYEA, DOUGLAS IRVING, b Rochester, NY, Sept 20, 30; m 57; c 4. ORGANIC CHEMISTRY. *Educ:* Clarkson Tech, BS, 51; Cornell Univ, MS, 53; Univ SC, PhD(chem), 54. *Prof Exp:* Proj assoc org chem, Univ Wis, 54-56; res scientist, Gen Labs, US Rubber Co, 56-57 & Res Ctr, 57-64, sr res scientist, 64-70; res assoc corp res & develop, Oxford Mgt & Res Ctr, 72-80, RES ASSOC CORP PROTECTION CHEM RES, UNIROYAL CHEM, 80- *Mem:* Am Chem Soc; Royal Soc Chem. *Res:* Chemistry of sulfur compounds; chemistry of nitrogen compounds; biological activity of organic compounds; chemical structure retrieval; molecular geometry and computer graphics. *Mailing Add:* Brookwood Rd Bethany New Haven CT 06525-3148

RELYEA, JOHN FRANKLIN, b Stuttgart, Ark, Oct 25, 47; m 74; c 1. SOIL SCIENCE. *Educ:* Univ Ark, BS, 69, MS, 72, PhD(agron), 78. *Prof Exp:* RES SCIENTIST ENVIRON CHEM, BATTELLE PAC NORTHWEST LABS, 77- *Mem:* AAAS; Am Soc Agron. *Res:* Soil chemistry and soil physics; environmental chemistry; radionuclide chemistry in geologic media. *Mailing Add:* 7702 W 13th Ave Kennewick WA 99337

RELYEA, KENNETH GEORGE, b New York, NY, Oct 24, 41; m 62; c 1. ICHTHYOLOGY, ECOLOGY. *Educ:* Fla State Univ, BA, 62, MS, 65; Tulane Univ, PhD(ichthyol), 67. *Prof Exp:* Asst prof biol, Jacksonville Univ, 67-72, assoc prof, 72-; MEM STAFF, DEPT BIOL, OHIO DOMINICAN COL. *Mem:* AAAS; Am Soc Ichthyol & Herpet; Ecol Soc Am. *Res:* Systematics, ecology and behavior of Ictalurid catfishes and killifishes. *Mailing Add:* Biol Dept Armstrong State Col Savannah GA 31419

REMAR, JOSEPH FRANCIS, b Bridgeport, Pa, Oct 2, 38; m 67; c 1. ORGANIC CHEMISTRY. *Educ:* Villanova Univ, BS, 60; Pa State Univ, PhD(chem), 66. *Prof Exp:* SR RES CHEMIST, ARMSTRONG CORK CO, 66- *Mem:* Am Chem Soc. *Res:* Urethane foams and coatings. *Mailing Add:* 127 Wellington Rd Lancaster PA 17603

REMBERT, DAVID HOPKINS, JR, b Columbia, SC, Jan 14, 37; m 60; c 4. PLANT PHYLOGENY. *Educ:* Univ SC, BS, 59, MS, 64; Univ Ky, PhD(biol), 67. *Prof Exp:* Instr biol, Converse Col, 64-65; instr bot, Univ Ky, 67; from asst prof to assoc prof, 67-81, PROF BIOL, UNIV SC, 81- *Concurrent Pos:* Asst dean, Col Sci & Math, Univ SC, 72-76, actg dean, 75, assoc chmn biol, 87- *Mem:* Bot Soc Am; Linnean Soc London. *Res:* Embryology and phylogeny in legumes; floristics in the southeastern United States; botanical history and garden history; eighteenth century botanical history. *Mailing Add:* Dept Biol Univ SC Columbia SC 29208

REMEDIOS, E(DWARD) C(HARLES), b Vengurla, India, Nov 17, 41; c 1. CHEMICAL ENGINEERING. *Educ:* Univ Edinburgh, BSc, 65, PhD(chem eng), 69; Univ Calif, Berkeley, MBA, 79. *Prof Exp:* Res engr, Chevron Res Co, 69-73; sr resource engr, 73-78, supv resource engr, 78-80, coordr, 80-86, MGR ECON & FORECASTING, PAC GAS & ELEC, 87- *Mem:* Pac Coast Gas Asn; Pac Coast Elec Asn. *Res:* Development, design and economic evaluation of energy supply projects. *Mailing Add:* 33 Toledo Way San Francisco CA 94123

REMENYIK, CARL JOHN, b Budapest, Hungary, May 5, 27; US citizen; m 68; c 1. FLUID MECHANICS. *Educ:* Swiss Fed Inst Technol, Dipl, 51; Johns Hopkins Univ, PhD(aeronaut), 62. *Prof Exp:* Asst fluid dynamics, Swiss Fed Inst Technol, 51-52, asst mach tools, 52-53; sr aeronaut engr, Convair Div, Gen Dynamics, Tex, 57; eng specialist, Martin-Marietta, Md, 61-64; asst prof fluid mech, 64-66, assoc prof fluid mech, 66-80, PROF DEPT ENG SCI & MECH, UNIV TENN, KNOXVILLE, 80- *Concurrent Pos:* Consult, Reactor Div, Oak Ridge Nat Labs, 64- & molecular anat prog, 72- *Mem:* Am Phys Soc; Sigma Xi. *Res:* Aerodynamically generated sound in boundary layers; heat transfer in hypersonic boundary layers; mechanics of biological fluids; centrifugation; dynamics of oscillating liquids and bubbles; magnetohydrodynamic vortex flow. *Mailing Add:* Dept Eng Sci & Mech Univ Tenn Knoxville TN 37996-2030

REMER, DONALD SHERWOOD, b Detroit, Mich, Feb 16, 43; m 69; c 3. ENGINEERING ECONOMICS & MANAGEMENT, CHEMICAL & BIO ENGINEERING. *Educ:* Univ Mich, Ann Arbor, BS, 65; Calif Inst Technol, MS, 66, PhD(chem eng, bus econ), 70. *Prof Exp:* Tech contact engr, Exxon Chem Co, USA, 70-71, chem raw mat div coordr, 72, startup engr new ethylene unit, 72-73, sr proj engr, 72-73, econ & forecast coordr, 73-75, task force mgr, 74-75; assoc prof, 75-80, dir energy inst, 81-83, PROF ENG, HARVEY MUDD COL ENG & SCI, CLAREMONT, 80-; COFOUNDER & PARTNER, CLAREMONT CONSULT GROUP, 79. *Concurrent Pos:* Sr eng consult, Caltech's Jet Propulsion Lab, 75-80, mem tech staff, 80-; Westinghouse Found grant eng econ & Shelby Cullum Found grant eng mgt, Harvey Mudd Col, 78-; case study ed, The Eng Economist 78-89; mem adv coun, Nat Energy Found, 81-86; dir, Eng Econ Div, Am Soc Eng Educ, 80-83, Am Soc Eng Mgt, 81-83; econ chairperson, Nat Tech Prog Comt, Fuels & Petrochemicals Div, Am Inst Chem Engrs. *Honors & Awards:* First Pl Nat Pub Rels Award, Am Inst Chem Engrs, 76; Outstanding Res Award, NASA, 83. *Mem:* Am Inst Chem Engrs; Am Soc Eng Mgt; Am Soc Eng Educ; Am Chem Soc Biochem Technol Div. *Res:* Industrial process and project cost estimation, venture, and risk analysis; engineering management and engineering economic analysis and optimization of capital projects, biotechnology, biochemical process economics; energy management and planning; air and water pollution abatement; solar energy and cogeneration; life cycle cost economic analysis. *Mailing Add:* Dept Eng Harvey Mudd Col Eng & Sci Claremont CA 91711-2834

REMERS, WILLIAM ALAN, b Cincinnati, Ohio, Oct 14, 32; m 61; c 2. ORGANIC CHEMISTRY, MEDICINAL CHEMISTRY. *Educ:* Mass Inst Technol, BS, 54; Univ Ill, PhD, 58. *Prof Exp:* USPHS res fel org chem, Oxford Univ, 58-59; org chemist, Lederle Labs, Am Cyanamid Co, NY, 59-70; assoc prof, 70-72, prof med chem, Sch Pharm & Pharmaceut Sci, Purdue Univ, 72-76, assoc head dept, 74-76; head dept, 80-85, PROF MED CHEM, COL PHARM, UNIV ARIZ, 76- *Concurrent Pos:* Res grants, Nat Cancer Inst, 71- & Bristol Labs, 75- *Mem:* Acad Pharmaceut Sci; Am Chem Soc; Am Pharmaceut Asn; Am Soc Microbiol. *Res:* Antibiotics; heterocycles; synthetic methods; cancer chemotherapeutic agents; molecular mechanics. *Mailing Add:* Col Pharm Univ Ariz Tucson AZ 85721

REMICK, FORREST J(EROME), b Lock Haven, Pa, Mar 16, 31; m 53; c 2. NUCLEAR ENGINEERING, MECHANICAL ENGINEERING. *Educ:* Pa State Univ, BS, 55, MS, 58, PhD, 63. *Prof Exp:* Design engr, Bell Tel Labs, Inc, 55; nuclear engr, Nuclear Reactor Facil, Pa State Univ, 56-59, dir, 59-65, assoc prof, 63-67, actg dir, Ctr Air Environ Studies, 76-78, dir, Inst Sci & Eng & asst to vpres res & grad studies, 67-79, prof nuclear eng, 67-81, asst vpres res & grad studies & dir inter-col res progs, 79-81; dir, Off Policy Eval, US Nuclear Regulatory Comn, 81-82, prof nuclear engr & assoc vpres res, 82-89, COMNR, US NUCLEAR REGULATORY COMN, 89- *Concurrent Pos:* Dir, Curtiss-Wright Nuclear Res Lab, 60-65; mem res reactor subcomt, Nat Acad Sci-Nat Res Coun, 63-65; mem, Atomic Safety & Licensing Bd Panel, 72-82; consult, US Nuclear Regulatory Comn, USAF, Dept Energy, Inst Nuclear Power Opers, Nat Nuclear Accrediting Bd, 82-88, Sci Adv Comt, Idaho Nat Eng Lab, 84-89 & Adv Comt Reator Safeguards, 82-89; chmn, Adv Comt Reactor Safeguards, 89 & Reactor Safety Adv Comt, Savannah River Site, 86-89. *Mem:* Fel Am Nuclear Soc; Am Soc Mech Engrs; Am Soc Eng Educ; Sigma Xi. *Res:* Reactor design and operation; heat transfer and fluid flow in reactor systems. *Mailing Add:* 305 E Hamilton Ave State College PA 16801

REMILLARD, STEPHEN PHILIP, CELL & MOLECULAR BIOLOGY. *Educ:* Princeton Univ, PhD(biol), 81. *Prof Exp:* RES ASSOC, BRANDEIS UNIV, 81- *Mem:* Am Soc Cell Biol. *Mailing Add:* Dept Biol Brandeis Univ 211 Bassine Waltham MA 02254

REMINE, WILLIAM HERVEY, b Richmond, Va, Oct 11, 18; m 43; c 4. SURGERY. *Educ:* Univ Richmond, BS, 40; Med Col Va, MD, 43; Univ Minn, MS, 52; Am Bd Surg, dipl. *Hon Degrees:* DSc, Univ Richmond, 65. *Prof Exp:* From asst prof to assoc prof, 59-69, PROF SURG, MAYO GRAD SCH MED, UNIV MINN, 69-, HEAD SECT SURG, MAYO FOUND, 52- *Concurrent Pos:* Consult, Surgeon Gen, 53- *Mem:* Fel Am Col Surgeons; Asn Mil Surgeons US; Soc Surg Alimentary Tract; hon fel Venezuelan Soc Surg; hon fel Colombian Col Surg; Sigma Xi; Am Surg Asn. *Res:* Surgery of the gastrointestinal tract; head and neck surgery. *Mailing Add:* 129 Island Dr Ponte Vedra Beach FL 32082

REMINGTON, C(HARLES) R(OY), JR, b Webster Groves, Mo, July 15, 24; m 46; c 2. MECHANICAL ENGINEERING. *Educ:* Mo Sch Mines, BSME, 49, MSME, 50. *Prof Exp:* From instr to prof mech eng, Univ Mo, Rolla, 50-89; RETIRED. *Concurrent Pos:* Asst dir indust res ctr, Univ Mo, 65-66. *Mem:* Am Soc Eng Educ; Soc Automotive Engrs; Am Soc Mech Engrs; Nat Soc Prof Engrs; Sigma Xi. *Res:* Heat transfer by conduction in solids, liquids and gases; thermal contact resistance; automotive emission studies and control. *Mailing Add:* Dept Mech Eng 649 Salem Ave Rolla MO 65401

REMINGTON, CHARLES LEE, b Reedville, Va, Jan 19, 22; m 88; c 3. EVOLUTIONARY GENETICS, PALEONTOLOGY. *Educ:* Principia Col, BS, 43; Harvard Univ, AM, 47, PhD(biol), 48. *Prof Exp:* From instr to assoc prof zool, 48-56, res assoc, Peabody Mus, 53-56, assoc cur entom, 56-74, assoc prof biol, 56-83, prof forest entom, 79-83, fel Pierson Col, Yale Univ, 50-, CUR ENTOM, PEABODY MUS, 74-, PROF BIOL, ENTOM & MUSEOLOGY, 83- *Concurrent Pos:* Ed, Jour Lepidopterists Soc, 47-64; entom ed, Conn Geol & Natural Hist Surv, 51-76; secy, Rocky Mountain Biol Lab, 55-59, trustee, 62-63, Guggenheim fel, 58-59; dir, Zero Pop Growth, 68-71, vpres, 69-71; dir, Coun Pop & Environ, 69-76, prog chmn, First Nat Cong; dir, Equil Fund, 70-76, pres, 73-75; res fel entom, CSIRO, Australia, 76; res fel zool, Campinas Univ, Brazil, 81; res assoc entom, Univ Calif, Berkeley, 81; mem, Survival Serv Comn, Int Union Conserv Nature & Natural Resources, 79- *Mem:* Soc Study Evolution; Am Soc Naturalists; Ecol Soc Am; Soc Syst Zool; Lepidopterists Soc (pres, 71). *Res:* Animal interspecific hybridization; island biology; genetics and biology of mimicry; systematics and caryology of Lepidoptera, Thysanura and Microcoryphia; biomed ethics; museology. *Mailing Add:* 440 Prospect St New Haven CT 06511-8112

REMINGTON, JACK SAMUEL, b Chicago, Ill, Jan 19, 31; div; c 2. INTERNAL MEDICINE. *Educ:* Univ Ill, BS, 54, MD, 56; Am Bd Internal Med, dipl, 65. *Prof Exp:* Intern, Univ Calif Serv, San Francisco County Hosp, 56; res assoc, Nat Inst Allergy & Infectious Dis, 57-59; asst resident med, Med Ctr, Univ Calif, San Francisco, 59-60; Nat Inst Allergy & Infectious Dis sr res fel infectious dis, Harvard Med Sch & Thorndike Mem Lab, 60-62; from instr to assoc prof, 62-74, PROF MED, SCH MED, STANFORD UNIV, 74-; CHMN, DEPT IMMUNOL & INFECTIOUS DIS, RES INST & CHIEF CONSULT INFECTIOUS DIS, MED CLIN, PALO ALTO MED FOUND, 62- *Concurrent Pos:* Scientist under US-Soviet Health Exchange, 66, 69; consult, Vet Admin Hosp, Palo Alto, 62-; spec consult, Proctor Found, Med Ctr, Univ Calif, San Francisco, 66-; infectious dis consult, WHO, 67-, Pan-Am Health Orgn, 67- & Dept Army, Ft Ord, Calif, 71-; mem bd, Gorgas Mem Inst, 72-78. *Honors & Awards:* Marcus A Krupp Res Chair, Palo Alto Med Found. *Mem:* Am Asn Immunol; Am Soc Microbiol; Infectious Dis Soc Am (pres, 87-88); Immunocompromised Host Soc (pres, 88-89). *Res:* Congenital infection; acquired toxoplasmosis; compromised host and infection; immunoglobulins in body secretions; defense mechanisms of the host; role of cellular immunity in resistance to infections with intracellular organisms; tumor immunology. *Mailing Add:* Res Inst Palo Alto Med Found 860 Bryant St Palo Alto CA 94303

REMINGTON, LLOYD DEAN, b Jackson, Mich, Dec 29, 19; m 44; c 4. ANALYTICAL CHEMISTRY. *Educ:* Univ Mich, BS, 42; Univ Fl, MEd, 54, PhD(anal chem), 66; Cornell Univ, MST, 62. *Prof Exp:* Chemist, Buick Motors Div, Gen Motors Corp, 42-45; teacher jr high schs & jr cols, Pinellas County, Fla, 46-66; from asst prof to prof, 65-86, EMER PROF CHEM, UNIV NC, ASHEVILLE, 86- *Concurrent Pos:* Ford Found Univ Chicago Proj sci adv, EPakistan Exten Ctr, 69-71. *Honors & Awards:* Am Chem Soc Award, 61. *Mem:* fel Am Inst Chemists. *Res:* Chemistry; comparison of British and North American practices in teaching science. *Mailing Add:* Dept Chem Univ NC Asheville NC 28804

REMINGTON, RICHARD DELLERAINE, b Nampa, Idaho, Aug 2, 31; m 52; c 2. BIOSTATISTICS. *Educ:* Univ Mont, BA, 52, MA, 54; Univ Mich, MPH, 57, PhD(pub health statist), 58. *Hon Degrees:* DSc, Univ Mont, 84. *Prof Exp:* Instr math, Univ Mont, 55-56; from asst prof to prof biostatist, Univ Mich, Ann Arbor, 58-69, dean, Sch Pub Health, 74-82; prof biomet & assoc dean, Sch Pub Health, Univ Tex, Houston, 69-74; vpres acad affairs & dean fac, 82-87, interim pres, 87-88, FOUND DISTINGUISHED PROF PREVENTIVE MED & ENVIRON HEALTH, UNIV IOWA, 82- *Concurrent Pos:* USPHS spec res fel, Dept Med, Statist & Epidemiol, London Sch Hyg & Trop Med, Eng, 66-67; vis prof, Univ Calif, Berkeley, 72; vpres res, Am Heart Asn, 74-75 & 78-80; consult, USPHS, US Vet Admin & Nat Res Coun, Mich State Dept Health. *Honors & Awards:* First Lowell Reed Lectr, Am Pub Health Asn, 78; Albert Lasker Spec Pub Health Award, 80; Louis A Conner Mem Lectr, Am Heart Asn, 82, Gold Heart Award, 83. *Mem:* Inst Med-Nat Acad Sci; fel Am Statist Asn; fel Royal Statist Soc; Biomet Soc; fel Am Pub Health Asn; Am Epidemiol Soc; Am Heart Asn; Int Soc & Fedn Cardiol; Soc Epidemiol Res. *Res:* Medical and health applications of statistical and mathematical models; epidemiology of hypertension and cardiovascular diseases. *Mailing Add:* Prev Med 2514 Steindler Bldg Univ Iowa Iowa City IA 52242

REMINGTON, WILLIAM ROSCOE, b Danville, Ohio, Nov 10, 18; m 42; c 6. ORGANIC CHEMISTRY. *Educ:* Univ Chicago, SB, 40, PhD(org chem), 44. *Prof Exp:* Asst, Nat Defense Res Comt Proj, Univ Chicago, 42-44, group leader, 44-45; res chemist, Jackson Lab, E I du Pont de Nemours & Co, 45-51, group leader, 51, div head, 51-71; prof, Nat Polytech Sch, Quito, Ecuador, 71-73; div head, E I du Pont de Nemours & Co, 73-76, res assoc, Jackson Lab, 76-82; RETIRED. *Mem:* Am Chem Soc. *Res:* Dyes; dyeing mechanisms; polymer permeability; organic synthesis. *Mailing Add:* 210 S Lindamere Wilmington DE 19809

REMLER, EDWARD A, b Vienna, Austria, Dec 26, 34; US citizen; m 61; c 2. PHYSICS. *Educ:* Mass Inst Technol, BS, 55; Columbia Univ, MS, 60; Univ NC, PhD(physics), 63. *Prof Exp:* Res assoc physics, Univ NC, 63-64; instr, Princeton Univ, 64-66; fel, Lawrence Livermore Lab, Univ Calif, 66-67; from asst prof to assoc prof, 67-77, PROF PHYSICS, COL WILLIAM & MARY, 77- *Mem:* Am Phys Soc. *Res:* Quantum mechanics; particle physics. *Mailing Add:* Dept Physics Col William & Mary Williamsburg VA 23185

REMLEY, MARLIN EUGENE, b Walcott, Ark, Apr 25, 21; m 43; c 3. PHYSICS. *Educ:* Southeast Mo State Col, AB, 41; Univ Ill, MS, 48, PhD(physics), 52. *Prof Exp:* Instr physics & math, Southeast Mo State Col, 46-47; asst physics, Univ Ill, 47-52; res engr exp physics, NAm Aviation, Inc, 52-55, supvr exp physics, Atomics Int Div, 55-56, group leader reactor kinetics, 56-58, actg chief reactor develop, 58, dir spec projs, 59-60, reactor physics & instrumentation, 60-61, dir health safety & radiation serv, Energy Systs Group, 67-84, DIR, NUCLEAR SAFETY & LICENSING, ROCKWELL INT, 84- *Mem:* Am Phys Soc; fel Am Nuclear Soc; Atomic Indust Forum. *Res:* Nuclear and reactor physics; nuclear scattering; scintillation counters; reactor design and development; reactor dynamics and safety; radiological safety; nuclear materials management and safeguards; author or coauthor of over 50 publications. *Mailing Add:* 19112 Halsted St Northridge CA 91324

REMMEL, RANDALL JAMES, b Peoria, Ill, Aug 23, 49. INORGANIC CHEMISTRY. *Educ:* Ill State Univ, BS, 71; Ohio State Univ, PhD(inorg chem), 75. *Prof Exp:* Res assoc, Mat Lab, Polymer Br, USAF, 75-76; asst prof chem, Univ Ala, Birmingham, 76-82; VIS LECTR, UNIV SFLA, TAMPA, 82- *Mem:* Am Chem Soc. *Res:* Synthesis and characterization of metallacarboranes, metallaboranes and boron-boron bonded polyhedcal boraves; nuclear magnetic resonance relaxation studies of boranes; borane derivatives and their anions. *Mailing Add:* 818 Chipaway Dr Apollo Beach FL 33570

REMMEL, RONALD SYLVESTER, b West Bend, Wis, July 18, 43; m 72. NEUROPHYSIOLOGY, NEUROANATOMY. *Educ:* Calif Inst Technol, BS, 65; Princeton Univ, PhD(physics), 71. *Prof Exp:* Res assoc physics, Princeton Univ, 71-72; fel physiol, Univ Calif, Berkeley, 72-74; fel ophthal, Med Sch, Johns Hopkins Univ, 74-75; from asst prof to assoc prof physiol, Univ Ark Med Sci, 83-83. *Concurrent Pos:* NIH fel, 72-75; fel, Fight for Sight Inc, NY, 75; NIH gen res grant, Univ Ark for Med Sci, 75-76; prin investr, Nat Eye Inst res grant, 76-79; grants, NIMH, 79-81 & NSF, 80- *Mem:* Soc Neurosci; Asn Res Vision & Ophthal; Am Physiol Soc; AAAS; Sigma Xi. *Res:* Study of cat brainstem neurons involved in the control of eye movements; vestibular reflexes and spinal motor mechanisms through use of electrophysiological techniques. *Mailing Add:* Remmel Labs 26 Bay Colony Dr Ashland MA 01721-1840

REMMENGA, ELMER EDWIN, b Douglas, Nebr, Jan 9, 27; m 53; c 5. STATISTICS. *Educ:* Univ Nebr, BS, 50; Purdue Univ, MS, 53, PhD, 55. *Prof Exp:* Asst statistician, Exp Sta, Purdue Univ, 50-55; from asst prof to assoc prof math, 55-62, sta statistician, 55-64, chief comput ctr, 57-62, PROF APPL STATIST, COLO STATE UNIV, 62- *Concurrent Pos:* Math consult, Bur Mines Res Ctr, Colo, 58-; vis biometrician, Waite Agr Res Inst, Univ Adelaide, 61-62; vis prof, Univ Colo, 62-65; statist consult, Nat Water Qual Lab, Environ Protection Agency, Minn, 70-; vis prof, Swiss Fed Forestry Res Inst. *Mem:* Biomet Soc; Am Statist Asn; Inst Math Statist. *Res:* Application of statistical methods to biological sciences; design sampling; computing. *Mailing Add:* Dept Statist Colo State Univ Ft Collins CO 80523

REMO, JOHN LUCIEN, b Brooklyn, NY, Dec 13, 41; c 2. LASER RESONATOR OPTICS, ENERGY EXPERT SYSTEMS. *Educ:* Manhattan Col, BS, 63; State Univ NY, Stony Brook, MS, 71; Polytech Inst NY, MS, 73, PhD(physics), 79. *Prof Exp:* Res scientist astrophysics, Copenhagen Univ Observ, 69-70 & Bartol Res Found Franklin Inst, 70-71; prof energy, Ctr Energy Policy & Res, Ny Inst Technol, 84-87, prof mech eng, Dept Eng, 85-87, PROF PHYSICS, NY INST TECHNOL, 86-; CHIEF SCIENTIST ERG SYSTS, LASER OPTIC SYSTS, INC, 82- *Concurrent Pos:* Adj assoc prof geol, Hofstra Univ, 73-84, physics, 80-84; sci-technol adv, Long Island Region Export Coun, 83-86; lectr mech eng, State Univ NY, Stony Brook, 84-85; sr scientist, dept areospace & appl mech, Polytech Inst NY, 85-86. *Honors & Awards:* Nininger Meteorite Award, 72-73. *Mem:* Am Phys Soc; Optical Soc Am; Meteoritical Soc; Am Geophys Union; Sigma Xi; Intrp Soc Optical Eng. *Res:* Development of integral operator methods to describe active laser resonator dynamics; utilization of solar energy for electricity generation and interior daylighting; development of expert system software for micro computers; analysis of imaging characteristics for extreme ultraviolet laser systems; optical computing systems designs; developed energy conservation & alternate energy software; developed laser resonator design optics; developed hermetic optics for xuv radiation; numerous publications in optics, astrophysics geophysics and energy systems. *Mailing Add:* Brackenwood Path Head of the Harbor St James NY 11780

REMOLD, HEINZ G, b Bad Reichenhall, Germany, May 27, 37. IMMUNOLOGY. *Educ:* Univ Munich, PhD(zool), 64. *Prof Exp:* Assoc prof, 75-83, PROF MED, SCH MED, HARVARD UNIV, 83- *Mem:* Am Asn Immunol; Am Fedn Clin Res; Am Asn Biol Chemists; Am Soc Pathologists. *Mailing Add:* Dept Med Seeley G Mudd Bldg Harvard Med Sch 250 Longwood Ave Boston MA 02115

REMOLE, ARNULF, b Melhus, Norway, July 1, 28; Can citizen; m 65; c 1. PHYSIOLOGICAL OPTICS, OPTOMETRY. *Educ:* Univ Man, BFA, 58; Ont Col Optom, OD, 62; Ind Univ, Bloomington, MS, 67, PhD(physiol optics), 69. *Prof Exp:* Instr optom, Sch Optom, Ont Col Optom, 62-66; teaching assoc physiol optics, Ind Univ, Bloomington, 68-69; asst prof, 69-74, assoc prof, 74-81, PROF OPTOM, UNIV WATERLOO, 81- *Mem:* Fel Am Acad Optom; Can Asn Optom; Optical Soc Am. *Res:* Psychophysics of vision; border effects; visual pattern responses arising from temporal modulations of the stimulus; binocular vision; visual performance evaluation; optics of the eye; optometrical instrumentation; aniseikonia. *Mailing Add:* Sch Optom Univ Waterloo Waterloo ON N2L 3G1 Can

REMONDINI, DAVID JOSEPH, b Deming, NMex, Dec 27, 31; m 52; c 6. GENETICS. *Educ:* Univ Calif, Santa Barbara, BA, 55; Univ Utah, MS, 64; Utah State Univ, PhD(zool-genetics), 67. *Prof Exp:* Asst prof biol, Gonzaga Univ, 67-74; assoc prof biol sci, Mich Technol Univ, 74-77; EXEC SECY GENETICS STUDY SECT, DIV RES GRANTS, NIH, DHHS, 77- *Concurrent Pos:* Jesuit Res Coun res grant, Gonzaga Univ, 67-71, dir summer sessions & spec progs, 71-73; consult genetics, Sacred Heart Med Ctr, Spokane, 73-74. *Mem:* AAAS; Genetics Soc Am; Am Soc Human Genetics; Genetics Soc Can. *Res:* Human cytogenetics; Drosophila genetics; temperature sensitivity; maternal effects. *Mailing Add:* Div Res Grants RRB NIH Bethesda MD 20892

REMPEL, ARTHUR GUSTAV, b Russia, Jan 5, 10; nat US; m 34; c 3. ZOOLOGY. *Educ:* Oberlin Col, AB, 34; Univ Calif, PhD(zool), 38. *Hon Degrees:* DSc, Whitman Col, 87. *Prof Exp:* Actg cur, Mus Zool & Anthrop, Oberlin Col, 31-34; custodian, Dept Zool, Univ Calif, 34-35, asst, 35-38; from instr to prof, 38-75, cur, Mus Natural Hist, 38-46 & 53-71, EMER PROF BIOL, WHITMAN COL, 75- *Mem:* Fel AAAS; Sigma Xi. *Res:* Embryology. *Mailing Add:* 635 University St Walla Walla WA 99362

REMPEL, GARRY LLEWELLYN, b Regina, Sask, Aug 20, 44. INORGANIC CHEMISTRY, CATALYSIS. *Educ:* Univ BC, BSc, 65, PhD(phys inorg chem), 68. *Prof Exp:* Nat Res Coun Can fel, Imp Col, Univ London, 68-69; asst prof, 69-73, assoc prof, 73-80, PROF CHEM ENG, UNIV WATERLOO, 80-, CHMN DEPT, 88- *Concurrent Pos:* Consult, Polysar Ltd, 81- & Chinook Chem, 81- *Mem:* NY Acad Sci; fel Chem Inst Can; Am Chem Soc; Can Inst Mining & Metall. *Res:* Coordination chemistry and homogeneous catalysis; organometallic chemistry; physical chemistry of hydrometallurgical processes; polymer supported and entrapped metal catalysts; catalyts for waste water treatment; chemical modification of polymers. *Mailing Add:* 532 Sandbrooke Ct Waterloo ON N2T 2H4 Can

REMPEL, HERMAN G, b Ukraine, Apr 8, 02; nat US; m 33; c 3. CHEMISTRY. *Educ:* Bethel Col, Kans, BA, 27. *Prof Exp:* Chemist, Inyo Chem Co, 27-29; chief chemist, Twining Labs, Calif, 29-71; agr consult, Resources Int, 71-81 RETIRED. *Mem:* Am Chem Soc; Nat Soc Prof Engrs; Int Food Technologists. *Res:* Agricultural chemistry; insecticides; toxicology; chemical engineering. *Mailing Add:* Monarch Pines Mobile Home Park 64 Briggs Ave Pacific Grove CA 93950

REMPEL, WILLIAM EWERT, b Man, Can, July 6, 21; nat US; m 48; c 2. ANIMAL BREEDING. *Educ:* Univ Man, BSA, 44, MSc, 46; Univ Minn, PhD(animal breeding), 52. *Prof Exp:* Instr animal husb, Univ Man, 46-47; agr rep, Man Dept Agr, 47-48; asst, 48-49, from instr to assoc prof, 50-64, PROF ANIMAL HUSB, UNIV MINN, ST PAUL, 64- *Concurrent Pos:* Dir genetics ctr, Univ Minn, 65-67. *Mem:* Fel AAAS; Am Soc Animal Sci; Genetics Soc Am; NY Acad Sci; Am Genetic Asn; Can Soc Animal Sci. *Res:* Genetics. *Mailing Add:* Dept Animal Sci Univ Minn 1404 Gortner Ave St Paul MN 55108

REMPFER, GERTRUDE FLEMING, b Seattle, Wash, Jan 30, 12; m 42; c 4. PHYSICS. *Educ:* Univ Wash, BS, 34, PhD(physics), 39. *Prof Exp:* Instr physics, Mt Holyoke Col, 39-40 & Russell Sage Col, 40-42; physicist, Naval Res Lab, 42-43 & SAM Lab, Columbia Univ, 44; proj engr, Farrand Optical Co, 45-51; assoc prof eng, Antioch Col, 51-52; assoc prof physics, Fisk Univ, 53-57; assoc prof, Pac Univ, 57-59; assoc prof, 59-68, prof, 68-77, EMER PROF PHYSICS, PORTLAND STATE UNIV, 77- *Concurrent Pos:* Consult, AMP, Inc, 51-57, Tektronix, 60-70 & Elektros, 70-75; NSF grant, 79-82; pres, E-scope, 75- *Mem:* Am Phys Soc; Am Asn Phys Teachers; Electron Micros Soc Am; Sigma Xi. *Res:* Electron physics; electron and ion optics; electron microscopy. *Mailing Add:* Dept Physics Portland State Univ PO Box 751 Portland OR 97207

REMPFER, ROBERT WEIR, b Parkston, SDak, Apr 14, 14; m 42; c 4. MATHEMATICS. *Educ:* Univ SDak, BA, 33; Northwestern Univ, MA, 34, PhD(math), 37. *Prof Exp:* Instr math, Rensselaer Polytech Inst, 37-44; physicist, SAM Labs, Columbia Univ, 44-45 & Farrand Optical Co, 45-50; prof math, Antioch Col, 50-53; assoc prof, Fisk Univ, 53-57; assoc prof, 57-58, PROF MATH, PORTLAND STATE UNIV, 58- *Concurrent Pos:* Chmn dept math, Portland State Univ, 58-67. *Mem:* Am Math Soc; Sigma Xi. *Res:* Gaseous diffusion; electron optical design; interference optics; probability; information theory; geometry. *Mailing Add:* Portland State Univ PO Box 751 Portland OR 97207

REMSBERG, ELLIS EDWARD, b Buckeystown, Md, Oct 24, 43; m 67; c 2. ATMOSPHERIC PHYSICS. *Educ:* Va Polytech Inst, BS, 66; Univ Wis-Madison, MS, 68, PhD(meteorol), 71. *Prof Exp:* Jr res asst, Nat Radio Astron Observ, 62-65; geophysicist, US Coast & Geod Surv, 66; res asst chem, Univ Minn, 69-70; lectr meteorol, Univ Wis-Madison, 71; res asst prof chem, Col William & Mary, 71-72; res asst prof & NASA res grant geophys, Old Dom Univ, 72-73; aerospace technologist, 73-80, SR RES SCIENTIST, LANGLEY RES CTR, NASA, 80-; fel, Univ Wash, Seattle, 83-84. *Concurrent Pos:* Fel, Univ Wash, Seattle, 83-84; prin investr, NASA, 85- *Honors & Awards:* Floyd Thompson fel, NASA, 83-84. *Mem:* Am Geophys Union; Am Meteorol Soc; Sigma Xi. *Res:* Air chemistry; remote sensing; satellite meteorology; processes in the stratosphere; laser applications. *Mailing Add:* Langley Res Ctr NASA M/S 401B Hampton VA 23665-5225

REMSBERG, LOUIS PHILIP, JR, b Rupert, Idaho, Sept 14, 33; m 57; c 3. NUCLEAR CHEMISTRY. *Educ:* Univ Idaho, BS, 55, MS, 56; Columbia Univ, PhD(phys chem), 61. *Prof Exp:* Actg instr chem, Univ Idaho, 55-56; CHEMIST, BROOKHAVEN NAT LAB, 61- *Mem:* Am Chem Soc. *Res:* Nuclear reactions and properties. *Mailing Add:* Brookhaven Nat Lab Upton NY 11973

REMSEN, CHARLES C, III, b Newark, NJ, May 16, 37; m 60, 76; c 4. MICROBIOLOGY, LIMNOLOGY. *Educ:* Nat Agr Col, BS, 60; Syracuse Univ, MS, 63, PhD(microbiol), 65. *Prof Exp:* NIH fel, 65-67; asst scientist biol, Woods Hole Oceanog Inst, 67-71, assoc scientist, 71-75; assoc prof zool & assoc scientist, Ctr Great Lake Studies, Univ Wis, Milwaukee, 75-83, coordr zool & microbiol, 76-84, prof & sr scientist, 83-87, actg dir, 87-89, DIR, CTR GREAT LAKE STUDIES, UNIV WIS-MILWAUKEE, 89- *Concurrent Pos:* Spec serv appointment, Grad Sch, Boston Univ; NSF rep, 2nd US-Japan Conf Microbiol; chmn, joint comt biol oceanog, Mass Inst Technol, WHOI PhD Prog, 71-74; deleg, Coun Ocean Affairs, 90- *Mem:* Int Asn Great Lakes Res; Am Soc Limnol Oceanog; Electron Micros Soc Am; Am Geophys Union. *Res:* Aquatic microbiology, microbial ecology, ultrastructure of autotrophic procaryotes, methane oxidations, biogeochemistry, freshwater hydrothermal vent communities. *Mailing Add:* Ctr Great Lakes Studies 600 E Greenfield Ave Milwaukee WI 53204

REMSEN, JAMES VANDERBEEK, JR, b Newark, NJ, Sept 21, 49; m. NEOTROPICAL BIRD BIOLOGY. *Educ:* Stanford Univ, BA & MA, 71; Univ Calif, Berkeley, PhD(zool), 78. *Prof Exp:* Asst prof, 78-83, assoc prof, 83-89, PROF DEPT ZOOL & PHYSIOL, LA STATE UNIV, 89-, CUR BIRDS, MUS ZOOL, 78- *Mem:* AAAS; Am Ornithologists Union; Ecol Soc Am; Am Soc Naturalists; Cooper Ornith Soc; Wilson Ornith Soc; Assoc Field Ornithologists. *Res:* Ecology and evolution of neotropical birds; aspects in which they differ from birds of the temperate zone; birds of the Andes and western Amazonia. *Mailing Add:* Mus Zool Foster Hall 119 La State Univ Baton Rouge LA 70803

REMSEN, JOYCE F, radiobiology, chemical carcinogensis, for more information see previous edition

REMSON, IRWIN, b New York, NY, Jan 23, 23; m 48; c 2. HYDROLOGY, ENVIRONMENTAL GEOLOGY. *Educ:* Columbia Univ, AB, 46, AM, 49, PhD, 54. *Prof Exp:* Asst geol, Columbia Univ, 47-49; geologist, US Geol Surv, 49-60; assoc prof civil eng, Drexel Inst, 60-65, prof & chief marshal fac, 65-68; chmn dept appl earth sci, 75-82, Barney & Estell Morris prof earth sci, 81, PROF GEOL, STANFORD UNIV, 68- *Concurrent Pos:* Lectr, Drexel Inst, 54-60; Lindback Found Award, 66. *Honors & Awards:* Birdsall lectr geol, Geol Soc Am. *Mem:* Geol Soc Am; Am Geophys Union; Soil Sci Soc Am; Sigma Xi. *Res:* Ground water geology; soil moisture movement; ground water recharge. *Mailing Add:* Dept Geol Stanford Univ Stanford CA 94305

REMY, CHARLES NICHOLAS, b Hudson, NY, May 31, 24; m 52; c 4. BIOCHEMISTRY. *Educ:* Syracuse Univ, PhD(biochem), 52. *Prof Exp:* Am Cancer Soc fel, Sch Med, Univ Pa, 52-53 & Div Biochem, Mass Inst Technol, 53-54; instr biochem, State Univ NY Upstate Med Ctr, 54-60, asst prof, 60-62; assoc prof, 62-68, PROF BIOCHEM, BOWMAN GRAY SCH MED, 68- *Concurrent Pos:* Biochemist, Med Res Div, Vet Admin Hosp, Syracuse, NY, 54-56, prin scientist, 56-62. *Mem:* AAAS; Am Soc Biol Chemists; Am Chem Soc; Soc Exp Biol & Med; Am Soc Microbiol; Sigma Xi. *Res:* Biomethylation of nucleic acids; biosynthesis of ribosomes; biological regulation of nucleic acids and protein synthesis; taurine biosynthesis and transport. *Mailing Add:* Dept Biochem Bowman Gray Sch Med Winston-Salem NC 27103

REMY, DAVID CARROLL, b Waco, Tex, July 17, 29; m 63; c 2. MEDICINAL CHEMISTRY. *Educ:* Univ Calif, Los Angeles, BS, 51, MS, 52; Univ Wis, PhD, 58. *Prof Exp:* Res chemist, Elastomer Chem Dept, E I du Pont de Nemours & Co, 58-60; fel oncol, McArdle Mem Lab, Med Sch, Univ Wis, 60-62; sr res fel, 62-80, SR INVEST, MERCK, SHARP & DOHME RES LABS, 80- *Mem:* Am Chem Soc. *Res:* Medicinal chemistry; CNS drugs; blood coagulation. *Mailing Add:* 607 Jenkins Lane MR 1 N Wales PA 19454

REN, PETER, b Macau, Mar 12, 48; US citizen; m 78; c 2. TECHNICAL MANAGEMENT. *Educ:* Adelphia Univ, BA, 71; Univ RI, PhD(biochem), 76, Rutgers Univ, MBA, 85. *Prof Exp:* Res asst, Univ RI, 71-76; res assoc biochem, Sch Med, Univ Md, 76-78; res biochem, Beecham Prod, 78-82, group leader, 82-85; mgr, Oral B Labs, 85-87, dir mfg, 87-89; SECT HEAD, COLGATE PALMOLIVE, 89- *Concurrent Pos:* Assoc, over the counter prod develop, Food & Drug Admin, Prod Scheduling, Clin Testing. *Mem:* AAAS; Am Asn Dental Res. *Res:* Mechanism of action of vitamin K; anticoagulant drugs; clotting proteins synthesis; basement membrane metabolism; collagen metabolism; oral hygiene; bacterial adhesion; caries formation; inflammations; chemotaxis. *Mailing Add:* Colgate Palmolive Technol Ctr 909 River Rd Piscataway NJ 08855

REN, SHANG YUAN, b Chongqin, Sichuan, China, Jan 10, 40; m 68; c 2. SEMICONDUCTOR PHYSICS, SEMICONDUCTOR DEVICE PHYSIC. *Educ:* Beijing Univ, BS, 63, PhD(physics), 66. *Prof Exp:* Engr, Beijing Second Semiconductor Fact, 68-73; teacher, Univ Sci & Tech, China, 73-78; vis scholar appl physics, Stanford Univ, 78-80; res assoc solid state physics, Univ Ill, 80-81; lect phys, 78-83, assoc prof physics, 83-85, PROF PHYSICS, UNIV SCI & TECH, CHINA, 85-; RES SCIENTIST, ARIZ STATE UNIV, 91- *Concurrent Pos:* Prin investr, Chinese Acad Sci, 83-85, Chinese Nat Educ Comt, 86-88, Chinese Nat Sci Found, 86-88; mem, Chinese Nat Acad Comt, Condensed Matter Theory & Statist Physics; vis prof, Univ Notre Dame, 86-90. *Mem:* Chinese Phys Soc; Am Phys Soc. *Res:* Theory condensed matter; electronic structure, optical, transport, vibrational and mechanical properties of semiconductors, semiconductor superlattices and possible device applications. *Mailing Add:* Dept Physics & Astron Ariz State Univ Tempe AZ 85287-1540

REN, SHANG-FEN, b Hunan, China. SEMICONDUCTORS, ELECTRONIC STATES. *Educ:* Beijing Univ, China, BS, 70; Tex A&M Univ, PhD(physics), 86. *Prof Exp:* Ed, Inst Sci & Technol Changsha, Hunan, China, 72-77; teaching & res asst physics, Univ Sci & Technol China, 77-81 & Tex A&M Univ, 83-86; res assoc, 81-83, RES PHYSICIST, UNIV ILL, URBANA-CHAMPAIGN, 86- *Concurrent Pos:* Vpres, Women Acad Prof Group, Univ Ill, Urbana-Champaign, 88-90. *Mem:* Am Phys Soc; Am Asn Women Sci. *Res:* Theoretical studies on electrons and phonons in semiconductors and their alloys, at semiconductor surfaces, in superlattices and other heterostructures; vibrational properties. *Mailing Add:* Dept Physics Univ Ill 1110 W Green St Urbana IL 61801

RENARD, KENNETH G, b Sturgeon Bay, Wis, May 5, 34; m 56; c 3. CIVIL ENGINEERING, HYDROLOGY & WATER RESOURCES. *Educ:* Univ Wis, BS, 57, MS, 59; Univ Ariz, PhD(civil eng), 72. *Prof Exp:* Hydraul engr, Agr Res Serv, USDA, Wis, 57-59, res hydraul engr, Ariz, 59-64, res hydraul engr, Southwest Watershed Res Ctr, 64-68, dir, 68-88, DIR, ARIDLAND WATERSHED MGT RES CTR, AGR RES SERV, USDA, 88-, HYDRAUL ENGR, 88- *Concurrent Pos:* Adj prof, Univ Ariz; ed, J Irrig & Drainage Eng, 83-85; mem exec comt, Irrig & Drainage Div, Am Soc Civil Engrs. *Mem:* Soil Conserv Soc Am (pres, 79); fel Am Soc Civil Engrs; AAAS; Am Soc Agr Engrs; Am Geophys Union; fel Soil & Water Conserv Soc. *Res:* Watershed hydrology relating land practices to water yields and peak rates of discharge; sediment transport phenomenon in ephemeral stream beds; erosion prediction from varying land use. *Mailing Add:* Arid Land Watershed Mgt Res Unit 2000 E Allen Rd Tucson AZ 85719

RENARD, ROBERT JOSEPH, b Green Bay, Wis, Dec 22, 23; m 47; c 4. METEOROLOGY. *Educ:* Univ Chicago, MS, 52; Fla State Univ, PhD(meteorol), 70. *Prof Exp:* Asst meteorol, Univ Chicago, 51-52; from asst prof to prof meteorol, 52-90, chmn dept, 80-90, EMER PROF, US NAVAL POSTGRAD SCH, 90- *Mem:* Fel Am Meteorol Soc; Am Geophys Union; Nat Weather Asn; Am Polar Soc. *Res:* Synoptic, polar and satellite meteorology; emphasis on observations, marine fog, visibility, Antarctic and model output statistics. *Mailing Add:* Dept Meteorol Naval Postgrad Sch Monterey CA 93943-5000

RENARDY, MICHAEL, b Stuttgart, Ger, Apr 9, 55; m 81; c 3. VISCOELASTIC FLUIDS. *Educ:* Univ Stuttgart, dipl, 77 & 78 & PhD(math), 80. *Prof Exp:* Res assoc, Univ Stuttgart, 78-80; post doc fel, Univ Wis, 80-81 & Univ Minn, 81-82; from asst prof to assoc prof math, Univ Wis, 82-86; assoc prof, 86-89, PROF MATH, VA POLYTECH INST & STATE UNIV, 89- *Honors & Awards:* Fed Victor, Fed Competition, Found Ger Sci. *Mem:* Am Math Soc; Soc Indust & Appl Math; Int Soc Interaction Mech & Math; Soc Natural Philos; Soc Rheology. *Res:* Problems in nonlinear partial different equations, in particular, equations modelling viscoelastic fluids. *Mailing Add:* Dept Math Va Polytech Inst & State Univ Blacksburg VA 24061-0123

RENARDY, YURIKO, b Sapporo, Japan, Jan 15, 55; m 81; c 3. FLUID DYNAMICS, COMPUTATIONAL FLUID DYNAMICS. *Educ:* Australian Nat Univ, BSc, 77; Univ Western Australia, PhD(math), 81. *Prof Exp:* Res assoc, Math Res Ctr, Univ Wis-Madison, 80-83, lectr, 82-83, prog coordr, 83-86; asst prof, 86-89, ASSOC PROF MATH, VA POLYTECH INST & STATE UNIV, 90- *Concurrent Pos:* Lectr math, Univ Minn, 81-82; vis fel, Australian Nat Univ, 84. *Mem:* Am Phys Soc. *Res:* Fluid dynamics, with emphasis on stability and numerical methods. *Mailing Add:* Dept Math Va Polytech Inst & State Univ 460 McBride Hall Blacksburg VA 24061-4097

RENAUD, LEO P, Can citizen. NEUROSCIENCE, NEUROENDOCRINOLOGY. *Educ:* Univ Ottawa, BA, 61, MD, 65; McGill Univ, PhD(physiol), 72. *Prof Exp:* From asst prof to assoc prof neurol, McGill Univ, 73-81; asst physician, 73-78, ASSOC PHYSICIAN MED, MONTREAL UNIV HOSP, 78-; PROF NEUROL, MCGILL UNIV, 81- *Concurrent Pos:* Scholar, Med Res Coun, Can, 73-78 & Found Health Res Quebec, 78-85. *Honors & Awards:* Gold Medal, Royal Col Physicians & Surgeons, Can, 85. *Mem:* Am Physiol Soc; Can Physiol Soc; Endocrine Soc; Soc Neurosci. *Res:* Electrophysiology of mammalian neurosecretory neurons; neurotransmitter regulation of their excitability and hormone (vasopressin and oxytocin) release using in-vivo and in-vitro approaches; central neural processing of cardiovascular inputs to the brain. *Mailing Add:* Div Neurol Ottawa Civic Hosp 1053 Carling Ave Ottawa ON K1Y 4E9 Can

RENAUD, SERGE, b Cartelegue, France, Nov 21, 27; Can citizen; m 55; c 1. EXPERIMENTAL PATHOLOGY. *Educ:* Univ Bordeaux, BA, 47; Univ Montreal, VMD, 57, PhD(exp med), 60, PhD(hematol) 78. *Prof Exp:* Res assoc, Montreal Heart Inst, 60-63, chief lab exp path, 63-73; DIR, UNIT 63, NIH & MED RES, FRANCE, 73-; PROF NUTRIT, UNIV MONTREAL, 75- *Concurrent Pos:* Med Res Coun Can & Que Heart Found grants, 61-; vis prof, Boston Univ, 71-72; prof path, Univ Montreal, 72-73, dir dept nutrit, 75-80. *Honors & Awards:* Borden Award for Nutrit, 67; Award of the Found Française de Nutrit, 83. *Mem:* Soc Exp Biol & Med; Am Heart Asn; Nutrit Soc Can; Am Soc Exp Path; Int Acad Path. *Res:* Influence of nutrition, stress, hormones on the pathogenesis of thrombosis, atherosclerosis and coronary heart disease. *Mailing Add:* Res Unit 63 NIH & Med Res 22 Doyen Lejune Ave Bron 69500 France

RENAULT, JACQUES ROLAND, b Alameda, Calif, July 26, 33; m 56; c 2. GEOCHEMISTRY, PETROLOGY. *Educ:* Stanford Univ, BS, 57; NMex Inst Mining & Technol, MS, 59; Univ Toronto, PhD(geol), 64. *Prof Exp:* Explor geologist, Bear Creek Mining Co, 59-60, Southwest Potash Corp, 61, F R Joubin & Assoc, 62-63; geologist, 64-80, SR GEOLOGIST, STATE BUR MINES & MINERAL RESOURCES, NMEX INST MINING & TECHNOL, 80- *Concurrent Pos:* Grant, Geol Surv Can, 62-63, NMex Energy Inst, 76, 78; exec bd, NMex Energy Inst, NMex State Univ, 78-80; adj prof, geosci dept, NMex Tech. *Mem:* AAAS; Geol Soc Am; Mineral Soc Am; Am Geophys Union; Sigma Xi. *Res:* Staratigraphic geochemistry of Espanola Basin, NMex; mineral physics, especially x-ray diffraction and thermoluminescence; ingeous petrology; x-ray flourescence spectroscopy. *Mailing Add:* 1210 South Dr Socorro NM 87801

RENCHER, ALVIN C, b St Johns, Ariz, Dec 21, 34; m 62; c 3. LINEAR MODELS, MULTIVARIATE ANALYSIS. *Educ:* Brigham Young Univ, BS, 59, MA, 62; Va Polytech Inst, PhD(statist), 68. *Prof Exp:* Statistician, Hercules Inc, 62-63; chmn dept, 80-85, PROF STATIST, BRIGHAM YOUNG UNIV, 63-, ASSOC DEAN, 85- *Concurrent Pos:* NSF fac fel, 67-68; epidemiol consult, Kennecott Copper Corp, 70-71. *Mem:* Am Statist Asn. *Res:* Best subset regression; discriminant analysis; effect of individual variables in multivariate analysis. *Mailing Add:* 206 TMCB Brigham Young Univ Provo UT 84602

RENCRICCA, NICHOLAS JOHN, b New York, NY, Mar 22, 41; m; c 6. HEMATOLOGY, MALARIOLOGY. *Educ:* St Francis Col, NY, BS, 62; St John's Univ, NY, MS, 64; Boston Col, PhD(physiol), 67; Univ Mass Med Sch, MD, 91. *Prof Exp:* Teaching asst physiol, anat & zool, St John's Univ, NY, 62-64; NIH-Nat Heart Inst res fel hematol, Sch Med, Tufts Univ, 67-70; from assoc prof to prof biol sci, Univ Lowell, 70-84, actg dean, Col Pure & Appl Sci, 84-86. *Concurrent Pos:* Res assoc, Dept Army & Univ Lowell, 73-75; ref ed, J Hematol, 75; NSF fel, Boston Univ, 78-80. *Mem:* AAAS; Am Soc Hemat; Int Soc Exp Hemat; NY Acad Sci; Soc Exp Biol & Med; Sigma Xi; AMA. *Res:* Control of hematopoiesis; stem cell proliferation and differentiation; erythropoiesis in rodent malaria; hyperbaric oxygen-induced toxicity; effects on stress of erythropoiesis. *Mailing Add:* 148 Princeton Blvd Lowell MA 01851

RENDA, FRANCIS JOSEPH, b Brooklyn, NY, June 16, 39; m 63; c 2. SOLID STATE PHYSICS. *Educ:* Brooklyn Col, BS, 62; Syracuse Univ, MS, 65, PhD(physics), 69. *Prof Exp:* MEM TECH STAFF PHYSICS, SANTA BARBARA RES CTR, HUGHES AIRCRAFT CO, 69- *Mem:* Am Inst Physics. *Res:* Photoconductivity. *Mailing Add:* 270 Savona Ave Goleta CA 93017

RENDELL, DAVID H, b St John's, Nfld, July 28, 35; m 61; c 3. THEORETICAL PHYSICS. *Educ:* Dalhousie Univ, BSc, 56, MSc, 57; Univ BC, PhD(physics), 62. *Prof Exp:* from asst prof to assoc prof, Mem Univ Nfld, 59-75, asst dean sci, 71-74, assoc dean sci, 74-83, head dept, 82-90, PROF PHYSICS, MEM UNIV NFLD, 75- *Mem:* Am Asn Physics Teachers; Can Asn Physicists; Sigma Xi. *Res:* Atomic and molecular physics; quantum mechanics. *Mailing Add:* Dept Physics Mem Univ Nfld St John's NF A1B 3X7 Can

RENDIG, VICTOR VERNON, b Wis, July 4, 19; m 44; c 2. SOIL FERTILITY, CROP QUALITY. *Educ:* Univ Wis, BS, 42, PhD(soil sci, biochem), 49. *Prof Exp:* From jr soil chemist to assoc prof, 49-63, prof, 63-88, PROF EMER SOILS & PLANT NUTRIT, UNIV CALIF, DAVIS, 88- *Mem:* Soil Sci Soc Am; Am Soc Agron; Am Soc Plant Physiologists; Am Chem Soc. *Res:* Effects of soil fertility and plant nutrition on plant composition and metabolism: nitrogen in cereal grain crops; sulfur(s) and S/Seienium in forages. *Mailing Add:* Dept Land Air & Water Res Univ Calif Davis CA 95616

RENDINA, GEORGE, b New York, NY, July 1, 23; m 48; c 4. BIOCHEMISTRY. *Educ:* NY Univ, AB, 49; Univ Kans, MA, 53, PhD, 55. *Prof Exp:* Instr biochem, Univ Kans, 54-55; Nat Found Infantile Paralysis fel, Univ Mich, 55-56, sr biochemist, 57; sr fel, E B Ford Inst Med Res, 57-58; instr physiol chem, Sch Med, Johns Hopkins Univ, 58-61; res biochemist, Training Sch, 62-63, chief biochemist, 64-66, dir, Isotope Lab, Cent Wis Colony, 66-67; from assoc prof to prof, 67-84, EMER PROF CHEM, BOWLING GREEN STATE UNIV, 84- *Concurrent Pos:* Adj assoc prof, Med Col Ohio, 70-77; grants, NSF & NIH. *Mem:* AAAS; Am Chem Soc; Sigma Xi. *Res:* Neurochemistry; protein synthesis; kinetics; enzymology. *Mailing Add:* 30 Brier Lane Brewster MA 02631

RENDTORFF, ROBERT CARLISLE, b Carlisle, Pa, Mar 22, 15; m 37; c 1. TROPICAL MEDICINE. *Educ:* Univ Ill, AB, 37, MS, 39; Johns Hopkins Univ, ScD(protozool), 44, MD, 49. *Prof Exp:* Instr epidemiol, Virol Lab, Mich, 42-44; med entomologist, Ministry Sanit & Social Assistance, Venezuela, 46; sr asst scientist, USPHS, 48, from asst surgeon to surgeon, 49-55; from asst prof to assoc prof, Col Med, 55-66, prof community med, 66-79, EMER PROF, UNIV TENN, MEMPHIS, 79- *Concurrent Pos:* Ed, Tenn Med Alumnus; consult epidemiologist infection control, 79-88. *Res:* Medical parasitology; epidemiology of human protozoan diseases, venereal diseases and respiratory viruses. *Mailing Add:* Four N Ashlawn Memphis TN 38112-4308

RENEAU, DANIEL DUGAN, JR, b Woodville, Miss, June 11, 40; m 61; c 2. CHEMICAL ENGINEERING. *Educ:* La Polytech Inst, BS, 63, MS, 64; Clemson Univ, PhD(chem eng), 66. *Prof Exp:* Res engr, Commercial Solvents Corp, 62-63; actg instr chem eng, La Polytech Inst, 63-64; res engr, Esso Res & Eng Co, 66-67; from asst prof to assoc prof chem eng, 67-77, prof biomed eng & head dept, 77-80, VPRES ACAD AFFAIRS, LA TECH UNIV, 80- *Concurrent Pos:* Res engr, Humble Oil & Refining Co, 64; NIH fel, 66. *Mem:* Am Chem Soc; Am Inst Chem Engrs. *Res:* Chemical and biomedical engineering, especially mathematical modeling, dynamic system behavior, transport phenomena, mass transport and oxygen diffusion in brain; iron metabolism; oxygen transport in placenta. *Mailing Add:* Pres Off La Tech Univ PO Box 3188 Ruston LA 71272

RENEAU, JOHN, b Beloit, Wis, May 1, 27; m 55; c 3. AUDIOLOGY, SPEECH PATHOLOGY. *Educ:* Univ Wis, BS, 51; Univ Denver, MS, 58, PhD(audiol, speech path), 60. *Prof Exp:* Dir speech & hearing, State Home & Training Sch, Denver, Colo, 56-60; DIR SPEECH & HEARING, CENT WIS COLONY, 64- *Concurrent Pos:* Fel med audiol, Med Sch, Univ Iowa, 60-63; NIH spec fel neurophysiol, Inst Med Physics, Utrecht, Neth, 63-64; lectr, Univ Wis-Madison, 71-; mem comt hearing, bioacoust & biomech, Nat Acad Sci, 71; mem sensory study sect, Social & Rehab Serv, Dept HEW, 71- *Honors & Awards:* Rosemary Dybwad Int Award, 67. *Mem:* Fel Am Speech & Hearing Asn. *Res:* Study of sensory electroneurophysiology using averaged evoked responses; research audiology. *Mailing Add:* 5410 Russett Rd Madison WI 53711

RENEAU, RAYMOND B, JR, b Burkesville, Ky, Sept 11, 41. SOIL CHEMISTRY. *Educ:* Berea Col, BS, 64; Univ Ky, MS, 66; Univ Fla, PhD(soil chem), 69. *Prof Exp:* Asst prof agron, Tex Tech Univ, 70-71; from asst prof to assoc prof soil pollution, 71-86, PROF SOIL ENVIRON QUAL, VA POLYTECH INST & STATE UNIV, 86- *Concurrent Pos:* Consult, Jamaica Sch Agr, 70- *Mem:* AAAS; Am Soc Agron. *Res:* Movement of septic pollutants through natural soil systems and the potential contamination of ground and surface waters. *Mailing Add:* Crop & Soil Environ Sci 339 Smyth Hall Va Polytech Inst & State Univ Blacksburg VA 24061

RENEKE, JAMES ALLEN, b Jacksonville, Fla, Sept 21, 37; m 61; c 4. MATHEMATICAL ANALYSIS. *Educ:* Univ Fla, BA, 58, MA, 60; Univ NC, Chapel Hill, PhD(math), 64. *Prof Exp:* Assoc prof math, Newberry Col, 64-66; asst prof, 66-71, ASSOC PROF MATH, CLEMSON UNIV, 71- *Concurrent Pos:* Vis assoc prof, Univ Houston, 72-73. *Mem:* Am Math Soc. *Res:* Mathematical system theory, in particular, system problems of realization, identification and control. *Mailing Add:* Dept Math Sci Clemson Univ Clemson SC 29631

RENEKER, DARRELL HYSON, b Birmingham, Iowa, Dec 5, 29; m 53; c 2. PHYSICS. *Educ:* Iowa State Univ, BS, 51; Univ Chicago, MS, 55, PhD, 59. *Prof Exp:* Mem tech staff, Bell Tel Labs, Inc, 51-53; physicist, Polychems Dept, Exp Sta, E I du Pont de Nemours & Co, 59-63, eng mat lab, 63-64, cent res dept, 64-69; chief polymer crystal physics sect, 75, dep chief polymer sci & standards div, 75-80, dep dir, Ctr Mat Sci, 80-85, POLYMERS DIV, NAT BUR STANDARDS, 85- *Concurrent Pos:* Exec secy, Comt Mat, Off Sci & Technol Policy, Off Pres, Wash, 85- *Mem:* Am Phys Soc; Soc Plastics Engrs; Electron Micro Soc Am; Mat Res Soc. *Res:* Ultra-sonic waves in metals; electronic properties of metals; physical properties of polymers and molecular solids. *Mailing Add:* Inst Mat Sci & Eng Nat Bur Standards Gaithersburg MD 20899

RENFREW, EDGAR EARL, b Colfax, Wash, Apr 8, 15; m 43; c 2. INDUSTRIAL ORGANIC CHEMISTRY. *Educ:* Univ Idaho, BS, 36; Univ Minn, PhD(org chem), 44. *Prof Exp:* Chemist, Bunker Hill & Sullivan Mining Co, Idaho, 36-38; asst, Univ Minn, 39-43; res chemist, Gen Aniline & Film Corp, Pa, 44-58; mgr dyestuffs res, Koppers Co, Inc, 58-62; res chemist, Minn Mining & Mfg Co, 62-66; dir res & develop, Am Aniline Prod, Inc, 66-72; vpres res & develop, Am Color & Chem Co, 72-80; RETIRED. *Concurrent Pos:* Consult, 80-83. *Mem:* Am Chem Soc; Am Asn Textile Chemists & Colorists. *Res:* Dyestuffs and intermediates; aromatic intermediates. *Mailing Add:* 1189 S Hillview St Lock Haven PA 17745

RENFREW, MALCOLM MACKENZIE, b Spokane, Wash, Oct 12, 10; m 38. SCIENCE EDITOR. *Educ:* Univ Idaho, BS, 32, MS, 34; Univ Minn, PhD(phys chem), 38. *Hon Degrees:* DSc, Univ of Idaho, 76. *Prof Exp:* Asst physics, Univ Idaho, 32-33, chem, 33-35; asst chem, Univ Minn, 35-37; res chemist, E I du Pont de Nemours & Co, 38-44, res supvr, 44-49; dir chem res, Gen Mills Inc, 49-54; dir res & develop, Spencer Kellog & Sons, Inc, 54-58; prof chem & head div phys sci, Univ Idaho, 59-67; staff assoc, Adv Coun Col Chem, Stanford Univ, 67-68; head dept, 68-73, prof chem, 68-76, EMER PROF, UNIV IDAHO, 76- *Concurrent Pos:* Consult, Mat Adv Bd, Nat Acad Sci-Nat Res Coun, 62-67; dir, Col Chem Consult Serv; exec vpres, Idaho Res Found, 77-78, patent mgr, 78-88, safety ed, J Chem Educ, 77-90. *Honors & Awards:* Harry & Carol Mosher Award, 86 & CHAS Award, Am Chem Soc, 85. *Mem:* AAAS; Am Chem Soc; Am Inst Chemists; Am Inst Chem Engrs; Soc Chem Indust. *Res:* Polymer chemistry; organic coatings and plastics. *Mailing Add:* Dept Chem Univ Idaho Moscow ID 83843

RENFREW, ROBERT MORRISON, b Glasgow, Scotland, Mar 11, 38; Can citizen; m 60; c 4. TRANSPORTATION SYSTEM DESIGN. *Educ:* Univ Toronto, Can, BASc, 60. *Prof Exp:* Chief engr, Husky Mfg & Tool Works, 69-73; sr prog engr, Govt Toronto, 73-75; prog dir, Urban Transp Develop Corp, 75-78, vpres eng, 78-80, sr vpres, 80-82, exec vpres, Res & Develop Ltd, 82-83; exec dir, Can Inst Guided Ground Transp, 83-86; SR VPRES, ENG & TECHNOL, UTDC INC, 86- *Res:* Transportation systems and related technologies; control systems; propulsion technology; urban transit market research; materials technology; civil infrastructure related to transportation. *Mailing Add:* 843 Safari Dr Kingston ON K7M 6W2 Can

RENFRO, J LARRY, RENAL PHYSIOLOGY, COMPARATIVE OSMO REGULATION. *Educ:* Univ Okla, PhD(zool), 70. *Prof Exp:* PROF PHYSIOL, UNIV CONN, 74- *Mailing Add:* Univ Conn 75 N Eagleville Rd Box U-42 Storrs CT 06268

RENFRO, WILLIAM CHARLES, b Hillsboro, Tex, Jan 16, 30; m 55; c 4. BIOLOGICAL OCEANOGRAPHY. *Educ:* Univ Tex, BA, 51, MA, 58; Ore State Univ, PhD(oceanog), 67. *Prof Exp:* Marine biologist, Tex Game & Fish Comn, 58-59; fishery res biologist, US Bur Com Fisheries, 59-64; asst prof oceanog, Ore State Univ, 67-71; chief radiobiol group, Int Atomic Energy Agency, Lab Marine Radioactivity, Monaco, 71-73; chief environ progs, 73-80, DIR ENVIRON PROGS, NORTHEAST UTILITIES CO, 80- *Mem:* AAAS; Am Nuclear Soc; Am Soc Limnol & Oceanog. *Res:* Marine and aquatic radioecology; estuarine ecology; marine pollution; radiochemistry; shrimp life history and ecology; fish physiology. *Mailing Add:* Environ Progs NE Utilities Co PO Box 270 Hartford CT 06141-0270

RENFROE, HARRIS BURT, b Meridian, Miss, Nov 22, 36; m 64; c 2. ORGANIC & PHARMACEUTICAL CHEMISTRY. *Educ:* Miss State Univ, BS, 58; Univ Ill, PhD(org chem), 61. *Prof Exp:* NIH fel org chem, Zurich, 61-62; chemist, Lederle Labs, Am Cyanamid Co, 63-65; res chemist, Geigy Res Labs, 65-69, proj leader, 69-71, group leader, Dept Org Chem, 72-81, dir chem res, 81-86, ASSOC DIR, DRUG DIS COORD, CIBA-GEIGY CORP, 86- *Mem:* Am Chem Soc; Sigma Xi. *Res:* Synthesis of pharmacologically active compounds; exploratory organic synthesis. *Mailing Add:* 12 Stonehedge Dr West Nyack NY 10994

RENGAN, KRISHNASWAMY, b Varalotti, India, Aug 9, 37; m 69; c 2. NUCLEAR CHEMISTRY, ANALYTICAL CHEMISTRY. *Educ:* Univ Kerala, BSc, 57; Univ Mich, PhD(chem), 66. *Prof Exp:* Sci off radiochem div, Bhabha Atomic Res Ctr, Bombay, 58-70; assoc prof, 70-80, PROF CHEM, EASTERN MICH UNIV, 80- *Honors & Awards:* Res Award, Sigma Xi. *Mem:* Am Phys Soc; Am Chem Soc; fel AAAS; Health Phys Soc; Sigma Xi. *Res:* Decay scheme studies; radiochemical separations; application of activation analysis to environmental problems. *Mailing Add:* Dept Chem Eastern Mich Univ Ypsilanti MI 48197

RENGSTORFF, GEORGE W(ILLARD) P(EPPER), b Seattle, Wash, Oct 27, 20; m 42; c 2. METALLURGY. *Educ:* Univ Wash, BS, 42; Mass Inst Technol, SM, 48, ScD, 50. *Prof Exp:* Res engr, Process Metall Div, Battelle Mem Inst, 49-51, asst chief, 51-55, asst chief, Phys Metall Div, 55-60, asst chief, Metals Sci Group, 60-61, res assoc, Process Metall Div, 61-66, tech adv, 66-69; assoc prof, 69-75, PROF MAT SCI, UNIV TOLEDO, 75- *Mem:* AAAS; Am Soc Metals; Am Inst Mining, Metall & Petrol Engrs; Nat Asn Corrosion Eng; Am Soc Eng Educ. *Res:* Chemical metallurgy; refractory metals; high purity metals; friction and wear of metals in corrosive aqueous solutions. *Mailing Add:* Dept Chem Eng Univ Toledo 2801 W Bancroft Toledo OH 43606

RENICH, PAUL WILLIAM, b La Junta, Colo, May 5, 19; m 43; c 3. CHEMISTRY. *Educ:* Bethel Col Kans, AB, 42; Univ Kans, MA, 44, PhD(phys chem), 49. *Prof Exp:* Prof chem, Kans Wesleyan Univ, 48-74, dean, 51-69, pres, 69-73; instr, Haskell Indian Jr Col, 74-84; ADJ PROF, BETHEL COL, 84- *Concurrent Pos:* Ford Found fel, Mass Inst Technol, 54-55; assoc, NCent Asn Cols, 59-60, consult-examr, 60-74. *Mem:* Am Chem Soc; AAAS. *Res:* Physical chemistry, including electrodeposition and physiochemical properties. *Mailing Add:* 720 W 17th St CT Newton KS 67114-1465

RENIS, HAROLD E, b Highland Park, Ill, Jan 1, 30; m 51; c 3. VIROLOGY, BIOCHEMISTRY. *Educ:* Elmhurst Col, BS, 51; Bradley Univ, MS, 54; Purdue Univ, PhD(biochem), 56. *Prof Exp:* Asst instr chem, Bradley Univ; res asst biochem, Purdue Univ, Nat Heart Inst fel microbiol; RES ASSOC VIROL, SR SCIENTIST, UPJOHN CO, 84- *Mem:* AAAS; Am Chem Soc; Am Microbiol Soc; Tissue Cult Asn; Soc Exp Biol Med; Sigma Xi; Int Soc Antiviral Res. *Res:* Tissue culture; nucleic acid antagonists; virus chemotherapy; animal virology; immunomodulators. *Mailing Add:* 6631 Trotwood Kalamazoo MI 49002

RENKA, ROBERT JOSEPH, b Summit, NJ, Dec 28, 47. NUMERICAL ANALYSIS, MATHEMATICAL SOFTWARE. *Educ:* Univ Tex, Austin, BA & BS, 76, MA, 79, PhD(computer sci), 81. *Prof Exp:* Asst prof, 84-89, ASSOC PROF COMPUTER SCI, UNIV NTEX, 89- *Concurrent Pos:* Prin investr, NSF grant, 90-91. *Mem:* Asn Comput Mach; Soc Indust & Appl Math. *Res:* Numerical analysis; mathematical software; curve and surface fitting; computational geometry; scattered data interpolation and smoothing. *Mailing Add:* Dept Computer Sci Univ NTex PO Box 13886 Denton TX 76203

RENKEN, JAMES HOWARD, b El Paso, Tex, July 1, 35; m 78; c 3. COMPUTATIONAL PHYSICS. *Educ:* Ohio State Univ, BSc & MSc, 58; Calif Inst Technol, PhD(physics), 63. *Prof Exp:* Staff mem, Theory & Analysis Div, Sandia Nat Labs, 64-67, supvr Theoret Div, 67-80, supvr, Hostile Environ Div, 80-83, MGR, RADIATION EFFECTS DEPT, SANDIA NAT LABS, 83- *Mem:* Am Phys Soc; Am Nuclear Soc. *Res:* Transport theory; interaction of radiation with matter; mathematical physics. *Mailing Add:* Org No 9350 Sandia Nat Labs PO Box 5800 Albuquerque NM 87185

RENKEY, EDMUND JOSEPH, JR, b Pittsburgh, Pa, May 19, 40. MECHANICAL ENGINEERING. *Educ:* Pa State Univ, BS, 63, MBA, 66. *Prof Exp:* Assoc engr, Wright Aero Div, Curtiss-Wright Corp, 63-65; sr contract adminr, Marvel-Schebler Div, Borg Warner Corp, 67-71; proposal adminr power generation, Babcock & Wilcox Co, 71-72; sr procurement specialist, 72-74, sect mgr, 74-78, sr engr, 78-81, PRIN ENGR, WESTINGHOUSE HANFORD CO, WESTINGHOUSE ELEC CO, 81- *Concurrent Pos:* Adj prof, Cent Wash Univ, 76 & 78. *Mem:* Am Soc Mech Engrs. *Res:* Diaphragm compressor development; auxiliary equipment/systems for testing space-based nuclear reactors. *Mailing Add:* 1963 Marshall Ave Richland WA 99352

RENKIN, EUGENE MARSHALL, b Boston, Mass, Oct 21, 26; m 55, 67; c 4. PHYSIOLOGY, PHARMACOLOGY. *Educ:* Tufts Col, BS, 48; Harvard Univ, PhD(med sci), 51. *Prof Exp:* Assoc biologist, Brookhaven Nat Lab, 51-55; sr asst scientist, Nat Heart Inst, 55-57; from asst prof to prof physiol & chmn dept, Sch Med, George Wash Univ, 57-63; prof pharmacol & head div, Sch Med, Duke Univ, 63-69, prof physiol, 69-74; PROF HUMAN PHYSIOL & CHMN DEPT, SCH MED, UNIV CALIF, DAVIS, 74- *Concurrent Pos:* NSF sr fel, 60-61; Bowditch lectr, Am Physiol Soc, 63; Wellcome vis prof physiol, 78. *Honors & Awards:* E M Landis Award, 77; B W Zweifach Award, 84; C J Wiggers Award, 85. *Mem:* Am Physiol Soc; Am Heart Asn; Microcirculatory Soc (pres, 75); Sigma Xi. *Res:* Peripheral circulation; capillary and membrane permeability; lymph circulation. *Mailing Add:* Dept Human Physiol Univ Calif Sch Med Davis CA 95616

RENN, DONALD WALTER, b East Rutherford, NJ, Mar 23, 32; m 55; c 3. BIOTECHNOLOGY, IMMUNOCHEMISTRY. *Educ:* Franklin & Marshall Col, BS, 53; Mich State Univ, PhD(org chem), 57. *Prof Exp:* Res chemist, John L Smith Mem Cancer Res, Pfizer, Inc, 57-65, asst dept head chem, 61-65, group leader immunochem cancer & viral chem, 62-65; sr chemist, Marine Colloids, Inc, 65-68, res scientist, 68-69, sr scientist, 69-79, dir spec proj res group, 70-72, dir res, MCI Biomed, 72-75; mgr, Biotechnol Venture, Marine Colloids Div, FMC Corp, 77-79, dir biomed res, 79-80, corp res fel, 79-81, dir corp explor lab, 83-85, SR RES FEL, FMC CORP, 81-, DIR EXPLOR TECHNOL GROUP, MARINE COLLOIDS, DIV, 87- *Mem:* Am Chem Soc; Sigma Xi; AAAS; fel Am Inst Chemists; Am Asn Clin Chem; Am Soc Microbiol; Tissue Cult Asn; NY Acad Sci; Int Plant Molecular Biol Asn. *Res:* Marine natural product chemistry; immunochemical and electrophoretic methods for disease detection; isolation and characterization of antibotics and anti-tumor agents from natural product sources; immunochemistry of cancer; viral and polysaccharide chemistry, microbial products, organic synthesis; biotechnology tools and techniques. *Mailing Add:* Brewster Pt Glen Cove ME 04846-0088

RENNAT, HARRY O(LAF), b Estonia, Aug 6, 22; nat US; m 51; c 3. MECHANICAL ENGINEERING. *Educ:* Univ Wis, BS, 53, MS, 54, PhD(mech eng), 56. *Prof Exp:* Res asst, Bjorksten Res Labs, Inc, Wis, 51-56; ASSOC PROF MECH ENG, COLO STATE UNIV, 56- *Mem:* Am Soc Mech Engrs. *Res:* Heat transfer; materials science. *Mailing Add:* 6015 S County Rd 11 Ft Collins CO 80525

RENNE, DAVID SMITH, b Harvey, Ill, Nov 27, 43; m 66; c 2. METEOROLOGY. *Educ:* Kalamazoo Col, BA, 66; Colo State Univ, MS, 69, PhD(earth resources), 75. *Prof Exp:* Meteorologist, US Weather Bur, Lansing, 66; res asst atmospheric sci, Colo State Univ, 66-71; meteorologist air pollution, Environ Qual Bd, San Juan, PR, 71-72; res asst earth resources, Colo State Univ, 72-74; RES SCIENTIST METEOROL, PAC NORTHWEST LABS, BATTELLE MEM INST, 75- *Concurrent Pos:* Consult air pollution, Marlatt & Assoc, Ft Collins, Colo, 72-75 & Inst Ecol-Urban Secondary Impacts Workshop, 74; consult hydrol, City of Ft Collins Planning Dept, 73-74. *Mem:* Am Meteorol Soc; Sigma Xi; Air Pollution Control Asn. *Res:* Research and project management in wind characteristics for wind energy utilization; long range transport and wet and dry deposition of pollutants. *Mailing Add:* 502 Bonifant St Silver Spring MD 20910-5530

RENNEKE, DAVID RICHARD, b Gaylord, Minn, June 29, 40; m 63; c 3. PHYSICS, COMPUTER SCIENCE. *Educ:* Gustavus Adolphus Col, BS, 62; Iowa State Univ, MS, 64; Univ Kans, PhD(physics), 70. *Prof Exp:* Asst prof, 68-77, assoc prof phys, 77-86, PROF PHYSICS, AUGUATANA COL, ILL, 86- *Concurrent Pos:* Consult comput graphics, John Deere Tech Ctr, Moline, 77-86. *Mem:* Am Asn Phys Teachers; Sigma Xi. *Res:* Computer applications in education; mathematical modeling of physical systems; holography; optical reflection from solids; photoconductivity in ionic crystals; atomic imperfections in solids; cryogenic gas storage systems; computer graphics; x-ray diffraction of titanium nitrate. *Mailing Add:* Dept Phys Augustana Col Rock Island IL 61201

RENNELS, MARSHALL L, b Marshall, Mo, Sept 2, 39; m 71. NEUROANATOMY. *Educ:* Eastern Ill Univ, BS, 61; Univ Tex Med Br Galveston, MA, 64, PhD(neuroendocrinol, neurocytol), 66. *Prof Exp:* Asst prof anat, 66-71, Univ Md, asst prof neurol, 69-71, assoc prof anat, 71-79, PROF ANAT & ASSOC PROF NEUROL, SCH MED, UNIV MD, BALTIMORE, 79-, DIR MD/PHD PROG, SCH MED & GRAD SCH, 89- *Mem:* AAAS; Electron Micros Soc Am; Am Asn Anatomists; Soc Neurosci; Int Soc Cerebral Blood Flow & Metab; Sigma Xi. *Res:* Ultrastructural and histochemical investigations of the cerebrovascular system and its innervation by autonomic and central neurons; related studies examine the characteristics of the cerebral microcirculatory system and its relationships to the cerebral parenchyma. *Mailing Add:* Sch Med Univ Md 655 W Baltimore St Baltimore MD 21201

RENNER, DARWIN S(PRATHARD), b Powell, Ohio, Oct 15, 10; m 44; c 3. ELECTRICAL ENGINEERING. *Educ:* Ohio State Univ, BEE, 32, MSc, 33. *Prof Exp:* Comput, Geophys Serv, Tex, 35-37, res engr, 37-42, supt electronics div, 42-45, sr engr, 45-47; vpres, 47-50, PRES, GEOTRONIC LABS, INC, 50- *Mem:* Am Phys Soc; Soc Explor Geophys; Inst Elec & Electronics Engrs; Am Geophys Union; Sigma Xi. *Res:* Automatic amplifiers; servomechanisms; filter systems; magnetic detectors; test sets-fault locators; mixing systems; noise meters; interference eliminators; cameras; recording oscillographs; inductive and capacitive components; computers. *Mailing Add:* 1314 Cedar Hill Ave Dallas TX 75208

RENNER, GERARD W, b Boston, Mass, Dec 23, 21. PHYSICS, ELECTRONICS. *Educ:* Harvard Univ, AB, 43. *Prof Exp:* Spec res assoc, Underwater Sound Lab, Harvard Univ, 43-45, spec res assoc, Systs Res Lab, 45; sect head transducers, Submarine Signal Co, 46-49; dept mgr, Submarine Signal Div, Raytheon Corp, 49-62; res assoc & consult, Parke Math Labs, Inc, 62; dept mgr acoust systs, Hazeltine Corp, 62-67; pres-treas, Appl Res Assocs, Inc, Uphams Corner, 67-77; dir advan develop, MASSA Corp, 77-78, mgr eng, 78; dir advanced acoustic technol, Hazeltine Corp, 78-81; pres, Appl Res Assoc, Inc, 81-88; RETIRED. *Concurrent Pos:* Group leader, Lowell Technol Inst Res Found, 67-69; consult-dir, Advan Technol Systs, Inc, Arlington, Va, 69- *Mem:* Acoust Soc Am; Am Phys Soc; Inst Elec & Electronics Engrs; NY Acad Sci. *Res:* Acoustics; magnetostriction, piezoelectric, ceramic transducer; underwater sound; ultrasonics; telecommunications; biomedical engineering. *Mailing Add:* 51 Bellevue St Dorchester MA 02125-2511

RENNER, JOHN WILSON, b De Smet, SDak, July 25, 24; m 48; c 2. SCIENCE EDUCATION, PHYSICS. *Educ:* Huron Col, BA, 45; Univ SDak, BA, 46, MA, 48; Univ Iowa, PhD(sci & math educ), 55. *Prof Exp:* Teacher high sch, SDak, 46-47; instr physics, Univ SDak, 47-52; instr physics & math, Minn State Teachers Col, 48-49; supvr & teacher phys sci, Univ High Sch, Iowa, 52-55; asst prof sci educ, Univ Ill, 55-56; dir radiol defense sch, Fed Civil Defense Admin, 56-58; asst prof physics & sci educ & actg head dept physics, Creighton Univ, 58-59; assoc exec secy, Nat Sci Teachers Asn, 59-62; PROF SCI EDUC, UNIV OKLA, 62 - *Concurrent Pos:* Consult, Educator's Progress Serv, Wis; dir teacher serv, Frontiers Sci Found Okla, 62-68; coordr, Okla Trial Ctr, Sci Curric Improv Study, Lawrence Hall Sci, Univ Calif, Berkeley, 65-73. *Mem:* Nat Asn Res Sci Teaching (pres, 79-80); Nat Educ Asn; Nat Sci Teacher's Asn; Sigma Xi; Am Asn Physics Teachers. *Res:* author of over 150 articles and publications. *Mailing Add:* Col Educ Univ Okla Norman OK 73019

RENNER, RUTH, b Lewistown, Mont, Nov 17, 25. NUTRITION. *Educ:* Univ Alta, BSc, 48, MSc, 50; Cornell Univ, PhD(animal nutrit), 60. *Prof Exp:* Res asst soils, Univ Alta, 50-52, res asst poultry, 52-55; res assoc nutrit & poultry, Cornell Univ, 55-58, asst, 58-60; from asst prof to assoc prof, 60-70, PROF NUTRIT, SCH HOUSEHOLD ECON, UNIV ALTA, 70- *Honors & Awards:* Borden Award, Nutrit Soc Can, 70. *Mem:* Am Inst Nutrit; Poultry Sci Asn; Nutrit Soc Can. *Res:* Nutritive value of proteins; energy and fat metabolism. *Mailing Add:* RR 1 Priddis AB T0L 1W0 Can

RENNER, TERRENCE ALAN, b Evergreen Park, Ill, Dec 1, 47. PHYSICAL CHEMISTRY. *Educ:* DePaul Univ, Chicago, BS, 69; Yale Univ, PhD(phys chem), 72. *Prof Exp:* ASST CHEMIST PHYS CHEM, CHEM ENG DIV, ARGONNE NAT LAB, ARGONNE, ILL, 73- *Concurrent Pos:* Fel, Argonne Nat Lab, Argonne, Ill, 72-73. *Mem:* Sigma Xi. *Res:* Transport properties of reacting gas mixtures; thermal conductivity; nucleation of superheated liquids; tritium and hydrogen transport in sodium-cooled fast breeder reactors; tritium permeation through reactor construction materials; sodium cold trap optimization. *Mailing Add:* 7941 S Toledo Tulsa OK 74136

RENNERT, JOSEPH, b Mannheim, Ger, July 26, 19; nat US; m 47; c 2. PHOTOCHEMISTRY. *Educ:* City Col New York, BS, 48; Syracuse Univ, MS, 52, PhD(chem), 53. *Prof Exp:* Asst anal chem, Syracuse Univ, 48-50 & phys chem, 50-52; res chemist, Ozalid div, Gen Aniline & Film Corp, 52-54 & Chas Bruning Co, 54-55; proj engr, Balco Res Labs, 55-56; from res scientist to sr res scientist, Inst Math Scis, NY Univ, 56-62; from asst prof to prof, 62-86, EMER PROF CHEM, CITY COL NEW YORK, 86- *Concurrent Pos:* Lectr, City Col New York, 56-; consult, Itek Corp, 60-62. *Mem:* AAAS; Am Chem Soc; fel Am Inst Chemists; NY Acad Sci; Am Soc Photobiol. *Res:* Photochemistry; mechanisms of photo-cyclo-addition, photo- scission, photo-redox and photosensitized reactions; photochemical and photophysical imaging and information storage systems; solid state photochemistry; photobiology. *Mailing Add:* 525 Fordham Pl Paramus NJ 07652

RENNERT, OWEN M, b New York, NY, Aug 8, 38; m 63; c 3. PEDIATRICS, BIOCHEMISTRY. *Educ:* Univ Chicago, BS & BA, 57, MD, 61, MS, 63; Am Bd Pediat, dipl, 67. *Prof Exp:* Res & clin assoc neurol, Nat Inst Neurol Dis & Blindness, Md, 64-66; instr pediat & chief resident, Univ Chicago, 66-67, asst prof pediat, 67-68; from assoc prof to prof pediat & biochem, Col Med, Univ Fla, 68-78, head inst div genetics, endocrinol & metab, 70-77; PROF PEDIAT & BIOCHEM, COL MED, UNIV OKLA, 77-, HEAD DEPT PEDIAT, 77- *Concurrent Pos:* Mem bd med examr, Ill, 61, Fla, 70 & Nat Bd Med Examr, 62; mem adv comt inborn errors of metab, Bd Health, Fla; NIH & Nat Cystic Fibrosis Fedn fels, 71-76; Fla Heart Asn fel, 72-73. *Mem:* AAAS; fel Am Acad Pediat; Soc Pediat Res; Pan-Am Med Asn. *Res:* Inborn errors of metabolism; human genetics. *Mailing Add:* Dept Pediat Georgetown Univ Hosp 3800 Reservoir Rd NW Washington DC 20007

RENNHARD, HANS HEINRICH, b Aarau, Switz, Sept 26, 28; nat US; m 55; c 2. ORGANIC CHEMISTRY, METABOLISM OF FOOD ADDITIVES. *Educ:* Swiss Fed Inst Technol, dipl, 52, DSc, 55. *Prof Exp:* Fel, Mass Inst Technol, 55-57; from res chemist to sr res chemist, 57-73, sr res investr, 74-79, PRIN RES INVESTR, PFIZER, INC, 80- *Mem:* Swiss Chem Soc; Am Chem Soc. *Res:* Antibiotics; tetracyclines; flavor enhancers; pharmacology of food additives; carbohydrate chemistry and metabolism; radiobiology; analytical methods development. *Mailing Add:* Pfizer Inc CTRL Res Groton CT 06340

RENNICK, BARBARA RUTH, b Ashtabula, Ohio, Oct 23, 19. PHYSIOLOGY. *Educ:* Wayne State Univ, BSc, 42, MSc, 44; Univ Mich, MD, 50. *Prof Exp:* Pharmacologist, Ciba Pharmaceut Prod, 44-46; instr physiol, State Univ NY Upstate Med Ctr, 50-54, asst prof, 54-61; assoc prof, Mt Holyoke Col, 61-64; prof pharmacol, Med Sch, State Univ NY Buffalo, 65-82; RETIRED. *Concurrent Pos:* USPHS spec fel, Oxford Univ, 54-55; Wellcome Found traveling fel, 58. *Mem:* AAAS; Am Physiol Soc; Am Soc Pharmacol & Exp Therapeut; Am Soc Nephrology. *Res:* Renal tubular function; membrane transport; autonomic pharmacology; newborn physiology; general physiology. *Mailing Add:* 3304 Dominica Ct Punta Gorda FL 33950

RENNIE, DONALD ANDREWS, b Medicine Hat, Alta, Apr 21, 22; m 48; c 3. SOIL CHEMISTRY. *Educ:* Univ Sask, BSA, 49; Univ Wis, PhD(soils), 52. *Prof Exp:* From asst prof to prof, 52-64, head dept, 64-81, DEAN COL AGR, UNIV SASK, 84- *Concurrent Pos:* Dir, Sask Inst Pedology, 65-81; head soils irrig & crop prod sect, Int Atomic Energy Agency, Vienna, Austria, 68-70. *Honors & Awards:* Award, Am Chem Soc, 67; Centennial Medal, Can, 67; Agr Hall Fame, 82. *Mem:* Fel Can Soc Soil Sci (secy-treas, 56-58, pres, 76-77); fel Agr Inst Can (vpres, 59-60); fel Soil Sci Soc Am; fel Am Soc Agron; fel Can Soc Soil Sci. *Res:* Soil chemistry and fertility. *Mailing Add:* Col Agr Univ Sask Saskatoon SK S7N 0W0 Can

RENNIE, DONALD WESLEY, b Seattle, Wash, Apr 2, 25; m 47; c 4. PHYSIOLOGY. *Educ:* Univ Wash, BS, 47; Univ Ore, MS & MD, 52. *Prof Exp:* Instr physiol, Univ Wis, 53; from asst prof to assoc prof, 59-66, chmn dept, 72-80, PROF PHYSIOL, SCH MED, STATE UNIV NY, BUFFALO, 66-, VPRES RES & GRAD EDUC, 80- *Concurrent Pos:* Res fel, Harvard Med Sch, 56-58; vis prof, Univ Milan, 66-67, Univ Toronto, 86. *Mem:* AAAS; Am Physiol Soc; Am Fedn Clin Res; Undersea Med Soc; Sigma Xi. *Res:* Hemodynamics; heart; environmental physiology; metabolism; kidneys. *Mailing Add:* Dept Physiol Med Sci Univ Buffalo Buffalo NY 14214

RENNIE, JAMES CLARENCE, b Ont, Sept 14, 26; m 50; c 2. ANIMAL BREEDING. *Educ:* Univ Toronto, BS, 47; Iowa State Col, MS, 50, PhD, 52. *Prof Exp:* Asst agr rep, Dept Agr, Ont, 47-49; tech off, Cent Exp Farm, Ottawa, 49; assoc prof animal husb, Ont Agr Col, Univ Guelph, 52-56, chmn dept, 65-71, prof, 56-74; exec dir educ & res, 74-78, asst dep minister technol & field serv, 85-90, ASST DEP MINISTER EDUC, RES & SPEC SERV, ONT MINISTRY AGR & FOOD, 78-, DIR, AGR RES INST ONT, 90- *Concurrent Pos:* Coordr exten, dept animal & poultry sci, Ont Agr Col, Univ Guelph, 71-73, actg dean res, 73-74. *Honors & Awards:* Award of Merit, Can Soc Animal Sci. *Mem:* Fel Agr Inst Can; Ont Inst Agrologists. *Mailing Add:* Ont Ministry Agr & Food Legis Bldg Queen's Park Toronto ON M7A 1A3 Can

RENNIE, PAUL STEVEN, b Toronto, Ont, Feb 9, 46; m 68; c 1. PROSTATE CANCER, ANDROGEN ACTION. *Educ:* Univ Western Ont, BSc, 69; Univ Alta, PhD(biochem), 73. *Prof Exp:* Res assoc, Univ Alta, 75-76, asst prof med, 76-79, assoc prof, 79; RES SCIENTIST, BC CANCER AGENCY, 79-; PROF SURG, UNIV BC, 86- *Concurrent Pos:* Med Res Coun res fel, Imperial Cancer Res Fund, 73-75; res scholar, Nat Cancer Inst Can, 76-79. *Mem:* Endocrine Soc; Can Soc Clin Invest; Biochem Soc. *Res:* Biochemical control of growth in androgen responsive organs and neoplasms; genetic markers in breast cancer. *Mailing Add:* BC Cancer Agency 600 W Tenth Ave Vancouver BC V5Z 4E6 Can

RENNIE, ROBERT JOHN, b Prince Albert, Sask, Sept 12, 49; m 70; c 2. SOIL MICROBIOLOGY. *Educ:* Univ Sask, BSA Hons, 71; Laval Univ MSc, 72; Univ Minn, PhD(soil microbiol), 75. *Prof Exp:* Res assoc dinitrogen fixation, Agr Res Coun, nit Nitrogen Fixation, UK, 75-76; assoc officer, Food & Agr Orgn, Int Atomic Energy Agency, Div Atomic Energy Food & Agr, UN, 76-78; res scientist dinitrogen fixation, Agr Can, 78-85; mgr, Agr Sci Progs, 85-90, AGR BIOL, EXXON CHEM, 90- *Concurrent Pos:* Nat Res Coun Can fel, Agr Res Coun Unit Nitrogen Fixation, UK, 75-76; mem, Can Comt Nitrogen Fixation, Agr Can, 78-85. *Mem:* Am Soc Microbiol; Am Soc Agron; Agr Inst Can; Can Soc Microbiologists. *Res:* Dinitrogen fixing bacteria associated with legumes and non-legumes such as spring wheat and temperate prairie grasses; isotope techniques and immunofluorescent procedures are applied throughout. *Mailing Add:* 31 Temple Crescent W Lethbridge AB T1K 4T3 Can

RENNIE, THOMAS HOWARD, b Coral Gables, Fla, Nov 26, 43; m 68; c 4. AQUATIC ECOLOGY, INVERTEBRATE ZOOLOGY. *Educ:* Univ Miami, BS, 65; Tex A&M Univ, MS, 67, PhD(aquatic biol), 75. *Prof Exp:* Instr zool, Ohio State Univ, 72-73; asst prof biol, Augustana Col, Rock Island, Ill, 74-79; sr biologist, Wapora, 79-80; environ specialist, Savannah, Ga, US Army Corps Engrs, 80-82, New Orleans, La, 82-83, Galveston, Tex, 83-88, chief, Environ & Anal Br, Chicago, Ill, 88-89, STAFF MGR, COASTAL PLANNING BR, US ARMY CENGR, GALVESTON, TEX, 89- *Mem:* Sigma Xi. *Res:* Freshwater and marine invertebrate ecology, mainly of zooplankton and benthos; copepod taxonomy and physiology. *Mailing Add:* 2703 Yorktown Dr Dickinson TX 77539-4424

RENNILSON, JUSTIN J, b Berkeley, Calif, Dec 10, 26; m 54; c 3. PHOTOMETRY, SPECTRORADIOMETRY. *Educ:* Univ Calif, AB, 50; Tech Univ, Berlin, 55. *Prof Exp:* Assoc engr, Visibility Lab, Univ Calif, San Diego, 55-61; sr scientist, Jet Propulsion Lab, Calif Inst Technol, 61-69, sr res fel, 69-74; VPRES RES & ENG, GAMMA SCI, INC, 74- *Concurrent Pos:* Instr, San Diego State Col, 56-58; co-investr, NASA Surveyor TV Exp, 63-68; consult, Cohu Electronics Co, 68-74 & Photo Res Co, 71-74; co-investr, NASA Apollo Geol Exp 11-17. *Mem:* Optical Soc Am; Sigma Xi; Am Soc Testing & Mat. *Res:* Photometry, colorimetry; optical instrument design; spectroradiometric standards; specroradiometry; retroreflection. *Mailing Add:* 4141 S Tropico Dr La Mesa CA 92041

RENO, FREDERICK EDMUND, b Reno, Nev, July 20, 39; m 64; c 2. PRODUCT SAFETY. *Educ:* Univ San Francisco, BS, 61; Univ Nev, Reno, MS, 63; Utah State Univ, PhD(toxicol), 67. *Prof Exp:* Lab instr human anat & physiol, Univ Nev, 63-64; res assoc emergency med servs, State of Nev, 64; assoc res coordr toxicol, Hazleton Labs, 67-68, proj mgr toxicol, 68-72, dir toxicol dept, 72-80, dir sci develop, 80-81, vpres, Hazleton Labs Am, Inc, 82-87; CONSULT, 88- *Mem:* AAAS; Soc Toxicol; Am Col Vet Toxicol; Europ Soc Toxicol; Am Col Toxicol. *Res:* Toxicological and teratological evaluation of new drugs, agricultural chemicals, food additives, cosmetics and industrial chemicals; reproductive physiology; consultant to industry on product safety issues, toxicology, and toxic tort cases. *Mailing Add:* Reno Assocs 3725 Ridgelea Dr Fairfax VA 22031

RENO, HARLEY W, b Oakland, Calif, Feb 13, 39; m; c 2. HAZARDOUS WASTES MANAGEMENT, ENVIRONMENTAL MANAGEMENT. *Educ:* Okla State Univ, BS, 61, MS, 63, PhD, 67. *Prof Exp:* Asst biol fishes, Okla State Univ, 62-67; from asst prof to assoc prof biol, Baylor Univ, 67-74; vis prof, Pan Am Univ, 74-75; tech coordr & supvr ecol sci, Williams Bros

Environ Serv, Williams Bros Eng Co, Resource Sci Corp, 75-78; environ mgr, Nuclear Waste Prog, 78-81, PRIN PROG PROJ ENGR & CONSULT, EG & G IDAHO, INC, 81-; PROF ENG & ENVIRON SCI, IDAHO NAT ENG LAB, UNIV IDAHO, 80- *Concurrent Pos:* Postdoctoral fel, Univ Okla, 74; adj prof environ sci, Univ Tulsa, 76-78. *Honors & Awards:* Stoye Award, Am Soc Ichthyol & Herpet, 64. *Mem:* AAAS; Am Soc Zoologists; Am Soc Study Evolution; Sigma Xi; Ecol Soc Am; Am Nuclear Soc. *Res:* Environmental transport of radioactive isotopes; hydrocarbon transport systems; surface coal mining; biological morphomechanics and systems evolution. *Mailing Add:* Waste Mgt Dept Box 1625 EG&G Idaho Inc Idaho Falls ID 83415

RENO, MARTIN A, b Erie Co, Pa, July 14, 36; m 55; c 3. CHEMICAL PHYSICS. *Educ:* Edinboro State Col, BS, 55; Harvard Univ, EdM, 58; Rensselaer Polytech Inst, MS, 60; Western Reserve Univ, PhD(chem), 66. *Prof Exp:* Teacher, High Sch, Ohio, 58-66; asst dean natural & social sci, 70-72, PROF PHYSICS, HEIDELBERG COL, 66-, DIR COMPUT CTR, 75- *Concurrent Pos:* Dir col sci improv prog, Heidelberg Col, 69-73. *Mem:* Am Chem Soc; Am Asn Physics Teachers; Nat Sci Teachers Asn. *Res:* Low temperature calorimetry; thermodynamic properties of fluorine containing gases; x-ray structure determination of biologically important molecules. *Mailing Add:* Dept Physics Heidelberg Col Tiffin OH 44883

RENO, ROBERT CHARLES, b New York, NY, Feb 26, 43; m 65; c 2. SOLID STATE PHYSICS, NUCLEAR SPECTROSCOPY. *Educ:* Manhattan Col, BS, 65; Brandeis Univ, MA, 67, PhD(physics), 71. *Prof Exp:* Nat Res Coun res assoc physics, Nat Bur Standards, 71-73; asst prof, 73-78, ASSOC PROF PHYSICS, UNIV MD, BALTIMORE COUNTY, 78- *Concurrent Pos:* Consult, Nat Bur Standards, 74-; prin investr, Petrol Res Fund grant, 77- *Mem:* Am Phys Soc. *Res:* Hyperfine interactions in solids; perturbed angular correlations; Mossbauer spectroscopy; physics of metals and alloys; positron annihilation; electron microscopy. *Mailing Add:* Dept Physics Univ Md Baltimore County Catonsville MD 21228

RENOLL, ELMO SMITH, b Glen Rock, Pa, Jan 25, 22; m 45; c 2. AGRICULTURAL ENGINEERING. *Educ:* Auburn Univ, BS, 47; Iowa State Univ, MS, 49. *Prof Exp:* Engr agr eng, USDA, 45-49; from asst prof to assoc prof, 49-72, prof agr eng, Auburn Univ, 72-81; RETIRED. *Honors & Awards:* Outstanding Serv Cert, Am Soc Agr Engrs, 64. *Mem:* Sr mem Am Soc Agr Engrs; Am Soc Eng Educ; AAAS. *Res:* Agricultural power and machinery; machinery use and selection; programming, modeling and simulation; hay and forage machinery utilization; alcohol and other alternative fuel utilization. *Mailing Add:* 939 S Gay St Auburn AL 36830

RENOLL, MARY WILHELMINE, b St Petersburg, Pa, June 26, 06. CHEMISTRY. *Educ:* Grove City Col, AB, 27; Ohio State Univ, MS, 30. *Prof Exp:* Chemist, Midgley Found, Ohio State Univ, 30-38; res chemist, Monsanto Chem Co, Ohio, 39-44; res assoc chem, Res Found, Ohio State Univ, 45-57, from asst supvr to assoc supvr, 58-67, res assoc, 68, independent researcher chem, food sci & nutrit, 69-73, res assoc, Res Found, 73-76; RETIRED. *Mem:* AAAS; Am Chem Soc; fel Am Inst Chemists; NY Acad Sci; Sigma Xi. *Res:* Organic fluorine compounds; nucleoproteins; organic and rubber chemistry; nucleic acids. *Mailing Add:* 886 W Tenth Ave Columbus OH 43212

RENSCHLER, CLIFFORD LYLE, b Evansville, Ind, Sept 20, 55; m 78; c 2. OPTICAL SPECTROSCOPY. *Educ:* Univ Evansville, BS, 77; Univ Ill, PhD(chem), 81. *Prof Exp:* Mem tech staff, 81-89, SUPVR, SANDIA NAT LABS, 89- *Mem:* Am Chem Soc; Sigma Xi; Mat Res Soc. *Res:* Radio luminescent light and scintillator development; photo degradation kinetics; photoresists and other materials for microlithography; carbon formation by organic pyrolysis. *Mailing Add:* Orgn 1812 Sandia Nat Labs Albuquerque NM 87185

RENSE, WILLIAM A, b Massillon, Ohio, Mar 11, 14; m 42; c 3. SPACE PHYSICS. *Educ:* Case Western Reserve Univ, BS, 35; Ohio State Univ, MS, 37, PhD(physics), 39. *Prof Exp:* Instr physics, La State Univ, 39-40; asst prof, Univ Miami, 40; vis asst prof, Rutgers Univ, 41-42; from asst prof to assoc prof, La State Univ, 43-49; from assoc prof to prof physics, 49-80, co-dir lab atmospheric & space physics, 56-78, EMER PROF, UNIV COLO, BOULDER, 80- *Mem:* Am Geophys Union; Am Phys Soc; Sigma Xi; Am Astron Soc. *Res:* Vacuum spectroscopy; solar ultraviolet and stellar spectroscopy; upper air physics; space science, rocket and satellite experiments. *Mailing Add:* 204 Birch Dr Lafayette LA 70506

RENSINK, MARVIN EDWARD, b Mason City, Iowa, Jan 29, 39; m 61; c 3. MAGNETIC FUSION ENERGY. *Educ:* St Johns Univ, BS, 60; Univ Calif, Los Angeles, MS, 62, PhD(physics), 67. *Prof Exp:* Physicist, Hughes Aircraft Co, 60-63; res asst, Univ Calif, Los Angeles, 65-67; PHYSICIST, LAWRENCE LIVERMORE NAT LAB, 67- *Mem:* Am Phys Soc. *Res:* Theoretical studies of plasma confinement in magnetic fusion devices. *Mailing Add:* Lawrence Livermore Nat Lab 7000 E Ave L-637 Livermore CA 94550

RENTHAL, ROBERT DAVID, b Chicago, Ill, Oct 29, 45; m 77; c 2. PROTEIN CHEMISTRY. *Educ:* Princeton Univ, BA, 67; Columbia Univ, PhD(biochem), 72. *Prof Exp:* NIH fel molecular biophys, Yale Univ, 72-74; Nat Res Coun assoc life sci, Ames Res Ctr, NASA, 74-75; asst prof, 75-80, assoc prof, 80-87, PROF BIOCHEM, UNIV TEX, SAN ANTONIO, 87- *Concurrent Pos:* Vis scientist, Cardiovasc Res Inst, Univ Calif, San Francisco, 79-80; NIH sr fel, Biochem Dept, Univ Tex Health Sci Ctr, San Antonio, 91-92. *Mem:* AAAS; Am Chem Soc; Biophys Soc; Am Soc Biochem & Molecular Biol; NY Acad Sci. *Res:* Structure and function of membrane proteins; mechanisms of energy transduction by photoreceptor membranes; rhodopsin and bacteriorhodopsin; structure and function of retinal rod cell connecting cilium. *Mailing Add:* Div Earth & Phys Sci Univ Tex San Antonio TX 78249

RENTMEESTER, KENNETH R, b Green Bay, Wis, Apr 26, 31; m 67; c 1. ORGANIC CHEMISTRY. *Educ:* St Norbert Col, BS, 52; Northwestern Univ, MS, 61. *Prof Exp:* Res assoc, Res Ctr, Am Can Co, 50-80, sr res assoc, 50-82, chemist, 52-82; CONSULT, INT EXEC SERV CORP, 82- *Mem:* AAAS; Am Chem Soc. *Res:* Protective organic coatings for glass, metal and plastic packages; high strength adhesive bonding systems; coil and sheet applied precoatings for deep prawn food containers with formulation and application. *Mailing Add:* Am Can Co 736 Highland Ave Barrington IL 60010

RENTON, JOHN JOHNSTON, b Pittsburgh, Pa, Nov 25, 34; m 61; c 2. GEOCHEMISTRY. *Educ:* Waynesburg Col, BS, 56; WVa Univ, MS, 59, PhD(geol), 65. *Prof Exp:* Mem staff solid state physics, Res & Develop Off, USAF, 60-63; from asst to assoc prof, 65-75, PROF GEOL GEOCHEM, WVA UNIV, 75-, COOP GEOCHEMIST, WVA GEOL & ECON SURV, 65- *Concurrent Pos:* NSF grant exp diagenesis, WVa Univ, 65-67. *Mem:* AAAS; Geol Soc Am; Am Asn Petrol Geol. *Res:* Geochemistry of coal; acid mine drainage. *Mailing Add:* WVa Geol Survey PO Box 879-White Hall Morgantown WV 26506

RENTON, KENNETH WILLIAM, b Galashiels, Scotland, Apr 6, 44; Can citizen; m 66; c 2. PHARMACOLOGY. *Educ:* Sir George Williams Univ, BSc, 72; McGill Univ, PhD(pharmacol), 75. *Prof Exp:* Fel pharmacol, Univ Minn, 75-77; from asst prof to assoc prof, 77-81, PROF PHARMACOL, 81-, HEAD DEPT, DALHOUSIE UNIV, 88- *Concurrent Pos:* Can Med Res Coun fel, 75-77, scholar, 77-82. *Honors & Awards:* Merck Award Pharmacol, 87. *Mem:* Can Fedn Biol Sci; Am Soc Pharmacol & Exp Therapeut; Can Soc Clin Pharamcol. *Res:* Drug biotransformation by cytochrome P-450; drug metabolism during infectious disease; drug interaction; drug toxicity; marine environ toxicol. *Mailing Add:* Dept Pharmacol Dalhousie Univ Halifax NS B3H 3J5 Can

RENTZ, DAVID CHARLES, b San Francisco, Calif, May 14, 42; m 63. SYSTEMATIC ENTOMOLOGY. *Educ:* Univ Calif, Berkeley, BS, 65, MS, 66, PhD(entom), 70. *Prof Exp:* Asst cur entom, Acad Natural Sci Philadelphia, 70-75; asst cur, Calif Acad Sci, 75-77; SR PRIN RES SCIENTIST, DIV ENTOM, COMMONWEALTH SCI & INDUST RES ORGN, 77- *Concurrent Pos:* Res assoc, Calif Acad Sci, 69-75. *Mem:* AAAS; Am Entom Soc (vpres, 71, pres, 72-74); Am Inst Biol Sci; Asn Trop Biol; Soc Syst Zool; Entom Am Soc (pres, 73-74); Int Asn Sound Arch Australia Secretag. *Res:* Classification and behavior of katydids of the subfamily Decticinae; origin and evolution of the Orthoptera of the Channel Islands, California; taxonomy and ecology of Australian orthoptera. *Mailing Add:* Div Entom CSIRO PO Box 1700 Canberra City ACT 2601 Australia

RENTZEPIS, PETER M, b Kalamata, Greece, Dec 11, 34; US citizen; m 60; c 2. CHEMICAL PHYSICS. *Educ:* Denison Univ, BS, 58; Syracuse Univ, MS, 60; Cambridge Univ, PhD(phys chem), 63. *Prof Exp:* Mem tech staff phys chem, Gen Elec Res Labs, 60-61; mem tech staff phys chem, Bell Labs, 63-73, head, Phys Chem Dept, 73-86; PROF CHEM, PRESIDENTIAL CHAIR, UNIV CALIF, IRVINE, 86- *Concurrent Pos:* Adj prof, Univ Pa, 69-; exec mem comt phys chem div, Am Chem Soc; prof, Yale Univ, 81. *Honors & Awards:* Langmuir Prize in Chem Physics, Am Phys Soc, 73; Peter Debye Award, Am Chem Soc, 79,; A Cressy Morrison Award, NY Acad Sci, 78. *Mem:* Nat Acad Sci; fel Am Phys Soc; fel NY Acad Sci; Royal Soc Chem. *Res:* Lasers; photochemistry; kinetics; picosecond spectroscopy. *Mailing Add:* Dept Chem Univ Calif Irvine CA 92717

RENUART, ADHEMAR WILLIAM, b Miami, Fla, Aug 10, 31; m 52; c 9. PEDIATRIC NEUROLOGY, NEUROCHEMISTRY. *Educ:* Duke Univ, BS, 52, MD, 56. *Prof Exp:* Assoc, Duke Univ, 61-71, from asst prof to assoc prof pediat, Med Ctr, 72-76, assoc clin prof pediat, 72-76; ASSOC CLIN PROF NEUROL, UNIV NC MED CTR, CHAPEL HILL, 76- *Concurrent Pos:* Dir res, Murdoch Ctr, 61-76. *Mailing Add:* Rte 1 Box 184A Franklinton NC 27525

RENWICK, J ALAN A, b Dundee, Scotland, May 7, 36; m 83; c 2. INSECT BEHAVIOR. *Educ:* Dundee Tech Col, HNC, 60; City Col New York, MA, 64; Univ Gottingen, DF, 70. *Prof Exp:* Lab asst, Scottish Hort Res Inst, 58-60; res asst, 60-66, from asst chemist to assoc chemist, 66-83, CHEMIST, BOYCE THOMPSON INST, CORNELL UNIV, 83- *Concurrent Pos:* Adj prof entom, Cornell Univ, 87- *Mem:* Am Chem Soc; Entom Soc Am; Int Soc Chem Ecol. *Res:* Chemical factors affecting oviposition behavior of phytophagous insects; positive stimuli (recognition of suitable host plants) and negative stimuli involved in host selection. *Mailing Add:* Boyce Thompson Inst Cornell Univ Ithaca NY 14853

RENZEMA, THEODORE SAMUEL, b Grand Rapids, Mich, July 12, 12; m 39, 68; c 1. EXPERIMENTAL SOLID STATE PHYSICS. *Educ:* Hope Col, AB, 34; Rutgers Univ, MS, 37; Purdue Univ, PhD(physics), 48. *Prof Exp:* Asst physics, Rutgers Univ, 34-37; asst physics, Purdue Univ, 37-42, instr, 42-47, Off Res & Inventions & Off Naval Res indust res fel, 46-48; assoc prof physics, Clarkson Col Technol, 48-50, prof & chmn dept, 50-65; vis prof, 65-66, chmn dept, 73-74, prof, 66-79, EMER PROF PHYSICS, STATE UNIV NY ALBANY, 79- *Concurrent Pos:* Res assoc, Dudley Observ, 65-66. *Mem:* Am Phys Soc; Am Crystallog Asn; Am Asn Physics Teachers; Microbeam Anal Soc. *Res:* Electron diffraction investigations of thin films of metals; semiconductor surfaces; corrosion; x-ray crystallography; electron microprobe analysis; x-ray spectroscopy. *Mailing Add:* Three Toll Lane Albany NY 12203-5517

RENZETTI, ATTILIO D, JR, b New York, NY, Nov 11, 20; m 47; c 4. PULMONARY DISEASES, PHYSIOLOGY. *Educ:* Columbia Univ, AB, 41, MD, 44. *Prof Exp:* Asst prof indust med, Postgrad Med Sch, NY Univ, 49-51; asst prof med, Sch Med, Univ Utah, 52-53; asst prof, Col Med, State Univ NY Upstate Med Ctr, Syracuse Univ, 53-59, assoc prof, 59-60; assoc prof, Johns Hopkins Univ & Univ Md, 60-61; chmn Pulmonary Dis Div, 61-87, from assoc prof to prof, 61-90, EMER PROF MED, UNIV UTAH, 90-

Concurrent Pos: Fel cardio-pulmonary physiol, Bellevue Hosp Chest Serv, 49-51; Nat Heart & Lung Inst grants, Univ Utah, 61-; mem subspecialty bd pulmonary dis, Am Bd Internal Med, 65-72, chmn, 70-72; mem epidemiol & biomet adv comt, Nat Heart & Lung Inst, 70-73, mem comt spec ctr res, 71; consult, Vet Admin Hosp, Salt Lake City, Utah, 70-85. *Mem:* Am Lung Asn; Am Fedn Clin Res; Am Thoracic Soc (pres-elect, 74-75, pres, 75-76); fel Am Col Physicians; NY Acad Sci; Sigma Xi. *Res:* Pulmonary function in disease; applied respiratory physiology. *Mailing Add:* 1801 London Pl Rd Salt Lake City UT 84124-3531

RENZETTI, NICHOLAS A, b New York, NY, Sept 30, 14; m 45; c 4. PHYSICS. *Educ:* Columbia Univ, AB, 35, AM, 36, PhD(physics), 40. *Prof Exp:* Asst physics, Columbia Univ, 37-40, res assoc, 40; physicist, Bur Ord, US Navy, 40-44, sci res administr, Naval Ord Test Sta, Calif, 44-54; sr physicist, Air Pollution Found, Calif, 54-59; sect chief telecommun, 59-63, mgr, Tracking & Data Systs for Planetary Projs, 64, tech mgr, eng, 64-80, PROG MGR, JET PROPULSION LAB, CALIF INST TECHNOL, 80- *Concurrent Pos:* Consult, US Naval Ord Test Sta, Pasadena, Calif, 54-62, Gen Motors Corp, Mich, 56-61 & Air Pollution Found, 59-61. *Honors & Awards:* NASA Outstanding Leadership Medal. *Mem:* Sigma Xi. *Res:* Telecommunications science and engineering with space vehicles; air pollution; underwater technology. *Mailing Add:* 1321 Virginia Rd San Marino CA 91108

RENZI, ALFRED ARTHUR, b Rochester, NY, July 20, 25; m 54; c 5. PHYSIOLOGY, ENDOCRINOLOGY. *Educ:* Fordham Univ, BS, 47; Syracuse Univ, MS, 49, PhD(zool), 52. *Prof Exp:* Asst endocrinol, Syracuse Univ, 47-51; from assoc endocrinologist to sr endocrinologist, Ciba Pharmaceut Co, NJ, 52-60, assoc dir physiol, 60-62, head endocrine-pharmacol sect, 62-67; head, Pharmacol Dept, Dow Human Health Res & Develop Labs, Ind, 67-71, assoc scientist, Dow Chem Co, 71-80, assoc scientist, Med Dept, 80-88, assoc scientist, Toxicol Dept, Merrell Dow Res Inst, 88-90; RETIRED. *Mem:* Am Physiol Soc; Am Soc Pharmacol & Exp Therapeut; Endocrine Soc. *Res:* Steroid hypertension; water metabolism; kidney function; adrenal cortex; inflammation; reproduction; atherosclerosis; lipid metabolism. *Mailing Add:* 40 Staten Pl Zionsville IN 46077

REPA, BRIAN STEPHEN, b Detroit, Mich, Apr 5, 42; m 70; c 3. BIOENGINEERING. *Educ:* Univ Mich, Dearborn Campus, BS, 65, Ann Arbor, MS, 66, PhD(bioeng), 72. *Prof Exp:* Co-op & res engr elec eng, Eng & Res Ctr, Ford Motor Co, 62-65; res asst manual-mach systs, Univ Mich, 68-70, lab instr analog comput, 70-71; sr prof engr impaired driver countermeasures, 72-73, proj mgr driver physiol, 73-74, staff res engr driver-vehicle performance, 74-82, prog mgr, safety res, 82-84, ACTIV HEAD, HUMAN FACTORS, PROJ TRILBY, GEN MOTORS RES LABS, 84- *Concurrent Pos:* Mem road user characteristics comt, Transp Res Bd, 74-, mem simulation & measurement of driving comt, 75-; mem passenger car safety comt, Soc Automotive Engrs, 75-, mem vehicle dynamics comt, 78- *Mem:* Soc Automotive Engrs. *Res:* Mathematical modeling of the driver-vehicle system; effects of vehicle characteristics on driver-vehicle performance; perceptual cues used by drivers in controlling their vehicles; driving simulators; driver impairment detection. *Mailing Add:* 22868 ShagBark Rd Birmingham MI 48010

REPAK, ARTHUR JACK, b New York, NY, Mar 19, 40; m 66; c 2. PROTOZOOLOGY. *Educ:* Univ Mich, Ann Arbor, BS, 61; LI Univ, MS, 64; NY Univ, PhD(biol), 67. *Prof Exp:* dir clin lab, 801st Med Group, Griffiss AFB, Rome, NY, 68-70; from asst prof to assoc prof, 70-79, PROF BIOL, QUINNIPIAC COL, 79- *Concurrent Pos:* Consult, Ecol Consults, 70-, Examr fac, Charter Oak Col, 90-93 & Int Bus Mach Corp; Yale-Lilly fel, 77-78; vis fac fel, Yale Univ, 78-82, res assoc, 71-82; mem Comn Water Pollution Control Authority, Cheshire, Conn, 84-; dir, Conn Jr Acad Sci & Eng, 84-; mem bd dirs, Conn Sci Fair Asn, 86; Northeast regional dir, Sigma Xi, 90-93. *Mem:* Am Micros Soc; Am Soc Microbiol; Sigma Xi; Soc Protozoologists. *Res:* Taxonomy and morphology of heterotrichous ciliates; encystment and excystment of protozoa; nutritional and physiological studies of marine and freshwater heterotrichous ciliates. *Mailing Add:* Dept Biol Sci Quinnipiac Col Box 158 Hamden CT 06518-0569

REPASKE, ROY, b Cleveland, Ohio, Mar 17, 25; m 50; c 3. MOLECULAR BIOLOGY, MICROBIAL BIOCHEMISTRY. *Educ:* Western Reserve Univ, BS, 48; Univ Mich, MS, 50; Univ Wis, PhD(bact, biochem), 54. *Prof Exp:* Res assoc bact metab, Univ Wis, 53; from instr to assoc prof bact, Ind Univ, 54-59; chemist, Lab Microbiol, 59-74 & Off Sci Dir, 74-80, CHEMIST, LAB MOLECULAR MICROBIOL, INST ALLERGY & INFECTIOUS DIS, NIH, 80-, ASST LAB CHIEF, 91- *Concurrent Pos:* Consult, NASA, 62-65 & Naval Med Res Inst, 90-; chmn gen div, Am Soc Microbiol, 63-64; instr, Found Advan Studies Sci, 66-73; prog dir biochem, NSF, 73-74. *Mem:* Fel AAAS; Am Soc Microbiol; Am Soc Biol Chem; Sigma Xi; Am Soc Virol. *Res:* Molecular biology of retroviruses, recombinant DNA. *Mailing Add:* Inst Allergy & Infectious Dis Bldg 4 Rm 303 NIH Bethesda MD 20892

REPINE, JOHN E, b Rock Island, Ill, Dec, 26, 44; m 69, 88; c 4. LUNG, PHAGOCYTE OXYGEN RADICAL RESEARCH. *Educ:* Univ Wis-Madison, BS, 67; Univ Minn, Minneapolis, MD, 71. *Prof Exp:* From instr to assoc prof internal med, Univ Minn, 74-79; asst dir & div exp med, 79-89, PROF MED & DIR, WEBB-WARINGA LUNG INST, 89-; ASSOC PROF MED, SCH MED, UNIV COLO, 79-, ASSOC PROF PEDIAT, 81- *Concurrent Pos:* Mem res comt & site vis, Am Lung Asn, NIH; young pulmonary investr grant, Nat Heart & Lung Inst, 74-75; Basil O'Connor starter res award, Nat Found March of Dimes, 75-77; estab investr award, Am Heart Asn, 76-81; co-chmn steering comt, Aspen Lung Conf, 80, chmn, 81. *Mem:* AAAS; Am Asn Immunologists; Am Fedn Clin Res; Am Heart Asn; Am Thoracic Soc; Am Soc Clin Invest; Asn Am Physicians. *Res:* Role of phagocytes and oxygen radicals in lung injury and host defense. *Mailing Add:* Webb-Waring Lung Inst Univ Colo Med Ctr 4200 E Ninth Ave Denver CO 80220

REPKA, BENJAMIN C, b Buffalo, NY, July 5, 27; m 53; c 4. PATENTS, POLYPROPLENE. *Educ:* Canisius Col, BS, 49, MS, 52; Purdue Univ, PhD(phys org chem), 57. *Prof Exp:* Chemist, Carborundum Co, 50-53; res chemist, Hercules, Inc, 56-71, sr res chemist, 71-73, res scientist, 73-78, res assoc & proj leader polypropylene, 78-83, mgr patent coord, 83-87 & new technol, 87- 89; RETIRED. *Mem:* Am Chem Soc; Sigma Xi. *Res:* Polyolefins; organometallics; heterogeneous catalysis; polymerization processes; technology acquisition. *Mailing Add:* 206 N Star Rd Newark DE 19711-2935

REPKO, WAYNE WILLIAM, b Detroit, Mich, Mar 21, 40; m 66; c 3. ELEMENTARY PARTICLE PHYSICS. *Educ:* Wayne State Univ, BS, 63, PhD(theoret physics), 67. *Prof Exp:* Res assoc theoret physics, Wayne State Univ, 67-68 & Johns Hopkins Univ, 68-70; from asst prof to assoc prof, 70-79, PROF THEORET PHYSICS, MICH STATE UNIV, 79- *Concurrent Pos:* Vis assoc prof, Johns Hopkins Univ, 76- *Mem:* Am Phys Soc. *Res:* Quantum field theory; quantum electrodynamics; elementary particle physics. *Mailing Add:* Dept Physics Mich State Univ East Lansing MI 48823

REPLOGLE, CLYDE R, b Detroit, Mich, Nov 13, 35; m 85; c 1. PHYSIOLOGY, OPERATIONS RESEARCH. *Educ:* Mich State Univ, BS, 58, MS, 60, PhD(physiol), 67. *Prof Exp:* Instr physiol, US Air Force Inst Technol, 62-64, asst prof bioeng & physiol, 64-68, chief, Environ Physiol Br, 68-72, adj prof bioeng, 68-77; chief, Environ med div, 72-77, Manned Systs Effectiveness Div, 77-79, CHIEF, SPEC PROJ BR, HUMAN ENG DIV, AEROSPACE MED RES LAB, 79- *Concurrent Pos:* Adj prof eng, Air Force Inst Technol, 68-; adj prof biomed sci, Wright State Univ, 79-; fel, Prints Maurits Lab, Hague, Neth. *Honors & Awards:* Barchi Prize, 86; Rist Prize, 88. *Res:* Chemical warfare defense analysis; man-machine system, human operator and manned weapon system performance; environmental medicine; all aspects of the impact of chemical warfare environments on Air Force operations; toxic effects of agents and their impact on human performance; casualty analysis; human factor analysis of the impact of protective equipment. *Mailing Add:* Human Eng Div Aerospace Med Res Lab Wright-Patterson AFB OH 45433

REPLOGLE, JOHN A(SHER), b Charleston, Ill, Jan 13, 34; m 57; c 2. HYDRAULICS, AGRICULTURAL ENGINEERING. *Educ:* Univ Ill, BS, 56, MS, 58, PhD(civil eng), 64. *Prof Exp:* Agr engr, Soil Conserv Serv, USDA, 56; res asst agr eng, Univ Ill, 56-58, instr, 58-63; res agr engr, 63-66, res hydraul engr, 66-75, RES LEADER, WATER CONSERV LAB, AGR RES SERV, USDA, 75 - *Concurrent Pos:* Consult, USAID, India, 82 & 83, Bangladesh, 85 & 87; chmn, Irrigation & Drainage Div, Am Soc Civil Engrs, 88-89. *Honors & Awards:* James R Croes Medal, Am Soc Civil Engrs, 77. *Mem:* AAAS; Am Soc Agr Engrs; Am Soc Civil Engrs. *Res:* Agricultural land drainage, irrigation and hydrology; fluid mechanics; hydraulic structures; water resources planning and development; flow measurement; basic physical science; irrigation hydraulics; irrigation-distribution system operations. *Mailing Add:* US Water Conserv Lab 4331 E Broadway Phoenix AZ 85040

REPLOGLE, LANNY LEE, b San Bernardino, Calif, Oct 30, 34; m 58; c 2. ORGANIC CHEMISTRY. *Educ:* Univ Calif, Berkeley, BS, 56; Univ Wash, PhD(org chem), 60. *Prof Exp:* Res instr chem, Univ Wash, 60-61; from asst prof to assoc prof, 61-69, PROF CHEM, CALIF STATE UNIV, SAN JOSE, 69- *Concurrent Pos:* NSF grants, San Jose State Col, 62-69; Nat Res Coun sr res associateship, Ames Res Ctr, NASA, 70-71; consult, Paul Masson Vineyards. *Mem:* Am Chem Soc. *Res:* Chemistry of azulene and its derivatives; nonbenzenoid aromatics; hetercyclic analogs of nonbenzenoid aromatic hydrocarbons; wine chemistry. *Mailing Add:* Dept Chem San Jose State Col San Jose CA 95192

REPLOGLE, ROBERT LEE, b Ottumwa, Iowa, Sept 30, 31; m 58; c 3. CARDIOVASCULAR SURGERY, THORACIC SURGERY. *Educ:* Cornell Col, BS, 56; Harvard Med Sch, MD, 60. *Hon Degrees:* DSc, Cornell Col, 72. *Prof Exp:* Asst resident surg, Peter Bent Brigham Hosp, Boston, Mass, 61-63; res fels, Children's Hosp Med Ctr, Boston & Harvard Med Sch, 63-64; sr resident, Children's Hosp Med Ctr, Boston, 64-65; asst resident, Mass Gen Hosp, Boston, 65-66; sr resident, Children's Hosp Med Ctr, 66, asst, 66-67; asst prof surg, Pritzker Sch Med, Univ Chicago, 67-70, assoc prof surg & chief pediat surg, 70-74, prof surg & chief cardiac surg, 74-79; CHIEF CARDIAC SURG, MICHAEL REESE HOSP, 79-; CHIEF CARDIAC SURG, COLUMBIA HOSP, 86-; CHIEF CARDIAC SURG, INGALLS HOSP, 89- *Concurrent Pos:* Asst surg, Harvard Med Sch, 66-67; dir cardiac surg, Michael Reese Hosp & Med Ctr, 77- *Mem:* Soc Univ Surg; Am Asn Thoracic Surg; Int Soc Surg; Am Surg Asn. *Res:* Cardiovascular physiology; rheology. *Mailing Add:* Michael Reese Hosp Chicago IL 60616

REPORTER, MINOCHER C, b Bombay, India, Feb 8, 28; US citizen; m 52; c 4. DEVELOPMENTAL BIOLOGY, MOLECULAR BIOLOGY. *Educ:* Johns Hopkins Univ, AB, 52; Mass Inst Technol, PhD(food technol), 59. *Prof Exp:* Res asst, Develop Dept, A D Witten Co, Mass, 59; res asst nutrit, Mass Inst Technol, 59; res fel, Geront Br, Nat Heart Inst, 59-62; asst investr develop biol, Carnegie Inst Dept Embryol, 62-65; staff scientist, 65-68, investr, Battelle-Charles F Kettering Res Lab, 68-86; vis prof, Bot Dept, Miami Univ, Oxford, Ohio, 87-89; PROF, DEPT BOT & PLANT PATH, ORE STATE UNIV, 89- *Concurrent Pos:* Adj assoc prof, Antioch Col, 68- & Wright State Univ, 79-; mem mantech environ serv, Environ Res Lab, US Environ Protection Agency, 89-90; affil, Ecol Planning & Toxicol Inc, 91- *Mem:* AAAS; Am Soc Microbiol; Am Soc Biol Chemists; Am Soc Cell Biol; Am Soc Plant Physiol; Biophys Soc. *Res:* Nitrogen fixation and control of energy supply; differentiation and cellular development; cell physiology; nitrogen fixation in culture; secondary metabolites from cultured cells; new tests for assessment of rhizophere ecology using plants, plant cell cultures and bacteria. *Mailing Add:* 1005 NW Alder Creek Dr Corvales OR 97330

REPPER, CHARLES JOHN, b Philadelphia, Pa, Nov 1, 34. SOLID STATE PHYSICS. *Educ:* St Joseph's Univ, Pa, BS, 56; Drexel Univ, MS, 65. *Prof Exp:* Physicist, Res Lab, 56-64 & Appl Res Lab, 64-66, res scientist & proj engr, Phys Electronics Dept, Philco Corp, 66-72; LEAD PROJ ENGR, GEN

ELEC CO, 72- *Mem:* Am Phys Soc. *Res:* Photoconductivity; electroluminescent gallium arsenide P-N junctions; thin film metal-oxide structures; solid state photo detectors; metallurgy and measurements of the electrical and optical properties of semiconductors; environmental effects on solid state devices and electro-optical sensors. *Mailing Add:* 7324 N 20th St Philadelphia PA 19138

REPPERGER, DANIEL WILLIAM, b Charleston, SC, Nov 24, 42; m 68, 88; c 2. ELECTRICAL ENGINEERING, MATHEMATICS. *Educ:* Rensselaer Polytech Inst, BSEE, 67, MSEE, 68; Purdue Univ, PhD(elec eng), 73, PE, 75. *Prof Exp:* Res asst elec eng, Rensselaer Polytech Inst, 67-68; teaching instr, Purdue Univ, 68-71, David Ross res fel, 71-73; Nat Res Coun appointment eng control theory, Aerospace Med Res Lab, 73-74, systs analyst elec eng, Systs Res Lab, 74-75, SYSTS ANALYST ELEC ENG, AEROSPACE MED RES LAB, WRIGHT PATTERSON AFB, 75- *Concurrent Pos:* Reviewer tech papers, Inst Elec & Electronics Engrs Trans Automatic Control, 72-, Inst Elec & Electronics Engrs Trans Systs, Man & Cybernet, 74- & Human Factors, 77-; adj prof, Sch Elec Eng, Wright State Univ, 83- *Honors & Awards:* H Schuck Award, 78; Armstrong Award, 80; F Russ Biomed Award, 89. *Mem:* Inst Elec & Electronics Engrs; Nat Soc Prof Engrs. *Res:* Modern control systems theory; modeling; identification; man-machine systems; numerical algorithms; applications and theoretical aspects of optimal control and estimation theory; human factors engineering; holder of eight patents. *Mailing Add:* Bldg 33 Aerospace Med Res Lab Wright Patterson AFB Dayton OH 45433-6573

REPPERT, STEVE MARION, b Sioux City, Iowa, Sept 4, 46; m 68; c 3. CIRCADIAN RHYTHM RESEARCH, NEUROPEPTIDES. *Educ:* Univ Nebr, Omaha, BS, 73, MD, 73. *Prof Exp:* Pediat resident, Mass Gen Hosp, 73-76; clin fel , Harvard Med Sch, 73-76; clin assoc, NIH, 76-79; from instr to asst prof, 79-85, ASSOC PEDIAT, HARVARD MED SCH, 85- *Concurrent Pos:* Res fel, Charles King Trust, Boston, 81-83; prin investr, NIH grants, 81, March Dimes grants 81-88 & estab investr, Am Heart Asn, 85; dir, Lab Develop Chronobiol, Mass Gen Hosp, 83. *Mem:* Am Heart Asn; Am Physiol Soc; Am Soc Clin Invest; Endocrine Soc; Soc Neurosci; Soc Pediat Res. *Res:* Neurobiology; neuropeptides; circadian rhythms; pineal gland. *Mailing Add:* Chronobiol Lab Harvard Univ Mass Gen Hosp Childrens Serv Boston MA 02114

REPPOND, KERMIT DALE, b Farmerville, La, Oct 31, 45. ORGANIC CHEMISTRY, PHYSICAL CHEMISTRY. *Educ:* Northeast La Univ, BS, 67. *Prof Exp:* CHEMIST, KODIAK LAB, NAT MARINE FISHERY SERV, 76- *Res:* Development of methods to enhance utilization of various seafood for human consumption. *Mailing Add:* Kodiak Nat Marine Fishery Serv 900 Trident Way Kodiak AK 99615-7401

REPPUCCI, NICHOLAS DICKON, b Boston, Mass, May 1, 41; m 67; c 3. PSYCHOLOGY. *Educ:* Univ NC, Chapel Hill, BA, 62; Harvard Univ, MA, 64, PhD(clin psychol), 68. *Prof Exp:* Lectr & res assoc psychol, Harvard Univ, 67-68; from asst prof to assoc prof psychol, Yale Univ, 68-76; PROF PSYCHOL, UNIV VA, 76- *Concurrent Pos:* Dir community psychol, Psychol Dept, Univ Va, 76-, dir grad studies, 86-, consult, Inst Law, Psychiat & Pub Policy, 78-; chair, Taskforce Psychol & Pub Policy, Am Psychol Assoc, 80-84; mem, Internal Review Comts, NIMH, 80-83 & 87-89; dir Yale Psychol Educ Clin, Yale Univ, 70-73, dir clin psychol, dept psychol, 69-70; dir clin psychol, dept psychol, Univ Va, 76-80; assoc ed, Law & Human Behav, 88-; scholar award psychol, Va Soc Sci Asn, 91. *Honors & Awards:* G Stanley Hall Lectr, Am Psychol Assoc, 84. *Mem:* Fel Am Psychol Assoc (pres, div 27 community psychol, 86). *Res:* Research children's competencies, child sexual abuse, juvenile delinquency, custody and other psycho-legal issues of children and families; preventive interventions relating to changing human service organizations. *Mailing Add:* 301 Gilmer Hall Dept Psychol Univ Va Charlottesville VA 22903

REPPY, JOHN DAVID, b Lakewood, NJ, Feb 16, 31; m 59; c 3. PHYSICS. *Educ:* Univ Conn, BA, 54, MS, 56; Yale Univ, PhD(physics), 61. *Prof Exp:* NSF fel physics, Oxford Univ, 61-62; asst prof, Yale Univ, 62-66; from assoc prof to prof physics, 66-87, JOHN L WETHERILL PROF PHYSICS, CORNELL UNIV, 87- *Concurrent Pos:* Guggenheim fel, 72-73; Fulbright-Hays fel, 78; Guggenheim fel & Sci Res Coun sr res fel, 79-80. *Honors & Awards:* Fritz London Award, 81. *Mem:* Nat Acad Sci; fel Am Phys Soc; fel AAAS. *Res:* Macroscopic quantum properties of superconductors and superfluid helium; cooperative phenomena. *Mailing Add:* Dept Physics Cornell Univ Clark Hall Ithaca NY 14853

REQUA, JOSEPH EARL, b Willits, Calif, Oct 17, 38; m 63; c 1. OPERATING SYSTEMS & NETWORK SOFTWARE, MANAGEMENT OF SOFTWARE DEVELOPMENT. *Educ:* Univ Calif, Berkeley, BS, 61, MA, 63; Univ Ill, Urbana, MS, 65. *Prof Exp:* Programmer, Lawrence Livermore Nat Lab, 65-81, researcher, 81-83, group leader, Computation Dept, 83-86, user systs div leader, 86-91, STAFF MEM, LIVERMORE COMPUTER CTR, LAWRENCE LIVERMORE NAT LAB, 91- *Res:* Advancement in the state of the art of distributed supercomputer networking technology; architecture; protocols; creating systems; graphic user interfaces; scientific visualization. *Mailing Add:* Lawrence Livermore Nat Lab PO Box 808 Livermore CA 94550

REQUARTH, WILLIAM H, b Charlotte, NC, Jan 23, 13; m 48; c 4. SURGERY. *Educ:* James Milliken Univ, BS, 34; Univ Ill, MD, 39, MSc, 40; Am Bd Surg, dipl, 47. *Prof Exp:* Intern, St Luke's Hosp, Chicago, 38-39; residency surg, Cook County Hosp, 39-46; asst prof, 47-66, PROF SURG, UNIV ILL COL MED, 66- *Concurrent Pos:* Mem attend staff, Macon County Hosp, St Mary's Hosp, Decatur, Ill & Ill Res Hosp. *Mem:* Am Soc Surg Hand; Am Asn Surg Trauma; fel Am Col Surgeons; Soc Surg Alimentary Tract. *Res:* General surgery; surgery of the hand. *Mailing Add:* 158 W Prairie Ave Decatur IL 62523

REQUE, PAUL GERHARD, b New York, NY, May 28, 07; m 36; c 2. MEDICINE. *Educ:* Duke Univ, MD, 34. *Prof Exp:* Assoc dermat & instr internal med, Sch Med, Duke Univ, 40-46; assoc prof dermat, Sch Med, Univ Ala, Birmingham, 46-81; ASSOC DIR DERMAT & SYPHIL, LLOYD NOLAND FOUND HOSP, 47- *Concurrent Pos:* Lectr, Univ NC, 40-42; consult, Vet Admin Hosp, 46-52. *Mem:* Soc Invest Dermat; Am Acad Dermat; Nat Asn Chain Drug Stores; AMA. *Res:* Borate absorption through the dermis and mucous membranes; chromatography; drug idiosyncrasy; antibody-antigen reactions; systemic sclerosis. *Mailing Add:* 801 Vestavia Lake Dr Birmingham AL 35216-2031

REQUICHA, ARISTIDES A G, b Monte Estoril, Portugal, Mar 18, 39; m 70. COMPUTER AIDED DESIGN & MANUFACTURING, GEOMETRIC MODELING. *Educ:* Univ Lisbon, EE, 62; Univ Rochester, PhD(elec eng), 70. *Prof Exp:* Lectr physics, Univ Lisbon, 61-63; res scientist, Saclant Res Ctr, NATA, 70-73; res assoc, Univ Rochester, 73-75, sr scientist & assoc dir, Prod Automation Proj, 75-85, dir, 85-86, assoc prof elec eng, 83-86; PROF COMPUTER SCI & ELEC ENG, UNIV SOUTHERN CALIF, 86- *Concurrent Pos:* Assoc ed, Asn Comput Mach Trans Graphics, 84-90; area ed, Graphic Models & Image Processing, 89- *Mem:* Inst Elec & Electronics Engrs; Asn Comput Mach; AAAS; Am Asn Artificial Intel; Soc Mfg Engrs; Sigma Xi. *Res:* Programmable automation; computer aided design and manufacturing systems for electromechanical products; concurrent design of products and processes; aritifical intelligence and computational geometry; spatial reasoning; geometric uncertainty; automatic planning for manufacturing of inspection; object-oriented geometric computation. *Mailing Add:* Computer Sci Dept Univ Southern Calif Los Angeles CA 90089-0782

RERICK, MARK NEWTON, b Syracuse, NY, Jan 31, 34; m 56; c 5. ORGANIC CHEMISTRY, SYSTEMS ANALYSIS. *Educ:* Le Moyne Col, BS, 55; Univ Notre Dame, PhD(chem), 58. *Prof Exp:* Res assoc, Univ Notre Dame, 58-59; res fel, Calif Inst Technol, 59-60; from asst prof to assoc prof, 60-68, PROF ORG CHEM, PROVIDENCE COL, 69- *Mem:* AAAS; Am Chem Soc. *Res:* Reductions of organic compounds with complex and mixed metal hydrides; conformational analysis of mobile cyclohexane systems; systems analysis of complex systems. *Mailing Add:* Dept Chem Providence Col Providence RI 02918-0000

RESCH, GEORGE MICHAEL, b Baltimore, Md, Mar 26, 40; m 63; c 1. PHYSICS, ASTRONOMY. *Educ:* Univ Md, BS, 63; Fla State Univ, MS, 65, PhD(physics), 74. *Prof Exp:* Res asst physics, Fla State Univ, 63-68; res assoc astron, Clark Lake Radio Observ, 68-71; res asst astron, Univ Md, 71-74; mem tech staff, Jet Propulsion Lab, 74-83, prog mgr, 83-90; CONSULT, 90- *Mem:* Am Inst Physics; Am Astron Soc; Am Geophys Union. *Res:* Instrumentation development and measurement systems with application to geodesy and astronomy. *Mailing Add:* Jet Propulsion Lab M-S 238-700 4800 Oak Grove Dr Pasadena CA 91109

RESCH, HELMUTH, b Vienna, Austria, May 22, 33; m 60; c 2. FOREST PRODUCTS. *Educ:* Agr Univ, Vienna, dipl eng, 56; Utah State Univ, MS, 57, PhD(wood technol), 60. *Prof Exp:* Res asst, Utah State Univ, 56-57, US Forest Serv, 57 & J Neils Lumber Co, 58; asst, Agr Univ, Vienna, 58-60; asst & assoc prof wood technol, Univ Calif, Berkeley, 62-70; prof & head dept, Forest Prod Dept, Sch Forestry, Ore State Univ, 70-87 CONSULT, 87- *Mem:* Soc Wood Sci & Technol (pres, 80-81); Forest Prod Res Soc; Int Acad Wood Sci. *Res:* Physical properties of wood, processing of timber into lumber, plywood and other manufactured products. *Mailing Add:* Dean Research SUNY Col Environ Sci & Forestry Syracuse NY 13210

RESCH, JOSEPH ANTHONY, b Milwaukee, Wis, Apr 29, 14; m 39; c 3. NEUROLOGY. *Educ:* Univ Wis, BS, 36, MD, 38; Am Bd Psychiat & Neurol, dipl, 49. *Prof Exp:* Rockefeller fel neurol, Univ Minn, 46-48; pvt pract, 48-62; from assoc prof to prof neurol, Univ Minn, 62-84, asst vpres health sci affairs, 70-77, head dept, 77-82, EMER PROF NEUROL, UNIV MINN, MINNEAPOLIS, 84- *Mem:* Fel Am Acad Neurol; Am Asn Neuropath; Am Electroencephalog Soc; Am Neurol Asn. *Res:* Cerebrovascular disease; geographic pathology. *Mailing Add:* 900 River Beach Rd The Sea Ranch CA 95497

RESCHER, NICHOLAS, b Hagen, Ger, July 15, 28; nat US; m; c 4. PHILOSOPHY OF SCIENCE, SYMBOLIC LOGIC. *Educ:* Queens Col NY, BS, 49; Princeton Univ, PhD(philos), 51. *Hon Degrees:* LHD, Loyola Univ, 70. *Prof Exp:* Instr philos, Princeton Univ, 51-52; res mathematician, Rand Corp, 54-57; assoc prof philos, Lehigh Univ, 57-61; prof, 61-70, UNIV PROF PHILOS, UNIV PITTSBURGH, 70- *Concurrent Pos:* Ford Found fel, 59; Guggenheim fel, 70-71; ed-in-chief, Am Philos Quart; secy gen, Int Union Hist & Philos Sci, 69-75; hon mem, Corpus Christi Col, Oxford, 78-; pres, eastern div, Am Philos Asn, 89-90. *Honors & Awards:* Alexander von Humboldt Prize, 83. *Mem:* Am Philos Asn; Philos Sci Soc. *Res:* Philosophy of science. *Mailing Add:* Dept Philos Univ Pittsburgh Pittsburgh PA 15260

RESCIGNO, ALDO, b Milan, Italy, Aug 27, 24; US citizen; m 50; c 1. MATHEMATICAL BIOLOGY. *Educ:* Univ Milan, Laurea in Physics, 48. *Prof Exp:* Res asst med physics, Tumor Ctr, Italy, 49-52; asst phys chem, Bracco Indust Chim, 52-55; tech dir nuclear instrumentation, Metalnova SpA, 55-59 & DISI, 59-61; res asst biophys, Donner Lab, Univ Calif, Berkeley, 61-62, lectr med physics, 62-63, asst prof, 63-64; fel phys biochem, Australian Nat Univ, 65-69; assoc prof physiol, Univ Minn, Minneapolis, 69-75; mathematician, Lab Theoret Biol, Nat Cancer Inst, NIH, 75-77; prof physiol, Univ Minn, Minneapolis, 77-79; prof biomath, Univ Witwatersrand, Johannesburg, 80-81; res prin, Sch Med, Yale Univ, New Haven, 82-87; prof pharmacokinetics, Univ Ancona, Italy, 87-88; PROF PHARMACOKINETICS, UNIV PARMA, ITALY, 88- *Concurrent Pos:* Vis prof, LADSEB, Ctr Nuclear Res, Padova, Italy, 75; Kiiliam Scholar, Univ Calgary, 79; vis scientist, Genentech Inc, San Francisco, 90- *Mem:* Soc Gen Systs Res; Brit Inst Physics; Soc Math Biol; Soc Pharmacokinetics & Biopharmaceut. *Res:* Mathematical models of biological system; theory of compartments; drug and tracer kinetics; general system theory; deterministic population dynamics. *Mailing Add:* 691 Panorama Dr San Francisco CA 94131-1226

RESCIGNO, THOMAS NICOLA, b New York, NY, Sept 10, 47; m 86. CHEMICAL PHYSICS. *Educ:* Columbia Col, BA, 69; Harvard Univ, MA, 71, PhD(chem physics), 73. *Prof Exp:* Res fel, Calif Inst Technol, 73-75; staff scientist, 75-79, group leader, 79-85, SR SCIENTIST, LAWRENCE LIVERMORE NAT LAB, 85- *Honors & Awards:* Am Inst Chemists Medal. *Mem:* Am Phys Soc; Am Chem Soc. *Res:* Theoretical atomic and molecular physics-low energy electron scattering; atomic and molecular photoabsorption; electronic structure of atoms and molecules; manybody theory. *Mailing Add:* Lawrence Livermore Nat Lab PO Box 5508 Livermore CA 94550

RESCONICH, EMIL CARL, b Portage, Pa, Oct 17, 23. PLANT VIROLOGY. *Educ:* St Francis Col, Pa, BA, 54; Univ Notre Dame, PhD(biol), 59. *Prof Exp:* NSF fel, Univ Calif, Berkeley, 59-60; asst prof biol, Col Steubenville, 60-62; assoc prof, 62-72, chmn dept, 62-72, PROF BIOL, ST FRANCIS COL, PA, 72- *Mem:* AAAS; Am Soc Plant Physiol; Bot Soc Am; Am Phytopath Soc; Sigma Xi. *Res:* Physiology of plant virus infection; cell physiology; plant virology. *Mailing Add:* Dept Biol St Francis Col Loretto PA 15940-0600

RESCONICH, SAMUEL, b Portage, Pa, July 31,33. ORGANIC CHEMISTRY. *Educ:* St Francis Col, Pa, BS, 54; Purdue Univ, PhD(org chem), 61. *Prof Exp:* Teaching asst gen chem, Purdue Univ, 54-56, Westinghouse Elec fel, 56-59, Naval ord res fel org chem, 59-60; from asst prof to assoc prof, 60-70, PROF CHEM, ST FRANCIS COL, PA, 70- *Mem:* Am Chem Soc; Sigma Xi; Am Asn Univ Professors. *Res:* Organometallic compounds; organic synthesis of fluorine containing compounds. *Mailing Add:* 1319 Gillespie Ave Portage PA 15946-1529

RESCORLA, ROBERT A, b Pittsburgh, Pa, May 9, 40; div; c 2. EXPERIMENTAL PSYCHOLOGY. *Educ:* Swarthmore Col, BA, 62; Univ Pa, PhD(psychol), 66. *Hon Degrees:* MA, Yale Univ, 75. *Prof Exp:* From asst prof to prof psychol, Yale Univ, 66-81; chmn dept, 85-88, PROF PSYCHOL, UNIV PA, 81-, JAMES SKINNER PROF SCI, 86- *Honors & Awards:* Distinguished Sci Contrib Award, Am Psychol Asn, 86. *Mem:* Nat Acad Sci; Am Psychol Asn; Psychonomic Soc; Soc Exp Psychologists; AAAS. *Res:* Elementary learning processes in nonhuman animals, especially pavlovian conditioning and instrumental learning. *Mailing Add:* Dept Psychol Univ Pa Philadelphia PA 19104

RESHKIN, MARK, b East Orange, NJ, May 31, 33; m 61; c 2. GEOMORPHOLOGY, GENERAL ENVIRONMENTAL SCIENCES. *Educ:* Rutgers Univ, AB, 55; Ind Univ, MA, 58, PhD(geol), 63. *Prof Exp:* Instr geol, Univ Maine, 63-64; from asst prof to assoc prof geol, Ind Univ Northwest, 64-71, chmn dept, 68-71, dir div pub & environ affairs, 72-75, assoc prof, 71-77, PROF PUB & ENVIRON AFFAIRS, IND UNIV NORTHWEST, 71-, PROF GEOL, 77-, ASST VICE CHANCELLOR ACAD AFFAIRS, 89- *Concurrent Pos:* Sr scientist, Nat Park Serv, 80-82. *Mem:* AAAS; Geol Soc Am; Glaciol Soc; Am Quaternary Asn; Am Pub Admin. *Res:* Geomorphology of northern Rocky Mountains; geomorphology and glacial geology of northern Indiana; environmental geology of northwest Indiana; environmental management; research administration. *Mailing Add:* Dept Geosci Ind Univ Northwest 3400 Broadway Gary IN 46408

RESHOTKO, ELI, b New York, NY, Nov 18, 30; m 53; c 3. AEROSPACE SCIENCES. *Educ:* Cooper Union, BME, 50; Cornell Univ, MME, 51; Calif Inst Technol, PhD(aeronaut, physics), 60. *Prof Exp:* Aeronaut res scientist, Lewis Lab, Nat Adv Comt Aeronaut, 51-55, head, Fluid Mech Sect, 56-57, head high temperature plasma sect, Lewis Res Ctr, NASA, 60-61, chief, Plasma Physics Br, 61-64; from assoc prof to prof eng, Case Western Reserve Univ, 64-88, head, Div Fluid Thermal & Aerospace Sci, 70-76, chmn, Dept Mech & Aerospace Eng, 76-79, KENT H SMITH PROF ENG, CASE WESTERN RESERVE UNIV, 88- *Concurrent Pos:* Mem res adv comt fluid mech, NASA, 61-64, aeronaut adv comt, 80- & chmn, adv subcomt aerodyn, 83-; mem, Plasma Physics Panel, Physics Study Comt, Nat Acad Sci, 64; Susman vis prof, Israel Inst Technol, 69-70; chmn, US Boundary Layer Transition Study Group, 70-; consult, Gould Corp, Arvin/Calspan, Dynamics Technol Inc, Boeing Co, United Technol Res Ctr; US mem, Fluid Dynamics Panel, Adv Group Aerospace Res & Develop, NATO, 81-88; chmn steering comt, Case/NASA-Lewis Inst Computational Mech in Propulsion, 85-; dean, Case Inst Technol, 86-87. *Honors & Awards:* Fluid & Plasmadynamics Award, Am Inst Aeronaut & Astronaut, 80. *Mem:* Nat Acad Eng; fel Am Acad Mech (pres, 86-87); fel Am Soc Mech Engrs; fel Am Inst Aeronaut & Astronaut; fel Am Phys Soc; fel AAAS. *Res:* Boundary layer theory and transition; aerodynamic heating; hydrodynamic stability; magnetohydrodynamics; advanced propulsion and power generation. *Mailing Add:* Sch Eng Case Western Reserve Univ Cleveland OH 44106

RESING, HENRY ANTON, b Chicago, Ill, May 23, 33; m 56; c 7. SURFACE CHEMISTRY, SOLID STATE CHEMISTRY. *Educ:* DePaul Univ, BS, 55; Univ Chicago, MS, 57, PhD(chem), 59. *Prof Exp:* Phys chemist, Solid State Div, 59-69, Sect Head, Surface & Solid Kinetics Sect, 69-76, Sect Head, Electroactive Solids Sect, 76-80, SR SCIENTIST, CHEM DIV, US NAVAL RES LAB, 80- *Concurrent Pos:* Nat Acad Sci-Nat Res Coun fel, 59-61; vis scientist, Mass Inst Technol, 71-72; vis lectr, Inst Earth Sci, Louvain Univ, 71. *Honors & Awards:* Sigma Xi Pure Res Award, US Naval Res Lab, 84. *Mem:* Am Chem Soc; Am Carbon Soc; Mat Res Soc. *Res:* Self diffusion in molecular crystals and motion of molecules adsorbed on solid surfaces by means of nuclear magnetic resonance; properties of conducting polymers and graphites; surface and polymer analysis. *Mailing Add:* ONR Lon Br Off PO Box 39 FPO New York NY 09510-7000

RESKO, JOHN A, b Patton, Pa, Oct 28, 32; m 62; c 2. REPRODUCTIVE PHYSIOLOGY. *Educ:* St Charles Sem, AB, 56; Marquette Univ, MS, 60; Univ Ill, PhD(physiol), 63. *Prof Exp:* USPHS fel steroid biochem, Col Med, Univ Utah, 63-64; asst scientist, 64-77, from asst prof to assoc prof, 65-77, PROF PHYSIOL, ORE HEALTH SCI UNIV, 77-, CHMN DEPT, 81-; SCIENTIST REPRODUCTIVE PHYSIOL, ORE REGIONAL PRIMATE RES CTR, 71- *Mem:* AAAS; Soc Study Reproduction; Am Physiol Soc; Endocrine Soc; Soc Neurosci. *Res:* Fetal endocrinology; influence of hormones on behavior; regulation of brain aromatase activity. *Mailing Add:* Dept Physiol Ore Health Sci Univ Portland OR 97201-3098

RESLER, E(DWIN) L(OUIS), JR, b Pittsburgh, Pa, Nov 20, 25; m 48; c 5. AERONAUTICAL ENGINEERING. *Educ:* Univ Notre Dame, BS, 47; Cornell Univ, PhD(aeronaut eng), 51. *Prof Exp:* Res assoc, Grad Sch Aeronaut Eng, Cornell Univ, 48-51, asst prof, 51-52; assoc res prof, Inst Fluid Dynamics & Appl Math, Univ Md, 52-56; assoc prof aerospace & elec eng & appl physics, 56-68, dir grad sch Aerospace Eng, 63-72, PROF AEROSPACE & ELEC ENG & APPL PHYSICS, CORNELL UNIV, 58-, JOSEPH NEWTON PEW, JR PROF ENG, 68- *Concurrent Pos:* Prin physicist, Avco Corp, 63; dir, Sibley Sch Mech & Aerospace Eng, 72-77; resident consult, Pratt & Whitney Aircraft, 80-81; vis prof, Case Western Reserve, Lewis Res Lab, NASA, Cleveland, Ohio, 88-89. *Mem:* Am Phys Soc; Am Inst Aeronaut & Astronaut; Int Acad Astronaut; Sigma Xi; Int Sci Radio Union. *Res:* Gas dynamics; aerodynamics; shock waves; magnetohydrodynamics; ferrohydrodynamics; pollution control; hypersonic engines. *Mailing Add:* Upson Hall Cornell Univ Ithaca NY 14853

RESNEKOV, LEON, b Cape Town, SAfrica, Mar 20, 28; m 55; c 2. MEDICINE. *Educ:* Univ Cape Town, MB, ChB, 51, MD, 65; FRCP, 72. *Prof Exp:* Intern med & surg, Groote Schuur Hosp, Univ Cape Town, 52; registr med, Kings Col Hosp, London, 54-59, registr cardiol, 59-61, sr registr, Inst Cardiol, Nat Heart Hosp, London, 61-67; from assoc prof to prof med, 67-82, dir myocardial infarction res unit, 67-81, joint dir sect cardiol, 71-81, RAWSON PROF MED, UNIV CHICAGO, 82- *Mem:* Fel Royal Soc Med; Brit Cardiac Soc; fel Am Col Cardiol. *Res:* Hemodynamics and cardiac physiology in humans and experimental animals; high energy electrical current on cardiac function; electrical control of cardiac rhythm disturbances. *Mailing Add:* Dept Med Cardiol Univ Chicago Med Ctr 5841 S Maryland Ave Chicago IL 60637

RESNICK, CHARLES A, b New York, NY, July 28, 39; m 63; c 3. PHARMACOLOGY, TOXICOLOGY. *Educ:* Brooklyn Col Pharm, BS, 62; Univ Pittsburgh, PhD(pharmacol), 70. *Prof Exp:* PHARMACOLOGIST, FOOD & DRUG ADMIN, 70- *Mem:* NY Acad Sci; Am Heart Asn. *Res:* Beta adrenoceptor blocking agents; cardiovascular applications and oncogenic potential. *Mailing Add:* 1733 Glastonberry Rd Rockville MD 20854

RESNICK, JOEL B, b Brooklyn, NY, Jan 24, 35; m 57; c 6. SYSTEMS ENGINEERING. *Educ:* City Col New York, BS, 57; Mass Inst Technol, MS, 62. *Prof Exp:* Staff engr, Lincoln Lab, Mass Inst Technol, 57-65, group leader, 65-70; systs engr & analyst, US Arms Control & Disarmament Agency, 70-72; dir strategic forces off, Off Asst Secy Defense for systs analysis, 72-78, dir, Prog Assessment Off, Intel Community Staff, 78-80; div mgr & asst vpres, Sci Applns Inc, 80-91; PROF STAFF, HOUSE ARMED SERV COMT, 91- *Res:* Analysis of national security issues related to planning for the US militiary posture. *Mailing Add:* 10604 Trotters Trail Potomac MD 20854

RESNICK, LAZER, b Montreal, Que, June 25, 38; m 64; c 3. PARTICLE PHYSICS, THEORETICAL PHYSICS. *Educ:* McGill Univ, BSc, 59; Cornell Univ, PhD(theoret physics), 65. *Prof Exp:* Res assoc physics, Brookhaven Nat Lab, 64-65; Nat Res Coun Can fel, Niels Bohr Inst, Copenhagen, Denmark, 65-66 & Tel-Aviv Univ, 66-67; asst prof, 67-71, ASSOC PROF PHYSICS, CARLETON UNIV, 71- *Mem:* Am Phys Soc; Can Asn Physicists. *Res:* High energy and elementary particle physics; quantum mechanics. *Mailing Add:* Dept Physics Carleton Univ Ottawa ON K1S 5B6 Can

RESNICK, MARTIN I, b Brooklyn, NY, Jan 12, 43; m 65; c 2. UROLOGIC ONCOLOGY, URORADIOLOGY. *Educ:* Alfred Univ, BA, 64; Bowman Gray Sch Med, MD, 69; Northwestern Univ, MS, 73. *Prof Exp:* Instr, 75-77, asst prof, 77-79, assoc prof, urol, Bowman Gray Sch Med, 79-81; PROF & CHMN, UROL, CASE WESTERN RESERVE UNIV SCH MED, 81-, PROF, ONCOL, 86- *Concurrent Pos:* consult, Organ Systs Coord Ctr Nat Cancer Inst, 83- *Honors & Awards:* Gold Cystoscope Award, Am Urol Asn. *Mem:* Am Urol Asn; Am Col Surgeons; Soc Univ Urol; Am Asn Genitourinary Surgeons; Clin Soc Genitourinary Surgeons. *Res:* Studies consist of identification of characterization of urinary macromolecules as they relate to urolithiasis Other areas of imaging modalities. *Mailing Add:* 2065 Adelbert Rd Cleveland OH 44106

RESNICK, OSCAR, b Bayonne, NJ, Apr 27, 24; m 49; c 2. PHYSIOLOGY, PHARMACOLOGY. *Educ:* Clark Univ, AB, 44; Harvard Univ, MA, 45; Boston Univ, PhD(biol), 55. *Prof Exp:* Instr biol, St Petersburg Jr Col, 46-47; instr physiol, Univ Minn, 49-50; ed asst, Biol Abstr, 50-51; res scientist, Nat Drug Co, Pa, 51-53; SR SCIENTIST, WORCESTER FOUND EXP BIOL, 53- *Concurrent Pos:* Lectr physiol, Grad Sch Dent, Boston Univ, 62-; lectr psychol, Clark Univ, 65-; res consult, Abraham Ribicoff Res Ctr, Norwich Hosp, Conn; mem, Psychodelic Res & Study Group. *Mem:* AAAS; fel Am Col Neuropsychopharmacol. *Res:* Metabolism of biogenic amines in man; clinical psychopharmacology; prenatal malnutrition and the developing nervous system. *Mailing Add:* Worcester Found Exp Biol 222 Maple Ave Shrewsbury MA 01545

RESNICK, PAUL R, b New York, NY, Apr 7, 34; m 66; c 1. FLUORINE CHEMISTRY. *Educ:* Swarthmore Col, BA, 55; Cornell Univ, PhD(org chem), 61. *Prof Exp:* Fel, Univ Calif, Berkeley, 60-62; from chemist to sr res chemist, 62-74, res assoc, 74-85, res fel, 85-88, SR RES FEL, E I DU PONT DE NEMOURS & CO, INC, 88- *Mem:* Am Chem Soc. *Mailing Add:* DuPont Polymers PO Box 80353 E I du Pont de Nemours & Co Wilmington DE 19880-0353

RESNICK, ROBERT, b Baltimore, Md, Jan 11, 23; m 45; c 3. THEORETICAL PHYSICS. *Educ:* Johns Hopkins Univ, AB, 43, PhD(physics), 49. *Prof Exp:* Physicist, Nat Adv Comt Aeronaut, 44-46; from asst prof to assoc prof physics, Univ Pittsburgh, 49-56; assoc prof, 56-57, PROF PHYSICS, RENSSELAER POLYTECH INST, 58-, EDWARD P HAMILTON DISTINGUISHED PROF SCI, 75- *Concurrent Pos:* Mem, Comn Col Physics, 60-68; hon res fel, Harvard Univ, 64-65; adv ed, John Wiley & Sons, Inc, 67-; Fulbright prof, Peru, 71; vis prof, People Repub China, 81 & 85; adv ed, Macmillan Publ, 90-; coun, Textbook Authors Asn, 90- *Honors & Awards:* Exxon Found Award, 53; Oersted Medal, Am Asn Physics Teachers, 75. *Mem:* Fel Am Phys Soc; Am Soc Eng Educ; Am Asn Physics Teachers (pres, 88-89); fel AAAS; Sigma Xi; Am Asn Univ Profs. *Res:* Aerodynamics; nuclear and atmospheric physics; instructional materials; educational research and development; relativity and quantum physics; history of physics; author or coauthor of 7 different textbooks in relativity, quantum physics and general physics. *Mailing Add:* Dept Physics Rensselaer Polytech Inst Troy NY 12180-3590

RESNICK, SIDNEY I, b New York, NY, Oct 27, 45; m 69; c 2. MATHEMATICS. *Educ:* Queens Col, NY, BA, 66; Purdue Univ, Lafayette, MS, 68, PhD(math statist), 70. *Prof Exp:* Lectr probability & statist, Israel Inst Technol, 69-72; asst prof statist, Stanford Univ, 72-77; assoc prof, Colo State Univ, 78-80, prof statist, 81-87; PROF OPERS RES, CORNELL UNIV 87- *Concurrent Pos:* Vis prof, Erasmus Univ, 81-82, Sussex Univ, 86-87. *Honors & Awards:* Lady Davis fel, 82; SERC fel, 86. *Mem:* Fel Inst Math Statist. *Res:* Probability and stochastic processes; extreme value theory; regular variation; weak convergence; stochastic models. *Mailing Add:* Dept Oper Res Cornell Univ Ithaca NY 14853

RESNICK, SOL DONALD, b Milwaukee, Wis, June 15, 18; m 81; c 2. HYDROLOGY. *Educ:* Univ Wis, BS, 41 & 42, MS, 49. *Prof Exp:* Asst hydrol engr, Tenn Valley Authority, 42-43; instr math, Carson-Newman Col, 43-44; asst prof hydrol, Colo State Univ, 49-52; irrig specialist, Int Coop Admin, India, 52-57; from assoc dir to dir, Water Resources Res Ctr, 66-84, EMER PROF HYDROL & HYDROLOGIST, UNIV ARIZ, 84- *Concurrent Pos:* Citizen ambassador prog, People to People, China & Tibet, 87; vis scientist, Ben Gurion Univ, Israel, 89. *Mem:* Am Soc Agr Eng; Am Soc Civil Engr. *Res:* Irrigation and urban hyrology. *Mailing Add:* Water Resources Res Ctr Univ Ariz Tucson AZ 85721

RESNIK, FRANK EDWARD, b Pleasant Unity, Pa, Oct 14, 28; m 52; c 3. ANALYTICAL CHEMISTRY. *Educ:* St Vincent Col, BS, 52; Univ Richmond, MS, 55. *Hon Degrees:* DSc, St Vincent Col, 85. *Prof Exp:* Res chemist, Philip Morris USA, 52-54, group leader, 54-56, from asst supvr to sr supvr, 56-60, tech asst to mgr res, 60-62, mgr, Analytical Serv Div, 62-67, dir com develop, 67-71, dir develop, 71-72, dir, Res Ctr Opers, 72-76, vpres opers admin, 76-78, vpres Tobacco Opers, 78-80, exec vpres, 80-82, pres, chief exec officer & mem bd dirs, 84-89, chmn, 89-90; RETIRED. *Mem:* AAAS; Am Chem Soc; NY Acad Sci; Am Inst Chemists. *Res:* Mass spectrometry; infrared, near-infrared and ultraviolet spectrometry; polarography; chromatography; paper, column and gas chromatography; ion exchange; analytical chemistry; instrumental and colorinetric methods development; microtechniques in spectroscopy and chromatography. *Mailing Add:* Philip Morris USA 120 Park Ave New York NY 10017

RESNIK, HARVEY LEWIS PAUL, b Buffalo, NY, Apr 6, 30; m 64; c 3. SUICIDOLOGY. *Educ:* Univ Buffalo, BA, 51; Columbia Univ, MD, 55; Univ Pa, certs med hypnosis & marriage counseling, 62; Del Valley Group Psychother Inst, cert group ther, 62; Philadelphia Psychoanalysis Inst, cert psychoanalysis, 67; Am Bd Psychiat & Neurol, dipl, 66. *Prof Exp:* Intern, Philadelphia Gen Hosp, 55, resident, 56; resident, Jackson Mem Hosp, Miami, 59-61; consult, Ment Health Asn Southeastern Pa, 65-67; assoc prof psychiat & assoc chmn dept, Sch Med, State Univ NY Buffalo, 67-68, prof psychiat & dep chmn dept, 68-69; chief ctr studies suicide prev, NIMH, 69-72, chief sect crisis intervention, Suicide & Ment Health Emergencies, 72-74; MED DIR, HUMAN BEHAV FOUND, 74- *Concurrent Pos:* Fel, Hosp Univ Pa, 62; fel, Reproductive Biol Res Found, St Louis, 71; clin prof, Sch Med, George Washington Univ, 69-; prof lectr, Sch Med, Johns Hopkins Univ, 69-; ed, Bull Suicidology, Am Psychiat Asn, 69-74; prof, Fed City Col, 71-72; consult, WHO, Am Red Cross, Nat Cancer Inst, Nat Naval Med Ctr, Dept Defense & Pub Defender's Off; clin prof, Sch Med, Uniformed Serv Univ Health Sci, 77-79; spec Nato fel Brussels, Belg, 86; vis prof psychiat, Sch Med & Law, Cath Univ Leuven, Belg, 86. *Honors & Awards:* Charles W Burr Res Prize, Philadelphia Gen Hosp, 56; Gold Medal, Am Psychiat Asn, 72. *Mem:* Fel Am Psychiat Asn; fel Am Col Psychiat; Am Acad Psychiat Law. *Res:* Crisis intervention; emergency mental health services; suicide, including prevention programs, diagnosis and management of suicidal and depressed individuals; treatment of sexual dysfunctions and criminal sexual offenders. *Mailing Add:* Air Rights Ctr 7315 Wisconsin Ave No 900E Bethesda MD 20814-3202

RESNIK, ROBERT ALAN, b New York, NY, Nov 11, 24; m 52. BIOCHEMISTRY. *Educ:* Purdue Univ, BS, 48, MS, 50, PhD(biophys, physiol), 52. *Prof Exp:* NIH fel, Northwestern Univ Ill, 52-53; chief biochemist, Ophthalmic Chem Sect, Nat Inst Neurol Dis & Blindness, NIH, 53-63, chemist, Lab Phys Biol, Nat Inst Arthritis & Metab Dis, 63-68, chief eval scientist biophys sci, Div Res Grants, NIH, 68-70, chief, Res Analysis & Eval Br, 70-71, prog planning off, Nat Eye Inst, 71-73, chief reports & eval br, Nat Heart & Lung Inst, 73-78; RETIRED. *Mem:* AAAS; Am Chem Soc; Am Soc Biol Chemists; Sigma Xi. *Res:* Interactions of metals with proteins; protein chemistry; nucleic acids and proteins; science administration. *Mailing Add:* 5508 Hoover St Bethesda MD 20817

RESNIK, ROBERT KENNETH, b Pleasant Unity, Pa, May 19, 36; m 64. INORGANIC CHEMISTRY. *Educ:* St Vincent Col, BS, 58; Univ Pittsburgh, PhD(inorg chem), 64. *Prof Exp:* Res chemist, Exxon Res & Eng Co, 64-65; sr inorg chemist, 65-66, SCIENTIST, RES LABS, J T BAKER CHEM CO, 66- *Mem:* Am Chem Soc; Sigma Xi. *Res:* Coordination and inorganic compounds; agriculture chemicals; industrial inorganic chemicals; properties of antacid chemicals. *Mailing Add:* 1709 Wynnwood Lane N Easton PA 18042

RESNIKOFF, GEORGE JOSEPH, b New York, NY, Mar 25, 15; m 43; c 1. MATHEMATICAL STATISTICS. *Educ:* Univ Chicago, SB, 50; Stanford Univ, MS, 52, PhD, 55. *Prof Exp:* Res assoc statist, Stanford Univ, 52-57; from assoc prof to prof indust eng, Ill Inst Technol, 57-64; staff mem, 64-70, dean grad studies, prof & chmn Dept Statist, 72-80, EMER PROF MATH & STATIST, CALIF STATE UNIV, HAYWARD, 80- *Concurrent Pos:* Chmn, Dept Indust Eng, Ill Inst Technol, 58-64; lectr, Univ Chicago, 60-64; vis scholar, Univ Calif, Berkeley, 62-63; NSF fel, 62-63. *Mem:* Fel AAAS; fel Inst Math Statist (treas, 64-72, exec secy, 73-78); fel Am Statist Asn. *Res:* Applications of mathematics and statistics to industrial problems. *Mailing Add:* Dept Statist Calif State Univ Hayward CA 94542

RESNIKOFF, HOWARD L, b New York, NY, May 13, 37; m 59; c 3. MATHEMATICS. *Educ:* Mass Inst Technol, BS, 57; Univ Calif, Berkeley, PhD(math), 63. *Prof Exp:* Mathematician, Electrodata Div, Burroughs Corp, 57-58; res scientist, Res Labs, Lockheed Missiles & Space Co, 62-64; mem, Inst Advan Study, 64-66; asst prof math, Rice Univ, 66-68, actg dir comput ctr, 71, assoc prof math, 68-73, prof, 73-75; chmn math, Univ Calif, Irvine, 75-78, prof math, 75-80; dir div info sci & technol, NSF, 78-80; mem fac, Harvard Univ, 80-; VP RES & DIR RES, THINKING MACH CORP. *Concurrent Pos:* NSF fel, Univ Munich, 62-63; partner, R&D Consult Co, Calif, 64- & R&D Press, Calif, 72-77; Alexander von Humboldt Found, US sr scientist award, Univ Muenster, 74-75. *Res:* Automorphic function theory; theory of Jordan algebras; linguistic structure of written language; information retrieval; library automation; history of mathematics. *Mailing Add:* Aware Inc University Pl Suite 310 124 Mt Auburn St Cambridge MA 02138

RESO, ANTHONY, b London, Eng, Aug 10, 31; nat US. STRATIGRAPHY, PETROLEUM GEOLOGY. *Educ:* Columbia Univ, AB, 54, MA, 55; Rice Univ, PhD(geol), 60. *Prof Exp:* Instr geol, Queens Col, NY, 54; asst, Columbia Univ, 54-55; geologist, Atlantic Richfield Corp, 55-56; asst, Univ Cincinnati, 56-57; asst prof geol, Amherst Col, 59-62; staff res geologist, Tenneco Oil Co, 62-86, GEOL MGR, PEAK PRODUCTION CO, 86-, VPRES 88- *Concurrent Pos:* Res consult, 60-61; Geol Soc Am grant & Am Asn Petrol Geologists grant, 58-59; NSF fel, 59; cur invert paleont, Pratt Mus, Amherst, Mass, 59-62; lectr, Univ Houston, 62-65; mem, Bd Advs, Gulf Univ Res Corp, 67-75, chmn, 68-69; vis prof geol, Rice Univ, Houston, 80; gen chmn, Am Asn Petrol Geologists Nat Conv, 79; consult, Gulf Coast Geol Library Inc, 87- *Honors & Awards:* Distinguished Serv Award, Am Asn Petrol Geologists, 85. *Mem:* Fel Geol Soc Am; fel AAAS; Am Asn Petrol Geologists (treas, 86-88); Paleont Soc; Soc Econ Paleontologists & Mineralogists; Sigma Xi. *Res:* Earth history and subdivision of the geological time scale; invertebrate paleontology; world cretaceous stratigraphy; petroleum geology; distribution of economic resources and international trade. *Mailing Add:* Peak Prod Co PO Box 130785 Houston TX 77219-0785

RESS, RUDYARD JOSEPH, b Bronx, NY, Oct 7, 50; m 81. PHARMACOLOGY. *Educ:* Univ Fla, BS, 74, PhD(physiol), 81. *Prof Exp:* Res fel pharmacol, div cardiol, Hershey Med Ctr, Pa State Univ, 81-83; SR RES PHARMACOLOGIST, HOECHST-ROUSSEL PHARMACEUT, INC, AM HOECHST CORP, 83- *Mem:* Am Physiol Soc. *Res:* Cardiovascular research and antihypertensive therapy; new drug development. *Mailing Add:* Dept Commun Schering Plough Inc K-6-1 G-11 2000 Gallopeng Hill Rd Kenilworth NJ 07033

RESSLER, CHARLOTTE, b West New York, NJ, July 21, 24. ORGANIC CHEMISTRY, BIOCHEMISTRY. *Educ:* NY Univ, BA, 44; Columbia Univ, MA, 46, PhD(org chem), 49. *Prof Exp:* Res assoc biochem, Med Col, Cornell Univ, 49-54, from asst prof to assoc prof, 55-74; PROF PHARMACOL, UNIV CONN HEALTH CTR, FARMINGTON, 74- *Concurrent Pos:* Am Heart Asn estab investr, Med Col, Cornell Univ, 57-59; assoc mem, Inst Muscle Dis, Inc, 59-63, mem, 63-, head div protein chem, 59-74; mem med chem study sect, NIH, 72-75. *Mem:* AAAS; Am Chem Soc; Am Soc Biol Chemists. *Res:* Peptide hormones; amino acid metabolism; natural and synthetic neurothyrogens; enzyme characterization; amide reactions; organic rearrangements. *Mailing Add:* 4G Talcott Glen Rd Farmington CT 06032

RESSLER, NEIL WILLIAM, b Columbus, Ohio, June 1, 39; m 81; c 3. ENGINEERING, LASERS. *Educ:* Gen Motors Inst, BSME, 62; Univ Mich, MS, 63, PhD(physics), 67. *Prof Exp:* Res scientist, Sci Res Staff, Ford Motor Co, 67-71, prin eng chassis, 71-73, supvr, 73-76, dept mgr, 76-78, exec engr, 78-81, chief engr, Climate Control Div, EXEC DIR, VEHICLE OPER, FORD MOTOR CO, 89- *Mem:* Soc Automotive Engrs. *Res:* Research development and production engineering of automotive climate control systems and heat exchangers. *Mailing Add:* 3420 Riverbend Dr Ann Arbor MI 48105

RESSLER, NEWTON, b Detroit, Mich, Sept 19, 23; m 54; c 2. BIOPHYSICS. *Educ:* Univ Mich, BS, 47; Univ Chicago, MS, 49; Wayne State Univ, PhD(biochem), 53. *Prof Exp:* Res biochemist, Wayne County Gen Hosp, 53-65; asst prof biochem, Univ Mich, 65-68; from assoc prof to prof, 68-90, EMER PROF BIOCHEM PATH, UNIV ILL MED CTR, 90- *Mem:* Am Chem Soc; Am Asn Clin Chemists; Biophys Soc; Sigma Xi; Nat Acad Clin Biochem. *Res:* Nature and control of enzymes; multiple enzyme forms and energy transduction. *Mailing Add:* 30 Red Haw Rd Northbrook IL 60062

REST, DAVID, b Chicago, Ill, Mar 27, 17; m 39; c 2. CHEMICAL ENGINEERING. *Educ:* Armour Inst Technol, BS, 37; Univ Calif, MS, 60. *Prof Exp:* Supt chem prod, G D Searle & Co, 38-47; chief res engr, Armed Forces Qm Food & Container Inst, 47-55, chief nuclear effects br, 55-58, dir food radiation preserv div, 58-62; FOOD PROCESSING CONSULT, ARTHUR D LITTLE, INC, 62- *Mem:* Am Chem Soc; Health Physics Soc; Inst Food Technol. *Res:* Radiation processing of biological materials; low pressure sublimation of water in food stuffs; process control of food production facilities; chemical and nuclear engineering design; technical and economic evaluation of international food processing enterprises. *Mailing Add:* 3450 N Lakeshore Dr No 2702 Chicago IL 60657

REST, RICHARD FRANKLIN, b Chicago, Ill. MICROBIOLOGY. *Educ:* Univ Mass, Amherst, BS, 70; Univ Kans, PhD(microbiol), 74. *Prof Exp:* Fel neutrophil functions, Dept Bact & Immunol, Sch Med, Univ NC, Chapel Hill, 74-77; asst prof phagocyte bact interactions, 77-83, assoc prof microbiol, Sch Med, Health Sci Ctr, Univ Ariz, 83; assoc prof, 83-90, PROF, DEPT MICROBIOL, HAHNEMANN UNIV SCH MED, PHILADELPHIA, PA, 90- *Concurrent Pos:* NIH fel, Dept Bact & Immunol, Sch Med, Univ NC, 75-77; Nat Inst Allergy & Infectious Dis grants, 78-94. *Mem:* Am Soc Microbiol. *Res:* Interaction of pathogenic Neisseria with human phagocytes and epithelial cells; biochemical and enzymatic analysis of human leukocyte phagolysosomes. *Mailing Add:* Dept Microbiol MS410 Hahnemann Univ Sch Med Philadelphia PA 19102-1192

RESTAINO, ALFRED JOSEPH, b Brooklyn, NY, Feb 18, 31; m 53; c 5. PHYSICAL CHEMISTRY, POLYMER CHEMISTRY. *Educ:* St Francis Col, NY, BS, 52; Polytech Inst Brooklyn, MS, 54, PhD(chem), 55. *Prof Exp:* AEC fel, 54-56; supvr radiation res, Martin Co, 56-58; mgr radiation chem sect, ICI Americas, Inc, 58-68, mgr polymer chem, 68-71, asst dir chem, res dept, 71-74, dir, corp res dept, 75-87; PRES, CREATIVE ASSETS & CONSULT CORP, 87- *Concurrent Pos:* Asst prof, St Francis Col, NY, 55-56; adj prof, Drexel Inst Technol, 57; mem Gov Sci Adv Coun, Del, 70-72; mem sci adv comt, AEC, 70-73; prof grad exten, Univ Del, 71-74. *Mem:* Am Chem Soc; NY Acad Sci; AAAS. *Res:* Organic chemistry; thermoset polymers, water soluble polymers, thermoplastics and photochemistry; novel reactions for synthesis of specialty products including agricultural chemicals; reaction injection molding; composites; advanced materials. *Mailing Add:* 615 Black Gates Rd Wilmington DE 19803

RESTAINO, FREDERICK A, b Brooklyn, NY, Dec 9, 34; m 58; c 3. PHYSICAL PHARMACY, PHARMACEUTICAL DEVELOPMENT. *Educ:* St John's Univ, NY, BS, 56; Rutgers Univ, MS, 58; Purdue Univ, PhD(phys pharm), 62. *Prof Exp:* From asst instr to instr org chem, Rutgers Univ, 56-58; unit head aerosol res, 62-65, unit head sterile prod, 65-70, unit head fluids-topicals, 69-70, unit head tablet & capsules & sr res fel, Res Labs, 70-84, DIR PHARMACEUT DEVELOP, MERCK SHARP & DOHME, 85- *Mem:* Am Pharmaceut Asn; Am Asn Pharmaceut Scientists. *Res:* Solubilization; aerosol research; sterile products research and development; fluids-topical development; tablet and capsule research and development; solid dosage form development. *Mailing Add:* Merck Sharp & Dohme Res Labs West Point PA 19486

RESTEMEYER, WILLIAM EDWARD, b Cincinnati, Ohio, Apr 28, 16; m 43; c 2. APPLIED MATHEMATICS, ELECTRICAL & COMPUTER ENGINEERING. *Educ:* Univ Cincinnati, EE, 38, MA, 39. *Hon Degrees:* DSc, Capitol Inst Technol, 76. *Prof Exp:* Officer, Naval Res Lab, 45; from instr to assoc prof math, 40-61, admin aide, Dept Math, 68-75, PROF ELEC & COMPUT ENG & MATH SCI, UNIV CINCINNATI, 61-, ADMIN ASSOC, 87- *Concurrent Pos:* Consult, Avco-Crosley, Kettering Lab, Fed Security Agency, Gen Elec Corp, Argonne Nat Lab, Math Asn Am, NSF, NATO US Off Educ, Nat Aero & Space Admin & Cincinnati Milacron Corp, 44-; accreditation bd, Eng & Technol, NY State Dept Educ; vis scientist, Ohio Acad Sci; asst dept head, Univ Cincinnati, 76- *Mem:* Am Soc Eng Educ; Math Asn Am; sr mem Inst Elec & Electronics Engrs; Int Math Union. *Res:* Signal and systems analysis. *Mailing Add:* Col Eng Univ Cincinnati Cincinnati OH 45221-0030

RESTER, ALFRED CARL, JR, b New Orleans, La, July 11, 40; m; c 2. NUCLEAR PHYSICS. *Educ:* Miss Col, BS, 62; Univ NMex, MS, 65; Vanderbilt Univ, PhD(physics), 69. *Prof Exp:* Fel, Delft Univ Technol, 69-70, scientist, 70-71; scientist, Oak Ridge Nat Lab, 71-72; guest prof physics, Inst Radiation & Nuclear Physics, Univ Bonn, 72-75; asst prof physics, Emory Univ, 75-76; assoc prof, Tenn Technol Univ, 76-77; vis assoc prof physics, 78-81, assoc res scientist, Space Astron Lab, 81-87, DIR INST ASTROPHYSICS & PLANETARY EXPLOR, UNIV FLA, 88- *Honors & Awards:* Antarctic Serv Medal, 88. *Mem:* Am Phys Soc; Sigma Xi; AAAS; Am Astron Soc. *Res:* Nuclear radiation detectors for use in space; man in space as experimenter; heavy ion reactions; nuclear structure; studies of medium mass nuclei; on-line isotope separators; computer analysis of nuclear data; nuclear astrophysics and astronomy; scientific ballooning in Antarctica; space experiments. *Mailing Add:* Inst Astrophy & Planetary Explor Univ Fla One Progress Blvd Box 33 Alachua FL 32615

RESTER, DAVID HAMPTON, b Bogalusa, La, June 14, 34; m 58. NUCLEAR PHYSICS. *Educ:* Tulane Univ, BS, 56; Rice Univ, MA, 58, PhD(physics), 60. *Prof Exp:* Asst prof physics, Tulane Univ, 60-63; sr scientist, Ling-Tempco-Vought Inc Res Ctr, 63-71; sr scientist, Advan Technol Ctr, Inc, 71-73, supvr electrophys, 73-79; MGR ANALYTICAL METHODS, MEAD OFF SYSTS, 79- *Mem:* Am Phys Soc; Soc Info Display. *Res:* Low energy studies of nuclear structure by method of internal conversion electron spectroscopy; study of electron scattering at intermediate energies observing bremsstrahlung production and scattered electrons. *Mailing Add:* 7027 DeLoache Ave Dallas TX 75225

RESTORFF, JAMES BRIAN, b Wytheville, Va, June 22, 49; m 73; c 1. SOLID STATE PHYSICS. *Educ:* Univ Md, BS, 71, MS, 75, PhD(physics), 76. *Prof Exp:* RES PHYSICIST, NAVAL SURFACE WARFARE CTR, 76- *Mem:* Am Phys Soc. *Res:* Magnetic materials. *Mailing Add:* Code R43 White Oak Lab Naval Surface Warfare Ctr 10901 New Hampshire Ave Silver Spring MD 20903-5000

RESTREPO, RODRIGO ALVARO, b Medellin, Colombia, Nov 6, 30. MATHEMATICS. *Educ:* Lehigh Univ, BA, 51; Calif Inst Technol, PhD(math), 55. *Prof Exp:* Res fel, Calif Inst Technol, 55-56; lectr, Univ BC, 56-58; vis asst prof, Stanford Univ, 58-59; from asst prof to assoc prof, 59-70, PROF MATH, UNIV BC, 70- *Concurrent Pos:* Vis prof, Interam Statist Training Ctr, Chile, 63-64 & COPPE, Fed Univ Rio de Janeiro, 70-71, 72 & 74. *Mem:* Am Math Soc; Math Asn Am; Can Math Cong. *Res:* Game theory; linear programming. *Mailing Add:* Dept Math Univ BC 2075 Westbrook Pl Vancouver BC V6T 1W5 Can

RESWICK, JAMES BIGELOW, b Ellwood City, Pa, Apr 16, 22; m 73; c 3. BIOMEDICAL ENGINEERING. *Educ:* Mass Inst Technol, SB, 43, SM, 48, ScD(mech eng), 54. *Hon Degrees:* DEng, Rose Polytech Inst, 68. *Prof Exp:* Instr mach design, Mass Inst Technol, 46-50, from asst prof to assoc prof mech eng, 50-59; prof mech eng & dir eng design, Case Inst Technol, 59-70; prof biomed eng & orthopaed, Univ Southern Calif, 70-81; assoc dir, Nat Inst Handicapped Res, US Dept Educ, Wash, DC, 81-83; dir, Va Rehab Res & Develop Eval Unit, Va Med Ctr, Wash, 84-88; ASSOC DIR, NAT INST DISABILITY & REHAB RES, US DEPT EDUC, WASH, DC, 88- *Concurrent Pos:* NSF sr fel, Imp Col, Univ London, 57-58; mem prosthetics res & develop comt, Nat Acad Sci, 60 & Inst Med, 72; dir, Rehab Eng Ctr, Rancho Los Amigos, 70-80. *Honors & Awards:* Isabel & Leonard Goldenson Award, United Cerebral Palsy Asn, 74. *Mem:* Inst Med-Nat Acad Sci; Nat Acad Eng; assoc Am Acad Orthop Surgeons; Biomed Eng Soc; fel Inst Elec & Electronics Engrs; Rehab Engr Soc NAm (founding pres). *Res:* Engineering design education; automatic control theory and application; dynamics; product design and development; biomedical engineering; administration of rehabilitation engineering. *Mailing Add:* 1003 Dead Run Dr McLean VA 22101

RETALLACK, GREGORY JOHN, b Hobart, Australia, Nov 8, 51; m 81; c 2. PALEOPEDOLOGY, TERRESTRIAL PALEOECOLOGY. *Educ:* Macquarie Univ, Australia, BA, 73; Univ New Eng, BSc Hons, 74, PhD(geol), 78. *Prof Exp:* Vis asst prof geol, Northern Ill Univ, 77-78; vis scholar biol, Ind Univ, 78-79 & proj co-dir, 79-81; asst prof, 81-85, ASSOC PROF GEOL, UNIV ORE, EUGENE, 86- *Honors & Awards:* Stillwell Medal, Geol Soc Australia. *Mem:* Sigma Xi; Geol Soc Am; Geol Soc Australia; Bot Soc Am; AAAS. *Res:* Paleopedological and paleoecological research into Gondwanan Triassic fossil plants; Cretaceous dispersal and rise to dominance of angiosperms; Tertiary development of grasslands; evolution of soils through geological time. *Mailing Add:* Dept Geol Univ Ore Eugene OR 97403

RETALLICK, WILLIAM BENNETT, b Yonkers, NY, Jan 16, 25; m 49; c 1. CHEMICAL ENGINEERING. *Educ:* Univ Mich, BSE, 48; Univ Ill, MS, 52, PhD(chem eng), 53. *Prof Exp:* Process engr, Phillips Petrol Co, 48-50 & Consol Coal Co, 53-64; res engr, Houdry Process & Chem Co, 64-71, sr chem engr, 71-74; vpres res & develop, Oxy-Catalyst, Inc, 74-80; CONSULT CHEM ENGR, 80- *Concurrent Pos:* Develop award, US Dept Energy, NSF. *Mem:* Am Chem Soc; Am Inst Chem Engrs; Sigma Xi. *Res:* Pilot scale research in coal; exploratory research on chemicals and catalysts; catalysts and catalytic processes; catalytic combustion; catalytic converters for automobiles. *Mailing Add:* 1432 Johnny's Way West Chester PA 19382

RETCOFSKY, HERBERT L, b Brownsville, Pa, Apr 1, 35; m 61; c 7. SPECTROSCOPY, COAL SCIENCE. *Educ:* Calif State Col, Pa, BS, 57; Univ Pittsburgh, MS, 65. *Prof Exp:* Res physicist phys chem, 58-76, chief molecular spectros, Br Analytical Chem Coal, 76-79, MGR ANALYTICAL CHEM DIV, PITTSBURGH ENERGY TECHNOL CTR, US DEPT ENERGY, 79- *Concurrent Pos:* US assoc ed, Fuel, 81- *Mem:* Am Chem Soc; Soc Appl Spectros. *Res:* Applications of spectral techniques in coal research. *Mailing Add:* Pittsburgh Energy Technol Ctr PO Box 10940 Pittsburgh PA 15236-0940

RETELLE, JOHN POWERS, JR, b Flushing, NY, Jan 1, 46; m 75; c 2. AEROSPACE ENGINEERING. *Educ:* US Air Force Acad, BS, 67; Univ Colo, MS, 69, PhD(aerospace eng), 78; Golden Gate Univ, MBA, 71. *Prof Exp:* Flight test engr aerospace eng, Air Force Flight Test Ctr, Edwards AFB, Calif, 69-72 & French Flight Test Ctr, Istres, France, 72-73; from instr to asst prof aeronaut, USAF Acad, 73-75, assoc prof aeronaut & dir labs & develop, 79-81; prog mgr sensor simulation, Air Force Human Resources Lab, Williams AFB, Ariz, 82-84; prog mgr, Defense Advan Res Proj Agency, Arlington, Va, 84-88; dir, advan appln, Lockheed Corp, Calabasis, Calif, 88-90; MGR, ADVAN COMPUTER LABS, LOCKHEED MISSILES & SPACE CO, PALO ALTO, CALIF, 90- *Concurrent Pos:* Flight test engr, Air Force Flight Test Ctr, Edwards AFB, Calif, 69-72 & French Flight Test Sch, Istres, Frances, 72-73. *Mem:* Assoc fel Am Inst Aeronaut & Astronaut; Inst Elec & Electronics Engrs; Am Asn Artificial Intel. *Res:* Wind tunnel experimentation on dynamic stall and unsteady flow separation using laser Doppler techniques and numerical simulation. *Mailing Add:* Lockheed Missiles & Space Co Res & Develop Div 3251 Hanover St Palo Alto CA 94304

RETHERFORD, JAMES RONALD, b Panama City, Fla, Oct 1, 37; m 61. MATHEMATICS. *Educ:* Fla State Univ, BS, 59, MS, 60, PhD(math), 63. *Prof Exp:* Asst prof math, Univ Chattanooga, 60-61; instr, Fla State Univ, 64; from asst prof to assoc prof, 64-70, PROF MATH, LA STATE UNIV, BATON ROUGE, 70- *Mem:* Am Math Soc; Math Asn Am. *Res:* Series and operators determined by series in Banach spaces. *Mailing Add:* La State Univ Baton Rouge LA 70803

RETHWISCH, DAVID GERARD, b LaCrosse, Wis, Mar 28, 59; m 81; c 3. HETEROGENEOUS CATALYSTS, MEMBRANE SCIENCE. *Educ:* Univ Iowa, BS, 79; Univ Wis Madison, PhD(chem eng), 85. *Prof Exp:* Asst prof, dept chem & mats eng, 85-90, ASSOC PROF CHEM ENG, DEPT CHEM & BIOCHEM ENG, UNIV IOWA, 90- *Concurrent Pos:* Consult, Oral B Labs & State La, 85, Rolscreen Inc, 86, Amana Corp, 88-90, Amoco Chem Co, 90- *Mem:* Am Chem Soc; Am Inst Chem Engr; Am Soc Eng Educ; Sigma Xi; Am Soc Eng Educators. *Res:* Use of photoresponsive polymers as membranes with capability of real time control; enzyme-facilitated transport through liquid membranes; enzyme production of saccharide based polymers. *Mailing Add:* Dept Chem & Biochem Eng Univ Iowa Iowa City IA 52242

RETIEF, DANIEL HUGO, b Winburg, OFS, Repub SAfrica, Apr 25, 22; US citizen; m 52; c 3. PREVENTIVE DENTISTRY, DENTAL MATERIALS. *Educ:* Univ Stellenborch, BSc, 42, MSc, 44, DSc, 84; Univ Witwatersrand, BDS, 53, PhD(dent), 75. *Prof Exp:* Prof res & dir res, Dent Res Unit, SAfrican Med Res Coun & Univ Witersrand, 70-76; PROF BIOMAT & SR SCIENTIST, INST DENT RES, SCH DENT, UNIV ALA, 77- *Concurrent*

Pos: Sr foreign dent scientist fel, Am Asn Dent Res, 75; assessor, Nat Health & Med Res Coun, Commonwealth Australia, 82-84; referee, Med Res Coun Can, 82-84; consult, dent serv, Vet Admin, 83- *Mem:* Int Asn Dent Res; Royal Soc SAfrica; Int Dent Fedn; Europ Orgn Caries Res; Soc Biomat; AAAS; Royal Soc Belgium Dent Med; Acad Dent Mat. *Res:* Polymeric dental resins; acid etch technique; preventive dentistry; effect of flouride and micronutrients on dental caries; dental caries epidemiology. *Mailing Add:* 3624 Bellemeade Way Birmingham AL 35223

RETNAKARAN, ARTHUR, b Trichy, India, Aug 28, 34; m 60; c 2. INSECT PHYSIOLOGY, BIOCHEMISTRY. *Educ:* Univ Madras, MA, 55; Univ Wis, MS, 64, PhD(entom), 67. *Prof Exp:* Lectr zool, Voorhees Col, Vellore, India, 55-58, prof, 58-62; fel, Univ Wis, 67-68; RES SCIENTIST, FOREST PEST MGT INST, 68- *Concurrent Pos:* Vis prof, Univ Louis Pasteur, Strasbourg, France, 74-75; Fulbright scholar, 62-67; vis scientist, Commonwealth Sci & Industrial Res Orgn, Canberra, Australia, 83-84. *Mem:* Entom Soc Am; Entom Soc Can; AAAS. *Res:* Use of juvenile hormone analogs and moult inhibitors in controlling forest insect pests; insect reproductive physiology; benzoyl ureas for insect control; antijuvenile hormones; chitin synthesis in insects. *Mailing Add:* 50 Ashgrove Ave Sault Ste Marie ON P6A 5M7 Can

RETSEMA, JAMES ALLAN, b Muskegon, Mich, Feb 27, 42; m 69; c 2. BIOCHEMISTRY, MICROBIOLOGY. *Educ:* Mich State Univ, BS, 64; Univ Iowa, MS, 67, PhD(biochem), 69. *Prof Exp:* Res asst biochem, Univ Iowa, 64-69; RES ADV IMMUNOL & INFECTIOUS DIS, CENT RES, PFIZER, INC, 69- *Concurrent Pos:* Fel, McArdle Lab Cancer Res, Univ Wis, 69. *Mem:* Am Soc Microbiol; Am Acad Microbiol. *Res:* Discovery and development of antibacterials and antiprotozoan and their spectrum of activity, mode of action, mechanism of destruction and structural activity relationships. *Mailing Add:* Cent Res Pfizer Inc Groton CT 06340

RETTALIATA, JOHN THEODORE, b Baltimore, Md, Aug 18, 11; m 70; c 3. MECHANICAL ENGINEERING. *Educ:* Johns Hopkins Univ, BE, 32, DEng, 36. *Hon Degrees:* DEng, Mich Col Mining & Technol, 56 & Rose Polytech Inst, 70; DSc, Valparaiso Univ, 59; LLD, DePaul Univ, 62, 62 & Chicago-Kent Col Law, 69; LHD, Loyola Univ, 70. *Prof Exp:* Instr & head dept, Baltimore Col Ctr, Md, 34-35; lab technician, USDA, 35; head calculation div, Allis-Chalmers Co, Wis, 36-44, mgr res & gas turbine develop div, 44-45; prof mech eng & head dept, Ill Inst Technol, 45-48, dean eng, 48-52, vpres, 50-52, pres, Inst, IIT Res Inst & Inst Gas Technol, 52-73; chmn bd, Banco Di Roma, Chicago, 73-87; RETIRED. *Concurrent Pos:* Mem bd vis, Air Univ, 55-58, chmn, 57-58; mem, Nat Aeronaut Space Coun, 59. *Honors & Awards:* Jr Award, Am Soc Mech Engrs, 41, Spec Award, 51. *Mem:* Fel AAAS; fel Am Soc Mech Engrs. *Res:* Super-Saturated steam; gas turbine engineering; jet engineering; science administration. *Mailing Add:* 8901 S Pleasant Chicago IL 60620

RETTENMEYER, CARL WILLIAM, b Meriden, Conn, Feb 10, 31; m 54; c 2. ENTOMOLOGY, ECOLOGY. *Educ:* Swarthmore Col, BA, 53; Univ Kans, PhD(entom), 62. *Prof Exp:* From asst prof to assoc prof entom, Kans State Univ, 60-71; head syst & evolutionary biol, 80-83, PROF BIOL, UNIV CONN, 71-; exec off, biol sci group, 83-85, DIR, CONN STATE MUS NATURAL HIST, 82- *Concurrent Pos:* NSF res grants, 62- & Orgn Trop Studies, 65, 67 & 69. *Mem:* Fel AAAS; Am Asn Mus; Asn Trop Biol; Animal Behav Soc. *Res:* Ecology and behavior of army ants and associated arthropods; taxonomy of Dorylinae; insect behavior, mimicry mutualism; biological photography; Neotropical insects. *Mailing Add:* Mus Natural Hist Univ Conn Rm 75 N Eagleville Rd U-23 Storrs CT 06269-3023

RETTENMIER, CARL WAYNE, b Erie, Pa, Oct 23, 52. MOLECULAR BIOLOGY, PATHOLOGY. *Educ:* Syracuse Univ, BS, 74; Rockefeller Univ, PhD(virol), 79; Cornell Univ, MD, 80. *Prof Exp:* Resident anat path, Lab Path, NIH, 80-82, jr staff pathologist, 82-83; res assoc tumor cell biol, 83-84, ASST MEM TUMOR CELL BIOL, ST JUDE CHILDREN'S RES HOSP, 84- *Mem:* AAAS; Am Soc Microbiol; Am Assoc Path. *Res:* Oncogene expression; tyrosine-specific protein kinases; mechanisms of cell surface receptor-mediated signal transduction; growth factor biosynthesis. *Mailing Add:* Dept Tumor Cell Biol St Jude Children's Res Hosp Memphis TN 38105

RETTORI, OVIDIO, b Parana, Entre Rios, Arg, June 16, 34; m 76; c 3. CANCER PHYSIOPATHOLOGY. *Educ:* Nat Col Parana, BSc, 51; Univ Buenos Aires, MD, 59. *Prof Exp:* Fel, NIH, 64-68; sr res scientist, Nat Res Coun, Arg, 68-76; assoc prof physiol respiratory & kidney physiol, Sch Med, Univ Buenos Aires, 68-76; physiol prof respiratory & kidney physiol, Sch Med, Univ Oriente Venezuela, 76-86, res coordr, 84-86; PHYSIOL PROF CANCER PHYSIOPATH, SCH MED, CAISM/UNICAMP, BRAZIL, 87. *Honors & Awards:* Augusto Pi-Suner Ann Award Best Physiol Exp Work, Catalan Soc, Caracas, Venezuela, 81. *Mem:* Am Physiol Soc. *Res:* Cancer physiopathology; experimental models; detection, identification and inhibition of humoral mediators involved in the remote lethal effects of cancer. *Mailing Add:* Lab Pesq Bioq Div Oncol CAISM/UNICAMP CP 6151 Campinas SP 13081 Brazil

RETZ, KONRAD CHARLES, b Oelwein, Iowa, Feb 19, 52. NEUROPHARMACOLOGY, NEUROCHEMISTRY. *Educ:* Augustana Col, Ill, BA, 74; Univ Iowa, PhD(pharmacol), 79. *Prof Exp:* Dept Pharmacol & Exp Therapeut, 79-81, fel, Dept Neurosci, Sch Med, Johns Hopkins Univ, Baltimore, 81-82, Long Island Res Inst, SUNY, Stony Brook, 82-83; ASST PROF PHARMACOL, TEX COL OSTEOP MED, 83- *Mem:* AAAS; Am Chem Soc; Soc Neurosci; NY Acad Sci. *Res:* Mechanisms of analgesia; mode of action of excitatory amino acid neurotransmitters; regulation of energy metabolism in the central nervous system and calcium involvement in neurotransmission; modulation of behavior in autoimmune mice; neurotransmission in aging. *Mailing Add:* Am Osteop Asn 142 E Ontario St Chicago IL 60611-2864

RETZER, KENNETH ALBERT, b Jacksonville, Ill, Nov 6, 33; m 53; c 3. MATHEMATICS EDUCATION. *Educ:* Ill Col, AB, 54; Univ Ill, EdM, 57, PhD(math educ), 67. *Prof Exp:* Instr, High Sch, Ill, 54-58, asst supt, 55-58; from asst prof to assoc prof, 59-74, asst head dept, 68-70, PROF MATH, ILL STATE UNIV, 74- *Concurrent Pos:* Partic, NSF Acad Year Inst Math, Univ Ill, 58-59. *Mem:* AAAS; Math Asn Am. *Res:* Effects of teaching logic on verbalization and on transfer of discovered mathematical generalizations; strategies for teaching mathematics. *Mailing Add:* Dept Math Stv 329D Ill State Univ Normal IL 61761

RETZLAFF, ERNEST (WALTER), neurophysiology, medical physiology; deceased, see previous edition for last biography

RETZLOFF, DAVID GEORGE, b Pittsburgh, Pa, Feb 19, 39; m 71; c 1. CATALYSIS, MODELING. *Educ:* Univ Pittsburgh, BS, 63, MS, 65, PhD(chem eng), 67. *Prof Exp:* Mem staff, Lab Chem, Technol & Tech High Sch, Delft, Neth, 67-68; res assoc, Univ Colo, 68-69; asst prof chem eng, Kans State Univ, 69-73; res engr, Exxon Res & Eng Co, 73-75; asst prof, 75-84, ASSOC PROF CHEM ENG, UNIV MO, COLUMBIA, 84- *Mem:* Am Inst Physics; Am Math Soc; Am Chem Soc; Soc Ind & Appl Math. *Res:* Development of catalysts to perform specific chemical transformations; mathematical analysis of models for chemical reactors. *Mailing Add:* Dept Chem Eng Univ Mo 1076 Columbia MO 65211

REUBEN, JACQUES, b Plovdiv, Bulgaria, Mar 17, 36; US citizen; m 63; c 3. POLYMER CHARACTERIZATION. *Educ:* Technion-Israel Inst Technol, BSc, 61, MSc, 65; Weizmann Inst Sci, PhD(phys chem), 69. *Prof Exp:* Teaching fel, Univ Pa, 69-71, vis lectr biophys, 76-77; res assoc, Weizmann Inst Sci, 71-72, sr scientist, 72-76; assoc prof chem, Univ Houston, 77-80; sr res chemist, 80-88, RES SCIENTIST, HERCULES INC, 88- *Concurrent Pos:* Co-ed, Biol Magnetic Resonance, 76- *Mem:* Am Chem Soc; Am Soc Biol Chemists. *Res:* Nuclear magnetic resonance studies of molecules of biological interest, and analysis and description of cellulose derivatives; carbon-13 nuclear magnetic resonance spectroscopy of carbohydrates; interaction of small molecules and ions with biological macromolecules; lanthanide shift reagents. *Mailing Add:* Res Ctr Hercules Inc Wilmington DE 19894

REUBEN, JOHN PHILIP, b Seattle, Wash, Mar 12, 30; m 55; c 3. BIOPHYSICS. *Educ:* Grinnell Col, BA, 54; Univ Rochester, MS, 56; Univ Fla, PhD(physiol), 59. *Prof Exp:* Res assoc, 59-60, NSF fel, 60-62, asst prof, 60-67, assoc prof, 67-78, PROF NEUROL, COL PHYSICIANS & SURGEONS, COLUMBIA UNIV, 78- *Mem:* Biophys Soc; Am Physiol Soc. *Res:* Neurophysiology; neuropharmacology; muscle; molecular studies of synaptic and electrically excitable membranes. *Mailing Add:* Fundamental & Explor Res Bldg 80-3B Merck Sharp & Dohme PO Box 2000 Rahway NJ 07065

REUBEN, RICHARD N, b New York, NY, June 21, 20; m 49; c 3. MEDICINE. *Educ:* Columbia Univ, AB, 40, MD, 43. *Prof Exp:* Assoc prof pediat neurol, 54-70, assoc prof, 70-85, PROF CLIN NEUROL, SCH MED, NY UNIV, 85- *Mem:* Am Acad Neurol; Child Neurol Soc; Am Acad Pediat. *Res:* Neurological disorders of childhood. *Mailing Add:* Dept Neurol NY Univ Sch Med 550 First Ave New York NY 10016

REUBEN, ROBERTA C, b Chicago, Ill, Jan 25, 36; m 55; c 4. PHARMACEUTICAL INDUSTRY. *Educ:* Columbia Univ, BA, 69, PhD(biochem), 73. *Prof Exp:* Postdoctoral molecular biol, Roche Inst Molecular Biol, 73-75; asst prof human genetics, Columbia Univ, 75-80; dir molecular biol, Merck & Co, 80-85; vpres molecular biol, Cistron Biotechnol, 85; dir virol, Schering Plough, 86-87, dir bus develop, 87-89; PRES, JOHNSTON REUBEN ASSOC, 89- *Mem:* Soc Biol Chemists; AAAS. *Res:* Biochemistry; molecular biology. *Mailing Add:* John Reuben Assoc PO Box 857 Far Hills NJ 07931

REUBER, MELVIN D, b Blakeman, Kans, Nov 10, 30. PATHOLOGY, MEDICINE. *Educ:* Univ Kans, AB, 52, MD, 58. *Prof Exp:* Intern path, Sch Med, Univ Md, 58-59, resident, 59-61; res fel, Beth Israel Hosp & Harvard Med Sch, 61-62, asst pathologist & asst instr, 62-63; med officer, Lab Path, Nat Cancer Inst, 63-65; from asst prof to assoc prof path, Sch Med, Univ Md, Baltimore City, 65-74; consult human & exp path, 75-76; pathologist, Frederick Cancer Res Ctr, Md, 76-81; CONSULT HUMAN & EXP PATH, 81- *Concurrent Pos:* Med officer, Lab Biol, Nat Cancer Inst, 65-69 & etiol, 69-71. *Mem:* Am Soc Exp Path; Am Asn Cancer Res; Am Asn Path & Bact; Soc Toxicol; Int Acad Path. *Res:* Hepatic carcinogenesis; toxicology. *Mailing Add:* 11014 Swansfield Rd Columbia MD 21044

REUCROFT, PHILIP J, b Leeds, Eng, Mar 29, 35; m 61; c 3. INDUSTRIAL & MANUFACTURING ENGINEERING, POLYMER PHYSICS. *Educ:* Univ London, BSc, 56, PhD(phys chem) & dipl, Imp Col, 59. *Prof Exp:* Fel phys chem, Nat Res Coun Can, 59-61; res chemist, Franklin Inst Res Labs, Pa, 61-63, sr res chemist, 63-65, sr staff chemist, 65-66, actg lab mgr, 66-67, lab mgr, 67-69; assoc prof, 69-75, PROF MAT SCI, UNIV KY, 75-, DIR, MAT CHARACTERIZATION FACIL, 88- *Concurrent Pos:* Consult, Franklin Inst Res Labs, Pa, 69-; Ashland Oil Found prof, Univ Ky, 70-74, Inst for Mining & Minerals Res fel, 78-; mem adv comt, Am Carbon Soc, 79-85. *Honors & Awards:* Fel Am Soc Metals, 87. *Mem:* Am Soc Metals; Am Chem Soc; Royal Soc Chem; Am Phys Soc; Am Carbon Soc. *Res:* Solid-gas interactions; adsorption; intermolecular forces; thermodynamics; solid state properties of polymers and molecular solids; conductivity; photoconductivity; molecular diffusion; crystallinity and crystal growth; coal science; heterogeneous catalysts. *Mailing Add:* Dept Mat Sci & Eng Univ Ky Lexington KY 40506-0046

REUCROFT, STEPHEN, b Leeds, Eng, May 17, 43; m 70. PARTICLE PHYSICS. *Educ:* Univ Liverpool, BSc, 65, PhD(physics), 69. *Prof Exp:* Demonstr, Univ Liverpool, 65-69; res fel, Europ Orgn Nuclear Res, Switz, 69-71; res assoc, Vanderbilt Univ, 71-73; asst prof res, Vanderbilt Univ, 73-78; STAFF PHYSICIST, EUROP ORGN NUCLEAR RES, GENEVA, SWITZ,

79- *Concurrent Pos:* Vis scientist, Europ Orgn Nuclear Res, 71-78, sci assoc, 78; vis physicist, Fermilab, Chicago, 75- & Brookhaven Nat Lab, 76-; staff scientist, Max-Planck Inst, Munich, 77; adj assoc prof, Vanderbilt Univ, 79-; sci assoc, Europ Coun Nuclear Res, Geneva, 78 & group leader, Europ Parliament; adj prof, Northeastern Univ, 84-85, prof, 86-, chmn, 88- *Mem:* Fel Inst Physics Eng; Am Phys Soc. *Res:* Fundamental structure and basic interactions of sub-nuclear particles. *Mailing Add:* Dept Physics Northeastern Univ 360 Huntington Ave Boston MA 02115

REUDINK, DOUGLAS O, b West Point, Nebr, May 6, 39; m 61. MATHEMATICS, PHYSICS. *Educ:* Linfield Col, BA, 61; Ore State Univ, PhD(math), 65. *Prof Exp:* Mem tech staff, 65-72, head, Satellite Systs Res Dept, 72-79, DIR, RADIO RES LAB, BELL LABS, 79- *Mem:* Am Inst Aeronaut & Astronaut; fel Inst Elec & Electronics Engrs. *Res:* Communications; satellite systems; mobile radio; integral transforms; wave propagation. *Mailing Add:* Boeing Aerospace & Elec PO Box 3999 MS 85-29 Seattle WA 98124

REULAND, DONALD JOHN, b Philadelphia, Pa, May 25, 37; m 60; c 2. FORENSIC CHEMISTRY. *Educ:* St Joseph's Univ, Pa, BS, 59; Carnegie-Mellon Univ, MS, 61, PhD(nuclear-inorg chem), 63. *Prof Exp:* Teaching asst chem, Carnegie-Mellon Univ, 59-61, proj chemist, 61-63; res chemist, Thomas A Edison Res Lab, 63-64; asst prof inorg-nuclear chem, 64-68, assoc prof chem, 68-75, PROF CHEM, IND STATE UNIV, TERRE HAUTE, 75- *Concurrent Pos:* Res chemist, Ames Nat Lab, 67. *Mem:* Sigma Xi. *Res:* Environmental chemistry; forensic chemistry. *Mailing Add:* Dept Chem Ind State Univ Terre Haute IN 47809

REULAND, ROBERT JOHN, b Philadelphia, Pa, Feb 9, 35; m 59; c 2. RADIOCHEMISTRY. *Educ:* St Joseph's Col, Philadelphia, BS, 56; Iowa State Univ, MS, 59, PhD(inorg chem), 63. *Prof Exp:* Instr chem, St Joseph's Col, Philadelphia, 59-60; mem tech staff mat res, Tex Instruments Inc, Dallas, 63-64; PROF CHEM, LORAS COL, IOWA, 64- *Mem:* Sigma Xi. *Res:* Preparation and structure determination of metal tungsten bronzes. *Mailing Add:* Dept Chem Loras Col Dubuque IA 52001

REUNING, RICHARD HENRY, b Wellsville, NY, Jan 3, 41; m 63; c 2. PHARMACOLOGY. *Educ:* State Univ NY, Buffalo, BS, 63, PhD(pharmaceut), 68. *Prof Exp:* USPHS fel pharmacol, Univ Mo-Kansas City, 68-69; from asst prof to assoc prof, 70-80, PROF PHARMACEUT, COL PHARM, OHIO STATE UNIV, 80- *Honors & Awards:* Lyman Award, Am Asn Cols Pharm, 75. *Mem:* AAAS; Acad Pharmaceut Sci; Am Asn Cols Pharm; Am Soc Pharmacol Exp Therapeut. *Res:* Biological drug transport; biopharmaceutics of digitalis glycosides; pharmacokinetics; assay, metabolism and pharmacokinetics of narcotic antagonists; alterations of membrane permeability. *Mailing Add:* Dept Pharm Col Med Ohio State Univ 500 W 12th Ave Columbus OH 43210

REUPKE, WILLIAM ALBERT, b Chicago, Ill, Jan 22, 40; m 83; c 2. SPACE SYSTEMS. *Educ:* Northwestern Univ, BA, 61; Ind Univ, MA, 67; Ga Inst Technol, MS, 73, PhD(nuclear eng), 77. *Prof Exp:* Physicist, NASA Lewis Res Ctr, 63-64; res engr, Lockheed Missiles & Space Co, 67-68; eng physicist, Stanford Linear Accelerator Ctr, 68-71; staff mem, Los Alamos Nat Lab, 77-82; SR ENGR, COMPUTER SCI CORP, 83- *Honors & Awards:* Group Achievement Award, NASA, 85. *Mem:* AAAS; Am Phys Soc; fel Brit Interplanetary Soc; Am Inst Aeronaut & Astronaut; Am Nuclear Soc; Inst Elec & Electronics Engrs. *Res:* Spacecraft ground support systems; interstellar transport and communication; nuclear rocket propulsion; space history. *Mailing Add:* Computer Sci Corp 10110 Aerospace Rd Lanham-Seabrook MD 20706

REUSCH, WILLIAM HENRY, b Carbondale, Ill, Dec 2, 31; m 56; c 3. SYNTHETIC ORGANIC CHEMISTRY. *Educ:* Univ Mich, BS, 53; Columbia Univ, PhD(chem), 57. *Prof Exp:* NSF fel, Imp Col, Univ London, 57-58; from asst prof to assoc prof, 58-68, PROF CHEM, MICH STATE UNIV, 68- *Concurrent Pos:* NIH spec fel, Stanford Univ, 65-66. *Mem:* Am Chem Soc; Royal Soc Chem. *Res:* Natural products and their rational synthesis; strained ring intermediates. *Mailing Add:* Dept Chem Mich State Univ East Lansing MI 48823

REUSS, ROBERT L, b New York, NY, May 31, 42; m 66. GEOLOGY. *Educ:* Ohio Wesleyan Univ, AB, 64; Univ Mich, Ann Arbor, MS, 67, PhD(geol), 70. *Prof Exp:* Asst prof, 69-75, ASSOC PROF GEOL, TUFTS UNIV, 75- *Mem:* Mineral Soc Am; Geol Soc Am; Soc Econ Geologists. *Res:* Igneous and metamorphic petrology; relationship and timing of igneous events, metamorphic reactions and structural deformation. *Mailing Add:* Dept Geol Tufts Univ Medford MA 02155

REUSS, RONALD MERL, b Buffalo, NY, Jan 29, 33; m 54, 79; c 4. HUMAN BIOLOGY, SCIENCE EDUCATION. *Educ:* State Univ NY, Albany, BA, 54, MA, 55; State Univ NY, Buffalo, DEd, 70. *Prof Exp:* Sci teacher gen sci & biol, Kenmore Pub Schs, 55-64; PROF BIOL, STATE UNIV NY BUFFALO, 64-, INSTR ANAT & PHYSIOL & PRE-HEALTH ADV, 78- *Concurrent Pos:* Sci consult, Carson City Schs, 68 & S-K Sci Co, Tonawanda, NY, 77-80 & Matte Polygraph, Buffalo, NY, 84-88; mem res staff, Lung Tumor Antigens, Roswell Park Inst, NY, 79. *Mem:* Nat Sci Teachers Asn; AAAS. *Res:* Individualized instruction in anatomy and physiology and muscle physiology; polygraph validity. *Mailing Add:* State Univ NY Buffalo 1300 Elmwood Ave Buffalo NY 14222

REUSSER, FRITZ, b Steffisburg, Switz, Dec 19, 28; US citizen; m 53; c 2. MOLECULAR BIOLOGY, MICROBIOLOGY. *Educ:* Swiss Fed Inst Technol, Dipl, 53, DSc(microbiol), 55. *Prof Exp:* Res asst microbiol, Swiss Fed Inst Technol, 53-55; Nat Res Coun Can fel, Prairie Regional Lab, Nat Res Ctr, Sask, 55-57; res assoc, Res Labs, 57-71, SR SCIENTIST, RES DIV, UPJOHN CO, KALAMAZOO, 71- *Concurrent Pos:* Mem, adv bd, Pohl Cancer Res Lab, Okla State Univ. *Mem:* Am Chem Soc; Am Soc Microbiol. *Res:* Aids fermentation; genetic engineering; antibiotics; mode of action. *Mailing Add:* 6548 Trotwood Portage MI 49002

REUSSNER, GEORGE HENRY, b Bethlehem, Pa, Dec 18, 18; m 50; c 2. DENTAL RESEARCH, NUTRITION. *Educ:* Lehigh Univ, BA, 40; Purdue Univ, MS, 50. *Prof Exp:* Chemist, Bethlehem Steel Corp, 40-42 & 46-48; assoc chemist, Gen Foods Corp, 49-54, proj leader biochem, 54-60, from chemist to sr chemist, Tech Ctr, 61-70, res specialist, 70-75, sr res specialist, 75-81, prin scientist, Cent Res, Gen Foods Tech Ctr, 81-84; RETIRED. *Mem:* Am Chem Soc; Am Asn Lab Animal Sci; Am Inst Nutrit; Soc Environ Geochem & Health; Int Asn Dental Res. *Res:* Mineral nutrition; dental health and diet; cereal nutrition. *Mailing Add:* 31 Nostrand Dr Toms River NJ 08757-5645

REUSZER, HERBERT WILLIAM, b Jamestown, Mo, Aug 4, 03; wid; c 2. SOIL MICROBIOLOGY. *Educ:* Univ Mo, BS, 25; Rutgers Univ, MS, 30, PhD(soil microbiol), 32. *Prof Exp:* Res specialist agron, NJ Exp Sta, 26-29, asst soil microbiol, 29-32, instr, 32-33; assoc bacteriologist, Exp Sta, Colo State Col, 33-40; coop agent, Soil Conserv Serv, USDA & Exp Sta, Auburn Univ, 40-47; assoc prof, 47-70, EMER PROF AGRON, PURDUE UNIV, 70- *Concurrent Pos:* Asst marine bact, Oceanog Inst Woods Hole, 30-31, jr marine bacteriologist, 32-33. *Mem:* Fel AAAS; Am Soc Microbiol; Soil Sci Soc Am; Am Soc Agron; Brit Biochem Soc; Sigma Xi. *Res:* Soil microbiology; origin and nature of soil organic matter; decomposition of cellulose; nonsymbiotic nitrogen fixation; relation of microorganisms to soil physical properties and liberation of plant nutrients in the soil; microbial decomposition of organotoxicants; axenic growth of plants. *Mailing Add:* Rte 3 Box 71-B Denton MD 21629

REUTER, GERALD LOUIS, b Providence, RI, Mar 16, 34; m 57; c 3. VETERINARY & AGRICULTURAL PHARMACY. *Educ:* RI Col Pharm, BS, 56. *Prof Exp:* Pharmaceut chemist, Hess & Clark Div, Richardson Merrell, Inc, 60-68; sect head vet formulations pharmaceut develop, 68-79, res assoc pharmaceut res & develop, 79-88, HEAD, PARENTERAL PROD, PHARMACEUT SCI, WYETH AYERST LABS, 88- *Mem:* Acad Pharmaceut Sci; Am Pharmaceut Asn; Inst Food Technol; Int Pharmaceut Fedn. *Res:* Veterinary pharmaceutical products, including animal health products and feed medication products; insecticide products research and development; aerosol technology; plant pathogen and piscicide formulations: liquid and parenteral formulation, research and development. *Mailing Add:* 23 Crescent Dr Plattsburgh NY 12901

REUTER, HARALD, b Düsseldorf, Ger, Mar 25, 34; Swiss citizen; m 60; c 3. PHARMACOLOGY. *Educ:* Univ Freiburg, Ger, Med, 59; Univ Mainz, Ger, Dr Med, 60. *Prof Exp:* Asst pharmacol, Univ Mainz, Ger, 60-65, privatdozent, 65-69; dean fac med, 83-85, PROF PHARMACOL, UNIV BERN, SWITZ, 69-, CHMN, 72- *Concurrent Pos:* Vis scientist, Mayo Clin, Rochester, Minn, 67-68; vis prof, Yale Univ, Conn, 78-79 & 86, Japan Soc Promotion Sci, 78 & Beijing Univ, China, 87. *Honors & Awards:* Award for Outstanding Res, Int Soc Heart Res, 84; Marcel-Benoist Prize, Swiss Govt, 85; Schmiedeberg Medal, Ger Pharmacol Soc, 87. *Mem:* Ger Soc Pharmacol & Toxicol; Physiol Soc Gt Brit; Am Physiol Soc. *Res:* Regulation and modulation of ion channels, ion transport and receptors in cell membranes; calcium and cell function; growth and differentiation of neurons. *Mailing Add:* Dept Pharmacol Univ Bern Friedbuehlstr 49 Bern CH-3010 Switzerland

REUTER, ROBERT A, b Dunkirk, NY, Aug 3, 28; m 56; c 3. MATERIAL SCIENCE, PHYSICAL CHEMISTRY. *Educ:* St Bonaventure Univ, BS, 49. *Prof Exp:* Exp chemist, Am Locomotive Co, 49-50; engr, Nat Carbon Co, 53-57; group leader nuclear fuel develop, Carbon Prod Div, Union Carbide Corp, 57-63, asst plant mgr, Nuclear Fuel Prod, Union Carbide Corp, Lawrenceburg, Tenn, 63-71, proj mgr, Polycrystalline Graphite Develop, Union Carbide Corp, Parma, Ohio, 71-73; PRES, DYLON INDUSTS INC, 73- *Mem:* Am Nuclear Soc; Am Ceramic Soc; Am Soc Lubrication Engrs. *Res:* Uranium carbide nuclear fuels; chemical vapor deposition coatings; high temperature materials processing; polycrystalline graphite production; refractory cements; solid state lubricants. *Mailing Add:* Dylon Industs Inc 7700 Clinton Rd Cleveland OH 44144-1045

REUTER, ROBERT CARL, JR, b Pittsburgh, Pa, Apr 30, 39; m 61; c 3. THEORETICAL & APPLIED MECHANICS. *Educ:* Univ Ill, BS, 64, MS, 65, PhD(appl mech), 67. *Prof Exp:* Mem tech staff struct mech, Martin Marietta Corp, 67-68; mem tech staff, 68-79, SUPVR APPL MECH DIV 1544, SANDIA NAT LABS, 79- *Res:* Development and application of the mechanics of composite materials with an emphasis on residual stresses in composites; mechanical and thermomechanical analysis of heterogeneous, orthotropic, wound structures; evaluation of the effects of processing parameters on mechanical states in wound structures; supervise technical staff in numerous areas of applied mechanics. *Mailing Add:* Sandia Nat Labs Div 1544 PO Box 5800 Albuquerque NM 87185

REUTER, STEWART R, b Detroit, Mich, Feb 14, 34; m 66. RADIOLOGY. *Educ:* Ohio Wesleyan Univ, AB, 55; Case Western Reserve Univ, MD, 59; Am Bd Radiol, dipl, 64; San Francisco Univ, JD, 80. *Prof Exp:* Intern med, Hosp, Univ Calif, 59-60, researcher, 60-63; instr radiol, Stanford Univ, 63-64; Picker res fel, 64-66; from asst prof to assoc prof, Univ Mich, 66-69; assoc prof, Univ Calif, San Diego, 69-72; prof radiol, Univ Mich, Ann Arbor, 72-75; prof radiol & vchmn radiol, Univ Calif, Davis, 76-80 & prof radiol, Univ Calif, San Francisco, 76-80; chief radiol, Martinez Vet Admin Hosp, 76-80; PROF RADIOL & CHMN DEPT, UNIV TEX HEALTH SCI CTR, SAN ANTONIO, 80- *Res:* Angiography, particularly visceral circulation and development of techniques to improve visceral angiography; therapeutic angiography. *Mailing Add:* 3923 Morgans Creek San Antonio TX 78230

REUTER, WILHAD, b Flensburg, Ger, Jan 2, 30; m 53; c 2. ANALYTICAL CHEMISTRY. *Educ:* Univ Mainz, BS, 54, MS, 56, PhD, 58. *Prof Exp:* Analytical chemist, O Hommel Co, 58 & Keystone Carbon Co, 59; anal chemist, Int Bus Mach Corp, NY, 60-70, ANALYTICAL CHEMIST, INT BUS MACH CORP RES CTR, 70- *Mem:* Am Chem Soc. *Res:* Instrumental analytical chemistry. *Mailing Add:* IBM Res Ctr PO Box 218 Yorktown Heights NY 10598

REUTER, WILLIAM L(EE), b Hartford, SDak, July 21, 34; m 59; c 2. ELECTRICAL ENGINEERING. *Educ:* SDak Sch Mines & Technol, BS, 56, MS, 58; Iowa State Univ, PhD(elec eng), 67. *Prof Exp:* From instr to assoc prof elec eng, SDak Sch Mines & Technol, 56-72; mgr res & develop, Dunham Assocs, Inc, 72-79; pres, Rapidata, Inc, 73-79; chief engr, 79-82, VPRES OPERS, PETE LIEN & SONS, INC, 82- *Concurrent Pos:* Vpres, Res Specialists Inc, 69-80, bd dir, 70-90. *Mem:* Inst Elec & Electronics Engrs; Sigma Xi; Nat Soc Prof Engrs. *Res:* Systems; network theory; numerical methods; computer programming; process control; signal processing. *Mailing Add:* 3402 Fairhaven Rapid City SD 57702

REUTHER, THEODORE CARL, JR, b Wheeling, WVa, Apr 16, 33; m 55; c 2. METALLURGICAL ENGINEERING, NUCLEAR MATERIALS. *Educ:* Carnegie Inst Technol, BS, 56, MS, 58; Cath Univ Am, DrEngr, 65. *Prof Exp:* Res metallurgist, US Naval Res Lab, 59-68; metall engr, Div Reactor Develop & Technol, USAEC. 68-75, metal engr, Off Fusion Energy, US Dept Energy, 75-90; SR DIR DEVELOP STAFF, OAK RIDGE NAT LAB, 90- *Mem:* Metall Soc; Am Soc Metals. *Res:* Diffusion; crystal growth and defects; refractory metals and alloys; creep; nuclear metallurgy; grain boundary energy and structure; irradiation effects; fusion reactor materials development. *Mailing Add:* Oak Ridge Nat Lab Washington Off 11 Clemson Ct Rockville MD 20850

REUTHER, WALTER, b Manganoui, NZ, Sept 21, 11; nat US; m 35; c 2. HORTICULTURE. *Educ:* Univ Fla, BS, 33; Cornell Univ, PhD(plant physiol), 40. *Prof Exp:* Instr res, Univ Fla, 33-36; asst horticulturist, Citrus Exp Sta, Univ Fla, 36-37; asst, Cornell Univ, 37-40, asst prof pomol, 40; from assoc horticulturist to prin horticulturist, USDA, 40-55; sr horticulturist in-chg, US Date Garden, Calif, 41-46; head dept hort, Univ Fla, 55-56; chmn dept, 56-66, prof, 56-77, EMER PROF HORT, UNIV CALIF, RIVERSIDE, 77- *Concurrent Pos:* Mem adv comt citrus & subtrop fruit res, USDA, 62-63 & hort crops res, 63-69; mem eval comt sr fels, Nat Acad Sci, 64-; mem comt trop studies; consult to USAID, FAO, UNDP, Rockefeller Found, World Bank and var other govts & pvt insts, 63-; chmn bd, Am Soc Hort Sci, 63-64. *Mem:* AAAS; fel Am Soc Hort Sci (vpres, 61-62, pres, 62-63); Am Soc Plant Physiol; Am Pomol Soc (vpres, 65-66). *Res:* Mineral nutrition of tree fruits; relation of leaf analysis to nutritional status of citrus; water relations and irrigation of tree fruits; toxicity of copper in citrus orchard soils; influence of climate on citrus. *Mailing Add:* Dept Bot & Plant Sci Univ Calif Riverside CA 92521

REUWER, JOSEPH FRANCIS, JR, b Harrisburg, Pa, May 31, 31; m 58; c 2. TRIBOLOGY, CERAMICS. *Educ:* Lehigh Univ, BS, 53; Mass Inst Technol, SM, 57; Univ NH, PhD(org chem), 62. *Prof Exp:* Res chemist, Armstrong World Industs, Inc, 62-65, res supvr, 65-69, sr scientist, 69-90, SR PRIN SCIENTIST, ARMSTRONG WORLD INDUSTS, INC, 90- *Concurrent Pos:* Adj prof, Franklin & Marshall Col, 64-66. *Mem:* Am Chem Soc. *Res:* Tribology of polymeric surfaces, plasma and corona chemistry, ceramics and reaction mechanisms. *Mailing Add:* 144 N School Lane Lancaster PA 17603

REVAY, ANDREW W, JR, b New Kensington, Pa, Oct 8, 33; m 57; c 2. ELECTROMAGNETICS, ENGINEERING EDUCATION. *Educ:* Univ Pittsburgh, BS, 55, MS, 56, PhD(elec eng), 63. *Prof Exp:* Asst prof elec eng, Univ Pittsburgh, 59-64, assoc res prof, 64-67; assoc prof elec eng, Fla Inst Technol, 67-69, head dept, 71-80, head dept mech eng, 72-78, assoc dean res, 77-80, dean sci & eng, 80-86, PROF ELEC ENG, FLA INST TECHNOL, 69-, VPRES, ACAD AFFAIRS, 86- *Concurrent Pos:* Consult adv bd hardened elec power systs, Nat Acad Sci, 63-70 & Harris Corp, 68-86; prin investr res contracts. *Honors & Awards:* Centennial Medal, Inst Elec & Electronics Engrs. *Mem:* Inst Elec & Electronics Engrs; Am Soc Eng Educ; Nat Soc Prof Eng; Sigma Xi. *Res:* Electromagnetic field theory and application; lightning and electromagnetic pulse protection. *Mailing Add:* FIT-VP Acad Affairs W Univ Blvd Melbourne FL 32901-6988

REVEAL, JAMES L, b Reno, Nev, Mar 29, 41; m 61, 78; c 3. PLANT TAXONOMY, SYSTEMATIC BOTANY. *Educ:* Utah State Univ, BS, 63, MS, 65; Brigham Young Univ, PhD(bot), 69. *Prof Exp:* Asst prof, 69-74, assoc prof, 74-81, PROF BOT, UNIV MD, COLLEGE PARK, 81- *Concurrent Pos:* Res assoc, Smithsonian Inst, 70-; secy-gen, Int Cong Syst & Evoluntionary Biol, 73-85; co-pres, Int Cong Syst & Evolutionary Biol, 86- *Mem:* AAAS; Am Inst Biol Sci; Bot Soc Am; Am Soc Plant Taxon; Int Asn Plant Taxon; Sigma Xi; fel Linnean Soc. *Res:* Floristic studies in intermountain West, vascular plants of North America, northern Mexico and state of Maryland; monographical studies in Eriogonum and related genera; botanical nomenclature. *Mailing Add:* Dept Bot Univ Md College Park MD 20742-5815

REVEL, JEAN PAUL, b Strasbourg, France, Dec 7, 30; nat US; m 57, 86; c 3. CELL BIOLOGY. *Educ:* Univ Strasbourg, France, BS, 49; Harvard Univ, PhD, 57. *Prof Exp:* Whitney fel anat, Med Col, Cornell Univ, 57-58, res assoc, 58-59; instr, Harvard Med Sch, 61, assoc, 61-63, from asst prof to prof, 63-71; prof, 71-78, ALBERT BILLINGS RUDDOCK PROF BIOL, CALIF INST TECHNOL, 78- *Concurrent Pos:* Mem, Molecular Biol Study Sect, NIH, 70-74 & mem, Nat Adv Coun, Div Res Resources, 84-89; mem, Develop Biol Panel, NSF, 76-80; mem bd sci adv, Nat Inst Aging,; chair sect G, AAAS, 91- *Mem:* AAAS; Am Asn Anat; Am Soc Cell Biol (pres, 72); Soc Develop Biol; Electron Micros Soc Am (pres, 88). *Res:* Correlation between structure and function; ultrastructural cytochemistry; investigation of cell to cell communication; structure of channel proteins by molecular biology and cytology; structure-function relationships; electron microscopy. *Mailing Add:* Div Biol 156-29 Calif Inst Technol Pasadena CA 91125

REVELANTE, NOELIA, b Rovinj, Croatia, Yugoslavia, Jan 17, 42; m 74; c 1. BIOLOGICAL OCEANOGRAPHY. *Educ:* Univ Zagreb, Yugoslavia, BSc, 66, MSc, 70, PhD, 74. *Prof Exp:* From res asst to sr assoc res scientist, Ctr Marine Res, Inst Rudjer Boskovic, Yugoslavia, 66-75; sr scientist, Australian Inst Marine Sci, 76-78; assoc res prof, 78-90, RES PROF, UNIV MAINE, 90- *Concurrent Pos:* Int Comn Sci Explor Mediter Sea; Int Atom Energy Fel, 74-75, Queen's Fel, Australia, 84-85. *Mem:* AAAS; Am Soc Limnol & Oceanog; Int Phycol Soc. *Res:* Pico- and nanoplankton ecology and taxonomy; primary and secondary aquatic production. *Mailing Add:* Dept Zool Univ Maine Orono ME 04469

REVELL, JAMES D(EWEY), b Toledo, Ohio, Feb 17, 29; m 55; c 2. AIRCRAFT NOISE CONTROL, AERODYNAMICS. *Educ:* Univ Calif, Los Angeles, BS, 52, MS, 58, PhD(eng), 66. *Prof Exp:* Thermodynamicist, Northrup Aircraft, Inc, Calif, 52-54, aerodynamicist, 54-57; sr tech specialist, NAm Aviation, Inc, 57-60; sr specialist dynamics & loads, Norair Div, Northrup Corp, 60-62, mem tech mgt, 62-65; res develop scientist, 65-75, RES & DEVELOP SCIENTIST ACOUST, LOCKHEED AIRCRAFT CORP, CALIF, 75- *Mem:* Assoc fel Am Inst Aeronaut & Astronaut; Acoust Soc Am. *Res:* Fluid mechanics; unsteady aerodynamics; acoustics of moving fluids; aerodynamic noise and jet noise theory; turbulence; boundary layer theory; gas dynamics; scattering refraction; aeroelasticity; noise transmission through structures. *Mailing Add:* 23853 Arroyo Park Dr Apt 311 Valencia CA 91355

REVELLE, CHARLES S, b Rochester, NY, Mar 26, 38; m 62; c 2. ENVIRONMENTAL SCIENCE, ARMS CONTROL POLICY. *Educ:* Cornell Univ, BChE, 61, PhD(civil eng), 67. *Prof Exp:* Res assoc, Ctr Environ Qual Mgt, Cornell Univ, 67, asst prof environ systs eng, 67-70; assoc prof, 70-75, PROF GEOG & ENVIRON ENG, JOHNS HOPKINS UNIV, 75-, COORDR OPERS RES GROUP, 78- *Concurrent Pos:* Vis asst prof, Johns Hopkins Univ, 68-69; Fulbright award, Erasmus Univ, Rotterdam, Neth, 75. *Mem:* Regional Sci Asn; Inst Mgt Sci; Opers Res Soc Am; Am Geophys Union; AAAS. *Res:* Applications of systems analysis and operations research to environmental and public problems such as water quality and water quanity management, public health systems; modeling of urban and regional problems, especially location systems such as ambulance and fire protectionn; modelling of arms control alternatives. *Mailing Add:* Ames Hall Johns Hopkins Univ Baltimore MD 21218

REVELLE, ROGER (RANDALL DOUGAN), oceanography, science policy; deceased, see previous edition for last biography

REVESZ, AKOS GEORGE, b Balassagyarmat, Hungary, July 25, 27; US citizen; m 56, 75; c 1. SOLID STATE CHEMISTRY. *Educ:* Budapest Tech Univ, Dipl Ing, 50, Dr Ing, 68. *Prof Exp:* Res asst x-ray crystallog, Phys Chem Inst, Budapest Tech Univ, 49; staff mem thermodyn iron metall, Iron & Metal Res Inst, Budapest, 50; staff mem semiconductors, Tungsram Co, 51-54, dept head, 55-56; staff mem anodic oxide films & solid state capacitors, Philips Co, Neth, 57-59 & growth & properties thin films, RCA Labs, 60-69; mem tech staff, Comsat Labs, 69-72, dept head, 72-74, sr staff scientist, 74-81, sr scientist, 81-84; CONSULT SOLID STATE & MAT SCI, REVESZ ASSOC, 84- *Concurrent Pos:* Vis res prof, Howard Univ, Wash, DC, 84- *Mem:* Fel Am Inst Chemists; Electrochem Soc. *Res:* Semiconductor and solid state devices; oxidation of semiconductors; insulator-semiconductor interfaces; properties of thin oxide films; noncrystalline solids. *Mailing Add:* 7910 Park Overlook Dr Bethesda MD 20817

REVESZ, GEORGE, b Budapest, Hungary, July 29, 23; nat US; m 48; c 2. RADIOLOGY, ELECTRICAL ENGINEERING. *Educ:* Swiss Fed Inst Technol, MS, 48; Univ Pa, PhD, 64. *Prof Exp:* Develop engr, Salford Labs, Gen Elec Co, Eng, 49-52; mem staff res, Brit Rayon Res Asn, 52-54; sr proj engr, 54-57, sect head, 57-59, tech dir, Robertshaw-Fulton Controls Co, 59-61; res sect mgr, Philco Corp, Pa, 61-62, mgr instrumentation, Microelectronics Div, 62-68; assoc prof, 66-76, PROF RADIOL, SCH MED, TEMPLE UNIV, 76-, ADJ PROF SCH ENG, 84- *Honors & Awards:* Bowen Award, Brit Inst Phys, 54; Stauffer Award, Asn Univ Radiologists, 82. *Mem:* Asn Univ Radiologists; Am Asn Physicists Med; Optical Soc Am. *Res:* Measuring and improving diagnostic accuracy in medicine; computer analysis of medical images; medical decision making and computers. *Mailing Add:* Dept Radiol Sch Med Temple Univ Philadelphia PA 19140

REVESZ, ZSOLT, b Tatabanya, Hungary, Mar 18, 50; Swiss citizen; m 88. COMPUTATIONAL EXPERIMENTS, ENGINEERING ANALYSIS. *Educ:* Voeresmarty Gymnazium, Budapest, BA, 68; Tech Univ Budapest, dipl eng, 74; Century Univ, DSc(eng), 80. *Prof Exp:* Res scientist & assoc prof computer sci, Tech Univ Budapest, 74-75; develop engr, Brown, Boveri & Cie Co, Switz, 75-78; eng analyst & proj mgr, Electrowatt Eng Serv, Switz, 79-89; GEN MGR, REVESZ & ASSOCS, SWITZ, 85- *Concurrent Pos:* Fac adv, Century Univ, Beverly Hills & Albuquerque, 80-; res scientist & sr asst energy technol, Swiss Fed Inst Technol, 89-91. *Mem:* Am Soc Mech Engrs; Am Nuclear Soc. *Res:* Engineering application in structural mechanics, piping analysis, computational fluid flow-heat transfer and energy systems; developments for analysis algorithms, computer graphics, nuclear and non-nuclear electricity generating devices; author of over 60 publications. *Mailing Add:* Assoc Consult Engrs Revesz & Assocs PO Box 1126 Baden CH-5401 Switzerland

REVETTA, FRANK ALEXANDER, b Monongahela, Pa, June 18, 28; m 61; c 2. GEOPHYSICS. *Educ:* Univ Pittsburgh, BS, 53; Ind Univ, MA, 57; Univ Rochester, PhD(geophys), 70. *Prof Exp:* Geophysicist, Geophys Serv Inc, 57-58; teacher earth sci, Elizabeth-Forward High Sch, 59-62; from instr to assoc prof, 62-77, PROF GEOL, STATE UNIV NY COL POTSDAM, 77- *Mem:* Am Geophys Union; Nat Asn Geol Teachers. *Res:* Gravity and magnetic surveys; interpretation of magnetic anomalies, seismology with emphasis on regional seismic networks, seismic and electrical methods of prospecting. *Mailing Add:* Dept Geol Sci State Univ NY Col Piierrpont Ave Potsdam NY 13676

REVILLARD, JEAN-PIERRE RÉMY, b Suresnes, France, Jan 12, 38; m 61; c 2. IMMUNOPHARMACOLOGY, MUCOSAL IMMUNITY. *Educ:* Claude Bernard Univ, Lyon, MD, 64, MSc, 62, cert immunol, 73. *Prof Exp:* Postdoctoral immunol, NIH, NY Univ Med Sch, 66-67; asst nephrology,

Lyon's Hosp, 68-73; asst prof immunol, 73-81, dean biol, Human Biol Fac, 77-84, COORDR, DEPT IMMUNOL, CLAUDE BERNARD UNIV, LYON, 74-, PROF, 81- *Concurrent Pos:* Spec adv on PhD biol, Minister Educ, 82-85; dir, Uro-Nephro Transplantation Clin Immunol, Res Unit, NIH & Med Res, Nat Ctr Sci Res, 86-; vpres, Educ Comt Int, Union Immunol Socs, 89- *Mem:* Am Asn Immunologists; Brit Soc Immunol; Transplantation Soc; Europ Dialysis & Transplant Asn. *Res:* Basic and clinical immunology; new treatments with monoclonal antibodies or cytokines, mucosal immunity, B and T lymphocyte activation. *Mailing Add:* Hop E Herriot Pavillon P Lyon 69437 France

REVOILE, SALLY GATES, b Pittsburgh, Pa. AUDIOLOGY. *Educ:* Univ Md, BA, 62, MA, 65, PhD(hearing sci), 70. *Prof Exp:* Res audiologist, Vet Admin Hosp, Wash, DC, 62-76; ASST PROF AUDIOL RES, DEPT HEARING RES, GALLAUDET COL, 76- *Mem:* Am Speech & Hearing Asn; Acoust Soc Am. *Res:* Use of hearing aids by the hearing impaired. *Mailing Add:* 4112 Culver St Kensington MD 20895

REVZIN, ALVIN MORTON, b Chicago, Ill, Nov 8, 26; m 56. PHARMACOLOGY, NEUROPHYSIOLOGY. *Educ:* Univ Chicago, SB, 47, SM, 48; Univ Colo, PhD(physiol), 57. *Prof Exp:* Instr physiol, Med Col SC, 48-49; asst, Col Dent, NY Univ, 49-50; asst, Child Res Coun, Denver, Colo, 51-55; med res assoc, Galesburg State Res Hosp, 57-60; pharmacologist, Nat Heart Inst, 60-63; PHARMACOLOGIST, CIVIL AEROMED INST, FED AVIATION ADMIN, 63- *Concurrent Pos:* Adj prof pharmacol, 63-71, prof pharmacol & psychiat, Univ Okla, 71- *Mem:* Am Soc Pharmacol & Exp Therapeut; Soc Neurosci; Asn Res Vision Ophthal; Aerospace Med Asn; Human Factors Soc. *Res:* Neuropharmacology of psychotomimetic compounds; neurotoxicity of pesticides and environmental pollutants; comparative neurology of the avian brain; bioeffects of nonionizing electromagnetic radiation; vision and performance; vision. *Mailing Add:* Civil Aeromed Inst AAM-621 Fed Aviation Admin PO Box 25082 Oklahoma City OK 73125

REVZIN, ARNOLD, B Chicago, Ill, Jan 23, 43; m 66; c 2. MOLECULAR BIOLOGY, BIOPHYSICAL CHEMISTRY. *Educ:* Univ Mich, Ann Arbor, BSc(chem eng) & BSc(eng math), 64; Univ Wis-Madison, PhD(chem), 69. *Prof Exp:* Res assoc, Enzyme Inst, Univ Wis, 69-70; res fel, polymer dept, Weizmann Inst, Israel, 70-72; res fel, Max Planck Inst Biophys Chem, Germany, 72-73; NIH fel & res assoc, Inst Molecular Biol, Univ Ore, Eugene, 73-75; asst prof, 75-81, assoc prof, 81-86, PROF BIOCHEM, MICH STATE UNIV, 86-, ASSOC DEAN, COL NAT SCI, 87- *Concurrent Pos:* Dir, biochem prog, NSF, 84-85. *Mem:* Biophys Soc; Am Soc Biol Chemists. *Res:* Physical and biochemical studies of nucleic acid-protein interactions involved in regulating transcription. *Mailing Add:* Dept Biochem Mich State Univ East Lansing MI 48824-1319

REW, ROBERT SHERRARD, biological chemistry, pharmacology, for more information see previous edition

REWCASTLE, NEILL BARRY, b Sunderland, Eng, Dec 12, 31; Can citizen; m 58; c 4. NEUROPATHOLOGY. *Educ:* St Andrews Univ, MB, ChB, 55; Univ Toronto, MA, 62; Royal Col Physicians & Surgeons, Can, cert, 62, FRCP(C), 68. *Prof Exp:* Lectr path, Univ Toronto, 64-68, actg head div neuropath, Banting Inst, 65-69, head div, 69-81, from assoc prof to prof neuropath, dept path, 69-81; pathologist, Toronto Gen Hosp, 64-81; PROF PATH & HEAD DEPT, UNIV CALGARY, 81-; DIR DEPT HISTOPATH, FOOT HILLS HOSP, CALGARY, 81- *Concurrent Pos:* Res fel path, Med Res Coun Can, 60-64; Med Res Coun Can & Muscular Dystrophy Asn Can res fels, 64-70. *Mem:* Am Asn Neuropath; Can Asn Path; Can Asn Neuropath (secy, 65-69, pres, 77-79); Can Med Asn. *Res:* Human nervous system diseases; skeletal muscle diseases; electron microscopy. *Mailing Add:* Dept Path Foothill Gen Hosp 1403 29th St NW Calgary AB T2N 2T9 Can

REWOLDT, GREGORY, b Ann Arbor, Mich, Apr 21, 48; m 84. PLASMA PHYSICS. *Educ:* Calif Inst Technol, BS, 70; Mass Inst Technol, PhD(physics), 74. *Prof Exp:* Physicist, Res Lab Electronics, Mass Inst Technol, 74-75; res physicist, 75-89, PRIN RES PHYSICIST, PLASMA PHYSICS LAB, PRINCETON UNIV, 89- *Concurrent Pos:* Lectr, dept astrophys sci, Princeton Univ, 79-84, assoc prof, 84-85. *Mem:* Am Phys Soc. *Res:* Theoretical plasma physics, especially drift and trapped particle instabilities, and stellarator MHD computations. *Mailing Add:* Plasma Physics Lab Princeton Univ PO Box 451 Princeton NJ 08543

REX, ROBERT WALTER, b New York, NY; m 52. EXPLORATION GEOLOGY. *Educ:* Univ Costa Rica, BS, 46; Harvard Univ, AB, 51; Stanford Univ, MS, 53; Univ Calif, PhD(oceanog), 58. *Prof Exp:* Geologist, US Geol Surv, 51-53; oceanogr, US Navy Electronics Lab, 53; geologist, Scripps Inst Oceanog, Univ Calif, San Diego, 54-57; sr res assoc geochem, Chevron Oil Field Res Co, Standard Oil Co Calif, 58-67; prof geol sci, Univ Calif, Riverside, 67-72; vpres explor, Pac Energy Corp, Hughes Aircraft Co, Marina Del Rey, Calif, 72-73; pres, Repub Geothermal Inc, Santa Fe Springs, 73-83, chmn bd, 83-85; CONSULT VENTURE CAPITAL & INVEST BANKING, 85- *Concurrent Pos:* Dir geothermal resources prog, Univ Calif, Riverside, 68-72; res geologist, Inst Geophys & Planetary Physics, 67-72, asst dir, 71-72; chmn, Geothermal Adv Bd, Univ Calif, 70-72; consult, Jet Propulsion Lab, Calif Inst Technol, 63-65; consult, US Bur Reclamation; res affil, Inst Geophys, Univ Hawaii, 69-; mem, Tech Adv Comt, Calif Geothermal Resources Bd, 71-; consult, Los Alamos Sci Lab, 71- & Oak Ridge Nat Lab, 72-; mem, President's Panel, Off Sci & Technol, 71-73 & Nat Adv Panel, Hawaii Geothermal Proj, 73-; dir, Geothermal Resources Coun, 76-86. *Mem:* Geol Soc Am; Am Geophys Union; AAAS; Geochem Soc; Geothermal Resources Coun; Mineral Soc Am. *Res:* Exploration and resource assessment of geothermal resources; oil and gas exploration; economic geology; mineral-water interactions; mineralogy of deep sea sediments and atmospheric dust; x-ray powder diffraction; geochemistry of geothermal systems; clays and clay minerals; exploration systems, strategy, economics and technology; energy resource and technology evaluation for venture capital and investment banking. *Mailing Add:* 2780 Casalero Dr La Habra Heights CA 90631

REXER, JOACHIM, b New York, NY, Nov 30, 28; m 62; c 2. METALLURGY. *Educ:* Brooklyn Col, BA, 52; Iowa State Univ, MS, 59, PhD(metall), 62. *Prof Exp:* Jr chemist, Ames Labs, Iowa State Univ, 52-54 & 57, asst metall, 57-62; metallurgist, Parma Tech Ctr, Union Carbide Corp, 62-71, sr res scientist, Carbon Prod Div, 71-82, mgr, Micros Dept, 82-88; CONSULT, NASA SPACE STA FREEDOM, ANALEX CORP, 89- *Honors & Awards:* Recipient of Certs Recognition Creative Develop Tech Innovations, NASA, 82. *Mem:* Am Soc Metals. *Res:* High temperature reaction kinetics; chemical vapor deposition; free space reactions; vacuum and pressure technology; powder metallurgy; high temperature processing in hydrogen atmospheres. *Mailing Add:* 9269 Highland Dr Cleveland OH 44141

REXFORD, DEAN R, b Orleans, Vt, Oct 1, 15; m 43; c 3. FLUORINE CHEMISTRY. *Educ:* Norwich Univ, BS, 37; Swiss Fed Inst Tech, Dr Tech, 43. *Prof Exp:* Chemist dyestuffs, Ciba Co, Inc, NY, 37-39; Rockefeller Res Asn fel, Princeton Univ, 45-46, instr chem, 46-50, asst prof, Frick Chem Lab, 50-53; from res chemist to sr patent chemist, E I Du Pont de Nemours & Co, Inc, 56-74, patent assoc, 74-80; PATENT CONSULT, UNIV DEL & INDUST, 80- *Mem:* Am Chem Soc; Am Inst Chemists. *Mailing Add:* 2323 W 16th St Wilmington DE 19806

REXFORD, EVEOLEEN NAOMI, psychiatry, psychoanalysis, for more information see previous edition

REXROAD, CAIRD EUGENE, JR, b Fairmont, WVa, Jan 6, 47; m 68; c 2. REPRODUCTIVE PHYSIOLOGY. *Educ:* WVa Univ, BS, 68; Univ Wis-Madison, MS, 72, PhD(reproductive physiol & endocrinol), 74. *Prof Exp:* RES PHYSIOLOGIST DAIRY CATTLE, REPRODUCTION LAB, ANIMAL SCI INST, BELTSVILLE AGR RES CTR, MD, 74- *Mem:* Am Soc Animal Sci; Soc Study Reproduction; Am Dairy Sci Asn. *Res:* Endocrine regulation of sperm transport and uterine motility; transfer of genes into farm animals. *Mailing Add:* 321 Old Line Ave Laurel MD 20724

REXROAD, CARL BUCKNER, b Columbus, Ohio, Apr 2, 25; m 51; c 1. GEOLOGY. *Educ:* Univ Mo, BA, 49, MS, 50; Univ Iowa, PhD(geol), 55. *Prof Exp:* Instr geol, La Tech Univ, 50-53; asst prof, Tex Tech Univ, 55-58; assoc prof, Univ Houston, 58-61; PALEONTOLOGIST, IND GEOL SURV, 61- *Concurrent Pos:* Adj prof, Ind Univ, 79-; chief, Pander Soc, 85-90; Ill State Geol Surv Res Affil, 57-61; Vis prof, Univ Iowa, 67-68. *Mem:* Soc Econ Paleontologists & Mineralogists; Am Asn Petrol Geol; Pander Soc; Paleontological Soc. *Res:* Carboniferous, Silurian and Ordovician conodonts and related stratigraphy. *Mailing Add:* Ind Geol Surv 611 N Walnut Grove Bloomington IN 47405

REY, CHARLES ALBERT, b Oklahoma City, Okla, Apr 8, 34; m 60, 89; c 2. CONTAINERLESS PROCESSING, MICROGRAVITY RESEARCH IN SPACE. *Educ:* Univ Chicago, AB, 56, BS, 57, MS, 59, PhD(physics), 64. *Prof Exp:* Res assoc physics, Enrico Fermi Inst Nuclear Studies, 63-64; physicist, Lawrence Radiation Lab, Univ Calif, 64-70; asst prof physics, Univ Notre Dame, 70-74; dir res, Interand Corp, 77-78; vpres, 78-86, PRES INTERSONICS, INC, 86- *Mem:* AAAS; Am Phys Soc; Sigma Xi; Am Ceramic Soc. *Res:* Acoustics; materials processing; communications engineering; high energy physics; elementary particle structure; particle detectors; electronics; acoustic levitation; containerless processing; microgravity research; materials properties management. *Mailing Add:* 1332 Woodland Lane Riverwoods IL 60015

REY, WILLIAM K(ENNETH), b New York, NY, Aug 11, 25; m 46; c 2. AEROSPACE ENGINEERING. *Educ:* Univ Ala, BSAE, 46, MSCE, 49. *Prof Exp:* Asst & instr math, Univ Ala, 46-47, from instr to asst prof eng mech, 47-52, from assoc prof to prof aeronaut eng, 52-60, PROF AEROSPACE ENG, UNIV ALA, 60-, ASST DEAN ENG, 76- *Concurrent Pos:* Proj dir res projs, NASA. *Mem:* Am Soc Eng Educ; assoc fel Am Inst Aeronaut & Astronaut; Nat Soc Prof Engrs. *Res:* Properties of materials under various environments, especially fatigue behavior at elevated temperatures; analytical and experimental determination of stress distributions in aircraft and missile components. *Mailing Add:* Dept Aerospace Eng Univ Ala Box 870200 Tuscaloosa AL 35487-0200

REYER, RANDALL WILLIAM, b Chicago, Ill, Jan 23, 17; m 43; c 2. DEVELOPMENTAL BIOLOGY. *Educ:* Cornell Univ, BA, 39, MA, 42; Yale Univ, PhD(zool), 47. *Prof Exp:* Lab asst biol, Yale Univ, 42-46, instr zool, 47-50; instr biol, Wesleyan Univ, 46-47; asst prof anat, Sch Med, Univ Pittsburgh, 50-57; from assoc prof to prof, 57-87, actg chmn dept, 77-78, EMER PROF ANAT, SCH MED, WVA UNIV, 87- *Concurrent Pos:* Prin investr res grants, NIH, 51-82. *Mem:* Am Soc Zoologists; Soc Develop Biol; Am Asn Anat; Int Soc Develop Biol; Am Inst Biol Sci; Differentiation Soc. *Res:* Embryonic development and regeneration of the crystalline lens and neural retina in Amphibia; microsurgical, light, electron microscopic and autoradiographic techniques; embryology; microscopic; regeneration. *Mailing Add:* Dept Anat WVa Univ Sch Med 4052 HSN Morgantown WV 26506

REYERO, CRISTINA, b Madrid, Spain, Mar 15, 49; Austrian citizen. CELL PHYSIOLOGY. *Educ:* Complutense Univ, Madrid, Spain, BS, 65, MS, 71, PhD(biochem), 74; Univ Vienna, PhD(biochem), 75. *Prof Exp:* Res assoc, L Boltzmann Inst Leukemia Res, Vienna, Austria, 74-75, dept path, div immunol, Sch Med, Yale Univ, 77-79 & dept dermat, 80, dept anat II, Univ Heidelberg, Fed Repub Germany, 81-83; asst prof biochem, Vienna Univ Vet Sci, 75-77; asst prof immunil & cell biol, Dept Biol, Northeastern Univ, 83-88; FOUNDER & PRES, INTERTECH VENTURES, LTD, 91- *Mem:* Am Soc Cell Biol; Am Asn Immunologists. *Res:* Regulation of cell growth and differentiation induced by low molecular weight products of tumor lymphoid cells. *Mailing Add:* 70 Walnut St Wellesley MA 02181

REYES, ANDRES ARENAS, b Bongabon, Philippines, Feb 26, 31; Can citizen; m 58; c 3. PLANT PATHOLOGY, AGRICULTURAL MICROBIOLOGY. *Educ:* Araneta Univ, Philippines, BSA, 54; Wash State Univ, Pullman, MSA, 58; Univ Wis-Madison, PhD(plant path), 61. *Prof Exp:* Instr plant path, Araneta Univ, Philippines, 54-55; plant quarantine officer, Bur Plant Indust, Philippines, 55-57; plant pathologist, Philippine Packing Corp, Bukidnon, 61-63; res scientist plant path, Nat Inst Sci & Technol, Philippines, 63-65; RES SCIENTIST PLANT PATH, AGR CAN, 67- *Concurrent Pos:* Res asst, Univ Wis-Madison, 57-61; lectr, Araneta Univ, Philippines, 64-65 & Adamson Univ, Philippines, 64-65; fel, Tex A&M Univ, 65-67. *Honors & Awards:* Spec Citation Plant Protection, Philippine Soc Advan Sci, 64. *Mem:* Am Phytopath Soc; Am Soc Hort Sci; Can Phytopath Soc. *Res:* Postharvest pathology of vegatable crops. *Mailing Add:* Agr Can Vineland Station ON L0R 2E0 Can

REYES, PHILIP, b Tulare, Calif, Sept 5, 36; m 62; c 3. NUCLEOTIDE METABOLISM, BIOCHEMISTRY. *Educ:* Univ Calif, Davis, 58, MS, 59, PhD(biochem), 63. *Prof Exp:* Res fel, McArdle Lab Cancer Res, Univ Wis-Madison, 63-65; USPHS fel enzymol, Scripps Clin & Res Found, 65-67; res assoc, Children's Cancer Res Found, 67-70; from asst prof to assoc prof, 70-83, PROF BIOCHEM, SCH MED, UNIV NMEX, 83- *Concurrent Pos:* Nat Adv Environ Health Sci Coun, NIH, 80-83, Cancer Res Manpower Rev Comt, 85-88; Scholar award, Leukemia Soc Am, 70-75. *Mem:* AAAS; Am Asn Cancer Res; Am Soc Biochem & Molecular Biol; Am Soc Trop Med Hyg. *Res:* Molecular basis for the mechanism of action of pyrimidines, purine and folic acid analogs; nucleotide synthesis and metabolism; cancer chemotherapy; biochemistry of malaria parasites. *Mailing Add:* Dept Biochem Univ NMex Sch Med Albuquerque NM 87131

REYES, VICTOR E, b Caracas, Venezuela, May 27, 59; m 82; c 1. PHARMACOLOGY. *Educ:* Tex Tech Univ, BS, 80, MS, 82; Univ Tex, PhD(microbiol immunol), 86. *Prof Exp:* Teaching asst med microbiol, Tex Tech Univ Health Sci Ctr, 80-82; asst, immunol, Univ Tex Med Br, 82-84, McLaughlin fel, 84-86; res assoc, 86-89, instr, 89-91, ASST PROF, IMMUNOL, UNIV MASS MED CTR, 91- *Mem:* Am Asn Immunologists. *Res:* Roles or invariant chains in antigen processing and presentation; antigen presentation mediated by class I MHC molecules; structural characteristics of antigenic peptides. *Mailing Add:* 30 Plantation St Worcester MA 01604

REYES, ZOILA, b Bucaramanga, Colombia, July 18, 20. ORGANIC CHEMISTRY. *Educ:* Univ WVa, AB & MS, 45; Johns Hopkins Univ, PhD(org chem), 48. *Prof Exp:* Res chemist, Gen Aniline & Film Corp, 48; sr org chemist, 56-86, CONSULT, SRI INT, 86- *Mem:* AAAS; Am Chem Soc; Sigma Xi; NY Acad Sci. *Res:* Proteins; amino acids; fatty acids and derivatives; heterocyclic compounds; photochemistry; photopolymers; polymer technology; microencapsulation; chemical and radiation induced graft copolymerization; biotechnology. *Mailing Add:* 317 Yale Menlo Park CA 94025

REYHNER, THEODORE ALISON, b Paterson, NJ, Nov 17, 40; m 68. FLUID MECHANICS. *Educ:* Stanford Univ, BS, 62, MS, 63, PhD(eng mech), 67. *Prof Exp:* Sr specialist engr, 60-80, PRIN ENGR, BOEING COM AIRPLANE CO, SEATTLE, 80- *Mem:* Assoc fel Am Inst Aeronaut & Astronaut. *Res:* Boundary layer and shock wave-boundary layer interactions computations; three-dimensional transonic potential flow; computational fluid mechanics; numerical analysis. *Mailing Add:* 2435 S 304th St Federal Way WA 98003

REYHNER, THEODORE O, b Paterson, NJ, Apr 19, 15; m 40; c 2. ENGINEERING. *Educ:* Newark Col Eng, BS, 37; Columbia Univ, AM, 38; NY Univ, PhD(educ admin), 50; Stanford Univ, MS, 63. *Prof Exp:* Struct draftsman, Robins Conveying Belt Co, NY, 37-38; instr math & physics, Newark Col Eng, 40-42; instr physics, Cooper Union, 42-43; instr civil eng, Lehigh Univ, 43-44; engr timber mech div, Forest Prods Lab, US Forest Serv, Wis, 44-46; assoc prof civil eng, Univ NDak, 46-47; assoc prof archit, Sch Archit & Allied Arts, Univ Ore, 47-49; assoc prof civil eng, Univ Denver, 49-53 & Mich Col Mining & Technol, 53-56; actg head dept eng, 57-58, head civil eng, 67-74 & 78, prof, 56-82, chmn div eng, 78-81, EMER PROF CIVIL ENG, CALIF STATE UNIV, CHICO, 82- *Concurrent Pos:* NSF sci fac fel, 62-63; consult civil eng, 82- *Mem:* Fel Am Soc Civil Engrs; Am Soc Eng Educ; Am Concrete Inst. *Res:* Reinforced concrete; structural engineering; statistical education. *Mailing Add:* 1325 Neal Dow Ave Chico CA 95926

REYNA, LUIS GUILLERMO, b Cordoba, Arg, Apr 6, 56; m 84; c 2. APPLIED MATHEMATICS. *Educ:* Univ Nat Cordoba, Licenciado, 78; Calif Inst Technol, PhD(appl math), 83. *Prof Exp:* Courant instr appl math, Courant Inst, NY Univ, 82-84; Humboldt res fel, Technische Hochschule Aachen, 84-86; RES STAFF MEM, INT BUS MACH CORP WATSON RES CTR, 86- *Mem:* Soc Indust & Appl Math. *Res:* Applied mathematics; scientific computation; computational fluid mechanics. *Mailing Add:* IBM Watson Res Ctr PO Box 218 Yorktown Heights NY 10598

REYNAFARJE, BALTAZAR, BIOENERGENICS, TRANSDUCTION. *Educ:* Univ St Marcos, Lima, Peru, MD, 53. *Prof Exp:* ASST PROF BIOL CHEM, SCH MED, JOHNS HOPKINS UNIV, 75- *Res:* Effects of oxidation concentration on energy. *Mailing Add:* Sch Med Johns Hopkins Univ 725 N Wolfe St Baltimore MD 21205

REYNARD, ALAN MARK, b Boston, Mass, Oct 11, 32; m 61; c 2. PHARMACOLOGY. *Educ:* George Washington Univ, BS, 53; Univ Minn, PhD(biochem), 60. *Prof Exp:* Fel biochem, Univ Wash, 60-62; fel pharmacol, Yale Univ, 62-64; asst prof, 64-68, ASSOC PROF PHARMACOL, SCH MED, STATE UNIV NY BUFFALO, 68- *Res:* Antibiotics; antibiotic-resistance in bacteria complement. *Mailing Add:* Dept Pharmacol 108 Farber Hall SUNY Health Sci Ctr 3435 Main St Buffalo NY 14214

REYNARD, KENNARD ANTHONY, b Philadelphia, Pa, Jan 13, 39; m 66; c 1. POLYMER CHEMISTRY, INORGANIC CHEMISTRY. *Educ:* St Louis Univ, BS, 60, MS, 64, PhD(chem), 67. *Prof Exp:* Proj engr, US Air Force, Fla, 66-69; group leader polymer chem, Horizons Res Inc, 69-72, head chem dept, 72-74, mgr contract res & develop, 74-76; tech mgr, 76-79, tech mgr fluids, emulsions & compounds, 79-88, DIR QUAL ASSURANCE, WACKER SILICONES CORP, 88- *Honors & Awards:* IR-100 Award, Indust Res Mag, 71. *Mem:* Am Chem Soc; Am Soc Qual Control. *Res:* Synthesis of monomers and polymers polysiloxanes; characterization of polyphosphazenes, inorganic chemistry; elastomers; phosphorus-nitrogen compounds, boron-nitrogen compounds, organo-silicon compounds, organometallic chemistry; waste utilization, electrostatics, inks and toners; research and development in chemistry and physics as related to materials and processes. *Mailing Add:* Wacker Silicones Corp Sutton Rd Adrian MI 49221

REYNES, ENRIQUE G, chemical engineering, chemistry, for more information see previous edition

REYNHOUT, JAMES KENNETH, b Mysore City, India, July 6, 42; US citizen; m 63; c 4. DEVELOPMENTAL BIOLOGY. *Educ:* Barrington Col, BS, 64; Brown Univ, MS, 68, PhD(develop biol), 71. *Prof Exp:* Res assoc biol, Purdue Univ, 70-73; asst prof biol, Holy Cross Col, 73-75; res scientist, Mich Cancer Found, 76-78; ASST PROF BIOL, OAKLAND UNIV, 78- *Mem:* Soc Develop Biol; Am Soc Zoologists; AAAS; Sigma Xi. *Res:* Developmental regulation, especially endocrine mechanisms stimulating meiosis; chromosomal interaction in hybrids; chromosomal proteins in developmental regulation. *Mailing Add:* Biol Dept Bethel Col 3900 Bethel Dr St Paul MN 55112

REYNIK, ROBERT JOHN, b Bayonne, NJ, Dec 25, 32; m 59; c 6. MATERIALS SCIENCE, RESEARCH ADMINISTRATION. *Educ:* Univ Detroit, BS, 56; Univ Cincinnati, MS, 60, PhD(theoret phys chem), 63. *Prof Exp:* Mem tech staff, Bell Tel Labs, Inc, NY, 57-58; res fel appl sci & elec eng, Univ Cincinnati, 58-59, lectr math & elec eng, 59-61; head physics dept, Ohio Col Appl Sci, 61-62; lectr math, Xavier Univ, Ohio, 62-63; fel metall eng, Univ Pa, 63-64; from asst prof to assoc prof, Drexel Univ, 64-70; assoc prog dir eng mat, Div Eng, NSF, 70-71, dir, eng mat prog, Div Mat Res, 71-74, dir metall prog, 74-84, sect head, Metall Polymers, Ceramics & Electronic Mat, 84-90, HEAD OFF SPEC PROGS MAT, NSF, 90- *Concurrent Pos:* Indust consult, 65-; res grants, NSF, 65-72 & Am Iron & Steel Inst, 67-71; NIH dent training grant, 67-72; chmn, Physics & Chem Mat Comt, Metall Soc, chmn Met Trans A Publ Comt, mem first deleg, People's Rep China, 78 & mem Prog Comt, 80-83; mem Planning Group & prog dir, Electo, Metall & Corrosion Work Group, US-USSR Sci & Technol Int Agreement, 73-; deleg to nat comn on mat policy, NSF liaison rep numerous comt, Nat Acad Sci; comt, Nat Metall Adv Bd, 71-; chmn, interagency mat group, Comt Mat/Off Sci & Technol Policy, 84-; vis prof mat sci & eng, Univ Pa, 82-83; liason rep, mat systs eng sub-panel, Eng Res Bd, Comt Eng & Tech Systs, NSF liaison rep comts, Nat Acad Sci & Nat Mats Adv Bd, 71-; chmn, Long Range Planning Comt, Electronic, Magnetic Photonic Mat Div, Metall Soc; mem, sr exec serv, Fed Govt, 85- *Mem:* Am Inst Mining Metall & Petrol Engrs; Metall Soc; Sigma Xi; Am Soc Metals; Am Chem Soc; Mat Res Soc; AAAS; Am Phys Soc. *Res:* Transport properties of liquid metals; biomaterials; electronic and magnetic behavior of materials; research management. *Mailing Add:* Div Mat Res NSF Washington DC 20550

REYNOLDS, ALBERT KEITH, pharmacology; deceased, see previous edition for last biography

REYNOLDS, BRIAN EDGAR, b Drogheda, Ireland, Mar 5, 36; m 60; c 2. MEDICINAL CHEMISTRY, ORGANIC CHEMISTRY. *Educ:* Queen's Univ, Belfast, 59, PhD(org chem), 62. *Prof Exp:* Res fel chem, Univ Rochester, 62-64; RES SCIENTIST, MED CHEM, MCNEIL LABS, INC, 64- *Mem:* Am Chem Soc; Royal Soc Chem. *Res:* Synthesis and biosynthesis of natural products; synthesis of organic chemicals for possible use as pharmaceuticals. *Mailing Add:* 1644 Arran Way Dresher PA 19025

REYNOLDS, BRUCE G, b Ft Myers, Fla, Jan 16, 37; m 61; c 2. NUCLEAR PHYSICS. *Educ:* Fla State Univ, BS, 61, PhD(physics), 66. *Prof Exp:* Asst prof physics, Fla State Univ, 66-68; asst prof physics & elec eng, Ohio Univ, 68-71; physicist & staff engr, Martin Marietta Corp, 71-74; physicist, Argonne Nat Lab, 74-80; mem tech staff, Bell Tel Labs, 80-81; prin engr, Stromberg Carlson, 81-; GEN MGR, OMEGA. *Mem:* Am Phys Soc; Am Nuclear Soc; Inst Elec & Electronics Engrs. *Res:* Experimental high energy particle physics; hardware/software design of data analysis systems; electrooptical instrumentation and systems engineering; operations research and economic planning; electromagnetic pulse phenomena and electromagnetic compatibility; analysis of telecommunication systems; hydrology and water resource analysis. *Mailing Add:* Gen Mgr Omega PO Box 1231 Taveres FL 32778

REYNOLDS, CHARLES ALBERT, b Colorado Springs, Colo, Apr 1, 23; m 53; c 6. CHEMISTRY. *Educ:* Stanford Univ, AB, 44, MA, 46, PhD(analytical chem), 47. *Prof Exp:* Asst instr chem, Stanford Univ, 44-47; from asst prof to assoc prof, 47-59, assoc chmn dept, 61-67, PROF CHEM, UNIV KANS, 60- *Concurrent Pos:* Mem opers res group, US Army Chem Corps, 51-53; tech dir, Edgewood Arsenal, 67-69. *Mem:* Am Chem Soc. *Res:* Organic functional group analysis; complex ion reactions in non-aqueous solvents; thermochemical methods of analysis. *Mailing Add:* 2209 Hill Ct Lawrence KS 66049

REYNOLDS, CHARLES C, b Webb City, Mo, July 17, 27; m 48; c 3. METALLURGY, CERAMICS. *Educ:* Mass Inst Technol, SB, 47, SM, 54, PhD(metall), 64. *Prof Exp:* Asst prof metall, Mass Inst Technol, 54-57; asst prof, Thayer Sch, Dartmouth Col, 57-60, asst dean, 57-59; fel, Mass Inst Technol, 60-62; assoc prof mech eng, Worcester Polytech Inst, 62-67, prof, 67-81, George F Fuller prof mech eng, 65-82; ADJ PROF & CONSULT, DEPT INDUST & MFG ENG, UNIV RI, 82- *Mailing Add:* Dept Indust Mfg Eng Worcester Polytech Inst Worcester MA 01609

REYNOLDS, CHARLES F, b 1947. SLEEP, AFFECTIVE DISORDERS. *Educ:* Univ Va, BA, 69; Yale Univ, MD, 73; Am Bd Psychiat & Neurol, dipl, 78. *Prof Exp:* ASSOC PROF PSYCHIAT & NEUROL, UNIV PITTSBURGH, 83-; DIR SLEEP EVAL CTR, WESTERN PSYCHIAT INST & CLIN, 84- *Concurrent Pos:* Dir Geopsychiat Clin Res Unit, Western Psychiat Inst & Clin, 87-; mem Psychopath & Clin Biol Res Rev Comt, NIMH, 83-87; prin investr, NIMH, 83-, res scientist develop award, 80- *Honors & Awards:* Marie Eldridge Award, Am Psychiat Asn, 82. *Mem:* Psychiat Res Soc; Am Col Psychiatrists. *Res:* Sleep, aging and mental illness; the effects of depression on sexual function in men and on maintenance therapies in late life depression. *Mailing Add:* 3811 OHara St Pittsburgh PA 15261

REYNOLDS, CHARLES WILLIAM, b Ala, Nov 30, 17; m 55; c 1. HORTICULTURE. *Educ:* Univ Ala, AB, 41; Ala Polytech Inst, BS, 47, MS, 49; Univ Md, PhD(hort), 54. *Prof Exp:* Instr hort, Ala Polytech Inst, 49-52, assoc prof, 52-53; from asst prof to assoc prof veg crops, Univ Md, College Park, 54-65, prof hort, 65-82; RETIRED. *Concurrent Pos:* Consult, Guatemala, 75, USSR, 78-79 & People's Repub China, 85. *Mem:* Am Soc Hort Sci. *Res:* Mineral nutrition of cucumbers; effects of supplemental irrigation on yield and quality of vegetable corps; cultural studies with cauliflower; vegetable seed production in the United States. *Mailing Add:* 705 Andrews Rd Opelika AL 36801

REYNOLDS, CLAUDE LEWIS, JR, b Roanoke, Va, Dec 16, 48; m 70; c 4. EXPERIMENTAL SOLID STATE PHYSICS, MATERIALS SCIENCE. *Educ:* Va Mil Inst, BS, 70; Univ Va, MS, 72, PhD(mat sci), 74. *Prof Exp:* Sr scientist, Dept Mat Sci, Univ Va, 74-75; res assoc physics, Univ Ill, Urbana, 75-77; sr proj engr, Union Carbide Corp, 77-80; MEM TECH STAFF, BELL LABS, 80- *Mem:* Am Phys Soc; Am Asn Physics Teachers; Sigma Xi; Inst Elec & Electronics Engrs; Mat Res Soc. *Res:* Molecular beam epitaxy; growth and properties of aluminum gallium arsenide; implant and activation of gallium arsenide; gallium arsenide integrated circuits; semiconductor injection lasers; amorphous materials; properties of materials at low temperatures. *Mailing Add:* Bell Labs 2525 N 12th St Reading PA 19612

REYNOLDS, DAVID B, B July 5, 49; m; c 3. BIOMEDICAL ENGINEERING, ARTIFICIAL ORGANS-IMPLANTS. *Educ:* Univ Va, PhD(biomed eng), 78. *Prof Exp:* ASSOC PROF BIOMED & HUMAN FACTORS ENG, WRIGHT STATE UNIV, 80- *Mem:* Biomed Eng Soc; Am Physiol Soc. *Res:* Biofluid mechanics; pulmonary mechanics; artificial urinary sphincter; computer modeling of cardiovascular system. *Mailing Add:* 139 Eng & Math Sci Bldg Wright State Univ Dayton OH 45435-0001

REYNOLDS, DAVID GEORGE, b South Chicago Heights, Ill, Nov 25, 33; m 87; c 2. PHYSIOLOGY. *Educ:* Knox Col, Ill, BA, 55; Univ Ill, Urbana, MS, 57; Univ Iowa, PhD(physiol), 63. *Prof Exp:* Med Serv Corps, US Army, 57-77, chief basic sci br, US Army Med Field Serv Sch, Tex, 58-60 & 63-65, actg chief dept gastroenterol, Walter Reed Army Inst Res, 65-68, asst chief, 68-72, dep dir div surg, 72-74, dir div surg, 74-77; from assoc prof to prof surg, Univ Iowa, 77-87, dir, Surg Labs & Res, Hosps & Clins, 77-87; PROF SURG & DIR RES, DEPT SURG, UNIV S FLA COL MED, 87- *Concurrent Pos:* Spec lectr, Sch Hosp Admin, Baylor Univ, 63-65. *Mem:* Am Physiol Soc; Soc Exp Biol & Med; Asn Acad Surg; Am Fedn Clin Res; NY Acad Sci; Shock Soc (pres, 86-87). *Res:* Gastrointestinal physiology and pharmacology; endogenous opioids; sphlanchnic blood flow; shock. *Mailing Add:* Dept Surg Box 16 Univ SFla Col Med Tampa FL 33612

REYNOLDS, DAVID STEPHEN, b Cincinnati, Ohio, Nov 12, 32; m 58; c 4. APPLIED STATISTICS, APPLIED MATHEMATICS. *Educ:* Univ Cincinnati, Mech Engr, 54; Fla State Univ, MS, 65, PhD(statist), 69. *Prof Exp:* Dept mgr mfg, 54-60, statist engr, 60-62, statist serv group leader, 63-73, DATA SERV MGR PROD DEVELOP, PROCTER & GAMBLE CO, 73- *Mem:* Am Statist Asn; Inst Math Statist. *Mailing Add:* 9636 Leebrook Dr Cincinnati OH 45231

REYNOLDS, DON RUPERT, b Shreveport, La, Aug 22, 38. MYCOLOGY. *Educ:* Tex A&M Univ, BS, 60; La State Univ, MS, 62; Univ Tex, Austin, PhD(bot), 70. *Prof Exp:* Mycologist, Univ Philippines, 63-67; CUR & HEAD BOT, NATURAL HIST MUS, LOS ANGELES, 75- *Concurrent Pos:* Adj prof, Univ Southern Calif, 75. *Mem:* Mycol Soc Am. *Res:* Systematics and evolution ascomycete fungi. *Mailing Add:* 1011 Beyer Way No 43 San Diego CA 92154

REYNOLDS, DONALD C, b Sioux City, Iowa, July 28, 20. GENERAL PHYSICS, SOLID STATE PHYSICS. *Educ:* Morningside Col, BS, 43; Univ Iowa, MS, 48. *Prof Exp:* Mat researcher, Battelle Mem Inst, 48-52; res engr, Aerospace Res Lab, 52-75, SR SCIENTIST, AVIONIS LAB, WRIGHT PATTERSON AFB, 75- *Honors & Awards:* Photo Voltaic Founders Award, Inst Elec & Electronics Engrs, 85. *Mem:* Fel Am Phys Soc. *Mailing Add:* 1603 S Fountain Ave Springfield OH 45506

REYNOLDS, DONALD KELLY, b Portland, Ore, Dec 9, 19; m 45; c 3. ELECTRICAL ENGINEERING. *Educ:* Stanford Univ, BA, 41, MA, 42; Harvard Univ, PhD(eng sci, appl physics), 48. *Prof Exp:* Res engr, Radio Res Lab, Harvard Univ, 42-45; sr res engr, Stanford Res Inst, 48-53; assoc prof elec eng, Tech Inst Aeronaut, Brazil, 53-56; prof & head dept, Seattle Univ, 56-59; prof, 59-, EMER PROF ELEC ENG, UNIV WASH. *Concurrent Pos:* Sci attache, US Dept State, Am Embassy, Brazil, 72-74. *Mem:* Fel Inst Elec & Electronics Engrs. *Res:* Electronic circuits; antennas; applied electromagnetic theory. *Mailing Add:* Dept Elec Eng Univ Wash Seattle WA 98195

REYNOLDS, ELBERT BRUNNER, JR, b Bryan, Tex, Sept 17, 24; m 64; c 2. MECHANICAL ENGINEERING. *Educ:* Tex A&M Univ, BS, 47; Pa State Univ, MS, 48; Univ Wis, PhD(mech eng), 57. *Prof Exp:* From instr to asst prof mech eng, Pa State Univ, 48-53; serv engr, E I du Pont de Nemours & Co, 57-61; assoc prof mech eng, Univ Va, 61-64; ASSOC PROF MECH ENG & TECHNOL, TEX TECH UNIV, 64- *Concurrent Pos:* Consult, US Naval Weapons Lab, Va, 63-64; NSF grant, 65-67. *Mem:* Am Soc Mech Engrs; Soc Automotive Engrs. *Res:* Thermodynamics; heat transfer; compressible fluid flow; optical instruments for study of heat transfer and compressible fluid flow; direct energy conversion. *Mailing Add:* Dept Technol Tex Tech Univ Lubbock TX 79406

REYNOLDS, GEORGE THOMAS, b Trenton, NJ, May 27, 17; m 44; c 4. BIOPHYSICS, HIGH ENERGY PHYSICS. *Educ:* Rutgers Univ, BS, 39; Princeton Univ, MA, 42, PhD(physics), 43. *Prof Exp:* Res physicist, Nat Defense Res Comt, 41-44, from asst prof to assoc prof, 46-58, dir, Ctr Environ Studies, 71-74, PROF PHYSICS, PRINCETON UNIV, 58- *Concurrent Pos:* Guggenheim fel, 55-56; consult radiation detection, blast effects & oceanog, 58-; fel, Churchill Found, Cambridge Univ, 73-74; mem bd trustees, Rutgers Univ, 75-; Sci & Eng Res Coun fel, Oxford Univ, 81-82, Royal Soc Res fel, 85. *Mem:* Fel Am Phys Soc; Am Geophys Union; Biophys Soc; fel AAAS. *Res:* Mass spectroscopy; fluid dynamics; cosmic ray; high energy nuclear physics; image intensification; bioluminescence; x-ray diffraction of biological structures. *Mailing Add:* Dept Physics Princeton Univ Princeton NJ 08544

REYNOLDS, GEORGE WARREN, b Decatur, Ill, May 10, 16; m 41; c 2. METEOROLOGICAL SUPPORT PROGRAMS & SYSTEMS. *Educ:* James Millikin Univ, BA, 39; St Louis Univ, MS, 50; Tex A&M Univ, PhD(meteorol & Oceanog), 62. *Prof Exp:* Res assoc & lectr meteorol, Univ Mich, 56-58; meteorologist, US Army Electronic Proving Ground, 58-62; design engr, Martin-Marietta Corp, 62-63; br chief, Geophys Environ, Us Army Res Off, Europe, 63-65; prof meteorol & oceanog, Utah State Univ, 66-72; mgr. Environ Off, Woodward Clyde Consults, 72-75; supvr, Air Qual Assessment, Tenn Valley Authority, 75-83; SR SCIENTIST, LATCO, 83- *Concurrent Pos:* Vis assoc prof meteorol, Univ Mo, Columbia, 65-66. *Mem:* Am Meteorol Soc; Air & Waste Mgt Asn. *Res:* Meteorological applications to practical problems; meteorological measurement systems such as Doppler weather radar, lightning detection systems, automatic weather observing stations, quality assurance; tornado forces; wind shear and thermodynamic stability of the lowest 100 meters of the atmosphere. *Mailing Add:* Rte 7 Box 402 Florence AL 35630

REYNOLDS, GEORGE WILLIAM, JR, b South Glens Falls, NY, Aug 18, 28; m 80; c 3. MATERIALS PHYSICS, CHEMICAL PHYSICS. *Educ:* Col Educ, Albany, BS, 53, MS, 58; Ohio State Univ, PhD(physics), 66. *Prof Exp:* Teacher, Del Acad & Cent Sch, 53-56; teacher & chmn sci dept, Fulton High Sch, 56-59; asst prof sci educ, Col Educ, Albany, 59-61; instr physics, Ohio State Univ, 62-63; from asst prof to assoc prof sci educ, State Univ NY, Albany, 63-65, assoc prof sci, 65-70, assoc prof physics, 70-; RETIRED. *Concurrent Pos:* Lectr, Naval Res Off Sch, 61-63 & 69-72; consult, Naval Res Lab, 79-88, small computer sys consult, 84- *Mem:* Am Asn Physics Teachers; Nat Sci Teachers Asn; Am Meteorol Soc; Mat Res Soc; Am Phys Soc. *Res:* Chemical kinetics; ion beam-solid interactions and analysis; design and development of instrumentation; ion beam metallurgy. *Mailing Add:* Reynolds Sci Enterprises 26951 Leport St SE Bonita Spring FL 33923

REYNOLDS, GLENN MYRON, b Alexandria, Minn, Jan 30, 36; m 58; c 3. NUCLEAR PHYSICS. *Educ:* Univ Minn, BS, 61, MS, 63, PhD(physics), 66. *Prof Exp:* Res assoc nuclear physics, Cyclotron Lab, Univ Mich, 66-68; staff assoc, Gulf Gen Atomic, Inc, 68-70; staff scientist nuclear physics, 70-84, VPRES, SCI APPLNS INT CORP, 85- *Mem:* Am Phys Soc; Am Soc Nondestructive Testing; Am Nuclear Soc. *Res:* Charged particle reactions; gamma ray spectroscopy; fission; neutron spectroscopy; nondestructive testing; radiation effects. *Mailing Add:* 1611 Calle Plumerias Encinitas CA 92024

REYNOLDS, HAROLD TRUMAN, b Manteca, Calif, Oct 28, 18; m 46; c 4. ENTOMOLOGY. *Educ:* Univ Calif, BS, 41, PhD, 49. *Prof Exp:* From asst entomologist to assoc entomologist, 48-62, chmn dept, 69-74, PROF ENTOM & ENTOMOLOGIST, UNIV CALIF, RIVERSIDE, 62-, HEAD, DIV BIOL CONTROL, 78- *Mem:* AAAS; Entom Soc Am. *Res:* Biology, ecology and pest management of pests affecting field crops, particularly cotton; selective insecticides. *Mailing Add:* Dept Entom Univ Calif 900 University Ave Riverside CA 92521

REYNOLDS, HARRY AARON, JR, b Lumberton, NJ, Feb 6, 28; m 59; c 2. VETERINARY PATHOLOGY. *Educ:* Gettysburg Col, AB, 52; Univ Pa, VMD, 56; Univ Ill, Urbana, MS, 63, PhD(vet med sci), 66. *Prof Exp:* Instr vet path, Univ Pa, 56-59; USPHS fel, 60-62,; from instr to assoc prof, 62-85, PROF VET PATH, UNIV ILL, URBANA, 85- *Concurrent Pos:* Chmn vet path, Univ Ill, Urbana, 68-70, 77-80, 82-84 & 86-90, asst head, Dept Vet Pathobiol, 88-90, mem, CL Davis Found Advan Vet Path. *Honors & Awards:* Norden Award, Norden Labs, 70. *Mem:* AAAS; Am Vet Med Asn; Int Acad Path; NY Acad Sci. *Res:* Viral cytopathology; small animal reproduction and diseases of reproductive organs; neoplastic diseases. *Mailing Add:* 2631 VMBSB 2001 S Lincoln Univ Ill Urbana IL 61801

REYNOLDS, HARRY LINCOLN, b Portchester, NY, Mar 31, 25; m 50; c 2. NUCLEAR ENGINEERING. *Educ:* Rennselaer Polytech Inst, BS, 47; Univ Rochester, PhD(physics), 51. *Prof Exp:* Sr scientist, Oak Ridge Nat Lab, 51-55; sr scientist & div leader, Lawrence Livermore Nat Lab, 55-64, assoc dir, 65-81; prog dir, NASA, Houston, 65; dep assoc dir, Los Alamos Nat Lab, 81-85; DIR ADVAN CONCEPTS, ROCKWELL INT CORP, 85- *Concurrent Pos:* Consult, Dept Defense & Dept Energy, 65-85, Oak Ridge Nat Lab, 90-; mem, Army Sci Bd, 82-88. *Mem:* Fel Am Phys Soc; Am Defense Preparedness Soc. *Res:* Energetic heavy ion nuclear physics; nuclear reactor studies; nuclear weapon physics; arms control; spacecraft design; strategic defense. *Mailing Add:* 801 Via Somonte Palos Verdes Estates CA 90274

REYNOLDS, HERBERT MCGAUGHEY, b Bryan, Tex, Aug 28, 42; m 67; c 2. BIOMECHANICS, ANTHROPOMETRICS. *Educ:* Southern Methodist Univ, BA, 64, MA, 71, PhD(phys anthrop), 74. *Prof Exp:* Anthropologist, Protection & Survival Lab, Civil Aeromed Inst, Fed Aviation Admin, 69-73, res anthrop, 73; res investr, Univ Mich, 73-74, asst res scientist phys anthrop, dept biomed, Hwy Safety Res Inst, 74-77; from asst prof to assoc prof, 77-91, PROF, DEPT BIOMECH & DEPT ANTHROP, MICH STATE UNIV, 91- *Concurrent Pos:* Consult, Biomed Sci Dept, Res Labs, Gen Motors Corp, 75-76, Franklin Inst, Philadelphia, 77-78, Haworth Inc, 79-81, Hoover Universal, 81-85 & Motor Wheel, 85; Johnson Controls, 86-87. *Mem:* Soc Study Human Biol; Human Biol Coun; Am Soc Biomech; Am Asn Phys Anthropologists; Sigma Xi. *Res:* Three-dimensional systems anthropometry; biomechanics; kinematics; human morphology; osteology; mathematical modeling of the human body; human factors. *Mailing Add:* Dept Biomech Mich State Univ East Lansing MI 48824-1316

REYNOLDS, JACK, b Norman, Okla, Jan 11, 29; m 64; c 2. ELECTRICAL & NUCLEAR ENGINEERING. *Educ:* Univ Okla, BSEE, 56, ME, 58; Univ Lund, PhD(physics), 64. *Prof Exp:* Engr, Radio Corp Am, 56; adv studies scientist, Lockheed Missile Systs Div, 58; instr elec eng, Univ Okla, 56-58, instr & engr comput proj, 59-61, res assoc, Res Inst, 61; consult, Standard Elec A/B, Denmark, 62-63; asst prof elec eng, Univ Okla & res assoc, Res Inst, 64-67, proj dir, 65-67; mem tech staff, Equip Res & Develop Lab, 67-78, SEMICONDUCTOR PROG MGR, TEX INSTRUMENTS INC, 78- *Concurrent Pos:* NSF res grant, 65-67; sr mem, tech staff, Sematech. *Mem:* Inst Elec & Electronics Engrs; Sigma Xi. *Res:* Solid state nuclear particle detectors; measurement of beta spectra; investigation of nonlinear effects in semiconductor devices; microwave integrated circuits; solid state microwave devices and components. *Mailing Add:* 13215 Roaring Springs Lane Dallas TX 75240

REYNOLDS, JACQUELINE ANN, b Los Angeles, Calif, Oct 19, 30; c 4. BIOCHEMISTRY, BIOPHYSICS. *Educ:* Pac Univ, BS, 51; Univ Wash, PhD(phys chem), 63. *Prof Exp:* Res asst prof microbiol, Wash Univ, 66-69; from asst prof to assoc prof biochem, 69-80, prof physiol, 80-88, PROF CELL BIOL, DUKE UNIV, 88- *Concurrent Pos:* John Simon Guggenheim fel, 77-78. *Mem:* Am Soc Biol Chemists; Biophys Soc; NY Acad Sci; Soc Gen Physiologists. *Res:* Lipid-proteins interactions; structure of serum lipoproteins and biological membranes; physical chemistry of amphiphiles; active transport. *Mailing Add:* Dept Cell Biol Duke Univ Med Ctr Durham NC 27710

REYNOLDS, JAMES BLAIR, b Ypsilanti, Mich, Nov 9, 39; m 60; c 6. FISHERY SCIENCE. *Educ:* Utah State Univ, BS, 61; Iowa State Univ, MS, 63, PhD(fishery biol), 66. *Prof Exp:* Fishery biologist, Great Lakes Fishery Lab, US Fish & Wildlife Serv, 66-72; asst leader, Mo Coop Fishery Unit, US Fish & Wildlife Serv & asst prof, Sch Forestry, Fisheries & Wildlife, Univ Mo-Columbia, 72-78; LEADER & ASSOC PROF, ALASKA COOP FISH RES UNIT, UNIV ALASKA, FAIRBANKS, 78- *Mem:* Am Fisheries Soc; NAm Benthological Soc; Am Inst Fishery Res Biol; Sigma Xi. *Res:* Fishery mensuration and dynamics, including effects of exploitation; aquatic habitat alteration; biometrics and sampling methodology. *Mailing Add:* Coop Fish Res Unit Univ Alaska 138 Arctic Health Fairbanks AK 99775

REYNOLDS, JAMES HAROLD, b Ogden, Utah, Feb 18, 45; m 67; c 6. ENVIRONMENTAL ENGINEERING. *Educ:* Utah State Univ, BS, 70, MS, 72, PhD(environ eng), 74. *Prof Exp:* Res engr, Utah Water Res Lab, 73-74; asst prof environ eng, Utah State Univ, 74-80, head div, 78-80; PRIN ENGR, JAMES M MONTGOMERY CONSULT ENGRS INC, 80- *Concurrent Pos:* Consult engr & vpres, Middlebrooks & Assoc, 72-80; pres, Intermountain Consults & Planners, 78-80. *Mem:* Water Pollution Control Fedn; Asn Environ Eng Profs; Govt Refuse Collection & Disposal Asn. *Res:* Water and wastewater treatment; biological kinetics; physical chemical waste treatment lagoons; small wastewater treatment systems toxicity. *Mailing Add:* James M Montgomery Consult Engrs 545 Indian Mound Wayzata MN 55391

REYNOLDS, JEFFERSON WAYNE, b Elizabethton, Tenn, Sept 11, 26; m 54; c 2. PHYSICAL CHEMISTRY. *Educ:* ETenn State Col, BS, 50; Ohio State Univ, MS, 54. *Prof Exp:* Res chemist, 51-52 & 54-63, SR RES CHEMIST, TENN EASTMAN CO, EASTMAN KODAK CO, 63- *Mem:* Am Chem Soc; fel Am Inst Chemists. *Res:* Surface chemistry-surface area and pore structure; transition elements-preparation of heterogeneous catalysts; aliphatic chemistry-evaluation of catalysts in hydrogenations, oxidations, condensations; measurement of design data; thermodynamics. *Mailing Add:* Rte 1 Box 224 Fall Branch TN 37656-9738

REYNOLDS, JOHN C, b Ill, June 1, 35; m 60; c 2. COMPUTER SCIENCES. *Educ:* Purdue Univ, BS, 56; Harvard Univ, AM, 57, PhD(physics), 61. *Prof Exp:* From asst physicist to assoc physicist, appl math div, Argonne Nat Lab, 61-70; PROF COMPUT & INFO SCI, SYRACUSE UNIV, 70- *Concurrent Pos:* Actg asst prof, Stanford Univ, 65-66; prof lectr comt info sci, Univ Chicago, 68; mem working group 2.3 prog methodology, Int Fedn Info Processing, 69-; sr res assoc, Queen Mary Col, Univ London, 70-71; vis res fel, Univ Edinburgh, 76-77; mem working group formal lang definition, Int Fedn Info Processing, 77-; researcher, Nat Inst Res Info & Automation, 83-84. *Honors & Awards:* Annual Prog Systs & Lang Paper Award, Asn Comput Mach, 71. *Mem:* Asn Comput Mach. *Res:* Design of computer programming languages; mathematical semantics; programming methodology. *Mailing Add:* Dept Computer Sci Carnegie Mellon Univ 88126 Wean Hall Pittsburgh PA 15213-3896

REYNOLDS, JOHN DICK, b Darby, Pa, June 6, 21; m 48. PLANT EMBRYOLOGY. *Educ:* Temple Univ, BS, 49, MEduc, 51; Univ SC, PhD(biol), 66. *Prof Exp:* Instr biol, Hampden-Sidney Col, 50-51; asst prof, Coker Col, 51-62; instr, Univ SC, 62-65; asst prof, Univ Southern Miss, 65-67; assoc prof biol, Va Commonwealth Univ, 67-87; RETIRED. *Concurrent Pos:* Instr, Univ SC, Lancaster Campus, 59-62. *Mem:* AAAS; Am Inst Biol Sci; Bot Soc Am; Int Soc Plant Morphol; Phytochem Soc. *Res:* Uses of infrared spectrophotometry in plant taxonomy; studies in cytoplasmic male sterility; callose distribution and function in the plant kingdom; effect of toxins on gametophyte development in plants. *Mailing Add:* Star Rte Box 124 Topping VA 23169

REYNOLDS, JOHN ELLIOTT, III, b Baltimore, Md, Nov 8, 52; m 75; c 1. MARINE MAMMALOGY. *Educ:* Western Md Col, BA, 74; Univ Miami, MS, 77, PhD(biol oceanog), 80. *Prof Exp:* Asst prof biol, 80-86, assoc prof biol & marine sci, 86-90, PROF BIOL & MARINE SCI, ECKERD COL, 90- *Concurrent Pos:* Adj fac prof natural sci, Dept Marine Sci, Univ SFla, 82-; prin investr, res grants, Fla Power & Light Co, 82-, US Marine Mammal Comn, 82-90, Save the Manatee Club, 87-, Fla Dept Educ, 85-90; coordr biol, 83-85, marine sci, 83-86, chmn natural sci, Eckerd Col, 86-; mem, Sci Adv Comt, Save the Manatee Club, 87-; mem, Comt Sci Adv on Marine Mammals, US Marine Mammal Comn, 89-, chair, 90-91, chmn, 91-; Res Award, Fla Wildlife Fedn, 77. *Mem:* Am Soc Mammalogists; Soc Marine Mammal. *Res:* Marine mammals, especially West Indian manatees and bottlenose dolphins; functional anatomical studies; population assessments; behavioral research; applications of geographic information systems to management of marine mammals and their habitats. *Mailing Add:* Eckerd Col PO Box 12560 St Petersburg FL 33733

REYNOLDS, JOHN HAMILTON, b Cambridge, Mass, Apr 3, 23; m 75; c 5. ISOTOPE GEOPHYSICS, COSMOCHEMISTRY. *Educ:* Harvard Univ, AB, 43; Univ Chicago, SM, 48, PhD(physics), 50. *Hon Degrees:* DSc, Coimbra Univ, Portugal, 87. *Prof Exp:* Asst, Electro-Acoustic Lab, Harvard Univ, 41-43; assoc physicist, Argonne Nat Lab, 50; from asst prof to assoc prof, 50-61, chmn dept, 84-86, EMER PROF PHYSICS & RES PHYSICIST RECALLED, UNIV CALIF, BERKELEY, 89- *Concurrent Pos:* Guggenheim fel, Bristol Univ, 56; NSF sr fel, Univ Sao Paulo, 63; Fulbright-Hays res award, Univ Coimbra, Portugal, 71; NSF US-Australia Cooperative Sci awardee, Univ Western Australia, 78; Guggenheim fel, Los Alamos Sci Lab, 87. *Honors & Awards:* John Price Wetherill Medal, Franklin Inst, 65; J Lawrence Smith Medal, Nat Acad Sci, 67; Leonard Medal, Meteoritical Soc, 73; Except Sci Achievement Medal, NASA, 73. *Mem:* Nat Acad Sci; AAAS; fel Am Phys Soc; Geochem Soc; Meteoritical Soc. *Res:* Mass spectrometry; meteoritics; lunar studies; origin and chronology of solar system; noble gas geochemistry. *Mailing Add:* Dept Physics Univ Calif Berkeley CA 94720

REYNOLDS, JOHN HORACE, b Darby, Pa, Aug 7, 37; m 63; c 2. AGRONOMY, PLANT PHYSIOLOGY. *Educ:* Univ Md, BS, 59; Univ Wis, MS, 61, PhD(forage crop physiol), 62. *Prof Exp:* Res asst agron, Univ Wis, 59-62; from asst prof to assoc prof agron, 62-78, PROF PLANT & SOIL SCI, UNIV TENN, KNOXVILLE, 78- *Mem:* Am Soc Agron; Soil & Water Conserv Soc; Crop Sci Soc Am. *Res:* Forage crop physiology; forage quality analysis. *Mailing Add:* Dept Plant & Soil Sci Univ Tenn Knoxville TN 37996

REYNOLDS, JOHN HUGHES, IV, b Rome, Ga, Sept 25, 40; m 63; c 1. PHYSICAL CHEMISTRY. *Educ:* Shorter Col, Ga, BA, 62; Clemson Univ, MS, 65, PhD(chem), 68. *Prof Exp:* Sr res chemist, R J Reynolds Tobacco Co, 68-76, res group leader, 76-80, mgr biobehav res, 80-90, PRIN SCIENTIST, R J REYNOLDS TOBACCO CO, 90- *Mem:* Sigma Xi; Asn Chemoreception Sci; AAAS; NY Acad Sci. *Res:* Biobehavioral aspects of tobacco product use. *Mailing Add:* Bowman Gray Tech Ctr R J Reynolds Tobacco Co Winston-Salem NC 27102

REYNOLDS, JOHN KEITH, b London, Ont, Sept 29, 19; m 45; c 3. WILDLIFE MANAGEMENT. *Educ:* Univ Western Ont, BSc, 49, MSc, 50, PhD, 52. *Prof Exp:* Wildlife biologist, Ont Dept Lands & Forests, 52-54, from asst dist forester to dist forester, 54-63, supvr fisheries, 63-64, chief exec off, Prime Minister's Dept, 64-69, secy cabinet, Off of the Prime Minister, 69-71, dep minister, Prime Minister's Dept, 71-72, dep prov secy resources develop, 72-74, dep minister, Ministry Natural Resources, 74-81; CHMN, TORONTO REGION CONSERV AUTHORITY, 81- *Concurrent Pos:* Mem adv comt sci policy, comt coordrs Can coun resource & environ ministers, Can forestry adv coun, fac adv bd, Fac Forestry, Univ Toronto & adv coun, Sch Admin Studies, York Univ; mem bd trustees, Can Nat Sportsmen's Fund; pres, J K Reynolds Consult Inc. *Mem:* Wildlife Soc; Can Can Soc Environ Biologists. *Res:* Life history studies on Canadian birds and mammals; management of fur-bearers and big game; administration of natural resources. *Mailing Add:* Five Castledene Crescent Scarborough ON M1T 1R9 Can

REYNOLDS, JOHN TERRENCE, b Savannah, Ga, Oct 26, 37; m 59; c 4. FLUID MECHANICS, NUCLEAR PHYSICS. *Educ:* Rice Univ, BA, 60; Duke Univ, PhD(physics), 64. *Prof Exp:* PHYSICIST, KNOLLS ATOMIC POWER LAB, GEN ELEC CO, 64- *Mem:* Am Phys Soc. *Res:* Turbulent fluid flow; neutron cross sections. *Mailing Add:* 25 Walden Glen Ballston Lake NY 12019

REYNOLDS, JOHN THEODORE, b Boston, Mass, Apr 27, 25; m 48; c 1. BACTERIOLOGY. *Educ:* Boston Col, BS, 51; Univ Mass, MS, 55, PhD(bact), 62. *Prof Exp:* Instr biol, Springfield Col, 52-54; instr bact, Smith Col, 54-56; from asst prof to prof bact, 56-74, PROF MICROBIOL, CLARK UNIV, 74- *Concurrent Pos:* NIH fel marine microbiol, Univ Miami, 63-64; res assoc, Inst Indust & Agr Microbiol, Univ Mass, 65-; Fulbright lectr, Univ Saigon, 66-67; adj prof, Univ Mass, Amherst, 69- *Mem:* Sigma Xi. *Res:* Aquatic bacteriology; microbiological cellulolytic activity; microbial physiology. *Mailing Add:* Dept Biol Clark Univ Worcester MA 01610

REYNOLDS, JOHN WESTON, b Portland, Ore, Aug 5, 30; m 54. PEDIATRICS. *Educ:* Reed Col, BA, 51; Univ Ore, MD, 56. *Prof Exp:* From intern to resident, Med Sch, Univ Minn, Minneapolis, 56-59, from instr to prof pediat, 61-77; PROF DEPT PEDIAT, ORE HEALTH SCI UNIV, 77- *Concurrent Pos:* NIH spec res fel, Med Sch, Univ Minn, Minneapolis, 59-61; NIH career develop award, 64-74. *Mem:* AAAS; Soc Pediat Res; Endocrine Soc; Am Pediat Soc. *Res:* Pediatric endocrinology; metabolism and nutrition; steroid metabolism in infants and children; neonatal medicine. *Mailing Add:* Dept Pediat Ore Health Sci Univ Portland OR 97201

REYNOLDS, JOHN Z, b Kansas City, Kans, July 31, 40; m 63; c 2. TECHNICAL MANAGEMENT. *Educ:* Kans State Univ, BS, 62; Univ Mich, MS & MPh 64, PhD(environ health), 66. *Prof Exp:* Consult, Resource Develop, James Calvert Consult Engr, 67; engr, Indust Conserv, Commonwealth Assoc Inc, 67-69; dir staff, Environ/Corp Planning Res, 69-91, DIR GAS FACIL, PLANNING & RES, CONSUMERS POWER CO, 90- *Concurrent Pos:* Tech adv, Edison Elec Inst, 70-80; mem, Lake Michigan Cooling Water Studies Panel, 73-75, Non-Radiation Environ Effects Comt, Am Nat Standards Inst, 75, Michigan Sea Grant External Adv Comt, 76; instr, Jackson Community Col, 77; proj mgr, Electric Power Res Inst, 78-79. *Mem:* AAAS. *Res:* Environmental effects of pumped storage hydroelectric development, cooling systems and water resource development. *Mailing Add:* Consumers Power Co 212 W Michigan Ave Jackson MI 49201

REYNOLDS, JOSEPH, b New York, NY, May 19, 35; m 73; c 2. CHEMICAL ENGINEERING, POLLUTION CONTROL. *Educ:* Cath Univ Am, BA, 57; Rensselaer Polytech Inst, PhD(chem eng), 64. *Prof Exp:* From asst prof to assoc prof, 64-77, chmn dept, 76-83, PROF CHEM ENG, MANHATTAN COL, 77- *Concurrent Pos:* Consult, Argonne Nat Lab, Consol Edison Co & private indust; chmn, Nat Conf Energy & Environ. *Mem:* Am Inst Chem Engrs; Air Pollution Control Asn; Am Soc Eng Educ. *Res:* Modelling of air pollution control equipment; electrostatic precipitators; filter bag houses; modelling and design of heat recovery equipment; hazardous waste incineration. *Mailing Add:* Manhattan Col Manhattan Col Pkwy Riverdale NY 10471

REYNOLDS, JOSEPH MELVIN, b Woodlawn, Tenn, June 16, 24; m 50; c 3. PHYSICS. *Educ:* Vanderbilt Univ, BA, 46; Yale Univ, MS, 47, PhD(physics), 50. *Prof Exp:* Instr physics, Conn Col, 48-49; from asst prof to prof, 50-62, head, Dept Physics & Astron, 62-65, vpres grad studies & res, 65-68, vpres instr & res, 68-81, actg vpres acad affairs, 66-68, vpres acad affairs, 81-85, Boyd prof, 62-85, VPRES & BOYD EMER PROF PHYSICS, LA STATE UNIV, BATON ROUGE, 85- *Concurrent Pos:* Guggenheim fel, Kamerlingh Onnes Lab, Univ Leiden, 58-59; mem nat sci bd, Nat Sci Found, 66-78; mem navig studies bd, Nat Acad Sci, 74-75, chmn, Panel Advan Navig Systs, 78-, Space Sci Bd, Nat Acad Sci, 88-, Gov Bd, Am Inst Physics, 87-; space sci bd, study "Major Directions of Space Sci", Nat Acad Sci, 84-; task force Scientific Uses Space Sta, NASA, 84-; chmn PHCE Sci Rev Bd, NASA, 83- *Mem:* Fel Am Phys Soc; fel AAAS; Sigma Xi; Am Inst Aeronaut & Astronaut. *Res:* Low temperature physics; liquid helium; superconductivity in pure metals and alloys; gravitational radiation; magnetic properties of metals; transport effects and nuclear magnetic resonance in metals; gravitational physics. *Mailing Add:* Dept Physics La State Univ Baton Rouge LA 70803

REYNOLDS, JOSHUA PAUL, b High Falls, NC, Oct 17, 06; m 37; c 2. ZOOLOGY. *Educ:* Guilford Col, BS, 28; Univ NC, MS, 29; Johns Hopkins Univ, PhD(zool), 34. *Hon Degrees:* DSc, Univ NC, Wilmington, 85. *Prof Exp:* Instr biol, Guilford Col, 29-31; actg asst prof zool, Univ NC, 32-33; from asst prof to to prof biol, Birmingham-Southern Col, 34-49; prof zool, Fla State Univ, 49-64, from asst dean to dean col arts & sci, 51-64; prof biol & dean fac, 64-71, vchancellor acad affairs, 71, ADJ PROF BIOL, UNIV NC, WILMINGTON, 71- *Concurrent Pos:* Gen Educ Bd grants, Univ Pa, 40-41 & 48-49. *Mem:* Fel AAAS; Am Soc Zoologists; Genetics Soc Am. *Res:* Genetics and cytology of Sciara; human genetics; chromosome behavior. *Mailing Add:* 1813 Azalea Dr Wilmington NC 28403

REYNOLDS, KEVIN A, b Oxford, Eng, Mar 18, 63. BIOSYNTHESIS, ENZYME MECHANISMS. *Educ:* Southampton Univ, BSc, 84, PhD(chem), 87. *Prof Exp:* Res assoc, Dept Chem, Univ Wash, Seattle, 87-89; ASST PROF BIOCHEM, SCH PHARM, UNIV MD, BALTIMORE, 89- *Mem:* Am Soc Pharmacog; Am Asn Cols Pharm. *Res:* Elucidation of biosynthetic pathways to antibiotics and other secondary metabolites in microorganisms; investigations of the origins of these pathways and mechanistic studies of the enzymes responsible for catalyzing the individual steps. *Mailing Add:* Univ Md 20 N Pine St Baltimore MD 21201

REYNOLDS, LARRY OWEN, b Norfolk, Va, Dec, 11, 40. ELECTROMAGNETICS, RADIATIVE TRANSPORT. *Educ:* Univ Wash, BSEE, 69, MSEE, 70, PhD(elec eng), 75. *Prof Exp:* Res asst electromagnetics, Ctr Bioeng, Univ Wash, 69-70, res assoc, 70-75, sr NIH fel, 75-76, fac res assoc, Dept Elec Eng, 76-79, adj asst res prof, Ctr Bioeng, 81-82, from asst res prof to assoc res prof, dept nuclear eng, 79-85; ASSOC PROF, DEPT BIOENG, UNIV ILL, CHICAGO, 85- *Concurrent Pos:* Prin investr, dept elec eng, Univ Wash, 76-79, dept nuclear eng, 78-85, Ctr Bioeng, Univ Ill, Chicago, 81-83, dept bioeng, 85-; dir, Biomat Eval, Int Biomat Consortium, Wash Technol Ctr, Univ Wash, Seattle, 85; mem, NIH Surg & Bioeng Study Sect, 85-89; NSF Bioeng Res Panel, 87. *Mem:* Sigma Xi; Optical Soc Am; Soc Biomat; Soc Indust & Appl Math; AAAS; Inst Elec & Electronics Engrs; Soc Artificial Organs; sr mem Biomed Eng Soc. *Res:* Elecromagnetic wave propagation with applications in the area of inverse and direct remote calculational techniques for characterizing dense scattering media such as found in the ocean, atmosphere and biological tissue. *Mailing Add:* Dept Bioeng MC1063 Univ Ill Chicago IL 60680

REYNOLDS, LESLIE BOUSH, JR, b Lakeland, Fla, Aug 16, 23; m 47; c 2. PHYSIOLOGY, CLINICAL MEDICINE. *Educ:* Randolph-Macon Col, BS, 49; Ga Inst Technol, MS, 51; Univ SC, PhD(physiol), 61; Northwestern Univ, MD, 66. *Prof Exp:* Engr textile fibers dept, Dacron Res Div, E I du Pont de Nemours & Co, 51-54, group leader analytical res & process control, 54-58; asst physiol, Med Col SC, 58-61; asst prof, Med Sch, Northwestern Univ, Chicago, 61-64, res assoc med, 64-67; actg chmn dept physiol, Univ Tenn Med Units, Memphis, 68-69, assoc prof physiol & med, 67-76; staff mem, Al-Med Pract Corp, Dresden, Tenn, 76-77; PROF PHYSIOL & FAMILY PRACT, QUILLEN-DISHNER COL MED, ETENN STATE UNIV, JOHNSON CITY, 77-, ASST DEAN & DIR MED EDUC, 77- *Concurrent Pos:* Dir, Memphis Emphysema Clin, 70-72; pres, Asseverator Enterprises, Inc, 76- *Mem:* Aerospace Med Asn; Am Physiol Soc; Am Chem Soc; Am Thoracic Soc; Am Col Chest Physicians; Am Acad Family Physicians. *Res:* Lung mechanics and reflexes; medical education; chest disease. *Mailing Add:* Col Med PO Box 924 Kingsport TN 37662-0924

REYNOLDS, MARION RUDOLPH, JR, b Salem, Va, Nov 1, 45; div; c 2. MATHEMATICAL STATISTICS. *Educ:* Va Polytech Inst & State Univ, BS, 68; Stanford Univ, MS, 71, PhD(oper res), 72. *Prof Exp:* Statistician, Hercules, Inc, 68; asst prof, 72-81, ASSOC PROF STATIST & FORESTRY, VA POLYTECH INST & STATE UNIV, 81- *Mem:* Am Statist Asn; Opers Res Soc Am; Am Soc Qual Control. *Res:* Sequential analysis; nonparametric statistics; quality control; validation of simulation models; applications of statistics and operations research to forestry. *Mailing Add:* Dept Statist Va Polytech Inst Blacksburg VA 24061

REYNOLDS, MARJORIE LAVERS, b Collingwood, Ont, Jan 10, 31; m 63; c 2. NUTRITION. *Educ:* Univ Toronto, BA, 53; Univ Minn, MS, 57; Univ Wis, PhD(nutrit, biochem), 64. *Prof Exp:* Res dietitian, Mayo Clinic, 57-59 & Cleveland Metrop Gen Hosp, 59-60; res asst nutrit, Univ Wis, 60-63; res assoc, Univ Tenn, 63-66; instr nutrit, Ft Sanders Hosp Sch Nursing, 67-76, renal dietitian, Ft Sanders Kidney Ctr, 78-79; INSTR, STATE TECH INST, KNOXVILLE, 82- *Mem:* Am Dietetic Asn. *Res:* Human nutrition; obesity; gastric secretion. *Mailing Add:* 7112 Stockton Dr Knoxville TN 37909

REYNOLDS, MICHAEL DAVID, b Jacksonville, Fla, Mar 30, 54; m 73; c 2. EDUCATION ADMINISTRATION, SCIENCE ADMINISTRATION. *Educ:* Thomas Edison State Col, BA, 79; Univ NFla, MEd, 83; Univ Fla, PhD(sci ed & astron), 90. *Prof Exp:* Field entomologist, USDA, 79; instr & dept chair physics, Fletcher Sr High Sch, 80-86; lectr & amb space sci, Fla Dept Educ, 86-88; dir astron, Brest Planetarium, Mus Sci & Hist, 88-91; EXEC DIR, CHABOT OBSERV & SCI CTR, 91- *Concurrent Pos:* Adj prof astron, Univ NFla, 83-91; mem, Space Sci Adv Bd, Nat Sci Teachers Asn, 87-90; consult, Am Col Testing Bd, 90-91; bd mem, Adams Environ Adv Bd, Jacksonville, Fla, 90-91. *Mem:* Am Astron Soc; Nat Sci Teachers Asn. *Res:* Chemistry of photographic emulsions for use in astronomical photography, particularly planetary; meteoritics; writing techniques include the above, along with space sciences and techniques in astronomy instruction. *Mailing Add:* Chabot Observ & Sci Ctr 4917 Mountain Blvd Oakland CA 94611

REYNOLDS, ORLAND BRUCE, b Mountain Home, Idaho, Feb 15, 22; m 54; c 1. ANIMAL PHYSIOLOGY. *Educ:* Idaho State Col, BS, 44; Boston Univ, AM, 55, PhD(biochem), 60. *Prof Exp:* Fel biol, Harvard Univ, 60-61; instr, Sch Med, Boston Univ, 61-62; asst prof, Middlebury Col, 62-68; from asst to prof, 68-88, EMER PROF BIOL, NORTHERN MICH UNIV, 88- *Mem:* AAAS; Am Chem Soc; Sigma Xi; Comt Sci Invest Claims Paranormal. *Res:* Comparative physiology; vision. *Mailing Add:* 225 E Michigan St Marquette MI 49855-3823

REYNOLDS, ORR ESREY, b Baltimore, Md, Mar 3, 20; m 42, 71; c 2. PHYSIOLOGY. *Educ:* Univ Md, BS, 41, MS, 43, PhD(physiol), 46. *Prof Exp:* Asst physiologist, NIH, Md, 43-44; physiologist, Marine Corps Air Sta, Va, 44-45; physiologist, Aviation Br, Res Div, Bur Med, US Dept Navy, 45-46, dir biol sci div, Off Naval Res, 47-57; dir off sci, US Dept Defense, 57-62; dir biosci prog, NASA, 62-70; educ officer, 70-72, EXEC SECY-TREAS, AM PHYSIOL SOC, BETHESDA, MD, 72- *Concurrent Pos:* Exec dir surv physiol, Am Physiol Soc, 52-53, exec secy-treas, Soc, 72-; lectr, Univ Md, 47-51. *Mem:* AAAS; Am Physiol Soc; Aerospace Med Asn; Am Soc Zoologists; Am Inst Biol Sci. *Res:* Respiratory physiology; physiology of pain perception. *Mailing Add:* 2701 N Ocena Blvd No 17A Ft Lauderdale FL 33308-7575

REYNOLDS, PETER HERBERT, b Toronto, Ont, Sept 28, 40; m 67; c 2. GEOCHRONOLOGY, GEOPHYSICS. *Educ:* Univ Toronto, BSc, 63; Univ BC, PhD(geochronology), 67. *Prof Exp:* Nat Res Coun Can fel, Australian Nat Univ, 68-69; asst prof physics, 69-76, ASSOC PROF PHYSICS & GEOL, DALHOUSIE UNIV, 76- *Mem:* Can Geophys Union; Geol Asn Can. *Res:* Potassium-argon geochronology; oxygen isotope abundance patterns. *Mailing Add:* Dept Geol Dalhousie Univ Halifax NS B3H 4H6 Can

REYNOLDS, PETER JAMES, b New York, NY, Nov 19, 49. STATISTICAL PHYSICS, PHASE TRANSITIONS. *Educ:* Univ Calif, Berkeley, AB, 71; Mass Inst Technol, PhD(physics), 79. *Prof Exp:* RES ASSOC-INSTR PHYSICS, BOSTON UNIV, 79-, ASST RES PROF, 79- *Concurrent Pos:* Vis scientist, Nat Res Comput Chem, Lawrence Berkeley Labs, 80-81. *Mem:* Am Phys Soc; Sigma Xi; NY Acad Sci. *Res:* Phase transitions in disordered systems, particularly the critical properties of generalized percolation models near their connectivity threshold, and of polymeric systems, especially by renormalization group and Monte Carlo simulation techniques; quantum Monte Carlo studies of molecules. *Mailing Add:* 3008 Cedarwood Lane Falls Church VA 22042

REYNOLDS, RAY THOMAS, b Lexington, Ky, Sept 2, 33; m 62; c 2. PLANETARY SCIENCES. *Educ:* Univ Ky, BS, 54, MS, 60. *Prof Exp:* Proj scientist, Am Geog Soc, Thule, Greenland, 60-61; res scientist, Theoret Studies Br, Space Div, 62-70, chief, 70-79, res scientist, 79-88, ASSOC THEORET & PLANETARY STUDIES BR, SPACE SCI DIV, AMES RES CTR, NASA, 88- *Honors & Awards:* Newcombe-Cleveland Award, AAAS, 79; Except Sci Achievement Medal, NASA, 80. *Mem:* AAAS; Am Astron Soc; fel Am Geophys Union; fel Meteoritical Soc; Am Inst Aeronaut & Astronaut. *Res:* Theoretical studies of the origin, evolution and present state of the solar system with emphasis upon the composition, structure and thermal history of the planets and their satellites. *Mailing Add:* Theoret Studies Br Space Sci Div NASA Ames Res Ctr Moffett Field CA 94035

REYNOLDS, RICHARD ALAN, b Los Angeles, Calif, Dec 14, 38; m 63; c 2. MATERIALS SCIENCE, SOLID STATE PHYSICS. *Educ:* Stanford Univ, BSc, 60, MSc, 63, PhD(mat sci), 66; Univ Sheffield, MSc, 61. *Prof Exp:* Mem tech staff math physics, Cent Res Labs, Tex Instruments, Inc, 65-75; DIR DEFENSE SCI OFF, DEFENSE ADVAN RES PROJ AGENCY, 75- *Mem:* Inst Elec & Electronics Engrs Electrochem Soc; Infrared Info Asn; Am Ceramic Soc. *Res:* Electron transport in semiconductors; photoconductive processes in infrared detector materials; materials preparation; processing of semiconductors; compound semiconductors; display technology; high temperature mechanical properties of metals. *Mailing Add:* 4500 S Four Mile Run Rd Arlington VA 22204

REYNOLDS, RICHARD CLYDE, b Saugerties, NY, Sept 2, 29; m 54; c 3. MEDICAL EDUCATION. *Educ:* Rutgers Univ, New Brunswick, BS, 49; Johns Hopkins Univ, MD, 53. *Hon Degrees:* DSc, Hahneman Univ, 89. *Prof Exp:* From intern to resident med, Johns Hopkins Hosp, 53-55, resident, 57-58, fel med, 58-59; pvt pract, 59-68; assoc prof med, 68-71, asst dean community health, 71-73, prof community health & family med & chmn dept, Col Med, Univ Fla, 71-78; dean & prof med, Rutgers Med Sch, 78-88, sr vpres acad affairs, 84-88; EXEC VPRES, ROBERT WOOD JOHNSON FOUND, 88- *Mem:* Am Col Physicians; AMA; Am Acad Family Physicians. *Res:* Rural health; problems of health care delivery; evaluation of medical education. *Mailing Add:* Robert Wood Johnson Found PO Box 2316 Princeton NJ 08543

REYNOLDS, RICHARD JOHNSON, organic chemistry; deceased, see previous edition for last biography

REYNOLDS, RICHARD TRUMAN, b Oakland, Calif, Dec 31, 42. ECOLOGY, ZOOLOGY. *Educ:* Ore State Univ, BS, 70, MS, 75, PhD(wildlife ecol), 79. *Prof Exp:* Res assoc, Dept Fisheries & Wildlife, Ore State Univ, 73-74; wildlife biologist res, Bur Land Mgt, 78-79, RES ANIMAL ECOLOGIST, FOREST SERV, ROCKY MOUNTAIN FOREST & RANGE EXP STA, USDA, 79 - *Concurrent Pos:* Consult res, Forest Serv Range & Wildlife Habitat Lab, 74 & Fish & Wildlife Serv, USDA, 75; affil fac, Dept Fisheries & Wildlife Biol, Colo State Univ, 81- *Mem:* Cooper's Ornith Soc; Am Ornith Union; Ecol Soc Am; Raptor Res Found; Asn Field Ornithologists. *Res:* Community and behavioral ecology of vertebrates with an emphasis on the morphological, behavioral and ecological underpinnings of their habitat requirements and preferences. *Mailing Add:* Rocky Mountain Forest & Range Exp Sta 222 S 22nd St Laramie WY 82070

REYNOLDS, ROBERT COLTART, JR, b Scranton, Pa, Oct 4, 27; m 50; c 3. PETROLOGY, GEOCHEMISTRY. *Educ:* Lafayette Col, BA, 51; Wash Univ, PhD(geol), 55. *Prof Exp:* Sr res engr, Res Ctr, Pan-Am Petrol Corp, 55-60; from asst prof to assoc prof, 60-69, PROF GEOL, DARTMOUTH COL, 69- *Concurrent Pos:* Instr, Benedictine Heights Col, 56-60; expert, US Army Cold Regions Res & Eng Lab, 64-; consult, Oak Ridge Nat Lab, 66. *Mem:* Geochem Soc; fel Geol Soc Am; fel Mineral Soc Am; Clay Minerals Soc. *Res:* Rates and types of chemical weathering and aqueous transport in extreme environments; clay mineralogy and ion exchange processes on clays; computer modeling of the chemistry of aquatic systems. *Mailing Add:* Eight Brook Rd Hanover NH 03755

REYNOLDS, ROBERT D, b Butler, Pa, Dec 11, 44. CARDIOVASCULAR CLINICAL RESEARCH. *Educ:* Clarion State, BA, 70; Univ Cincinnati, PhD(physiol), 74. *Prof Exp:* MEM STAFF, CLIN RES DEPT, MARION MERRELL DOW, INC. *Mem:* Am Heart Asn; Am Soc Pharmacol & Exp Therapeut. *Mailing Add:* Clin Res Dept Marion Merrell Dow Inc 5401 E 103rd St Baptiste Facil Kansas City MO 64137

REYNOLDS, ROBERT DAVID, b Mansfield, Ohio, June 25, 43; m 64; c 2. NUTRITION, BIOLOGICAL CHEMISTRY. *Educ:* Ohio State Univ, Columbus, BS, 65; Univ Wis-Madison, PhD(cancer res), 71. *Prof Exp:* Res fel cancer res, Biochem Inst, Univ Freiburg, WGer, 71-72; asst mem cancer res, Fred Hutchinson Cancer Res Ctr, Seattle, 72-73; res assoc vitamin D metabol, Dept Biochem, Univ Wis, 73-75; RES CHEMIST VITAMIN B6 NUTRIT, VITAMIN & MINERAL NUTRIT LAB, HUMAN NUTRIT RES CTR, USDA, 75- *Concurrent Pos:* Fel, Damon Runyon Mem Fund for Cancer Res, 71-72; adj prof, Dept Human Nutrit & Food Systs Univ Md, 80-; hon consult, Child's Health Ctr, Warsaw, Poland, 87. *Mem:* AAAS; Am Inst Nutrit; Am Soc Clin Nutrit. *Res:* Energy metabolism at high altitude; nutritional requirements of vitamin B-6. *Mailing Add:* Vitamin & Mineral Nutrit Lab Human Nutrit Res Ctr Agr Res Serv USDA Beltsville MD 20705

REYNOLDS, ROBERT EUGENE, b Dallas, Tex, Nov 25, 34; m 56, 70; c 3. PHYSICS. *Educ:* Univ Tex, BA(math) & BS(physics), 56, MA, 58, PhD(physics), 61. *Prof Exp:* From asst prof to assoc prof, 63-79, PROF PHYSICS, REED COL, 79- *Mem:* Am Phys Soc; Am Asn Physics Teachers; Am Geophys Union; Fedn Am Scientists. *Res:* Theoretical physics. *Mailing Add:* Dept Physics Reed Col 3203 SE Woodstock Blvd Portland OR 97202

REYNOLDS, ROBERT N, b Troy, NY, Feb 26, 22; m 51; c 3. ANESTHESIOLOGY. *Educ:* Yale Univ, BS, 44; Albany Med Col, MD, 46. *Prof Exp:* Instr surg anesthesia, 52-54, from asst prof to assoc prof anesthesia, 54-66, PROF ANESTHESIA, SCH MED, TUFTS UNIV, 66- *Concurrent Pos:* Asst anesthesiol, New Eng Med Ctr Hosp, Boston, Mass, 51-53, asst anesthetist, 54-55, anesthetist, 55-67, sr anesthetist, 67-; consult pediat anesthesiol, US Naval Hosp, Chelsea, Mass, 68- *Mem:* AMA; Am Soc Anesthesiol; Am Acad Pediat; Am Soc Pharmacol & Exp Therapeut; Asn Univ Anesthetists; Sigma Xi. *Res:* Respiratory physiology in infants; pediatric anesthesia. *Mailing Add:* 46 Homestead Park Needham MA 02194

REYNOLDS, ROBERT WARE, b Kingsport, Tenn, Sept 21, 42; m 65; c 2. SOLID STATE PHYSICS. *Educ:* Davidson Col, BS, 64; Vanderbilt Univ, PhD(physics), 69. *Prof Exp:* Res scientist, Advan Technol Ctr, Inc, 69-76; MEM TECH STAFF, GEN RES CORP, 76- *Mem:* Am Phys Soc; Sigma Xi. *Res:* Military systems analysis; high energy laser effects and propagation modeling; electron paramagnetic resonance spectroscopy; infrared detection. *Mailing Add:* Sparta Inc 4901 Corporate Dr NW-102 Huntsville AL 35805

REYNOLDS, ROBERT WILLIAMS, b Buffalo, NY, Feb 9, 27; m 51; c 2. PHYSIOLOGY, PSYCHOLOGY. *Educ:* Cornell Univ, AB, 49; Univ Buffalo, MA, 50, PhD(psychol), 57. *Prof Exp:* Res chemist, E I du Pont de Nemours & Co, 52-53; asst psychol, Univ Buffalo, 54-55; from instr to assoc prof psychol, 56-72, chmn dept, 73-76, PROF PSYCHOL, UNIV CALIF, SANTA BARBARA, 72- *Concurrent Pos:* Fel, Dept Physiol & Biophys, Univ Wash, 61-62. *Mem:* AAAS; Am Physiol Soc; Psychonomic Soc. *Res:* Influence of drugs on the nervous system and motivated behavior; nervous system and problems of learning and consciousness; physiological basis of motivations, emotions and attention. *Mailing Add:* Dept Psychol Univ Calif Santa Barbara CA 93106

REYNOLDS, ROGER SMITH, b Las Vegas, Nev, Oct 28, 43; m 65; c 2. NUCLEAR ENGINEERING. *Educ:* Univ Nev, Reno, BS, 65; Kans State Univ, MS, 69, PhD(nuclear eng), 71. *Prof Exp:* Res asst struct shielding res, Kans State Univ, 67-69, exp supvr, 69-70, proj dir, 70-71; prof nuclear eng, Miss State Univ, 71-87; SR ENGR, ADVAN NUCLEAR FUELS CORP, 87- *Mem:* Am Nuclear Soc; Sigma Xi; Nat Soc Prof Engr. *Res:* Radiation detection and measurement; neutron activation analysis; dosimetry; shielding; thermal hydraulic analysis. *Mailing Add:* 1211 Adair Dr Richland WA 99352

REYNOLDS, ROLLAND C, b Pomona, Calif, Feb 18, 25; m 46; c 1. PATHOLOGY. *Educ:* Southern Methodist Univ, BA, 49; Univ Tex, MD, 56; Am Bd Path, cert path anat & clin path. *Prof Exp:* Intern, Dallas Vet Admin Hosp, 56; fel path, 57-61, from asst prof to assoc prof, 61-70, PROF PATH, UNIV TEX HEALTH SCI CTR DALLAS, 70- *Concurrent Pos:* Resident, Parkland Mem Hosp, 57-61, pathologist & jr attend staff, 61-; dir labs chest div, Woodlawn Hosp, 62-74. *Mem:* Am Thoracic Soc; Am Soc Microbiol; Am Soc Cell Biol; Am Soc Clin Path; Electron Micros Soc. *Res:* Experimental pathology of nucleus and nucleolus in experimental carcinogenesis and inhibition of nucleic acid synthesis; experimental pulmonary disease. *Mailing Add:* Dept Path Univ Tex Southwestern Med Sch Dallas TX 75235

REYNOLDS, RONALD J, b Chicago Heights, Ill, May 17, 43; m 66; c 2. SPACE PHYSICS, ASTRONOMY. *Educ:* Univ Ill-Champaign, BS, 65; Univ Wis-Madison, MS, 67, PhD(physics), 71. *Prof Exp:* Res assoc, Nat Acad Sci-Nat Res Coun, Goddard Space Ctr, NASA, 71-73; res assoc & lectr physics, 73-76, asst scientist, 76-81, assoc scientist, 81-87, SR SCIENTIST, DEPT PHYSICS, UNIV WIS-MADISON, 87- *Mem:* Am Astron Soc; Am Inst Physics; Int Astron Union. *Res:* Detection and spectroscopic analysis of faint emission lines from the interstellar medium and the earth's upper atmosphere. *Mailing Add:* Dept Physics Univ Wis Madison WI 53706

REYNOLDS, ROSALIE DEAN (SIBERT), b Jacksonville, Ill, Mar 8, 26; m 48. ORGANIC CHEMISTRY. *Educ:* Ill Col, AB, 47; Univ Wyo, MS, 50, PhD(org chem), 53. *Prof Exp:* Instr chem, Univ Wyo, 52; res assoc, Univ Colo, 53-54; lectr & res assoc, Univ Southern Calif, 54-55; asst prof, Univ Wyo, 55-60; from assoc prof to prof chem, Northern Ill Univ, 60-81; CONSULT, ACAD PRESS, 73- *Mem:* Am Chem Soc; Sigma Xi. *Res:* Theoretical and synthetic organic chemistry. *Mailing Add:* 1852 Perry Ct Sycamore IL 60178

REYNOLDS, SAMUEL D, JR, b Upper Darby, Pa, Dec 19, 31; m 54, 84; c 4. WELDING CODES & STANDARDS, WRITING WELDING HANDBOOK CHAPTERS. *Educ:* Lehigh Univ, BSME, 53. *Prof Exp:* Engr/sr engr, Heat Transfer Div, Westinghouse Elec Corp, 57-68, mgr mat eng, Heat Transfer Div, 68-77, fel engr, Tampa Div, 77-80, lead reactor eng, US Nuclear Regulatory Comn, 80-86, FEL ENGR, PGBU, WESTINGHOUSE ELEC CORP, 86- *Concurrent Pos:* Fac mem, Temple Univ Eve Col, 57-62 & Drexel Inst Technol (Univ) Evening Col, 62-68; mem, Filler Metals Comn, Am Welding Soc, 70- & Sect IX, Am Soc Mech Engrs, 80- *Mem:* Fel Am Soc Metals; Am Welding Soc; Nat Asn Corrosion Engrs. *Res:* Development of specialized welding procedures for power plant heat exchangers, pressure vessels, steam turbines and electrical generators. *Mailing Add:* Westinghouse Elec Corp PGBU 4400 Alafaya Trail MC303 Orlando FL 32826-2399

REYNOLDS, STEPHEN EDWARD, b Decatur, Ill, Dec 11, 16; m 38; c 1. MECHANICAL ENGINEERING. *Educ:* Univ NMex, BS, 39. *Hon Degrees:* HDL, NMex State Univ, 77. *Prof Exp:* From apprentice engr to process engr, Phillips Petrol Co, Okla, 39-42; asst prof mech eng, Univ NMex, 42-43 & 46; proj supvr, Thunderstorm Res Prog, NMex Inst Mining & Technol, 46-55; STATE ENGR, NMEX, 55- *Concurrent Pos:* Consult, President's Adv Comt Weather Modification, 54-57; mem adv panel weather modification, NSF, 60-63; mem, President's Water Pollution Control Adv Bd, 67-70 & NMex Water Qual Control Comn, 67-; mem comt water qual policy, Nat Acad Sci, 73-; mem comt water resources, Natural Resources Coun, 75- *Mem:* AAAS; Am Phys Soc; Am Meteorol Soc; Sigma Xi. *Res:* Atmospheric electricity; cloud physics; properties of water; water resources development and management. *Mailing Add:* 1213 N Laguna Ave Farmington NM 87401-7071

REYNOLDS, TELFER BARKLEY, b Regina, Sask, July 30, 21; nat US; m 55; c 2. INTERNAL MEDICINE. *Educ:* Univ Calif, Los Angeles, AB, 41; Univ Southern Calif, MD, 45; Am Bd Internal Med, dipl, 53. *Hon Degrees:* Dr, Univ Montpellier, 85. *Prof Exp:* Chief resident physician, Los Angeles County Gen Hosp, 50-51; Giannini res fel med, Hammersmith Hosp, London, Eng, 52; res fel, 53, from asst prof to prof, 53-78, CLAYTON G LOOSLI PROF MED, SCH MED, UNIV SOUTHERN CALIF, 78- *Mem:* Am Soc Clin Invest; Asn Am Physicians; master Am Col Physicians. *Res:* Liver diseases. *Mailing Add:* Dept Internal Med Sch Med Univ Southern Calif Los Angeles CA 90033

REYNOLDS, THOMAS DE WITT, b Detroit, Mich, July 25, 29; m 51; c 3. PSYCHIATRY. *Educ:* Univ Chicago, BA, 47, MD, 55. *Prof Exp:* Intern med, George Washington Univ Hosp, 55-56; resident psychiat, 56-57 & 59-61, asst dir behav studies, 61-70, supvry med officer psychiat res, W A White Serv, 70-71, chief, 71-72, CLIN DIR, W A WHITE SERV, ST ELIZABETHS HOSP, WASH, DC, 72- *Concurrent Pos:* Assoc prof, George Washington Univ, 67- *Res:* Mathematical approaches to behavioral time series; stability characteristics of such time series in schizophrenic and other types of psychiatric patients. *Mailing Add:* Barton Hall St Elizabeths Hosp Martin L King Jr Ave Washington DC 20032

REYNOLDS, TOM DAVIDSON, b Gatesville, Tex, Apr 2, 29; m 54; c 2. ENVIRONMENTAL ENGINEERING. *Educ:* Tex A&M Univ, BSCE, 50; Univ Tex, MSSE, 61, PhD(civil eng), 63. *Prof Exp:* Proj engr, Lockwood, Andrews & Newnan, Consult Engrs, 50-51 & 53-59; asst prof civil eng, Univ Tex, 64-65; from asst prof to assoc prof, 65-77, PROF CIVIL ENG, TEX

A&M UNIV, 77- *Mem:* Am Soc Civil Engrs; Am Water Works Asn; Water Pollution Control Fedn. *Res:* Investigations concerning water and waste treatment, particularly industrial waste treatment using biological processes. *Mailing Add:* Dept Civil Eng Tex A&M Univ College Station TX 77843

REYNOLDS, VERNON H, b Oak Park, Ill, Dec 31, 26; c 2. SURGERY, ONCOLOGY. *Educ:* Vanderbilt Univ, BA, 52, MD, 55. *Prof Exp:* Res assoc microbiol & surg, 60-61, asst prof, 62-69, ASSOC PROF SURG, SCH MED, VANDERBILT UNIV, 69- *Concurrent Pos:* Intern & resident surg, Peter Bent Brigham Hosp, Boston, 55-62; NIH fel exp path, 57-58; Arthur Tracy Cabot teaching fel surg, Harvard Med Sch, 61-62; Am Cancer Soc adv clin fel, 62-65; Markle scholar med sci, 62- *Mem:* Am Col Surgeons; Sigma Xi. *Res:* Carbohydrate chemistry; tumor metabolism; cancer chemotherapy. *Mailing Add:* Dept Surg Vanderbilt Univ Med Sch Nashville TN 37232

REYNOLDS, WARREN LIND, b Gull Lake, Sask, Nov 29, 20; m 46; c 3. INORGANIC CHEMISTRY. *Educ:* Univ BC, BA, 49, MA, 50; Univ Minn, PhD, 55. *Prof Exp:* Asst, 52-54, lectr anal chem, 54-55, lectr inorg chem, 55-56, from asst prof to assoc prof, 56-67, PROF INORG CHEM, UNIV MINN, MINNEAPOLIS, 67- *Concurrent Pos:* NSF sr fel, 62-63; Fulbright-Hays res award, 72-73. *Mem:* Am Chem Soc; Am Phys Soc; Royal Soc Chem. *Res:* Kinetics of electron-transfer and substitution reactions; solvation numbers, labilities and contact shifts of metal ions in non-aqueous solvents; bonding in inorganic species. *Mailing Add:* Dept Chem Univ Minn Minneapolis MN 55455

REYNOLDS, WILLIAM CRAIG, b Berkeley, Calif, Mar 16, 33; m 53; c 3. MECHANICAL ENGINEERING. *Educ:* Stanford Univ, BS, 54, MS, 55, PhD(mech eng), 57. *Prof Exp:* Aeronaut res scientist, NASA-Ames Lab, 55; from asst prof to assoc prof mech eng, Stanford Univ, 57-66, chmn dept, 72-82, chmn, Inst Energy Studies, 74-82, PROF MECH ENG, STANFORD UNIV, 66-, CHMN DEPT, 89- *Concurrent Pos:* Nuclear engr, Aerojet-Gen Nucleonics Div, Gen Tire & Rubber Co, 57; NSF sr fel, Nat Phys Lab, UK, 64-65; vis prof, Pa State Univ, 72; Fairchild scholar, Calif Inst Technol, 84; co-chmn, Stanford Integrated Mfg Asn, 90-; consult fluid & appl mech. *Honors & Awards:* G Edwin Burks Award, Am Soc Eng Educ, 72; Fluids Eng Award, Am Soc Mech Engrs, 89. *Mem:* Nat Acad Eng; fel Am Soc Mech Engrs; fel Am Phys Soc (secy-treas, 84 & 85); Am Asn Univ Professors; Sigma Xi; Am Soc Eng Educ; Am Inst Aeronaut & Astronaut. *Res:* Blowdown thermodynamics; ignition of metals; non-isothermal heat transfer; zero-g fluid mechanics; turbulent boundary layer flow structure; turbulence-wall interactions; stability of gas films; stability of laminar and turbulent flows; boundary-layer calculation methods; surface-tension-driven flows; organized waves in turbulent shear flows; turbulence computation; unsteady turbulent boundary layers; internal combustion engine cylinder flows; unsteady jets and separating flows; turbulence modeling. *Mailing Add:* Dept Mech Eng Stanford Univ Stanford CA 94305-3030

REYNOLDS, WILLIAM FRANCIS, b Boston, Mass, Jan 31, 30; m 62; c 2. ALGEBRA. *Educ:* Col of the Holy Cross, AB, 50; Harvard Univ, AM, 51, PhD(math), 54. *Prof Exp:* Res fel math, Harvard Univ, 54-55; C L E Moore instr, Mass Inst Technol, 55-57; from asst prof to prof, 57-70, WALKER PROF MATH, TUFTS UNIV, 70- *Concurrent Pos:* Instr, Col Holy Cross, 54-55; prin investr, NSF grants, Tufts Univ, 65-74. *Mem:* AAAS; Am Math Soc; Math Asn Am. *Res:* Representation theory of finite groups, especially modular and projective representations with applications to structure of groups. *Mailing Add:* Three Preble Gardens Rd Belmont MA 02178

REYNOLDS, WILLIAM ROGER, b Chicago, Ill, Dec 27, 29; m 56; c 5. ZEOLITE PETROLOGY. *Educ:* Univ Wis, BS, 58; Fla State Univ, MS, 62, PhD(geol), 66. *Prof Exp:* Sr geologist, Pan Am Petrol Corp, 66-68; asst prof, 68-73, ASSOC PROF GEOL, UNIV MISS, 73- *Mem:* Soc Econ Paleontologists & Mineralogists; Nat Asn Geol Teachers; Sigma Xi; Clay Minerals Soc; Am Geol Soc. *Res:* Stratigraphy; clay mineralogy; sedimentation. *Mailing Add:* Dept Geol & Geol Eng Univ Miss University MS 38677

REYNOLDS, WILLIAM WALTER, b Pasadena, Calif, Jan 29, 25; m 53; c 3. PHYSICAL CHEMISTRY. *Educ:* Univ Calif, BS, 48. *Prof Exp:* Res chemist, Shell Oil Co, 48-55, group leader, 55-60, asst chief res chemist, 60-63, spec analyst, 63-65, mgr prod planning, Petrochem Div, Shell Chem Co, 65-67, mgr mkt res, 67-68, mgr econ coord, Chem Econ Dept, 68-75, mgr chem prod econ, 75-79, mgr chem stategic studies, 79-81, econ consult, Prod Econ Dept, 81-87; CONSULT, 87- *Concurrent Pos:* Mem adv bd, Corp Planner Roundtable, Duke Univ. *Honors & Awards:* Roon Award, 57, 58. *Mem:* Am Chem Soc; Chem Mkt Res Asn. *Res:* Petroleum solvents; surface coatings technology; polymer solutions; antioxidants; high temperature fluids; wear of internal combustion engines; petroleum derived additives; petrochemicals; market planning and decision theory; micro-economics; organizational effectiveness. *Mailing Add:* 23519 Creekview Dr Spring TX 77389

REYNOLDS, WYNETKA ANN KING, b Coffeyville, Kans, Nov 3, 37; m 58; c 2. EMBRYOLOGY. *Educ:* Kans State Teachers Col, BS, 58; Univ Iowa, MS, 60, PhD(zool), 62. *Hon Degrees:* DSc, Ind State Univ, 80, Ball State Univ, 85 & Emporia State Univ, 87; LHD, McKendree Col, 84 & Univ NC, Charlotte, 88; PhD, Fu Jen Cath Univ, Taiwan, 87. *Prof Exp:* Asst prof biol, Ball State Univ, 62-65; from asst prof to prof anat, Univ Ill Med Ctr, 65-79, res prof obstet & gynec, 73-79, assoc vchancellor res & dean, Grad Col, 77-79; prof anat, obstet, gynec & provost, Ohio State Univ, 79-82; CHANCELLOR, CALIF STATE UNIV, 82- *Concurrent Pos:* Mem biol bd, Grad Record Exam, Am Inst Biol Sci; mem comt nutrit mother & presch children, Nat Acad Sci; mem, Primate Adv Bd, Res Resources, NIH. *Honors & Awards:* Prize Award, Cent Asn Obstet & Gynec, 68. *Mem:* Am Asn Anat; Am Soc Zoologists; Soc Develop Biol; Soc Gynec Invest; Endocrine Soc; Sigma Xi. *Res:* Transplantation of endocrine pancreas; calcium metabolism in pregnancy; toxicity of methylmercury for fetus and neonate; nutrition during development of fetus. *Mailing Add:* City Univ NY 535 E 80th St New York NY 10021

REYNOLDS-WARNHOFF, PATRICIA, b Washington, DC, Feb 26, 33; m 56; c 3. ORGANIC CHEMISTRY. *Educ:* Trinity Col, DC, AB, 54; Mass Inst Technol, SM, 59; Univ Southern Calif, PhD(org chem), 62. *Prof Exp:* Res asst org chem, Nat Heart Inst, 55-57; res assoc org biochem, 62-63, instr, 63-64, sessional lectr org chem, 64-65, ASST PROF ORG CHEM, UNIV WESTERN ONT, 65- *Res:* Natural product structure determination; mechanisms of epoxide rearrangements; rearrangements of alpha-haloketones; hydride transfer reactions. *Mailing Add:* Dept Chem Univ Western Ont London ON N6A 5B7 Can

REYNOSO, GUSTAVO D, b Gomez Palacio, Mex, Sept 18, 32; US citizen; m 59; c 4. MEDICINE, PATHOLOGY. *Educ:* Ateneo Fuente Univ, Mex, BS, 50; Univ Nuevo Leon, MD, 58. *Prof Exp:* Resident path, St Luke's Hosp, Milwaukee, Wis & Marquette Univ, 61; M K Kellogg Found grant, prof path, Sch Med & dir clin labs, Univ Hosp, Univ Nuevo Leon, 61-65; asst clin prof path, Sch Med, Marquette Univ, 65-68; chief cancer res pathologist, Roswell Park Mem Inst, 68-72; CHIEF PATH, WILSON MEM HOSP, 72- *Concurrent Pos:* Assoc pathologist, St Luke's Hosp, Milwaukee, 65-68; asst prof exp path, State Univ NY Buffalo, 68-72; mem immunol subcomt, Nat Colorectal Cancer Prog, 71- *Mem:* AAAS; Am Soc Clin Pathologists; Col Am Pathologists; Asn Clin Scientists; Am Asn Clin Chemists. *Res:* Cancer immunology; biochemical diagnosis of cancer; hormonal interactions in the cancer patient. *Mailing Add:* Lept Path Wilson Mem Hosp 3557 Harnson St Johnson City NY 13790

REZA, FAZLOLLAH M, b Resht, Iran, Jan 1, 15; m 45; c 6. ELECTRICAL CIRCUIT THEORY, INFORMATION THEORY. *Educ:* Teheran Univ, BS & MS, 38; Columbia Univ, MS, 46; Polytech Inst NY, PhD(elec eng), 50. *Prof Exp:* Mem staff elec eng, Mass Inst Technol, 51-55 & Syracuse Univ, 55-68; ambassador, Univ Paris, 69-74; ADJ PROF SYSTS & COMMUN, MCGILL UNIV, CAN, 75- & CONCORDIA UNIV, CAN, 79- *Concurrent Pos:* Vis prof, Polytech Zurich, Swiss Fed Inst Technol, 62-63, Univ Colo, Boulder, 62 & 65 & Royal Tech Univ, Copenhagen, 63; pres, Tech Univ Sharif (Aryamehr), Iran, 67-68 & Tehran Univ, 68-69; ambassador Iran to Unesco, Paris, 69-74 & Iran to Can, 74-78; hon prof, Polytech Univ, NY, 75 & McGill Univ, Can, 78; hon mem bd, Atomic Energy Iran, 89. *Mem:* Fel Inst Elec & Electronics Engrs; emer mem Am Math Soc; fel AAAS; NY Acad Sci; Sigma Xi. *Mailing Add:* Five Sadhurst Ct Ottawa ON K1V 9W9 Can

REZAK, MICHAEL, b Feldafing, Ger, Sept 10, 48; US citizen. NEUROANATOMY, NEUROPHYSIOLOGY. *Educ:* Univ Wis, BA, 70; Bradley Univ, MA, 72; Univ Ill, PhD(anat-neuroanat), 76. *Prof Exp:* Assoc anat, 76-77, ASST PROF, DEPT ANAT, MED CTR, UNIV ILL, 77- *Mem:* Soc Neurosci; Asn Res Vision & Ophthal; AAAS; Am Soc Primatologists; Am Psychol Asn. *Res:* Neuroanatomical and neurophysiological organization of the mammalian central nervous system. *Mailing Add:* Anat Dept Univ Ill Med Ctr 1853 W Polk St Chicago IL 60680

REZAK, RICHARD, b Syracuse, NY, Apr 26, 20; m 65; c 1. GEOLOGICAL OCEANOGRAPHY. *Educ:* Syracuse Univ, AB, 47, PhD(geol), 57; Wash Univ, AM, 49. *Prof Exp:* Instr geol, St Lawrence Univ, 49-51; geologist, US Geol Surv, 52-58; res geologist, Shell Develop Co, 58-63, res assoc, 63-67; from assoc prof to prof, 67-90, VIS PROF OCEANOG, TEX A&M UNIV, 91- *Concurrent Pos:* Prin investr, US Minerals Mgt Serv, 74-83, mem outer continental shelf adv bd & regional tech working group, Gulf Mex region, 83-; consult, Espy-Houston, Inc & Racal Decca Surv, Inc, 81-84, Tech Disciplines, Inc, 89-90. *Mem:* Fel Geol Soc Am; Soc Econ Paleont & Mineral; Int Phycol Soc; fel AAAS; Am Asn Petrol Geologists. *Res:* Systematics and environmental significance of fossil algae; contributions of algae to carbonate sediments; carbonate sedimentology and diagenesis, especially cementation; seismic stratigraphy and salt tectonics. *Mailing Add:* 3600 Stillmeadow Dr Bryan TX 77802-3324

REZANKA, IVAN, b Prachatice, Czech, Sept 30, 31; m 59; c 2. NUMERICAL METHODS. *Educ:* Charles Univ, Prague, MS, 54; Czech Acad Sci, PhD(physics), 62. *Prof Exp:* Res asst aerodynamics, Res & Testing Inst Aerodynamics, Czech, 54-55; from physicist to sr physicist, Nuclear Res Inst, 55-68; res assoc nuclear physics, Res Inst Physics, Stockholm, 68-69; res assoc & lectr, Heavy Ion Accelerator Lab, Yale Univ, 69-74, asst dir, 71-74; scientist, 74-81, sr scientist, 81-83, MGR, XEROX CORP, 84- *Concurrent Pos:* Consult, Doll Res, Inc, 74-76. *Honors & Awards:* Czech Acad Sci Award, 62, Prize, 68. *Mem:* Am Phys Soc. *Res:* Nuclear physics, gamma ray and beta ray spectroscopy; computer hardware and software; activation analysis; medical physics; applied physics, fluid dynamics, continuum mechanics, acoustics; xerography; electrography; ink jet physics. *Mailing Add:* Six Squire Lane Pittsford NY 14534

REZEK, GEOFFREY ROBERT, b New York, NY, Nov 23, 41; m 72; c 2. COMPUTER INTEGRATED MANUFACTURING, PROJECT MANAGEMENT. *Educ:* Long Island Univ, BS, 64, MS, 65. *Prof Exp:* Indust engr, Components Div, Int Bus Mach Corp, 65-67, systs engr, 67-69, mfg indust specialist, 69-83, mfg indust mkt & consult, US Mkt & Serv, 83-91; COMPUTER INTEGRATED MFG CONSULT, G R REZEK & ASSOCS, 91- *Mem:* Inst Indust Engrs; Am Prod & Inventory Control Soc. *Res:* Computer integrated manufacturing to improve productivity and process simplification. *Mailing Add:* 110 Raymond St Darien CT 06820

REZNICEK, ANTON ALBERT, b Plochingen, Ger, June 11, 50; Can citizen; m 78. PLANT TAXONOMY, PHYTOGEOGRAPHY. *Educ:* Univ Guelph, BSc, 71; Univ Toronto, MSc, 73, PhD(bot), 78. *Prof Exp:* asst cur, 78-87, ASSOC CUR, UNIV MICH, ANN ARBOR, 87- *Concurrent Pos:* Dir, Matthaei Bot Gardens, Univ Mich, Ann Arbor, 87-89. *Mem:* Am Soc Plant Taxonomists; Int Asn Plant Taxon; Bot Soc Am. *Res:* Systematics of Cyperaceae wordwide, primarily Carex; geography of the North American flora, especially disjunct species; plant migrations, persistence of relict species, and dynamics of plant communities that harbor relict species; floristics of the Great Lakes region; history of botanical exploration. *Mailing Add:* Univ Mich Herbarium N Univ Bldg Ann Arbor MI 48109-1057

REZNICK, BRUCE ARIE, b New York, NY, Feb 3, 53. ANALYSIS, COMBINATORICS. *Educ:* Calif Inst Technol, BS, 73; Stanford Univ, PhD(math), 76. *Prof Exp:* Asst prof math, Duke Univ, 76-78; NSF fel, Univ Calif, Berkeley, 78-79; asst prof, 79-83, assoc prof, 83-89, PROF MATH, UNIV ILL, URBANA-CHAMPAIGN, 89- *Concurrent Pos:* Putnam Probs Comn, 82-85; Sloan fel, 83-87. *Mem:* Am Math Soc; Math Asn Am; Asn Women Math. *Res:* Algebra and number theory which are susceptible to combinational methods including polynomials, lattice points and inequalities. *Mailing Add:* Math Dept Univ Ill 1409 W Green St Urbana IL 61801

REZNIKOFF, WILLIAM STANTON, b New York, NY, Apr 29, 41; m 67; c 3. MOLECULAR GENETICS. *Educ:* Williams Col, Mass, BA, 63; Johns Hopkins Univ, PhD(biol), 67. *Prof Exp:* Fel bact genetics, Harvard Med Sch, 68-70; from asst prof to assoc prof biochem, 70-78, PROF BIOCHEM, COL AGR & LIFE SCI, UNIV WIS-MADISON, 78- *Concurrent Pos:* Nat Inst Gen Med Sci career develop award, 72-77; Harry & Evelyn Steenbock career develop award, Univ Wis, 74-78; mem, Adv Subcomt Genetic Biol, 76-79; mem, Rev Comt Cellular & Molecular Basis Dis, Nat Inst Gen Med Sci, 80-83; bd dirs, Promega, 80-91; consult, Biogen SA, 81. *Mem:* Am Soc Microbiol; Am Soc Biol Chemists. *Res:* Analysis of the structure and function of genetic regulatory regions, promoters, operators, associated with genes in Escherichia coli and its viruses; analysis of bacterial transposable elements. *Mailing Add:* Dept Biochem Univ Wis 420 Henry Mall Madison WI 53706

RHAMY, ROBERT KEITH, b Indianapolis, Ind, Nov 26, 27; m 51; c 4. UROLOGY, PHYSIOLOGY. *Educ:* Ind Univ, BS, 49, MS, 50, MD, 52; Am Bd Urol, dipl, 62. *Prof Exp:* Res assoc physiol, Ind Univ, 48-49, intern, 52-53, asst resident, 55-56, resident urol, 56-59, from instr to assoc prof, 59-64; PROF UROL & HEAD DEPT, SCH MED, VANDERBILT UNIV, 64-, CHIEF UROL SURG, UNIV HOSP, 64- *Concurrent Pos:* Chief urol, Nashville Gen Hosp, 64-; consult, US Vet Admin Hosp, 64-, Jr League Home Crippled Children, 65- & USPHS, 65-; prog consult, Nat Found, 65-; mem, Int Cong Nephrology, Nat Dialysis Comt & Nat Kidney Found. *Mem:* Am Urol Asn; Soc Pediat Urol; Am Acad Pediat; Am Soc Nephrology; Am Fertil Soc. *Res:* Renal hypertension; pediatric urology; ureteral physiology; chemical carcinogenesis. *Mailing Add:* Dept Urol Vanderbilt Univ T2106 Med Ctr Nashville TN 37232

RHEAD, WILLIAM JAMES, b Paris, France, Feb 20, 46; m 72; c 2. MEDICAL GENETICS, BIOCHEMICAL GENETICS. *Educ:* Univ Calif, San Diego, BA, 68, PhD(chem), 74, MD, 74; Yale Univ, MPh, 69. *Prof Exp:* From asst prof to assoc prof, 79-89, PROF PEDIAT, UNIV IOWA, 89-, ACTG DIR, DIV MED GENETICS, 91- *Concurrent Pos:* Fac partic, Inter-Dept PhD Prog in Genetics, Univ Iowa, 79-, Human Nutrit, 80-, Neurosci, 84-; pres & bd dirs, Asn Glycogen Storage Dis, 89-; bd dirs, Soc Inherited Metab Dis, 81-94. *Honors & Awards:* Noel Raine Award, Soc Study Inborn Errors Metab, 82. *Mem:* Am Soc Human Genetics; AAAS; Soc Study Inborn Errors Metab; Soc Pediat Res. *Res:* Enzymatic, molecular and genetic defects causing inherited diseases of fatty acid oxidation, mitochondrial function and energy metabolism in man. *Mailing Add:* Dept Pediat Iowa City IA 52242

RHEE, CHOON JAI, b Pyungyang, Korea, Sept 13, 35; m 64. TOPOLOGY. *Educ:* Univ of the South, BA, 60; Univ Ga, MA, 62, PhD(math), 65. *Prof Exp:* Asst prof math, Randolph-Macon Woman's Col, 65-66; asst prof, 66-71, assoc prof, 71-86, PROF MATH, WAYNE STATE UNIV, 87- *Mem:* Am Math Soc; Math Asn Am; Sigma Xi. *Res:* Homotopy functors; point set topology. *Mailing Add:* Dept Math Wayne State Univ 5950 Cass Ave Detroit MI 48202

RHEE, G-YULL, b Kyunggi-Do, Korea, Feb, 10, 39; US citizen; m 68; c 2. PHYSIOLOGICAL ECOLOGY OF ALGAE. *Educ:* Seoul Nat Univ, BS, 61; Northeastern Univ, Boston, MS, 67; Cornell Univ, Ithaca, NY, PhD(aquatic microbiol), 71. *Prof Exp:* From res scientist I to res scientist IV, 71-80, RES SCIENTIST V, CTR LABS & RES, NY DEPT HEALTH, ALBANY, 80- *Concurrent Pos:* Adj assoc prof, Cornell Univ, Ithaca, NY, 80- *Mem:* Am Soc Limnol & Oceanog; Phycol Asn Am; Am Soc Microbiol; AAAS. *Res:* Effects of environmental factors in algae physiology and ecology. *Mailing Add:* 213 Featherwood Ct Schenectady NY 12303

RHEE, HAEWUN, b Seoul, Korea, Sept 12, 37; m 65; c 2. MATHEMATICS. *Educ:* Johns Hopkins Univ, AB, 60; Univ Mass, Amherst, PhD(math), 68. *Prof Exp:* Instr math, Ohio Wesleyan Univ, 64-65; asst prof, Am Int Col, 67-68; assoc prof, 68-72, PROF MATH, STATE UNIV NY COL ONEONTA, 72- *Mem:* Am Math Soc. *Res:* Partial differential equations. *Mailing Add:* Dept Math State Univ NY Col Oneonta NY 13820

RHEE, JAY JEA-YONG, b Seoul, Korea, Oct 3, 37; US citizen; m 63; c 2. CHEMICAL PHYSICS, PHYSICAL CHEMISTRY. *Educ:* Univ La Verne, BA, 62; Univ NMex, MS, 66, PhD(chem), 67. *Prof Exp:* From asst prof to assoc prof, 68-77, PROF CHEM, UNIV LA VERNE, 77- *Concurrent Pos:* Vis assoc prof chem, Univ Calif, Davis, 75; consult, Synthane-Taylore Corp, La Verne, Calif; resident dir, Korea Telecommun Co, Ltd, Bell Tel, Antwerp, Belgium, 79-81. *Mem:* Am Chem Soc; Am Inst Physics; Korean Chem Soc. *Res:* Multiphoton spectroscopy and ODMR applications; laser applications on isotope separations; quantum theory; molecular spectroscopy. *Mailing Add:* 2803-B Blossom Lane Redondo CA 90278-2009

RHEE, MOON-JHONG, b Shinanchoo, Korea, Feb 19, 35; m 66; c 4. PLASMA PHYSICS. *Educ:* Seoul Nat Univ, BS, 58, MS, 60; Cath Univ Am, PhD(appl physics), 70. *Prof Exp:* Instr physics, Seoul Nat Univ, 64-66; res fel appl physics, Cath Univ Am, 66-70; from asst prof to assoc prof, 70-83, PROF ELEC ENG, UNIV MD, 83- *Concurrent Pos:* Vis prof, Seoul Nat Univ, 77-78 & Nat Univ Rosario, 85; consult, Naval Surface Weapons Ctr, 80- *Mem:* Am Phys Soc; Korean Phys Soc; Inst Elec & Electronics Engrs. *Res:* Collective ion acceleration; generation and measurement of charged particles in plasma focus; microwave generation; pulsed power system; picosecond optoelectronic switching. *Mailing Add:* 2305 Nees Lane Silver Spring MD 20904

RHEE, SEONG KWAN, b Mokpo, Korea, Aug 18, 36; US citizen; m 75; c 4. MATERIALS SCIENCE, METALLURGY. *Educ:* Chosun Univ, BS, 59; Univ Cincinnati, MS, 62, PhD(mat sci), 66. *Prof Exp:* Mem tech staff, Sci Res Inst, Korean Ministry Nat Defense, 59-60; DIR, MAT TECHNOL GROUP, ALLIED-SIGNAL AUTOMOTIVE TECH CTR, 82- *Concurrent Pos:* Chmn, int conf wear of mat, 81. *Mem:* Metall Soc; Am Soc Mech Engrs; fel Am Ceramic Soc; fel Am Inst Chemist; Nat Inst Ceramic Engrs. *Res:* Surface energy of solids; oxidation of metals; alloying behavior of metals; composite materials; tribology and friction materials; abrasive ceramics; grain growth in metals and ceramics; hard coatings; automotive sensors. *Mailing Add:* Mat Technol Group Allied Signal Automotive Tech Ctr 900 W Maple Rd Troy MI 48083

RHEE, SUE GOO, b Seoul, Korea, July 6, 43; m; c 2. ENZYMOLOGY. *Educ:* Seoul Nat Univ, BS, 65; Catholic Univ Am, PhD(org chem), 71. *Prof Exp:* Ordnance Maint Officer, Korean Army, 65-67; teaching asst, Phys Chem Lab, Catholic Univ Am, Wash, DC, 67-68, res asst, 68-71, postdoctoral fel, 71-72; postdoctoral fel, State Univ NY, Binghamton, 72-73; postdoctoral fel, 73-74, vis scientist, 74-75, staff fel, 75-79, sr biochem, 79-88, CHIEF, SECT SIGNAL TRANSDUCTION, LAB BIOCHEM, NAT HEART LUNG BLOOD INST, NIH, 88- *Mem:* Am Soc Biochem & Molecular Biol; Am Chem Soc. *Res:* Signal transduction. *Mailing Add:* Nat Heart Lung & Blood Inst Bldg 3 Rm 203 Bethesda MD 20205

RHEES, RAYMOND CHARLES, b Ogden, Utah, Jan 29, 14; m 38; c 5. INORGANIC CHEMISTRY, ANALYTICAL CHEMISTRY. *Educ:* Utah State Univ, BS, 40, MS, 44; Iowa State Univ, PhD(light scattering), 51. *Prof Exp:* Chemist, Kalunite Inc, 42-44; instr chem, Utah State Univ, 44-45 & Iowa State Univ, 45-51; sr chemist, Union Carbide Nuclear Corp, 51-52, sect head atomic energy, 52-53, dept head, 53-56; sect head anal chem, Am Potash & Chem Corp, 56-60, sect head boron hydrides, 60-63, sect head tech serv, 63-68; dir res & develop, 68-70, VPRES RES, PAC ENG & PROD CO NEV, 70- *Mem:* Am Chem Soc; Water Pollution Control Fedn; Electrochem Soc; Air Pollution Control Asn. *Res:* Light scattering; analytical and uranium chemistry; fine particle properties; boron hydrides; chlorates and perchlorates; borates; chemical specialties; electrolytic processes in pollution control. *Mailing Add:* 657 Sixth St Boulder City NV 89005-2941

RHEES, REUBEN WARD, b Ogden, Utah, Apr 1, 41; m 63; c 5. NEUROENDOCRINOLOGY. *Educ:* Univ Utah, BS, 67; Colo State Univ, PhD(physiol), 71. *Prof Exp:* Res asst physiol, Colo State Univ, 67-70, teaching asst physiol, 70-71; fel anat, Univ Utah Med Sch, 71-72; asst prof physiol, Weber State Col, 72-73; from asst prof to assoc prof zool, 73-84, PROF, ZOOL, BRIGHAM YOUNG UNIV, 84- *Concurrent Pos:* Vis prof, Anatomy, UCLA Med Sch. *Mem:* Am Physiol Soc. *Res:* Neuroendocrine mechanisms by which hormones exert regulatory and behavioral effects on the central nervous system; interrelationships between the endocrine and nervous systems. *Mailing Add:* Dept Zool Brigham Young Univ Provo UT 84602

RHEIN, ROBERT ALDEN, b San Francisco, Calif, Aug 18, 33; m 56; c 5. INORGANIC CHEMISTRY, POLYMER CHEMISTRY. *Educ:* Univ Calif, BS, 55; Univ Pittsburgh, MS, 58; Univ Wash, PhD(chem), 62. *Prof Exp:* Assoc engr, Westinghouse Elec Corp, 55-58; eng designer, Boeing Airplane Co, 58-60; sr engr chem, Jet Propulsion Lab, Calif Inst Technol, 62-79; SR ENGR CHEM, NAVAL WEAPONS CTR, 79- *Concurrent Pos:* Postdoctoral appointment, Univ Mass, 83-84. *Mem:* Am Chem Soc. *Res:* Ultrasonics in liquid systems; inorganic fluorine chemistry; powdered metals combustion; propellant chemistry; silicon chemistry; elastomers. *Mailing Add:* 424 Kendall Ave Ridgecrest CA 93555

RHEINBOLDT, WERNER CARL, b Berlin, Ger, Sept 18, 27; US citizen; m 59; c 2. NUMERICAL ANALYSIS. *Educ:* Heidelberg Univ, dipl, 52; Univ Freiburg, PhD(math), 55. *Prof Exp:* Mathematician aerodyn, Eng Bur Blume, Ger, 55-56; fel appl math, Inst Fluid Dynamics & Appl Math, Univ Md; 56-57; mathematician numerical anal, Comput Lab, Nat Bur Standards, 57-59; asst prof & dir comput ctr, Syracuse Univ, 59-62; dir comput sci ctr, Univ Md, College Park, 62-65, res assoc prof, Inst Fluid Dynamics & Appl Math, 62-63, res prof, 63-72, res prof comput sci ctr, 68-78, prof, Dept Math, 72-78, dir appl math prog, 74-78; ANDREW W MELLON PROF MATH, UNIV PITTSBURGH, 78- *Concurrent Pos:* Asn Comput Mach rep, Nat Acad Sci-Nat Res Coun, 65-67; ed, J Numerical Anal, Soc Indust & Appl Math, 65-, managing ed, 70-73; consult ed, Acad Press, 67-; vis prof, Soc Math Data Processing, Bonn, Ger, 69; consult, Div Comput Res, NSF, 72-75 & 81-89; mem, Adv Comt, Army Res Off, 74-78, exec comt, 81-84; chmn, Comt Appl Math, Nat Res Coun, 79- 85, mem, Bd Math Sci, 84-90; chmn, Bd Trustees, Soc Indust & Appl Math. *Honors & Awards:* Sr Alex V Humboldt Award, 88. *Mem:* Am Math Soc; Soc Indust & Appl Math (vpres, 76, pres, 77-78); fel AAAS. *Res:* Applied and computational mathematics; computer applications. *Mailing Add:* Dept Math & Statist Univ Pittsburgh Pittsburgh PA 15260

RHEINGOLD, ARNOLD L, b Chicago, Ill, Oct 6, 40; m 66; c 2. ORGANOMETALLIC CHEMISTRY, CRYSTALLOGRAPHY. *Educ:* Case Western Reserve Univ, AB, 62, MS, 63; Univ Md, PhD(inorg chem), 69. *Prof Exp:* Proj mgr organometall chem, Glidden-Durkee Div, SCM Corp, 63-65; res fel, Va Polytech Inst, 69-70; from asst prof to prof chem, State Univ NY Col Plattsburgh, 70-81; vis prof, 81-82, assoc prof, 82-86, PROF, CHEM, UNIV DEL, 86- *Mem:* AAAS; Am Chem Soc; Am Crystal Asn. *Res:* Transition metal/main group cluster synthesis; synthesis and characterization of main group compounds; preparative electrochemistry; main-group homoatomic ring and chain structures; crystallography. *Mailing Add:* Dept Chem Univ Del Newark DE 19716

RHEINLANDER, HAROLD F, b Ashland, Maine, June 10, 19; m 42. SURGERY. *Educ:* Univ Maine, BA, 41; Harvard Med Sch, MD, 44; Am Bd Surg, dipl, 52; Am Bd Thoracic Surg, dipl, 62. *Prof Exp:* Asst surg, Harvard Univ, 45-46, John Milton fel, 48-49; from instr to assoc prof, 49-66, PROF

SURG, SCH MED, TUFTS UNIV, 66-, VCHMN, DEPT SURG, 79- *Concurrent Pos:* Intern, Peter Bent Brigham Hosp, 44-45, resident, 45-46; resident, Childrens Hosp, Boston, 48-49; resident, New Eng Ctr Hosp, 49-51, asst surg, 50-52, asst surgeon, 52-58, surgeon & chief thoracic serv, 58-; consult, Boston Vet Admin Hosp, 66- *Mem:* Soc Thoracic Surg; Am Asn Thoracic Surg; Am Soc Artificial Internal Organs; Am Col Chest Physicians; Soc Vasc Surg; Am Col Surg. *Mailing Add:* 171 Harrison Ave Boston MA 02111

RHEINS, LAWRENCE A, b Cincinnati, Ohio, Mar 8, 55; m 77; c 2. IMMUNODERMATOTOXICOLOGY. *Educ:* Univ Cincinnati, BS, 78, MS, 80, PhD(biol sci), 84. *Prof Exp:* Postdoctoral fel, Dept Dermat, Univ Cincinnati, 84-86, asst prof dermat, 86-88; mgr, Skin Care Lab, Procter & Gamble, 88-90; ASST DIR CLIN SAFETY & TOXICOL, HILL TOP RES INC, 90- *Concurrent Pos:* Dermat Found res grant, 85. *Mem:* Am Asn Immunologists; Soc Investigative Dermat; Soc Pediat Dermat; NY Acad Sci; Soc Cosmetic Chemists; Cosmetic Toiletry & Fragrance Asn. *Res:* Effects of environmental toxicants on the immune cells of the skin, immunodermatotoxicology; mechanism of action of inflammatory mediations of the skin. *Mailing Add:* 343 Poage Farm Rd Cincinnati OH 45215

RHEINS, MELVIN S, b Cincinnati, Ohio, May 13, 20; m 48; c 2. MICROBIOLOGY, IMMUNOLOGY. *Educ:* Miami Univ, BA & MA, 46; Ohio State Univ, PhD(microbiol), 49. *Prof Exp:* From instr to assoc prof pathogenic microbiol, 49-59, actg chmn dept microbiol, 64-65, chmn, 65-67, PROF MICROBIOL, OHIO STATE UNIV, 59- *Concurrent Pos:* Spec consult, USPHS, 61-63. *Mem:* Fel Am Acad Microbiol; Am Soc Microbiol; NY Acad Sci; Sigma Xi. *Res:* Pathogenesis and immunology of pulmonary diseases, collagen diseases and ocular infection, diseases and malignancies. *Mailing Add:* Dept Microbiol 484 W 12th Ave Columbus OH 43210

RHEINSTEIN, JOHN, b Gardelegen, Ger, May 23, 30; US citizen; m 56; c 3. PHYSICS. *Educ:* Dartmouth Col, AB, 51; Univ Chicago, MS, 57; Munich Tech Univ, PhD(physics), 61. *Prof Exp:* Lectr physics, Munich Br, Univ Md, 58-60; staff mem systs analysis, Mass Inst Technol, 61-66, assoc group leader, 66-69, from asst site mgr to assoc site mgr, 69-73, group leader, Lincoln Lab, 73-90; CONSULT, 90- *Mem:* AAAS; Inst Elec & Electronics Engrs. *Res:* Medical physics; electromagnetic theory; reentry physics; systems analysis. *Mailing Add:* Ten Gould Rd Lexington MA 02173

RHEINSTEIN, PETER HOWARD, b Cleveland, Ohio, Sept 7, 43; m 69; c 1. PHARMACEUTICAL REGULATION, ADMINISTRATION OF HEALTH CARE DELIVERY. *Educ:* Mich State Univ, BA, 63, MS, 64; Johns Hopkins Univ, MD, 67; Univ Md, JD, 73; Am Bd Family Pract, dipl, 77. *Prof Exp:* Med intern, US Pub Health Serv Hosp, San Francisco, 67-68, med resident, Baltimore, 68-70; instr, Univ Md Sch Med, 70-73; dir, Div Drug Advertising & Labeling, US Food & Drug Admin, 74-82, actg dep dir, 83-84, actg dir, Off Drugs, 83-84, dir, Off Drug Standards, 84-90, DIR, MED STAFF, OFF HEALTH AFFAIRS, US FOOD & DRUG ADMIN, 90- *Concurrent Pos:* Med dir extended care facils, CHC Corp, 72-74; adj prof forensic med, George Washington Univ, 74-76; adv on essential drugs, Regional Off Southeast Asia, Manila, WHO, 81-; Food & Drugs Admin deleg to US Pharmacopeial Conv, Inc, 85-90; bd dirs, Drug Info Asn, 82-90; mem bd gov, Am Col Legal Med, 83-; chmn ann meeting, Drug Info Asn, 91. *Honors & Awards:* President's Award, Am Col Legal Med, 85, 86, 89 & 90; Outstanding Serv Award, Drug Info Asn, 90. *Mem:* Drug Info Asn (pres, 84-85, 88-89); fel Am Col Legal Med (treas, 85-88 & 90-); fel Am Col Family Physicians; AMA; Am Bar Asn; Fed Bar Asn. *Res:* Evaluation of medical technologies to determine safety, effectiveness and cost; mechanisms by which technologies become known to the professions and to the public and by which they are incorporated into medical practice. *Mailing Add:* 621 Holly Ridge Rd Severna Park MD 21146-3520

RHEINWALD, JAMES GEORGE, b Chicago, Ill, June 25, 48; m 68; c 1. CYTOLOGY, BIOCHEMISTRY. *Educ:* Univ Ill, Urbana, BS & MS, 70; Mass Inst Technol, PhD(cell biol), 75. *Prof Exp:* Res assoc cell biol, Mass Inst Technol, 76-78; asst scientist tumor biol, Dana-Farber Cancer Inst, 78-83, assoc prof physiol, 83-90; from asst prof to assoc prof physiol, Harvard Med Sch, 78-90; VPRES, RES & DEVELOP, BIOSURFACE TECH, INC, CAMBRIDGE, MA, 91- *Honors & Awards:* Am Cancer Soc Fac Res Award. *Mem:* Tissue Cult Asn; Am Soc Cell Biol; Int Soc Differentiation. *Res:* Growth control and differentiated function in human epithelial tissues and cultured cells; identification and study of tissue- and tumor-specific keratin proteins; cancer-related phenotypes expressed in cell culture. *Mailing Add:* Biosurface Tech Inc Univ Park 64 Sidney St Cambridge MA 02139

RHEMTULLA, AKBAR HUSSEIN, b Zanzibar, Tanzania, June 8, 39; Can citizen; m 67; c 3. GROUP THEORY, ORDERED STRUCTURES. *Educ:* Univ Cambridge, PhD(math), 67. *Prof Exp:* PROF MATH, UNIV ALTA, 68- *Concurrent Pos:* Vis prof, Univ Sao Paulo, Brazil, 82, Universita dejli Studi, Napoli, Italy, 88; mem bd, Can Math Soc, 85- *Mem:* Am Math Soc; Can Math Soc. *Res:* Author of 50 publications in algebra, mostly in group theory and ordered structures. *Mailing Add:* Dept Math Univ Alta Edmonton AB T6G 2G1 Can

RHIM, JOHNG SIK, b Korea, July 24, 30; US citizen; m 62; c 6. VIROLOGY, MEDICAL SCIENCES. *Educ:* Seoul Nat Univ, BS, 53, MD, 57. *Prof Exp:* Res fel poliovirus, Children's Hosp Res Found, Cincinnati, Ohio, 58-60; res fel reovirus, Baylor Col Med, 60-61; res fel Japanese B encephalitis, Grad Sch Pub Health, Univ Pittsburgh, 61-62; res assoc infant diarrhea, Sch Med, La State Univ, Costa Rica, 62-64; vis scientist arbovirus, Nat Inst Allergy & Infectious Dis, 64-66; proj dir career res, Microbiol Asn Inc, 66-78; virologist, 78-80, SR INVESTR, NAT CANCER INST, 80-; ADJ, PROF, GEORGETOWN UNIV, WASH, DC, 87- *Concurrent Pos:* Mem bd dirs, Winchester Sch, Silver Spring, Md. *Mem:* AAAS; Am Asn Cancer Res; Am Asn Immunologists; Soc Exp Biol & Med; Am Soc Microbiol; AMA; Int Assoc Comp Leukemia Res; Int Soc Prev Oncol. *Res:* In vitro chemical physical and viral cocarcinogenesis; immunoprevention of cancer; mechamism of carcinogenesis; isolation and characterization of oncogenes. *Mailing Add:* Nat Cancer Inst NIH 9000 Rockville Pike Bldg 37 1C03 Bethesda MD 20892

RHIM, WON-KYU, b Seoul, Korea, Oct 20, 37; US citizen; m 64; c 4. PHYSICS, PHYSICAL CHEMISTRY. *Educ:* Seoul Nat Univ, Korea, BS, 61, MS, 63; Univ NC, Chapel Hill, PhD(physics), 70. *Prof Exp:* Res assoc chem phys, Mass Inst Technol, 69-71; res fel chem phys, Dept Chem Eng, 71-73, MEM TECH STAFF PHYSICS, JET PROPULSION LAB, CALIF INST TECHNOL, 73- *Mem:* Am Phys Soc. *Res:* Solid state nuclear spin dynamics and spin thermodynamics; study of magnetic interactions in solids; instrumentation for magnetic resonance experiments. *Mailing Add:* 1800 San Pasqual St Pasadena CA 91106

RHINEHART, ROBERT RUSSELL, II, b Neptune, NJ, Jan 19, 46; m 67; c 2. CHEMICAL PROCESS CONTROL PLASMA ETCHING, FLUIDS DYNAMICS. *Educ:* Unv Md, Col Park, BS, 68, MS, 69; NC State Univ, Raleigh, PhD(chem eng), 85. *Prof Exp:* Engr, Fibers Tech Ctr, Celanese Corp, 69-72, engr, Celriver Plant, 72-73, sr engr, Fibers Mkt Co, 73-80, area supvr, Celriver Plant, 80-82; teaching asst introd comput prog, NC State Univ, 82-85; ASST PROF CHEM ENG, TEX TECH UNIV, 85- *Concurrent Pos:* Tech chmn, IND-Asn Nonwoven Indust; proj engr coal gasification, Pilot Plant, NC State Univ, 83-85, consult, process control, 87-88. *Honors & Awards:* Schoenborn Award, Am Inst Chem Engrs, 85. *Mem:* Am Inst Chem Engrs; Instrument Soc Am; Am Soc Eng Educ. *Res:* Modeling, optimization, and control of chemical processes of industrial nature; plasma etching and semiconductors. *Mailing Add:* Dept Chem Eng Tex Tech Univ PO Box 4679 Lubbock TX 79409

RHINES, FREDERICK N(IMS), metallurgy; deceased, see previous edition for last biography

RHINES, PETER BROOMELL, b Hartford, Conn, July, 23, 42; m 68. FLUID DYNAMICS. *Educ:* Mass Inst Technol, BSc & MSc, 64; Cambridge Univ, UK, PhD(appl math & theoret physics), 67. *Prof Exp:* Fel oceanog, Dept Meteorol, Mass Inst Technol, 67-68, asst prof, 68-71; res scientist, Cambridge Univ, UK, 71-72; assoc scientist oceanog, Woods Hole Oceanog Inst, 72-74, sr scientist, 74-84; PROF OCEANOG & ATMOSPHERIC SCI, SCH OCEANOG, UNIV WASH, SEATTLE, 84- *Concurrent Pos:* Vis prof, Nat Ctr Atmospheric Res, 69 & 72, Univ BC, 75, Univ Colo, 76, Calif Inst Tech & Princeton Univ, 78; fel, Christ's Col, Cambridge, UK, 79-80; Guggenheim fel, Dept Appl Math & Theoret Physics, Cambridge Univ, 79-80; Green scholar, Inst Geophys & Planetary Physics, Univ Calif, San Diego, 81; Queen's fel, marine sci, Australia, 88. *Mem:* Nat Acad Sci; fel Am Geophys Union. *Res:* Circulation of the oceans, waves, eddies and currents; climate and transport of natural and artificial trace chemicals in the seas. *Mailing Add:* Sch Oceanog W B 10 Univ Wash Seattle WA 98195

RHINESMITH, HERBERT SILAS, b Westtown, NY, Oct 25, 07; m 39; c 2. ORGANIC CHEMISTRY. *Educ:* Wesleyan Univ, AB, 29, AM, 30; Harvard Univ, AM, 31, PhD(chem), 33. *Prof Exp:* Teaching fel chem, Wesleyan Univ, 34-36, instr, 36-38; from instr to prof, 38-74, EMER PROF CHEM, ALLEGHENY COL, 74- *Concurrent Pos:* Mem staff, Keystone Ord Works; res assoc, Calif Inst Technol, 55-57. *Mem:* AAAS; Am Chem Soc; fel Am Inst Chemists; NY Acad Sci. *Res:* Action of Grignard reagent on acetylenic esters; epoxidations with peracetic acid; beta, gamma-oxido esters; amino acid; protein chemistry; normal and abnormal hemoglobins; chemistry of serotonin, especially quantitative determinations. *Mailing Add:* 617 David Dr Oxford OH 45056

RHO, JINNQUE, b Korea, Sept 15, 38; US citizen; m 70; c 1. MICROBIOLOGY. *Educ:* Seoul Nat Univ, BS, 61; Clark Univ, MS, 69; Univ Mass, PhD(microbiol), 72. *Prof Exp:* Res asst, Inst Agr & Indust Microbiol, Univ Mass, 68-71, teaching asst soil microbiol, 71-72, postdoc fel, Dept Environ Sci, 72-74, sr res assoc, Water Res Ctr, Mass Agr Exp Sta, 74-77, asst prof aquatic microbiol, Dept Environ Sci, 77-79; asst prof, Univ Bridgeport, 79-82, Yale vis fac fel, 82-83, assoc prof, 85-89, ELIPHALET REMINGTON PROF ENVIRON MICROBIOL, UNIV BRIDGEPORT, 82-, PROF MICROBIOL, BIOL DEPT, 90-, YALE VIS FAC FEL, 90- *Concurrent Pos:* Consult, USDA Forest Service, Conn, New Eng Res Inc, Mass, Protech Inc, Conn, Aquapura Corp, Conn; coordr, Olin Corp Environ Sci & Engr Prog, 88-, co-prin investr nitrogen transformation by heterotrophs, Mass Agr Exp Sta; affil asst prof, Dept Biol, Clark Univ, 77-79. *Mem:* Am Soc Microbiol; Am Soc Limnol & Oceanog; Sigma Xi. *Res:* Ecology and biochemistry of heterotrophic nitrifying bacteria; copper resistant bacteria in aquatic environments; stabilizer compositions for stabilizing aqueous systems. *Mailing Add:* 36 Whitney Lane Orange CT 06477

RHO, JOON H, b Pyongbuk, Korea, Jan 19, 22; m 47; c 5. BIOCHEMISTRY. *Educ:* Seoul Nat Univ, BS, 50; Duke Univ, MS, 56, PhD(biochem), 58. *Prof Exp:* Asst prof biol, Sung Kyun Kwan Univ, Korea, 53-56, from asst prof to prof, 57-59; Nat Acad Sci fel, Calif Inst Technol, 59-62, sr scientist, Jet Propulsion Lab, 62-66, tech staff mem, 67-73, sr biologist, Calif Inst Technol, 73-75; assoc prof med & pharm, 75-82, PROF MED & PHARM, SCHS MED & PHARM, UNIV SOUTHERN CALIF, 82- *Concurrent Pos:* Prin investr, Apollo Sample Analysis Porphyrin Compounds, 71-74; mem bd gov, State Bar Calif. *Mem:* Am Chem Soc; AAAS; Int Soc Study Origin Life; Soc Exp Biol Med & Neurosci. *Res:* Neurogenic hypertension; chemical carcinogenesis, metabolism of antihypertentive drugs; fluorometric analyses of biological and biomedicinal compounds. *Mailing Add:* Dept Pharm & Med Univ S Calif 1985 Zonal Ave Los Angeles CA 90033

RHOADES, BILLY EUGENE, b Lima, Ohio, Sept 27, 28; m 49; c 2. MATHEMATICS. *Educ:* Ohio Northern Univ, AB, 51; Rutgers Univ, MS, 53; Lehigh Univ, PhD, 58. *Prof Exp:* Asst math, Rutgers Univ, 52-53; from instr to prof, Lafayette Col, 53-65; assoc prof, 65-69, PROF MATH, IND UNIV, BLOOMINGTON, 69- *Mem:* Am Math Soc; Math Asn Am; Nat Coun Teachers Math. *Res:* Transformations in sequence spaces; fixed point theorems. *Mailing Add:* 3128 Coppertree Dr Bloomington IN 47401

RHOADES, EVERETT RONALD, b Lawton, Okla, Oct 24, 31; m 53; c 5. INTERNAL MEDICINE, MICROBIOLOGY. *Educ:* Univ Okla, MD, 56. *Prof Exp:* Intern, Gorgas Hosp, CZ, 56-57; resident med, Med Ctr, Univ Okla, 57-60, clin asst, 60-61; chief infectious dis, Wilford Hall, USAF Hosp, 61-66; from asst prof med & microbiol to prof med, Med Ctr, Univ Okla, 66-82, chief infectious dis, 68-82; DIR, INDIAN HEALTH SERV, 82- *Concurrent Pos:* Consult, Med Ctr, Univ Okla, 63-66, Surgeon Gen, 65-66 & Eastern & Western Okla Tuberc Sanatarium, 66-75; chief infectious dis, Vet Admin Hosp, Oklahoma City, 66-; mem adv coun, Nat Inst Allergy & Infectious Dis; mem health comt, Asn Am Indian Affairs; Markle scholar acad med, 67-72. *Mem:* Am Soc Microbiol; Am Fedn Clin Res; Am Col Physicians; Asn Am Indian Physicians (pres, 75); Infectious Dis Soc Am. *Res:* Various aspects of cryptococcosis, including the effect of antifungal compounds on organisms and humans; host-parasite factors in lower respiratory infections. *Mailing Add:* 921 N 13th St Univ Okla Med Ctr Oklahoma City OK 73104

RHOADES, HARLAN LEON, b Tuscola, Ill, Mar 7, 28; m 53; c 2. PLANT NEMATOLOGY. *Educ:* Univ Ill, BS, 52, MS, 57, PhD(plant path), 59. *Prof Exp:* Soil conservationist, Agr Res Serv, USDA, 52-55; asst plant nematologist, Univ Fla, 59-67, assoc prof & assoc nematologist, 67-73, prof nematol & nematologist, 73-90; RETIRED. *Mem:* Soc Nematol. *Res:* Control of plant nematodes attacking vegetables. *Mailing Add:* 106 Crystal View S Sanford FL 32773-4808

RHOADES, JAMES DAVID, b Tulare, Calif, May 13, 37; m 58; c 3. SOIL SCIENCE, CLAY MINERALOGY. *Educ:* Univ Calif, Davis, BS, 62, MS, 63; Univ Calif, Riverside, PhD, 66. *Prof Exp:* Res soil chemist, 65-74, res leader, 74-88, DIR, US SALINITY LAB, AGR RES SERV, USDA, 89- *Concurrent Pos:* adj prof soil sci, Univ Calif, Riverside. *Honors & Awards:* Appl Res Award, Soil Sci Soc Am, 89. *Mem:* Fel Am Soc Agron; fel Soil Sci Soc Am; Soil & Water Conserv Soc; Int Comn Irrig & Drainage. *Res:* Interactions between salts in waters and soils; assessment of soil salinity; fixation of mineral elements by clay minerals; water quality criteria; use of saline waters for irrigation. *Mailing Add:* 17065 Harlow Heights Dr Riverside CA 92503

RHOADES, JAMES LAWRENCE, b Mishawaka, Ind, Apr 24, 33; m 58; c 4. ENZYMOLOGY, PROTEIN CHEMISTRY. *Educ:* Purdue Univ, BS, 55, MS, 57, PhD(chem), 61. *Prof Exp:* From asst prof to assoc prof chem, Northwestern State Col, La, 60-70; prof chem, 70-80, WILLIAM E REID PROF CHEM, BERRY COL, 80-, CHMN DEPT, 70- *Concurrent Pos:* Vis assoc prof chem, Purdue Univ, 65, 67 & 69; res assoc, Univ Tenn, Memphis, 63, 75, 76, 77, 79 & 84 & Emory Univ, 88. *Mem:* Am Chem Soc; Sigma Xi. *Res:* Chemistry and mechanism of action of enzymes, especially the plant phenolase complex; biosynthesis of coenzyme A; B-protein assay for cancer; liver transglutaminase. *Mailing Add:* Dept Chem Berry Col Mt Berry GA 30149

RHOADES, MARCUS MORTON, b Graham, Mo, July 24, 03; m 31; c 2. GENETICS. *Educ:* Univ Mich, BS, 27, MS, 28; Cornell Univ, PhD(genetics), 32. *Hon Degrees:* DSc, Ind Univ, 82. *Prof Exp:* Researcher genetics, Cornell Univ, 28-29; plant breeding, 30-35; agent, USDA, 35-37, geneticist, Arlington Exp Farm, 37-40; from assoc prof to prof bot, Columbia Univ, 40-48; prof, Univ Ill, 48-58; prof & chmn dept, 58-68, distinguished prof, 68-74, EMER PROF BOT, IND UNIV, BLOOMINGTON, 74- *Concurrent Pos:* Managing ed, Genetics, 40-47; vis prof, Univ Sao Paulo, 47-48, NC State Col, 53 & Cornell Univ, 56; Jesup lectr, Columbia Univ, 58; mem vis comt biol, Brookhaven Nat Lab, 62-65; mem div comt biol & med, NSF, 63-66; mem genetics study sect, NIH, 63-66; mem exec comt, Div Biol & Agr, Nat Res Coun, 63-66; hon fel, Australian Nat Univ, 65-66; mem selection comt, Guggenheim Mem Found, 65-76. *Honors & Awards:* Thomas Hunt Morgan Medal, Genetics Soc Am. *Mem:* Nat Acad Sci; Am Soc Naturalists; Genetics Soc Am (vpres, 42, pres, 43); Bot Soc Am; Am Genetic Asn (pres, 50-53); Am Philos Soc; Am Acad Arts & Sci. *Res:* Cytogenetics of maize. *Mailing Add:* Dept Biol Ind Univ Bloomington IN 47401

RHOADES, RICHARD G, b Northampton, Mass, Aug 15, 38; m 67; c 3. CHEMICAL ENGINEERING, MATHEMATICS. *Educ:* Rensselaer Polytech Inst, BChE, 60, PhD(chem eng, math), 64; Mass Inst Technol, MS, 77. *Prof Exp:* Res chem engr, Propulsion Lab, US Army Missile Command, 63-66, prog mgr air breathing propulsion, 66-72, dir, Army Rocket Propulsion Technol & Mgt Ctr, 68-70, chief, Adv Res Projs Div, 69, group leader ballistic missile defense propulsion, 70-73, dir propulsion directorate, US Army Missile Res & Develop Command, 73-81, assoc dir technol, 81-88, ASSOC DIR SYSTS, US ARMY MISSILE COMMAND, 89- *Concurrent Pos:* Adj asst prof, Univ Ala, Huntsville, 65-69. *Honors & Awards:* Firepower Award, Am Defense Preparedness Asn, 89. *Mem:* Am Inst Chem Engrs; Sigma Xi. *Res:* Fluid dynamics of packed beds; gas generation and pressurization for missiles; analytical techniques and methodology for propulsion system selection and technology planning; air breathing propulsion system analysis, design and experimentation. *Mailing Add:* 3604 Lookout Dr Huntsville AL 35801

RHOADES, RODNEY A, b Greenville, Ohio, Jan 5, 39; m 61; c 2. PHYSIOLOGY. *Educ:* Miami Univ, BS, 61, MS, 63; Ohio State Univ, PhD(physiol), 66. *Prof Exp:* NASA fel, Ohio State Univ, 64-66; asst prof appl physiol, Pa State Univ, 66-72, assoc prof biol, 72-75; scientist, NIH, 75-76; assoc prof, 77-81, PROF PHYSIOL & CHMN DEPT, SCH MED, IND UNIV, 81- *Concurrent Pos:* NIH career develop award, 75-80. *Mem:* AAAS; Am Physiol Soc; Biophys Soc; Am Thoracic Soc; Soc Exp Biol & Med. *Res:* Pulmonary circulation. *Mailing Add:* Dept Physiol-MS 374-A Ind Univ Sch Med 635 Barnhill Dr Indianapolis IN 46223

RHOADS, ALLEN R, b Reading, Pa, Dec 19, 41; m 69. ENZYMOLOGY. *Educ:* Kutztown State Univ, BS, 66; Univ Md, PhD(biochem), 71. *Prof Exp:* Fel biochem & pharmacol, 71-72, asst prof, 72-77, ASSOC PROF BIOCHEM, COL MED, HOWARD UNIV, 77- *Concurrent Pos:* Prin investr regulation cyclic 3 & nucleotide phosphodiesterases, USPHS grant, 78-; investr, Cancer Ctr, Howard Univ, 79- *Mem:* Am Chem Soc; AAAS; NY Acad Sci. *Res:* Regulation of cyclic nucleotides and calcium; structure and function of cyclic 3',5'- nucleotide phosphodiesterase and calmodulin in cellular function of cardiac and nervous tissue. *Mailing Add:* Dept Biochem Col Med Howard Univ 520 W ST NW Washington DC 20059

RHOADS, DONALD CAVE, b Rockford, Ill, Feb 14, 38; m 59; c 2. PALEOECOLOGY. *Educ:* Cornell Col, BA, 60; Univ Iowa, MS, 63; Univ Chicago, PhD(paleo zool), 65. *Prof Exp:* From assoc prof to prof geol, Yale Univ, 65-85; SR SCIENTIST, SCI APPLNS INT CORP, 85- *Concurrent Pos:* adj prof, Boston Univ, 88- *Res:* Organism-sediment relationships in marine environment; biogenic sedimentary structures in marine sediments; environmental events recorded in skeletal structures of marine invertebrates; biogenic processes on the seafloor; ecology of low oxygen marine environments. *Mailing Add:* Sci Applns Int Corp 89 Water St Woods Hole MA 02543

RHOADS, FREDERICK MILTON, b New Site, Miss, Jan 12, 36; m 53; c 2. SOIL CHEMISTRY, SOIL PHYSICS. *Educ:* Miss State Univ, BS, 58, PhD(soil chem), 66; Tex A&M Univ, MS, 63. *Prof Exp:* Soil scientist, Soil Conserv Serv, USDA, 58-61; asst soil chemist, 66-72, assoc soil chemist, 72-78, PROF SOIL SCI, N FLA RES & EDUC CTR, UNIV FLA, 78- *Mem:* Soil Sci Soc Am. *Res:* Plant nutrition of field and vegetable crops; irrigation of field and vegetable crops; soil fertility and testing; soil-water. *Mailing Add:* Rte 3 Box 4370 Quincy FL 32351

RHOADS, GEORGE GRANT, b Philadelphia, Pa, Feb 11, 40; m 65; c 2. EPIDEMIOLOGY, INTERNAL MEDICINE. *Educ:* Haverford Col, BA, 61; Harvard Univ, MD, 65; Univ Hawaii, MPH, 70. *Prof Exp:* Intern, Univ Pa Hosp, 65-66, resident, 66-68; lt commander, Heart Dis Control Prog, USPHS, 68-70; asst dir, Honolulu Heart Study, 70-71 & 72-74; from assoc to prof pub health, Univ Hawaii, 74-82, chmn, Dept Pub Health Sci, 78-81; CHIEF, EPIDEMIOL BR, NAT INST CHILD HEALTH & HUMAN DEVELOP, NIH, 82- *Mem:* AAAS; Soc Epidemiol Res; Int Epidemiol Asn; Am Pub Health Asn; Am Epidemiol Soc. *Res:* Epidemiology of chronic disease; coronary heart disease and stroke in Japanese migrants and their descendants; epidemiology of low birth weight, congenial malformations and other problems in maternal and child health. *Mailing Add:* Dept Environ Comm Med 675 Hoes Lane Piscataway MD 08854

RHOADS, JOHN MCFARLANE, b Vineland, NJ, Jan 7, 19; m 44; c 4. PSYCHIATRY. *Educ:* Va Polytech Inst, BS, 40; Temple Univ, MD, 43, MSc, 52; Am Bd Psychiat & Neurol, dipl, 50. *Prof Exp:* Instr psychiat, Sch Med, Temple Univ, 49-53, assoc, 53-55, asst prof, 55-56; assoc prof, 56-62, PROF PSYCHIAT, SCH MED, DUKE UNIV, 62- *Concurrent Pos:* Consult, US Vet Admin Hosp, Philadelphia, 53-56 & Durham, 57-; lectr, Sch Theol, Temple Univ, 53-56; instr, Wash Psychoanal Inst, 58-65; fac, Univ NC/Duke Psychoanal Inst, 65-89. *Mem:* AMA; Am Psychiat Asn; Am Psychoanal Asn. *Res:* Psychotherapy and psychoanalysis; relationships between psychiatry and religion; integration of psychotherapies. *Mailing Add:* Dept Psychiat Duke Univ Sch Med Box 3903 Durham NC 27710

RHOADS, JONATHAN EVANS, b Philadelphia, Pa, May 9, 07; m 36, 90; c 6. SURGERY. *Educ:* Haverford Col, BA, 28; Johns Hopkins Univ, MD, 32; Univ Pa, DSc(med), 40. *Hon Degrees:* LLD, Univ Pa, 60, MA, 71; DSc, Haverford Col, 62, Swarthmore Col, 69, Med Col Pa, 74, Hahnemann Med Col, 78, Duke Univ, 79, Georgetown Univ, 79, Med Col Ohio, 85; DLitt, Jefferson Univ, 79; DMedSc, Yale Univ, 90. *Prof Exp:* Intern, Hosp Univ Pa, 32-34, asst chief resident, 34; asst instr surg, Sch Med, Univ Pa, 34-35, instr, 35-39, assoc, 39-47, assoc prof, 47-49, J William White prof surg res, 49-50, prof surg, Grad Sch Med, 50-64, prof surg, Sch Med, 51-57, provost, Univ, 56-59, from actg dir to dir, Harrison Dept Surg Res, 44-72, John Rhea Barton prof surg, 59-72, PROF SURG, SCH MED, UNIV PA, 72- *Concurrent Pos:* Mem, Franklin Inst, Pa; mem adv coun, Nat Inst Gen Med Sci, 63-67; vpres sci affairs, Inst Med Res, 64-76; mem, US Senate Panel Consults on Conquest of Cancer, 70; mem & chmn, Nat Cancer Adv Bd, 72-79. *Honors & Awards:* Sheen Award, Am Med Asn, 80; Ann Nat Award, Am Cancer Soc, 73; Prize, Soc Int Surg, 79; Goldberger Award, AMA; Medal, Nat Cancer Inst, 87; Swanberg Award, Am Med Writers Asn, 87; Medal of the Surgeon Gen of the US, 87; Benjamin Franklin Medal, Am Philos Soc, 88. *Mem:* Am Philos Soc (secy, 63-66, pres, 77-84); Int Fedn Surg Cols (vpres, 72-78, pres, 78-81); Soc Clin Surg (pres, 58-60); Am Surg Asn (pres, 72-73); Am Col Surgeons (pres, 71-72); Royal Col Surgeons Eng; Royal Col Surgeons Edinburg; Royal Col Physicians & Surgeons Can; Polish Asn Surgeons; Asn Surgeons India. *Res:* Nutrition of surgical patients; physiological factors regulating the level of prothrombin; factors affecting adhesion formation; clinical aspects of cancer. *Mailing Add:* Dept Surg Sch Med Univ Pa Philadelphia PA 19104

RHOADS, PAUL SPOTTSWOOD, medicine; deceased, see previous edition for last biography

RHOADS, ROBERT E, b San Antonio, Tex, Oct 14, 44; m 66; c 3. NUCLEIC ACID BIOCHEMISTRY, PLANT VIROLOGY. *Educ:* Rice Univ, BA, 66; George Washington Univ, PhD(biochem), 71. *Prof Exp:* NIH fel biochem, 66-68 & Roche Inst Molecular Biol, 68-70; fel pharmacol, dept pharmacol, Stanford Univ, 70-72, res assoc, 72-73; asst prof, 73-79, res prof, 84-85, ASSOC PROF BIOCHEM, DEPT BIOCHEM, UNIV KY, 80- *Concurrent Pos:* Assoc prof, Inst Molecular & Cellular Biol, Univ Louis Pasteur, Strasbourg, France, 80-81; guest prof, Inst Biochem, Univ Vienna, Austria, 85. *Mem:* Am Soc Biol Chemists; Am Soc Virol; AAAS. *Res:* The biochemistry of protein synthesis in eukaryotes and the mode of virus gene expression; chesnut blight disease and diseases caused by potyviruses; mRNA structure and function. *Mailing Add:* Dept Biochem Med Col Univ Ky Lexington KY 40536

RHOADS, SARA JANE, b Kansas City, Mo, June 1, 20. ORGANIC CHEMISTRY. *Educ:* Univ Chicago, BS, 41; Columbia Univ, PhD(chem), 49. *Prof Exp:* Instr chem, Hollins Col, 44-45; from instr to prof chem, Univ Wyo, 48-84; RETIRED. *Concurrent Pos:* NSF sr fel, Swiss Fed Inst Technol, 56-57; vis prof, Univ Wash, 65. *Honors & Awards:* Garvan Medal, Am Chem Soc, 82. *Mem:* Am Chem Soc. *Res:* Mechanisms of organic reactions; thermal rearrangements and decompositions. *Mailing Add:* 466 N Ninth St Laramie WY 82070

RHOADS, WILLIAM DENHAM, b Livingston, Mont, Dec 8, 34; m 59; c 3. ANALYTICAL CHEMISTRY, METABOLISM. *Educ:* Col Pac, BS, 59, MS, 60, PhD(chem), 68. *Prof Exp:* Res assoc analytical chem, Allergan Pharmaceut, Calif, 64-66; res asst, Diamond Walnut Growers, Calif, 66-68; sr analytical chemist, Abbott Labs, Ill, 68-70; pres & dir res, Analytical Develop Corp 71-75, DIR RHOADS SCI CO, 76-; dir, Rhoads Sci Co, 76-83; sr analytical chemist, Ciba-Geigy, NC, 83-84; PRES, COLO ANALYTICAL, 84- *Mem:* Am Chem Soc. *Res:* Development of analytical procedures for the pharmaceutical, veterinary, agricultural chemical and food and beverage industries, specializing in gas and high speed liquid chromatography. *Mailing Add:* Col Analytical Res & Develp Corp 4720 Forge Rd Unit 108 Colorado Springs CO 80907

RHODE, ALFRED S, transportation, logistics, for more information see previous edition

RHODE, EDWARD A, JR, b Amsterdam, NY, July 25, 26; m 55; c 5. VETERINARY MEDICINE, PHYSIOLOGY. *Educ:* Cornell Univ, DVM, 47. *Prof Exp:* Instr vet med, Kans State Col, 48-51; from asst prof to assoc prof vet med, 51-64, actg dean, Sch Vet Med, 77-78, assoc dean instr, 71-82, PROF VET MED, UNIV CALIF, DAVIS, 64- , DEAN, SCH VET MED, 82- *Concurrent Pos:* USPHS spec fel, 59-60 & 66-67; mem, basic sci coun, Am Heart Asn; actg dir, Vet Med Training Hosp, 68-69. *Mem:* AAAS; Am Vet Med Asn; Am Physiol Soc; Am Soc Vet Physiol & Pharmacol; Am Col Vet Internal Med; Sigma Xi; Asn Am Vet Med Cols. *Res:* Comparative mammalian cardiovascular physiology; veterinary cardiology; clinical medicine. *Mailing Add:* Sch Vet Med Univ Calif Davis CA 95616

RHODE, SOLON LAFAYETTE, III, b Reading, Pa, Dec 28, 38; m 65; c 2. VIROLOGY, CELL BIOLOGY. *Educ:* Princeton Univ, AB, 60; Thomas Jefferson Univ, MD, 64, PhD(exp path), 68. *Prof Exp:* Fel exp path, Jefferson Med Col, 68-69; med officer, US Navy, 69-70; assoc investr, Inst Med Res, Bennington, Vt, 70-80, dir, 80-84; assoc prof, 84-87, PROF, MED SCH, UNIV NEBR, 87- *Res:* Virology of parvoviruses and DNA viruses; DNA replication and repair; molecular genetics; cancer. *Mailing Add:* Eppley Inst 600 42nd St Omaha NE 68198

RHODE, WILLIAM STANLEY, b Chicago, Ill, Nov 4, 41; c 4. NEUROPHYSIOLOGY. *Educ:* Univ Wis, Madison, BS, 63, MS, 64, PhD(elec eng), 70. *Prof Exp:* Fel neurophysiol, 69-70, asst dir, 70-72, asst dir & asst prof, 72-77, PROF, COMPUT FACIL LAB, NEUROPHYSIOL, UNIV WIS, 83- *Honors & Awards:* Samuel Talbot Award, Inst Elec & Electronics Engrs, 70. *Mem:* Sigma Xi. *Res:* Auditory neurophysiology; investigation of cochlear mechanics; use of computers and instrumentation in neurophysiology; morphological-physiological correlations in the cochlear nucleus; neural circuits and complex signal processing in the auditory system. *Mailing Add:* Dept Neurophys Univ Wis Med Sch 1300 University Ave Madison WI 53706

RHODEN, RICHARD ALLAN, b Coatesville, Pa, May 8, 30; m 69; c 1. MAMMALIAN TOXICOLOGY, OCCUPATIONAL HEALTH & SAFETY. *Educ:* Lincoln Univ, AB, 51; Drexel Univ, MS, 67, PhD(environ toxicol), 71. *Prof Exp:* Chemist mil procurement, Defense Personnel Support Ctr, 51-56; chemist coatings develop, Naval Air Eng Ctr, 56-62, chemist aerospace safety & health, 62-66, res chemist, Naval Air Develop Ctr, 66-72; environ scientist environ health, Environ Protection Agency, 72-75; res pharmacologist occup health, Nat Inst Occup Safety & Health, NIH, 75-82, health scientist adminr, Nat Cancer Inst, 82-84, exec secy Safety & Occup Health Study Sect, 84-89; HEALTH SCIENTIST, AM PETROL INST, 89- *Concurrent Pos:* Lectr, Philadelphia Col Art, 71-72; lectr biol sci, Fed City Col, 73-74; fed exec, Develop Prog, 78-80. *Mem:* Am Chem Soc; Am Conf Govt Indust Hygienists; fel AAAS; Am Indust Hyg Asn; fel Am Inst Chemists; Am Col Toxicol. *Res:* Inhalation toxicology; occupational and environmental health effects. *Mailing Add:* PO Box 34472 Washington DC 20043-4472

RHODES, ALLEN FRANKLIN, b Estherville, Iowa, Oct 3, 24; m 62; c 2. SUBSEA PRODUCTION EQUIPMENT, DEEP OIL & GAS DRILLING. *Educ:* Villanova Univ, BSME, 47; Univ Houston, ML, 50. *Prof Exp:* Asst dir engr admin, Hughes Tool Co, 47-52; pres, McEvoy Co, 52-63; vpres engr & res, Rockwell Mfg Co, 63-70; vpres corp planning & develop, ACF Indust, 71-73; pres & chief exec officer, McEvoy Oilfield Equip Co, 74-79; exec vpres & chief exec officer, Goldrus Marine Drilling, 79-82; pres & chief exec officer, Warren Oilfield serv, 81-82; Anglo Energy, 83-86; pres & chief exec officer, Gripper Inc, 87-90; VPRES & CHIEF FINANCE OFFICER, HYDROTECH SYSTEMS, INC, 91- *Concurrent Pos:* chmn, Dept Trans Gas Pipeline Safety Standards comt, 69-73; consultant, 82- *Honors & Awards:* Robert Henry Thurston Award, Am Soc Mech Engrs, 78; Howard Conley Medal, Am Nat Standard Inst, 80; Charles Russ Richards Mem Award, Am Soc Mec Engrs, 87. *Mem:* Am Soc Mech Engrs; fel Inst Mech Engrs, Gr Brit; Nat Acad Eng; Soc Petroleum Engrs. *Res:* Pioneer in completion equipment for deep oil and gas wells; methods and equipment for subsea well completion. *Mailing Add:* 5449 John Dreaper Houston TX 77056

RHODES, ANDREW JAMES, b Inverness, Scotland, Sept 19, 11; Can citizen; c 4. MEDICAL MICROBIOLOGY, PUBLIC HEALTH. *Educ:* Univ Edinburgh, MB, ChB, 34, MD, 41; FRCP(E), 41; FRCP(C), 53; FFCM, 77. *Prof Exp:* House surgeon & asst bacteriologist, Royal Infirmary Edinburgh, 34-35; lectr bact, Univ Edinburgh, 35-41; pathologist, Emergency Med Serv, Shropshire, 41-45; lectr, Dept Bact & Immunol, London Sch Hyg & Trop Med, 45-47; prof virus infections, Sch Hyg, 47-56, PROF MICROBIOL, SCH HYG & FAC MED, UNIV TORONTO, 56- *Concurrent Pos:* Bacteriologist, Shropshire County Coun, 41-45; res assoc, Connaught Med Res Labs, Univ Toronto, 47-53; dir & virologist, Res Inst, Hosp Sick Children, Toronto, 53-56, active consult, 56-; chmn sect animal viruses, Int Cong Microbiol, Italy, 53; assoc prof pediat, Univ Toronto, 53-56, head dept microbiol, 56-70, dir sch hyg, 56-70; mem Exp Panel Virus Dis, WHO, Switz, 53-78, rapporteur, Exp Comt Poliomyelitis, Italy, 54, mem Sci Group Human Virus Vaccines, Switz, 65; Can del, Int Conf Health Educ, USA, 62; Can rep, Int Comt Virus Nomenclature, 64-78; from assoc med dir to med dir, Lab Br, Ont Ministry Health, 70-75; consult, Ministry Health, 75-80; chmn, Rabies Adv Comt, Ont Ministry Natural Resources, 79-87, mem, 88. *Mem:* Can Med Asn; Can Asn Med Bact; Can Pub Health Asn; fel Royal Soc Can. *Res:* Virology; public health; microbiology. *Mailing Add:* 79 Rochester Ave Toronto ON M4N 1N7 Can

RHODES, ASHBY MARSHALL, b Hinton, WVa, July 5, 23. PLANT BREEDING. *Educ:* WVa Univ, BS, 48; Mich State Col, PhD(farm crops), 51. *Prof Exp:* From instr to asst prof, 51-65, assoc prof, 65,79, PROF VEG CROPS, UNIV ILL, URBANA, 79- *Mem:* Am Soc Hort Sci; Am Genetic Asn; Soc Syst Zool; Sigma Xi. *Res:* Breeding of sweet corn, cucurbits and horse radish; taxonomy; economic botany. *Mailing Add:* 2101 Cherry St NE St Petersburg FL 33704

RHODES, BUCK AUSTIN, b LaUnion, NMex, Aug 30, 35; c 1. RADIOLOGY, PHARMACOLOGY. *Educ:* NMex State Univ, BS, 58; Johns Hopkins Univ, PhD(radiol sci), 68. *Prof Exp:* From asst prof radiol sci to assoc prof radiol & environ health, Sch Med, Hyg & Pub Health, Johns Hopkins Univ, 66-75; prof pharmacol & radiol, Univ Kans Med Ctr, 75-76; prof pharm & radiol, Univ NMex, 76-81; vpres sci affairs, Summa Med Corp, Albuquerque, NMex, 81-85; PRES, RHOMED, INC, ALBUQUERQUE, NMEX, 85- *Honors & Awards:* Hon dipl, Am Bd Sci Nuclear Med. *Mem:* Soc Nuclear Med. *Res:* Development of new radiopharmaceuticals and diagnostic tests for vascular diseases; immuno diagnostics and therapy. *Mailing Add:* 1104 Stanford Dr NE Albuquerque NM 87106

RHODES, CHARLES KIRKHAM, b NY, June 30, 39; m 64, 76; c 4. ATOMIC PHYSICS, LASERS. *Educ:* Cornell Univ, BEE, 63; Mass Inst Technol, MEE, 65, PhD(physics), 69. *Prof Exp:* Staff specialist, Control Data Corp, NY, 69-70; physicist, Lawrence Livermore Lab, Univ Calif, 70-75; sr physicist, Molecular Physics Ctr, SRI Int, 75-77; prof, 78-82, res prof physics, 82-87, ALBERT A MICHELSON PROF PHYSICS, UNIV ILL, CHICAGO, 87- *Concurrent Pos:* Lectr appl sci, Univ Calif, Davis, 71-75; mem, Adv Group Electron Devices; comt mem, Joint Coun Quantum Electronics; adj prof elec eng, Stanford Univ. *Mem:* Fel Am Phys Soc; fel Inst Elec & Electronics Engrs; Europ Physics Soc; fel Optical Soc Am; Sigma Xi; fel AAAS. *Res:* X-ray production, femto second lasers, atomic and molecular energy transfer, chemical processes and kinetics, coherent pulse propagation, saturation spectroscopy; nonlinear optics; collisional broadening of spectral lines, and high pressure electron-beam excited ultraviolet and visible lasers. *Mailing Add:* Dept Physics Univ Ill Box 4348 Chicago IL 60680

RHODES, DALLAS D, b El Dorado, Kans, Aug 8, 47; m 79. GEOMORPHOLOGY. *Educ:* Univ Mo-Columbia, BS, 69; Syracuse Univ, MA & PhD(geol), 73. *Prof Exp:* Asst prof geol, Univ Vt, 73-77; from asst prof to assoc prof geol, 77-86, chmn dept, 85-87, DIR, FAIRCHILD AERIAL PHOTOG COLLECTION, WHITTER COL, 81-, CHMN DEPT GEOL, 90-, PROF, 86- *Concurrent Pos:* Consult geologist, NY State Geol Surv, 75-76 & Jet Propulsion Lab, 80-84; resident dir, Denmark's Int Study Prog, 83; vis researcher, Univ Uppsala, Sweden, 84; vis prof geol, Univ Mo, Columbia, 90. *Mem:* Am Geophys Union; Geol Soc Am; Sigma Xi; Nat Asn Geol Teachers. *Res:* Detailing relationships of fluvial hydraulic geometry to the river's sedimentology, channel shape, and channel pattern; analysis of hydraulic geometry in terms of a most probable state; planetary geomorphology with emphasis on Mars. *Mailing Add:* Dept Geol Whittier Col Whittier CA 90608

RHODES, DAVID R, b Wichita, Kans, Oct 22, 36; m 54; c 4. ANALYTICAL CHEMISTRY, ELECTROCHEMISTRY. *Educ:* Friends Univ, BA, 57; Univ Ill, MS, 59, PhD(chem), 61. *Prof Exp:* Res chemist, 61-71, sr res chemist, 71-74, SR RES ASSOC, CHEVRON RES CO, 74- *Mem:* Am Chem Soc; Electrochem Soc. *Res:* Electroanalytical chemistry; pollution analysis; trace analysis; corrosion; fuel cells. *Mailing Add:* 1301 Quarry Ct No 112 Richmond CA 94801-4153

RHODES, DONALD FREDERICK, b Johnstown, Pa, July 1, 32; m 56. RADIOISOTOPE APPLICATIONS. *Educ:* Univ Pittsburgh, BS, 54, MLitt, 56; Pac Western Univ, PhD(physics), 82. *Prof Exp:* Res asst, Radiation Lab, Univ Pittsburgh, 53-54, instr elec measurements, physics dept, 55-56; engr, Westinghouse Elec Corp, 56-57; radio safety officer, Gulf Res & Develop Co, 68-83, res physicist, 58-86; INDEPENDENT CONSULT, 86- *Concurrent Pos:* Consult nuclear technol, US Govt, 73-74 & radiotracer tests, Exxon Co, 81-82; accident prev coun, Fed Aviation Admin, 81-85. *Honors & Awards:* IR-100 Award, 68. *Mem:* Am Nuclear Soc; Health Physics Soc; Inst Elec & Electronics Engrs. *Res:* Nuclear instrumentation development, radioactive tracer studies of chemical plant processes and oil well enhanced recovery applications; neutron activation analysis, radiation effects, nuclear well logging, applications of ultrasonics and flexible automation with robotics; early development of on-line data processing hardware; current applications of microcomputers. *Mailing Add:* Univ Pittsburgh Appl Res Ctr 345 William Pitt Way Pittsburgh PA 15238-1327

RHODES, DONALD R(OBERT), b Detroit, Mich, Dec 31, 23; div; c 4. MUSICOLOGY. *Educ:* Ohio State Univ, BEE, 45, MSc, 48, PhD(elec eng), 53. *Prof Exp:* Res assoc elec, Ohio State Univ, 44-54; res engr, Cornell Aeronaut Lab, Inc, 54-57; head basic res dept, Radiation, Inc, 57-61, sr scientist, 61-66; UNIV PROF ELEC ENG, NC STATE UNIV, 66-

Concurrent Pos: Instr, Ohio State Univ, 48-52. *Honors & Awards:* John T Bolljahn Award, Inst Elec & Electronics Engrs, 63; Benjamin G Lamme Medal, Ohio State Univ, 75. *Mem:* Fel AAAS; fel Inst Elec & Electronics Engrs. *Res:* Antenna synthesis. *Mailing Add:* Dept Elec Eng NC State Univ Raleigh NC 27695-7911

RHODES, E(DWARD), b Elland, Eng, Jan 31, 38; m 62; c 3. CHEMICAL ENGINEERING. *Educ:* Univ Manchester, BScTech, 60, MScTech, 61, PhD(chem eng), 64. *Prof Exp:* Asst lectr chem eng, Univ Manchester, 62-64; from asst prof to assoc prof, 64-74, PROF CHEM ENG, UNIV WATERLOO, 74-, CHMN DEPT, 76- *Mem:* Brit Inst Chem Engrs; Am Inst Chem Engrs; Can Soc Chem Eng. *Res:* Multiphase flow; mass transfer; boiling; condensation. *Mailing Add:* Dept Chem Eng-2500 Univ Calgary Calgary AB T2N 1N4 Can

RHODES, EDWARD JOSEPH, JR, b San Diego, Calif, June 1, 46; m 72; c 2. SOLAR ASTRONOMY, SPACE PHYSICS. *Educ:* Univ Calif, Los Angeles, BS, 68, MA, 71, PhD(astron), 77. *Prof Exp:* Fel researcher, Univ Calif, Los Angeles, 75-77, asst res astronomer, 78-79; asst prof, 79-85, ASSOC PROF ASTRON, UNIV SOUTHERN CALIF, 85- *Concurrent Pos:* Scientist, Jet Propulsion Lab, 70-77, sr scientist, 78-83, mem tech staff, 83-; res fel, Dept Physics, Calif Inst Technol, 77-78; adj asst prof astron, Univ Southern Calif, 78-79; mem, Adv Solar Observ Sci Study Team, NASA, 81-83, Solar Beacon Sci Study Team, 81-83, Solar Cycle & Dynamics Sci Working Group, 78-79 & Star Probe Imaging Comt, 79-80. *Mem:* Am Astron Soc; Am Geophys Union; Sigma Xi; Int Astron Union. *Res:* Observational and theoretical research into the internal structure of the sun using the tool of solar oscillations; spatial behavior of the solar wind; operator of the 60-foot solar tower telescope of the Mount Wilson observatory. *Mailing Add:* Dept Astron Univ Southern Calif Mail Code 1342 Los Angeles CA 90089

RHODES, FRANK HAROLD TREVOR, b Warwickshire, Eng, Oct 29, 26; m 52; c 4. GEOLOGY, PALEONTOLOGY. *Educ:* Univ Birmingham, BSc, 48, PhD, 50, DSc, 63. *Hon Degrees:* DSc, Univ Birmingham, 63, Bucknell Univ, 85; LLD, Wooster Col, 76, Nazareth Col Rochester, 79; LHD, Colgate Univ, 80, Univ Wales, 81, Johns Hopkins Univ, 82, Wagner Col, 82, Hope Col, 82, Rensselaer Polytech Inst, 82, LeMoyne Col, 84 ; D Litt, Univ Nev, 82,. *Prof Exp:* Fel & Fulbright scholar, Univ Ill, 50-51; lectr geol, Univ Durham, 51-54; asst prof, Univ Ill, 54-55, assoc prof, 55-56, dir, Wyo Field Sta, 56; prof geol & head dept, Univ Wales, Swansea, 56-68, dean fac sci, 67-68; prof geol & mineral, Univ Mich, Ann Arbor, 68-77, res assoc, Mus Paleont & dean, Col Lit Sci & Arts, 71-74, vpres acad affairs, 74-77; PRES, CORNELL UNIV, 77-, PROF GEOL & MINERAL, 77- *Concurrent Pos:* External examinerships, Univ Bristol, 58-61, Univ Belfast, 60-62, Oxford Univ & Univ Reading, 63-65; vis prof, Univ Ill, 59; Gurley lectr, Cornell Univ, 60; dir first int field studies conf, NSF-Am Geol Inst, 61; ed geol ser, Commonwealth & Int Libr, 62-; mem, Bd Geol Surv Gt Brit, 63-65; mem Australian vchancellor's comt vis, Australian Univs, 64; Brit Coun lectr univs & geol surveys, India, Pakistan, Thailand, Turkey & Iran, 64; NSF sr scientist fel, Ohio State Univ, 65-66; mem geol & geophys comt & subcomt postgrad awards, Nat Environ Res Coun, 65-68; Bownocker lectr, Ohio State Univ, 66; mem curric panel, Coun Educ Geol Sci, 70-71, chmn panel, 71; mem bd trustees, Carnegie Found Advan Teaching, 78-; vchmn, Am Coun Educ, 83-86, chmn, 85-86, mem bd dirs, 83- *Honors & Awards:* Daniel Pidgeon Fund Award, Geol Soc London, 53; Lyell Fund Award, 57; Bigsby Medal, 67. *Mem:* Geol Soc London; Brit Palaeont Asn (vpres, 63-68); Brit Asn Advan Sci; Geol Soc Am; Paleont Soc. *Res:* Stratigraphy; micropaleontology, especially conodonts; evolution; extinction; biogeochemistry; paleoecology; higher education; science and public policy. *Mailing Add:* Cornell Univ 300 Day Hall Ithaca NY 14853

RHODES, IAN BURTON, b Melbourne, Australia, May 29, 41; m 64; c 2. ELECTRICAL ENGINEERING, APPLIED MATHEMATICS. *Educ:* Melbourne, BE, 63, MEngSc, 65; Stanford Univ, PhD(elec eng), 68. *Prof Exp:* Res engr, Stanford Res Inst, 67; asst prof elec eng, Mass Inst Technol, 68-70; assoc prof eng & appl sci, Wash Univ, 70-76, prof, 76-80; PROF ELEC & COMPUT ENG, UNIV CALIF, SANTA BARBARA, 80- *Mem:* Soc Indust & Appl Math; Inst Elec & Electronics Engrs. *Res:* Decision and control sciences; system theory; control theory; estimation theory; optimization theory. *Mailing Add:* Eight Roe End Kingsbury London NW9 9BI England

RHODES, JACOB LESTER, b Linville, Va, Jan 13, 22; m 60; c 4. NUCLEAR PHYSICS. *Educ:* Lebanon Valley Col, BS, 43; Univ Pa, PhD, 58. *Prof Exp:* Asst res physicist, Johns Hopkins Univ, 43-46; asst instr physics, Univ Pa, 46-49, asst res physicist, 49-52; asst prof & chmn dept, Roanoke Col, 52-56; from assoc prof to prof, 57-85 EMER PROF PHYSICS, LEBANON VALLEY COL, 85- *Mem:* Am Phys Soc; Am Asn Physics Teachers. *Res:* Low energy nuclear physics, electronics; x-ray diffraction. *Mailing Add:* Dept Physics Lebanon Valley Col Annville PA 17003-1518

RHODES, JAMES B, b Kansas City, Mo, July 22, 28; m 60; c 3. GASTROENTEROLOGY, BIOCHEMISTRY. *Educ:* Univ Kans, AB, 54, MD, 58. *Prof Exp:* Intern med & surg, Univ Chicago Hosps, 58-59, resident med, 60-62; asst med, 64-66; from asst prof to assoc prof med & physiol, 66-79, PROF MED, UNIV KANS MED CTR, KANSAS CITY, 79- *Concurrent Pos:* Fel biochem, Ben May Lab Cancer Res, Univ Chicago, 59-60, trainee gastroenterol, Univ Hosps, 62-64; fel, Chicago Med Sch, 64-66, res assoc biochem, 64-66. *Mem:* AAAS; Am Physiol Soc; AMA; Am Gastroenterol Asn; Am Soc Gastrointestinal Endoscopy; fel Am Col Physicians. *Res:* Clinical gastroenterology; digestive biochemistry; digestion and absorption of the intestinal epithelial cell; physiology. *Mailing Add:* Dept Med Univ Kans Med Ctr Kansas City KS 66103

RHODES, JOHN LEWIS, b Columbus, Ohio, July 16, 37; m; c 4. MATHEMATICS. *Educ:* Mass Inst Technol, BS, 60, PhD(math), 62. *Prof Exp:* NSF fel, Paris, France, 62-63; from asst prof to assoc prof, 63-70, PROF MATH, UNIV CALIF, BERKELEY, 70- *Concurrent Pos:* Vpres, Krohn-Rhodes Res Inst, 64-68; USAF res grant, Univ Calif, Berkeley, 65-68; mem, Inst Advan Study, 66; Alfred P Sloan fel, 67; NSF grants, 63-; ed, J Pure & Appl Algebra, Publ NHolland; ed-in-chief, Int J Algebra & Computation World Sci Publ. *Res:* Algebraic theory of finite state machines, finite and infinite semi groups; finite physics from an algebraic viewpoint; neural nets; context free languages. *Mailing Add:* Dept Math Univ Calif Berkeley CA 94720

RHODES, JOHN MARSHELL, b Oakland, Calif, Mar 26, 26; m 67; c 2. NEUROPSYCHOLOGY. *Educ:* Univ Calif, Los Angeles, BA, 53; Los Angeles State Col, MA, 55; Univ Southern Calif, PhD, 59. *Prof Exp:* Instr, Whittier Col, 58-59; holder of USA table comp zool, Sta Zool, Naples, 59-60; USPHS fel neurophysiol, Fac Med, Univ Aix Marseille, 60-62; asst res anatomist & assoc mem brain res inst, Space Biol Lab, Univ Calif, Los Angeles, 62-65; assoc prof psychol, 65-71, prof psychol & neurol, 71-84, EMER PROF, UNIV NMEX, 84-; PVT PRACT, CLIN NEUROPSYCHOL, 84- *Mem:* Am Psychol Asn; Am EEG Soc. *Res:* Acute and chronic animal experiments with implanted electrodes; clinical electroencephalographic studies with neurological disorders in humans; multiple physiological recording and its computer application; clinical neuropsychology. *Mailing Add:* 5801 Osuna NE Apt 207 Albuquerque NM 87109

RHODES, JUDITH CAROL, b Tulsa, Okla, Jan 27, 49. MEDICAL MYCOLOGY, VIRULENCE MECHANISMS. *Educ:* Univ Okla, BS, 71, MS, 73: Univ Calif, Los Angeles, PhD(microbiol & immunol), 80. *Prof Exp:* Fel mycol sect, Lab Clin Invest, Nat Inst Allergy & Infectious Dis, NIH, 80-82; asst prof, 82-87, ASSOC PROF MYCOL, DEPT PATH & LAB MED, UNIV CINCINNATI COL MED, 87- *Concurrent Pos:* Dir, mycol lab, Univ Hosp, Univ Cincinnati, 82-, assoc dir, clin microbiol, 85-; prin investr, dept path & lab med, Univ Cincinnati Col Med, 84- *Mem:* Am Soc Microbiol; Mycol Soc Am; Med Mycol Soc Americas; Int Soc Human & Animal Mycol; Am Soc Clin Path. *Res:* Virulence mechanisms in pathogenic fungi, especially Cryptococcus neoformans and Aspergillus. *Mailing Add:* Dept Path & Lab Med Univ Cincinnati Col Med Cincinnati OH 45267-0529

RHODES, LANDON HARRISON, b Alton, Ill, Mar 15, 47; m 70; c 1. PLANT PATHOLOGY. *Educ:* Univ Ill, BS, 70, MS, 75, PhD(plant path), 77. *Prof Exp:* Asst prof, 76-82, ASSOC PROF PLANT PATH, OHIO STATE UNIV, 82- *Mem:* Am Phytopath Soc; Mycol Soc Am. *Res:* Forage crop pathology. *Mailing Add:* Dept Plant Path Ohio State Univ Columbus OH 43210

RHODES, MITCHELL LEE, b Chicago, Ill, Feb 12, 40; m 63; c 3. PULMONARY DISEASES. *Educ:* Univ Ill, Urbana, BS, 61, Chicago, MD, 65; Lake Forest Grad Sch Mgt, MBA, 91. *Prof Exp:* USPHS trainee pulmonary dis, Univ Chicago, 68-70, instr med, 70; clin instr med, Univ Calif, San Francisco, 71-72; from asst prof to assoc prof med, Col Med, Univ Iowa, 72-76; from assoc prof to prof med, Col Med, Ind Univ, 76-85; ASSOC DEAN CLIN AFFAIRS & PROF MED, CHICAGO MED SCH, 85- *Concurrent Pos:* Chief pulmonary dis, USPHS Hosp, San Francisco, 70-72; NIH pulmonary acad awardee, Nat Heart & Lung Inst, 74-76; gov, Am Col chest Physicians, 79-85, Bd Regents, 90- *Honors & Awards:* Cecile Lehman Mayer Res Award Pulmonary Dis, Am Col Chest Physicians, 74. *Mem:* Asn Am Med Col; Am Col Physicians; Am Col Chest Physicians (treas, 90-); AMA; Am Col Physician Execs. *Res:* Use of computer assisted instruction in medical education; correlation of ultrastructural and metabolic changes in lung tissue. *Mailing Add:* Chicago Med Sch 3333 Green Bay Rd North Chicago IL 60064

RHODES, RICHARD AYER, II, b Hartford, Conn, Feb 13, 22. PHYSICS. *Educ:* Bowdoin Col, AB, 43; Yale Univ, MS, 47; Brown Univ, PhD(physics), 61. *Prof Exp:* Instr, Bowdoin Col, 43-44; jr physicist, US Naval Res Lab, 44-45; asst prof, physics dept, Univ Conn, 47-62; asst prof, physics dept, Univ Fl, 62-66; assoc prof, physics, Fl Presbyterian/Eckerd Col, 66-73; vis assoc prof, physics dept, Randolf-Macon Col, 85-86; vis prof, Va Milit Inst, 86-87; VIS PROF, PHYSICS DEPT, ECKERD COL, 87- *Mem:* Am Physical Soc; Acoustical Soc Am; Optical Soc Am; AAAS. *Res:* Liquids; ionic collisions. *Mailing Add:* 205 NW Monroe Circle N St Petersburg FL 33702

RHODES, RICHARD KENT, connective tissue biochemistry, for more information see previous edition

RHODES, ROBERT ALLEN, b Harrisonburg, Va, May 10, 41; m 63; c 3. PHARMACEUTICAL CHEMISTRY. *Educ:* Bridgewater Col, BA, 63; Univ Md, Baltimore, PhD(pharmaceut chem), 68. *Prof Exp:* Asst biochem, Ahmadu Bello Univ, Nigeria, 68-70; from asst prof to assoc prof, 70-84, PROF CHEM, MID GA COL, 84- *Concurrent Pos:* Fel, Oak Ridge Assoc Univs, 76; fel, GD Searle, 83; res assoc, Ga State Univ, 84-90. *Mem:* AAAS; Am Chem Soc; Sigma Xi. *Res:* Indole and heterocyclic synthesis. *Mailing Add:* 224 Brookwood Dr Dublin GA 31021

RHODES, ROBERT CARL, b Detroit, Mich, Nov 14, 36; m 65; c 1. ORGANIC CHEMISTRY, ANALYTICAL CHEMISTRY. *Educ:* Univ Calif, Riverside, BA, 58; NMex Highlands Univ, MS, 59; Univ Wash, PhD(org chem), 63. *Prof Exp:* Res chemist, 63-70 & sr res chemist, Wilmington plant, 70-77, SUPVR ANALYTICAL CHEMISTS LAB, AGR PROD DEPT, E I DU PONT DE NEMOURS & CO, INC, LA PORTE, TX, 77- *Mem:* Am Chem Soc; Sigma Xi. *Res:* Behavior and movement of organic chemicals in soil. *Mailing Add:* Agr Prod Dept E I du Pont de Nemours & Co Inc La Porte TX 77571

RHODES, ROBERT SHAW, b Orangeburg, SC, Mar 3, 36; m 64; c 2. HEMATOLOGY, OCCUPATIONAL MEDICINE. *Educ:* SC State Col, BS, 58; Meharry Med Col, MD, 62. *Prof Exp:* Intern, Hubbard Hosp, Nashville, 62-63; med officer & aviation pathologist, Armed Forces Inst Path Aerospace Br, 67-70; fel hemat, Dept Internal Med, Vanderbilt Univ, 70-72; resident path, Meharry Med Col, 63-67, dir hemat & clin labs, 72-75, assoc prof internal med & path, 72-78; assoc med dir, 78-82, DIR HEALTH SERV,

HYDROMATIC DIV, GEN MOTORS, 82-; MEM STAFF, DEPT SURG, CASE WESTERN RESERVE UNIV. *Concurrent Pos:* Consult Aerospace Pathologists, Nat Transp Safety Bd, 67-70; assoc investr clin studies, Vanderbilt Univ Clin Res Ctr, 70-72; proj dir, Nat Heart & Lung Inst Contract, Clin Trials Vaso-occlusive Crisis Treatment, Meharry Med Col, 71-73; mem ad hoc sickle cell contracts rev comt, Nat Heart & Lung Inst, 74-76; consult, Nat Asn Sickle Cell Dis, Nat Educ Proj Sickle Cell Dis, 75- *Mem:* Am Soc Clin Path; Am Asn Blood Banks; Am Soc Hemat; AAAS. *Res:* Miscellaneous studies on the natural history of sickle cell disease and the sickling process. *Mailing Add:* Div Med Dir GMC HydraMatic M/C 540 Ypsilanti MI 48198

RHODES, RONDELL H, b Abbeville, SC, May 25, 18. DEVELOPMENTAL BIOLOGY, HISTOLOGY. *Educ:* Benedict Col, BS, 40; Univ Mich, MS, 50; NY Univ, PhD(biol), 60. *Prof Exp:* Instr biol, Lincoln Univ, Mo, 47-49; asst prof, Tuskegee Inst, 50-55; from asst prof to prof, 61-88, actg chmn dept, 66-67,chemn dept biol sci, 67-70, 73-76 & 74-82, EMER PROF BIOL, FAIRLEIGH DICKINSON UNIV, 88- *Mem:* AAAS; Nat Asn Biol Teachers; Am Soc Zool; Sigma Xi; Am Inst Biol Soc; NY Acad Sci; Am Asn Univ Prof. *Res:* Histochemical studies involving ribonucleic acid and alkaline phosphatase in developing pituitary glands of the amphibia. *Mailing Add:* 122 Ashland Pl Apt 5H Brooklyn NY 11201

RHODES, RUSSELL G, b St Louis, Mo, Aug 28, 39; m 61; c 1. PHYCOLOGY. *Educ:* Univ Mo-Kansas City, BS, 61; Univ Tenn, MS, 63, PhD(bot), 66. *Prof Exp:* Res assoc bot, Univ Tenn, 66; assoc prof bot, Kent State Univ, 66-77; PROF LIFE SCI & HEAD DEPT, SOUTHWEST MO STATE UNIV, 77- *Concurrent Pos:* Mem, Ohio Biol Surv, 67. *Mem:* Phycol Soc Am; Brit Phycol Soc; Int Phycol Soc; Sigma Xi. *Res:* Morphogenesis of brown algae; cultivation of acidophilic algae; isolation and cultivation of Chyrsophytan algae. *Mailing Add:* Life Sci Dept SW Missouri State Univ 901 S National Springfield MO 65802

RHODES, WILLIAM CLIFFORD, b Birmingham, Ala, Aug 8, 32; m 57; c 3. THEORETICAL CHEMISTRY, CHEMICAL DYNAMICS. *Educ:* Howard Col, AB, 54; Johns Hopkins Univ, PhD(biochem), 58. *Prof Exp:* Am Cancer Soc fel chem, Fla State Univ, 58-59, from instr to assoc prof, 59-70, exec dir, Inst Molecular Physics, 75-79, dir, 79-80, PROF CHEM, FLA STATE UNIV, 70- *Concurrent Pos:* Am Cancer Soc fel, 60-61; NSF sr fel, 64-65; NIH career develop award, 65-70. *Mem:* Sigma Xi. *Res:* Biophysics; quantum chemistry; dynamic aspects of molecular excitation and relaxation processes; selective excitation by laser and conventional light; energy channeling in molecular systems. *Mailing Add:* Dept Chem Fla State Univ Tallahassee FL 32306

RHODES, WILLIAM GALE, protein crystallography, for more information see previous edition

RHODES, WILLIAM HARKER, b Titusville, NJ, Apr 26, 25; m 58; c 5. VETERINARY MEDICINE. *Educ:* NY Univ, BA, 51; Univ Pa, VMD, 55, MMedSci, 58. *Prof Exp:* Instr, 55-58, assoc, 58-59, from asst prof to assoc prof, 59-70, PROF RADIOL, SCH VET MED, UNIV PA, 70-, CHIEF SECT, 64- *Concurrent Pos:* Chmn organizing comt, Am Bd Vet Radiol; ed, Jour Am Vet Radiol Soc. *Mem:* Am Vet Med Asn; Am Vet Radiol Soc (past pres); Educ Vet Radio Sci (past pres). *Res:* Veterinary clinical radiology; degenerative joint disease and radiology of the canine thorax. *Mailing Add:* 516 S 15th St Philadelphia PA 19146

RHODES, WILLIAM HOLMAN, b Oneonta, NY, Sept 13, 35; m 57; c 2. CERAMICS, METALLURGY. *Educ:* Alfred Univ, BS, 57; Mass Inst Technol, ScD(ceramics), 65. *Prof Exp:* Trainee, Gen Elec Co, 57-58, engr, 60-62; group leader ceramics, Mat Sci Dept, Avco Systs Div, 65-73; SR STAFF SCIENTIST, GTE LABS, INC, 73- *Concurrent Pos:* Co-chair, panel on ceramics & ceramic composites, Nat Acad Sci, 85. *Honors & Awards:* Ross Coffin Purdy Award, 83; Hobart M Kraner Award, 83; Leslie H Warner Award, 85. *Mem:* Fel Am Ceramic Soc (pres, 88-89); Sigma Xi. *Res:* Physical ceramics, especially mechanical and optical properties, kinetics of densification by sintering and pressure sintering and basic diffusion studies; fabrication of ceramics, composites and metals from powders. *Mailing Add:* GTE Labs Inc 40 Sylvan Rd Waltham MA 02254

RHODES, WILLIAM TERRILL, b Palo Alto, Calif, Apr 14, 43; div; c 3. ELECTRICAL ENGINEERING, OPTICS. *Educ:* Stanford Univ, BS, 66, MSEE, 68, PhD(elec eng), 72. *Prof Exp:* Res asst electronics labs, Stanford Univ, 69-71; from asst prof to assoc prof, 71-80, PROF ELEC ENG, GA INST TECHNOL, 81- *Concurrent Pos:* Consult, Naval Res Lab, 74-, Lockheed Electronics, 80- & Aerodyne Res, Inc, 81-; Humboldt res fel, Univ Erlangen, Nurnberg, 76; dir, Optical Soc Am, 84-86; gov, Soc Photo-Optical Instrumentation Engrs, 83-85; ed, Appl Optics. *Mem:* Fel Soc Photo-Optical Instrumentation Engrs; fel Optical Soc Am; Inst Elec & Electronics Engrs. *Res:* Synthetic aperture optics; image formation; hybrid optical-digital signal processing; numerical and algebraic optical processing; noncoherent optical processing; acousto-optic signal processing. *Mailing Add:* Sch Elec Eng Ga Inst Technol Atlanta GA 30332-0250

RHODES, YORKE EDWARD, b Elizabeth, NJ, Mar 25, 36; m 75; c 4. PHYSICAL ORGANIC & SPACE CHEMISTRY. *Educ:* Univ Del, BS, 57, MS, 59; Univ Ill, PhD(org chem), 64. *Prof Exp:* NIH fel org chem, Yale Univ, 64-65; asst prof, 65-71, ASSOC PROF ORG CHEM, NY UNIV, 71- *Concurrent Pos:* Guest prof, Freiburg Univ, 72-73 & Tech Univ Munich, 77-78; Alexander von Humboldt sr US scientist award, 78; assoc prof, Univ Grenoble, France, 87. *Mem:* Am Chem Soc; Royal Chem Soc. *Res:* Small ring chemistry; carbonium ion rearrangements; neighboring group participation, especially by alkyl and cyclopropane; solvolysis mechanisms; nonaqueous solvents; stereochemistry and reaction mechanisms; synthesis and reactions of strained polycyclic systems; organic chemistry of the interstellar media, including proposal and synthesis of new compounds. *Mailing Add:* Dept Chem NY Univ New York NY 10003

RHODIN, JOHANNES A G, b Lund, Sweden, Sept 30, 22; m 80; c 2. CELL BIOLOGY, ELECTRON MICROSCOPY. *Educ:* Karolinska Inst, Stockholm, MD, 50, PhD(anat), 54. *Prof Exp:* Instr anat, Karolinska Inst, 50-54, asst prof, 54-57; assoc prof, Sch Med, New York Univ, 58-60, prof, 60-64; prof & chmn, Dept Anat, New York Med Col, 64-74, Univ Mich, Ann Arbor, 74-77 & Med Sch, Karolinska Inst, 77-79; PROF & CHMN, DEPT ANAT, UNIV SFLA, 79- *Concurrent Pos:* Landis Res Award, Microcirulatory Soc, 70. *Mem:* Am Soc Cell Biol; Am Asn Anatomists; Microcirculatory Soc; Europ Soc Microcirculation; Europ Artery Club. *Res:* Functional and ultrastructural changes that precede and occur during the development of hypertension in the spontaneously hypertensive rat, using a combination of intravital microscopy and electron microscopy. *Mailing Add:* Dept Anat Col Med Univ SFla 12901 N 30th Tampa FL 33612

RHODIN, THOR NATHANIEL, JR, b Buenos Aires, Arg, Dec 9, 20; US citizen; m 48; c 4. CHEMICAL PHYSICS. *Educ:* Haverford Col, BS, 42; Princeton Univ, AM, 45, PhD(chem physics), 46. *Prof Exp:* Res assoc, Manhattan Proj, Princeton Univ, 42-47; mem staff, Inst Study Metals, Univ Chicago, 47-51; res assoc, Eng Res Lab, E I du Pont de Nemours & Co, 51-58; assoc prof appl physics, 58-68, PROF APPL PHYSICS & MEM FAC, SCH APPL & ENG PHYSICS, CORNELL UNIV, 68- *Concurrent Pos:* NSF fel, Cambridge Univ, 64-65; vis prof, Mass Inst Technol, Univ Munich & Univ Tokyo; adv ed, Surface Sci J; Humboldt sr scientist Award, WGermany, 85-86. *Mem:* Fel Am Phys Soc; Am Vacuum Soc; Am Chem Soc; Am Asn Univ Profs. *Res:* Physics and chemistry of metal and semiconductor surfaces and interfaces; synchrotron radiation spectroscopy; cluster chemistry; surface extended x-ray absorption fine structures; laser-solid interaction; time resolved surface processes; ion enhanced surface chemistry. *Mailing Add:* Sch Appl & Eng Physics Clark Hall G217 Cornell Univ Ithaca NY 14850

RHODINE, CHARLES NORMAN, b Denver, Colo, Jan 10, 31; m 57; c 3. ELECTRICAL & BIOLOGICAL ENGINEERING. *Educ:* Univ Wyo, BS, 57, MS, 59; Purdue Univ, PhD, 73. *Prof Exp:* Assoc prof, 59-80, PROF ELEC ENG, UNIV WYO, 80-, ASST HEAD DEPT, 83- *Concurrent Pos:* NSF fel, Purdue Univ, 67-69. *Mem:* Inst Elec & Electronics Engrs; Nat Soc Prof Engrs; Sigma Xi. *Res:* Communications and information theory, its application to biological systems; development of computer-aided teaching equipment. *Mailing Add:* Dept Elec Eng Univ Wyo Box 3295 Univ Sta Laramie WY 82070

RHOTEN, WILLIAM BLOCHER, b Orange, NJ, Feb 11, 43; m 69; c 3. CELLULAR BIOLOGY. *Educ:* Colo State Univ, BS, 65; Univ Ill, Urbana, MS, 68; Pa State Univ, PhD(anat), 71. *Prof Exp:* Res fel path, Sch Med, Wash Univ, 71-73; instr med, Div Endocrinol & Metab, Sch Med, Univ Ala, 73-74; asst prof anat, Sch Basic Med Sci, Col Med, Univ Ill, Urbana, 74-80; asst prof, 80-82, ASSOC PROF ANAT, NJ MED SCH, UNIV MED & DENT NJ, 82- *Concurrent Pos:* Reviewer, Diabetes, NSF & Sci Endocrinol; head, Histol & Cell Biol Unit, Univ Cape Town, 85-86. *Mem:* Am Asn Anat; Am Soc Cell Biol; AAAS; NY Acad Sci; Am Diabetes Asn; Electron Micros Soc Am. *Res:* Cellular biology; gene expression in experimental diabetes; in situ hybridization; structure-function relationships in the endocrine pancreas; in situ immunodetection; calcium-binding proteins. *Mailing Add:* Dept Anat NJ Med Sch 185 S Orange Ave Newark NJ 07103-2714

RHOTON, ALBERT LOREN, JR, b Parvin, Ky, Nov 18, 32; m 57; c 4. NEUROSURGERY. *Educ:* Ohio State Univ, BS, 54; Wash Univ, MD, 59. *Prof Exp:* Fel neurol surg, Sch Med, Wash Univ, 63-65, NIH spec fel neuroanat, 65-66; from instr to asst prof neurol surg, Mayo Found, Univ Minn, 66-72; prof neurol surg & chief div, 72-79, R D KEENE FAMILY PROF NEUROL SURG & CHMN DEPT, UNIV FLA, 79- *Concurrent Pos:* Consult, Gainesville Vet Admin Hosp; NIH travel award, 65; Krayenbuhl lectr & hon guest, Swiss Soc Neurol Surg, 75 & Japanese Neurosurg Soc, 77; vis fac mem & lectr, Harvard Univ, Duke Univ, Johns Hopkins Univ, Univ Pa, Univ Chicago & Univ Calif, San Francisco & Los Angeles; bd gov, Am Col Surgeons, 79-85. *Honors & Awards:* Billings Bronze Medal, AMA, 69; Jones Award, Am Asn Med Illustr, 69. *Mem:* Am Asn Neurol Surg (treas, 83-86, vpres, 87, pres elect, 88); Am Col Surgeons; Cong Neurol Surg (vpres, 73-74, pres, 78); Neurosurg Soc Am; Soc Neurol Surg (treas, 76-79); Am Surg Asn. *Res:* Microneurosurgery; microsurgical anatomy; neuroanatomy of the cranial nerves; microsurgery of cerebrovascular disease; developer of microsurgical instruments. *Mailing Add:* Box J265 Univ Fla Med Ctr Gainesville FL 32610

RHYKERD, CHARLES LOREN, b Cameron, Ill, Apr 7, 29; m 54; c 3. PLANT PHYSIOLOGY, AGRONOMY. *Educ:* Univ Ill, BS, 51, MS, 52; Purdue Univ, PhD(agron), 57. *Prof Exp:* Asst agron, Univ Ill, 51-52; technician, Producers Seed Co, Ill, 53-54; asst instr agron, Ohio State Univ, 54-55; asst, Purdue Univ, 55-56, instr, 56-57; soil scientist, US Regional Pasture Res Lab, Pa, 57-60; from asst prof to prof agron, 60-87, ASSOC DIR, INT PROG AGR, PURDUE UNIV, WEST LAFAYETTE, 87- *Concurrent Pos:* AID consult, Brazil, 65; vis prof, Univ Calif, Davis, 67-68; co-dir, Nat Corn & Sorghum Proj, Brazil, AID contract with Brazilian Ministry of Agr & Purdue Univ, 73-75; mem, Am Forage & Grassland Coun; short-term aid consult, Portugal, 81, 82 & 84. *Mem:* Am Soc Agron; Crop Sci Soc Am. *Res:* Physiology of forage crops; soil fertility. *Mailing Add:* Intern Progs, AG AD Bldg Purdue Univ West Lafayette IN 47907

RHYNE, A LEONARD, b Charlotte, NC, Dec 19, 34; m 56; c 4. MATHEMATICAL STATISTICS. *Educ:* Univ NC, AB, 57; NC State Univ, PhD(math statist), 64. *Prof Exp:* Prof math, Elon Col, 60-62; assoc prof biostatist, Bowman Gray Sch Med, 62-80, dir comput ctr, 65-77; DIR OPERS RES, LORILLARD INC, 80- *Mem:* Inst Math Statist; Biomet Soc; Am Statist Asn. *Res:* Applications of computer science and statistical methods in various areas of research. *Mailing Add:* 1837 Runnymeade Rd Winston-Salem NC 27104

RHYNE, JAMES JENNINGS, b Oklahoma City, Okla, Nov 14, 38; m 82, 90; c 2. SOLID STATE PHYSICS. *Educ:* Univ Okla, BS, 59; Univ Ill, Urbana, MS, 61; Iowa State Univ, PhD(physics), 65. *Prof Exp:* Res asst physics, Iowa State Univ, 63-65; res physicist, Solid State Div, US Naval Ord Lab, 65-75; PHYSICIST, NAT INST STANDARDS & TECHNOL, 75-; DIR, MU RES REACTOR & PROF PHYSICS, UNIV MO-COLUMBIA, 90- *Concurrent Pos:* Nat Acad Sci-Nat Res Coun associateship, US Naval Ord Lab, 65-66; adj prof, Am Univ, 75-80, co-ed, Proc Conf Magnetism & Magnetic Mat, 70-75. *Mem:* Fel Am Phys Soc. *Res:* Magnetic and transport properties of rare-earth metals and compounds; neutron scattering in magnetic materials. *Mailing Add:* Res Reactor Univ Mo Columbia MO 65211

RHYNE, THOMAS CROWELL, b Lincolnton, NC, Nov 8, 42; m 65; c 1. PHYSICAL INORGANIC CHEMISTRY. *Educ:* Appalachian State Univ, BS, 65, MA, 67; Va Polytech Inst & State Univ, PhD(chem), 71. *Prof Exp:* Res assoc chem, Aerospace Res Labs, Wright-Patterson AFB, Ohio, 71-72; asst prof, 72-80, PROF CHEM & ASST DEAN, GRAD SCH, APPALACHIAN STATE UNIV, 80- *Concurrent Pos:* Nat Acad Sci-Nat Res Coun fel, 71-72. *Mem:* Am Chem Soc; Am Soc Mass Spectrometry. *Res:* Negative ions; gas chromatography; mass spectrometry. *Mailing Add:* Grad Sch Appalachian State Univ Boone NC 28608

RHYNE, V(ERNON) THOMAS, b Gulfport, Miss, Feb 18, 42; m 61; c 2. ELECTRICAL ENGINEERING, COMPUTER SCIENCE. *Educ:* Miss State Univ, BS, 62; Univ Va, MEE, 64; Ga Inst Technol, PhD(elec eng), 67. *Prof Exp:* Aerospace technologist data systs, Langley Res Ctr, NASA, 62-65; instr elec eng, Ga Inst Technol, 65-67; from asst prof to prof elec eng, Tex A&M Univ, 67-83; DIR, CAD PROG, MCC, 84- *Concurrent Pos:* Consult, Tex Instruments Inc, Dallas, 68-69 & Elec Power Res Inst, 78-82; Darby & Darby expert witness, 79-80. *Honors & Awards:* Terman Award, Am Soc Eng Educ, 80; Spensley Horn Jubas & Lubitz, 83- *Mem:* Inst Elec & Electronics Engrs. *Res:* Digital systems design in various applications areas, especially satellite navigation systems and electric power distribution automation; computer design; arithmetic processes; computer software. *Mailing Add:* MCC 3500 W Balcones Ctr Dr Austin TX 78759

RHYNER, CHARLES R, b Wausau, Wis, Mar 25, 40; m 63; c 4. SOLID STATE PHYSICS. *Educ:* Univ Wis, BS, 62, MS, 64, PhD(physics), 68. *Prof Exp:* Asst prof physics, Univ Wis, Kenosha, 67-68; physicist, US Naval Radiol Defense Lab, Calif, 68; from asst prof to assoc prof, 68-85, PROF PHYSICS, UNIV WIS-GREEN BAY, 85- *Mem:* Am Phys Soc; Sigma Xi; Am Asn Physics Teachers. *Res:* Thermoluminescence; color centers in alkali halides; radiation dosimetry; solid waste management. *Mailing Add:* Dept Physics Univ Wis Green Bay WI 54305

RHYNSBURGER, ROBERT WHITMAN, b Kalamazoo, Mich, Dec 5, 25. ASTRONOMY. *Educ:* George Washington Univ, BA, 50; Georgetown Univ, MA, 55. *Prof Exp:* Astronomer, US Naval Observ, 52-80; RETIRED. *Honors & Awards:* Supt's Award, US Naval Observ, 75. *Res:* Meridian astrometry; proper motions; astronomical history. *Mailing Add:* Four Governors Pl Durham NC 27705-6427

RIAHI, DANIEL NOUROLLAH, b Shahrekord, Iran, Oct 30, 43; US citizen; m 84. GEOPHYSICS, MECHANICAL ENGINEERING. *Educ:* Tehran Univ, BM, 66; Fla State Univ, MS, 70, PhD(appl math), 74. *Prof Exp:* Teaching asst appl math, Fla State Univ, 70-72, res asst fluid mech, 72-74, postdoctoral fel fluid mech, 74-77; instr math, Winthrop Col, 77-78; sr researcher fluid mech, Univ Calif, Los Angeles, 78-80; vis asst prof, 80-82, asst prof, 82-85, ASSOC PROF MECH, UNIV ILL, URBANA-CHAMPAIGN, 85- *Concurrent Pos:* Consult math, Sch Social Work, Fla State Univ, 70-71; chmn, Appl Math Comt, Winthrop Col, 77-78; eng mech chief ad, Univ Ill, 85-86, eng mech coordr, 85-86; vis scholar, Cambridge Univ, Eng, 86; prin investr, Nat Ctr Supercomput Applns, 90-91 & NSF, 86-91; chmn, Awards Comt, Theory & Appl Mech Dept, Univ Ill, 90-91, Tech Session, Div Fluid Dynamics, Am Phys Soc, 88 & 89; chmn invited speakers, Mid-Western Mech Conf, 91 & tech session organizer, 91. *Mem:* Soc Indust & Appl Math; Am Phys Soc; assoc fel Am Inst Aeronaut & Astronaut; Am Acad Mech; Sigma Xi. *Res:* Author of over 93 publications including invited papers; fluid mechanics; applied mathematics; heat transfer; magneto hydrodynamics; instability and materials processing. *Mailing Add:* Dept Theoret & Appl Mech Univ Ill 216 Talbot Lab Urbana IL 61801

RIAZ, M(AHMOUD), b Paris, France, Feb 27, 25; nat US; m 64; c 1. ELECTRICAL ENGINEERING. *Educ:* Univ Paris, LLB, 44; Univ Cairo, Egypt, BSc, 46; Rensselaer Polytech Inst, MEE, 47; Mass Inst Technol, ScD, 55. *Prof Exp:* Asst elec eng, Mass Inst Technol, 52-54, from instr to asst prof, 54-59; assoc prof, 59-78, PROF ELEC ENG, UNIV MINN, MINNEAPOLIS, 78- *Honors & Awards:* Levy Medal, Franklin Inst, 72. *Mem:* Inst Elec & Electronics Engrs; Brit Inst Elec Eng; Int Solar Energy Soc. *Res:* Energy conversion and control; electromechanical systems; power systems; solar energy systems. *Mailing Add:* Dept Elec Eng Univ Minn Minneapolis MN 55455

RIBAK, CHARLES ERIC, b Albany, NY, July 19, 50; m 77; c 2. NEUROSCIENCE, NEUROANATOMY. *Educ:* State Univ NY, Albany, BS, 71; Boston Univ, PhD(neuroanat), 76. *Prof Exp:* Assoc res scientist neurosci, City Hope Nat Med Ctr, 75-78; from asst prof to assoc prof, 78-90, PROF ANAT, UNIV CALIF, IRVINE, 90- *Concurrent Pos:* Assoc ed, J Neurocytology, Brain Res, Epilepsy Res; Klingenstein fel neurosci. *Honors & Awards:* Jacob Javits Award, Am Psychiat Asn 90. *Mem:* Am Asn Anatomists; Soc Neurosci; AAAS; Int Brain Res Orgn. *Res:* Analysis of neurons in normal and epileptic cerebral cortex with the electron microscope; neurocytology of gamma-aminobutyric acid neurons with immunocytochemistry; electron microscopic studies of local circuit neurons in the hippocampus and neocortex. *Mailing Add:* Dept Anat & Neurobiol Univ Calif Irvine CA 92717

RIBAN, DAVID MICHAEL, b Chicago, Ill, May 10, 36; m 63; c 3. SCIENCE EDUCATION. *Educ:* Northern Ill Univ, BSEd, 57; Univ Mich, MA, 60; Purdue Univ, MS, 67, PhD(physics educ), 69. *Prof Exp:* Teacher physics, Luther High Sch South, Chicago, 57-60 & Leyden Sch, Northlake, Ill, 60-67; teaching asst physics, Purdue Univ, 68-69; assoc prof, 70-73, PROF PHYSICS, INDIANA UNIV PA, 74- *Concurrent Pos:* Dir, Intermediate Sci Curric Study Training Prog, Installed User Prog, NSF, 72-73 & Teacher Training Prog, 75-76, dir, Proj Physics Training Prog, 74-75 & 75-76. *Mem:* Nat Asn Res Sci Teaching; Am Asn Physics Teachers; AAAS; Nat Sci Teachers Asn; Nat Educ Asn. *Res:* Science curriculum implementation; learning theory; effects of field work in science in promoting learning; history of the development of scientific ideas. *Mailing Add:* Dept Physics Indiana Univ Pa Wey 59 Indiana PA 15701

RIBBE, PAUL HUBERT, b Bristol, Conn, Apr 2, 35; m 58; c 3. MINERALOGY. *Educ:* Wheaton Col, Ill, BS, 56; Univ Wis, MS, 58; Cambridge Univ, PhD(crystallog), 63. *Prof Exp:* Mineralogist, Corning Glass Works, 58-60; NSF fel, Univ Chicago, 63-64; asst prof geol, Univ Calif, Los Angeles, 64-66; assoc prof, 66-72, PROF MINERAL, VA POLYTECH INST & STATE UNIV, 72- *Concurrent Pos:* Ed, Reviews Mineral. *Mem:* Microbeam Analysis Soc; fel Mineral Soc Am (vpres, 85-86, pres, 86-); Mineral Asn Can; Asn Earth Sci Ed. *Res:* Crystal structure analysis and chemistry of rock-forming minerals, particularly feldspars and orthosilicates. *Mailing Add:* Dept Geol Sci Va Polytech Inst & State Univ Blacksburg VA 24061

RIBBENS, WILLIAM B(ENNETT), b Grand Rapids, Mich, May 26, 37; m 72. ELECTRICAL ENGINEERING. *Educ:* Univ Mich, Ann Arbor, BSEE, 60, MS, 61, PhD(elec eng), 65. *Prof Exp:* Design engr, Lear Inc, Mich, 60; res asst, 60-63, res engr, Cooley Elec Lab, 63-65, assoc res engr, 65-67, proj dir, 67-69, ASSOC PROF ELEC & COMPUT ENG, UNIV MICH, ANN ARBOR, 69- *Concurrent Pos:* Consult to various industs. *Mem:* Inst Elec & Electronics Engrs; Optical Soc Am. *Res:* Coherent optical data processing and optical metrology; instrumentation. *Mailing Add:* Dept Elec & Comput Eng Univ Mich Ann Arbor MI 48109

RIBBONS, DOUGLAS WILLIAM, biochemistry, microbiology, for more information see previous edition

RIBE, FRED LINDEN, b Laredo, Tex, Aug 14, 24; m 46; c 4. PHYSICS. *Educ:* Univ Tex, BS, 44; Univ Chicago, SM, 50, PhD(physics), 51. *Prof Exp:* Engr, Eng Res Assocs, Inc, 46-47; asst, Inst Nuclear Studies, Univ Chicago, 47-50; mem staff, Los Alamos Sci Lab, Univ Calif, 51-74, div leader, Controlled Thermonuclear Res Div, 74-77; prof, 77-90, EMER PROF NUCLEAR ENG, UNIV WASH, SEATTLE, 90- *Concurrent Pos:* Vis prof, Univ Iowa, 56; Guggenheim fel, Inst Plasma Physics, Munich, Ger, 63-64; adj prof physics, Univ Tex, Austin, 74-76. *Mem:* Fel Am Phys Soc; Am Nuclear Soc. *Res:* Nuclear reactions; fast neutron research; atomic collisions; plasma physics. *Mailing Add:* 1821 San Mountain Dr Santa Fe NM 87505

RIBE, M(ARSHALL) L(OUIS), b San Antonio, Tex, June 17, 19; m 43; c 2. ELECTRICAL ENGINEERING. *Educ:* Univ Tex, BSEE, 39, BSME, 40; Rutgers Univ, MSEE, 50. *Prof Exp:* Chief radio relay br, Signal Corps Eng Labs, US Dept Army, Ft Monmouth, NJ, 51-56; mgr systs eng projs, Int Div, Radio Corp Am, 56-59, mgr commun systs progs, Commun Lab, 59-64, mgr ground equip prog, Commun Systs Div, 64-66, mgr eastern region, Defense Electronic Prod, 67-69, mgr, Eastern Field Off Opers, Aerospace Systs Div, RCA Corp, 69-86; RETIRED. *Mem:* Sr mem Inst Elec & Electronics Engrs. *Res:* Management of electronic systems marketing, design and development. *Mailing Add:* 805 Bermuda St Austin TX 78734

RIBELIN, WILLIAM EUGENE, b Oxnard, Calif, Oct 1, 24; m 49; c 1. VETERINARY PATHOLOGY. *Educ:* Iowa State Col, DVM, 49; Wash State Col, MSc, 52; Univ Wis, PhD, 56. *Prof Exp:* Instr vet path, Wash State Col, 49-52; res vet, USDA, Holland, 52-54; prof vet path, Auburn Univ, 56-58; head path invests, Western Regional Lab, USDA, 58-61; sr scientist, Environ Health Lab, Am Cyanamid Co, 61-68; dir res, Animal Resource Ctr & prof vet path & med path, Univ Wis-Madison, 68-78; SR PATHOLOGIST, MONSANTO CO, 78- *Mem:* Am Vet Med Asn; Am Asn Pathologists; Soc Toxicol; Conf Res Workers Animal Dis; Am Col Vet Path. *Res:* Comparative and toxicologic pathology. *Mailing Add:* 12673 Tallow Hill Lane St Louis MO 63166

RIBENBOIM, PAULO, b Recife, Brazil, Mar 13, 28; m 51; c 2. MATHEMATICS. *Educ:* Univ Brazil, BS, 48; Univ Sao Paulo, PhD, 57. *Hon Degrees:* Dr, Univ Caen, France. *Prof Exp:* Asst, Cent Brazil Phys Res, 49-50; prof math, Army Tech Sch, Brazil, 52-53; res chief, Inst Pure & Appl Math, 57-59; vis assoc prof, Univ Ill, 59-62; PROF MATH, QUEEN'S UNIV, ONT, 62- *Concurrent Pos:* Vis prof, Northeastern Univ, Boston, 65, Univ Paris, 69-70 & Univ Ill, Urbana, 83. *Mem:* Fel Royal Soc Can; Am Math Soc; assoc Brazilian Acad Sci; Can Math Soc; Math Soc France. *Res:* Algebra; theory of ideals; commutative algebra; algebraic number theory; Fermat's last theorem. *Mailing Add:* Dept Math Queen's Univ Kingston ON K7L 3N6 Can

RIBES, LUIS, b Madrid, Spain, Sept 12, 40; m 64; c 2. ALGEBRA. *Educ:* Univ Madrid, Licenciado, 62; Univ Rochester, MA, 65, PhD(math), 67; Univ Madrid, DrCiencias, 69. *Prof Exp:* Asst prof math, Univ Ill, Urbana, 67-68; res assoc & asst prof math, Queen's Univ, 68-70; asst prof, 70-72, assoc prof, 72-79, PROF MATH, CARLETON UNIV, 79- *Mem:* Am Math Soc; Can Math Cong; Spanish Math Soc. *Res:* Cohomology of groups, discrete and profinite; structure of free profinite and free products of profinite groups; combinational group theory. *Mailing Add:* Dept Math Carleton Univ Ottawa ON K1S 5B6 Can

RIBI, EDGAR, b Zurich, Switz, Sept 5, 20; nat US; m 51; c 2. BIOPHYSICS. *Educ:* Univ Berne, PhD(chem, physics, mineral), 48. *Prof Exp:* Asst teacher chem, Univ Berne, 46-48; res assoc phys chem, Upsala Col, 49-50, res assoc biochem, 51; fel, NIH, 52, biophysicist, Rocky Mountain Lab, Nat Inst

Allergy & Infectious Dis, 52-57, HEAD MOLECULAR BIOL SECT, ROCKY MOUNTAIN LAB, NAT INST ALLERGY & INFECTIOUS DIS, 57- *Concurrent Pos:* NIH lectr, 62; mem tuberc panel, US-Japan Coop Med Serv Prog, 65- *Honors & Awards:* Product & Eng Master Design Award, 62; US Dept HEW Superior Serv Award, 67. *Mem:* AAAS; Am Thoracic Soc; Int Cong Microbiol; Am Venereal Dis Asn. *Res:* Structure and biological function of Enterobacteriaceae, mycobacteria and tumor cells; development of vaccines against tuberculosis and identification of mycobacterial adjuvants useful for immunotherapy of cancer. *Mailing Add:* 311 S Eighth St Hamilton MT 59840

RIBLET, GORDON POTTER, b Boston, Mass, Dec 12, 43; m 72; c 2. ELECTRICAL ENGINEERING, PHYSICS. *Educ:* Yale Univ, BS, 65; Univ Pa, MS, 66, PhD(physics), 70. *Prof Exp:* Res scientist physics, Univ Cologne, 70-72; RES SCIENTIST ENG, MICROWAVE DEVELOP LABS, INC, 72- *Concurrent Pos:* Dir, Parametric Indust, 76-80; pres, Fab-Braze Corp, 77- *Mem:* Am Phys Soc; Inst Elec & Electronics Engrs. *Res:* Solid state physics, especially effect of magnetic impurities on superconductivity; microwave technology, especially circuit properties of multiport networks. *Mailing Add:* Microwave Develop Labs Inc 11 Michigan Dr Natick MA 01760

RIBLET, HENRY B, b Clayton, NMex, May 20, 11. ELECTRICAL ENGINEERING. *Educ:* Friends Univ, AB, 34. *Prof Exp:* Br supvr, Data Control, Appl Physics Lab, Johns Hopkins Univ, 64-76; RETIRED. *Mem:* Fel Inst Elec & Electronics Engrs. *Mailing Add:* 10842 Deric Circle Tampa FL 33635

RIBLET, LESLIE ALFRED, b Wayne County, Ohio, Aug 10, 41; m 65; c 1. PHARMACOLOGY, PHYSIOLOGY. *Educ:* Ashland Col, BS, 63; Univ Mo-Kansas City, BS, 66, MS, 68; Univ Iowa, PhD(pharmacol), 71. *Prof Exp:* Sr scientist, Mead Johnson Res Ctr, 71-74, sr investr, Mead Johnson Pharmaceut Div, 74-76, sr res assoc pharmacol, 76-79, sect mgr, 79-; DIR, PRECLIN RES PHARMACEUT, RES & DEVELOP DIV, BRISTOL-MYERS CO. *Mem:* Soc Neurosci; Sigma Xi; AAAS; Soc Exp Biol Med; NY Acad Sci. *Res:* Central nervous system pharmacology with emphasis on quantitative electroencephalogram correlates of behavioral and neurochemical indices of brain function. *Mailing Add:* Bristol-Myers Pharmaceut Div Five Res Pkwy Wallingford CT 06492

RIBLET, ROY JOHNSON, b Charlotte, NC, Dec 24, 42; m 64; c 3. IMMUNOLOGY, GENETICS. *Educ:* Calif Inst Technol, BS, 64; Stanford Univ, PhD(genetics), 71. *Prof Exp:* Fel immunol, Salk Inst, 71-74; res assoc immunogenetics, Inst Cancer Res, 74-75, asst mem, 75-78, assoc mem immunogenetics, 78-; MEM STAFF, MED BIOL INST. *Concurrent Pos:* Fel, Damon Runyon Mem Fund Cancer Res, Inc, 71-73; spec fel, Leukemia Soc Am, 73-75; NIH res grant, 77-84. *Mem:* Genetics Soc Am; AAAS; Am Asn Immunologists. *Res:* Genetic control of the immune response; mouse antibody genetics; mitogen response genetics; antibody structure and function. *Mailing Add:* Med Biol Inst 11077 N Torrey Pines Rd La Jolla CA 92037

RIBNER, HERBERT SPENCER, b Seattle, Wash, Apr 9, 13; m 49; c 2. AERONAUTICAL & ASTRONAUTICAL ENGINEERING. *Educ:* Calif Inst Technol, BS, 35; Wash Univ, MS, 37, PhD(physics), 39. *Prof Exp:* From physicist to dir lab, Brown Geophys Co, Tex, 39-40; from physicist to head stability analysis sect, Langley Lab, Nat Adv Comt Aeronaut, Va, 40-49, from consult to head boundary layer sects, Lewis Lab, Ohio, 49-54; res assoc, Inst Aerospace Studies, Univ Toronto, 55-56, from asst prof to assoc prof, 56-59; staff scientist aeroacoustics, 75-76, DISTINGUISHED RES ASSOC, NASA LANGLEY RES CTR, 78-; EMER PROF AEROSPACE STUDIES, INST AEROSPACE STUDIES, UNIV TORONTO, 86- *Concurrent Pos:* Prof aerospace studies, Inst Aerospace Studies, Univ Toronto, 59-86; vis prof, Univ Southampton, 60-61; consult, De Havilland Aircraft, Ministry Transp, Can, 70-73 & Gen Elec Co, Cincinnati, 73-75; chmn Sonic Boom Panel, Int Civil Aviation Orgn, 70-71. *Honors & Awards:* Turnbull Lectr, Can Aeronaut & Space Inst, 68; Aeroacoustics Award, 76, Dryden lectr, Am Inst Aeronaut & Astronaut, 81. *Mem:* Fel Am Inst Aeronaut & Astronaut; fel Royal Soc Can; fel Am Phys Soc; fel Acoust Soc Am; fel Can Aeronaut & Space Inst; Can Acoust Asn. *Res:* X-rays; cosmic rays; development of gravity meter; aerodynamics; aeroacoustics; jet noise; sonic boom; acoustics of thunder; propellers. *Mailing Add:* Inst Aerospace Studies Univ Toronto 4925 Dufferin St Downsview ON M3H 5T6 Can

RICARDI, LEON J, b Brockton, Mass, Mar 21, 24; m 47; c 3. ELECTRICAL ENGINEERING. *Educ:* Northeastern Univ, BS, 49, MS, 52, PhD(elec eng), 69. *Prof Exp:* Engr, James L Waters, Inc, 49-50, Andrew Alford Consult Eng, 50-51, proj engr, Gabriel Labs, 51-54; staff mem, Mass Inst Technol, 54-55, asst group leader, radio frequency components & antennas, 55-57, group leader, antenna & commun syst, 57-85, head technol adv off, Lincoln Labs, 82-85; PRES-CONSULT ENGR, LJR, INC, DIV ELECTROMAGNETIC SCI, 85- *Concurrent Pos:* Lectr & teacher, Northeastern Univ, 69-80. *Mem:* Fel Inst Elec & Electronics Engrs. *Res:* Design and development of microwave components and antennas for use in ground, air and space communications systems and radars; investigation of electromagnetic wave propagation phenomena; communications systems analysis. *Mailing Add:* 750 W Sycamore Ave El Segundo CA 90245

RICCA, PAUL JOSEPH, b Brooklyn, NY, Apr 25, 39; m 61; c 2. ANALYTICAL CHEMISTRY. *Educ:* Syracuse Univ, AB, 61; Purdue Univ, PhD(analytical chem), 66. *Prof Exp:* Asst, Purdue Univ, 61-65; adv planning analyst, LTV Aerospace Corp, Dallas, 67-68, eng specialist, 68-71, dir labs & vpres, Anderson Labs, Inc, Ft Worth, 71-75; PRES, RICCA CHEM CO, 75- *Mem:* AAAS; Am Chem Soc; Am Ord Asn. *Res:* Electrochemical kinetics and reaction mechanisms; formulation and development of new missile system concepts; production of prepared chemical reagents and testing solutions. *Mailing Add:* 3315 Thorntree Ct Arlington TX 76016

RICCA, VINCENT THOMAS, b New York, NY, Dec 19, 35; m 57; c 3. HYDROLOGY, HYDRAULIC ENGINEERING. *Educ:* City Col New York, BS, 62; Purdue Univ, Lafayette, MSCE, 64, PhD(water resources eng), 66. *Prof Exp:* Teaching assoc civil eng, Purdue Univ, Lafayette, 62-63; from asst prof to assoc prof, 63-73, PROF CIVIL ENG, OHIO STATE UNIV, 73- *Concurrent Pos:* Hydrol & hydraul eng consult, 66- *Honors & Awards:* Raymond Q Armington Award, Ohio State Univ, 68 & Lichtenstien Mem Award, 72. *Mem:* Am Soc Civil Engrs; Am Soc Eng Educ; Am Water Resources Asn; Am Geophys Union. *Res:* Small watershed hydrology; flood plain management; streamflow simulation computer modeling; groundwater studies; stream surface profiles; acid mine drainage models. *Mailing Add:* Dept Civil Eng N470 Hitchcock Hall Ohio State Univ Columbus OH 43210

RICCI, BENJAMIN, b Cranston, RI, Apr 5, 23; m 44; c 3. APPLIED PHYSIOLOGY. *Educ:* Springfield Col, BPE, 49, MS, 50, PhD(appl physiol), 58. *Prof Exp:* Chmn dept, 71-73, PROF EXERCISE SCI, UNIV MASS, AMHERST, 66- *Concurrent Pos:* Fulbright fel, Inst Work Physiol, Oslo, Norway, 71 & Inst Human Physics, Rome, Italy, 78. *Mem:* Sigma Xi. *Res:* Adaptation of biochemical, biomechanical, heat regulatory, cardiopulmonary and neuromuscular systems to stress imposed by work or exercise. *Mailing Add:* 615 Bay Rd Amherst MA 01002

RICCI, ENZO, b Buenos Aires, Arg, Nov 8, 25; m 57; c 3. APPLIED PHYSICS, ATOMIC PHYSICS. *Educ:* Univ Buenos Aires, Lic chem sci, 52, PhD(chem), 54; Univ Tenn, MS, 71. *Prof Exp:* Staff mem nuclear chem res, Arg AEC, 53-61, head activation analysis group nuclear chem res & develop, 61-62; staff mem nuclear chem & physics res & develop, Oak Ridge Nat Lab, 62-80; mgr res & develop enrichment safeguards, Nuclear Div, Union Carbide Corp, 80-89; RETIRED. *Concurrent Pos:* Lab asst, Univ Buenos Aires, 50-57, lab demonstr, 58-59, prof, 61-62; Int Atomic Energy Agency fel, Chalk River Nuclear Labs, Can, 59-61. *Mem:* Fel Am Nuclear Soc; Am Chem Soc; Arg Chem Asn. *Res:* Nuclear methods of analysis; nuclear reaction cross sections; fusion plasma-wall interactions; x-ray and electron spectroscopy. *Mailing Add:* 996 W Outer Dr Oak Ridge TN 37830

RICCI, JOHN ETTORE, b New York, NY, Jan 1, 07. PHYSICAL CHEMISTRY. *Educ:* NY Univ, BS, 26, MS, 28, PhD(chem), 31. *Prof Exp:* From instr to prof, 31-77, EMER PROF CHEM, NY UNIV, 77- *Concurrent Pos:* Consult, Oak Ridge Nat Lab, 53- *Mem:* Am Chem Soc. *Res:* Phase rule; aqueous solubilities; solid solutions; non-aqueous solvents; ionization constants; hydrogen ion concentration; fused salts. *Mailing Add:* 17 Nolan Pl Yonkers NY 10704

RICCI, JOHN SILVIO, JR, b Springfield, Mass, Aug 27, 40. PHYSICAL CHEMISTRY, INORGANIC CHEMISTRY. *Educ:* Am Int Col, AB, 62; Columbia Univ, MA, 63; State Univ NY, Stony Brook, PhD(chem), 69. *Prof Exp:* Fel chem, Northwestern Univ, 69-70; prof, Windham Col, 70-77 & Williams Col, 77-81; PROF CHEM, UNIV SOUTHERN MAINE, 81- *Concurrent Pos:* Res collabr, Brookhaven Nat Lab, 71- *Mem:* Am Chem Soc; Am Crystallog Asn. *Res:* Molecular structure determination of transition metal complexes; organo-phosphorus and organo-sulfur compounds. *Mailing Add:* Univ Southern Maine 96 Falmouth St Portland MA 04103

RICCIARDI, ROBERT PAUL, b Quincy, Mass, June 18, 46; m 83. GENETIC ENGINEERING, MOLECULAR VIROLOGY. *Educ:* Boston Univ, BA, 68; Col William & Mary, MA, 73; Univ Ill, Urbana, PhD(cellular & molecular biol), 77. *Prof Exp:* Teacher chem, Lisbon High Sch, NH, 68-69; Am Cancer Soc fel, Brandeis Univ, 77-78, Nat Cancer Inst fel, Sch Med, Harvard Univ, 78-80, Charles A King Trust fel, 80-81; ASST PROF GENETICS & MICROBIOL, UNIV PA, 81-; ASST PROF CANCER RES, WISTAR INST, 81- *Concurrent Pos:* Consult, Dept Molecular Genetics, Smith Kline & Beckman, 85- , Workshop Oncogenic Human Polyomaviruses, Nat Cancer Inst, 85; vis prof, Sch Med, Univ Ferrara, Italy, 86. *Res:* Investigation of mechanisms that regulate both viral genes and oncogenes. *Mailing Add:* Hwy 507 Gouldsboro PA 18424

RICCIUTI, FLORENCE CHRISTINE, b New Haven, Conn, Aug 29, 44. HUMAN GENETICS. *Educ:* Albertus Magnus Col, BA, 66; Yale Univ, PhD(biol), 73. *Prof Exp:* Asst res med genetics, Yale Univ Med Sch, 66-68, fel, Dept Human Genetics, 72-75; PROF BIOL & CHMN DEPT BIOL, ALBERTUS MAGNUS COL, 75- *Mem:* Sigma Xi. *Res:* Human gene mapping using somatic cell genetics and studying differentiation and X chromosome inactivation in embryonic and adult tissues. *Mailing Add:* 42 Livingston New Haven CT 06511

RICCOBONO, PAUL XAVIER, b New York, NY, Jan 5, 39; div. DETERGENTS, TEXTILES. *Educ:* Brooklyn Col, BS, 59; NY Univ, MS, 63, PhD(org chem), 64. *Prof Exp:* Res chemist, Nat Biscuit Co, 59-60 & E I du Pont de Nemours & Co, Inc, 64-67; sr res chemist, Airco, 67-68; group leader, Cent Res Labs, J P Stevens & Co, Inc, 68-71, mgr, Cent Analysis Dept, 71-73, mgr, 73-77, dir, Mat Res & Eval Dept, 77-81; dir res, Congoleum Corp, 81-84; dept head, Colgate-Palmolive Co, 84-89; MGR, HOUSEHOLD PROF RES & DEVELOP, BLOCK DRUG CO, 90- *Concurrent Pos:* Assoc res dir, Indust Res Inst. *Mem:* Instrument Soc Am; Am Chem Soc; Am Soc Testing & Mat. *Res:* Resolution of optically active cyclooctatetraene derivatives; photochemistry of maleic anhydride derivatives; preparation and properties of graft copolymers; toxic chemicals legislation; textile chemistry; advanced materials characterization and analysis techniques; utilization of cold plasma and radiation for materials performance modification; detergent formulations; household products consumer research; household product research and development. *Mailing Add:* 72 Dogwood Lane Bedminster NJ 07921

RICE, BARBARA SLYDER, b Chambersburg, Pa, Dec 19, 37; m 63; c 4. MATHEMATICS. *Educ:* Clark Univ, AB, 59; Univ Va, MA, 61, PhD(math), 65. *Prof Exp:* Res asst math, Univ Va, 61-63; adj prof, Fla Inst Technol, 63-73; ASSOC PROF MATH, ALA AGR & MECH UNIV, 75- *Mem:* Am Math Soc; Math Asn Am. *Res:* Development of more effective teaching processes for use with mathematically inexperienced students. *Mailing Add:* 308 Flemington Rd SE Huntsville AL 35802

RICE, BERNARD, b Milwaukee, Wis, Dec 5, 14. PHYSICAL CHEMISTRY. *Educ:* George Washington Univ, BS, 37; Univ Chicago, PhD(chem), 48. *Prof Exp:* Chemist, Nat Bur Standards, 38-42; from asst prof to assoc prof, 49-61, PROF CHEM, ST LOUIS UNIV, 61- *Mem:* Am Chem Soc. *Res:* Molecular structure; spectroscopy; theoretical chemistry. *Mailing Add:* 7456 Parkdale Ave St Louis MO 63105

RICE, CHARLES EDWARD, b Seminole, Okla, Feb 13, 32; m 56; c 2. ENGINEERING, AGRICULTURE. *Educ:* Okla State Univ, BS, 60, MS, 61; Univ Minn, Minneapolis, PhD(agr eng), 72. *Prof Exp:* Res engr, Agr Res Serv, USDA, 61-66; asst prof, 66-74, ASSOC PROF AGR ENG, OKLA STATE UNIV, 74-; RES HYDRAULIC ENGR, AGR RES SERV, USDA, 78- *Mem:* Am Soc Agr Engrs; Sigma Xi. *Res:* Open channel hydraulics; hydraulics of conservation structures; overland flow. *Mailing Add:* Rte 5 Box 336 Stillwater OK 74074

RICE, CHARLES MERTON, b Whitmore Lake, Mich, Jan 26, 25; m 47; c 8. NUCLEAR ENGINEERING. *Educ:* Albion Col, AB, 48; Univ Mo-Rolla, MS, 49; Oak Ridge Reactor Sch, MS, 53. *Prof Exp:* Assoc prof physics, Oglethorpe Univ, 49-51; physicist, AEC, 51-54; proj supvr, Ford Instrument Co, 54-55; head atomic power eng group, Sargent & Lundy, 55-56; mgr reactor eng dept, Advanced Technol Lab, 56-59; prog mgr, Aerojet Gen Corp, 59-69; pres, Idaho Nuclear Corp, 69-70 & Aerojet Nuclear Co, 70-72; pres & chmn bd, Energy Inc, 72-81; prin, LRS Consult, 81-90; PRES, RICE INC, 78- *Honors & Awards:* Prod Eng Master Design Award, 63. *Mem:* Fel Am Nuclear Soc; Atomic Indust Forum. *Res:* Gas cooled reactor technology, nuclear rocket propulsion and water reactor safety. *Mailing Add:* Rice Inc 355 W 14th St Idaho Falls ID 83402

RICE, CHARLES MOEN, III, b Sacramento, Calif, Aug 25, 52. ANIMAL RNA VIRUSES. *Educ:* Univ Calif, Davis, BS, 74; Calif Inst Technol, PhD(biochem), 81. *Prof Exp:* Res fel, Calif Inst Technol, 81-84, staff biologist, 85; vis fel, Australian Nat Univ, 85; asst prof, 86-91, ASSOC PROF, SCH MED, WASH UNIV, 91- *Mem:* AAAS; Am Soc Virol; Am Soc Microbiol; Soc Gen Microbiol (Brit); Am Soc Biochem & Molecular Biol. *Res:* Molecular genetics of RNA virus replication, primarily alphaviruses, flaviviruses and Hepatitis C virus. *Mailing Add:* Dept Molecular Microbiol Wash Univ Sch Med Box 8230 660 S Euclid Ave St Louis MO 63110-1093

RICE, CHRISTINE E, BIOCHEMISTRY. *Educ:* Univ Toronto, Can, PhD(bact), 32. *Prof Exp:* Scientist, Dept Agr, Ottawa, Can, 46-67; RETIRED. *Mailing Add:* 16 Isabella St Perth ON K7H 2W6 Can

RICE, CLIFFORD PAUL, b San Diego, Calif, May 12, 40; m 74; c 3. DATA QUALITY THROUGH DATA MONITORING & RECORDING. *Educ:* Wash State Univ, BS, 62; Cornell Univ, PhD(bot & insect biochem), 72. *Prof Exp:* Res assoc environ chem, Syracuse Res Corp, 71-74; instr pesticide chem, Univ RI, 74-77, res assoc environ chem, 76-77; assoc res scientist environ chem, Great Lakes Res Div, Univ Mich, 78-85; res scientist environ chem, Large Lakes Res Sta, US Environ Protection Agency, 86; supvry chemist anal chem, 86-89, RES CHEMIST ENVIRON CHEM, PATUXENT WILDLIFE RES CTR, 89- *Concurrent Pos:* Adv, State NY Environ Comn, NY State Dept Environ Conserv, 78-; chief scientist, Activ of Joint US/USSR expeds to Bering & Pac Ocean, Patuxent Wildlife Res Ctr, 88-; liaison mem, TSCA's Interagency Testing Comt, 89- *Mem:* AAAS; Am Chem Soc. *Res:* Environmental chemistry and effects of organochlorine pollutants; long range atmospheric transport of toxaphenes; microlayer processes involving movement of PCBs in the Great Lakes; congener specific effects and occurrence of PCBs in avian systems. *Mailing Add:* Patuxent Wildlife Res Ctr Laurel MD 20708

RICE, DALE WARREN, b Grand Haven, Mich, Jan 21, 30. MARINE MAMMALOGY. *Educ:* Ind Univ, AB, 52; Univ Fla, MS, 55. *Prof Exp:* Wildlife res biologist, US Fish & Wildlife Serv, 55-58; WILDLIFE RES BIOLOGIST, NAT MARINE FISHERIES SERV, 58- *Concurrent Pos:* Mem cetacean specialist group, Survival Serv Comn, Int Union Conserv Nature & Natural Resources, 67- *Mem:* Am Soc Mammalogists; Wildlife Soc; Soc Syst Zool; Soc Marine Mammal; Am Ornithologist Union. *Res:* Life history, ecology, population dynamics and systematics of marine mammals, especially baleen whales and sperm whales. *Mailing Add:* 14334 Edgewater Lane Seattle WA 98125

RICE, DALE WILSON, b Jamestown, NY, Nov 21, 32; m 60; c 2. PHYSICAL CHEMISTRY. *Educ:* Mass Inst Technol, SB, 54, PhD(polymer chem), 61. *Prof Exp:* Res assoc, Corning Glass Works, 61-75; eng consult, Centorr Assocs, Inc, 75-78, tech dir, 78-81; PRES, DELTA LABS, 81-; TECH DIR, BEAUMAC CO, 88- *Mem:* AAAS; Am Chem Soc; Am Ceramic Soc; Am Asn Crystal Growth. *Res:* High temperature technology; engineering design; high temperature and vacuum technology. *Mailing Add:* Garland Rd RFD No 1 Barrington NH 03825

RICE, DAVID E, b Northfield, Minn, Dec 8, 33; m 65; c 3. ORGANIC CHEMISTRY, POLYMER CHEMISTRY. *Educ:* St Olaf Col, BA, 55; Univ Minn, PhD(org chem), 59. *Prof Exp:* Res specialist, 59-80, STAFF SCIENTIST, 3M CO, 80- *Mem:* AAAS; Am Chem Soc; Sigma Xi. *Res:* Fluorocarbon polymers; organic synthesis. *Mailing Add:* 3M Co Control Res Labs 3M Ctr St Paul MN 55101

RICE, DENNIS KEITH, b Newell, WVa, Dec 12, 39; m 70; c 3. LASER PHYSICS, STRATEGIC PLANNING. *Educ:* Cleveland State Univ, BEE, 64; Univ Southern Calif, MSEE, 66, PhD(elec eng), 69. *Prof Exp:* Prin engr, Laser Technol Labs, 71-74, dir eng, 74, mgr, Laser Lab, 74-76, prog mgr, 76-78, mgr laser-optical eng, Northrop Corp Res & Technol Ctr, 78-80, asst to gen mgr technol & planning, 80-82, mgr advan systs, 82-85, vpres advan systs, 85-86, vpres & chief scientist, Northrop Electro-Mech Div, 86-89, vpres transition team, Northrop Electronics Systs Div, 89-90, VPRES, SYSTS ENG & PLANNING, NORTHROP CORP, 90- *Mem:* Am Mgt Asn; Am Defense Preparedness Asn; Nat Security Indust Asn. *Res:* Optics; thin films; spectroscopy; solid state physics; atmospheric physics. *Mailing Add:* 650 S Scout Trail Anaheim CA 92807

RICE, DOROTHY PECHMAN, b Brooklyn, NY, June 11, 22; m 43; c 3. HEALTH STATISTICS, MEDICAL ECONOMICS. *Educ:* Univ Wis, BA, 41. *Hon Degrees:* DSc, Col Med & Dent. *Prof Exp:* Pub health analyst health serv res, Div Hosp & Med Facil, USPHS, 60-62; social sci analyst, Div Res & Statist, Social Security Admin, 62-64; pub health analyst econ, Div Community Health Serv, USPHS, 64-65; chief, Health Ins Res Br, Social Security Admin, 65-72, dep asst comnr res & statist, 72-76; dir health statist, Nat Ctr Health Statist, 76-82; PROF IN RESIDENCE, UNIV CALIF, SAN FRANCISCO, 82- *Honors & Awards:* Jack C Massey Award, 78. *Mem:* Inst Med-Nat Acad Sci; fel Am Statist Asn; fel Am Pub Health Asn; Am Econ Asn. *Res:* The organization, delivery and financing of health services, cost of illness, aging and chronic illness; disability. *Mailing Add:* Dept Social & Behav Sci Univ Calif San Francisco CA 94143-0612

RICE, ELMER HAROLD, biochemistry, for more information see previous edition

RICE, ELROY LEON, b Edmond, Okla, Jan 31, 17; m 45; c 2. ALLELOPATHY, CHEMICAL ECOLOGY. *Educ:* Cent State Col, Okla, BA, 38; Univ Okla, MS, 42; Univ Chicago, PhD(bot), 47. *Prof Exp:* From asst prof to prof bot, 48-66, David Ross Boyd prof, 67-81, DAVID ROSS BOYD EMER PROF BOT, UNIV OKLA, 81- *Concurrent Pos:* Vis prof biol, Purdue Univ, 62-63. *Mem:* Fel AAAS; Ecol Soc Am; Am Soc Plant Physiol. *Res:* Changes in microclimate, microorganisms, soil factors and allelochemicals during plant succession; writing of scientific monographs in chemical ecology. *Mailing Add:* Dept Bot & Microbiol Univ Okla Norman OK 73019

RICE, FRANK J, b Putnam, Conn, Oct 10, 24; m 52; c 7. EMBRYOLOGY. *Educ:* Colo State Univ, BS, 50; Univ Wyo, MS, 51; Univ Mo, PhD(genetics), 56. *Prof Exp:* Mgr beef cattle ranch, San Carlos Apache Tribal Enterprises, Ariz, 53-54; geneticist, USDA, Mont, 56-61; from asst prof to assoc prof biol, 61-79, chmn dept, 70-76, PROF BIOL, FAIRFIELD UNIV, 79- *Res:* Genetics; physiology of reproduction; natural family planning. *Mailing Add:* Dept Biol Fairfield Univ Fairfield CT 06430-7524

RICE, FREDERICK ANDERS HUDSON, b Bismarck, NDak, Feb 19, 17; m 49. ORGANIC CHEMISTRY. *Educ:* Dalhousie Univ, BA, 37, MSc, 45; Ohio State Univ, PhD(org chem), 48. *Prof Exp:* Res assoc bact, Johns Hopkins Univ, 48-50, asst prof microbiol, 51-54; res chemist, US Naval Powder Factory, 54-55, assoc head chem div, 55-56, head fundamental process div, 56-60; chief res, Off Qm Gen, 59-62; phys scientist, res div, Army Materiel Command, 62-63; prof, 63-82, EMER PROF CHEM, AM UNIV, 82- *Concurrent Pos:* Sr Fulbright award, Univ London, 52-53. *Honors & Awards:* Hillebrand Prize, Wash Chem Soc, 72. *Mem:* AAAS; Am Chem Soc; NY Acad Sci; Royal Soc Chem; Soc Exp Biol & Med. *Res:* Carbohydrate chemistry; chemistry of microorganisms; biochemistry of blood cell regulation; physics and chemistry of propellants; ultrasonics. *Mailing Add:* 484 Spinnaker Ft Lauderdale FL 33326-1632

RICE, JACK MORRIS, b Salina, Kans, Aug 30, 48; m 75; c 2. GEOLOGY. *Educ:* Dartmouth Col, AB, 70; Univ Wash, MS, 72, PhD(geol), 75. *Prof Exp:* Gibbs instr, Yale Univ, 75-77; ASST PROF GEOL, UNIV ORE, 77- *Mem:* Geol Soc Am; Mineral Soc Am; Am Geophys Union. *Res:* Field, analytical and theoretical studies bearing on the mineralogy and petrology of metamorphic rocks; thermodynamics of rock-forming silicate minerals. *Mailing Add:* 3580 W 20th Vancouver BC V6S 1E7 Can

RICE, JAMES K, b Pittsburgh, Pa, Mar 12, 23; m 46; c 2. WATER CHEMISTRY. *Educ:* Carnegie Inst Technol, BS, 46, MS, 47. *Prof Exp:* Res engr, Cyrus W Rice & Co, 47-52, sr engr, 52-59, pres, 59-67, pres & gen mgr, Rice Div, NUS Corp, 67-73, sr vpres, 73-76; CONSULT ENGR, 76- *Concurrent Pos:* Mem, Coal Slurry Adv Panel, Off Technol Assessment, US Cong, 76-78. *Honors & Awards:* Award of Merit, Am Soc Testing & Mat, 70. *Mem:* Fel Am Soc Testing & Mat; fel Am Inst Chemists; Am Chem Soc; Nat Asn Corrosion Engrs; fel Am Soc Mech Engrs. *Res:* Monitoring of microchemical contaminants in the aquatic environment and in industrial water and waste water; water technology of thermal power systems. *Mailing Add:* 17415 Batchellor's Forest Olney MD 20832

RICE, JAMES KINSEY, b Harvey, Ill, June 5, 41; m 63; c 2. CHEMICAL PHYSICS, LASERS. *Educ:* Ind Univ, Bloomington, BS, 63; Calif Inst Technol, PhD(chem), 68. *Prof Exp:* Res fel chem, Calif Inst Technol, 68-69; STAFF MEM CHEM PHYSICS, SANDIA LABS, 69- *Mem:* AAAS; Am Chem Soc; Am Phys Soc; Am Soc Mass Spectrom; Sigma Xi. *Res:* Chemical lasers; reaction dynamics; electron beam pumped gas laser. *Mailing Add:* 12428 Chelwood Trail NE Albuquerque NM 87112

RICE, JAMES R, b Frederick, Md, Dec 3, 40; m; c 2. GEOPHYSICS, ENGINEERING MECHANICS. *Educ:* Lehigh Univ, BS, 62, MSC, 63, PhD, 64. *Hon Degrees:* ScD, Lehigh Univ, 85. *Prof Exp:* Fel, Brown Univ, 64-65; from asst prof to prof eng, 65-81, Ballou prof theoret & appl mech, 73-81, MCKAY PROF ENG SCI & GEOPHYS, HARVARD UNIV, 81- *Mem:* Nat Acad Sci; Nat Acad Eng; fel Am Soc Mech Engrs; AAAS; fel Am Geophys Union. *Res:* Fracture mechanics in technology and geophysics; earthquake phenomena. *Mailing Add:* Dept Earth & Plantary Sci Div Appl Sci Harvard Univ Cambridge MA 02138

RICE, JAMES THOMAS, b Birmingham, Ala, Feb 7, 33; m 54, 75; c 4. SCIENCE & WOOD TECHNOLOGY. *Educ:* Auburn Univ, BS, 54; NC State Univ, MS, 60, PhD(wood technol), 64. *Prof Exp:* From instr to asst prof wood technol, NC State Univ, 59-65; assoc prof, Univ Ga, 65-69; mgr wood adhesives develop, Central Resin Develop Lab, Ga-Pac Corp, 69-70; ASSOC PROF FOREST RESOURCES, UNIV GA, 70- *Concurrent Pos:* Tech coordr, Adhesive & Sealant Coun, 66-69 & 72-78; chmn, Am Soc Testing & Mat comt D-14 on Adhesives, 86-90. *Mem:* Am Soc Testing & Mat; Forest Prod Res Soc; Soc Wood Sci & Technol. *Res:* Adhesives and adhesive bonded products, especially those with wood as an adherend. *Mailing Add:* Sch Forest Resources Univ Ga Athens GA 30602

RICE, JERRY MERCER, b Washington, DC, Oct 3, 40; m 69, 78; c 2. EXPERIMENTAL PATHOLOGY, BIOCHEMISTRY. *Educ:* Wesleyan Univ, BA, 62; Harvard Univ, PhD(biochem), 66. *Prof Exp:* Res scientist, Biol Br, 66-69, head perinatal carcinogenesis sect, Lab Exp Path, 73-80, SR SCIENTIST, LAB EXP PATH, DIV CANCER CAUSE & PREV, NAT CANCER INST, 69-, CHIEF, LAB COMP CARCINOGENESIS, 81- *Honors & Awards:* Outstanding Serv Medal, USPHS, 90. *Mem:* Sigma Xi; Am Chem Soc; Teratology Soc; Am Asn Cancer Res; Am Asn Pathologists. *Res:* Chemical carcinogenesis, especially transplacental carcinogenesis. *Mailing Add:* Lab Comp Carcinogenesis Nat Cancer Inst Ft Detrick Frederick MD 21702-1201

RICE, JOHN RISCHARD, b Tulsa, Okla, June 6, 34; m 54; c 2. APPLIED MATHEMATICS, COMPUTER SCIENCE. *Educ:* Okla State Univ, BS, 54, MS, 56; Calif Inst Technol, PhD(math), 59. *Prof Exp:* Nat Res Coun-Nat Bur Stand res fel math, Nat Bur Stand, 59-60; sr res mathematician, Gen Motors Res Labs, 60-64; prof math & comput sci, 64-89, HEAD, 83-, DISTINGUISHED PROF COMPUT SCI, PURDUE UNIV, 89- *Concurrent Pos:* Chmn, COSERS Panel on Numerical Computation, 74-78; ed-in-chief, Asn Comput Mach Trans Math Software, 74-; chmn, Signum, 77-79; Int Fed Info Processing Working Group 2.5. *Honors & Awards:* George E Forsythe Mem lectr, 75. *Mem:* Nat Acad Sci; Am Math Soc; Soc Indust & Appl Math; Asn Comput Mach; fel AAAS. *Res:* Approximation theory; numerical analysis; mathematical software supercomputing; author or coauthor of various publications. *Mailing Add:* Dept Comput Sci Purdue Univ West Lafayette IN 47907

RICE, JOHN T(HOMAS), b New London, Conn, Feb 4, 31. MECHANICAL ENGINEERING. *Educ:* Univ Conn, BSE, 52; Newark Col Eng, MSME, 54; Columbia Univ, EngScD(mech of solids), 62. *Prof Exp:* Engr propeller div, Curtiss-Wright Corp, 52-54, proj engr, Wright Aeronaut Div, 54-57; supvr struct mech, Elec Boat Div, Gen Dynamics Corp, 62-64; chmn, 82-90, PROF MECH ENG, PRATT INST, 64- *Mem:* Am Soc Mech Engrs; Am Soc Eng Educ. *Res:* Dynamics; vibrations; mechanical design; systems engineering. *Mailing Add:* Dept Mech Eng Pratt Inst Brooklyn NY 11205

RICE, KENNER CRALLE, b Rocky Mount, Va, May 14, 40. SYNTHETIC ORGANIC CHEMISTRY, MEDICINAL CHEMISTRY. *Educ:* Va Mil Inst, BS, 61; Ga Inst Technol, PhD(org chem), 66. *Prof Exp:* Capt, Walter Reed Army Inst Res, Wash, DC, 66-68; NIH fel, Ga Inst Technol, 68-69; sr scientist, Process Res, Ciba Pharmaceut Co, Ciba-Geigy Corp, 69-72; NIH sr staff fel, 72-76, res chemist, Nat Inst Arthritis, Metab & Digestive Dis, 77-86; CHIEF, SECT DRUG DESIGN & SYNTHESIS, NAT INST DIABETES DIGESTIVE & KIDNEY DIS, 87- *Concurrent Pos:* Adj prof pharmacol, dept pharmacol & exp therapeut, Sch Med, Univ Md, Baltimore. *Honors & Awards:* Sato Mem Int Award, 83. *Mem:* Am Chem Soc; Pytochemical Soc Europe; Am Pharmaceut Asn; Soc Neurosci. *Res:* The chemistry of analgesics, their antagonists and other drugs which act on the central nervous system; stereochemistry of drugs in relation to their mechanism of action and receptor interactions; isolation, structural elucidation and synthesis of natural products, especially alkaloids; positron emission tomography in the study of the central nervous system. *Mailing Add:* 9007 Kirkdale Rd Bethesda MD 20817-3331

RICE, MARION MCBURNEY, b Syracuse, NY, Feb 20, 23; m 52; c 2. BACTERIOLOGY, BOTANY. *Educ:* DePauw Univ, AB, 48, MA, 49; Univ Wis, PhD(bact), 53. *Prof Exp:* With Eli Lilly & Co, 41-44, asst, 45; asst, Ind Univ, 48; asst, Eli Lilly & Co, 49; asst bact, Univ Wis, 49-50; bacteriologist, Stuart Circle Hosp, Richmond, Va, 53-54; instr biol, Richmond Prof Inst, Col William & Mary, 55-58; asst prof bot & bact, Rockford Col, 58-60 & 63-66; instr biol, Beloit Col, 60; assoc prof bot, Univ Wis, Rock County Campus, 66-88; RETIRED. *Concurrent Pos:* Asst bact, Univ Wis, 59. *Mem:* Am Soc Microbiol. *Res:* Antibiotics; cytology of streptomyces; phytogeography; botulism. *Mailing Add:* 2514 W Memorial Dr Janesville WI 53545

RICE, MARY ESTHER, b Washington, DC, Aug 3, 26. INVERTEBRATE ZOOLOGY. *Educ:* Drew Univ, AB, 47; Oberlin Col, MA, 49; Univ Wash, PhD(zool), 66. *Prof Exp:* Instr zool, Drew Univ, 49-50; res assoc radiation biol, Col Physicians & Surgeons, Columbia Univ, 50-53; res biologist, NIH, 53-61; teaching asst, Univ Wash, 61-66; assoc cur invert zool, Mus Natural Hist, 66-74, scientist-in-chg, Smithsonian Marine Sta, Link Port, 81-89, CUR INVERT ZOOL, MUS NATURAL HIST, SMITHSONIAN INST, 74-, DIR, SMITHSONIAN MARINE STA, LINK PORT, 89- *Concurrent Pos:* Mem-at-large, Biol Sci Sect Comt, AAAS. *Mem:* Fel AAAS; Am Inst Biol Sci; Am Soc Zoologists (pres, 79). *Res:* Development, reproductive biology and systematics of marine worms of the phylum Sipuncula; life histories of marine invertebrates. *Mailing Add:* Dept Invert Zool Smithsonian Mus Natural Hist Washington DC 20560

RICE, MICHAEL JOHN, b Cowes, UK, Dec 25, 40; m 65; c 3. CONDENSED MATTER PHYSICS, THEORETICAL PHYSICS. *Educ:* Univ London, BSc, 62, PhD(theoret physics), 66. *Prof Exp:* Asst prof physics, Imp Col, Univ London, 65-68; staff physicist, Gen Elec Res & Develop Lab, 68-71; staff mem, Brown Boveri Res Ctr, Switz, 71-74; sr scientist, 74-84, PRIN SCIENTIST, XEROX WEBSTER RES, 84- *Concurrent Pos:* Vis asst prof physics, State Univ NY, Stony Brook, 68; Nordita prof physics, Nordisk Inst Theoret Atomic Physics, Denmark, 78-79; consult, Optical Spectros Prog, Xerox-Ohio State Univ, 78- *Mem:* Fel Am Phys Soc; Am Chem Soc; fel Inst Physics; fel Phys Soc UK; fel Swiss Phys Soc; Sigma Xi. *Res:* Microscopic theory of quantum fluids, metals and alloys, semiconductors, ionic conductors and organic radical-ion solids. *Mailing Add:* Xerox Webster Res Ctr 39D114 Webster NY 14580

RICE, NANCY REED, b Chicago, Ill, July 20, 40. ONCOGENIC RETROVIRUSES. *Educ:* Stanford Univ, BA, 61; Harvard Univ, MA, 63, PhD(biol), 69. *Prof Exp:* Fel molecular biol, 68-71, staff mem molecular biol, Carnegie Inst Wash Dept Terrestrial Magnetism, 72-76; staff, 76-80, sr scientist, 80-83, HEAD, MOLECULAR BIOL RETROVIRUSES SECT, FREDERICK CANCER RES DEVELOP CTR, 83- *Concurrent Pos:* Mem aging rev comt, Nat Inst Aging, 74-78; mem develop biol panel, NSF, 75. *Mem:* Am Soc Biol Chemists. *Res:* Molecular biology of retroviruses; rel oncogene. *Mailing Add:* ABL-Basic Res Prog Frederick Cancer Res Develop Ctr Frederick MD 21701

RICE, NOLAN ERNEST, zoology, for more information see previous edition

RICE, NORMAN MOLESWORTH, b Ottawa, Ont, Oct 13, 39; m 62; c 2. MATHEMATICS. *Educ:* Queen's Univ, Ont, BS, 62; Calif Inst Technol, PhD(math), 66. *Prof Exp:* ASSOC PROF MATH, QUEEN'S UNIV, ONT, 65- *Mem:* Am Math Soc; Math Asn Am; Can Math Cong. *Res:* Functional analysis; vector lattices. *Mailing Add:* Dept Math Queen's Univ Kingston ON K7L 3N6 Can

RICE, PAUL LAVERNE, b Bancroft, Nebr, Dec 28, 06; m 39; c 3. MEDICAL ENTOMOLOGY. *Educ:* Univ Idaho, BS, 31, MS, 32; Ohio State Univ, PhD(entom), 37. *Hon Degrees:* DSc, Alma Col, Mich, 67. *Prof Exp:* Instr entom & asst entomologist, Univ Idaho, 31-33; asst entomologist, Univ Del, 36-37; prof biol, Alma Col, 37-42; assoc entomologist & actg head dept, Univ Del, 42-45; prof biol, head dept & dean fac, Alma Col, 45-50; prof biol, Whittier Col, 50-58; assoc dir, Malaria Eradication training ctr, AID, Jamaica, 58-62; mem staff grants prog, NIH, 62-64; various pos, Vector-Borne Dis Training, Ctr Dis Control, USPHS, 64-70; contractor, Vector Borne Dis Training, USPHS Ctr Dis Control, 72-78; RETIRED. *Concurrent Pos:* In chg malaria team, USPHS, Int Coop Admin, Ethiopia, 55-57; Consult, 70-72. *Res:* Mosquito borne diseases. *Mailing Add:* 2373 Burnt Creek Rd Decatur GA 30033

RICE, PETER (FRANKLIN), b Toronto, Ont, May 18, 39; m 64; c 2. FOREST PATHOLOGY. *Educ:* Univ Toronto, BScF, 62, MScF, 64, PhD(forest path), 68. *Prof Exp:* Lectr forest path, Univ Toronto, 66-68; PATHOLOGIST, ROYAL BOT GARDENS, 68-, ASST DIR CONSERV, UNIV LETHBRIDGE. *Mem:* Can Phytopath Soc; Am Phytopath Soc; Can Inst Forestry. *Res:* Diseases of ornamental plants, especially woody plants. *Mailing Add:* Royal Bot Gardens PO Box 399 Hamilton ON L8N 3H8 Can

RICE, PETER MILTON, b Montclair, NJ, Oct 14, 37; m 62; c 2. OPERATIONS RESEARCH, SOFTWARE SYSTEMS. *Educ:* St John's Col, MD, AB, 59; Fla State Univ, PhD(math), 63. *Prof Exp:* From asst prof to assoc prof, 63-84, asst to dean, 77-82, PROF MATH, UNIV GA, 84- *Concurrent Pos:* Alexander von Humbolt res fel & Sarah Moss res fel, Univ Bonn, 66-67; Alexander von Humboldt res fel, Inst Math Econ, Univ Bielefeld & Inst Higher Educ, Vienna, 75-76. *Mem:* Math Assoc Am. *Res:* Software development for educational testing; artificial intelligence. *Mailing Add:* Dept Math Univ Ga Athens GA 30602

RICE, PHILIP A, b Ann Arbor, Mich, Aug 3, 36; m 59; c 3. CHEMICAL ENGINEERING. *Educ:* Univ Mich, BSE, 59, MSE, 60, PhD(chem eng), 63. *Prof Exp:* NATO fel, Inst Phys Chem, Univ Gottingen, 62-63; chem engr, Analytical Serv, Inc, Va, 63-65; from asst prof to assoc prof, 65-77, PROF CHEM ENG, SYRACUSE UNIV, 77- *Concurrent Pos:* Res assoc prof, Upstate Med Ctr, State Univ NY, 70-; prog mgr, Chem, Biochem & Thermal Eng Div, NSF, 84-85; chmn, Dept Chem Eng & Mat Sci, Syracuse Univ, 85-90. *Mem:* Am Inst Chem Engrs; Am Chem Soc; AAAS; Sigma Xi; Am Asn Univ Professors. *Res:* Transport and metabolism in animal cell systems; water renovation processes; vacuum spay stripping of dissolved and emulsified organics from water; heat transfer processes involving a change of phase. *Mailing Add:* Dept Chem Eng & Mat Sci Syracuse Univ Syracuse NY 13244

RICE, RANDALL GLENN, b Smithfield, Utah, Nov 13, 18; m 46. INFORMATION SCIENCE. *Educ:* Pomona Col, BA, 40; Stanford Univ, PhD(biochem), 44. *Prof Exp:* Asst biochem, Stanford Univ, 40-42, Off Sci Res & Develop Contract, 43-44; Corn Prod Co fel, Cornell Univ, 45-46, instr chem, 46-48; chemist, USDA, 48-54; from asst ed to assoc ed, Chem Abstr, 54-59, head biochem, Indexing Dept, 59-62; chief mat sci div, Dept Defense Doc Ctr, Cameron Sta, Va, 63-64; dep chief & ed officer, Study Support Br, Hq US Air Force, Pentagon, 65-68, chief, Study Support Br, 68-71, chief tech info serv br, 71-78; RETIRED. *Mem:* AAAS; Am Chem Soc. *Res:* Flavor chemistry of citrus products; chemistry of natural products; isolation and identification of natural constituents; chemical literature; classification and indexing vocabularies for military science. *Mailing Add:* 6404 21st Ave W Apt M506 Bradenton FL 34209-4663

RICE, RICHARD EUGENE, b Leominster, Mass, June 13, 43. GROUND WATER SYSTEMS, ARTIFICIAL & NATURAL MEMBRANES. *Educ:* Univ NH, BS, 65; Univ Mich, MS, 67; Univ Mont, MFA, 74; Mich State Univ, PhD(phys chem), 82. *Prof Exp:* Res asst, St Vincent Hosp, Worcester, Mass, 68-72; Int Res & Exchanges Bd fel, Inst Colloid & Water Chem, Ukr Acad Sci, Kiev, USSR, 79-80; Nat Res Coun res assoc, US Army Chem Res & Develop Ctr, 83-84; Res scientist, Holcomb Res Inst, Butler Univ, 84-89; Vis asst prof chem, phys & astron, Ind Univ NW, 88-90; ASST PROF, DIV MULTIDISCIPLINARY STUD, NC STATE UNIV, 90- *Mem:* AAAS; Am Chem Soc; Am Geophys Union; Hist Sci Soc. *Res:* Theoretical and mathematical description of the movement of ions and molecules in response to gradients of concentration, electric potential, pressure and temperature; theory of contaminant transport and reaction in ground water systems; theory of ion transport across artificial and natural membranes. *Mailing Add:* Div Multidisciplinary Studies NC State Univ Raleigh NC 27695-7101

RICE, RICHARD W, b Ainsworth, Nebr, Aug 10, 31; m 52; c 2. ANIMAL NUTRITION, BIOCHEMISTRY. *Educ:* Univ Nebr, BS, 53, MS, 58; Mich State Univ, PhD(animal nutrit), 60. *Prof Exp:* From asst prof to prof animal sci, Univ Wyo, 60-75; PROF ANIMAL SCI & HEAD DEPT, UNIV ARIZ, 75- *Mem:* Soc Range Mgt; Am Soc Animal Sci; Am Dairy Sci Asn. *Res:* Ruminant nutrition; forage evaluation; factors affecting feed intake; fat metabolism in the ruminant; applied animal ecology. *Mailing Add:* 1112 W Giaconda Way Tucson AZ 85704

RICE, RIP G, b New York, NY, Apr 19, 24; m 48; c 1. CHEMISTRY, ENVIRONMENTAL SCIENCES. *Educ:* George Washington Univ, BS, 47; Univ Md, PhD, 57. *Prof Exp:* Analytical chemist, Nat Bur Standards, DC, 47-50; chemist, US Naval Res Lab, 50-55; org chemist, US Naval Ord Lab, Md, 55-57; res chemist, Gen Dynamics/Convair, Tex, 57-59, staff scientist, Sci Res Lab, 59-60, tech dir adv prod dept, Calif, 60-62; sr chemist, W R Grace & Co, 62-63, res supvr, 63-64, mgr inorg chem res, 64-67, dir contract opers, 67-72; mgt consult-resident rep, 72-77; corp mgr gov relations, Jabcob Eng Group, 71-81, dir, environ systs, Adv Systs Div, 81-82; PRES, RICE INT CONSULT ENTERPRISES, 82- *Concurrent Pos:* Tech adv, Int Ozone Asn, 74-; ed-in-chief, Ozone Sci & Eng, 85- *Honors & Awards:* Founders Award, Int Ozone Asn, 79. *Mem:* Am Chem Soc; Water Pollution Control Fedn; Int Ozone Asn (pres-elect, 79-81, pres, 82-84); Am Water Works Asn; Am Inst Chem Engrs; Water Quality Asn; Int Asn Water Pollution Res; Int Bottled Water Asn; Nat Environ Health Asn; Nat Water Well Asn; Soc Soft Drink Technol. *Res:* Ozone technology; inorganic, polymer, organic chemistry. *Mailing Add:* 1331 Patuxent Dr Ashton MD 20861

RICE, ROBERT ARNOT, b San Francisco, Calif, Apr 4, 11; wid. SCIENCE EDUCATION. *Educ:* Univ Calif, BA, 34, MA, 47. *Prof Exp:* Teacher, Geyserville Union High Sch, Calif, 35-40, prin, 40-41; teacher, Berkeley High Sch, 41-61, chmn dept sci, 49-61; supvr sci & math, Berkeley Unified Sch Dist, 61-64, consult, 64-70, dir On Target Sch, 71-73, work experience coordr, 73-75; DIR NORTHERN CALIF-WESTERN NEV JR SCI & HUMANITIES SYMP, 62- *Concurrent Pos:* Exec dir, San Francisco Bay Area Sci Fair, 54-59, bd dirs, 60-; mem, Chem Comt, Nat Sci Teacher's Asn, 56-60, adv comt mem, 70; adminr, NSF Summer Insts Sci Teachers, Univ Calif, Berkeley, 57-65; coordr, Children's Area, US Sci Exhibit, Century 21 Expos, Seattle, 61; asst to dir, Lawrence Hall Sci, 64-69, coordr public progs, 69-75, liaison mem, 66-69 & 66-73; pres, Calif Sci Teachers Asn, 49-50, exec dir, 62-90; dir, 18th Int Sci Fair, San Francisco, 67; mem, Int Sci Fair Coun, Sci Serv Inc, 59-68; chmn, public educ comt, Alameda County Heart Asn, 67-69, bd dirs, 66-71; coordr, Indust Initiatives Sci & Math Educ, 85-87, dir acad, 87; mem adv & res comt, Alameda County TB & Health Asn, 65-69; bd dirs, Calif Heart Asn, 66-71; ed, Sci Teacher, 60-61. *Honors & Awards:* Armed Forces Chem Asn Award, 56; Distinguished Serv to Sci Educ, Nat Sci Teachers Asn, 86. *Mem:* Nat Sci Teachers Asn (pres, 60-61). *Res:* Author and co-author of several publications. *Mailing Add:* Lawrence Hall Sci Univ Calif Berkeley CA 94720

RICE, ROBERT BRUCE, earth sciences, for more information see previous edition

RICE, ROBERT HAFLING, b Birmingham, Ala, Dec 31, 44. TOXICOLOGY, CELL BIOLOGY. *Educ:* Mass Inst Technol, SB, 67; Univ Calif, Berkeley, PhD(molecular biol), 72. *Prof Exp:* Fel, Univ Calif, Davis, 72-75, Mass Inst Technol, 75-79; asst prof, Harvard Sch Pub Health, 79-85, asst prof toxicol, 85-; ASSOC PROF TOXICOL, UNIV CALIF, DAVIS. *Mem:* AAAS; Am Soc Cell Biol; Soc Toxicol. *Res:* Biochemical aspects of differentiation and toxicology of cultivated epithelial cells, with emphasis on mechanisms of cross-linked envelope formation and chronic effects of exposure to xenobiotics in the environment. *Mailing Add:* Dept Environ Toxicol Univ Calif Davis CA 95616-8588

RICE, ROBERT VERNON, b Barre, Mass, Aug 13, 24; div; c 2. BIOCHEMISTRY. *Educ:* Northeastern Univ, BS, 50; Univ Wis, MS, 52, PhD(biochem), 55. *Prof Exp:* Asst biochem, Univ Wis, 51-54; fel chem physics, sr fel independent res, 57-67, prof biol sci & chmn dept, Mellon Inst Sci, 71-77, prof biochem, 67-87, EMER PROF, BIOCHEM, CARNEGIE-MELLON UNIV, 87- *Concurrent Pos:* Vis prof, Med Ctr, Univ Calif, San Francisco, 63. *Mem:* AAAS; Am Soc Biol Chem; Am Soc Cell Biol; Biophys Soc; Am Soc Hort Sci. *Res:* Electron microscopy of muscle; physical biochemistry of macromolecules; cell biology; plant tissue culture. *Mailing Add:* Sippewissett Greenhouse PO Box 219 Woods Hole MA 02543

RICE, ROY WARREN, b Seattle, Wash, Aug 31, 34; m 64; c 2. CERAMICS, PHYSICS. *Educ:* Univ Wash, BS, 57, MS, 62. *Prof Exp:* Res engr, Boeing Co, 57-68; sect head, 68-74, head ceramics br, Naval Res Lab, 74-84; DIR MAT RES, W R GRACE, 84- *Mem:* Am Phys Soc; fel Am Ceramic Soc; Am Chem Soc; fel Am Soc Metals; AAAS. *Res:* Ceramics, especially the relationships between processing-microstructure and mechanical properties. *Mailing Add:* 5411 Hopark Dr Alexandria VA 22310

RICE, STANLEY ALAN, b Los Angeles, Calif, Mar 4, 47; m; c 2. POLLUTION BIOLOGY, INVERTEBRATE DEVELOPMENT. *Educ:* Calif State Univ, Long Beach, BS, 73, MS, 75; Univ SFla, PhD(invertebrate zool), 78. *Prof Exp:* Res asst, Calif State Univ, Long Beach, 73-75, Univ SFla, 75-78; fel, Harbor Br Found, 78-80; staff scientist, Mote Marine Lab, 80-84; asst prof biol, 84-87, ASSOC PROF BIOL, UNIV TAMPA, 87- *Mem:* Am Soc Zoologists; Am Micros Soc; Sigma Xi. *Res:* Invertebrate life histories; polychaete reproduction and development; population genetics; culture of marine invertebrates; pollution biology. *Mailing Add:* Div Sci & Math Univ Tampa Tampa FL 33606-1490

RICE, STANLEY DONALD, b Vallejo, Calif, Jan 20, 45; m 66; c 1. POLLUTION BIOLOGY, COMPARATIVE PHYSIOLOGY. *Educ:* Chico State Col, BA, 66, MA, 68; Kent State Univ, PhD(physiol), 71. *Prof Exp:* Res physiologist, Nat Marine Fisheries Serv, 71-86, PROG MGR, HABITAT RES ALASKA, DUKE BAY FISHERIES LAB, 87- *Concurrent Pos:* Invited contribr, Petrol Marine Environ, Nat Acad Sci; vis researcher, Beaufort Lab, 86, dept zool-physiol& La State Univ, Baton Rouge, 85 & 87. *Mem:* Am Soc Zoologist; Sigma Xi; AAAS. *Res:* Environmental research, lab and field, fish and crabs, oil effects, metabolism, uptake; TBT toxicity, uptake; author of 70 scientific publications; environmental physiology, smoltification, salinity tolerance; Alaskan fish and invertebrates; author of 70 publications. *Mailing Add:* PO Box 210155 Auke Bay Fisheries Lab Auke Bay AK 99821

RICE, STEPHEN LANDON, b Oakland, Calif, Nov 23, 41; m 65; c 2. MECHANICAL ENGINEERING. *Educ:* Univ Calif, Berkeley, BS, 64, MEng, 69, PhD(mech eng), 72. *Prof Exp:* Design engr, Lawrence Berkeley Lab, 64-69; from asst prof to prof mech eng, Univ Conn, 72-83; chair, Dept Mech Eng & Aerodyn Sci, 83-88, ASSOC DEAN ENG & DIR RES, CENT FLA UNIV, 88- *Concurrent Pos:* Dir, Automation, Robotics, Mfg Lab, Univ Conn, 80- & Design Proj Prog, 73-; prin investr, AFOSR Wear Proj, Univ Conn, 76-80, Dept Energy Wear Proj, 81-84, CAD/CAM Kinematics Proj, Control Data, 80-82, NSF Wear-Laser Speckle Proj, 86-88 & NASA/KSC Coop Agreement, 89-; ed, DELOS Lab Compendium, Am Soc Eng Educ, 76-82; evaluator, ASME/ABET Mech Eng Prog, 87- *Honors & Awards:* Teetor Award, Soc Automative Engrs, 75; Young Fac Award, Dow Chem Co/Am Soc Eng Educ, 75; Fulbright-Hays Award, 78-79. *Mem:* Am Soc Eng Educ; Am Soc Mech Engrs; Soc Tribologists & Lubrication Engrs; Soc Mfg Engrs. *Res:* Wear of materials, including role of mechanical parameters in formation of characteristic subsurface zones in sliding and repetitive impulsive loading in metals, polymers, composites and dental restoratives, laser speckle metrology, friction modeling, and tribodynamics; computer based education; evaluation of educational effectiveness; robotics, automation and intelligent manufacturing. *Mailing Add:* Cent Fla Univ PO Box 25000 Orlando FL 32816-0450

RICE, STUART ALAN, b New York, NY, Jan 6, 32; m 52; c 2. PHYSICAL CHEMISTRY. *Educ:* Brooklyn Col, BS, 52; Harvard Univ, AM, 54, PhD(chem), 55. *Hon Degrees:* DSc, Notre Dame Univ, 82; DSc, City Univ NY, 82. *Prof Exp:* Jr fel, Soc Fels, Harvard Univ, 55-57; from asst prof to prof, Univ Chicago, 57-60, dir, James Franck Inst, 62-68, chmn dept, 71-77, Louis Block prof, 69-77, FRANK P HIXON DISTINGUISHED SERV PROF CHEM, UNIV CHICAGO & JAMES FRANCK INST, 77-, MEM COMT MATH BIOL, 68-, DEAN, DIV PHYS SCI, 80- *Concurrent Pos:* Alfred P Sloane fel, 58-62; Guggenheim fel, 60-61; NSF sr fel & vis prof, Free Univ Brussels, 65-66; mem bd dirs, Bull Atomic Sci, 65-; co-ed Advan Chem Physics, 66-; mem adv bd inst statist mech & thermodyn, Univ, Tex, 67-; vis prof, H C Orsted Inst, Copenhagen Univ, 70-71.; King lectr, Johns Hopkins Univ, 63; Falk-Plaut lectr, Columbia Univ & Reilly lectr, Univ Notre Dame, 64; Farkas lectr, Hebrew Univ Jerusalem, 65; Venable lectr, Univ NC, 68; G K Rollefson lectr, Univ Calif, Berkeley, 68; Louderman lectr, Wash Univ, 68; W A Noyes lectr, Univ Tex, 75, A D Little lectr, Northeastern Univ, 76; Foster lectr, Univ Buffalo, 76; Liversidge lectr, Univ Sydney, 78; F T Gucker lectr, Univ Indiana, 76; A R Gordon distinguished lectr, Univ Toronto, 78; Baker lectr, 85-86; Centenary lectr, Royal Soc Chem, 86-87. *Honors & Awards:* A Cressy Morrison Prize, NY Acad Sci, 55; Award Pure Chem, Am Chem Soc, 62, L H Baekeland Award, 71; Marlowe Medal, Faraday Soc, 63; Medal of Free Univ Brussels, 66; Llewellyn John & Harriet Manchester Quantrell Award, 70; Sci Achievement Award Medal, City Univ NY, 78; Peter Debye Award, Am Chem Soc, 85; Hildebrand Award, Am Chem Soc, 86. *Mem:* Nat Acad Sci; AAAS; Am Chem Soc; Am Phys Soc; Chem Soc; Danish Royal Soc; Am Philos Soc; Am Acad Arts & Sci. *Res:* Statistical theory of matter; transport phenomena in dense media; electronic structure of liquids, solids and molecular crystals; statistical mechanics of simple systems; theory of phase transitions; photochemistry; properties of liquid surfaces and mono layers; optimal control of selectivity of chemical reactions. *Mailing Add:* James Franck Inst 5640 Ellis Ave Chicago IL 60637

RICE, THEODORE ROOSEVELT, b Ky, Jan 19, 19; m 41; c 3. MARINE ECOLOGY. *Educ:* Berea Col, AB, 41; Harvard Univ, MA & PhD(zool), 49. *Prof Exp:* From fishery res biologist to dir radiobiol lab, Bur Com Fisheries, US Fish & Wildlife Serv, 49-69, dir, Atlantic Estuarine Fisheries Ctr, 69-76, DIR BEAUFORT LAB, NAT MARINE FISHERIES SERV, 76- *Concurrent Pos:* Adj prof zool, NC State Univ, 63- *Honors & Awards:* Nat Oceanic & Atmospheric Admin Award, 72. *Mem:* Am Soc Limnol & Oceanog; Atlantic Estuarine Res Soc; Am Nuclear Soc; Am Fisheries Soc. *Res:* Estuarine productivity; pollution. *Mailing Add:* 410 Cathcart Dr Anderson SC 29624

RICE, THOMAS B, b Washington, DC, July 18, 46; m 68; c 2. AGRICULTURAL RESEARCH. *Educ:* Amherst Col, BA, 68; Yale Univ, PhD(biol), 73. *Prof Exp:* Postdoctoral plant tissue cult & genetics, Brookhaven Nat Lab, 73-74; res assoc, Mich State Univ, 74-76; sr res scientist, Cent Res, Pfizer, Inc, Groton, Conn, 76-78, proj leader, 78-80, mgr, 80-81, dir, 81-85; vpres & dir res, DeKalb-Pfizer Genetics, 85-86, exec vpres, dir res & dir US Prod Opers, 86-90, PRES & CHIEF OPERATING OFFICER, DEKALB PLANT GENETICS, 90- *Concurrent Pos:* Adj asst prof bot, Conn Col, 82, adj assoc prof, 84, adj prof, 85. *Mem:* AAAS; Crop Sci Soc Am; Int Asn Plant Cell & Tissue Cult; assoc mem Sigma Xi. *Mailing Add:* DeKalb Plant Genetics 3100 Sycamore Rd DeKalb IL 60115

RICE, THOMAS KENNETH, immunobiology, microbiology, for more information see previous edition

RICE, THOMAS MAURICE, b Dundalk, Ireland, Jan 26, 39; nat US; m 66; c 3. THEORETICAL SOLID-STATE PHYSICS. *Educ:* Univ Col Dublin, Ireland, BSc, 59, MSc, 60; Univ Cambridge, Eng, PhD(physics), 64. *Prof Exp:* Asst lectr physics, Univ Birmingham, 63-64; res assoc physics, Univ Calif, San Diego, 64-66; mem tech staff physics, Bell Labs, 66-75, res head theoret physics dept, 75-78, head surface physics dept, 78-81; PROF, FED TECH INST, ZURICH, SWITZ, 81- *Concurrent Pos:* Vis lectr physics, Fed Tech Inst, Zurich, Switzerland, 70-71; prof physics, Simon Fraser Univ, Burnaby, BC, 74-75. *Mem:* Fel Am Phys Soc; Europ Phys Soc; Swiss Phys Soc. *Mailing Add:* Theoretische Phyik ETH-Honggerberg Zurich 8093 Switzerland

RICE, W(ILLIAM) B(OTHWELL), b Montreal, Que, June 10, 18; m 44; c 3. MECHANICAL ENGINEERING. *Educ:* McGill Univ, BEng, 44, MEng, 56; Sir George Williams Univ, BSc, 50; Univ Montreal, DASc, 59. *Hon Degrees:* LLD, Concordia Univ, 87. *Prof Exp:* Asst prof mech eng, McGill Univ, 47-50; from assoc prof to prof, 50-84, EMER PROF MECH ENG, QUEEN'S UNIV, KINGSTON, ONT, 84- *Concurrent Pos:* Distinguished vis prof, Ariz State Univ, 82-83; pres, NAm Mfg Res Inst, Soc Mfg Engrs, 83-84.

Honors & Awards: Gold Medal, Soc Mfg Engrs, 82; Eng Medal, Asn Prof Engrs Ont, 85. *Mem:* Fel Am Soc Mech Engrs; fel Eng Inst Can (pres, 85-86); fel Soc Mfg Engrs; Int Inst Prod Eng Res; fel Can Soc Mech Eng (pres, 80-81). *Res:* Manufacturing processes, particularly the effect of friction on mechanics of cutting, extrusion and rolling; manufacturing systems for small organizations. *Mailing Add:* Dept Mech Eng Queen's Univ Kingston ON K7L 3N6 Can

RICE, WALTER WILBURN, b Harrogate, Tenn, Apr 30 18; m 43; c 1. ANALYTICAL CHEMISTRY. *Educ:* Lincoln Mem Univ, BS, 42. *Prof Exp:* Chemist & lab supvr, DOE, Union Carbide Nuclear Co, 46-83; RETIRED. *Mem:* AAAS; Am Chem Soc; Am Vacuum Soc; emer mem Am Soc Mass Spectrometry. *Res:* Mass spectrometry; vacuum technology; gamma ray scintillation. *Mailing Add:* 121 W Maiden Lane Oak Ridge TN 37830

RICE, WARREN, b Okla, Oct 11, 25; c 3. ENGINEERING. *Educ:* Agr & Mech Col, Tex, PhD(mech eng), 58. *Prof Exp:* Instr mech eng, Tex Tech Col, 49-50; from instr to assoc prof, Agr & Mech Col, Tex, 50-58; prof eng, 58-89, chmn, Dept Mech Eng, 67-74, EMER PROF ENG, ARIZ STATE UNIV, 89- *Concurrent Pos:* NSF grants, 56-57; indust res grants, 58-66 & 67-74. *Mem:* Am Soc Mech Engrs. *Res:* Engineering science; fluid mechanics; heat transfer, particularly boundary layer study and devices influenced by boundary layer phenomena. *Mailing Add:* 2042 E Balboa Dr Tempe AZ 85282

RICE, WENDELL ALFRED, b Saskatoon, Sask, Apr 24, 39; m 62; c 3. SYMBIOTIC DINITROGEN FIXATION. *Educ:* Univ Sask, BSA, 63, MSc, 66, PhD(soil microbiol), 70. *Prof Exp:* RES SCIENTIST SOIL MICROBIOL, RES STA, CAN DEPT AGR, 70- *Concurrent Pos:* Soils adv, Barani Agr Res & Develop Proj, Islamabad, Pakistan, 85-87. *Mem:* Can Soc Soil Sci; Int Soc Soil Sci; Can Soc Microbiologists. *Res:* Effect of environmental factors on Rhizobium growth and survival, nodulation, and nitrogen fixation; microbial transformations of soil nitrogen. *Mailing Add:* Soil Sect Res Sta PO Box 29 Beaverlodge AB T0H 0C0 Can

RICE, WILLIAM ABBOTT, b Delaware, Ohio, Dec 8, 12; m 46; c 2. PHYSICAL GEOLOGY. *Educ:* Ohio Wesleyan Univ, BA, 34; Yale Univ, PhD(geol), 40. *Prof Exp:* Instr geol, Yale Univ, 36-38; instr, Utah State Col, 38-40; geol trainee, Shell Oil Co, 40-41; asst prof geol, Univ NC, 41-42; chemist, Tenn Valley Authority, 42-46; from asst prof to prof geol, 47-75, EMER PROF GEOL, MT UNION COL, 76- *Concurrent Pos:* Instr, Ohio State Univ, 47, 49, 51, 55, 59 & 63-65, geologist, Exp Sta, 47; vis assoc prof, Univ Ill, 59-60. *Mem:* AAAS; Geol Soc Am. *Res:* Chemical industrial microscopy; huronian geology, geography, philosophy and implications of science. *Mailing Add:* 2241 S Seneca Alliance OH 44601

RICE, WILLIAM JAMES, b Whallonsburgh, NY, Aug 6, 27. THERMODYNAMICS. *Educ:* Worcester Polytech Inst, BS, 47, MS, 48; Princeton Univ, PhD(chem eng), 64. *Prof Exp:* Instr chem eng, Cath Univ Am, 48-53; from asst prof to assoc prof, 57-69, PROF CHEM ENG, VILLANOVA UNIV, 69- *Concurrent Pos:* Vis res prof, Inst Energy Conversion, Univ Del, 77-78, pres, Villanova Univ Sigma Xi (87-88). *Mem:* Am Inst Chem Engrs; Int Solar Energy Soc; Am Soc Eng Educ; Sigma Xi; Am Asn Univ Prof. *Res:* Fluid dynamics; transport properties; separation processes; thermodynamics. *Mailing Add:* Dept Chem Eng Villanova Univ Villanova PA 19085

RICH, ABBY M, b Newark, NJ, Mar 20, 50. ELECTRON MICROSCOPY, LEUKOCYTES. *Educ:* Oberlin Col, BA, 72; Univ NC, PhD(zool), 78. *Prof Exp:* Asst, Med Ctr, NY Univ, 78-81, res asst, 81-82, res assoc & fel, 82-85, asst prof cell biol, 85-86; RETIRED. *Mem:* Am Asn Cell Biol; Harvey Soc; NY Acad Sci. *Res:* Mechanisms of cell activation with an emphasis on stimulus response coupling; leukocytes and other phagocytic cells. *Mailing Add:* 161 W 16 St Apt 9B New York NY 10011

RICH, ALEXANDER, b Hartford, Conn, Nov 15, 24; m 52; c 4. MOLECULAR BIOPHYSICS, MOLECULAR BIOLOGY. *Educ:* Harvard Univ, AB, 47; Harvard Med Sch, MD, 49. *Hon Degrees:* DSc, Fed Univ Rio de Janeiro, 81. *Prof Exp:* Res fel chem, Calif Inst Technol, 49-54; chief, Sect Phys Chem, NIH, 54-58; from assoc prof to prof, 58-74, WILLIAM THOMPSON SEDGWICK PROF BIOPHYS, MASS INST TECHNOL, 74- *Concurrent Pos:* Vis scientist, Cavendish Lab, Cambridge, Eng, 55-56; mem postdoctoral fel bd, NIH, 55-58, mem career award comt, 64-67; Guggenheim Found fel, 63; mem vis comt, biol dept, Yale Univ, 63 & Weizmann Inst Sci, 65-66; mem, exobiol comt, Space Sci Bd, Nat Acad Sci, 64-65 & US Nat Comt, Int Orgn Pure & Appl Biophys, 65-67 & 79-83, chmn, Comt USSR & Eastern Europe Exchange Prog, 73-76, mem, adv bd, Acad Forum, 75-82, Gov-Univ-Indust Res Round Table, 84-; mem corp, Marine Biol Lab, Woods Hole, 65-77; mem, Lunar & Planetary Missions Bd, NASA, 68-71, biol team, Viking Mars Mission, 69-80 & life sci comt, 70-75; mem biol adv comt, Oak Ridge Nat Lab, 72-76; mem int res & exchanges bd, Am Coun Learned Socs, 73-76; mem sci adv bd, Stanford Synchroton Radiation Proj, 76-80; mem, Nat Sci Bd, 76-82; mem, US-USSR Joint Comn Sci & Technol, Dept State & sr consult, Off Sci & Technol Policy, Exec Off Pres, Wash, DC, 77-81; mem coun, Pugwash Conf Sci & World Affairs, Geneva, Switz, 77-82; mem sci rev comt, Howard Hughes Med Inst, Miami, Fla, 78-; mem bd dirs, Med Found Boston, Mass, 81-90; mem, Nat Adv Bd Physicians Social Responsibility, 83-; chmn, sci adv comt, Dept Molecular Biol, Mass Gen Hosp, Boston, Ma, 83-87; fel, Nat Res Coun, 49-51 & mem gov bd, 85-88; mem, Comt USSR & Eastern Europe, Nat Res Coun, Wash, DC, 86-, Nat Adv Comt Pew Scholar Prog, Pew Mem Trust, New Haven, Conn, 86-88. *Honors & Awards:* Skylab Achievement Award, NASA, 74; Theodore von Karmen Award, 76; Pres Award, NY Acad Sci, 77; Jabotinsky Medal, Jabotinsky Found, 80; Lewis S Rosenstiel Award, 83. *Mem:* Nat Acad Sci; sr mem Inst Med-Nat Acad Sci; Biophys Soc; Am Crystallog Asn; fel Am Acad Arts & Sci; fel AAAS; Am Soc Biol Chemists; Am Soc Microbiol; Am Philos Soc; hon mem Japanese Biochem Soc; Am Chem Soc. *Res:* Molecular structure of biological systems; x-ray crystallography; protein chemistry; nucleic acid chemistry; polymer molecular structure; information transfer in biological systems; mechanism of protein synthesis; origin of life; physical chemistry of nucleotides and polynucleotides; author or coauthor of over 350 publications. *Mailing Add:* Dept Biol Mass Inst Technol Cambridge MA 02139

RICH, ARTHUR, atomic physics; deceased, see previous edition for last biography

RICH, ARTHUR GILBERT, b Brooklyn, NY, Mar 21, 36; m 64; c 2. PHYSICAL PHARMACY, PHARMACEUTICS. *Educ:* Columbia Univ, BS, 57; Univ Iowa, MS, 59, PhD, 62. *Prof Exp:* Res chemist, Julius Schmid, Inc, 62-63, proj leader pharmaceut res & develop, 63-65; sr res scientist, Johnson & Johnson, New Brunswick, 65-69; group leader, Ortho Pharmaceut Corp, 69-72; prog mgr, Avon Prods, Inc, Suffern, NY, 72-88; VPRES, SCI, DE LAIRE, INC. NY, 88- *Concurrent Pos:* Union Carbide res fel, State Univ Iowa, 57-62; lectr, Ctr Prof Advan, 76-77. *Mem:* AAAS; Acad Pharmaceut Sci; Am Pharmaceut Asn; NY Acad Sci; fel Royal Soc Health; Sigma Xi; Soc Cosmetic Chemists. *Res:* Design and development of suitable pharmaceutical and cosmetic vehicles for maximum topical effect of drug and cosmetic agents. *Mailing Add:* Seven Craftwood Dr Spring Valley NY 10977

RICH, AVERY EDMUND, b Charleston, Maine, Apr 9, 15; m 38; c 2. PLANT PATHOLOGY. *Educ:* Univ Maine, BS, 37, MS, 39; Wash State Col, PhD, 50. *Prof Exp:* Asst supvr, Farm Security Admin, Maine, 39-40; instr high sch, Maine, 40-41; 4-H Club agent, Univ NH, 41-43; asst agronomist, Exp Sta, RI State Col, 43-47, exten agronomist, 46-47; asst plant pathologist, Exp Sta, State Col Wash, 47-51, from instr to asst prof, 47-51; from assoc prof to prof bot, 51-82, plant pathologist, 51-82, assoc dean life sci & agr, 72-82, EMER PROF PLANT PATH, UNIV NH, 82- *Res:* Potato diseases; fungicides; virus diseases; phytopathology. *Mailing Add:* Univ NH 13 Burnham Ave Durham NH 03824

RICH, BEN R, b Manila, Philippines, June 18, 25. AEROTHERMODYNAMICS. *Educ:* Univ Calif, Los Angeles, BA, 49, Berkeley, MS, 50; Harvard Univ, AMP, 68. *Prof Exp:* Vpres, fighter preliminary design, Lockheed Adv Aero Co, 72-79,vpres & gen mgr, adv develop proj, 75-84, corp vpres, 77, pres, 84-86, pres, Lockheed Aero Develop CO, 86-90; RETIRED. *Honors & Awards:* Collier Trophy, 90. *Mem:* Nat Acad Eng; hon fel Am Inst Aeronaut & Astronaut; Am Defense Preparedness Asn. *Mailing Add:* Lockheed Corp 2555 N Hollywood Way PO Box 551 Burbank CA 91520

RICH, CHARLES CLAYTON, b Cincinnati, Ohio, Dec 8, 22; m 66; c 2. GEOLOGY. *Educ:* Wittenberg Univ, AB, 45; Harvard Univ, MA, 50, PhD(geol), 60. *Prof Exp:* Lectr geol, Victoria Univ, NZ, 52-54; from inst to prof geol, 58-90, dir univ honors prog, 65-69, EMER PROF GEOL, BOWLING GREEN STATE UNIV, 90- *Mem:* Geol Soc Am; Nat Asn Geol Teachers; Am Quaternary Asn. *Res:* Glacial and Pleistocene geology. *Mailing Add:* Dept Geol Bowling Green State Univ Bowling Green OH 43403

RICH, CLAYTON, b New York, NY, May 21, 24; div; c 1. MEDICINE, ENDOCRINOLOGY. *Educ:* Cornell Univ, MD, 48. *Prof Exp:* Extern path, NY Hosp, Cornell Univ, 48, asst physician, 49-50; intern med, Albany Hosp & Med Col, Union Univ, NY, 48-49, asst resident & asst med, 50-51; asst, Rockefeller Inst & asst physician, Hosp, 53-58, asst prof, Inst & assoc physician, Hosp, 58-60; from asst prof to prof med, Sch Med, Univ Wash, 60-71, assoc dean, 68-71; chief staff, Stanford Univ Hosp, 71-77, dean & vpres med affairs, Stanford Univ, 71-78, Karl & Elizabeth Naumann prof med, 77-78; vis sr scholar, Inst Med-Nat Acad Sci, 79-80; exec dean, Col Med, 80-83, PROF MED, COL MED & PROVOST & EXEC OFFICER, OKLA UNIV, OKLAHOMA CITY, 80-, VPRES HEALTH SCI, 83-, PROF HEALTH ADMIN, COL PUB HEALTH, 85- *Concurrent Pos:* Chief radioisotope serv, Vet Admin Hosp, 60-70, assoc chief staff, 62-71, chief staff, 68-70; attend physician, Univ & King County Hosps, Seattle, 62-71. *Mem:* Inst Med-Nat Acad Sci; Am Soc Clin Invest; Endocrine Soc; Am Col Physicians; AMA; Asn Am Physicians. *Res:* Academic administration. *Mailing Add:* Okla Univ PO Box 26901 Oklahoma City OK 73190

RICH, DANIEL HULBERT, b Fairmont, Minn, Dec 12, 42; m 64; c 2. BIO-ORGANIC CHEMISTRY. *Educ:* Univ Minn, BS, 64; Cornell Univ, PhD(org chem), 68. *Prof Exp:* Res assoc org chem, Cornell Univ, 68; res chemist, Dow Chem Co, 68-69; fel org chem, Stanford Univ, 69-70; asst prof, 70-75, assoc prof, 75-81, PROF PHARMACEUT CHEM, UNIV WIS-MADISON, 81- *Concurrent Pos:* NIH fel, 68; mem, Bioorg Natural Prod Study Sect, NIH, 81-; consult, 80- *Honors & Awards:* H I Romnes Award, 80. *Mem:* AAAS; Am Chem Soc; Am Pharmaceut Asn. *Res:* Synthesis of peptides and hormones; inhibition of peptide receptors and proteases; characterization, synthesis, and mechanisms of action of peptide natural products. *Mailing Add:* Sch Pharm 425 N Charter St Madison WI 53705-1508

RICH, EARL ROBERT, b Marquette, Mich, Aug 30, 25; m 74; c 3. ECOLOGY. *Educ:* Univ Chicago, SB, 49, PhD(zool), 54. *Prof Exp:* NSF fel statist, Univ Calif, 54-55, lectr, 55-56, instr biostatist, 56-57; from asst prof to assoc prof zool, Univ Miami, 57-70, assoc dean col arts & sci, 68-70, prof biol, 70-85; PRES, RIO PALENQUE RES CORP, 72- *Mem:* Ecol Soc Am; AAAS; Am Soc Zool; Soc Wetland Scientists; Sigma Xi. *Res:* Population dynamics and ecology; wetland ecology; man's impact on environment; inshore marine biology. *Mailing Add:* PO Box 249118 Coral Gables FL 33124

RICH, ELLIOT, b Brigham City, Utah, May 27, 19; m 43; c 6. CIVIL ENGINEERING. *Educ:* Utah State Univ, BS, 43; Univ Utah, MS, 51; Univ Colo, PhD(civil eng), 68. *Prof Exp:* Hydraul engr, US Bur Reclamation, 46-47; instr, Weber Col, 47-50, head eng dept, 50-56; assoc prof, 56-67, head dept civil eng, 67-75, PROF CIVIL ENG, UTAH STATE UNIV, 67-, ASSOC DEAN, COL ENG, 75- *Concurrent Pos:* Consult, AEC, 59. *Mem:* Am Soc Eng Educ; Am Soc Civil Engrs (pres, 77-78). *Res:* Structures. *Mailing Add:* 1640 E 1140 N Logan UT 84321

RICH, HARRY LOUIS, b New York, NY, Apr 30, 17; m 41; c 2. SHIP PROTECTION, MARINE SCIENCE. *Educ:* Brooklyn Col, BA, 39. *Prof Exp:* Jr physicist, David Taylor Naval Ship Res & Develop Ctr, US Navy, 42-43, asst physicist, 43-44, physicist, 44-49, supvr, shock sect, 49-53, shock br, 53-65, coordr, shock res, 66-67, tech dir, oper dive-under, 67-68, head, engr facil div, 70-71, asst to tech dir, 72-74; lectr eng, Mass Inst Technol, 74-89; RETIRED. *Concurrent Pos:* Navy rep, Dept Defense Shock & Vibration Info Ctr, 57-71; US rep, Int Electrotech Comn, 64-74, Int Standards Orgn, 64-; sci adv, S Korean Navy, 71-72; consult, 74- *Mem:* Fel Acoust Soc Am; fel Inst Environ Sci; Soc Naval Architects & Marine Engrs. *Res:* Dynamics of explosions and their effects on mechanical systems and structures; instrumentation for shock and vibration; design of mechanical systems for shock loading and shock simulation. *Mailing Add:* 6765 Brigadoon Dr Bethesda MD 20817

RICH, JIMMY RAY, b Collins, Ga, Oct 29, 50. NEMATICIDES, NEMATODE RESISTANCE IN PLANTS. *Educ:* Univ Ga, BSA, 72, MS, 73; Univ Calif, Riverside, PhD(plant path), 76. *Prof Exp:* From asst prof to assoc prof, Univ Fla, 76-86, actg dir, Agr Res Ctr, 80-84, asst dir, 84-88, PROF, UNIV FLA, 86- *Concurrent Pos:* Chmn, Tobacco Dis Coun; pres, Orgn Trop Am Nematologists. *Honors & Awards:* Distinguished Serv Award, Orgn Trop Am Nematologists. *Mem:* Soc Nematologists; Am Phytopath Soc; Orgn Trop Am Nematologists. *Res:* Nematode management; chemical control, plant resistance and biocontrol; new crops and management procedures. *Mailing Add:* Agr Res & Educ Ctr Rte 2 Box 2181 Live Oak FL 32060

RICH, JOHN CHARLES, b Wichita, Kans, Oct 12, 37; m 63; c 3. ASTROPHYSICS, LASERS. *Educ:* Harvard Univ, AB, 59, AM, 60, PhD(astron), 67. *Prof Exp:* Physicist, Air Force Weapons Lab, 60-74, div chief, 74-77, comdr, 77-78, prog dir, 78-82; eng dir, Perkin-Elmer Corp, 83-85, gen mgr, 85-89; PRES, HUGHES DANBURY OPTICAL, 89- *Concurrent Pos:* Instr, Univ Va, 69-70. *Mem:* Am Astron Soc; Am Geophys Union; Am Inst Aeronaut & Astronaut; Sigma Xi. *Res:* Atomic and molecular physics; radiative processes; stellar atmospheres; optics. *Mailing Add:* 28 Sharp Hill Rd Ridgefield CT 06877

RICH, JOSEPH ANTHONY, b Hazardville, Conn, July 23, 16; m 45; c 2. PLASMA PHYSICS. *Educ:* Harvard Univ, BSc, 38; Brown Univ, MSc, 39; Yale Univ, PhD(physics), 50. *Prof Exp:* Physicist, Cent Res Lab, Monsanto Chem Co, 43-45, sr physicist, 45-47; physicist, Knolls Atomic Power Lab, Gen Elec Co, 49-52 & Electron Physics Dept, Res & Develop Ctr, 52-60, physicist, Gen Physics Lab, Res & Develop Ctr, 60-81; RETIRED. *Mem:* Am Phys Soc; Math Asn Am; sr mem Inst Elec & Electronics Engrs; Sigma Xi. *Res:* Microwave electronics; nuclear physics; high current arcs. *Mailing Add:* 1385 Ruffner Rd Schenectady NY 12309

RICH, JOSEPH WILLIAM, b New Orleans, La, Aug 6, 37; m 60; c 2. NONEQUILIBRIUM GAS DYNAMICS, CHEMICALLY REACTING FLOWS. *Educ:* Carnegie Inst Technol, BS, 59; Univ Va, MAE, 61; Princeton Univ, MA, 63, PhD(aerospace & mech sci), 65. *Prof Exp:* Scientific staff, Cornell Aeronautical Lab, 65-72; prin engr, Calspan Corp,72-82, head physics & chem sect, Arvin/Calspan Adv Technol Ctr, 82-86; PROF MECH ENG, OHIO STATE UNIV, 86-, PROF CHEM PHYSICS, 90- *Concurrent Pos:* Vis prof, Dept Mich Eng, Carnegie-Mellon Univ, 85, Fulbright fel, Ecole des Arts et Manufactures, Paris, 88. *Mem:* am Inst Aeronaut & Astronaut; AAAS. *Res:* Nonequilibrium gas dynamics; molecular energy transfer; development of new gas lasers; developed first electrically-excited supersonic flow carbon monoxide gas laser; isotope separation in vibrationally nonequilibrium gases. *Mailing Add:* 286 W South St Wothington OH 43085

RICH, KENNETH C, b Berkeley, Calif, Apr 7, 43; m; c 2. PEDIATRICS. *Educ:* Tulane Univ, MD, 70. *Prof Exp:* ASSOC PROF PEDIAT, COL MED, UNIV ILL, 82- *Mem:* Soc Pediat Res; Am Rheumatism Asn; Am Asn Immunologists. *Res:* Pediatric rheumatology and immunology. *Mailing Add:* Dept Pediat Univ Ill Col Med 840 S Wood St Chicago IL 60612

RICH, KENNETH EUGENE, b Alton, Ill, Nov 19, 43; m 68; c 2. SOFTWARE SYSTEMS, MEDICAL INSTRUMENTATION. *Educ:* Rose-Hulman Inst Technol, BS, 66; Univ Rochester, PhD(biophys), 71. *Prof Exp:* Res assoc biophys & instrumentation, Dept Biochem, Case Western Reserve Univ, 71-75; staff fel comput instrumentation & molecular biol, Nat Inst Neurologic Commun Dis & Stroke, 76-78; scientist comput instrumentation, Technico Instrument Corp, 78-81; sr comput appl scientist, 81-83, MGR, COMPUTERIZED SYST DEVELOP, CAPINTEC, INC, 83- *Concurrent Pos:* USPHS fel, Case Western Reserve Univ, 71-75. *Mem:* Biophys Soc; Am Chem Soc; Sigma Xi; AAAS; Inst Elec & Electronics Engrs; Soc Photo-Optical Instrumentation Engrs. *Res:* Management of systems group working on ambulatory cardiac function monitor and nuclear medicine department management systems; image processing. *Mailing Add:* Capintec Inc Six Arrow Rd Ramsey NJ 07644

RICH, LEONARD G, b New York, NY, Mar 28, 25; m 55; c 1. PHYSICS, ELECTRICAL ENGINEERING. *Educ:* St Lawrence Univ, BS, 45. *Prof Exp:* Res physicist, Crystal Res Labs, Inc, 46-47; pres, Norbert Photo Prod Co, 47-49; asst chief engr electronics, McMurdo Silver Co, 49-50, chief engr, 50-51; proj engr, New London Instrument Co, Inc, 51-53; proj leader res & develop, Andersen Labs, Inc, 53-56; sr proj engr, Roth Lab, 56-65; SCI DIR, GERBER SCI INC, 65-, VPRES RES GERBER SCI PROD, 82- *Mem:* Inst Elec & Electronics Engrs. *Res:* Electronic physics; ultrasonic delay lines; magnetic memory for signal-to-noise improvement through video integration; signal generators and test instruments for industry; computer-controlled digital and analog graphic output devices; robotic scanners; digital adaptive servos; computer generation of color images; Ink-Jet technology; large areas color sign generators; optical lens generators; eight patents. *Mailing Add:* Gerber Sci Prod 151 Batson Dr Manchester CT 06040

RICH, LINVIL G(ENE), b Pana, Ill, Mar 10, 21; m 44; c 2. ENVIRONMENTAL ENGINEERING. *Educ:* Va Polytech Inst, BS, 47, MS, 48, PhD(biochem), 51; Environ Eng Intersoc, dipl. *Prof Exp:* From instr to assoc prof sanit eng, Va Polytech Inst, 48-55; from assoc prof to prof, Ill Inst Technol, 56-61; dean eng, 61-72, prof 72-81, alumni prof, 82-87, EMER ALUMNI PROF, ENVIRON SYSTS ENG, CLEMSON UNIV, 87- *Concurrent Pos:* Consult, Environ Protection Agency. *Honors & Awards:* Hering Medal, Am Soc Civil Engrs, 83. *Mem:* Fel Am Soc Civil Engrs; Am Soc Eng Educ; Am Acad Environ Engrs; Asn Environ Eng Prof. *Res:* Environmental engineering. *Mailing Add:* Col Eng Clemson Univ Clemson SC 29632

RICH, MARK, b Chicago, Ill, Feb 1, 32; m 58; c 5. CARBONIFEROUS FORAMINIFERAL BIOSTRATIGRAPHY. *Educ:* Univ Calif, Los Angeles, AB, 54; Univ Southern Calif, MA, 56; Univ Ill, PhD(geol), 59. *Prof Exp:* Asst geol, Univ Ill, 57-59; asst prof, Univ NDak, 59-63; assoc prof, 63-70, PROF GEOL, UNIV GA, 70- *Mem:* Geol Soc Am; Nat Water Well Asn; Am Asn Petrol Geol. *Res:* Sedimentary petrology; stratigraphy; micropaleontology; foraminiferal biostratieraphy of carboniferous rocks in southern Appalachians. *Mailing Add:* Dept Geol Univ Ga Athens GA 30602

RICH, MARVIN A, b New York, NY, Apr 21, 31; m 66. MICROBIOLOGY, VIROLOGY. *Educ:* Brooklyn Col, BS, 52; Rutgers Univ, MS, 54, PhD(microbiol), 57. *Prof Exp:* Res asst biochem, Inst Appl Biol, 52-53; res asst microbiol, Med Ctr, Columbia Univ, 53; res assoc biophys, Sloan-Kettering Inst Cancer Res, 57-59, asst mem, 59-61, assoc, 61-62; asst mem, Albert Einstein Med Ctr, 61-62, assoc mem & dir lab cancer res, 62-65, mem & chmn dept cell biol, 65-72; chmn biol dept, 72-73, dir div biol sci, 73-74, EXEC VPRES & SCI DIR, MICH CANCER FOUND, 75- *Concurrent Pos:* Res collabr, Brookhaven Nat Lab, 61-62; USPHS career develop award, 64-72; consult, spec virus cancer prog, Nat Cancer Inst, 66-; mem leukosis team, WHO, 71-; mem bd gov, Nat Found Encephalitis Res. *Mem:* AAAS; Am Chem Soc; Am Asn Immunol; Am Asn Cancer Res; Am Soc Cell Biol. *Res:* Relationship between viruses and cancer, especially the characterization of virus-induced neoplasias in animal systems as models for leukemia and mammary carcinoma in man. *Mailing Add:* AMC Cancer Res Ctr 1600 Pierce St Lakewood CO 80214

RICH, MARVIN R, b Bronx, NY, Oct 29, 47. BIOPHYSICS. *Educ:* Rensselaer Polytech Inst, BS, 69; NY Univ, MS, 77, PhD(biophys), 81. *Prof Exp:* RES ASST, DEPT BIOL, NY UNIV, 81- *Mem:* AAAS; NY Acad Sci; Am Soc Photobiol. *Res:* Use of model membrane systems in the study of lipid peroxidation and free radical mechanisms. *Mailing Add:* 320 W 87th St New York NY 10024

RICH, MICHAEL, b Chicago, Ill, July 23, 40; m 63; c 4. MATHEMATICS. *Educ:* Roosevelt Univ, BS, 62; Ill Inst Technol, MS, 65, PhD(math), 69. *Prof Exp:* Asst prof math, Ind Univ, 67-69; from asst prof to prof math, Temple Univ, 69-90. *Concurrent Pos:* Vis assoc prof, Ben Gurion Univ Negev, 74-75. *Mem:* Am Math Soc. *Res:* Ring theory; nonassociative algebras; software systems. *Mailing Add:* Six Haerez St Ginot Shomron Israel

RICH, PETER HAMILTON, b Wellfleet, Mass, Nov 7, 39; div; c 1. LIMNOLOGY. *Educ:* Hunter Col, AB, 63; Mich State Univ, MS, 66, PhD, 70. *Prof Exp:* Res assoc biol, Brookhaven Nat Lab, 70-72; asst prof, 72-80, ASSOC PROF ECOL, UNIV CONN, 80- *Mem:* AAAS; Ecol Soc Am; Am Soc Limnol & Oceanog; Int Asn Theoret & Appl Limnol. *Res:* Measurement of the functional, community parameters of the aquatic ecosystem, especially benthos; aquatic ecosystem energetics. *Mailing Add:* Ecol & Evolutionary Biol U-42 Univ Conn Storrs CT 06269-3042

RICH, RICHARD DOUGLAS, b Chicago, Ill, Nov 30, 36; m 59; c 2. ORGANIC CHEMISTRY. *Educ:* Rensselaer Polytech Inst, BChE, 58; Univ Md, PhD(org chem), 68. *Prof Exp:* Res chemist, US Naval Ord Lab, Md, 58-68; sr scientist, Bickford Res Labs, Inc, Ensign-Bickford Co, Conn, 68-70; res chemist, 70-78, res assoc, 78-79, SCIENTIST, LOCTITE CORP, 79- *Concurrent Pos:* Adj prof, Univ Hartford, 69-72. *Mem:* Am Inst Chemists; Am Chem Soc; Adhesion Soc. *Res:* Organic chemistry of explosives; plastic binders for propellants and explosives; anaerobic and cyanoacrylate adhesives. *Mailing Add:* Loctite Corp 705 N Mountain Rd Newington CT 06111

RICH, ROBERT PETER, b Lowville, NY, Aug 28, 19; wid; c 2. MATHEMATICS, COMPUTER SCIENCES. *Educ:* Hamilton Col, AB, 41; Johns Hopkins Univ, PhD(math), 50. *Prof Exp:* Mathematician, Appl Physics Lab, Johns Hopkins Univ, 50-89, dir comput ctr, 56-85, assoc prof biomed eng, 69-89, oper analyst, 85-89; RETIRED. *Mem:* AAAS; Am Math Soc; Asn Comput Mach; Soc Indust & Appl Math. *Res:* Digital computing. *Mailing Add:* 1109 Schindler Dr Silver Spring MD 20903

RICH, ROBERT REGIER, b Newton, Kans, Mar 7, 41; m 74; c 2. IMMUNOBIOLOGY. *Educ:* Oberlin Col, AB, 62; Univ Kans, MD, 66. *Prof Exp:* Intern, Univ Wash, 66-67, asst resident, 67-68; clin assoc immunol, NIH, 68-71; NIH res fel immunol, Harvard Med Sch, 71-73; from asst prof to assoc prof, 73-78, HEAD, IMMUNOL SECT, BAYLOR COL MED, 77-, PROF MICROBIOL & IMMUNOL & CHIEF CLIN IMMUNOL, 78-, PROF MED, 79-, VPRES & DEAN RES, 90- *Concurrent Pos:* Asst med, Peter Bent Brigham Hosp, Boston, 72-73; attend physician, Vet Admin Hosp, Houston, 73-; adj asst prof, Grad Sch Biomed Sci, Univ Tex Health Sci Ctr, Houston, 75-79; prog dir, Gen Clin Res Ctr, Methodist Hosp, Houston, 75-77; NIH res career develop award, 75-77; investr, Howard Hughes Med Inst, 77-; mem immunobiol study sect, NIH, 77-81; assoc ed, J Immunol, 78-82 & J Infections Dis, 83-88; adj prof, Grad Sch Biomed Sci, Univ Tex Health Sci Ctr, Houston, 79-; adv ed. J Exp Med, 80-84; Transplantation Biol & Immunol Comt, Nat Inst Allergy & Infections Dis, 82-86, chmn, 84-86; Nat Ctr Grants Subcomt, Arthritis Found, 83-86, chmn, 84-86, Nat Res Comt, 84-, vchmn, 85-86, chmn, 86-89; bd dirs, Am Bd Allergy & Immunol, 88-, chmn, 91-; adv comt, res prog, Nat Mult Sclerosis Soc, 89-; bd gov, Am Med Internal Med,

90- *Mem:* Am Asn Immunologists; Am Asn Pathologists; fel Am Col Physicians; Am Soc Clin Invest; Asn Am Physicians; fel Am Acad Allergy & Immunol; fel Infections Dis Soc Am; Am Clin & Climat Soc; Clin Immunol Soc. *Res:* Major histocompatibility complex and T lymphogne genetics and function in mice and humans. *Mailing Add:* Dept Microbiol & Immunol Baylor Col Med Houston TX 77030

RICH, RONALD LEE, b Washington, Ill, Mar 29, 27; m 53; c 4. PERIODICITY. *Educ:* Bluffton Col, BS, 48; Univ Chicago, PhD(chem), 53. *Prof Exp:* Instr chem, Bethel Col, Kans, 50-51; mem staff, Los Alamos Sci Lab, 53; assoc prof chem, Bethel Col, Kans, 53-55; chemist, Nat Bur Stand, 55-56; prof chem, Bethel Col, Kans, 56-63; res fel, Harvard Univ, 63-64; prof chem, Bethel Col, Kans, 64-66; prof, Int Christian Univ, Tokyo, 66-69; prof chem & chmn div natural sci, Bethel Col, Kans, 69-71; prof chem, Int Christian Univ, Tokyo, 71-79; dean & prof chem, Bluffton Col, Ohio, 79-80, scholar in residence, 81-89; res fel, NC State Univ, 89-90; COMPUT CONSULT, 85- *Concurrent Pos:* Vis prof chem, Stanford Univ, 74, Univ Ill, 75 & Univ Oregon, 84. *Mem:* Am Chem Soc. *Res:* Correlation of chemical and physical properties with electron structure. *Mailing Add:* 112 S Spring St Bluffton OH 45817-1112

RICH, ROYAL ALLEN, b North Platte, Nebr, Aug 6, 34; m 56; c 4. REPRODUCTIVE PHYSIOLOGY. *Educ:* Univ Nebr, BS, 57; Utah State Univ, MS, 60, PhD(physiol), 65. *Prof Exp:* Instr zool & physiol, Utah State Univ, 62-64; assoc prof zool, 65-73, PROF ZOOL, UNIV NORTHERN COLO, 73- *Mem:* Soc Study Reproduction. *Res:* Mammalian physiology; influence of ovarian hormones on uterine biochemistry, especially deciduoma formation in the rat; post-partum involution. *Mailing Add:* Dept Biol Sci Univ Northern Colo Greeley CO 80631

RICH, SAUL, b Detroit, Mich, Nov 25, 17; m 46; c 2. PHYTOPATHOLOGY. *Educ:* Univ Calif, BS, 38, MS, 39; Ore State Col, PhD(plant path), 42. *Prof Exp:* Asst plant pathologist, Univ Wyo, 46; res assoc, Crop Protection Inst, 47-48, from asst plant pathologist to sr plant pathologist, 48-72, chief dept plant path & bot, 72-83, EMER PLANT PATHOLOGIST, CONN AGR EXP STA, 83- *Mem:* Fel AAAS; fel Am Phytopath Soc (treas, 52-58); Soc Indust Microbiol (pres, 66). *Res:* Air pollution; fungicides; vegetable diseases. *Mailing Add:* 65 Adla Dr Hamden New Haven CT 06514

RICH, SUSAN MARIE SOLLIDAY, b Rockford, Ill, June 17, 41; m; c 1. MICROBIOLOGY, IMMUNOLOGY. *Educ:* Beloit Col, Wis, BS, 63; Univ Wis-Madison, MS, 65; Baylor Col Med, PhD(microbiol & immunol), 79. *Prof Exp:* Res technician, Univ Chicago, 66-67; proj specialist, Dept Med Genetics, Univ Wis-Madison, 67-70; res asst, Dept Path, Med Sch, Harvard Univ, 70-73; from instr to assoc prof, 73-87, PROF MICROBIOL & IMMUNOL, BAYLOR COL MED, 87- *Concurrent Pos:* Prin investr, NIH, 80-83, 82-87 & 90-91; mem, Immunobiol Study Sect, Div Res Grants, NIH, 81-85 & Allergy & Clin Immunol Subcomt Allergy Immunol & Transplantation Res Comt, Nat Inst Allergy & Infectious Dis, 88-; co-dir, Med Scientist Training Prog, Baylor Col Med, 90- *Mem:* Am Asn Immunologists. *Res:* Regulation of Ts cell growth and differentiation; regulatory abnormalities in immunologic diseases; author of numerous technical publications. *Mailing Add:* Dept Microbiol & Immunol Baylor Col Med Houston TX 77030

RICH, TERRELL L, b Montpelier, Idaho, Sept 12, 39; m 63; c 5. CARDIOVASCULAR PHYSIOLOGY. *Educ:* Idaho State Univ, BS, 64; Univ Calif, Los Angeles, PhD(physiol), 71. *Prof Exp:* Res assoc, Univ Pa, 74-79; asst res physiologist, 71-74 & 79-82, assoc res physiologist, 82-87, RES PHYSIOLOGIST, SCH MED, UNIV CALIF, LOS ANGELES, 87- *Concurrent Pos:* Lectr, Univ Calif, Irvine, 72 & Univ Calif, Los Angeles Ext, 73. *Mem:* Am Physiol Soc; AAAS. *Res:* Excitation-contraction characteristics of heart; calcium binding, exchange and transport kinetics of myocardial sarcolemma and intracellular compartments using radioactive isotopes, fluorescent dyes; video monitoring of myocyte function; membrane patch clamp; myocardial metabolisim. *Mailing Add:* Dept Physiol A3-376 CHS UCLA Med Sch Los Angeles CA 90024-1760

RICH, THOMAS HEWITT, b Evanston, Ill, May 30, 41; Australian citizen; m 66; c 2. PALEOMAMMALOGY. *Educ:* Univ Calif, Berkeley, AB, 64, MA, 67; Columbia Univ, PhD(geol), 73. *Prof Exp:* CUR VERT PALEONT, MUSEUM VICTORIA, MELBOURNE, AUSTRALIA, 74- *Mem:* Soc Vert Paleont; AAAS; Geol Soc Australia; Australian Mammal Soc. *Res:* Documentation of the origin and evolution of the Australion mammal fauna; Australian dinosaurs; new fossil localities. *Mailing Add:* 2119 View Rd Glen Waverley Victoria 3150 Australia

RICH, TRAVIS DEAN, b Ryan, Okla, Oct 5, 40; m 64; c 2. REPRODUCTIVE PHYSIOLOGY. *Educ:* Okla State Univ, BS, 62, MS, 67; Purdue Univ, Lafayette, PhD(animal sci), 70. *Prof Exp:* Beef herdsman, Okla State Univ, 62-65, teaching asst animal sci, 65-67; res asst, Purdue Univ, Lafayette, 67-70; asst prof, SDak State Univ, 70-72; researcher beef cattle reproduction, Res & Exten Ctr, Tex A&M Univ, 72-73; exten beef cattle specialist, Dept Animal Sci, Okla State Univ, 73-78, assoc prof animal sci & indust, 74-78; BEEF NURTITIONIST, MOORMAN MFG CO. *Mem:* Am Soc Animal Sci; Soc Study Reproduction; Am Polled Hereford Asn (pres,81). *Res:* Beef cattle reproduction; endocrinology of postpartum period, puberty and superovulation; genetic improvement of beef cattle; physiology of reproduction. *Mailing Add:* 418 Brookside Liberty MO 64068

RICHARD, ALFRED JOSEPH, b Gardner, Mass, Mar 30, 28; m 54. PHYSICAL CHEMISTRY. *Educ:* Clark Univ, PhD(chem), 58. *Prof Exp:* Asst prof, 58-64, assoc prof chem, 64-76, PROF PHARMACEUT CHEM, MED COL VA, 76- *Mem:* Am Chem Soc. *Res:* Physical properties of proteins; compressibilities of pure liquids. *Mailing Add:* Med Col Va Va Commonwealth Univ Richmond VA 23298

RICHARD, BENJAMIN H, b Phoenixville, Pa, May 6, 29. STRUCTURAL GEOLOGY, GEOPHYSICS. *Educ:* Va Polytech Inst, BS, 58; Ind Univ, MA, 61, PhD(geol), 66. *Prof Exp:* From instr to asst prof geol, Wittenberg Univ, 62-66; asst prof, 66-70, ASSOC PROF GEOL, WRIGHT STATE UNIV, 70- *Mem:* Nat Asn Geol Teachers; Sigma Xi. *Res:* Structural geology and geophysics; use of gravity to locate large pockets of gravel within glacial debris. *Mailing Add:* Dept Geol Sci Wright State Univ Colonel Glenn Hwy Dayton OH 45435

RICHARD, CHRISTOPHER ALAN, b St Louis, Mo, June 28, 51; m 82. NEURAL CONTROL OF RESPIRATION, STATE-RELATED CONTROL OF RESPIRATION. *Educ:* Murray State Univ, BS, 79; Univ Louisville, PhD(physiol), 87. *Prof Exp:* Univ fel physiol, dept physiol & biophys, Univ Louisville, 82-87; POSTDOCTORAL RES, DEPT ANAT & CELL BIOL, UNIV CALIF, LOS ANGELES, 87-, POSTDOCTORAL FEL, BRAIN RES INST, 90- *Mem:* Am Physiol Soc; Soc Neurosci; Sleep Res Soc. *Res:* Modulatory activities of the amygdala on respiratory control especially during arousal, attention and defensive type reactions using single unit recordings in awake unrestrained animals during various sensory presentations. *Mailing Add:* Brain Res Inst Sch Med Univ Calif 10833 Le Conte Ave Los Angeles CA 90024-1761

RICHARD, CLAUDE, b Quebec City, Que, Mar 18, 44; m 67; c 2. BREEDING FOR DISEASE RESISTANCE, ROOT DISEASE. *Educ:* Laval Univ, BSc, 67, MSc, 69, PhD(phytopath), 73. *Prof Exp:* Prof bot, Laval Univ, 72-73; RES SCIENTIST AGR, AGR CAN, 73- *Concurrent Pos:* Ed, Phytoprotection, 82-85; asst ed, Can J Plant Path, 87-89, Can J Bot, 89- *Mem:* Can Phytopath Soc; Am Phytopath Soc. *Res:* Foliar and root diseases of forage legumes mainly alfalfa; selection of alfalfa for persistance and disease resistance; survey of alfalfa diseases; ice nucleation activity of bacteria and fungi. *Mailing Add:* Agr Can Sta Res 2560 Hochelaga Blvd Ste-Foy PQ G1V 2J3 Can

RICHARD, CLAUDE, b Trois-Rivieres, Que, Nov 28, 32; m 59; c 2. PLASMA PHYSICS, OPTICAL PHYSICS. *Educ:* Laval Univ, BASc, 59; Univ London, MSc & DIC, 61, PhD(microwave & plasma physics), 68. *Prof Exp:* Jr mem sci staff microwave & plasma physics, RCA Ltd, 61-65, mem sci staff optical & microwave physics, 65-68, sr mem sci staff space & plasma physics, 68-70; sr researcher laser plasma interaction, Inst Hydro Res, Que, 70-71, sci dir basic res, 71-80, dir res & testing & prod & conserv energy, 80-83; special assignment, Hydro-Quebec Int, Que, 83-84; vpres develop, 84-89, PRES, AGENCE QUEBECOISE DE VALORISATION INDUSTRIELLE DE LA RECHERCHE, 89- *Mem:* Can Asn Physicists; Am Phys Soc; Inst Elec & Electronics Engrs. *Res:* Microwave optics; interaction of electromagnetic wave with plasmas; plasma diagnostics; focused microwave systems; electrochemical properties of reentry plasmas; negative ion generation in gas discharge; high resolution-multiple imaging optical systems; gaseous lasers; laser plasma interaction. *Mailing Add:* Aqvir 300 Rue Leo Pariseau PO Box 1116 Montreal PQ H2W 2P4 Can

RICHARD, JEAN-PAUL, b Quebec, Que, June 10, 36; m 63; c 2. EXPERIMENTAL PHYSICS. *Educ:* Laval Univ, BA, 56, BS, 60; Univ Paris, DSpec(physics), 63, DSc(physics), 65. *Prof Exp:* Res attache physics, Nat Ctr Sci Res, France, 63-65; res assoc, 65-68, from asst prof to assoc prof, 68-81, PROF PHYSICS, UNIV MD, COLLEGE PARK, 81- *Mem:* Am Inst Physics; Can Asn Physicists. *Res:* Gravity; astronomy; relativity; earth physics. *Mailing Add:* Dept Physics & Astron Univ Md College Park MD 20742

RICHARD, JOHN L, b Melbourne, Iowa, May 19, 38; div; c 3. MYCOLOGY. *Educ:* Iowa State Univ, BS, 60, MS, 63, PhD(mycol), 68. *Prof Exp:* Microbiologist, Nat Animal Dis Ctr, 63-90, RES LEADER, MYCOTOXIN RES UNIT, NAT CTR AGR UTILIZATION RES, USDA, 90- *Mem:* Med Mycol Soc Americas; Am Soc Microbiol; Int Soc Human & Animal Mycol; Wildlife Dis Asn; Am Asn Off Analytical Chemists; Sigma Xi. *Res:* Fluorescent antibody, cultural techniques, transmission and electron microscopy of Dermatophilus congolensis; equine ringworm; mycotoxicoses; effects of mycotoxins on immunity; aerosol-toxins and infectious fungal agents; avian aspergillosis; interactions of toxins with infectious disease. *Mailing Add:* Nat Ctr Agr Utilization Res USDA Box 70 Peoria IL 61614

RICHARD, JOHN P, US citizen. BIOCHEMISTRY. *Educ:* Ohio State Univ, BS, 74, PhD(chem), 79. *Prof Exp:* Teaching fel, Brandeis Univ, 79-82; res assoc, Fox Chase Cancer Ctr, 82-84; ASST PROF CHEM, UNIV KY, 85- *Concurrent Pos:* Res fel, Cambridge Univ, Eng, 84-85. *Mem:* Am Chem Soc; Am Soc Biol Chemists; AAAS. *Res:* Mechanism of enzyme action; mechanism of reaction of small molecules in solution. *Mailing Add:* Dept Chem Univ Ky 125 Chem-Physics Bldg Lexington KY 40506-0055

RICHARD, PATRICK, b Crowley, La, Apr 28, 38; m 60; c 2. ELECTRON PHYSICS, ATOMIC PHYSICS. *Educ:* Univ Southwestern La, BS, 61; Fla State Univ, PhD(physics), 64. *Prof Exp:* Res asst prof nuclear physics, Univ Wash, 65-68; from asst prof to prof physics, Univ Tex, Austin, 68-72; PROF PHYSICS, KANS STATE UNIV, 72- *Concurrent Pos:* Consult, Columbia Sci Res Inst, 69-71. *Mem:* Am Phys Soc. *Res:* Characteristic x-rays and Auger electons produced in collisions of energetic heavy ions with heavy atoms. *Mailing Add:* Dept Physics Cardwell Hall Kans State Univ Manhattan KS 66506

RICHARD, PIERRE JOSEPH HERVE, b Montreal, Can, July 9, 46; m 70. PALYNOLOGY, PALEOECOLOGY. *Educ:* Univ Laval, BS, 67; Univ Paris, DEA, 68; Univ Montpellier, Dr Etat, 76. *Prof Exp:* Prof bot, Univ Que, Chicoutimi, 71-76; PROF GEOG, UNIV MONTREAL, 76- *Mem:* Palynology Asn Fr Lang; Bot Asn Can; Can Quaternary Asn; Int Soc Limnol; Can Asn Palynology. *Res:* Pollen analysis of late Pleistocene deposits, mainly in Quebec, for paleobiogeographic reconstruction; pollen morphology and methodology, organic sediments and other microfossils. *Mailing Add:* Dept Geog Univ Montreal CP 6128 Montreal PQ H3C 3J7 Can

RICHARD, RALPH MICHAEL, b South Bend, Ind, Dec 15, 30; m 61; c 3. STRUCTURAL ENGINEERING, SOLID MECHANICS. *Educ:* Univ Notre Dame, BSCE, 52; Wash Univ, MSCE, 56; Purdue Univ, PhD(civil eng), 61. *Prof Exp:* Instr civil eng, Wash Univ, 55-56; engr, McDonnell Aircraft, McDonnell Douglas Corp, 56-58; res asst, Purdue Univ, 59-61; asst prof civil eng, Univ Notre Dame, 61-63; assoc prof, 63-65, PROF CIVIL ENG, UNIV ARIZ, 65- *Concurrent Pos:* Consult, US Dept Defense, 63-65, General Dynamics Corp, 66-72, Kitt Peak Nat Observ, 68-70, City Investing Co, Los Angeles, 72-78, Welton Becket & Assoc, 74-76 & US Ballistics Lab, Md, 78. *Mem:* Am Soc Civil Engrs; Am Acad Mech; Sigma Xi. *Res:* Aseismic design; steel connection design; lightweight high resolution optical structural systems. *Mailing Add:* Dept Civil Eng Univ Ariz Tucson AZ 85721

RICHARD, RICHARD RAY, b Nederland, Tex, Sept 12, 27; m 55; c 2. ORBITAL EXPERIMENTS, THERMAL MANAGEMENT. *Educ:* Univ Tex, BS, 55. *Prof Exp:* Test engr instrumentation, Convair Div, Gen Dynamics Corp, 54-61; sr engr electronics, Brown Eng Co, Ala, 61-63; proj engr instrumentation, Manned Spacecraft Ctr, 63-65, head measurement sect, 65-67, head infrared sect, 67-81, HEAD ADV PROGS, JOHNSON SPACE CTR, NASA, 81-, EXP INTEGRATION MGR, 85- *Mem:* Instrument Soc Am. *Res:* Aircraft flutter prediction instrumentation; noncontacting vibration and measurement techniques; Saturn fuel measurement techniques; angular accelerometer and miscellaneous devices; development of digital sensing techniques; cryogenic refrigeration using molecular adsorption in zeolites for gas storage; raising operating temperature of photonic infrared detectors; one patent. *Mailing Add:* Johnson Space Ctr ID3 NASA Rd 1 Houston TX 77058

RICHARD, ROBERT H(ENRY), b Warrington, Fla, Sept 3, 27; m 53; c 6. ELECTRONIC & SYSTEMS ENGINEERING. *Educ:* Auburn Univ, BSEE, 50; Fla State Univ, MS, 51; Johns Hopkins Univ, MSE, 59, PhD(elec eng), 62. *Prof Exp:* Electronics aide, Nat Bur Standards, 49-50, electronic scientist, 50-51, physicist, 51-52; proj engr, Radiation Res Corp, 52-54; res staff asst electronics, Carlyle Barton Lab, Johns Hopkins Univ, 54-57, res assoc, 57-62, res scientist, 62-63; sr res engr, HRB-Singer, Inc, Pa, 63-65, staff engr, 65-68; mem prof staff, Systs Eval Group, Ctr Naval Analysis, 68-75, Opers Eval Group, 75-78, div dir, Opers Eval Group, 78-80, mem Naval Studies Group, 81-87; RETIRED. *Concurrent Pos:* Ctr Naval Analysis fel, Admiralty Surface Weapons Estab, Eng, 72-73. *Mem:* Sr mem Inst Elec & Electronics Engrs; Opers Res Soc Am; Mil Opers Res Soc (pres, 78); Sigma Xi. *Res:* Systems analysis; operations research; sensor systems; electronic warfare. *Mailing Add:* 6704C Lee Hwy Arlington VA 22205

RICHARD, TERRY GORDON, b Marshfield, Wis, Feb 25, 45; m 76; c 3. ENGINEERING MECHANICS, METALLURGY. *Educ:* Univ Wis-Madison, BS, 68, MS, 69, PhD(eng mech), 73. *Prof Exp:* Engr, Owen Ayers & Assoc, 66-67 & Naval Weapons Ctr, 68; teaching asst mech, Univ Wis-Madison, 68-72, fel, 73-75; prof eng mech, Ohio State Univ, 75-81; PROF ENG MECH, UNIV WIS-MADISON, 81- *Concurrent Pos:* Res scientist, Kimberly Clark Corp, 69-70; consult, Battelle Mem Inst-Columbus Div, 76-79, Al Lee Corp, 77-80, Columbia Gas Corp, 79, Joint Implant Surgeons Inc, 79-81 & Sensotec Inc, 79-81. *Mem:* Soc Exp Stress Analysis; Soc Adv Eng Educ; Sigma Xi; Am Soc Mech Engrs. *Res:* Photoelasticity, holography, fatigue, fracture mechanics and cryogenic materials characterization. *Mailing Add:* Dept Eng Mech 2348 Eng Bldg Univ Wis 1415 Johnson Dr Madison WI 53706

RICHARD, WILLIAM RALPH, JR, industrial organic chemistry, paper chemistry; deceased, see previous edition for last biography

RICHARDS, A(LVIN) M(AURER), b Akron, Ohio, Sept 24, 26; m 49; c 2. STRUCTURAL ENGINEERING. *Educ:* Univ Akron, BSCE, 48; Harvard Univ, MSCE, 49; Univ Cincinnati, PhD(struct), 68. *Prof Exp:* Designer, Barber & Magee, Ohio, 49; from asst prof to prof, 49-83, EMER PROF CIVIL ENG, UNIV AKRON, 83- *Concurrent Pos:* Consult. *Mem:* Am Soc Civil Engrs. *Res:* Computer applications to structural design and to the instructional process. *Mailing Add:* 4067 Wilshire Circle Sarasota FL 34238

RICHARDS, ADRIAN F, b Worcester, Mass, Apr 1, 29; div; c 3. MARINE SCIENCE & TECHNOLOGY MANAGEMENT. *Educ:* Univ NMex, BS, 51; Univ Calif, Los Angeles, PhD(oceanog), 57. *Prof Exp:* Res asst submarine geol, Scripps Inst, Univ Calif, 51-55, grad res geologist, 55-57; geol oceanogr, US Navy Hydrographic Off, 57-60; actg exec secy, Div Earth Sci, Nat Res Coun, 60; liaison scientist, London Br Off, Off Naval Res, 61-64; from assoc prof to prof geol & civil eng, Univ Ill, Urbana, 64-69; adj prof oceanog & ocean eng, Depts Geol Sci & Civil Eng, Lehigh Univ, 68-69, prof, 69-82, dir, Marine Geotech Lab, 70-82; sr consult, Fugro BV, Leidschendam, Holland, 82-83, vpres res & develop, 83-87; PRES, ADRIAN RICHARDS CO, 87- *Concurrent Pos:* Instr, USDA Grad Sch, 59-60; Nat Acad Sci-Nat Res Coun resident res assoc, US Navy Electronics Lab, 60-61; Royal Norweg Coun Sci & Indust Res fel, Norweg Geotech Inst, 63-64; mem, Joint Oceanog Insts Deep Explor Sampling, Sedimentary, Petrol & Phys Properties Panel, 70-82, chmn, 76-82, Europ Sci Found rep Shipboard Measurement Panel, 88-; mem, Downhole Measurements Panel, 75-82; mem, US deleg on training, educ & mutual assistance, Intergovt Oceanog Comn, UNESCO, 71-73; co-chmn panel on undersea sci & technol, Nat Acad Sci-Nat Acad Eng, 72-73; mem adv coun, Univ-Nat Oceanog Lab Syst, 72-76, chmn, Alvin Rev Comt, 75-77, vchmn, 78-79; vis prof oceanog, Lafayette Col, 73; mem seafloor eng comt, Marine Bd, Nat Res Coun, 73-76; consult marine sci & technol, Govt of Iran, UNESCO, 74; ed-in-chief, Marine Geotechnol, 74-83, assoc ed, 83-; mem vis comt, Dept Ocean Eng, Woods Hole Oceanog Inst, 74 & 76; Trans-Can lectr, Soil Mech Subcomt, Nat Res Coun Can, 74; rep, Eng Comt Oceanic Resources, Comt Training Educ & Mutual Assistance, Intergovt Oceanog Comn, UNESCO, 75-90, chmn, Eng Workshop, 82 & Workshop Marine Sci, 86 & 88; assoc ed, Ocean Eng, 75-; consult marine sci & technol, Adrian F Richards & Assocs, Inc, 75-, Marine Facil Panel, US-Japan Coop Prog Nat Resources, 72-82, Sea-Bottom Surv Panel, 77-82 & Off Ocean Eng, Nat Oceanic & Atmospheric Admin, 77-81; Nat Res Coun rep, Int Prog & Int Coop Oceans, Dept State, 77-79; mem ocean eng rev comt, Rosenstiel Sch Marine & Atmospheric Sci & Sch Eng & Environ Design, Univ Miami, 78; consult, United Nations, 79, Unesco Div Marine Sci, 79-; vis lectr, Nat Taiwan Univ, 90, vis researcher, Nat Res Inst Pollution Resources, Japan, 90; vis scientist, Woods Hole Oceanog Inst, 82; mem orgn comt, Seabed Mech Symp, Int Union Theoret & Appl Mech, 83-84; keynote speaker, Oceanog Int Conf, Eng, 84, mem comt, 84-88; prof eng geol, Delft Univ Technol, Holland, 88- *Honors & Awards:* C A Hogentogler Award, Am Soc Testing & Mat, 73, Spec Serv Award, 90. *Mem:* Fel AAAS; fel Marine Technol Soc; fel Geol Soc Am; fel Geol Soc London; fel Am Soc Civil Engrs; Royal Inst, Neth. *Res:* Geotechnical ocean engineering; Strategic management in marine science and technology for industry, governments, and international organizations; geotechnical properties and processes of sediments of the continental margins and ocean basins; marine science and technology education in developing nations. *Mailing Add:* Adrian Richards Co Uiterweg 309 Aalsmeer 1431 AJ Netherlands

RICHARDS, ALBERT GLENN, b Lake Forest, Ill, May 29, 09; m 35, 66; c 3. INSECT PHYSIOLOGY. *Educ:* Univ Ga, AB, 29; Cornell Univ, PhD(entom), 32. *Prof Exp:* Asst, Cornell Univ, 30-32; actg head dept entom, Ward's Nat Sci Estab, 32; asst zool, Univ Rochester, 33-36; asst, Am Mus Natural Hist, 36-37; instr biol, City Col New York, 37-39; from instr to asst prof zool, Univ Pa, 39-45; assoc prof entom & zool, 45-49, prof, 49-77, EMER PROF ENTOM & ZOOL, UNIV MINN, ST PAUL, 77- *Concurrent Pos:* Guggenheim fel & Fulbright res scholar, Max Planck Inst Biol, 57-58; vis lectr, Am Inst Biol Sci, 59-65; guest investr, Max Planck Inst Physiol Behav, 66-67; guest prof, Univ Munich, 66-67; mem, Marine Biol Lab, Woods Hole, Mass. *Mem:* Entom Soc Am (vpres, 49); Am Entom Soc (secy, 43-45); Am Soc Zool; Brit Soc Exp Biol; Electron Micros Soc Am. *Res:* Histology, ultrastructure and physiology of insects; arthropod cuticle; biological microscopy. *Mailing Add:* Dept Entom Fisheries & Wildlife Univ Minn St Paul MN 55108

RICHARDS, ALBERT GUSTAV, b Chicago, Ill, Jan 7, 17; m 42; c 5. DENTAL RADIOLOGY. *Educ:* Univ Mich, BS, 40, MS, 43. *Prof Exp:* From instr to prof dent, Univ Mich, Ann Arbor, 40-74, Marcus L Ward prof, 74-82; RETIRED. *Concurrent Pos:* Consult, Vet Admin Hosp, Ann Arbor, 54-, Nat Res Coun, 58, dent health proj, Dept HEW, 62- & comt x-ray protection dent off, Nat Comn Radiation Protection, 63-; mem, Nat Adv Environ Coun, USPHS, 69-71. *Mem:* Am Dent Asn; Am Acad Dent Radiol; Acad Oral Roentgenol (pres, 62-63). *Res:* Electron microscopy, radiation hygiene and dosimetry in dentistry; radiographic technics for pedodontists and exodontists; erythema; dental and x-ray machine designs; invented dynamic tomography and dental x-ray technique trainer. *Mailing Add:* 395 Rock Creek Dr Ann Arbor MI 48104

RICHARDS, BERT LORIN, JR, plant pathology, for more information see previous edition

RICHARDS, CHARLES DAVIS, b Cumberland, Md, May 14, 20; m 43; c 5. PLANT TAXONOMY. *Educ:* Wheaton Col, Ill, BA, 43; Univ Mich, MA, 47, PhD(bot), 52. *Prof Exp:* From instr to asst prof bot, Mich Col Mining & Technol, 47-50; instr, Univ Mich, 51; from instr to assoc prof, 52-63, PROF BOT, UNIV MAINE, ORONO, 63- *Res:* Flora and grasses of Maine; plant ecology of Mt Kathdin; aquatic flowering plants; plant geography; rare and endangered plants. *Mailing Add:* 22 Spencer Orono ME 04473

RICHARDS, CHARLES NORMAN, b Buffalo, NY, Mar 3, 42. PHYSICAL ORGANIC CHEMISTRY. *Educ:* Canisius Col, BS, 64; Univ Hawaii, PhD(phys org chem), 68. *Prof Exp:* SR RES CHEMIST, CORN PROD RES, ANHEUSER-BUSCH, INC, 68- *Mem:* Am Chem Soc. *Res:* Pyrolysis gas chromatography applied to synthetic and natural polymers; thermodynamic activation parameters in mixed aqueous systems; modified food starches. *Mailing Add:* 12215 Country Manor Ln St Louis MO 63141

RICHARDS, CLYDE RICH, b Paris, Idaho, June 9, 21; m 46; c 3. DAIRY HUSBANDRY. *Educ:* Utah State Agr Col, BS, 43; Cornell Univ, MS, 49, PhD(dairy husb), 50. *Prof Exp:* Asst animal husb, Cornell Univ, 46-50; lectr animal breeding, Super Sch Agr, Athens, Greece, 50-51; from asst prof to assoc prof animal indust, Univ Del, 51-61; dep asst adminr, Coop State Res Serv, USDA, Utah State Univ, 71-72, prin animal nutritionist, 61-83, dir, Int Feedstuffs Inst, Agr Res Serv, 83-85; RETIRED. *Mem:* AAAS; Am Inst Nutrit; Am Soc Animal Sci; Am Dairy Sci Asn; Sigma Xi. *Res:* Roughage digestibility using indicator techniques; nutrient content of lima bean silage; ketosis in dairy cattle; agricultural research programs at land grant colleges of 1890; research administration. *Mailing Add:* 1772 E 1400 N Logan UT 84321

RICHARDS, DALE OWEN, b Morgan, Utah, July 4, 27; m 55; c 3. STATISTICS. *Educ:* Utah State Univ, BS, 50; Iowa State Univ, MS, 57, PhD(statist, indust eng), 63. *Prof Exp:* Statistician, Gen Elec Co, 52-55; instr indust eng, Iowa State Univ, 55-59, asst prof, 59-63; assoc prof statist, 63-66, chmn dept, 66-69, PROF STATIST, BRIGHAM YOUNG UNIV, 67- *Concurrent Pos:* Consult, Am Can Co, 58; opers analyst, USAF contract, Iowa State Univ Standby Unit, 59-70; consult, Nuclear Div, Kaman Aircraft Corp, 60, CEIR, Inc, 65-70 & Andrulus Res Corp, 87-91. *Mem:* Am Statist Asn; Am Soc Qual Control. *Res:* Application of statistical and operations research techniques to industrial situations; SPC and TQM. *Mailing Add:* Dept Statist Brigham Young Univ Provo UT 84602

RICHARDS, EARL FREDERICK, b Detroit, Mich, Mar 11, 23; m 46; c 2. ELECTRICAL ENGINEERING. *Educ:* Wayne State Univ, BS, 51; Mo Sch Mines & Metall, MS, 61; Univ Mo, PhD(elec eng), 71. *Prof Exp:* Engr elec, Electronic Control Corp, 51-52, Pa Salt Mfg Co, 52-54; instr elec, Mo Sch Mines & Metall, 58-61; from asst prof to assoc prof, 62-80, PROF ELEC ENG, UNIV MO, ROLLA, 80- *Concurrent Pos:* Lectr elec, Univ Detroit, 56-68; res engr, Argonne Nat Lab, 63; consult, Ford Motor Co, 66-67, Emerson Elec Co, 78-80, Magnetic Peripherals, 80, Wanlass Corp, 81 & Public Serv Comn Mo, 82, US CEngr, 83, Asn of Mos, 80, Coop, 85. *Mem:*

Sigma Xi; Inst Elec & Electronics Engrs; Nat Soc Prof Engrs. *Res:* Linear and non-linear control systems theory; simulation and modelling techniques; digital filtering; power system analysis and stability; author of over 75 publications in the area of power and control systems. *Mailing Add:* Dept Elec Eng Univ Mo Rolla MO 65401

RICHARDS, EDMUND A, b Grand Rapids, Mich, Jan 25, 35; m 56; c 2. PHYSIOLOGY, PHARMACOLOGY. *Educ:* Purdue Univ, BS, 57; Univ Ill, MS, 59, PhD(physiol), 65; Univ Stockholm, MD, 67. *Prof Exp:* Assoc pharmacologist, Eli Lilly & Co, 59-61; res assoc physiol, Univ Ill, 63-65; guest scientist, Karolinska Inst, Sweden, 65-67; assoc prof physiol, Baylor Univ Col Med, 67-69; PROF PHYSIOL, UNIV NORTHERN COLO, 69- *Concurrent Pos:* Int lectr, Gt Brit & Scand, 72; Astra Pharmaceut Co, Gen Mills, Inc, WHO, Int Drug Control, Foreign Med Schs & AIDS Int. *Honors & Awards:* Int Pharmacol Award, 75; R A Gregory Award Med Res, 73; William S Merrill Award, 77. *Mem:* Am Gastroenterol Soc; Am Physiol Soc; Am Inst Biol Sci; fel Royal Soc; Int Pharmacol Cong; NY Acad Sci; fel Royal Brit Med Soc. *Res:* Gastroenterological research related to pancreatic and gastric secretion and peptic ulceration; the transfer of pharmacological active drugs across the human placental membranes; pharmacology of smooth muscle. *Mailing Add:* Dept Biol & Zool Univ Northern Colo Greeley CO 80639

RICHARDS, F PAUL, b Stoneham, Mass, Sept 22, 46; div; c 2. FISHERY BIOLOGY, IMPACT ASSESSMENT. *Educ:* Univ Mass, Amherst, BS, 68; Southeastern Mass Univ, MS, 74. *Prof Exp:* Asst marine fishery biologist, Mass Div Marine Fisheries, 69; fishery biologist, Essex Marine Lab, 69-72; fisheries lab mgr, environ sci div, NUS Corp, 74-76; aquatic ecol prog mgr, Ecol Analysts, Inc, 76-79; SUPVR, ECOL SCI & PLANNING DEPT, ENVIRON DIV, CHARLES T MAIN INC, 79- *Concurrent Pos:* Consult, US Army CEngr, 70-71. *Mem:* Am Fisheries Soc; Am Inst Fishery Res Biologists; Atlantic Fishery Biologists. *Res:* Identification, assessment and mitigation of the impacts of large capital project construction and operation on aquatic ecosystems; changes in distribution, behavior and population dynamics of fishes. *Mailing Add:* Environ Div Charles T Main Inc Prudential Ctr Boston MA 02199

RICHARDS, FRANCIS ASBURY, oceanography; deceased, see previous edition for last biography

RICHARDS, FRANCIS RUSSELL, b Biloxi, Miss, Jan 19, 44; m 65; c 2. OPERATION RESEARCH, APPLIED STATISTICS. *Educ:* La Polytech Univ, BS, 65; Clemson Univ, MS, 67, PhD(math sci), 71. *Prof Exp:* Chem engr, Olin Matheson, 62-63; oper res analyst, Naval Weapons Lab, 65-67; prof opers res, Naval Postgrad Sch, 70-87; PRES, EVAL TECHNOL INC, 87- *Mem:* Opers Res Soc Am; Sigma Xi; Am Statist Asn. *Res:* Applications of multi-attribute utility theory to multi-criteria decision making; test and evaluation; constrained multi-item inventory analysis; reliability, maintenance and availability; design of experiments; improving testing efficiency, expert systems. *Mailing Add:* 1008 Franklin Monterey CA 93940

RICHARDS, FRANK FREDERICK, b London, Eng, Nov 14, 28; m 58; c 3. BIOCHEMISTRY, MEDICINE. *Educ:* Cambridge Univ, BA, 53, MB, BChir, 56, MD, 63. *Prof Exp:* Intern surg, St Mary's Hosp, London, 57-58; intern med, Oxford Univ, 58; resident, Brompton Hosp, 59; sr resident invest med, St Mary's Hosp Med Sch, 59-60, lectr, 60-64; res assoc biochem, Harvard Med Sch, 64-66; assoc med, Mass Gen Hosp, 66-68; assoc prof med & microbiol, 68-74, PROF INTERNAL MED, SCH MED, 74-, DIR, YALE-MACARTHUR CTR MOLECULAR PARASITOL, YALE UNIV, 85- *Concurrent Pos:* Res fel biochem, St Mary's Hosp, London, 58-59; Am Heart Asn adv res fel, 64-66; estab investr, Am Heart Asn, 66-71. *Res:* Protein chemistry and molecular biology as applied to parasitology and virology; pulmonary disease. *Mailing Add:* Yale-MacArthur Ctr Molecular Parasitol 700 LEPH 60 College St New Haven CT 06510-8056

RICHARDS, FREDERIC MIDDLEBROOK, b New York, NY, Aug 19, 25; m 59; c 3. PROTEIN CHEMISTRY. *Educ:* Mass Inst Technol, SB, 48; Harvard Univ, PhD, 52. *Hon Degrees:* DSc, Univ New Haven. *Prof Exp:* Res fel phys chem, Harvard Univ, 52-53; from asst prof to assoc prof biochem, 54-62, chmn dept molecular biophys & biochem, 63-67 & 69-72, PROF BIOCHEM, YALE UNIV, 62-; DIR, JANE COFFIN CHILDS MEM FUND MED RES, 76- *Concurrent Pos:* Fel, Nat Res Coun, Carlsberg Lab, Copenhagen, Denmark, 54, NSF, Cambridge Univ, 55 & Guggenheim, 67. *Honors & Awards:* Pfizer-Paul Lewis Award Enzyme Chem, 65; Kai Linderstrom-Lang Award, 78; Merck Prize, Am Soc Biochem & Molecular Biol, 88; Stein & Moore Prize, Protein Soc, 88. *Mem:* Nat Acad Sci; AAAS; Biophys Soc (pres, 72); Am Crystallog Asn; Am Soc Biochem & Molecular Biol (pres, 79). *Res:* Proteins. *Mailing Add:* 69 Andrews Rd Guilford CT 06437

RICHARDS, FREDERICK, II, b Charleston, SC, Aug 28, 38; m 62; c 3. MEDICAL ONCOLOGY, HEMATOLOGY. *Educ:* Davidson Col, BS, 60; Med Univ SC, MD, 64. *Prof Exp:* Chief internal med, USAF Hosp, Lubbock, Tex, 66-68; instr med, 71-73, asst prof, 73-77, ASSOC PROF MED, BOWMAN GRAY SCH MED, 77- *Mem:* Am Soc Clin Oncol; Am Soc Hemat; AMA; fel Am Col Physicians. *Res:* Treatment of malignant disease and hematological conditions. *Mailing Add:* 300 S Hawthorne Rd Winston-Salem NC 27103

RICHARDS, GARY PAUL, b Springfield, Mass, June 4, 50; m 73; c 3. VIROLOGY, ANALYTICAL METHODS RESEARCH. *Educ:* Univ NH, BA, 73. *Prof Exp:* Dir qual control & res, Rockland Shrimp Corp, Maine, 73-75; food inspector, Northeast Inspection Off, Gloucester, Mass 75-77, microbiologist, College Park Lab, Md, 77-78, dir, Bio Res & Testing Lab, 78-88, RES MICROBIOLOGIST, CHARLESTON LAB, SC, NAT MARINE FISHERIES SERV, NAT OCEANIC & ATMOSPHERIC ADMIN, US DEPT COM, 78-; DIR, BIO RES & TESTING LAB, 78- *Mem:* Am Soc Microbiol; Sigma Xi. *Res:* Human enteric viruses in shellfish including hepatitis A, polio, Norwalk and other viruses of public health significance; extraction and assay of viruses from contaminated shellfish; evaluate cell cultures for virus propagation and assay; development and evaluate molecular biology methods for virus detection. *Mailing Add:* Nat Marine Fisheries Serv Charleston Lab PO Box 12607 Charleston SC 29412

RICHARDS, GEOFFREY NORMAN, b Eng, Apr, 3, 27; Australian citizen; m 50; c 3. WOOD CHEMISTRY. *Educ:* Birmingham Univ, Eng, BSc, 48, PhD(chem), 51, DSc, 64. *Prof Exp:* Org chemist, Brit Rayon Res Asn, 51-57, sr org chemist, 58-60; dep dir res, AMF Brit Res Lab, 60-64; Nevitt prof chem, James Cook Univ, N Queensland, 65-85, dean fac sci, 71-74; PROF & DIR, WOOD CHEM LAB, UNIV MONT, 85- *Concurrent Pos:* Asst prof biochem, Purdue Univ, 57-58; tech adv, Am Machine & Foundry Co, Europe, 64; vis prof, Univ Miami Med Sch, 69; counr, Australian Inst Nuclear Sci, 72-85; exec secy, Int Carbohydrate Orgn, 78-83. *Mem:* Fel Royal Soc Chem; fel Royal Australian Chem Inst. *Res:* Carbohydrate and polysaccharide chemistry; reaction mechanisms; digestion in ruminants; chemistry of marine mucins; sucrose chemistry and technology; wood chemistry and technology; biomass utilization, especially pyrolysis, gasification and combustion. *Mailing Add:* Wood Chem Lab Univ Mont Missoula MT 59812

RICHARDS, GRAYDON EDWARD, b Gilmer Co, WVa, July 2, 33; m 56; c 2. SOIL FERTILITY. *Educ:* WVa Univ, BS, 55; Ohio State Univ, MSc, 59, PhD(agron), 61. *Prof Exp:* Teacher high sch, WVa, 55-57; res asst agron, Ohio State Univ, 57-61; res agronomist, Int Minerals & Chem Corp, 61-67; sr res scientist & proj leader agron, Continental Oil Co, 67-70; regional agronomist, Agr Div, Olin Corp, 70-79; chief agronomist, Smith-Douglass Div, Borden Inc, 80; acct supvr fertilizer, Doane Agr Serv, 80-81; mgr, agr & tech serv, Vistron Corp, 81-83; assoc prof, Univ Ark, Monticello Campus, 84, head, Dept Agr & dir, Southewast Res & Exten Ctr, 84-89; PRES, REEDER AUTO PARTS, INC, 90- *Honors & Awards:* Jour Award, Am Soc Agron, 76. *Mem:* Am Soc Agron; Soil Sci Soc Am; Am Chem Soc. *Res:* Chemistry of soil potassium and essential micronutrient elements; plant growth regulators; micronutrient nutrition of plants; fertilization for optimum yields; efficient use of phosphatic fertilizer. *Mailing Add:* 914 Ann Marshall MO 65340

RICHARDS, HAROLD REX, b Timmins, Ont, Feb 9, 26; m 49; c 2. TEXTILE TECHNOLOGY. *Educ:* Univ Leeds, BSc, 49, PhD(textile sci), 54. *Prof Exp:* Sr phys chemist, Defence Res Bd, Can, 56-64; head dept textiles, clothing & design, Univ Guelph, 64-70; prof textiles/clothing & head dept, Colo State Univ, 71-85; RETIRED. *Concurrent Pos:* Consult, Mich Chem Corp, 70- *Honors & Awards:* Textile Sci Award, Textile Tech Fedn Can, 68. *Mem:* Can Inst Textile Sci (pres, 65-66); fel Brit Textile Inst; fel Brit Plastics Inst; Fiber Soc. *Res:* Textile and polymer science. *Mailing Add:* 305 W Magnolia St No 363 Ft Collins CO 80521

RICHARDS, HUGH TAYLOR, b Baca Co, Colo, Nov 7, 18; m 44; c 6. NUCLEAR PHYSICS, ION SOURCES. *Educ:* Park Col, BA, 39; Rice Inst, MA, 40, PhD(physics), 42. *Prof Exp:* Asst physics, Rice Univ, 41-42, res assoc uranium proj, Off Sci Res & Develop, 42; scientist, Univ Minn, 42-43 & Manhattan Dist, Los Alamos Sci Lab, 43-46; res assoc nuclear physics, Univ Wis-Madison, 46-47, from asst prof to prof, 47-52, chmn dept physics, 60-63, 66-69 & 85-88, assoc dean col lett & sci, 63-66, prof 52-88, EMER PROF, NUCLEAR PHYSICS, UNIV WIS-MADISON, 88- *Mem:* Fel Am Phys Soc. *Res:* Nuclear scattering cross sections; yields and angular distributions of nuclear reactions; nuclear energy levels and reaction energies; isospin forbidden reactions; negative ion sources. *Mailing Add:* Dept Physics Univ Wis Madison WI 53706

RICHARDS, JACK LESTER, b Apr 29, 40; US citizen; m 64; c 2. ORGANIC & PHOTOGRAPHIC CHEMISTRY. *Educ:* Rochester Inst Technol, BS, 65; Univ Rochester, PhD(org chem), 70. *Prof Exp:* Sr res chemist, 69-76, tech assoc, 77-81, RES ASSOC, RES LABS, EASTMAN KODAK CO, 81- *Mem:* Am Chem Soc. *Res:* Synthesis of organic compounds for applications in image formation systems; chemistry of organosulfur and organosulfur-nitrogen compounds; research and development work on color photographic systems; production and quality control of color photographic products; worldwide coordination of complex technical programs. *Mailing Add:* 2259 Latta Rd Rochester NY 14612

RICHARDS, JAMES AUSTIN, JR, b Boston, Mass, Apr 15, 16; m 39; c 4. PHYSICS. *Educ:* Oberlin Col, BA, 38; Duke Univ, PhD(physics), 42. *Prof Exp:* Instr physics, Bucknell Univ, 42-46; tutor physics & math, Olivet Col, 46-49; asst prof physics, Univ Minn, Duluth, 49-51; res physicist, Am Viscose Corp, 51-55; prof physics, Drexel Inst Technol, 55-65; dean instr, Community Col Philadelphia, 65-68; prof physics, State Univ NY Agr & Tech Col Delhi, 68-82, instrnl developer, 82-85; RETIRED. *Mem:* Soc Friends Kiwanis. *Res:* Atomic physics; nuclear science. *Mailing Add:* Box 162 Treadwell NY 13846

RICHARDS, JAMES FREDERICK, b Amherst Island, Ont, Mar 15, 27; div; c 3. BIOCHEMISTRY. *Educ:* Queen's Univ, Ont, BA, 49, MA, 52; Univ Western Ont, PhD(biochem), 58. *Prof Exp:* Nat Res Coun Can overseas fel biochem, Glasgow Univ, 59-60; from instr to assoc prof, 60-75, PROF BIOCHEM, UNIV BC, 75- *Mem:* Can Biochem Soc. *Res:* Polyamine metabolism and hormone function. *Mailing Add:* Dept Biochem Univ BC Vancouver BC V6T 1W5 Can

RICHARDS, JAMES L, b Kankakee, Ill, Dec 26, 46; m 69; c 2. OPERATIONS RESEARCH, COMPUTER SCIENCE. *Educ:* Ill State Univ, BS, 68; Univ Mo-Rolla, MS, 72, PhD(math), 76. *Prof Exp:* Asst prof math & comput sci, 76-81, assoc prof comput sci, 81-83, PROF COMPUT SCI, BEMIDJI STATE UNIV, 83- *Mem:* Asn Comput Mach. *Res:* Integer programming, optimization theory and programming languages. *Mailing Add:* Dept Math & Comput Sci Bemidji State Univ Bemidji MN 56601

RICHARDS, JOANNE S, b Exeter, NH, Apr 23, 45. REPRODUCTIVE ENDOCRINOLOGY. *Educ:* Oberlin Col, BA, 67; Brown Univ, MAT, 68, PhD(physiol chem), 70. *Prof Exp:* Asst prof biol, Univ NDak, 70-71; fel, 71-73, instr, 73-74, asst prof reprod endocrinol, Univ Mich, 74-; PROF, DEPT CELL BIOL, BAYLOR COL MED. *Concurrent Pos:* Prin investr, NIH grants, 76-79; Nat Inst Child Health & Human Develop res career develop award, 78. *Res:* Mammalian reproductive endocrinology; mechanisms of hormone action; ovarian cell physiology; molecular endocrinology. *Mailing Add:* Dept Cell Biol Baylor Col Med One Baylor Plaza Houston TX 77030

RICHARDS, JOHN HALL, b Berkeley, Calif, Mar 13, 30; m 54, 75; c 4. BIOCHEMISTRY. *Educ:* Univ Calif, BS, 51, PhD, 55; Oxford Univ, BSc, 53. *Prof Exp:* Instr chem, Harvard Univ, 55-57; from asst prof to assoc prof, 57-70, PROF CHEM, CALIF INST TECHNOL, 70- *Concurrent Pos:* Consult, Appl Biosysts, E I du Pont de Nemours & Co. *Honors & Awards:* Lalor Award, 56. *Mem:* Am Chem Soc; Protein Soc. *Res:* Mechanism of protein function; molecular immunology. *Mailing Add:* Chem Div 147-75 Calif Inst Technol Pasadena CA 91125

RICHARDS, JONATHAN IAN, b New York, NY, Dec 4, 36. MATHEMATICS. *Educ:* Univ Minn, BA, 57; Harvard Univ, MA, 59, PhD(math), 60. *Prof Exp:* Instr math, Mass Inst Technol, 60-62; asst prof, 62-67, assoc prof, 67-80, PROF MATH, UNIV MINN, MINNEAPOLIS, 80- *Mem:* Am Math Soc; Math Asn Am. *Res:* Functional analysis; functions of one complex variable; number theory. *Mailing Add:* Dept Math Univ Minn 206 Church St SE Minneapolis MN 55455

RICHARDS, JOSEPH DUDLEY, b Hanover, NH, Sept 13, 17; m 40; c 4. INDUSTRIAL CHEMISTRY. *Educ:* Dartmouth Col, AB, 39. *Prof Exp:* Analyst, Duraloy Co, Pa, 39-40 & Weirton Steel Co, WVa, 40-41; spectrogr, Am Steel & Wire Co, Mass, 40-44, US Bur Mines, 44-45 & Nat Bur Stand, Washington, DC, 45; res chemist in-chg instruments lab, Chem & Pigment Div, Baltimore, 45-50, res group leader chem pigment , Chem & Pigment Div, 50-57, asst to dir res pigments & color dept, 57-61, group leader new prod, 61-63, liaison mkt res, 63-66, mgr econ eval, Glidden-Durkee Div, SCM Corp, 66-67, mgr res serv, 67-70, mgr com develop, 70-80, mgr spec sales & com develop, pigments group, 80-83; RETIRED. *Mem:* Am Chem Soc; fel Am Inst Chemists. *Res:* Emission spectroscopy; electron microscopy; x-ray diffraction; pigment and inorganic chemistry; research planning; chemical marketing. *Mailing Add:* 113 Tenbury Rd Lutherville MD 21093

RICHARDS, KENNETH JULIAN, b Long Beach, Calif, Nov 29, 32; m 58; c 3. CHEMICAL METALLURGY, METALLURGICAL ENGINEERING. *Educ:* Univ Utah, Salt Lake City, BS, 56, PhD(metall eng), 62. *Prof Exp:* Process eng refining, Union Oil, 56; process eng fractionation, C F Braun, 57; develop eng rare earth separation, US Intel Agency, 57-59; group leader metals & ceramics res, Aerospace Res Labs, 62-67; sr scientist chem metall, Kennecott Copper Corp, 67-70, sect head refining res, 70-72, mgr process metall, 72-74, res dir, Metal Mining Div Res Ctr, 74-84; PRES TECH DIV, KERR-MCGEE CORP, 84- *Concurrent Pos:* Mat consult, Air Force Mat Lab, 65-67, Air Force Ballistics Missile Div, 65-67, & NASA, 65-67. *Mem:* Metall Soc; Am Inst Mining Metall & Petrol Eng; Am Soc Metals; Am Inst Chem Eng; Soc Mining Engrs. *Res:* Extractive metallurgy; process development; technical planning; research management. *Mailing Add:* PO Box 25861 Oklahoma City OK 73125

RICHARDS, L(ORENZO) WILLARD, b Logan, Utah, July 11, 32; m 66; c 2. PHYSICAL CHEMISTRY. *Educ:* Calif Inst Technol, BS, 54; Harvard Univ, AM, 56, PhD, 60. *Prof Exp:* From instr to asst prof chem, Amherst Col, 59-66; indust res fel, Cabot Corp, 66-68, mem tech staff, 68-74; mem tech staff, Environ Monitoring & Serv Ctr, Rockwell Int, 74-79; sr proj mgr, Meteorol Res, Inc, 79-82; VPRES, SONOMA TECH, INC, 82- *Concurrent Pos:* USPHS fel, Univ Calif, Berkeley, 64-65; adj prof, Rensselaer Polytech Inst, 71-73. *Mem:* Am Chem Soc; Am Phys Soc; Optical Soc Am; Combustion Inst; Air Pollution Control Asn. *Res:* Atmospheric chemistry; aerosols; light scattering by particles; multiple scattering of light; gas phase chemical kinetics; atmospheric sciences. *Mailing Add:* 5510 Skylane Blvd Suite 101 Santa Rosa CA 95403-1083

RICHARDS, LAWRENCE PHILLIPS, mammalogy, vertebrate paleontology, for more information see previous edition

RICHARDS, LORENZO ADOLPH, b Fielding, Utah, Apr 24, 04; m 30; c 3. EXPERIMENTAL PHYSICS. *Educ:* Utah State Univ, BS, 26, MA, 27; Cornell Univ, PhD(physics), 31. *Hon Degrees:* DTechSc, Israel Inst Technol, 52; DSc, Utah State Univ, 74. *Prof Exp:* Asst physics, Cornell Univ, 27-29, instr, 29-35; physicist, Battelle Mem Inst, 35; from asst prof to assoc prof physics, Iowa State Univ, 35-39; sr soil physicist salinity lab, Bur Plant Indust, USDA, 39-42; Nat Defense Res fel & group supvr, Calif Inst Technol, 42-45; chief physicst salinity lab, Bur Plant Indust, USDA, Calif, 45-66; CONSULT PHYSICIST IN-CHG RES & DEVELOP, LARK INSTRUMENTS, 66- *Concurrent Pos:* Consult, Ministry Agr, Egypt, 52; vpres & dir, Moistomatic, Inc, 54-62; lectr, Univ Alexandria, 62; prof in residence, Univ Calif, Riverside, 68-69; consult, Ford Found. *Honors & Awards:* Ord Develop Award, Dept Navy, 45; Super Serv Award, USDA, 59. *Mem:* Fel AAAS; Am Soc Agron (pres, 65); Am Phys Soc; Soil Sci Soc Am (pres, 52); Am Geophys Union. *Res:* Physics of soil water; automatic irrigation; diagnosis and improvement of saline and alkali soils; electronic sensors for soil water matric suction and salinity. *Mailing Add:* PO Box 3852 Carmel CA 93921

RICHARDS, MARK P, BIOCHEMISTRY, PHYSIOLOGY. *Educ:* Rutgers Univ, PhD(nutrit biochem), 77. *Prof Exp:* RES ANIMAL SCIENTIST, NONRUMINANT ANIMAL NUTRIT LAB, USDA, 79- *Mailing Add:* Nonruminant Animal Nutrit Lab USDA Bldg 200 Rm 201 Beltsville MD 20705

RICHARDS, MARVIN SHERRILL, b Somerville, NJ, Jan 27, 22. ORGANIC CHEMISTRY. *Educ:* Princeton Univ, BS, 43; Lehigh Univ, MS, 49; Rutgers Univ, PhD(chem), 53. *Prof Exp:* Res chemist, Johns-Manville Corp, 43; develop chemist, Am Cyanamid Co, 46-47; asst instr org chem, Lehigh Univ, 48-49; asst instr, Rutgers Univ, 49-52, instr, 52-54; from asst prof to prof phys chem, Drew Univ, 54-69; sr assoc prof chem, 69-73, PROF CHEM, MIAMI-DADE COMMUNITY COL, 73- *Mem:* Am Chem Soc. *Res:* Mechanism; diazoketones; halogenations; synthesis of pyrimidines. *Mailing Add:* Dept Chem Miami-Dade Community Col 11380 NW 27th Ave Miami FL 33167-3418

RICHARDS, NORMAN LEE, marine biology, environmental law, for more information see previous edition

RICHARDS, NORVAL RICHARD, b Ont, Can, July 2, 16; m 51; c 2. SOILS. *Educ:* Univ Toronto, BSA, 38; Mich State Univ, MS, 46. *Hon Degrees:* DSc, Laval Univ, 67. *Prof Exp:* Agr res officer soil classification, Agr Res Br, Can, 38-50; prof soils & head dept, Univ Guelph, 50-62, dean, 62-72, prof land resource sci, Ont Agr Col, 72-81; RETIRED. *Concurrent Pos:* Chmn, Can Agr Res Coun, 75- *Mem:* Int Soil Sci Soc; fel Soil Conserv Soc Am; Can Soc Soil Sci (pres, 56); fel Agr Inst Can (pres, 75). *Res:* Soil classification. *Mailing Add:* 59 Green Guelph ON N1H 2H4 Can

RICHARDS, OLIVER CHRISTOPHER, b Jamesburg, NJ, Jan 13, 33; m 63; c 2. BIOCHEMISTRY. *Educ:* Syracuse Univ, BS, 55; Univ Ill, PhD(biochem), 60. *Prof Exp:* NSF fel, Univ Minn, Minneapolis, 62-63; NSF fel, Univ Calif, Los Angeles, 63-64, res assoc, 64-65; from instr to asst prof, 65-72, ASSOC PROF BIOCHEM, COL MED, UNIV UTAH, 72- *Mem:* Am Soc Biol Chemists. *Res:* Structure, replication and function of extranuclear DNAs in eukaryotes; replication of animal viruses. *Mailing Add:* Dept Biochem Univ Utah Salt Lake City UT 84132

RICHARDS, OSCAR WHITE, environmental physiology, for more information see previous edition

RICHARDS, PAUL BLAND, b North Attleboro, Mass, Aug 14, 24; m 45; c 5. EMERGENCY MANAGEMENT, DISASTER MANAGEMENT. *Educ:* US Naval Acad, BS, 45; Harvard Univ, MS, 50; Case Inst Technol, MS, 53, PhD(math), 59. *Prof Exp:* Res engr, Babcock & Wilcox Res Ctr, 50-54; instr math & supvr res proj, Case Inst Technol, 54-57; mgr systs analysis, Thompson Ramo Wooldridge, Inc, 57-59; mathematician, Gen Elec Co, 59-62; mgr math, sci comput, mech & math physics, Aerospace Res Ctr, Gen Precision, Inc, 62-68; supt math & info sci div, 68-74, dir fleet med support proj, 74-87, OPERS RES ANALYSIS, NAVAL RES LAB, 87- *Mem:* Fel Am Astronaut Soc (pres, 71-73). *Res:* Pioneered research in the design and evaluation of emergency and disaster management systems; principal analytical tool is computer simulation; applications include natural disasters, man-made disasters and military medical operations. *Mailing Add:* 7617 Range Rd Alexandria VA 22306-2425

RICHARDS, PAUL GRANSTON, b Cirencester, Eng, Mar 31, 43; US citizen; m 68; c 2. SEISMIC WAVE PROPAGATION. *Educ:* Univ Cambridge, BA, 65; Calif Inst Technol, MS, 66, PhD(geophys), 70. *Prof Exp:* Asst res geophysicist, Univ Calif, San Diego, 70-71; from asst prof to assoc prof, 71-79, chmn dept, 80-83, assoc dir, Lamont-Doherty Geol Observ, 80-83, PROF GEOL SCI, COLUMBIA UNIV, 79-, MELLON PROF, 87- *Concurrent Pos:* Alfred P Sloan Found fel, 73-74; Am ed, Geophys J, Royal Astron Soc, 73-77; Guggenheim Found fel, 77-78; MacArthur Found fel, 81-86; William C Foster fel, US Arms Control & Disarmament Agency, 84-85; pres-elect, Seismol, Am Geol Union. *Honors & Awards:* James B Macelwane Award, Am Geophys Union, 77. *Mem:* Seismol Soc Am; fel Am Geophys Union; Royal Astron Soc; Arms Control Asn. *Res:* Co-author of 2 volumes text in quantitative seismology; interpretation of seismic signals to infer properties of the earth, and of earthquake and explosion sources; analysis of seismic data on nuclear explosions, and the relationship to nuclear test ban treaty verification. *Mailing Add:* Lamont-Doherty Geol Observ Palisades NY 10964

RICHARDS, PAUL LINFORD, b Ithaca, NY, June 4, 34; m 65; c 2. PHYSICS. *Educ:* Harvard Univ, AB, 56; Univ Calif, PhD(physics), 60. *Prof Exp:* NSF fel, Royal Soc Mond Lab, Cambridge Univ, 59-60; mem tech staff, Bell Labs, Inc, 60-66; PROF PHYSICS, UNIV CALIF, BERKELEY, 66- *Concurrent Pos:* Miller Inst fac fel, 70-71, 87-88; Guggenheim fel, Cambridge Univ, 73-74; Alexander von Humboldt sr scientist, Max Planck Inst, Stuttgart, 82; vis prof, Ecole Normale Superieure, Paris, 84. *Mem:* Nat Acad Sci; Am Phys Soc; Am Acad Arts & Sci. *Res:* Astrophysics; low temperature solid state physics; far infrared; superconductivity; magnetic resonance. *Mailing Add:* Dept Physics Univ Calif Berkeley CA 94720

RICHARDS, PETER MICHAEL, b San Jose, Calif, Dec 20, 34; m 59; c 3. HYDROGEN IN SOLIDS, KINETICS OF REACTIONS. *Educ:* Mass Inst Technol, BS, 57; Stanford Univ, PhD(physics), 62. *Prof Exp:* Imp Chem Industs res fel, Oxford Univ, 62-64; from asst prof to prof physics, Univ Kans, 64-71; mem staff, 71-74, div supvr, 74-80, MEM STAFF, SANDIA NAT LABS, 80- *Concurrent Pos:* Consult, US Army Missile Command, Ala, 70-72. *Mem:* Fel Am Phys Soc. *Res:* Magnetic properties of matter; ionic conductivity; metal hydrides. *Mailing Add:* Sandia Nat Labs Org 1112 Albuquerque NM 87185

RICHARDS, R RONALD, b Wenatchee, Wash, Nov 22, 37; m 61; c 3. PHYSICAL CHEMISTRY. *Educ:* Seattle Pac Col, BS, 59; Univ Wash, PhD(phys chem), 64. *Prof Exp:* Assoc prof, 64-77, PROF CHEM, GREENVILLE COL, 77- *Concurrent Pos:* Fel, NASA Langley Res Ctr, Old Dom Univ, 71-72; res fel, Argonne Nat Lab, 78-79. *Mem:* Am Chem Soc. *Res:* Thermodynamics; kinetics; analysis. *Mailing Add:* 619 E College Ave Greenville IL 62246

RICHARDS, RICHARD DAVISON, b Grand Haven, Mich, Mar 10, 27; m 50; c 3. OPHTHALMOLOGY. *Educ:* Univ Mich, AB, 48, MD, 51; Univ Iowa, MSc, 57; Am Bd Ophthal, dipl, 58. *Prof Exp:* Asst prof ophthal, Col Med, Univ Iowa, 58-60; PROF OPHTHAL & HEAD DEPT, SCH MED, UNIV MD, BALTIMORE CITY, 60- *Concurrent Pos:* Attend physician, Vet Admin Hosp, Iowa City, 58-60. *Mem:* AAAS; Asn Res Vision & Ophthal; Am Acad Ophthal & Otolaryngol; fel Am Col Surgeons; Am Ophthal Soc. *Res:* Radiation cataracts. *Mailing Add:* Dept Ophthal Univ Md Sch Med Baltimore MD 21201

RICHARDS, RICHARD EARL, exotic wildlife ecology, zoo biology; deceased, see previous edition for last biography

RICHARDS, ROBERTA LYNNE, b Salt Lake City, Utah, Apr 20, 45; m 76; c 2. BIOCHEMISTRY, IMMUNOLOGY. *Educ:* Bucknell Univ, BS, 67; Purdue Univ, PhD(biochem), 74. *Prof Exp:* Res chemist biochem & immunol lipids, Dept Immunol, 74-78, RES CHEMIST BIOCHEM & IMMUNOL LIPIDS, DEPT MEMBRANE BIOCHEM, WALTER REED ARMY INST RES, 78- *Mem:* AAAS; Am Oil Chemists Soc; Am Chem Soc; Am Asn Immunol. *Res:* Immunology of membrane lipids and liposomal model membranes; receptor functions of lipids; efficacy of liposomes as antigen carriers and adjuvants in vaccines. *Mailing Add:* Dept Membrane Biochem Walter Reed Army Inst Res Washington DC 20307-5100

RICHARDS, ROGER THOMAS, b Akron, Ohio, June 19, 42; m 86. ACOUSTICS, PHYSICS. *Educ:* Westminster Col, Pa, BS, 64; Ohio Univ, MS, 68; Pa State Univ, PhD(acoust), 80. *Prof Exp:* Assoc engr, Transducer Lab, Gen Dynamics & Electronics, 68-69, engr, Acoust Dept, 69-71; NASA trainee, Pa State Univ, 71-74; staff assoc acoust, Appl Res Lab, 76-80; sr scientist, Bolt, Beranek & Newman, 84-87; PHYSICIST, NAVAL UNDERWATER SYST CTR, 87- *Concurrent Pos:* Consult electro-acoust res, State College, Pa, 73-75; grad asst, Pa State Univ, 74-80. *Mem:* AAAS; Acoust Soc Am; Am Inst Aeronaut & Astronaut; Nat Speleol Soc; Am Cryptographic Asn; NY Acad Sci. *Res:* Acoustic propagation and scattering; design of sonar transducers and arrays; sociological and psychological effects of noise pollution. *Mailing Add:* Stonington Landing 34 Cove Side Lane Stonington CT 06378-9741

RICHARDS, THOMAS L, b Santa Monica, Calif, Feb 2, 42; m 65; c 4. MARINE BIOLOGY, ZOOLOGY. *Educ:* Calif State Univ, Long Beach, BS, 64, MA, 66; Univ Maine, PhD(zool), 69. *Prof Exp:* PROF BIOL, CALIF POLYTECH STATE UNIV, SAN LUIS OBISPO, 69- *Concurrent Pos:* Res assoc, Univ Malaysia, 76-77; res affil, Univ Hawaii Inst Marine Biol, 83-84; actg dir, Tropical Agr Progs, Bigham Young Univ, Hawaii, 84-85; sabbatical, Darling Marine Sci Ctr, Univ Maine, 90. *Mem:* Marine Biol Asn UK; Sigma Xi; Am Soc Zool; Western Soc Naturalists. *Res:* Physiological ecology of intertidal invertebrates, adults and larvae, oyster and abalone mariculture; developmental biology of invertebrate meroplankton; fresh water prawn aquaculture; marine environmental education. *Mailing Add:* Dept Biol Sci Calif Polytech State Univ San Luis Obispo CA 93407

RICHARDS, VICTOR, b Ft Worth, Tex, June 4, 18; m 41; c 4. SURGERY. *Educ:* Stanford Univ, AB, 35, MD, 39; Am Bd Surg, dipl, 45; Bd Thoracic Surg, dipl, 45. *Prof Exp:* From instr to prof surg, Sch Med, Stanford Univ, 42-59, chmn dept, 55-58, clin prof, 59-91; chief surg, Children's Hosp, 59-90; clin prof surg, Univ Calif, San Francisco, 65-91; RETIRED. *Concurrent Pos:* Commonwealth res fel, Harvard Univ, 50-51; mem spec comt, USPHS, 58-62; ed, Oncol, 68-76; mem surg study sect B, NIH; consult, USPHS, Letterman Gen, Oak Knoll Naval & Travis AFB Hosps. *Honors & Awards:* Gold-Headed Cane Lectr, Univ Calif Sch Med, 88. *Mem:* Soc Exp Biol & Med; Sigma Xi; Pan-Pac Surg Asn (vpres, 72-); Am Cancer Soc; Soc Surg Alimentary Tract (vpres, 72-73); Am Surg Soc; Am Thoracic Soc. *Res:* Cancer; cardiovascular surgery; transplantation and preservation of tissues and cells. *Mailing Add:* Children's Hosp 3838 California St San Francisco CA 94118

RICHARDS, WALTER BRUCE, b Cortland, Ohio, Jan 29, 41; m 62; c 2. PHYSICS. *Educ:* Oberlin Col, AB, 61; Univ Calif, Berkeley, PhD(physics), 66. *Prof Exp:* Physicist, Lawrence Radiation Lab, 65-66; res assoc & lectr physics, Tufts Univ, 66-67; from asst prof to assoc prof, 67-82, chmn physics dept, 86-90, PROF PHYSICS, OBERLIN COL, 82- *Concurrent Pos:* Vis assoc prof physics, Case Western Reserve Univ, 81-82; NASA summer fac fel, Lewis Res Ctr, 83, 84. *Mem:* Acoust Soc Am; Am Asn Physics Teachers. *Res:* Musical acoustics. *Mailing Add:* Dept Physics Oberlin Col Oberlin OH 44074-1088

RICHARDS, WILLIAM JOSEPH, b Scranton, Pa, Apr 7, 36; m 58; c 3. ICHTHYOLOGY. *Educ:* Wesleyan Univ, BA, 58; State Univ NY, MS, 60; Cornell Univ, PhD(vert zool), 63. *Prof Exp:* Fishery biologist, Biol Lab, US Bur Com Fisheries, DC, 63-65, res syst zoologist, 65, Trop Atlantic Biol Lab, 65-71; zoologist & prog mgr, Nat Marine Fisheries Serv, 71-77, dir, Miami Lab, 77-83; SR SCIENTIST, SOUTHEAST FISHERIES CTR, 83- *Concurrent Pos:* Mem working group, Food & Agr Orgn UN, 65-; adj asst prof, Inst Marine Sci, Univ Miami, 66-68, adj assoc prof, Rosenstiel Sch Marine & Atmospheric Sci, 68-74, adj prof, 74-; ed, Bull Marine Sci, 74-; sci ed, Nat Marine Fisheries Serv, 83-86. *Mem:* Am Soc Ichthyol & Herpet; Western Soc Naturalists; fel Am Inst Fishery Res Biologists; Sigma Xi. *Res:* Systematics of fishes, especially the study of larval pelagic fishes and the family Triglidae; recruitment mechanisms of fishes; larval fish ecology especially relating to physical oceanography and ocean climate. *Mailing Add:* Southeast Fisheries Ctr 75 Virginia Beach Dr Miami FL 33149

RICHARDS, WILLIAM REESE, b Springfield, Mo, June 27, 38; m 66; c 1. BIOCHEMISTRY, BIO-ORGANIC CHEMISTRY. *Educ:* Univ Calif, Riverside, AB, 61; Univ Calif, Berkeley, PhD(org chem), 66. *Prof Exp:* NIH fel chem path, St Mary's Hosp Med Sch, Eng, 66-67; fel bact, Univ Calif, Los Angeles, 67-68; asst prof, 68-83, ASSOC PROF CHEM, SIMON FRASER UNIV, 83- *Concurrent Pos:* Sabbatical leave microbiol, Univ Freiburg, WGermany, 77-78; res scientist biochem, Univ Bristol, UK, 83-84; sabbatical leave molecular biol, Univ Sheffield, UK, 90. *Mem:* Am Soc Photobiol. *Res:* Biosynthesis of chlorophylls: affinity chromatography, photoaffinity labeling, and kinetic mechanisms of enzymes of chlorophyll synthesis; labeling of membrane proteins; regulation of bacteriochlorophyll synthesis. *Mailing Add:* Dept Chem Simon Fraser Univ Burnaby BC V5A 1S6 Can

RICHARDS, WINSTON ASHTON, b Trinidad, WI; m 64; c 8. MATHEMATICAL STATISTICS. *Educ:* Marquette Univ, BS, 59, MS, 61; Univ Western Ont, MA, 66, PhD(math), 71. *Prof Exp:* Instr math, Aquinas Col, 60-61; chmn dept high sch, Mich, 64-65; ASSOC PROF MATH & STATIST, PA STATE UNIV, CAPITOL COL, 69- *Concurrent Pos:* Pa State Univ res grant, 72-73; statist expert, Orgn Am States, 75-, Pa Dept Agr, 89, 90 & 91; consult training govt statisticians, Repub Trinidad, Tobago & eastern Caribbean, 76-81; consult, UN, 79; vis sr lectr, Univ West Indies, Trinidad, 80-81 & 81-82. *Mem:* Int Asn Surv Statisticians; Inst Math Statist; Math Asn Am; Can Math Cong; Am Statist Asn. *Res:* Exact distribution theory; n-dimensional geometry; mathematical modeling; national income; applied statistics. *Mailing Add:* Dept Math Pa State Univ Capitol Col Middletown PA 17057

RICHARDSON, ALBERT EDWARD, b Lovelock, Nev, Feb 4, 29; m 59; c 7. NUCLEAR FISSION, GAMMA SPECTROMETRY. *Educ:* Univ Nev, BS, 50; Iowa State Univ, PhD(phys chem), 56. *Prof Exp:* Asst radiochem, Ames Lab, Iowa State Univ, 50-55; asst prof, 55-60, ASSOC PROF PHYS CHEM, NMEX STATE UNIV, 60- *Concurrent Pos:* Vis prof, Adams State Col, 63; consult, White Sands Missile Range, 65-71, contractor, 73-74, chemist, 81-88, res chemist, 88-; US AEC fel, Univ Colo, 68-89; vis staff mem, Los Alamos Nat Lab, 75-80 & Sandia Lab, 83. *Mem:* Am Chem Soc. *Res:* Nuclear chemistry; activation analysis; hot atom chemistry; neutron radiography. *Mailing Add:* NMex State Univ Box 30001 Dept 3C Las Cruces NM 88003-0001

RICHARDSON, ALFRED, JR, b Jersey City, NJ, Feb 18, 32; m 56; c 4. ORGANIC CHEMISTRY. *Educ:* Rutgers Univ, BS, 53; Lehigh Univ, MS, 55, PhD(chem), 58. *Prof Exp:* Asst chem, Lehigh Univ, 53-58; proj leader med chem res, William S Merrell Co, 58-67, sect head org res, 67-71, mgr res info, Merrell-Nat Labs, 71-76, dir com develop, Richardson-Merrell Inc, 76-81, dir sci & com develop, 81-84, ASSOC DIR DRUG REG, MERRELL DOW PHARMACEUT INC, 84- *Concurrent Pos:* Lectr eve col, Univ Cincinnati, 66-70. *Mem:* AAAS; Am Chem Soc; Drug Info Asn; Am Inst Chemists; NY Acad Sci. *Res:* Medicinal chemistry; synthetic organic chemistry; interdisciplinary product development; licensing; health care products; regulatory affairs. *Mailing Add:* Merrell Dow Pharmaceut Inc 2110 E Galbraith Rd Cincinnati OH 45215

RICHARDSON, ALLAN CHARLES BARBOUR, b Toronto, Ont, July 14, 32; US citizen; m 56, 86; c 3. RADIATION HEALTH, NUCLEAR PHYSICS. *Educ:* Col William & Mary, BS, 54; Univ Md, MS, 58. *Prof Exp:* Physicist, Nat Bur Stand, 58-69; exec secy radiol health sci training comt, Environ Control Admin & Radiol Health Study Sect, US Environ Protection Agency, 69-72, spec asst sci coord & eval, Off Radiation Progs, 72-73, asst standard develop, 73-77, chief fed guide br, 77-80, chief gen radiation standards br, 80-82, chief, guides & criteria br, Off Radiation Progs, 82-90, ASST DIR, CRITERIA & STANDARDS DIV, OFF RADIATION PROGS, US ENVIRON PROTECTION AGENCY, 90- *Concurrent Pos:* Consult, radiation protection policy, Nuclear Energy Agency, Orgn Econ Coop & Develop, Paris, 79- & Int AEC, 82- *Honors & Awards:* Bronze Medals, US Environ Protection Agency, 73 & 83. *Mem:* Am Nuclear Soc; Health Physics Soc; Am Phys Soc. *Res:* Radiation protection standards; fast neutron cross-sections; neutron age measurements. *Mailing Add:* Off Radiation Progs ANR 460 Environ Protection Agency 401 M St SW Washington DC 20460

RICHARDSON, ALLYN (ST CLAIR), b Edmonton, Alta, Nov 16, 18; nat US; wid; c 6. ENGINEERING. *Educ:* Univ BC, BASc, 41; Harvard Univ, SM, 49. *Prof Exp:* Asst chemist analytical & process control, BC Pulp & Paper Co, Can, 41-42; from asst engr to sr engr, Can Dept Nat Health & Welfare, 42-46, dist engr pub health, 46-50; res engr hydraul, Harvard Univ, 50-53, res engr bact aerosol viability study, 53-54, res engr soil stabilization res, 54-55; sr engr radar dept, Raytheon Co, 55-58; asst prof civil eng & dir fluid network lab, Tufts Univ, 58-59; proj dir instrumentation res & develop, United Res, Inc, 59-61 & Trans-Sonics, Inc, 61-65; res proj engr, WHO, 65-67; dir off res progs, Region I, Environ Protection Agency, 67-84; COMPUTILITY PERSONAL COMPUT & SMALL SYSTS SERV. *Concurrent Pos:* Consult & vis prof sanit sci, Cent Univ Venezuela; staff adv res & educ projs, Am Region Hq, Wash, DC; consult, Univ Tehran; chmn, Bd Water Commn, 82- *Mem:* Am Water Works Asn; Inst Elec & Electronics Engrs-Computer Asn. *Res:* Methods for control and improvement of air, water and land environment; environmental needs and standards, especially water hygiene, water pollution control and solid wastes management; application of microcomputers in management of environmental control processes; application of microcomputers in small water supply systems design, expansion and management. *Mailing Add:* Computility PO Box 257 West Groton MA 01472-0257

RICHARDSON, ARLAN GILBERT, b Beatrice, Nebr, Jan 23, 42; m 66. BIOCHEMISTRY, ORGANIC CHEMISTRY. *Educ:* Peru State Col, BA, 63; Okla State Univ, PhD(biochem), 68. *Prof Exp:* Teaching asst biochem, Okla State Univ, 67-68; asst prof chem, Fort Lewis Col, 68-69; NIH fel biochem, Univ Minn, St Paul, 69-71; asst prof chem, 71-75, ASSOC PROF CHEM & BIOL SCI, ILL STATE UNIV, 71- *Mem:* AAAS; Am Chem Soc; Am Soc Microbiol; Sigma Xi. *Res:* Bacterial transformation and genetics; protein synthesis, especially the effect of various diets upon polysome profiles and protein synthesis in rat liver. *Mailing Add:* Dept Chem Ill State Univ Normal IL 61761

RICHARDSON, ARTHUR JEROLD, b Aransas Pass, Tex, Mar 16, 38; m 65; c 1. AGRICULTURAL REMOTE SENSING, GEOGRAPHIC INFORMATION SYSTEMS. *Educ:* Texas Agr & Indust Col, BA, 65. *Prof Exp:* Mem staff, phys sci tech, 67-68, PHYSICIST, USDA, 68- *Mem:* Am Soc Photogram. *Res:* Agricultural remote sensing; published studies using spectral radiometric measurements of crop and soil conditions in the field, from aircraft, and satellite multispectral sensors. *Mailing Add:* USDA-ARS-ASRU 2413 E Hwy 83 Weslaco TX 78596

RICHARDSON, BILLY, b Channelview, Tex, June 20, 36; m 80. RESEARCH ADMINISTRATION. *Educ:* Tex A&M Univ, BS, 58, MS, 64, PhD(biol), 67. *Prof Exp:* Res technician, Tex A&M Univ, 58-59, instr floricult, 59-62; chief cellular physiol br, Environ Sci Div, USAF Sch Aerospace Med, 67-72, dep chief environ sci div, 72-73, chief crew environ br, 73-76, dep chief crew technol div, 76-78, chief biomet div, 78-79, dir spec proj, 79-80, dir chem defense prog off, 80-81, chief, Crew Technol Div, 81-82; tech dir, 82-87, prog exec, 87-89, DEP ASST TO SECY DEFENSE, CHEM MATTERS, CHEM CTR, PENTAGON, 89- *Concurrent Pos:* Consult, new fighter aircraft, Can Forces, 77-78. *Mem:* Fel Am Inst Chemists; assoc fel Aerospace Med Asn; Am Inst Biol Sci; Sigma Xi; Am Defense Preparedness Asn. *Res:* Steroid metabolism in plants; mechanism of oxygen toxicity; aerospace physiology; cellular and biochemical effects of environmental stresses; aircrew protection and life support systems; chemical warfare defense. *Mailing Add:* 4611 Kimby Lane Aberdeen MD 21001

RICHARDSON, BOBBIE L, b Detroit, Mich, May 24, 26; m 49; c 5. ENGINEERING. *Educ:* Purdue Univ, BS, 53, MS, 54, PhD(eng sci), 59. *Prof Exp:* Instr eng mech, Purdue Univ, 54-57; mech engr reactor eng div, Argonne Nat Lab, 57-58; from asst prof to assoc prof mech eng, 58-67, chmn dept, 59-67, PROF ELEC ENG, MARQUETTE UNIV, 67- *Mem:* Am Soc Mech Engrs; Am Nuclear Soc; Am Soc Eng Educ. *Res:* Nuclear engineering; digital systems engineering; two-phase flow; heat transfer. *Mailing Add:* Dept Elec Eng Marquette Univ Milwaukee WI 53233

RICHARDSON, CHARLES BONNER, b Dallas, Tex, Jan 18, 30; , 59; c 1. PHYSICS. *Educ:* Univ Pittsburgh, BS, 57, PhD(physics), 62. *Prof Exp:* Res asst prof physics, Univ Wash, 62-66; from asst prof to assoc prof, 66-74, PROF PHYSICS, UNIV ARK, FAYETTEVILLE, 74- *Concurrent Pos:* Vis scientist, Brookhaven Nat Labs, 83, Naval Res Labs, 87. *Mem:* Am Phys Soc. *Res:* Optics; atmospheric physics. *Mailing Add:* Dept Physics Univ Ark Fayetteville AR 72701

RICHARDSON, CHARLES CLIFTON, b Wilson, NC, May 7, 35; m 61; c 2. BIOCHEMISTRY. *Educ:* Duke Univ, BS, 59, MD, 60. *Hon Degrees:* AM, Harvard Univ, 67. *Prof Exp:* Intern med, Duke Univ, 60-61; res fel, dept biochem, Sch Med, Stanford Univ, 61-63; from asst prof to assoc prof, Harvard Med Sch, 64-69, chmn dept, 78-87, PROF BIOCHEM, HARVARD MED SCH, 69-, EDWARD S WOOD PROF BIOCHEM, 79- *Concurrent Pos:* Career Develop Award, NIH, 67-76; consult, physiol chem study sect, NIH, 70-74; assoc ed, Annual Rev Biochem, 72-83, ed, 83; mem, Nat Bd Med Examrs, 73-76, adv comt, nucleic acids & protein systhesis, Am Cancer Soc, 75-78, adv div, Max-Planck-Inst Moleculare Gentik, Berlin, 80-89, vis comt, Boston Biomed Res Found, 85- coun res & clin invest, Am Cancer Soc, 89-92; assoc, Helicon Found, San Diego, Ca, 83; bd dir, US Biochem Corp, Cleveland, 83-; mem sci adv comt, Genetics Inst, Cambridge, Mass, 86-; mem ed bd, J Biol Chem, 68-73, 84-88. *Honors & Awards:* Merit Award, NIH, 86. *Mem:* Nat Acad Sci-Inst Med; Am Chem Soc; Am Soc Biol Chemists; fel Am Acad Arts & Sci; Asn Med Sch Dept Biochem; Am Cancer Soc. *Res:* DNA metabolism. *Mailing Add:* Dept Biol Chem & Molecular Pharmacol Harvard Med Sch 25 Shattuck St Boston MA 02115

RICHARDSON, CLARENCE ROBERT, b Lovelock, Nev, Jan 10, 31; m 55; c 2. PHYSICS. *Educ:* Univ Nev, BS, 57; Johns Hopkins Univ, PhD(physics), 63. *Prof Exp:* Jr physicist, Naval Ord Test Sta, Calif, 57; physicist, Appl Physics Lab, Johns Hopkins Univ, 58-59; from asst physicist to assoc physicist, Brookhaven Nat Lab, 63-67; physicist high energy physics prog, Div Res, US AEC, 67-73, physicist nuclear sci prog, 73-75, dept mgr solar inst proj off, US Energy Res & Develop Admin, 75-76, dir prog planning div, 79, phys sci planning specialist, prog planning div, basic energy sci, 76-79, prog mgr medium energy nuclear physics, 79-90, DEP DIR NUCLEAR PHYSICS DIV, DEPT ENERGY, 90- *Concurrent Pos:* Vis physicist, Europ Orgn Nuclear Res, Switz, 70-71; co-chmn, Bubble Chamber Working Group, 73-74 & 78-79. *Mem:* Am Phys Soc. *Res:* Elementary particle research using bubble chambers; eta meson and omega hyperon. *Mailing Add:* 20513 Topridge Dr Boyds MD 20841

RICHARDSON, CLARENCE WADE, b Temple, Tex, Nov 15, 42; m 64; c 3. HYDROLOGY, CIVIL ENGINEERING. *Educ:* Tex A&M Univ, BS, 64, MS, 66; Colo State Univ, PhD(civil eng), 76. *Prof Exp:* Agr engr, Tex Agr Exp Sta, 64-65 & Agr Res Serv, 66-77, agr engr, Sci & Educ Admin-Fed Res, 78-80, res leader, 80-86, LAB DIR, AGR RES SERV, USDA, 86- *Honors & Awards:* Cert Merit, USDA, 81. *Mem:* Am Soc Agr Engrs; Soil Conserv Soc Am. *Res:* Deterministic hydrologic modeling; stochastic simulation of precipitation patterns; agricultural water quality. *Mailing Add:* Grassland Soil & Water Res Lab 808 E Blackland Rd Temple TX 76502

RICHARDSON, CURTIS JOHN, b Gouverneur, NY, July 27, 44; m 72; c 2. WETLAND ECOLOGY, PLANT PHYSIOLOGY. *Educ:* State Univ NY, Cortland, BS, 66; Univ Tenn, PhD(ecol), 72. *Prof Exp:* Asst prof resource ecol, Sch Natural Resources, Univ Mich, 72-77, asst prof plant ecol, Biol Sta, 73; assoc prof, 77-87, PROF RESOURCE ECOL, SCH FORESTRY & ENVIRON STUDIES & DEP DIR, ECOTOXICOL PROG, 80-, DIR, WETLAND CTR, 89- *Concurrent Pos:* Ecologist, AEC, 68; res fel ecol, Ecol Sci Div, Oak Ridge Nat Lab, 70-72; nat rep, Inst Ecol, 79- *Mem:* AAAS; Am Inst Biol Sci; Ecol Soc Am; Sigma Xi; Soil Sci Soc Am; Soc Wetland Scientists (pres, 87-88). *Res:* Ecosystem analysis of wetland and forest systems; linkages between terrestrial and aquatic ecosystems with an emphasis on phosphorus chemistry and biogeochemical cycles as influenced by man; plant stress physiology biomarker, and plant ecotoxicology. *Mailing Add:* Sch Environ Duke Univ Durham NC 27706

RICHARDSON, DANIEL RAY, b Martinsville, Ind, May 5, 39; m 59; c 2. PHYSIOLOGY. *Educ:* Ind Univ, Bloomington, BA, 65, Ind Univ, Indianapolis, PhD(physiol), 69. *Prof Exp:* Fel, Univ Calif, San Diego, 69-70; asst prof, 70-74, ASSOC PROF PHYSIOL & BIOPHYS, SCH MED, UNIV KY, 74- *Mem:* Microcirc Soc; NY Acad Sci; Soc Exp Biol & Med; Am Heart Asn; Am Physiol Soc. *Res:* Studies of peripheral vascular dynamics in man and laboratory animal models. *Mailing Add:* Dept Physiol & Biophys Univ Ky Med Sch Lexington KY 40536

RICHARDSON, DAVID LOUIS, b New York, NY, Sept 27, 48; c 2. ORBITAL MECHANICS, PLANETARY MOTION. *Educ:* Ind Univ, AB, 70; Cornell Univ, MS, 72, PhD(space mech), 77. *Prof Exp:* Struct analyst, US Naval Surface Weapons Lab, 73-74; orbital analyst, Comput Sci Corp, 74-77; from asst prof to assoc prof, 77-88, PROF DYNAMICS & ORBITAL MECH, UNIV CINCINNATI, 89- *Concurrent Pos:* Prin investr, NSF, 78-80, 84-88; res worker, Nat Bur Standards, 80- *Mem:* Am Astron Soc; Int Astron Union; Am Astronaut Soc. *Res:* Planetary motion analysis; orbital dynamics of artificial satellites; application of analytical and semi-analytical methods to the problems of space mechanics; dynamical systems. *Mailing Add:* Dept Aerospace Eng ML No 70 Univ Cincinnati Cincinnati OH 45221

RICHARDSON, DAVID W, b Nanking, China, Mar 22, 25; US citizen; m 48; c 4. CARDIOLOGY. *Educ:* Davidson Col, BS, 47; Harvard Med Sch, MD, 51. *Prof Exp:* Intern, Yale-New Haven Med Ctr, 51-52, asst resident, 52-53; from asst resident to resident, Med Col Va, 53-55, NIH fel cardiovasc physiol, 55-56; chief cardiovasc sect & assoc chief staff res, Vet Admin Hosp, 56-62; vis fel cardiovasc dis, Oxford Univ, 62-63; assoc prof med, 63-67, actg chmn dept, 73-74, chmn div cardiol, 72-86, PROF MED, MED COL VA, 67- *Concurrent Pos:* Fel coun clin cardiol, Am Heart Asn; consult, Vet Admin Coop Study Antihypertensive Agents, 62-82; Va Heart Asn Chair cardiovasc res, Med Col Va, 62-71; vis prof, Inst Cardiovasc Res, Univ Milan, 71-72. *Mem:* Am Fedn Clin Res; Am Clin & Climat Asn; Am Soc Clin Invest; fel Am Col Physicians; fel Am Col Cardiol. *Res:* Clinical hypertension and cardiology; prevention of myocardial infarction; neural control of circulation; cardiac arrhythmias. *Mailing Add:* Med Col Va Box 128 Richmond VA 23298-0128

RICHARDSON, DON ORLAND, b Auglaize Co, Ohio, May 12, 34; c 4. ANIMAL SCIENCE, DAIRY SCIENCE. *Educ:* Ohio State Univ, BS, 56, MS, 57 & PhD(animal breeding), 61. *Prof Exp:* Asst dairy breeding, Ohio State Univ, 56-58, instr, 60; dairy husbandman, Agr Res Serv, USDA, 58-61, dairy genetist, 61-63; asst prof dairy sci, 63-67, assoc prof dairying, 67-72, assoc prof, 72-75, PROF ANIMAL SCI, UNIV TENN, KNOXVILLE, 75- *Mem:* Am Dairy Sci Asn. *Res:* Evaluation of progress resulting from various selection schemes utilized with daity cattle, including an evaluation of correlated responses to single trait selection on milk yield. *Mailing Add:* Dean Agr Exp Sta Univ Tenn Knoxville TN 37901-1071

RICHARDSON, DONALD EDWARD, b Vicksburg, Miss, Oct 5, 31; div; c 5. NEUROSURGERY. *Educ:* Millsaps Col, BS, 53; Tulane Univ, MD, 57. *Prof Exp:* Assoc prof, 64-74, PROF & CHMN DEPT NEUROSURG, SCH MED, TULANE UNIV, 80-; DIR, PAIN TREAT CTR, HOTEL DIEU HOSPITAL, NEW ORLEANS, LA, 78- *Concurrent Pos:* Assoc clin prof neurosurg, La State Univ Med Ctr, New Orleans, 74-80. *Mem:* AAAS; AMA; Am Asn Neurol Surg; Am Col Surgeons; Int Soc Res Stereonencephalotomy. *Res:* Clinical and research neurosurgery; basic neurophysical research; electrophysiology of the sensory system of spinal cord and brain. *Mailing Add:* Sch Med Dept Neurosurg Tulane Univ 1430 Tulane Ave New Orleans LA 70112

RICHARDSON, EDWARD HENDERSON, JR, b Baltimore, Md, Dec 24, 11; m 48; c 3. OBSTETRICS & GYNECOLOGY. *Educ:* Princeton Univ, AB, 34; Johns Hopkins Univ, MD, 38; Am Bd Obstet & Gynec, dipl, 47. *Prof Exp:* From instr to asst prof, Sch Med, Johns Hopkins Univ, 43-59, emer prof gynec, 82; RETIRED. *Concurrent Pos:* Consult, Vet Admin Hosp, Baltimore, Md. *Mem:* Fel, Am Col Obstet & Gynec. *Res:* Female urology. *Mailing Add:* 304 Northwind Rd Baltimore MD 21204

RICHARDSON, ELISHA ROSCOE, b Monroe, La, Aug 15, 31; m 67; c 3. ORTHODONTICS, ANATOMY. *Educ:* Southern Univ, BS, 51; Meharry Med Col, DDS, 55; Univ Ill, MS, 63; Univ Mich, PhD(Human growth & develop), 88. *Prof Exp:* NIH fel, Univ Ill, 60-62; prof & chmn, dept orthod, Univ Colo, 85-88; from asst prof dent radiol to assoc prof orthod, 62-76, prof & dir postgrad educ, 67-85, assoc dean, 77-85, PROF & DEAN, SCH DENT, MEHARRY MED COL, 88- *Concurrent Pos:* Prin investr, Nat Inst Child Health & Human Develop, res grant, 65-68; guest lectr, John F Kennedy Ctr Res Educ & Human Develop, 66-68 & George Peabody Col, 66-68; prin investr, Nat Inst Dent Res grant, 68; consult, Vet Admin Hosp, Nashville, Tenn, 68; pres, Craniofacial Biol Group, Int Asn Dent Res, 78-79. *Mem:* Am Dent Asn; Am Asn Orthod; fel Am Col Dentists; Int Asn Dent Res; NY Acad Sci. *Res:* Craniofacial region; maxillary growth; periodontal membrane; uvula and tongue; tooth size and eruption, growth of face and jaws. *Mailing Add:* Sch Dent Meharry Med Col l005 DB Todd Blvd Nashville TN 37208

RICHARDSON, ERIC HARVEY, b Portland, Ore, Aug 14, 27; Can citizen; m 78. APPLIED OPTICS, ASTRONOMY. *Educ:* Univ BC, BA, 49, MA, 51; Univ Toronto, PhD(molecular spectros), 59. *Prof Exp:* Res asst, Gen Elec Co, Stanmore Labs, London, 52-53; res asst, Dominion Astrophys Observ, 54-57, res officer, 59-91; CONSULT, EHR OPTICAL SYSTS, 91- *Concurrent Pos:* Optical consult, NASA Lunar Laser Ranging Exp, McDonald Observ, 69, Can-France-Hawaii Telescope Corp, 73-78, Viking Satellite, 81-85 & NASA Astrometric Space Telescope proposal, 85; mem, high resolution camera instrument definition team, large space telescope, NASA, 73-75; co-investr, Space Shuttle, 85- *Honors & Awards:* N Copernicus Medal, Poland, 73. *Mem:* Int Astron Union; Soc Photo-Optical Instrumentation Engrs. *Res:* Optical design of space and ground based telescopes and associated instruments. *Mailing Add:* 1871 Elmhurst Pl Victoria BC V8N 1R1 Can

RICHARDSON, EVERETT V, b Scottsbluff, Nebr, Jan 5, 24; m 48; c 3. HYDRAULICS, HYDROLOGY. *Educ:* Colo State Univ, BS, 49, MS, 60, PhD(civil eng), 65. *Prof Exp:* Hydraul engr, Wyo Qual Water Br, US Geol Surv, 49-53 & Iowa Surface Water Br, 53-56, res hydraul engr, Water Resources Div, 56-68; assoc prof, Colo State Univ, 65-68, adminr, Eng Res Ctr, 68-83, dir, Hydraul Lab, 83-88, prof civil eng, 68-88, prof-in-chg, Hydraul Prog, 83-88, TRANS PROF CIVIL ENG, COLO STATE UNIV, 88-; SR ASSOC, RESOURCE CONSULTS INC, FT COLLINS, 88- *Concurrent Pos:* Consult, US Bur Pub Rd, 65-, US CEngr, 67-, World Bank Reconstruct & Develop, 70-, Colo Hwy Dept, 74-, US Bur Reclamation, 74- & USAID, 76-, US Transp Safety Bd, 87-88; dir, USAID Prog, Colo State Univ, 74-76, proj dir, Egypt Water Use & Mgt, 77-85 & Eygpt Irrigation Improv & Res Proj, 85-; mem joint workshop on res mgt, Nat Acad Sci/Egypt Nat Acad Sci, 75; mem, Int Comn Irrigation & Drainage, AAAS, NY State Bridge Safety Assurance Task Force, 88-90; chmn Task Force Bridges Scout, Am Soc Civil Engrs, 90- *Honors & Awards:* J C Stevens Award, Am Soc Civil Engrs, 61. *Mem:* Am Soc Civil Engrs; Sigma Xi. *Res:* Internal structure of turbulent shear flow; diffusion of waste in natural streams; measurement of fluid flow; erosion and sedimentation, river mechanics, irrigation and water management. *Mailing Add:* Eng Res Ctr Colo State Univ Ft Collins CO 80523

RICHARDSON, F C, b Whitehaven, Tenn, Sept 22, 36; m 60; c 2. BOTANY. *Educ:* Rust Col, BA, 60; Atlanta Univ, MSc, 64; Univ Calif, Santa Barbara, PhD(bot), 67. *Prof Exp:* Asst prof bot, 67-71, assoc prof bot & chmn dept, 71-72, prof bot, 82, dean Arts & Sci, Ind Univ Northwest, 72-84; vpres acad affairs, Jackson State Univ, 84-85; vpres acad affairs, Moorhead State Univ, 85-89; PRES, BUFFALO STATE COL, 89- *Mem:* Am Inst Biol Sci; Bot Soc Am; Int Soc Plant Morphol; Am Asn Higher Educ; Am Assoc State Cols & Univ. *Res:* Plant morphology; origin and evolution of the angiosperms using the anatomy and development of the flower, particularly the carpel, as the primary tool; nodal anatomy of elm species in connection with Dutch elm succeptibility in the family. *Mailing Add:* Buffalo State Col 1300 Elmwood Ave Buffalo NY 14222

RICHARDSON, FRANCES MARIAN, b Roanoke, Va, May 6, 22. CHEMICAL & BIOMEDICAL ENGINEERING. *Educ:* Roanoke Col, BS, 43; Univ Cincinnati, MS, 47. *Prof Exp:* Chemist, E I du Pont de Nemours & Co, 43-45; asst chem, Univ Cincinnati, 45-47; res chemist, Leas & McVitty, Inc, 48-49; res assoc, 51-60, assoc dir, eng oper prog, 80-85, RES ASSOC PROF ENG RES, NC STATE UNIV, 60-, DIR, EXTRADEPT DEGREE PROGS, COL ENG, 85- *Concurrent Pos:* Vis assoc prof, Case Western Reserve Univ, 67-68. *Mem:* AAAS; Royal Soc Health; Soc Women Engrs; Am Inst Chem Engrs; Am Soc Eng Educ; Sigma Xi. *Res:* Infrared imaging thermography; biomedical engineering; fluid flow; flow visualization. *Mailing Add:* NC State Univ 184 Weaver Labs Raleigh NC 27695-7625

RICHARDSON, FREDERICK S, b Carlisle, Pa, June 8, 39; m 59; c 4. THEORETICAL CHEMISTRY, PHYSICAL CHEMISTRY. *Educ:* Dickinson Col, BS, 61; Princeton Univ, PhD(chem), 66. *Prof Exp:* Instr chem, Princeton Univ, 65-66; officer, US Army CEngr, 66-68; fel, Univ Calif, San Diego, 68-69; from asst prof to assoc prof chem, 69-78, chmn dept, 84-87, PROF CHEM, UNIV VA, 78- *Concurrent Pos:* Dreyfus Found scholar, 72-78. *Mem:* AAAS; Am Chem Soc; Am Phys Soc. *Res:* Theoretical and experimental aspects of molecular electronic spectroscopy; optical properties of lanthanide ions and complexes; natural and magnetic optical activity in molecules and crystals; coupling of electronic states by molecular vibrations. *Mailing Add:* Dept Chem Univ Va Charlottesville VA 22901

RICHARDSON, GARY HAIGHT, b Grace, Idaho, Nov 30, 31; m 54; c 6. DAIRY CHEMISTRY, DAIRY MICROBIOLOGY. *Educ:* Utah State Agr Col, BS, 53; Univ Wis, PhD(dairy & food industs), 60. *Prof Exp:* Dairy chemist, Res Labs, Swift & Co, Ill, 59-61; res mgr, Dairyland Food Labs, Inc, Wis, 61-63, res dir, 63-67; prof dairy & food sci, 67-73, PROF NUTRIT & FOOD SCI, UTAH STATE UNIV, 73- *Concurrent Pos:* Dairyland Food Labs, Inc grant, 68-; USPHS grant, 69-72; Kellogg fel, Univ Col, Cork, Ireland, 81. *Mem:* Am Dairy Sci Asn; Inst Food Technol; Am Soc Microbiol; Inst Asn Milk, Food & Environ Sanit. *Res:* Dairy cultures; dehydration; accelerated cheese flavor development; enzyme utilization in production of dairy flavors; staphylococcal enterotoxin production in cheese products; prevention of defects in Swiss cheese; assay of milk constitutents. *Mailing Add:* Dept Nutrit & Food Sci Utah State Univ Logan UT 84322-8700

RICHARDSON, GEORGE S, b Boston, Mass, Dec 1, 21; m 58; c 3. ENDOCRINOLOGY, GYNECOLOGY. *Educ:* Harvard Univ, BA, 43, MD, 46. *Prof Exp:* From intern to resident surg, Mass Gen Hosp, 46-55; instr surg, Harvard Med Sch, 54-55; asst, Mass Gen Hosp, 55-59; clin assoc, 59-71, asst prof, 71-74, ASSOC PROF SURG, HARVARD MED SCH, 74- *Concurrent Pos:* Res fel physiol, Harvard Med Sch, 47-48; NIH res grants, 58-; assoc vis surgeon, Mass Gen Hosp, 64-75, vis surgeon, 76-, gynecologist, 77-; Nat Cancer Inst spec fel, Southwest Found Res & Educ, 69-71. *Mem:* Am Fertil Soc; Endocrine Soc; Soc Pelvic Surgeons; Am Col Surgeons. *Res:* Steroid hormones in relation to neoplasia; human endometrium. *Mailing Add:* Mass Gen Hosp Boston MA 02114

RICHARDSON, GERALD LAVERNE, b Ft Morgan, Colo, Sept 21, 28. CHEMICAL ENGINEERING. *Educ:* Univ Colo, BS, 50. *Prof Exp:* Jr engr, Hanford Atomic Prod Div, Gen Elec Co, 50-55, engr, 55-62, sr engr, 62-65; sr develop engr, Pac Northwest Lab, Battelle Mem Inst, 65-69, res assoc, 69-70; prin engr, Westinghouse Hanford Co, 70-79, fel engr, 79-84; RETIRED. *Res:* Separation and purification of radioactive isotopes from irradiated uranium; development of solvent extraction processes and equipment; nuclear fuel cycle waste management. *Mailing Add:* 1109 Pine St Richland WA 99352-2135

RICHARDSON, GRAHAM MCGAVOCK, b Emory, Va, Jan 10, 12; m 40; c 3. ORGANIC CHEMISTRY. *Educ:* Univ Tenn, BS, 34; Mass Inst Technol, PhD(org chem), 39. *Prof Exp:* Chemist, Acetate Yarn Div, Tenn Eastman Corp, 34-36; instr chem, Franklin Tech Inst, Boston, 38-39; chemist, E I Du Pont de Nemours & Co, Inc, 39-42 & 44-51, supvr, 42-44 & 51-62, specialist textile fibers, dyeing & finishing, 62-76; DIR RES & DEVELOP, LUTEX CHEM CORP, 76- *Mem:* Am Asn Textile Chem & Colorists; Am Asn Textile Technol; Am Soc Testing & Mat. *Res:* Process development of dyes; dyeing; organo sodium compounds; detergents; wetting agents; textile processing and finishing; cosmetics; vinyl polymers; flammability test methods for textile materials. *Mailing Add:* Box 3654 Greenville Wilmington DE 19807

RICHARDSON, GRANT LEE, b Safford, Ariz, June 20, 19; m 43; c 6. AGRONOMY. *Educ:* Univ Ariz, BS, 47, MS, 48; Ore State Col, PhD(farm crops), 50. *Prof Exp:* Asst prof agron, Purdue Univ, 50-53; assoc prof, 53-57, prof agron, Ariz State Univ, 57-83; RETIRED. *Concurrent Pos:* Team leader, Ariz State Univ-Kufra Agr Team, 73-77; res agron, Wash State Univ, 84-86; res dir, Univ Wyo, 88-89. *Mem:* Crop Sci Soc Am; Am Soc Agron. *Res:* Crop production in arid regions; crop physiology. *Mailing Add:* 1810 E Alameda Dr Tempe AZ 85282

RICHARDSON, HAROLD, b Ferryhill, Eng, Apr 13, 38; m 60; c 3. MEDICAL MICROBIOLOGY. *Educ:* Univ Durham, BSc, 59, MB, BS, 62; Univ Newcastle, Eng, MD, 68. *Prof Exp:* Demonstr bact, Univ Newcastle, Eng, 63-64, lectr, 64-69, sr lectr microbiol, 69-71; assoc prof, 71-76, PROF & DIR MICROBIOL, MED CTR, MCMASTER UNIV, 76- *Concurrent Pos:* Consult, United Newcastle Upon Tyne Hosps, 69-71. *Honors & Awards:* Comdr, Order of St John of Jerusalem, 75. *Mem:* Am Soc Clin Path; Am Soc Microbiol; Path Soc Gt Brit & Ireland. *Res:* Control of colicine production and role of colicinogeny in epidemiology of Escherichia coli infection. *Mailing Add:* Dept Microbiol Rm 2N30 Chedoke-McMaster Hosp Hamilton ON L8N 3Z5 Can

RICHARDSON, HENRY RUSSELL, b Pittsburgh, Pa, July 24, 38; m 60; c 2. SEARCH THEORY, STOCHASTIC PROCESSES. *Educ:* Univ Pittsburgh, BS, 60; Brown Univ, MS, 62, PhD(math), 65. *Prof Exp:* Sr vpres, Daniel H Wagner Assoc, 64-85; vpres, Ctr Naval Analysis, 85-87; prof math & oper res, US Naval Acad, 87-88; VPRES, METRON, 88- *Mem:* Am Math Soc; Oper Res Soc Am; Inst Math Statist. *Res:* Application of probability theory to problems in search for lost objects and in financial portfolio optimization. *Mailing Add:* 5911 Colfax Ave Alexandria VA 22311

RICHARDSON, HERBERT HEATH, b Lynn, Mass, Sept 24, 30; m 73; c 5. MECHANICAL ENGINEERING. *Educ:* Mass Inst Technol, SB & SM, 55, ScD(mech eng), 58. *Prof Exp:* Res engr, Dynamic Anal & Control Lab, Mass Inst Technol, 53-57, proj supvr, 57-58; ord officer, US Army, Ballistics Res Lab, Aberdeen, Md, 58-59; from asst prof to prof mech eng, Mass Inst Technol, 59-70, head systs & design div, dept mech eng & dir Analog-Hybrid Comput Facil, 67-70; chief scientist, US Dept Transp, 70-72; prof mech eng, Mass Inst Technol, 72-84, head dept, 74-82, assoc dean, 82-84; DEP CHANCELLOR & DEAN ENG, TEX A&M UNIV SYST, 84-, DIR, TEX ENG EXP STA, 85- *Concurrent Pos:* Sr consult, Foster-Miller Assocs, 58- & dir; mem tech adv bd, US Dept Energy, Dow Chem Co. *Honors & Awards:* Moody Award, 70; Centennial Medal, Am Soc Mech Engrs, 80, Rufus Oldenberger Medal, 84. *Mem:* Nat Acad Eng; Am Soc Eng Educ; NY Acad Sci; fel AAAS; hon mem Am Soc Mech Engrs; Sigma Xi; Soc Automotive Engrs. *Res:* Dynamic systems; automatic control; lubrication; transportation; fluid mechanics. *Mailing Add:* Eng Res Ctr Tex A&M Univ 301 Wisenbaker College Station TX 77843-3126

RICHARDSON, IRVIN WHALEY, b Tulsa, Okla, May 28, 34; m 58. BIOPHYSICS, BIOMATHEMATICS. *Educ:* Stanford Univ, BS, 56; Univ Calif, Berkeley, PhD(biophys), 67. *Prof Exp:* Asst prof physiol, Inst Physiol, Aarhus Univ, 67-68 & 69-71; vis assoc prof biophys, Sch Med, Univ Calif, San Francisco, 71-73; dir biophys sect, Med Fac, Dalhousie Univ, 74-75, from assoc prof to prof biophys, 73-82; CONSULT, 82- *Concurrent Pos:* Am Heart Asn vis scientist, Sch Med, Univ Calif, San Francisco, 71-72; assoc ed, Bull Math Biol, 73-79. *Mem:* Am Phys Soc. *Res:* Irreversible thermodynamics; membrane transport theory; mathematical modeling of physiological systems; theory of measurement, perception and representation. *Mailing Add:* 908 Middle Ave Apt O Menlo Park CA 94025

RICHARDSON, J MARK, b Duncan, Okla, Apr 27, 54. STOCHASTIC PROCESSES, MATHEMATICAL ANALYSIS. *Educ:* Okla State Univ, BS, 75, MS, 77, PhD(elec eng), 80. *Prof Exp:* Res assoc & teaching assoc circuit analysis, Okla State Univ, 75-80; MEM TECH STAFF, SANDIA NAT LABS, 80- *Concurrent Pos:* Software engr, Halliburton Co, 77; mem tech staff, Sandia Nat Labs, 78. *Mem:* Inst Elec & Electronics Engrs. *Res:* Stochastic integration and its relation to numerical integration; nuclear safety and safeguards at operating nuclear power plants and fuel cycle facilities; signal processing; estimation theory and techniques. *Mailing Add:* 2518 Virginia Duncan OK 73533

RICHARDSON, J(OHN) STEVEN, b London, Ont, Mar 24, 43; m 66; c 2. PSYCHONEUROPHARMACOLOGY, NEUROCHEMISTRY. *Educ:* Univ Toronto, BA, 65; Univ Vt, MA, 68, PhD(psycho-pharmacol), 72. *Prof Exp:* Postdoctoral neurochem & histopharmacol, Lab Clin Sci, NIMH, NIH, 71-73; from asst prof to assoc prof, 76-83, PROF PHARMACOL, UNIV SASK, 83- *Concurrent Pos:* Consult, Dannemara State Hosp, NY, 68-71; med consult, Royal Univ Hosp, Saskatoon, 79-; consult in forensic psychopharmacol; vis assoc prof, Univ Conn, 81-82. *Mem:* Can Col Neuropsychopharmacol; Am Soc Pharmacol & Exp Therapeut; Pharmacol Soc Can; Am Soc Neurochem; Int Soc Neurochem; Soc Neurosci; Can Asn Neurosci. *Res:* Neuropsychopharmacological and neurochemical analysis of brain function utilizing biochemical, histochemical, pharmacological and behavioral methodologies to elucidate the mechanisms whereby the limbic system and the basal ganglia exert homeostatic control over cognitive, emotional, hormonal, cardiovascular and motor activity. *Mailing Add:* Dept Pharmacol Col Med Univ Sask Saskatoon SK S7N 0W0 Can

RICHARDSON, JAMES ALBERT, b Schenectady, NY, Jan 27, 15; m 49; c 4. PHARMACOLOGY. *Educ:* Univ SC, BS, 36; Univ Miss, MS, 40; Univ Tenn, PhD(physiol), 49. *Prof Exp:* Instr pharmacy, Univ Miss, 40-41; instr physiol & pharmacol, 42-47, assoc pharmacol, 48, from asst prof to prof, 49-85, actg chmn dept, 71-72, EMER PROF PHARMACOL, MED UNIV SC, 85- *Mem:* AAAS; Am Soc Pharmacol & Exp Therapeut; Soc Exp Biol & Med; Am Fedn Clin Res; Sigma Xi. *Res:* Autonomic drugs; cardiovascular drugs; spinal anesthesia; toxicology of kerosene and decaborane; blood coagulation; catecholamines. *Mailing Add:* Dept Pharmacol Med Univ SC Charleston SC 29425

RICHARDSON, JAMES T(HOMAS), b Gillingham, Eng, Aug 5, 28; nat US; m 50; c 2. CHEMICAL ENGINEERING. *Educ:* Rice Inst, BA, 50, MA, 54, PhD(physics), 55. *Prof Exp:* Jr chemist, Pan-Am Refining Corp, 50-52; Welch Found fel, Rice Inst, 55-56; res physicist, Humble Oil & Refining Co, 56-65, sr res physicist, Esso Res & Eng Co, 59-64, res specialist, 64-66, res assoc, 66-69; assoc prof, 69-70, PROF CHEM ENG, UNIV HOUSTON, 70- *Mem:* Am Phys Soc; Am Chem Soc; Sigma Xi; Am Inst Chem Engrs. *Res:* Mass and infrared spectroscopy; x-ray and electron diffraction; electron microscopy; adsorption; magnetism; low temperature physics; adiabatic demagnetization; catalysis; defect solid state; electron spin resonance. *Mailing Add:* 5004 Arrowhead Baytown TX 77521-2902

RICHARDSON, JAMES WYMAN, b Sioux Falls, SDak, Aug 8, 30; m 52; c 4. QUANTUM CHEMISTRY. *Educ:* SDak Sch Mines & Technol, BS, 52; Iowa State Col, PhD(chem), 56. *Prof Exp:* Asst, Ames Lab, Iowa State Col, 53-56; res assoc physics, Univ Chicago, 56-57; from instr to assoc prof chem, 57-73, PROF CHEM, PURDUE UNIV, WEST LAFAYETTE, 73- *Concurrent Pos:* Visitor, Philips Res Labs, Neth, 67-68, Univ Groningen, Neth, 84, Univ Leiden, Neth, 84; assoc ed, J Solid State Chem, 84- *Mem:* Am Chem Soc; Am Phys Soc. *Res:* Theory of electronic properties of small molecules; transition-metal complex ions, and ionic solids. *Mailing Add:* Dept Chem 1393 Brown Bldg Purdue Univ West Lafayette IN 47907-1393

RICHARDSON, JANE S, b Teaneck, NJ, Jan 25, 41; m 63; c 2. BIOCHEMISTRY. *Educ:* Swarthmore Col, BA, 62; Harvard Univ, MA, 66, MAT, 66. *Hon Degrees:* DSc, Swarthmore Col, 86. *Prof Exp:* Tech asst, Dept Chem, Mass Inst Technol, 64-69; gen phys scientist, Lab Molecular Biol, Nat Inst Arthritis & Metab Dis, NIH, 69; assoc, Dept Anat, 70-84, med res assoc prof, Depts Biochem & Anat, 84-88, MED RES ASSOC PROF, DEPT BIOCHEM, DUKE UNIV, 88- *Concurrent Pos:* MacArthur Found Inst grant, 85-90; coun mem, Biophys Soc, 85-88, exec bd, 86-88; co-dir, Molecular Graphics & Modeling Shared Resource, Duke Comprehensive Cancer Ctr, 88-; Merck, Sharp & Dohme res grant, 88-91; coun mem, Protein Soc, 89-, Nat Ctr Res Resources, NIH, 90-; indust consult, Upjohn, Hoffman-LaRoche, Allied Chem, Becton Dickinson, Nutrasweet, Biosym & Tripos. *Mem:* Nat Acad Sci; Biophys Soc; Am Crystallog Asn; Protein Soc; Molecular Graphics Soc. *Res:* Comparison and classification of protein structures; design of new proteins for synthesis; protein crystallography; protein felding; representation of protein structures; conformational details in proteins; interpretation of electron density maps and evaluation of errors; structural information in reciprocal space; internal packing, subunit packing and crystal packing; concerted motions for protein modeling; comparison of nuclear magnetic resonance and x-ray structures; electronic publishing. *Mailing Add:* Dept Biochem Duke Univ Med Ctr Durham NC 27710

RICHARDSON, JASPER E, b Memphis, Tenn, Nov 8, 22; m 47; c 5. NUCLEAR PHYSICS, PETROLEUM ENGINEERING. *Educ:* Yale Univ, BS, 44; Rice Univ, MA, 48, PhD(physics), 50. *Prof Exp:* Instr physics, Univ Miss, 46-47; asst prof, Auburn Univ, 50-51; physicist, Med Div, Oak Ridge Inst Nuclear Studies, 51-53, Univ Tex MD Anderson Hosp & Tumor Inst, 53-55; res physicist, Bellaire Res Ctr, Shell Develop Co, Tex, 55-69, sr engr, Shell Oil Co, 69-72, staff engr, 72-86; RETIRED. *Mem:* Am Phys Soc; Soc Petrol Engrs. *Res:* Petrophysics; nuclear physics; medical physics; field testing new techniques for tertiary oil recovery. *Mailing Add:* 15015 Parkville Dr Houston TX 77068

RICHARDSON, JAY WILSON, JR, b Salt Lake City, Utah, Aug 1, 40. AQUATIC ENTOMOLOGY, AQUATIC ECOLOGY. *Educ:* Univ Utah, BS, 62, MS, 64. *Prof Exp:* Teaching asst invert zool, Univ Utah, 62-64; ENTOMOLOGIST, ACAD NAT SCI PHILADELPHIA, 64- *Concurrent Pos:* Biologist, Bur Com Fisheries, 62; asst to dir entom, Stroud Water Res Ctr, Acad Nat Sci Philadelphia, 69-73; entomologist, Coun Environ Qual, Exec Off President, 71-72, Savannah River Plant, E I du Pont de Nemours & Co, Inc, 71-75 & Nat Comn Water Qual, 74-75; proj mgr environ consult, Dept Limnol, Acad Nat Sci Philadelphia, 75- *Mem:* Sigma Xi; Entom Soc Am; Am Entom Soc; Ecol Soc. *Res:* Taxonomy and ecology of ephemeroptera; trichoptera; environmental pollution, especially fresh water biological monitoring. *Mailing Add:* 873 E Vine St Murray UT 84107

RICHARDSON, JEFFERY HOWARD, b Oakland, Calif, Nov 23, 48; m 78. PHYSICAL CHEMISTRY. *Educ:* Calif Inst Technol, BS, 70; Stanford Univ, PhD(chem), 74. *Prof Exp:* CHEMIST, LAWRENCE LIVERMORE LAB, 74-, CHEM SCI DIV LEADER, 88- *Concurrent Pos:* Consult, Mallinckrodt, 78. *Mem:* Am Chem Soc; Sigma Xi. *Res:* Applications of laser spectroscopy to analytical problems; oil shale chemistry; photoelectrochemistry; organic materials. *Mailing Add:* L-326 Lawrence Livermore Nat Lab Livermore CA 94550

RICHARDSON, JOHN CLIFFORD, neurology, for more information see previous edition

RICHARDSON, JOHN L(LOYD), b Ventura, Calif, Jan 29, 35; m 57; c 5. CHEMICAL ENGINEERING, BIOENGINEERING & BIOMEDICAL ENGINEERING. *Educ:* Stanford Univ, BS, 56, PhD(chem eng), 64. *Prof Exp:* Res & develop engr, Aeronutronic Div, Ford Motor Co, 61-62, sr engr, 62-64; sect supvr, chem Lab, Res Labs, Philco Corp, 64-65, mgr Appl Chem Dept Appl Res Labs, Aeronutronic Div, Newport Beach, 65-71, Mgr Res & Eng Dept, Liquid process Prod, 71-74; gen mgr, Recovery Systs, Oxy Metal Industs Corp, 74-78, mgr mkt & proj mgr, Voylite-Plating Systs Div, 78-81; pres & chief exec officer, Seagold Industs Corp, Burnaby, BC, 81-84; pres & chief exec office, Engenics Inc, Menlo Park, Calif, 84-88; MGR CONSULT, 88- *Mem:* Am Inst Chem Engrs. *Res:* Energy and mass transfer in chemically reacting systems; reverse osmosis and ultrafiltration, desalination, water purification, and effluent treatment; the development and growth of technology-based businesses, including the management of product commercializations and turnarounds. *Mailing Add:* 2356 Branner Dr Menlo Park CA 94025

RICHARDSON, JOHN MARSHALL, b Rock Island, Ill, Sept 5, 21; m 44; c 4. PHYSICS, TELECOMMUNICATIONS. *Educ:* Univ Colo, BA, 42; Harvard Univ, MA, 47, PhD(physics), 51. *Prof Exp:* Assoc head, Physics Div, Inst Indust Res, Univ Denver, 50-52; physicist, Nat Bur Stand, 52-60, chief, Radio Stand Lab, 60-67, dep dir, Inst Basic Stand, 66-67; dir, Off Stand Rev, US Dept Com, 67-68, actg dir, Off Telecommun, 69-70 & 72-76, dep dir, 70-72, dir, 76-78, chief scientist, Nat Telecommun & Info Admin, 78-80; PRIN STAFF OFFICER, NAT ACAD SCI, NAT RES COUN, 80- *Concurrent Pos:* Nat Bur Stand rep, Consult Comt Definition Second, Int Comt Weights & Measures, 61-67; conf chmn, Int Conf Precision Electromagnetic Measurements, 62; chmn UC Comn I, Int Union Radio Sci, 64-67, mem at large, US Nat Comt, 69-72; exec secy, Comt Telecommun, Nat Acad Eng, 68-69; US mem, Panel Comput & Commun, Orgn Econ Coop & Develop, 70-77; vchmn, Working Party Info, Comput & Communs Policy, 77-79; mem Comt Commun & Info Policy, Inst Elec & Electronics Engrs, 80. *Honors & Awards:* Gold Medal Award, US Dept Com, 64. *Mem:* Fel AAAS; fel Inst Elec & Electronics Engrs; fel Am Phys Soc. *Res:* Gaseous discharges; microwave spectroscopy; microwave physics; atomic time and frequency standards; precision electromagnetic measurements and standards; telecommunications technology, management and policy; science policy. *Mailing Add:* 7116 Arma T Dr Bethesda MD 20817

RICHARDSON, JOHN MEAD, b San Francisco, Calif, Nov 25, 18; m 69; c 1. THEORETICAL PHYSICS, INFORMATION SCIENCES. *Educ:* Calif Inst Technol, BS, 41; Cornell Univ, PhD(phys chem), 44. *Prof Exp:* Res assoc, Cornell Univ, 44-45; mem tech staff, Bell Tel Labs, 45-49; phys chemist, US Bur Mines, Pa, 49-50, head kinetics sect, 50-53; physicist, Hughes Aircraft Co, 53-54; sr staff physicist, Ramo Wooldridge Corp, 54-56; sr staff physicist, Hughes Res Labs, Calif, 56-57, asst head physics lab, 57-58, sr staff consult, 58-61, mgr theoret studies dept, 61-68; MEM STAFF, SCI CTR, ROCKWELL INT CORP, 68- *Concurrent Pos:* Instr, Stevens Inst Technol, 47-48; vis prof, Univ Calif, Los Angeles, 62; ed, J Cybernet, 73-78 & Info Sci, 68- *Mem:* Am Phys Soc; Soc Eng Sci. *Res:* Mathematical physics; underwater explosions and damage; combustion and detonation waves; hypersonic flow; dielectrics and ferroelectrics; liquids, plasmas and order-disorder phenomena; irreversible statistical mechanics; mathematical methods in statistical mechanics; automatic pattern recognition and estimation theory. *Mailing Add:* Sci Ctr Rockwell Int Corp 1049 Camino Dos Rios Thousand Oaks CA 91360

RICHARDSON, JOHN PAUL, b Pittsfield, Mass, June 27, 38; m 66; c 3. MOLECULAR BIOLOGY, GENE EXPRESSION. *Educ:* Amherst Col, BA, 60; Harvard Univ, PhD(biochem), 66. *Prof Exp:* NIH fel, Inst Biol Physio-Chimique, Paris, 65-67; Am Cancer Soc fel, Inst Molecular Biol, Geneva, Switz, 67-69; res assoc molecular genetics, Univ Wash, 69-70; from asst prof to assoc prof, 70-78, PROF CHEM, IND UNIV, BLOOMINGTON, 78- *Concurrent Pos:* NIH res grant, Ind Univ, Bloomington, 71-; career develop award, NIH, 72-77, pathobiol chem study sect, 75-78, Genetic Basis Dis rev comt, 81-85; ed, Gene, 83- *Mem:* Am Soc Biol Chemists; Am Soc Microbiol; fel AAAS. *Res:* Enzymology and regulation of RNA biosynthesis; control of virus development. *Mailing Add:* Dept Chem Ind Univ Bloomington IN 47405

RICHARDSON, JOHN REGINALD, b Edmonton, Alta, Oct 31, 12; nat US; m 38; c 2. NUCLEAR PHYSICS. *Educ:* Univ Calif, Los Angeles, AB, 33; Univ Calif, PhD(physics), 37. *Hon Degrees:* DSc, Univ Victoria, Can. *Prof Exp:* Nat Res Coun fel physics, Univ Mich, 37-38; asst prof, Univ Ill, 38-42; physicist, Manhattan Proj, Univ Calif, 42-46; assoc prof, 46-52, PROF PHYSICS, UNIV CALIF, LOS ANGELES, 52- *Concurrent Pos:* Sci liaison officer, Off Naval Res, London, 53-54 & 56-57; mem comt sr reviewers, US AEC, 52-71; dir, Tri Univ Meson Facility, Univ BC, 71-76; vis scientist, 76- *Mem:* Fel Am Phys Soc; Sigma Xi. *Res:* Nucleon-nucleon interaction; nuclear structure; cyclotrons. *Mailing Add:* Dept Physics Univ Calif Los Angeles CA 90024-1547

RICHARDSON, JONATHAN L, b Philadelphia, Pa, May 15, 35; m 63; c 2. LIMNOLOGY, PLANT ECOLOGY. *Educ:* Williams Col, BA, 57; Univ NZ, MA, 60; Duke Univ, PhD(zool), 65. *Prof Exp:* From asst prof to assoc prof, 66-80, PROF BIOL, FRANKLIN & MARSHALL COL, 80- *Concurrent Pos:* NSF res grant, EAfrican lakes & climatic hist, 69-72; US-Australia Coop sci fel, 80-81; interim dir, N Mus, Franklin & Marshall Col, 90-91. *Mem:* AAAS; Int Soc Limnol; Ecol Soc Am; Am Inst Biol Sci; Am Soc Limnol Oceanog; Int Soc Diatom Res. *Res:* Limnology and paleoecology of tropical lakes; ecology of diatoms; plant ecology of disturbed habitats; primary productivity of aquatic ecosystems; stream ecology. *Mailing Add:* 521 State St Lancaster PA 17603

RICHARDSON, JOSEPH GERALD, b Gulf, Tex, Oct 28, 23; m 71; c 4. ENGINEERING ADMINISTRATION. *Educ:* Tex A&M Univ, BS, 47; Mass Inst Technol, MS, 48. *Prof Exp:* Jr res eng, Humble Oil & Refining Co, 48-51, res supvr, 51-64; sect supvr, Exxon Prod Res Co, 64-80, res scientist, 80-82, sr eng scientist, 82-86; CONSULT, 86- *Concurrent Pos:* Adv bd, Exploitech, 87- *Honors & Awards:* Uren Award, Soc Petrol Engrs, 77, Degolyer Award, 78. *Mem:* Hon mem Nat Acad Engrs; hon mem Am Inst Mech Engrs; hon mem Soc Petrol Engrs. *Res:* Basic research on multiphase oil, gas, and water flow in porous rocks; developed standard core analysis procedures; developed theory for capillary inhibitions of water by reservoir rocks. *Mailing Add:* 11767 Katy Freeway Suite 225 Houston TX 77079

RICHARDSON, KATHLEEN SCHUELLER, b New York, NY, Sept 28, 38; m 70. ORGANIC CHEMISTRY. *Educ:* Bryn Mawr Col, BA, 60; Radcliffe Col, MA, 62; Harvard Univ, PhD(chem), 66. *Prof Exp:* Res scientist org chem, Bell Tel Labs, 66-68; asst prof, Vassar Col, 68-74; lectr, Ohio State Univ, 74-75; instr, 75-76, asst prof, 76-80, ASSOC PROF CHEM, CAPITAL UNIV, 80- *Mem:* AAAS; Am Chem Soc. *Res:* Cycloaddition reactions. *Mailing Add:* 415 Clinton Heights Ave Columbus OH 43202-1251

RICHARDSON, KEITH ERWIN, b Tucson, Ariz, Apr 22, 28; m 52; c 6. PHYSIOLOGICAL CHEMISTRY & TOXICOLOGY. *Educ:* Brigham Young Univ, BS, 52, MS, 55; Purdue Univ, PhD(biochem), 58. *Prof Exp:* Asst biochem, Purdue Univ, 55-58; fel agr chem, Mich State Univ, 58-60; from asst prof to assoc prof physiol chem, 60-67, PROF PHYSIOL CHEM & VCHMN DEPT, OHIO STATE UNIV, 67- *Mem:* AAAS; Am Chem Soc; Am Soc Plant Physiol; Am Soc Biol Chem; Sigma Xi; Am Inst Nutrit. *Res:* Enzymology; primary hyperoxaluria and oxalic acid metabolism; enzyme regulation; intermediate metabolism; metabolic inborn errors of metabolism. *Mailing Add:* Dept Physiol Chem 3133 Edgefield Rd Columbus OH 43221

RICHARDSON, LAVON PRESTON, b Ranger, Tex, July 6, 25; m 50; c 2. MICROBIOLOGY. *Educ:* Tex Christian Univ, AB, 48; NTex State Univ, MA, 49; Okla State Univ, EdD, 58. *Prof Exp:* From instr to asst prof biol, Cent State Univ, Okla, 49-57; from instr to asst prof bact, Okla State Univ, 57-59; adminr, DeLeon Munic Hosp, Tex, 59-60; asst prof biol, Tarleton State Col, 60-61; asst prof microbiol, 61-64, co-dir sci teaching ctr, 67, ASSOC PROF MICROBIOL, OKLA STATE UNIV, 65- *Concurrent Pos:* Asst dir spec proj, NSF, 66-67; coordr, This Atomic World Prog. *Mem:* Am Soc Microbiol; Nat Asn Biol Teachers; Am Pub Health Asn; Am Inst Biol Sci. *Res:* Colony movement in Bacillus alvei. *Mailing Add:* 1618 Chiquita Ct Stillwater OK 74075

RICHARDSON, LEE S(PENCER), b Syracuse, NY, Mar 17, 29; m 56; c 2. PHYSICAL METALLURGY, MATHEMATICS. *Educ:* Mass Inst Technol, SB, 50, SM, 51, ScD(phys metall), 56. *Prof Exp:* Res asst oxidation & nitriding titanium, Mass Inst Technol, 50-51, res asst creep rupture titanium, 53-55; jr metallurgist, Oak Ridge Nat Lab, 51-52; res metallurgist, Westinghouse Elec Corp, 55-56; indust staff mem, Los Alamos Sci Lab, 56-58; res metallurgist, Westinghouse Elec Corp, 58-60, supvry metallurgist, 60-63; mgr ceramics & metall res, Foote Mineral Co, 63-69, dir res & eng, Ferroalloy Div, 69-71, dir res & develop, 72-77; mgr mat sci, 77-80, mgr hot cell opers, 80-83, TECH LEADER PLASMA PROCESSING, EG&G IDAHO INC, 83- *Mem:* Am Soc Metals; Am Inst Mining, Metall & Petrol Engrs. *Res:* High temperature materials; alloy development; ferroalloys; inorganic chemistry of lithium and manganese; computer simulation; materials for energy production and conversion. *Mailing Add:* 210 N Garfield Moscow ID 83843

RICHARDSON, LEONARD FREDERICK, b Brooklyn, NY, Nov 23, 44; m 72; c 2. MATHEMATICAL ANALYSIS. *Educ:* Yale Univ, BA & MA, 65, PhD(math), 70. *Prof Exp:* Instr math, Yale Univ, 70-71; CLE Moore instr math, Mass Inst Technol, 71-73; from asst prof to assoc prof, 73-84, PROF MATH, LA STATE UNIV, 85- *Concurrent Pos:* NSF res grant, 74-84; NSF Int travel grant, 75, 76, 81; vis assoc prof math, Univ Conn, 77-78. *Mem:* Am Math Soc. *Res:* Harmonic analysis on manifolds; projections, measures, distributions and differential operators on nilmanifolds; representation theory. *Mailing Add:* Dept Math Lockett Hall La State Univ Baton Rouge LA 70803

RICHARDSON, MARY ELIZABETH, pathology, electron microscopy; deceased, see previous edition for last biography

RICHARDSON, MARY FRANCES, b Barbourville, Ky, Sept 3, 41. INORGANIC CHEMISTRY, CRYSTALLOGRAPHY. *Educ:* Univ Ky, BS, 62, PhD(chem), 67. *Prof Exp:* Contractor, Aerospace Res Labs, Wright-Patterson AFB, 67-71; from asst prof to assoc prof, 71-81, chmn dept, 79-82, PROF CHEM, BROCK UNIV, 81- *Mem:* Am Chem Soc; Am Crystallog Asn; fel Chem Inst Can. *Res:* Crystallographic packing; asymmetric syntheses and stereochemical control of reactions on single crystals; x-ray crystal structures; polymorphism in crystals. *Mailing Add:* Dept Chem Brock Univ St Catharines ON L2S 3A1 Can

RICHARDSON, MICHAEL LEWELLYN, b Kerrville, Tex, Dec 31, 50; m 76. MUSCULOSKELETAL RADIOLOGY. *Educ:* Tex A&M Univ, BS, 72; Baylor Col Med, MD, 75. *Prof Exp:* Staff internist, USAF, 76-78, radiol resident, 78-81, staff radiologist, 81-83; postdoctoral fel, musculoskeletal radiol, Univ Calif, San Francisco, 83-84; assoc prof radiol, 84-88, ASSOC PROF RADIOL, UNIV WASH, 88- *Concurrent Pos:* Manuscript referee, Radio Graphics, Univ Wash, 87-91 & dir musculoskelet radiol, 87-90; guest ed, Radiol Clin N Am, 86 & 88; manuscript reference, Am J Roentgenology, 86-91. *Mem:* Radiol Soc NAm; Am Roentgen Ray Soc; Asn Univ Professors; Am Col Radiol; Int Skeletal Soc. *Res:* Microcomputers and biostatistics in diagnostic radiology; musculoskeletal radiology. *Mailing Add:* Dept Radiol MS-5B-05 Univ Wash Sch Med 1956 NE Pacific St Seattle WA 98195

RICHARDSON, NEAL A(LLEN), b Casper, Wyo, Mar 14, 26; m 52; c 2. ENGINEERING. *Educ:* Univ Calif, Los Angeles, BS, 49, MS, 53, PhD(eng), 62. *Prof Exp:* Lectr eng, Univ Calif, Los Angeles, 49-62; mgr, TRW Systs Group, 62-76, mgr coal conversion progs, 76-78, dir advan systs eng, TRW Electronics Group, TRW Energy Systs Group, 78-86; RETIRED. *Concurrent Pos:* NSF sci fac fel, 61. *Mem:* Soc Automotive Engrs; Sigma Xi. *Res:* Advanced energy conversion processes including electrochemical systems; vehicle power train development and engineering including power plant emissions characterization and control; chemical processes based on coal; new catalytic process to produce high BTU gas; electronic controls and sensors for transportation. *Mailing Add:* 30823 Cartier Dr Rancho Palos Verdes CA 90274

RICHARDSON, PAUL ERNEST, b Covington, Ky, Dec 29, 34; m 58; c 1. PLANT ANATOMY & PATHOLOGY. *Educ:* Univ Ky, AB, 57; Univ Cincinnati, MEd, 62, MS, 66, PhD(bot), 68; Univ NC, Chapel Hill, MAT, 63. *Prof Exp:* Teacher, Lloyd Mem High Sch, Ky, 57-59 & Holmes High Sch, 59-62; from asst prof to assoc prof, 68-82, researcher, Okla Agr Exp Sta, 73-82, PROF BOT, OKLA STATE UNIV, 82-, RES PROF PLANT PATH, AGR RES STA, 82- *Mem:* Bot Soc Am; AAAS; Am Phytopath Soc; Torrey Bot Club; Am Inst Biol Sci; Sigma Xi. *Res:* Structure, especially ultrastructure of pathological and stress states in vascular (crop) plants; comparative structure of flowering plants. *Mailing Add:* Dept Bot Okla State Univ Stillwater OK 74078

RICHARDSON, PAUL NOEL, b Minneapolis, Minn, Mar 31, 25; m 46, 84; c 3. ORGANIC CHEMISTRY. *Educ:* Univ Minn, BS, 49, PhD(org chem), 52. *Prof Exp:* Res chemist, 52-57, tech rep, 57-64, consult, 64-65, supvr, 66-70, sr res chemist, Plastic Dept, 71-80, RES ASSOC, POLYMER PROD DEPT, E I DU PONT DE NEMOURS & CO, INC, 80- *Concurrent Pos:* Instr, Univ Del. *Honors & Awards:* Pres Cup, Soc Plastic Engrs, 72. *Mem:* Am Chem Soc; Soc Plastic Engrs (treas, 70). *Res:* Polymer chemistry; plastics engineering. *Mailing Add:* 11 N Wynwyd Dr Newark DE 19711-7424

RICHARDSON, PETER DAMIAN, b West Wickham, Eng, 35. BIOMEDICAL ENGINEERING. *Educ:* Univ London, BSc, 55, PhD(eng) & DIC, 58; City & Guilds of London Inst, ACGI, 55. *Hon Degrees:* MA, Brown Univ, 65; DSc(eng), Univ London, 74, DSc, 83. *Prof Exp:* Demonstr eng, Imp Col, Univ London, 55-58; res assoc, 58-60, from asst prof to prof eng, 60-84, PROF ENG & PHYSIOL, BROWN UNIV, 84- *Concurrent Pos:* Sci Res Coun Eng sr res fel, 67. *Honors & Awards:* Humbolt Found Award, 76; Prize Med, Jung Found, 87. *Mem:* Fel Am Soc Mech Engrs; Am Soc Eng Educ; Am Soc Artificial Internal Organs; Biomed Eng Soc; Europ Soc Artificial Organs. *Res:* Heat and mass transfer, fluid dynamics; theory and technology of artificial internal organs; blood flow. *Mailing Add:* Div Eng Box D Brown Univ Providence RI 02912

RICHARDSON, PHILIP LIVERMORE, b New York, NY, Oct 31, 40; m 66; c 2. PHYSICAL OCEANOGRAPHY, MARINE SCIENCE. *Educ:* Univ Calif, Berkeley, BS, 64; Univ RI, MS, 70, PhD(phys oceanog), 74. *Prof Exp:* Lieutenant jr grade, US Coast & Geod Surv, 64-66; asst, Sch Oceanog, Univ RI, 67-69, res asst, 69-73, asst prof, 73-74; from asst scientist to assoc scientist, 74-89, SR SCIENTIST, WOODS HOLE OCEANOG INST, 89- *Concurrent Pos:* Vis scientist, Nat Mus Natural Hist, Oceanog Phys Lab, Paris, 78-79, Oceanog Ctr Brittany, Brest, France, 83 & Scripp's Inst Oceanog, La Jolla, 86. *Mem:* AAAS; Am Geophys Union; Am Meteorol Soc. *Res:* General ocean circulation; gulf stream; oceanic eddies; equatorial currents; historical studies. *Mailing Add:* Woods Hole Oceanog Inst Woods Hole MA 02543

RICHARDSON, RALPH J, b Jamestown, NDak, Feb 28, 41; m 64; c 3. SOLID STATE PHYSICS. *Educ:* Rockhurst Col, AB, 62; St Louis Univ, MS, 64, PhD(physics), 69. *Prof Exp:* Scientist, McDonnell Douglas Res Labs, McDonnell Douglas Corp, 69-81; mgt dir, int technol, 81-83, source technol, 83-85, MGT DIR, AIR PROD & CHEM, SRI INT, 85- *Mem:* Am Phys Soc. *Res:* Electron spin resonance; combustion; chemical lasers; high purity gases for semiconductor applications. *Mailing Add:* 4976 Meadow Lane Macungie PA 18062

RICHARDSON, RANDALL MILLER, b Santa Monica, Calif, Dec 29, 48; m 77; c 2. GEOPHYSICS, EARTH SCIENCE. *Educ:* Univ Calif, San Diego, BA, 72; Mass Inst Technol, PhD(geophysics), 78. *Prof Exp:* asst prof, 78-83, ASSOC PROF GEOSCIENCES, UNIV ARIZ, 83- *Mem:* Am Geophys Union; Sigma Xi. *Res:* Intraplate deformation and driving mechanism for plate tectonics through observation and finite element modeling of stress; analysis of strain accumulation release at plate boundaries; inverse modeling. *Mailing Add:* Dept Geosci Univ Ariz Tucson AZ 85721

RICHARDSON, RAYMAN PAUL, b Piedmont, Mo, May 17, 39; m 66; c 2. SCIENCE EDUCATION. *Educ:* Cent Methodist Col, AB, 61; Univ Mo, MST, 64; Ohio State Univ, PhD(sci educ), 71. *Prof Exp:* Teacher chem & math, St Clair Pub Schs, 61-64; teacher phys sci, Antilles Sch Syst, 64-67; from instr to assoc prof, 71-78, PROF PHYS SCI & SCI EDUC, FAIRMONT STATE COL, 78- *Mem:* Am Chem Soc; Nat Asn Res Sci Teaching; Nat Sci Teachers Asn. *Res:* Measurement of scientific curiosity and interests of elementary school children. *Mailing Add:* Fairmont State Col Fairmont WV 26554

RICHARDSON, RAYMOND C(HARLES), b Junction City, Kans, Sept 26, 29; m 55; c 3. CHEMICAL ENGINEERING. *Educ:* Univ Colo, BS, 54; Kans State Univ, MS, 58; Iowa State Univ, PhD(chem eng), 63. *Prof Exp:* Sr tech serv rep, Spencer Chem Co, 54-56; instr chem eng, Iowa State Univ, 61-62; from asst prof to assoc prof, Univ Ariz, 62-67; hybrid simulation engr, Phillips Petrol Co, 67-69; mgr process br, 69-75, mgr admin & control div, 75-77, MGR PROCESS TECHNOL DIV, APPL AUTOMATION, INC, 77- *Mem:* Am Inst Chem Engrs; Simulation Coun; Sigma Xi. *Res:* Chemical and petroleum processes; modeling and simulation of processes; application of computer control. *Mailing Add:* Appl Automation Inc Pawhuska Rd Bartlesville OK 74004

RICHARDSON, RICHARD HARVEY, b Mexia, Tex, Mar 24, 38; m 70, 87; c 3. BIOLOGICAL PEST CONTROL, EVOLUTION. *Educ:* Tex A&M Univ, BS, 59; NC State Univ, MS, 62, PhD(genetics), 65. *Prof Exp:* Assoc res scientist, Genetics Found, 64-65, NIH fel zool, 65-67, from lectr to assoc prof, 65-79, PROF ZOOL, UNIV TEX, AUSTIN, 79- *Concurrent Pos:* USPHS career develop award, 70-75; on loan, Entom Dept, Univ Hawaii-Manoa. *Mem:* AAAS; Genetics Soc Am; Soc Study Evolution; Biomet Soc; Am Soc Naturalists. *Res:* Genetics of mating behavior and population structure; chromosomal, behavioral and biochemical changes during the evolution of the genus Drosophila; population genetics and ecology of natural and laboratory populations; genetics, evolutionary biology and biogeography of screwworms, Cochliomyia species; ticks; ecology of native prairies; water conservation practices utilizing native plants. *Mailing Add:* 600 Texas Ave Austin TX 78705

RICHARDSON, RICHARD LAUREL, b Chelan, Wash, Aug 12, 26; m 48; c 6. ELECTRIC POWER SYSTEM THEORY, ULTRASONICS. *Educ:* Univ Colo, Boulder, BS, 53; Univ Idaho, MS, 61. *Prof Exp:* Tech grad, Gen Elec Co, 53-55, engr, 55-60, mathematician, 60-65; sr res scientist, Pac Northwest Labs, Battelle Mem Inst, 65-80; sr engr, UNC Nuclear Indust Inc, 80-85; staff engr, Rockwell Hanf Co; prin engr, Westinghouse Hanford Co, 88-90; STAFF ENGR, PAC NORTHWEST LABS, BATTELLE NORTHWEST, 90- *Concurrent Pos:* Instr, Columbia Basin Col; lectr, Joint Ctr Grad Study, Richland, Wash; lectr IV, Wash State Univ, Tri Cities. *Honors & Awards:* Inst Elec Electronics Engrs Centennial Medal, 84. *Mem:* Inst Elec & Electronics Engrs; Sigma Xi. *Res:* Stress wave propagation, thermal diffusion; sphere packing and related molecular models, especially transmission and distribution of electrical energy and industrial control. *Mailing Add:* 4950 Dove Ct W Richland WA 99352

RICHARDSON, ROBERT COLEMAN, b Washington, DC, June 26, 37; m 62; c 2. LOW TEMPERATURE PHYSICS. *Educ:* Va Polytech Inst, MS, 60; Duke Univ, PhD(physics), 66. *Prof Exp:* From asst prof to prof, 67-87, F R NEWMAN PROF PHYSICS, CORNELL UNIV, 87-, DIR LAB ATOMIC & SOLID STATE PHYSICS, 90- *Concurrent Pos:* Vis scientist, Bell Lab, 84; mem, NRC Panel Basic Standards, 83- *Honors & Awards:* Simon Prize, Brit Phys Soc, 77; Buckley Prize, Am Phys Soc, 81. *Mem:* Nat Acad Sci; fel Am Phys Soc; fel AAAS. *Res:* Studies of thermal and magnetic properties of solid and liquid helium at very low temperatures. *Mailing Add:* Dept Physics Clark Hall Cornell Univ Ithaca NY 14853

RICHARDSON, ROBERT ESPLIN, b Alameda, Calif, Nov 9, 24; m 46; c 6. PHYSICS, ELECTRICAL ENGINEERING. *Educ:* Univ Okla, BS, 45; Univ Calif, PhD(physics), 51. *Prof Exp:* Physicist, Radiation Lab, Univ Calif, 46-52; mem staff, Lincoln Lab, Mass Inst Technol, 52-85; CONSULT, 85- *Mem:* AAAS; Am Phys Soc; Inst Elec & Electronics Eng. *Res:* Reentry physics; radar; electronics; computer systems and applications. *Mailing Add:* 159 Merriam St Weston MA 02193

RICHARDSON, ROBERT LLOYD, b Syracuse, NY, Oct 1, 29; m 57; c 6. ELECTRICAL ENGINEERING, NUCLEAR MEDICINE. *Educ:* Syracuse Univ, BEE, 51, MEE, 56, PhD(elec eng), 61. *Prof Exp:* Res asst elec eng, Syracuse Univ, 51-53, res assoc, 53-56, from instr to asst prof, 56-64; res engr, Syracuse Univ Res Corp, 64-70; ASSOC PROF RADIOL, STATE UNIV NY HEALTH SCI CTR, 70- *Concurrent Pos:* Lectr, Syracuse Univ, 71-76, adj prof, 76- *Mem:* Inst Elec & Electronics Engrs; Soc Nuclear Med. *Res:* Electronic circuits and electronics applied to medicine. *Mailing Add:* Div Nuclear Med State Univ NY Health Sci Ctr Syracuse NY 13210-2399

RICHARDSON, ROBERT LOUIS, b Lexington, Ky, Mar 19, 22; m 50; c 3. BACTERIOLOGY. *Educ:* Univ Louisville, DMD, 44; Univ Iowa, MS, 53. *Prof Exp:* Practicing dentist, Ky, 47-48; from instr to asst prof restorative dent, Univ Tex, 48-51; instr periodontia, 51-52, asst prof microbiol, Col Med, 53-64, ASSOC PROF MICROBIOL, COL MED, UNIV IOWA, 64- *Mem:* AAAS; Am Soc Microbiol; Int Asn Dent Res. *Res:* In vitro studies of dental caries; microorganisms in the mouth. *Mailing Add:* 1806 Jefferson Ave Des Moines IA 50314

RICHARDSON, ROBERT WILLIAM, b Sydney, Australia, Sept, 5, 35; US citizen; div; c 2. STATISTICAL MECHANICS, NUCLEAR PHYSICS. *Educ:* Univ Mich, BS & MA, 58, PhD(physics), 63. *Prof Exp:* Asst res scientist, Courant Inst Math Sci, 63-65, from asst prof to assoc prof physics, 65-75, chmn dept, 82-85, PROF PHYSICS, NY UNIV, 75- *Concurrent Pos:* Consult, Lawrence Berkeley Lab. *Mem:* Am Phys Soc; AAAS; Sigma Xi. *Res:* Many-body problem; nuclear models; low temperature physics; transport theory. *Mailing Add:* Dept Physics NY Univ New York NY 10003

RICHARDSON, RUDY JAMES, b Winfield, Kans, May 13, 45; m 70, 85; c 1. NEUROTOXICOLOGY. *Educ:* Wichita State Univ, BS, 67; Harvard Univ, ScM, 73, ScD(physiol-toxicol), 74. *Prof Exp:* NASA trainee chem, State Univ NY Stony Brook, 67-70; Nat Int Environ Health Sci trainee toxicol, Harvard Univ, 70-74; res biochem neurotoxicol, Med Res Coun, Eng, 74-75; from asst prof to assoc prof, 75-84, PROF TOXICOL, UNIV MICH, 84-, ASSOC PROF NEUROL, 87- *Concurrent Pos:* Consult, Environ Protection Agency, 76-; mem safe drinking water comt & toxicol subcomt, Nat Acad Sci, 78-79 & 84; vis scientist, Warner-Lambert/Parke-Davis Pharmaceut Res Div, 82-83; invited speaker, Second Int Meeting Cholinesterases, Bled, Yugoslavia & Gordon Conf Toxicol, 84; pres, Neurotoxicol specialty sect, Soc Toxicol, 87-88; mem sci adv panel, US Environ Protection Agency, 87-90; consult, Off Technol Assessment, US Cong, 88-90, NIOSH, 90. *Mem:* Am Chem Soc; AAAS; Soc Neurosci; Soc Toxicol; Sigma Xi; Am Soc Neurochem. *Res:* Delayed neurotoxicity of organophosphorus compounds; neurotoxic esterase, neuropathy target esterase; models of neurological disease; maintenance and plasticity of neurons; biomembranes; biological functions of glutathione; transport of heavy metals; leukocytes as biomonitors and models of certain neuronal functions; neuroimmunomodulation; connectivity of scientific fields; philosophy of science. *Mailing Add:* Toxicol Res Lab Sch Pub Health Univ Mich Ann Arbor MI 48109-2029

RICHARDSON, STEPHEN H, b Kalamazoo, Mich, June 30, 32; m 83; c 4. MICROBIOLOGY, BIOCHEMISTRY. *Educ:* Univ Calif, Los Angeles, BA, 55; Univ Southern Calif, MS, 59, PhD(bact), 60. *Prof Exp:* Lectr, Univ Southern Calif, 59-61; res assoc, Tobacco Industs Res Comt, 60-61; trainee, Enzyme Inst, Univ Wis, 61-63; from asst prof to assoc prof, 63-71, PROF MICROBIOL, BOWMAN GRAY SCH MED, 71- *Concurrent Pos:* Mem cholera adv comt, NIH, 69-73, US-Japan Cholera Panel, 69-76; adj prof biol, Wake Forest Univ, 71-; found lectr, Am Soc Microbiol, 71-72; guest scientist, Cholera Res Lab, Bangladesh, 72-73; consult, NIH Infectious Dis Comt, 77; pres NC br, Am Soc Microbiol, 77-78; mem, bacteriol/mycology study sect, NIH, 78-82; consult, Int Ctr Diarrheal Dis Res, Bangladesh, 83-, ICDDRB & UNICEF, Columbia, SAm, 84. *Mem:* AAAS; Am Soc Microbiol; Am Soc Trop Med & Hyg; Sigma Xi. *Res:* Bacteriology; microbial physiology; membrane systems and virulence; mechanisms of microbial pathogenesis; physiology of Vibrio cholerae; enterotoxin-induced diarrheal diseases; genetic mechanisms of host resistance; virulence factors of halophilic vibrios; international health. *Mailing Add:* Dept Microbiol Bowman Gray Sch Med 300 Hawthrone Rd SW Winston-Salem NC 27103

RICHARDSON, SUSAN D, b Brunswick, Ga, Oct 13, 62; m 84. MASS SPECTROMETRY, STRUCTURAL ELUCIDATION. *Educ:* Ga Col, Milledgeville, BS, 84; Emory Univ, Atlanta, PhD(chem), 89. *Prof Exp:* Postdoctoral assoc, 89, RES CHEMIST, US ENVIRON PROTECTION AGENCY, 89- *Mem:* Sigma Xi; Am Soc Mass Spectrometry; Am Chem Soc. *Res:* Mass spectrometric techniques, including low and high resolution EI and CI mass spectrometry and FAB mass spectrometry, to identify chemical pollutants of unknown structure that are found in the environment. *Mailing Add:* 5024 Logan's Run Loganville GA 30249

RICHARDSON, THOMAS, b Ft Lupton, Colo, Dec 4, 31; m 54; c 2. FOOD CHEMISTRY, BIOCHEMISTRY. *Educ:* Univ Colo, Boulder, BS, 54; Univ Wis-Madison, MS, 56, PhD(biochem), 60. *Prof Exp:* Fel food chem, Univ Calif, Davis, 60-62; from asst prof to assoc prof, 62-69, PROF FOOD CHEM, UNIV WIS-MADISON, 70- *Mem:* Am Chem Soc; Inst Food Technol; Am Dairy Sci Asn. *Res:* Application of insoluble enzymes to food processing, analysis and structure; applied enzymology in general. *Mailing Add:* Dept Food Sci & Tech Univ Calif Davis CA 95616

RICHARDSON, VERLIN HOMER, b Gage, Okla, July 5, 30; m 51; c 3. CHEMISTRY, SCIENCE EDUCATION. *Educ:* Northwestern State Col, Okla, BS, 52; Phillips Univ, MEd, 57; Okla State Univ, MS, 58; Univ Okla, PhD(sci educ), 69. *Prof Exp:* Teacher, Pub Sch, Okla, 52-56; instr chem, El Dorado Jr Col, Kans, 58-62; assoc prof, 62-76, PROF CHEM, CENT STATE UNIV, OKLA, 76- *Mem:* Am Chem Soc. *Res:* Inorganic chemistry. *Mailing Add:* 316 Ramblewood Terr Edmond OK 73034-4330

RICHARDSON, WALLACE LLOYD, b Santa Barbara, Calif, Sept 16, 27; m 51; c 4. ORGANIC CHEMISTRY. *Educ:* Univ Calif, BS, 51; Mass Inst Technol, PhD(org chem), 54. *Prof Exp:* Sr res assoc, Fuels, Asphalts & Spec Prods, 54-69, mgr, fuel chem div, MGR EXPLOR ADDITIVES DIV, PROD RES DEPT, CHEVRON RES CO, STANDARD OIL, CALIF, 84- *Mem:* Am Chem Soc; Sigma Xi. *Res:* Mechanism of combustion chemistry of knock and antiknock reactions; application of surfactants in hydrocarbon systems; wax crystal modification for improvement of low temperature flow. *Mailing Add:* 24 Prado Way Lafayette CA 94549-2332

RICHARDSON, WILLIAM C, b Passaic, NJ, May 11, 40; c 2. EDUCATION ADMINISTRATION. *Educ:* Trinity Col, Hartford, Conn, BA, 62; Univ Chicago, MBA, 64, PhD, 71. *Prof Exp:* From asst prof to prof health serv, Sch Pub Health & Community Med, Univ Wash, 71-84, dean grad sch & vprovost res, 81-84; exec vpres & provost, Pa State Univ, 84-90, prof family & community med, Milton S Hershey Med Ctr, 84-90; PRES, JOHNS HOPKINS UNIV, 90-, PROF HEALTH & POLICY MGT, SCH HYG & PUB HEALTH, 90- *Mem:* Inst Med-Nat Acad Sci; fel Am Pub Health Asn. *Res:* Financing of health care. *Mailing Add:* Off Pres 242 Garland Hall Johns Hopkins Univ Baltimore MD 21218

RICHARDSON, WILLIAM HARRY, b Los Angeles, Calif, Sept 15, 31; c 2. ORGANIC CHEMISTRY. *Educ:* Univ Calif, Los Angeles, BS, 55; Univ Ill, PhD(org chem), 58. *Prof Exp:* Fel, Univ Wash, 58-60; res chemist, Calif Res Corp, 60-63; PROF ORG CHEM, SAN DIEGO STATE UNIV, 63- *Concurrent Pos:* Petrol Res Fund grants, 63-64 & 69-71; US Army Res Off-Durham grants, 65-68, 71-74, 74-77, 77-80 & 80-83; NSF grant, 85-88, prog officer, 90-91; Int Bus Mach Corp, 88-91. *Mem:* Am Chem Soc; Royal Soc Chem. *Res:* Reaction mechanisms of organic peroxides and related chemiluminescence reactions; oxidation of organic compounds; neighboring group reactions; photoinitiators. *Mailing Add:* Dept Chem SDSU San Diego CA 92115

RICHART, F(RANK) E(DWIN), JR, b Urbana, Ill, Dec 6, 18; m 45; c 3. CIVIL ENGINEERING. *Educ:* Univ Ill, BS, 40, MS, 46, PhD(appl mech), 48. *Hon Degrees:* DSc, Univ Fla, 72, Northwestern Univ, 87. *Prof Exp:* Asst, Univ Ill, 46-47, spec res assoc, 47-48; asst prof mech eng, Harvard Univ, 48-52; assoc prof, Univ Fla, 52-54, prof, 54-62; chmn dept, Univ Mich, Ann Arbor, 62-69, prof, 62-67, W J Emmons distinguished prof civil eng, 77-86; RETIRED. *Concurrent Pos:* NSF fac fel, 59-60. *Honors & Awards:* Middlebrooks Award, Am Soc Civil Engrs, 56, 59, 60 & 66, Wellington Prize, 63; Karl Terzaghi lectr, 74; Karl Terzaghi Award, Am Soc Civil Engrs, 80. *Mem:* Nat Acad Eng; hon mem Am Soc Civil Engrs. *Res:* Soil dynamics; soil mechanics and foundations; stress analysis. *Mailing Add:* 2210 Hill St Ann Arbor MI 48104

RICHART, RALPH M, b Wilkes Barre, Pa, Dec 14, 33; c 2. PATHOLOGY, OBSTETRICS & GYNECOLOGY. *Educ:* Johns Hopkins Univ, BA, 54; Univ Rochester, MD, 58. *Prof Exp:* Teaching fel path, Harvard Med Sch, 59-60; instr path & obstet & gynec, Med Col Va, 61-63; from asst prof to assoc prof path, 63-69, PROF PATH, COL PHYSICIANS & SURGEONS, COLUMBIA UNIV, 69-, DIR PATH & CYTOL, SLOANE HOSP, PRESBY HOSP, NEW YORK, 63- *Concurrent Pos:* USPHS spec res fel, 61-63, career res develop award, 61-63 & 65-69; asst vis obstetrician & gynecologist, Harlem Hosp, New York, 63; from asst attend pathologist to assoc attend pathologist, Presby Hosp, 63-69, attend pathologist, 69-; consult, Ford Found Pop Off, 69. *Mem:* Am Soc Cytol; Soc Gynec Invest; assoc fel Am Col Obstetricians & Gynecologists; Int Acad Cytol; Am Asn Path & Bact. *Res:* Cervical neoplasia; human reproduction. *Mailing Add:* Dept Path Div OB/Gyn Path Columbia Univ 630 W 168th St New York NY 10032

RICHARZ, WERNER GUNTER, b Troisdorf, WGer, June 24, 48; Can citizen. AERO ACOUSTICS. *Educ:* Univ Toronto, BASc, 72, MASc, 74, PhD(aero acoust), 78. *Prof Exp:* ASST PROF, INST SPACE SCI, UNIV TORONTO, 78-; MEM STAFF DEPT MECH & AERONAUT ENG,

CARLETON UNIV. *Mem:* Am Inst Aeronaut & Astronaut; Acoust Soc Am; Can Aeronaut & Space Inst. *Res:* Generation of sound by unsteady flows; stability of shear flows; unsteady aerodynamics. *Mailing Add:* Dept Mech & Aeronaut Eng Carleton Univ Ottawa ON K1S 5B6 Can

RICHASON, BENJAMIN FRANKLIN, JR, b Logansport, Ind, July 24, 22. REMOTE SENSING, CARTOGRAPHY. *Educ:* Ind Univ, BA, 48, MA, 49; Univ Nebr, PhD(geog), 60. *Prof Exp:* Instr geog, Morton Jr Col, 49-51; PROF GEOG, CARROLL COL, 52- *Concurrent Pos:* Pres, Wis Coun Conserv Educ, 60-61, Nat Coun Geog Educ, 69-70 & Wis Coun Geog Educ, 78-79; ed, Remote Sensing Quart, 78-84 & Remote Sensing Feature, J Geog, 78-83. *Mem:* Asn Am Geographers; Am Soc Photogrammetry. *Res:* Remote sensing using infrared and radar to explore ore bodies and archaeological sites. *Mailing Add:* 308 E Roberta Ave Waukesha WI 53186

RICHASON, GEORGE R, JR, b Turners Falls, Mass, Apr 3, 16; m 40; c 1. NUCLEAR CHEMISTRY. *Educ:* Univ Mass, BS, 37, MS, 39. *Prof Exp:* Instr high sch, Mass, 39-42 & 46-47; from asst prof to assoc prof, 47-64, ASSOC HEAD DEPT, UNIV MASS, AMHERST, 61-, PROF CHEM, 64- *Mem:* Am Chem Soc; Sigma Xi. *Res:* Radiochemistry. *Mailing Add:* Dept Chem Univ Mass Amherst MA 01003-0035

RICHBERG, CARL GEORGE, b Syracuse, NY, July 10, 28; m 55. FOOD SCIENCE. *Educ:* Syracuse Univ, BS, 51, MS, 53, PhD(microbiol), 56. *Prof Exp:* Asst indust microbiol, bact & food tech, Syracuse Univ, 51-56, effects of radiation on food, Inst Indust Res, 56; res assoc, Res Ctr, 57-72, DEVELOP SCIENTIST, RES CTR, LEVER BROS CO, 72- *Mem:* AAAS; NY Acad Sci; Sigma Xi; Inst Food Technol; Fedn Am Scientists; Am Asn Cereal Chemists. *Res:* Oral microbiology; emulsions and protein chemistry; industrial microbiology; submerged culture methods; germicides; sterilization; dairy science; baking science and technology. *Mailing Add:* 344 Concord St Cressvill NJ 07626

RICHELSON, ELLIOTT, b Cambridge, Mass, Apr 3, 43; m 69; c 3. PSYCHOPHARMACOLOGY. *Educ:* Brandeis Univ, BA, 65; Johns Hopkins Univ, MD, 69; Am Bd Psychiat & Neurol, cert, 76. *Prof Exp:* Asst prof pharmacol & exp therapeut, Sch Med, Johns Hopkins Univ, 72-75; from asst prof to assoc prof psychiat & pharmacol, 75-81, PROF PSYCHIAT & PHARMACOL, MAYO MED SCH, 81-, CONSULT MAYO CLIN, 75-, DIR RES, 88- *Concurrent Pos:* Borden res award med, Sch Med, Johns Hopkins Univ, 69, NIMH res scientist develop award, 74; distinguished investr, Mayo Found, 90; asst secy, Soc Biol Psychiat, 90. *Honors & Awards:* A E Bennet Basic Sci Res Award, Soc Biol Psychiat, 77; Daniel H Efron Award, Am Col Neuropsychopharmacol, 85. *Mem:* Fel Am Psychiat Asn; Am Soc Neurochem; Am Soc Pharmacol & Exp Therapeut; Soc Biol Psychiat. *Res:* Psychiatry and pharmacology. *Mailing Add:* Mayo Clin 4500 San Pablo Rd Jacksonville FL 32224

RICHER, CLAUDE-LISE, b St Hyacinthe, Que, Nov 20, 28. MICROSCOPIC ANATOMY, ENDOCRINOLOGY. *Educ:* Univ Montreal, BA, 48, MD, 54, MS, 57. *Prof Exp:* Asst histol, 57-59, asst prof, 59-68, assoc prof, 68-80, PROF ANAT, UNIV MONTREAL, 80-, ASST DEAN FAC MED, 69- *Mem:* Endocrine Soc; Am Asn Anatomists; Can Asn Anat; Can Physiol Soc; Am Chem Soc. *Res:* Neuroendocrinology; magnesium deficiency and its effect on adrenal function. *Mailing Add:* Dept Chem CP 6128 Succ A Univ Montreal Montreal PQ H3C 3J7 Can

RICHER, HARVEY BRIAN, b Montreal, Que, Apr 7, 44; m 72; c 2. ASTRONOMY. *Educ:* McGill Univ, BS, 65; Univ Rochester, MS, 68, PhD(physics, astron), 70. *Prof Exp:* Assoc astron, Univ Rochester, 65-70; from instr to assoc prof, 70-83, PROF ASTRON, UNIV BC, 83- *Concurrent Pos:* Vis prof, Univ Uppsala, Sweden, 77-78; vis astron, Can-France Hawaii telescope, 84-85; NSERC Grant Selection Comt, 84-86; Killam sr fac fel, 91- *Mem:* AAAS; Am Astron Soc; Can Astron Soc; Royal Astron Soc. *Res:* Carbon stars; globular culsters. *Mailing Add:* Dept Geophys & Astron Univ BC 129 2219 Main Mall Vancouver BC V6T 1W5 Can

RICHER, JEAN-CLAUDE, b Montreal, Que, Feb 23, 33; m 55; c 4. ORGANIC CHEMISTRY. *Educ:* Univ Montreal, BSc, 54, MSc, 56, PhD(chem), 58. *Prof Exp:* Lectr chem, 57-58, from asst prof to assoc prof, 60-70, PROF CHEM, UNIV MONTREAL, 70- *Concurrent Pos:* Res assoc, Univ Notre Dame, 58-60; invited prof, Univ Toulouse, 68-69; sci adv, Sci Coun Can, 71-72. *Honors & Awards:* Herbert Lank Lectr, Univ Del, 66. *Mem:* Am Chem Soc; Chem Inst Can. *Res:* Organometallic derivatives; oxidation reactions; peptide synthesis; nuclear magnetic resonance; conformational analysis. *Mailing Add:* Dept Chem Univ Montreal Montreal PQ H3C 3J7 Can

RICHERSON, HAL BATES, b Phoenix, Ariz, Feb 16, 29; m 53; c 5. ALLERGY, IMMUNOLOGY. *Educ:* Univ Ariz, BS, 50; Northwestern Univ, MD, 54. *Prof Exp:* Resident internal med, Univ Iowa Hosps, 61-64, fel allergy, 64-66, FRP, ASST PROF TO PROF INTENAL MED & DIR, ALLERGY-IMMUNOL DIV, COL MED, UNIV IOWA, 66- *Concurrent Pos:* Consult, Vet Admin Hosp, Iowa City, 66-; vis lectr, Med Sch, Harvard Univ, 68-69; NIH spec fel immunol, Mass Gen Hosp, 68-69; mem, report rec comt allergy-immunol, Liason Comt on Grad Med Educ, 80-85; mem, Pulmonary Dis Adv Comt, Nat Heart, Lung & Blood Inst, NIH, 83-87, Gen Clin Res Ctr, Comn Div Res Resources, NIH, 89-93. *Mem:* Fel Am Col Physicians; fel Am Acad Allergy; Am Asn Immunol; Am Fedn Clin Res; Am Thoracic Soc. *Res:* Study of the lung as an immunologic target organ; animal models of hypersensitivity pneumonitis; pathogenesis of bronchial asthma. *Mailing Add:* Dept Internal Med Univ Iowa Hosps & Clins Iowa City IA 52242

RICHERSON, JIM VERNON, b Bossier City, La, Sept 22, 43; m 65; c 1. ENTOMOLOGY. *Educ:* Univ Mo-Columbia, BA, 65, MSc, 68; Simon Fraser Univ, PhD(biol sci), 72. *Prof Exp:* Res technician, Biol Control Insects Lab, USDA, 65-68; res assoc entom, Pa State Univ, University Park, 72-76; fel entom, Tex A&M Univ, College Station, 76-79; ASST PROF BIOL, SUL ROSS STATE UNIV, ALPINE, TEX, 79- *Mem:* Entom Soc Am. *Res:* Behavior of insects as it relates to pest management and control programs; host-parasite relationships and sex pheromone biology and behavior; medical-veterinary entomology; aquatic entomology; bio-control of range and weeds. *Mailing Add:* Dept Biol Sul Ross State Univ Alpine TX 79832

RICHERSON, PETER JAMES, b San Mateo, Calif, Oct 11, 43; m 78; c 2. LIMNOLOGY, HUMAN ECOLOGY. *Educ:* Univ Calif, Davis, BS, 65, PhD(zool), 69. *Prof Exp:* from asst prof to prof ecol, Div Environ Studies, Univ Calif, Davis, 71-89, dir Inst Ecol, 83-89. *Concurrent Pos:* Consult, Nat Water Comn, 70-71; co-investr, NSF grants, 72- *Honors & Awards:* Guggenheim Fel, 84; Stanley Prize, 89. *Mem:* AAAS; Am Soc Limnol & Oceanog; Ecol Soc Am; Soc Human Ecol; Am Soc Naturalists. *Res:* Human ecology; theory of cultural evolution; plankton community ecology. tropical limnology. *Mailing Add:* Div Environ Studies Univ Calif Davis CA 95616

RICHERT, ANTON STUART, b Newton, Kans, May 19, 35; m 60; c 2. PARTICLE PHYSICS, NUCLEAR PHYSICS. *Educ:* Caltech, BS, 57; Cornell Univ, PhD(exp physics), 62. *Prof Exp:* Res assoc high energy physics, Cornell Univ, 62-63; asst prof physics, Univ Pa, 63-69; assoc prof physics, Ore State Univ, 69-76; sr systs analyst, Sun Studs-Veneer, Roseburg, Ore, 76-86; SR SCIENTIST, BIO-DYNAMICS RES & DEVELOP CORP, EUGENE, ORE, 86- *Mem:* Am Phys Soc. *Res:* Photo production and neutral decays of pion resonances; cosmic ray muons; lepton conservation; pion-nucleus interactions. *Mailing Add:* 581 Brookside Eugene OR 97405

RICHERT, NANCY DEMBECK, b Pittsburgh, Pa, July 23, 45. MOLECULAR BIOLOGY. *Educ:* Univ Rochester, PhD(microbiol), 73. *Prof Exp:* Expert, 78-82, sr staff fel, 82-84, SR INVESTR, NAT CANCER INST, NIH, 84- *Mem:* Sigma Xi; AAAS; Endocrine Soc; Am Soc Cell Biol. *Mailing Add:* 4601 N Park Ave Apt 1702B Chevy Chase MD 20815

RICHES, DAVID WILLIAM HENRY, b Apr 8, 55; m; c 2. PEDIATRICS. *Educ:* Univ Birmingham, Eng, BSc Hons, 76, PhD(immunol), 79. *Prof Exp:* Res fel, Dept Immunol, Univ Birmingham, Eng, 79-83; res fel, 83-85, STAFF RESEARCHER, DEPT PEDIAT, NAT JEWISH CTR IMMUNOL & RESPIRATORY MED, DENVER, 85-; ASST PROF DEPT BIOCHEM, BIOPHYS & GENETICS, HEALTH SCI CTR, UNIV COLO, 85-, ASST PROF, DIIV PULMONARY SCI, DEPT MED, 89- *Concurrent Pos:* grants, Biomed Res Support, NJC, Colo Inst Res, R01 CA50107, & SCOR, 85-97; reviewer, J Am Rev Respiratory Dis, J Immunol, Clin Immunol & Immunopath, Immunol, Biochem Pharmacol, Substance & Alcohol Actions/Misuse & Lymphokine Res; grant reviewer, NSF, Vet Admin Career Develop Prog & NIH. *Mem:* Soc Leukocyte Biol; Am Soc Cell Biol. *Res:* Immunology; pediatrics. *Mailing Add:* Dept Pediat Nat Jewish Ctr Immunol & Respiratory Med 1400 Jackson St Denver CO 80206

RICHES, WESLEY WILLIAM, b Mt Pleasant, Mich, Feb 13, 14; m 41; c 2. CHEMISTRY. *Educ:* Cent Mich Univ, AB, 35; Univ Mich, MS, 36, PhD(chem), 41. *Hon Degrees:* ScD, Cent Mich Univ, 63. *Prof Exp:* Res chemist pigments dept, E I Du Pont De Nemours & Co, 41-55, salesman, 55-63, tech serv chemist, 63-64, group supvr, 64-66, mgr, Chem Dyes, 66-79; RETIRED. *Mem:* Am Chem Soc; Tech Asn Pulp & Paper Inst. *Res:* Pigments; surface chemistry. *Mailing Add:* 207 Cokesbury Village Hockessin DE 19707-1505

RICHEY, CLARENCE B(ENTLEY), b Winnipeg, Man, Dec 28, 10; m 36; c 2. AGRICULTURAL ENGINEERING. *Educ:* Iowa State Univ, BSAE, 33; Purdue Univ, BSME, 39. *Prof Exp:* Time study engr, David Bradley Mfg Works, Ill, 33-36; instr farm power-mach, Purdue Univ, 36-41; asst prof, Ohio State Univ, 41-43; supvr adv develop eng, Elec Wheel Co, Ill, 43-46; proj engr, Harry Ferguson, Inc, Mich, 46-47; res eng, Dearborn Motors Corp, 47-53; supvr tractor & implement div, Ford Motor Co, 53-57, chief res engr, 57-62; chief engr & partner, Five Mfg Co, Ohio, 62-64; chief engr, Fowler Div, Massey-Ferguson Inc, Calif, 64-69, prod mgt engr, Massey-Ferguson Ltd, 70; assoc prof, 70-76, EMER PROF AGR ENG, PURDUE UNIV, 76- *Honors & Awards:* Cyrus Hall McCormick Gold Medal, Am Soc Agr Engrs, 77. *Mem:* Am Soc Agr Engrs. *Res:* Farm equipment; field machinery and tractors; biomass energy & gasification. *Mailing Add:* 2217 Delaware Dr West Lafayette IN 47906

RICHEY, HERMAN GLENN, JR, b Chicago, Ill, May 25, 32; m 62; c 3. ORGANIC CHEMISTRY. *Educ:* Univ Chicago, BA, 52; Harvard Univ, MA, 55, PhD(chem), 59. *Prof Exp:* NSF fel chem, Yale Univ, 58-59; from asst prof to assoc prof, 59-69, asst head chem, 83-88, PROF CHEM, PA STATE UNIV, 69- *Concurrent Pos:* Sloan fel, 64-68; John Simon Guggenheim fel, 67-68; consult, Koppers Co, Inc, 63-88, INDSPEC Chem Corp, 89- *Mem:* Am Chem Soc; Royal Soc Chem; Sigma Xi. *Res:* Reaction mechanisms and intermediates; new synthetic methods; organometallic chemistry of polar main group elements. *Mailing Add:* Dept Chem Pa State Univ University Park PA 16802

RICHEY, WILLIS DALE, b Bedford, Ohio, July 26, 30; c 1. PHYSICAL CHEMISTRY. *Educ:* Hiram Col, BA, 52; Univ Rochester, PhD(chem), 58. *Prof Exp:* Res chemist, Diamond Alkali Co, 57-58; from asst prof to assoc prof chem, Bethany Col, 58-62; vis assoc prof, Colby Col, 62-63; from asst prof to assoc prof, 63-74, Buhl prof, 76-77, PROF CHEM, CHATHAM COL, 74- *Concurrent Pos:* Vis scholar, Freer Gallery of Art, Smithsonian Inst, 69-70; fac res participant, Pittsburgh Energy Technol Ctr, Dept Energy, 78-79. *Mem:* AAAS; Am Chem Soc; Am Phys Soc; Royal Soc Chem; Int Inst Conserv Hist & Artistic Works; Am Geophys Union; Mineral Soc Am. *Res:* Chemical aspects of the conservation of objects of historic and artistic value; thermodynamics of the conversions of inorganic constituents of coals during liquifaction and gasification processes; corosion of metals. *Mailing Add:* Dept Chem Chatham Col Pittsburgh PA 15232

RICHIE, JOHN PETER, JR, b Holden, Mass, Nov 11, 56; m 89. AGING, METABOLIC EPIDEMIOLOGY. *Educ:* Worcester Polytech Inst, BS, 78; Univ Louisville, MS, 83, PhD(biochem), 86. *Prof Exp:* Res assoc toxicol, Dept Pharm & Toxicol, Univ Louisville, 86-87; sr res fel, Div Nutrit & Endocrinol, 87-88, ASSOC RES SCIENTIST, NUTRIT BIOCHEM, DIV NUTRIT CARCINOGENESIS, AM HEALTH FOUND, 88- *Honors & Awards:* George A Sacher Award, Geront Soc Am, 86. *Mem:* Sigma Xi; Soc Exp Biol & Med; Am Soc Biochem & Molecular Biol; Oxygen Soc; Geront Soc Am. *Res:* Biochemistry of aging, with emphasis on glutathione and other redox systems; effects of aging and nutrition on host factors which regulate susceptibility of individuals to diseases and toxins; metabolic epidemiology of diseases of aging. *Mailing Add:* Am Health Found One Dana Rd Valhalla NY 10595

RICHLEY, E(DWARD) A(NTHONY), b Cleveland, Ohio, Sept 5, 28; m 50; c 2. MECHANICAL ENGINEERING, PHYSICS. *Educ:* Cleveland State Univ, BME, 59; Case Western Reserve Univ, MS, 63. *Prof Exp:* Res scientist, Lewis Res Ctr, 59-62, head propulsion components sect, 62-68, chief ion physics br, 68-70, mem dir staff, 70-72, chief off oper analysis & planning, 72-76, chief mgt opers officer, 76-80, DIR ADMIN, LEWIS RES CTR, NASA, 80- *Mem:* Assoc fel Am Inst Aeronaut & Astronaut. *Res:* Institutional and research and development operations analysis and planning. *Mailing Add:* 4801 W 229 St Cleveland OH 44126

RICHLIN, JACK, b New York, NY, Jan 17, 33; m 76; c 2. PHYSICAL CHEMISTRY. *Educ:* Brooklyn Col, BS, 54; Purdue Univ, MS, 57; Rutgers Univ, PhD(chem), 64. *Prof Exp:* Sr res chemist, Allied Chem Corp, 62-65; from asst prof to assoc prof, 65-86, PROF PHYS CHEM, MONMOUTH COL, NJ, 86- *Mem:* Am Chem Soc. *Res:* Polymer physics; computer application to education; solution and surface properties of detergents and surfactants. *Mailing Add:* Dept Chem Monmouth Col W Long Branch NJ 07764

RICHMAN, ALEX, b Winnipeg, Man, Jan 23, 29; m 52; c 4. PSYCHIATRY, EPIDEMIOLOGY. *Educ:* Univ Man, MD, 53; McGill Univ, dipl psychiat, 57; Johns Hopkins Univ, MPH, 60. *Prof Exp:* Staff asst comt ment health serv, Can Ment Health Asn, 56-60; asst prof psychiat, Univ BC, 60; proj dir, Can Royal Comn Health Serv, 62-63; head sect social psychiat, Univ BC, 63-66; med officer, WHO, Geneva, 66-67; assoc prof epidemiol & dir training prog psychiat epidemiol, Columbia Univ, 67-69; prof psychiat, Mt Sinai Sch Med & chief utilization rev psychiat, Beth Israel Med Ctr, 69-78; PROF DEPT PSYCHIAT & PREV MED & DIR TRAINING & RES UNIT PSYCHIAT EPIDEMIOL, DALHOUSIE UNIV, 78- *Concurrent Pos:* Nat Ment Health res award, Can Ment Health Asn, 64; assoc prof psychiat, Univ BC, 64-66; consult, WHO, Ministry Health, Jamaica, 64-66; consult, Southern NB Ment Health Planning Comt, 80-82 & Policy & Planning Unit, Ment Health Br, Ont Ministry of Health, 81-; Nat Health Scientist award, 78-82. *Mem:* Am Psychiat Asn; Biomet Soc; Am Statist Asn; Can Psychiat Asn; Royal Col Psychiat. *Res:* Social psychiatry; epidemiology; mental disorders; quality assurance; evaluation; planning of mental health services. *Mailing Add:* Dept Psych Dalhousie Univ Halifax NS B3H 4H7 Can

RICHMAN, DAVID M(ARTIN), b New York, NY, Mar 13, 32; m 60; c 2. CHEMICAL ENGINEERING, ENERGY SCIENCE. *Educ:* Columbia Univ, AB, 53, BA, 54, MS, 56. *Prof Exp:* Chem engr, Nuclear Eng Dept, Brookhaven Nat Lab, 55-58; radiation specialist, Div Isotope Develop, US AEC, 58-60, chemist, Div Res, 60-71, chief, Eng Chem & Isotope Prep Br, Div Phys Res, 71-72; head, Indust Appln & Chem Sect, Int Atomic Energy Agency, 72-74; chief, Chem Energy & Geosci Br, ERDA, 74-76, sr prog analyst, Off Asst Admin Solar & Geothermal Energy, 76-77; head Prog Planning & Implementation Off, Off Asst Secy Conserv & Solar Appln, Dept Energy, 77-78, sr prog analyst, Off Asst Secy Energy Technol, 79-80, actg dir, Res & Tech Assessment Div, 80, staff phys scientist, Off Basic Energy Sci, Energy Res, 80-90, SCI ADV TO DIR ENERGY RES, DEPT ENERGY, 90- *Mem:* Am Nuclear Soc; Am Chem Soc; Am Inst Chem Engrs. *Res:* Isotopic radiation source design; radiation chemistry; administration of basic research; isotope separations, transplutonium element production; separations chemistry; research materials distribution; solar energy. *Mailing Add:* Off Energy Res ER-1 Dept Energy Washington DC 20585

RICHMAN, DAVID PAUL, b Boston, Mass, June 9, 43; m 69; c 2. NEUROIMMUNOLOGY, EXPERIMENTAL NEUROPATHOLOGY. *Educ:* Princeton Univ, AB, 65; Johns Hopkins Univ, MD, 69. *Prof Exp:* Intern & asst resident, Albert Einstein Col Med, 69-71; asst resident neurol, Mass Gen Hosp, 71-73, chief resident, 73-74, clin & res fel, 74-76; from asst prof to assoc prof, 76-85, PROF, DEPT NEUROL & COMT IMMUNOL & NEURO BIOL, UNIV CHICAGO, 85-, MARJORIE & ROBERT E STRAUS PROF NEUROL SCI, 88- *Concurrent Pos:* Med Avd Bd, Myasthenia Gravis Found; instr neurol, Harvard Univ, 75-76; mem, aging rev comt, Nat Inst Aging, 84-85 & Immunol Sci Study Sect, Diag Related Group, NIH, 86-90. *Mem:* Sigma Xi; Am Acad Neurol; AAAS; Am Asn Immunol; Am Neurol Asn. *Res:* Cellular immunology of neurological diseases and structure and function of acetylcholine receptors; myasthenia gravis and experimental myasthenia; monocloual anti-acetylcholine receptor antibodies and anti -idiotypic antibodies; chimeric antibodies. *Mailing Add:* 4830 S Woodlawn Ave Chicago IL 60615

RICHMAN, DONALD, b Brooklyn, NY, Sept 15, 22. ELECTRICAL ENGINEERING. *Educ:* City Col New York, BEE, 43;Polytech Inst Brooklyn, MEE, 48. *Prof Exp:* PRES, RICHMAN RES CORP, 65- *Mem:* Fel Inst Elec & Electronics Engrs. *Mailing Add:* Richman Res Corp Four Astro Place Dix Hills NY 11746

RICHMAN, ISAAC, b Havana, Cuba, Apr 3, 32; US citizen; m 60; c 2. ELECTROOPTICS, SPECTROSCOPY. *Educ:* Univ Calif, Los Angeles, BA, 54, MA, 58, PhD(physics), 63. *Prof Exp:* Res engr, Elec Div, Nat Cash Register Co, 56-58; mem tech staff crystal physics, Lab Div, Aerospace Corp, 63-66; res physicist, Univ Calif, Los Angeles, 66; SR STAFF SCIENTIST, McDONNELL DOUGLAS ELECTRONIC SYSTS CO, 66- *Mem:* Am Phys Soc. *Res:* Research and development in the areas of infrared and visible detection, imaging, radiometry, and spectroscopy; lattice vibration studies; infrared photoconductor studies, spectroscopy of dielectrics. *Mailing Add:* 1842 Port Manleigh Pl Newport Beach CA 92660-6626

RICHMAN, JUSTIN LEWIS, b Providence, RI, Apr 12, 25; m 57; c 3. MEDICINE, CARDIOLOGY. *Educ:* Brown Univ, AB, 46; Tufts Univ, MD, 49; Am Bd Internal Med, dipl, 56. *Prof Exp:* Lectr, Harvard Univ, 52-53; lectr, Univ, 53-60, from instr to sr instr, 56-58, asst prof, 58-79, ASSOC PROF MED, SCH MED, TUFTS UNIV, 70- *Concurrent Pos:* USPHS fel cardiol, 51-53; physician-in-chg, Dept Med, Boston Dispensary, 56-68, chief electrocardiography lab, 57-68; consult, Mass Heart Asn, 57-60 & NH Heart Asn, 59-; physician-in-chief, Med Clin, New Eng Med Ctr Hosps, 56-68. *Mem:* AAAS; Am Heart Asn; Am Soc Internal Med; fel Am Col Cardiol. *Res:* Clinical cardiology; spatial vectorcardiography and electrocardiography. *Mailing Add:* 25 Boylston St Chestnut Hill MA 02167

RICHMAN, MARC H(ERBERT), b Boston, Mass, Oct 14, 36; m 63. METALLURGY, MATERIALS SCIENCE. *Educ:* Mass Inst Technol, BS, 57, ScD(metall), 63. *Prof Exp:* Instr metall, Mass Inst Technol, 57-60, res asst, 60-63; from asst prof to assoc prof eng, 63-67, dir, Electron Microscopy Facil, 70-86, PROF ENG, BROWN UNIV, 70- *Concurrent Pos:* Instr, Dept Educ, Commonwealth of Mass, 58-62; consult engr, 58 -; adj staff, Dept Med, Miriam Hosp, Providence, RI, 74-86; bioengr, Dept Orthop, RI Hosp, Providence, 79 -; pres, Marc H Richman, Inc, Consult Engrs, 81 - *Honors & Awards:* Outstanding Young Faculty Award, Am Soc Eng Educ, 69; Albert Sauveur Award, Am Soc Metals, 69. *Mem:* Am Soc Metals; Am Crystallog Asn; Am Inst Mining, Metall & Petrol Engrs; fel Am Inst Chem; Am Ceramic Soc. *Res:* Study of phase transformations by optical, electron and field ion microscopy; relation of properties to structure; development of ceramic materials by microstructural design; biomaterials in orthopaedics and cardiovascular systems; forensic engineering. *Mailing Add:* Div Eng Brown Univ Providence RI 02912

RICHMAN, ROBERT MICHAEL, b Pasadena, Calif, Apr 27, 50; m 76. TRANSITION METAL PHOTOCHEMISTRY. *Educ:* Occidental Col, AB, 71; Univ Ill, Urbana, MS, 72, PhD(inorg chem), 76. *Prof Exp:* NSF fel, Calif Inst Technol, 76-77; asst prof, Carnegie-Mellon Univ, 77-84, asst dept head chem, 84-87; ASSOC PROF & CHMN, SCI DEPT, MT ST MARY'S COL, 87- *Mem:* Am Chem Soc. *Res:* Transition metal photochemistry in homogeneous solution aimed at developing new strategies for solar energy conversion. *Mailing Add:* Sci Dept Mt St Mary's Col Emmitsburg MD 21727

RICHMAN, ROGER H, b Newark, NJ, May 4, 29. METALLURGY & PHYSICAL METALLURGICAL ENGINEERING. *Educ:* NMex Inst Mining Technol, BS, 50. *Prof Exp:* PRIN & TECH DIR, DAEDALUS ASSOCS, INC, 85- *Mem:* Fel Am Soc Metals Int; Am Inst Mining Metall & Petrol Engrs; Nat Asn Corrosion Engrs; Mat Res Soc; AAAS; Sigma Xi. *Mailing Add:* Daedalus Assocs Inc 1674 Stierlin Rd Mountain View CA 94043

RICHMAN, SUMNER, b Boston, Mass, Dec 15, 29; m 52; c 3. AQUATIC ECOLOGY. *Educ:* Hartwick Col, AB, 51; Univ Mass, MA, 53; Univ Mich, PhD(zool), 57. *Prof Exp:* From instr biol to assoc prof, 57-70, PROF BIOL, LAWRENCE UNIV, 70-, CHMN DEPT, 77- *Concurrent Pos:* Vis prof marine biol, Tel-Aviv Univ & Marine Lab, Eilat, Israel, 72 & Chesapeake Biol Lab, Univ Md, 74-75, Marine Biol Lab, Woods Hole, 88-89; col accreditation evaluator, N Cent Asn Col & Sec Schs, 72-; Smithsonian Inst Foreign Currency Grant, 72, sea grants, 78-, NSF-ROA grant, 88-89. *Mem:* AAAS; Ecol Soc Am; Am Soc Limnol & Oceanog; Sigma Xi; Int Cong Limnol. *Res:* Energy transformation in aquatic systems; secondary productivity and zooplankton feeding behavior. *Mailing Add:* Dept Biol Lawrence Univ Appleton WI 54912

RICHMOND, ARTHUR DEAN, b Long Beach, Calif, Mar 13, 44; m 71; c 2. ATMOSPHERIC PHYSICS, UPPER ATMOSPHERIC ELECTRODYNAMICS. *Educ:* Univ Calif, Los Angeles, BS, 65, PhD(meteorol), 70. *Prof Exp:* Asst res meteorologist, Univ Calif, Los Angeles, 70-71; Nat Acad Sci resident res assoc, Air Force Cambridge Res Labs, 71-72; res assoc upper atmospheric physics, High Altitude Observ, Nat Ctr Atmospheric Res, 72-76; res assoc, Coop Inst Res Environ Sci, Univ Colo, 76-77; res assoc, Nat Oceanic & Atmospheric Admin, 77-80, space scientist, 80-83; SCIENTIST, NAT CTR ATMOSPHERIC RES, 83- *Concurrent Pos:* Consult, Rand Corp, 66-69; NATO fel in sci, Lab Physique de l'Exosphere, Univ Paris, 73-74; comt mem, Comt Solar-Terrestrial Res, Nat Res Coun, 82-85; Japan Soc for Prom Sci Fel, 86; co-chmn, div 2, Int Assoc Geomagnetism & Aeronomy, 87- *Mem:* AAAS; Am Geophys Union; Am Meteorol Soc. *Res:* Upper atmospheric electric fields and currents; geomagnetism; atmospheric dynamics. *Mailing Add:* NCAR High Altitude Observ Boulder CO 80307-3000

RICHMOND, CHARLES WILLIAM, b New Martinsville, WVa, Jan 8, 38; m 66; c 3. ORGANIC CHEMISTRY. *Educ:* David Lipscomb Col, BA, 60; Univ Miss, PhD(org chem), 64. *Prof Exp:* Asst prof chem, David Lipscomb Col, 64-69; assoc prof, 69, PROF CHEM, UNIV NORTH ALA, 69- *Mem:* Am Chem Soc. *Res:* Preparation of heterocyclic compounds for use as potential drugs. *Mailing Add:* Dept Chem Univ N Ala Florence AL 35632

RICHMOND, CHESTER ROBERT, b South Amboy, NJ, May 29, 29; m 52; c 4. RADIOBIOLOGY, RISK ANALYSIS. *Educ:* NJ State Col, Montclair, BA, 52; Univ NMex, MS, 54, PhD(biol), 58. *Prof Exp:* Asst physiol, Univ NMex, 54-55; asst, Los Alamos Sci Lab, 55-57, mem staff, 57-68; mem staff, Div Biol & Med, US AEC, 68-71; leader biomed res group, Los Alamos Sci Lab, 71-73, alternate health div leader, 73-74; ASSOC DIR BIOMED & ENVIRON SCI, OAK RIDGE NAT LAB, 74-; PROF BIOMED SCI, UNIV TENN-OAK RIDGE GRAD SCH BIOMED SCI, 75- *Concurrent Pos:* Mem, Nat Coun Radiation Protection & Measurements, 74-; comt 2, Int

Comn Radiol Protection, 77-; bd dirs, Nat Coun Radiation Protection & Measurements & Inst Biomed Imaging, Univ Tenn, 88- *Honors & Awards:* E O Lawrence Award, US AEC, 74; G Failla Award & Lectr, Radiation Res Soc, 76; W H Langham Mem Lectr, Univ Ky, 87. *Mem:* Fel AAAS; Health Physics Soc; Sigma Xi; Soc Risk Analysis; Radiation Res Soc; NY Acad Sci. *Res:* Water and electrolyte metabolism; comparative metabolism of radionuclides; anthropometry; biological effects of internal emitters; health & environ effects of energy production; radiobiology of actinide elements. *Mailing Add:* Oak Ridge Nat Lab PO Box 2008 Oak Ridge TN 37830

RICHMOND, F(RANCIS) M(ARTIN), metallurgical engineering; deceased, see previous edition for last biography

RICHMOND, GERALD MARTIN, b Providence, RI, July 30, 14; m 41, 67; c 4. QUATERNARY GEOLOGY. *Educ:* Brown Univ, BA, 36; Harvard Univ, MA, 39; Univ Colo, PhD, 54. *Prof Exp:* Instr geol, Univ Conn, 40; geologist, Spec Eng Div, Panama Canal, 41-42; SR GEOLOGIST, BR CENT GEN GEOL, US GEOL SURV, 42- *Concurrent Pos:* NSF grant, Alps, 60-61; mem qual adv group, US Comn Stratig Nomenclature; mem stratig comn, Int Union Quaternary Res, 77-; mem, US Nat Comt for Int Geol Correlation Prog, 74-78; US working group, Quaternary Glaciation in Northern Hemisphere, 74- *Honors & Awards:* Kirk Bryan Award, Geol Soc Am, 65; Albrecht Penck Medal, Deutsche Quartarvereinigung, 78. *Mem:* AAAS; Int Union Quaternary Res (secy gen, 62-65, pres, 65-69); fel Geol Soc Am; Am Quaternary Asn; Sigma Xi. *Res:* Glacial and surficial geology; quaternary stratigraphy and correlation; fossil soils; geomorphology. *Mailing Add:* US Geol Surv Denver Fed Ctr Denver CO 80225

RICHMOND, ISABELLE LOUISE, NEUROBIOLOGY, NEUROSURGERY. *Educ:* Cornell Univ, PhD(neurobiol), 68. *Prof Exp:* ASSOC PROF NEUROSURG, EASTERN VA MED SCH, 84-; CLIN PROF SURG, UNIFORMED SERVS UNIV, 86- *Res:* Nerve repair; pituitary adenoma. *Mailing Add:* 408 Medical Tower Norfolk VA 23507

RICHMOND, J(ACK) H(UBERT), b Kalispell, Mont, July 30, 22; m 46; c 3. ELECTRICAL ENGINEERING. *Educ:* Lafayette Col, BS, 50; Ohio State Univ, MSc, 52, PhD(elec eng), 55. *Prof Exp:* Res assoc, 52-54, asst supvr, 54-56, from asst prof to assoc prof, 55-62, PROF ELEC ENG, OHIO STATE UNIV, 62-, ASSOC SUPVR, 56- *Mem:* Fel Inst Elec & Electronics Engrs. *Res:* Electromagnetic field theory; antennas; radomes; scattering. *Mailing Add:* 4678 Johnstown Rd Gahanna OH 43230

RICHMOND, JAMES KENNETH, b Chattanooga, Tenn, June 23, 20; m 44; c 4. PHYSICS. *Educ:* Ga Inst Technol, BS, 42; Univ Pittsburgh, MS, 43, MA, 58. *Prof Exp:* Res engr, Westinghouse Elec Corp, 42-44, 47-49; asst nuclear physics, Univ Pa, 46-47; physicist combustion, US Bur Mines, 49-59; res specialist advan propulsion, Sci Res Labs, Boeing Co, 59-69, sr basic res scientist, 69-71; res physicist, US Bur Mines, 71-80, suprvry res physicist, 80-83; CONSULT, MINING & INDUST CADRE, 83- *Concurrent Pos:* Instr, Carnegie Inst Technol, 55-59; consult, Comt Fire Res & Fire Res Conf, Nat Acad Sci-Nat Res Coun, 59; vis instr, Community Col, Allegheny County, PA, 84-87; instr, Bellevue & Edmond's Community Cols, 89- *Honors & Awards:* US Dept Interior Award, 59 & 75. *Mem:* Am Phys Soc; Inst Elec & Electronics Engrs. *Res:* Prevention of fires and explosions in coal mines, oil shale mines and other mines; supervision of group engaged in conducting research on full-scale explosions in experimental mines and the instrumentation thereof. *Mailing Add:* 12553 37th Ave NE Seattle WA 98125

RICHMOND, JAMES M, b Armstrong, Iowa, July 29, 41; m 63; c 1. NITROGEN CHEMISTRY OF FATTY ACIDS, SURFACTANTS. *Educ:* Iowa State Univ, BS, 63; Kans State Univ, PhD(org chem), 74. *Prof Exp:* Captain, USMC, 63-69; chemist, Procter & Gamble Co, 74-76; chemist, Armak Co, 76-78, sect head, 78-80, area mgr, Akzo Chemie Am, Akzo Am 80-83, dept head, 83-89, CHEM RES MGR, AKZO CHEMICALS, 89- *Concurrent Pos:* Edm Cationic Surfactants, Org Chem, 90. *Mem:* Am Chem Soc; Am Oil Chemists Soc; Sugar Indust Technol. *Res:* Process and product development of nitrogen derivatives of natural fats and oils, especially amines, amides, ethoxylates, nitriles and quaternary ammonium salts; author of one book and 14 patents. *Mailing Add:* 29 W 304 Hartman Dr Naperville IL 60564

RICHMOND, JONAS EDWARD, b Prentiss, Miss, July 17, 29; m 57; c 2. BIOCHEMISTRY. *Educ:* Univ Tenn, BS, 48; Univ Rochester, MS, 50, PhD(biochem, biophys), 53. *Prof Exp:* Res assoc biophys, Univ Rochester, 50-55, instr biophys & biochem, 56-57; estab investr, Am Heart Asn, Harvard Med Sch, 57-63; assoc biochemist, 63-69, BIOCHEMIST, UNIV CALIF, BERKELEY, 69- *Concurrent Pos:* NIH fel, Univ Rochester, 53-55; Commonwealth fel, Oxford Univ, 55-56; mem, Allergy & Immunol A Study Sect, NIH, 65-69, nutrit study sect, Marc Study Sect; consult, US HEW, 66- & NIH, 71, 74; pres, Alameda County Heart Asn, 74-; vpres, Am Heart Asn Calif affil, chmn, Res Comn. *Mem:* AAAS; Am Chem Soc; Radiation Res Soc; Am Soc Biol Chem; Biophys Soc; Sigma Xi; Am Physiol Soc; Am Soc Cell Biol. *Res:* Chemistry and biochemistry of proteins; protein and amino acid metabolism; transport and membrane function; intermediary metabolism; radiation chemistry; biochemistry and biophysics of growth; molecular biology; cell recognition and differentiation; cell surface chemistry. *Mailing Add:* Dept Nutrit Sci 219 Morgan Hall Univ Calif Berkeley CA 94720

RICHMOND, JONATHAN YOUNG, b Norwalk, Conn, Feb 10, 41; m 66; c 4. VIROLOGY, GENETICS. *Educ:* Univ Conn, BA, 62, MS, 64; Hahnemann Med Col, PhD(genetics), 68. *Prof Exp:* NSF-Nat Res Coun fel virol & cytol, US Dept Agr, 67-69, res microbiologist, 69-79, biol safety officer, Plum Island Animal Dis Ctr, 79-83; chief, Safety Opers Sect, Occup Safety & Health Br, Div Safety, NIH, 83-90; DIR, OFF HEALTH & SAFETY, CTR DIS CONTROL, 90- *Concurrent Pos:* Safety specialist, Disaster Med Assistance Team, PHHS, 83-88, AIDS video tech expert, 88. *Honors & Awards:* Silver Beaver Award, Bd Sci Affairs, 81; Cert Merit, USDA/Agr Res Serv, 83. *Mem:* Am Asn Lab Animal Sci; Am Biol Safety Asn (pres, 86); fel Am Acad Microbiol. *Res:* Virus/cell interrelationships; sterilization of biological materials by gamma irridation; development and presentation training programs in: occupational safety and health, biological safety, chemical safety, animal use and care, infectious waste management. *Mailing Add:* Off Health & Safety MS GO8 Ctr Dis Control Atlanta GA 30333

RICHMOND, JULIUS BENJAMIN, b Chicago, Ill, Sept 26, 16; m 37; c 2. MEDICINE. *Educ:* Univ Ill, BS, 37, MD & MS, 39; Am Bd Pediat, dipl. *Prof Exp:* Resident, Cook County Hosp, Chicago, Ill, 46; prof pediat, Col Med, Univ Ill, 46-53; prof & chmn dept, Col Med, State Univ NY Upstate Med Ctr, 53-71; prof child psychiat & human develop, Fac Pub Health & Fac Med, Harvard Med Sch, 71-73, prof prev & social med & chmn dept, 71-79, prof health policy, 81-88, dir, Div Health Policy Res & Educ & John D Macarthur prof mgt, 83-88, EMER PROF, HARVARD MED SCH, 88- *Concurrent Pos:* Markle Found scholar med sci, 48-53; supt inst juvenile res, Ill State Dept Pub Welfare, 52-53; dir, Proj Headstart, Off Econ Opportunity, 65-; dean med fac, State Univ NY Upstate Med Ctr, 65-71; psychiatrist in chief, Children's Hosp Med Ctr, Boston, 71-74; trustee, Child Welfare League Am; asst secy for Health & Surgeon Gen, USPHS, HEW, 77-81; adv child health policy, Children's Hosp Med Ctr, 81- *Honors & Awards:* Guestz Lienhart Award, Inst Med, Nat Acad Sci, 86. *Mem:* Inst Med-Nat Acad Sci; Health Policy Soc; Am Pub Health Asn; Soc Exp Biol & Med; Soc Pediat Res. *Res:* Pediatrics; psychological aspects of pediatrics; child development. *Mailing Add:* 79 Beverly Rd Chestnut Hill MA 02167

RICHMOND, MARTHA ELLIS, b Wilmington, Del, Sept 10, 41; m 69; c 2. BIOCHEMISTRY. *Educ:* Wellesley Col, Mass, AB, 62; Tufts Univ, PhD(biochem), 69; Harvard Univ, MPH, 88. *Prof Exp:* Res fel bact & immunol, Harvard Med Sch, 69-70; res assoc med, Sch Med, Tufts Univ, 70-73; lectr, 73-74, asst prof biol, Univ Mass, Boston, 74-75; from asst prof to assoc prof, 75-83, PROF CHEM, SUFFOLK UNIV, 83- *Concurrent Pos:* Consult staff scientist, Health Effects Inst, Cambridge, Mass, 89- *Mem:* Sigma Xi; Am Chem Soc. *Res:* Biosynthesis of complex carbohydrates; regulation of complex carbohydrate synthesis in mammalian systems. *Mailing Add:* Dept Chem Suffolk Univ Beacon Hill Boston MA 02114

RICHMOND, MILO EUGENE, b Cutler, Ill, Aug 29, 39. VERTEBRATE ZOOLOGY, REPRODUCTIVE BIOLOGY. *Educ:* Southern Ill Univ, Carbondale, BA & BS, 61; Univ Mo-Columbia, MS, 63, PhD(zool), 67. *Prof Exp:* Asst instr zool, Univ Mo-Columbia, 64-67; asst prof biol, ETenn State Univ, 67-68; ASST PROF WILDLIFE SCI, NY STATE COL AGR & LIFE SCI & ASST LEADER NY COOP WILDLIFE RES UNIT, CORNELL UNIV, 68- *Mem:* Am Soc Mammal; Wildlife Soc. *Res:* Ecology and physiology of reproduction of vertebrates; mammalian population dynamics, especially microtine rodents. *Mailing Add:* Dept Natural Resources 206-B Fernow Hall Cornell Univ Ithaca NY 14850

RICHMOND, PATRICIA ANN, b Salina, Kans, May 18, 47; m 82. DEVELOPMENT OF SPECIALTY PROTEIN & STARCH BASED PRODUCTS. *Educ:* Kans State Univ, BS, 70; Cornell Univ, MS, 72, PhD(food chem), 75. *Prof Exp:* Food technologist, A E Staley Mfg Co, 75-77, lab mgr new prod develop, 77-79, group mgr sweetner develop, 79-85, mgr food & indust starch res & develop, Staley Continental Inc, 85-88, DIR FOOD INGREDIENT RES & DEVELOP, A E STALEY MFG CO, 88- *Mem:* Inst Food Technologists; Am Asn Cereal Chemists. *Res:* Development of specialty protein and starch-based products for the food industry; development of high fructose corn syrup and other corn-based sweetners. *Mailing Add:* A E Staley Mfg Co 2200 E Eldorado Decatur IL 60525

RICHMOND, ROBERT CHAFFEE, b New York, NY, May 3, 43; m 68; c 1. RADIOBIOLOGY. *Educ:* Univ NH, BA, 66; Univ Tex, Austin, MA, 70, PhD(radiation biol), 72. *Prof Exp:* Teacher biol & chem, Chester High Sch, Vt, 66-68; res assoc radiation biol, Univ Kans, Lawrence, 73-75; vis scientist radiation biol, US Army Natick Develop Ctr, 75-77; instr radiol & radiation med, Boston Univ Med Ctr, 77-79; res asst prof, 79-87, RES ASSOC PROF, NORRIS COTTON CANCER CTR, DARTMOUTH-HITCHCOCK MED CTR, HANOVER, NH, 87- *Concurrent Pos:* Nat Res Coun fel, 75-77. *Mem:* Radiation Res Soc; Sigma Xi; AAAS; NY Acad Sci. *Res:* Chemical and thermal potentiation of cellular sensitivity and mutagenesis to radiation; chemistry and consequences of radiation-induced and antitumor drug-induced damage. *Mailing Add:* Norris Cotton Cancer Ctr Dartmouth Hitchcock Med Ctr Hanover NH 03755

RICHMOND, ROBERT H, b White Plains, NY, May 26, 54. INVERTEBRATE REPRODUCTIVE BIOLOGY, ENVIRONMENTAL BIOLOGY OF TROPICS. *Educ:* Univ Rochester, BS, 76; State Univ NY, MS, 82, PhD(biol sci), 83. *Prof Exp:* Res tech, Dept Radiation Biol & Biophys, Univ Rochester, 75-76; res, Mid Pac Res Lab, 80-82; fel, Smithsonian Trop Res Inst, 84-85, Smithsonian Inst, 85-86; from asst prof to assoc prof marine biol, 86-88, DIR, UNIV GUAM MARINE LAB, 88- *Concurrent Pos:* Adj grad fac, Dept Zool, Univ Hawaii, Manoa, 80- *Mem:* Am Soc Zoologists; Western Soc Naturalists; Int Soc Reef Studies. *Res:* Coral reef biology and ecology; reproductive biology and larval ecology of invertebrates; population biology and population genetics; environmental biology of tropical marine and island communities; evolutionary biology. *Mailing Add:* Marine Lab Univ Guam Univ Guam Sta Mangilao GU 96923

RICHMOND, ROLLIN CHARLES, b Nairobi, Kenya, May 31, 44; US citizen; m 75. POPULATION GENETICS, BIOCHEMICAL GENETICS. *Educ:* San Diego State Univ, AB, 66; Rockefeller Univ, PhD(genetics), 71. *Prof Exp:* From asst prof to assoc prof zool, Ind Univ, 70-75; assoc prof genetics, NC State Univ, 76; assoc prof, Ind Univ, 76-81, prof, 81-90; DEAN, COL ARTS & SCI, UNIV SFLA, 90- *Concurrent Pos:* Assoc ed, J Soc Study Evolution, 75-77, Genetica, 81- & J Heredity, 81- *Mem:* Sigma Xi; Genetics Soc Am; Soc Study Evolution; Am Soc Naturalists; Ecol Soc Am. *Res:* Population genetics of natural and artifical populations of Drosophila with particular emphasis on the adaptive significance of isozyme variants; behavioral genetics of Drosophila. *Mailing Add:* Col Arts & Sci Univ SFla Tampa FL 33620

RICHMOND, THOMAS G, b Buffalo, NY, Jan 4, 57; m 89. ACTIVATION OF C-F BONDS. *Educ:* Brown Univ, ScB, 79; Northwestern Univ, PhD(chem), 84. *Prof Exp:* Res fel, Calif Inst Technol, 83-85; asst prof, 85-91, ASSOC PROF CHEM, UNIV UTAH, 91- *Concurrent Pos:* Camille & Henry Dreyfus new fac fel, Dreyfus Found, 85; NSF presidential young investr, 89; Alfred P Sloan res fel, 91-93. *Mem:* Am Chem Soc. *Res:* Inorganic and organometallic chemistry; activation of C-F bonds; metal based molecular receptors; coordination chemistry and the design of new materials; environmental chemistry. *Mailing Add:* Dept Chem Univ Utah Salt Lake City UT 84112

RICHMOND, VIRGINIA, biochemistry, for more information see previous edition

RICHMOND, WILLIAM D, b Denver, Colo, July 19, 25; m 48; c 5. MECHANICAL ENGINEERING. *Educ:* Univ Wis, BSME, 46. *Prof Exp:* Engr, Bur Reclamation, 46-47, & Hanford Atomic Proj Opers, 47, Gen Elec Co, 47-64, proj engr, 47-56, supvr proj engr, 56-59, plant mgr reactor opers, 59-68; asst lab dir, 68-71, dir proj & facil, Pac Northwest Div, 71-79, dir, Hanford Proj, 79-85, DEP DIR, ENG TECHNOL, BATTELLE MEM INST. 85- *Res:* Nuclear reactors. *Mailing Add:* 1014 Cedar Ave Richland WA 99352

RICHSTONE, DOUGLAS ORANGE, b Alexandria, Va, Sept 20, 49. ASTRONOMY, ASTROPHYSICS. *Educ:* Calif Inst Technol, BS, 71; Princeton Univ, PhD(astrophys), 75. *Prof Exp:* Res fel astron, Calif Inst Technol, 74-76; asst prof physics, Univ Pittsburgh, 77-80; from asst prof to assoc prof astron, 80-88, chmn, Astron Dept, 85-90, PROF ASTRON, UNIV MICH, 88- *Mem:* Royal Astron Soc; Am Astron Soc; Int Astron Union. *Res:* Quasi-stellar objects; structure of galaxies and clusters of galaxies; stellar dynamics; cosmology. *Mailing Add:* Dept Astron Univ Mich Ann Arbor MI 48109

RICHTER, BURTON, b Brooklyn, NY, Mar 22, 31; m 60; c 2. PHYSICS. *Educ:* Mass Inst Technol, BS, 52, PhD(physics), 56. *Prof Exp:* Res assoc, High Energy Physics Lab, 56-60, asst prof, Physics Dept, 60-63, assoc prof, Stanford Linear Accelerator Ctr, 63-67, tech dir, 82-84, PROF, STANFORD LINEAR ACCELERATOR CTR, STANFORD UNIV, 67-, PAUL PIGOTT PROF PHYS SCI, 79-, DIR, 84- *Concurrent Pos:* Consult, Dept of Energy, NSF; mem Gen Motors sci adv comt, 81-87; bd dir, Teknowledge Inc, Middlefield Capital Corp; mem Accelerator Adv Comt, Desy, Hamburg, Ger; mem adv comt, Lincoln Lab, Mass Inst Technol; mem, Hist Sci Prog Comt, Stanford Univ; mem bd dirs, Varian Corp, Litel Instruments. *Honors & Awards:* Nobel Prize in Physics, 76; Loeb Lectr, Harvard Univ, 74; DeShalit Lectr, Weizmann Inst, 75; E O Lawrence Medal, US Dept Energy, 76. *Mem:* Nat Acad Sci; fel AAAS; fel Am Phys Soc; Europ Phys Soc; fel Am Acad Arts & Sci. *Res:* High energy physics; particle accelerators. *Mailing Add:* Stanford Linear Accelerator Ctr Stanford Univ PO Box 4349 Mail Bin 80 Stanford CA 94309

RICHTER, CURT PAUL, psychobiology; deceased, see previous edition for last biography

RICHTER, DONALD, b Brooklyn, NY, Sept 3, 30; m 67; c 2. STATISTICS, DATA ANALYSIS. *Educ:* Bowdoin Col, AB, 52; Univ NC, PhD(statist), 59. *Prof Exp:* Asst prof statist, Univ Minn, 59-61; mem tech staff, Bell Tel Labs, 61-64; assoc prof statist, 64-72, dir doctoral off, 74-76, PROF STATIST, GRAD SCH BUS, NY UNIV, 72- *Mem:* Am Statist Asn; Inst Math Statist; Int Asn Statist Comput. *Res:* Statistical methods; statistical software. *Mailing Add:* Dept Statist Grad Sch Bus New York Univ 100 Trinity Pl New York NY 10006

RICHTER, DOROTHY ANNE, b New Britain, Conn, June 26, 48. APPLIED, ECONOMIC & ENGINEERING GEOLOGY. *Educ:* Bates Col, BS, 70; Boston Col, MS, 73; Harvard Univ, 75-80. *Prof Exp:* Staff geologist, dept earth & planetary sci, Mass Inst Technol, 72-76; chief geologist, Rock of Ages Corp, 76-84; PRIN & SR GEOLOGIST, HAGER-RICHTER GEOSCI, INC, 84- *Mem:* Sigma Xi; Geol Soc Am; Am Inst Prof Geologists; Am Soc Testing & Mat. *Res:* Effects of microstructures on physical properties of terrestrial and lunar igneous rocks; geology of dimension stone resources and applications. *Mailing Add:* Hager-Richter Geosci Inc Eight Industrial Way-DIO Salem NH 03079

RICHTER, EDWARD EUGENE, b Hebron, Ill, Oct 22, 19; m 41; c 2. ORGANIC CHEMISTRY, INORGANIC CHEMISTRY. *Educ:* DePauw Univ, BA, 41. *Prof Exp:* Chemist, Jones-Dabney Co, 41-44 & Am-Marietta Co, 46-50; chief chemist, Kay & Ess Co, 50-51; tech dir & spec projs engr, Moran Paint Co, 51-60; tech dir & vpres, Blatz Paint Co Inc, 60-89; RETIRED. *Mem:* Am Chem Soc; Fedn Socs Coatings Technol. *Res:* Protective and decorative industrial type organic coatings. *Mailing Add:* 304 Old Farm Rd Louisville KY 40207-2308

RICHTER, ERWIN (WILLIAM), b Ironwood, Mich, Jan 29, 34. BIOCHEMISTRY. *Educ:* Northern Mich Univ, BS, 56; Univ Northern Iowa, MS, 63; Univ Iowa, PhD(biochem), 70. *Prof Exp:* High sch teacher, Mich, 56-62; instr phys sci, 63-67, asst prof chem, 70-72, ASSOC PROF CHEM, UNIV NORTHERN IOWA, 72- *Res:* Particle analysis of atmospheric pollutants; structure of proteins; pesticide residues in fish, phesants and rabbits in Iowa; science education. *Mailing Add:* Seven Mohawk Dr Action MA 01720

RICHTER, G PAUL, b Rahway, NJ, Aug 13, 37; m 61; c 2. CHEMICAL EDUCATION. *Educ:* Grinnell Col, BA, 59; Univ Minn, PhD(inorg chem), 68. *Prof Exp:* Asst prof, 65-59, assoc prof, 69-78, PROF CHEM, WVA WESLEYAN COL, 78- *Concurrent Pos:* Researcher, Univ Gottingen, W Germany, 76-77. *Mem:* AAAS; Am Chem Soc; Nat Wildlife Fedn; Nat Speleological Soc; Am Forestry Asn; Am Asn Univ Profs. *Res:* Sulfur-nitrogen chemistry; nonmetal compounds; inorganic aquatic chemistry. *Mailing Add:* c/o Darl Richter 16 Sutton Pl Groton CT 06340

RICHTER, GEORGE NEAL, b Denver, Colo, Mar 13, 30; m 60; c 5. GASIFICATION. *Educ:* Yale Univ, BE, 51; Calif Inst Technol, MS, 52, PhD(chem eng), 57. *Prof Exp:* Res fel, Calif Inst Technol, 57-58, asst prof chem eng, 58-65; sr res chem engr, Montebello Res Lab, 65-80, technologist, 80-83, res mgr, 83-89, RES FEL, TEXACO INC, MONTEBELLO RES LAB, 89- *Mem:* Am Inst Chem Engrs. *Res:* Synthesis gas and hydrogen generation by partial oxidation; coal gasification and conversion; alternate energy process development and commercialization; waste gasification and environmental impact measurements; research planning and coordination. *Mailing Add:* 1470 Granada Ave San Marino CA 91108

RICHTER, GOETZ WILFRIED, b Berlin, Germany, Dec 19, 22; nat US; wid; c 3. PATHOLOGY, CELL BIOLOGY. *Educ:* Williams Col, AB, 43; Johns Hopkins Univ, MD, 48; Am Bd Path, dipl, 53. *Prof Exp:* Instr path, Med Col, Cornell Univ, 50-51, res assoc, 51-53, from asst prof to assoc prof, 53-67; PROF PATH, UNIV ROCHESTER, 67- *Concurrent Pos:* Ledyard fel, NY Hosp & Cornell Univ, 51-53; Rockefeller Found grant, 56-61; vis investr, Rockefeller Inst, 56-57; Health Res Coun New York career scientist, 61-67; consult, NIH, 63-67, ad hoc, 68-; ed, Int Rev Exp Path; ed, Beitraege zur Pathologie, 71-78, Path Res & Pract, 78-90, Am J Path, 65-86. *Mem:* Electron Micros Soc Am; Soc Exp Biol & Med; Am Asn Path; Am Soc Cell Biol; Int Acad Path; AAAS; Sigma Xi; Harvey Soc. *Res:* Experimental pathology; cell biology; ferritin and iron metabolism; pathology of heart muscle; lead poisoning. *Mailing Add:* Univ Rochester Dept Path 601 Elmwood Ave Rochester NY 14642

RICHTER, HAROLD GENE, b Fontanet, Ind, Mar 5, 25; m 54; c 4. INORGANIC CHEMISTRY. *Educ:* Franklin Col, AB, 47; Mass Inst Technol, MS, 50, PhD(inorg chem), 52. *Prof Exp:* Jr physicist, Argonne Nat Lab, 47-48; instr chem, Univ Ore, 52-54; radiochemist, US Naval Radiol Defense Lab, Calif, 54-55; asst to pres, Nuclear Sci & Eng Corp, Pa, 55-58; chief radiochemist, Res Triangle Inst, 59-71; CHEMIST, OFF AIR PROGS, ENVIRON PROTECTION AGENCY, 71- *Mem:* AAAS; Am Chem Soc; Sigma Xi. *Res:* Radiochemistry of fission products; industrial applications of isotopes; analytical chemistry; biomedical instrumentation; atmospheric chemistry; air pollution control. *Mailing Add:* 1202 Willow Dr Chapel Hill NC 27514

RICHTER, HELEN WILKINSON, b Biloxi, Miss; c 1. RADIATION CHEMISTRY, FREE RADICAL REACTIONS. *Educ:* Woman's Col Ga, BA, 67; Ohio State Univ, MS, 70, PhD(phys chem), 74. *Prof Exp:* Res assoc radiation chem, Brookhaven Nat Lab, Upton, NY, 74-75; res chemist, Radiation Res Labs, 76, sr res chemist, dept chem, Carnegie-Mellon Univ, 76-84; asst prof, 84-89, ASSOC PROF DEPT CHEM, UNIV AKRON, 89- *Concurrent Pos:* Fel, Nat Defense Educ Act Title IV, 67-70; vis res assoc & vis scientist, Radiation Lab, Univ Notre Dame, 76-; Samuel & Emma Winters fel, 80. *Mem:* Am Asn Adv Sci; Am Chem Soc; Soc Free Res. *Res:* Application of physical chemistry and radiation chemistry to the study of biochemical reaction mechanisms; free radical reactions; oxy-radicals of biological interest; iron-catalyzed peroxide decompositions. *Mailing Add:* Dept Chem Univ Akron Akron OH 44325-3601

RICHTER, HERBERT PETER, b St Paul, Minn, Jan 22, 39; m 62; c 2. PHYSICAL CHEMISTRY, POLYMER CHEMISTRY. *Educ:* San Diego State Univ, BA, 65, MS, 67. *Prof Exp:* Teaching asst & res asst, Chem Dept, San Diego State Univ, 66-67; chemist, Electronic Warfare Dept, 67-70; res chemist, 70-79, res coordr, Polymer Chem, 79-86, HEAD, COMBUSTION & DETONATION RES BR, RES DEPT, NAVAL WEAPONS CTR, 86- *Mem:* Am Chem Soc; Sigma Xi. *Res:* Energetic materials; polymer and surface chemistry; physical and mechanical properties of polymers; photochemistry and photophysics; chemiluminescence. *Mailing Add:* Naval Weapons Ctr Code 3891 China Lake CA 93555

RICHTER, JOEL EDWARD, b Newport, RI, Oct 23, 49; m 71; c 3. INTERNAL MEDICINE, GASTROENTEROLOGY. *Educ:* Tex A&M Univ, BS, 70; Univ Tex Southwestern Med Sch, MD, 75. *Prof Exp:* Intern internal med, US Naval Hosp, Philadelphia, 75-76; resident, Nat Naval Med Ctr, Bethesda, Md, 76-78, fel gastroenterol, 78-80, staff, 80-82; from asst prof to assoc prof gastroenterol, Bowman Gray Med Sch, Winston-Salem, NC, 82-89; PROF MED DIR CLIN RES, UNIV ALA, BIRMINGHAM, 89- *Concurrent Pos:* Staff gastroenterologist, NC Baptist Hosp, Winston-Salem, 82- *Mem:* Am Gastroenterol Asn; fel Am Col Physicians; Am Soc Gastrointestinal Endoscopy; Am Asn Advan Liver Dis; fel Am Col Gastroenterol. *Res:* Gastroesophagal reflux; esophagal motility disorders; non-cardiac chest pain; functional gastrointestinal diseases. *Mailing Add:* Gastrointestinal Div Univ Ala Birmingham AL 35294

RICHTER, JOHN LEWIS, b Laredo, Tex, July 26, 33; m 58; c 2. NUCLEAR PHYSICS. *Educ:* Univ Tex, Austin, BS, 54, MS, 56, PhD(physics), 58. *Prof Exp:* STAFF MEM PHYSICS, LOS ALAMOS SCI LAB, 58-, GROUP LEADER, TD-4, 78- *Mem:* Am Phys Soc. *Res:* Nuclear weapon design. *Mailing Add:* Los Alamos Nat Lab MS 669 PO Box 1663 Los Alamos NM 87545

RICHTER, JUDITH ANNE, b Wilmington, Del, Mar 4, 42. NEUROPHARMACOLOGY, NEUROCHEMISTRY. *Educ:* Univ Colo, BA, 64; Stanford Univ, PhD(pharmacol), 69. *Prof Exp:* Wellcome Trust fel, Cambridge Univ, 69-70 & Inst Psychiat, Univ London, 70-71; from asst prof pharmacol to assoc prof pharmacol & neurobiol, 71-84, PROF PHARMACOL & NEUROBIOL, MED SCH, IND UNIV, INDIANAPOLIS, 84- *Concurrent Pos:* Vis assoc prof pharmacol, Univ Ariz Health Sci Ctr, Tucson, 83; mem, Biomed Res Rev Comn, Nat Inst Drug Abuse, 83-87. *Mem:* AAAS; Am Soc Neurochem; Int Soc Neurochem; Soc Neurosci; Am Soc Pharmacol Exp Therap; Sigma Xi. *Res:* Neurotransmission and drugs affecting transmitter systems; mechanism of action of barbiturates; weaver mutant mouse. *Mailing Add:* Inst Psychiat Res Ind Univ Med Sch 791 Union Dr Indianapolis IN 46202-4887

RICHTER, MAXWELL, b Montreal, Que, June 11, 33; m 58; c 2. IMMUNOLOGY, PATHOLOGY. *Educ:* McGill Univ, BSc, 54, PhD(biochem), 58, MD, 64. *Prof Exp:* From asst prof to assoc prof exp med, McGill Univ, 58-72; assoc prof immunol & allergy, 70-72; PROF PATH, UNIV OTTAWA, 72- *Concurrent Pos:* Mem staff, Dept Immunol & Allergy, Royal Victoria Hosp, Montreal, 68-; Med Res Coun scholar immunol, Univ Ottawa Clin, Victoria Hosp, 70- *Mem:* Am Acad Allergy; Am Asn Immunol; NY Acad Sci; Brit Soc Immunol; Can Soc Immunol. *Res:* Immunopathology. *Mailing Add:* Dept Path Fac Med Univ Ottawa 451 Smyth Rd Ottawa ON K1H 8M5 Can

RICHTER, RAYMOND C, b Riverside, Calif, Apr 17, 18; m 40; c 3. GEOLOGY. *Educ:* Univ Calif, BA, 40, BS, 42. *Prof Exp:* Supv eng geologist, Calif Dept Water Resources, 46-66, staff geologist, Resources Agency, 66-75; consult geologist, 76-82; RETIRED. *Mem:* Geol Soc Am; Asn Eng Geologists. *Res:* Engineering geology; ground water geology. *Mailing Add:* 2252 Camborne Dr Modesto CA 95356

RICHTER, REINHARD HANS, b Reinswalde, Ger, Oct 3, 37; m 65. ORGANIC CHEMISTRY. *Educ:* Univ Tuebingen, BS, 60; Stuttgart Tech Univ, MS, 62, PhD(org chem), 65. *Prof Exp:* NIH fel, Mellon Inst, 65-68; staff scientist, Donald S Gilmore Res Labs, UpJohn Co, 68-84; mgr chem res, 85-88, res mgr, 88-89, SR RES MGR, DOW CHEM, USA, 89- *Honors & Awards:* Chamberland Award, New Haven Sect, Am Chem Soc, 83. *Mem:* Am Chem Soc; Ger Chem Soc. *Res:* Organic synthesis; heterocycles; isocyanates; nitrile oxides; thermoplastic polyurethanes; polyurethane foams. *Mailing Add:* Dow Chem Co Freeport TX 77541

RICHTER, ROY, b Brooklyn, NY, Aug 25, 56; c 3. THEORETICAL PHYSICS. *Educ:* Rennselaer Polytech Inst, BS(math) & BS(physics), 75; Cornell Univ, MS, 79, PhD(theoret physics), 82. *Prof Exp:* Teaching asst physics, dept physics, Cornell Univ, 75-78; res asst theoret physics, Nordic Inst Theoret Atomic Physics, 79-81; teaching fel, Mont State Univ, 81-82; res fel, theoret physics, Gen Motors Res Labs, 82, sr res scientist, 82-85, STAFF RES SCIENTIST, GEN MOTORS RES LABS, 85-, SECT MGR, COMPUTATIONAL PHYSICS, 89- *Mem:* Am Phys Soc. *Res:* Theoretical solid state physics; metals; surfaces. *Mailing Add:* Dept Physics Gen Motors Res Labs 30500 Mound Rd Warren MI 48090-9055

RICHTER, STEPHEN L(AWRENCE), b Brooklyn, NY, Dec 29, 42; m 62; c 1. RADAR ELECTRONIC-COUNTER-COUNTER-MEASURES & SURVEILLANCE SYSTEMS, ALGORITHM DEVELOPMENT. *Educ:* Columbia Univ, BS, 63, MS, 64, PhD(electromagnetic theory), 67. *Prof Exp:* Res asst elec eng, Columbia Univ, 63-66, preceptor, 66-67, asst prof, 67-68, res scientist, 68; asst prof, City Col New York, 68-71; sr engr, 71-73, PRIN ENGR, RAYTHEON CO, 73- *Concurrent Pos:* Consult, 64-67 & 70; reviewer proceedings, Inst Elec & Electronics Engrs, 68-87. *Mem:* Sr mem Inst Elec & Electronics Engrs; Am Phys Soc; NY Acad Sci; Sigma Xi. *Res:* Systems analysis; adaptive signal processing and discrimination; communication and information theory; radar; electronic-counter-counter-measures; software requirements; optical and image processing; stochastic processes; computer and numerical methods; electromagnetic wave propagation; guided waves and antennas; applied mathematics and physics; system integration and testing. *Mailing Add:* Raytheon Co Mail Stop T3TF8 PO Box 1201 Tewksbury MA 01876-0901

RICHTER, THOMAS A, b Chicago, Ill, Feb 25, 38; div; c 1. CCDS, ELECTRO-OPTICS. *Educ:* Univ Wis-Madison, BS(metall eng) & BS(naut sci), 61, MS, 64; Univ Chicago, MBA, 69. *Prof Exp:* Metallurgist, Griffin Wheel Co, 64-67; chief metallurgist, Chicago Rawhide, 68; mgr, Optical Coating Lab, 69-75; dir, Optical Radiation, 76-79; vpres, Exotic Mat, 80; vpres & gen mgr, Magmum Technol, 81-83; OWNER, RICHTER ENTERPRISES, 84- *Concurrent Pos:* Consult, Jet Propulsion Lab, 87-88; mem liaison comt, Int Soc Optical Eng Indust Rels, 88-, prog chmn, Aerospace Sensing Symposa, 89-; panelist, Res Adv Bd, Aviation Week, 90-91. *Mem:* Int Soc Optical Eng. *Res:* Granted three patents; optical components. *Mailing Add:* 640 19th St Manhattan Beach CA 90266

RICHTER, WARD ROBERT, b Union Grove, Wis, Feb 27, 30; m 54; c 3. PATHOLOGY, ELECTRON MICROSCOPY. *Educ:* Iowa State Univ, DVM, 55, MS, 62. *Prof Exp:* Instr vet path, Iowa State Univ, 57-60; chief cellular path, US Army Med Res Lab, Ft Knox, Ky, 60-63; res assoc path, Univ Louisville, 61-63; head cellular path & electron micros, Abbott Labs, 63-68; dir, Path Div, Inst Res & Develop Corp, 68-88; PATH DEPT, CHEVRON ENVIRON HEALTH CTR INC, 88- *Concurrent Pos:* Assoc prof, Iowa State Univ, 64-; dep dir, A J Carlson Animal Res Facil, Univ Chicago; consult, Argonne Nat Lab, US Dept HEW & Univs Assoc for Res & Educ in Path. *Mem:* Am Vet Med Asn; Am Col Vet Path; Int Acad Pathologists. *Res:* Electron microscopy applied to study of drug toxicity, drug action and laboratory animal diseases; cell biology; pathology of trauma and subcellular injury produced by physical environmental alterations. *Mailing Add:* Kemdru Int Inc 14311 Honeysuckle Evansville IN 47711

RICHTER, WAYNE H, b New York, NY, June 7, 36; c 1. MATHEMATICS. *Educ:* Swarthmore Col, AB, 58; Princeton Univ, MA, 60, PhD(math), 63. *Prof Exp:* From instr to asst prof math, Rutgers Univ, 61-69; ASSOC PROF MATH, UNIV MINN, MINNEAPOLIS, 69- *Concurrent Pos:* Fac fel, Rutgers Univ, 64-65; NSF res grant, 66-68 & 70-75. *Mem:* Am Math Soc; Asn Symbolic Logic. *Res:* Theory of recursive functions; mathematical logic. *Mailing Add:* Dept Math Univ Minn Minneapolis MN 55455

RICHTERS, ARNIS, b Sauka, Latvia, Sept 23, 28; US citizen; c 1. EXPERIMENTAL PATHOLOGY. *Educ:* Univ Ariz, BS, 57, MS, 59; Univ Southern Calif, PhD(exp path), 67. *Prof Exp:* From instr to asst prof, 68-75, assoc prof path, 75-91, PROF, SCH MED, UNIV SOUTHERN CALIF, 91- *Concurrent Pos:* Site vis, Breast Cancer Task Force, Nat Cancer Inst, 72; reviewer, Monroe County Cancer & Leukemia Asn, 73- *Mem:* Tissue Cult Asn; Int Acad Path; AAAS; Sigma Xi. *Res:* Correlation of in vivo and in vitro behavior of cancer and the lymphocyte responses in cancer draining lymph nodes; ultrastructure of lymphocyte-target cell interactions; air pollution and health. *Mailing Add:* 2205 Tall Pine Dr Duarte CA 91010

RICHTOL, HERBERT H, b New York, NY, Aug 13, 32; m 56; c 4. ANALYTICAL CHEMISTRY, PHOTOCHEMISTRY. *Educ:* St Lawrence Univ, BS, 54; NY Univ, PhD(chem), 61. *Prof Exp:* Instr chem, NY Univ, 60-61; from asst prof to assoc prof, 61-74, PROF CHEM, RENSSELAER POLYTECH INST, 74-, DEAN, UNDERGRAD COL, 85- *Mem:* AAAS; Am Chem Soc. *Res:* Photoelectrochemistry; luminescence; liquid crystals. *Mailing Add:* Undergrad Col Rensselaer Polytech Inst Troy NY 12180-3590

RICK, CHARLES MADEIRA, JR, b Reading, Pa, Apr 30, 15; m 38; c 2. CYTOGENETICS, EVOLUTION. *Educ:* Pa State Col, BS, 37; Harvard Univ, AM, 38, PhD(genetics), 40. *Prof Exp:* Tech asst, Harvard Univ, 38; instr & jr geneticist, 40-44, asst prof & asst geneticist, 44-49, assoc prof & assoc geneticist, 49-55, PROF VEG CROPS & GENETICIST, UNIV CALIF, DAVIS, EXP STA, 55- *Concurrent Pos:* Guggenheim fel, 49, 51; Rockefeller res fel, 56-57; vis lectr, NC State Col, 56; fac res lectr, Univ Calif, 61; Carnegie vis prof, Univ Hawaii, 63; mem, Galapagos Int Sci Proj, 64; lectr, Univ Sao Paulo, 65; vis scientist, Univ PR, 68; centennial lectr, Ont Agr Col, Univ Guelph, 74; adj prof, Univ Rosario, Argentina, 80, Univ lectr, Cornell Univ, 88. *Honors & Awards:* Vaughan Award, Am Soc Hort Sci, 45; Campbell Award, AAAS, 59; M A Blake Award, Am Soc Hort Sci, 74; Frank N Meyer Medal, Am Genetics Asn, 82; Thomas Roland Medal, 83; Distinguished Econ Botanist,Soc Econ Bot, 87; Genetic & Plant Breeding Award, Nat Coun Com Plant Breeders, 87. *Mem:* Nat Acad Sci; AAAS; Genetics Soc Am; Soc Study Evolution; Am Soc Hort Sci. *Res:* Natural relationships amongst the tomato, Lycopersicon, species are investigated via cytology, incompatibility, hybrid fertility, and genetic variability, including isozymes. *Mailing Add:* Dept Veg Crops Univ Calif Davis CA 95616

RICK, CHRISTIAN E(DWARD), b Kansas City, Mo, Oct 9, 13; m 38. CHEMICAL ENGINEERING. *Educ:* Univ Kans, BS, 36; Univ Del, MS, 44. *Prof Exp:* Res chemist, Pigments Dept, E I du Pont de Nemours & Co, Inc, 36-58, res supvr, Staff Sect, 58-78; RETIRED. *Honors & Awards:* Citation, Off Sci Res & Develop, 45. *Mem:* Am Chem Soc. *Res:* Engineering economics; titanium metal, compounds and minerals; germanium, silicon and boron semiconductors; oxide ceramics and refractories; instruments; electronics; computers. *Mailing Add:* 1903 Greenhill Ave Wilmington DE 19806

RICK, PAUL DAVID, MICROBIAL PHYSIOLOGY, BIOCHEMISTRY. *Educ:* Univ Minn, PhD(biochem), 71. *Prof Exp:* ASSOC PROF MICROBIOL, UNIFORMED SERV HEALTH SCI UNIV, 83- *Res:* Genetics; biogenesis of microbial membranes. *Mailing Add:* Uniformed Serv Health Sci Univ 4301 Jones Bridge Rd Bethesda MD 20814

RICKABY, DAVID A, PULMONARY PHYSIOLOGY, CARDIOVASCULAR PHYSIOLOGY. *Educ:* Med Col Wis, PhD(physiol), 79. *Prof Exp:* RES PHYSIOLOGIST, VET ADMIN MED CTR, MILWAUKEE, 79- *Mailing Add:* W 159 N10764 Captains Dr Germantown WI 53022

RICKARD, CORWIN LLOYD, b Medina, Ohio, Sept, 26, 26; m 48; c 2. SCIENCE ADMINISTRATION, NUCLEAR ENGINEERING. *Educ:* Univ Rochester, BS, 47, MS, 49; Cornell Univ, PhD(eng, math), 61. *Prof Exp:* Asst eng, Univ Rochester, 47-48, instr, 48-49, asst prof, 49-52; instr, Cornell Univ, 52-54; nuclear engr, Brookhaven Nat Lab, 54-56; res staff mem nuclear eng, Gen Atomic Div, Gen Dynamics Corp, 56-66, vpres, Gulf Gen Atomic Inc, 67-77, exec vpres, Gen Atomic Co, 77-83; EXEC VPRES, SIBIA INC, 83- *Concurrent Pos:* Consult, Brookhaven Nat Lab, 49, Worthington Corp, NY, 50, Boeing Aircraft Co, Wash, 51, Adv Electronic Res Lab, Gen Elec Co, 53-54. *Mem:* Am Nuclear Soc (vpres, 80-81, pres, 81-82); Am Soc Mech Engrs; AAAS. *Res:* Thermodynamics; gas cooled nuclear reactor core and plant development, design and engineering; nuclear power plant economics and fuel cycles. *Mailing Add:* PO Box 472 Rancho Santa Fe CA 92067

RICKARD, EUGENE CLARK, b Wichita, Kans, Oct 19, 43; m 70; c 2. PROTEIN CHARACTERIZATION, CAPILLARY ELECTROPHORESIS. *Educ:* Wichita State Univ, BS, 65, MS, 67; Univ Wis-Madison, PhD(anal chem), 72. *Prof Exp:* Lectr anal chem, Univ Wis-Madison, 70-71; sr chemist, 71-76, res scientist, 77-84, SR RES SCIENTIST, ELI LILLY & CO, 85- *Concurrent Pos:* Mem, Fel Comt, Anal Div, Am Chem Soc, 90- *Mem:* Am Chem Soc; Sigma Xi. *Res:* Development of analytical methods to determine purity; impurities and stability of oncolytic drugs including vinca alkaloids, monoclonal antibody and drug conjugates; protein characterization; immunoassays. *Mailing Add:* Drop 1503 Eli Lilly & Co Lilly Corp Ctr Indianapolis IN 46285

RICKARD, JAMES ALEXANDER, b Austin, Tex, July 9, 26; m 52; c 2. PHYSICS. *Educ:* Tex A&I Univ, BS & MS, 48; Univ Tex, PhD(physics), 54. *Prof Exp:* Res engr, Houston Res Lab, Humble Oil & Ref Co, 53-62, res mgr, Esso Prod Res Co, Tex, 63-71, sr adv, Exxon Corp, 71-76, planning mgr, Exxon Prod Res Co, 76-84; res mgr, Standard Oil Prod Co, 84-87, vpres, 87-89; vpres, Brit Petrol, 89-90; RETIRED. *Mem:* Am Phys Soc; Marine Technol Soc (pres, 77-78); Am Inst Mining, Metall & Petrol Eng; Soc Naval Architects & Marine Engrs; Sigma Xi. *Res:* Petroleum; ocean engineering. *Mailing Add:* 3837 Del Montee Houston TX 77209

RICKARD, JAMES JOSEPH, b Seattle, Wash, Sept 20, 40; m 63. ASTRONOMY, COMPUTER SCIENCE. *Educ:* San Jose State Univ, BA, 62; Univ MD, MS, 65, PhD(astron), 68. *Prof Exp:* Teaching asst astron, Univ Md, 62-64, res asst radio astron, 64-67, res fel, 67-68; res fel astron, Calif Inst Technol-Hale Observ, 68-69; staff astronr, Europ Southern Observ, 69-76; res scientist astron, Univ Iowa, 77-80; PARTNER/OWNER, BORREGO SOLAR SYSTS, 80-; SOLAR ASTRONR, UNIV MD CLRO, 82- *Concurrent Pos:* Earth resources consult, Terra Inst, 76- *Mem:* Am Astron Soc; Int Astron Union. *Res:* Galactic structure; physics of the interstellar medium; instrumentation; computer systems design; extragalactic radio sources. *Mailing Add:* PO Box 777 Borrego Springs CA 92004

RICKARD, JOHN TERRELL, b Humboldt, Tenn, Sept 17, 47; m 71; c 3. SIGNAL PROCESSING, INFORMATION THEORY. *Educ:* Fla Inst Technol, BS, 69, MS, 71; Univ Calif, San Diego, PhD(eng physics), 75. *Prof Exp:* Electronics engr, Harris Corp, Melbourne, Fla, 69-71; sr prin engr, 75-80, SR VPRES, ORINCON CORP, 80- *Mem:* Inst Elec & Electronics Engrs. *Res:* Random processes; adaptive signal processing; detection theory. *Mailing Add:* Orincon Corp 9363 Towne Centre Dr San Diego CA 92122

RICKARD, LAWRENCE VROMAN, b Cobleskill, NY, July 10, 26; m 52; c 5. PALEONTOLOGY, STRATIGRAPHY. *Educ:* Cornell Univ, BA, 51, PhD(geol), 55; Univ Rochester, MS, 53. *Prof Exp:* Asst prof geol, St Lawrence Univ, 55-56; sr paleontologist, 56-73, assoc paleontologist, 73-77, PRIN PALEONTOLOGIST, MUS & SCI SERV, STATE GEOL SURV, NY, 77- *Concurrent Pos:* Mem, Paleont Res Inst. *Mem:* Geol Soc Am; Paleont Soc. *Res:* Stratigraphy and paleontology of the Silurian and Devonian rocks of New York. *Mailing Add:* RD 1 Kinderhook NY 12106

RICKARD, LEE J, b Miami, Fla, Dec 24, 49; m 72, 86; c 2. INFRARED ASTRONOMY, RADIO ASTRONOMY. *Educ:* Univ Miami, BS, 69; Univ Chicago, MS, 72, PhD(astrophys), 75. *Prof Exp:* Res assoc, Nat Radio Astron Observ, 75-77, asst scientist, 77-79, assoc scientist, 79-80; from asst prof to assoc prof, Dept Physics & Astron, Howard Univ, 84-87; ASTROPHYSICIST, NAVAL RES LAB, 87- *Concurrent Pos:* Radio astronr, Naval Res Lab, 83-84; res astronr, Sachs/Freeman Assoc, Inc, 84-86; res astronr, Appl Res Corp, 86-87. *Mem:* Am Astron Soc; Int Astron Union. *Res:* Radio spectroscopy, specifically interstellar molecules, molecular constituents of galaxies, interstellar masers and atomic recombination lines; problems of active galactic nuclei; infrared emission from galaxies; image processing and sensor enhancement. *Mailing Add:* Code 4213 3 Naval Res Lab Washington DC 20375-5000

RICKARD, WILLIAM HOWARD, JR, b Walsenburg, Colo, May 15, 26; m 53; c 1. BOTANY. *Educ:* Univ Colo, BA, 50, MA, 53; Wash State Univ, PhD(bot), 57. *Prof Exp:* Asst prof bot, N Mex Highlands Univ, 57-60; biol scientist, Hanford Atomic Prod Opers, Gen Elec Co, 60-65; SR RES SCIENTIST, ECOL DEPT, PAC NORTHWEST LAB, BATTELLE MEM INST, 65- *Mem:* AAAS; Ecol Soc Am; Soc Range Mgt; Northwest Sci Asn; Sigma Xi. *Res:* Plant ecology; fate and behavior of trace elements and radionuclides in terrestrial ecosystems; primary productivity and mineral cycling in shrub steppe ecosystems. *Mailing Add:* 1904 Lassen Ave Richland WA 99352

RICKART, CHARLES EARL, b Osage City, Kans, June 28, 13; m 42; c 3. MATHEMATICS. *Educ:* Univ Kans, BA, 37, MA, 38; Univ Mich, PhD(math), 41. *Hon Degrees:* MA, Yale Univ, 59. *Prof Exp:* Peirce instr & tutor math, Harvard Univ, 41-43; from instr to prof, 43-83, chmn dept, 59-65, EMER PROF MATH, YALE UNIV, 83- *Mem:* AAAS; Am Math Soc; Math Asn Am. *Res:* Functional analysis; theory of Banach algebras; function algebras; structuralism and structures. *Mailing Add:* Dept Math Yale Univ New Haven CT 06520

RICKBORN, BRUCE FREDERICK, b New Brunswick, NJ, Feb 23, 35; m 55; c 1. ORGANIC CHEMISTRY. *Educ:* Univ Calif, Riverside, BA, 56; Univ Calif, Los Angeles, PhD(chem), 60. *Prof Exp:* Asst prof chem, Univ Calif, Berkeley, 60-62; from asst prof to assoc prof, 62-71, provost, Col Creative Studies, 69-71, PROF CHEM, UNIV CALIF, SANTA BARBARA, 71- *Concurrent Pos:* NSF sr fel, 66-67; Alfred P Sloan fel, 67-69; Fulbright sr lectr, Bogata, Columbia, 70; assoc dean, Col Letters & Sci, Univ Calif, Santa Barbara, 71-73, dean, 73-78. *Mem:* Am Chem Soc; AAAS; Sigma Xi. *Res:* Mechanism and stereochemistry of small ring forming and opening reactions; organometallics; conformational analysis; photochemistry; hydride reduction; strong base induced reactions. *Mailing Add:* Dept Chem Univ Calif Santa Barbara CA 93106

RICKELS, KARL, b Wilhelmshaven, Ger, Aug 17, 24; US citizen; m 64; c 3. PSYCHIATRY, PSYCHOPHARMACOLOGY. *Educ:* Univ Munster, MD, 51. *Prof Exp:* PROF PSYCHIAT & PHARMACOL, UNIV PA, 69-, STUART & EMILY MUDD PROF HUMAN BEHAV REPRODUCTION, 77- *Concurrent Pos:* Dir psychopharmacol res, Univ Pa, 64-; chmn, OTC Sedatives, Tranquilizers and Sleep Aids Rev Panel, Food & Drug Admin, 72-75; chief psychiat, Phildelphia Gen Hosp, 75-77. *Mem:* Fel Acad Psychosom Med; fel Am Soc Clin Pharmacol & Therapeut; fel Am Col Neuropsychopharmacol; fel Am Psychiat Asn; fel Int Col Neuropsychopharmacol. *Res:* Clinical psychopharmacology; evaluation of psychotropic drugs in anxiety, depression and insomnia and tranquilizer dependency; study of the role of non-specific factors in drug and placebo response; assessment of emotional symptoms in various family practice and OB-GYN patients and their response to stress and hormonal treatment, i.e. premenstrual syndrome, in vitro fertilization and teenage pregnancy. *Mailing Add:* Dept Pharm Univ Pa Hosp 36th & Hamilton Wk Philadelphia PA 19104

RICKENBERG, HOWARD V, b Nuremberg, Ger, Feb 3, 22; nat US; m 53; c 3. DEVELOPMENTAL BIOLOGY. *Educ:* Cornell Univ, BS, 50; Yale Univ, PhD(microbiol), 54. *Prof Exp:* Am Cancer Soc fel, Pasteur Inst, Paris, 54-55 & Nat Inst Med Res, Eng, 55-56; from instr to asst prof microbiol, Univ Wash, 56-60; assoc prof bact, Ind Univ, 61-63, prof bact, 63-66; prof microbiol, Sch Med, Univ Colo, Denver, 66-71, prof biochem, biophys & genetics, 71-88; prof, Dept Molecular & Cellular Biol, Nat Jewish Ctr Immunol & Respiratory Med, 86-88; RETIRED. *Concurrent Pos:* Ida & Cecil Green investigatorship develop biochem, 75-88; Fulbright scholar, Univ Ivory Coast, Abiddan, 88-89; consult, Immunotech, Marseille, 90- *Mem:* Am Soc Microbiol; Am Soc Biol Chem; Am Soc Cell Biol. *Res:* Gene-enzyme relationships, metabolism and differentiation. *Mailing Add:* Immunotech SA Luminy Marseille 13288 France

RICKER, NEIL LAWRENCE, b Stambaugh, Mich, Oct 15, 48; m 71. CHEMICAL ENGINEERING. *Educ:* Univ Mich, Ann Arbor, BS, 70; Univ Calif, Berkeley, MS, 72, PhD(chem eng), 78. *Prof Exp:* Sci systs analyst, Air Prod & Chem, Inc, 72-75; ASSOC PROF CHEM ENG, UNIV WASH, 78- *Mem:* Am Inst Chem Engrs; Am Chem Soc. *Res:* Process dynamics and control; chemical process analysis and conceptual design; wastewater treatment by physical and chemical methods. *Mailing Add:* Dept Chem Eng BF-10 Univ Wash Seattle WA 98195

RICKER, RICHARD EDMOND, b Newport News, Va, Feb 26, 52; m; c 2. STRESS CORROSION CRACKING, HYDROGEN EMBRITTLEMENT. *Educ:* NC State Univ, BS, 75, MS, 78; Rensselaer Polytech Inst, PhD(mat eng), 83. *Prof Exp:* Sr engr, Lynchburg Res Ctr, Babcock & Wilcox Co, 77-79; asst prof mat sci & eng, Univ Notre Dame, 84-86; metallurgist, 86-89, GROUP LEADER, CORROSION GROUP, NAT INST STANDARDS & TECHNOL, 89- *Mem:* Am Soc Metals Int; Am Soc Testing & Mat; Electrochem Soc; Metall Soc Am Inst Mech Engrs; Nat Asn Corrosion Engrs; AAAS. *Res:* Corrosion measurement techniques, computerized systems for corrosion information and the corrosion behavior of new materials; mechanisms of environmentally induced fracture and environmentally induced fracture behavior of new materials. *Mailing Add:* Nat Inst Standards & Technol Corrosion Group 223-B254 Gaithersburg MD 20899

RICKER, RICHARD W(ILSON), b Galion, Ohio, Mar 30, 14; m 44; c 5. CERAMICS ENGINEERING. *Educ:* Alfred Univ, BS, 34; Pa State Col, MS, 50, PhD(ceramics), 52. *Prof Exp:* Mem staff res & develop, Libbey-Owens-Ford Glass Co, 34-48; res assoc, Pa State Col, 48-52; res engr, Alcoa Res Labs, Aluminum Co Am, 52-55, asst chief process metall div, 55-60; asst dir res, Ferro Corp, 60-66; tech dir ceramics, Harshaw Chem Co, 66-71; CERAMIC CONSULT, 71- *Concurrent Pos:* Consult refractories & whitewares projs, South Korea, Turkey, Mexico, Columbia, Brazil & Egypt. *Mem:* Fel Am Ceramic Soc; Sigma Xi. *Res:* Phase equilibria; refractories; porcelain enamel; ceramic color pigments; whitewares. *Mailing Add:* 1431 NE 55 St Ft Lauderdale FL 33334

RICKERT, DAVID A, b Trenton, NJ, Mar 14, 40; m 62; c 3. ENVIRONMENTAL SCIENCES. *Educ:* Rutgers Univ, BS, 62, MS, 65, PhD(environ sci), 69. *Prof Exp:* Chemist urban hydrol prog, 69-72, res hydrologist, chief, Willamette River Basin Study, 72-75, res hydrologist, chief, Land-Use-River Qual Study, Water Resources Div, 75-77, chief, Ore 208 Assessment, Ore Dept Environ Qual, 77-79, coordr, River Qual Assessment Prog, 79-80, sr staff scientist, 80-82, spec asst to chief hydrologist, 82-85, CHIEF, OFF WATER QUAL, WATER RESOURCES DIV, US GEOL SURV, 85- *Concurrent Pos:* Secy, Water Qual Comn, Int Asn Hydrol Sci, 81-87, vpres, 87- *Honors & Awards:* W R Boggess Award, Am Water Resources Asn, 74. *Mem:* AAAS; Int Asn Hydrological Sci; Am Water Resources Asn (vpres, 82-84, pres, 84). *Res:* Development and implementation of large-scale interdisciplinary programs for land and water-resources assessment; environmental chemistry and biology; applied hydrology; environmental geology; land-use analysis. *Mailing Add:* Off Water Qual US Geol Surv Mail Stop 412 Reston VA 22092

RICKERT, DOUGLAS EDWARD, b Sioux City, Iowa, Jan 27, 46; m 81. TOXICOLOGY, PHARMACOLOGY. *Educ:* Univ Iowa, BS, 68, MS, 72, PhD(pharmacol), 74; Am Bd Toxicol, dipl, 80. *Prof Exp:* Asst prof pharmacol, Mich State Univ, 74-77; ANALYTICAL BIOCHEMIST, CHEM INDUST INST TOXICOL, 77- *Mem:* AAAS; Soc Toxicol; Am Soc Pharmacol & Exp Therapeut; Am Soc Mass Spectros. *Res:* Biological disposition of foreign compounds and their toxicity. *Mailing Add:* Glaxo Inc Five Moore Dr Research Triangle Park NC 27709

RICKERT, NEIL WILLIAM, b Perth, Australia, June 2, 39; m 65; c 2. COMPUTER OPERATING SYSTEMS. *Educ:* Univ Western Australia, BSc, 62; Yale Univ, PhD(math), 65. *Prof Exp:* From instr to asst prof math, Yale Univ, 65-68; assoc prof math, Univ Ill, Chicago Circle, 68-85; PROF COMPUT SCI, NORTHERN ILL UNIV, 85- *Concurrent Pos:* Mem, Inst Advan Study, 66-67. *Mem:* Am Math Soc; Math Asn Am; Asn Comput Mach; AAAS; Inst Elec & Electronics Engrs. *Res:* Computer operating systems; distributed computer systems, including distributed databases and distributed computational algorithms. *Mailing Add:* Dept Comput Sci Northern Ill Univ De Kalb IL 60115

RICKERT, RUSSELL KENNETH, b Chalfont, Pa, Feb 6, 26; m 49; c 3. PHYSICS. *Educ:* West Chester State Col, BS, 50; Univ Del, MS, 53; NY Univ, EdD(sci ed), 61. *Prof Exp:* Teacher high sch, Md, 50-52 & Del, 52-55; instr phys sci, Salisbury State Col, 55-56; from asst prof to prof phys sci, 56-74, chmn dept sci, 64-68, dean, Sch Sci & Math, 69-79, PROF PHYSICS, WEST CHESTER STATE COL, 74- *Mem:* AAAS; Am Asn Physics Teachers. *Res:* Physics teaching, especially the development of physical science courses for students who are not science majors; aerospace science education; solar energy research and development. *Mailing Add:* Dept Physics West Chester State Col West Chester PA 19383

RICKETT, FREDERIC LAWRENCE, b Woodridge, NJ, Mar 11, 39; m 63; c 3. BIOCHEMISTRY. *Educ:* Pa State Univ, BS, 61, MS, 63, PhD, 66. *Prof Exp:* Instr biochem, Pa State Univ, 63-66; res assoc, 66-74, SR RES ASSOC ANALYTICAL BIOCHEM, DEPT RES & DEVELOP, AM TOBACCO CO, 74- *Mem:* Am Chem Soc; Sigma Xi. *Res:* Analytical methodology for lipids and other plant constituents, tobacco and tobacco smoke, flavoring agents; pesticide analysis, process development and quality control related to tobacco. *Mailing Add:* Dept Res & Develop Am Tobacco Co PO Box 899 Hopewell VA 23860

RICKETTS, GARY EUGENE, b Willard, Ohio, Aug 2, 35; m 58; c 3. ANIMAL SCIENCE. *Educ:* Ohio State Univ, BS, 57, MS, 60, PhD(animal sci), 63. *Prof Exp:* Livestock exten specialist sheep & beef cattle, 64-86, EXTEN SPECIALIST, SHEEP & EXTEN PROG LEADER, UNIV ILL, URBANA, 86- *Honors & Awards:* Am Soc Animal Sci Ext Award, 84. *Mem:*

Am Soc Animal Sci; Am Regist Prof Animal Scientists. *Res:* Authored or co-authored three journal articles, six abstracts, eleven sheep day report articles, three beef day report articles, two hundred sheep reports, a hundred and twenty beef performance testing letters, thirty circulars, bulletins and leaflets, and two hundred and eighty-six articles appearing in livestock publications. *Mailing Add:* 321 Mumford Hall Univ Ill 1301 W Gregory Dr Urbana IL 61801

RICKETTS, JOHN ADRIAN, b Lakewood, Ohio, Feb 29, 24; m 48; c 2. PHYSICAL CHEMISTRY. *Educ:* Ind Univ, BS, 48; Western Reserve Univ, MS, 50, PhD(chem), 53. *Prof Exp:* Lectr, Fenn Col, 51; from asst prof to assoc prof, 52-62, dir grad studies, 66-69, PROF CHEM, DEPAUW UNIV, 62- *Mem:* Am Chem Soc; Royal Soc Chem; Sigma Xi. *Res:* Electrochemistry; thermodynamics, chemical kinetics. *Mailing Add:* 702 Highridge Greencastle IN 46135

RICKEY, FRANK ATKINSON, JR, b Baton Rouge, La, Nov 21, 38; m 62; c 2. NUCLEAR PHYSICS. *Educ:* La State Univ, Baton Rouge, BS, 60; Fla State Univ, PhD(physics), 66. *Prof Exp:* Res assoc nuclear physics, Los Alamos Sci Lab, 66-68; asst prof, 68-76 ASSOC PROF NUCLEAR PHYSICS, PURDUE UNIV, LAFAYETTE, 76- *Mem:* Am Phys Soc. *Res:* Low energy nuclear physics; charged particle reactions; nuclear structure. *Mailing Add:* Dept Physics Purdue Univ West Lafayette IN 47907

RICKEY, MARTIN EUGENE, b Memphis, Tenn, Aug 4, 27; c 3. NUCLEAR PHYSICS, MUSICAL ACOUSTICS. *Educ:* Southwestern at Memphis, BS, 49; Univ Wash, PhD(physics), 58. *Prof Exp:* From asst prof to assoc prof physics, Univ Colo, 58-65; assoc prof, 65-68, PROF PHYSICS, IND UNIV, 68- *Concurrent Pos:* Vis staff mem, Los Alamos Sci Lab, 69-; consult, Brookhaven Nat Lab, 71-72 & Hahn-Meitner Inst, Berlin, 72- *Honors & Awards:* Sr Scientist Award, Alexander von Humboldt Found, WGer, 75. *Mem:* Fel Am Phys Soc; Sigma Xi. *Res:* Cyclotron design and development; nuclear reactions at low and intermediate energies; acoustical properties of musical instruments. *Mailing Add:* 3171 Picadilly St Bloomington IN 47401

RICKLEFS, ROBERT ERIC, b San Francisco, Calif, June 6, 43. ECOLOGY. *Educ:* Stanford Univ, AB, 63; Univ Pa, PhD(biol), 67. *Prof Exp:* Nat Res Coun vis res assoc, Smithsonian Trop Res Inst, 67-68; from asst prof to assoc prof, 68-80, PROF BIOL, UNIV PA, 80- *Mem:* Soc Study Evolution; Am Soc Naturalists; Ecol Soc Am; Am Ornithologists Union; Cooper Ornith Soc. *Res:* Evolutionary ecology; development and reproductive biology of birds; population and community ecology. *Mailing Add:* Dept Biol Univ Pa Philadelphia PA 19104

RICKLES, FREDERICK R, b Chicago, Ill, Sept 24, 42; m 64; c 2. HEMATOLOGY. *Educ:* Col Med, Univ Ill, MD, 67. *Prof Exp:* Intern & resident internal med, Univ Rochester-Strong Mem Hosp, 67-70, fel hemat, 70-71; dir, Coagulation Res Lab, Walter Reed Army Inst Res, 71-74; from asst prof to assoc prof, 74-84, co-chief, 81-85, PROF MED, UNIV CONN, 84-, CHIEF, DIV HEMAT-ONCOL, SCH MED, 85- *Concurrent Pos:* Clin asst prof med, George Washington Univ, 71-74; attend physician, Walter Reed Army Med Ctr, 71-74 & John Dempsey-Univ Conn Hosp, 74-; chief, Hemat Sect, Vet Admin Hosp, Newington, Conn, 74- & Hemat-Oncol Sect, 81-; consult, Artificial Kidney-Chronic Uremia Prog, Nat Inst Arthritis, Metab & Digestive Dis-NIH, 74-78; Vet Admin res grant, 74-; Nat Cancer Inst & Nat Heart, Lung & Blood Inst grants, NIH, 78; grants, Am Heart Asn, 83-86 & Am Cancer Soc, 85-; med res serv assoc chief staff, Vet Admin Hosp, Newington, Conn, 78-81; vpres, Med Sci Affairs, Nat Hemophilia Found, 87- *Honors & Awards:* Humanitarian Award, Nat Hemophilia Found, 90. *Mem:* AAAS; Am Fedn Clin Res; Am Asn Immunologists; Am Soc Hemat; Int Soc Thrombosis & Hemostasis. *Res:* Blood coagulation; cancer; delayed hypersensitivity; endotoxin biochemistry; membrane proteins. *Mailing Add:* Univ Conn Sch Med Health Ctr Farmington CT 06032

RICKLES, NORMAN HAROLD, b Seattle, Wash, May 8, 20; m 50; c 3. ORAL PATHOLOGY. *Educ:* Wash Univ, DDS, 47; Univ Calif, MS, 51. *Prof Exp:* Instr dent med, Col Dent, Univ Calif, 47-48, lectr, 48-52, asst clin prof, 52-56; from assoc prof to prof dent, 56-85, head dept oral path, 56-76, EMER PROF DENT, DENT SCH, UNIV ORE, 85- *Concurrent Pos:* Fel, Armed Forces Inst Path, 54-56; consult to Surgeon Gen, Madigan Army Hosp, Ft Lewis, Wash, 58-; Fulbright prof, Sch Dent Med, Hebrew Univ, Israel, 66-67; sabbatical vis prof & consult, Guys Hosp, London, 77-78. *Mem:* Am Soc Clin Path; Am Dent Asn; Am Acad Oral Med; Am Acad Oral Path (pres, 62-63); Int Acad Path. *Res:* Dental caries; allergy; methods of teaching dental students; screening tests for dental patients; histochemistry and methods of evaluation of oral disease; fluorescent and immunoperoxidase microscopy; transplantation of dental pulp. *Mailing Add:* Dept Dent Univ Ore Portland OR 97201

RICKLIN, SAUL, b New York, NY, Sept 5, 19; m 47; c 4. CHEMICAL ENGINEERING. *Educ:* Columbia Univ, BS, 39, ChE, 40. *Prof Exp:* Process engr, Metal & Thermit Corp, NJ, 40-46; consult engr, 46-47; asst prof chem, Brown Univ, 47-54; consult engr, Ricklin Res Assocs, 54-59; vpres, Dixon Industs Corp, 59-66, exec vpres, 66-70, pres, 70-77, chmn, 77-81, consult, 81-91; RETIRED. *Concurrent Pos:* Dir, NTN Rulon Industs Co, Ltd, Japan, 66-; vpres, Valflon, SpA, Italy, 71-81; vpres, G D Spencer Co, Ltd, Can, 71-; dir, Entwistle Corp, 76-90, EFD Corp; dir, 82-89. *Mem:* Am Chem Soc; Am Inst Chem Engrs. *Res:* Thermit reactions; aluminothermics; incendiaries; tracer ammunition; ceramic materials; metal organics; fine particle technology; electroplating; metal finishing; corrosion; friction and wear of dry bearings; fluorocarbon plastics. *Mailing Add:* PO Box 91 Bristol RI 02809

RICKMAN, RONALD WAYNE, b Pomeroy, Wash, June 28, 40; m 63; c 3. SOIL CONSERVATION. *Educ:* Wash State Univ, BS, 63; Univ Calif, Riverside, PhD(soil physics), 66. *Prof Exp:* SOIL SCIENTIST, SCI & EDUC ADMIN-AGR RES, USDA, 66- *Mem:* Am Soc Agron; Am Soc Agr Engrs; Sigma Xi. *Res:* Dryland small grain growth simulation; wheat root growth and water use; water conservation; wheat production. *Mailing Add:* 4515 SW Perkins Ave Pendleton OR 97801

RICKS, BEVERLY LEE, b Grand Chenier, La, Oct 10, 28; m 51; c 2. PHYSIOLOGY, ZOOLOGY. *Educ:* Miss State Univ, BS, 50, PhD(physiol), 69; Miss Col, MS, 53. *Prof Exp:* Teacher high schs, Miss, 51-65; asst prof, 65-70, ASSOC PROF BIOL, NORTHEAST LA UNIV, 70- *Honors & Awards:* OBTA Award, Nat Asn Biol Teachers, 63. *Mem:* Nat Asn Biol Teachers; Am Inst Biol Sci; Entom Soc Am. *Res:* Feeding preferences; seasonal changes in stored nutrients; distribution of digestive enzymes in the imported fire ant. *Mailing Add:* Dept Biol Northeast La Univ Monroe LA 71209

RICKSECKER, RALPH E, b Cleveland, Ohio, Sept 9, 12; m 38; c 2. CHEMICAL METALLURGY, METALLURGICAL ENGINEERING. *Educ:* Western Reserve Univ, BS, 44, MS, 50. *Prof Exp:* Anal chemist, Chase Brass & Copper Co, Inc, Cleveland, 30-35, asst chief chemist, 34-35, chief chemist, 35-45, process metallurgist, 45-48, chief metallurgist, 48-50, dir metall, 50-77; METALL CONSULT, 77- *Mem:* Am Chem Soc; Am Soc Test & Mat; Soc Automotive Eng; Am Inst Mining, Metall & Petrol Eng; Am Soc Metals. *Res:* Physical metallurgy of copper and copper alloys. *Mailing Add:* 130 E 196th St Euclid OH 44119

RICKTER, DONALD OSCAR, b Rio Dell, Calif, May 5, 31; m 59; c 2. INFORMATION SCIENCE. *Educ:* Univ Calif, Davis, AB, 52, MS, 55; Mich State Univ, PhD(chem), 64. *Prof Exp:* Instr chem, Santa Ana Col, 57-59; SCIENTIST, RES DIV, POLAROID CORP, CAMBRIDGE, 64- *Mem:* Am Chem Soc. *Res:* Photographic chemistry (developers, dyes, novel polymers, restrainers, synthesis); management of information systems; selective dissemination of information. *Mailing Add:* 88 Hemlock St Arlington MA 02174-2157

RICORD, LOUIS CHESTER, b Burbank, Calif, Aug 16, 51. BIOMEDICAL COMPUTING, INFORMATION SYSTEMS. *Educ:* Univ Utah, BS, 73, ME, 75, PhD(med biophys, comput), 78. *Prof Exp:* res scientist Biomed Info Systs, 78-83, MGR APPL INFO SYSTS, BATTELLE COLUMBUS LABS, 83- *Concurrent Pos:* Res assoc, NIH, 74-78; adj asst prof, Ohio State Univ, 80- *Mem:* Inst Elec & Electronics Engrs; Comput Soc; Soc Comput Med; Asn Comput Mach. *Res:* Computer assisted decision making in health sciences; medical data base design and construction; computer applications in laboratory animal toxicology; general information system applications development. *Mailing Add:* Columbus Labs Battelle Mem Inst 505 King Ave Columbus OH 43201

RIDDELL, JAMES, b Quill Lake, Sask, Sept 28, 33; m 57; c 3. MATHEMATICS. *Educ:* Univ Alta, BSc, 57, MSc, 61, PhD(math), 67. *Prof Exp:* Geophysicist, Shell Can Ltd, Calgary, 57-59; asst prof math, Eastern Mont Col, 62-64; asst prof, 67-73, ASSOC PROF MATH, UNIV VICTORIA, BC, 73- *Concurrent Pos:* Nat Res Coun grants, Univ Victoria, BC, 68- *Mem:* Math Asn Am; Can Math Cong; Am Math Soc. *Res:* Combinatorial number theory; combinatorics; additive prime number theory. *Mailing Add:* Dept Math Univ Victoria Box 1700 Victoria BC V8W 2Y2 Can

RIDDELL, JOHN EVANS, b Montreal, Que, May 21, 13; m 39; c 5. ECONOMIC GEOLOGY. *Educ:* McGill Univ, BEng, 35, MSc, 36, PhD(geol), 53. *Prof Exp:* Consult Univ Sask, 45-46, lectr geol, 47-49; lectr geol, McGill Univ, 49-50, from asst prof to assoc prof, 50-58; prof & chmn dept, Carleton Univ, Can, 58-62; pres & managing dir, Mt Pleasant Mines, Ltd, 62-65; pres & dir res, Int Geochem Assocs Ltd, 65-87; CONSULT, 74- *Concurrent Pos:* Pres & dir of various mining & mineral explor co; chmn, Coxheath Gold Holdings Ltd, 85- *Mem:* Am Inst Mining, Metall & Petrol Eng; fel Geol Soc Am; fel Royal Soc Can; fel Geol Asn Can; Can Inst Mining & Metall; Sigma Xi; Asn Explor Geochemists. *Res:* Mineral exploration, particularly use of geochemical techniques and development of integrated interpretation of field data using geotectonics, geophysics and geology. *Mailing Add:* PO Box 220 Annapolis Co Bridgetown NS B0S 1C0 Can

RIDDELL, ROBERT JAMES, JR, b US, June 25, 23; m 50; c 3. THEORETICAL PHYSICS. *Educ:* Carnegie Inst Technol, BS, 44; Univ Mich, MS, 47, PhD(physics), 51. *Prof Exp:* From instr to asst prof physics, Univ Calif, 51-55, physicist, Radiation Lab, 55-58; physicist, AEC, 58-60; physicist, Lawrence Berkeley Lab, 60-82; RETIRED. *Mem:* Am Phys Soc. *Res:* High energy physics and strong interactions. *Mailing Add:* 1095 Arlington Blvd El Cerrito CA 94530

RIDDICK, FRANK ADAMS, JR, b Memphis, Tenn, June 14, 29; m 52; c 3. ENDOCRINOLOGY. *Educ:* Vanderbilt Univ, BA, 51, MD, 54. *Prof Exp:* Instr med, Med Sch, Tulane Univ, New Orleans, 61-65, from clin asst prof to assoc prof, 65-77; from asst med dir to assoc med dir, 68-75, head sect endocrinol & metab dis, Dept Internal Med, 76-85, MED DIR, OCHSNER CLIN, 75-; CLIN PROF MED, MED SCH, TULANE UNIV, NEW ORLEANS, 77- *Mem:* Inst Med-Nat Acad Sci; fel Am Col Physicians Execs; Am Fedn Clin Res; AMA; Am Acad Med Dirs; Am Soc Internal Med. *Mailing Add:* Ochsner Clinic 1514 Jefferson Hwy New Orleans LA 70121

RIDDICK, JOHN ALLEN, b Greenville, Tex, Dec 20, 03; m 36. ANALYTICAL CHEMISTRY. *Educ:* Southwestern Univ, Tex, AB & AM, 26; Univ Iowa, PhD(analytical chem), 29. *Hon Degrees:* DSc, Southwestern Univ, Tex, 73. *Prof Exp:* Asst, Southwestern Univ, 24-26 & Univ Iowa, 26-28; asst prof chem, Univ Miss, 28-31; consult, 31-35; instr, Southern Methodist Univ, 35-36; analytical chemist, Commercial Solvents Corp, Ind, 36-42, chief analytical chemist, Res & Develop Div, 45-64, res chemist, 64-68. *Concurrent Pos:* Consult & author, 69- *Mem:* AAAS; Am Chem Soc. *Res:* Analysis in nonaqueous media; physical and thermodynamic properties of solvents; azeotropes. *Mailing Add:* 522 Centenary Dr Baton Rouge LA 70808

RIDDIFORD, LYNN MOORHEAD, b Knoxville, Tenn, Oct 18, 36; m 70. INSECT ENDOCRINOLOGY, DEVELOPMENTAL BIOLOGY. *Educ:* Radcliffe Col, AB, 58; Cornell Univ, PhD(develop biol, protein chem), 61. *Prof Exp:* NSF res fel biol, Harvard Univ, 61-63; instr, Wellesley Col, 63-65; res fel, Harvard Univ, 65-66, from asst prof to assoc prof, 66-73; assoc prof zool, 73-75, PROF ZOOL, UNIV WASH, 75- *Concurrent Pos:* NSF res grant, 64-; mem, trop med & parasitol study sect, NIH, 74-78, res grant, 75-; res grant, USDA, 78-82, 85-87 & 89-91, panel mem, Biol Stress on Plants, 79 & 89; mem, Int Comt for Symp on Comp Endocrinol, 78-; vis scholar, Stanford Univ, 79-80, Univ Cambridge, 86-87; Guggenheim fel, 79-80; panel mem, Regulatory Biol, NSF, 84-88; mem, gov bd, Int Ctr Insect Physiol & Ecol, Nairobi, Kenya, 85-, vis scientist, 87 & chmn, prog comt, 89-91; Guggenheim Fel, 79-80; Sr Int Fel, NIH, 86-87. *Mem:* Fel AAAS; Am Soc Cell Biol; Am Soc Zoologists (pres, 91); Soc Develop Biol; Am Soc Biochem & Molecular Biol; Entom Soc Am; fel Royal Entom Soc. *Res:* Hormonal control of insect development and behavior; insect embryogenesis; olfaction; pheromones. *Mailing Add:* Dept Zool Univ Wash Seattle WA 98195

RIDDLE, DONALD LEE, b Vancouver, Wash, July 26, 45; m 69; c 2. DEVELOPMENTAL GENETICS, NEMATOLOGY. *Educ:* Univ Calif, Davis, BS, 68, Berkeley, PhD(genetics), 71. *Prof Exp:* NIH trainee genetics, Univ Calif, Berkeley, 69-71, assoc molecular biol, Santa Barbara, 71-72; fel, Jane Coffin Childs Mem Fund, Med Res Coun Lab Molecular Biol, 73-75; from asst prof to assoc prof biol, 75-85, PROF BIOL, UNIV MO, COLUMBIA, 85-, DIR MOLECULAR BIOL PROG, 89- *Concurrent Pos:* Prin investr, Res Grants & Contracts, NIH, 77-, NSF, 83-86; dir, Caenorhabditis Genetics Ctr, 79-92; vis prof nemat, Commonwealth Sci & Indust Res Orgn, Australia, 83; mem, Div Res Grants, Genetics Study Sect, NIH, 85-; ed bd, Developmental Genetics, 86-88; assoc ed, J Nematol, 88-90; vis prof, Simon Fraser Univ, BC, Can, 89; honor lectr, Asn Big 8 Univs, 89-90. *Honors & Awards:* Chancellor's Award for Outstanding Fac Res, Univ Mo, 87. *Mem:* Genetics Soc Am; Soc Develop Biol; Soc Nematologists; AAAS. *Res:* Genetic analysis of nematode larval development and behavior using the formation of a dispersal stage called the dauer larva as a model system; molecular genetics of nematode RNA polymerase II. *Mailing Add:* Div Biol Sci Tucker Hall Univ Mo Columbia MO 65211

RIDDLE, GEORGE HERBERT NEEDHAM, b New York, NY, Mar 29, 40; m 65; c 3. PHYSICS, ELECTRON OPTICS & DIGITAL CONTROL. *Educ:* Princeton Univ, AB, 62; Univ Ill, MS, 64; Cornell Univ, PhD(appl physics), 71. *Prof Exp:* Mem tech staff, RCA Labs, 64-66; res physicist, Esso Res & Eng Co, 71-73; mem tech staff, RCA Labs, 73-84, sr mem tech staff, 84-87; SR MEM TECH STAFF, DAVID SARNOFF RES CTR, 87- *Mem:* Sigma Xi; Soc Info Display. *Res:* Surface physics; electron optics and electron beam instrumentation; video-disc stylus technology; display device technology. *Mailing Add:* 21 Grover Ave Princeton NJ 08540

RIDDLE, LAWRENCE H, b Jenkintown, Pa, Mar 1, 54; m 85. FUNCTIONAL ANALYSIS, MATHEMATICAL STATISTICS. *Educ:* Carnegie-Mellon Univ, BS, 76; Univ Ill, MS, 81, PhD(math), 82. *Prof Exp:* ASST PROF MATH, EMORY UNIV, 82- *Mem:* Am Math Soc; Math Asn Am; Sigma Xi. *Res:* Measure and integration in Banach spaces; functional analysis. *Mailing Add:* Dept Math & Comput Sci Emory Univ Atlanta GA 30322

RIDDLE, WAYNE ALLEN, b Madison, Wis, Nov 3, 45. INVERTEBRATE PHYSIOLOGY. *Educ:* Utah State Univ, BS, 68; Univ NMex, MS, 73, PhD(biol), 77. *Prof Exp:* Vis asst prof biol, Univ NMex, 76-77; asst prof, 77-83, ASSOC PROF PHYSIOL, ILL STATE UNIV, 83- *Mem:* Am Soc Zoologists; Ecol Soc Am. *Res:* Respiratory physiology; cold hardiness; water relations and osmoregulation of terrestrial arthropods and molluscs. *Mailing Add:* Dept Biol Sci Ill State Univ Normal IL 61761

RIDENER, FRED LOUIS, JR, b El Reno, Okla, Sept 19, 44; m 71. ELEMENTARY PARTICLE PHYSICS. *Educ:* NMex State Univ, BS, 68; Iowa State Univ, PhD(physics), 76. *Prof Exp:* Asst prof, 77-83, ASSOC PROF PHYSICS, PA STATE UNIV, 83- *Mem:* Am Phys Soc. *Res:* Elementary particle theory, especially polarization, electroweak form factors and relativistic wave equations; response functions for nonlinear systems. *Mailing Add:* Dept Physics Pa State Univ New Kensington PA 15068

RIDENHOUR, RICHARD LEWIS, b Santa Rosa, Calif, July 13, 32; m 54; c 5. FISH BIOLOGY. *Educ:* Humboldt State Col, BS, 54; Iowa State Col, MS, 55, PhD(fisheries), 58. *Prof Exp:* Aquatic biologist com fisheries, State Fish Comn, Ore, 58-60; from asst prof to assoc prof, Humboldt State Univ, 60-70, asst dean acad affairs, 67-69, dean acad planning, 69-81, PROF FISHERIES, HUMBOLDT STATE UNIV, 70-, DEAN, COL NAT RES, 81- *Concurrent Pos:* Consult, Water Develop Proj, Modesto & Turlock Irrig Dist, Int Eng Co, San Francisco. *Mem:* Am Fisheries Soc; Am Inst Fishery Res Biol. *Res:* Biometrics; ecology. *Mailing Add:* Col Nat Res Humboldt State Univ Arcata CA 95521

RIDENOUR, MARCELLA V, b New Martinsville, WVa, Nov 29, 45; m 73; c 2. MOTOR DEVELOPMENT. *Educ:* Miami Univ, BS, 67; Purdue Univ, MS, 68 & PhD(motor develop), 72. *Prof Exp:* PROF MOTOR DEVELOP, TEMPLE UNIV, 74- *Mem:* Am Soc Testing & Mat. *Mailing Add:* Dept Phys Educ Temple Univ Philadelphia PA 19122

RIDEOUT, DONALD ERIC, b Burlington, Nfld, May 15, 42; m 65; c 2. MATHEMATICS. *Educ:* Mem Univ Nfld, BA, 63, BSc, 64; McGill Univ, PhD(math), 70. *Prof Exp:* Lectr math, McGill Univ, 66-70; asst prof, 70-74, ASSOC PROF MATH, MEM UNIV NFLD, 74- *Concurrent Pos:* Nat Res Coun Can fel, 70-72 & 72-74. *Mem:* Math Soc Can. *Res:* Algebraic number theory. *Mailing Add:* Dept Math Mem Univ Nfld St John's NF A1C 5S7 Can

RIDEOUT, JANET LITSTER, b Bennington, Vt, Jan 6, 39; m 73. CHEMISTRY. *Educ:* Mt Holyoke Col, AB, 61, MA, 63; State Univ NY Buffalo, PhD(chem), 68. *Prof Exp:* Res chemist, Burroughs Wellcome Co, 68-70, sr res chemist, 70-79, group leader, Exp Ther Dept, 79-83, group leader, org chem div, 83-88, ASST DIV DIR, BURROUGHS WELLCOME CO, 88- *Mem:* Am Chem Soc; NY Acad Sci; fel Am Inst Chemists. *Res:* Organic synthesis; heterocyclic compounds; nucleosides; purines; pyrimidines; imidazoles; metabolites; cancer chemotherapy; antiviral chemotherapy. *Mailing Add:* Org Chem Dept Burroughs Wellcome Co Research Triangle Park NC 27709

RIDEOUT, SHELDON P, b Toronto, Ohio, Nov 5, 27; m 47; c 6. PHYSICAL METALLURGY. *Educ:* Mich Technol Univ, BS, 48; Univ Notre Dame, MS, 51. *Prof Exp:* Engr, Ladish Co, Wis, 48-49; works tech metall lab, Savannah River Plant, 51-63, process supvr, 53-55, sr supvr, 55-63, res supvr, Nuclear Mat Div, 63-78, CHIEF SUPVR, REACTOR MAT TECH, SAVANNAH RIVER LAB, E I DU PONT DE NEMOURS & CO, INC, 78- *Mem:* Am Soc Metals; Nat Asn Corrosion Engrs. *Res:* Metallurgy of materials for nuclear reactors; stress corrosion of stainless steel and titanium alloys; hydrogen effects in metals. *Mailing Add:* 245 Barnard Ave SE Aiken SC 29801

RIDEOUT, VINCENT C(HARLES), b Chinook, Alta, May 22, 24; nat US; m 39; c 4. ELECTRICAL ENGINEERING. *Educ:* Univ Alta, BSc, 38; Calif Inst Technol, MS, 40. *Prof Exp:* Mem tech staff elec eng, Bell Tel Labs, Inc, 39-46; asst prof, Univ Wis, 46-49, assoc prof, 49-54; vis prof, Indian Inst Sci, 54-55; PROF ELEC ENG, UNIV WIS-MADISON, 55- *Concurrent Pos:* Vis prof, Univ Colo, 63-64 & Inst Med Physics, Utrecht, Holland, 70-71; mem, Ball Mfrs Engrs Comt. *Mem:* AAAS; fel Inst Elec & Electronics Engrs. *Res:* Control systems, computing, bioengineering, socio-economics. *Mailing Add:* Univ Wis 543 Eng Res Bldg Madison WI 53706

RIDER, AGATHA ANN, b May 13, 19; m 48. PUBLIC HEALTH NUTRITION, FAT SOLUBLE VITAMINS. *Educ:* Johns Hopkins Univ, MS, 46. *Prof Exp:* asst prof biochem, Sch Hyg & Pub Health, Johns Hopkins Univ, 72-86; RETIRED. *Mem:* Am Chem Soc; Am Inst Nutrit; Sigma Xi; AAAS. *Res:* Vitamin A nutrition. *Mailing Add:* 5400 Vantage Point Rd No 1111 Columbia MD 21044

RIDER, BENJAMIN FRANKLIN, b Cleveland, Ohio, Dec 4, 21; m 44. ANALYTICAL CHEMISTRY. *Educ:* Mt Union Col, BS, 43; Purdue Univ, MS, 44, PhD(analytical chem), 47. *Prof Exp:* Res assoc, Knolls Atomic Power Lab, 47-57, RES ASSOC, VALLECITOS NUCLEAR CTR, GEN ELEC CO, 57- *Mem:* Am Chem Soc; fel Am Inst Chem; Am Soc Test & Mat. *Res:* Spectrophotometry; radiochemistry; medical radioisotopes processing; nuclear fuel burnup analysis; fission yields compilation. *Mailing Add:* 4137 Norris Rd Fremont CA 94536-5013

RIDER, DON KEITH, b Rockford, Ill, Feb 12, 18; m 47; c 3. ORGANIC POLYMER CHEMISTRY, MATERIALS ENGINEERING. *Educ:* Univ Mich, BS, 39, MS, 41. *Prof Exp:* Chemist resin develop, Rohm & Haas Co, 41-46, group leader ion exchange develop, 46-47; chemist binder res, Chicopee Mfg Corp Div, Johnson & Johnson, 47-48; chemist, Resins Laminates Develop, 48-59, head organic mat res & develop dept, 59-75, head organic mat eng dept, 75-78; RETIRED. *Concurrent Pos:* Mem tech panel, Mat Adv Bd, Nat Acad Sci-Nat Res Coun, 56-57, spec comt adhesive bonded struct components, Bldg Res Adv Bd, 63-65; consult, Archit Plastics Int, 64-66; mem ad hoc comt, Predictive Testing, Nat Acad Sci-Nat Res Coun, 70-72; consult, 78- *Mem:* Am Chem Soc. *Res:* Adhesives, bonded structures and casting resins; laminates; structural plastics; laminated thermoset materials; printed circuits; fibrous reinforcements. *Mailing Add:* 788 Park Shore Dr B24 Naples FL 33940

RIDER, JOSEPH ALFRED, b Chicago, Ill, Jan 30, 21; m 43; c 2. GASTROENTEROLOGY. *Educ:* Univ Chicago, SB, 42, MD, 44, PhD(pharmacol), 51; Am Bd Internal Med, dipl; Am Bd Gastroenterol, dipl. *Prof Exp:* Intern internal med, Presby Hosp, Chicago, 44-45; asst resident med, Univ Tex, 47-49; resident, Univ Chicago, 49-50, instr, 51-52; asst prof, Med Ctr, Univ Calif, San Francisco, 53-59, asst clin prof, 59-66, asst chief gastrointestinal clin, 53-61; DIR GASTROINTESTINAL RES LAB, FRANKLIN HOSP, 63- *Mem:* Am Soc Pharmacol & Exp Therapeut; Soc Exp Biol & Med; Am Geriat Soc; fel Am Gastroenterol Asn; fel Am Col Physicians. *Res:* Tolerance of organic phosphates in men; hypersensitivity factors in ulcerative colitis; cytology in the diagnosis of gastrointestinal malignancies; gastric secretion and motility; color television endoscopy. *Mailing Add:* Ralph K Davies Med Ctr Franklin Hosp San Francisco CA 94114

RIDER, PAUL EDWARD, SR, b Des Moines, Iowa, Nov 22, 40; m 63; c 3. PHYSICAL CHEMISTRY. *Educ:* Drake Univ, BA, 62; Iowa State Univ, MS, 64; Kans State Univ, PhD(phys chem), 69. *Prof Exp:* Instr chem, Drake Univ, 64-66; vis prof, Coe Col, 69; from asst prof to assoc prof, 69-79, asst provost, 82-83, PROF CHEM, UNIV NORTHERN IOWA, 79-, EXEC DIR, IOWA ACAD SCI, 88- *Concurrent Pos:* Consult, Shell Develop Co, Houston. *Mem:* Am Chem Soc; AAAS; Iowa Acad Sci. *Res:* Thermodynamic studies of weak hydrogen bonds; theoretical consideration of polymer solution formation; history and philosophy of science. *Mailing Add:* Dept Chem Univ Northern Iowa Cedar Falls IA 50613

RIDER, RONALD EDWARD, b Pasadena, Calif, June 11, 45; m 69; c 2. COMPUTER SCIENCE, PHYSICS. *Educ:* Occidental Col, AB, 67; Wash Univ, AM, 69, PhD(physics), 72. *Prof Exp:* Res scientist comput sci & physics, Xerox Corp, 72-77, mgr electronic subsyst develop, 77-86, co-mgr, Palo Alto Res Ctr, 86-89, VPRES SYSTS ARCHIT, XEROX CORP, 89- *Res:* Computer systems; word processing; image processing. *Mailing Add:* 3333 Coyote Hill Dr Palto Alto CA 94304-1314

RIDER, ROWLAND VANCE, b Syracuse, NY, Aug 23, 15; m 48. BIOSTATISTICS. *Educ:* Amherst Col, AB, 37; Syracuse Univ, MA, 38; Johns Hopkins Univ, ScD(biostatist), 47. *Prof Exp:* From asst prof to assoc prof pub health admin, 50-65, prof pop dynamics, 65-82, EMER PROF POP DYNAMICS, SCH HYG & PUB HEALTH, JOHNS HOPKINS UNIV, 82- *Mem:* AAAS; Pop Asn Am; Am Pub Health Asn; Am Statist Asn; Int Union Sci Study Pop. *Res:* Study prematures, maternal and child health and family planning administration evaluation. *Mailing Add:* 5400 Vantage Point Rd No 1111 Columbia MD 21044

RIDGE, DOUGLAS POLL, b Portland, Ore, Nov 9, 44; m 71; c 6. PHYSICAL CHEMISTRY. *Educ:* Harvard Col, AB, 68; Calif Inst Technol, PhD(chem), 72. *Prof Exp:* From asst prof to assoc prof, 72-85, PROF CHEM, UNIV DEL, 85- *Mem:* Am Phys Soc; Am Chem Soc. *Res:* Reactive and nonreactive ion molecule interactions in the gas phase; ion cyclotron resonance spectroscopy; gas phase organometallic chemistry. *Mailing Add:* Dept Chem Univ Del Newark DE 19716

RIDGE, JOHN CHARLES, b Bethlehem, Pa, Mar 9, 55; m 85. PLEISTOCENE PALEOMAGNETISM, GLACIAL SEDIMENTATION. *Educ:* Lehigh Univ, BS, 77, MS, 83; Syracuse Univ, PhD(geol), 85. *Prof Exp:* Teaching asst geol, Syracuse Univ, 80-85; ASST PROF GEOL, TUFTS UNIV, 85- *Concurrent Pos:* Geologist, NY State Geol Surv, 80-82 & 85- *Mem:* Geol Soc Am; Am Quaternary Asn; Am Geophys Union. *Res:* Paleomagnetism and secular variation of remanent magnetization as recorded in glacial sediments, varve and glacial stratigraphy and glacial processes; genesis of landscapes during interglacial and periglacial intervals; geomorphology. *Mailing Add:* Dept Geol Tufts Univ Medford MA 02155

RIDGE, JOHN DREW, b Cincinnati, Ohio, July 3, 09; m 41; c 2. APPLIED, ECONOMIC & ENGINEERING GEOLOGY. *Educ:* Univ Chicago, SB, 30, SM, 32, PhD(econ geol), 35. *Prof Exp:* Res petrographer, Universal-Atlas Cement Co, Ind, 35-36; petrologist, US Nat Park Serv, Washington, DC, 36-37; geologist, Cerro de Pasco Copper Corp, Peru, 37-40 & NJ Zinc Co, 46-47; assoc prof econ geol, Pa State Univ, 47-51, prof mineral econ, 51-64, asst dean col mineral indust, 53-64, prof econ geol & mineral econ, 64-75, head dept mineral econ, 51-75; RETIRED. *Concurrent Pos:* Mem, Earth Sci Div, Nat Res Coun, 67-70 & 68-71; mem panel mineral econ, Nat Acad Sci, 67-79; ed, Graton-Sales Volume, 68-72; Nat Acad Sci exchange scientist in Poland, 69 & 70, Romania, 73, USSR & Yugoslavia, 77; mem comt crit & strategic mat, Nat Mat Adv Bd, 69-71; exchange scientist, NSF, 69 & 73 & Nat Res Coun, 70, 72, 74 & 77; adj prof geol, Univ Fla, 75-80, actg chmn dept, 80-81, vis prof, 80-83. *Honors & Awards:* Henry Krumb Lectr, 71; Mineral Econ Award, 72. *Mem:* Int Asn Genesis Ore Deposits (pres, 76-80, past pres, 80-84); Soc Econ Geol; Am Soc Mining, Metall & Petrol Engrs; fel Am Mineral Soc; fel Geol Soc Am. *Res:* Geology and geochemistry of metallic ore deposits; chemistry of metasomatism, stable isotopes and ore genesis; politics and economics of mineral exploration and exploitation. *Mailing Add:* 400 Brentwood Rd Charlottesville VA 22901

RIDGEWAY, BILL TOM, b Columbia, Mo, Dec 23, 27; m 52; c 3. PARASITOLOGY, PROTOZOOLOGY. *Educ:* Friends Univ, AB, 52; Wichita State Univ, MS, 58; Univ Mo, PhD(zool), 66. *Prof Exp:* From instr to asst prof zool, Southwestern Col, 58-63; res asst parasitol, Univ Mo, 63-66, asst prof invert zool, 66; assoc prof, 66-71, PROF ZOOL & PARASITOL, EASTERN ILL UNIV, 71- *Concurrent Pos:* Vis prof, Inland Environ Lab, Univ Md, 74-75; prog officer, Ann Midwest Conf Parasitologists, 81, presiding officer, 85; contract scientist, Ill Dept Conserv, 76-80. *Mem:* Am Soc Parasitol; Am Micros Soc; Am Soc Protozoologists. *Res:* Protozoan ectosymbionts of freshwater invertebrates; helminth and protozoan parasites of wild rodents; systematics of nemotodes. *Mailing Add:* Dept Zool Eastern Ill Univ Charleston IL 61920

RIDGWAY, ELLIS BRANSON, b Philadelphia, Pa, May 14, 39; m 64, 80; c 3. PHYSIOLOGY, BIOPHYSICS. *Educ:* Mass Inst Technol, SB, 63; Univ Ore, PhD(biol), 68. *Prof Exp:* NATO fel, Univ Col, Univ London, 69; USPHS fel, Cambridge Univ, 69-70; USPHS fel, Friday Harbor Labs, Univ Wash, 70-72; asst prof, 72-77, ASSOC PROF PHYSIOL, MED COL VA, 77- *Mem:* AAAS; Soc Gen Physiol; Biophys Soc. *Res:* Role of calcium as an activator in muscle, nerve and synapse; biophysics of muscle contraction and membrane permeability. *Mailing Add:* Med Col Va Box 551 MCV Sta Richmond VA 23298

RIDGWAY, GEORGE JUNIOR, b Lincoln, Nebr, Aug 26, 22; m 47; c 3. MICROBIOLOGY, BIOCHEMISTRY. *Educ:* Univ Wash, Seattle, BS, 49, MS, 51, PhD(microbiol), 54. *Prof Exp:* Biochemist, US Bur Com Fisheries, 54-64; asst lab dir, Northeast Fisheries Ctr, Nat Marine Fisheries Serv, Woods Hole, 64-71, dir, Biol Lab, Maine, 71-73, asst ctr dir, 73-75; RETIRED. *Mem:* AAAS; Sigma Xi. *Res:* Comparative immunology, immunogenetics, immunochemistry and biochemical genetics of fishes and other animals as applied to discrimination of natural populations. *Mailing Add:* RR 5 Box 267 Gardiner ME 04345

RIDGWAY, HELEN JANE, b Ft Worth, Tex, Aug 10, 37. BIOCHEMISTRY. *Educ:* NTex State Col, BA, 59; Baylor Univ, MS, 63, PhD(chem), 68. *Prof Exp:* Chemist, 60-68, res investr biochem, 68-86, CHAIR CHEM DEPT, WADLEY INSTS MOLECULAR MED, 86- *Concurrent Pos:* Mem coun thrombosis, Am Heart Asn. *Mem:* Am Chem Soc; fel Int Soc Hemat; Am Heart Asn. *Res:* Biochemistry of blood coagulation; leukemia and cancer chemotherapy; platelet function. *Mailing Add:* 2844 Major Ft Worth TX 76112

RIDGWAY, JAMES STRATMAN, b Paintsville, Ky, July 27, 36; m 59; c 2. POLYMER CHEMISTRY, SYNTHETIC FIBERS. *Educ:* Univ Louisville, BS, 58, MS, 59, PhD(org chem), 61. *Prof Exp:* Technician, Girdler Co, 57; teaching asst gen chem & qual anal, Univ Louisville, 57-58, res asst org chem, 58- 61; res chemist, Chemstrand Res Ctr, Inc, 61-66, sr res chemist, 66-67; sr res chemist, 68-69, res specialist, 69-76, SR RES SPECIALIST, MONSANTO CO, 76- *Mem:* Am Chem Soc; fel Am Inst Chemists. *Res:* Polymer structure-property relationships; synthesis of polyamides and copolyamides; synthetic fiber applications for textile and tire cords; melt and solution polycondensation reactions; high modulus organic fibers; fiber flammability, textile colorfastness; anti-soil and stain resistant fibers; carpet fiber properties. *Mailing Add:* Monsanto Chem Co PO Box 12830 Pensacola FL 32575

RIDGWAY, RICHARD L, b Brownfield, Tex, Nov 9, 35; m 57; c 2. ENTOMOLOGY. *Educ:* Tex Tech Col, BS, 57; Cornell Univ, MS, 59, PhD(entom), 60. *Prof Exp:* Exten entomologist, Tex A&M Univ, 60-63, res entomologist, Entom Res Div, 63-70, entomologist-in-charge, Cotton Inst, 70-72, entomologist, Univ, 71-72, mem grad fac, 65-75, res leader, 72-75; staff scientist cotton & tobacco insects, Agr Res Serv, USDA, Washington, DC, 75-80; MEM FAC, ENTOM DEPT, TEX A&M UNIV, 80- *Honors & Awards:* Geigy Recognition Award, Entom Soc Am, 72. *Mem:* AAAS; Entom Soc Am; Int Orgn Biol Control. *Res:* Methods of application of systematic insecticides; behavior of insecticides in plants and soil; selective insecticides; biological control of insect pests. *Mailing Add:* 3788 Evans Trail Way Beltsville MD 20705

RIDGWAY, ROBERT WORRELL, b Hampton, Va, July 14, 39; m 60; c 1. ORGANIC CHEMISTRY. *Educ:* Drexel Univ, BS, 66; Univ NH, PhD(org chem), 70. *Prof Exp:* NSF fel & res assoc, Princeton Univ, 69-70; asst prof chem, J C Smith Univ, 70-72; asst prof chem, Rollins Col, 72-75, assoc prof, 75-80; MGR OFF COOP EDUC, AM CHEM SOC, 80- *Mem:* Am Chem Soc; Royal Soc Chem; Sigma Xi. *Res:* Organometallic stereochemistry, especially asymmetric reductions and biologically important systems; synthetic organic chemistry. *Mailing Add:* 11 Bon Aire Dr Olivette MO 63132

RIDGWAY, SAM H, b San Antonio, Tex, June 26, 36; m 60. ENVIRONMENTAL PHYSIOLOGY. *Educ:* Tex A&M Univ, BS, 58, DVM, 60; Cambridge Univ, PhD(neurobiol), 73. *Prof Exp:* Res vet, US Naval Missile Ctr, 62-63, Univ Southern Calif, 63-65 & Univ Calif, Santa Barbara, 65-66; res vet, 66-72, head, Biomed Div, Naval Undersea Ctr, 72-80; SR SCIENTIST, NAVAL OCEAN SYSTS CTR, 80- *Concurrent Pos:* USN res sponsor, 62- & Univ Calif, Santa Barbara & Inst Environ Stress, 65-; sr res fel, Cambridge Univ, 70-72; sci adv, Marine Mammal Comn, 75-77, chmn, 77-78; coun mem, Inst Lab Animal Resources, Nat Acad Sci-Nat Res Coun, 75-78; mem res comt, San Diego Zoo, 73- *Honors & Awards:* Gilbert Curl Sci Award, Naval Undersea Ctr, 73. *Mem:* NY Acad Sci; AAAS; Am Vet Med Asn; Sigma Xi; Explorers Club; Acoust Soc Am; Int Asn Aquatic Animal Med. *Res:* Marine mammal physiology; dolphin neurobiology; aquatic animal medicine; bioacoustics. *Mailing Add:* 1150 Anchorage Lane-608 Naval Ocean Systs Ctr San Diego CA 92106

RIDGWAY, STUART L, b Freeport, Ill, July 27, 22. ENERGY CONVERSION. *Educ:* Haverford Col, BS, 43; Princeton Univ, PhD(physics), 52. *Prof Exp:* Group leader fire control res, US Naval Res Lab, 43-46; instr physics, Princeton Univ, 50-52, res assoc, 52-53; res assoc, Brookhaven Nat Lab, 53-56; mem sr staff, TRW Inc, 56-62; mem sr tech staff, Gen Tech Corp, 62-66; sr physicist, Princeton Appl Res Corp, 66-73; sr res scientist, R & D Assocs, 73-89; sr mech engr, Pac Int Ctr for High Technol Res, 89-90; CONSULT, 90- *Mem:* Am Phys Soc; Combustion Inst; NY Acad Sci; AAAS. *Res:* Beta decay; high energy physics; heat transfer; combustion dynamics; motor vehicle exhaust control; ocean thermal energy conversion. *Mailing Add:* 537 Ninth St Santa Monica CA 90402

RIDGWAY, WILLIAM C(OMBS), III, b Orange, NJ, Apr 28, 36; m; c 6. COMPUTING. *Educ:* Princeton Univ, BSE, 57; NY Univ, BEE, 59; Stanford Univ, PhD(elec eng), 62. *Prof Exp:* Mem tech staff, 57-68, head mil data systs eng dept, 68-70, head ocean data systs dept, 70-72, dir, Par Software & Data Processing Ctr, 72-73, dir, Comput Technol & Prog Develop Ctr, Bell Tel Labs, 73-82; dir, Comput Telecommunications & Eng Info Ctr, 82-85; adminr, Telecommun & Info Systs, Dept Treas, NJ, 85-90; INDEPENDENT CONSULT, 90- *Concurrent Pos:* Mem resource & technol panel, Comput Sci & Eng Bd, Nat Acad Sci; mem tech staff, AT&T Info Systs, 82-85. *Mem:* Asn Comput Mach; Inst Elec & Electronics Engrs; Sigma Xi. *Res:* Development of advanced computing services. *Mailing Add:* 68 Neck Rd Old Lyme CT 06371

RIDHA, R(AOUF) A, b Karbala, Iraq, Aug 25, 37; US citizen. ENGINEERING MECHANICS, STRUCTURAL ENGINEERING. *Educ:* Univ Baghdad, BS, 59; Univ Ill, Urbana, MS, 63, PhD(struct), 66. *Prof Exp:* Supv engr struct, Govt Iraq, 59-62; anal specialist, Energy Controls Div, Bendix Corp, 66-73; res assoc, Cent Res Labs, Firestone Tire & Rubber Co, 73-78; head, Eng Mech Sect, Gen Tire & Rubber Co, 78-80, sect head, Physics & Eng Mech Sect, Res Div, 80-83; mgr tire physics & math, Akron, Ohio, 83-86, MGR TIRE-VEHICLE ENG TECHNOL, GOODYEAR TECH CTR, LUXENBOURG GOODYEAR TIRE & RUBBER CO, 86- *Concurrent Pos:* Lectr, Ind Univ, South Bend, 69-73 & Mich State Univ, 70-71; lectr IV, Univ Akron, 77-78. *Mem:* Assoc fel Am Inst Aeronaut & Astronaut. *Res:* Engineering mechanics; structural optimization; stability; nonlinear analysis; finite element methods; composite materials; aerospace structures; numerical methods; tire mechanics; tire stresses and deformation; analysis and design of composite structures. *Mailing Add:* Goodyear Tech Ctr Goodyear Tire & Rubber Co L-7550 Colmar Berg Luxembourg

RIDINGS, GUS RAY, b Arbyrd, Mo, Nov 22, 18; m 41; c 2. RADIOLOGY. *Educ:* Ark State Col, AB, 39; Vanderbilt Univ, MD, 50. *Prof Exp:* Instr radiol, Col Med, Vanderbilt Univ, 55-56; assoc prof, Sch Med, Univ Miss, 56-57; prof, Sch Med, Univ Okla, 57-62; prof, Sch Med, Univ Mo, Columbia, 63-67; clin prof radiother, Univ Tex Southwestern Med Sch Dallas & radiotherapist, St Paul Hosp, 67-71; dir, C J Williams Cancer Treat Ctr, Baptist Mem Hosp, 71-75; dir, Southeast Hosp Radiation Oncol Ctr, 75-; MEM STAFF, RIDGE RADIOL ONCOL ASN. *Concurrent Pos:* Consult, Vet Admin Hosp. *Mem:* Soc Nuclear Med; Radiol Soc NAm; Am Asn Cancer Educ; Am Soc Therapeut Radiologists; Sigma Xi. *Res:* Radiation therapy. *Mailing Add:* Ridings Radiation Oncol Asn 937 Broadway Cape Girardeau MO 63701

RIDLEN, SAMUEL FRANKLIN, b Marion, Ill, Apr 24, 16; m 46; c 3. POULTRY SCIENCE. *Educ:* Univ Ill, BS, 40; Mich State Univ, MS, 57. *Prof Exp:* Instr high sch, Ill, 40-43; asst prof poultry, Exten Div, Univ Ill, 46-53; gen mgr, Honegger Breeder Hatchery, Ill, 53-56; assoc prof poultry, Univ Conn, 57-58; from assoc prof to prof, 58-86, asst head dept animal sci, 78-86,

EMER PROF POULTRY, COOP EXTEN SERV, UNIV ILL, URBANA, 86- *Honors & Awards:* Poultry Sci Asn Exten Award, 65; Super Serv Award, USDA, 82. *Mem:* Fel Poultry Sci Asn; World Poultry Sci Asn. *Res:* Poultry management and production economics; effect of different cage densities and protein levels on the performance of laying hens. *Mailing Add:* 106 W Mumford Dr Urbana IL 61801-5818

RIDLEY, ESTHER JOANNE, b Pittsburgh, Pa, Sept 5, 24. PLANT PHYSIOLOGY. *Educ:* Fisk Univ, BA, 45; Univ Pittsburgh, MS, 50; Okla State Univ, PhD(bot), 67. *Prof Exp:* Asst biol, Univ Pittsburgh, 49-50, lab technician, Med Ctr, 50-54, res asst immunol, 54-56, lab supvr blood procurement, Cent Blood Bank, 56-59; instr biol, Morgan State Col, 59-60, instr sci educ, 60-61; instr bot, Baltimore Jr Col, 61-62; asst prof, Morgan State Col, 62-65, assoc prof, 67-69, chmn dept, 72-81, prof sci educ, 70-84, acting dean, Col Arts & Sci, 87-88, PROF BIOL, MORGAN STATE UNIV, 84- *Mem:* AAAS; NY Acad Sci; Mycol Soc Am; Am Bot Soc; Sigma Xi. *Res:* Developmental physiology; plant anatomy and morphology. *Mailing Add:* Dept Biol Morgan State Univ Baltimore MD 21239

RIDLEY, PETER TONE, b Meriden, Conn, Nov 11, 36; m 59; c 3. PHYSIOLOGY, PHARMACOLOGY. *Educ:* Rutgers Univ, BA, 59; Univ Pa, PhD(physiol), 64. *Prof Exp:* Jr pharmacologist, Smith Kline & French Labs, 59-60; USPHS fel, Karolinska Inst, Sweden, 64-65; asst prof physiol, George Washington Univ, 65-67; asst dir pharmacol, 67-71, dir pharmacol, 71-75, dep dir res, 75-81, VPRES RES & DEVELOP, ALLERGAN PHARMACEUT, DIV SMITH KLINE & FRENCH LABS, 81- *Mem:* Am Physiol Soc; Am Gastroenterol Asn. *Res:* Physiology and pharmacology of the gastrointestinal tract; central neural control of gastrointestinal function. *Mailing Add:* 865 Sandcastle Corona Del Mar CA 92625

RIDOLFO, ANTHONY SYLVESTER, b Montclair, NJ, Oct 27, 18; m 42; c 2. RHEUMATOLOGY, CLINICAL PHARMACOLOGY. *Educ:* Rutgers Univ, BS, 40; Ohio State Univ, MS, 42, PhD, 47, MD, 54; Am Bd Internal Med, dipl, 64. *Prof Exp:* Assoc prof pharm, Univ Toledo, 42-44; assoc prof, Ohio State Univ, 47-50; fel cardiol, Marion County Gen Hosp, Indianapolis, Ind, 57-58; from physician to sr physician, Eli Lilly & Co, 58-73, sr clin pharmacologist, 73-86; RETIRED, 86- *Concurrent Pos:* Instr, Sch Med, Ind Univ, Indianapolis, 57-60, from asst to assoc, 60-65, from asst prof to assoc prof, 65-75, clin prof med, 75-79, prof med, 79-; mem bd dirs, Ind Arthritis Found, 61-, pres, 71-72; chmn sect rheumatic dis & anti-inflammatory agent, Am Soc Clin Pharmacol & Therapeut, 74-77; mem, US Pharmacopeia Panel on Analgesics, Sedatives & Anti-Inflammatory Agents; mem, Comt Relationships Pharmaceut Houses, Nat Arthritis Found, 81- *Mem:* Fel Am Col Physicians; Am Fedn Clin Res; Am Rheumatism Asn; AMA; Am Soc Clin Pharmacol & Therapeut. *Mailing Add:* 5207 S State Rd No 421 Cionsville IN 46077

RIE, JOHN E, b New York, NY, Aug 26, 44; m 71. POLYMER CHEMISTRY, PHOTOCHEMISTRY. *Educ:* Univ Vt, BA, 66; Wayne State Univ, PhD(chem), 72. *Prof Exp:* Res chemist photopolymers, Kalle Aktiengesellschaft Div, Hoechst AG, 71-72, prod mgr photoresist, 72-74; sr res chemist photopolymers, Photopolymer Systs, W R Grace & Co, 74-78; group leader, Dynacure Printed Circuit Prod, Thiokol/Dynachem Corp, 78-80; mgr mat sci, PCK Technol Div, Kollmorgen Corp, 80-87; CONSULT, 91- *Mem:* Am Chem Soc; Sigma Xi. *Res:* Photopolymer research and development for coatings and photoresist applications. *Mailing Add:* 38 Tunxis Circle Meriden CT 06450-7401

RIEBESELL, JOHN F, b Oneida, NY, March 18, 48. PLANT ECOLOGY, POPULATION ECOLOGY. *Educ:* State Univ NY, Albany, BS, 70, MS, 71; Univ Chicago, PhD(biol), 75. *Prof Exp:* Vis asst prof biol, Col Wooster, 76-77; asst prof, 77-83, ASSOC PROF BIOL, UNIV MICH, DEARBORN, 83- *Concurrent Pos:* Dir, Adirondack Lab, 76- *Mem:* Ecol Soc Am; Sigma Xi; AAAS. *Res:* Population biology and physiological ecology; photosynthetic adaptations in plants; effects of environmental modification on population densities and community stability; land use planning. *Mailing Add:* Dept Natural Sci Univ Mich Dearborn MI 48128

RIEBMAN, LEON, b Coatesville, Pa, Apr 22, 20; m 42; c 2. ELECTRONICS ENGINEERING. *Educ:* Univ Pa, BSEE, 43, MSEE, 47, PhD(elec eng), 51. *Prof Exp:* Sr engr, Philco Corp, 45-46; res assoc & part-time instr, Univ Pa, 48-51; pres, Am Electronic Labs, Inc, 51-87, CHMN BD & CHIEF EXEC OFFICER, AEL INDUSTS INC, 87- *Concurrent Pos:* Mem bd eng educ, Univ Pa, 68; dir, Ampal, 70- *Mem:* Fel Inst Elec & Electronics Engrs. *Res:* Electronics, particularly antenna and microwaves; computers. *Mailing Add:* AEL Industs Inc Richardson Rd Box 552 Lansdale PA 19446

RIECHEL, THOMAS LESLIE, b Bakersfield, Calif, July 9, 50. ANALYTICAL CHEMISTRY, ELECTROCHEMISTRY. *Educ:* Univ Calif, Davis, BS, 72, Riverside, PhD(chem), 76. *Prof Exp:* Res assoc chem, State Univ NY Buffalo, 76-77 & Univ Del, 78; asst prof, 78-83, ASSOC PROF CHEM, MIAMI UNIV, 83- *Mem:* Am Chem Soc; Sigma Xi; Soc Electroanal Chem. *Res:* Electrochemical and spectroscopic studies of models for metalloenzymes; ion selective electrodes; chemically modified electrodes; room temperatures molten salt electrolytes. *Mailing Add:* Dept Chem Miami Univ Oxford OH 45056

RIECHERT, SUSAN ELISE, b Milwaukee, Wis, Oct 20, 45. ZOOLOGY. *Educ:* Univ Wis-Madison, BA, 67, MS, 70, PhD(zool), 73. *Prof Exp:* From asst prof to assoc prof, 73-82, PROF ZOOL, UNIV TENN, 82- *Concurrent Pos:* Assoc cur invert, Univ Wis Zool Mus, 71-77; asst ed, J Arachnology; pres, Am Arachnologist, 84-85; Fogarty Found Fel. *Mem:* Ecol Soc Am; Animal Behav Soc; Sigma Xi; Am Arachnological Asn; Entom Soc Am; AAAS. *Res:* Food-based spacing in spiders; underlying factors responsible for observed patterns of local animal distribution; game playing and interaction strategies in spiders; evolution of social behavior; generalist predator control of pests in agroecosystems. *Mailing Add:* Dept Zool Univ Tenn Knoxville TN 37996-0810

RIECK, H(ENRY) G(EORGE), b Eugene, Ore, Aug 30, 22; m 53; c 3. CHEMICAL ENGINEERING. *Educ:* Ore State Col, BS, 43, MS, 44. *Prof Exp:* Engr, Gen Elec Co, 47-65; SR RES SCIENTIST, PAC NORTHWEST LABS, BATTELLE MEM INST, 65- *Res:* Development and design of equipment used in air sampling (land, marine and aircraft based systems); water sampling (rivers, lakes and ocean based systems); collection and measurement of radionuclides in various sampling regimes; nuclear counting equipment and techniques in nondestructive assay of low level nuclear wastes for transuranic radionuclide content; remote handling equipment and techniques for manipulation of high level radionuclide sources. *Mailing Add:* Pac Northwest Lab PO Box 999 Richland WA 99352

RIECK, JAMES NELSON, b Wheeling, WVa, Aug 7, 39; c 2. POLYMER CHEMISTRY. *Educ:* West Liberty State Col, BS, 61; WVa Univ, PhD(org chem), 73. *Prof Exp:* Res chemist, 60-73, group leader polymer chem, 73-81, SECT MGR POLYURETHANE RES, MOBAY CHEM CORP, 81- *Mem:* Am Chem Soc. *Res:* Polyurethane textile coatings; polyurethane adhesives; prepolymers; modified isocyanates; bonding agents; rubber adhesive agents. *Mailing Add:* 26 Maple Ln Bethlehem Wheeling WV 26003

RIECK, NORMAN WILBUR, b Union City, NJ, Feb 16, 23; m 53; c 1. ANATOMY. *Educ:* Hope Col, AB, 53; Univ Mich, MS, 56, PhD(anat), 57. *Prof Exp:* Instr anat, Sch Med, Temple Univ, 57-59 & Sch Med, Univ Mich, 59-62; from assoc prof to prof, Hope Col, 62-86; RETIRED. *Mem:* Am Asn Anat; Am Mus Natural Hist; Am Inst Biol Sci. *Res:* Neuroanatomy; stimulation of occipital lobe of monkey. *Mailing Add:* 986 Laketown Dr Holland MI 49423

RIECKE, EDGAR ERICK, b Spencer, Iowa, Dec 29, 44; m 66; c 1. ORGANIC POLYMER CHEMISTRY. *Educ:* Univ SDak, BA, 67; Univ Mo-Kansas City, PhD(chem), 71. *Prof Exp:* Fel chem, Univ Rochester, 72; RES CHEMIST, EASTMAN KODAK CO, 72- *Mem:* Am Chem Soc. *Mailing Add:* 843 Independence Dr Webster NY 14580

RIECKER, ROBERT E, b Chicago, Ill, Oct 31, 36. GEOLOGY, GEOPHYSICS. *Educ:* Univ Colo, BA, 58, PhD(geol), 61. *Prof Exp:* Gen physicist, Air Force Cambridge Res Labs, 64-72, geophysicist, 72-76; alt group leader geol res, 76-77, group leader, 77-79, PROG MGR, BASIC RES GEOSCI, LOS ALAMOS NAT LAB, 76- *Concurrent Pos:* Lectr geol, Boston Col, 64-76; mem US Nat comn rock mech, Nat Acad Sci-Nat Res Coun, 70-73; adv panel geol-geophys, Los Alamos Sci Lab; adj prof geol, Grad Sch, Univ NMex, 77- & prof, Univ, 80-; geol dept, Univ NMex. *Mem:* AAAS; fel Geol Soc Am; Am Geophys Union; Sigma Xi. *Res:* Physical and mechanical properties and modes of deformation of rocks, minerals, and solids under geophysically realistic conditions of high pressure and temperature to gain insight into tectonophysical manifestations within the earth. *Mailing Add:* 147 Piedra Loop Los Alamos NM 87544

RIECKHOFF, KLAUS E, b Weimar, Ger, Feb 8, 28; Can citizen; m 49; c 3. CHEMICAL PHYSICS, SOLID STATE PHYSICS. *Educ:* Univ BC, BSc, 58, MSc, 59, PhD(physics), 62. *Prof Exp:* Mem res staff physics, Res Lab, Int Bus Mach Corp, Calif, 62-65; assoc prof, 65-66, actg dean sci, 66-67, assoc dean grad studies, 73-76, PROF PHYSICS, SIMON FRASER UNIV, 66- *Concurrent Pos:* Vis prof, Inst Appl Physics, Univ Karlsruhe, 69-70; vis scientist res lab, Int Bus Mach Corp, 76-77; vis prof, Univ PR, Mayaguez, 87 & 89, Univ Queensland, Brisbane, Queensland, Australia, 87, Univ Bayreuth, Bayreuth, Ger, 87 & 90; mem bd govs, Simon Fraser Univ, 78-84, 87-93. *Mem:* AAAS; Am Phys Soc; Can Asn Physicists. *Res:* Low temperature solid state; magneto-optics; spin-lattice relaxation; nonlinear optics; multiphoton processes in organic molecules; intermolecular and intramolecular energy transfer; spontaneous and stimulated Brillouin and Raman scattering; molecular luminescence; electrohydrodynamics. *Mailing Add:* Dept Physics Simon Fraser Univ Burnaby BC V5A 1S6 Can

RIEDEL, BERNARD EDWARD, b Provost, Alta, Sept 25, 19; m 44; c 3. PHARMACY. *Educ:* Univ Alta, BSc, 43, MSc, 49; Univ Western Ont, PhD(biochem), 53. *Hon Degrees:* DSc, Univ Alta, 90. *Prof Exp:* Assoc prof pharm, Univ Alta, 46-50, 52-58, prof, 58-62, exec asst to vpres, 62-67; prof pharmaceut sci & dean fac, Univ BC, 67-84, coodr health sci, 77-84; RETIRED. *Concurrent Pos:* Chmn, Bd Trustees, BC Organ Transplant Soc. *Mem:* AAAS; Am Chem Soc; Can Biochem Soc; Pharmacol Soc Can; hon mem Can Pharmaceut Asn. *Res:* Biochemistry; radioisotope technology. *Mailing Add:* 8394 Angus Dr Vancouver BC V6P 5L2 Can

RIEDEL, EBERHARD KARL, b Dresden, Ger, Dec 25, 39. STATISTICAL MECHANICS, THEORETICAL SOLID STATE PHYSICS. *Educ:* Univ Koln, Cologne, Ger, Physics Dipl, 64; Tech Univ Munchen, Munich, Ger, Dr rer nat, 66. *Prof Exp:* Res physicist, Max-Planck Inst Physics, Munich, Ger, 65-68 & Inst V Laue-Langevin, Munich, Ger & Grenoble, France, 69; res assoc physics, Cornell Univ, Ithaca, NY, 69-71; from asst prof to assoc prof physics, Duke Univ, Durham, NC, 71-75; assoc prof, 75-78, PROF PHYSICS, UNIV WASH, 78- *Concurrent Pos:* Nordita prof, Niels Bohr Inst, Copenhagen, Denmark, 78; vis scientist, Ctr Nuclear Studies, Grenoble, France, 83. *Mem:* Fel Am Phys Soc; Europ Physics Soc. *Res:* Theories of condensed matter, especially phase transitions and critical phenomena. *Mailing Add:* Dept Physics Univ Wash Seattle WA 98195

RIEDEL, ERNEST PAUL, b New York, NY, July 30, 31; m 56; c 2. LASERS, OPTICAL PHYSICS. *Educ:* Cornell Univ, BEngPhys, 54; Univ Wis, MS, 57; Mich State Univ, PhD(physics), 61. *Prof Exp:* Scientist, Lockheed Aircraft Corp, 56-58; sr res scientist, 61-69, mgr quantum electronics, 69-72, adv scientist, 72-74, mgr lasers & optical technol, 74-78, MGR APPL PHYSICS, RES & DEVELOP CTR, WESTINGHOUSE ELEC CORP, 78- *Mem:* AAAS; Am Phys Soc; Soc Photo-Optical Instrumentation Engrs. *Res:* Symmetry of antiferromagnetic crystals; spectroscopy of laser materials; electromagnetic wave propagation. *Mailing Add:* 4472 Kilmer Dr Murrysville PA 15668

RIEDEL, GERHARDT FREDERICK, b Santa Monica, Calif, July 8, 51; m 73; c 2. PHYTOPLANKTON PHYSIOLOGICAL ECOLOGY, TRACE ELEMENT GEOCHEMISTRY. *Educ:* Humboldt State Univ, BA & BS, 74; Ore State Univ, MS, 78, PhD(oceanog), 83. *Prof Exp:* Fel, Harbor Br Found, 83-85; investr, 85-87, SR SCIENTIST, ACAD NATURAL SCI, BENEDICT ESTUARINE RES LAB, 87- *Mem:* AAAS; Am Geophys Union; Am Soc Limnol Oceanog; Phycol Soc Am; Oceanog Soc. *Res:* Trace element and nutrient uptake by marine and estuarine organisms, particularly the effect of chemical forms on uptake; loss and regeneration of trace elements and nutrients involving chemical transformations. *Mailing Add:* Benedict Estuarine Res Lab Benedict MD 20612

RIEDEL, RICHARD ANTHONY, b Milwaukee, Wis, Feb 26, 22; m 45; c 4. DENTISTRY. *Educ:* Marquette Univ, DDS, 45; Northwestern Univ, MS, 48; Am Bd Orthod, dipl, 57. *Prof Exp:* Instr orthod, Northwestern Univ, 48-49; from instr to assoc prof, 49-71, actg dean, 80-81, prof, 71-90, EMER PROF, DEPT ORTHOD, SCH DENT, UNIV WASH, 90- *Concurrent Pos:* Pvt pract, 49-; dir, Am Bd Orthod, 72-, pres, 78-79; serve & rotate, "HOPE", Maceio, Brazil, 73; assoc dean acad affairs, Univ Wash, 77-80. *Honors & Awards:* Ketcham Award, Am Asn Orthod. *Mem:* Am Dent Asn; Am Asn Orthod; fel Am Col Dentists; Asn Dent Res; Sigma Xi. *Res:* Orthodontics; cephalometric evaluation of the skeletal interrelationships of the human face and esthetics as related to clinical orthodontics. *Mailing Add:* Dept Orthod Univ Wash Sch Dent Seattle WA 98195

RIEDEL, WILLIAM REX, b South Australia, Sept 5, 27; m 52, 63; c 3. MICROPALEONTOLOGY. *Educ:* Univ Adelaide, MSc, 52, DSc, 76. *Prof Exp:* Paleontologist, SAustralian Mus, 48-50; res fel, Oceanog Inst, Sweden, 50-51; paleontologist, SAustralian Mus, 54-55; asst res geologist, 56-62, assoc res geologist, 62-68, RES GEOLOGIST, SCRIPPS INST OCEANOG, UNIV CALIF, SAN DIEGO, 68- *Concurrent Pos:* With US Geol Surv, 59; chmn, Geol Res Div, Scripps Inst Oceanog, 78-83. *Mem:* Paleont Soc. *Res:* Systematic and stratigraphic investigations of Mesozoic to Quaternary Radiolaria; deep sea sediments; stratigraphy of microscopic fish skeletal debris; information-handling for stratigraphic and paleoenvironmental interpretations. *Mailing Add:* 553 Gravilla Pl La Jolla CA 92038

RIEDER, CONLY LEROY, b Orange, Calif, Nov 2, 50; m 79; c 2. CELL BIOLOGY, HIGH VOLTAGE ELECTRON MICROSCOPY. *Educ:* Univ Calif, Irvine, BS, 72; Univ Ore, Eugene, MS, 75, PhD(cell biol), 77. *Prof Exp:* Res fel cell biol, Univ Ore, Eugene, 75-77; res asst zool, Univ Wis, Madison, 77-79, NIH fel pathobiol, 79-80; RES SCIENTIST BIOPHYS, WADSWORTH CTR LABS & RES, ALBANY, 80-; ASSOC PROF, SCH PUB HEALTH, STATE UNIV NY, ALBANY, 85- *Concurrent Pos:* Res scientist IV, Wadsworth Ctr Labs & Res, NY State Dept Health, Albany, 80-; adj assoc prof biol, State Univ NY, Albany, 80-; adj assoc prof physiol, Albany Med Col. *Mem:* Am Soc Cell Biol; Entom Micros Soc Am; NY Acad Sci. *Res:* Develop and apply video light microscopic, high voltage electron microscopic and immunologic methods to study the mechanisms involved in the movement of cell organelles, especially chromosomes. *Mailing Add:* Wadsworth Ctr Labs & Res NY State Dept Health PO Box 509 Albany NY 12201

RIEDER, RONALD FREDERIC, b New York, NY, July 13, 33; m 64; c 2. MEDICINE, HEMATOLOGY. *Educ:* Swarthmore Col, BA, 54; NY Univ, MD, 58; Am Bd Internal Med, dipl & cert med & hemat. *Prof Exp:* Intern, III, NY Univ Med Div, Bellevue Hosp, 58-59, asst resident, 59-60, fel microbiol, NY Univ, 60-61; fel immunol, Pasteur Inst, Paris, 61-62; fel hemat, Sch Med, Johns Hopkins Univ, 62-64; asst resident, III & IV, NY Univ Med Div, 64-65; res assoc, Montefiore Hosp, 65-67; from asst prof to assoc prof, 67-76, dir hemat, 76-89, PROF MED, STATE UNIV NY DOWNSTATE MED CTR, 76- *Concurrent Pos:* Assoc med, Albert Einstein Col Med, 65-67; vis prof lectr, Nuffield Unit Med Genetics, Univ Liverpool, 68; WHO traveling fel, 68; Macy Found fac scholar, 75-76; vis fel, Wolfson Col, Oxford Univ, 75-76. *Mem:* Am Soc Clin Invest; Asn Am Physicians; Am Soc Hemat; fel Am Col Physicians; Am Physiol Soc. *Res:* Abnormal hemoglobins; hemoglobin synthesis; thalassemia. *Mailing Add:* Dept Med State Univ NY Health Sci Ctr Brooklyn NY 11203

RIEDER, RONALD OLRICH, b Wyandotte, Mich, Feb 20, 42; m 66; c 2. PSYCHIATRIC RESEARCH, PSYCHIATRIC EDUCATION. *Educ:* Harvard Univ, BA, 64, MD(med), 84. *Prof Exp:* Intern pediat, Johns Hopkins Hosp, 68-69; resident psychiat, Albert Einstein Col Med, 69-71; res assoc psychiat, 71-73, res psychiatrist, Intramural Res, NIMH, 73-79; DIR RESIDENCY TRAINING, DEPT PSYCHIAT, COLUMBIA UNIV, 79-, PROF, CLIN PSYCHIAT, 89- *Mem:* Am Psychopath Asn; Am Psychiat Asn; Am Asn Psychiat Residency Training; AAAS. *Res:* Genetics and neuroanatomical aspects of schizophrenia. *Mailing Add:* NY State Psychiat Inst 722 W 168th St New York NY 10032

RIEDER, SIDNEY VICTOR, b Philadelphia, Pa, Oct 22, 21; m 49; c 2. BIOCHEMISTRY. *Educ:* Philadelphia Col Pharm, BS, 43; Univ Pa, MS, 48, PhD(biochem), 53. *Prof Exp:* Instr biochem, Univ Pa, 52-53; from instr to asst prof, Sch Med, Yale Univ, 53-61. *Mem:* AAAS. *Res:* Intermediary metabolism of amino sugars; carbohydrates. *Mailing Add:* 118 Rockaway Ave Marblehead MA 01945

RIEDER, WILLIAM G(ARY), b Williston, NDak, Oct 28, 34; m 62; c 2. MECHANICAL ENGINEERING. *Educ:* NDak State Univ, BSME, 56; Ohio State Univ, MSc, 62; Univ Nebr, Lincoln, PhD(mech eng), 71. *Prof Exp:* Design engr, Atomic Energy Div, Phillips Petrol Co, 56-59; prin mech engr, Battelle Mem Inst, Ohio, 59-65; asst prof mech eng, NDak State Univ, 65-68; consult, Dept Econ Develop, State of Nebr, 68-69; from asst prof to assoc prof, 70-81, PROF MECH ENG, NDAK STATE UNIV, 81- *Mem:* AAAS; Am Soc Mech Engrs; Am Soc Eng Educ. *Res:* Energy conversion; thermal sciences; fluid mechanics; nuclear experiments; vacuum and processing phenomena; writing engineering modeling. *Mailing Add:* Dept Mech Eng NDak State Univ Fargo ND 58105

RIEDERER-HENDERSON, MARY ANN, b Buffalo, NY, July 21, 43; m 67. CELL BIOLOGY. *Educ:* Daemon Col, BS, 64; Univ Wis-Madison, MS, 66; Univ Ga, PhD(biochem), 71. *Prof Exp:* Instr microbiol, Univ Fla, 71-72; asst prof biol, Rollins Col, Winter Park, Fla, 72-76, assoc prof, 76-77; res asst prof, Sch Med, Univ Wash, 77-87; CONSULT, 87- *Mem:* AAAS; Am Soc Cell Biol. *Res:* Growth factors; collagen biosynthesis; ultrasound and tissues. *Mailing Add:* 1057 Summit Ave E Seattle WA 98102-4432

RIEDERS, FREDRIC, b Vienna, Austria, July 9, 22; US citizen. TOXICOLOGY, PHARMACOLOGY. *Educ:* NY Univ, AB, 48, MS, 49; Jefferson Med Col, PhD(pharmacol & toxicol), 51. *Prof Exp:* Prod chemist, Myer's 1890 Soda, Inc, New York, 40-42; qual control chemist, Penetone Corp, NJ, 42-43; jr toxicologist forensic toxicol, Lab Chief Med Examr New York, 46-49; from instr to assoc prof, 51-56, PROF PHARMACOL, JEFFERSON MED COL, 56-; PRES & LAB DIR, NAT MED SERV, INC, 70-; PRES, TOXICON ASSOC, LTD, 76- *Concurrent Pos:* Chief toxicologist forensic toxicol, Off Med Examr, Philadelphia, 56-70; ed bull, Int Asn Forensic Toxicol, 60-63; NIH fel, Jefferson Med Col, 63-68; NIH & Pa Health Dept fels, Off Med Examr, Philadelphia, 67-69; adj prof toxicol, Drexel Univ, 67-69; mem toxicol study sect, NIH, 69-70. *Mem:* Am Acad Forensic Sci; Am Bd Clin Chem; Am Soc Pharmacol & Exp Therapeut; Am Chem Soc; Int Asn Forensic Toxicol. *Res:* Bioanalytical and forensic toxicology; heavy metals; cyanogenetic and interactive mechanisms of toxic actions. *Mailing Add:* Nat Med Serv Inc PO Box 433-A Willow Grove PA 19090

RIEDESEL, CARL CLEMENT, b Ind, June 5, 10; m 39; c 2. PHARMACOLOGY. *Educ:* Univ Idaho, BS, 34; Univ Nebr, MS, 47; Univ Iowa, PhD(physiol), 52. *Prof Exp:* Retail pharmacist, 34-37; hosp pharmacist, 38-42; instr physiol, Univ Nebr, 47-48; assoc prof pharmacol, Idaho State Col, 48-56; chmn dept, 56-70, prof physiol & pharmacol, 56-80, EMER PROF PHYSIOL & PHARMACOL, UNIV PAC, 80-, ASST DEAN PHARMACEUT SCI, 70- *Mem:* Am Pub Health Asn; Am Pharmaceut Asn; Coun Med TV. *Res:* Toxicity of barbiturates; choline and carbohydrate metabolism; pharmacologic actions of natural products. *Mailing Add:* 5200 Adahmore Lane Stockton CA 95208

RIEDESEL, MARVIN LEROY, b Iowa City, Iowa, Nov 8, 25; m 49; c 1. PHYSIOLOGY. *Educ:* Cornell Col, BA, 49; Univ Iowa, MS, 53, PhD, 55. *Prof Exp:* Asst, Col Dent, 50-53, fel physiol, Univ Iowa, 53-55; res assoc occup health, Grad Sch Pub Health, Univ Pittsburgh, 55-59; from asst prof to assoc prof, 59-71, PROF BIOL, UNIV NMEX, 71- *Mem:* AAAS; Am Physiol Soc; NY Acad Sci; Int Soc Biometeorol; Soc Cryobiol. *Res:* Environmental and comparative physiology; mammalian hibernation; electrolyte metabolism; water balance. *Mailing Add:* Dept Biol Univ NMex Albuquerque NM 87131

RIEDHAMMER, THOMAS M, b Buffalo, NY. DRUG DELIVERY. *Educ:* State Univ NY, BA, 70, PhD(chem), 75. *Prof Exp:* Sr res chemist, Bausch & Lomb Inc, 75-77, dept mgr, 78-81, dir, 82-84; dir, 84-85, vpres, 85-86, PRES, PACO RES CORP, 87- *Mem:* Parenteral Drug Asn; Asn Advan Med Inst; Am Chem Soc. *Res:* Products and treatments for disorders and diseases of the anterior segment of the eye; transdermal drug delivery systems; awarded 12 US patents. *Mailing Add:* 1613 Chipmunk Ct Toms River NJ 08755

RIEDINGER, LEO LOUIS, b Brownwood, Tex, Nov 25, 44; m 66; c 1. NUCLEAR PHYSICS. *Educ:* Thomas More Col, AB, 64; Vanderbilt Univ, PhD(physics), 69. *Prof Exp:* NSF fel, Niels Bohr Inst, Copenhagen Univ, 68-69; res assoc nuclear physics, Univ Notre Dame, 69-71; asst prof, 71-76, ASSOC PROF PHYSICS, UNIV TENN, KNOXVILLE, 71- *Concurrent Pos:* Consult, Oak Ridge Nat Lab, 72-73. *Mem:* Am Phys Soc. *Res:* Low energy nuclear structure; radioactive decay experiments; in-beam coulomb-excitation and heavy-ion reaction experiments. *Mailing Add:* Dept Physics & Astron Univ Tenn Gill Physics Bldg Knoxville TN 37996-1200

RIEDL, H RAYMOND, b Colorado Springs, Colo, Aug 25, 35; m 59; c 7. PHYSICS. *Educ:* Creighton Univ, BS, 57. *Prof Exp:* PHYSICIST, SOLID STATE BR, NAVAL SURFACE WEAPONS CTR, 57- *Mem:* Am Phys Soc. *Res:* Solid state physics; semiconductors; epitaxial films; amorphous films. *Mailing Add:* White Oak Lab Naval Surface Weapons Ctr Silver Spring MD 20910

RIEDL, JOHN ORTH, JR, b Milwaukee, Wis, Dec 9, 37; m 61; c 5. MATHEMATICS. *Educ:* Marquette Univ, BS, 58; Univ Notre Dame, South Bend, MS, 60, PhD(math), 63. *Prof Exp:* Asst prof, 65-70, asst dean, 69-74, assoc dean, 74-87, actg dean, 85-87, ASSOC PROF MATH, OHIO STATE UNIV, 70-, DEAN & DIR, MANSFIELD REGIONAL CAMPUS. *Concurrent Pos:* Chmn, comt minicourses, Math Asn Am, 80-87. *Honors & Awards:* Grad Fel, NSF. *Mem:* Math Asn Am. *Res:* Functional analysis. *Mailing Add:* Ohio State Univ 1680 University Dr Mansfield OH 44906

RIEDMAN, RICHARD M, b Long Beach, Calif, Sept 9, 33; m 57; c 2. AUDIOLOGY. *Educ:* Univ Redlands, BA, 55, MA, 56; Univ Pittsburgh, PhD(audiol), 62. *Prof Exp:* From asst prof to assoc prof speech & hearing, 62-68, PROF SPEECH PATH & AUDIOL, SAN DIEGO STATE UNIV, 68- *Mem:* Acoust Soc Am; Am Speech & Hearing Asn. *Res:* Clinical and experimental audiology. *Mailing Add:* Dept Commun Dis San Diego State Univ 5300 Campanile San Diego CA 92182

RIEFFEL, MARC A, b New York, NY, Dec 22, 37; m 59; c 3. OPERATOR ALGEBRAS, GROUP REPRESENTATIONS. *Educ:* Harvard Univ, AB, 59; Columbia Univ, PhD(math), 63. *Prof Exp:* Lectr, 63-64, from actg asst prof to asst prof, 64-68, assoc prof, 68-73, PROF MATH, UNIV CALIF, BERKELEY, 73- *Mem:* Am Math Soc. *Res:* Functional analysis. *Mailing Add:* Dept Math Univ Calif Berkeley CA 94720

RIEG, LOUIS EUGENE, geology, for more information see previous edition

RIEGEL, GARLAND TAVNER, b Bowling Green, Mo, Aug 26, 14; m 41; c 4. ENTOMOLOGY. *Educ:* Univ Ill, BS, 38, MS, 40, PhD(entom), 47. *Prof Exp:* Asst entom, Ill Natural Hist Surv, 37-42; fel, Grad Col, Univ Ill, 47-48; from asst prof to prof, 48-78, head dept, 63-76, EMER PROF ZOOL, EASTERN ILL UNIV, 78- *Concurrent Pos:* Distinguished prof, Eastern Ill Univ, 78. *Mem:* AAAS; Entom Soc Am; Soc Syst Zool. *Res:* Classification of Braconidae, Alysiinae and Dacnusinae; insect ecology and morphology. *Mailing Add:* Dept Zool Eastern Ill Univ Charleston IL 61920

RIEGEL, ILSE LEERS, b Berlin, Germany, June 16, 16; m 40; c 2. CANCER RESEARCH. *Educ:* Univ Wis, Madison, BA, 41, MA, 49, PhD(endocrinol), 52. *Prof Exp:* Fel, 52-54, managing ed, Cancer Res, 54-64, SR SCIENTIST, MCARDLE LAB CANCER RES, UNIV WIS-MADISON, 64- *Mem:* Sigma Xi. *Mailing Add:* McArdle Lab Cancer Res Univ Wis Madison WI 53706

RIEGEL, KURT WETHERHOLD, b Lexington, Va, Feb 28, 39; m 74; c 3. RESEARCH ADMINISTRATION. *Educ:* Johns Hopkins Univ, AB, 61; Univ Md, PhD(astron), 66. *Prof Exp:* Res fel astron, Univ Md, 66; asst prof, Univ Calif, Los Angeles, 66-74; mgr energy conserv prog, Fed Energy Admin, 74-75, chief, technol & consumer prod energy conserv, 75-77, DIR, CONSUMER PROD DIV, DEPT ENERGY, 77-; HEAD, NAT ASTRON CTRS, NSF, 82-; DIR, ENVIRON PROF SAFETY & HEALTH, USN. *Concurrent Pos:* Consult, Aerospace Corp, 67-70; prof astron, Extens, Univ Calif, Los Angeles, 68-74; consult, Rand Corp, 73-74; vis fel, Univ Leiden, 72-73; dir environ eng & technol, Environ Protection Agency, 80-82. *Mem:* AAAS; Am Phys Soc; Am Astron Soc; Int Astron Union; Int Radio Sci Union. *Res:* Environmental pollution control technology; galactic radio astronomy; interstellar medium; energy technology. *Mailing Add:* Dept Navy Naval Sea Systs Command Sea-OOT Washington DC 20362-5101

RIEGER, ANNE LLOYD, b Philadelphia, Pa, Feb 6, 35; m 57; c 1. ORGANIC CHEMISTRY. *Educ:* Reed Col, BA, 56; Stanford Univ, MS, 59; Columbia Univ, PhD, 62. *Prof Exp:* Res assoc chem, 62-68, res assoc biol & med sci, 68-70, 71-74, res assoc chem, 74-77, ASST PROF RES, BROWN UNIV, 77- *Mem:* AAAS; Sigma Xi; Am Chem Soc. *Res:* Free radical chemistry and halogenating agents; synthesis of steroids; isolation and indentification of steroids from biological systems; electron transfer intermediates in photosynthetic systems; inorganic free radical reactions. *Mailing Add:* Dept Chem Brown Univ Box H Providence RI 02912

RIEGER, MARTIN MAX, b Braunschweig, Ger, Apr 12, 20; nat US; m 43; c 2. PHYSICAL ORGANIC CHEMISTRY. *Educ:* Univ Ill, BS, 41; Univ Minn, MS, 42; Univ Chicago, PhD(chem), 48. *Prof Exp:* Chemist, Transparent Package Co, 42-44; instr, DePaul Univ, 47; res chemist, Lever Bros Co, 48-55; sr res assoc, 55-60, dir toiletries & cosmetics res, 60-71, assoc dir chem-proprietary res, 71-75, dir chem-biol res, 75-77, dir chem res, 77-82, sr res fel, Consumer Prod Div, Warner Lambert Co, 77-86; CONSULT, 86- *Concurrent Pos:* Ed, J Soc Cosmetic Chem, 62-67. *Honors & Awards:* Cosmetic Indust Buyer & Suppliers Asn Award, 62; Medal Award, Soc Cosmetic Chemists, 74. *Mem:* Am Chem Soc; Soc Cosmetic Chem (pres, 72-73). *Res:* Cosmetics; proprietaries; aging; skin; hair; antacids; oral hygiene; cosmetic science with emphasis on safety and sterility. *Mailing Add:* 304 Mountain Way Morris Plains NJ 07950-1910

RIEGER, PHILIP HENRI, b Portland, Ore, June 24, 35; m 57; c 1. PHYSICAL CHEMISTRY. *Educ:* Reed Col, BA, 56; Columbia Univ, PhD(chem), 62. *Prof Exp:* Instr chem, Columbia Univ, 61-62; from instr to assoc prof, 62-77, PROF CHEM, BROWN UNIV, 77- *Concurrent Pos:* Vis prof, Univ Otago, Dunedin, NZ, 77-78. *Mem:* Am Chem Soc; Royal Soc Chem. *Res:* Solution physical chemistry of transition metal coordination and organometallic compounds; electron spin resonance; electrochemistry; mechanisms of reactions initiated by electron transfer. *Mailing Add:* Dept Chem Brown Univ Providence RI 02912

RIEGER, SAMUEL, b New York, NY, Sept 29, 21; m 47; c 3. SOIL MORPHOLOGY. *Educ:* Cornell Univ, BS, 43; Univ Wis, MS, 47; State Col Wash, PhD(soils), 52. *Prof Exp:* Soil surveyor, State Geol & Natural Hist Surv, Wis, 46, 47; soil scientist, US Bur Reclamation, 47-49; asst soils, State Col Wash, 49-52; soil scientist, Soil Conserv Serv, USDA, 52-55, state soil scientist, 55-78; RETIRED. *Mem:* Soil Sci Soc Am; Am Soc Agron; Sigma Xi. *Res:* Soil morphology, genesis, and classification, particularly of arctic and subarctic soils. *Mailing Add:* 5817 NE 181 St Seattle WA 98155

RIEGERT, PAUL WILLIAM, b Can, Dec 5, 23; m 48; c 4. INSECT PHYSIOLOGY. *Educ:* Univ Sask, BA, 44; Mont State Col, MSc, 48; Univ Ill, PhD(entom, physiol), 54. *Prof Exp:* Agr asst entom, Can Dept Agr, 44-47; lectr entom, Mont State Col, 47-48; entomologist, Entom Sect Res Lab, Can Dept Agr, 48-68; prof biol, Univ Sask, 68-72; head dept, Univ Regina, 72-79, prof biol, 72-86; RETIRED. *Mem:* Fel Entom Soc Can; Orthopterist Soc; fel Royal Entom Soc London. *Res:* Sensory and behavioral aspects of insect physiology; bioenergetics; history of entomology. *Mailing Add:* 103 Mayfair Crescent Regina SK S4S 5T9 Can

RIEGERT, R(ICHARD) P(AUL), computer science, materials engineering, for more information see previous edition

RIEGGER, OTTO K, b Howell, Mich, Dec 11, 35; m 60; c 3. METALLURGICAL ENGINEERING. *Educ:* Univ Mich, BSE, 58, MSE, 59, PhD(metall eng), 63. *Prof Exp:* Res engr appl res off, Ford Motor Co, 62-64; res engr, 64-67, DIR RES, RES LAB, TECUMSEH PROD CO, 67- *Mem:* Am Soc Metals; Am Foundrymen's Soc; Am Ceramic Soc; Soc Mfg Eng; Soc Die Casting Eng. *Res:* Applied research administration; computer aided product design and simulation; manufacturing engineering and process specification. *Mailing Add:* 2059 Delaware Ann Arbor MI 48103

RIEGLE, GAIL DANIEL, b De Soto, Iowa, Feb 19, 35; m 60; c 2. ENDOCRINOLOGY, REPRODUCTIVE PHYSIOLOGY. *Educ:* Iowa State Univ, BS, 57; Mich State Univ, MS, 60, PhD(physiol), 63. *Prof Exp:* From asst prof to assoc prof, 64-76, PROF PHYSIOL, MICH STATE UNIV, 76-, ASSOC DEAN ACAD AFFAIRS, COL OSTEOP MED, 80- *Concurrent Pos:* Develop of Undergrad & Grad Clin Educ Prog. *Mem:* Geront Soc; Am Physiol Soc; Soc Exp Biol & Med; Sigma Xi. *Res:* Effects of stress and aging on endocrine and reproductive control systems. *Mailing Add:* Col Osteop Med Mich State Univ East Lansing MI 48823

RIEHL, HERBERT, b Munich, Ger, Mar 30, 15; nat US; m 52; c 2. METEOROLOGY. *Educ:* NY Univ, MS, 42; Univ Chicago, PhD(meteorol), 47. *Prof Exp:* Instr meteorol, Univ Wash, 41-42; from instr to prof, Univ Chicago, 42-60; prof & head dept, Colo State Univ, 60-72; prof, Free Univ Berlin, 72-76; vis scientist, Nat Ctr Atmospheric Res, 76-79; prof, Coop Inst Res Environ Sci, 79-83; RETIRED. *Concurrent Pos:* Consult, USN & US Weather Bur; consult, Sen, Berlin, 74-75. *Honors & Awards:* Losey Award, Am Inst Aeronaut & Astronaut, 60; Andrew G Clark Award, Colo State Univ, 72; Meisinger Award, Am Meteorol Soc, 47, Rossby Award, 79. *Mem:* AAAS; Am Meteorol Soc; Am Geophys Union; Royal Meteorol Soc; Ger Meteorol Soc; Sigma Xi. *Res:* General circulation; jet streams; tropical meteorology; climatology; hydrometeorology. *Mailing Add:* 1200 Humboldt St Apt 706 Cheesman Towers W Ste H Denver CO 80218

RIEHL, JAMES PATRICK, b Toms River, NJ, Aug 6, 48; m 72; c 1. PHYSICAL CHEMISTRY, THEORETICAL CHEMISTRY. *Educ:* Villanova Univ, BS, 70; Purdue Univ, PhD(phys chem), 75. *Prof Exp:* Instr chem, Univ Va, 75-77; ASST PROF CHEM, UNIV MO, ST LOUIS, 77- *Concurrent Pos:* Res investr, Am Chem Soc Petrol Res Fund, 78-80 & Res Corp, 78. *Mem:* Am Chem Soc. *Res:* Molecular dynamics and structure of condensed phases; molecular spectroscopy; optical activity. *Mailing Add:* Dept Chem Univ Mo St Louis MO 63121

RIEHL, JERRY A, b Kalispell, Mont, July 25, 33; m 75; c 7. PHYSICS, CHEMISTRY. *Educ:* Seattle Univ, BS, 62; Wash State Univ, PhD(chem), 66. *Prof Exp:* Assoc prof physics, Seattle Univ, 66-74, chmn dept, 71-74; sr res specialist, Boeing Com Airline Co, 74-80; Scientist, Phys Dynamics, 80-83; RES PROF CHEM, SEATTLE UNIV, 78-; DIR TECH EDUC, S SEATTLE COMMUNITY COL, 84- *Concurrent Pos:* Res consult, 74- *Mem:* Am Chem Soc; Am Phys Soc; Am Soc Mech Engr; fel Inst Environ Sci; Soc Advan Educ; Am Soc Testing & Mat. *Res:* Air pollution control; gas-turbine emissions; nuclear analysis. *Mailing Add:* 9315 Fauntleroy Way SW Seattle WA 98136

RIEHL, MARY AGATHA, b Raleigh, NDak, Feb 17, 21. ORGANIC CHEMISTRY. *Educ:* Col St Scholastica, BA, 42; Inst Divi Thomae, MS, 45; Cath Univ, PhD(chem), 66. *Prof Exp:* PROF CHEM, COL ST SCHOLASTICA, 45- *Concurrent Pos:* NSF sci fac fel, 62-63; NSF res participation grant, Argonne Nat Lab, 69, Ill Inst Technol, 71. *Mem:* AAAS; Am Chem Soc; Am Inst Chem; Sigma Xi. *Res:* Biochemical studies in cancer, especially enzyme systems in cancerous and normal tissue; chromic acid oxidations of organic compounds; EPA laboratory. *Mailing Add:* Col St Scholastica Duluth MN 55811

RIEHL, ROBERT MICHAEL, b Borger, Tex, Aug 8, 51; m 78; c 1. IN VITRO FERTILIZATION & EMBRYO CULTURE. *Educ:* WTex State Univ, BS, 73, MS, 75; Univ Tex, San Antonio, PhD(physiol), 80. *Prof Exp:* Lab asst biol, WTex State Univ, 71-73, grad teaching asst, 73-75; grad teaching asst physiol, Grad Sch Biomed Sci, Univ Tex, San Antonio, 75-80; res fel, Dept Cell Biol, Mayo Clin-Found, 81-86; res assoc, 80-81, ASST PROF & DIR, HUMAN IN VITRO FERTIL & EMBRYO TRANSFER LAB, DEPT OBSTET-GYNEC, HEALTH SCI CTR, UNIV TEX, SAN ANTONIO, 86- *Mem:* Endocrine Soc; Soc Study Reproduction. *Res:* Biochemical endocrinology of reproduction; cell and tissue culture in vitro; comparative endocrinology of steroid receptors; epithelial cell biology. *Mailing Add:* Dept Obstet-Gynec Univ Tex Health Sci Ctr 7703 Floyd Curl Dr San Antonio TX 78284

RIEHLE, ROBERT ARTHUR, JR, b San Diego, Calif, Oct 24, 47; m. UROLITHIASIS, SHOCK WAVE LITHOTRIPSY. *Educ:* Yale Univ, BA, 69; Columbia Univ, MD, 73. *Prof Exp:* Asst prof urol, Wayne State Univ, 80-81; asst prof urol surg, Med Sch & asst attend surgeon, 81-86, assoc prof urol surg & assoc attend surgeon, 87-90, MED DIR, NY HOSP, CORNELL UNIV, 90- *Concurrent Pos:* Asst ed, Endourology, 86-90. *Mem:* Am Col Surgeons; Am Urol Asn; Am Col Physician Execs; Int Soc Urol; Endourol Soc; AMA. *Res:* Disintegration of kidney stones using extra-corporeal shock wave lithotripsy; use of percutaneous surgery for stone removal and intra-renal surgery; dose response curve analysis of urologic procedures. *Mailing Add:* Four Dutch Village Menands NY 12204

RIEHM, CARL RICHARD, b Kitchener, Ont, May 2, 35; m 58; c 2. ALGEBRA. *Educ:* Univ Toronto, BA, 58; Princeton Univ, PhD(math), 61. *Prof Exp:* From lectr to asst prof math, McGill Univ, 61-63; from asst prof to assoc prof, Univ Notre Dame, 63-73; chmn dept, 73-79, PROF MATH, MCMASTER UNIV, 73- *Concurrent Pos:* Mem, Inst Advan Study, 66-67; vis prof, Harvard Univ, 67-68. *Mem:* Am Math Soc; Can Math Soc. *Res:* Integral quadratic forms; classical groups; orthogonal representations of finite groups. *Mailing Add:* Dept Math McMaster Univ 1280 Main St W Hamilton ON L8S 4K1 Can

RIEHM, JOHN P, b Fergus, Ont, Mar 24, 35; m 63; c 1. BIOCHEMISTRY. *Educ:* Ont Agr Col, BSA, 56; Mich State Univ, PhD(biochem), 62. *Prof Exp:* Fel, Cornell Univ, 62-65; asst prof biochem, Univ Calif, Santa Barbara, 65-70; assoc prof, 70-77, PROF BIOL & CHMN DEPT, UNIV WEST FLA, 77- *Mem:* Am Chem Soc. *Res:* Chemical and physical properties of proteins. *Mailing Add:* Dept Biol Univ WFla Pensacola FL 32504

RIEKE, CAROL ANGER, b Milwaukee, Wis, Jan 17, 08; m 32; c 2. MATHEMATICS, ASTRONOMY. *Educ:* Northwestern Univ, BA, 28, MA, 29; Radcliffe Col, PhD(astron), 32. *Prof Exp:* Berliner fel, Radcliffe Col, 32-33, tutor astron, 33-36; instr, Johns Hopkins Univ, 37; asst physics, Univ Chicago, 38-42; computer, Mass Inst Tech, 42-46; instr, Purdue Univ, 47-52; instr math & astron, Thornton Community Col, 57-85; RETIRED. *Concurrent Pos:* Asst, Mass Inst Tech, 33-36, computer, 33-38. *Honors & Awards:* Caroline I Wilby Prize. *Mem:* Am Astron Soc. *Res:* Spectroscopic parallaxes; astronomical spectroscopy; molecular spectra; airplane propulsion; galactic clusters. *Mailing Add:* 2535 N Ave San Valle Tucson AZ 85715-3404

RIEKE, GARL KALMAN, b Seattle, Wash, June 30, 42. ANATOMY, NEUROSCIENCE. *Educ:* Univ Wash, BS, 65, BS, 66; La State Univ, PhD(anat), 71. *Prof Exp:* Asst prof anat, Hahnemann Med Col & Hosp, 73-78; asst prof anat, Col Med, Tex A&M Univ, 78-; MEM STAFF, DEPT ANAT SCI, MEHARRY MED COL. *Concurrent Pos:* Fel physiol, Univ Calif, Los Angeles, 71-73. *Mem:* AAAS; Sigma Xi; Soc Neurosci. *Res:* Neuronal interactions; local circuits; homing and magnetic fields. *Mailing Add:* Dept Anat Sci Meharry Med Col 1005 D B Todd Blvd Nashville TN 37208

RIEKE, GEORGE HENRY, b Boston, Mass, Jan 5, 43. ASTRONOMY. *Educ:* Oberlin Col, AB, 64; Harvard Univ, MA, 65, PhD(physics), 69. *Prof Exp:* Fel astron, Smithsonian Astrophys Observ, 69-70; res assoc, Lunar & Planetary Lab, 70-73, asst prof, Steward Observ & Lunar & Planetary Lab, 73-75, assoc prof, 75-80, PROF ASTRON, STEWARD OBSERV & LUNAR & PLANETARY LAB, UNIV ARIZ, 80- *Concurrent Pos:* Alfred P Sloan Found fel, 76-80; prin investr, Space Infrared Telescope Facil; dep dir, Steward Observ. *Res:* Infrared astronomy. *Mailing Add:* 5801 Paseo Ventoso Tucson AZ 85715

RIEKE, HERMAN HENRY, III, b Louisville, Ky, June 18, 37; m 64; c 6. PETROLEUM ENGINEERING, GEOLOGY. *Educ:* Univ Ky, BS, 59; Univ Southern Calif, MS, 64 & 65, PhD(petrol eng), 70. *Prof Exp:* Chief geologist, United Minerals, Inc, Los Angeles, 63-64; staff engr, Electro-Osmotics, Inc, Los Angeles, 64-66; lectr petrol eng, Univ Southern Calif, 66-68, res scientist, Res & Develop, Continental Oil Co, Ponca City, 69-71; asst prof petrol eng, Col Mineral & Energy Resources, WVa Univ, 71-81, assoc prof, 76-81; mem tech staff, TRW Inc, 77-81; prod mgr, 81-82, dir explor & geol, Poi Energy, Inc, Cleveland, Ohio, 82-83; sr vpres, Geofax, Inc, Cleveland, Ohio, 84-89; dir, WVa Energy & Environ Res Corp, Inc, Morgantown, WVa, 84-89; TECH STAFF, DIRECTORATE GEN MINERAL RESOURCES, JEDDAH, SAUDI ARABIA, 89- *Concurrent Pos:* Eng scientist, Res Eng Exp Sta, WVa Univ, 71-81. *Honors & Awards:* Crown Medal, Shah of Iran, 76. *Mem:* Soc Petrol Engrs; Geol Soc Am; Soc Prof Well Log Analysts; Soc Econ Paleont & Mineral; Am Inst Prof Geologists; Int Asn Study Clays. *Res:* Abnormal subsurface pressure detection; reservoir engineering; formation evaluation; compaction of sediments; geothermal energy; evaluation of mineral resources; oil and gas production; microcomputer based expert systems; artificial intelligence; expert witness, oil & gas. *Mailing Add:* 161 Poplar Dr Morgantown WV 26505

RIEKE, JAMES KIRK, b Barrington, Ill, Apr 15, 24; m 49; c 3. PHYSICAL CHEMISTRY. *Educ:* Univ Ill, BS, 49; Univ Wis, MS, 52, PhD(chem), 54. *Prof Exp:* Asst chem, Univ Wis, 49-54; res chemist, 54-57, proj leader, 57-62, group leader, 62-65, asst to dir plastics dept lab, 65-70, ASST TO DIR PHYS RES, DOW CHEM USA, 70- *Mem:* AAAS; Am Chem Soc; Soc Plastics Eng; Sigma Xi; Soc Rheology. *Res:* Luminescence properties induced in crystals by high energy radiation; physical properties of high polymers; graft copolymers. *Mailing Add:* Dow Chem Co-Bl 1702 Plastics Res Midland MI 48640

RIEKE, MARCIA JEAN, b Hillsdale, Mich, June 13, 51; div. ASTRONOMY. *Educ:* Mass Inst Technol, SB, 72, PhD(physics), 76. *Prof Exp:* Res assoc, Lunar & Planetary Lab, Univ Ariz, 76-78, res assoc infrared astron, 78-79, asst astronr, Steward Observ, 79-83, assoc astronr, 83-84, ASSOC PROF, UNIV ARIZ, 84- *Concurrent Pos:* Counr, Am Astron Soc, 90-92. *Honors & Awards:* Van Biesbroeck Prize, 80. *Mem:* Sigma Xi; Am Astron Soc. *Res:* Infrared observations of extragalactic objects; development of infrared detectors for astronomical use. *Mailing Add:* Steward Observ Univ Ariz Tucson AZ 85721

RIEKE, PAUL EUGENE, b Kankakee, Ill, May 13, 34; m 57; c 2. TURF MANAGEMENT, SOIL SCIENCE. *Educ:* Univ Ill, BS, 56, MS, 58; Mich State Univ, PhD(soil sci), 63. *Prof Exp:* From asst prof to assoc prof, 63-72, PROF CROP & SOIL SCI, MICH STATE UNIV, 72- *Mem:* Sigma Xi; Soil Sci Soc Am; fel Crop Sci Soc Am; Am Soc Agron; Am Soc Hort Sci. *Res:* Physical and chemical properties of soils affecting turf management. *Mailing Add:* Dept Crop & Soil Sci 584 Plant & Soil Sci Bldg Mich State Univ East Lansing MI 48824

RIEKE, REUBEN DENNIS, b Lucan, Minn, Mar 7, 39; m 62; c 2. ORGANIC CHEMISTRY. *Educ:* Univ Minn, Minneapolis, BCh, 61; Univ Wis-Madison, PhD(org chem), 66. *Prof Exp:* Assoc phys org chem, Univ Calif, Los Angeles, 65-66; from asst prof to prof phys org chem, Univ NC, Chapel Hill, 66-76; prof, NDak State Univ, 76-77; Interim chmn & chmn, dept chem, 81-85, PROF CHEM, UNIV NEBR, LINCOLN, 77- *Concurrent Pos:* NIH res fel, 65-66; participant, Am Chem Soc Course Molecular Orbital Theory, 68-70; fel, Alfred P Sloan Found, 73-77; fel, Alexander von Humboldt Found, 73-74; vis prof, Tech Univ, Munich, 73-74, Prince Univ, 85 & Stanford Univ, 90; mem adv bd, critical rev in surface chem, 88; app, Howard S Wilson Regents Prof Chem, 87. *Mem:* Am Chem Soc; Sigma Xi. *Res:* Preparation and study of chemistry of activated metals, electrochemical studies of organic and organometallic compounds; development of new synthetic methods using organometallics; preparation and study of organic metals. *Mailing Add:* Dept Chem Univ Nebr Lincoln NE 68588-0304

RIEKE, WILLIAM OLIVER, b Odessa, Wash, Apr 26, 31; m 54; c 3. MEDICAL ADMINISTRATION, ANATOMY. *Educ:* Pac Lutheran Univ, BA, 53; Univ Wash, MD, 58. *Prof Exp:* Instr anat, Sch Med, Univ Wash, 58-61, asst prof biol struct, 61-64, assoc prof, 64-66, admin officer, 63-66; prof anat, Univ Iowa, 66-71, chmn dept, 66-69, dean protem, 69-70; prof anat & exec vchancellor, Univ Kans Med Ctr, Kansas City, 71-76; PRES, DEPT ANAT, PAC LUTHERAN UNIV, 76- *Mem:* AAAS; Am Asn Anatomists; Am Soc Cell Biol; Sigma Xi. *Res:* Hematology; immunology; cell kinetics; metabolism. *Mailing Add:* 13611 Spanaway Loop Rd Tacoma WA 98444-1118

RIEKELS, JERALD WAYNE, b Muskegon, Mich, Oct 31, 32; m 58; c 3. PLANT PHYSIOLOGY, HORTICULTURE. *Educ:* Mich State Univ, BS, 59, MS, 60; Univ Calif, Davis, PhD(plant physiol), 64. *Prof Exp:* Asst prof, 64-68, ASSOC PROF HORT SCI, UNIV GUELPH, 68- *Concurrent Pos:* Grants, Nat Res Coun Can, 65-69, Ont Dept Univ Affairs, 69 & Potash Inst Can, 71. *Mem:* Am Soc Hort Sci; Am Soc Plant Physiol; Int Soc Hort Sci; Coun Agr Sci & Technol; Coun Soil Testing & Plant Anal; Sigma Xi. *Res:* Vegetable physiology, culture and production with emphasis on mineral nutrition. *Mailing Add:* Dept Hort Univ Guelph Guelph ON N1G 2W1 Can

RIEL, GORDON KIENZLE, b Columbus, Ohio, Oct 26, 34; m 54; c 3. HEALTH PHYSICS. *Educ:* Univ Fla, BChE, 56; Univ Md, MS, 61, PhD, 67. *Prof Exp:* Chem engr, 56-59, PHYSICIST, NAVAL ORD LAB, 59-, CONSULT, 69- *Mem:* Health Physics Soc; Instrument Soc Am; Nat Soc Prof Engrs. *Res:* Measurement of radiation in the ocean including cosmic rays, natural and artificial radioactive isotopes; neutron spectrometry and personnel dosimetry; radiation monitoring systems for nuclear reactors and environment. *Mailing Add:* 1210 Bayview Ct Edgewater MD 21037

RIEL, RENE ROSAIRE, b Sherrington, Que, Oct 21, 23; m 58; c 3. FOOD SCIENCE. *Educ:* Univ Montreal, BSA, 47, MSc, 49; Univ Wis, PhD(dairy indust, biochem), 52. *Prof Exp:* Prof dairy chem, Univ Montreal, 52-53; res officer, Chem Div, Sci Serv, 53-59; head chem sect, Dairy Tech Res Inst, Can Dept Agr, 59-62; dir dept food sci, Laval Univ, 62-71, prof food opers, 62-74; food res coordr, Can Dept Agr, 74-83; dir, Saint-Hyacinthe Ctr Food Sci Res, 83-88; FOOD CONSULT, 88- *Honors & Awards:* David Prize, 60; Berard Award, 74. *Mem:* Am Dairy Sci Asn; Inst Food Technologists; Can Inst Food Sci & Technol; Chem Inst Can; Fr-Can Asn Advan Sci. *Res:* Protein extraction; texturization; modified fats; freeze-drying. *Mailing Add:* 60 rue Therien Hull PQ J8Y 1J1 Can

RIEMANN, HANS, b Harte, Denmark, Mar 11, 20; m 43, 65; c 2. VETERINARY MEDICINE. *Educ:* Royal Vet & Agr Col, Denmark, DVM, 43; Copenhagen Univ, PhD(microbiol), 63, Am Col Vet Prev Med, dipl. *Prof Exp:* Pvt pract, 43; vet inspector food, Pub Health Serv, Denmark, 43-45; microbiologist, Tech Lab, Ministry Fisheries, 45-54; chief microbiologist, Danish Meat Res Inst, 54-57 & dir res planning, 60-64; res asst microbiol, Univ Ill, 57-60; res fel food microbiol, Univ Calif, Davis, 64-65, from lectr to assoc prof, 65-69, prof pub health, 69-80, chief investr, Field Res & Training Prog, 71-80, chairperson, Dept Epidemiol & Prev Med, 78-88, dir, Master Prev Vet Med Prog, 81-88, PROF, UNIV CALIF, DAVIS, 80- *Concurrent Pos:* Lectr, Royal Vet & Agr Col, Denmark, 60-64. *Honors & Awards:* Dr C O Jensen's Food Microbiol Reward, 53. *Mem:* Am Vet Med Asn; Am Soc Microbiol; Sigma Xi. *Res:* Veterinary epidemiology and preventive medicine. *Mailing Add:* Epidemiol & Prev Med Univ Calif Davis CA 95616

RIEMANN, JAMES MICHAEL, b Philadelphia, Ill, Oct 14, 40; m 63; c 2. ORGANIC CHEMISTRY. *Educ:* Berea Col, BA, 62; Univ Ohio, PhD(chem), 68. *Prof Exp:* PROF CHEM, PFEIFFER COL, 66-, DIR ACAD COMPUT, 79-, CHMN, DIV NATURAL & HEALTH SCI, 82- *Concurrent Pos:* Vis prof gen chem, Iowa State Univ, 73-74. *Mem:* Am Chem Soc; Sigma Xi. *Res:* Flash vacuum pyrolysis in synthesis; ozonolysis of aqueous organic mixtures. *Mailing Add:* PO Box 326 Misenheimer NC 28109

RIEMANN, JOHN G, b Gladstone, NMex, June 18, 28; m 61. ENTOMOLOGY. *Educ:* Tex Tech Col, BA, 51; Univ Tex, MA, 54, PhD(zool), 61. *Prof Exp:* Asst prof biol, West Tex State Col, 57-60; vis asst prof zool, Tulane Univ, 61-62; res entomologist, Man & Animal Br, 62-64, RES ENTOMOLOGIST, METAB & RADIATION LAB, ENTOM RES DIV, USDA, 64- *Mem:* Entom Soc Am; AAAS. *Res:* Radiation biology, cytology and reproductive physiology of insects. *Mailing Add:* 1006 Third Ave No 1 Fargo ND 58103

RIEMENSCHNEIDER, ALBERT LOUIS, b Cody, Nebr, May 18, 36; m 62; c 3. ELECTRICAL ENGINEERING. *Educ:* SDak Sch Mines & Technol, BSEE, 59, MSEE, 62; Univ Wyo, PhD(elec eng), 69. *Prof Exp:* Engr, Sperry Utah Co, 59-60; instr elec eng, Univ Wyo, 62-67; asst prof, SDak Sch Mines & Technol, 67-74; gen mgr, Syncom, Inc, 74-80; assoc prof, 80-82, PROF ELEC ENG, SDAK SCH MINES & TECHNOL, 82-, HEAD DEPT, 83- *Concurrent Pos:* Res engr, Natural Resources Res Inst, 64-67; consult, Respec Corp, 70-74, Durham Assoc, 80-82 & ALR Eng, 82- *Honors & Awards:* Benjamin Dasher Award, Am Soc Eng Educ/Inst Elec & Electronics Engrs, 82; John A Curtis Award, Am Soc Eng Educ, 83. *Mem:* Inst Elec & Electronics Engrs; Nat Soc Prof Engrs. *Res:* Control systems; instrumentation; hybrid computations; computer aided design; computer aided instruction; microprocessors; digital control. *Mailing Add:* 4051 Corral Dr Rapid City SD 57702

RIEMENSCHNEIDER, PAUL ARTHUR, b Cleveland, Ohio, Apr 17, 20; m 45; c 6. MEDICINE, RADIOLOGY. *Educ:* Baldwin-Wallace Col, BS, 41; Harvard Univ, MD, 44; Am Bd Radiol, dipl. *Prof Exp:* Asst radiol, Harvard Med Sch, 49; from asst prof to assoc prof, Med Col, Syracuse Univ, 50-51; from assoc prof to prof, Col Med, State Univ NY Upstate Med Ctr, 52-65; DIR DIAG RADIOL, SANTA BARBARA COTTAGE HOSP, 64- *Concurrent Pos:* Consult, Oak Ridge Inst; mem bd trustees, Am Bd Radiol. *Honors & Awards:* Gold Medal, Am Col Radiol, 82. *Mem:* Roentgen Ray Soc (pres, 80); Radiol Soc NAm; Am Col Radiol (pres, 75); Am Fedn Clin Res. *Res:* Diagnostic radiology with particular emphasis on cerebral angiography. *Mailing Add:* PO Box 689 Santa Barbara CA 93102

RIEMENSCHNEIDER, SHERMAN DELBERT, b Alliance, Ohio, Sept 11, 43; m 67; c 2. MATHEMATICS. *Educ:* Hiram Col, AB, 65; Syracuse Univ, MA, 67, PhD(math), 69. *Prof Exp:* Lectr math, Univ Wash, 69-70; Nat Res Coun grant & asst prof, 70-75, assoc prof, 75-82, PROF MATH, UNIV ALTA, 82- *Concurrent Pos:* Vis scholar, Univ Tex, 76-77; vis assoc prof, Univ SC, 80-81. *Mem:* Am Math Soc; Math Asn Am; Soc Indust & Appl Math; Can Math Soc; Can Appl Math Soc. *Res:* Interpolation of operators; approximation theory; splines. *Mailing Add:* Dept Math Univ Alta Edmonton AB T6G 2E2 Can

RIEMER, DONALD NEIL, b Newark, NJ, Feb 14, 34; m 56; c 2. AQUATIC BIOLOGY. *Educ:* Rutgers Univ, BS, 56, PhD(weed control), 66; Auburn Univ, MS, 60. *Prof Exp:* Fisheries biologist, NJ Div Fish & Game, 61-62; instr aquatic weed sci, 62-66, asst res prof, 66-69, chmn dept soil & crops, 75-82, ASSOC RES PROF AQUATIC WEED SCI, RUTGERS UNIV, NEW BRUNSWICK, 69- *Mem:* Soc Aquatic Plant Mgt; Asn Aquatic Vascular Plant Biologists. *Res:* Life histories, ecology and control of aquatic vegetation. *Mailing Add:* Dept Soils & Crops Rutgers Univ New Brunswick NJ 08903-0231

RIEMER, PAUL, b Poland, Mar 20, 24; nat US; m 49; c 3. ENGINEERING. *Educ:* Univ Sask, BE, 47. *Prof Exp:* From asst prof to prof civil eng, Univ Sask, 50-89; RETIRED. *Mem:* Fel Can Soc Civil Engrs; Eng Inst Can. *Res:* Structural engineering. *Mailing Add:* 34 Cambridge Crescent Saskatoon SK S7H 3P8 Can

RIEMER, ROBERT KIRK, c 1. HORMONAL REGULATION, UTERINE PHYSIOLOGY. *Educ:* Univ Calif, Santa Barbara, BA, 76; Univ Ark, PhD(pharmacol), 82. *Prof Exp:* TEACHING FEL, UNIV CALIF, SAN FRANCISCO, 82-, ASST RES PHARMACOLOGIST, 86- *Mem:* Am Soc Pharmacol & Exp Therapeut; AAAS. *Res:* Hormonal regulation of autonomic responses in uterus and myocardium. *Mailing Add:* Rm 1485M Univ Calif San Francisco CA 94143-0550

RIEMER, ROBERT LEE, b Sheboygan, Wis, June 11, 51; m 80. SCIENCE POLICY. *Educ:* Univ Wis-Madison, BS, 73; Univ Kans, Lawrence, MS, 78, PhD(physics), 80. *Prof Exp:* Postdoctoral res assoc, High-Energy Physics Group, Univ Kans, Lawrence, 80; proj geophysicist, Gulf Oil Co, 80-85; prog officer, 85-88, ASSOC STAFF DIR, BD PHYSICS & ASTRON, NAT ACAD SCI-NAT RES COUN, 88- *Mem:* Am Phys Soc; Sigma Xi. *Res:* Produce science policy reports in physics and astronomy for federal agencies. *Mailing Add:* Nat Res Coun 2101 Constitution Ave Washington DC 20418

RIEMER-RUBENSTEIN, DELILAH, b Brooklyn, NY, Aug 28, 10; m 37; c 3. MEDICAL ADMINISTRATION, NEUROPSYCHIATRY. *Educ:* Tufts Univ, BS, 31; Med Col Pa, MD, 36. *Prof Exp:* Intern, Univ Hosp, Boston, Mass, 36-37; jr physician, Boston Dispensary, 37-47; ward physician, Vet Admin Hosp, Bedford, Mass, 48-53, chief phys med & rehab serv, 53-63; dir, John T Berry Rehab Ctr, Mass Dept Ment Health, 63-79; med dir, Pentucket Chronic Hosp, 79-84; consult, Shaughnessy Rehab Hosp, 84-85; RETIRED. *Concurrent Pos:* Physician in chg, Am Red Blood Donor Ctr, New Eng, Am Red Cross, 42-45; asst, Sch Med, Tufts Univ, 57-63, instr, 63-66; asst physician, Boston Dispensary, Mass, 58-66; fel, Harvard Sch Pub Health. *Honors & Awards:* Bronze Medal, Am Cong Phys Med, 55. *Mem:* Am Cong Rehab Med; Am Psychiat Asn; Am Asn Ment Deficiency. *Res:* Mental retardation in all its aspects from developmental to results of habilitation and rehabilitation; psychiatry; geriatrics. *Mailing Add:* 164 Ward St Newton Center MA 02159

RIEMERSMA, H(ENRY), b Neth, Nov 30, 28; US citizen; m 56; c 4. ELECTRICAL ENGINEERING, PHYSICS. *Educ:* Univ Mich, BS, 57; Carnegie Inst Technol, MS, 65. *Prof Exp:* Intermediate res engr, 58-61, res engr, 61-64, sr engr, 65-70, res prog adminr, 70-75, SR ENGR, RES & DEVELOP CTR, WESTINGHOUSE ELEC CORP, 75- *Res:* Attainment of ultrahigh vacua; superconducting equipment, magnets, transformers, generators; electrical breakdown in various environment. *Mailing Add:* 617 Penny Drive Pittsburgh PA 15235

RIES, EDWARD RICHARD, b Freeman, SDak, Sept 18, 18; m 49, 64; c 2. PETROLEUM GEOLOGY, EXPLORATION GEOLOGY. *Educ:* Univ SDak, AB, 41; Univ Okla, MS, 43, PhD(geol), 51. *Prof Exp:* Geophys interpreter, Robert H Ray, Inc, 42; jr geologist, Carter Oil Co, 43-44, geologist, 46-51; sr geologist, Standard Vacuum Oil Co, India, 51-53, regional geologist, Standard Vacuum Petrol Mataschappij, Indonesia, 53-59, geol adv, 59-62; geol adv, Mobil Petrol Co, Inc, 62-65, staff geologist, Int Div, Mobil Oil Corp, 65-71, regional explorationist, Mobil Tech Serv, Inc, 71-73, sr regional explorationist, Mobil Explor & Prod Serv Inc, 73-76, sr geol adv, 76-79, assoc geol adv, 79-82, geol consult, 82-83 & consult to Mobil Oil Corp, 83-85; PETROL GEOLOGIST & INT PETROL CONSULT, EUROPE, SINO-SOVIET & SE ASIA, 85- *Concurrent Pos:* Vis lectr, Calcutta Univ, 52-53 & NY Univ, 65-70; assoc ed, Am Asn Petrol Geologists, 76-82. *Mem:* AAAS; Geol Soc Am; NY Acad Sci; Am Asn Petrol Geologists; Soc Explor Geophysicists. *Res:* Search for and evaluate the petroleum potential of hydrocarbon provenances and their specific prospects by use of applied geological and geophysical techniques; author of numerous proprietary and published domestic and international geological, geophysical, and geochemical reports and professional papers; spectrum covers generation, migration and entrapment of hydrocarbons; evaluation of production and reserves and estimates of future hydrocarbon potential. *Mailing Add:* 6009 Royal Crest Dr Dallas TX 75230-3434

RIES, HERMAN ELKAN, JR, b Scranton, Pa, May 6, 11; m 40, 81; c 2. MONOMOLECULAR FILMS, ELECTRON MICROSCOPY OF THIN FILMS. *Educ:* Univ Chicago, BS, 33, PhD(phys chem), 36. *Prof Exp:* Head lab instr phys chem, Univ Chicago, 35-36; head, Phys Chem Sect & assoc dir, Catalysis Div, Sinclair Res Labs, 36-51; res assoc surface chem, Standard Oil Co, Ind, 51-72; vis prof phys chem, Inst Chem Res, Kyoto Univ, 72-74; RES ASSOC BIOPHYS, DEPT MOLECULAR GENETICS & CELL BIOL, UNIV CHICAGO, 74- *Concurrent Pos:* Vis scientist, Cavendish Lab, Univ Cambridge, 64 & Argonne Nat Lab, 77-78; res assoc surface chem, Univ Paris, 85. *Honors & Awards:* Ipatieff Prize, Am Chem Soc, 50, Cert of Merit, 75 & 76; Welch Lectr, Univ Tex, 78 & 79. *Mem:* Am Chem Soc; fel AAAS; Int Union Pure & Appl Chem; Int Soc Colloid & Surface Sci. *Res:* Adsorption and thin-film studies as applied to catalysis, corrosion, lubrication, and waste-water treatment; structure of monolayers of compounds related to those in cell membranes using pressure-area isotherms and electron microscopy. *Mailing Add:* 5660 Blackstone Ave Chicago IL 60637

RIES, RICHARD RALPH, b New Ulm, Minn, Nov 16, 35; m 64; c 2. OPERATIONS MANAGEMENT. *Educ:* St Edward's Univ, BS, 57; Univ Minn, MS, 59, PhD(physics), 63. *Prof Exp:* Res assoc physics, Max Planck Inst Chem, 63-64; instr physics, Harvard Univ, 64-65; asst prog dir, US-Japan Coop Sci Prog, NSF, 65-66, staff assoc int sci activities, Tokyo Liaison Off, 66-70, prof assoc, Off Int Progs, 70-72, regional mgr Europ & Am sect, 72-74, dept head, Off Int Progs, 75-76, dir opers & anal, Tech & Int Affairs, 77-84, exec officer, 84-89, DIR DIV INT PROGS, NSF, 90- *Mem:* AAAS; Am Phys Soc. *Res:* Mass spectroscopy; positive ion optics; precise measurement of atomic masses; administration of international cooperative science programs; science policy; allocation of science resources; operations management; planning and evaluation. *Mailing Add:* 2500 Childs Lane Alexandria VA 22308

RIES, RONALD EDWARD, b Powell, Wyo, Feb 7, 44; m 66; c 2. GRASS SEEDLING MORPHOLOGY & ANATOMY. *Educ:* Univ Mont, BS, 66, MS, 68; Univ Wyo, PhD(range mgt), 73. *Prof Exp:* Teaching & res asst forest grazing, Univ Mont, 66-68; nat resource specialist forestry & range, Bur Land Mgt, 68; res asst arid land ecol, Univ Wyo, 70-73, res assoc, 73-74; RANGE SCIENTIST RECLAMATION & SEEDLING ESTAB RES, AGR RES SERV, USDA, 74- *Concurrent Pos:* Chmn, Energy Resources Div, Soil Conserv Soc Am, 86. *Honors & Awards:* Outstanding Reclamation Researcher, Am Soc Surface Mining & Reclamation, 84. *Mem:* Soc Range Mgt; Sci Res Soc NAm; Ecol Soc Am; Am Soc Surface Mining & Reclamation (pres, 85). *Res:* Principles and techniques important in revegetation of man-caused disturbed areas with emphasis on go-back cropland; effects of water on plant species and community establishment; prairie hay as a seed source for revegetation; grass seedling establishment, morphology and anatomy. *Mailing Add:* Northern Great Plains Res Ctr Box 459 Mandan ND 58554

RIES, STANLEY K, b Kenton, Ohio, Sept 6, 27; m 49; c 3. HORTICULTURE, PLANT PHYSIOLOGY. *Educ:* Mich State Col, BS, 50; Cornell Univ, MS, 51, PhD(veg crops), 54. *Prof Exp:* From asst prof to assoc prof, 53-65, PROF HORT, MICH STATE UNIV, 65- *Concurrent Pos:* Consult, Int Atomic Energy Agency, Vienna, 72-82; coop researcher, Rockefeller Found & Ford Found in Mex, Costa Rica, Turkey, Asia, various chem companies. *Honors & Awards:* Golden Key Outstanding Res Award, AAAS. *Mem:* Fel AAAS; fel Am Soc Hort Sci; Brit Plant Growth Regulation Soc; Am Soc Plant Physiol; Sigma Xi; Plant Growth Regulation Soc. *Res:* Plant growth regulation; isolation; identification and mode of action of new naturally occuring plant growth regulators. *Mailing Add:* Dept Hort Mich State Univ East Lansing MI 48824-1325

RIES, STEPHEN MICHAEL, b Watertown, SDak, Apr 4, 44; m 66; c 3. PLANT PATHOLOGY. *Educ:* SDak State Univ, BS, 66; Mont State Univ, PhD(microbiol), 71. *Prof Exp:* Res assoc plant path, Mont State Univ, 71; res assoc chem, Univ Colo, 71-72; res assoc plant physiol, Univ Calif, Riverside, 72-73; asst prof, 73-80, ASSOC PROF PLANT PATH, UNIV ILL, URBANA, 80- *Mem:* Am Phytopath Soc. *Res:* Identification and control of diseases of fruit crops and the motility and chemotaxis of bacterial pathogens of fruit crops. *Mailing Add:* N427 Turner Hall Univ Ill 1102 S Goodwin Urbana IL 61801

RIESE, RUSSELL L(OYD), b Kulm, NDak, June 20, 23; m 45; c 1. ELECTRICAL ENGINEERING. *Educ:* Univ Wash, BSEE, 46; Okla State Univ, MS, 50, PhD, 55. *Prof Exp:* With US Civil Serv, Sig Corps, US Dept Army, 42-43; res assoc, Phys Sci Lab, NMex State Univ, 46-47, from instr to assoc prof elec eng, 47-56; prof elec & comput eng & chmn depts, Ariz State Univ, 57-61; prof & assoc chmn, Sch Eng, San Fernando Valley State Col, 61-63; assoc dean acad planning, Off of Chancellor, Calif State Cols, 63-67; chief higher educ specialist & head sect acad plans & progs, Calif Postsecondary Educ Comn, 67-82; RETIRED. *Concurrent Pos:* Consult, Daley Elec Co, 58-59, Gen Elec Co, 59-61, Western Mgt Consults, 60-63, Marquardt Corp, 61-62, Electronics Assocs Inc, 65- & W N Samarzich & Assocs, 79-81, Higher Educ, 82- *Mem:* Am Soc Eng Educ; Inst Elec & Electronics Engrs; Sigma Xi. *Res:* Network synthesis; reliability; radar; controls; computers; power systems. *Mailing Add:* 4120 Wintercrest Ln Shingle Springs CA 95682

RIESE, WALTER CHARLES RUSTY, b Newport, RI, June 8, 51; m 73, 84; c 2. PETROLEUM EXPLORATION, GEOCHEMISTRY. *Educ:* NMex Tech, BS, 73; Univ NMex, MS, 77, PhD(geol), 80. *Prof Exp:* Asst geologist, Vanguard Explor, 71 & NMex Bur Mines & Mineral Resources, 72-73; geologist, Technol Appln Ctr, Univ NMex, 73-74; proj geologist, Gulf Mineral Resources Co, 74-81; proj geochemist, Anaconda Copper Co, 81-83, admin coordr, Anaconda Minerals, 83-84; sr geologist, Arco Explor, 84-85, area geologist, 85-89, DIST GEOLOGIST, ARCO OIL & GAS CO, 89- *Concurrent Pos:* Res asst, Univ NMex, 79-81; instr, Arapahoe Community Col, 83 & Univ Houston, 85-; affil fac, Colo State Univ, 85-; adj asst prof, Rice Univ, 85- *Mem:* Soc Econ Geologists; Am Inst Professional Geologists; Asn Explor Geochemists; Am Asn Petrol Geologists; Geol Soc Am; Sigma Xi; Int Asn Cosmochem & Geochem. *Res:* Applied geochemical exploration research in the biogeochemistry and geomicrobiology; seismic stratigraphy; scanning electron microscopic alterations studies; geochemical reservoir and formation damage studies. *Mailing Add:* 2009 Mountain Oak Rd Bakersfield CA 93311

RIESELBACH, RICHARD EDGAR, b Milwaukee, Wis, Dec 5, 33; m 56; c 3. INTERNAL MEDICINE, PHYSIOLOGY. *Educ:* Univ Wis-Madison, BS, 55; Harvard Med Sch, MD, 58. *Prof Exp:* Fel nephrology, Wash Univ, 62-64, instr internal med, Med Sch, 64-65; from instr to assoc prof internal med, 65-73, chmn dept, 73-77, PROF MED, MED SCH, UNIV WIS-MADISON, 73-, CHIEF MED & COORDR ACAD AFFAIRS, MT SINAI MED CTR, 73- *Concurrent Pos:* Markle Found scholar, 69. *Mem:* Am Soc Clin Invest; Am Fedn Clin Res; Am Soc Nephrology. *Res:* Renal physiology and pathophysiology. *Mailing Add:* Mt Sinai Hosp Univ Wis Med Sch Dept Med 850 N 12th Madison WI 53233

RIESEN, AUSTIN HERBERT, b Newton, Kans, July 1, 13; m 39; c 2. PHYSIOLOGICAL PSYCHOLOGY. *Educ:* Univ Ariz, AB, 35; Yale Univ, PhD(psychol), 39. *Hon Degrees:* DSc, Univ Ariz, 81. *Prof Exp:* Res assoc psychobiol, Yale Univ, 39-49; aviation physiologist, US Army Air Corps, 43-46; from assoc prof to prof psychol, Univ Chicago, 49-62; prof, 62-80, EMER PROF PSYCHOL, UNIV CALIF, RIVERSIDE, 80- *Concurrent Pos:* Vis res prof, Univ Rochester, 51-53; NIH res grant, 62-75; chmn dept psychol, Univ Calif, Riverside, 63-68; mem career develop rev comt, NIMH, 64-69; consult primate resources adv comt, Nat Inst Child Health & Human Develop, 71-75, res grant, 76-82; mem Div 3 & 6, Am Psychol Asn, pres Div 6, 66-67. *Mem:* Am Psychol Asn; Soc Neurosci; Int Soc Develop Psychobiol (pres, 70-71); Int Brain Res Orgn. *Res:* Sensory deprivation; brain development; behavioral development; sensitive periods; primate behavior; comparative psychology. *Mailing Add:* Dept Psychol Univ Calif Riverside CA 92521

RIESEN, JOHN WILLIAM, b Summit, NJ, Aug 18, 41; m 63; c 2. REPRODUCTIVE PHYSIOLOGY. *Educ:* Univ Mass, BS, 63; Univ Wis, MS, 65, PhD, 68. *Prof Exp:* Fel, Primate Ctr, Univ Wis, 68-70; from asst prof to assoc prof animal indust, 70-83, PROF ANIMAL SCI, UNIV CONN, 83- *Mem:* Am Soc Animal Sci; Soc Study Reproduction; Am Dairy Sci Asn. *Res:* Physiology of the postpartum cow; control of ovulation; physiology and endocrinology of spermatogenesis. *Mailing Add:* Animal Sci U-40 Univ Conn Storrs CT 06269-4040

RIESENFELD, PETER WILLIAM, b Minneapolis, Minn, Sept 6, 45; m 72; c 2. NUCLEAR SCIENCE. *Educ:* Univ Calif, Berkeley, BS, 67; Princeton Univ, MS, 69, PhD(chem & physics), 71. *Prof Exp:* Res asst, Inst Theoret Physics, Univ Frankfurt, 72-74; instr physics, Ind Univ, Bloomington, 74-75; RESEARCHER, RI NUCLEAR SCI CTR, 75- *Concurrent Pos:* Consult, Res Inst Nuclear Physics, Ger, 72-74. *Mem:* Sigma Xi; Am Phys Soc; AAAS. *Res:* Heavy ion physics, nuclear techniques for chemical analysis. *Mailing Add:* 71 Top Hill Rd PO Box 232 Saunderstown RI 02874

RIESENFELD, RICHARD F, b Milwaukee, Wis, Nov 26, 44; m 74; c 2. COMPUTER SCIENCE. *Educ:* Princeton Univ, AB, 66; Syracuse Univ, MA, 69, PhD(comput sci), 73. *Prof Exp:* Asst prof elec eng, 72-74, asst prof, 74-76, assoc prof, 77-81, PROF COMPUT SCI & CHMN DEPT, UNIV UTAH, 81- *Concurrent Pos:* Consult comput aided design; adj asst prof math, Univ Utah, 74-76. *Mem:* Asn Comput Mach; Soc Indust & Appl Math; Math Asn Am. *Res:* Developing mathematical models for representing geometric shape information in a computer; computer graphics; an experimental computer-aided design/computer-aided manufacturing system. *Mailing Add:* Dept Comput Sci Univ Utah Salt Lake City UT 84112

RIESER, LEONARD M, b Chicago, Ill, May 18, 22; m 44; c 3. ATOMIC PHYSICS, NUCLEAR PHYSICS. *Educ:* Univ Chicago, BS, 43; Stanford Univ, PhD(physics), 52. *Prof Exp:* Asst physics, Metall Lab, Univ Chicago, 44 & Los Alamos Sci Lab, 45-46; asst, Stanford Univ, 46-51, res assoc, 51-52; from instr to prof, Dartmouth Col, 52-81, dept provost, 59-64, dean fac arts & sci, 64-69 & 71-82, provost, 67-71, vpres, 71-82, FAIRCHILD PROF PHYSICS, DARTMOUTH COL, 81- *Concurrent Pos:* Pres, New Eng Conf Grad Educ, 65-66; chmn, AAAS, 74; chmn bd, Bull Atomic Scientists; mem, Fulbright Coun, Int Exchange Scholars, 88; vis scholar, Mac Arthur Found; mem bd trustees, Hampshire Col & Latin Am Scholar Prog Am Univ. *Mem:* AAAS (pres 73-); Interciencia Asn (vpres, 76-, pres, 80-84); Am Phys Soc; Biophys Soc; Am Asn Physics Teachers. *Res:* Reflection of x-rays; x-ray microscopy; proportional counters; experimental nuclear physics; biophysics. *Mailing Add:* Dept Physics Dartmouth Col Hanover NH 03755

RIESKE, JOHN SAMUEL, b Provo, Utah, Sept 25, 23; m 49; c 6. BIOCHEMISTRY. *Educ:* Brigham Young Univ, BA, 49; Univ Utah, PhD(biol chem), 56. *Prof Exp:* Biochemist, US Army Res & develop Command, 55-60; trainee respiratory enzymes, Inst Enzyme Res, Univ Wis, 60-64; asst prof, Inst enzyme Res, Univ Wis, 64-66; assoc prof, 66-75, PROF PHYSIOL CHEM, OHIO STATE UNIV, 75- *Concurrent Pos:* USPHS career develop award, 64-65. *Mem:* AAAS; Am Soc Biol Chemists. *Res:* Structure and function of respiratory enzymes; cholinesterase enzymes; photosynthesis. *Mailing Add:* Dept Physiol Chem Ohio State Univ 1645 Neil Ave Columbus OH 43210

RIESS, KARLEM, b New Orleans, La, Apr 17, 13. PHYSICS. *Educ:* Tulane Univ, BS, 33, MS, 35; Brown Univ, PhD(physics), 43. *Prof Exp:* Instr math, Tulane Univ, 33-35, reader, 35-36; prof, high sch, La, 36-42; jr chemist, USN Yard, Philadelphia, 42; instr physics, Brown Univ, 42-43; from asst prof to prof, 43-78, EMER PROF PHYSICS, TULANE UNIV, 78- *Mem:* Am Phys Soc; Am Chem Soc; Am Math Soc; Am Crystallog Asn. *Res:* Electromagnetic waves in a bent pipe of rectangular cross section; biophysics; mathematical physics. *Mailing Add:* 17 Audubon Blvd New Orleans LA 70118

RIESS, RONALD DEAN, b North English, Iowa, Sept 28, 40; m 62; c 1. MATHEMATICS. *Educ:* Iowa State Univ, BS, 63, MS, 65, PhD(numerical anal), 67. *Prof Exp:* Instr math, Iowa State Univ, 66-67; asst prof, 67-78, ASSOC PROF MATH, VA POLYTECH INST & STATE UNIV, 78- *Res:* Conditioning eigen value problems; numerical analysis of integration. *Mailing Add:* Dept Math Va Polytech Inst & State Univ Blacksburg VA 24061

RIESZ, PETER, b Vienna, Austria, Oct 2, 26; US citizen; m 53; c 3. PHYSICAL CHEMISTRY, RADIATION BIOLOGY. *Educ:* Oxford Univ, BA, 46, BSc, 47; Columbia Univ, PhD(phys chem), 53. *Prof Exp:* Res assoc phys chem, Pa State Univ, 53-54; res assoc, Brookhaven Nat Lab, 54-56; res assoc radiation chem, Argonne Nat Lab, 56-58; res chemist, 58-81, chief, Radiation Biol Sect, Lab Pathophysiol, 81-83, SR INVESTR, RADIATION ONCOL BR, NAT CANCER INST, 83- *Mem:* Am Chem Soc; Radiation Res Soc; Biophys Soc; Am Soc Photobiol. *Res:* Sonochemistry; radiation chemistry; effects of ionizing and ultraviolet radiation on nucleic acids and proteins; electron spin resonance; free radical mechanisms; spin trapping; photochemistry. *Mailing Add:* Nat Cancer Inst Bldg 10 Rm B1B50 Bethesda MD 20892

RIETHOF, THOMAS ROBERT, b Teplice, Czech, July 29, 27; nat US; m 49. PHYSICAL CHEMISTRY, ELECTROOPTICS. *Educ:* Manchester Col, AB, 49; Purdue Univ, MSc, 51, PhD(phys chem), 54. *Prof Exp:* Physicist, Aerophys Applns, 56-63, mgr reentry physics, 64-68, mgr sensor systs, 68-80, CONSULT PHYSICIST, SPACE SYSTS DIV, GEN ELEC CO, PHILADELPHIA, 80- *Mem:* AAAS; Am Chem Soc; Optical Soc Am; Am Inst Aeronaut & Astronaut. *Res:* Electrooptical sensors and experiments; remote sensing, optical and infrared, military and civilian; thermal radiation and its measurement. *Mailing Add:* 4124 Barberry Dr Lafayette Hill PA 19444

RIETVELD, WILLIS JAMES, b Harvey, Ill, Oct 30, 42. FOREST PHYSIOLOGY. *Educ:* Ore State Univ, BS, 65, MS, 67; Univ Ariz, PhD(plant physiol), 74. *Prof Exp:* Res plant physiologist, Rocky Mountain Forest & Range Exp Sta, US Forest Serv, Forestry Sci Lab, Northern Ariz Univ, 66-76; prin plant physiologist, North Cent Forest Exp Sta, Forestry Sci Lab, Southern Ill Univ, 76-; MEM STAFF, FORESTRY SCI LAB, RHINELANDER, WIS. *Mem:* Soc Am Foresters; Sigma Xi. *Res:* Physiological quality of tree planting stock and post-planting seedling growth; water relations; allelopathy. *Mailing Add:* Forestry Sci Lab PO Box 898 Rhinelander WI 54501

RIETZ, EDWARD GUSTAVE, chemistry; deceased, see previous edition for last biography

RIEWALD, PAUL GORDON, b E Grand Rapids, Mich, Aug 31, 41; m 68; c 2. MATERIALS SCIENCE, METALLURGICAL ENGINEERING. *Educ:* Univ Mich, BS, 63, MS, 64, PhD(metall eng), 68. *Prof Exp:* Res engr mat eng & develop, 68-74, sr res engr, 74-77, res assoc Mat Eng & Develop, 77-82, SR RES ASSOC, E I DU PONT DE NEMOURS & CO, INC, 82- *Res:* Characterization and study of fibrous materials; industrial applications, research and development of products from fibers including advanced composites; ropes and cables; ballistic armor; fiber reinforced cement/concrete, automotive uses. *Mailing Add:* Chestnut Run Labs Bldg 701 E I du Pont De Nemours & Co Inc Wilmington DE 19880

RIEWE, MARVIN EDMUND, plant-animal systems, grazing management, for more information see previous edition

RIFAS, LEONARD, b Sept 20, 46; m; c 2. METABOLISM, BONE RESEARCH. *Educ:* Univ Mo, Columbia, AB, 69, MS, 73. *Prof Exp:* Dir, Cell Cult Lab, Dept Biochem, Albert Einstein Col Med, Bronx, NY, 73-79; RES INSTR MED, JEWISH HOSP, SCH MED, WASH UNIV, ST LOUIS, 79- *Concurrent Pos:* Reviewer, J Calcified Tissue Int, J Bone & Mineral Res, J Cell & Tissue Res, Am J Physiol, J Clin Endocrinol & Metab. *Mem:* Am Soc Cell Biol; AAAS; Am Soc Bone & Mineral Res; NY Acad Sci; Fedn Am Soc Exp Biol. *Res:* Study of bone cell biology including effect of cytokines on bone cell physiology and function; relationship of immune system products on the pathophysiology of osteoporosis; study of metalloproteinase secretion by bone cells under the influence of cytokines; immunological assays; immunology; cell biology; bone cell matabolism. *Mailing Add:* Dept Med Div Endocrinol Jewish Hosp Wash Univ Med Ctr 216 S Kingshighway St Louis MO 63110

RIFE, DAVID CECIL, b Cedarville, Ohio, Jan 3, 01; m 58; c 1. GENETICS. *Educ:* Cedarville Col, BS, 22; Ohio State Univ, BS, 23, AM, 31, PhD(genetics), 33. *Prof Exp:* Instr sci & agr, Mission Sch, Sudan, 27-30; from instr to prof genetics, Ohio State Univ, 34-57; livestock adv, Int Coop Admin, Thailand, 57-59, dep sci attache, Am Embassy, New Delhi, India, 60-62; dir int rels, Am Inst Biol Sci, 62-63; sci adminr, Nat Inst Gen Med Sci, 63-65; vis prof genetics, Univ Fla, 66-71; CONSULT GENETICS, 71- *Concurrent Pos:* Consult geneticist, Ohio Agr Exp Sta, 44-52; Fulbright lectr, Univ Cairo, 51-52; Fulbright res scholar, Uganda, 55-56; mem panel selection awards, Fulbright Found, 56. *Mem:* Am Soc Human Genetics (secy, 55); Am Genetic Asn; Genetics Soc Am; AAAS; Int Soc Twin Studies. *Res:* Human population and behavioral genetics; dermatoglyphics; handedness; twins; genetics of coleus; bovine genetics. *Mailing Add:* 154 W Avenida del Rio Clewiston FL 33440

RIFE, WILLIAM C, b Chicago, Ill, Dec 29, 33; m 62, 70. ORGANIC CHEMISTRY. *Educ:* NCent Col, BA, 56; Univ Ill, PhD(chem), 60. *Prof Exp:* Assoc prof chem, Parsons Col, 60-62; patent chemist, Owens-Ill Glass Co, 62-64; prof chem, NCent Col, 64-72, chmn div humanities, 72-76; fel, Pa State Univ, 76-77; HEAD, DEPT CHEM, CALIF POLYTECH STATE UNIV, 77- *Mem:* AAAS; Am Chem Soc. *Res:* Organometallic compounds; isobenzofurans. *Mailing Add:* 1657 Southwood Dr San Luis Obispo CA 93401-6029

RIFFEE, WILLIAM HARVEY, b Steubenville, Ohio, Feb 17, 44; m 67; c 1. PHARMACOLOGY. *Educ:* WVa Univ, BS Pharm, 67; Ohio State Univ, PhD(pharmacol), 75. *Prof Exp:* Pharm officer, USPHS, 67-70; teaching & res assoc pharmacol, Col Pharm, Ohio State Univ, 71-75; ASST PROF PHARMACOL, UNIV TEX, AUSTIN, 75- *Mem:* Am Pharmaceut Asn. *Res:* Investigation of the effects of drugs on the central nervous systems with particular emphasis on neurotransmitter dynamics; the study of the mechanisms responsible for the development of tolerance to and physiological dependence on various drugs of abuse. *Mailing Add:* Col Pharm Univ Tex Austin TX 78712

RIFFENBURGH, ROBERT HARRY, b Blacksburg, Va, June 19, 31; m 52; c 5. STATISTICS, DECISION METHODOLOGY. *Educ:* Col William & Mary, BS, 51, MS, 53; Va Polytech Inst, PhD, 57. *Prof Exp:* Asst, Univ Hawaii, 51-52; asst prof math, Va Polytech Inst, 55-57; asst prof math, Univ Hawaii, 57-61; sr systs analyst, Lab Electronics, Inc, Calif, 61-62; prof statist & head dept, Univ Conn, 62-70; sr scientist, Naval Ocean Systs Ctr, 70-82; PROF STATIST & MGT, WEBSTER UNIV LEIDEN, LEIDEN, NETH, 82- *Concurrent Pos:* Math statistician, US Fish & Wildlife Serv, 58-61; res assoc, Scripps Inst Oceanog, Univ Calif, San Diego, 68-69; prof oceanog & mgt, San Diego State Univ, 68-74 & 78-82; lectr, Europ Div, Univ Md, Heidelburg, WGer, 74-77. *Mem:* Am Statist Asn; Am Soc Qual Control; Biomet Soc; fel Biomed Soc. *Res:* Data exploration; data salvage; time-dependent multivariate analysis; predicting human decision; classification methods. *Mailing Add:* Babcock Lake Rd Hoosick Falls NY 12090

RIFFER, RICHARD, b Chicago, Ill, Dec 3, 39. NATURAL PRODUCTS CHEMISTRY, AIR & WATER POLLUTION. *Educ:* Ind Univ, BS, 61; Univ Calif, Berkeley, MS, 63, PhD(agr chem), 67. *Prof Exp:* Asst specialist, natural prod chem, Forest Prod Lab, Univ Calif, Berkeley, 63-69; res chemist, US Forest Serv, 69-72; CHIEF CHEMIST, CALIF & HAWAIIAN SUGAR CO, 72- *Concurrent Pos:* Res chemist, Statewide Air Pollution Res Ctr, Univ Calif, 69-72; exec comt, US Nat Comt Sugar Analysis. *Mem:* Am Chem Soc; Am Asn Sugar Beet Technol; Asn Off Analytical Chem. *Res:* Structure elucidation of natural products; pyrolysis mechanisms; cellulose biosynthesis; electrokinetic properties of colloids; ion exchange and reverse osmosis; flavor and aroma chemistry; light scattering; polysaccharides; toxicology, cancer and forensic chemistry. *Mailing Add:* 1401 Walnut St Berkeley CA 94709

RIFFEY, MERIBETH M, b Waukegan, Ill, Mar 18, 24. BIOLOGY. *Educ:* Northwestern Univ, BS, 46, MS, 48; Wash State Univ, PhD, 59. *Prof Exp:* Instr biol, DC Teachers Col, 49-52, Everett Jr Col, 53-54 & Grays Harbor Col, 56-57; asst prof, 57-67, ASSOC PROF BIOL, WESTERN WASH STATE UNIV, 67- *Res:* Avian physiology and ecology; integration of sciences in nursing; avian population studies on all major habitats (including riparian) on the Arizona strip. *Mailing Add:* 8935 SW Washington Dr Portland OR 97223

RIFFLE, JERRY WILLIAM, b Mishawaka, Ind, Jan 7, 34; m 59; c 6. FOREST PATHOLOGY. *Educ:* Mich State Univ, BS, 57, MS, 59; Univ Wis-Madison, PhD(plant path), 62. *Prof Exp:* Res asst forest path, Mich State Univ, 57-58 & Univ Wis-Madison, 58-62; RES PLANT PATHOLOGIST, USDA FOREST SERV, ROCKY MOUNTAIN FOREST & RANGE EXP STA, 62- *Mem:* Am Phytopath Soc; Soc Nematologists; Mycol Soc Am. *Res:* Diseases of trees in plantings and natural stands; mycorrhizae of conifers; forest nematology. *Mailing Add:* Box 152 Rte 4 Syracuse IN 46567

RIFINO, CARL BIAGGIO, b New York, NY, Aug 21, 38; m 64; c 5. PHARMACEUTICAL CHEMISTRY. *Educ:* Fordham Univ, BS, 59; St John's Univ, NY, MS, 64; Purdue Univ, West Lafayette, PhD(med chem), 68. *Prof Exp:* Res investr, Olin Mathieson & Co, Inc, 67-68; res investr, Squibb-Beech Nut, Inc, 68-71, preformulations sect, Squibb Corp, 71-72, process develop mgr, Topical & Parenteral Trade Prod, Squibb Corp, 72-76; sect chief new prod develop, Morton-Norwich Corp, 76-77; SUPVR PROCESS DEVELOP, ICI AMERICAS, INC, 77- *Concurrent Pos:* Bd dir, Int Soc Pharmaceut Engrs, 82-86. *Mem:* Am Pharmaceut Asn; Acad Pharmaceut Sci; Sigma Xi; Am Mgt Asn; Parenteral Drug Asn; Int Soc Pharmaceut Engrs; Am Asn Pharmaceut Scientists. *Res:* Physical pharmacy, especially suspension technology and dissolution characteristics of solid solutions; scaleup activities in solids and liquids. pharmaceutical science; process validation activities. *Mailing Add:* ICI Pharmaceut Old Baltimore Pike PO Box 4520 Newark DE 19711

RIFKIN, ARTHUR, b New York, NY, Apr 7, 37; m 61; c 2. PSYCHOPHARMACOLOGY. *Educ:* Columbia Col, BA, 57; State Univ NY Downstate Med Ctr, MD, 61. *Prof Exp:* Staff psychiatrist, Hillside Hosp, 67-69, dir, Aftercare Clin, 69-76; res psychiatrist, NY State Psychiat Inst, 76-79; assoc clin prof psychiat, Col Physicians & Surgeons, 77-79; dir, div clin psychopharmacol res, Mt Sinai Med Ctr & assoc prof, 79-84, PROF PSYCHIAT, MT SINAI SCH MED, 84-; ASSOC DIR, DEPT PSYCHIAT, MT SINAI SERV, ELMHURST HOSP CTR, NY, 84- *Concurrent Pos:* Fel psychiat res, State Univ NY Downstate Med Ctr, 67-69. *Mem:* Fel Am Col Neuropsychopharmacol; Collegium Int Neuropsychopharmacologicicum; Am Psychopath Asn; Psychiat Res Soc; fel Am Psychiat Asn. *Res:* Drug treatment of schizophrenia, affective disorders, panic disorders and character disorders. *Mailing Add:* Queens Hosp Ctr 82-68 164th St Jamaica NY 11432

RIFKIN, BARRY RICHARD, b Trenton, NJ, Mar 30, 40; div; c 2. EXPERIMENTAL PATHOLOGY, ORAL MEDICINE. *Educ:* Ohio State Univ, BS, 61; Univ Ill, MS, 64; Temple Univ, DDS, 68; Univ Rochester, PhD(path), 74. *Prof Exp:* Assoc prof path & dent res, Med Sch, Univ Rochester, 73-80, assoc pathologist, Strong Mem Hosp, 74-80; ASSOC PROF ORAL MED, PATHOBIOL & ORAL PATH, COL DENT, NY UNIV, 80-, CHMN, DEPT ORAL MED, 80-, ASSOC PROF BIOMED SCI, GRAD FAC, COL ARTS & SCI, 81- *Concurrent Pos:* Prin investr, Pathogenesis of Bone Loss in Peridontal Disease, Nat Inst Dent Res, 76-80. *Mem:* Int Acad Path; Int Asn Dent Res; Am Soc Bone & Mineral Res; Sigma Xi. *Res:* Inflammation and bone resorption; mechanisms of localized bone loss; pathogenesis of bone loss in periodontal disease; vitro structure of bone resorption. *Mailing Add:* Dept Oral Med NY Univ Dent Ctr 421 First Ave New York NY 10010

RIFKIN, ERIK, b Brooklyn, NY, Sept 13, 40; m 64; c 2. PATHOBIOLOGY. *Educ:* Rutgers Univ, BA, 64; Univ Hawaii, MS, 67, PhD(zool), 69. *Prof Exp:* Res asst biol, Rutgers Univ, 64-65; res asst, Univ Hawaii, 66-67; Nat Res Coun assoc pathobiol, Naval Med Res Inst, 69-70; dir ecol, Antioch Col, 70-72; consult environ planning, Urban Life Ctr, Columbia, Md, 72-73; dir environ studies prog, New Col, 73-74; dir, Environ Planning/Res Inst, 74-78; PRES, RIFKIN & ASSOCS, INC, 79- *Mem:* Am Inst Biol Sci; AAAS; Am Soc Zoologists; Soc Invert Path. *Res:* Pathobiology of invertebrates, with emphasis on those organisms cultured for aquaculture and/or mariculture systems; surface mining and the environment; environmental policy; analysis and evaluation of the adverse environmental effects caused by coal and nonfuel mining operations. *Mailing Add:* 5950 Symphony Woods Rd Apt 518 Columbia MD 21044

RIFKIND, ARLEEN B, b New York, NY, June 29, 38; m 61; c 2. TOXICOLOGY, ENDOCRINOLOGY. *Educ:* Bryn Mawr Col, BA, 60; NY Univ, MD, 64. *Prof Exp:* Intern med, III & IV Med Div, Bellevue Hosp, 64-65, first year resident, 65; clin assoc endocrine br, Nat Cancer Inst, 65-68; res assoc endocrine pharmacol, Rockefeller Univ, 68-71; asst prof pediat, Cornell Univ, 71-75, asst prof med, 71-82, from asst prof to assoc prof pharmacol, 73-82, chmn Gen Fac Coun, 84-86, PROF PHARMACOL & ASSOC PROF MED, MED COL, CORNELL UNIV, 83- *Concurrent Pos:* Staff fel, Nat Inst Child Health & Human Develop, 65-68; USPHS spec fel, Rockefeller Univ, 68-70; USPHS spec fel, 71-72; adj asst prof, Rockefeller Univ, 71-74; prin investr grants, Nat Found, Am Cancer Soc, NY State Health Res Found, Nat Inst Environ Health Sci, 72-; mem, Environ Health Sci Review Comt, Nat Inst Environ Health Sci, 81-83 & 84-85, chmn, 85-, Superfund Basic Res Prog Review Group, 87, 88; mem, Toxicol Study Sect, 89-93; mem, Environ Health & Safety Coun, Am Health Found, 90-94. *Mem:* AAAS; Am Soc Pharmacol & Exp Therapeut; Am Soc Clin Pharmacol; Endocrine Soc; Am Soc Clin Invest; Toxicol Soc. *Res:* Heme, porphyrin and mixed function oxidase regulation; biochemical pharmacology; mechanisms of polychlorinated biphenyl and dioxin toxicity; role of arachidonic acid metabolism in dioxin toxicity. *Mailing Add:* Dept Pharmacol Med Col Cornell Univ 1300 York Ave New York NY 10021

RIFKIND, BASIL M, b Glasgow, Scotland, Sept 17, 34; US citizen; m; c 3. LIPID METABOLISM RESEARCH. *Educ:* Univ Glasgow, MB, ChB, 57, MD, 72; FRCP(G), 73. *Prof Exp:* House surgeon, County Hosp, Ormskirk, Eng, 57-58; house physician, Royal Infirmary, Glasgow, Scotland, 58, sr house officer internal med, 59-60, registrar, 60-63, registrar internal med to prof, E M McGirr Univ Dept Med, 63-65, sr registrar, Glasgow Teaching Hosp, 65-71; dep chief, Lipid Metab Br, 71-74, proj officer, Lipid Res Clin Prog, 71-80, CHIEF, LIPID METAB-ATHEROGENESIS BR, DIV HEART & VASCULAR DIS, NAT HEART, LUNG & BLOOD INST, 74-, DEP ASSOC DIR, HYPERTENSION, ARTERIOSCLEROSIS & LIPID METAB PROG, 79- *Concurrent Pos:* Sr house officer clin path, Crumpsall Hosp, Manchester, Eng, 58-59. *Honors & Awards:* Watson Prize Lectr, Royal Col Physicians & Surgeons, Glasgow, 70. *Res:* Hypertension; internal medicine; vascular diseases. *Mailing Add:* Nat Heart Lung & Blood Inst Div Heart & Vascular Dis Lipid Metab-Atherogenesis Br Fed Bldg Rm 4A14A 7550 Wisconsin Ave Bethesda MD 20892

RIFKIND, DAVID, b Los Angeles, Calif, Mar 11, 29; m 57; c 2. INTERNAL MEDICINE, INFECTIOUS DISEASES. *Educ:* Univ Calif, Los Angeles, AB, 50, PhD(microbiol), 53; Univ Chicago, MD, 57. *Prof Exp:* Res asst, Univ Chicago, 55-57; clin assoc, Nat Inst Allergy & Infectious Dis, 59-61; from instr to asst prof med, Univ Colo Med Ctr, Denver, 62-67, head sect infectious dis, 66-67; PROF MICROBIOL & HEAD DEPT, COL MED, UNIV ARIZ, 67-, PROF MED, 71-, HEAD SECT INFECTIOUS DIS, 71- , CLIN PROF, DEPT INT MED SECT INFECTIOUS DIS, 83- *Concurrent Pos:* Consult, Fitzsimons Gen Hosp, Denver, 64-; attend physician, Vet Admin Hosp, 66- *Mem:* Am Soc Microbiol; Am Fedn Clin Res; Infectious Dis Soc Am. *Res:* Infectious diseases complicating immunosuppressive drug therapy; mechanisms of action of endotoxin; respiratory, viral and mycoplasmal infections; viral latency and activation; antimicrobial drug therapy. *Mailing Add:* Dept Microbiol Univ Ariz Col Med Tucson AZ 85724

RIFKIND, JOSEPH MOSES, b New York, NY, Jan 13, 40; m 64; c 4. PHYSICAL BIOCHEMISTRY. *Educ:* Yeshiva Univ, BA, 61; Columbia Univ, MA, 62, PhD(phys chem), 66. *Prof Exp:* Res assoc biophys chem, Univ Minn, 65-67; from staff fel to sr staff fel molecular biol, Geront Nat Inst Child Health & Human Develop, 68-73; res chemist, 73-85, CHIEF SECT MOLECULAR DYNAMICS, NAT INST AGING, 85- *Concurrent Pos:* Pegram hon fel, 65-66; NIH fel, 67. *Mem:* AAAS; Am Chem Soc; Biophys Soc; Geront Soc. *Res:* Thermodynamics and kinetics of conformational transitions in polypeptides, proteins and nucleic acids; structure function relationships in proteins and nucleic acids; oxygenation of hemoglobin; interaction between proteins and nucleic acids; regulation of oxygen transport; hemolysis of the erythrocyte; the erythrocyte membrane; membrane fluidity; oxyradicals. *Mailing Add:* Nat Inst Aging 4940 Eastern Ave Baltimore MD 21224

RIFKIND, RICHARD A, b New York, NY, Oct 26, 30; m 56; c 2. MEDICINE, HEMATOLOGY. *Educ:* Yale Univ, BS, 52; Columbia Univ, MD, 55. *Prof Exp:* Intern med, Presby Hosp, New York, 55-56, resident, 56-57 & 60-61; assoc med, Col Physicians & Surgeons, Columbia Univ, 62-63, from asst prof to prof med & human genetics, 63-81; MEM, SLOAN-KETTERING INST & DIR, SLOAN-KETTERING DIV, GRAD SCH MED SCI, CORNELL UNIV, 81-, CHMN, SLOAN-KETTERING INST, 83- *Concurrent Pos:* Nat Found fel, 59-60; USPHS trainee hemat, 61-62; Guggenheim fel, 65-66. *Mem:* Am Soc Cell Biol; Asn Am Physicians; Am Soc Clin Invest; Electron Micros Soc Am; Am Soc Hemat. *Res:* Developmental biology; molecular biology; hematology. *Mailing Add:* Mem Sloan-Kettering Cancer Ctr 1275 York Ave New York NY 10021

RIGANATI, JOHN PHILIP, b Mt Vernon, NY, Apr 11, 44; m 66; c 3. ELECTRICAL ENGINEERING, APPLIED MATHEMATICS. *Educ:* Rensselaer Polytech Inst, BEE, 65, MEng, 66, PhD(elec eng), 69. *Prof Exp:* Co-op engr commun systs, Advan Systs Develop Div, IBM Corp, 61-63, co-op engr thin films & CPU design, Res Div, 63-65; engr mini-comput control systs, Syst Sales & Eng, Gen Elec Co, 67; instr elec eng, Rensselaer Polytech Inst, 68-69; mem tech staff pattern recognition, Electronics Res Ctr, Rockwell Int Corp, 69-77, chief scientist identification systs, Collins Commun Switching Systs Div, Com Telecommun Group, 77-80; chief comput syst component div, Nat Bur Standards, 80-85; DIR SYST RES,

SUPERCOMPUTING RES CTR, 85- *Concurrent Pos:* Lectr, Inst Elec & Electronics Engrs, 75-78; co-ed, J of Supercomputing, 85- *Honors & Awards:* Eng of Yr Award, Rockwell Int, 77. *Mem:* Inst Elec & Electronics Engrs; Pattern Recognition Soc; Sigma Xi; Asn Comput Mach. *Res:* Pattern recognition applied to identification systems and image and speech processing; queueing and statistical sampling theory; distributed computer systems; microfilm information retrieval; digital signal processing; coding and decoding; cryptography; data base design; music theory; high performance computer architecture, operating systems, compilers, languages and performance measures. *Mailing Add:* 13013 Brandon Way Gaithersburg MD 20878

RIGAS, ANTHONY L, b Andros, Greece, May 3, 31; US citizen; m 59; c 1. ELECTRICAL ENGINEERING. *Educ:* Univ Kans, BSEE, 58, MSEE, 62; Univ Beverly Hills, PhD(eng), 78. *Prof Exp:* Elec engr, US Naval Missile Ctr, Point Mugu, 58-61; teaching fel, Univ Kans, 61-63; sr res engr, Lockheed Missile & Space Div, 63-65; sr res engr, Delmo Victor Co, 65-66; from asst prof to assoc prof, 66-73, PROF ELEC ENG & DIR ENG EDUC, OUTREACH DIV, UNIV IDAHO, 73- *Concurrent Pos:* Asst prof, Grad Prog, San Jose State Col, 63-65; NASA-Stanford faculty fel, 67; NSF fel, Princeton Univ, 68; US Cong fel, AAAS, 75-76 & Inst Elec & Electronics Engrs; Cong fel, Inst Elec & Electronics Engrs. *Mem:* Am Soc Eng Educ; Simulation Coun; Sigma Xi; fel Inst Elec & Electronics Engrs. *Res:* Missiles and space vehicles guidance and control systems; biological and environmental systems analysis; computer simulation. *Mailing Add:* Off Dean Eng Mich State Univ 173 Eng Bldg East Lansing MI 48824

RIGAS, DEMETRIOS A, b Andros, Greece, Feb 2, 21; US citizen; m 55; c 2. BIOCHEMISTRY, BIOPHYSICS. *Educ:* Univ Eng Sci, Ahtens, ChE, PhD(phys chem), 41. *Prof Exp:* From res asst to res assoc biochem, 47-53, from asst prof to assoc prof, 53-63, PROF BIOCHEM, MED SCH, ORE HEALTH SCI UNIV, 63- *Concurrent Pos:* Vis prof, Med Sch, Univ Athens & Democritos Ctr Nuclear Res, 70-71. *Mem:* Sigma Xi; Am Asn Biol Chemists; Biophys Soc; Am Chem Soc; NY Acad Sci. *Res:* Biophysical chemistry of proteins; effects of ionizing radiations on mammalian cells; lymphocyte transformation and function; kinetics of cell proliferation. *Mailing Add:* Dept Biochem Ore Health Sci Univ Sch Med 3180 SWS Jackson Pk Rd Portland OR 97201

RIGAS, HARRIETT B, b Winnipeg, Man, Apr 30, 34; US citizen; m 59; c 1. ELECTRICAL ENGINEERING. *Educ:* Queen's Univ, Ont, BSc, 56; Univ Kans, MS, 59, PhD(elec eng), 63. *Prof Exp:* Engr, Mayo Clin, 56-57; instr physics, math & eng, Ventura Col, 59-60; sr res engr, Lockheed Missile & Space Co, 63-65; from asst prof to assoc prof info sci & elec eng, Wash State Univ, 65-76, mgr hybrid facility, 68-80, prof elec eng, 76-84; chmn elec eng, Naval Postgrad Sch, 80-84, prof & chmn, 84-87; PROF & CHMN, MICH STATE UNIV, 87- *Concurrent Pos:* Asst prof, San Jose State Col, 63-65 & Richland Grad Ctr, 66-67; prog dir, NSF, 75-76. *Mem:* Inst Elec & Electronics Engrs; Am Soc Eng Educ; Soc Comput Simulation; Soc Women Engrs. *Res:* Control system stability and optimization; computer systems; hybrid computer and automatic patching; logic design; parallel processing. *Mailing Add:* Dept Elec Eng 260 Engr Bldg Mich State Univ East Lansing MI 48824

RIGASSIO, JAMES LOUIS, b Union City, NJ, Aug 13, 23; m 59; c 4. ENGINEERING. *Educ:* Newark Col Eng, BS, 48; Yale Univ, ME, 49. *Prof Exp:* Develop engr, Johnson & Johnson, 49-52, indust engr, Ethicon Inc Div, 52-55, chief engr, 55-58; asst prof, 58-59, assoc prof, 59-65, PROF INDUST ENG, NEWARK COL ENG, 65-, CHMN DEPT, 69- *Concurrent Pos:* Adj prof, Newark Col Eng, 56-58; grants, Newark Col Eng Res Found, 62-63 & 68-69; consult, NJ Sch Bds Asn, 68- & Nat Asn Advan Colored People, 71- *Mem:* Am Soc Mech Engrs; Am Inst Indust Engrs; Am Soc Eng Educ; Nat Soc Prof Engrs; Indust Rels Res Asn. *Res:* Queuing theory; production process design and control; scheduling theory; work methods; job evaluation. *Mailing Add:* 23 Colony Dr Summit NJ 07901

RIGATTO, HENRIQUE, b Porpo Alegre, Brazil, Nov 23, 37. FETAL & NEWBORN RESPIRATION CONTROL, PEDIATRICS. *Educ:* Univ Rio Grande Do Sul, Brazil, MD, 63. *Prof Exp:* PROF PEDIAT, WOMEN'S HOSP-HEALTH SCI CTR, 80- *Mem:* Am Fedn Clin Res; Soc Pediat Res; NY Acad Sci. *Mailing Add:* Dept Pediat Child Health Univ Man 735 Notre Dame Ave Winnipeg MB R3E 0L8 Can

RIGAUD, MICHEL JEAN, b Paris, France, Oct 22, 39; Can citizen; m 63; c 3. MATERIALS SCIENCE. *Educ:* Polytech Sch, Montreal, BApplSc, 63, MApplSc, 64, Dr(metall), 66. *Prof Exp:* Asst prof metall, Polytech Sch, Montreal, 66-71, assoc prof & head dept, 71-74; assoc dir res, Sidbec-Dosco, 74-76; PROF METALL ENG & CHMN DEPT, POLYTECH SCH, MONTREAL, 76- *Concurrent Pos:* Consult, Metall & Mat. *Mem:* Can Metall Soc (treas, 68-); fel Can Inst Mining & Metall (pres, Metall Soc, 76); Am Soc Metals; Am Ceramic Soc. *Res:* Extractive metallurgy; steelmaking; direct reduction; refractories; ceramics. *Mailing Add:* Dept Metall Eng Campus Univ Montreal Montreal PQ H3C 3A7 Can

RIGBY, CHARLOTTE EDITH, b Winnipeg, Man, July 9, 40. MICROBIOLOGY, HEALTH SCIENCES. *Educ:* Univ Man, BSc, 60, MSc, 63; Univ Ottawa, PhD(microbiol), 71; Univ Sask, dipl vet microbiol, 75. *Prof Exp:* Lectr med microbiol, Univ Man, 66-68; res scientist vet microbiol, Inst Animal Sci, Havana, Cuba, 71-73; resident, Univ Sask, 74-75; res scientist vet microbiol, Animal Dis Res Inst, Agr Can, 75-81, Animal Path Lab, 81-83; RES SCIENTIST VET MICROBIOL, ANIMAL DIS RES INST, 83- *Mem:* Am Soc Microbiol; Can Soc Microbiologists. *Res:* Molecular virology of livestock diseases; brucellosis of cattle. *Mailing Add:* 2748 Howe Ottawa ON K2B 6W9 Can

RIGBY, DONALD W, b Anaheim, Calif, Feb 14, 29; m 50; c 1. PARASITOLOGY. *Educ:* La Sierra Col, BA, 50; Walla Walla Col, MA, 56; Loma Linda Univ, PhD(biol, parasitol), 67. *Prof Exp:* Med technologist, USPHS Hosp, Ft Worth, Tex, 53-54; PROF PARASITOL, WALLA WALLA COL, 58- *Mem:* AAAS; Am Soc Parasitol; Am Soc Zool; Sigma Xi. *Res:* General parasitology; host-parasite relationships; invertebrate zoology. *Mailing Add:* Dept Biol Walla Walla Col College Place WA 99324

RIGBY, E(UGENE) B(ERTRAND), b Provo, Utah, Apr 12, 30; m 55; c 4. CERAMIC ENGINEERING. *Educ:* Brigham Young Univ, BES, 56; Univ Utah, PhD(ceramic engr), 62. *Prof Exp:* Sr res engr, United Tech Ctr, United Aircraft Corp, 62-64; res chemist, E I du Pont de Nemours & Co, Inc, 64-69; staff engr, IBM Corp, 69-72, adv eng inorg res, 72-85, sr eng, 85-90; RETIRED. *Concurrent Pos:* Adj prof mat sci & eng, Univ Ariz, 83-86. *Mem:* Am Ceramic Soc. *Res:* Diffusion in ionic inorganic crystals; high strength ceramics; high wear resistant ceramics; magnetic materials for recording and computer technology. *Mailing Add:* 2415 N 1200 E Provo UT 84604

RIGBY, F LLOYD, b Calgary, Alta, Nov 10, 18; m 63; c 2. AGRICULTURAL CHEMISTRY, BIOCHEMISTRY. *Educ:* Univ Alta, BSc, 42, MSc, 44; McGill Univ, PhD(agr chem), 48. *Prof Exp:* Sr res chemist, Res Div, Can Breweries, 48-63, dir res, 63-65; vpres, 69-79, TECH DIR, JOHN I HAAS, INC, 65-, EXEC VPRES, 80- *Mem:* Am Soc Brewing Chemists; Master Brewers Asn Am; Brit Inst Brewing; Am Chem Soc. *Res:* Plant biochemistry; fermentation biochemistry; food flavors; development of hop concentration. *Mailing Add:* John I Haas Inc PO Box 1441 Yakima WA 98907

RIGBY, FRED DURNFORD, b Missoula, Mont, Sept 11, 14; m 37; c 2. MATHEMATICS. *Educ:* Reed Col, BA, 35; Univ Iowa, MS, 38, PhD(math), 40. *Prof Exp:* Statistician, Pac Northwest Regional Planning Comn, Ore, 37; asst math, Univ Iowa, 36-40; instr, Tex Tech Col, 40-43; mathematician, US Off Naval Res, 46-49, head logistics br, 49-58, dir math sci div, 58-62, dep res dir, 62-63; dean grad sch, 63-68, assoc vpres acad affairs, 68-73, prof math, statist & comput sci, 68-80, dir, Instnl Studies & Res, 73-78, EMER PROF, TEX TECH UNIV, 80- *Mem:* AAAS; Am Math Soc; Math Asn Am. *Res:* Mathematical theories of decision making; mathematical statistics; theory of partially ordered systems; self-organizing systems. *Mailing Add:* 3822 53rd St Lubbock TX 79413

RIGBY, J KEITH, b Fairview, Utah, Oct 8, 26; m 45; c 3. PALEONTOLOGY. *Educ:* Brigham Young Univ, BS, 48, MS, 49; Columbia Univ, PhD(geol), 52. *Prof Exp:* Geologist, Carter Oil Co, 47-48; geologist, Humble Oil & Refining Co, 52-53; from asst prof to prof geol, 53-90, ACTG DEAN GRAD SCH, BRIGHAM YOUNG UNIV, 90- *Concurrent Pos:* Consult geologist, Union Oil Co Can, 60-74 & Phillips Petrol Co, 75; vis prof, La State Univ, 65; ed, J Paleont, 82-85. *Mem:* Geol Soc Am; Paleont Soc; Am Asn Petrol Geologists; Geol Asn Can; Paleont Asn; Soc Econ Paleont Mineral. *Res:* Carbonate deposition and paleoecology of reefs; Upper Paleozoic paleontology; regional geology of Utah, Nevada and Western Canada; fossil sponges and reefs of the world; paleoecology. *Mailing Add:* 210 Page Sch Brigham Young Univ Provo UT 84601

RIGBY, MALCOLM, b Hartford, Wash, Oct 26, 09; m 33; c 2. METEOROLOGY. *Prof Exp:* Coop weather observer, US Weather Bur, Wash, 27-28, observer, 28-30, sr observer & first asst, Alaska, 30-31, officer in chg airport sta, Spokane, Wash, 32-38, North Head, 38-40, Mont, 41-42, climat res & reports, Washington, DC, 42-46 & climat res reports & abstr, 46-49; ed, Meteorol Abstr & Bibliog, 49-59; cur rare book collections, Nat Oceanic & Atmospheric Admin, 75-79; ED, METEOROL & GEOASTROPHYS ABSTR, AM METEOROL SOC, 60- *Concurrent Pos:* Librn, US Weather Bur, 54; mem, Nat Fedn Sci Abstracting & Indexing Serv, 58-70, vpres, 66-67; chmn subcomt mechanization of universal decimal classification, Int Fedn Document, 63-70, mem cent classification comn, 62-84, chmn comt geol-geophys, 62-65 & 70-; chmn comt universal decimal classification, US Nat Comt, Nat Res Coun-Nat Acad Sci, 62-65 & 80-, mem-at-large, 65-66; mem comn hydrol, World Meteorol Orgn, 64-67, reporter, 67-71, chmn comt universal decimal classification, 72-73, chmn working group bibliog prob, 73-87; tech info specialist, Sci Info & Document Div, Environ Sci Serv Admin, 66-70, historian, Environ Sci Info Ctr, Nat Oceanic & Atmospheric Agency, 70-73; consult, 74-; hon fel, Nat Fedn Abstracting & Info Serv, 82- *Honors & Awards:* Am Meteorol Soc Spec Award, 72. *Mem:* AAAS; Am Meteorol Soc; Am Soc Info Sci; Am Geophys Union; Royal Meteorol Soc. *Res:* Scientific information documentation; multilingual vocabularies; preservation of rare books; library and document classification & automation. *Mailing Add:* 2846 Lorcom Lane Arlington VA 22207

RIGBY, PAUL HERBERT, b Humboldt, Ariz, Aug 6, 24; m 54; c 2. MANAGEMENT SCIENCES, APPLIED STATISTICS. *Educ:* Univ Tex, Austin, BBA, 45, MBA, 48, PhD(statist), 52. *Prof Exp:* Sr price economist, Regional Off, Off Price Stabilization, Wash, 51-52; res assoc & asst prof mkt & regional econ, Bur Bus Res, Univ Ala, 52-54; assoc prof statist & dir ctr res bus & econ, Ga State Univ, 54-56; prof econ & dir ctr res bus & econ, Univ Houston, 56-62; assoc prof mgt & dir bus studies, Ctr Res, Univ Mo, Columbia, 62-64; PROF MGT SCI, DIR DIV RES & DEAN RES & GRAD PROGS, PA STATE UNIV, UNIVERSITY PARK, 64- *Concurrent Pos:* Fulbright fel, Nat Univ Mex, 62; vpres, Assoc Univs Bus & Econ Res, 66-67, pres, 67-68; consult, Instituto Politecnicol Nacional, Mexico City, 81; pres, Centre Community Hosp Corp, 87- *Mem:* Am Statist Asn; Inst Mgt Sci; Am Inst Decision Sci. *Res:* Cost benefit analysis; problem solving and decision making; program analysis and evaluation; managerial economics. *Mailing Add:* Res & Grad Progs Smeal Col Bus Pa State Univ University Park PA 16802

RIGBY, PERRY G, b East Liverpool, Ohio, July 1, 32; m 57; c 4. INTERNAL MEDICINE, HEMATOLOGY. *Educ:* Mt Union Col, BS, 53; Western Reserve Univ, MD, 57; Am Bd Internal Med, dipl, 64. *Prof Exp:* Intern med, Univ Va Hosp, 58, resident, 60; clin asst, Boston City Hosp, 61-62; fel hemat, Mass Mem Hosp, 62; from asst prof to assoc prof, Univ Nebr Med Ctr,

Omaha, 64-69, prof anat, 69-74, dir hemat, 68-74, asst dean curric, 71-72, assoc dean acad affairs, 72-74, prof internal med, 69-78, prof med educ, chmn dept med & educ admin & dean, Col Med, 74-78; dean, 81-85, PROF MED & ASSOC DEAN ACAD AFFAIRS, LA STATE UNIV, SHREVEPORT, 78-, CHANCELLOR, 85- *Concurrent Pos:* Head hemat, Eugene C Eppley Inst, 64-68; Markle scholar acad med, 65. *Mem:* Am Chem Soc; Am Soc Hemat; Am Asn Cancer Res; fel Am Col Physicians; Int Soc Hemat. *Res:* Immunology; cancer biology; RNA metabolism. *Mailing Add:* La State Univ Med Ctr 433 Bolivar St New Orleans LA 70112-2223

RIGDEN, JOHN SAXBY, b Painesville, Ohio, Jan 10, 34; m 53; c 6. ACOUSTICS. *Educ:* Eastern Nazarene Col, BS, 56; Johns Hopkins Univ, PhD(phys chem), 60. *Prof Exp:* Res fel chem physics, Harvard Univ, 60-61; asst prof physics, Eastern Nazarene Col, 61-64, assoc prof & head dept, 64-67; assoc prof, Middlebury Col, 67-68; assoc prof, 68-74, PROF PHYSICS UNIV MO, ST LOUIS, 74- *Concurrent Pos:* Res assoc, Harvard Univ, 66-67; Nat Endowment Humanities grant, 70; US rep, Int Sci Exhib, Burma, 70; Fulbright fel, Burma, 71, Uruguay, 75; ed, Am J Physics, 78- *Mem:* Am Phys Soc; Am Asn Physics Teachers; Hist Sci Soc. *Res:* History and philosophy of science; teaching of science. *Mailing Add:* Dept Physics Univ Mo St Louis MO 63121

RIGDON, ORVILLE WAYNE, b Ashland, La, Aug 25, 32; m 55; c 2. ORGANIC CHEMISTRY, PETROLEUM CHEMISTRY. *Educ:* Northwestern State Col, La, BS, 58; Univ Va, PhD(org photo oxidation), 66. *Prof Exp:* SR PROJ CHEMIST, RES & TECH DEPT, TEXACO INC, PORT ARTHUR, 72- *Mem:* AAAS; Am Chem Soc. *Res:* Petrochemicals applied research; additives and commodity chemicals. *Mailing Add:* 4618 Fountainhead Dr Houston TX 77066

RIGDON, RAYMOND HARRISON, Musella, Ga, Jul 30, 05. NEOPLASM, EXPERIMENTAL INFLAMMATION. *Educ:* Emory Univ, MD, 31. *Prof Exp:* Prof Path, Med Br, Univ Tex, Galveston; RETIRED. *Mailing Add:* S Main St PO Box 545 Madison GA 30650

RIGDON, ROBERT DAVID, b Louisville, Ky, Dec 11, 42. TOPOLOGY. *Educ:* Princeton Univ, AB, 65; Univ Calif, Berkeley, PhD(math), 70. *Prof Exp:* Asst prof math, Northwestern Univ, 70-72; vis asst prof, Univ Ky, 72-73; lectr, Calif State Univ, Dominguez Hills, 73-75; asst prof, 75-80, ASSOC PROF MATH, IND UNIV-PURDUE UNIV, INDIANAPOLIS, 80- *Mem:* Sigma Xi; Am Math Soc. *Res:* Obstruction theory in algebraic topology. *Mailing Add:* 2822 Lake Forest Indianapolis IN 46268

RIGERT, JAMES ALOYSIUS, b Beaverton, Ore, Feb 13, 35. ROCK MECHANICS. *Educ:* Univ Portland, BS, 57; Cornell Univ, MS, 60; Univ Ill, PhD(physics), 72; Tex A&M Univ, PhD(geophys), 80. *Prof Exp:* Asst prof, 73-79, ASSOC PROF GEOPHYS, UNIV NOTRE DAME, 79- *Mem:* AAAS; Am Geophys Union; Geol Soc Am; Sigma Xi. *Res:* Deformation processes in rock; internal deformation, rock strength, and frictional properties of surfaces at high pressures. *Mailing Add:* Dept Earth Sci Univ Notre Dame Notre Dame IN 46556

RIGGI, STEPHEN JOSEPH, b Pa, Oct 11, 37; m 57; c 3. GENERAL MANAGEMENT. *Educ:* Univ Scranton, BS, 59; Univ Tenn, MS, 61, PhD(psysiol), 63. *Prof Exp:* Group leader, Pennwalt Pharmaceut Div, Lederle Labs, 63-74, dir pharmacol, 74-76, dir biol res, 76-81, vpres res & develop, 81-85, exec vpres, 85-86, pres, 86-88; PRES & CHIEF EXEC OFFICER, TELOR OPHTHAL PHARMACEUT, INC, 89- *Mem:* Am Soc Pharmacol & Exp Therapeut; Asn Res Vision & Ophthal. *Res:* Development of pharmaceuticals for treatment of age-related diseases of the eye. *Mailing Add:* Telor Ophthal Pharmaceut Inc 500 W Cummings Park Suite 6950 Woburn MA 10801

RIGGLE, EVERETT C, b Spokane, Wash, Aug 4, 32; m 52; c 3. NUMERICAL ANALYSIS. *Educ:* Eastern Wash State Col, AB, 52; Ore State Univ, MS, 58. *Prof Exp:* PROF MATH, CALIF STATE UNIV, CHICO, 58-, CHMN DEPT, 73- *Concurrent Pos:* NSF fac fel. *Mem:* Asn Comput Mach; Math Asn Am. *Res:* Gradient methods for the solution of linear systems; Tchcbycheff approximation. *Mailing Add:* Dept Math Calif State Univ Chico CA 95929

RIGGLE, J(OHN) W(EBSTER), b Painesville, Ohio, Feb 6, 24; m 49. CHEMICAL & ELECTRICAL ENGINEERING. *Educ:* Carnegie Inst Technol, BSc, 45; Univ Mich, MS, 46; Univ Del, MSc, 61. *Prof Exp:* Res engr, E I du Pont de Nemours & Co, Inc, Wilmington, 46-64, staff engr, Tenn, 64-69, sr res engr, WVa, 69-73, process engr, Design Div, Del, 73-76, sr develop engr, 76-78, proj coordr, photo prod dept, NY, 78-82; RETIRED. *Mem:* Sigma Xi. *Res:* Fundamental engineering properties of materials; high vacuum fluid dynamics; mass transfer operations; photoelectric analysis; process dynamics; high polymer technology; water gel explosives technology; nitric acid technology; electronic thick film technology. *Mailing Add:* PO Box 370 Lewiston NY 14092

RIGGLE, JOHN H, b Avella, Pa, May 28, 26; m 51; c 2. MATHEMATICS. *Educ:* Washington & Jefferson Col, BA, 50; Univ Pittsburgh, MLitt, 52; Cent Mich Univ, MA, 64. *Prof Exp:* Teacher high sch, Pa, 50-64; ASSOC PROF MATH, CALIFORNIA STATE COL, PA, 64- *Mailing Add:* Dept Math Calif State Col Third St California PA 15419

RIGGLE, TIMOTHY A, b Coshocton, Ohio, Dec 24, 40; div; c 4. MATHEMATICS. *Educ:* Wittenberg Univ, AB, 62, MEd, 66; Ohio State Univ, PhD(math educ), 68; Case Western Res Univ, MS, 76. *Prof Exp:* Instr math, Lima Campus, Ohio State Univ, 66-68; assoc prof, 68-76, PROF MATH & GRAD FAC, BALDWIN-WALLACE COL, 76-, PROF & DEPT HEAD, MATH & COMPUTER SCI. *Mem:* Math Asn Am; Asn Comput Mach. *Res:* Mathematics education; operations research. *Mailing Add:* Dept Math Baldwin-Wallace Col 275 Eastland Rd Berea OH 44107

RIGGLEMAN, JAMES DALE, b Washington, DC, Feb 6, 33; m 54; c 2. HORTICULTURE, PLANT PHYSIOLOGY. *Educ:* Univ Md, BS, 55, MS, 61, PhD(hort, plant physiol), 64. *Prof Exp:* Res asst hort, Univ Md, 57-64; res biologist, Plant Res Lab, 64-67, sr sales res biologist, 67-72, prod develop mgr, 72-78, prod develop coordr, 78-79, mgr field stas, 79, mgr new prod develop, 79-81, mgr licensing & univ relations, 81-84, MGR LICENSING, ASIA PAC, E I DU PONT DE NEMOURS & CO, INC, 84- *Concurrent Pos:* Mem, N Cent Weed Control Conf; Coun Agr Sci & Technol. *Mem:* Fel Weed Sci Soc Am (pres); Sigma Xi; Int Weed Sci Soc; Weed Sci Soc Japan; Southern Weed Sci Soc; Northeastern Weed Sci Soc. *Res:* Experimental insecticides, fungicides and herbicides. *Mailing Add:* Du Pont Agr Prod Barley Mill Plaza Walkers Mill Bldg PO Box 80038 Wilmington DE 19880-0038

RIGGS, ARTHUR DALE, b Modesto, Calif, Aug 8, 39; m 60; c 3. MOLECULAR BIOLOGY. *Educ:* Univ Calif, Riverside, AB, 61; Calif Inst Technol, PhD(biochem), 66. *Prof Exp:* USPHS fel, Salk Inst Biol Studies, Calif, 66-69; SR RES SCIENTIST, CITY OF HOPE MED CTR, 69-, CHMN, BIOL DIV, 81- *Mem:* AAAS. *Res:* Chromosome structure; gene regulation; X chromosome inactivation; DNA methylation. *Mailing Add:* City Hope Med Ctr 1500 Duarte Rd Duarte CA 91010

RIGGS, AUSTEN FOX, II, b New York, NY, Nov 11, 24; m 52; c 3. BIOCHEMISTRY. *Educ:* Harvard Univ, AB, 48, AM, 49, PhD, 52. *Prof Exp:* Instr biol, Harvard Univ, 53-56; from asst prof to assoc prof zool, 56-65, PROF ZOOL, UNIV TEX, AUSTIN, 65- *Mem:* Sigma Xi. *Res:* Biochemistry of proteins. *Mailing Add:* Dept Zool Univ Tex Austin TX 78712

RIGGS, BENJAMIN C, b Stockbridge, Mass, May 11, 14; m 67; c 4. PSYCHIATRY, PSYCHOANALYSIS. *Educ:* Harvard Univ, AB, 36; Columbia Univ, MD, 40; Am Bd Psychiat & Neurol, cert psychiat, 51; Boston Psychoanal Inst, cert psychoanalyst, 64. *Prof Exp:* Intern med & surg, Bellevue Hosp, 40-41; resident gastroent, Hosp Univ Pa, 41-43; instr biochem, Univ Pa, 43-44, assoc, 44-45; resident psychiat, Baldpate Hosp, Mass, 45-47; resident, Metrop State Hosp, 47-48, asst supt, 49-50, clin dir, Adult Outpatient Clin, 49-50; asst instr, Dept Psychiat, Harvard Univ, 53-58, clin assoc, 58-66; asst clin prof, Emory Univ, 66-70; assoc clin prof, 68-70, actg chmn dept, 74-75, PROF, MED UNIV SC, 70-, EMER PROF PSYCHIAT, 86- *Concurrent Pos:* Intern clin path, Germantown Hosp, 41; intern obstet & gynec, Hosp Univ Pa, 42; fel gastroent, Univ Pa, 42-43; resident psychiat, Mass Gen Hosp, 46-47; attend, Vet Admin Hosp, Charleston, 70-74, consult, 74- *Mem:* AAAS; Am Psychoanal Soc; Int Psychoanal Soc; fel Am Psychiat Asn; Am Acad Psychoanal. *Res:* Biochemistry of oxygen poisoning; general systems theory in psychiatry; analysis of interpersonal process. *Mailing Add:* Dept Psychiat & Behav Sci Med Univ SC 171 Ashley Ave Charleston SC 29425

RIGGS, BYRON LAWRENCE, b Hot Springs, Ark, Mar 24, 31; m 55; c 2. INTERNAL MEDICINE. *Educ:* Univ Ark, BS, 51, MD, 55; Univ Minn, MS, 62. *Prof Exp:* Consult internal med, Mayo Clin, 62; from instr to prof med, Mayo Med Sch, Univ Minn, 62-85, chmn, Div Endocrinol, 74-85; RETIRED. *Concurrent Pos:* Royal Soc Med traveling fel, 73. *Mem:* Am Fedn Clin Res; Endocrine Soc; Am Col Physicians; Am Soc Clin Invest; Am Soc Bone & Mineral Res (pres, 85-86); Asn Am Physicians. *Res:* Bone and calcium metabolism. *Mailing Add:* 432 Tenth Ave SW Rochester MN 55901

RIGGS, CARL DANIEL, b Indianapolis, Ind, Dec 7, 20; m 54; c 4. ZOOLOGY. *Educ:* Univ Mich, BS, 44, MS, 46, PhD(zool), 53. *Prof Exp:* Asst, Univ Mich, 44-45, fel, 45-47; instr zool, Univ Okla, 48-49, asst prof & actg dir, Okla Biol Surv, 49-54, dir, 54-70, from assoc prof to prof zool, 54-71, dir, Univ Biol Sta, 50-69, cur, Mus Zool, 54-66, dean, Grad Col, 65-71, vpres grad studies, 66-71, actg provost, 70-71; vpres acad affairs & prof biol, 71-80, DEAN, GRAD SCH & COORDR UNIV RES, UNIV SFLA, 80- *Concurrent Pos:* Consult, Coun Grad Schs, 65- *Mem:* AAAS; Am Fisheries Soc; Am Soc Ichthyologists & Herpetologists; Sigma Xi. *Res:* Taxonomy, natural history and distribution of North American fresh water fishes. *Mailing Add:* Dir Math & Sci & Comput Univ SFla Tampa FL 33620

RIGGS, CHARLES LATHAN, b Bearden, Ark, Aug 13, 23; m 51. MATHEMATICS. *Educ:* Tex Christian Univ, BA, 44; Univ Mich, MA, 45; Univ Ky, PhD(math, statist), 49. *Prof Exp:* Instr math, Univ Ky, 46-49; asst prof, Kent State Univ, 49-51; asst prof, East Tex State Univ, 51-53; assoc prof, 53-60, PROF MATH, TEX TECH UNIV, 60- *Concurrent Pos:* Consult, NSF/USAID on assignment to India, 67 & 68. *Mem:* Am Math Soc; Math Asn Am. *Res:* Mathematical statistics; probability. *Mailing Add:* 3805 61st St Lubbock TX 79413

RIGGS, CHARLES LEE, b Clayton, NMex, Aug 21, 46; m 67; c 2. TEXTILES & DETERGENTS. *Educ:* Southwestern Okla State Univ, BS, 67; Okla State Univ, PhD(chem), 74. *Prof Exp:* Asst prof & detergency res coordr, 74-79, assoc prof, 79-84, PROF, TEX WOMAN'S UNIV RES INST, 84- *Mem:* Am Chem Soc; Am Asn Textile Chemists & Colorists; Sigma Xi; Am Oil Chemists Soc. *Res:* Interactions between surfactants, alkalis and other detergent components and auxiliaries with textile fibers and finishes. *Mailing Add:* PO Box 22509 Denton TX 76204-0509

RIGGS, DIXON L, b St Mary's, WVa, June 25, 24; m 52; c 2. HUMAN PHYSIOLOGY. *Educ:* Marietta Col, AB, 49; Univ Mich, MS, 50. *Prof Exp:* Res assoc ecol, Inst Human Biol, Univ Mich, 50; asst prof biol, Simpson Col, 51; assoc prof, Huron Col, 52-58; ASSOC PROF BIOL, UNIV NORTHERN IOWA, 58- *Mem:* AAAS; Nat Sci Teachers Asn; Am Soc Mammal. *Res:* Mammalian physiology and ecology; alcohol, particularly the ingestion, metabolism and effects on the body. *Mailing Add:* Dept Biol Univ Northern Iowa Cedar Falls IA 50614

RIGGS, HAMMOND GREENWALD, JR, b Drumright, Okla, July 30, 31; m 59; c 2. MICROBIOLOGY. *Educ:* Okla State Univ, BA, 55, MS, 65; Univ Tex Southwestern Med Sch Dallas, PhD(microbiol), 69. *Prof Exp:* Res technician & chemist oil prod anal, Jersey Prod Res Co, Okla, 55-62; FROM

ASST PROF TO ASSOC PROF MED MICROBIOL MED SCH, UNIV MO, COLUMBIA, 68- *Concurrent Pos:* Univ Mo assoc prof advan fel diag virol, Yale Univ, 69; vis prof, King Faisal Univ Col Med, Damman, Saudi Arabia, 83-84, Univ Autonoma Guadalajara, Mex, 87 & 88, Ross Univ Sch Vet Med, St Kitt's, WI, 88. *Honors & Awards:* O B Williams Award, Am Soc Microbiol. *Mem:* Am Soc Microbiol. *Res:* Bacterial genetics concerned with regulation of biosynthesis of cell wall and virulence factors; diagnostic medical microbiology. *Mailing Add:* Dept Microbiol M643 Med Sci Bldg Univ Mo Med Ctr Columbia MO 65212

RIGGS, JAMES W, JR, b Houston, Tex, Mar 15, 14; m 40; c 2. PHYSICS. *Educ:* Loma Linda Univ, BA, 47; Tex A&M Univ, MS, 53, PhD(physics), 58. *Prof Exp:* From instr to asst prof math & physics, Loma Linda Univ, 47-53, from assoc prof to prof physics, 53-82, head dept, 59-82, EMER PROF PHYSICS, LOMA LINDA UNIV, 82- *Mem:* Am Asn Physics Teachers; Optical Soc Am; Am Phys Soc; Am Geophys Union. *Res:* Molecular physics and spectroscopy. *Mailing Add:* Star Rte 759 Sourdough Rd West Point CA 95255

RIGGS, JOHN L, b Kingman, Kans, Nov 26, 26; m 51; c 2. VIROLOGY. *Educ:* Univ Calif, Berkeley, AB, 54; Univ Kans, MA, 57, PhD(bact), 59. *Prof Exp:* Microbiol trainee virol, Univ Mich, 59-61, asst prof epidemiol, 61-62; VIROLOGIST, VIRAL & RICKETTSIAL DIS LAB, STATE DEPT HEALTH, CALIF, 62- *Mem:* Am Soc Microbiol; Am Asn Cancer Res; Am Asn Immunol; Soc Exp Biol & Med. *Res:* Application of fluorescent antibody techniques to virology; oncogenic viruses. *Mailing Add:* 12982 Polvera Ct San Diego CA 92128

RIGGS, KARL A, JR, b Thomasville, Ga, Aug 12, 29; m 52; c 3. ECONOMIC GEOLOGY. *Educ:* Mich State Univ, BS, 51, MS, 52; Iowa State Univ, PhD(geol), 56. *Prof Exp:* Instr geol, Iowa State Univ, 52-56; sr res geologist, Mobil Field Res Lab, 56-59; consult geol, Tex, 59-66; asst prof, Western Mich Univ, 66-68; ASST PROF GEOL, MISS STATE UNIV, 68- *Concurrent Pos:* Consult, numerous firms, 52-91. *Mem:* Mineral Soc Am; Soc Sedimentary Geol; fel Geol Soc Am; Am Inst Prof Geologists; Asn Eng Geologists; Am Inst Mining, Metall & Petrol Engs. *Res:* Rock classification; mineralogy and petrology of serpentinite; carbonates; Pleistocene till and loess, stratigraphy and sedimentation; strategic minerals. *Mailing Add:* Dept Geol Miss State Univ Mississippi State MS 39762-5857

RIGGS, LORRIN ANDREWS, b Harput, Turkey, June 11, 12; US citizen; m 37; c 2. PHYSIOLOGICAL PSYCHOLOGY. *Educ:* Dartmouth Col, AB, 33; Clark Univ, AM, 34, PhD(psychol), 36. *Prof Exp:* Nat Res Coun fel biol sci, Johnson Found Med Physics, Univ Pa, 36-37; instr psychol, Univ Vt, 37-38 & 39-41; res assoc, Brown Univ, 38-39; res assoc, 41-45, from asst prof to prof psychol, 45-68, Edgar J Marston prof, 68-77, EMER PROF PSYCHOL, BROWN UNIV, 77- *Concurrent Pos:* Assoc ed, J Optical Soc Am, Vision Res & Sensory Processes, 62-71; Guggenheim fel, Cambridge Univ, 71-72; pres, Eastern Psychol Asn, 75-76; William James fel, Am Psychol Soc, 89. *Honors & Awards:* Warren Medal, 56; Friedenwald Award, 66; Edgar D. Tillyer Award, Optical Soc Am, 69, Ives Medal, 82; Prentice Medal, Am Acad Optom, 73; Distinguished Sci Contrib Award, Am Psychol Asn, 74; Kenneth Craik Award, St Johns Col, Cambridge Univ, 79. *Mem:* Nat Acad Sci; AAAS (vpres, 64); Asn Res Vision & Ophthal (pres, 77); Optical Soc Am; Soc Exp Psychol; Am Acad Arts & Sci. *Res:* Human vision; vision in animals. *Mailing Add:* Nine Diman Pl Providence RI 02906-2103

RIGGS, LOUIS WILLIAM, b Pearsau, Tex, June 29, 22; m 46; c 2. ENGINEERING. *Educ:* Univ Calif, Berkeley, BS, 48. *Prof Exp:* Civil engr, Div Bay Toll Xing, State Calif, 48-51; proj engr, Tudor Eng Co, 51-61, vpres, 61-63, pres, 63-84, chmn, 84-86, consult, 86-88; RETIRED. *Concurrent Pos:* Chmn, Building Futures Coun, 84-88. *Honors & Awards:* A P Greensfelder Award, Am Soc Civil Engrs, 67. *Mem:* Nat Acad Eng; Am Pub Works Asn; Soc Mil Engrs (vpres & pres, 80-82); Am Consult Engrs Coun (vpres, 79-81); Am Soc Civil Engrs. *Res:* Consulting civil engineering; application of physical and engineering principals to the design of bridges, highways and transit structures. *Mailing Add:* 3682 Happy Valley Rd Lafayette CA 94549

RIGGS, OLEN LONNIE, JR, b Bethany, Okla, Aug 25, 25; m 47; c 2. CORROSION, ELECTROCHEMISTRY. *Educ:* Eastern Nazarene Col, BS, 49. *Prof Exp:* Group supvr corrosion sci, Continental Oil Co, 52-68; res dir mat eng, Koch Industs, Inc, 68-69; sr res assoc mat, Getty Oil Co, 69-70; SR STAFF CHEMIST ELECTRO CHEM, KERR-MCGEE CORP, 71- *Concurrent Pos:* Mem adv bd, Univ Okla Continuing Educ Div, 73- *Mem:* Electrochem Soc; Nat Asn Corrosion Engrs; fel Am Inst Chemists; NY Acad Sci; Sigma Xi. *Res:* New electrolytic processes based on modern concepts; metal/solution interfaces, especially corrosion process and its control; design of electrolytic cells and their component parts; author of 50 papers. *Mailing Add:* Olran Corp 1216 Anita Ave Oklahoma City OK 73127

RIGGS, PHILIP SHAEFER, b Chicago, Ill, May 30, 06; m 39; c 2. ASTRONOMY. *Educ:* Carnegie Inst Technol, BS, 27; Univ Calif, PhD(astron), 44. *Prof Exp:* Asst prof astron, Washburn Univ, 37-38, from asst prof to prof physics & astron, 39-47; instr astron, Univ Ill, 38-39; from assoc prof to prof, 47-76, EMER PROF ASTRON, DRAKE UNIV, 76- *Concurrent Pos:* NSF sci fac fel, Univ Calif, 57-58; vis prof, Univ Mich, 54, Univ Ill, 60 & Univ Iowa, 64. *Mem:* Am Astron Soc; Sigma Xi. *Res:* Stellar and galactic astronomy. *Mailing Add:* 5921 Winwood Dr No 204 Johnston IA 50131

RIGGS, RICHARD, b Polo, Ill, Oct 8, 38; m 61; c 1. MATHEMATICS EDUCATION. *Educ:* Knox Col, Ill, AB, 60; Rutgers Univ, MA, 64, EdD(math educ), 68; Stevens Inst Technol, MS, 85. *Prof Exp:* High sch teacher, Mich, 60-61 & Tex, 61-63; from instr to assoc prof, 64-78, PROF MATH, JERSEY CITY STATE COL, 78- *Mem:* Math Asn Am. *Res:* Mathematics education at the secondary and undergraduate level. *Mailing Add:* Dept Math Jersey City State Col Jersey City NJ 07305

RIGGS, ROBERT D, b Pocahontas, Ark, June 15, 32; m 54; c 4. PHYTONEMATOLOGY. *Educ:* Univ Ark, BSA, 54, MS, 56; NC State Col, PhD(plant path), 58. *Prof Exp:* Asst plant path, Univ Ark, 54-55; asst, NC State Col, 55-58; from asst prof to assoc prof, 58-68, PROF PLANT PATH, UNIV ARK, FAYETTEVILLE, 68- *Mem:* Soc Nematologists; Sigma Xi. *Res:* Plant parasitic nematodes; variability, control and host range. *Mailing Add:* Dept Plant Path/217 Plant Sci Univ Ark Fayetteville AR 72701

RIGGS, RODERICK D, b Racine, Wis, Apr 15, 31; m 55; c 3. NUCLEAR PHYSICS. *Educ:* Dubuque Univ, BS, 55; Iowa State Univ, MS, 57; Mich State Univ, PhD, 71. *Prof Exp:* Instr physics & chem, Jackson Community Col, 58-61, dean men, 61-62, dean students, 62-65, chmn dept physics, 65-69, prof physics & eng & head dept, 69-87, prof physics, 73-89; PROF PHYSICS, SPRING ARBOR COL, MICH, 89- *Concurrent Pos:* Lectr & consult physics curriculum, Spring Arbor Col, 65-69, adj prof, 84-89; consult, Sci Assocs, 65-83. *Mem:* Am Asn Physics Teachers. *Res:* Theory and development of training programs in nuclear reactor technology; physics curriculum development; comparative European science education; low energy gamma ray spectroscopy. *Mailing Add:* 2605 S St Anthony Dr Jackson MI 49203

RIGGS, SCHULTZ, b Owensboro, Ky, Feb 10, 41. MATHEMATICS. *Educ:* Univ Ky, BS, 62, MS, 64, PhD(math), 70. *Prof Exp:* Instr math, Western Ky Univ, 67-69; ASST PROF MATH, JACKSON STATE COL, 71- *Mem:* Am Math Soc; Math Asn Am. *Res:* Analysis. *Mailing Add:* Dept Math Jackson State Col 1400 Lynch St Jackson MS 39217

RIGGS, STANLEY R, b Watertown, Wis, May 20, 38; m 60; c 2. GEOLOGY. *Educ:* Beloit Col, BS, 60; Dartmouth Col, MA, 62; Univ Mont, PhD(geol), 67. *Prof Exp:* Res & explor geologist, Int Minerals & Chem Corp, 62-67; from asst prof to assoc prof, 67-77, PROF GEOL, E CAROLINA UNIV, 77- *Mem:* Geol Soc Am; Soc Econ Paleontologists & Mineralogists; Sigma Si. *Res:* Modern nearshore and estuarine sediment studies in the southeastern United States; interpretation of the Atlantic Coastal Plain stratigraphy and sedimentary petrology. *Mailing Add:* Dept Geol ECarolina Univ PO Box 2751 Greenville NC 27834

RIGGS, STUART, b Port Arthur, Tex, Sept 23, 28; c 3. MEDICINE, MICROBIOLOGY. *Educ:* Rice Inst, 47-49; Univ Tex, MD, 53; Am Bd Internal Med, dipl, 61, cert, 74, cert infectious dis, 72. *Prof Exp:* Intern, Univ Iowa, 53-54, resident internal med, Univ Hosps, 54-57; instr med, Univ Tex Med Br Galveston, 59-61; res fel infectious dis, Univ Tex Southwestern Med Sch Dallas, 61-62; from asst prof to assoc prof, 62-69, clin assoc prof 69-84, ASST PROF MICROBIOL, BAYLOR COL MED, 66-, CLIN PROF MED, 84- *Concurrent Pos:* Attend physician, John Sealy Hosp, Galveston, Tex, 59-61 & Parkland Mem Hosp, Dallas, 61-62; attend physician, Ben Taub Gen Hosp, Houston, 62-69, head sect infectious dis med serv & supvr clin microbiol lab, 66-69; consult, Vet Admin Hosp, Houston, 64-66; attend physician internal med, Methodist Hosp, Tex, 67-; attend physician internal med & chief infectious dis, St Luke's Episcopal Hosp, 73-85. *Mem:* AAAS; Am Fedn Clin Res; Am Soc Microbiol; fel Am Col Physicians; AMA. *Res:* Viral respiratory and central nervous system infections; mycoplasma serology; infectious diseases. *Mailing Add:* 6624 Fannin Houston TX 77030

RIGGS, THOMAS ROWLAND, b Dallas, Ore, Oct 30, 21; m 58; c 2. BIOCHEMISTRY, NUTRITION. *Educ:* Ore State Col, BS, 44, MS, 45; Tufts Univ, PhD(biochem, nutrit), 50. *Prof Exp:* Instr biochem, Tufts Univ, 49-50, from instr to asst prof biochem & nutrit, Med & Dent Schs, 50-55; from asst prof to prof, 55-86, EMER PROF BIOL CHEM, UNIV MICH, ANN ARBOR, 86- *Mem:* Am Soc Biol Chemists; Am Inst Nutrit. *Res:* Amino acid transport, especially as altered by hormones and nutritional factors. *Mailing Add:* 11906 Dunmore Ann Arbor MI 48103

RIGGS, VICTORIA G, b Feb 14, 56; US citizen. ELECTRICAL ENGINEERING. *Educ:* State Univ NY, Binghamton, BS, 82; Rutgers State Univ, MS, 88. *Prof Exp:* MEM TECH STAFF, AT&T BELL LABS, 82- *Mem:* Inst Elec & Electronics Engrs; Asn Women Sci; Am Soc Qual Control. *Res:* Lightwave semiconductor research and development and manufacturing. *Mailing Add:* AT&T Bell Labs Rm 6F-415 600 Mountain Ave Murray Hill NJ 07974-0636

RIGGSBY, ERNEST DUWARD, b Nashville, Tenn, June 12, 25; m 62; c 1. SCIENCE EDUCATION, SCIENCE WRITING. *Educ:* Tenn Polytech Inst, BS, 48; George Peabody Col, BA, 55, MA, 56, EdS, 58, EdD, 64. *Prof Exp:* High sch teacher, 53-54; teacher math & sci, Univ of the South, 54-55; from instr to prof phys sci & sci educ, Troy State Col, 55-67; vis prof sci educ, Auburn Univ, 67-69; PROF SCI EDUC & PHYS SCI, COLUMBUS COL, 69- *Concurrent Pos:* Mem, Nat Aerospace Educ Adv Comt, 60-; educ consult, Ark Proj, Int Paper Co, 61 & US Steel Corp, 62; vis scientist, Ala Acad Sci & NSF Coop High Sch-Col Sci Proj, 64-66; spec consult, Ala Proj, US Off Educ, 65-66; vis prof, Fla Inst Technol, 70-74; vis grad prof, Univ PR, 75-76. *Honors & Awards:* Gen Aviation Mfg Asn Cert Merit in Aerospace Sci, 75. *Mem:* Fel AAAS; Nat Asn Res Sci. *Res:* Philosophy of science and scientific methodology; programmed instruction for use in science education; science teacher education through newer media; dimensional analysis as applied to teacher education in science; aerospace science. *Mailing Add:* Sch Educ Columbus Col Columbus GA 31993-2399

RIGGSBY, WILLIAM STUART, b Ashland, Ky, July 25, 36; m 63; c 3. MOLECULAR BIOLOGY, BIOPHYSICS. *Educ:* George Washington Univ, AB, 58, Yale Univ, MS, 60, PhD(molecular biol), 64. *Prof Exp:* USPHS fel nucleic acids, Oak Ridge Nat Lab, 65-68, biochemist, 68-69; from asst prof to assoc prof, 69-79, PROF MICROBIOL, UNIV TENN, KNOXVILLE, 79- *Concurrent Pos:* NIH career develop award, 72-79; vis prof molecular genetics, Univ PR, 81- *Mem:* AAAS; Am Soc Microbiol; Genetics Soc Am; Soc Study Evolution. *Res:* Nucleic acid sequence homology; yeast evolution; control of RNA synthesis in yeast morphogenesis; DNA probes for clinical identification. *Mailing Add:* Dept Microbiol Univ Tenn Knoxville TN 37916

RIGHTHAND, VERA FAY, b Pittsfield, Mass, Sept 4, 30. VIROLOGY, MICROBIAL BIOCHEMISTRY. *Educ:* Univ Rochester, BA, 52; Rutgers Univ, PhD(microbiol), 63. *Prof Exp:* Jr biologist, Am Cyanamid Co, 52-55; res asst virol, Rockefeller Inst Med Res, 55-59; from instr to asst prof, State Univ NY Buffalo, 63-68; asst prof, 68-74, ASSOC PROF VIROL, SCH MED, WAYNE STATE UNIV, 74- *Mem:* AAAS; Am Soc Microbiol; Sigma Xi; Am Soc Virol; fel Am Acad Microbiol. *Res:* Host cell-virus interrelationship; picornaviruses; measles virus; virus plaque mutants; oncogenic viruses; SV40 and Rous sarcoma virus studies; biochemical replication of viruses; factors influencing cell susceptibility to virus infections; persistent infections by normally cytocidal viruses. *Mailing Add:* Dept Immunol & Microbiol Wayne State Univ Sch Med Detroit MI 48201

RIGHTMIRE, GEORGE PHILIP, b Boston, Mass, Sept 15, 42; m 66; c 2. PALEO-ANTHROPOLOGY, SKELETAL BIOLOGY. *Educ:* Harvard Col, AB, 64; Univ Wis, MS, 66, PhD(human biol), 69. *Prof Exp:* Asst prof, 69-73, chmn dept, 76-78, assoc prof, 73-82, PROF ANTHROP, STATE UNIV NY, BINGHAMTON, 82- *Concurrent Pos:* Nat Inst Gen Med Sci spec res fel, Osteological Res Lab, Univ Stockholm, 73; vis scientist, Archeol Dept, Univ Cape Town, 75-76; assoc ed, Am J Phys Anthrop, 83-88, J Human Evol, 86-87; Contrib ed, Quart Rev Archaeol, 87-88, Rev Archaeol, 89- *Mem:* Am Asn Phys Anthrop; Human Biol Coun; Sigma Xi; Soc Syst Zool; fel AAAS. *Res:* Biometric studies of recent human populations in Africa; early humans in Africa and Asia; statistical methods in physical anthropology; hominid paleontology. *Mailing Add:* Dept Anthrop State Univ NY Binghamton NY 13902-6000

RIGHTMIRE, ROBERT, b Bedford, Ohio, Sept 28, 31; m 58; c 4. NUCLEAR CHEMISTRY. *Educ:* Hiram Col, BA, 53; Carnegie Inst Technol, MS, 56, PhD(chem), 57. *Prof Exp:* Tech specialist & electrochemist, Standard Oil Co, Ohio, 57-61, systs coordr, 61-64, res supvr, 64-69, mgr electrokinetics div, 69-71, petrol prod develop, 71-73, mgr petrol res & develop, 76-81, dir corp res, 81-85; PRES, SYTRAN ASN, 85- *Concurrent Pos:* Vis prof, Case Western Reserve Univ, 85. *Mem:* Am Chem Soc; Am Petrol Inst. *Res:* Energy conversion; battery research; coal and shale oil conversion process research; biochemistry; photochemistry; photovoltaic materials research; petroleum fuels and lubricants; geochemistry; catalysis research; surface science. *Mailing Add:* 220 Hiram College Dr Northfield OH 44067-2417

RIGHTSEL, WILTON ADAIR, b Terre Haute, Ind, July 21, 21; m 46; c 2. BACTERIOLOGY. *Educ:* Ind Univ, AB, 42; Ind State Teachers Col, MS, 47; Univ Cincinnati, PhD(bact), 51; Am Bd Med Microbiol, dipl. *Prof Exp:* Med technologist, St Anthony's Hosp, Terre Haute, 46-47; lab asst bact, Univ Cincinnati, 48-51; med bacteriologist, Biol Labs, Chem Corps, Camp Detrick, 51-52; assoc res virologist, Parke, Davis & Co, 52-53, res virologist, 53-58, sr res virologist, 58-61, lab dir virol, 62-63, dir virol, 63-66; ASSOC PROF MICROBIOL, MED UNITS, UNIV TENN, MEMPHIS, 66-; TECH DIR MICROBIOL, BAPTIST MEM HOSP, MEMPHIS, 66-, CLIN ASSOC PROF MICROBIOL, 76- *Mem:* Am Soc Microbiol; Am Soc Clin Path; Tissue Cult Asn; Brit Soc Gen Microbiol; fel Am Acad Microbiol; Signa Xi. *Res:* Medical bacteriology and immunology; tularemia; virology; chemotherapy of virus diseases and neoplasms; tissue culture; application of cell cultures in viruses and cancer. *Mailing Add:* 5886 Brierhedge Ave Memphis TN 38119

RIGLER, A KELLAM, b Lincoln, Nebr, June 8, 29; m 52; c 4. NUMERICAL ANALYSIS, OPTICS. *Educ:* Simpson Col, BA, 50; Univ Nebr, MA, 52; Univ Pittsburgh, PhD(math), 62. *Prof Exp:* Comput analyst, Douglas Aircraft Corp, 52-54; mathematician, Sandia Corp, 54-55; sr mathematician, Westinghouse Elec Corp, 59-69; PROF COMPUT SCI, UNIV MO, ROLLA, 69- *Concurrent Pos:* Lectr, Univ Conn, 58-59; sr lectr, Carnegie-Mellon Univ, 63-69; consult, Westinghouse Elec Corp, 69-, Environ Res Inst Mich, 87- *Mem:* Soc Indust & Appl Math; Math Prog Soc; Neural Network Soc. *Res:* Nonlinear programming applied to engineering design. *Mailing Add:* Rte 6 Box 517 Rolla MO 65401

RIGLER, NEIL EDWARD, b Waco, Tex, Nov 2, 08; m 34; c 2. ENVIRONMENTAL CHEMISTRY. *Educ:* Trinity Univ, Tex, BS, 30; Univ Tex, MA, 32, PhD(org chem), 35. *Prof Exp:* From asst to instr, Univ Tex, 30-35; agent, Bur Plant Indust, USDA, 35-37; res chemist, E R Squibb & Sons, 37; assoc agronomist, Exp Sta, NC State Col, 37-38; plant physiologist, Exp Sta, Agr & Mech Col, Tex, 38-43; org chemist & group leader antibiotic res, Heyden Chem Corp, 43-53; dir process develop antibiotics, Fine Chem Div, Am Cyanamid Co, 53-55, sr res chemist, Lederle Labs Div, 55-60, head, Anal Develop Dept, 60-73; consult, Havens & Emerson, Inc, 73-82; RETIRED. *Mem:* Fel AAAS; fel Am Inst Chemists; Am Chem Soc; NY Acad Sci. *Res:* Isolation, purification, identification and determination of natural products; analytical and chemical process development; pollution analysis; industrial waste treatment. *Mailing Add:* 1111 David Brainerd Dr Jamesburg NJ 08831-1696

RIGNEY, CARL JENNINGS, b Port Arthur, Tex, July 28, 25; m 48; c 5. PHYSICS. *Educ:* Univ Louisville, BS, 47; Northwestern Univ, MS, 48, PhD, 51. *Prof Exp:* Asst prof physics, Southern Ill Univ, 50-51; asst prof, Northern Ill Univ, 51-56; prof, Stephen F Austin State Col, 56-57; head dept, 57-78, PROF PHYSICS, LAMAR UNIV, 57- *Concurrent Pos:* Consult, Gen Elec Co, Ill, 54-55. *Mem:* Am Phys Soc; Am Asn Physics Teachers. *Res:* Measurements for thermal conductivity. *Mailing Add:* Dept Physics Lamar Univ Beaumont TX 77710

RIGNEY, DAVID ARTHUR, b Waterbury, Conn, Aug 8, 38; m 65; c 2. PHYSICAL METALLURGY. *Educ:* Harvard Univ, AB, 60, SM, 62; Cornell Univ, PhD(mat sci, eng), 66. *Prof Exp:* Fac mem, 67-75, PROF METALL ENG, OHIO STATE UNIV, 75- *Concurrent Pos:* Vis researcher, Cambridge Univ, Eng, 81; deleg, US/China Bilateral meeting, 81, chmn, US/China Wear meeting, 83. *Mem:* Fel Am Soc Metals; Am Inst Mining, Metall & Petrol Engr. *Res:* Solidification; liquid metals; magnetic resonance; electromigration; friction and wear (materials aspects, deformation, microstructure). *Mailing Add:* Dept Mat Sci & Eng Ohio State Univ Columbus OH 43210-1179

RIGNEY, DAVID ROTH, b Carbondale, Ill, Dec 27, 50. BIOPHYSICS, MOLECULAR BIOLOGY. *Educ:* Univ Tex, Austin, BA, 72, PhD(physics), 78. *Prof Exp:* Res scientist biophysics, Inst Cancer Res, 78-80, biophysicist, 81-86; RES ASSOC, MASS INST TECHNOL, 86-; ASST PROF MED, HARVARD MED SCH, 86- *Concurrent Pos:* NIH trainee, 78-80; assoc dir, Arrhythmia & Bioeng. *Mem:* AAAS; Biophys Soc; Cell Kinetics Soc; Tissue Cult Asn. *Res:* Theoretical biology; cell physiology; biochemical stochastics; biostatistics; cardiovascular physiology; electro-optical instrumentation; bioengineering; animal physiology. *Mailing Add:* 60 Aldworth St Jamaica Plain MA 02130-2755

RIGNEY, JAMES ARTHUR, b Flushing, NY, July 12, 31; m 58; c 4. ORGANIC CHEMISTRY, BIOCHEMISTRY. *Educ:* Fordham Univ, BS, 53; Va Polytech Inst, MS, 59, PhD(org chem), 61. *Prof Exp:* Res chemist, Am Cyanamid Co, 53-57; res chemist, Esso Res Labs, 61-66; sr chemist, Enjay Chem Co, La, 66-67; from asst prof to assoc prof, 67-78, chmn dept, 72-77, PROF CHEM, UNIV PRINCE EDWARD ISLAND, 78- *Mem:* Am Chem Soc; NY Acad Sci. *Res:* Organosulfur chemistry; chemistry of marine plants; catalysis. *Mailing Add:* Dept Chem Univ Prince Edward Island Charlottetown PE C1A 4P3 Can

RIGNEY, MARY MARGARET, b Albany, Mo, Nov 10, 26. MEDICAL MICROBIOLOGY. *Educ:* Northwest Mo State Col, BS, 50; Univ Mo-Columbia, MS, 60, PhD(microbiol), 68. *Prof Exp:* From asst prof to assoc prof, 68-78, PROF BIOL, UNIV WIS-OSHKOSH, 78- *Mem:* AAAS; Am Soc Microbiol; Sigma Xi. *Res:* Growth characteristics of aeromonas species; isolation and characterization of endotoxins and hemolysins of aeromonas species; pathogenicity of aeromonas species for warm and cold-blooded animals. *Mailing Add:* Halsey Sci Univ Wis Oshkosh WI 54902

RIGOR, BENJAMIN MORALES, SR, b Rizal, Philippines, Oct 13, 36; US citizen; m 61; c 3. ANESTHESIOLOGY. *Educ:* Univ Philippines, BS, 57; Univ of the East, Manila, MD, 62; Am Bd Anesthesiol, dipl, 70. *Prof Exp:* Instr pharmacol, Univ of the East Med Sch, 62-63; res assoc, Univ Ky, 65-66, resident anesthesiol, Med Ctr, 66-68, asst prof, 68-69; assoc prof, 69-71, prof anesthesiol & chmn dept, Col Med, NJ, 71-74; prof anesthesiol & chmn dept, Med Sch, Univ Tex, Houston, 74-81, prof & med dir nurse anesthesiol educ, 76-81; PROF ANESTHESIOL & CHMN DEPT, SCH MED, UNIV LOUISVILLE, 81- *Concurrent Pos:* USPHS grant, Univ Ky, 63; Am Heart Asn grant, 66; chief obstet anesthesia, Naval Hosp, Portsmouth, Va, 71-73; consult anesthesiol, Vet Admin Hosp, East Orange, Newark Beth Israel Med Ctr & St Barnabas Med Ctr, NJ, 71-74; consult, M D Anderson Hosp & Univ Tex Cancer Systs, 74-81; chief anesthesia, Hermann-Univ Hosp, Houston, 74- & Univ Hosp, Louisville 81- *Mem:* Am Soc Anesthesiol; Int Anesthesia Res Soc; fel Am Col Anesthesiol; AMA; Acad Anesthesiol. *Res:* Biological transport of non-electrolytes and electrolytes in the blood brain barrier; clinical pharmacology of drugs; fluid and parenteral therapy; clinical anesthesia. *Mailing Add:* Dept Anesthesiol Univ Louisville Sch Med Health Sci Ctr Louisville KY 40292

RIGROD, WILLIAM W, b New York, NY, Mar 29, 13; m 39. OPTICAL RESONATORS, PHYSICAL OPTICS. *Educ:* Cooper Union, BS, 34; Cornell Univ, MS, 41; Polytech Inst Brooklyn, DEE, 50. *Prof Exp:* Res scientist, Ignatyev Res Inst, USSR, 34-35, All-Union Electrotech Inst, USSR, 35-39; develop engr, Westinghouse Elec Corp, NJ, 40-51; mem tech staff, Electronics Res Lab, Bell Labs, Inc, 51-77; consult, Los Alamos Sci Lab, 79-88; RETIRED. *Mem:* Sr mem Inst Elec & Electronics Eng; emer mem Sigma Xi; emer mem Optical Soc Am. *Res:* Microwave electronics; physics of electron beams and gaseous discharges; lasers; physical optics. *Mailing Add:* Rte 19 Box 91-T Sunlit Hills Santa Fe NM 87505

RIGSBY, GEORGE PIERCE, b Wichita Falls, Tex, Nov 19, 15; m 43, 68; c 2. GEOLOGY. *Educ:* Calif Inst Technol, BS, 48, MS, 50, PhD(geol), 53. *Prof Exp:* Res scientist, Snow, Ice & Permafrost Res Estab, CEngrs, US Army, 53-56; res scientist, USN Electronics Lab, 56-59; staff scientist, Arctic Inst NAm, 59-68; assoc prof geol, US Int Univ, Elliott Campus, 68-74; geologist, Geothermal Surv, Inc, 74-85; RETIRED. *Mem:* Geol Soc Am; Am Mineral Soc; Am Geophys Union; Glaciol Soc; Sigma Xi. *Res:* Glaciology; field and laboratory investigation of ice; arctic field research; mineralogy; petrology; petrography; geothermal field research. *Mailing Add:* 1542 Alcala Pl San Diego CA 92111

RIHA, WILLIAM E, JR, b New Brunswick, NJ, Sept 15, 43; m 66; c 2. MICROBIOLOGY. *Educ:* Rutgers Univ, New Brunswick, BS, 65, MS, 69, PhD(food sci), 72. *Prof Exp:* Res asst food sci, Rutgers Univ, New Brunswick, 65-72; food scientist, Hunt-Wesson Foods, Inc, 72-74, group leader, 74-76, sect head, 76; dir tech serv, Cadbury North Am, Peter Paul Cadbury, 76-78, mgr food technol, 78-80; with res & tech serv, Pepsico, Inc, 80-83, group mgr, US Prod Develop & dir, Int Prod Develop, 83-88; VPRES RES & DEVELOP, JOSEPH E SEAGRAM & SONS INC, 88- *Concurrent Pos:* Consult food indust, 68-72. *Mem:* Inst Food Technologists; Sigma Xi. *Res:* Food product development and research; food microbiology. *Mailing Add:* Joseph E Seagram & Sons Inc Three S Corporate Park Dr White Plains NY 10604

RIHM, ALEXANDER, JR, b New York, NY, May 18, 16; m 40; c 2. SANITARY ENGINEERING. *Educ:* NY Univ, BS, 36, MS, 39. *Prof Exp:* Mem field party, Brader Construct Corp, 37-39; dist sanit engr, NY State Dept Health, 39-44, water supply engr, 47-49, water pollution control engr, 49-51, chief radiol health & air sanit sect, 52-57, exec secy air pollution control bd, 57-66, asst comnr health, 66-70; dir div air resources, NY State Dept Environ Conserv, 70-76; CONSULT, 76- *Concurrent Pos:* Adj prof, Rensselaer Polytech Inst; mem Nat Air Qual Criteria Adv Comt. *Honors & Awards:* J Smith Griswold Award, Air Pollution Control Asn, 80. *Mem:* Am Pub Health Asn; hon mem Air Pollution Control Asn (pres, 73-74). *Res:* Air pollution control. *Mailing Add:* 28 Euclid Ave Delmar NY 12054

RIJKE, ARIE MARIE, b Velsen, Neth, Apr 6, 34; US citizen; m 74. BIOMATERIALS, POLYMER SCIENCE. *Educ:* State Univ Leiden, BS, 56, MS & PhD(phys chem), 61; State Univ NY, MS, 60; Univ Amsterdam, MD, 78. *Prof Exp:* Res officer surface chem, Nat Defense Res Orgn, Neth, 61-64; res officer polymer, Coun Sci & Indust Res, 64-66; lectr chem, Univ Cape Town, 66-67; res assoc polymer physics, Inst Molecular Biophys, Fla State Univ, 67-69; lectr, Univ Witwatersrand, 69-70; SR SCIENTIST MAT SCI, UNIV VA, 70- *Mem:* Am Chem Soc; Opers Res Soc; AAAS. *Res:* Implant tissue response; development of new biomate rials for soft and hard tissue replacement; dental composites. *Mailing Add:* Dept Mat Sci Sch Eng & Appl Sci Univ Va Charlottesville VA 22901

RIKANS, LORA ELIZABETH, b Grand Rapids, Mich, Feb 7, 40; m 62; c 2. BIOCHEMICAL PHARMACOLOGY. *Educ:* Mich State Univ, BS, 61, MS, 62; Univ Mich, Ann Arbor, PhD(pharmacol), 75. *Prof Exp:* Res assoc nutrit, Mich State Univ, 62-63; clin chemist, St Marys Hosp, Saginaw, Mich, 64-67; res asst pharmacol, Dow Chem Co, Midland, Mich, 67-69; fel, Univ Mich, 70-75, scholar pharmacol, 75-77; from asst prof to assoc prof, 77-90, PROF PHARMACOL, UNIV OKLA, 90- *Concurrent Pos:* Masua Hon Lectr, 89-90. *Mem:* Am Soc Pharmacol & Exp Therapeut; Sigma Xi; Int Soc Study Xenobiotics. *Res:* Aging modification of drug metabolism; aging modification of drug toxicity; mechanisms of chemically induced toxicity. *Mailing Add:* Dept Pharmacol PO Box 26901 Oklahoma City OK 73190

RIKE, PAUL MILLER, b Duquesne, Pa, Feb 6, 13; m 45. CARDIOLOGY, INTERNAL MEDICINE. *Educ:* Univ Pittsburgh, BS, 36, MD, 38; Thiel Col, DSc, 71. *Prof Exp:* ASST PROF MED, SCH MED, UNIV PITTSBURGH, 48- *Concurrent Pos:* Active staff, Magee Womens Hosp & Presby Univ Hosp; consult, Western Psychiat Inst & Clin. *Mem:* Am Heart Asn; fel Am Col Physicians; fel Am Col Cardiol; fel Am Col Angiol. *Mailing Add:* Magee Womens Hosp Forbes & Halket Sts Pittsburgh PA 15213

RIKER, DONALD KAY, b New York, NY, Oct 22, 45; m 65; c 2. PHARMACEUTICAL RESEARCH & DEVELOPMENT, NEUROPHARMACOLOGY. *Educ:* Univ Kans, BA, 69; Cornell Univ, PhD(neurobiol behav), 77. *Prof Exp:* Fel, Rockefeller Univ, 68-70; fel pharmacol, 76-79, res assoc, Yale Univ, 80-82; sr res investr, Richardson-Vicks Inc, 82-84, prin res investr, 84-86, asst dir, Appl Clin Res, 86-90, ASSOC DIR, CLIN DEVELOP, RICHARDSON-VICKS/PROCTER & GAMBLE, 90- *Concurrent Pos:* Mem, US Antarctic Res Exped, 68; res health scientist, Vet Admin, 80-81. *Honors & Awards:* US Antarctic Serv Medal, 68. *Mem:* AAAS; Soc Neurosci; Int Soc Neurochem; Sigma Xi; Am Soc Pharmacol & Exp Therapeut; Am Soc Neurochem. *Res:* Pharmaceutical research management; design and execution of clinical programs to evaluate drug safety and efficacy; over-the-counter development of new technology; respiratory drug development; Rx/OTC switch. *Mailing Add:* Richardson-Vicks USA Proctor & Gamble One Far Mill Crossing Shelton CT 06484

RIKER, WALTER FRANKLYN, JR, b Bronx, NY, Mar 8, 16; m 41; c 3. NEUROPHARMACOLOGY, CLINICAL PHARMACOLOGY. *Educ:* Columbia Univ, BS, 39, Cornell Univ, MD, 43. *Hon Degrees:* DSc, Med Col Ohio, 80. *Prof Exp:* Res fel pharmacol, Med Col, Cornell Univ, 41-44, from instr to prof, 44-83, instr med, 45-46, chmn pharmacol dept, 56-83, Revlon prof pharmacol, 79-86, EMER PROF PHARMACOL, MED COL, CORNELL UNIV, 86- *Concurrent Pos:* Traveling fel, Am Physiol Soc Cong Oxford, Eng, 47; vis prof, Univ Kans, 53-54; mem pharmacol comt, Nat Bd Med Exam, 56-59; mem pharmacol study sect, USPHS, 56-59, mem pharmacol training grant comt, 58-61, chmn, 61-63; mem adv coun, Nat Inst Gen Med Sci, 63; mem toxicol panel, President's Sci Adv Comt, 65; mem pharmacol-toxicol rev comt, NIH, 65-68; mem adv comts, Pharmaceut Mfrs Asn Found, 65-87; vis scientist, Roche Inst Molecular Biol, 71-72, adj mem, 72-75; mem adv coun, Nat Inst Environ Health Sci, 72-75; dir, Richardson Vicks Inc, 79-85; mem Unitarian Serv Med Exchange Prog, Japan, 56; assoc ed, J Pharmacol & Exp Therapeut, 50-57; med adv drugs, Nat Football League, 73-85; sci adv bd, Sterling Winthrop Res Inst, 73-76; chmn sci adv comt, Irma T Hirschal Trust, 73-83, mem consult comt, 73-; Sterling Drug Vis Prof, Cornell Univ Med Col, 79, mem bd overseers, 81-86; hon mem med staff NY Hosp, 80- *Honors & Awards:* John J Abel Prize, Am Pharmacol Soc, 51; Torald Sollmann Award, Am Soc Pharmacol & Exp Ther, 86; Oscar B Hunter Award in Clin Pharmacol, Am Soc Clin Pharmacol & Ther, 90. *Mem:* Fel AAAS; Am Acad Neurol; Am Soc Pharmacol & Exp Therapeut; Sigma Xi; Am Soc Clin Pharmacol & Therapeut; Am Col Clin Pharmacol; Soc Neurosci; hon fel Am Col Clin Pharmacol. *Res:* Neuromuscular transmission; neuropharmacology; general pharmacology; neuromuscular pharmacology; delineation and reactivities of mammalian motor nerve endings; laboratory and clinical studies of anti-curare drugs, muscle relaxants and myasthenia gravis; neurotoxicity and pharamacologic changes during denervation; drug abuse. *Mailing Add:* Dept Pharmacol Cornell Univ Med Col 1300 York Ave New York NY 10021

RIKER, WILLIAM KAY, b New York, NY, Aug 31, 25; m 47, 83; c 3. PHARMACOLOGY. *Educ:* Columbia Univ, BA, 49; Cornell Univ, MD, 53. *Prof Exp:* Intern, II Med Div, Bellevue Hosp, New York, 53-54; instr pharmacol, Sch Med, Univ Pa, 54-57, assoc, 57-59, asst prof, 59-61; Nat Inst Neurol Dis & Blindness spec fel physiol, Sch Med, Univ Utah, 61-64; assoc prof, Woman's Med Col Pa, 64-68, prof & chmn dept, 68-69; prof & chmn dept, 69-91, EMER PROF PHARMACOL, MED SCH, ORE HEALTH SCI UNIV, 91- *Concurrent Pos:* Pa Plan scholar, 58-61; field ed neuropharmacol, J Pharmacol & Exp Therapeut, 68-; mem, Pharmaceut Toxicol Prog Comn, 68-72 & Neurol Dis Prog Proj Rev Comt B, 75-79; mem pharmacol-morphol adv comt, Pharmaceut Mfrs Asn Found, 70-85, sci adv coun, 78- *Mem:* Am Soc Pharmacol & Exp Therapeut (secy-treas, 78-79, pres, 85-86); Japan Pharmacol Soc; Am Epilepsy Soc; NY Acad Sci. *Res:* Physiology and pharmacology of synaptic transmission. *Mailing Add:* Dept Pharmacol Sch Med Ore Health Sci Univ Portland OR 97201

RIKIHISA, YASUKO, RICKETTSIAL DISEASE, IMMUNOLOGY. *Educ:* Univ Tokyo, Japan, PhD(pharmacol), 77. *Prof Exp:* Assoc prof pathobiol, Va Polytech Inst & State Univ, 81-86; ASSOC PROF PATHOBIOL, COL VET MED, OHIO STATE UNIV, 86- *Mailing Add:* Dept Vet Pathobiol Col Vet Med Ohio State Univ 1925 Coffey Rd Columbus OH 43210

RIKMENSPOEL, ROBERT, biophysics, instrumentation; deceased, see previous edition for last biography

RIKOSKI, RICHARD ANTHONY, b Kingston, Pa, Aug 13, 41; div; c 2. ELECTRICAL & ELECTRONICS ENGINEERING. *Educ:* Univ Detroit, BEE, 64; Carnegie-Mellon Univ, MSEE, 65, PhD(elec eng, appl space sci), 68; Case Western Reserve Univ, post-doctoral, 71. *Prof Exp:* Solid state engr, Electronic Defense Lab, Int Tel & Tel Corp, NJ, 62, solid state engr, Space Commun Lab, 63; guid engr, AC Electronics Div, Gen Motors Corp, Wis, 64; instr elec eng, Carnegie-Mellon Univ, 67-68; asst prof, Univ Pa, 68-74; assoc prof elec eng, Ill Inst Technol, 74-80; PRES, TECH ANALYSIS CORP, CHICAGO, 78- *Concurrent Pos:* Engr, Hazeltine Res, Ill, 69; consult metroliner vehicle dynamics, Ensco Inc, Va, 70; NASA-Am Soc Eng Educ fel, Case Western Reserve Univ, 71; mem educ activ bd, Inst Elec & Electronics Engrs, 79-83; mem Energy Comt, Inst Elec & Electronics Engrs, 84-85; eng consult. *Mem:* Sr mem Inst Elec & Electronics Engrs; Sigma Xi; Nat Fire Protection Asn. *Res:* Plasma physics; magnetohydrodynamics; circuit theory, simulations; thick film; microelectronics; fluid mechanics; energy conversion; product liability; evaluation of new technological concepts; analysis of patent claims; engineering problem solving. *Mailing Add:* PO Box 444 Beverly Shores IN 46301

RIKVOLD, PER ARNE, b Hadsel, Norway, Oct 4, 48. STATISTICAL PHYSICS, COMPUTATION PHYSICS. *Educ:* Univ Oslo, Norway, BSc, 71, MSc, 76; Osaka Univ Foreign Studies, Japan, dipl, 77; Temple Univ, Philadelphia, Pa, PhD(physics), 83. *Prof Exp:* Res fel physics, Japanese Ministry Educ, Kyushu Univ, Fukuoka, Japan, 77-78 & Norweg Res Coun Sci & Humanities, Temple Univ, 81-83; res assoc physics, Univ Oslo, Norway, 78-81; postdoctoral res assoc mech eng, State Univ NY, Stony Brook, 83-85; sr res chemist prod develop indust water treatment, ChemLink Indust & Petrol Chem, Subsid Atlantic Richfield Co, 85-87; ASSOC PROF PHYSICS, FLA STATE UNIV, 87- *Concurrent Pos:* Vis scientist, Dept Physics, Kyushu Univ, Fukuoka, Japan, 79, Dept Theoret Physics, Univ Geneva, Switz, 81-82, Inst Solid State Physics, Julich, WGer, 82; vis scholar, Ctr Advan Computational Sci, Temple Univ, 86-87; fac assoc, Supercomputer Computations Res Inst & Ctr Mat Res & Technol, Fla State Univ, 87-; vis scientist, Tohwa Inst Sci & Kyushu Univ, Fukuoka, Japan, 91. *Mem:* Am Phys Soc; Norweg Phys Soc. *Res:* Computationally oriented, statistical-mechanics based research in materials science and chemical physics; phase transitions in low-dimensional systems and their applications to surface science, electrochemistry and high-temperature superconductors. *Mailing Add:* Dept Physics B-159 Fla State Univ Tallahassee FL 32306

RILA, CHARLES CLINTON, b Pittsburgh, Pa, Aug 1, 28; m 50; c 2. INORGANIC CHEMISTRY, ORGANIC CHEMISTRY. *Educ:* Col Wooster, BA, 50; Ill Inst Technol, PhD(chem), 55; Univ Evansville, MS, 85. *Prof Exp:* From instr to asst prof chem, Ohio Wesleyan Univ, 55-62; assoc prof, Parsons Col, 62-65; chmn div natural sci, 67-79, PROF CHEM & HEAD DEPT, IOWA WESLEYAN COL, 65- *Concurrent Pos:* Dir, Res & Develop, SAI Corp, 79-81. *Mem:* Am Chem Soc. *Res:* Transition metal complexes; chemical education; computer applications to chemical education. *Mailing Add:* Dept Chem Iowa Wesleyan Col Mt Pleasant IA 52641

RILES, JAMES BYRUM, b Dexter, Iowa, Feb 16, 38; m 60; c 3. ALGEBRA. *Educ:* Reed Col, BA, 59; Univ London, PhD(algebra), 67. *Prof Exp:* From asst prof to assoc prof, 67-76, PROF MATH, ST LOUIS UNIV, 76-, CONSULT, COLLEGIATE ASSISTANCE PROG, 70- *Mem:* Am Math Soc; London Math Soc. *Res:* Infinite group theory. *Mailing Add:* 750 Harvard Ave St Louis MO 63130

RILEY, BERNARD JEROME, b Eau Claire, Wis, Feb 15, 28; m 51; c 8. REGULATION LIAISON. *Educ:* St Mary's Col, Minn, BS, 48; Univ Detroit, MS, 50, MBA, 61. *Prof Exp:* Radioceramicist, Glass Div Res Labs, Pittsburgh Plate Glass Co, 54-55; sr res chemist, Isotope Lab, Res Labs, Gen Motors Tech Ctr, 55-65, sect chief, Mil Vehicles Oper, Detroit Diesel Allison Div, 65-77, staff develop engr, Auto Safety Eng, 77-88; RETIRED. *Mem:* Am Chem Soc; Int Asn Auto-Theft Investr. *Res:* Use of radioactive tracers in solution of research problems, principally in fields of electrochemistry and surface chemistry. *Mailing Add:* 365 Willowtree Lane Rochester Hills MI 48306-4254

RILEY, CHARLES MARSHALL, b Chicago, Ill, Aug 17, 20; m 46; c 4. ECONOMIC GEOLOGY. *Educ:* Univ Chicago, BS, 42; Univ Minn, MS, 48, PhD(geol), 50. *Prof Exp:* Asst geol, Univ Minn, 47-50; asst prof, Univ Nebr, 50-57; res geologist, Exxon Prod Res Co, 57-83; res geologist, Aramco, 83-86; RETIRED. *Mem:* Fel Geol Soc Am; Am Asn Petrol Geol. *Res:* Sedimentary petrography; organic geochemistry; dolomite; petroleum; economic minerals. *Mailing Add:* 11938 Wink Dr Houston TX 77024

RILEY, CHARLES VICTOR, biology, ecology; deceased, see previous edition for last biography

RILEY, CLAUDE FRANK, JR, b Milledgeville, Ga, Apr 24, 22; m 47; c 3. AERONAUTICAL & MECHANICAL ENGINEERING. *Educ:* Ga Inst Tech, BS, 43; Univ Mich, MS, 49. *Prof Exp:* Propulsion engr, Bell Aircraft Corp, 46-47; assoc prof, Univ Mich, 47-50; sr operating vpres, Booz, Allen & Hamilton, 50-71; corp & group vpres & mem exec comt, Auerbach Corp Sci & Technol, 71-72, gen mgr, Auerback Asn, Inc, 71-72; vpres & gen mgr, Computing & Software Corp, 72-74; vpres, Tracor, Inc, mem bd dirs, Tracor-Jitco, 74-84; VPRES & MEM BD, COMPREHENSIVE HEALTH SERV, INC, 84-, PROMANA, INC, 84- & M A BAHETH CO, INC, 89-; OWNER, RILEY ASSOCS, 84- *Concurrent Pos:* Mem bd dir, Am Fedn Info Processing

Socs; exec off, Ga Mil Col; chmn distinguished lect series, Nat Bd Trade; lectr, Int Telemetry Conf; chmn, Proj Aristolle, Nat Security Indust Asn. *Mem:* Assoc fel Am Inst Aeronaut & Astronaut; Prof Engrs Soc. *Res:* Aeronautics; propulsion; instrumentation; computer technology; scientific and general management and marketing. *Mailing Add:* 10825 Foxhunt Lane Potomac MD 20854

RILEY, CLYDE, b Niagara Falls, NY, Feb 19, 39; m 61. PHYSICAL CHEMISTRY. *Educ:* Univ Rochester, BS, 60; Fla State Univ, PhD. *Prof Exp:* Proj assoc, Univ Wis-Madison, 65-67, asst prof, 67-68, assoc prof, 68-79, PROF CHEM, UNIV ALA, HUNTSVILLE, 79-, CHMN DEPT, 72- *Mem:* Am Chem Soc. *Res:* Laser induced chemistry; reactive scattering from crossed molecular beams; pyrolysis decomposition mechanisms by modulated molecular beam velocity; analysis mass spectrometry; electrodeposition in low gravity. *Mailing Add:* 3713 Crestmore Ave Huntsville AL 35805

RILEY, DANNY ARTHUR, b Rhinelander, Wis, Nov 18, 44; m 70; c 1. ANATOMY. *Educ:* Univ Wis, BS, 66, PhD(anat), 71. *Prof Exp:* Muscular Dystrophy Asn Am fel, NIH, 72-73; asst prof anat, Univ Calif, San Francisco, 73-; MEM STAFF, DEPT ANAT, MED COL WIS. *Mem:* Int Soc Electromyographic Kinesiology; Am Asn Anatomists. *Res:* Skeletal muscle, differentiation of fiber types; regeneration, neural dependence, hormonal dependence as studied histochemically, electronmicroscopically and physiologically. *Mailing Add:* 2705 Clearwater Dr Brookfield WI 53005

RILEY, DAVID, b New York, NY, Sept 6, 42; m; c 2. PULMONARY DISEASE. *Educ:* Univ Md, MD, 68. *Prof Exp:* From asst prof to assoc prof, 73-87, PROF MED, ROBERT WOOD JOHNSON MED SCH, 87- *Concurrent Pos:* Adj prof physiol & biophys, Robert Wood Johnson Med Sch. *Mem:* Am Thoracic Soc; Am Physiologic Soc; Am Col Chest Physicians; Am Col Physicians. *Res:* Study of mechanism of lung injury with particular emphasis on connective tissue; genetics of interstitial lung disease. *Mailing Add:* Dept Med UMDNJ-Robert Wood Johnson Med Sch CN 19 New Brunswick NJ 08903-0019

RILEY, DAVID WAEGAR, b Winchester, Mass, May 7, 21; div; c 3. RHEOLOGY, MATERIAL SCIENCE ENGINEERING. *Educ:* Tufts Col, BS, 43; Ohio State Univ, MSc, 49, PhD(chem), 51. *Prof Exp:* Res chemist, Res Ctr, Goodyear Tire & Rubber Co, 43-44 & 46; asst gen chem, Ohio State Univ, 46 & 49-51; res chemist, Polychem Dept, E I du Pont de Nemours & Co, 51-54; res chemist, Silicones Div, Union Carbide Corp, 54-60; sr develop engr, Western Elec Co, 60-67; res adv & group mgr technol, Gen Cable Res Ctr, 67-76; dir plastics eng, Sci Process & Res, 76-78; mat res specialist, Tenneco Chemicals, Inc, 78-81; CONSULT, EXTRUSION ENGRS, 81- *Mem:* Fel AAAS; fel Am Inst Chem; Am Chem Soc; fel Soc Plastics Engrs; Inst Elec & Electronics Engrs. *Res:* Polyethylene; silicone polymers; rheology of polyvinyl chloride compounds; thermal stability and extrusion of polymers; computer simulation of extrusion; coefficient of friction of plastics and solids conveying; calendering; theory of lubrication of polyvinyl chloride compounds; instruments for polymer melt analysis. *Mailing Add:* Extrusion Eng 858 Princeton Ct Neshanic Station NJ 08853-9686

RILEY, DENNIS PATRICK, b Tiffin, Ohio, Jan 22, 47; m 73; c 1. CATALYTIC OXIDATION CHEMISTRY. *Educ:* Heidelberg Col, BS, 69; Ohio State Univ, PhD(chem), 75. *Prof Exp:* Staff indexer chem, Chem Abstr Serv, 69-71; fel chem, Univ Chicago, 75-76; RES CHEMIST, TECHNOL DIV, PROCTER & GAMBLE CO, 76- *Mem:* Am Chem Soc; AAAS; Sigma Xi. *Res:* Catalytic asymmetric hydrogenations; better catalysts for molecular oxygen activation. *Mailing Add:* 800 Chancellor Heights Dr Ballwin MO 63011

RILEY, DONALD RAY, b Goshen, Ind, Mar 6, 47; m 68; c 2. COMPUTER GRAPHICS & COMPUTER AIDED DESIGN, INTELLIGENT SYSTEMS. *Educ:* Purdue Univ, BS, 69, MS, 70, PhD(mech eng), 76. *Prof Exp:* From asst prof to assoc prof, 76-88, PROF MECH ENG, UNIV MINN, 88- *Concurrent Pos:* Dir, Comput Graphics & Comput Aided Design Lab, Productivity Ctr, Univ Minn, 79-, assoc dir, Productivity Ctr, 87-; contrib ed, Comput Mech Eng, Am Soc Mech Engrs, 82-, assoc ed, J Mech & Mach Theory, 84-; co-founder & vpres, Minn Techno Transfer, Inc, 84-; mem tech adv bd, Aries Technol Inc, 86-; co-founder & bd dirs, Digital Dent Systs, Inc, 88; chmn, Computers in Eng Div, Am Soc Mech Engrs, 90-91. *Honors & Awards:* Ralph R Teetor Educ Award, Soc Automotive Engrs, 85; AT&T Found Excellence Educ Award, Am Soc Eng Educ, 85. *Mem:* Am Soc Mech Engrs; Inst Elec & Electronics Engrs Comput Soc; Asn Comput Mach; Soc Mfg Engrs; Nat Comput Graphics Asn; Am Soc Eng Educ. *Res:* Applications of computer graphics technology to mechanical engineering design and manufacturing; computer-aided design; computer-aided manufacturing; knowledge-based systems; bioengineering; computer-aided mechanism design; over 80 referred publications resulting from research. *Mailing Add:* Mech Eng Dept Univ Minn 111 Church St SE Minneapolis MN 55455

RILEY, EDGAR FRANCIS, JR, b Platteville, Wis, Nov 26, 14; m 43; c 4. RADIOBIOLOGY. *Educ:* Univ Wis, AB, 38; Univ Iowa, PhD(bot), 53. *Prof Exp:* Assoc biologist, Manhattan Proj, Oak Ridge, Tenn, 43-48; asst bot, 50-53, res assoc, Radiation Res Lab, 53-55, from asst prof to prof, 55-85, EMER PROF RADIOBIOL, RADIATION RES LAB, UNIV IOWA, 85- *Concurrent Pos:* Consult, AEC. *Mem:* AAAS; Radiation Res Soc; Am Inst Biol Sci; NY Acad Sci; Am Soc Plant Physiol; Cell Kinetics Soc. *Res:* Comparative biological effects of different ionizing radiations; effect of x-radiation on plant growth; radiation cataracts. *Mailing Add:* 540 S Summit Iowa City IA 52240-5632

RILEY, GENE ALDEN, b Wheeling, WVa, July 7, 30; m 53; c 3. PHARMACOLOGY. *Educ:* Duquesne Univ, BS, 52; Case Western Reserve Univ, PhD(pharmacol), 61. *Prof Exp:* From instr to asst prof pharmacol, Western Reserve Univ, 61-66; assoc prof, 66-72, chmn dept pharmacol, 70-87, PROF PHARMACOL, SCH PHARM, DUQUESNE UNIV, 72- *Concurrent Pos:* NIH res grant, 64-69; mem teaching staff, St Francis Hosp, Pittsburgh, 67-87. *Mem:* Am Pharmaceut Asn; Acad Pharmaceut Sci; Am Soc Consult Pharmacists; Nat Asn Retail Druggists. *Res:* Hormonal control of intermediary metabolism; drug mechanisms leading to intracellular variations in adenosine-phosphate. *Mailing Add:* Dept Pharmacol & Toxicol Duquesne Univ Sch Pharm Pittsburgh PA 15282

RILEY, HARRIS D, JR, b Clarksdale, Miss, Nov 12, 25; m 50; c 3. PEDIATRICS. *Educ:* Vanderbilt Univ, BA, 45, MD, 48. *Prof Exp:* Instr pediat, Sch Med, Vanderbilt Univ, 53-57; PROF PEDIAT, SCH MED, UNIV OKLA, 58- *Mem:* Soc Pediat Res; Am Acad Pediat; Am Pediat Soc; Infectious Dis Soc Am. *Res:* Infectious diseases, immunology and renal disease. *Mailing Add:* Children's Mem Hosp Dept Pediat Univ Okla Med Ctr Oklahoma City OK 73190

RILEY, JAMES A, b Minneapolis, Minn, May 26, 37; m 62; c 2. PHYSICS. *Educ:* Univ Minn, BS, 60; Temple Univ, MA, 64; Univ Minn, PhD(physics), 69. *Prof Exp:* Pub sch teacher, Mich, 60-63; instr physics, Mankato State Col, 64-65; from asst prof to assoc prof, 69-79, PROF PHYSICS, DRURY COL, 79- *Concurrent Pos:* Res Corp Fredrick Gardner Cottrell grant, 70. *Mem:* Am Asn Physics Teachers; Am Asn Univ Prof. *Res:* Interaction of atomic oxygen with solid surfaces; analysis of causes of failure of carbonated beverage bottles and closures. *Mailing Add:* Dept Physics Drury Col Springfield MO 65802

RILEY, JAMES DANIEL, b Tuscola, Ill, June 25, 20; m 52; c 2. PURE MATHEMATICS. *Educ:* Park Col, AB, 42; Univ Kans, MA, 48, PhD(math), 52. *Prof Exp:* Asst instr math, Park Col, 46; asst, Univ Kans, 46-48; mathematician, Naval Res Lab, 48-49; asst, Univ Md, 49-50; mathematician, Naval Ord Lab, 52-54; asst prof math, Univ Ky, 54-55; asst prof, Iowa State Univ, 55-58; mem tech staff, Space Tech Labs, Inc Div, Thompson Ramo Wooldridge, Inc, 58-61; mem tech staff, Aerospace Corp, 61-66; staff mathematician, Hughes Aircraft Co, Calif, 66-67; sr exec adv, Western Div, McDonnell Douglas Astronautics Co, Huntington Beach, 67-72, sr staff scientist, 72-74; mem tech staff, TRW, Inc, 74-75; lectr math, Univ Kebangsaan, Malaysia, 75-77; mem staff, Abacus Prog Corp, 77-80, Honeywell, Inc, 80-85; mem staff, Lockheed Aircraft Corp, 85-90; INSTR, MOORPARK COL, 88- *Mem:* Am Math Soc; Soc Indust & Appl Math; Math Asn Am. *Res:* Complex variables; numerical analysis. *Mailing Add:* 6195 Sylvan Dr Simi Valley CA 93063-4754

RILEY, JOHN PAUL, b Celista, BC, June 27, 27; m 52; c 4. AGRICULTURAL ENGINEERING. *Educ:* Univ BC, BASc, 50; Utah State Univ, CE, 53, PhD(civil eng), 67. *Prof Exp:* Res fel civil eng, Utah State Univ, 51-52; asst hydraul engr, Water Rights Br, BC Prov Govt, 52-54; instr agr eng, Ore State Univ, 54-57; dist eng, Water Rights Br, BC Prov Govt, 57-62, proj eng, Water Invests Br, 62-63; res asst, 63-67, assoc prof, 67-71, PROF CIVIL ENG, UTAH STATE UNIV, 71- *Honors & Awards:* Gov Medal for Sci & Technol. *Mem:* Am Soc Civil Engrs; Sigma Xi; Am Soc Eng Educ; fel Am Water Resources Asn; Am Geophys Union. *Res:* Water rights and computer simulation of water resource systems. *Mailing Add:* Dept Civil Eng Utah State Univ Logan UT 84322-4110

RILEY, JOHN THOMAS, b Bardstown, Ky, Apr 2, 42; m 63; c 2. COAL CHEMISTRY, ANALYTICAL INSTRUMENTATION. *Educ:* Western Ky Univ, BS, 64; Univ Ky, PhD(inorg & anal chem), 68. *Prof Exp:* From asst prof to assoc prof, 68-81, actg head dept, 81, MGR, COAL & FUEL CHARACTERIZATION LAB, WESTERN KY UNIV, 81-, DIR CTR COAL SCI, 85- *Concurrent Pos:* Prin investr, twenty-nine coal chem grants, 79-90; consult, industs, 70- *Mem:* Am Chem Soc; Sigma Xi; Am Soc Testing & Mat. *Res:* Coal desulfurization, self-heating and quality deterioration of coal and chemistry of micronized coal; determination of major, minor and trace elements in coal, coal ash and coal-derived materials. *Mailing Add:* Dept Chem Western Ky Univ Bowling Green KY 42101

RILEY, KENNETH LLOYD, b New Orleans, La, Feb 25, 41; m 61; c 2. CHEMICAL ENGINEERING. *Educ:* La State Univ, BS, 63, MS, 65, PhD(chem eng), 67. *Prof Exp:* Staff engr, 67-80, SR STAFF ENGR, EXXON RES & DEVELOP LABS, EXXON CO, USA, 80- *Concurrent Pos:* Adj prof chem eng, La State Univ, 78- *Mem:* Am Inst Chem Engrs; Am Chem Soc; Catalysis Soc. *Res:* Heterogeneous catalysis; development of petroleum and petrochemical processing catalysts and processes; catalyst characterization; process modelling; catalyst preparation; catalyst deactivation. *Mailing Add:* 1289 Rodney Dr Baton Rouge LA 70808-5874

RILEY, LEE HUNTER, JR, b St Louis, Mo, May 21, 32; m 57; c 2. ORTHOPEDIC SURGERY. *Educ:* Univ Okla, BS, 54, MD, 57. *Prof Exp:* From instr to assoc prof, 63-73, PROF ORTHOP SURG, SCH MED, JOHNS HOPKINS UNIV, 73-, CHMN DEPT, 79- *Concurrent Pos:* Fel orthop surg, Armed Forces Inst Path, 61; consult, Perry Point Vet Admin Hosp, 63- & Loch Raven Vet Admin Hosp, 68-; mem subcomt rehab & related health serv personnel, Nat Res Coun, 68-71; orthop surgeon-in-chief, Johns Hopkins Hosp, 79- *Mem:* Am Acad Orthop Surg; Am Col Surgeons; Orthop Res Soc; Am Orthop Asn; Asn Acad Surg. *Res:* Intracellular calcification; orthopedic pathology; total joint replacement; cervical spine surgery. *Mailing Add:* Dept Orthop Surg Johns Hopkins Univ 720 Rutland Ave Baltimore MD 21205

RILEY, MARK ANTHONY, b Salford, UK, June 10, 59; m 85; c 2. HIGH SPIN STATES IN NUCLEI, GAMMA-RAY SPECTROSCOPY. *Educ:* Univ Liverpool, BSc Hons, 81, PhD(nuclear struct), 85. *Prof Exp:* Res assoc nuclear physics, Niels Bohr Inst, 85-87, Oak Ridge Nat Lab & Univ Tenn, 87-88; advan fel, Univ Liverpool, 88-90; ASST PROF PHYSICS, FLA STATE UNIV, 91- *Mem:* Am Phys Soc; Inst Physics UK. *Res:* Effect of rapid rotation upon the unique quantum system of the atomic nucleus using the techniques of gamma-ray spectroscopy. *Mailing Add:* Dept Physics Fla State Univ Tallahassee FL 32306

RILEY, MICHAEL VERITY, b Bradford, Eng, Dec 27, 33; m 63; c 4. BIOCHEMISTRY, OPHTHALMOLOGY. *Educ:* Cambridge Univ, BA, 55, MA, 60; Univ Liverpool, PhD(biochem), 61. *Prof Exp:* USPHS fel, Sch Med, Johns Hopkins Univ, 61-62; sr lectr biochem, Inst Ophthal, Univ London, 62-69; assoc prof, 69-78, PROF BIOMED SCI, INST BIOL SCI, OAKLAND UNIV, 78- *Concurrent Pos:* Lister travel fel, Royal Col Surgeons Eng, 67; vis prof, Sch Med, Wash Univ, 67-68; NIH career develop award, 71; mem vision res & training comt, Nat Eye Inst, 71-75; vis prof, Welsh Nat Sch Med, Univ Wales, Cardiff, 75. *Honors & Awards:* Alcon Research Inst Award, 87. *Mem:* Asn Res Vision & Ophthal; Sigma Xi; NY Acad Sci. *Res:* Transport processes and metabolism that relate to control of hydration and transparency of the cornea; other ocular transport and metabolism. *Mailing Add:* Eye Res Inst Oakland Univ Rochester MI 48309-4401

RILEY, MICHAEL WALTERMIER, b Sedalia, Mo, Feb 11, 46; m 71; c 2. INDUSTRIAL ENGINEERING, HUMAN FACTORS. *Educ:* Univ Mo-Rolla, BS, 68; NMex State Univ, MS, 73; Tex Tech Univ, PhD(indust eng), 75. *Prof Exp:* Field engr oil prod, Shell Oil Co, 68-70; data analyst missile testing, US Army, 70-72; res asst measurement studies, NMex State Univ, 72-73; instr indust eng, Tex Tech Univ, 73-75; ASST PROF INDUST ENG, UNIV NEBR, 75- *Concurrent Pos:* NSF traineeship, 74-75. *Honors & Awards:* R R Teetor Award, Soc Automotive Engrs, 77. *Mem:* Am Inst Indust Engrs; Am Soc Eng Educ; Sigma Xi; Human Factors Soc; Soc Automotive Engrs. *Res:* Applied human factors; applied operations research. *Mailing Add:* Indust & Mgt Syst Eng Dept Univ Nebr 175 Nebr Hall Lincoln NE 68588

RILEY, MONICA, b New Orleans, La, Oct 4, 26; div; c 3. MOLECULAR GENETICS, GENOME EVOLUTION. *Educ:* Smith Col, AB, 47; Univ Calif, Berkeley, PhD(biochem), 60. *Prof Exp:* Asst prof bact, Univ Calif, Davis, 60-66; assoc prof biol, 66-75, PROF BIOCHEM, STATE UNIV NY STONY BROOK, 75- *Concurrent Pos:* USPHS fel, Stanford Univ, 61-62; vis scientist, Univ Brussels, 81; chmn, Comt Molecular & Genetic Microbiol Pub & Sci Affairs Bd, Am Soc Microbiol & rep Basic Biol Bd Nat Acad Sci, 84-87; mem, working groups of recombonant DNA adv comt & NIH, 86-88; mem organizer, conf on agn bact chromosome, sponsored by NSF & Am Soc Microbiol, 88; chmn & orgn, Gordon Conf Pop Biol & Evolution Microorganisms, 89. *Mem:* Fedn Am Socs Exp Biol; Am Soc Microbiol; Sigma Xi. *Res:* Molecular mechanisms of genome evolution in enterobacteria; evolution of the bacterial genome is studied by comparative analysis of genetic maps and by analysis of nucleotide sequences of related genes and sequences of their gene products. *Mailing Add:* Dept Biol State Univ NY Stony Brook NY 11790

RILEY, MONICA, b New Orleans, La, Oct 24, 26; m 49; c 3. DATABASE CONSTRUCTION. *Educ:* Smith Col, BA, 47; Univ Calif, Berkeley, PhD(biochem), 60. *Prof Exp:* Asst prof bact, Univ Calif, Davis, 60-66; from asst prof to prof, 66-89, EMER PROF BIOCHEM, STATE UNIV NY, STONY BROOK, 89-; ASSOC SCIENTIST, MARINE BIOL LAB, WOODS HOLE, 89- *Concurrent Pos:* Chair, Comt Genetics & Molecular Microbiol, Pub & Sci Affairs, 84-90; mem, Recombinant DNA Adv Comt, NIH, 88-91; vis prof, Univ Paris-Sud, Orsay, 91; chair, Comt Genomics, Am Soc Microbiol, 91- *Mem:* Sigma Xi; Am Soc Microbiol. *Res:* Molecular biology and genetics of E coli; database construction; E coli genes and metabolism. *Mailing Add:* Marine Biol Lab Woods Hole MA 02543

RILEY, N ALLEN, earth sciences, research administration; deceased, see previous edition for last biography

RILEY, PATRICK EUGENE, b Baltimore, Md, June 23, 49; m 72; c 2. ELECTRONICS ENGINEERING. *Educ:* Va Polytechnic Inst, BS, 72; Loyola Col Baltimore, MES, 79. *Prof Exp:* Engr, AAI Corp, 72-76 & Westinghouse Elec Corp, 76-79; design engr, Johnston Labs (BD), 79-80; proj mgr, 80-85, dir field serv, 85-90, ASSOC DIR ENG, BECTON DICKINSON, 90- *Mem:* Am Soc Microbiol. *Res:* Design and development of in vitro medical diagnostic instruments for microbiology and immunochemistry. *Mailing Add:* Becton Dickinson Seven Loveton Circle Sparks MD 21152

RILEY, PETER JULIAN, b Kamloops, BC, July 6, 33; m 59; c 4. EXPERIMENTAL NUCLEAR PHYSICS. *Educ:* Univ BC, BASc, 56, MASc, 58; Univ Alta, PhD(nuclear physics), 62. *Prof Exp:* Res asst, Fla State Univ, 58-59; from asst prof to assoc prof, 62-76, PROF PHYSICS, UNIV TEX, AUSTIN, 76- *Concurrent Pos:* Res partic, Oak Ridge Nat Lab, 66; prog dir, Intermediate Energy Nuclear Physics, NSF, Washington, DC, 84-85. *Mem:* Am Phys Soc. *Res:* Nuclear spectroscopy; reactions and scattering; direct nuclear reactions; nucleon-nucleon interactions, with emphasis on elastic and inelastic polarization measurements at medium energies; measurements of rare decays. *Mailing Add:* Dept Physics Univ Tex Austin TX 78712

RILEY, REED FARRAR, b Chicago, Ill, Aug 5, 27; m 51; c 6. INDUSTRIAL CHEMISTRY. *Educ:* Univ Ill, BS, 49; Mich State Univ, PhD(chem), 54; Columbia Univ, JD, 71. *Prof Exp:* Asst prof, Bucknell Univ, 54-57; from asst prof to assoc prof, Polytech Inst Brooklyn, 57-65; eng specialist, Bayside Labs, Gen Tel & Electronics, Inc, 65-68; patent attorney, Standard Oil Co, Inc, 71-74, sr patent attorney, 74-77, dir tech liaison, 77-85; PATENT ATTY, 85- *Concurrent Pos:* Res Corp grant, 56-57; NSF res grant, 63-65; AEC contract, 63-65; adj prof, NY Inst Technol, 69-71. *Mem:* Sigma Xi. *Res:* Activated carbons; catalysed processes. *Mailing Add:* 2411 N Burling Chicago IL 60614

RILEY, RICHARD FOWBLE, b South Pasadena, Calif, Mar 23, 17; wid. PHYSIOLOGICAL CHEMISTRY. *Educ:* Pomona Col, BA, 39; Univ Rochester, PhD(biochem), 42. *Prof Exp:* Asst biochem & pharmacol, Sch Med & Dent, Univ Rochester, 39-42, instr pharmacol, 42-45; fel biochem, Sch Med, Buffalo, 45-47; asst prof pharmacol, Univ Rochester, 47-49; fel clin radiol, Sch Med, Univ Calif, 49-53, fel physiol chem & radiol, 53-55, assoc prof, 55-82; RETIRED. *Concurrent Pos:* Rep from Univ Calif, Los Angeles, to tech info div, AEC, DC. *Mem:* AAAS; Am Chem Soc; Am Soc Pharmacol & Exp Med; Radiation Res Soc; Soc Exp Biol & Med. *Res:* Radiation biology; radioactive pharmaceuticals for nuclear medicine. *Mailing Add:* 24055 Paseo Del Lago Apt 257 Laguna Hills CA 92653-2638

RILEY, RICHARD KING, b Marshalltown, Iowa, June 4, 36; m 61; c 3. MECHANICAL ENGINEERING. *Educ:* State Univ Iowa, BS, 61, MS, 65; Univ Mo-Rolla, PhD(mech eng), 70. *Prof Exp:* Develop engr, Western Elec Corp, 65-66; from instr to asst prof mech eng, Univ Mo-Rolla, 66-76; prin engr, Dravo Corp, 76-78; res engr, 78-80, SR RES ENGR, PHILLIPS PETROL CO, 80- *Mem:* Am Soc Mech Engrs; Sigma Xi; Soc Automotive Engrs. *Res:* Petroleum fuels and combustion research. *Mailing Add:* 6600 SE Baylor Dr Bartlesville OK 74003

RILEY, RICHARD LORD, b North Plainfield, NJ, July 10, 11; m 47; c 3. MEDICINE. *Educ:* Harvard Univ, BS, 33, MD, 37. *Prof Exp:* Intern, Chas V Chapin Hosp, Providence, RI, 37-38, St Luke's Hosp, NY, 38-40; asst resident, Chest Serv, Bellevue Hosp, New York, 40-42; researcher, Off Sci Res & Develop, 42-43; from assoc to asst prof med, Columbia Univ, 47-50; assoc prof, Inst Indust Med, NY Univ, 49-50; from assoc prof to prof med, 50-77, from assoc prof to prof environ med, Sch Hyg & Pub Health, 50-77, EMER PROF MED, JOHNS HOPKINS UNIV, 77-, EMER PROF ENVIRON HEALTH SCI, 77- *Concurrent Pos:* Physiol Res Sect, Sch Aviation Med, USN, Pensacola, Fla; prof honorario, Facultad de Medicina, Universidad Autonoma de Puebla, Mex. *Honors & Awards:* Trudeau Medal, Am Lung Asn, 70. *Mem:* Assoc Am Physiol Soc; assoc Am Thoracic Soc (pres, 77-78); assoc Am Soc Clin Invest; assoc Asn Am Physicians; Sigma Xi; hon mem, Can Thoracic Soc; hon mem, Thoracic Soc UK. *Res:* Respiratory physiology; cardiovascular physiology; airborne infection. *Mailing Add:* Petersham MA 01366-0066

RILEY, ROBERT C, b Brooklyn, NY, Aug 14, 28; m 53; c 2. BIOCHEMISTRY. *Educ:* Hobart Col, BA, 51; Clemson Univ, MS, 59; Rutgers Univ, New Brunswick, PhD(entom), 64. *Prof Exp:* Qual control & formulation chemist, Geigy Agr Chem Div, Geigy Chem Corp, 51-55; analytical chemist, Clemson Univ, 57-59; from res asst to res assoc entom, Rutgers Univ, 59-64, from asst res prof to assoc res prof entom, 64-68; PRIN ENTOMOLOGIST, COOP STATE RES SERV, USDA, 68- *Mem:* Entom Soc Am; AAAS. *Res:* Integrated control of insects and pest management; economic entomology; pesticides; research information and retrieval systems. *Mailing Add:* 5013 Cedar Lane West Bethesda MD 20814

RILEY, ROBERT GENE, b Oakland, Calif, June 9, 46. ANALYTICAL CHEMISTRY. *Educ:* Calif State Univ, Hayward, BS, 69; State Univ NY, PhD(org chem), 74. *Prof Exp:* Res assoc agr chem, Wash State Univ, 74-76; res scientist, Battelle Mem Inst, 76-78, sr res scientist, 78-82, mgr, Environ Chem Sect, Earth Sci Dept, 82-87, mgr, Terrestrial Sci Sect, Environ Sci Dept, 87-88, MGR, GEOCHEM SECT, ENVIRON SCI DEPT, PAC NORTHWEST LABS, BATTELLE MEM INST, 88- *Mem:* Am Chem Soc. *Res:* Transport, fate and effects of energy related contaminants in aquatic, terrestrial and subsurface environments; distribution and fate of anthropogenic pollutants discharged to estuarine water bodies; synthetic, organic and natural products chemistry. *Mailing Add:* Battelle Northwest PO Box 999 Richland WA 99352

RILEY, ROBERT LEE, b Iola, Kans, Jan 8, 35; m 58; c 1. POLYMER CHEMISTRY, ORGANIC CHEMISTRY. *Educ:* Regis Col, Colo, BS, 56. *Prof Exp:* Chemist, Convair Div, Gen Dynamics Corp, Tex, 57-58, res chemist, Gen Dynamics Sci Res Lab, Calif, 58-62, from staff assoc to staff mem, Gen Atomic Div, Gen Dynamics Corp, 62-67, staff mem, Gulf Gen Atomic, Inc, 67-72, mgr membrane res & develop, Gulf Environ Systs Co, 72-74; dir res & develop dept, fluid systs div, Universal Oil Prod Inc, 74-84; dir, Technol Fluid Systs Div, Allied Signal Corp, 85-86; PRES, SEPARATION SYSTS INT, 86- *Concurrent Pos:* Guest scientist, Max Planck Inst Biophys, 71. *Honors & Awards:* Indust Res, Inc Award, 72. *Mem:* AAAS; Am Chem Soc; Nat Water Supply Improv Asn. *Res:* Environmental science and technology; biomedical engineering; membrane research and development for desalination by reverse osmosis; seawater desalination; gas separation processes; membrane transport, structure and separations technology. *Mailing Add:* 5803 Cactus Way La Jolla CA 92037

RILEY, STEPHEN JAMES, b Washington, DC, July 17, 43. PHYSICAL CHEMISTRY. *Educ:* Oberlin Col, Ohio, BA, 65; Harvard Univ, MA, 67, PhD(chem), 70. *Prof Exp:* NIH fel chem, Univ Calif, San Diego, 70-72, res asst chem, 72-73; asst prof chem, Yale Univ, 73-80; chemist, 80-90, SR CHEMIST, ARGONNE NAT LAB, 90- *Mem:* Am Phys Soc. *Res:* Chemical and physical properties of isolated metal clusters. *Mailing Add:* Chem Div Argonne Nat Lab 9700 S Cass Ave Argonne IL 60439

RILEY, THOMAS N, b Mishawaka, Ind, Dec 2, 39; m 60; c 2. MEDICINAL CHEMISTRY. *Educ:* Univ Ky, BS, 63; Univ Minn, Minneapolis, PhD(med chem), 69. *Prof Exp:* Teaching asst med chem, Col Pharm, Univ Minn, 63-64, res asst, 64-67, PHS trainee, 67-69; from asst prof to prof, Sch Pharm, Univ Miss, 69-82; head, 82-90, acting dean, 87-88, PROF MED CHEM, DEPT PHARM SCI, SCH PHARM, AUBURN UNIV, 90- *Mem:* AAAS; Am Chem Soc; Am Asn Cols Pharm; Sigma Xi. *Res:* Design, synthesis and evaluation of organic medicinal agents in an attempt to elucidate the molecular mechanisms of drug action; medicinal chemistry of drugs affecting the nervous system including analgesics, anticonvulsants, antihistaminics and bronchodilators. *Mailing Add:* Sch Pharm Sci Auburn Univ Auburn AL 36849

RILEY, WILLIAM F(RANKLIN), b Allenport, Pa, Mar 1, 25; m 45; c 2. ENGINEERING MECHANICS, MECHANICAL ENGINEERING. *Educ:* Carnegie Inst Technol, BS, 51; Ill Inst Technol, MS, 58. *Prof Exp:* Mech engr, Mesta Mach Co, 51-54; assoc engr, Armour Res Found, 54-58, res engr, 58-61, sr res engr, 61, sect mgr, IIT Res Inst, 61-64, sci adv, 64-66; from assoc prof to prof, 66-78, DISTINGUISHED PROF ENG MECH,

IOWA STATE UNIV, 78- *Concurrent Pos:* Am consult, US Agency Int Develop Summer Inst Prog, Bihar Inst Technol, Sindri, India, 66, Indian Inst Technol, Kanpur, India, 70. *Honors & Awards:* M M Frocht Award, 77. *Mem:* Fel & Soc Exp Mech. *Res:* Photoelasticity, especially the solution of three-dimensional and dynamic stress problems. *Mailing Add:* 1518 Meadowlane Ave Ames IA 50010

RILEY, WILLIAM ROBERT, b Bellaire, Ohio, July 31, 22; m 49; c 2. PHYSICS. *Educ:* Hiram Col, AB, 44; Ohio State Univ, BSc, 51, MA, 52, PhD(sci ed, physics), 59. *Prof Exp:* Instr math & physics, Hiram Col, 46-48; asst physics, Ohio State Univ, 48-50, instr, 51-53, instr in chg demonstrations, 53-59, from asst prof to assoc prof, 59-87, EMER ASSOC PROF PHYSICS, OHIO STATE UNIV, 87- *Concurrent Pos:* Consult, Bur Educ Res, Ohio State Univ, 57-58, res found mobile lab, 59-60; dir, In-serv Insts in Physics, NSF-Ohio State Univ, 60-67, 68-73 & 78-82; consult, North Bengal Univ, India, 64; mem, NSF sci liaison staff, New Delhi, India, 67-68. *Mem:* Sigma Xi; AAAS; Nat Sci Teachers Asn; Am Asn Physics Teachers. *Res:* Science education. *Mailing Add:* Dept Physics Ohio State Univ 174 W 18th Ave Columbus OH 43210-1106

RILL, RANDOLPH LYNN, b Canton, Ohio, Oct 19, 44; m 66; c 4. PHYSICAL BIOCHEMISTRY. *Educ:* Franklin & Marshall Col, BA, 66; Northwestern Univ, Evanston, PhD(phys chem), 71. *Prof Exp:* USPHS fel biophys, Dept Biochem & Biophys, Ore State Univ, Corvallis, 70-72; from asst prof to assoc prof, 72-83, PROF CHEM, FLA STATE UNIV, TALLAHASSEE, 84- *Honors & Awards:* Career Develop Award, USPHS, 75. *Mem:* Biophys Soc; Am Soc Biol Chemists. *Res:* Physical and chemical studies of DNA-protein interactions; DNA drug and small molecule interactions; structure of chromatin subunits; conformational properties of DNA; biological NMR. *Mailing Add:* Dept Chem Fla State Univ Tallahassee FL 32306

RILLEMA, JAMES ALAN, b Grand Rapids, Mich, Nov 6, 42; m 68; c 3. ENDOCRINOLOGY, PHYSIOLOGY. *Educ:* Calvin Col, BS, 64; Mich State Univ, MS, 66, PhD(physiol), 68. *Prof Exp:* NIH fel, Emory Univ, 68-70, res assoc physiol, 70-71; asst prof, 71-75, assoc prof, 75-79, PROF PHYSIOL, SCH MED, WAYNE STATE UNIV, 79- *Concurrent Pos:* Fogarty Sr Int Fel, VanLeeuwenholtz Cancer Inst, Amsterdam, Neth, 80-81. *Mem:* AAAS; Endocrine Soc; Am Physiol Soc; Soc Exp Biol & Med; Sigma Xi. *Res:* Hormones, especially mechanism of action. *Mailing Add:* Dept Physiol 5374 Scott Hall Sch Med Wayne State Univ 540 E Canfield Detroit MI 48201

RILLING, HANS CHRISTOPHER, b Cleveland, Ohio, June 6, 33; m 55; c 3. BIOCHEMISTRY. *Educ:* Oberlin Col, BA, 55; Harvard Univ, PhD(biochem), 60. *Prof Exp:* Res assoc biochem, Univ Mich, 58-60, from instr to assoc prof, 60-71, PROF BIOCHEM, UNIV UTAH, 71- *Concurrent Pos:* USPHS res career develop award, 63-73. *Mem:* Am Soc Biol Chemists. *Res:* Enzymology and mechanism of the biosynthesis of terpenes. *Mailing Add:* Dept Biochem Univ Utah Sch Med 410 Chipeta Way Rm 215 Salt Lake City UT 84108

RILLINGS, JAMES H, b Mineola, NY, June 5, 42; m 78. AUTOMATIC CONTROL SYSTEMS, INTELLIGENT CONTROL. *Educ:* Rensselaer Polytech Inst, BSEE, 64, MSEE, 66, DEng, 68. *Prof Exp:* Instr elec eng, Rensselaer Polytech Inst, 66-68; engr, Electronics Res Ctr, NASA, 68-70; PRIN RES ENGR, GEN MOTORS RES LABS, 70- *Mem:* Inst Elec & Electronics Engrs; Soc Automotive Engrs; Sigma Xi. *Res:* Computer applications to automatic control; intelligent systems; systems engineering. *Mailing Add:* 668 Ardmoor Dr Birmingham MI 48010-1716

RIM, DOCK SANG, mathematics; deceased, see previous edition for last biography

RIM, KWAN, b Korea, Nov 7, 34; m 62; c 3. ENGINEERING MECHANICS, BIOMEDICAL ENGINEERING. *Educ:* Tri-State Col, BS, 55; Northwestern Univ, MS, 58, PhD(theoret & appl mech), 60. *Prof Exp:* Engr, Int Bus Mach Corp, NY, 56; from asst prof to assoc prof, 60-68, PROF MECH, UNIV IOWA, 68, CHMN DEPT MECH & HYDRAUL, 72-, ASSOC DEAN ENG, 74-, ADJ PROF MED, DEPT ORTHOP SURG, 70- *Concurrent Pos:* Consult, Deere & Co, Ill, 64-66; actg chmn dept mech & hydraul, Univ Iowa, 71-72. *Honors & Awards:* Teetor Educ Fund Award, Soc Automotive Eng, 65. *Mem:* Soc Indust & Appl Math; Am Soc Mech Engrs; Am Soc Eng Educ; Sigma Xi. *Res:* Classical mechanics of deformable bodies, such as the theory of elasticity and viscoelasticity; optimal design; biomechanics. *Mailing Add:* 604 Granada Ct Iowa City IA 52240

RIM, YONG SUNG, b Changryun, Korea, Mar 15, 35; m 59; c 2. ORGANIC CHEMISTRY. *Educ:* Yonsei Univ, Korea, BS, 57; Univ Tex, Austin, PhD(chem), 67. *Prof Exp:* Res chemist, 68-70, res scientist, 70-75, sr res scientist, Oxford Res Ctr, 75-78, sr group leader elastomer res, 78-80, MGR ETHYLENE-PROPYLENE-DIENE MONOMER RES & DEVELOP, UNIROYAL CHEM, UNIROYAL INC, 80- *Mem:* Am Chem Soc. *Res:* Nonbenzenoid aromatic chemistry; free radical chemistry; fire retardant polymer chemistry; elastomer syntheses and evaluation of elastomers. *Mailing Add:* Yukong Ltd PO Box 400 Naugatuck CT 06770-0400

RIMAI, DONALD SAUL, b New York City, NY, Oct 17, 49; m 77. SOLID STATE PHYSICS. *Educ:* Rensselear Polytech Inst, BS, 71; Univ Chicago, MS, 73, PhD(physics), 77. *Prof Exp:* Res assoc, Purdue Univ, 77-79; SR RES SCIENTIST, EASTMAN KODAK, 79- *Mem:* Am Phys Soc; NY Acad Sci; Sigma Xi; AAAS. *Res:* Electrical and photoconducting properties of organic insulators; ultrasonic properties; adhesion & surface forces. *Mailing Add:* PO Box 505 Webster NY 14580

RIMAI, LAJOS, b Budapest, Hungary, Apr 10, 30; US citizen; m 54; c 3. NONLINEAR OPTICS, SPECTROSCOPY. *Educ:* Univ San Paulo, Brazil, EE, 52, BS, 53; Harvard Univ, PhD(appl physcis), 59. *Prof Exp:* Instr elec eng, Inst Aeronaut Technol, San Jose Campus, San Paulo, Brazil, 53-55; scientist physics, Res Div, Raytheon Co, 59-64; STAFF SCIENTIST, PHYSICS DEPT, FORD MOTOR CO, 54- *Mem:* Am Phys Soc. *Res:* Spectroscopic probes of molecular interactions, structure and combustion; nonlinear laser spectroscopic diagnostics of combustion; resonance effects in Raman spectroscopy; resonance Raman spectroscopy of biologically active chromophores (visual pigments); laser velocimetry. *Mailing Add:* 22364 Long Blvd Dearborn MI 48124

RIMBEY, PETER RAYMOND, b LaGrande, Ore, Aug 27, 47. THEORETICAL SOLID STATE PHYSICS. *Educ:* Eastern Ore State Col, BA, 69; Univ Ore, MS, 71, PhD(physics), 74. *Prof Exp:* Res assoc, Ind Univ, 73-74; fel, Ames Lab, Energy Res & Develop Admin, Iowa State Univ, 75-78; fel, Univ Wis-Milwaukee, 78-80; mem staff, Boeing Aerospace, 80-84; physics dept, Eastern Ore State Col, 83-84; PHYSICS DEPT, UNIV SEATTLE, 84- *Concurrent Pos:* Contract engr & appl math, Vector Res, Seattle. *Mem:* Am Phys Soc; Sigma Xi; AAAS; Am Asn Physics Teachers. *Res:* Many-body theory; condensed matter; optical properties of semiconductors, molecular crystals and metals; photoemission; surface physics; photo-electro-chemistry; excitons and polarons; transport in semiconductors; electro- and thermo-transport; atomic diffusion; applied mathematics. *Mailing Add:* Dept Physics Univ Seattle Seattle WA 98122

RIMES, WILLIAM JOHN, b Lake Providence, La, Mar 31, 18. ANALYTICAL CHEMISTRY. *Educ:* Spring Hill Col, BS, 43; Cath Univ, MS, 45; La State Univ, PhD(chem), 57. *Prof Exp:* Assoc prof, 56-73, chmn dept, 58-66, PROF CHEM, SPRING HILL COL, 73-, PRES, 66-, DIR CHEM, 77- *Mem:* Am Chem Soc. *Res:* Instrumental methods of analysis. *Mailing Add:* St Charles Borroneo Church PO Box A118 Grand Cotau LA 70541

RIMLAND, DAVID, b Havana, Cuba, Aug 8, 44; US citizen; wid; c 2. INFECTIOUS DISEASES. *Educ:* Emory Univ, BS, 66 MD, 70; Am Bd Internal Med, dipl & cert infectious dis. *Prof Exp:* Intern & resident internal med, Barnes Hosp, 70-72; epidemic intel serv officer, Ctr Dis Control, 72-74; resident & fel, Emory Univ Affil Hosps, 74-77; asst prof, 77-81, ASSOC PROF MED, INFECTIOUS DIS, SCH MED, EMORY UNIV, 81- *Concurrent Pos:* Staff physician, Vet Admin Med Ctr, Atlanta, 77-90, chief, infectious dis, 90- *Mem:* Am Soc Microbiol; Am Fedn Clin Res; Infectious Dis Soc Am. *Res:* HIV disease; epidemiology of hospital infections. *Mailing Add:* Med Serv Vet Admin Med Ctr 1670 Clairmont Rd Decatur GA 30033

RIMM, ALFRED A, b Atlantic City, NJ, Apr 13, 34; m 58; c 4. GENETICS, STATISTICS. *Educ:* Rutgers Univ, 56, MS, 58, PhD(dairy sci), 62. *Prof Exp:* Sr res scientist, Roswell Park Mem Inst, 63-66; PROF BIOSTATIST, MED COL WIS, 66- *Mem:* Am Statist Asn; Biomet Soc; Am Pub Health Asn. *Res:* Controlled clinical trials; epidemiology studies in obesity and heart disease; public health statistics; teaching biostatistics in medical school; computers for medical research; immunobiology; disease registeries. *Mailing Add:* Med Col Wis 8701 Watertown Plank Rd Milwaukee WI 53226

RIMOIN, DAVID (LAWRENCE), b Montreal, Que, Nov 9, 36; m 62, 80; c 3. MEDICAL GENETICS. *Educ:* McGill Univ, BSc, 57, MSc, & MD, CM, 61; Johns Hopkins Univ, PhD(human genetics), 67. *Prof Exp:* Asst prof med & pediat, Wash Univ, 67-70; chief med genetics, Harbor Gen Hosp, 70-86; assoc prof, 70-73, PROF MED & PEDIAT, SCH MED, UNIV CALIF, LOS ANGELES, 73-; CHMN, DEPT PEDIAT, CEDARS SINAI MED CTR, LOS ANGELES, 86- *Concurrent Pos:* Lectr med, Sch Med, Johns Hopkins Univ, 67-71; consult, Orthop Hosp, Los Angeles, 70-, Fairview State Hosp, Costa Mesa, 71- & Cedars-Sinai Med Ctr, Los Angeles, 71- *Honors & Awards:* E Mead Johnson Award, 76; Ross Outstanding Young Investr, 76. *Mem:* Am Soc Clin Invest; Am Fedn Clin Res (secy-treas, 73-76); Soc Pediat Res; Am Soc Human Genetics; fel Am Col Physicians; Asn Am Physicians; Am Pediat Soc. *Res:* Dwarfism; birth defects; genetic disorders of the endocrine glands. *Mailing Add:* 8700 Beverly Blvd Los Angeles CA 90048

RIMROTT, F(RIEDRICH) P(AUL) J(OHANNES), b Halle, Ger, Aug 4, 27; m 55; c 4. ENGINEERING MECHANICS. *Educ:* Univ Karlsruhe, Dipl Ing, 51; Univ Toronto, MASc, 55; Pa State Univ, PhD(eng mech), 58; Tech Univ Darmstadt, DrIng, 61. *Prof Exp:* Design engr locomotives, Henschel-Werke Kassel, Ger, 51-52; instr mech eng, Univ Toronto, 53-55; from instr to asst prof eng mech, Pa State Univ, 55-60; from asst prof to assoc prof mech eng, 60-67, PROF MECH ENG, UNIV TORONTO, 67- *Concurrent Pos:* Vis prof, Wien Tech Univ, 69, 86, Hannover Tech Univ, 70, Ruhr Univ, 71 & Univ Wuppertal, 78 & Hamburg-Harburg Tech Univ, 87; pres, 15th Int Cong Theoret & Appl Mech, 80. *Honors & Awards:* Queen's Silver Anniversary Medal, 77. *Mem:* Fel Can Soc Mech Engrs; fel Eng Inst Can; Sigma Xi; Can Aeronaut & Space Inst; fel Am Soc Mech Engrs; fel Inst Mech Engrs. *Res:* Plasticity; creep; fatigue; strain and stress measurement; vibrations; strength of materials; elasticity; machine design; gyrodynamics. *Mailing Add:* Dept Mech Eng Univ Toronto Toronto ON M5S 1A4 Can

RINALDI, LEONARD DANIEL, b Torrington, Conn, Feb 15, 24; m 50; c 3. MATHEMATICS. *Educ:* Univ Conn, BS, 45; Univ Buffalo, MS, 55. *Prof Exp:* Res mathematician, Cornell Aeronaut Labs, Inc, 45-55; dir, Arithmetion Consults, 55-58; PRES, RINALDI DATA PROCESSING CONSULTS, INC, 58- *Mem:* Math Asn Am; Asn Comput Mach; Am Mkt Asn; Am Math Soc; Inst Mgt Sci. *Res:* Probability and statistics; theory and practice of computation; mathematics of resource use; aeronautical engineering. *Mailing Add:* 506 Springfield Ave Cranford NJ 07016

RINALDO, CHARLES R, ACQUIRED IMMUNE DEFICIENCY SYNDROME. *Educ:* Univ Utah, PhD(microbiol), 73. *Prof Exp:* ASSOC PROF PATH, SCH MED, UNIV PITTSBURGH, 85- *Mailing Add:* Dept Path Infectious Dis & Microbiol Univ Pittsburgh A 417 Crabtree Hall Pittsburgh PA 15261

RINARD, GILBERT ALLEN, b Denver, Colo, Dec 16, 39; m 61; c 4. PHYSIOLOGY, ENDOCRINOLOGY. *Educ:* George Fox Col, BA, 61; Ore State Univ, MS, 63; Cornell Univ, PhD(endocrinol), 66. *Prof Exp:* NIH fel pharmacol, Case Western Reserve Univ, 66-68; from instr to asst prof Physiol, 68-76, ASSOC PROF PHYSIOL, SCH MED, EMORY UNIV, 76- *Mem:* Am Physiol Soc. *Res:* Mechanism of action of drugs used in asthma therapy; cyclic nucleotides; smooth muscle; cellular mechanisms of asthma. *Mailing Add:* Dept Physiol Woodruff McAdam Bldg Emory Univ 1364 Clifton Rd NE Atlanta GA 30322

RINCHIK, EUGENE M, b Troy, NY, May 29, 57. GENETICS. *Educ:* Cornell Univ, BS, 79; Duke Univ, PhD(genetics, microbiol & immunol), 83. *Prof Exp:* SR STAFF SCIENTIST, BIOL DIV, OAK RIDGE NAT LAB, 85- *Concurrent Pos:* Adj asst prof, Univ Tenn, 86- *Mem:* Int Mammalian Genome Soc. *Res:* Experimental germ-line mutagenesis and developmental genetics in the mouse; mouse genome analysis. *Mailing Add:* Biol Div Oak Ridge Nat Lab PO Box 2009 Oak Ridge TN 37831-8077

RIND, DAVID HAROLD, b New York, NY, May 1, 48. CIRCULATION MODELING, CLIMATE DYNAMICS. *Educ:* City Col, City Univ New York, BS, 69; Columbia Univ, MA, 73, PhD(meteorol & geophysics), 76. *Prof Exp:* Res assoc, Lamont Doherty Geol Observ, 76-81; SPACE SCIENTIST, GODDARD INST SPACE STUDIES, NASA, 81- *Concurrent Pos:* Lectr, Columbia Univ, 78-82, adj asst prof, 82-; consult, Goddard Inst Space Studies, NASA, 79-81. *Mem:* Am Meteorol Soc; Am Geophys Union. *Res:* Climate and upper atmosphere research using computer models of the atmosphere; observations of upper atmosphere parameters and geophysicsal wave propagation. *Mailing Add:* 201 W 70th St New York NY 10023

RINDERER, THOMAS EARL, b Dubuque, Iowa, Sept 16, 43; m 67, 80. GENETICS, INSECT BEHAVIOR. *Educ:* Loras Col, BS, 66; Ohio State Univ, MSc, 68, PhD(insect path), 75. *Prof Exp:* RES GENETICIST & RES LEADER, BEE BREEDING & STOCK CTR LAB, AGR RES SERV, USDA, 75- *Mem:* Entom Soc Am; Bee Res Asn; Soc Invert Path; Am Behav Soc. *Res:* Epigenetics of disease events, including breeding for resistance and susceptibility in insects and understanding the influence of environmental factors; behavior, behavior genetics, and pathogenetics of honey bees; nector foraging and defensive behavior of both European and Africanized honeybees. *Mailing Add:* 5034 Alvin Dark Ave Baton Rouge LA 70820

RINDERKNECHT, HEINRICH, b Zurich, Switz, Jan 21, 13; US citizen; m 39; c 4. BIOCHEMISTRY, ORGANIC CHEMISTRY. *Educ:* Swiss Fed Inst Technol, MSc, 36; Univ London, PhD(biochem), 39. *Hon Degrees:* DSc, Univ London, 80. *Prof Exp:* Res chemist, Roche Prod, Ltd, Eng, 39-47; assoc dir biochem res, Aligena Co, Switz, 47-49; res fel chem, Calif Inst Technol, 49-54; res dir org chem, Crookes Labs, London, Eng, 54-55; dir res & develop, Calbiochem, Calif, 55-62; sr res fel chem, Calif Inst Technol, 62-70; from assoc prof to prof, Sch Med, Univ Southern Calif, 64-70; chief med biochem, Vet Admin Hosp, Sepulveda, 70-85; PROF MED, UNIV CALIF, LOS ANGELES, 70- *Concurrent Pos:* Consult, Cilag A G, Switz, 49-54, Geigy Pharmaceut, 55-57 & Stuart Co, Calif, 59-61; mem ed comt, Int J Pancreatology, 86- *Mem:* AAAS; Am Chem Soc; Am Fedn Clin Res; Brit Biochem Soc; fel Royal Soc Health; NY Acad Sci. *Res:* Development of synthetic analgesics and local anesthetics; synthesis of nucleotides, enzyme substrates and metabolites; ultrarapid method for fluorescent labeling of proteins; role of proteolytic and lysosomal enzymes in acute pancreatitis and cancer of the pancreas; radioimmunoassay of peptide hormones. *Mailing Add:* 1971 Cielito Lane Santa Barbara CA 93105

RINDLER, WOLFGANG, b Vienna, Austria, May 18, 24; m 59, 77; c 3. RELATIVITY THEORY, COSMOLOGY. *Educ:* Univ Liverpool, BSc, 45, MSc, 47; Univ London, PhD, 56. *Prof Exp:* Asst lectr pure math, Univ Liverpool, 47-49; lectr math, Univ London, 49-56; from instr to asst prof, Cornell Univ, 56-63; from assoc prof to prof, Southwest Ctr Advan Studies, 63-69; prof math, 69-80, PROF PHYSICS, UNIV TEX, DALLAS, 80- *Concurrent Pos:* NSF fel, Hamburg & King's Col, Univ London, 61-62; Nat Res Coun Italy fel, Univ Rome, 68-69; vis prof, Univ Vienna, 75 & 87; vis fel, Churchill Coll, Cambridge, 90. *Mem:* Fel Am Phys Soc; fel Royal Astron Soc; Int Astron Union; Int Soc Gen Relativity & Gravitation. *Res:* Relativity; cosmology; spinors. *Mailing Add:* Dept Physics Univ Tex Dallas Richardson TX 75083-0688

RINDONE, GUY E(DWARD), b Buffalo, NY, Aug 11, 22; m 43; c 2. CERAMICS. *Educ:* Alfred Univ, BS, 43; Pa State Univ, MS, 46, PhD(ceramics), 48. *Prof Exp:* Glass technologist, Sylvania Elec Prod, Inc, 43-45; glass technologist, Glass Sci Inc, 45-47; asst, 47-48, from asst prof to prof, 49-81, chmn sect, 69-80, EMER PROF CERAMIC SCI, PA STATE UNIV, UNIVERSITY PARK, 81-; PRES, MAT RES CONSULTS, INC, 81- *Concurrent Pos:* Prog chmn, Int Cong Glass, DC, 62; mem coun, Int Comn Glass; mem, Univs Space Res Asn, NASA, 72- *Honors & Awards:* Forrest Award, Am Ceramic Soc, 63; Founders Award, Am Ceramic Soc, 84; Bleininger Award, 88; Toledo Glass Award, Am Ceramic Soc, 84. *Mem:* Fel Am Ceramic Soc (vpres, 77-78); fel Brit Soc Glass Technol. *Res:* Ceramic and glass science; solarization; gas evolution anelasticity, nucleation and crystallization processes and electrochemical behavior of glass; small angle x-ray scattering of glass; relaxation processes; glass strength; fluorescence; microstructure, laser light scattering; nuclear waste encapsulation; optical glass; gradient index glasses. *Mailing Add:* 116 Steidle Bldg Penn State Univ University Park PA 16802

RINE, DAVID C, b Bloomington, Ill, Nov 16, 41; m; c 2. COMPUTER SCIENCE, INFORMATION SYSTEMS. *Educ:* Ill State Univ, BS, 60; Okla State Univ, MS, 66; Univ Iowa, PhD(math), 70. *Prof Exp:* PROF & CHMN COMPUT & INFO SCI, GEORGE MASON UNIV, 85- *Concurrent Pos:* Consult, govt & indust contracts. *Honors & Awards:* Honor Roll Award, Inst Elec & Electronics Engrs Comput Soc, 77 & 81, Spec Award, 77; Centennial Award, Inst Elec & Electronics Engrs, 84, Pioneer Award, Computer Soc, 88. *Mem:* Inst Elec & Electronics Engrs; Computer Soc; Am Fedn Info Processing Soc; Int Coun Comput Educ; Asn Comput Mach. *Res:* Database systems; knowledge based systems with an emphasis on logic programming, logic design and software engineering; software engineering; systems analysis. *Mailing Add:* Dept Comput Sci George Mason Univ Fairfax VA 22030

RINEHART, EDGAR A, b Guthrie, Okla, Oct 16, 28; m 59; c 2. MOLECULAR SPECTROSCOPY, LASERS. *Educ:* Cent State Col, Okla, BS, 52; Univ Okla, MS, 55, PhD(physics), 61. *Prof Exp:* Asst prof physics, Univ Idaho, 60-61; adj asst prof, Univ Okla, 61-64; asst prof, 64-67, assoc prof, 67-77, PROF PHYSICS & ASTRON, UNIV WYO, 77- *Concurrent Pos:* Consult, Lawrence Livermore Lab, 66- & Martin Marietta Corp, 78-79; Fac res partic, US DOE Oak Ridge Assoc Univ, 87. *Mem:* Am Phys Soc; Am Chem Soc; Soc Appl Spectros; Optical Soc Am; Sigma Xi; Inst Elec & Electronics Engrs. *Res:* Microwave spectroscopy, line widths and intensities; free radicals; laser physics; optics; astronomy. *Mailing Add:* Dept Physics & Astron Univ Wyo PO Box 3905 Laramie WY 82071

RINEHART, FRANK PALMER, b Washington, DC, Mar 1, 44. BIOPHYSICAL CHEMISTRY. *Educ:* Western Md Col, BA, 66; Univ Calif, Berkeley, PhD(chem), 71. *Prof Exp:* Lectr chem, Univ Ife, Nigeria, 71-74; researcher biophys chem, Univ Calif, Davis, 74-77, lectr chem, 77; from asst prof to assoc prof, 77-87, PROF CHEM, UNIV VI, 88- *Mem:* Am Chem Soc. *Res:* Analysis of repetitive sequences in eucaryotic DNA; design of microcomputer software for chemistry courses. *Mailing Add:* Dept Chem Univ VI Charlotte Amalie VI 00802

RINEHART, JAY KENT, b Xenia, Ohio, Apr 13, 40; m 61; c 2. CHEMISTRY. *Educ:* Univ Cincinnati, BS, 62; Univ Minn, Minneapolis, PhD(org chem), 67. *Prof Exp:* Sr res chemist, 67-85, RES ASSOC, PPG INDUSTS, INC, 85- *Mem:* Am Chem Soc; AAAS. *Res:* Synthesis of new pesticides; structure-activity correlations as a tool for new pesticide design; carbene chemistry; small ring chemistry; bridged aromatic compounds; organic sulfur chemistry; heterocyclic chemistry; agricultural chemistry; synthesis of isotopically labelled compounds; radiochemistry. *Mailing Add:* PO Box 4000 Princeton NJ 68543

RINEHART, JOHN SARGENT, b Kirksville, Mo, Feb 8, 15; m 40; c 2. PHYSICS. *Educ:* Northeastern Mo State Teachers Col, BS, 34, AB, 35; Calif Inst Technol, MS, 37; Univ Iowa, PhD(physics), 40. *Prof Exp:* Asst physics, Calif Inst Technol, 35-37; asst, Univ Iowa, 37-40, Kans State Col, 40 & Wayne State Univ, 41-42; assoc physicist, Nat Bur Stand, 42; sect tech aide, Nat Defense Res Comt, 42-45; head physicist & supvr exp range, Res & Develop Div, NMex Sch Mines, 45-48; head terminal ballistics br, US Naval Ord Testing Sta, Inyokern, 49-50, mech br, 50-51 res physicist, 51-55; asst dir astrophys lab, Smithsonian Inst, 55-58; prof mining eng & dir mining res lab, Colo Sch Mines, 58-64; dir res, US Coast & Geod Surv, 64-65; dir sci & eng, Environ Sci Serv Admin, 65-66, sr res fel & dir univ rels, Inst Environ Res, 66-68, sr res fel, Nat Oceanic & Atmospheric Admin, 68-73; RETIRED. *Concurrent Pos:* Adj prof, Univ Colo, Boulder, 68-73. *Honors & Awards:* Presidential Cert Merit. *Mem:* Am Geophys Union. *Res:* Interior, exterior and terminal ballistics; failure of metals under impulsive loading; hypersonics; rock physics; geysers; geothermal areas. *Mailing Add:* Hyperdynamics PO Box 392 Santa Fe NM 87501-0392

RINEHART, KENNETH LLOYD, b Chillicothe, Mo, Mar 17, 29; m 61; c 3. ORGANIC CHEMISTRY. *Educ:* Yale Univ, BS, 50; Univ Calif, PhD(chem), 54. *Prof Exp:* From instr to assoc prof, 54-64, univ scholar, 89-92, PROF ORG CHEM, UNIV ILL, URBANA, 64- *Concurrent Pos:* Fels, Orgn Europ Econ Coop, 60, Guggenheim, 62, Fulbright, 66 & Erskine, 83; distinguished vis lectr, Tex A&M Univ, 80; consult, PharmaMar & NIH. *Honors & Awards:* Squibb Lectr, Rutgers Univ, 61 & 83; Werner Lectr, Univ Kans, 65; A D Little Lectr, Mass Inst Technol, 75; Barnett Lectr, Northeastern Univ, 81; Andrews Lectr, New South Wales, 82; Smith Lectr, Okla State Univ, 83; Res Achievement Award, Am Soc Pharmacog, 89; Karcher lectr, Univ Okla, 91. *Mem:* Am Chem Soc; Royal Soc Chem; Am Soc Biol Chem; Am Soc Microbiol; fel AAAS; Am Soc Mass Spectrometry. *Res:* Structure, biosynthesis and synthesis of natural products; antibiotics and marine natural products; mass spectrometry. *Mailing Add:* 454 Roger Adams Lab Univ Ill 1209 W California St Urbana IL 61801

RINEHART, ROBERT FROSS, mathematics, for more information see previous edition

RINEHART, ROBERT R, b Shenandoah, Iowa, Apr 5, 32; m 55; c 3. GENETICS, RADIATION BIOLOGY. *Educ:* San Diego State Col, AB, 58; Univ Tex, PhD(zool), 62. *Prof Exp:* USPHS fel genetics, Oak Ridge Nat Lab, 62-63; res assoc, Ind Univ, 63-64; assoc prof biol, 64-70, PROF BIOL, SAN DIEGO STATE UNIV, 70- *Concurrent Pos:* AEC res grant, 64-; res assoc, State Univ Leiden, Neth, 68-69. *Mem:* AAAS; Genetics Soc Am. *Res:* Radiation induced mutation repair of radiation damage; chromosome mechanics in Drosophila. *Mailing Add:* Dept Biol San Diego State Univ 5300 Campanilel Dr San Diego CA 92182

RINEHART, WALTER ARLEY, b Peoria, Ill, June 1, 36; m 59; c 2. ENGINEERING MANAGEMENT, AEROSPACE ENGINEERING. *Educ:* Univ Mo-Rolla, BS, 59. *Prof Exp:* Assoc engr thermodyn, McDonnell Douglas Corp, 59, from res asst to res assoc arc heaters, 59-64, res scientist, 64-69, group engr, 69-70, sr group engr gas dynamics, 70-72, sect mgr, 72-77, br chief, 77-86, chief tech engr, 86-90, DIR, MCDONNELL DOUGLAS CORP, 90- *Concurrent Pos:* Nat Tech Comt Ground Testing & Simulation, 75-78. *Mem:* Am Inst Aeronaut & Astronaut. *Res:* Hypervelocity vehicle technology development with emphasis toward aerodynamics; aerothermodynamics, propulsion, thermal management; structures and materials, CFD and vehicle synthesis for future hypersonic vehicles. *Mailing Add:* 16232 Lone Cabin Dr Chesterfield MO 63005

RINEHART, WILBUR ALLAN, b Mansfield, Ohio, Aug 17, 30; m 63; c 3. GEOPHYSICS, COMPUTER SCIENCE. *Educ:* Bowling Green State Univ, BS, 59; Univ Utah, MS, 62. *Prof Exp:* Geophysicist, US Coast & Geod Surv, 64-73; geophysicist, US Geol Surv, 74-75; geophysicist, Nat Oceanic & Atmospheric Admin, 76-86; RETIRED. *Concurrent Pos:* Publ, Seismol Res Letts, 84-90. *Mem:* Seismol Soc Am; Am Geophys Union. *Res:* Solid earth geophysics data management; earthquake loss estimation to structures. *Mailing Add:* 9119 Chapel Valley Rd Dallas TX 75220

RINER, JOHN WILLIAM, b Kansas City, Mo, July 29, 24; m 47; c 5. MATHEMATICS. *Educ:* Rockhurst Col, BS, 47; Univ Notre Dame, MS, 49, PhD(math), 53. *Prof Exp:* Instr math, Univ Notre Dame, 49-50 & St Peters Col, 53-56; asst prof, St Louis Univ, 56-59; from asst prof to assoc prof, 59-69, assoc, Systs Res Group, 60-64, PROF MATH, OHIO STATE UNIV, 69-, VCHMN DEPT, 70- *Mem:* Am Math Soc. *Res:* Topology; organization theory. *Mailing Add:* Math Dept Ohio State Univ Columbus OH 43210

RINES, HOWARD WAYNE, b Portland, Ind, Feb 19, 42; m 65; c 3. PLANT GENETICS, PLANT BREEDING. *Educ:* Purdue Univ, BS, 64, MS, 66; Yale Univ, PhD(genetics), 69. *Prof Exp:* Captain, US Army Reserves, 69-71; asst prof bot & genetics, dept bot, Univ Ga, Athens, 71-76; RES GENTICIST, AGR RES SERV, USDA, 76-; ADJ ASSOC PROF, DEPT AGRON & PLANT GENETICS, UNIV MINN, ST PAUL, 76- *Mem:* Am Soc Agron; Int Asn Plant Tissue Cult. *Res:* Plant genetics; application of genetics, tissue culture, and molecular biology to crop improvement; introgression of useful genes from wild species and other exotic sources into cultivated species by conventional and nonconventional techniques. *Mailing Add:* Dept Agron & Plant Genetics Univ Minn 504 Borlaug St Paul MN 55108

RING, B ALBERT, medicine, for more information see previous edition

RING, DENNIS RANDALL, b Texarkana, Ark, Aug 6, 52; m 76; c 3. ECONOMIC ENTOMOLOGY. *Educ:* Baylor Univ, BS, 74; Tex A&M Univ, MS, 78, PhD(entom), 81. *Prof Exp:* Res technician entom, Tex Agr Exp Sta Pecan Insects, 74; res asst pecans & teaching asst hort & floral entom, Tex A&M Univ, 77-81, postdoctoral pecans & insects, 82-84, modeling tick develop & survival, 84-85; res scientist, Screwworm Res Lab, Agr Res Serv, USDA, 81-82 & Southeastern Fruit & Tree Nut Res Lab, 85-87; postdoctoral pecan insects, Coastal Plains Exp Sta, Univ Ga; postdoctoral sugarcane insects & mex rice borer, Waslaco, Tex, 87-88, POSTDOCTORAL COTTON INSECTS, ECON INJURY LEVELS, SIMULATION MODELING & FIELD TESTING, TEX AGR RES & EXT CTR, CORPUS CHRISTI, TEX, 88- *Concurrent Pos:* Instr introductory entom, Tex A&M Univ, Kingsville, Tex, 90. *Mem:* AAAS; Entom Soc Am; Am Registry Prof Entomologists. *Res:* Biology, ecology, bionomics, integrated pest management, economic injury levels, and modeling of arthropods attacking cotton, sugarcane, pecan, ticks, and screwworms; evaluation of genetically engineered cotton. *Mailing Add:* Tex Agr Res & Exten Ctr Rte 2 Box 589 Corpus Christi TX 78406-9704

RING, JAMES GEORGE, b Chicago, Ill, July 26, 38. DATABASE ADMINISTRATION, NUCLEAR MEASUREMENTS. *Educ:* Ill Inst Technol, BS, 60, PhD(physics), 66. *Prof Exp:* Res assoc physics, Univ Ill, Urbana, 65-68; vis asst prof, Univ Ill, Chicago Circle, 68-71; res physicist, Packard Instrument Co, 72-74, sr res physicist, lab, 74-82; prin tech analyst, 82-90, SR SYST ANALYST, COMMONWEALTH EDISON, 91- *Mem:* Am Phys Soc; Sigma Xi. *Res:* Optical properties of color centers in alkali halides; Borrmann effect in germanium and silicon; laser Raman scattering; applied research on radiation detectors; computer assisted spectral analysis. *Mailing Add:* 10626 S Albany Ave Chicago IL 60655

RING, JAMES WALTER, b Worcester, NY, Feb 24, 29; m 59; c 1. SOLAR ENERGY, INDOOR AIR QUALITY. *Educ:* Hamilton Col, AB, 51; Univ Rochester, PhD(physics), 58. *Prof Exp:* From asst prof to assoc prof, 57-69, chmn dept, 68-80, PROF PHYSICS, HAMILTON COL, 69-, WINSLOW PROF PHYSICS, 75- *Concurrent Pos:* NSF sci fac fel, AERE Harwell, UK, 65-66; attached physicist, Phys Chem Lab, Oxford Univ, UK, 73; vis fel, Princeton Univ, 81; guest reseacher, Tech Univ Denmark, 87; radiation officer, 64-84, Hamilton Col, 64-84, eng liaison officer, 75-, pres, Am Asn Univ Prof chap, 87-, chmn dept, 91- *Mem:* Am Phys Soc; Am Asn Physics Teachers; Int Solar Energy Soc; Am Asn Univ Profs; Sigma Xi. *Res:* Dielectrics; inelastic neutron scattering; H-bonded liquids; vibrational modes; solar energy; energy conservation; passive solar heating; infiltration and ventilation; computer energy management; thermal comfort indoors. *Mailing Add:* Dept Physics Hamilton Col Clinton NY 13323

RING, JOHN ROBERT, b Warsaw, Ind, Nov 19, 15; m 48; c 3. ANATOMY. *Educ:* Univ Ill, AB, 39; Brown Univ, ScM, 41, PhD(biol), 43. *Prof Exp:* Asst biol, Brown Univ, 39-42; instr anat, Sch Med, St Louis Univ, 43-47; from asst prof to assoc prof, 47-62, asst dean pre-clin instr & res, 67-72, chmn anat dept, 74-77, PROF ANAT, SCH DENT MED, WASH UNIV, 62-, DIR ADMIS, 72-, ASST DEAN BIOMED SCI & CHMN DEPT, 77- *Mem:* Am Asn Dent Schs; AAAS; Am Asn Anat; Int Asn Dent Res; Sigma Xi. *Res:* Sex endocrinology; gerontology; histology and histochemistry of endocrine organs and oral tissues. *Mailing Add:* Wash Univ Sch Dent Med 755 Catalpa Ave St Louis MO 63119

RING, MOREY ABRAHAM, b Detroit, Mich, Aug 31, 32; m 60; c 4. INORGANIC CHEMISTRY. *Educ:* Univ Calif, Los Angeles, BS, 54; Univ Wash, PhD(chem), 60. *Prof Exp:* Fel chem, Johns Hopkins Univ, 60-61; sr engr, Rocketdyne Div, NAm Aviation, Inc, 61-62; from asst prof to assoc prof chem, 62-69, PROF CHEM, SAN DIEGO STATE UNIV, 69- *Mem:* Am Chem Soc. *Res:* Chemistry of group IV compounds, especially silicon hydrides. *Mailing Add:* Dept Chem San Diego State Univ San Diego CA 92182-0002

RING, PAUL JOSEPH, b Winthrop, Mass, Dec 1, 28; m 66; c 4. PHYSICS & RADIOLOGICAL SCIENCE. *Educ:* Boston Col, BS, 50; Rensselaer Polytech Inst, MS, 54; Brown Univ, PhD(physics), 63. *Prof Exp:* Scientist physics, Nuclear Magnetics Corp, Perkin-Elmer Corp, 56-57; res asst, Brown Univ, 57-60; consult, Metals & Controls, Inc, Mass, 60-61; res asst, Brown Univ, 61-63; scientist semiconductors, Transitron Electronics Corp, Mass, 63-64; scientist electrooptics, RCA, Mass, 64-67; ASSOC PROF PHYSICS, UNIV LOWELL 67- *Concurrent Pos:* Vis prof, Inst Educ Technol, Univ Surrey, Eng. *Mem:* Health Physics Soc; Sigma Xi. *Res:* Solid state and infrared physics; cathodoluminescence; minerals; nuclear magnetic resonance; education; ultrasonics; dosimeters; spectral measurements of dosimeters. *Mailing Add:* Dept Physics & Appl Sci Univ Lowell Lowell MA 01854

RING, RICHARD ALEXANDER, b Glasgow, Scotland, Sept 24, 38; m 60; c 2. ENTOMOLOGY. *Educ:* Glasgow Univ, BSc, 61, PhD(entom), 65. *Prof Exp:* Asst prof zool, Univ BC, 64-65; Nat Res Coun Can fel, Entom Res Inst, Can Dept Agr, 65-66; from asst prof to assoc prof, 66-88, PROF ENTOM, UNIV VICTORIA, BC, 88- *Mem:* Soc Cryobiol; Can Soc Zool; Can Entom Soc; Arctic Inst NAm. *Res:* Insect ecology and physiology; extrinsic and intrinsic control of diapause and coldhardiness in insects; intertidal insects. *Mailing Add:* Dept Biol Univ Victoria Box 1700 Victoria BC V8W 2Y2 Can

RING, ROBERT E, b Ragsdale, Ind, Aug 6, 22; m 44; c 2. AERONAUTICAL ENGINEERING. *Educ:* Purdue Univ, BS, 50; Denver Univ, MBA, 56. *Prof Exp:* Asst prof, Western Mich Univ, 51-62; assoc prof, 62-80, PROF AERONAUT, SAN JOSE STATE UNIV, 80- *Res:* Aircraft accident research and investigation; man power requirements in the aeronautics and aerospace industry. *Mailing Add:* 3646 Debra Way San Jose CA 95117

RINGEISEN, RICHARD DELOSE, b Kokomo, Ind, Mar 18, 44; m 65; c 2. MATHEMATICS, GRAPH THEORY & APPLICATIONS. *Educ:* Manchester Col, BS, 66; Mich State Univ, MS, 68, PhD(math), 70. *Prof Exp:* Asst prof math, Colgate Univ, 70-74; from asst prof to assoc prof math, Ind Univ-Purdue Univ, Ft Wayne, 74-79; from asst prof to assoc prof math, 79-86, prof & dir grad studies math, 86-88, HEAD MATH SCI, CLEMSON, 88- *Concurrent Pos:* Vis scientist human eng, Aerospace Med Res Lab, Wright-Patterson AFB, 78-79; sci officer discrete math, Off Naval Res, 85-86; actg dir, math sci, 86. *Mem:* Am Math Soc; Math Asn Am; Soc Indust & Appl Math; Soc Indust Appl Math. *Res:* Graph theory, with special emphasis in applications, modelling and topological problems; network models for social sciences. *Mailing Add:* Dept Math Sci Clemson Univ Clemson SC 29631

RINGEL, SAMUEL MORRIS, b New York, NY, Nov 29, 24; m 51; c 3. CLINICAL PHARMACOLOGY, INFECTIOUS DISEASES. *Educ:* Hunter Col, BA, 50; Univ Mich, MS, 51; Mich State Univ, PhD(mycol), 56. *Prof Exp:* Jr microbiologist, Hoffmann-La Roche, 51-52; asst mycol, Mich State Univ, 53-56; from assoc plant pathologist to plant pathologist, Agr Mkt Serv, USDA, 56-61; from sr scientist to sr res assoc, Dept Microbiol & Immunol, Warner-Lambert Res Inst, 61-78; from asst dir to assoc dir clin pharmacol, Revlon Health Care Group, 79-84; PHARMACEUT CONSULT, RINGEL ASSOCS, 85- *Concurrent Pos:* Vis scientist, NJ Acad Sci, 65-67; consult numerous univs, industs & insts, 78-85. *Mem:* Int Soc Human & Animal Mycoses; Soc Indust Microbiol; Am Soc Clin Pharmacol & Therapeut; Am Soc Microbiol; NY Acad Sci. *Res:* Fungal enzymes; antifungal agents; post-harvest diseases of fruits and vegetables; medical and pharmaceutical microbiology; antibiotics; cardiovascular research. *Mailing Add:* PO Box 26625 Tamarac FL 33320-6625

RINGEL, STEVEN ADAM, b New Brunswick, NJ, July 21, 62; m 85. ELECTRONIC MATERIALS, MATERIALS CHARACTERIZATION. *Educ:* Pa State Univ, BS, 84, MS, 86; Ga Inst Technol, PhD(elec eng), 91. *Prof Exp:* Res asst semiconductors, Dept Eng Sci, Pa State Univ, 84-86; RES ASST, MICROELECTRONICS RES CTR & SCH ELEC ENG, GA INST TECHNOL, 86- *Mem:* Inst Elec & Electronics Engrs; Am Phys Soc. *Mailing Add:* 3402 Seven Pines Ct Atlanta GA 30339

RINGENBERG, LAWRENCE ALBERT, mathematics; deceased, see previous edition for last biography

RINGER, DAVID P, CHEMICAL CARCINOGENESIS, GENE EXPRESSION. *Educ:* Wayne State Univ, Detroit, PhD(biochem), 73. *Prof Exp:* SECT HEAD, BIOCHEM PHARMACOL, SAMUEL ROBERTS NOBLE FOUND, INC, 75- *Res:* Carcinogenic metabolism. *Mailing Add:* Dept Biochem Pharmacol PO Box 2180 Samuel Roberts Noble Found Inc Hwy 199 Ardmore OK 73402

RINGER, LARRY JOEL, b Cedar Rapids, Iowa, Sept 24, 37; m 60; c 3. STATISTICS. *Educ:* Iowa State Univ, BS, 59, MS, 62; Tex A&M Univ, PhD(statist), 66. *Prof Exp:* Res asst statist, Iowa State Univ, 59-61; from instr to assoc prof, 65-75, res scientist, Data Processing Ctr, 74-76, PROF STATIST & ASST HEAD DEPT, TEX A&M UNIV, 75- *Concurrent Pos:* Assoc res statistician, Tex Transportation Inst, 70-79; Statistical consult, Data Processing Ctr, 81-82. *Mem:* Am Statist Asn; Am Soc Qual Control; Sigma Xi. *Res:* Statistical methods and techniques as applied to problems in engineering and physical sciences; agriculture. *Mailing Add:* Dept Statist Tex A&M Univ College Station TX 77843

RINGER, ROBERT KOSEL, b Ringoes, NJ, Feb 21, 29; m 51; c 1. PHYSIOLOGY. *Educ:* Rutgers Univ, BS, 50, MS, 52, PhD(physiol), 55. *Prof Exp:* Asst prof avian physiol, Rutgers Univ, 55-57; from asst prof to assoc prof, 57-64, PROF AVIAN PHYSIOL, MICH STATE UNIV, 64-, PROF PHYSIOL, 65- *Concurrent Pos:* On leave, Unilever Res Lab, Eng, 66-67 & Environ Protection Agency Res Lab, Corvallis, Ore, 84-85; trustee, Am Asn Accreditation Lab Animal Care. *Mem:* Am Asn Avian Path; Am Physiol Soc; fel Poultry Sci Asn (2nd vpres); World Poultry Sci Asn; Soc Environ Toxicol & Chem. *Res:* Cardiovascular research; endocrinology. *Mailing Add:* Poultry Sci Dept Mich State Univ East Lansing MI 48823

RINGHAM, GARY LEWIS, b Boonville, Ind, Nov 11, 41; m 64; c 2. PHYSIOLOGY, PHARMACOLOGY. *Educ:* Butler Univ, BS, 63, MS, 66; Univ Utah, PhD(pharmacol), 71. *Prof Exp:* Fel physiol, Univ Colo Med Ctr, Denver, 71-73; res instr, Univ Utah, 73-76, asst prof physiol, 76-; CLIN RES ASSOC, ABBOTT LABS. *Res:* Physiology and pharmacology of synaptic transmission; excitation-secretion coupling; impulse origin and conduction. *Mailing Add:* Abbott Labs Dept 49J APGA-2 Abbott Park IL 60061

RINGLE, DAVID ALLAN, b Wausau, Wis, Sept 28, 24; m 49. PHYSIOLOGY, IMMUNOLOGIC TEST METHODS. *Educ:* Univ Wis, BS, 49; Columbia Univ, MA, 52; NY Univ, PhD(cellular physiol), 60. *Prof Exp:* Scientist, Warner-Lambert Pharmaceut Co, NY, 52-53; biologist, Am Cyanamid Co, Conn, 53-54; USPHS fel, Col Physicians & Surgeons, Columbia Univ, 60-62; prin physiologist, Midwest Res Inst, 62-76; CONSULT IMMUNOPHYSIOL, 76- *Mem:* Harvey Soc; Am Soc Zoologists; Reticuloendothelial Soc; Sigma Xi. *Res:* Tissue metabolism; reticuloendothelial system function; plasma protein alterations; amphibian yolk; shock; structure and function of cell membranes; cellular immunology; lymphocyte physiology; antilymphocyte serum; immunologic effects of morphine; immunologic detection methods. *Mailing Add:* PO Box 8013 Prairie Village KS 66208

RINGLE, JOHN CLAYTON, b Kokomo, Ind, Aug 28, 35; m 60; c 4. NUCLEAR ENGINEERING. *Educ:* Case Inst Technol, BS, 57, MS, 59; Univ Calif, Berkeley, PhD(nuclear eng), 64. *Prof Exp:* Res physicist, Cambridge Res Lab, 63-66; assoc prof, 66-84, PROF NUCLEAR ENG, RADIATION CTR, ORE STATE UNIV, 84-, ASSOC DEAN GRAD SCH, 80- *Concurrent Pos:* Vis scientist, Cadarache Nuclear Lab, France, 75. *Mem:* Am Phys Soc; Am Nuclear Soc; Sigma Xi. *Res:* Safety analysis and radiological engineering calculations for nuclear power plants, environmental effects of nuclear power, energy system analysis; radioactive waste management. *Mailing Add:* Radiation Ctr Ore State Univ Corvallis OR 97331

RINGLEE, ROBERT J, b Sacramento, Calif, Apr 23, 26; m 49; c 3. MECHANICS, ELECTRICAL ENGINEERING. *Educ:* Univ Wash, BS, 46, MS, 48; Rensselaer Polytech Inst, PhD(mech), 64. *Prof Exp:* Engr, Gen Elec Co, 48-55, supvr design, 55-60, sr engr, 60-67, mgr systs reliability, Elec Utility Eng Oper, 67-69; prin engr & dir, 69-86, PRIN CONSULT, POWER TECHNOL INC, 86- *Concurrent Pos:* Mem, Expert Adv Study Comt 38, Int Conf Large Elec Systs, 53-; adj prof, Polytech Inst Brooklyn, 65-66; Atwood Assoc, Int Conf Large High Voltage Elec Systs. *Mem:* Fel Inst Elec & Electronics Engrs; fel AAAS. *Res:* Systems sciences; vibration, noise; process modelling; system reliability; control theory. *Mailing Add:* Power Technol Inc PO Box 1058 Schenectady NY 12301

RINGLER, DANIEL HOWARD, b Oberlin, Ohio, Aug 19, 41; m 63; c 2. LABORATORY ANIMAL MEDICINE. *Educ:* Ohio State Univ, DVM, 65; Univ Mich, MS, 69; Am Col Lab Animal Med, dipl, 71. *Prof Exp:* Fel, 67-69, instr, 69-71, asst prof, 71-75, assoc prof, 75-79, PROF LAB ANIMAL MED, UNIV MICH, 79- *Concurrent Pos:* Consult, Animal Resources Br, Div Res Resources, NIH, 73-; mem, Am Asn Accreditation Lab Animal Care, 74-, Adv Coun, Inst Lab Animal Resource, Nat Res Coun-Nat Nat Acad Sci, 75-78. *Honors & Awards:* Griffin Award, Am Asn Lab Animal Sci. *Mem:* Am Asn Lab Animal Sci; Am Vet Med Asn. *Res:* Spontaneous diseases of laboratory animals; use of animals in biomedical research; diseases and husbandry of amphibians; pathogenic bacteriology. *Mailing Add:* Unit Lab Animal Med Univ Mich Ann Arbor MI 48109

RINGLER, IRA, b Brooklyn, NY, Feb 11, 28; m 54; c 3. BIOCHEMISTRY. *Educ:* Ohio State Univ, BS, 51; Cornell Univ, MNutrS, 53, PhD(biochem), 55. *Prof Exp:* Res fel endocrinol, Harvard Univ, 55-57; group leader, Exp Therapeut Res Sect, Lederle Labs, Am Cyanamid Co, 57-61, head dept metab chemother, 61-66, dir exp therapeut res sect, 66-69, dir res, 69-75; vpres res mgt & corp develop, 75-76; vpres pharmaceut prod res & develop, Abbott Labs, 76-83; PRES, TAP PHARMACEUT, 83- *Mem:* Am Chem Soc; Endocrine Soc; Am Soc Pharmacol & Exp Therapeut; Am Soc Biol Chem. *Res:* Biochemical aspects of endocrinology. *Mailing Add:* Tap Pharmaceut 1400 Sheridan Rds North Chicago IL 60064

RINGLER, NEIL HARRISON, b Long Beach, Calif, Nov 12, 45; m 68. AQUATIC ECOLOGY, FISHERY BIOLOGY. *Educ:* Calif State Univ, Long Beach, BA, 67; Ore State Univ, MS, 70; Univ Mich, PhD(fisheries biol), 75. *Prof Exp:* Res asst, Ore Game Comn, Ore State Univ, 68-69; teaching fel ichthyol & aquatic entom, Sch Natural Resources, Univ Mich, 71-74; from asst prof to prof zool, 75-89, ASSOC CHMN & PROF ENVIRON FOREST & BIOL, STATE UNIV NY COL ENVIRON SCI & FOREST BIOL, 89- *Concurrent Pos:* Res biologist, Pac Biol Sta, Nanaimo, BC, 85. *Mem:* Sigma Xi; Am Fisheries Soc; Am Soc Ichthyologists & Herpetologists; Ecol Soc Am; NAm Benthological Soc. *Res:* Foraging tactics and behavior of fishes; salmonid biology; population ecology of fishes and aquatic invertebrates; role of predation in structering aquatic communities; stream ecology; effects of forest practices on streambeds; restoration of perturbed fish communities. *Mailing Add:* Dept Envir & Forest Biol SUNY Col Environ Sci & Forestry Syracuse NY 13210

RINGLER, ROBERT L, biochemistry; deceased, see previous edition for last biography

RINGO, GEORGE ROY, b Minot, NDak, Jan 19, 17; m 41; c 3. EXPERIMENTAL PHYSICS. *Educ:* Univ Chicago, BS, 36, PhD(physics), 40. *Prof Exp:* Physicist, US Rubber Co, RI, 41; physicist, Naval Res Lab, DC, 41-48; PHYSICIST, ARGONNE NAT LAB, 48- *Mem:* Am Phys Soc; Sigma Xi; fel Am Phys Soc. *Res:* Neutron physics; ion microscopy. *Mailing Add:* 16 W 220 97th St Hinsdale IL 60521-6898

RINGO, JAMES LEWIS, b La Grange, Ill, May 17, 51; m 81; c 3. LEARNING & MEMORY, VISION. *Educ:* Mass Inst Technol, BS, 73; Duke Univ, PhD(physiol), 79. *Prof Exp:* Fulbright fel med physics, Univ Amsterdam, 81-82; res fel neurosci, Ctr Brain Res, Univ Rochester, 82-85, asst prof neurosci, 85-87, asst prof, Dept Physiol, 87-91, ASSOC PROF, DEPT PHYSIOL, UNIV ROCHESTER, 91- *Mem:* AAAS; NY Acad Sci; Neurosci Soc. *Res:* Neurophysiologic description of vision and visual memory; the inter hemispheric transfer of information and the split brain. *Mailing Add:* 122 Irvington Rd Rochester NY 14620

RINGO, JOHN ALAN, b Spokane, Wash, Dec 29, 41; m 64; c 3. BIOENGINEERING, BIOMEDICAL ENGINEERING. *Educ:* Wash State Univ, BS, 64; Univ Wash, MS, 67, PhD(elec eng), 71. *Prof Exp:* Assoc engr, Douglas Aircraft Co, 64; assoc res engr, Boeing Co, 66-67; from asst prof to assoc prof, 77-83, PROF ELEC ENG, WASH STATE UNIV, 83- *Mem:* Sigma Xi. *Res:* Application of engineering principles for understanding the interactive heart-artery systems; specialized instrumentation for measurement in physiological systems; solid state device properties and their application in linear circuit design. *Mailing Add:* Dept Computer Sci Wash State Univ Pullman WA 99164-1210

RINGO, JOHN MOYER, b Columbia, Mo, Nov 25, 43; m 69, 90; c 2. BEHAVIORAL GENETICS. *Educ:* Univ Calif, Berkeley, AB, 69; Univ Calif, Davis, PhD(behav genetics), 73. *Prof Exp:* From asst prof to assoc prof, 74-88, PROF ZOOL, UNIV MAINE, 88- *Concurrent Pos:* Assoc ed, Behav Genetics, 78-81, J, Evolution, 89-91. *Mem:* Behav Genetics Asn; Soc Study Evolution; Genetics Soc Am; Am Soc Naturalists; Sigma Xi. *Res:* Genetic analysis of stereotyped behavior, biological clocks, and reproductive behavior in Drosophilia. *Mailing Add:* Dept Zool Univ Maine Orono ME 04469

RINGS, ROY WILSON, b Columbus, Ohio, Aug 15, 16; m 42; c 3. ENTOMOLOGY. *Educ:* Ohio State Univ, BSc, 38, MSc, 40, PhD(entom), 46. *Prof Exp:* Asst, Ohio Exp Sta, 41-42; specialist in chg, Div Plant Indust, State Dept Agr, 46-47; asst entomologist, 47-53; assoc prof entom, Ohio Agr Exp Sta, 54-61; assoc chmn ctr, 61-73, prof entom, 61-77, EMER PROF ENTOM, OHIO STATE UNIV 77- *Concurrent Pos:* US Army, med entom, 43-46. *Mem:* Entom Soc Am; Am Mosquito Control Asn; fel, AAAS. *Res:* Noctuidae of Ohio. *Mailing Add:* Ohio Agr Res & Develop Ctr Wooster OH 44691

RINGSDORF, WARREN MARSHALL, JR, b Elba, Ala, May 2, 30; m 55; c 2. DENTISTRY, NUTRITION. *Educ:* Asbury Col, AB, 51; Univ Ala, MS, 56; Univ Ala, Birmingham, DMD, 56; Am Bd Oral Med, dipl, 63. *Prof Exp:* Pvt dent pract, Ala, 58-59; asst prof, 59-64, assoc prof, 64-82, PROF CLIN DENT, UNIV ALA, BIRMINGHAM, 82- *Concurrent Pos:* Fel, Univ Ala, Birmingham, 60-62; attend oral med, Vet Admin Hosp, Birmingham, Ala, 60-; consult, Hq US Army Infantry Ctr, Ft Benning, Ga, 64-74 & Nutritech Corp, Santa Barbara, 80-; pres, Seven, Inc. *Honors & Awards:* Chicago Dent Soc Res Award, 66 & 68; Honors Achievement Award, Angiol Res Found, Inc, 68. *Mem:* Am Dent Asn; Am Acad Oral Med; Acad Orthomolecular Psychiat. *Res:* Role of diet and nutrition in host resistance and susceptibility and its relationship to prevention of disease occurrence and recurrence. *Mailing Add:* Dept Oral Med Univ Ala Birmingham AL 35294

RINI, FRANK JOHN, b New York, NY, Feb 18, 52; m 84; c 1. OPHTHALMOLOGY, RADIATION PHYSICS & BIOLOGY. *Educ:* Columbia Univ, BA, 74, MA, 76, MPh, 77, PhD(physics), 78, MD, 82. *Prof Exp:* Res assoc radio physics, Columbia Univ, 76-78; res assoc radiobiol, Lawrence Berkeley Lab, 79; res assoc ocular radio, Eye Inst, Columbia Univ, 81-86. *Concurrent Pos:* Ophthal resident, Eye Inst, Columbia, 83-86. *Mem:* Am Acad Ophthal; Asn Res Vision & Ophthal. *Res:* Effects of different radiation modalities on ocular tissues, particulary the lens. *Mailing Add:* 15 Council Crest Rd Sloatsburg NY 10974-1802

RINK, GEORGE, b Riga, Latvia, Jan 17, 42; US citizen; m 66; c 3. BLACK WALNUT GENETICS. *Educ:* NY Univ, BA, 63; Univ Tenn, MS, 71, PhD(forest genetics), 74. *Prof Exp:* Asst prof forestry, Stephen F Austin State Univ, 75-80; RES GENETICIST, N CENT FOREST EXP STA, USDA FOREST SERV, 80- *Concurrent Pos:* Adj asst prof, Dept Forestry, Southern Ill Univ, 81- *Mem:* Sigma Xi; Soc Am Foresters; Walnut Coun. *Res:* Genetic variation in black walnut, white oak and southern pines; genetic resistance to stress during establishment. *Mailing Add:* Forestry Sci Lab USDA Forest Serv Southern Ill Univ Carbondale IL 62901

RINK, RICHARD DONALD, b Chicago, Ill, Mar 29, 41; m 62; c 2. HUMAN ANATOMY. *Educ:* Beloit Col, BA, 63; Tulane Univ, PhD(anat), 67. *Prof Exp:* From instr to assoc prof, 67-80, PROF ANAT, SCH MED, UNIV LOUISVILLE, 80-, RES PROF SURG ANAT, 81- *Mem:* Am Physiol Soc; Am Asn Anatomists; Shock Soc. *Res:* Oxygen supply to tissue in relation to shock; injury mechanisms. *Mailing Add:* Dept Anat Health Sci Ctr Univ Louisville Sch Med Louisville KY 40292

RINKEMA, LYNN ELLEN, LEUKOTRIENE RESEARCH. *Educ:* Loyola Univ, MS, 81. *Prof Exp:* PHARMACOLOGIST, ELI LIILY & CO, 81- *Mailing Add:* Lilly Res Labs MC931 Eli Lilly & Co Indianapolis IN 46285

RINKER, GEORGE ALBERT, JR, b Lubbock, Tex, Feb 7, 45; c 3. ELECTROMAGNETIC MATTER, SELF-ORGANIZING INFORMATION NETWORKS. *Educ:* Franklin Col, BA, 67; Univ Calif, Irvine, MA, 70, PhD(physics), 71. *Prof Exp:* Fel, 71-73, STAFF MEM PHYSICS, LOS ALAMOS SCI LAB, 73- *Concurrent Pos:* Vis scientist, Nuclear Energy Res Inst, J lich, WGer, 76-77; vis lectr, Univ Fribourg, Switz, 77, 85. *Mem:* Am Phys Soc. *Res:* Advanced computational methods applied to fundamental problems in many body-quantum-mechanical electromagnetic systems. *Mailing Add:* T-1 MSB221 Los Alamos Nat Lab Los Alamos NM 87545

RINKER, GEORGE CLARK, b Hamilton, Kans, Apr 8, 22; m 44; c 3. ANATOMY. *Educ:* Univ Kans, AB, 46; Univ Mich, MS, 48, PhD(zool), 51. *Prof Exp:* From instr to asst prof anat, Univ Mich, 50-62; assoc prof, 62-70, asst dean sch med, 70-73, PROF ANAT, SCH MED, UNIV SDAK, VERMILLION, 70-, ASSOC DEAN SCH MED, 73- *Mem:* Am Asn Anat. *Res:* Comparative mammalian and human gross anatomy. *Mailing Add:* Dept Anat Univ SDak Sch Med Vermillion SD 57069

RINKER, ROBERT G(ENE), b Vincennes, Ind, Dec 31, 29; m 63; c 3. CHEMICAL ENGINEERING. *Educ:* Rose Polytech Inst, BS, 51; Calif Inst Technol, MS, 55, PhD(chem eng), 59. *Prof Exp:* Res fel, Calif Inst Technol, 59-60, asst prof, 60-67; assoc prof, 67-73, chmn dept, 73-78, PROF CHEM ENG, UNIV CALIF, SANTA BARBARA, 73- *Concurrent Pos:* Consult, NAm Instrument Co, 55, Electro Optical Systs, Inc, 60, Dow Chem Co, 64-65, MHD Res, Inc, 65-67, Sci Appln, Inc, 70-87, JRB Assocs, 70-87 & Omnia Res, 85-; fel, Am Inst Chem Engrs. *Mem:* AAAS; Am Inst Chem Engrs; Sigma Xi. *Res:* Chemical kinetics; reactor design; catalysis. *Mailing Add:* Chem Engr Univ Calif Santa Barbara CA 93106

RINNE, JOHN NORMAN, b Pawnee City, Nebr, Mar 19, 44; m 66; c 1. FISHERY BIOLOGY, AQUATIC ECOLOGY. *Educ:* Peru State Col, BSE, 66; Ariz State Univ, MS, 69, PhD(zool), 73. *Prof Exp:* Chief field invest res, Ariz State Univ, 72-73; fishery res biologist, EAfrican Community, 73-75; fishery res biologist, US Fish Wildlife Serv, 76; FISHERY RES BIOLOGIST, US FOREST SERV, 76- *Concurrent Pos:* Consult, Squawfish Recovery Team, 76-77, Ariz Trout Recovery Team & Gila Trout Recovery Team, 76-; area coordr, Desert Fishes Coun, 77- *Mem:* Am Fisheries Soc; Am Inst Fishery Res Biologists. *Res:* Habitat, biology and distribution of endangered fish in the southwestern United States; fisheries and ecology of desert and tropical resources. *Mailing Add:* 1025 E Balboa Circle Tempe AZ 85282

RINNE, ROBERT W, b Hammond, Ind, Jan 6, 32; m 56; c 4. PLANT PHYSIOLOGY. *Educ:* DePauw Univ, BA, 55; Purdue Univ, BS, 57, MS, 59, PhD(plant physiol), 61. *Prof Exp:* Asst plant physiol, Purdue Univ, 57-61; res assoc microbiol, Dartmouth Med Sch, 61-62; NIH fel biochem, Wayne State Univ, 62-64; PROF PLANT PHYSIOL, DEPT AGRON, UNIV ILL, URBANA, 64- *Mem:* Am Soc Plant Physiol; Crop Sci Soc Am. *Res:* Metabolism of the developing soybean seed. *Mailing Add:* Dept Agron Univ Ill S 312 Turner Hall Urbana IL 61801

RINNE, VERNON WILMER, b Pawnee City, Nebr, Dec 15, 25; m 48; c 5. DENTISTRY. *Educ:* Univ Nebr, BS & DDS, 53. *Prof Exp:* From instr to assoc prof oper dent, 53-68, chmn dept restorative dent, 69-72, PROF ADULT RESTORATION, COL DENT, UNIV NEBR-LINCOLN, 69- *Concurrent Pos:* Consult, USPHS, 60. *Mem:* Am Asn Dent Res. *Res:* Dental materials. *Mailing Add:* Dept Restorative Dent Univ Nebr Col Dent Lincoln NE 68583-0740

RINSE, JACOBUS, b Amsterdam, Neth, Apr 4, 00; m 31; c 3. PHYSICAL CHEMISTRY, BIOCHEMISTRY. *Educ:* Univ Amsterdam, PhD(chem), 27. *Prof Exp:* Instr phys chem, Univ Amsterdam, 23-27; fel, John Hopkins Univ, 27; res chemist, Int Paper Co, 28; chief chemist & asst dir, N V Pieter Schoen a Zn Zaandam, Neth, 28-38; founder & dir, Chem Tech Advice Bur, 38-49; dir & owner, Chem Res Assocs, 50-80; RETIRED. *Mem:* Am Chem Soc; Am Oil Chemists Soc; Asn Consult Chemists & Chem Engrs (treas); NY Acad Sci; Brit Oil & Colour Chemists Asn; Tell Oil Asn. *Res:* Allotropy of mercury sulfide and vapor pressure measurements of mercuric iodide; tung oil; alkyd resins; styrenated resins; vinyl esters; paints; degumming of fibers; exchange esterifications; pigments; thixotropy; organic metal compounds; greases; antiacids; metallic resins; atherosclerosis and aging. *Mailing Add:* East Dorset VT 05253

RINSLEY, DONALD BRENDAN, psychoanalytic research, medical education; deceased, see previous edition for last biography

RINZEL, JOHN MATTHEW, b Milwaukee, Wis, July 18, 44; m 67; c 2. APPLIED MATHEMATICS. *Educ:* Univ Fla, BS, 67; NY Univ, MS, 68, PhD(math), 73. *Prof Exp:* Mathematician, Div Comput Res & Technol, NIH, 68-70 & 73-75, RES MATHEMATICIAN & CHIEF, MATH RES BR, NAT INST DIABETES, DIGESTIVE & KIDNEY DIS, NIH, 75- *Concurrent Pos:* Vis instr, Dept Math, Univ Md, College Park, 75- *Mem:* Soc Indust & Appl Math; Am Math Soc; Soc Neurosci. *Res:* Biomathematics; mathematical models for electrical signaling in biological cells; theoretical neurophysiology; numerical analysis. *Mailing Add:* Bldg 31 Rm 4B-54A Nat Inst Health Math Res Br Bethesda MD 20892

RIO, DONALD C, b New Britain, Conn, July 7, 57. EUKARYOTIC GENE EXPRESSION. *Educ:* Univ Colo, BA, 79; Univ Calif, Berkeley, PhD(biochem), 83. *Prof Exp:* Res asst, 79-83, fel biochem, Univ Calif, Berkeley, 84-86; asst prof, 87-, ASSOC PROF, DEPT BIOL, MASS INST TECHNOL. *Concurrent Pos:* Assoc mem, Whitehead Inst Biomed Res, 87- *Honors & Awards:* Presidential Young Investr Award, NSF, 88. *Mem:* AAAS. *Res:* Regulation of eukaryotic gene expression; transcriptional control mechanisms; eukaryotic transposable elements; differential gene expression during development of the fruit fly, Drosophila melanogaster; RNA transcription and processing. *Mailing Add:* Whitehead Inst Biomed Res Mass Inst Technol Nine Cambridge Ctr Cambridge MA 02142

RIO, MARÍA ESTHER, b González Chaves, Buenos Aires, Apr 5, 36; m 67; c 2. NUTRITIONAL RECOVERY, EVALUATION NUTRITIONAL STATUS. *Educ:* Tandil Nat Sch, BS, 54; Univ Buenos Aires, Pharmacyst, 60, Biochemist, 62, Biochemist PhD(nutrit), 69. *Hon Degrees:* Univ Buenos Aires, Degree with honors, 62. *Prof Exp:* Minor positions basic nutrit, Sch Pharm & Biochem, 61-70, from asst prof to assoc prof, 70-85, PROF NUTRIT, SCH PHARM & BIOCHEM, 85-, HEAD DEPT HEALTH SCI, 91- *Concurrent Pos:* Career investr nutrit, Nat Res Coun, CONICET, 70-; exec secy, Nat Prog Food Res Sci & Technol, 83-86; prin investr, PID-CONICET Proj, 84-, IDRC N Degree 3-)-83- 0085, 85-88, SANCOR-CONICET-CERELA Proj, 88-91 & SANCOR-CMN Proj, 90-91; consult, Centro Munic de Nutrición (Off Nutrit Ctr Buenos Aires City), 90- & Master on Human Nutrit, Univ San Luis, 90- *Mem:* Am Inst Nutrit; Am Soc Clin Nutrit; Int Union Nutrit Sci. *Res:* Nutritional recovery projects conducted to evaluate proficiency of formulations for nutritional support and recovery; biochemical evaluation of nutritional status: basic and operative research. *Mailing Add:* Junin 956 2nd p Buenos Aires 1113 Argentina

RIO, SHELDON T, b Raymond, Mont, May 9, 27; m 50; c 2. MATHEMATICS. *Educ:* Westmar Col, BA, 50; Mont State Univ, MA, 54; Ore State Univ, PhD(anal), 59. *Prof Exp:* Chmn, Dept Math, Pac Univ, 55-57; assoc prof, Western Wash State Col, 59-63; chmn dept math, Southern Ore Col, 63-72, prof, 63-89, dir, Sch Sci & Math, 79-89; RETIRED. *Concurrent Pos:* Prof, Univ Delhi, 67. *Mem:* Am Math Soc; Math Asn Am; Nat Coun Teachers Math. *Res:* Analysis; topology. *Mailing Add:* 570 Taylor St Ashland OR 97520

RIOCH, DAVID MCKENZIE, neuropsychiatry; deceased, see previous edition for last biography

RIOPEL, JAMES L, b Kittery, Maine, May 24, 34; m 56; c 3. PLANT MORPHOLOGY. *Educ:* Bates Col, AB, 56; Harvard Univ, MS, 58, PhD(biol), 60. *Prof Exp:* Asst prof, 60-66, assoc dean grad sch arts & sci, 68-69, ASSOC PROF BIOL, UNIV VA, 66-, DIR MT LAKE BIOL STA, 60- *Concurrent Pos:* NSF Instnl grant, 62-63; Am Cancer Soc Instnl grant, 63-64; NSF grant, 65-67. *Mem:* Bot Soc Am; Torrey Bot Club; Soc Develop Biol; Int Soc Plant Morphol. *Res:* Developmental plant anatomy and morphogenesis; regulation on cell differentiation and organ origin and determination. *Mailing Add:* Dept Biol Univ Va Gilmer Hall 278 Charlottesville VA 22903

RIOPELLE, ARTHUR J, b Thorpe, Wis, Apr 22, 20; m 42; c 3. SENSATION & PERCEPTION, BRAIN FUNCTION. *Educ:* Univ Wis-Madison, PhD(psychol), 50. *Prof Exp:* Boyd prof, 79-89, EMER BOYD PROF PSYCHOL, LA STATE UNIV, 89- *Mem:* Am Physiol Soc; Am Psychol Asn. *Res:* Primate behavior. *Mailing Add:* Dept Psychol La State Univ Baton Rouge LA 70803-5501

RIORDAN, H(UGH) E(RNEST), mechanical engineering, for more information see previous edition

RIORDAN, JAMES F, b New Haven, Conn, Feb 6, 34; m 70; c 4. BIOLOGICAL CHEMISTRY. *Educ:* Fairfield Univ, BS, 55; Fordham Univ, MS, 57, PhD(enzym), 61. *Hon Degrees:* MA, Harvard Univ, 87. *Prof Exp:* Instr chem, US Merchant Marine Acad, 57-58 & Fordham Univ, 58-61; res fel biol chem, Harvard Med Sch, 61-64, assoc, 66-68, from asst prof to assoc prof, 68-87, res assoc, 64-65, Brigham & Women's Hosp; asst dir, Clin Chem Lab, 66-86, ASSOC BIOCHEMIST, BRIGHAM & WOMEN'S HOSP, 61-, MED DIR, CLIN CHEM LAB, 86-; PROF BIOCHEM, HARVARD MED SCH, 87- *Concurrent Pos:* Nat Found fels, 62-63 & 64-65; NIH fel, 63-64; ed, J Inorg Biochem, 79-; exec ed, Analytical Biochem, 79-; mem adv bd, Clin Chem Standards, Nat Bur Standards, 71-74, Med Chem Study Sect, NIH, 74-77, chmn, Bioanalytical & Metallobiochem Study Sect, 79-82. *Mem:* Am Chem Soc; Am Soc Biochem & Molecular Biol; AAAS. *Res:* Chemical modification of proteins; proteolytic enzymes; angiogenesis; zinc metalloenzymes; angiotensin converting enzyme. *Mailing Add:* Seeley G Mudd Bldg 250 Longwood Ave Boston MA 02115

RIORDAN, JOHN RICHARD, b St Stephen, NB, Sept 2, 43; m 70. BIOCHEMISTRY. *Educ:* Univ Toronto, BSc, 66, PhD(biochem), 70. *Prof Exp:* Fel, Max Planck Inst Biophys, Frankfurt, 70-73; INVESTR BIOCHEM, RES INST, HOSP SICK CHILDREN, 73-; PROF, DEPTS BIOCHEM & CLIN BIOCHEM, UNIV TORONTO, 74- *Concurrent Pos:* Can Cystic Fibrosis Found fel, 70; res grants, Med Res Coun Can, 81- & Can Cystic Fibrosis Found, 82- *Mem:* Can Fedn Biol Soc; Am Soc Biol Chem. *Res:* Studies of mammalian cell plasma membrane glycoproteins, particularly normal structure, function and aberrations thereof in genetic disease; including cystic fibrosis, disorders of myclimation and cancer. *Mailing Add:* Hosp Sick Children 555 University Ave Toronto ON M5G 1X8 Can

RIORDAN, MICHAEL DAVITT, b Willimantic, Conn, Oct 21, 21; m 45; c 1. PETROLEUM CHEMISTRY. *Educ:* Col Holy Cross, BS, 43, MS, 44. *Prof Exp:* Analytical chemist, Texaco Inc, 43-45, chemist, 45-54, proj leader fuels res, 54-57, asst supvr, 57-61, asst supvr process res, 61- 67, res supvr chem res, 67-76, staff coordr, 76-79, mgr petrol prod res staff, 79-82; RETIRED. *Mem:* Am Chem Soc. *Res:* Petroleum products; petroleum processing; petrochemicals. *Mailing Add:* PO Box 1755 Anna Maria FL 33501

RIORDON, J(OHN) SPRUCE, b Springs, SAfrica, June 28, 36; Can citizen; m 63; c 3. COMPUTER SYSTEMS ENGINEERING, COMMUNICATIONS. *Educ:* McGill Univ, BEng, 57, MEng, 61; Univ London, PhD(automatic control eng), 67. *Prof Exp:* Res officer, Nat Res Coun Can, 57-68; sessional lectr, 67-68, asst prof eng, 68-70, assoc prof systs eng, 70-77, chmn dept systs eng & comput sci, 70-75, chmn dept systs eng & comput sci, 78-81, PROF SYSTS ENG, CARLETON UNIV, 77-, DEAN ENG, 81- *Concurrent Pos:* Nat Res Coun Can res grant, 68-; consult, Dept Energy, Mines & Resources, 71-73, Ministry State Urban Affairs, 75-76 & Dept Commun, 80- *Mem:* Inst Elec & Electronics Engrs; Can Info Processing Soc. *Res:* Mobile communications; computer networks; distributed databases; modelling and simulation; information systems design; application of digital computers to on-line process control; optimum control of stochastic processes; adaptive control systems; modelling of dynamic systems; information storage and retrieval; management information systems. *Mailing Add:* 69 Promenade Ave Ottawa ON K2E 5X9 Can

RIOS, PEDRO AGUSTIN, b Havana, Cuba, Apr 26, 38; US citizen; m 60; c 2. CRYOGENICS. *Educ:* Mass Inst Technol, BS, 59 & 60, MS, 67, ScD(mech eng), 69. *Prof Exp:* Plant engr, Airco Indust Gases, 60-62, asst plant supt, 63-65; res assoc, Mass Inst Technol, 69-70; mech & proj engr, Gen Elec Co, 70-73, mgr, Rotating Mach Unit, 73-77, mgr, Electro-Mech Br, Res & Develop Ctr, 77-87; VPRES, GUTIERREZ CO, 87- *Honors & Awards:* IR-100 Award Indust Res & Develop, 80. *Mem:* Am Soc Mech Engrs; Sigma Xi. *Res:* Application of cryogenics and superconductivity to rotating electrical machinery, electrical apparatus and magnets; computer aided engineering tools for electromagnetic and electromechanical devices and fluid flow. *Mailing Add:* 11 Wright Farm Concord MA 01742

RIOUX, CLAUDE, b Mont-Joli, Que, June 4, 53; m 76; c 2. SPACE PHYSICS. *Educ:* Univ Laval, BAC, 75, MSc, 78, PhD(nuclear physics), 82. *Prof Exp:* Natural sci & eng res coun fel, Lawrence Berkeley Lab, 82-84; res fel, 84-89, RES ASSOC, UNIV LAVAL, 89- *Res:* Microgravity and fractal aggregates. *Mailing Add:* Dept Physics Univ Laval Pavillon Vachon St Foy PQ G1K 7P4 Can

RIOUX, ROBERT LESTER, b Natick, Mass, June 11, 27; m 58; c 6. GEOLOGY. *Educ:* Univ NH, BA, 53; Univ Ill, MS, 55, PhD(geol), 58. *Prof Exp:* Asst geol, Univ Ill, 54-55; GEOLOGIST, US GEOL SURV, 56- *Mem:* Geol Soc Am; Am Asn Petrol Geologists. *Res:* Economic geology of mineral fuels and fertilizers; conservation of mineral lands; economic geology of mineral fuels and fertilizaers; conservation of mineral lands. *Mailing Add:* US Geol Surv Geol Div 12201 Sunrise Valley Dr MS 915 Reston VA 22092

RIPARBELLI, CARLO, b Rome, Italy, Nov 15, 10; nat US; m 58. AERONAUTICAL ENGINEERING. *Educ:* Univ Rome, Italy, DSc(civil eng), 33, DSc(aeronaut eng), 34, libero docente, 40. *Prof Exp:* Design engr aircraft, S A Caproni, Italy, 35-37, chief designer, 41-43; asst prof aeronaut eng, Univ Rome, 37-41; res assoc, Princeton Univ, 47-48; from asst prof to assoc prof, Cornell Univ, 49-55; design specialist aircraft, Convair Div, Gen Dynamics Corp, 55-59, mem res staff space craft, Gen Atomic Div, 60-65, eng staff specialist, Pomona Div, 65-72; eng consult, 73-85; RETIRED. *Concurrent Pos:* Designer, Italian Air Ministry, 41-43; consult, Princeton Univ, 49-50, Aeronaut Macchi, Italy, 49-53, Cornell Aeronaut Lab, NY, 51, Bur Ships, USN, 53-55 & Aerospace Corp, Sci Applns Inc, 73-80. *Mem:* assoc fel Am Inst Aeronaut & Astronaut; Italian Aerotechnol Asn. *Res:* Dynamics of structures; impact problems; theoretical and experimental stress analysis; design of aircraft and space craft structures. *Mailing Add:* 4429 Arista Dr San Diego CA 92103

RIPIN, BARRETT HOWARD, b Troy, NY, Oct 27, 42. PLASMA PHYSICS. *Educ:* Rensselaer Polytech Inst, BS, 64; Univ Md, PhD(physics), 71. *Prof Exp:* Res asst plasma physics, Univ Md, 65-70; res assoc controlled thermonuclear res, Plasma Physics Lab, Princeton Univ, 70-71; asst prof physics, Univ Calif, Los Angeles, 71-73; RES PHYSICIST & BR HEAD, NAVAL RES LAB, 73- *Concurrent Pos:* Assoc ed, Phys Rev Lett, Trans Plasma Sci, Lasers & Particle Beams & Phys Rev A; mem, Fusion Policy adv comt, Dept Energy; chmn, task force on financing journals, Am Phys Soc, counr-at-large, 91-94. *Honors & Awards:* Appl Sci Award, Sigma Xi, 88. *Mem:* Sr mem Inst Elec & Electronics Engrs; fel Am Phys Soc; AAAS. *Res:* Experimental investigations of laser-produced plasmas, laser fusion, controlled thermonuclear research and space plasmas; laser light scattering; self-generated magnetic fields; nonlinear wave interactions; hydrodynamic interactions. *Mailing Add:* Code 4780 Space Plasma Br Naval Res Lab Washington DC 20375-5000

RIPKA, WILLIAM CHARLES, b Los Angeles, Calif, June 2, 39; m 67; c 1. ORGANIC CHEMISTRY, PHARMACEUTICAL CHEMISTRY. *Educ:* Calif Inst Technol, BS, 61; Univ Ill, PhD(chem), 66. *Prof Exp:* RES FEL, MED PROD DEPT, E I DU PONT DE NEMOURS & CO, INC, 65-; VPRES, PHARMACEUT RES, CORVAS, INC, 90- *Mem:* Am Chem Soc. *Res:* Computer graphics and molecular modeling in drug design. *Mailing Add:* 10819 Red Rock Dr San Diego CA 92131

RIPLEY, DENNIS L(EON), b Joplin, Mo, Aug 30, 38; m 58; c 3. CHEMISTRY, ENGINEERING. *Educ:* Kans State Univ, BS, 59; Univ Tex, PhD(phys chem), 67. *Prof Exp:* Chem engr, Dow Chem Co, 59-61 & US Bur Mines, 61-63; from res chemist to mgr, planning & econ, Phillips Petrol Co, 66-88; MGR, PROCESSING & THERMO, IIT RES INST, NIPER, 88- *Mem:* Am Chem Soc. *Res:* Synthetic fuels; catalysts and catalytic processes; petroleum processes; fuel stability. *Mailing Add:* Po Box 315 Bartlesville OK 74005

RIPLEY, EARLE ALLISON, b Sydney, NS, June 29, 33; m 67; c 1. BIOMETEOROLOGY, AGROMETEOROLOGY. *Educ:* Dalhousie Univ, BSc, 53; Univ Toronto, MA, 55. *Prof Exp:* Meteorologist, Meteorol Br, Can, 55-60 & Nigerian Meteorol Serv, 60-62; agrometeorologist, EAfrican Agr & Forestry Res Orgn, 62-67; micrometeorologist, Matador Proj, 68-74, from assoc prof to prof, plant ecol dept, 74-84, PROF, CROP SCI & PLANT ECOL DEPT, UNIV SASK, 84- *Mem:* Fel Can Meteorol Soc; fel Royal Meteorol Soc; Am Meteorol Soc. *Res:* Micrometeorology; agricultural meteorology; environmental impact analysis; drought climatology. *Mailing Add:* Dept Crop Sci & Plant Ecol Univ Sask Saskatoon SK S7N 0W0 Can

RIPLEY, ROBERT CLARENCE, b Attleboro, Mass, Oct 24, 40. ANATOMY, CELL BIOLOGY. *Educ:* Brown Univ, AB, 62; Univ Calif, Los Angeles, PhD(anat), 66. *Prof Exp:* From instr to asst prof biol med sci, 67-74, asst dean, 74-78, ASSOC DEAN HEALTH CAREERS, BROWN UNIV, 78- *Mem:* Am Asn Med Cols. *Res:* Electron microscopy. *Mailing Add:* Dean Col Box G-A 124 Brown Univ Providence RI 02912

RIPLEY, SIDNEY DILLON, II, b New York, NY, Sept 20, 13; m 49; c 3. ZOOLOGY. *Educ:* Yale Univ, BA, 36; Harvard Univ, PhD(zool), 43. *Hon Degrees:* MA, Yale Univ, 61; DHL, Marlboro Col, 65 & Williams Col, 72; DSc, George Washington Univ, 66, Cath Univ Am, 68, Univ Md, 70, Cambridge Univ, 74, Brown Univ, 75 & Trinity Col, 77; LLD, Dickinson Col, 67, Hofstra Univ, 68 & Yale Univ, 75; DEng, Stevens Inst Technol, 77, Gallaudet Col, 81, Johns Hopkins Univ, 84 & Harvard Univ, 84. *Prof Exp:* Zool collector, Acad Natural Sci Philadelphia, 36-39; vol asst, Am Mus Natural Hist, NY, 39-40; asst, Harvard Univ, 41-42; asst cur birds, Smithsonian Inst, 42-43; lectr, Yale Univ, 46-49, from asst prof to assoc prof zool, 49-61, prof biol, 61-64; secy, 64-84, EMER SECY, SMITHSONIAN INST, 84- *Concurrent Pos:* Fulbright fel, Northeast Assam, 50; Guggenheim fel, Yale Univ & NSF fel, Indonesia, 54; from assoc cur to cur, Peabody Mus Natural Hist, Yale Univ, 46-64, dir, 59-64; dir, Pac War Mem, 46-50; trustee, White Mem Found; pres, Int Coun Bird Preserv, 58-82, emer pres, 82-; mem int comt, Int Ornith Cong, 62; bd trustees, World Wildlife Fund, 62-, Leader exped, Yale Univ & Smithsonian Inst, India & Nepal, 46-47, Nat Geog Soc, Yale Univ & Smithsonian Inst, Nepal, 48-49, Neth New Guinea, 60 & Bhutan & India, 67-77; deleg, UN Sci Conf Conserv Utilization Resources, 49; deleg, Int Union Preserv Nature, Caracas, 52, mem exec bd, 64; deleg, Stockholm Environ Conf, 72. *Honors & Awards:* Gold Medals, NY Zool Soc, 66 & Zool Soc Belg, 70; Tata Mem Lectr, 75; Arthur A Medal Ornith, Cornell Univ, 84; James Smithson Medal, Smithsonian Inst, 84; Presidential Medal of Freedom, 85; Thomas Jefferson Award, Am Soc Interior Decorators; Delacour Medal, Int Coun Bird Preserv, 82; Onassis Medal, Athens, 84; Distinguished Serv Award, Am Asn Mus, 85; Padma Bhushan Award, Govt of India, 86; Charles Reed Bishop Medal, 90. *Mem:* Nat Acad Sci; fel AAAS; fel Am Ornithologists Union; Soc Study Evolution; Soc Syst Zool; Am Philos Soc; Am Acad Arts Sci; Sigma Xi; fel Am Acad Arts & Lett. *Res:* Speciation and evolution in vertebrate zoology, primarily ornithology. *Mailing Add:* Smithsonian Inst NHB Rm 336 Washington DC 20560

RIPLEY, THOMAS H, b Bennington, Vt, Nov 18, 27; m 48; c 3. BIOLOGY. *Educ:* Va Polytech Inst, BS, 51, PhD, 58; Univ Mass, MS, 54. *Prof Exp:* Wildlife consult res admin, Dept Fish & Game, Mass, 53-56; instr biol, Va Polytech Inst, 56-57; wildlife consult res admin, Dept Fish & Game, Mass, 57; res biologist, Comn Game & Inland Fisheries, Va, 57-58; asst dir, Southeast Forest Exp Sta, asst to dep chief for res & chief range & wildlife res, US Forest Serv, 58-69; dir forestry, fisheries & wildlife develop, 69-78, MGR, OFF NATURAL RESOURCES, TENN VALLEY AUTHORITY, 79- *Mem:* Wildlife Soc; Soc Am Foresters; Am Forestry Asn (pres, 81-82); Am Inst Biol Sci; Int Union Forestry Res Orgn. *Res:* Forest land management; wildlife, range, watershed and timber resources. *Mailing Add:* Conserv Dept 701 Broadway Nashville TN 37203

RIPLEY, WILLIAM ELLIS, b Turlock, Calif, Nov 4, 17; m 40; c 3. MARINE RESOURCES DEVELOPMENT. *Educ:* Univ Wash, BS, 39. *Prof Exp:* Fish biologist, Calif Dept Fish & Game, 40-53, asst & chief, Marine Resources Br, 53-63; chief, tech asst, US Bur Com Fisheries, 63-65; dir fisheries develop, Brazil, 65-69, fisheries officer, NY, 70-82, CONSULT FISHERIES DEVELOP, UN DEVELOP PROG, 82- *Concurrent Pos:* Fisheries adv United Nation Develop Prog, Brazil, Indonesia, Colombia, Malaysia, Sri Lanka, Senegal, E & W Africa, Indian Ocean, Carribbean, Singapore, Hong Kong, Japan, 54-55 & 70-84; spec adv to US Embassy, Brazil, 68-69. *Mem:* Am Fisheries Soc; Int Fisheries Soc; Am Inst Fisheries Res Biologists. *Res:* Marine fisheries, shark, sardine, demersal and trawl fisheries; development of marine resources; formulator and administrator of 125 international marine and freshwater fisheries programs for the United Nations and United Nations Development Program. *Mailing Add:* Paradise Park 34 395 Hiram Rd Santa Cruz CA 95060

RIPLING, E(DWARD) J, b Lewistown, Pa, Feb 25, 21; m 43; c 3. METALLURGICAL ENGINEERING. *Educ:* Pa State Univ, BS, 42; Case Inst Technol, MS, 48, PhD(phys metall), 52. *Prof Exp:* Metallurgist, Westinghouse Elec Co, 42-43 & Copperweld Steel Co, 43-44; asst, Case Inst Technol, 46-52, asst prof & res dir, 52-55; lab dir mech metall, Continental Can Co, Inc, 55-60; PRES & DIR RES, MAT RES LAB, INC, 60- *Honors & Awards:* David Ford Mc Farland Award, Penn State Chap, Am Soc Metals, 84. *Mem:* Fel Am Soc Metals; Am Inst Mining, Metall & Petrol Engrs. *Res:* Metal forming; mechanical properties of materials. *Mailing Add:* Mat Res Lab Inc One Science Rd Glenwood IL 60425

RIPMEESTER, JOHN ADRIAN, b Voorburg, Neth, Feb 11, 44; Can citizen; m 67; c 2. PHYSICAL CHEMISTRY. *Educ:* Univ BC, 65, PhD(chem), 70. *Prof Exp:* Res assoc chem, Univ Ill, Urbana-Champaign, 70-72; fel, Nat Res Coun Can, 72-74, from asst res officer to assoc res officer, 74-82, sr res officer chem, 82-90, SR RES OFFICER, STEACIE INST MOLECULAR SCI, NAT RES COUN CAN, 90- *Concurrent Pos:* Vis scientist, Univ BC, 82-83; sect head, Colloid & Clathrate Chem Sect, NRC, 86-90; adj prof, Carleton Univ, 87- *Mem:* Chem Inst Can. *Res:* Molecular motion in solids; nuclear magnetic resonance; supramolecular chemistry. *Mailing Add:* 29 Delong Dr Ottawa ON K1J 7E5 Can

RIPPEN, ALVIN LEONARD, b Campbell, Nebr, Nov 6, 17; m 43; c 3. DAIRY SCIENCE. *Educ:* Univ Nebr, BSc, 40; Ohio State Univ, MSc, 41. *Prof Exp:* Sales engr, Creamery Package Mfg Co, 45-50; plant supt dairy processing, Kegle Dairy Co, Mich, 50-57; asst prof agr eng, Mich State Univ, 57-64, assoc prof food sci, 64-69, prof, 69-80, exten specialist, 57-80; RETIRED. *Mem:* Am Dairy Sci Asn. *Res:* Dairy products processing; dairy plant engineering. *Mailing Add:* Dept Food Sci & Human Nutrit Mich State Univ East Lansing MI 48824

RIPPEN, THOMAS EDWARD, b Lansing, Mich, Apr 19, 53; m 82; c 1. SEAFOOD TECHNOLOGY & MARKETING. *Educ:* Mich State Univ, BS, 75, MS, 81. *Prof Exp:* Asst food sci, Mich State Univ, 78-81; Marine exten agt seafood, 81-88, DIR SEAFOOD EXTEN UNIT, VA COOP EXTEN SERV, 89- *Mem:* Inst Food Technologists. *Res:* Assist seafood industry with processing, marketing product development and quality concerns specialize in crabmeat pastureurization; methods include evaluation of processing procedures, training of managers computer applications and applied research. *Mailing Add:* Va Tech Seafood Exten Unit PO Box 369 Hampton VA 23669-0369

RIPPERGER, EUGENE ARMAN, b Stover, Mo, July 7, 14; m 40; c 3. ENGINEERING MECHANICS. *Educ:* Kans State Col, BS, 39; Univ Tex, MS, 50; Stanford Univ, PhD(eng mech), 52. *Prof Exp:* Asst res engr, Portland Cement Asn, 39-42; asst engr, US War Dept, 42-43; from instr to asst prof, Univ Tex, 46-50; asst, Stanford Univ, 50-52; assoc dir eng mech res lab, 52-64, dir, 64-85, prof aerospace eng & eng mech, 52-82, EMER PROF ENG MECH, UNIV TEX, AUSTIN, 82- *Mem:* Fel Am Soc Mech Engrs; Soc Exp Mech; Inst Elec & Electronics Engrs. *Res:* Impact; experimental mechanics; bioengineering. *Mailing Add:* 3700 Highland View Dr Austin TX 78731

RIPPIE, EDWARD GRANT, b Beloit, Wis, May 29, 31; m 55; c 1. PHARMACEUTICS. *Educ:* Univ Wis, BS, 53, MS, 56, PhD(pharm), 59. *Prof Exp:* Asst pharm, Univ Wis, 56; from asst prof to assoc prof, 59-66, head dept, 66-74, dir grad studies, 74-81, head dept, 86-88, PROF PHARMACEUT, UNIV MINN, MINNEAPOLIS, 66- *Concurrent Pos:* Mem comt rev, US Pharmacopoeia, 70-80; res consult, Minn Mining & Mfg Co, 74- *Honors & Awards:* Ebert Prize, Acad Pharmaceut Sci Am Pharaceut Asn, 82, Advan Indust Pharm Award, 85. *Mem:* Am Chem Soc; Am Pharmaceut Asn; Am Asn Cols Pharm; fel Acad Pharmaceut Sci; fel Am Inst Chemists; fel AAAS. *Res:* Pharmaceutics; physical chemical behavior of physiologically active chemical species within anisotropic solvents; mechanisms of mass transport within beds of particulate solids; viscoelasticity of pharmaceutical tablets during and after compression. *Mailing Add:* Univ Minn Col Pharm Minneapolis MN 55455

RIPPLE, WILLIAM JOHN, b Yankton, SDak, Mar 10, 52. REMOTE SENSING, GEOGRAPHIC INFORMATION SYSTEMS. *Educ:* SDak State Univ, BS, 74; Univ Idaho, MS, 78; Ore State Univ, PhD(geog), 84. *Prof Exp:* Data analyst, SDak State Planning Bur, 77-81; res assoc, 84-88, ASST PROF, DEPT FOREST RESOURCES & DIR, ENVIRON REMOTE SENSING APPLICATIONS LAB, ORE STATE UNIV, 88- *Concurrent Pos:* Prin investr, Earth Commercialization Applications Prog, NASA, 88-; assoc ed, Photogram Eng & Remote Sensing J, 88- *Honors & Awards:* Presidential Citation for Meritorious Serv, Am Soc Photogram & Remote Sensing, 90. *Mem:* Am Soc Photogram & Remote Sensing; Soc Am Foresters; Int Asn Landscape Ecol. *Res:* Remote sensing and geographic information systems for the study of forest ecosystems, landscape ecology and wildlife habitat; author and co-author of over 35 technical publications. *Mailing Add:* 24549 Evergreen Rd Philomath OR 97370

RIPPON, JOHN WILLARD, b Toledo, Ohio, May 19, 32. MEDICAL MYCOLOGY. *Educ:* Univ Toledo, BS, 53; Univ Ill, MS, 57, PhD(microbiol), 59. *Prof Exp:* Res asst biochem, Univ Ill, 57-59; res assoc, Northwestern Univ, 59-60; instr biol, Loyola Univ, Ill, 60-63; asst prof, 63-69, ASSOC PROF MED, UNIV CHICAGO, 70- *Concurrent Pos:* Res bacteriologist, Vet Admin Hosp, Hines, Ill, 59-60, consult biochemist, 60-62, consult mycologist, 73-; ed-in-chief Mycopathologia, 74- *Mem:* Am Soc Microbiol; Mycol Soc Am; Int Soc Human & Animal Mycol; Sigma Xi. *Res:* Mechanisms of fungal pathogenicity; physiology of dimorphism in pathogenic fungi. *Mailing Add:* RR 1 - 375 Sawyer MI 49125-9801

RIPPS, HARRIS, b New York, NY, Mar 9, 27; m 49; c 3. PHYSIOLOGY. *Educ:* Columbia Univ, BS, 50, MS, 53, MA, 56, PhD(physiol psychol), 59. *Prof Exp:* Assoc optom, Columbia Univ, 51-56; from asst prof to assoc prof ophthal, 59-67, PROF OPHTHAL & PHYSIOL, SCH MED, NY UNIV, 67- *Concurrent Pos:* Nat Inst Neurol Dis & Blindness spec fels, 62 & 63; USPHS career develop award, 63- *Mem:* AAAS; Biophys Soc; Harvey Soc; Am Asn Res Vision & Ophthal; NY Acad Sci. *Res:* Visual physiology, especially visual pigments; electrical activity of retina. *Mailing Add:* Dept Ophthal Univ Ill Col Med PO Box 6998 Chicago IL 60680

RIPS, E(RVINE) M(ILTON), b Tulsa, Okla, Mar 7, 21; m 48; c 3. ELECTRICAL ENGINEERING. *Educ:* Mass Inst Technol, BS, 42; Carnegie Inst Technol, MS, 47. *Prof Exp:* Asst exp historadiography, Sloan-Kettering Inst Cancer Res, 48-50; instr elec eng, Polytech Inst Brooklyn, 50-52; asst chief engr, Cent Transformer Co, Ill, 52-56; chief engr, Hamner Electronics Co, NJ, 56-58; from asst prof to assoc prof elec eng, NJ Inst Technol, 58-84; RETIRED. *Concurrent Pos:* Consult, forensic eng, 65- *Mem:* Sr mem Inst Elec & Electronics Engrs. *Res:* Circuit design by digital computers; regulated direct-current power supplies. *Mailing Add:* PO Box 1534 New Brunswick NJ 08903

RIPY, SARA LOUISE, b Lawrenceburg, Ky, July 22, 24. MATHEMATICAL ANALYSIS. *Educ:* Randolph-Macon Woman's Col, BA, 46; Univ Ky, MA, 49, PhD(math), 57. *Prof Exp:* Instr math, Univ Ky, 46-54, Randolph-Macon Woman's Col, 54-56, Univ Ky, 56-57 & Vassar Col, 57-58; from asst prof to assoc prof, 58-67, chmn dept, 70-86, PROF MATH, AGNES SCOTT COL, 67- *Mem:* Math Asn Am; Am Math Soc. *Res:* Summability theory; analytic continuation. *Mailing Add:* 143 Winnona Dr Decatur GA 30030

RIRIE, DAVID, b Ririe, Idaho, Mar 20, 22; m 46; c 5. AGRONOMY. *Educ:* Brigham Young Univ, BS, 48; Rutgers Univ, PhD, 51. *Prof Exp:* Agronomist, Sugar Beet Proj, Univ Calif, 51-55; chmn dept agr, Church Col NZ, 55-63; soils & irrig farm adv, 63-80, DIR, MONTEREY COUNTY, AGR EXTEN SERV, UNIV CALIF, 81- *Concurrent Pos:* Consult, FMC Int, Eastern Europe. *Mem:* Am Soc Agron; Soil Sci Soc Am; Am Soc Hort Sci. *Res:* Nutrition studies with sugar beets; effects of growth regulators on sugar beets; peat land reclamation; vegetable crop culture; cereal crop fertilization. *Mailing Add:* 1061 University Ave Salinas CA 93901

RIS, HANS, b Bern, Switz, June 15, 14; nat US; m 80; c 2. CELL BIOLOGY. *Educ:* Columbia Univ, PhD(cytol), 42. *Prof Exp:* Asst zool, Columbia Univ, 39-40, lectr, 41-42; Seessel fel, Yale Univ, 42; instr biol, Johns Hopkins Univ, 42-44; asst physiol, Rockefeller Inst, 44-47, assoc, 47-49; assoc prof, 49-53, prof, 53-84, EMER PROF ZOOL, UNIV WIS-MADISON, 84- *Mem:* Nat Acad Sci; fel Am Acad Arts & Sci; Am Soc Naturalists; Genetics Soc Am; Am Soc Cell Biol; fel AAAS. *Res:* Chromosome structure and chemistry; physiology of cell nucleus; mitosis; cell ultrastructure. *Mailing Add:* Dept Zool Univ Wis Madison WI 53706

RISBUD, SUBHASH HANAMANT, b New Delhi, India, Aug 3, 47; m 74; c 1. GLASS SCIENCE, REFRACTORY MATERIALS. *Educ:* Indian Inst Technol, BS, 69; Univ Calif, Berkeley, MS, 71, PhD(ceramic eng), 76. *Prof Exp:* Eng assoc mat res, Stanford Univ, 71-73; ceramic engr, GTE-WESGO Corp, Calif, 73-74; asst prof mech eng, Univ Nebr-Lincoln, 76-78; asst prof mat sci & eng, Lehigh Univ, 78-79; ASST PROF CERAMIC ENG, UNIV ILL, URBANA, 79- *Concurrent Pos:* Prin investr, Mat Res Lab, Univ Ill, Urbana, 79- *Honors & Awards:* Ross Coffin Purdy Award, Am Ceramic Soc, 79. *Mem:* Am Ceramic Soc; Am Soc Eng Educ. *Res:* Glasses and glass-ceramics: developing the scientific framework for new and unusual glasses in non-oxide ceramic systems, crystallization behavior, microstructure, and properties. *Mailing Add:* Dept Mat Sci 153 Mines Bldg Univ Ariz Tucson AZ 85721

RISBY, EDWARD LOUIS, b Clarksdale, Miss, Sept 14, 33; m 57; c 3. PARASITOLOGY, CELL BIOLOGY. *Educ:* Lane Col, BS, 56; Southern Ill Univ, Carbondale, MA, 59; Tulane Univ, PhD(parasitol, cell biol), 68. *Prof Exp:* Asst prof biol, Lane Col, 58-61 & Southern Univ, 61-66; asst prof microbiol, Meharry Med Col, 68-78, asst dean grad studies, 71-78; prof & head, Dept Biol Sci, 78-88, DEAN, GRAD STUDIES & RES, TENN STATE UNIV, 88- *Concurrent Pos:* Consult premed prog, United Negro Col Fund, Fisk Univ, 71- & Dillard Univ, 72. *Mem:* Am Soc Parasitol; Soc Protozool; Am Soc Microbiol. *Res:* Comparative study of trypanosomal physiology and pathobiology observed in experimental trypanosomiasis. *Mailing Add:* Sch Grad Studies & Res Tenn State Univ 3500 J A Merritt Blvd Nashville TN 37209-1561

RISBY, TERENCE HUMPHREY, b Essex, Eng, June 9, 47; m 83; c 1. ANALYTICAL CHEMISTRY. *Educ:* Imperial Col, Univ London, DIC, 68, PhD(chem), 70; Royal Inst Chem, MRIC Chem, 75; FRSC, 80. *Prof Exp:* Res fel chem, Univ Madrid, 70-71; res assoc, Univ NC, Chapel Hill, 71-72; asst prof chem, Pa State Univ, 72-80; assoc prof, 79-88, PROF CHEM, JOHNS HOPKINS UNIV, 88-, DIR, DIV ENVIRON CHEM & BIOL, 89- *Concurrent Pos:* Europ fel, Royal Soc, 70-71; consult, Appl Sci Labs, 74-78, Sci Res Instruments Corp, 75-77. *Mem:* Royal Inst Chem; Royal Inst Gt Brit; Am Chem Soc; Soc Appl Spectros; Am Soc Mass Spectros; fel assoc, Inst Chem; fel Royal Soc Chem; Sigma Xi. *Res:* Mechanism of fragmentation and excitation of molecules in electrical discharges; ion-molecule reactions in various ionization sources for mass spectrometry; thermodynamics of solute-solvent interaction in chromatography. *Mailing Add:* Johns Hopkins Univ 615 N Wolfe St Baltimore MD 21205

RISCH, STEPHEN JOHN, ecology, entomology; deceased, see previous edition for last biography

RISDON, THOMAS JOSEPH, b Detroit, Mich, Sept 22, 39; m 79; c 1. LUBRICANTS, TRIBOLOGY. *Educ:* Univ Detroit, BSChE, 62. *Prof Exp:* Res asst, 63-67, sr res asst, 67-70, res assoc, 70-75, sr res assoc, 75-78, SR RES SPECIALIST, CLIMAX MOLYBDENUM CO, AMAX, INC, 78- *Mem:* Am Soc Lubrication Engrs. *Res:* New uses and applications for molybdenum disulfide and other molybdenum compounds as lubricants or lubricant additives. *Mailing Add:* 4871 Dexter-Pinckney Rd Dexter MI 48130

RISEBERG, LESLIE ALLEN, b Malden, Mass, July 23, 43; m 64; c 2. PHYSICS. *Educ:* Harvard Univ, AB, 64; Johns Hopkins Univ, PhD(physics), 68. *Prof Exp:* Mem tech staff, Bell Tel Labs, 68; guest lectr physics, Hebrew Univ Jerusalem, 68-69; mem tech staff, Corp Res & Eng, Tex Instruments Inc, 70-72; mem tech staff, 72-75, res mgr, 75-81, dir, Lighting Technol Ctr, 81-82, DIR, COMPONENTS RES LAB, GTE LABS, INC, 82- *Concurrent Pos:* Consult, Raytheon Co, 70- & Lawrence Livermore Lab, 74. *Mem:* Am Phys Soc; Inst Elec & Electronics Engrs; Sigma Xi; Optical Soc Am. *Res:* Quantum electronics; spectroscopy; solid state physics; optics; optical pumping; gas discharges; phosphors; solid state devices; lighting technology; materials science; integrated circuits; fiber optics; semiconductor technology. *Mailing Add:* GTE Labs Inc 40 Sylvan Rd Waltham MA 02254

RISEMAN, EDWARD M, b Washington, DC, Aug 15, 42. COMPUTER VISION. *Educ:* Clarkson Col Technol, BS, 64; Cornell Univ, MS, 66, PhD(elec eng), 69. *Prof Exp:* From asst prof to assoc prof, 69-78, chmn dept comput & info sci, 81-85, PROF COMPUT & DIR COMPUT VISION LAB, DEPT COMPUT & INFO SCI, UNIV MASS, AMHERST, 78- *Concurrent Pos:* Co-prin investr, equip grants, NSF, 77-79, 82-84, Digital Equipment Corp, 77; res grants, NSF, 79-, Army Res Inst, 80-82, Defense Advan Res Proj Agency, 82-87, Air Force Off Sci Res, 83-, Defense Mapping Agency & Off Naval Res, 84-86, Honeywell, 85, Gen Dynamics, 85-86 & Naval Res Lab, 85; consult & inst observer, SRI, Menlo Park, 78; prin investr res grant, Digital Equip Corp, 80-81 & NSF, 84-87. *Mem:* Inst Elec & Electronics Engrs; Asn Comput Mach; Pattern Recognition Soc; Am Asn Artificial Intel. *Res:* Visual perception by machine; development of real-time computers. *Mailing Add:* Coins Dept Grad Res Ctr Univ Mass Amherst MA 01003

RISEN, WILLIAM MAURICE, JR, b St Louis, Mo, July 22, 40; m 64; c 2. PHYSICAL CHEMISTRY, INORGANIC CHEMISTRY. *Educ:* Georgetown Univ, ScB, 62; Purdue Univ, PhD(phys chem), 67. *Prof Exp:* From instr to assoc prof, 66-75, chmn dept, 72-80, PROF CHEM, BROWN UNIV, 75- *Concurrent Pos:* Consult, NSF; mem, Coun Chem Res; consult, Indust. *Mem:* Am Chem Soc; Am Phys Soc; Am Ceramic Soc. *Res:* Molecular spectroscopy; studies of glasses and polymers; metal-metal bonding; far infrared and laser Raman spectra of ionic polymers and glasses; electron-delocalized materials. *Mailing Add:* Dept Chem Brown Univ Providence RI 02912

RISER, MARY ELIZABETH, b Richland, Wash, Aug 1, 45; m 78. SOMATIC CELL GENETICS, ENDOCRINOLOGY. *Educ:* Newcomb Col, Tulane Univ, BS, 67; Univ Tex Grad Sch Biomed Sci, MS, 70, PhD(genetics & cell biol), 73. *Prof Exp:* Fel endocrinol, 74-77, ASST PROF CELL BIOL, BAYLOR COL MED, 77- *Concurrent Pos:* Fel, NIH, 74-77. *Mem:* Am Soc Cell Biol; Tissue Cult Asn; Sigma Xi. *Res:* Genetic controls in cells: in vitro techniques somatic cell hybridization, and chromosome banding; malignant characteristics of cells; hormone regulation of functions; controls involved in peptide hormone synthesis. *Mailing Add:* Dept Cell Biol Baylor Col Med One Baylor Plaza Houston TX 77030

RISER, NATHAN WENDELL, b Salt Lake City, Utah, Apr 11, 20; m 43; c 3. INVERTEBRATE ZOOLOGY. *Educ:* Univ Ill, AB, 41; Stanford Univ, AM, 47, PhD, 49. *Prof Exp:* Actg instr biol, Stanford Univ, 49; instr zool, Univ Pa, 49-50; from assoc prof to prof biol, Fisk Univ, 50-56; res assoc marine biol, Woods Hole Oceanog Inst, 56-57; prof biol, 57-79, dir, Marine Sci Inst, 66-79, prof, 79-85, EMER PROF MARINE BIOL, NORTHEASTERN UNIV, 85- *Concurrent Pos:* Vis prof, Univ NH, 50-52 & 53-58; instr, Marine Biol Lab, Woods Hole, 52; res assoc, Woods Hole Oceanog Inst, 57-59; assoc, Mus Comp Zool, Harvard Univ, 57-62. *Mem:* Am Micros Soc. *Res:* Nemertine morphology, systematics; invertebrate systematics, morphology, histology and embryology; soft bodied interstitial fauna of the Northwest Atlantic and New Zealand. *Mailing Add:* Marine Sci Inst East Point Nahant MA 01908-1696

RISHEL, RAYMOND WARREN, b Phillips, Wis, June 27, 30; m 57; c 2. MATHEMATICS. *Educ:* Univ Wis, BS, 52, MS, 53, PhD(math), 59. *Prof Exp:* Instr math, Brown Univ, 59-60; res specialist, Boeing Co, Wash, 60-68; assoc prof math, Wash State Univ, 68-69; mathematician, Bell Tel Labs, 69-72; PROF MATH, UNIV KY, 72- *Mem:* Soc Indust & Appl Math. *Res:* Optimal control theory; probability. *Mailing Add:* Dept Math Univ Ky Lexington KY 40506

RISHELL, WILLIAM ARTHUR, b Lock Haven, Pa, Mar 1, 40; m 60; c 2. POULTRY BREEDING. *Educ:* Univ Md, BS, 62; Iowa State Univ, MS, 65, PhD(poultry breeding), 68. *Prof Exp:* Geneticist & dir breeding res, Indian River Int, Tex, 67-75; res geneticist, 75-76, DIR RES, ARBOR ACRES FARM, INC, 76- *Mem:* Poultry Sci Asn; World Poultry Sci Asn; Am Genetic Asn. *Res:* Applied poultry breeding and research; statistical analyses of experiments; estimation of genetic parameters. *Mailing Add:* Arbor Acres Farm Inc Marlborough Rd Glastonbury CT 06033

RISING, EDWARD JAMES, b Troy, NY, Nov 10, 26; m 49; c 3. INDUSTRIAL ENGINEERING. *Educ:* Rensselaer Polytech Inst, BME, 50; Syracuse Univ, MME, 53; Univ Iowa, PhD(indust eng), 59. *Prof Exp:* Instr mech eng, Syracuse Univ, 51-54; asst prof, Kans State Univ, 54-56; instr mech & hydraul, Univ Iowa, 56-60; assoc prof indust eng & asst dean, Univ Mass, Amherst, 60-71, prof indust eng & opers res, 71-89; RETIRED. *Concurrent Pos:* Consult, Mo River Div, US Corps Engrs, 58, Educ Testing Serv, NJ, 60, Package Mach Corp, Mass, 60, City of Gardner, 61, Franklin County Hosp, Greenfield, 64-66, Sprague & Carlton, NH, 66, Paper Serv Corp, Mass, 70 & US Pub Health Serv, 72; Joseph Lucas vis prof, Univ Birmingham. *Mem:* Am Soc Eng Educ; Am Inst Indust Engrs; Am Hosp Asn; Opers Res Soc Am. *Res:* Hospital systems; engineering education. *Mailing Add:* Dept Indust Eng Univ Mass Amherst MA 01003

RISING, JAMES DAVID, b Kansas City, Mo, Aug 10, 42; m 65; c 1. EVOLUTIONARY BIOLOGY, SYSTEMATIC ZOOLOGY. *Educ:* Univ Kans, BA, 64, PhD(zool), 68. *Prof Exp:* Fel, Cornell Univ, 68-69; ASSOC PROF ZOOL, UNIV TORONTO, 69- *Mem:* Am Ornithologists Union; Am Soc Naturalists; Soc Study Evolution; Ecol Soc Am; Soc Syst Zool. *Res:* Biology of birds; distribution, abundance, ecology, systematics, behavior and physiology of birds, especially systematic theory; interpopulational variation of vertebrate animals. *Mailing Add:* Dept Zool Univ Toronto Toronto ON M5S 1A1 Can

RISINGER, GERALD E, b Pekin, Ill, Nov 13, 33; m 53; c 2. CHEMISTRY. *Educ:* Bradley Univ, BS, 55; Iowa State Univ, PhD(org chem), 60. *Prof Exp:* Asst prof org chem, Arlington State Col, 60-62 & La Polytech Inst, 62-63; asst prof biochem, 63-71, ASSOC PROF BIOCHEM, LA STATE UNIV, BATON ROUGE, 71- *Concurrent Pos:* Petrol Res Fund grant, 63-66; USPHS res grant, 64-66. *Mem:* Am Chem Soc; Sigma Xi. *Res:* Bio-organic chemistry; characterization and synthesis of natural products; biogenesis of alkaloids and terpenes; biochemical mechanisms of coenzymatic and enzymatic reactions. *Mailing Add:* 650 Seyburn Dr Baton Rouge LA 70803

RISIUS, MARVIN LEROY, b Buffalo Center, Iowa, July 20, 31; m 59; c 2. PLANT BREEDING. *Educ:* Iowa State Univ, BS, 58; Cornell Univ, MS, 62, PhD(plant breeding), 64. *Prof Exp:* Res assoc corn breeding, Cornell Univ, 64-66; from asst prof to assoc prof, 66-77, prof forage breeding, 77-79, PROF SMALL GRAIN BREEDING, PA STATE UNIV, UNIVERSITY PARK, 79- *Mem:* Am Soc Agron; Crop Sci Soc Am. *Res:* Plant breeding and genetics with small grains. *Mailing Add:* Dept Agron Pa State Univ University Park PA 16802

RISK, MICHAEL JOHN, b Toronto, Ont, Feb 17, 40; m 65; c 2. MARINE ECOLOGY, PALEOECOLOGY. *Educ:* Univ Toronto, BSc, 62; Univ Western Ont, MSc, 64; Univ Southern Calif, PhD(biol), 71. *Prof Exp:* ASST PROF GEOL, McMASTER UNIV, 71-; STAFF MEM, DEPT GEOL SCI, BROCK UNIV. *Mem:* AAAS; Soc Econ Paleontologists & Mineralogists; Geol Asn Can. *Res:* Species diversity and substrate complexity; coral reef diversity; trace fossils; early history of the invertebrate phyla; animal-sediment relationships. *Mailing Add:* Dept Geol Sci McMaster Univ 1280 Main St W Hamilton ON L8S 4L8 Can

RISKA, DAN OLOF, nuclear physics, theoretical physics, for more information see previous edition

RISKIN, JULES, b Oakland, Calif, Aug 25, 26; m 60; c 2. PSYCHIATRY. *Educ:* Univ Chicago, PhB, 48, MD, 54. *Prof Exp:* Intern, Col Med, Univ Ill, 54-55; intern psychiat res, Cincinnati Gen Hosp, Ohio, 55-58; actg dir, 58, assoc dir, 59-76, DIR, MENT RES INST, 76- *Concurrent Pos:* NIMH res grant co-prin investr, Ment Res Inst, 61-65, prin investr, 66-69. *Mem:* Am Psychiat Asn; AMA. *Res:* Developing a methodology for studying whole family interaction; studying technique and theory of family therapy. *Mailing Add:* 2345 Byron Palo Alto CA 94301

RISLEY, JOHN STETLER, b Seattle, Wash, Mar 3, 42; m 65; c 3. ATOMIC PHYSICS, PHYSICS COURSEWARE. *Educ:* Univ Wash, BS, 65, MS, 66, PhD(physics), 73. *Prof Exp:* Teaching asst & res assoc physics, Univ Wash, 65-75; vis asst prof, Univ Nebr, 76; from asst prof to assoc prof, 76-84, PROF PHYSICS, NC STATE UNIV, 84- *Concurrent Pos:* Secy, Int Conf Physics of Electronic & Atomic Collisions, 77-; mem prog comt electronic & atomic physics, Am Phys Soc, 78- 79 & 84-87; column ed, Physics Teacher, 83-88; ed, Physics Acad Software, 87-; co-chmn, Conf Computers in Physics Instr, 88-; dir, Acad Software Libr, 88- *Mem:* Am Phys Soc; Sigma Xi; AAAS; Am Asn Physics Teachers. *Res:* Atomic collisions physics; negative ions; autodetaching states; vacuum ultraviolet radiation; synchrotron radiation; density matrix; electron collisions; physics courseware; computers in teaching physics. *Mailing Add:* Dept PhysicS NC State Univ Raleigh NC 27695-8202

RISLEY, MICHAEL SAMUEL, REPRODUCTIVE BIOLOGY. *Educ:* City Univ New York, PhD(biol), 77. *Prof Exp:* ASSOC PROF CELL BIOL & HISTOL, MED COL, CORNELL UNIV, 84- *Res:* Reproductive genetic toxicology; chromosome structure; spermatogenesis. *Mailing Add:* Dept Biol Fordham Univ Bronx NY 10458

RISLOVE, DAVID JOEL, b Rushford, Minn, Nov 16, 40; m 63; c 2. ORGANIC CHEMISTRY. *Educ:* Winona State Col, BA, 62; NDak State Univ, PhD(chem), 68. *Prof Exp:* Am Petrol Inst res asst, NDak State Univ, 65-68; assoc prof, 68-77, PROF CHEM, WINONA STATE UNIV, 77- *Concurrent Pos:* Vis scientist, 3M Co, St Paul, Minn, 83-84, Univ Minn, 76. *Mem:* Am Chem Soc. *Res:* Pyrrole and porphyrin chemistry; synthetic organic chemistry; kinetics and mechanisms of organic reactions; acid-base chemistry in non-aqueous media; polyanilines and N-substituted-2,4-dinitroaniline compounds. *Mailing Add:* Dept Chem Winona State Univ Winona MN 55987

RISS, WALTER, b New Britain, Conn, Jan 1, 25; m 51; c 4. NEUROANATOMY, PSYCHOLOGY. *Educ:* Univ Conn, BA, 49; Univ Rochester, PhD(psychol), 53. *Prof Exp:* Res assoc anat, Univ Kans, 52-53, USPHS fel, 53-54; from instr to assoc prof anat, 54-69, asst dean grad studies, 70-73, dir biol psychol, 71-75, PROF ANAT, STATE UNIV NY DOWNSTATE MED CTR, 70- *Concurrent Pos:* Prin founder & ed-in-chief, Brain, Behav & Evolution, 68-86. *Mem:* Am Asn Anatomists; Soc Neurosci. *Res:* Evolution of the nervous system and behavior; brain functions and behavior; testing brain function with use of computers. *Mailing Add:* Three Addison Lane Greenvale NY 11548

RISSANEN, JORMA JOHANNES, b Finland, Oct 20, 32; m 57; c 2. MATHEMATICS. *Educ:* Tech Univ Helsinki, dipl eng, 56, Techn Lic, 60, Techn Dr(control theory, math), 65. *Prof Exp:* MEM RES STAFF, INFO THEORY & MATH, IBM RES, 60- *Honors & Awards:* Outstanding Innovation Award, IBM Res, 81. *Res:* Information theory; estimation and statistics. *Mailing Add:* IBM Res K52/802 650 Harry Rd San Jose CA 95120-6099

RISSE, GUENTER BERNHARD, b Buenos Aires, Arg, Apr 28, 32; US citizen; m; c 3. HISTORY OF MEDICINE. *Educ:* Univ Buenos Aires, MD, 58; Univ Chicago, MA, 66, PhD(hist), 71. *Prof Exp:* Asst med, Med Sch, Univ Chicago, 63-67; asst prof hist med, Med Sch, Univ Minn, 69-71; assoc prof hist med, 71-76, prof hist health sci, 76-85, prof health sci, Univ Wis-Madison, 80-85; PROF & CHMN HIST HEALTH SCI, UNIV CALIF-SAN FRANCISCO, 85- *Concurrent Pos:* Am Philos Soc fel, 72, 77 & 82; NIH grant, Univ Wis, 72-74, 82-83; Univ Calif grant, 87. *Honors & Awards:* William H Welch Medal, Am Asn Hist Med, 88. *Mem:* Am Asn Hist Med; Ger Soc Hist Med, Sci & Technol; Hist Sci Soc; Int Acad Hist Med; Brit Soc Social Hist Med. *Res:* History of modern european medicine; history of the hospital, history of epidemics, twentieth century health sciences. *Mailing Add:* Dept Hist Health Sci Univ Calif San Francisco 533 Parnassus Ave San Francisco CA 94143-0726

RISSER, ARTHUR CRANE, JR, b Blackwell, Okla, July 8, 38; m 78; c 3. ORNITHOLOGY. *Educ:* Grinnell Col, BA, 60; Univ Ariz, MS, 63; Univ Calif, Davis, PhD(zool), 70. *Prof Exp:* Mus technician, Smithsonian Inst, US Nat Mus, 63-64; res assoc med ecol, Int Ctr Med Res & Training, Univ Md, 64-65; asst zool, Univ Calif, Davis, 65-67, lab technician, 67-70; asst prof biol, Univ Nev, Reno, 70-74; asst cur birds, 74-76, cur birds, 76-81, gen cur birds, 81-86, GEN MGR, SAN DIEGO ZOO, 86- *Concurrent Pos:* Adj prof zool, San Diego State Univ, 77-; mem, Calif condor recovery team; co-chmn, working group on Calif condor captive reproduction & re-introduction; dir, Int Fedn Conserv Birds, 80- *Mem:* Am Asn Zool Parks & Aquariums; Am Pheasant & Waterfowl Soc; Cooper Ornith Soc; Am Fedn Aviculture; Int Fedn Conserv Birds. *Res:* Avian reproduction. *Mailing Add:* 2015 Wedgemere Rd El Cajon CA 92020

RISSER, PAUL GILLAN, b Blackwell, Okla, Sept 14, 39; m 61; c 4. ECOSYSTEMS. *Educ:* Grinnell Col, BA, 61; Univ Wis-Madison, MS, 65, PhD(bot & soils), 67. *Prof Exp:* Res asst, Jackson Lab, 61-63; asst prof bot, Univ Okla, Norman, 67-72, assoc prof, 72-77, chmn, Dept Bot & Microbiol & prog dir ecosyst studies, 77-81, prof, 78-81; chief, Ill Natural Hist Surv, 81-; VPRES, DEPT RES, UNIV NMEX. *Mem:* Ecol Soc Am (secy, 78-82); Sigma Xi; Brit Ecol Soc; Soc Range Mgt. *Res:* Systems analysis of grassland ecosystems, particularly, dynamics of energy and material storage and transfer; studies of vegetation structure; natural resource planning. *Mailing Add:* Dept Res Univ NMex 108 Scholes Hall Alberqueque NM 87131

RISSLER, JANE FRANCINA, b Martinsburg, WVa, Jan 1, 46. PLANT PATHOLOGY. *Educ:* Shepherd Col, BA, 66; WVa Univ, MA, 68; Cornell Univ, PhD(plant path), 77. *Prof Exp:* Fel fungal physiol, Boyce Thompson Inst, 77-78; ASST PROF PLANT PATH, UNIV MD, 78- *Mem:* Am Phytopath Soc. *Res:* Phytopathogenic bacteria; diseases of ornamental plants and turf grass. *Mailing Add:* Six Plateau Pl No V Greenbelt MD 20770

RISTENBATT, MARLIN P, b Lebanon, Pa, Oct 12, 28; m 57; c 1. COMMUNICATION SYSTEMS. *Educ:* Pa State Univ, BS, 52, MS, 54; Univ Mich, PhD(elec eng), 61. *Prof Exp:* Instr elec eng, Pa State Univ, 52-54; sr engr, HRB-Singer, Inc, 54-56; assoc res engr, 56-61, res engr, 61-65, GROUP LEADER COMMUN, UNIV MICH, ANN ARBOR, 65- *Mem:* AAAS; sr mem Inst Elec & Electronics Engrs. *Res:* Application of communications theory, decision theory, estimation theory and computer methods to devise and evaluate new communication techniques and systems. *Mailing Add:* Dept Elec Comput Eng Univ Mich Ann Arbor MI 48109

RISTEY, WILLIAM J, b Cresco, Iowa, Aug 1, 38; m 66; c 4. POLYMER PHYSICS. *Educ:* Memphis State Univ, BS, 60; Vanderbilt Univ, PhD(phys chem), 66; Northwestern Univ, MM, 79. *Prof Exp:* Res assoc crystallog, Vanderbilt Univ, 65-66; chemist struct chem, Esso Res & Eng , 66-69; sr scientist polymer physics, Chemplex Co, 69-78; prin scientist mat res, Arco Chem Co, 78-86; RES SCI, HERCULES INC, 87- *Mem:* Am Chem Soc; Am Phys Soc. *Res:* Polymer and solid state research; structure-property relationships; polymer orientation in the solid state; polymer structure via fractionation and thermal analysis; transmission and scanning electron microscopy with elemental analysis. *Mailing Add:* 605 Hickory Lane Berwyn PA 19312

RISTIC, MIODRAG, b Serbia, Yugoslavia, May 16, 18; nat US; m 50; c 1. VETERINARY MEDICINE. *Educ:* Univ Munich, dipl, 50; Col Vet Med, Ger, Dr Vet Med, 50; Univ Wis, MS, 53; Univ Ill, PhD, 59. *Prof Exp:* Asst prof microbiol, Col Vet Med, Ger, 50-51; proj asst vet sci, Univ Wis, 51-53; from assoc pathologist to pathologist, Univ Fla, 53-60; PROF VET PATH & HYG, UNIV ILL, URBANA, 60-, PROF VET RES, 65- *Concurrent Pos:* Mem comt anaplasmosis & transmissible dis of swine, US Animal Health Asn; Anglo-Am fel, Europe; mem comt on rickettsia, Bergey's Manual; consult, Agency Int Develop & Rockefeller Found. *Mem:* Soc Immunol; Conf Res Workers Animal Dis; Am Vet Med Asn; Am Soc Trop Med & Hyg; Soc Protozool. *Res:* Infectious diseases of domestic animals, with special emphasis on blood diseases of man and animals. *Mailing Add:* Dept Vet Path, Hyg & Vet Res Col Vet Med Univ Ill 2001 S Lincoln Ave Urbana IL 61801

RISTIC, VELIMIR MIHAILO, b Skopje, Yugoslavia, Oct 10, 36; m 64; c 1. PLASMA PHYSICS, ELECTRICAL ENGINEERING. *Educ:* Univ Belgrade, BS, 60, MS, 64; Stanford Univ, MS, 66, PhD(elec eng), 69. *Prof Exp:* Res asst elec eng, Boris Kidric Inst Nuclear Sci, Belgrade, 61-62; lectr, Univ Belgrade, 62-65; res asst plasma physics, Stanford Univ, 66-68; asst prof, 68-72, assoc prof, 72-80, PROF ELEC ENG, UNIV TORONTO, 80- *Concurrent Pos:* Res assoc, Inst Geomagnetic Sci & lectr, Univ Nis, 62-65; Nat Res Coun Can & Defence Res Bd Can res grants, Univ Toronto, 68-72; Ont Dept Univ Affairs res grant, 69-70. *Mem:* Inst Elec & Electronics Engrs. *Res:* Microwave acoustics; acousto-optic signal processing; wave-wave and wave-particle interactions in acoustics, optics and electromagnetics; real-time signal processing. *Mailing Add:* Dept Elec Eng Univ Toronto Toronto ON M5S 1A1 Can

RISTOW, BRUCE W, b Chicago, Ill, June 24, 40; m 66; c 3. PHYSICAL CHEMISTRY. *Educ:* Northwestern Univ, BA, 62; Cornell Univ, PhD, 66. *Prof Exp:* NIH fel, Yale Univ, 66-67; from asst prof to assoc prof chem, 67-72, dean grad studies, 72-75, asst vpres, 75-86, ASSOC VPRES ACAD AFFAIRS, STATE UNIV NY, GENESEO, 86- *Mem:* Am Chem Soc; Sigma Xi. *Res:* Electron spin resonance; electrochemistry. *Mailing Add:* 540 Antlers Dr State Univ NY Rochester NY 14618

RISTROPH, JOHN HEARD, b New Orleans, La, Nov 12, 46; m 69; c 3. FORECASTING, ECONOMIC ANALYSIS. *Educ:* La State Univ, BS, 69, MS, 70; Va Polytech Inst & State Univ, PhD(indust eng & opers res), 75. *Prof Exp:* From instr to asst prof indust eng, La State Univ, 73-78; dir econ develop & dir policy & planning, La Dept Natural Resources, 78-82; ASSOC PROF ENG MGT, UNIV SOUTHWESTERN LA, 82- *Concurrent Pos:* Lectr, La State Univ & Univ Southwestern La, 78-82. *Mem:* Inst Indust Engrs; Opers Res Soc Am; Am Soc Eng Educ; Inst Mgt Sci. *Res:* Forecasting, economic analysis, operations management and computer applications; author of 10 journal articles. *Mailing Add:* 819 Dafney Dr Lafayette LA 70503

RISTVET, BYRON LEO, b Tacoma, Wash, Aug 22, 47; m 75. ENVIRONMENTAL SCIENCES. *Educ:* Univ Puget Sound, BS, 69; Northwestern Univ, PhD(geol), 76. *Prof Exp:* sci adv geol, Air Force Weapons Lab, 73-77, geoscientist, Defense Nuclear Agency, 77-83, S-CUBED DIV, MAXWELL LABS, 83- *Honors & Awards:* Secy Air Force Res & Develop Award, 75. *Mem:* Geochem Soc; Soc Econ Paleontologists & Mineralogists; Clays & Clay Minerals Soc; Am Mineral Soc. *Res:* Quaternary geology of atolls; reverse weathering reactions and their global implications; explosion effects phenomena on the solid earth; environmental impact analysis. *Mailing Add:* 1505 Soplo Rd SE Albuquerque NM 87123

RITCEY, GORDON M, b Halifax, NS, May 17, 30; m 55; c 1. INORGANIC CHEMISTRY, ORGANIC CHEMISTRY. *Educ:* Dalhousie Univ, BSc, 52. *Prof Exp:* Chemist, Radioactivity Div, Mines Br, Dept Energy, Mines & Resources, Can, 52; chief chemist, Eldorado Mining & Refining Co, 52-57, chief chemist res & develop & res chemist, 57, chief chemist, 59-60, head chem res group, 60-67; res scientist & head hydrometall sect, Mineral Sci Div, Dept Energy Mines & Resources, 67-80, res scientist & head, Process Metall Sect, Extrn Metall Labs, Canmet, 80-88; CONSULT, HYDROMETALL & WASTE MGT, 89- *Concurrent Pos:* Mem, UN & CESO projs, Africa, Egypt & Brazil; lectr & consult. *Honors & Awards:* Hydrometall Medal, Chem Inst Can, Alcan Award. *Mem:* Am Inst Mining, Metall & Petrol Eng; Am Chem Soc; Chem Inst Can; Can Soc Chem Engrs; Can Inst Mining & Metall; Soc Chem Indust. *Res:* Solution chemistry relating to hydrometallurgy; recovery and separation of metals from leach solutions resulting from work on solvent extraction has resulted in plants for the recovery of uranium, cobalt, nickel, zirconium and hafnium; rare earth separations. *Mailing Add:* 258 Grandview Rd Nepean ON K2H 8A9 Can

RITCH, ROBERT, b New Haven, Conn, May 14, 44. OPTHALMOLOGY. *Educ:* Harvard Univ, BA, 67; Albert Einstein Univ, MD, 72; Am Bd Ophthal, dipl, 77; Am Bd Laser Surg, dipl, 85. *Prof Exp:* Lab instr biol, Rice Univ, Houston, Tex, 67-68; intern, St Vincent's Med Ctr, New York, NY, 72-73; resident, Mt Sinai Sch Med, NY, 73-75, chief resident, 75-76, fel, Heed Opthal Found, 76-77, NIH fel, 77-78, from instr to assoc prof ophthal, 77-82; PROF OPHTHAL, NY MED COL, VALHALLA, 83-; CHIEF, GLAUCOMA SERV, NY EYE & EAR INFIRMARY, 83- *Concurrent Pos:* Grants, numerous insts, 64-; asst clin ophthalmologist, City Hosp Ctr Elmherst, 76-78, dir, Glaucoma Serv, 78-82; consult ophthalmologist, Vet Admin Hosp, Bronx, NY, 77-82; dir, Glaucoma Clin, Beth Israel Med Ctr, 77-78; rep, Nat Comt Glaucoma Educ, Washington, DC, 78; mem, Med Serv Plan Adv Coun, Mt Sinai, 79-81; actg dir, City Hosp Ctr Elmhurst, 79-81, dir, 81-82; vis prof, Univ Fla, Gainesville, 83, Scheie Eye Inst, Pa, 85 & Health Sci Ctr, State Univ NY, 87; chmn sci adv bd, Glaucoma Found, 84-; mem, Glaucoma Adv Comt, Nat Soc Prevent Blindness, 86-, Glaucoma Screening Comt, 87-; consult ophthal, Manhattan Eye, Ear & Throat Hosp, 88-; mem spec study sect, NIH, 91; bd dirs, Dooley Found & UN Vol Develop Coun, 91- *Honors & Awards:* Honor Award, Am Acad Ophthal, 85; John Roche Mem Lectr, Winthrop Univ Med Ctr, NY, 85; Charimet Kanchararanya Mem Lectr, Fifth Ann Bangkok Ophthal cong, 89; Spec Honoree, Glaucoma Found, 90; Charles H May Mem Lectr, NY Acad Med, 91. *Mem:* Fel Am Acad Ophthal; fel Am Col Surgeons; fel Int Col Surgeons; fel Am Soc Laser Surg & Med; Ophthalmic Laser Surg Soc (secy-treas, 82-); Am Med Asn; Int Glaucoma Cong; AAAS; Am Soc Cell Biol; Asn Res Vision & Ophthal. *Mailing Add:* NY Eye & Ear Infirmary 310 E 14th St New York NY 10003

RITCHEY, JOHN ARTHUR, industrial engineering, for more information see previous edition

RITCHEY, JOHN MICHAEL, b Wichita, Kans, Dec 14, 40; m 64; c 1. INORGANIC CHEMISTRY, ORGANOMETALLIC CHEMISTRY. *Educ:* Univ Colo, PhD(inorganic chem), 68. *Prof Exp:* Clin chemist, Wesley Med Ctr, 62-63; asst prof chem, Furman Univ, 68-70 & Northern Ariz Univ, 70-72; asst prof, 72-73, chmn dept, 73-77, assoc prof, 74-77, PROF CHEM, FT LEWIS COL, 78- *Concurrent Pos:* Consult, Four Corners Environ Res Inst, 72-, dir, 73-; collabr, Los Alamos Nat Lab, 80-; counr, Coun Undergrad Res, 84-86, 90- *Mem:* AAAS; Am Chem Soc; Sigma Xi. *Res:* Organometallic chemistry, especially heterobimetallic synthesis, zero valent metal cluster compounds; trace metal analysis, especially in natural systems; synthetic inorganic and organic chemistry, especially ligand design. *Mailing Add:* Dept Chem Ft Lewis Col Durango CO 81301

RITCHEY, KENNETH DALE, b Washington, DC, Oct 24, 44; m 71; c 2. SOIL FERTILITY. *Educ:* Carnegie Inst Technol, BS, 65; Cornell Univ, MS, 67, PhD(agron), 73. *Prof Exp:* Soil fertil specialist int agr progs, Univ Wis, 73-74; RES ASSOC SOIL FERTIL, CORNELL UNIV, 74- *Mem:* Int Soil Sci Soc; Am Soc Agron; Am Soil Sci Soc; Brazilian Soil Sci Soc. *Res:* Nitrogen, magnesium, micronutrient and potassium responses in highly weathered tropical soils; amelioration of subsoil acidity by leaching of calcium sulfate. *Mailing Add:* Trop Agr Res Sta Box 70 Mayaguez PR 00709

RITCHEY, SANFORD JEWELL, b Columbia, Miss, Feb 6, 30; m 57; c 3. NUTRITION. *Educ:* La State Univ, BS, 51; Univ Ill, MS, 56, PhD(animal nutrit), 57. *Prof Exp:* Fel biochem, Tex A&M Univ, 57-59, asst prof food & nutrit, 59-63; assoc prof, 63-66, head dept nutrit & foods, 66-73, assoc dean, Col Home Econ, 73-80, PROF HUMAN NUTRIT, VA POLYTECH INST & STATE UNIV, 69-, DEAN, COL HUMAN RESOURCES, 80- *Honors & Awards:* Borden Award, 79. *Mem:* AAAS; Am Chem Soc; Inst Food Technologists; Am Inst Nutrit. *Res:* Food science; nutritional relationships in growing children. *Mailing Add:* Col Human Resources Va Polytech Inst & State Univ Blacksburg VA 24061

RITCHEY, WILLIAM MICHAEL, b Mt Vernon, Ohio, June 2, 25; m 47; c 3. PHYSICAL CHEMISTRY. *Educ:* Ohio State Univ, BS, 50, MS, 53, PhD(phys chem), 55. *Prof Exp:* Chemist, Battelle Mem Inst, 52-55; sr res chemist, Res Dept, Standard Oil Co, Ohio, 55-66, group leader, 66-68; assoc prof phys chem, 68-79, ASSOC PROF MACROMOLECULAR SCI, CASE WESTERN RESERVE UNIV, 69-, PROF CHEM & MACROMOLECULAR SCI, 79- *Concurrent Pos:* Asst, Ohio State Univ, 50-54. *Mem:* AAAS; Am Chem Soc; Soc Appl Spectros. *Res:* Applications of nuclear magentic resonance in the solution and solid state, polymer characterization in structures, motion and morphology; x-ray spectroscopy and differential thermal analysis to petroleum chemistry, polymers and fossil fuels. *Mailing Add:* Dept Chem Case Western Reserve Univ Cleveland OH 44106

RITCHIE, ADAM BURKE, b Waynesboro, Va, Sept 8, 39; m 61; c 2. THEORETICAL CHEMISTRY. *Educ:* Univ Va, BA, 60, MA, 61, PhD(chem), 68. *Prof Exp:* Air Force Off Sci Res fel chem, Harvard Univ, 68-69; Nat Acad Sci-Nat Res Coun-NASA resident res assoc atomic physics, Goddard Space Flight Ctr, 69-71; from asst prof to prof chem, Univ Ala, 71-86; STAFF MEM, TEST PROG, LAWRENCE LIVERMORE NAT LAB, 86- *Mem:* Am Chem Soc. *Res:* Faraday effect; perturbation theory; correlation energies of molecules; atomic and molecular collision processes. *Mailing Add:* 1951 Creek Rd Livermore CA 94550

RITCHIE, ALEXANDER CHARLES, b Auckland, NZ, Apr 2, 21; m 56. PATHOLOGY. *Educ:* Univ NZ, MB, ChB, 44; Oxford Univ, DPhil(path), 50; Royal Col Physicians Can, cert specialist gen path, 55; Am Bd Path, dipl, 56; FRCP(C), 64; FRCPath, 70; FRCP(Australasia), 72. *Prof Exp:* Mem, Brit Empire Cancer Campaign Res Unit, Oxford Univ, 47-49, Walker studentship path; vis fel oncol, Chicago Med Sch, 51-52; lectr path, McGill Univ, 54-55, Douglas res fel, 55-56, asst prof, 55-58, Miranda Fraser assoc prof comp path, 58-61; head dept, 61-74, prof, 61-89, EMER PROF PATH, UNIV TORONTO, 86- *Concurrent Pos:* Mem consult panel, Can Tumour Registry, 58-75; pathologist-in-chief, Toronto Gen Hosp, 62-75; consult, Wellesley Hosp, 62-74, Hosp Sick Children, 62-, Women's Col Hosp, 63-74, Ont Cancer

Inst, 64-74, Toronto Western Hosp, 75-, Toronto Gen Hosp, 86- & Mt Sinai Hosp, 72-75 & 88-; chief examr lab med, Royal Col Physicians & Surgeons, Can, 73-77; secy, World Asn Soc Path, 75-81, pres, 81-85, pres, World Path Found, 80-84. *Honors & Awards:* Centennial Medal, Can, 67, Jubilee Medal, 77. *Mem:* Am Asn Cancer Res; fel Col Am Path; hon mem Can Asn Path (pres, 67-69); Int Acad Path; hon mem Asn Clin Pathologists; Can Asn Pathologists. *Res:* Oncology; pathology of tumors and cancer; occupational lung disease. *Mailing Add:* 625 Avenue Rd Suite 304 Toronto ON M4V 2K7 Can

RITCHIE, AUSTIN E, b Van Wert, Ohio, Feb 3, 18; m 42; c 4. AGRICULTURE, AGRICULTURAL EDUCATION. *Educ:* Ohio State Univ, BSc, 46, MSc, 51, PhD(agr educ), 55. *Prof Exp:* Teacher, Gibsonburg High Sch, 47-48 & Hilliard High Sch, 48-50; from instr to assoc prof, Ohio State Univ, 48-62, asst dean & secy agr, 57-63 & 64-65, actg exec dean spec serv, 63-64, prof agr educ, 62-81, asst dean acad affairs agr, 65-81, EMER PROF AGR EDUC, OHIO STATE UNIV, 81- *Mem:* AAAS. *Res:* Administration college of agriculture and home economics; agriculture curricula; general education and teacher education. *Mailing Add:* 5609 Hayden Run Rd Amlin OH 43002

RITCHIE, BETTY CARAWAY, b Dyersburg, Tenn, June 16, 29; m 66. AUDIOLOGY. *Educ:* La State Univ, BA, 50, MA, 51; Northwestern Univ, PhD(audiol), 64. *Prof Exp:* Speech correctionist, La Pub Schs, 51-53; dir, WTenn Hearing & Speech Ctr, 53-60; asst prof speech & dir hearing eval ctr, Univ Wis-Milwaukee, 63-70; assoc prof, Southern Ill Univ, 70-72; ASSOC PROF SPEECH PATH & AUDIOL, UNIV WIS-MILWAUKEE, 72-, CHMN DEPT, 82- *Concurrent Pos:* Audiological consult, Interstate Dropforge, 78- *Mem:* Am Audiol Soc; Am Speech-Lang-Hearing Asn; Acad Rehab Audiol. *Res:* Effects of compression amplication on speech intelligibility; evaluation of the efficiency of the verbal auditory screening for children; aural rehabilitation for older adults. *Mailing Add:* Dept Speech Path & Audiol Univ Wis, PO Box 413 Milwaukee WI 53201

RITCHIE, BRENDA RACHEL (BIGLAND), b Jordans, Eng, Sept 23, 27; m 51; c 2. PHYSIOLOGY. *Educ:* Univ London, BSc, 49, PhD(physiol), 68. *Hon Degrees:* DSc, Univ London, 87. *Prof Exp:* Res asst physiol, Univ Col, Univ London, 49-51, asst lectr, 51-53; lectr pharmacol, Royal Free Hosp for Women, London, Eng, 55-56; res asst, Albert Einstein Col Med, 61-64; lectr biol, Lehman Col, 66-67; asst prof, Marymount Col, NY, 67-70; assoc prof, 70-73, PROF BIOL, QUINNIPIAC COL, 73-, NIH RES GRANT, 71- *Concurrent Pos:* Prin investr, NIH & Marking Device Asn grants; NIH res grant, 71-; assoc fel, John B Pierce Found, Conn, 79-85, fel, 85- *Mem:* Brit Physiol Soc; Am Physiol Soc Neurosci; fel Am Col Sports Med; Sigma Xi. *Res:* Human motor control and mechanisms of fatigue. *Mailing Add:* John B Pierce Found 290 Congress Ave New Haven CT 06519

RITCHIE, CALVIN DONALD, b Arlington, Va, Jan 30, 30; m 52; c 4. ORGANIC CHEMISTRY. *Educ:* George Washington Univ, BS, 54, PhD(phys org chem), 60. *Prof Exp:* Org chemist, Food & Drug Admin, 56-60; Welch fel chem, Rice Univ, 60-61; PROF CHEM, STATE UNIV NY BUFFALO, 61- *Concurrent Pos:* Vis scholar, Stanford Univ, 76. *Honors & Awards:* Schoellkopf Medal, Western NY Sect Am Chem Soc, 70. *Mem:* Am Chem Soc; Royal Soc Chem. *Res:* Physical organic chemistry, particularly dealing with solvent effects and substituent effects in organic chemistry. *Mailing Add:* Dept Chem State Univ NY Buffalo NY 14214

RITCHIE, DAVID MALCOLM, b Woodbury, NJ, April 13, 50; m 72; c 2. PULMONARY PHARMACOLOGY, HYPERSENSITIVITY DISEASES. *Educ:* Rutgers Univ, BA, 72; Hahnemann Med Col, MS, 74, PhD(pharmacol), 76. *Prof Exp:* Fel pharmacol, Hahnemann Med Col, 74-76; res assoc, Med Col Pa, 76-78; scientist, 78-80, sr scientist, 80-85, PRIN SCIENTIST, R W JOHNSON PHARMACEUT RES INST, 85- *Mem:* Am Chem Soc; Int Soc Immunopharmacol; Am Soc Pharmacol Exp Ther; Am Heart Asn. *Res:* Leukotriene and lipoxygenase pathway of arachidonic metabolism as they relate to hypersensitivity disease; role of leukotrienes and their management in asthma and allergic disorders. *Mailing Add:* R W Johnson Pharmaceut Res Inst Rte 202 P O Box 300 Raritan NJ 08869-0602

RITCHIE, GARY ALAN, b Washington, DC, Aug 23, 41. PHYSIOLOGICAL ECOLOGY. *Educ:* Univ Ga, BS, 64; Univ Wash, MF, 66, PhD(forest ecol), 71. *Prof Exp:* Environ engr, US Army Corps Engrs, 71-73; environ impact analyst, 73-74, tech planner forestry & raw mat res develop, 74-76, RES PROJ LEADER PLANT PHYSIOL, WEYERHAEUSER CO, 77- *Mem:* Sigma Xi; Soc Am Foresters. *Res:* Financial analysis of long and short-term research investments in forestry, forest regeneration, forest management and forest genetics; reforestation and tree seedling production technology. *Mailing Add:* 8026 61 Ave NE Olympia WA 98506

RITCHIE, HARLAN, b Albert City, Iowa, Aug 3, 35; m 57; c 3. ANIMAL HUSBANDRY. *Educ:* Iowa State Univ, BS, 57; Mich State Univ, PhD(animal husb), 64. *Prof Exp:* Asst instr, 57-64, from asst prof to assoc prof, 64-71, PROF ANIMAL HUSB, MICH STATE UNIV, 71- *Honors & Awards:* Exten Award, Am Soc Animal Sci. *Mem:* Am Soc Animal Sci. *Res:* Trace elements in swine nutrition; beef cattle management. *Mailing Add:* Dept Animal Husb 104 Anthony Mich State Univ East Lansing MI 48823

RITCHIE, JAMES CUNNINGHAM, b Aberdeen, Scotland, July 20, 29; m 54; c 3. BOTANY, ECOLOGY. *Educ:* Aberdeen Univ, BSc, 51, DSc(bot), 62; Sheffield Univ, PhD, 55. *Prof Exp:* Sr res demonstr, Univ Sheffield, 52-53; Royal Comn Exhib 1851 sr res scholar, Montreal Bot Gardens & Man, 54-55; Nat Res Coun Can fel, Man, 55-56, from asst prof to assoc prof bot, 56-65; prof biol, Trent Univ, 65-69; prof, Dalhousie Univ, 69-70; chmn dept, 70-75, PROF LIFE SCI DIV, SCARBOROUGH COL, UNIV TORONTO, 72- *Concurrent Pos:* Nat Res Coun Can exchange scientist, USSR, 61; Sr Killam res scholar, 77-78; actg assoc dir, Lab Palynol, CNRS, Montpellier, France, 82. *Honors & Awards:* Lawson Medal, Can Bot Asn, 85; Cooper Award, Ecol Soc Am, 90. *Mem:* Ecol Soc Am; fel Arctic Inst NAm; Can Bot Soc (pres, 69); Brit Ecol Soc; Bot Soc Brit Isles; fel Royal Soc Canada, 88. *Res:* Quaternary ecology, especially Holocene and Pleistocene vegetation and ecology of North West America and North Africa. *Mailing Add:* Life Sci Div Scarborough Col Univ Toronto Scarborough ON M1C 1A4 Can

RITCHIE, JERRY CARLYLE, b Richfield, NC, Dec 13, 37; m 66; c 2. ECOLOGY. *Educ:* Pfeiffer Col, BA, 60; Univ Tenn, Knoxville, MS, 62; Univ Ga, PhD(bot), 67. *Prof Exp:* Res asst ecol, Oak Ridge Nat Lab, 62; fel plant sci, Southeastern Watershed Res Ctr, Univ Ga, 67-68; botanist, US Sedimentation Lab, Agr Res Serv, USDA, Oxford, Miss, 68-78; SOIL SCIENTIST, HYDROL LAB, USDA, BELTSVILLE, 78- *Mem:* Ecol Soc Am; Brit Ecol Soc; Am Inst Biol Sci; Am Soc Agron. *Res:* Radioecology, limnology, sedimentation; remote sensing. *Mailing Add:* ARS Hydrol Lab BARC-W Bldg 007 Beltsville MD 20705

RITCHIE, JOE T, b Palestine, Tex, June 2, 37; m 59; c 1. SOIL PHYSICS, PHYSICAL CHEMISTRY. *Educ:* Abilene Christian Col, BS, 59; Tex Tech Col, MS, 61; Iowa State Univ, PhD(soil physics), 64. *Prof Exp:* Lab asst agr, Abilene Christian Col, 57-59; res asst, Tex Agr Exp Sta, 59-61; asst agron, Iowa State Univ, 61-64; physicist, Tex Res Found, 64-66; res soil scientist, Soil & Water Conserv Res Div, Agr Res Serv, USDA, 66-; STAFF MEM, INST WATER RES, MICH STATE UNIV. *Concurrent Pos:* Consult, Tex Instruments, Inc, 65. *Mem:* Am Soc Agron; Soil Sci Soc Am; Am Chem Soc; Am Statist Asn; Am Geophys Union; Sigma Xi. *Res:* Application of gas-solid chromatography for analysis of soil gases; measurement of microclimate as related to evapotransportation; soil moisture estimation under row crops. *Mailing Add:* Plant & Soil Sci Bldg Mich State Univ East Lansing MI 48824-1325

RITCHIE, JOSEPH MURDOCH, b Scotland, June 10, 25; m 51; c 2. PHARMACOLOGY, PHYSIOLOGY. *Educ:* Aberdeen Univ, BSc, 44; Univ London, BSc, 49, PhD(physiol), 52, DSc, 60. *Hon Degrees:* MA, Yale Univ, 68; DSc, Aberdeen Univ, 87. *Prof Exp:* Lectr physiol, Univ London, 49-51; mem staff, Nat Inst Med Res, Eng, 51-56; from asst prof to prof pharmacol, Albert Einstein Col Med, 56-68; chmn dept pharmacol, 68-74, dir div biol sci, 75-78, EUGENE HIGGINS PROF PHARMACOL, SCH MED, YALE UNIV, 68- *Concurrent Pos:* Overseas fel, Churchill Col, Cambridge Univ, 64-65. *Honors & Awards:* Van Dyke Mem Award, 83. *Mem:* Am Soc Pharmacol; Am Physiol Soc; Brit Physiol Soc; fel Royal Soc; Brit Pharmacol Soc. *Res:* Biophysics of muscle and nerve. *Mailing Add:* Dept Pharmacol Sch Med Yale Univ New Haven CT 06510

RITCHIE, KIM, b Korea, Apr 13, 36; US citizen; m 59; c 1. BIOCHEMISTRY. *Educ:* ETenn State Col, BA, 59; Univ Tenn, MS, 61; Ariz State Univ, PhD(chem), 67. *Prof Exp:* Res chemist, Tenn Eastman Co, 61-62; res biochemist, Parke-Davis Co, 62-64; res biochemist, Barrow Neurol Inst, St Joseph's Hosp, Phoenix, Ariz, 64-68; sr chemist, Motorola Inc, 68-76, lab mgr, Process Technol Lab, Semiconductor Res & Develop Lab, Semiconductor Prod Div, 76-80; STAFF MEM, AUX CERAMICS, 80- *Mem:* Electrochem Soc; AAAS; Am Chem Soc; NY Acad Sci. *Res:* Neurochemistry; copper metabolism in central nervous system; organometallic interaction with biopolymers; solid state chemistry; polymer surface chemistry. *Mailing Add:* 409 Lafayette Rd Myrtle Beach SC 29577

RITCHIE, ROBERT OLIVER, b Plymouth, Eng, Jan 2, 48; div; c 1. MECHANICAL ENGINEERING, FRACTURE MECHANICS. *Educ:* Cambridge Univ, BA, 69, MA & PhD(metall, mat sci), 73, ScD, 89. *Prof Exp:* Res assoc metall & mat sci, Churchill Col, Cambridge Univ, 72-74; lectr mat sci & eng, Univ Calif, Berkeley, 74-76, res metallurgist mat, Lawrence Berkeley Lab, 76; asst prof mech eng, Mass Inst Technol, 77-78, assoc prof, 78-81; assoc prof metall, 81-82, PROF MAT SCI, UNIV CALIF, BERKELEY, 82-; DIR, CTR ADV MAT, LAWRENCE BERKELEY LAB, 87-, DEP DIR, MAT SCI DIV, 90- *Concurrent Pos:* Goldsmith's res fel, Churchill Col, Cambridge Univ, 72-74; Miller res fel, Univ Calif, Berkeley, 74-76. *Honors & Awards:* Most Outstanding Sci Accomplishment Award, Dept Energy, 82, 89; G R Irwin Medal, Am Soc Testing & Mat, 85; C W McGraw Res Award, Am Soc Eng Educ, 87; C H Mathewson Gold Medal, Am Inst Mech Engrs, 85; M A Grossmann Award, Am Soc Metals, 80. *Mem:* Fel Am Soc Metals; Am Inst Mining, Metall & Petrol Engrs; fel Brit Inst Metals; Am Soc Testing & Mat; Mat Res Soc; fel Inst Metals UK. *Res:* Deformation and failure of engineering materials, especially metallurgy, toughness, fatigue and environmentally-assisted failure and wear of metals and alloys; ceramics and composites. *Mailing Add:* Dept Mat Sci & Mineral Eng Univ Calif Berkeley CA 94720

RITCHIE, ROBERT WELLS, b Alameda, Calif, Sept 21, 35; m 57; c 2. COMPUTER SCIENCE, MATHEMATICS. *Educ:* Reed Col, BA, 57; Princeton Univ, MA, 59, PhD(math), 61. *Prof Exp:* Res instr, Dartmouth Col, 60-62; from asst prof to assoc prof math, Univ Wash, 62-69, assoc dean grad sch, 66-69, assoc math & comput sci & vprovost acad admin, 69-72, prof comput sci & vprovost & asst vpres acad affairs, 72-76, prof & chmn comput sci, 77-83; mgr, Comput Sci Lab, Palo Alto Res Ctr, 83-88; vpres, Univ Affairs, Xerox Corp, 88; dir, Computer Systs Ctr, Hewlett-Packard Labs, 88-90, DIR, UNIV AFFAIRS, HEWLETT-PACKARD CO, 90- *Concurrent Pos:* Vis scientist, Xerox Palo Alto Res Ctr, 81-82. *Mem:* AAAS; Am Math Soc; Asn Symbolic Logic; Asn Comput Mach; Inst Elec & Electronics Engrs. *Res:* Mathematical logic and linguistics; computability theory; theory of algorithms; complexity theory. *Mailing Add:* Hewlett-Packard Co 1501 Page Mill Rd-MS 1U Palo Alto CA 94304

RITCHIE, RUFUS HAYNES, b Blue Diamond, Ky, Sept 24, 24; m 44; c 2. RADIATION PHYSICS. *Educ:* Univ Ky, BS, 47, MS, 49; Univ Tenn, PhD, 59. *Prof Exp:* Instr physics, Univ Ky, 48-49; corp fel, 85-90, PHYSICIST, HEALTH PHYSICS DIV, OAK RIDGE NAT LAB, 49-, DISTINGUISHED RES STAFF MEM, 78-, SR CORP FEL, 90- *Concurrent Pos:* Vis res prof, Inst Physics, Aarhus Univ, 61-62; Ford Found prof physics, Univ Tenn, 65- prof, Univ Ky, 68-69; sr vis fel, Cavendish Lab, Cambridge

Univ, 75-76, overseas fel, Churchill Col, 75-76; vis prof, Bhzba Atomic Res Ctr, Bombay, Incdia, 73, New York Univ, 72; vis prof, Inst Physics, Odense Univ, Denmark, 80-81; bd dir, Pellissippi Int Inc, 87; exec comt, Southeastern Sect, Am Phys Soc. *Honors & Awards:* Jesse W Beams Award. *Mem:* AAAS; fel Am Phys Soc; Radiation Res Soc. *Res:* Radiopharmaceutical development; collective interactions in condensed matter. *Mailing Add:* Health & Safety Res Div Oak Ridge Nat Lab PO Box 2008 Oak Ridge TN 37830-6123

RITCHIE, STEPHEN G, b Melbourne, Australia, Nov 14, 54. TRANSPORTATION ENGINEERING, ARTIFICIAL INTELLIGENCE. *Educ:* Monash Univ, BE, 77, M Eng Sc, 81; Cornell Univ, PhD(civil & environ eng), 83. *Prof Exp:* Asst prof transp eng, Univ Wash, 83-85; asst prof, 85-88, ASSOC PROF TRANSP ENG, UNIV CALIF, IRVINE, 88-, VCHAIR, DEPT CIVIL ENG, 88- *Concurrent Pos:* Res engr, Inst Transp Studies, Univ Calif, Irvine, 85-; NSF presidential young investr award, 87. *Mem:* Am Soc Civil Engrs; Inst Transp Engrs; Am Asn Artificial Intel; Transp Res Bd. *Res:* Transportation and traffic systems engineering; advanced technology development and application; artificial intelligence. *Mailing Add:* Inst Transp Studies Univ Calif 330 Berkeley Pl Irvine CA 92717

RITCHIE, WALLACE PARKS, JR, b St Paul, Minn, Nov 4, 35; m 60; c 3. SURGERY. *Educ:* Yale Univ, BA, 57; Johns Hopkins Univ, MD, 61; Univ Minn, PhD(surg), 71. *Prof Exp:* From intern to resident surg, Yale-New Haven Med Ctr, 61-63; from resident to chief resident, Sch Med, Univ Minn, 64-69, instr, 69-70; chief dept surg gastroenterol, Div Surg, Walter Reed Army Inst Res, 70-73; from asst prof to assoc prof surg, Sch Med, Univ VA, 73-76, prof, 76-; STAFF MEM, DEPT SURG, TEMPLE UNIV. *Concurrent Pos:* Consult, Vet Admin Hosp, Roanoke, Va, 73-; Am Heart Asn estab investr, Univ Va, 74- *Mem:* Am Fedn Clin Res; Am Gastroenterol Asn; Asn Acad Surg (pres, 76-77); Soc Univ Surgeons; Am Surg Asn. *Res:* Gastric mucosal resistance as a factor in ulcerative disease of the upper gastrointestinal tract. *Mailing Add:* Dept Surg Temple Univ Broad Ontario St Philadelphia PA 19140

RITENOUR, GARY LEE, b Warsaw, Ind, Oct 5, 38; m 63; c 2. PLANT PROTECTION, PLANT PHYSIOLOGY. *Educ:* Purdue Univ, West Lafayette, BS, 60; Univ Calif, Davis, MS, 62, PhD(plant physiol), 64. *Prof Exp:* Univ Ill res fel, Univ Ill, Urbana, 64-66 & farm adv agron crops, Agr Exten Serv, Univ Calif, 66-69; assoc prof, 69-74, PROF AGRON, CALIF STATE UNIV, FRESNO, 74- *Mem:* Am Soc Agron; Weed Sci Soc Am. *Res:* Use of agricultural chemicals in crop production. *Mailing Add:* Dept Plant Sci Calif State Univ 6241 N Maple Ave Fresno CA 93740-0072

RITER, JOHN RANDOLPH, JR, b Denver, Colo, Apr 18, 33; m 55; c 4. PHYSICAL CHEMISTRY. *Educ:* Colo Sch Mines, PRE, 56; Univ Wash, PhD(chem), 62. *Prof Exp:* Jr engr, Boeing Co, 56-58, assoc res engr, 58-60; instr chem, Univ Wash, 61-62; from asst prof to assoc prof, 62-71, PROF CHEM, UNIV DENVER, 71- *Concurrent Pos:* Assoc Western Univ fac fel physics, Univ Calif, Berkeley, 68; vis prof, Math Inst, Oxford Univ, 71; consult, Colo Pathologists Regional Lab, 76- *Mem:* Am Chem Soc; Am Phys Soc. *Res:* Thermodynamics; high-temperature spectroscopy. *Mailing Add:* 2507 S Kearney St Denver CO 80222-6325

RITER, STEPHEN, b Providence, RI, Mar 7, 40; m 64; c 2. ELECTRICAL ENGINEERING. *Educ:* Rice Univ, BA, 61, BSEE, 62; Univ Houston, MS, 67, PhD(elec eng), 68. *Prof Exp:* From asst prof to prof elec eng, Tex A&M Univ, 68-79, asst dir, Ctr Urban Progs, 72-76, assoc dir, Ctr Energy & Mineral Resources, 76-77, dir, Tex Energy Exten Serv, 77-79; prof elec eng & chmn dept, Comput Sci Dept, 80-89, DEAN ENG, UNIV TEX, EL PASO. *Concurrent Pos:* Ed, Trans Geosci, Inst Elec & Electronics Engrs, 72; prin investr, res prog image enhancement using artificial intel, Nat Inst Justice, 84-; proj dir, Develop Meteorol Sensor RPVS, US Army, 86-; chmn, El Paso Pub Util Regulatory Bd, 86- *Mem:* Inst Elec & Electronics Engrs; Marine Technol Soc; Sigma Xi; Technol Transfer Soc. *Res:* Image processing; fiber optics radar systems environmental resource management. *Mailing Add:* Elec Eng Tex Univ El Paso TX 79968

RITLAND, RICHARD MARTIN, b Grants Pass, Ore, July 3, 25; m 46; c 5. VERTEBRATE ZOOLOGY, PALEONTOLOGY. *Educ:* Walla Walla Col, BA, 46; Ore State Col, MS, 50; Harvard Univ, PhD, 54. *Prof Exp:* From instr to asst prof biol, Atlantic Union Col, 47-52; from instr to asst prof, Loma Linda Univ, 54-60; from assoc prof to prof paleont, Andrews Univs, 60-80, PROF GEOL, 77- *Mem:* AAAS; Soc Study Evolution; Soc Vert Paleont; Paleont Soc; Geol Soc Am. *Res:* Tertiary and cretaceous paleo-ecology. *Mailing Add:* PO Box 263 Berrien Springs MI 49103

RITSCHEL, WOLFGANG ADOLF, b Trautenau, Czech, Jan 10, 33; m 60; c 2. PHARMACOKINETICS. *Educ:* Innsbruck Univ, MPharm, 55; Univ Strasbourg, DPharm, 60; Univ Vienna, DPhil, 65; Univ Villareal, MD, 89. *Prof Exp:* Chief pharmacist, Girol SA, Zurich, Switz, 58-59; head pharmaceut res, Biochemie AG, Kundl, Austria, 59-61; prof chem & pharmaceut, Notre Dame Col, Dacca, EPakistan, 61-64, head dept pharm, 62-64; head technol pharmaceut res, Siegfried AG S-ckingen, Ger, 65-68; assoc prof, 69-72, prof biopharmaceut, 72-77, prof pharmacokinetics & biopharmaceut, 77-, PROF PHARMACOL, MED COL, UNIV CINCINNATI, 81-, HEAD DIV PHARMACEUT & DRUG DELIVERY SYSTS, 85- *Concurrent Pos:* Asst prof, Teaching Hosp, Kufstein, Austria, 59-61; adj prof, Univ Basel, 65-68; Nat Pharmaceut Coun grant biopharmaceut, Univ Cincinnati, 69-71; dir pharm serv-res, Cincinnati Gen Hosp, Med Ctr, Univ Cincinnati, 75; prof, Nat Univ Mayor San Marcos, Lima, Peru, 73. *Honors & Awards:* Theodore Koerner Prize, Pres of Repub Austria, 62, Cross of Honor for Sci & Arts, 75; Hertha Heinemann Mem Prize, Ger Pharmaceut Indust, 65. *Mem:* Acad Pharmaceut Sci; Int Pharmaceut Fedn; Int Soc Chemother; Ger Asn Trop Med; fel Am Clin Pharmacol. *Res:* Pathways of absorption of drugs; bioavailability of drugs; development of testing procedures; pharmacokinetics of drugs; clinical pharmacokinetics; geriatrics. *Mailing Add:* Col Pharm Univ Cincinnati Med Ctr Cincinnati OH 45267-0004

RITSKO, JOHN JAMES, b Pittston, Pa, July 8, 45; m 67; c 2. EXPERIMENTAL SOLID STATE PHYSICS. *Educ:* Mass Inst Technol, BS, 67; Princeton Univ, MS, 69, PhD(physics), 74. *Prof Exp:* SCIENTIST PHYSICS, XEROX CORP, 74- *Mem:* Am Phys Soc; Sigma Xi. *Res:* Study of electronic states and elementary excitations of organic and molecular solids by means of high energy inelastic electron scattering and ultraviolet photoemission spectroscopy. *Mailing Add:* 70 High Ridge Rd Mt Kisco NY 10549

RITSON, DAVID MARK, b London, Eng, Nov 10, 24; m 52; c 4. PHYSICS. *Educ:* Oxford Univ, BA, 44, DPhil(physics), 48. *Prof Exp:* Res fel physics, Dublin Inst Advan Studies, Ireland, 48-49; asst & instr, Univ Rochester, 49-52; lectr physics & res physicist, Mass Inst Technol, 52-64; assoc prof, 64-71, PROF PHYSICS, STANFORD UNIV, 71- *Res:* Cosmic rays; high energy accelerator physics; particularly properties of fundamental particles. *Mailing Add:* Stanford-Linear Acc Ctr Sandhill Rd Stanford CA 94305

RITT, PAUL EDWARD, JR, b Baltimore, Md, Mar 3, 28; m 50; c 6. CHEMISTRY. *Educ:* Loyola Col, Md, BS, 50; Georgetown Univ, MS, 52, PhD(chem), 54. *Prof Exp:* Lab asst, Loyola Col, Md, 50; res asst, Harris Res Labs, Inc, 50-52; chemist, Melpar, Inc, Westinghouse Air Brake Co, 52-54, proj chemist, 54-56, chief chemist, 56-57, mgr chem lab, 57-59, mgr phys sci lab, 59-60, dir res, 60-62, vpres res, 62-67; pres, Appl Sci Div & Appl Technol Div, Litton Industs, 67-68; vpres & dir res, GTE Labs, Inc, 68-86; DIR ACAD AFFAIRS, BABSON COL, 86- *Concurrent Pos:* Lectr, Univ Va, 57-59 & Am Univ, 58-; vpres & gen mgr, Training Corp Am, Melpar, Inc, 65-66, pres, 66-67. *Mem:* Am Chem Soc; Electrochem Soc; Am Ceramic Soc; Am Inst Chemists; NY Acad Sci; Inst Elec & Electronics Engrs; AAAS; Royal Soc Chemists. *Res:* Organometallic synthesis; rare earth ceramics; solid state phenomenon; thin films; integrated circuits; optical communications; plasma physics; high temperature measurements; space instrumentation; special purpose data processing; electron systems; systems integration; telecommunications; robotry; operations research. *Mailing Add:* 36 Sylvan Lane Weston MA 02193

RITT, ROBERT KING, b New York, NY, Dec 30, 24; m 50; c 4. APPLIED MATHEMATICS. *Educ:* Columbia Univ, AB, 44, PhD(math), 53. *Prof Exp:* Lectr math, Columbia Univ, 46-48; from instr to assoc prof, Univ Mich, 48-62; div mgr, Conductron Corp, 62-68; pres, Ritt Labs, Inc, 68-71; chmn dept, 71-76, PROF MATH, ILL STATE UNIV, 71- *Mem:* Am Math Soc; Math Asn Am. *Res:* Electromagnetic theory; foundations of statistical mechanics; perturbation theory of symetric operators. *Mailing Add:* Dept Math Ill State Univ Normal IL 61761

RITTELMEYER, LOUIS FREDERICK, JR, b Mobile, Ala, Dec 23, 24; m 49; c 8. PSYCHIATRY. *Educ:* Spring Hill Col, BS, 45; Med Col Ala, MD, 47. *Prof Exp:* Instr prev med & asst dir dept gen pract, Col Med, Univ Tenn, 54-55; assoc prof & dir post-grad educ & student health serv, Med Ctr, Univ Miss, 55-59; assoc med dir, Mead Johnson & Co, 59-60, vpres & med dir, 60-63; from asst prof to assoc prof, 66-77, PROF PSYCHIAT, SCH MED, GEORGETOWN UNIV, 77- *Concurrent Pos:* Consult, Surgeon Gen, US Army, 57-65. *Mem:* Fel Am Psychiat Asn; Asn Acad Psychiat. *Res:* Psychiatry in primary care medicine; continuing education. *Mailing Add:* Dept Psychiat Georgetown Univ Hosp Washington DC 20007

RITTENBAUGH, CHERYL K, b Pittsburgh, Pa, Nov 26, 46. NUTRITIONAL ANTHROPOLOGY & EPIDEMIOLOGY. *Educ:* Rice Univ, BA, 68; Univ Calif Los Angeles, MA, 71; Univ Mich Ann Arbor, MPH, 79. *Prof Exp:* Teaching asst Dept Anthrop, Univ Calif Los Angeles, 70-71; asst prof Dept Anthrop, Col Human Med, Mich State Univ, 75-79; from instr to asst prof, Dept Anthrop, Univ Ariz, 72-75, res assoc, Dept Family & Community Med, 80-81, adj asst prof, 81-83, from res asst prof to res assoc prof, 83-87, ASSOC PROF, DEPT FAMILY & COMMUNITY MED, UNIV ARIZ, 87- *Concurrent Pos:* Numerous grants, var orgn, 71-91; fac grad prog nutrit sci, Univ Ariz, 81-, actg chief, Nutrit Sect, Dept Family & Community Med, 89-90; US rep comt II/4, Nutrit & Anthrop, Int Union Nutrit Sci, 86-90; grant reviewer, NIH, Nat Cancer Inst & NSF. *Mem:* Fel AAAS; Am Anthrop Asn; Am Asn Phys Anthropologists; Am Pub Health Asn; Soc Med Anthrop; Am Inst Nutrit; fel Human Biol Coun. *Res:* Human nutrition and food consumption; epidemiology; diet and cancer; diabetes; obesity; nutritional and medical anthropology; contemporary US, Native Americans, Egypt. *Mailing Add:* Dept Family & Community Med Univ Ariz Health Sci Ctr Tucson AZ 85724

RITTENBERG, ALAN, particle physics; deceased, see previous edition for last biography

RITTENBERG, MARVIN BARRY, b Los Angeles, Calif, Sept 10, 31; m 54; c 2. IMMUNOLOGY, MICROBIOLOGY. *Educ:* Univ Calif, Los Angeles, AB, 54, MA, 57, PhD(microbiol), 61. *Prof Exp:* Inst Microbiol fel, Rutgers Univ, 61-63; NIH fel immunochem, Calif Inst Technol, 63-66; from asst prof to assoc prof med & microbiol, 66-73, PROF MICROBIOL & IMMUNOL, MED SCH, ORE HEALTH SCI UNIV, 73- *Concurrent Pos:* Leukemia Soc Am scholar, 67-71; consult diag immunol, United Med Labs, Portland, 68-73; mem exec comt & bd dirs, Ore Comprehensive Cancer Ctr Prog, 72-76; assoc ed, J Immunol, 75-81; consult, Crime Detection Lab Syst, Ore State Police, 75-77, NIH Immunobiol Study Sect, 77-81 & Zymogenetics, Inc, 84-90; vis prof zool, Univ Col London, 72-73 & 80-81; mem, Adv Comt Immunol & Immunother, Am Cancer Soc, 78-82, scholar, 80-81; vis scientist, Imp Cancer Res Fund Labs, London, 87-88. *Mem:* AAAS; Am Asn Immunologists; Am Soc Microbiol; Sigma Xi. *Res:* Development and control of immunological memory through differentiation of B lymphocytes; evolution of molecular recognition by antibodies. *Mailing Add:* Dept Microbiol & Immunol Ore Health Sci Univ Portland OR 97201

RITTENBERG, SYDNEY CHARLES, b Chicago, Ill, Dec 19, 14; m 41; c 2. BACTERIOLOGY. *Educ:* Univ Calif, Los Angeles, AB, 35, MA, 37; Univ Calif, PhD(bact), 41. *Prof Exp:* Asst chem, Univ Calif, Los Angeles, 35-36; asst marine bact, Scripps Inst, Univ Calif, 37-41; res chemist, Technicolor Motion Picture Corp, 41-42; chief bacteriologist, Comt Infectious Wounds, Nat Res Coun, Tulane Univ, 42-43; chief microbiologist, Res Div, S B Penick & Co, NY, 43-47; from asst prof to prof bact, Univ Southern Calif, 47-62; prof bact & chmn dept, 62-85, EMER PROF, UNIV CALIF, LOS ANGELES, 85- *Honors & Awards:* Carski Found Distinguished Teaching Award, Am Soc Microbiol, 69. *Mem:* Am Soc Microbiol; Am Chem Soc; Brit Soc Gen Microbiol; Am Soc Biol Chemists. *Res:* Physiology and metabolism of bacteria Chemolithotrophy; physiology of the Bdellovibrio. *Mailing Add:* 17236 Village 17 Camarillo CA 93012-7404

RITTENBURY, MAX SANFORD, b Bailey, NC, Dec 16, 28; m 50; c 2. SURGERY, BIOCHEMISTRY. *Educ:* The Citadel, 46-49; Med Col Va, MD, 53. *Prof Exp:* Resident gen surg, Med Col Va, 56-62, from instr to asst prof, 62-66; from asst prof to assoc prof, 66-72, PROF GEN SURG, MED UNIV SC, 72- *Concurrent Pos:* Surg res fel, Med Col Va, 59-62, USPHS res grants, 59-, fel, 60-62, spec fel, 63-64. *Mem:* AAAS; Am Col Surgeons; Soc Surg Alimentary Tract; Am Asn Surg Trauma; Am Fedn Clin Res. *Res:* Disease of the pancreas; immune and enzymatic response of the body to stress and thermal injury; surgical bacteriology. *Mailing Add:* Dept Surg Med Univ SC 171 Ashley Ave Charleston SC 29401

RITTENHOUSE, HARRY GEORGE, b Spokane, Wash, Oct 19, 42. BIOLOGICAL CHEMISTRY. *Educ:* Univ Puget Sound, BS, 68; Wash State Univ, PhD(biochem), 72. *Prof Exp:* Am Cancer Soc fel biochem, Univ Calif, Los Angeles, 73-74; asst prof biochem, 75-78, sr res assoc, Ment Health Res Inst, Univ Mich, Ann Arbor, 78-; HEAD TUMOR MAKERS RES LAB, ABBOTT LAB. *Res:* Isolation and characterization of surface glycoproteins from cultured animal cells. *Mailing Add:* Specialty Labs Inc/Cytometrics Inc 2211 Michigan Blvd Santa Monica CA 90404-3900

RITTENHOUSE, LARRY RONALD, b Lewellen, Nebr; c 4. RANGE SCIENCE. *Educ:* Utah State Univ, BS, 62; Univ Nebr, MS, 66, PhD, 69. *Prof Exp:* Asst, Univ Nebr, 64-69; from asst prof to assoc prof, Eastern Ore Agr Res Ctr, 69-74; assoc prof, Tex A&M Res & Exten Ctr, 75-81; PROF RANGE SCI, COLO STATE UNIV, 81- *Mem:* Am Soc Animal Sci; Soc Range Mgt. *Mailing Add:* Dept Range Sci Colo State Univ Ft Collins CO 80523

RITTENHOUSE, SUSAN E, b New York, NY, Mar 18, 45. LIPID METABOLISM. *Educ:* Harvard Univ, PhD(biochem), 72. *Prof Exp:* From asst prof to assoc prof biochem, Sch Med, Boston Univ, 76-84; ASSOC PROF BIOCHEM, BIRMINGHAM WOMEN'S HOSP, HARVARD UNIV, 84- *Mem:* Am Soc Cell Biol; Am Soc Biol Chemists; Sigma Xi; Am Heart Asn; Int Soc Thrombosis & Hematosis. *Mailing Add:* Dept Biochem B-411 Health Sci Complex Univ Vermont Burlington VT 05405

RITTER, A(LFRED), b Brooklyn, NY, Mar 15, 23; m 47; c 3. AERODYNAMICS. *Educ:* Ga Inst Technol, BS, 43, MS, 47; Cornell Univ, PhD(aerodyn), 51. *Prof Exp:* Aerodynamicst, Glenn L Martin Co, 46; asst, Ga Inst Technol, 47-48; aeronaut res engr, Off Naval Res, 51-54; supvr aerophys, Armour Res Found, Ill Inst Technol, 54-58; vpres & dir res, Adv Res Div, Therm Inc, 58-64, pres, Therm Advan Res, Inc, NY, 64-68; asst head, Appl Mech Dept, 68-70, asst head, 70-78, head, Aerodyn Res Dept, Calspan Corp, 78-80; dir technol, Calspan Corp, Arnold Eng Develop Ctr, Arnold AFB, 80-86; sr staff, Booz-Allen & Hamilton, 86-88; PRES, RITTER INC, 88- *Concurrent Pos:* Instr, Eve Div, Ill Inst Technol, 56-58; vis lectr, Cornell Univ, 65; mem Nat Adv Bd, Univ Tenn Space Inst, chmn, 87-90; mem, Int Coun Aeronaut Sci prog comt, Nat Res Coun comt on Earthquake Eng, 84, on Assessment Nat Aerospace Wind Tunnel Facil, 87-88, SDIO Blue Ribbon Panel Rev ARROW Missile Prog, 89; adj prof aerospace eng, Univ Ala, Huntsville, 88- *Mem:* AAAS; fel Am Inst Aeronaut & Astronaut (tech vpres, 81-85); NY Acad Sci; Sigma Xi. *Res:* Shock wave theory; high temperature gas dynamics; transonic aerodynamics; technical management; strategic defense missile systems; aero-optics. *Mailing Add:* 10148 Dunbarton Dr Huntsville AL 35803

RITTER, CARL A, b Confluence, Pa, Jan 23, 32; m 59; c 1. IMMUNOPHARMACOLOGY. *Educ:* Syracuse Univ, AB, 55; State Univ NY, PhD(pharmacol), 64; Univ Pa, AM, 72. *Prof Exp:* Asst prof, 66-71, ASSOC PROF PHARMACOL, SCH VET MED, UNIV PA, 72- *Concurrent Pos:* Spec fel, Inst Path, Karolinska Inst, Stockholm, 67. *Mem:* Am Soc Pharmacol & Exp Therapeut; NY Acad Sci; affil Am Col Vet Toxicologists; Am Soc Microbiol; Am Asn Cancer Res. *Res:* Immunogenetics; biological response; relation in pathogenesis. *Mailing Add:* 27 N Elm Ave Aldan PA 19018

RITTER, DALE FRANKLIN, b Allentown, Pa, Nov 13, 32; m 53; c 4. GEOLOGY. *Educ:* Franklin & Marshall Col, AB, 55, BS, 59; Princeton Univ, MA, 63, PhD(geol), 64. *Prof Exp:* Asst prof geol, Franklin & Marshall Col, 64-72; prof geol, Southern Ill Univ, Carbondale, 72-90; EXEC DIR, QUAT SCI CTR, DESERT RES INST, RENO, 90- *Concurrent Pos:* Chmn, Quat Geol, Geomorphol Div, Geol Soc Am, 88-89; RIS rep, Int Asn Geomorphol, 91- *Mem:* AAAS; Geol Soc Am; Sigma Xi. *Res:* Geomorphology and Pleistocene geology specifically in analysis of processes. *Mailing Add:* 3628 Hillsdale Ct Sparks NV 89434

RITTER, E GENE, b N Kansas City, Mo, Apr 2, 28; m 53; c 2. SPEECH-LANGUAGE PATHOLOGY. *Educ:* William Jewell Col, AB, 50; Univ Mo, MA, 56, PhD(speech path), 62. *Prof Exp:* Teacher speech & English, Lathrop Pub Schs, 51-53; supvr, Lab Schs, Univ Mo, 53-56; instr speech, Univ Hawaii, Hilo, 56-58; clinician speech path, Univ Mo, 58-62; assoc prof speech & hearing clin, Univ Hawaii, 62-69; ASSOC PROF SPEECH & HEARING CTR, IND UNIV, BLOOMINGTON, 69 - *Concurrent Pos:* Consult speech path, var hosps, Honolulu, 62-69, Gov of Guam, 67 & Bloomington Hosp & Bloomington Convalescent Ctr, 69 -; fel, Mayo Grad Sch Med, 66. *Mem:* Fel Am Speech-Lang-Hearing Asn. *Res:* Diagnosis and treatment of aphasia and apraxia of speech. *Mailing Add:* Speech & Hearing Ctr Ind Univ Bloomington IN 47405

RITTER, EDMOND JEAN, b Cleveland, Ohio, Dec 11, 15; c 4. BIOCHEMISTRY. *Educ:* Ohio State Univ, BChE, 37; Univ Toledo, MS, 41; Univ Cincinnati, PhD, 70. *Prof Exp:* Chemist & chem engr, Sun Oil Co, 37-43; org res chemist, Sharples Chem, Inc, Pa Salt Mfg Co, 43-49; dir, Cimcool Lab, Cincinnati Milling Mach Co, 49-65; dir biol res, Laser Lab, 65-68, assoc prof, 68-85, biochemist, Div Basic Sci Res,Inst Develop Res, 68-87, EMER PROF RES PEDIAT, CHILDREN'S HOSP RES FOUND, UNIV CINCINNATI, 85- *Mem:* AAAS; Am Chem Soc; Teratology Soc; Soc Develop Biol. *Res:* Protein chemistry; experimental teratology and cytology; mental retardation; developmental biology; relationship of inhibition of DNA and ATP synthesis to death and differentiation of proliferating cells. *Mailing Add:* 432 Evanswood Pl Cincinnati OH 45220

RITTER, ENLOE THOMAS, b Memphis, Tenn, June 21, 39; div. PHYSICS. *Educ:* Southwestern at Memphis, BS, 61; Johns Hopkins Univ, PhD(physics), 66. *Prof Exp:* Staff mem, P Div, Los Alamos Sci Lab, 66-68; physicist, 68-80, DIR, DIV NUCLEAR PHYSICS, US DEPT ENERGY, 80- *Mem:* Am Phys Soc. *Res:* Nuclear science. *Mailing Add:* ER-221-6TN US Dept Energy Washington DC 20585

RITTER, GARRY LEE, b Michigan City, Ind, Nov 14, 49; m 71. ANALYTICAL CHEMISTRY. *Educ:* Wabash Col, BA, 70; Univ NC, PhD(analytical chem), 76. *Prof Exp:* Res assoc, Nat Bur Standards, 76-77; sr scientist, Schering Corp, 77-80; proj mgr, Ciba Geigy Pharmaceut, 83; SR SOFTWARE ENGR NICOLET INSTRUMENT CORP, 83- *Concurrent Pos:* Nat Acad Sci-Nat Res Coun fel, Nat Bur Standards, 76-77. *Mem:* Am Chem Soc; Am Statist Asn. *Res:* Interpretation of chemical data using optimization techniques and robust statistical techniques; numerical solutions to mixture problems; automation of laboratory instrumentation with active control and real time data acquisition and interpretation. *Mailing Add:* Nicolet Instrument Corp PO Box 4451 Madison WI 53744-4451

RITTER, GERHARD X, b Bochum, WGer, Oct 27, 36. MATHEMATICS & COMPUTER VISION RESEARCH. *Educ:* Univ Wis-Madison, PhD(math), 71. *Prof Exp:* PROF COMPUT SCI & MATH, UNIV FLA, 83-, DIR, CTR COMPUT VISION RES, 87- *Mem:* Math Asn Am; Asn Comput Mach; Inst Elec & Electronics Engrs. *Res:* Computer vision research. *Mailing Add:* Computer & Info Sci Dept Univ Fla Gainesville FL 32611

RITTER, HARTIEN SHARP, b Iola, Kans, Oct 13, 18; m 41; c 2. PHYSICAL CHEMISTRY. *Educ:* Univ Kans, AB, 41; Univ Akron, MS, 54, PhD, 64. *Prof Exp:* Res chemist, Olin Industs, Ill, 41-46; asst dir res, Calcium Carbonate Co, 46-48; sr res chemist, Chem Div, Pittsburgh Plate Glass Co, 48-59, supvr inorg chem, 60-64; sr supvr inorg chem, Chem Div, PPG Industs, Inc, 64-81; CONSULT COATINGS, 81- *Mem:* Am Chem Soc; Fedn Socs Plant Technol. *Res:* Colloid chemistry of pigments in paints and related products; application research in protective coatings. *Mailing Add:* 1495 Shanabrook Dr Akron OH 44313

RITTER, HOPE THOMAS MARTIN, JR, b Allentown, Pa, Sept 24, 19; m 46, 70; c 5. CELL BIOLOGY. *Educ:* Cornell Univ, AB, 43; Lehigh Univ, MS, 47, PhD, 55. *Prof Exp:* Instr zool & gen biol, Lehigh Univ, 48-55, asst prof zool, gen biol & comp physiol, 55-57; res fel, Biol Labs, Harvard Univ, 57-59, lectr biol, 59-61; asst prof, State Univ NY Buffalo, 61-66; prof, Univ Ga, 66-87; RETIRED. *Concurrent Pos:* Lectr cell div mech, 87- *Mem:* AAAS; Soc Protozoologists; Am Soc Cell Biologists. *Res:* Biology of Cryptocercus and termite protozoa; insect blood cells; tissue culture; anaerobic metabolism. *Mailing Add:* 775 Riverhill Dr Athens GA 30606

RITTER, HUBERT AUGUST, b St Louis, Mo, Aug 30, 24; m 49; c 1. OBSTETRICS & GYNECOLOGY. *Educ:* Westminster Col, Mo, AB, 45; St Louis Univ, MD, 48. *Prof Exp:* Actg chmn dept obstet & gynec, St Louis Univ, 76-78; PRES EDUC & RES FOUND, AMA, 77- *Concurrent Pos:* Trustee, Am Med Asn, 76-; comnr, Nat Joint Pract Comn, 77- & Joint Comn on Accreditation of Hosps, 82-85. *Honors & Awards:* Robert Schlueter Award, 85. *Mem:* Am Fertil Soc; Am Col Surgeons; Am Col Obstet & Gynec; AMA (secy-treas, 82-85). *Res:* Incompetence of uterine cervix. *Mailing Add:* AMA 1035 Bellevue St Louis MO 63117

RITTER, JAMES CARROLL, b Denver, Colo, Apr 12, 35; m 59; c 4. NUCLEAR PHYSICS. *Educ:* Univ Colo, AB, 57; Purdue Univ, MS, 62. *Prof Exp:* Physicist, Radiation & Nucleonics Lab, Westinghouse Elec Corp, 59-60 & Nat Bur Standards, 61-62; res physicist, 62-73, head, Satellite Survivability Sect, 73-83, HEAD, RADIATION EFFECTS BR, US NAVAL RES LAB, 83- *Mem:* Am Phys Soc; Sigma Xi; AAAS; Inst Elec & Electronics Engrs. *Res:* Nuclear reactions and spectroscopy; decay schemes; resonance neutron-capture gamma-ray studies; radiation hardening of electronic devices; satellite survivability; space radiation; single event upset; space dosimetry; Combined Release and Radiation Effects Satellite microelectronics experiment; High Temperature Supeconductivity Space experiment. *Mailing Add:* Code 4610 US Naval Res Lab Washington DC 20390

RITTER, JOHN EARL, JR, b Baton Rouge, La, July 17, 39; m; c 4. MATERIALS SCIENCE, METALLURGY. *Educ:* Mass Inst Technol, BS, 61, MS, 62; Cornell Univ, PhD(metall), 66. *Prof Exp:* From asst Prof to assoc prof, 65-76 PROF MECH ENG, UNIV MASS, AMHERST, 76. *Mem:* Am Ceramic Soc. *Res:* Mechanical behavior of materials; adhesion between dissimilar materials. *Mailing Add:* Dept Mech Eng Univ Mass Amherst MA 01003

RITTER, JOSEPH JOHN, US citizen. SYNTHETIC INORGANIC CHEMISTRY. *Educ:* Siena Col, BS, 60; Univ Hawaii, MS, 63; Univ Md, PhD(inorg chem), 71. *Prof Exp:* RES CHEMIST, NAT BUR STANDARDS, 63- *Mem:* Am Chem Soc. *Res:* Reactivity of volatile boranes and silanes; inorganic chemistry of corrosive reactions on ferrous metals; nature of passive films on metals; electrochemistry. *Mailing Add:* NIST Ceramics Div Bldg 223 RA 25B Gaithersburg MD 20899

RITTER, KARLA SCHWENSEN, b Detroit, Mich, Oct 30, 50; m 74. BIOCHEMISTRY. *Educ:* Ohio State Univ, BS, 71; Univ Calif, Berkeley, PhD(entom), 76. *Prof Exp:* Res fel, Harvard Univ, 76-77; vis asst prof, 78-79, asst prof, 79-85, ASSOC PROF BIOL & ENTOM, DREXEL UNIV, 85- *Mem:* Sigma Xi; Entom Soc Am; Am Entom Soc; Soc Invertebrate Path; Am Registry Prof Entomologists. *Res:* Function and metabolism of sterols in insects; physiological and histopathological effects of insect disease. *Mailing Add:* 5870 Timber Ridge Trail Madison WI 53711

RITTER, MARK ALFRED, b San Francisco, Calif, May 11, 48; m 74. RADIATION BIOLOGY. *Educ:* Univ San Francisco, BS, 70; Univ Calif, MS, 72, PhD(nuclear eng), 76; Univ Miami, MD, 84. *Prof Exp:* Res assoc, Sch Pub Health, Harvard Univ, 76-77; asst prof, Univ Pa, 77-82, resident radiation ther, 85-; ASST PROF, CLIN CANCER CTR, UNIV WIS. *Mem:* Radiation Res; Am Soc Photobiol; AAAS. *Res:* Mechanisms of DNA damage and repair; their relationship to cellular inactivation, mutagenesis, and transformation in normal or mutant human cells. *Mailing Add:* 600 Highland Ave Madison WI 53792

RITTER, NADINE MARIE, b New Orleans, La, June 20, 58; m 80; c 1. GENERAL CELL BIOLOGY, PROTEIN BIOCHEMISTRY. *Educ:* Univ Houston-Clear Lake, BS, 84; Rice Univ-Houston, MA, 88, PhD(cell biol), 88. *Prof Exp:* Res asst, Dent Sci Inst, Univ Tex, 78-80, sr res asst, 80-84; grad fel, Dept Molecular Biol & Biochem, Rice Univ, 84-88; POSTDOCTORAL FEL, DEPT BIOCHEM, DENT BR, UNIV TEX, 88- *Concurrent Pos:* Chem coordr, Houston Mus Natural Sci, 85-89; young investr award, Tex Mineralized Tissue Soc, 89; vol scientist, Sci-By-Mail, Boston Mus Sci, 89-91; sci consult, Children's Mus Houston, 90-91. *Mem:* Sigma Xi; Asn Women Sci; Am Asn Dent Res. *Res:* Cell and molecular mechanisms of bone formation; biology of the extracellular matrix; developmental biology of the skeleton; evolutionary mechanisms of limb regeneration; science education; science and children; women scientists' issues. *Mailing Add:* Dept Biochem Dent Br Univ Tex PO Box 20068 Rm DB 3-108 Houston TX 77225

RITTER, PRESTON PECK OTTO, b Memphis, Tenn, Mar 29, 41; m 61; c 2. BIOCHEMISTRY. *Educ:* Univ Calif, Berkeley, BS, 63; Univ Wis-Madison, MS, 65, PhD(biochem), 67. *Prof Exp:* Investr biochem, Biol Div, Oak Ridge Nat Lab, 67-69; res asst prof, Col Med, Baylor Univ, 69-70; asst prof, 70-78, PROF CHEM, EASTERN WASH UNIV, 78- *Concurrent Pos:* Vis staff, Health Res Lab, Los Alamos Sci Lab, NMex, 75-76. *Res:* Absorption and metabolism of antibiotics; protein and nucleic acid biochemistry. *Mailing Add:* Dept Chem & Biochem Eastern Wash Univ Cheney WA 99004

RITTER, R(OBERT) BROWN, b Winchester, Va, Jan 12, 21; m 45; c 2. CHEMICAL ENGINEERING. *Educ:* Ohio State Univ, BChE & MS, 50. *Prof Exp:* Chem engr, E I du Pont de Nemours & Co, 50-53; sr chem engr, C F Braun & Co, 53-59 & Chemet Engrs, Inc, 59-62; chief process engr, Aetron Div, Aerojet Gen Corp, 62-67; asst tech dir, Heat Transfer Res, Inc, Alhambra, 67-80; sr tech specialist, Fluor Technol, Inc, Irvine, Calif, 80-87; vpres engr, Spottwood Eng, Inc, Seal Beach, Calif, 87-89; CONSULT ENGR, 89- *Concurrent Pos:* Lectr, Long Beach State Univ, 80-81, Calif Poly State Univ, 83-84. *Honors & Awards:* Am Soc Eng Educ Div Eng Graphics Convair Award, 59. *Mem:* Am Inst Chem Engrs. *Res:* Heat transfer, condensation, fouling and rating methods; thermodynamics, chemical process, cryogenics , nuclear waste treatment and disposal; process plant design, process equipment, instrumentation and controls. *Mailing Add:* 595 Old Ranch Rd Seal Beach CA 90740-2836

RITTER, ROGERS C, b Pleasanton, Nebr, Oct 27, 29; m 50; c 3. EXPERIMENTAL GRAVITATIONAL PHYSICS, BIOPHYSICS. *Educ:* Univ Nebr, BSc, 52; Univ Tenn, PhD(physics), 61. *Prof Exp:* Inst eng, Oak Ridge Gaseous Diffusion Plant, 52-59, Oak Ridge Inst Nuclear Studies fel physics, Oak Ridge Nat Lab, 59-61; from asst prof to assoc prof, 61-70, PROF PHYSICS, UNIV VA, 70- *Concurrent Pos:* Mem, Fundamental Constants Comt, Nat Res Coun-Nat Acad Sci, 79-84 & Panel Basic Standards, Nat Res Coun, 82- *Res:* Fundamental constants; precision measurement; biophysics; urology and cardiovascular medical physics; experimental gravitation. *Mailing Add:* Dept Physics Univ Va Charlottesville VA 22903

RITTER, WALTER PAUL, b Brooklyn, NY, Oct 8, 29; m 56; c 1. NEUROPSYCHOLOGY. *Educ:* City Col New York, BA, 53; Columbia Univ, PhD(psychol), 63. *Prof Exp:* Psychologist, Suffolk County Ment Health Bd, 62-64; supvr clin psychologist, NY Univ Med Ctr, Goldwater Hosp, 64-65; fel neurol, 65-68, from asst prof neurol to vis asst prof anat, Albert Einstein Col Med, 68-74; assoc prof, 71-80, PROF PSYCHOL, LEHMAN COL, 80- *Concurrent Pos:* Vis asst prof neurosci, Albert Einstein Col Med, 74- *Mem:* Am Psychol Asn; AAAS; Psychonomic Soc; Int Neuropsychol Soc. *Res:* Electrophysiological correlates of information processing. *Mailing Add:* Dept Psychol CUNY Lehman Col Bedford Park Blvd W Bronx NY 10468

RITTER, WILLIAM FREDERICK, b Stratford, Ont, Mar 25, 42; m 66; c 2. WATER RESOURCES. *Educ:* Univ Guelph, BSA, 65; Univ Toronto, BAS, 66; Iowa State Univ, MS, 68, PhD(agr eng, sanit eng), 71. *Prof Exp:* Res assoc agr eng, Iowa State Univ, 66-71; from asst prof to assoc prof, 71-82, PROF AGR ENG, UNIV DEL, 82- *Concurrent Pos:* Res grants, Hercules, Inc, 72-76, Del Water Resources Ctr, 74-76 & 78-80, 82-87, E I du Pont de Nemours & Co, Inc, 74-75 & 76-77, Univ Del Res Found, 74-75, 208 Prog, Environ Protection Agency, 76-81, Allied Chem Co, 77-78, State of Del, 79-81, 85-, Environ Protection Agency, 80-86, Clean Lakes Prog, Environ Protection Agency, 81-83, USDA, Del, 83-87, 86-89; exec comt, Irrigation & Drainage Div, Am Soc Civil Engrs, 87-92; bd dir, USDA, 89-90, Am Soc Agr Engs, 90-92, US Geol Surv, 90-93. *Honors & Awards:* Young Engr of Year, Am Soc Agr Eng, Gunlogoson; Water Resources Award, New Castle Co; Super Achievement Award, Environ Protection Asn. *Mem:* Am Soc Civil Engrs; Am Soc Agr Engrs; Water Pollution Control Fedn; Am Water Works Asn. *Res:* Agricultural waste management; water quality modeling; land disposal of industrial and municipal wastes; irrigation; groundwater pollution and lake eutrophication; non point source pollution. *Mailing Add:* Dept Agr Eng Univ Del Newark DE 19711

RITTERHOFF, ROBERT J, medicine; deceased, see previous edition for last biography

RITTERMAN, MURRAY B, b New York, NY, Oct 19, 14; m 44; c 4. APPLIED MATHEMATICS. *Educ:* NY Univ, PhD(math), 55. *Prof Exp:* Instr math, Long Island Univ, 47-52; engr, Sylvania Elec Prod Inc Div, Gen Tel & Electronics Corp, 52-59, eng specialist, GTE Labs, 59-72; asst prof math, York Col, NY, 72-77; ASST PROF MATH, HOFSTRA UNIV, 77- *Mem:* Am Math Soc; Math Asn Am. *Res:* Differential and difference equations; communication and information theory; electron optics. *Mailing Add:* 576 Marion Dr East Meadow NY 11554

RITTERSON, ALBERT L, b New Brunswick, NJ, Mar 13, 24; m 57; c 1. MEDICAL PARASITOLOGY. *Educ:* Rutgers Univ, BS, 45, MS, 48; Univ Calif, Los Angeles, PhD(zool), 52. *Prof Exp:* From instr to asst prof, 52-68, assoc prof, 68-86, EMER PROF MICROBIOL & PARASITOL, SCH MED, UNIV ROCHESTER, 86- *Concurrent Pos:* China Med Bd fel, Cent Am, 57. *Mem:* AAAS; Am Soc Parasitologists; Am Soc Trop Med & Hyg; Sigma Xi. *Res:* Host parasite relationships; leishmaniasis, trichinosis and malaria in the golden and Chinese hamsters; innate resistance and parasite invasion; microbiology; immunology. *Mailing Add:* Sch Med & Dent Univ Rochester Box 672 Rochester NY 14642

RITTMANN, BRUCE EDWARD, b St Louis, Mo, Nov 17, 50; m 86. ENVIRONMENTAL BIOTECHNOLOGY, BIOLOGICAL PROCESSES. *Educ:* Wash Univ, St Louis, BS & MS, 74; Stanford Univ, Phd(environ eng), 79. *Prof Exp:* Environ engr, Sverdrup & Parcel & Assocs, Inc, St Louis, MO, 74-75; grad res asst, Stanford Univ, 75-79, scholar & lectr, 79; lectr, San Jose State Univ, 79; from asst prof to assoc prof, 80-88, PROF ENVIRON ENG, UNIV ILL, URBANA-CHAMPAIGN, 88- *Concurrent Pos:* Presidential young investr award, NSF, 84; Xerox fac res award, Col Eng, Univ Ill, Urbana-Champaign, 85; mem adv comt, Crit Eng Systs Div, NSF, 86-90; Univ Scholar, Univ Ill, 87; mem, Comt Simulation Contaminant Transport in Groundwater, Nat Res Coun, Nat Acad Sci, 87-89; bd dirs, Asn Environ Eng Professors, 88-92; managing ed, Biodegradation, 90-95. *Honors & Awards:* Eng Sci Award, Asn Environ Eng Professors, 79 & CH2M-Hill Award, 89; Acad Achievement Award, Am Water Works Asn, 90; Walter L Huber Res Prize, Am Soc Civil Engrs, 90. *Mem:* Water Pollution Control Fedn; Am Soc Civil Engrs; AAAS; Int Asn Water Pollution Res & Control; Asn Environ Eng Professors (vpres, 89-90, pres, 90-91); Am Water Works Asn; Am Soc Microbiol. *Res:* Application of biotechnology for environmental control; use of molecular tools in the study of biological processes, biofilm kinetics, biodegradation of low-concentration and hazardous organic chemicals, and in situ bioremediation of contaminated aquifers. *Mailing Add:* Dept Civil Eng Univ Ill 205 N Mathews Ave Urbana IL 61801-2397

RITTNER, EDMUND SIDNEY, b Boston, Mass, May 29, 19; m 42; c 1. APPLIED PHYSICS, ENERGY CONVERSION. *Educ:* Mass Inst Technol, SB, 39, PhD(chem), 41. *Prof Exp:* Little fel chem, Mass Inst Technol, 41-42, res assoc, Div Indust Coop, 42-46; res physicist & sect chief, Philips Labs Div, NAm Philips Co, Inc, 46-62, dir, Dept Physics, 62-69; mgr physics lab, Comsat Labs, 69-81, exec dir phys sci, 81-84; CONSULT SCIENTIST, 84- *Honors & Awards:* Inst Elec & Electronics Engrs Photovoltaic Founders Award, 85. *Mem:* Fel Inst Elec & Electronics Engrs; fel Am Phys Soc. *Res:* Semiconductors; photoconductivity; infrared; thermionic emission; solid state devices; solar cells. *Mailing Add:* 700 New Hampshire Ave Washington DC 20037-2406

RITTS, ROY ELLOT, JR, b St Petersburg, Fla, Jan 16, 29; m 53; c 3. MICROBIOLOGY, IMMUNOLOGY. *Educ:* George Washington Univ, AB, 48, MD, 51; Am Bd Microbiol, dipl; Am Bd Med Lab Immunol, Dipl. *Prof Exp:* Intern, DC Gen Hosp, 51-52; fel med, Sch Med, George Washington Univ, 52-53, resident, George Washington Univ Hosp, 53-54; res fel, Harvard Med Sch, 54-55; vis investr microbiol & path, Rockefeller Inst, 55-57, res assoc immunol, 57-58; from assoc prof microbiol to prof microbiol & trop med, Sch Med, Georgetown Univ, 58-64, chmn dept microbiol & trop med & prof lectr med, 59-64; dir inst biomed res, educ & res found, AMA, 64-68, dir med res, 66-68; chmn dept, Mayo Clin, 68-80, PROF MICROBIOL, MAYO MED SCH, UNIV MINN, 68-, PROF ONCOL & HEAD, MICROBIOL RES LAB, 80- *Concurrent Pos:* Life Ins Med Res Fund fel, 54-57; prof lectr, Sch Med, Univ Chicago, 64-68; mem, Am Bd Microbiol, 65-67; chmn ad hoc sci adv comt, USPHS-Food & Drug Admin, 70-71, consult, 71-; chmn adv comt diag prod, Food & Drug Admin, 72-75, Nat Food & Drug Adv Comt, 75-78; mem carcinogenesis comt, Nat Comt Clin Lab Standards, Nat Cancer Inst, 80-; mem, Am Bd Med Lab Immunol, 76-83 & 88-92; mem bd dirs, Int Assoc Study Lung Cancer, 74-76, secy, 76-78; sect ed, J Immunol, 84-88; mem exp comt immunol, WHO, 84-93; mem Exp Immunol Study Sect, NIH, 85-89. *Mem:* Fel Am Col Physicians; fel Asn Clin Sci; fel Royal Soc Health; fel Am Col Chest Physicians; fel Am Acad Microbiol; fel Infectious Dis Soc Am; Am Asn Immunologists; Am Asn Cancer Res; Am Soc Clin Oncol. *Res:* Immunotherapy; tumor immunology. *Mailing Add:* Microbiol Res Lab Mayo Clin Rochester MN 55905

RITTSCHOF, DANIEL, b Morenci, Ariz, Feb 26, 46; m 80; c 2. BEHAVIORAL ECOLOGY, CHEMICAL ECOLOGY. *Educ:* Univ Mich, BS, 68, MA, 70, PhD, 75. *Prof Exp:* Teaching asst, Univ Mich, 68-74, lectr zool, 74-75; fel, Univ Calif, Riverside, 75-78, res physiologist, Los Angeles, 78-80; marine scientist, Univ Del, 80-88; ASST PROF ZOOL, MARINE

STUDIES, DUKE UNIV, 88- *Concurrent Pos:* Res dir, Biosponge Aquaculture Prod Co; lectr, Univ Del. *Mem:* Am Soc Zoologists; Asn Chemoreception Sci; Int Soc Chem Ecol. *Res:* Marine chemical sensing; anti fouling; molecules and mechanisms of chemoattraction of Urosalpinx cinerea; functions of chemical sensing in the integration of resource utilization in marine environments and in the gastropod shell habitat web; chemical camouflaging and inhibition facilitation of chemoresponses, soluble pollutants; teratogenic effects; assays of behavioral toxicity. *Mailing Add:* Marine Lab Duke Univ Pivers Island Beaufort NC 28516

RITVO, EDWARD R, b Boston, Mass, June 1, 30; m 61; c 4. PSYCHIATRY. *Educ:* Harvard Col, BA, 51; Boston Univ, MD, 55. *Prof Exp:* From asst prof to assoc prof, 64-77, PROF IN RESIDENCE PSYCHIAT, SCH MED, UNIV CALIF, LOS ANGELES, 77- *Res:* Adult and child psychiatry, including psychoanalysis and neurophysiology. *Mailing Add:* Neuropsych Inst 760 Westwood Plaza Los Angeles CA 90024

RITZ, VICTOR HENRY, b New York, NY, Apr 4, 34. SURFACE PHYSICS, RADIATION PHYSICS. *Educ:* Polytech Inst Brooklyn, BS, 55; Univ Md, Col Park, MS, 62; Univ Sao Paulo, PhD(physics), 67. *Prof Exp:* Physicist, Nat Bur Standards, 56-59; res physicist, US Naval Res Lab, 59-89; CONSULT, B-K SYSTS, 89- *Mem:* Am Phys Soc; Am Vacuum Soc; Sigma Xi. *Res:* Electron emission from solids; radiation effects in solids; radiation dosimetry. *Mailing Add:* 114 Princess St Alexandria VA 22314

RITZERT, ROGER WILLIAM, b Aurora, Ill, Jan 24, 36; m 62; c 3. BIOCHEMISTRY. *Educ:* NCent Col, Ill, BA, 58; Mich State Univ, MS, 61, PhD(biochem), 66. *Prof Exp:* Asst biochem, Mich State Univ, 58-65; res biochemist, Div Biochem & Microbiol, Battelle Mem Inst, 65-67; SR BIOCHEMIST, TECH CTR, OWENS-ILL, INC, 67- *Concurrent Pos:* Adj asst prof, Med Col Ohio, Toledo, 71- *Mem:* Am Soc Microbiol; Am Chem Soc; Tissue Cult Asn. *Res:* Biochemistry of plant growth and development; cellular responses responses to toxicants in vitro. *Mailing Add:* 6009 Jeffrey Lane Sylvania OH 43560

RITZ-GOLD, CAROLINE JOYCE, b Cleveland, Ohio; m 78. NONEQUILIBRIUM THERMODYNAMICS. *Educ:* Calif State Univ, Long Beach, BS, 66, MS, 69; Univ Southern Calif, PhD(biochem), 78. *Prof Exp:* Fel, Cardiovasc Res Inst, Univ Calif, San Francisco, 78-82; NSF vis prof, San Francisco State Univ, 84-86; RES SCIENTIST, BIOMOLECULAR SCI, 86- *Concurrent Pos:* Advan res fel, Am Heart Asn, 80-81. *Mem:* Biophys Soc; Am Phys Soc; Sigma Xi; Molecular Graphics Soc; Protein Soc. *Res:* application of fractal concepts and condensed matter theory to the structure and dynamics of biomolecules and assemblies; mechanisms of free energy conversion, transfer and storage; analysis and modeling of biological systems using dynamical systems methods. *Mailing Add:* Biomolecular Sci 38451 Timpanogas Fremont CA 94536

RITZI, EARL MICHAEL, b Worcester, Mass, June 4, 46; m 69; c 1. TUMOR IMMUNOLOGY, RETROVIROLOGY. *Educ:* Univ Mass, BS, 68; Princeton Univ, MA, 70, PhD(virol), 72. *Prof Exp:* Res asst virol, Princeton Univ, 72; res assoc, Worcester Found Exp Biol, 72-75; sr staff assoc virol, Inst Cancer Res, Columbia Univ, 75-77; asst prof virol, Univ Tenn, Ctr Health Sci, 77-82; assoc prof virol, State Okla Organized Res Prog, OCOMS, Tulsa, 82-84; ASSOC PROF VIROL, HEALTH SCI CTR, TEX TECH UNIV, 84- *Concurrent Pos:* Chmn, Immunol Sect, Nat Cancer Inst Site Rev Panel, 78-, mem virol sect, 78-; prin investr, Am Cancer Soc Willie Mae Darwin Mem Grant, 78-81 & NIH Cancer Res Grant, 81-86; chairperson, Memphis Am Cancer Soc Fund Raising Camp, 80-; sci fair judge, Am Soc Microbiol, 80-; virol lectr, Miami Int Med Sch Rev, 89- *Mem:* Am Asn Cancer Res; Am Asn Immunologists; Am Soc Microbiol. *Res:* Expression of retroviral proteins and human tumor-associated antigens, studied to determine their utility as signals for the presence of solid tumors. *Mailing Add:* Dept Microbiol Health Sci Ctr Tex Tech Univ Lubbock TX 79430

RITZMAN, ROBERT L, b Peoria, Ill, Nov 19, 32; m 59; c 1. INORGANIC CHEMISTRY, RADIOCHEMISTRY. *Educ:* Bradley Univ, BS, 55; Rensselaer Polytech Inst, PhD(phys chem), 61. *Prof Exp:* Sr scientist phys chem, Battelle-Columbus Labs, 59-75; sr scientist, 75-80, vpres, Sci Applns Int Corp, 80-84; TECH ADV, ELEC POWER RES INST, 84- *Mem:* Am Chem Soc; fel Am Nuclear Soc. *Res:* Chemical separations; nuclear reactor safety and accident analysis; nuclear radiation effects; fission-product chemistry; fission-gas release; environmental impact analysis; nuclear fuel cycle risk analysis; nuclear reactor severe accident phenomena. *Mailing Add:* 829 Henderson Ave Sunnyvale CA 94086

RITZMAN, THOMAS A, b Mayaguez, PR, Feb 18, 14; m 41; c 3. TREATMENT ANXIETY, TREATMENT OBESITY. *Educ:* Yale Univ, BA, 36; Harvard Med Sch, MD, 40. *Prof Exp:* Pvt pract obstet & gynecol, 46-81; PVT PRACT, MED HYPNOANALYSIS, 73- *Honors & Awards:* Achievement Award, Am Acad Med Hypnoanalysis, 88. *Mem:* Am Col Obstet & Gynecol; Am Acad Med Hypnoanalysis; NY Acad Sci. *Res:* The use of guided analysis to effect permanent cure of emotion and systemic physical illness. *Mailing Add:* 279 Pleasant St Concord NH 03301

RITZMANN, LEONARD W, b South Bend, Ind, Sept 8, 21; m 42; c 3. INTERNAL MEDICINE, CARDIOLOGY. *Educ:* Valparaiso Univ, AB, 42; Wash Univ, MD, 45; Am Bd Internal Med, dipl. *Prof Exp:* Intern internal med, Barnes Hosp, St Louis, 45-46; asst resident med, Salt Lake County Gen Hosp, 47-48, fel cardiol, 48-49, chief resident, 49-50; Am Heart Asn fel cardiol, Postgrad Med Sch London, Eng, 50-51; sect chief cardiol, Vet Admin Hosp, Portland, 53-56, chief med, 56-60, sect chief cardiol, 60-70, staff cardiol, 70-90; prof med, Sch Med, Univ Ore, 63-90; RETIRED. *Mem:* Am Heart Asn; Christian Med Soc (pres, 78-80). *Res:* Electrocardiography; vectorcardiography; cardiac arrhythmias; cor pulmonale; coronary heart disease; pacemakers. *Mailing Add:* 4230 SW Sixth Ave Portland OR 97201

RITZMANN, RONALD FRED, b Cicero, Ill, June 16, 43. PSYCHOPHARMACOLOGY, DRUG ABUSE. *Educ:* Northern Ill Univ, BA, 65, MA, 68, PhD(neurosci), 73. *Prof Exp:* Res assoc, 77-78, ASST PROF PHYSIOL, DEPT PHYSIOL & BIOPHYS, MED CTR, UNIV ILL, 78-, ASSOC FAC, DRUG & ALCOHOL ABUSE RES & TRAINING PROG, 80-; RES PHARMACOLOGIST, WESTSIDE VET ADMIN MED CTR, 84- *Mem:* Neurosci Soc; Am Soc Neurochem; Int Soc Biomed Res Alcoholism. *Res:* Neurochemical basis for the development of tolerence and physical dependence on psychoactive drugs with particular interest in the modification of the addictive states by peptides. *Mailing Add:* Dept Psychiat Bldg 1 Olive View Med Ctr 14445 Olive View Dr Sylmar CA 91342-1493

RIVA, JOHN F, b Digohan (Belluno), Italy, June 17, 29; US citizen; div; c 3. PALEONTOLOGY, BIOSTRATIGRAPHY. *Educ:* Univ Nev, BA, 50, MSc, 57; Columbia Univ, PhD(geol), 62. *Prof Exp:* Res assoc geol, McGill Univ, 61-63; asst prof, Villanova Univ, 63-66; from asst prof to assoc prof, 66-77, TITULAR PROF LAVAL UNIV, 77- *Concurrent Pos:* Soc Sigma Xi res awards, 64- 65, 67 & 70; NSF res grant, 64-67; res assoc, Columbia Univ, 66-68 & 70-71; Nat Res Coun Can yearly grants, 67-; prin investr, res on graptolites, ordovician biostratigraphy. *Mem:* Fel Geol Soc Am; Palaeont Asn. *Res:* Ordovician and Silurian graptolites, their taxonomy and biostratigraphy. *Mailing Add:* 1238 de Rouville Ave Ste Foy PQ G1W 3T7 Can

RIVA, JOSEPH PETER, JR, b Chicago, Ill, Oct 31, 35; m 63; c 2. GEOLOGY, STRUCTURAL GEOLOGY. *Educ:* Carleton Col, BA, 57; Univ Wyo, MS, 59. *Prof Exp:* Geologist, Tenn Gas & Oil Co, 59; consult geologist, G H Otto Co, Chicago, 61-65 & Earth Sci Labs, Ohio, 65-66; geologist, Sci Info Exchange, Smithsonian Inst, 66-67, actg chief earth sci br, 67-69, chief, 69-74; asst chief, 85-86, SPECIALIST EARTH SCI, SCI POLICY RES DIV, CONG RES SERV, LIBR CONG, WASH, DC, 74- *Concurrent Pos:* Geol consult, 66-; detail, US geol surv, 80; mem Comt Offshore Hydrocarbon Resource Estimation Methodology, Nat Res Coun, Nat Acad Sci, Comt Undiscovered Oil & Gas Res. *Honors & Awards:* Reeve & Noyes Acad Prize, Carleton Col; Fel Award, NSF. *Mem:* Am Asn Petrol Geologists; Am Inst Prof Geologists; Sigma Xi. *Res:* Petroleum geology and resources; natural gas; underground gas storage; engineering geology; science policy; energy policy; geological information and retrieval; environmental geology, geothermal energy, editing and writing; author of over 150 publications in energy and the earth sciences including World Petroleum Resources and Reserves, Westview Press, Fossil Fuels, Encyclopedia Britannica. *Mailing Add:* 9705 Mill Run Dr Great Falls VA 22066

RIVARD, JEROME G, b Hudson, Wis, Nov 21, 32; m 55; c 3. FLUID MECHANICS. *Educ:* Univ Wis, BSME, 55. *Prof Exp:* vpres, Allied Signal, Inc, 86-88. *Concurrent Pos:* Dir eng, Bendix, 62-76; chief eng, Ford Motor Co, 76-86. *Mem:* Nat Acad Eng; fel Inst Elec & Electronics Engrs; fel Soc Automotive Engrs. *Res:* Application of electronics to automotive systems. *Mailing Add:* Global Technol Bus Develop 31078 Rivers Edge Ct Birmingham MI 48010

RIVAS, MARIAN LUCY, b New York, NY, May 6, 43. MEDICAL GENETICS, COMPUTER SYSTEMS. *Educ:* Marian Col, BS, 64; Ind Univ, MS, 67, PhD(med genetics), 69; Am Bd Med Genetics, cert, 82. *Prof Exp:* Fel med genetics, Dept Med, Johns Hopkins Hosp, 69-71; asst prof biol, Douglas Col, Rutgers Univ, 71-75; assoc prof, 75-82, dir genetic coun, Hemophilia Ctr, 77-79, PROF MED GENETICS, ORE HEALTH SCI UNIV, 82-; ASSOC SCIENTIST, NEUROL SCI INST, GOOD SAMARITAN HOSP, 78-; ASSOC PROF PEDIAT, UNIV TENN, MEMPHIS. *Concurrent Pos:* Lectr Biol, Marian Col, 66-68; dir, Genetic Counseling Grad Prog, Rutgers Univ, 71-75; adj asst prof med genetics, Ind Univ, 71-75; mem, Mammalian Cell Lines Comt, Nat Inst Gen Med Sci, NIH, 75-76, adv comt, 76-80; consult, Interregional Cytogenetics Register Syst & Nat Mutant Cell Bank, 75-; staff mem, Emmanuel Hosp, Portland, 81-; mem, NIH-Venezuela Comn Huntington's Chorea, 82- *Mem:* Am Soc Human Genetics; Am Epilepsy Soc; NY Acad Sci; Sigma Xi. *Res:* Human gene mapping; genetic aspects of epilepsy; human pedigree and segregation analyses; computer applications in clinical genetics and genetic counseling; population genetics; ethnic distribution of genetic disease. *Mailing Add:* Univ Tenn 711 Jefferson Suite 523 Memphis TN 38165

RIVELA, LOUIS JOHN, b Brooklyn, NY, Feb 24, 42; m 64; c 1. GENERAL CHEMISTRY, ANALYTICAL CHEMISTRY. *Educ:* Rutgers Univ, New Brunswick, BS, 63; Univ NC, Chapel Hill, MS, 67, PhD(inorg chem), 70. *Prof Exp:* From asst prof to assoc prof inorg chem, 77-89, CHMN, DEPT INORG CHEM, WILLIAM PATERSON COL, 69-, PROF, 89- *Concurrent Pos:* Princeton fac fel, 87. *Mem:* Am Chem Soc; Sigma Xi; Am Inst Chem. *Res:* Synthesis and characterization of coordination compounds containing organophophines. *Mailing Add:* 115 Andover Dr Wayne NJ 07470

RIVELAND, A(RVIN) R(OY), b Buxton, NDak, July 29, 23; m 52; c 3. CIVIL ENGINEERING. *Educ:* Univ NDak, BS, 45; Univ Nebr, MS, 54. *Prof Exp:* Engr, Lium & Burdick, Engrs, 45-57; from instr to assoc prof, 49-80, PROF CIVIL ENG, UNIV NEBR, LINCOLN, 80- *Mem:* Am Soc Civil Engrs; Am Soc Eng Educ; Am Concrete Inst. *Res:* Structures; ultimate strength design in reinfored concrete. *Mailing Add:* Dept Civil Eng Univ Nebr Lincoln NE 68588-0531

RIVELLO, ROBERT MATTHEW, b Washington, DC, May 20, 21; m 50; c 4. STRUCTURES, MATERIALS. *Educ:* Univ Md, BS, 43, MS, 48. *Prof Exp:* From instr to asst prof mech eng, 46-49, from asst prof to prof aeronaut eng, 49-79, EMER PROF AEROSPACE ENG, UNIV MD, COLLEGE PARK, 79-; ENGR, PRIN PROF STAFF, APPL PHYSICS LAB, JOHN HOPKINS UNIV, 79- *Concurrent Pos:* Engr, Fairchild Aircraft Co, 48; aeronaut eng officer, USAF, 51-53; assoc engr, Appl Physics Lab, Johns Hopkins Univ, 53-54, consult, 54-79, sr engr, 56-79; mem spec struct subcomt, Panel Piloted Aircraft, Res & Develop Bd, 52-53; mem comt elevated temperature struct test facil, USAF; consult, Smithsonian Inst, 57-58; struct subcomt, Navy

Aeroballistics Comt, 80- *Mem:* Am Soc Mech Engrs; Am Inst Aeronaut & Astronaut. *Res:* Applied Mechanics; aircraft and missile structures; materials, dynamics, elasticity and thermal stresses. *Mailing Add:* 8502 Hunter Creek Trail Potomac MD 20854

RIVENSON, ABRAHAM S, b Roman, Rumania, Nov 22, 26; US citizen; m 49; c 1. PATHOLOGY OF CANCER, CARCINOGENESIS. *Educ:* Univ Bucharest, MD, 54. *Prof Exp:* RES PROF PATH, NY MED COL, VALHALLA, 77-, RES PROF ANAT, 80- *Concurrent Pos:* Head Histopathology, Am Health Found, 77- *Mem:* NY Acad Med; Am Asn Cancer Res; Europ Asn Cancer Res. *Res:* Experimental pathology; carcinogenesis. *Mailing Add:* Am Health Found Naylor Dana Inst One Dana Rd Valhalla NY 10595-1599

RIVERA, AMERICO, JR, b New York, NY, Aug 22, 28; m 58; c 4. BIOCHEMISTRY, NEUROCHEMISTRY. *Educ:* Inter-Am Univ PR, San German, 52; Fordham Univ, MS, 56; Columbia Univ, PhD(biochem), 63. *Prof Exp:* USPHS fel, Univ Wis-Madison, 63-65; res chemist, NIH, San Juan, PR, 65-70 & Md, 70-75; HEALTH SCIENTIST ADMINR, NAT INST GEN MED SCI, 75- *Concurrent Pos:* Lectr, Med Sch, Univ PR, 71-72. *Mem:* Brit Biochem Soc. *Res:* Effects of energy deprivation on the carbohydrate metabolism of the monkey brain; intermediary metabolism and mechanisms of mitosis in microorganisms and cell cultures; administration of biomedical engineering research grants. *Mailing Add:* PO Box 583 Olney MD 20830-0583

RIVERA, EVELYN MARGARET, b Hollister, Calif, Nov 10, 29; div. ENDOCRINOLOGY, CANCER. *Educ:* Univ Calif, Berkeley, AB, 52, MA, 60, PhD(zool), 63. *Prof Exp:* Am Cancer Soc fel biochem, Nat Inst Res Dairying, Reading, Eng, 63-65; from asst prof to assoc prof, 65-72, PROF ZOOL, MICH STATE UNIV, 72- *Concurrent Pos:* Sabbatical leave, Cancer Res Lab, 71-72; mem, Exp Biol Comt, Breast Cancer Task Force, Nat Cancer Inst, 73-76, Carcinogenesis Comt, 76-79, Cancer Cause & Prev Comt, 79-81, Reproductive Endocrinol Comt, 85-87; sabbatical leave, Transplantation Biol Sect, Clin Res Ctr, Harrow, Eng, 78-79; ed, Tissue Cult Asn Report, 86-90; reviewing ed, In Vitro Cellular & Develop Biol, 90- *Honors & Awards:* UNESCO Award, Int Cell Res Orgn, 65; Res Career Develop Award, NIH, 67-72 & Res Fel Award, 78-79. *Mem:* Fel AAAS; Am Asn Can Res; Soc Exp Biol & Med; Tissue Cult Asn; Brit Soc Endocrinol. *Res:* Biology of mammary tumors. *Mailing Add:* Dept Zool Mich State Univ East Lansing MI 48824-1115

RIVERA, EZEQUIEL RAMIREZ, b Alpine, Tex, Oct 17, 42; m 70; c 1. ELECTRON MICROSCOPY, ULTRASTRUCTURE. *Educ:* Sul Ross State Col, Tex, BS, 64; Purdue Univ, MS, 67; Univ Tex, Austin, PhD(biol sci & bot), 73. *Prof Exp:* Res technician bot & plant pathol, Purdue Univ, 66-67; clin lab technician clin chem, US Army, Ft Dix, NJ, 67; asst chief biochem & toxicol, 6th US Army Med Lab, Ft Baker, Calif, 68; chief biochem, 376th Med Lab, US Army Support, Thailand, 69-70; instr bot, Univ Tex, Austin, 72; asst prof biol, Univ Notre Dame, Ind, 73-74; from asst prof to assoc prof, 74-85, PROF BIOL SCI, UNIV LOWELL, MASS, 85- *Concurrent Pos:* Consult, Pathol Dept, Bon Secours Hosp, Mass, 80-; referee, Scanning Electron Micros, 80-, Protoplasma Int J Cell Biol, 82-; guest assoc prof trop med, Mahidol Univ, Bangkok, Thailand, 81; dir biol sci, New Eng Soc Electron Micros, 82, from vpres to pres, 83-84, dir, 85; testing, Univ Lowell Res Found, 83-; referee, Physiol, Cellular & Molecular Biol Comt, NSF, 83-; guest prof animal biol, Univ Perpignan, Perpignan, France, 86; pres, Univ Lowell Club, Sigma Xi, 89-91. *Mem:* Bot Soc Am; Am Soc Plant Physiologists; Electron Micros Soc Am; Sigma Xi; Microbeam Analysis Soc. *Res:* Ultrastructure of overwintering and desert xerophytic plants; cytochemistry of secretory components in marine snails; cytology of schistosome development; plant physiology of carbon 13/carbon 12 fractionation and calcium oxalate; spermagenesis in snails, fish. *Mailing Add:* Dept Biol Univ Lowell Lowell MA 01854

RIVERA, WILLIAM HENRY, b El Paso, Tex, Jan 26, 31; m 58; c 5. ANALYTICAL CHEMISTRY. *Educ:* Univ Louisville, BS, 53, PhD(chem), 62. *Prof Exp:* Nuclear engr, Gen Dynamics/Ft Worth, 56-58; asst chem, Univ Louisville, 58-62; asst prof, 62-63, asst grad dean, 73-76, ASSOC PROF CHEM, UNIV TEX, EL PASO, 63- *Concurrent Pos:* Consult, Nuclear Effects Directorate, White Sands Missile Range, NMex, 64- *Mem:* Am Chem Soc; Am Soc Testing & Mat; Sigma Xi. *Res:* Radiation effects on organic materials, radiation induced polymerization; photon and neutron activation analysis; dosimetry; environmental analysis; radiation protection. *Mailing Add:* 805 E Blacker Ave El Paso TX 79902

RIVERA-CALIMLIM, LEONOR, NEUROPSYCHOPHARMACOLOGY. *Educ:* Univ St Tomas, Manila, Phillipines, MD, 53. *Prof Exp:* ASSOC PROF PHARMACOL, DEPT PHARMACOL, SCH MED & DENT, UNIV ROCHESTER, 76- *Res:* Drugs in Parkinsonism; drugs in psychiatry. *Mailing Add:* Dept Pharmacol & Med Sch Med & Dent Univ Rochester 601 Elmwood Ave Rochester NY 14642

RIVERO, JUAN ARTURO, b Santurce, PR, Mar 5, 23; c 2. HERPETOLOGY. *Educ:* Univ PR, BS, 45; Harvard Univ, MA, 51, PhD(biol), 52. *Prof Exp:* Asst plant physiologist, Inst Trop Agr, PR, 45; from instr to assoc prof, 43-58, dir inst marine biol & zool garden, 54-63, dir dept biol, Univ, 59-60, actg dean arts & sci, 62-63, dean, 63-66, PROF ZOOL, UNIV PR, MAYAGUEZ, 58- *Concurrent Pos:* Assoc, Mus Comp Zool, Harvard Univ, 68; Guggenheim Found fel, 70; temporary investr, Venezuelan Inst Sci Invests; Herpet League fel. *Honors & Awards:* distinguished prof, Univ PR, 87- *Mem:* AAAS; Asn Island Marine Labs Caribbean (pres, 58-60); corresp mem Soc Venezolana de Ciencias Naturales; Soc Syst Zool; PR Acad Arts & Sci; Soc La Salle Ciencias Naturales. *Mailing Add:* Dept Biol Univ PR Mayaguez PR 00708

RIVERS, DOUGLAS BERNARD, b Beatrice, Nebr, May 24, 51; m 83; c 2. FERMENTATION MICROBIOLOGY, ENVIRONMENTAL BIOREMEDIATION. *Educ:* Kans State Univ, BS, 74; Auburn Univ, MS, 76; Univ Ark, PhD(microbiol), 83. *Prof Exp:* Biochemist, Gulf Oil Corp, 76-79, group leader, 79; res assoc, Univ Ark, 79-83, asst prof microbiol, 84; sr microbiologist, Archer Daniels Midland Co, 84-86; sr biotechnologist, Southern Res Inst, 86-87; HEAD, FERMENTATION & BIOPROCESS TECHNOL SECT, SOUTHERN RES INST, 87- *Concurrent Pos:* Adj prof, dept biology, Univ Ala, Birmingham. *Mem:* Am Soc Microbiol; Soc Indust Microbiol. *Res:* Environmental bioremediation; process development through biochemical engineering in foods, specialty chemicals and commodity chemicals; fermentations and separations to produce enzymes, ethanol and monoclonal antibodies; diagonostics based on Enzyme-Linked Immunosorbent Assay technology; technology assessments. *Mailing Add:* Southern Res Inst PO Box 55305 Birmingham AL 35255-5305

RIVERS, JERRY MARGARET, b Bogota, Tex, Sept 29, 29. NUTRITION. *Educ:* Tex Tech Univ, BS, 51, MS, 58; Pa State Univ, PhD(nutrit, biochem), 61. *Prof Exp:* Dietitian, USPHS, 51-53 & Methodist Hosp, 53-57; from asst prof to prof, 62-84, EMER PROF NUTRIT, CORNELL UNIV, 84- *Concurrent Pos:* Res grants, NIH, 63-, Nutrit Found, 66- & USDA, 78-; fels, Mary Swantz Rose, General Foods, Mead Johnson; adj prof, Univ Tex, 85- *Mem:* Am Dietetic Asn; Am Inst Nutrit; Sigma Xi; Am Home Econ Asn. *Res:* Ascorbic acid metabolism. *Mailing Add:* 10819 Crown Colony Dr 37 The Greens Austin TX 78747

RIVERS, JESSIE MARKERT, b Elizabeth City, NC, July 9, 49; m 71. ORGANIC CHEMISTRY, NATURAL PRODUCTS CHEMISTRY. *Educ:* Meredith Col, AB, 71; NC State Univ, PhD(org chem), 78. *Prof Exp:* SR RES CHEMIST TOBACCO CHEM, R J REYNOLDS TOBACCO CO, 77- *Mem:* Am Chem Soc; Sigma Xi. *Res:* Isolation, characterization and synthesis of naturally occurring tobacco constituents; relationship of tobacco chemistry to tobacco utilization. *Mailing Add:* 4837 River Ridge Rd Pfafftown NC 27040

RIVERS, PAUL MICHAEL, b Schenectady, NY, July 18, 44; m 66; c 2. ORGANIC CHEMISTRY, ANALYTICAL CHEMISTRY. *Educ:* LeMoyne Col, BS, 66; Univ Notre Dame, Ind, PhD(org chem), 70. *Prof Exp:* Sr chemist, 70-71, DIR, QUAL CONTROL & DIR, ANALYSIS DEPT, REILLY TAR & CHEM CORP, 71- *Mem:* Am Chem Soc; Am Soc Testing Mats. *Res:* Gas and liquid chromatographic separation of pyridine derivatives; analytical methods in air and water pollution control; isolation and identification of organic chemicals. *Mailing Add:* 8244 Colt Dr Plainfield IN 46168-9755

RIVERS, WILLIAM J(ONES), b Lakeland, Fla, May 28, 36; m 58; c 2. FLUID MECHANICS, HYDRAULICS. *Educ:* Univ Fla, BME, 59; Purdue Univ, MSME, 61, PhD(mech eng), 64; Univ Southern Calif, MSAE, 70. *Prof Exp:* Sr res engr, Rocketdyne Div, NAm Aviation, Inc, 64-65; PROF MECH ENG, CALIF STATE UNIV, NORTHRIDGE, 65- *Concurrent Pos:* Res engr, Rocketdyne Div, NAm Aviation, Inc, 65-66; develop engr, Marquardt Corp, 66-67; thermodynamics engr, Lockheed Calif Co, 68-69; NSF fac fel, Univ Southern Calif, 69-70; tech consult, Peerless Pump Co, 71-72. *Honors & Awards:* Ralph R Teetor Award, Soc Automotive Engrs, 70. *Mem:* Am Soc Mech Engrs. *Res:* Centrifugal pump performance and axial thrust; fluid power systems; experimental methods. *Mailing Add:* Mech Eng Dept Calif State Univ Northridge CA 91330

RIVES, JOHN EDGAR, b Birmingham, Ala, 1933; m 56; c 6. SPECTROSCOPY & SPECTROMETRY. *Educ:* Auburn Univ, BS, 55; Duke Univ, PhD(physics), 62. *Prof Exp:* Asst prof physics, Univ Ga, 59-60; instr & res assoc, Duke Univ, 61-63; from asst prof to assoc prof, 63-76, PROF PHYSICS, UNIV GA, 76- *Concurrent Pos:* Consult, Oak Ridge Nat Lab, 67-68. *Mem:* Fel Am Phys Soc; Am Asn Univ Professors; Sigma Xi. *Res:* Optical properties of solids; laser induced phonon physics; heat transport in magnetic insulators. *Mailing Add:* Dept Physics & Astron Univ Ga Athens GA 30602

RIVEST, BRIAN ROGER, b Schenectady, NY, May 16, 50; m 75; c 2. INVERTEBRATE REPRODUCTIVE BIOLOGY, BIOLOGY OF GASTROPODS. *Educ:* Cornell Univ, BS, 72; Univ NH, MS, 75; Univ Wash, PhD(zool), 81. *Prof Exp:* Actg assoc dir invert zool, Fri Harbor Labs, Univ Wash, 80-81; asst prof, 81-88, ASSOC PROF INVERT ZOOL, ZOOL MARINE BIOL & INTROD BIOL, STATE UNIV NY, CORTLAND, 88- *Concurrent Pos:* Vis assoc prof, Shoals Marine Lab, Cornell Univ, 82- *Mem:* Am Soc Zoologists; Western Soc Naturalists; AAAS; Am Malacological Union. *Res:* Developmental patterns and reproductive biology of gastropod molluscs, with special interest in extra embryonic sources of nutrition. *Mailing Add:* Biol Dept State Univ NY Cortland NY 13045

RIVEST, ROLAND, chemistry, for more information see previous edition

RIVEST, RONALD L, CRYPTOGRAPHY. *Educ:* Yale Univ, BS, 69; Stanford Univ, PhD(computer sci), 74. *Prof Exp:* PROF COMPUTER SCI, MASS INST TECHNOL, 74- *Mem:* Nat Acad Eng; Inst Elec & Electronics Engrs; Asn Comput Mach. *Mailing Add:* Mass Inst Technol 545 Technology Sq Rm 324 Cambridge MA 02139

RIVETT, ROBERT WYMAN, b Omaha, Nebr, Jan 20, 21; m 40; c 3. BIOCHEMISTRY. *Educ:* Univ Nebr, BS, 42, MS, 43; Univ Wis, PhD(biochem), 46. *Prof Exp:* Res microbiologist, Abbott Labs, 46-48, group leader antibiotic develop, 48-57, asst to dir develop, 57-58, mgr develop, 58-59, dir, 59-64, dir, Sci Admin & Serv, 64-71; dir, Corp Qual Assurance Standards & Audits, 71-76, dir, Qual Assurance Agr & Vet Div, 76-77, Sci Prod Div, 77-78, vpres, Alpha Therapeut Corp, Qual Assurance, 78-82; CONSULT, PHARMACEUT & RELATED FIELDS, 82- *Mem:* Am Chem Soc; Am Inst Chem EngrS; Am Soc Qual Control. *Res:* Organic chemistry; nutrition of bacteria; natural products; streptolin; fermentation equipment. *Mailing Add:* 3303 Taos Ct Deming NM 88030-9601

RIVETTI, HENRY CONRAD, b Feb 21, 24; US citizen; m 72; c 3. DENTISTRY, PROSTHODONTICS. *Prof Exp:* Assoc prof, Sch Dent, Fairleigh Dickinson Univ, 69-75, prof prosthodont & actg chmn dept, 75-90, chmn removable partial denture sect, 72-90; RETIRED. *Concurrent Pos:* Grant, Sch Dent, Fairleigh Dickinson Univ, 71- *Mem:* Fel Am Col Dentists; Am Dent Asn; Am Prosthodont Soc. *Res:* Dental and psychological aspects involved in temporo-mandibular joint dysfunction syndrome. *Mailing Add:* 80A Long Beach Blvd North Beach NJ 08008

RIVIER, CATHERINE L, b Vaud, Switzerland, June 21, 43; m 67; c 2. REPRODUCTIVE ENDOCRINOLOGY. *Educ:* Univ Lausanne, Switzerland, Lic es Sci, 68, PhD, 72. *Prof Exp:* Fel, 72-74, sr res assoc, 74-79, ASST RES PROF, SALK INST, 79-; ASSOC RES PROF, CLAYTON FEDN LABS, PEPTIDE BIOL. *Mem:* Endocrine Soc; Am Physiol Soc; Soc Neurosci; Soc Study Reproduction; Res Soc Alcoholism. *Res:* Mechanism of control of prolactin, gonadotropin and adrenocorticotrophic hormone secretion. *Mailing Add:* Salk Inst 10010 Torrey Pine Rd La Jolla CA 92037

RIVIER, JEAN E F, b Casablanca, Morocco, July 14, 41; Swiss citizen; m 67; c 2. NEUROENDOCRINOLOGY. *Educ:* Univ Lausanne, PhD, 68. *Prof Exp:* Fel, Univ Lausanne, 69; fel, Rice Univ, Houston, Tex, 69-70; res assoc, Saik Inst, 70-73, from asst prof to assoc prof, 73-83, assoc prof & sr res mem, 84-89, PROF, SAIK INST, 89- *Mem:* Am Chem Soc; Endocrine Soc; Protein Soc; Am Peptide Soc; Am Soc Biochem & Molecular Biol. *Res:* Neuroendocrinology involved in isolation and analysis of new peptide hormones, their total synthesis and pharmacology. *Mailing Add:* Salk Inst PO Box 85800 San Diego CA 92186-5800

RIVIER, NICOLAS YVES, b Lausanne, Switz, Aug 5, 41; m 71; c 3. GLASS, MANY-BODY THEORY. *Educ:* Univ Lausanne, dipl physics, 65; Cambridge Univ, PhD(physics), 68. *Prof Exp:* Asst prof physics, Univ Calif, Los Angeles, 68-69; lectr, Univ Calif, Riverside, 69-70; lectr, 70-86, READER THEORET SOLID STATE PHYSICS, IMP COL, 86- *Concurrent Pos:* Vis prof, Univ Fed Da Paraiba, Brazil, 73-77, Univ Provence, France, 79-82, Univ Porto, Portugal, 82, Univ Lausanne & Geneva, Switz, 85 & Lausanne Univ, 90; vis scientist, Inst Theoret Physics, Univ Calif, Santa Barbara, 83, Los Alamos Nat Lab, 84 & Argonne Nat Lab, 87-90. *Mem:* Am Phys Soc; Europ Phys Soc; Inst Physics. *Res:* Glass structure and properties; gauge aspects of condensed matter; geometry and topology of disorder; cellular networks, structure and evolution; maximum entropy inference; quasicrystals; grain boundaries. *Mailing Add:* Blackett Lab Imp Col London SW7 2BZ England

RIVIERE, GEORGE ROBERT, b Decatur, Ill, Feb 26, 43; m 71; c 2. IMMUNOLOGY, DENTISTRY. *Educ:* Drake Univ, BA, 66; Univ Ill, Chicago Med Ctr, BSD, 66, DDS, 68, MS, 70; Univ Calif, Los Angeles, PhD(immunol), 73. *Prof Exp:* Researcher, Dent Res Inst, USN, 73-75; asst prof, 75-77, assoc prof, 77-81, prof, Sch Med & Dent, Dent Res Inst, Univ Calif, Los Angeles, 82-87; PROF & CHMN DEPT ORAL BIOL, SCH DENT, UNIV MO, KANS CITY, 87- *Concurrent Pos:* USPHS fel, Nat Inst Dent Res, 73-75; prin investr, Nat Inst Dent res grant, 76-83 & NIH career res develop award, 77-82. *Mem:* Int Asn Dent Res; Sigma Xi; AAAS; Am Asn Immunologists. *Res:* Transplantation immunology; immunogenetics; regulation of immune responses, especially immunity to enteric microorganisms and role of immunity in development. *Mailing Add:* 1119 SW Comus St Portland OR 97219-6493

RIVIERE, JIM EDMOND, b New Bedford, Mass, Mar 3, 53; m 76; c 3. PHARMACOKINETICS, TOXICOLOGY. *Educ:* Boston Col, BS, 76, MS, 76; Purdue Univ, DVM & PhD(pharmacol), 80. *Prof Exp:* Postdoctoral assoc pharmacol, Purdue Univ, 80-81; from asst prof to assoc prof, NC State Univ, 81-88; PROF TOXICOL, NC STATE UNIV, 88- *Concurrent Pos:* Prin investr, NIH, US Army Med Res & Develop Command, B Dickenson Res ctr & USDA, 84-; expert panelist, Dermal Toxicol Worshops, Environ Protection Agency, 88; dir, Cutaneous Pharmacol & Toxicol Ctr, NC State Univ, Batelle, 88-; ed, J Vet Pharmacol Therapeut, 89- *Honors & Awards:* Beecham Award Res Excellence, 86. *Mem:* Soc Toxicol; Am Vet Med Asn; Am Acad Vet Pharmacol & Therapeut; Sigma Xi; Am Asn Pharmaceut Scientists. *Res:* Dermatopharmacology; cutaneous toxicology; toxicokinetics of drug and chemical percutaneous absorption using invivo and invitro animal models; author of two books and 152 manuscripts. *Mailing Add:* Col Vet Med Cutaneous Pharmacol Toxicol Ctr NC State Univ 4700 Hillborough St Raleigh NC 27606

RIVIN, DONALD, b Brooklyn, NY, Oct 5, 34; m 56; c 4. SURFACE CHEMISTRY. *Educ:* Columbia Univ, BA, 55, MA, 57, PhD(chem), 62. *Prof Exp:* Res chemist, Cabot Corp, Billerica, 59-61, group leader org chem, 61-69, sr res assoc, Fine Particle Technol Dept, 69-74, dir, 74-80, corp res fel & dir environ health, 80-88; CHIEF MAT SECT, US ARMY NATICK RD&E CTR, 88- *Concurrent Pos:* Vchmn, Resource Recovery Coun, 80-84. *Mem:* AAAS; Am Chem Soc; Catalysis Soc; Am Carbon Soc; Sigma Xi. *Res:* Adsorption and reactions on carbon and oxide surfaces; heterogeneous catalysis; organic reaction mechanisms; polymer reinforcement by carbon black; environmental chemistry; adsorption properties of colloidal and microporous solids. *Mailing Add:* US Army Natick RD&E Ctr Natick MA 01760-5019

RIVKIN, ISRAEL, b Rochester, NY, Feb 14, 38; m 65; c 4. IMMUNOLOGY. *Educ:* Yeshiva Univ, BA, 59; NY Univ, MS, 68; Univ Conn, PhD(immunol), 74. *Prof Exp:* Res asst rheumatology, State Univ NY Downstate Med Ctr, 64-65; res assoc biochem pharmacol, 68-70, sr res investr immunol, Squibb Inst Med Res, 73-78; PRIN SCIENTIST, DEPT INFLAMMATION & ALLERGY, SCHERING CORP, 78- *Concurrent Pos:* Squibb sabbatical fel, 70. *Mem:* NY Acad Sci; AAAS. *Res:* The study of the effects of drugs on in vitro and in vivo neutrophil and monocycle chemotaxis; design biological systems for testing new anti-inflammatory drugs. *Mailing Add:* One Opatut Ct Edison NJ 08817

RIVKIN, MAXCY, b Columbia, SC, Mar 31, 37; m 59; c 2. CONTROL SYSTEMS, PAPER SCIENCE. *Educ:* Univ SC, BS, 59. *Prof Exp:* Group leader tech serv papermaking systs, Kraft Div, 63-70, group leader process systs, Covington Res Ctr, 70-78, DIR RES, LAUREL RES CTR, WESTVACO CORP, 78- *Mem:* Tech Asn Pulp & Paper Indust; Instrument Soc Am; AAAS; NY Acad Sci; Soc Rheology. *Res:* Development and implementation of experimental or prototype process control systems in the manufacture of pulp and paper and of allied products; development and implementation of systems for the management of research and innovation; properties of coated papers. *Mailing Add:* c/o Westvaco Corp 299 Park Ave Attn: June Arsenault New York NY 10171

RIVKIN, RICHARD BOB, b Brooklyn, NY, Nov 17, 49; div; c 3. BIOLOGICAL OCEANOGRAPHY. *Educ:* City Col New York, BS, 72, MS, 75; Univ RI, PhD(biol sci), 79. *Prof Exp:* Assoc res scientist phytoplankton ecol, Johns Hopkins Univ, 78; STAFF MEM, HORN PT ENVIRON LAB, UNIV MD. *Mem:* Am Soc Limnol & Oceanog; Phycol Soc Am; Am Soc Plant Physiologists; Sigma Xi. *Res:* Phytoplankton nutrient physiology and biochemistry; phytoplankton ecology; interactions of water motion and phytoplankton ecology. *Mailing Add:* Horn Pt Environ Lab Univ Md Box 775 Cambridge MD 21613

RIVLIN, RICHARD S, b Forest Hills, NY, May 15, 34; m 60, 76; c 4. DRUG-NUTRIENT INTERACTIONS, VITAMIN METABOLISM. *Educ:* Harvard Univ, AB, 55, MD, 59; Am Bd Internal Med, dipl, 69. *Prof Exp:* Attend physician med serv, Baltimore City Hosps, 64-66; assoc med, Col Physicians & Surgeons, 66-67, from asst prof to assoc prof med, 67-79; chief & asst physician endocrinol, Francis Delafield Hosp, 66-75; PROF MED, CORNELL UNIV MED COL, 79-; CHIEF NUTRIT DIV, NY HOSP-CORNELL MED CTR, 79-; CHIEF NUTRIT SERV, MEM SLOAN-KETTERING CANCER CTR, 79- *Concurrent Pos:* Lectr clin med, Johns Hopkins Univ Sch Med, 65-66; asst attend physician med, Columbia Presby Hosp, 66-73 & assoc attend physician med, 73-79; prin investr, Clin Nutrit Res Unit, 80-; ser ed, Contemp Issues Clin Nutrit, 81-; vis prof, Univ Ariz, Tucson, 84 & Wright State Univ, Dayton, Ohio, 85; consult diet, nutrit & cancer, Am Cancer Soc, 85-; mem, Nat Adv Comt Colorectal Cancer, Am Cancer Soc, 85-; mem, Nat Sci Adv Coun, Am Fedn Aging, 85-; mem Black-White Survival Study, NIH Sci Adv Comt, Nat Cancer Inst, 85-; mem coun, Am Soc Clin Nutrit, 89. *Honors & Awards:* Grace A Goldsmith Lect Award, Am Col Nutrit, 81; Scroll of Appreciation, Am Fedn Aging Res, 85; Virgil Sydenstricker Lectr, Med Col Georgia, 89. *Mem:* Am Soc Clin Nutrit; Am Inst Nutrit; Am Soc Clin Invest; Soc Exp Biol & Med; fel Am Col Physicians; Am Physiol Soc. *Res:* Inhibition of flavin biosynthesis by Adriamycin in rat skeletal muscle and mechanisms underlying the differential effects of ethanol on the bio-availability of riboflavin and flavin-adenine dinucleotide; chlorpromazine and quinacrine inhibit flavin-adenine dinucleotide biosynthesis in skeletal muscle and alteration in age-related decline of beta-adrenergic receptor binding in adipocytes during riboflavin deficiency. *Mailing Add:* Mem Sloan-Kettering Cancer Ctr Box 140 1275 York Ave New York NY 10021

RIVLIN, RONALD SAMUEL, b London, Eng, May 6, 15; m 48; c 1. APPLIED MATHEMATICS. *Educ:* Cambridge Univ, BA, 37, MA, 39, ScD(math), 52. *Hon Degrees:* DSc, Univ Ireland & Univ Nottingham, Eng, 80, Tulane Univ, 82, Univ Thessaloniki, 84. *Prof Exp:* Res scientist, Res Labs, Gen Elec Co, Ltd, 37-42; sci officer, Ministry Aircraft Prod, Telecommun Res Estab, 42-44; from physicist to supt res, Brit Rubber Producers Res Asn, 44-52; consult, US Naval Res Lab, 52-53; prof appl math, Brown Univ, 53-63, chmn div appl math, 58-63, L Herbert Ballou Univ Prof, 63-67; centennial univ prof & dir, Ctr Appln Math, 67-80, ADJ PROF, LEHIGH UNIV, 80- *Concurrent Pos:* Guest scientist, Nat Bur Standards, 46-47; res scientist, Davy-Faraday Lab, Royal Inst, London, 47-52; vis lectr, Calif Inst Technol, 53; Guggenheim fel, Univ Rome, 61-62; mem mech adv comt, Nat Bur Standards, 65-70; mem comt appln math, Nat Res Coun, 65-68; vis prof, Univ Paris, 66-67; mem, Nat Comt Theoret & Appl Mech, 72-80, chmn, 76-78; Alexander von Humboldt sr scientist award, 81; fel, Inst Advan Study, Berlin, 84-85; distiguished vis prof, Univ Delaware, 85-86. *Honors & Awards:* Bingham Medal, Soc Rheology, 58; Panetti Prize, Acad Sci of Turin, 75; Timoshenko Medal, Am Soc Mech Engrs, 87. *Mem:* Nat Acad Eng; fel Am Acad Arts & Sci; Soc Rheology; fel Am Soc Mech Engrs; hon mem Mexican Soc Rheology; fel Soc Eng Sci; fel Am Phys Soc; fel Acad Mech; hon mem Royal Irish Acad; hon mem Accad Naz d Lincei. *Res:* Finite elasticity; physics of rubber; continuum mechanics of viscoelastic solids and fluids. *Mailing Add:* Dept Mech Eng Lehigh Univ Bethlehem PA 18015

RIVLIN, THEODORE J, b Brooklyn, NY, Sept 11, 26. MATHEMATICS. *Educ:* Brooklyn Col, BA, 48; Harvard Univ, MA, 50, PhD(math), 53. *Prof Exp:* Instr math, Johns Hopkins Univ, 52-55; asst, Inst Math Sci, NY Univ, 55-56; sr math analyst, Engine Div, Fairchild Engine & Aircraft Corp, 56-59; MATHEMATICIAN, IBM CORP, 59- *Mem:* Am Math Soc; Soc Indust & Appl Math; Math Asn Am. *Res:* Approximation theory; function theory; numerical analysis. *Mailing Add:* TJ Watson Res Ctr PO Box 218 IBM Yorktown Heights NY 10598

RIXON, RAYMOND HARWOOD, b Vancouver, BC, July 17, 26; m 57; c 4. LIVER REGENERATION, ENDOCRINOLOGY. *Educ:* Univ BC, BA, 48, MA, 50; Univ Western Ont, PhD(physiol), 55. *Prof Exp:* Fel physiol, Fac Med, Univ Western Ont, 55-56; asst res officer, Atomic Energy Can Ltd, 56-62, assoc res officer, 62-68; assoc res officer, 68-70, SR RES OFFICER, DIV BIOL, NAT RES COUN CAN, 70- *Res:* The physiological and biochemical control of cell proliferation, especially liver regeneration. *Mailing Add:* Nine Parklane Ct Gloucester ON K1B 3H3 Can

RIZACK, MARTIN A, b New York, NY, Nov 19, 26; m 64, 87; c 5. BIOCHEMISTRY, PHARMACOLOGY. *Educ:* Columbia Univ, MD, 50; Rockefeller Univ, PhD, 60; Am Bd Internal Med, dipl, 57, recertified, 74. *Prof Exp:* Intern, Cornell Med Div, Bellevue Hosp, 50-51, asst resident physician, 51; asst resident physician, St Luke's Hosp, 53-55, chief resident physician,

55-56, res assoc, 56-57; asst physician, Rockefeller Univ, 57-60, asst prof med & assoc physician, 60-65, assoc dean grad studies, 68-74, assoc prof, head lab cellular biochem & pharmacol & physician, 65-90, EMER PROF, ROCKEFELLER UNIV, 90- *Concurrent Pos:* Fel, Rockefeller Univ, 57-60; asst attend physician, St Luke's Hosp, 60-71, assoc attend physician, 71-80, adj sr attend physician, 80-; consult ed, Med Lett, 74- *Mem:* Fel Am Col Physicians; Am Soc Pharmacol & Exp Therapeut; Am Soc Biol Chemists. *Res:* Biochemistry and physiology of hormone action; lipolytic enzymes; biochemical pharmacology. *Mailing Add:* Rockefeller Univ 1230 York Ave Box 74 New York NY 10021-6399

RIZKALLA, SAMI H, b Alexandria, Egypt, Feb 5, 45; Can citizen; m 70; c 2. STRUCTURAL ENGINEERING, BRIDGES. *Educ:* Alexandria Univ, Egypt, BSc, 65; NC State Univ, MSc, 74, PhD(civil), 76. *Prof Exp:* struct engr, Vibro Cast-in-place, Piles, Egypt, 65-66 & Govt Egypt, 66-68; instr civil, Alexandria Univ, Egypt, 68-71; grad teaching civil, NC State Univ, 71-76; postdoctoral fel civil, Univ Alta, Can, 76-78; from asst prof to assoc prof, 79-88, PROF STRUCT UNIV MAN, CAN, 88-; CONSULT ENGR STRUCT, 80- *Concurrent Pos:* Consult engr, Gonda, Egypt, F Saleh, USA & Con Force St, Can, 81-84, Penner & Keeler Partners, Reid Crowther, Can Nat, 84-88 & Wardrop Eng, Dillon Eng & ID Eng, 89-; vis prof, New South Wales, Sydney, Australia & Nagaoka Tech Univ, Yamanashi Univ, Japan, 88; vpres, Prairie Region, Can Soc Civil Eng, 89-91, chmn, Struct Div, 91- *Mem:* Can Soc Civil Engrs; Am Concrete Inst; fel Am Soc Civil Eng; Prestressed Concrete Inst; Japanese Soc Civil Eng; Can Standard Asn. *Res:* Behaviour of reinforced concrete and prestressed concrete structures; laboratory and field testing of structures; durability of concrete; wooden poles for hydro transmission; use of advanced composite material in civil engineering application. *Mailing Add:* Seven Prestwood Pl Winnipeg MB R3T 4Y9 Can

RIZKI, TAHIR MIRZA, b Hyderabad, India, Jan 8, 24; nat US; m 49; c 3. DEVELOPMENTAL GENETICS, CELL BIOLOGY. *Educ:* Osmania Univ, India, BSc, 44; Muslim Univ, MSc, 46; Columbia Univ, PhD(zool), 53. *Prof Exp:* Demonstr zool, Muslim Univ, 45-46, lectr, 46-48; asst, Columbia Univ, 51-52; Cramer res fel genetics, Dartmouth Col, 52-54; Am Cancer Soc fel, Yale Univ, 54-56; from asst prof to assoc prof biol, Reed Col, 56-61; assoc prof, 61-64, PROF BIOL, UNIV MICH, ANN ARBOR, 64- *Mem:* AAAS; Am Soc Zoologists; Am Soc Naturalists; Sigma Xi; Soc Invert Path. *Res:* Developmental genetics of Drosophila; biology of hemocytes and melanotic tumors in Drosophila; parasitoids of Drosophila; cell death. *Mailing Add:* Div Biol Sci Univ Mich Ann Arbor MI 48109

RIZZA, PAUL FREDERICK, b New Britain, Conn, Dec 15, 38; m 79; c 3. CARTOGRAPHY. *Educ:* Cent Conn State Col, BS, 65, MS, 68; Univ Ga, PhD(geog), 73. *Prof Exp:* Teacher geog & hist, Haddam Sch, Conn, 65-67; teacher world human geog, Univ Ga, 68-70; asst prof geog, 72-75, chmn geog & cartog, 72-78, assoc prof, 75-78, actg dean, Sch Social & Behav Sci, 78-80, PROF & CHMN GEOG & ENVIRON STUDIES DEPT, SLIPPERY ROCK UNIV, 81- *Concurrent Pos:* Sr Fulbright scholar, Coun Int Exchange Scholars, Finland, 76; vis prof, Helsinki Sch Econ, Finland, 81. *Mem:* Asn Am Geogr; Am Cong Surv & Mapping; Am Geog Soc; NAm Cartog Info Soc. *Res:* Rural land use in the United States and Scandinavia; thematic cartography. *Mailing Add:* Dept Geog & Environ Studies Slippery Rock Univ Slippery Rock PA 16057

RIZZI, GEORGE PETER, b Middletown, Conn, Sept 25, 37; m 59; c 2. ORGANIC CHEMISTRY. *Educ:* Worcester Polytech Inst, BS, 59, MS, 61, PhD(org chem), 63. *Prof Exp:* Res assoc chem, Stanford Univ, 63-64; RES CHEMIST, MIAMI VALLEY LABS, PROCTER & GAMBLE CO, 64- *Mem:* Am Chem Soc; Sigma Xi. *Res:* Organic synthesis; natural products; flavors; organic reactions involved in biochemical processes. *Mailing Add:* 542 Blossom Hill Lane Cincinnati OH 45224

RIZZO, ANTHONY AUGUSTINE, b Birmingham, Ala, June 8, 28; m 55, 73; c 5. DENTISTRY, MICROBIOLOGY. *Educ:* Birmingham-Southern Col, AB, 52; Univ Ala, DMD & MS, 56. *Prof Exp:* Instr chem, Lab, Univ Ala, 50-51; fel periodont, Sch Dent, 56-57, prin investr periodont dis, Lab, 57-68, chief mat sci & spec clin studies prog, 68-71, chief periodont dis prog, 70-73, spec asst prog coordr, 73-84, deputy assoc dir, 84-85, CHIEF PERIODONT DIS PROG, NAT INST DENT RES, 85- *Concurrent Pos:* Ed-in-chief, Ala Dent Rev, 55-56. *Honors & Awards:* Commendation Medal, US Pub Health Serv, 77, Outstanding Serv Medal, 88. *Mem:* Int Asn Dent Res. *Res:* Microbiology and immunopathology of periodontal disease; periodontology. *Mailing Add:* Extramural Progs Nat Inst Dent Res WB-509 Bethesda MD 20892

RIZZO, DONALD CHARLES, b Boston, Mass, June 10, 45. ENTOMOLOGY, PARASITOLOGY. *Educ:* Boston State Col, AB, 68; Cornell Univ, MS, 70, PhD(entom), 73. *Prof Exp:* Res asst insect path, Cornell Univ, 68-73; instr biol, Siena Heights Col, 73-74; asst prof, 74-78, CHMN DIV NAT SCI & MATH, MARYGROVE COL, 75-, ASSOC PROF BIOL, 78- *Concurrent Pos:* Dir, Minority Inst Sci Improv Prog, NSF, inst grants, 84-85, special project grants, 85-87. *Mem:* Nat Sci Teachers Asn; Am Asn Univ Prof; Nat Geog Soc; Am Inst Biol Sci. *Res:* Fungal pathogens of medically and economically important insects, especially dipterans. *Mailing Add:* Dept Biol Marygrove Col 8425 W McNichols Rd Detroit MI 48221

RIZZO, PETER JACOB, b Gary, Ind, Dec 10, 40; m; c 2. PLANT PHYSIOLOGY, CELL BIOLOGY. *Educ:* Ind Univ, AB, 67, MA, 68; Univ Mich, PhD(plant physiol), 72. *Prof Exp:* Res assoc plant physiol, Purdue Univ, 72-75; from lectr to asst prof, 75-82, ASSOC PROF CELL BIOL, TEX A&M UNIV, 82- *Concurrent Pos:* NIH fel, 72. *Mem:* Soc Protozoologists; AAAS; Sigma Xi; Soc Cell Biol. *Res:* Histone occurance in lower eukaryotes; histone-like proteins in dinoflagellates; evolution of nucleosomes. *Mailing Add:* Biol Dept Tex A&M Univ College Station TX 77843

RIZZO, THOMAS GERARD, b New York, NY, June 22, 55; m 85. GRAND UNIFIED THEORIES, PHENOMENOLOGY. *Educ:* Fordham Univ, BS, 74; Columbia Univ, MA, 75; Univ Rochester, PhD(physics), 78. *Prof Exp:* Res assoc, Physics Dept, Brookhaven Nat Lab, 79-81; asst prof & assoc physicist, 81-83, assoc prof physics & physicist, 83-87, PROF PHYSICS & SR PHYSICIST, AMES LAB, IOWA STATE UNIV, 87- *Concurrent Pos:* Vis scientist, Technion, 86, Univ Tex-Austin, 87, Triumf, 87. *Mem:* Am Phys Soc; AAAS; Europ Phys Soc. *Res:* Phenomenological implications of unified field theories; how low energy experiments can be used to distinguish between various schemes; superstring-inspired E6 models; fourth generation particles. *Mailing Add:* Dept Physics Iowa State Univ Ames IA 50011

RIZZUTO, ANTHONY B, b Baton Rouge, La, Sept 11, 30; m 57; c 4. MICROBIOLOGY, BIOCHEMISTRY. *Educ:* La State Univ, BS, 52; Miss State Univ, MS, 55. *Prof Exp:* Plant microbiologist, Am Sugar Co, NY, 56-59, res microbiologist, 59-63, sr res scientist, Res & Develop Div, 63-69, asst dir, 69-74, DIR, AMSTAR CORP, 74- *Concurrent Pos:* Mem, Nat Comn Uniform Methods Sugar Analysis, 60- & Int Comn, 62- *Mem:* Nat Acad Sci; Soc Indust Microbiol; Inst Food Technol; Am Soc Microbiol. *Res:* Industrial, agricultural, food, dairy and sanitation microbiology; new product development; new process development. *Mailing Add:* Amstar Corp 266 Kent Ave Brooklyn NY 11211

ROACH, ARCHIBALD WILSON KILBOURNE, b Omaha, Nebr, Sept 15, 20; m 42; c 3. BOTANY. *Educ:* Univ Colo, BA, 46, MA, 48; Ore State Univ, PhD, 51. *Prof Exp:* From asst prof to assoc prof, 50-57, prof bot, 57-77, PROF BIOL SCI, NTEX STATE UNIV, 77- *Concurrent Pos:* Mem, Int Bur Plant Taxon & Nomenclature. *Mem:* Ecol Soc Am. *Res:* Phytosociology; southwestern aquatic actinomycetes. *Mailing Add:* 2319 Fowler Denton TX 76201

ROACH, DAVID MICHAEL, b Detroit Lakes, Minn, Oct 10, 39; m 61; c 2. OCEANOGRAPHY. *Educ:* SDak Sch Mines & Technol, BS, 61, MS, 63; Ore State Univ, PhD(oceanog), 74. *Prof Exp:* Instr physics, SDak Sch Mines & Technol, 63-64; asst prof, Wis State Univ-Whitewater, 64-66; PROF PHYSICS, CALIF POLYTECH STATE UNIV, SAN LUIS OBISPO, 66- *Res:* Determination of refractive index distributions of oceanic particulates. *Mailing Add:* Dept Physics Calif Polytech State Univ San Luis Obispo CA 93407

ROACH, DON, b Bono, Ark, Dec 10, 36; m 57; c 1. ANALYTICAL CHEMISTRY. *Educ:* Ark State Univ, BS, 59; Univ Ark, Fayetteville, MS, 63; Univ Mo-Columbia, PhD(anal biochem), 70. *Prof Exp:* Technician, Univ Ark, Fayetteville, 63-64; asst prof chem, Miami-Dade Jr Col, 64-67; chemist, Exp Sta Chem Labs, Univ Mo-Columbia, 67-68; assoc prof, 69-80, PROF CHEM & CHMN DEPT, MIAMI-DADE COMMUNITY COL, 80- *Concurrent Pos:* Consult. *Honors & Awards:* William Henry Hatch Fel. *Mem:* Am Chem Soc. *Res:* Gas-liquid chromatography of biologically important compounds. *Mailing Add:* 12190 SW 93 Ave Miami FL 33176-5002

ROACH, DONALD VINCENT, b Oak Grove, Mo, Jan 18, 32. PHYSICAL CHEMISTRY. *Educ:* Univ Mo, BS, 54, PhD(phys chem), 62. *Prof Exp:* Res assoc, Univ Calif, Berkeley, 62-63 & Univ Mo-Columbia, 63-64; res chemist, US Naval Weapons Lab, 64-65; asst prof, 65-72, ASSOC PROF CHEM, UNIV MO-ROLLA, 72- *Mem:* Am Chem Soc. *Res:* Gas-solid surface interactions; adsorption of gases on solids; exchange of energy between gases and solids. *Mailing Add:* Dept Chem Univ Mo Rolla MO 65401

ROACH, FRANCIS AUBRA, b Coleman, Tex, Apr 5, 35; m 56; c 2. MATHEMATICAL ANALYSIS. *Educ:* Univ Tex, BA, 59, MA, 60, PhD(math), 66. *Prof Exp:* Chemist, Tex Butadiene & Chem Corp, 57-58; programmer math, Shell Oil Co, 60-61; instr, San Jacinto Col, 61-62 & Univ Tex, 62-66; asst prof, Univ Ga, 66-69; from asst prof to assoc prof math, Univ Houston, 69-81; res mathematician, Western Geophys Co, Houston, 84-86; SR MATHEMATICIAN, FORMAL SYSTS DESIGN & DEVELOP, 87- *Mem:* Am Math Soc. *Res:* Mathematical analysis, especially continued fractions. *Mailing Add:* 6219 Fawnwood Spring TX 77389

ROACH, J ROBERT, b Stockton, Ill, Nov 24, 13; m 50; c 3. ORGANIC CHEMISTRY, FOOD CHEMISTRY. *Educ:* Iowa State Col, BS, 36; Purdue Univ, PhD(org chem), 42. *Prof Exp:* Pittsburgh Plate Glass Co fel, Northwestern Univ, 41-43; proj leader org chem, Gen Mills, Inc, 43-50, sect leader, Food Develop Dept, 50-56, dept head, 56-61, dir food develop activity, 61-69, vpres & dir res & develop food group, 69-77; RETIRED. *Mem:* Am Chem Soc; Am Asn Cereal Chemists; Inst Food Technologists. *Res:* Food development; convenience foods. *Mailing Add:* 24 Luverne Minneapolis MN 55419

ROACH, JOHN FAUNCE, b Boston, Mass, July 21, 12; m 39; c 1. RADIOLOGY. *Educ:* Harvard Univ, AB, 35, MD, 39. *Prof Exp:* From asst prof to assoc prof radiol, Sch Med, Johns Hopkins Univ, 47-50; chmn dept, 50-77, PROF RADIOL, ALBANY MED COL, 50-; CONSULT, NY STATE EDUC DEPT, 81- *Concurrent Pos:* Consult, Vet Admin Hosp, Albany; mem pub health coun, State Health Dept, NY; trustee & pres, Am Bd Radiol; pres, Am Bd Med Specialties; exec secy, NY State Bd for Med, 77-81. *Mem:* Radiol Soc NAm; Am Roentgen Ray Soc (past pres); fel Am Col Radiol. *Mailing Add:* Dept Radiol Albany Med Col Albany NY 12208

ROACH, MARGOT RUTH, b Moncton, NB, Dec 24, 34. BIOPHYSICS, BIOENGINEERING. *Educ:* Univ NB, BSc, 55; McGill Univ, MD, CM, 59; Univ Western Ont, PhD(biophys), 63; FRCP(C), 65. *Hon Degrees:* DSc, Univ New Brunswick, 81. *Prof Exp:* Jr intern, Victoria Hosp, London, Ont, 59-60; Med Res Coun Can fel, Univ Western Ont, 60-62, Ont Heart Found fel cardiol, Victoria Hosp, 62-63, asst resident med, 63-64; asst resident, Toronto Gen Hosp, Ont, 64-65; from asst prof to assoc prof med, 65-78, from asst prof to assoc prof biophys, 65-71, chmn dept, 70-78, PROF BIOPHYS, UNIV WESTERN ONT, 71-, PROF MED, 78- *Concurrent Pos:* Med Res Coun Can fel, Nuffield Inst Med Res, Oxford, Eng, 65-67; Med Res Coun Can scholar biophys & med, Univ Western Ont, 67-70; res grants, Med Res Coun Can, 68-70 & 71-, Ont Heart Found, 68-, Can Tuberc Asn, 69-70, Ont Thoracic Soc, 71- & Picker Found, 71-73; Young investr's award, Am Col Cardiol, 63; mem active teaching staff med, Victoria Hosp, 67-72; Can rep, Adv Group Aeronaut Res & Develop, NATO, 69-71; mem comt on scholar, Med Res Coun Can, 71-72; res consult, Westminster Hosp, 71-73; mem adv comt biophys, Nat Res Coun, 71-78; consult, Univ Hosp, 72-; assoc ed, Can J Physiol Pharmacol, 73-79; Commonwealth vis scientist apppl math & theoret physics, Cambridge, 75. *Honors & Awards:* Ciba Found Award, 59. *Mem:* Can Physiol Soc; Can Cardiovasc Soc; Can Soc Clin Invest; Can Biophys Soc. *Res:* Hemodynamics; elastic properties of tissues; effects of vibration on tissues; arterial elasticity; atherosclerosis; the role of hemodynamic factors in localized arterial disease, such as poststenotic dilatation, atherosclerosis, brain aneurysms, dissecting aneurysms. *Mailing Add:* Dept Med Biophys Univ Western Ont London ON N6A 5C1 Can

ROACH, PETER JOHN, b Rangeworthy, UK, June 8, 48; m 75. GLYCOGEN METABOLISM, MECHANISMS OF HORMONE ACTION. *Educ:* Univ Glasgow, BSc, 69, PhD(biochem), 72. *Prof Exp:* Fel, Univ Calif, Los Angeles, 72-74, Univ Va, 74-75, Univ Pisa, Italy, 75-77; instr pharmacol, Univ Va, 77-79; from asst prof to assoc prof, 79-85, PROF BIOCHEM, SCH MED, IND UNIV, 85- *Mem:* Am Soc Biol Chem. *Res:* Mechanisms by which hormones regulate enzyme activity, with the main focus on covalent phosphorylation in the control of glycogen metabolism. *Mailing Add:* Dept Biochem Sch Med Ind Univ 635 Barnhill Dr Indianapolis IN 46223

ROACH, WILLIAM KENNEY, b Cincinnati, Ohio, Feb 2, 42; m 64; c 2. ENTOMOLOGY. *Educ:* St Louis Univ, BS, 64; Ohio State Univ, PhD(entom), 71. *Prof Exp:* Surv entomologist, Ohio State Univ, 69-70; SPECIALIST-IN-CHARGE, PLANT PEST CONTROL SECT, OHIO DEPT AGR, 72- *Mem:* Entom Soc Am. *Res:* Taxonomy of Empidid flies. *Mailing Add:* Plant Pest Control Sect Ohio Dept Agr 8995 E Main Reynoldsburg OH 43068

ROACHE, LEWIE CALVIN, b Dalzell, SC, Oct 31, 25; m 59. ZOOLOGY. *Educ:* SC State Col, BS, 47, MS, 54; Cath Univ, PhD(zool), 60. *Prof Exp:* From instr to assoc prof, 47-69, from actg head to head dept, 56-69, chmn dept natural sci, 69-77, PROF BIOL, SC STATE COL, 69-, DEAN, SCH ARTS & SCI, 77- *Mem:* AAAS; Ecol Soc Am; Nat Inst Sci. *Res:* Systematics of the freshwater Cyclopoid Copepods. *Mailing Add:* Off Dean Sch Arts & Sci SC State Col PO Box 1746 Orangeburg SC 29117

ROADS, JOHN OWEN, b Boulder, Colo, Jan 20, 50. METEOROLOGY, CLIMATOLOGY. *Educ:* Univ Colo, BA, 72; Mass Inst Technol, PhD(meteorol), 77. *Prof Exp:* Res asst, Nat Ctr Atmospheric Res, 70-72; NSF fel, 72-77, ASST RES METEOROLOGIST, SCRIPPS INST OCEANOG, UNIV CALIF, SAN DIEGO, 77- *Concurrent Pos:* Res asst, Mass Inst Technol, 75-77. *Mem:* Am Meteorol Soc. *Res:* Numerical modelling; dynamic meteorology; climate modelling. *Mailing Add:* Scripps Inst Oceanog Univ Calif San Diego La Jolla CA 92093

ROADSTRUM, WILLIAM H(ENRY), b Chicago, Ill, June 22, 15; m 43; c 2. ENGINEERING, ENGINEERING ECONOMICS. *Educ:* Lehigh Univ, BS, 38; Carnegie Inst Technol, MS, 48, PhD(elec eng), 55. *Prof Exp:* Elec engr, US Bur Mines, Pa, 39-41, 45-47; asst prof electronics, US Naval Postgrad Sch, 48-50, 52-53; systs engr, adv electronics ctr, Gen Elec Co, NY, 55-57, mgr missile systs eng, Light Mil Dept, 57-59, mgr detection & surveillance eng, 59-60, mgr info storage & retrieval, Adv Tech Labs, 60-63; from adj prof to prof elec eng, 63-80, EMER PROF ELEC ENG, WORCESTER POLYTECH INST, 80- *Concurrent Pos:* Consult, eng practices & eng mgt, William H Roadstrum Inc, 64- *Mem:* Sigma Xi; sr mem Inst Elec & Electronics Engrs. *Res:* Systems engineering; engineering economy; management theory and techniques in engineering and development projects; education and development of engineers. *Mailing Add:* Nine Juniper Lane Holden MA 01520

ROAKE, WILLIAM EARL, b Oregon City, Ore, Sept 30, 19; m 42; c 2. NUCLEAR ENGINEERING, MATERIALS SCIENCE ENGINEERING. *Educ:* Ore State Col, BS, 41, MS, 42; Northwestern Univ, PhD, 49. *Prof Exp:* Chemist, Charleton Labs, Portland, Ore, 41; res chemist, Nat Defense Res Comt Contract, Cent Labs, Northwestern Univ, 42-45 & Calif Inst Technol, 45; res chemist, Hanford Atomic Prods Oper, Gen Elec Co, 48-58, sr engr, 58-62, mgr, Fuels Testing & Analysis Unit, 62-65; mgr, Ceramics Res Unit, Pac Northwest Labs, Batelle Mem Inst, 65-66, mgr, Fast Reactor Fuels Sect, 65-70; mgr, FFTF Fuel Develop Sect, Wadco Corp, 66-68, dep mgr, 68-70, mgr, Core Components Subdiv, 70-73; mgr, Fuels Subdiv, Westinghouse-Hanford Co, 73-76, mgr, Fuels & Controls Subdept, 76-79, asst mgr, Appl Systs Develop Dept, 79-84. *Concurrent Pos:* Consult, 84- *Mem:* Am Chem Soc; Sigma Xi; fel Am Nuclear Soc; Am Inst Aeronaut & Astronaut. *Res:* Nuclear reactor fuels development, testing, design, fabrication and reprocessing. *Mailing Add:* 2336 Harris Ave Richland WA 99352

ROALES, ROBERT R, b New York, NY, July 17, 44; m 69; c 1. ENVIRONMENTAL TOXICOLOGY, ENVIRONMENTAL PHYSIOLOGY. *Educ:* Iona Col, BS, 66; NY Univ, MS, 69, PhD(biol), 73. *Prof Exp:* Lectr biol, Bor Manhattan Community Col, 69-70; teaching fel biol, NY Univ, 70-73, asst res scientist, 73-74; from asst prof to assoc prof anat & physiol, Natural Sci Dept, 74-87, actg dean arts & sci, 90, CHMN, DEPT BIOL & PHYS SCI, IND UNIV, KOKOMO, 87-, COORDR ALLIED HEALTH SCI, 87- *Concurrent Pos:* Consult, T F H Publ, 73; adj asst prof biol, LaGuardia Community Col & NY Univ, 73-74. *Honors & Awards:* Founder's Award, NY Univ, 74. *Mem:* AAAS; Am Fisheries Soc; Am Inst Biol Sci; Am Soc Ichthyologists & Herpetologists; NY Acad Sci; Sigma Xi; Nat Asn Adv Health Professions. *Res:* Establishment of tolerance limits of fish and fish embryos to various heavy metals and organic phosphorous pesticides; effects of these compounds on inhibiting the immune systems of fish as evidenced by decreased antibody levels when exposed to viral and bacterial antigens; effects of heavy metal pollutants on lipid content of fish. *Mailing Add:* Dept Biol & Phys Sci Ind Univ PO Box 9003 2300 S Washington St Kokomo IN 46904-9003

ROALSVIG, JAN PER, b Stavanger, Norway, Mar 23, 28; m 54; c 5. NUCLEAR PHYSICS, HIGH ENERGY PHYSICS. *Educ:* Univ Oslo, BSc, 52, MSc, 55; Norweg Inst Pedag, BEd, 53; Univ Sask, PhD(physics), 59. *Prof Exp:* Asst prof physics, St John's Univ, NY, 59-62; PROF PHYSICS, STATE UNIV NY, BUFFALO, 62- *Concurrent Pos:* Res prof, Chalmers Univ Technol, Sweden, 69-70, Univ BC, Can, 79; docent, Norweg Tech Univ, 72-73. *Mem:* Am Phys Soc; Am Asn Physics Teachers; Can Asn Physicists; Norweg Phys Soc. *Res:* Pair-production; absolute beta-counting; photo-alpha and photo-neutron reactions; nuclear spectroscopy; gamma ray detectors; hyperfragments. *Mailing Add:* Physics/Rm 259 Fronczak Hall SUNY Buffalo-N Campus Buffalo NY 14260

ROAN, VERNON P, b Ft Myers, Fla, Nov 19, 35; c 2. MECHANICAL ENGINEERING, AEROSPACE ENGINEERING. *Educ:* Univ Fla, BS, 58, MS, 59; Univ Ill, PhD(aeronaut & mech eng), 66. *Prof Exp:* Assoc prof mech eng, 71-73, grad coordr, 85-90, PROF MECH ENG, UNIV FLA, 73-, DIR, CTR ADVAN STUDIES ENG, 89- *Concurrent Pos:* Consult, E I du Pont de Nemours & Co, Inc, 73-; chmn, Subcomt Univ Activities Elec & Hybrid Elec Vehicles, Inst Elec & Electronics Engrs, 77-; consult, Jet Propulsion Lab, 80-; mem staff, Brunel Univ, Uxbridge, Eng, 79-80; consult, Pratt & Whitney Aircraft, 85-, Georgetown Univ, 88- *Honors & Awards:* Nat Winner Urban Vehicle Design, Student Competition Relevant Eng, 72, Nat Winner Era II Wind Systs Design, 77, Nat Winner Energy Efficient Vehicle Design, 79; Ralph R Tector Award, Soc Automotive Engrs, 75. *Mem:* Soc Automotive Engrs; Sigma Xi; Am Soc Mech Engrs. *Res:* Gasdynamics; propulsion systems; wind systems; electric and hybrid electric vehicles; development of first hybrid electric bus in US; alternative fuels. *Mailing Add:* Dept Mech Eng Univ Fla Gainesville FL 32611

ROANE, CURTIS WOODARD, b Norfolk, Va, Apr 19, 21; m 47; c 2. PLANT PATHOLOGY. *Educ:* Va Polytech Inst, BS, 43, MS, 44; Univ Minn, PhD(plant path), 53. *Prof Exp:* From asst plant pathologist to assoc plant pathologist, Exp Sta, Va Polytech Inst & State Univ, 47-68, from assoc prof to prof, 48-86, EMER PROF PLANT PATH, VA POLYTECH INST & STATE UNIV, 86- *Concurrent Pos:* Assoc ed, Phytopathology, 69-71, Plant Dis Reporter, 78-81; chmn, Southern Small Grain Worker's Conf, 70-72, Eastern Wheat Workers Conf, 78-81; chmn, NAm Barley Res Workers, 81-84; pres, Potomac Div, Am Phytopath Soc. *Mem:* Sigma Xi; fel Am Phytopath Soc. *Res:* Control of cereal crop and soybean diseases through resistance; genetics of host-parasite interaction; education. *Mailing Add:* 607 Lucas Dr Blacksburg VA 24060

ROANE, MARTHA KOTILA, b Munising, Mich, Nov 1, 21; m 47; c 2. MYCOLOGY, TAXONOMIC BOTANY. *Educ:* Mich State Col, BS, 44; Univ Minn, MS, 46; Va Polytech Inst & State Univ, PhD(bot), 71. *Prof Exp:* Eng aide advan develop, Pratt & Whitney Aircraft Corp, 44-45; technician soils, Agr Eng, Va Polytech Inst & State Univ, 47-48, instr math, 56-63; from instr to asst prof, Radford Col, 63-68, asst dir, Off Instnl Res, 68-69; coordr, Gen Biol Lab, 71-72, cur fungi, 72-75, adj asst prof bot, 75-77, ADJ PROF PLANT PATH, VA POLYTECH INST & STATE UNIV, 77- *Mem:* AAAS; Bot Soc Am; Mycol Soc Am; Am Phytopathological Soc; Sigma Xi; Am Rhododendron Soc. *Res:* Monographic studies of Endothia and North American species of Rhododendron; morphology, taxonomy and development of the Chytridiales; role of fungal pigments, proteins and enzymes in taxonomy; diseases of rhododendrons & azaleas, grasses of Virginia; polymyxa graminis and wheat soil born mosaic and wheat spindle streak viruses. *Mailing Add:* Dept Plant Path & Physiol Va Polytech Inst & State Univ Blacksburg VA 24061

ROANE, PHILIP RANSOM, JR, b Baltimore, Md, Nov 20, 27. VIROLOGY, IMMUNOLOGY. *Educ:* Morgan State Col, BS, 52; Johns Hopkins Univ, MS, 60; Univ Md, PhD, 70. *Prof Exp:* Asst microbiol, Sch Med, Johns Hopkins Univ, 60-64; virologist, Microbiol Assocs, Inc, 64-72, dir qual control, 67-72; asst prof, 72-77, ASSOC PROF MICROBIOL, HOWARD UNIV, 78- *Concurrent Pos:* Mem, Virol Study Sect, NIH, 76-80, Viral & Rickettsial Rev Subcomt, 79-; consult, Hem Res Inc, Rockville Md, 81- *Mem:* AAAS; Am Soc Microbiol; Am Asn Immunologists. *Res:* Biophysical and immunological properties of viruses; biochemistry of tumor and transformed cells induced by a common virus. *Mailing Add:* Dept Microbiol Howard Univ 2400 Sixth St NW Washington DC 20059

ROANTREE, ROBERT JOSEPH, b Elko, Nev, Sept 11, 24. MEDICAL MICROBIOLOGY. *Educ:* Stanford Univ, AB, 45, MD, 48. *Prof Exp:* From instr to asst prof, 56-63, ASSOC PROF MED MICROBIOL, SCH MED, STANFORD UNIV, 63- *Concurrent Pos:* Bank Am-Giannini fel, 55-56; Lederle med fac award, 57-60; vis fel, Lister Inst, 63-64. *Mem:* AAAS; Am Soc Microbiol; NY Acad Sci; Sigma Xi. *Res:* Genetic changes affecting the bacterial cell wall and their effects upon virulence and susceptibility to antibiotics. *Mailing Add:* Dept Med Microbiol Stanford Univ Stanford CA 94305

ROARK, BRUCE (ARCHIBALD), b New York, NY, Jan 19, 20; m 46, 77; c 7. PLANT PHYSIOLOGY, PLANT BREEDING. *Educ:* Univ Western Australia, BSc, 49 & 50; Univ Adelaide, PhD, 56. *Prof Exp:* Plant physiologist, Sci & Educ Admin, Agr Res, USDA, Mayaguez, PR, 56-58 & Miss, 58-70, supvry plant physiologist, 70-79; plant breeder, Northrup King Seed Co, 81-83; JTPA PLANNER, S DELTA PLANNING & DEVELOP DIST, 84- *Mem:* Am Soc Plant Physiologists; Am Soc Agron; Crop Sci Soc Am; Sigma Xi. *Res:* Water relations; stress resistance genotype and environment interactions; upland cotton; plant breeding. *Mailing Add:* 1304 Newport Pl Greenville MS 38701

ROARK, JAMES L, b Kansas City, Mo, Mar 27, 43; m 67; c 2. ORGANIC CHEMISTRY, CHEMISTRY EDUCATION. *Educ:* Nebr Wesleyan Univ, BA, 65; Tex Christian Univ, PhD(chem), 69. *Prof Exp:* From asst prof to assoc prof chem, Kearney State Col, 69-75, prof & chmn dept, 76-81. *Concurrent Pos:* Robert Welch res fel, Tex Christian Univ, 67-69; NSF fel, Tufts Univ, 71; dir water analysis lab, Kearney State Col, 72-76, proj dir water qual study,

73-75, high ability high sch students, 74-77,sec chem teachers, 77-; vis prof, Univ Va, 81-82. *Mem:* Am Chem Soc; Nat Sci Teachers Asn; Sigma Xi. *Res:* Non-traditional methods of instruction; computer based methods of instruction; chemistry curriculums for secondary education; natural products chemistry. *Mailing Add:* 824 W 27th St Kearney NE 68847

ROARK, TERRY P, b Okeene, Okla, June 11, 38; m 63; c 1. ASTRONOMY, ASTROPHYSICS. *Educ:* Oklahoma City Univ, BA, 60; Rensselaer Polytech Inst, MS, 62, PhD(astron), 66. *Prof Exp:* Asst prof astron, Ohio State Univ, 66-76, prof, 76, asst provost curric, 77-79, assoc provost instr, 79-83; prof physics & vpres student affairs, Kent State, 83-87; PRES, UNIV WYO, 87- *Concurrent Pos:* Comnr, Western Interstate comn higher educ, 87- *Mem:* Am Astron Soc; Int Astron Union; Astron Soc Pac; Sigma Xi. *Res:* Observational and theoretical investigation of the solid interstellar medium, binary and white dwarf stars. *Mailing Add:* Univ Wyo 1306 Ivinson Laramie WY 82070

ROATH, WILLIAM WESLEY, b Torrington, Wyo, Dec 7, 34; m 55; c 7. PLANT BREEDING. *Educ:* Mont State Univ, BS, 57, PhD(genetics), 69. *Prof Exp:* Res agronomist, Dekalb Agr Res, Inc, 71; res geneticist, Agr Res Serv, USDA, 78-82; agronomist, Zamare Proj, Zambia, 82-84; RES AGRONOMIST, AGR RES SERV, USDA, 85- *Mem:* Am Soc Agron; Crop Sci Soc Am. *Res:* Germ plasm (oilseeds and new crop) evaluation and enhancement. *Mailing Add:* Rte 1 Box 161 Madrid IA 50156

ROB, CHARLES G, b Weybridge, Eng, May 4, 13; nat US; m 41; c 4. SURGERY. *Educ:* Cambridge Univ, MA, 34, MB, BCh, 37, MCh, 41, MD, 60; FRCS, 39. *Hon Degrees:* MCh, Trinity Col, Dublin, 61. *Prof Exp:* Reader surg, Univ London, 46-50, prof, 50-60; prof surg & chmn dept, Sch Med, Univ Rochester, 60-78; PROF SURG, E CAROLINA SCH MED, GREENVILLE, NC, 78- *Concurrent Pos:* Consult, Brit Army, Royal Nat Orthop Hosp, London; mem court examr, Royal Col Surgeons. *Mem:* Asn Surg Gt Brit & Ireland (hon secy); Royal Soc Med; Int Soc Cardiovasc Surg (pres, 61); hon fel Venezuelan Surg Soc. *Mailing Add:* Uniformed Serv Univ Health 4301 Jones Bridge Rd Bethesda MD 20814

ROBACK, SELWYN, aquatic pollution biology, taxonomy of chironomidae; deceased, see previous edition for last biography

ROBAUGH, DAVID ALLAN, b Uniontown, Pa, Mar 28, 54. HIGH TEMPERATURE REACTION MECHANISMS, KINETICS. *Educ:* Calif State Univ, BS, 76; WVa Univ, PhD(phys chem), 81. *Prof Exp:* Res fel, Chem Kinetics Div, Nat Bur Standards, 81-85; ASSOC CHEMIST, MIDWEST RES INST, KANSAS CITY, MO, 85- *Mem:* Am Chem Soc. *Res:* Kinetics and mechanisms of high-temperature organic reactions via very low-pressure pyrolysis and shock tube experiments. *Mailing Add:* Midwest Res Inst 425 Volker Blvd Kansas City MO 64110-2241

ROBB, ERNEST WILLARD, b Dodge City, Kans, Sept 4, 31; m 60; c 3. ORGANIC CHEMISTRY. *Educ:* Kans State Univ, BS, 52; Harvard Univ, MS, 54, PhD(chem), 56. *Prof Exp:* Fels, Iowa State Col, 56-57, Mass Inst Technol, 57-58 & Harvard Univ, 58-59; res chemist, Fritzsche Bros, Inc, 59-60; res scientist, Philip Morris Inc, 60-63, sr scientist, 63-65; assoc prof, 65-75, PROF ORG CHEM, STEVENS INST TECHNOL, 75- *Mem:* AAAS; Am Chem Soc. *Res:* Mechanism of organic reactions; organic stereochemistry. *Mailing Add:* Dept Chem Stevens Inst Technol Castle Pt Sta Hoboken NJ 07030-5991

ROBB, JAMES ARTHUR, b Pueblo, Colo, Nov 13, 38; m 62; c 4. PATHOLOGY, VIROLOGY. *Educ:* Univ Colo, Boulder, BA, 60; Univ Colo, Denver, MD, 65; Am Bd Anat Path, dipl, Am Bd Dermatopath, dipl; Am Bd Cytopath, dipl. *Prof Exp:* Intern anat path, Yale Med Ctr, 65-66, resident, 66-68; res assoc molecular virol, Nat Inst Arthritis & Metab Dis, 68-71; from asst prof to assoc prof path, Univ Calif, San Diego, 71-78, VCHMN & STAFF PATHOLOGIST, GREEN HOSP SCRIPPS CLIN, LA JOLLA, CALIF, 78- *Concurrent Pos:* Adj prof path, Sch Med, Univ Calif, 78-; fel Col Am Pathol. *Mem:* AAAS; Am Soc Cell Biol; Am Soc Microbiol; Am Asn Pathologists; Am Soc Clin Pathologists. *Res:* Molecular biology of latent human viruses; study of the animal virus-mammalian cell interactions resulting in various human diseases; detection of antigens in formalin-fixed, paraffin-embedded human tissues. *Mailing Add:* Path Dept Cedars Med Ctr 1400 NW 12th Ave Miami FL 33136

ROBB, RICHARD A, b Price, Utah, Dec 2, 42. MEDICAL IMAGING, COMPUTED TOMOGRAPHY. *Educ:* Univ Utah, BA, 65, MS, 68, PhD(comput sci), 71. *Prof Exp:* DIR, RES COMPUT FACIL, MAYO FOUND, 72-, PROF BIOPHYSICS, GRAD SCH MED, 84- *Concurrent Pos:* Prin investr & staff consult, Res Comput Facil, Mayo Found, 76- *Mem:* Am Physiol Soc; Biomed Eng Soc; AAAS; Asn Advan Technol Biomed Sci; Sigma Xi. *Res:* Computurized processing, display and analysis of biomedical imagery; x-ray computed tomography; three dimensional image display. *Mailing Add:* Biophys Mayo Found Mayo Grad Sch Med Rochester MN 55901

ROBB, RICHARD JOHN, b Detroit, Mich, Nov 1, 50; m 71; c 2. MOLECULAR IMMUNOLOGY, LYMPHOKINES. *Educ:* Mich State Univ, BS, 72; Harvard Univ, PhD(biochem & molecular biol), 78. *Prof Exp:* Jane Coffin Child mem Fund fel, Uppsala Univ, Sweden, 78-79 & Darthmouth Univ, 79-80; res assoc, Dartmouth Univ, 80-81; prin scientist, E I Du Pont de Nemours Co, Inc, 81-84, res leader, 84-86, group leader, Cent Res & Develop Dept, 86-88, res fel, Med Prod Dept, 88-90, RES LEADER, INFLAMMATORY DIS, DUPONT MERCK PHARMACEUTICALS, 91- *Concurrent Pos:* Assoc ed, J Immunol, 85-; adj assoc prof, Univ Pa, 85-; group leader, Corp Res & Develop, Dupont, 86- *Mem:* Am Asn Immunologists. *Res:* Mechanistic action of lymphokines; purification and molecular characterization of lymphokines and their receptors; signal transduction and differentiation events following lymphokine-receptor interaction. *Mailing Add:* Glenolden Lab Rm 205 DuPont Merck Pharmaceuticals 500 S Ridgeway Ave Glenolden PA 19036

ROBB, THOMAS WILBERN, b Marshall, Mo, Apr 25, 53; m; c 2. RUMINANT NUTRITION, FORAGE EVALUATION. *Educ:* Central Mo State Univ, BS, 73; Univ Mo, Columbia, MS, 75; Univ Ky, PhD(animal sci), 80. *Prof Exp:* Asst prof animal sci, NC State Univ, 81-84, stationed in NE Brazil, 81-83; res assoc, Virginia Tech Univ, 84-85; intl tech serv rep, 85-89, PROJ MGT, PITMAN-MOORE, 89- *Mem:* Am Soc Animal Sci. *Res:* Ruminant nutrition research, primarily forages in northeast Brazil, meat producing sheep and goats and with dairy goats. *Mailing Add:* Pitman-Moore PO Box 207 Terre Haute IN 47808

ROBB, WALTER L(EE), b Harrisburg, Pa, Apr 25, 28; m 54; c 3. CHEMICAL ENGINEERING. *Educ:* Pa State Col, BS, 48; Univ Ill, MS, 49, PhD(chem eng), 51. *Hon Degrees:* DSc, Worcester Polytech Inst. *Prof Exp:* Engr, Knolls Atomic Power Lab,Gen Elec Co, 51-56, res engr, Res Lab, 56-62, mgr chem process sect, 62-66, mgr res & develop, Silicone Prod Dept, 66-68, mgr med developer, Schenectady, 68-71, gen mgr silicone prod dept, Waterford, 71-73, gen mgr chem & metall div, 73, sr vpres & group exec, Med Systs Group, 73-86, SR VPRES CORP RES & DEVELOP, GEN ELEC CO, SCHENECTADY, 86- *Concurrent Pos:* Mem, Health Adv Comt-OTA, Exec Comt & Bd Health Indust Mfrs Asn. *Honors & Awards:* Pioneer Award, Imaging Div, Nat Elec Mfg Asn. *Mem:* Nat Acad Eng. *Res:* Permeable membranes; diagnostic imaging equipment. *Mailing Add:* 1358 Ruffner Rd Schenectady NY 12309-2506

ROBBAT, ALBERT, JR, b Boston, Mass, Apr 11, 54; m; c 2. ELECTROCHEMISTRY, CHROMATOGRAPHY. *Educ:* Univ Mass, Boston, BA, 76; Pa State Univ, PhD(chem), 80. *Prof Exp:* ASSOC PROF CHEM, TUFTS UNIV, 80- *Concurrent Pos:* Res assoc anal chem, US Dept Energy, Pittsburgh, 81-82, prin investr, Anal Div. *Mem:* Am Chem Soc; Electrochem Soc. *Res:* Development of GC-MS for on-site detection of organic compounds; Electroreactivity and affinity of biologically important species: iron, sulfur, molybdenum proteins, and condensed thiophenes; application of radio frequency, static and magnetic fields in chromatography. *Mailing Add:* Dept Chem Tufts Univ Medford MA 02155

ROBBEN, FRANKLIN ARTHUR, combustion, diagnostics, for more information see previous edition

ROBBERS, JAMES EARL, b Everett, Wash, Oct 18, 34; m 57. PHARMACOGNOSY. *Educ:* Wash State Univ, BS & BPhar, 57, MS, 61; Univ Wash, PhD(pharmacog), 64. *Prof Exp:* Asst prof pharmacog, Univ Houston, 64-66; from asst prof to assoc prof, 66-75, PROF PHARMACOG, PURDUE UNIV, W LAFAYETTE, 75- *Concurrent Pos:* Ed, J Nat Prod, 84-93. *Honors & Awards:* Edwin Leigh Newcomb Award, 63; Hon Mem, Asn Francaise pour l'Enseignement et la Recherche en Pharmacognosie. *Mem:* Am Soc Pharmacog. *Res:* Isolation, physiology, biosynthesis, genetics and metabolic control of fungal metabolites. *Mailing Add:* Sch Pharm & Pharmacol Sci Purdue Univ West Lafayette IN 47907-0702

ROBBERSON, DONALD LEWIS, b Shawnee, Okla, Sept 10, 41; div; c 3. GENE STRUCTURE, GENE TRANSPOSITION. *Educ:* Okla Baptist Univ, BS, 63; Calif Inst Technol, PhD(biophys), 71. *Prof Exp:* Res fel mitochondrial DNA replication, dept path, Stanford Univ Med Ctr, 71-72; res fel polyoma virus genetics, Imp Cancer Res Fund, 72-73; asst biologist, 73-79, from asst prof to assoc prof molecular biol, 73-84, actg head dept, 79-83, PROF GENETICS & GENETICIST, UNIV TEX SYST CANCER CTR MD ANDERSON HOSP & TUMOR INST, 84-, CHIEF, SECT MOLECULAR GENETICS, 83- *Concurrent Pos:* Mem, molecular biol study sect, NIH, 80- *Mem:* Am Soc Cell Biol; Am Soc Biol Chemists. *Res:* Structure and replication of mitochondrial DNA sequences in normal and malignat animal cells; electron microscopy of nucleic acids; recombinant DNA techniques. *Mailing Add:* Dept Genetics MD Anderson Hosp 6723 Bertner Ave Houston TX 77030

ROBBIN, JOEL W, b Chicago, Ill, May 28, 41; m 66. MATHEMATICS. *Educ:* Univ Ill, BS, 62; Princeton Univ, MA, 64, PhD(math), 65. *Prof Exp:* Instr math, Princeton Univ, 65-67; asst prof, 67-73, PROF MATH, UNIV WIS-MADISON, 73- *Mem:* Am Math Soc; Asn Symbolic Logic. *Res:* Logic; differential equations. *Mailing Add:* Dept Math Univ Wis 480 Lincoln Dr Madison WI 53706

ROBBINS, ALLEN BISHOP, b New Brunswick, NJ, Mar 31, 30; m 79; c 4. PHYSICS. *Educ:* Rutgers Univ, BSc, 52; Yale Univ, MS, 53, PhD(physics), 56. *Prof Exp:* From instr to assoc prof, 56-68, PROF PHYSICS, RUTGERS UNIV, NEW BRUNSWICK, 68-, CHMN DEPT, 79- *Concurrent Pos:* Imp Chem Industs res fel, Univ Birmingham, 57-58, lectr, 60-61. *Mem:* AAAS; fel Am Phys Soc; Am Asn Physics Teachers. *Res:* Nuclear physics. *Mailing Add:* Dept Physics Rutgers Univ New Brunswick NJ 08855-0849

ROBBINS, APRIL RUTH, BIOCHEMICAL GENETICS, CELL BIOLOGY. *Educ:* Brown Univ, PhD(biochem genetics), 74. *Prof Exp:* SR INVESTR, NAT INST DIABETES, DIGESTIVE & KIDNEY DIS, NIH, 83- *Mailing Add:* Genetics & Biochem Br NIH Bldg 10 Rm 9015 Bethesda MD 20892

ROBBINS, CHANDLER SEYMOUR, b Belmont, Mass, July 17, 18; m 48; c 4. ORNITHOLOGY. *Educ:* Harvard Univ, AB, 40; George Washington Univ, MA, 50. *Prof Exp:* Chief sect, 61-76, WILDLIFE RES BIOLOGIST, MIGRATORY NON-GAME BIRD STUDIES, US FISH & WILDLIFE SERV, 45- *Concurrent Pos:* Ed, Md Birdlife, 47-; mem, Int Bird Census Comt; secy, Int Bird Ringing Comt, 71-87; tech ed, Audubon Field Notes, 52-70 & Am Birds, 71-74 & 79-89. *Honors & Awards:* Arthur A Allen Award; Ludlow Griscom Award; Paul Bartsch Award; Eugene Eisenmann Award; Chuck Yeager Award. *Mem:* Fel Am Ornith Union; Wilson Ornith Soc; Cooper Ornith Soc; Am Meteorol Soc; Nat Audubon Soc. *Res:* Distribution, migration, abundance and habitat requirements of North American birds; monitoring of bird population levels; analysis of records of banded birds; breeding bird atlas methodology; field identification of birds. *Mailing Add:* US Fish & Wildlife Serv Laurel MD 20708-9619

ROBBINS, CLARENCE RALPH, b Point Marion, Pa, Aug 25, 38; m 71; c 2. COSMETIC CHEMISTRY, ORGANIC CHEMISTRY. *Educ:* WVa Wesleyan Col, BS, 60; Purdue Univ, PhD(org chem), 64. *Prof Exp:* Res chemist hair res, 64-66; sect head, 66-72, sect head toiletries, 72-74, sr sect head hair & soap prod, 74-77, sr res assoc skin & hair res, 77-81, sr scientist hair res, 81-86, RES FEL HAIR RES COLGATE PALMOLIVE CO, 86- *Concurrent Pos:* Asst prof dermal pharmacol, Farleigh Dickinson Univ, 83-84. *Honors & Awards:* Maison de Nevarree Medal Award, 89. *Mem:* Soc Cosmetic Chemists. *Res:* Chemical and physical properties of human hair and skin, especially physical property changes of hair and skin to toiletry and surfactant treatments and to other environmental influences. *Mailing Add:* Colgate Palmolive Co 909 River Rd Piscataway NJ 08854

ROBBINS, D(ELMAR) HURLEY, b Elmira, NY, Dec 5, 37; m 68; c 1. ENGINEERING MECHANICS. *Educ:* Univ Mich, BSE(eng mech) & BSE(math), 60, MSE, 61, PhD(eng mech), 65. *Prof Exp:* Asst prof eng mech, Ohio State Univ, 65-67; res engr, Hwy Safety Res Inst, Univ Mich, Ann Arbor, 67-88; PROJ MGR & RES ENG, TRW, INC, 88- *Honors & Awards:* Metrop Life Award, Nat Safety Coun, 70. *Mem:* Am Soc Mech Engrs; Acoust Soc Am; Soc Exp Stress Analysis; Sigma Xi. *Res:* Theory of shells; mechanical properties of engineering and biological materials; analytical and experimental simulation of automotive crash impact; acoustics of musical instruments. *Mailing Add:* 1469 Biggers Dr Rochester Hills MI 48309

ROBBINS, DAVID ALVIN, b May 11, 47. BANACH MODULES, BANACH BUNDLES. *Educ:* Dartmouth Col, AB, 67; Bucknell Univ, MA, 68; Duke Univ, MA, 70, PhD(math), 72; Rensselaer Polytech Inst, MS, 83. *Prof Exp:* From asst prof to assoc prof, Trinity Col, 72-84, chmn dept, 78-84 & 88-90, spec asst to pres, 88-90, PROF MATH, TRINITY COL, 84- *Concurrent Pos:* Ace fel academ adm, 87-88. *Mem:* Am Math Soc; Math Asn Am; Am Asn Higher Educ. *Res:* Fields of Banach spaces. *Mailing Add:* Dept Math Trinity Col Hartford CT 06106

ROBBINS, DAVID O, b Bryn Mawr, Pa, July 1, 43; m 71; c 2. VISION, ELECTROPHYSIOLOGY. *Educ:* Lycoming Col, BA, 65; Univ Del, MA, 68, PhD(physiol psychol), 70. *Prof Exp:* Fel vision, Eye Res Found, 69-70, prin investr, Dept Physiol, 70-73, dir res, 72-73; asst prof, 73-78, ASSOC PROF, DEPT PSYCHOL, OHIO WESLEYAN UNIV, 78- *Concurrent Pos:* Prin investr contracts, US Army Res & Develop Command, 70- *Mem:* Asn Res Vision & Ophthal; Sigma Xi; Am Psychol Soc. *Res:* Electrophysiological bases of color vision with emphasis on single cell receptive field organization to color and movement in reptiles; adverse effects of intense, coherent (laser) light on retinal physiology and function in rhesus monkeys. *Mailing Add:* Dept Psychol Ohio Wesleyan Univ Delaware OH 43015

ROBBINS, DONALD EUGENE, b San Saba, Tex, July 4, 37; m 56; c 3. PHYSICS. *Educ:* Tex Christian Univ, BA, 60; Univ Houston, PhD(physics), 69. *Prof Exp:* Nuclear engr, Gen Dynamics/Ft Worth, 60-62; sr nuclear engr, Ling Temco Vought, Dallas, 62-63; PHYSICIST, JOHNSON SPACE CTR, NASA, 63- *Concurrent Pos:* Lectr, San Jacinto Col, Tex, 69-70; lectr, Univ Houston, 72-78. *Mem:* Am Phys Soc. *Res:* Atmospheric physics; catalytic attack of minor species on stratospheric ozone; institute measurements of ozone and other stratospheric species; measurement of vacuum ultraviolet photoabsorption cross section for gases of interest to atmospheric physics. *Mailing Add:* 15022 St Cloud Houston TX 77062

ROBBINS, EDITH SCHULTZ, b Galveston, Tex, July 4, 41; m 66; c 1. CELL BIOLOGY, ELECTRON MICROSCOPY. *Educ:* Barnard Col, AB, 62; NY Univ, MS, 65, PhD(biol), 70. *Prof Exp:* Instr, 68-71, from asst prof to assoc prof, 72-80, PROF BIOL, MANHATTAN COMMUNITY COL, CITY UNIV NY, 81- *Concurrent Pos:* Adj instr, NY Univ Sch Med, 70-72, adj asst prof, 72-75, adj assoc prof, 75- *Mem:* Am Soc Cell Biol; AAAS; Sigma Xi; NY Acad Sci; Electron Micros Soc Am. *Res:* Studies on epithelial polarity using cultured cells; examination of cells with various electron microscopy techniques including freeze-fracture and scanning; trypanosome structure/function correlations; cellular morphometry of bronchi. *Mailing Add:* Dept Cell Biol NY Univ Sch Med 550 First Ave New York NY 10016

ROBBINS, ELEANORA IBERALL, b Wash, DC, July 20, 42; m 72. BIOGEOLOGY, PALEOECOLOGY. *Educ:* Ohio State Univ, BS, 64; Univ Ariz, MS, 72; Pa State Univ, PhD(geosci), 82. *Prof Exp:* Volunteer, US Peace Corps Geol Surv, Tanzania, 64-66, GEOLOGIST, US GEOL SURV, 67- *Concurrent Pos:* Mem bd, Am Asn Stratig Palynologists, 89-91. *Mem:* AAAS; Am Asn Stratig Palynologists; Asn Women Geoscientists; Soc Wetland Scientists. *Res:* Relationship between coal, petroleum, and ore deposits; publications on organic tissues and minerals in Precambrian iron formation, in Mn pisolites and nodules, and in carbonaceous gold deposits. *Mailing Add:* US Geol Surv Nat Ctr MS 956 Reston VA 22092

ROBBINS, ERNEST ALECK, b Boy River, Minn, Mar 26, 26; m 46; c 6. BIOCHEMISTRY. *Educ:* Univ Minn, BChem, 51, PhD(biochem), 57. *Prof Exp:* Eng aide, State Hwy Dept, Minn, 49-51; asst, Univ Minn, 51-56; res chemist, Rohm and Haas Co, Pa, 56-71; head food biochem, 71-75, ASSOC DIR YEAST PROD RES, ANHEUSER-BUSCH INC, 75- *Mem:* Am Chem Soc; Am Asn Cereal Chem; Inst Food Technol; Sigma Xi. *Res:* Production, properties and utilization of enzymes from microorganisms; development of food and fermentation products. *Mailing Add:* 4429 Meadow Dr High Ridge MO 63049

ROBBINS, FREDERICK CHAPMAN, b Auburn, Ala, Aug 25, 16; m 48; c 2. PEDIATRICS. *Educ:* Univ Mo, BA, 36, BS, 38; Harvard Med Sch, MD, 40; Am Bd Pediat, dipl, 51. *Hon Degrees:* DSc, John Carroll Univ, 55, Univ Mo, 58; LLD, Univ NMex, 68; DSc, Univ NC, 79, Tufts Univ, 83, Med Col Ohio, 83, Albert Einstein Col Med, 84, Med Col Wis, 84; DMedSc, Med Col Pa, 84; LLD, Univ Ala, Birmingham, 85. *Prof Exp:* Resident bact, Children's Hosp, 40-41, intern med, 41-42, asst resident, 46-47, resident, 47-48; Nat Res Coun fel virus dis, Children's Hosp, Boston, 48-50; instr & assoc pediat, Harvard Med Sch, 50-52; prof pediat, Sch Med, Case Western Reserve Univ, 50-80, dean, Sch Med, 66-80, prof community health, 73-80; pres, Inst Med-Nat Acad Sci, 81-85; UNIV PROF, CASE WESTERN RESERVE UNIV, 80- & EMER DEAN, SCH MED, 80- *Concurrent Pos:* Asst, Children's Med Serv, Mass Gen Hosp, Boston, 50-52; dir, Dept Pediat & Contagious Dis, Cleveland Metrop Gen Hosp, 52-66; consult, Communicable Dis Ctr, USPHS, Ga, 54-60 & adv comt, Bur Biologics, Food & Drug Admin, 84- *Honors & Awards:* Nobel Prize in Physiol & Med, 54; First Mead Johnson Award, 53; Award for Distinguished Achievement Modern Med, 63; Med Mutual Hon Award, 69. *Mem:* Nat Acad Sci; Am Philos Soc; Am Pediat Soc (pres, 73-74); emer mem Ad Soc Clin Invest; Soc Exp Biol & Med; Soc Pediat Res (pres, 61-62); Am Acad Arts & Sci; emer mem Am Acad Pediat; fel AAAS; hon mem Asn Am Physicians. *Res:* Recognition and epidemiology of Q fever; immunology of mumps; tissue culture-poliomyelitis virus. *Mailing Add:* Dept Epidemiol & Biostatist Case Western Res Univ Sch Med Cleveland OH 44106

ROBBINS, GORDON DANIEL, b Rocky Mount, NC, Feb 8, 40; m 67; c 2. ELECTROCHEMISTRY. *Educ:* Univ NC, BS, 62; Princeton Univ, MA, 64, PhD(electrochem), 66. *Prof Exp:* Fulbright fel, Molten Salts, Tech Univ Norway, 66-67; res chemist, 67-72, dir, Off Prof Univ Rels, 72-77, DIR, INFO DIV, OAK RIDGE NAT LAB, 78- *Mem:* Am Soc Info Sci. *Res:* Aqueous electrochemistry; thermodynamics, transport properties, electrochemistry and electrical conductivity of molten salts. *Mailing Add:* 14 Brookside Dr Oak Ridge TN 37830

ROBBINS, HERBERT ELLIS, b New Castle, Pa, Jan 12, 15; m 43, 66; c 5. MATHEMATICAL STATISTICS. *Educ:* Harvard Univ, AB, 35, AM, 36, PhD(math), 38. *Hon Degrees:* ScD, Purdue Univ, 74. *Prof Exp:* Instr & tutor math, Harvard Univ, 36-38; asst, Inst Advan Study, 38-39; instr math, NY Univ, 39-42; assoc prof math statist, Univ NC, 46-49, prof, 50-53; prof math statist, Columbia Univ, 53-85; PROF STATIST, RUTGERS UNIV, 86- *Concurrent Pos:* Guggenheim fel, 52-53 & 75-76. *Mem:* Nat Acad Sci; fel Inst Math Statist (pres, 65-66); Am Acad Arts & Sci. *Res:* Mathematical statistics, especially sequential experimentation; theory of probability. *Mailing Add:* Dept Statist Rutgers Univ New Brunswick NJ 08903

ROBBINS, JACKIE WAYNE DARMON, b Spartanburg, SC, Feb 6, 40; m 63; c 2. AGRICULTURAL ENGINEERING, DRIP IRRIGATION. *Educ:* Clemson Univ, BS, 61, MS, 65; NC State Univ, PhD(biol & agr eng), 70. *Prof Exp:* E C McArthur fel, Clemson Univ, 62-63; asst prof agr eng, La State Univ, 63-65; res asst, NC State Univ, 65-68, res assoc biol & agr eng, 68-70; assoc prof agr eng, Univ Mo, 70-71; prof agr eng & head dept, 71-87, PROF CIVIL ENG, LA TECH UNIV, 87- *Concurrent Pos:* NSF grant & vis prof, Univ Hawaii, 77-78; consult, dept tech serv, Repub SAfrica, 79, Water Res Comn, 81 & Comn Int du Genic Rural, 86; owner, Robbins Asn, Irrigation-Mart; pres, La Blueberry Asn Inc. *Mem:* Am Soc Agr Engrs; Am Soc Eng Educ; Sigma Xi; Irrigation Asn; Nat Soc Prof Engrs. *Res:* Agricultural waste and waste water management; soil and water conservation; groundwater hydrology; irrigation. *Mailing Add:* Rte 6 Box 1241 Ruston LA 71270

ROBBINS, JACOB, b Yonkers, NY, Sept 1, 22; m 49; c 3. ENDOCRINOLOGY, MEDICAL RESEARCH. *Educ:* Cornell Univ, AB, 44, MD, 47; Am Bd Internal Med, dipl. *Prof Exp:* Intern med, NY Hosp, 47-48; res, Mem Hosp, 48-50; instr, Med Col, Cornell Univ, 50-54; res scientist, 54-62, CHIEF CLIN ENDOCRINOL BR, NAT INST ARTHRITIS, DIABETES, DIGESTIVE & KIDNEY DIS, 62- *Concurrent Pos:* Res fel, Sloan-Kettering Inst Cancer Res, 50-53, asst, 53-54, clin asst, Mem Ctr, 51-53, asst attend physician, 53-54; vis scientist, Calsberg Lab, Copenhagen, Denmark, 59-60. *Honors & Awards:* Van Meter Award, Am Thyroid Asn, 55. *Mem:* AAAS; Endocrine Soc; Am Soc Clin Invest; Am Physiol Soc; Am Thyroid Asn (pres, 74-75); Asn Am Physicians; Europ Thyroid Asn; hon mem Japan Endocrine Soc. *Res:* Thyroid physiology; biochemistry and disease. *Mailing Add:* 7203 Bradley Blvd Bethesda MD 20817

ROBBINS, JAMES CLIFFORD, LIPID SYNTHESIS, ATHEROSCLEROSIS. *Educ:* Univ Mich, PhD(biochem), 73. *Prof Exp:* RES FEL, MERCK & CO, INC, 84- *Mailing Add:* Dept Biochem Regulation Merck Sharp & Dohme PO Box 2000 Rahway NJ 07065

ROBBINS, JAY HOWARD, b New York, NY, Feb 10, 34; m 61; c 3. CELL BIOLOGY, MEDICINE. *Educ:* Harvard Univ, AB, 56; Columbia Univ, MD, 60. *Prof Exp:* Intern, Mt Sinai Hosp, NY, 60-61, asst resident, 61-62, fel med, 63; res assoc, NIMH, 63-65, SR INVESTR, NAT CANCER INST, 65- *Mem:* Am Fedn Clin Res. *Res:* DNA repair and carcinogenesis; cell growth and differentiation; study of effect of DNA-damaging agents on cultured cells from patients with degenerative neurological disease and/or with cancer. *Mailing Add:* Nat Cancer Inst Bldg 10 Rm 12 N-238 Bethesda MD 20892

ROBBINS, JOHN ALAN, b Syracuse, NY, Jan 19, 38; m 66; c 2. LIMNOLOGY, PHYSICS. *Educ:* Swarthmore Col, BA, 59; Univ Rochester, PhD(nuclear physics), 67. *Prof Exp:* Res assoc nuclear physics, Univ Rochester, 68-69; res assoc air chem, dept atmospheric sci, Univ Mich, Ann Arbor, 69-70; asst res physicist, Great Lakes Res Div, Univ Mich, Ann Arbor, 70-73, assoc res physicist, 73-76, sr res scientist limnol, 76-80; PHYSICIST, GREAT LAKES ENVIRON RES LAB, NAT OCEANIC & ATMOSPHERIC ADMIN, ANN ARBOR, MICH, 80-; ADJ PROF, UNIV MICH, SCH NATURAL RESOURCES, 84- *Concurrent Pos:* Co-founder & consult, Environ Res Group, Inc, Ann Arbor, 70-; consult, Wildlife Supply Co, Saginaw, Mich, 74- & Radiol & Environ Res Div, Argonne Nat Lab, Ill, 78-; guest ed, Chem Geol, 84. *Honors & Awards:* RPI Sci Award, 85; Bausch & Lomb Sci Achievement Award, 55. *Mem:* Am Soc Limnol & Oceanog; Int Asn Great Lakes; Sigma Xi; Soc Int Limnol. *Res:* Physics and chemistry of lakes; radiolimnology and sediment geochronology; radiotracer studies of animal-sediment interactions; mathematical modeling of limnological processes. *Mailing Add:* 1115 Spring St Ann Arbor MI 48103

ROBBINS, JOHN B, b Brooklyn, NY, Dec 1, 32; m. HUMAN DEVELOPMENT. *Educ:* New York Univ, BA, 56, MD, 59. *Hon Degrees:* MD, Univ Goteborg, Sweden, 76. *Prof Exp:* Student investr, New York Univ, Col Med, 57-59; intern/resident, Children's Med Serv, Mass Gen Hosp, Boston, 59-60; postdoctoral res fel, Dept Pediat, Univ Fla, 61-64; guest scientist, Dept Chem Immunol, Weizmann Inst Sci, Rehovot, Israel, 65-66; asst prof pediat & microbiol, Univ Fla, Gainesville, 64-67; from asst prof to assoc prof pediat, Albert Einstein Col Med, 67-70; clin dir, Nat Inst Child Health & Human Develop, NIH, 70-72, chief, Develop Immunol Br, 71-74; dir, Div Bact Prod, Bur Biologics, Food & Drug Admin, 74-83; CHIEF LAB DEVELOP & MOLECULAR IMMUNITY, NAT INST CHILD HEALTH & HUMAN DEVELOP, NIH, 83- *Concurrent Pos:* Assoc mem, Armed Forces Epidemiol Bd, 70; assoc ed, J Immunol, 74; consult, WHO, Nat Inst Vaccines & Sera & Shanghai Inst Vaccines & Sera, Beijing, People's Repub China, Dept Pub Health & Inst Prev Med, Nat Taiwan Univ. *Honors & Awards:* E Mead Johnson Award, Am Acad Pediat, 75; Henry Bale Mem Lectr, Nat Inst Biol Standards & Control, 79; Erwin Neter Mem Lectr, Univ Buffalo, 84; Henry L Barnett Lectr, Albert Einstein Col Med, 85; Maxwell Finland Lectr, Infectious Dis Soc Am, 89; Louis Weinstein Lectr, Tufts Univ, 89. *Mem:* Am Asn Immunologists; Soc Pediat Res; Soc Infectious Dis; Am Soc Clin Invest; Asn Am Physicians. *Res:* Vaccines. *Mailing Add:* NIH Nat Inst Child Health & Human Develop Lab Develop & Molecular Immunity Bldg 6 Rm 141 Bethesda MD 20892

ROBBINS, JOHN EDWARD, b Chamberlin, SDak, Apr 2, 35; m 58; c 1. BIOCHEMISTRY. *Educ:* Carroll Col, Mont, BA, 58; Mont State Univ, MS, 61, PhD(chem), 63. *Prof Exp:* NIH fel, Univ Ore, 63-66; asst prof, 67-70, assoc prof chem, 70-77, ASSOC PROF BIOCHEM, MONT STATE UNIV, 77- *Mem:* AAAS; Am Chem Soc; Sigma Xi. *Res:* Subunit structure of enzymes as related to their functional roles in enzyme functions of catalysis and control. *Mailing Add:* Dept Chem Mont State Univ Bozeman MT 59715

ROBBINS, KELLY ROY, STATISTICS, BIOCHEMISTRY. *Educ:* Univ Ill, PhD(nutrit), 79. *Prof Exp:* ASSOC PROF NUTRIT, UNIV TENN, KNOXVILLE, 79- *Mailing Add:* Dept Animal Sci Univ Tenn 208 Brehm Hall PO Box 1071 Knoxville TN 37901-1071

ROBBINS, KENNETH CARL, b Chicago, Ill, Sept 1, 17; m 46; c 2. BIOCHEMISTRY. *Educ:* Univ Ill, BS, 39, MS, 40, PhD(biol chem), 44. *Prof Exp:* Instr biol chem, Univ Ill, 44-47; asst prof dept path, Western Reserve Univ Sch Med, 47-51; head protein sect, Biochem Res Dept, Res Labs, Armour & Co, 51-56, sect head, Biochem Dept, Cent Labs, Res Div, 56-58; dir biochem res & develop dept, Michael Reese Res Found, 58-73, sci dir, 73-84; pro, 70-87, EMER PROF MED & PATH, PRITZKER SCH MED, UNIV CHICAGO, 87-; RES PROF, NORTHWESTERN UNIV, 88- *Concurrent Pos:* Mem hemat study sect, Div Res Grants, NIH, 71-75 & 78-82; chmn, Gordon Conf Hemostasis, 75; mem, Blood Diseases and Resources Adv Comt, Nat Heart, Lung & Blood Inst, NIH, 76-80; co-chmn, Subcomt on Fibrinolysis, Int Comt on Thrombosis & Hemostasis, 77-80, chmn, 80-82; mem, Int Comt Thrombosis & Hemostasis, 80-86. *Honors & Awards:* Elwood A Sharp Lect Award, Sch Med, Wayne State Univ, 71; Prix Servier Medal & Prize, Fifth Int Congress Fibrinolysis, Malmo, Sweden, 80. *Mem:* Soc Exp Biol & Med; Am Chem Soc; Am Asn Immunol; Am Soc Biol Chem; Am Soc Hemat; Int Soc Thrombosis & Hemostasis. *Res:* Blood coagulation and fibrinolysis; animal proteins and enzymes. *Mailing Add:* 6101 N Sheridan Rd E Apt 36C Chicago IL 60660

ROBBINS, LANNY ARNOLD, b Wahoo, Nebr, Apr 3, 40; m 62; c 2. CHEMICAL ENGINEERING. *Educ:* Iowa State Univ, BS, 61, MS, 63, PhD(chem eng), 66. *Prof Exp:* Chem engr, 66-67, res engr, 67-70, sr pilot plant engr, 70-72, res specialist, 72-73, sr res specialist, 73-76, assoc scientist, 76-79, sr assoc scientist, 79-83, res scientist, 83-88, SR RES SCIENTIST, DOW CHEM CO, 88- *Concurrent Pos:* Adj prof, Mich State Univ, 82. *Mem:* Am Inst Chem Engrs; Sigma Xi. *Res:* Commercial separations and purification processes, with emphasis on melt crystallization, liquid-liquid extraction, packed tower strippers, absorbers and distillation; miniplant design and operation; process scale liquid chromatography and pressure swing adsorption. *Mailing Add:* 4101 Old Pine Trail Midland MI 48640

ROBBINS, LEONARD GILBERT, b Brooklyn, NY, Aug 10, 45; m 83; c 2. GENETICS, MOLECULAR BIOLOGY. *Educ:* Brooklyn Col, BS, 65; Univ Wash, PhD(genetics), 70. *Prof Exp:* NIH trainee zool, Univ Tex, Austin, 70-72; from asst prof to assoc prof, 72-82, PROF ZOOL, MICH STATE UNIV, 83- *Concurrent Pos:* NIH-Fogarty sr int fel, Madrid, Spain, 80-81, Bari, Italy, 91-92. *Mem:* Genetics Soc Am; Fedn Am Scientist. *Res:* Formal and molecular genetics of higher organisms; maternal and zygotic gene action; genetic regulation of chromosome behavior in Drosophila; molecular and functional organization of ribosanal RNA genes. *Mailing Add:* Genetics Prog S308 Molecular Biol Addn Mich State Univ East Lansing MI 48824-1312

ROBBINS, LOUISE MARIE, physical anthropology; deceased, see previous edition for last biography

ROBBINS, MARION LERON, b Chesnee, SC, Aug 18, 41; m 65; c 4. HORTICULTURE, PLANT BREEDING. *Educ:* Clemson Univ, BS, 64; La State Univ, MS, 66; Univ Md, PhD(hort breeding), 68. *Prof Exp:* Asst prof hort, Iowa State Univ, 68-72; assoc prof, 72-79, resident dir, Sweet Potato Res Sta, 84-88, PROF HORT, COASTAL EXP STA, CLEMSON UNIV, 79-, RESIDENT DIR, CALHOUN RES STA, 88- *Concurrent Pos:* Assoc ed, Crop Res; contrib ed & columnist, Am Veg Grower & Greenhouse Grower; consult on veg prod & tea. *Mem:* Am Soc Hort Sci; Int Soc Hort Sci; Res Ctr Adminr Soc. *Res:* Vegetable production research; breeding and physiology; influence of cultural practices on yield and quality of vegetables; tea as a crop for the United States; sweet potato research and development. *Mailing Add:* Calhoun Res Sta PO Box 539 Calhoun LA 71225

ROBBINS, MURRAY, b Brooklyn, NY, Mar 9, 31; m 54; c 1. INORGANIC CHEMISTRY. *Educ:* Brooklyn Col, BS, 53, MA, 54; Polytech Inst Brooklyn, PhD(inorg chem), 62. *Prof Exp:* Res chemist, Columbia Univ Mineral Beneficiation Labs, 53-54; instr inorg chem, Polytech Inst Brooklyn, 60-62; res chemist, David Sarnoff Res Labs, Radio Corp Am, 62-67; MEM TECH STAFF, BELL TEL LABS, INC, 67- *Mem:* Am Chem Soc; Am Phys Soc; Sigma Xi. *Res:* Synthesis and properties of solid state materials and crystal growth. *Mailing Add:* 110 Kent Dr Berkeley Heights NJ 07922

ROBBINS, NAOMI BOGRAD, b Paterson, NJ, May 8, 37; m 62; c 2. STATISTICS, RELIABILITY. *Educ:* Bryn Mawr Col, AB, 58; Cornell Univ, MA, 62; Columbia Univ, PhD(math statist), 71. *Prof Exp:* MEM TECH STAFF, BELL LABS, 60- *Mem:* Am Statist Asn; Sigma Xi. *Mailing Add:* 11 Christine Ct Wayne NJ 07470

ROBBINS, NORMAN, b Brooklyn, NY, Apr 15, 35; m 68; c 2. NEUROPHYSIOLOGY. *Educ:* Columbia Col, AB, 55; Harvard Univ, MD, 59; Rockefeller Univ, PhD(biol), 66. *Prof Exp:* Intern med, 2nd Div, Bellevue Hosp, 59-60; jr asst resident med, Peter Bent Brigham Hosp, 60-61; fel neurobiol, Rockefeller Univ, 61-66; res assoc neurophysiol, Spinal Cord Div, Nat Inst Neurol Dis & Blindness, 66-69; vis scientist, Kyoto Prefectural Univ Med, 69-70; from asst prof to assoc prof anat, 70-84, prof develop genetics & anat, 84-89, PROF NEUROSCI, SCH MED, CASE WESTERN UNIV, 89- *Mem:* AAAS; Soc Neurosci. *Res:* Neurobiology; physiology of development, plasticity and aging of the neuromuscular junction. *Mailing Add:* Dept Neurosci Sch Med Case Western Reserve Univ Cleveland OH 44106

ROBBINS, OMER ELLSWORTH, JR, physical chemistry, for more information see previous edition

ROBBINS, PAUL EDWARD, b Camden, NJ, Apr 4, 28; m 54; c 3. ORGANIC CHEMISTRY. *Educ:* Univ Pa, BS, 52; Ga Inst Technol, PhD(chem), 56. *Prof Exp:* Res chemist, E I du Pont de Nemours & Co, 56-60; assoc prof textile chem, Clemson Univ, 60-66; assoc prof, 66-75, prof, 75-86, EMER PROF CHEM, ARMSTRONG STATE COL, 86- *Mem:* Am Chem Soc; AAAS; Sigma Xi. *Res:* Catalytic hydrogenation; physical and synthetic organic chemistry; polymer chemistry and rheology; textile chemistry; films. *Mailing Add:* 14 Keystone Dr Savannah GA 31406

ROBBINS, PHILLIPS WESLEY, b Barre, Mass, Aug 10, 30; m 53; c 2. CELL SURFACE PROTEINS & CARBOHYDRATES. *Educ:* DePauw Univ, AB, 52; Univ Ill, PhD(biochem), 55. *Prof Exp:* Res assoc, Mass Gen Hosp, 55-57; asst prof, Rockefeller Inst, 57-59; from asst prof to prof biochem, 59-77, AM CANCER SOC PROF BIOCHEM, MASS INST TECHNOL, 77- *Honors & Awards:* Eli Lilly Award Biol Chem, 66. *Mem:* Nat Acad Sci; Am Soc Biochem & Molecular Biol; Am Acad Arts & Sci; Am Soc Microbiol; Sigma Xi. *Res:* Sulfate activation; synthesis of complex polysaccharides; cell-virus relationships. *Mailing Add:* Dept Biol Ctr Cancer Res Mass Inst Technol Cambridge MA 02139

ROBBINS, R(OGER) W(ELLINGTON), b Belmont, Mass, July 25, 20; m 46; c 2. ELECTRICAL ENGINEERING. *Educ:* Univ Wis, BS, 42. *Prof Exp:* Elec draftsman, Jackson & Moreland, Mass, 42-43; field engr, Submarine Signal Co, 43, supvr tech div, Equip Dept, 43-45, asst admin engr, 45-46; res engr, Operadio Mfg Co, 46-51, proj engr, 51-53; mgr govt contracts div, Dukane Corp, 53-65, tech dir, spec prod div, 65-70, eng mgr, 70-86; RETIRED. *Mem:* AAAS; Inst Elec & Electronics Engrs; Ultrasonic Indust Asn (pres, 75-77). *Res:* Devices which require the use of moving mechanical parts in conjunction with electrical circuits; optics; ultrasonics. *Mailing Add:* 1341 South St Geneva IL 60134

ROBBINS, RALPH COMPTON, b Spurrier, Tenn, Feb 7, 21; m 51. NUTRITION, PHYSIOLOGY. *Educ:* Tenn Technol Univ, BS, 52; Iowa State Univ, MS, 55; Univ Ill, PhD(nutrit), 58. *Prof Exp:* Res fel nutrit & biochem, Med Col SC, 58-59; asst prof nutrit, 59-66, ASSOC PROF NUTRIT, INST FOOD & AGR SCI, UNIV FLA, 66- *Mem:* Am Chem Soc. *Res:* Effectiveness in the diet of naturally occurring blood cell antiadhesive compounds against certain types of circulating dysfunction. *Mailing Add:* 1410 NW 28th St Gainesville FL 32605

ROBBINS, RALPH ROBERT, b Wichita, Kans, Sept 2, 38. ASTROPHYSICS. *Educ:* Yale Univ, BA, 60; Univ Calif, Berkeley, PhD(astron), 66. *Prof Exp:* McDonald Observ fel, Univ Tex, Austin, 66-67; asst prof physics, Univ Houston, 67-68; asst prof astron, Univ Tex, 68-72, ASSOC PROF ASTRON, UNIV TEX, AUSTIN, 72- *Concurrent Pos:* Prin investr, NSF grant, Univ Tex Austin, 70-72; sci educ consult, AID. *Mem:* Am Astron Soc; Int Astron Union; fel Royal Astron Soc; Am Asn Physics Teachers. *Res:* Theoretical and observational astrophysics of gas nebulae and Seyfert galaxies; radiative transfer; atomic and plasma physics. *Mailing Add:* Dept Astron Univ Tex Austin TX 78712

ROBBINS, RICHARD J, UNDERGROUND EXCAVATION SYSTEMS. *Prof Exp:* PRES, ROBBINS CO. *Mem:* Nat Acad Eng. *Mailing Add:* Off Pres Robbins Co 2245 76th Ave S Box 97027 Kent WA 98031-0427

ROBBINS, ROBERT, medicine; deceased, see previous edition for last biography

ROBBINS, ROBERT JOHN, b Niles, Mich, May 10, 44; m 84. COMPUTER SCIENCE, DATABASE SYSTEMS & THEORY MANAGEMENT. *Educ:* Stanford Univ, AB, 66; Mich State Univ, BS, 73, MS, 74, PhD(zool), 77. *Prof Exp:* Instr genetics, Mich State UNiv, 76-77, from asst prof to assoc prof, 77-87; staff assoc, 87-89, PROG DIR, NSF, 90- *Concurrent Pos:* Fel, Univ Calif, Davis, 78-79. *Mem:* Am Inst Biol Sci; Am Soc Microbiol; Genetics Soc Am; Inst Elec & Electronics Engrs Computer Soc; Asn Comput Mach. *Res:* Computer applications in biology; general genetics; database systems & theory. *Mailing Add:* Div Instrumentation & Resources NSF 1800 G St Washington DC 20550

ROBBINS, ROBERT KANNER, b New York, NY, Oct 26, 47. BUTTERFLY EVOLUTION. *Educ:* Brown Univ, AB, 69; Tufts Univ, PhD(biol), 78. *Prof Exp:* Fel biol, Smithsonian Trop Res Inst, 78-79; fel biol, US Nat Mus, 81-82; ASSOC CUR, DEPT ENTOM, SMITHSONIAN INST, 83, RES ENTOMOLOGIST. *Concurrent Pos:* Exec coun, Lepidopterists' Soc. *Mem:* Soc Study Evolution; Soc Am Naturalists; AAAS; Lepidopterists' Soc; Soc Syst Zool. *Res:* Inferring butterfly phynology; using phylogenies to study the evolution of butterfly predator avoidance mechanisms, foodplant use, courtship/mating behavior, biogeography and species diversity; hereditability of behavior; statistical estimation of demographic parameters from field data; systematics of lepidoptera. *Mailing Add:* 2326 Huidekoper Pl Washington DC 20007

ROBBINS, ROBERT RAYMOND, b Des Moines, Iowa, May 28, 46. BOTANY, ULTRASTRUCTURE. *Educ:* Iowa State Univ, BS, 68; Univ Ill, Urbana, MS, 73. PhD(bot), 77. *Prof Exp:* Instr human anat, US Army Med Field Serv Sch, San Antonio, Tex, 69-71; lectr bot, Univ Ill, Urbana, 76, fel, 76-77; asst prof bot, Univ Wis-Milwaukee, 77; MEM STAFF, IDAHO STATE UNIV. *Mem:* AAAS; Bot Soc Am; Am Bryol & Lichenological Soc; Am Inst Biol Sci; Am Fern Soc. *Res:* Ultrastructure of plant reproductive cells, especially bryophyte and lower vascular plant spermatogenesis; ragweed pollen development. *Mailing Add:* Idaho State Univ Box 8007 Pocatello ID 83209

ROBBINS, STANLEY L, b Portland, Maine, Feb 27, 15; m 40; c 3. PATHOLOGY. *Educ:* Mass Inst Technol, BS, 36; Tufts Univ, MD, 40. *Prof Exp:* From instr to prof path, Boston Univ Sch Med, 41-72, chmn dept, 72-80; VIS PROF PATH, HARVARD UNIV SCH MED, 80- *Concurrent Pos:* Vis prof, Univ Glasgow, 60 & Hebrew Univ, Israel, 75; consult, Framingham Union Hosp, 65-80 & Vet Admin Hosp, Roxbury, Mass, 70-80; dir, Mallory Inst Path, 70-76; assoc ed, Human Path, 75- *Mem:* Am Asn Pathologists; Int Acad Path; AAAS; Am Soc Exp Path. *Res:* Pathology. *Mailing Add:* Dept Path Brigham & Womens Hosp 75 Francis St Cambridge MA 02115

ROBBINS, WAYNE BRIAN, b Dayton, Ohio, Oct 12, 51; m 72; c 3. ANALYTICAL CHEMISTRY. *Educ:* Univ Cincinnati, BS, 75, MS, 77, PhD(anal chem), 78. *Prof Exp:* Asst prof chem, Utah State Univ, 78-79; res scientist, Union Camp Corp, 79-85; MGR, S D WARREN CO, 85- *Concurrent Pos:* Adj prof, Dept Chem Eng, Univ Maine, Orono, Maine. *Mem:* Am Chem Soc; Soc Appl Spectros; Tech Asn Pulp & Paper Indust. *Res:* Process development for pulp and paper. *Mailing Add:* S D Warren Co Res Lab PO Box 5000 Westbrook ME 04092

ROBBINS, WILLIAM PERRY, b Atlanta, Ga, May 29, 41; m 65. ELECTRICAL ENGINEERING. *Educ:* Mass Inst Technol, BSEE, 63, MSEE & Elec Engr, 65; Univ Wash, PhD(elec eng), 71. *Prof Exp:* Res engr, Boeing Co, 65-69; asst prof, 69-75, ASSOC PROF ELEC ENG, UNIV MINN, MINNEAPOLIS, 75- *Mem:* Am Inst Physics; Inst Elec & Electronics Engrs; Sigma Xi. *Res:* Acoustic surface wave properties and devices; active circuit design. *Mailing Add:* Dept Elec Eng Univ Minn Minneapolis MN 55455

ROBBLEE, ALEXANDER (ROBINSON), b Calgary, Alta, Jan 21, 19; m 44; c 4. POULTRY NUTRITION. *Educ:* Univ Alta, BSc, 44, MSc, 46; Univ Wis, PhD(poultry biochem), 48. *Prof Exp:* Govt agr fieldman, Dept Agr, Alta, 40-44; instr poultry nutrit, Univ Alta, 45-46 & Univ Wis, 47-48; from asst prof to prof, 48-84, EMER PROF POULTRY NUTRIT, UNIV ALTA, 84- *Honors & Awards:* Golden Award, Can Feed Indust Asn. *Mem:* Fel Poultry Sci Asn; fel Agr Inst Can; Worlds' Poultry Sci Asn; Can Soc Nutrit Sci. *Res:* Studies of B-vitamins (riboflavin, B12, biotin, pyridoxine) and investigation of the nutritive value of rapeseed and canola meals for poultry. *Mailing Add:* Dept Animal Sci Univ Alta Edmonton AB T6G 2P5 Can

ROBE, THURLOW RICHARD, b Petersburg, Ohio, Jan 25, 34; m 55; c 4. ENGINEERING MECHANICS. *Educ:* Ohio Univ, BSCE, 55, MS, 62; Stanford Univ, PhD(eng), 66. *Prof Exp:* Engr, Lamp Div, Gen Elec Co, Ohio, 55, Locomotive & Car Equip Dept, Pa, 55-56, Flight Propulsion Lab, Ohio, 59-60; from asst prof to assoc prof eng mech, Univ KY, 65-75, prof, 75-80, assoc dean, Col Eng, 76-80; DEAN, COL ENG & TECHNOL, OHIO UNIV, 80- *Concurrent Pos:* Eng consult, Westinghouse Air Brake Co, 67-69, Int Bus Mach Corp, 69-70 & QED Assocs, 75-; fel, Am Coun Educ, 70-71. *Mem:* Am Soc Mech Engrs; Am Soc Eng Educ; Nat Soc Prof Engrs. *Res:* Engineering mechanics with emphasis on analysis of dynamical systems. *Mailing Add:* Col Eng Technol Ohio Univ Athens OH 45701

ROBECK, GORDON G, b Denver, Colo, Feb 3, 23; m 51; c 3. CIVIL ENGINEERING. *Educ:* Univ Wis, BS, 44; Mass Inst Technol, MS, 50. *Hon Degrees:* DSc, Univ Cincinnati, 85. *Prof Exp:* Comn officer, USPHS, 44-74; dir, Drinking Water Res Div, US Environ Protection Agency, 70-85; RETIRED. *Concurrent Pos:* Consult, water treat eng, 85-; mem, Water Sci & Technol Bd, Nat Res Coun, 86-89. *Honors & Awards:* Huber Res Prize, Am Soc Civil Engrs, 65; 10 Awards from Am Waterworks Asn; Gold Medal, US Environ Protection Agency, 81. *Mem:* Nat Acad Eng; hon mem Am Water Works Asn; life mem Water Pollution Control Fedn; Int Water Supply Asn; hon mem Am Soc Civil Engrs. *Res:* Environmental engineering. *Mailing Add:* 614 Q Avenida Sevilla Laguna Hills CA 92653

ROBEL, GREGORY FRANK, b Yakima Wash, May 5, 55. CONTROL THEORY, FUNCTIONAL ANALYSIS. *Educ:* Harvey Mudd Col, BS, 77; Univ Mich, Ann Arbor, PhD(math), 82. *Prof Exp:* Asst prof math, Iowa State Univ, Ames, 82-86; SR SPECIALIST ENGR, BOEING CO, SEATTLE, WASH, 86- *Concurrent Pos:* Lectr, Cogswell Col North, Kirkland, Wash, 87- *Mem:* Am Inst Aeronaut & Astronaut; Am Math Soc; Inst Elec and Electronics Engrs; Soc Indust & Appl Math. *Res:* Functional analysis and operator theory, and their applications to the engineering disciplines of control systems and signal processing. *Mailing Add:* Boeing M/S 7W-78 PO Box 3707 Seattle WA 98124

ROBEL, ROBERT JOSEPH, b Lansing, Mich, May 21, 33; m 60. APPLIED ECOLOGY, FISH & WILDLIFE SCIENCE. *Educ:* Mich State Univ, BS, 56; Univ Idaho, MS, 58; Utah State Univ, PhD(ecol), 61. *Prof Exp:* Biologist aide, Idaho Fish & Game Dept, 57-58; res asst ecol, Utah State Univ, 58-61; from asst prof to assoc prof biol, 61-71, PROF BIOL, KANS STATE UNIV, 71- *Concurrent Pos:* Aquatic biologist, Bear River Club Co, 58-61; governor's sci adv, 69-80; chmn, Gov Energy & Natural Resources Coun; proj leader, US Congress Off Technol Assessment; distinguished vis lectr, numerous countries; Fulbright res fel. *Mem:* Fel AAAS; Wildlife Soc; Ecol Soc Am; Am Soc Mammal; Animal Behavior Soc; Sigma Xi; Brit Ecol Soc; Am Ornithologists Union; Am Inst Biol Sci. *Res:* Animal ecology with emphasis on avian population dynamics, comparative avian behavior, avian dispersal and movement patterns and avian bioenergetics; natural resources management; energy and environmental considerations, environmental assessments. *Mailing Add:* Div Biol Ackert Hall Kansas State Univ Manhattan KS 66506

ROBEL, STEPHEN B(ERNARD), b Selah, Wash, Jan 29, 23; m 54; c 5. MECHANICAL ENGINEERING. *Educ:* Seattle Univ, BS, 48; Univ Notre Dame, MS, 51. *Prof Exp:* Instr math & physics, 48-49, from asst prof to assoc prof, 50-77, PROF MECH ENG, SEATTLE UNIV, 77- *Mem:* Am Soc Mech Engrs; Am Soc Eng Educ; Am Soc Artificial Internal Organs. *Res:* Thermodynamics; heat transfer; applied mechanics; cardiovascular research. *Mailing Add:* Dept Mech Eng Seattle Univ 900 12th & E Columbia Seattle WA 98122

ROBENS, JANE FLORENCE, b Utica, NY, July 23, 31. VETERINARY MEDICINE, TOXICOLOGY. *Educ:* Cornell Univ, DVM, 55; Am Bd Vet Toxicol, dipl. *Prof Exp:* Vet clinician, Ambassador Animal Hosp, 58-61; vet med officer, US Food & Drug Admin, 61-65, res vet, Div Toxicol Eval, 65-68; asst dir drug regulatory affairs, Hoffmann-La Roche Inc, 68-75; toxicologist, Cancer Bioassay Prog, Tracor Jitco, Inc, 75-79; chief, Animal Feed Safety Br, Food & Drug Admin, 79-80; PROG OFFICER FOOD SAFETY & HEALTH, AGR RES SERV, USDA, 81- *Concurrent Pos:* Assoc ed, Toxicol & Appl Pharmacol, 78- *Mem:* Am Vet Med Asn; Women's Vet Med Asn; Am Asn Lab Animal Sci; Soc Toxicol; Am Col Vet Toxicol. *Res:* Evaluation of the safety and efficacy of drugs used in veterinary medicine; teratological and carcinogenic potential of pesticides, drugs and other chemicals. *Mailing Add:* 5713 Lone Oak Dr Bethesda MD 20814

ROBERGE, ANDREE GROLEAU, b Quebec, Que, Aug 5, 38; m 60; c 4. BIOCHEMISTRY. *Educ:* Laval Univ, BSc, 60, DSc(biochem), 69. *Prof Exp:* Med Res Coun Can fel, McGill Univ, 69-70; Med Res Coun Can fel, Laval Univ, 70-71, prof neurochem, Labs Neurobiol, Fac Med, 71-89; EXEC DIR & SCI DIR, INST ARMAND-FRAPPIER, QUEBEC UNIV, 89- *Concurrent Pos:* Med Res Coun Can grants, 71-76 & scholar, 73-78. *Mem:* AAAS; Can Biochem Soc; Am Soc Neurochem; Brit Biochem Soc; Soc Neurosci. *Res:* Catecholaminergic and serotoninergic metabolisms related to locomotor activity to stressfull situation and to nutrition and drugs inhibiting or activating these metabolisms. *Mailing Add:* Dept Nutrit Humaine Laval Univ FSAA Quebec PQ G1K 7P4 Can

ROBERGE, FERNAND ADRIEN, b Thetford Mines, Que, June 11, 35; m 58; c 4. BIOMEDICAL ENGINEERING, CARDIOLOGY. *Educ:* Polytech Sch Montreal, BAS & Engr, 59, MScA, 60; McGill Univ, PhD(control eng, biomed eng), 64. *Prof Exp:* Develop engr numerical control, Sperry Gyroscope Co, Montreal, 60-61; from asst prof to prof physiol, Fac Med, 65-78, dir biomed eng, Inst Ecole Polytech, 78-88, PROF BIOMED ENG, UNIV MONTREAL, 78-, DIR RES GROUP BIOMED MODELING, 88- *Concurrent Pos:* Mem res group neurol sci, Med Res Coun Can, Univ Montreal, 67-75; mem, Sci Coun Can, 71-74; mem grant comt biomed eng, Med Res Coun Can, 71-76; mem sci comt, Can Heart Found, 74-77; mem, Killam Prog Can Coun, 74-77; mem elec eng comt, Nat Sci Eng Res Coun, Can, 81-83, chmn, 85-88. *Honors & Awards:* D W Ambridge Award, 64; Rousseau Award, Asn Can-France Advan Sci, 86; Leon Lortie Award, 87. *Mem:* Inst Elec & Electronics Engrs; Can Physiol Soc; Can Med & Biol Eng Soc (vpres, 74-76); Int Fedn Med Electronics & Biol Eng; Biomed Eng Soc. *Res:* Membrane biophysics; cardiovascular regulation and control, cardiac arrhythmias; assessment of medical technologies. *Mailing Add:* Inst de Genie Biomed Univ Montreal PO Box 6128 Sta A Montreal PQ H3C 3J7 Can

ROBERGE, JAMES KERR, b Jersey City, NJ, June 13, 38; m 61; c 2. ELECTRICAL ENGINEERING. *Educ:* Mass Inst Technol, SB, 60, SM, 62, ScD(elec eng), 66. *Prof Exp:* Res asst, Mass Inst Technol, 60-62 & 63-66, staff mem, Lincoln Lab, 62-63, res assoc, 66-67, from asst prof to assoc prof, 67-76, PROF ELEC ENG, MASS INST TECHNOL, 76- *Concurrent Pos:* Consult, 60- *Mem:* Inst Elec & Electronics Engrs; Sigma Xi. *Res:* Electronic circuit design, particularly low power, high performance designs for difficult environments; control system design. *Mailing Add:* Dept Elec Eng Mass Inst Technol 77 Mass Ave Cambridge MA 02139

ROBERGE, MARCIEN ROMEO, b Quebec, PQ, Jan 19, 34; m 59; c 5. MICROBIOLOGY. *Educ:* Laval Univ, BSc, 59, MSc, 61, PhD(microbiol), 65. *Prof Exp:* Res scientist microbiol forest soils, Can Forestry Serv, Dept Environ, 59-89; RETIRED. *Mem:* Am Soc Agron; Can Soc Soil Sci; Can Soc Microbiol; Can Inst Forestry. *Res:* Effects of soil microbes and on their activities of various soil and forest treatments, such as fertilization, thinning, scarification and plantation. *Mailing Add:* 712 Carré Bon-Accueil Ste-Foy PQ G1V 4C7 Can

ROBERSON, EDWARD LEE, b Tarboro, NC, June 10, 35; m 60; c 2. VETERINARY ANTHELMINTICS. *Educ:* Duke Univ, AB, 57, MAT, 65; Univ Ga, DVM, 61, PhD(parasitol), 72. *Prof Exp:* Teacher, Raleigh Pub Schs, NC, 63-66; NSF fel parasitol, NC State Univ, 66-67; res assoc path & parasitol, 67-70, from instr to assoc prof, 70-82, PROF PARASITOL, COL VET MED, UNIV GA, 82- *Mem:* Am Soc Parasitologists; Am Asn Vet Parasitologists (vpres, 77-79, pres, 79-81); Am Vet Med Asn; World Asn Advan Vet Parasitol. *Res:* Lactogenic and prenatal transmission and chemotheraphy of veterinary helminths. *Mailing Add:* Univ Ga Col Vet Med Athens GA 30602

ROBERSON, HERMAN ELLIS, b Texarkana, Tex, Apr 27, 34; m 62; c 1. GEOLOGY, MINERALOGY. *Educ:* Univ Tex, BS, 55, MA, 57; Univ Ill, PhD(geol), 59. *Prof Exp:* From instr to assoc prof, 59-74, PROF GEOL & ENVIRON STUDIES, STATE UNIV NY BINGHAMTON, 74- *Mem:* AAAS; Mineral Soc Am; Soc Econ Paleont & Mineral; Mineral Soc Gt Brit. *Res:* Clay mineralogy; sedimentation. *Mailing Add:* Dept Geol Sci & Environ Studies SUNY Binghamton NY 13901

ROBERSON, JOHN A(RTHUR), b Woodland, Wash, June 4, 25; m 47; c 3. HYDRAULIC ENGINEERING. *Educ:* Wash State Univ, BS, 48; Univ Wis, MS, 50; Univ Iowa, PhD, 61. *Prof Exp:* Asst hydraul engr, Wash State Univ, 50-54; physicist hydrodyn res, US Naval Mine Defense Lab, 55-56; asst prof civil eng, Wash State Univ, 56-59, assoc prof, 59-66; assoc prof, Seato Grad Sch Eng, Bangkok, 63-65; prof civil eng, Wash State Univ, 66-88, assoc dean res & grad studies, Col Eng, 81-83; RETIRED. *Concurrent Pos:* Faculty dept civil & environ eng. *Mem:* Am Soc Civil Engrs; Am Soc Eng Educ; Int Asn Hydraulic Res; Sigma Xi. *Res:* Fluid mechanics and hydraulics; author of two books. *Mailing Add:* NW 405 Orion Dr Pullman WA 99163

ROBERSON, NATHAN RUSSELL, b Robersonville, NC, Dec 13, 30; m 54; c 3. NUCLEAR PHYSICS. *Educ:* Univ NC, BS, 54, MS, 55; Johns Hopkins Univ, PhD(nuclear physics), 60. *Prof Exp:* Jr instr physics, Johns Hopkins Univ, 55-60; res assoc, Princeton Univ, 60-63; from asst prof to assoc prof, 63-74, PROF PHYSICS, DUKE UNIV, 74-; DEP DIR, TRIANGLE UNIVS NUCLEAR LAB, 88- *Concurrent Pos:* Mem bd dirs, Triangle Univs Comput Ctr, 75-80; mem, Network Steering Comt, Dept Energy, 86-89, Nuclear Phys Panel Computer, 88- *Mem:* Fel Am Phys Soc. *Res:* Nuclear spectroscopy; radiative capture reactions; use of electronic computers for data acquisition; studies of symmetry violating interactions. *Mailing Add:* Dept Physics Duke Univ Durham NC 27706

ROBERSON, ROBERT ERROL, dynamics, systems science; deceased, see previous edition for last biography

ROBERSON, ROBERT H, b Tuckerman, Ark, July 3, 28; m 51; c 6. POULTRY NUTRITION, BIOCHEMISTRY. *Educ:* Okla State Univ, BS, 51; Univ Ark, MS, 56; Mich State Univ, PhD(poultry nutrit, biochem), 59. *Prof Exp:* FROM ASST PROF TO PROF POULTRY NUTRIT, NMEX STATE UNIV, 59- *Mem:* Poultry Sci Asn; Am Inst Nutrit; World Poultry Sci Asn. *Res:* Mineral-zinc, calcium, protein and amino acids. *Mailing Add:* Dept 3I NMex State Univ Box 30003 Las Cruces NM 88001

ROBERSON, WARD BRYCE, b Hamilton, Ala, Jan 17, 39; m 68; c 1. PLANT PHYSIOLOGY. *Educ:* Harding Col, BA, 61; Utah State Univ, MS, 64, PhD(plant physiol), 67. *Prof Exp:* From asst prof to assoc prof, 66-78, PROF BIOL, HARDING UNIV, 78- *Mem:* AAAS. *Res:* Membrane permeability; plant water relations. *Mailing Add:* Box 932 Sta A Searcy AR 72143-5591

ROBERT, ANDRE, b Montreal, Que, Oct 6, 26; m 51; c 4. ENDOCRINOLOGY. *Educ:* Univ Montreal, BA, 44, MD, 50, PhD(exp med), 58. *Prof Exp:* Asst, Inst Exp Med & Surg, Univ Montreal, 50-52, asst prof endocrinol, Sch Med, 52-55; SR SCIENTIST, UPJOHN CO, 55- *Concurrent Pos:* Guest scientist, NIH, 60-61; vis scientist, Ctr Ulcer Res & Educ, Los Angeles, 80-81. *Mem:* Soc Exp Biol & Med; Am Gastroenterol Asn; Am Physiol Soc. *Res:* Peptic ulcer; hormones and inflammation; gastric secretion; prostaglandins; gastroenterology. *Mailing Add:* Upjohn Co Kalamazoo MI 49001

ROBERT, KEARNY QUINN, JR, b Liberty, Tex, June 12, 43; m 68; c 3. PHYSICS, ENGINEERING. *Educ:* Tulane Univ, BA, 65, BS, 65, PhD(physics), 70. *Prof Exp:* Asst physics, Tulane Univ, 66-70; res physicist, Gulf South Res Inst, 70-76; RES PHYSICIST, USDA SOUTHERN REGIONAL RES CTR, 77- *Concurrent Pos:* Lectr physics, St Mary's Dominican Col, 66; res consult, Physics Dept, Tulane Univ, 72-76, adj asst prof, Sch Eng, 76- *Mem:* Am Phys Soc; Am Asn Physics Teachers; Health Physics Soc; Am Soc Nondestructive Testing; Sigma Xi. *Res:* Environmental health sciences; radiological health and health physics; industrial hygiene; medical diagnostics; nondestructive testing; computer simulation; textile engineering; dust control; nuclear physics. *Mailing Add:* 4501 S Tonti St New Orleans LA 70125

ROBERT, SUZANNE, b Montreal, Que, Can, Feb 18, 57; m 83; c 2. ANIMAL STRESS, STRAY VOLTAGE. *Educ:* Univ Montreal, DVM, 79; Univ Claude Bernard, Lyon, France, DEA, 80. *Prof Exp:* Postdoctoral fel swine ethology, Univ Montreal, Fac Vet Med, 83-86; postdoctoral fel, 86-87, RES SCIENTIST SWINE ETHOLOGY, RES STA AGR CAN, LENNOXVILLE, 87- *Concurrent Pos:* Pres, Lennoxville Res Sta Animal Care Comt, 87; mem, Comn Agr Biol CPAQ, 89. *Mem:* Can Soc Animal Sci; Soc Vet Ethology. *Res:* Effects of stray voltage on the welfare, health and productivity of pigs; effects of feed restriction and develop alternative feeding practices that allow to achieve a high standard of animal performance and health while addressing animal welfare concern over confinement rearing. *Mailing Add:* Res Sta Agr Ctr 2000 Rd 108 E PO Box 90 Lennoxville PQ J1M 1Z3 Can

ROBERTO, JAMES BLAIR, b Portland, ME, Sept 4, 46; m 70; c 2. PARTICLE SOLID INTERACTIONS. *Educ:* Mass Instit Technol, SB, 68; Cornell Univ, MS, 70, PhD(appl physics), 74. *Prof Exp:* Group leader, 80-84, RES STAFF MEM, SOLID STATE DIV, OAK RIDGE NAT LAB, 74-, SECT LEADER, 86- *Concurrent Pos:* vis scientist, Kernforschungsanlage, Juelich, Fed Rep Ger, 77, 78; co-ed, Plasma Surface Interactions In Controlled Fusion Devices, 82; co-chmn, V Int Conf on Plasma surface interactions in controlled fusion devices, 82; vis scientist, Max-Planck-Institut für Plasmaphysik, Garching Fed Rep Ger, 83; co-ed, Advanced photon and particle techniques for the characterization of defects in solids, 85; Gen co-chmn, 1986 fall meeting Mat Res Soc, 86; co-chmn, sematech workshop on advanced ion implantation technol, 87; co-ed, Ion Beam Processing of Adv Electronic Mat, 89; co-chmn, Wash Mat Forum, 91. *Mem:* Mat Res Soc (vpres, 89, 1st vpres, 90, pres, 91); Am Phys Soc. *Res:* Interaction of particle beams with materials including ion-solid interactions; ion implantation, plasma-surface interactions; fusion materials research. *Mailing Add:* Solid State Div PO Box 2008 Oak Ridge Nat Lab Bldg 3025 Oakridge TN 37831-6024

ROBERTS, A(LBERT) S(IDNEY), JR, b Washington, NC, Sept 16, 35; m 57; c 4. MECHANICAL & NUCLEAR ENGINEERING. *Educ:* NC State Univ, BS, 57, PhD(nuclear eng, plasma physics), 65; Univ Pittsburgh, MS(mech eng), 59. *Prof Exp:* Assoc engr, Bettis Atomic Power Lab, Westinghouse Elec Corp, 57-60; res asst exp plasma physics, Plasma Physics Lab, NC State Univ, 61-65; from asst prof to assoc prof thermal eng, chmn & grad prog dir, thermal eng group, 72, assoc dean eng, 74-77, PROF MECH ENG, OLD DOM UNIV, 72. *Concurrent Pos:* Consult, NASA, 66-80; guest res engr, AB Atomenergi, Studsvik, Sweden, 68-69; guest lectr, Nat Acad Sci, Roumania, 68. *Mem:* Sigma Xi; Am Soc Eng Educ; Am Soc Mech Engrs; Am Soc Heating, Refrig & Air-Conditioning Engrs. *Res:* Heat transfer; physical gas dynamics; plasma energy conversion methods; nuclear power reactors; solar energy conversion; building heating, ventilating and air-conditioning energy systems; thermodynamics. *Mailing Add:* 5437 Glenhaven Crescent Norfolk VA 23508

ROBERTS, A WAYNE, b Chicago, Ill, Aug 29, 34; m 56; c 3. MATHEMATICS. *Educ:* Ill Inst Technol, BS, 56; Univ Wis, MS, 58, PhD(math), 65. *Prof Exp:* Instr math, Morton Jr Col, 58-62; teaching asst, Univ Wis-Madison, 62-65; PROF MATH, MACALESTER COL, 65- *Mem:* Math Asn Am. *Res:* Mathematical analysis; convex functions; optimization theory. *Mailing Add:* Dept Math Macalester Col St Paul MN 55105

ROBERTS, ALFRED NATHAN, b Welsh, La, Nov 6, 17; m 39; c 3. HORTICULTURE. *Educ:* Ore State Univ, BS, 39, MS, 41; Mich State Univ, PhD(hort), 53. *Prof Exp:* Asst, 39-41, from instr to assoc prof, 41-57, prof, 57-80, EMER PROF HORT, ORE STATE UNIV, 80- *Honors & Awards:* Alex Laurie Award, Am Soc Hort Sci, 65, Stark Award, 68, Colman Award, 72, J H Gourley Medal, 77. *Mem:* Fel Am Soc Hort Sci; Int Plant Propagator's Soc; Int Hort Soc; Scand Soc Plant Physiol. *Res:* Ornamental plant physiology, growth and development; bulb crop physiology; root regeneration physiology. *Mailing Add:* 3107 NW Firwood Pl Corvallis OR 97330

ROBERTS, AMMARETTE, chemistry, for more information see previous edition

ROBERTS, ANITA BAUER, b Pittsburgh, Pa, Mar 4, 42; m 64; c 2. GROWTH FACTORS, ONCOGENES. *Educ:* Oberlin Col, BA, 64; Univ Wis-Madison, PhD(biochem), 68. *Prof Exp:* Fel, dept pharmacol, Harvard Med Sch, 68-69; asst prof biochem, dept chem, Ind Univ, 73-75; SR SCIENTIST, LAB CHEMOPREV, NIH, NAT CANCER INST, 76-, DEPUTY DIR, 90- *Mem:* Am Soc Biol Chemists. *Res:* Study of peptide growth factors, particularly transforming growth factors, and their role in control of both normal and neoplastic cellular physiology. *Mailing Add:* Lab Chemoprev NIH Nat Cancer Inst Bldg 41 Rm C629 Bethesda MD 20892

ROBERTS, ARTHUR, b New York, NY, July 6, 12; m 35; c 2. NEUTRINO ASTRONOMY, PHYSICS INSTRUMENTATION. *Educ:* City Col NY, BS, 31; Columbia Univ, MA, 33; NY Univ, PhD(physics), 36. *Prof Exp:* Res assoc physics, Mass Inst Technol, 37-42, group leader, Radiation Lab, 42-45; assoc prof physics, Univ Iowa, 46-50; from assoc prof to prof, Univ Rochester, 50-60; sr physicist, Argonne Nat Lab, 60-67; sr physicist, Fermi Nat Lab, 67-79; vis prof physics, Univ Hawaii, 80-82, consult, 82-85; RETIRED. *Concurrent Pos:* Physicist, London Br, Off Naval Res, 53-54; vis physicist, CERN, Geneva, 61-62, Harwell Lab, Didcot, Eng, 65, Rutherford Lab, Eng, 73-74; adj prof, Univ Chicago, 70-73; music dir & pres, Light Opera Co, Chicago, 63-74; bd dirs, Contemporary Concerts, Chicago, 62-79. *Mem:* Am Physics Soc; Comput Music Asn; Sigma Xi. *Res:* Radioactivity; use of radioisotopes in medicine; nuclear physics; microwave spectroscopy; high energy physics; cosmic rays and neutrino astronomy. *Mailing Add:* 15176 Rochdale Circle Lombard IL 60148

ROBERTS, AUDREY NADINE, b Lawrence, Kans, Jan 12, 35. IMMUNOLOGY, VIROLOGY. *Educ:* Univ Kans, AB, 55, MA, 57, PhD(microbiol biol, biochem), 59. *Prof Exp:* Asst microbiol, Univ Kans, 55-59; USPHS res assoc, Ind Univ, 59-61; instr immunol & virol, Col Med, 61-62, asst prof prev med, 62-66, assoc prof microbiol, 66-73, PROF MICROBIOL, UNIV TENN MED UNITS, MEMPHIS, 73-, DEP CHMN, 77- *Concurrent Pos:* Prin investr, Nat Inst Allergy & Infectious Dis grants, 61-, Nat Inst Allergy & Infectious Dis career develop award, 64- *Mem:* AAAS; Am Soc Microbiol; Am Pub Health Asn; Am Asn Immunol; NY Acad Sci. *Res:* Quantitation of mechanisms of immune responses in vivo and in vitro utilizing autoradiography, immunochemistry, radiochemistry, tissue culture, cytochemistry and fluorescence microscopy; enterovirus and respiratory virus diseases, cellular immunity and cytochemistry of virus infected cells. *Mailing Add:* 365 Sequoia Univ Tenn Health Sci Ctr 800 Madison Ave Memphis TN 38117

ROBERTS, BRADLEY LEE, b Bristol, Va, Aug 11, 46. INTERMEDIATE ENERGY PARTICLE PHYSICS. *Educ:* Univ Va, BS, 68; Col William & Mary, MS, 70, PhD(physics), 74. *Prof Exp:* Res assoc medium energy nuclear physics, Sci Res Coun, Rutherford Lab, Chilton, Eng, 74-76 & Lab Nuclear Sci, Mass Inst Technol, 76-77; from asst prof to assoc prof, 77-89, PROF & ASSOC CHMN PHYSICS DEPT, BOSTON UNIV, 89- *Concurrent Pos:* Guest asst physicist, Brookhaven Nat Lab, 76-83; res affil, Lab Nuclear Sci, Mass Inst Technol, 78-; co prin investr, NSF grant Kaon & Hyperon Physics; guest assoc physicist, Brookhaven Nat Lab, 83-; mem, prog comt, Brookhaven Nat Lab AGS, 88-91, chmn, User's Exec Comt, 89-91. *Mem:* Am Phys Soc; Sigma Xi. *Res:* Intermediate energy nuclear and particle physics; exotic atoms; muon physics; kaon physics; hyperon physics; weak decays and CP violation. *Mailing Add:* Dept Physics 590 Commonwealth Ave Boston MA 02215

ROBERTS, BRUCE R, b Leonia, NJ, May 19, 33; m 56; c 2. PLANT PHYSIOLOGY. *Educ:* Gettysburg Col, AB, 56; Duke Univ, MF, 60, PhD(plant physiol), 63. *Prof Exp:* Res plant physiologist, Nursery Crops Res Lab, USDA, 63-89; CONSULT, 89- *Concurrent Pos:* Adj prof, Ohio Wesleyan Univ. *Honors & Awards:* Res Award, Int Soc Arboricult, 75, Authors Citation, 81. *Mem:* Am Soc Plant Physiol; Sigma Xi; Am Soc Hort Sci; hon life mem Int Soc Arboricult. *Res:* Plant and tree physiology; water relations; physiological response to plant stress. *Mailing Add:* Dept Bot Ohio Wesleyan Univ Delaware OH 43015

ROBERTS, BRYAN WILSON, b Pinehurst, NC, Feb 12, 38; m 60; c 2. ORGANIC CHEMISTRY. *Educ:* Univ NC, BS, 60; Stanford Univ, PhD(org chem), 64. *Prof Exp:* Nat Acad Sci-Nat Res Coun fel, Calif Inst Technol, 63-64; asst prof chem, Univ Southern Calif, 64-67; asst prof, 67-70, assoc prof, 70-81, asst chmn dept chem, 73-77, PROF CHEM, UNIV PA, 81-, CHMN DEPT, 81- *Mem:* Am Chem Soc; Royal Soc Chem. *Res:* Synthetic organic chemistry; synthesis of natural products and of molecular systems of theoretical interest. *Mailing Add:* Dept Chem D5 Univ Pa Philadelphia PA 19104-6323

ROBERTS, C SHELDON, b Rupert, Vt, Oct 27, 26; m 50; c 3. MATERIALS SCIENCE & ENGINEERING. *Educ:* Rensselaer Polytech Inst, BMetE, 48; Mass Inst Technol, SM, 49, ScD(metall), 52. *Hon Degrees:* DEng, Rensselaer Polytech Inst, 88. *Prof Exp:* Res metallurgist, Dow Chem Co, Mich, 51-56; mem sr staff, Semiconductor Lab, Beckman Instrument Corp, Calif, 56-57; head mat res & develop, Fairchild Semiconductor Corp, 57-61; head mat, Amelco Semiconductor Div, Teledyne, Inc, 61-63; CONSULT MAT & PROCESSES, 63- *Concurrent Pos:* Pres, Timelapse Inc, 71-74; trustee, Rensselaer Polytech Inst, 72- & Am Soc Metals, 84-87. *Honors & Awards:* Alfred Noble Award, 54. *Mem:* Am Soc Metals; Electrochem Soc; Mat Res Soc; Inst Elec & Electronics Engrs; Sigma Xi; Soc Air Safety Investrs; Metall Soc. *Res:* Deformation and fracture of metals; materials processing technology; magnesium and its alloys; semiconductor materials development; material processing technology; behavior of solid state electronic devices; failure analysis of engineering materials. *Mailing Add:* PO Box 4548 Sunriver OR 97707

ROBERTS, CARLYLE JONES, b Philadelphia, Pa, Apr 13, 28; m 50; c 4. HEALTH PHYSICS. *Educ:* Lehigh Univ, BS, 50; Univ Rochester, PhD(biophys), 54. *Prof Exp:* Asst, Univ Rochester, 50-51, res assoc, 53-54; health physicist, Nuclear Power Dept, Curtiss-Wright Corp, 54-57, chief res reactor div, 57-58; res assoc prof appl biol & assoc dir reactor proj, Ga Inst Technol, 58-63, head nuclear res ctr, 63-65, prof nuclear eng, dir sch & chief nuclear sci div, Eng Exp Sta, 65-71, chief biol sci div, 70-71; chief training sect, Int AEC, 71-73; prof nuclear eng, Ga Inst Technol, 73-77; MEM STAFF, ARGONNE NAT LAB, 77-; MEM STAFF, DAMES & MOORE. *Concurrent Pos:* Mem, Am Bd Health Physics, 75- *Mem:* AAAS; Health Physics Soc; Am Nuclear Soc. *Res:* Radiobiology; nuclear engineering. *Mailing Add:* Three Fawn Terr Orchard Park NY 14171

ROBERTS, CARMEL MONTGOMERY, b Atherton, Australia, Apr 5, 28; m 60; c 2. PHARMACOLOGY. *Educ:* Univ Queensland, BS, 48; Univ Southern Calif, MS, 50, PhD, 56. *Prof Exp:* Lectr biochem, Univ Melbourne, 57; asst prof pharmacol, 58-65, ASSOC PROF PHARMACOL, SCH MED, UNIV SOUTHERN CALIF, 65- *Concurrent Pos:* Consult, Los Angeles County Hosp; mem pharmacol-toxicol res prog comt, NIH, 77-81. *Mem:* Am Soc Cell Biol. *Res:* Pharmacology in relationship to mechanism of drug action at the cellular level; metabolism of fetal heart and drug effects on fetal heart metabolism; ionic basis of the electrical potential of cardiac cells; metabolism of the cardiac cell membrane; topography of receptor groups for acetylcholine on cardiac cell. *Mailing Add:* Dept Pharmacol & Nutrit Univ Southern Calif 2025 Zonal Ave Los Angeles CA 90033

ROBERTS, CATHERINE HARRISON, b Wareham, Miss, Apr 21, 58; m; c 3. PROTEIN CHEMISTRY, PHYSICAL BIOCHEMISTRY. *Educ:* Va Tech, BS, 80; Va Commonwealth Univ, PhD(biochem), 84. *Prof Exp:* Postdoctoral, 85-91, AFFIL RES ASSOC BIOCHEM, VA COMMONWEALTH UNIV, 87-, RES ASSOC MICROBIOL, 91- *Concurrent Pos:* Lectr, Va Commonwealth Univ, 88- *Mem:* Am Soc Biochem & Molecular Biol. *Res:* Characterization of the structure-function relationships of proteins; protein production and purification; characterizing of the protein using spectroscopy, calorimetry, immunology, co-factor association and biological function. *Mailing Add:* MCV Sta Box 678 Richmond VA 23298

ROBERTS, CHARLES A, JR, b Changsha, China, Oct 21, 25; US citizen; m; c 5. THEORETICAL PHYSICS. *Educ:* Univ Calif, Los Angeles, BS, 49; Univ Southern Calif, MS, 51; Univ Md, PhD(physics), 56. *Prof Exp:* Physicist, Physics Naval Res Lab, Wash, DC, 52-53; chmn dept, 56-70, PROF PHYSICS, CALIF STATE UNIV, LONG BEACH, 56- *Concurrent Pos:* Res Corp grant theoret physics, 58-60; NSF grant, 62-64; NSF fel, Brussels, 64-65; consult, Naval Ord Lab, Calif, Douglas Aircraft Co & Aeroneutronics; sr Fulbright award, Dept State, 74- *Mem:* Am Phys Soc. *Res:* Nonequilibrium statistical mechanics applied to the field of plasma physics; equilibrium statistical mechanics; interaction of electromagnetic waves with plasmas; elastic surface waves in solids. *Mailing Add:* Dept Physics Calif State Univ 1250 Bellflower Blvd Long Beach CA 90840

ROBERTS, CHARLES BROCKWAY, b Kansas City, Mo, July 31, 18; m 46; c 3. INORGANIC CHEMISTRY, ANALYTICAL CHEMISTRY. *Educ:* Univ Alta, BSc, 40; Univ Ark, MS, 50; Univ Ill, PhD(anal chem), 56. *Prof Exp:* Explosives chemist, Kankakee Ord Works, US Civil Serv, 41-43; chemist, Ethyl Corp, 43-46; teacher & prin pub schs, Ark, 47-53; asst analytical chem, Univ Ill, 53-56; analytical chemist, Dow Chem Co, 56-59, analytical specialist, 59-63, sr res chemist, 63-72, sr res specialist, 72-79, quality control supvr, 79-82; RETIRED. *Mem:* Am Chem Soc; Sigma Xi. *Res:* Analytical methods development. *Mailing Add:* 713 Columbia Rd Midland MI 48640-3430

ROBERTS, CHARLES SHELDON, b Newark, NJ, Sept 25, 37; m 59; c 2. SOFTWARE ENGINEERING, OPERATING SYSTEMS. *Educ:* Carnegie-Mellon Univ, BS, 59; Mass Inst Technol, PhD(physics), 63. *Prof Exp:* Mem tech staff, AT&T Bell Labs, 63-68, head, Comput Technol Dept, 68-70, Info Processing Res Dept, 70-73, Interactive Comput Systs Res Dept, 73-82 & Advan Systs Dept, 82-88; MGR, SYST ARCHIT LAB, HEWLETT-PACKARD CO, CUPERTINO, CALIF, 88- *Mem:* Am Phys Soc; Asn Comput Mach; Inst Elec & Electronics Engrs; Sigma Xi. *Res:* Computer operating systems and software; database management; software and system architecture; theory of molecular scattering; theory of plasmas; physics of The Van Allen belts and the earth's magnetosphere; experiments aboard earth satellites to measure particles and electromagnetic waves. *Mailing Add:* 210 Manresa Ct Los Altos CA 94022-4646

ROBERTS, CLARENCE RICHARD, b Cushing, Okla, May 4, 26; m 51; c 4. HORTICULTURE, PLANT PHYSIOLOGY. *Educ:* Okla State Univ, BS, 49, MS, 51; Tex A&M Univ, PhD(hort), 64. *Prof Exp:* Asst county agt, Okla State Univ, 50-54; exten horticulturist, Kans State Univ, 54-67; EXTEN HORTICULTURIST, UNIV KY, 67- *Concurrent Pos:* Indonesia training teachers, 89. *Mem:* Am Soc Hort Sci. *Res:* Nutrition of vegetable crops; post harvest physiology studies. *Mailing Add:* Dept Hort Univ Ky Lexington KY 40546

ROBERTS, DANIEL ALTMAN, b Micanopy, Fla, Jan 8, 22; m 44; c 3. PLANT VIROLOGY, TEACHING. *Educ:* Univ Fla, BS, 43, MS, 48; Cornell Univ, PhD(plant path), 51. *Prof Exp:* Asst plant path, Cornell Univ, 48-51, from asst prof to assoc prof, 51-59; assoc prof, 59-64, PROF PLANT PATH, UNIV FLA, 64- *Concurrent Pos:* Guggenheim fel, 58; consult ed, Encycl Sci & Technol, McGraw-Hill, 79- *Mem:* Am Phytopath Soc; Int Soc Plant Pathologists. *Res:* Physiology of plant viral infections; diseases of alfalfa. *Mailing Add:* Dept Plant Path Univ Fla Gainesville FL 32611

ROBERTS, DARRELL LYNN, safety & human engineering, energy management; deceased, see previous edition for last biography

ROBERTS, DAVID CRAIG, b Madison, Wis, Feb 21, 48; m 79; c 1. CHEMICAL INFORMATION SCIENCE, CHEMOMETRICS. *Educ:* Univ Wis, BA, 70; Mass Inst Technol, PhD(org chem), 75. *Prof Exp:* Res fel org chem, Univ Calif, Los Angeles, 75-77; asst prof chem, Rutgers Univ, 77-83; assoc prof chem, Fordham Univ, 83-85; SYSTS SCIENTIST & MEM TECH STAFF, MITRE CORP, 85- *Concurrent Pos:* Prin investr, Petrol Res Fund grant, 78-80 & NIH grant, 83-85. *Mem:* Am Chem Soc; NY Acad Sci. *Res:* Organic polymer and industrial chemistry; chemical estimation and modeling methods; information science; structure-activity relationships; computational chemistry. *Mailing Add:* Mitre Corp MS W-759 7525 Colshire Dr McLean VA 22102-3481

ROBERTS, DAVID DUNCAN, b Indiana, Pa, July 1, 54; m 76; c 2. CARBOHYDRATE BIOCHEMISTRY, CELL ADHESION. *Educ:* Mass Inst Tech, BS, 76; Univ Mich, PhD(bio chem), 83. *Prof Exp:* Postdoctoral scholar, Univ Mich, 83-84; staff fel, Nat Inst Diabetes & Digestive & Kidney Dis, NIH, 84-86, sr staff fel, 86-87, res chemist, 87-88, CHIEF BIOCHEM PATH SECT, NAT CANCER INST, NIH, 88- *Mem:* Am Soc Biochem & Molecular Biol; Soc Complex Carbohydrates; AAAS. *Res:* Role of thrombospondin in cell adhesion and migration and tumor metastasis; biochemistry of cell surface carbohydrates and their function in adhesion and host-pathogen interactions. *Mailing Add:* Lab Path Bldg 10 Rm 2A33 NIH Bethesda MD 20892

ROBERTS, DAVID HALL, b Washington, DC, Feb 4, 47; m 74; c 3. RADIO ASTRONOMY, THEORETICAL ASTROPHYSICS. *Educ:* Amherst Col, AB, 69; Stanford Univ, PhD(physics), 73. *Prof Exp:* Res assoc physics, Univ Ill, Urbana, 73-75; res physicist, Univ Calif, San Diego, 75-78; res scientist, Res Lab Electronics, Mass Inst Technol, 78-79; from asst prof to assoc prof, 84-89, PROF ASTROPHYS, BRANDEIS UNIV, 89- *Concurrent Pos:* Vis scientist, Res Lab Electronics, Mass Inst Technol, 79-82, Ctr Space Res, 82-87; Vis Assoc Radio Astron, Calif Inst Technol, 87-88; Consult, Jet Propulsion Lab, 87-88. *Honors & Awards:* Group Achievement Award, NASA, 88. *Mem:* Am Phys Soc; Am Astron Soc; Sigma Xi; Int Astron Union. *Res:* Non-thermal phenomena in galactic and extragalactic objects; radio astronomy; very-long baseline interferometry; pulsars, quasars, radio sources; radio astronomy from space. *Mailing Add:* Dept Physics Brandeis Univ Waltham MA 02254

ROBERTS, DAVID WILFRED ALAN, b Yeadon, Eng, Sept 21, 21; m 60; c 1. PLANT PHYSIOLOGY, BIOCHEMISTRY. *Educ:* Univ Toronto, BA, 42, PhD(plant physiol), 48. *Prof Exp:* Res assoc, Nat Cancer Inst Can, Toronto, 48-49; PLANT PHYSIOLOGIST, RES STA, CAN DEPT AGR, 49- *Mem:* AAAS; Am Soc Plant Physiol; Soc Cryobiol; Can Soc Plant Physiol. *Res:* Genetics of cold resistance of wheat and traits that correlate with cold hardiness; cold resistance. *Mailing Add:* 2129 18th Ave S Lethbridge AB P1K 1C7 Can

ROBERTS, DEAN WINN, JR, b Jan 9, 45; c 3. IMMUNOLOGY, TOXICOLOGY. *Educ:* Hahnemann Med Col, PhD(microbiol, immunol), 77. *Prof Exp:* RES SCIENTIST, UNIV ARK MED SCI, 79- *Concurrent Pos:* Rosenstadt vis prof, fac pharm, Univ Toronto, 88. *Honors & Awards:* FDA Comnr Spec Citation, 86. *Mem:* Am Asn Immunologists; Soc Toxicol. *Res:* Immunotoxicology; immunochemical detection of carcinogen-DNA adducts and drug-protein adducts. *Mailing Add:* Nat Ctr Toxicol Resources Jefferson AR 72079

ROBERTS, DEWAYNE, b Okla, Sept 7, 27; m 51; c 3. BIOCHEMICAL PHARMACOLOGY. *Educ:* Okla State Univ, BS; Wash Univ, PhD. *Prof Exp:* Res scientist, Roswell Park Mem Inst, 57-62; asst chief biochem pharmacol, Children's Cancer Res Found, Boston, 62-68; mem, Dept Preclin & Clin Pharmacol, St Jude Children's Res Hosp, 68-88; RETIRED. *Concurrent Pos:* Res assoc, Grad Dept Biochem, Brandeis Univ, 62-65; res assoc, Harvard Med Sch, 62-68; adj prof biol, Northeastern Univ, 66-67; assoc prof pharmacol, Univ Tenn, Memphis, 68- *Mem:* Am Asn Cancer Res; Am Soc Pharmacol & Exp Therapeut; Am Fedn Clin Res; Sigma Xi. *Res:* Biochemistry of nucleic acid precursors and the effect of drugs on their biosyntheses in relation to cancer chemotherapy; molecular basis of acquired resistance to oncolytic drugs. *Mailing Add:* 6193 Ivanhoe Rd Bartlett TN 38134

ROBERTS, DONALD DUANE, b Jamestown, NDak, Feb 18, 29; m 52; c 5. PHYSICAL ORGANIC CHEMISTRY. *Educ:* Jamestown Col, BS, 50; Loyola Univ, MS, 57, PhD(org chem), 62. *Prof Exp:* Res chemist, Sherwin Williams Co, 52-54; group leader polymer chem, Borg-Warner Corp, 54-58; asst proj chemist, Am Oil Co, 58-60; res assoc org chem, Plastics Div, Allied Chem Corp, 62-63; assoc prof chem, 63-74, PROF CHEM, LA TECH UNIV, 74- *Concurrent Pos:* Res Corp grant, 64-65; Petrol Res Fund grant, 66-68; NSF grant, 69-71. *Mem:* Am Chem Soc; Sigma Xi. *Res:* Cyclopropylcarbinyl system; solvent effects; medium size rings; linear free energy relationships; solvent effects in solvolysis reactions. *Mailing Add:* Dept Chem La Tech Univ Ruston LA 71272

ROBERTS, DONALD RAY, b Purcell, Okla, Mar 26, 42; m 67; c 2. MEDICAL ENTOMOLOGY, POPULATION DYNAMICS. *Educ:* Univ Mo-Columbia, BA, 65, MS, 66; Univ Tex, PhD(entom), 73. *Prof Exp:* Staff entomologist, USARSUPTHAI Command, Thailand, 68-69; med entomologist, US Army Environ Hyg Agency, Md, 69-70; med entomologist, US Army Med Res Unit, Belem, Brazil, 73-75, chief med res, 75-78, entomologist, US Army Med Res Unit, Brasilia, Brazil, 78-80; CHIEF, DEPT ENTOM, WALTER REED ARMY INST, 80- *Concurrent Pos:* Consult med entom, Pan Am Health Orgn, 73-78; prof med entom, Univ Brasilia, 78- *Mem:* Am Soc Trop Med & Hyg; Entom Soc Am; Am Mosquito Control Asn; Soc Brasileira Medicina Tropical; Soc Brasiliera Entomologia. *Mailing Add:* 12833 SW 146th Terr Miami FL 33186

ROBERTS, DONALD RAY, b Trenton, Tenn, Dec 21, 29; m 57; c 3. PLANT PHYSIOLOGY, WEED SCIENCE. *Educ:* Univ Tenn, BSA, 52, MS, 59; Auburn Univ, PhD(bot), 66. *Prof Exp:* Instr agron, Univ Tenn, 57-59; asst bot, Auburn Univ, 60-63; plant physiologist, Southeastern Forest Exp Sta, US Forest Serv, 63-; RETIRED. *Mem:* Am Soc Plant Physiologists; Soc Am Foresters. *Res:* Lightwood induction research; herbicide physiology, especially metabolism of atrazine; biosynthesis of oleoresin in slash pine and physiology of pine oleoresin extraction. *Mailing Add:* Rte 2 Box 317 Newberry FL 32669

ROBERTS, DONALD WILSON, b Phoenix, Ariz, Jan 20, 33; m 59; c 2. INSECT PATHOLOGY. *Educ:* Brigham Young Univ, BS, 57; Iowa State Univ, MS, 59; Univ Calif, Berkeley, PhD(entom), 64. *Prof Exp:* NSF fel, Swiss Fed Inst Technol, 64-65; asst entomologist, 65-69, assoc insect pathologist, 70-73, INSECT PATHOLOGIST, BOYCE THOMPSON INST PLANT RES, 74-, COORDR, INSECT PATH RESOURCE CTR, 79- *Concurrent Pos:* USPHS res grants, 66-76; consult, WHO, 74-78, res grants, 75-84; USDA res grants, 76-92; US-India exchange scientist, 78; consult, Brazil gov, 78-82, US Nat Acad Sci-Nat Res Coun, 76-77, US Agency Int Develop grants, 81-92; Fulbright sr res scholar, Australia, 85. *Honors & Awards:* L O Howard Distinguished Achievement Award, Etom Soc Am, 89. *Mem:* AAAS; Entom Soc Am; Am Soc Microbiol; Mycol Soc Am; Soc Invert Path (vpres, 86-88, pres, 88-90); Soc Entom Brazil. *Res:* Insect mycoses including molecular biology of mechanisms used by fungi to overcome insects; toxins produced by insect-infecting fungi; pathogens of mosquito larvae; pox-like viruses of insects; integration of pathogens into pest management programs in developing nations. *Mailing Add:* Boyce Thompson Inst Plant Res Cornell Univ Tower Rd Ithaca NY 14853

ROBERTS, DORIS EMMA, b Toledo, Ohio, Dec 28, 15. PUBLIC HEALTH NURSING. *Educ:* Peter Bent Brigham Sch Nursing, dipl nursing, 38; Geneva Col, BS, 44; Univ Minn, MPH, 58; Univ NC, PhD, 67. *Prof Exp:* Staff nurse, Vis Nurse Asn, New Haven, 38-40; sr nurse, Neighborhood House, Millburn, NJ, 42-45; supvr tuberc, Baltimore County Dept Health, Towson, Md, 45-46; tuberc consult, Md State Dept Health, 46-50; consult & chief nurse tuberc prog, USPHS, 50-57, consult div nursing, 58-63; chief nursing pract br, Health Resources Admin, HEW, 66-75, consult nursing, 75-83; RETIRED. *Concurrent Pos:* Comt officer, USPHS, 45-75; adj prof pub health nursing, Univ NC, Chapel Hill, 75-89; consult, WHO, 61-83; consult prof health prof training prog, US Vet Admin, 83-85. *Honors & Awards:* Sedgwick Mem Award, Am Pub Health Asn, 79; Meritorious Serv Award, NIH & USPHS, 71. *Mem:* Inst Med-Nat Acad Sci; fel Am Pub Health Asn. *Mailing Add:* 6111 Kennedy Dr Chevy Chase MD 20815

ROBERTS, DURWARD THOMAS, JR, b Chattanooga, Tenn, Jan 1, 42; m 62; c 3. ORGANIC POLYMER CHEMISTRY. *Educ:* Univ NC, Chapel Hill, BA, 64, PhD(org chem), 68. *Prof Exp:* Chemist, Esso Res Labs, Humble Oil & Refining Co, La, 67-70; sr res scientist, Cent Res Labs, Firestone Tire & Rubber Co, 70-71, group leader, 71-76, mgr, 76-90, MGR RES, BRIDGESTONE-FIRESTONE INC, 90- *Mem:* Am Chem Soc. *Res:* Catalytic polymerization of monomers; polymer chemistry polyurethane polymer synthesis; urethane coatings; tire research; catalytic reactions of petroleum hydrocarbons; rubber processing. *Mailing Add:* Res Labs Bridgestone-Firestone Inc Akron OH 44317

ROBERTS, EARL C(HAMPION), physical metallurgy; deceased, see previous edition for last biography

ROBERTS, EARL JOHN, b Magee, Miss, May 14, 13; m 44; c 2. CHEMISTRY. *Educ:* Miss Col, BA, 39; La State Univ, MS, 42. *Prof Exp:* High sch teacher, Miss, 39-40; analyst, Miss Testing Lab, 40-41; asst chem, La State Univ, 41-42; jr chemist, Southern Utilization Res Br, 42-44, from asst chemist to chemist, 44-62, res chemist, Southern Res & Develop Div, 62-72, SR CHEMIST, CANE SUGAR REFINING RES PROJ, SCI & EDUC ADMIN-AGR RES, USDA, 72- *Concurrent Pos:* Teaching fel, Tulane Univ, 48- *Res:* Organic chemistry of the by-products of the sugar industry; composition of sugar cane juice; structure of modified cellulose; identification and determination of minor constituents in refined cane sugar. *Mailing Add:* 6748 Orleans Ave New Orleans LA 70124

ROBERTS, EDGAR D, b Odessa, Tex, Mar 23, 31; m 50; c 2. VETERINARY PATHOLOGY. *Educ:* Colo State Univ, BS, 57, DVM, 59; Iowa State Univ, MS, 62, PhD(vet path), 65. *Prof Exp:* Res asst vet path, Vet Med Res Inst, 59-62; pathologist, Nat Animal Dis Lab, 62-63; asst prof vet path, Iowa State Univ, 63-65; vet pathologist, Rockefeller Found, 65-69; head animal health, Int Ctr Trop Agr, Colombia, 69-71; PROF VET PATH & HEAD DEPT, SCH VET MED, LA STATE UNIV, BATON ROUGE, 71- *Mem:* Am Vet Med Asn; Am Col Vet Path; Int Acad Path. *Res:* Pathogenesis of nutritional and infectious diseases of the swine and equine species; naturally occurring diseases of the tropics. *Mailing Add:* Dept Path-2301 Vet Med Bldg La State Univ Baton Rouge LA 70803

ROBERTS, EDWIN KIRK, b Marlton, NJ, Nov 15, 22; m 49; c 2. PHYSICAL CHEMISTRY. *Educ:* Earlham Col, BA, 49; Harvard Univ, PhD(chem physics), 53. *Prof Exp:* From instr to prof chem, Middlebury Col, 52-81; DIR LABS & LECTR CHEM, STANFORD UNIV, 82- *Concurrent Pos:* NSF fac fel, Stanford Univ, 69-70. *Mem:* Am Chem Soc. *Res:* Free radical kinetics; chemistry of coordination complexes. *Mailing Add:* Dept Chem Stanford Univ Stanford CA 94305

ROBERTS, ELLIOTT JOHN, physical inorganic chemistry, for more information see previous edition

ROBERTS, EUGENE, b Krasnodar, Russia, Jan 19, 20; US citizen; m 77; c 3. BIOCHEMISTRY. *Educ:* Wayne State Univ, BS, 40; Univ Mich, MS, 41, PhD(biochem), 43. *Hon Degrees:* Laurea Dr, Univ Florence. *Prof Exp:* Asst head inhalation sect uranium compounds res, Univ Rochester, 43-46; res assoc, Barnard Free Skin & Cancer Hosp, St Louis & Med Sch, Wash Univ, 46-54; chmn, Dept Biochem, 54-68, chmn, Div Neurosci, 68-83, DIR, DEPT NEUROBIOCHEM, BECKMAN RES INST CITY HOPE, DUARTE, CALIF, 83- *Concurrent Pos:* Res prof biochem, Sch Med, Univ Southern Calif, 70-, res prof neurol, 85-; fel, Ctr Advan Study Behav Sci, Stanford, Calif, 78; mem USA nat comt, Int Brain Res Orgn, Nat Res Coun, Nat Acad Sci, Wash, DC, 76-79; mem bd sci adv, La Jolla Cancer Res Found, Calif, 79-; mem bd trustees, Calif Found Biochem Res, La Jolla Calif, 79-, chmn, 81-82; mem bd sci counselors, Nat Inst Neurol & Comomun Dis & Stroke, 80-84; mem sci adv bd, Stallone Fund Autism Res, Culver City, Calif, & John Douglas French Found Alzheimer's Dis, Los Angeles, Calif, 83-86; Louis & Bert Freedman Found award res in biochem, NY Acad Sci, 86; adj sr scientist, Hal B Wallis Res Facil, Eisenhower Med Ctr, Rancho Mirage, Calif, 89- *Honors & Awards:* Distinguished Serv Award, Wayne State Univ, 66; Distinguished Scientist Award, Beckman Res Inst City Hope, 83. *Mem:* Nat Acad Sci; Am Asn Cancer Res; Am Chem Soc; Am Inst Chemists; Am Soc Biol Chemists; Am Soc Neurochem; Am Soc Pharmacol & Exp Therapeut; fel NY Acad Sci; Int Soc Neurochem; Sigma Xi. *Res:* Neurochemistry; biochemistry of cancer; comparative biochemistry; general metabolism. *Mailing Add:* Dept Neurobiochem Beckman Res Inst City Hope Duarte CA 91010

ROBERTS, FLOYD EDWARD, JR, b Philadelphia, Pa, Dec 2, 34; m 56; c 3. ORGANIC CHEMISTRY. *Educ:* Franklin Col, AB, 56; Pa State Univ, MS, 58; Purdue Univ, PhD(org chem), 63. *Prof Exp:* Chemist, Merck & Co, Inc, 62-64, sr chemist, 64-69, res fel, 69-79, SR RES FEL, MERCK SHARPE & DOHME RES LABS, 79- *Mem:* Am Chem Soc. *Res:* Steroids and other medicinals; antibiotics; natural product isolation. *Mailing Add:* 220 Valley Rd Princeton NJ 08540-3473

ROBERTS, FRANCIS DONALD, b Utica, NY, Jan 18, 38; m 64; c 4. ORGANIC CHEMISTRY. *Educ:* Syracuse Univ, BA, 59; Cornell Univ, PhD(org chem), 64. *Prof Exp:* Res chemist, New York, 64-66, sect head, 66-70, tech coordr toilet articles, Pharmaceut Div, 70-72, asst mgr oral res, 72-73, dir purchasing raw mat, 73-76, dir res & develop, 76-81, vpres res & develop, 81-86, VPRES TECHNOL & PLANNING, KENDALL CO DIV, COLGATE-PALMOLIVE CO, 86- *Mem:* AAAS; Am Chem Soc. *Res:* Synthesis of new organo-sulfur compounds; condensed aromatics; oral research, especially treatment of caries and periodontal disease; nonwoven fabrics, acrylic and rubber based pressure sensitive adhesives, medical devices, surgical dressings, urological products; anti-embolism compression therapy; IV therapy. *Mailing Add:* Two Bridle Path Circle Dover MA 02030-2400

ROBERTS, FRANKLIN LEWIS, b Waterboro, Maine, Feb 21, 34; wid; c 5. CYTOLOGY, GENETICS. *Educ:* Univ Maine, BS, 57; NC State Univ, PhD(genetics), 64. *Prof Exp:* Instr sci, Gorham State Teachers Col, 58-59; instr zool, NC State Univ, 59-64; from asst prof to assoc prof, 64-74, Coe Fund res grant, 65-66, chmn dept, 75-81, PROF ZOOL, UNIV MAINE, ORONO, 74- *Concurrent Pos:* NSF res grant, 66-69 & City Hope Med Ctr, Duarte, Calif, 68, Tilapice aquacult, Haiti, 80- *Mem:* Am Soc Nat; AAAS; Am Soc Ichthyol & Herpet; Genetics Soc Am. *Res:* Cytotaxonomy and cytogenetics of cold blooded vertebrates, particularly fishes; cell culture of cold blooded vertebrate tissues; genetics of parthenogenesis. *Mailing Add:* Dept Zool Murray Hall Univ Maine Orono ME 04469

ROBERTS, FRED STEPHEN, b New York, NY, June 19, 43; m 72; c 2. DISCRETE MATHEMATICS, MATHEMATICAL MODELING. *Educ:* Dartmouth Col, AB, 64; Stanford Univ, MS, 67, PhD(math), 68. *Prof Exp:* NIH traineeship math psychol, Univ Pa, 68; mathematician, Rand Corp, 68-72; assoc prof, 72-76, PROF MATH, RUTGERS UNIV, 76- *Concurrent Pos:* Daniel Webster Nat Scholar, 61-64; Woodrow Wilson fel, 64-65; mem, Inst Advan Study, Princeton Univ, 71-72; prin investr, NSF awards, 72-79, 83-85; vis prof oper res, Cornell Univ, 79-80; sci res awards, Air Force, 79-83, 85-; Alexander von Humboldt fel, 84; vis, AT&T Bell Labs, 86-88; consult, Rand Corp, Orgn Econ Coop & Develop, Contruct Eng Res Lab, Inst Gas

Technol; Robert G Stone vis prof, Northeastern Univ, 90- *Mem:* Am Math Soc; Math Asn Am; Soc Indust & Appl Math; Oper Res Soc Am; Soc Math Psychol; Consortium Math & Applns. *Res:* Mathematical models in the social, biological and environmental sciences; combinatorial mathematics and graph theory; theory of measurement; operations research. *Mailing Add:* Dept Math Rutgers Univ New Brunswick NJ 08903

ROBERTS, FREDDY LEE, b Dermott, Ark, Dec 29, 41; m 64; c 2. CIVIL ENGINEERING. *Educ:* Univ Ark, Fayetteville, BSCE, 64, MSCE, 66; Univ Tex, Austin, PhD(civil eng), 70. *Prof Exp:* Soils engr, Grubbs Consult Engrs, Inc, Ark, 65; res engr, Univ Ark, Fayetteville, 64-66; consult piles, Hudson, Matlock, Dawkins & Panak Res Engrs, 66-67; res engr & teaching asst hwy roughness, Univ Tex, Austin, 66-69; from asst prof to assoc prof civil eng, Clemson Univ, 69-75; vis assoc prof, Univ Tex, Austin, 75-78; chief engr, Austin Res Engrs, Inc, 78-81; RES ENGR, UNIV TEX, AUSTIN, 81-; MEM STAFF, DEPT CIVIL ENG, TEX A&M UNIV. *Concurrent Pos:* Mem, Hwy Res Bd, Nat Acad Sci-Nat Res Coun; mat & design consult, Rawhut Eng, Austin, 81- *Mem:* Am Soc Civil Engrs; Am Soc Testing & Mat. *Res:* Highway roughness, its measure and effect on highway user and vehicle; pavement material evaluation and use in design; highway geometric design; development of pavement evaluation and management systems; design and evaluation design procedure development; recycled mixture design; evaluation of stripping test methods. *Mailing Add:* Dept Civil Eng Auburn Univ Auburn AL 36849

ROBERTS, GEORGE A(DAM), b Uniontown, Pa, Feb 18, 19; m 71; c 3. METALLURGY. *Educ:* Carnegie Inst Technol, BSc, 39, MSc, 41, DSc, 42. *Prof Exp:* Res metallurgist, Vanadium-Alloys Steel Co, 41-45, chief metallurgist, 45-53, vpres, 53-61, pres, Vasco Metals Corp, Pa, 61-66; pres, 66-90, vchmn, 90-91, CHMN, TELEDYNE, INC, 91- *Mem:* Nat Acad Eng; fel Am Soc Metals (vpres, 53, pres, 54); Soc Mfg Eng; Am Chem Soc; fel Am Inst Mining, Metall & Petrol Engrs. *Res:* Metallurgy of tool steels; heat treatment of hardenable steels; powder metallurgy of ferrous alloys; mechanical properties of steels. *Mailing Add:* Teledyne Inc 1901 Ave of the Stars Los Angeles CA 90067

ROBERTS, GEORGE E(DWARD), b Portsmouth, Eng, Sept 25, 26; m 50; c 1. SYSTEMS ANALYSIS. *Educ:* Portsmouth Naval Col Eng, BSc, 45. *Prof Exp:* Apprentice elec eng, Portsmouth Dockyard, Brit Admiralty, 42-47, design draftsman, 47-51; design draftsman, Can Vickers Ltd, 51-52; elec engr, Tech Prods Dept, RCA Victor Co, Ltd, 52-56 & Defense Dept, 56-58, sr mem sci staff, Res Labs, 58-66; mem tech staff, Gen Res Corp, 66-74; systs analyst, Litton Mellonics, 74-76; mem prof staff, Geodynamics Corp. 76-80, mgr, 80-; RETIRED. *Res:* Probability and statistics; antennas; mathematical analysis. *Mailing Add:* 494 Gibralter Dr Anacortes WA 98221

ROBERTS, GEORGE P, b Barton, Vt, Dec 25, 37; m 59; c 4. POLYMER CHEMISTRY. *Educ:* Univ Vt, BS, 59; Northwestern Univ, PhD(phys chem), 64. *Prof Exp:* Res scientist, Corp Res Ctr, NJ, 64-69, group leader acrylonitrile-butadiene-styrene res, Chem Div, Conn, 69-74, mgr polymer appl res, Corp Res & Develop, 74-84, dir new technol, Uniroyal, Inc, 84-86; MEM STAFF, UNIROYAL CHEM CO, 86- *Mem:* Am Chem Soc; Soc Plastics Engrs. *Res:* Polymer characterization, relationship of properties of polymers to morphology and molecular structure; process development and polymer applications. *Mailing Add:* World Hq Uniroyal Chem Co Benson Rd Middlebury CT 06749

ROBERTS, GEORGE W(ILLARD), b Bridgeport, Conn, Aug 9, 38; m 63; c 2. CHEMICAL REACTION ENGINEERING, HETEROGENEOUS CATALYSIS. *Educ:* Cornell Univ, BChE, 61; Mass Inst Technol, ScD, 65. *Prof Exp:* Instr heat transfer, Mass Inst Technol, 63; chem engr, Rohm & Haas Co, 65-67, projs supvr, 68-69; assoc prof chem eng, Wash Univ, 69-71; mgr chem & chem eng, Engelhard Industs Div, Engelhard Minerals & Chem Corp, 72-77; dir, Corp Res & Development Dept, Air Prod & Chem, Inc, 77-81, gen mgr, Commercial Develop Div, Process Systs Group, 81-89; PROF & HEAD DEPT CHEM ENG, NC STATE UNIV, 89- *Mem:* Am Chem Soc; Am Inst Chem Engrs. *Res:* Chemical reaction engineering; heterogeneous catalysis; separations science; research management. *Mailing Add:* 1610 St Mary's St Raleigh NC 27608-2219

ROBERTS, GLENN DALE, b Gilmer, Tex, Apr 9, 43; m 73; c 3. CLINICAL MICROBIOLOGY, CLINICAL MYCOLOGY. *Educ:* NTex State Univ, BS, 67; Univ Okla, MS, 69, PhD(med mycol), 71; Am Bd Med Microbiol, cert, 79. *Prof Exp:* Fel clin mycol, Col Med, Univ Ky, 71-72; instr microbiol & lab med, 72-77, from asst prof to assoc prof, 77-86, PROF MICROBIOL & LAB MED, MAYO MED SCH, 87- *Concurrent Pos:* Consult, Mayo Clin & Mayo Found, 72- *Mem:* Med Mycol Soc of the Americas; Int Soc Human & Animal Mycol; Am Soc Clin Path; Am Soc Microbiol; Am Thoracic Soc; Am Acad Microbiol. *Res:* Diagnostic and clinical mycobacteriology and mycology; mycotic serology. *Mailing Add:* Sect of Clin Microbiol Mayo Clin Rochester MN 55905

ROBERTS, HAROLD R, b Four Oaks, NC, Jan 4, 30; m 58; c 2. INTERNAL MEDICINE, HEMATOLOGY. *Educ:* Univ NC, BS, 52, MD, 55. *Prof Exp:* Instr path, 60-62, assoc prof path & med, 62-70, chief div hemat, dept med, 68-77, PROF PATH & MED, SCH MED, UNIV NC, CHAPEL HILL, 70- *Res:* Blood coagulation; hemorrhage; thrombosis. *Mailing Add:* Dept Med & Path Hematol CB No 7035 Univ NC 416 Burnett Womack Chapel Hill NC 27599-7035

ROBERTS, HARRY HEIL, b Huntington, WVa, Feb 2, 40; m 63; c 1. GEOLOGY. *Educ:* Marshall Univ, BS, 62; La State Univ, Baton Rouge, MS, 66, PhD(geol), 69. *Prof Exp:* Asst prof geol res, 69-74, assoc prof marine sci, 74-78, PROF MARINE SCI, COASTAL STUDIES INST, LA STATE UNIV, BATON ROUGE, 78- *Mem:* Soc Econ Paleontologists & Mineralogists; Am Geophys Union; Coastal Soc; Geol Soc Am; Int Asn Sedimentol. *Res:* Sedimentology associated with reef and deltaic environments; dynamics and ecology of reefs. *Mailing Add:* Dept Geol La State Univ 103 Coastal Studies Bldg Baton Rouge LA 70803-4101

ROBERTS, HARRY VIVIAN, b Peoria, Ill, May 1, 23; m 43; c 2. STATISTICS. *Educ:* Univ Chicago, BA, 43, MBA, 47, PhD(bus), 55. *Prof Exp:* Mkt res analyst, McCann-Erickson, Inc, 46-49; from instr to assoc prof, 49-59, PROF STATIST, UNIV CHICAGO, 59- *Concurrent Pos:* Assoc ed, Am Statistician News, 53-73 & J Am Statist Asn, 76-; Ford Found fel, 59-60. *Mem:* AAAS; Am Econ Asn; fel Am Statist Asn; Inst Math Statist; Royal Statist Soc. *Res:* Statistical theory and applications, especially interactive data analysis; statistical decision theory and applications, especially to medical diagnosis and treatment. *Mailing Add:* Grad Sch Bus Univ Chicago 1101 E 58th St Chicago IL 60637

ROBERTS, HOWARD C(REIGHTON), b Wayne Co, Ill, Nov 1, 10; m 37; c 2. ENGINEERING. *Educ:* Univ Ill, AB, 33, EE, 44. *Prof Exp:* Asst physics, Univ Ill, 30-32; asst, Div Physics, State Geol Surv, Ill, 32-37; asst comput, Gen Geophys Co, Tex, 38-39; asst, res staff, Asn Am Railroads, Ill, 39-46; assoc prof civil eng, Univ Ill, Urbana, 46-53, assoc prof, Eng Exp Sta, 53-54, prof eng, 54-68, prof elec eng, 68-69; RETIRED. *Concurrent Pos:* Spec res assoc, Off Sci Res & Develop, 42-44; independent consult physicist, 68-; vis prof archit & acoust, Univ Ill, Urbana, 72-77; vis lectr, Univ Colo, Denver, 78-84. *Mem:* Fel AAAS; fel Instrument Soc Am; Soc Noise Control Eng. *Res:* Industrial measurements; automatic weighing and proportioning; noise and vibration control; architectural acoustics; industrial accident investigation; environmental engineering; industrial instrumentation. *Mailing Add:* 7199 S Vine Circle W Littleton CO 80122

ROBERTS, HOWARD RADCLYFFE, entomology; deceased, see previous edition for last biography

ROBERTS, IRVING, b Brooklyn, NY, Jan 9, 15; m 38, 66; c 2. GENETIC ENGINEERING. *Educ:* City Col New York, BS, 34; Columbia Univ, MS, 35, PhD(phys chem), 37. *Prof Exp:* Asst chem, Columbia Univ, 35-39; chem engr, Weiss & Downs, Inc, NY, 39-43; group leader, Manhattan Proj, 43-45; div engr, Process Div, Res & Develop Dept, Elliott Co, 45-50; consult, 50-56; dir planning, Reynolds Metals Co, 56-61, vpres, 61-78; consult engr, 78-82; RES PROF, DEPT MICROBIOL & IMMUNOL, MED COL VA, 82- *Mem:* Am Chem Soc; Am Inst Chem Engrs; Am Soc Mech Engrs. *Res:* Long-range planning and economic evaluation; the aluminum industry; oxygen and other low temperature plants; new chemical processes; development of DNA shuttle plasmids for anaerobic bacteria. *Mailing Add:* Three Westwick Rd Richmond VA 23233

ROBERTS, J(ASPER) KENT, b Ryan, Okla, Jan 15, 22; m 43; c 2. ENGINEERING. *Educ:* Univ Okla, BSCE, 47; Univ Mo, MSCE, 50. *Prof Exp:* From instr to assoc prof, 47-57, asst dean eng, 70-80, PROF CIVIL ENG, UNIV MO-ROLLA, 57- *Mem:* Am Soc Civil Engrs; Nat Soc Prof Engrs; Am Soc Eng Educ. *Mailing Add:* 906 Murry Lane Rolla MO 65401

ROBERTS, J T ADRIAN, b Northwich, Cheshire, Eng, May 21, 44; US citizen; m 65; c 2. MECHANICAL PROPERTIES, CORROSION. *Educ:* Manchester Univ, UK, BSc, 65, MSc, 66, PhD(metall), 68. *Prof Exp:* Group leader ceramic properties, Argonne Nat Lab, 68-74; sr prog mgr fuels & mat, nuclear power div, Elec Power & Res Inst, 74-85; mgr mat sci & technol, 85-87, Dep Dir Res, 87-89, DIR LAB PROGS, BATTELLE PAC NORTHWEST LABS, 89- *Concurrent Pos:* Vis assoc prof mat sci, Cornell Univ, 79; vis lectr mat sci, Univ Calif, Berkeley, 80-85 & Stanford Univ, 81-85; trustee, Am Ceramic Soc, 82-85; exec comt mem, mat sci & technol div, Am Nuclear Soc, 83-90, chmn, 89-90. *Mem:* Fel Am Ceramic Soc (vpres, 86-87); fel Am Nuclear Soc; Am Inst Metall Eng. *Res:* Materials science in support of energy production; ceramic and metal structural behavior for use in nuclear reactors (LWR, LMFBR) and advanced systems (fusion, MHD and batteries); environmental restoration and waste management. *Mailing Add:* Lab Progs Battelle Pac Northwest Labs Richland WA 99352

ROBERTS, JAMES ALLEN, b Beach, NDak, May 31, 34; m 86. UROLOGY. *Educ:* Univ Chicago, BS, 55, MD, 59. *Prof Exp:* Asst prof, 67-71, assoc prof, urol, Tulane Univ Sch Med, 71-75; res assoc, 67-72, res scientist, 72-78, SR RES SCIENTIST & HEAD, UROL, DELTA REGIONAL PRIMATE CTR, 78-; PROF & ASSOC CHMN, RES, TULANE UNIV SCH MED, 87- *Concurrent Pos:* Fogarty sr int fel, 84. *Honors & Awards:* Original Res Award, Southern Med Asn, 70. *Mem:* Sigma Xi; Am Urol Asn; Soc Univ Urologists; Urodynamics Soc. *Res:* Ureteral and renal physiology and studies on etiology of vesicoureteral reflux; urinary tract infection and the pathophysiology of pyelonephritis, etiology of renal damage from bacterial infection; clinical studies of urinary tract infection. *Mailing Add:* Dept Urol Delta Regional Primate Ctr 18703 Three Rivers Rd Covington LA 70433

ROBERTS, JAMES C, JR, b New York, NY, Feb 13, 26; m 49; c 4. PATHOLOGY. *Educ:* Wesleyan Univ, AB, 45; Univ Rochester, MD, 49. *Prof Exp:* Asst path, Wash Univ, 49-51; instr, Univ Pittsburgh, 53-55, asst prof, 55-58; pathologist & dir med educ, ETenn Baptist Hosp, 58-60; PATHOLOGIST & DIR RES, LITTLE CO MARY HOSP, 60-, DIR, CETACEAN RES LAB, 62- *Concurrent Pos:* Hartford fel, Arteriosclerosis Res, 55-58; prin investr, Res Grants, 57-; consult, Los Angeles County Harbor Hosp, 62-; co-chmn conf comp arteriosclerosis, Nat Heart Inst, Calif, 64-65; Nuffield traveling fel, Zool Soc London, 65. *Mem:* Am Heart Asn; Am Soc Mammal; AMA; Am Asn Path & Bact; fel Col Am Path. *Res:* Immunopathology; arteriosclerosis; comparative arteriosclerosis and morphology. *Mailing Add:* 15 Portuguese Bend Rd Rolling Hills CA 90274

ROBERTS, JAMES ERNEST, SR, b Newport, Ark, May 17, 24; m 46; c 3. ECONOMIC ENTOMOLOGY. *Educ:* Univ Ark, BSA, 54, MS, 55; Kans State Univ, PhD(entom), 63. *Prof Exp:* High sch teacher, Ark, 47-52; jr entomologist, Univ Ark, 55-56; asst entomologist, Ga Exp Sta, Univ Ga, 56-65; exten entomologist, Agr Exten Serv, Univ Ark, Fayetteville, 65-69; exten entomologist, Col Agr & Life Sci, Va Polytech Inst & State Univ, 69-88; RETIRED. *Concurrent Pos:* Consult, Nat Park Serv, 75. *Mem:* Entom Soc Am; Agr & Life Sci Fac Asn. *Res:* Field testing of insecticides for the control of insects affecting crops and livestock in Virginia. *Mailing Add:* 1007 Kentwood Dr Blacksburg VA 24060

ROBERTS, JAMES HERBERT, b Tucson, Ariz, Oct 27, 15; m 42; c 4. EXPERIMENTAL NUCLEAR PHYSICS. *Educ:* Univ Ariz, BS, 37, MS, 38; Univ Chicago, PhD(physics), 46. *Prof Exp:* Lab asst physics, Univ Chicago, 38-39, asst, Metall Lab, 42-43, res assoc, 43-44; demonstr, Mus Sci & Indust, 39-42; res scientist, Los Alamos Sci Lab, Univ Calif, 44-46, actg group leader, 46-48; from asst prof to prof physics, Northwestern Univ, 48-69; prof physics, Macalester Col, 69-79, chmn, Dept Physics & Astron, 75-76; consult, Westinghouse-Hanford Co, 76-82, fel scientist, 83-85; pres, Metrol Control Corp, 86-88, chief sci adv, 88-90; RETIRED. *Concurrent Pos:* Vis prof physics, Pahlavi Univ, Iran, 72. *Mem:* Fel Am Phys Soc; Am Asn Physics Teachers; Am Nuclear Soc. *Res:* Neutron physics involving fissionable material; neutron yields of light elements bombarded with alpha particles; neutron energy measurements with nuclear research emulsions; hyper fragments; detection of fission fragments and other charged particles with solid state track recorders; thermal annealing of mica and quartz solid state track recorders; absolute neutron spectra-fluence measurements in and out of nuclear reactors. *Mailing Add:* 2002 Howell Ave Richland WA 99352

ROBERTS, JAMES LEWIS, b Lima, Peru, Oct 23, 51; US citizen; m. BIOCHEMISTRY, MOLECULAR BIOLOGY. *Educ:* Colo State Univ, BS, 73; Univ Ore, PhD(chem), 78. *Prof Exp:* Res technician biochem, Colo State Univ, 70-73; teaching asst, Univ Ore, 73-74; fel biochem, Univ Calif, 78-79; MEM STAFF, DEPT BIOCHEM, COL PHYSICIANS & SURGEONS, COLUMBIA UNIV, 79- *Honors & Awards:* Golden Lamport Award, Excellence Basic Sci. *Mem:* AAAS; Am Chem Soc; Soc Neurosci; Int Neuroendocrine Soc. *Res:* Biosynthesis and regulation of the adrenocorticotropin-endorphin precursor; recombinant DNA cloning of pituitary and brain adrenocorticotropin-endorphin; glucocorticoid and thyroid hormone regulation of gene expression; gene structure. *Mailing Add:* Mt Sinai Med Ctr Neurobiology Box 1065 100th St & Fifth Ave New York NY 10029

ROBERTS, JAMES M, b Taylor, Mich, Mar 22, 41. OBSTETRICS & GYNECOLOGY. *Educ:* Univ Mich, MD, 66; Am Bd Obstet & Gynec, cert fetal & maternal med. *Prof Exp:* Resident obstet & gynec, Univ Mich Hosp, 67-71; postdoctoral res fel, Cardiovasc Res Inst, Univ Calif, San Francisco, 73-75, from asst prof to assoc prof, Dept Obstet, Gynec & Reproductive Sci, 75-84, assoc staff mem, 80-84, SR STAFF MEM, CARDIOVASC RES INST & PROF OBSTET, GYNEC & REPRODUCTIVE SCI, UNIV CALIF, SAN FRANCISCO, 84- *Concurrent Pos:* NIH res grants, 78-93; NIH res career develop awards, 79-84; perinatal consult, Marin Gen Hosp, 83-; mem coun, Perinatal Res Soc, 84-87, Comt Treat Hypertension During Pregnancy, NIH & 5 Yr Plan Comt, Nat Inst Child Health & Human Develop, 89; chmn, Fetal Mat Med Network, Nat Inst Child Health & Human Develop, 91-; prof fac mem, Molecular Med Prog, Univ Calif, San Francisco, 91- *Honors & Awards:* Gordon Jimenssen Lectr, Univ Okla, 90. *Mem:* Am Fedn Clin Res; Am Col Obstet & Gynec; Soc Gynec Invest; Endocrine Soc; Neurosci Soc; Soc Perinatal Obstetricians; Perinatal Res Soc (pres-elect, 90); NY Acad Sci; Am Soc Pharmacol & Exp Therapeut; Am Gynec & Obstet Soc. *Res:* Modulation of adrenergic response and receptors by steroid hormones; role of endothelial injury in preclampsia; author of more than 100 technical publications. *Mailing Add:* Dept Obstet Gynec & Reproductive Sci Sch Med Univ Calif Rm M-1489 San Francisco CA 94143

ROBERTS, JAMES RICHARD, b Flint, Mich, July 1, 37; m 60; c 3. PHYSICS. *Educ:* Univ Mich, BS, 59, MS, 61, PHD(physics), 64. *Prof Exp:* PHYSICIST, NAT BUR STAND, 64- *Concurrent Pos:* Nat Bur Stand training grant & vis scientist, Aarbus Univ, Denmark, 70-71, Los Alamos Nat Lab, Wroclaw, Poland. *Mem:* Am Inst Physics. *Res:* Plasma spectroscopy; experimental determination of atomic transition probabilities; stark broadening parameters and other atomic parameters of ionized plasmas. *Mailing Add:* 12512 W Old Baltimore Rd Boyds MD 20841

ROBERTS, JANE CAROLYN, b Malden, Mass, Nov 14, 32. PHYSIOLOGY. *Educ:* Univ Mass, Amherst, BS, 54; Univ Calif, Los Angeles, MA, 56, Univ Calif, Santa Barbara, PhD(physiol), 71. *Prof Exp:* Grad res physiologist, Med Ctr, Univ Calif, Los Angeles, 57-63, asst specialist physiol, 63-64, asst res physiologist, 64-67; assoc specialist, Dept Ergonomics & Phys Educ, Univ Calif, Santa Barbara, 67-71; res worker, Dept Ophthal Res, Columbia Univ, 71-72; asst prof, 72-79, ASSOC PROF, DEPT BIOL, CREIGHTON UNIV, 79- *Mem:* Am Physiol Soc; Int Hibernation Soc; Am Soc Zoologists; Soc Exp Biol & Med; Sigma Xi. *Res:* Metabolic and biochemical changes in hibernation and temperature acclimation. *Mailing Add:* Dept Biol Creighton Univ Omaha NE 68178

ROBERTS, JAY, b New York, NY, July 15, 27; m 50; c 2. PHARMACOLOGY. *Educ:* Long Island Univ, BS, 49; Cornell Univ, PhD(pharmacol), 53. *Prof Exp:* Instr physiol, Hunter Col, 53; res fel pharmacol, Med Col, Cornell Univ, 53-54, from instr to assoc prof, 54-66; prof, Univ Pittsburgh, 66-70; PROF PHARMACOL & CHMN DEPT, MED COL PA, 70- *Concurrent Pos:* Asst ed, Biol Abstr, 56-72; Lederle fac award, 57-60; mem study sect pharmacol, USPHS, 68-72; mem, Geriat Adv Panel, US Pharmocopeia, 80-90; sr investr award, SE Heart Asn; pres, Midatlantic Pharmacol Soc, 89-92. *Mem:* AAAS; Am Soc Pharmacol & Exp Therapeut; fel Am Col Cardiol; Soc Exp Biol & Med; Cardiac Muscle Soc (secy-treas, 65-67, pres, 67-69); Geront Soc Am. *Res:* Neuromuscular and cardiovascular pharmacology; aging. *Mailing Add:* Benson House T03 930 Montgomery Ave Rosemont PA 19010

ROBERTS, JEFFREY WARREN, b Flint, Mich, Feb 16, 44; m 68; c 2. MOLECULAR BIOLOGY, BIOCHEMISTRY. *Educ:* Univ Tex, BA, 64; Harvard Univ, PhD(biophys), 70. *Prof Exp:* NSF fel, Lab Molecular Biol, Med Res Ctr, 70-71; jr fel, Harvard Soc Fels, 71-73; asst prof, 74-80, assoc prof biochem, 80-85 PROF, BIOCHEM CORNELL UNIV, 86- *Res:* Regulation of gene expression, especially mechanism of lysogenic induction and the mechanism of positive control in the life cycle of bacteriophage lambda. *Mailing Add:* Biochem Molecular & Cell Biol Cornell Univ Ithaca NY 14853

ROBERTS, JERRY ALLAN, b Landis, NC, July 31, 31; m 55; c 4. APPLIED MATHEMATICS. *Educ:* NC State Univ, BEngPhys, 58, MS, 60, PhD(appl math), 64. *Prof Exp:* Res asst, NC State Univ, 57-63, instr, 63-65; asst prof, 65-69, ASSOC PROF MATH, DAVIDSON COL, 69- *Mem:* Math Asn Am; Soc Indust & Appl Math; Asn Comput Mach. *Res:* Use of numerical analysis and digital computers to obtain approximate solutions of boundary-value problems arising from problems in fracture mechanics; experimental statistics. *Mailing Add:* PO Box 1659 Davidson NC 28036

ROBERTS, JOAN MARIE, b Wall, Pa, Sept 28, 32; m 66. RADIOBIOLOGY, EXPERIMENTAL PSYCHOLOGY. *Educ:* Wayne State Univ, BS, 54, MA, 57, PhD(psychol), 62. *Prof Exp:* Exp psychologist, Henry Ford Hosp, Detroit, 54-55; from res asst to res assoc animal behav, Lafayette Clin, 56-63; instr radiobiol, Sch Med, Wayne State Univ, 63-66; from asst prof to assoc prof physiol & pharmacol, Univ Detroit, 66-83. *Concurrent Pos:* Consult, Detroit Mem Hosp, 63-66; USPHS res grant, Wayne State Univ & Univ Detroit, 64-71; Seed grant, Univ Detroit, 72-75. *Mem:* AAAS; Sigma Xi. *Res:* Effects of whole-body radiation on prenatal organisms, especially postnatal development and behavior; protection of the prenatal animal against irradiation by chemical agents administered to the pregnant animal. *Mailing Add:* 1154 Ashover Dr Bloomfield Hills MI 48013-1823

ROBERTS, JOEL LAURENCE, b Denver, Colo, Sept 5, 40; m 64; c 3. MATHEMATICS. *Educ:* Mass Inst Technol, BS, 63; Harvard Univ, MA, 64, PhD(math), 69. *Prof Exp:* Asst prof math, Purdue Univ, Lafayette, 68-72; from asst prof to assoc prof, 72-80, PROF MATH, UNIV MINN, MINNEAPOLIS, 80- *Concurrent Pos:* Vis lectr, Nat Univ Mex, 69; NSF res grant, Purdue Univ, Lafayette, 70-72, Univ Minn, 72-; vis scholar, Univ Calif, Berkeley, 80; vis prof, Nat Univ Mex, 87. *Mem:* Am Math Soc. *Res:* Algebraic geometry; properties of algebraic varieties in projective spaces; intersection theory and enumerative geometry; commutative algebra. *Mailing Add:* Dept Math Univ Minn Minneapolis MN 55455

ROBERTS, JOHN BURNHAM, b Manchester, NH, Apr 29, 13; m 45; c 2. CHEMICAL ENGINEERING. *Educ:* Dartmouth Col, AB, 34; Mass Inst Technol, MS, 36. *Prof Exp:* Chem engr, E I du Pont de Nemours & Co, Inc, 36-40, gen group supvr, Jackson Lab, 40-48, res engr, Eng Res Lab, 48-50, res supvr, 50-52, staff asst, 53, res mgr, 54-60, asst dir, 60-62, sect mgr, Eng Res Div, 63-65, eng fel, Eng Develop Lab, 66-69, eng fel, Eng Technol Lab, 70-74, sr eng fel, 74-78; CONSULT, 78- *Concurrent Pos:* Mem, Franklin Inst. *Mem:* Am Chem Soc; Am Inst Chem Engrs; Inst Elec & Electronics Engrs; NY Acad Sci. *Res:* Pollution abatement; environmental engineering; solar energy; coal conversion and feedstocks technology. *Mailing Add:* 2302 W 11th St Wilmington DE 19805-2606

ROBERTS, JOHN D, b Los Angeles, Calif, June 8, 18; m 42; c 4. ORGANIC CHEMISTRY. *Educ:* Univ Calif, Los Angeles, BA, 41, PhD(chem), 44. *Hon Degrees:* Dr rer nat, Univ Munich, 62; DrSci, Temple Univ, 64. *Prof Exp:* Instr chem, Univ Calif, Los Angeles, 44-45; Nat Res Coun fel, Harvard Univ, 45-46, instr, 46; from instr to assoc prof, Mass Inst Technol, 46-53; prof org chem, Calif Inst Technol, 53-72, chmn div chem & chem eng, 63-68, actg chmn, 72-73, vpres, provost & dean fac, 80-83, inst prof chem, 72-78, EMER INST PROF CHEM, CALIF INST TECHNOL, 88- *Concurrent Pos:* Consult, E I Du Pont de Nemours & Co, 49- & Union Carbide, 49-62; Guggenheim fel, 52-53 & 55; lectureships and professorships at var cols & univs, US & abroad, 56-; mem, adv panel chem, NSF, 57-60, chmn, 59-60, mem, adv comt math, phys & eng sci, 61-64, chmn, 62-64, mem, adv comt math & phys sci, 64-66; mem, chem adv panel, Off Sci Res, Air Force, 59-61; dir & consult ed, W A Benjamin, Inc, 60-67; chmn, Chem Sect, Nat Acad Sci, 68-71, chmn, Class I, 76-79; trustee, L B Leakey Found, 83-; mem bd dirs, Huntington Med Res Inst, 84-, treas, 89- *Honors & Awards:* Pure Chem Award, Am Chem Soc, 54, Roger Adams Award, 67, James Flack Norris Award, 79; Harrison Howe Award, 57; Nichols Medal, 71; Tolman Medal, 75; Linus Pauling Award, 80; Richards Medal, 82; Willard Gibbs Gold Medal, 83; Priestley Medal, 87; Madison Marshall Award, 89; Robert A Welch Award, 90; Nat Medal Sci, 90. *Mem:* Nat Acad Sci; Am Philos Soc; AAAS; Am Chem Soc; Am Acad Arts & Sci. *Res:* Small-ring organic compounds; relation between structure and reactivity; nuclear magnetic resonance spectroscopy of nitrogen and carbon; applications of nuclear magnetic resonance spectroscopy to biology and medicine. *Mailing Add:* Div Chem & Chem Eng Mail Code 164-30 Calif Inst Technol Pasadena CA 91125

ROBERTS, JOHN EDWIN, b Laconia, NH, Mar 6, 20; m 44; c 2. CHEMISTRY. *Educ:* Univ NH, BS, 42, MS, 44; Cornell Univ, PhD(inorg chem), 47. *Prof Exp:* Asst chem, Univ NH, 42-44 & Cornell Univ, 44-46; from asst prof to assoc prof, 46-62, PROF CHEM, UNIV MASS, AMHERST, 62- *Concurrent Pos:* Fel fluorine chem, Univ Wash, 58-59; vis prof, Univ Cairo, 61-62; sabbatical, Analytical Inst, Univ Vienna, 71. *Mem:* Am Chem Soc. *Res:* Molar refraction; indium fluoride systems; rare earth chemistry; fluorides of less familiar elements; the system indium trifluoride-water and the tendency of indium to form fluoanions; peroxydisulfuryldifluoride; fluorosulfonates; trifluoracetates; microchemistry and chemical microscopy; chemistry of art and archaeology. *Mailing Add:* 12 W Diane Dr Keene NH 03431

ROBERTS, JOHN ENGLAND, b Los Angeles, Calif, Dec 27, 22; m 57; c 2. PHYSICS. *Educ:* Univ Calif, BS, 44, PhD(chem), 50. *Prof Exp:* Tech asst chem, Univ Calif, 43-44, jr scientist, Manhattan Dist, 44-46, asst, Dept Chem, 46-50, physicist, Lawrence Radiation Lab, 50-70; CHMN DEPT SCI & MATH, PARKS COL, ST LOUIS UNIV, 70- *Concurrent Pos:* Lectr physics, St Mary's Col, Calif, 67-70. *Mem:* Am Phys Soc; Am Asn Physics Teachers. *Res:* Physics and mathematics. *Mailing Add:* Dept Sci & Math Parks Col St Louis Univ Cahokia IL 62206-1998

ROBERTS, JOHN FREDRICK, b Gallup, NMex, Oct 12, 28; m 51; c 4. ZOOLOGY, CELL BIOLOGY. *Educ:* Univ Ariz, BS, 56, PhD(zool), 64. *Prof Exp:* Asst prof biol, NMex Highlands Univ, 61-65; from asst prof to assoc prof zool, 65-74, PROF ZOOL, NC STATE UNIV, 74- *Concurrent Pos:* USPHS res grants, 64-65 & 69-72. *Mem:* Sigma Xi; Soc Protozool; Am Soc Cell Biol. *Res:* Mitochondrial nucleic acids; morphogenesis; trypanosomes. *Mailing Add:* Dept Zool NC State Univ Box 7617 Raleigh NC 27695-7617

ROBERTS, JOHN HENDERSON, b Raywood, Tex, Sept 2, 06; wid; c 1. MATHEMATICS. *Educ:* Univ Tex, AB, 27, PhD(math), 29. *Prof Exp:* Nat Res Coun fel, Univ Pa, 29-30; adj prof pure math, Univ Tex, 30-31; from asst prof to prof math, Duke Univ, 31-71, chmn dept, 66-68, emer prof, 71; RETIRED. *Concurrent Pos:* Lectr, Princeton Univ, 37-38. *Mem:* AAAS; Am Math Soc (secy, 54); Math Asn Am. *Res:* Point-set topology; integral equations. *Mailing Add:* Bldg 3-203 Carolina Meadows Chapel Hill NC 27514

ROBERTS, JOHN LEWIS, b Waukesha, Wis, May 23, 22; m 50; c 2. COMPARATIVE PHYSIOLOGY. *Educ:* Univ Wis, BS, 47, MS, 48; Univ Calif, Los Angeles, PhD, 53. *Prof Exp:* Asst zool, Univ Wis, 47-48 & Univ Calif, Los Angeles, 48-51; from instr to assoc prof physiol, 52-69, PROF PHYSIOL, UNIV MASS, AMHERST, 70- *Concurrent Pos:* With zool inst & mus, Univ Kiel, 59-60; USPHS spec fel, 66-67; guest lectr, Dept Zool, Univ Bristol, 66-67; sr res assoc, Nat Res Coun-Nat Oceanog & Atmospheric Admin, La Jolla, Calif, 73-75; panelist, NSF, 77-80; vis prof zool, Univ Hawaii, Manoa, Honolulu, 80; vis investr, Physiol Res Lab, Scripps Inst Oceanog, La Jolla, Calif, 84. *Mem:* Fel AAAS; Soc Gen Physiol; Am Fisheries Soc; Am Soc Zool (secy, 76-79); Sigma Xi. *Res:* Physiological adaptations of poikilotherms; respiratory and cardiac physiology of fish; neural control of respiration and swimming in fish. *Mailing Add:* Dept Zool Univ Mass Amherst MA 01003

ROBERTS, JOHN MELVILLE, physical metallurgy, materials science; deceased, see previous edition for last biography

ROBERTS, JOHN STEPHEN, b Apr 9, 37; m 59; c 2. PHYSIOLOGY, ENDOCRINOLOGY. *Educ:* Dartmouth Col, BA, 59; Duke Univ, PhD(zool), 65. *Prof Exp:* NIH fel, Sch Med, Case Western Reserve Univ, 64-67, Pop Coun res grant, 68-71, from sr instr to assoc prof physiol, 72-75; vis scientist, 73-75, SR SCIENTIST, WORCESTER FOUND EXP BIOL, 75- *Concurrent Pos:* NIH res grant, 71- *Mem:* AAAS; Am Physiol Soc; Endocrine Soc; Soc Study Reproduction. *Res:* Neuropharmacological studies of interactions among reproductive neuroendocrine control functions in intact animals using the systems analysis approach. *Mailing Add:* 653 South St Shrewsbury MA 01545

ROBERTS, JOSEPH, biochemistry, for more information see previous edition

ROBERTS, JOSEPH BUFFINGTON, b Albany, NY, Sept 9, 23; m 44; c 3. MATHEMATICS. *Educ:* Case Inst Technol, BS, 44; Univ Colo, MA, 50; Univ Minn, PhD(math), 55. *Prof Exp:* Jr scientist radiochem, Manhattan Proj, Univ Calif, 45-46; asst math biol, Univ Chicago, 46-48; instr chem & math, Univ Wyo, 48-49; asst math, Univ Minn, 50-52; from instr to assoc prof, 52-70, PROF MATH, REED COL, 70- *Concurrent Pos:* Vis asst prof, Wesleyan Univ, 56-57; NSF res grant, 58-67; res assoc, Univ London, 62-63; Agency Int Develop vis prof, Univ Col, Dar es Salaam, 65-67; vis prof, Dalhousie Univ, 69-70; consult, Bonneville Power Admin, 58-61. *Mem:* Am Math Soc; Math Asn Am; Fedn Am Sci; Indian Math Soc; London Math Soc. *Res:* Number theory; summability; orthogonal functions. *Mailing Add:* Dept Math Reed Col Portland OR 97202

ROBERTS, JOSEPH LINTON, b Atlanta, Ga, Nov 13, 29; m 53; c 3. PHYSICAL CHEMISTRY, BIOCHEMISTRY. *Educ:* Oglethorpe Univ, BS, 53; Univ SDak, MA, 55; Univ Cincinnati, PhD(theoret chem), 64. *Prof Exp:* Res asst systs analytical, Biostatist Dept, Kettering Lab, Col Med, Univ Cincinnati, 59, res asst analytical chem, Dept Toxicol, 59-61; res chemist, Dept Med Res, Vet Hosp, Cincinnati, 61-66; asst prof biochem & exp med, Col Med, Univ Cincinnati, 65-66; from asst prof to assoc prof chem, 66-74, PROF CHEM, MARSHALL UNIV, 74- *Mem:* Am Chem Soc. *Res:* Chemistry of fire supression; computer modeling of fire supression systems; models for effect of solvent or reactions. *Mailing Add:* Dept Chem Marshall Univ 400 Halgereer Blvd Huntington WV 25701

ROBERTS, JULIAN LEE, JR, b Columbia, Mo, June 15, 35; m 65. ANALYTICAL CHEMISTRY. *Educ:* Univ Southern Calif, BA, 57; Northwestern Univ, PhD(analytical chem), 62. *Prof Exp:* Instr chem, Univ Redlands, 61-63; res chemist & vis lectr, Univ Calif, Riverside, 63-64; from asst prof to assoc prof chem, 64-73, PROF CHEM, UNIV REDLANDS, 73- *Concurrent Pos:* NSF sci fac fel, Calif Inst Technol, 68-69; vis scientist, Lab Org & Analytical Electrochem, Ctr Nuclear Studies, Grenoble, France, 75-76. *Mem:* AAAS; Am Chem Soc; NY Acad Sci. *Res:* Structural and electroanalytical chemistry; mechanisms of electrochemical reactions; electrochemistry of dissolved gases and coordination compounds. *Mailing Add:* Dept Chem Univ Redlands Redlands CA 92373-0999

ROBERTS, KENNETH DAVID, b Montreal, Que, Nov 20, 31; m 60; c 1. BIOCHEMISTRY, ENDOCRINOLOGY. *Educ:* George Williams Col, BSc, 55; McGill Univ, PhD(biochem), 60. *Prof Exp:* Chemist, Charles E Frosst & Co, 49-55; Jane C Childs Mem Fund fel, Univ Basel, 61-63; asst prof biochem, Columbia Univ, 63-69; assoc prof, 69-77, PROF BIOCHEM, UNIV MONTREAL, 77- *Concurrent Pos:* Corresp ed, Steroids, 63-; consult, Endocrine Lab, Maisonneuve Hosp, Montreal, 69- mem grants comt, Med Res Coun Can, 71- *Mem:* AAAS; NY Acad Sci; Can Fertility Soc; Endocrine Soc; Can Biochem Soc. *Res:* Biochemistry of steroids and steroid conjugates; reproduction; biochemistry of fertilization. *Mailing Add:* 2084 de la Regence St Bruno de Montarville PQ J3V 4B6 Can

ROBERTS, LARRY SPURGEON, b Texon, Tex, June 30, 35; m 62; c 4. ZOOLOGY, PARASITOLOGY. *Educ:* Southern Methodist Univ, BSc, 56; Univ Ill, MSc, 58; Johns Hopkins Univ, ScD(parasitol), 61. *Prof Exp:* USPHS-NIH trainee, McGill Univ, 61-62; trainee, Univ Mass, Amherst, 62-63, from asst prof zool to prof 62-79; prof & chairperson, 79-84, PROF BIOL SCI, TEX TECH UNIV, LUBBOCK, 84- *Concurrent Pos:* USPHS-NIH spec fel, Johns Hopkins Univ, 69-70. *Honors & Awards:* Henry Baldwin Ward Medal, Am Soc Parasitol, 71. *Mem:* Am Soc Parasitol (vpres, 85); Soc Protozoologists; Am Soc Trop Med Hyg; Am Micros Soc (vpres, 75); Crustacean Soc; Int Soc Reef Studies; Wildlife Dis Asn. *Res:* Developmental biochemistry of helminth parasites, especially cestodes; regulation of carbohydrate and energy metabolism in cestodes; systematics and morphology of parasitic copepods. *Mailing Add:* Dept Biol Sci Tex Tech Univ Lubbock TX 79409-4149

ROBERTS, LAWRENCE G, b Norwalk, Conn, Dec 21, 37; c 2. DATA COMMUNICATIONS, PACKET SWITCHING. *Educ:* Mass Inst Technol, BS, 59, MS, 60, PhD(elec eng), 63. *Prof Exp:* Mem staff, Lincoln Lab, Mass Inst Technol, 63-67; dir, Info Process Techniques Defense Advan Res Projs Agency, 67-73; pres, Telenet Commun Corp, 73-79, GTE Telenet Corp, 80-82 & GTE Subscriber Network, 80-82; PRES, DHL CORP, 82-; CHMN, NET EXPRESS INC, 82- *Honors & Awards:* L M Erickson Award Commun Res, 81. *Mem:* Nat Acad Eng; Inst Elec & Electronics Engrs; Sigma Xi; Asn Comput Mach. *Mailing Add:* 170 Sunrise Dr Woodside CA 94062

ROBERTS, LEE KNIGHT, b Salt Lake City, Utah, Sept 9, 49; m 69; c 2. DERMATOLOGY, PHOTOBIOLOGY. *Educ:* Univ Utah, BS, 72, PhD(anat), 80. *Prof Exp:* Postdoctoral fel immunol, Univ NMex, 80-82; from instr to asst prof immunol & dermat, Univ Utah Sch Med, 82-89; SR PRIN SCIENTIST, SCHERING-PLOUGH INC, 89- *Mem:* Soc Investigative Dermat; Am Asn Immunologists; AAAS. *Res:* Photoimmunology and photodermatology; first to clone tumor-antigen-specific suppressor T-cells from uvb-exposed mice; demonstrating common tumor antigens being shared by uvb-induced tumors and uvb-exposed skin; characterization of antigen-presenting cell repopulation of uvb-exposed mouse skin. *Mailing Add:* Skin Biol Res Schering-Plough Inc 3030 Jackson Ave Memphis TN 38151

ROBERTS, LEIGH M, b Jacksonville, Ill, June 9, 25; m 46; c 4. PSYCHIATRY. *Educ:* Univ Ill, BS, 45, MD, 47. *Prof Exp:* Intern med, St Francis Hosp, Peoria, Ill, 47-48; pvt pract, 48-50; resident, Univ Hosps, 53-56, clin instr, Sch Med, 56-58, from asst prof to assoc prof, 59-71, actg chmn dept, 72-75, PROF PSYCHIAT, SCH MED, UNIV WIS-MADISON, 71- *Concurrent Pos:* Staff psychiatrist, Mendota State Hosp, Madison, 56-58, consult, 60-; consult, Wis Child Ctr, Sparta, 56-58, Cent State Hosp, 62-70 & State Ment Health Planning, 63-; mem spec rev bd, Wis Div Corrections, 63-; mem bd, Methodist Hosp, Madison, 65- & Goodwill Industs, 71- *Mem:* AAAS; fel Am Psychiat Asn; AMA. *Res:* Community psychiatry; psychiatry and law. *Mailing Add:* 600 Highland Ave Madison WI 53792

ROBERTS, LEONARD, b Prestatyn, North Wales, UK, Sept 27, 29; US citizen; m 55; c 2. APPLIED MATHEMATICS, AERODYNAMICS. *Educ:* Univ Manchester, BSc, 52, MSc, 54, PhD, 55. *Prof Exp:* Res assoc, Mass Inst Technol, 55-57; aerospace res engr theoret, Mech Div, Langley Res Ctr, NASA, 57-59, head math physics br, Dynamic Loads Div, 59-66, dir mission anal div, NASA Hq, 66-69; Stanford Sloan fel, 69-70; DIR AERONAUT & FLIGHT SYSTS, AMES RES CTR, NASA, 70-; MEM STAFF, DEPT ASTRONAUT ENG, STANFORD UNIV. *Concurrent Pos:* Consult prof aeronaut & astronaut, Stanford Univ, 75- *Mem:* Am Inst Aeronaut & Astronaut; Am Helicopter Soc. *Res:* Aerodynamics; flight dynamics; guidance and control; aeronautical vehicles; aviation systems; research using wind tunnels, simulators and experimental aircraft; management of aeronautical research and development. *Mailing Add:* Aeronaut & Astronaut Eng Durano 269 Stanford Univ Stanford CA 94305

ROBERTS, LEONIDAS HOWARD, b Garard's Ft, Pa, Feb 27, 21; m 45; c 2. PHYSICAL SCIENCES, ASTRONOMY. *Educ:* Waynesburg Col, BS, 48; WVa Univ, MS, 49; Univ Fla, PhD(appl math), 58. *Prof Exp:* From instr to asst prof phys sci, 49-61, assoc prof phys sci & astron, 61-71, PROF PHYS SCI & ASTRON, UNIV FLA, 71- *Mem:* Am Astron Soc. *Res:* Astronomy. *Mailing Add:* Dept Astron Univ Fla Gainesville FL 32611

ROBERTS, LESLIE GORDON, b Flin Flon, Man, Aug 16, 41; m 67. MATHEMATICS. *Educ:* Univ Man, BSc, 63; Harvard Univ, PhD(math), 68. *Prof Exp:* From asst prof to assoc prof, 68-83, PROF MATH, QUEEN'S UNIV, ONT, 83- *Mem:* Am Math Soc; Can Math Cong. *Res:* Algebraic K-theory; algebraic geometry. *Mailing Add:* Dept Math Queen's Univ Kingston ON K7L 3N6 Can

ROBERTS, LORIN WATSON, b Clarksdale, Mo, June 28, 23; m 67; c 3. PLANT PHYSIOLOGY. *Educ:* Univ Mo, AB, 48, MA, 50, PhD(bot), 52. *Prof Exp:* Asst bot, Univ Mo, 49-52; asst prof biol, Agnes Scott Col, 52-56, assoc prof, 56-57; from asst prof to assoc prof bot, 57-67, PROF BOT, UNIV IDAHO, 67- *Concurrent Pos:* Vis asst prof, Emory Univ, 52-55; pres bot sect, Int Cong Histochem & Cytochem, Paris, 60; Fulbright res prof, Kyoto Univ, 67-68; vis prof, Bot Inst, Univ Bari, 68; Maria Moors Cabot res fel, Harvard Univ, 74; Fulbright vis lectr, NEastern Hill Univ, Shillong, Meghalaya, India, 77; Fulbright sr scholar & vis fel, Australian Nat Univ, Canberra, 80; fel, Univ London, 84. *Honors & Awards:* Chevalier de l'Ordre du Merite Agricole, France, 61. *Mem:* AAAS; Bot Soc Am; Am Soc Plant Physiol; Int Asn Plant Tissue Cult. *Res:* Physiology of vascular differentiation; physiological action of plant hormones. *Mailing Add:* Dept Biol Sci Univ Idaho Moscow ID 83843

ROBERTS, LOUIS DOUGLAS, b Charleston, SC, Jan 27, 18; m 42; c 1. SOLID STATE PHYSICS, THEORETICAL PHYSICS. *Educ:* Howard Col, AB, 38; Columbia Univ, PhD(phys chem), 41. *Prof Exp:* Nat Res Coun fel, Cornell Univ, 41-42; res physicist, Gen Elec Co, NY, 42-46; physicist, Oak Ridge Nat Lab, 46-68; PROF PHYSICS, UNIV NC, CHAPEL HILL, 68- *Concurrent Pos:* Vchair, Southeastern Sect, Am Phys Soc, 54-55, chmn,

55-56; Fulbright fel & Guggenheim fel, Oxford Univ, 58-59; prof, Univ Tenn, 63-68. *Mem:* Fel Am Phys Soc. *Res:* Semiconductors; vacuum tube design and ion optics; neutron diffusion theory and measurement; low temperature physics; low energy nuclear and solid state physics; high pressure physics; electron many-body studies. *Mailing Add:* Dept Physics & Astron Univ NC Chapel Hill NC 27599-3255

ROBERTS, LOUIS REED, b Wray, Colo, July 8, 23; m 48; c 3. CHEMICAL ENGINEERING. *Educ:* Univ Colo, BS, 49; Rice Univ, MA, 51; Univ Tex, PhD(chem eng), 63. *Prof Exp:* Jr chemist, Gulf Oil Corp, 51-53, chemist, 53-56; chem engr, Southwest Res Inst, 56-58; sr chem engr, Union Tex Petrol Div, Allied Chem Corp, 62-63, engr, Cent Res Lab, 63-70, res chem engr, Chem Res Lab, 70-76; DIV DIR, STATE TEX, 76- *Mem:* Fel Am Inst Chem Engrs. *Res:* Vapor-liquid equilibria of hydrocarbon fixed-gas systems; high temperature chemistry; petrochemical and polymer processes. *Mailing Add:* 8611 Honeysuckle Austin TX 78759

ROBERTS, LOUIS W, b Jamestown, NY, Sept 1, 13; m 38; c 2. PHYSICS. *Educ:* Fisk Univ, AB, 35; Univ Mich, MS, 37. *Hon Degrees:* LLD, Fisk Univ, 85. *Prof Exp:* Teaching asst physics & math, Fisk Univ, 35-36; instr, St Augustine's Col, 37-40, assoc prof physics, 41-43; asst prof, Howard Univ, 43-44; sr engr, Sylvania Elec Prod, Inc, 44-46, sect head tubes, 46-47, mgr tube develop, 47-50; pres, dir & founder, Microwave Assocs, 50-51, vpres, 51-55; eng specialist, Bomac Labs, Inc Div, Varian Inc, 55-59; vpres, dir & founder, Metcom Inc, 59-67; consult optics & microwaves, Electronic Res Ctr, NASA, 67, chief, Microwave Lab, 67-68, chief, Optics & Microwave Lab, 68-70; from dep dir to dir technol, 70-77, dir energy & environ, 77-79, dep dir, 79, dir data systs & technol, 80-82, dir off admin, 82-83, assoc dir off opers eng, 83-84, actg dep dir, 84, actg dir, 84-85, DIR TRANSP SYSTS CTR, US DEPT TRANSP, 85- *Concurrent Pos:* Res assoc, Stand Oil NJ, 35-36; prof, A&T State Univ, NC, 41-43; instr, US Signal Corps Sch, 42; prof, Shaw Univ, 42-43 & Army Specialized Training Prog, Wash, DC, 43-44; staff mem, Res Lab Electronics, Mass Inst Technol, 50-51; pres, Elcon Lab, Inc, Metcom Inc & consult, Addison-Wesley Press, 63-67; mem, US-Japan Natural Resources Comt, US Dept Interior, 69-78, adv group aerospace res & develop, Nato, 73-79; mem adv bd Col Eng, Univ Mass, 72-; mem bd trustees, Univ Hosp, Boston Univ, 73-; mem pres adv bd, Bentley Col, 74-; vis engr & lectr, Mass Inst Technol, 79-80. *Mem:* AAAS; fel Inst Elec & Electronics Engrs; Am Phys Soc; Am Math Soc; Am Inst Aeronaut & Astronaut. *Res:* Microwave and optical techniques and components; plasma research and solid state component and circuit development; all aspects of transportation and logistics, including applications of artificial intelligence. *Mailing Add:* Five Michael Rd Wakefield MA 01880

ROBERTS, LYMAN JACKSON, CLINICAL PHARMACOLOGY. *Educ:* Univ Iowa, MD, 69. *Prof Exp:* PROF PHARMACOL & MED, VANDERBILT UNIV, 76- *Res:* Prostaglandin research; internal medicine. *Mailing Add:* Dept Pharmacol Vanderbilt Univ Nashville TN 37232-6602

ROBERTS, MARTIN, b Brooklyn, NY, May 8, 20; m 49; c 4. BIOCHEMISTRY. *Educ:* City Col New York, BS, 42; Brooklyn Col, MA, 47; Univ Southern Calif, PhD(biochem), 51. *Prof Exp:* Res org chem, Schwarz Labs, Inc, 43-47; sr clin res biochemist, Don Baxter, Inc, 51-66; mgt & med dir, Artificial Kidney Supply Div, Sweden Freezer Mfg Co, Seattle, 66-71; dir clin invest, 71-74, vpres mkt, CCI Life Systs Inc, Van Nuys, 74-76; vpres, 76-80, dir & med & regulatory affairs, Redy Labs, Organon Teknika, 80-85; CONSULT, ROBERTS ENTERPRISES, 85- *Mem:* Am Soc Artificial Internal Organs; Int Soc Artificial Internal Orgns; Am Soc Nephrology; Int Cong Nephrology. *Res:* Nucleic acid derivatives; parenteral and nutritional solutions, peritoneal and hemodialysis. *Mailing Add:* 16022 Parthenia St Sepulveda CA 91343

ROBERTS, MARY FEDARKO, b Pittsburgh, Pa, July 11, 47; m 74; c 1. BIOCHEMISTRY, PHYSICAL CHEMISTRY. *Educ:* Bryn Mawr Col, AB, 69; Stanford Univ, PhD(chem), 74. *Prof Exp:* Res assoc biochem, Univ Ill, Urbana-Champaign, 74-75; NIH trainee biol, Univ Calif, San Diego, 75-77, NIH fel chem, 77-78; asst prof chem, Mass Inst Technol, 78-86; assoc prof, 87-91, PROF CHEM, BOSTON COL, 91- *Mem:* Am Chem Soc. *Mailing Add:* Dept Chem Boston Col Chestnut Hill MA 02167

ROBERTS, MERVIN FRANCIS, b New York, NY, June 7, 22; m 49; c 4. ANIMAL BEHAVIOR, ICHTHYOLOGY. *Educ:* Alfred Univ, BS, 47. *Prof Exp:* Sr inspector, New York Port Authority, 47-56; lab mgr, De Lackner Helicopters, NY, 56-58 & Irco Corp, 58-59; assoc ed, McGraw Hill Publ Co, 59-60; ed supvr, Anco Tech Writing Serv, 60-61; mgr res & develop, T F H Publ, Inc, 61-67, asst to pres res & develop, 67-68; MGR RES & DEVELOP, WIDGET CO OLD LYME, 68- *Concurrent Pos:* Fish behav consult, Northeast Utilities Serv Corp, 70-; mem, Govr's Coun Marine Resources, Conn, 70-73, chmn, 71, shellfish comnr, 82-; consult, Salmonoid Aquacult, Nordic Enterprises Inc, 85-88, Tilopia Aquacult, Diocese Alleppey, Kerala, India, 87- *Mem:* Am Fisheries Soc; Am Littoral Soc; Estuarine Res Soc. *Res:* Photography of high speed animal movements; small animal maintenance; fish culture; tidemarsh ecosystems; author or coauthor of over 30 publications on care of caged animals and on tidemarsh life. *Mailing Add:* Duck River Lane Old Lyme CT 06371

ROBERTS, MICHAEL FOSTER, b Guatemala, Aug 8, 43; US citizen; m 66; c 2. CARDIOVASCULAR PHYSIOLOGY, THERMAL PHYSIOLOGY. *Educ:* Univ Calif, Berkeley, BA, 66; Univ Wis-Madison, MA, 68, PhD(zool), 72. *Prof Exp:* Asst fel & asst prof epidemiol, John B Pierce Found Lab, Yale Univ, 72-81; from asst prof to assoc prof, 81-90, PROF BIOL, LINFIELD COL, 90- *Concurrent Pos:* Prin investr, NIH, 82-85 & Am Heart Asn, 85-86. *Mem:* AAAS; Sigma Xi; Am Physiol Soc. *Res:* Peripheral circulation and temperature regulation; contractile mechanisms in peripheral blood vessels; influence temperature has on the process of transmitter release, uptake, degradation and receptor binding. *Mailing Add:* Linfield Col McMinnville OR 97128

ROBERTS, MORTON SPITZ, b New York, NY, Nov 5, 26; m 51; c 1. RADIO ASTRONOMY. *Educ:* Pomona Col, BA, 48; Calif Inst Technol, MS, 50; Univ Calif, PhD(astron), 58. *Hon Degrees:* DSc, Pomona Col, 79. *Prof Exp:* Asst prof physics, Occidental Col, 49-52; physicist underwater ord, US Naval Ord Testing Sta, 52-53; jr res astronr, Univ Calif, 57-58, NSF fel, 58-59, lectr astron & asst res astronr, Radio Astron Lab, 59-60; lectr astron & res assoc, Observ, Harvard Univ, 60-64; scientist, 64-78, from asst dir to dir, 69-84, SR SCIENTIST, NAT RADIO ASTRON OBSERV, 78- *Concurrent Pos:* Vis prof, Univ Calif, Berkeley, 68, State Univ NY, Stony Brook, 68, Inst Theoret Astron, Cambridge, 72, 86-87 & Univ Groningen, 72; Sigma Xi nat lectr, 70-71; assoc ed, Astron J, 77-79. *Mem:* Nat Acad Sci; Int Union Radio Sci; Am Astron Soc (vpres, 71-72); Int Astron Union (vpres, 88-); AAAS (vpres, 88-). *Res:* Galaxies; galactic structure; interstellar and extragalactic matter. *Mailing Add:* Nat Radio Astron Observ Edgemont Rd Charlottesville VA 22903

ROBERTS, NORBERT JOSEPH, b Alabama, NY, June 6, 16; m 43; c 5. MEDICINE. *Educ:* Canisius Col, 34-36; Univ Buffalo, MD, 40; Univ Minn, MS, 49; Am Bd Internal Med, dipl, 50; Am Bd Prev Med, dipl, 55. *Prof Exp:* Physician, Exxon Corp, 49-52; med dir, Pa RR Co, 52-55; assoc med dir, Exxon Corp, 35-73, med dir, 73-81, vpres med & environ health, 81; assoc clin prof, Mt Sinai Sch Med, 81-83; RETIRED. *Concurrent Pos:* Asst prof, Sch Med, Univ Pa, 53-73; mem US deleg, Permanent Comn & Int Asn Occup Health, 57-81; lectr, Inst Environ Med, NY Univ, 58-83. *Honors & Awards:* Knudsen Award, 76; Robert Kehoe Award, 80; Plaque, NY Acad Med, 1987. *Mem:* Fel Am Occup Med Asn (pres, 71-72); fel Am Col Prev Med; fel Am Col Physicians; fel Am Acad Occup Med; NY Acad Med (pres, 81-82). *Res:* Internal and occupational medicine; preventive medicine. *Mailing Add:* 688 Key Royale Dr Holmes Beach FL 33510

ROBERTS, NORMAN HAILSTONE, b Seattle, Wash, Mar 3, 22; div; c 3. APPLIED STATISTICS. *Educ:* Univ Wash, BS, 46, PhD(physics), 57. *Prof Exp:* Asst physicist, Appl Physics Lab, 55, instr mech eng, Univ, 56-57, from assoc physicist to physicist, Appl Physics Lab, 57-65, actg assoc prof mech eng, Univ, 66-68, ASSOC PROF MECH ENG, UNIV WASH, 68- *Concurrent Pos:* Consult to Nuclear Regulatory Comn, 74- *Mem:* AAAS; NY Acad Sci; Syst Safety Soc. *Res:* Electron energy loss; ferroelectricity; statistics, reliability and systems analysis, fault tree analysis; physics of the oceans. *Mailing Add:* Dept Mech Eng Univ Wash Seattle WA 98195

ROBERTS, P ELAINE, b Mt Clements, Mich, Feb 18, 44. HORMONAL REGULATION OF INSECT REPRODUCTION & GENE EXPRESSION. *Educ:* Western Mich Univ, BA, 67; Univ Ill, PhD(cell biol), 76. *Prof Exp:* Res asst, Western Mich Univ, 67-70; postdoctoral fel, Dept Molecular Med, Mayo Clin, 76-78 & Queen's Univ, Ont, 78-79; from asst prof to assoc prof cell biol & insect physiol, 86-91, PROF ENTOM, COLO STATE UNIV, 91- *Concurrent Pos:* NIH grant, 83; NSF grant, 87; vis scientist, USDA, 87; Colo Agr Exp Sta grant, 88. *Mem:* Asn Cell Biol; Entom Soc Am; Am Soc Zoologists; AAAS; Sigma Xi; Asn Women Sci. *Res:* Mechanism by which juvenile hormone regulates gene expression in insects. *Mailing Add:* Dept Entom Colo State Univ Ft Collins CO 80523

ROBERTS, PAUL ALFRED, b Chicago, Ill, Dec 22, 31; m 57, 71; c 5. GENETICS. *Educ:* Univ Ill, Urbana, BS, 53, Univ Ill, Chicago, MD, 57; Univ Chicago, PhD(zool), 62; Chicago Teachers Col, MEd, 62. *Prof Exp:* Res assoc biol, Oak Ridge Nat Lab, 62-63, biologist, 63-66; assoc prof zool, 66-73, PROF ZOOL, ORE STATE UNIV, 73- *Mem:* Genetics Soc Am. *Res:* Chromosome behavior; developmental biology; cytogenetics; aging in Drosophila. *Mailing Add:* Dept Zool Ore State Univ Corvallis OR 97331

ROBERTS, PAUL HARRY, b Aberystwyth, Wales, UK, Sept 9, 29; m 89. MAGNETOHYDRODYNAMICS. *Educ:* Cambridge Univ, Eng, BA, 51, MA & PhD(math), 54, ScD, 67. *Prof Exp:* Res assoc, Univ Chicago, 54-55; sci officer, Awre, Aldermaston, Eng, 55-56; ICI res fel, Univ Durham, Eng, 56-59, lectr physics, 59-61; assoc prof astron, Univ Chicago, 61-63; prof math, Univ Newcastle upon Tyne, UK, 63-85; PROF MATH & PROF GEOPHYS SCI, UNIV CALIF, LOS ANGELES, 86- *Mem:* Fel Royal Soc London. *Res:* Magnetohydrodynamics and applications to earths core; fluid mechanics, including two phase flow and superfluid mechanics. *Mailing Add:* Math Dept Univ Calif Los Angeles CA 90024

ROBERTS, PAUL OSBORNE, JR, b Memphis, Tenn, Mar 6, 33; m 57; c 3. TRANSPORTATION MANAGEMENT. *Educ:* Tex A&M Univ, BS, 55; Mass Inst Technol, SM, 57; Northwestern Univ, PhD(transp eng), 65. *Prof Exp:* Engr in charge electronic comput ctr, Michael Baker, Jr, Inc, Pa, 57-60; asst prof civil eng, Mass Inst Technol, 60-66; lectr transp, Harvard Univ, 66-69, assoc prof transp & logistics, Bus Sch, Harvard Univ, 69-72; prof transp eng & sr lectr, Sloan Sch Mgt, Mass Inst Technol, 72-76, prof civil eng & dir, ctr Transp Studies, 76-80; prin, Transp Consult Div, Booz Allen & Hamilton, Inc, 80-83; PRES, TRANSMODE CONSULT, INC, 83- *Concurrent Pos:* Mem, Hwy Res Bd, Nat Acad Sci-Nat Res Coun. *Mem:* Opers Res Soc Am; Inst Traffic Engrs. *Res:* Economic and engineering analysis of transportation systems, particularly as related to economic development; engineering project analyses; decision theory; business logistics; industry location; urban growth; trucking. *Mailing Add:* 3301 39th St NW Washington DC 20016

ROBERTS, PETER MORSE, b Pottstown, Pa, Sept 14, 55; m 85; c 2. SEISMOLOGY, ULTRASONIC MATERIAL TESTING. *Educ:* Mass Inst Technol, BS, 79, PhD(geophysics), 89. *Prof Exp:* Tech asst, Seismic Instrument Lab, Mass Inst Technol, 79-82, res asst earth, atomospheric & planetary sci, 82-85; res fel geol sci, Univ Southern Calif, 85-89; FEL GEOENG, LOS ALAMOS NAT LAB, 89- *Mem:* Am Geophys Union; Seismol Soc Am. *Res:* Teleseismic waveform modeling used to determine crustal structure beneath volcanic systems; laboratory development of new methods for crustal imaging using controlled ultrasonics test experiments; operating and designing equipment systems for recording seismic and ultrasonic data. *Mailing Add:* Los Alamos Nat Lab EES-4 Mail Stop D443 Los Alamos NM 87545

ROBERTS, R MICHAEL, REPRODUCTION, MOLECULAR BIOLOGY. *Educ:* Oxford Univ, PhD(plant biochem), 65. *Prof Exp:* PROF ANIMAL SCI & BIOCHEM, STATE UNIV MO, 86- *Mailing Add:* Biochem Molecular Biol J245 Jhmhc Univ Fla Gainesville FL 32611

ROBERTS, RALPH, b Bridgeton, NJ, May 31, 15; m 39; c 2. PHYSICAL CHEMISTRY. *Educ:* Cath Univ Am, BS, 36, MS, 38, PhD(phys chem), 40. *Prof Exp:* Chemist, US Naval Eng Exp Sta, 40-46, sci adminr phys chem, Off Naval Res, 46-55, sci liaison officer, London, 55-56, res coordr, 56-58, head propulsion chem br, 58-60, dir power prog, 60-74; mem, tech staff, Mitre Corp, 75-84; PRES, ROBERTS CONSULTS, INC, 84- *Concurrent Pos:* Guest chemist, Brookhaven Nat Lab, 51; vis scientist, Yale Univ, 71-72. *Mem:* Am Chem Soc; AAAS; Combustion Inst. *Res:* Chemical propulsion; energy conversion systems; electrochemical power sources. *Mailing Add:* 3308 Camalier Dr Chevy Chase MD 20815

ROBERTS, RALPH JACKSON, b Rosalia, Wash, Jan 31, 11; m 42; c 3. GEOLOGY. *Educ:* Univ Wash, BS, 35, MS, 37; Yale Univ, PhD, 49. *Prof Exp:* Geologist, US Geol Surv, 39-42, Cent Am, 42-45 & Nev-Utah, 45-71, tech adv econ geol, US Geol Surv, Saudi Arabia, 72-78, Nev, 78-80; CONSULT, 81- *Honors & Awards:* Distinguished Serv Award, Dept Interior, 83. *Mem:* Geol Soc Am; Soc Econ Geologists. *Res:* Economic and structural geology of Nevada; geology of Cordilleran fold-belt; volcanogenic ore deposits in Arabian Shield; economic geology, western United States and Central America. *Mailing Add:* 104 W Kinnear Pl Seattle WA 98119

ROBERTS, REGINALD FRANCIS, b Baton Rouge, La, Feb 14, 23. PHYSICAL CHEMISTRY. *Educ:* La State Univ, AB, 42, BS, 47, MS, 50. *Prof Exp:* From asst res chemist to sr res chemist, Kaiser Aluminum & Chem Corp, 50-58, sr develop chemist, 59-63; sr res chemist, Dow Chem Co, Plaquemine, 64-84, proj leader tech & patent liaison, 84-86; PVT REGIST PATENT AGENT. *Res:* Chemical kinetics; reaction mechanisms; physical organic chemistry; structural theory; effect of structure on reactivity; relationship between physical properties and molecular structure; thermodynamics. *Mailing Add:* PO Box 515 Baton Rouge LA 70821

ROBERTS, RICHARD, b Atlantic City, NJ, Feb 16, 38; m 60; c 3. MECHANICAL ENGINEERING. *Educ:* Drexel Univ, BS, 61; Lehigh Univ, MS, 62, PhD(mech eng), 64. *Prof Exp:* From asst prof to assoc prof, 64-77, PROF MECH ENG, LEHIGH UNIV, 77- *Honors & Awards:* Spraragen Award, Am Welding Soc, 72. *Mem:* Am Soc Mech Engrs; Am Soc Testing & Mat; Am Soc Eng Educ; Am Welding Soc; Am Soc Metals. *Res:* Fracture mechanics; material behavior; experimental stress analysis. *Mailing Add:* 524 S Lynn St Bethlehem PA 18015

ROBERTS, RICHARD A, electrical engineering; deceased, see previous edition for last biography

ROBERTS, RICHARD B, b New Haven, Conn, Feb 26, 33. INFECTIOUS DISEASES. *Educ:* Temple Univ, MD, 59. *Prof Exp:* PROF MED, MED COL, CORNELL UNIV, 70- *Mailing Add:* Dept Med Rm A-423 Med Col Cornell Univ 1300 York Ave New York NY 10021

ROBERTS, RICHARD CALVIN, b Akron, Ohio, May 26, 25; m 49; c 1. APPLIED MATHEMATICS. *Educ:* Kenyon Col, AB, 46; Brown Univ, ScM, 46, PhD(appl math), 49. *Prof Exp:* Asst appl math, Brown Univ, 46-48, res assoc, 48-50; fel, Univ Md, 50-51; mathematician, US Naval Ord Lab, 51-66, chief math dept, 60-66; div chmn, 66-74, dean div math & physics, 74-78, chmn dept math & comput sci, 82-85, PROF MATH, UNIV MD BALTIMORE COUNTY, 78- *Mem:* Am Math Soc; Math Asn Am; Am Comput Mach. *Res:* Fluid dynamics; numerical methods; finite differences. *Mailing Add:* Dept Math & Statist Univ Md Baltimore County Baltimore MD 21228

ROBERTS, RICHARD HARRIS, b Buffalo, NY, Oct 24, 24; m 54; c 3. MEDICAL ENTOMOLOGY, VETERINARY ENTOMOLOGY. *Educ:* Univ Buffalo, BA, 50, MA, 52; Univ Wis, PhD(med & vet entom), 56. *Prof Exp:* Asst invert zool, Univ Buffalo, 50-52; vet entom, Univ Wis, 52-56; res entomologist, Sci & Educ Admin-Fed Res, USDA, 56-84; RES ASSOC, FLA STATE COL ARTHOPODS, 88- *Concurrent Pos:* Adj assoc prof entom, Miss State Univ, 65-75; adj prof entom, Univ Fla, 76- *Mem:* Entom Soc Am; Am Mosquito Control Asn; Am Soc Trop Med & Hyg. *Res:* Livestock-affecting insects, especially biology, control and taxonomy of Tabanidae and the vectors of bovine anaplasmosis; pesticide development on mosquitoes, houseflies, stableflies, ticks and chiggers. *Mailing Add:* 2241 NW 49th Terr Gainesville FL 32605

ROBERTS, RICHARD JOHN, b Derby, Eng, Sept 6, 43; m 65, 86; c 4. MOLECULAR BIOLOGY. *Educ:* Sheffield Univ, BSc, 65, PhD(chem), 68. *Prof Exp:* Fel chem, Sheffield Univ, 68-69; fel biochem, Harvard Univ, 69-70; res assoc, Cold Spring Harbor Lab, 71-72, sr staff investr, 72-86, ASST DIR RES, COLD SPRING HARBOR LAB, 86- *Concurrent Pos:* Consult, New Eng Biolabs, 75- & Genex Corp, 78-; John Simon Guggenheim fel, 79-80. *Mem:* AAAS; Am Soc Microbiol; Fedn Am Soc Exp Biol. *Res:* Restriction endonucleases and their application for DNA sequence analysis and genetic engineering; RNA splicing; computer applications in molecular biology; DNA methylases. *Mailing Add:* Cold Spring Harbor Lab Cold Spring Harbor NY 11724

ROBERTS, RICHARD NORMAN, b Lockport, NY, Sept 3, 30; m 61; c 2. ANALYTICAL CHEMISTRY. *Educ:* Univ Buffalo, BA, 57, MA, 59, PhD(biochem), 62. *Prof Exp:* Fel biochem, Rutgers Univ, 62-64; res assoc, Cornell Univ, 64-65; res biochemist, Gen Elec Co, Syracuse, 65-69, analytical chemist, 69-90; ENVIRON CHEM, STATE UNIV NY, OSWEGO, 90- *Mem:* AAAS; Am Chem Soc; Sigma Xi. *Res:* Intermediary metabolism; metabolic pathways; products of microbial metabolism on synthetic and natural substrates; related organic-analytical techniques especially gas chromatography and infrared spectrophotometry; gas chromatographic identification of micro-organisms; environment analysis. *Mailing Add:* 117 North Way Camillus NY 13031-1221

ROBERTS, RICHARD W, b Milwaukee, Wis, Dec 9, 21; m 43; c 2. ORGANIC CHEMISTRY. *Educ:* Marquette Univ, BS, 43; Lawrence Univ, MS, 49, PhD, 51. *Prof Exp:* Instr chem, Marquette Univ, 47; res chemist, Int Paper Co, 51-52 & Marathon Corp, 52-63; res supvr & Computer mgr, Wausau Paper Mills Co, 63-85; RETIRED. *Mem:* Am Chem Soc; Tech Asn Pulp & Paper Indust; Am Soc Qual Control. *Res:* Computer process control; pulp bleaching; product development; statistical techniques. *Mailing Add:* 845 Everest Dr Rothschild WI 54474-1020

ROBERTS, ROBERT ABRAM, b Iowa City, Iowa, Oct 28, 23; m 49; c 3. MATHEMATICS. *Educ:* WVa Wesleyan Col, BS, 45; Univ WVa, MS, 48; Univ Mich, PhD, 54. *Prof Exp:* Asst prof math, Univ WVa, 52-53; asst prof, Univ Miami, 53-58; sr mathematician, Bettis Plant, Westinghouse Elec Corp, Pa, 58-59; assoc prof math, Ohio Wesleyan Univ, 59-61; from assoc prof to prof, Denison Univ, 61-72; chmn dept, 72-77, PROF MATH, WASH & LEE UNIV, 72- *Concurrent Pos:* NSF fel, 65-66. *Mem:* AAAS; Am Math Soc; Math Asn Am; Soc Indust & Appl Math; Sigma Xi. *Res:* Mathematical physics; atomic structure; numerical analysis; computing machinery; educational use of machines. *Mailing Add:* Rte 5 Box 126 Lexington VA 24450

ROBERTS, ROBERT RUSSELL, b Fitchburg, Mass, Mar 4, 31; m 53; c 3. MICROBIOLOGY, IMMUNOLOGY. *Educ:* Brigham Young Univ, BS, 70, MS, 72, PhD(microbiol), 75. *Prof Exp:* MICROBIOLOGIST, HAWAIIAN SUGAR PLANTERS ASN, 76- *Mem:* AAAS; Am Soc Microbiol; Am Chem Soc. *Res:* Biomass energy; immunology; microbial and yeast genetics; industrial fermentations; general applied microbiology; wastewater treatment. *Mailing Add:* Hawaiian Sugar Planters Asn Aiea HI 96701

ROBERTS, ROBERT WILLIAM, b Riverside, Ill, July 24, 23; m 46; c 3. CHEMICAL ENGINEERING. *Educ:* Wash Univ, St Louis, BSChE, 48; Univ Iowa, MS, 60, PhD(chem eng), 62. *Prof Exp:* Develop engr, Aluminum Ore Co, 48, prod engr, 49-51; foreman extrusion, Cryovac Div, W R Grace & Co, 51-52, plant engr, 52-57, tech dir, Western Div, 57-59; group supvr process res, Allegany Ballistics Lab, Hercules, Inc, 62-64; dir res & develop, Cadillac Plastics & Chem Co, 64-66; from assoc prof to prof, Univ Akron, 66-77, head dept, 70-77, Robert Iredell prof chem eng & res assoc, Inst Polymer Sci, 77-88; RETIRED. *Concurrent Pos:* Dean, Plastics Eng Prog, Algerian Inst Petrol, 79-80. *Mem:* Am Inst Chem Engrs; Soc Plastics Engrs; Am Soc Eng Educ. *Res:* Process design and economics; plastic processing; thermodynamics; research administration. *Mailing Add:* 2352 Chatham Rd Akron OH 44313

ROBERTS, RONALD C, b Meadville, Pa, Jan 28, 36; m 60, 85; c 2. PHYSICAL BIOCHEMISTRY. *Educ:* Pa State Univ, BS, 57; Univ Minn, PhD(biochem), 64. *Prof Exp:* Res asst biochem, Univ Minn, 57-64, NIH fel & res assoc, 64-66; RES BIOCHEMIST, MARSHFIELD MED FOUND, 66- *Concurrent Pos:* Adj assoc prof chem, Univ Wis-Stevens Pt, 78-83; adj prof, Biol Dept, Univ Wis-Oshkosh, 80-82; vis scientist, Ctr Res & Develop Life Sci, E I Du Pont, Wilmington, Del, 83-84. *Mem:* Am Chem Soc; Am Soc Biochemists & Molecular Biol; AAAS; Sigma Xi; Am Diabetes Asn. *Res:* Physical and chemical characterization of proteins, particularly blood serum proteins; chemical and immunological studies on the antigens of hypersensitivity pneumonitis. *Mailing Add:* Marshfield Med Found 1000 N Oak Ave Marshfield WI 54449-5790

ROBERTS, RONALD FREDERICK, b Brooklyn, NY, Mar 26, 44; m 66; c 2. SURFACE CHEMISTRY. *Educ:* St John's Univ, BS, 65; Long Island Univ, MS, 67; NY Univ, PhD(phys chem), 72. *Prof Exp:* MEM TECH STAFF RES, BELL LABS, 71- *Mem:* AAAS; Am Chem Soc; Soc Appl Spectroscopy. *Res:* Electron spectroscopic investigation of the adsorption of chemical species onto inorganic surfaces as related to corrosion inhibition, and the chemical modification of polymer surfaces as related to adhesion and other phenomena. *Mailing Add:* AT&T Erc Princeton PO Box 900 Princeton NJ 08540

ROBERTS, RONNIE SPENCER, b Pascagoula, Miss, June 5, 43; m 72; c 1. CHEMICAL ENGINEERING, BIOCHEMICAL ENGINEERING. *Educ:* Univ Miss, BSChE, 66; Univ Tenn, Knoxville, MS, 72, PhD(chem eng), 76. *Prof Exp:* Tech serv engr, Monsanto Co, 66-69; asst prof, 76-81, ASSOC PROF CHEM ENG, GA INST TECHNOL, 81- *Concurrent Pos:* Consult, Milliken & Co, 77-78, Stake Technol Ltd, 80-83, FTI, 89- *Mem:* Am Inst Chem Engrs; Am Soc Eng Educ; AAAS. *Res:* Vigorous stationary phase fermentation; microbial processes; reactor design; solvent delignification of biomass. *Mailing Add:* Sch Chem Eng Ga Inst Technol Atlanta GA 30332-0100

ROBERTS, ROYSTON MURPHY, b Sherman, Tex, June 11, 18; m 43; c 4. ORGANIC CHEMISTRY. *Educ:* Austin Col, BA, 40; Univ Ill, MA, 41, PhD(org chem), 44. *Hon Degrees:* DSc, Austin Col, 65. *Prof Exp:* Asst, Comt Med Res, Off Sci Res & Develop, Univ Ill, 43-45; res chemist, Merck & Co, 45-46; fel, Univ Calif, Los Angeles, 46-47; from asst prof to assoc prof, 47-61, PROF CHEM, UNIV TEX, AUSTIN, 61- *Concurrent Pos:* Fel, Petrol Res Fund, Zurich, 59-60; vis prof, Phillipps Univ, Marburgl, 67, Bucharest Polytech, 76; vis lectr, various Univ, Egypt, 82. *Mem:* Am Chem Soc. *Res:* Organic synthesis; reaction mechanisms; reactions of ortho esters; imidic esters and amidines; new Friedel-Crafts chemistry; thermal rearrangement reactions; author of three books and more than 100 sci articles, patents and reviews. *Mailing Add:* Dept Chem Univ Tex Austin TX 78712-1167

ROBERTS, RUFUS WINSTON, ophthalmology, for more information see previous edition

ROBERTS, SANFORD B(ERNARD), b New York, NY, Feb 20, 34; m 55; c 3. BIOMECHANICS, STRUCTURAL MECHANICS. *Educ:* City Col New York, BCE, 56; Univ Southern Calif, MSCE, 59; Univ Calif, Los Angeles, PhD(eng), 65. *Prof Exp:* Engr, Struct Space Div, N Am Aviation, Inc, 56-57; asst prof civil eng, Univ Southern Calif, 57-61; assoc eng, Univ

Calif, Los Angeles, 61-63; design specialist, Rocketdyne Div, N Am Aviation, Inc, 63-65; asst prof eng, 65-71, ASSOC PROF ENG, UNIV CALIF, LOS ANGELES, 71- *Concurrent Pos:* Pres, Asn Sci Adv, 71-; consult, Civil & Aerospace Eng Co. *Mem:* Am Soc Civil Engrs. *Res:* Biomechanics, especially the structural dynamic behavior of the human body under normal and traumatic conditions; structural mechanics, theory of plates and shells. *Mailing Add:* PO Box 970 Sanford FL 32771

ROBERTS, SHEPHERD (KNAPP DE FOREST), b Princeton, NJ, Mar 15, 32; m 55; c 4. COMPARATIVE PHYSIOLOGY. *Educ:* Princeton Univ, AB, 54, MA, 57, PhD(biol), 59. *Prof Exp:* Instr biol, Princeton Univ, 59-61; from asst prof to assoc prof, 61-75, PROF BIOL, TEMPLE UNIV, 75- *Concurrent Pos:* Prin investr, Bermuda Biol Sta, Off Naval Res, 59-60. *Mem:* AAAS; Animal Behav Soc; Int Soc Chromobiol. *Res:* Biological chronometry; diurnal rhythmic activities in animals. *Mailing Add:* Dept Biol Temple Univ Broad & Montgomery Philadelphia PA 19122

ROBERTS, SIDNEY, b Boston, Mass, Mar 11, 18; m 43. ENDOCRINOLOGY, NEUROCHEMISTRY. *Educ:* Mass Inst Technol, SB, 39; Univ Minn, MS, 42, PhD(biochem), 43. *Prof Exp:* Instr physiol, Sch Med, Univ Minn, 43-44; instr, Sch Med, George Washington Univ, 44-45; res assoc, Worcester Found Exp Biol, 45-47; asst prof physiol chem, Sch Med, Yale Univ, 47-48; from asst prof to assoc prof, 48-57, PROF BIOL CHEM, SCH MED, UNIV CALIF, LOS ANGELES, 57- *Concurrent Pos:* Guggenheim fel, Univ London, 57; consult, Vet Admin Hosp, Long Beach, Calif, 51-55, NSF, 55-59, Vet Admin Hosp, Los Angeles, 58-62, Los Angeles Co Heart Asn, 58-60, 62-63 & NIH, 60-63; exec comn, Am Chem Soc, 56-59; ed, Brain Res, Neurochem Res, Am J Physiol & Proc Soc Exp Biol & Med; chair, academic senate, Univ Calif, Los Angeles, 89-90. *Honors & Awards:* Ciba Award, Endocrine Soc, 53. *Mem:* AAAS; Endocrine Soc (vpres, 68-69); Am Soc Biol Chemists; Am Soc Neurochem; Am Physiol Soc; Am Chem Soc; Soc Neurosci; Biochem Soc Gt Brit; Int Soc Neurochem; Sigma Xi (pres, 59-60). *Res:* Regulation and role of protein synthesis and protein phosphorylations in endocrine and neural function. *Mailing Add:* Dept Biol Chem Sch Med Univ Calif Los Angeles CA 90024

ROBERTS, STEPHEN D, b Warsaw, Ind, Sept 3, 42. COMPUTER SCIENCE, COMPUTER SIMULATION. *Educ:* Purdue Univ, BS, 65, MS, 66, PhD(indust eng), 68. *Prof Exp:* From assoc prof to prof indust eng, Purdue Univ, 72-90; PROF INDUST ENG & HEAD DEPT, NC STATE UNIV, 90- *Concurrent Pos:* From assoc prof to prof internal med, Sch Med, Ind Univ, 72-90; head health systs, Regenstrief Inst, 74-90. *Mem:* Asn Comput Mach; Inst Mgt Sci; Inst Indust Engrs. *Mailing Add:* Dept Indust Eng NC State Univ Box 7906 Raleigh NC 27695-7906

ROBERTS, STEPHEN WINSTON, b Chicago, Ill, May 4, 41. SYSTEMS DESIGN & SYSTEMS SCIENCE. *Educ:* San Diego State Univ, BS, 72, MS, 74; Duke Univ PhD(plant ecol), 78. *Prof Exp:* Res asst, Dept Biol, San Diego State Univ, 72-74; teaching & res asst, Dept Forestry, Duke Univ, 74-77; adj prof biol, San Diego State Univ, Calif & res prof, Systs Ecol Res Group, 78-84; assoc prof, Fac Ciencias Biol, Pontificia Univ Catolica Chile, Santiago, 84-85; PRES, DATA DESIGN GROUP, LA JOLLA, CALIF, 85- *Concurrent Pos:* Grad fel, Duke Univ, 74-77; Res Award, Duke Univ, 76 & 77; co-prin investr, NSF grant, 77-81 & 81-84; environ consult, Chambers Consults & Planners, Stanton, Calif, 78-84; prin investr, US Dept Agr Grant, 81-84; researcher, photosynthesis in tropical plants, Orgn Tropical Studies Res Sta, Costa Rica, 84-85; Coop Res Projs, USDA Forest Serv, Riverside Fire Lab, 89- *Mem:* AAAS; Ecol Soc Am; Sigma Xi. *Res:* Ecophysiology research in plants, primarily gas exchange and water relations; measuring instruments associated with ecophysiology research; author of various publications. *Mailing Add:* Data Design Group PO Box 3318 La Jolla CA 92038

ROBERTS, SUSAN JEAN, b Plainfield, NJ, Feb 7, 58. DEVELOPMENTAL BIOLOGY, CELL BIOLOGY. *Educ:* Duke Univ, BS, 80; Univ Calif, San Diego, PhD(marine biol), 86. *Prof Exp:* Postdoctoral, Univ Calif, San Diego, 86-87; NIH postdoctoral fel, Univ Calif, Berkeley, 87-91; ASST PROF BIOL, UNIV HOUSTON, 92- *Mem:* AAAS; Soc Develop Biol; Am Soc Cell Biol; Am Soc Zoologists. *Res:* Early amphibian development and cell differentiation; regulation of protein secretion during the cell cycle and in embryonic epithelial cells; role of secreted proteins in pattern formation. *Mailing Add:* Dept Biol Univ Houston Houston TX 77204-5513

ROBERTS, THEODORE S, b Waukesha, Wis, July 29, 26; m 53; c 4. NEUROSURGERY. *Educ:* Univ Wis, BS, 50, MS, 52, MD, 55. *Prof Exp:* Instr neurosurg & neuroanat, Sch Med, Univ Wis, 60-61; from instr to prof neurosurg, Col Med, Univ Utah, 61-85, chmn, div neurol surg, 64-85; PROF NEUROL SURG, UNIV HOSP DEPT NEUROSURG, UNIV WASH, 85- *Concurrent Pos:* Clin investr, Vet Admin Hosp, 61-63. *Mem:* Am Asn Neurol Surg; fel Am Col Surgeons; AMA. *Res:* Neuroanatomy and neurophysiology; techniques of sterotoxic surgery; tumors and epilepsy disorders. *Mailing Add:* Rm 20 Univ Hosp Univ Wash 1959 Pacific NE Seattle WA 98195

ROBERTS, THOMAS D, b New York, NY, June 21, 35; m 57; c 3. ELECTROPHYSICS. *Educ:* Ore State Univ, PhD(physics), 65. *Prof Exp:* Physicist, US Bur Mines, Ore, 59-64 & Nat Bur Stand, Colo, 65-66; assoc prof physics, 66-74, prof physics & elec eng, 74-88, EMER PROF PHYSICS & ELEC ENG, UNIV ALASKA, 87- *Mem:* Am Phys Soc; Inst Elec & Electronics Engrs; Nat Asn Physics Teachers. *Res:* Propagation of electromagnetic radiation; communications systems. *Mailing Add:* Dept Elec Eng Univ Alaska Fairbanks AK 99701

ROBERTS, THOMAS DAVID, b Quanah, Tex, July 11, 38; m 57; c 3. ORGANIC CHEMISTRY, POLYMER CHEMISTRY. *Educ:* Abilene Christian Col, BS, 59; Ohio State Univ, PhD(chem), 67. *Prof Exp:* Res chemist, Explor Group, Pittsburgh Plate Glass Co, Ohio, 59-62; from asst prof to assoc prof chem, Univ Ark, Fayetteville, 67-77, prof, 77-79; chemist, 79-84, sr chem, 84-89, PRIN CHEMIST, TEX EASTMAN CO, 89- *Concurrent Pos:* Petrol Res Fund grant, 67-69, 71-73 & 75-79; Res Corp grant, 69; vis assoc prof, Univ Fla, 73. *Res:* Synthesis of strained small ring and pseudo-aromatic compounds; new polymers; new synthetic reactions. *Mailing Add:* Res Labs Tex Eastman Co Longview TX 75602

ROBERTS, THOMAS GEORGE, b Ft Smith, Ark, Apr 27, 29; m 58; c 2. LASERS, PLASMA PHYSICS. *Educ:* Univ Ga, BS, 56, MS, 57; NC State Univ, PhD(physics), 67. *Prof Exp:* Instr physics, Univ Ga, 56-57; physicist, Army Rocket & Guided Missile Agency, 57-62 & US Army Missile Command, 62-85; staff scientist, Phys Dynamics, Inc, 86-89; RES SCIENTIST, TECHNOCO, HUNTSVILLE, ALA, 85- *Concurrent Pos:* Consult, Ballistic Missile Defense Technol Ctr & Southeastern Inst Technol, 75-85, SAIC, Celtic Res & BDM, 85- *Mem:* Am Phys Soc; fel Optical Soc Am; sr mem Inst Elec & Electronics Engrs. *Res:* Behavior of very intense relativistic electron beams in plasmas; effects of short duration pulses of high energy on materials; high power laser technology where the active medium is a plasma or aerodynamic gases; the physics of high energy partice beams. *Mailing Add:* Technoco PO Box 4723 Huntsville AL 35815-4723

ROBERTS, THOMAS GLASDIR, b Clintonville, Wis, Aug 3, 18. MICROPALEONTOLOGY, STRATIGRAPHY. *Educ:* Univ Wis, PhB, 40; Columbia Univ, PhD(geol), 49. *Prof Exp:* Stratigrapher, Shell Oil Co, 41-46; instr geol, Univ Kans, 46-47; geologist & lab chief, Arctic Coastal Area Subsurface Invests, US Geol Surv, Alaska, 49-52, subsurface stratigrapher, Four Corners area, NMex, 52-55; from asst prof to prof, 56-85, EMER PROF GEOL, UNIV KY, 85- *Concurrent Pos:* Lectr, Univ Tulsa, 44-45; Fulbright & AID lectr, Ecuador, 71. *Mem:* Geol Soc Am; Am Asn Petrol Geologists; Sigma Xi. *Res:* Subsurface stratigraphy; Fusulinidae; phylogeny of Mississippian Archaeodiscidae of western Kentucky. *Mailing Add:* 612 Stratford Dr Lexington KY 40503

ROBERTS, THOMAS L, b Key West, Fla, Apr 17, 32; c 3. MICROBIAL BIOCHEMISTRY, GENETICS. *Educ:* Talladega Col, AB, 57; Trinity Univ, Tex, MS, 61; Clark Univ, PhD(microbiol, biochem), 65. *Prof Exp:* Res asst physiol, Med Br, Univ Tex, Galveston, 57-58; microbiologist, USAF Sch Aerospace Med, 58-63; from asst prof to prof biol, Worcester State Col, 65-69, chmn dept biol, 72-76, prof biol, 69-88, educ dir, Nuclear Med Technol Prog, Mass Med Sch, 76-88; RETIRED. *Concurrent Pos:* Res assoc, Clark Univ, 65-67, res affiliate, 67-68; trustee, Rehab Ctr Worcester County, Inc, 75-81, pres, 77-81. *Mem:* Fel Sigma Xi; AAAS; NY Acad Sci; Am Soc Microbiol; fel Am Inst Chemists; Sigma Xi. *Res:* Immunochemistry related to learning under physical stresses. *Mailing Add:* 321 Catherine St Key West FL 33040-7504

ROBERTS, THOMAS M, DIRECT & INVERSE SCATTERING, ELECTROMAGNETIC MEASUREMENTS. *Educ:* Univ Conn, BS, 81; Ind Univ, MS, 83, MA, 85, PhD(math physics), 87. *Prof Exp:* Res & teaching asst, Ind Univ, 81-85; sr tech analyst, Sabbagh Assocs, 85-89; postdoctoral fel, Ames Lab, US Dept Energy, 89-91; RESEARCH SCHOLAR, SCH AEROSPACE MED, BROOKS AFB, 91- *Mem:* Am Math Soc; Am Phys Soc. *Res:* Applied mathematics and mathematical physics; direct and inverse scattering for electromagnetically dispersive materials; quantum inverse scattering; direct scattering for anisotropic conductors. *Mailing Add:* Radiation Analysis Br Sch Aerospace Med Brooks AFB TX 78235-5301

ROBERTS, TIMOTHY R, b Middletown, Ohio, Mar 12, 54; m 74; c 1. ELECTROPLATING, PAINT FORMULATION. *Educ:* Univ Cincinnati, BS, 80, MS, 86. *Prof Exp:* Sr res engr corrosion & non-metallic coatings res, 80-87, sr staff supvr, 87-90, MGR METALLIC COATINGS RES, ARMCO RES & TECHNOL, 90- *Concurrent Pos:* Thesis adv, Univ Cincinnati, 86-88; vis scholar, Dept Educ, State Ohio, 90-91. *Res:* Develops new products and processes dealing with metallic coatings for steel substrates; electroplated steels; hot-dipped galvanized and vapor-deposited coatings; author of 13 publications; granted two patents. *Mailing Add:* 614 Quail Run Rd Middletown OH 45042

ROBERTS, VERNE LOUIS, b Kansas City, Mo, Aug 11, 39; m 57; c 3. BIOMECHANICS, MECHANICAL ENGINEERING. *Educ:* Univ Kans, BS, 60; Univ Ill, MS, 61, PhD(eng mech), 64. *Prof Exp:* From asst prof to assoc prof eng mech, Wayne State Univ, 63-66, assoc prof neurosurg, 66; head dept biomech, Univ Mich, Ann Arbor, 66-73; ADJ PROF, DEPT MECH ENG, DUKE UNIV, 73-, DIR, INST PROD SAFETY, 79- *Concurrent Pos:* Assoc urol, Wayne State Univ, 64-66, fac res fel, 65; consult, Vet Admin, 64-66; mem engrs Joint Coun Comt Interaction Eng with Med & Biol, 65-66; co-ed in chief, J Biomech, 67-; mem, Comt Head Protection, Unified Space Appln Mission Standards Inst, 68-; mem, Adv Comt, Stapp Car Crash Conf, 68-; dir, Nat Driving Ctr, 73-78. *Honors & Awards:* Res Award, Am Soc Testing & Mat, 66. *Mem:* Am Soc Mech Engrs; Am Soc Safety Engrs; Syst Safety Soc; Am Soc Biomech; Europ Soc Biomech. *Res:* Safe product design. *Mailing Add:* PO Box 1931 Durham NC 27702

ROBERTS, W(ILLIAM) NEIL, b Prince Albert, Sask, May 27, 31; m 58; c 1. METALLURGY. *Educ:* Univ Sask, BA, 52, Hons, 53, MA, 54; Univ Leeds, PhD(metall), 61. *Prof Exp:* Patent examr, Can Govt Patent & Copyright Off, 55-56; sci officer, Can Dept Mines & Tech Surv, 57-58, res scientist, Metal Physics Sect, 61-72, HEAD METAL PHYSICS SECT, PHYS METALL RES LABS, CAN DEPT ENERGY, MINES & RESOURCES, 72- *Mem:* Am Soc Metals; Am Inst Mining, Metall & Petrol Engrs; Am Soc Testing & Mat. *Res:* Development of improved sutures for microsurgery; evaluation of steels for line pipe; transmission electron microscopy of austenitic steels and chromium molybdenum steels for fuel conversion; transmission electron microscopy of dual phase steels for automotive applications. *Mailing Add:* 74 Rothwell Dr Ottawa ON K1J 7G6 Can

ROBERTS, WALDEN KAY, b Independence, Mo, July 1, 34; m 66; c 3. MICROBIOLOGY, BIOCHEMISTRY. *Educ:* Iowa State Univ, BS, 56; Univ Calif, Berkeley, PhD(biochem), 60. *Prof Exp:* NSF fel, Univ Newcastle, Eng, 60-62; asst prof molecular biol, Univ Calif, Berkeley, 62-67; PROF MICROBIOL, MED SCH, UNIV COLO, DENVER, 76- *Mem:* Am Soc Biol Chem. *Res:* Antifungal proteins from plants; cell-specific toxins; interferon. *Mailing Add:* Dept Microbiol & Immunol Univ Colo Med Ctr 4200 E Ninth Ave Denver CO 80262

ROBERTS, WALTER HERBERT B, b Field, BC, Jan 24, 15; US citizen; c 3. ANATOMY. *Educ:* Loma Linda Univ, MD, 39. *Prof Exp:* Med dir, Rest Haven Hosp, Sidney, BC, 40-53; PROF ANAT, LOMA LINDA UNIV, 56- *Mem:* Am Asn Anatomists. *Res:* Gross anatomy. *Mailing Add:* Dept Anat Loma Linda Univ Loma Linda CA 92350

ROBERTS, WALTER ORR, solar physics; deceased, see previous edition for last biography

ROBERTS, WARREN WILCOX, b Lincoln, Nebr, Oct 22, 26; m 67; c 2. NEUROSCIENCES. *Educ:* Stanford Univ, BA, 48, MA, 53; Yale Univ, PhD(psychol), 56. *Prof Exp:* Fel psychol, Yale Univ, 56-57; asst res anatomist, Univ Calif, Los Angeles, 57-59; asst prof psychol, Syracuse Univ, 59-62; from assoc prof to prof, 62-89, EMER PROF PSYCHOL, UNIV MINN, MINNEAPOLIS, 89- *Mem:* Soc Neurosci. *Res:* Brain mechanisms of motivation; thermoregulation. *Mailing Add:* N218 Elliott Hall Univ Minn Minneapolis MN 55455

ROBERTS, WENDELL LEE, b Orange City, Iowa, July 26, 38; div; c 4. INSECT PHEROMONES. *Educ:* Cent Col, Iowa, BA, 60; Univ Ind, PhD(org chem), 64. *Prof Exp:* NIH fel org chem, Mass Inst Technol, 64-65; from asst prof to assoc prof chem, 65-75, PROF INSECT BIOCHEM, STATE UNIV NY COL AGR, CORNELL UNIV, 76- *Honors & Awards:* J Everett Bossert Entom Award, 73; Alexander Von Humboldt Agr Award, 77. *Mem:* Nat Acad Sci; Am Chem Soc; Entom Soc Am; AAAS; Sigma Xi; NY Acad Sci. *Res:* Synthetic organic chemistry; isolation and characterization of sex, food and oviposition attractants of insects. *Mailing Add:* Dept Entom NY State Agr Exp Sta Geneva NY 14456

ROBERTS, WILLARD LEWIS, b Milton Junction, Wis, Mar 24, 04; m 28; c 3. NUTRITIONAL BIOCHEMISTRY. *Educ:* Milton Col, BA, 27; Univ Wis, MA, 32, PhD(biochem), 37. *Prof Exp:* Chemist, State Regulatory Dept, NDak, 28-30; asst chem, Univ Wis, 30-32; chemist, State Regulatory Dept, NDak, 32-35; asst biochem, Univ Wis, 35-37; dir cereal res, Post Prod Div, Gen Foods Corp, 37-41, dir biochem & analytical chem res, Cent Labs, 42-45, tech dir, Gaines Div, 45-47; gen mgr & secy, Fed Foods, Inc, Wis, 47-70; mgr res, agr-prod, Beatrice Foods Res Ctr, 70-; RETIRED. *Concurrent Pos:* Consult. *Mem:* Am Chem Soc. *Res:* Research, development and production of human foods and pet and fur bearing animal foods. *Mailing Add:* PO Box 62 Thiensville WI 53092-0062

ROBERTS, WILLIAM C, PATHOLOGY. *Prof Exp:* MEM STAFF, PATH BR, NIH. *Mailing Add:* Pathology Branch NIH Bldg 10A, 9000 Rockville Pike Bethesda MD 20892

ROBERTS, WILLIAM JOHN, b Philadelphia, Pa, June 5, 18; m 48; c 2. CHEMISTRY, FIBER TECHNOLOGY. *Educ:* Univ Pa, AB, 42, MS, 44, PhD(org chem), 47. *Prof Exp:* Asst chemist, United Gas Improv Co, Pa, 36-41, res chemist, 41-44, asst to mgr, 44-45; asst to dir, Pa Indust Chem Corp, 46-47, from asst dir res to dir res, 48-57; dir res, Summit Res Labs, Celanese Corp, 57-64, vpres & tech dir, Celanese Fibers Co, 65-74; vpres & tech dir, Fiber Div, FMC Corp, 74-76; consult, 76-86; RETIRED. *Concurrent Pos:* Asst instr, Univ Pa, 42-44. *Mem:* Am Chem Soc; Asn Res Dirs. *Res:* Organic synthesis; hydrocarbons; polymerization; isomerization; pyrolysis and petrochemicals; polymers; plastics; fibers, especially man-made fibers; oxygenated chemicals; research management. *Mailing Add:* 65 Peachcroft Rd Bernardsville NJ 07924

ROBERTS, WILLIAM KENNETH, b Provo, Utah, Dec 10, 28; m 56; c 3. RUMINANT NUTRITION. *Educ:* Calif State Polytech Col, BS, 52; Wash State Univ, MS, 57, PhD(animal nutrit), 59. *Prof Exp:* Res asst animal sci, Wash State Univ, 55-59; fel nutrit & biochem, Grad Sch Pub Health, Univ Pittsburgh, 59-60; from asst prof to assoc prof, animal nutrit, Univ Manitoba, 60-66; res specialist, Kern County Land Co, Calif, 66-68; NUTRITIONIST, NUTRITION-LABORATORY SERV, INC, 68- *Concurrent Pos:* Nat Res Coun Can res grants, 61-66. *Mem:* Am Soc Animal Sci; Can Soc Animal Sci; Sigma Xi; AAAS. *Res:* Ruminant nutrition concerning fat deposition and volatile fatty acid metabolism; vitamin A utilization; the role of potassium in ruminant nutrition. *Mailing Add:* Nutrit-Lab Servs Inc Box 237 Tolleson AZ 85353

ROBERTS, WILLIAM WOODRUFF, JR, b Huntington, WVa, Oct 8, 42; m 67; c 2. APPLIED MATHEMATICS, ASTRONOMY. *Educ:* Mass Inst Technol, SB, 64, PhD(appl math), 69. *Prof Exp:* From asst prof to assoc prof, 69-82, PROF APPL MATH, UNIV VA, 82- *Concurrent Pos:* Consult, Du Pont, 86-, Union Carbide Corp, 81; vis scientist, Inst Hautes Etudes Scientifiques, Bures-sur-Yvette, France, 74, Kapteyn Astron Inst, Univ Groningen, Neth, 74, Huygens Lab, Univ Leiden, Neth, 75, Int Bus Mach, T J Watson Res Ctr, Yorktown Heights, NY, 80, Inst Comput Appln Sci & Eng, NASA Langley Res Ctr, Hampton, Va, 80-81, Nat Radio Astron Observ, Va, 88; Nordita guest prof, Stockholms Observ, Saltsjobaden, Sweden, 74-75; mem, Va Inst Theoretical Astron. *Mem:* Am Astron Soc; Soc Indust & Appl Math; Int Astron Union; Sigma Xi; Am Inst Aeronaut & Astronaut. *Res:* Fluid mechanics; dynamics of galaxies of gas and stars; shock waves; star formation; nonlinear wave motion; computational mathematics; computational fluid dynamics; rarefied gas dynamics. *Mailing Add:* 309 Westminster Rd Charlottesville VA 22901

ROBERTSEN, JOHN ALAN, b Kenosha, Wis, June 21, 26; m 57; c 2. MEDICAL MICROBIOLOGY, ENVIRONMENTAL AND BIOHAZARD CONTROL. *Educ:* Univ Wis, BS, 52, MS, 56, PhD(med microbiol), 58. *Prof Exp:* Chief, Microbiol Sect, Leprosy Res, USPHS Hosp, Carville, La, 58-62; leader, Infectious Diseases Group, Pitman-Moore Div, Dow Chem Co, 62-65, leader, microbiol group, Biohazards Dept, Dow Chem Co, 65-69; dir microbiol, St Francis Hosp & Thornton-Haymond Labs, Indianapolis, 69-71; dir tech mkt, Kallestad Labs, Inc, Minneapolis, 71-73; VPRES & CHIEF OPERATING OFFICER, EXTENSOR CORP, 73- *Concurrent Pos:* Former US contrib ed, Int J Leprosy; founder & pres, Robertsen Biomed Consults; prin, Robertsen & Assocs Mgt Consults; consult, Zinpro Corp, Sci Int Res, Inc & others. *Mem:* AAAS; Int Leprosy Asn; Am Asn Contamination Control; Am Soc Trop Med & Hyg; Am Asn Lab Animal Sci; Am Soc Performance Improv; Am Radio Relay League; Nat Disaster Med Syst. *Res:* Immunodiagnostics; host-parasite relationships in tropical diseases, especially leprosy; control and containment of biological hazards; hospital environmental control; time allocation, behavior and task analyses of persons in health-care-related and administrative or professional positions; work measurement methodology at the managerial level; 8 patent disclosures; food processing industry sanitation. *Mailing Add:* 17273 Hampton Ct Minnetonka MN 55345-2517

ROBERTSHAW, JOSEPH EARL, b Providence, RI, Apr 19, 34. SYSTEMS ANALYSIS, ENERGY MANAGEMENT. *Educ:* Providence Col, BS, 56; Mass Inst Technol, MS, 58, PhD(physics), 61. *Prof Exp:* Assoc prof, 61-72, PROF PHYSICS, PROVIDENCE COL, 72- *Concurrent Pos:* Vis prof systs eng, Ga Inst Technol, 70-71; consult, GTE Labs, Waltham, Mass, 81-82. *Mem:* Am Phys Soc; Nat Asn Physics Teachers; Inst Elec & Electronics Engrs; Asn Energy Eng; Am Soc Eng Educ. *Res:* Systems analysis and engineering; systems approach to problem solving in engineering designs; socio-technical fields; energy management; factory productivity improvement; control theory; optimization. *Mailing Add:* Dept Physics Eng Systs Providence Col Providence RI 02918

ROBERTS-MARCUS, HELEN MIRIAM, b Panama City, Panama, June 22, 43; US citizen. BIOSTATISTICS. *Educ:* City Col New York, BS, 64; Johns Hopkins Univ, PhD(biostatist), 70. *Prof Exp:* Asst prof statist, Univ Calif, Riverside, 70-72; asst prof, Montclair State Col, 72-77, assoc prof math, 77-80. *Concurrent Pos:* Consult, Pac State Hosp, Pomona, Calif, 70-72, Loma Linda Univ Hosp, Calif, 70-72, Hosp Res Assoc, Fanwood, NJ, 77-79 & Boyle-Midway, Cranford, NJ, 81-; vis assoc prof, Oper Res & Indust Eng, Cornell Univ, 79-80. *Mem:* Am Statist Asn; Math Asn Am; Soc Indust Appl Math. *Res:* Stochastic models in biology; mathematical genetics. *Mailing Add:* Dept Math Comput Sci Montclair State Col Upper Montclair NJ 07043

ROBERTSON, A(LEXANDER) F(RANCIS), b Vancouver, BC, Aug 31, 12; nat US; m 46; c 3. MECHANICAL ENGINEERING. *Educ:* Univ Wis, BS, 35, MS, 38, PhD(mech eng), 40. *Prof Exp:* Engr, Fairbanks, Morse & Co, Wis, 35-37, res engr, 40-41; asst, Univ Wis, 37-38; engr, US Naval Ord Lab, DC, 41-46 & Manhattan Dist, Chicago, 46; res engr, Battelle Mem Inst, 46-47; res assoc, Inst Textile Technol, 47-50; chief fire res sect, 50-68, tech asst dir inst appl technol, 68-70, tech asst chief fire technol div, 70-73, sr scientist, 73-85, GUEST WORKER, CTR FIRE RES, NAT BUR STANDARDS, 85- *Concurrent Pos:* N Am ed, Fire & Mat, 76-85. *Honors & Awards:* Gold Medal Award, US Dept Com, 75; Edward Bennett Rosa Award, Nat Bur Standards, 78; Charles P Dudly Award, Am Soc Testing Mat, 86. *Mem:* Am Phys Soc; Combustion Inst; AAAS. *Res:* Diesel and gaseous combustion; heat transfer; electromechanical transducers; problems relating to unwanted fires; national and international fire standards. *Mailing Add:* 4228 Butterworth Pl NW Washington DC 20016

ROBERTSON, ABEL ALFRED LAZZARINI, JR, b Argentina, July 21, 26; m 58. PALEO PATHOLOGY. *Educ:* Cambridge Col, Arg, BA, 42; Univ Buenos Aires, MD, 51; Cornell Univ, PhD, 59. *Prof Exp:* Instr, Inst Parasitol, Med Sch, Univ Buenos Aires, 48-50, resident, 49-51, lectr, 50-51; from asst prof to assoc prof res surg, Postgrad Med Sch, NY Univ, 55-60, assoc prof path, 60-63; prof exp path, Cleveland Clin Found & mem staff, Cleveland Clin, 63-73; prof path & dir, Interdisciplinary Cardiovasc Res, Inst Path Case Western Reserve Univ, 73-82; PROF PATH & HEAD DEPT, UNIV ILL, 82- *Concurrent Pos:* Instr, Navy Hosp, Buenos Aires, Arg, 47-49; dir tissue bank, Buenos Aires Dept Pub Health, 48-51; asst vis surgeon, Bellevue Hosp, NY, 55 & NY Univ Hosp, 55; Am Heart Asn estab investr, 56-; estab investr, Am Heart Asn, 56-61; res career develop award, NIH, 61-63, mem path study sect, 75-; lectr, Univs Caracas, Rio de Janeiro, Sao Paulo, Riberao Preto, Buenos Aires & La Plata; assoc clin prof, Sch Med, Case Western Reserve Univ, 69-73; mem exec comt, Inter-Am Asn Atherosclerosis; deleg, Int Cong Hemat & Pan-Am Cong Am Col Surgeons; chmn, Gordon Res Conf Arteriosclerosis, 73; chmn & host, Hugh Lofland Conf Arterial Wall Metab, 78 & 88; counr angiol, Pan-Am Med Asn. *Honors & Awards:* Hon Achievement Award, Angiol Found. *Mem:* AAAS; Tissue Cult Asn; Am Asn Path; Am Soc Cell Biol; Am Col Cardiol. *Res:* Cardiovascular pathology, cell and organ culture; tissue transplantation immunity; molecular biology. *Mailing Add:* Dept Path MC 847 Univ Ill Med Ctr Chicago 1853 W Polk St Chicago IL 60612

ROBERTSON, ALAN ROBERT, b Wakefield, Eng, Sept 3, 40; m 64; c 2. COLORIMETRY, COLOR SCIENCE. *Educ:* Univ London, BSc, 62, PhD(physics), 65. *Prof Exp:* RES OFFICER PHYSICS, NAT RES COUN CAN, 65- *Concurrent Pos:* Assoc ed, Color Res & Appln, 78-; pres, Can Nat Comt, Int Comn Illum, 84-; pres, Int Color Asn, 90- *Mem:* Can Soc Color (pres, 79-81); Inter-Soc Color Coun; Optical Soc Am; Int Color Asn (vpres, 86-89). *Res:* Colour-difference evaluation, colour measurment, spectrophotometry, colour rendering, reflectance standards. *Mailing Add:* Nat Res Coun Montreal Rd Ottawa ON K1A 0R6 Can

ROBERTSON, ALEX F, b Staunton, Va, Dec 5, 32. PEDIATRICS, BIOCHEMISTRY. *Educ:* Univ Va, BA, 53, MD, 57; Univ Mich, MA, 62. *Prof Exp:* From instr to asst prof pediat, Univ Mich, 63-65; from asst prof to assoc prof, Ohio State Univ, 65-71, dir div neonatology, 65-71; chmn dept, 71-81, PROF PEDIAT, MED COL GA, 71- *Mem:* Soc Pediat Res; Am Pediat Soc; Am Acad Pediat; Sigma Xi; AOA. *Res:* Neonatology; bilirubin-albumin binding in neonates. *Mailing Add:* Dept Pediat Med Col Ga Augusta GA 30912

ROBERTSON, ALEXANDER ALLEN, b Edmonton, Alta, Jan 6, 19; m 49. PHYSICAL CHEMISTRY. *Educ:* Univ Alta, BSc, 40; McGill Univ, PhD(chem), 49. *Prof Exp:* Prin scientist, Pulp & Paper Res Inst Can, 49-80, dir appl chem div, 80-84; res assoc, 56-84, VIS SCIENTIST, MCGILL UNIV,

84- *Concurrent Pos:* Phys chemist, Swed Forest Prod Res Lab, 53-54. *Mem:* Can Pulp & Paper Asn; Sigma Xi. *Res:* Pulp and paper technology; physical properties of fibers and suspensions; polymer cellulose interactions; latexes. *Mailing Add:* Dept Chem McGill Univ Pulp & Paper Bldg Montreal PQ H3A 2T5 Can

ROBERTSON, ANDREW, b Port Huron, Mich, Sept 15, 36; m 65; c 2. AQUATIC ECOLOGY. *Educ:* Univ Toledo, BS, 58; Univ Mich, MA, 61, PhD(zool), 64. *Prof Exp:* Asst res limnologist, Univ Mich, Ann Arbor, 64-67, assoc res limnologist, 67-68; assoc prof zool, Univ Okla, 68-71; fishery biologist, IFYGL Proj Off, Nat Oceanic & Atmospheric Admin, 71-74, head biol chem group, Great Lakes Environ Res Lab, 74-81, dep dir, Off Marine Pollution Assessment, 81-82, dir, Nat Marine Pollution Prog Off, 82-86, CHIEF, OCEAN ASSESSMENTS DIV, NAT OCEANIC & ATMOSPHERIC ADMIN, 86- *Mem:* Am Soc Limnol & Oceanog; Ecol Soc Am; Crustanean Soc; Int Asn Great Lakes Res; Estuarine Res Fedn. *Res:* Great Lakes ecology; systematics and distribution of calanoid copepods; modeling of Great Lakes ecosystems; marine pollution. *Mailing Add:* NOAA/NOS N/OMA3 Rm 323 WSC-1 6001 Executive Blvd Rockville MD 20852

ROBERTSON, BALDWIN, b Los Angeles, Calif, Sept 26, 34; m 62; c 3. THEORETICAL PHYSICS. *Educ:* Stanford Univ, BS, 56, MS, 57, PhD(physics), 65. *Prof Exp:* Instr & res assoc, Cornell Univ, 64-66; res assoc, Nat Bur Standards, 66-68, Physicist, 68-89, PHYSICIST, NAT INST STANDARDS & TECHNOL, 89- *Mem:* Am Phys Soc. *Res:* Statistical mechanics; experimental and theoretical biophysical and chemical relaxation kinetics, membrane enzymes. *Mailing Add:* Biotechnol Div Nat Inst Standards & Technol Gaithersburg MD 20899

ROBERTSON, BEVERLY ELLIS, b Fredericton, NB, Feb 5, 39; m 58; c 2. CRYSTALLOGRAPHY, MOLECULAR BIOLOGY. *Educ:* Univ NB, BSc, 61; McMaster Univ, MSc, 65, PhD(physics), 67. *Prof Exp:* Assoc chem, Cornell Univ, 67-69; from asst prof to assoc prof physics & astron, 69-76, PROF PHYSICS & ASTRON, UNIV REGINA, 76- *Concurrent Pos:* Vis prof, Univ Stuttgart, 76-77; mem, Sci Coun Can, 78-84; adj prof, Univ Sask, 72- *Mem:* Am Crystallog Asn; Am Chem Soc; fel Humboldt Found; Health Physics Soc. *Res:* Crystal structure determination; environmental radon; deformation densities of nucleosidi analogs; crystallographic software; health physics. *Mailing Add:* Dept Physics Univ Regina Regina SK S4S 0A2 Can

ROBERTSON, BOBBY KEN, b Reed, Okla, June 20, 38. PHYSICAL CHEMISTRY. *Educ:* WTex State Univ, BA, 60; Tex A&M Univ, PhD, 65. *Prof Exp:* Asst prof, 65-74, ASSOC PROF PHYS CHEM, UNIV MO, ROLLA, 74-, DEAN STUDENTS, 78- *Mem:* Am Chem Soc; Am Crystallog Asn. *Res:* Interpretation of molecular structure by use of x-ray crystallography. *Mailing Add:* Dept Chem Univ Mo PO Box 249 Rolla MO 65401

ROBERTSON, CHARLES WILLIAM, JR, b Memphis, Tenn, Mar 2, 43; m 89. PHYSICS. *Educ:* Southwestern at Memphis, BS, 65; Fla State Univ, PhD(physics), 69. *Prof Exp:* Res assoc physics, Kans State Univ, 70-73; RES ASSOC, E I DU PONT DE NEMOURS & CO, INC, 73- *Mem:* Optical Soc Am; Sigma Xi. *Res:* Application of optical and spectroscopic techniques to the measurement of material properties in the infrared, visible and ultraviolet. *Mailing Add:* E I DuPont Exp Sta E357 Wilmington DE 19880-0357

ROBERTSON, CLYDE HENRY, b Heath Springs, SC, Aug 8, 29; m 51; c 2. ENTOMOLOGY. *Educ:* Wofford Col, BS, 50; Duke Univ, MA, 52, PhD(zool), 55. *Prof Exp:* From asst prof to prof biol, 56-90, head dept, 56-74, chmn div natural sci, 74-90, EMER PROF BIOL, PFEIFFER COL, 90- *Res:* Vertebrate morphology; development and metamorphosis of insectan respiratory systems. *Mailing Add:* Dept Biol Pfeiffer Col Misenheimer NC 28109

ROBERTSON, DALE NORMAN, b Whittier, Calif, Jan 8, 25; m 51; c 1. ORGANIC CHEMISTRY. *Educ:* Pomona Col, BA, 49; Univ Wis, MS, 51, PhD(biochem), 53. *Prof Exp:* Asst biochem, Univ Wis, 49-53; fel, Mellon Inst, 53-54; res chemist biochem dept, Dow Chem Co, 54-60; res chemist, Arapahoe Chem, Inc, 60-66, asst dir res, Arapahoe Chem Div, Syntex Corp, 66-72; from staff scientist to scientist, 72-82, SR SCIENTIST, POP COUN, ROCKEFELLER UNIV, 82- *Mem:* AAAS; Sigma Xi; Am Chem Soc; NY Acad Sci. *Res:* Porphyrins; 4-hydroxycoumarins; isolation, characterization and synthesis of porphyrins in bitumens and petroleum; agricultural and medicinal chemicals; process development research on organic and medicinal chemicals; analytical chemistry; liquid scintillation spectrometry; contraceptive development; clinical studies; regulatory affairs. *Mailing Add:* Pop Coun Rockefeller Univ York Ave & 66th St New York NY 10021

ROBERTSON, DAVID, b Dickson Co, Tenn, May 23, 47; m 76; c 1. AUTONOMIC DISORDERS, SPARE MEDICINE & PHYSIOLOGY. *Educ:* Vanderbilt Univ, BA, 69, MD, 73. *Prof Exp:* Instr med, Johns Hopkins Hosp, 77-78; from asst prof to assoc prof, 78-86, PROF MED & PHARM, VANDERBILT UNIV, 86- *Concurrent Pos:* Teaching & res scholar award, Am Col Physicians, 78-81 & res career develop award, 81-86; vis prof, Dept Molecular Endocrinol, All-Union Cardiol Res Ctr, Moscow, USSR, 84 & Dept Anat & Embryol, Univ Col London, UK, 85; Scholar, Burroughs Wellcome, 86-91; dir, Clin Res Ctr, Vanderbilt Univ, 87- & Ctr Space Physiol & Med, 89-; William N Creasy Vis Prof Clin Pharm, 90. *Mem:* Am Soc Clin Invest; Aerospace Med Asn; Am Fedn Clin Res. *Res:* How the brain controls the heart and blood pressure; acute blood pressure disturbances and chronic acquired or genetic deficiency and carried out. *Mailing Add:* AA3228 Med Ctr N Clin Res Ctr Vanderbilt Univ Nashville TN 37232-2195

ROBERTSON, DAVID C, b Scotia, Calif, Nov 13, 42; m 67; c 1. ATMOSPHERIC CHEMISTRY, THEORETICAL PHYSICS. *Educ:* Stanford Univ BS, 66; Univ Calif, Santa Barbara, PhD(theoret physics), 70. *Prof Exp:* PRIN SCIENTIST, SPECTRAL SCI INC. *Mem:* Am Phys Soc; Am Meteorol Soc. *Res:* IR spectroscopy; atmospheric radiative transport. *Mailing Add:* Spectral Sci Inc 99 S Bedford St Burlington MA 01803-5169

ROBERTSON, DAVID G C, b Dublin, Ireland, Dec 29, 41; m 70; c 2. METALLURGY & PHYSICAL METALLURGICAL ENGINEERING, RESEARCH ADMINISTRATION. *Educ:* Imp Col, London, UK, BSc, 63; Univ NSW, Sydney, Australia, PhD(metall), 68. *Prof Exp:* Lectr metall, Imp Col, 69-82, reader metall, 82-85; PROF METALL, UNIV MO, ROLLA, 85- *Concurrent Pos:* Consult, var co, 69-; dir, Generic Mineral Technol Ctr Pyrometall, 85- *Mem:* Metall Soc. *Mailing Add:* 215 Fulton Hall Univ Mo Rolla MO 65401-0249

ROBERTSON, DAVID MURRAY, b Melville, Sask, May 4, 32; m 56; c 3. NEUROPATHOLOGY, PATHOLOGY. *Educ:* Queen's Univ, Ont, MD, CM, 55, MSc, 60; Royal Col Physicians, cert, 60, fel neuropath, 68; Am Bd Path, cert anatomic path & neuropath, 77. *Prof Exp:* From asst prof to assoc prof, 62-69, PROF PATH, QUEEN'S UNIV, ONT, 69-, HEAD DEPT, 79- *Concurrent Pos:* Neuropathologist, Kingston Gen Hosp, 62-, chief pathologist, 79-87. *Mem:* Am Asn Neuropath; Can Asn Path; Can Asn Neuropath; Int Acad Path. *Res:* Nutritional diseases of nervous system; diabetic neuropathy; cerebrovascular disease. *Mailing Add:* Richardson Lab Dept Path Queen's Univ Kingston ON K7L 3N6 Can

ROBERTSON, DAVID WAYNE, b Dumas, Tex, July 30, 55; m 81; c 1. SEROTONIN RECEPTOR LIGANDS, POSITIVE INOTROPES. *Educ:* Stephen Austin State Univ, BS, 77; Univ Ill, MS, 78, PhD (org chem), 81. *Prof Exp:* Sr med chemist, 81-84, res scientist, 85-87, sr res scientist & group leader, 88-90, DIR, CNS RES, LILLY RES LABS, 90- *Mem:* Am Soc Pharmacol & Exp Therapeut; Am Chem Soc; AAAS; Am Heart Asn; Int Union Pure & Appl Chem. *Res:* Chemical tools to probe biochemical, physiological and pathophysiological processes; development of selective ligands to probe serotonin receptor subtypes and selective inhibitors of the multiple molecular forms of phosphodresterase. *Mailing Add:* Lilly Res Labs Lilly Corp Ctr Indianapolis IN 46285

ROBERTSON, DONALD, b Baltimore, Md, Nov 29, 10; m 55; c 1. MECHANICAL ENGINEERING. *Educ:* Cornell Univ, ME, 32. *Prof Exp:* Foreman copper furnace, Am Smelting & Refining Co, 33-36; eng specification writer, Leeds & Northrup Co, 36-37, engr, 37-45, head pyrometer sect, 45-55, chief res & develop, pyromet group, 55-62, head temperature measurements sect, Res & Develop Ctr, 62-68, mgr standard prod eng, Sensor Div, 68-69, mgr develop & eng, 69-73; temperature measurement expert, 73-76; INDEPENDENT TEMPERATURE CONSULT, 76- *Res:* Temperature primary elements. *Mailing Add:* 77 Middle Rd Apt 366 Bryn Mawr PA 19010

ROBERTSON, DONALD CLAUS, b Rockford, Ill, Mar 5, 40; m 62; c 2. BIOCHEMISTRY, MICROBIOLOGY. *Educ:* Univ Dubuque, BS, 62; Iowa State Univ, PhD(biochem), 67. *Prof Exp:* Chemist, Nat Animal Dis Lab, Ames, Iowa, 62-67; res assoc biochem, Mich State Univ, 67-70; from asst prof to assoc prof, 70-80, PROF MICROBIOL, UNIV KANS, 80- *Concurrent Pos:* NIH trainee, 67-70; mem bact & mycol study sect, NIH, 79-83; hon lectr, Mid-Am State Univs Asn, 81-82. *Mem:* Am Soc Microbiol; AAAS; Sigma Xi; Am Asn Univ Professors; Am Acad Microbiol; Am Soc Biochem Molecular Biol. *Res:* Microbial physiology; protein chemistry; biochemistry of host-parasite relationships. *Mailing Add:* Dept Microbiol Univ Kans Lawrence KS 66045

ROBERTSON, DONALD EDWIN, b Edmonton, Alta, Jan 8, 29; nat US; m 52; c 1. MOLECULAR DESIGN, PROCESS DEVELOPMENT. *Educ:* Univ Alta, BSc, 51; Univ Utah, PhD(chem), 59. *Prof Exp:* Res org chemist, US Gypsum Co, 60-62; res org chemist, Interchem Corp, 62-65; res org chemist, Com Solvents Corp & Int Minerals & Chem Corp, 65-86; TECH CONSULT, 86- *Mem:* Am Chem Soc; Chem Inst Can; fel Am Inst Chemists. *Res:* Design and synthesis of bioactive compounds; synthesis of natural products; carbohydrate and polymer chemistry; process and product development; technical writing. *Mailing Add:* 4951 Dixie Bee Rd Apt 6 Terre Haute IN 47802

ROBERTSON, DONALD HUBERT, b Monmouth, Maine, June 6, 34. ANALYTICAL CHEMISTRY. *Educ:* Bates Col, BS, 56; Univ Glasgow, PhD(chem), 73. *Prof Exp:* Researcher, Gen Foods Corp Res Ctr, 56-57, Natick Labs, US Army, 57-59 & Gen Foods Corp Res Ctr, 59-62; RESEARCHER, NATICK LABS, US ARMY, 62- *Mem:* AAAS; Am Chem Soc; Pattern Recognition Soc. *Res:* Chromatography; mass spectrometry; computer processing as related to natural products, especially foodstuffs. *Mailing Add:* PO Box 18 Natick MA 01760

ROBERTSON, DONALD SAGE, b Oakland, Calif, June 27, 21; m 42; c 3. MAIZE GENETICS. *Educ:* Stanford Univ, AB, 47; Calif Inst Technol, PhD, 51. *Prof Exp:* Head dept sci, Biola Bible Col, 51-57; from asst prof to assoc prof, 57-63, chmn dept, 75-80, PROF GENETICS, IOWA STATE UNIV, 63- *Honors & Awards:* Gov Sci Medal, Iowa, 84. *Mem:* Genetics Soc Am; AAAS; Am Genetic Asn; Sigma Xi. *Res:* Genetic control of mutation in maize; discoverer of the mutator transposable element system of maize. *Mailing Add:* Dept Zool & Genetics Iowa State Univ Ames IA 50011

ROBERTSON, DOUGLAS REED, b Buffalo, NY, Sept 11, 38; m 67; c 2. ANATOMY, ENDOCRINOLOGY. *Educ:* Univ Buffalo, BA, 61; State Univ NY Buffalo, MA, 63; State Univ NY Upstate Med Ctr, PhD(anat), 66. *Prof Exp:* Asst prof anat, Col Med, Univ Fla, 66-70; assoc prof, 70-76, PROF ANAT, STATE UNIV NY UPSTATE MED CTR, 76- *Concurrent Pos:* NIH res grant, Inst Arthritis, Metab & Digestive Dis, 67-70; NSF grant, Div Biol & Med Sci, 82-85. *Mem:* Am Asn Anatomists; Am Soc Cell Biol; Am Soc Zoologists; Sigma Xi. *Res:* Secretory mechanisms of the ultimobranchial body; endocrinology and physiology of the hormones, calcitonin and parathormone in amphibians; annual and diurnal rhythms of calcitonin secretion and plasma calcium; enzyme immunoassay for frog calcitonin; immunohistochemistry of calcitonin; autonomic nervous system. *Mailing Add:* Dept Anat & Cell Biol State Univ NY Health Sci Ctr Syracuse NY 13210

ROBERTSON, DOUGLAS SCOTT, b Three Rivers, Mich, Dec 29, 45; m 72; c 1. GEOPHYSICS, ASTRONOMY. *Educ:* Principia Col, BS, 68; Mass Inst Technol, PhD(geophys), 75. *Prof Exp:* Mem tech staff, Comput Sci Corp, 75-77; GEODESIST, NAT GEODETIC SURV, NAT OCEAN SURV, NAT OCEANIC & ATMOSPHERIC ADMIN, 77- *Mem:* Am Geophys Union; Am Astron Soc; AAAS; Int Astron Union. *Res:* Use of very-long-baseline radio interferometry to study polar motion, earth rotation, precession, nutation, earth tides, tectonic crustal deformations and astrometry. *Mailing Add:* 15 Pavilion Dr Gaithersburg MD 20878

ROBERTSON, DOUGLAS WELBY, b Crawford, Ga, June 13, 24; m 49; c 3. ELECTRONICS. *Educ:* Ga Inst Technol, BS, 51, MS, 57. *Prof Exp:* Electronic technician, US Civil Serv, 42-44 & 46-47; res asst eng exp sta, Ga Inst Technol, 50-53, res engr, 53-57; mem tech staff, ITT Labs, Int Tel & Tel Corp, 57-59; res engr, Ga Inst Technol, 59-62, head commun br, 62-72, chief commun div, 72-75, dir, Electronics Technol Lab, Eng Exp Sta, 75-80; CONSULT, 80- *Mem:* AAAS; sr mem Inst Elec & Electronics Engrs; Acoust Soc Am. *Res:* Voice intelligibility; speech communication systems; electromagnetic compatibility; piezoelectric crystals and oscillators; antennas; electronic measurements; solid state components; radar systems; research management. *Mailing Add:* 2937 Henderson Rd Tucker GA 30084

ROBERTSON, EDWARD L, b St Paul, Minn, July 16, 44; m 69; c 2. THEORY OF COMPUTATION, COMPUTIONAL COMPLEXITY. *Educ:* Calif Inst Technol, BS, 66; Univ Wis, MS, 68, PhD(comput sci), 70. *Prof Exp:* Asst prof comput sci, Univ Wis, 70-71; res fel, Univ Ghana, Africa, 71-72; asst prof, Univ Waterloo, 72-74, Pa State Univ, 74-78; assoc prof, 78-84, PROF COMPUT SCI, IND UNIV, 84- *Concurrent Pos:* Ed, Info Technol for Develop, 85-; vchmn, Comn Informatics in Develop, Int Fedn Info Processing, 85- *Mem:* Asn Comput Mach; Inst Elec & Electronics Engrs; Soc Indust & Appl Math. *Res:* Theory of computation and computational complexity; design and analysis of algorithms; software engineering; computers in economic development; database systems. *Mailing Add:* Dept Comput Sci Ind Univ Bloomington IN 47402

ROBERTSON, EUGENE CORLEY, b Tucumcari, NMex, Apr 9, 15; m 72; c 2. PHYSICAL PROPERTIES OF ROCKS. *Educ:* Univ Ill, BS, 36; Harvard Univ, MA, 48, PhD(geol), 52. *Prof Exp:* Mining geologist, Anaconda Co, 36-42; geophysicist, US Geol Surv, 49-91; RETIRED. *Honors & Awards:* Meritorius Serv Award, US Geol Surv, 81. *Mem:* Am Inst Mining, Metall & Petrol Eng; Am Geophys Union; Geol Soc Am; Mineral Soc Am. *Res:* Deformation of rocks; experimental geology; strength, elastic and thermal properties of rocks; fault characteristics in the earth and applications to overthrust faulting. *Mailing Add:* US Geol Surv 922 Nat Ctr Reston VA 22092

ROBERTSON, FORBES, b New Haven, Conn, May 24, 15; m 43; c 3. ECONOMIC GEOLOGY. *Educ:* Principia Col, BA, 38; Wash Univ, MS, 40; Univ Wash, PhD, 56. *Prof Exp:* Geologist, Reynolds Mining Co & Reynolds Metals Co, 42-45; sr geologist, Stand Oil Co NJ, 45-46; econ geologist, Mo State Geol Surv, 46-47; assoc prof geol, Mont Sch Mines, 47-55; lectr, Univ Wash, 55-57; prof petrol, Petrobras Sch Geol, Brazil, 57-59; prof geol & head dept, Principia Col, 59-78; pres, Western Minerals Inc, 78-80; RETIRED. *Concurrent Pos:* Geologist, Mont Bur Mines & Geol, 47-55. *Mem:* Geol Soc Am; Am Asn Petrol Geol; Mineral Soc Am. *Res:* Petrology and geology of mineral deposits. *Mailing Add:* 33400 Kuhio Hwy No A 107 Lihue-Kauai HI 96766-1050

ROBERTSON, FREDERICK JOHN, veterinary medicine, serology; deceased, see previous edition for last biography

ROBERTSON, FREDERICK NOEL, b Akron, Ohio, Oct 12, 35. HYDROLOGY & WATER RESOURCES, GEOCHEMISTRY. *Educ:* Kent State Univ, BA, 62; Ariz State Univ, MS, 75. *Prof Exp:* Shift supvr, Controls for Radiation, Inc, 62-63; res assoc, Horizons Res, Inc, 64-70; res hydrologist, Water Resources Ctr, Univ Del, 75-77; geochemist, US Geol Surv, Dept Interior, 78-86. *Concurrent Pos:* Prin investr Delaware Water Use, 74-75; consult hydrologist, 77-78. *Mem:* Soc Geochem & Health; Geol Soc Am; Nat Water Well Asn. *Res:* Evaluation of water quality and regional geochemistry in the Southwest; development of geochemical models for defining chemical reactions in ground-water systems. *Mailing Add:* 8520 N Rancho Catalina Ave Tucson AZ 85704

ROBERTSON, G PHILIP, b Houston, Tex, Oct 12, 53. ECOSYSTEM DYNAMICS, SOIL MICROBIOLOGY. *Educ:* Hampshire Col, Amherst, BA, 76; Ind Univ, Bloomington, PhD(biol), 80. *Prof Exp:* Andrew H Mellon Found postdoctoral fel, 80-81; asst prof, 85-90, ASSOC PROF, CROP & SOIL SCI DEPT, W K KELLOGG BIOL STA, MICH STATE UNIV, 90- *Concurrent Pos:* Ed, Plant & Soil, 84-89 & Ecol & Ecol Monogr, 88-91. *Mem:* Ecol Soc Am; Soil Sci Soc Am; AAAS. *Res:* Ecosystem ecology; nutrient dynamics in disturbed terrestrial communities, including agricultural; nitrogen fluxes; spatial dynamics; microbial transformations in soil. *Mailing Add:* Kellogg Biol Sta Mich State Univ Hickory Corners MI 49060

ROBERTSON, GEORGE GORDON, b St John, NB, Jan 30, 16; nat; m 46; c 2. ANATOMY. *Educ:* Acadia Univ, BSc, 36, BA, 37; Yale Univ, PhD(zool), 41. *Prof Exp:* Asst biol, Yale Univ, 38-41; fel anat, Sch Med, Univ Ga, 41-42; instr anat, La State Univ, 42-43; from instr to assoc prof, Baylor Univ Col Med, 43-52; prof, 52-80, chmn dept, 61-80, EMER PROF ANAT, UNIV TENN COL MED, 80- *Concurrent Pos:* Vis assoc prof anat, Col Basic Med Sci, Univ Tenn, Memphis, 48 & 51. *Mem:* Am Asn Anatomists. *Res:* Developmental genetics; ovarian transplantation; human embryology; experimental teratology. *Mailing Add:* 71 Wagon Trail Rd Black Mountain NC 28711-2564

ROBERTSON, GEORGE HARCOURT, b Evergreen Park, Ill, Jan 30, 43; m 67; c 2. BIOCHEMICAL ENGINEERING. *Educ:* Univ Ill, Urbana, BS, 65; Univ Calif, Berkeley, PhD(chem eng), 70. *Prof Exp:* RES CHEM ENGR AGR & FOOD ENG, WESTERN REGIONAL RES LAB, SCI & EDUC ADMIN-AGR RES, USDA, 69- *Mem:* AAAS; Inst Food Technologists; Am Inst Chem Engrs; Am Chem Soc; Nat Sweetcorn Breeders Asn; Int Asn Plant Tissue Cult. *Res:* Textile (wool) and food (sweetcorn) process modifications to reduce energy use, waste or pollutant generation; ethylalcohol fuels separation and generation; chemicals from plant tissue culture. *Mailing Add:* Western Regional Res Ctr USDA 800 Buchanan St Berkeley CA 94710

ROBERTSON, GEORGE LEVEN, b Alexandria, La, Feb 7, 21; m 43; c 3. ANIMAL SCIENCE. *Educ:* La State Univ, BS, 41; Tex A&M Univ, MS, 47; Univ Wis, PhD(animal husb, genetics), 51. *Prof Exp:* Asst animal husb, Tex A&M Univ, 41-42, instr, 46-48, from asst prof to assoc prof, 48-55; prof animal sci & head dept, 55-77; EXEC DIR, HONOR SOC PHI KAPPA PHI, LA STATE UNIV, BATON ROUGE, 77- *Mem:* Fel AAAS; Am Soc Animal Sci; Sigma Xi. *Res:* Physiology of reproduction in farm animals. *Mailing Add:* 7017 Perkins Rd Baton Rouge LA 70808

ROBERTSON, GEORGE WILBER, agricultural meteorology, for more information see previous edition

ROBERTSON, GLENN D(AVID), JR, b Los Angeles, Calif, July 10, 24; m 51; c 3. APPLIED PHYSICS. *Educ:* Rice Univ, BS, 49; Calif Inst Technol, MS, 50, ChE, 53. *Prof Exp:* Engr mat res & develop, Hughes Aircraft Co, 53-61, sr staff engr, 61-88; RETIRED. *Concurrent Pos:* Adj fac physics & geol, Pepperdine Univ, 77- *Mem:* Sigma Xi. *Res:* Applications of materials in electronics, missiles and aircraft; plastics; ceramics; radiation detection and dosimetry; chemical kinetics; mass and energy transport; infrared detection systems; fiber-optic waveguides; high purity silicon. *Mailing Add:* 6202 Frondosa Dr Malibu CA 90265

ROBERTSON, H THOMAS, II, b Abilene, Tex. RESPIRATORY PHYSIOLOGY, ANIMAL PHYSIOLOGY. *Educ:* Harvard Univ, MD, 68. *Prof Exp:* ASSOC PROF MED, UNIV HOSP, UNIV WASH, 82- *Honors & Awards:* Res Career Develop Award, NIH. *Mem:* Fel Am Col Physicians; Am Physiol Soc. *Res:* Respiratory physiology. *Mailing Add:* Dept Med Univ Hosp Rm 12 Univ Wash Seattle WA 98195

ROBERTSON, HARRY S(TROUD), b Montgomery, Ala, Sept 26, 21; m 43; c 3. PLASMA PHYSICS, STATISTICAL MECHANICS. *Educ:* Univ NC, BS, 42; Johns Hopkins Univ, PhD(physics), 49. *Prof Exp:* Radio engr, US Naval Res Lab, 42-46; asst physics, Johns Hopkins Univ, 47-49; from asst prof to assoc prof, 49-56, chmn dept, 52-62, 72-74, PROF PHYSICS, UNIV MIAMI, FLA, 56- *Concurrent Pos:* Vis lectr, Johns Hopkins Univ, 56-57; consult, Oak Ridge Nat Lab, 58-69 & Princeton Plasma Physics Lab, 61; vis prof, Univ Edinburgh, 71. *Mem:* AAAS; Fedn Am Scientists; Am Phys Soc; Sigma Xi. *Res:* Plasma transport properties; plasma stability and oscillations; cesium plasmas; moving striations; physics and society; statistical thermodynamics; solar energy conversion. *Mailing Add:* Dept Physics Univ Miami Univ Sta Box 248046 Coral Gables FL 33124

ROBERTSON, HUGH ELBURN, b Sask, Can, Oct 2, 19; m 42; c 5. BIOCHEMISTRY, MICROBIOLOGY. *Educ:* Univ Sask, MA, 43; Univ Minn, PhD(biochem), 53. *Prof Exp:* Res chemist, Int Nickel Co, Can, 40-46; provincial analyst, Sask Dept Pub Health, 46-49, dir provincial labs, 50-52; instr bact & immunol, Univ Minn, 52-53, asst prof, 54; dir provincial labs, Sask Dept Health, 55-86; RETIRED. *Concurrent Pos:* Adj prof dept chem, Univ Sask, Regina. *Mem:* Fel Chem Inst Can; Eng Inst Can; Can Pub Health Asn. *Res:* Evaluation of clinical diagnostic laboratory procedures for metabolic and infectious diseases; economics of delivery systems for automated laboratory procedures with particular emphasis on clinical chemistry. *Mailing Add:* Provincial Labs Sask Dept Health Regina SK S4S 5W6 Can

ROBERTSON, HUGH MERETH, b Johannesburg, SAfrica, Dec 19, 55. GENETICS OF TRANSPOSONS, ALTERNATIVE MATING BEHAVIOR STRATEGIES. *Educ:* Univ Witwatersrand, SAfrica, BSc, Hons, 78, PhD(zool), 82. *Prof Exp:* Guyer fel zool, Univ Wis-Madison, 82-84, lectr, 85, proj assoc genetics, 85-87; ASST PROF, DEPT ENTOM, UNIV, ILL, 87- *Mem:* Soc Study Evolution; Animal Behav Soc; Genetics Soc Am; AAAS. *Res:* Evolution of mating behavior and relevance to speciation; alternative behavioral strategies of insects; genetics of transposable elements. *Mailing Add:* Dept Entom Univ Ill 505 S Goodwin Urbana IL 61801

ROBERTSON, J(OHN) A(RCHIBALD) L(AW), b Dundee, Scotland, July 4, 25; Can citizen; m 54; c 3. MATERIALS SCIENCE. *Educ:* Cambridge Univ, BA, 50, MA, 53. *Prof Exp:* Sci officer metall, Atomic Energy Res Estab, UK Atomic Energy Authority, 50-55, sect leader, 55-57; res officer, Chalk River Nuclear Labs, Atomic Energy Can Ltd, 57-63, head reactor mat br, 63-70, dir fuels & mat div, 70-75, asst to vpres & gen mgr, Chalk River Nuclear Labs, 75-82, dir program planning, 82-85; CONSULT, 85- *Concurrent Pos:* Co-ed, J Nuclear Mat, 67-71. *Honors & Awards:* W B Lewis Medal, Can Nuclear Asn, 87. *Mem:* Royal Soc Can. *Res:* Research and development programs for nuclear reactor systems; irradiation effects in nuclear fuels. *Mailing Add:* One Kelvin Crescent Deep River ON K0J 1P0 Can

ROBERTSON, JACK M, b Clovis, NMex, Sept 26, 37; m 57; c 3. MATHEMATICS. *Educ:* Eastern NMex Univ, BS, 59; Univ Utah, MS, 61, PhD(math), 64. *Prof Exp:* Teaching asst math, Univ Utah, 59-64; asst prof, 64-70, assoc prof, 70-77, PROF MATH, WASH STATE UNIV, 77- *Concurrent Pos:* Assoc dir, NSF Summer Inst Sec Teachers, Wash State, 66-67, dir, 68-69. *Mem:* Am Math Soc; Math Asn Am. *Res:* Topology; analysis; mathematics education. *Mailing Add:* Dept Math Applied-Pure Wash State Univ Pullman WA 99164-2930

ROBERTSON, JACQUELINE LEE, b Petaluma, Calif, July 9, 47. INSECT PHYSIOLOGY, INSECT TOXICOLOGY. *Educ:* Univ Calif, Berkeley, BA, 69, PhD(entom), 73. *Prof Exp:* RES ENTOMOLOGIST, PAC SOUTHWEST FOREST & RANGE EXP STA, INSECT BIOCHEM & GENETICS PROJ, FOREST SERV, BERKELEY, CALIF, 79- *Concurrent Pos:* Teaching asst, Univ Calif, Berkeley, 70-72, res asst, 72-73; ed, J Econ Entom, 82- *Honors & Awards:* Outstanding Scientist, Int Union of Forestry Related Orgn, 86. *Mem:* Entom Soc Can; Entom Soc Am; AAAS; Am Soc Zoologists. *Res:* Influence of population genetics and biochemical markers on host-insect interactions; toxicology of insecticides to forest insect pests; toxicological biostatistics; interactions of phytophagous insects and their hosts in the context of population genetics and biochemical diversity. *Mailing Add:* Pac Southwest Forest & Range Sta 1960 Addison St Berkeley CA 94701

ROBERTSON, JAMES ALDRED, b Knoxville, Tenn, July 13, 31; m 54; c 2. CHEMISTRY, FRUIT QUALITY. *Educ:* Univ Tenn, BS, 53, MS, 57; Ohio State Univ, PhD(dairy technol), 62. *Prof Exp:* Chemist, M&R Dietetic Labs, Inc, 62-63; res chemist, Southern Utilization Res & Develop Div, New Orleans, La, 63-69, RES LEADER, RICHARD B RUSSELL AGR RES CTR, AGR RES SERV, USDA, ATHENS, GA, 69- *Mem:* Inst Food Technol. *Res:* Purification and specificity of lipase in milk; distribution of fatty acids in blood lipids; toxic fungal metabolites; composition, flavor and oxidative stability of sunflower seed and oil; methods of analysis of sunflower seed; field and storage damage of soybeans and sunflower; extending shelf life and reducing portharvest losser of fruits; postharvest factons influencing quality of stone fruit. *Mailing Add:* Richard B Russell Agr Res Ctr Agr Res Serv USDA PO Box 5677 Athens GA 30613

ROBERTSON, JAMES ALEXANDER, b Basswood, Man, Apr 15, 31; m 57; c 2. SOIL FERTILITY, SOIL MANAGEMENT. *Educ:* Univ Man, BSA, 53, MSc, 55; Purdue Univ, PhD, 63. *Prof Exp:* Assoc prof soil sci, 55-71, PROF SOIL SCI, UNIV ALTA, 71-, CHAIR, 89- *Concurrent Pos:* Mem, Alta Inst Agrology. *Mem:* Fel Can Soc Soil Sci (secy-treas, 58-60, pres, 72-73); Am Soc Agron; Sigma Xi. *Res:* Phosphorus sorption by Alberta soils; plant uptake of phosphorus from various soil horizons; methods of measuring phosphorus availability; soil management on long-term Breton plots; potassium status of Alberta soils. *Mailing Add:* Dept Soil Sci Univ Alta Edmonton AB T6G 2E3 Can

ROBERTSON, JAMES BYRON, b Spiceland, Ind, Mar 29, 37; m 61; c 2. MATHEMATICS. *Educ:* Mass Inst Technol, SB, 59; Ind Univ, PhD(math), 64. *Prof Exp:* From instr to asst prof math, Cornell Univ, 63-66; from asst prof to assoc prof, 66-77, PROF MATH, UNIV CALIF, SANTA BARBARA, 77- *Concurrent Pos:* NSF grant, 70-73; Fulbright lectr, Tbilisi, USSR, 77. *Mem:* Am Math Soc; Math Asn Am. *Res:* Ergodic theory; prediction theory. *Mailing Add:* Dept Math Univ Calif Santa Barbara CA 93106

ROBERTSON, JAMES DAVID, b Tuscaloosa, Ala, Oct 13, 22; m 46; c 3. ANATOMY. *Educ:* Univ Ala, BS, 42; Harvard Med Sch, MD, 45; Mass Inst Technol, PhD(biochem), 52. *Prof Exp:* Asst biol, Univ Ala, 40-42; intern, Boston City Hosp, 45-46; asst physician, Vet Admin Hosp, Ala, 47-48; asst physician, Med Dept, Mass Inst Technol, 48-52; asst prof path & oncol, Univ Kans Med Ctr, Kansas City, 52-55; hon res assoc anat, Univ Col, Univ London, 55-60; from asst prof to assoc prof neuropath, Harvard Med Sch, 60-66; prof anat, 66-75, James B Duke Prof Anat, Sch Med, 75-, chmn dept, 66-88, JAMES B DUKE PROF NEUROBIOL & EMER CHMN, DEPT ANAT, DUKE UNIV, 88- *Concurrent Pos:* Assoc biophysicist, McLean Hosp, Boston, 60-63, biophysicist, 64-67; mem cell biol study sect, NIH, 67-71; vis prof physiol chem inst, Univ Wurzburg, Germany, 78-79; consult, Dept Defense; chmn search comt anat dept, Uniformed Servs Univ Health Sci, 75; mem, bd trustees, NC Sch Math & Sci, 85. *Honors & Awards:* First Pomerat Mem Lectr, Univ Tex, 66; Damon Lectr, Univ Conn, 75; Ferris Lectr, Yale Univ, 77; Otto Mortenson Lectr, Univ Wis, Madison, 77; Alexander von Humboldt sr scientist award, Fed Repub Ger, 78; Cummings Mem lectr, Tulane Univ, 84; Pleanry lectr, First Beijing Conf & Exhib Instrumental Anal, China, 85; George Harvey Miller Distinguished lectr anat, Chicago Col Med, Univ Ill. *Mem:* Electron Micros Soc Am (pres, 84); Physiol Soc Gt Brit & Ireland; Am Asn Anatomists; Am Soc Cell Biol; Int Soc Neurochem; Soc Neurosci (chpt pres, 87-88). *Res:* Tissue ultrastructure, especially electron microscope studies of nerve junctional tissues and cell membranes. *Mailing Add:* Dept Anat Duke Univ Med Ctr Box 3011 Durham NC 27706

ROBERTSON, JAMES DOUGLAS, b New Rochelle, NY, Feb 15, 48; m 75. SEISMIC INTERPRETATION. *Educ:* Princeton Univ, BSE, 70; Univ Wis, MS, 72, PhD(geophys), 75. *Prof Exp:* Sr res geophysicist, 75-79, res dir geophys, 79-83, dep gen mgr, geol res, 83-85, OFFSHORE MGR, GEOPHYS DIV, ARCO, 85- *Concurrent Pos:* Vis lectr, Dept Geosci, Univ Tex, Dallas, 80-; lectr, continuing educ, Soc Explor Geophysicists, 83- *Mem:* Soc Explor Geophysicists; Am Asn Petrol Geologists; Am Geophys Union; Soc Petrol Engrs. *Res:* Shear wave seismic technology; seismic modeling; stratigraphic interpretation methods; interactive computer graphics for seismic analysis. *Mailing Add:* Arco Int Oil & Gas Co PAI-82409 2300 West Plano Pkwy Plano TX 75075

ROBERTSON, JAMES E(VANS), b Fairfax, Okla, Mar 30, 24; m 56; c 4. ELECTRICAL ENGINEERING. *Educ:* Okla Agr & Mech Col, BS, 47; Univ Ill, MS, 48, PhD(elec eng), 52. *Prof Exp:* Res asst prof elec eng, 52-56, res assoc prof, 56-59, PROF ELEC ENG & COMPUT SCI, UNIV ILL, URBANA, 59- *Concurrent Pos:* Fulbright scholar, Univ Sydney, 63. *Mem:* Inst Elec & Electronics Engrs. *Res:* Design of electronic digital computers. *Mailing Add:* Dept Comput Sci 271 DCI Univ Ill 1304 W Springfield Urbana IL 61801

ROBERTSON, JAMES MCDONALD, b Edinburgh, Scotland, Feb 27, 40; Can citizen; m 76; c 4. OCCUPATIONAL EPIDEMIOLOGY, ENVIROMENTAL EPIDEMIOLOGY. *Educ:* Univ toronto, DVM, 61; Univ Pa, MSc, 66. *Prof Exp:* Res fel epidemiol, Res Coun Can, 62-65; from asst prof to assoc prof, Univ Sask, 66-71; asst prof, 71-75, ASSOC PROF EPIDEMIOL, UNIV WESTERN ONT, 75- *Concurrent Pos:* Int cancer fel epidemiol, Harvard Sch Pub Health, 71-72; sessional lectr, Univ Sask, 72-78; assoc prof epidemiol, Fac Grad Studies, Univ Western Ont, 73-, actg dir, Lab Animal Serv, 82, assoc mem, Health Care Res Unit, 86-89, dir, Occup Health & Safety Resource Ctr, 86-; consult, Environ Health Asn Carbon Black Indust, 74- & Can Portland Cement Asn, 88- *Mem:* AAAS; Am Pub Health Asn; Am Col Epidemiol; Soc Epidemiol Res; Am Vet Med Asn; Can Vet Med Asn. *Res:* Morbidity and mortality studies in the carbon black industry; antioxidant vitamins in cataract prevention; environmental effects on reproduction. *Mailing Add:* 829 Hickory Rd London ON N6H 2V3 Can

ROBERTSON, JAMES MAGRUDER, b Port Clinton, Ohio, Sept 24, 43; m 70. ECONOMIC GEOLOGY. *Educ:* Carleton Col, BA, 65; Univ Mich, Ann Arbor, MS, 68, PhD(econ geol), 72. *Prof Exp:* Asst prof geol, Mich Technol Univ, 72-74; mining geologist, 74-86, sr econ geologist, 86-88, ASSOC DIR, NMEX BUR MINES & MINERAL RESOURCES, 88- *Mem:* Geochem Soc; Sigma Xi; Geol Soc Am; Soc Econ Geol. *Res:* Evaluating the geology, petrology and mineral resource potential of the Precambrian rocks of New Mexico. *Mailing Add:* NM Bur Mines Socorro NM 87801

ROBERTSON, JAMES SYDNOR, b Richmond, Va, Nov 27, 20; m 44; c 3. MEDICAL PHYSICS, NUCLEAR MEDICINE. *Educ:* Univ Minn, BS, 43, MB, 44, MD, 45; Univ Calif, PhD(physiol), 49. *Prof Exp:* From asst physiologist to assoc physiologist, Univ Calif, 46-50; biophysicist & asst physician, Brookhaven Nat Lab, 50-51, head med physics div & physician, 51-74; consult nuclear med, Mayo Clin, 75-84; DIR, HUMAN HEALTH DIV, US DEPT ENERGY, 84- *Mem:* AAAS; Am Physiol Soc; Radiation Res Soc; Soc Nuclear Med; Health Physics Soc; Am Math Soc. *Res:* Electrolyte metabolism, neutron capture therapy; tracer theory; radiation dosimetry; positron emission tomography. *Mailing Add:* ER 73 (GTN) US Dept Energy Washington DC 20585

ROBERTSON, JAMES THOMAS, b McComb, Miss, Apr 5, 31; m 52; c 6. NEUROSURGERY. *Educ:* Univ Tenn, Memphis, MD, 54; Am Bd Neurol Surg, cert, 62. *Prof Exp:* Teaching fel surg, Harvard Med Sch, 59-60; from instr to assoc prof neurosurg, 64-73, PROG DIR & PROF & CHMN, DEPT NEUROSURG, COL MED, UNIV TENN, 73-, ACTG CHAIR, 89-, PROF ANAT & NEUROBIOL, 89- *Concurrent Pos:* Chief neurosurg, USAF, Travis AFB, 60-63, Regional Med Ctr, Memphis, Veteran's Admin Med Ctr; consult, Methodist Hosp, Memphis, LeBonheur Children's Hosp, St Joseph's Hosp, Calif Med Facil, 60-63, Jackson-Madison County Hosp, Tenn & Semmes-Murphey Clin, 77-; mem exec comt, Am Med Asn Stroke Coun, 79-, chmn sci prog, 83-86, deleg, 85-87, vchmn, 91; deleg, Cong Neurol Surgeons, AMA, 75-80; chmn, Cerebral Vasc Sect, Am Asn Neurol Surgeons, 82, mem exec comt, 84; mem, Cardiovasc Comt, Am Col Surgeons, 85-88, bd gov, 87-, adv coun vasc surg, 88-91; vis prof various univs, 74-86; mem, Tech Rev Comt on Stroke, NIH, 81. *Mem:* Cong Neurol Surg (treas, 69-74, pres, 74-75); Am Asn Neurol Surgeons (treas, 86-89, vpres, 89-90, pres-elect, 90-91); fel Am Col Surgeons; Asn Acad Surg; Soc Univ Neurosurg (pres, 65); Am Acad Neurol Surg (treas, 81-83, secy, 83-86, pres-elect, 87, pres, 88); AMA. *Res:* Profound hypothermia; work with prostaglandins and the vasospasm phenomenon; platelet activity in experimental subarachnoid hemorrhage; pathogenesis and treatment of stoke; studies on cerebral vasospasm; author or co-author of 112 publications and one book. *Mailing Add:* 956 Court St Room A202 Memphis TN 38163

ROBERTSON, JEROLD C, b Provo, Utah, Mar 20, 33; m 53; c 3. PHYSICAL ORGANIC CHEMISTRY. *Educ:* Brigham Young Univ, BS, 58, PhD(org chem), 62. *Prof Exp:* From instr to asst prof, 61-74, ASSOC PROF ORG CHEM, COLO STATE UNIV, 74- *Concurrent Pos:* Assoc chmn, Chem Dept, Colo State Univ, 73- *Res:* Mechanisms of organic reactions, specifically aromatic electrophilic substitution, Baeyer-Villiger oxidation and Schmidt reaction with olefins; molecular orbital calculations on chemisorption of small molecules on metal surfaces. *Mailing Add:* Dept Chem Colo State Univ Ft Collins CO 80523

ROBERTSON, JERRY EARL, b Detroit, Mich, Oct 25, 32; m 55; c 3. ORGANIC CHEMISTRY, MEDICINAL CHEMISTRY. *Educ:* Miami Univ, BS, 54; Univ Mich, MS, 56, PhD(org chem), 59. *Prof Exp:* Sr chemist, Lakeside Labs, Colgate-Palmolive Co, Wis, 59-60, group leader cardiovasc med chem, 60-61, sect chief, 61-63; sr chemist, Riker Labs, Inc, 63-64, supvr synthetic med res, Cent Res Labs, 64-67, mgr, Synthetic Med Res Sect, 67-70, dir tech planning and coord, 70-71, dir, Chem Res Dept, 71-73, mgr, Surg Prod Dept, 74-75, gen mgr, Surg Prod Div, 75-79, div vpres surg, 79-80. *Mem:* Am Chem Soc. *Res:* Medicinal chemistry of cardiovascular and psychopharmacologic agents. *Mailing Add:* 3M Ctr Bldg 220-13S St Paul MN 55144-1000

ROBERTSON, JERRY L(EWIS), b Tulsa, Okla, Oct 25, 33; m 56; c 2. PROCESS MODELING-OPTIMIZATION. *Educ:* Okla State Univ, BS, 55; Northwestern Univ, PhD(chem eng), 62. *Prof Exp:* Chem engr, Esso Res & Eng Co, 55-65, sr engr, 65-69, eng assoc, 69-71, sect head, 71-74, mgr process & systs eng, Centrifuge Enrichment, Exxon Nuclear Co, Inc, 74-78, sr eng assoc, 78-82, ENG ADV, EXXON RES & ENG CO, 82- *Concurrent Pos:* Chmn, Heat Transfer & Energy Conversion Div, 86. *Mem:* Am Inst Chem Engrs; Am Chem Soc; AAAS; Am Petroleum Inst; NY Acad Sci. *Res:* Mass transfer in porous media; process design and economic optimization; plant start up; liquified natural gas processing; uranium enrichment; energy conservation/efficiency. *Mailing Add:* Exxon Res & Eng Co PO Box 101 Florham Park NJ 07932

ROBERTSON, JOHN CONNELL, b Carrollton, Ky, Nov 24, 31; m 56; c 4. ANIMAL NUTRITION. *Educ:* Univ Ky, BS, 53, MS, 57, PhD(animal nutrit), 60. *Prof Exp:* Area livestock specialist, 60-63, state exten livestock specialist, 63-66, PROF ANIMAL SCI, UNIV KY, 66-, ASSOC DEAN COL AGR, 69- *Mem:* Am Soc Animal Sci. *Res:* Interrelationships of certain minerals, mainly calcium, phosphorous, zinc, and amino acids. *Mailing Add:* Rm N6 Agr Sci Bldg Univ of Ky Lexington KY 40506

ROBERTSON, JOHN DAVID, b Poplar Bluff, Mo, Aug 5, 60; m 83; c 2. RADIOANALYTICAL CHEMISTRY, ION BEAM ANALYSIS. *Educ:* Univ Mo, BS, 82; Univ Md, PhD(chem), 86. *Prof Exp:* Postdoctoral fel, Lawrence Berkeley Lab, 87-89; ASST PROF CHEM, UNIV KY, 89- *Concurrent Pos:* Lectr, Univ Calif, Berkeley, 87-89. *Mem:* Am Chem Soc; Am Phys Soc. *Res:* Development and application of surface and trace element ion-beam analysis techniques; applications include fuel science and clean fuels, trace elements in neurological disorders and thin-films and thin-film devices. *Mailing Add:* Dept Chem Univ Ky Lexington KY 40506-0055

ROBERTSON, JOHN HARVEY, b Cheyenne, Wyo, Dec 6, 41; m 66; c 2. PROCESS CONTROL, MICROBIOLOGY. *Educ:* Univ Wyo, BS, 68. *Prof Exp:* Design draftsman, Dynaelectron Corp, 63 & Wyott Mfg, Wyo, 64-65; design engr & draftsman, State Wyo Engrs Off, 65-66; res microbiologist, Upjohn Co, 68-78, bioeng, 78-81, sr res microbiologist, 81-90, SR MICROBIOLOGIST, UPJOHN CO, 90- *Mem:* Am Chem Soc; Int Soc Pharmaceut Engrs; Qual Assurance Acad; Inst Environ Sci. *Res:* Development of microbiological assay methods for new product candidates or products; development of new sterilization and process methods for production; particulate control (clean room technology). *Mailing Add:* 7802 Pickering Portage MI 49081

ROBERTSON, JOSEPH HENRY, b Carrington, NDak, Jan 10, 06; m 33; c 5. RANGE CONSERVATION. *Educ:* Nebr State Teachers Col, Peru, AB, 28; Univ Nebr, MS, 32, PhD(bot), 39. *Prof Exp:* Teacher, pub schs, Nebr, 25-27 & Idaho, 28-30; instr biol, plant anat, bot & zool, Wis State Teachers Col, River Falls, 32-35; asst instr plant ecol, Univ Nebr, 36-39; jr range examr, Range Res, US Forest Serv, 40-42, forest ecologist, Range Exp Sta, 42-47; from assoc prof to prof agron & range mgt, 47-67, chmn dept, 52-64, range ecologist, 51-71, head Div Plant Sci, 59-65, prof range sci, 67-71, actg assoc dir, Agr Exp Sta, 75, actg assoc dean, Col Agr, 76, EMER PROF RANGE SCI, UNIV NEV, RENO, 71- *Concurrent Pos:* Lectr range mgt & chief of party, WVa Univ-USAID contract team & head, Dept Range Mgt, Egerton Col, Kenya, 65-67; consult watershed revegetation, Develop & Resources Corp, Khorramabad, Iran, 71-73; prin investr native shrub proj, Foresta Inst Ocean & Mountain Studies, 74-76, mem bd dirs, 75-79; leader, watershed veg surv, Nev Div Forestry, 76 & agr collection, Arch, Nev Univ, 78; consult, Desert Res Inst, Univ Nev, 80 & Res Mgt Co, 81. *Honors & Awards:* Frederic Renner Award, 77. *Mem:* Soc Range Mgt. *Res:* Artificial revegetation of range land; ecology of sagebrush-grass zone; domestication of native shrubs. *Mailing Add:* 920 Evans Ave Reno NV 89512-2805

ROBERTSON, KENNETH RAY, b Detroit, Mich, July 26, 41; c 1. TAXONOMIC BOTANY. *Educ:* Univ Kans, BS, 64, MA, 66; Wash Univ, PhD(bot), 71. *Prof Exp:* Teaching asst bot & biol, Univ Kans, 64-66; instr biol, Forest Park Community Col, 69-70; asst cur, Arnold Arboretum, Harvard Univ, 71-76; from asst scientist to assoc scientist, 76-84, SCIENTIST, ILL NAT HIST SUV, 84-, CUR HERBARIUM, 76- *Mem:* Am Soc Plant Taxonomists; New Eng Bot Club. *Res:* Classification and evolution of the Rosaceae; fruits and seeds, especially form, structure and dispersal; systematics of Jacquemontia (Convolvulaceae); flora of the southeastern United States. *Mailing Add:* 1202 Alton Dr Champaign IL 61820

ROBERTSON, LESLIE EARL, b Los Angeles, Calif, Feb 12, 28; m 82; c 4. STRUCTURAL ENGINEERING. *Educ:* Univ Calif, Berkeley, BS, 52. *Hon Degrees:* DSc, Univ Western Ont, 83; DEng, Rensselaer Polytech Inst, 86, Lehigh Univ, 90. *Prof Exp:* Struct engr, Kaiser Engrs, Oakland, Calif, 52-54, John A Blume, San Francisco, 54-57 & Raymond Int Co, NY, 57-58; DIR DESIGN & CONSTRUCT, LESLIE ROBERTSON ASSOC, 58- *Concurrent Pos:* Chmn comt risks & liabilities, Kinetic Energies Resource Coun, 75-; chair, Coun Tall Bldgs & Urban Habitat; steering comt, Int Coun Wind Eng; mem, Cornell Eng Col Coun; invited lectr, various nat & int univs. *Honors & Awards:* Raymond C Reese Res Prize, Am Soc Civil Engrs, 74; Richard J Carrol lectr, Johns Hopkins Univ, 85; Inst Honor, Am Inst Archit, 89. *Mem:* Nat Acad Eng; fel Am Soc Civil Engrs; Am Concrete Inst; Am Soc Testing & Mat; Am Soc Concrete Construction; Int Asn Bridge & Struct Eng; NY Acad Sci; Am Inst Archit; Univs Coun Consult Engrs. *Res:* Author of 100 articles. *Mailing Add:* 211 East 46th St New York NY 10017-2989

ROBERTSON, LYLE PURMAL, b Vancouver, BC, Aug 15, 33; m 55; c 2. NUCLEAR PHYSICS, INTERMEDIATE ENERGY PHYSICS. *Educ:* Univ BC, BA, 55, MA, 58, PhD(nuclear physics), 63. *Prof Exp:* Res officer reactor physics, Atomic Energy Can Ltd, 57-60; Nat Res Coun Can overseas fel, 63-65; sr res officer nuclear physics, Rutherford High Energy Lab, Eng, 65-66; assoc prof physics, 66-72, PROF PHYSICS, UNIV VICTORIA, BC, 72- *Concurrent Pos:* Royal Soc Can Rutherford Mem fel, 63-64. *Mem:* Can Asn Physicists; Am Phys Soc. *Res:* Intermediate energy nuclear and particle physics, associated with triumf; nucleon-nucleon interaction at intermediate energy. *Mailing Add:* Dept Physics & Astron Univ Victoria Box 3055 Victoria BC V8W 3P6 Can

ROBERTSON, LYNN SHELBY, JR, b Ind, Sept 19, 16; m 41; c 3. SOIL SCIENCE. *Educ:* Purdue Univ, BS, 40, MS, 41; Mich State Univ, PhD, 55. *Prof Exp:* Asst soil sci, Mich State Univ, 41-43, soil surveyor, 43, asst instr, 44-46, asst prof, 46-52; asst prof, Nat Univ Colombia, 52-53; assoc prof, 58-62, EXTEN SPECIALIST SOIL SCI, MICH STATE UNIV, 54-, PROF, 62- *Concurrent Pos:* Consult micronutrients, Taiwan, 71. *Mem:* AAAS; fel Soil Sci Soc Am; fel Soil Conserv Soc Am; fel Am Soc Agron; Int Soil Sci Soc. *Res:* Soil management, especially as related to systems of farming, tillage and economic use of commercial fertilizer. *Mailing Add:* Crop & Soil Sci Mich State Univ East Lansing MI 48823

ROBERTSON, MALCOLM SLINGSBY, b Brantford, Ont, July 18, 06; US citizen; m 34; c 2. MATHEMATICS. *Educ:* Univ Toronto, BA, 29, MA, 30; Princeton Univ, PhD(math), 34. *Prof Exp:* Nat Res Coun fel, Univ Chicago, 34-35; instr math, Yale Univ, 35-37; from instr to prof math, Rutgers Univ, 37-66; prof 66-72, EMER PROF MATH, UNIV DEL, 72- *Mem:* Math Asn Am; Am Math Soc; Sigma Xi. *Res:* Theory of functions of complex variable; conformal mapping; univalent functions; multivalent and typically real functions. *Mailing Add:* 107 29-18th St Dawson Creek BC V1G 4N5 Can

ROBERTSON, MERTON M, b Scobey, Mont, Aug 16, 24; m 57; c 1. EXPERIMENTAL PHYSICS. *Educ:* Univ Mont, BA, 51; Univ Wis, MS, 56, PhD(physics), 60. *Prof Exp:* Proj assoc physics, Univ Wis, 60-61; mem tech staff, Sandia Lab, AEC, 61-66, tech div supvr, 66-72, PROJ LEADER, SANDIA LAB, DEPT ENERGY, 72- *Concurrent Pos:* Fulbright fel, Netherlands, 51-52; sci comdr, Sandia Lab Airborn Solar Eclipse Exped, Argentina, 66. *Mem:* Am Phys Soc. *Res:* Optical spectroscopy and instrumentation; interferometry; hyperfine structure; nuclear moments of radioactive atoms; optical studies of missile reentries; solar physics; plasma physics; high speed radiometry, photography and photometry; fiber optics; opto electronics. *Mailing Add:* 6608 Natalie NE Albuquerque NM 87110

ROBERTSON, NAT CLIFTON, b Atlanta, Ga, July 23, 19; m 46; c 3. PHYSICAL CHEMISTRY, RESOURCE MANAGEMENT. *Educ:* Emory Univ, AB, 39; Princeton Univ, PhD(phys chem), 42. *Hon Degrees:* ScD, Emory Univ, 70. *Prof Exp:* Asst chem, Princeton Univ, 40-41; res assoc, Nat Defense Res Comt, 42-43; res chemist, Standard Oil Develop Co, 43-47; group leader, Celanese Corp Am, 47-51; dir petrochem dept, Nat Res Corp, 51-55; vpres & dir res, Escambia Chem Corp, 55-58; vpres res & develop, Spencer Chem Co & Spencer Chem Div, Gulf Oil Corp, 58-66; vpres res, Air Prod & Chem, Inc, 66-69, sr vpres & dir, 69-77; dir & sci adv, Marion Labs, Inc, 77; dir, C H Kline & Co, 77-86; RETIRED. *Mem:* Fel AAAS. *Res:* Kinetics of gas reactions; catalysis; physical methods of separation; free radicals. *Mailing Add:* 156 Philip Dr Princeton NJ 08540

ROBERTSON, PHILIP ALAN, b Sept 9, 38; US citizen; m 67; c 2. PLANT ECOLOGY. *Educ:* Colo State Univ, BS, 62, MS, 64, PhD(plant ecol), 68. *Prof Exp:* Instr range sci, Colo State Univ, 67-68; asst prof biol, State Univ NY Col Oneonta, 68-70; asst prof bot, 70-77, ASSOC PROF BOT, SOUTHERN ILL UNIV, CARBONDALE, 77- *Mem:* Ecol Soc Am; AAAS; Am Inst Biol Sci; Soc Range Mgt. *Res:* Analysis of structure and function of terrestrial plant communities. *Mailing Add:* Dept Bot Life Sci Li-0431 Southern Ill Univ Carbondale IL 62901

ROBERTSON, RALEIGH JOHN, b Reinbeck, Iowa, Nov 8, 42; m 66; c 2. ORNITHOLOGY. *Educ:* Grinnell Col, BA, 65; Univ Iowa, MSc, 67; Yale Univ, PhD(ecol), 71. *Prof Exp:* DIR, BIOL STA, QUEENS UNIV, 72-, PROF BIOL, 83- *Mem:* Am Ornithologists Union; Ecol Soc Am; Can Soc Zoologists. *Res:* Behavioral ecology of reproduction in birds including mating systems, sexual selection, parental investment and competition. *Mailing Add:* Biol Dept Queens Univ Kingston ON K7L 3N6 Can

ROBERTSON, RANDAL MCGAVOCK, b Tampa, Fla, Mar 12, 11; m 39; c 3. PHYSICS. *Educ:* Glasgow Univ, MA, 32; Mass Inst Technol, PhD(physics), 36. *Prof Exp:* Asst, Columbia Univ, 36-37; res assoc, Norton Co, Mass, 37-42; staff mem radiation lab, Mass Inst Technol, 42-46; head mech & mat br, US Off Naval Res, DC, 46-48, dir phys sci div, 48-51, dep natural sci, 51-52, sci dir, 52-58; asst dir math, phys & eng sci, NSF, 58-61, assoc dir res, 61-70; dean res div & prof physics, 70-76, EMER DEAN RES DIV, VA POLYTECH INST & STATE UNIV, 76- *Mem:* Fel AAAS; fel Am Phys Soc; Soc Am Foresters. *Res:* Arc cathode phenomena; nuclear magnetic moments; physics of solid materials, especially abrasives and refractories; microwave linear array scanning antennas; radar; research administration. *Mailing Add:* 1404 Highland Circle SE Blacksburg VA 24060

ROBERTSON, RAYMOND E(LIOT), b St Louis, Mo, Aug 17, 40; m 77; c 1. ORGANIC CHEMISTRY, PETROLEUM CHEMISTRY. *Educ:* Cent Mo State Col, BS, 62; Colo State Univ, MS, 71; Univ Wyo, PhD(chem), 76. *Prof Exp:* Res chemist org synthesis, Tretolite Co, Petrolite Corp, Webster Groves, Mo, 63-69; instr chem, Eastern Wyo Col, 75-76; res chemist phys & org chem, 76-83, SR RES SCIENTIST & MGR EXPLOR & NEW PROD DIV, WESTERN RES INST, UNIV WYOMING RES CORP, 83- *Mem:* Am Chem Soc; Sigma Xi. *Res:* Micellar catalysis; petroleum recovery and demulsification chemistry; asphalt chemistry and relationships between asphalt physical and chemical properties; acyloin condensation chemistry. *Mailing Add:* Western Res Inst Box 3395 Univ Sta Laramie WY 82071

ROBERTSON, RICHARD EARL, b Long Beach, Calif, Nov 12, 33; m 55, 74; c 2. MATERIALS SCIENCE & ENGINEERING. *Educ:* Occidental Col, BA, 55; Calif Inst Technol, PhD(chem), 60. *Prof Exp:* NSF fel, Wash Univ, 59-60; phys chemist, Gen Elec Res & Develop Ctr, 60-70; staff scientist, Ford Motor Co, 70-86; PROF MAT SCI & ENG, UNIV MICH, 86- *Mem:* AAAS; fel Am Phys Soc; Am Chem Soc; Sigma Xi. *Res:* Mechanical properties and structure of polymers; adhesion; structure and mechanical properties of polymers; behavior of fiber composites. *Mailing Add:* Dept Mat Sci & Eng Univ Mich 2300 Hayward St Ann Arbor MI 48109-2136

ROBERTSON, RICHARD THOMAS, b Spokane, Wash, July 25, 45. NEUROBIOLOGY. *Educ:* Wash State Univ, BS, 67; Univ Calif, Irvine, MS, 68, PhD(biol sci), 72. *Prof Exp:* Res assoc neuroanat, Univ Oslo, 71-72; res scientist neurobiol, Fels Res Inst, 72-76; ASST ASSOC PROF ANAT & BIOL SCI, UNIV CALIF IRVINE, 76- *Concurrent Pos:* Adj mem, Psychol Dept, Antioch Col, 73-76. *Mem:* Soc Neurosci; Am Asn Anatomists; Psychonomic Soc. *Res:* Neuroanatomical and neurophysiological studies of nonspecific sensory systems of the brainstem, thalamus and cerebral cortex. *Mailing Add:* Dept Anat & Biol Sci Univ Calif Irvine Col Med Irvine CA 92717

ROBERTSON, ROBERT, b Suffolk, Eng, Nov 14, 34; c 1. MALACOLOGY. *Educ:* Stanford Univ, AB, 56; Harvard Univ, PhD(biol), 60. *Prof Exp:* Asst cur mollusks, Acad Natural Sci Philadelphia, 60-65, assoc cur mollusks, 65-76, cur malacol, 76-88, chmn malacol, 69-72, HON CUR MALACOL, ACAD NATURAL SCI PHILADELPHIA, 88- *Concurrent Pos:* Secy, Inst Malacol, 65-70, pres-elect, 70-73, pres, 73-74, co-ed-in-chief, Malacologia, 73-88. *Mem:* Fel AAAS; Am Malacol Union (pres, 83-84); Marine Biol Asn UK; Malacol Soc Japan; Australia Soc Malacol. *Res:* Marine gastropods; systematics; larvae; anatomy; life histories; ecology, especially foods and reproduction of gastropods; paleontology; marine zoogeography. *Mailing Add:* Acad Natural Sci Philadelphia Dept Malacol 19th & Pkwy Philadelphia PA 19103

ROBERTSON, ROBERT GRAHAM HAMISH, b Ottawa, Ont, Oct 3, 43; m 80; c 1. NEUTRINO PHYSICS. *Educ:* Oxford Univ, BA, 65; McMaster Univ, PhD(nuclear physics), 71. *Prof Exp:* Res assoc, Cyclotron Lab, Mich State Univ, 71-72, asst prof nuclear physics, 72-73, from asst prof to prof physics, 73-83; staff mem, 81-88, FEL, LOS ALAMOS NAT LAB, 88- *Concurrent Pos:* Res assoc, Princeton Univ, 75-76; Alfred P Sloan fel, 76-78. *Mem:* Brit Inst Physics; Can Asn Physicists; Am Phys Soc; fel Am Phys Soc. *Res:* Weak interactions; atomic beam magnetic resonance; nuclear astrophysics; isobaric multiplets; nuclei far from stability; neutrino mass. *Mailing Add:* MS-D449 Los Alamos Nat Lab Los Alamos NM 87545

ROBERTSON, ROBERT JAMES, b Hazleton, Pa, Oct 7, 43; m 67; c 1. EXERCISE PHYSIOLOGY. *Educ:* West Chester State Col, BS, 66; Univ Pittsburgh, MA, 67, PhD(health & phys educ), 73. *Prof Exp:* Asst dir, Phys Fitness Res Lab, Univ Health Ctr, 71-73; asst prof & dir, Phys Fitness Res Lab, Dept Health Educ, Nebr Ctr Health Educ, Univ Nebr, 73-76; ASSOC PROF & DIR, HUMAN ENERGY RES LAB, DEPT HEALTH & PHYS EDUC, UNIV PITTSBURGH, 76- *Concurrent Pos:* Coordr, Health & Phys Educ Prog, Univ Pittsburgh. *Mem:* Am Alliance Health Phys Educ & Recreation; Am Heart Asn; Am Col Sports Med. *Res:* Physiological and perceptual correlates of exercise stress; exercise as a therapeutic modality in coronary heart disease; energy cost of load carriage; effect of red blood cell reinfusion and bicarbonate ingestion on physical working capacity; carbohydrate metabolism. *Mailing Add:* Dept Health & PE Recreation Rm 242 Trees Hall Univ Pittsburgh Pittsburgh PA 15261

ROBERTSON, ROBERT L, b Blountsville, Ala, July 20, 25; m 65; c 1. ENTOMOLOGY. *Educ:* Auburn Univ, BS, 50, MS, 54. *Prof Exp:* Asst county agr agent, Auburn Univ, 50-52, res asst entom, 52-54, asst prof, 54-57; entomologist, Am Cyanamid Co, 57-58; exten entomologist, Univ Ga, 58-60; assoc prof, 60-69, exten prof,69-84, EMER PROF ENTOM, NC STATE UNIV, 84- *Concurrent Pos:* Consult, Dow Chem USA, Am Agr Serv, Inc, Ciba-Geigy. *Mem:* Entom Soc Am; Int Turfgrass Asn. *Res:* Insects of economic importance on tobacco, cotton, peanuts, ornamentals and turf. *Mailing Add:* 409 Holly Circle Cary NC 27511

ROBERTSON, ROSS ELMORE, b Kennetcook, NS, Oct 5, 15; m 45; c 4. PHYSICAL CHEMISTRY, ORGANIC CHEMISTRY. *Educ:* Mt Allison Univ, BSc, 41, MSc, 42; McGill Univ, PhD(chem), 44. *Prof Exp:* From mem staff to prin res officer, Nat Res Coun Can, 44-69; prof chem, 69-81, AOSTRA res prof, 80-82, EMER PROF CHEM, UNIV CALGARY, 81- *Mem:* Am Chem Soc; fel Chem Inst Can; fel Royal Soc Can; Royal Soc Chem. *Res:* Detailed mechanisms of solvolysis; solvent isotope effects in kinetics and equilibria; secondary deuterium isotope effects; water-oil emulsions; water purification. *Mailing Add:* Dept Chem Univ Calgary Calgary AB T2N 1N4 Can

ROBERTSON, SCOTT HARRISON, b Washington, DC, Nov 6, 45; m 73. PLASMA PHYSICS. *Educ:* Cornell Univ, BS, 68, PhD(appl physics), 72. *Prof Exp:* Sr res assoc plasma physics, Columbia Univ, 72-74; asst res physicist, 75-80, ASSOC RES PHYSICIST PLASMA PHYSICS, UNIV CALIF, IRVINE, 80- *Mem:* Am Phys Soc; AAAS; Inst Elec & Electronics Engrs. *Res:* Experimental investigations of the propagation of intense ion and electron beams, and their interaction with magnetically confined target plasma, particularly heating, focusing, and microwave emission due to collective processes. *Mailing Add:* Astro-Geophys/Campus Box 391 Univ Colo Boulder CO 80309

ROBERTSON, STELLA M, MONOCLONAL ANTIBODIES, IMMUNE REGULATION. *Educ:* Johns Hopkins Univ, PhD(biol), 78. *Prof Exp:* ASST DIR THERAPEUT RES, ALCON LABS INC, 81- *Mailing Add:* Alcon Labs Inc 6201 S Freeway Ft Worth TX 76134

ROBERTSON, STUART DONALD TREADGOLD, electrical engineering; deceased, see previous edition for last biography

ROBERTSON, T(HOMAS) M(ILLS), b Tallula, Ill, Oct 1, 22; m 43; c 2. ELECTRICAL ENGINEERING. *Educ:* Univ Ill, BS, 43. *Prof Exp:* Electronic engr, Farnsworth TV & Radio, 46-47 & Kellex Corp, 47-48; sect leader labs, Vitro Corp Am, 48-53, group leader, 53-54, asst dept head, 54-57, dept head, 57-69, dept head, Vitro Labs Div, Automation Industs Inc, 69-76, dir customer rels, 76-80, vpres, Vitro Corp, 80-86; PRES TM ROBERTSON INC, 86- *Concurrent Pos:* Chmn anti-submarine warfare adv comt, Nat Security Indust Asn. *Mem:* Sr mem Inst Elec & Electronics Engrs; Nat Soc Prof Engrs. *Res:* Ordnance equipment; weapons and systems for undersea warfare. *Mailing Add:* TM Robertson Inc 11404 Rouen Dr Potomac MD 20854

ROBERTSON, THOMAS N, b St Andrews, Scotland, Oct 22, 31; US citizen; m 58; c 2. MATHEMATICS. *Educ:* Univ St Andrews, BSc, 53; Col Aeronaut, MSc, 55; Univ Southern Calif, MA, 60. *Prof Exp:* Aerodynamicist, Eng Elec Co, 55-57; lectr math, Univ Southern Calif, 58-60; from instr to assoc prof math, 60-79, chmn dept, 69-75 & 78-83, PROF MATH, OCCIDENTAL COL, 79- *Concurrent Pos:* Danforth teacher grant, 63-64; Fulbright lectr, Turkey, 73-74. *Mem:* Math Asn Am. *Res:* Supersonic aerodynamics; numerical analysis. *Mailing Add:* Occidental Col Los Angeles CA 90041

ROBERTSON, TIM, b Denver, Colo, Oct 4, 37; m 59; c 4. MATHEMATICAL STATISTICS. *Educ:* Univ Mo, BA, 59, MA, 61, PhD(statist), 66. *Prof Exp:* Asst prof math, Cornell Col, 61-63; from asst prof to assoc prof statist, 65-74, PROF STATIST, UNIV IOWA, 74- *Concurrent Pos:* Vis prof, Univ NC, 74-75 & Univ Calif, Davis, 83-84. *Mem:* Fel Inst Math Statist; fel Am Statist Asn; Math Asn Am; Int Statist Inst. *Res:* Mathematical statistics with particular interests in the theory and applications of order restricted estimates, hypothesis tests and related problems. *Mailing Add:* Dept Statist & Actuarial Sci Univ Iowa Iowa City IA 52242

ROBERTSON, W(ILLIAM) D(ONALD), b Montreal, Que, Dec 23, 13; nat US; m 38, 74; c 2. MATERIALS SCIENCE. *Educ:* Mass Inst Technol, BSc, 42, DSc(metall), 48; MA, Yale Univ, 57. *Prof Exp:* Res metallurgist, Aluminum Labs, Ltd, 42-45; res assoc, Inst Study Metals, Chicago, 48-50; from asst prof to prof metall, 50-63, prof, 63-78, EMER PROF APPL SCI, YALE UNIV, 78- *Concurrent Pos:* Fulbright sr res scholar, Cambridge Univ & overseas fel, Churchill Col, 64-65; vis prof, Univ Sussex, 71; sr res fel, Univ Warwick, Eng, 74-75; sr scientist in residence, Univ Va, 78- *Honors & Awards:* Willis R Whitney Award, Nat Asn Corrosion Engrs, 65. *Mem:* Int Inst Conserv Hist & Artistic Works Eng. *Res:* Physical metallurgy; corrosion of metals; crystal plasticity and fracture; structure and properties of surfaces by low energy electron diffraction techniques and electron spectroscopy; preservation of architectural monuments (cathedrals); physical techniques for evaluating durability of limestone masonry used in restoration of Westminster Abbey; identification of preservatives for architectural masonry. *Mailing Add:* 2107 Minor Rd Charlottesville VA 22903

ROBERTSON, WALTER VOLLEY, b Malakoff, Tex, Apr 6, 31; m 57; c 5. VERTEBRATE ZOOLOGY. *Educ:* Stephen F Austin State Col, BS, 51; Tex A&M Univ, MS, 59, PhD(zool), 64. *Prof Exp:* Instr zool, Tex A&M Univ, 60-64; from asst prof to assoc prof, 64-73, PROF BIOL & ADMIN ASST, STEPHEN F AUSTIN STATE UNIV, 73- *Mem:* Am Soc Zoologists. *Res:* Comparative vertebrate anatomy, osteology of North American clupeid fishes and brain morphology of rodents. *Mailing Add:* Dept Biol Stephen F Austin State Univ Austin Bldg North St Nacogdoches TX 75962

ROBERTSON, WILBERT JOSEPH, JR, b Washington, DC, Mar 28, 28; m 55; c 4. INORGANIC CHEMISTRY. *Educ:* George Washington Univ, BS, 50; Univ Wis, PhD(inorg chem), 55. *Prof Exp:* Chemist, Uranium Div, Mallinckrodt Chem Works, 55-60, res supvr uranium processing, 60-66, chemist, Opers Div, 66-67; sr staff engr, Nuclear Div, 67-70, RES PROJ CHEMIST, TECH DIV, KERR-McGEE CORP, 70- *Mem:* Am Chem Soc; Am Nuclear Soc; Sigma Xi. *Res:* Hydrometallurgy; solvent extraction; mineral processing and purification; extractive metallurgy. *Mailing Add:* 7320 Hammond Ave Kerr-McGee Corp Oklahoma City OK 73132

ROBERTSON, WILLIAM, IV, b Glen Ridge, NJ, Sept 12, 43; m 71; c 2. SCIENCE POLICY, ENVIRONMENTAL MANAGEMENT. *Educ:* Parsons Col, BS, 66; Sam Houston State Univ, MA, 69. *Prof Exp:* Tutor biol, Parsons Col, 66-67; res technician biochem, Med Ctr, NY Univ, 67-68; sci secy, Comt Water Qual Criteria, Nat Acad Sci, 71-72, staff officer, Environ Studies Bd, 72-73, tech asst, Comn Natural Resources, 73-78, exec secy, Int Environ Progs Comt, 74-78; PROG DIR, ANDREW W MELLON FOUND, 79- *Mem:* Am Soc Limnol & Oceanog; Ecol Soc Am. *Res:* Science policy. *Mailing Add:* Andrew W Mellon Found 140 E 62nd St New York NY 10021

ROBERTSON, WILLIAM G, b Bethesda, Md, Dec 2, 29; m 55; c 3. MEDICAL PHYSIOLOGY. *Educ:* Shepherd Col, BS, 54; WVa Univ, MS, 55; State Univ NY Buffalo, PhD(physiol), 63. *Prof Exp:* Instr zool, WVa Univ, 55-57; instr aviation physiol, Sch Aerospace Med, USAF, 57-62, res physiologist respiratory physiol, 62-63, chief sealed environ sect, 63-66; proj scientist, Garrett Corp, Calif, 66-70; sr staff scientist, Inhalation Toxicol Dept, Hazelton Labs, Inc, 70-71; chief environ physiol br, USAF Aerospace Med, 71-74; chmn, Med Biol Div, 74-78, prof physiol, Okla Col Osteop Med & Surg, 78-89, PROF PHYSIOL, COL OSTEOP MED, OKLA STATE UNIV, 89- *Mem:* AAAS; Am Physiol Soc; Aerospace Med Asn. *Res:* Respiratory and environmental physiology. *Mailing Add:* 3223 Riverside Dr Apt 148 Tulsa OK 74105

ROBERTSON, WILLIAM O, b New York, NY, Nov 24, 25; m 52; c 5. PEDIATRICS, MEDICAL ADMINISTRATION. *Educ:* Univ Rochester, BA, 46, MD, 49. *Prof Exp:* From instr to assoc prof pediat, Col Med, Ohio State Univ, 56-63, asst dean, 62-63; from asst dean to assoc dean, 63-72, assoc prof, 63-72, PROF PEDIAT, SCH MED, UNIV WASH, 72- *Concurrent Pos:* Consult, US Army Madigen Hosp, 64-72. *Mem:* AAAS; Am Med Writers' Asn; Am Asn Poison Control Ctr (treas, 63-64). *Res:* Medical malpractice and risk management; nutrition; accidental poisoning; education. *Mailing Add:* Dept Pediat Univ Wash Seattle WA 98195

ROBERTSON, WILLIAM VAN BOGAERT, b New York, NY, Sept 15, 14; m 41, 68; c 3. BIOCHEMISTRY, NUTRITION. *Educ:* Stevens Inst Technol, ME, 34; Univ Freiburg, PhD(chem), 37. *Prof Exp:* Res chemist, Mass Gen Hosp, Boston, 38-41; res fel, Nat Cancer Inst, 41-44; res assoc, Univ Chicago, 44-45; asst prof exp med, Univ Vt, 45-48, assoc prof biochem & exp med, 48-52, prof biochem, 52-61; from assoc prof to prof, 61-79, emer prof biochem, stanford univ, 79-81; prog dir metab biol, NSF, 81-87; RETIRED. *Concurrent Pos:* Vis prof biochem, Univ del Valle, Cali, Colombia, 67-68 & Univ Saigon, Viet-Nam, 74; consult nutrit, Nat Inst Child Health & Human Develop, Bethesda, Md, 78; dir res & educ, Children's Hosp, Stanford, 62-79; prog mgr human nutrit, NSF, 79-81. *Mem:* AAAS; Am Chem Soc; Sigma Xi; Am Soc Biol Chemists; Am Soc Clin Nutrit. *Res:* Biochemistry of connective tissue; nutritional biochemistry. *Mailing Add:* 329 Millicent Way Shreveport LA 71106

ROBERTS-PICHETTE, PATRICIA RUTH, b Hamilton, NZ, Dec 22, 30; m 67; c 2. ECOLOGY. *Educ:* Univ NZ, BSc, 53, MSc, 54; Duke Univ, PhD, 57. *Prof Exp:* Asst prof biol, Pfeiffer Col, 57-58; from asst prof to assoc prof, Univ NB, 58-67; consult & pvt researcher, 67-73; exec secy, Man & Biosphere Prog Can, liaison & coord directorate, Environ Can, 73-79; dep secy to JAC, CGIAR, Rome, Italy, 82-89; sr prog officer, Multilateral Br, 79-82, SR POLICY ANALYST, AMERICAS BR, CEDA, 89- *Concurrent Pos:* Consult dept natural resources, Prov NB, 66-68 & dept agr, 68-71, Can Dept Forestry, 68 & Environ Can, 71-72. *Mem:* AAAS; Ecol Soc Am; Can Bot Asn. *Res:* Community ecology; vegetation and floras of specific regions; plant succession and distribution; pollution biology. *Mailing Add:* 430 Besseret St Ottawa ON K1N 6C1 Can

ROBERTSTAD, GORDON WESLEY, b Madison, Wis, Sept 29, 23; m 48; c 4. BACTERIOLOGY. *Educ:* Lniv Wis, BS, 49, MS, 51; Colo State Univ, PhD(bact), 59. *Prof Exp:* Instr bact, Univ Wyo, 49-50; proj asst, Univ Wis, 51-52; instr bact, Univ Wyo, 52-57, asst prof, 57-59, assoc prof microbiol, 59-64; prof bact & head dept, SDak State Univ, 64-68; dir health related prog, 75-81, prof microbiol, 68-88, EMER PROF, UNIV TEX, EL PASO, 88- *Concurrent Pos:* NIH fel, Commun Dis Ctr, Atlanta, Ga, 62-63. *Mem:* Fel Am Acad Microbiol; Am Soc Allied Health Professions; Sigma Xi; Mycol Soc Am; Int Soc Human & Animal Mycol. *Res:* Antigenic studies of Vibrio fetus; taxonomy of microaerophilic Actionomycetes; epidemiology of dermatophytic and systemic fungi; aeromycology. *Mailing Add:* Dept Biol Sci Univ Tex El Paso TX 79968

ROBEY, FRANK A, INFLAMMATION, PEPTIDE SYNTHESIS. *Educ:* Catholic Univ, PhD(phys chem), 77. *Prof Exp:* RES CHEMIST, MOLECULAR PHARMACOL LAB, DIV BIOCHEM & BIOPHYSICS, CTR DRUGS & BIOLOGICS, FOOD & DRUG ADMIN, 79- *Mailing Add:* 8729 Ridge Rd Bethesda MD 20817

ROBEY, PAMELA GEHRON, b Oct 31, 52; m; c 2. BONE CELL BIOCHEMISTRY, CELL BIOLOGY. *Educ:* Cath Univ Am, PhD(biol), 79. *Prof Exp:* Sr staff fel, 83-87, BIOLOGIST, NAT INST DENT HEALTH, NIH, 87- *Mem:* Am Soc Bone & Mineral Res; Am Soc Biochem & Molecular Biol; Am Soc Cell Biol. *Res:* Matrix protein biochemistry. *Mailing Add:* Bone Res Br Bldg 30 Rm 214 NIH 9000 Rockville Pike Bethesda MD 20205

ROBEY, ROGER LEWIS, b Fairmont, WVa, June 18, 46; m 68; c 2. SYNTHETIC ORGANIC CHEMISTRY. *Educ:* Marietta Col, BS, 68; Princeton Univ, MS, 71, PhD(org chem), 72. *Prof Exp:* Assoc, Ohio State Univ, 72-74; SR ORG CHEMIST, ELI LILLY & CO, 74- *Mem:* Am Chem Soc. *Res:* Organic synthesis; synthesis of heterocycles. *Mailing Add:* 4831 Brentridge Pkwy Greenwood IN 46143

ROBIE, NORMAN WILLIAM, b Washington, DC, Jan 21, 42; m. PHARMACOLOGY. *Educ:* Auburn Univ, BS, 64, MS, 69; Med Univ SC, PhD(pharmacol), 72. *Prof Exp:* Fel clin pharmacol, Sch Med, Emory Univ, 72-75; asst prof pharmacol, Univ Tex Health Sci Ctr, San Antonio, 75-77; asst prof, 77-79, ASSOC PROF PHARMACOL, LA STATE UNIV MED CTR, 79- *Mem:* Am Soc Pharmacol & Exp Therapeut. *Res:* Autonomic cardiovascular pharmacology. *Mailing Add:* Dept Pharmacol La State Univ Med Ctr 1100 Florida Ave New Orleans LA 70119

ROBIE, RICHARD ALLEN, b Winchendon, Mass, Oct 13, 28; m 68. MINERALOGY. *Educ:* Dartmouth Col, AB, 50; Univ Chicago, MS, 53, PhD(geochem), 57. *Prof Exp:* Chemist, Univ Chicago, 54-56; GEOPHYSICIST, US GEOL SURV, 57- *Concurrent Pos:* Prin investr, Lunar Samples Prog. *Mem:* AAAS; Mineral Soc Am; Am Chem Soc; Am Geophys Union; Geochem Soc. *Res:* Thermodynamic properties of minerals; low temperature heat capacities and aqueous solution calorimetry of carbonates and silicates; elastic constants of single crystals; specific heats of lunar soils. *Mailing Add:* US Geol Surv Stop 959 12201 Sunrise Valley Dr Reston VA 22092

ROBILLARD, GEOFFREY, b Niagara Falls, NY, Feb 25, 23; m 63; c 2. CHEMICAL ENGINEERING. *Educ:* Mass Inst Technol, BS, 44, MS, 47. *Prof Exp:* Chemist, Carbide & Carbon Chem Co Div, Union Carbide Corp, 47-52; res engr, Calif Inst Technol, 52-53, sect chief, 53-59, div chief, 59-63, dep proj mgr, 63-68, mgr, Eng Mech Div, 68-73, dep asst lab dir, 73-76, asst lab dir planning & rev, 76-78, asst lab dir for energy & technol appln, Jet Propulision Lab, 78-85, asst lab dir for eng & review, 85-88; RETIRED. *Mem:* Am Inst Aeronaut & Astronaut; Sigma Xi. *Res:* Propulsion; structures; materials; system design and development; Ranger, Mariner & Voyager projects; energy, safety, reliability and quality assurance. *Mailing Add:* PO Box 842 Bodega Bay CA 94923

ROBILLIARD, GORDON ALLAN, b Victoria, BC, May 19, 43; m 72; c 2. MARINE ECOLOGY, RESOURCE DAMAGE ASSESSMENT METHODOLOGY. *Educ:* Univ Victoria, BSc, 65; Univ Wash, MS, 67, PhD(zool), 71. *Prof Exp:* Sr aquatic ecologist, Woodward-Clyde Consults, 71-85; SR CONSULT & VPRES, ENTRIX INC, 85- *Concurrent Pos:* Mem comt on appl ecol theory to environ problems, Nat Acad Comn Life Sci, 84-86. *Mem:* Ecol Soc Am; Sigma Xi. *Res:* Ecology and feeding habits of opisthobranch molluscs; ecological consequences of predation by fish and large motile invertebrates in marine benthic communities; polar marine ecology; methods to evaluate the value of natural resources damaged by man's activities. *Mailing Add:* Entrix Inc 3214 Kirby Lane Walnut Creek CA 94598

ROBIN, ALLEN MAURICE, b Chicago, Ill, Jul 10, 34; m; c 2. PROJECT MANAGEMENT. *Educ:* Univ Ill, BS, 57; Univ Southern Calif, MS, 66. *Prof Exp:* Sr chem engr, VELSICOL Chem Corp, 56-62; sr thermo engr, General Dynamics Corp, Pomona, Calif, 62-65; RES MGR, TEXACO INC, MONTEBELLO RES LAB, 65- *Mem:* Am Inst Chem Eng. *Res:* Coal gasification; high temperature desulfurization/filtration of fuel gas; waste gasification. *Mailing Add:* 2517 E Gelid Ave Anaheim CA 92806

ROBIN, BURTON HOWARD, b Chicago, Ill, Mar 19, 26; m 47, 87; c 3. ORGANIC CHEMISTRY. *Educ:* Roosevelt Univ, BS, 48; Univ Chicago, MS, 49. *Prof Exp:* Res chemist, Corn Prod Co, 49-51; res chemist, Swift & Co, 51-57; res chemist, Visking Corp, 58-59; res chemist, Nalco Chem Co, 59-63; chmn dept phys sci, 69-74, assoc prof, 63-76, PROF CHEM, KENNEDY-KING COL, 76- *Mem:* AAAS; Am Chem Soc; Sigma Xi. *Res:* Coagulants; surfactants; corrosion and rust inhibitors; flotation agents; vinyl polymers; emulsion polymerization; fats and fatty acids and derivatives; qualitative analysis; physical science. *Mailing Add:* Dept Phys Sci Kennedy-King Col 6800 S Wentworth Ave Chicago IL 60621

ROBIN, EUGENE DEBS, b Detroit, Mich, Aug 23, 19; m; c 2. PHYSIOLOGY. *Educ:* George Washington Univ, SB, 46, SM, 47, MD, 51. *Prof Exp:* Res fel med, Harvard Med Sch, 52-53; sr asst resident med serv, Peter Bent Brigham Hosp, Boston, 53-54; asst med, Harvard Med Sch, 54-55, instr, 55-58, assoc, 58-59; from assoc prof to prof, Sch Med, Univ Pittsburgh, 59-70; prof med & physiol, Sch Med, Stanford Univ, 70-88, actg chmn, Dept Med, 71-73, actg chmn & curric & acad consult, Dept Physiol, 80-88, EMER PROF MED & PHYSIOL, SCH MED, STANFORD UNIV, 88- *Concurrent Pos:* Asst, Peter Bent Brigham Hosp, 52-53, chief med resident, 54-55, jr assoc, 55-57, assoc, 57-59, assoc dir, Cardiovasc Training Prog, 58-59; chmn pulmonary adv comt, Nat Heart & Lung Inst, 71-74. *Mem:* Am Physiol Soc; Am Soc Clin Invest; Am Thoracic Soc (pres, 70-71); Am Col Physicians. *Res:* Clinical physiology; intracellular acid-base metabolism; intracellular gas exchange; comparative physiology and biochemistry. *Mailing Add:* PO Box 1185 Trinidad CA 95570

ROBIN, MICHAEL, b New York, NY, May 17, 19; m 43; c 2. ORGANIC CHEMISTRY. *Educ:* City Col New York, BS, 40; NY Univ, ChE, 44; Brooklyn Col, MS, 50. *Prof Exp:* Sr org chemist org & pharmaceut res & develop, Nepera Chem Co, Inc, NY, 46-51; res & develop chemist metallic soap, Nuodex Prod Co, Inc, NJ, 51-53; chief chemist org & pharmaceut, Simpson Labs, Simpson Coal & Chem Co, 53-56; group leader chem org res & develop, Catalin Corp Div, 56-68, mgr chem prod div lab, 68-73, plant mgr fine chem dept, 73-74, mgr, Mfg Servs, Ashland Chem Co, Ashland Oil & Refining Co, 74-76; tech dir, 77-86, VPRES TECH DIR & DIR INPLANT TRAINING WORLD WIDE, CTR PROF ADVAN, EAST BRUNSWICK, NJ, 86- *Mem:* AAAS; Am Chem Soc; Am Inst Chem; Soc Plastics Eng; NY Acad Sci; Am Inst Chem Engrs. *Res:* Fine organic and pharmaceutical product and process research and development; organic synthesis; antioxidants and stabilizers for organic materials. *Mailing Add:* 1508 Ashbrook Dr Scotch Plains NJ 07076

ROBINETTE, CHARLES DENNIS, b Conway, Ark, June 23, 35; div; c 1. RADIOBIOLOGY, STATISTICS. *Educ:* State Col Ark, BS, 57; Colo State Univ, MS, 65, PhD(radiation biol), 71. *Prof Exp:* Statistician med, Southern Res Support Ctr, Vet Admin, 68-72; SR PROF OFFICER, MED FOLLOW-UP AGENCY, NAT ACAD SCI, 72- *Concurrent Pos:* Instr div biometry med ctr, Univ Ark, Little Rock, 68-72; statistician, Radiation Effects Res Found, Hiroshima, Japan, 77-79. *Mem:* Am Statist Asn; Radiation Res Soc. *Res:* Application of statistics methods to medical research. *Mailing Add:* Nat Acad Sci Washington DC 20418

ROBINETTE, HILLARY, JR, b Wilmington, Del, Jan 27, 13; m 34; c 2. CHEMISTRY. *Educ:* Temple Univ, AB, 34. *Prof Exp:* Res chemist, Rohm and Haas Co, Pa, 33-39; pres, W H & F Jordon Jr Co, 39-41; mem res staff, Com Solvents Corp, Ind, 41-42; mgr mkt develop, Publicker Industs, Inc, Pa, 45-48; res dir, Amalgamated Chem Corp, 48-52; PRES, ROBINETTE RES LABS, INC, 52- *Mem:* AAAS; Am Asn Textile Chem & Colorists; Am Soc Testing & Mat; fel Am Inst Chemists. *Res:* Organic chemistry; textile chemicals. *Mailing Add:* 10333 Campana Dr Sun City AZ 85351-1098

ROBINETTE, MARTIN SMITH, b Sacramento, Calif, Sept 18, 39; m 65; c 3. AUDIOLOGY. *Educ:* Univ Utah, BS, 65, MS, 67; Wayne State Univ, PhD(audiol), 70. *Prof Exp:* Asst prof audiol, Univ Wyo, 70-74; assoc prof audiol, 74-80, ASSOC PROF COMMUN, UNIV UTAH, 80- *Mem:* Acoust Soc Am; Int Soc Audiol; Am Speech & Hearing Asn. *Res:* Lateralization of sound image from intensity cues; binaural detection ability at large interaural intensity differences; test for functional hearing loss; diplacusis. *Mailing Add:* Dept Audiol Mayo Clin 200 First St SW Rochester MN 55905

ROBINOVITCH, MURRAY R, b Brandon, Man, Jan 17, 39; m 64. ORAL BIOLOGY, EXPERIMENTAL PATHOLOGY. *Educ:* Univ Minn, BS, 59, DDS, 61; Univ Wash, PhD(salivary gland protein synthesis), 67. *Prof Exp:* From instr to assoc prof, 66-75, PROF ORAL BIOL, UNIV WASH, 75-, MEM, CTR RES ORAL BIOL, 71-, ACTG CHMN DEPT ORAL BIOL, 73- *Mem:* AAAS; Am Soc Cell Biol; Am Dent Asn; Int Asn Dent Res. *Res:* Protein synthesis in salivary and other exocrine glands; secretion and the secretory product; oral histology and pathology. *Mailing Add:* Dept Oral Biol Univ Wash Seattle WA 98195

ROBINOW, CARL FRANZ, b Hamburg, Ger, Apr 10, 09; m 38; c 2. MICROBIOLOGY. *Educ:* Univ Hamburg, MD, 35. *Prof Exp:* Researcher, Copenhagen, Denmark, 35-37, St Bartholomew's Hosp, London, 37-40, Strangeways Lab, Cambridge Univ, 40-47; vis lectr, US univs, 47-49; assoc prof, 49-56, PROF BACT & IMMUNOL, UNIV WESTERN ONT, 56- *Mem:* Am Soc Microbiol; Bot Soc Am; fel Royal Soc Can. *Res:* Cytology of bacteria and fungi. *Mailing Add:* 1161 Beechwood Pl London ON N6A 5C1 Can

ROBINOW, MEINHARD, b Hamburg, Ger, May 19, 09; US citizen; m 44; c 3. NUTRITION, PEDIATRICS. *Educ:* Hamburg, Ger, MD, 35. *Prof Exp:* Asst prof pediat, Med Col, Augusta, Ga, 38-39 & 42-43; dir phys growth, Fels Res Inst, 39-42; prof pediat, Sch Med, Univ Va, 75-79; prof, 79-81, clin prof, 81-87 EMER CLIN PROF PEDIAT, SCH MED, WRIGHT STATE UNIV, 82- *Concurrent Pos:* From 1st lieutenant to Major, US Army Med Corps, 43-46. *Mem:* Am Acad Pediat; Am Pediat Soc; Am Soc Human Genetics; Am Nutrit Inst; AAAS; AMA. *Res:* Dysmorphology and genetics - delineation of new malformation syndromes; nutritional anthropometry. *Mailing Add:* Dept Pediat Wright State Univ Dayton OH 45431

ROBINS, CHARLES RICHARD, b Harrisburg, Pa, Nov 25, 28; m 65; c 3. ICHTHYOLOGY. *Educ:* Cornell Univ, BA, 50, PhD(ichthyol), 54. *Prof Exp:* Asst gen zool, Cornell Univ, 50-51, ichthyol & taxon, 51-54; from asst prof to prof, 56-69, chmn dept marine sci, 61-63, MAYTAG PROF MARINE BIOL, SCH MARINE & ATMOSPHERIC SCI, UNIV MIAMI, 69-, CUR FISHES, 66-, PROF ICHTHYOLOGY, 74-, CHMN, DIV BIOL & LIVING RESOURCES, 78-, ACTG DEAN, ROSENTIEL SCH MARINE & ATMOSPHERIC SCI, 81- *Concurrent Pos:* Res assoc, Cornell Univ, 52; ed,

Bull Marine Sci & Gulf & Caribbean, 61-62; mem panel syst biol, NSF, 66-69; proj & prog rev comn, Marine Lab, Duke Univ, 68-70; comt inshore & estuarine pollution, Hoover Found, 69; mem ecol adv comt, Environ Protection Agency, 75-78. *Mem:* AAAS; Am Soc Ichthyologists & Herpetologists (vpres, 64 & pres-elect, 82); Am Soc Syst Zool; Am Soc Zool; Am Inst Fishery Res Biol. *Res:* Taxonomy, morphology, ecology and behavior of fishes. *Mailing Add:* 9190 SW 61st Ct Miami FL 33156

ROBINS, ELI, b Houston, Tex, Feb 22, 21; m 46; c 4. PSYCHIATRY, NEUROLOGY. *Educ:* Rice Inst, BA, 40; Harvard Med Sch, MD, 43. *Prof Exp:* Asst psychiat, Harvard Med Sch, 44-45; asst neurol, Sch Med, Boston Univ, 48; USPHS fel Sch Med, 49-51, instr neuropsychiat, 51-53, from asst prof to prof psychiat, 53-66, head dept, 63-75, WALLACE RENARD PROF PSYCHIAT, SCH MED, WASH UNIV, 66- *Concurrent Pos:* Fel psychiat, New Eng Ctr Hosp, 48. *Honors & Awards:* Salmon Medalist, NY, 81. *Mem:* Am Soc Clin Invest; Histochem Soc; Am Soc Biol Chemists; Am Col Nuclear Physicians; Am Psychiat Asn. *Res:* Biochemistry of the nervous system; psychiatric disease. *Mailing Add:* 4940 Audubon St St Louis MO 63110

ROBINS, JACK, b Roselle, NJ, Feb 17, 19; m 49; c 2. PHYSICAL CHEMISTRY, POLYMER CHEMISTRY. *Educ:* City Col New York, BS, 40; Univ Buffalo, MA, 48; Polytech Inst Brooklyn, PhD(phys chem), 59. *Prof Exp:* Chemist, Vandium Corp Am, 44-48; chemist, Am Electrometal Corp, 48-49; chemist, Bd Transport, NY, 49-54; chemist, Wilmot & Cassidy, Inc, 55-56; res chemist, ICI Am, Inc, 59-75, res chemist, 75-82, COMPUT CONSULT, ATLAS POWDER CO, 82 - *Mem:* Am Chem Soc. *Res:* Analytical chemistry; inorganic solutions; gas chromatography; thermodynamics and material properties; microcomputers, software, hardware and interfacing with instruments; scientific programmer. *Mailing Add:* 139-B Cistus Plaza Cranbury NJ 08512

ROBINS, JANIS, b Riga, Latvia, Aug 3, 25; nat US; m 51; c 4. POLYMER CHEMISTRY, CATALYSIS. *Educ:* Univ Wash, Seattle, BS, 52, PhD(chem), 57. *Prof Exp:* Analytical chemist, Wash Farmers Coop, 52; analytical chemist, State Dept Agr, Wash, 52-53; analytical chemist, Am-Marietta Co, 54, res chemist, 55-57; res chemist, Minn Mining & Mfg Co, 57-65; res assoc, Archer Daniels Midland Co, 65-67; res assoc, Ashland Chem Co, 67-72; sr res specialist, 3M Co, 72-80, div scientist, 80-90; PRES, CATALYTIC CROSSLINK, INC, 90- *Concurrent Pos:* Asst prof chem, Macalester Col, 60-65. *Mem:* Am Chem Soc; Am Foundrymens Soc. *Res:* Physical organic chemistry; polymer synthesis; metal ion catalysis; polyurethane, furan, epoxy and phenolic resin technology; foundry binder technology; adhesives technology. *Mailing Add:* Catalytic Crosslink Inc 11 Ludlow Ave St Paul MN 55108

ROBINS, MORRIS JOSEPH, b Nephi, Utah, Sept 28, 39; m 60, 73; c 8. BIO-ORGANIC CHEMISTRY, NUCLEIC ACID COMPONENTS-ANALOGUES. *Educ:* Univ Utah, BA, 61; Ariz State Univ, PhD(org chem), 65. *Prof Exp:* Cancer res scientist biochem, Roswell Park Mem Inst, 65-66; res assoc org chem, Univ Utah, 66-69; from asst prof to prof chem, Univ Alta, 69-88; prof, 87-89, J REX GOATES PROF CHEM, BRIGHAM YOUNG UNIV, 89- *Concurrent Pos:* Mem adv comt chemother & hemat, Am Cancer Soc, 77-80; vis prof med chem, Univ Utah, 81-82; mem grant panel, Nat Cancer Inst Can, 83-86; adj prof med microbiol infectious Dis, Univ Alta, 88- *Mem:* Am Chem Soc; Am Asn Cancer Res. *Res:* Chemistry of nucleic acid components, nucleoside analogues and related biomolecules; transformations of natural product nucleosides, mechanism-based enzyme inhibitors, anticancer and antiviral agents. *Mailing Add:* Chem Dept Brigham Young Univ Provo UT 84602-1049

ROBINS, NORMAN ALAN, b Chicago, Ill, Nov 19, 34; m 56; c 2. CHEMICAL ENGINEERING, MATHEMATICS. *Educ:* Mass Inst Technol, BS, 55, MS, 56; Ill Inst Technol, PhD, 72. *Prof Exp:* Metallurgist, 56-60, res metallurgist, 60-62, asst mgr, Res Dept, 62-67, assoc mgr, 67-72, dir process res, 72-77, vpres res, 77-84, vpres technol assessment, 84-86, vpres tech assessment & strategic planning, 86-87, VPRES STRATEGIC PLANNING, INLAND STEEL CO, 87- *Mem:* Am Inst Mining, Metall & Petrol Engrs; Am Inst Chem Engrs; Math Asn Am. *Res:* Process research and computer process control. *Mailing Add:* Inland Steel Co Mail Code 8-129 3210 Watling St East Chicago IN 46312

ROBINS, RICHARD DEAN, b North Manchester, Ind, Nov 19, 42; m 69. ORGANIC CHEMISTRY. *Educ:* Manchester Col, BA, 64; Ohio State Univ, MS, 66, PhD(org chem), 68. *Prof Exp:* RES CHEMIST, LUBRIZOL CORP, 69- *Mem:* Am Chem Soc. *Res:* Physical organic and polymer chemistry. *Mailing Add:* 29400 Lakeland Blvd Wickliss OH 44092

ROBINS, ROLAND KENITH, b Scipio, Utah, Dec 13, 26; m 48; c 6. ORGANIC CHEMISTRY. *Educ:* Brigham Young Univ, AB, 48, MA, 49; Ore State Col, PhD(chem), 52. *Prof Exp:* Res assoc, Wellcome Res Labs, Burroughs Wellcome & Co, Inc, NY, 52-53; from asst prof chem to assoc prof chem, NMex Highlands Univ, 53-57; assoc prof, Ariz State Univ, 57-60, prof, 60-64; prof, Univ Utah, 65-69; vpres res & develop & dir, Nucleic Acid Res Inst, Int Chem & Nuclear Corp, 69-74, sr vpres res & develop, 74-77; PROF CHEM & BIOCHEM, BRIGHAM YOUNG UNIV, 77-, DIR, CANCER RES CTR, 78- *Concurrent Pos:* Consult, Parke, Davis & Co, Mich, 55-60, Midwest Res Inst, Mo, 58-64, Nat Cancer Inst, 59-69 & Merck Sharp & Dohme Div, Merck & Co Inc, 57-65; mem chem panel, Cancer Chemother, Nat Serv Ctr, NIH, 60-62. *Mem:* Am Chem Soc; Am Asn Cancer Res. *Res:* Synthesis of purines and pyrimidines; purine nucleosides; condensed pyrimidine systems, especially in cancer research. *Mailing Add:* Nucleic Acid Res Inst 3300 Hyland Ave Costa Mesa CA 92626-1503

ROBINSON, A(UGUST) R(OBERT), agricultural & hydraulic engineering; deceased, see previous edition for last biography

ROBINSON, ALBERT DEAN, b Sherman Mills, Maine, Mar 28, 39; m 61; c 2. GENETICS, MYCOLOGY. *Educ:* Univ Maine, BA, 61; Johns Hopkins Univ, MAT, 62; Univ Iowa, PhD(bot), 68. *Prof Exp:* Teacher high sch, Nyack, NY, 62-65; from asst prof to assoc prof genetics, 68-80, assoc prof, 80-83, PROF BIOL, STATE UNIV NY CPL POTSDAM, 83- *Concurrent Pos:* Dept chmn, State Univ NY Col Potsdam, 86. *Mem:* Am Genetics Asn; Am Inst Biol Sci; Am Soc Human Genetics; Nat Asn Biol Teachers; Sigma Xi. *Res:* meiosis in chives. *Mailing Add:* Dept Biol State Univ NY Col Potsdam Potsdam NY 13676

ROBINSON, ALFRED GREEN, b Thomasville, Ga, Feb 19, 28; m 51; c 4. PETROLEUM CHEMISTRY. *Educ:* Emory Univ, AB, 49, MS, 51, PhD(chem), 55. *Prof Exp:* Chemist, Hercules Powder Co, 51-52; chemist, Tenn Eastman Co Div, 55-58, chemist, Tex Eastman Co Div, 58-73, develop assoc, 73-74, head develop div, 74-77, dir res div, 77-84, dir res & develop 84-89, VPRES, RES & DEVELOP, TEX EASTMAN CO, EASTMAN KODAK CO, 89- *Mem:* Am Chem Soc. *Res:* Chemical properties of aliphatic carbonyl compounds; synthesis of polymers by condensation polymerization. *Mailing Add:* 501 Terrace Dr Longview TX 75601

ROBINSON, ALIX IDA, b Ft Worth, Tex, Oct 26, 37; div. CELLULAR DIFFERENTIATION, NUCLEIC ACID STRUCTURE. *Educ:* Univ Tex, Austin, BA, 59, PhD(zool), 64. *Prof Exp:* Res assoc bot, Univ Ill, Urbana, 64; NIH fel path, Med Sch, Univ Gothenburg, Sweden, 64-65; instr zool, Univ Ill, Urbana, 65-66; instr anat, State Univ NY Health Sci Ctr, 66-68, res assoc, 68-71, asst prof, 71-75, ASSOC PROF MICROBIOL, STATE UNIV NY HEALTH SCI CTR, 75- *Concurrent Pos:* Career develop award, NIH, 68-73; prin investr, NIH grant, 73-76, NSF grants, 81-88; Blinker fel acad admin, State Univ NY, 89. *Mem:* Am Soc Cell Biol; Soc Develop Biol; Am Soc Microbiol. *Res:* Role of asymmetric cell division in cellular differentiation including the source of polar development and the role of microtubules in nuclear migration during fern spore germination. *Mailing Add:* Dept Microbiol State Univ NY Health Sci Ctr Syracuse NY 13210

ROBINSON, ALLAN RICHARD, b Lynn, Mass, Oct 17, 32; m 55; c 3. PHYSICAL OCEANOGRAPHY. *Educ:* Harvard Univ, BA, 54, MA, 56, PhD(physics), 59. *Prof Exp:* NSF fel meteorol & oceanog, Cambridge Univ, 59-60; from asst prof to assoc prof, 60-68, dir, Ctr Earth & Planetary Physics, 72-75, GORDON McKAY PROF GEOPHYS FLUID DYNAMICS & MEM CTR EARTH & PLANETARY PHYSICS, HARVARD UNIV, 68-, CHMN, COMT OCEANOG, 72-; ASSOC PHYS OCEANOGR WOODS HOLE OCEANOG INST, 60- *Concurrent Pos:* Co-chmn, Mid-Ocean Dynamics Exp I Sci Coun, NSF, 71-74; Guggenheim fel, Cambridge Univ, Eng, 72-73; Co-ed-in-chief, Dynamics of Atmospheres & Oceans, 76- *Mem:* Fel Am Acad Arts & Sci. *Res:* Oceanography; dynamics of oceanic motions and geophysical fluid dynamics. *Mailing Add:* Dept Earth Sci Harvard Univ Pierce Hall Cambridge MA 02138

ROBINSON, ARTHUR, b New York, NY, Jan 12, 14; m 41; c 2. PEDIATRICS, GENETICS. *Educ:* Columbia Univ, AB, 34; Univ Chicago, MD, 38. *Prof Exp:* Assoc prof pediat & biophys, 63-66, prof biophys & genetics & chmn dept, 71-74, DISTINGUISHED PROF PEDIAT, SCH MED, UNIV COLO, DENVER, 66- *Concurrent Pos:* Dir prof serv, Nat Jewish Hosp & Res Ctr, 75-83. *Mem:* AAAS; Am Pediat Soc; Am Acad Pediat; Am Soc Human Genetics. *Res:* Cytogenetics. *Mailing Add:* 4101 E Ellsworth Ave Denver CO 80222

ROBINSON, ARTHUR B, b Chicago, Ill, Mar 24, 42; m 72; c 6. MOLECULAR BIOLOGY OF AGING, PROTEIN CHEMISTRY. *Educ:* Calif Inst Technol, BS, 63; Univ Calif, San Diego, PhD(chem), 67. *Prof Exp:* Asst prof chem, Biol Dept, Univ Calif, San Diego, 68-73; res prof, Linus Pauling Inst, 73-79, vpres, 73-75, pres, 75-78; PRES & RES PROF, ORE INST SCI & MED, 81- *Res:* Molecular biology of aging; nutrition and preventive medicine; protein chemistry; civil defense engineering. *Mailing Add:* Ore Inst Sci & Med 2251 Dick George Rd Cave Junction OR 97523

ROBINSON, ARTHUR GRANT, b Wadena, Sask, July 7, 16; m 42; c 1. ENTOMOLOGY. *Educ:* Univ Man, BSA, 50, PhD(entom), 61; McGill Univ, MSc, 52. *Prof Exp:* Agr Res officer, Can Dept Agr, 50-53; from assoc prof to emer prof entom, Univ Man, 53-82, head dept, 77-81; RETIRED. *Mem:* Entom Soc Am; fel & hon mem Entom Soc Can; Agr Inst Can. *Res:* Biology and taxonomy of aphids. *Mailing Add:* Dept Entom Univ Man Winnipeg MB R3T 2N2 Can

ROBINSON, ARTHUR R(ICHARD), b Brooklyn, NY, Oct 28, 29. STRUCTURAL MECHANICS. *Educ:* Cooper Union, BCE, 51; Univ Ill, MS, 53, PhD(civil eng), 56. *Prof Exp:* Res assoc mech & mat, Univ Minn, 55-57, asst prof, 57-58, asst prof aeronaut eng, 58-60; assoc prof civil eng, 60-63, PROF CIVIL ENG, UNIV ILL, URBANA, 63- *Honors & Awards:* Moisseiff Award, 70; Walter L Huber Civil Eng Res Award, Am Soc Civil Engrs, 69. *Mem:* Am Soc Civil Engrs; Am Soc Mech Engrs. *Res:* Numerical methods; stress waves in solids; analysis of structural systems; earthquake engineering. *Mailing Add:* 2129 Newmark Civil Eng Lab 205 N Mathews Ave Urbana IL 61801

ROBINSON, ARTHUR ROBIN, b Montreal, Que, May 26, 43; m 68; c 4. AGRICULTURAL CHEMISTRY, ENDOCRINOLOGY. *Educ:* McGill Univ, BSc, 67, MSc, 70, PhD(agr chem), 75. *Prof Exp:* PROF CHEM, NS AGR COL 75- *Concurrent Pos:* Proj leader res, NS Dept Agr & Mkt, 76-88, Can NS Livestock Feed initiative Agreement, 87- *Res:* Animal reproduction with particular reference to hormones and plant estrogens; the application of sewage sludge to agricultural land; determination of lead and other metal ion content of Eastern Canada maple products; cervical mucus peroxidase and heat detection in dairy cattle; effect of Omega three fatty acids in poultry. *Mailing Add:* PO Box 550 Truro NS B2N 5E3 Can

ROBINSON, ARTHUR S, b New York, NY, Sept 26, 25; m; c 6. TECHNICAL MANAGEMENT. *Educ:* Columbia Univ, BS, 48; NY Univ, MS, 51; Columbia Univ, DSc, 57. *Prof Exp:* PRES, SYSTS TECHNOL DEVELOP CORP, 80- *Mem:* Fel Inst Elec & Electronics Engrs. *Mailing Add:* Syst-Technol Develop Corp 1035 Sterling Rd No 101 Herndon VA 22070

ROBINSON, BARRY WESLEY, engineering physics, for more information see previous edition

ROBINSON, BEATRICE LETTERMAN, b Bloomsburg, Pa, June 6, 41; m 63; c 2. PHYCOLOGY. *Educ:* Bloomsburg State Col, BS, 63; Syracuse Univ, PhD(bot), 68. *Prof Exp:* Adj asst prof, Syracuse Univ, 68-75; from adj asst prof to asst prof, 68-80, assoc acad dean, 80-82, ASSOC PROF, LE MOYNE COL, 82- *Mem:* AAAS; Phycol Soc Am. *Res:* Development in blue-green algae, specifically, photomorphogenesis in Nostoc. *Mailing Add:* Dept Biol Le Moyne Col Le Moyne Heights Syracuse NY 13214

ROBINSON, BEROL (LEE), b Highland Park, Mich, June 25, 24; m 48; c 3. NUCLEAR PHYSICS, SCIENCE EDUCATION. *Educ:* Harvard Univ, AB, 48; Johns Hopkins Univ, PhD(physics), 53. *Prof Exp:* Asst prof physics, Univ Ark, 52-56; assoc prof, Western Reserve Univ, 60-67; assoc prof, Case Western Reserve Univ, 67-71; prog specialist, Univ Sci Educ, Unesco, 71-84, consult, 85-90; RETIRED. *Concurrent Pos:* Res fel, Israel AEC, 61-62; vis prof, Rensselaer Polytech Inst, 64; asst to dir, Educ Res Ctr, Mass Inst Technol, 69-71; dir educ component, US Metric Study, 70-71; mem adv comt educ, Europ Phys Soc, 75-80; assoc mem int comn physics educ, Int Union Pure & Appl Physics, 74-80. *Honors & Awards:* Sigma Xi. *Mem:* AAAS; Am Phys Soc; Am Asn Physics Teachers. *Res:* Nuclear and x-ray spectroscopy; Mossbauer effect; physics laboratory instruction equipment; science education; international science administration. *Mailing Add:* One rue du General Gouraud Meudon F-92190 France

ROBINSON, BRIAN HOWARD, b Derby, UK, Sept 24, 44. BIOCHEMISTRY, GENETICS. *Educ:* Bristol Univ, BSc, 65, PhD(biochem), 68. *Prof Exp:* Can Med Res Coun fel biochem, 68-70; lectr, Univ Sheffield, 70-73; SCIENTIST BIOCHEM & GENETICS, RES INST, HOSP SICK CHILDREN, TORONTO, 73- *Concurrent Pos:* Prof, Dept Pediat, Univ Toronto, 73-, Dept Biochem, 74-; Can Med Res Coun grants, 75-80. *Mem:* Brit Biochem Soc; Can Biochem Soc. *Res:* Mitochondrial metabolite transport; keto acid dehydrogenases; hereditary disorders of metabolism leading to lactic acidosis and keto acidosis; branched-chain amino acid metabolism; mitochondrial respiratory chain; mitochondrial DNA. *Mailing Add:* Hosp Sick Children 555 University Ave Toronto ON M5S 1A8 Can

ROBINSON, BRUCE B, b Chester, Pa, Oct 13, 33; m 60; c 2. PLASMA PHYSICS. *Educ:* Drexel Inst, 56; Princeton Univ, MA, 58, PhD(plasma transport properties), 61. *Prof Exp:* Res asst gen relativity, Yerkes Observ, Univ Chicago, 61; res asst plasma stability, Univ Calif, San Diego, 61-63; mem tech staff atomic physics & solid state plasmas, RCA Labs, 63-75; asst dir prog strategies, US Energy Res & Develop Admin, 75-77; dir technol implementation & policy integration, US Dept Energy, 77-81; sr adv to vpres res, Exxon Res & Develop Corp, 81-84; dep dir, 84-87, DIR RES PROGS, OFF NAVAL RES, 87- *Concurrent Pos:* Exec dir com tech adv bd, US Dept Com, 73-75. *Honors & Awards:* Presidential Meritorious Award, 89. *Mem:* Inst Elec & Electronics Engrs; Am Phys Soc. *Res:* Relativistic cosmology; plasma transport theory; atomic collision theory; plasma stability theory; solid-state device physics; energy research and development planning. *Mailing Add:* 3437 N Emerson St Arlington VA 22217

ROBINSON, C PAUL, b Detroit, Mich, Oct 9, 41; m 63; c 2. LASERS, CHEMICAL PHYSICS. *Educ:* Christian Bros Col, BS, 63; Fla State Univ, Tallahassee, PhD(physics), 67. *Hon Degrees:* PhD, Christian Bros Univ, 89. *Prof Exp:* Chief test operator nuclear reactor tests, Nev, Los Alamos Nat Lab, 67-70, res physicist, Advan Concepts Group, NMex, 70-71, alt group leader chem laser res & develop, 71-73, proj dir laser isotope separation, 73-76, appl photochem div leader, 76-80, assoc dir, 80-85; sr vpres, Ebasco Corp, 85-88; ambassador, US Nuclear Testing Talks, Geneva, Switz, 88-90; MEM STAFF, SANDIA NAT LABS, 90- *Mem:* Am Phys Soc; AAAS; Am Nuclear Soc. *Res:* Laser isotope separation research including work on uranium enrichment, laser spectroscopy, laser induced chemistry and tunable lasers; weapons physics; arms control issues. *Mailing Add:* Sandia Nat Lab Org 9400 Albuquerque NM 87185

ROBINSON, CAMPBELL WILLIAM, b Edmonton, Alta, June 22, 33; c 3. CHEMICAL ENGINEERING. *Educ:* Univ BC, BASc, 61; Univ Calif, Berkeley, PhD(chem eng), 71. *Prof Exp:* Process engr, Gulf Oil Can Ltd, 57-62, process unit supt, 62-64, sr process engr, 64-65, tech serv supvr, 65-66; from asst prof to assoc prof, 71-81, PROF CHEM ENG, UNIV WATERLOO, 81- *Concurrent Pos:* Vis prof, Univ Calif, Berkeley, 78-79; assoc ed, Can J Chem Eng, 80-84; co-ed, Comprehensive Biotechnol, vol 4, 85; vis prof, EPF, Lausanne, Switz, 85-86; ed, Can J Chem Eng, 89- *Honors & Awards:* J W T Spinks lectr, Univ Sask, 91. *Mem:* Can Soc Chem Eng; Chem Inst Can; Am Chem Soc. *Res:* Mass transfer; biochemical engineering, fermentation and waste treatment process design. *Mailing Add:* Dept Chem Eng Univ Waterloo Waterloo ON N2L 3G1 Can

ROBINSON, CASEY PERRY, b Idabel, Okla, Oct 10, 32; m 55; c 3. PHARMACOLOGY. *Educ:* Univ Okla, BS, 54, MS, 67; Vanderbilt Univ, PhD(pharmacol), 70. *Prof Exp:* Sr sci investr pharmacol, Am Heart Asn, Med Sch, Univ Calif, Los Angeles, 70-71; assoc prof pharmacol, Col Med, 71-80, from asst prof to assoc prof pharmacodynamics, Col Pharm, 71-81, PROF PHARMACOL, COL MED, UNIV OKLA, 81-, PROF PHARMACODYNAMICS, COL PHARM, 81- *Concurrent Pos:* Co-prin investr, NSF grant, 74-76; prin investr, EPA grant, 77-80, DOD grant, 85- *Mem:* AAAS; Soc Exp Biol & Med; Am Soc Pharmacol & Exp Therapeut; Soc Toxicol; NY Acad Sci. *Res:* Pesticide and drug effects on cardiovascular responses; neuroeffector mechanisms of blood vessels. *Mailing Add:* Col Pharm Health Sci Ctr Univ Okla PO Box 26901 Oklahoma City OK 73190

ROBINSON, CECIL HOWARD, b London, Eng, Nov 5, 28; m 56; c 5. ORGANIC CHEMISTRY, PHARMACOLOGY. *Educ:* Univ London, BSc, 50, PhD(chem), 54. *Prof Exp:* NSF fel, Wayne State Univ, 54-55; chemist, Glaxo Labs Ltd, 55-56; from chemist to sr chemist, Schering Corp, 56-63; from asst prof to assoc prof, 63-77, PROF PHARMACOL, SCH MED, JOHNS HOPKINS UNIV, 77- *Mem:* AAAS; Am Chem Soc; Royal Soc Chem; Am Soc Pharmacol & Exp Therapeut. *Res:* Synthetic steroids; enzyme inhibitors; design and synthesis of chemotherapeutic agents; enzyme mechanism; biochemistry. *Mailing Add:* Dept Pharmacol Johns Hopkins Univ Sch Med Baltimore MD 21205

ROBINSON, CHARLES ALBERT, b Newton, Mass, July 22, 21; m 54; c 1. ORGANIC CHEMISTRY. *Educ:* Brown Univ, ScB, 43; Mass Inst Technol, PhD(org chem), 50. *Prof Exp:* Develop res chemist, Merck & Co, 43-47; asst chem, Mass Inst Technol, 47-50; res chemist, Arnold, Hoffman & Co, 50-53, res sect leader, 53-57, mem staff develop dept, 57-59, sect leader tech serv, 59-63; sr res scientist, Wyeth Labs, Inc, 63-70, asst to dir, Biol & Chem Develop Div, 70-77, assoc dir chem develop, Res & Develop Dept, 77-81; RETIRED. *Mem:* Am Chem Soc. *Res:* Synthesis of organic compounds; process development; pharmaceuticals. *Mailing Add:* 31 Lonsdale Lane Kenneth Square PA 19348

ROBINSON, CHARLES C(ANFIELD), b East Orange, NJ, Oct 18, 32; m 61; c 2. COLLOIDAL SCIENCE, PHOTOCONDUCTIVITY. *Educ:* Miami Univ, BS, 55; Mass Inst Technol, BS, 55, MS, 57, EE, 59, PhD(elec eng), 60. *Prof Exp:* Asst prof elec eng, Mass Inst Technol, 61-62; sr physicist, Res Ctr, Am Optical Corp, 62-77; TECH SPECIALIST & PROJ MGR, XEROX CORP, 77- *Mem:* Inst Elec & Electronics Engrs; Optical Soc Am. *Res:* Colloidal systems photoconductivity; optical properties of materials; xerographic instrumentation; optical thin films; rare earth studies in glass; magneto-optical effects. *Mailing Add:* Xerox Corp Phillips Rd Webster NY 14580

ROBINSON, CHARLES DEE, b Dallas, Tex, July 16, 32; m 54. MATHEMATICS. *Educ:* Hardin-Simmons Univ, BA, 56; Univ Tex-Austin, MA, 61, PhD(math), 64. *Prof Exp:* Instr math, Hardin-Simmons Univ, 56-59; asst prof, Ariz State Univ, 64-65; assoc prof, Univ Miss, 65-68; chmn, Div Sci, 69-80, PROF MATH & HEAD DEPT, HARDIN-SIMMONS UNIV, 68-, DIR INSTNL RES, 80- *Concurrent Pos:* Dir student sci training projs math, NSF, 67-68, 70-71 & 74-80; pres, Tex Asn Acad Adminrs Math Scis, 88-89; dir, Univ Info Systm, 84- *Mem:* Am Math Soc; Math Asn Am. *Res:* Functional analysis; computational mathematics. *Mailing Add:* Dept Math Hardin-Simmons Univ 2200 Hickory Box 1170 Abilene TX 79698

ROBINSON, CHARLES J, b Wheeling, WVa, July 16, 47; m 69; c 5. REHABILITATION ENGINEERING. *Educ:* Col Steubenville, BS, 69; Ohio State Univ, MS, 71, Wash Univ, DSc, 79. *Prof Exp:* Mem tech staff, Bell Tel Labs, Columbus, Ohio, 69-74; postdoctoral assoc anesthesiol, Yale Univ, 79-81; BIOMED ENGR, REHAB RES & DEVELOP CTR, VET ADMIN HINES HOSP, 81-; PROF NEUROL, STRITCH SCH MED, LOYOLA UNIV, 89- *Concurrent Pos:* Adj asst prof, Physiol Dept, Stritch Sch Med, Loyola Univ, 82-89; assoc dir, Rehab Res & Develop Ctr, Vet Admin Hines Hosp, 83-, prog dir rehab neurosci, 84-; prin investr, merit rev, US Dept Vet Affairs, 83-; assoc ed, Inst Elec & Electronics Engrs Trans Biomed Eng, 87-90; vis lectr, Bioeng Prog, Univ Ill, Chicago, 90- *Mem:* Fel Inst Elec & Electronics Engrs; Inst Elec & Electronics Engrs Eng Med & Biol Soc (vpres 88 & 89, pres 90 & 91); Rehab Eng Soc NAm; Soc Neurosci. *Res:* Rehabilitative neuroscience with a particular emphasis on spinal cord injury; micturition dysfunction; lower limb spasticity and motor control following spinal cord injury; exercise physiology in wheelchair bound individuals. *Mailing Add:* Vet Admin Hines Hosp Rehab Res & Develop Ctr PO Box 20 Hines IL 60141

ROBINSON, CHARLES NELSON, b Fayetteville, Tenn, Nov 18, 28. ORGANIC CHEMISTRY. *Educ:* Maryville Col, BS, 49; Univ Tenn, MS, 51, PhD(org chem), 53. *Prof Exp:* Asst, Univ Tenn, 49-50; asst chemist org res, Oak Ridge Nat Lab, 52; fel & asst, Univ Ill, 53-54; from asst prof to assoc prof chem, La Polytech Inst, 56-61; assoc prof, 61-66, PROF CHEM, MEMPHIS STATE UNIV, 66- *Mem:* Am Chem Soc; Sigma Xi. *Res:* Structure proof and synthesis of alkaloids and related compounds; organophosphorus chemistry. *Mailing Add:* 547 Wild Cherry Cove Memphis TN 38117

ROBINSON, CLARK, b Seattle, Wash, Dec 29, 43; m 66. MATHEMATICAL ANALYSIS. *Educ:* Univ Wash, BS, 66; Univ Calif, Berkeley, PhD(math), 69. *Prof Exp:* Vis prof math, Inst Pure & Appl Math, Rio de Janeiro, Brazil, 70-71; from asst prof to assoc prof, 69-78, chmn dept, 85-87, PROF MATH, NORTHWESTERN UNIV, 78- *Mem:* Am Math Soc; Math Asn Am; Soc Indust Appl Math. *Res:* Differential dynamical systems; global analysis. *Mailing Add:* 1431 Noyes Evanston IL 60201

ROBINSON, CLARK SHOVE, JR, b Reading, Mass, May 13, 17; m 42; c 2. PHYSICS. *Educ:* Mass Inst Technol, SB, 38, PhD(physics), 42. *Prof Exp:* Res assoc radiation lab, Mass Inst Technol, 41-43; res asst prof physics, Univ Ill, Urbana, 46-51, res assoc prof, 51-55, prof, 55-76; translr & ed, Soviet J Nuclear Physics, 76-91; RETIRED. *Concurrent Pos:* Adj prof physics, Mont State Univ, Bozeman, 76- *Mem:* Am Phys Soc. *Res:* X-ray study of glass structure; crystal growth in tungsten wire; microwave and other vacuum tubes; development of betatrons; nuclear physics; mesons; Russian translation. *Mailing Add:* PO Box 1765 Bozeman MT 59771

ROBINSON, CURTIS, b Wilmington, NC, May 12, 34; m 68; c 3. PLANT PHYSIOLOGY. *Educ:* Morgan State Col, BS, 60; Howard Univ, MS, 68; Univ Md, PhD(bot), 73. *Prof Exp:* Teacher sci & math, Lincoln High Sch, Leland, NC, 60-61 & Williston Sr High Sch, Wilmington, 61-62; res biologist, Radiation Biol Lab, Smithsonian Inst, 62-69; res technician immunol, Microbiol Assocs, 65-67; ASSOC PROF BIOL, EDINBORO STATE COL, 73- *Mem:* Am Soc Plant Physiologists. *Res:* Studies of light, hormonal, and mineral interactions in etiolated mung bean seedlings. *Mailing Add:* Dept Biol/Health Servs Edinboro Univ Pa Edinboro PA 16444

ROBINSON, D(ENIS) M(ORRELL), b London, Eng, Nov 19, 07; nat US; m 32; c 2. ENGINEERING. *Educ:* Univ London, BS, 27, PhD(elec eng), 29; Mass Inst Technol, MS, 31. *Prof Exp:* Mem staff eng res, Callender's Cable Co, 31-35; mem staff TV develop, Scophony, Ltd, 39; radar develop assignments, Ministry Aircraft Prod, 39-45; prof elec eng, Univ Birmingham, 45-46; pres, High Voltage Eng Corp, Burlington, 46-70, chmn bd, 70-80; RETIRED. *Concurrent Pos:* Chmn bd trustees, Marine Biol Lab, Woods Hole, 71-76, hon chmn bd trustees, 76- *Honors & Awards:* Order of the Brit Empire; US Medal of Freedom, 47. *Mem:* Nat Acad Eng; Am Acad Arts & Sci (secy, 70-73); Am Phys Soc; Brit Inst Elec Engrs. *Res:* High-voltage behavior of solids; television; radar; particle accelerators. *Mailing Add:* 44 Gosnold Rd Woods Hole MA 02543

ROBINSON, DAN D, forestry, for more information see previous edition

ROBINSON, DANIEL ALFRED, b Schenectady, NY, Apr 9, 32; m 58; c 3. ALGEBRA. *Educ:* NY State Teachers Col Albany, BA, 53; Rensselaer Polytech Inst, MS, 54; Univ Wis-Madison, PhD(math), 64. *Prof Exp:* From asst prof to assoc prof math, 59-76, PROF MATH, GA INST TECHNOL, 76- *Mem:* Am Math Soc; Sigma Xi. *Res:* General algebraic structures; loop theory. *Mailing Add:* Sch Math Ga Inst Technol Atlanta GA 30332

ROBINSON, DANIEL E, instrumentation, for more information see previous edition

ROBINSON, DANIEL OWEN, b Colonia Dublan, Mex, Jan 28, 18; m 42; c 6. SOIL SCIENCE. *Educ:* Brigham Young Univ, AB, 42; Univ Ariz, MS, 47; Ohio State Univ, PhD(soil physics), 49. *Prof Exp:* Instr agron, Ohio State Univ, 49-50; assoc prof agron & head dept agr, 50-55, prof agron & dir div agr, 55-70, PROF AGR, ARIZ STATE UNIV, 70- *Mem:* AAAS; Soil Sci Soc Am; Am Soc Agron; Sigma Xi. *Res:* Soils; soil physics; plant nutrition. *Mailing Add:* Box 484 Thatcher AZ 85552

ROBINSON, DAVID, b Larne, Ireland, Jan 13, 29; US citizen; m 59; c 4. PLASMA PHYSICS. *Educ:* Queen's Univ, Belfast, BSc, 51, MSc, 53; Univ Western Ont, PhD(physics), 57. *Prof Exp:* Asst lectr physics, Queen's Univ, Belfast, 52-54; res assoc, Univ Western Ont, 54-58; res assoc, Univ Southern Calif, 58-60, asst prof elec eng, 62-64; assoc prof physics, Univ Windsor, 64-68; assoc prof, 68-70, chmn dept, 68-86, PROF PHYSICS, DRAKE UNIV, 70- *Concurrent Pos:* Consult, Northrop Space Labs, 62-64, Electro-Optical Systems, Calif, 65. *Mem:* Am Asn Physics Teachers. *Res:* Molecular spectroscopy; plasmas in hypersonic flow; diagnostic techniques in high-current plasmas and plasma accelerators for space applications. *Mailing Add:* Dept Physics Drake Univ Des Moines IA 50311

ROBINSON, DAVID ADAIR, b Boston, Mass, Dec 9, 25; m 80. NEUROPHYSIOLOGY. *Educ:* Brown Univ, BA, 47; Johns Hopkins Univ, MSc, 56, DrEng(elec eng), 59. *Prof Exp:* From proj engr to vpres res, Airpax Electronics, Inc, Md, 51-56, Fla, 58-61; from instr to assoc prof med, 61-70, from asst prof to assoc prof elec eng, 63-75, assoc prof biomed eng, 70-75, assoc prof ophthal, 71-75, PROF OPHTHAL & PROF BIOMED ENG, SCH MED, JOHNS HOPKINS UNIV, 75- *Concurrent Pos:* Consult visual sci study sect, Div Res Grants, NIH, 66-70, consult eng in biol & med training comt, 71-73; consult comt vision, Nat Res Ctr, Nat Acad Sci, 77-81 & consult planning comt, Nat Eye Inst, NIH, 76 & 81; NIH res grant, Sch Med, Johns Hopkins Univ, 69- *Mem:* Asn Res Vision & Ophthal; Soc Neurosci. *Res:* Neurophysiology of the eye movement control system. *Mailing Add:* Dept Ophthal Johns Hopkins Univ Sch Med Baltimore MD 21205

ROBINSON, DAVID BANCROFT, b Bellefonte, Pa, Feb 4, 24; m 45; c 5. PSYCHIATRY. *Educ:* Pa State Univ, BS, 44; Univ Pa, MD, 49; Univ Minn, MS, 57. *Prof Exp:* Intern, Geisinger Mem Hosp Foss Clin, Danville, Pa, 49-50, resident med, 50-51; staff psychiatrist, Rochester State Hosp, Minn, 55-56; instr psychiat, Mayo Clin, Mayo Grad Sch Med, Univ Minn, 57-58; from asst prof psychiat, to assoc prof psychiat, 58-68, actg chmn dept psychiat, 64-68 & 77-78, dir Adult Psychiat In-Patient Unit, 69-81, PROF PSYCHIAT, STATE UNIV NY UPSTATE MED CTR, 68-, CONSULT-LIAISON PSYCHIAT, 81- *Concurrent Pos:* Fel med, Mayo Found, Sch Med, Univ Minn, 52-55; sr psychiatrist, Syracuse Vet Admin Hosp & Crouse-Irving Mem Hosp, Syracuse, 58-61; coordr, Psychiat State Hosp Residents, NY State Dept Ment Hyg, 61-64; coordr, Grad Training Prof, Dept Psychiat, State Univ NY Upstate Med Ctr, 61-64; attend psychiatrist, Crouse-Irving Mem Hosp, Syracuse, 61-71 & Syracuse Vet Admin Hosp, 61-; proj dir, NIMH Training Grants, State Univ NY Upstate Med Ctr, 64-68, pres, Med Col Assembly, 71-73. *Honors & Awards:* Silver Award, Psychiat Asn, 67. *Mem:* Fel Am Psychiat Asn. *Res:* Psychotherapy; psychosomatic medicine; determinants of deviant behavior and psychosis; methods of teaching psychotherapy. *Mailing Add:* 2870 W Lake Rd Skaneateles NY 13152

ROBINSON, DAVID LEE, b St Louis, Mo, May 2, 43; div; c 2. NEUROPHYSIOLOGY. *Educ:* Springfield Col, BS, 65; Wake Forest Univ, MS, 68; Univ Rochester, PhD(neurosci), 72. *Prof Exp:* Fel neurophysiol, Lab Neurobiol, NIMH, 71-74; res physiologist, Neurobiol Dept, Armed Forces Radiobiol Res Inst, 74-78; RES PHYSIOLOGIST, LAB SENSORIMOTOR RES, NAT EYE INST, 78- *Concurrent Pos:* Fel psychiat, NIMH, 73-74, guest scientist, Lab Neurobiol, 74-78. *Mem:* Soc Neurosci; Asn Res Vision & Ophthal. *Res:* Neural control of visual attention; information processing the visual system; influence of eye movements on the visual system. *Mailing Add:* Lab Sensorimotor Res Nat Eye Inst 10/10C101 Bethesda MD 20892

ROBINSON, DAVID MASON, b Eccles, Eng, July 7, 32; m 65; c 2. CELL BIOLOGY, CRYOBIOLOGY. *Educ:* Durham Univ, BSc, 55, PhD(zool), 58. *Prof Exp:* Sect head entom, Cotton Res Sta, Uganda, 59-61; res officer, Hope Dept Zool, Oxford Univ, 61-63; mem sci staff, Med Res Coun Radiobiol Res Unit, Atomic Energy Res Estab, Eng, 63-66; prin sci officer & head cell biol, Microbiol Res Estab, Eng, 66-69; asst res dir, Blood Res Lab, Am Red Cross, 69-74; prof biol, Georgetown Univ, 74-80, adj prof anat & cell biol, 81-89; expert consult, 80-83, sr sci adv, 83-86, dep chief, Hypertension & Kidney Dis Br, 86-88, ASSOC DIR SCI PROG, DIV HEART & VASCULAR DIS, NAT HEART, LUNG & BLOOD INST, NIH, 88-; PROF LECTR, LIB STUDIES, GEORGETOWN UNIV, 89- *Mem:* Soc Cryobiol (secy, 75-76); Biophys Soc; Brit Asn Cancer Res; Am Soc Cell Biol; Brit Soc Low Temperature Biol (treas, 68). *Res:* Functional role of water in living systems; nature and repair of freezing injury in mammalian cells; nature and role of animal cell surface macromolecules. *Mailing Add:* Nat Heart Lung-Blood Inst NIH Fed Bldg Rm 416 Bethesda MD 20892

ROBINSON, DAVID NELSON, b Malden, Mass, July 14, 33. APPLIED MECHANICS, RHEOLOGY. *Educ:* Northeastern Univ, BSci, 61; Brown Univ, MSci, 63, PhD(appl mech), 66. *Prof Exp:* Prof appl mech, Dept Theoret & Appl Mech, Cornell Univ, 66-74; res staff mem appl mech, Oak Ridge Nat Lab, 74-82; PROF CIVIL ENG, UNIV AKRON, 82-; RESIDENT RES ASSOC, NASA, LERE, 82- *Concurrent Pos:* Consult appl mech, Int Bus Mach Corp, 68-72. *Mem:* Am Soc Mech Eng; Am Soc Civil Eng; Sigma Xi. *Res:* Constitutive equations for materials at elevated temperature; mechanics of composite materials; plasticity; viscoplasticity. *Mailing Add:* Dept Civil Eng Univ Akron Akron OH 44325-3905

ROBINSON, DAVID WEAVER, b Kansas City, Mo, Nov 15, 14; m 40; c 4. SURGERY. *Educ:* Univ Kans, AB, 35, MS, 47; Univ Pa, MD, 38; Am Bd Plastic Surg, dipl. *Prof Exp:* Instr anat, 35-36, from instr to assoc prof surg, 41-54, chmn, Sect Plastic Surg, 46-72, prof surg, 54-78, DISTINGUISHED PROF SURG, COL HEALTH SCI, UNIV KANS, KANSAS CITY, 74- *Concurrent Pos:* VChancellor clin affairs-actg exec vchancellor, Am Bd Plastic Surg, 61-67, chmn, 67. *Mem:* Am Soc Plastic & Reconstruct Surg (pres, 66-67); Am Surg Asn (2nd vpres, 65-66); Am Asn Plastic Surg; Am Col Surgeons; Sigma Xi. *Mailing Add:* 5516 Groveland Ave Baltimore MD 21215

ROBINSON, DAVID ZAV, b Montreal, Que, Sept 29, 27; nat US; m 54; c 2. OPTICS. *Educ:* Harvard Univ, AB, 46, AM, 47, PhD(chem physics), 50. *Prof Exp:* Physicist, Baird-Atomic, Inc, 49-52, asst dir res, 52-59; sci liaison officer, US Off Naval Res, London, Eng, 59-60; asst dir res, Baird-Atomic, Inc, 60-61; tech asst, Off Sci & Technol, Exec Off of President, Wash, DC, 61-62, tech specialist, 62-67; vpres acad affairs, NY Univ, 67-70; vpres, 70-80, exec vpres, Carnegie Corp New York, 80-88; EXEC DIR, CARNEGIE COMM SCI, TECHNOL & GOVT, 88- *Concurrent Pos:* US del, Int Comn Optics, 59-63, chmn, 61-63; mem, Comt Physics & Soc, Am Inst Physics, 67-; consult, President's Sci Adv Comt, 67-74; mem, NY State Energy Res & Develop Auth, 71-76; consult, NSF, 71-74 & Nat Acad Sci, 73- *Mem:* Optical Soc Am; Sigma Xi. *Res:* Optics; government and science. *Mailing Add:* Carnegie Corp 437 Madison Ave New York NY 10022

ROBINSON, DEAN WENTWORTH, b Boston, Mass, July 22, 29; m 83; c 3. CHEMICAL LASERS, SPECTROSCOPY. *Educ:* Univ NH, BS, 51, MS, 52; Mass Inst Technol, PhD(infra-red spectros), 55. *Prof Exp:* From asst prof to assoc prof chem, 55-66, chmn dept, 76-83, PROF CHEM, JOHNS HOPKINS UNIV, 66- *Concurrent Pos:* Fulbright res grant & Guggenheim fel, 66-67; vis prof, Univ Lille, 91. *Mem:* Am Chem Soc; fel AAAS. *Res:* Far infrared and low temperature spectroscopy; electronic spectra of small molecules; non-linear optical properties of materials; far infrared chemically-pumped lasers. *Mailing Add:* Dept Chem Johns Hopkins Univ Baltimore MD 21218

ROBINSON, DEREK JOHN SCOTT, b Montrose, Scotland, Sept 25, 38; m; c 3. MATHEMATICS. *Educ:* Univ Edinburgh, BSc, 60; Cambridge Univ, PhD(math), 63. *Prof Exp:* Lectr math, Queen Mary Col, Univ London, 65-68; from asst prof to assoc prof, 68-74, PROF MATH, UNIV ILL, URBANA, 74- *Honors & Awards:* Sir Edmund Whitaker Mem Prize, Edinburgh Math Soc, 71; Alexander von Humboldt Prize, 79. *Mem:* Am Math Soc; London Math Soc. *Res:* Theory of groups. *Mailing Add:* Dept Math 332 Illini Hall Univ Ill Urbana IL 61801

ROBINSON, DONALD ALONZO, b Joliet, Ill, Aug 4, 20; m 49; c 6. ORGANIC CHEMISTRY. *Educ:* Iowa State Col, BS, 42; Univ Wis, PhD(org chem), 48. *Prof Exp:* Res chemist, Naugatuck Chem Div, US Rubber Co, 42-46; res chemist, Mallinckrodt Chem Works, Mallinckrodt, Inc, 48-51, mfg improvement supvr, Brentwood Assocs, Inc, 51-56, asst to opers mgr, 56-62, mkt res mgr, 62-65, asst dir com develop, 65-68, asst to pres, St Louis, 68-72, asst to vchmn, 72-79, asst to pres, 79-82; PRES, BRENTWOOD ASSOCS, INC, 83- *Mem:* Am Chem Soc; Chem Mkt Res Asn. *Res:* Halogenation of amines; catalytic dehydrogenation; claisen condensations. *Mailing Add:* 2301 St Clair Ave Brentwood MO 63144

ROBINSON, DONALD KEITH, b Truro, NS, Aug 29, 32; m 65. PARTICLE PHYSICS. *Educ:* Dalhousie Univ, BSc, 54, MSc, 56; Oxford Univ, DPhil(physics), 60. *Prof Exp:* Res assoc high energy physics, Brookhaven Nat Lab, 60-62; from asst physicist to assoc physicist, 62-66; assoc prof high energy physics, 66-71, PROF HIGH ENERGY PHYSICS, CASE WESTERN RESERVE UNIV, 71- *Mem:* Am Phys Soc. *Res:* High energy particle interactions in bubble chambers; resonance production and decay; exchange processes; counter experiments; K-meson interactions; anti-proton interactions; radiative production and decay of hyperons. *Mailing Add:* Dept Physics Case Western Reserve Univ Cleveland OH 44106

ROBINSON, DONALD NELLIS, b New Brunswick, NJ, Nov 21, 33; m 60; c 2. ORGANIC CHEMISTRY, POLYMER CHEMISTRY. *Educ:* Cornell Univ, AB, 55; Univ Minn, PhD(org chem), 59. *Prof Exp:* Res chemist elastomers dept, E I du Pont de Nemours & Co, Del, 59-64, develop chemist, Ky, 64-67; asst prof chem, Ky Southern Col, 67-69, actg chmn dept, 68-69; SR RES CHEMIST, PENNWALT CO & ATOCHEM NAM, 69- *Mem:* Am Chem Soc. *Res:* Thermoplastics (synthesis, blends, physical evaluation); elastomers; indoles. *Mailing Add:* Atochem NAm 900 First Ave PO Box 1536 King of Prussia PA 19406-0018

ROBINSON, DONALD STETSON, b Pittsfield, Mass, Aug 14, 28; m 55; c 4. PHARMACOLOGY, MEDICINE. *Educ:* Rensselaer Polytech Inst, BChemEng, 49; Univ Pa, MD, 59; Univ Vt, MS, 66; Am Bd Internal Med, dipl, 66. *Prof Exp:* Sr investr, Nat Heart Inst, 66-68; from asst prof to assoc prof med & pharmacol, Col Med, Univ Vt, 70-77; PROF PHARMACOL & MED & CHMN DEPT PHARMACOL, SCH MED, MARSHALL UNIV, 77- *Concurrent Pos:* Pharmaceut Mfrs Asn fac develop award, 68-70; Burroughs Wellcome Fund scholar clin pharmacol, 71. *Mem:* Am Soc Pharmacol & Exp Therapeut; Am Fedn Clin Res. *Res:* Clinical and biochemical pharmacology; drug interactions; psychopharmacology; narcotic antagonists. *Mailing Add:* CNS Clin Res Box 5100 Wallingford CT 06492

ROBINSON, DONALD W(ALLACE), JR, b Minocqua, Wis, Sept 28, 21; m 47; c 4. MECHANICAL & AEROSPACE ENGINEERING. *Educ:* Northeastern Univ, BS, 47; Rensselaer Polytech Inst, MS, 68. *Prof Exp:* Aerodynamicist, Chance Vought Aircraft, 47-50; aerodynamicist, Kaman Aircraft Corp, 50-52, flight test engr, 52-53, chief test & develop, 53-56, proj engr, 56-58, proj mgr res & develop, 58-60, chief res engr, 60-69, dir res & develop, 69-76, vpres eng, 76-78, vpres planing & mkt, Kaman Aerospace Corp, 78-87; CONSULT, 87- *Mem:* Am Inst Aeronaut & Astronaut; Am Helicopter Soc; Sigma Xi. *Res:* Fluid mechanics; vibrations; systems analysis; statistical forecasting; management sciences. *Mailing Add:* 43 Tamara Circle Avon CT 06001

ROBINSON, DONALD WILFORD, b Salt Lake City, Utah, Feb 29, 28; m 52; c 7. MATHEMATICS. *Educ:* Univ Utah, BS, 48, MA, 52; Case Western Reserve Univ, PhD(math), 56. *Prof Exp:* From asst prof to assoc prof math, 56-62, PROF MATH, BRIGHAM YOUNG UNIV, 62- *Concurrent Pos:* NSF fac sr res fel, Calif Inst Technol, 62-63; vis prof, Naval Postgrad Sch, 69-70; Fulbright-Hays Lectureship, Univ Carabobo, Valencia, Venezuela, 76-77; vis prof, Rijksunivergiteit Gent, Ghent, Belig, 86. *Mem:* Am Math Soc; Math Asn Am; Sigma Xi; Soc Indust Appl Math. *Res:* Linear algebra and matrix theory. *Mailing Add:* Brigham Young Univ 290 TMCB Provo UT 84602

ROBINSON, DOUGLAS WALTER, b Niagara Falls, NY, Oct 22, 34; m 63; c 2. ANALYTICAL CHEMISTRY. *Educ:* Hamilton Col, AB, 56; Cornell Univ, MS, 59; Univ RI, PhD(anal chem), 64. *Prof Exp:* Chemist, Cadet Chem Corp, 58-59; chemist, Hooker Chem Corp, 59-61; res chemist, E I du Pont de Nemours & Co, 64-66; proj leader, Pennwalt Corp, King of Prussia, 66-87; SR CHEMIST, JOHNSON MATTHEY, WAYNE, 88- *Mem:* Soc Appl Spectros; Am Chem Soc. *Res:* Mass spectroscopy; gas chromatography. *Mailing Add:* 401 Riverview Ave Swarthmore PA 19081

ROBINSON, E ARTHUR, JR, ERGODIC THEORY, DYNAMICAL SYSTEMS. *Educ:* Tufts Univ, BS, 77; Univ Md, MA, 81, PhD(math), 83. *Prof Exp:* Mem math, Math Sci Res Inst, 83-84; lectr math, Univ Pa, 84-86; asst prof, 86-90, ASSOC PROF MATH, GEORGE WASHINGTON UNIV, 90- *Concurrent Pos:* Prin investr, NSF, 85-; mem math, Inst Advan Study, Princeton Univ, 86-87. *Mem:* Am Math Soc; Math Asn Am. *Res:* Spectral multiplicity in ergodic theory; theory of extensions and the theory of joinings; ergodic theory and topological dynamics of tilings. *Mailing Add:* Dept Math George Washington Univ Washington DC 20052

ROBINSON, EDWARD J, b New York, NY, June 16, 36; m 59; c 2. PHYSICS. *Educ:* Queens Col, NY, BS, 57; NY Univ, PhD(physics), 64. *Prof Exp:* Substitute in physics, Queens Col, NY, 58-59; lectr, City Col New York, 59-61; instr, NY Univ, 61-63; res assoc, Joint Inst Lab Astrophys, Nat Bur Standards & Univ Colo, 64-65; from asst prof to assoc prof, 65-81, PROF PHYSICS, NY UNIV, 82- *Mem:* Am Phys Soc. *Res:* Theoretical atomic physics, including atomic structure, scattering, and the interaction of laser radiation with atoms. *Mailing Add:* Univ Dept Physics Four Wash Pl New York NY 10003

ROBINSON, EDWARD LEE, b Clanton, Ala, Nov 6, 33; m 54; c 3. NUCLEAR PHYSICS. *Educ:* Samford Univ, AB, 54; Purdue Univ, MS, 58, PhD(physics), 62. *Prof Exp:* From asst prof to prof, Samford Univ, 61-67, head dept, 61-67; dir, Radiation Biol Lab, 67-80, assoc prof, 67-77, PROF PHYSICS, UNIV ALA, BIRMINGHAM, 77- *Concurrent Pos:* Consult, Hayes Int Corp, 63-68, Accident Reconstruction & Applied Physics Problems. *Mem:* AAAS; Am Phys Soc; Am Asn Physics Teachers; Soc Automotive Eng. *Res:* Nuclear spectroscopy; applied physics. *Mailing Add:* Dept Physics Univ Ala Birmingham AL 35294

ROBINSON, EDWARD LEWIS, b Pittsburgh, Pa, Aug 29, 45; m; c 1. ASTRONOMY. *Educ:* Univ Ariz, BA, 69; Univ Tex, Austin, PhD(astron), 73. *Prof Exp:* Astronomer, Lick Observ, Univ Calif, Santa Cruz, 73-74; asst prof, 74-80, assoc prof, 80-85, PROF ASTRON, UNIV TEX, AUSTIN, 85- *Concurrent Pos:* Alfred P Sloan Found fel, 78-80. *Mem:* Am Astron Soc; Int Astron Union. *Res:* Observational astrophysics, especially white dwarfs, neutron stars and black holes; interacting binary stars; accretion. *Mailing Add:* Dept Astron Univ Tex Austin TX 78712

ROBINSON, EDWIN ALLIN, b Denver, Colo, Nov 17, 07; m 33; c 3. CHEMISTRY. *Educ:* Univ Denver, AB, 28, MS, 29; Columbia Univ, PhD(org chem), 33. *Prof Exp:* Instr chem, Univ Denver, 28-29; asst, Columbia Univ, 30-32; res chemist, Tenn Eastman Corp, Tenn, 33-36; tech dir indust div, Nopco Chem Co, 36-48; asst vpres, 48-53, vpres, 53-67, dir, 58-67, vpres, Nopco Chem Div, Diamond Shamrock Corp, 67-69; INT CHEM CONSULT, 69- *Mem:* Am Chem Soc. *Res:* Textile fiber processing lubricants; surface active chemical production and evaluation; female sex hormones; substituted benzocinchoninic acids; polyurethane foams; metallic soaps and vinyl stabilizers; defoamers; water soluble polymers. *Mailing Add:* 105 Fairmount Ave Chatham NJ 07928-2315

ROBINSON, EDWIN HOLLIS, b Florence, Ala, Dec 16, 42; m 65; c 2. NUTRITION, FISHERIES. *Educ:* Samford Univ, BS, 69; Auburn Univ, MS, 72, PhD(nutrit), 77. *Prof Exp:* Res assoc biochem, Miss State Univ, 77-81; asst prof wildlife & fisheries sci, Tex A&M Univ, 81-; FISHERY BIOLOGIST, DELTA RES & EXTEN CTR, MISS STATE UNIV. *Mem:* Am Inst Nutrit; Am Fisheries Soc; Sigma Xi. *Res:* Nutritional requirements of various aquatic animals. *Mailing Add:* Delta Res & Exten Ctr PO Box 197 Stoneville MS 38776

ROBINSON, EDWIN JAMES, JR, b Wilkes-Barre, Pa, Feb 7, 16; m 48; c 6. PARASITOLOGY. *Educ:* Dartmouth Univ, AB, 39; NY Univ, MS, 41, PhD(biol), 49. *Prof Exp:* Asst biol, NY Univ, 39-48; instr parasitol med col, Cornell Univ, 48-51; sr asst scientist, USPHS, 51-54; from asst prof to assoc prof biol, Kenyon Col, 54-60, prof, 60; prof biol, Macalester Col, 63-84; RETIRED. *Mem:* AAAS; Am Micros Soc; Am Soc Parasitol; Soc Protozool; Sigma Xi. *Res:* Ecology and life cycles of trematodes and filarial nematodes. *Mailing Add:* 5928 Halifax Ave S Edina MN 55424

ROBINSON, EDWIN S, b Saginaw, Mich, Apr 29, 35; m 62; c 2. GEOPHYSICS, GEOLOGY. *Educ:* Univ Mich, BS, 57, MS, 59; Univ Wis, PhD(geophys, geol), 64. *Prof Exp:* Res asst geophys, Willow Run Labs, Univ Mich, 56-57; asst geophysicist, Arctic Inst NAm, 57-58; res asst geophys, Willow Run Labs, Univ Mich, 58-59; proj assoc geophys & polar res ctr, Univ Wis, 59-64; asst prof geophys, Univ Utah, 64-67; assoc prof, 67-72, PROF GEOPHYS, VA POLYTECH INST & STATE UNIV, 72- *Mem:* Am Geophys Union; Soc Explor Geophys; Glaciol Soc; fel Geol Soc Am. *Res:* Studies of seismic waves; exploration seismology in Antarctica; regional gravity and magnetic surveys interaction of earth tides and ocean tides; characteristics of seismic waves from nuclear explosions. *Mailing Add:* Dept Geol Sci Va Polytech Inst Blacksburg VA 24060

ROBINSON, ENDERS ANTHONY, b Boston, Mass, Mar 18, 30. MATHEMATICAL PHYSICS. *Educ:* Mass Inst Tech, SB, 50, SM, 52 , PhD(geophys), 54. *Prof Exp:* Geophysicist, Gulf Oil Corp, 54-55; instr math, Mass Inst Tech, 55-56; petrol economist, Exxon Corp, 56-57; asst prof statist, Mich State Univ, 58; assoc prof math, Univ Wis, 58-62; dep prof statist, Uppsala Univ, Sweden, 60-64; vpres & dir, Geoscience Inc, 64-65; vpres & dir, Digicon Inc, 65-70; pres, Robinson Res, 70-82; vis prof mech, Cornell Univ, 81-82; McMANN PROF GEOPHYS, UNIV TULSA, 83- *Concurrent Pos:* TRW distinguished lectr, Univ Southern Calif, 80; vis fel, Int Bus Mach Sci Ctr, Rome, Italy, 83. *Honors & Awards:* Donald G Fink Prize, Inst Elec & ctronics Engr, 84; Soc Explor Geophys Medal, 69-; Conrad Schlumberger Award, Europ Asn Explor Geophysicists, 67. *Mem:* Nat Acad Eng; Soc Explor Geophys; Europ Asn Explor Geophysicists. *Res:* History of science and engineering; popular scientific exposition; seismic wave propagation; signal analysis and digital signal processing; imaging systems and technology; digital spectral analysis and statistical time series analysis; electrical engineering systems and noise analysis; classical mathematical physics and partial differential equations; probability theory and mathematical statistics. *Mailing Add:* Dept Geosci Univ Tulsa Tulsa OK 74104-3189

ROBINSON, FARREL RICHARD, b Wellington, Kans, Mar 23, 27; m 49; c 4. VETERINARY PATHOLOGY, TOXICOLOGY. *Educ:* Kans State Univ, BS, 50, BS, DVM & MS, 58; Tex A&M Univ, PhD(vet path), 65, Am Bd Vet Toxicol, dipl; Am Col Vet Pathol, dipl. *Prof Exp:* Res vet, Aerospace Med Res Lab, Wright-Patterson AFB, USAF, 58-62, chief pat br, 64-68, mem staff, Armed Forces Inst Path, 68-71, chief vet path, 72-74, registrar, Am Registries Vet & Comp Path, 72-74; dir, animal dis diag lab, 78-85, interim head, dept vet pathobiol, 86-88, PROF TOXICOL-PATH, PURDUE UNIV, 74-, CHIEF TOXICOL SERV, ANIMAL DIS DIAG LAB, 85- *Concurrent Pos:* Vpres, Am Bd Vet Toxicol, 71-73, pres, 76-79; head, Int Ref Ctr Comp Oncol, WHO, 72-74, collab, Tumors Eye, 72-74; vpres, Am Asn Vet Lab Diag, 86, pres, 87. *Mem:* Am Vet Med Asn; Am Col Vet Path; Soc Toxicol; Am Bd Vet Toxicol; Am Asn Vet Lab Diag. *Res:* Pathology of the respiratory system; pathology of laboratory animals; oxygen toxicity; beryllium toxicity; pathology of toxicologic diseases; oncology. *Mailing Add:* Sch Vet Med Purdue Univ West Lafayette IN 47907

ROBINSON, FRANK ERNEST, b Oaklyn, NJ, Oct 29, 30; m 83; c 5. AGRONOMY, IRRIGATION. *Educ:* Rutgers Univ, BS, 52; Purdue Univ, PhD(soil physics), 58. *Prof Exp:* Asst, Purdue Univ, 55-58; assoc agronomist exp sta, Hawaiian Sugar Planters Asn, 58-64; from asst water scientist to assoc water scientist, 64-73, WATER SCIENTIST, DEPT LAND, AIR & WATER RESOURCES, UNIV CALIF, 73- *Concurrent Pos:* Mem, Int Comn Irrig & Drainage. *Mem:* Int Soc Soil Sci; Am Soil Sci Soc; Am Soc Agr Eng; Am Geophys Union; Am Soc Agron; Am Soc Hort Sci; hon mem Irrig Asn. *Res:* Irrigation management and salinity control; nitrate mobility, soil drainage, sprinkler and drip irrigation of vegetable crops; growth of plants with geothermal water. *Mailing Add:* 1004 E Holton Rd El Centro CA 92243

ROBINSON, GENE CONRAD, b Hurricane, La, July 31, 28; m 59; c 2. ORGANIC CHEMISTRY. *Educ:* Univ Chicago, PhB, 47, MS, 49; Univ Ill, PhD(org chem), 52. *Prof Exp:* Res chemist, Univ Calif, Los Angeles, 52-54; res chemist & res assoc, 54-73, SUPVR, ETHYL CORP, 73- *Mem:* Am Chem Soc; Oceanog Soc; Int Oceanog Found. *Res:* Physical organic chemistry; organometallic chemistry. *Mailing Add:* 1064 N Leighton Dr Baton Rouge LA 70806-1835

ROBINSON, GEORGE DAVID, b Leeds, Eng, June 8, 13; m 48; c 2. ATMOSPHERIC PHYSICS. *Educ:* Leeds Univ, BSc, 33, PhD(physics), 36. *Prof Exp:* Res asst chem, Leeds Univ, 35-36; tech officer, UK Meteorol Off, 37-44; sci officer atmospheric physics, Kew Observ, Eng, 46-57; dep dir, UK Meteorol Off, 57-68; RES FEL ATMOSPHERIC PHYSICS, CTR ENVIRON & MAN, INC, 68- *Honors & Awards:* Buchan Prize, Royal Meteorol Soc, 52. *Mem:* Fel Brit Inst Physics; hon mem Royal Meteorol Soc (pres, 65-67); fel Am Meteorol Soc. *Res:* Radiative transfer in the earth's atmosphere; predictability of weather and climate; artificial modification of climate. *Mailing Add:* 676 Fern St West Hartford CT 06107

ROBINSON, GEORGE EDWARD, JR, b Cary, Miss, May 4, 16; m 51; c 3. PHYSIOLOGY. *Educ:* Alcorn Agr & Mech Col, BS, 42; Univ Ill, MS, 45, PhD(animal sci), 50. *Prof Exp:* Head animal husb dept, Alcorn Agr & Mech Col, 45-47; assoc prof animal husb, 50-63, PROF ANIMAL SCI, SOUTH UNIV, BATON ROUGE, 63-, CHMN DEPT, 74- *Mem:* AAAS; Am Soc Animal Sci; Sigma Xi. *Res:* Physiology of reproduction; poultry and swine nutrition. *Mailing Add:* 2574 79th Ave Baton Rouge LA 70807

ROBINSON, GEORGE H(ENRY), metallurgy, for more information see previous edition

ROBINSON, GEORGE WALLER, b Winchester, Ky, Mar 29, 41; m 78; c 2. ANALYTICAL CHEMISTRY, BIOCHEMISTRY. *Educ:* Centre Col Ky, BA, 63; Duke Univ, PhD(biochem), 68. *Prof Exp:* Res assoc protein chem, Rockefeller Univ, 67-70; asst prof biochem, Univ Ky, 70-77; assoc prof chem, Centre Col Ky, 78-81; asst prof chem, Southern Tech Inst, 84-88; ASSOC PROF CHEM, SOUTHERN COL TECHNOL, 88- *Concurrent Pos:* Vis asst prof, Ga Inst Technol, 81-83, res assoc, 83-84. *Mem:* Am Chem Soc; Sigma Xi. *Res:* Structure-function relationships in proteins and enzymes; electrochemical detectors for high performance liquid chromatography, amino acid analysis, polysaccharide analysis. *Mailing Add:* 2886 Cherokee St Kennesaw GA 30144

ROBINSON, GEORGE WILSE, b Kansas City, Mo, July 27, 24; m 50. PICOSECOND SPECTROSCOPY. *Educ:* Ga Inst Technol, BS, 47, MS, 49; Univ Iowa, PhD(chem), 52. *Prof Exp:* Asst phys chem, Univ Iowa, 50-52; fel, Univ Rochester, 52-54; asst prof, Johns Hopkins Univ, 54-59; from assoc prof chem to prof phys chem, Calif Inst Technol, 59-75; ROBERT A WELCH PROF PHYS CHEM, TEX TECH UNIV, 76- *Concurrent Pos:* Prof phys chem & chmn dept, Univ Melbourne, Australia, 75-76. *Honors & Awards:* Alexander von Humboldt Award, 84. *Mem:* fel Am Phys Soc. *Res:* Molecular structure; liquid water; electron/proton hydration dynamics; solvent effects on chemical reactions; ultrafast molecular processes. *Mailing Add:* Dept Chem PO Box 4260 Lubbock TX 79409

ROBINSON, GERALD ARTHUR, b Hamilton, Ont, Apr 8, 29; m 53; c 6. RADIOBIOLOGY. *Educ:* Univ Western Ontario, BSc, 52, MSc, 54, PhD(biochem), 58. *Prof Exp:* Res asst biochem med sch, Univ Western Ont, 52-57; lectr, 57-58, from asst prof to assoc prof biomed sci, 58-70, PROF BIOMED SCI, ONT VET COL, UNIV GUELPH, 70- *Res:* Avian endocrinology and mineral metabolism. *Mailing Add:* Dept Biomed Sci Univ Guelph Guelph ON N1G 2W1 Can

ROBINSON, GERALD GARLAND, b St Louis Co, Minn, May 8, 33; m 60; c 2. ZOOLOGY, PHYSIOLOGY. *Educ:* Univ Minn, BS, 55, PhD(zool), 60. *Prof Exp:* From instr to assoc prof, 60-77, PROF BIOL SCI, UNIV SFLA, 77- *Mem:* Am Soc Pharmacog. *Mailing Add:* Dept Biol Sci Univ SFla 4202 Fowler Ave Tampa FL 33620

ROBINSON, GERSHON DUVALL, b Tulsa, Okla, Apr 2, 18; m 76; c 6. GEOLOGY. *Educ:* Northwestern Univ, BS, 39; Univ Calif, MA, 41. *Prof Exp:* Asst geol, Univ Calif, 39-41; petrol geologist, Tide Water Assoc Oil Co, Tex, 41-42; geologist field geol, Alaska, US Geol Surv, 42-45, in charge volcano invests, 45-48, asst chief gen geol br, Colo, 48-51, actg chief, 51-52, field geologist, Mont, 52-60, chief, Northern Rocky Mt Br, 60-64, res geologist, Rocky Mt Environ Geol Br, Colo, 64-72, res geologist, Geol Div, 72-84; RETIRED. *Concurrent Pos:* Consult, Res & Develop Bd, 47-50, Earth Sci Adv Comt, NSF, 76-80; assoc ed, Geol Soc Am, 67-74; chmn, Fed Geothermal Environ Adv Panel, Dept Interior, 78-84; consult, 84- *Honors & Awards:* Meritorious Serv Award, Dept Interior, 82. *Mem:* Fel Mineral Soc Am; Soc Econ Geol; fel Geol Soc Am. *Res:* Ore deposits; volcanology; structural geology; geology applied to urban problems; disposal of radioactive waste. *Mailing Add:* 2830 Somass Dr Victoria BC V8R 1R8 Can

ROBINSON, GERTRUDE EDITH, b Peoria, Ill, Apr 28, 23. MATHEMATICS EDUCATION. *Educ:* Ill State Univ, BS, 45; Univ Wis-Madison, MS, 51, PhD(math educ), 64. *Prof Exp:* Teacher high sch, Ill, 45-57 & Wis, 57-60; asst prof, Univ Ga, 63-69, assoc prof math, 69-77; examiner, Educ Testing Serv, Princeton, NJ, 77-84, sr examiner, 84-87; RETIRED. *Concurrent Pos:* Consult, Ga Educ TV & Ga State Dept Educ, 64-77; Training Teacher Trainers Proj fel, NY Univ, 70-71. *Mem:* Math Asn Am; Res Coun Diagnostic & Prescriptive Math; Nat Coun Teachers Math; NY Acad Sci. *Res:* Diagnostic testing in mathematics for elementary school children; improving spatial ability of girls. *Mailing Add:* 102 Macon Ave Asheville NC 28801

ROBINSON, GILBERT C(HASE), b Lykens, Pa, July 19, 19; m 47. CERAMICS ENGINEERING. *Educ:* NC State Col, BCerE, 40. *Hon Degrees:* DSc, Alfred Univ, 70. *Prof Exp:* Geol aide, Tenn Valley Authority, 40-41, jr chemist, 41, jr chem engr, 41-42, asst chem engr, Wilson Dam, Ala, 42-44; PROF CERAMICS ENG & HEAD DEPT, CLEMSON UNIV, 46- *Honors & Awards:* Wilson Award, Am Ceramic Soc, 57. *Mem:* Fel Am Ceramic Soc; Am Soc Testing & Mat; Nat Inst Ceramic Engrs; Sigma Xi. *Res:* Acoustic emission; analysis of clay minerals; ceramic raw materials; fundamentals of drying and firing; factory systems for preassembled masonry buildings; ceramic forming processes. *Mailing Add:* Dept Ceramic Eng Clemson Univ Clemson SC 29631

ROBINSON, GILBERT DE BEAUREGARD, b Toronto, Ont, June 3, 06; c 2. MATHEMATICS. *Educ:* Univ Toronto, BA, 27; Cambridge Univ, PhD, 31. *Prof Exp:* Lectr math, Univ Toronto, 31-34, from asst prof to assoc prof, 34-54, vpres res, 65-71, prof math, 54-71, EMER PROF, UNIV TORONTO, 71- *Concurrent Pos:* Nat Res Coun Can, 41-45; vis prof, Mich State Univ, 53-54, Univ BC, 63 & Univ NZ, 68. *Mem:* Am Math Soc; Math Asn Am; Royal Soc Can; Can Math Cong (pres, 53-57); Sigma Xi. *Res:* Theory of groups, especially representation theory; author of 3 books in mathematics. *Mailing Add:* 185 Univ Col Univ Toronto Toronto ON M5S 1A1 Can

ROBINSON, GLEN MOORE, III, b El Dorado, Ark, July 23, 43; m 67; c 2. PHYSICAL CHEMISTRY. *Educ:* La Polytech Inst, BS, 65; Tulane Univ, La, PhD(phys chem), 70. *Prof Exp:* Chemist, E I du Pont de Nemours & Co, Inc, 66; sr chemist, 69-76, SR RES SPECIALIST, MEMORY TECHNOLOGIES RES LAB, 3M CO, 76- *Concurrent Pos:* Assoc prof, Bethel Col, 72. *Honors & Awards:* Harlan Vergin Award, 3M Co, 74; Heinrich Hertz Award, Inst Electronic & RadioEngrs, 85. *Mem:* Am Chem Soc; Optical Soc Am; Sigma Xi. *Res:* Magnetic tape; ceramics; surface chemistry; molecular spectroscopy; natural and magnetically induced optical activity; applied statistics; computerization of laboratory instruments; computer analysis of micrographs, optical and electron; interferometry. *Mailing Add:* Memory Technol Group Lab Bldg 236-3C-89 3M Co 3M Ctr St Paul MN 55144

ROBINSON, GLENN HUGH, b Rosedale, Ind, May 20, 12; m 76; c 3. SOILS, LAND USE PLANNING. *Educ:* Purdue Univ, BS, 38, MS, 40; Univ Wis, PhD(soils), 50. *Prof Exp:* Asst agron, Purdue Univ, 38-42; assoc soil surveyor, US Forest Serv, 42-43, assoc soil surveyor, Bur Plant Indust, Soils & Agr Eng, USDA, NC, 43-46 & Wis, 46-51, soil correlator, 51-54, sr soil correlator, Soil Conserv Serv, 54-61; sr soil scientist, Food & Agr Orgn, UN, Brit Guiana, 61-64, sr soil scientist & proj mgr, Wad Medani, Sudan, 64-69, Land Develop Dept, Thailand, 69-73 & Indonesia, 73-75; sr scientist, econ res sect, USDA, Saudi Arabia, 76-80; CONSULT SOILS & LAND USE PLANNING, 80- *Concurrent Pos:* Consult, Food & Agr Orgn, Malawi, 65, Sudan, 70, Calif, 81 & Jordan, 83. *Mem:* Am Soc Agron; Soil Sci Soc Am; Sigma Xi; Int Soc Soil Sci. *Res:* Soil and land classification, mapping and utilization; agriculture development in new areas; classification and evaluation of soils for Wetland Rice; soil properties in relation to cotton production. *Mailing Add:* RR 1 Box 210A Carbon IN 47837

ROBINSON, GORDON HEATH, b Detroit, Mich, Oct 23, 31. HUMAN FACTORS ENGINEERING. *Educ:* Wayne State Univ, BS, 54; Univ Mich, Ann Arbor, MS, 55, PhD(instrumentation, eng, psychol), 62. *Prof Exp:* Asst res engr & lectr, Dept Indust Eng & Opers Res, Univ Calif, Berkeley, 61-66; from asst prof to assoc prof indust eng, 66-75, prof indust eng, univ wis, madison, 75-; RETIRED. *Mem:* Am Psychol Asn; Am Asn Univ Professors; Human Factors Soc; Brit Ergonomics Res Soc. *Res:* Human performance models; accident causation; sociotechnical systems; quality of working life. *Mailing Add:* PO Box 370 Guerneville CA 95446-0370

ROBINSON, GUNER SUZEK, b Nazilli, Turkey, Feb 5, 37; US citizen; div; c 2. ELECTRICAL ENGINEERING. *Educ:* Istanbul Tech Univ, BS, 60, MS, 61; Polytech Inst Brooklyn, PhD(elec eng), 66. *Prof Exp:* Asst prof elec eng, Middle East Tech Univ, Ankara, Turkey, 66-68; mem tech staff, Comsat Labs, Commun Satellite Corp, 68-73; res scientist image processing, Image Processing Inst, Univ Southern Calif, 73-76; mem res & tech staff, 76-77, mgr signal processing lab, Northrop Res & Technol Ctr, 77-84, DIR, ADVAN TECHNOL LAB, ELECTRO-MECH DIV, NORTHROP CORP, 84- *Concurrent Pos:* Consult, Re-transfer Technol to Turkey, UN Develop Prog, Turkey, 77-78 & 81. *Mem:* Inst Elec & Electronics Engrs; Soc Photo Instrumentation Engrs; Am Asn Artificial Intel; Am Defense Preparedness Asn; Asn Comput Mach; Am Asn Univ Women. *Res:* Education in electrical engineering and computer science; digital signal processing; image signal processing, compression and transmission; applications of image processing to military problems and industrial automation. *Mailing Add:* Advan Technol Lab Northrop Electro Mech Div 500 E Orangethorpe Ave Anaheim CA 92801

ROBINSON, HAMILTON BURROWS GREAVES, b Philadelphia, Pa, Feb 16, 10; m 29; c 3. PATHOLOGY, HISTOLOGY. *Educ:* Univ Pa, DDS, 34; Univ Rochester, MS, 36. *Hon Degrees:* ScD, Georgetown Univ, 75. *Prof Exp:* Rockefeller & Carnegie fel, Univ Rochester, 34-37; from asst prof to assoc prof oral histol & path, Wash Univ, 37-44; prof dent, Ohio State Univ, 44-58; prof dent & dean sch dent, 58-75, EMER PROF DENT & EMER DEAN SCH DENT, UNIV MO, KANSAS CITY, 75- *Concurrent Pos:* Ed, J Dent Res, 36-58 & Dent World, 75-84; dir post-grad div, Col Dent, Ohio State Univ, 47-53, assoc dean, 53-58; pres, Am Asn Dent Schs, 67; actg chancellor, Univ Mo-Kansas City, 67-68; former consult to Surgeon Gen, USPHS, US Army, USN & Vet Admin Hosps, Kansas City & Wadsworth, Kans; vis prof dent, Univ Calif, Los Angeles, 75-83. *Honors & Awards:* Fauchard Medal, 51; Callahan Gold Medal, 64; Jarvis-Burkhart Medal, 67; Hinman Distinguished Serv Medal, 80; Distinguished Serv Award, Am Acad Oral Path, 85. *Mem:* AAAS; Am Dent Asn (vpres, 71-72); fel Am Acad Oral Path (past pres); hon mem Am Acad Oral Med; hon mem Am Soc Oral Surg; Int Asn Dent Res (pres, 59-60). *Res:* Oral tumors and cysts; dental caries; relationship of oral and systemic diseases; diseases of dental pulp. *Mailing Add:* 3243 1B San Amadeo Laguna Hills CA 92653

ROBINSON, HAROLD ERNEST, b Syracuse, NY, May 22, 32. BOTANY, ENTOMOLOGY. *Educ:* Ohio Univ, BS, 55; Univ Tenn, MA, 57; Duke Univ, PhD(bot), 60. *Prof Exp:* Asst prof biol, Wofford Col, 60-62; CUR BOT, SMITHSONIAN INST, 62- *Mem:* Bot Soc Am; Am Soc Plant Taxon; Am Bryol & Lichenological Soc; Sigma Xi. *Res:* Bryophytes of Latin American and India; Asteraceae; Dolichopodidae. *Mailing Add:* Dept Bot Smithsonian Inst Washington DC 20560

ROBINSON, HAROLD FRANK, quantitative genetics, plant breeding; deceased, see previous edition for last biography

ROBINSON, HARRY, b Oldham, Lancashire, Eng, Apr 7, 25; US citizen; m 46; c 2. BIOSTATISTICS. *Educ:* Univ Manchester, BSc, 51; Univ Pittsburgh, MS, 64, ScD(biostatist), 68. *Prof Exp:* Biostatistician, Hillsboro County, Fla, 59-61; from asst prof to assoc prof dept prev & community med, 68-75, prog dir arthritis res, Sect Rheumatol, 68-75, prof dept community med, Col Med, Univ Tenn, Memphis, 75-76; MED SYSTS DIR, MED DEPT, NY TEL CO, 76- *Concurrent Pos:* Mem criteria for scleroderma subcomt, Diag & Therapeut Criteria Comt, Arthritis Found, 70-, mem subcomt data on nat needs, 72-, mem comt criteria govt subcomt, 72-; consult arthritis suppl to Nat Health Interview Surv, HEW; chief sect biostatist, 71-76, assoc prin investr,

Arthritis Res Prog, Sect Rheumatol, 72-76. *Mem:* Am Statist Asn; Am Pub Health Asn; Biomet Soc; Data Processing Mgrs Asn. *Res:* Multi-disciplinary studies; evaluation of medical care delivery systems; federal study to develop an experiment to assess the influence of variables on total cancer mortality; federal study of panencephalitis; federal study of transient ischemic attacks in the University of Tennessee Cerebral Vascular Research Center; computer systems in occupational medicine; stress factors in industry. *Mailing Add:* Med Dept NY Tel Co 1095 Avenue of Americas Rm 2561 New York NY 10036

ROBINSON, HENRY WILLIAM, b Chicago, Ill, Apr 26, 24; m 50; c 5. TROPICAL PUBLIC HEALTH. *Educ:* Brigham Young Univ, BA, 47; Univ Southern Calif, MA, 49; Stanford Univ, PhD(trop pub health), 62. *Prof Exp:* Instr zool, 52-55, from asst prof to assoc prof, 55-66, PROF MICROBIOL & ZOOL, SAN JOSE STATE UNIV, 66- *Mem:* Am Soc Parasitol; Royal Soc Trop Med & Hyg; Am Soc Trop Med & Hyg. *Res:* Antigenic analysis of blood and tissue flagellates, insect tissue culture-parasite models; epidemiology of parasitic diseases. *Mailing Add:* Dept Biol Sci San Jose Univ Wash Sq San Jose CA 95192

ROBINSON, HOWARD ADDISON, b Rotterdam, NY, July 30, 09; m 35; c 3. ATOMIC SPECTROSCOPY, PHYSICS OF GLASS. *Educ:* Mass Inst Technol, SB, 30, PhD(physics), 35. *Prof Exp:* Chief physicist, Armstrong Cork Co, 36-51; first secy, US Embassy Paris, US Dept State, 51-57; chmn & prof, Dept Physics, Adelphi Univ, 57-74; RETIRED. *Concurrent Pos:* Attache, US Embassy Stockholm, US Dept State, 47-48; ed, Soviet J Optical Technol, 64- *Mem:* Fel Am Phys Soc; Optical Soc Am. *Res:* Atomic spectroscopy; physics of glass. *Mailing Add:* Four Walnut St Gloucester MA 01930

ROBINSON, HUGH GETTYS, b New Orleans, La, Oct 30, 28. PRECISION MEASUREMENT & ATOMIC PHYSICS, ATOMIC CLOCKS & QUANTUM OPTICS. *Educ:* Emory Univ, AB, 50; Duke Univ, PhD(physics), 54. *Prof Exp:* Res assoc physics, Duke Univ, 54-55; res assoc, Univ Md, 55-56; res assoc, Univ Wash, Seattle, 56-57; instr, Yale Univ, 57-58, asst prof, 58-63; lectr, Harvard Univ, 63-64; assoc prof, 64-70, PROF PHYSICS, DUKE UNIV, 70- *Mem:* Fel AAAS; Fel Am Phys Soc; Sigma Xi. *Res:* Atomic physics; precision measurements in atomic and molecular physics, including fundamental constants; design and construction of appropriate apparatus; atomic clock; diode laser use and control. *Mailing Add:* 240 Physics Dept Duke Univ Durham NC 27706

ROBINSON, IVOR, b Liverpool, Eng, Oct 7, 23; m 63; c 3. MATHEMATICS, THEORETICAL PHYSICS. *Educ:* Cambridge Univ, BA, 47. *Prof Exp:* Lectr math, Univ Col Wales, 50-58; res assoc physics, Univ NC, 59-60; res assoc, Syracuse Univ, 60-61, 62-63; vis prof physics & astron, Cornell Univ, 61-62; prof, Southwest Ctr Advan Studies, 63-69; prof physics, 69-73, PROF MATH SCI, UNIV TEX, DALLAS, 73- *Concurrent Pos:* Res assoc, King's Col, Univ London, 59; vis lectr, Inst Theoret Physics, Polish Acad Sci, 59; vis prof, Univ Sydney, 65-66 & Tel-Aviv Univ, 66 & 70; state chair reserved for foreign scholars, Col France, 70. *Mem:* Int Astron Union; Am Astron Soc; Am Math Soc; Am Phys Soc. *Res:* Gravitational radiation. *Mailing Add:* Dept Math Sci Univ Tex PO Box 830688 Sta Jo 4-2 Richardson TX 75083-0688

ROBINSON, J(AMES) MICHAEL, b Shreveport, La, Oct 13, 43; m 70; c 5. ORGANIC CHEMISTRY, MEDICINAL CHEMISTRY. *Educ:* La Tech Univ, BS, 67, MS, 69; La State Univ, Baton Rouge, PhD(org chem), 73. *Prof Exp:* Res assoc med chem, Purdue Univ, 73-74, NIH res fel, 74-75; res assoc org chem, Tulane Univ, 75-76; asst prof, 76-79, ASSOC PROF CHEM, UNIV TEX PERMIAN BASIN, 80- *Concurrent Pos:* Consult, steel foundarys and precious metals extraction. *Mem:* Am Chem Soc. *Res:* Organic, medicinal and analytical chemistry of enamines, imines, pyridines and thiazyls; biomass to liquid fuels, energy research; mechanisms and drug design. *Mailing Add:* Dept Chem Univ Tex the Permian Basin Odessa TX 79762

ROBINSON, JACK LANDY, b Durant, Okla, Jan 6, 40; m 61; c 3. ANALYTICAL CHEMISTRY. *Educ:* Southeastern State Col, BA, 62; Univ Okla, PhD(chem), 66. *Prof Exp:* From asst prof to assoc prof, 66-77, PROF CHEM, SOUTHEAST STATE UNIV, 77- *Concurrent Pos:* NIH. *Mem:* Am Chem Soc. *Res:* Analysis of polycyclic aromatic hydrocarbons and microbes by chromatographic methods; analysis of oxalic acid in body fluids; chromatography of polymeric adsorbents. *Mailing Add:* PO Box 238 Mead OK 73449-0238

ROBINSON, JAMES LAWRENCE, b Boston, Mass, Feb 23, 42; m 63; c 3. BIOCHEMISTRY, NUTRITION. *Educ:* Univ Redlands, BS, 64; Univ Calif, Los Angeles, PhD(biochem), 68. *Prof Exp:* Teaching asst analytical chem & biochem, Univ Calif, Los Angeles, 65-66; NIH res fel, Inst Cancer Res, Philadelphia, 68-70; from asst prof to assoc prof biochem, 76-85, PROF BIOCHEM, UNIV ILL, URBANA, 85- *Concurrent Pos:* Mem nutrit sci fac, Univ Ill, 72; researcher, Nutrit Res Ctr, Meudon, France, 78-79, Biochem Dept, Univ Nijmegen, Neth, 87-88. *Honors & Awards:* Nutrit Res Award, 81. *Mem:* Am Soc Biol Chemists; Am Dairy Sci Asn; Am Inst Nutrit. *Res:* Enzyme mechanisms; metabolic regulation; specific interests in biochemistry, nutrition and physiology associated with borine inherited disorders (deficiency of uridine monophosphate synthase, citrullinemia, and factor XI deficiency). *Mailing Add:* Dept Animal Sci Univ Ill 1207 W Gregory Dr Urbana IL 61801

ROBINSON, JAMES MCOMBER, b Petoskey, Mich, Aug 22, 20; m 44; c 4. CLINICAL BIOCHEMISTRY. *Educ:* DePauw Univ, AB, 42; Ind Univ, PhD(org chem), 44. *Prof Exp:* Res assoc, Off Sci Res & Develop, Ind Univ, 44-45; res assoc res found, Ohio State Univ, 46; org chemist, Merck & Co, Inc, 47-55; org chemist, Aero-Jet-Gen Corp Div, Gen Tire & Rubber Co, 55-65; org chemist autonetics div, NAm Aviation, Inc, Anaheim, 65-70; org chemist clin biochem, Los Angeles County, Univ Southern Calif Med Ctr, 71-86; RETIRED. *Mem:* Fel AAAS; fel Am Inst Chem; Am Chem Soc; Am Asn Clin Chemists. *Mailing Add:* 525 Baughman Ave Claremont CA 91711

ROBINSON, JAMES VANCE, b Corsicana, Tex, July 27, 43; m 62; c 3. ENTOMOLOGY. *Educ:* Tex A&M Univ, BS, 67, MS, 71; Miss State Univ, PhD(entom), 75. *Prof Exp:* Surv entomologist, Miss Agr & Forestry Exp Sta, 71-75; EXTEN ENTOMOLOGIST, TEX AGR EXTEN SERV, 75- *Mem:* Entom Soc Am. *Mailing Add:* Dept Biol Univ Tex Box 19088 Uta Sta Arlington TX 76019

ROBINSON, JAMES WILLIAM, b Syracuse, NY, Apr 4, 38; m 64; c 4. ELECTRICAL ENGINEERING. *Educ:* Univ Mich, Ann Arbor, BSE, 59, MSE, 61, PhD(elec eng), 65. *Prof Exp:* Res asst elec eng, Univ Mich, Ann Arbor, 64-65, res engr, 65-66; asst prof, 66-70, assoc prof, 70-80, PROF ENG, PA STATE UNIV, 80- *Mem:* Inst Elec & Electronics Engrs; Am Soc Eng Educ. *Res:* Electron and ion beams; beam-surface interactions; electrical discharges. *Mailing Add:* Dept Elec & Computer Eng Pa State Univ University Park PA 16802

ROBINSON, JAMES WILLIAM, b Kidderminster, Eng, July 12, 23; nat US; m 46; c 3. ANALYTICAL CHEMISTRY. *Educ:* Univ Birmingham, BSc, 49, PhD(analytical chem), 52, DSc, 77. *Prof Exp:* Sr sci officer, Brit Civil Serv, 52-55; res assoc, La State Univ, 55-56; sr chemist res labs, Esso Standard Oil Co, 56-63; tech adv, Ethyl Corp, 63-64; assoc prof chem, 64-66, asst dir, Environ Inst, 71-86, PROF CHEM, LA STATE UNIV, BATON ROUGE, 66- *Concurrent Pos:* Ed, Spectros Letters, Environ Sci & Health & CRC Handbook Spectros; asst ed, Anal Chimica Acta & Appl Spectros Reviews; Guggenheim fel, 74; chmn, Analytical Gordon Conf, 74. *Mem:* Am Chem Soc; Chem Soc; Soc Appl Spectros; fel Royal Chem Soc. *Res:* Speciation analysis of trace metals; atomic absorption; molecular spectroscopy; light structure; air quality control; air pollution analysis; acid rain. *Mailing Add:* Dept Chem 440 Chopping Hall La State Univ Baton Rouge LA 70803

ROBINSON, JEROME DAVID, b Stamford, Conn, Jan 30, 41; m 64; c 2. CHEMICAL ENGINEERING. *Educ:* City Univ New York, BChE, 63; Univ Del, MChE, 66, PhD(chem eng), 68. *Prof Exp:* Res & develop chem engr, E I du Pont de Nemours & Co, 63-64; instr chem eng, Univ Del, 67; DIR ENG & TECHNOL, AM CYANAMID CO, 68- *Concurrent Pos:* Adj prof, Newark Col Eng, 69-70. *Mem:* Am Inst Chem Engrs. *Res:* Chemical and environmental process and systems design, analysis and control. *Mailing Add:* Am Cyanamid Co Wayne NJ 07470

ROBINSON, JERRY ALLEN, b Danville, Ill, Dec 18, 39; m 69; c 2. REPRODUCTIVE PHYSIOLOGY, ENDOCRINOLOGY. *Educ:* Wabash Col, BA, 63; Univ Cincinnati, MS, 66, PhD(zool), 70. *Prof Exp:* Trainee, Wis Regional Primate Res Ctr, Univ Wis-Madison, 70-72, res assoc, 72-73, asst scientist, 73-78, assoc scientist, 78-86; grants assoc, NIH, Bethesda, Md, 86-87; HEALTH SCIENTIST ADMINR, NAT INST ENVIRON HEALTH SCI, RES TRIANGLE PARK, NC, 87- *Concurrent Pos:* Asst dir, Inst Aging, Univ Wis-Madison, 82-85; asst ed, Biol Reproduction, Soc Study Reproduction, 85-86. *Mem:* Am Soc Primatologists; Soc Study Reproduction; Am Soc Zool; Endocrine Soc; Sigma Xi. *Res:* Steroid hormone secretion, metabolism and mechanism of action. *Mailing Add:* Nat Inst Environ Health MD3-02 PO Box 12233 Research Triangle Park NC 27709

ROBINSON, JOHN, b Vancouver, BC, Apr 25, 22; m 46; c 4. BACTERIOLOGY. *Educ:* Univ BC, BSA, 44; McGill Univ, MSc, 45, PhD(bact), 50. *Prof Exp:* Res officer, Nat Res Coun Can, 46-48 & 49-50; res officer, Can Dept Agr, 50-56; bacteriologist, HEW, 56-65; assoc prof, microbiol & immunol, Univ Western Ont, 65-73, Ont, 73-; RETIRED. *Honors & Awards:* Can Silver Jubilee Medal, 77. *Mem:* Asn Adv Sci Can; Am Soc Microbiol; Can Soc Microbiol (secy-treas, 59-61, 72-74, first vpres, 76-77, pres, 77-78). *Res:* Production of antibiotics by fungi; metabolism of halophilic bacteria; isolation and characterization of toxins produced by Staphylococcus aureus; mode of action of lysis of specific bacteria by the predaceous bacterium Bdellovibrio; Bdellovibrio and the ecology of polluted water. *Mailing Add:* J & J Robinson Christmas Tree Farm RR3 Mouth of Keswick NB E0H 1N0 Can

ROBINSON, JOHN E, JR, dentistry; deceased, see previous edition for last biography

ROBINSON, JOHN MITCHELL, LEUKOCYTE ACTIVATION, ENDOCYTOSIS. *Educ:* Vanderbilt Univ, PhD(biol), 76. *Prof Exp:* ASSOC PROF PATH, HARVARD MED SCH, 86- *Mailing Add:* Dept Cell Biol Neurobiol & Anat Ohio State Univ 4072 Graves Hall 333 W Tenth Ave Columbus OH 43210

ROBINSON, JOHN MURRELL, b Lecompte, La, Mar 26, 45. ELECTRONIC & MAGNETIC PROPERTIES OF SOLIDS. *Educ:* La State Univ, BS, 67; Fla State Univ, MS, 70, PhD(physics), 72. *Prof Exp:* Tech asst, Univ Munich, 70-71; res assoc, Fla State Univ, 72-73; asst prof, 73-78, ASSOC PROF PHYSICS, IND UNIV-PURDUE UNIV, 78-, CHMN DEPT, 79- *Mem:* Am Phys Soc; Am Asn Physics Teachers. *Res:* Theoretical models of electronic and magnetic properties of rare earth and actinide materials, particularly mixed valence materials. *Mailing Add:* Dept Physics Ind Univ-Purdue Univ Ft Wayne IN 46805

ROBINSON, JOHN PAUL, b Providence, RI, Jan 1, 39; m 60; c 3. COMMUNICATION ENGINEERING. *Educ:* Iowa State Univ, BSEE, 60; Princeton Univ, MSE, 62, PhD(elec eng), 66. *Prof Exp:* Mem tech staff, RCA Labs, 60-62 & Int Bus Mach Labs, NY, 63-65; from asst prof to assoc prof, 65-72, PROF INFO ENG, UNIV IOWA, 72- *Mem:* Inst Elec & Electronics Engrs; Sigma Xi. *Res:* Digital systems; switching theory; privacy and data security; codes for error detection and correction. *Mailing Add:* 5408 EB Dept EE Univ Iowa Iowa City IA 52242

ROBINSON, JOHN PRICE, b Charlotte, Tenn, Dec 1, 27; m 60; c 2. MICROBIOLOGY. *Educ:* Univ Tenn, BS, 54; Vanderbilt Univ, PhD(microbiol), 61. *Prof Exp:* Instr, 61-63, asst prof, 63-68, ASSOC PROF MICROBIOL, SCH MED, VANDERBILT UNIV, 68- *Concurrent Pos:*

USPHS sci res grants, 64-65 & NSF, 66-67. *Mem:* Electron Micros Soc; Am Soc Microbiol. *Res:* Mechanism of antibody-antigen complex formation; biochemical approach and visualization in the electron microscopy. *Mailing Add:* Dept Microbiol Vanderbilt Univ Nashville TN 37232

ROBINSON, JOSEPH DEWEY, b Ottumwa, Iowa, Mar 22, 28. PHYSICAL CHEMISTRY, PHYSICS. *Educ:* Drake Univ, AB, 49; Univ NMex, MSc, 52; Wash Univ, PhD, 56. *Prof Exp:* Chemist, Shell Develop Co Div, Shell Oil Co, 56-60, res assoc physics, 60-71, sr res assoc, 71-72, sr staff res physicist, 72-78, proj leader, borehole physics, 78-85; CONSULT, 85- *Mem:* Soc Petrol Eng; Am Phys Soc. *Res:* Molecular physics; theoretical chemistry; petrophysics. *Mailing Add:* 3026 Stanton Houston TX 77025

ROBINSON, JOSEPH DOUGLASS, b Asheville, NC, Nov 28, 34; m 58; c 2. PHARMACOLOGY. *Educ:* Yale Univ, MD, 59. *Prof Exp:* Intern med, Stanford Univ Hosp, 59-60; res assoc neurochem, NIH, 60-62; fel pharmacol, Sch Med, Yale Univ, 62-64; from asst prof to assoc prof pharmacol, 64-72, PROF PHARMACOL, STATE UNIV NY HEALTH SCI CTR, 72- *Concurrent Pos:* NSF sr fel pharmacol, Cambridge Univ, 71-72; mem pharmacol study sect, NIH, 80-84. *Honors & Awards:* Javits Neurosci Investr Award, 86. *Mem:* AAAS; Am Soc Pharmacol & Exp Therapeut; NY Acad Sci; Am Soc Neurochem; Biophys Soc. *Res:* Membrane structure, permeability and transport; neurochemistry; history and philosophy of science. *Mailing Add:* Dept Pharmacol State Univ NY Health Sci Ctr Syracuse NY 13210

ROBINSON, JOSEPH EDWARD, b Regina, Sask, June 25, 25; div; c 3. PETROLEUM EXPLORATION. *Educ:* McGill Univ, BEng, 50, MSc, 51, PhD(geol), 68. *Prof Exp:* Geophysist, Imp Oil Ltd, 51-66; sr geologist, Union Oil Can, 68-76; PROF GEOL, SYRACUSE UNIV, 76- *Concurrent Pos:* Vis indust assoc, Kans Geol Surv, 70-72; vis prof, Syracuse Univ, 74; asst ed, Int Asn Math Geol, 76-80; prin investr, US Dept Energy, 79-81; geol consult, J E Robinson & Assoc, 76- *Mem:* Soc Explor Geophysists; Am Asn Petrol Geologists; Can Soc Petrol Geologists; Int Asn Math Geol; Soc Independent Prof Earth Scientists. *Res:* Geologic data base construction; management and computer applications in exploration for petroleum, natural gas and other economic minerals; environmental geology and geophysics. *Mailing Add:* Dept Geol Heroy Geol Lab Syracuse Univ Syracuse NY 13244-1070

ROBINSON, JOSEPH ROBERT, b New York, NY, Feb 16, 39; m 59; c 3. PHARMACEUTICS. *Educ:* Columbia Univ, BS, 61, MS, 63; Univ Wis, PhD(pharm), 66. *Prof Exp:* From asst prof to assoc prof, 66-74, PROF PHARM, UNIV WIS-MADISON, 74- *Concurrent Pos:* Vis prof, Johann Goethe Univ, Frankfurt, WGer, 89. *Honors & Awards:* Ebert Prize & Res Achievement Award, APhA Found; Janot Medal. *Mem:* Am Pharmaceut Asn; Am Chem Soc; fel Acad Pharmaceut Sci; fel AAAS; NY Acad Sci; Controlled Release Soc; fel Am Asn Pharmaceut Scientists. *Res:* Biopharmaceutics; ophthalmic pharmacology; mechanisms of drug transport and activity in the eye; controlled drug delivery. *Mailing Add:* Sch Pharm Univ Wis Madison WI 53706

ROBINSON, KENNETH ROBERT, b Akron, Ohio, Nov 17, 21; m 47; c 4. ORGANIC CHEMISTRY. *Educ:* Ohio Northern Univ, BA, 43; Purdue Univ, MS, 46; Mich State Univ, PhD(chem), 50. *Prof Exp:* Res chemist, E I du Pont de Nemours & Co, 50-55; sr chemist, Koppers Co, Inc, 55-61; res chemist, Maumee Chem Co, 61-63; group leader, Ashland Oil, Inc, 64-71, res assoc res & develop dept, 71-74, mgr govt contracts, 74, mgr cent coding control, 74-87; RETIRED. *Res:* Petroleum chemistry; hydrocarbon oxidation; ozonation; custom chemical synthesis; coal; coal tar chemicals; aromatic chemicals; cellulose and viscose chemistry; synthetic monomer and polymer chemistry; research planning; fuel science; computer sciences. *Mailing Add:* 800 Mission Hills Lane Worthington OH 43235

ROBINSON, KENT, b Reese, NC, June 22, 24; m 48; c 2. BIOLOGY, SCIENCE EDUCATION. *Educ:* Appalachian State Teachers Col, BS, 50, MA, 52; Ohio State Univ, PhD(sci educ, biol), 66. *Prof Exp:* Teacher pub schs, NC, 50-56; PROF BIOL, APPALACHIAN STATE UNIV, 56- *Res:* Botany. *Mailing Add:* Dept Biol Appalachian State Univ Boone NC 28608

ROBINSON, LAWRENCE BAYLOR, b Tappahannock, Va, Sept 14, 19; m 56; c 3. PHYSICS. *Educ:* Va Union Univ, BS, 39; Harvard Univ, MA, 41, PhD(chem physics), 46. *Prof Exp:* Instr math & physics, Va Union Univ, 41-42; teacher math, USSignal Corps Schs, Md, 42; tester radio parts, Victor Div, Radio Corp Am, NJ, 43; asst prof chem & physics, Va Union Univ, 44; asst prof physics, Howard Univ, 46-47, instr phys sci, 47-48, assoc prof physics, 48-51; res physicist, Atomic Energy Res Div, US Naval Res Lab, 53-54; asst prof physics, Brooklyn Col, 54-56; mem tech staff, Space Tech Labs, Inc Div, Thompson Ramo Wooldridge, Inc, 56-60; lectr, 57-60, from assoc prof to prof eng, 60-74, asst dean, Sch Eng & Appl Sci, 69-74, PROF ENG & APPL SCI, UNIV CALIF, LOS ANGELES, 74- *Concurrent Pos:* NSF fel, 66-67; guest prof, Aachen Tech Univ, 66-67. *Mem:* Am Phys Soc; Am Asn Physics Teachers. *Res:* Zeta potentials of solutions of electrolytes; surface tension of electrolytes; neutron physics; reactor theory; collision between electrons and atoms; interatomic forces and collisions; magnetic properties of solids; nonequilibrium thermodynamics. *Mailing Add:* Chem Eng 5405 Boelter Univ Calif 405 Hilgard Ave Los Angeles CA 90024

ROBINSON, LEWIS HOWE, b Cody, Wyo, Sept 11, 30; m 71. AIR QUALITY MODELING, APPLIED MATHEMATICS. *Educ:* San Jose State Col, BA, 52; Univ Calif, Berkeley, MA, 59. *Prof Exp:* Meteorologist, Pac Southwest Forest & Range Exp Sta, US Forest Serv, 57-60; meteorologist, Pac Gas & Elec Co, 61-66; meteorologist, WeatherMeasure Corp, 66; meteorologist, Aerojet-Gen Corp Div, Gen Tire & Rubber Co, 66-67; pres, Robinson Assocs, 68-71; air pollution meteorologist, 72-73, sr air pollution meteorologist, 73-76, chief res & planning sect, 76-79, dir planning div, Bay Area Air Quality Mgt Dist, 79-87; RETIRED. *Mem:* Am Meteorol Soc; Math Asn Am; Asn Comput Mach. *Res:* Air pollution; diffusion; atmospheric turbulence; statistical prediction. *Mailing Add:* 1100 Gough St, Apt 14 A San Francisco CA 94109

ROBINSON, LLOYD BURDETTE, b Gravelburg, Sask, Aug 28, 29; US citizen; m; c 2. ELECTRONIC INSTRUMENTATION, ASTRONOMY. *Educ:* Univ Sask, BA, 53, MA, 54; Univ BC, PhD(physics), 57. *Prof Exp:* Res officer electronics, Atomic Energy Can Ltd, 57-62; electronics engr, Lawrence Radiation Lab, Univ Calif, 62-69; RES PHYSICIST & ASTRONR INSTRUMENTATION & ASTRON, LICK OBSERV, UNIV CALIF, SANTA CRUZ, 69- *Concurrent Pos:* NSF grants astron, 71- *Mem:* Soc Photo-Optical Instrumentation Engrs; Int Astron Union; Am Astron Soc; Astron Soc Pac. *Res:* Development of electronic and optical instruments for observational optical astronomy; use of small computers as an aid in control and data acquisition; development of low light level sensors. *Mailing Add:* Lick Observ NS-2 Univ Calif Santa Cruz CA 95064

ROBINSON, LOUIS, mathematics, computer sciences; deceased, see previous edition for last biography

ROBINSON, M JOHN, b Monroe, Mich, June 19, 38; m 60; c 4. NUCLEAR ENGINEERING. *Educ:* Univ Mich, BS, 60, MS, 62, PhD(nuclear eng), 65. *Prof Exp:* Res asst nuclear eng, Univ Mich, 60-63, asst res engr, 63-64, res assoc nuclear eng, 64-65, lectr, 65-69; assoc prof nuclear eng, Kans State Univ, 69-72; nuclear engr, 72-77, proj mgr, 77-78, PARTNER, BLACK & VEATCH ENGRS-ARCHITECTS, 79- *Concurrent Pos:* Inst Sci & Technol res fel, 65-66; Int Atomic Energy Agency tech asst expert, heat transfer adv Govt Brazil, 70-71. *Mem:* Am Nuclear Soc. *Res:* Fluid flow and heat transfer; nuclear power systems; reactor physics; shielding; fuel management; radiological effects. *Mailing Add:* Black & Veatch Engrs-Architects Power Div 11401 Lamar Overland Park KS 66211

ROBINSON, MARGARET CHISOLM, b Savannah, Ga, Oct 31, 30; m 52; c 2. BIOLOGY. *Educ:* Savannah State Col, BS, 52; Univ Mich, MS, 55; Wash Univ, PhD(plant physiol), 69. *Prof Exp:* Teacher natural sci, Jefferson County Training High Sch, 52-54; instr biol, Ft Valley State Col, 57-59; teaching fel molecular biol, Wash Univ, 65-66; prof biol, 66-69, head biol dept, 69-80, dept chmn natural sci, 71-80, actg dean, 80-81, DEAN, SCH SCI & TECH, SAVANNAH STATE COL, 81- *Concurrent Pos:* Consult, Sol C Johnson High Sch, 73-77 & 84-88, Herschel V Jenkins High Sch, 75-76, Frank W Spencer Elementary 74 & Richard Arnold High Sch, 76. *Mem:* Am Soc Cell Biol; Am Asn Plant Physiologists; Am Coun Educ; Nat Tech Asn; AAAS; Am Inst Biol Sci. *Res:* Inherited changes in Euglena gracilis induced by ultracentrifugation; antibiotics; antihistamines; micronutrients dificiencies. *Mailing Add:* 4317 Whatley Ave Savannah GA 31404

ROBINSON, MARK TABOR, b Oak Park, Ill, June 23, 26; m 47; c 2. RADIATION EFFECTS, COMPUTATIONAL PHYSICS. *Educ:* Univ Ill, BS, 46; Okla State Univ, MS, 49, PhD(chem), 51. *Prof Exp:* Instr chem, Okla State Univ, 47-49, res assoc, 49-51; RES STAFF MEM, SOLID STATE DIV, OAK RIDGE NAT LAB, 51- *Concurrent Pos:* Vis scientist, Metall Div, Atomic Energy Res Estab, Eng, 64-65, Inst Solid State Res, Nuclear Res Estab, Ger, 71-72 & Max-Planck Inst for Plasma Physics, Ger, 83; western hemisphere regional ed, Radiation Effects, 80-84. *Mem:* Am Chem Soc; Am Phys Soc. *Res:* Theory of radiation effects in solids; atomic collisions in solids; digital computer applications. *Mailing Add:* 112 Miramar Circle Oak Ridge TN 37830

ROBINSON, MARTIN ALVIN, b New York, NY, Sept 12, 30; m 56; c 3. INORGANIC CHEMISTRY. *Educ:* NY Univ, BA, 52; Univ Buffalo, MS, 54; Ohio State Univ, PhD(inorg chem), 61. *Prof Exp:* Sr chemist, Battelle Mem Inst, 56-61; group supvr inorg res & develop, Olin Corp, 61-69; dept mgr indust fine chem, J T Baker Chem Co, NJ, 69-71; MGR RES & DEVELOP, SPECIALTY CHEM DIV, ALLIED CHEM CORP, 72-, DIR, BUFFALO RES LAB, 85- *Concurrent Pos:* Asst prof, Southern Conn State Col, 62-69. *Mem:* Am Chem Soc; Sigma Xi; fel Am Inst Chemists. *Res:* Catalysis; transition metal complexes. *Mailing Add:* 167 Wood Acre Dr East Amherst NY 14051-1758

ROBINSON, MERTON ARNOLD, b Los Angeles, Calif, Sept 13, 25; m 49; c 3. SCIENTIFIC INSTRUMENTATION, ANALYTICAL CHEMISTRY. *Educ:* Univ Calif, Los Angeles, BS, 49. *Prof Exp:* Head anal lab, Riker Labs, Inc, 49-53 & Carnation Res Lab, 53-59; mgr prod assurance, Bechman Instruments, Inc, Anaheim, 59-83; mgr regulatory affairs, Sensormedics Corp, Anaheim, 83-87; RETIRED. *Res:* Optical instruments for satellites; gas chromatography for trace contaminant analysis. *Mailing Add:* 1041 Brookwood Dr La Habra CA 90631

ROBINSON, MICHAEL HILL, b Preston, Eng, Jan 7, 29; m 55. ZOOLOGY. *Educ:* Univ Wales, BSc; Oxford Univ, Dphil, 66. *Prof Exp:* Teacher sci, UK Sec Schs, 63-70; biologist, Tropical Res Inst, 66-84, asst dir, 80, actg dir, 80-81, dep dir, 81-84, DIR NAT ZOOL PARK, SMITHSONIAN INST, 84- *Concurrent Pos:* Vis lectr, Univ Pa, 69; reader biol, New Univ Ulster, 71; adj prof zool, Univ Miami, Coral Gables, 81-; dir, Am Arachnological Soc, 82-; guest lectr, Univ Papua, New Guinea, 74. *Mem:* Fel Linnaean Soc; fel Royal entom Soc London; fel Inst Biol; Zool Soc UK; Am Arachnological Soc; Soc Study Animal Behav. *Res:* Tropical ecology, and behavior of predators; anti-predator adaptations; courtship and mating behavior in invertebrates (insects and spiders; evolutionary implications of behavior. *Mailing Add:* Nat Zool Park Smithsonian Inst Washington DC 20008

ROBINSON, MICHAEL K, b Toledo, Ohio, Jan 4, 51; m 76; c 5. ALLERGIC HYPERSENSITIVITY, TUMOR IMMUNOLOGY. *Educ:* Univ Notre Dame, BSc, 73, MSc, 75; State Univ NY, PhD(microbiol), 79. *Prof Exp:* Sr immunologist, Dept Immunochem Res, Evanston Hosp, 82-85; STAFF SCIENTIST GROUP LEADER, PROCTOR & GAMBLE CO, 85- *Concurrent Pos:* Consult, Immunotoxicol Working Group, Task Force Environ Cancer, Heart & Lung Dis, 88-89. *Mem:* Am Asn Immunologists; AAAS; Am Soc Cancer Res; Soc Investigative Dermat. *Res:* Immune mechanisms in cancer and hypersensitivity; recently working on immunology of skin and respiratory hypersensitivity supporting both product safety and drug development activities. *Mailing Add:* Procter & Gamble Co Miami Valley Labs Cincinnati OH 45239-8707

ROBINSON, MYRON, b Bronx, NY, Mar 4, 28; m 63; c 4. AIR POLLUTION, ENVIRONMENTAL SCIENCE. *Educ:* City Col New York, BS, 49; NY Univ, MS, 58; Cooper Union, PhD(physics), 75. *Prof Exp:* Physicist, Nat Bur Stand, 50-53; physicist, US Army Biol Warfare Labs, 54-56; electronic scientist, US Navy Appl Sci Lab, 53-54 & 56-58; asst dir res, Res-Cottrell, Inc, 58-68; aerosol physicist, Health & Safety Lab, US AEC, NY, 68-75; vis prof, Hebrew Univ, Jerusalem, 75-76; prin sci assoc, Dart Indust, 78-80; prof, Queensborough Community Col, 81-82; ADJ PROF, LONG ISLAND UNIV, 82-; SAFETY & HEALTH MGR, DEFENSE LOGISTICS AGENCY, 87- *Concurrent Pos:* Consult air pollution control, Res-Cottrell, Inc, Precipitair Pollution Control, Seversky Electronatom, Environ Protect Agency, Energy Res Co, FluiDyne Eng Corp, Israel Environ Protection Agency, Argonne & Brookhaven Nat Labs, India Ministry Energy, Kerr-McGee Corp, UN Indust Develop Agency, 68-; adj prof, Cooper Union, 70-75; vis lectr, Univ Western Ont, 70-71 & Nehru, Madras & Madurai Univs, 76, Israel Inst Technol, 76, Univ Wash & Univ Wis, 80; Fulbright-Hayes fel, 75. *Mem:* Inst Elec & Electronics Engrs; Air Pollution Control Asn; Electrostatic Soc Am (pres, 71-73). *Res:* Particulate air pollution control mechanisms, particularly the extension of electrostatic precipitation to untried areas of application; aerosol technology in industrial hazard evaluation. *Mailing Add:* 73-32 136th St Flushing NY 11367

ROBINSON, MYRTLE TONNE, b Cincinnati, Ohio, Jan 12, 29; m 51; c 3. ORGANIC CHEMISTRY. *Educ:* Berea Col, BA, 50; Purdue Univ, MS, 52. *Prof Exp:* Lab mgr, 69-74, LAB DIR, LINDAU CHEM, INC, 74- *Mailing Add:* PO Box 13565 Columbia SC 29201

ROBINSON, NEAL CLARK, b Seattle, Wash, March 6,42; m 68; c 1. MEMBRANE BIOCHEMISTRY. *Educ:* Univ Wash, BS, 64, PhD(biochem), 71. *Prof Exp:* Res assoc & fel, Duke Univ, 71-74; res assoc, Univ Ore, 74-75; lectr biochem, Univ Calif, Davis, 75-77; from asst prof to assoc prof, 77-89, PROF BIOCHEM, UNIV TEX HEALTH SCI CTR, 89- *Mem:* Biophys Soc; Am Chem Soc; Sigma Xi; Am Soc Biol Chemists. *Res:* Protein-phospholipid and protein-detergent interactions with solubilized membrane protein complexes isolated from the inner mitochondrial membrane, especially cytochrome, oxiclase and cardiolipm. *Mailing Add:* Dept Biochem Univ Tex Health Sci Ctr 7703 Floyd Curl Dr San Antonio TX 78284-7760

ROBINSON, NORMAN EDWARD, b Tadley, Eng, Oct 12, 42; m 67; c 2. PULMONARY PATHOPHYSIOLOGY. *Educ:* Univ London, Eng, BVetMed, 65; Univ Calif, Davis, PhD(physiol), 72. *Prof Exp:* Pvt pract vet surg, Eng, 65-66; intern large animal med, Univ Pa, 66-67; assoc vet med, Univ Calif, Davis, 67-70; assoc prof, 72-78, PROF PHYSIOL & LARGE ANIMAL MED, MICH STATE UNIV, 78-, MATILDA R WILSON PROF LARGE ANIMAL CLIN SERV, 88- *Concurrent Pos:* Prin investr, develop collateral ventilation, 75-82 & model bronchial hypereactivity, 81- *Mem:* Am Physiol Soc; Am Thoracic Soc; Am Vet Med Asn. *Res:* Pathophysiology of lung disease in the large domestic mammals, particularly chronic airway disease in the horse and pneumonia in cattle; species variations in collateral ventilation. *Mailing Add:* A10 Vet Clin Ctr Mich State Univ East Lansing MI 48824

ROBINSON, PAUL RONALD, b Philadelphia, Pa, Aug 6, 50. INORGANIC CHEMISTRY, COORDINATION CHEMISTRY. *Educ:* Univ Mo-Columbia, BS, 72, AM, 73; Univ Calif, San Diego, PhD(chem), 77. *Prof Exp:* Res assoc chem kinetics, Univ Ill, Urbana-Champaign, 76-77, vis asst prof chem, 77-78; staff res assoc inorg chem, Oak Ridge Nat Labs, 78-81; TECHNOL SALES REP & RES CHEMIST, UNOCAL CORP, 81- *Mem:* AAAS; Sigma Xi; Am Chem Soc. *Res:* Preperation, characterization and testing of catalysts for hydrorefining and synthesis gas conversion; synthesis and reactions of transition metal complexes; chemical kinetics; environmental chemistry. *Mailing Add:* Unocal Sci & Technol Div PO Box 76 Brea CA 92621

ROBINSON, PETER, b New York, NY, July 19, 32; m 54, 84; c 2. PALEONTOLOGY, GEOLOGY. *Educ:* Yale Univ, BS, 54, MS, 58, PhD(geol), 60. *Prof Exp:* Instr geol, Harpur Col, 55-57; res assoc paleont, Yale Peabody Mus, 60-61; cur mus, 61-71, asst prof, 61-71, PROF NATURAL HIST, UNIV COLO, BOULDER, 71-, CUR GEOL, 61- *Concurrent Pos:* NSF grants Eocene Insectivora res, 60-61 & co-recipient for salvage archaeol, Aswan Reservoir, Sudan, 64-67; res assoc, Carnegie Mus Natural Hist, 66-; Smithsonian grant paleont res, Tunisia, 67-81; dir, Colo Paleont Exped to Tunisia, 67-87. *Mem:* AAAS; Soc Vert Paleont; Australian Soc Mammal; Paleont Soc. *Res:* Vertebrate paleontology, especially fossil mammals. *Mailing Add:* Univ Colo Mus Boulder CO 80309-0315

ROBINSON, PETER, b Hanover, NH, July 9, 32; m 57; c 2. STRUCTURAL GEOLOGY, PETROLOGY. *Educ:* Dartmouth Col, AB, 54; Univ Otago, NZ, MSc, 58; Harvard Univ, PhD(geol), 63. *Prof Exp:* Raw mat engr, Columbia Iron Mining Co, US Steel Corp, 56-58; from instr to asst prof, 62-69, assoc prof, 69-76, PROF GEOL, UNIV MASS, AMHERST, 76- *Concurrent Pos:* Petrologists Club lectr, Carnegie Inst Geophys Lab, 65; Am Geol Inst vis scientist, Maine & Dalhousie Univs, 66; NSF res grants & joint res grants with H W Jaffe, 66-74, & with J M Rhodes, 80-86; mem exped, Metamorphic Rocks, Chatham Island, NZ Plateau, 68; subsurface mapping, Northfield Mountain Pumped Storage Hydroelec Proj, 68-70; geol consult, Metrop Dist Comt, Boston, 72-80; co-compiler, US Geol Surv bedrock geol map of Mass, 75-83, DNAG Continent-Ocean Transect E-1 Adirondacks-Georges Bank, 82-86; partic, compiler, field trip leader, Int Geol Correlations Prog, Caledonide Orogen Proj, US, Scand, Can, Gt Brit, 79-84. *Mem:* Fel, Geol Soc Am; fel Mineral Soc Am; Am Geophys Union; Mineral Asn Can. *Res:* Structural geology and stratigraphy of metamorphic rocks in New England and New Zealand; metamorphic mineral facies; crystal chemistry and exsolution in amphiboles and pyroxenes. *Mailing Add:* Dept Geol Univ Mass Amherst MA 01003

ROBINSON, PETER JOHN, b Kinston, UK, July 4, 44; m 68; c 2. CLIMATOLOGY, ATMOSPHERIC RADIATION. *Educ:* Univ London, BSc, 65, MPhil, 68; McMaster Univ, PhD(climat), 72. *Prof Exp:* Asst prof, 71-76, ASSOC PROF CLIMAT, UNIV NC, CHAPEL HILL, 76- *Concurrent Pos:* Asst dir, Nat Climate Prog Off, Nat Oceanic & Atmospheric Admin, 80-82; dir, NC Climate Prog, 82- *Mem:* Am Meteorol Soc; Am Asn State Climatologists (secy-treas, 79-80). *Res:* Techniques and benefits of using climate data and information in solution of operational planning problems of commercial enterprises; studies of social impacts of climatic change. *Mailing Add:* Dept Geog Univ NC Chapel Hill NC 27599-3220

ROBINSON, PRESS L, b Florence, SC, Aug 2, 37; m 64; c 2. PHYSICAL CHEMISTRY. *Educ:* Morehouse Col, BS, 59; Howard Univ, MS, 62, PhD(phys chem), 63. *Prof Exp:* Chemist, NIH, 61; from asst prof to assoc prof, 63-68, PROF PHYS CHEM, SOUTHERN UNIV, BATON ROUGE, 68- *Concurrent Pos:* Res grants, 64-68. *Mem:* Am Chem Soc; Sigma Xi. *Res:* Molten salt chemistry; explosive metal forming of liquid alloys. *Mailing Add:* Dept Chem Southern Univ PO Box 10155 Baton Rouge LA 70813

ROBINSON, RALPH M(YER), b Terre Haute, Ind, Aug 17, 26; m 56; c 2. CHEMICAL ENGINEERING, ORGANIC CHEMISTRY. *Educ:* Univ Ill, BS, 49; Univ Mich, MSE, 50. *Prof Exp:* Chem engr res & develop, Argonne Nat Labs, 51-53; chem engr, Abbott Labs, 53-60, group leader develop, High Pressure Lab, 60-62, mgr eng develop dept, 62-68, opers mgr oral & topicals prod, Pharmaceut Mfg Div, 68-87, PHARMACEUT CONSULT, ABBOTT LABS, 87- *Mem:* Am Chem Soc; Sigma Xi; fel Am Inst Chem Engrs. *Res:* Administration and evaluation of research and development; economic and financial analysis; process development in fermentation, pharmaceutical and chemical areas. *Mailing Add:* 705 Colville Pl Waukegan IL 60087

ROBINSON, RAPHAEL MITCHEL, b National City, Calif, Nov 2, 11; wid. MATHEMATICS. *Educ:* Univ Calif, AB, 32, MA, 33, PhD(math), 35. *Prof Exp:* Instr math, Brown Univ, 35-37; from instr to prof, 37-73, EMER PROF MATH, UNIV CALIF, BERKELEY, 73- *Mem:* Am Math Soc; Math Asn Am. *Res:* Functions of a complex variable; theory of numbers; foundations of mathematics; geometry. *Mailing Add:* Dept Math Univ Calif Berkeley CA 94720

ROBINSON, RAYMOND FRANCIS, b Albany, Ore, Sept 25, 14; m; c 9. GEOLOGY. *Educ:* Union Col, NY, BSc, 36; McGill Univ, MSc, 38. *Prof Exp:* Eng asst, NY State Dept Pub Works, Utica, 39-40; underground sampler, Anaconda Copper Mining Co, Butte, Mont, 41-43, asst mining engr, 43-45; resident geologist & engr, Am Smelting & Refining Co & Fed Mining & Smelting Co, Wallace, Idaho, 45-47; chief geologist, Sunshine Mining Co, Kellogg, Idaho, 47-53; actg dist supvr, Tucson Off Surv, US Geol Surv, 53; sr explor geologist, Bear Creek Mining Co, 53-56, supv geologists, 56-58, sr explor geologist, Develop Div, 58-64; field geologist, Phelps Dodge Corp, Douglas, Ariz, 64-65; dist mgr, Northwest Explor Div Duval Corp, 65-69, dist mgr, Holt McPhar, 69-71; consult geol eng, Raymond F Robinson, Inc, Reno, Nev, 71-90; RETIRED. *Mem:* Am Inst Mining, Metall & Petrol Engrs; Soc Econ Geologists; Asn Explor Geologists. *Res:* Metal exploration, evaluation and development; surface and underground; regional to specific mining properties; surface and underground; all phases of reconnaissance, detailed examination, evaluation and development; planning, administration and field execution; world-wide operation. *Mailing Add:* 180 W Laramie Dr Reno NV 89511

ROBINSON, REX JULIAN, b Maxwell, Ind, Nov 15, 04; m 32; c 3. ANALYTICAL CHEMISTRY. *Educ:* DePauw Univ, AB, 25; Univ Wis, MA, 27, PhD(chem), 29. *Prof Exp:* Asst, Univ Wis, 26-29; from instr to prof, 29-71, from instr to assoc prof, Oceanog Labs, 32-41, EMER PROF CHEM, UNIV WASH, 71- *Mem:* Am Chem Soc. *Res:* Colorimetric and micro quantitative methods of analysis. *Mailing Add:* 7324 16th NE Seattle WA 98115

ROBINSON, RICHARD ALAN, b Monroe, Mich, Oct 4, 42; c 1. NUCLEAR ENGINEERING, MECHANICAL ENGINEERING. *Educ:* Univ Mich, BSE, 64, MSE, 66. *Prof Exp:* Res scientist nuclear technol, 67-78, dept mgr nuclear waste isolation, proj mgt div, 78-90, PROG MGR, ENVIRON TECHNOL, BATTELLE COLUMBUS LABS, 90- *Mem:* Am Nuclear Soc. *Res:* Nuclear technology; advanced nuclear fuel development; experimental reactor engineering; PWR and BWR LOCA analysis and experimental studies; nuclear material shipping and transportation; nuclear waste disposal and isolation. *Mailing Add:* Battelle-Environ Technol 505 King Ave Columbus OH 43201

ROBINSON, RICHARD C(LARK), b Seattle, Wash, Nov 4, 37; m 60; c 3. CHEMICAL ENGINEERING. *Educ:* Univ Wash, BS, 59; Univ Wis-Madison, MS, 61, PhD(chem eng), 65. *Prof Exp:* Res engr, Chevron Res Co, Standard Oil Co, Calif, 65-70, sr res engr, 70-72; asst prof, Colo Sch Mines, 72-74; SR RES ENGR, CHEVRON RES CO, 74- *Mem:* Am Inst Chem Engrs; Am Chem Soc; Sigma Xi. *Res:* Diffusion in compressed fluids; diffusion coefficient correlation by corresponding states; chemical reaction kinetics; catalytic reforming of naphtha; hydrofining of oils; fluidized catalytic cracking. *Mailing Add:* Chevron Res Co 576 Standard Ave Richmond CA 94802

ROBINSON, RICHARD CARLETON, JR, b Walton, NY, Aug 29, 27; m 49; c 4. RISK & RELIABILITY, CUMPUTER SCIENCES. *Educ:* Alfred Univ, BA, 50; Kent State Univ, MA, 52; Am Univ, ABD(math statist), 70. *Prof Exp:* Physicist, Argonne Nat Lab, 52-53 & Savannah River Lab, E I du Pont de Nemours & Co, 53-57; analyst, Opers Res Off, Johns Hopkins Univ, 58-61; sr opers res analyst, Res Analysis Corp, Va, 61-72 & Gen Res Corp, Va, 72-75; sr opers res analyst, Ketron Inc, Va, 75-77; SR OPERS RES ANALYST, US NUCLEAR REGULATORY COMN, WASH, DC, 77- *Concurrent Pos:* Prin investr, Res Analysis Corp, 66-72. *Mem:* Am Asn Physics Teachers; Opers Res Soc Am. *Res:* Nuclear reactors and exponential assemblies; electronic instrumentation design for nondestructive testing;

programming and systems analysis; computer simulation and gaining; mathematical modeling, applied probability; nuclear safeguards; probabilistic risk analysis of nuclear power plants. *Mailing Add:* 4013 Cleveland St Kensington MD 20895-3806

ROBINSON, RICHARD DAVID, JR, stellar astronomy, for more information see previous edition

ROBINSON, RICHARD WARREN, b Los Angeles, Calif, Apr 14, 30; m 63; c 2. HORTICULTURE. *Educ:* Univ Calif, Davis, BS, 52, MS, 53; Cornell Univ, PhD(veg crops), 62. *Prof Exp:* PROF HORT SCI, NY STATE AGR EXP STA, 61- *Mem:* Am Soc Hort Sci. *Res:* Vegetable breeding, genetics and physiology. *Mailing Add:* Agr Exp Sta Dept Hort Sci Cornell Univ Geneva NY 14456

ROBINSON, ROBERT EARL, b Covington, Ky, Aug 3, 27; m 51; c 3. SYNTHETIC ORGANIC CHEMISTRY, ORGANOMETALLIC CHEMISTRY. *Educ:* Berea Col, BA, 49; Purdue Univ, MS, 51, PhD, 53. *Prof Exp:* Proj leader, Nat Distillers & Chem Corp, 53-64; group leader, Stauffer Chem Co, 64-66; dir res & develop, Cardinal Chem Co, 66-67; exec vpres & dir, 67-87, PRES, LINDAU CHEM INC, 87- *Concurrent Pos:* Consult synthetic organic & organometallic chemistry. *Mem:* Am Chem Soc; fel Am Inst Chemists; NY Acad Sci; AAAS. *Res:* Organometallic compounds; transition metal catalysts; synthetic resins; resin additives; epoxy curing agents; alkali metals; pharmaceutical intermediates; quate rnary ammonium compounds. *Mailing Add:* Lindau Chem Inc PO Box 13565 Columbia SC 29201

ROBINSON, ROBERT EUGENE, b Provo, Utah, Jan 9, 27; m 51; c 1. ANALYTICAL CHEMISTRY. *Educ:* Univ Okla, BS, 49, MS, 52; Univ Mich, PhD(chem), 59. *Prof Exp:* Chemist asphalt prod develop, Kerr-McGee Oil Co, 49-50; chemist oil prod res, Sun Oil Co, 52-54; res chemist, Shell Develop Co, 58-69, sr res chemist, 69-75, staff res chemist, 75-90; RETIRED. *Concurrent Pos:* Vis res fel, Dept Chem, Leeds, UK, 70-91. *Mem:* Am Chem Soc. *Res:* Raman, infrared and molecular spectroscopy; Raman spectra of fluorocarbon derivatives; catalytic process research and development; separation sciences; gas and supercritical chromatography. *Mailing Add:* 11631 Jaycreek PO Box 820764 Houston TX 77070

ROBINSON, ROBERT GEORGE, b Beacon, NY, Aug 13, 37; m 60; c 3. CHEMICAL ENGINEERING. *Educ:* Clarkson Tech Univ, BChE, 58, MChE, 60; Pa State Univ, PhD(chem eng), 64. *Prof Exp:* Scientist, Upjohn Co, 64-77, sr scientist, 77-80, res mgr, 80-84, assoc dir, chem process res & develop, 84-90, ASSOC DIR, CHEM PROD, UPJOHN CO, 90- *Concurrent Pos:* Chmn, Food, Pharmaceut & Bioeng Div, Am Inst Chem Engrs, 87-88. *Mem:* Am Inst Chem Engrs; Am Chem Soc; AAAS. *Res:* Controlled cycling mass transfer; countercurrent crystallization; separation and purification; hazard evaluation. *Mailing Add:* Dept Chem Eng 2385 Woody Noll Kalamazoo MI 49002-7666

ROBINSON, ROBERT GEORGE, b Minneapolis, Minn, Jan 26, 20. AGRONOMY. *Educ:* Iowa State Univ, BS, 41; Univ Minn, MS, 46, PhD(agron, soils), 48. *Prof Exp:* Teacher high sch, Iowa, 41-42; asst, 42-47, from asst prof to assoc prof, 48-73, prof, 73-86, EMER PROF AGRON, UNIV MINN, ST PAUL, 86- *Mem:* Am Soc Agron; Weed Sci Soc Am; Sigma Xi; Soc Econ Bot; Crop Sci Soc Am. *Res:* Field crop production; new and special field crops; farm management. *Mailing Add:* Dept Agron Univ Minn St Paul MN 55108

ROBINSON, ROBERT L(OUIS), JR, b Muskogee, Okla, June 14, 37; m 58. CHEMICAL ENGINEERING. *Educ:* Okla State Univ, BS, 59, MS, 62, PhD(chem eng), 64. *Prof Exp:* Sr res engr, Pan Am Petrol Corp, Okla, 64-65; from asst prof to assoc prof chem eng, 65-72, REGENTS PROF CHEM ENG, OKLA STATE UNIV, 84- *Mem:* Am Inst Chem Engrs; Am Chem Soc; Am Soc Eng Educ; Soc Petrol Engrs. *Res:* Thermodynamic and transport properties. *Mailing Add:* 2002 Crescent Dr Stillwater OK 74074

ROBINSON, ROBERT LEO, b Kansas City, Mo, Mar 14, 26; m 55; c 2. PHARMACOLOGY. *Educ:* Univ Kans, MA, 54, PhD(physiol), 58. *Prof Exp:* From instr to prof pharmacol, Med Ctr, WVa Univ, 59-87; RETIRED. *Mem:* Am Soc Pharmacol & Exp Therapeut. *Res:* Adrenal medullary physiology; pharmacology of autonomic drugs. *Mailing Add:* 1126 Valley View Ave Morgantown WV 26505

ROBINSON, ROBERT W, b Atlanta, Ga, Nov 23, 41. COMPUTER SCIENCE. *Educ:* Dartmouth Col, AB & AM, 63; Cornell Univ, PhD(math), 66. *Prof Exp:* Prof math, Southern Ill Univ, 82-84; PROF & HEAD DEPT COMPUTERR SCI, UNIV GA, 84- *Mem:* Asn Computer Mach; Inst Elec & Electronics Engrs; Am Math Soc; Australian Math Soc. *Mailing Add:* Dept Computer Sci Univ Ga 415 Boyd Grad Studies Bldg Athens GA 30802

ROBINSON, ROSCOE ROSS, b Oklahoma City, Okla, Aug 21, 29; m 52; c 2. INTERNAL MEDICINE, NEPHROLOGY. *Educ:* Cent State Univ, BS, 49; Univ Okla, MD, 54; Am Bd Internal Med, dipl, 62. *Prof Exp:* Intern med, Duke Univ, 54-55, asst resident, 55-56; Am Heart Asn res fel, Columbia-Presby Med Ctr, 56-57; instr, Duke Univ, 57-58, assoc, 60-62, from asst prof to assoc prof, 62-69, dir Div Nepthrol, 62-80, prof med, 69-, chief, Div Nephrology, 80-; NEPHROLOGIST, VANDERBILT UNIV. *Concurrent Pos:* Chief resident, Vet Admin Hosp, Durham, NC, 57-58, clin investr, 60-62, attend physician, 62-; sr investr, NC Heart Asn, 62-; mem exec comt, Coun Kidney & Cardiovasc Dis, Am Heart Asn; nat consult internal med, USAF Surgeon Gen; mem sci adv bd, Nat Kidney Found; ed, Kidney Int. *Mem:* Am Physiol Soc; Am Clin & Climat Asn; Europ Dialysis & Transplant Soc; Am Soc Clin Invest; Am Soc Artificial Internal Organs. *Res:* Renal disease and physiology. *Mailing Add:* Dept Med Vanderbilt Univ Med Ctr D 3300 Med Ctr N Nashville TN 37232

ROBINSON, ROSS UTLEY, b Minneapolis, Minn, July 30, 28; m 53; c 6. ANALYTICAL BIOCHEMISTRY. *Educ:* Colgate Univ, BA, 49; Wesleyan Univ, MA, 51; Mass Inst Technol, MS, 53. *Prof Exp:* Phys chemist, Abbott Labs, 53-58, sci instrumentation group leader, 58-62, sect mgr, 62-67, res & develop coordr, 67-69, advan technol mgr, 69-71, prod planning & develop, 71-73, dir contract res & develop, 73-75, dir advan systs res, 75-79; assoc dir, Corp Res & Develop, Boehringer Mannheim Corp, 80-81; vpres, Res & Develop, ICL Scientific, 81-84; pres & chief exec officer, Mesa Diagnostics, 84-85; exec vpres, Los Alamos Diagnostics, 85-89; PRES, CARDINAL ASSOCS, 85- *Mem:* AAAS; Am Chem Soc; Sigma Xi. *Res:* Physical, analytical and medical instrumentation; technological forecasting. *Mailing Add:* 2393 Botulph Rd Santa Fe NM 87505

ROBINSON, ROY GARLAND, JR, b Arkansas City, Kans, Mar 14, 21; m 74; c 1. PHYSIOLOGY, HISTOLOGY. *Educ:* Univ Ariz, BS, 48; Univ Southern Calif, MS, 54, PhD(biol), 65. *Prof Exp:* Instr histol, Sch Dent, Univ Southern Calif, 53-65, asst prof physiol, 65-69, chmn dept, 60-69; PROF ZOOL, MCNEESE STATE UNIV, 69- *Mem:* AAAS; Am Soc Mammal. *Res:* Metabolic effects of thyroxin; carbohydrate metabolism of Ascaris lumbricoides; tooth development of lepisosteus. *Mailing Add:* 1501 S Greenfield Circle McNeese State Univ 4100 Ryan St Lake Charles LA 70605

ROBINSON, RUSSELL LEE, b Louisville, Ky, July 30, 31; m 53; c 3. NUCLEAR PHYSICS. *Educ:* Univ Louisville, BA, 53; Ind Univ, MS, 55, PhD(physics), 58. *Prof Exp:* Asst physics, Ind Univ, 53-58; physicist, Oak Ridge Nat Lab, 58-83; SCI DIR, HOLIFIELD HEAVY ION RES FACIL, 83- *Mem:* Fel Am Phys Soc. *Res:* Gamma-ray spectroscopy; heavy-ion induced reactions. *Mailing Add:* Oak Ridge Nat Lab PO Box 2008 Oak Ridge TN 37831-6368

ROBINSON, STEPHEN MICHAEL, b Columbus, Ohio, Apr 12, 42; m 68; c 2. MATHEMATICAL PROGRAMMING, SYSTEMS ANALYSIS. *Educ:* Univ Wis, Madison, BA, 62, PhD(computer sci), 71; NY Univ, MS, 63. *Prof Exp:* Mem staff, Computer & Numerical Analysis Div, Sandia Corp, 62; instr math, US Mil Acad, 68-69; asst dir, Math Res Ctr, 71-74, from asst prof to assoc prof computer sci, 72-76, assoc prof, 76-79, PROF INDUST ENG & COMPUTER SCI, UNIV WIS, MADISON, 79- *Concurrent Pos:* Assoc ed, Opers Res, 74-86, Math Opers Res, 75-81, Math Opers forschung & Statist, 77-83; assoc ed, Math Prog, 86-, ed, Math Opers Res, 81-86, adv ed, Math Opers Res, 87- *Mem:* Math Prog Soc; Opers Res Soc Am; Inst Mgt Sci; Inst Indust Engrs. *Res:* Mathematical programming; operations research. *Mailing Add:* Dept Indust Eng 1513 University Ave Madison WI 53706-1572

ROBINSON, STEWART MARSHALL, b Schenectady, NY, Jan 7, 34; m 60; c 3. MATHEMATICS. *Educ:* Union Col, BS, 55; Duke Univ, PhD(math), 59. *Prof Exp:* Asst prof math, Univ RI, 59-61; asst prof, Smith Col, 61-64; asst prof, Union Col, NY, 64-66, assoc prof, 66-68; ASSOC PROF MATH, CLEVELAND STATE UNIV, 68- *Mem:* Am Math Soc; Math Asn Am. *Res:* Partial differential equations; topology. *Mailing Add:* 3334 Berkeley Ave Cleveland Heights OH 44118

ROBINSON, SUSAN ESTES, b Radford, Va, Apr 26, 50; m 81; c 2. NEUROPHARMACOLOGY. *Educ:* Vanderbilt Univ, BA, 72, PhD(pharmacol), 76. *Prof Exp:* Staff fel, Lab Preclin Pharmacol, NIMH, 76-79; asst prof, dept med pharmacol & toxicol, Tex A&M Univ, 79-81; asst prof, 81-87, ASSOC PROF PHARMACOL, DEPT PHARMACOL & TOXICOL, MED COL VA, VA COMMONWEALTH UNIV, 87- *Honors & Awards:* Lyndon Baines Johnson Res Award, Am Heart Asn, 80. *Mem:* AAAS; Am Soc Pharmacol & Exp Therapeut; Soc Neurosci; Am Soc Neurochem; Int Soc Psychoneuroendocrinol. *Res:* Interactions between neurotransmitters in the nervous system and the physiological relevance of the interactions. *Mailing Add:* Dept Pharmacol & Toxicol Med Col Va Box 613 MCV Sta Richmond VA 23298-0613

ROBINSON, TERENCE LEE, b El Paso, Tex, June 29, 55; m 76; c 6. POMOLOGY, CULTURAL SYSTEMS & PHYSIOLOGY. *Educ:* Brigham Young Univ, BS, 78; Wash State Univ, MS, 82, PhD(hort), 84. *Prof Exp:* Res asst hort, Wash State Univ, 78-84; asst prof, 84-90, ASSOC PROF POMOL, NY STATE AGR EXP STA, CORNELL UNIV, 90- *Mem:* Am Soc Hort Sci. *Res:* Environmental and cultural limitations of yield in tree fruits; development of orchard production systems that integrate cultivar, soil, climatic, economic and cultural factors to maximize returns. *Mailing Add:* Dept Hort Sci NY State Agr Exp Sta Geneva NY 14456

ROBINSON, TERRANCE EARL, b Rochester, NY, May 22, 49. PHYSIOLOGICAL PSYCHOLOGY, NEUROSCIENCE. *Educ:* Univ Lethbridge, BA, 72; Univ Sask, MA, 74; Univ Western Ont, PhD(psychol), 78. *Prof Exp:* Lectr psychol, Univ Western Ont, 76-77; fel psychobiol, Univ Calif, Irvine, 77-78; from asst prof to assoc prof, 78-89, PROF PSYCHOL, UNIV MICH, ANN ARBOR, 89- *Concurrent Pos:* Nat Res Coun Can fel, Univ Calif, Irvine, 77-78. *Honors & Awards:* Res Career Develop Award, Nat Inst Neurol & Commun Dis & Stroke. *Mem:* Fel AAAS; Soc Neurosci; fel Am Psychol Asn; NY Acad Sci. *Res:* Neuropsychopharmacology; brain-behavior relations; neuroplasticity. *Mailing Add:* Neurosci Lab Bldg Univ Mich 1103 E Huron Ann Arbor MI 48109

ROBINSON, THANE SPARKS, b Kansas City, Kans, Apr 8, 28; m 54, 67; c 2. ECOLOGY. *Educ:* Univ Kans, AB, 51, PhD(zool), 56. *Prof Exp:* Asst state biol survr, Univ Kans, 51-54, from asst instr to instr zool, 54-57; from asst prof to assoc prof, Western Mich Univ, 57-63; assoc prof, 63-66, chmn dept biol, 66-68, assoc dean, Col Arts & Sci, 68-72, PROF BIOL, UNIV LOUISVILLE, 66- *Concurrent Pos:* Dir, Adams Ctr Ecol Studies, 59-63. *Mem:* AAAS; Wilson Ornith Union; Am Ornith Union; Am Soc Zoologists. *Res:* Terrestial and micro-environmental ecology; microclimatology and bioclimatology of homoiotherms. *Mailing Add:* Dept Biol Univ Louisville Louisville KY 40292

ROBINSON, THOMAS B, b Kansas City, Mo, Feb 28, 17. CIVIL ENGINEERING, ENVIRONMENTAL ENGINEERING. *Educ:* Kans Univ, BS, 39; Columbia Univ, MS, 40. *Prof Exp:* Proj mgr & proj engr, Black & Veatch, 40-56, exec partner, 56-65, asst managing partner, 65-73, managing partner, 73-83; RETIRED. *Mem:* Nat Acad Eng; Am Waterworks Asn; Water Pollution Control Fedn; Nat Soc Prof Engrs; Am Soc Civil Engrs. *Mailing Add:* Black & Veatch 1500 Meadow Lake Pkwy Kansas City MO 64114

ROBINSON, THOMAS FRANK, biophysics, muscular physiology; deceased, see previous edition for last biography

ROBINSON, THOMAS JOHN, b Volga, Iowa, May 7, 35; m 59; c 2. MATHEMATICS. *Educ:* Luther Col, BA, 56; Iowa State Univ, MS, 58, PhD(math), 63. *Prof Exp:* Instr math, Univ NDak, 58-60 & Iowa State Univ, 60-63; from asst prof to assoc prof, 63-72, PROF MATH, UNIV N DAK, 72- *Concurrent Pos:* Fel, Sch Behav Studies, Univ NDak, 71-72. *Mem:* Am Math Soc; Math Asn Am. *Res:* Topology; algebra. *Mailing Add:* PO Box 8162 Grand Forks ND 58202

ROBINSON, TREVOR, b Springfield, Mass, Feb 20, 29; m 52; c 3. BIOCHEMISTRY. *Educ:* Harvard Univ, AB, 50, AM, 51; Univ Mass, MS, 53; Cornell Univ, PhD(biochem), 56. *Prof Exp:* Res assoc bact & bot, Syracuse Univ, 56-60; from asst prof to assoc prof chem, 61-66, assoc prof, 66-82, PROF BIOCHEM, UNIV MASS, AMHERST, 82- *Mem:* Am Soc Plant Physiol; Phytochem Soc NAm; AAAS; Am Soc Pharmacognosy; Sigma Xi; Early Am Industs Asn. *Res:* Plant biochemistry; alkaloids; tannins; history of science. *Mailing Add:* Dept Biochem Univ Mass Amherst MA 01003

ROBINSON, WILBUR EUGENE, b Viola, Kans, Aug 27, 19; m 42; c 2. CHEMISTRY, FUEL SCIENCE. *Educ:* Sterling Col, BA, 41. *Prof Exp:* Teacher high sch, 41-42; chemist, US Bur Mines, Nev, 42-44, supvry chemist, Wyo, 47-64; sect supvr, Laramie Energy Technol Ctr, US Dept Energy, 64-80; RETIRED. *Mem:* Am Chem Soc; Sigma Xi. *Res:* Constitution and properties of oil-shale kerogen. *Mailing Add:* 1516 Sheridan Laramie WY 82070

ROBINSON, WILLIAM COURTNEY, JR, b Weatherford, Tex, July 4, 37; m 64; c 2. METALLURGY, CHEMICAL ENGINEERING. *Educ:* Univ Tex, BS, 59; Iowa State Univ, PhD(metall), 64. *Prof Exp:* Jr chem engr, Humble Oil & Refining Co, 59; asst, Ames Lab, 60-64; metallurgist, Oak Ridge Nat Lab, 64-69; sr res scientist, Lockheed-Calif Co, 69-70; group leader, 70-85, PLANT MGR, KEMET ELECTRONICS, 85- *Mem:* Am Soc Metals. *Res:* Vapor deposition of refractory metals; corrosion and adhesive bonding problems on aircraft; high temperature sinterins and powder metallurgical fabrication; anodization; capacitor manufacture. *Mailing Add:* Kemet Corp PO Box 5928 Greenville SC 29606

ROBINSON, WILLIAM EDWARD, b Nashua, NH, Dec 27, 48. BIVALVE TOXICOLOGY, INORGANIC BIOCHEMISTRY. *Educ:* Boston Univ, AB, 70; Northeastern Univ, MS, 77, PhD(biol), 81. *Prof Exp:* Instr, Histol Tech, Northeastern Univ, 77, res fel, Dept Biol & Marine Sci Inst, 77-81, instr, Quant Fluorescense Microscopy, 81; postdoctoral fel, Chem Dept, Brandeis Univ, 81-83; assoc scientist, 83-88, RADIATION SAFETY OFFICER, EDGERTON RES LAB, NEW ENG AQUARIUM, 83-, SR SCIENTIST, 88- *Concurrent Pos:* Numerous grants from many orgns, 78-92; vis investr, Egerton Res Lab, New Eng Aquarium, 81-83; adj res assoc, Marine Sci & Maritime Studies Ctr, Northeastern Univ, 83-, chem dept, Brandeis Univ, 85- *Mem:* Am Soc Zoologists; AAAS; Am Micros Soc; Am Soc Limnol & Oceanog; Nat Shellfish Asn. *Res:* Metal uptake, transport and detoxification in bivalve molluscs; function of vanadium and tunichrome in ascidians; feeding and digestion in bivalve molluscs-adult and larval; chlorophyll breakdown and utilization by invertebrates; effects of turbidity on invertebrate feeding and digestion; molluscan energetics; malacology; numerous publications. *Mailing Add:* Edgerton Res Lab New Eng Aquarium Cent Wharf Boston MA 02110-3399

ROBINSON, WILLIAM H, b Philadelphia, Pa, Feb 15, 43; m 64; c 3. ENTOMOLOGY. *Educ:* Kent State Univ, BA, 64, MA, 66; Iowa State Univ, PhD(entom), 70. *Prof Exp:* Asst prof, 70-80, ASSOC PROF ENTOM, VA POLYTECH INST & STATE UNIV, 80- *Mem:* Entom Soc Am; Entom Soc Can; Am Entom Soc. *Res:* Biology and immature stages of Phoridae; Diptera biology and taxonomy; insects associated with thermal water. *Mailing Add:* Dept Entom Vet Admin Polytech Inst 215 Price Hall Blacksburg VA 24061-0319

ROBINSON, WILLIAM JAMES, b Erie, Pa, Feb 19, 29; m 57; c 2. DENDROCHRONOLOGY. *Educ:* Univ Ariz, BA, 57, MA, 59, PhD(anthrop), 67. *Prof Exp:* From res asst to assoc prof, 63-76, asst dir lab, 72-81, dir lab, 82-86, PROF DENDROCHRONOLOGY, LAB TREE-RING RES, UNIV ARIZ, 76- *Mem:* Soc Am Archaeol; Am Quaternary Asn; Tree-Ring Soc. *Res:* Application of special techniques of dendrochronology to archaeology, especially the non-chronological aspect which views the material as an artifact, and reconstruction of past environments. *Mailing Add:* 960 E Foothills Dr Univ Ariz Tucson AZ 85718

ROBINSON, WILLIAM KIRLEY, b Syracuse, NY, Apr 1, 25; m 50; c 4. PHYSICS. *Educ:* Univ Mo, BS, 45; Carnegie Inst Technol, MS, 54, PhD(physics), 59. *Prof Exp:* Instr math, Brevard Col, 48-51; assoc prof, 59-71, prof physics, 71-80, HENRY PRIEST PROF PHYSICS, ST LAWRENCE UNIV, 80- *Mem:* Am Phys Soc; Am Asn Physics Teachers. *Res:* Nuclear measurements; astronomy. *Mailing Add:* 19 College St Canton NY 13617

ROBINSON, WILLIAM LAUGHLIN, b Ironwood, Mich, Mar 29, 33; m 59; c 2. WILDLIFE ECOLOGY. *Educ:* Mich State Univ, BS, 54; Univ Maine, MS, 59; Univ Toronto, PhD(zool, ecol), 63. *Prof Exp:* Asst leader, Maine Coop Wildlife Res Unit, Maine, 63; asst prof biol, Middlebury Col, 63-64; asst prof, 64-69, PROF BIOL, NORTHERN MICH UNIV, 69- *Concurrent Pos:* NSF res grants, 66-69; sci fac fel, San Diego State Col, 71-72; US Forest Serv res contracts, 72-79; Nat Audubon Soc res grant, 73-75; US Fish & Wildlife Serv contracts, 78-81; Ruffed Grouse Soc res grant, 82-91; Earthwatch grants, 85-91. *Mem:* Sigma Xi; Wildlife Soc; Am Soc Mammal. *Res:* Winter shelter requirements, social behavior and populations of white-tailed deer; homing behavior of meadow mice; ecology of spruce grouse; ecology of wolves; ecology of woodcock; ecology of loons. *Mailing Add:* 410 E Crescent St Marquette MI 49855

ROBINSON, WILLIAM ROBERT, b Longview, Tex, May 30, 39; m 62; c 3. SOLID STATE CHEMISTRY, X-RAY CRYSTALLOGRAPHY. *Educ:* Tex Technol Col, BS, 61, MS, 62; Mass Inst Technol, PhD(chem), 66. *Prof Exp:* NSF fel, Univ Sheffield, 66-67; from asst prof to assoc prof, 67-79, PROF CHEM, PURDUE UNIV, LAFAYETTE, 79- *Concurrent Pos:* Adj assoc prof, Dept Earth & Space Sci, State Univ NY Stony Brook, 73. *Mem:* AAAS; Am Chem Soc; Am Crystallog Asn. *Res:* X-ray crystallographic studies of inorganic compounds; solid state chemistry; synthesis and characterization of transition metal oxides, sulfides, and related compounds. *Mailing Add:* Dept Chem Purdue Univ West Lafayette IN 47907

ROBINSON, WILLIAM SIDNEY, b Bloomington, Ind, Nov 24, 33; m 65. INTERNAL MEDICINE, MOLECULAR BIOLOGY. *Educ:* Ind Univ, AB, 56; Univ Chicago, MS & MD, 60. *Prof Exp:* Intern internal med, Columbia-Presby Med Ctr, NY, 60-61, jr asst resident, 61-62; sr asst resident, Univ Chicago Hosps, 62-63, resident, 63-64; NIH spec fel, Univ Calif, Berkeley, 64-65, asst prof molecular biol & res biologist, Virus Lab, 65-67; from asst prof to assoc prof med, 67-76, PROF MED, DIV INFECTIOUS DIS, SCH MED, STANFORD UNIV, 76- *Concurrent Pos:* Res fel biochem, Argonne Cancer Res Hosp & Univ Chicago, 62-64; mem cancer res training comt, NIH, 71- *Mem:* AAAS; Am Soc Microbiol; Am Soc Clin Invest. *Res:* Biochemistry of virus infection and replication; malignant transformation of cells by tumor viruses; nucleic acid metabolism; infectious diseases. *Mailing Add:* Dept Med Div Infectious Dis Stanford Univ Sch Med Stanford CA 94305

ROBINSON-WHITE, AUDREY JEAN, b Houston, Tex, June 14, 43; m 87; c 1. ANESTHESIOLOGY, VASCULAR CELL PHYSIOLOGY. *Educ:* Spelman Col, AB, 65; Boston Col, MS, 75; Boston Univ, PhD(cell biol), 80. *Prof Exp:* Postdoctoral fel, Nat Heart, Lung & Blood Inst, NIH, 80-82, Pratt fel, Nat Inst Gen Med Sci, 82-84; RES ASST PROF, DEPT ANESTHESIOL, UNIFORMED SERV UNIV HEALTH SCI, BETHESDA, MD, 84-, DEPT PHYSIOL, 84- *Concurrent Pos:* Nat Res Serv Award, NIH, 80-82, First Award, 87-91, small instrumentation grant, 89; prin investr, Competitive Award, Uniformed Serv Univ Health Sci, 85-88. *Honors & Awards:* Res Award, Shiley Biomed, 87. *Mem:* AAAS; Am Physiol Soc; Am Soc Pharmacol & Exp Therapeut; NY Acad Sci. *Res:* Biochemistry of the vascular endothelial cell which includes a study of the uptake and metabolism of biogenic amines by endothelial cells from different species and vascular regions; study of the action of three classes of anesthetics on endothelial cell biochemistry, signal transduction systems, and function. *Mailing Add:* Dept Anesthesiol Uniformed Serv Univ Health Sci 4301 Jones Bridge Rd Bethesda MD 20814

ROBINTON, ELIZABETH DOROTHY, b Woburn, Mass, June 27, 10. MICROBIOLOGY, PUBLIC HEALTH. *Educ:* Columbia Univ, BS, 38; Smith Col, MA, 42; Yale Univ, PhD(pub health), 50. *Prof Exp:* Microbiologist, Div Labs, State Dept Health, Conn, 31-42; instr bact, Woman's Col NC, 42-43; instr, Goucher Col, 43-44; from instr to asst prof bact, 44-56, assoc prof bact & pub health, 56-62, prof microbiol & pub health, 62-65, prof biol sci, 65-73, chmn dept, 66-69, EMER PROF BIOL SCI, SMITH COL, 73- *Concurrent Pos:* WHO fel, 65; ed-in-chief, Health Lab Sci, 66-74; chmn, lab sect, Am Pub Health Asn. *Mem:* Med Mycol Soc Americas; Conf State & Prov Pub Health Lab Dirs; Am Soc Microbiol; fel Am Pub Health Asn; fel Am Acad Microbiol. *Res:* Environmental and public health microbiology; medical mycology. *Mailing Add:* 242 Kimball Farms 193 Walker St Lenox MA 01240

ROBISHAW, JANET D, b Midland, Mich, June 18, 56; m. PHARMACOLOGY, PHYSIOLOGY. *Educ:* Cent Mich Univ, BS, 79; Pa State Univ, PhD(physiol), 83. *Prof Exp:* Teaching asst, Pa State Univ, 79-81; NIH postdoctoral fel pharmacol, Health Sci Ctr, Univ Tex, 83-86; MEM STAFF, WEIS CTR RES, GEISINGER CLIN, 86- *Concurrent Pos:* Lectr, Dept Pharmacol, Wash Univ, Case Western Univ & Univ Pa, 86, Dept Neurobiol, Columbia Univ & Dept Physiol, Pa State Univ, 87, Am Heart Asn, 88 & Dept Pharmacol, Mayo Clin, 89; ad hoc mem, Biochem Study Sect, Am Cancer Soc, 87-90; ad hoc reviewer, Nat Heart, Lung & Blood Inst, NIH. *Mem:* Am Soc Biochem & Molecular Biol. *Res:* Author of 23 technical publications. *Mailing Add:* Weis Ctr Res 26-14 Geisinger Clin Danville PA 17822

ROBISON, D(ELBERT) E(ARL), b Weiser, Idaho, June 11, 20; m 41; c 1. MECHANICAL ENGINEERING. *Educ:* Univ Idaho, BS, 50; Purdue Univ, MS, 52, PhD(jet propulsion), 55. *Prof Exp:* Instr mech eng & res engr, Exp Sta, Univ Idaho, 46-50; asst, Purdue Univ, 50-51, instr & res assoc, 51-55, asst prof, 55-56; gas dynamicist & actg supvr rocket engine sect, Gen Elec Co, 56-57, mgr combustion unit, 57-58, supvr rocket systs & processes, 58-61; mgr thrust chamber tech staff, Aerojet Gen Corp, 61-63, asst div mgr, 63-64, prog mgr, 64-70, vpres eng, Aerojet Liquid Rocket Co, Sacramento, 70-72, dir advan prod develop, Aerojet Energy Conversion Co, 72-76; PROF MECH ENG, CALIF POLYTECH STATE UNIV, SAN LUIS OBISPO, 76- *Mem:* Am Inst Aeronaut & Astronaut; Sigma Xi. *Res:* Rocket and space flight research and development; theory and procedures for evaluating rocket propulsion systems; comparison of liquid and solid propellant missile systems. *Mailing Add:* 6462 N Briarwood Fresno CA 93711

ROBISON, GEORGE ALAN, b Lethbridge, Alta, Nov 4, 34; m 56; c 2. BIOCHEMICAL PHARMACOLOGY, ENDOCRINOLOGY. *Educ:* Univ Alta, BSc, 57; Tulane Univ, MS, 60, PhD(pharmacol), 62. *Prof Exp:* Res fel pharmacol, Sch Med, Western Reserve Univ, 62-63; res assoc physiol, Sch

Med, Vanderbilt Univ, 63-64, instr physiol & pharmacol, 64-66, asst prof pharmacol, 66-69, assoc prof pharmacol & physiol, 69-72; PROF PHARMACOL & CHMN DEPT, UNIV TEX MED SCH HOUSTON, 72- *Concurrent Pos:* Investr, Howard Hughes Med Inst, 70-72; co-ed, Advances Cyclic Nucleotide Res, 72- *Honors & Awards:* J Murray Luck Award, Nat Acad Sci, 79. *Mem:* Am Chem Soc; Am Soc Pharmacol & Exp Therapeut; Endocrine Soc; NY Acad Sci; Soc Neurosci. *Res:* Biochemical basis of hormone action; biochemical basis of animal behavior. *Mailing Add:* 250 Stoney Creek Dr Houston TX 77024-6209

ROBISON, HENRY WELBORN, b Albany, Ga, Mar 24, 45; m 66; c 2. ICHTHYOLOGY. *Educ:* Ark State Univ, BS, 67, MS, 68; Okla State Univ, PhD(zool), 71. *Prof Exp:* Res asst fish social behav, Okla State Univ, 70-71; asst prof zool, Southern Ill Univ, Carbondale, 71; from assoc prof to prof biol, 71-86, DEAN SCI & TECHNOL, SOUTHERN ARK UNIV, 87- *Concurrent Pos:* Ark Wildlife Conservationist, 78; pres, Ark Acad Sci, 80-81; hon prof, Southern Ark Univ, 81. *Mem:* Am Soc Ichthyologists & Herpetologists; Sigma Xi. *Res:* Taxonomy; ecology and behavior of cyprinid and perciform fishes; social behavior of anabantoid fishes; fishes of Arkansas; evolution of reproductive behavior in fishes. *Mailing Add:* Dept of Biol Southern Ark Univ Magnolia AR 71753

ROBISON, LAREN R, b Georgetown, Idaho, Mar 25, 31; m 55; c 6. PLANT GENETICS, WEED SCIENCE. *Educ:* Brigham Young Univ, BS, 57, MS, 58; Univ Minn, PhD(plant genetics), 62. *Prof Exp:* Res leader new crops, Univ Nebr-Lincoln, 62-65, res leader exten weed control, vchmn & exten leader, Agron Dept, 65-71; prof agron & hort & chmn dept, 71-82, ASSOC DEAN, COL BIOL & AGR SCI, BRIGHAM YOUNG UNIV, 82-, DIR, BENSON AGR FOOD INST. *Mem:* Am Soc Agron; Crop Sci Soc Am; Weed Sci Soc Am; Sigma Xi; Coun Agr Sci & Technol. *Res:* Plant and soil relationships; new and potentail crops breeding; weed control methods, including herbicides used, dissipation and crop tolerance; international agriculture development. *Mailing Add:* Dept Agron & Hort Brigham Young Univ Provo UT 84601

ROBISON, NORMAN GLENN, b Littlefield, Tex, Oct 5, 38; m 63; c 3. GENETICS, AGRONOMY. *Educ:* Tex Tech Col, BS, 61; Univ Nebr, MS, 63, PhD(agron), 67. *Prof Exp:* TROP MAIZE RES DIR, DEKALB-PFIZER GENETICS, 65- *Mem:* Am Soc Agron. *Res:* Direct tropical maize breeding programs worldwide. *Mailing Add:* 110 Thornbrook DeKalb IL 60115

ROBISON, ODIS WAYNE, b Lawton, Okla, Aug 23, 34; m 56; c 3. GENETICS, REPRODUCTIVE PHYSIOLOGY. *Educ:* Okla State Univ, BS, 55; Univ Wis, MS, 57, PhD(genetics, animal husb), 59. *Prof Exp:* From asst prof to assoc prof, 59-74, PROF ANIMAL SCI & GENETICS, NC STATE UNIV, 74- *Concurrent Pos:* NSF travel grant, NATO Conf, 62 & 68; mem, AID Mission to Peru; lectr several foreign countries; sect ed, J Animal Sci; dir, Nat Swine Improv Fedn; ed, J Animal Sci Appl Sect; Fulbright Scholar. *Honors & Awards:* Rockefeller Prentice Mem Award, Am Soc Animal Sci; Serv Award, Nat Swine Improv Fedn. *Mem:* AAAS; Am Soc Animal Sci; Biomet Soc; Sigma Xi; Am Genetic Asn. *Res:* Genetic control of developmental and physiological processes; formulation of selection schemes and breeding systems for efficient manipulation of populations; interaction of genetics and maternal influence on developing processes. *Mailing Add:* 226 Polk Hall NC State Univ Box 7621 Raleigh NC 27695-7621

ROBISON, RICHARD ASHBY, b Fillmore, Utah, Jan 10, 33; m 53; c 3. GEOLOGY, PALEONTOLOGY. *Educ:* Brigham Young Univ, BS, 57, MS, 58; Univ Tex, PhD(geol), 62. *Prof Exp:* Geologist, US Geol Surv, 59-60; asst prof geol, Univ Utah, 62-66; assoc cur invert paleont, Smithsonian Inst, 66-67; from assoc prof to prof geol, Univ Utah, 67-74; HEDBERG PROF GEOL, UNIV KANS, 74- *Mem:* Int Paleont Asn; Palaeont Asn London; Geol Soc Am; Paleont Soc; Soc Econ Paleont & Mineral. *Res:* Paleontology, particularly Cambrian trilobites and stratigraphy. *Mailing Add:* Dept Geol 120 Lindley Univ Kans Lawrence KS 66045-2124

ROBISON, WENDALL C(LOYD), b Des Moines, Iowa, July 31, 23; m 47; c 3. ELECTRICAL ENGINEERING. *Educ:* Iowa State Univ, BSc, 47, MSc, 48, PhD(elec eng), 57. *Prof Exp:* Engr, Gen Elec Co, 48-49; from instr to assoc prof elec eng, 49-71, PROF ELEC ENG, UNIV NEBR, LINCOLN, 71- *Mem:* AAAS; Am Soc Eng Educ; Inst Elec & Electronics Engrs. *Res:* Electric network theory including both linear and nonlinear networks; system theory. *Mailing Add:* Dept Elec Eng Univ Nebr Lincoln NE 68588

ROBISON, WILBUR GERALD, JR, b Cheyenne, Wyo, Dec 27, 33; m 57; c 4. CELL BIOLOGY, EXPERIMENTAL PATHOLOGY. *Educ:* Brigham Young Univ, AB, 58, MA, 61; Univ Calif, Berkeley, PhD(genetics), 65. *Prof Exp:* Res geneticist, Univ Calif, Berkeley, 63-65; res fel anat, Harvard Med Sch, 65-66; asst prof biol, Univ Va, 66-72; sr staff fel, 72-76, geneticist & cell biologist, lab vision res, 76-82, chief sect exp anat, 82-85, CHIEF, SECT PATHOPHYSIOL, NAT EYE INST, 85- *Concurrent Pos:* US Air Force Off Sci Res-Nat Acad Sci res fel, 65-66. *Honors & Awards:* Spec Achievement Award, Nat Eye Inst, NIH; hon mem Argentine Med Asn. *Mem:* Am Soc Cell Biol; Asn Res Vision & Ophthal. *Res:* Experimental pathology of the eye; ultrastructural and functional interrelationships between the pigment epithelium and the visual cells of the retina; prevention of visual complications of diabetes using inhibitors of aldose reductase; vitamin A, vitamin E, and aging pigments. *Mailing Add:* Bldg 10 Rm 10N-105 Nat Eye Inst Lab Mechanism of Oscular Dis NIH Bethesda MD 20892

ROBISON, WILLIAM LEWIS, b Grinnell, Iowa, June 18, 38; m 59; c 3. ECOLOGY, RADIOBIOLOGY. *Educ:* Cornell Col, AB, 56; Univ Calif, Berkeley, MS, 62, PhD(biophysics), 66. *Prof Exp:* SR RES SCIENTIST ENVIRON SCI, UNIV CALIF, LAWRENCE LIVERMORE LAB, 65- *Mem:* AAAS; Health Physics Soc; Nat Coun Radiation Protection & Measurement. *Res:* Environmental science; radionuclide and stable element transport and fate; radiation biology; uptake, retention, dose assessment to populations via food chains. *Mailing Add:* 774 Canterbury Ave Livermore CA 94550

ROBITAILLE, HENRY ARTHUR, b Washington, DC, Sept 2, 43; m 68; c 3. DIRECTOR, SCIENCE & TECHNOLOGY. *Educ:* Univ Md, BS, 66; Mich State Univ, MS, 67, PhD(hort), 70. *Prof Exp:* Prof hort, Okla State Univ, 70-72; prof hort, Purdue Univ, 72-81; DIR SCI & TECHNOL, EPCOT CTR, 81- *Concurrent Pos:* Chmn, Adminr WG & vpres elect Indust Sect, Am Soc Hort Sci; adj prof, Univ Vicosa, Brazil, 73-75, Univ Ariz & Univ Fla, 81-; plant physiologist, USAID, MG, Brazil, 73-75; adv bd, Valencia Community Col, 83- *Honors & Awards:* Cert of Appreciation, USDA, 87; Tribute of Appreciation, US Environ Protection Agency, 88. *Mem:* Am Soc Hort Sci; Sigma Xi; AAAS. *Res:* Administer a department doing research and engineering in many areas related to both agriculture and oceanography-marine biology; special expertise in science communications and technical pavilion support. *Mailing Add:* Land-Epcot Ctr PO Box 10000 Lake Buena Vista FL 32830

ROBKIN, MAURICE, b New York, NY, Apr 25, 31; m 62; c 3. NUCLEAR ENGINEERING, BIOENGINEERING. *Educ:* Calif Inst Technol, BS, 53; Oak Ridge Sch Reactor Technol, dipl, 54; Mass Inst Technol, PhD(nuclear eng), 61. *Prof Exp:* Physicist, Bettis Atomic Power Lab, Westinghouse Elec Co, 54-56 & Valecitos Atomic Lab, Gen Elec Co, 61-67; assoc prof, 67-79, PROF NUCLEAR ENG, UNIV WASH, 79-, PROF ENVIRON HEALTH, 81- *Concurrent Pos:* Vis scientist, Cambridge Univ, 76; consult indust nuclear eng & health physics. *Mem:* AAAS; NY Acad Sci; Am Nuclear Soc; Radiation Res Soc; Health Physics Soc. *Res:* Health physics; environmental radioactivity. *Mailing Add:* Radiol Sci, SB-75 Univ Wash Seattle WA 98195

ROBL, HERMANN R, b Vienna, Austria, Aug 7, 19; nat US; m 42; c 1. THEORETICAL PHYSICS. *Educ:* Univ Vienna, PhD, 48, Dr habil, 52. *Prof Exp:* Asst, Inst Theoret Physics, Vienna, 48-52, asst prof, 52-55; asst, Phys Sci Div, Off Ord Res, 55-56, assoc dir, Physics Div, 56-57, dir, 57-62; dep chief scientist, US Army Res Off, 62-73, chief scientist, 73-75, tech dir, 75-85; RETIRED. *Concurrent Pos:* From vis asst prof to vis assoc prof, Duke Univ, 59-65, adj prof, 66-; Army Res Off-Durham res & study proj scholar, 65. *Honors & Awards:* Korner Award, 54. *Mem:* Am Phys Soc. *Res:* Quantum optics and mechanics. *Mailing Add:* 2215 Elmwood Ave Durham NC 27707

ROBL, ROBERT F(REDRICK), JR, electromagnetic casting, aluminum electrolysis, for more information see previous edition

ROBLES, LAURA JEANNE, BIOLOGY OF PHOTORECEPTORS. *Educ:* Univ Calif, PhD(biol), 75. *Prof Exp:* PROF BIOL, CALIF STATE UNIV, 75- *Res:* Electron microscopy; vitamin A cycling. *Mailing Add:* Dept Biol Calif State Univ 1000 E Victoria St Carson CA 90747

ROBLIN, JOHN M, engineering management, chemical engineering, for more information see previous edition

ROBLIN, JOHN M, b Sagada, Philippines, Feb 20, 31; m 66; c 3. ECONOMIC ANALYSIS, STRATEGIC PLANNING. *Educ:* Princeton Univ, BS, 53; Mass Inst Technol, MS, 55; Case Univ, PhD(chem eng), 62. *Prof Exp:* Chmn coal & coal chem res, Repub Steel Corp, 63-65, head new prod res, 65-74, mgr, Process & Mat Pkg, 74-78, dir, Res & Develop, 78-81, dir, Strategic Planning, 81-85; dir corp planning, LTV Steel Corp, 85-87; dir eng res, 87-89, DIR APPL RES, UNIV NC, CHARLOTTE, 89- *Concurrent Pos:* Affil, Sr Consult Network, 86- *Honors & Awards:* F C Zeisberg Award, Am Inst Chem Engrs, 53; Regional Tech Award, Am Iron & Steel Indust, 65. *Mem:* Am Iron & Steel Inst; Am Soc Eng Educ; Sigma Xi; Am Inst Chem Engrs. *Res:* Process analysis and development, especially in natural resources, metallurgical industries, coal and coke; management of technology; technology transfer; intellectual capital. *Mailing Add:* 130 Appl Res Facil Univ NC Charlotte NC 28223

ROBOCK, ALAN, b Boston, Mass, Sept 7, 49; m 80. CLIMATE DYNAMICS, CLIMATOLOGICAL DATA ANALYSIS. *Educ:* Univ Wis, Madison, BA, 70; Mass Inst Technol, SM, 74, PhD(meteorol), 77. *Prof Exp:* Volunteer meteorol, US Peace Corps, Philippines, 70-72; asst prof, 77-82, ASSOC PROF, DEPT METEOROL, UNIV MD, 82- *Concurrent Pos:* Fel & coun mem, Coop Inst Climate Studies. *Mem:* Am Meteorol Soc; AAAS; Am Geophys Union; Fedn Am Scientists. *Res:* Numerical modeling of the climate system; causes of climate change, especially volcanic eruptions, carbon dioxide and natural variability; snow and ice-albedo feedback; climatological data analysis-surface temperature and snow cover; nuclear winter; soil moisture. *Mailing Add:* Dept Meteorol Univ Md College Park MD 20742

ROBOLD, ALICE ILENE, b Daleville, Ind, Feb 7, 28; m 55; c 1. MATHEMATICS, TEACHER EDUCATION. *Educ:* Ball State Univ, BS, 55, MA, 60, EdD, 65. *Prof Exp:* Asst prof math, 64-69, assoc prof, 69-76, PROF MATH SCI, BALL STATE UNIV, 76- *Mem:* Nat Coun Teachers Math; Sch Sci & Math Asn. *Res:* Background of college instructors of mathematics for prospective elementary school teachers. *Mailing Add:* Dept Math Sci Ball State Univ Muncie IN 47306

ROBOZ, JOHN, b Budapest, Hungary, Oct 14, 31; US citizen; m 61; c 2. ANALYTICAL BIOCHEMISTRY, CHEMOTHERAPY. *Educ:* Eotvos Lorand, Budapest, BS, 55; NY Univ, MS, 60, PhD(phys chem), 62. *Prof Exp:* Sr engr, Gen Tel & Electronics Res Labs, NY, 57-63; group leader gas analysis res, Cent Res Labs, Air Reduction Co, 63-69; res assoc prof clin chem, 69-74, assoc prof, 74-81, PROF NEOPLASTIC DIS, MT SINAI SCH MED, 81- *Mem:* Am Chem Soc; Am Soc Mass Spectrometry; Am Asn Cancer Res; NY Acad Sci; Fedn Am Soc Exp Biol. *Res:* Identification and quantification of antineoplastic agents and metabolites in body fluids and tissues; biological markers of cancer; biochemical diagnosis of opportunistic infections; mass spectrometry, high performance liquid chromatography and other instrumental techniques in clinical chemistry. *Mailing Add:* Mt Sinai Sch Med 11 E 100th St New York NY 10029-5291

ROBROCK, RICHARD BARKER, II, b Cleveland, Ohio, Dec 29, 41; m 70; c 2. INTELLIGENT NETWORK SYSTEMS, NETWORK SERVICES. *Educ:* Case Inst Technol, BS, 63, MS, 65, PhD(elec eng), 67. *Prof Exp:* Mem tech staff, Bell Labs, 67-70, supvr, 70-79, dept head, 79-83, dir, 83-84, ASST VPRES, BELL COMMUN RES, 84- *Mem:* Fel Inst Elec & Electronics Engrs; AAAS; Sigma Xi. *Res:* Software systems for telecommunications networks; real-time fault-tolerant data bases; telecommunications services; programming languages for rapid service creation. *Mailing Add:* Six Timothy Lane Bedminster NJ 07921

ROBSON, ANTHONY EMERSON, b London, Eng, March 29, 32; US citizen. EXPERIMENTAL PLASMA PHYSICS. *Educ:* Oxford Univ, BA, 52, MA, 56, DPhil, 56. *Prof Exp:* Sci officer, UK Atomic Energy Authority, Harwell & Culham Labs, 56-66; res scientist, Univ Tex, Austin, 66-72; head, exp plasma physics br, 72-89, SR SCIENTIST, CONTROLLED FUSION RES & APPLICATIONS, NAVAL RES LAB, WASH, DC, 89- *Concurrent Pos:* Vis scientist, Imp Col Sci & Technol, London, UK, 85-86. *Mem:* Fel Am Phys Soc. *Res:* Arc discharges; plasma solid interaction; mirror machines; collisionless shock waves; turbulent plasma heating; high magnetic field generation; homopolar generators; relativistic electron beams; high density Z-pinches; fusion reactor studies. *Mailing Add:* Code 4708 Naval Res Lab Washington DC 20375-5000

ROBSON, DONALD, b Leeds, Eng, Mar 19, 37; m 60, 71; c 3. NUCLEAR & PARTICLE PHYSICS. *Educ:* Univ Melbourne, BSc, 59, MSc, 61, PhD(nuclear physics), 63. *Prof Exp:* Res assoc nuclear physics, 63-64, from asst prof to assoc prof physics, 64-67, PROF PHYSICS, FLA STATE UNIV, 67-, CHAIRPERSON, 85- *Concurrent Pos:* Fulbright scholar, 63-64; A P Sloan fel, 66-72; vis prof, Princeton Univ, 71-72, Univ München, WGer, 76-77; Alexander von Humboldt sr scientist award, 76-77. *Honors & Awards:* Tom W Bonner Prize, Am Phys Soc, 72. *Mem:* Fel Am Phys Soc. *Res:* Theoretical nuclear and particle physics. *Mailing Add:* Dept Physics Fla State Univ Tallahassee FL 32306

ROBSON, DOUGLAS SHERMAN, b St John, NDak, July 30, 25; m 49; c 3. BIOMETRICS. *Educ:* Iowa State Col, BS, 49; Cornell Univ, MS, 51, PhD(statist), 55. *Prof Exp:* Biometrician, Cornell Univ, 49-53, res assoc plant breeding, 54-55, from asst prof to assoc prof, 55-62, prof, 62-87, EMER PROF PLANT BREEDING & BIOMET, CORNELL UNIV, 87- *Concurrent Pos:* NIH career develop award, Cornell Univ, 62-72; pres, Eastern NAm Region Biomet Soc, 70; consult, 87- *Mem:* Fel Am Statist Asn; fel Am Inst Fishery Res Scientists; Biomet Soc. *Res:* Biological statsitics; sampling theory and applications. *Mailing Add:* 150 MacLaren St PH6 Ottawa ON K2P 0L2 Can

ROBSON, HARRY EDWIN, b Kans, July 19, 27; m 50; c 3. PHYSICAL CHEMISTRY. *Educ:* Univ Kans, BS, 49, PhD(chem), 59. *Prof Exp:* Res assoc, Esso Res Labs, Humble Oil & Refining Co, 57-72, res assoc, 72-78, SR RES ASSOC, EXXON RES & DEVELOP LABS, 78- *Res:* Petroleum process catalysts; inorganic synthesis. *Mailing Add:* 3131 Congress Blvd Baton Rouge LA 70808-3139

ROBSON, HOPE HOWETH, soil microbiology, for more information see previous edition

ROBSON, JOHN HOWARD, b East Liberty, Ohio, July 26, 40; m 61; c 2. POLYMER CHEMISTRY, ORGANIC CHEMISTRY. *Educ:* Ohio Northern Univ, BS, 62; Ohio State Univ, PhD(org chem), 67. *Prof Exp:* Proj chemist, 66-77, mgr tech recruiting & mgr tech & managerial educ, 77-80, res scientist, 80-88, TECHNOL MGR/GROUP LEADER, RES & DEVELOP, UNION CARBIDE CORP, 88- *Mem:* Am Chem Soc. *Res:* Development of intermediates and application technology for flexible, high-resiliency and rigid polyurethane foams; chemical process innovation of development; prices and project research and development for ethylene oxide, ethylene glycol and formulated products. *Mailing Add:* Union Carbide Tech Cent 770-436 PO Box 8361 South Charleston WV 25303

ROBSON, JOHN MICHAEL, b London, Eng, Mar 26, 20; m 50; c 3. NUCLEAR PHYSICS. *Educ:* Cambridge Univ, BA, 42, MA, 46, ScD, 63. *Prof Exp:* Physicist, Radar Res & Develop Estab, Eng, 42-45, Atomic Energy Res Estab, 45-50 & Atomic Energy Can, Ltd, 50-60; prof physics, Univ Ottawa, 60-69; prof physics, Sultan Qaboos Univ, Oman, 86-88; prof, 69-85, EMER PROF PHYSICS, MCGILL UNIV, 85- *Honors & Awards:* Gold Medal, Can Asn Physicists, 78. *Mem:* Fel Am Phys Soc; Royal Soc Can; Can Asn Physicists (past pres). *Res:* Radioactive decay of the neutron; inelastic scattering of fast neutrons; shielding of nuclear reactors; ultra cold neutrons; applied mechanics. *Mailing Add:* Dept Physics McGill Univ 3600 Univ St Montreal PQ H3A 2T8 Can

ROBSON, JOHN ROBERT KEITH, b Darlington, Eng, Oct 9, 25; c 4. NUTRITION, PUBLIC HEALTH. *Educ:* Univ Durham, MB, BS, 48; Univ Edinburgh, DTM&H, 56; Univ London, DPH, 59; Univ Newcastle, MD, 68. *Prof Exp:* Physician, hosp & pvt med pract, 48-52; spec grade med officer, Tanganyika Govt, 52-62; nutrit adv, WHO, Philippines, 62-64 & Egypt, 64-67; from assoc prof to prof human nutrit, Sch Pub Health, Univ Mich, Ann Arbor, 67-75; PROF NUTRIT, DEPT FAMILY PRACT, MED UNIV SC, 75- *Concurrent Pos:* Ed, Ecol of Food & Nutrit, 70. *Honors & Awards:* World Hunger Media Award, 82; Golden Amaranth Award, 77. *Mem:* Fel Royal Soc Trop Med & Hyg; Am Bd Nutrit. *Res:* Evaluation of nutritional status in communities and individuals; ethno-nutrition and nutritive value of prehistoric foods. *Mailing Add:* 330 Middle St Mt Pleasant SC 29464

ROBSON, JOHN WILLIAM, b Coshocton, Ohio, Sept 6, 23; m 48; c 2. PHYSICS. *Educ:* Oberlin Col, BA, 49; Case Inst Technol, MS, 52, PhD(physics), 54. *Prof Exp:* Asst physics, Case Inst Technol, 49-51, instr, 51-54; asst prof, 54-58, assoc prof, 58-74, prof, 74-83, EMER PROF PHYSICS, UNIV ARIZ, 83- *Mem:* Am Asn Physics Teachers; Sigma Xi. *Res:* Low energy nuclear physics; applied optics and acoustics. *Mailing Add:* 5671 E Whittier Dr Tucson AZ 85711

ROBSON, RICHARD MORRIS, b Atlantic, Iowa, Dec 9, 41; m; c 4. BIOCHEMISTRY, ANIMAL SCIENCE. *Educ:* Iowa State Univ, BS, 64, MS, 66, PhD(biochem), 69. *Prof Exp:* NIH fel biochem, Iowa State Univ, 65-69; asst prof biochem & animal sci, Univ Ill, 69-72; assoc prof, 72-77, PROF BIOCHEM & ANIMAL SCI, IOWA STATE UNIV, 77- *Honors & Awards:* Meats Res Award, Am Meat Sci Asn, 84; Meat Res Award, Am Soc Animal Sci, 85; Vis Sci Award, WGer Cancer Res Ctr, 86-87. *Mem:* AAAS; Am Heart Asn; Am Soc Animal Sci; Biophys Soc; Am Soc Cell Biol; Am Soc Biochem & Molecular Biol. *Res:* Biochemistry of muscle tissue with emphasis on the chemistry, structure, function and turnover of the myofibrillar/cytoskeletal. *Mailing Add:* Molecular Biol Bldg Iowa State Univ Ames IA 50011

ROBSON, RONALD D, b Leicester, Eng, Oct 21, 33; m 57; c 2. PHARMACOLOGY. *Educ:* Univ Leeds, BSc, 54 & 55, PhD(pharmacol), 58. *Prof Exp:* Mem staff pharmacol, Beecham Res Labs, UK, 58-61; mem staff pharmacol, Wellcome Res Labs, 61-66; group leader, Merck-Frosst Labs, Can, 66-67; sect head & sr scientist, Warner-Lambert Res Inst, 68-71; mgr cardiovasc res, 71-72, dir pharmacol, Pharmaceut Div, 72-78, exec dir biol res, 78-81, SCIENTIFIC ADV, CIBA-GEIGY CORP, 81- *Mem:* Am Soc Pharmacol & Exp Therapeut; Brit Pharmacol Soc; Am Pharm Asn; NY Acad Sci. *Res:* Pharmacology applied to the search for improved therapies for cardiovascular and psychotic disorders. *Mailing Add:* Pharmaceut Div Res Dept Ciba-Geigy Corp Summit NJ 07901

ROBUSTO, C CARL, b Bridgeport, Conn, Nov 29, 16; m 44; c 2. MATHEMATICS, PHYSICS. *Educ:* St John's Univ, NY, BS, 39; Columbia Univ, MA, 46; NY Univ, MS, 50; Fordham Univ, PhD, 54. *Prof Exp:* Assoc prof, 46-56, acad vpres, Queens, 78-80, PROF MATH & PHYSICS, ST JOHN'S UNIV, NY, 56-, EXEC VPRES, 80- *Concurrent Pos:* Dean jr col, St John's Univ, 62-67, dean gen studies, 68-71; acad vpres, Staten Island & dean, Notre Dame Col, 71-78. *Mailing Add:* St John's Univ Jamaica NY 11439

ROBYT, JOHN F, b Moline, Ill, Feb 17, 35; m 58; c 2. BIOCHEMISTRY. *Educ:* St Louis Univ, BS, 58; Iowa State Univ, PhD(biochem), 62. *Prof Exp:* Asst prof biochem, La State Univ, 62-63; NIH fel, Lister Inst Prev Med, London, Eng, 63-64; res assoc, 64-67, asst prof, 67-73, assoc prof, 73-83, PROF BIOCHEM, IOWA STATE UNIV, 83- *Concurrent Pos:* Consult, E I du Pont de Nemours & Co, 70- *Mem:* Am Soc Biol Chemists; Am Chem Soc. *Res:* Study of the mechanisms of carbohydrase action, the mode of substrate binding, the sequence of catalytic events, the types of groups involved, especially with the polysaccharide synthesizing and degrading enzymes; study of the mechanisms of polysaccharide synthesizing and degrading enzymes; chemical modification of carbohydrates to form new products, especially from starch and sucrose, and for inhibitors of carbohydrate enzymes. *Mailing Add:* RR 4 Ames IA 50010

ROCCI, MARIO LOUIS, JR, b Utica, NY, Nov 6, 52. PHARMACOKINETICS, BIOANALYTICAL CHEMISTRY. *Educ:* Syracuse Univ NY, Buffalo, BS, 76 PhD(pharmaceut), 81. *Prof Exp:* Asst prof pharm & dir, Pharmacokinetics Res Lab, Philadelphia Col Pharm & Sci, 80-83; res assoc prof med & head lab investigative med, Div Clin Pharmacol, Dept Med, Jefferson Med Col, 83-88; DIR PHARMACEUT RES, ONEIDA RES SERV INC, 88- *Concurrent Pos:* Consult, Muck Sharp & Dome Res Lab, 85-88, Nat Asn Clin Res, 85-; clin assoc prof pharm, Philadelphia Col Pharm & Sci, 86- *Mem:* Sigma Xi; fel Am Col Clin Pharmacol; Am Asn Pharmaceut Scientist; Am Soc Clin Pharmacol & Therapeut; Am Fedn Clin Res. *Res:* Evaluating the clinical pharmacokinetics and pharmacodynamics of established as well as investigational drugs; author of over 150 manuscripts, book chapters and abstracts. *Mailing Add:* 27 The Hills Dr Utica NY 13501-5513

ROCCO, GREGORY GABRIEL, b Lawrence, Mass, Sept 16, 26; m 50; c 3. RADIOCHEMISTRY. *Educ:* Boston Univ, BA, 49; Univ Mich, MS, 50. *Prof Exp:* Radiochemist, Tracerlab, Inc, 49-63; staff chemist, Wentworth Inst, 63-66; mgr radiochem, New Eng Nuclear Corp, North Billercia, Mass, 66-72, mgr Nuclides & Sources Div, 72-81; site mgr, Dupont NEN Prod, 81-85; RETIRED. *Res:* Development of radiochemical procedures for the separation and decontamination of reactor and cyclotron produced isotopes; development of the use of isotopes for medicine, industry and research. *Mailing Add:* 1149 Hillsborough Mile Apt 602 N Hillsborough Beach FL 33062

ROCEK, JAN, b Prague, Czech, Mar 24, 24; US citizen; m 47; c 2. PHYSICAL ORGANIC CHEMISTRY. *Educ:* Prague Tech Univ, ChemE, 49, PhD(chem), 53. *Prof Exp:* Chemist, Inst Chem, Czech Acad Sci, Prague, 53-60; res fel chem, Harvard Univ, 60-62; from assoc prof to prof, Cath Univ, 62-66; dean, grad sch, 70-79, actg head dept, 80-81, PROF CHEM, UNIV ILL, CHICAGO, 66-, HEAD DEPT, 81- *Concurrent Pos:* Actgdean, grad sch, Univ Ill, 69-70; vis scholar, Stanford Univ & Univ Cambridge, Eng, 79-80. *Mem:* Am Chem Soc; Sigma Xi. *Res:* Mechanisms of oxidation reactions. *Mailing Add:* Dept Chem Univ Ill Box 4348 Chicago IL 60680

ROCHBERG, RICHARD HOWARD, b Baltimore, Md, May 15, 43; m 68; c 2. MATHEMATICS. *Educ:* Princeton Univ, AB, 64; Harvard Univ, MA, 66, PhD(math), 70. *Prof Exp:* Res assoc, Inst Future, 69-70; from asst prof to assoc prof, 70-81, PROF MATH, WASH UNIV, 81- *Concurrent Pos:* Sr vis fel, Univ Col, London, 78-79. *Mem:* Am Math Soc. *Res:* Function theory, spaces of analytic functions, operator theory. *Mailing Add:* Dept Math Wash Univ St Louis MO 63130

ROCHE, ALEXANDER F, b Melbourne, Australia, Oct 17, 21; m 45; c 3. CHILD GROWTH, ANTHROPOMETRICS. *Educ:* Univ Melbourne, MB, BS, 46, PhD(anat), 54, DSc(child growth), 66, MD, 69; FRACP, 80. *Prof Exp:* Intern med, St Vincent's Hosp, Melbourne, 46-48, asst to outpatients surgeon, 48-50; lectr anat, Univ Melbourne, 50-52, sr lectr, 52-62, reader, 62-68; chmn dept growth genetics, 68-71, chief sect phys growth & genetics, Sect Fels

Longitudinal Study & Families, Sect Measurement Growth & Maturity, Fels Res Inst, Ohio, 71-77; prof obstet & gynec, 77-90, PROF PEDIAT, WRIGHT STATE UNIV, 77-, UNIV PROF & PROF COM HEALTH, 90- *Concurrent Pos:* Demonstr, Univ Melbourne, 48-50; Smith-Mundt & Fulbright fels, 52-53; teaching fel, Western Reserve Univ, 52-53; Rockefeller traveling grant, 52-; consult, Royal Children's Hosp, Melbourne, 67-68, Children's Med Ctr, Dayton, Ohio, 69-, Hamilton County Diag Clin Ment Retarded, 69-, Univ Cincinnati Affil Prog Ment Retarded, 69-, USAF, Pan-Am Health Orgn, WHO, Inst Nutrit Cent Am & Panama, Nat Health & Nutrit Exam Surv, 71- & Nat Pituitary Agency, 72-; vis prof, Ohio State Univ, 76, Univ Md, 78; mem, Pediat Adv Sub-comt, Food & Drug Admin, 78; fels prof pediat, fels prof obstet & gynecol, Wright State Univ Sch Med; consult, Nat Health & Nutrit Exam Serv, Dept Health & Human Serv, 71. *Mem:* Am Asn Phys Anthrop; Soc Res Child Develop; Soc Study Human Biol; Anat Soc Gt Brit & Ireland; fel Human Biol Coun. *Mailing Add:* Div Human Biol 1005 Xenia Ave Yellow Springs OH 45387-1695

ROCHE, BEN F, JR, b Winona, Miss, Feb 2, 24; m 50; c 2. RANGE MANAGEMENT. *Educ:* Univ Calif, Davis BS, 51; Wash State Univ, MS, 60; Univ Idaho, PhD(range ecol), 65. *Prof Exp:* County exten agent land develop, 51-54, veg mgt, 54-57, weed specialist, 58-65, asst prof range ecol, 65-66, assoc prof, 66-71, PROF RANGE ECOL, WASH STATE UNIV, 71- *Concurrent Pos:* Coordr res, Colockum Multiple Use Res Ctr, 66- *Mem:* Weed Sci Soc Am; Soc Range Mgt. *Res:* Ecology of secondary succession as created by man's disorder of the primary, particularly the exotic species that seem preadapted to the site. *Mailing Add:* Rte 3 Box 544 Pullman WA 99163

ROCHE, EDWARD BROWINING, b Stamford, Conn, Apr 29, 38; c 2. MEDICINAL CHEMISTRY, CHEMICAL PHARMACOLOGY. *Educ:* Butler Univ, BS, 61, MS, 63; Ohio State Univ, PhD(med chem), 66. *Prof Exp:* Asst prof, 66-71, ASSOC PROF BIOMED CHEM, COL PHARM, UNIV NEBR MED CTR, OMAHA, 71-, ASST DEAN, 80- *Mem:* AAAS; Am Pharmaceut Asn; Acad Pharmaceut Sci; Am Chem Soc; Sigma Xi. *Res:* Design of compounds for analgesic drug-receptor interaction studies; synthetic organic medicinal chemistry; the application of physical organic chemistry to the study of mechanism of biological activity. *Mailing Add:* Col Pharm Univ Nebr Med Ctr 600 S 42nd St Omaha NE 68198-6000

ROCHE, EDWARD TOWNE, b Buenos Aires, Arg, Mar 8, 25; nat US; m 52; c 3. INVERTEBRATE ZOOLOGY. *Educ:* San Diego State Col, AB, 48; Univ Southern Calif, MS, 52, PhD(zool), 57. *Prof Exp:* Asst & assoc, Univ Southern Calif, 53-57; instr life sci, Compton Jr Col, 57-59; from asst prof to assoc prof, 59-69, PROF BIOL SCI, CALIF STATE POLYTECH UNIV, 69- *Res:* Histology; parasitology; biological education; venomous fishes. *Mailing Add:* Dept Biol Sci Calif State Polytech Univ Pomona CA 91766

ROCHE, GEORGE WILLIAM, b San Francisco, Calif, May 27, 21; m 54; c 2. FORENSIC SCIENCE. *Educ:* Univ Calif, Berkeley, AB, 42; Univ Minn, Minneapolis, MS, 52. *Prof Exp:* Crime lab analyst, Bur Criminal Apprehension, State Minn, 46-54, lab dir, 54-62; criminalist, Dept Justice, State Calif, 62-64, supvy criminalist, 64-69; assoc prof, 69-71, PROF DEPT CRIMINAL JUSTICE, CALIF STATE UNIV, SACRAMENTO, 71- *Mem:* Am Acad Forensic Sci; Am Chem Soc; Soc Appl Spectros; Am Soc Criminol; Inst Asn Identification. *Res:* Recognition, individualization and evaluation of physical evidence by application of the natural sciences to law-science matters. *Mailing Add:* 7233 Milford St Sacramento CA 95822

ROCHE, JAMES NORMAN, b Ithaca, NY, Oct 14, 1898; m 37; c 1. CHEMISTRY. *Educ:* Univ Pittsburgh, BS, 24, PhD(chem), 27. *Prof Exp:* Asst, Univ Pittsburgh, 24-27, instr, 27-30; res chemist, Am Tar Prod Co, Pa, 30-44, mgr, Tech Sect, Tar Prod Div, Koppers Co, Inc, 44-49, mgr creosote & pitch sales, 49-52, mgr sales develop, 53-63, tech sales consult, 63-75; RETIRED. *Concurrent Pos:* Dir, Am Wood Preservers Inst, 50-63. *Honors & Awards:* Award Merit, Am Wood Preservers Asn, 72. *Mem:* Emer mem Am Chem Soc; Forest Prod Res Soc; hon mem Am Wood Preservers Asn; Am Chem Soc. *Res:* Glycerides; disinfectants; coal tar chemicals; creosote; pitches. *Mailing Add:* 250 Pantops Mountain Rd No 206 Charlottesville VA 22901

ROCHE, LIDIA ALICIA, b Havana, Cuba, May 9, 39; m 61; c 3. NUCLEAR WASTE DISPOSAL, HEALTH PHYSICS. *Educ:* The Am Univ, BS, 69, PhD(phys chem), 75. *Prof Exp:* Res technician biomed, Georgetown Univ, 63-69; chemist, Gillette Res Inst, 70-72, res chemist, 75-78; RES ANALYST, 78-, US NUCLEAR REGULATORY COMN, 87, PROG MGR, 86-, TECH ASST TO EXEC DIR OPERS 41,. *Mem:* Am Chem Soc. *Res:* Medical uses of radio isotopes; fuel production through waste management and disposal. *Mailing Add:* 623 Warfield Dr Rockville MD 20850-1921

ROCHE, MARCEL, endocrinology, parasitology, for more information see previous edition

ROCHE, RODNEY SYLVESTER, b Oxford, Eng, July 9, 34; div; c 3. BIOPHYSICAL CHEMISTRY, POLYMER CHEMISTRY. *Educ:* Univ Glasgow, BSc, 57, PhD(polymer chem), 65. *Prof Exp:* Sci officer, Chem Div, UK Atomic Energy Authority Exp Reactor Estab, Scotland, 57-61; asst lectr chem, Univ Glasgow, 62-65; from asst prof to prof chem, 65-78, PROF BIOCHEM & CHEM, UNIV CALGARY, 87-, ASSOC DEAN RES, FAC GRAD STUDIES, 89- *Concurrent Pos:* Vis scientist, Weizmann Inst Sci, 71-72, 76 & 80. *Mem:* AAAS; Am Chem Soc; Chem Soc; fel Royal Soc Chem; fel Chem Inst Can; Sigma Xi; NY Acad Sci. *Res:* Physical chemistry of macromolecules; conformational studies of polypeptides and proteins; calcium binding proteins; calmodulin; protein engineering and the protein folding problem. *Mailing Add:* Dept Biol Sci Univ Calgary Calgary AB T2N 1N4 Can

ROCHE, THOMAS EDWARD, b Denver, Colo, Feb 17, 44; m 66; c 2. BIOCHEMISTRY. *Educ:* Regis Col, Colo, BS, 66; Wash State Univ, PhD(chem), 70. *Prof Exp:* NIH res fel, Clayton Found Biochem Inst, Univ Tex, Austin, 70-72, res assoc, 72-74; from asst prof to assoc prof, 74-82, PROF BIOCHEM, KANS STATE UNIV, 82-, HEAD DEPT, 90- *Mem:* Am Chem Soc; Fedn Am Socs Exp Biol; AAAS. *Res:* Structure and function of 2-ketoacid dehydrogenase complexes; regulation of mammalian pyruvate dehydrogenase complex by enzymatic interconversion; cellular organization. *Mailing Add:* Dept Biochem Willard Hall Kans State Univ Manhattan KS 66506

ROCHE, THOMAS STEPHEN, b New York, NY, Apr 9, 46; m 70; c 1. INORGANIC CHEMISTRY, PHYSICAL CHEMISTRY. *Educ:* Manhattan Col, BS, 67; State Univ NY, Buffalo, PhD(inorganic chem), 72. *Prof Exp:* Res assoc inorganic chem, Wayne State Univ, 72-73; res assoc organometallic chem, Univ Chicago, 73-75; res chemist, Pullman Kellogg Res & Develop Lab, 75-80; RES ASSOC, OLIN CHEM, 80- *Mem:* Am Chem Soc; Sigma Xi. *Res:* Inorganic and organometallic chemistry; reaction mechanisms; catalysis; new process research. *Mailing Add:* 8602 E Cheryl Dr Scottsdale AZ 85258

ROCHEFORT, JOHN S, b Boston, Mass, June 15, 24; m 52; c 5. ELECTRICAL ENGINEERING. *Educ:* Northeastern Univ, BS, 48; Mass Inst Technol, SM, 51. *Prof Exp:* Asst, Servomech Lab, Mass Inst Technol, 48-49, staff engr, 49; res assoc elec eng, 49-52, from asst prof to assoc prof commun, 52-62, actg chmn dept elec eng, 72-73, chmn, 77-82, PROF ELEC ENG, NORTHEASTERN UNIV, 62, PROF COMP ENG, 82-, DIR ELECTRONIC, RES LAB, 87- *Concurrent Pos:* Vis prof elec eng, Univ Alaska, 74-75; vis engr, Air Force Geophys Lab, Bedford, Mass, 82-83. *Mem:* Am Soc Eng Educ; sr mem Inst Elec & Electronics Engrs. *Res:* Analysis and instrumentation in information theory, networks and radio telemetry; development of airborne instrumentation and radio telemetry systems for high altitude balloons, sounding rockets, shuttle and orbital vehicles; systems incorporating microprocessor control to change experiment in flight as function of experimental data received. *Mailing Add:* Dept Elec Eng 360 Huntington Ave Boston MA 02115

ROCHEFORT, JOSEPH GUY, b Astorville, Ont, July 5, 29; m 54; c 2. BIOCHEMISTRY, ENDOCRINOLOGY. *Educ:* Laurentian Univ, BA, 51; McGill Univ, BSc, 54, MSc, 56, PhD(biochem), 58. *Prof Exp:* Biochemist, Regional Labs, Dept Health & Welfare, NB, 54-55; Nat Mutiple Sclerosis Soc fel, 58-60; sr scientist, Dept Pharmacol, Abbott Labs, 60-65 & Dept Biochem, 66-69, asst dir, Dept Clin Pharmacol, 69-76, dir dept Clin Res, Can, 76-83; res projects coordinator, Sanofi Recherche, Montpellier, France, 83-87; PRES, PHARMEDICAL CONSULTS INC, 88- *Concurrent Pos:* Lectr, Concordia Univ, 64-79. *Mem:* Endocrine Soc; Can Physiol Soc; Can Fertil Soc; Am Fertil Soc; Can Soc Clin Invest; Soc Toxicol Can. *Res:* Anterior pituitary-adrenocorticotrophic hormone distribution and release; adrenal responses to stress; bioassay of synthetic and natural steroid hormones; adrenal steriodogenesis; biochemistry of inflammation; bioavailability; pharmacokinetics; clinical investigation of new drugs. *Mailing Add:* 3450 King Edward Ave Montreal PQ H4B 2H2 Can

ROCHELEAU, ROBERT, environmental sciences; deceased, see previous edition for last biography

ROCHELLE, ROBERT W(HITE), b Nashville, Tenn, June 23, 23; m 49; c 4. ELECTRICAL ENGINEERING. *Educ:* Univ Tenn, BS, 47; Yale Univ, ME, 49; Univ Md, PhD(elec eng), 63. *Prof Exp:* Electronic scientist, US Naval Res Lab, 49-55, head magnetic amplifier sect, 55-58; br head flight data systs br, Goddard Space Flight Ctr, NASA, 58-71, assoc chief commun & navig div, 71-73; prof elec eng, Univ Tenn, 73-89; DIR, RES & DEVELOP, EMPRUVE, INC, 89- *Concurrent Pos:* Lectr, Univ Md, 57-59. *Honors & Awards:* Medaille du CNES, France, 65. *Mem:* AAAS; fel Inst Elec & Electronics Engrs; Am Soc Eng Educ; Sigma Xi. *Res:* Application of microcomputers in space and industrial instrumentation. *Mailing Add:* Empruve Inc 1016 Weisgarber Rd Suite 260 Knoxville TN 37909

ROCHESTER, DUDLEY FORTESCUE, b Bennington, Vt, May 21, 28; m 50; c 2. PULMONOLOGY, RESPIRATORY MUSCLE PHYSIOLOGY. *Educ:* Columbia Col, AB, 50; Columbia Univ, MD, 55. *Prof Exp:* Instr med, Col Physicians & Surgeons, Columbia Univ, 62-63, assoc, 63-65, from asst prof to assoc prof, 65-76; PROF MED & HEAD PULMONARY DIV, DEPT INTERNAL MED, SCH MED, UNIV VA, 76- *Concurrent Pos:* Physician, Univ Va Hosp, 76- *Mem:* Fel Am Col Physicians; fel Am Col Chest Physicians; Am Physiol Soc; Am Thoracic Soc; Am Fedn Clin Res. *Res:* Function of respiratory muscles and diaphragm with regard to strength, endurance and susceptibility to fatigue in normal humans and patients with chronic pulmonary disease. *Mailing Add:* 103 Shawnee Crt Charlottesville VA 22901

ROCHESTER, EUGENE WALLACE, b Greenville, SC, July 15, 43; m 68; c 2. AGRICULTURAL ENGINEERING. *Educ:* Clemson Univ, BS, 65; NC State Univ, MS, 68, PhD(biol & agr eng), 70. *Prof Exp:* ASSOC PROF AGR ENG, AUBURN UNIV, 70- *Concurrent Pos:* Consult, Irrig Syst Design, 78-; William Howard Smith fac fel award, Sch Agr & Agr Exp Sta, Auburn Univ, 76. *Mem:* Am Soc Agr Engrs; Irrig Asn; Am Soc Civil Engrs. *Res:* Field crop irrigation, especially machinery types, energy requirements and systems for irregular fields. *Mailing Add:* Dept Agr Eng Auburn Univ Auburn AL 36849

ROCHESTER, MICHAEL GRANT, b Toronto, Ont, Nov 22, 32; m 58; c 3. GEOPHYSICS, ASTRONOMY. *Educ:* Univ Toronto, BA, 54, MA, 56; Univ Utah, PhD(physics), 59. *Prof Exp:* Lectr physics, Univ Toronto, 59-60, asst prof, 60-61; from asst prof to assoc prof, Univ Waterloo, 61-67; assoc prof, Mem Univ Nfld, 67-70, prof physics, 70, prof earth sci, 82, UNIV RES PROF, MEM UNIV NFLD, 86- *Concurrent Pos:* Mem, Working Group Physical Processes in Earth's Interior, Int Geodynamics Proj, 71-79, Comn Rotation of Earth, Int Astron Union, 73- & Can Subcomt Geodynamics, 74-80; vis prof,

York Univ, 74-75, 82-83 & Univ Queensland, 77; mem, Can Nat Comt Int Union Geodesy & Geophys, 74-75, 84-88; mem, Nat Sci Eng Res Coun Earth Sci Grant Selection Comt, 79-82; mem, Bd Dir Lithoprobe Proj, 87- *Honors & Awards:* Tuzo Wilson medal, Can Geophysical Union, 86. *Mem:* AAAS; Am Geophys Union; Can Asn Physicists; Royal Astron Soc; Can Geophys Union; Sigma Xi; fel Royal Soc Can. *Res:* Rotation of the earth; earth tides; dynamics of the earth's core; geomagnetism; planetary physics. *Mailing Add:* Dept Earth Sci Mem Univ Nfld St John's NF A1B 3X5 Can

ROCHLIN, PHILLIP, b New York, NY, Mar 24, 23; div; c 2. CHEMISTRY. *Educ:* City Col New York, BS, 43; NY Univ, MS, 49; Rutgers Univ, MLS, 60. *Prof Exp:* Analytical chemist, var cos, 43-49; res chemist, Picatinny Arsenal, NJ, 50-63; sci analyst, Nat Referral Ctr Sci & Technol, Libr Cong, DC, 63; supvry chemist & mgr tech libr, Naval Propellant Plant, Indian Head, 63-68; chief accessions & indexing br, Nat Hwy Safety Inst Doc Ctr, 68-69; supvry chemist & dir, Tech Info Div, Naval Ord Sta, Indian Head, 69-84, chemist & tech info specialist, Environ & Energy Off, 79-84; EXEC DIR, ASN RECORDED SOUND COLLECTIONS, 85- *Concurrent Pos:* Instr library Sci, Charles County Community Col. *Mem:* Emer mem Am Chem Soc; Asn Recorded Sound Collections. *Res:* Explosives and propellants; information storage and retrieval; missiles and rockets. *Mailing Add:* Asn Recorded Sound Collections PO Box 10162 Silver Spring MD 20914

ROCHLIN, ROBERT SUMNER, b Yonkers, NY, June 25, 22; m 50; c 2. NUCLEAR PHYSICS, ARMS CONTROL. *Educ:* Cornell Univ, BEE, 44, PhD(physics), 52. *Prof Exp:* Radio engr, US Naval Res Lab, 44-45; physicist, Gen Elec Co, 51-63; MEM STAFF, US ARMS CONTROL & DISARMAMENT AGENCY, 63-, CHIEF SCIENTIST, 82- *Concurrent Pos:* Mem US del, US-Soviet Strategic Arms Limitation Talks, Vienna, Austria, 70 & Nuclear Fuel Cycle Eval, 78-80. *Res:* Arms control; negotiations, research and policy formulation. *Mailing Add:* 2709 Ross Rd Chevy Chase MD 20815

RO-CHOI, TAE SUK, b Seoul, Korea, May 8, 37; m 67; c 2. BIOCHEMISTRY, PHARMACOLOGY. *Educ:* Soo Do Med Col, Korea, MD, 62; Baylor Col Med, MS, 64, PhD(pharmacol), 68. *Prof Exp:* Teaching asst, 62-67, from instr to asst prof, 67-72, RES ASSOC PROF PHARMACOL, BAYLOR COL MED, 72- *Concurrent Pos:* On leave, Baylor Col Med, 78- *Mem:* AAAS; Am Soc Biol Chemists; Am Soc Pharmacol & Exp Therapeut; Am Asn Cancer Res. *Res:* Nuclear RNA of cancer and normal cells; primary sequence of nuclear low molecular weight RNA; nuclear and nucleolar RNA transcription and control mechanism of gene expression; ribonucleoprotein complex and their functions. *Mailing Add:* 4147 Martinshire Houston TX 77025

ROCHOVANSKY, OLGA MARIA, b New York, NY. BIOCHEMISTRY, VIROLOGY. *Educ:* Queens Col, BS, 50; NY Univ, PhD(biochem), 60. *Prof Exp:* From asst to assoc biochem, Pub Health Res Inst New York, Inc, 60-71, assoc virol, 71-75, assoc mem virol, 75-77; mem, Christ Hosp Inst Med Res, 77-87; RETIRED. *Concurrent Pos:* Fel, Univ Calif, Berkeley, 61-62; Nat Inst Allergy & Infectious Dis grant, 79. *Mem:* Am Soc Microbiol; AAAS; Am Soc Biol Chemists; Harvey Soc. *Res:* Studies on influenza viral RNA and protein synthesis in vitro. *Mailing Add:* 3918 Chesterwood Dr Silver Spring MD 20906

ROCHOW, EUGENE GEORGE, b Newark, NJ, Oct 4, 09; m 35, 52; c 3. CHEMISTRY. *Educ:* Cornell Univ, BChem, 31, PhD(chem), 35. *Hon Degrees:* MA, Harvard Univ, 48; Dr rer nat, Brunswick Tech Univ, 66. *Prof Exp:* Res chemist, Halowax Corp, NJ, 31-32; asst chem, Cornell Univ, 32-35; chemist, Res Lab, Gen Elec Co, 35-48; from assoc prof to prof, 48-70, EMER PROF INORG CHEM, HARVARD UNIV, 70- *Concurrent Pos:* Mem, Nat Res Coun, 48. *Honors & Awards:* Baekeland Medal, 49; Matiello Award, 58; Perkin Medal, 62; Kipping Award, 65; Norris Award, Am Chem Soc, 74; Stock Prize, German Chem Soc, 83. *Mem:* AAAS; Am Chem Soc; Am Inst Chemists; French Soc Indust Chemists; Int Acad Law & Sci. *Res:* Organosilicon chemistry and silicones; inorganic chemistry. *Mailing Add:* Myerlee Manor 107 1499 Brandywine Circle Ft Myers FL 33919-6764

ROCHOW, THEODORE GEORGE, b Newark, NJ, July 8, 07; m 58; c 1. CHEMICAL MICROSCOPY, RESINOGRAPHY. *Educ:* Cornell Univ, BChem, 29, PhD(chem), 34. *Prof Exp:* Group leader micros, Am Cyanamid Co, 34-55, projs mgr chem, 55-56, res fel, 56-69; assoc prof, 69-74, EMER ASSOC PROF TEXTILE TECHNOL, NC STATE UNIV, RALEIGH, 74- *Honors & Awards:* Templin Award, Am Soc Testing & Mat, 73. *Mem:* Am Chem Soc; Am Soc Testing & Mat; Electron Micros Soc Am; Optical Soc Am; Am Inst Physics. *Res:* Chemical microscopy; resinography. *Mailing Add:* Wedgwood Apts No 33 704 Smallwood Dr Raleigh NC 27605-1346

ROCHOW, WILLIAM FRANTZ, b Lancaster, Pa, Mar 12, 27; m 53; c 2. PLANT PATHOLOGY. *Educ:* Franklin & Marshall Col, BS, 50; Cornell Univ, PhD(plant path), 54. *Prof Exp:* Asst plant path, Cornell Univ, 50-54; from asst prof to prof plant path, Cornell Univ, 55-87; RETIRED. *Concurrent Pos:* Nat Found fel, Univ Calif, 54-55; plant pathologist, USDA, 55-86. *Honors & Awards:* Superior Serv Award, USDA, 66; Ruth Allen Award, Am Phytopath Soc, 85. *Mem:* Fel AAAS; fel Am Phytopath Soc; Am Inst Biol Sci; Am Soc Virol. *Res:* Plant virology, especially virus-vector relationships. *Mailing Add:* 48 Woodcrest Ave Ithaca NY 14850

ROCK, BARRETT NELSON, b Warren, Ohio, Sept 8, 42; m 67; c 2. PLANT ANATOMY. *Educ:* Univ Vt, BA, 66; Univ Md, MS, 70, PhD(bot), 72. *Prof Exp:* Asst prof, 72-78, ASSOC PROF BIOL, ALFRED UNIV, 78- *Concurrent Pos:* Field Botanist, Columbia Gas Corp, Ohio, 78-; sabbatical leave, Dept Bot, Univ Calif, Davis, 80. *Mem:* Bot Soc Am; Int Asn Wood Anatomists. *Res:* Anatomical study of vegetative plant tissue, including wood, leaves, and stem tips; megaphytic members of the Asteraceae (compositae); remote sensed vegetation data. *Mailing Add:* 20 Pinecrest Lane Durham NH 03824-3113

ROCK, CHET A, b Vancouver, Wash, Dec 8, 44; m. WATER POLLUTION, WATER QUALITY. *Educ:* Wash State Univ, BS, 68; Stanford Univ, MS, 71; Univ Wash, PhD(environ eng), 74. *Prof Exp:* Sanitary eng, Environ Sanitation Prog, USPHS, 68-70; sanitary eng, Lake Restoration Sect, Water Quality Div, Dept Ecol, State Wash, 74-76 & Indust Waste Div, 76-79, asst prof, 79-84; asst chmn, 84, ASSOC PROF ENVIRON ENG, DEPT CIVIL ENG, UNIV MAINE, 85- *Mem:* Am Soc Civil Engrs (pres, Maine Sect, 85-86); Water Pollution Control Fedn; Asn Environ Eng Prof; Int Peat Soc; NAm Lake Mgt Soc. *Res:* Onsite wastewater treatment; treatment of industrial wastes; ecological effects of wastewater; restoration of eutrophic lakes; constructed welands. *Mailing Add:* Dept Civil Eng Univ Maine 457 Aubert Hall Orono ME 04469

ROCK, ELIZABETH JANE, b Plattsburgh, NY, Dec 14, 24. PHYSICAL CHEMISTRY. *Educ:* Col Mt St Vincent, BS, 46; Smith Col, MA, 48; Pa State Col, PhD(chem), 51. *Prof Exp:* Asst, Cryogenic Lab, Pa State Col, 48-51, res assoc, Solid State Lab, 51-52; instr chem, Vassar Col, 52-55; from assoc prof to prof textiles, Univ Tenn, 55-59; lectr, 59-61, from assoc prof to prof, 61-70, chmn dept, 67-70, dir sci ctr, 73-75 & 76-78, ARTHUR J & NELLIE Z COHEN PROF CHEM, WELLESLEY COL, 70- *Concurrent Pos:* Textile chemist, Exp Sta, Univ Tenn, 55-59; NSF sci fac fel thermochem, Oxford Univ, 66-67; extramural assoc, NIH, 79; vis res prof, Tufts Univ, 81-82. *Mem:* Am Chem Soc. *Res:* Physical chemistry of conservation of stone in monuments; infrared laser induced reactions. *Mailing Add:* Dept Chem Wellesley Col Wellesley Boston MA 02181-8249

ROCK, GAIL ANN, b Winnipeg, Man, Oct 20, 40; m 66; c 2. TRANSFUSION MEDICINE, BLEEDING DISORDERS. *Educ:* St Patricks Col, BSc, 62; Univ Ottawa, PhD(biochem), 66 & MD, 72. *Prof Exp:* Postdoctoral biophys, Nat Res Coun Can, 65-67; fel biochem, Univ Ottawa, 68-69; med intern, Ottawa Gen Hosp, 72-73, physician emergency med, 73-74; MED DIR, CAN RED CROSS BLOOD TRANSFUSION SERV, OTTAWA, 74- *Concurrent Pos:* Mem matching grants review comt & prog site review comt, Am Red Cross, 83, subcomt factor eight & von Willebrand factor, Int Soc Thrombosis & Hemostasis, prog comt, Haemonetics Res Inst, 81-84, 86 & 87, ad hoc comt, Estab World Asn Apheresis, 85, fractionator, supplier & user working group, Can Red Cross, planning prog comt, First Meeting Int Soc Apheresis, 86, subcomt transfusion med, Am Soc Hemat, 87-; chmn, estab investr award, initial review group, Am Red Cross, 84-88, sci prog comt, Second Int Meeting, 86-88, sci prog comt ann meeting, Am Asn Blood Banks, 85-; invited consult, blood transfusion prog, Cuba, & Int Red Cross, Brazil, 84; consult, dept med & lab med, Univ Ottawa, 78-; adj prof, dept biochem, Univ Ottawa, 77-, dept lab med, 78- clin assoc prof, dept med, 86-; prin investr of more than twenty-five grants contracts from various study groups & health & welfare found in Can & Am, 77-88. *Mem:* World Apherris Asn (pres, 88-90); Am Asn Blood Banks; Am Soc Apheresis; Can Aphersis Study Group; In Soc Blood Transfusion. *Res:* Biochemistry of factor VIII and other blood clotting factors and their interaction with platelets; improving yields of fractionated coagulation products; optimization of platelet storage conditions; toxicology of plasticizers in blood products; role of vascular endothelium in hemostasis and thrombosis and apheresis applications. *Mailing Add:* 270 Sandridge Rd Ottawa ON K1L 5A2 Can

ROCK, GEORGE CALVERT, b Roanoke, Va, May 26, 34; m 70. ENTOMOLOGY. *Educ:* Bob Jones Univ, BS, 57; Va Polytech Inst & State Univ, MS, 60; Cornell Univ, PhD(entom), 63. *Prof Exp:* Asst prof entom, Va Poltech Inst & State Univ, 64-67; from asst prof to assoc prof, 67-76, PROF ENTOM, NC STATE UNIV, 76- *Mem:* Entom Soc Am; Sigma Xi. *Res:* Nurtitional requirements of insects; pest population management in apple orchards; insect resistance to insecticides. *Mailing Add:* Dept Entom NC State Univ Box 7626 Raleigh NC 27695

ROCK, MICHAEL KEITH, b Milwaukee, Wis, Aug 26, 51; m 78. NEUROSCIENCE. *Educ:* Univ Dallas, BA, 73; Univ Tex Med Br, Galveston, PhD(physiol), 77. *Prof Exp:* Res assoc neurophysiol, Sch Med, Washington Univ, 77-79 & Univ Va, 79-80; assoc mem neurophysiol, Marine Biomed Inst, 80-81; asst prof, 81-85, ASSOC PROF, SCH ALLIED HEALTH SCI, UNIV TEX MED BR GALVESTON, 85- *Concurrent Pos:* Adj mem, Marine Biomed Inst, 81- *Mem:* Sigma Xi; Soc Neurosci. *Res:* Neurophysiology of simple nervous systems. *Mailing Add:* Sch Allied Health Sci Univ Tex Med Br Galveston TX 77550

ROCK, PAUL BERNARD, b Kansas City, Mo, Dec 15, 45. PHYSIOLOGY, MEDICINE. *Educ:* Univ Colo, BA, 69, MA, 72, PhD(biol), 80; Chicago Col Ostheo Med, MD, 80. *Prof Exp:* Investr, US Army Res Inst Environ Med, 81-87; med resident, Fitzsimons Army Med Ctr, 87-90; INVESTR, ALTITUDE RES DIV, US ARMY RES INST ENVIRON MED, 90- *Mem:* Sigma Xi; Am Soc Mammalogists; Am Col Physicians. *Res:* Investigation of environmental medical problems and physiology with an emphasis in medical problems in high terrestrial altitude. *Mailing Add:* Altitude Res Div US Army Res Inst Environ Med Kansas St Natick MA 01760

ROCK, PETER ALFRED, b New Haven, Conn, Sept 29, 39; m 59; c 3. ELECTROCHEMISTRY, COMBUSTION PROCESSES. *Educ:* Boston Univ, AB, 61; Univ Calif, Berkeley, PhD(chem), 64. *Prof Exp:* From asst prof to assoc prof, 64-75, chmn dept, 80-85, PROF CHEM, UNIV CALIF, DAVIS, 75- *Concurrent Pos:* Nat Inst Neurol Dis & Stroke fel, Ind Univ, 70-71; consult, Dorland Med Dictionaries & World Book Encycl. *Mem:* AAAS; Am Chem Soc; Sigma Xi. *Res:* Chemical thermodynamics; isotope effects; oscillatory phenomena; energy utilization and conversion; combustion reactions. *Mailing Add:* Dept Chem Univ Calif Davis CA 95616

ROCKAFELLAR, RALPH TYRRELL, b Milwaukee, Wis, Feb 10, 35; m 64; c 3. MATHEMATICS. *Educ:* Harvard Univ, AB, 57, PhD(math), 63. *Prof Exp:* Teaching fel math, Harvard Univ, 60-62; mem res staff, Mass Inst Technol, 62-63; asst prof, Univ Tex, 63-65; vis asst prof, Princeton Univ, 65-66; from asst prof to assoc prof, 66-73, PROF MATH & ADJ PROF COMPUT SCI, UNIV WASH, 73- *Concurrent Pos:* Univ Tex res grant, Math

Inst, Copenhagen, 64; Air Force Off Sci Res grants, Princeton Univ, 66 & Univ Wash, 66-68. *Honors & Awards:* Dantzig Prize, 82. *Mem:* Soc Indust & Appl Math. *Res:* Optimization; variational analysis; convexity. *Mailing Add:* Dept Math Univ Wash-GN-50 Seattle WA 98195

ROCKAWAY, JOHN D, JR, b Cincinnati, Ohio, Mar 7, 38; m 62; c 4. GEOLOGICAL ENGINEERING. *Educ:* Colo Sch Mines, BS, 61; Purdue Univ, MSE, 63, PhD(civil eng), 68. *Prof Exp:* Instr geol, Purdue Univ, 65-68; from asst prof to assoc prof, 68-76, chmn dept, 81-87, PROF GEOL ENG, UNIV MO-ROLLA, 76- *Concurrent Pos:* Eng geologist, US Army CEngr, 76- *Mem:* Asn Eng Geol; Am Inst Mining, Metall & Petrol Engrs; Int Asn Eng Geologists; Geol Soc Am. *Res:* Geotechnical studies of the physical and engineering properties of earth materials; the impact of land-use and development upon the geological environment. *Mailing Add:* Dept Geol Eng Univ Mo Rolla MO 65401

ROCKCASTLE, VERNE NORTON, b Rochester, NY, Jan 1, 20; m 43; c 2. ECOLOGY. *Educ:* Syracuse Univ, AB, 42; Mass Inst Technol, SM, 44; Cornell Univ, PhD, 55. *Prof Exp:* Instr meteorol, Mass Inst Technol, 43-44; from asst prof biol to assoc prof biol, State Univ NY, Brockport, 47-56; from assoc prof to prof, 59-86, EMER PROF SCI & ENVIRON EDUC, CORNELL UNIV, 86- *Honors & Awards:* Carleton Award, Nat Sci Teachers Asn, 90. *Mem:* Am Nature Study Soc (pres, 65); fel AAAS. *Res:* Elementary science education; concept development in science. *Mailing Add:* 102 Sunset Dr Ithaca NY 14850

ROCKE, DAVID M, b Chicago, Ill, June 4, 46; m 71; c 2. ROBUST STATISTICS, QUALITY CONTROL. *Educ:* Shimer Col, AB, 66; Univ Ill, Chicago Circle, MA, 68, PhD(math), 72. *Prof Exp:* Vis lectr math, Univ Ill, Chicago Circle, 72-74; univ prof bus admin, Col Bus & Pub Serv, Goveners State Univ, 74-80; assoc prof, grad sch mgt, 80-86, PROF GRAD SCH MGT, UNIV CALIF, DAVIS, 86- *Honors & Awards:* Youden Prize, 82 & Shewell Award, Chem Div, Am Soc Qual Control, 86; Interlab Testing Award, Am Statist Asn, 85. *Mem:* Math Asn Am; Am Math Soc; Am Statist Asn; AAAS; Inst Math Statist; Royal Statist Soc; Am Soc Qual Control; Soc Indust & Appl Math. *Res:* Robust statistical methods; quality control and inprovement; statistical computing; statistics and public policy. *Mailing Add:* Grad Sch Mgt Univ Calif Davis CA 95616

ROCKETT, JOHN A, b Philadelphia, Pa, Aug 6, 22; m 56; c 2. FLUID MECHANICS. *Educ:* Mass Inst Technol, BS, 44; Brown Univ, MS, 51; Harvard Univ, PhD(appl physics), 57. *Prof Exp:* Res engr, Nat Adv Comt Aeronaut, 47-49; res engr, Dept Aeronaut, Mass Inst Technol, 50-53; res engr, United Aircraft Corp res proj, Harvard Univ, 53-57; chief fuel cell technol, United Aircraft Corp, 57-65; dir basic res, Factory Mutual Eng Corp, Mass, 65-68; spec asst to dir, Inst Appl Technol, 68, chief, Off Fire Res & Safety, 68-72, chief fire physics res, 73-80, sr scientist, Ctr Fire Res, Nat Bur Standards, 81-86; CONSULT, 86- *Concurrent Pos:* Vis prof, Tokyo Sci Univ, 82, 84, vis fel, Nat Bldg Technol Ctr, Sydney, 87; vis scientist, VTT, Espoo Helsinki, Finland, 91. *Honors & Awards:* Silver Medal, Dept Commerce, 77. *Mem:* Fire Protection Engrs. *Res:* Internal and external aerodynamics; hydrodynamics; combustion and flames; solid state physics; electrochemistry; mass-heat transfer; fire behavior. *Mailing Add:* 4701 Alton Pl NW Washington DC 20016-2041

ROCKETT, THOMAS JOHN, b Medford, Mass, June 4, 34; m 64; c 3. MINERALOGY, CERAMICS, METALS & POLYMERS. *Educ:* Tufts Univ, BS, 56; Boston Col, MS, 58; Ohio State Univ, PhD(mineral eng), 63. *Prof Exp:* Res mineralogist, Ohio State Univ, 59-61; res ceramist, Wright-Patterson AFB, 61-65; sr res ceramist, Monsanto Res Co, Mass, 65-67, scientist, New Enterprises Div, Monsanto Corp, 67-72; assoc prof, 72-78, PROF MAT & CHEM ENG, UNIV RI, 78- *Concurrent Pos:* NSF fel, 63; lectr, Univ Dayton, 63-65 & Boston Col, 65-71; chmn, Univ RI, 81-87. *Mem:* Mineral Soc Am; Am Ceramic Soc; Am Chem Soc; Am Soc Metals. *Res:* High temperature phase equilibria; composite materials; water/polymer interactions; glass systems and phase separation in glasses; ceramic composites; dental cement chemistry; oxide fiber; laser-glass interactions; polymer stability; mineralogy; petrology. *Mailing Add:* Dept Chem Eng Univ RI Kingston RI 02881-0805

ROCKETTE, HOWARD EARL, JR, b Baltimore, Md, Feb 6, 44; m 68; c 3. BIOSTATISTICS. *Educ:* Franklin & Marshall Col, BA, 65; Pa State Univ, MA & PhD(statist), 72. *Prof Exp:* PROF BIOSTATIST, UNIV PITTSBURGH, 83-; DIR OPERS RES, INST INT NATURAL FAMILY PLANNING, 85- *Concurrent Pos:* Dir, Statist Unit, Otititis Media Res Ctr, 81-, co-dir, Statist Ctr, Nat Surg Adjuvant Breast Proj, 82- *Mem:* Am Statist Asn; Biomet Soc; Inst Math Statist; Am Pub Health Asn; Soc Occup & Environ Health. *Res:* Development of methodological techniques and the evaluation and collection of data in the fields of biology, medicine and health, particularly in the areas of clinical trial evluation and occupational health. *Mailing Add:* Dept Biostatist 318C Parron Hall Grad Sch Pub Health Univ Pittsburgh Pittsburgh PA 15261

ROCKEY, JOHN HENRY, b Madison, Wis, Feb 2, 31. OPHTHALMOLOGY, IMMUNOCHEMISTRY. *Educ:* Univ Wis, BS, 52, MD, 55; Univ Pa, PhD(molecular biol), 68. *Prof Exp:* Intern, Hosp Univ Pa, 55-56; asst resident path, Cornell Med Sch-New York Hosp, 58-59; res assoc & asst physician, Rockefeller Inst Hosp, 59-62; asst prof microbiol, 62-69, assoc med, 66-70, from asst prof to assoc prof ophthal, 69-73, PROF OPHTHAL, SCH MED, UNIV PA, 73- *Concurrent Pos:* Prin investr, USPHS-NIH grants, 62-75; mem grad group molecular biol, Univ Pa, 69-, mem grad group path, 71- & mem grad group immunol, 72-; mem sci staff med res coun molecular pharmacol unit, Med Sch, Cambridge Univ, 70; co-investr, NSF grant, 70-72. *Honors & Awards:* William J Bleckwenn Award, 55. *Mem:* Am Asn Immunol. *Res:* Reaginic antibodies; primary structural studies of visual pigments; multiple molecular forms of antibodies. *Mailing Add:* Dept Ophthal Univ Pa Scheie Eye Inst 51 N 39th St Philadelphia PA 19104

ROCKHILL, THERON D, b Malone, NY, Feb 9, 37; m 58; c 2. MATHEMATICS. *Educ:* Houghton Col, BA, 59; Syracuse Univ, MS, 62; State Univ NY Buffalo, EdD(math & educ), 69. *Prof Exp:* Teacher math, Newfield Cent Sch, 59-61; from asst prof to assoc prof, 62-72, chmn dept, 77-81, PROF MATH, STATE UNIV COL BROCKPORT, 72- *Concurrent Pos:* Res assoc, Ctr Res Col Instr Sci & Math, Fla State Univ, 71; pres, Asn Math Teachers NY, 85-86; NSF sci fac fel. *Mem:* Am Math Soc; Math Asn Am; Nat Coun Teachers Math; Soc Indust & Appl Math. *Res:* Numerical analysis and the use of computers in teaching mathematics; author of one book. *Mailing Add:* 15 Sherwood Dr Brockport NY 14420

ROCKHOLD, ROBIN WILLIAM, b Dayton, Ohio, Sept 29, 51; m 78; c 1. HYPERTENSION, NEUROPHARMACOLOGY. *Educ:* Kenyon Col, AB; Univ Tenn, PhD(pharmacol), 78. *Prof Exp:* Res fel pharmacol, Pharmacol Inst Univ Heidelberg, WGermany, 78-79; res fel physiol, dept physiol & biophysics, Univ Tenn Ctr Health Sci, 79-82, asst prof, 82-83; asst prof, 83-88, ASSOC PROF PHARMACOL, DEPT PHARMACOL & TOXICOL, UNIV MISS MED CTR, 88- *Mem:* Am Heart Asn; Soc Neurosci; Am Soc Pharmacol & Exp Therapeut. *Res:* Regulation of circulatory homeostasis by the central nervous system in normal and pathophysiological conditions such as hypertension; role of neuropeptides and amino acids in specific hypothalamic nuclei. *Mailing Add:* Dept Pharmacol & Toxicol Univ Miss Med Ctr 2500 State St Jackson MS 39216

ROCKLAND, LOUIS B, b NY, July 14, 19; m 43; c 3. AGRICULTURAL CHEMISTRY, FOOD SCIENCE. *Educ:* Univ Calif, Los Angeles, BA, 40, MA, 47, PhD(phys & biol sci, chem), 48. *Prof Exp:* Asst chem, Univ Calif, Los Angeles, 40-47, res assoc filter paper chromatog, 49-50; res chemist, Fruit & Veg Chem Lab, Western Utilization Res & Develop Div, Agr Res Serv, USDA, Calif, 49-70, res chemist, Western Regional Res Ctr, 70-80; prof & chmn, Dept Food Sci & Nutrit, 80-83, dir, Food Sci Res Ctr, 83-88, EMER DIR, FOOD SCI RES CTR, CHAPMAN COL, 88-; PRES, FOODTECH RES & DEVELOP, 88- *Concurrent Pos:* Chmn, Second Int Symp Properties of Water, Osaka, Japan, 78. *Honors & Awards:* Special Award, Inst Food Technologists, 83. *Mem:* Fel AAAS; Am Chem Soc; fel Inst Food Technologists; Am Soc Biol Chemists; Am Asn Cereal Chemists. *Res:* Chemical, biological and physical properties of dry legume seeds; chemical properties of amino acids and proteins; methods for filter paper, thin layer and gas chromatography; citrus fruits and walnuts; lemon oil; moisture sorption; water activity-stability relationships in foods and natural products; food product developement. *Mailing Add:* 45 Corliss Dr Moraga CA 94556

ROCKLIN, ALBERT LOUIS, b Toronto, Ont, May 28, 21; nat US; m 46; c 2. PHYSICAL CHEMISTRY, INDUSTRIAL CHEMISTRY. *Educ:* Univ Toronto, BA, 43, MA, 44, PhD(phys chem), 46. *Prof Exp:* Asst prof chem, exten, Purdue Univ, 46-47, univ, 47-50; res chemist, Western Div, Dow Chem Co, 51-58; chemist, 58-72, sr res chemist, Shell Develop Co, 72-86; RETIRED. *Concurrent Pos:* Tech writing consult, 87- *Honors & Awards:* Roon Found Awards Coatings Technol, 76, 78, 80 & 85. *Mem:* Fel AAAS; fel Am Chem Soc; Fedn Socs Coatings Technol. *Res:* Applied solution theory relating to surface coatings, solubility, solvent evaporation from thin films, and solution viscosity; development of performance materials. *Mailing Add:* PO Box 3581 Laguna Hills CA 92654-3581

ROCKLIN, ISADORE J, b Chicago, Ill, Nov 9, 07; m 39; c 1. ROCKLINIZING, ELECTRONIC DISCHARGE MACHINING. *Educ:* Univ Iowa, BS, 30. *Prof Exp:* Mfg engr, Western Elec Co, 30-32; head, Indust Eng, Jensen Radio Corp, 31-32; asst chief engr, Grigsby Grunow Co, 32; chief engr, Sioux City Foundry & Boiler Co, 32-34; GEN MGR, ROCKLIN MFG CO, 34- *Mem:* Nat Soc Prof Engrs; Am Soc Mech Engrs; Soc Mfg Engrs; Am Inst Indust Engrs. *Res:* Electronic application of spark discharge machining. *Mailing Add:* 110 S Jennings St Sioux City IA 51104

ROCKLIN, ROSS E, LYMPHOKINES, IMMUNO REGULATION. *Educ:* Howard Univ, MD, 66. *Prof Exp:* Chief, Div Allergy, New Eng Med Ctr Hosp, 78-89; prof med allergy, Sch Med, Tufts Univ, 82-89; SR ASSOC DIR, CLIN RES, BOEHRINGER INGELHEIM CORP, 89- *Mailing Add:* Boehringer Ingelheim Corp 90 E Ridge PO Box 368 Ridgefield CT 06877

ROCKLIN, ROY DAVID, b San Francisco, Calif, Aug 3, 53. ELECTROANALYTICAL CHEMISTRY, ION CHROMATOGRAPHY. *Educ:* Univ Calif, Santa Cruz, AB, 75; Univ NC, Chapel Hill, PhD(analytical chem), 80. *Prof Exp:* Res chemist, 80-82, SR CHEMIST, MKT APPLNS, DIONEX CORP, 82- *Mem:* Am Chem Soc. *Res:* Ion chromatography; electrochemical detectors for liquid and ion chromatography; electroanalytical instrumentation. *Mailing Add:* Dionex Corp PO Box 3603 Sunnyvale CA 94088-3603

ROCKMORE, RONALD MARSHALL, b New York, NY, Aug 10, 30; m 60; c 2. THEORETICAL MEDIUM ENERGY PHYSICS. *Educ:* Brooklyn Col, BS, 51; Columbia Univ, PhD(physics), 57. *Prof Exp:* Asst physics, Columbia Univ, 52-55; NSF fel, 57-58; res assoc, Brookhaven Nat Lab, 58-60; asst prof, Brandeis Univ, 60-63; assoc prof, 63-79, PROF PHYSICS, RUTGERS UNIV, 79- *Concurrent Pos:* Mem, Inst Advan Study, 57-58; vis lectr, Univ Minn, 58; consult, Repub Aviation Corp, 59-60, Rand Corp, 61-72 & Inst Defense Anal, 62-64; vis physicist, Brookhaven Nat Lab, 64 & 71; visitor, Neils Bohr Inst, Copenhagen, Denmark, 66 & Stanford Linear Accelerator Ctr, 67; vis staff mem, Los Alamos Sci Lab, 67, 72-75 & 88, Argonne Nat Lab, 67 & Ctr Theoret Physics, Trieste, 68; vis mem staff, Theoret Physics Inst, Univ Alta, 72-84; vis theorist, SIN, Villigen, 78; Rutgers Univ fac fel, Imp Col, Univ London, 68-69; Rutgers FASP fel, State Univ NY, Stony Brook, 78 & 88; vis theorist, Saclay, Rutgers FASP fel, 88, Gif-sur-Yvette, 79, 81, 82, 83, 84, 85, 87 & 88. *Mem:* Fel Am Phys Soc. *Res:* Field theory; pion physics; many-body problem; strong and weak interactions; medium-energy physics. *Mailing Add:* Dept Physics Rutgers Univ Piscataway NJ 08855

ROCKOFF, MAXINE LIEBERMAN, b Gary, Ind, July 15, 38; m 56; c 3. MATHEMATICS. *Educ:* George Washington Univ, BS, 58; Univ Pa, MA, 60, PhD(math), 64. *Prof Exp:* Programmer, Univ Pa, 58-60, res fel physiol, 60-61; mathematician, Comput Lab, Nat Bur Standards, 61-64; res assoc, Comput Sci Ctr & Inst Fluid Dynamics, Univ Md, 64-65; res assoc epidemiol & pub health, Yale Univ, 65-68; asst prof, Appl Math, Comput & Biomed Comput Lab, Washington Univ, 68-71; health scientist adminr, Nat Ctr Health Serv Res 71-75, prog analyst, Off Planning, Eval & Legis, 75-76, health scientist adminr, Nat Ctr Health Serv Res, 76-78, prog analyst, Off Planning & Eval, Dept Energy, 78-79; vpres planning & res, Corp Pub Broadcasting, 79-80; MGR MKT TECHNOL, MERRILL LYNCH, PEIRCE, FENNER & SMITH, 80- *Concurrent Pos:* Mem, Panel Impact Comput, Math Curriculum, Comt Undergrad Prog Math, 71-72; mem bd trustees, Soc Indust & Appl Math, 76-78, chmn bd, 78; mem, Biotechnol resources Adv Comt, NIH, 78-81. *Mem:* AAAS; Asn Comput Mach; Soc Indust & Appl Math. *Res:* Development of mathematical models for physiological systems; numerical analysis; analysis of health care systems; application of technology and manpower innovations in health care delivery; evaluation of public programs; application of emerging electronic technologies to financial services delivery. *Mailing Add:* Clark Rockoff & Assoc One Sherman Sq New York NY 10023

ROCKOFF, SEYMOUR DAVID, b Utica, NY, July 21, 31; c 3. RADIOLOGY. *Educ:* Syracuse Univ, AB, 51; Albany Med Col, MD, 55; Univ Pa, MSc, 61. *Prof Exp:* Staff radiologist, Clin Ctr, NIH, 61-65; from asst prof to assoc prof radiol, Sch Med, Yale Univ, 65-68; assoc prof, Mallinckrodt Inst Radiol, Sch Med, Wash Univ, 68-71; chmn dept, 71-77, PROF RADIOL, GEORGE WASHINGTON UNIV MED CTR, 71-, HEAD SECT CHEST RADIOL, 77- *Concurrent Pos:* Asst attend radiologist, Yale-New Haven Med Ctr, Conn, 65-68; ed-in-chief, Investigative Radiol, 66-76, emer ed-in chief, 76-; asst radiologist, Barnes & Allied Hosps, St Louis, Mo; consult radiologist, Homer G Phillips Hosp, St Louis & Vet Admin Hosp, 69-71, NIH, Nat Naval Med Ctr, Bethesda, Md & Washington Vet Admin Hosp, DC, 73-, immunodiagnosis, Nat Cancer Inst, NIH, Bethesda, Md, 77 & asbestosis res, 83-; exec dir, Soc Thoracic Radiol, 83- *Mem:* Fel Am Col Radiol; Am Fedn Clin Res; Asn Univ Radiol; AMA; Radiol Soc NAm; Soc Thoracic Radiol (pres, 83-84). *Res:* Contrast media toxicity; image analysis of radiographs; asbestosis imaging. *Mailing Add:* Dept Radiol George Washington Univ Med Ctr 901 23rd St NW Washington DC 20037

ROCKOWER, EDWARD BRANDT, b Philadelphia, Pa, May 29, 43; m 64, 82; c 2. PHYSICS, OPERATIONS RESEARCH. *Educ:* Univ Calif, Los Angeles, BS, 64; Brandeis Univ, MA, 67, PhD(physics), 75. *Prof Exp:* Res assoc physics, LTV Res Ctr, Western Div, 64-65; vpres, Univ Home Serv, Inc, 71-74; sr opers analyst oper res, Gen Dynamics Corp, 75-77, Ketron, Inc, 77-79; physicist, Laser Isotope Separation Prog, Lawrence Livermore Nat Lab, 79-82; instr physics & computer sci, Univ Md, 82-84; ASSOC PROF OPERS RES, NAVAL POSTGRAD SCH, 84- *Concurrent Pos:* Adj prof mgt sci, MBA prog, LaSalle Col, 77-78. *Mem:* Am Phys Soc; Opers Res Soc Am. *Res:* Generalized Cherenkov radiation of massive fields in plasmas or by faster-than-light particles; applied stochastic processes in lasers and queueing systems; economic and financial analysis; laser propagation, relativity, reliability, search theory. *Mailing Add:* 49 Alta Mesa Circle Monterey CA 93940

ROCKS, LAWRENCE, b New York, NY, Aug 27, 33. CHEMISTRY. *Educ:* Queens Col, NY, BS, 55; Purdue Univ, MS, 57; Vienna Tech Univ, Dr Tech, 64. *Prof Exp:* Asst prof, 58-72, ASSOC PROF CHEM, C W POST COL, LONG ISLAND UNIV, 58- *Mem:* Am Chem Soc. *Mailing Add:* 29 Seville Lane Stony Brook NY 11790-1799

ROCKSTAD, HOWARD KENT, b Ada, Minn, Aug 5, 35. SOLID STATE PHYSICS. *Educ:* St Olaf Col, BA, 57; Univ Ill, Urbana, MS, 59, PhD(physics), 63. *Prof Exp:* Res physicist, Corning Glass Works, NY, 63-70; res physicist, Energy Conversion Devices, Inc, 71-73; sr proj engr, Micro-Bit Div, Control Data Corp, 74-80; sr res scientist, 81-86, SR SYSTS ENGR, JET PROPULSION LAB, ATLANTIC RICHFIELD CO, 87- *Concurrent Pos:* Lectr, Elmira Col, 67. *Mem:* Am Phys Soc; Inst Elec & Electronics Engrs. *Res:* Physics of electron-beam-accessed metal-oxide-semiconductor memories; metal-oxide-semiconductor charge storage; semiconductor device physics; high-electric-field transport; physics of amorphous semiconductors; optical and electronic properties of semiconductors and insulators; color centers; thermoelectricity; microprobe analysis; silicon micromachining. *Mailing Add:* 1227 Tierra Dr Thousand Oaks CA 91362

ROCKSTEIN, MORRIS, b Toronto, Ont, Jan 8, 16; nat US; wid; c 2. PHYSIOLOGY. *Educ:* Brooklyn Col, BA, 38; Columbia Univ, MA, 41; Univ Minn, PhD(insect physiol & biochem), 48. *Prof Exp:* Asst entom, Univ Minn, 41-42; from asst prof to assoc prof zoophysiol, Wash State Univ, 48-53; from asst prof to assoc prof physiol, Sch Med, NY Univ, 53-61; prof physiol & biophys, Univ Miami, 61-81, chmn dept, Sch Med, 67-71, prof nursing, 81-82; pres, Cortisol Med Res, Inc, 83-85; EMER PROF PHYSIOL & BIOL, SCH MED, UNIV MIAMI, 81- *Concurrent Pos:* Nat Res Coun fel in natural sci, 46-48; instr, Marine Biol Lab, Woods Hole, 54-60, trustee, 62-65; mem sci coun, Geront Res Found, 60-63; adv, Nat Inst Aging, 61; consult, Am Pub Health Asn, 61-74; adv study sect trop med, NIH, 62-66; abstractor, Excerpta Medica & Chem Abstr; ed, Biol Sci Bannerstone Lectr Ser Geriat & Geront, 64-72; ed biol sci, Acad Press, 72-; vpres, Int Asn Prolong Human Lifespan; mem sci adv coun, Am Comt, Weizmann Inst, 74-88. *Mem:* Fel AAAS; fel Entom Soc Am; Sigma Xi; Soc Gen Physiol; fel Geront Soc (pres, 65-66); Am Physiol Soc; Sigma Xi; Am Soc Zoologists. *Res:* Insect physiology and enzymology; biochemistry of flight; physiology of aging; radiobiology; marine biology. *Mailing Add:* 600 Biltmore Way Apt 805 Coral Gables FL 33134

ROCKSTROH, TODD JAY, b Terre Haute, Ind, Sept 1, 56; m 80; c 2. HIGH POWER SOLID STATE LASERS. *Educ:* Purdue Univ, BS, 78, MS, 80; Univ Ill, PhD(mech eng), 86. *Prof Exp:* Mem tech staff, Bell Labs, 80-83; mfg engr, 86-88, proj engr, 88-90, STAFF ENGR, GEN ELEC AIRCRAFT ENGINES, 91- *Mem:* Laser Inst Am; Optical Soc Am. *Res:* High power (industrial) solid state lasers, primarily slab-base crystal resonator optimization; non-conventional optical design and high power fiber optic transmission. *Mailing Add:* 3691 Spring Mill Way Maineville OH 45039

ROCKWELL, DAVID ALAN, b Chicago, Ill, Nov 11, 45; m 86; c 1. OPTICAL PHYSICS. *Educ:* Univ Ill, Urbana, BS, 67; Mass Inst Technol, PhD(physics), 73. *Prof Exp:* Res assoc laser mat, Physics Dept, Univ Southern Calif, 73-75; head, Laser Physics Sect, Laser Div, Hughes Aircraft Co, 75-81, sr staff physicist, Hughes res Labs, 81-85, HEAD, ADVAN LASER SOURCES SECT, HUGHES RES LABS, 85- *Mem:* Am Phys Soc; Inst Elec & Electronics Engrs Laser & Elec & Optics Soc; Optical Soc Am. *Res:* Research and development of solid state laser devices; nonlinear and electrooptic devices; high energy laser sources; phase-conjugate lasers. *Mailing Add:* Dept Optical Physics Hughes Res Lab 3011 Malibu Canyon Rd Malibu CA 90265

ROCKWELL, DONALD O, b Canton, Pa, Oct 2, 42; m 72. FLUID MECHANICS. *Educ:* Bucknell Univ, BS, 64; Lehigh Univ, MS, 65, PhD(mech eng), 68. *Prof Exp:* NSF trainee mech eng, Lehigh Univ, 64-67; group leader fluid syst, Harry Diamond Labs, 68-70; asst prof mech eng, Lehigh Univ, 70-72, assoc prof, 72-74; guest prof & Von Humboldt fel, Univ Karlsruhe, Ger, 74-75; PROF MECH ENG, LEHIGH UNIV, 76- *Concurrent Pos:* Co-dir, Volkswagen Found Prog, Univ Karlsruhe & Lehigh Univ, 77-82; NSF grant, Lehigh Univ, 78-83, NASA grant, 78-81; overseas fel, Churchill Col, Cambridge Univ, 81; consult fluid induced vibration & noise generation, var firms in US, Ger, France, Austria & Switz. *Mem:* Am Phys Soc; Am Inst Aeronaut & Astronaut; Am Soc Mech Engrs; Ger Soc Air & Space Travel. *Res:* Unsteady fluid mechanics. *Mailing Add:* Dept Mech Eng Lehigh Univ Packard Lab Bldg 19 Bethlehem PA 18015

ROCKWELL, HARRIET ESTHER, therapeutics, for more information see previous edition

ROCKWELL, JULIUS, JR, b Taunton, Mass, July 25, 18; m 64; c 5. RIPARIAN DESIGN & CONSTRUCTION & RESTORATION, ENVIRONMENTAL MANAGEMENT. *Educ:* Univ Mich, BS, 40; Univ Wash, PhD(fisheries), 56. *Prof Exp:* Sci asst, Int Pac Halibut Fisheries Comn, 46; from jr res asst to res assoc, Fisheries Res Inst, Univ Wash, 46-54; fisheries res biologist & proj leader, Biol Lab Seattle, Bur Com Fisheries, US Fish & Wildlife Serv, 54-59, chief, Fish Counting Prog, 59-62, chief, Oceanog Instrumentation Unit, Wash, DC, 62-67, spec asst, Br Marine Fisheries, 67-70; staff fishery biologist, Alaska Pipeline Off & Off Special Proj, Bur Land Mgt, US Dept Interior, Anchorage, 70-82; ADJ PROF BIOL, ALASKA PAC UNIV, 79- *Concurrent Pos:* Consult riparian restoration speleology. *Honors & Awards:* Superior Performance Award, Bur Com Fisheries, 62. *Mem:* AAAS; Am Fisheries Soc; Am Inst Fishery Res Biologists; Marine Technol Soc; fel Nat Speleology Soc; Soc Ecol Restoration & Mgt. *Res:* Developing methods for riparian habitat protection, evaluation and restoration; data handling and processing systems; operations research; statistical analysis; cave exploration, conservation and management. *Mailing Add:* 2944 Emory St Anchorage AK 99508-4466

ROCKWELL, KENNETH H, b Huntington, Pa, Jan 27, 36; m 58; c 3. ZOOLOGY. *Educ:* Juniata Col, BS, 57; Brown Univ, MS, 60; Pa State Univ, PhD(zool), 67. *Prof Exp:* From instr to assoc prof, 60-72, chmn dept, 69-74, & 79-86, PROF BIOL, JUNIATA COL, 72- *Concurrent Pos:* Vis scholar, Stanford Univ, 74-75 & Univ Calif, Davis, 87. *Mem:* AAAS; Am Inst Biol Sci. *Res:* Histology of endocrine organs; physiology of altitude exposure. *Mailing Add:* Dept Biol Juniata Col 1700 Moore St Huntingdon PA 16652

ROCKWELL, ROBERT FRANKLIN, b Dayton, Ohio, Dec 19, 46. POPULATION GENETICS, POPULATION ECOLOGY. *Educ:* Wright State Univ, BS, 69, MS, 71; Queen's Univ, Ont, PhD(biol), 75. *Prof Exp:* Fel, City Univ New York, 75-76; from asst prof to assoc prof, 76-85, PROF BIOL, CITY COL NEW YORK, 86- *Concurrent Pos:* Res assoc, Am Mus Natural Hist, NY, 86- *Mem:* Genetic Soc Am; Soc Study Evolution; Biomet Soc; Behav Genetics Soc; Am Soc Genetics; Sigma Xi. *Res:* Genetic structure of natural populations; biostatistical methods; population biology of migratory waterfowl; fitness estimation in natural populations. *Mailing Add:* Dept Biol City Col Convent Ave at 138th St New York NY 10031

ROCKWELL, ROBERT LAWRENCE, b Portsmouth, NH, Nov 18, 35; m 60; c 3. AEROSOL FLUID DYNAMICS. *Educ:* Univ Calif, Berkeley, AB, 59; Stanford Univ, MS, 64, PhD(aeronaut & astronaut), 70. *Prof Exp:* Physicist, 58-70, AERONAUT ENGR, NAVAL WEAPONS CTR, CHINA LAKE, 70- *Concurrent Pos:* Lectr mech eng, Univ Southern Calif, 72-74, elec eng, Calif State Univ, Northridge, 80- *Res:* Nonlinear modeling of arterial blood flow, with emphasis on viscous and viscoelastic effects; mechanics of aerosols, particularly fluid dynamic effects on coagulation of chain aggregates. *Mailing Add:* 607 Randall St Ridgecrest CA 93555-3307

ROCKWELL, SARA CAMPBELL, b Somerset, Pa, Sept 9, 43; c 2. RADIATION BIOLOGY, EXPERIMENTAL CANCER THERAPY. *Educ:* Pa State Univ, BS, 65; Stanford Univ, PhD(biophys), 71. *Hon Degrees:* MA, Yale Univ, 90. *Prof Exp:* Res fel radiol, Sch Med, Stanford Univ, 71, 72, 74; res fel, Institut de Recherche Radiobiologie, Clinique Institut Gustave Roussy, 73; from asst prof to assoc prof, 74-84, PROF THERAPEUT RADIOL, SCH MED, YALE UNIV, 84-, PROF, YALE COMPREHENSIVE CANCER CTR, 89- *Concurrent Pos:* prin investr, NIH grants & Am Cancer Soc grants, 79-; mem, Exp Therapeut Study Sect, Div Res Grants, NIH, 82-86, Yale Comprehensive Cancer Ctr, 83-, vis comt, dept med, Brookhaven Nat Lab, 87-91 & Sci Adv Bd Prev Diag, & Ther, Am Cancer Soc, 90. *Mem:* Radiation Res Soc; Cell Kinetics Soc (pres, 82-83, vpres, 81-82); Am Soc Therapeut Radiol & Oncol; Am Asn Cancer Res; Am Asn Univ Women; Bioelectromagnetics Soc. *Res:* The biology of solid cancers and the development and testing of new agents and regimens which may improve the treatment of these malignancies. *Mailing Add:* Dept Therapeut Radiol Yale Sch Med PO Box 3333 New Haven CT 06510-8040

ROCKWELL, THEODORE, b Chicago, Ill, June 26, 22; m 47; c 4. NUCLEAR & CHEMICAL ENGINEERING. *Educ:* Princeton Univ, BSE, 43, MS, 45. *Hon Degrees:* ScD, Tri-State Univ, 60. *Prof Exp:* Engr, Electromagnetic Separation Plant, Clinton Engr Works, Tenn, 44-45; head shield eng group, Oak Ridge Nat Lab, 45-49; nuclear engr naval reactors, AEC & nuclear propulsion, Bur Ships, 49-53, dir nuclear technol div, 53-54, tech dir, 54-64; prin officer & dir, 64-87, DIR, MPR ASSOCS, INC, WASH, DC, 87- *Concurrent Pos:* Chmn nat shield eng group, AEC, 48-49; mem reactor safety panel, & chmn, Reactor Safety Task Force, Atomic Indust forum, 66-68; mem adv group, NIH, 66; mem adv comt, Dept Chem Eng, Princeton Univ, 67-72; res assoc, Wash Ctr Foreign Policy Res, Johns Hopkins Univ, 65-67, consult, Joint Cong Comt Atomic Energy, 67; pvt pract. *Honors & Awards:* AEC Distinguished Serv Medal & Navy Distinguished Civilian Serv Medal, 60. *Mem:* AAAS; Sigma Xi; Am Soc Psych Res; Parapsychol Asn. *Res:* Criteria, procedures and facilities for safe operation of naval and central station nuclear power plants; nuclear and marine engineering. *Mailing Add:* 3403 Woolsey Dr Chevy Chase MD 20815-3924

ROCKWELL, THOMAS H, b Loma Linda, Calif, May 2, 29; m 53; c 5. INDUSTRIAL ENGINEERING. *Educ:* Stanford Univ, BS, 51; Ohio State Univ, MS, 53, PhD(indust eng), 57. *Prof Exp:* Design engr, Standard Oil Co Calif, 53-54; res assoc opers res, 55-57, from asst prof to assoc prof indust eng, 57-65, prof indust eng, Ohio State Univ, 65-88; RETIRED. *Concurrent Pos:* USPHS grants, accident prev res, 62-, transp res, 66-; mem road user characteristics & traffic safety comt, Hwy Res Bd, 64-, chmn road user characteristics comt, 72; consult, Nationwide Ins, 64-; mem, Gov Traffic Comt; consult indust environ. *Mem:* Fel AAAS; Am Soc Eng Educ; Am Soc Safety Engrs; Inst Elec & Electronics Engrs; fel Human Factors Soc. *Res:* Operations research; human factors engineering; transportation accident prevention; human performance experimental design. *Mailing Add:* Indust Eng Ohio State Univ 1971 Neil Ave Columbus OH 43210

ROCKWOOD, STEPHEN DELL, b Ft Scott, Kans, Apr 8, 43; m 64; c 2. LASERS. *Educ:* Grinnell Col, BA, 65; Calif Inst Technol, MS, 67, PhD(physics), 69. *Prof Exp:* Res officer laser develop, Air Force Weapons Lab, 70-72; staff mem laser develop, 72-75, group leader tunable lasers, 75-80, dep assoc dir 80-86, ASSOC DIR, INERTIAL FUSION, LOS ALAMOS NAT LAB, 86- *Concurrent Pos:* Adj prof, Univ NMex Inst Modern Optics; vpres, Advan Technol Assocs; dir, Los Alamos Laser Fusion Prog; consult problems in laser develop & indust applications; sr Vpres, Sci Appln Corp 86- *Honors & Awards:* USAF Sci Achievement Award, 71. *Mem:* Am Phys Soc. *Res:* Development of tunable ir lasers for isotope separation; sulfur, boron, carbon and silicon isotopes. *Mailing Add:* 13036 Decant Dr Poway CA 92064

ROCKWOOD, WILLIAM PHILIP, b Albany, NY, Dec 7, 30; m 53; c 5. PSYCHO-PHARMACOLOGY OF DRUGS & ALCOHOL, PHYSIOLOGY. *Educ:* Boston Col, BS, 57; Syracuse Univ, MS, 59; NY Univ, PhD(endocrine physiol), 68. *Prof Exp:* From instr to assoc prof, 61-76, PROF BIOL, RUSSELL SAGE COL, 76-, CHMN BIOL DEPT, 87- *Concurrent Pos:* Expert witness, alcoholism court trials. *Mem:* AAAS; Sigma Xi; NY Acad Sci. *Res:* Pateral effects of alcohol on newborn children. *Mailing Add:* 15 Milner Ave Albany NY 12203

ROD, DAVID LAWRENCE, b Gardner, Mass, Apr 23, 38; m 66. MATHEMATICS. *Educ:* Mass Inst Technol, BS, 60; Univ Wis-Madison, MS, 62, PhD(math), 71. *Prof Exp:* From asst prof to assoc prof, 66-82, PROF MATH, UNIV CALGARY, 82- *Mem:* Am Math Soc. *Res:* Hamiltonian systems of differential equations and dynamical systems theory. *Mailing Add:* Univ Calgary Calgary AB T2N 1N4 Can

RODABAUGH, DAVID JOSEPH, b Kansas City, Mo, Jan 14, 38; m 59; c 3. ALGEBRA, COMPUTER SCIENCE. *Educ:* Univ Chicago, SB, 59, SM, 60; Ill Inst Technol, PhD(math), 63. *Prof Exp:* Instr math, Ill Inst Technol, 62-63; asst prof, Vanderbilt Univ, 63-65; from asst prof to assoc prof, Univ Mo-Columbia, 65-78, prof math, 78-; MATHEMATICIAN, LOCKHEED CORP, CALIF. *Concurrent Pos:* Consult, NASA, 65-66; NSF grant, Univ Mo-Columbia, 67-68; prof, Calif State Univ, Northridge; staff scientist & engr, Lockheed Corp, 81- *Mem:* Asn Comput Mach; Inst Elec & Electronics Engrs; Soc Indust & Appl Math; Sigma Xi. *Mailing Add:* 2760 Bitternut Circle Lochead Bldg 311 Simi Valley CA 93065

RODAHL, KAARE, b Bronnoysund, Norway, Aug 17, 17; nat US; m 46; c 2. WORK PHYSIOLOGY. *Educ:* Univ Oslo, MD, 48, Dr med, 50. *Prof Exp:* Spec consult, US Dept Air Force, 49, chief dept physiol, Arctic Aeromed Lab, Ladd AFB, Alaska, 50-52, dir res, 54-57; asst prof physiol, Univ Oslo, 52-54; dir res, Lankenau Hosp, 57-65; dir, Inst Work Physiol, 65-87; prof physiol, Norweg Col Phys Educ, 66-87; RETIRED. *Concurrent Pos:* Hon mem staff & fac, Command Gen Staff Col, US Army. *Mem:* Am Physiol Soc. *Res:* Environmental physiology and medicine; work physiology; metabolism nutrition; vitamins. *Mailing Add:* Maltrostveien 40 Oslo 3 0390 Norway

RODAN, GIDEON ALFRED, b Rumania, June 14, 34; m 72; c 2. CELL BIOLOGY. *Educ:* Hebrew Univ, Hadassah, MD, 64; Weizmann Inst Sci, PhD(chem), 70. *Prof Exp:* From asst prof to prof oral biol, 70-85, head, 78-85, EXEC DIR BONE BIOL & OSTEOPOROSIS, MERCK SHARP & DOHME RES LABS, 85- *Concurrent Pos:* coun, NIDR, 88- *Mem:* Am Soc Bone & Mineral Res (pres, 88); Am Soc Cell Biol. *Res:* Cell biology of hard tissues; hormonal control of growth and differentiation in bone-derived cells. *Mailing Add:* Merck Sharp & Dohme West Point PA 19486

RODARTE, JOSEPH ROBERT, b Temple, Tex, Apr 1, 38. PULMONARY PHYSIOLOGY, PATHOPHYSIOLOGY. *Educ:* Harvard Univ, MD, 64. *Prof Exp:* PROF MED, MAYO MED SCH, 81- *Mailing Add:* Baylor Col Med Smith Tower Suite 1225 6550 Fannin Houston TX 77030

RODBARD, DAVID, b Chicago, Ill, July 6, 41; m 77. MEDICINE, BIOPHYSICS. *Educ:* Univ Buffalo, BA, 60; Western Reserve Univ, MD, 64. *Prof Exp:* Intern med, King County Hosp, Seattle, 64-65; resident, Hahnemann Hosp, Philadelphia, 65-66; clin assoc med res, NIH, 66-69, sr investr, 69-78, SECT HEAD MED RES, BIOPHYS ENDOCRINOL SECT, ENDOCRINOL & REPRODUCTION RES BR, NAT INST CHILD HEALTH & HUMAN DEVELOP, 79- *Concurrent Pos:* Consult, Int Atomic Energy Agency, 70- & WHO, 74-; assoc ed, Am J Physiol, 76- *Honors & Awards:* Young Investr Award, Clin Radioassay Soc, 79; Ayerst Award, Endocrine Soc, 81. *Mem:* Endocrine Soc; Am Physiol Soc; Am Soc Biol Chemists; Biomet Soc; Am Soc Clin Invest. *Res:* Endocrinology; physiology; biomathematics; biochemistry; radio immunoassay; physical-chemistry of proteins; neurotransmitters. *Mailing Add:* NIH Bldg 10 Rm 6C101 9000 Rockville Blvd Bethesda MD 20892

RODBELL, MARTIN, b Baltimore, Md, Dec 1, 25; m 40; c 4. BIOCHEMISTRY. *Educ:* Johns Hopkins Univ, BA, 49; Univ Wash, PhD(biochem), 54. *Prof Exp:* Res assoc biochem, Univ Ill, 54-56; biochemist, Nat Heart Inst, 56-61 & Nat Inst Arthritis & Metab Dis, 61-70, chief lab nurtit & endocrinol, Nat Inst Arthritis, Metab & Digestive Dis, 70-85; SCI DIR, NAT INST EVIRON HEALTH SCI, 85- *Honors & Awards:* Gairdner Award, 84; Sci Merit Award, Nat Inst Arthritis, Diabetes, Digestive & Kidney Dis, 85; Distinguished Service Award, NIH, 73; Lounsbery Award; Jacobers Award. *Mem:* Nat Acad Sci; AAAS; Am Soc Biol Chem. *Res:* Mechanisms of hormone action; structure and function of membrane lipoproteins. *Mailing Add:* Nat Inst Environ Health Sci PO Box 12210 Res Triangle Park NC 22709

RODDA, BRUCE EDWARD, b Schenectady, NY, June 21, 42; m 62, 89; c 1. MEDICAL STATISTICS, CLINICAL PHARMACOLOGY. *Educ:* Alfred Univ, BA, 65; Tulane Univ, MS, 67, PhD(biostatist), 69; Fairleigh Dickinson Univ, MBA, 82. *Prof Exp:* Res scientist med statist, Eli Lilly & Co, 69-76; sr dir, Cbards Int, Merck & Co, Rahway, NJ, 76-88; VPRES, BRISTOL MYERS SQUIBB PHARM RES INST, PRINCETON, NJ, 88- *Concurrent Pos:* Asst prof pharmacol, Ind Univ Med Ctr, 72-76; asst prof biostatist, Rockford Med Sch, 78-; guest investr, Rockefeller Univ, 77-; consult, Vet Admin, NIH. *Mem:* Am Statist Asn; Am Soc Clin Pharmacol & Therapeut; Biomet Soc; Sigma Xi; fel Royal Statist Soc. *Res:* Pharmacokinetic modeling; general statistical methodology; design of clinical trials. *Mailing Add:* Bristol Myers Squibb Pharm Res Facil PO Box 4000 Princeton NJ 08540

RODDA, ERROL DAVID, b Platteville, Wis, June 3, 28; m 55; c 2. AGRICULTURAL & STRUCTURAL ENGINEERING. *Educ:* Univ Ill, Urbana, BS, 51, MSCE, 60, MS, 64; Purdue Univ, PhD(agr eng), 65. *Prof Exp:* Engr, Caterpillar Tractor Co, Ill, 51-58; res assoc agr eng, Univ Ill, Urbana-Champaign, 58-62; asst prof, Univ Calif, Davis, 64-68; assoc prof, 68-75, PROF AGR ENG & FOOD ENG, UNIV ILL, URBANA-CHAMPAIGN, 75- *Concurrent Pos:* USAID-Univ Ill adv agr eng, Uttar Pradesh Agr Univ, India, 68-70. *Mem:* Am Soc Agr Engrs; Am Soc Eng Educ; AAAS. *Res:* Grain processing and storage; agricultural structures; engineering systems design; food engineering; fuel alcohol. *Mailing Add:* 1304 W Pennsylvania Ave Urbana IL 61801

RODDA, PETER ULISSE, b Albuquerque, NMex, Nov 18, 29; div. GEOLOGY. *Educ:* Univ Calif, Los Angeles, AB, 52, PhD(geol), 60. *Prof Exp:* Asst geol, Univ Calif, Los Angeles, 54-57, instr, 57; lectr geol, Univ Tex, Austin, 63-66, assoc prof, 67-71, res scientist, Bur Econ Geol, 58-71; CUR DEPTS INVERT ZOOL & GEOL, CALIF ACAD SCI, 71- *Mem:* Geol Soc Am; Paleont Soc; Am Asn Petrol Geol; AAAS; Soc Econ Paleont Mineral. *Res:* Stratigraphy and invertebrate paleontology of the Cretaceous and Cenozoic; evolutionary history of the gastropoda; Mesozoic and Cenozoic geology of California. *Mailing Add:* Dept Geol Calif Acad Sci San Francisco CA 94118

RODDEN, ROBERT MORRIS, b Roswell, NMex, June 28, 22; m 46; c 2. PHYSICS, BIOLOGY. *Educ:* US Mil Acad, BS, 44; Univ Calif, Berkeley, MS, 54. *Prof Exp:* Chief weapons develop div, Defense Nuclear Agency, Dept Defense, 44-64; asst dir opers eval, SRI Int, 64-72, dir opers eval, 72-74, dir, Ctr Resources & Environ, 74-77; prin consult, 77-87, pres, RMR Assocs, 80-88; RETIRED. *Concurrent Pos:* Dir, Systs eng task group, Nat Indust Pollution Control Coun, 70-72. *Mem:* Asn Energy Engrs; Soc Petrol Engrs; Am Cogeneration Soc. *Res:* Energy analysis; planning; environmental analysis; information systems; research and development management; energy, environment, nuclear and technology assessment. *Mailing Add:* PO Box 223039 Carmel CA 93922-3039

RODDICK, JAMES ARCHIBALD, b New Westminster, BC, Feb 23, 25; m 63; c 2. GEOLOGY. *Educ:* Univ BC, BASc, 48; Calif Inst Technol, MS, 50; Univ Wash, PhD, 55. *Prof Exp:* GEOLOGIST, DEPT ENERGY, MINES & RESOURCES GEOL SURV, CAN, 50- *Mem:* Fel Geol Soc Am; fel Geol Asn Can; Can Inst Mining & Metall. *Res:* Cordilleran geology; granitic rocks. *Mailing Add:* Dept Energy Mines & Resources Geol Surv Can 100 W Pender St Vancouver BC V6B 1R8 Can

RODDICK, JOHN WILLIAM, JR, b Dallas, Tex, June 13, 26; m 49; c 3. OBSTETRICS & GYNECOLOGY. *Educ:* Northwestern Univ, BS, 47, BM, 49, MD, 50, MS, 56. *Prof Exp:* Resident obstet & gynec, Chicago Wesley Mem Hosp, 52-56; clin asst, Northwestern Univ, 56-57, instr, 57-59, assoc, 59-62, asst prof, 62-64; from assoc prof to prof, Univ Ky, 64-71; PROF OBSTET & GYNEC & CHMN DEPT, SCH MED, SOUTHERN ILL UNIV, 72- *Concurrent Pos:* Fel surg, Mem Ctr Cancer & Allied Dis, 61-62. *Mem:* AMA; Am Col Obstet & Gynec; Soc Gynec Invest; Am Col Surg; Soc Gynec Oncol. *Res:* Gynecologic oncology, clinical and laboratory. *Mailing Add:* Dept Obstet & Gynec Southern Ill Univ Sch Med PO Box 3926 Springfield IL 62708

RODDIS, LOUIS HARRY, JR, marine & energy engineering; deceased, see previous edition for last biography

RODDIS, WINIFRED MARY KIM, COMPUTER-AIDED DESIGN, STRUCTURAL DESIGN. *Educ:* Mass Inst Technol, BS, 77, MS, 87, PhD(civil eng), 89. *Prof Exp:* Struct engr, Stone & Webster Eng Corp, 77-81, Souza & Truf Eng, 81-84 & A G Lichtenstein Engrs, 86; ASST PROF CIVIL ENG, DEPT CIVIL ENG, UNIV KANS, 88- *Mem:* Am Soc Civil Engrs; Am Soc Eng Educ; Am Asn Artificial Intel; Soc Women Engrs. *Res:* Expert systems; knowledge based systems; artificial intelligence; structural design; infrastructure maintenance; nondestructive evaluation; bridge fatigue; engineering ethics. *Mailing Add:* Dept Civil Eng Univ Kans 2008 Learned Lawrence KS 66045-2225

RODDY, DAVID JOHN, b Springfield, Ohio, May 27, 32; c 3. GEOLOGY, GEOPHYSICS. *Educ:* Miami Univ, AB, 55, MS, 57; Calif Inst Technol, PhD(physics, geol), 66. *Prof Exp:* Instr geol, Miami Univ, 54-57; geologist, Jet Propulsion Lab, Calif Inst Technol, 60-64; GEOLOGIST, US GEOL SURV, 66- *Concurrent Pos:* Geologist, Calif Oil Co, 55; instr, USAF Inst Technol, 66. *Mem:* Geol Soc Am; Am Geophys Union; Mineral Soc Am; Sigma Xi. *Res:* Large-scale impact cratering mechanics related to the earth and planets; shock metamorphic studies of very high pressure shock wave deformed natural materials. *Mailing Add:* US Geol Surv 2255 N Gemini Dr Flagstaff AZ 86001

RODDY, MARTIN THOMAS, b Washington, DC, Dec 17, 46; m 73. CELL BIOLOGY. *Educ:* St Ambrose Col, BS, 69; Cath Univ Am, MS, 72, PhD(cell biol), 75; Am Bd Toxicol, dipl 80 & 85. *Prof Exp:* Instr human anat, Cath Univ Am, 73-75; med rev officer, Gillette Med Eval Lab, 75-81, sr toxicologist, 81-88; SECT HEAD, HUMAN SAFETY, NOXEL, INC, 88- *Mem:* Sigma Xi; AAAS; Am Soc Cell Biol; Tissue Cult Asn Am; Am Bd Toxicol; Am Col Toxicol. *Res:* The potential of various organic metal compounds to cause sister chromated exchanges in vitro. *Mailing Add:* Noxell Inc 11050 York Rd Hunt Valley MD 21030

RODE, DANIEL LEON, b Delphos, Ohio, Aug 10, 42; m 67; c 2. SOLID STATE PHYSICS, PLASMA PHYSICS. *Educ:* Univ Dayton, BS, 64; Case Western Reserve Univ, MS, 66, PhD(appl physics), 68. *Prof Exp:* Mem tech staff semiconductor microwave devices, Bell Tel Labs, 68-70, supvr semiconductor mat, 70-80; PROF, WASH UNIV, 80- *Concurrent Pos:* Fel, Max-Planck Inst, 77. *Mem:* Am Phys Soc; AAAS; Inst Elec & Electronics Engrs; Sigma Xi. *Res:* Electron theory of crystals; band structure of solids; semiconductor crystal growth; thermonuclear plasmas. *Mailing Add:* Campus Box 1127 Wash Univ St Louis MO 63130

RODE, JONATHAN PACE, b Worcester Mass, Oct 2, 48; m 69. ELECTRICAL ENGINEERING, INFRARED IMAGING. *Educ:* Univ Ore, BA, 70, PhD(physics), 76. *Prof Exp:* MEM TECH STAFF INFRARED, ROCKWELL INT SCI CTR, 76- *Mem:* Am Phys Soc. *Res:* Development of high performance; two-dimensional arrays of infrared detectors and their incorporation into infrared imaging systems. *Mailing Add:* Rockwell Sci Ctr 1049 Camino Dos Rios Thousand Oaks CA 91360

RODEBACK, GEORGE WAYNE, b Soda Springs, Idaho, Aug 12, 21; m 57; c 2. PHYSICS. *Educ:* Univ Idaho, BS, 43; Univ Ill, MS, 47, PhD(physics), 51. *Prof Exp:* Mem staff radiation lab, Mass Inst Technol, 42-45; from res engr to sr tech specialist, Atomics Int Div, NAm Aviation Inc, 51-60; ASSOC PROF PHYSICS, NAVAL POSTGRAD SCH, 60- *Mem:* Am Phys Soc; Am Asn Physics Teachers; Sigma Xi. *Res:* Nuclear and reactor physics; analytical mechanics. *Mailing Add:* Dept Physics USN Postgrad Sch Monterey CA 93943

RODELL, CHARLES FRANKLIN, b La Crosse, Wis, Aug 1, 42; m 67; c 1. POPULATION GENETICS. *Educ:* Univ Wis, BS, 65; Univ Minn, MS, 67, PhD(genetics), 72. *Prof Exp:* Fel systs ecol, Natural Resource Ecol Lab, Colo State Univ, 72-74; asst prof biol, Vanderbilt Univ, 74-80; asst prof, 79-81, ASSOC PROF BIOL, COL ST BENEDICT, ST JOHN'S UNIV, 82- *Mem:* AAAS; Genetics Soc Am; Soc Study Evolution; Sigma Xi. *Res:* Development and analysis of life history parameters and demographic patterns on the genetic structure of populations. *Mailing Add:* 709 13th Ave St Cloud MN 56301

RODELL, MICHAEL BYRON, b Brooklyn, NY, Sept 4, 32; m 80; c 2. PHARMACY, ANALYTICAL CHEMISTRY. *Educ:* Univ Md, BS, 58; Univ Tex, MS, 64, PhD(pharm), 66. *Prof Exp:* Res assoc, Dorsey Labs, Wander Co, 66-67, mgr pharmaceut & anal res, Dorsey Labs, Sandoz-Wander, Inc, 67-72; mgr regulatory affairs, 72-74, dir regulatory affairs & clin develop, Hyland Div, 74-81, vpres, Regulatory Affairs, Hyland Therapeut Div, Travenol Labs, Inc, 82-; MEM STAFF, ETHICAL PROD DIV, TARRYTOWN. *Mem:* Am Pharmaceut Asn; Am Chem Soc; Acad Pharmaceut Sci; NY Acad Sci; AAAS. *Res:* Development of pharmaceutical dosage forms; design of stability protocols; analytical methodology. *Mailing Add:* Armour Pharmaceut 920 Harvest Dr Suite 200 Blue Bill PA 19422

RODEMEYER, STEPHEN A, b Freeport, Ill, Oct 2, 40; m 65; c 2. PHYSICAL ORGANIC CHEMISTRY. *Educ:* Col St Thomas, BS, 62; Univ Calif, Berkeley, PhD(chem), 66. *Prof Exp:* NSF res assoc chem, Radiation Lab, Univ Notre Dame, 66-67; from asst prof to assoc prof, 67-74, chmn dept, 71-77, PROF CHEM, CALIF STATE UNIV, FRESNO, 74- *Mem:* Am Chem Soc. *Res:* Kinetics and mechanism; radiation and photochemistry. *Mailing Add:* Dept Chem Calif State Univ Fresno CA 93740-0070

RODEMS, JAMES D, b Springfield, Ill, Apr 30, 26; m 47; c 3. ELECTRICAL ENGINEERING, ELECTRONICS. *Educ:* Univ Louisville, BEE, 47; Univ Ill, MS, 50. *Prof Exp:* Test engr, Gen Elec Co, 47-48; asst elec eng, Univ Ill, 48-50; instr, Ohio State Univ, 50-51; control systs engr, Bell Aircraft Corp, 51-53; res assoc radar & control systs, Univ Ill, 53-57; res engr radar & control systs, Defense Systs Lab, Syracuse Univ Res Corp, 57-78, dir, 61-78; PRES, JDR SYST CORP, 78- *Concurrent Pos:* Consult, USN Dept, 62- *Mem:* Inst Elec & Electronics Engrs. *Res:* Radar systems; detection; tracking; fire control; feedback; control systems; hybrid real time analog-digital control; maintenance and test engineering; production and process control engineering. *Mailing Add:* 701 Nottingham Rd Syracuse NY 13224

RODEN, GUNNAR IVO, b Tallinn, Estonia, Dec 27, 28; m 58; c 3. PHYSICAL OCEANOGRAPHY, CLIMATOLOGY. *Educ:* Univ Calif, Los Angeles, MS, 56. *Prof Exp:* Res oceanographer, Univ Calif, San Diego, 56-64, asst specialist oceanog, 65-66; res assoc phys oceanog, Univ Wash, 66-68, sr res assoc, 68-82, prin res assoc, 82-91, RES PROF PHYS OCEANOG, UNIV WASH, 91- *Concurrent Pos:* Assoc ed, Am Geophys Union, 87-; tech rev consult, Transls Bd, Am Geophys Union, 65-; Off Naval Res grants, 68- *Mem:* Am Geophys Union; Am Meteorol Soc. *Res:* Oceanic fronts; meso-scale and large-scale thermohaline structure and circulation; flow over topography; remote sensing of the ocean environment; regional oceanography; sea level and climatic change. *Mailing Add:* Sch Oceanog WB-10 Univ Wash Seattle WA 98195

RODENBERGER, CHARLES ALVARD, b Muskogee, Okla, Sept 11, 26; m 49; c 2. ENGINEERING, FORENSIC ENGINEERING. *Educ:* Okla State Univ, BS, 48; Southern Methodist Univ, MS, 59; Univ Tex, Austin, PhD(aerospace eng), 68. *Prof Exp:* Jr petrol engr, Amoco Prod Co, 48-51; chief engr, McGregor Bros, Inc, 53-54; petrol engr, Gen Crude Oil Co, 54; sr design engr aircraft struct, Gen Dynamics/FW, 54-60; Halliburton prof eng, 66-78, asst dean, Col Eng, 77-80, prof, 60-82, emer prof aerospace eng, Tex A&M Univ, 82; CONSULT, GEN DYNAMICS/FW, 82- *Concurrent Pos:* Consult, Gen Motors Defense Res Labs, 65-66, Meiller Res, Inc, 66-82 & Southwest Res Inst, 67; expert witness, McMahon, Smart, Surovilc, Suttle, Bohrman & Cobb, 83-85, Mullen, Mac Innes, Redding & Grove, 87-88, Glandon, Erwin, Scarborough, Baker, Choate & Arnot, 87-88; nat pres, JETS Inc, 78-80, Tex State coordr, 75-82, chmn, Adv Comt, 88-89. *Mem:* Am Inst Aeronaut & Astronaut; Am Soc Eng Educ; Sigma Xi; Am Soc Mech Engrs; Nat Soc Prof Engrs (vpres, 80 & 81). *Res:* Design engineering productivity; engineering innovation; oil spill containment; bioengineering orthotic devices; hypervelocity devices; composite structures; problem definition in systems engineering; two patents. *Mailing Add:* HC 85 Box 60 Baird TX 79504-9603

RODENHUIS, DAVID ROY, b Michigan City, Ind, Oct 5, 36; m 58; c 2. DYNAMIC METEOROLOGY. *Educ:* Univ Calif, Berkeley, BS, 59; Pa State Univ, BS, 60; Univ Wash, PhD(atmospheric sci), 67. *Prof Exp:* From asst prof to assoc prof fluid dynamics & appl math, 72-76, ASSOC PROF METEOROL, UNIV MD, COLLEGE PARK, 76- *Concurrent Pos:* Exec scientist, US Comt Global Atmospheric Res Prog, Nat Acad Sci, 72; sci officer, World Meteorol Orgn, 75-; US-USSR exchange scientist, 80. *Mem:* Am Geophys Union; Am Meteorol Soc. *Res:* Tropical meteorology; convection models; dynamic climate models. *Mailing Add:* Climate Analysis Ctr World Weather Bldg 5200 Auth Rd Camp Springs MD 20746

RODER, HANS MARTIN, b Schenectady, NY, June 30, 30; m 51; c 3. FLUID THERMODYNAMICS. *Educ:* Univ Colo, BA, 55. *Prof Exp:* Physicist cryog, 55-66, supvr physicist data compilation, 66-75, EXP PHYSICIST, THERMOPHYSICS DIV, NAT BUR STANDARDS, 75- *Concurrent Pos:* Mem quantum fluids panel, Int Union Pure & Appl Chem, Thermodyn Tables Proj Ctr, 67- *Honors & Awards:* Gold Medal, US Dept Com, 66; Russell B Scott Mem Award, Cryogenic Eng Conf, 70. *Mem:* Am Phys Soc; Res Soc Am. *Res:* Experimental research on the equilibrium and transport properties of simple fluids; correlation and derivation of thermodynamic functions from the experimental results. *Mailing Add:* 720 S 42nd Boulder CO 80302

RODERICK, GILBERT LEROY, b Waukon, Iowa, Aug 18, 33; m 62. SOIL ENGINEERING. *Educ:* Iowa State Univ, BS, 60, MS, 63, PhD(soil eng), 65. *Prof Exp:* Engr, US Bur Reclamation, 60-61; res assoc soil eng, Iowa Eng Exp Sta, Iowa State Univ, 63-65; asst prof, Univ RI, 65-68; asst prof, 68-70, ASSOC PROF CIVIL ENG, UNIV WIS-MILWAUKEE, 70- *Concurrent Pos:* Mem, Hwy Res Bd, Nat Acad Sci-Nat Res Coun. *Mem:* Am Soc Civil Engrs. *Res:* Soil stabilization with chemicals; physicochemical properties of soils by x-ray diffraction and adsorption isotherm studies; frost action in soils; ground water flow; lake bottom sediments; disposal of dredging spoil. *Mailing Add:* 4326 N Ardmore Ave Shorewood WI 53211

RODERICK, THOMAS HUSTON, b Grand Rapids, Mich, May 10, 30; m 58; c 2. GENETICS. *Educ:* Univ Mich, AB, 52, BS, 53; Univ Calif, Berkeley, PhD(genetics), 59. *Prof Exp:* Asst psychol, Univ Calif, 55-58; from assoc staff scientist to sr staff scientist, 59-73, SR STAFF SCIENTIST, JACKSON LAB, 75- *Concurrent Pos:* Lectr, Univ Maine, 65-; vis lectr, Univ Calif, Berkeley, 68; staff, Ctr Human Genetics, 69-, pres bd dirs, 74-77, 89-; adj prof, Univ RI, 70-73; geneticist, Energy Res & Develop Admin, 73-75. *Mem:* AAAS; Genetics Soc Am; Am Soc Human Genetics; AAAS. *Res:* Mammalian genetics; mutagenesis in mammals; population and cytogenetics. *Mailing Add:* Jackson Lab 600 Main St Bar Harbor ME 04609-0800

RODERICK, WILLIAM RODNEY, b Chicago, Ill, Aug 6, 33; m 65. ORGANIC CHEMISTRY, MEDICINAL CHEMISTRY. *Educ:* Northwestern Univ, BS, 54; Univ Chicago, SM, 55, PhD(chem), 57. *Prof Exp:* Res fel chem, Harvard Univ, 57-58; asst prof, Univ Fla, 58-62; sr res chemist, Abbott Labs, 62-70, assoc res fel, 70-71; asst prof natural sci, 72-73, assoc prof natural sci & chem, 73-85, PROF NATURAL SCI & CHEM, ROOSEVELT UNIV, 85- *Mem:* Am Chem Soc; AAAS; Sigma Xi. *Res:* Structural and synthetic organic chemistry; synthesis of antiviral agents. *Mailing Add:* 15193 W Redwood Lane Libertyville IL 60048-1447

RODERUCK, CHARLOTTE ELIZABETH, b Walkersville, Md, Dec 2, 19. NUTRITION. *Educ:* Univ Pittsburgh, BS, 40; State Col Wash, MS, 42; Univ Iowa, PhD(biochem), 49. *Prof Exp:* Res chemist, Children's Fund Lab, Mich, 42-46; from asst prof to prof nutrition, Iowa State Univ, 48-73, asst dean grad col, 71-73, assoc dean col home econ, 73-77, Mary B Welch Distinguished prof home econ, 72-88, dir, World Food Inst, 77-88; RETIRED. *Concurrent Pos:* Vis prof, Univ Baroda, 64-66. *Honors & Awards:* Garst Mem Award, UNA-USA, 88. *Mem:* Soc Exp Biol & Med; Am Chem Soc; Am Inst Nutrit; Sigma Xi; Soc Nutrit Educ; Am Home Econ Asn. *Res:* Nutrition education; nutritional status; intermediary metabolism. *Mailing Add:* 228 Parkridge Circle Ames IA 50010-3645

RODESILER, PAUL FREDERICK, b Adrian, Mich, Sept 10, 41; m 65; c 1. INORGANIC CHEMISTRY, X-RAY CRYSTALLOGRAPHY. *Educ:* Capital Univ, BS, 63; Univ Western Ont, MSc, 66; Queen Mary Col, Univ London, PhD(chem), 68. *Prof Exp:* Res asst chem & crystallog, Univ SC, 69-71, asst prof chem, 71-72, res asst chem & crystallog, 72-73; asst prof math & sci, 73-77, ASSOC PROF CHEM, COLUMBIA COL, SC, 77- *Concurrent Pos:* Instr, Midland Tech Sch, 72-73. *Mem:* Am Chem Soc; Am Crystallog Asn. *Res:* Preparation, structure and bonding of metallo-organic compounds, including the use of x-ray diffraction techniques. *Mailing Add:* 6137 Rutledge Hill Rd Columbia SC 29209

RODEWALD, LYNN B, b Norcatur, Kans, Nov 15, 39; m 63; c 2. ORGANIC CHEMISTRY. *Educ:* Whittier Col, BA, 61; Iowa State Univ, PhD(org chem), 64. *Prof Exp:* NIH fel chem, Princeton Univ, 64-65; asst prof, Univ Tex, Austin, 65-72; vis asst prof, Univ Okla, 72-75; ASST PROF CHEM, TOWSON STATE UNIV, 75- *Concurrent Pos:* Petrol Res fund grant, 65-66. *Mem:* Am Chem Soc; Sigma Xi. *Res:* Organic reaction mechanisms; electroorganic chemistry; small ring chemistry. *Mailing Add:* 8215 Rider Ave Baltimore MD 21204

RODEWALD, PAUL GERHARD, JR, b Pittsburgh, Pa, May 15, 36; m 58; c 4. CATALYSIS. *Educ:* Haverford Col, BA, 58; Pa State Univ, PhD(organosilicon chem), 62. *Prof Exp:* Res chemist, Cent Res Div Lab, Mobil Oil Corp, 63-66, sr res chemist, 66-78, assoc, 76-91, RES ASSOC, CENT RES DIV LAB, MOBIL OIL CORP, 91- *Mem:* Am Chem Soc; Am Chem Soc Petrol Chem. *Res:* Organosilicon chemistry; electrophilic aromatic substitution; organic synthesis; catalysis. *Mailing Add:* Mobil Res & Develop Corp CTRL Res Div Lab PO Box 1025 Princeton NJ 08540

RODEWALD, RICHARD DAVID, b Nyack, NY, Mar 20, 44; m 69. CELL BIOLOGY. *Educ:* Harvard Univ, BA, 66; Univ Pa, PhD(biochem), 70. *Prof Exp:* NSF fel, 70; fel biol, Univ Calif, San Diego, 70-71; fel path, Harvard Med Sch, 71-73; ASST PROF BIOL, UNIV VA, 73- *Concurrent Pos:* Instr physiol, Marine Biol Lab, Woods Hole, 74. *Mem:* Am Soc Cell Biol; AAAS. *Res:* Selective transport of immunoglobulins across the small intestine; glomerular permeability to macromolecules. *Mailing Add:* Dept Biol Gilmer Hall-72 Univ Va Charlottesville VA 22901

RODEY, GLENN EUGENE, b Mansfield, Ohio, Mar 25, 36; m 58; c 4. HUMAN HISTOCOMPATIBILITY, TRANSPLANTATION IMMUNOLOGY. *Educ:* Ohio Univ, BS, 57; Ohio State Univ, MD, 61. *Prof Exp:* Resident internal med, Milwaukee County Gen Hosp & Med Col, Wis, 62-63 & 65-67; fel immunol, Univ Minn, 67-69; dir histocompatibility, Milwaukee Blood Ctr, 70-74; assoc prof path & med, Med Col, Wis, 74-76; assoc prof, 76-81, PROF MED & PATH, SCH MED, WASH UNIV, MO, 81- *Concurrent Pos:* NIH mem, Transplantation Immunol Comt, Nat Inst Allergy & Infectious Dis, 74-78; mem, Transplantation Comt, Am Nat Red Cross, 79-; managing ed, Human Immunol, 79- *Mem:* Am Soc Histocompatibility & Immunogenetics (Am Asn Clin Histocompatibility Testing, pres, 82-83); Am Asn Immunologists; Am Soc Hemat; Cent Soc Clin Res; Transplantation Soc. *Res:* Structure and function of human histocompatibility locus antigens, genetic complex and gene products; how these factors contribute to disease susceptibility and their functions in generating effective immune responses. *Mailing Add:* 1916 N Decatur Rd Atlanta GA 30307

RODGER, WALTON A, chemical engineering, for more information see previous edition

RODGERS, ALAN SHORTRIDGE, b St Louis, Mo, Oct 23, 31; div; c 5. PHYSICAL CHEMISTRY. *Educ:* Princeton Univ, AB, 53; Univ Colo, PhD(phys chem), 60. *Prof Exp:* Sr chemist, Minn Mining & Mfg Co, 60-65; phys chemist, Stanford Res Inst, 65-67; ASSOC PROF CHEM, TEX A&M UNIV, 67- *Mem:* Am Chem Soc. *Res:* Chemical kinetics; free radical thermochemistry; structure and bond dissociation energy. *Mailing Add:* Dept Chem Tex A&M Univ College Station TX 77843-5000

RODGERS, AUBREY, b Lexington, Miss, June 11, 29; m 55; c 3. PHYSICS, MATHEMATICS. *Educ:* Miss Col, BS, 57; Rensselaer Polytech Inst, dipl, 67; Univ Calif, Los Angeles, dipl 68. *Prof Exp:* Physicist electronics, Naval Coastal Syst Lab, Fla, 57-60; physicist, mech, 60-62 & gen, 62-64, PHYSICIST RES, US ARMY MISSILE COMMAND, ALA, 64- *Res:* Investigates revolutionary concepts and techniques of gyroscopic instrumention used in Army inertial guidance, stabilization and navigation systems. *Mailing Add:* 216 Creek Trail Madison AL 35758-8514

RODGERS, BILLY RUSSELL, b Fitzgerald, Ga, Sept 5, 36; m 75; c 4. HYDROCARBON FUELS & COAL CONVERSION, HAZARDOUS & RADIOACTIVE WASTE TECHNOLOGY. *Educ:* Univ Fla, BSChE, 66, MSE, 67; Univ Tenn, PhD(chem eng), 80. *Prof Exp:* Task leader petrol res, Shell Develop Co, 68-72; mgr filter develop, Fluid Handling Div, Keene Corp, 72-74; mgr coal conversion, 74-83, mgr waste technols, 83-86, MGR MODELING, SIMULATION & ARTIFICIAL INTEL, OAK RIDGE NAT LAB, 86- *Concurrent Pos:* Lectr, Univ Tenn, 80-; mem, panel environ protection, safety & hazardous mat, Nat Acad Sci/Nat Res Coun, 84-; dir, Fuels & Petrochemical Div, Am Inst Chem Engrs, 85-88. *Mem:* Am Inst Chem Engrs. *Res:* Conversion of heavy petroleum residues to lighter products followed by conversion of coal to lighter products; physical chemistry of seperating solids from liquids in heavy media; chemically hazardous and radioactive waste technology; artificial intelligence applied to modeling and relational databases. *Mailing Add:* PO Box 2008 Bldg 4500 N Oak Ridge Nat Lab Oak Ridge TN 37831-6234

RODGERS, BRADLEY MORELAND, b Montclair, NJ, Jan 16, 42; m 69; c 2. PEDIATRIC SURGERY. *Educ:* Dartmouth Col, BA, 63, Dartmouth Med Sch, BS, 64; Johns Hopkins Univ, MD, 66. *Prof Exp:* Asst prof surg & pediat, Med Ctr, Univ Fla, 74-76, assoc prof, 76-78, prof & assoc chief, Div Pediat Surg, 78-81; PROF SURG & PEDIAT & CHIEF, DIV PEDIAT SURG, MED SCH, UNIV VA, 81-, CHIEF CHILDREN'S SURG, UNIV VA HOSP, 81- *Mem:* Am Col Surgeons; Asn Acad Surg. *Res:* Neonatal gastric physiology. *Mailing Add:* Univ Va Med Ctr Box 181 Charlottesville VA 22908

RODGERS, CHARLES H, b Sept 5, 32; US citizen; m 66; c 3. PHYSIOLOGICAL PSYCHOLOGY. *Educ:* Los Angeles State Col Arts & Sci, BA, 58, MS, 61; Claremont Grad Sch, PhD(psychol), 66. *Prof Exp:* Nat Inst Child Health & Human Develop fel physiol & neuroendocrinol, Sch Med, Stanford Univ, 66-68; asst prof psychol, physiol & pharmacol, Iowa State Univ, 68-70; asst prof psychol, physiol & pharmacol, Col Med, Univ Ill, 70-72, assoc prof, 72-81; assoc grant, 81-82 DIR, UROL & MANPOWER PROG, NAT INST DIABETES, DIGESTIVE & KIDNEY DIS, NIH, 82- *Concurrent Pos:* Nat Inst Child Health & Human Develop grant, Iowa State Univ & Univ Ill Col Med, 68-71; Vet Admin grant, West Side Vet Admin Hosp, Chicago, 70-73, psychologist, 70-74; consult, Cook County Hosp, Ill, 74-81. *Mem:* Am Physiol Soc; Endocrine Soc. *Res:* Reproductive physiology in the male and female. *Mailing Add:* Nat Inst Digestive Diabetes & Kidney Dis Manpower Prog 5333 Westboard Ave Rm 621 Bethesda MD 20892-4500

RODGERS, CHARLES LELAND, botany, for more information see previous edition

RODGERS, EARL GILBERT, b Trenton, Fla, Jan 27, 21; m 43; c 2. AGRONOMY, WEED SCIENCE. *Educ:* Univ Fla, BS, 43, MS, 49; Iowa State Univ, PhD(plant physiol), 51. *Prof Exp:* Asst county agent, Wauchula, Fla, 46-47; from instr to assoc prof, 47-59, grad coordr agron, 69-81, PROF AGRON, UNIV FLA, 59-, COORDR AGRON TEACHING, 65- *Concurrent Pos:* Consult, USAF, Eglin AFB, Fla, 57. *Mem:* Fel Weed Sci Soc Am (pres, 73); fel Am Soc Agron. *Res:* Crop production; weed science. *Mailing Add:* 611 S W 16th Pl Gainesville FL 32601

RODGERS, EDWARD J(OHN), engineering, dynamics; deceased, see previous edition for last biography

RODGERS, FRANK GERALD, b Belfast, N Ireland, UK, Oct 5, 46; m 70; c 2. PATHOGENIC MECHANISMS, MICROBIAL ULTRA STRUCTURE. *Educ:* Univ Surrey, Eng, BSc, 69, PhD(virol & electron micros), 77; Inst Biol, London, Eng, MIBiol, 77. *Prof Exp:* Basic microbiologist virol & electron micros, Pub Health Lab Serv, Virus Ref Lab, London, 69-74, sr microbiologist, 74-75; sr microbiologist, clin microbiol, Pub Health Lab Serv, Nottingham, Eng, 75-83, prin microbiologist, 83-85; asst prof, 85-87, ASSOC PROF MICROBIOL, UNIV NH, 87- *Concurrent Pos:* Assoc lectr, Univ Surrey, Eng, 71-75, Trent Polytech, Eng, 78-85; med sci teacher, Med Sch, Univ Nottingham, Eng, 78-85; spec prof, Fac Sci, Univ Nottingham, Eng, 78-85; sect ed, Manual Clin Microbiol, 5th ed, Am Soc Microbiol; bd dirs, Northeast Asn Clin Microbiol & Infectious Dis, 86-91, pres, 89-90. *Mem:* Royal Micros Soc Eng; Soc Gen Microbiol Eng; Inst Biol Eng; Asn Clin Pathologists Eng; Am Soc Microbiol; AAAS; Sigma Xi. *Res:* Mechanisms by which intracellular infectious agents initiate disease at the cellular and molecular levels; pathogenic mechanisms in legionellosis and listeriosis and Campylobacter infections; ultrastructure of infectious agents and their response to antimicrobials; genetic regulation of virulence. *Mailing Add:* Dept Microbiol Univ NH Durham NH 03824-3544

RODGERS, GLEN ERNEST, b Farmington, Maine, Dec 27, 44; m 66; c 3. INORGANIC CHEMISTRY. *Educ:* Tufts Univ, BS, 66; Cornell Univ, PhD(chem), 71. *Prof Exp:* Asst prof chem, Muskingum Col, 70-75; asst prof, 75-81, ASSOC PROF CHEM, ALLEGHENY COL, 81- *Concurrent Pos:* Vis prof chem, Univ Cincinnati, 77 & Boston Univ, 78; NSF fac develop fel, Univ BC, 81-82. *Mem:* Am Chem Soc. *Res:* Infrared analysis of the interaction between amino acids and heavy metal cations; synthesis and characterization of metal complexes of homoanular ferrocene derivatives; topics in chemical education and interdisciplinary studies. *Mailing Add:* Dept Chem Allegheny Col Meadville PA 16335-3902

RODGERS, JAMES EARL, b Los Angeles, Calif, Aug 19, 43; m 82; c 2. MEDICAL PHYSICS. *Educ:* Calif State Univ, Long Beach, BS, 66; Univ Calif, Riverside, PhD(physics), 72; Am Bd Radiol, cert therapeut radiol physics, 83. *Prof Exp:* Res physicist, Naval Weapons Ctr-Corona, 67-68; res assoc theoret physics, State Univ NY, Albany, 72-76; asst prof therapeut radiol, Tufts Univ Sch Med, radiation oncol res fel & spec & sci staff, 76-80; asst prof, 81-86, ASSOC PROF RADIATION MED, SCH MED, GEORGETOWN UNIV, 87-; DIR RADIATION PHYSICS DIV, DEPT RADIATION MED, GEORGETOWN UNIV HOSP, 80-; DIR, RADIATION SCI DEPT, GEORGETOWN UNIV, 90- *Mem:* Am Phys Soc; Am Col Radiol; Health Physics Soc; Am Asn Phys Med; Am Col Med Physics; Am Soc Therapeut Radiol & Oncol. *Res:* Computer applications in radiation physics and dosimetry; Monte Carlo simulation of radiation energy deposition; digital image processing in radiation therapy. *Mailing Add:* Dept Radiation Med Georgetown Univ Hosp Washington DC 20007

RODGERS, JAMES EDWARD, b Boise, Idaho, Jan 13, 38; m 58; c 2. PHYSICAL ORGANIC CHEMISTRY. *Educ:* Westmont Col, BA, 60; Univ Calif, Berkeley, PhD(chem), 64. *Prof Exp:* Res asst chem, Univ Calif, Berkeley, 60-64; from instr to asst prof, North Park Col, 64-66; from asst prof to assoc prof, Bethel Col, 66-73; vis prof, Westmont Col, 73-74; prof, Bethel Col, 74-75; assoc prof, 75-78, PROF CHEM, AZUSA PAC COL, 78- *Concurrent Pos:* NSF res grant, North Park Col, 65-66. *Mem:* Am Chem Soc; Royal Soc Chem; Royal Inst Chem; Sigma Xi. *Res:* Organic synthesis; photochemistry of allylic systems; reaction mechanisms of the saturated carbon; free radical chemistry. *Mailing Add:* 552 W Comstock Glendora CA 91740

RODGERS, JOHN, b Albany, NY, July 11, 14. FIELD GEOLOGY. *Educ:* Cornell Univ, BA, 36, MS, 37; Yale Univ, PhD(geol), 44. *Prof Exp:* Asst geol, Cornell Univ, 35-36, instr, 36-37; field geologist, US Geol Surv, 40-46; from instr to prof, 46-62, Silliman prof, 62-85, EMER SILLIMAN PROF GEOL,

YALE UNIV, 85- *Concurrent Pos:* Asst ed, Am J Sci, 48-54, ed, 54-; secy, Comn Stratig, Int Geol Cong, 52-60; sr fel, NSF, France, 59-60; vis lectr, Col France, 60; comnr, Conn Geol & Natural Hist Surv, 60-71; Nat Acad Sci exchange scholar, USSR, 67; Guggenheim fel, Australia, 73-74; vpres, Societe Geologique de France, 60. *Honors & Awards:* Medal of Freedom, US Army, 47; Penrose Medal, Geol Soc Am, 81; Gadry Prize, Geol Soc France, 87; Fourmanier Medal, Royal Acad Sci, Fine Arts Letts Belg, 87. *Mem:* Nat Acad Sci; AAAS; Geol Soc Am (pres, 70); Am Asn Petrol Geologists; Am Geophys Union; Am Philos Soc. *Res:* Field geology in deformed sedimentary rocks; stratigraphy and structural geology of Appalachian Mountains; comparative anatomy of mountain ranges. *Mailing Add:* Yale Univ Dept Geol PO Box 6666 New Haven CT 06511-8130

RODGERS, JOHN BARCLAY, JR, b Cleveland, Ohio, Jan 5, 33; m 55; c 3. INTERNAL MEDICINE, GASTROENTEROLOGY. *Educ:* Denison Univ, BA, 55; Harvard Med Sch, MD, 59. *Prof Exp:* Clin & res fel gastroenterol, Mass Gen Hosp, 64-66; from asst prof to assoc prof med, 66-74, PROF MED, ALBANY MED COL, 74-, CHIEF SECT GASTROENTEROL, 81- *Mem:* Am Fedn Clin Res; Am Gastrointestinal Asn; Am Asn Study Liver Dis; Am Soc Clin Invest. *Res:* Small bowel function and lipid absorption; factors influencing sterol absorption. *Mailing Add:* Dept Med Albany Med Col 43 New Scotland Ave Albany NY 12208

RODGERS, JOHN JAMES, b Glasgow, Scotland, Mar 31, 30; US citizen; m 53; c 1. LUBRICATION ENGINEERING, RESEARCH ENGINEERING. *Educ:* Wayne State Univ, BS, 52, MS, 53. *Prof Exp:* Res engr, Res Labs, Gen Motors Corp, 52-59, sr res engr, 59-81, staff res engr, 81-89; CONSULT, 89- *Honors & Awards:* Henry Ford Mem Award, Soc Automotive Engrs, 61. *Mem:* Soc Automotive Engrs; Am Soc Lubrication Engrs. *Res:* Friction in lubricated sliding systems, primarily clutch plate systems in automatic transmissions and controlled-slip rear axles; lubricant-seal compatibility. *Mailing Add:* 31239 W Lyons Circle Warren MI 48090

RODGERS, LAWRENCE RODNEY, SR, b Clovis, NMex, Mar 9, 20; m 43; c 3. MEDICINE. *Educ:* WTex State Col, BS, 40; Univ Tex, MD, 43; Am Bd Internal Med, dipl, 57, cert, 74. *Prof Exp:* Intern, Philadelphia Gen Hosp, Pa, 43-44, resident internal med, 46-49; asst prof, 49-72, ASSOC PROF CLIN MED, BAYLOR COL MED, 72-; PROF CLIN MED, UNIV TEX MED SCH HOUSTON, 72- *Concurrent Pos:* Attend physician, Hermann Hosp, 49-66, chmn dept med, 66-71; assoc internist, Univ Tex M D Anderson Hosp & Tumor Inst, 49- *Mem:* AMA; Am Col Physicians; fel Royal Soc Health. *Res:* Internal medicine; inheritable disorders. *Mailing Add:* 5508 Briar Dr Houston TX 77056

RODGERS, MICHAEL A J, b Chesterfield, Eng, Oct 10, 36. RADIATION CHEMISTRY, PHOTOCHEMISTRY. *Educ:* Univ Manchester, Eng, MSc, 64, PhD(chem), 66. *Prof Exp:* Fel, Lawrence Berkeley Lab, 66-67; sr res assoc, Univ Manchester, Eng, 68-69, lectr chem, 69-76; RES COORDR, CTR FAST KINETICS RES, UNIV TEX, AUSTIN, 76- *Mem:* Am Chem Soc; Royal Soc Chem; Am Soc Photobiol; Radiation Res Soc. *Res:* Nature and properties of unstable, short-lived reaction intermediates in chemistry; application of time-resolved techniques in chemistry and biology. *Mailing Add:* Ctr Photochemical Scis Bowling Green State Univ Bowling Green OH 43403-0002

RODGERS, NELSON EARL, b Fredonia, Pa, May 18, 15; m 37; c 1. MICROBIOLOGY, BIOCHEMISTRY. *Educ:* Allegheny Col, AB, 37; Univ Wis, AM, 40, PhD(bact), 42. *Prof Exp:* Asst biol, Allegheny Col, 36-37 & bact, Univ Wis, 37-42; chief bacteriologist, Western Condensing Co, 42-46, res mgr, 46-56, assoc dir res, Foremost Dairies, Inc, Calif, 56-64; res assoc, Pillsbury Co, 64-76; CONSULT MICROBIOL & BIOCHEM, 76- *Concurrent Pos:* Bacteriologist, Natural Hist Surv, Wis, 38-39; Nutrit Coun, Am Feed Mfrs Asn, 51-54; ed bd, Appl Microbiol, 53-56; AAAS fel, 67. *Honors & Awards:* Indust Achievement Award, Inst Food Technologists, 67. *Mem:* AAAS; Am Soc Microbiol; Am Chem Soc. *Res:* Industrial fermentations; vitamin synthesis by microorganisms; unidentified growth factors in animal nutrition; biochemistry of milk and whey products; microbial polysaccharides; cereal products. *Mailing Add:* 4262 Circle Dr Wayzata MN 55391

RODGERS, RICHARD MICHAEL, b Scranton, Pa, Nov 12, 45. BIOCHEMISTRY. *Educ:* Univ Scranton, BS, 67; Columbia Univ, PhD(biochem), 71. *Prof Exp:* Fel virol, Div Infectious Dis, Stanford Univ, 71-73; sr chemist biochem, Syva Res Inst, 73-78; group leader, 78-81, group leader, Syntex Med Diag, 81-85, GROUP LEADER, SYVA MICROBIOL RES & DEVELOP, SYVA CO, 85- *Mem:* AAAS; Am Chem Soc; Sigma Xi. *Res:* Mechanism of enzyme action; use of enzymes in immunoassays; protein purification; fluorescence immunoassays. *Mailing Add:* 900 Arastradero Rd Palo Alto CA 94304

RODGERS, ROBERT STANLEIGH, b Kew Gardens, NY, Jan 5, 45; m 75; c 2. ANALYTICAL CHEMISTRY. *Educ:* Polytech Inst Brooklyn, BS & MS, 66; Clarkson Col Technol, PhD(chem), 71. *Prof Exp:* Res asst chem, Calif Inst Technol, 70-72; asst prof, Mich State Univ, 72-73; asst prof chem, Lehigh Univ, 73-79; MEM STAFF, PRINCETON APPL RES, EG&G, INC, 79- *Mem:* Am Chem Soc; Electrochem Soc. *Res:* Electrode kinetics; laboratory microprocessors. *Mailing Add:* EG&G Princeton Appl Res PO Box 2565 Princeton NJ 08543-2565

RODGERS, SHERIDAN JOSEPH, b Ellwood City, Pa, Mar 26, 29; m 50; c 4. ANALYTICAL CHEMISTRY, ENVIRONMENTAL HEALTH. *Educ:* Geneva Col, BS, 54. *Prof Exp:* Res chemist, 52-60, SECT HEAD ANALYTICAL CHEM, MSA RES CORP, 60- *Mem:* Am Nuclear Soc; Am Soc Test & Mat; Am Indust Hyg Asn. *Res:* Aerosol generation and sampling; air pollution monitoring; measurement and control of dust and toxic fumes in underground mines and vehicular tunnels; development and testing of protective clothing; monitoring and control of workplace health and safety hazards. *Mailing Add:* RD 2 Box 4655 Ellwood City PA 16117

RODGMAN, ALAN, b Aberdare, Wales, Feb 7, 24; nat US; m 47; c 3. ORGANIC CHEMISTRY. *Educ:* Univ Toronto, BA, 49, MS, 51, PhD(org chem), 53. *Prof Exp:* Asst, Banting & Best, Dept Med Res, Toronto, 47-53, res assoc, 53-54; sr res chemist, R J Reynolds Tobacco Co, 54-65, head, Natural Prod Chem Sect, 65-72 & 74-75, actg mgr chem res, 73, mgr anal res, 75-76, dir res, 76-80, dir fund res, 80-87; CONSULT, 87- *Mem:* Am Chem Soc; NY Acad Sci; Chem Inst Can. *Res:* Composition of tobacco smoke; tobacco smoke and health. *Mailing Add:* 2828 Birchwood Dr Winston-Salem NC 27103

RODIA, JACOB STEPHEN, b Chicago, Ill, Apr 7, 23; m 49; c 5. BIO-ORGANIC CHEMISTRY. *Educ:* Loyola Univ, Ill, BS, 47; Univ Ill, MS, 48, PhD(org chem), 52. *Prof Exp:* Res chemist, Int Minerals & Chem Corp, Ill, 52-54; assoc prof org chem, Drake Univ, 54-56; res chemist, Westinghouse Labs, 56-60 & 3M Res Labs, 60-63; assoc prof, 63-70, PROF CHEM, ST JOSEPH'S COL, IND, 70-, CHMN DEPT CHEM, 81- *Mem:* Am Chem Soc. *Res:* Polymer chemistry; synthesis, mechanism of formation, properties and reactions of new organic and polymeric materials. *Mailing Add:* 1170 Carnaby East Bloomington IN 47401-8730

RODIECK, ROBERT WILLIAM, b Highland Falls, NY, Apr 17, 37; m 61. VISION. *Educ:* Mass Inst Technol, BS, 58, MS, 61; Univ Sydney, PhD(physiol), 65. *Prof Exp:* Lectr physiol, Univ Sydney, 62-67, sr lectr, 67-72, reader physiol, 72-78; BISHOP PROF, UNIV WASH, 78- *Mem:* Neurosci Soc. *Res:* Neurophysiology; retinal neurophysiology. *Mailing Add:* Dept Ophthal RJ-10 Univ Wash Seattle WA 98195

RODIG, OSCAR RUDOLF, b Rahway, NJ, Nov 15, 29; m 64; c 2. ENZYME CHEMISTRY, BIOSYNTHESIS. *Educ:* Rutgers Univ, BS, 51; Univ Wis, PhD(chem), 54. *Prof Exp:* Fulbright fel, Univ Manchester, 54-55; USPHS fel, Swiss Fed Inst Technol, 55-56; asst prof chem, 56-61, ASSOC PROF CHEM, UNIV VA, 61- *Concurrent Pos:* Ramsay Mem fel, 54-55; vis assoc prof, Univ Strasbourg, 61-62, Swiss Fed Inst Technol, 73 & Max Planck Inst Biochem, Munich, 74. *Honors & Awards:* J Shelton Horsley Award, Va Acad Sci, 67. *Mem:* Am Chem Soc; Sigma Xi; Int Soc Herocyclic Chem. *Res:* Chemistry of natural products, especially steroids and antibiotics; mechanisms of enzymatic reactions. *Mailing Add:* Dept Chem Univ Va Charlottesville VA 22901

RODIN, ALVIN E, b Winnipeg, Man, Mar 25, 26; m 51, 74; c 4. PATHOLOGY, MEDICAL EDUCATION. *Educ:* Univ Man, MD, 50, MSc, 59; FRCP(C), 59. *Prof Exp:* Teaching fel path, Queen's Univ, Ont, 56-57; res assoc, Univ Man, 57-59; assoc dir path, Royal Alexandra Hosp, Edmonton, Alta, 59-60; dir path, Misericordia Hosp, 61-63; from asst prof to prof path, Univ Tex Med Br Galveston, 63-75; PROF PATH & CHMN POSTGRAD MED & CONTINUING EDUC, SCH MED, WRIGHT STATE UNIV, 75- *Concurrent Pos:* Consult pathologist, Med-Surg Res Inst, Univ Alta, 59-63. *Mem:* Fel Col Am Path; fel Royal Col Med (Can.); Am Osler Soc; Int Acad Path; fel Royal Soc Med. *Res:* Mercury nephrotoxicity; relationship of pineal to tumor growth; ultrastructure of pineal; radiation induced tumors; congenital heart disease; perinatal disease; medical education; medical history; medical humanism. *Mailing Add:* 3041 Maginn Dr Beaver Creek OH 45385

RODIN, BURTON, b St Louis, Mo, June 19, 33; m 62; c 2. MATHEMATICS. *Educ:* Univ Calif, Los Angeles, BA, 55, PhD(math), 61; Univ Chicago, MS, 58. *Prof Exp:* Asst prof math, Harvard Univ, 61-63, Univ Minn, 63-64, Univ Calif, San Diego, 64-65 & Stanford Univ, 65-66; from asst prof to assoc prof math, 66-72, chmn dept, 77-81, PROF MATH, UNIV CALIF, SAN DIEGO, 72- *Mem:* Am Math Soc. *Res:* Complex analysis. *Mailing Add:* Dept Math C-012 Univ Calif San Diego La Jolla CA 92093

RODIN, ERVIN Y, b Budapest, Hungary, Jan 17, 32; US citizen; m 56; c 3. APPLIED MATHEMATICS. *Educ:* Univ Tex, Austin, BA, 60, PhD(math), 64. *Prof Exp:* Spec instr math, Univ Tex, Austin, 60-64; sr mathematician, Wyle Labs, 64-66; assoc prof, 66-77, PROF APPL MATH, WASH UNIV, 77- *Concurrent Pos:* Organizer & Ed Proceedings, Symp Apollo Appln, 65; mem, Adv Comt Data Processing Systs Antiballistic Missile Defense, Nat Acad Sci, 68-71 & Adv Bd, RETA Consult Environ Engrs, 71-; organizer & gen chmn, Symp Eng Sci in Biomed, 69; chmn, Aleph Found, 71-; organizer & gen chmn, Int Meeting Pollution, Eng & Sci Solutions, Israel, 72; organizer & dir, Appl Finite Element Technol in RR Indust, 75; ed-in-chief, Int J Comput & Math with Appln, Int Series Monographs on Modern Appl Math & Comput Sci, Int J Math Modelling & Appl Math Lett; pres, Inst Appl Sci, Inc. *Mem:* AAAS; Am Math Soc; Math Asn Am; Soc Indust & Appl Math; Soc Eng Sci; Int Asn Math Modeling; Am Inst Aeronaut & Astronaut. *Res:* Applied mathematics, particularly applications of artificial intelligence to differential game theory, to population studies and to the growth of tumors; nonlinear partial differential and similar equations. *Mailing Add:* Dept Systs Sci & Math Wash Univ Box 1040 St Louis MO 63130

RODIN, JUDITH, b Philadelphia, Pa, Sept 9, 44; m 78; c 1. PSYCHIATRY. *Educ:* Univ Pa, Philadelphia, AB, 66; Columbia Univ, PhD(psychol), 70. *Prof Exp:* Asst prof psychol, New York Univ, 70-72; from asst prof to prof psychol, 72-83, dir grad studies, 78-82, PHILIP R ALLEN PROF PSYCHOL, YALE UNIV, 84-, PROF MED & PSYCHIAT, 85-, CHAIR DEPT PSYCHOL, 89-, DEAN, 91- *Concurrent Pos:* Woodrow Wilson fel, 66-68; NSF postdoctoral fel, Univ Calif, Irvine, 71; prin investr, NSF, 73-75 & 77-82, NIH, 81-86, 84-89 & 90-, MacArthur Found, 83-88; mem, Comn Pvt & Pub Needs, Cong Subcomt, 74-75, Comt Substance Abuse & Habitual Behav, Nat Acad Sci, 76-82, Bd Sci Affairs, Am Psychol Asn, 79-82 & Clin Appln & Prev Adv Comt, Nat Heart, Lung & Blood Inst, 80-84; chair, Fogarty Ctr Task Force Res Obesity, NIH, 76-78 & John D & Catherine T MacArthur Found, 83-; assoc ed, Personality & Social Psychol Bull, 76-79, chief ed, Appetite, 79-; vchair, Panel Health & Behav, Inst Med, 79; John Simon Guggenheim Found fel, 86-87. *Honors & Awards:* Katherine D McCormack Distinguished Lectr, Stanford Univ, 82; Katz-Newcomb Lectr, Univ Mich, 84; Distinguished Creative Contrib Award, Gerontol Soc Am, 87; Robert E Miller Lectr, Univ

Pittsburgh, 87; Bishop Robert F Joyce Lectr, Univ Vt, 89; Esther & Isadore Kesten Mem Lectr, Univ Southern Calif, 91. *Mem:* Inst Med-Nat Acad Sci; fel Am Psychol Asn; fel Acad Behav Med Res; fel AAAS; NY Acad Sci; Sigma Xi; fel Am Acad Arts & Sci; fel Soc Behav Med (pres, 90-91); Soc Exp Social Psychol. *Res:* Health-promoting and health-damaging bahavior; author of numerous publications in medical journals. *Mailing Add:* Psychol Dept Yale Univ Two Hillhouse Ave Box 11A New Haven CT 06520

RODIN, MARTHA KINSCHER, b New York, NY, May 18, 29; m 51; c 3. ANATOMY, NEUROPHYSIOLOGY. *Educ:* Wagner Col, BS, 52; Wayne State Univ, MS, 65, PhD(anat), 67. *Prof Exp:* From instr to asst prof, 67-73, ASSOC PROF ANAT, COL MED, WAYNE STATE UNIV, 73- *Concurrent Pos:* Anat consult & assoc psychiat, Lafayette Clin, 72- *Honors & Awards:* Sigma Xi Res Award, 67. *Mem:* AAAS; Am Asn Anat. *Res:* Autonomic nerve endings; mechanisms of convulsive disorders; fluorescent mapping of catacholamines in brain. *Mailing Add:* Three Mountainwood Lane Sandy UT 84092

RODIN, MIRIAM BETH, epidemiology, medical anthropology, for more information see previous edition

RODINE, ROBERT HENRY, b West Pittston, Pa, Nov 9, 29; m 53; c 3. APPLIED MATHEMATICS, PROBABILITY. *Educ:* Mansfield State Col, BSEd, 52; Purdue Univ, MS, 57, PhD, 64. *Prof Exp:* Chemist, Westinghouse Elec Corp, 52-55; instr math & statist, Purdue Univ, 57-64; asst prof statist, State Univ NY Buffalo, 64-68; ASSOC PROF MATH, NORTHERN ILL UNIV, 68- *Mem:* Int Asn Math Modeling; Inst Math Statist. *Res:* Applied probability; applications of probability to the natural sciences and engineering. *Mailing Add:* Dept Math Northern Ill Univ DeKalb IL 60115

RODINI, BENJAMIN THOMAS, JR, b Philadelphia, Pa, Feb 26, 47; m 75; c 2. COMPOSITE MATERIALS. *Educ:* Drexel Univ, BS, 70, MS, 72, PhD(appl mech), 75. *Prof Exp:* Sr structures engr, Fort Worth Div, Gen Dynamics, 75-79; struc eng, Space Syst Div, 79-86, staff eng mat eng, 86-88, MGR, MAT ANALYSIS & DEVELOP, GEN ELEC, 88- *Concurrent Pos:* Adj prof, Drexel Univ, 75. *Res:* Characterization, development, analysis and design of advanced composite structures (polymer and metal matrix). *Mailing Add:* Space Syst Div Gen Elec Co PO Box 8555 U4019 Philadelphia PA 19101

RODKEY, FREDERICK LEE, b Limon, Colo, Apr 13, 19; m 43; c 2. BIOCHEMISTRY. *Educ:* Whitworth Col, BS, 42; Univ Idaho, MS, 43; Harvard Univ, PhD(biol chem), 48. *Prof Exp:* Asst, Univ Idaho, 42-43; tutor biochem sci, Harvard Univ, 44-48, assoc, 49-51, asst prof, 51-58; CHEMIST, US NAVAL MED RES INST, 58- *Concurrent Pos:* Instr biochem, Harvard Univ, 47-50; consult, New Eng Deaconess Hosp, 48-58. *Mem:* Am Chem Soc; Am Soc Biol Chem; Am Asn Clin Chem. *Res:* Biological oxidation reduction; enzymology; clinical chemistry. *Mailing Add:* 3908 Dresden St Kensington MD 20895

RODKEY, LEO SCOTT, b Topeka, Kans, Jan 18, 41; m 63; c 1. IMMUNOLOGY. *Educ:* Univ Kans, BA, 64, PhD(microbiol), 68. *Prof Exp:* NIH fel immunochem, Col Med, Univ Ill, 68-70; asst prof, 70-75, ASSOC PROF BIOL, KANS STATE UNIV, 75- *Concurrent Pos:* Mem, Basel Inst Immunol, 77-78; NIH res career develop award. *Mem:* AAAS; Am Asn Immunologists; Am Asn Pathologists; NY Acad Sci; Am Soc Microbiol. *Res:* Isotypic, allotypic and idiotypic determinants of antibodies; autoantiidiotypic regulation of immune processes; immunopathology. *Mailing Add:* Dept Path Lab Med Univ Tex Health Sci Ctr PO Box 20708 Houston TX 77225-0708

RODKIEWICZ, CZESLAW MATEUSZ, b Turka, Poland. MECHANICAL ENGINEERING. *Educ:* Polish Univ Col, London, dipl ing, 50; Univ Ill, MSc, 63; Case Inst Technol, PhD(mech eng), 67. *Prof Exp:* Res engr transsonic wind tunnel, English Elec Co, 52-54; tech asst mech eng, Dowty Equip Ltd, Can, 54-55; lectr, Ryerson Inst Technol, Toronto, 55-58; from asst prof to prof, 58-84, EMER PROF, DEPT MECH ENG, UNIV ALTA, 84- *Mem:* Fel NY Acad Sci; fel Am Soc Mech Engrs; Can Soc Mech Engrs; Sigma Xi. *Res:* Fluid mechanics and heat transfer, especially hypersonic flight, lubrication, blood flow and ice formations. *Mailing Add:* Dept Mech Eng Univ Alta Edmonton AB T6G 2G7 Can

RODMAN, CHARLES WILLIAM, b Delaware, Ohio, Mar 27, 28; m 48; c 1. ENVIRONMENTAL NOISE, ACOUSTIC INSTRUMENTATION. *Educ:* Ohio State Univ, BSc, 71. *Prof Exp:* Res technician mech eng, 51-59, prin res engr acoust, 59-86, SR NOISE ADV, ENERGY SYSTS DIV, BATTELLE MEM INST, 86- *Concurrent Pos:* Chair, Comt E-33 & Comt Standards, Am Soc Testing & Mat. *Honors & Awards:* Wallace Waterfall Award, Am Soc Testing & Mat. *Mem:* Acoust Soc Am; Am Soc Testing & Mat; Inst Noise Control Eng. *Res:* Community noise; environmental noise; architectural acoustics; machinery noise control; dynamics instrumentation and analysis. *Mailing Add:* Battelle Mem Inst 505 King Ave Columbus OH 43201

RODMAN, HARVEY MEYER, b New York, NY, Sept, 8, 40. ENDOCRINOLOGY, DIABETES. *Educ:* Columbia Col, BA, 62; Univ Chicago, MD, 66. *Prof Exp:* Asst prof, 72-80, ASSOC PROF MED, SCH MED, CASE WESTERN RESERVE UNIV, 80-, RES FEL & DIR, CLIN RES CTR, 72- *Mem:* Am Diabetes Asn; Endocrine Soc; Am Fedn Clin Res. *Res:* Immunology of diabetes; diabetes in pregnancy; secondary causes of diabetes. *Mailing Add:* Univ Hosp 2074 Abington Rd Cleveland OH 44120

RODMAN, JAMES PURCELL, b Alliance, Ohio, Nov 11, 26; m 50; c 4. PHYSICAL MATHEMATICS, PHYSICS GENERAL. *Educ:* Mt Union Col, BS, 49; Wash Univ, MA, 51; Yale Univ, PhD(astrophysics), 63. *Prof Exp:* Instr, dept physics & math, 51-59, assoc prof physics & astron, 62-65, chmn dept, 63-74, dir Comput Ctr, 66-74 & 77, chmn dept physics & astron, 77-85, dir tech eng, Radio Sta WRMU-FM, 70-76, DIR CLARKE OBSERV, 51-, PROF APPL PHYSICS & ASTROPHYS, MT UNION COL, 66- *Concurrent Pos:* From vpres to pres, Alliance Tool Co, 51-58; physicist, Alliance Ware Inc, 51-55, corp secy, 54-55, Alliance Mach Co, 57-72; res assoc, dept astron, Yale Univ, 62-68; dir, Mt Union Bank, 67-71; dir, United National Bank & Trust Co, 71-; trustee, Western Reserve Acad, 70-; vis fel, Yale Univ, 82. *Mem:* Am Phys Soc; Am Astron Soc; Am Geophys Union; fel Royal Astron Soc; Am Optical Soc; fel AAAS; Astron Soc Pac; Am Asn Physics Teachers; Am Inst Physics; Sigma Xi. *Res:* Observational astrophysics of unstable stellar atmospheres; astrophysical-physical instrumentation design; x-ray and beta-spectroscopy. *Mailing Add:* Dept Physics & Astron Mt Union Col 1972 Clark Ave Alliance OH 44601

RODMAN, MORTON JOSEPH, b Boston, Mass, Jan 28, 18; m 43; c 5. PHARMACOLOGY. *Educ:* Mass Col Pharm, BS, 39; Georgetown Univ, PhD, 50. *Prof Exp:* Asst prof biol sci, Rutgers Univ, Piscataway, 50-53, chmn dept, 53-71, from assoc prof to prof pharmacol, Col Pharm, 53-85, chmn dept, 71-81, EMER PROF PHARMACOL, COL PHARM, RUTGERS UNIV, PISCATAWAY, 85- *Concurrent Pos:* Consult accident prev prog, USPHS, 57-64; consult, Dept Neurol & Psychiat, US Vet Admin, 58-80, Bur Maternal & Child Welfare, NJ State Dept Health, 60-64, New York City Dept Health, 60-64 & Dept Drugs, Coun Drugs, AMA, 63-; NSF vis scientist, Am Asn Cols Pharm, 63-; comnr, Narcotic Drug Study Comn, NJ Legis, 63-70. *Honors & Awards:* Lindback Found Award, 62. *Mem:* Fel AAAS; Am Pharmaceut Asn; sci affil AMA; Am Pub Health Asn; Walter Reed Soc. *Res:* Physiology and pharmacology of thermoregulation; toxicology of plant products; organic chemicals in household products; neuropharmacology; psychopharmacology; evaluation of therapeutic efficacy and safety of drugs in clinical use. *Mailing Add:* 41 Wayland Dr Verona NJ 07044-2330

RODMAN, NATHANIEL FULFORD, JR, b Norfolk, Va, July 24, 26; m 51, 70; c 4. PATHOLOGY. *Educ:* Princeton Univ, AB, 47; Univ Pa, MD, 51. *Prof Exp:* Intern, Lankenau Hosp, Philadelphia, 51-52; fel & resident path, Sch Med, Univ NC, Chapel Hill, 52-53 & 55-58, from instr to assoc prof, 58-70; prof, Col Med, Univ Iowa, 70-74; chmn dept, 74-89, PROF PATH, MED CTR, WVA UNIV, 74- *Concurrent Pos:* Nat Heart & Lung Inst res career develop award, 67-70; mem coun thrombosis, Am Heart Asn. *Mem:* AAAS; Electron Micros Soc Am; AMA; Int Acad Path; Soc Exp Biol & Med; Sigma Xi. *Res:* Ultrastructural aspects of problems in thrombosis and atherosclerosis; cytopathology and surgical pathology. *Mailing Add:* Dept Path Sch Med WVa Univ Med Ctr Morgantown WV 26506

RODMAN, TOBY C, b Philadelphia, Pa, May 30, 18; m 43; c 2. CHROMOSOME STRUCTURES, IMMUNOLOGY. *Educ:* Philadelphia Col Pharm & Sci, BSc, 37; NY Univ, MS, 61, PhD(biol), 64. *Prof Exp:* USPHS fel, Columbia Univ, 64-67, res assoc biol sci, 67-69; from instr to assoc prof anat, 69-83, PROF CELL BIOL & ANAT, MED COL CORNELL UNIV, 83- *Concurrent Pos:* Vis assoc prof, Rockefeller Univ, 76-77, adj assoc prof, 77-83, adj prof, 83- *Mem:* AAAS; Am Soc Cell Biol; Am Asn Anat; Soc Study Reproduction. *Res:* Molecular organization of mammalian gametes and fertilized eggs; immunology of sperm. *Mailing Add:* Dept Anat Cornell Univ Med Col New York NY 10021

RODNAN, GERALD PAUL, medicine, for more information see previous edition

RODNEY, DAVID ROSS, b Jane, Mo, May 15, 19; m 42; c 3. HORTICULTURE. *Educ:* Univ Mo, BS, 40; Ohio State Univ, MSc, 46, PhD(hort), 50. *Prof Exp:* Asst prof pomol, State Univ NY Col Agr, Cornell, 48-53; asst horticulturist, Citrus Exp Sta, Univ Calif, 53-57; from assoc horticulturist to horticulturist, Exp Sta, 57-83, EMER PROF PLANT SCI, UNIV ARIZ, 83- *Mem:* Am Soc Hort Sci. *Res:* Citrus rootstocks; tree physiology; plant nutrition. *Mailing Add:* 2025 Cotton Tail Ave Yuma AZ 85364

RODNEY, EARNEST ABRAM, b Scranton, Pa, Mar 30, 17; m 43; c 2. METEOROLOGY, MATHEMATICS. *Educ:* East Stroudsburg State Col, Pa, BS, 40. *Prof Exp:* Meteorologist, Com US Weather Bur, Wash, DC, 45-46; air traffic control meteorol, USAF, 46, meteorologist, Pope Field, NC, 46-48; meteorologist, Com US Weather Bur, Jacksonville, Fla, 48-49, prin asst, Asheville, NC, 49-51, prin asst, Greensboro, NC, 51-56; meteorologist hurricane res & radiomarine, Nat Weather Serv, Wash, DC, 56-61, meteorologist in chg, Com Nat Ocean & Atmospheric Admin, Fletcher, 61-79; RETIRED. *Concurrent Pos:* Teacher math & sci, Pender County Bd Educ, Burgaw, NC, 40-42; meteorologist & opers officer, USAF 42-45. *Mem:* Am Meteorol Soc. *Res:* Hurricane research and quantitative precipitation research; operational field. *Mailing Add:* 28 Westridge Dr Asheville NC 28803

RODNEY, WILLIAM STANLEY, b Scranton, Pa, Dec 20, 26; m 49; c 4. NUCLEAR PHYSICS, ASTROPHYSICS. *Educ:* Univ Scranton, BS, 49; Cath Univ, MS, 53, PhD, 55. *Prof Exp:* Physicist optics, Nat Bur Stand, 49-56; Guggenheim fel, Royal Inst Technol, Sweden, 56-57; physicist optics, Nat Bur Stand, 57-58; physicist, Off Sci Res, USAF, 58-62; prog dir, Nuclear Physics Prog, NSF, 62-87; PHYSICS DEPT, GEORGETOWN UNIV, 87- *Concurrent Pos:* Lectr, Cath Univ, 53; vis prof, Calif Inst Technol, 73-74; adj prof, Georgetown Univ, Wash, DC, 81; Fulbright sr scholar, 83 & 89. *Mem:* Fel AAAS; fel Am Phys Soc; fel Optical Soc Am. *Res:* Nuclear astrophysics. *Mailing Add:* 10707 Muirfield Dr Potamac MD 20854

RODNING, CHARLES BERNARD, b Pipestone, Minn, Aug 4, 43; m 68; c 3. FLEXIBLE ENDOSCOPY. *Educ:* Gustavus Adolphus Col, BS, 65; Univ Rochester, MD, 70; Univ Minn, PhD(anat), 79. *Prof Exp:* Comdr med corps, US Naval Regional Med Ctr, Okinawa, Japan, 79-81; asst prof, 81-84, ASSOC PROF, DEPTS SURG & ANAT, COL MED, UNIV S ALA, 84-, ACTG CHMN, 88- *Concurrent Pos:* Mem SW Oncol Group, 88- *Honors & Awards:* Bacaner Basic Sci Award, Minn Med Found, 79. *Mem:* Asn Acad Surg; Am Col Surgeons; Am Asn Surg Trauma; Soc Am Gastrointestinal Endoscopic Surgeons. *Res:* Gastrointestinal immunology and transplantation; flexible fiberoptic endoscopy. *Mailing Add:* 2451 Fillingim St Mobile AL 36617

RODOLFO, KELVIN S, b Manila, Philippines, Dec 20, 36; US citizen; m 73; c 2. MARINE GEOLOGY, VOLCANOLOGY. *Educ:* Univ Philippines, BS, 58; Univ Southern Calif, MS, 64, PhD, 67. *Prof Exp:* Geologist, San Jose Oil Co, 59-60; asst, Univ Southern Calif, 61-64, res assoc, 64-66; asst prof, 66-70, ASSOC PROF GEOL SCI, UNIV ILL, CHICAGO, 70- *Concurrent Pos:* Geologist, Tidewater Oil Co, 62-63; geol oceanogr, Int Indian Ocean Exped, 64; oceanogr, Environ Sci Serv Admin Res Vessel Oceanogr Global Cruise, 67; lithologist, Leg 16, Deep Sea Drilling Proj, 71, Leg 26, 72 & Leg 59, 78, Leg 126, Ocean Drilling Prog, 89; NSF res grants 67, 69, 70, 71, 78, 85 & 88; Danforth Found Assoc, 80-86; sr scientist, Philippine Inst Volcanology & Seismology, 82- *Mem:* Fel AAAS; fel Geol Soc Am; Soc Econ Paleont & Mineral; Am Geophys Union. *Res:* Regional tectonics of Southeast Asia; deep-sea sedimentology; volcanology; volcaniclastic sedimentology. *Mailing Add:* Dept Geol Sci M/C 186 Univ Ill Chicago Chicago IL 60680

RODOWSKAS, CHRISTOPHER A, JR, b Baltimore, Md, July 26, 39. PHARMACY, PHARMACY ADMINISTRATION. *Educ:* Fordham Univ, BS, 61; Purdue Univ, MS, 63, PhD(pharm), 68. *Prof Exp:* Asst prof pharm, Univ Conn, 64-68; from asst prof to assoc prof, Ohio State Univ, 68-73; vis dir, Pharm Manpower Info Proj, Am Asn Cols Pharm, 72, dir, Off Res & Develop, 73-74, exec dir, 75-81; PROF & CHMN, DEPT PHARM PRACT, HOWARD UNIV, 81- *Concurrent Pos:* Mem, Nat Adv Coun Health Professions Educ, HEW, 76-79 & Bd Dirs, Am Found Pharmaceut Educ, 75-81. *Mem:* AAAS; Am Pharmaceut Asn; Am Pub Health Asn; Am Soc Hosp Pharmacists. *Res:* Application of administrative sciences to health care delivery systems with emphasis on drugs and pharmaceutical services. *Mailing Add:* 4630 Montgomery Ave Suite 201 Bethesda MD 20814

RODRICK, GARY EUGENE, b McPherson, Kans, Oct 5, 43; c 2. PARASITOLOGY, SEAFOOD PROCESSING IRRADIATION & DEPURATION. *Educ:* Kans State Col Pittsburg, BA, 66, MS, 67; Univ Okla, PhD(zool), 71. *Prof Exp:* NIH fel parasitol, Univ Mass, Amherst, 71-73; res assoc pathobiol, Lehigh Univ, 73-75; asst prof biol, Dodge City Col, 75-78; from asst prof to assoc prof comp med, Col Pub Health, Univ SFla, 82-87, vchmn dept comp med & assoc prof, 82-87; ASSOC PROF DEPT FOOD SCI, UNIV FLA, 87- *Mem:* Sigma Xi; Am Soc Parasitol; Am Micros Soc; Am Soc Trop Med & Hyg; Am Soc Zoologists; Am Microbiol Soc; Inst Food Technologists. *Res:* Public health aspects of edible shellfish and fish; seafood processing; schistosomiasis. *Mailing Add:* 8406 SW 103rd Ave Gainesville FL 32608

RODRICK, MARY LOFY, IMMUNOREGULATION, THERMAL INJURY. *Educ:* Univ Calif, Berkeley, PhD(immunol), 81. *Prof Exp:* ASST PROF IMMUNOL, HARVARD MED SCH, 82- *Mailing Add:* Brigham & Women's Hosp Harvard Med Sch 75 Francis St Boston MA 02115

RODRICKS, JOSEPH VICTOR, b Brockton, Mass, Feb 25, 38; m 75; c 2. RISK ASSESSMENT. *Educ:* Mass Inst Technol, BS, 60; Univ Md, PhD(biochem), 68; Am Bd Toxicol, dipl, 81. *Prof Exp:* Res chemist, Food & Drug Admin, 65-69; fel, chem Univ Calif, Berkeley, 69-70; chief, Biochem Br, Food & Drug Admin, 70-72, assoc dir, Div Toxicol, 73-77, dep assoc comnr sci, 78-80; vpres & dir life sci, Clement Assocs, 80-82; ASSOC PROF CHEM, UNIV MD, 65-; PRIN, ENVIRON CORP, 82- *Concurrent Pos:* Mem, Comt Toxicol, Nat Acad Sci-Nat Res Coun, 78-82 & Comt Risk Assessment, 81-83. *Mem:* Acad Toxicol Sci; Soc Risk Analysis; Am Col Toxicol; Am Chem Soc. *Res:* Review and evaluation for toxicity and exposure data for purposes of estimating the risk to public health associated with exposures to environmental pollutants, occupational hazards, food additives, pesticides, industrial chemicals and natural toxins. *Mailing Add:* Environ Corp 4350 N Fairfax Arlington VA 22203

RODRIGO, RUSSELL GODFREY, Can citizen. HETEROCYCLIC SYNTHESIS, ANIONIC REARRANGEMENTS & CYCLISATIONS. *Educ:* Univ Ceylon, Sri Lanka, BSc Hons, 58; Univ Nottingham, Eng, PhD(chem), 63. *Prof Exp:* Asst prof chem, Univ Ceylon, Sri Lanka, 63-67; postdoctoral res asst, Univ Waterloo, Can, 68-69; asst prof, Waterloo Lutheran Univ, Can, 69-76; assoc prof, 73-77, PROF CHEM, WILFRID LAURIER UNIV, CAN, 77-, CHMN DEPT, 87- *Concurrent Pos:* Adj prof, Univ Waterloo, 77-; vis prof, Univ Auckland, NZ, 78. *Mem:* Can Inst Chem. *Res:* Development of novel synthetic pathways to benzylisoquinoline, indole, acridone alkaloids and Aspergillus mycotoxins; syntheses using isobenzofurans as reactive intermediates and leading to lignans, anthracyclinones and aromatic antibiotics. *Mailing Add:* Dept Chem Wilfrid Laurier Univ Waterloo ON N2L 3C5 Can

RODRIGUE, GEORGE PIERRE, b Paincourtville, La, June 19, 31; m 55; c 6. ELECTRICAL ENGINEERING. *Educ:* La State Univ, BS, 52, MS, 54; Harvard Univ, PhD(appl physics), 58. *Prof Exp:* Sr staff engr, Sperry-Rand Corp, 58-61, res consult, 61-68; prof, 68-77, REGENTS PROF ELEC ENG, GA INST TECHNOL, 77- *Concurrent Pos:* Consult, Sperry-Rand Corp, 68-72, US Army Missle Command, 70-73, Airtron, Inc, 76-82, Los Alamos Sci Lab, 81-85; bd dir, Electromagnetic Sci Inc, 69-73. *Mem:* Fel Inst Elec & Electronics Engrs (vpres, 82-83). *Res:* Ferrites; application of ferrites to microwave devices; microwave acoustic phenomena and devices; near field antenna measurements. *Mailing Add:* Dept Elec Eng Ga Inst Tech 225 North Ave NW Atlanta GA 30332

RODRIGUES, MERLYN M, b Lucknow, India, Aug 24, 38; US citizen; m 67; c 1. OPHTHALMIC PATHOLOGY, CELL BIOLOGY. *Educ:* Agra Univ, BS, 55; Madras Univ, India, MD, 59; New Delhi Univ, India, MS, 64; George Washington Univ, PhD(path), 71. *Prof Exp:* Dir path serv, Eye Path, Wills Eye Hosp, 71-76; sect chief clin eye path, Eye Res, Nat Eye Inst, NIH, 77-87; DIR EYE PATH, LAB EYE PATH, UNIV MD, BALTIMORE, 87- *Concurrent Pos:* Chmn, Anat & Path Prog Comt, Asn Res Vision & Opthal, 84-85; mem, Rev Comt, Fight for Sight, 86-88. *Honors & Awards:* Hon Award, Am Acad Ophthal, 84; Sr Sci Investr, Res Prevent Blindness, 88. *Mem:* Am Asn Pathologists; Int Soc Eye Res; Electron Micros Soc Am; Am Acad Opthal; Int Acad Eye Path. *Res:* Characterization of histopathologic changes in corneal genetic diseases; retinoblastoma; retinal degenerations; glaucoma; ocular infections. *Mailing Add:* Ophthalmic Path Lab MSTF Rm 5-00B Univ Md Ten S Pine St Baltimore MD 21201

RODRIGUEZ, ANDRES F, b Havana, Cuba, July 20, 29; US citizen; m 56; c 3. PHYSICS EDUCATION. *Educ:* Univ Havana, Dr Sci(phys chem), 55. *Prof Exp:* Instr, Havana Univ, Cuba, 55-58, asst prof physics, 58-61; lectr physics, Univ PR, Humacao Regional Col, 63-64; from asst prof to prof physics, 64-86, chmn, Dept Physics, 74-80, PROF PHYSICS & ENG-PHYSICS, UNIV PAC, 86- *Concurrent Pos:* Orgn Am States scholar, Inst Physics, Bariloche, Arg, 60, Nat Univ, Mex, 61-63; res assoc, Spec Training Div, Oak Ridge Assoc Univs, Tenn, 67; consult, Int Latin Am Prog, Oak Ridge Assoc Univs, 67-70; expert physics, UNESCO, Univ Antioquia, Colombia, 71-73; sr Fulbright lectr, Colombia SAm, 73 & 74. *Mem:* Am Soc Eng Educ; Am Asn Physics Teachers; Am Phys Soc. *Res:* Physics education; applied physics; engineering physics. *Mailing Add:* Dept Physics Univ Pac 3601 Pac Ave Stockton CA 95211

RODRIGUEZ, ARGELIA VELEZ, b Havana, Cuba, Nov 23, 36; US citizen; m 54; c 2. MATHEMATICS, PHYSICS. *Educ:* Marianao Inst, Cuba, BS, 55; Univ Havana, PhD(math), 60. *Prof Exp:* Asst prof math, Marianao Inst, Cuba, 58-61; asst prof math & physics & head dept, Tex Col, 62-64; from asst prof to assoc prof, 64-72, PROF MATH & DEPT HEAD, BISHOP COL, 72-, CHAIRPERSON, DIV NATURAL & MATH SCI, 75- *Concurrent Pos:* Lectr mod math, NSF In-Serv Inst Sec Teachers, Bishop Col, 65-67; fel comput assisted instruct, Tex Christian Univ, 67-68; dir & coordr, US Off Educ Proj Elem Math Teachers & Teacher Trainers, 70-73; NSF Coop Col Sch Sci Prog Jr High Sch Teachers Math, 72-73; dir & coordr, NSF Math Instrnl Improv Implementation Prog for High Sch Teachers, Dallas Independent Sch Dist, 73-76; liaison officer, Argonne Nat Lab, 75- *Mem:* AAAS; Am Math Soc; Nat Coun Teachers Math; Math Asn Am. *Res:* Classical analysis; solutions of partial differential equations; teaching strategies in mathematics and curriculum development for disadvantaged college students. *Mailing Add:* 1246 W Villaret Blvd San Antonio TX 78224

RODRIGUEZ, AUGUSTO, b New York, NY, Oct 5, 54; m 76; c 2. RESEARCH SCIENCE. *Educ:* Univ PR, BS, 76, PhD(chem), 80. *Prof Exp:* Fel org chem, Emory Univ, 80-82; res chemist polymer chem, E I DuPont de Nemours, 82-85; sr res scientist cellulose chem, Kimberly Clark Corp, 85-88; ASSOC PROF POLYMER & PEROXIDE CHEM, CLARK ATLANTA UNIV, 88- *Concurrent Pos:* Fel, NIH, 80. *Mem:* Am Chem Soc. *Res:* Polysaccharides; use natures polymers to produce thermoplastic materials, superabsorbent products and consumer goods; peroxides, heterocyclic polymers. *Mailing Add:* 665 Wood Valley Trace Roswell GA 30076-9704

RODRIGUEZ, CARLOS EDUARDO, b San Antonio, Tex, Apr 23, 41. COMPUTER SCIENCE. *Educ:* Univ Tex, BS, 62, PhD(chem), 66. *Prof Exp:* Res assoc, Univ Tex, 66-68; staff asst comput animated films, Adv Coun Col Chem, 67-68; asst prof, 68-73, ASSOC PROF COMPUT SCI, E TEX STATE UNIV, 73- *Concurrent Pos:* Coun Libr Resources fel, 74-75. *Mem:* Asn Comput Mach; Asn Develop Comput Based Instrnl Systs. *Res:* Computer-assisted instruction; computer science curriculum development. *Mailing Add:* Dept Comput Sci ETex State Univ Commerce TX 75428

RODRIGUEZ, CHARLES F, b San Antonio, Tex, July 1, 38; m 62; c 4. ANALYTICAL CHEMISTRY, EVIRONMENTAL SCIENCE. *Educ:* St Mary's Univ, Tex, BS, 61. *Prof Exp:* Asst chemist, 61-64, res chemist, 64-74, sr res chemist, Southwest Res Inst, 74-83; CONSULT, 83- *Mem:* Am Chem Soc (treas, 68); Sigma Xi. *Res:* Analytical characterization of petroleum distillate fuels; development of analytical methods for applications in energy; development of analytical methods for environmental; characterization of water, wastewater, soil, air pollutants. *Mailing Add:* 1803 S Zarzamora St PO Box 7386 San Antonio TX 78207-0386

RODRIGUEZ, DENNIS MILTON, b Tampa, Fla, July 21, 43; m 70. MATHEMATICS. *Educ:* Univ S Fla, BA, 65; Univ Calif, Riverside, MA, 66, PhD(math), 69. *Prof Exp:* Asst prof, 69-76, ASSOC PROF MATH, UNIV HOUSTON DOWNTOWN COL, 76- *Mem:* Math Asn Am. *Res:* Probability theory; stochastic processes. *Mailing Add:* Dept Math Univ Houston Downtown Col One Main St Houston TX 77002

RODRIGUEZ, EUGENE, b New York, NY, Mar 26, 33; m 58, 84; c 4. IMMUNOLOGY. *Educ:* Queen's Col, NY, BS, 54; Johns Hopkins Univ, ScM, 59, ScD(microbiol), 66. *Prof Exp:* Instr microbiol, Med Sch, Johns Hopkins Univ, 66-67; mem field staff, Rockefeller Found, 67-70; IMMUNOLOGIST, DIV EXP PATH, DEPT LAB MED, ST JOHN'S MERCY MED CTR, 70- *Concurrent Pos:* Vis prof, Mahidol Univ, Bangkok, Thailand, 67-69. *Res:* Mechanism of antigen-antibody reactions; biological consequences of antigen-antibody reactions such as immediate and delayed hypersensitivity and complement fixation; immunopath of glomerular diseases. *Mailing Add:* 3322 Colard Lane PO Box 913 Lyons CO 80540-0931

RODRIGUEZ, FERDINAND, b Cleveland, Ohio, July 8, 28; m 51; c 2. CHEMICAL ENGINEERING, POLYMER SCIENCE. *Educ:* Case Inst Technol, BS, 50, MS, 54; Cornell Univ, PhD(chem eng), 58. *Prof Exp:* Develop engr, Ferro Chem Corp, Ohio, 50-54; from asst prof to assoc prof chem eng, 58-71, PROF CHEM ENG, CORNELL UNIV, 71- *Concurrent Pos:* Consult, Union Carbide Corp, 60-69. *Mem:* Am Chem Soc; fel Am Inst Chem Engrs; Soc Hisp Prof Engrs; Soc Plastics Engrs. *Res:* Formation, fabrication and evaluation of polymeric materials. *Mailing Add:* Sch Chem Eng Olin Hall Cornell Univ Ithaca NY 14853

RODRIGUEZ, GILBERTO, b Caracas, Venezuela, May 12, 29; m 59; c 6. BIOLOGICAL OCEANOGRAPHY, ECOLOGY. *Educ:* Cent Univ Venezuela, BSc, 55; Univ Miami, MSc, 57; Univ Wales, PhD(zool), 70. *Prof Exp:* Investr marine biol, Univ of the Orient, 59-60; assoc prof biol, Cent Univ Venezuela, 67-77; head, ctr ecol, Venezuelan Inst Sci Res, 81-83; INVESTR BIOL, VENEZUELAN INST SCI RES, 67-; CONSULT, VENEZUELAN INST PETROL TECHNOL, 91- *Concurrent Pos:* Mem, Venezuelan Coun Agr Res, 67-68. *Mem:* Brit Soc Exp Biol; Venezuelan Soc Natural Sci; Crustacean Soc. *Res:* Ecology of estuaries; taxonomy of freshwater crabs; environmental impact. *Mailing Add:* Venezuelan Inst Sci Res PO Box 21827 Caracas Venezuela

RODRIGUEZ, HAROLD VERNON, b New Orleans, La, Aug 30, 32; m 56; c 4. CHEMICAL ENGINEERING. *Educ:* La State Univ, BS, 54, MS, 58, PhD(chem eng), 62. *Prof Exp:* Engr trainee, Natural Gas Dept, Magnolia Petrol Co, 54-55; instr chem eng, La State Univ, 60-61; sr res technologist, Mobil Oil Corp, 61-69; assoc prof chem eng, 69-71, dir div eng, 71-76, DEAN, COL ENG, UNIV SALA, 76- *Mem:* Am Inst Chem Engrs; Sigma Xi; Am Soc Eng Educ; Soc Am Military Engrs; Nat Soc Prof Engrs. *Res:* Application of heat to underground formations for oil recovery. *Mailing Add:* Admin 300 Univ SAla Mobile AL 36688-0002

RODRIGUEZ, JACINTO, b Havana, Cuba, Jan 24, 32; m 61; c 2. HIGH-VOLTAGE & HIGH POWER ELECTRICAL TESTING. *Educ:* Havana Univ, MSEE, 55; Ill Inst Technol, MS, 67. *Prof Exp:* MGR ELEC LABS, S&C ELEC CO, CHICAGO, ILL, 60- *Concurrent Pos:* Elec engr, Cuban Elec Co, 56-60. *Mem:* Inst Elec & Electronics Engrs. *Res:* High voltage and high power electrical testing. *Mailing Add:* S&C Elec Co 6601 N Ridge Blvd Chicago IL 60626

RODRIGUEZ, JOAQUIN, b New York, NY, Jan 9, 34; m 66. GEOLOGY. *Educ:* Hunter Col, BA, 55; Ohio State Univ, MSc, 57; Ind Univ, PhD(geol), 60. *Prof Exp:* Lectr, 59-62, from instr to asst prof, 62-70, assoc prof, 70-79, PROF GEOL, HUNTER COL, 79- *Concurrent Pos:* Comt Examr, Grad Record Exam Geol, 86-91; eval panels, Grad Fel Earth Sci & Instrumentation & Lab Improv, NSF, 89, 90. *Mem:* AAAS; fel Geol Soc Am; Paleont Soc; Nat Asn Geol Teachers; Int Paleont Asn. *Res:* Invertebrate paleontology; paleoecology; palichnology; Devonian-Mississippian brachiopods; biostratigraphy; microcomputer applications to geology. *Mailing Add:* Dept Geol & Geog Hunter Col 695 Park Ave New York NY 10021

RODRIGUEZ, JORGE LUIS, b Yauco, PR, June 6, 40. SEED PRODUCTION, CROP PRODUCTION & MANAGEMENT. *Educ:* Univ PR, Mayaguez, BS, 62; La State Univ, Baton Rouge, MS, 67; Univ Ark, Fayetteville, PhD(agron), 82. *Prof Exp:* Div agronomist, sugarcane prod, C Brewer PR Inc, 62-63; asst field mgr, Sugar Corp PR, 74-76; res asst sugarcane breeding, Gurabo Substa, Univ PR, 63-69 & 77-86, asst to dean, Mayaguez, 86-87, dean & dir, Col Agr, 88-90, ASSOC PLANT BREEDER, GURABO SUBSTA, UNIV PR, 90- *Concurrent Pos:* Agr consult, Sugar Corp PR, 82-, mem, Biomass Comt, 87-; sugarcane consult, Hacienda La Caharta, El Salvador, 85- *Mem:* Caribbean Food Crops Soc. *Res:* New sugarcane varieties adapted to mechanization, high sucrose content or fermentable solids resistant to diseases mainly rust and smut, and with desirable agronomic characters. *Mailing Add:* Agr Exp Sta PO Box 306 Gurabo PR 00658

RODRIGUEZ, JOSE ENRIQUE, b San Juan, PR, Oct 16, 33; m 61; c 2. VIROLOGY. *Educ:* Yale Univ, BS, 55; Univ Pa, PhD(microbiol), 63. *Prof Exp:* Res asst virol, Children's Hosp, Philadelphia, 63-65; NIH fel, Inst Virol, Univ Wurzburg, 65-68; asst prof, 68-74, ASSOC PROF MICROBIOL, UNIV IOWA, 74- *Mem:* Am Soc Microbiol; Sigma Xi. *Res:* Interferons. *Mailing Add:* Dept Microbiol Univ Iowa Iowa City IA 52242

RODRIGUEZ, JUAN GUADALUPE, b Espanola, NMex, Dec 23, 20; m 48; c 4. ENTOMOLOGY. *Educ:* NMex State Univ, BS, 43; Ohio State Univ, MS, 46, PhD(entom), 49. *Prof Exp:* Asst, Ohio State Univ, 46-47, Exp Sta, 48-49; from asst entomologist to assoc entomologist, 49-59, assoc prof, 59-61, PROF ENTOM, UNIV KY, 61- *Concurrent Pos:* Ed, Insect & Mite Nutrit, 72 & Recent Advances Acarol, 79; gov bd, Entom Soc Am, 85-88; chair, centennial comt, Entom Soc Am, 87-89. *Honors & Awards:* Outstanding Award Acarology, Am Reg Prof Entom, 84; J E Russart Mem Award, Entom Soc Am, 86. *Mem:* AAAS; Am Inst Biol Sci; Entom Soc Can; Sigma Xi; Acarol Soc Am; Entom Soc Am. *Res:* Axenic culture of arthropods; insect/mite nutrition; insect pest management; host-plant resistance to insects/mites; nutritional ecology of arthropods. *Mailing Add:* Dept Entom Univ Ky Lexington KY 40546-0091

RODRIGUEZ, LORRAINE DITZLER, b Ava, Ill, July 4, 20; m 48; c 4. MICROBIOLOGY, TOXICOLOGY. *Educ:* Southern Ill Norm Univ, BEduc, 43; Ohio State Univ, MS, 44; Univ Ky, PhD(microbiol), 73. *Prof Exp:* Asst nutritionist, Ohio Agr Res & Develop Ctr, 44-49; fel, Dept Entom, Univ Ky, 73-74; consult, 74-79; EXTENSION SPECIALIST, DEPT ENTOM, UNIV KY, 79- *Mem:* Am Chem Soc; Soc Environ Toxicol & Chem. *Res:* Acaricide resistance; relation of fungi to acarines; biodegradation of pesticides; pesticide impact assessment. *Mailing Add:* 1550 Beacon Hill Rd Lexington KY 40504

RODRIGUEZ, LUIS F, b Arecibo, PR, Dec 30, 47; US citizen; m 72; c 2. NUCLEAR & HAZARDOUS WASTE TREATMENT, MEDICAL DEVICES MANUFACTURING. *Educ:* Univ Puerto Rico, BS, 68; Univ Cincinnati, MS, 70, PhD(chem & nuclear eng), 72, Univ Santa Clara, MBA, 82. *Prof Exp:* Sr engr, Monsanto Res Corp, 73-74; sr engr, Gen Elec Co, Wilmington, NC, 74-78; Prod mgr, San Jose, Calif, 78-88; dir eng, Permutit Co, Zurn Ind, 88-89; DIR ENG, MEDTRONIC, INC, 89- *Mem:* Am Nuclear Soc; Am Inst Chem Engrs; Am Soc Qual Control. *Res:* Interested in hazardous waste treatment, water treatment and management of nuclear waste. *Mailing Add:* C7 Zeus St Monte Olimpo Guaynabo PR 00657

RODRIGUEZ, ROCIO DEL PILAR, b Rio Piedras, PR. PHYTOPATHOLOGY. *Educ:* Univ PR, BSc, 67, MSc, 72; Pa State Univ, PhD(plant path), 88. *Prof Exp:* Instr biol, 72-73, instr plant path, 79-83, ASST PROF PLANT PATH, UNIV PR, 88- *Concurrent Pos:* Consult plant dis, Agr Exp Sta, 81-, prin investr & co-leader, 82-; chairperson, Thesis Comt, Univ PR, 89-; adv, Org Agr Proj, 91- *Mem:* Am Phytopath Soc. *Res:* Chemical control of fungal and bacterial diseases of beans, cowpeas, pigeon peas, citron, vegetables and coffee; host-parasite relations of knot disease of citron and root-rot of alfalfa; screened for resistance to coffee rust and pigeon pea foliar diseases. *Mailing Add:* Dept Plant Protection Col Agr Sci Univ PR Box 5000 Mayaquez PR 00709-5000

RODRIGUEZ, SERGIO, b Lautaro, Chile, Dec 12, 30; m 59; c 2. SOLID STATE PHYSICS. *Educ:* Univ Calif, Berkeley, AB, 55, MA, 56, PhD, 58. *Prof Exp:* Asst prof physics, Univ Wash, 58-59; res asst prof, Univ Ill, 59-60; asst prof, Purdue Univ, 60-61 & Princeton Univ, 61-62; asst prof, Princeton Univ, 61-62; assoc prof, 62-64, PROF PHYSICS, PURDUE UNIV, LAFAYETTE, 64- *Concurrent Pos:* Consult, Ford Motor Co, Mich, Int Bus Mach Corp & Argonne Nat Lab; John Simon Guggenheim Mem fel, 67-68. *Honors & Awards:* Alexander von Humboldt Sr US Scientist Award, Alexander von Humboldt Stiftung, Bonn, Fed Repub Ger, 74. *Mem:* Fel Am Phys Soc. *Res:* Solid state theory; statistical mechanics. *Mailing Add:* Dept Physics Purdue Univ Lafayette IN 47907

RODRIGUEZ-PARADA, JOSE MANUEL, b Orense, Spain, May 4, 53; Venezuelan citizen; m 78; c 1. POLYMER LIQUID CRYSTALS, POLYMERIC SURFACES & THIN FILMS. *Educ:* Universidad Metropolitana, Caracas, BS, 75; Case Western Reserve Univ, MS, 83, PhD(polymer sci), 86. *Prof Exp:* Instr polymer chem, Universidad Metropolitana, Caracas, Venezuela, 75-78; postdoctoral fel polymer sci, Max-Planck-Inst Für PolymerForschung, Mainz, Ger, 87-88; RES STAFF POLYMER SCI, DU PONT CENT RES & DEVELOP, 88- *Mem:* Am Chem Soc. *Res:* Polymer synthesis and characterization; cationic ring-opening polymerization; liquid crystalline polymers; amphiphilic polymers; polymeric surfaces and thin films. *Mailing Add:* Du Pont Cent Res & Develop Exp Sta PO Box 80328 Wilmington DE 19880-0328

RODRIGUEZ-SIERRA, JORGE F, b Havana, Cuba, Sept 18, 45; m 85; c 3. NEUROENDOCRINOLOGY, NEUROSCIENCE. *Educ:* Calif State Col, Los Angeles, BA, 70; Calif State Univ, Los Angeles, MA, 72; Rutgers Univ, PhD(psychobiol), 76. *Prof Exp:* Fel neuroendocrinol, Wis Regional Primate Ctr, 76-77; asst prof, 78-82, assoc prof, 82-89, PROF ANAT, MED CTR, UNIV NEBR, 89- *Concurrent Pos:* Fel anat, Med Ctr, Univ Nebr, 78; asst prof, dept psychol, Univ Nebr, Omaha, 79-82, assoc prof, 82-89; vis scientist, NIMH, 85, prof, 89- *Mem:* Endocrine Soc; Soc Neuroscience; Soc Study Reproduction; Int Soc Neuroendocrinol; Electron Micros Soc Am; Soc Exp Biol Med. *Res:* Control of pituitary gland hormonal release by the brain; control of neural development by hormones; development of contraceptives; role of prostaglandins in behavior and pituitary function; neuroendocrinology of behavior. *Mailing Add:* Dept Anat Med Ctr Univ Nebr Omaha NE 68198-6395

RODRIQUEZ, MILDRED SHEPHERD, b Sterling, Okla, Mar 31, 23; m 43; c 1. NUTRITION, BIOCHEMISTRY. *Educ:* Okla State Univ, BS, 43; Univ Ariz, MS, 63, PhD(agr biochem, nutrit), 69. *Prof Exp:* Res assoc biol chem, Georgetown Univ, 68-69; res nutritionist, Human Nutrit Inst, Agr Res Serv, USDA, 69-72; prof nutrit, Calif State Univ, Northridge, 73-74; PROF NUTRIT, CALIF STATE UNIV, LONG BEACH, 74- *Mem:* AAAS; Am Chem Soc; Am Dietetic Asn; Latin Am Nutrit Soc; Nutrit Ed Soc. *Res:* Human nutrition; metabolism and requirements for proteins, vitamin A and folic acid; developing sensitive criteria for evaluating nutritional status; evaluating nutrition intervention programs; obesity. *Mailing Add:* 2275 Legion Dr Signal Hill CA 90806

RODWELL, JOHN DENNIS, b Boston, Mass, Oct 9, 46; m 71; c 2. MOLECULAR IMMUNOLOGY, MONOCLONAL ANTIBODIES. *Educ:* Univ Mass, Amherst, BA, 68; Lowell Tech Inst, MS, 71; Univ Calif, Los Angeles, PhD(biochem), 76. *Prof Exp:* Fel, Univ Pa Sch Med, 76-80, res asst prof microbiol, 80-81; sr scientist, 81-82, group leader, 82-84, dir chem res, vpres, Discovery Res, 87-89, VPRES, RES & DEVELOP, CYTOGEN CORP, 89- *Concurrent Pos:* Adj asst prof microbiol, Univ Pa Sch Med, 81-89, adj assoc prof, 89- *Mem:* Am Asn Immunologists; Am Chem Soc; AAAS. *Res:* Applications of monoclonal antibodies in in vivo diagnosis and therapy. *Mailing Add:* Cytogen Corp 201 Col Rd E Princeton Forrestal Ctr Princeton NJ 08540

RODWELL, VICTOR WILLIAM, b London, Eng, Sept 10, 29; US citizen; m 52; c 4. BIOCHEMISTRY. *Educ:* Wilson Teachers Col, BS, 51; George Washington Univ, MS, 52; Univ Kans, PhD(biochem), 56. *Prof Exp:* USPHS fel biochem, Univ Calif, Berkeley, 56-58; asst prof, Med Ctr, Univ San Francisco, 58-65; assoc prof, 65-71, PROF BIOCHEM, PURDUE UNIV, WEST LAFAYETTE, 72- *Mem:* Am Soc Biochem & Molecular Biol. *Res:* Regulation of 3-hydroxy-3-methylglutaryl-coenzyme A reductase and cholesterol biosynthesis. *Mailing Add:* Dept Biochem Purdue Univ West Lafayette IN 47907

ROE, ARNOLD, b New York, NY, June 13, 25; m 61; c 5. ENGINEERING. *Educ:* NY Univ, BS, 47; Univ Calif, Los Angeles, MS, 59, PhD(eng), 64. *Prof Exp:* Engr, Am Creosoting Co, 47-48; supt eng, Am & Foreign Power, 48-53; consult, TAMS-Turkish Govt, 53-58; assoc, Univ Calif, Los Angeles, 58-64, proj dir teaching systs lab, 59-64; chmn dept mech & mat, 64-67, PROF ENG, CALIF STATE UNIV, NORTHRIDGE, 67- *Concurrent Pos:* UNESCO consult higher educ, 65-; consult, Technomics, 66. *Honors & Awards:* Award, Am Soc Eng Educ, 59. *Mem:* Am Soc Mech Engrs; Am Soc Eng Educ. *Res:* Programmed learning; computer-aided instruction; adaptive learning systems; power engineering; hydroelectric dams; engineering education. *Mailing Add:* Dept Eng Calif State Univ 18111 Nordhoff St Northridge CA 91330

ROE, BENSON BERTHEAU, b Los Angeles, Calif, July 7, 18; m 45; c 2. CARDIO-THORACIC SURGERY. *Educ:* Univ Calif, AB, 39; Harvard Univ, MD, 43; Am Bd Surg, dipl; Am Bd Thoracic Surg, dipl. *Prof Exp:* Intern surg, Mass Gen Hosp, 43-44, from asst resident to resident, 46-50; asst clin prof, 52-59, assoc prof, 59-68, chief thoracic & cardiac surg, 66, chief cardiac surg, 58-76, co-chief cardiothoracic surg, 77-87, prof surg, 65-89, EMER PROF SURG, SCH MED, UNIV CALIF, SAN FRANCISCO, 89- *Concurrent Pos:* Nat Res Coun fel, Harvard Med Sch, 47-48; Moseley traveling fel, Univ Edinburgh, 51; vis thoracic surgeon, Vet Admin Hosp, Ft Miley; consult thoracic surgeon, St Luke's Hosp; vis surgeon, San Francisco Gen Hosp; dir & chmn, Am Bd Thoracic Surg; vis prof, univs in US, Poland,

Spain, England, Nigeria; sr mem, Cardiovasc Inst, Sch Med, Univ Calif, San Francisco, 58- *Honors & Awards:* Silver Medal, Am Heart Asn. *Mem:* Soc Vascular Surg (vpres); Soc Univ Surg; Am Asn Thoracic Surg; fel Am Col Surg; Am Surg Asn; Soc Thoracic Surgeons (vpres & pres, 72-73); hon mem Polish Surg Asn. *Res:* Cardiopulmonary physiology; author 159 articles including 2 textbooks, 18 textbook chapters, 1 Presidential Address. *Mailing Add:* Dept Surg 593 Univ Calif Sch Med San Francisco CA 94143-0118

ROE, BRUCE ALLAN, b New York, NY, Jan 01, 42; m 63; c 2. BIOCHEMISTRY, MOLECULAR BIOLOGY. *Educ:* Hope Col, BA, 63; Western Mich Univ, MA, 67, PhD(chem), 70. *Prof Exp:* Teacher chem & physics, Marshall Pub Sch, Mich, 63-68; grad teaching fel chem & biochem, Western Mich Univ, 68-70; postdoctoral res fel biochem, State Univ NY, Stony Brook, 70-73; from asst prof to assoc prof chem, Kent State Univ, 73-81; PROF CHEM, UNIV OKLA, 81-, ADJ PROF, BIOCHEM & MOLECULAR BIOL, HEALTH SCI CTR. *Concurrent Pos:* NIH fel, State Univ NY, Stony Brook, 71-72; NIH res grants, Kent State Univ, 74-81, Univ Okla, 81-, NIH res career develop fel, 76-81; res assoc prof, Col Med, Northeastern Ohio Univ, 77-81; Sabbatical res assoc, Med Res Coun, Cambridge, Eng, 78-79; sabbatical res, Med Res Coun, Cambridge, Eng, 90. *Mem:* AAAS; Am Soc Biol Chemists. *Res:* The role of modified nucleotides in mammalian transfer ribonucleic acids; structure of mammalian transfer ribonucleic acids; structure of mammalian oncogenes and transfer ribonucleic acid genes; structure and function of tumor transfer ribonucleic acids. *Mailing Add:* Dept Chem Univ Okla Norman OK 73019

ROE, BYRON PAUL, b St Louis, Mo, Apr 4, 34; m 61; c 2. PHYSICS. *Educ:* Wash Univ, AB, 54; Cornell Univ, PhD, 59. *Prof Exp:* From instr to assoc prof, 59-69, PROF PHYSICS, UNIV MICH, ANN ARBOR, 69- *Concurrent Pos:* Brit sci res fel, 79-80. *Mem:* Fel Am Phys Soc. *Res:* High energy physics; fundamental particles; weak interactions. *Mailing Add:* Dept Physics Univ Mich Ann Arbor MI 48109

ROE, DAPHNE A, b London, Eng, Jan 4, 23; US citizen; m 54; c 3. NUTRITION, MEDICINE. *Educ:* Univ London, MB, BS, 46, MD, 50. *Prof Exp:* Asst pathologist, Royal Free Hosp, London, Eng, 45-46, house physician, 46-47, A M Bird scholar path, 47-48; registr, St John's Hosp Dis Skin, 48-52, first asst, 53-54; res assoc, Univ Pa, 54-57; res assoc, 61-63, asst prof, 63-70, assoc prof clin nutrit, 70-76, PROF NUTRIT, CORNELL UNIV, 76- *Concurrent Pos:* Res fel, Harvard Univ, 53-54; proj dir health & nutrit eval & rehab men & women poverty groups, Dept Labor. *Mem:* AMA; Am Acad Dermat; Soc Invest Dermat; AAAS; Am Heart Asn; Am Public Health Asn; fel Royal Col Physician; Fel Royal Soc Med. *Res:* Keratinization of skin in health and disease; drug induced malnutrition; effects of oral contraceptives on folate status; prenatal malnutrition caused by drugs; cutaneous losses of nutrients; sulfur metabolism with special reference to sulfur requirements for detoxication. *Mailing Add:* Div Nutrit Sci Cornell Univ 108 Savage Hall Ithaca NY 14853

ROE, DAVID CHRISTOPHER, b Toronto, Ont, July 9, 48; m 79. NUCLEAR MAGNETIC RESONANCE. *Educ:* McGill Univ, BSc, 69; Univ Calif-Santa Barbara, PhD(chem), 74. *Prof Exp:* Teaching fel chem, Univ BC, 75-79; RES CHEMIST, E I DU PONT DE NEMOURS & CO, 79- *Mem:* Am Chem Soc; Am Phys Soc; AAAS. *Res:* Nuclear magnetic resonance studies of slow exchange phenomena followed by magnetization transfer; two-dimensional nuclear magnetic resonance methods for the determination of structure and dynamics in solution. *Mailing Add:* Cent Res & Develop Du Pont Exp Sta Wilmington DE 19880-0356

ROE, DAVID KELMER, b Gig Harbor, Wash, Jan 9, 33; div; c 3. ELECTROCHEMISTRY, ANALYTICAL CHEMISTRY. *Educ:* Pac Lutheran Col, BA, 54; State Col Wash, MS, 56; Univ Ill, PhD, 59. *Prof Exp:* NSF fel, 59-60; chemist, Shell Develop Co, 60-62; asst prof, Mass Inst Technol, 62-68; assoc prof, Ore Grad Ctr, 68-72; assoc prof, 72-77, PROF CHEM, PORTLAND STATE UNIV, 77- *Mem:* AAAS; Am Chem Soc; Electrochem Soc. *Res:* Electroanalytical methods; chemical instrumentation. *Mailing Add:* Dept Chem Portland State Univ Portland OR 97207

ROE, GLENN DANA, b Danbury, Conn, Mar 5, 31; m 53; c 2. GEOCHEMISTRY. *Educ:* Tex Christian Univ, BA, 59, MA, 61; Mass Inst Technol, PhD(geochem), 65. *Prof Exp:* Asst astron, Mass Inst Technol, 62-63, asst geol, 63, asst geochronology, 63-64; sr res geologist, 64-71, supvr, Geochem Sect, 71-73, dir, 73-78, DIR MINERALS & GEOCHEM SECTS, ATLANTIC RICHFIELD CO, 78- *Mem:* Am Asn Petrol Geologists. *Res:* Isotopic study of earth mantle rocks, origin and age; geochemical study of sedimentary rocks and their correlation. *Mailing Add:* 7746 El Santo Lane Dallas TX 75248

ROE, JAMES MAURICE, JR, b Clarksville, Tenn, June 22, 42. COMPUTER SUPPORT & ANALYSIS. *Educ:* Austin Peay State Univ, BS, 65; Univ Ala, MBA, 87. *Prof Exp:* Power plant design engr, Chrysler Corp, 66-68; reliability engr, Fed Elec Corp, 68-70; mem tech staff, Computer Sci Corp, 70-74; sr engr, Intergraph Corp, 74-78; dir, computer serv & instr computer sci, Athens State Col, 78-81; proj mgr, Intermetrics Inc, 81; mgr, Macy Ctr Acad Res, 81-87, MGR PLANNING & STAFF SERV, HEALTH INFO SYSTS, UNIV ALA, BIRMINGHAM, 87- *Concurrent Pos:* Instr, Athens State Col, 77-78. *Mem:* Digital Equip Computer Users Soc; Healthcare Info & Mgt Systs Soc. *Res:* Computer support and analysis for the Lipid Research Project and for coronary heart disease risk factor in young adults; systems management; control software for space experiments. *Mailing Add:* Health Info Systs Univ Ala Hosp 619 19th St S Birmingham AL 35233

ROE, KENNETH A, industrial engineering, naval architecture; deceased, see previous edition for last biography

ROE, PAMELA, b San Angelo, Tex, Oct 18, 42. ZOOLOGY. *Educ:* Univ Tex, Austin, BA, 65; Univ Wash, MS, 67, PhD(zool), 71. *Prof Exp:* PROF BIOL SCI, CALIF STATE UNIV, STANISLAUS, 71- *Mem:* AAAS; Ecol Soc Am; Am Soc Zoologists; Am Soc Naturalists. *Res:* Marine invertebrate natural history and ecology; invertebrate zoology. *Mailing Add:* Dept Biol Sci Calif State Univ Stanislaus Turlock CA 95380

ROE, PETER HUGH O'NEIL, b Birmingham, Eng, May 18, 34; m 58; c 2. GRAPH THEORETIC SYSTEMS MODELLING & DESIGN, COMPUTER NETWORK DEVELOPMENT. *Educ:* Univ Toronto, BASc, 59; Univ Waterloo, MSc, 60, PhD(elec eng), 63 , Univ Technol, Compeigne, France, Dipl, 76. *Prof Exp:* Lectr math, 60, lectr elec eng, 60-63, asst prof, 63-64, asst prof design & elec eng, 65, assoc prof, 65-69, assoc dean, eng, 77, 78 & 80-86, PROF SYSTS DESIGN, UNIV WATERLOO, 69- *Concurrent Pos:* Vis asst prof, Thayer Sch Eng, Dartmouth Col, 63-64; consult, Inst Design, Univ Waterloo, 65-; vis assoc prof, NS Tech Col, 68-69; vis prof, Technol Univ Compiegne & Advan Sch Eng Marseille, France, 74-75. *Mem:* Inst Elec & Electronics Engrs. *Res:* Systems theory (modelling and simulation), bond graphic, graph theory, matroids; computer operating systems, local area networks, theory of engineering design. *Mailing Add:* Dept Systs Design Univ Waterloo Waterloo ON N2L 3G1 Can

ROE, RYONG-JOON, b Pyongyang, Korea, Mar 22, 29; US citizen; m 61; c 3. POLYMER SCIENCE, PHYSICAL CHEMISTRY. *Educ:* Seoul Nat Univ, BS, 52, MS, 55; Univ Manchester, PhD(polymer chem), 57. *Prof Exp:* Res chemist, Arthur D Little Res Inst, Inveresk, Scotland, 57-60; res assoc polymer chem, Duke Univ, 60-63; res chemist, E I du Pont de Nemours & Co, Inc, 63-68; mem tech staff, Bell Labs, Inc, 68-75; assoc prof, 75-80, PROF MAT SCI & METALL ENG, UNIV CINCINNATI, 80- *Concurrent Pos:* Vis scholar, Centre Recherche Macromolecule, Strasbourg, France, 83-84. *Mem:* Am Phys Soc; Am Chem Soc. *Res:* Physics and chemistry of polymers; thermodynamics of polymers; surface properties of polymers; application of x-ray diffraction to study of polymers; small-angle x-ray scattering; computer simulation of polymers. *Mailing Add:* Univ Cincinnati Mail Location 12 Cincinnati OH 45221-0012

ROE, WILLIAM P(RICE), b Dover, Del, Aug 25, 23; m 43; c 2. METALLURGY. *Educ:* Vanderbilt Univ, AB, 47, MS, 48, PhD(metall, inorg chem), 52. *Prof Exp:* Process develop chemist, Carbide & Carbon Chems Co Div, Union Carbide Corp, 48-49; res metallurgist, Titanium Div, Nat Lead Co, 51-56; sr metallurgist, Southern Res Inst, 56-57; sect leader, Cent Res Labs, 57-60, res supt, 60-63, mgr, 63-69, dir, 69-74, vpres, Asarco, Inc, 74-86; dir, Int Lead Zinc Res Orgn, 72-86; RETIRED. *Concurrent Pos:* Chmn, Int Lead Res Orgn, 83-85; mem, comt vis, Sch Eng, Vanderbilt Univ, 84-; dir, ASA Ltd, 81- *Mem:* Am Soc Metals; Am Inst Mining, Metall & Petrol Engrs; Sigma Xi. *Res:* Metallurgy of nonferrous metals. *Mailing Add:* 320 San Juan Dr Ponte Vedra Beach FL 32082

ROEBBER, JOHN LEONARD, b Bonne Terre, Mo, Mar 23, 31; m 55; c 3. PHYSICAL CHEMISTRY. *Educ:* Wash Univ, AB, 53; Univ Calif, PhD(chem), 57. *Prof Exp:* Sr chemist, Res Ctr, Texaco, Inc, NY, 57-64; asst prof, 64-70, ASSOC PROF CHEM, NORTHEASTERN UNIV, 70-, EXEC OFFICER, 76- *Mem:* Am Chem Soc. *Res:* Photochemistry; reaction kinetics; free radicals; infrared and ultraviolet spectroscopy; low temperature chemistry. *Mailing Add:* 77 Old Post Rd East Walpole MA 02032-1417

ROEBUCK, ISAAC FIELD, b Graham, Tex, Sept 6, 30; m 57; c 1. PETROLEUM ENGINEERING. *Educ:* Univ Tex, BS, 53, MS, 55. *Prof Exp:* From instr to asst prof petrol eng, Univ Tex, 54-57; petrol reservoir engr, Core Lab, Inc, 57-58, sr reservoir engr, 58-59, supv petrol engr, 59-61, asst mgr eng, 61-70; vpres & gen mgr, Eng Numerics Corp, 71-73; mgr educ serv, Core Labs, Inc, 74-78; pres, Roebuck-Walton, Inc, 79-88; PRES, ROEBUCK ASSOCS, INC, 88- *Concurrent Pos:* Chmn continuing educ, Soc Petrol Engrs, 84-85; adj prof geol sci, Southern Methodist Univ, 82- *Mem:* Soc Petrol Engrs; Am Asn Petrol Geologists; AAAS; Soc Independent Prof Engrs. *Res:* Flow through porous media; improved oil recovery; computer science; petroleum economics and risk analysis. *Mailing Add:* PO Box 25024 Roebuck-Walton Inc Dallas TX 75225

ROECKER, ROBERT MAAR, b US, Nov 30, 22; m 58. VERTEBRATE ZOOLOGY. *Educ:* Cornell Univ, BS, 47, MS, 48, PhD(zool), 51. *Prof Exp:* Biologist, NY State Dept Conserv, 50-62; assoc prof, 62-70, prof biol, State Univ NY Col Geneseo, 70-85; RETIRED. *Mem:* Am Soc Mammal; Wildlife Soc; Am Soc Ichthyologists & Herpetologists. *Res:* Life history of gray squirrel; life history and management work on New York waters; mammalogy; ichthyology; herpetology; wildlife and natural resource conservation; vertebrate taxonomy. *Mailing Add:* 4309 Reservoir Rd Geneseo NY 14454

ROEDDER, EDWIN WOODS, b Monsey, NY, July 30, 19; m 45; c 2. GEOCHEMISTRY. *Educ:* Lehigh Univ, BA, 41; Columbia Univ, AM, 47, PhD(geol), 50. *Hon Degrees:* DSc, Lehigh Univ, 76. *Prof Exp:* Res engr, Bethlehem Steel Co, Pa, 41-46; asst geol, Columbia Univ, 46-49; from asst prof to assoc prof mineral, Univ Utah, 50-55; chief solid state group, Geochemical & Petrol Br, 55-60, staff geologist, 60-62, geologist, US Geol Surv, 62-87; ASSOC, DEPT EARTH & PLANETARY SCI, HARVARD UNIV, 87- *Concurrent Pos:* Mem comt geochemical res, NSF, 54-55. *Honors & Awards:* Except Sci Achievement Medal, NASA, 73; Werner Medal, German Mineral Asn, 85; Roebling Medal, Mineral Soc Am, 86; Penrose Medal, Soc Econ Geologists, 88. *Mem:* Nat Acad Sci; Soc Econ Geol; Mineral Soc Am (vpres, 81-82, pres, 82-83); Am Geophys Union; Geochem Soc (pres, 76-77). *Res:* Ore deposition; fluid inclusions in minerals; studies of lunar materials; nuclear waste storage problems; volcanology. *Mailing Add:* Dept Earth & Planetary Sci Harvard Univ Cambridge MA 02138

ROEDEL, GEORGE FREDERICK, b Saginaw, Mich, July 1, 16; m 42; c 3. POLYMER CHEMISTRY. *Educ:* Valparaiso Univ, BS, 38; Purdue Univ, MS, 40, PhD(agr chem), 42. *Prof Exp:* Asst, Purdue Univ, 42; assoc, Res Lab, Gen Elec Co, 43-50; vpres, Tewes-Roedel Plastics Corp, 50-55; prod engr, Chem Mat Dept, 55-56, mgr chem & insulation, AC Motor & Generator Lab, 56-60, specialist silicone resin chem, Silicone Prod Dept, Gen Elec Co, 60-83; RETIRED. *Concurrent Pos:* Assoc prof, Carroll Col, Wis, 50-55. *Mem:* Am Chem Soc. *Res:* Composition and properties of oils and fats; methyl silicone resins; silicone rubber; copolymerization of vinyl silicon compounds; boron chemistry; suspension polymerization of chlorotrifluoroethylene; research and development on silicone resins. *Mailing Add:* 2178 Apple Tree Lane Schenectady NY 12309-4739

ROEDER, CHARLES WILLIAM, b Hershey, Pa, Oct 12, 42; m 69; c 1. STEEL STRUCTURES, SEISMIC DESIGN. *Educ:* Univ Colo, Boulder, BS, 69; Univ Ill, Urbana, MS, 71; Univ Calif, Berkeley, PhD(civil eng), 77. *Prof Exp:* Struct engr, J Ray McDermott, Inc, 71-74; res asst, Univ Calif, 74-77; PROF CIVIL ENG, UNIV WASH, 77- *Concurrent Pos:* Prin investr, NSF, Nat Acad Sci, & Am Inst Steel Construct, 77-; chmn, tech comts, Am Soc Civil Engrs, 83- & Comt Steel Bridges, Transp Res Bd, 90- *Honors & Awards:* J James Croes Medal, Am Soc Civil Engrs, 79, Raymond C Reese Prize, 84. *Mem:* Am Soc Civil Engrs; Earthquake Eng Res Inst; Am Inst Steel Construct; Am Welding Soc. *Res:* Structural engineering; seismic behavior of steel structures; thermal movements in bridges; behavior of bridge bearings; heat straightening and curving of steel; inelastic behavior of steel structures; behavior of elastomeric bearings. *Mailing Add:* Univ Wash 233 B More Hall FX-10 Seattle WA 98195

ROEDER, DAVID WILLIAM, b Philadelphia, Pa, June 19, 39; m 66; c 2. PURE MATHEMATICS. *Educ:* Univ NMex, BS, 60; Univ Calif, Berkeley, MA, 62, Univ Calif, Santa Barbara, PhD(math), 68. *Prof Exp:* John Wesley Young Res Instr Math, Dartmouth Col, 68-70; from asst prof to assoc prof, 70-85, chmn dept, 75-85, PROF MATH, COLO COL, 85- *Mem:* Am Math Soc; Math Asn Am. *Res:* Duality theory of locally compact groups and topological groups; number theory and quadratic forms. *Mailing Add:* Dept Math Colo Col Colorado Springs CO 80903

ROEDER, EDWARD A, b Sellersville, Pa, Apr 25, 39. SOLID STATE PHYSICS. *Educ:* Lafayette Col, BS, 61; Lehigh Univ, MS, 63, PhD(physics), 67. *Prof Exp:* Asst prof, 67-74, ASSOC PROF PHYSICS, MORAVIAN COL, 74- *Mem:* Am Asn Physics Teachers. *Res:* Transport number measurements in silver chloride at high temperatures. *Mailing Add:* Dept Physics Moravian Col Main St & Elizabeth Ave Bethlehem PA 18018

ROEDER, LOIS M, b 1932. PEDIATRICS. *Educ:* Johns Hopkins Univ, ScD, 71. *Prof Exp:* ASSOC PROF PEDIAT, DEPT PEDIAT, SCH PHARM & SCH MED, UNIV MD, 73- *Mem:* Am Soc Neurochemistry; Am Inst Nutrit. *Res:* Metabolic aspects of brain development; etiology of sudden infant death syndrome. *Mailing Add:* Dept Pediat Sch Med Univ Md 655 W Baltimore St Baltimore MD 21201

ROEDER, MARTIN, b Long Branch, NJ, Aug 19, 25; m 57; c 2. PHYSIOLOGY, BIOCHEMISTRY. *Educ:* Queens Col, NY, BS, 48; Univ NMex, MS, 51; Univ NC, PhD(zool), 54. *Prof Exp:* Asst biol, Univ NMex, 49-51; US AEC fel, Univ NC, 51-53, asst zool, 53-54; asst prof chem, Woman's Col NC, 54-56, biol, 56-59, assoc prof, 59-64; from asst dean to actg dean arts & sci, 66-74, assoc prof, 64-89, PROF BIOL SCI, FLA STATE UNIV, 90- *Concurrent Pos:* Vpres finance & admin, Fla State Univ, 90- *Mem:* Fel AAAS; Am Soc Zoologists; Soc Gen Physiol; Asn Southeastern Biologists; Sigma Xi. *Res:* Cellular physiology; enzyme induction; gene-enzyme relationships; respiratory activity of tumor cells; mineral nutrition; bioluminescence and evolutionary significance of bioluminescence. *Mailing Add:* Dept Biol Sci Fla State Univ Tallahassee FL 32306-2043

ROEDER, PETER LUDWIG, b Medford, Mass, Mar 6, 32; m 53; c 3. GEOCHEMISTRY, PETROLOGY. *Educ:* Tufts Univ, BS, 54; Pa State Univ, PhD(geochem), 60. *Prof Exp:* Fel geochemistry, Pa State Univ, 60-61 & NMex Inst Mining & Technol, 61-62; assoc prof, 62-70, PROF GEOCHEM & HEAD GEOL DEPT, QUEEN'S UNIV, ONT, 70- *Honors & Awards:* Past Pres Medal, Mineral Asn Can. *Mem:* Mineral Soc Am; Mineral Asn Can. *Res:* High temperature phase equilibrium studies in systems analogous to basic igneous rocks; analytical problems in petrology. *Mailing Add:* Miller Hall Queen's Univ Kingston ON K7L 3N6 Can

ROEDER, ROBERT CHARLES, b Stratford, Ont, Oct 7, 37; m 61; c 2. ASTRONOMY, PHYSICS. *Educ:* McMaster Univ, BSc, 59, MSc, 60; Univ Ill, Urbana, PhD(astron), 63. *Prof Exp:* Asst prof astron, Univ Ill, Urbana, 62-63; lectr physics, Queen's Univ, Ont, 63-64; from asst prof to prof astron, Univ Toronto, 64-83; PROF PHYSICS, SOUTHWESTERN UNIV, 83- *Concurrent Pos:* Consult, Kitt Peak Nat Observ, 71-72; vis prof physics, Univ Tex, 79-80. *Mem:* Am Astron Soc; Am Physical Soc; Int Astron Union; Int Soc Gen Relativity & Gravitation; Am Asn Physics Teachers. *Res:* Cosmology-determination of models of universe; quasars-investigations of the nature of redshifts; relativistic optics. *Mailing Add:* Dept Physics Southwestern Univ Georgetown TX 78626

ROEDER, ROBERT GAYLE, b Boonville, Ind, June 3, 42; m 90; c 2. NUCLEIC ACID BIOCHEMISTRY. *Educ:* Wabash Col, Ind, BA, 64; Univ Ill, Urbana, MS, 65; Univ Wash, Seattle, PhD(biochem), 69. *Hon Degrees:* DSc, Wabash Col, 90. *Prof Exp:* Pre-doctoral fel, US Pub Health Serv, 65-69; from asst prof to prof biol chem, Sch Med, Wash Univ, 71-78, prof genetics, 78-82; PROF BIOCHEM & MOLECULAR BIOL, ROCKEFELLER UNIV, 82- *Concurrent Pos:* Gilbert Scholar, Wabash Col, 60-64; NIH res career develop award, 73-78; fel, Am Cancer Soc, Carnegie Inst Wash, 69-71; mem, Molecular Biol Study Sect, NIH, 75; James S McDonnell prof biochem genetics, 79-82, Arnold & Mabel Beckman prof biochem & molecular biol, 85-; consult, Am Cancer Soc, 83-86; chairperson, Gordon Res Conf on Nucleic Acids, 82. *Honors & Awards:* Eli Lilly Award Biol Chem, Am Chem Soc, 77; US Steel Found Award Molecular Biol, Nat Acad Sci, 86. *Mem:* Nat Acad Sci; Am Soc Biol Chemists; Am Soc Microbiologists; AAAS; Am Chem Soc; NY Acad Sci; Harvey Soc. *Res:* Regulation of gene expression during cellular growth, differentiation, and transformation; isolation and mechanistic analysis of cellular and viral factors involved in transcriptional control. *Mailing Add:* Rockefeller Univ 1230 York Ave New York NY 10021-6399

ROEDER, STEPHEN BERNHARD WALTER, b Dover, NJ, Aug 26, 39; m 69; c 2. NUCLEAR MAGNETIC RESONANCE, SCIENTIFIC INSTRUMENTATION. *Educ:* Dartmouth Col, BA, 61; Univ Wis, PhD(chem), 65. *Prof Exp:* Mem tech staff, Bell Labs, Inc, 65-66; instr physics, Univ Ore, 66-68; from asst prof to assoc prof, San Diego State Univ, 68-74, chmn dept, 75-78, chmn dept chem, 79-86, dir, MLA Prog, 86-87 & 89-90, PROF PHYSICS & CHEM, SAN DIEGO STATE UNIV, 74- *Concurrent Pos:* Fel chem, Univ Ore, 66-68; vis prof, Univ BC, 74-75 & Tex A&M Univ, 82; vis staff mem, Los Alamos Sci Lab, 74-; consult, Lovelace Med Found, 86- *Mem:* AAAS; Am Chem Soc; Am Phys Soc. *Res:* Pulsed nuclear magnetic resonance; instrumentation; magnetoencephalography. *Mailing Add:* Dept Physics San Diego State Univ San Diego CA 92182

ROEDERER, JUAN GUALTERIO, b Trieste, Italy, Sept 2, 29; US citizen; m 52; c 4. SPACE PHYSICS, PSYCHOACOUSTICS. *Educ:* Univ Buenos Aires, DSc(physics), 52. *Prof Exp:* Teaching asst physics, Univ Buenos Aires, 52-53; res asst high energy physics, Max Planck Inst, 53-55; prof physics, Univ Buenos Aires, 56-67; prof, Univ Denver, 67-77; dir, Geophys Inst, 77-86, PROF, UNIV ALASKA, 86- *Concurrent Pos:* Vis staff mem, Dept Energy, Los Alamos Sci Lab, 68-86; pres, Sci Comn Solar Terrestrial Physics, Int Coun Sci Unions, 86-90; mem, polar res bd, Nat Acad Sci, 80-84, atmosphere sci & climate bd, 83-85; chmn, US Arctic Res Comn, 87- *Mem:* Fel Am Geophys Union; Arg Geophys Soc; Acoustical Soc Am; Int Asn Geomag & Aeronomy (pres, 75-79); fel AAAS; corresp Acad Sci Arg; corresp Acad Sci Austria. *Res:* Physics of the Magnetosphere, particularly radiation belts, diffusion of trapped particles; psychol physics of music; perception of musical sounds. *Mailing Add:* Geophys Inst Univ Alaska Fairbanks AK 99775-0800

ROEDERER, MARIO, b Buenos Aires, Arg, Apr 30, 63; US citizen. GENETICS. *Educ:* Harvey Mudd Col, BS, 83; Carnegie-Mellon Univ, PhD(biol sci), 88. *Prof Exp:* Res asst, Carnegie-Mellon Univ, 83-88; POSTDOCTORAL FEL GENETICS, STANFORD UNIV, 88- *Concurrent Pos:* Spec fel, Leukemia Soc Am, Inc, 91- *Mem:* Am Soc Cell Biol; AAAS; Sigma Xi. *Res:* Regulation of mammalian gene expression by inflammatory cytokines and intracellular thiols, especially that of the human immunodeficiency virus; advanced flow cytometric tools for measurement of gene expression during development, especially in the immune system. *Mailing Add:* Dept Genetics Stanford Univ Stanford CA 94305

ROEGER, ANTON, III, b Philadelphia, Pa, Oct 19, 35; m 67; c 2. TECHNICAL MANAGEMENT, FUEL TECHNOLOGY & CHEMICAL ENGINEERING. *Educ:* Lehigh Univ, BS, 57; Pa State Univ, MS, 59; Univ Va, DSc(chem eng), 63. *Prof Exp:* Res engr, Indust Chem Div, Res & Develop Lab, Shell Chem Co, Deer Park, 62-71; sr engr, Tex Eastern Transmission Corp, 71-77, res adv, 77-78, coordr, Res & Technol Div, 78-80, tech mgr, Synfuels Div, 80-84, tech mgr, Res & Bus Develop Div, 84-86, prog mgr energy utilization, 86-89, tech mgr, Corp Environ Div, 89; RETIRED. *Concurrent Pos:* Proj adv, Cogeneration & Power Systs & res adv, Gas Process Res, Gas Res Inst. *Mem:* Am Inst Chem Engrs; Sigma Xi. *Res:* Synthetic gaseous and liquid fuels; chemicals; two phase flow; radiation chemistry; process research and development; cogeneration of steam and electricity. *Mailing Add:* 4618 Shatner Dr Houston TX 77066

ROEHL, PERRY OWEN, b Detroit, Mich, Jan 2, 25; m 52; c 4. PETROLEUM GEOLOGY. *Educ:* Ohio State Univ, BS, 50; Stanford Univ, MS, 52; Univ Wis, PhD(geol), 55. *Prof Exp:* Exploitation engr, Shell Oil Co, 55-57, from prod geologist to sr prod geologist, 57-62, sr prod geologist, Shell Oil Co Can, Ltd, 62, res geologist, Shell Develop Co, 62-66; res assoc, Union Oil Co Calif, 66-75; consult, 75-81; HERNDON DISTINGUISHED PROF GEOL, TRINITY UNIV, SAN ANTONIO, TEX, 81- *Concurrent Pos:* Chmn carbonate adv comt, Am Petrol Inst, 70-75; Esso Distinguished lectr, Australia, 86-87. *Mem:* Fel Geol Soc Am; Am Asn Petrol Geol; Soc Econ Paleont & Mineral; Am Inst Prof Geol; Am Inst Mining, Metall & Petrol Geol; Sigma Xi. *Res:* Carbonate geology, recent sedimentation; petroleum geology; paleoecology; sedimentary petrography; petrophysics; invertebrate paleontology; stratigraphy; marine geology; geological engineering. *Mailing Add:* Dept Geol Trinity Univ 715 Stadium Dr San Antonio TX 78284

ROEHRIG, FREDERICK KARL, b Peoria, Ill, June 25, 42; m 67. PHYSICAL METALLURGY, METALLURGICAL ENGINEERING. *Educ:* Bradley Univ, BS, 65; Univ Ill, MS, 67; Ohio State Univ, PhD(metall eng), 76. *Prof Exp:* Res metallurgist, Battelle-Columbus Lab, 67-72; res assoc metallurgy, Owens-Corning Fiberglas Tech Ctr, 76-87; VPRES, GELLES LABS, INC, 87- *Mem:* Sigma Xi; Am Soc Metals; Am Inst Mining, Metall & Petrol Eng; Am Powder Metall Inst; Nat Soc Prof Eng. *Res:* Physical metallurgy of high temperature alloys, particularly powder metallurgy; failure analysis of ferrous-non-ferrous alloy systems; field-freezing and electrotransport; patents. *Mailing Add:* 4800 Hayden Blvd Columbus OH 43221

ROEHRIG, GERALD RALPH, b Aurora, Ill, Nov 2, 41; m 63; c 3. ORGANIC CHEMISTRY, PHYSICAL CHEMISTRY. *Educ:* Aurora Col, BS, 63; Univ Ky, MS, 65; Ind Univ, PhD(org chem), 70. *Prof Exp:* Asst prof chem, Aurora Col, 65-75, prof, 75-79; ASST PROF, ORAL ROBERTS UNIV, 79- *Concurrent Pos:* Chmn, Dept Chem, Aurora Col, 70-79, chmn, Div Natural Sci, 74-79, res dir, Res Corp, 76-79; educ consult, Argonne Nat Lab, 71-77; consult, Process Systs Div, John Zink Co, 80-82. *Mem:* AAAS; Am Chem Soc. *Res:* Synthesis of potential antitumor agents. *Mailing Add:* 2306 Stonebrook Carrollton TX 75007-5726

ROEHRIG, JIMMY RICHARD, b Yokohama, Japan, Oct 12, 49. COLLIDING BEAM PHYSICS. *Educ:* Univ Chicago, BA, 71, PhD(physics), 77. *Prof Exp:* Res assoc, Univ Chicago, 77-78; RES ASSOC, STANFORD LINEAR ACCELERATOR CTR, 78- *Mailing Add:* 28 Roosevelt Circle Palo Alto CA 94306

ROEHRIG, KARLA LOUISE, b Sycamore, Ill, Aug 18, 46; m 67. BIOCHEMISTRY, NUTRITION. *Educ:* Univ Ill, BS, 67; Ohio State Univ, PhD(phys chem), 77. *Prof Exp:* Res asst nutrit biochem, Inst Nutrit, Ohio State Univ, 67-71 & Dept Food Sci & Nutrit, 71-75; res assoc biochem, Sch Med, Ind Univ, 77-78; asst prof, 78-83, ASSOC PROF NUTRIT BIOCHEM, DEPT FOOD SCI & NUTRIT, OHIO STATE UNIV, 83- *Concurrent Pos:* Showalter fel, Dept Biochem, Sch Med, Ind Univ, 77-78; bd mem, Columbus

Zool Asn. *Mem:* AAAS; Am Diabetes Asn; Am Heart Asn; Biochem Soc; NY Acad Sci; Am Inst Nutrit; Am Soc Biochem & Molecular Biol; Sigma Xi (pres, 90-91). *Res:* Metabolic regulation, especially dietary and hormonal control of enzymes involved in carbohydrate and lipid metabolism; mechanisms of insulin action; enzyme control mechanisms; mitochondrial malicenzymes. *Mailing Add:* Dept Food Sci & Nutrit 2121 Fyffe Rd Columbus OH 43210

ROEL, LAWRENCE EDMUND, b Brooklyn, NY, Aug 19, 49. NEUROCHEMISTRY, NEUROPHARMACOLOGY. *Educ:* Princeton Univ, AB, 71; Mass Inst Technol, PhD(nutrit biochem), 76. *Prof Exp:* Fel neurosci, Univ Calif, San Diego, 76-78; ASST PROF ANAT, MED SCH, NORTHWESTERN UNIV, 78- *Mem:* Soc Neuroscience; Am Chem Soc; NY Acad Sci; Sigma Xi. *Res:* Regulation of protein synthesis in neurons and glia; effects of neurotransmitter release; neuronal recognition and synaptogenesis. *Mailing Add:* Nine Hillcrest Lane Woodbury NY 11797

ROELFS, ALAN PAUL, b Stockton, Kans, Nov 18, 36; m 83; c 5. PLANT PATHOLOGY. *Educ:* Kans State Univ, BS, 59, MS, 64; Univ Minn, St Paul, PhD(plant path), 70. *Prof Exp:* Plant pathologist, Coop Rust Lab, Plant Protection Div, Agr Res Serv, 65-69, res plant pathologist, Cereal Rust Lab, Plant Protection Progs, Animal & Plant Health Inspection Serv, 69-75, SUPVRY RES PLANT PATHOLOGIST, CEREAL RUST LAB, AGR RES SERV, USDA, 75- *Concurrent Pos:* Prof, Univ Minn. *Mem:* Fel Am Phytopathological Soc; AAAS; Int Asn Aeriobiol; US Fedn Culture Collections; Int Soc Plant Path; hon mem Can Phytopathological Soc. *Res:* Cereal rust epidemiology and physiological race distribution; resistance to the cereal rusts. *Mailing Add:* Cereal Rust Lab Univ Minn St Paul MN 55108

ROELLIG, HAROLD FREDERICK, b Detroit, Mich, Apr 23, 30; m 59; c 4. PALEONTOLOGY. *Educ:* Concordia Col, Mo, BA, 54; Concordia Sem, dipl theol, 57; Columbia Univ, PhD(geol), 67. *Prof Exp:* Campus chaplain, Lutheran Church, Mo Synod, 60-69; chmn earth sci dept, Adelphi Univ, 78-81, assoc prof geol, 69-83; RETIRED. *Mem:* Geol Soc Am; Paleont Soc; Soc Vert Paleont; Sigma Xi. *Res:* Philosophical and theological interpretation of evolutionary phenomena; evolution and paleoecological interpretation of marine faunas. *Mailing Add:* 14520 Ferns Corner Rd Monmouth OR 97361-9707

ROELLIG, LEONARD OSCAR, b Detroit, Mich, May 17, 27; m 52; c 3. POSITRON ANNIHILATION, SURFACE & ATOMIC PHYSICS. *Educ:* Univ Mich, AB, 50, MS, 55, PhD(physics), 59. *Prof Exp:* Asst, Univ Mich, 53-58; from asst prof to prof physics, Wayne State Univ, 58-78, dean acad admin, 71-72, assoc provost physics, 72-76; vchancellor acad affairs & prof physics, City Univ New York, 78-83, PROF PHYSICS, CITY COL NEW YORK, 78- *Concurrent Pos:* Consult, High Energy Physics Div, Argonne Nat Lab, 59-62, Space Tech Lab, 62-63 & Gen Motors Res Lab, 69-71; dir sci res prog, Inner City High Sch Students, 67-71. vis prof, Univ Col, Univ London, 68-69 & Tata Inst Fundamental Res, Bombay, India, 73; pres, Cent Solar Energy Res Corp, 77. *Mem:* Am Phys Soc; NY Acad Sci. *Res:* Solid state studies using position and positronium beams, materials research; atomic physics. *Mailing Add:* Dept Physics City Col New York Convent & 138th St New York NY 10031

ROELOFS, TERRY DEAN, b Manistique, Mich, Nov 3, 42. FISHERIES. *Educ:* Mich State Univ, BS, 65; Univ Wash, MS, 67; Ore State Univ, PhD(fisheries), 71. *Prof Exp:* Asst prof, 71-74, assoc prof, 74-78, PROF FISHERIES, HUMBOLDT STATE UNIV, 78- *Mem:* AAAS; Am Fisheries Soc; Am Soc Limnol & Oceanog. *Res:* Water pollution biology; fisheries ecology and limnology; anadromous salmonid ecology and management techniques designed to increase the natural production of these fishes through habitat rehabilitation and enhancement. *Mailing Add:* Dept Fisheries Humboldt State Univ Arcata CA 95521

ROELOFS, THOMAS HARWOOD, b Arlington, Va, Oct 10, 37; m 62; c 2. ELECTRICAL ENGINEERING. *Educ:* Cornell Univ, BEE, 60, MEE, 61, PhD(elec eng), 64. *Prof Exp:* Asst prof, 64-70, assoc prof, 68-72, PROF ELEC ENG, UNIV HAWAII, 72- *Concurrent Pos:* Mem staff, Adtech Inc, Honolulu, 69- *Mem:* Am Geophys Union. *Res:* Clear air radar echoes; ionospheric total electron content. *Mailing Add:* Dept Elec Eng HOL 441 Univ Hawaii Manoa 2540 Dole St Honolulu HI 96822

ROELOFS, WENDELL L, b Orange City, Iowa, July 26, 38; m 89; c 4. INSECT COMMUNICATIONS SYSTEM. *Educ:* Cent Col, Iowa, BA, 60; Ind Univ, Bloomington, PhD(org chem), 64, MIT Cambridge, MA, 65. *Hon Degrees:* DSc Cent Col, 85, Hobart & William Smith Cols, 88, Ind Univ, 88, Univ Lund, Sweden, 89, Free Univ Brussels, Belg, 89. *Prof Exp:* Fel, NIH 62-64; from asst prof to prof, 65-78, LIBERTY HYDE BAILEY PROF INSECT BIOCHEM, NY STATE AGR EXP STA, DEPT ENTOMOL, CORNELL UNIV, 78- *Concurrent Pos:* Deleg insect control, People's Repub China, 76; comt biol pest species, Nat Res Coun, 76; US-USSR Sci Exchange Conf, Tashent, USSA; sci lectr, Univ Md, 85. *Honors & Awards:* J Everett Bussart Award, Entomol Soc Am, 73; Alexander von Humboldt Award, 77; Am Boyce Mem lectr, Univ Calif, Riverside; Wolf Prize, 82; Nat Medal Science, 83; Silver Medal, Int Soc Chem Ecol, 90. *Mem:* Nat Acad Sci; fel AAAS; fel Entomol Soc Am; Am Chem Soc; Sigma Xi; Am Acad Arts & Sci. *Res:* The insect communication system; insect monitoring and control programs; author of over 250 scientific journals; defining sex pheromone blends that are biosynthesized and genetically control by female moths. *Mailing Add:* Dept Entomol NY State Agri Exp Sta Geneva NY 14456

ROELS, OSWALD A, b Temse, Bel, Sept 16, 21; US citizen; m 50; c 1. BIOCHEMISTRY, MARINE BIOLOGY. *Educ:* Cath Univ Louvain, BS, 40, MS, 42, PhD(org chem), 44. *Prof Exp:* Head agr biochem dept, Nat Inst Agron Res Belgian Congo, 45-49; tech officer, org chem res dept, Imp Chem Industs, Eng, 49-53; dir, Produits Chimiques des Flandres, Belg, 53-55; head nutrit biochem div, Inst Sci Res Cent Africa, Belg Congo, 55-60; assoc prof nutrit biochem, Columbia Univ, 60-65, sr res assoc, Lamont-Doherty Geol Observ, 65-76, assoc chmn marine biol prog, 67-69, chmn, 69-75; prof marine studies, Univ Tex, 76-80; CHMN, MARITEK CORP, 80- *Concurrent Pos:* Consult, Sorbonne Univ, 57 & Vanderbilt Univ, 59; mem, Int Conf Biochem Probs Lipids; prof biol & biochem, Univ Inst Oceanog, City Col New York, 69-76, vis prof, 76-80; adj prof, Rockefeller Univ, 69-76; chmn, UN Indust Develop Orgn-Food & Agr Orgn exp group mkt on fish protein concentrate, Morocco, 69; vis res prof, Laval Univ, Que, 71-79. *Mem:* AAAS; Am Soc Biol Chem; Am Soc Limnol & Oceanog; Inst Environ Sci; Marine Technol Soc. *Res:* N-transfer in aquatic food chains; artificial upwelling, mariculture; effluent aquaculture. *Mailing Add:* Maritek Corp PO Box 6755 Corpus Christi TX 78466

ROEMER, ELIZABETH, b Oakland, Calif, Sept 4, 29. ASTROMETRY, SOLAR SYSTEM ASTRONOMY. *Educ:* Univ Calif, BA, 50, PhD, 55. *Prof Exp:* Asst astron, 50-52, lab technician, Lick Observ, 54-55, res astronr, Univ Calif, 55-56; res assoc, Yerkes Observ, Univ Chicago, 56; astronr, Flagstaff Sta, US Naval Observ, 57-66, actg dir, 65; assoc prof, 66-69, PROF ASTRONR, LUNAR & PLANETARY LAB, UNIV ARIZ, 69-, ASTRONR, STEWARD OBSERV, 80- *Concurrent Pos:* Partic, prog vis profs in astron, Am Astron Soc & NSF, 60-75; mem panel planetary astron, Nat Acad Sci, 67; mem adv comt astron, Off Naval Res, 69-70; chmn comt, Dept Planetary Sci, Univ Ariz, 72-73; mem, Space Sci Rev Panel Associateship Prog, Off Sci Personnel, Nat Res Coun, 73-75, chmn, 75; mem subcomt space telescope, Space Sci Steering Comt, NASA, 77-78; vchmn, Div Dynamical Astron, Am Astron Sci, 73, chmn, 74; chmn working group comets, Comn 20 Positions & Motions Minor Planets, Comets & Satellites, Int Astron Union, 64-79 & 85-88, vpres, 79-82, pres, Comn 20, 82-85, vpres, 73-76, 85-88, pres, Comn 6 Astron Telegrams, 76-79 & 88-; mem coun, Am Astron Soc, 67-70 & Comet Medal Comt, Astron Soc Pac, 68-74; comt man at large, Sect D, AAAS, 66-69 & 75-78, coun mem, 66-69 & 72-73. *Honors & Awards:* Dorothea Klumpke Roberts Prize, 50; Donohoe lectr, Astron Soc Pac, 62; Benjamin Apthorp Gould Prize, Nat Acad Sci, 71. *Mem:* Fel AAAS; Am Astron Soc; Am Geophys Union; Astron Soc Pac; Sigma Xi; Int Astron Union. *Res:* Comets and minor planets; astrometry and practical astronomy; computation of orbits; astrometric and astrophysical investigations of comets, minor planets and satellites; dynamical astronomy. *Mailing Add:* Lunar & Planetary Lab Univ Ariz Tucson AZ 85721

ROEMER, LOUIS EDWARD, b Washington, DC, July 5, 34; m 58; c 3. ELECTRICAL ENGINEERING, PHYSICS. *Educ:* Univ Del, BS, 55, MS, 63, PhD(appl sci), 67. *Prof Exp:* Assoc engr, Sperry Gyroscope Co, Sperry Rand Corp, 60-63; from instr to asst prof elec eng, Univ Del, 63-68; PROF ELEC ENG, UNIV AKRON, 68- *Mem:* Inst Elec & Electronics Engrs; Sigma Xi. *Res:* Wave propagation in plasma; time domain reflectometry. *Mailing Add:* Elec Eng Dept La Tech Univ Ruston LA 71212

ROEMER, MILTON IRWIN, b Paterson, NJ, Mar 24, 16; m 39; c 2. MEDICAL CARE SYSTEMS. *Educ:* Cornell Univ, BA, 36, MA, 39; NY Univ, MD, 40; Univ Mich, MPH, 43. *Prof Exp:* Med officer venereal dis, NJ State Dept Health, 41-42; med officer med care, USPHS, 43-49; assoc prof pub health, Yale Univ, 49-51; sect chief social med, WHO, Geneva, 51-53; dir bur med care, Sask Dept Health, 53-56; prof admin med, Cornell Univ, 57-61; PROF PUB HEALTH, UNIV CALIF, LOS ANGELES, 62- *Concurrent Pos:* World Health Orgn, USAID. *Honors & Awards:* Henry E Sigerist Lectr, Yale Univ, 80; Sedgwick Medalist, Am Pub Health Asn, 83; Stubenbord Prof, Cornell, 84; Rosenstadt Prof, Univ Toronto, 88. *Mem:* Inst Med-Nat Acad Sci; Am Pub Health Asn. *Res:* Organization of medical care; health insurance; rural health; international health care systems; health manpower. *Mailing Add:* Sch Pub Health Univ Calif Los Angeles CA 90024

ROEMER, RICHARD ARTHUR, b Minneapolis, Minn, Sept 12, 39; m 72; c 6. NEUROSCIENCE, COMPUTER SCIENCE. *Educ:* Calif State Univ, Northbridge, BA, 68; Univ Calif, Irvine, PhD(psychobiol), 73; Univ Bellgrade, DMS, 84. *Prof Exp:* Sr eng aide digital comput, Litton Data Systs, 62-64; comput specialist reentry systs, Gen Elec Co, 64-66; sr elec engr digital comput, Calif State Univ, Northridge, 66-67; analyst mgt systs, Syst Develop Corp, 67-69; comput syst consult, Enki Res Inst, 69-73; res asst & res assoc, Univ Calif, Irvine, 69-73; res fel psychol, Harvard Univ, 73-75; med res scientist, 75-80, assoc prof, Dept Psychiat, 81-85, PROF, DEPT PSYCHIAT & NEUROL, TEMPLE UNIV, 85- *Concurrent Pos:* Analyst mgt systs, 69; clin asst prof, Dept Psychiat, Temple Univ, 75-77, res assoc prof, 77-81, assoc prof, 81-85; lectr psychol, Univ Pa & Rutgers Univ, 77-; sr Fulbright scholar, 81, 83 & 84. *Mem:* Soc Neurosci; Inst Elec & Electronics Engrs; Nat Asn On-Line Comput Psychol; Soc Systs, Man & Cybernet; Soc Eng Med & Biol; Am Psychopath Asn; Soc Biol Psychiat. *Res:* Assessment of neuropsychological and neuropharmacological relationships in mentally ill patients; development of multivariate statistical applications in neurobiology. *Mailing Add:* Dept Psychiat Temple Univ Sch Med 3400 Broad St Philadelphia PA 19140

ROENIGK, WILLIAM J, b Cleveland, Ohio, Jan 26, 29; m 53; c 4. VETERINARY RADIOLOGY. *Educ:* Ohio State Univ, DVM, 54; Baylor Univ, MSc, 58; Am Col Vet Radiol, ACVR, 66. *Prof Exp:* From asst prof to prof vet radiol, Vet Clin, Ohio State Univ, 58-67; assoc prof comp radiol & lab animal med, Col Med & dir, Div Vet Med, Children's Hosp Res Found, Univ Cincinnati, 67-72; prof vet radiol, NY State Vet Col, Cornell Univ, 72-75; PROF VET RADIOL & HEAD DEPT, COL VET MED, TEX A&M UNIV, 75- *Concurrent Pos:* NIH grant, 62-65; consult, five major co. *Mem:* Am Vet Med Asn; Am Col Vet Radiol (pres, 70); Educ Vet Radiol Sci (pres, 66); Radiol Soc NAm; Am Vet Radiol Soc. *Res:* Diagnostic radiology; radiation therapy; nuclear medicine. *Mailing Add:* Col Vet Med Tex A&M Univ College Station TX 77843

ROEPE, PAUL DAVID, b Suffern, NY, June 20, 60; m 90. BIOLOGICAL MEMBRANE TRANSPORT, VIBRATIONAL SPECTROSCOPY. *Educ:* Boston Univ, BA, 82, MA & PhD(chem & math), 87. *Prof Exp:* Mem molecular biol, Roche Inst Molecular Biol, Nutley, NJ, 87-89; res assoc, Molecular Biol Inst, Univ Calif, Los Angeles, 89-90; ASST MEM PHARMACOL, MEM SLOAN-KETTERING CANCER CTR, 90-; ASST PROF PHARMACOL, CORNELL UNIV MED COL, 90- *Concurrent Pos:*

Sackler scholar, Raymond & Beverly Sackler Found, 90. *Honors & Awards:* Young Investr Award, Biophys Soc, 88. *Mem:* AAAS; Biophys Soc; NY Acad Sci. *Res:* Molecular level studies of chemotherapeutic drug transport in normal and tumor cells. *Mailing Add:* Dept 6012 Mem Sloan-Kettering Cancer Ctr New York NY 10021

ROEPKE, HARLAN HUGH, b Rochester, Minn, Nov 14, 30; m 58; c 2. SEDIMENTARY PETROLOGY, ENVIRONMENTAL GEOLOGY. *Educ:* Univ Minn, Minneapolis, BA, 53, MS, 58; Univ Tex, Austin, PhD(geol), 70; Purdue Univ, MS, 76. *Prof Exp:* Res asst geol, Minn Geol Surv, 56-58; geologist, Paleont & Stratig Br, US Geol Surv, 58-60, Alaskan Br, summer 60; from asst prof to assoc prof, 65-77, PROF GEOL, BALL STATE UNIV, 77- *Concurrent Pos:* Geol consult, 72- *Mem:* Geol Soc Am; Soc Econ Paleont & Mineral; Sigma Xi; Nat Asn Geol Teachers; Nat Asn Coeducational Teachers; Soc Econ Paleont & Mineral. *Res:* Petrology of carbonate rocks; x-ray florescence study of Indiana chert. *Mailing Add:* 4806 W University Muncie IN 47304

ROEPKE, RAYMOND ROLLIN, b Bodaville, Kans, Jan 5, 11; m 42; c 1. BIOCHEMISTRY. *Educ:* Kans State Col, BS, 33, MS, 34; Univ Minn, PhD(biophys), 38. *Prof Exp:* Asst, Agr Exp Sta, Kans State Col, 33-35; asst physiol, Univ Minn, 39-40; asst chem, Lafayette Col, 40-41; res biochemist, Am Cyanamid Co, 42-56, group leader pharmacol res, Lederle Labs, 56-76; RETIRED. *Mem:* AAAS; Am Chem Soc; Am Soc Pharmacol & Exp Therapeut; NY Acad Sci. *Res:* Osmotic pressures of biologic fluids; intestinal absorption of electrolytes; rheology of the blood; bacterial mutation and metabolism; chemotherapy; toxicology and safety evaluation; biochemical pharmacology. *Mailing Add:* 1080 Arcadia Ave No 104 Vista CA 92084

ROER, ROBERT DAVID, b New York, NY, Oct 15, 52; m 76; c 1. BIOMINERALIZATION. *Educ:* Brown Univ, ScB, 74; Duke Univ, PhD(zool), 79. *Prof Exp:* From asst prof to assoc prof, 79-90, PROF BIOL SCI, UNIV NC, WILMINGTON, 90- *Mem:* Am Physiol Soc; Am Soc Zoologists; Am Soc Bone & Mineral Res; AAAS; Crustacean Soc. *Res:* Mechanisms of membrane transport in relation to osmoregulation and in relation to biomineralization; mineral nucleation in crustacean cuticle; fluid dynamics in relation to bone physiology. *Mailing Add:* Ctr Marine Sci Res 7205 Wrightsville Ave Wilmington NC 28403

ROERIG, DAVID L, PULMONARY-DRUG UPTAKE, NARCOTIC ANALGESICS. *Educ:* Kans State Univ, PhD(biochem), 70. *Prof Exp:* TOXICOL-RES HEALTH SCIENTIST, WOOD VET ADMIN MED CTR, 72-; ASST PROF ANESTHESIOL, DEPT ANESTHESIOL & PHARMACOL, MED COL WIS, 74- *Res:* Drug metabolism and distribution. *Mailing Add:* Dept Anesthesiol Pharmacol & Toxicol Med Col Wash Vet Admin Med Ctr Milwaukee WI 53295

ROERIG, SANDRA CHARLENE, OPIATES ACTION, MECHANISMS TO ANALGESIA. *Educ:* Med Col Wis, MA, 76. *Prof Exp:* RES ASSOC PHARMACOL, MED COL WIS, 76- *Mailing Add:* Dept Pharmacol Univ Minn Med Sch 3-260 Millard Hall 435 Delware St SE Minneapolis MN 55455

ROESCH, WILLIAM CARL, b Saginaw, Mich, Nov 11, 23; m 46; c 3. PHYSICS. *Educ:* Miami Univ, Ohio, AB, 45; Calif Inst Technol, PhD(physics), 49. *Prof Exp:* Mgr, Radiol Physics, Gen Elec Co, 49-64; mgr radiol physics, Battelle-Northwest, 65, sr res assoc, 66-70, staff scientist, 70-85; RETIRED. *Concurrent Pos:* Mem, Panel Reassessment on Atomic Bomb Dosimetry, Nat Acad Sci, 82-87 & Biol Effects Ionozing Radiation-4 Comt, 85-88; chmn, Sci Comt No 80 Radiabiol Skin, Nat Coun Radiation Protection & Measurements, 83-90. *Mem:* Am Phys Soc; Radiation Res Soc; Health Phys Soc; Am Asn Physicists Med. *Res:* Radiological physics; instrumentation and dosimetry methods for alpha, beta and gamma rays and neutrons; physics of radiobiology and radiation protection. *Mailing Add:* 1646 Butternut Richland WA 99352

ROESEL, CATHERINE ELIZABETH, b Augusta, Ga, Feb 6, 20. IMMUNOLOGY. *Educ:* Vanderbilt Univ, BA, 41; Wash Univ, PhD(bact), 51. *Prof Exp:* Fel, Carnegie Inst, 50-51; from instr to assoc prof, 51-76, PROF CELL & MOLECULAR BIOL, MED COL GA, 76- *Mem:* AAAS; Am Soc Microbiol; Asn Am Med Cols. *Res:* Antibody formation; Rubella virus; tissue culture. *Mailing Add:* 2722 Cherry Lane Augusta GA 30909

ROESER, ROSS JOSEPH, b Louisville, Ky, Nov 14, 42; m 63; c 3. AUDIOLOGY. *Educ:* Western Ill Univ, BS, 66; Northern Ill Univ, MA, 67; Fla State Univ, PhD(audiol), 72. *Prof Exp:* DIR, CALLIER CTR, UNIV TEX, DALLAS, 72- *Concurrent Pos:* Clin assoc prof otolaryngol, Med Sch, Univ Tex Southwestern, 73-; prof grad prog, Univ Tex, Dallas, 73-; Health Educ & Welfare grants, cent auditory processing children, 77-79 & tactile aids, 80-82; NASA grant, Hearing Aid Malfunction Detection Unit. *Mem:* Fel Am Speech & Hearing Asn; Am Auditory Soc (secy-treas, 73-); fel Soc Ear, Nose & Throat Advan Children. *Res:* Central auditory processing in children, tactile aids and other audiology related areas. *Mailing Add:* Callier Ctr Commun Dis Univ Tex 1966 Inwood Dallas TX 75235

ROESIJADI, GURITNO, b Tokyo, Japan, Apr 4, 48; US citizen; m 70; c 3. COMPARATIVE PHYSIOLOGY, AQUATIC TOXICOLOGY. *Educ:* Univ Wash, BS, 70; Humboldt State Univ, MS, 73; Tex A&M Univ, PhD(biol), 76. *Prof Exp:* Res scientist, Battelle Northwest Labs, 76-80, sr res scientist, 80-84; assoc prof biol, Dept Biol, Pa State Univ, 84-86; ASSOC PROF TOXICOL, CHESAPEAKE BIOL LAB, CTR ENVIRON & ESTUARINE STUDIES, UNIV MD, 86- *Concurrent Pos:* US ed, Marine Environ Res, 80-86, ed, 86- *Mem:* AAAS; Am Soc Zoologists; Sigma Xi; Soc Environ Toxicol & Chem. *Res:* Adaptive mechanisms to environmental conditions; physiological, biochemical and molecular mechanisms for the regulation of metals in aquatic animals. *Mailing Add:* Chesapeake Biol Lab Univ Md Box 38 Solomons MD 20688

ROESING, TIMOTHY GEORGE, b Abington, Pa, May 14, 47; m 69; c 2. VIROLOGY, VACCINES. *Educ:* Univ Md, BS, 69; Hahnemann Med Col, MS, 73, PhD(microbiol & immunol), 76. *Prof Exp:* Sr res scientist microbiol, Smith Kline Diag Div, Smith Kline Corp, 76-77; sr proj develop biologist microbiol, Merck Sharp & Dohme Div, Merck & Co, 77-82, sr proj microbiologist, 82-85; mgr, biol qual control tech serv, 86-87, mgr, bact vaccines & blood prod, Recombivax, 88-90, MGR DIPLOID CELL VACCINES, MERCK PHARMACEUT MFG DIV, 91- *Concurrent Pos:* Fel, Smith Kline Diag, 76-77. *Mem:* Am Soc Microbiol; Tissue Cult Asn. *Res:* Virus-cell interactions with coxsackieviruses; role of coxsackieviruses in heart disease and pancreatitis; enzyme linked immunoassays and fluorescent immunoassays for human viruses. *Mailing Add:* Merck Pharmaceut Mfg Div W28-1 West Point PA 19486

ROESKE, ROGER WILLIAM, b Valders, Wis, July 30, 27; m 55; c 2. BIO-ORGANIC CHEMISTRY. *Educ:* Univ Wis, BA, 48; Univ Ill, PhD(org chem), 51. *Prof Exp:* Merck fel, Swiss Fed Inst Tech, 51-52; res assoc biochem, Med Col, Cornell Univ, 52-55; sr chemist, Eli Lilly & Co, Ind, 55-61; from asst prof to assoc prof, 62-66, PROF BIOCHEM, SCH MED, IND UNIV, 77- *Concurrent Pos:* Res Career Develop Award, USPHS, 62-71. *Mem:* AAAS; Am Chem Soc. *Res:* Mechanism of enzyme action; synthesis of peptides; membrane-active peptides. *Mailing Add:* Dept Biochem MS 425A Ind Univ Sch Med 635 Barnhill Dr Indianapolis IN 46223

ROESLER, FREDERICK LEWIS, b Milwaukee, Wis, Feb 26, 34; m 57, 68; c 5. ATOMIC PHYSICS, OPTICAL SPECTROSCOPY. *Educ:* St Olaf Col, BA, 56; Univ Wis, MS, 58, PhD(physics), 62. *Prof Exp:* Res assoc physics, Univ Wis, 62-63; NSF fel, Lab Aime-Cotton, France, 63-64; from asst prof to assoc prof, 64-70, PROF PHYSICS, UNIV WIS-MADISON, 70- *Concurrent Pos:* Consult, Los Alamos Sci Lab, 65-69 & Argonne Nat Lab, 77-; mem comt line spectra of elements, Nat Acad Sci-Nat Res Coun, 67-72; vis assoc prof physics, Univ Ariz, 70; von Humboldt Found Sr US Scientist Award, 75-; vis prof physics, Univ Munich, 76; sr res assoc, Nat Acad Sci, Nat Res Coun, Space Flight Ctr, 82-83; NSF Prog dir aeronomy, 91. *Honors & Awards:* von Humboldt Sr Scientist, 75; Group Achievement Award, NASA, 81. *Mem:* Optical Soc Am; Am Geophys Union. *Res:* Interference spectroscopy; astronomy; aeronomy; astrophysics. *Mailing Add:* Dept Physics B621 Sterling Hall Univ Wis Madison WI 53706

ROESLER, JOSEPH FRANK, b Chicago, Ill, Dec 15, 30; m 61; c 2. ENVIRONMENTAL ENGINEERING, PHYSICAL CHEMISTRY. *Educ:* Roosevelt Univ, BS, 54; Okla State Univ, MS, 61; Univ Cincinnati, MS, 70. *Prof Exp:* Res chemist, Ill Inst Technol, 56-59, Okla State Univ, 60-61 & Rauland Corp, 61-62; res chemist, Div Air Pollution, USPHS, 62-67; sanit engr & mgr instrumentation & automation of waste systs, Nat Environ Res Ctr, 67-78, regional liaison officer, US Environ Protection Agency,78-82, environ engr, Off Res & Develop, 78-88, chief, sampling & field measurements, 82-88; pres, Cincinnati Engrs Inc, 88-90; SUPT LABS, METRO WATER, NASHVILLE, 90- *Concurrent Pos:* Dir, Water & Waste Water Indust Div, Instrument Soc Am, 86- *Honors & Awards:* Water & Waste Water Div Award, Instrument Soc Am, 81, Kermit Kischer Award, 83. *Mem:* Instrument Soc Am; Water Pollution Control Fedn; Am Soc Civil Engrs; Sigma Xi (pres, 86). *Res:* Mathematical modeling of advanced waste water treatment processes; environmental health engineering; aerosols and chemical instrumentation pertaining to air pollution; kinetics and photoemissive surfaces; instrumentation for water quality; automation of wastewater treatment plants. *Mailing Add:* 5630 Valley View Rd Brentwood TN 37027-4687

ROESMER, JOSEF, b Konigsberg, Germany, Aug 29, 28; m 59; c 1. NUCLEAR SCIENCE. *Educ:* Univ Mainz, BS, 52, MS, 55; Clark Univ, PhD(nuclear chem), 64. *Prof Exp:* Adv scientist, Nuclear Sci & Eng Corp, 60-64; sr scientist, Astronuclear Lab, 64-71, FEL SCIENTIST, NUCLEAR ENERGY SYSTS, WESTINGHOUSE ELEC CORP, 71- *Mem:* NY Acad Sci; Am Chem Soc. *Res:* Nuclear reactions; application of neutron-prompt gamma reactions to chemical analysis; formation, activation & transport etc.; radiochemical studies; low-level counting; fission product diffusion; applications of computers; high-temperature chemistry; composition of PWR reactor coolants near the operating temperature; primary and secondary coolant systems; minimization of occupational radiation; whole PWR coolant loop determination; waste management; compatibility of decontamination with construction materials of nuclear reactor circuits. *Mailing Add:* 969 Holly Lynne Dr Pittsburgh PA 15236

ROESNER, LARRY A, RESEARCH ADMINISTRATION. *Educ:* Valparaiso Univ, Ind, BS, 63; Colo State Univ, MS, 65; Univ Wash, PhD(sanit eng), 69. *Prof Exp:* Grad res asst, Colo State Univ, 63-65, Univ Wash, 65-67; prin engr, Water Resources Engrs, Inc, 68-75; assoc, 75-83, vpres, Water Resources & Environ Sci S Region, 83-88, TECH DIR, WATER RESOURCES & ENVIRON SCI SOUTH REGION, CAMP DRESSER & MCKEE INC, 83-, SR VPRES, 88- *Concurrent Pos:* Chmn, Eng Found Conf, 88; vchmn, Non-Point Sources Pollution Comt, Water Pollution Control Fedn; mem, Comt Wastewater Mgt Coastal Urban Areas, Nat Res Coun. *Honors & Awards:* Walter L Huber Civil Eng Res Prize, 75. *Mem:* Nat Acad Eng; fel Am Soc Civil Engrs; Am Water Resources Asn; Water Pollution Control Fedn; Am Inst Hydrol; Nat Soc Prof Engrs. *Res:* Author of various publications. *Mailing Add:* Camp Dresser & McKee Inc 555 Winderley Pl Suite 200 PO Box 945375 Maitland FL 32794-5375

ROESS, WILLIAM B, b Evanston, Ill, Sept 8, 38; m 57; c 4. GENETICS, MOLECULAR BIOLOGY. *Educ:* Blackburn Col, BA, 61; Fla State Univ, PhD(genetics), 66. *Prof Exp:* From asst prof to assoc prof, 66-74, chmn, Collegium, 76-86, actg provost, 79, PROF BIOL, COLLEGIUM NATURAL SCI, ECKERD COL, 75- *Concurrent Pos:* NIH spec fel, Oak Ridge Nat Labs, 71. *Mem:* AAAS; Sigma Xi. *Res:* Mechanisms and genetic control of amino acid transport in human tissue culture cells; genetic control of membrane synthesis; carcinogenesis in human tissue culture cells. *Mailing Add:* Collegium Natural Sci Eckerd Col St Petersburg FL 33733

ROESSLER, BARTON, metallurgy, engineering; deceased, see previous edition for last biography

ROESSLER, CHARLES ERVIN, b Elysian, Minn, May 1, 34; m 56; c 7. HEALTH PHYSICS, RADIOLOGICAL HEALTH. *Educ:* Mankato State Col, AB, 55; Univ Rochester, MS, 56; Univ Pittsburgh, MPH, 59; Univ Fla, PhD(environ eng), 67; Am Bd Health Physics, cert, 61. *Prof Exp:* Health physicist, Nuclear Power Dept, Res Div, Curtiss-Wright Corp, 56-58; radiol physicist, Fla State Bd Health, 59-65; asst prof radiation biophys, 67-72, asst prof environ eng, 72-73, assoc prof, 73-79, PROF ENVIRON ENG, UNIV FLA, 79- *Concurrent Pos:* Consult natural radioactivity to phosphate indust; lectr, radiation emergency planning. *Mem:* AAAS; Health Physics Soc; Am Indust Hyg Asn; Am Conf Govt Indust Hygienists. *Res:* Environmental radiation, particularly naturally occurring radioactivity. *Mailing Add:* RR 1 Box 139H Elysian MN 56028-0139

ROESSLER, DAVID MARTYN, b London, Eng, Apr 29, 40; m 83; c 2. PHYSICS, SPECTROSCOPY. *Educ:* Univ London, BSc, 61, PhD(physics), 66. *Prof Exp:* Fel, Univ Calif, Santa Barbara, 66-68; mem, tech staff spectroscopy, Bell Labs, 68-70; STAFF RES SCIENTIST, SPECTROSCOPY, GEN MOTORS RES LABS, 70- *Concurrent Pos:* Fac mem, Physics Dept, Wayne State Univ, 86-, Lawrence Technol Univ Mo, 90-; consult & invited speaker on laser mat processing, optical properties mat. *Mem:* Brit Inst Physics; Sigma Xi; Optical Soc Am; Int Solar Energy Soc; Laser Inst Am; Soc Photo-optical Instrumentation Engrs; Am Inst Physics. *Res:* Optical Properties of solids and aerosols; reflection; luminescence; photoacoustic spectroscopy; laser processing of materials; solar energy materials; author of numerous publications. *Mailing Add:* Dept Physics Gen Motors Res Labs Warren MI 48090-9055

ROESSLER, MARTIN A, b Hempstead, NY, Apr 7, 39. FISH BIOLOGY. *Educ:* Univ Miami, BS, 61, MS, 64, PhD(marine sci), 67. *Prof Exp:* Instr fisheries biol, Inst Marine Sci, Univ Miami, 66-69, from asst prof to assoc prof, Rosenstiel Sch Marine & Atmospheric Sci, 69-73; PROG DIR, TROP BIOINDUST DEVELOP CO, 73- *Mem:* Am Fisheries Soc; Am Soc Ichthyologists & Herpetologists; Am Soc Limnol & Oceanog. *Res:* Ecology of estuaries in Florida; biology of pink shrimp Penaeus duorarum; ecology of fishes; power plant siting and environmental effects of coastal zone development. *Mailing Add:* 7821 SW 114th St Miami FL 33156

ROESSLER, ROBERT L, b Neillsville, Wis, Sept 2, 21; m 46; c 3. PSYCHIATRY. *Educ:* Univ Wis, PhB, 42; Columbia Univ, MD, 45; Am Bd Psychiat & Neurol, dipl, 51. *Prof Exp:* Intern, Englewood Hosp, NJ, 45-46; resident psychiat, Vet Admin Hosp, Madison, Wis, 46-48; resident, Strong Mem Hosp, Rochester, NY, 48-49; instr, Sch Med, Univ Rochester, 49-50; from asst prof to prof, Sch Med, Univ Wis, 50-63, from actg chmn dept to chmn dept, 56-61, dir psychiat inst, 60-61; PROF PSYCHIAT, BAYLOR COL MED, 63- *Concurrent Pos:* Consult, US Info Agency, 59-61 & Vet Admin Hosp, Madison, 59-61. *Mem:* AAAS; fel Am Psychiat Asn; Am Psychosom Soc; Soc Psychophysiol Res; NY Acad Sci. *Res:* Psychophysiology; psychosomatic medicine. *Mailing Add:* Dept Psychiat Baylor Col Med 1200 Moursund Ave Houston TX 77030

ROESSMANN, UROS, b Vevce, Slovenia, Yugoslavia, Sept 9, 25; US citizen; m 57; c 4. NEUROPATHOLOGY. *Educ:* Ohio State Univ, BSc, 51; Western Reserve Univ, MD, 57. *Prof Exp:* Intern med, Michael Reese Hosp, 57-58; resident neurol, Univ Hosps Cleveland, 58-59; resident anat path, Inst Path, Western Reserve Univ, 59-60, resident neuropath, 61-62; resident anat path, Cleveland Clin, 60-61; captain & asst pathologist, Neuropath Br, Armed Forces Inst Path, 62-64; res assoc, Ment Health Res Inst, Univ Mich, 64-65; from instr to asst prof, 65-82, PROF INST PATH, CASE WESTERN RESERVE UNIV, 82- *Concurrent Pos:* Consult, Vet Admin Hosp, 74-, Mt Sinai Hosp, 77- *Mem:* Am Asn Neuropathologists. *Res:* Morphological aspects of central nervous system pathology; growth and development of the central nervous system; skeletal muscle response to stimulation. *Mailing Add:* Inst Path Case Western Reserve Univ 2085 Adelbert Rd Cleveland OH 44106

ROEST, ARYAN INGOMAR, b Chicago, Ill, June 13, 25; m 50; c 4. VERTEBRATE ZOOLOGY, TAXONOMY. *Educ:* Univ Va, BS, 45; Ore State Col, BS, 48, MS, 49, PhD(zool), 54. *Prof Exp:* Asst, Ore State Col, 49; instr biol & math, Cent Ore Col, 52-55; prof biol, Calif Polytech State Univ, San Luis Obispo, 55-90; RETIRED. *Mem:* Am Soc Mammal; Soc Marine Mammal. *Res:* Vertebrate field zoology, including mammals, birds, reptiles and amphibians; systematic studies of sea otter, red fox, kit fox & kangaroo rat; mammal studies. *Mailing Add:* 1197 10th St Los Osos CA 93402

ROETH, FREDERICK WARREN, b Houston, Ohio, Aug 21, 41; m 68; c 3. WEED SCIENCE, CROP PRODUCTION. *Educ:* Ohio State Univ, BS, 64; Univ Nebr, Lincoln, MS, 67, PhD(agron), 70. *Prof Exp:* Res asst weed sci, Univ Nebr, Lincoln, 64-69; asst prof, Purdue Univ, W Lafayette, 69-75; from asst prof to assoc prof, 75-83, PROF AGRON, UNIV NEBR, LINCOLN, 83- *Concurrent Pos:* Vis botanist, Univ Calif, Davis, 84-85. *Mem:* Weed Sci Soc Am; Am Soc Agron; Coun Agr Sci & Technol; Sigma Xi. *Res:* Weed biology and control in agronomic crops, herbicide dissipation. *Mailing Add:* Box 66 S Cent Ctr Clay Center NE 68933

ROETHEL, DAVID ALBERT HILL, b Milwaukee, Wis, Feb 17, 26; m 53; c 2. SCIENCE ADMINISTRATION, SCIENCE COMMUNICATIONS. *Educ:* Marquette Univ, BS, 50, MS, 52; Oak Ridge Sch Reactor Technol, cert, 53. *Prof Exp:* Tech specialist, AEC, 52-57; asst to exec secy & mgr, Prof Rel Off, Am Chem Soc, 57-72; exec dir, Nat Registry Clin Chem, 67-72, Am Asn Clin Chem, 68-70, Am Orthotic & Prosthetic Asn, Am Acad Orthotists & Prosthetists, Am Bd Cert Orthotics & Prosthesis, 72-76; EXEC DIR, NAT CERT COMN CHEM & CHEM ENG, AM INST CHEMISTS, 77- *Concurrent Pos:* Mem, Gov Md Comn Sci Develop, 69; secy-gen, Seventh Int Cong Orthotics & Prosthetics, 75-76, Second World Congress, 75-77; secy-treas, Comn Profs in Sci & Technol, 79-82, comnr, 77-; dir, Coun Eng & Sci Soc Execs, 82-86; ed, Chemist, 77-; dir, Am Inst Chem, China-US scientific exchange, 85-, trustee, 82- *Honors & Awards:* Nat Six Award, Nat Registry Clin Chem, 72. *Mem:* Am Inst Chemists; Am Chem Soc; AAAS; Coun Eng & Sci Soc Execs. *Res:* Matters of professional interest to chemist and chemical engineers, such as ethics, economic patterns in compensation, pensions, professionsl liability insurance and legislative and regulatory concerns which affect the profession. *Mailing Add:* 13218 Bregman Rd 7315 Wisconsin Ave Silver Spring MD 20904

ROETLING, PAUL G, b Buffalo, NY, Sept 12, 33; m 60. PHYSICS, OPTICS. *Educ:* Univ Buffalo, BA, 55, PhD(physics), 60. *Prof Exp:* Consult, Cornell Aeronaut Lab, 59-60, physicist, 60-63, sect head optics, 63-68; prin scientist, 68-70, mgr optics res area, 70-74, mgr image processing area, 74-78, RES FEL, XEROX CORP, 78- *Concurrent Pos:* Hon mem, Rochester Sect, Optical Soc Am. *Mem:* Am Phys Soc; fel Optical Soc Am; Inst Elec & Electronics Engrs; fel Soc Photographic Scientists & Engrs. *Res:* Optical image formation and image processing. *Mailing Add:* 468 Ontario Dr Ontario NY 14519

ROETMAN, ERNEST LEVANE, b Chandler, Minn, Sept 18, 36; m 62; c 2. APPLIED MATHEMATICS, ANALYTICAL MECHANICS. *Educ:* Univ Minn, BA, 57; Ore State Univ, PhD(math & mech eng), 63. *Prof Exp:* Asst math, Aachen Tech Univ, 63; tech specialist, Bell Tel Labs, 63-65; asst prof, Stevens Inst Technol, 65-68; assoc prof math, Univ Mo-Columbia, 68-75, prof, 75-80; PROF ENGR, BOEING COMPUT SERV, 80- *Concurrent Pos:* Guest prof, Aachen Tech Univ, 75-76; von Humboldt fel, 76; guest prof, Ore State Univ, 79-80. *Mem:* Soc Indust & Appl Math; Soc Natural Philos; Int Soc Oxygen Transp Tissue. *Res:* Biofluid mechanics; partial differential equations; irregular boundary value problems; numerical analysis; fluid flow. *Mailing Add:* 3016 67th Ave SE Mercer Island WA 98040

ROFFLER-TARLOV, SUZANNE K, b Missoula, Mont, Apr 4, 38. NEUROCHEMISTRY, NEUROGENETICS. *Educ:* Univ Chicago, PhD(biopsychol), 68. *Prof Exp:* Asst prof, 80-86, ASSOC PROF NEUROSCI, SCH MED, TUFTS UNIV, 86- *Mailing Add:* Prog Neurosci Tufts Univ Sch Med 136 Harrison Ave Boston MA 02111

ROFFMAN, STEVEN, b New York, NY, Apr 29, 44. BIOCHEMICAL PHARMACOLOGY. *Educ:* Queens Col, NY, BA, 65; New York Univ, MS, 68, PhD(biochem), 74. *Prof Exp:* Fel biochem, Albert Einstein Col Med, NY, 73-75; res assoc, 75-78, assoc, 78-79, asst prof pharmacol, 79-81, RES ASSOC MED, COL PHYSICIANS & SURGEONS, COLUMBIA UNIV, 81- *Concurrent Pos:* Info syst mgr, Dept Allergy & Infectious Dis, Roosevelt/St Luke's Hosp Ctr, 87- *Mem:* AAAS. *Res:* Role of prokolytic enzymes in the pathophysiology of inflammation and carcinogenesis; role of vasoactive peptides released in inflammatory disease of lung and skin. *Mailing Add:* Dept Allergy & Infectious Dis Roosevelt/St Luke's Hosp Ctr 428 W 59th St New York NY 10019

ROFFWARG, HOWARD PHILIP, b New York, NY, June 9, 32; m 55; c 2. SLEEP-WAKE PHYSIOLOGY, SLEEP DISORDERS. *Educ:* Columbia Univ, AB, 54, MD, 58. *Prof Exp:* Res assoc, Mt Sinai Hosp, 61-62; instr psychiat, Columbia Univ, 62-66; from asst prof to assoc prof psychiat, Albert Einstein Col Med, 66-77; PROF PSYCHIAT, UNIV TEX HEALTH SCI CTR, DALLAS, 77-, DIR RES, DEPT PSYCHIAT & DIR, SLEEP RES LAB, 77- *Concurrent Pos:* NIMH career res scientist awards & res proj grants, 62-76; dir sleep EEG lab, NY State Psychiat Inst, 62-66, asst attend psychiatrist, 63-66; adj attend psychiatrist, Montefiore Hosp & Med Ctr, 66-71, assoc attend psychiatrist, 71-77; ed-in-chief, Sleep Rev Brain Info Serv, 74-75; mem, clin projs res rev comt, NIMH, 76-80; chmn, comt diag classification, Asn Sleep Dis Ctr, 76-79, continuing chair, 76-; consult dir, Sleep & Wake Dis Ctr, dept psychiat, Presby Hosp-Univ Tex Health Sci Ctr, Dallas, 82- *Honors & Awards:* Pioneer in Sleep Res, Asn Psychophysiol Study Sleep, 77; Nathaniel Kleitman Prize, Asn Sleep Dis Ctr, 81. *Mem:* Sleep Res Soc (vpres, 80-87, pres, 85-87); fel Am Psychiat Asn; fel Am Col Psychiatrists; Am Sleep Dis Asn (pres elect & pres, 88-90). *Res:* Physiology and psychophysiology of sleep and sleep disorders; clinical psychiatry. *Mailing Add:* Univ Tex Southwestern Med Ctr 5323 Harry Hines Blvd J4 126 Dallas TX 75235

ROGALSKI-WILK, ADRIAN ALICE, b Chicago, Ill, Aug 2, 53; m 91. CELL MEMBRANE CYTOSKELETON INTERACTIONS. *Educ:* Univ Chicago, BA, 75; Univ Ill, PhD(cell biol), 81. *Prof Exp:* Postdoctoral fel Nat Res Serv Award cell biol, Univ Calif, San Diego, 81-83, NIH postdoctoral fel, 83-85; asst prof, 85-91, ASSOC PROF CELL BIOL, DEPT ANAT & CELL BIOL, UNIV ILL, CHICAGO, 91- *Concurrent Pos:* Prin investr, NIH grant, Dept Anat & Cell Biol, Univ Ill, Chicago, 86-, PEW scholar, 87-93. *Mem:* Am Soc Cell Biol; AAAS; NY Acad Sci. *Res:* Molecular cell biology of membrane-cytoskeleton interactions. *Mailing Add:* Dept Anat & Cell Biol Univ Ill 808 S Wood St Chicago IL 60612

ROGAN, ELEANOR GROENIGER, b Cincinnati, Ohio, Nov 25, 42; div; c 1. BIOCHEMISTRY. *Educ:* Mt Holyoke Col, AB, 63; Johns Hopkins Univ, PhD(biochem), 68. *Prof Exp:* Lectr biol sci, Goucher Col, 68-69; fel biochem, Univ Tenn, 69-71, res assoc cancer res, 71-73; res assoc, 73-76, from asst prof to assoc prof, 76-90, TENURE PROF CHEM CARCINOGENESIS, EPPLEY INST & DEPT PHARM SCI, MED CTR, UNIV NEBR, 90- *Mem:* Am Asn Cancer Res; AAAS; Am Soc Biochem & Molecular Biol. *Res:* Mechanism of carcinogenesis by polycyclic aromatic hydrocarbons; nuclear monooxygenase enzyme activities especially those catalyzing activation of hydrocarbons; binding of hydrocarbons to DNA and structure of hydrocarbon-nucleic acid base derivatives. *Mailing Add:* Eppley Inst Univ Nebr Med Ctr Omaha NE 68198-6805

ROGAN, JOHN B, b Kansas City, Kans, Sept 3, 30; m 66. ORGANIC CHEMISTRY, POLYMER CHEMISTRY. *Educ:* Univ Wyo, BS, 52; Univ Calif, Berkeley, PhD(chem), 55. *Prof Exp:* Res chemist, Elastomer Chem Dept, E I du Pont de Nemours & Co, 55-59; res assoc org chem, Univ Wyo, 59; asst prof chem, Colo State Univ, 59-62; assoc prof, Univ Nev, Reno, 62-66; res chemist, 66-69, group leader, Res & Develop Dept, 69-73, RES ASSOC, RES & DEVELOP DEPT, AMOCO CHEM CORP, 73- *Concurrent Pos:* Res grants, NSF, 60-62, Am Chem Soc Petrol Res Fund, 64-66 & Desert Res Inst, Univ Nev, 66. *Mem:* AAAS; Am Chem Soc; Am Inst Chem Eng. *Res:* Polymer synthesis; property-structure relationships. *Mailing Add:* 319 Lake St W Tarpon Springs FL 34689-2599

ROGAN, WALTER J, b Bridgeport, Pa, May 27, 49. PUBLIC HEALTH & EPIDEMIOLOGY. *Educ:* LaSalle Univ, BA, 71; Univ Calif, Berkeley, MPH, 75, San Francisco, MD, 75. *Prof Exp:* Med officer, 78-86, chief, Epidemiol Br, 86-91, ASSOC DIR, DIV BIOMET & RISK ASSESSMENT, NAT INST ENVIRON HEALTH SCI, NIH, 91- *Concurrent Pos:* Adj assoc prof, Sch Pub Health, Univ NC, 88- *Mem:* Am Epidemiol Soc; Soc Pediat Res; AAAS; Soc Epidemiol Res; Am Col Prev Med. *Res:* Effects of perinatal exposure to toxic chemicals on childhood growth and development. *Mailing Add:* Nat Inst Environ Health Sci Epidemiol Br PO Box 12233 Research Triangle Park NC 27709

ROGATZ, PETER, b New York, NY, Aug 5, 26; m 49; c 2. PUBLIC HEALTH. *Educ:* Columbia Univ, BA, 46, MPH, 56; Cornell Univ, MD, 49. *Prof Exp:* Dir study home care prog, Hosp Coun Greater New York, 53-55; Commonwealth Fund fel, Columbia Univ, 55-56; assoc med dir, Health Ins Plan Greater New York, 56-57; med adminr, East Nassau Med Group, 57-58; assoc dir hosp & health agency study, Fedn Jewish Philanthropies NY, 58-59; dep dir, Montefiore Hosp, New York, 60-63; dir, Long Island Jewish Hosp, 64-68; dir univ hosp, State Univ NY, Stony Brook, 68-71, co-dir grad prog health care admin, 69-71; sr vpres, Blue Cross & Blue Shield Greater New York, 71-76; pres & co-founder, RMR Health & Hosp Mgt Consults, Inc, 76-84; PROF CLIN COMMUNITY MED, SCH MED, STATE UNIV NY, STONY BROOK, 68-; VPRES MED AFFAIRS, VIS NURSE SERV, NY, 84-; LECTR, SCH PUB HEALTH, COLUMBIA UNIV, 58- *Concurrent Pos:* Dir, Study Home Care Progs, Hosp Coun Greater New York, 53-55; Commonwealth Fund fel, Columbia Univ, 55-56; assoc med dir, Health Ins Plan Greater New York, 56-57; med adminr, ENassau Med Group, 57-58; assoc dir, Hosp & Health Agency Study, Fedn Jewish Philanthropies New York, 58-59; lectr, Sch Pub Health, Columbia Univ, 58- *Honors & Awards:* Dean Conley Award, Am Col Hosp Adminrs, 75. *Mem:* Fel Am Col Physicians; fel Am Col Hosp Adminr; fel Am Col Prev Med; fel Am Pub Health Asn; Asn Teachers Prev Med. *Res:* Medical care organization and delivery of health services; community health planning; medical care administration. *Mailing Add:* 76 Oakdale Lane Roslyn Heights NY 11577-1535

ROGAWSKI, MICHAEL ANDREW, b Los Angeles, Calif, April 8, 52; c 2. NEUROLOGY, NEUROSCIENCE. *Educ:* Amherst Col, BA, 74; Sch Med, Yale Univ, MD & PhD(pharmacol), 80. *Prof Exp:* Med staff fel, lab neurophysiol, Nat Inst Neurol Dis & Stroke, NIH, 81-86; ASST PROF NEUROL, SCH MED, JOHNS HOPKINS UNIV, 86-; CHIEF, NEURONAL EXCITABILITY SECT, NAT INST NEUROL DIS & STROKE, NIH, 90- *Concurrent Pos:* Resident & fel neurol, Johns Hopkins Hosp, 82-85; active staff clin ctr, NIH, 87- *Mem:* Sigma Xi; Soc Neurosci; Am Soc Pharmacol & Exp Therapeut; Am Acad Neurol; Am Epilepsy Asn. *Res:* Cellular neurophysiology and neuropharmacology; anticonvulsant drugs. *Mailing Add:* Neuronal Excitability Sect NINDS NIH Bldg 10 Rm 5N-248 Bethesda MD 20892

ROGER, WILLIAM ALEXANDER, b Toronto, Ont, Apr 6, 47; m 69; c 2. ELECTRONICS, COMPUTER SCIENCES. *Educ:* Univ Toronto, BASc, 68, MSc, 70; Univ Alta, PhD(physics), 74. *Prof Exp:* Killam fel physics, Dalhousie Univ, 74-77; DEFENSE SCIENTIST ELECTRONICS & SIGNAL PROCESSING, DEFENSE RES ESTAB ATLANTIC, 77- *Res:* Design and fabrication of electronic instruments for the armed forces; signal processing and analysis of oceanographic acoustic data. *Mailing Add:* Defense Res Estab Atlantic PO Box 1012 Dartmouth NS B2Y 3Z7 Can

ROGERS, ADRIANNE ELLEFSON, b Aberdeen, Wash, Feb 18, 33; m 54; c 3. EXPERIMENTAL PATHOLOGY. *Educ:* Radcliffe Col, AB, 54; Harvard Med Sch, MD, 58. *Prof Exp:* Intern med, Beth Israel Hosp, Boston, 58-59; res fel path, Boston City Hosp & Harvard Med Sch, 60-62; USPHS res fel, Mallory Inst Path, 62-64, res fel, 64-65; res assoc, Mass Inst Technol, 66-72, sr res scientist nutrit & food sci, 72-85; PROF & ASSOC CHMN PATH, BOSTON UNIV SCH MED, 84- *Concurrent Pos:* Asst, Harvard Med Sch, 64-67, instr, 68-73; sr resident path, Peter Bent Brigham Hosp, Boston, Mass, 76-77, clin fel, 77-78. *Mem:* Am Asn Cancer Res; Am Soc Exp Path; Am Inst Nutrit. *Res:* Interaction between diet and chemical carcinogenesis; nutritional liver disease, including fatty liver and cirrhosis; toxicity. *Mailing Add:* Dept Path L804 Boston Univ Sch Med 80 E Concord St Boston MA 02118

ROGERS, ALAN BARDE, b Sergeant Bluffs, Iowa, Nov 11, 18; m 41; c 3. CHEMISTRY. *Educ:* Iowa State Col, BSc, 42. *Prof Exp:* Res chemist, Armour & Co, 41-42, chem process develop, 45-48, chem mkt develop, 48-53, head dairy, poultry & specialty prods res, 53-60, assoc tech dir, Food Res Div, 60-64, asst dir, Food Res Div, 64-67, asst dir res, Food Res Div, 67-80; RETIRED. *Concurrent Pos:* Mem, Nat Turkey Fedn Res Comt; dir & pres, Inst Am Poul Ind Res Coun; fel, Inst Food Technologists. *Mem:* Poultry Sci Asn; Inst Food Technologists. *Res:* Poultry; frozen foods; dehydrated and dairy products. *Mailing Add:* 6135 Gold Dust Ave Scottsdale AZ 85253

ROGERS, ALAN ERNEST EXEL, b Harare, Zimbabwe, Oct 3, 41; US citizen; m 68; c 2. RADIO ASTRONOMY. *Educ:* Univ Col Rhodesia & Nyasaland, BSc, 62; Mass Inst Technol, SM, 64, PhD(elec eng), 67. *Prof Exp:* ASST DIR, HAYSTACK OBSERV, MASS INST TECHNOL, 67- *Honors & Awards:* Rumford Medal, Am Acad Arts & Sci, 71. *Mem:* AAAS; Inst Elec & Electronics Engrs; Am Astron Soc; Am Geophys Union; Am Cong Surv & Mapping. *Res:* Emission and absorption of microwave radiation by interstellar hydroxl radical; radar mapping of Venus; very long baseline interferometry; radiometric instrumentation. *Mailing Add:* Haystack Observ Westford MA 01886

ROGERS, ALVIN LEE, b Houston, Tex, Jan 18, 29. MEDICAL MYCOLOGY, MEDICAL MICROBIOLOGY. *Educ:* Southeastern State Univ, BS, 48; Mich State Univ, PhD(med mycol, mycol), 67. *Prof Exp:* Teacher high sch, Tex, 48-59; instr biol, comp anat & microbiol, Bay City Jr Col, 59; instr biol bot, zool & microbiol, Port Huron Jr Col, 59-61; from instr to assoc prof, 65-84, PROF MED MYCOL, MICH STATE UNIV, 84- *Concurrent Pos:* Res assoc, Belo Horizonte Vet Sch, Brazil, 60, Bot Inst Sao Paulo, Brazil, 61 & Nat Univ Colombia, 68; co-dir, Mycol Sect Health Ctr & Animal Diag Lab, Mich State Univ, 77- *Honors & Awards:* Bessey Res Award, 64. *Mem:* Int Soc Human & Animal Mycol; Med Mycol Soc Americas; Am Soc Microbiol; Mycol Soc Am; Soc Gen Microbiol; Sigma Xi; Med Mycol Soc Am. *Res:* Medical mycology; pathogenicity of fungi pathogenic in humans and animals and metabolic products produced by these fungi; adherance of Candida albicans to biological surfaces; development of animal models with mycoses for use in antifungal treatment, specifically within liposomes. *Mailing Add:* Dept Microbiol & Publ Health & Med Technol Prog Mich State Univ Plant Biol Bldg East Lansing MI 48824-1312

ROGERS, BEVERLY JANE, b Pasadena, Tex, Sept 24, 43; m 72; c 1. BIOCHEMISTRY, REPRODUCTIVE BIOLOGY. *Educ:* Univ Tex, Austin, BA, 64; Univ Wis-Madison, MS, 70; Univ Hawaii, Manoa, PhD(biochem), 72. *Prof Exp:* Researcher, Pasteur Inst, 72-73; instr sci, Phoenix Study Group, Vietnam, 73-74; NIH fel reproductive biol, Pac Biomed Res Ctr, 74-76, asst researcher, 76-77, from asst prof to assoc prof obstet-gynec, 77-83; ASSOC PROF OBSTET-GYNEC, VANDERBILT UNIV, 83- *Honors & Awards:* Squibb Prize, Am Fertility Soc, 78. *Mem:* Am Soc Cell Biol; Soc Study Reproduction; Am Soc Andrology; Am Fertility Soc. *Res:* Biochemistry of capacitation of mammalian spermatozoa; metabolism of spermatozoa and occurrence of acrosome reaction in the presence of various energy sources and inhibitors; in vitro fertilization; male infertility testing and treatment. *Mailing Add:* 194 Carnavon Pkwy Nashville TN 37205

ROGERS, BRUCE G(EORGE), b Houston, Tex, Feb 20, 25. CIVIL ENGINEERING. *Educ:* Univ Houston, BS, 57; Univ Ill, MS, 58, PhD(civil eng), 61. *Prof Exp:* Engr geophys, Robert H Ray Co, Tex, 47-52; engr asst, Tex Hwy Dept, 53-54, sr draftsman, 54-56, assoc design engr, 56; consult civil eng, Turner & Collie, 56-57; teaching asst theoret & appl mech, Univ Ill, 57-59, instr, 59-61; from asst prof to assoc prof, 61-67, PROF CIVIL ENG, LAMAR UNIV, 67- *Mem:* Am Soc Civil Engrs; Am Soc Mech Engrs; Nat Soc Prof Engs; Masonry Soc. *Res:* Behavior of tapered beam-columns under ultimate loads; plastic design and analysis of non-prismatic members; masonry specifications. *Mailing Add:* Dept Civil Eng Lamar Univ Box 10024 Beaumont TX 77710-0024

ROGERS, BRUCE JOSEPH, b Pasadena, Calif, June 21, 24; m 56; c 4. PLANT PHYSIOLOGY, SCIENCE WRITING. *Educ:* Univ Calif, BSF, 49, MSF, 50; Calif Inst Technol, PhD(plant physiol), 55. *Prof Exp:* From asst prof to assoc prof plant physiol, Purdue Univ, 54-61; assoc plant physiologist, Univ Hawaii, 61-66; prof bot, NDak State Univ, 66-67; plant physiologist, Agr Res Serv, USDA, Pasadena, Calif, 67-69, Riverside, 69-71; consult & sci writer, 71-74; HEAD CLIN RES & DOC, LEE PHARMACEUT, 73- *Mem:* Am Inst Biol Sci; Soc Am Foresters; Sigma Xi. *Res:* Abscission; ethylene enzymology; herbicide action; ecology; dental and medical materials and devices. *Mailing Add:* 17372 Lido Lane Huntington Beach CA 92647

ROGERS, CHARLES C, b Crawfordsville, Ind, Jan 27, 31; m 54; c 3. ELECTRICAL ENGINEERING. *Educ:* Purdue Univ, BSEE, 53, MSEE, 57, PhD(elec eng), 60. *Prof Exp:* Res asst elec eng, Mass Inst Technol, 55; supt elec power, Crawfordsville Elec Light & Power Co, Ind, 56; teaching asst elec eng, Purdue Univ, 56-57, instr, 57-60; res engr electromagnetics, Collins Radio Co, 60-61; from asst prof to prof elec eng, 61-70, chmn dept, 65-70, PROF PHYSICS & ELEC ENG & CHMN DIV, 72-, Rose-Hulman Inst Technol. *Concurrent Pos:* Mem adv bd, Aerospace Res Appl Ctr, Indiana Univ, 65-; pvt consult engr. *Mem:* Inst Elec & Electronics Engrs; Am Soc Eng Educ. *Res:* Electromagnetic theory. *Mailing Add:* RR 32 Box 874 Terre Haute IN 47803

ROGERS, CHARLES EDWIN, b Rochester, NY, Dec 29, 29; m 54; c 3. PHYSICAL CHEMISTRY, MEMBRANE MATERIALS. *Educ:* Univ Syracuse, BS, 54; Univ Syracuse & State Univ NY, PhD(phys chem), 57. *Prof Exp:* Goodyear res fel & res assoc chem, Princeton Univ, 57-59; mem tech staff polymer chem, Bell Labs, 59-65; PROF MACROMOLECULAR SCI, CASE WESTERN RESERVE UNIV, 65-; CO-DIR, INTERUNIV CTR ADHESIVES, SEALANTS & COATINGS, 89- *Concurrent Pos:* Chmn comt, Consortium Univs Estab Overseas Educ Insts, 74-79; SRC sr vis fel, Imp Col, London, 71; assoc dir, Ctr Adhesives, Sealants & Coatings, 84-89. *Mem:* Am Chem Soc; Am Phys Soc; Adhesion Soc; NAm Membrane Soc. *Res:* Polymer science; solubility and diffusion in polymers; kinetics and mechanism of polymerization; polymer rheology and properties; polymer degradation; environmental effects on polymers; membrane separation processes and materials; adhesion and adhesives, sealants and coatings; surface science and technology. *Mailing Add:* Dept Macromolecular Sci Case Western Reserve Univ Cleveland OH 44106

ROGERS, CHARLES GRAHAM, b Summerside, PEI, Mar 13, 29; m 61; c 1. BIOCHEMISTRY, MICROBIOLOGY. *Educ:* McGill Univ, BSc, 52, MSc, 54; Univ Wis, PhD(microbiol), 63. *Prof Exp:* Chemist, 54-64, RES SCIENTIST, DEPT NAT HEALTH & WELFARE, CAN, 65- *Mem:* Am Soc Microbiol; Can Biochem Soc; Sigma Xi; Nutrit Soc Can; Can Inst Food Sci & Technol. *Res:* Dietary stress in relation to lipid composition and metabolism in rat liver tissue and cardiac muscle cells in tissue culture; enzymes of lipid metabolism in cultured heart cells. *Mailing Add:* Dept Health & Welfare Health Protection Br Ottawa ON K1A 0L2 Can

ROGERS, CHARLIE ELLIC, b Booneville, Ark, Aug 13, 38; m 71; c 2. ENTOMOLOGY. *Educ:* Northern Ariz Univ, BS, 64; Univ Ky, MS, 67; Okla State Univ, PhD(entom), 70. *Prof Exp:* Teacher biol & social studies, Dysart Pub Schs, Ariz, 64-65; res assoc entom, Okla State Univ, 70-71; asst prof, Agr Exp Sta, Tex A&M Univ, 71-75; res entomologist, Southwestern Great Plains Res Ctr, Sci & Educ Admin-Fed Res, USDA, 75-80; supvr res entomologist & res leader, Conserv & Prod Res Ctr, 80-83, LAB DIR, INSECT BIOL & POP MGT RES LAB, AGR RES SERV, USDA, TIFTON, GA, 83- *Concurrent Pos:* Assoc ed, Insect World Dig, Biol Sci Res Inst, 73-77; ed, Biol Control, 90- *Honors & Awards:* Prof Excellence Award, Am Agr Econ Soc,

79. *Mem:* Entom Soc Am; Am Registry Prof Entom; Southwest Entom Soc; Sigma Xi. *Res:* Biology, ecology and control of insect pests of sunflower and guar; biological control of agricultural pests; insect migration. *Mailing Add:* USDA/ARS-Insect Biol & Pop Mgt Res Lab PO Box 748 Tifton GA 31793

ROGERS, CLAUDE MARVIN, b Bloomington, Ind, Sept 22, 19; m 44; c 4. BOTANY, TAXONOMY. *Educ:* Ind Univ, AB, 40; Univ Mich, MA, 47, PhD, 50. *Prof Exp:* Stockroom custodian, Univ Okla, 43-44; instr biol, Univ Tex, 48-49; from asst prof to prof biol, Wayne State Univ, 49-85; RETIRED. *Mem:* AAAS; Am Soc Plant Taxon; Bot Soc Am; Int Asn Plant Taxon. *Res:* Taxonomy of flowering plants, especially family Linaceae. *Mailing Add:* Dept Biol Sci 210 Sci Hall Wayne State Univ 5950 Cass Ave Detroit MI 48202

ROGERS, DAVID ELLIOTT, b New York, NY, Mar 17, 26; m 72; c 3. INTERNAL MEDICINE. *Educ:* Cornell Univ, MD, 48. *Hon Degrees:* ScD, Thomas Jefferson Univ, 73, Tufts Univ, 82, Moorehouse Sch Med, 85; LHD, Rush Univ, 85; LLD, Univ Pa, 85, Albany Med Col, 89. *Prof Exp:* Intern & asst resident, Osler Med Serv, Johns Hopkins Hosp, 48-50; NSPHS fel, New York Hosp, 50-51; chief resident med, New York Hosp, 51-52, chief div infectious dis, 55-59; from asst prof to assoc prof med, Cornell Univ, 54-59; prof, chmn dept & physician-in-chief, Vanderbilt Univ, 59-68; vpres med & dean med sch, prof med & med dir, Johns Hopkins Univ, 68-71; PRES, ROBERT WOOD JOHNSON FOUND, 72-; CLIN PROF MED, RUTGERS UNIV, 75-; PROF, WALSH MCDERMOTT UNIV, 86- *Concurrent Pos:* Vis investr, Rockefeller Inst Med Res, 54-55; adj prof med, Cornell Univ, 74-; mem sci adv bd, Scripps Clinic & Res Found, 72-74, 77-; mem, Nat Vis Coun, Columbia Univ, 79- *Mem:* Inst Med Nat Acad Sci; Am Soc Clin Invest; Asn Am Physicians(treas, 64-69, pres,74-75); Am Clin & Climat Asn; Fel Am Col Physicians. *Res:* Host-parasite relationships in infectious diseases. *Mailing Add:* Cornell Univ Med Col 1300 York Ave New York NY 10021

ROGERS, DAVID FREEMAN, b Theresa, NY, Sept 3, 37. FLUID DYNAMICS, AEROSPACE ENGINEERING. *Educ:* Rensselaer Polytech Inst, BAeroE, 59, MS, 60, PhD(aeronaut, astronaut), 67. *Prof Exp:* Res assoc aeronaut eng, Rensselaer Polytech Inst, 62-64; from asst prof to assoc prof, 64-73, Dir CAD/IG, 75-85, PROF AEROSPACE ENG, US NAVAL ACAD, 74-, OFF NAVAL RES PROF, 71- *Concurrent Pos:* Sr consult, Cadcom, Inc, 70-; hon res fel, Univ Col London, 77-78; vis prof, Univ New Southwales, 80; Fujitsu fel Royal Melbourne, Inst Tech, Melbourne, Australia, 87. *Mem:* Am Inst Aeronaut & Astronaut; Soc Naval Archit & Marine Engrs; Asn Comput Mach. *Res:* Computer graphics; computer aided manufacturing; curve and surface description; compressible boundary layers; aerodynamics of nonrigid airfoils; computer aided design and interactive graphics; numerical methods; nonlinear two point asymptotic boundary value problems. *Mailing Add:* Dept Aerospace Eng US Naval Acad Annapolis MD 21402

ROGERS, DAVID PETER, b Barri, Wales, Mar 20, 57; m 82; c 2. MARINE METEOROLOGY, AIR POLLUTION. *Educ:* Univ E Anglia, UK, BSc, 80; Univ Southampton, PhD(oceanog), 83. *Prof Exp:* From asst res prof to assoc res prof boundary layer meteorol, Desert Res Inst, 83-87; from asst prof to assoc prof physics, Univ Nev, Reno, 85-87 & 87-88; asst res meteorologist, 88-89, ASSOC RES OCEANOGR, SCRIPPS INST OCEANOG, 89- *Mem:* Fel Royal Meteorol Soc; Am Meteorol Soc; AAAS. *Res:* Observational and numerical studies of the marine atmospheric boundary layer and problems of long range transport of pollutants. *Mailing Add:* Scripps Inst Oceanog La Jolla CA 92093-0230

ROGERS, DAVID T, JR, b Foley, Ala, Apr 10, 35; m 58; c 2. ECOLOGY. *Educ:* Huntington Col, AB, 58; Univ Ga, MS, 63, PhD(bird migration), 65. *Prof Exp:* Teacher, Marbury High Sch, 59-61 & Southern Union Jr Col, 61-62; from asst prof to assoc prof, 65-77, PROF BIOL, UNIV ALA, TUSCALOOSA, 77- *Concurrent Pos:* US Dept Interior grant, 70-71; NSF grant, 71; US Corps Eng grant, 72-73. *Mem:* Am Ornithologists Union. *Res:* Ornithology, especially population and community aspects; bird migration in Latin America. *Mailing Add:* Box 1927 Tuscaloosa AL 35487

ROGERS, DAVID WILLIAM OLIVER, b Toronto, Ont, Aug 11, 45; div; c 2. RADIATION DOSIMETRY, MONTE CARLO TRANSPORT. *Educ:* Univ Toronto, BSc, 68, MSc, 69, PhD(physics), 72. *Prof Exp:* Res assoc, Oxford Nuclear Physics Lab, UK, 72-73; RES OFFICER, RADIATION DOSIMETRY, NAT RES COUN CAN, 73-, HEAD, IONIZING RADIATION STANDARDS, 85- *Concurrent Pos:* Chmn, Med & Biol Physics Div, Can Asn Physicists, 83-84; assoc ed, Med Physics, 87-; mem, ICRU comt on Absorbed Dose Standards, 88-, CCEMRI(1) of the BIPM, 88. *Honors & Awards:* Sylvia Fedoruk Prize, 89. *Mem:* Can Asn Physicists; Health Physics Soc; Am Asn Physicists Med; Can Radiation Protection Asn. *Res:* Radiation dosimetry for medical and radiation protection purposes; development and use of coupled electron-photon transport; Monte Carlo codes for medical physics applications; national primary standards for radiation measurement. *Mailing Add:* Nat Res Coun Can Inst Nat Measurement Standards Montreal Rd Ottawa ON K1A 0R6 Can

ROGERS, DEXTER, b Kyoto, Japan, Dec 14, 21; m 45; c 4. BIOCHEMISTRY. *Educ:* Univ Mich, BS, 44, MS, 46; Ore State Univ, PhD(biochem), 54. *Prof Exp:* Res chemist, Western Condensing Co, Appleton, Wis, 45-51; NSF fel, Stanford Univ, 54-55; instr biochem, Univ Mich, 55-58; instr & res assoc chem, Univ Ore, 58-60; asst prof food & nutrit, Utah State Univ, 60-62, chem, 62-65; asst prof, Univ Mont, 65-66; USPHS spec fel, Ore State Univ, 66-68; prof, State Univ NY Col Cortland, 68-69; lectr, Portland State Univ, 69-70; prof & chmn dept, William Paterson Col NJ, 70-75; prof chem, Bloomfield Col, 75-77; applns & develop chemist, KONTES, 77-79; CONSULT TECH MGT & ENVIRON REGULATIONS, 80- *Mem:* AAAS; Am Chem Soc; Am Soc Microbiol. *Res:* Biochemistry of microorganisms; sugar transport mechanisms. *Mailing Add:* 535 Harding Hwy Mays Landing NJ 08330

ROGERS, DONALD B, b Moulton, Ala, Mar 2, 36; m 57; c 4. SOLID STATE INORGANIC CHEMISTRY. *Educ:* Vanderbilt Univ, BA, 58; Mass Inst Tech, PhD(inorg chem), 62. *Prof Exp:* Res scientist, Lincoln Lab, Mass Inst Tech, 62-65 & E I DuPont de Nemours, 65-69; res scientist, 65-69, res supvr, Ctr Res, 69-72, res mgr, Photo Prods, 72-75, lab dir, DuPont Deutschland, 75-78, dir mkt, Printing Systs, 78-82, dir res & develop, Textile Fibers, 82-84, DIR RES & DEVELOP ELECTRONICS, E I DUPONT DE NEMOURS, 84- *Mem:* Indust Res Inst. *Res:* ceramics and polymers photosensitive coatings; film casting and coating; microelectronic packaging and interconnection; OPTO electronics. *Mailing Add:* Dupont Electronic Barley Mill Plaza P-21 Wilmington DE 19880-0021

ROGERS, DONALD EUGENE, b Los Angeles, Calif, Aug 27, 32; m 55; c 2. FISH BIOLOGY. *Educ:* Calif State Polytech, San Luis Obispo, BS, 58; Univ Wash, MS, 61, PhD(fisheries), 67. *Prof Exp:* Fishery biologist, 60-68, mem fac fisheries, 69-77, RES ASSOC PROF FISHERIES, FISHERIES RES INST, UNIV WASH, 77- *Mem:* Am Inst Fishery Res Biologists; Am Fisheries Soc; Sigma Xi. *Res:* Fish population dynamics; biology of sockeye salmon. *Mailing Add:* Fisheries Res Inst Univ Wash Seattle WA 98195

ROGERS, DONALD RICHARD, b Richmond, Ky, Apr 27, 32; m 61. INORGANIC CHEMISTRY. *Educ:* Univ Ky, BS, 59, MS, 61, PhD(chem), 68. *Prof Exp:* Res chemist, 61-64, group leader chem, 64-65 & 68-71, sr res specialist, 72-79, FEL NUCLEAR OPERS DEPT, MOUND LAB, MONSANTO RES CORP, MIAMISBURG, 79- *Mem:* AAAS; Am Chem Soc; Sigma Xi. *Res:* Coordination chemistry of lanthanide and actinide elements; solvent extraction equilibria; radioisotopic fuels; environmental chemistry of plutonium; nuclear safeguards; uranium and plutonium measurement methods. *Mailing Add:* 975 Fernshire Dr Centerville OH 45459

ROGERS, DONALD WARREN, b Hackensack, NJ, Sept 10, 32; m 56; c 2. PHYSICAL CHEMISTRY, ANALYTICAL CHEMISTRY. *Educ:* Princeton Univ, BA, 54; Wesleyan Univ, MA, 56; NC Univ, PhD(chem, math), 60. *Prof Exp:* Asst prof chem, Robert Col, Istanbul, 60-63 & Long Island Univ, 63-64; master teacher chem & physics, Am Madrid, Spain, 64-65; from asst prof to assoc prof, 65-71, PROF CHEM, LONG ISLAND UNIV, 71- *Concurrent Pos:* NIH res grants, 71-73, 74-76, 77 & 81-, Cottrell res grant, 83-85 & PRF res grants, 85-87, 88-89 & 89-91; vis prof chem, Univ Ga, 76-77 & Barnard Col, 80-81. *Mem:* Am Chem Soc. *Res:* Enthalpies of hydrogenation; thermochemistry; molecular structure; enthalpimetry; solution theory. *Mailing Add:* Dept Chem Long Island Univ Brooklyn NY 11201

ROGERS, DOUGLAS HERBERT, b Wolfville, NS, June 2, 26; m 51; c 3. OPTICS, METAL PHYSICS. *Educ:* Dalhousie Univ, BSc, 47, MSc, 49; Mass Inst Technol, PhD(physics), 53. *Prof Exp:* Fel low temperature & solid state physics, Nat Res Coun, Can, 52-54; from asst prof to assoc prof, 54-65, PROF PHYSICS, ROYAL MIL COL CAN, 65- *Mem:* Optical Soc Am; Am Phys Soc; Am Asn Physics Teachers; Can Asn Physicists. *Res:* Attenuation of ultra sound; dislocations; acoustic emission; liquid crystals; holographic interferometry. *Mailing Add:* RR 1 Tunstall Bay C17 Bowen Island BC V0N 1G0 Can

ROGERS, EDWIN HENRY, b Newton, Mass, Nov 5, 36; m 60; c 3. COMPUTER SCIENCES, SYSTEMS DESIGN. *Educ:* Carnegie-Mellon Univ, BS, 58, MS, 60, PhD(math), 62. *Prof Exp:* Res technician, Woods Hole Oceanog Inst, 56-58; teaching asst & instr math, Carnegie Inst Technol, 58-62; Leverhulme vis fel, Univ Strathclyde, 62-63; US Army, 63-65; PROF COMPUTER SCI & MATH, RENSSELAER POLYTECH INST, 65- *Concurrent Pos:* Managing ed, Siam News, Soc Indust & Appl Math, 75-82; vis res prof, Univ Waterloo, 76-77; vis scientist, Gen Electric Res & Develop Ctr, 83. *Mem:* AAAS; Inst Elec & Electronics Engrs Comput Soc; Soc Indust & Appl Math; Asn Comput Mach; Inst Elec & Electronics Engrs. *Res:* Very-large-scale integration wafer scale systems; yield analysis and fault tolerance; parallel algorithms; nonlinear and multiparameter eigen problems; computer-integrated education. *Mailing Add:* Dept Comput Sci Rensselaer Polytech Inst Troy NY 12180-3590

ROGERS, EMERY HERMAN, b Los Angeles, Calif, Mar 31, 21; c 3. PHYSICS. *Educ:* Stanford Univ, AB, 43, PhD(physics), 51. *Prof Exp:* Asst physics, Stanford Univ, 42-43; physicist, US Naval Res Lab, 43-45; engr nuclear magnetic resonance, Varian Assocs, 49-53, mgr instrument field eng, 53-60, vpres & mgr instrument div, 60-63, vpres instrument group, 63-67; gen mgr, 67-75, gen mgr, Anal Instrument Group, Hewlett-Packard Co, 75-79, EXEC DIR, HEWLETT-PACKARD CO FOUND, 79- *Mem:* AAAS; Am Phys Soc; Instrument Soc Am; Sci Apparatus Makers Asn. *Res:* Nuclear magnetic resonance; electron paramagnetic resonance; magnetism; analytical instrumentation. *Mailing Add:* 218 Lowell Ave Palo Alto CA 94301

ROGERS, ERIC MALCOLM, physics; deceased, see previous edition for last biography

ROGERS, FRANCES ARLENE, b Northfield, Minn, Jan 24, 23; m 56; c 2. PROTOZOOLOGY & VERTEBRATE ZOOLOGY, ANATOMY. *Educ:* Drake Univ, BA, 44; Univ Chicago, MS, 46; Univ Iowa, PhD(zool), 53. *Prof Exp:* Instr biol, Earlham Col, 46-49; asst prof biol, Cornell Col, 54 & Shimer Col, 55-56; from asst prof to assoc prof, 66-85, PROF BIOL, DRAKE UNIV, 85- *Res:* Parasitic protozoology. *Mailing Add:* Dept Biol Drake Univ Des Moines IA 50311

ROGERS, FRED BAKER, b Trenton, NJ, Aug 25, 26. PREVENTIVE MEDICINE, FAMILY PRACTICE. *Educ:* Temple Univ, MD, 48; Univ Pa, MS, 54; Columbia Univ, MPH, 57; Am Bd Prev Med, dipl, 57. *Prof Exp:* Rotating intern, Univ Hosp, 48-49, chief resident med, 53-54, USPHS fel, Univ, 54-55, from asst prof to assoc prof prev med, Sch Med, 56-60, chmn dept family pract & community health, 70-77, PROF PREV MED, SCH MED, TEMPLE UNIV, 60- *Concurrent Pos:* Lectr epidemiol, Sch Pub

Health, Columbia Univ, 57-68; lectr pub health, Sch Nursing, Univ Pa, 64-67; consult med, US Naval Hosp, Philadelphia, 64-73; hon mem, Royal Soc Med, London, Eng, 90. *Mem:* AMA; fel Am Col Physicians; Am Osler Soc. *Res:* Epidemiology; immunization. *Mailing Add:* Dept Family Pract & Community Health Sch Med Temple Univ Philadelphia PA 19140

ROGERS, GARY ALLEN, b Compton, Calif, Jan 22, 45. BIOCHEMISTRY. *Educ:* Univ Calif, Los Angeles, BS, 68, Univ Calif, Santa Barbara, PhD(org chem), 73. *Prof Exp:* Fel biochem, Univ Calif, Los Angeles, 73-76; ASST PROF CHEM, UNIV TEX, DALLAS, 76- *Concurrent Pos:* NIH fel, Molecular Biol Inst, Univ Calif, Los Angeles, 73-75. *Mem:* Am Chem Soc. *Res:* Mechanisms of enzymic catalysis; protein modification reagents; mechanism of uncouplers of oxidative phosphorylation. *Mailing Add:* RR 2 No 8 Cushing TX 75760

ROGERS, GERALD STANLEY, b Reading, Pa, Feb 29, 28. MATHEMATICAL STATISTICS. *Educ:* Muhlenberg Col, BS, 48; Univ Wash, MA, 51; Univ Iowa, PhD(math), 58. *Prof Exp:* Instr, Lafayette Col, 54-55; asst prof math, Univ Ariz, 58-65; assoc prof, 65-76, PROF MATH, NMEX STATE UNIV, 76- *Mem:* Am Math Soc; Math Asn Am; Am Statist Asn; Inst Math Statist. *Res:* Statistical distribution theory. *Mailing Add:* Dept Math Sci NMex State Univ Las Cruces NM 88003

ROGERS, GIFFORD EUGENE, b Grand Island, Nebr, May 22, 20; m 42; c 4. REGIONAL DEVELOPMENT, IRRIGATION & DRAINAGE DESIGN. *Educ:* Univ Nebr, Lincoln, BSCE, 43; Purdue Univ, West Lafayette, Ind, MSCE, 48. *Prof Exp:* Chief engr & proj engr, United Fruit Co, Panama, Costa Rica, Honduras, Guatemala, Dominican Repub, 48-62; Consult, Int Develop Serv, 62-65, Develop & Resources Corp, 65-68, Robert R Nathan Assocs, 68-70, Eng Consults, Inc, Denver, Colo, 70-83, consult, Pub Admin Serv, 83-86 & Wilbur Smith Assocs, Columbia, SC, 86-91; CONSULT, 91- *Mem:* Fel Am Soc Civil Engrs; Nat Soc Prof Engrs; Am Soc Agr Engrs; assoc mem Sigma Xi; Am Acad Environ Engrs. *Res:* Asphalt pavement design; regional water resources development in Third-World countries; roads and highways construction and maintenance in Third-World countries; irrigation and drainage improvement in Honduras. *Mailing Add:* Leisure World Apt 129 G 13361 St Andrews Dr Seal Beach CA 90740

ROGERS, H(ARRY) C(ARTON), JR, b Patchogue, NY, Oct 3, 23; m 47; c 3. PHYSICAL & MECHANICAL METALLURGY, METALWORKING. *Educ:* Cornell Univ, AB, 47; Rensselaer Polytech Inst, PhD(metall), 56. *Prof Exp:* Res metallurgist, Gen Elec Co, 48-69; prof, 69-84, head, Dept Mat Eng, 87-90, AW GROSVENOR PROF MAT ENG, DREXEL UNIV, 84- *Concurrent Pos:* Vis res scientist, Melbourne Res Labs, BHP Steel Co, Australia, 81 & Mat & Mech Res Ctr, Army, Watertown, MA, 83; vis prof, Monash Univ, Australia, 81. *Honors & Awards:* McKay-Helm Award, Am Welding Soc, 76 & Adams Mem Award. *Mem:* Fel Am Soc Metals; Am Welding Soc; Am Inst Mining, Metall & Petrol Engrs; fel Am Inst Chemists. *Res:* Materials processing; mechanical behavior and fracture of solids; hydrogen embrittlement of metals; high pressure processing of metals; material behavior under high rates of loading. *Mailing Add:* Dept Mat Eng Drexel Univ Philadelphia PA 19104

ROGERS, HARTLEY, JR, b Buffalo, NY, July 6, 26; m 54; c 3. MATHEMATICS, PROBABILITY. *Educ:* Yale Univ, AB, 46, MS, 50; Princeton Univ, PhD(math), 52; Cambridge Univ, MA, 68. *Prof Exp:* Asst instr math, Princeton Univ, 50-52; Benjamin Pierce instr, Harvard Univ, 52-55; vis lectr, 55-56, from asst prof to assoc prof, 56-64, chmn faculty, 71-73, assoc provost, 74-80, PROF MATH, MASS INST TECHNOL, 64- *Concurrent Pos:* Guggenheim fel, 60-61; vis fel, Clare Hall, Cambridge Univ & NSF sr fel, 67-68; ed, Annals Math Logic & J Comput & Syst Sci, 69- & J Symbolic Logic, 63-67; trustee, Buckingham, Browne & Nichols Sch, 75-81 & Kingsley Trust Assoc, 76-78; consult, NSF, 80- *Honors & Awards:* Ford Award, Math Asn Am. *Mem:* Am Math Soc; Math Asn Am; Asn Symbolic Logic (vpres, 64-67). *Res:* Mathematical logic and recursion theory; probability and statistics, mathematical models; mathematics education; physics and applied mechanics. *Mailing Add:* Dept Math Mass Inst Technol 77 Massachusetts Ave Cambridge MA 02139

ROGERS, HARVEY WILBUR, b Annapolis, Md, Feb 20, 45; m 70; c 2. INCINERATION RESEARCH, INFECTIOUS WASTE MANAGEMENT. *Educ:* WVa Univ, BS, 67, MS, 68. *Prof Exp:* Sanit engr, off solid waste mgt prog, 68-71, chief systs sect, Off Solid Waste Mgt Prog, USPHS, 71-73; sanit engr, Off Solid Waste Mgt Prog, US Environ Protect Agency, 73-74; environ engr, div res serv, NIH, 74-80, chief environ systs sect, 80-85, chief pollution control sect, 85-86, chief environ protect br, Div Safety, 86-89; ENVIRON ENG, INCINERATION SPECIALIST, AGENCY TOXIC SUBSTANCES DIS REGISTRY, 89- *Concurrent Pos:* Lectv univ prof, 75-; advisor, Nat Acad Sci, 82-83, WHO, Hosp Waste mgt, 84-, US Environ Protect Agency, 85-86, NSF, 88; guest fac, Am Soc Hosp Engrs, 85-88. *Mem:* Air & Waste Mgt Asn; Am Soc Mech Engrs. *Res:* Author of several articles on special waste management; hazardous chemical waste management and incineration technology. *Mailing Add:* Agency Toxic Substances & Dis Registry MS E32 1600 Clifton Rd NE Atlanta GA 30333

ROGERS, HORACE ELTON, b Philadelphia, Pa, Dec 5, 02; m 26; c 2. PHYSICAL CHEMISTRY, ANALYTICAL CHEMISTRY. *Educ:* Dickinson Col, BSc, 24; Lafayette Col, MSc, 25; Princeton Univ, PhD(chem), 30. *Prof Exp:* Instr sci, Dickinson Col, 25-27; asst chem, Princeton Univ, 27-29; assoc prof, 29-41, prof anal chem, 41-52, Alfred Victor duPont prof, 52-71, chmn dept, 58-65, EMER ALFRED VICTOR DUPONT PROF ANALYTICAL CHEM, DICKINSON COL, 71- *Concurrent Pos:* Head dept, Pa Area Col, Harrisburg, 46-48. *Mem:* Emer mem Am Chem Soc. *Res:* Gas chromatography of metal chelates; metal analysis by thin-layer chromatography. *Mailing Add:* 900 W South St Carlisle PA 17013

ROGERS, HOWARD GARDNER, b Houghton, Mich, June 21, 15; m 40; c 5. PHYSICS, CHEMISTRY. *Prof Exp:* Res asst new light polarizers, Polaroid Corp, 37-41, Nat Defense Res Coun contract, 41-44. res mgr optical plastics, 44-47, sr scientist, 47-54, dept mgr color photog res, 54-66, dir, 66-68, vpres, 68-78, from assoc dir to dir res, 75-85, sr vpres, 78-85, EMER DIR RES, POLAROID CORP, 85- *Concurrent Pos:* Sr res fel, Polaroid Corp, 68-; consult, 85- *Honors & Awards:* Wetherill Medal, Franklin Inst, 66; Achievement Award, Indust Res Inst, 87. *Mem:* AAAS; hon mem Soc Photog Scientists & Engrs; fel Optical Soc Am; fel Am Inst Chemists; fel Am Acad Arts & Sci. *Res:* New one step color photographic processes; new light polarizing materials; new optical plastics; visual 3-D systems; acknowledgement of 144 various patents. *Mailing Add:* 20 Newton St Weston MA 02193-2311

ROGERS, HOWARD H, b New York, Dec 26, 26; m 78; c 3. ELECTROCHEMISTRY, COMPUTER PROGRAMMING. *Educ:* Univ Ill, BS, 49; Mass Inst Technol, PhD(inorg chem), 53. *Prof Exp:* Res asst chem, Mass Inst Technol, 51-52; from res chemist to res group leader inorg chem, Res Div, Allis-Chalmers Mfg Co, 52-61; from res specialist to sr tech specialist, Res Div, Rocketdyne Div, NAm Rockwell Corp, 61-70; chief res scientist, Martek Instruments, Inc, 70-73; SR SCIENTIST & ENGR, HUGHES AIRCRAFT CO, 73- *Mem:* Am Chem Soc; Electrochem Soc; Sigma Xi; Forth Interest Group. *Res:* Batteries, especially nickel-hydrogen; electrochemistry; electronics; fluorine chemistry; high temperature chemistry; advanced wastewater treatment; chemical instrumentation; battery (computer) modelling. *Mailing Add:* 18361 Van Ness Ave Torrance CA 90504

ROGERS, HOWARD TOPPING, b Savedge, Va, July 22, 08; m 37; c 1. SOIL SCIENCE, AGRONOMY. *Educ:* Va Polytech Inst, BS, 30; Mich State Col, MS, 36; Iowa State Col, PhD(soil fertility), 42. *Prof Exp:* Instr high schs, Va, 30-34; asst soil surv, Mich State Col, 35; asst soil technologist, Exp Sta, Va Polytech Inst, 36-41; assoc soil chemist, Ala Polytech Inst, 42-46; agronomist & chief, Soil & Fertilizer Res Br, Tenn Valley Authority, 47-51; head dept, 51-66, prof, 66-76, EMER PROF AGRON & SOILS, AUBURN UNIV, 76- *Concurrent Pos:* Consult, Tenn Valley Authority, 80. *Mem:* Soil Sci Soc Am; fel Am Soc Agron; Sigma Xi. *Res:* Exo-enzyme systems of plant roots and availability of organic phosphorus to plants; soil and water conservation; effect of lime on physico-chemical properties of soils and crop yields; boron requirements of legumes and boron fixation in soils; summary and analysis of data on crop response to phosphates; agri-chemicals; new fertilizer practices. *Mailing Add:* Dept Agron & Soils Funchess Hall Auburn Univ Auburn AL 36849

ROGERS, HOWELL WADE, b Ripley, Miss, Nov 26, 43; m 63; c 3. MEDICAL MICROBIOLOGY, VIROLOGY. *Educ:* Univ Miss, BA, 65, MS, 67; Univ Okla, PhD(med microbiol), 70. *Prof Exp:* Instr microbiol, Med Ctr, Univ Miss, 70-72, actg asst prof, 72; asst prof, 72-83, ASSOC PROF MICROBIOL & IMMUNOL, IND UNIV SCH MED EVANSVILLE CTR, 83- *Mem:* AAAS; Am Soc Microbiol; Sigma Xi. *Res:* Viral enzymes; Herpes virus immunology; cell-mediated immunity to Herpes simplex virus. *Mailing Add:* Ind Univ Sch Med-Evansville 8600 Univ Blvd Evansville IN 47712

ROGERS, HUGO H, JR, b Atmore, Ala, Aug 2, 47; m 70. AIR POLLUTION, PLANT PHYSIOLOGY. *Educ:* Auburn Univ, BS, 69, MS, 71; Univ NC, Chapel Hill, PhD(air pollution), 75. *Prof Exp:* Environ engr air pollution & veg, Res Triangle Inst, 75-76; from asst to assoc prof bot, NC State Univ, 76-84; PLANT PHYSIOLOGIST, AGR RES SERV, USDA, 76-; PROF AGRON, AUBURN UNIV, 84- *Concurrent Pos:* Res assoc air pollutant uptake by veg, NC State Univ, 75-76; adj assoc prof, Dept Environ Sci & Eng, Sch Pub Health, Univ NC, Chapel Hill, 78- *Mem:* Air Pollution Control Asn; Am Soc Plant Physiologists; Am Soc Agr Eng; Agron Soc Am; Crop Sci Soc Am. *Res:* Interaction of vegetation with atmospheric chemicals, especially plant gas exchange and effects of air pollutants on plants; carbon dioxide and field studies; design of research equipment for such work; global environmental change. *Mailing Add:* Nat Soil Dynamics Lab USDA-Agr Res Serv PO Box 792 Auburn AL 36831

ROGERS, JACK DAVID, b Point Pleasant, WVa, Sept 3, 37; m 58; c 2. MYCOLOGY, PLANT PATHOLOGY. *Educ:* Davis & Elkins Col, BS, 60; Duke Univ, MF, 60; Univ Wis, PhD(plant path), 63. *Prof Exp:* Asst prof forestry & asst plant pathologist, 63-68, assoc prof plant path, 68-72, PROF PLANT PATH, WASH STATE UNIV, 72- *Mem:* Mycol Soc Am (vpres, 75-76, pres-elect, 76-77, pres, 77-78); Am Phytopath Soc; Bot Soc Am. *Res:* Forest pathology; botany; cytology; genetics; evolution of Ascomycetes. *Mailing Add:* Dept Plant Path Wash State Univ Pullman WA 99164-6430

ROGERS, JACK WYNDALL, JR, b Austin, Tex, Jan 13, 43; m 62; c 2. TOPOLOGY. *Educ:* Univ Tex, BA, 63, MS, 65, PhD(math), 66. *Prof Exp:* From asst prof to assoc prof math, Emory Univ, 66-73; assoc prof, 73-76, PROF MATH, AUBURN UNIV, 76- *Mem:* Am Math Soc; Math Asn Am. *Mailing Add:* Dept Math Auburn Univ 218 Parker Hall Auburn AL 36849

ROGERS, JAMES ALBERT, b Sault Ste Marie, Ont, July 4, 40; m 66; c 2. PHARMACEUTICS, PHYSICAL PHARMACY. *Educ:* Univ Toronto, BS, 63, MS, 66; Univ Strathclyde, Scotland, PhD(pharmaceut technol), 69. *Prof Exp:* Pharmacist, Tamblyn Drug Co, Ltd, 63; asst prof, 69-74, assoc prof pharm, 74-88, PROF, UNIV ALTA, 88- *Concurrent Pos:* bd dir, Vexco Labs Inc, Edmonton, AB. *Mem:* Asn Faculties Pharm Can; Am Pharmaceut Asn; Am Asn Pharmaceut Scientists; NY Acad Sci. *Res:* Dosage form design and formulation of drug delivery systems; drug-biomembrane interactions; partition coefficient basis of quantitative structure activity relationships; liposome approaches to drug dissolution, drug stability and oral drug delivery. *Mailing Add:* Fac Pharm & Pharmaceut Sci Univ Alta Edmonton AB T6G 2N8 Can

ROGERS, JAMES EDWIN, b Waco, Tex, Feb 24, 29; m 57; c 1. GROUND WATER HYDROLOGY, GEOLOGIC MAPPING. *Educ:* Univ Tex, BS, 55, MA, 61. *Prof Exp:* Supv hydrologist, US Geol Surv, Alexandria, LA, 56-85; GEO-HYDROL CONSULT, 85- *Mem:* Fel Geol Soc Am; Nat Waterwell Asn. *Res:* Ground-water flow. *Mailing Add:* 4008 Innis Dr Alexandria LA 71303

ROGERS, JAMES JOSEPH, b Salem, NJ, Oct 30, 42; m 65; c 3. HYDROLOGY, SYSTEM MODELING. *Educ:* Utah State Univ, BS, 64; Univ Ariz, MS, 71, PhD(hydrol), 73. *Prof Exp:* Forester mgt, Apache Nat Forest, 65-71; forester watershed, Rocky Mountain Forest & Range Exp Sta, 71-72, hydrologist, 72-74, hydrologist watershed, 76-83; COMPUT PROG ANALYST, PAC NORTHWEST RES STA, 83- *Mem:* Sigma Xi; Am Geophys Union. *Res:* Development of computer systems for understanding and predicting effects of land management activities on resource outputs and productivity of southwestern forest ecosystems. *Mailing Add:* Pac Northwest Res Sta 2770 Sherwood Lane Suite 2A Juneau AK 99801

ROGERS, JAMES SAMUEL, b Cassville, Ga, Aug 20, 34; m 56, 89; c 1. AGRICULTURAL ENGINEERING, SOIL PHYSICS. *Educ:* Univ Ga, BSAE, 55, MS, 62; Univ Ill, Urbana, PhD(soil physics), 69. *Prof Exp:* Agr engr, Agr Res Serv, USDA, Univ Ga, 57-64 & Univ Ill, 64-68; agr engr sci & educ admin-agr, USDA, Univ Fla, 68-83, asst prof agr eng, 77-83; AGR ENGR AGR RES SERV, USDA, 83-, ASSOC PROF AGR ENG, LA STATE UNIV, 83- *Mem:* Am Soc Agr Engrs; Am Soc Agron; Soil Sci Soc Am; Soil Conserv Soc Am. *Res:* Water management in wet soils to include drainage, irrigation, water quality, climate modification and more efficient use of water in agriculture. *Mailing Add:* La State Univ PO Box 25071 Baton Rouge LA 70894-5071

ROGERS, JAMES STEWART, b Santa Maria, Calif, Apr 5, 32; Can citizen; m 56; c 4. EXPERIMENTAL SOLID STATE PHYSICS. *Educ:* Univ Sask, BEng, 54; Univ Alta, MSc, 62, PhD(physics), 64. *Prof Exp:* Fel, Commonwealth Sci & Indust Res Orgn, Australia, 64-66; ASSOC PROF PHYSICS, UNIV ALTA, 68- *Mem:* Australian Inst Physics. *Res:* Electron tunneling into superconductors and ferromagnetic materials. *Mailing Add:* Dept Physics Univ Alta Edmonton AB T6G 2E2 Can

ROGERS, JAMES TED, JR, b Statesville, NC, July 26, 42; m 66; c 2. MATHEMATICS. *Educ:* Univ NC, Chapel Hill, BS, 64; Univ Calif, Riverside, MA, 66, PhD(math), 68. *Prof Exp:* From asst prof to assoc prof, 68-77, PROF MATH, TULANE UNIV, 77- *Mem:* AAAS; Am Math Soc. *Res:* Point set topology. *Mailing Add:* Dept Math Tulane Univ La 6823 St Charles Ave New Orleans LA 70118

ROGERS, JAMES TERENCE, (JR), b Montreal, Que, Nov 19, 26; m 57; c 4. MECHANICAL ENGINEERING. *Educ:* McGill Univ, BEng, 48, MEng, 50, PhD(mech eng), 53. *Prof Exp:* Res engr, Can Dept Mines & Tech Surv, 50-51; asst prof mech eng, Royal Mil Col Can, 53-55; engr & design specialist, Nuclear Div, Canadair Ltd, 55-59; res engr, Gen Atomic Div, Gen Dynamics Corp, 59-60; tech counsr, Atomic Power Dept, Can Gen Elec, 60-70; PROF MECH ENG, CARLETON UNIV, 70- *Concurrent Pos:* Mem assoc comt heat transfer, Nat Res Coun Can, 66-71; Can deleg, Assembly Int Heat Transfer Conf, 67; mem comt energy sci policies, Sci Coun Can, 76-79, Can Nat Comt Heat Transfer, 79-; mem sci coun, Int Ctr Heat & Mass Transfer, Yugoslavia, 68; consult, Atomic Energy Can Ltd, 70- *Honors & Awards:* Robert W Angus Medal, Eng Inst Can, 64. *Mem:* Can Soc Mech Engrs; Eng Inst Can; Am Soc Mech Engrs; Can Nuclear Soc. *Res:* Coolant mixing in nuclear reactor fuel bundles; two-phase flow and heat transfer problems in nuclear reactors; nuclear reactor safety; combined-purpose use of nuclear reactors; effective utilization of energy. *Mailing Add:* Mech & Aerospace Eng Carleton Univ Ottawa ON K1S 5B6 Can

ROGERS, JAMES VIRGIL, JR, b Johnson City, Tenn, Oct 7, 22; m 45; c 4. RADIOLOGY. *Educ:* Emory Univ, BS, 43, MD, 45. *Prof Exp:* PROF RADIOL, SCH MED, 65-, DIR, DIAG RADIOL, EMORY CLIN, 85- *Concurrent Pos:* Vis radiologist, Grady Mem Hosp, 51-; attend radiologist, Atlanta Vet Admin Hosp, 51- *Mem:* Radiol Soc NAm; Am Roentgen Ray Soc; AMA; fel Am Col Radiol. *Res:* Pulmonary vasculature in heart and lung diseases; mammography. *Mailing Add:* Sect Diag Radiol Emory Clin 1365 Clifton Rd NE Atlanta GA 30322

ROGERS, JERRY DALE, b Nebo, WVa, Mar 27, 54. PHYSICAL CHEMISTRY. *Educ:* WVa Univ, BS, 76; Univ Fla, PhD(chem), 80. *Prof Exp:* Nat res coun fel, Goddard Space Flight Ctr, NASA, 80-82,; STAFF RES SCIENTIST, GEN MOTORS RES LABS, 82- *Mem:* Coblentz Soc; Am Chem Soc; Am Geophys Union; Sigma Xi. *Res:* Infrared spectroscopy; atmospheric chemistry and physics; urban air pollution and global changes; chlorofluorocarbons. *Mailing Add:* Environ Sci Dept GM Res Labs Warren MI 48090

ROGERS, JESSE WALLACE, b Littlefield, Tex, June 8, 41; m 62; c 1. PHYSICAL CHEMISTRY. *Educ:* Univ Tex, Arlington, BS, 63; Tex Christian Univ, PhD(phys chem), 68. *Prof Exp:* From asst prof to assoc prof, 67-77, chmn dept, 69-78, actg vpres acad affairs, 78-80, PROF CHEM, MIDWEST UNIV, 77-, PRES ACAD AFFAIRS, 80- *Mem:* Am Chem Soc. *Res:* Electrochemical kinetics. *Mailing Add:* Dept Chem Midwestern State Univ 3400 Taft Blvd Wichita Falls TX 76308

ROGERS, JOHN ERNEST, b Ames, Iowa, May 15, 47; c 1. MICROBIOLOGY, BIOCHEMISTRY. *Educ:* Univ Ill, BS, 69; Wash State Univ, PhD(biochem), 74. *Prof Exp:* NIH fel, Nat Inst Environ Health Sci, 75; res scientist microbiol, Pac Northwest Labs, Battelle Mem Inst, 77-85; RES SCIENTIST MICROBIOL, ENVIRON PROTECTION AGENCY, ATHENS, GA, 85- *Mem:* Am Soc Microbiol. *Res:* Microbiology of fates and effects of trace metals, radionuclides and organic pollutants in the environment. *Mailing Add:* 150 Richmond Way Bogart GA 30622

ROGERS, JOHN GILBERT, JR, b New York, NY, Sept 13, 41; m 68; c 2. WILDLIFE ECOLOGY. *Educ:* Cornell Univ, BS, 63; NMex State Univ, MS, 66; NC State Univ, PhD(wildlife ecol), 71. *Prof Exp:* Wildlife specialist, NC Agr Exten Serv, 66-67; wildlife biologist, 69-78, STAFF BIOLOGIST, DIV WILDLIFE RES, US FISH & WILDLIFE SERV, 78- *Mem:* Am Soc Mammalogists; Ecol Soc Am; Wildlife Soc. *Res:* The chemical senses in animal depredations control. *Mailing Add:* US Fish & Wildlife Serv 1011 E Tudor Rd Anchorage AK 99503

ROGERS, JOHN JAMES WILLIAM, b Chicago, Ill, June 27, 30; m 56; c 2. GEOLOGY. *Educ:* Calif Inst Technol, BS, 52, PhD(geol), 55; Univ Minn, MS, 52. *Prof Exp:* From instr to prof geol, Rice Univ, 54-74, chmn dept, 71-74; W R KENAN, JR PROF GEOL, UNIV NC, CHAPEL HILL, 75- *Mem:* Am Asn Petrol Geologists; Geol Soc Am; Soc Econ Paleont & Mineral; Geophys Union; hon fel Geol Soc Africa; Geol Soc India. *Res:* Regional geology and studies of crustal evolution in southern India, northeastern Africa, the Middle East, and the Caribbean. *Mailing Add:* Dept Geol CB No 3315 Univ NC Chapel Hill NC 27599-3315

ROGERS, JOHN PATRICK, migratory bird management, wildlife ecology, for more information see previous edition

ROGERS, JOSEPH WOOD, b Jamaica, NY, Mar 7, 37; m 60; c 2. ELECTRICAL ENGINEERING. *Educ:* Cornell Univ, BEE, 59; Univ Mich, PhD(elec eng), 65. *Prof Exp:* Asst prof, Bucknell Univ, 65-71, assoc prof elec eng, 65-83; INDEPENDENT CONTRACTOR, 83- *Res:* Electrical circuit theory; mathematics; education. *Mailing Add:* 17226 Beach Dr NE Seattle WA 98155

ROGERS, KENNETH CANNICOTT, b Teaneck, NJ, Mar 21, 29; m 56; c 3. PHYSICS. *Educ:* St Lawrence Univ, BS, 50; Columbia Univ, MA, 52, PhD(physics), 56. *Hon Degrees:* LHD, St Lawrence Univ, 83; MEng, Stevens Inst Technol, 64, DEng, 87, LHD, St Lawrence Univ, 83. *Prof Exp:* Res assoc, Lab Nuclear Studies, Cornell Univ, 55-57; from asst prof to assoc prof, 57-64, head, dept physics, 68-72, actg provost & dean fac, 72, pres, 72-87, PROF PHYSICS, STEVENS INST TECHNOL, 64-, EMER PRES, 87- *Concurrent Pos:* Vis prof, City Col New York, 65-66; consult, Stanford Res Inst, 59-62, Grumman Aircraft Eng Corp, 62-63 & Vitro Labs, 62-72; comnr, US Nuclear Regulatory Comn, 87- *Mem:* Fel Royal Soc Arts; sr mem Inst Elec & Electronics Engrs; NY Acad Sci; Newcomen Soc; Sigma Xi; fel AAAS. *Res:* Plasma, particle accelerator and high energy particle physics; physical electronics. *Mailing Add:* US Nuclear Regulatory Comn Washington DC 20555

ROGERS, KENNETH D, b Cincinnati, Ohio, May 23, 21. PREVENTIVE MEDICINE. *Educ:* DePauw Univ, AB, 42; Univ Cincinnati, MD, 45; Univ Pittsburgh, MPH, 52. *Prof Exp:* From asst prof to assoc prof maternal child health, Grad Sch Pub Health, 52-60, PROF COMMUNITY MED, SCH MED, UNIV PITTSBURGH, 60-, CHMN, 80- *Res:* Public health; pediatrics. *Mailing Add:* Dept Community Med/Scaife Hall M200 Sch Med Univ Pittsburgh 4200 Fifth Ave Pittsburgh PA 15261

ROGERS, KENNETH SCIPIO, b Lafayette, Ind, Oct 8, 35; m 90; c 2. BIOCHEMISTRY. *Educ:* Purdue Univ, BSA, 57, MS, 59, PhD(biochem), 62. *Prof Exp:* USPHS fel, Sch Med, Johns Hopkins Univ, 61-64; from asst prof to assoc prof, 64-76, PROF BIOCHEM, MED COL VA, VA COMMONWEALTH UNIV, 77- *Concurrent Pos:* Vis prof pharmacol & nutrit, Univ SC Sch Med, 88- *Mem:* AAAS; Am Soc Biol Chem; Soc Exp Biol & Med; Sigma Xi; Int Soc Quantum Biol; Am Inst Nutrit. *Res:* Physical biochemistry; mechanisms of enzyme inhibition and glutamate dehydrogenase; molecular interactions of hydrophobic cations and anions with mitochondria, DNA and proteins; experimental diabetes, hepatocyte, phosphate and vitamin B6 metabolism. *Mailing Add:* Dept Biochem Box 614 Med Col Va Richmond VA 23298

ROGERS, LEE EDWARD, b Haxtun, Colo, Sept 12, 37; m 58; c 3. ECOLOGY, ENTOMOLOGY. *Educ:* Univ Nebr, BS, 67, MS, 69; Univ Wyo, PhD(entom), 72. *Prof Exp:* Teaching asst biol, Univ Nebr, Omaha, 66-69, instr, 69; STAFF SCIENTIST ECOL, PAC NORTHWEST LAB, BATTELLE MEM INST, 72- *Concurrent Pos:* Lectr, Wash State Univ, 77- *Mem:* Entom Soc Am. *Res:* Ecology of semi-arid regions; insect ecology; bioenergetics; effects of electric fields on the environment. *Mailing Add:* Environ Sci Dept Battelle Mem Inst Richland WA 99352

ROGERS, LEWIS HENRY, b DeFuniak Springs, Fla, Oct 1, 10; m 34; c 2. ANALYTICAL CHEMISTRY. *Educ:* Univ Fla, BS, 32, MS, 34; Cornell Univ, PhD(chem), 41. *Prof Exp:* Spectrochem analyst, Exp Sta, Univ Fla, 34-39, from assoc chemist to chemist, 39-48; res chemist, Carbide Nuclear Co Div, Union Carbide Corp, 48-52; leader, Analytical Div, Kraftco Res Lab, 52-54; sr chemist, Air Pollution Found, 54-58; from assoc dir to dir, West Orange Lab, Automation Industs, Inc, NJ, 58-69; dir corp res & develop, Automation Industs, Inc, Calif, 69-71; exec vpres, Air Pollution Control Asn, 71-78; PRIN ENGR ENVIRON SCI & ENGR, 78- *Mem:* AAAS; Am Chem Soc; hon mem Air Pollution Control Asn. *Res:* Society administration; spectrochemistry; soil and nuclear chemistry; photochemistry; trace elements in agriculture; air pollution; high temperature reactions; research management. *Mailing Add:* 2607 NW 22nd Ave Gainesville FL 32605

ROGERS, LLOYD SLOAN, b Waukegan, Ill, Apr 23, 14. SURGERY. *Educ:* Trinity Col, Conn, BS, 36; Univ Rochester, MD, 41; Am Bd Surg, dipl, 51. *Prof Exp:* Instr chem, Trinity Col, Conn, 36-37; intern surg, Strong Mem Hosp, Rochester, NY, 41-42; asst resident surgeon & resident, Strong Mem & Genesee Hosps, 46-50; asst chief surg serv, Crile Vet Admin Hosp, Cleveland, Ohio, 51-53; from asst prof to assoc prof surg, State Univ NY Health Sci Ctr, 53-65, actg chmn dept, 67-70, head div gen surg, Univ Hosp, 67-77, vchmn, dept surg, 78-84, PROF SURG COL MED, STATE UNIV NY HEALTH SCI CTR, 65- *Concurrent Pos:* Chief surgeon, Vet Admin Hosp, 53-81; attend surgeon, Univ Hosp, State Univ NY Health Sci Ctr, 54-;

attend surgeon, Crouse-Mem Hosp, 58-; chmn surg adjuvant cancer chemother infustion study group, Vet Admin, 61-80, nat partic surg consult comt, 65-69, mem surg res prog comt, 65-69; mem surg study sect, NIH, 62-67; consult, St Joseph, 66- *Mem:* Am Soc Colon & Rectal Surgeons; Am Col Surg; Soc Surg Alimentary Tract; Cent Surg Asn; Asn Vet Admin Surgeons (pres, 67); Int Soc Surg. *Res:* Cancer surgery, research and chemotherapy; hyperbaric medicine; gastrointestinal physiology and surgery. *Mailing Add:* Dept Surg State Univ NY Health Science Ctr 750 E Adams St Syracuse NY 13210

ROGERS, LOCKHART BURGESS, b Manchester, Conn, July 16, 17; m 52; c 2. ANALYTICAL CHEMISTRY. *Educ:* Wesleyan Univ, BA, 39; Princeton Univ, MA, 40, PhD(analytical chem), 42. *Prof Exp:* Instr analytical chem, Stanford Univ, 42-43, asst prof, 43-46; group leader, Oak Ridge Nat Lab, 46-48; from asst prof to prof, Mass Inst Technol, 48-61; prof, Purdue Univ, West Lafayette, 61-74; Graham Perdue prof, 74-86, EMER PROF CHEM, UNIV GA, 87- *Concurrent Pos:* Mem adv comt anal chem div, Oak Ridge Nat Lab; mem, Anal Chem Comt, Nat Res Coun; consult, Lawrence Livermore Lab, Oak Ridge Nat Lab, USAF Off Sci Res, NSF, Nat Bur Standards, US Army Res Off-Durham, US Environ Protection Agency & NIH; mem, Comn Equilibrium Data, Int Union Pure & Appl Chem, 67-71 & Comn Nomenclature, 75-78. *Honors & Awards:* Analytical Chem Award, Am Chem Soc, 68, Chromatography Award, 74; Stephen dal Nogare Award, Chromatography Forum, 72; 100th Anniversary Award, Asn Official Anal Chemists, 84; Benedetti-Pichler Award, Am Microchemical Soc, 85; Herty Medal, Am Chem Soc, 86. *Mem:* Am Chem Soc; Electrochem Soc; Am Acad Arts & Sci; Instrument Soc Am; Sigma Xi. *Res:* Instrumental methods of chemical analysis; separation processes; trace analysis. *Mailing Add:* 219 Rolling Wood Dr Athens GA 30605-3329

ROGERS, LORENE LANE, b Prosper, Tex, Apr 3, 14; m 35. BIOCHEMISTRY. *Educ:* NTex State Col, BA, 34; Univ Tex, MA, 46, PhD(chem), 48. *Hon Degrees:* DSc, Oakland Univ, 72; LLD, Austin Col, 77. *Prof Exp:* Instr high sch, Tex, 34-35; asst, Clayton Found Biochem Inst, Univ Tex, 46-47; prof chem, Sam Houston State Col, 47-49; Eli Lilly & Co fel, Univ Tex, Austin, 49-50, res scientist, Clayton Found Biochem Inst, 50-64, exec asst, 51-57, asst dir, 57-64, assoc dean grad sch, 64-71, vpres univ, 71-74, pres ad interim, 74-75, prof nutrit, 62-79, pres univ, 75-79, EMER PRES, UNIV TEX, AUSTIN, 79- *Mem:* AAAS; Am Chem Soc; Am Inst Nutrit; fel Am Inst Chem. *Res:* Synthesis of hydantoins; metabolic interrelationships of vitamins and amino acids; metabolic patterns; alcoholism; mental retardation; congenital malformations. *Mailing Add:* Four Nob Hill Circle Austin TX 78746

ROGERS, LYNN LEROY, b Grand Rapids, Mich, Apr 9, 39; c 4. ETHOLOGY, ECOLOGY. *Educ:* Mich State Univ, BS, 68; Univ Minn, MS, 70, PhD(ecol), 77. *Prof Exp:* Res asst black bears, Bell Mus Natural Hist, Univ Minn, 68-72; dir, Wildlife Res Inst, 72-76; WILDLIFE RES BIOLOGIST, USDA FOREST SERV, N CENT FOREST EXP STA, 76- *Honors & Awards:* Anna M Jackson Award, Am Soc Mammalogists, 74; US Forest Serv Qual Res Award, 88. *Mem:* Wildlife Soc; Am Soc Mammalogists; Soc Am Foresters. *Res:* Determining habitat use, food habits, social behavior, travels, physiology and population dynamics of black bears, white-tailed deer and moose in the upper midwest; filming and photographing animal communication and behavior. *Mailing Add:* NCent Forest Exp Sta Star Rte 1 Box 7200 Ely MN 55731

ROGERS, MARION ALAN, b Columbus, Ind, Nov 4, 36; div; c 4. GEOCHEMISTRY, PETROLEUM GEOLOGY. *Educ:* Earlham Col, AB, 58; Univ Minn, MS, 62, PhD(geol), 65. *Prof Exp:* Res geologist, Exxon Prod Res Co, Tex, 65-68; res specialist, Serv & Res Labs, Imp Oil, Ltd, Alta, 68-72; res supvr petrol geochem, Exxon Prod Res Co, 73-78, sr supvr geologist, Lafayette Prod Dist, Exxon Co, 78-80, dist geologist, Cent Dist, SE Explor Div, New Orleans, 80-81, DIV MGR, RESERVOIR EVAL DIV, EXXON PROD RES CO, 81- *Mem:* Fel Geol Soc Am; Am Asn Petrol Geologists; Am Chem Soc; Geochem Soc; Can Soc Petrol Geologists. *Res:* Petroleum geochemistry; light hydrocarbons; carbon isotopes; biodegradation and water-washing; deasphalting; reservoir bitumens and asphalts; sulfur in oils. *Mailing Add:* 23 Golden Sunset Circle The Woodlands TX 77381-4156

ROGERS, MARK CHARLES, CARDIOPULMONARY RESUSCITATION, CEREBRAL HYPOXIA. *Educ:* State Univ NY, MD, 69. *Prof Exp:* DIR PEDIAT INTENSIVE CARE UNIT, JOHNS HOPKINS UNIV MED SCH, 77-, PROF PEDIAT, 80-, PROF & CHMN DEPT ANESTHESIOL & CRITICAL CARE MED, 80- *Mailing Add:* Dept Anethesiol John Hopkins Hosp Blalock 1415 600 N Wolfe St Baltimore MD 21205

ROGERS, MARLIN NORBERT, b Mexico, Mo, Dec 18, 23; m 56; c 2. FLORICULTURE. *Educ:* Univ Mo, BS, 48, MS, 51; Cornell Univ, PhD(plant path), 56. *Prof Exp:* From assoc prof to prof, 60-87, EMER PROF HORT, UNIV MO-COLUMBIA, 87- *Concurrent Pos:* Vis prof, Univ Fla, Gainesville, 81-82. *Mem:* Fel Am Soc Hort Sci. *Res:* Plant growth regulators; physiological responses of herbaceous ornamentals to environment; diseases of ornamental plants. *Mailing Add:* 1322 Weaver Dr Columbia MO 65203

ROGERS, MICHAEL JOSEPH, biochemistry, immunology; deceased, see previous edition for last biography

ROGERS, MORRIS RALPH, b Poughkeepsie, NY, Feb 28, 24; m 54; c 2. INDUSTRIAL MICROBIOLOGY. *Educ:* Syracuse Univ, BS, 50; Hofstra Univ, MA, 52. *Prof Exp:* Bacteriologist, Nat Dairy Res Labs, Inc, NY, 50-52; microbiologist, Appl Microbiol Group, Pioneering Res Div, Sci & Adv Technol Lab, US Dept Army Natick Labs, 52-70, res microbiologist, Environ Protection Group, 70-83, chief, Mat Protection Br, 83-89; CONSULT, 89- *Concurrent Pos:* Gov bd, Am Inst Biol Sci, 72-75. *Honors & Awards:* Award, US Dept Army Natick Labs, 58; Charles Porter Award, Soc Indust Microbiol, 75. *Mem:* Soc Indust Microbiol (pres, 69-70); Sigma Xi; Am Soc Microbiol; Am Inst Biol Sci. *Res:* Germicides and fungicides, including inter-disciplinary microbiological and chemical problems; interaction of fungicides with base materials; synergism; effects of microorganisms and insects on prevention of deterioration of materials; biodegradable detergents; biodegration of organic compounds; biodegradation of petroleum products; water purification; sanitation systems; registration of disinfectants/fungicides with Environmental Protection Agency. *Mailing Add:* 47 Hadley Rd Framingham MA 01701

ROGERS, NELSON K, b Flushing, NY, May 17, 28; m 53; c 3. INDUSTRIAL & MARINE ENGINEERING. *Educ:* US Naval Acad, BS, 50; Ga Inst Technol, MS, 56. *Prof Exp:* Instr indust eng, Ga Inst Technol, 54-56; proj engr, Waterman Steamship Corp, 56-58; mgr, Pan-Atlantic Steamship Corp, 58-59; asst to pres, Sea-Land Serv, Inc, 59-60, vpres marine opers, 60-61, vpres construct opers, 61-63; lectr indust eng, 65-69, ASSOC PROF & ASSOC DIR INDUST ENG, GA INST TECHNOL, 69- *Concurrent Pos:* Sr mem, N K Rogers, Inc, 63- *Mem:* Am Inst Indust Engrs; Am Soc Eng Educ; Soc Am Mil Engrs; Soc Adv Mgt. *Res:* Design of transportation systems; design and simulation of economic systems; organizational and physical development of transportation enterprises; financial and economic analysis. *Mailing Add:* Dept Systs & Indust Eng Ga Inst Technol 225 N Ave Northwest Atlanta GA 30332

ROGERS, OWEN MAURICE, b Worcester, Mass, July 4, 30; m 56; c 2. GENETICS, PLANT BREEDING. *Educ:* Univ Mass, BVA, 52; Cornell Univ, MS, 54; Pa State Univ, PhD(genetics, plant breeding), 59. *Prof Exp:* Asst prof hort, Univ NH, 59-65, assoc prof plant sci, 65-72, actg chmn dept, 78-79, chmn dept, 79-90, PROF PLANT BIOL, UNIV NH, 72- *Mem:* Am Soc Hort Sci; Genetics Soc Am; Sigma Xi. *Res:* Genetics and plant breeding of Syringa and Pelargonium. *Mailing Add:* Dept Plant Biol Univ NH Durham NH 03824-3587

ROGERS, PALMER, JR, b New York, NY, Sept 7, 27; m 51, 81; c 4. BIOCHEMISTRY, MICROBIOLOGY. *Educ:* Johns Hopkins Univ, PhD(biol), 57. *Prof Exp:* Am Cancer Soc fel biochem, Biol Div, Oak Ridge Nat Lab, 57-59; asst prof agr biochem, Ohio State Univ, 59-63; assoc prof, 63-68, PROF MICROBIOL, UNIV MINN, MINNEAPOLIS, 68- *Concurrent Pos:* NIH spec fel, Max Planck Inst Med Res, Heidelberg, Ger, 72-73. *Mem:* AAAS; Am Soc Microbiol; Am Soc Biol Chemists; Soc Indust Microbiol. *Res:* Mechanisms of biosynthesis of enzymes in bacteria; regulation of bacterial protein and enzyme synthesis in Escherichia coli and Clostridium acetobutylicum. *Mailing Add:* Dept Microbiol Univ Minn Minneapolis MN 55455-0312

ROGERS, PETER H, b New York, NY, Jan 8, 45; m 66; c 3. PHYSICS. *Educ:* Mass Inst Technol, SB, 65; Brown Univ, PhD(physics), 70. *Prof Exp:* Res physicist acoust, Naval Res Lab, Washington, DC, 69-75, supvry res physicist, Underwater Sound Res Div, SCI OFFICER, OFF NAVAL RES, 81-; PROF, SCH MECH ENG, GA INST TECHNOL, ATLANTA. *Honors & Awards:* A B Wood Award & Prize, Inst Acoust, 79; Biennial Award, Acoust Soc Am, 80. *Mem:* Fel Acoust Soc Am. *Res:* Nonlinear acoustics; acoustic radiation theory; acoustic measurement theory; marine bioacoustics. *Mailing Add:* 205 Colewood Way Atlanta GA 30328

ROGERS, PHIL H, b McKinney, Tex, Apr 13, 24; m 46; c 4. ELECTRICAL ENGINEERING. *Educ:* Univ Tex, BS, 44, MS, 47; Univ Mich, PhD, 56. *Prof Exp:* Design engr, Westinghouse Elec Co, 47-48; instr elec eng, Univ Mich, 48-53, assoc res engr, Eng Res Inst, 53-55, asst prof elec eng, 55-56; prof, Univ Ariz, 56-59; asst dir res, Collins Radio Co, Iowa, 59-62; dir prog software develop, State of Tex, 63-67; mgr comput systs, 67-76, dir, 76-80, VPRES ELECTRONIC SYSTS SOFTWARE, E-SYSTS, 80- *Mem:* Inst Elec & Electronics Engrs; Sigma Xi. *Res:* Circuit theory; computer, computer systems and programming systems design; electronic systems embedded software development. *Mailing Add:* 7230 Cliffbrook Dallas TX 75240

ROGERS, PHILIP VIRGILIUS, b Utica, NY, Feb 7, 07; m 32; c 3. ENDOCRINOLOGY, EMBRYOLOGY. *Educ:* Hamilton Col, AB, 30, MA, 34; Yale Univ, PhD(anat), 37. *Prof Exp:* Asst biol, Hamilton Col, 32-34; lab asst anat, Yale Univ Sch Med, 34-37; from instr to asst prof biol, Hamilton Col, 37-45; fel, Johns Hopkins Univ, 45-46; from assoc prof to prof, 47-72, chmn dept, 60-72, EMER PROF BIOL, HAMILTON COL, 72- *Concurrent Pos:* Ford Found fel & res assoc, Stanford Univ, 53-54; trustee, Kirkland Col, 69- *Mem:* AAAS; Soc Endocrinol; assoc Am Soc Zool; Am Asn Anat; Sigma Xi. *Res:* Electrical potential changes during the reproductive cycle; comparative studies of the adrenal gland; effect of sulfa drugs on reproduction. *Mailing Add:* 29 Bristol Rd Clinton NY 13323

ROGERS, QUINTON RAY, b Palco, Kans, Nov 24, 36; m 56; c 4. NUTRITION, BIOCHEMISTRY. *Educ:* Univ Idaho, BS, 58; Univ Wis, MS, 60, PhD(biochem), 63. *Prof Exp:* Res assoc with Dr A E Harper, Mass Inst Technol, 62-63, instr nutrit, 63-64, asst prof physiol chem, 64-66; from asst prof to assoc prof, 66-74, PROF PHYSIOL CHEM, SCH VET MED, UNIV CALIF, DAVIS, 74- *Mem:* AAAS; Am Inst Nutrit; Am Physiol Soc; Sigma Xi; Am Acad Vet Nutrit; Soc Neurosci. *Res:* Amino acid nutrition and metabolism; feline nutrition; food intake regulation. *Mailing Add:* 1918 Alpine Pl Davis CA 95616

ROGERS, RALPH LOUCKS, b Wilkensburg, Pa, Feb 2, 22; m 47; c 4. INDUSTRIAL ORGANIC CHEMISTRY. *Educ:* Juniata Col, BS, 46; Univ Pa, MS, 48, PhD(org chem), 56; Del County Community Col, AAS, 86. *Prof Exp:* From assoc chemist to res chemist, Atlantic Ref Co, 51-69, sr res chemist, 69-83; CO-AD INSTR, DEL COUNTY COMMUNITY COL, 86- *Res:* Unnatural amino acids; effects of ionizing radiation on chemicals; physical testing of polymers; multi-component vapor liquid equilbrium; liquid-liquid distribution coefficients; accelerating rate calormetry. *Mailing Add:* 22 Ridgeway Ave Norwood PA 19074

ROGERS, RAYMOND N, b Albuquerque, NMex, July 21, 27; m 48; c 2. ANALYTICAL CHEMISTRY. *Educ:* Univ Ariz, BS, 48, MS, 51. *Prof Exp:* Chemist, Infilco, Inc, Ariz, 51-52; staff mem anal, 52-63, sect leader anal & stability, 63-64, alternate group leader, 64-74, GROUP LEADER, EXPLOSIVES RES, LOS ALAMOS SCI LAB, UNIV CALIF, 74- *Concurrent Pos:* Consult, Petrol Technol Corp, 72- *Mem:* Am Chem Soc. *Res:* Isotope dilution methods as applied to agricultural chemical problems; analytical chemistry, physical chemistry and thermal stability of organic high explosives, polymers and adhesives; analysis of archeological samples. *Mailing Add:* Los Alamos Nat Lab Univ Calif Box 1663 MST-7 Los Alamos NM 87545

ROGERS, RICHARD BREWER, b Paris, Tex, Oct 20, 44; m 66; c 2. AGRICULTURAL CHEMISTRY. *Educ:* Tulane Univ, BS, 66; Univ Wis, MS, 68; Univ Ala, PhD(org chem), 72. *Prof Exp:* Fel, Ind Univ, 71-73; sr res chemist org chem, Mich, 73-77, res specialist, Dow Chem Co, 77-91, DIR GLOBAL FORMULATIONS & ENVIRON CHEM, DOW ELANCO, 91- *Mem:* Am Chem Soc. *Res:* Heterocyclic chemistry; specifically biologically active derivatives of pyridine and fused pyridines, naphtheridine, quinazoline and benzo thiophene. *Mailing Add:* Dow Elanco 2001 W Main St PO Box 708 Greenfield IN 46140-0708

ROGERS, RICHARD C, b Burbank, Calif, June 29, 53; m 75; c 1. NEUROSCIENCE, AUTONOMIC NERVOUS SYSTEM. *Educ:* Univ Calif, Los Angeles, BA, 74, PhD(neurosci), 79. *Prof Exp:* Fel gastroenterol, Ctr Ulcer Res & Educ, 79-80; asst prof physiol, Sch Med, Northwestern Univ, 80-; PROF PHYSIOL, COL MED, OHIO STATE UNIV, 86- *Concurrent Pos:* Vis scientist, Dept Physiol, Sch Med, Fukuoka Univ, Japan, 80; prin investr, NIH, 81- *Mem:* Soc Neurosci. *Res:* Central neural elaboration of the autonomic nervous system; physiological and anatomical details of visceral afferent control over ingestive behavior, gastrointestinal function and neuroendocrine function. *Mailing Add:* Dept Physiol Col Med Ohio State Univ Graves Hall 333 W Tenth Ave Columbus OH 43210-1238

ROGERS, ROBERT LARRY, b Lawrence Co, Miss, Feb 10, 42; m 64; c 2. BIOCHEMISTRY, AGRONOMY. *Educ:* Miss State Univ, BS, 64; Auburn Univ, PhD(plant physiol, biochem), 68. *Prof Exp:* From asst prof to assoc prof, 67-74, PROF PLANT PHYSIOL, PLANT PATH DEPT, LA STATE UNIV, BATON ROUGE, 80-, SUPT, NORTHEAST RES STA, 74- *Mem:* Am Chem Soc; Am Soc Plant Physiol; Weed Sci Soc Am; Am Soc Agron. *Res:* Chemical weed control in agronomic crops; metabolism and mode of action of herbicides; vigor tests of planting seed. *Mailing Add:* Northeast Res Sta La State Univ PO Box 438 St Joseph LA 71366

ROGERS, ROBERT M, PULMONARY MEDICINE. *Educ:* Univ Pa, MD, 60. *Prof Exp:* PROF MED & ANESTHESIOL & CHIEF PULMONARY MED, UNIV PITTSBURGH MED SCH, 80- *Mailing Add:* Med/Scaife Hall 444 Univ Pittsburgh 3550 Terrace St Pittsburgh PA 15261

ROGERS, ROBERT N, b San Francisco, Calif, Nov 4, 33; m 55; c 4. SOLID STATE PHYSICS, ACADEMIC ADMINISTRATION. *Educ:* Stanford Univ, BS, 56, PhD(physics), 62. *Prof Exp:* Res assoc appl sci, Yale Univ, 65-66; asst prof physics, Wesleyan Univ, 60-67; from assoc prof to prof, Univ Colo, Boulder, 67-76, spec asst to pres, 73-75; assoc dean grad sch & coordr res, Univ Colo, Denver, 76-; MEM STAFF, DEPT PHYSICS, SAN FRANCISCO UNIV. *Concurrent Pos:* Consult, Sandia Lab, 65-66, 67-70, tech staff mem, 66-67; admin fel, Am Coun Educ, 73-74. *Mem:* AAAS; Am Phys Soc; Am Asn Physics Teachers. *Res:* Magnetic interactions in solids; paramagnetic resonance in solids and liquids; optical pumping; ion implantation. *Mailing Add:* Dean Grad Studies San Francisco State Univ 1600 Holloway St San Francisco CA 94132

ROGERS, ROBERT WAYNE, b Russellville, Ky, Aug 6, 38; m 61; c 3. ANIMAL SCIENCE, FISH & WILDLIFE SCIENCES. *Educ:* Univ Ky, BS, 60, MS, 62, PhD(animal sci), 64. *Prof Exp:* Instr, Univ Ky, 63-64; from asst prof to assoc prof, 64-74, PROF MEAT SCI, MISS STATE UNIV, 74- *Concurrent Pos:* Lectr, USDA Meat Inspection Training Ctr, 80-81, dir, Miss State Univ & USDA Meat Grading Training Sch, 81-86; dir tech serv, Southern Belle Foods Inc, 85. *Mem:* Am Soc Animal Sci; Am Meat Sci Asn; Inst Food Technol. *Res:* Curing and processing of pork and beef; methods of extending shelf-life of fresh meats; beef tenderness; efficient production of meat; live animal and carcass evaluation methods. *Mailing Add:* Dept Animal Sci PO Drawer 5228 Mississippi State MS 39762-5228

ROGERS, ROBIN DON, b Ft Lauderdale, Fla, Mar 4, 57; m 79; c 2. X-RAY CRYSTALLOGRAPHY. *Educ:* Univ Ala, BS, 78, PhD(inorg chem), 82. *Prof Exp:* Asst prof chem, 82-87, ASSOC PROF CHEM, NORTHERN ILL UNIV, 87- *Concurrent Pos:* Chmn, Rock River Sect, Am Chem Soc, 84-86. *Mem:* Am Chem Soc; Am Crystallog Soc; Sigma Xi. *Res:* Exploration of the synthetic and structural chemistry of the f-elements; f-element separations. *Mailing Add:* Dept Chem Northern Ill Univ DeKalb IL 60115

ROGERS, RODDY R, b Baytown, Tex, Jan 19, 34; m 64; c 2. METEOROLOGY. *Educ:* Univ Tex, BS, 55; Mass Inst Technol, SM, 57; NY Univ, PhD(meteorol), 64. *Prof Exp:* Asst res scientist, Nat Adv Comt Aeronaut, 57-58; asst physicist, Cornell Aeronaut Lab, 59-60, assoc physicist, 61-62, res physicist, 63-66; assoc prof, 66-74, PROF METEOROL, MCGILL UNIV, 74- *Concurrent Pos:* Chmn dept, McGill Univ, 78-87; counr, Am Meteorol Soc, 90- *Res:* Am Meteorol Soc. Res: Radar meteorology; cloud physics; precipitation physics. *Mailing Add:* Dept Meteorol Burnside Hall McGill Univ Montreal PQ H3A 2K6 Can

ROGERS, RODNEY ALBERT, b Lucas, Iowa, Aug 24, 26; m 56; c 2. BIOLOGY. *Educ:* Drake Univ, BA, 49, MA, 51; Univ Iowa, PhD(zool), 55. *Prof Exp:* From asst prof to assoc prof, 55-65, PROF BIOL, DRAKE UNIV, 65-, CHMN DEPT, 66- *Concurrent Pos:* Assoc prog dir, NSF, 67-68. *Mem:* AAAS; Am Soc Parasitol; Soc Protozool; Am Soc Trop Med & Hyg; Am Soc Zool. *Res:* Parasitology, especially helminthology; cytology; immunology; microbiology. *Mailing Add:* 4203 40th St Des Moines IA 50310

ROGERS, SAMUEL JOHN, b Florence, Ariz, Nov 5, 34; m 61; c 2. BIOCHEMISTRY. *Educ:* Univ Calif, Davis, BS, 56, DVM, 58; Univ Calif, Berkeley, PhD(biochem), 64. *Prof Exp:* Fel chem, Univ Ore, 64-66; asst prof, 66-71, ASSOC PROF BIOCHEM, MONT STATE UNIV, 71- *Mem:* Am Chem Soc. *Res:* Chemical modification of the ribiosome; dye-sensitized photooxidation mechanism of action; environmental mutagenesis, metal in mutagenesis. *Mailing Add:* Dept Chem Mont State Univ Bozeman MT 59717

ROGERS, SENTA S(TEPHANIE), b Braila, Romania, May 30, 23; US citizen. ENVIRONMENTAL HEALTH. *Educ:* Hunter Col, City Univ NY, AB, 45; Purdue Univ, MS, 48; George Washington Univ, PhD(org phys chem), 67. *Prof Exp:* Res fel analytical instrumentation, Georgetown Univ, Washington, DC, 66-67; assoc prof chem forensic sci, Northern Va Community Col, 67-69; res chemist, Bur Customs, 70; sr staff scientist, George Washington Univ, 71-75, asst res prof, Univ & dir, Occup Cancer Proj, Med Ctr, 75-78; COORDR, TOXIC SUBSTANCES PROJ, NAT WILDLIFE FEDN, WASHINGTON, DC, 80- *Mem:* Am Chem Soc; AAAS; NY Acad Sci; fel Am Inst Chemists; Sigma Xi. *Res:* The relationship between environmental pollutants and health; correlations of chemical structure-reactivity-physiological properties in toxic substances (simple and complex). *Mailing Add:* 3001 Veazey Terr NW Washington DC 20008

ROGERS, SPENCER LEE, b Topeka, Kans, Mar 28, 05; m 35; c 3. PHYSICAL ANTHROPOLOGY, ETHNOLOGY. *Educ:* San Diego State Univ, BA, 27; Claremont Grad Sch, MA, 30; Univ Southern Calif, PhD(anthrop), 37. *Prof Exp:* From instr to prof, 30-71, EMER PROF ANTHROP, SAN DIEGO STATE UNIV, 71- *Concurrent Pos:* Sci dir, San Diego Mus Man, 71-77, res anthropologist, 77- *Mem:* Am Asn Phys Anthrop. *Res:* Aboriginal medicine and surgery; ancient California populations; paleopathology; forensic anthropology. *Mailing Add:* San Diego Mus Man 1350 El Prado Balboa Park San Diego CA 92101

ROGERS, STANFIELD, pathology; deceased, see previous edition for last biography

ROGERS, STEARNS WALTER, b Alva, Okla, July 28, 34; m 55; c 2. BIOCHEMISTRY, ORGANIC CHEMISTRY. *Educ:* Northwestern State Col, BS, 56; Okla State Univ, PhD(biochem), 61. *Prof Exp:* Prof chem, Northwestern State Col, Okla, 61-76; ASSOC PROF CHEM, MCNEESE STATE UNIV, 76- *Mem:* Am Chem Soc; Am Inst Chemists. *Res:* Utilization of amino acid analogs in the synthesis of bacterial proteins; synthesis of mannans in Pseudonomas auregenosia; topic sporulation in Bacillus cereus organization; interaction of blood platlets with basement membrane and insoluble collagen; chemical modification of basement membranes. *Mailing Add:* Dept Chem McNeese State Univ 4100 Ryan St Lake Charles LA 70609

ROGERS, STEFFEN HAROLD, b Madison, Wis, Apr 17, 41; m 64. CELL BIOLOGY, PARASITOLOGY. *Educ:* Ga Southern Col, BS, 65; Vanderbilt Univ, PhD(biol), 68. *Prof Exp:* NIH training grant, Sch Pub Health & Hyg, Johns Hopkins Univ, 68-70; from asst prof to assoc prof, 70-85, assoc dean, 81-90, actg dean, 90, PROF ZOOL, UNIV TULSA, 85- *Res:* Cellular biology. *Mailing Add:* Dept Zool Univ Tulsa 600 S Col Ave Tulsa OK 74104

ROGERS, TERENCE ARTHUR, b London, Eng, Oct 8, 24; nat US; m 45; c 4. PHYSIOLOGY. *Educ:* Univ BC, BS, 52; Univ Calif, PhD, 55. *Prof Exp:* Instr physiol, Sch Med, Univ Rochester, 55-59; asst prof, Sch Med, Stanford Univ, 59-63; chmn dept physiol & assoc dean sch med, 67-71, PROF PHYSIOL, SCH MED, UNIV HAWAII, MANOA, 63-, DEAN SCH MED, 71- *Concurrent Pos:* Lederle med fac award, 58-59; USPHS sr res fel, 59; dir res & anal, Presidential Comn World Hunger, 79-80. *Mem:* AAAS; Am Physiol Soc; Aerospace Med Asn. *Res:* Fluid and electrolyte metabolism; starvation; arctic survival. *Mailing Add:* John A Burns Sch Med Univ Hawaii 1960 E West Rd Honolulu HI 96822

ROGERS, THOMAS EDWIN, b Mt Vernon, Iowa, Mar 19, 17; m 41; c 2. ZOOLOGY. *Educ:* Cornell Col, BA, 39; Univ Okla, MS, 41, PhD(zool, plant sci), 51. *Prof Exp:* Instr zool sci, Univ Okla, 46-49; assoc prof biol, Baylor Univ, 49-55; prof & chmn dept, Cornell Col, 55-64; mem Rockefeller Found spec field staff, Univ Valle, Colombia, 64-66; prof biol, Cornell Col, 66-82; RETIRED. *Res:* Cardiovascular studies on lower vertebrates; arthropod soil fauna. *Mailing Add:* 808 Summit Ave Mt Vernon IA 52314

ROGERS, THOMAS F, b Providence, RI, Aug 11, 23; m 46; c 3. PHYSICS, ELECTRONICS. *Educ:* Providence Col, BSc, 45; Boston Univ, AM, 49. *Prof Exp:* Res assoc, Radio Res Lab, Harvard Univ, 44-45; TV proj engr, Bell & Howell Co, 45-46; electronic scientist, Air Force Cambridge Res Ctr, Mass, 46-54, supvry physicist, 54-59; assoc head radio physics div, Lincoln Lab, Mass Inst Technol, 59-63, head commun div & mem steering comt, 63-64; asst dir defense res & eng, Off Secy Defense, US Dept Defense, 64-65, dep dir, 65-67; dir urban technol & res, US Dept Housing & Urban Develop, 67-69; vpres urban affairs, Mitre Corp, 69-72; dir, US Cong Space Study, 82-84; CHMN & BD DIR, LUNACORP, 90-; BD DIR, CHIEF SCIENTIST & ENGR, INT RADIO SATELLITE CORP, 90- *Concurrent Pos:* Assoc group leader, Lincoln Lab, Mass Inst Technol, 51-53; mem & later panel chmn, Commun Comt, Dept Navy Polaris Command, 60-64; mem, President's Sci Adv Comt Commun Satellite, 61-63, Commun Satellite Panel, Inst Defense Anal, 61-63, Dept Defense-NASA Tech Comt Commun Satellite, 61-64, US Nat Comt, UN Conf Appl Sci & Technol to Lesser Developed Nations, 63, Aeronaut & Astronaut Coord Bd, 65-67, Space Appln Bd, Nat Res Coun, 74-85, Regional Emergency Med Commun Comt, Nat Acad Sci, 76-79 & Voice of AM Adv Comt, Nat Res Coun, 84-88; chmn, NASA Space & Terrestrial Appl Adv Comt, Space Progs Adv Coun, 71-74; pres, Sophron Found, 80-; founder, Space Phoenix Prog & dir, External Tanks Corp, 86-; Aerospace Res & Develop Comt, Inst Elec & Electronics Engrs, 89- *Honors & Awards:* Space Pioneer Award, Nat Space Soc. *Mem:* Fel Inst Elec & Electronics Engrs; Am Phys Soc; Am Geophys Union; Sigma Xi; Brit Inst Physics. *Res:* Research, development and engineering in electronics; communications; command control; intelligence; reconnaissance; radio wave propagation; electronic memory devices; ultrasonics; molecular physics; housing; city planning and administration. *Mailing Add:* 7404 Colshire Dr McLean VA 22102

ROGERS, THOMAS HARDIN, b Washington, DC, June 10, 32; m 71; c 2. HAZARDOUS WASTE SITE CHARACTERIZATION, EARTHQUAKE HAZARD EVALUATION. *Educ:* Wesleyan Univ, BA, 54; Univ Calif, Berkeley, MA, 57. *Prof Exp:* Explor geologist, Standard Oil Co, 57-60; geologist, Calif Div Mines & Geol, 63-74; ENG GEOLOGIST, WOODWARD-CLYDE CONSULTS, 74- *Concurrent Pos:* Instr, Monterey Peninsula Col, 69-73, Univ Calif, Santa Cruz, 70-73, Univ Calif-Berkeley, 72 & Calif State Univ, Hayward, 76. *Mem:* Fel Geol Soc Am; Asn Eng Geologists; Seismological Soc Am; Am Asn Petrol Geologists; Earthquake Eng Res Inst; Sigma Xi. *Res:* Surface and subsurface geologic investigations; geologic hazard and waste distribution evaluations for various facilities; nuclear waste repository; class I hazardous waste disposal facility; petroleum refinery; superfund sites; nuclear power plants. *Mailing Add:* Woodward-Clyde Consults Suite 100 500 12th St Oakland CA 94607-4014

ROGERS, VERN CHILD, b Salt Lake City, Utah, Aug 28, 41; m 62; c 6. NUCLEAR PHYSICS, NUCLEAR ENGINEERING. *Educ:* Univ Utah, BS & MS, 65; Mass Inst Technol, PhD(nuclear eng), 69. *Prof Exp:* Nuclear engr reactor physics, Argonne Nat Lab, 68-69; vis assoc prof nuclear eng, Lowell Technol Inst, 70-71; assoc prof chem eng & physics, Brigham Young Univ, 69-73; mgr res & develop, IRT Corp, 73-76; vpres res & develop, Ford, Bacon & Davis Utah, Inc, 76-80; PRES RES & DEVELOP, ROGERS & ASSOCS ENG CORP, 80- *Concurrent Pos:* Adj prof nuclear eng, Univ Utah, 73-; consult to indust & govt or radiation, transp, health effects and disposal. *Mem:* Am Nuclear Soc; Am Phys Soc; Am Chem Soc; Am Health Physics Soc; Am Soc Prof Engrs. *Res:* Management of products related to nuclear fuel cycle and decontamination of nuclear facilities; low energy nuclear physics; radiation pathway and risk analysis. *Mailing Add:* 747 W 3500 S Bountiful UT 84010

ROGERS, WAID, b New York, NY, Sept 7, 27; m 51; c 3. SURGERY. *Educ:* Yale Univ, BA, 50; Cornell Univ, MD, 57; Univ Minn, MS, 66, PhD(surg & oncol), 67. *Prof Exp:* Instr surg, Univ Minn, 67; from asst prof to assoc prof, 67-74, PROF SURG, UNIV TEX HEALTH SCI CTR, 74-; CHIEF SURG, AUDIE MURPHY VET ADMIN HOSP, 73- *Concurrent Pos:* NIH spec fel, 63-; consult gen surg, San Antonio State Tuberc Hosp, 70- *Mem:* Sigma Xi; Am Soc Transplant Surgeons; Am Col Angiol; Transplantation Soc; AAAS. *Res:* Tumor blood flow; transplantation; parenteral nutrition. *Mailing Add:* 923 Serenade San Antonio TX 78213

ROGERS, WALTER RUSSELL, b Newark, NJ, Aug 11, 45; m 67; c 2. NEUROENDOCRINOLOGY OF CIRCADIAN RHYTHMS, NEUROTOXICOLOGY. *Educ:* Col Wooster, BA, 67; Univ Iowa, MA, 70, PhD(psychol), 72; Am Bd Toxicol, dipl, 88. *Prof Exp:* Res fel psychobiol, Univ Calif, Irvine, 71-72; asst prof psychol, Univ Northern Iowa, 72-74; SR RES PSYCHOLOGIST, BEHAV SCI SECT, DEPT BIOENG, SOUTHWEST RES INST, SAN ANTONIO, TEX, 74-, ACTG MGR, SECT, 83-, MGR BIOSCI, 89- *Concurrent Pos:* Adj assoc scientist, dept med & physiol, Southwest Found Biomed Res, 80-85; staff scientist, dept bioeng, Southwest Res Inst, 86-89; adj scientist, Dept Med & Physiol, Southwest Found Biomed Res, 85- *Mem:* AAAS; Am Heart Asn; Am Primatological Soc; Animal Behav Soc; Bioelectromagnetics Soc; Sigma Xi; Soc Toxicol; Behav Toxicol Soc; Int Neurotoxicol Soc. *Res:* Nonhuman primates as models for biomedical research where behavior is a variable; health effects of cigarette smoking, exercise and stress; behavioral toxicology of combustion products and electromagnetic fields; neurobehavioral toxicology; electromagnetic field bioeffects. *Mailing Add:* Dept Biosci & Bioeng Southwest Res Inst 6220 Culebra Rd San Antonio TX 78284

ROGERS, WILLARD L(EWIS), mechanical engineering; deceased, see previous edition for last biography

ROGERS, WILLIAM ALAN, b Trenton, NJ, Feb 22, 21; m 47; c 4. PHYSICS. *Educ:* Oberlin Col, AB, 47; Univ NMex, MS, 51; Univ Pittsburgh, PhD(physics), 58. *Prof Exp:* Res physicist, Res Lab, Westinghouse Elec Corp, 51-60; from assoc prof to prof physics, Thiel Col, 60-66; prof physics, Univ Petrol & Minerals, Dhahran, Saudi Arabia, 66-80, chmn dept, 66-70 & 71-74; MEM FAC, DEPT PHYSICS, TRINITY UNIV, 80- *Concurrent Pos:* Vis fel, Sch Eng & Appl Sci, Princeton Univ, 75-76. *Mem:* Am Asn Physics Teachers. *Res:* Atomic physics; microwave discharge and plasmas; undergraduate teaching. *Mailing Add:* 7422 Saddlewood Ave San Antonio TX 78204

ROGERS, WILLIAM EDWIN, b Carlisle, Pa, Apr 5, 36; m 60; c 3. ZOOLOGY, PARASITOLOGY. *Educ:* Dickinson Col, BS, 58; Pa State Univ, MS, 61; Univ Minn, PhD(zool), 69. *Prof Exp:* Asst prof biol, Shippensburg State Col, 64-65; asst prof, Lycoming Col, 65-70; assoc prof, 70-77, PROF BIOL, SHIPPENSBURG STATE COL, 77- *Mem:* Soc Protozool; Am Soc Parasitol. *Res:* Morphogenesis and ultrastructure of trypanosomatids found in muscoid flies. *Mailing Add:* Dept Biol Shippensburg Univ N Prince St Shippensburg PA 17257

ROGERS, WILLIAM IRVINE, b Brooklyn, NY, Dec 10, 27; m 54; c 2. RESEARCH MANAGEMENT DEVELOPMENT. *Educ:* Adelphi Col, BA, 49; Univ Vt, MS, 52; Univ Iowa, PhD(biochem), 56. *Prof Exp:* Asst, Univ Vt, 50-52 & Univ Iowa, 52-54; assoc technologist, Res Ctr, Gen Foods Corp, 56-59, proj leader, 59; consult biochem & sect leader, Life Sci Div, 59-70; sr staff mem, org develop sect, Arthur D Little, Inc, 69-84; PRES, CHADWICK ROGERS INC, 84- *Concurrent Pos:* Vis lectr, Boston Univ, Brandeis Univ, 86- *Mem:* Am Chem Soc; Sigma Xi. *Res:* Application of organic, analytical and physical chemistry to the problems of biochemistry and the physiological disposition and mechanisms of action of drugs; isolation and characterization of natural products; nutrition; organization and management development at interfaces of technical and other operations in academic, industrial and governmental agencies. *Mailing Add:* 53 Turkey Shore Rd Ipswich MA 01938

ROGERS, WILLIAM LESLIE, b Boston, Mass, Mar 27, 34; m 62; c 3. MEDICINE, MEDICAL PHYSICS. *Educ:* Ohio Wesleyan Univ, AB, 55; Case Western Reserve, MS, 61, PhD(physics), 67. *Prof Exp:* Engr, Hughes Aircraft Co, 55-58; instr physics, Univ Wyo, 61-62; fel, Case Western Reserve, 67-68; staff engr, Bendix Aerospace Systs, 68-70; res assoc, 70-72, ASST PROF INTERNAL MED, UNIV MICH, ANN ARBOR, 72- *Mem:* Soc Nuclear Med; Sigma Xi. *Res:* Improved nuclear medicine imaging techniques and instrumentation. *Mailing Add:* 1425 Cambridge Rd Ann Arbor MI 48104

ROGERS, WILMER ALEXANDER, b Mt Dora, Fla, Aug 17, 33; m 61; c 4. FISH PATHOLOGY, FISHERIES MANAGEMENT. *Educ:* Univ Southern Miss, BS, 58; Auburn Univ, MS, 60, PhD(fish mgt), 67. *Prof Exp:* Biologist aide, Miss Game & Fish Comn, 57-58; fishery biologist, Ala Dept Conserv, 60-62, US Fish & Wildlife Serv, 62-64; from instr to assoc prof, 64-77, PROF FISHERIES, AUBURN UNIV, 77- *Concurrent Pos:* Leader, Southeastern Coop Fish Dis Proj, 68-; ed, Southeastern Game and Fish Proc, 74-77; ed, J Aquatic Animal Health, 88- *Mem:* Am Fisheries Soc (pres, Southern Div, 77-78 & Fish Health Sect, 86-87); Am Soc Parasitol; Wildlife Dis Asn; Am Micros Soc. *Res:* General parasites and diseases of fish, especially taxonomy of monogenea; intensive culture of fish; fish immunology. *Mailing Add:* Dept Fisheries & Allied Aquacult Auburn Univ Auburn AL 36849-5419

ROGERSON, ALLEN COLLINGWOOD, b Stoke-on-Trent, Eng, Dec 9, 40; US citizen; m 60; c 2. NITROGEN FIXATION, MICROBIAL PHYSIOLOGY. *Educ:* Haverford Col, BA, 64; Dartmouth Col, PhD(molecular biol), 69. *Prof Exp:* NIH fel molecular biol, Albert Einstein Med Ctr, Philadelphia, Pa, 68-70; asst prof biol, Bryn Mawr Col, 70-76; res assoc & head, Biol Nitrogen Fixation Group, Div Agr, Fort Valley State Col, Ga, 76-79; assoc prof, 79-83, chmn dept, 79-86, PROF BIOL, ST LAWRENCE UNIV, 83- *Concurrent Pos:* NIH grant, 71-74; vis res prof, Univ Copenhagen, Denmark, 73-74, Univ Dundee, Scotland, 77 & Univ Rochester, 86-87. *Mem:* AAAS; Am Soc Microbiol; Sigma Xi. *Res:* Chromosome structure; bacterial conjugation regulation of molecular synthesis and growth in microorganisms; environmental microbiology; computer applications. *Mailing Add:* Biol Dept St Lawrence Univ Canton NY 13617

ROGERSON, ASA BENJAMIN, b Williamston, NC, Oct 24, 39; m 65; c 2. PLANT PHYSIOLOGY, AGRONOMY. *Educ:* NC State Univ, BS, 62; Va Polytech Inst, MS, 65, PhD(plant physiol), 68. *Prof Exp:* Asst prof crop sci, NC State Univ, 68-70; REGIONAL MGR RES & DEVELOP, UNIROYAL CHEM, 70- *Mem:* Weed Sci Soc Am. *Res:* Agricultural chemical products involving tobacco, soybeans, cotton, fruits, ornamentals and growth regulants. *Mailing Add:* Uniroyal Chem Co 158 Wind Chime Ct Raleigh NC 27615

ROGERSON, CLARK THOMAS, mycology, for more information see previous edition

ROGERSON, JOHN BERNARD, JR, b Cleveland, Ohio, Sept 3, 22; m 43; c 3. ASTROPHYSICS. *Educ:* Case Univ, BS, 51; Princeton Univ, PhD(astron), 54. *Prof Exp:* Carnegie fel astron, Mt Wilson Observ, 54-56; from res assoc & lectr to assoc prof, 56-67, PROF ASTRON, PRINCETON UNIV, 67- *Mem:* Am Astron Soc; Int Astron Union. *Res:* Solar atmosphere; photoelectric spectrophotometry; high altitude astronomy with balloon and satellite borne telescopes and equipment. *Mailing Add:* 277 Moore St Princeton NJ 08540

ROGERSON, PETER FREEMAN, b Stoke-on-Trent, Eng, Mar 12, 44; US citizen; m 66; c 2. ANALYTICAL CHEMISTRY. *Educ:* Univ Vt, BS, 66; Univ NC, Chapel Hill, PhD(analytical chem), 71. *Prof Exp:* Res assoc, Cornell Univ, 70-71; res chemist, Environ Res Lab, Environ Protection Agency, 71-86; ANALYTICAL CHEMIST, NAT WATER QUAL LAB, US GEOL SURV, 86- *Mem:* Am Soc Mass Spectrometry; Am Chem Soc; AAAS. *Res:* Analytical chemistry as applied to pollution control research. *Mailing Add:* US Geol Surv 5293 Ward Rd Arvada CO 80002-1811

ROGERSON, ROBERT JAMES, b Lancaster, Eng, July 4, 43; Can citizen; m 68; c 2. GLACIER MASS BALANCE, QUATERNARY HISTORY. *Educ:* Liverpool Univ, BA, 65; McGill Univ, MSc, 67; Macquarie Univ, Australia, PhD(earth sci), 79. *Prof Exp:* Sci officer, Glaciology Div, Inland Waters Br, Ottawa, 67-69; lectr, Mem Univ Nfld, 69-72, from asst prof to assoc prof, 72-86, prof & head geog dept, 86-88; PROF, DEPT GEOG, UNIV LETHBRIDGE, 88- *Concurrent Pos:* Consult, Westfield Minerals, 79 & Ice Eng, 81-82; secy coun, Sch Grad Studies Mem Univ, 85-86; asst dir, Labrador Inst Northern Studies, 85-88. *Mem:* Int Glaciol Soc; Geol Asn Can(pres, Nfld sect, 86-87); Sigma Xi (secy, 87-); Asn Am Geographers; Brit Geomorphol Res Group; Can Asn Geographers. *Res:* Mass-balance and dynamics of cirque glaciers in Torngat Mountains of Labrador and the Yoho region of Canadian Rockies; glacial geology and quaternary history of Labrador, Newfoundland and the Arctic. *Mailing Add:* Dept Geog Univ Lethbridge 4401 University Dr Lethbridge AB T1K 3M4 Can

ROGERSON, THOMAS DEAN, b Salt Lake City, Utah, Oct 31, 46; m 68; c 2. ORGANIC CHEMISTRY, AGRICULTURAL CHEMISTRY. *Educ:* Mich State Univ, BS, 68; Cornell Univ, MS, 71, PhD(org chem), 74. *Prof Exp:* Sr chemist, 74-81, REGISTRATION MGR, ROHM & HAAS CO, 81- *Concurrent Pos:* NIH fel, Dept Entom & Limnol, Cornell Univ, 73-74. *Res:* Synthesis and structure-activity relationships of biologically-active compounds; terpenes, alkaloids and toxins; pesticide metabolism and residue analysis; pesticide registration. *Mailing Add:* 136 Britany Dr Chalfont PA 18914

ROGGE, THOMAS RAY, b Oelwein, Iowa, Oct 29, 35; m 60; c 3. ENGINEERING MECHANICS, APPLIED MATHEMATICS. *Educ:* Iowa State Univ, BS, 58, MS, 61, PhD(appl math), 64. *Prof Exp:* Asst prof math, Univ Ariz, 64-65; from asst prof to assoc prof eng mech, 65-76, PROF ENG SCI & MECH, IOWA STATE UNIV, 76- *Mem:* Soc Indust & Appl Math; Soc Eng Sci. *Res:* Elastic wave propagation in bounded media; finite element analysis; blood flow in human arterial system. *Mailing Add:* Dept Eng Mech Iowa State Univ SCM 2019 Me/ESM Bldg Ames IA 50011

ROGGENKAMP, PAUL LEONARD, b Jefferson Co, Ky, May 3, 27; m 50; c 5. REACTOR PHYSICS, NUCLEAR ENGINEERING. *Educ:* Louisville Univ, BA, 49; Univ Ind, MS, 51, PhD(physics), 53. *Prof Exp:* Res physicist, 52-57, sr supvr, 57-59, sr physicist 59-61, chief supvr, 61-64, res mgr, 64-85, sr consult, E I du Pont de Nemours & Co Inc, Aiken, SC, 85-87; SELF EMPLOYED, 88- *Mem:* Am Phys Soc; fel Am Nuclear Soc. *Res:* Beta decay; operational planning; reactor engineering. *Mailing Add:* 1418 Socastee Dr North Augusta SC 29841

ROGIC, MILORAD MIHAILO, b Belgrade, Yugoslavia, July 23, 31; US citizen; m 56; c 3. ORGANIC CHEMISTRY, ENZYMOLOGY. *Educ:* Univ Belgrade, BS, 56, PhD(org chem), 61. *Prof Exp:* Asst prof chem, Univ Belgrade, 58-61; res assoc steroidal chem, Worcester Found Exp Biol, 61-62, org chem & stereochem, Univ Notre Dame, 62-64 & org chem, Univ Sask, 65-66; asst prof organoborane chem, Purdue Univ, 66-69; res group leader, Allied Corp, 69-77, res supvr, 77-80, sr res assoc org chem, 80-86; res mgr, 86-89, ASSOC RES DIR ORG CHEM, MALLINCKRODT MED INC, 89- *Mem:* Am Chem Soc. *Res:* Enzyme chemistry; organic reaction mechanism; metal catalyzed oxidants; chemistry of clinical reagents; organic chemistry of sulfur dioxide; organoborane chemistry; stereochemistry and conformational analysis; electrophilic additions; nitrogen containing organic compounds; x-ray contrast media agents; technetium based radiopharmaceuticals; molecular modeling; pharmaceutical chemistry. *Mailing Add:* Mallinckrodt Med Inc 675 McDonnell Blvd Box 5840 St Louis MO 63134

ROGLER, JOHN CHARLES, b Providence, RI, Sept 21, 27; m 51; c 2. ANIMAL NUTRITION. *Educ:* Univ RI, BS, 51; Purdue Univ, MS, 53, PhD(poultry nutrit), 58. *Prof Exp:* Res asst, 51-53 & 55-57, from asst prof to assoc prof, 57-66, PROF ANIMAL NUTRIT, PURDUE UNIV, WEST LAFAYETTE, 66- *Honors & Awards:* Am Feed Indust Assoc Nutrit Res Award, 85. *Mem:* Fel AAAS; Am Inst Nutrit; fel Poultry Sci Asn; Soc Exp Biol & Med. *Res:* Study requirements, interactions and biochemical functions of nutrients. *Mailing Add:* Dept Animal Sci Purdue Univ West Lafayette IN 47907

ROGNLIE, DALE MURRAY, b Grand Forks, NDak, July 12, 33; m 55; c 4. MATHEMATICS. *Educ:* Concordia Col, Moorhead, Minn, BA, 55; Univ NDak, MS, 58; Iowa State Univ, PhD, 69. *Prof Exp:* Instr math, Univ NDak, 56-60; res engr, Boeing Co, Wash, 60-66, supvr math group, 66; assoc prof, 69-77, PROF MATH, SDAK SCH MINES & TECHNOL, 77- *Concurrent Pos:* Eve lectr, Seattle Univ, 61 & Pac Lutheran Univ, 61-63. *Mem:* Soc Indust & Appl Math; Sigma Xi; Math Asn Am. *Res:* Special functions; integral transforms; numerical analysis. *Mailing Add:* 284 Westberry Ct N Rapid City SD 57702

ROGNLIEN, THOMAS DALE, b June 2, 45; US citizen; m 71; c 1. PLASMA PHYSICS, IONOSPHERIC PHYSICS. *Educ:* Univ Minn, BEE, 67; Stanford Univ, MS, 69, PhD(elec eng), 73. *Prof Exp:* Res assoc ionospheric plasma physics, Nat Oceanic & Atmospheric Admin, 72-74; res assoc plasma physics, Univ Colo, 74-75; PHYSICIST PLASMA PHYSICS, LAWRENCE LIVERMORE LAB, 75- *Concurrent Pos:* Nat Res Coun fel, 72-74. *Mem:* Am Phys Soc. *Res:* Linear and nonlinear waves in plasmas; real space and velocity space transport of particles in plasmas via theoretical and computational models. *Mailing Add:* Lawrence Livermore Lab MS L-630 PO Box 5511 Livermore CA 94550

ROGOFF, GERALD LEE, b New Haven, Conn, June 7, 39; m 62. PLASMA PHYSICS. *Educ:* Yale Univ, BA, 61; Mass Inst Technol, PhD(physics), 69. *Prof Exp:* Res asst physics, Mass Inst Technol, 61-66; sr scientist, 69-78, fel scientist, Res & Develop Ctr, Westinghouse Elec Corp, 78-82; PRIN MEM TECH STAFF, GTE LABS, INC, 82- *Concurrent Pos:* Lectr, Carnegie-Mellon Univ, 73-75; assoc ed, Inst Elec & Electronics Engrs Trans Plasma Sci, 85-90. *Mem:* AAAS; Am Phys Soc; Sigma Xi; Inst Elec & Electronics Engrs. *Res:* Physics of electrical discharges in gases; gaseous electronics; optical diagnostics of ionized gases. *Mailing Add:* PO Box 2973 Framingham MA 01701

ROGOFF, JOSEPH BERNARD, medicine; deceased, see previous edition for last biography

ROGOFF, MARTIN HAROLD, b New York City, NY, May 10, 26; m 53; c 2. PESTICIDE RESIDUE CHEMISTRY, ENVIRONMENTAL CHEMISTRY. *Educ:* Hunter Col, AB, 50; Pa State Univ, MS, 52, PhD(microbiol), 54. *Prof Exp:* Supv microbiologist, US Dept Interior, Bur Mines, 54-61; mgr microbiol, Int Minerals & Chem Corp, 61-67, mgr microbiol chem & molecular biol, 67-73; mgr biosci res, Sandoz Inc, 73-74; assoc dir sci, Pesticide Regist Div, US Environ Protection Agency, 74-79; prog coordr basic res, prog planning staff, Sci & Educ Admin, 79-83, nat prog leader biotechnol, nat prog staff, 83-86, dir, Western Region Res Ctr, 86-90, DIR STRATEGIC PLANNING, AGR RES SERV, USDA, 90- *Concurrent Pos:* Div chmn, Am Soc Microbiol, 78-80, pub affairs comt, 85- *Mem:* Am Soc Microbiol; Am Acad Microbiol. *Res:* Biotransformations and fermentations in bioprocessing for aromatic ring products, sulfur removal from coal, flavorous compound production; biological pest control agents; conversion of bulk agricultural materials to value-added products. *Mailing Add:* Western Regional Res Ctr 800 Buchanan St Albany CA 94710

ROGOFF, WILLIAM MILTON, b New York, NY, Mar 15, 16; m 47; c 2. VETERINARY ENTOMOLOGY. *Educ:* Univ Conn, BS, 37; Cornell Univ, PhD(insect morphol), 43. *Prof Exp:* Entomologist, US Navy, 43-46; assoc, Exp Sta, Univ Calif, Riverside, 46-47; asst prof entom & zool & asst entomologist, Exp Sta, SDak State Univ, 47-49, assoc prof & assoc entomologist, 49-53, prof & entomologist, 53-62; res entomologist, Agr Res Serv, Ore, 62-68, res leader western insects affecting man & animals, 68-78, res entomologist & prof entom, 78-80, emer prof entom, Livestock Insect Res Unit, Agr Res Serv, Univ Nebr, USDA, 80-83; ASSOC, EXP STA, UNIV CALIF, DAVIS, 84- *Concurrent Pos:* Fulbright res scholar, Commonwealth Sci & Indust Res Orgn, Australia, 55-56. *Mem:* Fel AAAS; Entom Soc Am; Am Mosquito Control Asn. *Res:* Insect morphology and behavior; pheromones; veterinary entomology; systemic insecticides in livestock; insect vectors of disease organisms. *Mailing Add:* 908 Hacienda Ave Davis CA 95616

ROGOL, ALAN DAVID, b New Haven, Conn, March 9, 41; m 68; c 2. PEDIATRICS, ENDOCRINOLOGY. *Educ:* Mass Inst Technol, BS, 63; Duke Univ, PhD(physiol), 70, MD, 70. *Prof Exp:* Intern pediat, Johns Hopkins Hosp, 70-71, resident, 71-72 & 74-75; fel endocrinol, NIH, 72-74; from asst prof to assoc prof pediat, 75-84, asst prof pharmacol, 77-87, PROF PEDIAT, SCH MED, UNIV VA, 84-, PROF PHARMACOL, 87- *Concurrent Pos:* Co-dir, Blue Ridge Poison Control Ctr, 81-87. *Mem:* Am Acad Pediat; Am Fedn Clin Res; Endocrinol Soc; Soc Pediat Res; Sigma Xi; fel Am Col Sports Med. *Res:* Neuroendocrinology; mechanism of growth hormone secretion mechanism of insulin action; mechanism of insulin action; endocrinology of endurance training. *Mailing Add:* 685 Explorers Rd Charlottesville VA 22901-8441

ROGOLSKY, MARVIN, b Passaic, NJ, Apr 17, 39; m 90; c 3. MICROBIOLOGY. *Educ:* Rutgers Univ, BA, 60; Northwestern Univ, MS, 62; Syracuse Univ, PhD(microbiol), 65. *Prof Exp:* Asst biol, Northwestern Univ, 60-62; NIH fel microbiol, Scripps Clin & Res Found, 65-67; from instr to assoc prof microbiol, Col Med, Univ Utah, 67-76; assoc prof biol & med, 77-80, PROF BASIC LIFE SCI, UNIV MO, KANSAS CITY, 80- *Concurrent Pos:* Gen Res Support Fund grant & Am Cancer Soc inst grant, 67-68; NIH res grants, 68-71, 72-75 & 76-79; Gen Res Support Fund grant, Univ Utah, 71-72; Weldon Springs Endowment Fund grants, 80-85; Sarah Morrison bequest basic res internal med grant, 85-87, fac research grant, 89-91. *Mem:* Am Soc Microbiol; fel Am Acad Microbiol. *Res:* Genetic regulation of bacterial sporulation; genetic mapping and gene transfer in staphylococcus aureus; genetic control of staphylococcal exfoliative toxin production; staphylococcal plasmids. *Mailing Add:* Biol Sci Bldg Univ Mo Kansas City MO 64110

ROGOSA, GEORGE LEON, b Lynn, Mass, Jan 16, 24; m 50. PHYSICS. *Educ:* Johns Hopkins Univ, AB, 44, PhD(physics), 49. *Prof Exp:* From jr instr to asst prof physics, Johns Hopkins Univ, 44-49; from asst prof to assoc prof physics, Fla State Univ, 49-56; physicist, Eng Develop Br, Div Reactor Develop, USAEC, 56-57 & Physics & Math Br, 57-73, asst dir nuclear sci, Div Phys Res, 73-75; asst dir nuclear sci, Div Phys Res, US Energy Res & Develop Admin, US Dept Energy, 75-77, dir, Div Nuclear Physics, 77-79; RETIRED. *Concurrent Pos:* Adj prof physics, Duke Univ, 80- *Honors & Awards:* Arthur S Flemming Award, 63. *Mem:* AAAS; Am Phys Soc. *Res:* Transuranic x-ray spectra; anamalous transmission of x-rays in Laue diffraction; x-ray absorption edges; technical administration of atomic and nuclear physics research programs. *Mailing Add:* Five Berwick Ct Durham NC 27707

ROGOWSKI, A(UGUSTUS) R(UDOLPH), b Shelton, Conn, June 18, 05; m 29; c 1. ENGINEERING. *Educ:* Yale Univ, BS, 27; Mass Inst Technol, SM, 28. *Prof Exp:* Engr, Curtiss Aeroplane & Motor Co, 28-34 & Fleetwings, Inc, 34-36; res assoc, Mass Inst Technol, 36-40, from asst prof to, prof 40-68, emer prof mech eng, 70-73; RETIRED. *Mem:* Soc Automotive Engrs; fel Am Soc Mech Engrs; Am Soc Eng Educ; Sigma Xi. *Res:* Alcohol fuel blends; two-stroke engine development; combustion of fuel sprays; laboratory equipment. *Mailing Add:* Box 92 Needham MA 02192

ROGOWSKI, ROBERT STEPHEN, b Batavia, NY, Oct 7, 38; m 58; c 3. PHYSICAL CHEMISTRY, MOLECULAR SPECTROSCOPY. *Educ:* Canisius Col, BS, 60; Mich State Univ, PhD(phys chem), 68. *Prof Exp:* SR RES SCIENTIST, NASA LANGLEY RES CTR, 68- *Concurrent Pos:* Adj prof chem, Christopher Newport Col, 69-79. *Mem:* AAAS; Am Chem Soc. *Res:* Metrology, optics, optical fiber sensors and properties of composite materials such as graphite/epoxy. *Mailing Add:* 232 Chickamauga Pike Hampton VA 23669

ROGUSKA-KYTS, JADWIGA, b Warsaw, Poland, May 11, 32; m 60. INTERNAL MEDICINE, NEPHROLOGY. *Educ:* Univ Poznan, Poland, MA, 52; Poznan Med Sch, MD, 58. *Prof Exp:* Intern, 58-59, resident, 59-63, NIH fel, 64-66, asst prof, 70-74, ASSOC PROF MED, MED SCH, NORTHWESTERN UNIV, CHICAGO, 74- *Concurrent Pos:* Attend physician, Cook County Hosp, Chicago, 64-68, Passavant Mem Hosp, 66-, Med Sch, Northwestern Univ, Chicago, 66- & Vet Admin Res Hosp, 70-; assoc chief div med, Northwestern Mem Hosp, Chicago. *Mem:* AAAS; Am Fedn Clin Res; Am Soc Nephrology; Am Heart Asn; fel Am Col Physicians. *Res:* Hypertension; renal pressor system; uremia. *Mailing Add:* Northwestern Univ Med Sch 707 N Fairbanks Ct Chicago IL 60611

ROHA, MAX EUGENE, b Meadville, Pa, Apr 14, 23; m 46; c 3. ORGANIC CHEMISTRY. *Educ:* Allegheny Col, BS, 44; Harvard Univ, AM, 47, PhD, 49. *Prof Exp:* Sr tech man, 49-59, mgr plastics res, 60-62, dir sci liaison, 62-78, dir explor new prods, 78-81, dir new spec polymers & chem, res & develop, Res Ctr, B F Goodrich Co, 81-84; INNOVATIVE SOLUTIONS, DR MAX ROHA ASSOCS, BRECKSVILLE, OH, 84- *Concurrent Pos:* Res, Case Western Reserve Univ, 84-, Univ Akron, 90- *Mem:* Am Chem Soc. *Res:* Polymers; plastics; research and development planning; project evaluation and conceptualization. *Mailing Add:* 8205-A Parkview Rd Brecksville OH 44141

ROHACH, ALFRED F(RANKLIN), b Toledo, Iowa, Jan 30, 34; m 61; c 4. NUCLEAR ENGINEERING. *Educ:* Iowa State Univ, BS, 59, MS, 61, PhD(nuclear eng), 63. *Prof Exp:* From asst prof, to assocassoc prof, 63-79, PROF NUCLEAR ENG, IOWA STATE UNIV, 79 - *Concurrent Pos:* AEC fel, 64-66; fac improvement leave, Southern Calif Edison Co, 77-78. *Mem:* Am Nuclear Soc. *Res:* Radiation shielding; numerical analysis in nuclear reactor theory; nuclear power economics; nuclear fuel management. *Mailing Add:* Dept Nuclear Eng Iowa State Univ 261 Sweeney Ames IA 50011

ROHAN, PAUL E(DWARD), b Evergreen Park, Ill, May 5, 43; m 70. NUCLEAR ENGINEERING, SOFTWARE DEVELOPMENT. *Educ:* Univ Detroit, BS, 65; Univ Ill, Urbana, MS, 66, PhD(nuclear eng), 70. *Prof Exp:* Sr staff physicist, Combustion Eng, Inc, 70-72, supvr nuclear eng, 72-74, sect mgr, 74-77, task area mgr physics, 77-79, mgr computer analysis, 79-90; SR CONSULT, 90- *Mem:* Sigma Xi; Am Nuclear Soc. *Res:* Computational methods for nuclear reactor analyses; reactor physics; numerical analysis. *Mailing Add:* 76 Overlook Dr Windsor CT 06095

ROHATGI, PRADEEP KUMAR, b Kanpur, India, Feb 18, 43; c 2. MATERIALS SCIENCE. *Educ:* Banaras Hindu Univ, India, BS, 61; Mass Inst Technol, MS, 63, DSc, 64. *Prof Exp:* Res Metallurgist, Int Nickel Co, 64-68; res engr mat, Homer Res Lab, Bethelehem Steel, 69-72; prof mat sci, Indian Inst Sci, Bangalore, India, 72-77; dir res, Coun Sci & Indust Res, Indian, 77-85; PROF MAT, COL ENG, UNIV WIS-MILWAUKEE, 85- *Concurrent Pos:* Vis prof, Indian Inst Technol, New Dehli, 82-83 & San Jose Univ, 83. *Mem:* Fel Am Soc Metals; Am Inst Metall Engrs; Am Foundrymen Soc. *Res:* Dendritic solidification of solutions; synthesis of cast metal ceramic composites and characterization of their mechanical, physical and tribological properties; structure and properties of natural fibers and synthesis of composites using these fibers; alloy development. *Mailing Add:* Dept Mat Univ Wis PO Box 413 Milwaukee WI 53201

ROHATGI, UPENDRA SINGH, b Kanpur, India, Mar 3, 49; m 75; c 1. MECHANICAL ENGINEERING. *Educ:* Indian Inst Technol, Kanpur, BTech, 70; Case Western Reserve Univ, MS, 72, PhD(fluid & thermal sci), 75. *Prof Exp:* MECH ENGR THERMOHYDRAUL REACTORS, BROOKHAVEN NAT LAB, 75- *Concurrent Pos:* Engr aircraft fuel pumps, TRW Inc, Cleveland, 74 & Chandler-Evans, 81; consult, UN Deveop Prog, Defence Nuclear Facil Safety Bd; vis prof, Indian Inst Technol; adj prof, Cooper Union, NY. *Mem:* Am Soc Mech Engrs; Sci Res Soc NAm. *Res:* Nonequilibrium two-phase flow; thermohydraulic of reactors and turbomachinery. *Mailing Add:* Brookhaven Nat Lab Bldg 475 B Upton NY 11973

ROHATGI, VIJAY, b Delhi, India, Feb 1, 39; m 71; c 1. MATHEMATICAL STATISTICS, PROBABILITY. *Educ:* Univ Delhi, BSc, 58, MA, 60; Univ Alta, MS, 64; Mich State Univ, PhD(math statist), 67. *Prof Exp:* From asst prof to assoc prof math, Cath Univ Am, 67-72; assoc prof, 72-73, chmn, dept math & statist, 83-85, PROF MATH, BOWLING GREEN STATE UNIV, 73- *Concurrent Pos:* Vis prof, Ohio State Univ, 79-80; investment analyst & consult, 87- *Mem:* Am Statist Asn; Int Statist Inst; Inst Math Statist. *Res:* Probability limit laws; statistical inference; sequential methods; stochastic processes; order statistics. *Mailing Add:* Dept Math Bowling Green State Univ Bowling Green OH 43403

ROHDE, CHARLES RAYMOND, b Glasgow, Mont, Sept 3, 22; m 45; c 5. PLANT BREEDING. *Educ:* Mont State Col, BS, 47; Univ Minn, PhD(plant genetics), 53. *Prof Exp:* Asst prof agron & small grain breeder, Univ Wyo, 50-51; assoc prof, 52-70, wheat breeder, Pendleton Exp Sta, 52-76,70-76, prof agron, Ore State Univ, 70-76; prof agron, Columbia Basin Agr Res Ctr, 76-87, emer prof, 87-; RETIRED. *Mem:* Am Soc Agron. *Res:* Wheat breeding and genetics. *Mailing Add:* 2230 SW Ladow Pendleton OR 97801

ROHDE, FLORENCE VIRGINIA, b Davenport, Iowa, May 15, 18. ENGINEERING MATHEMATICS, ACTUARIAL MATHEMATICS. *Educ:* Univ Northern Iowa, AB, 39; Univ Rochester, MM, 40; Miami Univ, AM, 45; Univ Ky, PhD(math), 50. *Prof Exp:* Teacher pub schs, Iowa, 40-42; instr music & math, Miami Univ, 42-45; asst math, Ohio State Univ, 45-46; instr math & astron, Univ Ky, 46-50; from instr to assoc prof math & astron, Univ Fla, 50-57; prof math, Univ Chattanooga, 57-66; prof, 66-83, EMER PROF MATH, MISS STATE UNIV 83- *Concurrent Pos:* Mathematician, US Naval Weapons Lab, 56 & 57 & Tenn Valley Authority, 59-62. *Mem:* Soc Indust & Appl Math; Am Math Soc; Math Asn Am. *Res:* Engineering and actuarial mathematics; numerical analysis; number theory. *Mailing Add:* Box 5172 Mississippi State MS 39762-5172

ROHDE, RICHARD ALLEN, b Peekskill, NY, Sept 28, 29; m 55; c 4. PLANT PATHOLOGY. *Educ:* Drew Univ, AB, 51; Univ Md, MS, 56, PhD(plant path), 58. *Prof Exp:* Asst plant path, Univ Md, 58-59; from asst prof to assoc prof, 59-68, head dept, 68-81, PROF PLANT PATH, UNIV MASS, AMHERST, 68-, ASSOC DEAN, FOOD & NAT RESOURCES & ASSOC DIR, MASS AGR EXP STA, 84- *Concurrent Pos:* Vis scientist, Rothamsted Exp Sta, Eng, 67; vis prof, Univ Ariz, 74; nematologist, Coop State Res Serv, USDA, 81. *Mem:* AAAS; Am Phytopathological Soc; Soc Nematol. *Res:* Ecology and physiology of plant parasitic nematodes; plant diseases caused by nematodes. *Mailing Add:* 112 Stockbridge Hall Univ Mass Amherst MA 01003

ROHDE, RICHARD WHITNEY, b Salt Lake City, Utah, Dec 23, 40; m 61; c 5. MATERIALS SCIENCE, TRIBOLOGY. *Educ:* Univ Utah, BS, 63, PhD(metall), 67. *Prof Exp:* Staff mem shock wave physics div, 67-69, supvr phys & mech metall div, 69-82, mgr mat dept, 83-90; MGR ENVIRON, HEALTH & SAFETY DEPT, SANDIA NAT LABS, 90- *Concurrent Pos:* Ed, Am Soc Mech Engrs Trans, J Eng Mat & Technol, 87- *Mem:* Metall Soc; Am Inst Mining, Metall & Petrol Engrs; Am Soc Metals; Am Soc Testing & Mat; Am Soc Mech Engrs. *Res:* Dislocation dynamics in shock loaded materials; shock loading equation of state; phase transformation in metals; dislocation dynamics; stress corrosion cracking; friction, lubrication and wear; inleastic deformation modeling. *Mailing Add:* Environ Health & Safety Dept 8540 Sandia Nat Lab Livermore CA 94551-0969

ROHDE, STEVE MARK, b Newark, NJ, May 18, 46; m; c 2. APPLIED MATHEMATICS, TRIBOLOGY. *Educ:* NJ Inst Technol, BS, 67; Lehigh Univ, MS, 69, PhD(math), 70. *Prof Exp:* Res scientist appl math, Gen Motors Res Labs, 70-79, head systs eng activ, Proj Trilby, 85-88, MGR, SYSTS ANALYSIS & INFO MGT DEPT, GEN MOTORS SYSTS ENERGY CTR, GEN MOTORS RES LABS, 88- *Honors & Awards:* Henry Hess Award, Am Soc Mech Engrs, 73; Harry Kummer Mem Award, Am Soc Testing & Mat, 76; Burt L Newkirk Tribology Award, Am Soc Mech Engrs, 77; Clifford Steadman Award, Inst Mech Engrs, 84. *Mem:* Fel Am Soc Mech Engrs; Soc Automotive Engrs. *Res:* Advanced vehicle systems; Tribology; computer aided engineering; control of mechanical systems. *Mailing Add:* Gen Motors Systs Eng Ctr 1151 Crooks Rd Troy MI 48084

ROHDE, SUZANNE LOUISE, b Denver, Colo, Feb 13, 63; m 85. THIN FILMS, VACUUM SCIENCE. *Educ:* Iowa State Univ, BS, 85; Northwestern Univ, MS, 88, PhD(mat sci & eng), 91. *Prof Exp:* Eng technician, Spangler Lab, Iowa State Univ, 85-86; res asst eng, Dept Mat Sci & Eng, Northwestern Univ, 86-89; dir res, Biophotonics, Inc, Milwaukee, 89-91; ASST PROF MECH ENG, UNIV NEBR-LINCOLN, 92- *Concurrent Pos:* Res scientist, Birl Indust Res Lab, Evanston, 89-91. *Mem:* Mat Res Soc; Am Vacuum Soc; Am Soc Metals; Metall Soc. *Res:* Thin film deposition equipment and processes; sputter deposition and molecular beam epitaxy based processes; characterization of thin film couples by sem, tem, xtem, auger, edax and x-ray. *Mailing Add:* Univ Nebr-Lincoln 2850 Sheridan Blvd Lincoln NE 68502

ROHEIM, PAUL SAMUEL, b Kiskunhalas, Hungary, July 11, 25; US citizen; m 57; c 1. PHYSIOLOGY, MEDICINE. *Educ:* Med Sch Budapest, MD, 51. *Prof Exp:* Demonstr physiol, Med Sch Budapest, 47-51, asst prof, 51-56; res assoc, Hahnemann Med Col, 57-58; instr med, Albert Einstein Col Med, 58-62, assoc in med, 62-63, from asst prof physiol to prof physiol & med, 63-77; PROF PHYSIOL, MED & PATH, LA STATE UNIV MED CTR, 77- *Concurrent Pos:* Intern, Univ Budapest Clins, 51-52; NIH fel, Albert Einstein Col Med, 58-62, New York Health Res Coun career scientist award, 68-73; estab investr, Am Heart Asn, 63-68, mem coun on arteriosclerosis. *Mem:* Am Physiol Soc; Fedn Am Socs Exp Biol; Hungarian Physiol Soc. *Res:* Lipid and lipoprotein metabolism; arteriosclerosis. *Mailing Add:* Dept Physiol & Path La State Univ Med Ctr 1542 Tulane Ave Rm 638 New Orleans LA 70112

ROHL, ARTHUR N, b Brooklyn, NY, May 11, 30. MINERALOGY. *Educ:* Columbia Univ, BS, 58, MA, 60, PhD(geol), 72. *Prof Exp:* Asst prof environ & occup med, Mt Sinai Sch Med, 69-85; PRIN, ARTHUR ROHL ASSOCS INC, 85- *Concurrent Pos:* Expert, New York Bd Educ, 72-85, US Environ Protection Agency & Occup Safety & Health Admin, 73-88, Repub Ireland, 74-85, Govt Bermuda & Nat Inst Occup Safety & Health, 75-86; consult, Philadelphia Bd Educ, 74-89. *Mem:* Sigma Xi; AAAS; Mineral Soc Am. *Res:* Identification and quantification of asbestos minerals in tissue specimens, water, air and in bulk samples; characterization of mineral dusts which cause disease, such as talc, quartz and asbestos. *Mailing Add:* Box 455 Durham PA 18039

ROHLF, F JAMES, b Blythe, Calif, Oct 24, 36; m 59; c 2. BIOMETRY, POPULATION BIOLOGY. *Educ:* San Diego State Col, BS, 58; Univ Kans, PhD(entom), 62. *Prof Exp:* Asst prof biol, Univ Calif, Santa Barbara, 62-65; assoc prof statist biol, Univ Kans, 65-69; assoc prof biol, 69-72, chmn, Dept Ecol & Evolution, 75-80 & 90-91, PROF BIOL, STATE UNIV NY STONY BROOK, 72- *Concurrent Pos:* Statist consult, NY Pub Serv Comn, 75-, IBM, 77-, US Environ Protection Agency, 78-80 & Appl Biomath Inc, 84-; vis scientist, IBM, Yorktown Heights, NY, 76-77 & 80-81. *Mem:* Biomet Soc; Asn Comput Mach; Soc Syst Zool; Classification Soc. *Res:* Applications of multivariate analysis, cluster and factor analysis to systematics morphometrics, and population biology. *Mailing Add:* Three Heritage Lane Setauket NY 11733

ROHLF, MARVIN EUGUENE, b Ida Grove, Iowa, Mar 10, 27; m 54; c 5. ANIMAL FEED PRODUCTION, BEEF CATTLE NUTRITION. *Educ:* Iowa State Univ, BS, 50; MS, 54. *Prof Exp:* Asst nutritionist, Golden Sun Feeds Inc, 54-56, nutritionist, 56-70; vpres, Res & Develop, Golden Sun Feeds Div United Brands, 70-80, sr vp, 80-87; SR VPRES RES & DEVELOP, GOLDEN SUN FEEDS INC, 87- *Concurrent Pos:* Chmn, Iowa Feed & Nutrit Service, 60-61; chmn various comt, Nat Feed Ingredient Asn, 76-81, asst chmn bd, 80-81, chmn bd, 81-82; bd gov, Advan Regist Prof Animal Scientists. *Mem:* Am Soc Animal Sci; Am Soc Poultry Sci; Am Soc Dairy Sci; Advan Regist Prof Animal Scientists. *Res:* Nutritonal conduction with swine of all ages; beef cattle, dairy beef and cage layers. *Mailing Add:* Golden Sun Feeds Inc Hwy 4 S Estherville IA 51334

ROHLFING, DUANE L, b Cape Girardeau, Mo, Nov 18, 33; m 56; c 4. BIOCHEMISTRY, EVOLUTION. *Educ:* Drury Col, BS, 55; Fla State Univ, MS, 60, PhD(biochem), 64. *Prof Exp:* Nat Acad Sci res assoc enzyme models, Ames Res Ctr, NASA, Calif, 64-66; Med Found Boston sr res fel enzymes, Mass Inst Technol, 66-68; from asst prof to assoc prof, 68-76, PROF BIOL, UNIV SC, 76- *Mem:* AAAS; Am Chem Soc; Am Soc Biol Chem; Int Soc Study Origins Life. *Res:* Molecular and cellular evolution; polyamino acids; protocells; catalysis. *Mailing Add:* Dept Biol Univ SC Columbia SC 29208

ROHLFING, STEPHEN ROY, b Toledo, Ohio, July 25, 36; m 59; c 2. MICROBIOLOGY, INFECTIOUS DISEASES. *Educ:* Bowling Green State Univ, BS, 58; Miami Univ, MS, 60; Western Reserve Univ, PhD(microbiol), 65. *Prof Exp:* Assoc microbiol, Chicago Med Sch, 65-66, asst prof, 66-70; from instr to asst prof pharmacog, Col Pharm, Univ Minn, Minneapolis, 70-80, assoc prof pharmacol, 80-90; SR SPECIALIST CLIN RES, RIKER LABS INC, 91- *Concurrent Pos:* Sr scientist, Riker Res Labs, Minn Mining & Mfg Co, 70-72, res specialist, Riker Labs, Inc, 72-75, supvr microbiol, 75-78, proj leader, Antiinfective Drug Res, 78-85, sr res specialist, 85-90, sr specialist clin res, 91- *Mem:* Can Soc Microbiol; Am Soc Microbiol; Brit Soc Antimicrobial Chemother; Europ Soc Clin Microbiol. *Res:* Gene-enzyme relationships; molecular characteristics of lymphomas; anti-infective research and development; asepsis and surgical wound infection. *Mailing Add:* 1081 Cedarwood Dr Woodbury MN 55125

ROHM, C E TAPIE, JR, b Long Beach, Calif, June 2, 47; m 72; c 8. INFORMATION MANAGEMENT, INFORMATION STRATEGIES & PLANNING. *Educ:* Brigham Young Univ, BS, 73, MA, 74; Ohio Univ, PhD(psychol, commun & mgt), 77. *Prof Exp:* Asst to dean, Ohio Univ, 77-79;

assoc prof bus & computers, Whittier Col, 81-83; asst prof mgt & computers, 79-81, PROF INFO MGT, CALIF STATE UNIV, SAN BERNARDINO, 83- *Concurrent Pos:* Corp treas, OU Med Assocs, Inc, 78-79; vis prof, Univ Dar es Salaam, 88-89; Fulbright sr scholar, 88-89; assoc ed, J Int Info Mgt, 90-93. *Mem:* Int Info Mgt Asn (pres, 86-91); AAAS. *Res:* Focus on decision-making within individuals to increase productivity. *Mailing Add:* Dept Info & Decision Sci Calif State Univ San Bernardino CA 92407

ROHMAN, MICHAEL, b New York, NY, Nov 21, 25; m 48; c 2. THORACIC & TRAUMA SURGERY. *Educ:* Boston Univ, MD, 50. *Prof Exp:* From intern to chief resident surg, Mass Mem Hosp, 50-55; asst & chief resident thoracic surg, Bronx Munic Hosp Ctr, 55-58; from instr to asst prof cardio-thoracic surg, Albert Einstein Col Med, 58-69, from asst clin prof to assoc clin prof, 69-70; PROF SURG, NY MED COL, 70-; CHIEF CARDIOTHORACIC SURG, LINCOLN MED & MENT HEALTH CTR, 79- *Concurrent Pos:* Nat Tuberc Asn fel, Albert Einstein Col Med, 57-59; consult cardio-thoracic surg, Northern Westchester Hosp, Yonkers Gen Hosp, United Hosp & Montrose Vet Admin Hosp, 66-79; dir surg, Westchester County Med Ctr, 64-79. *Mem:* Am Col Chest Physicians; Am Heart Asn; fel Am Col Surg; Soc Thoracic Surg; Am Asn Surg Trauma. *Res:* Small arterial anastamoses; pulmonary hemodynamics, particularly bronchial artery circulation. *Mailing Add:* Dept Surg Lincoln Med & Ment Health Ctr 234 E 149th St Bronx NY 10451

ROHN, ROBERT JONES, b Lima, Ohio, June 16, 18; m 42; c 4. MEDICINE. *Educ:* DePauw Univ, AB, 40; Ohio State Univ, MD, 43. *Prof Exp:* Intern, Ohio State Univ Hosp, 43-44, resident internal med, 44-46, resident hemat, 48-50; from instr to asst prof med, 50-55, assoc med & coordr cancer educ, 55-68, prof med, 68-74, cancer coordr, 68-85, Bruce Kenneth Wiseman prof med,74-85, DISTINGUISHED EMER PROF MED, IND UNIV, INDIANAPOLIS, 85- *Concurrent Pos:* Consult, Vet Admin & Army Hosps, 52-85; chmn, Nonprofit Div, Soc Retired Exec, 88- *Mem:* Am Soc Hemat; Reticuloendothelial Soc; fel Am Col Physicians; Am Fedn Clin Res; Int Soc Hemat. *Res:* Cytological and clinical hematology. *Mailing Add:* 7334 A King George Dr Indianapolis IN 46260

ROHNER, THOMAS JOHN, b Trenton, NJ, Jan 1, 36; c 2. UROLOGY. *Educ:* Yale Univ, BA, 57; Univ Pa, MD, 61. *Prof Exp:* Resident gen surg, Hosp, Univ Pa, 62-64, resident urol, 64-67, USPHS spec fel, Dept Pharm, Sch Med, 69-70; from asst prof to assoc prof, 70-75, PROF SURG, MILTON S HERSHEY MED CTR, PA STATE UNIV, 75-, CHIEF DIV UROL, 70- *Concurrent Pos:* Consult urol, Vet Admin Hosp, Lebanon, Pa, 70- & Elizabethtown Hosp Children & Youth, Pa, 70- *Mem:* Am Urol Asn; Am Col Surgeons; Soc Pediat Urol; Asn Acad Surg. *Res:* In vitro contractile mechanisms of bladder and ureteral smooth muscle. *Mailing Add:* Div Urol M S Hershey Med Ctr Pa State Univ 500 University Dr Box 850 Hershey PA 17033

ROHOLT, OLIVER A, JR, b Preston, Idaho, May 30, 16; m 44; c 2. IMMUNOLOGY. *Educ:* Univ Mont, BA, 39; Univ Calif, PhD(biochem), 53. *Prof Exp:* Asst, Utah State Agr Col, 40-41; asst, Univ Calif, 47-50; instr, State Col Wash, 50-56; sr cancer res scientist, Roswell Park Mem Inst, 56-63, assoc cancer res scientist, 63-65, prin cancer res scientist, 65-79, cancer res scientist vi, 79-83, actg dept chmn, molecular immunol, Roswell Park Mem Inst, 80-82; RETIRED. *Concurrent Pos:* Assoc res prof, Roswell Park Div, Grad Sch, State Univ NY Buffalo, 66-83; mem grad fac, Niagara Univ, 68-83. *Mem:* Am Chem Soc; Am Soc Biochem & Molecular Biol; Sigma Xi. *Mailing Add:* 2233 E Behrend Dr No 78 Phoenix AZ 85024-1853

ROHOVSKY, MICHAEL WILLIAM, b Youngstown, Ohio, Feb 26, 37; m 65; c 2. PATHOLOGY. *Educ:* Ohio State Univ, DVM, 60, MSc, 65, PhD(vet path), 67. *Prof Exp:* NIH fel vet path, Ohio State Univ, 62-67, asst prof, 67-69; dir path labs, Merrell Nat Labs, 69-72; dir path, Arthur D Little Inc, 72-77; vpres res & develop, Pittman-Moore Inc, 77-81; dir, med affairs & orthop res & develop, Johnson & Johnson Products Inc, 81-86, VPRES RES & DEVELOP, JOHNSON & JOHNSON ORTHOP INC, 86- *Concurrent Pos:* Adj asst prof path, Sch Med Univ Cincinnati, 69-72; indust rep, FDA adv panel, gen, med & personal use, 86- *Mem:* Am Vet Med Asn; fel Am Col Vet Pharmacol & Therapeut; Int Acad Path; Am Col Vet Pathologists. *Res:* Infectious diseases of animals; neoplastic diseases; antiviral therapy; toxicology and drug safety. *Mailing Add:* Johnson & Johnson Orthop Inc 501 Goerge St New Brunswick NJ 08903

ROHR, DAVID M, b Portsmouth, Va, Sept 9, 47. TAXONOMY, BIOGEOGRAPHY. *Educ:* Col William & Mary, BS, 69; Ore State Univ, PhD(geol), 77. *Prof Exp:* Asst prof geol, Univ Ore, 77-79; asst prof geol, Univ Wash, 79-80; from asst prof to assoc prof, 80-89, PROF GEOL, SUL ROSS STATE UNIV, 89- *Mem:* Paleont Soc; fel Geol Soc Am; Paleont Res Inst; Soc Econ Paleontologists & Mineralogists; Int Paleont Asn. *Res:* Lower Paleozoic (Ordovician through Lower Devonian) gastropods; paleoecology; lower Paleozoic stratigraphy. *Mailing Add:* Dept Geol Sul Ross State Univ Alpine TX 79832

ROHR, ROBERT CHARLES, b Leonia, NJ, Feb 27, 22; m 46; c 3. PHYSICS. *Educ:* Guilford Col, BS, 43; Univ Wis, MS, 47; Univ Tenn, PhD(physics), 55. *Prof Exp:* Physicist, Union Carbide Nuclear Co, div, Union Carbide Corp, 47-53; physicist, Gen Elec Co, 55-59, mgr cold water assembly, 59-66, mgr cold water assembly/adv test reactor, 66-72, mgr chem & radiol controls navy training support, 72-74, sr proj engr, Prototype Opers & Eng, 74-77, mgr Reactor Eng Oper Ship & Prototype Testing, Knolls Atomic Power Lab, Gen Elec Co, 77-87; ADJ PROF, NUCLEAR ENG & ENG PHYSICS DEPT, RPI, 89- *Mem:* Am Phys Soc; Am Nuclear Soc. *Res:* Critical assembly experiments; beta ray spectroscopy; radiation and particle detectors. *Mailing Add:* RD 3 249 Gordon Rd Schenectady NY 12306

ROHRBACH, KENNETH G, b Ashland, Ohio, Oct 16, 40; m 62; c 3. PLANT PATHOLOGY. *Educ:* Ohio State Univ, BS, 62; Univ Idaho, MS, 64; Colo State Univ, PhD(plant path), 67. *Prof Exp:* Res asst plant dis, Univ Idaho, 62-64 & Colo State Univ, 64-67; plant pathologist, Dole Co, 67-70 & Pineapple Res Inst, 70-73; assoc plant pathologist, 73-77, plant pathologist, 78-85, ASST DIR, HAWAII INST TROP AGR & HUMAN RESOURCES, UNIV HAWAII, MANOA, 85- *Mem:* Am Phytopath Soc. *Res:* Diseases of tropical crops. *Mailing Add:* Hawaii Inst Trop Agr & Human Resources Univ Hawaii-Manoa 3050 Maile Way Honolulu HI 96822

ROHRBACH, MICHAEL STEVEN, PULMONARY RESEARCH, BIOCHEMISTRY. *Educ:* Univ NC, PhD(biochem), 72. *Prof Exp:* ASSOC PROF BIOCHEM & MED, MAYO MED SCH, 77-; CONSULT, THORACIC RES & CELL BIOL, MAYO CLIN, 81- *Res:* Immunology. *Mailing Add:* Mayo Clinic 613 Plummer Bldg Rochester MN 55905

ROHRBACH, ROGER P(HILLIP), b Canton, Ohio, Oct 12, 42; m 65; c 3. AGRICULTURAL ENGINEERING. *Educ:* Ohio State Univ, PhD(agr eng), 68. *Prof Exp:* From asst prof to assoc prof, 68-78, PROF BIOL & AGR ENG, AGR EXP STA, NC STATE UNIV, 78- *Honors & Awards:* FMC Corp Young Designer Award, Am Soc Agr Engrs, 81. *Mem:* Am Soc Agr Engrs. *Res:* Fruit production mechanization; blueberries and grapes; irradiation processes applied to agriculture; process engineering in agriculture. *Mailing Add:* Dept Biol & Agr Eng NC State Univ Box 5906 Raleigh NC 27695-7625

ROHRER, DOUGLAS C, b Cleveland, Ohio, Dec 24, 42; m 66; c 2. X-RAY CRYSTALLOGRAPHY. *Educ:* Western Reserve Univ, BA, 66; Case Western Reserve Univ, PhD(chem), 70. *Prof Exp:* Res assoc crystallog, Crystallog Dept, Univ Pittsburgh, 70-73; res assoc, 73-74, res scientist cyrstallog, 74-78, ASSOC RES SCIENTIST, MOLECULAR BIOPHYS DEPT, MED FOUND BUFFALO, 78- *Concurrent Pos:* Vis prof, Sch Pharm, Oregon State Univ, 79. *Mem:* Am Chem Soc; Am Crystallog Asn; Am Inst Physics. *Mailing Add:* Upjohn Co 301 Henrietta St Kalamazoo MI 49001-0199

ROHRER, HEINRICH, b Buschs, Switzerland, June 6, 33. PHYSICS. *Educ:* Swiss Inst Technol, Dipl, 55, PhD(physics), 60. *Prof Exp:* Res asst, ETH, Zurich, Switz, 60-61; fel, Rutgers Univ, 61-63; MGR PHYSIC DEPT, IBM ZURICH RES LAB, 86- *Concurrent Pos:* Sabbatical, Univ Calif, Santa Barbara, 74-75; IBM fel, 86. *Honors & Awards:* Nobel Prize in Physics, 86. *Mailing Add:* IBM Zurich Res Lab Saemerstr Four 8803 Rueschlikon Zurich Switzerland

ROHRER, JAMES WILLIAM, CELLULAR IMMUNOLOGY, IMMUNE REGULATION. *Educ:* Univ Kans, PhD(microbiol), 75. *Prof Exp:* ASSOC PROF MICROBIOL & IMMUNOL, COL MED, UNIV S ALA, 85- *Mailing Add:* Dept Microbiol & Immunol 841 Col Med Univ S Ala 2086 Med Sci Bldg Univ S Ala 2086 Med Sci Bldg 307 University Blvd Mobile AL 36688

ROHRER, ROBERT HARRY, b Philadelphia, Pa, Sept 20, 18; m 42; c 4. PHYSICS. *Educ:* Emory Univ, BS, 39, MS, 42; Duke Univ, PhD(physics), 54. *Prof Exp:* From instr to asst prof eng, 40-51, from asst prof to assoc prof physics, 51-61, assoc prof radiol, 56-62, chmn dept physics, 63-69, PROF PHYSICS, EMORY UNIV, 61-, PROF RADIOL, SCH MED, 62- *Mem:* Am Phys Soc; Health Physics Soc; Soc Nuclear Med (vpres, 71-72); Am Asn Physicists Med; Am Asn Physics Teachers; Sigma Xi. *Res:* Nuclear radiation spectroscopy; radiation dosimetry; low energy neutron spectroscopy. *Mailing Add:* 457 Princeton Way NE Atlanta GA 30307

ROHRER, RONALD A, ELECTRICAL ENGINEERING, COMPUTER ENGINEERING. *Prof Exp:* HOWARD M WILKOFF UNIV PROF ELEC & COMPUTER ENG & DIR, CTR COMPUTER AIDED DESIGN, CARNEGIE-MELLON UNIV, 90- *Concurrent Pos:* Assoc ed, Inst Elec & Electronics Engrs Trans Circuit Theory, 66-69, Computer-Aided Design, Trans Circuits & Systs, 81-84; prog chmn, Int Conf Circuit Theory, Inst Elec & Electronic Engrs, 69, Long-Range Planning Comt, 69-70, chmn, Circuit Theory Group Computer-Aided Design Subcomt, 70-71; Alexander von Humboldt prize, 72; co-prin investr, Elec Power Res Inst, 77-80; prin investr, NSF grant, Univ Maine, 77-79, Southern Methodist Univ, 80-81, Carnegie-Mellon Univ, 88-90 & Semiconductor res Corp grant, 90; mem, Rev Eng & Eng Technol Studies Comt, Am Soc Eng Educ, 78-80. *Honors & Awards:* Frederick Emmons Terman Award, Am Soc Elec Eng, 78; Circuit & Systs Soc Award, Inst Elec & Electronics Engrs, 90. *Mem:* Nat Acad Eng; fel Inst Elec & Electronics Engrs; Inst Elec & Electronics Engrs, Circuits & Systs Soc (pres elect, 86, pres, 87). *Res:* Author of various publications. *Mailing Add:* Dept Elec & Computer Eng Carnegie Mellon Univ Pittsburgh PA 15213-3890

ROHRER, WESLEY M, JR, b Johntown, Pa, Sept 12, 21; m 43; c 3. MECHANICAL ENGINEERING, THERMODYNAMICS. *Educ:* Univ Pittsburgh, BS, 47; Mass Inst Technol, SM, 61. *Prof Exp:* Instr eng, Johnstown Ctr, 47-49, from instr to asst prof, 49-55, ASSOC PROF MECH ENG, UNIV PITTSBURGH, 55- *Concurrent Pos:* Consult, IBM Corp, 61-63, Adv Bd Hardened Elec Systs, Nat Acad Sci, 64-67 & World Bank, 80; vpres eng & natural sci, Univ Sci Ctr, Inc, 68-71, eng consult, 71-; vpres, Hosp Utility Mgt, Inc, 80-81. *Mem:* Combustion Inst (treas, 64-82); Am Soc Eng Educ; AAAS; Sigma Xi. *Res:* Radiative heat transfer; air pollution abatement; combustion; two-phase flow; boiling heat transfer; efficient energy utilization. *Mailing Add:* 220 Summit Dr Pittsburgh PA 15238

ROHRIG, IGNATIUS A, b Detroit, Mich, Jan 10, 10. METALLURGY. *Educ:* Univ Detroit, BSMet E, 33. *Prof Exp:* Metall engr, Detroit Edison, 33-59, asst supvr, Eng Lab, 40-53, asst div head mat eng dept, 59-75; CONSULT METALL APPLN, 75- *Concurrent Pos:* Instr metall, Detroit Tech, 47-50; chmn comt, Am Soc Mech Eng, sub-comts, 62-75, Am Soc Testing & Mat. *Mem:* Fel Am Soc Testing & Mat; fel Am Soc Metals. *Res:* Twenty publications in the field of applications of metals at elevated temperatures. *Mailing Add:* 15131 Newburgh Rd Livonia MI 48154

ROHRIG, NORMAN, b Creston, Iowa, Oct 28, 44; m 67; c 2. RADIOLOGICAL PHYSICS. *Educ:* Iowa State Univ, BS, 66; Univ Wis-Madison, MS, 68, PhD(nuclear physics), 73. *Prof Exp:* Res scientist, Radiol Res Labs, Col Physicians & Surgeons, Columbia Univ, 72-74; ASSOC SCIENTIST, MED DEPT, BROOKHAVEN NAT LAB, 74- *Mem:* Radiation Res Soc; Am Phys Soc; AAAS; Sigma Xi. *Res:* Measurements of W, energy required to create an ion pair, in gases relevant to neutron dosimetry; development of hardware for irradiation of mammalian cells with charged particles. *Mailing Add:* Bldg 535-A Brookhaven Nat Lab Upton NY 11973

ROHRINGER, ROLAND, b Nurnberg, Ger, Aug 12, 29; nat Can; m 58; c 2. PLANT PATHOLOGY. *Educ:* Univ Göttingen, PhD(agr), 56. *Prof Exp:* Fel plant path, Univ Göttingen, 56, res asst, 58-59; res asst biochem, Univ Wis, 56-58; res officer agr, Pesticide Res Inst, Can Dept Agr, 59, head, Cereal Rust Sect, 63-73, head cereal dis sect, 74-91; RETIRED. *Mem:* Phytochem Soc NAm. *Res:* Physiology of parasitism in plants; molecular biology of cereal rust diseases. *Mailing Add:* Agr Can Res St 195 Dafoe Rd Winnipeg MB R3T 2N9 Can

ROHRL, HELMUT, b Straubing, Ger, Mar 22, 27. MATHEMATICS. *Educ:* Univ Munich, DSc(math), 49, Habilitation, 53. *Prof Exp:* Asst prof math, Univ Würzburg, 49-51; asst prof, Univ Munich, 51-53, 54-55, docent, 55-58; res assoc, Univ Chicago, 58-59; from assoc prof to prof, Univ Minn, 59-64; vis prof, Harvard Univ, 62-63; chmn dept, 68-71, PROF MATH, UNIV CALIF, SAN DIEGO, 64- *Concurrent Pos:* Asst prof, Univ Munster, 53-54; vis prof, Princeton Univ, 67-68, Univ Munich, 72-73 & Univ Nagoya, 76. *Mem:* Am Math Soc; Ger Math Asn. *Res:* Pure mathematics. *Mailing Add:* Math Dept Univ Calif 0112 La Jolla CA 92093

ROHRLICH, FRITZ, b Vienna, Austria, May 12, 21; nat US; m 51; c 2. FOUNDATIONS OF PHYSICS, PHILOSOPHY OF SCIENCE. *Educ:* Inst Technol, Israel, ChemE, 43; Harvard Univ, MA, 47, PhD(physics), 48. *Prof Exp:* Mem staff, Inst Adv Study, 48-49; res assoc physics, Nuclear Studies Lab, Cornell Univ, 49-51; lectr, Princeton Univ, 51-53; from assoc prof to prof, Univ Iowa, 53-63; PROF PHYSICS, SYRACUSE UNIV, 63- *Concurrent Pos:* Vis prof, Johns Hopkins Univ, 58-59, Tel-Aviv Univ, 67, Univ Warsaw, 77, Acad Sci Bulgaria, 78, Univ Calif, Irvine, 81 & Stanford Univ, 81; consult, Nat Bur Standards, 58-61. *Honors & Awards:* Fulbright Lectr, US Dept State, Cent Europ, 74. *Mem:* NY Acad Sci; fel Am Phys Soc; Fedn Am Sci; Philos Sci Asn. *Res:* Quantum field theory; quantum electrodynamics; relativistic particle dynamics; interpretation of quantum mechanics; philosophy of science. *Mailing Add:* Dept Physics Syracuse Univ Syracuse NY 13244-1130

ROHRMANN, CHARLES A(LBERT), b Pendleton, Ore, Dec 17, 11; m 39; c 4. CHEMICAL ENGINEERING. *Educ:* Ore State Univ, BS, 34; Ohio State Univ, PhD(chem eng), 39. *Prof Exp:* Res chem engr, Grasselli Chem Dept, E I du Pont de Nemours & Co, Ohio, 39-43; plant process supvr, Pa, 43-48; chem engr, Gen Elec Co, 48-64; sr res assoc, Battelle Northwest, Wash, 65-68, consult, Chem & Metall Div, 68-70, staff engr, Chem Tech Dept, 70-77, RESIDENT CONSULT, PAC NORTHWEST LABS, BATTELLE MEM INST, 77- *Honors & Awards:* I R 100 Award, Am Inst Chem Engrs, 84. *Mem:* Am Chem Soc; Am Inst Chem Engrs; Sigma Xi. *Res:* Heavy inorganic and organic chemicals; sulfur-nitrogen chemicals; applications in the atomic energy industry; production recovery and applications of nuclear energy by-products and special radioisotopes; air pollution control; biomass conversion. *Mailing Add:* 4707 W Seventh Ave Kennewick WA 99336

ROHRS, HAROLD CLARK, b Alexandria, Ky, Sept 23, 40; m 62; c 2. CELL PHYSIOLOGY. *Educ:* Transylvania Col, AB, 61; Univ Ky, PhD(physiol), 66. *Prof Exp:* Asst prof, 66-72, assoc prof, 72-79, PROF ZOOL, DREW UNIV, 79- *Concurrent Pos:* Vis res prof, Univ Oslo, 73-74. *Mem:* Am Physiol Soc; Am Soc Zoologists. *Res:* Transport-process in blood and bone marrow cells. *Mailing Add:* Dept Zool Drew Univ 36 Madison Ave Madison NJ 07940

ROHRSCHNEIDER, LARRY RAY, b Minneapolis, Minn, Oct 2, 44; m 71; c 2. VIROLOGY, ONCOLOGY. *Educ:* Univ Minn, BS, 67; Univ Wis, PhD(oncol), 73. *Prof Exp:* Res fel, Nat Cancer Inst, NIH, 73-76; asst mem tumor virol, 76-80, ASSOC MEM TUMOR VIROL, FRED HUTCHINSON CANCER RES CTR, 81- *Concurrent Pos:* Res grant, HEW, USPHS, 77-; res asst prof microbiol, Univ Wash, 78- *Mem:* AAAS; Sigma Xi; Am Soc Cell Biol. *Res:* Investigation of the mechanism of transformation by avian RNA tumor virus; biochemistry of transforming proteins and expression of tumor specific cell surface antigens. *Mailing Add:* F Hutchinson Cancer Res 1124 Columbia St Seattle WA 98104

ROHSENOW, WARREN M(AX), b Chicago, Ill, Feb 12, 21; m 46; c 5. HEAT TRANSFER. *Educ:* Northwestern Univ, BS, 41; Yale Univ, MEng, 43, DEng(heat power), 44. *Prof Exp:* Lab asst mech eng, Yale Univ, 41-43, instr, 43-44; from asst prof to assoc prof, 45-56, prof, 56-86, EMER PROF MECH ENG, MASS INST TECHNOL, 86- *Concurrent Pos:* Consult engr, 46-; dir, Dynatech Corp, 57- *Honors & Awards:* Gold Medal, Am Soc Mech Engrs, 51, Jr Award, 52; Yale Eng Asn Award, 52; Heat Transfer Div Mem Award, Am Soc Mech Engrs, 67, Max Jakob Mem Award, 71. *Mem:* Fel Nat Acad Eng; hon mem Am Soc Mech Engrs; fel Am Acad Arts & Sci. *Res:* Gas turbines; heat transfer; thermodynamics. *Mailing Add:* Dept Mech Eng Mass Inst Technol 77 Massachusetts Ave Cambridge MA 02139

ROHWEDDER, WILLIAM KENNETH, b Williston Park, NY, June 15, 32. MASS SPECTROMETRY. *Educ:* Lehigh Univ, BS, 54, MS, 57, PhD(chem), 60. *Prof Exp:* Sr engr, Bendix Corp, 60-61; CHEMIST, NORTHERN REGIONAL RES LAB, AGR RES SERV, USDA, 62- *Mem:* Am Chem Soc; Am Oil Chemists Soc; Am Soc Mass Spectrometry. *Res:* Mass spectrometry of natural and isotopically labeled blood lipids, chemical derivatives from oilseeds and other natural products. *Mailing Add:* 3900 N Stable Ct Apt 205 Peoria IL 61614-6950

ROHWEDER, DWAYNE A, b Marshalltown, Iowa, Aug 12, 26; m 48; c 2. AGRONOMY. *Educ:* Iowa State Univ, BS, 48, MS, 56, PhD(crop prod & soil mgt), 63. *Prof Exp:* Co Exten dir, Exten Div, Iowa State Univ, 48-54, area exten agronomist, 55-59, exten agronomist, 59-63; prof forage crops & exten agronomist, 63-87, EMER PROF FORAGE CROPS, EXTEN DIV, UNIV WIS-MADISON, 88- *Concurrent Pos:* Agronomist & chief of party, Univ Wis-US Agency Int Develop univ develop contract, Univ Rio Grande do Sul, Brazil, 67-69; agr enten educr, Am Soc Agron, 79; prof agronomist, 88- *Mem:* Am Forage & Grassland Coun; fel Am Soc Agron; fel Crop Sci Soc; Soil Conserv Soc Am. *Res:* Forage crop production; forage quality and evaluation. *Mailing Add:* 26425 S Ribbonwood Sun Lakes AZ 85248

ROHWER, ROBERT G, b Ft Calhoun, Nebr, Nov 20, 20; m 43; c 4. CEREAL CHEMISTRY, CORN & OTHER GRAINS. *Educ:* Univ Nebr, BS, 43. *Prof Exp:* Chemist, Am Cyanamid Co, 43-48; plant supt, KrimKo Corp, 48-49; supr sweetner res, Clinton Corn processing Co, Standard brands, Inc, 49-59; res chemist, Grain Processing Corp, 59-62, sales mgr, 62-66, vpres, 66-78, sr vpres & dir mkt, 78-86; CONSULT, GRAIN PROD & MKT, 86- *Concurrent Pos:* Trustee Paper Technol Found, Western Mich Univ; arranging chair, Starch Round Table, 85- *Mem:* Am Chem Soc; Tech Asn Pulp & Paper Indust; Inst Food Technol; Am Asn Cereal Chem. *Res:* Corn wet milling; starch, sugars and syrups; ion exchange. *Mailing Add:* PO Box 4122 Jackson WY 83001-4122

ROHY, DAVID ALAN, b Santa Barbara, Calif, July 30, 40; m 64; c 3. SOLID STATE PHYSICS. *Educ:* Univ Calif, Santa Barbara, BA, 62; Cornell Univ, PhD(physics), 68. *Prof Exp:* Physicist instrumentation, Lawrence Livermore Labs, Univ Calif, 67-70; res engr instrumentation & energy storage, Solar Turbines, 70-76, prog mgr, energy storage, hydrogen & instrumentation Int Harvester, 76-80; mgr applied sci, 81-87, MGR PROCESS DEVELOP & DIR PROD DEVELOP, SOLAR TURBINE INC, CATERPILLAR, 87- *Mem:* Am Phys Soc; Am Defense Prepardness Asn. *Res:* Development of waste heat recovery; thermal energy storage and thermal energy conversion systems using metal hydrides; hydrogen energy systems; instrumentation for gas turbine engines, particularly gas and solid temperature measurement; development of gas turbine engines. *Mailing Add:* 8639 Warmwell Dr San Diego CA 92119

ROIA, FRANK COSTA, JR, b New Bedford, Mass, June 5, 36; m 60; c 3. ECONOMIC BOTANY, MICROBIOLOGY. *Educ:* Mass Col Pharm, BS, 58, MS, 60, PhD(biol sci), 67. *Prof Exp:* Assoc prof, 67-77, PROF BIOL, PHILA COL PHARM & SCI, 77- *Honors & Awards:* Award, Soc Cosmetic Chem, 67. *Mem:* Soc Cosmetic Chem; Sigma Xi; Am Soc Microbiol. *Res:* Antimicrobial activity of plants; microbial flora and its relationship to dandruff. *Mailing Add:* 413 Walnut Hill Rd West Chester PA 19382

ROISTACHER, SEYMOUR LESTER, b New York, NY, May 21, 22; m 46; c 3. DENTISTRY. *Educ:* City Col NY, BS, 41; NY Univ, DDS, 44. *Prof Exp:* DIR DENT, QUEENS HOSP CTR, LONG ISLAND JEWISH-HILLSIDE MED CTR, 64-, ATTEND DENTIST, 60-; PROF DENT MED, STATE UNIV NY, STONY BROOK, 70- *Concurrent Pos:* Mem, Sect Oral Surg, Anesthesia & Hosp Dent, Am Asn Dent Schs, 68-, chmn, Coun Hosp Dent Serv, 69-, vpres, Coun Hosp, 70- *Mem:* Fel Am Col Dent; Am Asn Hosp Dent; fel Int Col Dent. *Res:* Myofascial pain in the facial areas. *Mailing Add:* Dept Dent Med Queens Hosp Ctr 82-68 164th St Jamaica NY 11432

ROITMAN, JAMES NATHANIEL, b Providence, RI, June 29, 41; m 72; c 1. ORGANIC CHEMISTRY. *Educ:* Brown Univ, BA, 63; Univ Calif, PhD(org chem), 69. *Prof Exp:* RES CHEMIST, WESTERN REGIONAL LAB, USDA, 69- *Mem:* Am Chem Soc; NY Acad Sci; Am Soc Pharmacog. *Res:* Isolation, characterization, and synthesis of naturally occurring compounds possessing biological activity; pyrrolizidine alkaloids; flavones; antifungal metabolites from microorganisms. *Mailing Add:* Western Regional Res Ctr USDA Albany CA 94710

ROITMAN, JUDY, b New York, NY, Nov 12, 45; m 78; c 1. TOPOLOGY, BOOLEAR ALGEBRA. *Educ:* Sarah Lawrence Col, BA, 66; Univ Calif, Berkeley, MA, 72, PhD(math), 74. *Prof Exp:* Asst prof math, Wellesley Col, 74-77; mem, Inst Advan Study, 77; from asst prof math to assoc prof math, 78-86, PROF MATH, UNIV KANS, 86- *Mem:* Am Math Soc; Asn Symbolic Logic; Asn Women Math. *Res:* Applications of set theory to topology and Boolean algebra; consistency results. *Mailing Add:* Dept Math Univ Kans Lawrence KS 66045

ROITMAN, PETER, b Boston, Mass, Aug 30, 49. ELECTRONICS, SOLID STATE PHYSICS. *Educ:* Dartmouth Col, BA, 72; Princeton Univ, PhD(elec eng), 76. *Prof Exp:* NAT RES COUN FEL PHYSICS, NAT BUR STANDARDS, 76- *Mem:* Am Phys Soc; Am Vacuum Soc; Inst Elec & Electronics Engrs; Sigma Xi. *Mailing Add:* Nat Bur Standards Bldg 225 Rm 8310 Washington DC 20234

ROIZIN, LEON, neuropathology; deceased, see previous edition for last biography

ROIZMAN, BERNARD, b Chisinau, Romania, Apr 17, 29; nat US; m 50; c 2. MOLECULAR GENETICS, VIROLOGY. *Educ:* Temple Univ, BA, 52, MS, 54; Johns Hopkins Univ, ScD(virol), 56. *Hon Degrees:* LHD, Governors State Univ, Ill, 84; MD, Univ Ferrara, Italy. *Prof Exp:* Instr microbiol, Sch Med & Sch Hyg & Pub Health, Johns Hopkins Univ, 56-57, res assoc, 57-58, asst prof, 58-65; prof biophys & theoret biol, Univ Chicago, 70-84, prof virol, 81-84, chmn, Dept Molecular Genetics & Cell Biol, 85-88, PROF MICROBIOL & CHMN, COMT VIROL, UNIV CHICAGO, 69-84, 88-, JOSEPH REGENSTEIN DISTINGUISHED SERV PROF, 84- *Concurrent Pos:* Lederle med fac award, 60; Am Cancer Soc scholar, Pasteur Inst, France, 61-62; USPHS career develop award, 63-65; Am Cancer Soc fac res assoc award, 66-71; mem develop res working group, Spec Virus Cancer Prog, Nat Cancer Inst, 67-71, consult, 67-73; Int Agency Res Against Cancer traveling fel, Karolinska Inst, Sweden, 70; mem steering coun, Human Cell Biol Prog,

NSF, 71-74; mem sci adv coun, NY Cancer Inst, 71-; chmn herpesvirus study group, Int Comn Taxon of Viruses, 71-; consult, Nat Sci Fedn, 72-74; mem med adv bd, Leukemia Res Found, 72-77; lectr, Am Found Microbiol, 74-75; mem bd sci consult, Sloan Kettering Inst, NY, 75-81; mem, Task Force Virol, Nat Inst Allergy & Infectious Dis, 76-77; mem, Study Sect Virol & Res Grants Rev Bd, NIH, 76-80; mem, bd trustees, Goodwin Inst Cancer Res, 77-; mem adv bd, Northwestern Univ Cancer Ctr, 79-88; chmn sci adv bd, Teikeo Showa Univ Ctr, Tampa Bay Inst Biomed Res, 83-; mem, Nat Inst Med Comt for Establishing Vaccine Priorities, 83-85; mem, Breifing Panel on Prev & Treatment of Viral Dis, Off Pres Sci Adv, 86. *Honors & Awards:* Esther Langer Award Achievement Cancer Res, 74; ICN Int Prize Virol, 88. *Mem:* Nat Acad Sci; Am Asn Immunol; Soc Exp Biol & Med; Am Soc Microbiol; Am Soc Biol Chemists; Am Soc Virol. *Res:* Biochemistry and genetics of viral replication; viral oncology; biochemistry and molecular biology of herpes viruses, especially genosome structure, function and regulation of gene expression. *Mailing Add:* Univ Chicago 910 E 58th St Chicago IL 60637

ROJAS, RICHARD RAIMOND, b New York, NY, Sept 25, 31; m 54; c 2. UNDERWATER ACOUSTICS, SIGNAL PROCESSING. *Educ:* Col City New York, BEE, 52; Drexel Inst Technol, Philadelphia, MEE, 61. *Prof Exp:* Proj engr guided missile fuzing, Philco Corp, 52-60; dept mgr undersea warfare, Magnovox Gen Atronic Corp, 60-69; assoc div head undersea surveillance, 69-76, ASSOC DIR RES, NAVAL RES LAB, 76- *Concurrent Pos:* Lectr, Pa State Univ, 63-69. *Honors & Awards:* Fel, Acoust Soc Am. *Mem:* Sigma Xi; Acoust Soc Am; Marine Technol Soc; Inst Elec & Electronics Engrs; Asn Old Crows. *Res:* Ocean engineering; physical oceanography; radar; electronic warfare; application of signal processing and statistical techniques to sonar. *Mailing Add:* Naval Res Lab Code 5000 Washington DC 20375-5000

ROJIANI, KAMAL B, b Bombay, India, Oct 22, 48; Pakistan citizen. STRUCTURAL RELIABILITY. *Educ:* Univ Karachi, BE, 71; Univ Ill, Urbana, MS, 73, PhD(civil eng), 77. *Prof Exp:* Comput programmer, Dept Finance, State Ill, 74; teaching & res asst, Dept Civil Eng, Univ Ill, 73-77; ASST PROF STRUCT, DEPT CIVIL ENG, VA POLYTECH INST & STATE UNIV, 78- *Concurrent Pos:* Struct engr, Bechtel Power Corp, 80; co-prin investr, US Dept Interior, Bur Surface Mines, 80-82; prin investr, Res Initiation Grant, NSF, 81-83; mem, Comt Load Resistance Factor Design, Am Soc Civil Engrs, 82- *Mem:* Am Soc Civil Engrs. *Res:* Application of probabilistic concepts to the solution of civil engineering problems, and specifically in the area of structural reliability. *Mailing Add:* Dept Civil Eng Va Polytech Inst State Univ 200 Palton Hall Blacksburg VA 24061

ROJO, ALFONSO, b Burgos, Spain, Jan 22, 21; Can citizen; m 56; c 3. ICHTHYOLOGY. *Educ:* Univ Valladolid, BSc & BA, 43; Univ Madrid, MSc, 53, PhD(ichthyol), 56. *Prof Exp:* Biologist, Spanish Dept Fisheries, 54-58; asst scientist, Fisheries Res Bd Can, 58-61; from asst prof to assoc prof, comp anat & ichthyol, St Mary's Univ, NS, 61-68; assoc prof expert fisheries & oceanog, Food & Agr Orgn, UN, 68-70; assoc prof, 70-72, PROF ICHTHYOL, ST MARY'S UNIV, NS, 72- *Concurrent Pos:* Sci adv for Spain, Int Comn Northwest Atlantic Fisheries, 54-58; sci adv for Can, Int Comn Great Lakes, 58-61. *Mem:* Am Fisheries Soc; Can Soc Zool; Soc Syst Zool. *Res:* Problems of systematics at the species level in relation to larval forms of commerical fishes; study of a possibility of linking sciences and humanities through bioethics. *Mailing Add:* Dept Biol St Mary's Univ Robie St Halifax NS B3H 3C3 Can

ROKACH, JOSHUA, b Cairo, Egypt, Sept 17, 35; Can citizen; m. LEUKOTRIENES, PROSTAGLANDINS. *Educ:* Hebrew Univ, MSC, 62; Weizmann Inst Sci, Israel, PhD, 64. *Prof Exp:* Fel, Weizmann Inst Sci, 64-65, Max-Planck Inst, 65-66; sr res chemist, Merck Frosst Labs, Merck Frosst Can Inc, 66-74, group leader, 74-77, sr res fel, 77-79, dir med chem, 79-81, exec dir res, 81-88, distinguished sr scientist, 88-89; PROF CHEM, FLA INST TECHNOL, 89-, DIR, CLAUDE PEPPER INST AGING & THERAPEUT RES, 89- *Concurrent Pos:* Ed, Prostaglandins and J Lipid Mediators. *Honors & Awards:* Prix Urgel Archambault, 86; John LaBatt Ltd Award, Can Soc Chem, 88; Xerox Lectr, 88; Chinese Acad Sci Taiwan Award, 88; Prix Paul Ehrlich, Paris, France, 90. *Mem:* Am Chem Soc; Chem Inst Can; Asn Res Vision & Ophthal. *Res:* Syntheses of various metabolites of the lipoxygenase pathway of arachidonic acid and the biological properties of these metabolites with special emphasis on their relation to allergic and respiratory diseases. *Mailing Add:* Claude Pepper Inst Aging & Therapeut Res Fla Inst Technol 150 W University Blvd Melbourne FL 32901-6988

ROKEACH, LUIS ALBERTO, b Buenos Aires, Arg, Nov 1, 51; m 73; c 2. AUTOIMMUNE DISEASE. *Educ:* Hebrew Univ Jerusalem, Israel, BS, 74; Univ Libre de Bruxelles, Belg, MS, 81, PhD(molecular biol), 84. *Prof Exp:* Res asst, Dept Clin Path, Clin Olivos, Buenos Aires, Arg, 77-80; postdoctoral res assoc, Dept Biol, San Diego State Univ, 84-86; assoc res scientist, 86-87, RES SCIENTIST, AGOURON INST, LA JOLLA, CALIF, 87- *Concurrent Pos:* Prin investr, Agouron Inst, La Jolla, Calif, 90- *Mem:* Sigma Xi. *Res:* Role of autoantigens in autoimmune disease; cellular function of autoantigens. *Mailing Add:* Agouron Inst 505 Coast Blvd S La Jolla CA 92037

ROKEBY, THOMAS R(UPERT) C(OLLINSON), b Port Rowan, Ont, May 9, 21; nat US; m 46; c 3. AGRICULTURAL ENGINEERING. *Educ:* Ont Agr Col, BSA, 48; Univ Toronto, MSA, 50; Okla State Univ, PhD, 68. *Prof Exp:* Asst agr eng, Ont Agr Col, 48-50, asst res, 50-51; asst prof, SDak State Col, 51-55; from asst prof to assoc prof, 55-85, EMER PROF AGR ENG, UNIV ARK, FAYETTEVILLE, 85- *Concurrent Pos:* NSF fel, Okla State Univ, 64-65. *Mem:* Am Soc Agr Engrs; Int Solar Energy Soc. *Res:* Solar energy for heating farm and residential buildings; design and construction of farm buildings, particularly houses and poultry housing; design of grain and bulk materials storage; materials and energy research. *Mailing Add:* Dept Biol & Agr Eng Univ Ark Fayetteville AR 72701

ROKITKA, MARY ANNE, ENVIRONMENTAL & HYPERBARIC PHYSIOLOGY. *Educ:* State Univ NY, Buffalo, PhD(biol), 73. *Prof Exp:* ASST PROF PHYSIOL, STATE UNIV NY, BUFFALO, 76- *Res:* Aerospace physiology. *Mailing Add:* 120 Sherman Hall Health Sci Ctr State Univ NY 3435 Main St Buffalo NY 14214

ROKNI, MOHAMMAD ALI, b Tehran, Iran, May 23, 58. CHAOTIC DYNAMICS. *Educ:* WVa Univ, BSME, 79; Univ Md, MS, 81, PhD(mech eng), 86. *Prof Exp:* Instr mech eng, Univ Md, 82-87; adj prof mech eng, Univ DC, 87-88; MEM TECH STAFF, COMPUTER SCI CORP, 88- *Mem:* Am Soc Mech Engrs; Am Astron Soc. *Res:* Application of concepts and techniques from nonlinear dynamics and chaotic motion to the study of engineering problems; Kalman filtering and estimation techniques; spacecraft attitude determination and control. *Mailing Add:* 4901 Seminary Rd No 802 Alexandria VA 22311

ROKOP, DONALD J, b Aurora, Ill, May 15, 39; m 64; c 3. SPECIFIC ISOTOPE IONIZATION ENHANCEMENT, MEASUREMENT OF ATOM ABUNDANCES OF ISOTOPES. *Educ:* Lake Forest Col, BA, 61. *Prof Exp:* Sci assoc, Argonne Nat Lab, 61-78; STAFF SCIENTIST, LOS ALAMOS NAT LAB, 78- *Concurrent Pos:* Group leader anal mass spectrometry, Argonne Nat Lab, 68-78; founder & co-chair Isotope Ratio Measurements Comt, Am Soc Mass Spectrometry, 80-83; sect leader mass spectrometry, Los Alamos Nat Lab, 86-88, asst group leader, 88-89, tech coordr mass spectrometry develop, 89-; organizer, A O Nier Symposium Inorg Mass Spectrometry, 91. *Honors & Awards:* Award of Excellence, Dept Energy, 85; Antarctic Serv Medal, US Congress & NSF, 86. *Mem:* AAAS. *Res:* Thermal ionization and electron bombardment analytical mass spectrometric techniques for application to problems in nuclear physics, chemistry, geology, geophysics, and atmospheric dynamics; determining low abundance isotopes. *Mailing Add:* Los Alamos Nat Lab MS J-514 Los Alamos NM 87545

ROKOSKE, THOMAS LEO, b Danville, Ill, Feb 25, 39; m 61; c 3. SOLID STATE PHYSICS. *Educ:* Loyola Univ, La, BS, 61; Fla State Univ, MS, 63; Auburn Univ, PhD(physics), 73. *Prof Exp:* Instr physics, Fla State Univ, 64; from instr to asst prof, 64-67 & 71-74, ASSOC PROF PHYSICS, APPALACHIAN STATE UNIV, 74- *Concurrent Pos:* Vchmn, NC Adv Coun Metrication, 74-75, chmn, 75-76; dir, Indust Metric Educ Conf, 74-75. *Mem:* Am Phys Soc; Am Asn Physics Teachers. *Res:* Electrical conduction in thin films. *Mailing Add:* Dept Physics Appalachian State Univ Boone NC 28608

ROL, PIETER KLAAS, b Neth, Nov 22, 27; m 52; c 3. MATERIALS SCIENCE. *Educ:* Univ Amsterdam, Drs, 53, PhD(physics), 60. *Prof Exp:* Group leader isotope separation, Inst Atomic & Molecular Physics, Found Fundamental Res in Matter, Amsterdam, 53-60; fel, Space Sci Lab, Gen Dynamics/Convair, Calif, 60-62; res leader atomic collisions, Inst Atomic & Molecular Physics, Found Fundamental Res in Matter, 62-66; staff scientist, Convair Div, Gen Dynamics, 66-69; dir res, Inst Eng Sci, 74-76, assoc dean, Col Eng, 76-78, assoc provost acad planning, 78-81, PROF CHEM & METALL ENG, COL ENG, WAYNE STATE UNIV, 69-, INTERIM DEAN, 85- *Concurrent Pos:* Consult, Appl Res Labs, 68-69 & La Jolla Inst, 74- *Mem:* Am Phys Soc; Am Vacuum Soc; AAAS; Sigma Xi. *Res:* Atomic and molecular physics; surface physics; ion implantation in metals. *Mailing Add:* Col Eng 239 Physics Wayne State Univ 5950 Cass Ave Detroit MI 48202

ROLAND, ALEX, b Providence, RI, Apr 7, 44; m 79; c 3. HISTORY & PHILOSOPHY OF SCIENCE. *Educ:* US Naval Acad, BS, 66; Univ Hawaii, MA, 70; Duke Univ, PhD(hist), 74. *Prof Exp:* Historian, NASA, 73-81; assoc prof, 81-87, PROF MIL HIST & HIST TECHNOL, DUKE UNIV, 87- *Concurrent Pos:* Harold K Johnson vis prof mil hist, US Army War Col. *Mem:* Soc Hist Technol (secy, 84-); Am Mil Inst; Hist Sci Soc. *Res:* History of technology and war in the West from earliest times to the present, with specialization in aerospace history. *Mailing Add:* Dept Hist Duke Univ Durham NC 27706

ROLAND, CHARLES GORDON, b Winnipeg, Man, Jan 25, 33; m 53, 79; c 4. HISTORY OF MEDICINE, COMMUNICATION SCIENCE. *Educ:* Univ Man, BSc & MD, 58. *Prof Exp:* Physician pvt pract, 59-64; sr ed jour, AMA, 64-69; prof biomed commun & chmn dept & assoc prof hist med, Mayo Med Sch & Mayo Found, 69-77; JASON A HANNAH PROF HIST MED, MCMASTER UNIV, 77- *Concurrent Pos:* Sid W Richardson vis prof med humanities, Univ Tex Med Br, Galveston, 84; Can deleg, Int Soc His Med, 85- *Mem:* Am Med Writers Asn (pres, 69-70); Coun Biol Ed; Am Asn Hist Med (secy-treas, 76-79); Am Osler Soc (secy-treas, 75-85, pres, 86-87); Can Soc Hist Med; Hannah Inst Hist Med. *Res:* Variety of investigations into the sociology and history of scientific communication; history of medicine in the nineteenth century; health and medical care in Axis prisoner of war camps during World War II. *Mailing Add:* 3N10-HSC McMaster Univ Hamilton ON L8N 3Z5 Can

ROLAND, DAVID ALFRED, SR, b Cochran, Ga, Jan 2, 43; m 65; c 2. NUTRITIONAL BIOCHEMISTRY. *Educ:* Univ Ga, BS, 66, PhD(nutrit), 70. *Prof Exp:* Res asst nutrit, Univ Ga, 66-67; assoc prof mgt & poultry & asst poultry scientist, Univ Fla, 70-76; ALUMNI PROF POULTRY NUTRIT, AUBURN UNIV, 76- *Honors & Awards:* Research Award, Poultry Sci Asn, 73 & Egg Sci Award, 74. *Mem:* AAAS; Poultry Sci Asn; World Poultry Sci Asn. *Res:* Mineral metabolism; egg shell quality; poultry management; lipid metabolism; reproductive physiology. *Mailing Add:* Dept Poultry Sci Auburn Univ Auburn AL 36830

ROLAND, DENNIS MICHAEL, b New Castle, Ind, July 16, 49; m 71; c 1. PHARMACEUTICAL CHEMISTRY. *Educ:* Ball State Univ, BS, 71; Univ Vt, PhD(org chem), 77. *Prof Exp:* Fel org synthesis, Colo State Univ, 77-79; sr scientist, 79-84, sr res scientist, 84-89, MGR, INFLAMMATION/OSTEOARTHRITIS CHEM, PHARAMACEUT DIV, CIBA-GEIGY CORP, 89- *Concurrent Pos:* Nat Res Serv Award, NIH, 77-79. *Mem:* Am Chem Soc; NY Acad Sci. *Res:* Design and synthesis of new biologically active molecules for use as theraputic agents. *Mailing Add:* 20 Heather Lane Basking Ridge NJ 07920

ROLA-PLESZCZYNSKI, MAREK, b Fermo, Italy, Aug 9, 47; Can citizen; m 71; c 2. IMMUNOLOGY. *Educ:* Séminaire de Sherbrooke, BA, 66; Univ Sherbrooke, Que, MD, 70; Nat Bd Med Examr, dipl med, 71; FRCP(C), 78. *Prof Exp:* Res fel immunol, Sch Med, Georgetown Univ, 73-75 & Harvard Med Sch, 75-76; from asst prof to assoc prof, 76-85, RES SCHOLAR IMMUNOL, FAC MED, UNIV SHERBROOKE, 81-, PROF, 85- *Concurrent Pos:* Chmn, Exam Comt Clin Immunol, Royal Col Physicians, Canada, 84- *Honors & Awards:* Frosst Medal, Frosst Co & Univ Sherbrooke, 71. *Mem:* Am Asn Immunologists; Soc Pediat Res; Can Immunol Soc; Can Soc Clin Invest. *Res:* Discovery and characterization of a distinct subset of human peripheral blood lymphocytes; modulation of cytotoxic and immunoregulatory activities of lymphocytes by leukotrienes, neuropeptides and hormones; immunopathology of pulmonary diseases. *Mailing Add:* Dept Pediat Univ Sherbrooke Fac Med 3001 12th Ave N Sherbrooke PQ J1H 5N4 Can

ROLD, JOHN W, b Kirkman, Iowa, May 23, 27; m 50; c 4. GEOLOGY. *Educ:* Univ Coloro, Boulder, AB, 48, MS, 50. *Prof Exp:* Field geologist, Magnolia Petrol, (Mobil Oil), 48; dist geologist, Chevron Oil, 50-69; area geologist, La Gulf Coast, New Orleans, 53-55, Denver Basin Colo, 55-58; dist staff geologist, Colo, Nebr & Kans, 58-63, & Plains dist, Eastern Colo, Nebr & Kans, 63-65; dist geologist, Montana-Dalcotas Dist, Billings, 65-69; State GEOLOGIST & DIR, COLO, SURV, 69- *Concurrent Pos:* Crew geophysicist, various Rocky Mt provinces & Williston Basin, Wyo, Utah, Mont & N & S Dak, 51-52; chmn, Ges Soc Am, 80; mem bd gov, Am Geol Inst, 80-81; comt chmn, Nat Res Coun, 82-83. *Mem:* Am Asn Petrol Geol; Geol Soc Am; Am Geol Inst; Asn Am State Geologist. *Mailing Add:* 2534 S Balsom St Lakewood CO 80227

ROLDAN, LUIS GONZALEZ, b Garafia, Tenerifee Spain, Aug 8, 25; m 61; c 4. CRYSTALLOGRAPHY, TECHNICAL LAB ASSESSMENT. *Educ:* Univ Sevilla, Lic Chem, 50, DSc, 57. *Prof Exp:* Adj prof exp physics, Univ Sevilla, Spain, 53-57; from res physicist to sr res physicist, Brit Rayon Res Asn, Manchester, Eng, 57-61; sr res chemist, Cent Res Lab, Allied Chem Corp, 61-63, from scientist to sr scientist, 63-68; res assoc & head, Phys Chem Sect, J P Stevens & Co, Inc, Greenville, SC, 68-76, mgr physics & micros dept, Tech Ctr, 76-86; DIR, LGR MICRORES, 86- *Concurrent Pos:* Vis prof, Univ Beira Interior, Covilha, Portugal, 87; adj prof, NC State Univ, Raleigh, NC, 88-90; tech expert, Nat Voluntary Lab Accreditation Prog, Nat Inst Standards & Technol, 88- *Mem:* Fiber Soc; Am Chem Soc; Am Crystallog Asn; Royal Span Soc Phys & Chem. *Res:* Structure of organic compounds, macromolecules and polymers by x-ray diffraction techniques; morphology of these compounds by electron microscopy; relationship of structure and morphology with physical properties; metallography; ceramics structure; fiber physics and textiles: research and development. *Mailing Add:* 124 Becky Don Dr Greer SC 29651-1213

ROLETT, ELLIS LAWRENCE, b New York, NY, July 10, 30; m 56; c 3. INTERNAL MEDICINE, CARDIOLOGY. *Educ:* Yale Univ, BS, 52; Harvard Univ, MD, 55. *Prof Exp:* Intern med, Mass Gen Hosp, Boston, 55-56; asst resident, New York Hosp, 56-57; resident, Mass Gen Hosp, 59-60, clin fel cardiol, 60-61; asst med, Peter Bent Brigham Hosp, 61-63; from asst prof to prof, Univ NC, Chapel Hill, 63-74; prof med, Univ Calif, Los Angeles, 74-77; PROF MED, DARTMOUTH MED SCH, 77-; CHIEF CARDIOL, DARTMOUTH-HITCHCOCK MED CTR, 77- *Concurrent Pos:* Am Heart Asn res fel, 61-63; res grant, USPHS, 64-77, career develop award, 67-72; Lederle med fac award, 65-67; chief cardiol, Vet Admin Wadsworth Hosp Ctr, 74-77; mem, Merit Rev Bd Cardiovasc Studies, Vet Admin, chmn, 76-79. *Mem:* AAAS; Am Col Cardiol; Am Physiol Soc. *Res:* Cardiodynamics; influence of catecholamines on cardiac muscle function; contractile behavior of isolated myocytes. *Mailing Add:* Dept Med Dartmouth Col Maynard St Hanover NH 03755

ROLF, CLYDE NORMAN, b Dayton, Ky, June 26, 37; c 2. HYPERTENSION, NEPHROLOGY. *Educ:* Univ Ky, AB, 63, MD, 67. *Prof Exp:* Pvt pract nephrology, 72-77; assoc group dir, 77-79, group dir, 79-81, SR MED SPECIALIST, MERRELL RES CTR, 81-, ASSOC DIR MED RES, 85- *Concurrent Pos:* Fel med nephrologyk Good Samaritan Hosp, Cincinnati, 72; asst clin prof med, Col Med, Univ Cincinnati, 75- *Mem:* Am Nephrological Soc; Int Nephrological Soc; Am Col Physicians; Am Soc Artificial Internal Organs; Am Soc Clin Pharmacol & Therapeut. *Res:* Development, conduct and analysis of clinical trials with investigational antihypertensive agents and other investigational drugs. *Mailing Add:* 7475 Algonquin Dr Cincinnati OH 45215

ROLF, HOWARD LEROY, b Laverne, Okla, Nov 25, 28; m 61; c 4. MATHEMATICS. *Educ:* Okla Baptist Univ, BS, 51; Vanderbilt Univ, MA, 53, PhD(math), 56. *Prof Exp:* Instr math, Vanderbilt Univ, 54-56; asst prof, Baylor Univ, 56-57; assoc prof, Georgetown Col, 57-59; asst prof & dir comput cent, Vanderbilt Univ, 59-64; PROF MATH, BAYLOR UNIV, 64-, CHMN DEPT, 71- *Concurrent Pos:* Vis assoc, Calif Inst Technol, 67-68. *Mem:* Am Math Soc; Math Asn Am. *Res:* Lattice theory; abstract algebra. *Mailing Add:* Dept Math Baylor Univ Waco TX 76798-7328

ROLF, LESTER LEO, JR, b San Antonio, Tex, Nov 30, 40; m 64; c 3. COMPARATIVE PHARMACOLOGY, RENAL PHYSIOLOGY. *Educ:* St Mary's Univ, BA, 64; Tex A&M Univ, MS, 67, PhD(physiol), 69; Okla State Univ, DVM, 88. *Prof Exp:* NIH fel, Col Med, Univ Fla, 69-71, from instr to asst prof pharmacol, 71-74; asst prof physiol sci, Col Vet Med, Okla State Univ, 74-86; CONSULT, 86- *Mem:* Am Physiol Soc; Sigma Xi; Soc Exp Biol & Med; Am Asn Vet Physiologists & Pharmacologists; Am Vet Med Cols. *Res:* Pharmacokinetics of chemotherapeutic agents in fish. *Mailing Add:* 22-2 Willow W Stillwater OK 74075

ROLF, RICHARD L(AWRENCE), b Milwaukee, Wis, Nov 4, 35; m 61; c 1. STRUCTURAL MECHANICS, STRUCTURAL ENGINEERING. *Educ:* Marquette Univ, BS, 58; Univ Ill, Urbana, MS, 60. *Prof Exp:* Res asst civil eng, Univ Ill, Urbana, 58-60; res engr, Eng Design Div, Res Labs, Aluminum Co Am, 60-75, sr engr, 75-81, staff engr, Eng Properties & Design Div, 81-83, tech specialist, Prod Eng Div, 83-86, sr tech specialist, Prod Eng Div, 86-87, Prod Design & Mech Div, 87-90, TECH CONSULT, PROD DESIGN & MECH DIV, ALCOA LABS, ALUMINUM CO AM, 90- *Honors & Awards:* J James R Croes Medal, Am Soc Civil Engrs, 66. *Mem:* Sigma Xi; Am Acad Mech; Am Soc Civil Engrs. *Res:* Design analysis and testing of various structural members; development of design rules for aluminum, brittle materials and composites; damage evolution in materials. *Mailing Add:* Prod Design & Mech Div Alcoa Labs Bldg D Alcoa Center PA 15069

ROLFE, GARY LAVELLE, b Paducah, Ky, Sept 5, 46; m 68. ECOLOGY, BIOLOGY. *Educ:* Univ Ill, BS, 68, MS, 69, PhD(ecol), 71. *Prof Exp:* Dir metals task force, 71-77, asst prof, 72-75, assoc prof forest ecol, 75-80, PROF FORESTRY & CHMN DEPT, UNIV ILL, 80- *Concurrent Pos:* Asst dir, Ill Agr Exp Sta, 76-80. *Mem:* Ecol Soc Am; Soil Sci Soc Am; Soc Am Foresters. *Res:* Nutrient cycling in forest ecosystems; water quality; trace contaminants in the environment. *Mailing Add:* Dept Forestry Univ Ill 110 Mumford Hall Urbana IL 61801

ROLFE, JOHN, b London, Eng, Feb 15, 27; Can citizen; m 53; c 3. SOLID STATE PHYSICS. *Educ:* Univ London, BSc, 50, PhD(physics), 53. *Prof Exp:* Res physicist, A E I Res labs, Aldermaston, Eng, 53-55; fel, Radio & Elec Eng Div, Nat Res Coun Can, 55-57, res officer solid state physics, 57-75, sr res officer, Physics Div, 75-82; SCI ADV, BANK CAN, 82- *Mem:* Can Asn Physicists. *Res:* Electrical and optical properties of insulating crystals, especially alkali halide crystals; color centers in alkali halide crystals. *Mailing Add:* 11 Appleford St Ottawa ON K1J 6V1 Can

ROLFE, RIAL DEWITT, b St Louis, Mo, Feb 25, 52; m; c 2. ANAEROBIC BACTERIOLOGY. *Educ:* Univ Mo, Columbia, BA, 74, MS, 76, PhD(microbiol), 78. *Prof Exp:* Teaching fel, Sch Med, Univ Calif, Los Angeles, 78-79; dir, Infectious Dis Sect Res Lab, Wadsworth Va Med Ctr, Los Angeles, 79-81; asst prof, 81-86, ASSOC PROF, DEPT MICROBIOL, TEX TECH UNIV HEALTH SCI CTR, 86-, ASSOC CHMN, 89- *Concurrent Pos:* Adj asst prof, Sch Med, Univ Calif, Los Angeles, 79-81. *Mem:* Am Soc Microbiol; AAAS; Soc Intestinal Microbial Ecol & Dis. *Res:* Anaerobic bacteria in health and disease. *Mailing Add:* Dept Microbiol Tex Tech Health Sci Ctr Lubbock TX 79430

ROLFE, STANLEY THEODORE, b Chicago, Ill, July 7, 34; m 56; c 3. FRACTURE MECHANICS. *Educ:* Univ Ill, Urbana, BS, 56, MS, 58, PhD(civil eng), 62. *Prof Exp:* Res assoc civil eng, Univ Ill, Urbana, 56-62; sect supvr & div chief, US Steel Appl Res Lab, 62-69; Ross H Forney prof, 69-89, A P LEARNED PROF CIVIL ENG, UNIV KANS, 89-, CHMN DEPT, 75- *Concurrent Pos:* Chmn, Metall Studies Panel, Nat Acad Sci, 68-71; chmn low cycle fatigue comt, Pressure Vessel Res Coun, 68-70. *Honors & Awards:* Sam Tour Award, Am Soc Testing & Mat, 71. *Mem:* Nat Acad Eng; Am Soc Mech Engrs; Soc Exp Stress Anal; Am Soc Eng Educ; Am Soc Testing & Mat; Am Soc Civil Engrs. *Res:* Fracture mechanics; failure analysis; fatigue and fracture of structural materials as related to design; experimental stress analysis. *Mailing Add:* Dept Civil Eng Univ Kans Lawrence KS 66045

ROLL, BARBARA HONEYMAN HEATH, b Portland, Ore, Apr 4, 10; m 77. PHYSICAL ANTHROPOLOGY. *Educ:* Smith Col, BA, 32. *Hon Degrees:* LHD, Smith Col, 89. *Prof Exp:* Res assoc & exec dir, Constitution Lab, Col Physicians & Surgeons, NY, 48-53; instr phys anthrop, Monterey Peninsula Col, 66-74; RES ASSOC, DEPT ANTHROP, UNIV PA, 75- *Mem:* Fel AAAS; Am Asn Phys Anthrop; fel Am Anthrop Asn; Brit Soc Study Human Biol; Int Asn Human Biol. *Res:* Child development; somatotype methodology and interpretation of somatotype data. *Mailing Add:* 26030 Rotunda Dr Carmel CA 93923

ROLL, DAVID BYRON, b Miles City, Mont, Mar 16, 40; m 70; c 2. MEDICINAL CHEMISTRY. *Educ:* Univ Mont, BS, 62; Univ Wash, PhD(med chem), 66. *Prof Exp:* Res chemist, Pesticides Res Lab, Dept HEW, 66-67; assoc prof med chem, 67-76, PROF MED CHEM, COL PHARM, UNIV UTAH, 76-, ASSOC DEAN ACAD AFFAIRS, 77- *Mem:* Am Pharmaceut Asn. *Res:* Nuclear magnetic resonance and its applications to stereochemical and biochemical problems; medicinal chemistry. *Mailing Add:* Pharm/258 HLS Skaggs Hall Univ Utah Salt Lake City UT 84112

ROLL, DAVID E, b Baltimore, Md, June 4, 48; m 76; c 4. STRUCTURE & FUNCTION OF TOPOISOMERASES, REGULATION OF TOPOISOMERASES BY PHOSPHORYLATION. *Educ:* Harding Col, Searcy, Ark, BS, 70; Univ Ill, MS, 72, PhD(biochem), 76. *Prof Exp:* Res assoc biochem, Univ Pittsburgh Med Sch, 76-78; PROF BIOCHEM, ROBERTS WESLEYAN COL, 78- *Concurrent Pos:* Vis asst prof, Baylor Col Med, Houston, 79, 82 & 83, fel/sabbatical, 85-86; vis asst prof, Dept Biol, Univ Rochester, 84, vis prof, Path Dept, 89. *Mem:* Am Chem Soc; Fedn Am Scientists; Sigma Xi; Nat Asn Adv Health Prof. *Res:* Structure and function of topoisomerase enzymes and their regulation by phosphorylation; type I topoisomerase has been isolated and purified from Novikoff ascites cells; this topoisomerase has been activated by protein phosphorylation using purified serine Kinase enzymes. *Mailing Add:* Roberts Wesleyan Col 2301 Westside Dr Rochester NY 14624

ROLL, FREDERIC, b New York, NY, Sept 10, 21; m 50. STRUCTURAL ENGINEERING. *Educ:* City Col New York, BCE, 44; Columbia Univ, MS, 49, PhD(civil eng), 57; Univ Pa, MA, 71. *Prof Exp:* Civil engr, W S Briggs, NY, 50; stress analyst, Chance Vought Aircraft Corp, Conn, 44; civil engr, Andrews & Clarke, NY, 44-45; asst civil eng, Columbia Univ, 45-46, instr, 46-50; assoc, 50-57, from assoc prof to prof, 57-86, EMER PROF CIVIL ENG, UNIV PA, 86- *Concurrent Pos:* NSF fel, Eng & Port, 63-64, grant, 66-; mem res staff, Cement & Concrete Asn, 70-71; eng consult, 57-; pres, Del Valley Chap, Am Concrete Inst, 76; pres, Philadelphia Sect, Am Soc Civil Engrs, 81-82. *Honors & Awards:* Cert Appreciation, Am Soc Civil Engrs, 66, State-of-the-Art of Civil Eng Award, 74, Outstanding Serv Award, 75, Raymond Reese Res Award, 76, Struct Engr of the Year Award, 87; Cert

Appreciation, Am Concrete Inst, 90. *Mem:* Fel Am Concrete Inst; Am Soc Civil Engrs; Am Soc Testing & Mat; Soc Exp Stress Anal; Sigma Xi; Prestressed Concrete Inst. *Res:* Creep of plain concrete; shear and diagonal tension in reinforced concrete beams and slabs; structural model analysis; structural and materials testing; reinforced and prestressed concrete; fiber reinforced plastic structural systems. *Mailing Add:* 1928 Pine St Philadelphia PA 19103-6626

ROLL, JAMES ROBERT, b Chilton, Wis, Aug 31, 58; c 1. CONCEPTUAL DESIGN MODELING, SYSTEMS DESIGN. *Educ:* Univ Wis-Madison, BS, 80, MS, 81, PhD(mech eng), 84. *Prof Exp:* Sr res engr, Gen Motors Res Labs, 83-86, staff res engr, 86-88, SECT HEAD CONCEPTUAL DESIGN, GEN MOTORS-SYSTS ENG CTR, 88- *Mem:* Soc Automotive Eng; assoc mem Am Soc Mech Engrs; Sigma Xi. *Res:* Design technique and software to assist in the conceptual design of automotive vehicles including mechanisms, powertrains, optimization, trade off analysis techniques and advanced reliability techniques. *Mailing Add:* 1305 Baldwin Ann Arbor MI 48104

ROLL, JOHN DONALD, nuclear physics; deceased, see previous edition for last biography

ROLL, PAUL M, BIOCHEMISTRY. *Educ:* Stanford Univ, PhD(biochem), 46. *Prof Exp:* Prof biochem, Med Col Wis, 46-54; RETIRED. *Mailing Add:* 611 Seacliff Dr Aptos CA 95003-3538

ROLL, PETER GUY, b Detroit, Mich, Apr 13, 33; m 55; c 3. PHYSICS. *Educ:* Yale Univ, BS, 54, MS, 58, PhD(physics), 60. *Prof Exp:* Jr scientist nuclear reactor design, Westinghouse Atomic Power Div, 54-56; res asst physics, Yale Univ, 56-58, instr, 59-60; from instr to asst prof, Princeton Univ, 60-65; staff physicist, Comn Col Physics, Univ Mich, 65-66; spec asst to vpres acad admin, 71-84, ASSOC PROF PHYSICS, UNIV MINN, MINNEAPOLIS, 66-; VPRES INFO SERV, NORTHWESTERN UNIV, EVANSTON, IL, 84- *Mem:* AAAS; Am Asn Physics Teachers; Asn Comput Mach. *Res:* Low energy nuclear physics; gravity experiments; cosmic background radiation measurements; musical acoustics. *Mailing Add:* 633 Clark St Northwestern Univ Evanston IL 60201

ROLLAND, WILLIAM WOODY, b Asheville, NC, June 8, 31; m 50; c 4. NUCLEAR PHYSICS. *Educ:* King Col, BA, 53; Duke Univ, PhD, 63. *Prof Exp:* Instr math, King Col, 55-56, assoc prof physics, 59-68; assoc prof comput sci & dir comput ctr, St Andrews Presby Col, 68-83; PRES, ROLLAND MGT SYST, 83- *Mem:* Am Asn Physics Teachers. *Res:* Nuclear spectroscopy, particularly direct nuclear interactions; studies of variable stars, both eclipsing and intrinsic variables. *Mailing Add:* Rte 5 PO Box 135 Laurinberg NC 28352

ROLLASON, GRACE SAUNDERS, b New York, NY, Sept 14, 19; m 44; c 2. EMBRYOLOGY. *Educ:* Hunter Col, AB, 40; NY Univ, MS, 42, PhD(exp embryol), 48. *Prof Exp:* Asst, Harvard Univ, 44-45; assoc, Amherst Col, 46-47; res assoc, Univ Mass, Amherst, 47-56, from instr to assoc prof zool, 56-85; RETIRED. *Concurrent Pos:* Hunter Col scholar, Woods Hole Marine Biol Lab. *Res:* Mammalian and amphibian embryology. *Mailing Add:* 34 Red Gate Lane Amherst MA 01002

ROLLASON, HERBERT DUNCAN, b Beverly, Mass, Mar 20, 17; m 44; c 2. HISTOLOGY. *Educ:* Middlebury Col, AB, 39; Williams Col, AM, 41; Harvard Univ, MA, 43, PhD(biol), 49. *Prof Exp:* Instr anat, L I Col Med, 45-46; instr biol, Amherst Col, 46-48; from asst prof to prof, 48-84, asst dean, 65-71, assoc dean, Col Arts & Sci, 72-73, assoc chmn dept, 76-84, EMER PROF ZOOL, UNIV MASS, AMHERST, 84- *Mem:* AAAS; Am Soc Zool; Am Inst Biol Sci. *Res:* Kidney histology and cytology; compensatory hypertrophy; cellular ultrastructure. *Mailing Add:* 34 Red Gate Lane Amherst MA 01003

ROLLE, F ROBERT, b Jamaica, NY, May 21, 39; m 74; c 1. ANALYTICAL CHEMISTRY. *Educ:* Pratt Inst, BS, 61; Purdue Univ, MS, 65, PhD(phys org chem), 66. *Prof Exp:* Res fel kinetics, Univ London, 66-67; res chemist, 67-68, group leader, 68-71, asst mgr, 71-75, mgr anal chem, 75-79, asst dir, Cent Lab, 79-80, from asst dir to dir, Tech & Adv Serv, 80-84, RES FEL, CHICOPEE DIV, JOHNSON & JOHNSON, 84- *Honors & Awards:* P B Hofmann Res Scientist Award, 73. *Mem:* The Chem Soc; Am Chem Soc. *Res:* Analytical aspects of nonwoven fabrics; operating room gowns. *Mailing Add:* One Lafayette Rd W Princeton NJ 08540-2428

ROLLEFSON, AIMAR ANDRE, b Houston, Tex, Apr 26, 40; m 64; c 3. NUCLEAR STRUCTURE, PHYSICS. *Educ:* Rice Univ, Ba, 60, MA, 62, PhD(physics), 64. *Prof Exp:* Postdoctoral res assoc physics, Univ Pittsburgh, 64-66; postdoctoral res assoc physics, Univ Notre Dame, 66-68, asst prof, 68-75; asst prof, 75-80, PROF PHYSICS, UNIV ARK LITTLE ROCK, 80-, DEPT CHAIR PHYSICS & ASTRON, 89- *Concurrent Pos:* Vis prof, Univ Notre Dame, 80-81. *Mem:* Am Phys Soc; Am Asn Physics Teachers; Sigma Xi. *Res:* Experimental nuclear physics precision energy measurements using a broad-range magnetic spectrograph; study of nuclear reactions of astrophysical importance. *Mailing Add:* Physics & Astron Dept Univ Ark-Little Rock 2801 S University Little Rock AR 72204-3275

ROLLEFSON, RAGNAR, b Chicago, Ill, Aug 23, 06; m 36; c 4. PHYSICS. *Educ:* Univ Wis, BA, 26, MA, 27, PhD(physics), 30. *Prof Exp:* Asst, physics 27-30, from instr to assoc prof, 30-46, chmn dept, 47-51, 52-56 & 57-61, prof, 46-76, EMER PROF PHYSICS, UNIV WIS-MADISON, 76- *Concurrent Pos:* Mem staff, Radiation Lab, Mass Inst Technol, 42-46; chief scientist, Naval Res Lab, Boston, Mass, 46; mem staff, Proj Charles, Mass Inst Technol, 51, Lincoln Lab, 51-52; tech capabilities panel, Off Defense Mobilization, 54-55; chief scientist, US Army, 56-57; actg dir lab, Midwestern Univs Res Asn, 57-60; dir int sci affairs, Dept State, Washington, DC, 62-64. *Mem:* Fel Am Phys Soc. *Res:* Continuous spectrum of mercury vapor; radar; infrared dispersion of gases. *Mailing Add:* Dept Physics Univ Wis Madison WI 53706

ROLLEFSON, ROBERT JOHN, b Madison, Wis, Sept 9, 41; m 65; c 2. LOW TEMPERATURE PHYSICS, SURFACE PHYSICS. *Educ:* Univ Wis, BA, 63; Cornell Univ, PhD(physics), 70. *Prof Exp:* Res assoc physics, Univ Wash, 70-73; from asst prof to assoc prof, 73-86, PROF PHYSICS, WESLEYAN UNIV, 86- *Mem:* Am Phys Soc; Sigma Xi. *Res:* Properties of absorbed monolayer gas films; use of nuclear magnetic resonance spectroscopy to study the phases existing at various densities and dynamics of phase changes; investigations of the effects of surface structure on film properties. *Mailing Add:* Dept Physics Wesleyan Univ Middletown CT 06457

ROLLER, DUANE HENRY DUBOSE, b Eagle Pass, Tex, Mar 14, 20; m 42; c 1. HISTORY OF SCIENCE. *Educ:* Columbia Univ, AB, 41; Purdue Univ, MS, 49; Harvard Univ, PhD(hist sci & learning), 54. *Prof Exp:* Asst physics, Purdue Univ, 46-49; vis fel gen educ, Harvard Univ, 49-50, teaching fel, 51-54; asst prof hist sci & cur DeGolyer Collection, 54-58, assoc prof hist sci, 58-63, asst dir, Univ Okla Libr, 71-79, CUR HIST SCI COLLECTIONS, UNIV OKLA LIBR, 59-, MCCASLAND PROF, UNIV OKLA, 63-, DAVID ROSS BOYD PROF, 81- *Concurrent Pos:* Vis prof, Univ Colo, 60 & Univ Ore, 66; NSF sr fel, 61-62; mem, Nat Comt Hist & Philos Sci, 60-61, 63-68, chmn, 64-65; res fel, Am Sch Classical Studies, 69-70 & 77-78; Sigma Xi nat lectr, 77-78 & 78-82; NSF Chautauqua lectr, 78-79. *Mem:* Fel AAAS; Hist Sci Soc; corresp mem Int Acad Hist Sci; Sigma Xi. *Res:* Bibliography of the history of science; history of Greek and Renaissance science. *Mailing Add:* Hist Sci Collections Univ Okla 401 W Brooks Norman OK 73019

ROLLER, HERBERT ALFRED, b Magdeburg, Ger, Aug 2, 27; m 57. ZOOLOGY, BIOLOGICAL CHEMISTRY. *Educ:* Georg August Univ, Gottingen, PhD(zool), 62. *Prof Exp:* Proj assoc zool, Univ Wis-Madison, 62-65, asst prof pharmacol, 65-66, res assoc zool, 66-67, assoc prof, 67-68; prof biol, 68-77, head invert res, Inst Life Sci, 68-73, alumni prof, 80-85, dir inst develop biol, 73-83, prof biochem & biophys, 74-83, DISTINGUISHED PROF BIOL, TEX A&M UNIV, 77-; CHIEF SCIENTIST, ZOECON RES INST, PALO ALTO, CALIF, 85- *Concurrent Pos:* Vpres & dir res, Zoecon Corp, Calif, 68-72, sci adv, 72-85; mem adv panel regulatory biol, NSF, 69-72; res dir, Int Ctr Insect Physiol & Ecol, Nairobi, Kenya, 70-75; consult, Syntex Res Div, Calif, 67-68. *Mem:* AAAS; Am Soc Zool; Entom Soc Am; Soc Develop Biol; Sigma Xi; Int Ctr Insat Physiol & Ecol. *Res:* Physiology and biochemistry of morphogenetic hormone and neuropeptide systems in insects. *Mailing Add:* 824 N Rosemary Dr Bryan TX 77802

ROLLER, MICHAEL HARRIS, b Soldier, Kans, Apr 20, 22; m 44; c 3. ANIMAL PHYSIOLOGY. *Educ:* Kans State Univ, BS & DVM, 50, PhD(physiol), 66; cert biophys, Baylor Univ, 63. *Prof Exp:* Private practice, 50-61; Nat Defense Educ Act fel, 61-64; instr surg & med, Col Vet Med, Kans State Univ, 64, instr physiol, 65-66; USPHS res fel, 64-65; assoc prof, 66-73, prof entom & zool, 73-79, prof vet sci, SDak State Univ, 79-86; RETIRED. *Concurrent Pos:* Moorman res grant zool, 68-69. *Mem:* Am Vet Med Asn; Sigma Xi. *Res:* Ammonia intoxication in cattle, sheep and rabbits, especially blood and tissue changes and reproductive performance. *Mailing Add:* 1011 Forest St Brookings SD 57006

ROLLER, PAUL S, chemistry, for more information see previous edition

ROLLER, PETER PAUL, b Debrecen, Hungary, Nov 16, 40; US citizen; m 67; c 3. PEPTIDE SYNTHESIS, ANTI-TUMOR DRUG DEVELOPMENT. *Educ:* Univ BC, BSc, 63, MSc, 65; Stanford Univ, PhD(org chem), 69. *Prof Exp:* NIH fel, Univ Hawaii, 70 & Univ Va, 71-72; RES CHEMIST, NAT CANCER INST, NIH, 72- *Mem:* Am Chem Soc; AAAS; Protein Soc. *Res:* Design and chemical synthesis of peptides, drugs, growth factors, antitumor and antiviral agents; conformational analysis of proteins; spectroscopic analysis of peptides, carcinogens, natural products and hormones; mass spectrometry applications. *Mailing Add:* Lab Medicinal Chem NIH Bldg 37 Rm 5C-02 9000 Rockville Pike Bethesda MD 20892

ROLLER, WARREN L(EON), b Logansport, Ind, May 31, 29; m 51; c 5. AGRICULTURAL ENGINEERING. *Educ:* Purdue Univ, BS, 51, MS, 55, PhD(agr eng), 61. *Prof Exp:* Res fel agr eng, Purdue Univ, 53-54; jr agr engr, Agr Res Serv, USDA, 54-55; from instr to asst prof, Ohio Agr Exp Sta, 55-63, assoc prof, 63-68, assoc chmn dept, 68-81, chmn dept, 81-87, coord prog develop, 87-88, PROF AGR ENG, OHIO AGR RES & DEVELOP, 68-, ASSOC CHMN DEPT, 88- *Concurrent Pos:* Nat Acad Sci-Nat Res Coun sr vis scientist, US Army Res Inst Environ Med, 67- *Honors & Awards:* Jour Award, Am Soc Agr Engrs, 64, 70, 78 & 84, Nat Award, 75. *Mem:* Sigma Xi; Am Soc Agr Engrs. *Res:* Automatic feeding systems for animal production; environmental control for animal production; effect of thermal environment upon animal reproduction; energy efficiencies in agricultural production systems; biomass production for fuel. *Mailing Add:* 873 Ashwood Dr Wooster OH 44691

ROLLESTON, FRANCIS STOPFORD, b Montreal, Que, June 1, 40; m 64; c 3. BIOCHEMISTRY. *Educ:* Queen's Univ, Ont, BSc, 62; Oxford Univ, DPhil(biochem), 66. *Prof Exp:* Fel physiol, Univ Chicago, 66-67, res assoc biochem, 67-68; asst prof, Banting & Best Dept Med Res, Univ Toronto, 68-75; asst dir grants prog, 75-77, Dir Sp Prog, 77-83, Dir Pub Affairs, 83-86, DIR SCI EVAL, MED RES COUN CAN, 86- *Concurrent Pos:* Med Res Coun Can oper grants, 68-75. *Mem:* Can Biochem Soc. *Res:* Operation of peer review processes; guidelines for ethical research involving human subjects and animals. *Mailing Add:* Med Res Coun 20th Fl Jeanne Mance Bldg Tunney's Pasture Ottawa ON K1A 0W9 Can

ROLLETSCHEK, HEINRICH FRANZ, recursive function theory, for more information see previous edition

ROLLINGER, CHARLES N(ICHOLAS), b Chicago Heights, Ill, Aug 5, 34; m 57; c 5. COMPUTER SYSTEMS, TECHNICAL MANAGEMENT. *Educ:* Univ Detroit, BME, 57; Northwestern Univ, MS, 59, PhD(mech eng), 61. *Prof Exp:* Res engr, Roy C Ingersoll Res Ctr, Borg-Warner Corp, 60-61; from instr to asst prof math, USAF Acad, 61-63; res engr, Frank J Seiler Res

Lab, Off Aerospace Res, USAF, 64; sr res engr, Res Labs, Whirlpool Corp, 64-66, corp mgr eng & sci comput, 66-68, dir comput sci & technol, 68-72, dir eng res, 72-74; DIR COMPUT SERV, BERRIEN COUNTY GOVT, 75- *Concurrent Pos:* Lectr, Univ Denver, 64; lectr, Univ Notre Dame, 66, adj asst prof mgt, 78-82; sr lectr, Mich State Univ, 64-66; consult, R C Ingersoll Res Ctr, Borg-Warner Corp, 61-64; adj assoc prof indust eng, Univ Mich, 70-71. *Mem:* Opers Res Soc Am; Inst Mgt Sci; Am Soc Mech Engrs. *Res:* Information systems design; management science. *Mailing Add:* 867 Tucker Dr St Joseph MI 49085

ROLLINO, JOHN, b Brooklyn, NY, Oct 11, 44; m 70; c 2. PHYSICAL CHEMISTRY. *Educ:* St Francis Col, NY, BS, 66; Mass Inst Technol, PhD(chem), 69. *Prof Exp:* Asst prof chem & physics, St Francis Col, NY, 69-84; PROF CHEM & PHYSICS, UPSALA COL, 84- *Concurrent Pos:* Consult, PIC Corp, Orange, NJ. *Mem:* Am Chem Soc; Am Inst Physics; Am Asn Physics Teachers. *Res:* Low temperature thermodynamics; solid state charge transfer complexes. *Mailing Add:* 45 Wells Ct Bloomfield NJ 07003-3042

ROLLINS, HAROLD BERT, b Hamilton, NY, Feb 1, 39; m 60; c 1. INVERTEBRATE PALEONTOLOGY, PALEOECOLOGY. *Educ:* Colgate Univ, BA, 60; Univ Wis-Madison, MA, 63; Columbia Univ, PhD(geol, invert paleont), 67. *Prof Exp:* NSF-Great Lakes Col Asn teaching intern earth sci & biol, Antioch Col, 67-68, asst prof, 68-69; asst prof, 69-73, ASSOC PROF EARTH SCI, UNIV PITTSBURGH, 73- *Concurrent Pos:* Res assoc, Am Mus Natural Hist, NY & Carnegie Mus, Pa. *Mem:* Paleont Soc; Soc Syst Zool; Geol Soc Am; Brit Paleont Asn; Sigma Xi; Paleont Res Inst. *Res:* Phylogeny and functional morphology of Paleozoic Gastropoda; Devonian paleontology and stratigraphy; Paleozoic Monoplacophora; Pennsylvanian peleoecology. *Mailing Add:* Dept Earth Sci 318 Old Engin Hall Univ Pittsburgh Pittsburgh PA 15260

ROLLINS, HOWARD A, JR, b Dover, NH, July 12, 27; m 52; c 4. HORTICULTURE. *Educ:* Univ Conn, BS, 50; Univ NH, MS, 51; Ohio State Univ, PhD, 54. *Prof Exp:* Assoc prof hort, Winchester Fruit Res Lab, Va Polytech Inst & State Univ, 54-56, prof, 56-67, prof & head dept, 67-70; PROF HORT & CHMN DEPT, OHIO STATE UNIV & OHIO AGR RES & DEVELOP CTR, 70- *Mem:* Am Soc Hort Sci. *Res:* Winter hardiness of apple; apple production, harvest efficiency and tree fruit culture. *Mailing Add:* 1289 Shockey Dr Winchester VA 22601

ROLLINS, ORVILLE WOODROW, b Sybial, WVa, Dec 4, 23; m 47; c 1. INORGANIC CHEMISTRY, ANALYTICAL CHEMISTRY. *Educ:* Univ WVa, MS, 50; Georgetown Univ, PhD(chem), 66. *Prof Exp:* Instr & asst prof chem, Moravian Col Men, 49-51; from asst prof to assoc prof, 51-65, PROF CHEM, US NAVAL ACAD, 65- *Concurrent Pos:* Consult, Chem Div, Air Force Off Sci Res. *Mem:* Am Chem Soc. *Res:* Isopoly and heteropoly molybdates and tungstates; analytical methods. *Mailing Add:* RR 2 No 278 Queenstown MD 21658

ROLLINS, REED CLARK, b Lyman, Wyo, Dec 7, 11; m 39, 78; c 2. BOTANY, GENETICS. *Educ:* Univ Wyo, AB, 33; State Col Wash, SM, 36; Harvard Univ, PhD(bot), 41. *Prof Exp:* From instr to assoc prof biol, Stanford Univ, 40-48; assoc prof bot, 48-54, chmn inst res plant morphol, 55-66, inst plant sci, 65-69, dir Gray Herbarium, 48-78, supvr, Bussey Inst, 67-78, chmn admin comt, Farlow Libr & Herbarium, 74-78, Asa Gray prof syst bot, 54-82, EMER PROF SYST BOT, HARVARD UNIV, 82- *Concurrent Pos:* Asst cur, Dudley Herbarium, Stanford Univ, 40-41, cur, 41-48; assoc geneticist, Guayule Res Proj, USDA, 43-45, geneticist, Div Rubber Plant Invests, 47-48; prin geneticist, Stanford Res Inst, 46-47; ed-in-chief, Rhodora, 50-61; pres, Orgn Trop Studies, Inc, 64-65. *Honors & Awards:* Centenary Medal, French Bot Soc, 54; Cert of Merit, Bot Soc Am, 60; Congress Medal, XI Int Bot Cong, Seattle, 69, XII Int Bot Cong, Leningrad, 75; Gold Seal, Nat Coun State Garden Clubs, 81; Asa Gray Award, Am Soc Plant Taxonomists, 87; 25th Anniversary Medal, Orgn Trop Studies, 88. *Mem:* Nat Acad Sci; Am Acad Arts & Sci; Am Soc Nat (vpres, 60, pres, 66); Genetics Soc Am; Int Asn Plant Taxon (vpres, 50-54, pres, 54-59). *Res:* Cytology and systematics of the Cruciferae; cytogenetics of the guayule rubber plant and related species of Parthenium. *Mailing Add:* Gray Herbarium Harvard Univ 22 Divinity Ave Cambridge MA 02138

ROLLINS, ROGER WILLIAM, b Columbia City, Ind, Jan 23, 39; m 61; c 2. NON-LINEAR DYNAMICS & CHAOS, EXPERIMENTAL SOLID STATE PHYSICS. *Educ:* Purdue Univ, BS, 61; Cornell Univ, PhD(appl physics), 67. *Prof Exp:* From asst prof to assoc prof, 66-78, PROF PHYSICS, OHIO UNIV, 78- *Concurrent Pos:* Vis scientist, Inst Exp Nuclear Physics, Univ Karlsruhe, Ger, 72-73. *Mem:* AAAS; Am Phys Soc; Am Asn Physics Teachers. *Res:* Computer simulation and graphics; non-linear systems; superconductivity-hysteresis effects in type II superconductors, alternating current losses, bulk and surface effects; specific heat and magnetic properties of superconductors. *Mailing Add:* Dept Physics Ohio Univ Athens OH 45701

ROLLINS, RONALD ROY, b Tooele, Utah, Oct 2, 30; m 57; c 5. PHYSICAL CHEMISTRY, EXPLOSIVES. *Educ:* Univ Utah, BS, 59, PhD(metall), 62. *Prof Exp:* Staff mem direct energy conversion, Vallecitos Atomic Lab, Gen Elec Co, Calif, 62-64; from asst prof to assoc prof theory high explosives, Univ Mo-Rolla, 64-79; chmn mineral processing eng dept, 81-84, PROF EXPLOSIVES/MINING, WVA UNIV, 79- *Concurrent Pos:* Sr investr, Rock Mech & Explosives Res Ctr, 64-79. *Mem:* AAAS; Am Inst Mining, Metall & Petrol Engrs; Am Soc Eng Educ; Soc Explosive Engrs. *Res:* Hot wire initiation of secondary explosives; factors that sensitize primary and secondary explosives; theory of high explosives; shaped charge explosive effects; high pressure water jets; underground methane explosions; explosive blast casting. *Mailing Add:* Comer Bldg WVa Univ Morgantown WV 26506-6070

ROLLINS, WADE CUTHBERT, b Jersey City, NJ, Feb 12, 12; m 41; c 1. BIOLOGY. *Educ:* Univ Calif, AB, 33, MA, 35, PhD(genetics), 48. *Prof Exp:* Asst, 45-48, instr & jr animal husbandryman, 48-50, asst prof & asst animal husbandryman, 50-56, assoc prof & assoc animal husbandryman, 56-64, prof & geneticist, 64-78, EMER PROF ANIMAL SCI, UNIV CALIF, DAVIS, 78- *Mem:* Biomet Soc; Am Soc Animal Sci. *Res:* Application of genetics to livestock breeding; population genetics. *Mailing Add:* 442 University Ave Davis CA 95616

ROLLINS-PAGE, EARL ARTHUR, b Portland, Ore, Mar 6, 40; m 84; c 4. DEVELOPMENTAL BIOLOGY. *Educ:* Willamette Univ, BS, 61; Purdue Univ, MS, 64; State Univ NY Buffalo, PhD(develop biol), 67. *Prof Exp:* Fel biochem, C F Kettering Res Lab, 66-67; asst prof biol sci, State Univ NY Albany, 67-72; assoc prof biol, Sangamon State Univ, 72-78; MEM FAC, DEPT BIOL, AQUINAS COL, 78- *Concurrent Pos:* State NY Res Found fac fel, 68, grant-in-aid, 68-69; Am Cancer Soc res grant, 68-70; adj assoc prof, Sch Med, Southern Ill Univ, 73-78. *Mem:* AAAS; Am Soc Cell Biol; Soc Develop Biol; Int Soc Develop Biol; NY Acad Sci; Sigma Xi. *Res:* Structure of chromosomes; Chemical origins of life. *Mailing Add:* Dept Biol Aquinas Col Grand Rapids MI 49506

ROLLMAN, GARY BERNARD, b New York, NY, Nov 9, 41; m 67; c 2. EXPERIMENTAL PSYCHOLOGY. *Educ:* Univ Rochester, BA, 62; Univ Penn, MA, 63, PhD(Psychol), 67. *Prof Exp:* Fel, Princeton Univ, 67-69; PROF PSYCHOL, UNIV WESTERN ONT, 69- *Concurrent Pos:* Vis lectr psychol, Princeton Univ, 68-69; prin investr res grants, Natural Sci & Eng, Res Coun Can, 69-; vis scholar psychol, Univ Stockholm, 75-76; vis prof psychol, Univ St Andrews, Scotland, 82-83; mem, People to People Pain Specialists Deleg, People's Repub China, 84; steering comt, London Pain Interest Group, 85-90; exec comt, Can Pain Soc, 85-90, chmn, nominating comt, 86-88, chmn, local arrangements comt, 89-90, sci prog comt, 86-88 & 89-90; mem, task force special pain problems related women, Int Asn Study Pain, 90- *Mem:* Int Asn Study Pain; Can Pain Soc; Int Soc Psychophysics; Am Psychol Asn; fel Can Psychol Asn; Psychonomic Soc; fel Am Psychol Soc. *Res:* Pain measurement in laboratory and clinical settings; human experimental psychology, sensation and perception, particularly involving the somatosensory system, psychophysics. *Mailing Add:* Dept Psychol Univ Western Ont London ON N6A 5C2 Can

ROLLMAN, WALTER F(UHRMANN), chemical engineering, for more information see previous edition

ROLLMANN, LOUIS DEANE, b Kingman, Kans, Apr 26, 39; m 67; c 3. INORGANIC CHEMISTRY, PETROLEUM. *Educ:* Univ Kans, BA, 60, PhD(inorg chem), 67. *Prof Exp:* NIH fel phys chem, Calif Inst Technol, 67-68; from res chemist to sr res chemist, Mobil Res & Develop Corp, 68-75, assoc chem & prof leader, 75-77, res assoc, 77-80, group mgr, 80-85, mgr, Enhanced Oil Recovery, 85-90, MGR, HEAVY OIL RECOVERY, MOBIL RES & DEVELOP CORP, 90- *Mem:* Am Chem Soc; Soc Petrol Eng. *Res:* Zeolite synthesis; catalysis; metals in petroleum; petroleum processing; petroleum production; enhanced and heavy oil recovery; petroleum recovery; research management. *Mailing Add:* 6235 Pineview Rd Dallas TX 75248-3933

ROLLO, FRANK DAVID, b Endicott, NY, Apr 15, 39; m 60; c 1. PHYSICS, MEDICINE. *Educ:* Harpur Col, BS, 59; Univ Miami, MSc, 65; Johns Hopkins Univ, PhD(physics), 68; State Univ NY Upstate Med Ctr, MD, 72; Am Bd Nuclear Med, dipl, 74. *Prof Exp:* Res physicist, Dept Appl Res, Int Bus Mach Corp, NY, 59-60; assoc prof math & physics, Dept Eng Physics, Broome Tech Community Col, 60-64; radiol consult, Sinai Hosp, Greater Baltimore Med Ctr & Md Gen Hosp, Baltimore, 65-68; res assoc med physics & nuclear med, State Univ NY Upstate Med Ctr, 68-72; resident radiol & nuclear med, Med Ctr, Univ Calif, San Francisco, 72-74, asst prof med & radiol, 74-77; ASSOC PROF RADIOL & DIR NUCLEAR MED, UNIV HOSP, VANDERBILT UNIV, 77- *Concurrent Pos:* Sci lectr, Radiol Health Ctr, Rockville, Md, 66-68; res consult radiol, Duke Univ, 68-72; res consult appl physics lab, Johns Hopkins Univ, 68-; res consult, Univ Tex, Galveston, 73-; assoc dir nuclear med, Vet Admin Hosp, San Francisco, 74-77 & Vet Admin Hosp, Nashville, 77- *Honors & Awards:* Achievement Award, Am Math Soc, 63; Bronze Medal, Soc Nuclear Med, 75. *Mem:* Asn Physicists Med; Am Math Soc; Soc Nuclear Med; Asn Univ Radiologists; Radiol Soc NAm. *Res:* Calorimetry of Grenz rays; pulse height selection in scintigraphic imaging systems; frequency response analysis of imaging systems; determination of organ volumes with imaging systems; depth correction in organ imaging; thyroid uptake methodology; radionuclide evaluation of renal and cardiac function. *Mailing Add:* 500 W Main St PO Box 1438 Louisville KY 40201

ROLLO, IAN MCINTOSH, b Aberdeen, Scotland, May 28, 26; m 49; c 1. PHARMACOLOGY, MICROBIOLOGY. *Educ:* Aberdeen Univ, BSc, 45; Univ Man, PhD(pharmacol), 68. *Prof Exp:* Exp officer chem, Ministry of Food, Brit Civil Serv, 46-48; res asst chemother, Sch Trop Med, Univ Liverpool, 48-49; res scientist, Wellcome Labs Trop Med, London, 49-58; res scientist, Distillers Co, Ltd, 58-61; PROF PHARMACOL, FAC MED, UNIV MAN, 61- *Mem:* Am Soc Pharmacol & Exp Therapeut; Pharmacol Soc Can; Am Soc Trop Med & Hyg. *Res:* Chemosensitivity testing of human tumors; optimizing production of single-cell suspensions and increasing cloning efficiency in the culture of clonogenic tumor cells. *Mailing Add:* Dept Oral Biol & Pharmacol Univ Man Winnipeg MB R3T 2N2

ROLLOSSON, GEORGE WILLIAM, b Lake Charles, La, Oct 13, 23; m 58; c 2. PHYSICS. *Educ:* Univ Southwestern La, BS, 45; Mass Inst Technol, MS, 47; Univ NMex, PhD(physics), 52. *Prof Exp:* Asst prof physics, Univ Southwestern La, 47-48; staff mem physics instrumentation, Sandia Corp, NMex, 51-55, div supvr instrumentation, 55-65; group head, Field Exp Dept, Stanford Res Inst, 65-66, sr res engr, 67-69; chmn, dept sci & eng, 71-74, dean sci & math, 74-78, dean letters & sci, 78-81, PROF PHYSICS, MENLO COL, 69- *Concurrent Pos:* Lectr, Univ NMex, 60, part-time prof, 61; consult,

Thomas Bede Found, 70, Opers Res Assocs, 73; fel, Naval Res Lab, 84 & 85. *Mem:* Am Asn Physics Teachers. *Res:* Instrumentation; field experimentation; computers in education. *Mailing Add:* 709 San Conrado Terr No 4 Sunnyvale CA 94086

ROLLWITZ, WILLIAM LLOYD, b Dooley, Mont, Apr 28, 22; m 43; c 2. MAGNETIC RESONANCE, ELECTRONIC INSTRUMENTATION. *Educ:* Mass Inst Technol, BS, 50, MS, 52. *Prof Exp:* Res engr, Philco Corp, 49-52; res physicist, Dept Instrumentation Res, Southwest Res Inst, 51-59, mgr, Electronic Instrumentation Sect, 59-70, staff scientist, 70-74, INST SCIENTIST, SOUTHWEST RES INST, 74- *Concurrent Pos:* Lectr, San Antonio Col, 54- *Honors & Awards:* Swearingen Award, NASA, 54; Imagineer Award, Mind Sci Found, San Antonio, 87. *Mem:* Instrument Soc Am; Inst Elec & Electronics Engrs; Am Phys Soc; Am Soc Agr Engrs. *Res:* Biomedical electronics; nuclear magnetic, electron paramagnetic, nuclear quadrupole resonance and zero field nuclear magnetic resonance; transistor electronics; quality control instrumentation; nondestructive testing with magabsorption; resonance detection of hidden explosives; nuclear magnetic resonance and electron magnetic resonance in live animals; magnetic resonance imaging; developing nuclear magnetic resonance devices to measure moisture in materials, to detect explosives and illegal drugs hidden in letters, packages and baggage and nuclear magnetic resonance devices to screen persons for cancer in the breast, colon, prostate, rectum, uterus, cervix and urethra; receipt of 15 patents and 4 patent applications pending. *Mailing Add:* 213 Halbart Dr San Antonio TX 78213

ROLNICK, WILLIAM BARNETT, b Brooklyn, NY, Aug 20, 36; m 62; c 2. PHYSICS & ACOUSTICS, MATHEMATICS. *Educ:* Brooklyn Col, BS, 56; Columbia Univ, AM, 60, PhD(physics), 63. *Prof Exp:* Asst prof physics, US Merchant Marine Acad, 63-64; res assoc, Case Inst Technol, 64-66; asst prof, 66-70, assoc prof, 71-80, PROF PHYSICS, WAYNE STATE UNIV, 81- *Concurrent Pos:* Dir, Res Careers Minority Scholars, Wayne State Univ; Dept Energy res grant, 79-90; res consult, Physics Dept, Univ Mich, 79-89. *Mem:* Am Asn Physics Teachers. *Res:* Elementary particle theory; scattering theory; electromagnetic theory; quantum electrodynamics; tachyons. *Mailing Add:* Dept Physics Wayne State Univ Detroit MI 48202

ROLOFF, MARSTON VAL, b Charles City, Iowa, June 28, 43. TOXICOLOGY. *Educ:* Iowa State Univ, PhD(physiol & pharmacol), 74. *Prof Exp:* Sr scientist, 78-81, sr res group leader, 81-86, CONSULT, INHALATION TOXICOL, MONSANTO CO, 81- *Concurrent Pos:* Cmndg officer, Fleet Hosp Univ, Navy. *Mem:* Am Indust Health Coun; Soc Toxicol; Sigma Xi; Am Physiol Soc; Christian Educators Soc. *Res:* Aerosol research. *Mailing Add:* Two El Caballo St Charles MO 63301

ROLOFSON, GEORGE LAWRENCE, b Lincoln, Nebr, July 16, 38. TOXICOLOGY, ENTOMOLOGY. *Educ:* Univ Nebr, BSc, 61, MSc, 64; Va Polytech Inst, PhD(entom & toxicol), 68. *Prof Exp:* Staff specialist insecticide develop, Ciba-Geigy Corp, 68-70, group leader plant protectants, 70-72, toxicologist, 72-75, sr toxicologist, 75-78, mgr toxicol, 78-79; CONSULT TOXICOL & REGULATORY AFFAIRS, 79-; MGR, FED GOVT RELS, CIBA GEIGY CORP. *Mem:* Entom Soc Am. *Res:* Toxicology required for Federal Insecticide, Fungicide and Rodenticide Act and Toxic Substance Control Act. *Mailing Add:* Agr Div Ciba-Geigy Corp PO Box 18300 Greensboro NC 27419

ROLSTON, CHARLES HOPKINS, b Harrisonburg, Va, July 25, 27; m 53; c 3. INDUSTRIAL ORGANIC CHEMISTRY. *Educ:* Hampden-Sydney Col, BS, 48; Univ Md, MS, 53. *Prof Exp:* Res org chemist, Westvaco Chlorine-Alkali Div, Food Mach & Chem Corp, 52-54; res chem, Eastern Lab, Gibbstown, 56-72, sr res chemist, Exp Sta, 72-81, SR CHEMIST, JACKSON LAB, E I DU PONT DE NEMOURS & CO, INC, 81- *Mem:* Am Chem Soc; AAAS; Sigma Xi. *Res:* Synthetic organic chemistry; developmental research on organic intermediates; application of catalytic processes. *Mailing Add:* Jackson Lab E I du Pont de Nemours & Co Inc Deepwater NJ 08023

ROLSTON, DENNIS EUGENE, b Burke, SDak, June 20, 43; m 69; c 2. SOIL PHYSICS. *Educ:* SDak State Univ, BS, 65; Iowa State Univ, MS, 67; Univ Calif, Davis, PhD(soils), 70. *Prof Exp:* Lab technician, 68-70, asst prof soils, 70-77, assoc prof, 77-81, assoc dean, 81-86, PROF SOILS, UNIV CALIF, DAVIS, 81- *Mem:* Am Soc Agron; fel Soil Sci Soc Am; Int Soc Soil Sci. *Res:* Water movement in soils; diffusion and displacement of ions and gas in soil; denitrification; drip irrigation; contaminant transport in soil. *Mailing Add:* Dept Land Air & Water Univ Calif Davis CA 95616

ROLSTON, KENNETH VIJAYKUMAR ISSAC, b Etah, India, Apr 23, 51; US citizen; m 84; c 1. INFECTIOUS DISEASES. *Educ:* Christian Med Col, India, MB & BS, 72. *Prof Exp:* Intern, Christian Med Col, India, 73-74; Physician, Luteran World Relief Team Cambodia & Vietnam, 74-75; resident internal med, Franklin Square Hosp, Baltimore, 76-78, chief resident, 78-79; staff physician internal med, N Charles Gen Hosp, Baltimore, 79-81; fel infectious dis, Hahnemann Univ, 81-83; instr med, 83-84, ASST PROF MED, UNIV TEX MD ANDERSON HOSP & TUMOR INST, 84- *Concurrent Pos:* Co-investr, Nat Inst Allergy & Infectious Dis, 86- & various corp grants, 86- *Mem:* Am Col Physicians; Am Soc Microbiol; Am Venereal Dis Asn; Asn Gnotobiotics; Am Fed Clin Res; AMA; Infectious Dis Soc Am. *Res:* Evaluation of newer antimicrobial agents in-vitro and clinically in immuno suppressed patients; management of infectious complications in AIDS patients; the recognition and description of new infectious agents in AIDS patients. *Mailing Add:* Infectious Dis Box 47 Univ Tex MD Anderson Cancer Ctr 1515 Holcom Blvd Houston TX 77030

ROLSTON, LAWRENCE H, b Parkersburg, WVa, Apr 14, 22; m 41; c 4. ENTOMOLOGY. *Educ:* Marietta Col, AB, 49; Ohio State Univ, MS, 50, PhD, 55. *Prof Exp:* Jr entomologist & cur, Entom Mus, Univ Ark, 52-55; asst prof entom, Exp Sta, Ohio State Univ, 55-58; from assoc prof to prof, Univ Ark, 58-66, mem exp sta, 58-66; entom specialist, Tex A&M Univ, 66-68; PROF ENTOM, LA STATE UNIV, BATON ROUGE, 68- *Res:* Truck insects; agricultural entomology; taxonomy of pentatomidae. *Mailing Add:* Dept Entom La State Univ 410 Life Sci Bldg Baton Rouge LA 70803

ROLWING, RAYMOND H, b Toledo, Ohio, Mar 22, 31; m 56; c 4. MATHEMATICS. *Educ:* Christian Bros Col, BS, 55; Univ Notre Dame, MS, 58; Univ Cincinnati, PhD(math), 63. *Prof Exp:* Instr math, Christian Bros Col, Tenn, 58-60; from instr to asst prof, 58-70, asst dean, McMicken Col Arts & Sci, 68-71, ASSOC PROF MATH, UNIV CINCINNATI, 70- *Concurrent Pos:* Dir acad year in-serv summer inst sec teachers math & sci, NSF, 65-73; dir leadership develop proj, NSF, 73-75. *Mem:* Am Math Soc; Math Asn Am; Nat Coun Teachers Math. *Res:* Ordinary differential equations; existence and uniqueness theorems; quadratic nonlinear integral equations; calculus of variations; isoperimetric problems; econometrics; learning theories in mathematics; history of mathematics. *Mailing Add:* Dept Math 25-831 Old Chem Univ Cincinnati Cincinnati OH 45221

ROM, ROY CURT, b Milwaukee, Wis, Jan 29, 22; m 50; c 4. HORTICULTURE, POMOLOGY. *Educ:* Univ Wis-Madison, BS, 48, PhD(hort soils), 58. *Prof Exp:* Asst hort, Univ Wis-Madison, 54-58; PROF HORT, UNIV ARK, FAYETTEVILLE, 58- *Concurrent Pos:* Consult, Corp Farms, 70- & USAID prog; vis scientist, Nat Acad Sci, Poland & Czech, 75; sabbatical study, France, 79, Yugoslavia, 83. *Honors & Awards:* Marshall Wilder Medal, Am Pomol Soc, 85. *Mem:* Am Pomol Soc (secy, 80); Int Dwarf Fruit Tree Asn; fel Am Soc Hort Sci. *Res:* Nutrition, physiology, pruning and cultural practices in fruit production; weed control; breeding of apples and peaches; rootstock growth and development. *Mailing Add:* Dept Hort Univ Ark Fayetteville AR 72701

ROMACK, FRANK ELDON, b Jennings, Okla, Dec 26, 24; m 44; c 4. CARDIOVASCULAR DISEASES, ANIMAL PHYSIOLOGY. *Educ:* Univ Mo, BS, 51, MS, 61, PhD, 63. *Prof Exp:* Asst animal husb, Univ Mo, 59-61, from instr to asst prof vet anat, 60-65; asst prof, 65-67, ASSOC PROF VET BIOSCI & VET PROGS AGR, COL VET MED, UNIV ILL, URBANA, 67- *Concurrent Pos:* USAEC contract, 65- *Mem:* Am Asn Vet Anat; World Asn Vet Anat; Am Asn Anat; Am Soc Animal Sci; Soc Study Reproduction; Sigma Xi. *Res:* Endocrine metabolism and function; surgical anesthesia; circulatory dysfunction and tissue culture systems; atherosclerosis. *Mailing Add:* RR 2 Box 234 St Joseph IL 61873

ROMAGNANI, SERGIO, b Grosseto, Italy, Apr 7, 39; m 68; c 2. INTERNAL MEDICINE. *Educ:* Univ Florence, Dr, 64. *Prof Exp:* From asst prof to assoc prof clin immunol, 66-75, prof internal med, 86-89, PROF CLIN IMMUNOL, UNIV FLORENCE, 89- *Mem:* Col Int Allergol; Am Asn Immunologists; Ital Asn Immunol & Immunopath; Ital Asn Allergol & Clin Immunol. *Res:* Allergy immunodeficiency. *Mailing Add:* Via Guido Banti 20/G Florence 50139 Italy

ROMAGNOLI, ROBERT JOSEPH, b Chicago, Ill, Aug 16, 31. ELECTROMAGNETIC WAVES. *Educ:* Ill Inst Technol, BS, 53, MS, 54, PhD(physics), 57. *Prof Exp:* Instr physics, Ill Inst Technol, 57-59; lectr, El Camino Col, 59-60; from asst prof to assoc prof, 60-72, PROF PHYSICS, CALIF STATE UNIV, NORTHRIDGE, 72- *Concurrent Pos:* Consult to indust. *Mem:* Am Phys Soc; Am Asn Physics Teachers. *Res:* Electromagnetic radiation; optical propagation; photovoltaic devices; electron optics; electromagnetic surface plasma waves; nonlinear magneto-optics. *Mailing Add:* Dept Physics & Astron Calif State Univ Northridge CA 91330

ROMAN, ANN, b Tampa, Fla, Sept 8, 45; m 74; c 1. VIROLOGY, MOLECULAR GENETICS. *Educ:* Reed Col, BA, 67; Univ Calif, San Diego, PhD(biol), 73. *Prof Exp:* Fel immunol, Ore Health Sci Ctr, 73-75; from asst prof to assoc prof microbiol & immunol, 75-88, PROF MICROBIOL & IMMUNOL, SCH MED, IND UNIV, 88- *Concurrent Pos:* Vis scientist, dept virol, Weizmann Inst, Israel, 82-83; chmn elect, DNA viruses div, Am Soc Microbiol, 86-87, chmn, 87-88; mem, Cancer Biol & Immunol Contract Rev Comt, 88- *Mem:* AAAS; Am Soc Microbiol; Am Soc Virol. *Res:* Association of human papillomavirus DNA with benign and malignant lesions; interaction of human papillomavirus with keratinocytes; factor determining the intracellular fate of viral DNA. *Mailing Add:* Dept Microbiol & Immunol Ind Univ Sch Med 635 Barnhill Dr Indianapolis IN 46202

ROMAN, BERNARD JOHN, b Kingston, Pa, June 26, 40; m 65; c 3. PHYSICS. *Educ:* Carnegie-Mellon Univ, BSc, 62; Northwestern Univ, PhD(physics), 69. *Prof Exp:* Jr engr electronics, Radio Corp Am, 61, 62 & 63; mem staff magnetic devices, Bell Labs, 69-; MEM STAFF, MICROWAVE SEMICONDUCTOR CORP. *Mem:* Am Phys Soc. *Res:* Fabrication techniques for magnetic domain (bubble) devices. *Mailing Add:* Motorola Aprdl 3501 Ed Bluestein Blvd MS:K-10 Austin TX 78721

ROMAN, GUSTAVO CAMPOS, b Bogota, Columbia, Sept 7, 46; US citizen; m 72; c 2. NEUROEPIDEMIOLOGY. *Educ:* Col Emmanuel d'Alzon, BA, 64; Nat Univ Columbia, MD, 71; Am Bd Psychiat & Neurol, dipl neurol, 83. *Prof Exp:* Neurol resident, Med Ctr Hosp, Univ Vt, Col Med, 75-77, chief resident, 77-78, spec fel Neurol, 78-79; asst prof neurol, Dept Internal Med, Nat Univ Columbia & San Juan de Rios Univ Hosp, 80-81, Univ El Rosario Sch Med & San Jose Univ Hosp, 82-83; from asst prof to prof neurol, Dept Med & Surg Neurol, Tex Tech Univ Health Sci Ctr Sch Med, 83-88, actg chmn dept, 85-88; CHIEF, NEUROEPIDEMIOL BR, NAT INST NEUROL DIS & STROKE, NIH, BETHESDA, MD, 89- *Concurrent Pos:* Asst instr neuroanat, Dept Anat, Nat Univ Columbia, 72-73, asst prof neurol, 80-81; asst prof neurol, Univ El Rosario Sch Med & San Jose Univ Hosp, Bogota, 82-83; staff neurologist, Vet Admin Outpatient Clin, Lubbock, Tex, 83-85; prof staff mem, Lubbock Gen Hosp, 83-89; founding dir, Ctr Sleep Dis, Lubbock Gen Hosp & Tex Tech Univ, 85-88; dir, Electromyography Lab, Dept Med & Surg Neurol, 85-88. *Mem:* Fel Am Acad Neurol; NY Acad Sci; fel Am Col Physicians; Am EEG Soc; Soc Neurosci; AAAS; fel Royal Soc Trop Med & Hyg; Royal Soc Med. *Res:* Neurology; sleep disorders. *Mailing Add:* Neuroepidemiol Br NIH Fed Bldg Rm 714 7550 Wisconsin Ave Bethesda MD 20892

ROMAN, HERSCHEL LEWIS, genetics; deceased, see previous edition for last biography

ROMAN, JESSE, b Cabo Rojo, PR, June 18, 31; m 56; c 1. PLANT NEMATOLOGY. *Educ:* Univ PR, BS, 56; Auburn Univ, MS, 59; NC State Univ, PhD(nematol), 68. *Prof Exp:* Asst nematologist, Univ PR, Rio Piedras, 56-73, nematologist & head dept entom & nematol, 73-77, tech asst to dir, Agr Exp Sta, 77-; RETIRED. *Mem:* Soc Nematol; Orgn Trop Am Nematol (pres, 71); Caribbean Food Crop Soc; Am Soc Agr Sci (pres, 74). *Res:* Taxonomy; morphology; cytology; biology; population dynamics; reproduction and control of plant parasitic nematodes. *Mailing Add:* Three Hortencia No 242 Round Hill Tiugillolto PR 00976

ROMAN, LAURA M, b Chicago, Ill, Aug, 14, 55. MOLECULAR BIOLOGY, NEUROSCIENCES. *Educ:* Smith Col, BA, 77; Yale Univ, PhD(cell biol), 83. *Prof Exp:* Postdoctoral fel, Europ Molecular Biol Lab, 83-86 & Howard Hughes Med Inst, Univ Tex Southwestern, 86-91; ASST PROF BIOL, SCH MED, YALE UNIV, 91- *Mem:* Sigma Xi; Am Soc Cell Biol; AAAS; Fedn Am Soc Exp Biol. *Res:* Effect of retinouic acid on the migration and differentiation of murine neural crest cells. *Mailing Add:* Dept Cellular & Molecular Physiol Sch Med Yale Univ New Haven CT 06510

ROMAN, NANCY GRACE, b Nashville, Tenn, May 16, 25. ASTRONOMY. *Educ:* Swarthmore Col, BA, 46; Univ Chicago, PhD(astron), 49. *Hon Degrees:* DSc, Russell Sage Col, 66, Hood Col, 69, Bates Col, 71 & Swarthmore Col, 76. *Prof Exp:* Asst astron & astrophys, Univ Chicago, 46-48, res assoc, 49-52, instr, 52-54, asst prof, 54-55; astronr, Radio Astron Br, US Naval Res Lab, 55-56, head, Microwave Spectros Sect, 56-57, consult, 58-59; head observational astron prog, NASA Hq, 59-60, chief astron & astrophys progs, 60-61, chief astron & solar physics, 61-63, chief astron, 63-72, chief astron & relativity, 72-79; SR SCIENTIST, ASTRON DATA CTR, GODDARD SPACE FLIGHT CTR, SYSTS TECHNOL CORP, 80- *Concurrent Pos:* Consult, 79-90; trustee, Russel Sage Col, 73-78 & Swathmore Col, 80-84; prof, McDonnell Space Systs Div, 89- *Honors & Awards:* Fed Woman's Award, 62; Except Sci Achievement Award, NASA, 69 & Outstanding Sci Leadership Award, 78; William Randolf Lovelace II Award, Am Astron Soc, 80. *Mem:* Int Union Radio Sci; Int Astron Union; fel Am Astronaut Soc; Am Astron Soc; fel AAAS. *Res:* Spectral classification; stellar motions; photoelectric photometry; space research; astronomy data. *Mailing Add:* 4620 N Park Ave Apt 306W Chevy Chase MD 20815

ROMAN, PAUL, b Budapest, Hungary, Aug 20, 25; nat US; m 47, 62; c 4. THEORETICAL PHYSICS, MATHEMATICAL PHYSICS. *Educ:* Eotvos Lorand Univ, MSc, 47, PhD(physics), 48; Hungarian Acad Sci, DSc, 56. *Prof Exp:* Asst lectr physics, E-tv-s Lorand Univ, Budapest, 47-48, lectr, 48-50, sr lectr, 50-51, sr res worker, 54-56; dept chmn, Tech Pedag Col, Budapest, 51-52; res fel, Moscow State Univ, 52-53; dept chmn, Agr Univ, Budapest, 53-54; lectr, Univ Manchester, 57-61; assoc prof physics, Boston Univ, 60-62, prof, 62-78; DEAN GRAD STUDIES & RES, STATE UNIV NY PLATTSBURGH, 78- *Concurrent Pos:* Vis prof, Mex, 72, 74, 75 & 77 & Max Planck Inst, 76, 78, 80 & 82; assoc ed, J Math Physics, 78-81. *Mem:* Fel Am Phys Soc; Am Asn Physics Teachers; Int Asn Math Physicists; Nat Coun Univ Res Adminrs; Soc Res Admin. *Res:* Theory of elementary particles and quantum field theory; mathematical physics; science education and research administration. *Mailing Add:* 230 W Highland Ave Philadelphia PA 19118

ROMAN, RICHARD J, b Dec 19, 51; m 76; c 2. ROLE OF KIDNEY IN HYPERTENSION. *Educ:* Univ Tenn, PhD(pharmacol), 77. *Prof Exp:* From asst prof to assoc prof, 81-90, PROF PHYSIOL, MED COL WIS, 90- *Mem:* Am Physiol Soc; Am Soc Nephrology. *Res:* Renal physiology; hypertension; renal microcirculation. *Mailing Add:* Dept Physiol Med Col Wis Milwaukee WI 53226

ROMAN, STANFORD A, JR, b New York, NY, Nov 19, 42; div; c 2. EDUCATION ADMINISTRATION, MEDICINE. *Educ:* Dartmouth Col, AB, 64; Columbia Univ, MD, 68; Univ Mich, MPH, 75. *Prof Exp:* Assoc dir ambulatory care, Harlem Hosp Ctr, 72-73; dir clin serv, Healthco Inc, 73-74; asst prof med, Sch Med, Univ NC, 73-74; dir ambulatory care, Boston City Hosp, 74-78; asst prof med & sociomed sci, Sch Med, Boston Univ, 74-78, asst dean, 75-78; med dir, DC Gen Hosp, 78-81; assoc dean, Dartmouth Med Sch, 81-86, dep dean, 86-87; dean & vpres acad affairs, Mosehoabe Sch Med, 87-89; sr vpres med & prof affil, NY Hosp Corp, 89-90; DEAN & PROF, CITY UNIV NEW YORK MED SCH, 90- *Concurrent Pos:* Lectr, Sch Pub Health, Harvard Univ, 77-78; prin investr, Boston Comprehensive Sickle Cell Ctr, 75-78; proj dir, Southeast Washington, DC Primary Care Network, 78-91; bd mem, Nat Bd Med Engrs, 88-92; grant rev comt, Nat Health Prom & Dis Prev Ctr, 88-; nat adv group, Assessory Charge Med Educ, 89-; comn mem, NY State Coun Grad Med Educ, 91- *Mailing Add:* 101 W 87th St New York NY 10024

ROMANI, ROGER JOSEPH, b Sacramento, Calif, Dec 17, 19; m 59; c 4. PLANT PHYSIOLOGY, BIOCHEMISTRY. *Educ:* Univ Calif, BS, 51, PhD(plant physiol, biochem), 55. *Hon Degrees:* Dr, Nat Polytech Inst, Toulevre, France. *Prof Exp:* Asst, Univ Calif, Los Angeles, 51-55, asst food scientist, 57-59, lectr pomol, 70-74, ASSOC POMOLOGIST, UNIV CALIF, DAVIS, 59-, PROF POMOL, 74- *Concurrent Pos:* Mem food irradiation adv comt, AEC-Am Inst Biol Sci, 61-63; chmn, Gorden Res Conf Postharvest Physiol, 82. *Mem:* AAAS; Am Soc Plant Physiol; Am Soc Hort Sci. *Res:* Cellular aspects of maturation and senescence; radiation biochemistry; mitochondrial physiology. *Mailing Add:* 2727 Russell Blvd Davis CA 95616

ROMANKIW, LUBOMYR TARAS, b Zhowkwa, Ukraine, Apr 17, 31; Can citizen. MATERIALS ENGINEERING, ELECTROCHEMISTRY. *Educ:* Univ Alta, BSc, 55; Mass Inst Technol, MSc & PhD(metall), 62. *Prof Exp:* Mem res staff electrochem, Thomas J Watson Res Ctr, 62-63, mgr magnetic mat group, Components Div, 63-64, mem res staff mat & processes, 65-68, mgr magnetic mat & devices, 68-78, consult, IBM E Fishkill Develop Lab & Mfg, 78-80, MGR MAT & PROCESS STUDIES, THOMAS J WATSON RES CTR, IBM CORP, 81- *Concurrent Pos:* Instr, Mass Inst Technol, 59-61. *Mem:* Electrochem Soc (secy-tres, 79-80); Am Electroplaters Soc; Inst Elec & Electronics Engrs; Sigma Xi. *Res:* Magnetic thin films; deposition of thin films; dielectrics; magnetic device design; material selection and fabrication; electrodeposition; magnetic materials; electronic and magnetic device fabrication; chemical engineering; metallurgy. *Mailing Add:* IBM Res Corp Old Orchard Rd Armonk NY 10504

ROMANKO, RICHARD ROBERT, b Cortland, NY, Dec 18, 25; m 54; c 3. PLANT PATHOLOGY. *Educ:* Univ NH, BS, 53; Univ Del, MS, 55; La State Univ, PhD(plant path), 57. *Prof Exp:* Asst fruit fungicides, Univ Del, 53-55; asst plant pathologist, Exp Sta, Univ Idaho, 57-65, assoc plant pathologist, 65-84; EXT AGRONOMIST, IDACES, 84. *Concurrent Pos:* Sabbatical leave, Univ Calif, Davis, 67-68; mem, Western Alfalfa Improvement Conf. *Mem:* AAAS; Potato Asn Am; Am Phytopath Soc; Sigma Xi. *Res:* Physiology of parasitism; microbial genetics; hop diseases; verticillium diseases; alfalfa management; hop breeding; verticillium physiology. *Mailing Add:* Box 864 Parma ID 83660

ROMANO, ALBERT, b New York, NY, Feb 2, 27. MATHEMATICAL STATISTICS. *Educ:* Brooklyn Col, BA, 50; Wash Univ, St Louis, MA, 54; Va Polytech Univ, PhD, 61. *Prof Exp:* Asst prof math, Ariz State Univ, 58-60; staff statistician, Semiconductor Prod Div, Motorola Inc, 60-63; NSF res assoc, Nat Bur Standards, 62-63; assoc prof, 63-78, PROF MATH, SAN DIEGO STATE UNIV, 78- *Mem:* AAAS; Am Math Soc; Am Statist Soc; Inst Math Statist. *Res:* Evolution and population genetics; statistical models. *Mailing Add:* Dept Math San Diego State Univ San Diego CA 92182

ROMANO, ANTONIO HAROLD, b Penns Grove, NJ, Mar 6, 29; m 53; c 3. MICROBIOLOGY. *Educ:* Rutgers Univ, BS, 49, PhD(microbiol), 52. *Prof Exp:* Assoc microbiologist, Ortho Res Found Div, Johnson & Johnson, 52-54; instr microbial biochem, Rutgers Univ, 54-56; from sr asst scientist to sr scientist, Taft Sanit Eng Ctr, USPHS, Ohio, 56-59; from assoc prof to prof bact, Univ Cincinnati, 59-71; head microbiol sect, 74-84, PROF BIOL, UNIV CONN, STORRS, 71- *Concurrent Pos:* NSF sr fel, Univ Leicester, 67-68; vis fel, Cambridge Univ, 79; prog dir cell biol, NSF, 84-85. *Mem:* AAAS; Am Soc Microbiol. *Res:* Microbial physiology and biochemistry; regulation of sugar uptake in microorganisms and mammalian cells. *Mailing Add:* Dept Molecular & Cell Biol Univ Conn Storrs CT 06269-3044

ROMANO, JOHN, b Milwaukee, Wis, Nov 20, 08; m 33; c 1. NEUROLOGY. *Educ:* Marquette Univ, BS, 32, MD, 34. *Hon Degrees:* DSc, Med Col Wis, 71, Hahnemann Med Col, 74 & Univ Cincinnati, 79. *Prof Exp:* Extern psychiat, Milwaukee County Asylum Ment Dis, 32-33; intern med, Milwaukee County Hosp, 33-34; asst psychiat, Sch Med, Yale Univ, 34-35; Commonwealth Fund fel, Sch Med, Univ Colo, 35-38; Rockefeller fel neurol, Harvard Med Sch, 38-39, Freud fel, 39-42, asst med, 39-40, instr, 40-42; prof psychiat, Col Med, Univ Cincinnati, 42-46; prof, 46-68, chmn dept, 46-71, distinguished univ prof, 68-79, EMER DISTINGUISHED UNIV PROF PSYCHIAT, SCH MED & DENT, UNIV ROCHESTER, 79- *Concurrent Pos:* Intern & asst resident, New Haven Hosp, 34-35; asst psychiatrist, Colo Psychopath Hosp, Denver, 35-38; fel, Boston City Hosp, 38-39; assoc, Peter Bent Brigham Hosp, 39-42; dir dept psychiat, Cincinnati Gen Hosp, 42-46; mem nat adv ment health coun, USPHS, 46-49, chmn ment health career investr selection comt, 56-61; consult, NIMH, 46-61; psychiatrist-in-chief, Strong Mem Hosp, 46-71; mem adv comt, Behav Sci Div, Ford Found, 55-58; Commonwealth Fund advan fel Europ study, 59-60. *Mem:* Sr mem Inst Med-Nat Acad Sci; Am Soc Clin Invest; Am Epilepsy Soc; fel Am Psychiat Asn; Am Neurol Asn; Psychosom Soc. *Res:* Medical and psychiatric education; studies of schizophrenic patients and their families; general clinical psychiatry; psychosomatic medicine. *Mailing Add:* Dept Psychiat Sch Med & Dent Univ Rochester Rochester NY 14642

ROMANO, PAULA JOSEPHINE, b Rochester, NY, Mar 19, 40. IMMUNOLOGY, MICROBIOLOGY. *Educ:* Cath Univ Am, AB, 61; Duke Univ, PhD(microbiol, immunol), 74. *Prof Exp:* USPHS fel immunol, Nat Cancer Inst, 75-76; instr, Georgetown Univ, 76-79; dir, Histocompatibility Lab, Found Blood Res, 79-81; DIR, HISTOCOMPATIBILITY LAB, MILTON S HERSHEY MED CTR, 81- *Mem:* Am Soc Histocompatibility & Immunogenetics; Am Soc Microbiol; AAAS; Transplantation Soc. *Res:* Understanding of mechanisms of cellular immune reactions; role of histomcompatibility cell surface antigens in disease processes. *Mailing Add:* Dept Path & Pharmacol Milton S Hershey Med Ctr PO Box 850 Hershey PA 17033

ROMANO, SALVATORE JAMES, b Highland Park, NJ, June 2, 41; m 63; c 2. ANALYTICAL CHEMISTRY. *Educ:* Mt St Mary's Col, BS, 63; Rutgers Univ, PhD(analytical chem), 68. *Prof Exp:* Res chemist, Colgate-Palmolive Res Lab, 68-71; prin scientist & supvr, Ethicon Inc, 71-74, mgr, 74-78, assoc dir res, 78-79; vpres res & develop, Devro Inc, 79-82; dir qual anal, Ortho Pharm, 81-84; VPRES CORP QUAL ANALYSIS, JOHNSON & JOHNSON INC, 84- *Mem:* Am Chem Soc; Sigma Xi; Am Soc Qual Control. *Res:* Chemical separations, particularly optical isomer separations; high pressure liquid chromatography; pharmaceutical and biotechnical quality control; quality assurance. *Mailing Add:* 1595 Kearney Dr N Brunswick NJ 08902

ROMANOFF, ELIJAH BRAVMAN, b Clinton, Mass, Feb 15, 13; m 42. PHYSIOLOGY. *Educ:* Worcester Polytech Inst, BS, 34, MS, 36; Tufts Univ, PhD, 52. *Prof Exp:* Jr chemist, Commonwealth Mass, 35-38; jr chemist, Worcester State Hosp, 38-40, asst chemist, 40-41, dir lab, 46-47; res assoc, Worcester Found Exp Biol, 47-60, scientist, 60-62, assoc dir training prog physiol reproduction, 62-68, sr scientist, 62-69; prog dir metab biol, NSF, 69-; CONSULT. *Concurrent Pos:* Chemist, Texol Chem Works, 35-41; instr, Univ Exten, State Dept Educ, Mass, 38-42; vis prof, Med Sch, Univ PR, 57-60; lectr, Brown Univ, 65-69. *Mem:* AAAS; Am Physiol Soc; NY Acad Sci; Brit Soc Endocrinol. *Res:* Steroid metabolism; reproduction. *Mailing Add:* 70 Prospect St Shrewsbury MA 01545

ROMANOVICZ, DWIGHT KEITH, b Newport News, Va, Sept 1, 48. CELL BIOLOGY, ENZYME CYTOCHEMISTRY. *Educ:* Dickinson Col, BS, 70; Univ NC, Chapel Hill, PhD(bot), 75. *Prof Exp:* Res assoc cytochem, Dent Res Ctr, Sch Dent, Univ NC, 75-78; asst prof, 78-82, ASSOC PROF BIOL, WGA COL, 82- *Mem:* Am Soc Cell Biol; Electromicro Soc Am; Am Soc Plant Physiol; Phycological Soc Am. *Res:* Cytochemical localization of plant carbonic anhydrase; ultrastructural investigation of cell wall formation in algae; identification and distribution of scaled algae. *Mailing Add:* Dept Biol WGa Col Carrollton GA 30118

ROMANOW, LOUISE ROZAK, b Boston, Mass, July 15, 50. VECTOR-PLANT VIRUS RELATIONSHIPS, PLANT RESISTANCE. *Educ:* Univ Mass, Amherst, BS, 76; NC State Univ, MS, 80, PhD(entom), 84. *Prof Exp:* Teaching asst ecol, dept zool, NC State Univ, 76-77, res asst, dept entom, 77-84; ED PROG DEVELOP, NC EXTEN SERV, 81- *Concurrent Pos:* Vis fel, Inst Hort Plant Breeding, Netherlands, 84-85. *Mem:* Entom Soc Am; Am Phytopathological Soc. *Res:* Relationship of aphid vectors to nonpersistently transmitted viruses and their plant hosts; effects of plant resistance to vector and virus epidemiology; modeling virus epidemiology based on vector movement and characteristics. *Mailing Add:* 1010 Reedy Creek Rd Cary NC 21513

ROMANOWSKI, CHRISTOPHER ANDREW, b London, Eng, July 23, 53. ALUMINUM ALLOY PROCESSING. *Educ:* Univ Surrey, Eng, BS Hons, 75, PhD (metall), 81. *Prof Exp:* Res assoc mat eng, 79-80, asst prof, Drexel Univ, 80-82; res engr, Pechiney/Howmet, 82-83; tech mgr, Alumax, 83-84; mgr res & develop, 84-86, tech dir, 86-88, SALES MGR, HUNTER ENG, 88- *Concurrent Pos:* Consult, Alcoa Tech Ctr, 79-82. *Mem:* Metall Soc; Am Soc Metals; Inst Metallurgists. *Res:* Melting, degassing, filtration, continuous strip casting, rolling and annealing of aluminum alloys; development and application of strip cast and powder metallurgy aluminum alloys; design and engineering of aluminum plants. *Mailing Add:* Hunter Eng PO Box 5677 Riverside CA 92517

ROMANOWSKI, ROBERT DAVID, b Chicago, Ill, Oct 4, 31; m 51, 74; c 7. BIOCHEMISTRY. *Educ:* SDak State Univ, BS, 56, MS, 57; Purdue Univ, PhD(biochem), 61. *Prof Exp:* Asst, Purdue Univ, 57-61; asst prof vet biochem, Vet Res Lab, Mont State Univ, 61-66; res chemist, Beltsville Parasitol Lab, USDA, 66-72, res biochemist, Animal Parasitol Inst, Nat Agr Res Ctr, 72-88; STAFF MEM, LIVESTOCK POULTRY SCI INST, 88- *Mem:* Am Soc Parasitol. *Res:* Disease resistance of plants; urinary calculi; biochemistry, enzymology and immunology of nematodes; separation and isolation of antigens from nematodes. *Mailing Add:* Livestock Poultry Sci Inst Agr Res Serv USDA Agr Res Serv Bldy 1040 Rm 5 Beltsville MD 20705

ROMANOWSKI, THOMAS ANDREW, b Warsaw, Poland, Apr 17, 25; nat US; m 52; c 2. EXPERIMENTAL HIGH ENERGY PHYSICS. *Educ:* Mass Inst Technol, BS, 52; Case Inst Technol, MS, 55, PhD(physics), 57. *Prof Exp:* Physicist, Nat Bur Standards, Washington, DC, 52; instr, Case Inst Technol, 55-56; res physicist, Carnegie Inst Technol, 57-60; physicist, High Energy Physics Div, Argonne Nat Lab, 60-78; PROF PHYSICS, OHIO STATE UNIV, 64- *Mem:* Fel Am Phys Soc. *Res:* Nuclear physics; photonuclear reactions; radio propagation and upper atmospheric physics. *Mailing Add:* Smith Lab Ohio State Univ 174 W 18th Ave Columbus OH 43210

ROMANS, JAMES BOND, b Monroe, Iowa, Jan 24, 14; m 39; c 3. PHYSICAL CHEMISTRY. *Educ:* Cent Col, Iowa, BA, 35. *Prof Exp:* Teacher sci, Beaver Consol Sch, Iowa, 36-38; chemist oils & fuels, Nat Bur Standards, 38-46; chemist phys & surface chem, Naval Res Lab, Dept Navy, 46-80; consult, fire suppressive fluids, chem agent protection & fire extinguishants, 80-88; RETIRED. *Mem:* Am Chem Soc; Sigma Xi; AAAS. *Res:* Surface analysis; wettability of surfaces; lubricants; dielectric liquids; friction and wear; fatigue resistance of reinforced plastics in water; desensitization of explosives; frictional electrification of polymers; gas adsorption properties of charcoal; fire suppressive fluids; chemical agent protection; powder-type fire extinguishing agents. *Mailing Add:* 9111 Louis Ave Silver Spring MD 20910

ROMANS, JOHN RICHARD, b Montevideo, Minn, Mar 4, 33; m 56; c 4. ANIMAL SCIENCE, BIOCHEMISTRY. *Educ:* Iowa State Univ, BS, 55; SDak State Univ, MS, 64, PhD(animal sci), 67. *Prof Exp:* Asst mgr agr serv dept, John Morrell & Co, 55-56, hog buyer, 58-62; res asst meat & animal sci, SDak State Univ, 62-67; asst prof animal sci, Univ Ill, Urbana, 67-73, assoc prof, 73-; MEM STAFF, ANIMAL SCI DEPT, SDAK STATE UNIV. *Mem:* Am Soc Animal Sci; Am Meat Sci Asn; Inst Food Technol. *Res:* Increase efficiency of high-quality nutritious food production via meat animals. *Mailing Add:* Animal Sci Bldg SDak State Univ Brookings SD 57006

ROMANS, ROBERT CHARLES, b Hawthorne, Wis, Oct 12, 37; m 63; c 1. BOTANY. *Educ:* Univ Wis-Superior, BS, 65, MST, 66; Ariz State Univ, PhD(bot), 69. *Prof Exp:* Asst biol, Univ Wis-Superior, 65-66; asst prof, 69-75, ASSOC PROF BIOL, BOWLING GREEN STATE UNIV, 75- *Concurrent Pos:* Ohio Biol Surv grant, 71-72. *Mem:* Bot Soc Am; Am Asn Stratig Palynologists; Int Soc Plant Morphologists. *Res:* Palynology; paleobotany; plant anatomy and morphology. *Mailing Add:* Dept Biol Sci Bowling Green State Univ Bowling Green OH 43403

ROMANS, ROBERT GORDON, b Waverley, NS, Aug 19, 09; wid; c 3. PROTEIN CHEMISTRY. *Educ:* Univ Toronto, BA, 33, MA, 34, PhD(chem), 42. *Prof Exp:* Asst chem, Univ Toronto, 33-38; tech asst, Connaught Med Res Labs, 38-39, res asst, 39-46, res assoc, 46-61, res mem, 61-73, dir, Insulin Div, 73-74; consult, 74-78; RETIRED. *Mem:* Fel Chem Inst Can. *Res:* Raman effect; photochemistry; insulin. *Mailing Add:* Nine Tresillian Rd Downsview ON M3H 1L5 Can

ROMANS-HESS, ALICE YVONNE, b Wayne, WVa, June 27, 47. BIOPHYSICAL CHEMISTRY. *Educ:* Marshall Univ, BS, 69; Duke Univ, PhD(biophys chem), 74. *Prof Exp:* Instr biochem, Med Ctr, Univ Ala, 74-77; res scientist biochem, 77-81, prod develop scientist, 81-85, SR RES SCIENTIST ADVAN PROD DEVELOP, KIMBERLY-CLARK CORP, 85 - *Mem:* Am Chem Soc; Biophys Soc; Sigma Xi. *Res:* Biological membranes and protein-lipid interactions; plasma lipoproteins; physical chemistry of biological macromolecules. *Mailing Add:* Kimberly-Clark Corp PO Box 999 Neenah WI 54957-0999

ROMANSKY, MONROE JAMES, b Hartford, Conn, Mar 16, 11; m 43; c 4. INTERNAL MEDICINE. *Educ:* Univ Maine, AB, 33; Univ Rochester, MD, 37. *Prof Exp:* Intern med, Strong Mem Hosp, NY, 37-38, asst resident, 38-39; Gleason res fel, Univ Rochester, 39-40; chief resident, Strong Mem Hosp, 40-41; instr med, Off Sci Res & Develop Proj, Univ Rochester, 41-42; assoc prof, 46-57, Chief Univ Med Div, DC Gen Hosp, 50-69, PROF MED, SCH MED, GEORGE WASHINGTON UNIV, 57- *Concurrent Pos:* Consult, Walter Reed Army Hosp & Cent Off, US Vet Admin, 46-78, USPHS, 46-72, Clin Ctr, NIH, 53-72& Wash Hosp Ctr, 58-78. *Mem:* Am Fedn Clin Res; Soc Med Consults Armed Forces; Soc Exp Biol & Med; Am Soc Microbiol; Infectious Dis Soc Am; AMA; Am Col Physicians. *Res:* Infectious diseases; chemotherapy and antibiotics; immunology; nutrition in obesity. *Mailing Add:* 5600 Wisconsin Ave Chevy Chase MD 20815

ROMARY, JOHN KIRK, b Topeka, Kans, July 26, 34; div; c 3. PHYSICAL CHEMISTRY. *Educ:* Washburn Univ, BS, 56; Kans State Univ, PhD(phys chem), 61. *Prof Exp:* Asst prof phys chem, Univ Nev, 61-63; from asst prof to assoc prof, Washburn Univ, 63-68; PROF CHEM & CHMN DEPT, UNIV WIS-WHITEWATER, 68- *Concurrent Pos:* Asst prog dir, NSF, 67-68; mem, Gov Comn Educ, State Wis, 69-70. *Mem:* Am Chem Soc; Royal Soc Chem. *Res:* Polarographic, potentiometric, thermodynamic and calorimetric studies of coordination compound formation; synthesis of pyridyl polyamine ligands; development of computer programs for analysis of coordination data. *Mailing Add:* Dept Chem Univ Wis 800 W Main St Whitewater WI 53190

ROMBACH, HANS DIETER, b Herbolzheim, Ger, June 6, 53; m 78; c 2. SOFTWARE MEASUREMENT, DEVELOPMENT PROCESS MODELING. *Educ:* Univ Karlsvuhe, Ger, MS, 78; Univ Kaiserslautern, Ger, PhD(computer sci), 84. *Prof Exp:* Res scientist, Nuclear Res Ctr, Karlsvuhe, Ger, 78-79; res assoc software eng, computer Sci Dept, Univ Kaiserslautern, Ger, 79-84; vis asst prof, 84-86, ASST PROF SOFTWARE ENG, COMPUTER SCI DEPT, UNIV MD, COLLEGE PARK, 86-, ASST PROF SOFTWARE ENG, UMIACS, 86- *Concurrent Pos:* Prin investr, SEL NASA/Goddard, Md, 86-, NSF, Wash, DC, 90-; vis researcher, Software Eng Inst, Carnegie Mellon Univ, 87-88; consult, Inst Defense Anal, Fairfax Va & Europ projs, 88-, Northrop Corp, Calif, 89-; lectr, DEC Ger, 89-; NSF presidential young investr award res/teaching software eng, 90. *Mem:* Asn Comput Mach; Inst Elec & Electronics Engrs Computer Soc. *Res:* Formalization of software processes; integration of software measurement; exploration of effective mechanisms for learning and reuse of all kinds of software related experience. *Mailing Add:* Dept Computer Sci Univ Md College Park MD 20742

ROMBEAU, JOHN LEE, b Los Angeles, Calif, May 8, 39; m 70; c 2. OTHER MEDICAL & HEALTH SCIENCES. *Educ:* La Sierra Col, Calif, BA, 62; Loma Linda Univ, MD, 67. *Prof Exp:* Asst prof surg, Univ Calif Hosp, Davis, 77-79; asst prof surg, 79-86, ASSOC PROF DEPT SURG, HOSP UNIV PA, 86-, SURGEON, 79- *Concurrent Pos:* Doctor surg & nutrit support, Philadelphia Vet Admin Med Ctr, 79-; doctor surg & nutrit, Martinez Vet Admin Med Ctr, Calif, 77-79. *Mem:* Fel Am Col Surgeons; AMA; Am Soc Clin Nutrit; Am Soc Parenteral & Enteral Nutrit (vpres, 86, pres, 88); Asn Acad Surg; Soc Surg Alimentary Tract; Colombian Surg Soc; Colombian Soc Clin Nutrit; SAfrican Soc Parenteral & Enteral Nutrit. *Mailing Add:* Dept Surg Univ Pa Presby Med Ctr 3400 Spruce St Philadelphia PA 19104

ROMBERGER, JOHN ALBERT, b Northumberland Co, Pa, Dec 25, 25; m 51; c 2. PLANT PHYSIOLOGY. *Educ:* Swarthmore Col, BA, 51; Pa State Univ, MS, 53; Univ Mich, PhD(bot), 58. *Prof Exp:* Res fel plant physiol, Calif Inst Technol, 58-59; plant physiologist, US Forest Serv, 59-82; RETIRED. *Concurrent Pos:* Vis scholar, Univ Silesia, Poland, 81, 83; independent scholar-writer, 82- *Mem:* Am Soc Plant Physiol; Bot Soc Am; Scand Soc Plant Physiol; AAAS; Acad Independent Scholars. *Res:* Tree physiology; growth and development in woody plants; developmental biology; morphogenesis; wood anatomy; morphogenesis and development in higher organisms. *Mailing Add:* 741 S Lombard Ave Oak Park IL 60304

ROMBERGER, KARL ARTHUR, b Orwin, Pa, Sept 13, 34; m 58; c 4. ANALYTICAL CHEMISTRY, PHYSICAL CHEMISTRY. *Educ:* Lebanon Valley Col, BS, 56; Pa State Univ, PhD(analytical chem), 67. *Prof Exp:* Res chemist, Oak Ridge Nat Lab, 64-70; group leader chem, Kawecki-Berylco Industs, 70-79; RES ASSOC, CABOT CORP, 79- *Mem:* AAAS; Am Chem Soc; Sigma Xi. *Res:* Tantalum and columbium extraction, purification and metallurgy. *Mailing Add:* 100 Martin Ave Gilbertsville PA 19525

ROMBERGER, SAMUEL B, b Harrisburg, Pa, June 12, 39; m 66; c 1. GEOCHEMISTRY, ECONOMIC GEOLOGY. *Educ:* Pa State Univ, BS, 62, PhD(geochem), 68. *Prof Exp:* Asst prof geol, Mich State Univ, 66-69 & Univ Wis-Madison, 69-75; MEM FAC, COLO SCH MINES, 75- *Concurrent Pos:* Assoc prof geol, Off Water Resources grants, 66-69. *Mem:* Mineral Soc Am; Soc Econ Geologists; Geol Soc Am; Sigma Xi. Mineral Asn Can. *Res:* Determination of the nature, particularly the chemistry, of the processes responsible for the formation of metallic, and uranium ore deposits. *Mailing Add:* Dept Geol Colo Sch Mines Golden CO 80401

ROME, DORIS SPECTOR, b Albany, NY, Feb 10, 26; m 47; c 3. CYTOLOGY, PATHOLOGY. *Educ:* Skidmore Col, BA, 47; Albany Med Col, MD, 50; Am Bd Path, dipl, 72. *Prof Exp:* Dir, Cytol Lab, Albany Med Col, 51-80, from instr to assoc prof med, 53-80, dir, Cytol Lab, 77-80; DIR, ALBANY CYTOPATH LABS, INC, 81- *Concurrent Pos:* Am Cancer Soc fel, Albany Med Ctr, 51-52, univ fel, 52-53, univ fel path, 70-72, attend physician, Albany Med Ctr Hosp, 62-72, attend pathologist, 75-81. *Mem:* Am Soc Cytol; fel Am Col Path; Int Acad Path; AMA; fel Int Acad Cytol. *Res:* Diagnostic procedures for early detection of cancer by cytologic methods. *Mailing Add:* 214 Central Ave Albany NY 12206

ROME, JAMES ALAN, b New York, NY, Oct 12, 42. PLASMA PHYSICS. *Educ:* Mass Inst Technol, SB, 65, SM, 67, ScD, 71. *Prof Exp:* Instr elec eng, Mass Inst Technol, 67-71; PHYSICIST, FUSION ENERGY DIV, OAK RIDGE NAT LAB, 71-; PRES, SCI ENDEAVORS CORP, KINGSTON, TENN, 84- *Concurrent Pos:* Assoc ed, Physics Fluids, 79-82. *Mem:* Fel Phys Soc; Inst Elec & Electronics Engrs. *Res:* Particle containment in magnetic fusion devices; optimization of magnetic configurations; theory of neutral beam heating of tokamaks. *Mailing Add:* Bldg 9201-2 Fusion Energy Div Oak Ridge Nat Lab PO Box 2009 Oak Ridge TN 37831-8071

ROME, LEONARD H, b Youngstown, Ohio, Feb 2, 49; m 77; c 2. BIOCHEMISTRY, NEUROSCIENCES. *Educ:* Univ Mich, BS, 72, MS, 73, PhD(biol chem), 75. *Prof Exp:* Staff fel, NIH, 75-79; from asst prof to assoc prof, 79-88, PROF BIOL CHEM, SCH MED, UNIV CALIF, LOS ANGELES, 88-, VCHMN, BIOL CHEM, 89- *Mem:* Am Soc Biochem & Molecular Biol; Am Soc Cell Biol; Am Soc Neurochemistry; AAAS. *Res:* Myelination in the central nervous system utilizing primary cultures of oligodendroglial cells; structure & function of vaults, novel ribonucleoprotein found in cytoplasm; macromolecular transport-organelle biogenesis. *Mailing Add:* Dept Biol Chem 33-257 CHS Sch Med Univ Calif Los Angeles CA 90024

ROME, MARTIN, b New York, NY, June 20, 25; m 50; c 2. PHYSICAL CHEMISTRY. *Educ:* Brooklyn Col, AB, 47; Columbia Univ, AM, 49; Polytech Inst Brooklyn, PhD(phys chem), 51. *Prof Exp:* Sr engr, Electronic Tube Div, Westinghouse Elec Corp, 51-55; chief engr, Photosensitive Tube Div, Machlett Labs, Inc, 55-63; dir res & develop, Electro-Mech Res, Inc, 63-68; vpres & gen mgr, EMR Div, 68-71, VPRES & GEN MGR, EMR PHOTOELEC, WESTON INSTRUMENTS, INC, 71- *Mem:* AAAS; Optical Soc Am; Inst Elec & Electronics Engrs. *Res:* Photoelectric television camera tubes; semiconductor photoemissive and photoconducting layers; vacuum technology; x-ray image intensifying devices; ultraviolet and visible image converter and intensifier tubes; multiplier phototubes. *Mailing Add:* 55 Linwood Circle Princeton NJ 08540

ROMEO, JOHN THOMAS, b Plattsburgh, NY, July 4, 40; m 66; c 1. PLANT CHEMISTRY, CHEMICAL ECOLOGY. *Educ:* Hamilton Col, AB, 62; Univ Idaho, MS, 70; Univ Tex, Austin, PhD(bot), 73. *Prof Exp:* Teacher biol, East Syracuse-Minoa High Sch, 66-69; res assoc phytochem, Inst Biomed Res, Univ Tex, 73; chemist, Tex Air Control Bd, 74; asst prof biol, Oakland Univ, 74-77; asst prof, 77-79, assoc prof, 80-88, PROF BIOL, UNIV SFLA, TAMPA, 88- *Concurrent Pos:* NSF res grant, 80-86; vis res scientist, Kew Gardens, Eng, 86-87; hon res fel, Birkbeck Col, Univ London, 86-87. *Mem:* Sigma Xi; Phytochem Soc NAm (treas, 78-82, pres, 87-88); Int Soc Chem Ecol; Latin Am Bot Soc. *Res:* Non-protein amino acids, and alkaloids of tropical legumes, application to taxonomic and ecological problems. *Mailing Add:* Dept Biol Univ SFla Tampa FL 33620

ROMEO, TONY, b Batesville, Ark, Dec 20, 56; m 83. MICROBIOLOGY. *Educ:* Univ Fla, BS, 79, MS, 81, PhD(microbiol), 86. *Prof Exp:* Postdoctoral, Dept Biochem, Mich State Univ, 86-89; ASST PROF, DEPT MICROBIOL & IMMUNOL, TEX COL OSTEOP MED, UNIV NTEX, 89- *Concurrent Pos:* Adj asst prof, Dept Biochem & Molecular Biol, Tex Col Osteop Med, Univ NTex, 89- *Mem:* Am Soc Microbiol; Am Soc Biochem & Molecular Biol; AAAS. *Res:* Biochemical, macromolecular and cellular processes that are unique to the stationary phase in bacteria; study of glycogen biosynthesis genes of Escherichia coli as a model system to identify novel regulatory factors; author of 17 technical publications. *Mailing Add:* Dept Microbiol Tex Col Osteop Med 3500 Camp Bowie Blvd Ft Worth TX 76107

ROMER, ALFRED, b Pleasantville, NY, Aug 9, 06; m 33; c 3. PHYSICS, SCIENCE EDUCATION. *Educ:* Williams Col, BA, 28; Calif Inst Technol, PhD(physics), 35. *Hon Degrees:* ScD, St Lawrence Univ, 79. *Prof Exp:* Asst chem, Williams Col, 28-29; from asst prof to prof physics, Whittier Col, 33-43; assoc prof, Vassar Col, 43-46; from assoc prof to prof, 46-73, EMER PROF PHYSICS, ST LAWRENCE UNIV, 73- *Concurrent Pos:* Fel, Harvard Univ, 40-41; vis fel, Princeton Univ, 55-56; actg ed, Am J Physics, 62; vis staff mem, Educ Serv Inc, 62-63; consult, AID & Univ Grants Comn, India, 64 & 65; vis prof hist, Univ Calif, Santa Barbara, 75. *Mem:* AAAS; Am Phys Soc; Hist Sci Soc; Soc Hist Technol; Am Asn Physics Teachers (secy, 66-70). *Res:* History of physics; history of technology. *Mailing Add:* Dept Physics St Lawrence Univ Canton NY 13617

ROMER, I(RVING) CARL, JR, mechanical engineering; deceased, see previous edition for last biography

ROMER, ROBERT HORTON, b Chicago, Ill, Apr 15, 31; m 53; c 3. LOW TEMPERATURE PHYSICS. *Educ:* Amherst Col, BA, 52; Princeton Univ, PhD(physics), 55. *Prof Exp:* From instr to assoc prof, 55-66, PROF PHYSICS, AMHERST COL, 66- *Concurrent Pos:* Res assoc, Duke Univ, 58-59; vis physicist, Brookhaven Nat Lab, 63-; NSF fel, Univ Grenoble, 64-65; vis prof, Voorhees Col, 69-70; assoc ed, Am J Physics, 68-74, bk rev ed, 82-88, ed , 88- *Mem:* AAAS; Am Phys Soc; Am Asn Physics Teachers; Solar Energy Soc. *Res:* Nuclear magnetic resonance; environmental physics; energy statistics; physics for non-scientists. *Mailing Add:* Dept Physics Amherst Col Amherst MA 01002

ROMERO, ALEJANDRO F, b Mascota, Mex, Nov 16, 37; m 67; c 2. EMISSIONS RESEARCH, INTERNAL COMBUSTION ENGINES. *Educ:* Nat Univ Mex, BS, 66, MS, 73, PhD(mech eng), 83. *Prof Exp:* Propositions engr, Combustion Eng, 64-67; eng dept chief, Babcock Wilcox Co, 67-71; FULL PROF THERMAL SCI, FAC ENG, NAT UNIV MEX, 71- *Concurrent Pos:* Consult, Fibras Sinteticas, 70-71 & var ins co, 86-; vis prof, Tech Univ Munich, Ger, 78-80 & Univ Mich, Ann Arbor, 83-85; orgn dir, Nat Acad Eng, 81-83, cong reviewer, 85- *Mem:* Am Inst Aeronaut & Astronaut. *Res:* Radiation heat transfer and diffusion both of energy and mass; reduction of air pollution both from mobile and stationary sources. *Mailing Add:* Fac Eng Nat Univ Mex Apartado Postal 70-270 Mexico City 04510 DF Mexico

ROMERO, JACOB B, b Las Vegas, NMex, July 10, 32; m 61; c 2. CHEMICAL & NUCLEAR ENGINEERING. *Educ:* Univ NMex, BS, 54; Univ Wash, MS, 57, PhD(chem eng), 59. *Prof Exp:* Res specialist, Aerospace Group, Boeing Co, Wash, 59-66; assoc prof chem & nuclear eng, Univ Idaho, 66-69; specialist engr aerospace group, Boeing Co, Wash, 69-72; mem fac, 72-80, NATURAL SCI INSTR, EVERGREEN STATE COL, 80- *Mem:* Sigma Xi. *Res:* Fluid flow in fluidized beds; analysis of advanced nuclear reactor concepts; cryogenic fluid storage using superinsulations; transient behavior of heat pipes; lasers. *Mailing Add:* 2101 Beverly Bch Dr Nw Olympia WA 98502

ROMERO, JUAN CARLOS, b Mendoza, Arg, Sept, 15, 37; US citizen; m 63; c 2. ANIMAL PHYSIOLOGY, COMPUTER SCIENCE. *Educ:* San Jose Col, Mendoza, Arg, BS, 55; Univ Mendoza, MD, 64. *Prof Exp:* Res asst physiol, Inst Path Physiol, Mendoza, 62-66; res assoc physiol, Univ Mich, 68-73; assoc consult nephrol, 73-76, from asst prof to assoc prof physiol & med, 74-80, PROF PHYSIOL, MAYO SCH MED, MAYO FOUND, 81-, DIR HYPERTENSION RES LAB, 82- *Concurrent Pos:* Exec secy, Pan Am Coun High Blood Pressure, 76-78; estab investr, Am Heart Asn, 76-81, mem, med adv bd, Coun High Blood Pressure, 77-; mem, grant rev comt, Nat Heart, Lung & Blood Inst, NIH, 81-85; mem, extramural adv bd, dept physiol, Univ Mich, 84-; mem, adv bd rev res projs submitted to NASA, Am Inst Biol Sci, 84-85. *Honors & Awards:* Sibley-Hoobler lectr, Univ Mich, 85. *Mem:* Am Heart Asn; Am Fedn Clin Res; Am Inst Biol Sci; Am Physiol Soc; Am Soc Nephrology; Int Soc Hypertension. *Res:* Mechanisms by which the kidney can produce high blood pressure; identification of the hormones that control renal circulation and excretion of salt; manner in which renal circulation is altered; relationship between renal circulation and the function of the kidney to control salt equilibrium. *Mailing Add:* Physiol Dept Mayo Clin Rochester MN 55901

ROMERO-SIERRA, CESAR AURELIO, b Madrid, Spain, Nov 29, 31; Can citizen; c 5. ANATOMY. *Educ:* Univ Granada, BA & BSc, 49, MD, 59; Univ Zaragoza, MD, 60, DSc, 61. *Prof Exp:* Intern dermat, Univ Granada, 59; asst prof anat, Univ Zaragoza, 59-61; Swed Ministry Foreign Affairs fel, 61-62; scholar, Swed Inst, 62; NIH fel, Univ Calif, Los Angeles, 64-65; asst prof anat, Univ Ottawa, 65-67; assoc prof, 67-76, PROF ANAT, QUEEN'S UNIV, ONT, 76- *Concurrent Pos:* Researcher, Nat Health & Welfare, Ont, 68. *Mem:* AAAS; Am Soc Cell Biol; Can Asn Lab Animal Sci; Can Fedn Biol Socs; Pan-Am Asn Anat. *Res:* Interaction of electromagnetic fields with living organisms; a modeling of neural behavior; biomedical instrumentation. *Mailing Add:* Dept Anat Queen's Univ Kingston ON K7L 3N6 Can

ROMESBERG, FLOYD EUGENE, b Garrett, Pa, Jan 31, 27; div; c 3. CHEMICAL ENGINEERING, PHYSICAL CHEMISTRY. *Educ:* Pa State Univ, BS, 49; Bucknell Univ, MS, 50; Univ Cincinnati, PhD(phys chem), 53. *Prof Exp:* Chemist, 53-56, proj leader fibers, 56-59, group leader, 59-63, lab dir films, 63-65, asst to tech dir films res & develop, 65-78, ASSOC SCIENTIST, FOAMS & FILMS PROD DEVELOP, DOW CHEM CO, 78- *Mem:* Am Chem Soc. *Res:* Technology support of manufacturing and research and development for the industrial polymeric films, including polyethylene, polystyrene, saran, copolymer and multilayer and for the consumer products, including Handiwrap and Saranwrap films and Ziploc bags; product development for cellular foams, heat transfer studies, long range product development for films, foam and composites. *Mailing Add:* 9039 Mt Vernon Rd St Louisville OH 43071

ROMEY, WILLIAM DOWDEN, b Richmond, Ind, Oct 26, 30; m 55; c 3. GEOLOGY, SCIENCE EDUCATION. *Educ:* Ind Univ, AB, 52; Univ Calif, Berkeley, PhD(geol), 62. *Prof Exp:* From asst prof to assoc prof geol & sci teaching, Syracuse Univ, 62-69, exec dir, Earth Sci Educ Prog, 69-72; chmn dept geol & geog, 71-76, PROF GEOL, ST LAWRENCE UNIV, 71-, PROF & CHMN GEOG, 83- *Concurrent Pos:* Consult, Earth Sci Curriculum Proj, 63-72; vis geol scientist, Am Geol Inst, 64-72 & NY State Educ Dept, 65-69; NSF sci fac fel, Univ Oslo Geol Mus, 67-68; earth sci consult, Compton's Encycl, 71; assoc ed, Geol Soc Am Bulletin, 79-84 & J Geol Educ, 80- *Mem:* Fel AAAS; fel Geol Soc Am; Nat Asn Geol Teachers (vpres, 71-72, pres, 72-73); Asn Am Geographers; Am Geophys Union; Can Asn Geographers. *Res:* Igneous and metamorphic petrology; structural, Precambrian and general field geology; humanistic learning theory and human potential; regional geography; volcanology and natural hazards. *Mailing Add:* Dept Geog St Lawrence Univ Canton NY 13617

ROMICK, GERALD J, b Ennis, Tex, Jan 26, 32; m 54; c 6. AERONOMY. *Educ:* Univ Alaska, BS, 52, PhD(geophys), 64; Univ Calif, Los Angeles, MS, 54. *Prof Exp:* Spectroscopist, US Naval Ord Lab, Calif, 54-56; res asst auroral studies, 56-58, asst geophysicist, 58-64, from asst prof to prof, 64-82, EMER PROF GEOPHYS, GEOPHYS INST, UNIV ALASKA, 84-; SR RES CONSULT, KIA CONSULTS, INC, 82- *Concurrent Pos:* Consult, Lockheed Missiles & Space Co, 66-67, Johns Hopkins Univ, Appl Physics Lab, 86- *Mem:* Am Geophys Union; Int Asn Geomag & Aeronomy. *Res:* Physics of the upper atmosphere; studies of the excitation of the constituents by energetic particle, solar radiation and chemical processes. *Mailing Add:* Geophys Inst Univ Alaska Fairbanks AK 99775-0800

ROMIG, ALTON DALE, JR, b Bethlehem, Pa, Oct 6, 53. ELECTRON MICROSCOPY. *Educ:* Lehigh Univ, BS, 75, MS, 77, PhD(metall & mat sci), 79. *Prof Exp:* MEM TECH STAFF, AT&T TECHNOL INC, SANDIA NAT LABS, 79-; ASSOC PROF METALL, NMEX INST MINING & TECHNOL, 81- *Concurrent Pos:* Key reader, Metall Transactions, 80-; lectr, Lehigh Univ, 78-; supvr phys metall, AT&T Technol Inc, Sandia Nat Labs, 88- *Honors & Awards:* Castaing Award, Microbeam Anal Soc, 79; Burton Medal, Electron Micros Soc, 88. *Mem:* Am Inst Mining Metall & Petrol Engrs; Am Soc Metals; Microbeam Anal Soc (treas); Electron Micros Soc Am; Sigma Xi; Mat Res Soc. *Res:* Diffusion controlled phase transformations and phase equilibrium in multicomponent metallic alloy systems theoretically and by several experimental techniques including analytical electron microscopy. *Mailing Add:* 4923 Calle De Lune NE Albuquerque NM 87111

ROMIG, PHILLIP RICHARDSON, b Dennison, Ohio, July 24, 38; m 62; c 2. GEOPHYSICS, SEISMOLOGY. *Educ:* Univ Notre Dame, BS, 60; Colo Sch Mines, MS, 67, PhD(geophys), 69. *Prof Exp:* System engr, A C Electronics Div, Gen Motors Corp, 63-64; res assoc, earthquake seismol, 69-74, from asst prof to assoc prof, 74-82, PROF, COLO SCH MINES, 82-, DEPT HEAD, 83- *Concurrent Pos:* Sr geophysicist, Westinghouse Geores Lab, 69-74; staff seismologist, E D'Appolonia Consult Engrs, 75-80; consult, Sandia Nat Labs, 80-81. *Mem:* Seismol Soc Am; Am Geophys Union; Inst Elec & Electronics Engrs; Soc Explor Geophysicists; Earthquake Eng Res Inst. *Res:* Engineering geophysics and evaluation of seismic hazards, groundwater exploration and toxic waste mapping. *Mailing Add:* 2032 Goldenvue Golden CO 80401

ROMIG, ROBERT P, b Eugene, Ore, May 1, 36; m 65; c 2. CHEMICAL ENGINEERING. *Educ:* Ore State Univ, BS, 59, MS, 61; Carnegie Inst Technol, PhD(chem eng, fluid flow), 64. *Prof Exp:* ASSOC PROF CHEM ENG, SAN JOSE STATE UNIV, 64- *Mem:* Am Inst Chem Engrs; Am Comput Mach. *Res:* Applied mathematics; computer simulation of chemical processes. *Mailing Add:* Dept Chem Eng San Jose State Univ Washington Sq San Jose CA 95192

ROMIG, ROBERT WILLIAM MCCLELLAND, b Cornwall, Ont, July 17, 29; m 52; c 2. PLANT PATHOLOGY. *Educ:* Drew Univ, BA, 53; Purdue Univ, MS, 55, PhD(plant path), 57. *Prof Exp:* Asst geneticist, Agr Prog, Rockefeller Found, Colombia, 57-59, assoc geneticist, Chile, 59-62; res plant pathologist, USDA, Univ Minn, St Paul, 62-69; dir, wheat res, 78-79, GENETICIST & WHEAT PROJ LEADER, NORTHRUP KING & CO, 69-, VPRES RES, 79- *Mem:* AAAS; Am Phytopathological Soc; Am Soc Agron. *Res:* Small grains; disease resistance; epidemiology; wheat breeding. *Mailing Add:* Northrop-King Co 7500 Olson Mem Hwy Minneapolis MN 55427

ROMIG, WILLIAM D(AVIS), b Victor, Iowa, Aug 19, 14; m 39; c 2. CIVIL ENGINEERING, WATER RESOURCES. *Educ:* Univ Colo, BS, 36. *Prof Exp:* Levelman, US Bur Reclamation, Colo, 36, jr hydraul engr, State Water Conserv Bd, Colo, 37-41; from jr engr to engr, US Bur Reclamation, Colo, 41-45, Washington, DC, 45-53, planning engr, US Foreign Opers, Costa Rica, 53-55 & Washington, DC, 55-63; water resources engr, Near East & SAsia, USAID, Washington, DC, 63-72; CONSULT, 72- *Concurrent Pos:* Mem, Int Comn Irrig & Drainage. *Mem:* Am Soc Civil Engrs. *Res:* Hydrology; multi-purpose water projects, including municipal water and groundwater; agricultural water management. *Mailing Add:* 100 Inca Pkwy Boulder CO 80303

ROMIG, WILLIAM ROBERT, b Hope, Ark, Mar 4, 26; m 58; c 1. BACTERIOLOGY. *Educ:* Southwestern State Col, BA, 48; Univ Okla, MS, 54; Univ Tex, PhD(bact), 57. *Prof Exp:* From instr to asst prof, 57-70, prof, 70-, EMER PROF BACT, UNIV CALIF, LOS ANGELES. *Mem:* Am Soc Microbiol; Brit Soc Gen Microbiol. *Res:* Bacteriophage; radiation effects on bacteria; bacterial genetics and endospore formation; transformation. *Mailing Add:* Dept Microbiol 5304 Life Sci Univ Calif 405 Hilgard Ave Los Angeles CA 90024

ROMINGER, JAMES MCDONALD, b Charleston, Ill, May 19, 28; m 52; c 3. SYSTEMATIC BOTANY. *Educ:* Eastern Ill Univ, BS, 50; Univ NMex, MS, 55; Univ Ill, PhD(bot), 59. *Prof Exp:* Asst biol, Univ NMex, 50-52; instr, Univ Jacksonville, 55-56; asst bot, Univ Ill, 56-59; instr biol, Black Hills State Col, 59-61; asst prof, Western State Col Colo, 61-63; ranger-naturalist, Grand Teton Nat Park, 63; assoc prof bot, 63-74, prof biol, Northern Ariz Univ, 74-89, cur, Deaver Herbarium, 63-89, EMER PROF BOT, NORTHERN ARIZ UNIV, 90- *Concurrent Pos:* Range technician, Yellowstone Nat Park, 74 & 81. *Mem:* Sigma Xi. *Res:* Agrostology; vascular flora of northern Arizona, especially grass flora; genus Setaria in North America. *Mailing Add:* Dept Biol Box 5640 Northern Ariz Univ Flagstaff AZ 86011

ROMINGER, MICHAEL COLLINS, control systems, chemical engineering, for more information see previous edition

ROMMEL, FREDERICK ALLEN, medical microbiology, allergy & clinical immunology, for more information see previous edition

ROMMEL, MARJORIE ANN, analytical chemistry; deceased, see previous edition for last biography

ROMNEY, CARL FREDRICK, b Salt Lake City, Utah, June 5, 24; m 46; c 2. SEISMOLOGY. *Educ:* Calif Inst Technol, BS, 45; Univ Calif, PhD(geophys), 56. *Prof Exp:* Asst seismol, Univ Calif, 47-49; from seismologist to chief seismologist, Beers & Heroy, 49-54, consult, 54-55; supvry geophysicist, Air Force Off Atomic Energy, 55-58, asst tech dir, Air Force Tech Applns Ctr, 58-73; dep dir, 73-75, dir, Nuclear Monitoring Res Off, 75-79, dep dir res, Defense Advan Res Proj Agency, 79-83, dir, Ctr Seismic Studies, 83-91; VPRES, SCI APPLN INT CORP, 84- *Concurrent Pos:* Mem panel seismic improv, Off Spec Asst to President for Sci & Tech, 59; mem US deleg, Negotiations for Threshold Nuclear Test Ban Treaty Limiting Peaceful Nuclear Explosions, Moscow, 74-75; comprehensive nuclear test ban negotiations, Geneva, 77-78. *Mem:* Soc Explor Geophys; Seismol Soc Am; Am Geophys Union. *Res:* Explosion seismology; seismic detection methods and instruments; seismicity; differences between earthquake and explosion signals. *Mailing Add:* 4105 Sulgrave Dr Alexandria VA 22309

ROMNEY, EVAN M, b Duncan, Ariz, Jan 13, 25; m 48; c 4. PLANT NUTRITION, SOILS. *Educ:* Brigham Young Univ, BS, 50; Rutgers Univ, PhD(soils), 53. *Prof Exp:* Asst res soil scientist, Atomic Energy Proj, 53-60, assoc res soil scientist, 60-65, RES SOIL SCIENTIST, LAB BIOMED & ENVIRON SCI, UNIV CALIF, LOS ANGELES, 66-, ASSOC DIR, LAB BIOMED & ENVIRON SCI, 79-; ELEMENT MGR, NEV APPL ECOL GROUP, US DEPT ENERGY, 72- *Mem:* Soil Sci Soc Am; Am Soc Agron. *Res:* Cycling of radioactive materials in plants and soils; trace elements in plants and soils; soil chemistry of rare earth elements; fate and persistence of radioactive fallout in natural environment. *Mailing Add:* 3585 Grand View Blvd Mar Vista CA 90066

ROMNEY, SEYMOUR L, b New York, NY, June 8, 17; m 45; c 3. GYNECOLOGY, ONCOLOGY. *Educ:* Johns Hopkins Univ, AB, 38; NY Univ, MD, 42. *Prof Exp:* Resident gynecologist, Free Hosp Women, 49; 49; resident obstetrician, Boston Lying-in-Hosp, 50; instr gynec & obstet, Harvard Med Sch, 50-51, assoc, 52-56; prof & chair obstet & gynec, 57-72, DIR GYNEC CANCER RES, ALBERT EINSTEIN COL MED, 72- *Concurrent Pos:* Consult, Nat Cancer Inst, WHO, Planned Parenthood-World Pop, NY State Family Planning Asn & Ari* Res Comn. *Mem:* AAAS; Am Gynec & Obstet Soc; Soc Gynec Invest; Am Soc Clin Cancer Res; NY Acad Sci; NY Acad Med. *Res:* Human reproduction; population and fertility regulation; medical education; nutrition and female genital tract malignancy; cancer prevention. *Mailing Add:* 1081 Orienta Ave Mamaroneck NY 10543

ROMO, WILLIAM JOSEPH, b Oregon City, Ore, Jan 17, 34; m 67; c 1. PHYSICS. *Educ:* Univ Ore, 60, MS, 61; Univ Wis-Madison, PhD(theoret physics), 67. *Prof Exp:* Res assoc theoret physics, Univ Wis-Madison, 67-68 & Physics Div, Argonne Nat Lab, 68-70; from asst prof to assoc prof, 70-84, PROF PHYSICS, CARLETON UNIV, 84- *Mem:* Am Phys Soc; Can Asn Physicists; Sigma Xi. *Res:* Nuclear reaction theory; continuum shell model; resonances; nuclear structure. *Mailing Add:* Dept Physics Carleton Univ Ottawa ON K1S 5B6 Can

ROMOSER, WILLIAM SHERBURNE, b Columbus, Ohio, Oct 18, 40; m 64, 73; c 3. ENTOMOLOGY. *Educ:* Ohio State Univ, BSc, 62, PhD(entom), 64. *Prof Exp:* Res assoc entom, Ohio State Univ, 65; from asst prof to assoc prof entom, 65-76, PROF ZOOL, OHIO UNIV, 76- *Concurrent Pos:* Resident res assoc, Nat Res Coun, US Army Med Res Inst Infectious Dis, Ft Detrick, 84-85. *Mem:* AAAS; Entom Soc Am; Am Mosquito Control Asn; Am Soc Trop Med & Hyg. *Res:* Alimentary morphology and physiology of blood-feeding insects, particularly mosquitoes; development and metamorphosis of insects, particularly mosquitoes; arboviruses; medical entomology. *Mailing Add:* Dept Zool & Biomed Sci Ohio Univ Athens OH 45701

ROMOVACEK, GEORGE R, b Prague, Czech, May 17, 23; m 84; c 2. CHEMISTRY, FUEL TECHNOLOGY. *Educ:* Univ Chem Technol, Prague, MSc, 49, PhD(fuel technol), 55. *Prof Exp:* Sr lectr fuel technol, Univ Chem Technol, Prague, 51-62, assoc prof, 62-69; sr scientist, Coal Tar & Pitch, Koppers Co, Inc, 69-72, mgr, Carbon & Pitch Res, 72-90; RETIRED. *Concurrent Pos:* Brit Coun res fel, Univ Strathclyde, 65-66; NSF res assoc, Pa State Univ, 67-68. *Mem:* Am Chem Soc; Am Inst Metall Engrs. *Res:* Chemistry and chemical technology of petroleum, coal, coal tar and pitch; air pollution; carbon, natural and manufactured gas. *Mailing Add:* 111 Rydal Lane Pittsburgh PA 15237

ROMRELL, LYNN JOHN, b Idaho Falls, Idaho, Oct 20, 44; m 80; c 6. REPRODUCTIVE BIOLOGY, HUMAN ANATOMY. *Educ:* Idaho State Univ, BS, 67; Utah State Univ, PhD(zool), 71. *Prof Exp:* NIH res fel anat, Harvard Med Sch, 71-73, instr anat & mem lab human reproduction & reproductive biol, 73-75; from asst prof to assoc prof, 79-87, PROF ANAT & CELL BIOL & ASSOC DEAN EDUC, UNIV FLA COL MED, 87-; EXEC DIR, FLA STATE ANAT BD, 83- *Mem:* Sigma Xi; Soc Study Reproduction; Am Asn Anat; Am Soc Cell Biol. *Res:* Reproductive biology; cytology; cell biology; ultrastructure and function of isolated cells; genetic, biochemical and physiological factors which influence sperm development. *Mailing Add:* Dept Anat & Cell Biol Univ Fla Col Med Gainesville FL 32610

ROMSDAHL, MARVIN MAGNUS, b Havti, SDak, Apr 2, 30; m 58, 83; c 2. SURGERY, BIOLOGY. *Educ:* Univ SDak, AB, 52, BS, 54; Univ Ill, MD, 56; Univ Tex, PhD(biomed sci), 68. *Prof Exp:* Intern, Res & Educ Hosps, Univ Ill, 56-57; resident surg, Vet Admin Hosp, Hines, Ill, 57-58; clin assoc surg br, Nat Cancer Inst, 58-60; resident, Res & Educ Hosps, Univ Ill, 60-63, instr, 63-64; from asst prof to assoc prof, 67-75, PROF SURG, UNIV TEX M D ANDERSON HOSP & TUMOR INST, 75- *Concurrent Pos:* USPHS trainee, Univ Ill, 63-64; univ fel, Univ Tex, 64-67; assoc dir sci opers, Nat Large Bowel Cancer Proj, Houston, 77-79. *Honors & Awards:* Mead Johnson Award, Am Col Surg, 66-69. *Mem:* AMA; Am Asn Cancer Res; Am Col Surg; Soc Surg Alimentary Tract; Am Radium Soc. *Res:* Management and surgical treatment of cancer; dissemination of cancer and metastases in experimental systems; immunology of human solid tumors; management of breast, colon and rectal cancers; sarcomas, bone and soft tissue. *Mailing Add:* 4910 Wigton Houston TX 77096-4228

ROMSOS, DALE RICHARD, b Rice Lake, Wis, Nov 19, 41; m; c 1. NUTRITION. *Educ:* Univ Wis, BS, 64; Iowa State Univ, PhD(nutrit), 70. *Prof Exp:* Fel, Univ Ill, Urbana, 70-71; from asst prof to assoc prof, 71-79, PROF NUTRIT, MICH STATE UNIV, 79- *Concurrent Pos:* NIH res career develop award. *Honors & Awards:* Mead Johnson Res Award, Am Inst Nutrit. *Mem:* Brit Nutrit Soc; Soc Exp Biol & Med; Am Inst Nutrit; NY Acad Sci; Inst Food Technologists. *Res:* Lipid and carbohydrate metabolism; obesity; cardiovascular disease. *Mailing Add:* Dept Food Sci & Human Nutrit Mich State Univ 106 Food Sci Bldg E Lansing MI 48824-1224

ROMUALDI, JAMES P, b New York, NY, June 30, 29; m 58; c 3. CIVIL ENGINEERING. *Educ:* Carnegie Inst Technol, BS, 51, MS, 53, PhD, 54. *Prof Exp:* Fulbright fel, Karlsruhe Tech Univ, 54-55; from asst prof to assoc prof civil eng, Carnegie-Mellon Univ, 55-65, prof, 65-79, dir, Trans Res Inst, 66-79; VPRES, GAI CONSULTS, INC & PRES, FORENSIC CONSULTS & ENGRS, INC, 79- *Mem:* Am Soc Civil Engrs; Am Soc Testing & Mat; Am Concrete Inst; Am Soc Eng Educ. *Res:* Fracture mechanics, especially design and applications to reinforced concrete; applied mechanics and structural analysis; transportation engineering and planning. *Mailing Add:* 5737 Wilkins Ave Pittsburgh PA 15217-1403

RONA, DONNA C, b Jacksonville, Fla; m 74; c 1. COASTAL ENGINEERING, ENVIRONMENTAL ENGINEERING. *Educ:* Fla Atlantic Univ, BS, 73; Univ Miami, MS, 76. *Prof Exp:* Oceanogr, Nat Oceanic & Atmospheric Admin, 73-75; res asst, Univ Miami, 75-76; coastal engr,

Metrop Dade County, 76-77; proj engr, Connell, Metcalf & Eddy, 77-80; PRES, RONA COASTAL CONSULT, 79- *Concurrent Pos:* Adj prof coastal eng, Nova Univ, 80-90; ed, J Coastal Res, 83. *Mem:* Am Soc Civil Engrs; Acoust Soc Am; Am Geophys Union. *Res:* Marine acoustics and coastal engineering; physics; mathematics; engineering; biology; ecology; author of scientific publications and one book. *Mailing Add:* Rona Coastal Consult PO Box 490102 Key Biscayne FL 33149

RONA, GEORGE, b Budapest, Hungary, Mar 8, 24; Can citizen; m 54; c 2. PATHOLOGY. *Educ:* Med Univ Budapest, MD, 49; Hungarian Acad Sci, PhD(diabetes), 53; FRCP(C), 60; FRCPath. *Prof Exp:* From asst prof to assoc prof path, Med Univ Budapest, 50-54, dep dir, 54-56; resident, St Mary's Hosp, Montreal, 57-59; sr pathologist, Ayerst Res Labs, 59-61; from asst pathologist to assoc pathologist, St Mary's Hosp, 61-64; assoc prof, 67-71, PROF PATH, MCGILL UNIV, 71-; DIR & PRES, LAKESHORE DIAG SERV, 76- *Concurrent Pos:* Med Res Coun Can grant, 67-; pres Am sect, Int Study Group Res Cardiac Metab; vis prof, Univ Geneva, 73-74; pathologist & dir labs, Lakeshore Gen Hosp, 65-79, sr consult pathologist, 79- *Mem:* Am Asn Pathologists; Int Acad Path; fel Col Am Path; Can Asn Path. *Res:* Tuberculous meningitis; diabetic vascular lesions; experimental and human myocardial infarction. *Mailing Add:* 160 Stillview Rd Pointe Claire PQ H9R 2Y2 Can

RONA, MEHMET, b Adana, Turkey, Oct 18, 39; m 68; c 2. ELECTRICAL ENGINEERING. *Educ:* Robert Col, Istanbul, Turkey, BSc, 61; Princeton Univ, PhD(elec eng & physics), 66; Middle E Tech Univ, Dozent, 74. *Prof Exp:* Asst prof physics, Robert Col, 68-69; from asst prof to assoc prof physics, Middle E Tech Univ, 69-79; sr staff, 79-89, DIR PHYSICS, ARTHUR D LITTLE, INC, 89- *Mem:* Am Phys Soc; Europ Phys Soc; Am Math Asn. *Res:* Physics of reduced dimensions; electron-phonon interactions; superconductivity. *Mailing Add:* Arthur D Little Inc Acorn Park Boston MA 02140-2390

RONA, PETER ARNOLD, b Trenton, NJ, Aug 17, 34; m 74; c 1. MARINE GEOLOGY, GEOPHYSICS. *Educ:* Brown Univ, AB, 56; Yale Univ, MS, 57, PhD(marine geol, geophys), 67. *Prof Exp:* Explor geologist, Standard Oil Co, NJ, 57-59; res asst Hudson Labs, Columbia Univ, 60-61, marine geologist, 61-67, res assoc marine geol & geophys & prin investr, Ocean Bottom Studies, 67-69; prin investr, Metallogenesis Proj, 75-80, sr geophysicist & chief scientist, Trans-Atlantic Geotraverse, Atlantic Oceanog & Meteorol Labs, 69-80, prin investr, Marine Minerals Proj, 80-84, SR RES GEOPHYSICIST & PRIN INVESTR, VENTS PROG, NAT OCEANIC & ATMOSPHERIC ADMIN, 84- *Concurrent Pos:* Consult sea floor resources, UN, 70-; adj prof, Univ Miami, 74-; mem vis comt geol sci, Brown Univ, 74-77; trustee, Mus Sci, Miami, Fla, 74-, gov & vpres, 77-, chmn, 79-80 & trustee, Int Oceanog Found, 81-; assoc ed, Geol Soc Am, 75-82; mem, Sierra Club Adv Comt, Ocean Environ, 75-80; mem, Nat Oceanic & Atmospheric Admin, Marine Minerals Task Force, 76-84, Penrose Medal Comt, Geol Soc Am, 81-; chmn metallogenesis panel, Manganese Proj, Int Geol Correlation Prog, 76-85, chmn, NATO Advan Res Inst, 80-83; tech expert sea floor resources, US Dept State, 77, adv, Tectonic Map NAm Proj, 81-85; adj prof, Fla Int Univ, 78-79; mem, Vents Prog Coun, Nat Oceanic & Atmospheric Admin, 84-; distinguished lectr, Am Asn Petrol Geologists, 83-84; assoc ed, Am Geophys Union, 88-; lectr, Distinguished Lect Colloquium, Univ Colo, 91. *Honors & Awards:* Bruce Heezen Mem lectr, NY Acad Sci, 85; US Dept Com Gold Medal, 87; Francis P Shepard Medal for Excellence Marine Geol, 86. *Mem:* Fel Geol Soc Am; Am Geophys Union; Soc Explor Geophysicists; fel AAAS; Am Asn Petrol Geologists; Soc Econ Geologists; Acoust Soc Am; Sigma Xi. *Res:* Structure and development of continental margins and ocean basins; marine energy and mineral resources; author or coauthor of over 150 publications. *Mailing Add:* Nat Oceanic & Atmospheric Admin 4301 Rickenbacker Causeway Miami FL 33149

RONALD, ALLAN ROSS, b Portage la Prairie, Man, Aug 24, 38; m 62; c 3. INFECTIOUS DISEASES, MICROBIOLOGY. *Educ:* Univ Man, BSc & MD, 61; FRCP(C), 67; Am Bd Microbiol, dipl, 70. *Prof Exp:* Fel infectious dis, Univ Wash, 65-67, fel microbiol, 67-68; from asst prof to assoc prof microbiol & internal med, 68-77, prof med microbiol & head dept, 77-85, PROF INTERNAL MED & HEAD DEPT, UNIV MAN, 85-; PHYSICIAN-IN-CHIEF, HEALTH SCI CTR, WINNIPEG, 85- *Mem:* Fel Am Col Physicians; Infectious Dis Soc Am; Can Soc Clin Invest; Am Soc Microbiol; Am Soc Clin Invest. *Res:* Pathogenesis of recurrent urinary infection; antimicrobial susceptibility testing and resistance; chancroid, and hemophilus ducreyi; international health; AIDS. *Mailing Add:* Sect Infectious Dis St Boniface Hosp 409 Tache Ave Winnipeg MB R2H 2A6 Can

RONALD, BRUCE PENDER, b Chicago, Ill, Nov 22, 39; m 67. ORGANIC CHEMISTRY. *Educ:* Portland State Col, BS, 62; Univ Wash, PhD(chem), 68. *Prof Exp:* Asst prof, 68-77, ASSOC PROF CHEM, IDAHO STATE UNIV, 77- *Mem:* Am Chem Soc; Royal Soc Chem. *Res:* Carbonium ion reactions; stereochemistry of reactions in asymmetric environments; chemical kinetic applications of nuclear magnetic resonance spectrometry. *Mailing Add:* Dept Chem Idaho State Univ Pocatello ID 83209

RONALD, KEITH, b Llandaff, Wales, Aug 24, 28; m 54; c 1. ZOOLOGY. *Educ:* McGill Univ, BSc, 53, MSc, 56, PhD(parasitol), 58. *Prof Exp:* Marine pathologist, Que Dept Fisheries, 54-58, sr biologist & actg dir, 58; prof parasitol, Ont Agr Col, 58-62; sr scientist, Fisheries Res Bd Can, 62-64; prof zool & head dept, 64-71, DEAN COL BIOL SCI, UNIV GUELPH, 71- *Concurrent Pos:* Pres & chmn, Huntsman Marine Lab, 67-73; mem Nat Res Coun Grants Comt (Animal), 70-72; mem adv bd, Atlantic Regional Lab, Nat Res Coun, 70-72; mem, Fisheries Res Bd Can, 72-77; chmn, Sci Adv Comt, World Wildlife Fund, Can, 72-76; chmn seal group, Int Union Conserv Nature & Natural Resources, 77- *Honors & Awards:* Fry Medal Res, Can Soc Zoologists, 81. *Mem:* Can Soc Zool (pres, 72); fel Royal Geog Soc; fel Inst Biol; Sigma Xi. *Mailing Add:* Dept Zool Univ Guelph Guelph ON N1G 2W1 Can

RONALD, ROBERT CHARLES, b Blue Island, Ill, Apr 22, 44. ORGANIC CHEMISTRY. *Educ:* Portland State Col, BS, 66; Stanford Univ, PhD(org chem), 70. *Prof Exp:* Vis prof chem, Univ Sao Paulo, 70-72; sr res chemist & sect mgr chem, Syva Res Inst, Calif, 72-74; ASST PROF CHEM, WASH STATE UNIV, 74- *Concurrent Pos:* Nat Acad Sci overseas fel, 70-72. *Mem:* Am Chem Soc. *Res:* Synthetic organic chemistry; structure and synthesis of natural products and other biologically significant molecules; development of new methods and reagents for synthetic purposes. *Mailing Add:* Dept Chem Wash State Univ Pullman WA 99164-4630

RONAN, MICHAEL THOMAS, b Fall River, Mass, Jan 15, 49; m 76; c 2. ELEMENTARY PARTICLE PHYSICS. *Educ:* Southeastern Mass Univ, BS, 70; Northeastern Univ, MS, 73, PhD(physics), 76. *Prof Exp:* RES ASSOC EXP HIGH ENERGY PHYSICS, LAWRENCE BERKELEY LAB, 76- *Mem:* Am Phys Soc. *Mailing Add:* Lawrence Berkeley Lab B50B-5239 Univ Calif Berkeley CA 94720

RONCA, LUCIANO BRUNO, b Trieste, Italy, Apr 26, 35; US citizen; m 58, 89; c 3. RADAR GEOLOGY, GROUND WATER. *Educ:* Univ Kans, MS, 59, PhD(geol), 63. *Prof Exp:* Res assoc geochem, Univ Kans, 63-64; res scientist, Air Force Cambridge Res Labs, 64-67; scientist, Boeing Sci Res Labs, 67-70; assoc prof, 70-74, PROF GEOL, WAYNE STATE UNIV, 74- *Concurrent Pos:* Vis scientist, Lunar Sci Inst, Houston, Vernadsky Inst Acad Sci, USSR, Univ Bern, Switz, Observ de Paris, France; exchange scientist, Acad Sci USSR-USA, 76, 79-81 & 84-85. *Honors & Awards:* Mt Ronca Antarctica Award, 66. *Mem:* AAAS; fel Geol Soc Am; Am Geophys Union; Am Polar Soc. *Res:* Thermoluminescence; radiation damage in geological material; lunar and planetary geology; geostatistics; radar interpretation; ground water. *Mailing Add:* Dept Geol Wayne State Univ Detroit MI 48202

RONCADORI, RONALD WAYNE, b Centerville, Pa, Nov 19, 35; m 59; c 3. PLANT PATHOLOGY. *Educ:* Waynesburg Col, BS, 57; Univ WVa, MS, 59, PhD(plant path), 62. *Prof Exp:* Plant pathologist, US Forest Serv, 62-66; from asst prof to assoc prof 66-80, PROF PLANT PATH, UNIV GA, 80- *Mem:* Am Phytopath Soc. *Res:* General plant pathology, especially tree diseases; mycorrhizae; study of shade tree and woody ornamentals diseases with emphasis on dogwood anthracose. *Mailing Add:* Dept Plant Path Univ Ga Athens GA 30602

RONCO, FRANK, JR, b Pueblo, Colo, Sept 23, 26; m 48; c 2. FOREST MANAGEMENT. *Educ:* Colo State Univ, BS, 51, MS, 60; Duke Univ, DF, 68. *Prof Exp:* Admin forester, Rocky Mountain Forest & Range Exp Sta, US Forest Serv, USDA, 51-56, res forester, 57-86; CONSULT, 87- *Mem:* Soc Am Foresters; Sigma Xi. *Res:* Artificial regeneration and silviculture of Rocky Mountain conifers; tree physiology. *Mailing Add:* 1500 Edgewood St Flagstaff AZ 86004

RONDESTVEDT, CHRISTIAN SCRIVER, JR, b Minneapolis, Minn, July 13, 23; m 44; c 2. INDUSTRIAL ORGANIC CHEMISTRY. *Educ:* Univ Minn, BS, 43; Northwestern Univ, PhD(org chem), 48. *Prof Exp:* Asst, Northwestern Univ, 43-44; instr chem, Univ Mich, 47-52, asst prof, 52-56; Guggenheim fel, Univ Munich, 56-57; res chemist, 57-63, sr res chemist, 63-68, res assoc, 68-85, CONSULT, E I DU PONT DE NEMOURS & CO, 85- *Concurrent Pos:* Consult, US Rubber Co, 52-56; assoc prof, Univ Del, 60-61; counr, Am Chem Soc, 68-87. *Mem:* Am Chem Soc. *Res:* New low-cost synthesis of industrial organic chemicals, especially acid chlorides, amines, and halogenated compounds; mechanisms of organic reactions; organic sulfur compounds; olefins; free radicals; organic fluorine compounds; organometallic chemistry; polysaccharides. *Mailing Add:* 2547 Deepwood Dr Wilmington DE 19810-3632

RONEL, SAMUEL HANAN, b Metz, France. BIOMEDICAL ENGINEERING. *Educ:* Israel Inst Technol, BSc, 64, MSc, 66, DSc(polymers), 69. *Prof Exp:* Asst prof, Israel Inst Technol, 66-69; NIH fel, Clarkson Col Technol, 69-70; res chemist, Hydron Labs, Nat Patent Develop Corp, 70-71; group leader biomat, Hydro Med Sci, Inc, 71-74, dir res & develop, 74-75; VPRES RES & DEVELOP, NAT PATENT DEVELOP CORP, 75-; PRES, INTERFERON SCI, INC, 80- *Mem:* Am Chem Soc; Soc Biomat; Int Soc Interferon Res; NY Acad Sci. *Res:* Biocompatible synthetic materials useful for implantation; non-thrombogenic materials for use in the vascular system; prosthetic devices; production and clinical testing of interferon-drug release clinics. *Mailing Add:* Nat Patent Develop Corp Hydro Med Sci New Brunswick NJ 08901

RONEY, ROBERT K(ENNETH), b Newton, Iowa, Aug 5, 22; m 51; c 2. ELECTRICAL ENGINEERING, SPACE SYSTEMS. *Educ:* Univ Mo, BS, 44; Calif Inst Technol, MS, 47, PhD(elec eng), 50. *Prof Exp:* Res engr, Calif Inst Technol, 48-50; mgr systs anal, Hughes Aircraft Co, 50-60, tech dir res & develop labs, 60-61, from assoc mgr to mgr, Space Systs Div, 61-70, asst group exec, Space & Commun Group, 70, vpres, 73-85, sr vpres, 85-88; RETIRED. *Concurrent Pos:* Mem, US Dept Transp Com Space Transp Adv Comt, 84-87. *Mem:* Nat Acad Eng; Am Inst Aeronaut & Astronaut; fel Inst Elec & Electronics Engrs. *Res:* Dynamic systems analysis and feedback control. *Mailing Add:* 1105 Georgina Ave Santa Monica CA 90402

RONGONE, EDWARD LAUREL, b Cuyahoga Falls, Ohio, June 17, 26; m 58; c 4. BIOCHEMISTRY. *Educ:* Kent State Univ, BS, 50; St Louis Univ, PhD(biochem), 56. *Prof Exp:* Asst prof biochem, Sch Med, Tulane Univ, 56-63; assoc prof, 63-68, PROF BIOCHEM, SCH MED, CREIGHTON UNIV, 68- *Concurrent Pos:* Res biochemist, Ochsner Med Found, 56-63; consult, Baker Maid, Inc, La, 57-63. *Mem:* Am Chem Soc; Brit Soc Endocrinol; Am Soc Biol Chemists; Soc Exp Biol & Med; NY Acad Sci. *Res:* Steroid metabolism by in vitro and vivo methods; mechanism of steroid metabolism and steroid isolation; steroid metabolism by skin; protein isolation. *Mailing Add:* Dept Biol Chem Creighton Univ 1633 Holling Dr Omaha NE 68144

RONGSTAD, ORRIN JAMES, b Northfield, Wis, Apr 22, 31; m 62; c 3. WILDLIFE ECOLOGY. *Educ:* Univ Minn, BS, 59; Univ Wis, Madison, MS, 63, PhD(wildlife ecol & zool), 65. *Prof Exp:* Fel mammal res, Univ Minn, 65-67; asst prof & wildlife exten specialist, 67-71, assoc prof, 71-78, PROF WILDLIFE ECOL, UNIV WIS-MADISON, 78- *Mem:* Am Soc Mammal; Wildlife Soc. *Res:* Mammalian ecology, especially hares, rabbits and deer. *Mailing Add:* Dept Wildlife Ecol Univ Wis 226 Russell Labs 1630 Linden Dr Madison WI 53706

RONIS, MAX LEE, b May 8, 30; US citizen; m 54; c 3. OTOLARYNGOLOGY. *Educ:* Muhlenberg Col, BS, 52; Temple Univ, MD & MS, 56. *Prof Exp:* PROF OTORHINOL & CHMN DEPT, HEALTH SCI CTR, TEMPLE UNIV, 69-; DIR OTORHINOL, ST CHRISTOPHER'S HOSP CHILDREN, 69- *Mem:* Am Otol Soc; Am Acad Ophthal & Otolaryngol; Am Laryngol, Rhinology & Otol Soc. *Res:* Anatomy and pathophysiology of diseases of the cochlea-vestibular apparatus. *Mailing Add:* Dept Otorhinol Sch Med Temple Univ 3400 N Broad St Philadelphia PA 19140

RONKIN, R(APHAEL) R(OOSER), b Los Angeles, Calif, July 8, 19; m 49; c 2. PHYSIOLOGY. *Educ:* Stanford Univ, AB, 39; Univ Calif, MA, 41, PhD(zool), 49. *Prof Exp:* Biol sci, Univ Del, 49-67; sci liaison staff, New Delhi, 67-69, sci educ prog, 63-64 & 69-70, Int Progs, Nat Sci Found, 70-85; RETIRED. *Concurrent Pos:* Merck sr fel & guest investr, Biol Inst, Carlsberg Found, Denmark, 57-58; mem corp, Marine Biol Lab, Woods Hole. *Mem:* Fel AAAS; Soc Gen Physiol; Sigma Xi; Am Inst Biol Sci. *Res:* International cooperation in science and engineering; science education; cell biology. *Mailing Add:* 3212 McKinley St NW Washington DC 20015-1635

RONN, AVIGDOR MEIR, b Tel-Aviv, Israel, Nov 17, 38; m 63; c 2. CHEMICAL PHYSICS. *Educ:* Univ Calif, Berkeley, BSc, 63; Harvard Univ, AM, 64, PhD(phys chem), 66. *Prof Exp:* Res chemist, Nat Bur Standards, 66-68; from asst prof to assoc prof phys chem, Polytech Inst Brooklyn, 68-72; assoc prof chem, 73-76, PROF CHEM, BROOKLYN COL, 76- *Concurrent Pos:* Vis prof, Tel-Aviv Univ Israel, 71-72; Alfred P Sloan fel, 71-73; vis prof, Univ Sao Paulo, 73; vpres & gen mgr, LIC Indust, Inc, 79-81. *Mem:* AAAS; Am Phys Soc; Am Chem Soc. *Res:* Double resonance and energy transfer in rotational and vibrational spectra; infrared lasers; relaxation phenomena in gas lasers; laser catalyzed chemical reactions and laser induced isotope separation. *Mailing Add:* Dept Chem Brooklyn Col Bedford Ave & Ave H Brooklyn NY 11210

RONNINGEN, REGINALD MARTIN, b Frederic, Wis, Aug 19, 47. NUCLEAR PHYSICS. *Educ:* Univ Wis-River Falls, BS, 69; Vanderbilt Univ, PhD(physics), 75. *Prof Exp:* Res assoc physics, Vanderbilt Univ & Oak Ridge Nat Lab, 74-77; res assoc physics, Max-Planck für Kernphysik, Heidelberg, 77-78; res asst prof physics, 78-81, SPECIALIST, MICH STATE UNIV, EAST LANSING, 81- *Mem:* Sigma Xi; Am Phys Soc. *Res:* In-beam gamma ray spectroscopy, light and heavy ion Coulomb excitation, in elastic scattering. *Mailing Add:* Nat Superconducting Cyclotron Lab Mich State Univ East Lansing MI 48824

RONNINGEN, THOMAS SPOONER, b Hammond, Wis, Oct 3, 18; m 45; c 3. AGRONOMY, CROP BREEDING. *Educ:* Univ Wis, River Falls, BS, 39; Univ Wis, Madison, MS, 47, PhD(agron), 49. *Prof Exp:* Teacher high sch, Ind, 39-41; chemist, E I du Pont de Nemours & Co, Ind, 41-43; asst agron, Univ Wis, 46-49; from asst prof to assoc prof, Univ Md, 49-56; prin agronomist, Northeast State Agr Exp Stas, USDA, 56-63, asst to adminstr, 63-65, asst adminstr, 65-73, assoc adminr, Coop State Res Serv, 73-77, assoc dep dir coop res, Sci & Educ Admin, 77-79, dir-at-large, 79-84; RETIRED. *Concurrent Pos:* Partic, Fed Exec Inst, Va, 70; dir, Turkish Exec Mgt Seminar, 70-71. *Mem:* Fel AAAS; fel Am Soc Agron; Am Inst Biol Sci. *Res:* Forage breeding and management; microclimatology. *Mailing Add:* 1919 Blackbriar St Silver Spring MD 20903

RONY, PETER R(OLAND), b Paris, France, June 29, 39; US citizen; m 61; c 5. PHYSICAL CHEMISTRY, CHEMICAL ENGINEERING. *Educ:* Calif Inst Technol, BS, 60; Univ Calif, Berkeley, PhD(chem eng), 65. *Prof Exp:* Res specialist, Monsanto Co, 65-70; sr res engr, Esso Res & Develop Co, 70-71; assoc prof chem eng, 71-76, PROF CHEM ENG, VA POLYTECH INST & STATE UNIV, 76- *Concurrent Pos:* Dreyfuss Found Teacher-Scholar grant, 74; ed, Inst Elec & Electronics Engrs Micro, Inst Elec & Electronics Engrs Comput Soc, 83-84. *Honors & Awards:* Delox-Tektronix Award, 88. *Mem:* Am Chem Soc; Am Inst Chem Engrs; Inst Elec & Electronics Engrs; Am Soc Eng Educ. *Res:* Microcomputers; process instrumentation; chemical microengineering; process controls; engineering education laboratories; hardware systems. *Mailing Add:* Dept Chem Eng Va Polytech Inst & State Univ Blacksburg VA 24061

RONZIO, ROBERT A, b Boulder, Colo, Jan 24, 38. BIOCHEMISTRY, NUTRITION. *Educ:* Reed Col, BA, 60; Univ Calif, Berkeley, PhD(biochem), 66. *Prof Exp:* NIH fel biochem, Med Sch, Tufts Univ, 65-67; Am Cancer Soc fel, Univ Wash, 67-69; asst prof biochem, Mich State Univ, 69-74, assoc prof, 74-77; PROF & CHAIR BASIC MED SCI, BASTYR COL NATURAL HEALTH SCI, 80- *Concurrent Pos:* USPHS res career develop award, 72-77; resource fac prof, Evergreen State Col, 80-86; consult, Diagnos-Techs, Inc, Kent, Wash, 90- *Mem:* Am Chem Soc; fel Am Inst Chemists; Sigma Xi. *Res:* Formation and function of mammalian cell membranes; clinical nutrition. *Mailing Add:* 1712 Markham Ave NE Tacoma WA 98422

ROOBOL, NORMAN R, b Grand Rapids, Mich, Aug 19, 34; m 55; c 4. ORGANIC CHEMISTRY, MATERIALS SCIENCE. *Educ:* Calvin Col, BS, 58; Mich State Univ, PhD(org chem), 62. *Prof Exp:* Chemist, Shell Develop Co, 62-65; asst prof org chem, 65-68, assoc prof mat sci, 68-71, PROF MAT SCI, GEN MOTORS INST, 71- *Concurrent Pos:* Consult coatings appln, 79-; Rodes prof, Russelsheim, WGer; Dow fel. *Mem:* Am Inst Chemists; Soc Mfg Engrs; Soc Automotive Engrs. *Res:* paints and coatings application processes. *Mailing Add:* 114 Kenton Pl Peachtree City GA 30269

ROOD, JOSEPH LLOYD, b May 2, 22; m 49; c 4. PHYSICS. *Educ:* Univ Calif, AB, 43, MA, 47, PhD(physics), 48. *Prof Exp:* Asst physics, Univ Calif, 43-48; from instr to assoc prof, Univ San Francisco, 48-56; head mat physics lab, Bausch & Lomb, Inc, 56-67; prof physics, Univ Lethbridge, 67-87; RETIRED. *Mem:* AAAS; Am Phys Soc; Optical Soc Am; Am Ceramic Soc; Am Asn Physics Teachers. *Res:* Optical and solid state properties of materials, especially glass and other ceramics. *Mailing Add:* Box 1569 Ft MacLeod AB T0L 0Z0 Can

ROOD, ROBERT THOMAS, b Raleigh, NC, Mar 30, 42; m 78; c 2. ASTROPHYSICS. *Educ:* NC State Univ, BS, 64; Mass Inst Technol, PhD(physics), 69. *Prof Exp:* Res assoc physics, Mass Inst Technol, 69-71; res fel, Kellogg Lab, Calif Inst Technol, 71-73; asst prof, 73-78, ASSOC PROF ASTRON, UNIV VA, 78- *Mem:* AAAS; Am Astron Soc; Int Astron Union; Royal Astron Soc. *Res:* Stellar interiors; radio astronomy. *Mailing Add:* Dept Astron Univ Va Math-Astron Bldg Rm 314 Charlottesville VA 22903-0818

ROODMAN, STANFORD TRENT, b St Louis, Mo, Sept 17, 39; m 61; c 2. IMMUNOLOGY, BIOCHEMISTRY. *Educ:* Purdue Univ, BS, 61; Univ Mich, PhD(biochem), 68. *Prof Exp:* NIH Fel, Univ Calif, San Diego, 68-70; asst prof biochem, 70-74, asst prof, 74-85, ASSOC PROF PATH, MED SCH, ST LOUIS UNIV, 85- & DIR, GRAD PROG PATH, 78- *Concurrent Pos:* Vis asst prof, Med Sch, Wash Univ, 81-82, Howard Hughes doctoral fel, Selection Comt in Immunol, 88-92; consult, Vet Admin, 84-86, Sigma Chem, 90- *Mem:* AAAS; Am Asn Immunologists. *Res:* AIDS cellular immunology; pathology; characterization and role of Interleukin-6 in growth control of retroperitoneal fibromatosis cells from SRV-2 infected macaques with AIDS as an animal model for Kaposi's Sarcoma in human AIDS; flow cytometry. *Mailing Add:* Dept Path St Louis Univ Med Sch St Louis MO 63104-1028

ROOF, BETTY SAMS, b Columbia, SC, Apr 13, 26; div; c 4. INTERNAL MEDICINE, ENDOCRINOLOGY METABOLISM. *Educ:* Univ SC, BS, 44; Duke Univ, MD, 49. *Prof Exp:* Vol vis investr, Rockefeller Inst, NY, 49-50; from intern to asst resident med, Presby Hosp, NY, 50-53; vis fel, Col Physicians & Surgeons, Columbia Univ, 53-55; clin & res fel, Mass Gen Hosp, Boston, 55-56; asst, Rockefeller Inst, 56-57; res fel, Mass Gen Hosp, 57-59; asst res physician, Cancer Res Inst, Univ Calif, San Francisco, 62-63, assoc res physician, 67-71, lectr med, Dept Med & assoc res physician, Cancer Res Inst, 71-74, assoc clin prof med, Dept Med, 74; assoc prof, 74-80, asst dean student progress, 89-92, PROF MED, MED UNIV SC, 80- *Honors & Awards:* Pres Award, Am Col Obstet-Gynec, 75. *Mem:* Endocrine Soc; Int Endocrine Soc; Am Soc Bone & Mineral Res; Am Asn Cancer Res; fel Am Col Physicians; Am Fedn Clin Res. *Res:* Parathyroid hormone produced by non-endocrine tumors; metabolic bone disease; post-menopausal osteoporosis; breast cancer therapy with hormone; hyperparathyroidism. *Mailing Add:* Med Univ SC 171 Ashley Ave Charleston SC 29425

ROOF, JACK GLYNDON, b Cleburne, Tex, June 17, 13; m 41; c 3. PHYSICAL CHEMISTRY. *Educ:* Univ Calif, Los Angeles, BA, 34, MA, 35; Univ Wis, PhD(phys chem), 38. *Prof Exp:* From instr to asst prof chem, Ore State Col, 38-46; sr chemist, Shell Develop Co, 46-63, res assoc, 63-70; INSTR CHEM, GALVESTON COL, 70- *Concurrent Pos:* Chemist, Nat Defense Res Comt, Northwestern Univ, 42-45; consult, Univ Calif, 46; mem, Joint Task Force One, Bikini, 46. *Mailing Add:* 2017 46th St Galveston TX 77550

ROOF, RAYMOND BRADLEY, JR, b Battle Creek, Mich, Mar 3, 29; m 51; c 2. CRYSTALLOGRAPHY. *Educ:* Univ Mich, BS(chem eng) & BS(metall eng), 51, MS, 52, PhD(mineral, crystallog), 55. *Prof Exp:* Sr engr, Atomic Power Div, Westinghouse Elec Corp, 55-57; CRYSTALLOGR, LOS ALAMOS SCI LAB, 57- *Concurrent Pos:* Vis lectr, Univ Western Australia, 70; adj prof, Univ NMex, 71-72; mem, Joint Comt on Powder Diffraction Stands, 75- *Mem:* Am Crystallog Asn; Sigma Xi. *Res:* Crystallographic structure analysis; powder patterns; computer programming; metallurgical identifications; synthetic minerals. *Mailing Add:* Los Alamos Sci Lab MS 740 PO Box 1663 Los Alamos NM 87544

ROOK, HARRY LORENZ, b Middletown, Conn, May 31, 40; m 63; c 4. ANALYTICAL CHEMISTRY. *Educ:* Worcester Polytech Inst, BS, 62, MS, 67; Tex A&M Univ, PhD(analytical chem), 69. *Prof Exp:* Analytical chemist, Monsanto Res Corp, 62-65; res chemist, 69-75, sect chief neutron activation anal, 75-78, CHIEF GAS & PARTICULATE SCI DIV, CTR ANALYTICAL CHEM, NAT BUR STANDARDS, 78- *Concurrent Pos:* Sci consult to subcomt on environ & atmosphere, Comt Sci & Technol, US House of Representatives, 75; chmn radioactivity, Off Water Data Coord, 75-; chmn, D-22.03, Am Soc Testing & Mat. *Mem:* Am Chem Soc; Sigma Xi; Am Soc Testing & Mat. *Res:* Analytical methodology using nuclear techniques; studies into proper methods of sampling, storage and preservation of analytical samples; environmental analysis. *Mailing Add:* 312 E Main St Middletown CT 21769

ROOKS, H CORBYN, b Grand Rapids, Mich, Feb 9, 10; m 37; c 2. ENERGY, HEAT TRANSFER. *Educ:* Univ Mich, BS, 32, MS, 33. *Prof Exp:* Engr, Trane Co, 34-35, mgr coil sales, 35-37, mgr heat transfer sales, 37-51, vpres, 51-59, vpres eng, 59-68, vpres eng & res, 68-75; ENERGY CONSULT, US DEPT ENERGY, 75- *Concurrent Pos:* Consult, Dept Appl Sci, Brookhaven Nat Lab, 76-84, Off Technol Assessment, US Cong, 76-78 & Energy & Environ Div, Lawrence Berkeley Lab, 77- *Mem:* Am Soc Heating, Refrig & Air-Conditioning Engrs; Int Solar Energy Soc; Am Soc Eng Educ; Am Soc Mech Engrs; AAAS. *Res:* Solar energy. *Mailing Add:* 2607 Cass St La Crosse WI 54601

ROOKS, WENDELL HOFMA, II, b Ann Arbor, Mich, Oct 2, 31; m 55; c 5. PHARMACOLOGY. *Educ:* Calvin Col, AB, 53; Univ Mich, MS, 54. *Prof Exp:* Staff scientist, Worcester Found Exp Biol, Shrewsbury, Mass, 56-64; asst dept head bioassay, Inst Hormone Biol, Syntex Corp, 64-65, head dept bioassay, 65-80, asst dir inst biol sci, Res Div, 73-86, HEAD SCI INFO DEPT, SYNTEX USA INC, 87- *Mem:* Am Soc Pharmacol & Exp Therapeut. *Res:* Steroid and prostaglandin bioassay; anti-inflammatory, analgesic and immuno-suppressive pharmacology; reproductive physiology; acne vulgaris. *Mailing Add:* Syntex Corp 3401 Hillview Ave Palo Alto CA 94304

ROOM, ROBIN GERALD WALDEN, b Sidney, New South Wales, Australia, Dec 28, 39; m 87; c 3. ALCOHOL & DRUG EPIDEMIOLOGY, ALCOHOL SOCIAL SCIENCE. *Educ:* Princeton Univ, AB, 60; Univ Calif, Berkeley, MA, 62, MA, 67, PhD(sociol), 78. *Prof Exp:* Res assoc, Drinking Practices Study, Ment Res Inst, 65-69; res scientist, Social Res Group, Western Off, George Washington Univ, 69-71; lectr & prin admin analyst behav sci, Sch Pub Health, Univ Calif, Berkeley, 72-82, sci dir alcohol epidemiol, Social Res Group, 77-81, adj prof social & admin health sci, Sch Pub Health, 82-91; DIR & SR SCIENTIST ALCOHOL EPIDEMIOL, ALCOHOL RES GROUP & INST EPIDEMIOL & BEHAV MED, MED RES INST SAN FRANCISCO, 81- *Concurrent Pos:* Chair, Drinking & Drugs Div, Soc Study Social Probs, 75-77; prin investr & sci dir, Nat Alcohol Res Ctr grant, 77-91; panel mem, Nat Acad Sci-Inst Med, 78-82 & 86-87; chair & mem, Alcohol Abuse Prev Rev Comt, Nat Inst Alcohol Abuse & Alcoholism, 79-82; mem & adv, Expert Adv Panel on Drug Dependence & Alcohol Probs, WHO, 79-91. *Mailing Add:* Alcohol Res Group 2000 Hearst Ave Berkeley CA 94709-2176

ROON, ROBERT JACK, b Grand Rapids, Mich, Nov 3, 43; m 66; c 1. BIOCHEMISTRY, MICROBIOLOGY. *Educ:* Calvin Col, BS, 65; Univ Mich, Ann Arbor, MS, 67, PhD(biochem), 69. *Prof Exp:* Teaching asst biochem, Univ Mich, Ann Arbor, 65-69; Am Cancer Soc fel, Univ Calif, Berkeley, 70-71; asst prof, 71-77, ASSOC PROF BIOCHEM, UNIV MINN, MINNEAPOLIS, 77- *Concurrent Pos:* NSF fel, Univ Calif, Berkeley, 78. *Res:* Regulation of nitrogen metabolism in saccharomyces mechanism of amino acid transport in saccharomyces; biochemistry of protein secretion in saccharomyces. *Mailing Add:* Dept Biochem 4-225 Millard Univ Minn Sch Med 435 Delaware St SE Minneapolis MN 55455

ROONEY, JAMES ARTHUR, b Springfield, Vt, Sept 15, 43; m 67; c 1. PHYSICS. *Educ:* Clark Univ, Mass, BA, 65; Univ Vt, MS, 67, PhD(physics), 70. *Prof Exp:* Res assoc physics, Univ Vt, 71-72; asst prof physics, Univ Maine, 72-76, assoc prof, 76-; MEM STAFF, JET PROPULSION LAB, PASADENA, CALIF. *Mem:* Am Asn Advan Med Instrumentation; Am Inst Ultrasound Med; Acoust Soc Am; AAAS. *Res:* Biomedical ultrasonics concerning biological effects of ultrasound on membranes and enzymes as well as applications of sound and ultrasound to dispersal of cell aggregates; blood pressure measurements and coagulation studies. *Mailing Add:* Jet Propulsion Lab 4800 Oak Grove Dr Pasadena CA 91109

ROONEY, LAWRENCE FREDERICK, b Conrad, Mont, Nov 21, 26; m 56; c 3. ECONOMIC GEOLOGY. *Educ:* Univ Mont, BA, 48, MA, 50; Ind Univ, PhD(geol), 56. *Prof Exp:* Jr geologist, Mobil Oil Can, 56-58; asst prof geol, Univ Tex, 58-59; geologist, Humble Oil & Refining Co, 60-62 & Ind Geol Surv, 62-70; prof geol, Flathead Valley Community Col, 70-75; GEOLOGIST, US GEOL SURV, 75- *Mem:* Geol Soc Am. *Res:* Geology of industrial minerals. *Mailing Add:* PO Box 3245 Evergreen CO 80439

ROONEY, LLOYD WILLIAM, b Atwood, Kans, July 17, 39; m 63; c 3. FOOD SCIENCE. *Educ:* Kans State Univ, BS, 61, PhD(cereal chem), 66. *Prof Exp:* Res asst grain sci, Kans State Univ, 63-65; from asst prof to assoc prof cereal chem, 65-77, PROF FOOD SCI NUTRIT, TEX A&M UNIV, 77- *Mem:* AAAS; Am Asn Cereal Chemists; Inst Food Technologists; Am Chem Soc; Am Soc Agron. *Res:* Research program on determination of physical, chemical, nutritional and processing properties of cereal grains, especially sorghum, maize and wheat, involving close cooperation with plant breeders to use genetic material for cereal improvement. *Mailing Add:* Cereal Qual Lab Soil & Crop Sci Tex A&M Univ College Station TX 77843

ROONEY, PAUL GEORGE, b New York, NY, July 14, 25; Can citizen; m 50; c 5. MATHEMATICS. *Educ:* Univ Alta, BSc, 49; Calif Inst Technol, PhD(math), 52. *Prof Exp:* Lectr math, Univ Alta, 52-54, asst prof, 54-55; from asst prof to assoc prof, 55-62, PROF MATH, UNIV TORONTO, 62- *Concurrent Pos:* Ed-in-chief, Can J Math, 71-75. *Mem:* Am Math Soc; Math Asn Am; Can Math Soc (vpres, 79-81, pres, 81-83); fel Royal Soc Can. *Res:* Functional analysis. *Mailing Add:* Dept Math Univ Toronto Toronto ON M5S 1A1 Can

ROONEY, SEAMUS AUGUSTINE, b Cork, Ireland, Dec 19, 43; US citizen; m 68; c 3. LIPID BIOCHEMISTRY, LUNG BIOCHEMISTRY. *Educ:* Nat Univ, Ireland, BSc, 64, MSc, 66; Dublin Univ, PhD(biochem), 69, ScD(biochem), 90. *Prof Exp:* Fel lipid chem, Dept Microbiol, Univ Pa, 69-72; res assoc 72-77, sr res assoc, 77-81, sr res scientist, 81-90, RES PROF PEDIAT, DEPT PEDIAT, SCH MED, YALE UNIV, 90- *Mem:* Am Soc Biochem & Molecular Biol; Soc Pediat Res; Perinatal Res Soc; Am Thoracic Soc; Am Physiol Soc; Am Soc Cell Biol. *Res:* Hormonal control of pulmonary surfactant production during fetal development; regulation of surfactant secretion. *Mailing Add:* Dept Pediat Sch Med Yale Univ PO Box 3333 New Haven CT 06510-8064

ROONEY, THOMAS PETER, b New York, NY, June 29, 32; m 65; c 2. GEOLOGY. *Educ:* City Col New York, BS, 59; Columbia Univ, MA, 62, PhD(geol), 65. *Prof Exp:* GEOPHYSICIST, TERRESTRIAL SCI DIV, AIR FORCE GEOPHYS LAB, 66- *Mem:* Geol Soc Am; Am Geophys Union. *Res:* Satellite altimetry; remote sensing; rock deformation. *Mailing Add:* Earth Sci Div Phillips Lab Gravity Br Hanscom AFB MA 01731

ROONEY, VICTOR MARTIN, b Paris, Ill, Oct 9, 37; m 60; c 1. ELECTRONICS ENGINEERING. *Educ:* Univ Dayton, BEE, 65; Ohio State Univ, MSc, 70. *Prof Exp:* From instr to assoc prof, 70-78, PROF ELECTRONIC ENG, UNIV DAYTON, 78- *Concurrent Pos:* Bioeng consult, Miami Valley Hosp, 73-86; ad hoc visitor, Engrs Coun Prof, 73-81; regist eng, State Ohio. *Mem:* Inst Elec & Electronics Engrs. *Res:* Microprocessors; programmable logic controllers bioengineering; analysis of linear circuits; passive and active components. *Mailing Add:* Dept Electronic Eng Technol 300 College Park Dayton OH 45469

ROOP, RICHARD ALLAN, b Decatur, Ind, Jan 5, 55; m 74; c 3. TECHNICAL SERVICE ADMISTRATION, FOOD REGULATIONS. *Educ:* Purdue Univ, BS, 77, PhD(food sci), 81. *Prof Exp:* Group leader, Central Soya Co, Inc, 81-85; SCI DIR, HOLLY FARMS FOODS, INC, 85- *Mem:* Inst Food Technologists; Sigma Xi. *Res:* Develoment and improvement of products for the frozen food business; quality assurance; technical service; microbiology; production; process engineering. *Mailing Add:* Three Grandview Acres-A Millers Creek NC 28651-9770

ROOP, ROBERT DICKINSON, b Plainfield, NJ, Sept 23, 49; m 81; c 2. BIOLOGY, ECOLOGY. *Educ:* Hiram Col, BA, 71; State Univ NY Stony Brook, MA, 75. *Prof Exp:* Staff ecologist, Inst Ecol, 74-76; res assoc, Oak Ridge Nat Lab, 76-89; PROJ DIR, LABAT-ANDERSON INC, 89- *Mem:* AAAS; Am Inst Biol Sci; Ecol Soc Am; Water Pollution Control Fedn; Nat Asn Environ Prof. *Res:* Environmental impact assessment for energy programs and facilities; waste management; risk assessment for hazardous site clean-up; quality criteria for aquatic sediments. *Mailing Add:* Labat-Anderson Inc 575 Oak Ridge Turnpike Oak Ridge TN 37830-7173

ROORDA, JOHN, b Tzummarum, Neth, June 22, 39; Can citizen; m 62; c 4. CIVIL ENGINEERING, APPLIED MECHANICS. *Educ:* Univ Waterloo, BASc, 62; Univ London, PhD(struct eng), 65. *Prof Exp:* Fel, civil eng, Northwestern Univ, 65-66; from asst prof to assoc prof, 66-74, PROF CIVIL ENG, UNIV WATERLOO, 74- *Concurrent Pos:* Sr vis fel, Sci Res Coun, Eng, 68; vis prof, Cranfield Inst Technol, Eng, 71 & Univ Col, London, 74 & 78. *Res:* Structural analysis; stability of structures; dynamic response of structures; vibration problems; theory of elasticity. *Mailing Add:* Dept Civil Eng Univ Waterloo Waterloo ON N2L 3G1 Can

ROOS, ALBERT, b Leyden, Neth, Nov 22, 14; nat US; m 46; c 2. PHYSIOLOGY. *Educ:* State Univ Groningen, MD, 40. *Prof Exp:* Fel cardiol, 46-47, from instr to asst prof physiol, 47-54, assoc prof physiol & surg, 54-61, assoc prof physiol, 61-70, PROF PHYSIOL & BIOPHYS, SCH MED, WASH UNIV, 70-, RES PROF ANESTHESIOL, 61- *Mem:* AAAS; Am Physiol Soc. *Res:* Physiology of respiration; membrane transport. *Mailing Add:* Dept Cell Biol & Physiol Sch Med Wash Univ 660 S Euclid Ave St Louis MO 63110

ROOS, C(HARLES) WILLIAM, b Cairo, Ill, July 2, 27; m 50; c 3. CHEMICAL ENGINEERING, BIOCHEMICAL ENGINEERING. *Educ:* Wash Univ, BS, 48, MS, 49, DSc(chem eng), 51. *Prof Exp:* Asst chem eng, Wash Univ, 50-51; res chem eng, Monsanto Co, 51-56, group leader chem eng res, 56-59, sect leader, 59-63, technologist, 63-66, mgr res & develop pioneering res, 66-68, mgr petrol additives res, 68-70, dir technol planning & eval, 70-75, gen mgr, New Enterprise Div, 75-77, dir, Corp Res Labs, 77-79, dir, Technol Admin, 79-82; ASSOC PROF, CHEM ENG, AUBURN UNIV, 83- *Concurrent Pos:* Lectr, Wash Univ, 51- & St Louis Univ, 59- *Mem:* Am Inst Chem Engrs; Nat Soc Prof Engrs; Am Chem Soc. *Res:* Catalysis; biochemistry; chemical process development; reaction kinetics; research project evaluation; resource allocation; computer aided design. *Mailing Add:* 915 Cherokee Rd Auburn AL 36830

ROOS, CHARLES EDWIN, b Chicago, Ill, Apr 23, 27; m 52; c 4. ELEMENTARY PARTICLE PHYSICS. *Educ:* Univ Tex, BS, 48; Johns Hopkins Univ, PhD(physics), 53. *Prof Exp:* Assoc res staff, Johns Hopkins Univ, 53; from instr to asst prof physics, Univ Calif, Riverside, 54-59; from assoc prof to prof, 59-89, EMER PROF PHYSICS, VANDERBUILT UNIV, 89-, PROF RADIOL SCI, MED SCH, 84- *Concurrent Pos:* Res fel, Calif Inst Technol, 56-67; guest physicist, Brookhaven Nat Lab, CERN, Fermi lab; pres & chmn, Nat Recovery Technol, 83-; prof radiol sci, Vanderbilt Univ Med Sch, 84; treas & dir, Cryomagnetic, Inc, 66-69; dir, Am Magnetics, Inc, 69-85. *Mem:* Fel Am Phys Soc; NY Acad Sci; Sigma Xi. *Res:* Auger transitions; dibaryons, studies of charm and beauty guards, neutrino actrophysics (deep underwater muon and neutrino detection); nonferrons metal sorting-processing of municipal solid waste; photo mesonic reactions; superconductivity; hyperon magnetic moments; over 150 publications and holder of 16 patents. *Mailing Add:* Vanderbilt Univ Box 1807 Nashville TN 37235

ROOS, FREDERICK WILLIAM, b Sault Ste Marie, Mich, June 4, 40; m 63; c 1. AEROSPACE ENGINEERING. *Educ:* Univ Mich, BSE, 63, MSE, 65, PhD(aerospace eng), 68. *Prof Exp:* SCIENTIST, FLIGHT SCI DEPT, MCDONNELL DOUGLAS RES LABS, MCDONNELL DOUGLAS CORP, 68- *Mem:* Am Inst Aeronaut & Astronaut. *Res:* Unsteady separated fluid flows with emphasis on aircraft buffeting at low and transonic speed. *Mailing Add:* 11311 Clayton Rd St Louis MO 63131

ROOS, HENRY, b Jersey City, NJ, Dec 25, 21; m 54; c 1. MICROSCOPIC ANATOMY, PHYSIOLOGY. *Educ:* City Col New York, BS, 43; Columbia Univ, MA, 46, EdD(sci educ), 61; Univ NC, MPH, 52. *Prof Exp:* Asst prof biol, Westfield State Teachers Col, 47-48; instr, Springfield Col, 48-49; asst prof, ECarolina Col, 49-50; health educator, Hudson County Tuberc & Health Asn, 52-54; instr biol, Westchester Community Col, 55-61; assoc prof, 61-68, PROF BIOL, EASTERN CONN STATE COL, 68- *Mem:* AAAS; Nat Asn Biol Teachers. *Res:* Hematopoiesis in mouse spleen; energy potential of Connecticut farm wastes. *Mailing Add:* 247 Highland View Dr South Windham CT 06226

ROOS, JOHN FRANCIS, b Seattle, Wash, Jan 18, 32; m 56; c 3. FISHERIES MANAGEMENT. *Educ:* Univ Wash, BS, 55. *Prof Exp:* Proj leader chignik sockeye salmon studies, Fisheries Res Inst, Univ Wash, 55-60; chief biologist, Int Pac Salmon Fisheries Comn, 68-71, asst dir, 71-82, dir, 82-85; VPRES, PAC SEAFOOD PROCESSORS ASN, 88- *Mem:* Am Fisheries Soc; Am Inst Fishery Res Biologists. *Mailing Add:* Pac Seafood Processors Asn 4019 21st Ave W Suite 201 Seattle WA 98199

ROOS, LEO, b Amsterdam, Neth, Nov 10, 37; US citizen; m 61; c 3. PHYSICAL ORGANIC CHEMISTRY, PHOTOCHEMISTRY. *Educ:* City Col New York, BS, 61; Univ Cincinnati, PhD(phys org chem), 65. *Prof Exp:* Res chemist, Photoproducts Dept, E I du Pont de Nemours & Co, Inc, 65-71, sr res chemist, 71-76; dir new imaging systs, Xidex Corp, 76-84, TECH DIR, DYNACHEM CORP, 84- *Concurrent Pos:* Tech dir, Exxon Epid Div, 84; consult, coating technol, Photopolymer Mft. *Mem:* AAAS; Am Chem Soc; Soc Photog Sci & Eng. *Res:* Photopolymerization; homogeneous catalysis, specifically reaction of cobalt carbonyls with olefins; low energy visible light catalyzed; photopolymer reactions; adhesion between metal substrates and polymer under extreme conditions; electroless plating; catalytic reactions; coating Rheology pertaining to coating large scale planar devices. *Mailing Add:* Dynachem Corp 2631 Michelle Dr Tustin CA 92680

ROOS, PHILIP G, b Wauseon, Ohio, May 16, 38; m 63; c 1. NUCLEAR PHYSICS. *Educ:* Ohio Wesleyan Univ, BA, 60; Mass Inst Technol, PhD(physics), 64. *Prof Exp:* Vis asst prof physics, Univ Md, 64-65; AEC fel nuclear physics, Oak Ridge Nat Lab, 65-70; from asst prof to assoc prof nuclear physics, 71-75, PROF PHYSICS & ASTRON, UNIV MD, COLLEGE PARK, 75- *Res:* Study of nuclear reactions and nuclear structure using particle accelerators. *Mailing Add:* Dept Physics & Astron Univ Md College Park MD 20742

ROOS, RAYMOND PHILIP, b Brooklyn, NY, Apr 5, 44; m 67; c 1. NEUROLOGY, VIROLOGY. *Educ:* Columbia Col, BA, 64; State Univ NY Downstate Med Ctr, MD, 68. *Prof Exp:* Intern med, State Univ NY, Kings County Hosp, Brooklyn, 68-69; staff assoc spec chronic dis study sect, Nat Inst Neurol Dis & Stroke, 69-71; resident neurol, Johns Hopkins Univ, 71-74, instr neurol & NIH fel neurovirol, 74-78; MEM STAFF, DEPT NEUROL, MED CLINS, UNIV CHICAGO, 78- *Concurrent Pos:* Consult neurol, Moore Genetics Clin, Johns Hopkins Univ, 75. *Mem:* Am Acad Neurol. *Res:* Relationship of virus infections, especially defective, slow or unconventional infections to neurological degenerative diseases; virological etiology to certain neurological heritable diseases. *Mailing Add:* Dept Neurol Univ Chicago Box 425 Chicago IL 60637

ROOS, THOMAS BLOOM, b Peoria, Ill, Mar 19, 30; m 53; c 2. ZOOLOGY. *Educ:* Harvard Univ, AB, 51; Univ Wis, MS, 53, PhD(zool), 60. *Hon Degrees:* AM, Dartmouth, 71. *Prof Exp:* Instr zool, Univ Wis, 60; from instr to prof zool, 60-71, chmn dept, 69-70, 78-83, PROF BIOL, DARTMOUTH COL, 71- *Concurrent Pos:* USPHS spec fel, 66-67; mem comn undergrad educ biol sci, NSF, 66-70; hon res fel, Univ Col, London, 74-75; Fulbright fel, 88-89; Fulbright fel, Bhabba Atomic Res Ctr (BASC), CCMB, IICB, Univ Delhi, India, 88-89. *Mem:* AAAS; Am Soc Zool. *Res:* Phylogenetic and ontogenetic development and genetic control of hormonal secretory function and responsivity; theoretical biology; use of computers in biology. and teaching of biology; automation of data collection and analysis; invention, master scan interpretive densitometer. *Mailing Add:* Dept Biol Sci Dartmouth Col Hanover NH 03755

ROOSA, ROBERT ANDREW, b Manila, Philippines, June 18, 25; US citizen; m 89; c 3. MEDICAL MICROBIOLOGY, GENETICS. *Educ:* Univ Conn, BA, 50; Univ Pa, PhD(med microbiol), 57. *Prof Exp:* Res asst, Sch Med, Yale Univ, 50-52; fel, Nat Cancer Inst, 57-60; res assoc, 60-64, dep dir sci serv, 69-74, ASSOC PROF, WISTAR INST ANAT & BIOL, 64-, CUR MUS, 69-, SCI ADMINR, 75- *Concurrent Pos:* NSF fel, 57; Eleanor Roosevelt Int Cancer Res fel, Med Res Coun Exp Virus Res Unit, Univ Glasgow, 67-68; ed, Info Newsletter Somatic Cell Genetics, 69-73. *Mem:* AAAS; Nat Coun Univ Res Admin; Soc Res Admin; Tissue Cult Asn; Am Asn Cancer Res; Radiation Res Soc; Am Assoc Lab Animal Sci; Am Soc Microbiol. *Res:* Nuclear cytology; cancer chemotherapy; mechanisms of drug resistance; nutritional requirements of cells in culture; somatic cell genetics; tumor transplantation; role of the thymus in immunobiology; slow virus diseases of mammals. *Mailing Add:* Wistar Inst Anat & Biol 36th & Spruce Sts Philadelphia PA 19104

ROOSENRAAD, CRIS THOMAS, b Lansing, Mich, July 28, 41; m 64; c 1. MATHEMATICS. *Educ:* Univ Mich, BSc, 63, MSc, 64; Univ Wis, PhD(math), 69. *Prof Exp:* Asst prof math, Williams Col, 69-75, assoc dean & lectr, 75-83; DEAN STUDENTS & LECTR, CARLETON COL, 83- *Mem:* Am Math Soc; Math Asn Am. *Res:* Orthogonal polynomials; special functions. *Mailing Add:* Dean Students Carleton Col Northfield MN 55057-4020

ROOSEVELT, C(ORNELIUS) V(AN) S(CHAAK), b New York, NY, Oct 23, 15. MINING ENGINEERING, ELECTRONICS. *Educ:* Mass Inst Technol, BS, 38. *Prof Exp:* Engr, Am Smelting & Refining Co, Mex, 38-41; mgr, Mining Div, William Hunt & Co, China, 46-49, pres & dir, 49-50; res administr, Off Naval Res, 52-62; pres & dir, Linderman Eng Co, 54-68; vpres, US Banknote Co, Philadelphia, 65-70; DIR, COLUMBIA RES CORP, DC, 68- *Concurrent Pos:* Pres & dir, Int Industs, Inc, Hong Kong, 49-50; vpres, Security Banknote Co, Philadelphia, 49-65; consult, US Intel Bd, 64-74; mem bd trustees, Aerospace Corp, Los Angeles, 78-88. *Mem:* AAAS; Soc Indust Archaeol; NY Acad Sci. *Res:* Industrial electronic control equipment; industrial archaeology; naval electronic equipment; scuba equipment and underwater photography. *Mailing Add:* 2500 Que St NW Apt 604 Washington DC 20007-4320

ROOT, ALLEN WILLIAM, b Philadelphia, Pa, Sept 24, 33; m 58; c 3. PEDIATRIC ENDOCRINOLOGY. *Educ:* Dartmouth Col, AB, 55; Harvard Univ, MD, 58; Am Bd Pediat, dipl, 64, dipl endocrinol, 78, 86. *Prof Exp:* Intern, Strong Mem Hosp, Rochester, NY, 58-60; resident pediat, Hosp Univ Pa, Philadelphia, 60-62; fel pediat endocrinol, Children's Hosp Philadelphia, 62-65; assoc physician pediat, Sch Med, Univ Pa, 64-66, asst prof, 66-69; from assoc prof to prof, Sch Med, Temple Univ, 69-73; dir univ serv, All Children's Hosp, St Petersburg, 73-89; PROF PEDIAT & HEAD SECT PEDIAT ENDOCRINOL, COL MED, UNIV SFLA, 73- *Concurrent Pos:* Asst physician endocrinol, Children's Hosp Philadelphia, 65-66; USPHS career develop award, 68-69; chmn, Div Pediat, Albert Einstein Med Ctr, Philadelphia, 69-73; consult ed, J Pediat, 73-81 & J Adolescent Health Care, 79-; mem med adv bd, Nat Pituitary Agency, 74-89. *Mem:* Endocrine Soc; AAAS; Am Pediat Soc; Am Acad Pediat; Soc Pediat Res; Lawson Wilkins Pediat Endocrinol Soc (pres, 88-89). *Res:* Investigation of factors which regulate function of the hypothalamic-pituitary unit and the mechanisms of pubertal maturation and the effects of prenatal events upon development of the hypothalamic-pituitary axis. *Mailing Add:* All Children's Hosp Med Serv 801 Sixth St S St Petersburg FL 33731

ROOT, CHARLES ARTHUR, b Rochester, NY, Aug 25, 38; m 63; c 2. INORGANIC CHEMISTRY. *Educ:* Ohio Wesleyan Univ, BA, 60; Ohio State Univ, MSc, 62, PhD(chem), 65. *Prof Exp:* From instr to assoc prof, 65-82, PROF CHEM, BUCKNELL UNIV, 82- *Concurrent Pos:* NSF fac fel, Calif Inst Technol, 71-72, vis assoc prof, 79-80; res fel, Univ Mich, 90-91. *Mem:* AAAS; Am Chem Soc; Soc Environ Geochem & Health; Sigma Xi. *Res:* Thermal and photochemistry of vanadium complexes; complexes of oxazolidines. *Mailing Add:* Dept Chem Bucknell Univ Lewisburg PA 17837

ROOT, DAVID HARLEY, b Columbus, Ohio, Oct 10, 37; m 67; c 2. MATHEMATICS, ENERGY RESOURCES. *Educ:* Mass Inst Technol, BS, 59; Univ Wash, PhD(math), 68. *Prof Exp:* Asst prof math & statist, Purdue Univ, 68-74; MATHEMATICIAN, US GEOL SURV, 74- *Mem:* AAAS. *Res:* Study of the methods of extimation of the remnants of fossil fuels and of the potentials of various non-fossil fuel energy resources. *Mailing Add:* 12227 Quorn Lane Reston VA 22090

ROOT, ELIZABETH JEAN, b Lewiston, Maine, Sept 27, 31; m 52, 75; c 3. ELECTRON MICROSCOPY, NEUROSCIENCE. *Educ:* Univ Kans, BA, 53; Univ Tex, Austin, MA, 72, PhD(biol sci), 80. *Prof Exp:* Intern dietetics, Univ Kans Med Ctr, 54; lab & teaching asst, 72-78, asst instr, 78-79, instr 80-81, LECTR NUTRIT, UNIV TEX, AUSTIN, 81- *Mem:* Electron Micros Soc Am; Soc Neurosci; assoc Am Inst Nutrit; AAAS. *Res:* Electron microscopy and light microscopy of brain and other mammalian and avian tissues in order to study effects of nutrient deficiencies, interactions and metabolism; evaluation of human nutrient intakes. *Mailing Add:* Dept Home Econ Univ Tex GEA 115 Austin TX 78712

ROOT, HARLAN D, b Riders Mills, NY, Feb 16, 26; m 53; c 4. SURGERY. *Educ:* Cornell Univ, AB, 50, MD, 53; Univ Minn, PhD(surg), 61, Am Bd Surg, dipl, 61. *Prof Exp:* Univ fel, Univ Minn, 54-60, Nat Cancer Soc fel, 55-59, Am Cancer Soc fel, 59-61, instr surg, Med Sch, 60-66, instr surg, Med Sch, 61, asst prof, 61-66; assoc dir surg, Ancker Hosp, St Paul, Minn; assoc prof, 66-67, PROF SURG, UNIV TEX HEALTH SCI CTR, SAN ANTONIO, 67-, ASST CHMN, 66- *Concurrent Pos:* Consult, San Antonio State Chest Hosp, 66-, Audie L Murphy Mem Vet Admin Hosp, San Antonio, 73-; surg test comt, Nat Bd Med Examiners, 77-; dist comt applicants, 77- & comt trauma, Am Col Surg, 77-80; gov, Am Col Surgeons, 89- *Mem:* AMA; fel Am Col Surgeons; Soc Univ Surgeons; Am Asn Surg Trauma; Am Surg Asn. *Res:* Peripheral vasular surgery. *Mailing Add:* Dept Surg 7703 Floyd Curl Dr San Antonio TX 78284

ROOT, JOHN WALTER, b Kansas City, Mo, Oct 5, 35; m 60; c 2. PHYSICAL CHEMISTRY, RADIOCHEMISTRY. *Educ:* Univ Kans, BA, 58, PhD(chem), 64; Univ Calif, Los Angeles, cert, 66. *Prof Exp:* From asst prof to assoc prof chem, 66-75, assoc dir, Crocker Nuclear Lab, 74-77, PROF CHEM, UNIV CALIF, DAVIS, 75- *Concurrent Pos:* Consult, Gen Elec Co, 68-70; John Simon Guggenheim fel, 72-73. *Mem:* Am Chem Soc; Am Phys Soc; Am Nuclear Soc; AAAS; Sigma Xi. *Res:* Radiochemistry; hot-atom chemistry; chemical kinetics; radiation, isotope and physical chemistry. *Mailing Add:* 1333 Lincoln St Dept 386 Bellingham WA 98226

ROOT, L(EONARD) EUGENE, b Lewiston, Idaho, July 4, 10; m 35; c 3. AEROSPACE SCIENCE. *Educ:* Univ Pac, BA, 32; Calif Inst Technol, MS, 33 & 34. *Hon Degrees:* ScD, Col Pac, 58. *Prof Exp:* Asst chief aerodyn sect, Douglas Aircraft Co, Inc, Calif, 34-39, chief sect, 39-46, mem staff, Spec Eng Proj, 46-48; chief aircraft div, Rand Corp, 48-53; dir develop planning dept, Lockheed Aircraft Corp, 53-56, gen mgr missile & space div & vpres corp, 56-59, group vpres missiles & electronics, 59-61, pres, Lockheed Missiles & Space Co, 61-69, group vpres, Lockheed Aircraft Corp, 69-70; RETIRED. *Concurrent Pos:* Lectr, Calif Inst Technol, 37-38; consult, Hughes Aircraft Co, 42-43; mem aerodyn comt, Nat Adv Comt Aeronaut, 44-50; chmn aerodyn adv panel, AEC, Sandia Corp, 48-50; mem sci adv bd, Chief Staff, USAF, 48-59; adv, USAF Inst Air Weapons Res, Univ Chicago, 50-52; spec asst to dep chief staff, Develop Hq, USAF, 51-52; mem defense sci bd, Dir Defense, Res & Eng, 57-70. *Mem:* Nat Acad Eng; fel Am Inst Aeronaut & Astronaut (vpres, 58, pres, 62); fel Am Astronaut Soc; fel Royal Aeronaut Soc. *Res:* Aircraft design; aerodynamics; dynamic longitudinal stability; empennage design; flying qualities; systems and operational analyses in military sciences; development planning methods; missiles and spacecraft; aerospace management. *Mailing Add:* 1340 Hillview Dr Menlo Park CA 94025

ROOT, MARY AVERY, b Hartford, Conn, Oct 28, 18. PHARMACOLOGY. *Educ:* Oberlin Col, AB, 40; Radcliffe Col, MA, 49, PhD, 50. *Prof Exp:* Pharmacologist, Res Labs, Eli Lilly & Co, 50-63, res assoc, 63-85; RETIRED. *Mem:* Endocrine Soc; Am Soc Pharmacol & Exp Therapeut; Soc Exp Biol & Med; Am Diabetes Asn; NY Acad Sci. *Res:* Carbohydrate metabolism; insulin, diabetes. *Mailing Add:* 4425 N Emerson Indianapolis IN 46226

ROOT, PAUL JOHN, b Pittsburgh, Pa, Apr 1, 29; m 56; c 4. OIL & GAS RESERVOIR ENGINEERING, NATURAL GAS ENGINEERING. *Educ:* Pa State Univ, Bs, 52, MS, 54; Univ Tex, Austin, PhD(petrol eng), 61. *Prof Exp:* Engr, Consolidated Gas Supply Corp, 53-55; instr petrol eng, Pa State Univ, 55-56; asst prof, Univ Tex, Austin, 56-61; res engr, Gulf Res & Develop Co, 61-65; assoc prof petrol eng, Univ Okla, 65-70, prof, 73-75; tech dir, Natural Gas Surv, Fed Energy Regulatory Comn, 70-73; dir, Educ Progs, H Zinder & Assocs, 75-77; PRES, PETROL & GEOL ENG, INC, 77-

Concurrent Pos: Lectr, Carnegie-Mellon Univ, 64-65; res engr, Bur Bus & Econ Res, 68-69; consult engr, Fed Energy Regulatory Comn, 73-74; res fel, Off Sci & Pub Policy, 74-75; energy consult, Fed Energy Admin, 78-79 & Pertamina, Indonesian Nat Oil Co. 82-83. *Mem:* Soc Petrol Engrs.; Am Soc Eng Educ; AAAS. *Res:* Natural gas engineering and operations; reservoir engineering; enhanced oil recovery; energy supplies and utilization. *Mailing Add:* 1839 Rolling Hills Norman OK 73072

ROOT, RICHARD BRUCE, b Dearborn, Mich, Sept 7, 36; div; c 2. ECOLOGY. *Educ:* Univ Mich, BS, 58; Univ Calif, Berkeley, PhD(zool), 64. *Prof Exp:* Assoc zool, Univ Calif, Berkeley, 60-62; from asst prof to assoc prof, 64-79, PROF ECOL, CORNELL UNIV, 79- *Concurrent Pos:* Mem field staff, Rockefeller Found, Colombia, 70-71; vis prof, Univ Valle, Colombia, 70-71; ed, Ecol & Ecol Monogr, 71-73; res assoc, Mus Vert Zool, Univ Calif, Berkeley, 75; bd dir, Orgn Trop Studies, 76-; vis scientist, Oxford Univ, 83. *Honors & Awards:* Howell Award, Cooper Ornith Soc, 63. *Mem:* Ecol Soc Am (vpres, 79-80, pres, 85-86); Asn Trop Biol; Brit Ecol Soc; Am Ornith Union; Entom Soc Am. *Res:* Comparative ecology of insects; adaptive syndromes of birds and insects; plant/herbivore interactions; differences in the structure of natural and agricultural ecosystems; functional classifications. *Mailing Add:* Sect Ecol & Systematics Cornell Univ Cornell Hall Ithaca NY 14853

ROOT, RICHARD KAY, INTERNAL MEDICINE, INFECTIOUS DISEASES. *Educ:* Johns Hopkins Univ, MD, 63. *Prof Exp:* PROF MED & CHMN DEPT, UNIV CALIF, SAN FRANCISCO, 85- *Mailing Add:* Dept Med Univ Calif Box 0120 997 M San Francisco CA 94143-0120

ROOT, SAMUEL I, b Winnipeg, Man, Mar 1, 30; m 52; c 3. GEOLOGY. *Educ:* Univ Man, BSc, 52, MSc, 56; Ohio State Univ, PhD(geol), 58. *Prof Exp:* Geologist, Seaboard Oil Co, Can, 52-53; sr geologist, Int Petrol Co Ltd, Colombia, Peru, 57-63; staff geologist, Pa State Geol Surv, 63-65, chief field geologist, 66-78; mem staff, Esso Prospeccao Ltd, Brazil, 78-82; COL WOOSTER, 83- *Concurrent Pos:* Consult, World Bank; Keck grant, 90 & 91, Hughes grant, 89-91 & NSF grant, 90. *Mem:* Geol Soc Am; Am Asn Petrol Geologists; Soc Econ Geologists. *Res:* Areal and structural geology; stratigraphy; nonmetallic mineral deposits. *Mailing Add:* Dept Geol Col Wooster Wooster OH 44691

ROOT, WILLIAM L(UCAS), b Des Moines, Iowa, Oct 6, 19; m 40; c 2. ENGINEERING. *Educ:* Mass Inst Technol, PhD(math), 52. *Prof Exp:* Instr math, Mass Inst Technol, 47-51, mem staff, Instrumentation Lab, 51-52 & Lincoln Lab, 52-56, asst leader, Systs Res Group, 57-59, leader anal group, 59-61; prof, 61-87, EMER PROF AEROSPACE ENG, ELEC ENG & COMPUTER SCI, UNIV MICH, ANN ARBOR, 88- *Concurrent Pos:* Vis lectr, Harvard Univ, 58-59; vis prof, Math Res Ctr, Univ Wis, 63-64, Univ Calif, Berkeley, 66-67, Mich State Univ, 66 & 68; NSF sr fel & vis fel, Clare Hall, Univ Cambridge, 70; mem, Army Sci Bd, 79-83. *Honors & Awards:* Com Con Conf Bd Career Achievement Award, 87; Shannon lectr, Info Theory Soc, Inst Elec & Electronics Engrs, 86. *Mem:* Fel Inst Elec & Electronics Engrs; Am Math Soc. *Res:* Stochastic processes, statistical theory of communications and general system theory; system modelling and identification; estimation and detection theory; information theory. *Mailing Add:* Dept Aerospace Eng Univ Mich Ann Arbor MI 48109

ROOTARE, HILLAR MUIDAR, b Tallinn, Estonia, Apr 26, 28; US citizen; m 59; c 6. DENTAL MATERIALS, SURFACE CHEMISTRY. *Educ:* Wagner Col, BS, 52; Univ Mich, PhD(dent mat & pharmaceut chem), 73. *Prof Exp:* Chemist & res assoc, Bone Char Res Proj, Inc, Nat Bur Stand, 57-63; dir, Mat Technol Lab, Am Instrument Co, Inc, div Travenol Labs, 63-66; NIH trainee & res asst, Univ Mich, Ann Arbor, 66-73; dir tech res, L D Caulk Co, div Dentsply Int Inc, 73-75; sr res assoc, Dept Dent Mat, Sch Dent, Univ Mich, Ann Arbor, 75-77; RES SCIENTIST, MICROMERITICS INSTRUMENT CORP, 78- *Mem:* Int Asn Dent Res; Am Chem Soc; Am Soc Metals; Fine Particle Soc; Sigma Xi. *Res:* Physical and surface chemistry applied to use of hydroxyapatite as synthetic bone, in characterizing of surfaces, compaction and sintering of powders to reproduce the porous structures of bone for possible use as implants; design and construction of microprocessor controlled instrumentation for automatic measurements and analysis of pore size distributions of porous materials. *Mailing Add:* Adv Composite Mat Corp 1525 S Buncombe Rd Greer SC 29651-9208

ROOT-BERNSTEIN, ROBERT SCOTT, b Washington, DC, Aug 7, 53; m 78; c 2. BIOLOGY, IMMUNOLOGY. *Educ:* Princeton Univ, AB, 75, PhD(hist sci), 80. *Prof Exp:* Postdoctoral fel theoret biol, Salk Inst Biol Studies, 81-83, res assoc, 83-84; res assoc neurobiochem, Vet Admin Hosp, Brentwood, Calif, 85-87; asst prof natural sci, 88-90, ASSOC PROF PHYSIOL, DEPT PHYSIOL, MICH STATE UNIV, 90- *Concurrent Pos:* mem adv bd, J Theoret Biol, 84-91; vis assoc prof sci & art, Univ Calif, Los Angeles, 87; consult, Calif Mus Sci & Indust, 87-88, Biolark, Inc, 88-89, Res Div, Parke-Davis Pharmaceut, 90-91; mem bd dirs, Impression 5 Sci Mus, 89-90; contrib ed, Sciences, 89-91. *Honors & Awards:* MacArthur Prize, J&CD MacArthur Found, 81. *Res:* Theoretical and experimental investigation of autoimmunity; theory and applications of molecular complementarity; peptide, neurotransmitter and drug interactions; historical and philosophical studies of discovery process; research management. *Mailing Add:* Dept Physiol Mich State Univ East Lansing MI 48824

ROOTENBERG, JACOB, b Afula, Israel, Mar 23, 36; m 68; c 2. ELECTRICAL ENGINEERING. *Educ:* Israel Inst Technol, BSc, 60, MSc, 62, PhD(elec eng), 67. *Prof Exp:* Asst elec eng, Israel Inst Technol, 60-62, instr, 62-65, lectr, 65-67; asst prof, Columbia Univ, 67-72, assoc prof, 72-76; PROF DEPT COMPUT SCI, QUEENS COL, NY, 76- *Mem:* Inst Elec & Electronics Engrs; Sigma Xi. *Res:* Control systems; optimal control; stability of nonlinear systems; simulation and bioengineering problems. *Mailing Add:* 127 Falmouth St Brooklyn NY 11235

ROOTHAAN, CLEMENS CAREL JOHANNES, b Nymegen, Neth, Aug 29, 18; US citizen; m 50; c 5. PHYSICS. *Educ:* Inst Technol, Delft, MS, 45; Univ Chicago, PhD(physics), 50. *Prof Exp:* Instr, Cath Univ Am, 47-49; res assoc, 49-50, from instr to prof, 50-58, prof commun & info sci, 65-68, dir comput ctr, 62-68, LOUIS BLOCK PROF PHYSICS & CHEM, UNIV CHICAGO, 68- *Concurrent Pos:* Guggenheim fel, Cambridge Univ, 57; consult, Argonne Nat Lab, 58-66, Lockheed Missiles & Space Co, 60-65, Union Carbide Corp, 65- & IBM Corp, 65-; guest professorships, Ohio State, Columbus, Ohio, 76, Tech Univ, Lyngby, Denmark, 83, & Univ Del, 87-88. *Mem:* Fel Am Phys Soc; Int Acad Quantum Chem; corresp mem, Royal Dutch Acad Sci. *Res:* Theory of atomic and molecular structure; application of digital computers to scientific problems. *Mailing Add:* Dept Chem Univ Chicago 5735 S Ellis Ave Chicago IL 60637

ROOTS, BETTY IDA, b South Croydon, Eng. ZOOLOGY, NEUROBIOLOGY. *Educ:* Univ Col, Univ London, BSc, 49, dipl educ, 50, PhD(zool), 53, DSc, 81. *Prof Exp:* Asst lectr biol, Royal Free Hosp Sch Med, Univ London, 53-59; vis asst prof physiol, Univ Ill, Urbana, 59-61; asst lectr biol, Royal Free Hosp Sch Med, Univ London, 61-62, lectr anat, Univ Col, 62-66; vis scientist physiol, Univ Ill, Urbana, 66-67; res neuroscientist, Univ Calif, San Diego, 68-69; assoc prof zool, 69-72, PROF ZOOL, UNIV TORONTO, 72-, CHMN, 84- *Concurrent Pos:* Rose Sidgwick Mem fel, Univ Ill, Urbana, 59-60. *Mem:* Am Soc Neurochem; Am Oil Chem Soc; Soc Neurosci; Can Soc Zool; Int Soc Develop Neurosci; Int Soc Neurochem. *Res:* Structural and chemical changes in nervous system in relation to environmental factors; cell isolation techniques; structure and function of glial cells. *Mailing Add:* Dept Zool Univ Toronto 25 Harbord St Toronto ON M5S 1A1 Can

ROOTS, ERNEST FREDERICK, b Salmon Arm, BC, July 5, 23; m 55; c 5. GEOLOGY. *Educ:* Univ BC, MASc, 47; Princeton Univ, PhD(geol), 49. *Hon Degrees:* DSc, Univ Victoria, 83. *Prof Exp:* Surveyor, Nat Parks Serv, Can, 41-42; asst, Geol Surv Can, 43-44, tech officer, 45-47, geologist, 48-49; asst prof geol, Princeton Univ, 52-54; geologist, Geol Surv Can, 55-58, coordr, Polar Continetal Shelf Proj, 58-72; sr adv, Dept Energy Mines & Resources, 72-73; sci adv, 73-89, EMER SCI ADV, DEPT ENVIRON, 89- *Concurrent Pos:* Sr Geologist, Norweg-Brit-Swed Antarctic Exped, 49-52, 54-55; polar res bd, US Nat Acad Sci, 69-82; pres, Int Comn Snow & Ice, 79-84; mem coun, Comt Arctique, 80-; co-chmn, Can-USSR coop Arctic Sci, 84-91; gov & chmn rev comt, Arctic Inst NAm, 80-86; chmn comt, Int Arctic Sci, 90- *Honors & Awards:* Queen Elizabeth II Polar Medal, UK, 56; Patron's Medal, Royal Geog Soc, 65; Massey Medal, Royal Can Geog Soc, 79. *Mem:* Geol Soc Am; Am Geophys Union; Glaciol Soc; Arctic Inst NAm; fel Royal Soc Can. *Res:* Tectonics and mineral deposits of Canadian cordillera; geology of Antarctica and the Himalayas; glaciology; geology and geophysics of Arctic North America and the Arctic Ocean basin; environmental policy; climate change of polar regions; international arctic science policies. *Mailing Add:* Dept Environ Fontaine Bldg Ottawa ON K1A 0H3 Can

ROOVERS, J, b Deurne, Belg, June 3, 37; m 63; c 3. POLYMER CHEMISTRY, PHYSICAL CHEMISTRY. *Educ:* Cath Univ Louvain, BSc, 59, PhD(polymer chem), 62. *Prof Exp:* Nat Res Coun Can fel, 63-64; assoc pharmacol, Cath Univ Louvain, 65-66; assoc res officer, 67-77, SR RES OFFICER POLYMER CHEM, NAT RES COUN CAN, 77- *Mem:* Am Chem Soc; Can High Polymer Forum (secy, 77). *Mailing Add:* 21 Wren Rd Gloucester ON K1J 7H5 Can

ROOZEN, KENNETH JAMES, b Milwaukee, Wis, Jan 17, 43; m 81; c 3. GENETICS, MOLECULAR BIOLOGY. *Educ:* Lakeland Col, BS, 66; Univ SDak, MA, 68; Univ Tenn, Oak Ridge, PhD(microbial genetics), 71. *Prof Exp:* NIH fel pediat, Med Sch, Wash Univ, 71-74; from asst prof to assoc prof, Univ Ala, Birmingham, 81-87, vchmn, dept microbiol, 77-81, dean & co-dir, Grad Sch, 81-88, chmn, 84-87, asst vpres health affairs, 86-88, exec asst to pres, 88-89, vpres univ affairs, 89-90, PROF, UNIV ALA, BIRMINGHAM, 88-, VPRES RES & UNIV AFFAIRS, 90- *Concurrent Pos:* Policy fel, Robert Wood Johnson-Nat Acad Sci Health, 83-84. *Mem:* AAAS; Am Soc Microbiol; Sigma Xi; Am Soc Cell Biologists. *Res:* Mammalian cell genetics and biochemistry, specifically mutant isolation, correction of genetic defects, gene mapping and nucleic acid metabolism. *Mailing Add:* Vpres Res & Univ Affairs Univ Ala Birmingham AL 35226

ROPER, CLYDE FORREST EUGENE, b Ipswich, Mass, Oct 1, 37; m 58; c 2. BIOLOGICAL OCEANOGRAPHY, SYSTEMATIC ZOOLOGY. *Educ:* Transylvania Col, AB, 59; Univ Miami, Fla, MS, 62, PhD(marine sci), 67. *Prof Exp:* Sr res asst oceanog, Inst Marine Sci, Univ Miami, 64-66; assoc cur, 66-72, supvr mollusks, 68-70, CUR, SMITHSONIAN INST, 72- *Concurrent Pos:* adj lectr, George Washington Univ, 68-; affil grad fac, Dept Oceanog, Univ Hawaii, 75-76; chmn, Dept Invertebrate Zool, Nat Mus Natural Hist, Smithsonian Inst, 80-85; founding mem, 81, co-chmn, 81-83, coun mem, 81-86, int workshop coordr, Cephaloped Int Adv Coun, 87-88; Smithsonian Pub Coun; Nat Mus Natural Hist rep, Am Inst Biol Sci Coun. *Mem:* Soc Syst Zool; Australian Malacol Soc; Am Malacol Union (pres-elect, 78-80); Marine Biol Asn UK; Inst Malacol; Sigma Xi. *Res:* Systematics, distribution behavior and ecology of recent Cephalopoda (squid, octopus, cuttlefish) of the world, particularly oceanic forms; phylogenetic relationship of families and orders; functional anatomy of bioluminescent organs, skin structures; fisheries research utilization of cephalopods. *Mailing Add:* Dept Invert Zool NHB-118 Nat Mus Nat Hist Smithsonian Inst Washington DC 20560

ROPER, GERALD C, b Tewksbury, Mass, Dec 18, 33; m 57; c 3. PHYSICAL CHEMISTRY. *Educ:* Univ Boston, AB, 56, PhD(phys chem), 66. *Prof Exp:* From asst prof to assoc prof phys chem, 62-74, PROF CHEM, DICKINSON COL, 74-, CHMN DEPT, 71- *Mem:* Am Chem Soc. *Res:* Inorganic synthesis; transition metal chemistry. *Mailing Add:* Dickinson Col Carlisle PA 17013

ROPER, L(EON) DAVID, b Shattuck, Okla, Dec 13, 35; m 55; c 2. THEORETICAL PARTICLE PHYSICS, BIOPHYSICS. *Educ:* Okla Baptist Univ, AB, 58; Mass Inst Technol, PhD(pion-nucleon interaction), 63. *Prof Exp:* Asst, Mass Inst Technol, 58-63; fel, Lawrence Radiation Lab, Univ Calif, 63-65; asst prof physics, Ky Southern Col, 65-67; from asst prof to assoc prof, 67-74, actg head dept, 77-78, PROF PHYSICS, VA POLYTECH INST & STATE UNIV, 74- *Concurrent Pos:* Instr, Eastern Nazarene Col, 62-63; mem staff, KEK Nat Lab High Energy Physics, Japan, 80-81. *Mem:* Fel Am Phys Soc; Am Asn Physics Teachers; World Future Soc. *Res:* Particle scattering phenomenology; nonrenewable resource depletion; computational physics. *Mailing Add:* Dept Physics Va Polytech Inst & State Univ Blacksburg VA 24061-0435

ROPER, MARYANN, b Bayonne, NJ, July 17, 49. CANCER RESEARCH. *Educ:* Univ Del, BA, 71; Pa State Univ, MD, 75; Am Bd Pediat, dipl, 83. *Prof Exp:* Intern & resident pediat, Univ Colo Med Ctr, 75-77; fel pediat immunol, Univ Ala, Birmingham, 77-79, fel Pediat hemat/oncol, 80-81, asst prof med & pediat & asst dir, Bone Marrow Transplant Unit, 82; interim fac, Div Pediat, Hemat/Oncol, Georgetown Univ, Washington, DC, 82-83; pvt pract pediat, Birmingham, Ala, 83-84; sr investr, Biologics Eval Sect, Cancer Ther Eval Prog, Nat Cancer Inst, 85-86, spec asst to dir, 86-87, actg dep dir, 87-89, dep dir, 89-90; SCI CONSULT, CARTER CTR, EMORY UNIV, ATLANTA, 90- *Concurrent Pos:* Mem, Pediat Oncol Group: Leukemia Markers Ref Lab Coordr, Immunol & Marrow Transplant Comt; reviewer, J Blood, J Clin Oncol, Cancer Ther Reports; mem, Children's Cancer Study Group, Biologics Protocol Adv Comt, Ad Hoc Subcomt Prev & Control Res. *Mem:* Am Acad Pediat. *Res:* Childhood leukemia; immunologic cell markers and their influence on disease prognosis and on the use of experimental biologic therapy for the treatment of childhood malignancies. *Mailing Add:* Carter Ctr Emory Univ One Copenhill Atlanta GA 30307

ROPER, PAUL JAMES, b Detroit, Mich, June 29, 39; div. STRUCTURAL GEOLOGY, GEOLOGICAL EXPLORATION FOR OIL & GAS. *Educ:* Univ Mich, Ann Arbor, BS, 62; Univ Nebr, Lincoln, MS, 64; Univ NC, Chapel Hill, PhD(geol), 70. *Prof Exp:* From instr to asst prof geol, Lafayette Col, 69-76; lectr, Univ Wis, Oshkosh, 76-77; assoc prof, Univ SW La, 77-78; prod geologist, Superior Oil Co, 78-80; explor geologist, Ramco Explor, 80-81; pres, Aalpha Explor Co, 81-83; explor geologist, Celeron Oil & Gas Co, 83-85; GEO CONSULT, 85- *Concurrent Pos:* Explor geologist, Phillip D Parker, 78; consult, explor geol. *Mem:* Geol Soc Am; Am Geophys Union; Am Geol Inst; Am Asn Petrol Geologists; Earthquake Eng Res Inst. *Res:* Geology and tectonics of Brevard zone, Southern Appalachian Mountains, Motagua fault zone and Sierra de las Minas Mountains, Guatemala; proposed theory of plastic plate tectonics; post-Jurassic tectomism in eastern North America; exploration for oil & gas. *Mailing Add:* 1835 Weldon Houston TX 77073

ROPER, ROBERT, b Vienna, Austria, Jan 1, 28; US citizen; m 58; c 2. POLYMER CHEMISTRY. *Educ:* City Col New York, BS, 51; NY Univ, PhD(org chem), 57. *Prof Exp:* Chemist, Esso Res & Eng Co, 57-60, sr chemist, 60-65, res assoc surface coatings, Rubbers, Adhesives & Sealants, 65-78, RES ASSOC INFO ANALYSIS, EXXON RES & ENG CO, 78- *Mem:* Am Chem Soc. *Res:* Development of polymers for elastomers, sealants and coatings; effect of polymer structure on properties; polymer synthesis; information retrieval and analysis; patent and technical literature. *Mailing Add:* 22 Oakland Pl Summit NJ 07901

ROPER, ROBERT GEORGE, b Adelaide, Australia, Apr 30, 33; m 58; c 4. ATMOSPHERIC PHYSICS. *Educ:* Univ Adelaide, BSc, 57 & 58, PhD(physics), 63. *Prof Exp:* Demonstr physics, Univ Adelaide, 58, Radio Res Bd grant & res officer upper atmosphere, 62-63; Nat Acad Sci-Nat Res Coun resident res assoc, NASA, Goddard Space Flight Ctr, Md, 64-65; assoc prof aerospace eng, Ga Inst Technol, 65-66; sr res scientist, Australian Defence Sci Serv, 66-69; assoc prof to prof aerospace eng, 69-76, prof aerospace eng & geophys sci, 77-80, PROF ATMOSPHERIC SCI, GA INST TECHNOL, 80- *Concurrent Pos:* Pres, Int Comt Meteorol Upper Atmosphere, Int Asn Meteorol & Atmospheric Physics, 79-87; Middle Atmosphere Prog Steering Comt, Int Coun Sci Unions, 82-89; co-chmn, Global Meteor Observations Systs, Int Astron Union, Int Union Geod & Geophys, 82- *Mem:* Am Geophys Union; Am Meteorol Soc; fel Australian Inst Physics; Int Asn Geomagnetism & Aeronomy; fel Royal Meteorol Soc. *Res:* Design and construction of instrumentation for the measurement of lower and upper atmosphere winds and shears using radio and acoustic techniques; analysis and interpretation of wind data from various sources. *Mailing Add:* Sch Earth & Atmospheric Sci Ga Inst Technol Atlanta GA 30332-0340

ROPER, STEPHEN DAVID, b Rock Island, Ill, May 30, 45; m 85; c 2. DEVELOPMENTAL NEUROBIOLOGY, NEUROPLASTICITY. *Educ:* Harvard Col, BA, 67; Univ Col, London, PhD(physiol), 70. *Prof Exp:* Fel neurobiol, Harvard Med Sch, 70-73; asst prof anat, 73-79, asst prof physiol, 76-79, assoc prof anat & physiol, 79-85, PROF ANAT & PHYSIOL, MED SCH, UNIV COLO, 85-; PROF ANAT & NEUROBIOL, COLO STATE UNIV, 85- *Concurrent Pos:* Fulbright fel, 67-69; instr neurobiol, Harvard Med Sch, 71-73; NIH res career develop award; consult, NIH Site Visit Teams, 77-; mem, NIH Neurobiol Study Sect, 81-86; chmn, Dept Anat & Neurobiol, Colo State Univ, 82-89; counr, Asn Chem Reception Sci, 88-91. *Mem:* Am Asn Anat; Soc Neurosci; Am Physiol Soc. *Res:* Chemosensory transduction mechanisms in taste; development and maintenance of synaptic connections in the vertebrate, focusing on trophic interactions between neurons and their targets and upon the regeneration of neural connections after damage. *Mailing Add:* Dept Anat & Nerobiol Colo State Univ Ft Collins CO 80523

ROPER, WILLIAM L, b Birmingham, Ala, July 6, 48; m; c 1. MEDICINE. *Educ:* Univ Ala, BS, 70, MD, 74, MPH, 81. *Prof Exp:* Resident pediat, Med Ctr, Univ Colo, Denver, 74-77; health officer, Jefferson County Dept Health, Birmingham, Ala, 77-83; spec asst to Pres, health policy, White House, Washington, DC, 83-86; adminr, Health Care Financing Admin, Washington, DC, 86-89; dep asst to Pres, domestic policy & dir, White House Off Policy Develop, White House, 89-90; DIR, CTRS DIS CONTROL, ATLANTA, GA, 90-; ADMINR, AGENCY TOXIC SUBSTANCES & DIS REGISTRY, ATLANTA, GA, 90- *Concurrent Pos:* Community fel, Dept Health & Hosps, City & County, Denver, Colo, 75-76; asst state health officer, Ala Dept Pub Health, Birmingham, 81-83; White House fel, 82-83. *Mem:* Inst Med-Nat Acad Sci; AMA; Am Acad Pediat; Am Pub Health Asn. *Res:* Author or co-author of over 25 publications. *Mailing Add:* Ctrs Dis Control 1600 Clifton Rd NE Atlanta GA 30333

ROPES, JOHN WARREN, fish & wildlife sciences; deceased, see previous edition for last biography

ROPP, GUS ANDERSON, b Columbia, SC, July 31, 18; wid. ORGANIC CHEMISTRY. *Educ:* Univ SC, BS, 40; Univ Tenn, PhD(org chem), 49. *Prof Exp:* Prod chemist, Gen Chem Co, 41-43; res chemist, Rohm & Haas Co, Pa, 43-45; res chemist, Oak Ridge Nat Lab, 48-61; res scientist & group leader, Union Carbide Corp, 61-64; prof, 65-84, EMER PROF CHEM, COKER COL, 85- *Concurrent Pos:* USPHS res fel, Phys Chem Lab, Oxford Univ, 55-56. *Mem:* Am Inst Chemists; Am Chem Soc; Am Asn Univ Prof. *Res:* Effect of isotopes on reaction rates; isotopic tracers and organic reaction mechanisms; Diels-Alder reactions and synthesis of dienes; applications of mass spectrometry; gas chromatography; electroorganic chemistry. *Mailing Add:* 301 Park Ave Hartsville SC 29550-3237

ROPP, RICHARD C, b Detroit, Mich, Mar 26, 27; m 52; c 4. SOLID STATE CHEMISTRY. *Educ:* Franklin Col, AB, 50; Purdue Univ, West Lafayette, MS, 52; Rutgers Univ, Newark, PhD(phys chem), 71. *Prof Exp:* Adv develop engr, Sylvania Elec Prod, Pa, 52-63; mgr luminescence, Westinghouse Elec, NJ, 63-71; pres, Luminescence Technol, 71-73; consult, Allied Chem Corp, 72-73, staff scientist solid state chem, 73-77; dir technol, 78-79, vpres, Petrex Corp, 80-83; vpres, Enhanced Oil Recovery, Can, 84-87; environ consult, 87-88; RES SPECIALIST CHEM, RUTGERS UNIV, NEWARK, 71- *Concurrent Pos:* Vpres, Int Superconductor, New York, 89-90. *Mem:* Am Chem Soc; Am Inst Chemists; AAAS; fel Royal Soc Chem. *Res:* Luminescent materials; laser hosts; phosphate glass; nuclear waste encapsulation; petroleum recovery. *Mailing Add:* 138 Mountain Ave Warren NJ 07060

ROPP, WALTER SHADE, b Lakeland, Fla, Oct 15, 22; m 49; c 3. CHEMISTRY. *Educ:* Fla Southern Col, BS, 43; Pa State Col, MS, 44, PhD, 48. *Prof Exp:* Asst, Pa State Col, 43-44 & 45-47; res chemist, Exp Sta, 47-52, res supvr, 52-55, sr tech rep, 56-57, sr res chemist, Coatings Div, Res Ctr, 58-61, supvr coatings develop, Polymers Dept, 61-65, sr res chemist, Mat Res Div, 66-68, res assoc, 68-79, SR RES ASSOC, MAT SCI DIV, HERCULES INC, 79- *Concurrent Pos:* Instr, Pa State Col, 46-47. *Mem:* Am Chem Soc. *Res:* Plastics, elastomers, organic coatings and water soluble polymers. *Mailing Add:* 440 S Gulfview Blvd No 1503 N Clearwater Beach FL 33515-2518

ROQUEMORE, LEROY, b Columbia, La, Apr 6, 35. MATHEMATICAL MODELLING, COMPUTER SIMULATION. *Educ:* Southern Univ, BS, 57; La State Univ, MS, 63; Univ Southwest La, PhD(comput sci), 84. *Prof Exp:* Instr math, Southern Univ, 57-62; assoc res engr, Boeing Co, 63-65; asst prof math, 65-68, assoc prof comput sci, 68-72 & 74-84, dir comput ctr, 72-74, PROF COMPUT SCI, SOUTHERN UNIV, 84-, CHMN DEPT, 74- *Concurrent Pos:* Software engr, Western Elec, 68; assoc prof systs sci, La State Univ, 78-; lectr, IBM In-House Educ, 83-; prin investr, NSF, 79-81, Raytheon Data Syst, 81-82, IBM & NASA, 84- *Mem:* Asn Comput Mach; Inst Elec & Electronics Engrs; Asn Educ Data Systs; Data Processing Mgt Asn. *Res:* Design of automated auction mechanisms; effects of computer assisted instruction in science at developing institutions of higher education; technological information transfer. *Mailing Add:* Dept Comput Sci Southern Univ & A&M Col Baton Rouge LA 70813

ROQUES, ALBAN JOSEPH, b Paulina, La, Feb 3, 41; m 66; c 2. MATHEMATICAL ANALYSIS. *Educ:* Nicholls State Univ, BS, 63; La State Univ, MS, 65, PhD(math), 74. *Prof Exp:* Math analyst comput sci, Space Div, Chrysler Corp, 65-68; instr math, Southeastern La Univ, 68-69; instr, 71-72 & 74-76, asst prof math & comput sci, 76-79, assoc prof, 79-85, PROF MATH, LA STATE UNIV, EUNICE, 85-, COORDR DATA PROCESSING, 81- *Mem:* Math Asn Am. *Res:* Evolution equations in general Banach spaces. *Mailing Add:* 1471 W Elm St Eunice LA 70535

ROQUITTE, BIMAL C, b Calcutta, India, Sept 29, 31; m 61; c 2. PHYSICAL CHEMISTRY. *Educ:* Univ Calcutta, BSc, 52, MSc, 55; Univ Rochester, PhD(phys chem), 61. *Prof Exp:* Res assoc chem, Res Found, Ohio State Univ, 61-62; vis fel phys chem, Mellon Inst, 62; fel, Nat Res Coun Can, 62-64 & Mellon Inst, 64-66; ASSOC PROF PHYS CHEM, UNIV MINN, MORRIS, 66- *Mem:* Am Chem Soc. *Res:* Unimolecular decomposition of cyclobutane carboxaldehyde; photochemistry of bicyclic hydrocarbons; energy transfer in cyclopentanone both in the gas and solid phase; flash photolysis of hydrocarbon in the far ultraviolet. *Mailing Add:* Dept Math & Sci Univ Minn Morris MN 56267

RORABACHER, DAVID BRUCE, b Ypsilanti, Mich, June 8, 35; m 58; c 4. ANALYTICAL CHEMISTRY, INORGANIC CHEMISTRY. *Educ:* Univ Mich, BS, 57; Purdue Univ, PhD(anal chem), 63. *Prof Exp:* Res engr, Ford Motor Co, 57-59; PROF ANALYTICAL CHEM, WAYNE STATE UNIV, 63- *Concurrent Pos:* NIH fel, Max Planck Inst Phys Chem, 64-65; assoc dean, Wayne State Univ, 84-85. *Mem:* Am Chem Soc. *Res:* Kinetics and mechanisms of coordination reactions and electron transfer reactions; nonaqueous solvation effects; macrocyclic ligand complexes. *Mailing Add:* Dept Chem Wayne State Univ Detroit MI 48202

RORABAUGH, DONALD T, b Phoenixville, Pa, Sept 8, 44; c 1. MATERIALS SCIENCE ENGINEERING. *Educ:* Drexel Inst Technol, Philadelphia, Pa, BS, 67, MS, 69; Fla Inst Technol, MS, 86, MBA, 89. *Prof Exp:* Proj engr armament res & develop, US Army, 68-78, team leader, 78-84, sr proj engr, 84-87, PROJ MGR ARMAMENT RES & DEVELOP, US ARMY, 87- *Concurrent Pos:* Asst instr, Drexel Inst Technol, 67-69; boy scout leader, Boy Scouts Am, 69- *Mem:* Am Soc Metals; Am Inst Mining & Metall Eng; NY Acad Sci; Res Soc NAm; Am Defense Preparedness Asn; Fed Bus Asn. *Res:* Coordinate the efforts of government and industry on research and development efforts aimed at improving armor and anti-armor monitions; author of over 100 publications; granted three patents. *Mailing Add:* PO Box 477 Netcong NJ 07828

RORER, DAVID COOKE, b Darby, Pa, Oct 25, 37; m 61; c 3. NUCLEAR ENGINEERING, RESEARCH DIRECTION. *Educ:* Mass Inst Technol, BS, 59; Univ Ill, MS, 61; Duke Univ, PhD(physics), 64. *Prof Exp:* Res assoc physics, Duke Univ, 63-65; assoc physicist, Brookhaven Nat Lab, 65-72; reactor engr nuclear eng, Long Island Lighting Co, 72-75; DEP DIV MGR, REACTOR DIV, BROOKHAVEN NAT LAB, 80- *Concurrent Pos:* Adj prof, Polytech Inst New York, 76-84. *Mem:* Am Phys Soc; Am Nuclear Soc. *Res:* Neutron physics; nuclear cryogenics. *Mailing Add:* Ten Pine Path Port Jefferson NY 11777

RORIG, KURT JOACHIM, b Bremerhaven, Ger, Dec 1, 20; US citizen; m 49; c 3. MEDICINAL CHEMISTRY, PHARMACOLOGY. *Educ:* Univ Chicago, BS, 42; Carleton Col, MA, 44; Univ Wis, PhD(chem), 47. *Prof Exp:* Chemist, J Seagram & Sons, 42-43; res chemist, G D Searle & Co, 47-60, asst dir, 60-74, assoc dir chem res, 74-79, sect head cardiovasc & renal res, 74-86; PRES, CHEMI-DELPHIC CONSULT, 86-; ADJ PROF MED CHEM, UNIV ILL, CHICAGO, 89- *Concurrent Pos:* Lectr, Loyola Univ Chicago, 50-60 & 87-88. *Mem:* AAAS; Am Chem Soc; NY Acad Sci; Am Soc Pharmacol & Exp Therapeut. *Res:* Cardiovascular and psychotropic drugs; aldol condensations; synthesis of aliphatic disulfonic acids, pyrimidines, imidazoles, oxazolines, steroids and steroid analogs; drug toxicology. *Mailing Add:* 337 Hager Lane Glenview IL 60025

RORK, EUGENE WALLACE, b Beatrice, Nebr, Mar 22, 40; div; c 4. SPACE SURVEILLANCE, ELECTRO OPTICS. *Educ:* Ohio State Univ, BS, 62, MS, 65, PhD(physics), 71. *Prof Exp:* Vis res assoc & lectr physics, dept physics, Ohio State Univ, 71-73; physicist, USAF Avionics Lab, 73-75; MEM STAFF, LINCOLN LAB, MASS INST TECHNOL, 75- *Res:* Ground-based electro optical sensor systems and techniques for detection of artificial satellites in space from reflected sunlight at night and in daytime; photoelectronic imaging devices; computer-controlled telescopes; atmospheric optical phenomena; Mossbauer-effect spectroscopy of gadolinium and dysprosium nuclei. *Mailing Add:* Lincoln Lab PO Box 73 Lexington MA 02173

RORK, GERALD STEPHEN, b Horton, Kans, Feb 12, 47; m 69; c 3. PHARMACEUTICAL CHEMISTRY. *Educ:* Univ Kans, BS, 69, MS, 73, PhD(pharmaceut chem), 74. *Prof Exp:* Jr chemist, Cook Paint & Varnish Co, 69-71; sr res scientist pharmaceut res & develop, Wyeth labs, Inc, 74-80; with Riker Labs Inc, 78-81; ASSOC DIR PHARMACEUT CHEM, INTERX RES CORP, 81- *Mem:* Am Chem Soc; Am Asn Pharmaceut Scientists. *Res:* Improvement of drug bioavailability and dosage from stability; physical organic chemistry; mechanisms of elimination reactions and nucleophilic addition reactions involving slow proton transfer steps; design and development of drug delivery systems. *Mailing Add:* Interx Res Corp 2201 W 21st St Lawrence KS 66044

RORKE, LUCY BALIAN, b St Paul, Minn, June 22, 29; m 60. NEUROPATHOLOGY. *Educ:* Univ Minn, BA, 51, MA, 52, BS, 55, MD, 57. *Prof Exp:* Intern med, Philadelphia Gen Hosp, 57-58, resident physician path, 58-61, NIH fels neuropath, 61-62 & neonatal brain path, 63-69, asst neuropathologist & pediat pathologist, 62-68, chief neuropathologist, 68-69, chmn, dept anat path, 69-73, pres med staff, 73-75, chmn, dept path, 73-77; PROF PATH, SCH MED, UNIV PA, 73-, CLIN PROF NEUROL, 79- *Concurrent Pos:* Consult neuropathologist, Wyeth Res Labs, 62-87, Wistar Inst Anat & Biol, 67- & Inst Merieux, Lyons, France, 69-70; neuropathologist, Children's Hosp Philadelphia, 65-, pres med staff, 86-88. *Mem:* Am Asn Neuropath (pres, 81-82); Am Neurol Asn; Am Acad Neurol; Col Am Path. *Res:* Pediatric neuropathology and viral diseases of the nervous system; health sciences. *Mailing Add:* Dept Path Children's Hosp Philadelphia Philadelphia PA 19104-4303

RORRES, CHRIS, b Philadelphia, Pa, Jan 2, 41. MATHEMATICS. *Educ:* Drexel Univ, BS, 63; NY Univ, MS, 65, PhD(math), 69. *Prof Exp:* Asst prof, 70-80, ASSOC PROF MATH, DREXEL UNIV, 80- *Mem:* AAAS; Sigma Xi; Soc Indust & Applied Math; Am Math Soc; Math Asn Am. *Res:* Population dynamics; harvesting of renewable resources; solar energy; acoustic scattering. *Mailing Add:* Dept Math & Comput Sci Drexel Univ Philadelphia PA 19104

RORSCHACH, HAROLD EMIL, JR, b Tulsa, Okla, Nov 5, 26; m 51; c 2. PHYSICS. *Educ:* Mass Inst Technol, BS, 49, MS, 50, PhD(physics), 52. *Prof Exp:* From instr to prof, 52-81, chmn dept, 66-73, SAM & HELEN WORDEN PROF PHYSICS, RICE UNIV, 81- *Concurrent Pos:* Guggenheim fel, 61; vis prof, Baylor Col Med. *Mem:* Fel Am Phys Soc. *Res:* Low temperature and solid state physics; biophysics. *Mailing Add:* Dept Physics Rice Univ PO Box 1892 Houston TX 77251

RORSTAD, OTTO PEDER, b Alesund, Norway, Mar 26, 47; Can citizen; m 74; c 2. ENDOCRINOLOGY. *Educ:* Univ BC, BSc, 69, MD, 72; McGill Univ, PhD(exp med), 80. *Prof Exp:* Instr med, Harvard Univ, 79-80; from asst prof to assoc prof, 80-88, PROF MED, UNIV CALGARY, 88- *Concurrent Pos:* Head, Div Endocrinol & Metab, Dept Med, Univ Calgary, 90-; assoc ed, Clin & Investigative Med, 90- *Mem:* Endocrine Soc; Royal Col Physicians & Surgeons Can; Can Soc Clin Invest; Can Soc Endocrinol & Metab; Can Hypertension Soc. *Res:* Regulation of the vasculature by neurotransmitters and hormones with particular reference to vasoactive intestinal peptide; vascular receptor for vasoactive intestinal peptide and the mechanism of action of vasoactive intestinal peptide in normal and hypertensive models. *Mailing Add:* Dept Med Univ Calgary 3330 Hospital Dr NW Calgary AB T2N 4N1 Can

ROSA, CASIMIR JOSEPH, b Poland, 1933; US citizen; m. OXIDATION, DIFFUSION. *Educ:* Acad Mining & Metall, Cracow, Poland, BS & MS, 56; McMaster Univ, Hamilton, Can, PhD(metall), 65. *Prof Exp:* Res metallurgist, Atlas Steels Ltd, Can, 60-62; sr res metallurgist, Westinghouse Co, Can, 65-66; postdoctoral fel, Univ Denver, 66-69; PROF METALL, UNIV CINCINNATI, OHIO, 68- *Concurrent Pos:* Alexander von Humboldt-Stiftung, US Sr Scientist, Ger, 78; res scientist, General Electric Co, Cincinnati, Ohio, 85-86; vis scientist, Wright-Patterson AFB, Ohio, 82. *Honors & Awards:* Fulbright Lectr, Fulbright-Hays Comn & Korean Ministry Educ, 80. *Mem:* Am Inst Mining, Metall & Petrol Engrs; Electrochem Soc; Sigma Xi. *Res:* High temperature oxidation of metals, diffusion of gases in metals; surface properties as related to initial stages of oxidation; thermodynamics of solutions; environmental effects on metals. *Mailing Add:* Dept Mat Sci & Eng Univ Cincinnati Cincinnati OH 45221

ROSA, EUGENE JOHN, b Sacramento, Calif, May 25, 37; m 59. CHEMICAL PHYSICS. *Educ:* Univ Calif, Berkeley, BS, 59; Univ Wash, PhD(chem), 64. *Prof Exp:* Chemist, Shell Develop Co, 64-67, supvr appl physics, 67-72; gen mgr, Veekay Ltd, 72-78; vpres & gen mgr, Ondyne Inc, 78-88; PRES & GEN MGR, NYAD INC, 88- *Res:* Development of analytical and process control instrumentation; development, manufacture and sales of process instrumentation. *Mailing Add:* 913 Tarvan East Dr Martinez CA 94553

ROSA, NESTOR, b Myrnam, Alta, Jan 15, 36; m 57; c 4. PLANT PHYSIOLOGY, BIOCHEMISTRY. *Educ:* Univ Alta, BSc, 58, MSc, 60; Dalhousie Univ, PhD(biol), 66. *Prof Exp:* Horticulturist, 60-63, PLANT PHYSIOLOGIST, RES BR, RES STA, CAN DEPT AGR, 66- *Mem:* Can Soc Plant Physiol; Am Soc Plant Physiol; Am Soc Agron; Phytochemistry Soc NAm. *Res:* Physiological and biochemical studies related to growth and development and chemical changes in Nicotiana tabacum and other plants. *Mailing Add:* Agr Can Res Sta PO 186 Delhi ON N4B 2W9 Can

ROSA, RICHARD JOHN, b Detroit, Mich, Mar 19, 27; m 50; c 3. MAGNETOHYDRODYNAMICS. *Educ:* Cornell Univ, BEP, 53, PhD(eng physics), 56. *Prof Exp:* Res assoc, Cornell Univ, 55-56; prin res scientist, Avco-Everett Res Lab Div, Avco Corp, 56-71, chief scientist, MHD Generator Proj, 71-75; PROF MECH ENG, MONT STATE UNIV, 75- *Concurrent Pos:* Vis lectr, Stanford Univ, 66; magnetohydrodyn ed, J Adv Energy Conversion; vis prof, Univ Sydney, Australia, 78, Tokyo Inst Technol, Japan, 81; coordr, US/Japan Coop Res in Magnetohydrodynamics, 82-86. *Honors & Awards:* Wiley Award for Res; Faraday Mem Award, Int Liaison Group Magnetohydrodynamics; Rosa-Kantrowitz Award, Magnetohydrodynamcis Indust Forum. *Mem:* AAAS; sr mem Inst Elec & Electronics Engrs; sr mem Am Inst Aeronaut & Astronaut; Am Soc Mech Eng. *Res:* Applied physics; magnetohydrodynamics, particularly the development of magnetohydrodynamic generators for large-scale production of electric power from chemical or nuclear fission heat sources. *Mailing Add:* Dept Mech Eng Mont State Univ Bozeman MT 59717

ROSADO, JOHN ALLEN, nuclear survivability, high power microwaves; deceased, see previous edition for last biography

ROSALES-SHARP, MARIA CONSOLACION, b Manila, Philippines, Jan 1, 27; m 50; c 2. DEVELOPMENTAL BIOLOGY, TISSUE CULTURE. *Educ:* Univ Philippines, BS, 50; Univ Ill, MS, 66, PhD(develop biol, insect bionomics), 68. *Prof Exp:* Instr zool, Univ of the East, Manila, 51-55, asst prof embryol, comp anat & gen entom, 56-63; asst embryol, Univ Ill, Urbana, 63-65, asst entom, 65-68, res assoc insect bionomics & develop biol, 68-69, AID contract res assoc zool, 69-72, AID contract res assoc biologist, clin prof & res scientist, Univ NMex, 72-75; VIS ASST PROF ENTOM, UNIV ILL, URBANA, 75- *Concurrent Pos:* NIH fel entom, Univ Ill, Urbana, 68-69; mem, Smithsonian Inst. *Mem:* Sigma Xi; Am Inst Biol Scientist; NY Acad Sci; Entom Soc Am; Soc Develop Biol. *Res:* Establishment of primary tissue culture of anopheline mosquitoes; cultivation of the sporogonic forms of the malarial parasites in vitro; cell movement in vitro using mosquito tissues; histochemistry of mosquito tissues in vivo and in vitro; postembryonic development of aedine mosquitoes; application of tissue culture to problems in malariology. *Mailing Add:* Dept Entom Univ Ill 505 S Goodwin Ave Urbana IL 61801

ROSAN, ALAN MARK, b Buffalo, NY, Oct 6, 48; m 69; c 2. CATALYSIS. *Educ:* Earlham Col, BS, 70; Brandeis Univ, PhD(chem), 75. *Prof Exp:* Fel chem, Yale Univ, 75-77; sr res chemist, Corp Res & Technol, Allied Signal, 77-86; ASSOC PROF CHEM, DREW UNIV, 86- *Mem:* Am Chem Soc; Catalysis Soc; Sigma Xi; NY Acad Sci; AAAS; Int Union Pure & Appl Chem; Planetary Soc. *Res:* Applied and exploratory organometallic chemistry with application to catalysis, photochemistry, reaction mechanism, and organic synthesis. *Mailing Add:* Dept Chem Drew Univ Madison NJ 07940

ROSAN, BURTON, b New York, Aug 18, 28; m 51; c 3. MEDICAL MICROBIOLOGY. *Educ:* City Col New York, BS, 50; Univ Pa, DDS, 57, MSc, 62. *Prof Exp:* NIH fel, 57-59; res assoc microbiol, 59-62, from asst prof to assoc prof periodont, 62-68, from asst prof to assoc prof microbiol, 62-75, PROF MICROBIOL, SCH DENT MED, UNIV PA, 75- *Concurrent Pos:* Partic, World Conf Periodont, Mich, 66; co-prin investr, NIH grant, 59-68 & Ctr Oral Health Res, 68-, prin investr, 70-; vis assoc prof, State Univ NY Downstate Med Ctr, 71-72; mem study sect oral biol & med, NIH, Div Res Grants, 75-79; assessor, Australian Med Res Coun, 77-; vis scientist, Nat Inst Dent Res, Sydney, Australia. *Mem:* AAAS; Int Asn Dent Res; Am Soc Microbiol; NY Acad Sci; Sigma Xi; fel Am Acad Microbiol. *Res:* Serology and immunochemistry of the oral streptococci and relationship to other adherence to oral tissues; extracellular polysaccharides of the actinomycetes. *Mailing Add:* Sch Dent Med-Microbiol Univ Pa Philadelphia PA 19104

ROSANO, HENRI LOUIS, b Nice, France, Feb 29, 24; US citizen; m 53; c 3. CHEMISTRY. *Educ:* Sorbonne Univ, Lic es phys sc, 46, Dr Sc Eng, 51. *Prof Exp:* Fel, Columbia Univ, 54-55; sr res assoc, Lever Brothers Co, NJ, 55-59; res assoc mineral eng, Columbia Univ, 59-62; PROF CHEM, CITY COL NEW YORK, 62- *Honors & Awards:* Prize, Fatty Acid Mat Inst Paris, 51; Prize, Soc Cosmetic Chem, 73. *Mem:* Am Chem Soc; Sigma Xi. *Res:* Surface and colloid chemistry. *Mailing Add:* Dept Chem City Univ NY 138th St-Convent Ave New York NY 10031

ROSANO, THOMAS GERARD, b Albany, NY, Oct 22, 48; m 70; c 1. CLINICAL CHEMISTRY, BIOCHEMISTRY. *Educ:* State Univ NY Albany, BS, 70; Albany Med Col, PhD(biochem), 75. *Prof Exp:* Fel clin chem, Dept Lab Med, Univ Wash, 74-76; assoc dir clin chem, 76-86, DIR CLIN CHEM & HEAD LAB MED, ALBANY MED CTR, 86-; ASSOC PROF PUB HEALTH SCI, STATE UNIV NY, ALBANY, 85- *Concurrent Pos:* Assoc prof biochem & path, 76-89, prof path & lab med, Albany Med Col, 90- *Mem:* Am Asn Clin Chem; Acad Clin Lab Physicians & Scientists; NY Acad Sci. *Res:* Development of new methodology and techniques in the area of clinical biochemistry clinical toxicology; chromatography; radioimmunoassay; enzymology; endocrine testing; metatabolism of cyclosporine; tumor marker testing. *Mailing Add:* Div Lab Med Albany Med Ctr New Scotland Ave Albany NY 12208

ROSAR, MADELEINE E, b Alface, France, July 15, 55. MATHEMATICS, PHYSICS. *Educ:* Fordham Univ, BS, 77, MS, 80. *Prof Exp:* RES ASSOC, PHILLIP LAB, 78- *Mem:* Am Phys Soc. *Mailing Add:* 98 Somerset Dr Suffern NY 10901-6903

ROSATI, ROBERT LOUIS, b Providence, RI, Mar 3, 42; m 66; c 2. MEDICINAL CHEMISTRY. *Educ:* Providence Col, BS, 64; Mass Inst Technol, PhD(org chem), 69. *Prof Exp:* NIH fel chem, Harvard Univ, 69; res chemist, 70-80, SR RES INVESTR, MED RES LAB, PFIZER, INC, 80- *Mem:* Am Chem Soc. *Mailing Add:* RD 3 Box 66A Stonington CT 06378-9719

ROSATO, FRANK JOSEPH, b Somerville, Mass, Feb 28, 25; m 50; c 5. APPLIED PHYSICS. *Educ:* Northeastern Univ, BS, 47; Tufts Univ, MS, 49; Harvard Univ, SM, 50, PhD(appl physics), 53. *Prof Exp:* Engr electronics, Polaroid Corp, Mass, 48-49; mathematician, Snow & Schule, Inc, 50; tech dir, GTE Sylvania Inc, Gen Tel & Electronics Corp, 53-73, chief scientist, 73-89; RETIRED. *Concurrent Pos:* Teaching fel, Tufts Univ, 48-49; fel, Harvard Univ, 53; vis lectr, Lowell Univ, 56-61; consult, Inst Naval Studies, 61-65 & Inst Defense Analysis, 62; dir naval commun, US Navy, 64-69. *Mem:* Acoust Soc Am; Am Asn Physics Teachers; sr mem Inst Elec & Electronics Engrs. *Res:* Communication satellite systems; defense communication systems; electronic countermeasures; electromechanical transducers; electroacoustics. *Mailing Add:* 12 Blueberry Lane Lexington MA 02173

ROSAUER, ELMER AUGUSTINE, materials science; deceased, see previous edition for last biography

ROSAZZA, JOHN N, b Torrington, Conn, Dec 25, 40; m 62; c 3. PHARMACOGNOSY, BIO-ORGANIC CHEMISTRY. *Educ:* Univ Conn, BS, 62, MS, 66, PhD(org pharmacog), 68. *Prof Exp:* NIH trainee natural prod res, Univ Conn, 65-68; fel pharmaceut biochem, Univ Wis, 68-69; from asst prof to assoc prof pharmacog, Univ Iowa, 69-73, assoc prof pharm, 73-77, head med chem, Natural Prod Div, Col Pharm, 77-85, PROF PHARM, UNIV IOWA, 77-, HEAD MED CHEM, NATURAL PROD DIV, COL PHARM, 89- *Honors & Awards:* Res Achievement Award in Natural Prod, Am Pharmaceut Assoc, Acad Pharmaceut Sci. *Mem:* Am Chem Soc; Am Soc Microbiologists; Am Soc Pharmacog; Am Soc Biol Chemist; Soc Indust Microbiol. *Res:* Microbial and enzymatic transformations of organic compounds including natural products, agricultural chemicals, antibiotics, alkaloids, terpenes; microbiology, enzymology, microbial chemistry and biochemistry; all aspects of biocatalysis. *Mailing Add:* Col Pharm Univ Iowa Iowa City IA 52242

ROSBERG, DAVID WILLIAM, b Superior, Wis, Jan 3, 19; m; c 2. PLANT PATHOLOGY. *Educ:* St Olaf Col, BS, 40; Ohio State Univ, MS, 47, PhD(plant path), 49. *Prof Exp:* Asst bot & plant path, Ohio State Univ, 46-47, res found, 48-49; from asst prof to prof plant path, Tex A&M Univ, 49-60, prof plant physiol, path & head dept plant sci, 60-74, prof plant path, 74-81, emer prof, 81-; RETIRED. *Concurrent Pos:* Mem, President's Cabinet Comt on Environ, Subcomt on Pesticides, Task Group on Training Objectives & Stand as Resource Contact, 71- *Mem:* Fel AAAS; Am Phytopath Soc. *Mailing Add:* Rte 3 Box 262 College Station TX 77845

ROSBERG, ZVI, b Kassel, WGer, July 25, 47; m 72; c 2. COMMUNICATION NETWORKS, PROBABILITY. *Educ:* Hebrew Univ Jerusalem, BSc, 71, MA, 74, PhD(oper res), 78. *Prof Exp:* Vis asst prof, Univ Ill, 79-80; vis scientist IBM, TJ Watson Res Lab, 85-87; SR LECTR, ISRAEL INST TECHNOL, 80- *Res:* Optical control and performance analysis of stochastic models of communication networks. *Mailing Add:* 40 Pine Ave Ossining NY 10562

ROSBOROUGH, JOHN PAUL, b Chicago, Ill, June 23, 30; m 54; c 3. ANIMAL PHYSIOLOGY. *Educ:* Univ Ill, Urbana, BS, 51, MS, 53, BS, 54, DVM, 56, PhD(vet med sci), 69. *Prof Exp:* Res asst physiol, Univ Ill, Urbana, 61-64; NIH fel biophys, Baylor Col Med, 69-70, from asst prof to assoc prof physiol, 71-87; ADJ PROF PHYSIOL, DENT BR, UNIV TEX HEALTH SCI CTR, 86- *Mem:* Am Physiol Soc; Am Soc Vet Physiol & Pharmacol. *Res:* Cardiopulmonary resuscitation. *Mailing Add:* 7910 Mobud Dr Houston TX 77036

ROSCHER, DAVID MOORE, b Mt Vernon, NY, Mar 28, 37; m 64. PHOTOCHEMISTRY, PHYSICAL ORGANIC CHEMISTRY. *Educ:* Rutgers Univ, BS, 59; Purdue Univ, PhD(org chem), 66. *Prof Exp:* Robert A Welch fel, Univ Tex, 65-67; res chemist, Celanese Res Co, 67-70; sci instr, Matawan High Sch, 70-75; SCI INSTR, ALEXANDRIA CITY SCH DIST, 75- *Mem:* AAAS; Am Chem Soc; Nat Sci Teachers Asn. *Res:* Gas-phase photochemical processes; photochemical polymerization and solvolysis reactions. *Mailing Add:* 10400 Hunter Ridge Dr Oakton VA 22124

ROSCHER, NINA MATHENY, b Uniontown, Pa, Dec 8, 38; m 64. PHYSICAL ORGANIC CHEMISTRY. *Educ:* Univ Del, BS, 60; Purdue Univ, PhD, 64. *Prof Exp:* Eli Lilly fel & instr chem, Purdue Univ, 64-65; instr, Univ Tex, 65-67; sr staff chemist, Coca-Cola Export Corp, 68; asst prof chem, Douglass Col, Rutgers Univ, 68-74, asst dean col, 71-74; dir acad admin, 74-76, assoc prof chem, 74-79, assoc dean grad affairs & res, 76-79, vprovost acad serv, 79-85, dean fac affairs, 81-85, PROF CHEM, AM UNIV, 79- *Concurrent Pos:* Mem, Sci Manpower Comn, 79-84, pres, 81-82; bd dir, Am Inst Chemists, 81-86. *Mem:* Am Chem Soc; fel Am Inst Chemists; NY Acad Sci; Soc Appl Spectros; fel AAAS. *Res:* Reaction mechanisms in organic chemistry, particularly in inorganic ion, free radical and light catalysis; structures of organic molecules. *Mailing Add:* Dept Chem Am Univ Washington DC 20016-8014

ROSCHLAU, WALTER HANS ERNEST, b Sonneberg, Ger, Feb 14, 24; Can citizen; m 51; c 1. PHARMACOLOGY. *Educ:* Univ Heidelberg, MD, 51. *Prof Exp:* Res asst exp med, W P Caven Mem Res Found, Toronto, Ont, 51-55; res asst, Gardiner Med Res Found, 55-60; sr res asst pharmacol, Connaught Med Res Labs, Univ Toronto, 60-62, res assoc, 62-66, assoc prof, Fac Med, 66-69, PROF PHARMACOL, FAC MED, UNIV TORONTO, 69- *Concurrent Pos:* Mem coun thrombosis, Am Heart Asn. *Mem:* AAAS; Pharmacol Soc Can; Int Soc Thrombosis & Haemostasis. *Res:* Blood coagulation; fibrinolytic enzymes; clinical pharmacology. *Mailing Add:* Dept Pharmacol Univ Toronto Fac Med One Kings College Circle Toronto ON M5S 1A8 Can

ROSCOE, CHARLES WILLIAM, b Pocatello, Idaho, Nov 22, 24; m 55; c 1. MOLECULAR PHARMACOLOGY. *Educ:* Idaho State Col, BS, 48; Univ Wash, MS, 54, PhD(pharmaceut chem), 58. *Prof Exp:* Asst prof pharmaceut chem, Univ Mont, 58-62, res supvr, NSF Prog, 60-61; assoc prof, Univ Pac, 62-71, prof pharmaceut chem, 71-85; RETIRED. *Concurrent Pos:* Vis scholar, Univ Wash, 70. *Mem:* AAAS; Am Chem Soc; Am Pharmaceut Asn. *Res:* Medicinal chemistry; organic syntheses; relationships between chemical constitution and biologic activity. *Mailing Add:* Box 404 Valley Springs CA 95252

ROSCOE, HENRY GEORGE, b Bridgeport, Conn, Nov 24, 30; m 55; c 3. BIOCHEMISTRY, SCIENCE ADMINISTRATION. *Educ:* Columbia Univ, AB, 52; Cornell Univ, PhD(biochem), 60. *Prof Exp:* Sr res biochemist, Lederle Labs, 60-72; assoc, 72-73, health scientist adminr, Grants Assoc Off, 73-74 & Nat Inst Neurol & Commun Disorders & Stroke, 74-79, CHIEF, REFERRAL BR, DIV RES GRANTS, NIH, 79- *Concurrent Pos:* Exec secy, Res Rev Comt B, Nat Heart, Lung & Blood Inst, 79- *Res:* Lipid biochemistry and cardiovascular biochemistry. *Mailing Add:* Div Extramural Affairs Deputy Dir Nat Heart Blood & Lung Inst Rm 7A171 Bldg WB Bethesda MD 20205

ROSCOE, JOHN MINER, b Halifax, NS, Dec 31, 43; m 68; c 2. CHEMICAL KINETICS, GAS PHASE PHOTOCHEMISTRY. *Educ:* Acadia Univ, BSc, 65, MSc, 66; McGill Univ, PhD(chem), 70. *Prof Exp:* Fel appl physics lab, Johns Hopkins Univ, 69-70; from asst prof to assoc prof, 70-85, PROF CHEM, ACADIA UNIV, 85- *Concurrent Pos:* Distinguished vis scientist, Nat Resources Coun, Ottawa, Can, 83-84; Consult, Sohio. *Mem:* Chem Inst Can; Royal Soc Chem; InterAm Photochemistry Soc. *Res:* Chemical kinetics of reactions of atoms, molecules, free radicals and excited molecules in the gas phase; kinetic modelling; gas phase photochemistry. *Mailing Add:* PO Box 878 Wolfville NS B0P 1X0 Can

ROSCOE, JOHN STANLEY, JR, b Dakota Co, Minn, Oct 12, 22; m 46; c 6. PHYSICAL INORGANIC CHEMISTRY. *Educ:* Univ Chicago, PhB, 47, MS, 51; St Louis Univ, PhD(chem), 54. *Prof Exp:* Res assoc, St Louis Univ, 52-53; mem staff, Res Dept, Mathieson Chem Corp, 53-59, mem staff, Energy Div, Olin Mathieson Chem Corp, 59-61, res assoc, Chem Div, 61-68; dir res, Quantum Inc, 68-71; SR PARTNER, J S ROSCOE ASSOCS, 72- *Mem:* Fel AAAS; fel Am Inst Chemists; Am Chem Soc. *Res:* Synthetic and physical chemistry of the hydrides of aluminum and boron; organometallics; compounds of phosphorus, nitrogen and sulfur; high performance epoxy, urethane and polyimide coatings; urethane and silicone elastomers; glass reinforced plastics. *Mailing Add:* 267 Lanyon Dr Cheshire CT 06410

ROSE, AARON, chemical engineering; deceased, see previous edition for last biography

ROSE, ALBERT, solid state physics, human vision; deceased, see previous edition for last biography

ROSE, ARTHUR L, b Cracow, Poland, July 21, 32; US citizen; c 2. NEUROLOGY, PEDIATRICS. *Educ:* Univ Bristol, MB ChB, 57; Royal Col Physicians & Surgeons, dipl child health, 59; Am Bd Pediat, dipl, 63; Am Bd Psychiat & Neurol, dipl & cert neurol, 69, cert child neurol, 73. *Prof Exp:* Res fel pediat, Med Sch, Harvard Univ, 61-63; instr neuropath, Col Physicians & Surgeons, Columbia Univ, 66-67; from asst prof to assoc prof neurol & pediat, Albert Einstein Col Med, 67-75; assoc prof, 75-80, PROF NEUROL, STATE UNIV NY DOWNSTATE MED CTR, 80-, DIR, DIV PEDIAT NEUROL, 75- *Concurrent Pos:* Fel behav & neurol sci, Albert Einstein Col Med, 67-69, Nat Inst Neurol Dis & Stroke fels, 69-77; assoc attend neurologist, Bronx Munic Hosp & Montefiore Hosp, 74-; assoc attend neurologist & pediatrician, Albert Einstein Col Hosp, 74- *Mem:* Am Acad Neurol; Child Neurol Soc. *Res:* Investigation of neurotoxic substances on the development of the nervous system using pathological techniques. *Mailing Add:* Dept Neurol State Univ NY Downstate Med Ctr 450 Clarkson Ave Brooklyn NY 11203

ROSE, ARTHUR WILLIAM, b Bellefonte, Pa, Aug 8, 31; m 71; c 3. GEOCHEMICAL EXPLORATION, ORE DEPOSITS. *Educ:* Antioch Col, BS, 53; Calif Inst Technol, MS, 55, PhD(geol & geochem), 58. *Prof Exp:* From geologist to sr geologist mineral explor, Bear Creek Mining Co, Kennecott Copper Co, 57-64; mining geologist geol mapping, Div Mines & Minerals, State of Alaska, 64-67; from asst prof to assoc prof, 67-75, dir, mineral censerv sect, 78-85, PROF GEOCHEM, PA STATE UNIV, 75- *Concurrent Pos:* Mem, Nat Comt Geochem, Nat Res Coun, 78-81, mem, Bd Energy & Mineral Resources, 84-86. *Mem:* Asn Explor Geochemists (pres, 80-81); Geochem Soc; Soc Econ Geologist (vpres, 89); Geol Soc Am; Soc Mining Engrs. *Res:* Geochemical exploration; geology and geochemistry of metallic ore deposits; environmental geochemistry; economics of mineral resources; geochemistry of sedimentary rocks. *Mailing Add:* Pa State Univ Dept Geosci 218 Deike Bldg University Park PA 16802

ROSE, BIRGIT, b Tegernsee, WGer, Aug 21, 43; m 71. CELL PHYSIOLOGY, MEMBRANE CHANNELS. *Educ:* Univ Munich, PhD(natural sci), 70. *Prof Exp:* Lab technician, Med Sch, Stanford Univ, 62-63; lab technician, Stanford Res Inst, 63-64; from res asst to res assoc physiol, Columbia Univ, 67-71; res asst prof, 71-77, res assoc prof, 77-78, RES PROF PHYSIOL & BIOPHYS, MED SCH, UNIV MIAMI, 89- *Mem:* AAAS; Biophys Soc; Soc Gen Physiol; Am Soc Cell Biol. *Res:* Intercellular communication, its basis of mechanism and role in differentiation; role of calcium in membrane physiology; membrane structure; membrane channels. *Mailing Add:* Dept Physiol & Biophys PO Box 016430 Miami FL 33101

ROSE, BRAM, b Montreal, Que, Apr 21, 07; m 41; c 3. MEDICINE. *Educ:* McGill Univ, BA, 29, MD, 33, MSc, 37, PhD, 39. *Prof Exp:* Jr intern med, Royal Victoria Hosp, Montreal, Que, 33-34, asst med resident, 34-35; asst in med, Guy's Hosp, London, 35-36; physician & assoc in med, Royal Victoria Hosp, 41, from asst prof to assoc prof med, Univ, 45-63, prof exp med & dir, div immunochem & allergy, Royal Victoria Hosp, 63-75, EMER PROF MED, MCGILL UNIV, 75- *Mem:* Am Soc Clin Invest; fel Aerospace Med Asn; fel Am Col Physicians; Am Asn Immunologists; fel Am Acad Allergy (past pres). *Res:* Investigations of allergy; metabolism of histamine; naturally occurring antihistaminics; immunologlobulins and antigen antibody mechanisms. *Mailing Add:* RR 3 1944 CH Langlois Magog PQ J1X 3W6 Can

ROSE, CARL MARTIN, JR, b Macon, Ga, Aug 31, 36; m 60; c 2. PHYSICS. *Educ:* Yale Univ, BS, 58; Univ Chicago, SM, 62, PhD(physics), 67. *Prof Exp:* Res assoc physics, Duke Univ, 66-67, asst prof, 67-74; mem staff, 74-78, SUPVR, BELL LABS, 78- *Mem:* Am Phys Soc; Asn Comput Mach; Inst Elec & Electronics Engrs. *Res:* Electron beam lithography; real-time computing; pattern recognition; numerical computing techniques. *Mailing Add:* 44 Plymouth Dr Berkeley Heights NJ 07922

ROSE, CHARLES BUCKLEY, b Washington, DC, Feb 8, 38; m 61; c 3. ORGANIC CHEMISTRY. *Educ:* Brigham Young Univ, BS, 60; Harvard Univ, AM, 63, PhD(org chem), 66. *Prof Exp:* Asst prof chem, 66-73, ASSOC PROF CHEM, UNIV NEV, RENO, 73- *Mem:* Am Chem Soc; Chem Soc; fel Am Inst Chemists; Sigma Xi. *Res:* Structure elucidation of natural products; development of new synthetic methods; model systems of physiologically active compounds. *Mailing Add:* Dept Chem Univ Nev Reno NV 89557

ROSE, CHARLES WILLIAM, b Columbus, Ohio, May 20, 40; m 84; c 2. COMPUTER ENGINEERING, INFORMATION SCIENCE. *Educ:* Duke Univ, BS, 62, MS, 63; Case Western Reserve Univ, PhD(comput & info sci), 70. *Prof Exp:* Sr engr, Tex Instruments, Inc, 63-70; from asst prof to prof & chmn, Dept Comput Eng & Sci, Case Western Reserve Univ, 70-84, asst to pres univ comput servs, 78-79; PRES, ENDOT INC, 84- *Concurrent Pos:* Chmn spec interest group on design automation, Asn Comput Mach, 75-77; consult, US Govt, Nat Eng Consortium & var indust corp. *Mem:* Inst Elec & Electronics Engrs; Asn Comput Mach. *Res:* Computer system architecture; design automation of electronic systems. *Mailing Add:* Dept Computer Sci Case Western Reserve Univ 2040 Adelbert Rd Cleveland OH 44106

ROSE, DAVID, b Chicago, Ill, Nov 13, 21; m 44; c 2. PHYSICS. *Educ:* Univ NDak, BS, 42; Carnegie Inst Technol, DSc, 52. *Prof Exp:* Control chemist, Chicago Sanit Dist, 42; jr physicist, Metall Lab, Chicago, 44; jr engr, Manhattan Engrs, Tenn, 44-45; jr scientist, Los Alamos Sci Lab, Univ Calif, 45-46, asst engr, 46; asst res physicist, Carnegie Inst Technol, 47-52; assoc physicist, Argonne Nat Lab, 52-56; physicist, Gen Atomic Div, Gen Dynamics Corp, 56-66; consult engr, United Engrs & Constructors, 66-67; SR NUCLEAR ENGR, ARGONNE NAT LAB, 67- *Res:* Nuclear physics; power reactors; reactor safety. *Mailing Add:* Argonne Nat Lab 9700 S Cass Ave Argonne IL 60439

ROSE, DONALD CLAYTON, b Clearmont, Mo, Apr 30, 20; m 42; c 5. MATHEMATICS. *Educ:* Transylvania Col, AB, 45; Univ Ky, MA, 48, PhD(math), 54. *Prof Exp:* Instr astron & math, Univ Ky, 45-54; chmn Dept Math, Transylvania Col, 54-60; from assoc prof to prof, Univ SFla, 60-75, chmn dept, 61-75; prof math, Hillsborough Community Col, 75-77; PROF MATH, UNIV S FLA, 77- *Mem:* Am Math Soc; Math Asn Am. *Mailing Add:* 10716 Lake Carroll Way Tampa FL 33618

ROSE, EARL FORREST, b Isabel, SDak, Sept 23, 26; m 51; c 6. PATHOLOGY. *Educ:* Yankton Col, BA, 49; Univ SDak, BSM, 51; Univ Nebr, MD, 53; Southern Methodist Univ, LLB, 67. *Prof Exp:* Resident path, Med Ctr, Baylor Univ, 56-58; resident DePaul Hosp, St Louis, 58-60; fel forensic path, Med Col Va, 61, lectr forensic path, 61-63; prof path, Univ Tex Southwestern Med Sch, Dallas, 63-68; PROF PATH, COL MED, UNIV IOWA, 68- *Concurrent Pos:* Dep chief med examr, Commonwealth Va, 61-63; mem, Dallas County Med Examrs, 63-68; lectr, Col Law, Univ Iowa, 71. *Mem:* Fel Col Am Pathologists; fel Am Soc Clin Pathologists; Am Acad Forensic Sci; AMA. *Res:* Forensic pathology; surgical and autopsy pathology; application of pathology to law. *Mailing Add:* Dept Path Univ Iowa Col Med Iowa City IA 52240

ROSE, FRANCIS L, b Augusta, Ga, Dec 20, 35; m 55; c 4. ZOOLOGY. *Educ:* Univ Ga, BS, 60, MS, 62; Tulane Univ, PhD, 65. *Prof Exp:* NIH fel, Fla State Mus, 65-66; asst prof, 66-74, PROF BIOL, TEX TECH UNIV, 74- *Concurrent Pos:* Am Philos Soc grant, 65-66. *Mem:* Am Soc Ichthyologists & Herpetologists. *Res:* Anatomy, ecology, behavior and systematics. *Mailing Add:* Dept Biol Tex Tech Univ Lubbock TX 79409

ROSE, FRANK EDWARD, b Junction City, Kans, Mar 11, 27; m 48; c 4. SOLID STATE PHYSICS. *Educ:* Greenville Col, BS, 49; Univ Mich, AM, 57; Cornell Univ, PhD(exp physics), 65. *Prof Exp:* Teacher high sch, 49-54; instr physics & electronics, Gen Motors Inst, 54-58; res asst physics, Cornell Univ, 60-62; from lectr to asst prof, 63-68, chmn, Dept Physics & Astron, 78-80, ASSOC PROF PHYSICS, UNIV MICH, FLINT, 68- *Mem:* Am Phys Soc; Am Asn Physics Teachers; Am Sci Affil. *Res:* Research on electron properties of metals; magnetoresistance of alkali metals; helicon mode of wave propagation; computer assisted instruction; astronomy instructions; forensic physics; computer text processing; science museum exhibits. *Mailing Add:* Dept Physics & Astron Univ Mich Flint MI 48502-2186

ROSE, GENE FUERST, b Erie, Pa, Mar 15, 18; m 40; c 1. MATHEMATICS, COMPUTER SCIENCE. *Educ:* Case Inst Technol, BS, 38; Univ Wis, MA, 47, PhD(math), 52. *Prof Exp:* Chemist, Copperweld Corp, Ohio & Pa, 40-42; res engr, Res Labs, Westinghouse Elec Corp, 42-45; res assoc, Allegany Ballistics Lab, George Washington Univ, 45; asst & actg instr math, Univ Wis, 45-52; mem staff, Sandia Corp, 52-55; mem tech staff, Space Tech Labs, Inc, Thompson Ramo Wooldridge, Inc, 55-60; sr scientist, Syst Develop Corp, Calif, 60-68; prof math & info sci, Case Western Reserve Univ, 68-77; prof, dept comput sci, Calif State Univ, Fullerton, 77-85; RETIRED. *Concurrent Pos:* Guest prof, Munich Tech Univ, 66-67; guest lectr, Imp Col Sci & Technol, Univ London, 66; Alexander von Humboldt Found grant, Ges für Math und Datenverarbeitung MBH Bonn, 72-73. *Mem:* Am Math Soc; Math Asn Am; Asn Symbolic Logic. *Res:* Recursive function theory; foundations of mathematics; formal language theory; mathematical machine theory; relational data bases. *Mailing Add:* 1565 Sherwood Village Circle Placentia CA 92670

ROSE, GEORGE DAVID, b Chicago, Ill, Aug 28, 39; m; c 2. BIOPHYSICAL CHEMISTRY, PROTEIN FOLDING. *Educ:* Bard Col, BS, 63; Ore State Univ, MS, 72, PhD(biochem & biophys), 76. *Prof Exp:* Asst prof & asst res prof chem, Univ Del, 75-80; prof, 80-89, chmn dept, 88-89, DISTINGUISHED PROF BIOL CHEM, HERSHEY MED CTR, PA STATE UNIV, 89- *Concurrent Pos:* Res Career Develop Award, NIH, 80-85, prin investr grants; mem, bd ed adv, Biopolymers, 87-; exec ed, Proteins: Struct, Function & Genetics, 89-90. *Honors & Awards:* Hinkle Award & Lectureship, 85. *Mem:* Am Soc Biol Chemists; Am Chem Soc; Biophys Soc; AAAS; NY Acad Sci; Protein Soc. *Res:* Structure, function, dynamics and self-assembly of macromolecules of biological interest. *Mailing Add:* Dept Biol Chem Hershey Med Ctr Pa State Univ Hershey PA 17033

ROSE, GEORGE G, medicine; deceased, see previous edition for last biography

ROSE, GORDON WILSON, b Elmira, NY, Apr 25, 24; m 51; c 3. EPIDEMIOLOGY, CLINICAL MICROBIOLOGY. *Educ:* Wayne State Univ, AB, 50; Univ Detroit, MS, 54; Univ Mich, PhD(epidemiol sci), 65. *Prof Exp:* Asst chem, Wayne State Univ, 47-48, chem, gross anat & histol, Dept Mortuary Sci, 50-54, instr bact, histol & chem, 54-56; spec instr bact, Univ Detroit, 56-57; from asst prof to assoc prof, 57-80, PROF BACT, HISTOL & CHEM, WAYNE STATE UNIV, 80- *Concurrent Pos:* Spec instr, Providence Hosp, Detroit, 57-58; assoc dir dept mortuary sci, Wayne State Univ, 57-65, spec instr, Sch Med, 65; asst dir dept, Deaconess Hosp, 65-66. *Mem:* Am Asn Bioanalysts; Am Soc Clin Pathologists; Am Pub Health Asn; Am Soc Microbiol; Sigma Xi. *Res:* Clinical and post-mortem microbiology; histopathogenesis of infectious diseases; hospital epidemiology. *Mailing Add:* 1205 Devonshire Rd Grosse Pointe Park MI 48230

ROSE, HAROLD WAYNE, b Telluride, Colo, Jan 11, 40; m 64; c 4. ELECTRONICS ENGINEERING, COMPUTER SCIENCE. *Educ:* Univ Colo, Boulder, BS, 62, MS, 64; Ohio State Univ, PhD(elec eng), 72. *Prof Exp:* Res engr, 66-71, ELECTRONICS ENGR, AVIONICS LAB, AIR FORCE SYSTS COMMAND, WRIGHT-PATTERSON AFB, 71- *Mem:* Optical Soc Am; Am Inst Physics. *Res:* Holography; holographic optical elements; coherent optics; lasers; missile guidance techniques; electro-optical trackers; laser trackers. *Mailing Add:* 636 N Galloway St Xenia OH 45385

ROSE, HARVEY ARNOLD, b New York, NY, Nov 9, 47. FLUID DYNAMICS. *Educ:* City Col New York, BS, 68; Harvard Univ, MA, 69, PhD(physics), 75. *Prof Exp:* Health serv officer radiol health, USPHS, 70-72, fel fluid turbulence, Nat Ctr Atmospheric Res, 75-76; vis scientist turbulence res, Observ Nice, France, 76-77; fel, 77-79, MEM STAFF, LOS ALAMOS NAT LAB, 79- *Mem:* Sigma Xi. *Res:* Calculate properties of turbulence, using methods of quantum field theory in high temperature plasmas and in fluids. *Mailing Add:* 3095 Arizona Ave Los Alamos NM 87544

ROSE, HERBERT G, b Chicago, Ill, Feb 21, 30. MEDICINE. *Educ:* Univ Ill, MD, 54. *Prof Exp:* ASSOC CHIEF STAFF, RES & DEVELOP, BRONX VET ADMIN MED CTR, 74- *Mem:* Am Heart Asn; Am Soc Clin Nutrit; Am Fedn Clin Res. *Res:* Metabolic diseases; lipo protein metabolism. *Mailing Add:* Vet Admin Hosp Bronx NY 10468

ROSE, IRA MARVIN, b Brooklyn, NY, Feb 22, 21; m 54; c 1. ORGANIC CHEMISTRY, ANALYTICAL CHEMISTRY. *Educ:* Brooklyn Col, BA, 41; Columbia Univ, AM, 49, PhD(org chem), 52. *Prof Exp:* Analytical chemist, Wallerstein Labs, NY, 41-42; analytical chemist, War Dept, Edgewood Arsenal, Md, 42-44, org chemist, Chem Corps, 46-47; sr org chemist, US Vitamin Corp, NY, 52-58; group leader res & develop indust org chem, Nopco Chem Co, 58-71 & Nopco Div, 71-74; sr res assoc, Process Chem Div, Diamond Shamrock Chem Co, Morristown, 74-82; RETIRED. *Concurrent*

Pos: Consult, 82- *Mem:* Am Chem Soc. *Res:* Fine chemical synthesis; industrial organic synthesis, research, product and process development and general analytical chemistry; synthesis, research, development and analysis of chemical warfare agents. *Mailing Add:* 55 Greenwood Dr Millburn NJ 07041

ROSE, IRWIN ALLAN, b Brooklyn, NY, July 16, 26; m 55; c 4. BIOCHEMISTRY. *Educ:* Univ Chicago, PhD(biochem), 52. *Prof Exp:* Postdoctoral fel, Dept Med, Western Reserve Univ, 52-53 & Dept Pharmacol, NY Univ, 53-54; from instr to assoc prof, Dept Biochem, Sch Med, Yale Univ, 54-63; SR MEM, BIOCHEM DIV, INST CANCER RES, 63- *Concurrent Pos:* Affil phys biochem, Grad Sch, Univ Pa, 63-70, prof phys biochem, Univ, 63-76; Guggenheim fel, Univ Oxford & Hebrew Univ, 72. *Mem:* Nat Acad Sci; AAAS; Am Chem Soc; Soc Biol Chemists; Am Acad Arts & Sci. *Res:* Mechanisms of transfer enzyme; protein degradation; regulation of metabolism; author of numerous publications. *Mailing Add:* Inst Cancer Res 7701 Burholme Ave Philadelphia PA 19111

ROSE, ISRAEL HAROLD, b New Britain, Conn, May 17, 17. MATHEMATICS. *Educ:* Brooklyn Col, AB, 38, AM, 41; Harvard Univ, PhD(math), 51. *Prof Exp:* Tutor & instr math, Brooklyn Col, 38-41; instr, Pa State Univ, 42-46; from asst prof to assoc prof, Univ Mass, 48-60; from assoc prof to prof math, Hunter Col, 60-68, chmn dept, 66-68; chmn dept, 68-72 & 80-82, prof math, 86-82, EMER PROF & RES PROF, LEHMAN COL, 83- *Concurrent Pos:* Vis asst prof, Mt Holyoke Col, 51-52, vis assoc prof, 54-55 & 58-59; fel, Ford Found, 52-53. *Mem:* Am Math Soc; Math Asn Am. *Res:* Abstract algebra. *Mailing Add:* 18 Floral Dr Hastings-on Hudson NY 10706

ROSE, JAMES A, REGULATION OF VIRUS MACROMELECULAR SYNTHESIS. *Educ:* Harvard Univ, MD, 56. *Prof Exp:* CHIEF MOLECULAR STRUCT SECT, LAB BIOL VIRUSES, NAT INST ALLERGY & INFECTIOUS DIS, NIH, 72- *Mailing Add:* Nat Inst Allergy & Infectious Dis Bldg 5 Rm 309 NIH Twinbrook II 12441 Parklawn Dr Rockville MD 20852

ROSE, JAMES C, PERIONATAL ENDOCRINOLOGY. *Educ:* Med Col Va, PhD(physiol), 74. *Prof Exp:* ASSOC PROF PHYSIOL & PHARMACOL, BOWMAN GRAY SCH MED, 82-, ASSOC PROF OBSTET, 83- *Mailing Add:* Bowman Gray Sch Med Wake Forest Univ 300 Hawthorne Rd SW Winston-Salem NC 27103

ROSE, JAMES DAVID, b Ann Arbor, Mich, Mar 31, 42; m 66; c 2. NEUROSCIENCE, BIOPSYCHOLOGY. *Educ:* Cent Mich Univ, BS, 64; Ind Univ, PhD(psychol), 70. *Prof Exp:* Fel, Sch Med, Emory Univ, 69-71; asst prof psychol, Dartmouth Col, 71-74; asst prof neurophysiol & anat, Sch Med, Emory Univ, 74-76; assoc prof, 76-80, PROF PSYCHOL, UNIV WYO, 80- *Concurrent Pos:* Prin investr, NIH res grants, 72- *Mem:* Soc Neurosci; Am Asn Anatomists; Soc Psychoneuroendocrinol. *Res:* Neurological bases of behavior; neurophysiology; neuroanatomy. *Mailing Add:* Dept Psychol Univ Wyo Laramie WY 82071

ROSE, JAMES STEPHENSON, b Halifax, NS, Can, July 9, 26. ORGANIC CHEMISTRY. *Educ:* Dalhousie Univ, BSc, 48, MSc, 50; Yale Univ, PhD(chem), 55. *Prof Exp:* Asst prof chem, NS Tech Univ, 54-55; sr res chemist, Olin-Mathieson Chem Corp, 55-62; sr res chemist, Naval Res Estab, NS, 62-64; SR RES CHEMIST, UPJOHN CO, 64- *Res:* Organic reaction mechanisms; synthetic organic chemistry; structure-spectra correlation of organic compounds. *Mailing Add:* 1820 Durham Rd RFD 2 Guilford CT 06437

ROSE, JERZY EDWIN, b Buczacz, Poland, Mar 5, 09; nat US. NEUROPHYSIOLOGY. *Educ:* Jagiellonian Univ, MD, 34. *Prof Exp:* Asst neurol & psychiat, Stefan Batory Univ, Poland, 34-36; fel, Emperor William Inst Brain Res, Ger, 36-37 & Ger Inst Brain Res, 37-38; sr instr neurol & psychiat, Stefan Batory Univ, 38-39; res assoc neuropath, Johns Hopkins Univ, 40-43, from asst prof to assoc prof physiol & psychiat, 46-60; prof, 60-80, EMER PROF NEUROPHYSIOL, UNIV WIS-MADISON, 80- *Mem:* Nat Acad Sci; Soc Exp Biol & Med; Am Physiol Soc; Am Asn Anatomists; Am Neurol Asn. *Res:* Anatomy and physiology of the mammalian central nervous system; development of the organ of corti in situ. *Mailing Add:* Dept Neurophysiol Univ Wis Madison WI 53706

ROSE, JOHN CHARLES, b New York, NY, Dec 13, 24; m 48; c 5. PHYSIOLOGY. *Educ:* Fordham Univ, BS, 46; Georgetown Univ, MD, 50. *Hon Degrees:* ScD, Georgetown Univ; LLD, Mt St Mary's Col. *Prof Exp:* From instr to asst prof med, 54-58, assoc prof physiol & biophys, 58-60, chmn dept biophys, 59-63, dean, Sch Med, 63-73 & 78-79, PROF PHYSIOL & BIOPHYS, SCH MED, GEORGETOWN UNIV, 60-, PROF MED, 73- *Concurrent Pos:* Estab investr, Am Heart Asn, 54-57, coordr med educ, 57-58; med ed, Am Family Physician, 62-88; vchancellor, Med Ctr, 84-87. *Mem:* Am Physiol Soc; Biophys Soc; Soc Exp Biol & Med; Am Col Physicians; Am Fedn Clin Res. *Res:* Cardiovascular physiology; medical education. *Mailing Add:* 5710 Surrey St Chevy Chase MD 20815

ROSE, JOHN CREIGHTON, b Milwaukee, Wis, July 27, 22; div; c 4. GEOPHYSICS. *Educ:* Univ Wis, BS, 48, MS, 50, PhD(geol), 55. *Prof Exp:* Res assoc, Woods Hole Oceanog Inst, 50-55; from instr to asst prof geol & geophys, Univ Wis, 55-64; assoc geophysicist, 64-68, prof geoscience, 68-74, PROF GEOPHYS, UNIV HAWAII, 74-, GEOPHYSICIST, 76- *Concurrent Pos:* Consult, Aero Div, Minneapolis-Honeywell Regulator Co. *Mem:* Am Geophys Union; Europ Asn Explor Geophys. *Res:* International pendulum gravity reference standard; absolute gravity measurements by pulse recycling and laser interferometer; geodesy; explosion seismology; marine gravity. *Mailing Add:* Dept Geol & Geophys Marine Ctr Univ Hawaii-Manoa 2500 Campus Rd Honolulu HI 96822

ROSE, JOSEPH LAWRENCE, b Philadelphia, Pa, July 5, 42; m 63; c 3. MECHANICS, MECHANICAL ENGINEERING. *Educ:* Drexel Inst Technol, BSME, 65, MS, 67; Drexel Univ, PhD(appl mech), 70. *Prof Exp:* Engr, Hale Fire Pump Co, 61-62 & SKF Industs, Inc, 63-64; instr mech eng, 65-69, asst prof, 70-73, assoc prof, 73-78, PROF MECH ENG, DREXEL UNIV, 78- *Mem:* Am Soc Mech Engrs; Soc Exp Stress Analysis; Am Soc Nondestruct Test; Am Soc Testing & Mat; Acoust Soc Am; Sigma Xi. *Res:* Nondestructive testing; experimental mechanics; stress analysis; wave propagation; composite materials; biomechanics. *Mailing Add:* Dept Mech Eng & Mech 32nd & Chestnut Sts Drexel Univ Philadelphia PA 19104

ROSE, KATHLEEN MARY, b St Paul, Minn, Sept 5, 45. NUCLEIC ACID SYNTHESIS, ENZYMOLOGY. *Educ:* Mich State Univ, BS, 66, MS, 69; Pa State Univ, PhD(genetics), 77. *Prof Exp:* NIH fel biochem, Mich State Univ, 67-69; res asst, Pa State Univ, 72-78, asst prof, 78-82, assoc prof pharmacol, M S Hershey Med Ctr, 82-83; assoc prof, 83-87, PROF, DEPT PHARMACOL, MED SCH, UNIV TEX, 87- *Concurrent Pos:* Consult, Alcoa Found Award, Grad Prof Opportunities Prog, US Dept Educ, 81-82 & USPHS Res Career Develop Award, Nat Cancer Inst, 80-; prin investr, USPHS & Nat Inst Gen Med Sci, 79- *Mem:* Am Soc Biol Chemists; Am Soc Cell Biol; Sigma Xi; Am Soc Pharmacol & Exp Therapeut. *Res:* Regulation of ribosomal RNA synthesis; role of protein phosphorylation in gene expression; control of cell growth and transformation. *Mailing Add:* Univ Tex Med Sch PO Box 20708 Houston TX 77225

ROSE, KENNETH, b Bloomington, Ind, Apr 21, 35; m 59; c 2. ELECTRICAL ENGINEERING, PHYSICS. *Educ:* Univ Ill, BS, 55, MS, 57, PhD(elec eng), 61. *Prof Exp:* Physicist, Gen Elec Res Lab, 61-65; assoc prof elec eng, 65-71, PROF ELEC ENG, RENSSELAER POLYTECH INST, 71- *Concurrent Pos:* Consult, US Naval Res Lab, 66-69 & Gen Elec Co, 76-78; mem, Ctr Integrated Electronics, Rensselaer Polytech Inst. *Mem:* AAAS; Inst Elec & Electronics Engrs; Am Phys Soc. *Res:* Very-large-scale integration fabrication, design and testing; yield enhancement; thin film insulators; semiconductors; superconductors. *Mailing Add:* Elec Comput & Systs Eng Dept Rensselaer Polytech Inst Troy NY 12180-3590

ROSE, KENNETH DAVID, b Newark, NJ, June 21, 49; m 81; c 2. MAMMALIAN EVOLUTION. *Educ:* Yale Univ, BS, 72; Harvard Univ, MA, 74; Univ Mich, PhD(geol & paleont), 79. *Prof Exp:* Fel paleobiol, Smithsonian Nat Mus Natural Hist, 79-80; from asst prof to assoc prof, 80-90, PROF ANAT, SCH MED, JOHNS HOPKINS UNIV, 90- *Concurrent Pos:* Res collab, 81-89, Dept Paleobiol, Smithsonian Nat Mus Natural Hist, res assoc, 89-; co-ed, J Vert Paleont, 87-90; res assoc, Carnegie Mus Natural Hist, 90- *Mem:* Soc Vert Paleont; Paleont Soc; Am Soc Mammalogists; Soc Syst Zool; Sigma Xi; Am Asn Phys Anthrop; Soc Study Evolution. *Res:* Early Cenozoic mammals, with emphasis on systematics and evolution, functional anatomy of the teeth and skeleton, biostratigraphy and biochronology (use of fossil mammals for stratigraphic correlation and determination of relative age of strata), and species diversity. *Mailing Add:* Dept Cell Biol & Anat Sch Med Johns Hopkins Univ Baltimore MD 21205

ROSE, KENNETH E(UGENE), b Winfield, Kans, Oct 20, 15; m 39; c 2. METALLURGY, CORROSION. *Educ:* Colo Sch Mines, EMet, 39; Cornell Univ, MS, 43. *Prof Exp:* Jr engr, US Govt, 36-37; engr trainee, Caterpillar Tractor Co, 39-41; instr mech & metall, Cornell Univ, 41-43; res engr, Battelle Mem Inst, 43-46; asst prof mech & metall, Univ Okla, 46-47; from assoc prof to prof metall eng, Univ Kans, 47-75, chmn dept metall eng, 47-68, assoc dean eng, 68-75, prof mech eng, 75-84, EMER PROF MECH ENG, UNIV KANS, 84- *Concurrent Pos:* Consult, 48-; mem, State Bd Eng Examr, 56-59; Fulbright lectr, Nat Univ Eng, Peru, 60. *Mem:* Fel Am Soc Metals; Am Inst Mining, Metall & Petrol Engrs; Nat Asn Corrosion Engrs. *Res:* Metallurgy of cast metals; science of materials and materials processing; service failures; engineering education; corrosion. *Mailing Add:* Dept Mech Eng Univ Kans Lawrence KS 66045

ROSE, MICHAEL ROBERTSON, b Iserlohn, Ger, July 25, 55; Can citizen; m 76, 85; c 1. EVOLUTIONARY BIOLOGY, GERONTOLOGY. *Educ:* Queen's Univ, BSc, 75, MSc, 76; Univ, Sussex, PhD, 79. *Prof Exp:* NATO Sci fel, Univ Wis, Madison, 79-81; from asst prof to assoc prof biol, Dalhousie Univ, 81-88; assoc prof, 87-90, PROF BIOL, UNIV CALIF IRVINE, 90- *Concurrent Pos:* Univ res fel, Dalhousie Univ, 81-88. *Mem:* Genetics Soc Am; Soc Study Evolution. *Res:* Drosophila life-history evolution, including fitness-components and senescence; evolution with antagonistic pleiotropy; evolution of sex; human evolution. *Mailing Add:* Dept Ecol & Evolutionary Biol Univ Calif Irvine CA 92717

ROSE, MILTON EDWARD, b Newark, NJ, May 22, 25; m 48; c 3. MATHEMATICS, ENGINEERING. *Educ:* NY Univ, AB, 47, MS, 51, PhD, 53. *Prof Exp:* Res assoc, Inst Math Sci, NY Univ, 50-55; mathematician, Off Naval Res, 55-57; head appl math div, Brookhaven Nat Lab, 57-60; mathematician, Lawrence Radiation Lab, Univ Calif, 60-63; head math sci sect, NSF, 63-67, head off comput activities, 67-69; head dept math & statist, Colo State Univ, 69-70; vis prof, Univ Denver, 70-71; chief math & comput prog, US Energy Res & Develop Admin, 71-77; dir, Inst Comput Applns Sci & Eng, NASA Langley Res Ctr, 77-87; adj prof, NC AT&T, 87-89; RETIRED. *Concurrent Pos:* Consult, NSF. *Mem:* Soc Indust & Appl Math. *Res:* Numerical methods for partial differential equations; stefan problems; navier-stokes and euler equations for fluids; elastic behavior of composite materials. *Mailing Add:* 4505 Tower Rd Greensboro NC 27410

ROSE, MITCHELL, b Cleveland, Ohio, Mar 10, 51; m 74; c 5. GAS PHASE PHOTOCHEMISTRY, X-RAY FLUORESCENCE. *Educ:* Maimonides Col, BA, 75; Cleveland State Univ, MS, 77; Case Western Univ, PhD(chem), 79. *Prof Exp:* Anal lab mgr, Mogul Corp, Div Dexter Corp, 79-81; res adv, Master Builders Co, 81-85, PRES, TECH INNOVATIONS, DIV MARTIN MARIETTA CORP, 85- *Res:* Gas phase reaction kinetics; x-ray diffraction, and x-ray fluorescence spectroscopy; computer interfacing for laboratories. *Mailing Add:* 3788 Shannon Dr Cleveland Heights OH 44118

ROSE, NICHOLAS JOHN, b Ossining, NY, Apr 21, 24; m 46; c 4. MATHEMATICS. *Educ:* Stevens Inst Technol, ME, 44; NY Univ, MS, 49, PhD(math), 56. *Prof Exp:* From instr to prof math, Stevens Inst Technol, 46-68, head dept, 60-68, consult, 48-54; head dept, 68-77, prof math, 68-89, EMER PROF MATH, NC STATE UNIV, 89- *Concurrent Pos:* Consult, Bell Tel Labs, Inc, 54-60. *Mem:* Am Math Soc; Math Asn Am; Soc Indust & Appl Math. *Res:* Differential equations; matrix theory. *Mailing Add:* Dept Math NC State Univ PO Box 8205 Raleigh NC 27695-8205

ROSE, NOEL RICHARD, b Stamford, Conn, Dec 3, 27; m 51; c 4. MICROBIOLOGY, IMMUNOLOGY. *Educ:* Yale Univ, BS, 48; Univ Pa, AM, 49, PhD(microbiol), 51; State Univ NY Buffalo, MD, 64; Am Bd Microbiol, dipl; Am Bd Path, dipl; Am Bd Lab Immunol, dipl. *Hon Degrees:* Dr, Univ Calgiari, Italy, 90. *Prof Exp:* From instr to assoc prof bact & immunol, State Univ NY Buffalo, 51-66, prof microbiol, Sch Med, 66-73; prof immunol & microbiol & chmn dept, Sch Med, Wayne State Univ, 73-81; PROF & CHMN DEPT IMMUNOL & INFECTIOUS DIS, SCH HYG & PUB HEALTH & PROF MED, SCH MED, JOHNS HOPKINS UNIV, 81- *Concurrent Pos:* Consult, Niagara Sanatorium, NY, 53-56 & Edward J Meyer Mem Hosp, Buffalo, 56-73, Vet Admin Hosp, Oak Park, Mich & Sinai Hosp, Detroit, 73-81; from assoc dir to dir, Erie County Lab, 64-70; asst prof med, State Univ NY Buffalo, 64-73, dir ctr immunol, 70-73. *Mem:* Fel Am Pub Health Asn; Tissue Cult Asn; fel Am Acad Allergy; fel Col Am Path; fel Am Acad Microbiol; Am Asn Immunologist; Clin Immunol Soc; Soc Exp Biol & Med. *Res:* Autoimmunity; cellular immunology; cancer; clinical immunology; immunopathology. *Mailing Add:* Dept Immunol & Infectious Dis Johns Hopkins Univ 615 N Wolfe St Baltimore MD 21205

ROSE, NORMAN CARL, b Seattle, Wash, Mar 15, 29; m 54; c 4. ORGANIC CHEMISTRY. *Educ:* Univ Calif, BS, 50; Univ Kans, PhD(chem), 57. *Prof Exp:* From asst prof to assoc prof chem, Tex A&M Univ, 56-66; from assoc prof to prof, 66-78, ASST DEAN, COL LIB ARTS & SCI, PORTLAND STATE UNIV, 78- *Mem:* Am Chem Soc. *Res:* Learning theories. *Mailing Add:* Col Lib Arts & Sci Portland State Univ Portland OR 97207-0751

ROSE, PETER HENRY, b Lincoln, Eng, Jan 16, 25; m 52; c 2. PHYSICS. *Educ:* Univ London, BSc, 45, PhD(physics), 55. *Prof Exp:* Sci officer, Nat Gas Turbine Estab, 45-47; asst lectr physics, Univ Leicester, 47-48; res assoc, Mass Inst Technol, 51-52, physicist, Proj Lincoln, 52-53; lectr physics, Birmingham Univ, 55-56; vpres & dir res, High Voltage Eng Corp, 56-70; pres, Ion Physics Corp, 70-71; pres, Extrion Corp, 71-75; gen mgr, Varian Extrion Div, Varian Assocs, 75-77; pres, Nova Assoc, Inc, 78-80; PRES, ION IMPLANTATION DIV, EATON CORP, 80- *Mem:* Fel Am Phys Soc; fel Brit Inst Physics. *Res:* Low energy nuclear physics; accelerators; theoretical and experimental ion optics; plasma and atomic physics as related to the accelerator; ion implantation. *Mailing Add:* 85 Phillips Ave Rockport MA 01966

ROSE, PETER R, b Austin, Tex, July 3, 35; m 56, 78; c 3. PETROLEUM GEOLOGY, STRATIGRAPHY. *Educ:* Univ Tex, BS, 57, MA, 59, PhD(geol), 68. *Prof Exp:* Geologist, Shell Oil Co, 59-66; asst prof geol, State Univ NY Stony Brook, 68-69; staff geologist, Shell Oil Co, 69-73; chief Br Oil & Gas Resources, US Geol Surv, 73-76; chief geol, Energy Reserves Group, Inc, 76-80; PRES, TEL EXPLOR, 80- *Concurrent Pos:* Vis prof geol, Univ Tex, Austin, 83; distinguished lectr, Am Asn Petrol Geologists, 85; distinguished vis prof geol, Kans State Univ, 88-89. *Mem:* Am Asn Petrol Geologists; fel Geol Soc Am; Soc Econ Paleontologists & Mineralogists; AAAS; Am Inst Prof Geologists; Soc Independent Earth Scientists. *Res:* Petrology and paleoecology of carbonate rocks; carbonate reservoir rocks; analysis of petroleum basins; petroleum resource prediction; exploration risk and uncertainity. *Mailing Add:* Tel Explor 711 W 14th St Austin TX 78701

ROSE, PHILIP I, b New York, NY, Apr 3, 39; m 59; c 2. PHYSICAL CHEMISTRY. *Educ:* Univ Ariz, BS, 60; Purdue Univ, PhD(phys chem), 64. *Prof Exp:* Sr res chemist, Res Labs, 64-73, RES ASSOC, RES LABS, EASTMAN KODAK CO, 73- *Mem:* AAAS; Am Chem Soc; Sigma Xi. *Res:* Molecular characterization and physical chemistry of biopolymers. *Mailing Add:* 125 Summit Dr Rochester NY 14620

ROSE, RAYMOND EDWARD, b Canton, Ohio, July 17, 26; div; c 4. AERODYNAMICS, MATHEMATICS. *Educ:* Univ Kans, BSAE, 51; Univ Minn, MSAE, 56, PhD(aerodyn), 66. *Prof Exp:* From jr engr to scientist aerodyn, Rosemount Aero Labs, Univ Minn, 51-59, scientist, 59-62, res fel, Univ Minn, 62-66; prin res scientist to proj staff engr & supvr, Aerodyn, Fluid Mech & Control Sci Sect, Res Dept, Systs & Res Ctr, Honeywell, Inc, Minneapolis, Minn, 66-76; prog mgr, Aerodyn & Active Controls, Aircraft Energy Efficiency Prog Off, 76-79, prog mgr, Gen Aviation, Subsonic Aircraft Technol Off, 79-84, MGR, GEN AVIATION & COMMUTER AERODYN & COORDR, TECHNOL TRANSFER CONTROL, AERODYN DIV, OFF AERONAUT & SPACE TECHNOL, NASA HQ, WASHINGTON, DC, 84- *Concurrent Pos:* Consult, EDO Corp, NY, 62, Pioneer Parachute Co, 65 & Pillsbury Co, Minn, 65; adj prof, math dept, Southeastern Univ, Washington, DC, 81-; indust specialist, NASA Spec Assignment, Capitol Goods & Prod Mat Div, Off Export Admin, Int Trade Admin, US Dept Com, Washington, DC, 83-84; speaker & presenter aerospace sci & math, prof educator's workshops & student groups. *Honors & Awards:* Space Ship Earth Award, NASA, 80. *Mem:* Am Inst Aeronaut & Astronaut. *Res:* Shock swallowing and control concepts for air data sensing; jet-flap aerodynamics; natural laminar flow for general aviation aircraft viscous drag reduction; general aviation crash dynamics (structural design for occupant survivability); single pilot instrument flight rules operations (pilot workload reduction); general aviation aircraft design for stall/spin reduction/elimination. *Mailing Add:* Dept Comput Sci Southeastern Univ 501 Eye St SW Washington DC 20024

ROSE, RAYMOND WESLEY, JR, b Cleveland, Ohio, July 5, 41; m 65; c 2. MOLECULAR GENETICS. *Educ:* Bucknell Univ, BS, 63, MS, 65; Temple Univ, PhD(biol), 70. *Prof Exp:* Asst prof, 70-77, ASSOC PROF BIOL, BEAVER COL, 77- CHMN DEPT, 74- *Mem:* Genetics Soc Am; AAAS; Am Soc Zoologists. *Res:* Genetic control of protein synthesis and nucleic acid synthesis in Drosophila; protein and nucleic acid synthesis during early development in Xenopus; effect of environmental pollutants on mammalian development. *Mailing Add:* Dept Biol Beaver Col Glenside PA 19038

ROSE, RICHARD CARROL, b Minneapolis, Minn, Jan 2, 40. PHYSIOLOGY. *Educ:* Augsburg Col, 65; Mich State Univ, MS, 67, PhD(physiol), 69. *Prof Exp:* USPHS fel, Sch Med, Univ Pittsburgh, 70-71; asst prof, 71-74, assoc prof, 74-80, PROF PHYSIOL & SURG, COL MED, PA STATE UNIV, 81-, MEM STAFF, DEPT SURG. *Concurrent Pos:* Fogarty sr int fel, 81. *Mem:* Am Physiol Soc; Am Gastroenterol Asn; Biophys Soc; Sigma Xi. *Res:* Membrane physiology; electrophysiology of epithelial tissues; transport in intestine and gallbladder; vitamin absorption/metabolism. *Mailing Add:* Dept Physiol Univ NDak Grand Forks ND 58202

ROSE, ROBERT LEON, b San Francisco, Calif, Sept 3, 20; m 42; c 3. GEOLOGY. *Educ:* Univ Calif, AB, 48, MA, 49, PhD(geol), 57. *Prof Exp:* Geologist, Shell Oil Co, 49-53; assoc geol, Univ Calif, 53-56; actg asst prof geol, Stanford Univ, 56-57; asst econ geologist, Nev Bur Mines, 57-59; from asst prof to prof, 59-72, prof, 72-84, EMER PROF GEOL, SAN JOSE STATE UNIV, 84- *Concurrent Pos:* Consult, Chevron Resources, 73-79. *Mem:* Geol Soc Am; Mineral Soc Am; Am Asn Petrol Geologists. *Res:* Petrology and petrography of igneous and metamorphic rocks; geology of California and the Franciscan Formation. *Mailing Add:* 1080 Los Viboras Rd Hollister CA 95023

ROSE, ROBERT M(ICHAEL), b New York, NY, Apr 15, 37; m 61; c 3. PHYSICAL METALLURGY, SOLID STATE PHYSICS. *Educ:* Mass Inst Technol, SB, 58, ScD(phys metall), 61. *Prof Exp:* Ford Found fel eng, 61-63, from asst prof to assoc prof metall, 61-72, PROF MAT SCI & ENG, MASS INST TECHNOL, 72- *Concurrent Pos:* Prof, Mat Sci & Eng, Harvard Med Sch-MIT Div Health Sci & Technol, 78-; mem, Comt Space Biol & Med, Nat Res Coun, Nat Acad Sci, 84-87. *Honors & Awards:* Bradley Stoughton Award, Am Soc Metals, 68; Kappa Delta Award, Am Acad Orthop Surgeons, 73. *Mem:* AAAS; Am Phys Soc; Am Inst Mining, Metall & Petrol Engrs; Am Soc Metals; Orthop Res Soc; NY Acad Sci. *Res:* Electrical, magnetic, mechanical and thermodynamic properties of metals at cryogenic and high temperatures; theoretical and practical solid state physics; statistical thermodynamics; surgical implant materials; mammalian bone. *Mailing Add:* Rm 4-132 Mass Inst Technol Cambridge MA 02139

ROSE, SETH DAVID, b Dayton, Ohio, Nov 11, 48. DNA PHOTOCHEMISTRY, BIO-ORGANIC CHEMISTRY. *Educ:* Univ Calif, Berkeley, BS, 70; Univ Calif, San Diego, PhD(chem), 74. *Prof Exp:* NIH fel biophys, Johns Hopkins Univ, 74-76; asst prof, 76-80, ASSOC PROF CHEM, ARIZ STATE UNIV, 81- *Mem:* Am Chem Soc; Am Soc Photobiol. *Res:* Nucleic acid photochemistry; physical organic chemistry. *Mailing Add:* Dept Chem Ariz State Univ Tempe AZ 85287-1604

ROSE, STUART ALAN, b Dayton, Ohio, Sept 17, 42; m 65; c 2. ANALYTICAL CHEMISTRY. *Educ:* Univ Cincinnati, BA, 64; Wayne State Univ, PhD(analytical chem), 69. *Prof Exp:* Finished prod chemist, Wm S Merrell Div, Richardson-Merrell, Inc, 64-66; res chemist, Cent Res Div, 69-73, dept head-group leader, Lederle Labs Div, 73-76, asst to dir qual control, 76-77, mgr, Pearl River Qual Control, 77-81, DIR PHARMACEUT CONTROL, LEDERLE LABS DIV, AM CYANAMID CO, PEARL RIVER, 81- *Mem:* Am Chem Soc; Water Pollution Control Fedn. *Res:* General analytical problem solving; technical management. *Mailing Add:* Lederle Parenterals PO Box AC Pueblo Station Carolina PR 00628-8001

ROSE, TIMOTHY LAURENCE, b Cleveland, Ohio, July 6, 41; m 63; c 3. PHOTOELECTROCHEMISTRY, THIN FILMS. *Educ:* Haverford Col, BA, 63; Yale Univ, MS, 64, PhD(phys chem), 67. *Prof Exp:* NATO fel physics, Univ Freiburg, 68; from asst prof to assoc prof chem, Tex A&M Univ, 69-78; Nat Res Coun sr res assoc, Air Force Geophys Lab, 76-78; sr scientist, 78-83, GROUP LEADER, EIC LABS, INC, 84- *Mem:* AAAS; Am Chem Soc; Am Phys Soc; Am Vacuum Soc; Electrochem Soc; Mat Res Soc. *Res:* Gas phase methylene reactions and chemical activation; molecular and ion beam photodissociation processes; photochemical and photovoltaic solar energy conversion; electrical and optical properties of thin films; photochemical deposition; photoelectrochemistry on semiconductors; chemical sensors. *Mailing Add:* 97 Bartlett Hill Rd Concord MA 01742-1801

ROSE, VINCENT C(ELMER), b Fall River, Mass, July 31, 30; m 59; c 3. WATER POLLUTION, MARINE CORROSION. *Educ:* Univ RI, BS, 52, MS, 58; Univ Mo, PhD(chem eng), 64. *Prof Exp:* Pilot plant supvr, Lindsay Chem Co, 54-56; asst prof chem eng, 63-67, from asst prof to assoc prof, 67-82, PROF NUCLEAR & OCEAN ENG, UNIV RI, 83-, ASSOC DEAN GRAD SCH, 71- *Mem:* Am Inst Chem Engrs; Am Soc Eng Educ; Sigma Xi. *Res:* Plant siting; marine corrosion; water pollution control. *Mailing Add:* Grad Sch Univ RI Kingston RI 02881

ROSE, WALTER DEANE, b Liberty, Ind, Jan 10, 20; m 39; c 6. CAPILLARITY, TRANSPORT PHENOMENA. *Educ:* Univ Chicago, BS, 45. *Prof Exp:* Chemist geochem & petrol eng, Standard Oil NJ, 45-47; engr petrol eng, Gulf Res, 48-50; lectr petrol eng, Univ Tex, 50-52; scientist petrol eng, Continental, 52-54; prof petrol eng, Univ Ill, 56-68; prof civil eng, Technion, Israel, 68; prof petrol eng, Mid E Tech Univ, Turkey, 68-69; prof mech eng, Purdue Univ, 69-70; dean appl sci, Abadan Inst Technol, Iran, 71-73; prof petrol eng, Univ Ibadan, Nigeria, 73-75, NMex Tech, 75-77 & Univ Wyo, 78-79; scientist gas technol, Inst Gas Technol, 80-82; prof mech eng, Calif State Univ, Long Beach, 84-86; RETIRED. *Concurrent Pos:* Sci adv, French Petrol Inst, Paris, 62-63; lectr, Petroleos Mexicanos, Mexico

City, 65; consult, UNESCO, Pakistan, 70; vis prof, Univ Dacca, 70; vis fel, Commonwealth Sci & Indust Res Orgn, Canberra, Australia, 89; vis lectr, Inst Porous Flow, Langfang (Beijing), Peoples Repub China, 90. *Mem:* Soc Petrol Engrs. *Res:* Flow of fluids through porous media, including theoretical and experimental investigations of coupled transport processes, and of the action of capillary forces that control the distributions of interstitial fluids. *Mailing Add:* PO Box 2430 Sta A Champaign IL 61825

ROSE, WAYNE BURL, b Lamar, Colo, Dec 23, 32; m 56; c 4. PHYSICAL CHEMISTRY, PHARMACY. *Educ:* Adams State Col, BA, 55, MEd, 57; Kans State Univ, MS, 61, PhD(phys chem), 64. *Prof Exp:* Assoc chemist, Midwest Res Inst, 63-66; res chemist, Chemagro Corp, 66-68, sr res chemist, 68-69, mgr formulations res, 69-74; MGR CHEM RES, ANIMAL HEALTH DIV, MOBAY, 74- *Mem:* Sigma Xi; Royal Soc Chem; Am Chem Soc. *Res:* Oxygen fluoride and nitrogen fluoride chemistry; metal ligand complexes in deuterium oxide; histamine heparin interactions in aqueous solutions; surface chemistry; kinetics of decomposition; preparation of pharmaceutical products and their registration; analytical methods and packaging. *Mailing Add:* Mobay Animal Health Div 9009 W 67th Bldg 1 Merriam KS 66203

ROSE, WILLIAM DAKE, b Nashville, Tenn, Mar 15, 28; m 52; c 2. PETROLEUM GEOLOGY. *Educ:* Vanderbilt Univ, AB, 50, MS, 53. *Prof Exp:* Geologist, Gulf Oil Corp, 53-56 & Ky Geol Surv, 56-70; geologist-ed, Okla Geol Surv, 70-85; PUBL SUPVR, OCEAN DRILLING PROG, TEX A&M UNIV, 85- *Concurrent Pos:* Exec Comt, Am Geol Inst, 87-88. *Mem:* Geol Soc Am; Am Asn Petrol Geol; Am Inst Prof Geologists; Asn Earth Sci Ed (secy-treas, 68-71, pres, 83); Am Geophys Union. *Res:* Areal geology; stratigraphy. *Mailing Add:* Ocean Drilling Prog Tex A&M Univ PO Drawer GK College Station TX 77841

ROSE, WILLIAM INGERSOLL, JR, b Detroit, Mich, Oct 4, 44; m 67; c 2. VOLCANOLOGY. *Educ:* Dartmouth Col, AB, 66, PhD(geol), 70. *Prof Exp:* Res asst geol, Dartmouth Col, 66-68; res asst geochronology, Univ Ariz, 68-69; res asst geol, Dartmouth Col, 69-70; from asst prof to assoc prof, 70-79, PROF GEOL, MICH TECHNOL UNIV, 79-, HEAD DEPT GEOL ENG, GEOL & GEOPHYS, 90- *Concurrent Pos:* NSF grants, Guatemala, El Salvador & Nicaragua, 70-72, Guatemalan volcanoes, 73-74 & 78-80, volcanic ash, 75-77 & 90-91, Toba Tuff, 82-85 & volcanic gases, 85-87, 87-89 & 90-91, satellites & eruption clouds, 87-89; vis scientist, Nat Ctr Atmospheric Res, 77-78, Cascade Volcano observ, US Geol Surv, 81- & Los Alamos Nat Lab, 85-86. *Mem:* AAAS; Int Asn Volcanology & Chem Earth's Interior; Am Geophys Union; fel Geol Soc. *Res:* Volcanic gas geochemistry; igneous petrology; volcanic domes; active Central American volcanoes; volcanic ash. *Mailing Add:* Dept Geol Eng & Geol & Geophys Mich Technol Univ Houghton MI 49931

ROSE, WILLIAM K, b Ossining, NY, Aug 10, 35; m 61; c 3. ASTROPHYSICS. *Educ:* Columbia Col, AB, 57; Columbia Univ, PhD(physics), 63. *Prof Exp:* Res staff astrophys, Princeton Univ, 63-67; asst prof, Mass Inst Technol, 67-70, assoc prof, 71; PROF ASTROPHYS, UNIV MD, 76- *Mem:* Am Astron Soc; AAAS; Am Asn Univ Prof; Int Astron Union. *Res:* Stellar evolution; extragalactic astrophysics; radio astronomy; physical processes in stars; cosmology. *Mailing Add:* Dept Physics & Astron Univ Md College Park MD 20742

ROSE, ZELDA B, BIOPHOSPATE COMPOUNDS. *Educ:* Yale Univ, PhD(biochem), 55. *Prof Exp:* MEM, INST CANCER RES, 78- *Mailing Add:* 1862 Old Orchard Rd Abington PA 19001

ROSEBERRY, JOHN L, b Riverton, Ill, Sept 24, 36. WILDLIFE RESEARCH. *Educ:* Univ Ill, Urbana, BS, 58; Southern Ill Univ, Carbondale, MA, 61. *Prof Exp:* Res asst zool, Southern Ill Univ, 58-60, researcher zool, 61-84, assoc scientist, 84-89, SR SCIENTIST, COOP WILDLIFE RES LAB, SOUTHERN ILL UNIV, 89- *Concurrent Pos:* Consult controlled deer hunt, Crab Orchard Nat Wildlife Refuge, 66, 78-; biol adv deer permit task force, Ill Dept Conserv, 77 & 78; Deer Tech Adv Comt, 80- *Mem:* Wildlife Soc; Am Ornithologists' Union; Am Soc Mammalogists; Sigma Xi. *Res:* Vertebrate ecology and population dynamics, especially regulation, exploitation, computer simulation, habitat inventory and analysis. *Mailing Add:* Coop Wildlife Res Lab Southern Ill Univ Carbondale IL 62901

ROSEBERY, DEAN ARLO, b Stahl, Mo, Sept 23, 19; m 43; c 2. FISHERIES. *Educ:* Northeast Mo State Teachers Col, BS, 41; Va Polytech Inst, PhD(zool), 50. *Prof Exp:* Instr biol, Va Polytech Inst, 46-48; asst chief fish div, State Comn Game & Inland Fisheries, Va, 48-52; prof sci, 52-72, prof & head div sci, 70-85, EMER PROF BIOL & EMER HEAD DIV SCI, NORTHEAST MO STATE UNIV, 85- *Concurrent Pos:* Mem coun, AAAS, 85-88. *Mem:* Fel AAAS; Am Fisheries Soc; Wildlife Soc; Am Soc Limnol & Oceanog; Nat Asn Acad Sci (pres, 84-85). *Res:* Freshwater fish management; fisheries biology. *Mailing Add:* Div Sci Northeast Mo State Univ Kirksville MO 63501

ROSEBOOM, EUGENE HOLLOWAY, JR, b Columbus, Ohio, Sept 21, 26; m 58; c 4. NUCLEAR WASTE DISPOSAL, EXPERIMENTAL PETROLOGY. *Educ:* Ohio State Univ, BS, 49, MS, 51; Harvard Univ, PhD(geol), 58. *Prof Exp:* Res fel, Geophys Lab, Carnegie Inst, 56-59; geologist, Washington, DC, 59-74, asst chief geologist, Eastern Region, 74-79, prog coordr nuclear waste disposal, Geol Div, 80-84, chief, Off Regional Geol, 84-88, DIR STAFF, GEOL SURV, 89- *Mem:* Mineral Soc Am; Soc Econ Geol. *Res:* Phase equilibria among arsenides, sulfides, silicates; geologic applications; theory of phase equilibria diagrams. *Mailing Add:* US Geol Surv Stop 106 Reston VA 22092

ROSECRANS, JOHN A, b Brooklyn, NY, July 30, 35; m 58; c 3. PHARMACOLOGY. *Educ:* St John's Univ, BS, 57; Univ RI, MS, 60, PhD(pharmacol), 63. *Prof Exp:* NIMH fel, Univ Mich, Ann Arbor, 63-64; res asst prof pharmacol, Univ Pittsburgh, 64-65; trainee, Yale Univ, from asst prof to assoc prof, 67-78, PROF PHARMACOL, MED COL VA, 78- *Mem:* AAAS; Am Soc Pharmacol & Exp Therapeut; Soc Neurosci. *Res:* Behavioral and biochemical basis of drug dependence. *Mailing Add:* Commonwealth Univ Box 613 MCV Sci Richmond VA 23298

ROSEGAY, AVERY, b New York, NY, Sept 3, 29. TRITIUM RADIOCHEMISTRY, ISOTOPIC ORGANIC SYNTHESIS. *Educ:* Columbia Univ, BS, 50, MS, 53; NY Univ, PhD(org chem), 60. *Prof Exp:* SR RES FEL, MERCK, SHARP & DOHME RES LABS, RAHWAY, 60- *Mem:* Am Chem Soc. *Res:* Synthesis of tritium-labeled compounds; hydrogen-tritium exchange reactions. *Mailing Add:* Merck Res Labs PO Box 2000 Rahway NJ 07065-0900

ROSEHART, ROBERT GEORGE, b Owen Sound, Ont, July 29, 43; m 67; c 3. CHEMICAL ENGINEERING. *Educ:* Univ Waterloo, BASc, 67, MASc, 68, PhD(chem eng), 70. *Prof Exp:* Fluids & heat transfer engr, Atomic Energy Can, 66 & Can Gen Elec, 67; coordr, Chem Eng Group, 70-77, dean, Univ Schs, 77-84, PRES, LAKEHEAD UNIV, 84- *Concurrent Pos:* Consult, Atomic Energy Can, 67-84; Ont Forestry Coun, 84-; chmn, Ont Adv Comt on Resource Dependent Communities, 85-86; Thunder Bay Wafer Bd Study Govt Ont, 86-87; Ont Premiers Coun, 86-91; chmn, Ont Forest Resources Inventory Comt, 87-88; Ont chief, NAN Land Negotiation, 91- *Res:* Environmental engineering; heat transfer; two-phase flow; water treatment processes; northern development. *Mailing Add:* Lakehead Univ Thunder Bay ON P7B 5E1 Can

ROSELAND, CRAIG R, b Long Beach, Calif, Oct 15, 49; m 87; c 1. INSECT ENDOCRINOLOGY. *Educ:* Univ Calif, Irvine, BS, 72, PhD(insect develop), 76. *Prof Exp:* Res assoc, Univ Wash, 76-81; res assoc, Tree Fruit Res Sta, Wash State Univ, 81-82; res assoc, Kans State Univ, 83-85; ASST PROF ENTOM, NDAK STATE UNIV, 85- *Concurrent Pos:* NIH fel, 77-79. *Mem:* Entom Soc Am; AAAS. *Res:* Analyze the control of synthesis of catecholamines destined for the insect cuticle; investigate the insect attractants that are found in sunflower; study chemical defenses of sunflower against coleopteran pests. *Mailing Add:* Dept Entom NDak State Univ PO Box 5346 Univ Sta Fargo ND 58105

ROSELL, KARL-GUNNAR, analytical carbohydrate chemistry, polymer chemistry, for more information see previous edition

ROSELLE, DAVID PAUL, b Vandergrift, Pa, May 30, 39; m 67. MATHEMATICS. *Educ:* West Chester State Col, BS, 61; Duke Univ, PhD(number theory), 65. *Prof Exp:* Asst prof math, Univ Md, 65-68; from assoc prof to prof, La State Univ, Baton Rouge, 68-74; prof math, VA Polytech Inst & State Univ, 74-79, dean grad sch, 79-81, dean res & grad studies, 81-83, univ provost, 83-87; pres, Univ Ky, 87-90; PRES, UNIV DEL, 90- *Concurrent Pos:* Chief investr, NSF grants, 66-71 & 73-75; Nat Res Coun Can fel, 68; assoc ed, Am Math Monthly, 73-75. *Mem:* Am Math Soc; Math Asn Am (secy, 75-); Nat Coun Teachers Math. *Res:* Combinatorial analysis; number theory. *Mailing Add:* 132 Hullihen Hall Univ Del Newark DE 19716

ROSELLI, CHARLES EUGENE, b Harrisburg, Pa, Dec 26, 52; m 89; c 1. NEUROENDOCRINOLOGY, REPRODUCTIVE BIOLOGY. *Educ:* Franklin & Marshall Col, BA, 74; Hahnemann Univ, PhD(physiol), 81. *Prof Exp:* Postdoctoral assoc reprod physiol, Ore Regional Primate Res Ctr, 81-83; from instr to asst prof, 83-90, ASSOC PROF PHYSIOL, ORE HEALTH SCI UNIV, 90-; COLLAB SCIENTIST, DEPT REPRODUCTIVE BIOL, ORE REGIONAL PRIMATE RES CTR, 85- *Concurrent Pos:* Prin investr, Nat Inst Child Health & Human Develop, 87-92. *Honors & Awards:* Young Investr Award, Soc Study Reproduction, 83. *Mem:* Am Physiol Soc; Endocrine Soc; Soc Neurosci; Soc Study Reproduction. *Res:* Cellular mechanisms of androgen action in brain; effects of androgens on neuropeptide content and neurotransmitter receptor concentrations relative to neuroendocrine functions. *Mailing Add:* Dept Physiol L-334 Ore Health Sci Univ Portland OR 97201-3098

ROSELLI, ROBERT J, b Jan 21, 47; m; c 3. LUNG FLUID & SOLUTE EXCHANGE. *Educ:* Univ Calif, Berkeley, PhD(eng sci), 75. *Prof Exp:* PROF, BIOMED ENG, VANDERBILT UNIV, 85- *Mem:* Biomed Eng Soc; Am Physiol Soc; Microcirculatory Soc. *Res:* Cardiovascular system; biomechanics. *Mailing Add:* Dept Biomed Eng Box 36 Sta B Vanderbilt Univ Nashville TN 37235

ROSEMAN, ARNOLD S(AUL), b Boston, Mass, 1930; m 55; c 2. FOOD SCIENCE & TECHNOLOGY. *Educ:* Northeastern Univ, BS, 52; Univ Mass, MS, 54, PhD(food technol), 56. *Prof Exp:* Biochemist rice, Southern Regional Lab, Agr Res Serv, USDA, 56-60; sr scientist, Res & Develop Div, Kraft Co, 60-68; res mgr meats, Res & Develop Dept, John Morrell & Co, 68-74; dir res & develop food technol, CFS Continental, Inc, Chicago, 74-82; vpres, Food Beverages, Jerrico Inc, Lexington, KY, 82-85; dir res, Develop QC & QA, Horace W Longacre, Inc, Franconia, PA, 86-88; PRES, QUAL CONSULT INC, WAYNE, PA, 85-; SR RES ASSOC, DEPT FOOD SCI, BERKS CAMPUS, PA STATE UNIV, READING, 89- *Concurrent Pos:* fel Am Inst Chem; chmn, Food Serv Div, Inst Food Technologists, 76-78. *Mem:* Inst Food Technologists; Am Chem Soc; Am Asn Cereal Chemists; Am Soc Qual Control; Int Asn Milk Food & Environ Sanitarians. *Res:* Food and menu product and process development and improvement, especially institutional items and poultry products; organoleptic evaluation of foods; utilization of products by food service; effects of processing and storage on the chemical, physical and organoleptic properties; research management; management and policy; food law and regulatory compliance; various publications and patents. *Mailing Add:* 39 Militia Hill Dr Wayne PA 19087

ROSEMAN, JOSEPH JACOB, b Brooklyn, NY, Nov 11, 35. MATHEMATICS. *Educ:* Mass Inst Technol, BS, 57; Polytech Inst Brooklyn, MS, 61; NY Univ, PhD(math), 65. *Prof Exp:* Chem engr, Gen Elec Co, 57-59; asst prof math, Univ Wis-Madison, 65-67; assoc prof, Polytech Inst Brooklyn, 67-74; sr lectr, 74-79, ASSOC PROF, TEL AVIV UNIV, 79- *Mem:* Am Math Soc; Math Asn Am; Soc Natural Philos. *Res:* Mathematical elasticity; fluid dynamics; applied mathematics. *Mailing Add:* Tel Aviv Univ Sch Math Sci Tel Aviv Israel

ROSEMAN, SAUL, b New York, NY, Mar 9, 21; m 41; c 3. BIOCHEMISTRY. *Educ:* City Col NY, BS, 41; Univ Wis, MS, 44, PhD(biochem, org chem), 47. *Hon Degrees:* DM, Univ Lund, Sweden, 84. *Prof Exp:* Res assoc biochem & pediat, Univ Chicago, 48-51, asst prof, 51-53; res assoc, Rackham Arthritis Res Unit, Univ Mich Hosp, 53-54, from asst prof to prof biol chem, 54-65; chmn dept biol & dir, McCollum-Pratt Inst, 69-73 & 88-90, prof biol, 65-75, RALPH S O'CONNOR PROF CHAIR BIOL, JOHNS HOPKINS UNIV, 75- *Concurrent Pos:* Sci counr, Nat Cancer Inst, 72- *Honors & Awards:* T Duckett Jones Mem Award, Helen Hay Whitney Found, 73; Rosenstiel Award, Brandeis Univ, 74; Gairdner Found Int Award, 81; Townsend Harris Medal, City Col NY, 87; Merit Award, NIH, 87. *Mem:* Nat Acad Sci; Am Acad Arts & Sci; Am Chem Soc; Am Soc Biol Chemists; hon mem Biochem Soc Japan; Am Soc Microbiol; AAAS; Biophys Soc; Am Asn Univ Professors. *Res:* Chemistry and metabolism of complex carbohydrates; polysaccharides; glycoproteins; glycolipids. *Mailing Add:* Dept Biol Johns Hopkins Univ Charles & 34th Sts Baltimore MD 21218

ROSEMAN, THEODORE JONAS, b Chicago, Ill, Aug 2, 41; m 63; c 2. PHARMACEUTICAL CHEMISTRY. *Educ:* Univ Ill, BS, 63; Univ Mich, MS, 65, PhD(pharmaceut chem), 67. *Prof Exp:* Pharmacist, Westridge Med Ctr, 63; teaching asst pharm, Univ Mich, 63-64; res scientist, 67-80, SR SCIENTIST PHARM, UPJOHN CO, 80-; DIR, BAXTER HEALTHCARE CORP, DEERFIELD, ILL. *Concurrent Pos:* Past pres, Control Release Soc; adj prof, Philadelphia Col Pharm. *Mem:* Am Pharmaceut Asn; Am Chem Soc; Control Release Soc; Parenteral Drug Asn; fel Acad Pharmaceut Sci; USP Comt Rev; Am Soc Hosp Pharmacists; Am Asn Pharmaceut Scientists. *Res:* Physical-chemical properties of therapeutic agents; controlled release delivery systems; package-product interactions; drug release mechanisms; product research and development. *Mailing Add:* 16 Nottingham Dr Lincolnshire IL 60069

ROSEMARK, PETER JAY, b Los Angeles, Calif, July 28, 55. RADIATION THERAPY PHYSICS. *Educ:* Mass Inst Technol, Cambridge, ScB, 77; Univ Calif, Los Angeles, MSc, 79, PhD(med physics), 82; Am Bd Radiol, cert therapeut radiol physics, 84. *Prof Exp:* RADIATION PHYSICIST, DEPT RADIATION THER, CEDARS-SINAI MED CTR, 77- *Concurrent Pos:* Vis lectr, dept radiation oncol, Univ Calif, Los Angeles, 82-; consult physicist, radiation ther serv, Wadsworth Vet Admin, 84- *Mem:* Am Asn Physicists Med; Hosp Physicists Asn; Am Col Radiol; Am Soc Therapeut Radiol & Oncol. *Res:* Dose distribution, calculation and reporting of radioactive implants and other radiation therapy physics problems. *Mailing Add:* Dept Radiation Ther Cedars-Sinai Med Ctr 8700 Beverly Blvd Los Angeles CA 90048-0750

ROSEMBERG, EUGENIA, b Buenos Aires, Argentina. REPRODUCTIVE ENDOCRINOLOGY, PHARMACOLOGY OF REPRODUCTION. *Educ:* Nat Lyceum Women, Buenos Aires, BS, 36; Univ Buenos Aires, MD, 44. *Prof Exp:* Instr, dept anat, Univ Buenos Aires Med Sch, 40-46, instr pediat, 46-48, res fel, Johns Hopkins Univ Med Sch, endocrinol dept, 48-50, staff mem, Worcester Found Experimental Biol, Mass, 53-62; mem med staff, Worcester City Hosp, 55-85, pediat, Univ Hosp, Univ Md, 70-72; dir med res, Worcester City Hosp, Mass, 73-85; RES DIR, MED RES INST WORCESTER INC, MASS, 62-, PROF MED, UNIV MASS MED SCH, WORCESTER, MASS, 71- *Concurrent Pos:* Assoc pediat & res asst, pediat dept, Univ Hosp, Univ Buenos Aires, 43-48; vis scientist, Univ Montevideo Med Sch, Uruguay, 50; res fel, Pub Health Serv, NIH, Nat Inst Arthritis & Metab Dis, Endocrinol Sect, Bethesda, Md, 51-53; res fel, Med Res Inst & Hosp, Okla, 53; Nat Health, 65-85, Nat Inst Child Health & Human Develop, 69-82. *Mem:* Hon mem Arg Steril; corresp mem Fertil Soc Peru; Am Soc Androl; Am Fertil Soc; fel AAAS; Am Heart Asn. *Res:* Immunological and biological properties of glycoprotein hormones; study of the hypothalamic-pituitary-ovarian and testicular axis as it relates to female and male infertility. *Mailing Add:* Fifth floor Med Res Inst Worcester Inc Eight Portland St Worcester MA 01608

ROSEMOND, GEORGE P, b Hillsboro, NC, Aug 23, 10; m 37; c 1. SURGERY. *Educ:* Univ NC, BS, 32; Temple Univ, MD, 34, MS, 39; Am Bd Surg, dipl, 42; Am Bd Thoracic Surg, dipl, 60. *Prof Exp:* From instr to assoc prof surg, Sch Med, Temple Univ, 39-47, clin prof, 47-50, prof clin surg, 50-60, co-chmn, Dept Surg, 60-63, chmn, Div Surg, 63-73, prof surg & chmn dept, 63-73, EMER PROF SURG, SCH MED, TEMPLE UNIV, 79- *Concurrent Pos:* Mem adv comt & consult physician, Wilkes-Barre Vet Admin Hosp. *Mem:* Am Asn Thoracic Surg; Am Col Chest Physicians; Am Surg Asn; fel Am Col Surg; Am Cancer Soc (pres, 74). *Res:* Cancer, especially cancer of the breast. *Mailing Add:* 825 Old Hickory Rd Apt 100 Lancaster PA 17601

ROSEN, ALAN, b Tel Aviv, Israel, Aug 19, 27; nat US; m 56; c 1. SPACE PHYSICS, BIOMEDICAL ENGINEERING. *Educ:* Univ Southern Calif, BA, 51, MA, 54, PhD(nuclear physics), 58. *Prof Exp:* Asst, Univ Southern Calif, 51-55, lectr physics, 55-58; geophysicist, Space Tech Lab, Inc, Thompson Ramo Wooldridge, Inc, 58-68, DIR SPACE SCI LAB, TRW SYSTS GROUP, 68- *Mem:* Am Phys Soc; Am Asn Physics Teachers; Am Geophys Union; Am Inst Aeronaut & Astronaut; Sigma Xi. *Res:* Space physics, especially terrestrial radiation belt studies by means of earth satellites and space probes; studies of the magnetic characteristics of man; magnetocardiographic research; spacecraft charging by magnetospheric plasmas. *Mailing Add:* 1917 Clark Lane-B Redondo Beach CA 90278

ROSEN, ARTHUR LEONARD, b Chicago, Ill, Apr 30, 34; m 56; c 2. BIOPHYSICS, APPLIED MATHEMATICS. *Educ:* Roosevelt Univ, BS, 57; Univ Chicago, MS, 64, PhD(physics), 71. *Prof Exp:* Res assoc, Dept Cardiovasc Res, Michael Reese Hosp, Chicago, 55-58, res assoc, Dept Surg Res, 60-64; asst dir dept surg res, Hektoen Inst Med Res, 64-78; res assoc, 81-83, ASST PROF, DEP SURG, UNIV CHICAGO, 83-; RES ASSOC, DEPT SURG, MICHAEL REESE HOSP CHICAGO, 78- *Concurrent Pos:* Asst prof biophys, Dept Surg, Col Med, Univ Ill, 67-78; asst prof bioeng, Univ ILL, 80-81. *Mem:* Am Phys Soc; Biophys Soc; Soc Rheology; Inst Elec & Electronics Engrs; Am Asn Physicists in Med. *Res:* Red blood cell substitutes; control of erythropoiesis. *Mailing Add:* Dept Surg Michael Reese Hosp 2929 S Ellis Chicago IL 60616

ROSEN, ARTHUR ZELIG, b Oil City, Pa, Feb 5, 20; m 41. PHYSICS. *Educ:* Univ Calif, BA, 41, PhD(physics), 52. *Prof Exp:* Engr, Permanente Shipyards, 41-42; jr physicist radiation lab, Univ Calif, 42-45, asst physics, 46-51; engr, Berkeley Sci Co, 46; lectr physics, Santa Barbara Col, 51-53; from instr to assoc prof, 53-64, PROF PHYSICS, CALIF POLYTECH STATE UNIV, SAN LUIS OBISPO, 64- *Concurrent Pos:* NSF sci fac fel, 59. *Mem:* Am Phys Soc; Am Asn Physics Teachers. *Res:* Mass spectrograph development; cosmic ray studies at sea level and mountain altitudes; experimental nuclear physics; environmental radioactivity. *Mailing Add:* Dept Physics Calif Polytech State Univ San Luis Obispo CA 93401

ROSEN, BARRY PHILIP, b Hartford, Conn, June 18, 44. BIOCHEMISTRY. *Educ:* Trinity Col, BS, 65; Univ Conn, MS, 68, PhD(biochem), 69. *Prof Exp:* USPHS fel, Cornell Univ, 69-71; ASSOC PROF BIOCHEM, SCH MED, UNIV MD, BALTIMORE CITY, 71- *Mem:* Am Soc Microbiol; Am Soc Biol Chemists. *Res:* Structure and function of biological membranes; energy transduction and active transport in microorganisms. *Mailing Add:* Dept Biochem Wayne State Univ Sch Med 540 E Canfield Ave Detroit MI 48201

ROSEN, BERNARD, b New York, NY, June 6, 30; m 62; c 2. PLASMA PHYSICS. *Educ:* NY Univ, AB, 50, PhD(physics), 59. *Prof Exp:* Engr, Fed Tel & Radio Co, 51-52 & Bendix Aviation Co, 52-53; instr physics, NY Univ, 55-58; engr, Radio Corp Am, 58-60; from asst prof to assoc prof physics, 60-73, PROF PHYSICS, STEVENS INST TECHNOL, 73- *Concurrent Pos:* Consult, Plasma Physics Lab, Princeton Univ, 72-75. *Mem:* Am Phys Soc. *Res:* Computational plasma physics. *Mailing Add:* Dept Physics Stevens Inst Technol Castle Pt Sta Hoboken NJ 07030

ROSEN, BRUCE IRWIN, b Chicago, Ill, July 8, 52. ORGANIC CHEMISTRY. *Educ:* Northwestern Univ, BA, 74; Univ Southern Calif, PhD(org chem), 77. *Prof Exp:* Sr res chemist org chem, 3M Co, 77-79; res specialist, org chem, UOP Inc, 79-84; SR RES SCIENTIST, ORG CHEM, AMOCO CHEM, 84- *Mem:* Am Chem Soc; AAAS. *Res:* Synthetic organic chemistry, organometallic chemistry (heterogeneous and homogeneous catalysis) and polymer chemistry. *Mailing Add:* Amoco Chem PO Box 3011 Naperville IL 60566

ROSEN, C(HARLES) A(BRAHAM), b Toronto, Ont, Dec 7, 17; nat US; m 41; c 4. ELECTRICAL ENGINEERING. *Educ:* Cooper Union, BEE, 40; McGill Univ, MEng, 50; Syracuse Univ, PhD(elec eng), 56. *Prof Exp:* Sr examr, Brit Air Comn, NY, 40-43; proj engr, Fairchild Aircraft, Ltd, Can, 43-47; co-owner, Electrolabs Regist, 47-50; consult engr, Gen Elec Co, 50-57; mgr, Appl Physics Lab, SRI Int, 57-70, Artificial Intel Group, Info Sci Lab, 70-78, consult, 78-89; CONSULT, RICOH RES, 90- *Concurrent Pos:* Co-founder & pres, Machine Intel Co, 78-80, chmn, 80-87; dir, Ridge Vineyards Co, 62-86, Microbot Co, 80-81, Packet Tech Co, 80-89, Cochlea Co, 81-, Telebit Co, 83-89, Picodyne Co, 86-89; mem, Adv Comt Tech Innov, Nat Acad Sci, 72-77; mem, Oversight Comt, Nat Bur Standards, 75-78, Comt Army Robotics & Artificial Intel, Nat Res Coun, 81-83, 86-87, Space Syst & Tech Adv Comt, NASA, 86-90, Rehab Res Adv Comt, Vet Admin, 89-, US Army STAR Comt, Nat Res Coun, 89-; mem Space Syst & Tech Adv Comt, NASA, 86- *Honors & Awards:* Taylor Award, Inst Elec & Electronics Engrs, 75. *Mem:* Am Phys Soc; Sigma Xi; fel Inst Elec & Electronics Engrs; sr mem Soc Mfg Engrs; fel Am Asn Artificial Intel. *Res:* Learning machines; pattern recognition; artificial intelligence; electron-beam-activated micromachining processes; ferroelectric and piezoelectric devices; programmable automation; robotics. *Mailing Add:* 139 Tuscaloosa Ave Atherton CA 94025

ROSEN, CAROL ZWICK, b New York, NY; m 62; c 2. ENGINEERING PHYSICS, MATERIAL SCIENCE ENGINEERING. *Educ:* Brooklyn Col, BS, 53; NY Univ, MS, 55; Stevens Inst Technol, PhD(physics), 65. *Prof Exp:* Res assoc, HE & nuclear magnetic resonance, IBMWatson Lab, Columbia Univ, 65-67; res assoc, Stevens Inst Technol, 67-70; consult, Am Inst Physics, 70-82; asst prof physics, York Col, NY, 74-75; staff physicist, Am Phys Soc, 75-76; scientist non-woven web technol, Johnson & Johnson Co, 76-79; proj mgr superconductivity, Airco Superconductors, 79-80; scientist microelectronics, Advan Technol Systs Inc Div, Austin Co, 80-83; ENG SPECIALIST, GEC-MARCONI ELECTRONIC SYSTS CORP, 84- *Concurrent Pos:* adj prof physics, Queens Col, NY, 75-; index ed, Am Inst Physics, 70-73, consult ed, 73-82; instr chem, Barnard Col; instr physics, NY Univ, Fairleigh Dickinson Univ, Hunter Col & York Col-City Univ NY; eng specialist, Singer Electronic Systs Div, 84- *Mem:* Am Phys Soc; NY Acad Sci; Am Inst Metall Eng; Am Electroplating Soc; Sigma Xi; Soc Women Engrs (vpres, 73-74 & pres, 74-75). *Res:* Fabrication of superconducting wire and braid; research and product development of piezoelectric transducers including mathematical modeling of mechanical-electrical-thermal properties along with measurements of the material properties of novel and existing piezoelectrics. *Mailing Add:* 934 Red Rd Teaneck NJ 07666

ROSEN, DAVID, b New Haven, Conn, Jan 26, 21; m 46; c 4. MATHEMATICS. *Educ:* NY Univ, AB, 42; Univ Pa, AM, 49, PhD(math), 52. *Prof Exp:* Asst instr math, Univ Pa, 49-52; from instr to prof, 52-87, EMER PROF MATH, SWARTHMORE COL, 87- *Concurrent Pos:* NSF fac fel, 61-62; Fulbright-Hays lectr, Ireland, 71-72. *Mem:* Am Math Soc; Math Asn Am; Sigma Xi. *Res:* Automorphic functions; continued fractions. *Mailing Add:* Dept Math Swarthmore Col Swarthmore PA 19081

ROSEN, DAVID A, b Montreal, Que, Apr 18, 26; m 47; c 2. OPHTHALMOLOGY. *Educ:* McGill Univ, BSc, 47, MD, CM, 49; FRCS(C), 54. *Prof Exp:* From asst prof to assoc prof ophthal, Queen's Univ, Ont, 54-57, prof, 57-78; PROF OPHTHAL, STATE UNIV NY, STONY BROOK, 78- *Concurrent Pos:* Markle scholar, 54-59; chmn dept ophthal, Kingston Gen Hosp, 54-73; chmn, Dept Ophthal, Long Island Jewish Med Ctr, 78- *Mem:* Asn Res Vision & Ophthal; Am Acad Ophthal; fel Am Col Surgeons; Can Soc Ophthal; Can Med Asn. *Res:* Ophthalmic pathology; retinal vessel physiology and angiology; photocoagulation; health care delivery; medical education. *Mailing Add:* L1 Jewish Med Ctr New Hyde Park NY 11042

ROSEN, DIANNE L, VIROIDS. *Educ:* Dartmouth Col, PhD(biol), 84. *Prof Exp:* Fel, Genetics Lab, Rockefeller Univ, 83-86; CONSULT, 86- *Mailing Add:* 963 Hillside Ave Plainfield NJ 07060

ROSEN, EDWARD M(ARSHALL), b Chicago, Ill, Jan 28, 30; m 65; c 2. CHEMICAL ENGINEERING. *Educ:* Ill Inst Technol, BS, 51, MS, 53; Univ Ill, PhD(chem eng), 59. *Prof Exp:* SR SCI FEL CHEM ENG, MONSANTO CO, 59- *Concurrent Pos:* Monsanto acad leave chem eng, Stanford Univ, 62-63; lectr, Univ Mo, 72; affil prof, Wash Univ, 74; secy, Cache Corp, 84-85; chmn, Computing & Systs Technol Div, Am Inst Chem Engrs, 85. *Mem:* Am Inst Chem Engrs. *Res:* Process simulation and design. *Mailing Add:* Monsanto Co, F2WK 800 N Lindbergh St Louis MO 63167

ROSEN, FRED SAUL, b Newark, NJ, May 26, 30. IMMUNOLOGY. *Educ:* Lafayette Col, AB, 51; Western Reserve Univ, MD, 55. *Prof Exp:* Intern path, Children's Hosp, Boston, 55-56, jr asst res med, 56-59, sr asst resident, 59-60, res fel, 60-62, asst, 62-63, assoc med, 63-68, chief, Immunol Div, 68-85; teaching fel pediat, Harvard Med Sch, 59-60, res fel, 60-62, instr, 62-63, assoc, 64-65, from asst prof to assoc prof pediat, 66-72, JAMES L GAMBLE PROF PEDIAT, HARVARD MED SCH, 72-; SR ASSOC MED, CHILDREN'S HOSP, BOSTON, 68- *Concurrent Pos:* USPHS res officer, Lab Chem Pharmacol, Nat Cancer Inst, NIH, 57-59; fel, John Simon Guggenheim Mem Found, 74; vis prof, Dept Immunol, Royal Postgrad Med Sch, Univ London, Eng, 74-75; chmn, Sci Adv Comt, Ctr for Blood Res, Boston, 76-87; prog dir, Clin Res Ctr, Children's Hosp, Boston, 77-91; chmn, Comt Disadvantaged Scientists, Am Asn Immunologists, 78; vpres, Ctr for Blood Res, Boston, 81-87, pres, 87-; distinguished vis scholar, Christ's Col, Cambridge Univ, Eng, 83-; assoc ed, J Clin Invest, 85-87; mem, Med Adv Comt, Immune Deficiency Found, 86-; chmn, Expert Sci Comt on Immunodeficiency, WHO, 88; ed, Immunodeficiency Rev, 88-; vis prof pediat, Sch Med, Washington Univ, St Louis, Mo, 90-; adj prof, Dept Vet Pathobiol, Sch Vet Med, Purdue Univ, 91- *Honors & Awards:* E Mead Johnson Award for Pediat Res, Am Acad Pediat, 70. *Mem:* Inst Med-Nat Acad Sci; emer mem Am Soc Clin Invest; Am Pediat Soc; Am Asn Immunologists; Asn Am Physicians; emer mem Soc Pediat Res. *Mailing Add:* Ctr Blood Res Inc 800 Huntington Ave Boston MA 02115

ROSEN, GERALD HARRIS, b Mt Vernon, NY, Aug 10, 33; m 63; c 2. THEORETICAL PHYSICS. *Educ:* Princeton Univ, BSE, 55, MA, 56, PhD(physics), 58. *Prof Exp:* Res assoc dept aeronaut eng, Princeton Univ, 58-59; NSF fel, Inst Theoret Physics, Stockholm, Sweden, 59-60; prin scientist, Martin-Marietta Aerospace Div, 60-63; consult basic & appl res, Southwest Res Inst, 63-66; prof physics, 66-73, M RUSSELL WEHR PROF PHYSICS & ATMOSPHERIC SCI, DREXEL UNIV, 73- *Concurrent Pos:* Mem tech staff, Weapons Systs Eval Div, The Pentagon, 60. *Mem:* Fel AAAS; fel Am Phys Soc. *Res:* Theories of relativistic quantum and classical fields; theories of compressible, viscous and combustible fluid flows; turbulence phenomena; nonlinear reaction and transport processes; space-charge-limited currents; biomathematics; nonlinear partial differential equations. *Mailing Add:* 415 Charles Lane Wynnewood PA 19096

ROSEN, GERALD M, FREE RADICAL BIOLOGY. *Educ:* Clarkson Col Technol, PhD(chem), 69; Duke Univ Law Sch, JD, 79. *Prof Exp:* Assoc prof, 78-85, PROF PHARMACOL, DUKE UNIV MED SCH, 85- *Mailing Add:* Dept Pharmacol Univ Md Sch Pharm 20 N Pine St Baltimore MD 21201

ROSEN, HAROLD A, b New Orleans, La, Mar 20, 26. ASTRONAUTICS. *Educ:* Tulane Univ, BE, 47; Calif Inst Technol, ME, 48, PhD, 51. *Hon Degrees:* DSc, Tulane Univ, 75. *Prof Exp:* Vpres eng & mgr, commun systs div, Space & Commun Group, 56-75, VPRES, HUGHES AIRCRAFT CO, 75- *Concurrent Pos:* Astronaut engr, Nat Space Coun, 64. *Honors & Awards:* Commun Award, Am Inst Aeronautics & Astronautics, 68; Mervin J Kelly Award, Inst Elec & Electronics Engrs, 72, Alexander Graham Bell Medal, 82; L M Ericsson Int Prize, Sweden, 76; Lloyd V Berkner Award, Am Astronautical Soc, 76; Nat Medal Technol, President of US, 85; C & C Found Prize, Tokyo, 85. *Mem:* Nat Acad Eng; fel Am Inst Aeronautics & Astronautics; fel Inst Elec & Electronics Engrs. *Res:* Conceived spin stabilized synchronous communication satellite; development of advanced communication and satellite systems. *Mailing Add:* Hughes Aircraft Co Bldg S64 Mail Sta A402 Airport Sta PO Box 92919 Los Angeles CA 90009

ROSEN, HARRY MARK, b Philadelphia, Pa, Mar 19, 46. MEDICAL ADMINISTRATION. *Educ:* Univ Pa, BS, 68; Columbia Univ, MS, 70; Cornell Univ, PhD, 76. *Prof Exp:* Dir hosp students, New York City Health Serv Admin, 72-73; from asst prof to assoc prof, 73-80, PROF, MT SINAI SCH MED, 80-, VCHMN, DEPT HEALTH CARE MGT, 79- *Mem:* Am Pub Health Asn; Am Hosp Asn; Am Col Hosp Adminr. *Res:* Evaluation of health care services; quantitative methods in health care management; the quality assurance function. *Mailing Add:* Dept Mgt CUNY Bernard Baruch 17 Lexington Ave New York NY 10010

ROSEN, HENRY, b Bad Reichenhall, Ger, Nov 26, 46; US citizen; m 79; c 2. INFECTIOUS DISEASES, MICROBIAL BIOCHEMISTRY. *Educ:* Yale Col, AB, 68; Univ Rochester, MD, 72. *Prof Exp:* Intern internal med, Univ Wash Affil Hosp, 72-73, resident, 73-74; fel infectious dis, 74-77, from asst prof to assoc prof, 77-91, PROF MED, UNIV WASH, 91- *Concurrent Pos:* Consult, NIH, 81-85. *Mem:* Fel Infectious Dis Soc Am; Am Fedn Clin Res; Am Col Physicians. *Res:* Understanding the contribution of neutrophils, a subpopulation of circulating white blood cells, towards protecting host organisms from infections. *Mailing Add:* ZD-09 Univ Wash Seattle WA 98195

ROSEN, HOWARD, b Chicago, Ill, July 25, 39; m 64; c 2. DNA REPAIR, GENETIC RECOMBINATION. *Educ:* Univ Calif, Los Angeles, BS, 61, MS, 65, PhD(bot sci), 68. *Prof Exp:* Fel, Dept Human Genetics, Univ Mich, 69-70; PROF BIOL, CALIF STATE UNIV, LOS ANGELES, 70- *Concurrent Pos:* Res assoc, Carnegie-Mellon Univ, 72; vis prof zool, Duke Univ, 88-89. *Res:* Mechanism of repair of ultraviolet induced damage in chlamydomonas reinhardtii; mechanism of uniparental inheritance of chloroplast DNA in chlamydomonas reinhardtii. *Mailing Add:* Calif State Univ Dept Biol 5151 State Univ Dr Los Angeles CA 90032

ROSEN, HOWARD NEAL, b Takoma Park, Md, June 25, 42; m 69; c 2. CHEMICAL ENGINEERING, WOOD TECHNOLOGY. *Educ:* Univ Md, BS, 64; Northwestern Univ, MS, 66, PhD(chem eng), 69. *Prof Exp:* Develop engr petrol res, Shell Develop Co, 69-70; res chem engr, NCent Forest Exp Sta, 70-77, proj leader wood utilization, 77-85. *Concurrent Pos:* Adj asst prof, Forestry Dept, Southern Ill Univ, 71-84. *Mem:* Am Inst Chem Engrs; Forest Prod Res Soc; Soc Wood Sci & Technol; Int Union Forest Res Orgn; Hardwood Res Coun. *Res:* Drying, energy conservation, wood utilization, treated wood products, wood physics, crystallization. *Mailing Add:* USDA Forest Serv Forest Prod & Harvesting Res PO Box 96090 Washington DC 20090-6090

ROSEN, IRVING, b New York, NY, Apr 3, 24; m 48; c 2. POLYMER CHEMISTRY. *Educ:* Brooklyn Col, BA, 47; Ind Univ, MS, 49, PhD(chem), 51. *Prof Exp:* Res phys chemist, Reaction Motors, Inc, 51-52; sr chemist, Diamond Alkali Co, 52-55, res group leader polymer & radiation chem, 55-66, sr group leader polymer chem, 66-67; res supvr, 67-81, ADV TO DIR RES, STANDARD OIL CO, OHIO, 81- *Mem:* Am Chem Soc; Soc Plastics Engrs. *Res:* Free radical and ionic polymerization; thermoplastic and thermoset polymers; structure-property relations; monomer synthesis; catalysis; chemicals; synfuels. *Mailing Add:* 2657 Green Rd Cleveland OH 44122-1564

ROSEN, IRWIN GARY, b Long Beach, NY, Feb 19, 54. NUMERICAL ANALYSIS, CONTROL THEORY. *Educ:* Brown Univ, ScB, 75, ScM, 76, PhD(appl math), 80. *Prof Exp:* ASST PROF MATH, BOWDOIN COL, 80- *Concurrent Pos:* Vis scientist, Inst Comput Appln Sci & Eng, 81, consult, 81- *Mem:* Am Math Soc; Soc Indust & Appl Math. *Res:* Numerical approximation methods for the solution of parameter identification and optimal control problems for distributed parameter systems. *Mailing Add:* Univ Southern Calif Los Angeles CA 90089

ROSEN, JAMES CARL, b Los Angeles, Calif, July 30, 49; m 76; c 2. CLINICAL PSYCHOLOGY. *Educ:* Univ Calif, Berkeley, BA, 71; Univ Nev, PhD(psych), 76. *Prof Exp:* Asst prof, 76-82, ASSOC PROF PSYCHOL, UNIV VT, 82-, DIR CLIN PSYCHOL PROG, 84- *Res:* Eating disorders; bulimia nervosa; body image disturbance; weight reducing in adolescents; psychology of disability. *Mailing Add:* Dept Psychol Univ Vt Dewey Hall Burlington VT 05405

ROSEN, JAMES MARTIN, b Waseca, Minn, Mar 9, 39; m 67; c 2. PHYSICS. *Educ:* Univ Minn, BS, 61, MS, 63, PhD(physics), 67. *Prof Exp:* Res assoc, Univ Minn, 67-68; from asst prof to assoc prof, 68-78, PROF PHYSICS, UNIV WYO, 78- *Mem:* Am Geophys Union; Optical Soc Am. *Res:* Stratospheric constituents; atmospheric research, especially aerosols, ozone & water vapor; atmospheric electricity. *Mailing Add:* Dept Physics & Astron Univ Wyo Laramie WY 82071

ROSEN, JEFFREY KENNETH, b Middletown, NY, Dec 27, 41; m 70; c 2. DATA PROCESSING. *Educ:* Dartmouth Col, BA, 63; Brown Univ, MA & PhD(physiol), 72. *Prof Exp:* Teacher gen sci, Middletown Bd Educ, NY, 66-68; partner biol consult, L M Kraft Assocs, Goshen, NY, 68-69; asst prof physiol, Univ Dar es Salaam, Tanzania, 72-74; res investr & admin asst to dir, Animal Physiol & Husb, Amphibian Facil, 74-77, sr procedures analyst, data processing, Med Serv Plan Off, 78-80, PROGRAMMER & ANALYST, HOSP DATA SYSTS CTR, UNIV MICH HOSP, 80- *Concurrent Pos:* Fel, Brown Univ, 72. *Mem:* Am Asn Lab Animal Sci; AAAS. *Mailing Add:* 2129 Newport Rd Ann Arbor MI 48103

ROSEN, JEFFREY MARK, b New York, NY, Jan 5, 45; m 70; c 1. BIOCHEMISTRY, ENDOCRINOLOGY. *Educ:* Williams Col, BA, 66; State Univ NY Buffalo, PhD(biochem), 71. *Prof Exp:* Res assoc obstet & gynec, Sch Med, Vanderbilt Univ, 72-73; from asst prof to assoc prof, 73-82, PROF CELL BIOL, BAYLOR COL MED, 82- *Concurrent Pos:* NIH career develop award, 75-80; mem molecular cytol study sect, NIH, 79-83; vis scientist, Imperial Cancer Res Fund, London, 87-88; Am Cancer Soc Scholar grant, 87-88; mem, Am Chem Soc Sci Adv Comt biochem & chem carcinogenesis, 87-; exec ed, Nucleic Acids Res. *Mem:* AAAS; Am Chem Soc; Endocrine Soc; Am Soc Microbiol; Develop Biol Soc; Am Soc Biol Chemists; Am Soc Cell Biol. *Res:* Mechanism of steroid and peptide hormone action; hormonal regulation of mammary gland growth and differentiation; hormonal regulation of breast cancer. *Mailing Add:* Dept Cell Biol Baylor Col Med One Baylor Plaza Houston TX 77030-3498

ROSEN, JOHN FRIESNER, b New York, NY, June 3, 35; m 63; c 2. BIOCHEMISTRY, METABOLISM. *Educ:* Harvard Col, AB, 57; Columbia Univ, MD, 61. *Prof Exp:* Intern med, Montefiore Hosp, Bronx, NY, 61-62; resident pediat, Columbia-Presby Med Ctr 62-65; guest investr, Rockefeller Univ, 65-66, res assoc pediat endocrinol, 66-69, asst physician, Univ Hosp, 65-69; asst prof, 69-76, ASSOC PROF PEDIAT, ALBERT EINSTEIN COL MED & MONTEFIORE HOSP & MED CTR, 76- DIR PEDIAT METAB SERV & LABS, HOSP & CTR, 69- *Mem:* AAAS; Am Chem Soc; Am Fedn Clin Res; fel Am Acad Pediat. *Res:* Biochemical research in mineral metabolism; extraction and purification of human thyrocalcitonin; clinical research in pediatric mineral metabolism; vitamin D dependent rickets; basic and clinical research in mechanisms of lead's interaction with bone. *Mailing Add:* Dept Pediat Albert Einstein Col Med Moses Bldg Rm 404 111 E 210th St Bronx NY 10467

ROSEN, JOSEPH DAVID, b New York, NY, Feb 26, 35; m 62; c 4. FOOD CHEMISTRY, PESTICIDE CHEMISTRY. *Educ:* City Col New York, BS, 56; Rutgers Univ, PhD(org chem), 63. *Prof Exp:* Res chemist, E I du Pont de Nemours & Co, Inc, 63-65; from asst res prof to assoc res prof food sci, 65-74, res prof, 74-88, RES PROF II FOOD SCI, RUTGERS UNIV, NEW BRUNSWICK, 88- *Concurrent Pos:* Sci adv, Food & Drug Admin, 74-78, 81-87; adv bd, J Agr & Food Chem, 76-78; vis prof pesticide chem, Univ Calif, Berkeley, 78-79; co ed, J Food Safety, 78- 89; NJ Pesticide Rev Comt, 87- *Mem:* Am Chem Soc; Am Soc Mass Spectronomy; Inst Food Technologists.

Res: Analysis of pesticides, mycotoxins, and industrial chemicals in food by mass spectrometry; metabolism of xenobiotics; photochemistry of pesticides. *Mailing Add:* Dept Food Sci Cook Col Rutgers Univ New Brunswick NJ 08903

ROSEN, JUDAH BEN, b Philadelphia, Pa, May 5, 22; div; c 2. COMPUTER SCIENCE. *Educ:* Johns Hopkins Univ, BS, 43; Columbia Univ, PhD(appl math), 52. *Prof Exp:* Jr engr, Gen Elec Co, 43-44; develop engr, Manhattan Proj, 44-47; develop engr, Brookhaven Nat Lab, 47-48; res assoc, Princeton Univ, 52-54; head appl math dept, Shell Develop Co, 55-62; vis prof comput sci, Stanford Univ, 62-64; prof & chmn dept, Univ Wis-Madison, 65-71, prof math res ctr, 64-71; head dept, 71-80, PROF COMPUT SCI, UNIV MINN, 71- *Concurrent Pos:* Consult, Argonne Nat Labs, 76-; Lady Davis vis prof, Technion, Israel, 80; prin investr, NSF res grants, 79-82, 82-85 & 85-; invited lectr, Chinese Acad Sci Peking, 80; fel, Supercomput Inst, 85. *Mem:* Comput Sci Bd; Soc Indust & Appl Math; Asn Comput Mach; Math Prog Soc. *Res:* Computer algorithms and software for nonlinear programming and large-scale optimization problems; constrained global optimization and combinatorial optimization methods; parallel algorithms for vector/multiprocessor supercomputers. *Mailing Add:* 10305 N 28th Ave Plymouth MN 55441

ROSEN, LEON, b Los Angeles, Calif, Oct 4, 26; m 52; c 3. EPIDEMIOLOGY. *Educ:* Univ Calif, AB, 45, MD, 48, MPH, 50; Johns Hopkins Univ, DrPH, 53. *Prof Exp:* Intern, Gorgas Hosp, Panama, CZ, 48-49; med dir & head Pac Res Sect, Nat Inst Allergy & Infectious Dis, 50-78, dir, Pac Res Univ, Res Corp, 78-80, DIR ARBOVIRUS PROG, PAC BIOMED RES CTR, UNIV HAWAII, 80- *Honors & Awards:* Laveran Medal, Societe de Pathologie Exotique; Ashford Medal, Am Soc Trop Med & Hyg. *Mem:* Am Epidemiol Soc; Am Soc Trop Med & Hyg; Royal Soc Trop Med & Hyg; Infectious Dis Soc Am. *Res:* Virology; arthropod-borne viruses; nematode infections; medical entomology. *Mailing Add:* Arbovirus Prog Atherton Bldg 2nd Fl Univ Hawaii 3675 Kilauea Ave Honolulu HI 96816

ROSEN, LEONARD CRAIG, b New York, NY, Apr 14, 36; div; c 2. ENVIRONMENTAL PHYSICS, ENERGY CONVERSION. *Educ:* Cornell Univ, AB, 57; Columbia Univ, MBA, 58, MA, 64, PhD(physics), 68. *Prof Exp:* Asst prof physics & astron, Dartmouth Col, 69-76; MEM STAFF, LAWRENCE LIVERMORE NAT LAB, UNIV CALIF, 76- *Concurrent Pos:* Consult, Lawrence Livermore Lab, Univ Calif, 68-76 & Arthur D Little, Inc, 75-76; lectr, San Jose State Univ, 78-79, Univ Calif, Davis, 82-87, San Francisco State Univ, 90- *Mem:* Am Phys Soc. *Res:* Nucleosynthesis; mathematical models of air pollution; development of energy conversion systems; laser interactions in the atmosphere; radiative transport models. *Mailing Add:* Atmospheric Geophys Sci L-262 PO Box 808 Livermore CA 94550

ROSEN, LOUIS, b New York, NY, June 10, 18; m 41; c 1. PHYSICS. *Educ:* Univ Ala, BA, 39, MS, 41; Pa State Univ, PhD(physics), 44. *Hon Degrees:* DSc, Univ NMex, Univ Colo. *Prof Exp:* Asst physics, Univ Ala, 39-40, instr, 40-41; from asst to instr, Pa State Univ, 41-44; mem staff, Los Alamos Sci Lab, 44-46, alt group leader cyclotron group, 46-49, group leader, Nuclear Plate Lab, 49-65, alt div leader, 62-65, div leader, 65-85, dir, Los Alamos Meson Physics Fac, 65-85, sr fel, 85-90, EMER SR FEL, LOS ALAMOS SCI LAB, 90- *Concurrent Pos:* Guggenheim fel, 59-60; mem nuclear sci panel & chmn subpanel accelerators, President's Off Sci & Technol & Nat Acad Sci, 70-; mem, Gov Comt Tech Excellence in NMex, 70-; ad hoc comt nuclear sci, Nat Acad Sci, 75-77; mem, NMex Cancer Control Adv Comn, NIH; sesesquicentennial hon prof, Univ Ala, 81. *Honors & Awards:* E O Lawrence Award, 63; Golden Plate Award, Nat Acad Achievement, 64. *Mem:* Fel AAAS; fel Am Phys Soc. *Res:* High hydrostatic pressures; x-ray; cosmic rays; nuclear physics; particle accelerators; nuclear and particle physics; accelerators; cancer treatment. *Mailing Add:* Los Alamos Nat Lab PO Box 1663 Los Alamos NM 87545

ROSEN, MARC ALLEN, b Toronto, Ont, July 23, 58. ENERGY, THERMODYNAMICS & HEAT TRANSFER. *Educ:* Univ Toronto, BASc, 81, MASc, 83, PhD(mech eng), 87. *Prof Exp:* Res & teaching asst mech eng, Univ Toronto, 81-83; res assoc energy, Inst Hydrogen Systs, 83-86; PROF MECH ENG, RYERSON POLYTECH INST, 86- *Concurrent Pos:* Res assoc, Argonne Nat Lab, 87; adv Dynawatt Energy Res Corp, 88- *Res:* Thermodynamics and heat transfer especially second law analysis; energy systems analysis; thermal energy storage and solar energy; computer-aided process simulation. *Mailing Add:* Dept Mech Eng Ryerson Polytech Inst Toronto ON M5B 2K3 Can

ROSEN, MARVIN, b Newark, NJ, Jan 28, 27; m; c 2. ORGANIC CHEMISTRY, POLYMER CHEMISTRY. *Educ:* NY Univ, BA, 48, PhD(chem), 53. *Prof Exp:* Res chemist, Otto B May, Inc, 53-58; chief chem prod lab, Nuodex Chem Prod Co, 58-66, mgr res, Nuodex Div, Tenneco Chem, Inc, 66-69, vpres res & develop, 69-70, mgr res & develop, Tenneco Intermediates Div, 70-71, dir res & develop, Tenneco Organics & Polymers Div, 71-75, mgr polymer appln, 75-77; mgr org res, 77-83, SR RES ASSOC, LONZA INC, 83- *Mem:* Am Chem Soc; Soc Plastics Engrs; Am Soc Microbiol. *Res:* Terpenes; dyes and intermediates; protective coatings; vinyls; fungicides; polyvinyl chloride applications and development; hydantoin chemistry; preservatives. *Mailing Add:* 24 Mandon Dr Wayne NJ 07470

ROSEN, MICHAEL IRA, b Brooklyn, NY, Mar 7, 38; m 60; c 1. MATHEMATICS, NUMBER THEORY. *Educ:* Brandeis Univ, BA, 59; Princeton Univ, PhD(math), 63. *Prof Exp:* From instr to assoc prof math, 62-73, PROF MATH, BROWN UNIV, 73- *Concurrent Pos:* NSF res grants, 63-78 & 80-88; Off Naval Res fel, 65-66. *Mem:* Am Math Soc; Math Asn Am. *Res:* Algebraic numbers, algebraic functions, arithmetic algebraic geometry, cohomology of groups; Drinfeld modules, the Hilbert class field, formal groups, and special values of L-functions; class number identities and generalizations. *Mailing Add:* 157 Lancaster St Providence RI 02906

ROSEN, MILTON JACQUES, b Brooklyn, NY, Feb 11, 20; m 48; c 3. SURFACTANTS, SURFACE CHEMISTRY. *Educ:* City Col New York, BS, 39; Univ Md, MS, 41; Polytech Inst Brooklyn, PhD(org chem), 49. *Prof Exp:* Chemist, Jewish Hosp, Brooklyn, 40-42, Glyco Prod Co, 42-44 & Publicker Com Alcohol Co, Pa, 44; tutor, Brooklyn Col, 46-50, from instr to assoc prof, 50-66, PROF CHEM, BROOKLYN COL CUNY, 66-, DIR, SURFACTANT RES INST, 87- *Concurrent Pos:* Fel, Lenox Hill Hosp, 41; chem consult to indust & US govt, 46-; vis prof, Hebrew Univ, Israel, 58-59, 64-65, 71-72 & 80; grants, CUNY, 62-64 & 85-89, NSF, 63-71 & 79-90, Pilot Chem Co Calif, 76-78, Yamada Sci Found, Japan, 79, Lever Bros, 84-85, Vista Chem Co, 84-85, Alcon Labs, 86, GAF Corp, 87-89, Exxon Res Eng, 87-88 & Dow Chem Co, 88-91; NIH spec fel, 64-65; assoc ed, J Am Oil Chemists Soc, 81-, ed comn, J Dispersion Sci & Technol, 83-, adv bd, J Colloid Interface Sci, 90-; consult, Chinese Univ Develop Proj Two, US Nat Acad Sci, 86. *Honors & Awards:* Plenary lectr, World Surfactants Cong, Munich, 84. *Mem:* Am Chem Soc; Am Oil Chem Soc. *Res:* Surface active agents--correlations between structure and properties, utilization, synthesis, analysis; reactions at interfaces. *Mailing Add:* Surfactant Res Inst Brooklyn Col CUNY Brooklyn NY 11210

ROSEN, MILTON W(ILLIAM), b Philadelphia, Pa, July 25, 15; m 48; c 3. PROPULSION GUIDANCE & CONTROL. *Educ:* Univ Pa, BS, 37. *Prof Exp:* Radio engr guided missiles, Naval Res Lab, 40-45, head rocket sect, Rocket-Sonde Br, 47-52, head, Rocket Develop Br, 53-55, tech dir, Proj Vanguard, 55-58; chief, Rocket Vehicle Develop, NASA, 58-60, dep dir, Off Launch Vehicle Prog, 60-61, dir, Launch Vehicles & Propulsion, Off Manned Space Flight, 61-63, sr scientist, Off Dept Defense & Interagency Affairs, 63-72, dep assoc, Admin Space Sci, 72-74; exec secy, Space Sci Bd, 74-78, Comt Impacts Stratospheric Change, 78-80, exec secy, Comt Underground Coal Mine Safety, 80-83, exec dir, Space Appl Bd, NAS, 83-85; STUDY LEADER, INST LEARNING RETIREMENT, AM UNIV, 87- *Concurrent Pos:* Physicist, Liquid Rocket Sect, Jet Propulism Lab, Calif Inst Technol, 46-47. *Honors & Awards:* James H Wyld Award for Propulsion, Am Inst Aeronaut & Astronaut, 54. *Mem:* Fel Am Inst Aeronaut & Astronaut. *Res:* Radio and radar systems for control of guided missiles; ceramic liners for rocket combustion chambers; Viking high-altitude sounding rocket; conception of Vanguard earth satellite vehicle. *Mailing Add:* 5610 Alta Vista Rd Bethesda MD 20817

ROSEN, MORDECAI DAVID, b Brooklyn, NY, Nov 23, 51; m 73; c 3. PLASMA PHYSICS, LASER-PLASMA INTERACTIONS. *Educ:* Hebrew Univ, Jerusalem, BSc, 72; Princeton Univ, PhD(astrophys), 76. *Prof Exp:* From res asst to res assoc, Plasma Physics Lab, Princeton Univ, 72-76; after proj mgr shiva physics, 78-82, staff physicist plasma physics & laser fusion target design, 76-84, assoc div leader, laser target design, 84-90, DIV LEADER, LASER TARGET DESIGN, UNIV CALIF, LAWRENCE LIVERMORE LAB, 90- *Concurrent Pos:* Lectr, Dept Appl Sci, Univ Calif, Davis. *Honors & Awards:* Fel, Am Phys Soc; "Top 100 Innovators" Sci Dig, 85; Excellence in Plasma Physics Award, Am Phys Soc, 90. *Mem:* Am Phys Soc; Soc Photo-Optical Instrumentation Engrs. *Res:* Laser fusion target design; laser-plasma interactions; magnetohydrodynamic stability of tokamaks, shock waves in plasmas; radiation transport; suprathernal electron transport; physics of high energy density; x-ray laser design. *Mailing Add:* L-477 Lawrence Livermore Lab Livermore CA 94550

ROSEN, MORTIMER GILBERT, b Brooklyn, NY, Dec 31, 31; m 55; c 2. OBSTETRICS & GYNECOLOGY. *Educ:* Univ Wis, BS, 52; NY Univ, MD, 55. *Prof Exp:* Intern, Bellevue Hosp, New York, 55-56, resident med & surg, 56-57; resident obstet & gynec, Genesee Hosp, Rochester, NY, 59-62; from instr to assoc prof obstet & gynec, Sch Med & Dent, Univ Rochester, 62-73, dir res, 68-73; prof reproductive biol & dir, Prenatal Res Unit, Sch Med, Case Western Reserve Univ, 73-; Dir, Dept Obstet & Gynec & Maternity & Infant Care Proj, Cleveland Metrop Gen Hosp, 73-; WILLARD C RAPPLEYE PROF & CHMN, DEPT OBSTET & GYNEC, COL PHYSICIANS & SURGEONS, COLUMBIA UNIV; DIR, DEPT OBSTET & GYNEC, PRESBY HOSP, NEW YORK. *Mem:* Am Col Obstetricians & Gynecologists; Royal Soc Med; Soc Neurosci; Soc Psychophysiol Res. *Res:* Neurologic function and development of the fetus and newborn infant; antepartum and intrapartum human fetal monitoring systems; delivery of improved health care systems to obstetrics patients. *Mailing Add:* Columbia Presby Med Ctr 622 W 168th St New York NY 10032

ROSEN, NATHAN, b Brooklyn, NY, Mar 22, 09; m 32; c 2. THEORETICAL PHYSICS. *Educ:* Mass Inst Technol, SB, 29, SM, 30, ScD(physics), 32. *Prof Exp:* Nat Res Coun fel, Univ Mich, 32-33 & Princeton Univ, 33-34; res worker, Inst Advan Study, 34-36; prof theoret physics, Kiev State Univ, 36-38; assoc spectros res, Mass Inst Technol, 38-40; asst prof physics, Black Mountain Col, 40-41; from asst prof to prof, Univ NC, 41-52; prof physics, 52-77, res prof, 77-79, EMER PROF PHYSICS, ISRAEL INST TECHNOL, 79- *Honors & Awards:* Weizmann Res Prize, Tel-Aviv Munic, 68; Landau Res Prize, 75. *Mem:* Fel AAAS; fel Am Phys Soc; Am Asn Physics Teachers; Israel Acad Sci; Phys Soc Israel (vpres, 54-56, pres, 56-58). *Res:* Relativity and unified field theory; quantum theory; gravitation and cosmology. *Mailing Add:* Dept Physics Technion Israel Inst Technol Haifa 32000 Israel

ROSEN, NORMAN CHARLES, b Cleveland, Ohio, Oct 8, 41; m 64; c 2. PETROLEUM GEOLOGY, SEDIMENTOLOGY. *Educ:* Ohio State Univ, BSc, 63, MSc, 64; La State Univ, PhD(geol), 68. *Prof Exp:* Ed publs, Geol Surv Iran, 68-69; explor geologist, Texaco Inc, 69-74; sr geologist, Deminex A G, 74-78; proj geologist, Tenneco Oil Co, 78-80, sr geol specialist, 80-81; exec vpres, Robertson Res, 81-83; SR GEOLOGIST CONSULT, B P EXPLOR, 84- *Mem:* Am Asn Petrol Geologists; Soc Econ Paleontologists & Mineralogists; Sigma Xi. *Res:* Carbonate depositional environments; lithologic interpretation from wire-line logging techniques. *Mailing Add:* BP Explor PO Box 4587 Houston TX 77210

ROSEN, ORA MENDELSOHN, enzymology, endocrinology; deceased, see previous edition for last biography

ROSEN, PAUL, b New York, NY, Apr 2, 28; m 51; c 4. PHYSICS, ELECTRONICS. *Educ:* City Col New York, BS, 53; NY Univ, MS, 66. *Prof Exp:* Res physicist, Patterson Moos Div, Universal Coil Winding Mach, Inc, 53-56; res engr, Fairchild Camera & Instrument Corp, 56-58; res assoc, 58-70, AFFIL BIOPHYS, ROCKEFELLER UNIV, 70- *Concurrent Pos:* Instr, Brooklyn Polytech Inst, 62-71; mem adj fac, Fairleigh Dickinson Univ, 71- *Mem:* AAAS; Am Phys Soc; Sigma Xi. *Res:* Biomedical instrumentation; electro-optics; analytical chemistry instrumentation; digital computer interfacing and programming. *Mailing Add:* Rockefeller Univ New York NY 10021

ROSEN, PERRY, b Bronx, NY, Oct 2, 30; m 60; c 1. ORGANIC CHEMISTRY. *Educ:* City Col New York, BS, 53; Columbia Univ, MS & PhD(chem), 60. *Prof Exp:* Fel, Columbia Univ, 61-62; sr chemist, 62-70, res fel, 70-72, group chief, 72-76, sect chief, 76-80, assoc dir, 80-83, DIR, HOFFMANN-LA ROCHE, INC, 83- *Concurrent Pos:* Vis prof chem, Mass Inst Technol, 72- *Mem:* Am Chem Soc; NY Acad Sci; Am Inst Chemists; Int Soc Immunopharmacol; Int Union Pure & Appl Chem; AAAS. *Res:* Synthetic organic chemistry; development of new chemical reactions; development of new drugs primarily in the allergy/inflammation and cardiovascular areas. *Mailing Add:* Hoffman-La Roche Inc 340 Kingsland St Bldg 76/14 Nutley NJ 07110-1199

ROSEN, PHILIP, b New York, NY, Oct 31, 22; m 46, 66; c 3. BIOPHYSICS. *Educ:* Yale Univ, MS, 46, PhD(physics), 49. *Prof Exp:* Physicist, Nat Adv Comt Aeronaut, 44; asst high polymers, Polytech Inst Brooklyn, 45; lab asst, Yale Univ, 46-48; from instr to asst prof physics, Rensselaer Polytech Inst, 48-51; sr theoret physicist, Appl Physics Lab, Johns Hopkins Univ, 51-54; asst prof physics, Univ Conn, 54-57; PROF PHYSICS, UNIV MASS, AMHERST, 57- *Concurrent Pos:* Consult, Gen Elec Co, 56-57, appl physics lab, Johns Hopkins Univ, 58, 62-63 & Boeing Airplane Co, 59; vis prof, Univ Wis, 65; vis fel, Yale Univ, 65-66; Nat Cancer Inst spec fel biol, Oak Ridge Nat Lab, 72-73; guest scientist, Biol Dept, Brookhaven Nat Lab, 79-80. *Mem:* Fel Am Phys Soc; Biophys Soc; Am Soc Photobiol; Sigma Xi; AAAS; NY Acad Sci. *Res:* Microwaves; molecular quantum mechanics; kinetic theory; irreversible thermodynamics; plasma physics; electronic structure of nucleic acids and proteins, particularly the energy band structure; theory of radiation carcinogenesis; theory of ageing. *Mailing Add:* Dept Physics & Astron Univ Mass Amherst MA 01003

ROSEN, RICHARD DAVID, b Brooklyn, NY, Feb 23, 48; m 73. CLIMATE DYNAMICS. *Educ:* Mass Inst Technol, SB, 70, SM, 70, PhD(meteorol), 74. *Prof Exp:* Staff scientist res, Environ Res & Technol, Inc, 74-82; PRIN SCIENTIST & VPRES, BASIC RES, ATMOSPHERIC & ENVIRON RES, INC, 82- *Concurrent Pos:* Sr lectr, dept earth, atmospheric & planetary sci, Mass Inst Technol, 74-; mem, Comt Earth Sci, Nat Acad Sci, 82-84 & Spec Study Group, Int Asn Geodesy, 83- & Comt Southern Hemisphere, Am Meteorol Soc, 91; ed, Am Meteorol Soc Journals, Monthly Weather Rev, 86-87 & J Climate, 88-89. *Mem:* Am Meteorol Soc; Am Geophys Union. *Res:* Dynamics of general circulation of the atmosphere; atmospheric excitation of earth rotation and polar motion; interannual variability of large-scale circulation; diagnostic studies of quality of modern data assimilation techniques. *Mailing Add:* Atmospheric & Environ Res Inc 840 Mem Dr Cambridge MA 02139-3794

ROSEN, ROBERT, b Brooklyn, NY, June 27, 34; m 58; c 3. MATHEMATICAL BIOLOGY. *Educ:* Brooklyn Col, BS, 55; Columbia Univ, MA, 56; Univ Chicago, PhD(math biol), 59. *Prof Exp:* Asst analyst, Chicago Midway Labs, Ill, 56-57; asst math biol, Univ Chicago, 57-60, res assoc, 60-62, asst prof, 62-67; from assoc prof to prof math & biophys, State Univ NY Buffalo, 67-75, vis assoc prof, Ctr Theoret Biol, 66-67, assoc dir, 69-75; I W Killam prof physiol & biophys, 75-80, PROF PHYSIOL, DALHOUSIE UNIV, 80- *Concurrent Pos:* Vis fel, Ctr Study Democratic Insts, 71-72. *Mem:* Am Math Soc; Am Soc Cell Biol; Biophys Soc. *Res:* Relational aspects of general biological systems; quantum-theoretic mechanisms involved in the utilization of primary genetic information; morphogenesis and stability in biological systems. *Mailing Add:* Dalhousie Univ Sir Chas Tupper Bldg Halifax NS B3H 4H7 Can

ROSEN, SAMUEL, b New York, NY, Apr 14, 23; m 54; c 2. MICROBIOLOGY. *Educ:* Brooklyn Col, BA, 48; Univ Ill, MS, 49; Mich State Univ, PhD(microbiol), 53. *Prof Exp:* Technician microbiol, Pyridium Corp, 49-50; res assoc, Mich State Univ, 52-61; PROF MICROBIOL, OHIO STATE UNIV, 61- *Concurrent Pos:* Consult, Battelle Mem Inst, 62-, Procter & Gamble Co, 70-, Dent Res Inst, Great Lakes, 70, Southern Ill Univ, 72- & Nat Inst Dent Res, 75-; NIH career develop award, Ohio State Univ, 67-72. *Mem:* AAAS; Am Soc Microbiol; Int Asn Dent Res; Am Asn Dent Schs; Gnotobiotics Asn; Sigma Xi. *Res:* Experimental dental caries; experimental periodontal disease; microbial taxonomy. *Mailing Add:* 305 W 12th Columbus OH 43210

ROSEN, SAUL, b Port Chester, NY, Feb 8, 22; m 50; c 4. COMPUTER SCIENCE. *Educ:* City Col, BS, 41; Univ Cincinnati, MA, 42; Univ Pa, PhD, 50. *Prof Exp:* Instr math, Univ Del, 46-47; lectr, Univ Calif, Los Angeles, 48-49; asst prof, Drexel Inst Technol, 49-51; assoc res engr, Burroughs Corp, 51-52, mgr eastern appl math sect, Electrodata Div, 56-58; asst prof elec eng, Univ Pa, 52-54; assoc prof math, Comput Lab, Wayne Univ, 54-56; mgr comput prog & serv, Philco Corp, 58-60, consult, Comput & Prog Systs, 60-62; prof comput sci, Purdue Univ, 62-66; prof eng & assoc, Comput Ctr, State Univ NY Stony Brook, 66-67; dir, Univ Comput Ctr, 68-87, PROF MATH & COMPUT SCI, PURDUE UNIV, 67- *Honors & Awards:* Distinguished Serv Award, Asn Comput Mach, 84. *Mem:* Asn Comput Mach; Computer Soc; Inst Elec & Electronics Eng. *Res:* Operating systems, computer performance measurement and evaluation; history of computing; applications of electronic digital computers, programming systems and automatic programming. *Mailing Add:* Dept Comput Sci Purdue Univ West Lafayette IN 47907

ROSEN, SAUL W, b Boston, Mass, July 29, 28; div; c 3. INTERNAL MEDICINE, ENDOCRINOLOGY. *Educ:* Harvard Univ, AB, 47; Northwestern Univ, PhD(chem), 55; Harvard Med Sch, MD, 56; Am Bd Internal Med, cert, 63, dipl endocrinol, 65. *Prof Exp:* From intern med to asst resident, Univ Calif Med Ctr, San Francisco, 56-58; clin assoc, Nat Inst Arthritis & Metab Dis, 58-60; sr resident med, Med Ctr, Univ Calif, 60-61; sr investr, Clin Endocrinol Br, Nat Inst Arthritis, Metab & Digestive Dis, 61-84, DEP DIR, CLIN CTR, NIH, 84- *Mem:* Am Chem Soc; Am Fedn Clin Res; fel Am Col Physicians; Endocrine Soc; Sigma Xi; Asn Am Physicians. *Res:* Clinical endocrinology; ectopic tumor gonadotropins. *Mailing Add:* 7401 Westlake Terr Apt 1104 Bethesda MD 20817

ROSEN, SEYMOUR, b Chicago, Ill, Aug 21, 35; m 63; c 2. KIDNEY DISEASE. *Educ:* Univ Ill, MD, 59. *Prof Exp:* Assoc pathologist, 74-82, DIR, SURG PATH, BETH ISRAEL HOSP, 82-; ASSOC PROF PATH, SCH MED, HARVARD UNIV, 73- *Mem:* Fedn Am Soc Exp Biol; Int Acad Pathologists. *Mailing Add:* Dept Path Beth Israel Hosp 330 Brookline Boston MA 02215

ROSEN, SIDNEY, b Boston, Mass, June 5, 16; m 44; c 1. HISTORY OF SCIENCE. *Educ:* Univ Mass, AB, 39; Harvard Univ, MAT, 52, PhD(phys sci), 55. *Prof Exp:* From instr to asst prof phys sci, Brandeis Univ, 53-58; assoc prof sci educ, 58-60, from assoc prof to prof phys sci, 60-82, EMER PROF ASTRON, UNIV ILL, URBANA, 82- *Concurrent Pos:* Consult, Comnr Educ, Mass, 57-58 & Am Humanities Sem, 58; vis lectr, Harvard Univ, 58; spec sci consult, Ford Found, Colombia, SAm, 63-64 & 66; Fulbright fel, UK, 63; consult, Encycl Britannica Films, Inc, 64- *Honors & Awards:* Clara Ingram Judson Mem Award, 70. *Mem:* Fel AAAS; Nat Asn Res Sci Teaching; Hist Sci Soc; Nat Sci Teachers Asn; Am Asn Physics Teachers. *Res:* History of learning; problems of science teaching. *Mailing Add:* 113 Astron Univ Ill 1002 W Green St Urbana IL 61801

ROSEN, SIMON PETER, b London, Eng, Aug 4, 33; m 58, 87; c 4. PHYSICS. *Educ:* Oxford Univ, BA, 54, PhD(physics), 57. *Prof Exp:* Res assoc physics, Washington Univ, 57-59; scientist, Midwestern Univ Res Asn, 59-61; NATO fel, Oxford Univ, 61-62; from asst prof to prof physics, Purdue Univ, 62-83; staff mem, Los Alamos Nat Lab, 83-90; DEAN SCI, UNIV TEX, ARLINGTON, 90- *Concurrent Pos:* Tutorial fel, Univ Sussex, 69; sr theoret physicist, High Energy Physics Prog, Div Phys Res, US Energy Res & Develop Admin, Washington, DC, 75-77; prog assoc theoret physics, NSF, 81- *Mem:* Fel Am Phys Soc. *Res:* Symmetry theories of elementary particles; theory of weak interactions; high energy physics. *Mailing Add:* Col Sci Univ Tex Arlington TX 76019-0047

ROSEN, SOL, b New York, NY, Oct 12, 32; m 55; c 3. METALLURGY. *Educ:* City Col New York, BChE, 54; Columbia Univ, MS, 57; NY Univ, EngScDr, 62. *Prof Exp:* Scientist nuclear metall, Tech Res Corp, 56-58; engr, Sylvania-Corning Nuclear Corp, 58-59; asst scientist actinide alloy chem, Argonne Nat Lab, 59-62; sr res assoc struct mat, x-ray diffraction & electron micros, Pratt & Whitney Aircraft Div, United Aircraft Corp, Conn, 63-66, group leader, 66-67; metallurgist nuclear fuels & mat, US AEC, 67-73, US AEC & Environ Res Develop Admin sci rep, US Mission to Europ Communities, Belg, 73-76; US Dept Energy sci adv energy, US Mission to Orgn Econ Coop & Develop, France, 76-78; int tech adv, Off Nuc Progs, 78-79, DIR, INT NUCLEAR PROGS DIV, DEPT ENERGY, 79- *Res:* Nuclear and alternative energy technology; international relations. *Mailing Add:* Advan Reactor Progs Off Nuclear Energy Dept Energy Washington DC 20585

ROSEN, STEPHEN, b New York, NY, May 3, 34; m 84; c 2. HUMAN BIOMETEOROLOGY, TECHNOLOGY VENTURES. *Educ:* Queens Col, NY, BS, 55; Bryn Mawr Col, MA, 58; Adelphi Univ, PhD(physics), 66. *Prof Exp:* Physicist, Int Bus Mach Corp, 58-60; asst prof physics, State Univ NY Maritime Col, 60-67; res adv, Gen Res Labs, NJ, 67-69; mem prof staff, Hudson Inst, 69-70; dir res, Mkt & Planning Group, Inc, 72-84; mem staff technol investments, R S Enrlich & Co, 84-88; CHMN, SCI & TECHNOL ADV BD, NYANA & CELIA PAUL ASSOCS, INC, NY, 88- *Concurrent Pos:* Sr engr & scientist, Ford Instrument Co Div, Sperry Rand Corp, 63; res assoc, Astrophys Inst, Paris & Nuclear Studies Ctr, Saclay, France, 68; consult, Xerox Corp, Gen Tel & Electronics Corp & Carnegie Corp, 70-; consult, NSF, 73-74 & Fed Energy Admin, 74-75. *Mem:* AAAS; Am Phys Soc; Am Astron Soc; NY Acad Sci. *Res:* Beta and gamma ray spectroscopy; nuclear reactors; radiation effects; operational tactics; high energy interactions; origin and astrophysics of cosmic rays; technological and social forecasting; long-range planning; market planning for high-technology products and services; electronic and print journalism; technology investments; science-careers management and outplacement. *Mailing Add:* 35 W 81st St Apt 1D New York NY 10024-6045

ROSEN, STEPHEN L(OUIS), b New York, NY, Nov 25, 37. POLYMER SCIENCE & ENGINEERING. *Educ:* Cornell Univ, BChE, 60, PhD(chem eng), 64; Princeton Univ, MS, 61. *Prof Exp:* From asst prof to prof chem eng, Carnegie-Mellon Univ, 64-81; PROF CHEM ENG & CHMN DEPT, UNIV TOLEDO, 81 - *Mem:* Am Chem Soc; Am Inst Chem Engrs; Soc Plastics Engrs. *Res:* Polymeric materials; rheology; polymerization. *Mailing Add:* Dept Chem Eng Univ Mo Rolla MO 65401-0249

ROSEN, STEVEN DAVID, b New York, NY, Oct 20, 43. CELL BIOLOGY. *Educ:* Univ Calif, Berkeley, AB, 66; Cornell Univ, PhD(neurobiol), 72. *Prof Exp:* Fel cell biol, Dept Psychiat, Sch Med, Univ Calif, San Diego, 72-76; ASST PROF CELL BIOL, DEPT ANAT, UNIV CALIF, SAN FRANCISCO, 76- *Concurrent Pos:* Am Cancer Soc fel, Dept Psychiat, Univ Calif, San Diego, 72-74. *Mem:* AAAS; Am Soc Cell Biol. *Res:* Molecular basis of specific cell adhesion in cellular slime molds and higher systems. *Mailing Add:* Dept Anat Univ Calif Box 0452 San Francisco CA 94143

ROSEN, STEVEN TERRY, b Brooklyn, NY, Feb 18, 52; m 76; c 2. ONCOLOGY, HEMATOLOGY. *Educ:* Northwestern Univ, BM, 72, MD, 76. *Prof Exp:* GENEVIEVE TEUTON PROF, MED SCH, NORTHWESTERN UNIV, 89-, DIR CANCER CTR, 89- *Concurrent Pos:* Dir clin progs, Northwestern Mem Hosp, 89-; ed chief, J Northwestern Univ Cancer Ctr, 89- & Contemp Oncol, 90-; chmn, Exhibs/Indust Comt, Am Soc Clin Oncol. *Mem:* Am Asn Cancer Res; AAAS; Am Col Physicians; AMA; Am Soc Clin Oncol; Am soc Hemat; Cent Soc Clin Res. *Res:* Cutaneous T-cell lymphomas; biology of lung cancer; biologic therapies; hormone receptors. *Mailing Add:* 303 E Chicago Ave Chicago IL 60611

ROSEN, WILLIAM EDWARD, b New York, NY, Jan 29, 27; m 53; c 5. ORGANIC CHEMISTRY, MICROBIOLOGY. *Educ:* NY Univ, BA, 48; Harvard Univ, MA, 50, PhD(chem), 52. *Prof Exp:* Fel, Univ Southern Calif, 52 & Yale Univ, 52-53; sr res chemist, Ciba Pharmaceut Prod, Inc, 53-64; vpres res, Cambridge Res, Inc, 64-68; exec vpres, 68-89, PRES, SUTTON LABS, DIV GAF CHEM CORP, 89- *Concurrent Pos:* Adj prof, Rutgers Univ. *Mem:* Am Chem Soc; Soc Cosmetic Chem. *Res:* Cosmetic preservation; pharmaceuticals and fine chemicals; process research and development; organic synthesis; steroids; alkaloids. *Mailing Add:* 116 Summit Ave Chatham NJ 07928

ROSEN, WILLIAM G, b Portsmouth, NH, May 13, 21; m 47, 75; c 2. MATHEMATICS. *Educ:* Univ Ill, MS, 47, PhD(math), 54. *Prof Exp:* Instr math, Univ Md, 54-56, asst prof, 56-61; from asst prof dir to assoc prof dir sci educ, NSF, 61-64, staff assoc sci develop, 64-66, spec asst to dir, 66-70, prog dir, Mod Analytical & Probability Prog, 70-79, head, Math Sci Sect, 79-83; dep exec dir, US-Israel Binat Res Found, 83-86; consult, Boardeu Math Sci, Nat Acad Sci, 86-89; DEP EXEC DIR, US-ISRAEL BINAT SCI FOUND, 89- *Res:* Science administration. *Mailing Add:* Binat Sci Found PO Box 7677 Jerusalem 91076 Israel

ROSEN, WILLIAM M, b Lynn, Mass, Dec 27, 41; m 64; c 2. CHEMISTRY. *Educ:* Univ Calif, Los Angeles, BS, 63; Univ Calif, Riverside, PhD(chem), 67. *Prof Exp:* Teaching asst chem, Univ Calif, Riverside, 63-65, res assoc, 65-66, assoc, 66-67; res assoc, Ohio State Univ, 67-69; asst prof, Purdue Univ, 69-70; from asst prof to assoc prof, 70-82, PROF CHEM, UNIV RI, 82- *Concurrent Pos:* NIH fel, 67-69; assoc res chemist, Scripps Inst oceanog, 76-77. *Mem:* Am Chem Soc; Chem Soc; Am Inst Chem; AAAS. *Res:* Synthesis of interesting organic and inorganic chemical systems. *Mailing Add:* Dept Chem Univ RI Kingston RI 02881-0809

ROSENAU, JOHN (RUDOLPH), b Sheboygan, Wis, Feb 25, 43; m 65; c 2. AGRICULTURAL & FOOD ENGINEERING. *Educ:* Univ Wis-Madison, BS, 65, BSME, 66; Mich State Univ, PhD(agr eng), 70. *Prof Exp:* Asst prof food sci & industs, Univ Minn, St Paul, 70-73; ASSOC PROF FOOD ENG, UNIV MASS, AMHERST, 73-, HEAD DEPT, 85- *Mem:* Am Soc Agr Engrs; Inst Food Technol; Sigma Xi. *Res:* Utilization of protein components; new systems for cheese production, soy protein utilization. *Mailing Add:* Dept Food Eng Univ Mass Amherst MA 01003

ROSENAU, WERNER, b June 28, 29; US citizen. PATHOLOGY, IMMUNOLOGY. *Educ:* Univ Calif, San Francisco, MD, 56. *Prof Exp:* From asst prof to assoc prof, 61-72, PROF PATH, UNIV CALIF, SAN FRANCISCO, 72- *Mem:* Am Soc Exp Path; AAAS. *Mailing Add:* Dept Path Univ Calif San Francisco CA 94143

ROSENAU, WILLIAM ALLISON, b Redding, Conn, Nov 25, 26; m 62; c 3. PLANT NUTRITION. *Educ:* Yale Univ, BS, 48; Univ Conn, MS, 50; Pa State Univ, PhD(agron), 60. *Prof Exp:* Res asst forage crops, Eastern State Farmers Exchange, 50-51 & 54-56; asst prof, 60-69, res asst prof soils, Waltham Field Sta, 60-66, assoc prof, 69-77, PROF PLANT & SOIL SCI, UNIV MASS, AMHERST, 77- *Mem:* Am Soc Agron. *Res:* Nitrogen fertilization and calcium nutrition of corn; calcium-boron relationships in soils and crops. *Mailing Add:* Dept Plant & Soil Sci Univ Mass Amherst MA 01003

ROSENBAUM, DAVID MARK, b Boston, Mass, Feb 11, 35; m 64. RISK ANALYSIS, SYSTEMS ANALYSIS. *Educ:* Brown Univ, ScB, 56; Rensselaer Polytech Inst, MS, 58; Brandeis Univ, PhD(physics), 64. *Prof Exp:* Mem staff, Mitre Corp, 60-64; asst res prof physics, Boston Univ, 64-65; mem staff, Inst Defense Analysis, 65-67; expert commun & network analysis, Off Emergency Planning, Exec Off President, 67-68; assoc prof elec eng, Polytech Inst Brooklyn, 68-69; sr staff mem, Mitre Corp, 70-72; asst dir analysis & asst dir admin, systs & computerization, Off Nat Narcotics Intel, US Dept Justice, 72-73; consult, US AEC, 73-74; consult, Perm Subcomt Invests, US Senate, 74; sr staff mem, Mitre Corp, 74-76; consult to the comptroller gen, US Gen Accounting Off, 76-79; dep asst adminr, Radiation Progs, US Environ Protection Agency, 79-81; PRES, TECH ANALYSIS CORP, 81- *Concurrent Pos:* Consult, Off Emergency Planning, Exec Off President, 68-69; pres, Network Anal Corp, 68-70; mem, BD, Beta Instrument Corp, 68-72; chmn, Eng Found Conf Vulnerability Urban Areas Subversive Disruption, 72; consult, Perm Subcomt Invests, US Sen, 74- *Mem:* Am Phys Soc; Soc Indust & Appl Math. *Res:* Plasma physics; history of guerilla warfare; network analysis; elementary particles and mathematical foundations of quantum mechanics; energy and environmental policy studies; nuclear safeguards; nuclear proliferation; epidemiology; statistics. *Mailing Add:* 4620 Dittmar Rd Arlington VA 22207

ROSENBAUM, EUGENE JOSEPH, b New York, NY, July 22, 07; m 32; c 2. PHYSICAL CHEMISTRY. *Educ:* Univ Chicago, SB, 29, PhD(chem), 33. *Prof Exp:* Instr chem, Univ Chicago, 31-40; Lalor Found fel, Harvard Univ, 40-41; res chemist, Sun Oil Co, 41-58; prof chem, 58-72, EMER PROF CHEM, DREXEL UNIV, 72- *Mem:* AAAS; Am Chem Soc; Am Soc Testing & Mat; Soc Appl Spectros (pres, 52). *Res:* Applied spectroscopy; Raman spectra and molecular structure. *Mailing Add:* 11 Martins Run Apt B-105 Media PA 19063-1057

ROSENBAUM, FRED J(EROME), b Chicago, Ill, Feb 15, 37; m 60; c 2. ELECTRICAL ENGINEERING, MICROWAVE ENGINEERING. *Educ:* Univ Ill, Urbana, BS, 59, MS, 60, PhD(elec eng), 63. *Prof Exp:* Res asst ultramicrowave group, Dept Elec Eng, Univ Ill, Urbana, 59-63; res scientist, McDonnell Aircraft Corp, Mo, 63-65; chief scientist, Cent Microwave Co, 83-85; from asst prof to assoc prof elec eng, 65-71, PROF ELEC ENG, WASH UNIV, 71- *Concurrent Pos:* Ed, Trans Microwave Theory & Tech, 72-74; pres, Microwave Theory & Tech Soc, Inst Elec & Electronics Engrs, 81. *Honors & Awards:* Centennial Medal, Inst Elec & Electronics Engrs, 84. *Mem:* Fel Inst Elec & Electronics Engrs. *Res:* Electromagnetic theory; microwave devices; microwave ferrite devices; nonlinear circuit analysis; microwave solid state devices and circuits; millimeter wave solid state devices and circuits. *Mailing Add:* Dept Elec Eng Box 1127 Wash Univ St Louis MO 63130

ROSENBAUM, H(ERMAN) S(OLOMON), b Philadelphia, Pa, Oct 24, 32; m 53; c 3. METALLURGY. *Educ:* Univ Pa, BS, 53, MS, 56; Rensselaer Polytech Inst, PhD(metall), 59. *Prof Exp:* Lab asst, Sam Tour & Co, 52 & Franklin Inst, 53-56; metallurgist, Res Lab, 56-64, mgr mat struct & properties, Nucleonics Lab, Vallecitos Nuclear Ctr, 64-76, prin engr & prog mgr, 76-78, SR PROG MGR, NUCLEAR ENERGY ENG DIV, GEN ELEC CO, 78- *Honors & Awards:* IR-100, 65 & 83. *Mem:* Am Soc Metals; Am Inst Mining, Metall & Petrol Engrs; Am Nuclear Soc. *Res:* Crystal imperfections; precipitation kinetics in solids; radiation damage; microstructures and properties of irradiated materials; physical metallurgy. *Mailing Add:* 917 Kensington Dr Fremont CA 94539

ROSENBAUM, HAROLD DENNIS, b Fairplay, Ky, Aug 17, 21; m 70; c 6. MEDICINE. *Educ:* Berea Col, AB, 41; Harvard Univ, MD, 44. *Prof Exp:* Instr med, Sch Med, Univ Colo, 47-48; res fel pediat & radiol, Harvard Med Sch & Children's Med Ctr, Boston, 52; radiologist, John Graves Ford Hosp, Georgetown, Ky, 53-60; PROF DIAG RADIOL & CHMN DEPT, MED CTR, UNIV KY, 60- *Concurrent Pos:* Radiologist, Clark County Hosp, Winchester, Ky, 53-55, Eastern State Hosp, Lexington, 54-60, Woodford County Mem Hosp, Versailles, 55-60 & Proj Hope, Corinto, Nicaragua, 66-; consult cardiorentgenologist, St Joseph's Hosp, Lexington, Ky, 53-; US Dept HEW grant, Univ Ky, 70-; CARE-Medico prof, Honduras, 71- *Mem:* Am Roentgen Ray Soc; Radiol Soc NAm; AMA; Am Heart Asn; fel Am Col Radiol. *Res:* Congenital heart disease; radiology; medical and radiological education and clinical service. *Mailing Add:* Dept Diag Radiol Univ Ky Med Ctr Lexington KY 40506

ROSENBAUM, IRA JOEL, b New York, NY, June 5, 41; c 2. PHYSICS. *Educ:* Queen's Col, NY, BS, 62; Am Univ, MS, 67, PhD(physics), 71. *Prof Exp:* Physicist, Nat Bur Standards, 64-66; PHYSICIST, WHITE OAK LAB, NAVAL SURFACE WEAPONS CTR, 66- *Mem:* Am Phys Soc; Acoust Soc Am. *Res:* Acoustic properties of liquid metals at high pressures. *Mailing Add:* 3714 Woodbine St Chevy-Chase MD 20815

ROSENBAUM, JOEL L, b Massena, NY, Oct 4, 33; c 3. CELL & MOLECULAR BIOLOGY. *Educ:* Syracuse Univ, BS, 55, MS, 59, PhD(biol), 63; St Lawrence Univ, MSc, 57. *Prof Exp:* Fel, Univ Chicago, 63-68; from asst prof to assoc prof, 68-79, PROF BIOL, YALE UNIV, 79- *Honors & Awards:* Newcomb-Cleveland Award, AAAS, 68. *Mem:* Am Soc Cell Biol. *Res:* Cell and molecular biology of microtibiles; control of flagellan growth in chlamydenonas. *Mailing Add:* Dept Biol Yale Univ New Haven CT 06511

ROSENBAUM, JOSEPH HANS, b Hannover, Ger, July 1, 25; nat US; m 59; c 1. PHYSICS. *Educ:* Lowell Technol Inst, BS, 47; Clark Univ, PhD(chem), 50. *Prof Exp:* Phys chemist, US Naval Ord Lab, 50-53; asst prof textile chem, Lowell Technol Inst, 53-54; phys chemist, US Naval Ord Lab, 54-56; PHYSICIST, SHELL DEVELOP CO, 56- *Mem:* Am Chem Soc; Am Asn Textile Chemists & Colorists; Acoust Soc Am; Am Geophys Union; Soc Explor Geophys. *Res:* Elastic waves; thermodynamics; hydrodynamics; electrochemistry; applied mathematics. *Mailing Add:* 1308 Castle Court Blvd Houston TX 77006

ROSENBAUM, MANUEL, b Detroit, Mich, Sept 13, 29; m 55; c 3. GRANT ADMINISTRATION. *Educ:* Univ Mich, BS, 51, MS, 53, PhD(bact), 56. *Prof Exp:* Asst microbiol, Univ Mich, 53-55; instr med col, Cornell Univ, 56-58; res assoc, Wistar Inst Anat & Biol, Univ Pa, 58-60; chief microbiologist, Wayne County Gen Hosp, Mich, 60-63; res microbiologist, Parke, Davis & Co, 64-71; res assoc, Child Res Ctr, Mich, 71-76; grant & contract adminr, Mich Cancer Found, 76-82; grant & contract adminr, Wayne State Univ, 82-83; SOFTWARE SPECIALIST, CHRYSLER CORP, 84- *Mem:* AAAS; Am Asn Artificial Intel. *Res:* Protein synthesis in development of bacteriophage; nucleic acid metabolism in normal and virus-infected animal cells; viral oncogenesis. *Mailing Add:* 24111 Stratford Oak Park MI 48237-1927

ROSENBAUM, MARCOS, b Mexico, DF, Feb 26, 35; m 65; c 1. MATHEMATICAL PHYSICS. *Educ:* Nat Univ Mex, BS, 57; Univ Mich, MSE, 59, PhD(nuclear sci), 63. *Prof Exp:* From res asst to res assoc neutron physics, Univ Mich, Ann Arbor, 61-64; mem res staff physics & appl math, Ctr Advan Studies, G E Tempo, 64-71; head dept physics & math, 74-76, RES PROF MATH PHYSICS, NAT UNIV MEX, 71-, DIR, INST NUCLEAR SCI, 76- *Concurrent Pos:* Consult, Nat Inst Nuclear Energy, 71-86; mem scholar comt nuclear sci, Nat Coun Sci Technol, 75-; pres, Comt Int Rels, Acad Sci Res Mex. *Mem:* Sigma Xi; Am Phys Soc; AAAS; Acad Sci Res Mex; NY Acad Sci. *Res:* General relativity and gauge theories. *Mailing Add:* Instituto de Ciencias Nucleares UNAM Circuito Exterior Ciudad Universitaria Mexico 20 DF Mexico

ROSENBAUM, ROBERT ABRAHAM, b New Haven, Conn, Nov 14, 15; m 42; c 3. MATHEMATICS. *Educ:* Yale Univ, BA, 36, PhD(math), 47. *Hon Degrees:* MA, Wesleyan Univ, 54, LHD, 81; LHD, St Josephs Col, Conn, 70. *Prof Exp:* From instr to prof math, Reed Col, 40-53; prof math, 53-85, univ prof math & sci, 77-85, EMER PROF MATH & SCI, WESLEYAN UNIV,

85- *Concurrent Pos:* NSF fel, Math Inst, Oxford Univ, 58-59; dean sci, Wesleyan Univ, 63-65, provost, 65-67, acad vpres, 67-69, actg pres, 69-70, chancellor, 70-73; vis scholar, Univ Calif, Berkeley, 62; ed, Math Asn Am, 66-68; vis prof, Univ Mass, 73-74, Col St Thomas, St Paul Minn, 83; mem adv coun, Braitmayer Found, 74-, dir, Proj Increase Mastery Math Sci, 79- *Honors & Awards:* Baldwin Medal, 85. *Mem:* Fel AAAS; Am Math Soc; Math Asn Am (2nd vpres, 61-62). *Res:* Classical analysis; geometry; subadditive functions. *Mailing Add:* Wesleyan Univ Middletown CT 06457

ROSENBERG, AARON E(DWARD), b Malden, Mass, Apr 9, 37; m 61; c 3. SIGNAL PROCESSING, PATTERN RECOGNITION. *Educ:* Mass Inst Technol, SB & SM, 60; Univ Pa, PhD(elec eng), 64. *Prof Exp:* MEM TECH STAFF, AT&T BELL LABS, 64- *Honors & Awards:* Senior Award, Acoust, Speech & Signal Processing Soc, Inst Elec & Electronics Engrs, 86. *Mem:* Fel Inst Elec & Electronics Engrs; fel Acoust Soc Am. *Res:* Speech signal processing and pattern recognition; automatic speech; speaker recognition. *Mailing Add:* AT&T Bell Labs 600 Mountain Ave Rm 2d-528 New Providence NJ 07974

ROSENBERG, ABRAHAM, b New York, NY, Aug 12, 24; m 48; c 2. BIOCHEMISTRY. *Educ:* Univ Ill, BS, 47; Polytech Inst Brooklyn, MS, 52; Columbia Univ, PhD(biochem), 56. *Prof Exp:* Jr biochemist, Jewish Hosp Brooklyn, 47, res asst, 48-49, supvr clin chem, 49-51, res biochemist, 51-53; res assoc biochem, Columbia Univ, 57-61, asst prof, 61-68; from assoc prof to prof biol chem, Col Med, Pa State Univ, Hershey Med Ctr, 68-82; prof & chmn, Dept Biochem & Biophys, Loyola Univ Med Ctr, 83-88; prof psychiat, NY Univ, 88-90; PROF PSYCHIAT, UNIV CALIF, LOS ANGELES, 89- *Concurrent Pos:* USPHS fel, 56; NY Heart Asn res fel, 57-58, sr fel, 59-61; Health Res Coun New York career investr, 61-66; Fulbright prof & res scholar, 74; master res, Nat Inst Health & Med Res, France, 75; vis prof, Univ Heidelberg, 90. *Mem:* Am Soc Biol Chemists; Am Chem Soc; Am Soc Neurochem; Soc Protozool; fel Am Inst Chemists; Sigma Xi. *Res:* Fat-soluble vitamin; analysis, absorption, transport and conversion of provitamin; structure and isolation of complex glycolipids of nervous tissues; lipids in photosynthesis; neurochemistry; synaptic structure-function; cell surface enzymes; nerve development. *Mailing Add:* UCLA Neuropsychiat Inst 760 Westwood Plaza Los Angeles CA 90024-1759

ROSENBERG, ALBURT M, b Hollywood, Calif, Oct 26, 27; m 58; c 4. BIOPHYSICS. *Educ:* Harvard Univ, AB, 49; Univ Fla, MS, 51; Univ Pa, PhD(gen physiol), 58. *Prof Exp:* Fel, Basic Res Div, Eastern Pa Psychiat Inst, 58-59; instr biol, 59-60, asst prof natural sci, 60-66, ASSOC PROF NATURAL SCI, SWARTHMORE COL, 66- *Mem:* Am Asn Physics Teachers. *Res:* Marine egg physiology; role of cellular water; x-ray sensitivity of yeast; growing giant chromosomes; general education in physics; computers in physics teaching; divalent ions and cell division; sunflower seed order. *Mailing Add:* Dept Physics Swarthmore Col Swarthmore PA 19081

ROSENBERG, ALEX, b Berlin, Ger, Dec 5, 26; nat US; m 52, 85; c 3. MATHEMATICS. *Educ:* Univ Toronto, BA, 48, MA, 49; Univ Chicago, PhD(math), 51. *Prof Exp:* Res assoc math, Univ Mich, 51-52; from instr to assoc prof, Northwestern Univ, 52-61; chmn dept math, 66-69, prof math, Cornell Univ, 61-88; chmn dept, 86-87, PROF MATH, UNIV CALIF, SANTA BARBARA, 86- *Concurrent Pos:* Mem, Inst Advan Study, 55-57; algebra ed, Proc, Am Math Soc, 60-65, trustee, 74-83; vis prof, Univ Calif, Berkeley, 61-79; researcher, Queen Mary Col, Univ London, 63-64; vis researcher, Univ Calif, Los Angeles, 69-70, 82; ed, Math Asn Am, 74-78; vis prof, ETH Zurich, 76; mem Math Forschungs Inst, ETH Zurich, 76; DFG grant, Univ Dortmund, 84-85. *Honors & Awards:* Alexander v Humboldt-Stiftung sr US scientist award, Univ Munich, 75-76. *Mem:* Math Asn Am. *Res:* Homological algebra; structure theory of rings; quadratic forms; Witt rings. *Mailing Add:* Dept Math Univ Calif Santa Barbara CA 93106

ROSENBERG, ALEXANDER F, b Frankfurt am Main, Mar 8, 27; nat US; m 54; c 3. ANALYTICAL CHEMISTRY. *Educ:* City Col New York, BS, 51; Duke Univ, AM, 53, PhD(chem), 55. *Prof Exp:* Asst, Duke Univ, 51-54; res chemist, Shell Oil Co, 54-59; sr engr, Aircraft Nuclear Propulsion Dept-Nuclear Mat & Propulsion Opers, 59-63, prin chemist, 63-68, chemist, Major Appliance Labs, Major Appliance Bus Group, 68-71, mgr chem analysis, Major Appliance Labs, 71-79, mgr chem analysis, Appl Sci & Technol Lab, 79-82, MGR CHEM ANALYSIS & PROCESS, TECHNOL DEVELOP & APPLNS LAB, MAJOR APPLIANCE BUS GROUP, GEN ELEC CO, 82- *Concurrent Pos:* Instr, Univ Houston, 55-59. *Mem:* Chem Soc. *Res:* Development of analytical methods; gas chromatography; mass spectrometry; thermal analysis. *Mailing Add:* 10712 Sunderland Pl Louisville KY 40243

ROSENBERG, ALLAN (HERBERT), b Brooklyn, NY, Dec 18, 38; m 62; c 3. PHYSICAL CHEMISTRY. *Educ:* City Col NY, BS, 59; Yale Univ, MS, 62, PhD(chem), 64. *Prof Exp:* Res chemist, Allied Chem Corp, 64-69; SR RES INVESTR, BRISTOL-MYERS CORP, 69- *Mem:* AAAS; Am Chem Soc; Am Pharmaceut Asn; Royal Soc Chem. *Res:* Kinetics of drug decomposition; foam stabilization; wetting and adsorption phenomena with respect to human skin; dissolution and absorption of drugs. *Mailing Add:* 129 Irving Ave South Orange NJ 07079-2308

ROSENBERG, ANDREAS, b Tartu, Astomia, Nov 3, 24. PROTEIN CHEMISTRY. *Educ:* Univ Upsala, Sweden, PhD(biochem), 60. *Prof Exp:* PROF PATH, BIOCHEM & BIOPHYS, MED CTR, UNIV MINN, 64- *Mailing Add:* Univ Minn Box 198 Mayo Minneapolis MN 55455-0100

ROSENBERG, ARNOLD LEONARD, b Boston, Mass, Feb 11, 41; m 64; c 2. APPLIED GRAPH THEORY. *Educ:* Harvard Col, AB, 62; Harvard Univ, AM, 63, PhD(appl math), 66. *Prof Exp:* Res staff mem, T J Watson Res Ctr, IBM Corp, 65-81; prof Computer Sci, Duke Univ, 81-86; DISTINGUISHED UNIV PROF COMPUTER & INFO SCI, UNIV MASS, 86- *Concurrent Pos:* Vis asst prof, Polytech Inst Brooklyn, 67-69; adj assoc prof math, NY Univ, 70-73, adj prof comput sci, 80-81; vis lectr, Yale Univ, 78-79; vis prof comput sci, Univ Toronto, 79-80; ed-in-chief, Math Systs Theory. *Honors & Awards:* Cert Appreciation, Inst Elec & Electronics Engrs Computer Soc, 86. *Mem:* Asn Comput Mach; Soc Indust & Appl Math; Sigma Xi; Inst Elec & Electronics Engrs. *Res:* Mathematical theory of computation, with emphasis on computational structures and complexity of computation; concrete complexity; applications to parallel algorithms and arthitectures. *Mailing Add:* A267 Lederle Univ Mass Amherst MA 01003

ROSENBERG, ARNOLD MORRY, b Boston, Mass, Mar 18, 34; m 55; c 3. CONSTRUCTION CHEMICALS & MATERIALS. *Educ:* Boston Univ, AB, 55; Purdue Univ, PhD(phys chem), 60. *Prof Exp:* Chemist, Dewey & Almy Div, W R Grace & Co, 60-62, group leader, Construct Prod Div, 63-66; sr supvr, Polaroid Corp, 66-69; res mgr, Construction Prod Div, 69-71, res dir, 71-74, RES MGR, RES DIV, W R GRACE & CO, 74- *Honors & Awards:* IR-100 Award, Indust Res Mag, 80. *Mem:* Am Soc Testing & Mat; Am Concrete Inst; Am Chem Soc; Nat Asn Corrosion Engrs; Transp Res Bd. *Res:* Product development in concrete admixtures, fire proof coatings, insulation, roofing and corrosion control; author or coauthor of 40 publications. *Mailing Add:* 11836 Goya Dr Potomac MD 20854-3307

ROSENBERG, BARBARA HATCH, molecular biology, for more information see previous edition

ROSENBERG, CHARLES E, b Nov 11, 36, US citizen; c 2. HISTORY OF MEDICINE & BEHAVIORAL SCIENCE. *Educ:* Univ Wis, BA, 56; Columbia Univ, MA, 57, PhD, 61. *Prof Exp:* lectr, 61-62, res asst prof, Dept Hist & Hist Sci, Univ Wis, 62-63; from asst prof to assoc prof, 63-68, chmn dept, 74-75 & 79-83, PROF, DEPT HIST, UNIV PA, 68- *Concurrent Pos:* fel, Johns Hopkins Univ, Inst Hist Med, 60-61, Guggenheim fel, 65-66 & 88-90, NIH Res Grant, 64-70, sr fel, Nat Endow Humanities, 72-73, Rockefeller Humanities fel, 76-77, fel, Inst for Advan Study, 79-80, fel, Ctr Adv Study in Behav Sci, 83-84; mem coun, Hist Sci Soc, 72-, Hist Life Scis Study Sect, NIH, 72-75, Hist & Philos of Sci Panel, NSF, 81-82; Hist Sci Soc lectr, 82. *Honors & Awards:* William H Welch Medal, Am Asn Hist Med, 69; Fielding H Garrison lectr, Am Asn Hist Med, 82. *Mem:* Nat Acad Sci; Inst Med-Nat Acad Sci; Am Acad Arts & Sci; Am Asn Hist Med (vpres & pres-elect, 90-); Soc Social Hist Med (vpres 80, pres 81); Orgn Am Historians; Am Antique Soc. *Res:* History of American medicine and its role in society; author of numerous books and articles. *Mailing Add:* Dept Hist Sci Univ Pa Philadelphia PA 19104-6310

ROSENBERG, DAN YALE, b Stockton, Calif, Jan 8, 22; m 54; c 1. PHYTOPATHOLOGY. *Educ:* Col of the Pac, AB, 49; Univ Calif, Davis, MS, 52. *Prof Exp:* Jr plant pathologist, Bur Plant Path, Calif Dept Food & Agr, 52-55, asst plant pathologist, 55-59, plant pathologist, 59-63, prog supvr dis detection, 63-72, chief exclusion & detection, 72-76, chief nursery & seed serv, 76-82, spec asst, Div Plant Indust, 82-87; AGR CONSULT SPECIALIZING/REGULATORY AGR, 88- *Concurrent Pos:* Lectr, Univ Calif exten, Int Training & Educ, Univ Calif, Davis. *Mem:* Am Phytopath Soc. *Res:* Seed-borne plant diseases; regulatory plant pathology involving detection of new or rarely occurring plant diseases to California; regulatory agricultural pests including plant pathology, entomology, weeds occurring in nurseries; vertebrates; seed law enforcement; certification programs for fruit trees, citrus, avocado, and so on; gene resource conservation; recombinant DNA (biotechnology). *Mailing Add:* Agr Consult 2328 Swarthmore Dr Sacramento CA 95825

ROSENBERG, DAVID MICHAEL, b Edmonton, Alta, Aug 24, 43; m 65; c 2. FRESH WATERS. *Educ:* Univ Alta, BSc, 65, PhD(entom), 73. *Prof Exp:* RES SCIENTIST, FRESHWATER INST, CAN DEPT FISHERIES & OCEANS, 71- *Concurrent Pos:* Adj prof dept entom, Univ Man, 78- *Mem:* NAm Benthological Soc (pres, 86-87); Rawson Acad Aquatic Sci. *Res:* Environmental assessment using benthic microvertebrates; environmental effects of reservoirs and water diversions; ecology of Chironomidae; aquatic insects of freshwater wetlands; freshwater wetlands. *Mailing Add:* Dept Fisheries & Oceans Freshwater Inst 501 University Crescent Winnipeg MB R3T 2N6 Can

ROSENBERG, DENNIS MELVILLE LEO, b Johannesburg, SAfrica, Jan 27, 21; nat US; m 47. THORACIC SURGERY, CARDIOVASCULAR SURGERY. *Educ:* Univ Witwatersrand, BSc, 41 & 44, MB, BCh, 45; Am Bd Surg, dipl, 53; Bd Thoracic Surg, dipl, 54. *Prof Exp:* Asst surg, Tulane Univ, 46-47, instr, 48-52; surgeon, Biggs Hosp, Ithaca, NY, 53-54; ASSOC PROF CLIN SURG, SCH MED, TULANE UNIV, 56- *Concurrent Pos:* Resident, Ochsner Found, 48-52; registr, Children's Hosp, Johannesburg, SAfrica, 52; sr vis physician, Charity Hosp, 57-; prin investr, Touro Res Inst. *Mem:* Asn Thoracic Surg; fel Am Col Surg; fel Am Col Chest Physicians; Am Surg Asn; Soc Vascular Surg. *Res:* Development of open heart surgical apparatus; embolism; assisted circulation; arterial heterografts. *Mailing Add:* 3525 Prytania St New Orleans LA 70115

ROSENBERG, EDITH E, b Berlin, Ger, Jan 24, 28; Can citizen; div. PULMONARY PHYSIOLOGY. *Educ:* Univ Toronto, MA, 52; Univ Pa, PhD(physiol), 59. *Prof Exp:* Asst biophys, Univ Western Ont, 52-55; asst instr physiol, Sch Med, Univ Pa, 56-57; asst prof, Med Sch, Univ Montreal, 59-63; asst prof exp surg & lectr physiol, McGill Univ, 63-68; ASSOC PROF PHYSIOL, COL MED, HOWARD UNIV, 68 - *Concurrent Pos:* Univ res fel physiol, Sch Med, Univ Pa, 59; PI on grants, Nat Res Coun Can, Univ Montreal, 60-63 & McGill Univ, 64-68, Wash Heart Asn, 70-71; asst instr, Sch Med, Hebrew Univ, Israel, 54-55; NIH grant, 72-75. *Mem:* Am Physiol Soc; Biophys Soc; Can Physiol Soc; NY Acad Sci; Am Thoracic Soc. *Res:* Respiration and circulation, particularly pulmonary circulation and Va/Q distribution; alveolar-arterial tension differences; pulmonary surfactant; pulmonary diffusing capacity; effect of hyperbaric oxygen; lung elasticity. *Mailing Add:* Dept Physiol & Biophys Howard Univ Col Med Washington DC 20059

ROSENBERG, ELI IRA, b Brooklyn, NY, Feb 19, 43; m; c 1. ELEMENTARY PARTICLE PHYSICS. *Educ:* City Col New York, BS, 64; Univ Ill, Urbana, MS, 66, PhD(physics), 71. *Prof Exp:* Enrico Fermi fel physics, Enrico Fermi Inst, Univ Chicago, 71-72, res assoc, 72, instr, 72-74, asst prof, 74-79; from asst prof to assoc prof, 79-87, PROF PHYSICS, IOWA STATE UNIV, 87- *Concurrent Pos:* assoc physicist, Ames Lab, US Dept Energy, physicist, 81-87, sr physicist, 87-, prog dir high energy physics, 88- *Honors & Awards:* US Dept Energy Jr Investr, 79-80. *Mem:* AAAS; Sigma Xi; Am Phys Soc. *Res:* Experimental high energy physics; study of asymptotic behavior of scattering processes with counter techniques; direct production of leptons in hadron interactions; electron-positron annihilation; applications of microprocessors to experiments. *Mailing Add:* Dept Physics Iowa State Univ Ames IA 50011

ROSENBERG, FRANKLIN J, pharmacology, for more information see previous edition

ROSENBERG, FRED A, b Berlin, Germany, Mar 19, 32; nat US; m 57; c 1. MICROBIOLOGY. *Educ:* NY Univ, AB, 53; Rutgers Univ, PhD(sanit), 60. *Prof Exp:* Asst bact, Univ Fla, 54-57; asst sanit, Rutgers Univ, 57-60; res assoc physiol aspects water qual, Grad Sch Pub Health, Univ Pittsburgh, 60-61; from asst prof to assoc prof, 61-77, PROF MICROBIOL, NORTHEASTERN UNIV, 77- *Concurrent Pos:* Books ed, J Col Sci Teaching, 83-; vis prof, Inst Microbiol, Univ Hannover, WGer, 85, 87-88 & 91. *Mem:* Fel Am Pub Health Asn; Am Soc Microbiol; fel Am Acad Microbiol; Biodeterioration Soc Eng. *Res:* Microbiology of bottled water; marine and freshwater microbiology. *Mailing Add:* Dept Biol Northeastern Univ Boston MA 02115

ROSENBERG, GARY, b New Rochelle, NY, Oct 16, 59. MALACOLOGY, SYSTEMATICS. *Educ:* Princeton Univ, AB, 81; Harvard Univ, PhD(biol), 89. *Prof Exp:* ASST CUR, ACAD NATURAL SCI PHILADELPHIA, 89- *Mem:* AAAS; Am Malacol Union; Am Soc Zoologists; Geol Soc Am; Sigma Xi. *Res:* Evolution, biogeography and phylogenetic systematics of mollusks, particularly prosobranch gastropods; integrating information from comparative anatomy; allozyme electrophoresis; DNA sequencing and the fossil record. *Mailing Add:* Acad Natural Sci 1900 Benjamin Franklin Pkwy Philadelphia PA 19103

ROSENBERG, GARY DAVID, b Milwaukee, Wis, Aug 2, 44. BIOMINERALIZATION. *Educ:* Univ Wis, BS, 66; Univ Calif, Los Angeles, PhD(geol), 72. *Prof Exp:* Sr res assoc, geophys dept, Univ Newcastle-upon-Tyne, Eng, 72-76; res assoc orthop surg, Med Sch, Wash Univ, St Louis, 76-78; vis asst prof geol, Mich State Univ, 78-79; asst prof, 79-83, ASSOC PROF GEOL & PALEONT, IND-PURDUE UNIV, 83- *Mem:* Paleont Soc; UK Paleont Asn; AAAS; Am Malacol Union; Paleont Res Inst. *Res:* Structural and compositional growth patterns in skeletons of living and fossil organisms; paleoecological, medical, geophysical and environmental implications. *Mailing Add:* Dept Geol Ind-Purdue Univ 425 University Blvd Indianapolis IN 46202-5140

ROSENBERG, GILBERT MORTIMER, b Montreal, Que, Oct 26, 22; m 47; c 2. GERIATRIC MEDICINE, LONG TERM CARE. *Educ:* McGill Univ, BSc, 42, MDCM, 49, MSc & dipl internal med, 56. *Prof Exp:* Med dir & physician-in-chief, Maimonides Hosp & Home Aged, Montreal, Que, 56-77; prof med & dir, Fanning Ctr, Univ Calgary, 77-80; PROF MED GERIAT, QUEEN'S UNIV, KINGSTON, ONT, 80-; MED DIR & MEM, DEPT GERIAT & CONTINUING CARE MED, ST MARY'S LAKE HOSP, KINGSTON, ONT, 80- *Concurrent Pos:* Chmn, med adv bd, Can Geriat Res Soc, 86-90. *Mem:* Geront Soc Am (vpres, 77-78); Can Asn Geront (pres, 77-81); Am Col Physicians; Royal Col Physicians; Royal Col Physicians & Surgeons Can; Am Fedn Clin Res; Am Geriat Soc; Am Col Chest Physicians. *Res:* Involvement of chronic care patients in medical undergraduate teaching; calcium metabolism in older women. *Mailing Add:* 604-185 Ontario Kingston ON K7L 2Y7 Can

ROSENBERG, HARRY, b Feb 14, 40; Can citizen; m 64; c 2. BIOCHEMISTRY, IMMUNOLOGY. *Educ:* Univ Toronto, BSc, 61; Univ Mich, BS & PharmD, 68, MS, 70, PhD(pharmacog), 72. *Prof Exp:* Res asst pharmacog, Col Pharm, Univ Mich, 65-69; asst prof, Univ Pittsburgh, 71-72; assoc prof, 72-80, PROF BIOMED CHEM, UNIV NEBR MED CTR, 80-; PHARMACIST, LINCOLN GEN HOSP, 74- *Concurrent Pos:* Chmn, Dept Pharmaceut Sci, Campbell Univ. *Mem:* Am Asn Cols Pharm; Am Pharmaceut Asn; AAAS; Sigma Xi; Am Asn Col Pharm; Am Asn Pharmaceut Scientists. *Res:* Altered drug metabolism in diabetic state; synthesis and screening of antiarrythmic agents; enzymatic and non-enzymatic glucosylation of biological macromolecules. *Mailing Add:* Col Pharm NDak State Univ Fargo ND 58105

ROSENBERG, HENRY, b New York, NY, Sept 26, 41; c 3. MALIGNANT HYPERTHERMIA, ANESTHESIOLOGY. *Educ:* Albert Einstein Col Med, MD, 67. *Prof Exp:* PROF & CHMN DEPT ANESTHESIOL, HAHNEMANN UNIV, 81- *Mem:* Am Soc Anesthesiologists; AAAS; NY Acad Sci; Am Soc Pharmacol & Exp Therapeut; Bd Dir, Malignant Hyperthermia Asn USA. *Res:* Malignant hyperthermia, neuromuscular pharmacology. *Mailing Add:* Dept Anesthesiol Hahnemann Univ 230 N Broad St Philadelphia PA 19102

ROSENBERG, HENRY MARK, b Chicago, Ill, Jan 29, 14; m 39; c 1. RADIOLOGY, PERIODONTOLOGY. *Educ:* Northwestern Univ, DDS, 36. *Prof Exp:* From res asst to res assoc dent, 54-57, asst prof mat med & therapeut, 58-59, from asst prof to assoc prof radiol, 59-65, assoc head dept, 63-65, actg head dept, 65-67, PROF RADIOL, COL DENT, UNIV ILL MED CTR, 66-, HEAD DEPT, 67- *Concurrent Pos:* Consult, Ill State Psychiat Inst, 60- & div biol & med res, Argonne Nat Lab, 61-; co-prin investr, Nat Inst Dent Res, 62-65. *Mem:* Fel Am Col Dent; Am Dent Asn; Am Acad Periodont; Int Asn Dent Res; Am Acad Dent Radiol; Sigma Xi. *Res:* Aging of bone in human masticatory apparatus; dosimetry in radiology. *Mailing Add:* 763 La Crosse Wilmette IL 60091

ROSENBERG, HERBERT IRVING, b Brooklyn, NY, Oct 2, 39; m 77; c 2. VERTEBRATE MORPHOLOGY. *Educ:* City Col New York, BS, 61; State Univ NY Buffalo, PhD(biol), 68. *Prof Exp:* Fel insect behav, Cornell Univ, 67-69; asst prof vert zool, 69-73, asst dean fac arts & sci, 71-74, ASSOC PROF VERT ZOOL, UNIV CALGARY, 73- *Concurrent Pos:* Vis assoc prof zool, Univ Mich, 75-76 & Tel Aviv Univ, 82- 83, 90. *Mem:* Am Soc Zoologists; Am Soc Ichthyologists & Herpetologists; Can Soc Zool; Sigma Xi. *Res:* Comparative anatomy and histology of vertebrate organ systems. *Mailing Add:* Dept Biol Sci Univ Calgary Calgary AB T2N 1N4 Can

ROSENBERG, HERMAN, b Jersey City, NJ, Aug 27, 20. MATHEMATICS. *Educ:* NY Univ, BA, 39, MA, 48, PhD(math), 55. *Prof Exp:* Instr math, Jersey City Pub Sch Syst, 40-60; assoc prof, 60-61, chmn dept, 76-82, PROF MATH, JERSEY CITY STATE COL, 61-, COORDR GRAD STUDIES MATH, 66- *Concurrent Pos:* Assoc dir, Esso-Educ Found, Math Inst, NY Univ, 59-60; NSF lectr, Math Inst, Montclair State Col, 63- & Rutgers Univ, 70- *Mem:* Math Asn Am. *Res:* The impact of modern mathematics on trigonometry; modern applications of exponential and logarithmic functions; alternative structures for trigonometry and geometry. *Mailing Add:* Dept Math Jersey City State Col Jersey City NJ 07305-1597

ROSENBERG, HOWARD C, b Atlantic City, NJ, Apr 17, 47; m 69; c 2. NEUROPHARMACOLOGY, DRUG ABUSE. *Educ:* Ithaca Col, BA, 69; Cornell Univ, PhD(pharmacol), 75, MD, 76. *Prof Exp:* Fel pharmacol, Med Col, Cornell Univ, 76-77; from asst prof to assoc prof, 77-89, PROF PHARMACOL, MED COL OHIO, 90- *Mem:* Am Soc Pharmacol & Exp Therapeut; Soc Neurosci; Sigma Xi; AAAS. *Res:* Mechanisms for drug tolerance and dependence; effects of chronically administered benzodiazepines; anticonvulsant drugs. *Mailing Add:* Dept Pharmacol Med Col Ohio PO Box 10008 Toledo OH 43699-0008

ROSENBERG, IRA EDWARD, b New York, NY, Sept 25, 41; m 83; c 2. PHOTOCHEMISTRY, PHYSICAL ORGANIC CHEMISTRY. *Educ:* Hunter Col, BA, 63; Univ Md, College Park, MS, 66; George Washington Univ, PhD(org chem), 70. *Prof Exp:* Res fel photchem, Mich State Univ, 69-70; prin res scientist, 70-76, sect head, 76-78, MGR ADVAN INSTRUMENTATION, CLAIROL INC, 78- *Mem:* Am Chem Soc; Soc Cosmetic Chemists; Sigma Xi. *Mailing Add:* PO Box 10213 Stamford CT 06904

ROSENBERG, IRWIN HAROLD, b Madison, Wis, Jan 6, 35; m 64; c 2. GASTROENTEROLOGY, NUTRITION. *Educ:* Univ Wis-Madison, BS, 56; Harvard Univ, MD, 59. *Prof Exp:* Instr med, Harvard Med Sch, 65, assoc med, 67; res assoc, Thorndike Mem Lab-Boston City Hosp, 67; vis scientist, Dept Biophys, Weizmann Inst Sci, Israel, 68-69; asst prof med, Harvard Med Sch, 69-70; assoc prof med, Univ Chicago, 70, chief, sect gastroenterol, 71-86, prof med, 75-86, dir, Clin Nutrit Res Ctr, 79-86; PROF PHYSIOL, NUTRIT & MED, TUFTS UNIV, 86- *Concurrent Pos:* NIH career develop award, 68-74; Josiah Macy fac scholar award, 74; chmn Food Drug Admin Panel OTC Vitamin, Mineral & Hematinics Consult Nutrit Prog, US Govt, 75-; mem, Training Grants Study Sect, Gen Med Study Sect, 76-; chmn, Food & Nutrit Bd, Nat Res Coun. *Mem:* Am Gastroenterol Asn; Am Soc Clin Invest; Am Soc Clin Nutrit. *Res:* Study of intestinal absorption and malabsorption; human nutrition with emphasis on metabolic and nutritional aspects of aging. *Mailing Add:* USDA Human Nutrit Ctr Aging Tufts Univ 711 Washington St Boston MA 02111

ROSENBERG, ISADORE NATHAN, b Boston, Mass, May 19, 19; m 54; c 2. MEDICINE. *Educ:* Harvard Univ, AB, 40, MD, 43. *Prof Exp:* Intern med, Boston City Hosp, 44, asst res & resident, 47-49; asst, Med Sch, Tufts Col, 49-51, instr, 51-54; from asst prof to assoc prof, 54-71, PROF, DEPT MED, SCH MED, BOSTON UNIV, 71- *Concurrent Pos:* Asst med, Boston Univ, 49; res assoc, New Eng Ctr Hosp, 49-54; physician-in-chg endocrine unit, 5th & 6th Med Servs, Boston City Hosp, 54-72, physician-in-chief, 6th Med Serv, 63-72; chief med, Framingham Union Hosp, 72- *Honors & Awards:* Van Meter Prize, Am Goiter Soc, 51; Ciba Award, Endocrine Soc, 54. *Mem:* Am Soc Clin Invest; Endocrine Soc; AMA; Am Thyroid Asn; NY Acad Sci. *Res:* Chemical and metabolic studies of thyroid, pituitary and adrenal. *Mailing Add:* Dept Med Framingham Union Hosp 115 Lincoln St Framingham MA 01701

ROSENBERG, IVO GEORGE, b Brno, Czech, Dec 13, 34; m 66; c 2. MATHEMATICS. *Educ:* Purkyne Univ, Brno, MSc, 58, CandSc, 65, Dr rer nat(math), 66; Brno Tech Univ, Habil Dozent, 66. *Prof Exp:* From lectr to dozent math, Brno Tech Univ, 58-66; from lectr to sr lectr, Univ Khartoum, 66-68; assoc prof, Univ Sask, 68-71; ASSOC MEM, CTR MATH RES, UNIV MONTREAL, 71-, PROF MATH, 83- *Mem:* Can Math Soc; Am Math Soc. *Res:* Zero to one programming; universal algebra. *Mailing Add:* Math Univ Montreal 6128 Succ Ave Montreal PQ H3C 3J7 Can

ROSENBERG, JEROME LAIB, b Harrisburg, Pa, June 20, 21; m 46; c 2. BIOPHYSICAL CHEMISTRY. *Educ:* Dickinson Col, AB, 41; Columbia Univ, MA, 44, PhD(phys chem), 48. *Prof Exp:* Lectr chem, Columbia Univ, 42-44, res scientist, SAM Labs, 44-46, instr chem, Univ, 46-48; AEC fel, Univ Chicago, 48-50, res assoc, 50-53; from asst prof to prof chem, Univ Pittsburgh, 53-59, chmn dept biophys & microbiol, 69-71, prof molecular biol, 69-76, dean, Fac Arts & Sci, 69-86, vprovost, 78-89, chmn dept biol sci, 89-90, chmn dept commun, 90, PROF CHEM & BIOL SCI, UNIV PITTSBURGH, 76- *Concurrent Pos:* NSF sr fel, 62-63. *Mem:* Am Chem Soc; AAAS; Sigma Xi. *Res:* Photochemistry; photosynthesis; luminescence; molecular spectroscopy. *Mailing Add:* Dept Biol Sci Univ Pittsburgh Pittsburgh PA 15260

ROSENBERG, JERRY C, b New York, NY, Apr 23, 29; m 55; c 2. SURGERY, TRANSPLANTATION IMMUNOLOGY. *Educ:* Wagner Col, BS, 50; Chicago Med Sch, MD, 54; Univ Minn, Minneapolis, PhD(surg), 63; Mich State Univ, MBA. *Prof Exp:* Fulbright fel, Univ Vienna, 55-56; univ fel, Univ Minn, Minneapolis, 60-61; asst chief surg, USPHS Hosp, Staten Island, 61-63; instr, Univ Ky, 64-65; adj asst prof pharm, Univ Toledo, 67-68; dir surg, Maumee Valley Hosp, 65-68; assoc prof, 68-72, PROF SURG, SCH MED, WAYNE STATE UNIV, 72- , VPRES SURG SERV, HUTZEL HOSP. *Concurrent Pos:* Consult, Vet Admin Hosp, 68-; chief surg, Hutzel Hosp. *Mem:* Transplantation Soc; Am Asn Hist Med; Am Soc Artificial Internal Organs; Soc Univ Surg; Soc Surg Oncol. *Res:* Mechanisms of immune damage to transplanted organs; biochemical and physiological responses to shock and trauma; breast & thoracic oncology; oncology. *Mailing Add:* Dept Surg Hutzel H Sch Med Wayne State Univ 4707 St Antoine Detroit MI 48201

ROSENBERG, JONATHAN MICAH, b Chicago, Ill, Dec 30, 51. OPERATOR ALGEBRAS, GROUP REPRESENTATIONS. *Educ:* Harvard Univ, AB, 72; Univ Cambridge, MS, 73; Univ Calif, Berkeley, PhD(math), 76. *Prof Exp:* Asst prof math, Univ Pa, 77-81; assoc prof, 81-85, PROF MATH, UNIV MD, 85- *Concurrent Pos:* Sloan fel, 81-84. *Mem:* Am Math Soc; Math Asn Am. *Res:* Relations between operator algebras and geometry and topology; unitary representation theory of Lie groups. *Mailing Add:* Dept Math Univ Md College Park MD 20742

ROSENBERG, JONATHAN S, flow cytometry, lymphokine research, for more information see previous edition

ROSENBERG, JOSEPH, b New York, NY, Sept 8, 26; m 48; c 3. ORGANIC CHEMISTRY. *Educ:* City Col New York, BS, 48; Kans State Univ, MS, 49; Wayne State Univ, PhD(org chem), 51. *Prof Exp:* Polymer chemist, Gen Elec Co, 51-60; head org chem dept, Tracerlab, Inc, 60-62, mgr, Tracerlab Tech Prod Div, Lab Electronics, Inc, Mass, 62-69; group vpres, Int Chem & Nuclear Corp, 70-71; PRES, INTEREX CORP, 71- *Mem:* Am Chem Soc. *Res:* Synthetic organic chemistry; pyrolysis of esters; monomer synthesis; polymer chemistry; epoxy resins; polyurethanes; polyesters; addition and condensation polymers; acrylonitrile copolymers; synthesis of radiochemicals. *Mailing Add:* 46 Maugus Hill Rd Wellesley MA 02181

ROSENBERG, LAWSON LAWRENCE, b Hagerstown, Md, Apr 3, 20. PHYSIOLOGICAL CHEMISTRY. *Educ:* Johns Hopkins Univ, AB, 40, PhD, 51. *Prof Exp:* Asst physiol chem, Sch Med, Johns Hopkins Univ, 51-52, instr pediat, 52-53; jr res biochemist, Univ Calif, Berkeley, 53-56, asst res biochemist, 56-65, from assoc prof to prof physiol, 65-88, PROF MOLECULAR & CELLULAR BIOL, UNIV CALIF, BERKELEY, 88- *Mem:* Am Soc Biol Chemists; Endocrine Soc; Am Thyroid Asn. *Res:* Binding of metalloporphyrins to protein; enzymes; pituitary hormones and target organ relationships. *Mailing Add:* Dept Molecular Cellular Biol Univ Calif Berkeley CA 94720

ROSENBERG, LEON EMANUEL, b Madison, Wis, Mar 3, 33; m 54; c 4. GENETICS. *Educ:* Univ Wis, BA, 54, MD, 57. *Hon Degrees:* DSc, Univ Wis, 89. *Prof Exp:* From intern to resident med, Columbia-Presby Hosp, 57-59; clin assoc metab, Nat Cancer Inst, 59-61, sr invest, 61-62 & 63-65; resident med, Yale-New Haven Hosp, 62-63; from asst prof to prof human genetics, med & pediat, 65-84, DEAN, SCH MED, YALE UNIV, 84-; PRES, BRISTOL-MYERS SQUIBB PHARMACEUT RES INST, 91- *Concurrent Pos:* John Hartford Found grant, 65-68; NIH res grant, 65-81; Nat Inst Arthritis & Metab Dis res career develop award, 65-70; Guggenheim fel, 72; mem coun, Inst Med-Nat Acad Sci; bd trustees, Yale-New Haven Hosp. *Honors & Awards:* Borden Award, Am Acad Pediat, 73. *Mem:* Nat Acad Sci; Inst Med; Am Fedn Clin Res; Am Soc Human Genetics; Am Soc Clin Invest; Am Asn Med Clins; fel AAAS; fel Am Acad Arts & Sci. *Res:* Medical genetics; membrane function; biogenesis of mitochondrial enzymes; inherited disorders of amino acid metabolism; mechanism of vitamin transport and coenzyme synthesis. *Mailing Add:* Yale Univ Sch Med 333 Cedar St New Haven CT 06510

ROSENBERG, LEON T, b New York, NY, Feb 11, 28; m 50; c 2. IMMUNOLOGY. *Educ:* City Col New York, BSc, 46; Ohio State Univ, MSc, 48; NY Univ, PhD(biol), 58. *Prof Exp:* Res bacteriologist, Bellevue Hosp, 53-55; res assoc immunol, Bronx Hosp, 55-59; from asst prof to assoc prof immunol, 61-81, PROF MICROBIOL & IMMUNOL, STANFORD UNIV, 81- *Concurrent Pos:* NIH fel immunol, Stanford Univ, 59-60, Giannini Found fel, 60-61; Eleanor Roosevelt fel, Int Union Against Cancer, Karolinska Inst, Stockholm, 69-70; vis prof, La Trobe Univ, Melbourne, Australia, 86-87. *Mem:* AAAS; Am Asn Immunologists; Am Soc Microbiol. *Res:* Biological functions of antibody and complement; microbiology. *Mailing Add:* Dept Microbiol & Immunol Stanford Univ Stanford CA 94305

ROSENBERG, LEONARD, b New York, NY, Mar 11, 31. PHYSICS. *Educ:* City Col New York, BS, 52; NY Univ, MS, 54, PhD(physics), 59. *Prof Exp:* Assoc res scientist, NY Univ, 59-61 & Univ Pa, 61-63; from asst prof to assoc prof, 63-70, PROF PHYSICS, NY UNIV, 70- *Mem:* Am Phys Soc. *Res:* Scattering theory; low-energy atomic and nuclear physics. *Mailing Add:* Dept Physics NY Univ New York NY 10003

ROSENBERG, MARTIN, b Bridgeport, Conn, Feb 10, 46. GENE REGULATION. *Educ:* Purdue Univ, PhD(biochem), 70. *Prof Exp:* DIR MOLECULAR GENETICS, SMITHKLINE & FRENCH LABS, 82-, VPRES BIOPHARMACEUT, RES & DEVELOP, 85- *Concurrent Pos:* Adj prof, Depts Human Genetics & Microbiol, Univ Pa, biol sci, State Univ NY & biochem, UMD, NJ. *Honors & Awards:* Arthur S Fleming Award, 82. *Mem:* Am Soc Biol Chemists; Am Microbiol Soc. *Mailing Add:* Smith Kline Beecham Pharmaceuticals 709 Swedeland Rd PO Box 1539 King of Prussia PA 19406

ROSENBERG, MARVIN J, b New York, NY, Aug 27, 31; m 54; c 2. BIOLOGY, MOLECULAR GENETICS. *Educ:* City Col New York, BS, 52; Cornell Univ, MS, 54; Columbia Univ, PhD(molecular biol), 67. *Prof Exp:* Teacher sec sch, NY, 54-60; asst prof biol, State Univ NY Stony Brook, 60-68; assoc prof, 68-74, PROF BIOL, CALIF STATE UNIV, FULLERTON, 74-, CHMN DEPT BIOL SCI, 76- *Concurrent Pos:* State of NY Res Found grant in aid, 67-69. *Mem:* AAAS; Genetics Soc Am; Bot Soc Am. *Res:* Chromosome structure; radiomimetic effects on chromosome breakage and reunion; density gradient studies of DNA replication after treatment with thymidine analogs; exogenous DNA uptake by tumor cells. *Mailing Add:* Dept Biol Calif State Univ Fullerton CA 92634

ROSENBERG, MURRAY DAVID, b Boston, Mass, Jan 7, 25; m 47; c 3. CELL BIOLOGY, MEDICINE. *Educ:* Harvard Univ, AB, 47, MA, 48, MES, 50, PhD(appl physics), 52, MD, 56; Am Bd Family Physicians, dipl, 74. *Prof Exp:* Res assoc acoust, Harvard Univ, 52-54; consult biophys, NIH, 57-59; res assoc & asst prof develop biol, Rockefeller Inst, 59-65; PROF CELL BIOL, UNIV MINN, ST PAUL, 65-, PROF GENETICS, 77- *Concurrent Pos:* Res assoc, Mass Inst Technol, 58-59; consult div biol & med sci, NSF, 61-67, Off Sci & Technol, Exec Off President, 61-67 & med adv comt, State of Minn, 69-; Health Res Coun New York career scientist, 62-65; MacAulay nonresident fel inst animal genetics, Univ Edinburg, 63-66; mem exec comt & coun, Int Cell Res Orgn, 62; fel comt div biol & agr, Nat Acad Sci-Nat Res Coun, 63-67; mem cell biol study sect, NIH, 65-69; prog dir human cell biol, NSF, 71-72. *Mem:* Am Phys Soc; Acoust Soc Am; Biophys Soc; fel Am Acad Family Physicians; Asn Am Vol Physicians. *Res:* Physics and chemistry of cell in growth and development; cell contact relations; cell membranes; surface physics; reproductive biology; membrane enzymes. *Mailing Add:* Dept Genetics & Cell Biol 250 Bio Sci Univ Minn St Paul MN 55108

ROSENBERG, MURRAY DAVID, b Philadelphia, Pa, Feb 9, 40; m 63; c 2. DATABASE MANAGEMENT. *Educ:* Temple Univ, BA, 64, MBA, 71. *Prof Exp:* Res chemist synthetic org chem, Hoffmann-LaRoche Inc, 64-67; med chemist, Smith Kline Corp, 67-69; info scientist, Inst Sci Info, 69-70, sr info scientist, 70-72; systs & standards coordr qual control & info systs, Dutch Boys Paint Div, NL Industs, 72-73, mgr qual assurance & systs develop, 73-75; dir res info systs & mkt, Randex Corp, 75-77; facil leader, 77-81, SECT LEADER TECH INFO, PHILIP MORRIS USA, 81- *Concurrent Pos:* Mem bd, Doc Abstr, Inc, 84-85, pres, 86-87; mem, Va Sect ACS Publ Comt, 88-; chmn, Legislative Issues Comt, 88-; mem Exec Comt - Indust Tech Information Mgrs Group, 87- *Mem:* Am Chem Soc; Am Inst Chemists; Am Soc Info Sci; NY Acad Sci; Sigma Xi; Indust Tech Info Mgrs' Group. *Res:* Synthetic organic chemistry in the pharmaceutical industry; scientific and technical information storage and retrieval systems. *Mailing Add:* Res Ctr Philip Morris USA PO Box 26583 Richmond VA 23261

ROSENBERG, NORMAN J, b Brooklyn, NY, Feb 22, 30; m 50; c 2. MICROMETEOROLOGY, CLIMATOLOGY. *Educ:* Mich State Univ, BS, 51; Okla State Univ, MS, 58; Rutgers Univ, PhD(soil physics), 61. *Prof Exp:* Soil scientist, Israel Soil Conserv Serv, 53-55 & Israel Water Authority, 55-57; res asst soil physics, Okla State Univ, 57-58; from asst prof to assoc prof, Univ Nebr, Lincoln, 61-67, prof agr climat, 67-87, leader agr meteorol sect, Inst Agr & Natural Resources, 74-87, dir, ctr Agr Meteorol & Climat, 79-87, prof agr meteorol, George Holmes, 81-87, emer prof, 87; SR FEL & DIR, CLIMATE RESOURCES PROG, RESOURCES FOR FUTURE, 87- *Concurrent Pos:* Consult, Nat Oceanic & Atmospheric Admin, 62-, Great Western Sugar Co, 64-68, Water Resources Res Inst, US Dept Interior, 65- & US AID; vis prof, Israel Inst Technol, 68; NATO sr fel sci, 68; NSF grant, 71-; NASA grant, 72-; mem comt atmospheric sci, Nat Res Coun, 75-78, bd atmospheric sci & climate, 82-85; mem, Comt atmospheric Sci, Oak Ridge Assoc Univ, 80-87, Sandia Nat Labs, 90. *Honors & Awards:* Centennial Medal, Nat Oceanic & Atmospheric Admin, 70; Award Outstanding Achievement Biometeorol, Am Meteorol Soc, 78. *Mem:* Fel AAAS; fel Am Meteorol Soc; fel Am Soc Agron. *Res:* Microclimatology; ground level micrometeorology; evapotranspiration and windbreak influences on crop growth and development; global carbon dioxide balance and its interaction with plant growth; remote sensing of evapotranspiration in large regions; impact of drought on social, political and physical environment, development of strategies to cope with extended drought; impacts of climatic change. *Mailing Add:* Resources for Future 1616 P St NW Washington DC 20036

ROSENBERG, PAUL, b New York, NY, Mar 31, 10; m 43; c 1. NAVIGATION, PHOTOGRAMMETRY. *Educ:* Columbia Univ, AB, 30, AM, 33, PhD, 41. *Prof Exp:* Chemist, Hawthorne Paint & Varnish Corp, NJ, 30-33; asst physics, Columbia Univ, 34-39; instr, Hunter Col, 39-41; res assoc, Mass Inst Technol, 41, mem staff, Radiation Lab, 41-45; PRES, PAUL ROSENBERG ASSOCS, 45- *Concurrent Pos:* Lectr, Columbia Univ, 40-41; mem maritime res adv panel, Nat Acad Sci-Nat Res Coun, 59-60, chmn navig & traffic control panel, Space Appln Study, 68, chmn cartog & mapping panel, Comt Remote Sensing Progs for Earth Resources Surv, 73-77; chmn, Nat Conf Clear Air Turbulence, 66; mem navig adv comt, NASA, 69-70; mem bd dirs, Ctr Environ & Man, 76-85; gen chmn, Joint Conf, Radio Tech Comn Aeronaut, Radio Tech Comn Marine & Inst Navig; mem bd dir, Universal High Technol Corp, 81-85. *Honors & Awards:* Abrams Grand Award, Am Soc Photogram, 55; Cogswell Award, US Defense Dept, 86. *Mem:* Nat Acad Eng; fel AAAS (vpres, 66-69); Am Inst Navig (pres, 50-51); fel Inst Elec & Electronics Engrs; NY Acad Sci; fel Am Inst Chemists; fel Explorers Club. *Res:* Molecular beams; kinetic theory of gases; geometric and physical optics; ultrasonics; radar; navigation of land, marine, air and space vehicles; industrial electronics; photogrammetry; space technology; earth satellites; aeronautics; electrophotography; remote sensing. *Mailing Add:* 53 Fernwood Rd Larchmont NY 10538

ROSENBERG, PHILIP, b Philadelphia, Pa, July 28, 31; m 56; c 3. TOXINOLOGY, NEUROCHEMISTRY. *Educ:* Temple Univ, BS, 53; Univ Kans, MS, 55; Jefferson Med Col, PhD(pharmacol), 57. *Prof Exp:* Instr pharmacol, Jefferson Med Col, 57-58; res asst neurol & biochem, Col Physicians & Surgeons, Columbia Univ, 58-62, res assoc, 63, asst prof neurol, 63-68; asst dean grad studies, 71-75, 86-87, chmn sect pharmacol & toxicol, 68-88, PROF PHARMACOL, SCH PHARM, UNIV CONN, 68- *Concurrent Pos:* USPHS spec fel, Nat Inst Neurol Dis & Blindness, 60-62 & career develop award, 64-68; ed, Toxicon, 70-91; WHO spec consult & vis prof, Sch Med, Tel Aviv Univ, 74-75. *Honors & Awards:* Redi Award, 82; Javits Neurosci Investr Award, 87-94. *Mem:* Am Pharmaceut Asn; Am Chem Soc; Am Soc Pharmacol; Int Soc Toxinology (pres, 88-91); Am Soc Neurochem; Soc Toxicol. *Res:* Pharmacodynamics of drugs affecting the nervous system; actions of phospholipases; venom action on biological tissue; membranal permeability; actions of organophosphorus anticholinesterases; toxins and enzymes on membrane organization and asymmetry. *Mailing Add:* Sect Pharmacol & Toxicol Univ Conn Sch Pharm Storrs CT 06269

ROSENBERG, REINHARDT M, b Tubingen, Ger, Dec 17, 12; nat US; m 37; c 1. MECHANICS. *Educ:* Univ Pittsburgh, BS, 41; Purdue Univ, MS, 47. *Hon Degrees:* Dr, Univ Besancon, 62. *Prof Exp:* Engr, Flutter Group, Bell Aircraft Co, 42-44; design specialist, Consol Vultee Aircraft Corp, 44-46; from instr to asst prof aeronaut eng, Purdue Univ, 46-48; assoc prof, Univ Wash, 48-51; design specialist, Boeing Airplane Co, 51-53; prof appl mech, Univ Toledo, 53-58; prof eng mech, 58-64, Miller res prof, 64-76, prof, 76-82, EMER PROF ENG MECH, UNIV CALIF, BERKELEY, 82- *Concurrent Pos:* Guggenheim fel, Fulbright fel, 60; ed, Int J Nonlinear Mech & J Franklin Inst. *Res:* Nonlinear oscillations; vibration theory; biomechanics; dynamics; applied mathematics; theoretical and applied mechanics. *Mailing Add:* PO Box 678 Diablo CA 94528

ROSENBERG, RICHARD CARL, b Chicago, Ill, Mar 14, 43; m 69; c 3. MECHANICAL ENGINEERING, MATERIAL SCIENCE. *Educ:* Gen Motors Inst, BS, 66; Rensselaer Polytech Inst, MS, 67. *Prof Exp:* Res engr, Gen Motors Res Labs, 67-75, sr res engr, 75-80, staff res engr, 80-85, sr staff res engr, 85-90, SECT HEAD, SPECIF & VALIDATION, GEN MOTORS SYSTS ENG CTR, 90- *Mem:* Am Soc Mech Engrs; Soc Automotive Engrs. *Res:* Friction and wear mechanisms, especially bearing alloy development, rolling element bearings, lubricant additive effects and engine friction; vehicle systems engineering. *Mailing Add:* Requirements Eng GM Systs Eng Ctr Troy MI 48084

ROSENBERG, RICHARD MARTIN, b New York, NY, Jan 13, 33; m 55; c 3. INORGANIC CHEMISTRY, RESEARCH ADMINISTRATION. *Educ:* Brooklyn Col, BS, 54; Pa State Univ PhD(chem), 59. *Prof Exp:* Res chemist, Cent Res Dept, E I du Pont de Nemours & Co, Inc, 59-69, & Electrochem Dept, 69-70, res supvr, 70-75, res mgr, 75-80, prod mkt mgr, 80-81, prod mkt & planning mgr, 81-83, technical & planning mgr, 83-85, task force mgr, electronic mat div, 85-86, dir res, electronics, Electronics Dept, 86-88, dir, planning & new bus develop, 88, VPRES & GEN MGR, ELECTRONIC MAT, E I DU PONT DE NEMOURS & CO, INC, 89- *Mem:* Am Chem Soc; fel Am Ceramic Soc. *Res:* Solid state materials; hybrid microelectronics. *Mailing Add:* DuPont Electronics E I du Pont de Nemours & Co Inc Wilmington DE 19880-0021

ROSENBERG, RICHARD STUART, b Toronto, Ont, Aug 12, 39; m 78; c 3. SOCIAL ISSUES. *Educ:* Univ Toronto, BASc, 61, MASc, 64; Univ Mich, PhD(comput sci), 67. *Prof Exp:* Lab instr physics, Univ Toronto, 61-62; res asst elec eng, 62; asst in res eng, Univ Mich, 62-65, asst math, 65-68, asst prof comput sci, 67-68; from asst prof to assoc prof comput sci, Univ BC, 68-84; dir, Comput Sci Div, dept math, statist & comput sci, Dalhousie Univ, 84-86; ASSOC PROF COMPUT SCI, UNIV BC, 86- *Mem:* Am Asn Artificial Intel; Asn Comput Mach; Can Soc Computational Studies Intel (pres, 76-78); Comput Prof Soc. *Res:* Artificial intelligence, natural language understanding by computer; dialogue; reference problems; question-answering systems; bibliographic information; retrieval; computers and society. *Mailing Add:* Dept Comput Sci Univ BC Vancouver BC V6T 1W5 Can

ROSENBERG, ROBERT, physical metallurgy, for more information see previous edition

ROSENBERG, ROBERT, b Brooklyn, NY, Jan 3, 30; m 56; c 2. VISION. *Educ:* Columbia Univ, BS, 51, MS, 52, State Univ NY, OD, 73. *Prof Exp:* Chmn, 71-87, MEM FAC DEPT VISION SCI, STATE UNIV NY COL OPTOM, 74- *Concurrent Pos:* Mem 280 comt, Am Nat Standards Inst. *Mem:* Fel Am Acad Optom; Optom Soc Am; Am Optom Asn. *Res:* Optics; low vision. *Mailing Add:* Dept Vision State Univ NY Col Optomet New York NY 10010

ROSENBERG, ROBERT CHARLES, b New York, NY, Apr 25, 45. DBH INHIBITION ANTIHYPERTENSIVE & OTHER COMPOUNDS. *Educ:* Calif Inst Technol, PhD(bioinorg chem), 73. *Prof Exp:* Asst prof, 77-82, ASSOC PROF CHEM, HOWARD UNIV, 83- *Mem:* Am Chem Soc; Am Soc Biochem & Molecular Biol. *Res:* Enzyme structure and mechanics. *Mailing Add:* Dept Chem Howard Univ 2400 Sixth St NW Washington DC 20059

ROSENBERG, ROBERT D, THROMBOSIS, ATHEROSCLEROSIS. *Educ:* Mass Inst Technol, PhD(biophys), 69. *Prof Exp:* PROF MED, HARVARD MED SCH, 70-; PROF BIOCHEM, MASS INST TECHNOL, 81- *Mailing Add:* Dept Biochem Mass Inst Technol Bldg E 25 Rm 229 77 Mass Ave Cambridge MA 02139

ROSENBERG, ROBERT MELVIN, b Hartford, Conn, Mar 9, 26; m 51; c 4. PHYSICAL BIOCHEMISTRY. *Educ:* Trinity Col, Conn, BS, 47; Northwestern Univ, PhD(chem), 51. *Prof Exp:* Res assoc chem, Cath Univ Am, 50-51; asst dermat, Harvard Univ, 51-53; asst prof chem, Wesleyan Univ, 53-56; from asst prof to assoc prof chem, 56-67, chmn dept, 61-62, 66-67 & 79-81, assoc dean, Lawrence & Downer Cols, 68-75, PROF CHEM, LAWRENCE UNIV, 67- *Concurrent Pos:* NSF sci fac fel, Oxford Univ, 62-63; Am Chem Soc vis scientist, 63-74; resident dir, Assoc Cols Midwest Argonne Sem Prog, Argonne Nat Lab, 67-68; res assoc chem, Univ Wis-Madison, 75-76 & 81-82; res dir, Oak Ridge Sci Semester, 88; res assoc, Oak Ridge Nat Lab, 88-89. *Mem:* AAAS; Am Chem Soc; Royal Soc Chem. *Res:* Physical chemistry of proteins. *Mailing Add:* Dept Chem Lawrence Univ Appleton WI 54912

ROSENBERG, RONALD C(ARL), b Philadelphia, Pa, Dec 15, 37; m 59; c 4. SYSTEM DYNAMICS, COMPUTER-AIDED ENGINEERING. *Educ:* Mass Inst Technol, BSc & MSc, 60, PhD(mech eng), 65. *Prof Exp:* Asst prof mech eng, Mass Inst Technol, 66-69; assoc prof, 69-73, PROF MECH ENG, MICH STATE UNIV, 73- *Concurrent Pos:* Pres & consult, Rosencode Assoc, Inc. *Honors & Awards:* Teetor Award, Soc Automotive Engrs, 88. *Mem:* Am Soc Mech Engrs; Inst Elec & Electronics Engrs; Soc Comput Simulation; Am Soc Eng Educ. *Res:* Dynamic system modeling and behavior; computer-aided design; software development. *Mailing Add:* Dept Mech Eng Mich State Univ Eng Bldg East Lansing MI 48824-1226

ROSENBERG, SANDERS DAVID, b New York, NY, Dec 21, 26; m 46; c 2. FUELS SCIENCE, HIGH TEMPERATURE MATERIALS. *Educ:* Middlebury Col, AB, 48; Iowa State Univ, PhD(org chem), 52. *Prof Exp:* Sr chemist & group leader org res dept, Rahway Res Lab, Metal & Thermal Corp, NJ, 53-58; prin chemist & mgr fuels & combustion res dept, Chem Prod Div, Aerojet Gen Corp, 58-68, mgr fuels & combustion res, 68-70, mgr chem processing & mat, Aerojet Liquid Rocket Co, 70-73; scientist, Res & Technol, 80-89, CHIEF SCIENTIST, AEROJET PROPULSION DIV, AEROJET TECH SYSTS CO, 89- *Mem:* Assoc fel Am Inst Aeronaut & Astronaut. *Res:* Development of high energy liquid propellants; research on nature of combustion in chemical rocket engines, development of materials for advanced propulsion applications, and development and production of liquid rocket engine systems; chemical vapor deposition. *Mailing Add:* 628 Commons Dr Sacramento CA 95825

ROSENBERG, SAUL ALLEN, b Cleveland, Ohio, Aug 2, 27; c 2. INTERNAL MEDICINE, ONCOLOGY. *Educ:* Western Reserve Univ, BS, 48, MD, 53; Am Bd Internal Med, dipl. *Prof Exp:* Spec fel, Med Neoplasia, Mem Ctr Cancer & Allied Dis, NY, 57-58; from asst prof to assoc prof med & radiol, 61-70, PROF MED & RADIATION ONCOL, SCH MED, STANFORD UNIV, 70-, CHIEF DIV ONCOL, 65-, MAUREEN LYLES D'AMBROGIO PROF MED & RADIOL, 85-, ASSOC DEAN CLIN PROG PLANNING & DEVELOP, 89- *Concurrent Pos:* Chief resident physician, Peter Bent Brigham Hosp, Boston, 60-61; chief med serv, Stanford Univ Hosp, 69-71; Eleanor Roosevelt int fel, Am Cancer Soc, 71-72; chmn, Comt Med Oncol, Am Bd Internal Med, 78-80; assoc ed, J Clin Oncol, 82-88; K P Stephen Chang vis prof, Univ Hong Kong, 86. *Honors & Awards:* William Lister Rogers Award, 77; Walter Albion Hewlett Award, 83; David Karnofsky Medal, Am Soc Clin Oncol, 84; Jan Waldenstrom Medal, Swed Acad Oncol, 84; C Chester Stock Award, Mem Sloan Kettering Inst, 90. *Mem:* Inst Med-Nat Acad Sci; master Am Col Physicians; Radiation Res Soc; Am Asn Cancer Res; Am Soc Therapeut Radiol & Oncol; Am Fedn Clin Res; Asn Am Physicians; Am Soc Clin Oncol. *Res:* Clinical investigation of malignant lymphomas; cancer chemotherapy. *Mailing Add:* Div Oncol M211 Stanford Univ Med Ctr Stanford CA 94305-5306

ROSENBERG, SAUL H, b Philadelphia, Pa; c 1. EPIDEMIOLOGY. *Educ:* Univ Pa, AB, 61; Cath Univ Am, MA, 65; Johns Hopkins Univ, PhD(biostatist), 70. *Prof Exp:* Instr biostatist, Med Sch, Johns Hopkins Univ, 69-70; math statistician, Nat Ctr Health Statist, 70-72; asst prof biostatist, Med Sch, Temple Univ, 72-73; from asst prof to assoc prof biostatist, Sch Pub Health, Univ Ill, 73-80; chmn biostatist, Am Health Found, 80-82; math statistician, Environ Protection Agency, 82-85; assoc prof biostatist, East Carolina Univ Allied Health, 85-86; CHMN BIOSTATIST, ARMED FORCES INST PATH, 86-; ASSOC PROF BIOSTATIST, UNIV MD, 90- *Mem:* Biomet Soc; Am Statist Asn. *Res:* Statistical techniques to evaluate prognostic and diagnostic factors; estimation of a cure rate using censored data. *Mailing Add:* 10713 Lady Slipper Terr North Bethesda MD 20852-3403

ROSENBERG, SAUL HOWARD, b Boston, Mass, Feb 20, 57; m 81. PEPTIDE CHEMISTRY, ORGANIC CHEMISTRY. *Educ:* Mass Inst Technol, BS, 79; Univ Calif, PhD(org chem), 84. *Prof Exp:* RES INVESTR, ABBOTT LABS, 84- *Mem:* Am Chem Soc; Sigma Xi. *Res:* Development of novel peptide structures as antihypertensive agents; preparation of new peptide surrogate and the examination of the fate of peptide related compound in vivo. *Mailing Add:* D-47B AP10 Abott Lab Abbott Park IL 60064

ROSENBERG, STEVEN A, b New York, NY, Aug 2, 40. CANCER RESEARCH. *Educ:* Johns Hopkins Univ, MD, 64; Harvard Univ, PhD(biophys), 69. *Prof Exp:* Intern, 64-65, Peter Bent Brigham Hosp, Surg, 68-69 & 72-74; CHIEF SURG, NAT CANCER INST, 74- *Concurrent Pos:* USPHS, 70-72. *Mem:* Inst Med-Nat Acad Sci; Halsted Soc; Am Asn Immunologist; Am Asn Cancer Res. *Mailing Add:* Dept Health & Human Serv NIH Bldg 10 Rm 2B42 Bethesda MD 20205

ROSENBERG, STEVEN LOREN, b Oakland, Calif, Sept 27, 41; m 68. MICROBIOLOGY. *Educ:* Univ Calif, Berkeley, AB, 63, PhD(bact), 70. *Prof Exp:* Fel bact, Univ Mass, Amherst, 70-71, NIH fel, 71-72; microbiologist, Phys Chem Lab, Gen Elec Res & Develop Ctr, 72-76; mem staff, Lawrence Berkeley Lab, Univ Calif, 76-81; sr microbiologist, SRI Int, 81-86; CONSULT, 86- *Mem:* Am Chem Soc; Mycol Soc Am; Am Soc Microbiol. *Res:* Experimental study of bacterial evolution; enzymology of lignocellulose degradation; production of fuels and chemicals from cellulosic materials by fermentation; biodegradation of xanthan gum and other oil field chemicals; microbial enhanced oil recovery. *Mailing Add:* 5555 Greenridge Rd Castro Valley CA 94552

ROSENBERG, STUART A, b Petersburg, Va, Aug 22, 47; m 73; c 2. INTERNAL MEDICINE, RHEUMATOLOGY. *Educ:* Univ Va, MD, 73. *Prof Exp:* Asst prof med, Univ Va, 80-85; STAFF PHYSICIAN, OCHSNER CLIN, NEW ORLEANS, 85- *Mem:* Am Col Physicians; Am Asn Immunologists; Am Fedn Clin Res; Am Rheumatism Asn. *Mailing Add:* Ochsner Clin 1514 Jefferson Hwy New Orleans LA 70121

ROSENBERG, THEODORE JAY, b New York, NY, May 24, 37; m 60; c 2. SPACE PHYSICS. *Educ:* City Col New York, BEE, 60; Univ Calif, Berkeley, PhD(physics), 65. *Prof Exp:* Royal Norweg Coun Sci & Indust Res grant, Univ Bergen, 65-66; res assoc & lectr space sci, Rice Univ, 66-68; from res asst prof to res assoc prof, 71-75, res prof, Inst Fluid Dynamics & Appl Math, 75-76, RES PROF, INST PHYS SCI & TECHNOL, UNIV MD, COLLEGE PARK, 76- *Concurrent Pos:* Royal Norweg Coun Sci & Indust Res fel, Norweg Inst Cosmic Physics, 75-76. *Mem:* Am Geophys Union; Am Phys Soc. *Res:* Magnetosphere and ionosphere of earth; experimental investigations of magnetospheric plasma and wave-particle interaction processes including studies of the aurora, radiowave propagation, magnetic substorms and the ionospheric effects of particle energy deposition. *Mailing Add:* Inst Phys Sci & Technol Univ Md College Park MD 20742-2431

ROSENBERG, WARREN L, b Brooklyn, NY, Nov 30, 54; m 80; c 3. ANIMAL PHYSIOLOGY, CYTOLOGY. *Educ:* City Univ New York, BA, 76; New York Univ, MS, 79, PhD(biol), 82. *Prof Exp:* Fel physiol & biophys, Mt Sinai Sch Med, New York, 82-83; asst prof, 83-87, ASSOC PROF PHYSIOL, IONA COL, 87- *Concurrent Pos:* Lectr, Iona Col, 81-83; biol photogr, Biol Photo Serv, 81-; res assoc, Cellular Biol Lab, New York Univ, 82- *Mem:* AAAS; NY Acad Sci; Am Soc Cell Biol. *Res:* Role of renal glomerulus in the initiation and maintenance of hypertension; morphology and function of the glomerular filtration barrier. *Mailing Add:* 715 North Ave New Rochelle NY 10801-1890

ROSENBERGER, ALBERT THOMAS, b Butte, Mont, Jan 27, 50. OPTICAL PHYSICS, NONLINEAR DYNAMICS. *Educ:* Whitman Col, AB, 71; Univ Chicago, MS, 72; Univ Ill, Champaign-Urbana, PhD(physics), 79. *Prof Exp:* Vis asst prof physics, Drexel Univ, 79-80; asst prof physics, Western Ill Univ, 80-82; lectr & res assoc, Univ Tex, Austin, 82-85; asst prof physics, Southern Methodist Univ, 85-89; ASSOC PROF PHYSICS, UNIV ALA, HUNTSVILLE, 89- *Concurrent Pos:* Consult, Battelle Columbus Labs, 76-79. *Mem:* Am Phys Soc; Am Asn Physics Teachers; Optical Soc Am; Inst Elec & Electronics Engrs. *Res:* Optical physics and nonlinear dynamical phenomena in optical system including optical bistability and instability; coherent transient effects especially supperradiance and effects of losses in optical resonator. *Mailing Add:* Dept Physics Univ Ala Huntsville Huntsville AL 35899

ROSENBERGER, ALFRED L, b New York, NY, Nov 30, 49; m 79. PRIMATOLOGY, EVOLUTIONARY BIOLOGY. *Educ:* City Col New York, BA, 72; City Univ New York, PhD(anthrop), 79. *Prof Exp:* Lectr & fel, Dept Anat Sci, State Univ NY Stony Brook, 79-81; ASST PROF, DEPT ANTHROP, UNIV ILL, CHICAGO, 81- *Concurrent Pos:* Fulbright fel, Brazil, 82-83. *Mem:* Am Asn Phys Anthropologists; Am Asn Primatologists; Int Soc Primatology; Soc Syst Zool; Soc Vert Paleont. *Res:* Primate evolution, particularly evolution and adaptation of South and Central American monkeys and higher primates as a whole. *Mailing Add:* Dept Anthrop Univ Ill Box 4348 Chicago IL 60680

ROSENBERGER, FRANZ, b Salzburg, Austria, May 31, 33; m 59; c 3. CHEMICAL PHYSICS, THERMAL PHYSICS. *Educ:* Stuttgart Univ, MS, 64; Univ Utah, PhD, 69. *Prof Exp:* Res scientist, 69-70, asst res prof, 70-73, asst prof, 73- 78, assoc prof, 78-81, prof physics, 81-86, dir crystal growth lab, Univ Utah, 66-86; PROF PHYSICS & DIR, CTR MICROGRAVITY & MAT RES, UNIV ALA, HUNTSVILLE, 86- *Concurrent Pos:* Adj prof mat sci & eng, Univ Utah, 81-88. *Mem:* Am Asn Crystal Growth; Am Phys Soc. *Res:* Mass and heat transfer phenomena in crystal growth and materials processing; thermodynamics and kinetics of phase transitions; mass spectroscopy of high temperature vapors; fluid dynamics. *Mailing Add:* Ctr Microgravity & Mat Res Univ Ala Huntsville AL 35899

ROSENBERGER, JOHN KNOX, b Wilmington, Del, Dec 8, 42; m 64; c 2. ANIMAL VIROLOGY, AVIAN PATHOLOGY. *Educ:* Univ Del, BS, 64, MS, 66; Univ Wis-Madison, PhD(virol, immunol), 72. *Prof Exp:* Virologist, US Army Biol Labs, 67-69; from asst prof to assoc prof, 72-81, PROF VIROL & IMMUNOL, COL AGR SCI, UNIV DEL, 81-, CHAIRPERSON, DEPT ANIMAL SCI & AGR BIOCHEM, 78- *Mem:* Wildlife Dis Asn; Am Soc Microbiol; assoc Am Asn Avian Pathologists; AAAS; Sigma Xi. *Res:* Viral arthritis; virus induced immuno suppression; characterization of avian respiratory agents. *Mailing Add:* Dept Animal Sci & Agr Biochem Col Agr Sci Univ Del Newark DE 19716

ROSENBERRY, TERRONE LEE, b Ft Wayne, Ind, Mar 16, 43; m 65; c 2. BIOCHEMISTRY, NEUROCHEMISTRY. *Educ:* Oberlin Col, AB, 65; Univ Ore, PhD(biochem), 69. *Prof Exp:* Res assoc neurol, Columbia Univ, 69-72, asst prof, 72-79; assoc prof, 79-85, PROF PHARMACOL, CASE WESTERN RESERVE UNIV, 85- *Concurrent Pos:* Prin investr, NSF grant, 72-, NIH grant, 73- & Muscular Dystrophy Asn grant, 79-; Jacob Javits Neurosci Investr, NIH, 89- *Mem:* Am Chem Soc; Am Soc Biol Chem. *Res:* Membrane proteins. *Mailing Add:* Dept Pharmacol Case Western Reserve Univ 2119 Abington Rd Cleveland OH 44106

ROSENBLATT, CHARLES STEVEN, b Brooklyn, NY, Aug, 23, 52; m; c 2. LIGHT SCATTERING & OPTICS. *Educ:* Mass Inst Technol, SB, 74; Harvard Univ, PhD, 78. *Prof Exp:* Res fel physics, Lawrence Berkeley Lab, Univ Calif, 78-80; res staff physics, Francis Bitter Nat Magnet Lab, Mass Inst Technol, 80-87; ASSOC PROF, CASE WESTERN RESERVE UNIV, 87- *Mem:* Am Phys Soc; Sigma Xi. *Res:* Optical, magnetic, and mechanical studies of liquid crystals, liquids, and colloids; phase transitions; light scattering from small biological organisms. *Mailing Add:* Dept Physics Case Western Reserve Univ Cleveland OH 44106

ROSENBLATT, DANIEL BERNARD, b Bloomington, Ind, Apr 28, 56. EARTHQUAKE STATISTICS, GEODESY. *Educ:* Univ Calif, Berkeley, BA, 77; Stanford Univ, MA, 78; Univ Calif, Los Angeles, PhD(geophys), 89. *Prof Exp:* Programmer, US Geol Surv, 79-80; GEODESIST, NIED, JAPAN, 91- *Mem:* Am Geophys Union; Am Inst Physics. *Res:* Mathematical statistics; modeling of seismic risk; geodetic studies of fault movement; nonlinear science, investigation of the behavior of nonlinear and delay differential equations. *Mailing Add:* PO Box 2066 La Jolla CA 92038

ROSENBLATT, DAVID, b New York, NY, Sept 5, 19; m 50. MATHEMATICAL STATISTICS. *Educ:* City Col New York, BS, 40. *Prof Exp:* Assoc statistician, Off Price Admin, DC, 41-44; sr economist, Div Statist Stand, US Bur Budget, 44-47, 48-49; asst prof econ, Carnegie Inst Technol, 49-51; consult, 51-53; assoc prof statist, Am Univ, 53-55; prin investr, Off Naval Res, 53-57; res consult, Industs, 55-61; consult, Info Tech Div, Nat Bur Stand, 61-63, mathematician, 63-67; RES CONSULT, 68- *Concurrent Pos:* Littauer fel, Harvard Univ, 47-48; statist consult, Div Statist Standards, US Bur Budget, 49-53; consult, George Washington Univ, 55, 59-61, 68-; adj prof math & statist, Am Univ, 59-61. *Mem:* Fel AAAS; Am Math Soc; fel Am Statist Asn. *Res:* Applied stochastic processes; theory of graphs and theory of relations; mathematical theory of organizations and complex systems; design of statistical information systems; history of mathematical and symbolic methods in resource and social sciences; relation and Boolean algebraic methods; biological structures and relations. *Mailing Add:* 2939 Van Ness St NW Washington DC 20008

ROSENBLATT, DAVID HIRSCH, b Trenton, NJ, July 24, 27; m 49; c 3. POLLUTANT LIMIT VALUES, CHEMICAL PROPERTY ESTIMATION. *Educ:* Johns Hopkins Univ, BA, 46; Univ Conn, PhD(chem), 50. *Prof Exp:* Res & develop chemist, Baltimore Paint & Color Works, 50-51; chemist, Chem Warfare Labs, US Dept Army, 51-56; org chemist, Johns Hopkins Univ, 56-57; chemist, Chem Res & Develop Labs, US Dept Army, 57-63, chief decontamination res sect, Chem Lab, Edgewood Arsenal, 63-72, res chemist, Health Effects Res Div, US Army Biomed Res & Develop Lab, Ft Detrick, Md, 72-89; INDEPENDENT CONSULT & ENVIRON SCIENTIST, ARGONNE NAT LAB, ILL, 90- *Concurrent Pos:* Instr eve col, Johns Hopkins Univ, 57-70. *Mem:* AAAS; Am Chem Soc; Sigma Xi; NY Acad Sci; Soc Environ Toxicol Chem; Air & Waste Mgt Asn. *Res:* Nucleophilic displacements; mechanisms of oxidation of amines and other organics in aqueous solution; halogens and halogen oxides; decomposition and complexing of toxic chemical agents; specific analytical methods; risk assessment of environmental pollutants; chemistry and applications of chlorine dioxide; estimation of physico-chemical properties of environmental pollutants. *Mailing Add:* 3316 Forest Rd Baltimore MD 21208-3101

ROSENBLATT, DAVID SIDNEY, b Montreal, Que, July 14, 46; m 69; c 2. BIOCHEMICAL GENETICS, PEDIATRICS. *Educ:* McGill Univ, BSc, 68, MDCM, 70. *Prof Exp:* Intern pediat med, Montreal Children's Hosp, 70-71; clin & res fel pediat & genetics, Mass Gen Hosp, Harvard Med Sch, 71-73; fel biol, Mass Inst Technol, 73-74; asst resident pediat, Children's Hosp Med Ctr, Harvard Med Sch, 74-75; asst prof, 75-80, assoc prof, Ctr Human Genetics, 80, asoc prof pediat, 80-87, PROF MED, PEDIAT, CTR HUMAN GENETICS, DIR DIV MED GENETICS, MCGILL UNIV, MONTREAL GENERAL & ROYAL VICTORIA HOSPS, 87-, AUX PROF BIOL, 78- *Concurrent Pos:* Prin investr, Med Res Coun Can, Genetics Group, McGill Univ, 75-; mem genetics comt, Med Res Coun Can, 78-81, 88-; vchmn res comt, McGill Univ, Montreal Children's Hosp Res Inst, 80-84. *Honors & Awards:* Prix d'excellence for Res into Dis of Children, 86. *Mem:* Am Soc Human Genetics; Can Soc Clin Invest (secy-treas, 87-90); Soc Pediat Res; Can Col Med Genetics. *Res:* Inborn errors of folate metabolism in cultured human cells; methotrexate metabolism; inborn errors of cobalamin metabolism. *Mailing Add:* Dept Pediat & Biol McGill Univ 853 Sherbrooke St W Montreal PQ H3A 2T6 Can

ROSENBLATT, GERD MATTHEW, b Leipzig, Ger, July 6, 33; US citizen; m; c 2. PHYSICAL CHEMISTRY. *Educ:* Swarthmore Col, BA, 55; Princeton Univ, PhD(phys chem), 60. *Hon Degrees:* Dr, Vrije Univ Brussel, 89. *Prof Exp:* Chemist, Inorg Mat Res Div, Lawrence Radiation Lab, Univ Calif, 60-63; from asst prof to assoc prof chem, Pa State Univ, University Park, 63-70, prof, 70-81; assoc div leader, Chem-Mat Sci Div, Los Alamos Nat Lab, 81-82, chem div leader, 82-85; dep dir, 85-89, SR CHEMIST, LAWRENCE BERKELEY LAB, UNIV CALIF, 89- *Concurrent Pos:* Lectr, Univ Calif, Berkeley, 62-63; guest scientist & consult, Inorg Mat Res Div, Lawrence Berkeley Lab, Univ Calif, 68-84; mem, Comt High Temperature Sci & Technol, Nat Acad Sci-Nat Res Coun, 70-79 & 84-85, chmn , 77-79 & 84-85, chmn, Workshop High Temperature Sci, Nat Acad-Res Coun-NSF, 79, Panel Solid-State Sci, Acad-coun, 81-82, Numerical Data Adv Bd, 84-90, chmn, 86-90, & mem Res Briefing Panel Ceramics & Ceramic Compos, 85; vchmn, Gordon Res Conf High Temperature Chem, 72, chmn, 74; vis prof, Vrije Univ & fel, Solvay Inst, Univ Libre Burssels, Belg, 73; mem rev comt, Chem Eng Div, Argonne Nat Lab, 74-80, chmn, 77 & 78, rev comn, High Temperature Mat Lab, Oak Ridge Nat Lab, 78-81; consult, Hooker Chem Co, 76-78, Xerox Corp, 77-78, Los Alamos Sci Lab, 78, Aerospace Corp, 79-85 & Solar Energy Res Inst, 80-81; ed, Progress in Solid State Chem, 77-; mem, adv rev bd, Joint Army-Navy-Air Force Thermochem Tables, 77-; US nat rep, Comn High Temperature & Solid State Chem, 78-85, assoc mem, 85-, mem, US Nat Comt, 86-; vis fel, Southhampton Univ, & King's Col, Cambridge, 80; adj prof chem, Univ NMex, 81-85; mem, rev comt, Chem Sci Progs, Lawrence Berkeley Lab, 84 & chmn, rev comt, Chem & Mat Sci Dept, Lawrence Livermore Nat Lab, 85-87; mem, adv bd, Sandia Nat Lab, Livermore & Lockheed Missiles & Space Co, 85-89; mem, external rev comt, Chem Div, Los Alamos Nat Lab, 85 & external adv comt, Ctr Mat Sci, 85-, chmn, 88-; mem, rev team, Comt Army Chem Stockpile Disposal Prog, Nat Res Coun, Nat Acad Sci, 89-, Solid State Sci Comt, 88-, Sci & Tech Info Bd, 90; chmn, Western Regional Mat Sci & Eng Meeting, Nat Res Coun, 90; mem, Nat Lab Task Force, Indust Res Inst, 87-89, Basic Energy Sci Lab Prog Panel, Dept Energy, 85-89; mem, US Nat Comt Data Sci & Technol, 86- *Mem:* Am Phys Soc; Am Chem Soc; fel AAAS; Electrochem Soc. *Res:* Properties and behavior of high-temperature materials and high-temperature gases; thermodynamics; dynamics of gas-surface reactions and of vaporization; Raman scattering as a probe of high-temperature systems, solid-state materials, and surface processes. *Mailing Add:* 1177 Miller Ave Berkeley CA 94708

ROSENBLATT, JOAN RAUP, b New York, NY, Apr 15, 26; m 50. MATHEMATICAL STATISTICS. *Educ:* Columbia Univ, AB, 46; Univ NC, PhD(statist), 56. *Prof Exp:* Statist analyst, US Bur Budget, 48; asst statist, Univ NC, 53-54; mathematician, Nat Bur Standards, 55-69, chief statist eng lab, 69-78; DEP DIR COMPUT & APPL MATH, NAT INST STANDARDS & TECHNOL, 78- *Concurrent Pos:* Sect U statist chair, AAAS, 82; mem, Comt Appl & Theoret Statist, Nat Res Coun, 85-88. *Honors & Awards:* Fed Woman's Award, 71; Gold Medal, Dept Commerce, 76. *Mem:* Fel AAAS; Am Math Soc; Int Statist Inst; fel Am Statist Asn (vpres, 81-83); fel Inst Math Statist; AAAS (secy, 87-91); Bernouli Soc. *Res:* Nonparametric statistical theory; applications of statistical techniques in physical and engineering sciences; reliability of complex systems. *Mailing Add:* 2939 Van Ness St NW Apt 702 Washington DC 20008

ROSENBLATT, JUDAH ISSER, b Baltimore, Md, Feb 12, 31; m 56; c 1. BIOMETRICS, BIOSTATISTICS. *Educ:* Johns Hopkins Univ, BA, 51; Columbia Univ, PhD, 59. *Prof Exp:* Asst prof math & statist, Purdue Univ, 56-60; assoc prof math, Univ NMex, 60-68, dir math comput lab, 67-68; prof math, 68-76, PROF BIOMET, CASE WESTERN RESERVE UNIV, 76- *Concurrent Pos:* Consult, Sandia Corp, Albuquerque, NMex, 60-68, Gen Elec Co, 77- *Mem:* Am Math Soc; Inst Math Statist; Math Asn Am; Soc Critical Care Med. *Res:* Stochastic processes, numerical analysis, biostatistical modelling and analysis. *Mailing Add:* Med Branch Biomath Off Univ Tex 7131 Shearn Moody Plaza Galveston TX 77550

ROSENBLATT, KARIN ANN, b Chicago, Ill, Apr 22, 54. CANCER EPIDEMIOLOGY, EPIDEMIOLOGIC METHODS. *Educ:* Univ Calif, Santa Cruz, BA, 75; Univ Mich, MPH, 77; Johns Hopkins Univ, PhD(epidemiol), 88. *Prof Exp:* Postdoctoral fel, Univ Wash, 87-89; staff scientist, Fred Hutchinson Cancer Res Ctr, 89-91; ASST PROF HEALTH & SAFETY STUDIES, UNIV ILL, 91- *Mem:* Am Col Epidemiol; Soc Epidemiol Res; Am Pub Health Asn. *Res:* Etiology of male breast cancer, ovarian cancer and chronic lymphocytic leukemia; incidence of fallopian tube tumors; relationship between contraceptive hormones and liver and endometrial cancer. *Mailing Add:* 120 Huff Hall Univ Ill 1206 S Fourth St Champaign IL 61820

ROSENBLATT, MICHAEL, b Lund, Sweden, Nov, 27, 47; m 69; c 2. ENDOCRINOLOGY, BIOCHEMISTRY. *Educ:* Columbia Col, AB, 69; Harvard Med Sch, MD, 73. *Prof Exp:* From instr to asst prof med, Harvard Med Sch, Mass Gen Hosp, 76-85; vpres biol res & molecular biol, 84-89, SR VPRES RES, MERCK SHARP & DOHME RES LABS, MERCK & CO, INC, 89- *Concurrent Pos:* Chief endocrine unit, Mass Gen Hosp, 81-84, consult med, 84-; lectr med, Harvard Univ Sch Med, 85-; adj prof, Univ Pa Sch Med, 88- *Honors & Awards:* Fuller Albright Award, Am Soc Bone & Mineral Res, 86; Vincent du Vigneaud Award, Chem & Biol of Reptides. *Mem:* Endocrine Soc; Am Soc Bone & Mineral Res; Am Soc Biochem & Molecular Biol; Am Soc Clin Invest; Asn Am Physicians. *Res:* Interaction of parathyroid hormone with receptors; peptide hormone isolation of receptors molecular biology of parathyroid hormone secretion tumor-secreted hypercalcemia factor (paralthyroid hormone-related protein). *Mailing Add:* Merck Sharp & Dohme Res Labs West Point PA 19486

ROSENBLATT, MURRAY, b New York, NY, Sept 7, 26; m 49; c 2. STOCHASTIC PROCESSES, NON-PARAMETRIC METHODS. *Educ:* City Col New York, BS, 46; Cornell Univ, MS, 47, PhD(math), 49. *Prof Exp:* Res assoc math, Cornell Univ, 49-50; from instr to asst prof statist, Univ Chicago, 50-55; assoc prof math, Ind Univ, 56-59; prof appl math, Brown Univ, 59-64; PROF MATH, UNIV CALIF, SAN DIEGO, 64- *Concurrent Pos:* Off Naval Res grant, 49-50; Guggenheim fels, 65-66, 71-72; fel, Univ Col London, 66, 72 & Australian Nat Univ, 76; overseas fel, Churchill Col, 79. *Mem:* Nat Acad Sci; Am Math Soc; fel Inst Math Statist; fel AAAS; Int Statist Inst. *Res:* Probability theory; stochastic processes; time series analysis; turbulence. *Mailing Add:* Dept Math Univ Calif San Diego La Jolla CA 92093

ROSENBLATT, RICHARD HEINRICH, b Kansas City, Mo, Dec 21, 30; m 52; c 2. ZOOLOGY. *Educ:* Univ Calif, Los Angeles, AB, 53, MA, 54, PhD, 59. *Prof Exp:* Asst zool, Univ Calif, Los Angeles, 53-56; asst res zoologist, Inst, 58-65, from asst prof to assoc prof, 65-73, PROF MARINE BIOL, SCRIPPS INST OCEANOG, PROF, GRAD DEPT & RES ZOOLOGIST, SCI SUPPORT DIV, UNIV CALIF, 73-, VCHMN DEPT, 70-, CUR MARINE VERT, 58-, CHMN, GRAD DEPT, 80- *Mem:* AAAS; Soc Syst Zool; Am Soc Ichthyol & Herpet; Soc Study Evolution; Am Soc Zool; Sigma Xi. *Res:* Systematics, evolution, ecology and zoogeography of fishes. *Mailing Add:* 5160 Middleton Rd San Diego CA 92109

ROSENBLATT, ROGER ALAN, b Denver, Colo Aug 8, 45; m; c 2. FAMILY MEDICINE. *Educ:* Harvard Univ, BA, 67, MD & MPH, 71; Am Bd Family Pract, dipl, 74. *Prof Exp:* Intern, Affil Hosp, Univ Wash, 71-72, resident, 72-74, from clin instr to assoc prof, Sch Med, 74-85, PROF & VCHMN, DEPT FAMILY MED, SCH MED, UNIV WASH, 85- *Concurrent Pos:* Sr surg, USPHS, 74-77; regional med consult, Nat Health Serv Corps, 74-76, dir, 76-77; prin investr, USPHS, 77-83, NSF, 83-85 & NIH, 83-84; adj asst prof, Sch Pub Health, 78-81, adj assoc prof, 81-85; med staff mem, Hosp, Univ Wash, Seattle, 78-; vis lectr, Ben-Gurion Univ Negev, Israel, 79, Royal Australian Col Gen Practitioners, 84 & Univ Queensland, Australia, 87; consult, USDA, 79-80, HEW, 79, Off Sci & Technol, 80 & State Alaska, 81, Nat Rural Health Care Asn, 83-85; NIH sr int fel award, 83-84; vis prof, Univ Auckland, NZ, 83-84 & Univ Calgary, 88; co-founder & mem, Citizens for Qual Pub Schs & Parents Educ Union Seattle, 89- *Mem:* Inst Med-Nat Acad Sci; Am Acad Family Pysicians; Am Pub Health Asn. *Res:* Public health; author of numerous technical publications. *Mailing Add:* Dept Family Med Univ Wash HQ 3 Seattle WA 98195

ROSENBLITH, WALTER ALTER, b Vienna, Austria, Sept 21, 13; nat US; m 41; c 2. BIOPHYSICS. *Educ:* Univ Bordeaux, ing radiotelegraphiste, 36; Ecole Superieure d'Elec, Paris, ing radioelectricien, 37. *Hon Degrees:* ScD, Univ Pa, 76 & SDak Sch Mines & Technol, 80; Dr, Fed Univ Rio de Janeiro, 76; ScD, Brandeis Univ, 88. *Prof Exp:* Res engr, France, 37-39; res asst physics, NY Univ, 39-40; from asst prof to assoc prof, SDak Sch Mines, 43-47 & actg head dept physics; res fel, Psychoacoust Lab, Harvard Univ, 47-51; from assoc prof to prof commun biophys, Mass Inst Technol, 51-84, staff mem res lab electronics, 51-69, chmn fac, 67-69, from assoc provost to provost, 69-80, INST PROF, MASS INST TECHNOL, 75- *Concurrent Pos:* Lectr otol, Harvard Med Sch & Mass Eye & Ear Infirmary, 57-; inaugural lectr, Tata Inst Fundamental Res, Bombay, 62; mem sci adv bd, US Air Force, 61-62; consult, Life Sci Panel, President's Sci Adv Comt, 61-66; consult, WHO, 64-65; chmn comt electronic comput in life sci, Nat Acad Sci/Nat Res Coun, 60-64, mem brain sci comt, 65-68, chmn, 66-67; mem cent coun exec comt, Int Brain Res Orgn, 60-68, hon treas, 62-67; mem coun, Int Union Pure & Appl Biophys, 61-69 & pres comn biophys commun & control processes, 64-69; mem bd med, Nat Acad Sci, 67-70; mem, President's Comn Urban Housing, 67-68, Selection Comt John & Alice Tyler Prize Environ Achievement, 73- & Coun Foreign Rels, 83-; chmn sci adv coun, Callier Ctr Commun Disorders, 68-85; mem bd gov, Weizmann Inst Sci, 73-86; dir, Kaiser Industs, 68-76; mem adv comt dir, NIH, 70-74; mem comt on scholarly commun, People's Repub China, 77-86; mem, Bd Foreign Scholar, 78-81, chmn, 80-81; co-chmn comt for study on saccharin & food safety policy, Nat Res Coun-Inst Med, 78-79; mem bd trustees, Brandeis Univ, 79-; chmn res comt, Health Effects Inst, 81-89, mem bd, 89-; mem, Adv Panel Int Educ Exchange, US Info Agency, 82-86; vpres, Int Coun Sci Unions, 84-88; chmn, int adv panel of Chinese Univ Develop Project, 86-; consult, Carnegie Corp NY, 86- *Honors & Awards:* Weizmann lectr, Weizmann Inst Sci, Israel, 62. *Mem:* Nat Acad Sci (foreign secy, 82-86); Nat Acad Eng; Inst Med-Nat Acad Sci; Inst Elec & Electronics Engrs; Biophys Soc; AAAS; Am Acad Arts & Sci; Acoust Soc Am; World Acad Arts & Sci. *Res:* Electrical activity of the nervous system and brain function; sensory communication; science and technology in the university and society. *Mailing Add:* Rm E51-211 Mass Inst Technol Cambridge MA 02139

ROSENBLOOM, ALFRED A, JR, b Pittsburgh, Pa, Apr 5, 21; m; c 2. OPTOMETRY. *Educ:* Pa State Col, BA, 42; Ill Col Optom, OD, 48; Univ Chicago, MA, 53. *Hon Degrees:* DOS, Ill Col Optom, 54. *Prof Exp:* Instr neural physiol, 46-48, dir contact lens & subnorm vision clins, 47-48, lectr, 49-54, dean, 55-72, PRES, ILL COL OPTOM, 72-, PROF OPTOM, 77- *Concurrent Pos:* Am Optom Found fel, 52-54; reading clinician & res assoc, Dept Educ, Univ Chicago, 53-54; consult, Chicago Lighthouse for Blind, 55-; chmn adv res coun, Am Optom Found, 69-76; lectr, Brit Optical Asn, 70; mem, Optom Exten Prog; mem optom rev comt, Bur Health Prof Educ & Manpower, US Dept Health, Educ & Welfare, Region III; mem state contract task force, Am Optom Found. *Mem:* AAAS; fel Am Acad Optom; Am Optom Asn; Nat Soc Study Educ; Asn Schs & Cols Optom. *Res:* Visual problems of children and youth and the partially sighted; reading problems. *Mailing Add:* Chicago Lighthouse for Blind 1850 W Roosevelt Rd Chicago IL 60608

ROSENBLOOM, ARLAN LEE, b Milwaukee, Wis, Apr 15, 34; m 58; c 4. PEDIATRICS, ENDOCRINOLOGY. *Educ:* Univ Wis-Madison, BA, 55, MD, 58. *Prof Exp:* Intern, Los Angeles County Gen Hosp, 58-59; resident, Ventura County Hosp, Calif, 59-60; physician & chief, Medico Hosp, Kratie, Cambodia, 60-61, med officer, Medico, Inc, Pahang, Malaysia, 61-62; resident pediat, Univ Wis-Madison, 62-63, chief resident, 64-65; tech adv epidemiol, Commun Dis Ctr, USPHS, 66-68; dir clin res ctr, 74-80, from asst prof to assoc prof, 68-74, PROF PEDIAT ENDOCRINOL, COL MED, UNIV FLA, 74-; DIR DIABETES RES EDUC & TREAT CTR, 79- *Concurrent Pos:* USPHS training grants pediat & endocrinol, Univ Wis-Madison, 63-64, 65-66. *Mem:* Am Diabetes Asn; Am Acad Pediat; Pediat Endocrine Soc; Soc Pediat Res; Endocrine Soc; Am Pediat Soc; fel Am Col Epidemiol. *Res:* Natural history of diabetes mellitus, including acute and long term complications; growth hormone receptor deficiency in Ecuador, clinical genetic and metabolic studies. *Mailing Add:* Dept Pediat J296JHMHC Univ Fla Col Med Gainesville FL 32610

ROSENBLOOM, ARNOLD, engineering, for more information see previous edition

ROSENBLOOM, JOEL, b Denver, Colo, July 18, 35; m 58; c 2. BIOCHEMISTRY, BIOPHYSICS. *Educ:* Harvard Univ, AB, 57; Univ Pa, MD, 62, PhD(biochem), 65. *Prof Exp:* Asst prof biochem & med, Sch Med, 65-71, assoc prof biochem, Sch Dent Med, 71-74, PROF EMBRYOL & HISTOL & CHMN DEPT, SCH DENT MED, UNIV PA, 74- *Concurrent Pos:* Biochemist, Clin Res Ctr, Philadelphia Gen Hosp, 65-73; dir, Ctr Oral Health Res, 78- *Mem:* AAAS; NY Acad Sci; Am Chem Soc; Biophys Soc; Am Soc Biol Chemists. *Res:* Structure of macromolecules; ultracentrifugation; structure and biosynthesis of macromolecules, particularly collagen and elastin. *Mailing Add:* Dept Anat Dent Al Univ Pa Philadelphia PA 19104-6002

ROSENBLOOM, PAUL CHARLES, b Portsmouth, Va, Mar 31, 20; m 48, 55; c 2. MATHEMATICS. *Educ:* Univ Pa, AB, 41; Stanford Univ, PhD(math), 44. *Prof Exp:* Asst math, Stanford Univ, 41-43; from instr to asst prof, Brown Univ, 43-46; from asst prof to assoc prof, Univ Syracuse, 46-51; from assoc prof to prof, Univ Minn, 51-65; PROF MATH, TEACHERS COL, COLUMBIA UNIV, 65- *Concurrent Pos:* Dir math sect, Minn Nat Lab, State Dept Educ & Minn Sch Math & Sci Ctr, 58-65; Guggenheim fel & asst prof, Univ Lund, 47-48; vis prof, Columbia Univ, 52, Univ Kans, 53 & Harvard Univ, 56; mem, Inst Advan Study, 53; mem, Adv Panel Math Sci, NSF, 54-57; mem, Div Math, Nat Res Coun, 58-; vis prof, Univ Denver, 70, Sir George Williams Univ, Montreal, 72-74 & Israel Inst Technol, 78-79. *Honors & Awards:* Frechet Prize, Math Soc France, 50. *Mem:* Am Math Soc; Inst Advan Study; Math Asn Am. *Res:* Mathematical logic; function of complex variables; absolutely monotonic functions; Banach spaces; differential equations; mathematical education. *Mailing Add:* Dept Math & Sci Ed Teachers Col Columbia Univ 525 W 121st St Box 210 New York NY 10027

ROSENBLUM, ANNETTE TANNENHOLZ, b Brooklyn, NY, Oct 3, 42; m 66; c 2. SCIENCE POLICY, CHEMICAL SCIENCES. *Educ:* Queens Col, BS, 64; Univ Rochester, MS, 67; Ohio State Univ, PhD(org chem), 71. *Prof Exp:* Teaching asst chem, Univ Rochester, 64-66; teaching asst, Ohio State Univ, 66-67, res asst, 67-71; staff writer chem sci, News Serv, Am Chem Soc, Washington, Dc, 72-74, asst for res & coordr govt affairs, 74-79, asst to dir, Dept Pub Affairs, 79-84, MGR, OFF SCI POLICY ANALYSIS, AM CHEM SOC, WASHINGTON, DC, 84- *Concurrent Pos:* NY State Col teaching fel, Univ Rochester, 64-66; res chemist, Naval Ord Lab, Dept Navy, 72; interim co-adminr govt affairs, Dept Chem & Pub Affairs, Am Chem Soc, 77-78. *Mem:* AAAS; Am Chem Soc. *Res:* Science policy; legislation; federal regulations; chemical safety and health; environmental improvement. *Mailing Add:* 12008 Trailridge Dr Potomac MD 20854

ROSENBLUM, ARTHUR H, pediatrics, allergy; deceased, see previous edition for last biography

ROSENBLUM, BRUCE, b New York, NY, May 20, 26; m 82; c 3. PHYSICS. *Educ:* NY Univ, BS, 49; Columbia Univ, PhD(physics), 59. *Prof Exp:* Res physicist, Dept Physics, Univ Calif, Berkeley, 57-58; mem tech staff, Radio Corp Am Labs, 58-65, head, Gen Res Group, 65-66; dep provost, Stevenson Col, 68-73, assoc dir, Ctr Innovation & Entrepreneural Develop, 78-83, PROF PHYSICS, UNIV CALIF, SANTA CRUZ, 66-, CHMN DEPT, 66-70, 89- *Concurrent Pos:* VPres, Rev & Critique, 80-; consult. *Mem:* Am Phys Soc; Am Asn Physics Teachers; AAAS. *Res:* Molecular physics; microwave spectroscopy; plasmas and transport in semiconductors; superconductivity; biophysics; fundamentals of quantum mechanics. *Mailing Add:* Dept Physics Univ Calif Santa Cruz CA 95064

ROSENBLUM, CHARLES, b Brooklyn, NY, Sept 19, 05; m 33; c 1. PHYSICAL CHEMISTRY. *Educ:* Univ Rochester, BS, 27; Univ Minn, PhD(phys chem), 31. *Prof Exp:* Asst chem, Univ Minn, 27-30, assoc, 31-35; fel, Comn Relief Belg Educ Found Louvain, 35-37; instr chem, Princeton Univ, 37-41; sr chemist, Merck & Co, Inc, 41-42; sect head, Kellex Corp, NY, 44; sect head, Merck & Co, Inc, 46-50, head radioactivity lab, 50-65, sr investr & dir, 65-70; VIS SR RES BIOCHEMIST, PRINCETON UNIV, 70- *Honors & Awards:* Levy Medal, Franklin Inst, 40. *Mem:* Fel AAAS; Am Chem Soc; Soc Exp Biol & Med; fel Am Inst Chemists; Belg Am Educ Found. *Res:* Photochemistry; radiochemistry; radioactive indicators; activation analysis; adsorption; physical methods of analysis; structure of precipitates; stability of pharmaceuticals. *Mailing Add:* Pennewood Village Apt F114 Rte 413 Newtown PA 18940

ROSENBLUM, EUGENE DAVID, b Brooklyn, NY, Oct 13, 20; m 56; c 2. MICROBIOLOGY, GENETICS. *Educ:* Brooklyn Col, BA, 41; Univ Wis, MS, 48, PhD(bact), 50. *Prof Exp:* Bacteriologist, Biol Lab, Cold Spring Harbor, 50-52; res assoc microbiol, May Inst Med Res, Cincinnati, 52-53; from asst prof to assoc prof, 53-74, dir microbiol grad prog, 72-86, PROF MICROBIOL, UNIV TEX HEALTH SCI CTR, DALLAS, 53- *Mem:* AAAS; Am Soc Microbiol; Sigma Xi. *Res:* Genetics of pathogenic bacteria; antibiotic resistance; extrachromosomal inheritance; bacteriophage. *Mailing Add:* Dept Microbiol Univ Tex Southwestern Med Ctr Dallas TX 75235

ROSENBLUM, HAROLD, b Paterson, NJ, Mar 30, 18; m 41; c 3. ELECTRONICS ENGINEERING. *Educ:* Cooper Union, BChE, 43; NY Univ, MEE, 51. *Prof Exp:* Head, Radar Systs Sect, NY Naval Shipyard, 47-54; head, Flight Trainers Br, Naval Training Systs Ctr, 54-57, Air Tactics Br, 57-60, Strike-Air Defense Systs Trainers Div, 60-65 & Aerospace Systs Trainers Dept, 65-67, asst tech dir eng, 67-69, dep dir eng, 69-74; dir tech sales, Appl Devices Corp, Kissimmee, Fla, 77-78; sr staff consult, Link Simulation Systs Div, Singer Co, Silver Spring, Md, 80-87; MGT ENG CONSULT, 74- *Mem:* Sr mem Inst Elec & Electronics Engrs; NY Acad Sci; Sigma Xi. *Res:* Training devices, research techniques and instructional technology for simulation of air, surface underwater and land warfare weapon systems. *Mailing Add:* 1325 Classic Dr Longwood FL 32779-5816

ROSENBLUM, HOWARD EDWIN, b Brooklyn, NY, Apr 28, 28; m 50; c 4. COMMUNICATIONS ENGINEERING, ACOUSTICS. *Educ:* City Col New York, BS, 50. *Prof Exp:* Chief proj engr commun systs, Sanders Assocs, 61-62; div chief res & develop secure speech systs, Nat Security Agency, Ft George Meade, 63-71, asst dir res & develop, 71-73, asst dep dir res & eng, 73-74, dep dir, 74-78, dep dir commun security, 78-83; VPRES & GEN MGR, UNISYS DEFENSE SYSTS, 83- *Concurrent Pos:* Consult, Defense Sci Bd, 76-77; mem, Mil Commun Electronics Bd & Defense Telecommun Coun, 78- *Honors & Awards:* Except Civilian Serv Award, Nat Security Agency, 72. *Mem:* AAAS; Armed Forces Commun-Electronics Asn. *Res:* Cryptography; speech compression and digitalization; microelectronics; computer and information science. *Mailing Add:* 1809 Franwell Ave Silver Spring MD 20902

ROSENBLUM, IRWIN YALE, b Youngstown, Ohio, Nov 26, 42; m 78; c 1. INSULIN-GROWTH FACTOR. *Educ:* Pa State Univ, BS, 64; Univ Wis, PhD(biochem), 69. *Prof Exp:* Res fel, Univ Hawaii, 69-70; joint res assoc, Med Ctr & Univ Ariz, 75-76; sr res scientist, Miami Valley Labs, Proctor & Gamble Co, 71-78; asst prof toxicol, Albany Med Col, Union Univ, 78-81; res dir, Large Primate Div, White Sands Res Ctr, 81; ASSOC PROF PHARMACOL & TOXICOL, PHILADELPHIA COL PHARM & SCI, 82- *Concurrent Pos:* Dir clin chem, Int Ctr Environ Safety, 78-80; adj prof, Park Col, 79-81; consult, Coulston Int Corp, 81; adj prof, Drexel Univ, 85. *Mem:* Soc Toxicol; Am Soc Primatologists; Am Diabetes Asn; Int Primatol Soc; Sigma Xi. *Res:* Identification and characterization of insulin and insulin-like growth factor receptors in early preimplantation embryos; chemical-induced genotoxicity in male germ cells. *Mailing Add:* Philadelphia Col Pharm & Sci 43rd St & Kingsessing Mall Philadelphia PA 19104

ROSENBLUM, LAWRENCE JAY, b New York, NY, Jan 25, 44; m 66; c 2. MATHEMATICAL MODELING. *Educ:* Queens Col, NY, BA, 64; Ohio State UniV, MS, 66, PhD(math), 71. *Prof Exp:* Mathematician, Dept Navy, 71-73, prog mgr ocean environ, 73-77; MATHEMATICIAN, NAVAL RES LAB, 77- *Concurrent Pos:* Ed, newsletter, Comt Oceanic Eng & Technol, Inst Elec & Electronics Engrs, 85- *Mem:* Inst Elec & Electronics Engrs Computer Soc; Am Geophys; Asn Comput Mach. *Res:* Real-time computer systems in oceanography; mathematical modeling of oceanographic phenomena; digital imaging of oceanographic data. *Mailing Add:* Naval Res Lab Code 5170 4555 Overlook Ave SW Washington DC 20375

ROSENBLUM, LEONARD ALLEN, b Brooklyn, NY, May 18, 36; m 56; c 2. ETHOLOGY. *Educ:* Brooklyn Col, BA, 56, MA, 58; Univ Wis-Madison, PhD, 61. *Prof Exp:* Res asst psychol, Brooklyn Col, 56-58; res asst psychol, Univ Wis, 58-61, teaching fel, 61; from instr to assoc prof, 61-72, PROF PSYCHIAT, STATE UNIV NY DOWNSTATE MED CTR, 72- *Concurrent Pos:* NIMH career develop award, 64-71; reviewer small grants, NIMH, 74-77, basic behavioral sci, 78-82. *Mem:* Am Psychol Asn; Am Soc Primatol; Int Acad Sex Res; Int Primatol Soc; Int Soc Develop Psychobiol (pres, 80); Sigma Xi. *Res:* Mother-infant relations and effects of early experience on development in primates; sexual behavior and its development and control. *Mailing Add:* Dept Psychiat 450 Clarkson Ave Brooklyn NY 11203

ROSENBLUM, MARTIN JACOB, b Stamford, Conn, Dec 6, 28; m 55; c 2. COMPUTER SCIENCE, APPLIED MATHEMATICS. *Educ:* Yale Univ, BA, 50, MA, 51; State Univ NY Stony Brook, MS, 75; Stevens Inst Technol, MS, 81. *Prof Exp:* Asst physics, Yale Univ, 54-59, res assoc, 59-61; from assoc physicist to physicist, Brookhaven Nat Lab, 61-75; mem tech staff, Bell Labs, 75-80; MEM STAFF INFO SYSTS, WESTERN ELEC CO, 80- *Concurrent Pos:* Ford fel, Europ Orgn Nuclear Res, Switz, 63-64. *Mem:* AAAS; Asn Comput Mach; Inst Elec & Electronics Engrs. *Res:* Computer sciences; communication science; digital systems; data analysis. *Mailing Add:* AT&T LC 4W-H16 184 Liberty Corner Rd Warren NJ 07060

ROSENBLUM, MARVIN, b Brooklyn, NY, June 30, 26; m 59; c 5. MATHEMATICS. *Educ:* Univ Calif, BS, 49, MA, 51, PhD(math), 55. *Prof Exp:* Actg instr math, Univ Calif, 54-55; from asst prof to assoc prof, 55-65, chmn dept, 69-72, prof math, 65-78, COMMONWEALTH PROF MATH, UNIV VA, 78- *Concurrent Pos:* Mem, Inst Advan Study, 59-60. *Mem:* Am Math Soc; Math Asn Am; Soc Indust & Appl Math. *Res:* Hilbert space; harmonic analysis. *Mailing Add:* Dept Math Univ Va Math Astr Bldg 206 Charlottesville VA 22903

ROSENBLUM, MYRON, b New York, NY, Oct 20, 25; m 58. ORGANIC CHEMISTRY. *Educ:* Columbia Univ, AB, 49; Harvard Univ, AM, 50, PhD(chem), 54. *Prof Exp:* Res assoc, Columbia Univ, 53-55; asst prof chem, Ill Inst Technol, 55-58; from asst prof to assoc prof, 58-66, PROF CHEM, BRANDEIS UNIV, 66- *Concurrent Pos:* Guggenheim fel, 65-66; vis prof, Israel Inst Technol, 66. *Mem:* Am Chem Soc; Royal Soc Chem. *Res:* Organometallic chemistry of the transition elements; reaction mechanisms; synthesis. *Mailing Add:* Dept Chem Brandeis Univ Waltham MA 02154

ROSENBLUM, SAM, b New York, NY, Jan 25, 23; m 47; c 2. GEOCHEMISTRY. *Educ:* City Col New York, BS, 49; Stanford Univ, MS, 51. *Prof Exp:* Geologist, US Bur Reclamation, 51; geologist, Br Mineral Deposits, 52-57, Br Foreign Geol, Taiwan, 57-61, Geochem Census Unit, 61-63, Br Mil Geol, 63-65, Br Foreign Geol, Bolivia, 65-67, Off Int Geol, Liberia, 67-72, mineralogist, Br Explor Res, US Geol Surv, 72-80; CONSULT GEOL, 81- *Mem:* Geol Soc Am; Mineral Soc Am; Int Asn Geochem & Cosmochem; Soc Environ Geochem & Health; Asn Explor Geochemists. *Res:* Mineralogy; petrography; geochemistry; economic and exploration geology; rare-earth element mineralogy; application of geochemistry to health studies. *Mailing Add:* 12165 W Ohio Pl Lakewood CO 80228

ROSENBLUM, STEPHEN SAUL, b Brooklyn, NY, Sept 26, 42; m 72; c 2. EXPERIMENTAL SOLID STATE PHYSICS, PARTICLE ACCELERATOR DEVELOPMENT. *Educ:* Columbia Univ, AB, 63; Univ Calif, Berkeley, PhD(chem), 69. *Prof Exp:* Res fel physics, Calif Inst Technol, 69-70; guest lectr physics, Freie Univ, Berlin, 70-72; vis staff mem, Hahn-Meitner Inst, Berlin, 72-74; staff mem cryogenics, Los Alamos Sci Lab, 74-77; staff scientist physics, Lawrence Berkeley Lab, 77-78, staff scientist elec eng, 78-85; sr engr, Varian Assocs, 85-91; SR RESEARCHER, KOBE DEVELOP CORP, 91- *Mem:* Am Phys Soc; Am Chem Soc; affil Inst Elec & Electronics Engrs; fel Am Inst Chemists; Am Vacuum Soc. *Res:* Investigation of plasma chemistry and physics relevant to semiconductor processing; design of particle accelerators, particularly ion sources; study of magnetic mats. *Mailing Add:* Kobe Develop Corp 777 California Ave Suite 200 Palo Alto CA 94304

ROSENBLUM, WILLIAM I, b New York, NY, July 6, 35; m 58; c 3. NEUROPATHOLOGY, PATHOLOGY. *Educ:* Swarthmore Col, BA, 57; NY Univ, MD, 61. *Prof Exp:* USPHS fel path & neuropath & intern & resident path, Sch Med, NY Univ, 61-66; assoc prof neuropath, Northwestern Univ, 68-69; PROF PATH & CHMN, DIV NEUROPATH, MED COL VA, 69-, VCHMN, DEPT PATH, 79- *Concurrent Pos:* Res assoc, Nat Inst Neurol & Commun Disorders & Stroke & assoc pathologist, Clin Ctr, NIH, 66-68; asst pathologist, Passavant Mem Hosp, Chicago, 68-69; consult neuropath, Evanston Hosp, 68-69; mem exec comt, Coun on Stroke, Am Heart Asn. *Mem:* Am Asn Path & Bact; Am Asn Neuropath; Am Soc Exp Path; Am Physiol Soc; Microcirc Soc; Sigma Xi. *Res:* Cerebral circulation in health and disease. *Mailing Add:* Box 17 Div Neuropath Med Col Va Richmond VA 23298-0017

ROSENBLUM, WILLIAM M, US citizen. PHYSICS, OPTOMETRY. *Educ:* Univ Miami, BS, 58; Fla State Univ, MS, 60; Tufts Univ, PhD(physics), 67. *Prof Exp:* Asst, Fla State Univ, 58-60; instr physics, Univ Miami, 60-62 & Tufts Univ, 62-66; sr scientist, Phys Optics Dept, Tech Opers, Inc, 66-68, Optical Eng Dept, Polaroid Corp, Mass, 68-69 & NASA Electronics Res Ctr, 69-70; ASSOC PROF PHYSICS & OPTOM, UNIV ALA, BIRMINGHAM, 70- *Concurrent Pos:* Instr, Harvard Exten Ser, 67-68 & Northeastern Univ, 69-70. *Mem:* Am Phys Soc; Optical Soc Am. *Res:* Optical design and fabrication. *Mailing Add:* Dept Optom Univ Ala Med Ctr Birmingham AL 35294

ROSENBLUTH, JACK, b New York, NY, Nov 8, 30; m 60; c 3. CYTOLOGY. *Educ:* Columbia Univ, AB, 52; NY Univ, MD, 56. *Prof Exp:* Intern & asst resident med, Bellevue Hosp, New York, 56-58; Nat Found fel, Nat Inst Neurol Dis & Blindness, 58-59, sr asst surgeon, 59-61; USPHS spec fel, Med Ctr, Univ Calif, San Francisco, 61-62; instr anat, Harvard Med Sch, 62-63; asst prof, Albert Einstein Col Med, 63-66; assoc prof, 66-71, PROF PHYSIOL, SCH MED, NY UNIV, 71- *Concurrent Pos:* Mem, Neurol Study Sect B, NIH, 79-; vis prof, Univ London, 81-82. *Honors & Awards:* Javits Neurosci Invest Award, 85. *Mem:* AAAS; Am Asn Anatomists; Soc Neurosci; Am Soc Cell Biol. *Res:* Comparative cytology and physiology of nerve and muscle tissues; neurobiology. *Mailing Add:* Dept Physiol Sch Med NY Univ 550 First Ave New York NY 10016

ROSENBLUTH, MARSHALL N, b Albany, NY, Feb 5, 27; m 51; c 2. THEORETICAL PHYSICS. *Educ:* Harvard Univ, BS, 45; Univ Chicago, PhD, 49. *Prof Exp:* Instr physics, Stanford Univ, 49-50; mem staff, Los Alamos Sci Lab, Univ Calif, 50-56; sr res adv, Gen Dynamics Corp, 56-60; prof physics, Univ Calif, San Diego, 60-67; vis prof astrophys sci, Sch Natural

Sci, Inst Advan Study, Princeton Univ, 67-80, vis res physicist, Plasma Physics Lab, 67-80; with Inst Fusion Studies, Univ Tex, 80-; at dept physics, Princeton Univ; PROF PHYSICS, UNIV CALIF, SAN DIEGO, 87- *Mem:* Nat Acad Sci; fel Am Phys Soc. *Res:* Physics of plasmas. *Mailing Add:* Dept Physics Univ Calif La Jolla CA 92093

ROSENBLUTH, SIDNEY ALAN, b Deport, Tex, Nov 20, 33; m 62; c 2. PHARMACY. *Educ:* Univ Okla, BS, 55; Univ Tex, MS, 62, PhD(pharm), 66. *Prof Exp:* Pharmacist, Swindle Pharm, Tex, 55-56; hosp pharm resident, Med Ctr Pharm Serv, Univ Ark, 60-61; exchange pharmacist, Univ Hosp Pharm, Copenhagen, 61-62; res assoc, Drug-Plastic Res & Toxicol Lab, Univ Tex, 62-66; res fel, Univ Bath, 66; from asst prof to assoc prof pharmaceut, Col Pharm, Univ Tenn, Memphis, 66-71, prof, 71-75, asst dean clin affairs & chief clin pharm, 73-75, asst dean student affairs, 75-79, assoc dean, 79-81; PROF & DEAN, SCH PHARM, WVA UNIV, 81- *Concurrent Pos:* Mead Johnson Labs res grant pharm, 68-69; NIMH res grant & prin investr, 71-73, 76-79, 78-82; Am Cancer Soc, Tenn Div & Regional Med Prog, res grant & prin investr, 73-75; Vet Admin, training grant & prin investr, 74-78; NIMH, training grant & prin investr, 74-78; NIH Bur Health Manpower Educ, res grant & prin investr, 75-78; training grant & prin investr, NIMH, 76-80 & 78-82 & Dept Educ, 80-83. *Mem:* AAAS; Am Acad Pharmaceut Sci; Tissue Cult Asn; NY Acad Sci. *Res:* Use of tissue cultures in investigations of pharmacological and toxicological actions of drugs and chemicals; development and evaluation of new roles and education units for pharmacists in health care delivery. *Mailing Add:* 141 Poplar Dr Morgantown WV 26505

ROSENBROOK, WILLIAM, JR, b Omaha, Nebr, Mar 28, 38; m 69; c 2. ORGANIC CHEMISTRY. *Educ:* Univ Omaha, BA, 60; Mont State Univ, PhD(org chem), 64. *Prof Exp:* Fel, Univ Calif, Berkeley, 64-65; sr res chemist, Dept Biochem Res, 65-70, Dept Microbiol Chem, 70-74, Dept Chem Res, 74-82, SR RES CHEMIST, ANTI-INFECTIVE DIV, ABBOTT LABS, 82- *Mem:* AAAS; Am Chem Soc; Sigma Xi. *Res:* Natural products chemistry; isolation, characterization and modification of biologically active microbial products; synthetic organic chemistry. *Mailing Add:* Anti-Infective Res Div Abbott Labs North Chicago IL 60064

ROSENBURG, DALE WEAVER, b Hannibal, Mo, Dec 2, 27; m 54; c 3. ORGANIC CHEMISTRY. *Educ:* Culver-Stockton Col, AB, 50; Univ Mo, AM, 51, PhD(org chem), 58. *Prof Exp:* Jr chemist, Merck & Co, Inc, NJ, 51-55, chemist, Pa, 58-60, sr chemist & sect leader process develop org chem, Va, 60-64, sect leader process res org chem, NJ, 64-66; from asst prof to assoc prof, 66-73, PROF CHEM, NORTHWEST MO STATE UNIV, 73- *Mem:* Am Chem Soc; Sigma Xi. *Res:* Process research and development in the synthesis and manufacture of organic compounds of biological interest. *Mailing Add:* 617 Arkansas Ave Ordway CO 81063-1141

ROSENCRANS, STEVEN I, b Brooklyn, NY, Mar 13, 38; m 67; c 5. MATHEMATICS. *Educ:* Mass Inst Technol, SB, 60, PhD(math), 64. *Prof Exp:* Instr math, Mass Inst Technol, 64-65; from asst prof to assoc prof, 65-75, PROF MATH, TULANE UNIV, 75- *Concurrent Pos:* Vis assoc prof, Univ NMex, 71-72. *Res:* Mathematical physics; partial differential equations; stochastic processes. *Mailing Add:* Dept Math Tulane Univ 410 Gibson Hall 6823 St Charles Ave New Orleans LA 70118

ROSENCWAIG, ALLAN, b Poland, Jan 1, 41; US citizen. PHYSICS. *Educ:* Univ Toronto, BASc, 63, MA, 65, PhD(physics), 69. *Prof Exp:* Mem tech staff solid state res, Bell Labs, 69-76; sr scientist, Gilford Instrument Labs, 76-77; physicist, Lawrence Livermore Labs, 77-82; PRES, THERMA-WAVE INC, 82- *Concurrent Pos:* Ed, J Photoacoust. *Mem:* Am Chem Soc; Am Optical Soc; Am Phys Soc. *Res:* Thermal-wave imaging; photoacoustics. *Mailing Add:* 134 Timberlane Ct Oanville CA 94526-4210

ROSENDAHL, BRUCE RAY, b Jamestown, NY, Dec 28, 46; m 69; c 2. CONTINENTAL RIFTS, ANALYSIS SEISMIC DATA STRUCTURE & STRATIGRAPHY. *Educ:* Univ Hawaii, BS, 70, MS, 72; Univ Calif, San Diego, PhD(earth sci), 76. *Prof Exp:* Postodctoral fel geophys, Scripps Inst Oceanog, 76; from asst prof to prof geophys, Dept Geol, Duke Univ, 76-89; DIR, PROJ PROBE, GEOPHYS, INC, 82-; WEEKS PROF GEOPHYS, ROSENSTEIL SCH MARINE & ATMOSPHERIC SCI, UNIV MIAMI, 89-, DEAN, 89- *Concurrent Pos:* Consult, World Bank, Maj Petrol Cos, 76-; adv, Nat Geol Mag, 87-89, PBS Nova TV Spec, 88-89 & Time-Life Bks, 88-89; vpres & chief operating officer, Int Oceanog Found, 89-; tech ed, Sea Frontiers Mag, 89-; dean, Rosentiel Sch Marine & Atmospheric Sci, Univ Miami, 89-; bd dirs, Cono-Sur, Inc, 89-; bd gov, Joint Oceanog Insts, 89-; bd contributors, Miami Herald, 90- *Mem:* Joint Oceanog Insts; Am Geophys Union; Am Asn Petrol Geologists. *Res:* Rifted plate margins and continental rifting using exploration seismic methods; authored more than 100 scientific publications on subjects like Tectono-Stratigraphy, seismic models, and the geometry of rifting; continental rifts and paleo-rift margins-project Probe. *Mailing Add:* 4600 Rickenbacker Causeway Miami FL 33149

ROSENDAHL, GOTTFRIED R, b Borna, Ger, Mar 25, 11; nat US; m 39; c 4. OPTICAL SYSTEMS & LENS DESIGN. *Educ:* Dresden Tech, Dipl Eng, 35, Dr Eng, 38; Johns Hopkins Univ, MA, 36. *Prof Exp:* Res asst, Dresden Tech, 38-40; head optics lab, E Leitz, Inc, Ger, 40-42, sr scientist, Peenemuende, Ger, 43-46; head optics lab, E Leitz, Inc, 46-53; consult physics, Air Defense Command, Holloman AFB, NMex, 53-54; tech dir, E Leitz, Inc, NY, 54-57; sr scientist, Ball Bros Res Corp, Colo, 58-60, Gen Elec Co, 60-62, Fed Lab, Int Tel & Tel Corp, 63-64 & Link Group, Gen Precision, Inc, NY, 64-67; res physicist, Naval Training Equip Ctr, 67-78; RETIRED. *Concurrent Pos:* Lectr, Univ Pa, 61-62, Exten Ctr, Purdue Univ, 63-64, State Univ NY Binghamton, 66-67 & Univ Cent Fla, 79, 80 & 87. *Mem:* Fel Optical Soc Am. *Res:* Physical and geometrical optics; optical systems and instruments; photometry and radiometry; metrology. *Mailing Add:* 2079 Penguin Ct Oviedo FL 32765-8549

ROSENE, ROBERT BERNARD, organic chemistry, for more information see previous edition

ROSENE, WALTER, JR, b Ogden, Iowa, May 15, 12; m 37; c 2. WILDLIFE ECOLOGY, WILDLIFE MANAGEMENT. *Educ:* Iowa State Univ, BS, 34; Auburn Univ, MS, 37. *Prof Exp:* Game technician, Iowa State Planning Bd, 34-35; technician, Ala State Conserv Comn, 38; proj biologist, Soil Conserv Serv, USDA, 38-39, area biologist, 39-40, conservationist, 40-46; biologist, US Fish & Wildlife Serv, 46-69; CONSULT WILDLIFE MGT, 69- *Concurrent Pos:* Spec writing proj, NAm Wildlife Found, 64-70; vis prof, Miss State Univ, 78. *Mem:* Wildlife Soc. *Res:* Ecology and population dynamics of the bobwhite quail and mourning dove in the southeast; evaluation of introduced plants in wildlife management; evaluation of effects of insecticides on wildlife. *Mailing Add:* 127 Oak Circle Gadsden AL 35901

ROSENFELD, ARTHUR H, b Birmingham, Ala, June 22, 26; m 55; c 3. NUCLEAR PHYSICS. *Educ:* Va Polytech Univ, BS, 44; Univ Chicago, PhD(physics), 54. *Prof Exp:* Res assoc, Inst Nuclear Studies, Univ Chicago, 54-55; from asst prof to assoc prof, 57-63, actg chmn dept comput sci, 67-68, dir particle data group, 64-75, leader re group A, 71-73, PROF PHYSICS, UNIV CALIF, BERKELEY, 63-, RES ASSOC, LAWRENCE BERKELEY LAB, 56- *Concurrent Pos:* Mem comt uses comput, Nat Acad Sci-Nat Res Coun, 62-65 & statist data panel, Physics Surv Comt, 70-; mem panel univ comput facil, NSF, 63-66, chmn, 65-66, chief investr, Univ Comput Facil Grant, 70-; mem subpanel B, High Energy Physics Adv Panel, AEC, 67-69; mem physics info comt, Am Inst Physics, 69-71. *Mem:* Fel Am Phys Soc; Fedn Am Sci. *Res:* Physics of elementary particles; use of digital computers to process data. *Mailing Add:* Lawrence Rad Lab Univ Calif Bldg 90 Rm 3028 Berkeley CA 94720

ROSENFELD, AZRIEL, b New York, NY, Feb 19, 31; m 59; c 3. COMPUTER VISION, IMAGE PROCESSING. *Educ:* Yeshiva Univ, BA, 50, MHL, 53, MS, 54, DHL, 55; Columbia Univ, MA, 51, PhD(math), 57. *Hon Degrees:* DTech, Linköping Univ, Sweden, 80. *Prof Exp:* Physicist, Fairchild Controls Corp, 54-56; engr, Ford Instrument Co, 56-59; mgr res, Budd Electronics, Inc, 59-64; RES PROF COMPUT SCI, UNIV MD, COLLEGE PARK, 64-, DIR, CTR AUTOMATION RES, 83- *Concurrent Pos:* Vis asst prof, Grad Sch Math, Yeshiva Univ, 58-63; pres, Asn Orthodox Jewish Scientists, 63-65; pres, IM Tech, Inc, 75-; dir, Mach Vision Asn, 85-88. *Honors & Awards:* Piore Award, Inst Elec & Electronics Engrs, 85; Fu Award, Int Asn Pattern Recognition, 88. *Mem:* Math Asn Am; fel Inst Elec & Electronics Engrs; Asn Comput Mach; Int Asn Pattern Recognition (pres, 80-82); fel Am Asn Artificial Intel. *Res:* Computer processing of pictorial information. *Mailing Add:* 847 Loxford Terr Silver Spring MD 20901

ROSENFELD, CARL, b Baltimore, Md, Dec 25, 44. EXPERIMENTAL HIGH ENERGY PHYSICS. *Educ:* Mass Inst Technol, BS, 66; Calif Inst Technol, PhD(physic), 77. *Prof Exp:* Res assoc, Univ Rochester, 77-81, sr res assoc, 81-84; asst prof, res, La State Univ, 84-85; ASSOC PROF, UNIV SC, 86- *Mem:* Am Phys Soc. *Res:* Elementary particles; electron positron annihilation at high energy. *Mailing Add:* Dept Physics & Astron Univ SC Columbia SC 29208

ROSENFELD, CHARLES RICHARD, b Atlanta, Ga, Aug 25, 41; m 71; c 3. NEONATAL-PERINATAL MEDICINE, DEVELOPMENTAL BIOLOGY. *Educ:* Emory Univ, MD, 66. *Prof Exp:* Intern, Yale-New Haven Hosp, 66-67; from resident to chief resident, Albert Einstein Col Med, 67-71; from asst prof to PROF, UNIV TEX SOUTHWESTERN MED SCH, 73- *Concurrent Pos:* Assoc ed, Early Human Develop. *Mem:* Soc Pediat Res; Am Pediat Soc; Am Physiol Soc; Endocrine Soc. *Res:* Cardiovascular adaptation in pregnancy. *Mailing Add:* Dept Pediat Univ Tex Southwestern Med Ctr 5323 Harry Hines Blvd Dallas TX 75235

ROSENFELD, DANIEL DAVID, b Brooklyn, NY, May 7, 33; m 57; c 2. PETROLEUM CHEMISTRY. *Educ:* Brooklyn Col, BS, 55; Univ Pa, MS, 57, PhD(org chem), 61. *Prof Exp:* Res chemist, Esso Res & Eng Co, 60-66, sr res chemist, 66-71, res assoc, 71-72, head aromatics tech serv, 72-79, HEAD, AROMATICS LAB, EXXON CHEM CO, 79- *Mem:* Am Chem Soc. *Res:* Synthesis of solid rocket propellants; reactions in aprotic solvents; synthesis and screening of agricultural pesticides; process development; ketone solvents; hydotreating desulphurization studies; catalyst evaluations; zeolite adsorption. *Mailing Add:* Exxon Chem Co PO Box 4900 Baytown TX 77522

ROSENFELD, GEORGE, b Cambridge, Mass, Nov 6, 19; m 59. MEDICAL RESEARCH, ENDOCRINOLOGY. *Educ:* Mass Inst Technol, BS, 40, MS, 41; Geogetown Univ, PhD(biochem), 52. *Prof Exp:* Asst chemist, Mass State Dept Pub Health, 41-42; asst toxicologist, Med Res Div, US War Dept, 42; mem staff, Med Sch, Cornell Univ, 43-46; officer-in-chg, Navy Chem Warfare Sch, Calif, 46-50 & Naval Med Res Inst, 50-57; assoc res physiologist, Univ Calif, Berkeley, 57-60, res biochemist, 60-63; BIOMED CONSULT, 64- *Honors & Awards:* Res Commendation, US Navy, 56. *Mem:* Endocrine Soc; Am Physiol Soc; Sigma Xi; fel AAAS. *Res:* Stress physiology; endocrinopathy; acute and chronic alcoholism; neurohormones; radiation medicine. *Mailing Add:* 50 Vista Del Mar Ct Oakland CA 94611

ROSENFELD, ISADORE, b Montreal, Que, Sept 7, 26; US citizen; m 56; c 4. MEDICINE. *Educ:* McGill Univ, BSc, 47, MD & CM, 51; FRCPS(C). *Prof Exp:* Clin asst prof, 64-71, CLIN PROF MED, CORNELL UNIV COL MED, 79-; ATTEND PHYSICIAN, NY HOSP, 78- *Concurrent Pos:* Hatch Cummings vis prof med, Methodist Hosp, Baylor Col Med, 82; mem, bd visitors, Sch Med, Univ Calif, Davis, 82; chmn, Found Biomed Res; fel Coun Epidemiol, Am Heart Asn. *Mem:* Fel Am Col Physicians; fel Am Col Chest Physicians; fel Am Col Cardiol; NY Acad Sci. *Res:* Clinical cardiology, author of various medical publications. *Mailing Add:* 125 E 72nd St New York NY 10021

ROSENFELD, JACK LEE, b Pittsburgh, Pa, June 6, 35; m 69; c 2. COMPUTER SCIENCE, TECHNICAL EDITING. *Educ:* Mass Inst Technol, SB & SM, 57, ScD(elec eng), 61. *Prof Exp:* Res staff mem comput sci, T J Watson Res Ctr, 61-89, ASSOC ED, IBM J RES & DEVELOP, IBM CORP, 89- *Concurrent Pos:* Adj prof, Columbia Univ, 65-72. *Honors &*

Awards: Silver Core, Int Fedn Info Processing, 77. *Mem:* Asn Computer Mach; Inst Elec & Electronics Engrs. *Res:* Computer architecture, office automation, microprocessor system design; parallel processing; distributed systems; object-oriented programming. *Mailing Add:* IBM J Res & Develop Rm 2B-71 500 Columbus Ave Thornwood NY 10594

ROSENFELD, JEROLD CHARLES, b New Haven, Conn, Apr 13, 43; m 67; c 2. ORGANIC CHEMISTRY, POLYMER CHEMISTRY. *Educ:* Clark Univ, BA, 65; Yale Univ, PhD(chem), 70. *Prof Exp:* SR RES CHEMIST, OCCIDENTAL PETROL CORP, GRAND ISLAND, 70- *Mem:* Am Chem Soc. *Res:* Preparation of non-burning and high temperature plastics. *Mailing Add:* 18 Willow Green Dr Amherst NY 14228-3420

ROSENFELD, JOHN L, b Portland, Ore, July 14, 20; m 43; c 2. GEOLOGY. *Educ:* Dartmouth Col, AB, 42; Harvard Univ, AM, 49, PhD(geol), 54. *Prof Exp:* Asst prof geol, Mo Sch Mines, 49-55; vis asst prof, Wesleyan Univ, 55-57; from asst prof to assoc prof, 57-70, PROF GEOL, UNIV CALIF, LOS ANGELES, 70- *Concurrent Pos:* Guggenheim fel, 63; res assoc, Harvard Univ, 71-72. *Mem:* AAAS; fel Geol Soc Am; fel Mineral Soc Am; Am Geophys Union; Sigma Xi. *Res:* Structure, petrology and stratigraphy of metamorphic rocks in western New England; application of physical and chemical properties of minerals to petrology; structural petrology; solid inclusion piezothermometry. *Mailing Add:* 2401 Arbutus Dr Los Angeles CA 90049

ROSENFELD, LEONARD M, b Philadelphia, Pa, June 28, 38; m 62; c 3. PHYSIOLOGY, BIOCHEMISTRY. *Educ:* Univ Pa, AB, 59; Jefferson Med Col, PhD(physiol), 64. *Prof Exp:* Instr physiol, Jefferson Med Col, 64-68, physiol coordr, Sch Nursing, Col Allied Health Sci, 74-75, Jefferson Med Col, 75-80, coordr, Struct & Function Course Anat & Physiol, 77-79, ASST PROF PHYSIOL, JEFFERSON MED COL, THOMAS JEFFERSON UNIV, 68-, CLIN PROF, COL ALLIED HEALTH SCI, 84- *Concurrent Pos:* Vis lectr, Pa State Univ, Ogontz Campus, 69-71, vis prof, 71-74. *Mem:* AAAS; Soc Exp Biol & Med; Am Physiol Soc; Sigma Xi. *Res:* Medical/science education; history of medicine; electrolyte interactions in metabolic systems; factors influencing the phosphatase enzyme system; environmental biology; biological effects of air pollution; acute biochemical changes following myocardial infarction; intestinal integrity; alteration in intestinal blood flow; malnutrition. *Mailing Add:* 1030 Kipling Rd Rydal PA 19046

ROSENFELD, LEONARD SIDNEY, medicine, public health; deceased, see previous edition for last biography

ROSENFELD, LOUIS, b Brooklyn, NY, Apr 8, 25. BIOCHEMISTRY, CLINICAL CHEMISTRY. *Educ:* City Col New York, BS, 46; Ohio State Univ, MS, 48, PhD(physiol chem), 52. *Prof Exp:* Asst instr physiol chem, Ohio State Univ, 52; res assoc biochem, Univ Va, 52-53; biochemist, Wayne County Gen Hosp, 54-56 & Beth-El Hosp, 56-61; asst prof, 61-67, dir chem, Univ Hosp, 61-73, ASSOC PROF CLIN PATH, MED CTR, NY UNIV, 67-, DIR SPEC CHEM, UNIV HOSP, 73- *Concurrent Pos:* Instr sci, Rutgers Univ, Newark, 57-63. *Mem:* AAAS; Sigma Xi; Asn Clin Sci; Nat Acad Clin Biochem; Am Asn Clin Chem; NY Acad Sci. *Res:* Fibrinogen; plasma coagulation; heparin; electrophoresis; protein analysis, glycosaminoglycans, thyroid hormones; history of chemistry and medicine. *Mailing Add:* 1417 E 52nd St Brooklyn NY 11234

ROSENFELD, MARTIN HERBERT, b Rockaway Beach, NY, May 30, 26; m 52; c 2. MICROBIOLOGY, CLINICAL BIOCHEMISTRY. *Educ:* Brooklyn Col, BA, 50; St John's Univ, MS, 62, PhD(microbiol), 72. *Prof Exp:* Supvr labs, A Angrist, 50-55 & Horace Harding Hosp, 55-60; lab adminr, Montefiore-Morrisania Hosp Affil, 63-70; assoc prof, 70-74, chmn div diag progs, 70-77, PROF MED TECHNOL, SCH ALLIED HEALTH PROF, STATE UNIV NY, STONY BROOK, 74-, CHMN DEPT MED TECHNOL, 70-, ASST DEAN GRAD PROGS, 73- *Concurrent Pos:* Mem & chmn nat comt instnl orgn, Asn Schs Allied Health Prof, 71-74; mem sci lab technol adv comt, State Univ NY Agr & Tech Col, Cobleskill, 71-; mem bd adv, NY City Dept Lab Supvrs, 79-83; med technologist, Am Soc Clin Pathologists, 52-; clin lab scientist, Nat Cert Agency Med Lab Personnel, 79-; consult med technol educ, Marshall Univ, WVa, 85- *Mem:* Am Asn Clin Chem; NY Acad Sci; Am Soc Microbiol; Am Asn Bioanalysts; Am Inst Biol Sci; fel Nat Acad Clin Biochem; fel Asn Clin Scientists. *Res:* Detection of microorganisms and quantitative assay of blood constituents by automated chemiluminescent technique; ratios and analysis of lipids by infra-red spectroscopy. *Mailing Add:* Div Med Technol State Univ NY Health Sci Ctr Stony Brook NY 11794

ROSENFELD, MELVIN, b New York, NY, Apr 19, 34; m 57; c 1. MATHEMATICS. *Educ:* Univ Calif, Los Angeles, BA, 56, MA, 62, PhD(math), 63. *Prof Exp:* Lectr math, Univ Calif, Los Angeles, 63-64; from lectr to asst prof, 64-70, ASSOC PROF MATH, UNIV CALIF, SANTA BARBARA, 70- *Concurrent Pos:* NSF grant, 66-67. *Mem:* Am Math Soc; Math Asn Am. *Res:* Functional analysis. *Mailing Add:* 401 Yankee Farm Rd Santa Barbara CA 93109

ROSENFELD, MELVIN ARTHUR, b Chicago, Ill, Dec 26, 18; m 46; c 2. SEDIMENTOLOGY, ACADEMIC ADMINISTRATION. *Educ:* Univ Chicago, BS, 40; Pa State Univ, MS, 50, PhD(sedimentology, statist), 53. *Prof Exp:* Instr & map draftsman, Univ Chicago, 40-42, instr map interpretation, 46; petrol geologist, H D Hadley & I E Stewart, Consult, 46-47; from res asst to res assoc sedimentary petrog, Pa State Univ, 47-53; sr res geologist, Field Res Labs, Magnolia Petrol Co, 53-57; dir explor res div, Pure Oil Co, 57-65, leader corp model opers res team, 64-65; sr scientist & mgr info processing ctr, Woods Hole Oceanog Inst, 65-81; RETIRED. *Concurrent Pos:* Mem data mgt panel, Comt Oceanog, Nat Acad Sci, 67-71; chmn panel info handling, Joint Oceanog Inst Deep Earth Sampling Proj, 68-; mem US deleg, Working Group Int Exchange Oceanog Data, Int Oceanog Comn, Rome, 71. *Mem:* AAAS; Sigma Xi. *Res:* Application of statistics and computers to geology and oceanography. *Mailing Add:* PO Box 758 Damariscotta ME 04543

ROSENFELD, NORMAN SAMUEL, b New York, NY, July 21, 34; m 78; c 3. MATHEMATICS. *Educ:* Yeshiva Univ, BA, 54; Syracuse Univ, MA, 56; Yale Univ, PhD(math), 59. *Prof Exp:* Res asst, Yale Univ, 57-58; lectr, City Col New York, 58-59; asst res scientist, Courant Inst Math Sci, NY Univ, 59-61; from asst prof to assoc prof, NY Univ, 61-68; assoc prof, 68-79, PROF MATH, YESHIVA UNIV, 79-, DEAN COL, 80- *Concurrent Pos:* Vis asst prof, Belfer Grad Sch Sci, Yeshiva Univ, 60-61. *Mem:* Am Math Soc; Math Asn Am. *Res:* Functional analysis. *Mailing Add:* Yeshiva Univ 500 W 185th St New York NY 10033

ROSENFELD, ROBERT L, b New York, NY, Aug 6, 37; m 59; c 2. APPLIED MECHANICS, COMPUTER SCIENCE. *Educ:* Mass Inst Technol, SB, 59; Calif Inst Technol, PhD(appl mech), 62. *Prof Exp:* Engr, Components Div, IBM Corp, 62-64; mem tech staff, RCA Labs, 64-69; mgr applns dept, Appl Logic Corp, 69-70; staff analyst, Computer Servs Dept, Consumers Power Co, 70-74, reliability & performance adminr elec prod, 74-81, nuclear planning adminr nuclear opers, 81-84; prin eng, Tech Analysis Corp, 84-85; PROG MGR, DEFENSE ADVAN RES PROG, 85- *Mem:* Am Soc Mech Engrs; Inst Elec & Electronics Engrs; Asn Comput Mach. *Mailing Add:* Defense Advan Res Proj 1400 Wilson Blvd Arlington VA 22209-2308

ROSENFELD, ROBERT SAMSON, b Richmond, Va, June 24, 21; m 44; c 3. BIOCHEMISTRY. *Educ:* Univ Pittsburgh, PhD(chem), 50. *Prof Exp:* Asst prof biochem, Sloan-Kettering Div, Med Col, Cornell Univ, 50-63; INVESTR, INST STEROID RES, MONTEFIORE HOSP & MED CTR, 63-; PROF BIOCHEM, ALBERT EINSTEIN COL MED, 72- *Concurrent Pos:* Asst, Sloan-Kettering Inst Cancer Res, 50-56, assoc mem, 56-63. *Mem:* Am Chem Soc; Endocrine Soc; Am Heart Asn. *Res:* Steroid chemistry. *Mailing Add:* 5956 Kirkwall Ct E Dublin OH 43017-9001

ROSENFELD, RON GERSHON, b Brooklyn, NY, June 22, 46; m 68; c 2. ENDOCRINOLOGY, BIOLOGY. *Educ:* Columbia Col, 68; Stanford Univ, MD, 73. *Prof Exp:* Resident pediatrics, Stanford Univ Med Ctr, 73-76; from asst prof to assoc prof, 77-89, PROF, STANFORD UNIV, 89- *Mem:* Am Diabetes Assoc; AAAS. *Res:* Growth factors and receptors in normal and malignant cells. *Mailing Add:* Dept Pediat Stanford Univ Med Ctr Stanford CA 94305

ROSENFELD, SHELDON, b New York, NY, Dec 28, 21; m 53; c 4. PHYSIOLOGY. *Educ:* Middlesex Sch Vet Med, DVM, 45; Brooklyn Col, BA, 48; Univ Calif, MS, 55; Univ Southern Calif, PhD(physiol), 64. *Prof Exp:* Asst med, Med Col, Cornell Univ, 46-48; res assoc aviation med, Sch Med, Univ Southern Calif, 48-50, instr & res assoc physiol, 50-53, asst prof, 50-53; sr res assoc, Cedars Sinai Med Res Inst, Mt Sinai Hosp, 53-86; RETIRED. *Mem:* Am Physiol Soc; Am Heart Asn; Soc Exp Biol & Med; Int Soc Nephrology; Am Soc Nephrology. *Res:* Renal physiology; hypertension. *Mailing Add:* 1231 N Ogden Dr Los Angeles CA 90046

ROSENFELD, STEPHEN I, b New York, NY, May 8, 39; m 60; c 2. IMMUNOLOGY. *Educ:* Univ Rochester, BA, 59, MD, 63. *Prof Exp:* Instr med, Sch Med, Boston Univ, 70-72; from asst prof to assoc prof, 72-87, PROF MED, SCH MED, UNIV ROCHESTER, 87- *Mem:* Am Acad Allergy; Am Fedn Clin Res; Am Coll Rheumatalogy; Am Asn Immunol; Clin Immunol Soc. *Res:* Biological functions and genetics of the human complement system; clinical immunology; Fc receptors. *Mailing Add:* Box 695 Med Ctr Univ Rochester 601 Elmwood Ave Rochester NY 14642

ROSENFELD, STUART MICHAEL, b New Haven, Conn, Jan 28, 48. SYNTHETIC CHEMISTRY. *Educ:* Colby Col, BA, 69; Brown Univ, PhD(org chem), 73. *Prof Exp:* Fel chem, Dept Chem, Brandeis Univ, 72-73 & Dyson Perrins Lab, Oxford Univ, 73-74; asst prof, Dept Chem, Univ RI, 75-77 & Univ Mass, 77-78; asst prof chem, Brandeis Univ, 78-79 & Wellesley Col, 79-82; asst prof, 82-88, ASSOC PROF CHEM, SMITH COL, 88- *Mem:* Am Chem Soc; Sigma Xi. *Res:* observation and study of transient intermediates formed during chemical reactions; synthesis and structural studies of spirocyclopropyl compounds, beta-keto acids and strained aromatic compounds. *Mailing Add:* Dept Chem Smith Col Northampton MA 01063

ROSENFIELD, ALAN R(OBERT), b Chelsea, Mass, Sept 7, 31; m 60; c 2. FRACTURE MECHANICS. *Educ:* Mass Inst Technol, SB, 53, SM, 55, ScD, 59. *Prof Exp:* Metallurgist, Mass Inst Technol, 59-61; fel metall, Univ Liverpool, 61-62; RES LEADER, BATTELLE MEM INST, 62- *Concurrent Pos:* Consult, Open Univ, 74-85. *Mem:* Am Inst Mining, Metall & Petrol Engrs; Iron & Steel Inst Japan; Am Soc Testing & Mat; fel Am Soc Metals Int. *Res:* Fracture mechanics; advanced techniques to measure strength and toughness; relations between strength and microstructure; physical metallurgy; experimental and analytical studies of fracture resistance of engineering materials, principally steels and ceramics. *Mailing Add:* Battelle Mem Inst 505 King Ave Columbus OH 43201-2693

ROSENFIELD, ARTHUR TED, b Waterbury, Ct, Dec 7, 42; m 68; c 3. DIAGNOSTIC RADIOLOGY, URINARY TRACT IMAGING. *Educ:* Brandeis Univ, BA, 64; NY Univ, MD, 68. *Hon Degrees:* MA, Yale Univ, 84. *Prof Exp:* Intern med, Montefiore Hosp, Pittsburgh, 68-69; surgeon, USPHS, Baltimore, 69-70, sr surgeon, 70-71; resident, Beth Israel Hosp, Boston, 71-73, chief resident, 73-74; instr radiol, Mass Gen Hosp, 74-75; from asst prof to assoc prof, 75-83, PROF DIAG RADIOL, SCH MED, YALE UNIV, 83-, PROF UROL, 87- *Concurrent Pos:* Instr med, Univ Pittsburgh, 68-69; clin fel, Harvard Med Sch, 71-74. *Mem:* Radiol Soc N Am; Am Roentgen Ray Soc; Soc Uroradiol; Am Col Radiol; Am Inst Ultrasound Med; Asn Univ Radiologists. *Res:* The evaluation of urinary tract disease by imaging techniques, particularly sectional imaging; author of 150 publications. *Mailing Add:* Dept Diag Radiol Yale Univ Sch Med 333 Cedar St New Haven CT 06510

ROSENFIELD, DANIEL, b Philadelphia, Pa, Jan 3, 32; m 71. NUTRITION. *Educ:* Univ Mass, BS, 53; Rutgers Univ, MS, 55, PhD(food sci, biochem), 59. *Prof Exp:* Dir quality control & res, Engelhorn Meat Packing Co, 59-60; sr food technologist, Gen Foods Tech Ctr, 60-62, Off Tech Coop & Res, USAID, 65-67; group leader food proteins & plant physiol, Union Carbide Res Inst, 62-65, 67-68; dep dir nutrit & agribus group, USDA, 68-71, dir nutrit & tech serv, staff, Food & Nutrit Serv, 71-73; dir nutrit affairs, Miles Labs, Inc, 73-79; DIR SCI AFFAIRS, M&M/MARS, 79- *Concurrent Pos:* US deleg workshop food, US Nat Acad Sci-Indonesian Inst Sci, 68; mem new foods panel, White House Conf Food, Nutrit & Health, 69; partic workshop prepare guidelines nat prog food enrichment & fortification, Pan Am Health Orgn, 73; mem, World Soy Protein Conf, 73 & Xth Int Cong Nutrit, 75; adj prof nutrit, Sch Med, Ind Univ, 78- *Mem:* AAAS; Inst Food Technologists; Am Chem Soc; Am Asn Cereal Chemists; Soc Nutrit Educ. *Res:* Nutritional equivalency of new food analogs; technology of fortification and fabrication of protein foods; availability of food nutrients effectiveness of dietary supplements; relationship of diets to oral health and to exercise. *Mailing Add:* M&M/Mars High St Hackettstown NJ 07840

ROSENFIELD, JOAN SAMOUR, b Boston, Mass, Aug 12, 39. ATMOSPHERIC RADIATIVE TRANSFER, STRATOSPHERIC CLIMATE STUDIES. *Educ:* Brandeis Univ, AB, 61; Univ Minn, Minneapolis, PhD(phys chem), 69. *Prof Exp:* Res asst chem, Children's Cancer Res Found, Boston, 61-62; res specialist phys chem, Univ Minn, Minneapolis, 69-71, teaching fel, 71-72; staff fel physics, Nat Inst Arthritis, Metab & Digestive Dis, 73-77; SCI PROGRAMMER & ANALYST, SIGMA DATA SERV, NASA GODDARD SPACE FLIGHT CTR, 77- *Concurrent Pos:* Eastman Kodak sci award, 65-66. *Mem:* Am Meteorol Soc; Am Geophys Union. *Res:* Radiation in troposphere-stratosphere climate models; theoretical chemistry; electronic structure of molecules; Atmospheric remote sensing. *Mailing Add:* 7101 Ora Glen Ct Greenbelt MD 20770

ROSENFIELD, RICHARD ERNEST, b Pittsburgh, Pa, Apr 7, 15; m 44; c 2. HEMATOLOGY, IMMUNOLOGY. *Educ:* Univ Pittsburgh, BS, 36, MD, 40. *Prof Exp:* Resident hemat, Mt Sinai Hosp, 47-48, res asst, 48-51, clin asst, Outpatient Dept, Med Div, Hemat Clin, 48-53, res asst med, Blood Bank, 51-53, from asst to assoc hematologist, 53-61, dir, Blood Bank, 57-80, prof med, Mt Sinai Sch Med, 66-71, prof path, 72-85, prof med, 79-85, ATTEND HEMATOLOGIST, MT SINAI HOSP, 61-, EMER DIR, DEPT BLOOD BANK & CLIN MICROS, 81-, EMER PROF MED, MT SINAI SCH MED, 85- *Concurrent Pos:* Asst physician, Willard Parker Hosp, New York, 50-56; NIH res fel ABO erythroblastosis, Mt Sinai Hosp, New York, 59-70, NIH blood bank dirs training fel, 62-77, NIH contract hepatitis, New York & Boston, 72-73; ed-in-chief, Transfusion, Am Asn Blood Banks, 67-71. *Honors & Awards:* Landsteiner Award, Am Asn Blood Banks; Philip Levine Award, Am Soc Clin Pathologists, 75. *Mem:* NY ACad Sci; Am Soc Clin Pathologists; Int Soc Hemat; Int Soc Blood Transfusion; Am Asn Immunologists. *Res:* Study, largely immunochemical, of human blood types and the clinical significance of allogeneic immune responses. *Mailing Add:* Rm 263 Berg-Atran Mt Sinai Med Ctr Box 1079 New York NY 10029

ROSENFIELD, ROBERT LEE, b Robinson, Ill, Dec 16, 34; m; c 3. PEDIATRIC ENDOCRINOLOGY. *Educ:* Northwestern Univ, Evanston, BA, 56; Northwestern Univ, Chicago, MD, 60. *Prof Exp:* Intern, Philadelphia Gen Hosp, 60-61; resident pediat, Children's Hosp Philadelphia, 61-63; instr, Med Sch, Univ Pa, 65-68; from asst prof to assoc prof, 68-78, PROF PEDIAT & MED SCH, UNIV CHICAGO, 78- *Concurrent Pos:* USPHS trainee, Children's Hosp Philadelphia, 65-68; Forgarty Sr Int fel, 77-78. *Mem:* Endocrine Soc; Soc Pediat Res; Am Pediat Soc; Soc Gynec Invest; Lawson Wilkins Pediat Endocrine Soc. *Res:* Androgen physiology; reproductive endocrinology; growth disorders. *Mailing Add:* Dept Pediat Med Sch Univ Chicago Box 118 Chicago IL 60637

ROSENGREEN, THEODORE E, b Tooele, Utah, Feb 19, 37; m; c 4. HAZARDOUS & TOXIC WASTE DISPOSAL, ENERGY RESOURCE DEVELOPMENT & ANALYSIS. *Educ:* Univ Wash, BS, 62, MS, 65; Ohio State Univ, PhD(geol), 70. *Prof Exp:* Explor geologist, Humble Oil Refining Co, 65-67; sr geologist, assoc Dames & Moore, 70-85; GEOL CONSULT, 87- *Mem:* Fel Geol Soc Am. *Res:* Quaternary geology, nuclear and thermal power plant siting studies, environmental impact studies of energy resource developments; investigation of hazardous and toxic waste disposal. *Mailing Add:* 4412 143rd Ave SE Bellevue WA 98006

ROSENGREN, JACK WHITEHEAD, b Chula Vista, Calif, Oct 2, 26; m 49; c 2. PHYSICS. *Educ:* Univ Calif, AB, 48, PhD(physics), 52. *Prof Exp:* Physicist, Radiation Lab, Univ Calif, 48-52; from instr to asst prof physics, Mass Inst Technol, 52-55; head reactor physics sect, Missile Systs Div, Lockheed Aircraft Corp, 55-57; staff physicist, Aeronutronic Systs, Inc, Ford Motor Co, 57; physicist, Lawrence Livermore Lab, Univ Calif, 57-63, head large weapons physics div, 63-64, assoc dir nuclear design, 64-67 & spec projs, 67-72; dep dir sci & technol, Defense Nuclear Agency, 72-74; sr physicist, R&D Assocs, 74-90; ASST ASSOC DIR AT LARGE, LAWRENCE LIVERMORE NAT LAB, 90- *Concurrent Pos:* Mem, Polaris Re-entry Body Coord Comt, 58-60; mem Mark 11 re-entry vehicle joint working group, US Air Force, 60-63, penetration prog panel, Ballistic Systs Div, 62-64 & ballistic missile re-entry systs consult group, 66-68; mem sci adv comt, Defense Intel Agency, 65-71; mem sci adv panel, US Army, 66-68, consult, 68-70; assoc ed, J Defense Res, 67-70; mem defense sci bd, Nuclear Test Detection Task Force, 67-72; mem Bethe Panel, Foreign Weapons Eval Group, 67-72; mem comt sr reviewers, AEC, 69-72 & Energy Res & Develop Admin, 76-77; consult, Tech Eval Panel (classification), Dept Energy, 77-83. *Mem:* Am Phys Soc. *Res:* Experimental nuclear physics; nuclear weapon design; nuclear weapon effects; nuclear weapon security; nuclear materials safeguards; nuclear weapon requirements studies; nuclear test ban verification; arms control studies. *Mailing Add:* Lawrence Livermore Nat Lab L-01 Univ Calif PO Box 808 Livermore CA 94550

ROSENGREN, JOHN, b Wooster, Ohio, Sept 29, 28; m 55; c 3. FRESHWATER BIOLOGY. *Educ:* Col Wooster, BA, 49; Columbia Univ, MA, 55, EdD, 58. *Prof Exp:* Instr biol, Am Univ Beirut, 49-53; instr, High Sch, 55-58; assoc prof, 59-63, PROF BIOL, WILLIAM PATERSON COL NJ, 64- *Mem:* AAAS; Nat Sci Teachers Asn; Nat Asn Biol Teachers. *Res:* Freshwater Porifera. *Mailing Add:* Dept Biol William Paterson Col 300 Pompton Rd Wayne NJ 07470

ROSENHAN, A KIRK, b Jefferson City, Mo, May 3, 40; m 72; c 1. FIRE PROTECTION. *Educ:* Univ Mo-Columbia, BS, 62; Miss State Univ, MS, 65. *Prof Exp:* Engr, Eastman Kodak Co, 62-63; instr mech eng, Miss State Univ, 63-68, bus, 70-72 & indust eng, 74-85; grad asst mech eng, Okla State Univ, 68-69; engr, Int Asn Fire Chiefs, 69-70; head fire sci, Vo-Tec, Racine, Wis, 72-74; ENGR, 85- *Concurrent Pos:* Fire coordr, Oktibbeha County, Miss, 88-; mem, Safety Comt, Am Soc Mech Engrs; mem, 1901 Comt, Nat Fire Protection Asn. *Mem:* Soc Fire Protection Engrs; Am Soc Mech Engrs; Int Asn Arson Investigators. *Res:* Fluid dynamics in fire equipment and appliances. *Mailing Add:* Drawer KJ Mississippi State MS 39762

ROSENHEIM, D(ONALD) E(DWIN), b New York, NY, Mar 23, 26; m 58; c 2. ELECTRICAL ENGINEERING. *Educ:* Polytech Inst Brooklyn, BS, 49; Columbia Univ, MS, 57. *Prof Exp:* Develop engr, Servo Corp Am , 49-51; engr, Develop Labs, IBM Corp, 51-53, res staff mem, Thomas J Watson Res Ctr, 53-60, mgr circuits & systs, 60-63, mgr solid state electronics, 63-66, dir appl res, 66-72, asst IBM dir res, 72-73, dir, San Jose Res Lab, 73-83, ASST DIR, IBM ALMADEN RES CTR, IBM CORP, 83- *Concurrent Pos:* Lectr, City Col New York, 55-56 & Columbia Univ, 57-58. *Mem:* Sr mem Inst Elec & Electronics Engrs. *Res:* Digital systems and technology, especially solid state materials, devices and circuits for logic, storage and input/output applications in data processing systems. *Mailing Add:* Dept K01/802 IBM Corp Almaden Res Ctr 650 Harry Rd San Jose CA 95120

ROSENHOLTZ, IRA N, b New York, NY, May 14, 45; c 2. TOPOLOGY. *Educ:* Brandeis Univ, BA, 67; Univ Wis, MA & PhD(math), 72. *Prof Exp:* Asst prof math, Grinnell Col, Iowa, 72-73; ASST PROF MATH, UNIV WYO, 73- *Concurrent Pos:* NDEA Title IV fel, Univ Wis, 71-72. *Mem:* Math Asn Am; Am Math Soc. *Res:* Continua; fixed point problems; remetrization. *Mailing Add:* Eastern Ill Univ Charleston IL 61920

ROSENKILDE, CARL EDWARD, b Yakima, Wash, Mar 16, 37; div; c 2. FLUID DYNAMICS, NONLINEAR WAVE PROPAGATION. *Educ:* Wash State Univ, BS, 59; Univ Chicago, MS, 60, PhD(physics), 66. *Prof Exp:* Fel physics, Argonne Nat Lab, 66-68; asst prof math, NY Univ, 68-70; asst prof physics, 70-76; assoc prof physics, Kans State Univ, 76-79; PHYSICIST, LAWRENCE LIVERMORE NAT LAB, 79- *Concurrent Pos:* Vis scientist, Lawrence Livermore Lab, 77-78. *Mem:* AAAS; Am Astron Soc; Am Phys Soc; Soc Indust & Appl Math; Am Geophys Union; Acoust Soc Am. *Res:* Fluid dynamics; nonlinear wave propagation in complex media. *Mailing Add:* Lawrence Livermore Nat Lab L-84 PO Box 808 Livermore CA 94551-0808

ROSENKRANTZ, BARBARA G, b New York, NY, Jan 11, 23. HISTORY OF SCIENCE. *Educ:* Radcliff Col, AB, 42; Clark Univ, PhD, 70. *Prof Exp:* Assoc prof, 73-75, PROF HIST SCI, HARVARD UNIV, 75-, CHMN DEPT, 84- *Mem:* Inst Med-Nat Acad Sci; Am Hist Asn; Hist Sci Soc; Am Asn Hist Sci. *Mailing Add:* Hist Sci Dept Harvard Univ Sci Ctr 235 Cambridge MA 02138

ROSENKRANTZ, DANIEL J, b Brooklyn, NY, Mar 5, 43; m 69; c 4. REPRESENTATIONS OF DIGITAL CIRCUITS, DATABASE CONCURRENCY CONTROL. *Educ:* Columbia Univ, BS, 63, MS, 64 & PhD(elec eng), 67. *Prof Exp:* Info scientist, Bell Telephone Labs, 66-67, Gen Elec Res & Develop Ctr, 67-77; PROF COMPUT SCI, STATE UNIV NY, ALBANY, 77- *Concurrent Pos:* Area ed, formal languages & models of computation, J Asn for Comput Mach, 81-86; prin comput scientist, Phoenix Data Syst, 83-85; ed chief, J Asn for Computer Mach, 86- *Mem:* Asn Computer Mach; Inst Elec & Electronics Engrs Comput Soc; EATCS. *Res:* Computer aided design, VLSI theory, database systems, algorithms, complexity theory, compiler design theory, automata theory, and software engineering. *Mailing Add:* Comput Sci Dept State Univ NY 1400 Washington Ave Albany NY 12222

ROSENKRANTZ, HARRIS, b Brooklyn, NY, Mar 23, 22; m 51; c 2. BIOCHEMICAL PHARMACOLOGY. *Educ:* Brooklyn Col, AB, 43; NY Univ, MS, 46; Cornell Univ, MS, 48; Tufts Univ, PhD(physiol), 52. *Prof Exp:* Asst infrared anal of steroids, NY Hosp, 43-46, res assoc biochem, 48-51; res biochemist, Worcester Found Exp Biol, 52-60; dir biochem & vpres, Mason Res Inst, EG&G, Inc, 60-85; CONSULT, 85- *Concurrent Pos:* Affil prof, Clark Univ, 59-87; adj prof, Sch Vet Med, Tufts Univ, 81-87. *Mem:* Am Chem Soc; Am Soc Biol Chemists; Endocrine Soc; Am Soc Pharmacol & Exp Therapeut; Soc Toxicol; Am Physiol Soc. *Res:* Biochemistry of steroid hormones; vitamin E and muscular dystrophy; infrared spectroscopic analysis of biologically important compounds; biochemistry of the prostate; biochemical pharmacology, toxicology and endocrinology of marijuana, narcotic antagonists, antineoplastic, contraceptive agents and iron chelators. *Mailing Add:* 136 S Flagg St Worcester MA 01602-1831

ROSENKRANTZ, JACOB ALVIN, b New York, NY, Feb 12, 14; m 36; c 4. MEDICINE, HOSPITAL ADMINISTRATION. *Educ:* City Col New York, BS, 33; Columbia Univ, MA, 34, MD, 38. *Prof Exp:* Res asst, Columbia Univ, 41-42, 46; chief outpatient serv, Bronx Vet Admin Hosp, 47-49, asst chief prof servs, 49-52, attend, Med Serv, 52; chief prof servs, Vet Admin Hosp, East Orange, NJ, 52-56; adminr, Southern Div, Albert Einstein Med Ctr, 56-59; exec dir, Newark Beth Israel Hosp, 59-68; chief, Outpatient Clin, Vet Admin, 71-82; RETIRED. *Concurrent Pos:* Fel, Postgrad Med Sch, NY Univ, 41-42 & 46, asst physician, 41-, instr, 54-; clin assoc prof community & prev med & exec dir & assoc dean hosp admin, New York Med Col Flower & Fifth Ave Hosps, 68-71. *Mem:* Fel Am Geriat Soc; fel Am Pub Health Asn; fel Am Col Physicians; fel Am Col Prev Med; fel Am Col Healthcare Execs; Sigma Xi;

fel Royal Soc Health; fel NY Acad Sci. *Res:* Hypertension; galactose and glucose tolerance; arteriosclerosis; infectious mononucleosis; bilateral nephrectomy in rats; potassium and electrocardiograms; thyroid diseases; benign paroxysmal peritonitis; Paget's disease; puperperal sepsis. *Mailing Add:* 5367 A Privet Pl Delray Beach FL 33484

ROSENKRANTZ, MARCY ELLEN, b New York, NY, June 15, 48; m 79. COMPUTATIONAL CHEMISTRY. *Educ:* Harpur Col, BA, 69; State Univ NY, Binghamton, MA, 76, PhD(theoret chem), 84. *Prof Exp:* Teacher chem, Binghamton City Schs, 69-80; res assoc physics, Smithsonian Inst Ctr Astrophys, 84-85; resident res assoc chem, Nat Res Coun, 85-86; adj asst prof chem, State Univ NY, Binghampton, 86-87; RES SCIENTIST CHEM, UNIV DAYTON RES INST, 87- *Mem:* Am Chem Soc; Am Phys Soc; Sigma Xi. *Res:* Electronic structure calculations on systems which promise to possess high energy density; dispersion energies and polarizabilities of small molecules. *Mailing Add:* OLAC PL/RFE Edwards AFB CA 93523-5000

ROSENKRANZ, EUGEN EMIL, b Wilno, Poland, July 9, 31; US citizen. PLANT PATHOLOGY. *Educ:* Univ Wis, BS, 56, MS, 57, PhD(plant path), 61. *Prof Exp:* Fel microbial genetics, Univ Wis, 61-62, res assoc plant path, 63-64; RES PLANT PATHOLOGIST, AGR RES STA, USDA, 64- *Concurrent Pos:* Adj prof plant path, Miss State Univ, 71- *Mem:* Am Phytopath Soc. *Res:* Diseases of maize caused by viruses and mycoplasma-like organisms; insect transmission of mycoplasma-like organisms and viruses affecting corn and other graminaceous plants. *Mailing Add:* Dept Plant Path Miss State Univ PO Drawer PG Mississippi State MS 39762

ROSENKRANZ, HERBERT S, b Vienna, Austria, Sept 27, 33; US citizen; m 59; c 8. CANCER RESEARCH, RISK ASSESSMENT. *Educ:* City Col New York, BS, 54; Cornell Univ, PhD(biochem), 59. *Prof Exp:* Res assoc biochem, Univ Pa, 60-61; from asst prof to prof microbiol, Col Physicians & Surgeons, Columbia Univ, 61-76; prof microbiol & chmn, dept microbiol, NY Med Col, 76-81; prof & chmn, Dept Environ Health Sci & prof biochem, pediat, oncol & radiol, Case Western Res Univ, 81-90; PROF & CHMN, DEPT ENVIRON & OCCUP HEALTH, UNIV PITTSBURGH, 90- *Concurrent Pos:* Nat Cancer Inst fel, Sloan-Kettering Inst, 59-60; res career develop award, NIH, 65-75; vis prof, Hebrew Univ, Med Sch, Jerusalem, 71-72; Int Comn for Protection Against Environ Mutagens & Carcinogens, 77-; del, Fifth Soviet-Am Symp Environ Mutagenesis & Carcinogenesis, 78; chmn DNA panel, Gene & Toxicol Prog, Environ Protection Agency, 79-, scientific review panel health effects, 80-; guest investr, Nat Coun Res Ctr Japan, 80 & Toxicol Study Sect, NIH, 84. *Mem:* Soc Toxicol; Am Asn Cancer Res; Am Chem Soc; Am Soc Biol Chemists; Am Soc Microbiol; AAAS; Environ Mutagen Soc; Genetic Soc Am; Sigma Xi. *Res:* Chemistry and biology of nucleic acids; biochemical basis of mutagenicity and carcinogenicity; causes and prevention of cancer; risk assessment. *Mailing Add:* Dept Environ Occup Health Grad Sch Pub Health Univ Pa Pittsburgh PA 15261

ROSENKRANZ, PHILIP WILLIAM, b Buffalo, NY, Oct 30, 45. REMOTE SENSING. *Educ:* Mass Inst Technol, SB, 67, SM, 68, PhD(elec eng), 71. *Prof Exp:* Resident res assoc, Jet Propulsion Lab, Calif Inst Technol, 71-73; res assoc, 73-89, PRIN RES SCIENTIST, MASS INST TECHNOL, 89- *Concurrent Pos:* Mem, comt F, Int Union Radio Sci. *Mem:* AAAS; Am Geophys Union; Inst Elec & Electronics Engrs; Int Union Radio Sci. *Res:* Remote sensing of earth; radio astronomy. *Mailing Add:* Rm 26-343 Mass Inst Technol Cambridge MA 02139

ROSENLICHT, MAXWELL, b Brooklyn, NY, Apr 15, 24; m 53; c 3. MATHEMATICS. *Educ:* Columbia Univ, AB, 47; Harvard Univ, PhD(math), 50. *Prof Exp:* Nat Res Coun fel, Univ Chicago & Princeton Univ, 50-52; from asst prof to assoc prof math, Northwestern Univ, 52-59; PROF MATH, UNIV CALIF, BERKELEY, 59- *Concurrent Pos:* Fulbright fel, Univ Rome, 54-55; Guggenheim fel, 57-58; mem, Inst Advan Sci Studies, France, 62-63. *Honors & Awards:* Cole Prize, Am Math Soc, 60. *Mem:* Am Math Soc. *Res:* Algebraic groups; differential algebra. *Mailing Add:* Dept Math Univ Calif Berkeley CA 94720

ROSENMAN, IRWIN DAVID, b Brooklyn, NY, Apr 9, 23; m 50; c 3. ORGANIC CHEMISTRY, TRANSPORTATION. *Educ:* Univ Pa, BA, 45; Univ Conn, MS, 48; PhD(biochem), 51. *Prof Exp:* Asst instr anal chem, Univ Conn, 48-50; chemist, Ames Aromatics, Inc, 51-53; chemist & pres, Ames Labs, Inc, 53-85; CONSULT, 85- *Mem:* Sigma Xi; Am Chem Soc. *Res:* Organic synthesis of aliphatic and aromatic amines, oximes and other nitrogen containing compounds; antiskinning agents; electrophoresis; antioxidants and additives for paint and inks; amines; transportation consultant (hazardous materials). *Mailing Add:* Ames Labs Box 3024 Milford CT 06460

ROSENMAN, RAY HAROLD, b Akron, Ohio, Nov 17, 20; m 45, 78; c 2. MEDICINE. *Educ:* Univ Mich, AB, 41, MD, 44. *Prof Exp:* Intern, Michael Reese Hosp, Chicago, 44-45; resident path, Wayne County Gen Hosp, Eloise, Mich, 45-46; resident internal med & cardiovasc dis, Michael Reese Hosp, 48-50; ASSOC CHIEF DEPT MED, MT ZION HOSP & MED CTR, SAN FRANCISCO, CALIF, 51-; DIR CARDIOVASC RES, STANFORD RES INST, SRI INT, MENLO PARK, CALIF, 78- *Concurrent Pos:* Fel Coun Epidemiol, Am Heart Asn; consult, Sch Aerospace Med, San Antonio, Brooks AFB, USAF, 74-79; mem, Nat Task Force Epidemiol of Heart Dis, Nat Heart, Lung & Blood Inst, 78; Am Heart Asn fel, 59-60, assoc chief, Harold Brunn Inst, Mt Zion Hosp & Med Ctr, San Francisco, Calif, 51-78; mem study sect, Behav Med, NIH, 81-; mem adv bd coun, High Blood Pressure Res, Am Heart Asn, 66- & expert adv coun , Health Econ, Basel, 81-; trustee, Am Inst Stress, Yonkers, NY, 81-; co-chmn, Int Symp Psychophysiol Risk Factors, Carlsbad, Czech, 81; mem rev panel, Coronary-Prone Behav, Jacksonville, Fla & Workshop, Cholesterol & Non-Cardiovasc Mortality, Nat Heart, Lung & Blood Inst, Bethesda, Md, 81-; guest lectr, Australia, 74, 78 & 80; guest lectr, Klinik Hohenreid, Munich, Ger & Ciba Symp, Stratford, Eng, 78, Charing Cross Hosp, London, Eng, 79, Ciba Symp, Montsoult, France, Netherlands Inst Adv Study Humanities & Social Sci, Amsterdam, Holland, conf, Coronary-Prone Behav, Univ Freiburg, Ger & Univ SFla, Tampa, 80, conf, Psychol Factors Before & After Myocardial Infarction, Europ Soc Cardiol, Nice, France, Ger Conf, Coronary-Prone Behav, Altenberg, Ger & Switzerland, Univ Hosps, Basel, Zurich, Bern, Geneva & Lausanne, 81. *Mem:* Am Col Cardiol; Am Soc Clin Invest; Am Soc Internal Med; Am Physiol Soc. *Res:* Lipid metabolism; pathogenesis of coronary heart disease; hypertension and the predictive role of risk factors; role of Type-A behavior pattern in coronary heart disease. *Mailing Add:* SRI Int 333 Ravenswood Ave BN110 Menlo Park CA 94025

ROSENMANN, EDWARD A, b Santiago, Chile, Nov 11, 40; m 76; c 2. ENZYMOLOGY, PROTEIN CHEMISTRY. *Educ:* Univ Chile, BS, 60. *Prof Exp:* Biochemist, Univ Chile, 68, asst prof chem, Med Sch, 68-69; asst prof biochem, Fac Sci, Inst Biochem, Univ Austral, Chile, 69-70 & 71-75; Alcohol & Drug Addiction Res Found fel pharmacol, dept pharmacol, Univ Toronto, 70-71; Muscular Dystrophy Asn Can res fel, 76-79, RES ASSOC BIOCHEM, DEPT BIOCHEM, UNIV MAN, 79- *Res:* The study of protein structure and function; enzyme activity and regulation; sex hormone receptors; abnormal proteins in muscular dystrophy. *Mailing Add:* Dept Biochem Univ Man Winnipeg MB R3E 0W3 Can

ROSENOW, EDWARD CARL, JR, b Chicago, Ill, Apr 7, 09; m 31; c 2. MEDICINE. *Educ:* Carleton Col, BA, 31; Harvard Med Sch, MD, 35; Univ Minn, MSc, 39; Am Bd Internal Med, dipl, 46; FRCP, 68. *Hon Degrees:* DSc, Carleton Col, 67 & MacMurray Col, 73. *Prof Exp:* Exec dir, Los Angeles County Med Asn, 57-59; exec vpres, 60-77, EMER EXEC VPRES, AM COL PHYSICIANS, 77- *Concurrent Pos:* Clin prof, Sch Med, Univ Pa, 60-66, emer clin prof, 66-; trustee, Carleton Col, 68-; dir med educ, Grad Hosp, Philadelphia, 78- *Honors & Awards:* Stengel Award, Am Col Physicians, 76. *Mem:* Int Soc Internal Med (pres elec, 78-); Am Clin & Climat Asn; Am Fedn Clin Res; hon fel Am Col Chest Physicians; Am Med Writers Asn (pres, 65). *Res:* Postgraduate education; medical administration. *Mailing Add:* 4200 Pine St Philadelphia PA 19104

ROSENQUIST, BRUCE DAVID, b Chicago, Ill, June 19, 34; m 56; c 2. VETERINARY VIROLOGY. *Educ:* Iowa State Univ, DVM, 58; Univ Mo-Columbia, PhD(microbiol), 68. *Prof Exp:* Staff, USPHS, Atlanta, Ga & Haddonfield, NJ, 58-60; vet, Morton Grove Animal Hosp, Ill, 60-61 & Depster Animal Clin, Skokie, Ill, 61-64; res assoc vet microbiol, Sch Vet Med, 64-68, Space Sci Res Ctr, 65-66, from asst prof to assoc prof, 68-73, PROF VET MICROBIOL, UNIV MO-COLUMBIA, 73- *Concurrent Pos:* NIH fel, 66-68; mem working team for bovine & equine picornaviruses, WHO-Food & Agr Orgn Prog Comp Virol, 73- *Mem:* Conf Res Workers Animal Dis; Am Vet Med Asn; Int Soc Interferon Res; Am Col Vet Microbiol; Am Soc Virology. *Res:* Bovine viral respiratory disease; interferon; bovine rhinoviruses; nutrition and bovine viral disease. *Mailing Add:* Dept Vet Microbiol Univ Mo 100A Connaway Columbia MO 65211

ROSENQUIST, EDWARD P, b Dayton, Ohio, Feb 28, 38; m 64; c 3. ORGANIC CHEMISTRY, PHYSICAL CHEMISTRY. *Educ:* Denison Univ, BS, 60; Purdue Univ, PhD(org chem), 65. *Prof Exp:* Res chemist, Shell Oil Co, Ill, 65-69, chemist, Head Off Mfg Res, NY, 69-70, sr engr, Mkt Lubricants, Tex, 70-73, supvr staff, Shell Develop Co, Wood River, Ill & Houston, Tex, 73-78, mgr res recruitment, 78-80, mgr res & develop coord, Oil Prods, Shell Oil Co, 80-83, sr staff res engr, 83-85, RES MGR, SHELL DEVELOP CO, HOUSTON, TEX, 85- *Mem:* Am Chem Soc. *Res:* Organic synthesis; lubricants and detergents research. *Mailing Add:* 5614 Bermuda Dunes Lane Houston TX 77069

ROSENQUIST, GLENN CARL, b Lincoln, Nebr, Aug 29, 31; m 53; c 6. MEDICINE. *Educ:* Univ Nebr, BA, 53, MD, 57. *Prof Exp:* USPHS fel embryol, Carnegie Inst Washington, 63-65; fel pediat cardiol, Johns Hopkins Univ, 65-67, asst prof pediat, 67-71, assoc prof, 71-76; prof pediat & chmn, Col Med, Univ Nebr, 76-80; PROF CHILD HEALTH & DEVELOP, GEORGE WASHINGTON UNIV, 80-; DIR RES, CHILDREN'S HOSP, NAT MED CTR, 80- *Concurrent Pos:* Prof path, Johns Hopkins Univ, 72-76; mem, Human Embryol & Develop Study Sect, 75-79. *Mem:* Am Asn Anatomists; Soc Develop Biol; Teratol Soc; Am Col Cardiol; Am Acad Pediat. *Res:* Early development of organ systems in the chick embryo; embryology of the heart and lung; pathology of congenital heart disease. *Mailing Add:* Dept Child Health Develop Childrens Hosp Nat Med Ctr 111 Michigan Ave NW Washington DC 20010

ROSENQUIST, GRACE LINK, b Los Angeles, Calif; m 60; c 2. PEPTIDES PURIFICATION & ASSAY. *Educ:* Willamette Univ, BA, 54; Univ Wis-Madison, MS, 58, PhD(zool), 61. *Prof Exp:* Fel, Calif Inst Technol, 61-64; USPHS fel, Sch Med, Wash Univ, 64-66; instr med microbiol, 66-68, res assoc, 68-70, res assoc med & microbiol, Sch Med, Stanford Univ, 70-74; lectr, 75-80, PROF ANIMAL PHYSIOL, UNIV CALIF, DAVIS, 80- *Concurrent Pos:* Vis scientist, Inst Animal Physiol, Cambridge, UK, 72- 73, Ctr Ulcer Res & Educ, Los Angeles, 74-; vis scientist, Nat Inst Med Res, London, 81-82. *Mem:* AAAS; Am Asn Immunol. *Res:* Function and structure of cholecystokinin; evolution of gastrointestinal hormones; radioimmunoassay of gastrin, cholecystokinin; peptide purification; enzymatic sulfation of tyrosine in peptides and proteins. *Mailing Add:* Dept Animal Physiol Univ Calif Davis CA 95616

ROSENSHEIN, JOSEPH SAMUEL, b Kimball, WVa, Apr 19, 29; m 51; c 3. HYDROLOGY, GEOHYDROLOGY. *Educ:* Univ Conn, BA, 52; Johns Hopkins Univ, MA, 53; Univ Ill, Urbana, PhD(geol), 67. *Prof Exp:* Geologist, Ind Dist, US Geol Surv, 53-63, subdist chief, RI & NY Dist, 64-66, RI & Cent New Eng Dist, 66-67 & Tampa Subdist, Fla Dist, 67-75, chief Kans Dist, 75-87, DEP ASST CHIEF HYDROL, PROG COORD & TECH SUPPORT & INSTALLATION RESTORATION PROG COORD, WATER RESOURCES DIV, US GEOL SURV, 87- *Honors & Awards:* Meritorious Serv Award, Dept Interior, 86. *Mem:* Am Geophys Union; fel Geol Soc Am; Int Asn Hydrogeologists; Asn Groundwater Scientists & Engrs; Am Inst Hydrol. *Res:* Hydrology of aquifer systems; estuarine hydrology; environmental hydrology; subsurface storage of wastes; hydraulic characteristics and modeling of aquifer systems; remote sensing of the environment. *Mailing Add:* US Geol Surv 414 Nat Ctr Reston VA 22092

ROSENSHINE, MATTHEW, b New York, NY; m; c 2. OPERATIONS RESEARCH, MATHEMATICS. *Educ:* Columbia Univ, AB, 52, MA, 53; Univ Ill, MS, 56; State Univ NY Buffalo, PhD(opers res), 66. *Prof Exp:* Engr aerodyn, Bell Aircraft Corp, 53-56; instr, Univ Ala, Huntsville, 57; prin math, Aeronaut Lab, Cornell Univ, 58-68; PROF INDUST ENG, PA STATE UNIV, 68- *Concurrent Pos:* Instr, Univ Buffalo, 53-55 & NMex State Univ, 58; lectr, State Univ NY Buffalo, 66-68; consult, Aeronaut Lab, Cornell Univ, 69-70, Xerox Corp, 72, Fed Energy Admin, 73-76 & Fed RR Admin, 78-80. *Mem:* Opers Res Soc Am; Inst Mgt Sci; Am Inst Indust Engrs. *Res:* Queueing theory and control applied to large scale systems, especially air traffic and railroads. *Mailing Add:* 207 Hammond Bldg Pa State Univ University Park PA 16802

ROSENSON, LAWRENCE, b Brooklyn, NY, May 20, 31; m 63; c 2. ELEMENTARY PARTICLE PHYSICS, EXPERIMENTAL PHYSICS. *Educ:* Univ Chicago, AB, 50, MS, 53, PhD(physics), 56. *Prof Exp:* Res assoc physics, Univ Chicago, 56-58; from instr to assoc prof, 58-67, PROF PHYSICS, MASS INST TECHNOL, 67- *Mem:* Fel Am Phys Soc. *Res:* Experimental elementary particle physics; particle spectroscopy; weak and strong interactions. *Mailing Add:* Dept Physics Mass Inst Technol 77 Massachusetts Ave Cambridge MA 02139

ROSENSPIRE, ALLEN JAY, b New York, NY, May 25, 49; m 77; c 2. MOLECULAR BIOLOGY. *Educ:* State Univ NY, Stony Brook, BS, 70; State Univ NY, Buffalo, PhD(biophys sci), 78. *Prof Exp:* Res fel, dept microbiol, State Univ NY, Buffalo, 78-79; res fel lab immunobiol, Sloan-Kettering Inst Cancer Res, 79-82, res assoc, 83-85; sr res scientist, Bio Magnetech Corp, New York, 86; asst prof immunol & biophys, dept biol, 86-87, ASSOC, DEPT IMMUNOL & MICROBIOL, WAYNE STATE UNIV, DETROIT, 87- *Concurrent Pos:* Consult, Bio Magnetech Corp, New York, 86- *Mem:* Am Soc Immunologists; Sigma Xi. *Res:* Molecular biology of membrane receptors and the mechanisms of receptor mediated cell activation in the lymphocyte; the application of magnetic substrates in cell separation and immuno-assay techniques. *Mailing Add:* Dept Biol Sci Wayne State Univ Detroit MI 48202

ROSENSTARK, SOL(OMON), b Poland, May 13, 36; US citizen; m 60; c 2. ELECTRICAL ENGINEERING. *Educ:* City Col New York, BEE, 58; NY Univ, MEE, 61, PhD(elec eng), 66. *Prof Exp:* Jr engr, Polarad Electronics Corp, 58-59; assoc develop engr, Norden Labs, Div United Aircraft Corp, 59-61; lectr, City Col New York, 61-63; teaching fel, NY Univ, 64-65, res asst, 65-66; mem tech staff, Bell Tel Labs, 66-68; PROF ELEC & COMP ENG, NJ INST TECHNOL, 68- *Mem:* Inst Elec & Electronics Engrs; Sigma Xi. *Res:* Communication theory; electronics; microprocessors. *Mailing Add:* 17 Whitman St West Orange NJ 07052

ROSENSTEEL, GEORGE T, b Baltimore, Md, Sept 30, 47. THEORETICAL PHYSICS, NUCLEAR STRUCTURE. *Educ:* Univ Toronto, BS, 73; MS, 74, PhD, 75. *Prof Exp:* Nat Res Coun Can postdoc fel, McMaster Univ, 76-78; asst prof math, Ariz State Univ, 81-82; from asst prof to assoc prof, 78-84, PROF PHYSICS, TULANE UNIV, 84-, CHMN DEPT PHYSICS, 85- *Concurrent Pos:* Vis prof, La State Univ, 85, Yale Univ, 86; Brit Sci & Eng Univ Coun vis fel, Univ Sussex, 86; NSF grant, 79-; Nat Sci Found fel, 85-87, 87-90. *Mem:* Am Phys Soc; Am Math Soc; Sigma Xi; Am Asn Univ Prof. *Res:* Theoretical nuclear physics; mathematical physics; applied mathematics; author of 75 publications. *Mailing Add:* Tulane Univ Dept Physics New Orleans LA 70118

ROSENSTEIN, A(LLEN) B, b Baltimore, Md, Aug 25, 20; m; c 3. ENGINEERING. *Educ:* Univ Ariz, BS, 40; Univ Calif, Los Angeles, MS, 50, PhD, 58. *Prof Exp:* Elec engr, Convair Div, Gen Dynamics Corp, 40-41; sr elec engr, Lockheed Aircraft Corp, 41-42; chief plant engr, Utility Appliance Co, 42-44; lectr eng, 46-58, PROF ENG & APPL SCI, UNIV CALIF, LOS ANGELES, 58- *Concurrent Pos:* Consult, Atomic Energy Comn, US Air Force, US Corps Eng, Douglas Aircraft Co, Beckman Instruments & Marquardt Aircraft Co, 46-; chmn bd, Inet, Inc, 47-53, consult engr, 54-; dir, Pioneer Magnetics, Inc, Int Transformer Co, Inc & Foreign Resource Serv. *Mem:* AAAS; Am Soc Eng Educ; Inst Elec & Electronics Engrs; NY Acad Sci; Sigma Xi. *Res:* Magnetic amplifiers; ferromagnetic systems; automatic controls; system design; engineering organization; education; computer aided design; educational planning. *Mailing Add:* 314 S Rockingham Los Angeles CA 90049

ROSENSTEIN, ALAN HERBERT, b Baltimore, Md, July 4, 36; c 2. METALLURGY. *Educ:* Drexel Inst Technol, BS, 59; George Washington Univ, MA, 63; Colo Sch Mines, MS, 65, DSc, 66. *Hon Degrees:* DSc, Colo Sch Mine, 66. *Prof Exp:* Physical metallurgist ferrous & nonferrous, Naval Ship Res & Develop Ctr, Annapolis, 59-68, br chief ferrous metall, 68-71; prog mgr struct mat, 71-85, DEP DIR ELECTRONIC & MAT SCI, AIR FORCE OFF SCI RES, BOLLING AFB, WASHINGTON, DC, 85- *Mem:* Am Soc Metals; Sigma Xi; Am Welding Soc; Am Soc Testing & Mat; Am Inst Mining, Metall & Petrol Engrs; Mat Res Soc. *Res:* Structural materials research and development, especially metallurgy and mechanics of ferrous, titanium, aluminum, nickel, niobium and copper alloys and composite materials; administration of federal contracts and grants program in industry and universities dealing with above. *Mailing Add:* 9472 Turnberry Dr Potomac MD 20854

ROSENSTEIN, BARRY SHELDON, b Roanoke, Va, June 17, 51; m 73; c 2. PHOTOBIOLOGY, RADIATION BIOLOGY. *Educ:* Univ Rochester, BA, 73, MS, 76, PhD(radiation biol), 78. *Prof Exp:* Res fel, Brookhaven Nat Lab, 78-80; asst prof radiol, Univ Tex Health Sci Ctr, Dallas, 80-86; ASSOC PROF RADIOL MED, BROWN UNIV, 86- *Mem:* AAAS; Am Inst Biol Sci; Am Soc Photobiol; Environ Mutagen Soc; Sigma Xi; Radiation Res Soc. *Res:* Induction and repair of DNA damages induced by solar ultra-violet radiation. *Mailing Add:* Brown Univ Box 6-8093 Providence RI 02912

ROSENSTEIN, GEORGE MORRIS, JR, b Philadelphia, Pa, Feb 27, 37; m 59; c 1. MATHEMATICS, HISTORY OF MATHEMATICS. *Educ:* Oberlin Col, AB, 59; Duke Univ, MA, 62, PhD(math), 63. *Prof Exp:* Asst prof math, Western Reserve Univ, 63-67; from asst prof to assoc prof, 67-80, PROF MATH, FRANKLIN & MARSHALL COL, 80- *Concurrent Pos:* Chief reader, Advan Placement Exam, 88-91. *Mem:* Math Asn Am; Am Math Soc; Hist Sci Soc; Sigma Xi; Nat Coun Teachers Math. *Res:* History of 19th Century mathematics, focusing on the evolution of calculus; point-set topology; metric topology. *Mailing Add:* Dept Math Franklin & Marshall Col Box 3003 Lancaster PA 17604-3003

ROSENSTEIN, JOSEPH GEOFFREY, b London, Eng, Feb 8, 41; US citizen; m 69; c 5. MATHEMATICAL LOGIC, MATHEMATICS EDUCATION. *Educ:* Columbia Univ, BA, 61; Cornell Univ, PhD(math logic), 66. *Prof Exp:* Instr & res assoc math, Cornell Univ, 66; asst prof, Univ Minn, Minneapolis, 66-69; from asst prof to assoc prof, 69-79, vchmn dept & dir Undergrad Prog, 81-85, PROF MATH, RUTGERS UNIV, NEW BRUNSWICK, 79- *Concurrent Pos:* Mem Sch Math, Inst Advan Study, 76, 77. *Mem:* Am Math Soc; Math Asn Am; Asn Symbolic Logic. *Res:* Recursion theory; model theory; linear orderings. *Mailing Add:* Dept Math Rutgers Univ New Brunswick NJ 08903

ROSENSTEIN, LAURENCE S, b Philadelphia, Pa, Aug 19, 41; m 62; c 2. TOXICOLOGY, CARCINOGENESIS. *Educ:* Drexel Univ, BS, 64, MS, 65; Univ Cincinnati, PhD(toxicol), 70. *Prof Exp:* Toxicologist, Ins Inst Hwy Safety, 71-72; sr toxicologist, Environ Protection Agency, 72-77; sr toxicologist, SRI Int, 77-79; SR TOXICOLOGIST, ENVIRON PROTECTION AGENCY, 79-; CHIEF, SCI POLICY INTEGRATION STAFF, DELEG, INTERAGENCY TESTING COMT, INT PROG CHEM SAFETY TECH LEAD, 85-; COORD DIR, CLEARING HOUSE ON PHTHALATA ESTER, NAT TOXICOL PROG ENVIRON PROTECTION AGENCY, 81-; ASST DIR, DIV ANTIVIRAL DRUG PROD, US FOOD & DRUG ADMIN, 90- *Concurrent Pos:* Consult, Environ Defense Fund, 71-; vis scientist, Environ Protection Agency, 71-72, chief, Sci Policy Integration Staff, deleg, Interagency Testing Comt, Int Prog Chen Safety Tech Lead; adj prof, Univ Miami, 72-73 & NC State Univ, 73-; Pub Health Serv fel; chief, Risk Anal Br & prog dir existing chem prog, Environ Protection Agency, 87-90; coord dir, Clearing House Phthalata Ester, Nat Toxicol Prog, Environ Protection Agency, 90- *Honors & Awards:* Bronze Medals, US Environ Protection Agency. *Mem:* AAAS; Soc Toxicol; Am Asn Lab Animal Sci; Am Pub Health Asn; Am Col Toxicol. *Res:* Regulatory toxicology and risk assessment; chemical carcinogenesis; microscopic calcium metabolism. *Mailing Add:* 10222 Yearling Dr Rockville MD 20850

ROSENSTEIN, ROBERT, b New York, NY, Aug 7, 33; m 65; c 2. PHARMACOLOGY. *Educ:* Columbia Univ, BS, 55, MS, 57; Univ Utah, PhD(pharmacol), 62. *Prof Exp:* Teaching asst biol, Col Pharm, Columbia Univ, 55-57; res asst pharmacol, Col Med, Univ Utah, 57-58; res instr, Sch Med, Univ Okla, 62-65, asst res prof, 65; from instr to asst prof, 65-77, ASSOC PROF PHARMACOL, DARTMOUTH MED SCH, 77-, PHARMACOLOGIST, RES SERV, VET ADMIN CTR, 65- *Concurrent Pos:* Res pharmacologist, Civil Aeromed Res Inst, Fed Aviation Agency, 62-65. *Mem:* AAAS; Am Physiol Soc; Am Soc Pharmacol & Exp Therapeut; Soc Neurosci. *Res:* Respiratory pharmacology and physiology; actions of drugs upon brain stem; blood platelet pharmacology; pathology. *Mailing Add:* Vet Admin Hosp Dartmouth-Hitchcock Med Ctr White River Junction VT 05001

ROSENSTEIN, ROBERT WILLIAM, b New York, NY, Mar 26, 44; m 66; c 2. INFECTIOUS DISEASES, ANTIGEN BIOCHEMISTRY. *Educ:* Univ Wis, BS, 64; Yale Univ, PhD(biochem), 69. *Prof Exp:* Group mgr, Diag Res & Develop, Ortho Diag, 78-79; assoc investr immunol, Howard Hughes Med Inst, 79-82; mgr, 82-88, TECH DIR, INDUST RES & DEVELOP, BECTON DICKINSON ADVAN DIAG, 88- *Mem:* Am Chem Soc; Am Soc Microbiologists; Am Asn Immunologists. *Res:* Rapid diagnostics for infectious diseases and hormones. *Mailing Add:* 4273 Bright Bay Way Ellicott City MD 21043

ROSENSTEIN, SHELDON WILLIAM, b Chicago, Ill, Oct 16, 27; m 55; c 3. ORTHODONTICS. *Educ:* Northwestern Univ, DDS, 51, MSD, 55. *Prof Exp:* Attend orthodontist, Children's Mem Hosp, Chicago, 55-74; PROF ORTHOD, DENT SCH, NORTHWESTERN UNIV, 74- *Concurrent Pos:* Prof orthod, Sch Med, St Louis Univ, 69-74. *Mem:* Sigma Xi. *Res:* Growth and development of orthodontics; clinical cleft lip and palate. *Mailing Add:* 4801 W Peterson-515 Chicago IL 60646

ROSENSTOCK, HERBERT BERNHARD, b Vienna, Austria, Dec 5, 24; nat US; m 50; c 3. SOLID STATE PHYSICS. *Educ:* Clemson Univ, BS, 44; Univ NC, MS, 50, PhD(physics), 52. *Prof Exp:* Theoretical physicist, US Naval Res Lab, 51-80; res physicist, Sachs & Freeman Assocs, Landover, Md, 80-87; INDEPENDENT CONSULT, 87- *Concurrent Pos:* Res contract adv, Off Naval Res, 65-66; vis prof, Univ Utah, 70-71. *Mem:* AAAS; Am Phys Soc; fel Sigma Xi; Fedn Am Sci. *Res:* Solid state theory, especially lattice dynamics, energy transfer, color centers, multiphonon absorption, amorphous solids and dislocations; atmospheric physics and light propagation; probability theory; heat conduction; radiation damage to electronic devices. *Mailing Add:* 5715 Janice Lane Temple Hills MD 20748-4714

ROSENSTOCK, PAUL DANIEL, b Brooklyn, NY, Apr 1, 35; m 56; c 2. ORGANIC CHEMISTRY. *Educ:* Polytech Inst Brooklyn, BS, 56; Pa State Univ, PhD(chem), 60. *Prof Exp:* Chemist, Ethyl Corp, Mich, 60; med chemist, Nat Drug Co, Pa, 63-64; chemist, Rohm & Haas Co, 64-67, group leader, Plant Trouble-Shooting Lab, 67-70, head, Chem Process Develop Lab, 70-74, mgr, Process Develop Dept, 74-81, mgr develop & environ control, Rohn Haas Del Valley, Inc, 81-87, mgr regulatory affairs, 88-90; VPRES, BCM ENGRS, INC, 90- *Concurrent Pos:* Mem, bd examiners, Hazard Control Mgr Cert Bd, 85-89. *Mem:* Water Pollution Control Asn; Air Pollution & Waste Mgt Asn; Am Inst Indust Hyg Asn. *Res:* Process development and design; predominantly on polymers and extrusion; environmental sciences; sanitary and environmental engineering. *Mailing Add:* 3598 Neshaminy Valley Dr Bensalem PA 19020

ROSENSTRAUS, MAURICE JAY, b Brooklyn, NY, Mar 13, 51; m 77; c 1. BIOLOGY. *Educ:* Rensselaer Polytech Inst, BS, 72; Columbia Univ, PhD(biol), 77. *Prof Exp:* Fel develop genetics, Princeton Univ, 76-78; Asst Prof Biol, Rutgers Univ, New Brunswick, 78-85; sr res scientist, Enzo Biochem, NY, 85-89; PRIN RES SCIENTIST, CYTOGEN CORP, PRINCETON, 89- *Mem:* AAAS; Genetics Soc Am; Sigma Xi. *Res:* Monoclonal antibodies for targeted delivery to tumors. *Mailing Add:* 40 Smith Rd Somerset NJ 08873

ROSENSTREICH, DAVID LEON, b New York, NY, Nov 16, 42; m 65; c 3. CELLULAR IMMUNOLOGY, ALLERGY. *Educ:* City Col New York, BS, 63; Sch Med, NY Univ, MD, 67. *Prof Exp:* Clin assoc, Nat Inst Allergy & Infectious Dis, 69-72; sr investr, Nat Inst Dent Res, 72-79; PROF, DEPT MED, ALBERT EINSTEIN COL MED, 80-, DIR, DIV ALLERGY & IMMUNOL, 81- *Concurrent Pos:* Vis assoc prof, Rockefeller Univ, 78-79. *Mem:* Am Asn Immunologists; Am Soc Clin Invest; Am Fedn Clin Res; Am Acad Allergy; Am Asn Physicians. *Res:* Genetic regulation of the immune response and resistance to infection. *Mailing Add:* Dept Med Albert Einstein Col Med 1300 Morris Park Ave Bronx NY 10461

ROSENSWEIG, JACOB, b Montreal, Que, Dec 1, 30; m 52; c 3. THORACIC SURGERY, CARDIOVASCULAR SURGERY. *Educ:* McGill Univ, BSc, 51, MD, CM, 55, PhD(exp surg), 60. *Prof Exp:* Fel cardio-thoracic surg, Jewish Gen Hosp, Montreal, 61-62, jr asst, 62-66, assoc surgeon, 66-71, dir clin teaching unit surg, 69-71; assoc prof surg, Sch Med, Univ Conn, 71-74; clin assoc prof surg, Med Sch, Univ Tenn, Memphis, 74-77; clin assoc prof surg, Univ Ark Health Sci, 77-84; CHAPMAN PROF VASCULAR RES & DIR, CHAPMAN VASCULAR LAB, SOUTHERN COL OPTOM, MEMPHIS, 84- *Concurrent Pos:* Clin fel cardiac surg, Royal Victoria Hosp, 61-63; consult thoracic surg, Mt Sinai Hosp, Ste Agathe, Que, 62-71 & thoracic & cardiovasc surg, Maimonides Hosp, Montreal, 66-71; Que Heart Found grant, Jewish Gen Hosp, Montreal, 63-70; Med Res Coun Can grant, Lady Davis Inst Med Res, Montreal, 67-73, head surg div, Exp Surg, 67-68, sr investr, 68-71; vis consult, Nat Heart Inst, Md, 68; lectr surg, McGill Univ, 70; chief surg, Mt Sinai Hosp, Hartford, Conn, 71-74; consult cardiovasc surg, Vet Admin Hosp, Newington, Conn, 72-74; mem staff, Baptist Mem, St Joseph & Methodist Hosps, Memphis, 74-, St Francis Hosp, Memphis, 76-, chmn, Cardiovasc Surg & Thoracic Surg, 81-82; mem coun thrombosis & fel coun cardiovasc surg, Am Heart Asn. *Mem:* Am Col Surgeons; Am Asn Thoracic Surg; Soc Thoracic Surg; fel Am Col Cardiol. *Res:* Circulatory and pulmonary support systems; transplantation, specifically role of platelets in rejection; cerebrovascular and occular vascular disease. *Mailing Add:* Rosenweig Clin 3960 Knight Arnold Rd No 202 Memphis TN 38118-3035

ROSENSWEIG, NORTON S, b Brooklyn, NY, Aug 1, 35; m 63; c 3. GASTROENTEROLOGY, CLINICAL NUTRITION. *Educ:* Princeton Univ, AB, 57; NY Univ, MD, 61. *Prof Exp:* Instr med, Sch Med, Johns Hopkins, 64-66; res internist, US Army Med Res & Nutrit Lab, 66-69; asst prof med, Sch Med, Colo Univ, 67-69; assoc prof med, Med Sch, Cornell Univ, 78-79; from asst prof to assoc prof clin med, 69-78, ASSOC CLIN PROF MED, COL PHYSICIANS & SURGEONS, COLUMBIA UNIV, 79- *Concurrent Pos:* Dir clin GI, St Luke's Hosp, 69-78; attend physician, St Luke's-Roosevelt Hosp Ctr, 69-78, 79-; prin investr, NIH & US Army Res Grants, 70-75; gov, Am Col Gastroenterol, 78-80; chief gastroenterol & nutrit, N Shore Univ Hosp, 78-79. *Mem:* Am Gastroenterol Asn; Am Soc Clin Nutrit; Am Fedn Clin Res; Am Col Physicians; Am Col Gastroenterol; Am Asn Study Liver Dis. *Res:* Difference between blacks and whites in the incidence of lactase dificiency; series of papers on dietary, vitamin and hormone regulation of human intestinal enzymes; non-cardiac chest pain. *Mailing Add:* 125 E 74th St New York NY 10021

ROSENSWEIG, RONALD E(LLIS), b Hamilton, Ohio, Nov 8, 32; m 54; c 3. CHEMICAL ENGINEERING, FLUID MECHANICS. *Educ:* Univ Cincinnati, ChE, 55; Mass Inst Technol, SM, 56, ScD(chem eng), 59. *Prof Exp:* Asst prof chem eng, Mass Inst Technol, 59-62; sect chief, Avco Corp, Mass, 62-69; pres & tech dir, 69-72, chmn, treas & tech dir, 72-73, mem bd dirs, Ferrofluidics Corp, 69-83; res assoc, Corp Res Labs, 73-76, sr res assoc, 76-85, SCI ADV, EXXON RES & ENG CO, 85- *Concurrent Pos:* Indust consult, Nat Res Corp, Linde, Dynatech, 59-62; vis prof chem eng, Univ Minn, 80, vis prof physics, Univ Chicago, 90; steering comt chmn, ICMF, 77- *Honors & Awards:* IR-100 Awards, 65, 68, 70 & 72; Alpha Chi Sigma Award, Am Inst Chem Engrs, 85. *Mem:* Nat Acad Eng; Am Inst Chem Engrs; Inst Elec & Electronics Engrs; Am Phys Soc. *Res:* Magnetic fluids; turbulent mixing; reentry ablation; sensors; fluidized beds; ferrohydrodynamics. *Mailing Add:* 34 Gloucester Rd Summit NJ 07901

ROSENTHAL, ALEX, b Scollard, Alta, Oct 16, 14; m 47; c 4. ORGANIC CHEMISTRY. *Educ:* Univ Alta, BSc, 43, BEd, 47, MSc, 49; Ohio State Univ, PhD(chem), 52. *Prof Exp:* Teacher pub sch, Can, 34-37; prin & teacher high sch, 38-41, 44-46; instr, War Vet Sch, 46-47; fel, Univ Utah, 52-53; from asst prof to assoc prof, 53-64, PROF CHEM, UNIV BC, 65- *Concurrent Pos:* Vis prof, Cambridge Univ, 63-64; Killam sr res fel, 75-76; vis scientist, Salk Inst, Sloan Kettering Inst, 75-76. *Mem:* Am Chem Soc; fel Chem Inst Can. *Res:* Photoamidation studies; branched-chain sugar nucleosides; glycosyl amino acids; synthesis of analogs of nucleoside antibiotics. *Mailing Add:* Dept Chem Univ BC Vancouver BC V6T 1W5 Can

ROSENTHAL, ALLAN LAWRENCE, b Montreal, Que, Feb 16, 48; m 71; c 2. ANTIBIOTICS, ANTI-INFLAMMATORIES. *Educ:* McGill Univ, BS, 69; Fla State Univ, MS, 72; State Univ NY Upstate Med Ctr, PhD(microbiol), 75. *Prof Exp:* Fel, Brookhaven Nat Lab, 75-78; asst prof biol, Tex Christian Univ, 78-80; scientist, Alcon Labs, Inc, 80-91; VPRES CLIN RES, DIV PHARMACEUT, ALLERGAN INC, 91- *Concurrent Pos:* Adj prof, Tex Christian Univ, 80- *Res:* Genetic transformation of streptococcus pneumoniae and the acholeplasmas, with specific reference to the role of nucleuses; ocular anti-infective and anti-inflammatory agents. *Mailing Add:* Allergan Inc Pharmaceut Div 2525 Dupont PO Box 19534 Irvine CA 92713-9534

ROSENTHAL, ARNOLD JOSEPH, b New York, NY, July 9, 22; m 47; c 4. POLYMER CHEMISTRY, ORGANIC CHEMISTRY. *Educ:* Polytech Inst Brooklyn, PhD, 58. *Prof Exp:* Sr res chemist, Celanese Corp, Cumberland, Md, 41-44, 47, sr res assoc, Celanese Res Co, Summit, NJ, 47-82; vpres, Exec Sci Inst, 55-87, PARTNER, ESI ASSOCS, WHIPPANY, NJ, 87- *Concurrent Pos:* Ed, Qual Control & Appl Statist, 56-87 & Opers Res & Mgt Sci, 61-87. *Mem:* Am Chem Soc; Opers Res Soc Am; Am Soc Qual Control; Am Statist Asn. *Res:* Synthetic fibers; mechanisms of chemical reactions; physical chemistry of polymers; statistical design and analysis of experiments; operations research. *Mailing Add:* Eight Ford Hill Rd Whippany NJ 07981

ROSENTHAL, ARTHUR FREDERICK, b Brooklyn, NY, Aug 3, 31; m 67; c 3. BIOCHEMISTRY. *Educ:* Antioch Col, BS, 54; Harvard Univ, AM, 56, PhD(biochem), 60; Am Bd Clin Chem, dipl, 74. *Prof Exp:* Fel chem, Univ Birmingham, 61-62 & Auburn Univ, 62; CHIEF BIOCHEM, LONG ISLAND JEWISH-HILLSIDE MED CTR, 62-; ASSOC PROF PATH, STATE UNIV NY, STONY BROOK, 72- *Concurrent Pos:* Prin investr, NIH Proj Grant, 63-; adj prof biochem, PhD Prog Biochem, City Univ New York, 74-; adj prof, City Univ NY; exten assoc prof, Univ Puerto Rico, 76- *Mem:* Am Chem Soc; The Chem Soc; Am Asn Clin Chemists; Am Soc Biol Chemists. *Res:* Phospholipid chemistry and biochemistry; synthetic organophosphorus chemistry; lipid chemistry and metabolism; clinical chemistry; protein modification. *Mailing Add:* 21 Radnor Rd Great Neck NY 11023-1998

ROSENTHAL, DONALD, b Princeton, NJ, July 16, 26; m 55; c 4. ANALYTICAL CHEMISTRY, PHYSICAL CHEMISTRY. *Educ:* Princeton Univ, AB, 49; Columbia Univ, AM, 50, PhD(chem), 55. *Prof Exp:* Instr chem, Columbia Univ, 54-55 & Univ Minn, 55-56; from instr to asst prof, Univ Chicago, 56-61; assoc prof, 61-66, exec officer, Dept Chem, 81-88, PROF CHEM, CLARKSON COL, 66- *Concurrent Pos:* Sr res assoc, Brandeis Univ, 70-71; ed, Comput Chem Educ Newslett, 81-88; mem, comt comput chem educ, Div Chem Educ, Am Chem Soc. *Mem:* AAAS; Am Chem Soc; NY Acad Sci. *Res:* Solution chemistry; equilibria, acidity and kinetics in aqueous and non-aqueous solutions; potentiometry and spectrophotometry; instrumental methods of analysis; chromatography; computer simulation. *Mailing Add:* Dept Chem Clarkson Univ Potsdam NY 13676

ROSENTHAL, F(ELIX), b Munich, Ger, Feb 6, 25; nat US; c 3. MECHANICS. *Educ:* Ill Inst Technol, BS, 47, MS, 48, PhD(appl mech), 52. *Prof Exp:* Instr, Ill Inst Technol, 47-50, res engr, Armour Res Found, 48-54; proj engr & head, Math Sect, Res Ctr, Clevite Corp, 54-58; staff scientist, Raytheon Co, 58-61; adv engr, Fed Systs Div, IBM Corp, 61-68; vpres res & develop, Oceanog, Inc, 68-69; independent consult mech design, underwater acoust & signal processing, 69-71; head appl mech br, 71-78, RES MECH ENG, NAVAL RES LAB, 78- *Mem:* Am Soc Mech Engrs; Fedn Am Sci; sr mem Inst Elec & Electronics Engrs; World Federalist Asn. *Res:* Applied mechanics; elasticity, dynamics; acoustics; high speed automatic machines; piezoelectric transducers; wave propagation; oceanography; acoustic and seismic signal processing; theory & application of noise cancelling; structural dynamics; fluid mechanics; underwater engineering; computer structural design analysis methods; underwater acoustics; noise cancellation. *Mailing Add:* Code 4220 Naval Res Lab Washington DC 20375

ROSENTHAL, FRED, b Breslau, Ger, Feb 26, 31; US citizen; m 57; c 2. NEUROPHYSIOLOGY. *Educ:* Univ Calif, Berkeley, BA, 52, PhD(psychol), 56; Stanford Univ, MD, 60. *Prof Exp:* Res physiologist, Univ Calif, Berkeley, 60-62; fel neurophysiol, Univ Wash, 62-64; asst prof, New York Med Col, Flower & Fifth Ave Hosps, 64-69, assoc prof, 69-70; ASSOC RES PHYSIOLOGIST, CARDIOVASC RES INST, MED SCH, UNIV CALIF, SAN FRANCISCO, 70- *Mem:* Am Physiol Soc; Biophys Soc. *Res:* Description of the extracellular electric field around cortical pyramidal tract neurons; effects of radiation on brain function; relations between somatic motor systems and cardiovascular changes. *Mailing Add:* 2340 Sutter St San Francisco CA 94115

ROSENTHAL, FRITZ, b Wuerzburg, Ger, July 4, 11; nat US; m 38; c 2. INDUSTRIAL CHEMISTRY. *Educ:* Univ Bern, PhD(chem), 35. *Prof Exp:* Res chemist, Gen Elec Co, 36-38 & Forest Prod Chem Co, 38-39; plastics technologist, Univ Tenn, 39-42; chem engr, Radio Corp Am, 42-46; dir res, Nat Plastics, Inc, 46-49; sr res chemist, Nashua Corp, 49-54; proj leader, Armour Res Found, Ill Inst Technol, 54-56; dir prod develop, Knowlton Bros, Inc, NY, 56-68; sr assoc res & develop dept, Riegel Paper Corp, NJ, 68-70; SR ASSOC, SCM CORP, 70- *Mem:* Am Chem Soc; Tech Asn Pulp & Paper Indust; Am Inst Chemists; Sigma Xi. *Res:* Plastics and paper technology; high polymer literature; electrofax; xerography. *Mailing Add:* Cherry Hill Apt 402 E 2151 Rte 38 Cherry Hill NJ 08002

ROSENTHAL, GERALD A, b New York, NY, Jan 9, 39; m 60; c 3. PLANT BIOCHEMISTRY, CHEMICAL ECOLOGY. *Educ:* Syracuse Univ, BS, 62; Duke Univ, MF, 63, PhD(plant physiol, biochem), 66. *Prof Exp:* NIH fel biochem, Med Col, Cornell Univ, 66-69; asst prof biol, Case Western Reserve Univ, 69-72; from asst prof to assoc prof, 73-80, PROF BIOL, UNIV KY, 80- *Concurrent Pos:* NIH grant, 70-73, 74-78, 79-82 & 82-87; NSF grants, 73-75, 76-79, 80-83 & 89-93; Lady Davis prof agr entom, Hebrew Univ Jerusalem, 79; Fulbright-Hays res scholar, Univ Louis Pasteur, 85. *Mem:* Int Soc Chem Ecol. *Res:* Plant amino acid metabolism and enzymology; biochemistry of plant-insect interaction; chemical-ecology. *Mailing Add:* T H Morgan Sch Biol Sci Univ Ky Lexington KY 40506

ROSENTHAL, GERSON MAX, JR, b Pittsfield, Mass, Mar 13, 22; m 54; c 1. ECOLOGY, ZOOLOGY. *Educ:* Dartmouth Col, AB, 43; Univ Calif, MS, 51, PhD(zool), 54. *Prof Exp:* Asst zool, Univ Chicago, 46-48; asst, Univ Calif, 49-51, assoc, 52-53; Ford fel, Univ Chicago, 54, instr nat sci, 54-56, from asst prof to assoc prof, 59-76, prof biol, social sci & geog, 76-89, EMER PROF BIOL, UNIV CHICAGO, 87- *Concurrent Pos:* Consult ecol, Oak Ridge Inst Nuclear Res, 63-76; vis scientist, Argonnne Nat Lab, 69-70, consult ecol,

70-74. *Honors & Awards:* Quantrell Award, 63. *Mem:* AAAS; Ecol Soc Am; Am Inst Biol Sci; Soc Study Evolution. *Res:* Physiological ecology; energetics of ecosystems; radioecology; biogeography; relationship of organisms to microclimates; animal populations. *Mailing Add:* Univ Chicago 17 Gates Blake Chicago IL 60612

ROSENTHAL, HAROLD LESLIE, b Elizabeth, NJ, Mar 26, 22; m 47; c 2. BIOCHEMISTRY, NEUROSCIENCES. *Educ:* Univ NMex, BSc, 44; Rutgers Univ, PhD(biochem), 51. *Prof Exp:* Res biochemist, Rutgers Univ, 49-51; res biochemist, Philadelphia Gen Hosp, 51; instr med, Tulane Univ, 51-53; chief biochem, Rochester Gen Hosp, 53-58; from asst prof to assoc prof, 58-64, chmn dept, 58-74, prof, 65-87, EMER PROF MED SCI, SCH DENT, WASHINGTON UNIV, 87- *Concurrent Pos:* Res assoc, Minerva Found Med Res, Helsinki, Finland, 66; Nat Acad Sci exchange scientist, Inst Nutrit, Budapest, Hungary, 73-74. *Mem:* Am Chem Soc; Am Inst Nutrit; Am Inst Chemists; Am Soc Biol Chemists; fel AAAS; Sigma Xi. *Res:* Intermediary metabolism; vitamins and radioactive isotopes; neurochemistry; brain research. *Mailing Add:* 7541 Teasdale University City MO 63130

ROSENTHAL, HENRY BERNARD, b Philadelphia, Pa, Dec 31, 17; m 45; c 2. NUCLEAR SAFETY ENGINEERING, RISK ASSESSMENT. *Educ:* Trenton State Col, BS, 40; Rutgers Univ, MS, 51. *Prof Exp:* Group leader, critical nuclear exps, ANP Dept, Gen Elec Co, 51-56; supvr, critical reactor exps, Nuclear Div, Martin Marietta , 57-68, mgr, nuclear survivability, Orlando Div, 68-75; sr nuclear safety engr, US Dept Energy, 76-84; RETIRED. *Concurrent Pos:* Self-employed consult, nuclear safety, 84- *Mem:* Am Nuclear Soc; AAAS; Am Soc Safety Engrs; sr mem Syst Safety Soc; Soc Risk Analysis; Sigma Xi. *Res:* Safety analysis review, appraisal and risk assessment systems; missile system survivability and hardness to nuclear weapons effects; developmental experimental radiation, nuclear criticality and nuclear reactor physics. *Mailing Add:* 107 Water Oak Dr Sanford FL 32773

ROSENTHAL, HOWARD, b Brooklyn, NY, Sept 15, 24; m 51; c 2. CHEMICAL ENGINEERING. *Educ:* City Col New York, BChE, 44; NY Univ, MChE, 49. *Prof Exp:* Res engr, Sunray Elec Co, 46-47; asst chem eng, NY Univ, 48-49; mem tech staff, RCA Labs, 49-59, indust hyg & safety engr, RCA Corp, 59-63, sr tech adminr, 63-66, mgr tech admin, 66-68, adminr staff serv, Res & Eng, 68-70, dir, 70-72, staff vpres eng, 72-86; RETIRED. *Mem:* Inst Elec & Electronics Engrs. *Res:* Engineering management. *Mailing Add:* 3600 Conshohocken Ave Philadelphia PA 19131

ROSENTHAL, IRA MAURICE, b New York, NY, June 11, 20; m 43; c 2. PEDIATRICS. *Educ:* Univ Ind, AB, 40, MD, 43. *Prof Exp:* From asst prof to prof, 53-90, head dept, 74-82, EMER PROF PEDIAT, ABRAHAM LINCOLN SCH MED, UNIV ILL COL MED, 90- *Concurrent Pos:* Clin prof, Loyola Univ, Chicago, 90- *Mem:* Soc Pediat Res; Endocrine Soc; AMA; Am Acad Pediat; Am Fedn Clin Res; Am Pediat Soc; Sigma Xi. *Res:* Pediatric endocrinology and metabolism. *Mailing Add:* Dept Pediat Univ Chicago 5841 S Maryland Ave Box 118 Chicago IL 60637

ROSENTHAL, J WILLIAM, b New Orleans, La, Oct 30, 22; m 45; c 2. OPHTHALMOLOGY, SURGERY. *Educ:* Tulane Univ, BS, 43, MD, 45; Univ Pa, MSc, 52, Am Bd Ophthal, dipl, 54. *Hon Degrees:* DSc, Univ Pa 62. *Prof Exp:* From instr to assoc clin prof, 52-71, CLIN PROF OPHTHAL, MED SCH, TULANE UNIV, 71- *Concurrent Pos:* Sr ophthal surg, Touro Infirmary & Eye, Ear, Nose & Throat Hosp, 52-; chief ophthal, United Med Ctr Hosp & Smithsoman Inst; cur, Am Acad Ophthal Found Mus; trustee, Am Optical Co Mus. *Honors & Awards:* Serv Award, Am Acad Ophthal, 85, Honor Award, 90. *Mem:* French Ophthal Soc; fel Am Col Surgeons; Am Acad Ophthal & Otolaryngol; fel Royal Soc Med; fel Int Col Surgeons. *Res:* Tularemia; pterygia; glaucoma with brachydactyly and spherophakia; trachoma; accident prone children; microsurgery. *Mailing Add:* 3715 Prytania St New Orleans LA 70115

ROSENTHAL, JEFFREY, b El Paso, Tex, Sept 18, 53; m 75; c 3. PHOTOELECTROCHEMISTRY. *Educ:* Univ Minn, Duluth, BS, 75; Purdue Univ, PhD(anal chem), 80. *Prof Exp:* Asst prof, Oakland Univ, 80-82; from asst prof to assoc prof, 82-90, PROF ANALYTICAL CHEM, UNIV WIS, RIVER FALLS, 90- *Mem:* Am Chem Soc. *Res:* Semiconductor photoelectrochemistry; design of instrumentation (optimized for transient measurements); computer controlled instrumentation and chemical software in CTT. *Mailing Add:* Dept Chem Univ Wis River Falls WI 54022

ROSENTHAL, JENNY EUGENIE, electro-optics, spectroscopy, for more information see previous edition

ROSENTHAL, JOHN WILLIAM, b New York, NY, Sept 6, 45; m 68. MATHEMATICS. *Educ:* Mass Inst Technol, BS, 65, PhD(math), 68. *Prof Exp:* Asst prof math, State Univ NY Stony Brook, 68-71, NSF res contract, 69-71; asst prof, 71-74, ASSOC PROF MATH, ITHACA COL, 74- *Concurrent Pos:* Vis assoc prof, Mich State Univ, 77-78; vis lectr, Univ Sydney, Australia, 78. *Mem:* Math Asn Am; Am Math Soc; Asn Symbolic Logic. *Res:* Mathematical logic; model theory; categoricity and stability; infinitary languages; set theory; computational complexity; algebraically closed fields; asymptotic methods in combinatorics. *Mailing Add:* Dept Math Ithaca Col Danby Rd Ithaca NY 14850

ROSENTHAL, JUDITH WOLDER, b New York, NY, May 12, 45; c 1. SCIENCE FOR LIMITED ENGLISH PROFICIENT STUDENTS. *Educ:* Brown Univ, BA, 67, PhD(physiol chem), 71. *Prof Exp:* Fel, Univ Toronto, 71-72 & McMaster Univ, 72-73; PROF BIOL, KEAN COL NJ, 74- *Mem:* Nat Asn Sci Technol & Soc; AAAS; Nat Asn Bilingual Educ. *Res:* Science, technology and society; science for limited English proficient students. *Mailing Add:* Dept Biol Kean Col NJ Union NJ 07083

ROSENTHAL, KENNETH LEE, b Chicago, Ill, Nov 29, 50; m 77; c 2. CELL MEDIATED IMMUNITY, VIROLOGY. *Educ:* Univ Ill, Urbana, BSc, 72, MSc, 74; McMaster Univ, Can, PhD(med sci), 78. *Prof Exp:* Fel immunol, Scripps Clin Res Found, Calif, 78-80 & Univ Zurich, Switz, 80-81; asst prof, 81-86, ASSOC PROF IMMUNOL, MCMASTER UNIV, CAN, 86- *Concurrent Pos:* Damon Runyon Fel, Walter Winchell Cancer Found, 78 & Swiss Nat Sci Found, 80; scholar, Med Res Coun Can, 82-87. *Mem:* Am Asn Immunologists; Can Soc Immunol; Am Soc Microbiol; Am Soc Virol. *Res:* Generation, specificity and role of anti-viral cytotoxic T lymphocytes; acquired immunodeficiency syndrome (AIDS); virus-lymphocyte interactions. *Mailing Add:* Dept Path McMaster Univ Health Sci Ctr Hamilton ON L8N 3Z5 Can

ROSENTHAL, LEE, b Brooklyn, NY, Nov 28, 37. ELECTRICAL & SYSTEMS ENGINEERING. *Educ:* Polytech Inst Brooklyn, BEE, 58, PhD(elec eng), 67; Calif Inst Technol, MS, 59. *Prof Exp:* Engr, Hughes Aircraft Co, 58-59; lectr elec eng, City Col New York, 62-66; asst prof, Stevens Inst Technol, 66-70; asst prof, Hofstra Univ, 70-72; assoc prof, 72-80, PROF, FAIRLEIGH DICKINSON UNIV, 80- *Mem:* Am Soc Eng Educ; NY Acad Sci; Sigma Xi. *Res:* Linear integrated circuits; process control. *Mailing Add:* Dept Eng Technol Fairleigh Dickinson Univ Teaneck NJ 07666

ROSENTHAL, LEONARD JASON, b Boston, Mass, June 23, 42; m 76; c 2. VIROLOGY, MOLECULAR BIOLOGY. *Educ:* Univ Vermont, BA, 64; Southern Ill Univ, MA, 66; Kans State Univ, PhD(bact), 69. *Prof Exp:* Res fel, Harvard Med Sch, 69-71, instr, 72-73; Europ Molecular Biol Orgn fel, Univ Geneva, 73-74 & Leukemia Soc Am Spec fel, NIH, 74-75; asst prof microbiol, 75-82, ASSOC PROF MICROBIOL & PEDIAT, GEORGETOWN UNIV SCH MED & DENT, 82- *Concurrent Pos:* Mem, sci adv bd, Int Biotechnol, Inc; mem, Vincent L Lombardi Cancer Ctr, 75-; panelist, Res Initiation & Support Prog, NIH, 75-76, reviewer, 80-81; contract reviewer, Nat Inst Child Health & Human Develop, 79-81; Leukemia Soc Am Scholar, 80-85; sci adv bd, Int Biotechnol, Inc, 83-87; mem special study sect, Nat Heart, Lung, & Blood Inst, 83; hon prof virol, Wuhan Univ, 88-, hon prof microbiol, Jinan Univ, 88- *Mem:* Am Soc Microbiol; Sigma Xi; Am Soc Virol. *Res:* Determining the association of human herpes viruses with cancer; author of numerous books, chapters, and papers. *Mailing Add:* Dept Microbiol Georgetown Univ 3900 Reservoir Rd Washington DC 20007

ROSENTHAL, LOIS C, b Boston, Mass, Mar 20, 46; c 3. PHYSICAL CHEMISTRY. *Educ:* Simmons Col, BS, 68; Univ Calif, Berkeley, PhD(chem), 75. *Prof Exp:* Instr chem, Diablo Valley Col, 75-77; asst prof, 77-84, ASSOC PROF CHEM, UNIV SANTA CLARA, 84- *Mem:* Am Chem Soc. *Res:* Molecular motion in fluids; infrared and Raman spectroscopy. *Mailing Add:* Univ Santa Clara Santa Clara CA 95053

ROSENTHAL, LOUIS AARON, b New York, NY, Aug 16, 22; wid; c 3. ELECTRONICS ENGINEERING. *Educ:* City Col New York, BEE, 43; Polytech Inst Brooklyn, MEE, 47. *Prof Exp:* Elec engr, Star Elec Motor Co, NJ, 43; instr elec eng, Lehigh Univ, 44; from assoc prof to prof, 44-81, EMER PROF ELEC ENG, RUTGERS UNIV, NEW BRUNSWICK, 81- *Concurrent Pos:* Consult, US Naval Surface Weapons Ctr, 46-, Union Carbide Corp, 50-86 & var chemcorps. *Mem:* Sr mem Inst Elec & Electronics Engrs; Sigma Xi. *Res:* Corona discharge processing; static electricity hazards and control; electroexplosive devices. *Mailing Add:* 384B Stirling Dr Cranbury NJ 08512-3922

ROSENTHAL, MARCIA WHITE, radiation biology, for more information see previous edition

ROSENTHAL, MICHAEL DAVID, b Brooklyn, NY, Dec 30, 43; m 81; c 2. SOLID STATE PHYSICS. *Educ:* Wesleyan Univ, BA, 65; Cornell Univ, PhD(physics), 72. *Prof Exp:* Res assoc sci policy, Prog Sci, Technol & Soc, Cornell Univ, 71-73; instr physics, Dept Physics, Rutgers Univ, 73-74; asst prof physics, Swarthmore Col, 74-77; phys sci officer, Non Proliferation Bur, US Arms Control & Disarmament Agency, 77-81; SR OFFICER, INT ATOMIC ENERGY AGENCY, 81- *Concurrent Pos:* Vis scientist, Cornell Univ, 74; consult, math policy res, 76-77. *Mem:* Sigma Xi; Am Phys Soc; Am Nuclear Soc; Fedn Am Scientists. *Res:* Radiofrequency size effects in metals and electrical properties of metal-ammonia compounds and solutions; international safeguards and non-proliferation. *Mailing Add:* 8809 Gallant Green Dr McLean VA 22102

ROSENTHAL, MICHAEL R, b Youngstown, Ohio, Dec 2, 39; m 63; c 3. INORGANIC CHEMISTRY, ENVIRONMENTAL SCIENCES. *Educ:* Western Reserve Univ, AB, 61; Univ Ill, MS, 63, PhD(chem), 65. *Prof Exp:* From asst prof to assoc prof, 65-73, prof chem, Bard Col, 73-84, assoc dean acad affairs, 80-84; vpres acad affairs & dean, St Mary's Col Md, 84-89, prof chem, 84-89; PROVOST & DEAN FAC, PROF CHEM, SOUTHWESTERN UNIV, 89- *Mem:* AAAS; Am Chem Soc; Royal Soc Chem. *Res:* Coordination chemistry; water chemistry in the natural environment; water pollution. *Mailing Add:* 2914 Gabriel View Dr Georgetown TX 78628

ROSENTHAL, MIRIAM DICK, LIPID METABOLISM, CELL BIOLOGY. *Educ:* Brandeis Univ, PhD(develop biol), 74. *Prof Exp:* ASSOC PROF BIOCHEM, EASTERN VA MED SCH, 77- *Mailing Add:* Dept Biochem Med Sch Eastern Va PO Box 1980 Norfolk VA 23501

ROSENTHAL, MURRAY WILFORD, b Greenville, Miss, Feb 25, 26; m 49; c 2. CHEMICAL ENGINEERING. *Educ:* La State Univ, BS, 49; Mass Inst Technol, PhD(chem eng), 53. *Prof Exp:* Develop engr, 53-55, lectr, Oak Ridge Sch Reactor Technol, 55-56, leader reactor anal group, 55-59, group leader, Anal Advan Reactor Div, 59-61, proj engr, Pebble Bld Reactor Exp, 61-63, head long range planning sect, 63-65, dir, planning & eval, 65, tech asst to US Atomic Energy Comn asst gen mgr reactors, 65-66, dir, Molten Salt Reactor Prog, 66-73, actg dep lab dir, 73-74, assoc lab dir, Advan Energy Systs, 74-89, DEP DIR, OAK RIDGE NAT LAB, 89- *Concurrent Pos:* Vis prof, Mass Inst Technol, 61; tech asst to asst gen mgr reactors, Atomic Energy Comn, 65-66; US deleg, Panel Utilization Thorium in Power Reactors, Int Atomic Energy Agency, Vienna, 65, Manila Conf Probs & Prospects of Nuclear Power Appln Develop Countries, 66 & Int Conf on Nuclear Energy, Geneva, 71; adv panel fusion energy, Energy Res & Prod Subcomt, Comt Space Sci & Technol, US House Rep, 81-86; adv ed, Eng Sci & Technol News, 80-82; magnetic fusion adv comt, 82-84, Japan-US Magnetic Fusion Coor Comt, 82-88, US-USSR Joint Fusion Power Coor Comt, 85-88; mem adv panel, Off Technol Assessment Magnetic Fusion Res & Demonstration, 86-88. *Mem:* Nat Acad Eng; fel Am Nuclear Soc; Sigma Xi; AAAS. *Res:* Energy research and development. *Mailing Add:* Oak Ridge Nat Lab Bldg 4500N M/S 6241 PO Box 2008 Oak Ridge TN 37831-6241

ROSENTHAL, MURRAY WILLIAM, b Stamford, Conn, Aug 15, 18; m 47; c 2. PHARMACEUTICAL CHEMISTRY. *Educ:* Pa State Univ, BS, 41, MS, 42. *Prof Exp:* Chemist, Philadelphia Qm Depot, 42-43 & Edcan Labs, 46-48; asst res dir, McKesson-Robbins, 48-51; from asst res dir to res dir, Block Drug Co, Inc, 51-62, vpres res & develop, 62-83; CONSULT, PHARMACEUT INDUST, 83- *Concurrent Pos:* Dir, Therapeut Res Found. *Mem:* AAAS; Am Chem Soc; Am Pharmaceut Asn. *Res:* Development of proprietary pharmaceuticals and toiletries. *Mailing Add:* 688B Old Nassau Rd Jamesburg NJ 08831

ROSENTHAL, MYRON, BRAIN METABOLISM, ELECTROPHYSIOLOGY. *Educ:* Duke Univ, PhD(physiol & pharmacol), 69. *Prof Exp:* PROF NEUROL & PHYSIOL BIOPHYS, SCH MED, UNIV MIAMI, 77- *Mailing Add:* Sch Med Univ Miami Miami FL 33101

ROSENTHAL, NATHAN RAYMOND, b Washington, DC, Oct 29, 25; m 55; c 4. BIOCHEMISTRY. *Educ:* Georgetown Univ, BS, 49, MS, 51, PhD(biochem), 57. *Prof Exp:* Chemist, NIH, 51-54; biochemist, Walter Reed Army Inst Res, 54-58; chemist, Bur Drugs, Food & Drug Admin, 58-90; RETIRED. *Honors & Awards:* Meritorious Achievement Award, US Dept Army, 57. *Mem:* Am Chem Soc; Asn Off Anal Chem; Sigma Xi. *Res:* Biochemistry of endocrine glands and their regulation and metabolism; biological activity of organophosphorous pesticides; manufacturing controls on radiopharmaceutical, antineoplastic, cardio-renal, psychopharmacologic, new drugs and investigational new drugs. *Mailing Add:* 18708 Bloomfield Rd Olney MD 20832

ROSENTHAL, PETER (MICHAEL), b New York, NY, June 1, 41; m 60, 85; c 5. MATHEMATICS. *Educ:* Queens Col, NY, BS, 62; Univ Mich, Ann Arbor, MA, 63, PhD(math), 67. *Prof Exp:* Instr math, Univ Toledo, 66-67; from asst prof to assoc prof, 67-76, PROF MATH, UNIV TORONTO, 76- *Mem:* Am Math Soc; Can Math Soc. *Res:* Operators on Hilbert and Banach spaces. *Mailing Add:* Dept Math Univ Toronto Toronto ON M5S 1A1 Can

ROSENTHAL, RICHARD ALAN, b Newark, NJ, Aug 29, 36; div; c 2. MECHANICAL & OPTICAL ENGINEERING. *Educ:* Mass Inst Technol, BSME, 58, MSME, 59; Worcester Polytech Inst, MBA, 87. *Prof Exp:* Mem tech staff, Bell Tel Labs, Inc, 59-64; sr engr, Itek Corp, 64-65, staff engr, 65-68; sr optical engr, 68-71, sect mgr, 71-79, TECH MGR, INSTRUMENTATION & ELECTRONICS ENG DIV, POLAROID CORP, 79- *Mem:* Sr mem Am Soc Mech Engrs. *Res:* Electrooptical and electromechanical instruments and instrumentation systems. *Mailing Add:* Eng Instrumentation Develop Polaroid Corp Cambridge MA 02139

ROSENTHAL, RUDOLPH, b Atlantic City, NJ, Mar 24, 23; m 60; c 2. ORGANIC CHEMISTRY. *Educ:* Temple Univ, AB, 44, MA, 48; Univ Southern Calif, PhD(chem), 51. *Prof Exp:* Res chemist, Allied Chem Corp, 51-61; sr res chemist, Glenolden, 61-81, SR RES CHEMIST, ATLANTIC RICHFIELD CO, NEWTOWN SQUARE, 81- *Mem:* Am Chem Soc. *Res:* Olefin oxides; catalytic oxidation; nitration of olefins; isocyanates. *Mailing Add:* 484 Hilldale Ave Broomall PA 19008-3104

ROSENTHAL, SAUL HASKELL, b Brooklyn, NY, Nov 14, 36; m 64; c 2. PSYCHIATRY. *Educ:* Harvard Univ, BA, 58, Harvard Med Sch, MD, 62; Am Bd Psychiat & Neurol, dipl, 69. *Prof Exp:* Intern med, Boston City Hosp, Harvard Med Sch, 62-63, resident psychiat, Mass Ment Health Ctr, 63-66; psychiatrist, US Air Force, 66-68; asst prof, 68-69, coordr residency training, 68-72, assoc prof psychiat, Univ Tex Med Sch San Antonio, 69-73; PRIV PRACT, 73- *Concurrent Pos:* Assoc clin prof psychiat, Univ Tex Med Sch, San Antonio, 73- *Mem:* Fel Am Psychiat Asn; AMA. *Res:* Depression; electrosleep. *Mailing Add:* Oak Hills Med Bldg 7711 Louis Pasteur Dr San Antonio TX 78229

ROSENTHAL, SAUL W, b New York, NY, Aug 13, 18; m 54; c 2. ELECTROPHYSICS, ELECTRICAL ENGINEERING. *Educ:* Polytech Inst Brooklyn, BEE, 48, MEE, 52. *Prof Exp:* ASSOC PROF ELECTROPHYS, POLYTECH INST NEW YORK, 61-, ASST DIR, MICROWAVE RES INST, 78- *Concurrent Pos:* Chmn, Comt C-95 RF Radiation Hazards, Am Nat Standards Inst, 68-; chmn, Int Working Group Measurements Related Interaction Electromagnetic Fields Biol Systs, Int Union Radio Sci, 76-; mem, Satellite Power Syst Microwave Health & Ecol Rev Group, Environ Protection Agency, 78- *Mem:* Sigma Xi; NY Acad Sci; Inst Elec & Electronics Engrs; Bioelectromagnetics Soc; Int Union Radio Sci. *Res:* Interaction of electromagnetic fields with biological systems and microwave measurements. *Mailing Add:* 12 Milford Lane Melville NY 11747

ROSENTHAL, SOL ROY, b Russia; nat US; m 50, 72; c 2. PATHOLOGY, IMMUNOLOGY. *Educ:* Univ Ill, BS, MD, MS, PhD(path). *Prof Exp:* Intern, Cook County Hosp, 27-29, resident path; instr, Univ Ill Col Med, 30-32, instr bact & pub health, 34-36, assoc, 36-40, asst prof, 40-49, assoc prof prev med, 49-65, prof prev med & community health, 65-72; dir, Inst Tuberc Res, 48-72; med dir, Res Found, Chicago, 48-75, dir, 75-82; CHMN, PEOPLE TO PEOPLE FOR PEACE, 82-; FOUND, ACADEMIA FOR ARTS, 87- *Concurrent Pos:* Lectr, SAm, Cuba, China, 80, Japan, 80; Pasteur Inst Paris consult, USPHS, 47-53. *Mem:* Soc Exp Biol & Med; Am Soc Exp Pathologists; Am Physiol Soc; Am Thoracic Soc; Am Asn Path & Bact. *Res:* Experimental pathology and medicine; immunology; bacillus-Calmette-Guerin vaccination against tuberculosis and neoplasia; chemical mediator of pain; atherosclerosis; thermal and radiation injury; competetin; risk exercise; inventor tuberculin tine test; developed the BCG substrain, (Tice-Rosenthal) produces up to 94% cures after intra vesicular installation. *Mailing Add:* Box F Rancho Santa Fe CA 92067

ROSENTHAL, STANLEY ARTHUR, b Paterson, NJ, July 9, 26; m 48; c 2. MEDICAL MYCOLOGY, MICROBIOLOGY. *Educ:* Rutgers Univ, BS, 48; Univ Maine, MS, 50; Pa State Col, PhD(bact), 52. *Prof Exp:* USPHS spec fel, Prince Leopold Inst Trop Med, Antwerp, Belg, 60-61; from asst microbiol to asst prof dermat, 52-71, ASSOC PROF EXP DERMAT, MED CTR, NY UNIV, 61- *Mem:* Am Soc Microbiol; Med Mycol Soc Am; Int Soc Human & Animal Mycol. *Res:* Dermatophytosis; transmission of ringworm; identification of fungi. *Mailing Add:* Dept Dermat NY Univ Med Sch 550 First Ave New York NY 10016

ROSENTHAL, STANLEY LAWRENCE, b Brooklyn, NY, Dec 6, 29; m 53; c 3. METEOROLOGY. *Educ:* Fla State Univ, MS, 53, PhD(meteorol), 58. *Prof Exp:* Instr meteorol, Fla State Univ, 57-58; mem staff, Los Alamos Sci Lab, 58-59; asst prof meteorol, Fla State Univ, 59-60; mem staff, 60-75, chief modeling group, 75-78, DIR, NAT HURRICANE RES LAB, NAT OCEANIC & ATMOSPHERIC ADMIN, 78- *Concurrent Pos:* Adj prof, Univ Miami; chmn, Oceans & Atmospheres Sect, Fla Acad Sci, 78-79; mem coun, Am Meteorol Soc, 81-83, comt on hurricanes & trop meteorol, 81-84, chmn, 83-84. *Honors & Awards:* Gold Medal, Dept Com, 70. *Mem:* Fel AAAS; fel Am Meteorol Soc. *Res:* Application of dynamic meteorology ot the tropical portions of the atmosphere; application of numerical weather prediction techniques to the tropics; dynamics and numerical simulation of hurricanes. *Mailing Add:* 13301 SW 99th Pl Miami FL 33176

ROSENTHAL, THEODORE BERNARD, b New Haven, Conn, June 12, 14. MICROSCOPIC ANATOMY, PHYSIOLOGY. *Educ:* Yale Univ, BS, 35, PhD(physiol), 41. *Prof Exp:* Res assoc cancer, Barnard Free Skin & Cancer Hosp, St Louis, 46-47; res assoc & instr anat, Sch Med, Washington Univ, 47-54; assoc prof, Sch Med, Univ Ind, 54-55 & Sch Med, Emory Univ, 55-56; assoc prof anat, Sch Med, Univ Pittsburgh, 56-83; RETIRED. *Concurrent Pos:* Vis fel, Carlsberg Lab, Copenhagen Univ, 48. *Mem:* AAAS; Am Asn Anatomists. *Res:* Aging; arteriosclerosis; cancer. *Mailing Add:* 272 Northbellfield Pittsburgh PA 15213

ROSENTHAL, WALDEMAR ARTHUR, clinical chemistry, for more information see previous edition

ROSENTHAL, WILLIAM S, b New York, NY, June 21, 25; m 54; c 2. GASTROENTEROLOGY, PHYSIOLOGY. *Educ:* Univ Minn, BA, 46; State Univ NY, 50. *Prof Exp:* Intern, Michael Reese Hosp, Chicago, 50-51; resident med, Vet Admin Hosp, Brooklyn, 51-52 & 53-55; res asst liver dis, Postgrad Med Sch, Univ London, 55-56; res asst gastroenterol, Mt Sinai Hosp, New York, 57-61; from asst prof to assoc prof med, 61-69, PROF MED & CHIEF SECT, NY MED COL, 69-, RES PROF PHYSIOL, 73- *Concurrent Pos:* Chief sect gastroenterol, Westchester County Med Ctr, 77- *Mem:* Am Gastroenterol Asn; Am Asn Study Liver Dis; Am Col Physicians; Am Col Gastroenterol (pres, 77-78); Am Soc Human Genetics; Am Physiol Soc. *Res:* Clinical and laboratory investigation of the pathological physiology of gastrointestinal and liver disease. *Mailing Add:* Dept Med NY Med Col Westchester County Med Ctr Munger Pavilion Rm 206 Valhalla NY 10595

ROSENTHALE, MARVIN E, b Philadelphia, Pa, Dec 13, 33; m 59; c 2. PHARMACOLOGY. *Educ:* Philadelphia Col Pharm & Sci, BSc, 56, MSc, 57; Hahnemann Med Col, PhD(pharmacol), 60. *Prof Exp:* Pharmacologist, Lawall & Harrison Res & Control Labs, 51-60; pharmacologist, Wyeth Labs, Inc, 60-69, mgr immunoinflammatory pharmacol, 69-77; dir div pharmacol, 77-79, dir div biol res, 79-82, GROUP DIR DRUG DISCOVERY, ORTHO PHARMACEUT CORP, 82- *Concurrent Pos:* Vis assoc prof, Hahnemann Med Col, 63-; instr, Univ Pa, 64-67; adj prof, Philadelphia Col Pharm & Sci, 65- *Mem:* Am Soc Pharmacol & Exp Therapeut; Soc Exp Biol & Med; Am Rheumatism Asn. *Res:* Inflammation and allergy, particularly autoimmune diseases; cardiovascular and renal pharmacology. *Mailing Add:* 71 Bertrand Dr Princeton NJ 08540

ROSENWALD, GARY W, b Manhattan, Kans, Jan 10, 41; m 65; c 4. CHEMICAL ENGINEERING, PETROLEUM ENGINEERING. *Educ:* Univ Kans, BS, 64; Kans State Univ, MS, 66; Univ Kans, PhD(chem eng), 72. *Prof Exp:* High sch teacher math, Dickinson County Community High Sch, 64-66; res engr, 70-73, Sr res engr petrol prod, Cities Serv Co, 73-; AT ENG DEPT, UNIV WYO. *Mem:* Am Inst Mining, Metall & Petrol Engrs; Am Chem Soc; Am Inst Chem Engrs; AAAS. *Res:* Petroleum production, especially oil recovery by thermal and chemical flooding methods, reservoir performance and numerical simulation. *Mailing Add:* US Dept Energy Richland Opers PO Box 550 Richland WA 99352-0550

ROSENWASSER, LANNY JEFFERY, MEDICINE. *Educ:* New York Univ, MD, 72. *Prof Exp:* ASSOC PROF MED, SCH MED, TUFTS UNIV, 79- *Mailing Add:* Dept Med Nat Jewish Ctr Immunol & Resp Med 1400 Jackson St Denver CO 80206

ROSENZWEIG, CARL, b Brooklyn, NY, June 10, 46; m 76; c 3. QUANTUM CHROMODYNAMICS. *Educ:* Polytechnic Univ, BS, 67, MS, 67; Harvard Univ, PhD(physics), 72. *Prof Exp:* Res assoc, Univ Calif, 72-73; Weizman Inst Sci, 73-74, Univ Calif, 74-75; res asst prof, Univ Pittsburg, 75-76; from asst prof to assoc prof, 76-86, PROF PHYSICS, SYRACUSE UNIV, 86- *Concurrent Pos:* Vis scientist Inst Theoretical Physics, 88. *Mem:* Am Phys Soc. *Mailing Add:* Dept Physics Syracuse Univ Syracuse NY 13244-1130

ROSENZWEIG, DAVID YATES, b Detroit, Mich, June 9, 33; m 58; c 5. INTERNAL MEDICINE, PULMONARY DISEASE. *Educ:* Wayne State Univ, BS, 54, MD, 57. *Prof Exp:* Intern, Sinai Hosp, Detroit, 57-58; resident internal med, Vet Admin Hosp, Denver, 58-60; fel pulmonary dis, Univ Colo, 60-62; from instr to asst prof, 62-72, ASSOC PROF MED, MED COL WIS, 72- *Concurrent Pos:* Consult, Vet Admin Hosp, Wood, Wis, 66-; fel pulmonary physiol, Postgrad Med Sch, Univ London, 67-68. *Mem:* Am Thoracic Soc; Am Fedn Clin Res; Int Union Against Tuberculosis; fel Am Col Physicians; Am Physiol Soc; fel Am Col Chest Physicians. *Res:* Pulmonary disease and physiology; distribution of pulmonary ventilation and perfusion; exercise physiology; mycobacterial infections. *Mailing Add:* Dept Med Med Col Wis 8700 W Wisconsin Ave Milwaukee WI 53226

ROSENZWEIG, MARK RICHARD, b Rochester, NY, Sept 12, 22; m 47; c 3. NEUROSCIENCES. *Educ:* Univ Rochester, BA, 43, MA, 44; Harvard Univ, PhD(psychol), 49. *Hon Degrees:* Dr, Univ Rene Descartes, Sorbonne, 80. *Prof Exp:* Res assoc, Psycho-Acoust Lab, Harvard Univ, 49-51; from asst prof to assoc prof, 50-60, res prof, Miller Inst Basic Res in Sci, 58-59, 65-66, PROF PSYCHOL, UNIV CALIF, BERKELEY, 60- *Concurrent Pos:* Fulbright res fel & Soc Sci Res Coun fel, Paris, 60-61; ed, Ann Rev Psychol, 68-; Am Psychol Asn rep, Int Union Psychol Sci, 70-72; vis prof, Univ Paris, 73- 74, mem exec comt, 72-; pres, Int Union Psychol Sci, 88-92; rep div physiol & comp psychol, Coun Am Psychol Sci, 83-86; chmn US Nat Comt, Int Union Psychol Sci, Nat Acad Sci-Nat Res Coun, 85-88. *Mem:* Nat Acad Sci; fel Am Psychol Asn; Int Brain Res Orgn; Am Physiol Soc; Soc Neurosci; Soc Franscaise Psychol. *Res:* Neurophysiology and behavior; neural processes in learning and memory; history of psychology. *Mailing Add:* Dept Psychol Tolman Hall Univ Calif Berkeley CA 94720

ROSENZWEIG, MICHAEL LEO, b Philadelphia, Pa, June 25, 41; m 61; c 3. POPULATION ECOLOGY, MAMMALOGY. *Educ:* Univ Pa, AB, 62, PhD(zool), 66. *Prof Exp:* Asst prof biol, Bucknell Univ, 65-69 & State Univ NY Albany, 69-71; assoc prof, Univ NMex, 71-75; dept head, 75-76, PROF ECOL & EVOLUTIONARY BIOL, UNIV ARIZ, 75- *Concurrent Pos:* Ed in chief, Evolutionary Ecol, 86-; chair, Res Support Liaison Comt Ecol, Evolution Syst, 86-88 & Div Ecol, Am Soc Zoologists, 85-86. *Mem:* Ecol Soc Am; Soc Study Evolution; Am Soc Naturalists; Japanese Soc Pop Ecol; Am Soc Zoologists; Brit Ecol Soc. *Res:* Theory of predation; theory of habitat selection; speciation dynamics; population ecology of desert rodent communities. *Mailing Add:* Dept Ecol & Evolutionary Biol Univ Ariz Tucson AZ 85721

ROSENZWEIG, NORMAN, b New York, NY, Feb 28, 24; m 45; c 1. PSYCHIATRY, EDUCATION & RESEARCH ADMINISTRATION. *Educ:* Univ Chicago, MB, 47, MD, 48; Univ Mich, MS, 54; Am Bd Psychiat & Neurol, dipl, 54. *Prof Exp:* From instr to asst prof psychiat & interim chmn, dept psychiat, Med Sch, Univ Mich, 53-61; from asst prof to assoc prof, 62-73, PROF PSYCHIAT, WAYNE STATE UNIV, 73-; ASSOC PROF PSYCHIAT MICH STATE UNIV, 72- *Concurrent Pos:* Dir joint res proj, Univ Mich & Ypsilanti State Hosp, 57-59; Chmn dept, Sinai Hosp of Detroit, 61-; interim chmn, dept psychiat, Wayne State Univ, 87- *Honors & Awards:* Rush Gold Medal Award, Am Psychiat Asn; Presidential Award, Puerto Rico Med Asn, 81; Warren Williams Speaker's Award, Assembly of Am Psychiat Asn, 86. *Mem:* NY Acad Sci; fel Am Col Psychiat; AMA; fel Am Psychiat Asn; AAAS; hon mem, Royal Australia New Zealand Col Psychiat; hon mem, Indian Psychiat Soc. *Res:* Schizophrenia; mind-brain relationships; perceptual integration and isolation; psychological consequences of physical illness; social attitudes; primary preventive intervention to reduce untoward outcomes in teen-age pregnancy. *Mailing Add:* 14800 W McNichols Detroit MI 48235

ROSENZWEIG, WALTER, b Vienna, Austria, Sept 23, 27; US citizen; m 49; c 3. PHYSICS. *Educ:* Rutgers Univ, BS, 50; Univ Rochester, MS, 52; Columbia Univ, PhD(physics), 60. *Prof Exp:* Assoc health physicist, Brookhaven Nat Lab, 51-53; res physicist, Radiol Res Lab, Columbia Univ, 53-60; mem tech staff semiconductor physics, 60-72, supvr semiconductor memories, 72-85, SUPVR PROD SUPPORT, AT&T BELL LABS, 85- *Mem:* Radiation Res Soc; Inst Elec & Electronics Engrs. *Res:* Radiation dosimetry; radiation damage to semiconductors and devices; semiconductor memory design, large scale integrated circuit design; semiconductor device physics. *Mailing Add:* 8431 Glen View Ct Orlando FL 32819

ROSENZWEIG, WILLIAM DAVID, b New York, NY, Feb 6, 46. MICROBIOL ECOLOGY, ENVIRONMENTAL MICROBIOLOGY. *Educ:* St John's Univ, BS, 69; Long Island Univ, MS, 71; NY Univ, PhD(biol), 78. *Prof Exp:* Instr microbiol, NY Univ, 74-78; res fel microbiol, Rutgers Univ, 78-80; asst prof microbiol, Drexel Univ, 80-81; ASST PROF MICROBIOL, WEST CHESTER UNIV, 87- *Mem:* Am Soc Microbiol; Sigma Xi; Am Water Works Asn. *Res:* Environmental biology; ecology and physiology of nematode-trapping fungi; microbiological quality of potable water. *Mailing Add:* Dept Biol West Chester Univ West Chester PA 19383

ROSEVEAR, JOHN WILLIAM, b Glenns Ferry, Idaho, Feb 19, 27; m 49; c 4. BIOCHEMISTRY. *Educ:* Ore State Col, BS, 49; Northwestern Univ, MD, 53; Univ Utah, PhD(biochem), 58. *Prof Exp:* Intern, Evanston Hosp Asn, 54; asst to staff, Sect Biochem, Mayo Clin, 58-59, consult, 59-69; assoc prof, Div Health Comput Sci, Univ Minn, Minneapolis, 69-71; med dir & dir clin chem, Kallestad Labs Inc, 71-74, dir labs, 74-75, vpres labs, 75-76, vpres sci affairs, 76-79; PRES, BIO-METRIC SYSTS, INC, 79- *Concurrent Pos:* Clin assoc prof, Univ Minn, Minneapolis, 71- *Mem:* AAAS; Am Chem Soc; AMA. *Res:* Carbohydrate groups of proteins; diabetes mellitus; biochemical monitoring of patients; chromatographic analyses of organic acids; cancer detection tests; radioimmunoassay systems. *Mailing Add:* 6316 Barrie Rd Apt 1A Minneapolis MN 55435

ROSHAL, JAY YEHUDIE, b Chicago, Ill, Aug 27, 22; m 89. GENETICS. *Educ:* Univ Chicago, PhB, 48, SB, 49, SM, 50, PhD(bot), 53. *Prof Exp:* Instr bot, Oberlin Col, 53-55; res assoc, Ben May Lab, Univ Chicago, 55-56; asst prof bot, Eastern Ill Univ, 57-58; from instr to assoc prof, 58-63, chmn div sci & math, 62-64, prof, 63-84, EMER PROF BOT, UNIV MINN, MORRIS, 84-; CONSULT, GENDERM CORP, 90- *Concurrent Pos:* Mem secretariat, Space Sci Bd, Nat Acad Sci, 62-63; lectr, Univ Tex, 65-66; hon res assoc, Harvard Univ, 68-69; vpres, Genderm Corp, 84-90. *Mem:* AAAS; Mycol Soc Am; Am Soc Plant Physiol. *Res:* Mycology; physiology; microbial genetics. *Mailing Add:* 2124 N Monroe St Arlington VA 22207

ROSHKO, ALEXANA, US citizen. ANALYTICAL ELECTRON SPECTROSCOPY, ELECTRON MICROSCOPY. *Educ:* Mass Inst Technol, SB, 81, PhD(ceramics sci), 87. *Prof Exp:* Postdoctoral, Mass Inst Technol, 87-88; MAT SCIENTIST, NAT INST STANDARDS & TECHNOL, 88- *Mem:* Mat Res Soc; Am Ceramic Soc; Electron Micros Soc Am. *Res:* Structure-property relationships in electronic and magnetic ceramics; microstructure, grain boundary chemistry; point defect chemistry and diffusion, surface, grain boundary and bulk, in these materials and how they affect material behavior. *Mailing Add:* Nat Inst Standards & Technol 325 Broadway Boulder CO 80303

ROSHKO, ANATOL, b Bellevue, Alta, July 15, 23; nat US; m 57; c 2. TURBULENT SHEER FLOW. *Educ:* Univ Alta, BSc, 45; Calif Inst Technol, MS, 47, PhD(aeronaut), 52. *Prof Exp:* Instr math, Univ Alta, 45-46, lectr eng, 49-50; res fel, Calif Inst Technol, 52-55, from asst prof to prof, 55-85, actg dir, Grad Aeronaut Labs, 85-87, THEODORE VON KARMEN PROF AERONAUT, CALIF INST TECHNOL, 85- *Concurrent Pos:* Sci liaison officer, Off Naval Res, London, 61-62; consult, McDonnell Douglas Corp, 54-90 & Rocketdyne Corp Div, Rockwell Int, 84-90; founding dir, Wind Eng Res Inc, 70; mem, Aeronaut & Space Eng Bd. *Honors & Awards:* Dryden Res Lectr, Am Inst Aeronaut & Astronaut, 76; Dynamics Prize, Am Phys Soc, 87. *Mem:* Nat Acad Eng; fel AAAS; fel Am Inst Aeronaut & Astronaut; Am Phys Soc; fel Can Aeronaut & Space Inst. *Res:* Gas dynamics; separated flow; turbulence; transonic and supersonic aerodynamics. *Mailing Add:* Calif Inst Technol Mail Sta 105-50 1201 E California Blvd Pasadena CA 91125

ROSHWALB, IRVING, b New York, NY, Jan 24, 24; m 47; c 4. STATISTICS. *Educ:* City Col New York, BSS, 43; Columbia Univ, MA, 46. *Prof Exp:* Chief statistician, Opinion Res Corp, 50-55; SR VPRES, AUDITS & SURV, INC, 55- *Concurrent Pos:* Adj prof statist, Grad Sch Bus, City Univ New York, 46- *Mem:* Am Asn Pub Opinion Res; fel Am Statist Asn; Am Mkt Asn; AAAS. *Res:* Survey sampling procedures; applications to areas of marketing and public opinion research; data analysis. *Mailing Add:* Audits & Surv Inc 650 Ave of Americas New York NY 10011

ROSI, DAVID, b Utica, NY, Apr 22, 32; m 56; c 4. BIOLOGY, MUTASYNTHESIS. *Educ:* Utica Col, BS, 54; Univ NH, MS, 56. *Prof Exp:* Jr bacteriologist & jr sanit chemist, Div Labs & Res Water Pollution, NY State Dept Health, 56-58, sanit chemist, 58-60; from asst res biologist to res biologist, 60-70, SR RES BIOLOGIST, STERLING WINTHROP RES INST, 70-, GROUP LEADER, 66-, SR RES INVESTR, 89- *Mem:* Am Soc Microbiol; Sigma Xi. *Res:* Microbial conversion, isolation and characterization of synthetic organic compounds of pharmaceutical interest; in vivo and in vitro drug metabolism studies; new antibiotics via mutagenesis; natural products; pharmacokinetics. *Mailing Add:* Two Boncroft Dr E Greenbush NY 12061

ROSI, FRED, b Jan 13, 21. ELECTRONIC MATERIALS. *Educ:* Yale Univ, BE, 42, ME, 47, MS, 48, PhD(physics), 49. *Prof Exp:* Engr specialist, Sylvania Elec Prod, 49-54; mem tech staff, RCA Labs, dir, Mat Res Lab, vpres, Mat & Device Res, 54-72; dir, Res & Develop, Am Can Co, 73-75; gen dir, Res & Develop, Reynolds Metals Co, 75-78; RES PROF, DEPT MAT SCI, SCH ENG & APPL SCI, UNIV VA, CHARLOTTESVILLE, 78- *Concurrent Pos:* Consult, Cent Intel Agency, 72-73, 80, NASA, 79-81, Energy Conversion Devices, 83-, Gen Elec Co, 84-; mem Awards Comt, Am Inst Mettall Engrs, 58-67; mem, Res Adv Comt, NASA, 60-65; adv comt, Dept Physics, Univ Ky, 68-71; comt elec Mat Devices, Nat Acad Sci, 70-71; adv bd, Gould Incorp, 79-83. *Honors & Awards:* David Saonoff Medal in Sci. *Mem:* AAAS; fel Am Phys Soc; fel Am Inst Metall Engr; Sigma Xi. *Res:* Author of numerous articles in various publications. *Mailing Add:* Dept Mat Sci Sch Eng Appl Sci Univ Va Charlottesville VA 22901

ROSIER, RONALD CROSBY, b Baltimore, Md, Jan 23, 43; m 65; c 3. MATHEMATICS. *Educ:* Boston Col, AB, 65; Univ Md, PhD(math), 70. *Prof Exp:* ASST PROF MATH, GEORGETOWN UNIV, 70- *Mem:* Math Asn Am; Am Math Soc. *Res:* Functional analysis. *Mailing Add:* Dept Math 252 Reiss Georgetown Univ 37th & O Sts NW Washington DC 20057

ROSIN, ROBERT FISHER, b Chicago, Ill, Mar 19, 36; m 63; c 2. COMPUTER SCIENCE. *Educ:* Mass Inst Technol, BS, 57; Univ Mich, MS, 60, PhD(commun sci), 64. *Prof Exp:* From asst prof to assoc prof eng & appl sci, Yale Univ, 64-68; assoc prof comput sci & vchmn dept, State Univ NY Buffalo, 68-70, prof, 70-74; prof comput sci, Iowa State Univ, 74-76; mem tech staff, Bell Labs, Holmdel, NJ, 76-78, consult, 78-83; consulting mem tech staff, Syntex, Inc, Eatontown, NJ, 83-86; dist mgr, Bell Commun Res, Red Bank, NJ, 86-89; VPRES & PRIN ARCHITECT, ENHANCED SERV PROVIDERS, INC, SHREWSBURY, NJ, 89- *Concurrent Pos:* Nat Sci Found res grant, 70-72; NATO res grant, 72-73; vis prof, Univ Aarhus, 72-74. *Mem:* Asn Comput Mach; Comput Soc Inst Elec & Electronics Engrs; Sigma Xi. *Res:* Software and communication systems; software system architecture; realization of computer systems via software, hardware and firmware. *Mailing Add:* 23 Haddon Park Fair Haven NJ 07704

ROSING, WAYNE C, b Kenosha, Wis, Oct 6, 47. FUNGAL ULTRASTRUCTURE. *Educ:* Univ Wis, Madison, BS, 69; Univ Tex, Austin, PhD(bot & mycol), 75. *Prof Exp:* Asst prof, George Peabody Col, 76-80; asst prof, 80-84, ASSOC PROF BIOL, MIDDLE TENN STATE UNIV, 84- *Mem:* Mycol Soc Am; Sigma Xi; Bot Soc Am. *Res:* Ultrastructural studies of fungi; ascomycete spore formation. *Mailing Add:* PO Box 333 Fairview TN 37062

ROSINGER, EVA L J, b July 21, 41; Can citizen; m 69. ENVIRONMENTAL POLICY, ENVIRONMENTAL AFFAIRS. *Educ:* Tech Univ, Prague, MSc, 63, PhD(chem), 68. *Prof Exp:* Researcher indust res, 67-68; fel chem eng, Univ Toronto, 68-70; res librarian tech info, Tech Univ, Aachen, Ger, 71-72; res officer environ comput modelling, Atomic Energy Can Ltd, 73-79, head, Systs Assessment Sect, 79-84, sci asst dir, Waste Mgt Div, 80-84, head, Environ & Safety Assessment Br, 85-86, exec asst to pres, 86-87, dir Waste Mgt Concept Rev, 87-90; DIR GEN, CAN COUN MINISTERS ENVIRONMENT, 90- *Concurrent Pos:* Vpres, Radioactive Waste Mgt Comt, Orgn Econ Coop & Develop/Nuclear Energy Agency, Paris; bd dir, Radioactive Waste Mgt, Can Mem Eng Found, 89; adv coun, Univ Waterloo. *Mem:* Nat Acad Sci; Can Nuclear Soc (pres, 89-90); Women Sci & Eng. *Res:* Environmental policies; evaluation and analysis of environmental systems; scientific and organizational strategies related to environmental issues, documenting options and providing scientific advice. *Mailing Add:* Can Coun Ministers Environ 326 Broadway Suite 400 Winnipeg MB R3C 0S5 Can

ROSINGER, HERBERT EUGENE, b Ger, Apr 18, 42; Can citizen; m 69. SMALL REACTOR TECHNOLOGY. *Educ:* Univ Toronto, BASc, 66, MASc, 67, PhD(metall & mats sci), 70. *Prof Exp:* Humboldt fel, Inst Metal Physics & Metall, Aachen, WGer, 70-72; res officer metall, 72-78, head, Fuel Model Verification & Assessment Sect, 79-80, head, Fuel Transient Behav Sect, 80-81, Severe Fuel Damage Sect, 81-83, Systs Br, 84-85, dir technol, Maple Reactor Technol Div, 85-86, HEAD, SYSTS ANALYSIS BR, ATOMIC ENERGY CAN, LTD, 86- *Concurrent Pos:* Vis scientist at Kernforschungszentrum Karlsruhe, WGer, 79. *Mem:* Can Nuclear Soc. *Res:* Research and development management of small nuclear reactors for electricity and/or high quality steam production; safety research of nuclear power reactors. *Mailing Add:* Atomic Energy Can Ltd Pinawa MB R0E 1L0 Can

ROSINSKI, JAN, b Warsaw, Poland, Feb 10, 17; nat US; m 45; c 1. ATMOSPHERIC CHEMISTRY. *Educ:* Warsaw Inst Technol, dipl, 49; Univ Warsaw, PhD, Dr habil, 80; Univ Bologna, Libera Docenza, 70. *Prof Exp:* Res chemist, Brit Celanese, Ltd, 48-51; chief chemist, Poray, Inc, Ill, 51-52; sr engr, IIT Res Inst, 52-62; scientist, Nat Ctr Atmospheric Res, 62-89; RETIRED. *Res:* Fine particles; radiochemistry; colloids; geochemistry, atmospheric chemistry and physics; ice nucleation in marine, mixed marine-continental and continental air masses; chemistry of ice-forming and cloud condensation nuclei. *Mailing Add:* 590 Ord Dr Boulder CO 80303-4732

ROSINSKI, JOANNE, b Milwaukee, Wis. ELECTRON MICROSCOPY. *Educ:* Marquette Univ, BS, 66; State Univ NY, Buffalo, PhD(biol), 70. *Prof Exp:* Res fel, Harvard Univ, 71-73; asst prof biol, Williams Col, 73-78; res assoc, Brookhaven Nat Lab, 78-80; asst prof, 80-84, chmn dept, 82-84 ASSOC PROF BIOL, SWEET BRIAR COL, 84- *Concurrent Pos:* NIH fel, Harvard Univ, 71-73. *Mem:* Am Soc Plant Physiologists; Bot Soc Am; Am Inst Biol Sci. *Res:* Chloroplast development; ultrastructural and biochemical characterization of pigment-protein complexes in cyanobacteria and higher plants. *Mailing Add:* Biol Dept Sweet Briar Col Sweet Briar VA 24595

ROSINSKI, MICHAEL A, b Detroit, Mich, Oct 7, 62; m 91. AUTOMOTIVE PROTOTYPE BODY SHEETMETAL DEVELOPMENT. *Educ:* Oakland Univ, BSE, 85, MSME, 88. *Prof Exp:* Sr mfg engr, Prototype Develop, Gen Motors, 85-89; SR PROJ ENGR, NISSAN RES & DEVELOP, 89- *Mem:* Am Soc Mech Engrs; Am Welding Soc. *Res:* Automotive prototype sheetmetal development; reducing development cycle; developing alternative sheetmetal development methods and strategies and implementing these ideas on future car models. *Mailing Add:* 6842 Ellinwood White Lake MI 48383

ROSKA, FRED JAMES, b Sheboygan, Wis, May 7, 54. DIELECTRICS, ELECTROSTATICS. *Educ:* Univ Wis, BS, 76, PhD(chem), 81. *Prof Exp:* SPECIALIST, MINN MINING & MFG CO, 81- *Res:* High field dielectric properties of polymers; structure-property relationships in semi-crystalline polymers; electrostatic processes; electrical coronas. *Mailing Add:* Bldg 236 GB-03 3M Ctr St Paul MN 55144

ROSKAM, JAN, b The Hague, Neth, Feb 22, 30; US citizen; m 56. AEROSPACE ENGINEERING. *Educ:* Delft Technol Univ, MS, 54; Univ Wash, PhD(aeronaut, astronaut), 65. *Prof Exp:* Asst chief designer, Aviolanda Aircraft Co, Neth, 54-56; aerodyn engr, Cessna Aircraft Co, 57-59; sr group engr, Boeing Co, 59-67; from assoc prof to prof aerospace eng, 67-76, chmn dept, 72-76, ACKERS PROF AEROSPACE ENG, UNIV KANS, 76- *Concurrent Pos:* Consult, Gates Learjet Corp. *Mem:* Assoc fel Am Inst Aeronaut & Astronaut; fel Royal Aeronaut Soc; Neth Royal Inst Engrs; Sigma Xi. *Res:* Airplane stability, control and design; effects of aeroelasticity on stability, control and design; automatic flight controls. *Mailing Add:* Rte 4 Box 274 Ottawa KS 66067

ROSKAMP, GORDON KEITH, b Quincy, Ill, July 4, 50; m 74; c 2. WEED SCIENCE, AGRONOMY. *Educ:* Western Ill Univ, BS, 71; Univ Mo-Columbia, MS, 73, PhD(agron), 75. *Prof Exp:* Res asst weed sci, Western Ill Univ, 70-71 & Univ Mo-Columbia, 71-75; asst prof agron, 75-81, ASSOC PROF AGR, WESTERN ILL UNIV, 81- *Concurrent Pos:* Grants from var chem co, Ill Soybean Operating Bd & Nat Crop Ins Asn. *Mem:* Weed Sci Soc Am; Am Soc Agron. *Res:* Weed control, especially in corn, soybeans and pastures; hail damage to corn; control of multiflora rose; herbicide damage to soybeans. *Mailing Add:* Dept Agr Western Ill Univ Adams St Macomb IL 61455

ROSKES, GERALD J, b Edgewood, Md, June 27, 43. APPLIED MATHEMATICS. *Educ:* Mass Inst Technol, BS, 65, PhD(math), 70. *Prof Exp:* Mem tech staff, Bell Tel Labs, 69-70; ASSOC PROF MATH, QUEENS COL, CITY UNIV NY, 70- *Mem:* Am Math Soc; Soc Indust & Appl Math; Math Asn Am. *Res:* Nonlinear wave theory; singular perturbation theory. *Mailing Add:* Dept Math Queens Col Flushing NY 11367

ROSKIES, ETHEL, b Montreal, Can, Nov 27, 33; m 53, 81; c 2. HEALTH PSYCHOLOGY, BEHAVIORAL MEDICINE. *Educ:* McGill Univ, Can, BA, 54, MA, 61; Univ Montreal, MA, 63, PhD(psychol), 69. *Prof Exp:* Staff psychologist, Montreal Childrens Hosp, 66-68, lectr psychol, 68-69, asst prof psychol, 69-74, assoc prof psychol, 74-80; PROF PSYCHOL, UNIV MONTREAL, CAN, 80- *Concurrent Pos:* Vis scholar, Univ Calif, Berkeley, 77-78, Inst Soc Res, Mich, 85-86, MRC-ESRC Soc & Appl Psychol Unit, Univ Sheffield, UK, 86-; res assoc, Behavior Therapy Serv, Montreal Gen Hosp, 76-77; prin invest, Montreal Type A Intervention Proj, 82-85. *Mem:* Fel Can Psychol Asn; Am Psychol Asn; fel Soc Behav Med; fel Am Inst Stress; Can Register Health Serv Providers; Assoc Adv Behav Ther. *Res:* Management of work stress; behavioral interventions to reduce illness risk. *Mailing Add:* Dept Psychol Univ Montreal PO Box 6128 Br A Montreal PQ H3C 3J7 Can

ROSKIES, RALPH ZVI, b Montreal, Que, Nov 24, 40; m 63; c 3. ELEMENTARY PARTICLE PHYSICS. *Educ:* McGill Univ, BSc, 61; Princeton Univ, MA, 63, PhD(math), 66. *Prof Exp:* Instr physics, Yale Univ, 65-67; I Meyer Segals fel, Weizmann Inst, 67-68; asst prof, Yale Univ, 68-72; assoc prof, 72-78, PROF PHYSICS, UNIV PITTSBURGH, 79-; SCI DIR, PITTSBURGH SUPERCOMPUT CTR, 86- *Concurrent Pos:* Yale jr fac fel, Stanford Linear Accelerator Ctr, 71-72; Alfred P Sloan fel, 73-77. *Mem:* Am Phys Soc; Asn Comput Mach. *Res:* High energy theoretical physics; quantum electrodynamics; Yang-Mills fields; lattice field theory; high performance computing. *Mailing Add:* Dept Physics Univ Pittsburgh Pittsburgh PA 15260

ROSKOS, ROLAND R, b Arcadia, Wis, July 9, 40; m 66; c 2. PHYSICAL CHEMISTRY. *Educ:* Wis State Univ, La Crosse, BS, 62; Iowa State Univ, MS, 64, PhD(phys chem), 66. *Prof Exp:* Asst prof, 66-69, assoc prof, 69-76, PROF CHEM, UNIV WIS-LA CROSSE, 76- *Concurrent Pos:* Consult, Gunderson Clin, Ltd, La Crosse, Wis, 67- *Mem:* Am Chem Soc; Sigma Xi. *Res:* Chemical and medical instrumentation; physical and chemical properties as related to structure; applications and theory of lasers; isotope effects. *Mailing Add:* 1220 Seiler Lane La Crosse WI 54601

ROSKOSKI, JOANN PEARL, b Paterson, NJ, June 18, 47. ECOLOGY, MICROBIOLOGY. *Educ:* Rutgers Univ, BA, 69, MS, 72; Yale Univ, PhD(forest ecol), 77. *Prof Exp:* Res asst microbiol, E R Squibb Corp, 69-71; Rockefeller found fel, Res Inst Biotic Resources, 77-80; mem fac, plant sci dept, Univ Ariz, 80-83; assoc soil scientist, 83-85, ASSOC DIR, NIFTAL PROJ, UNIV HAWAII, 85- *Mem:* Ecol Soc Am; Sigma Xi; Int Soc Trop Ecol. *Res:* Nutrient cycling in tropical ecosystems, especially the role of nitrogen fixation in tropical agroecosystems; agroforestry; nitrogen fixation in agriculture. *Mailing Add:* Nat Res Coun 2101 Constitution Ave NW Washington DC 20418

ROSKOSKI, ROBERT, JR, b Elyria, Ohio, Dec 10, 39; m 75. BIOCHEMISTRY. *Educ:* Bowling Green State Univ, BS, 61; Univ Chicago, MD, 64, PhD(biochem), 68. *Prof Exp:* USPHS fel biochem, Univ Chicago, 64-66; sr investr, US Air Force Sch Aerospace Med, 67-69; USPHS spec fel, Rockefeller Univ, 69-72; from asst prof to assoc prof biochem, Univ Iowa, 72-79; PROF & HEAD BIOCHEM DEPT, LA STATE UNIV MED CTR, 79- *Concurrent Pos:* Mem, Biochem Test Comt, Nat Bd Med Examnrs, 81-84. *Mem:* Am Chem Soc; NY Acad Sci; Am Soc Neurochem; Am Soc Biol Chemists; Soc Neurosci; Am Soc Pharmacol & Exp Therapeut. *Res:* Characterization of cyclic adenosine monophosphate dependent protein kinase and tyrosine hydroxylase activity in the central nervous system; characterization of the acetylcholine receptor in the heart. *Mailing Add:* Dept Biochem La State Univ Med Ctr 1900 Perdido St New Orleans LA 70112

ROSLER, LAWRENCE, b Brooklyn, NY, Feb 20, 34; m 65; c 2. COMPUTER SCIENCES. *Educ:* Cornell Univ, BA, 53; Yale Univ, MS, 54, PhD(physics), 58. *Prof Exp:* Jr physicist, Cornell Aeronaut Labs, 53, jr mathematician, 54; mem tech staff, AT&T Bell Labs, Murray Hill, 57-67, supvr comput graphics systs group, 67-76, head, lang eng dept, 76-87; DIR, COMPUTER LANG LAB, HEWLETT PACKARD, PALO ALTO, CALIF, 87- *Mem:* AAAS; Sigma Xi; Asn Comput Mach; Am Phys Soc. *Res:* Man-machine graphical interaction; small-computer systems; computer languages. *Mailing Add:* Hewlett-Packard Co 3155 Porter Dr 28AB Palo Alto CA 94304

ROSLER, RICHARD S(TEPHEN), b Spokane, Wash, Jan 29, 37; m 61; c 4. CHEMICAL ENGINEERING. *Educ:* Gonzaga Univ, BS, 59; Northwestern Univ, PhD(chem eng), 62. *Prof Exp:* Sr aerothermo engr, United Aircraft Corp, 62-64; from asst prof to assoc prof chem eng, Gonzaga Univ, 64-68; process engr, Shell Develop Co, Calif, 68-69; dir res & develop, Applied Mat, Inc, 69-77; mgr semiconductor mat res, Motorola, Inc, 77-79; corp dir res & develop, Advan Semiconductor Mat Am, 79-86; mgr process eng, Spectrum CVD, Subsid Motorola, Inc, 86-88; SR MEM TECH STAFF, APPLIED MAT, INC, 88- *Mem:* Am Vacuum Soc; Electrochem Soc. *Res:* Technical management of development of equipment and processes on low pressure chemical vapor deposition reactors and plasma reactors for the processing of semiconductor wafers for the semiconductor industry. *Mailing Add:* 7500 E McCormick Pkwy No 78 Scottsdale AZ 85258

ROSLINSKI, LAWRENCE MICHAEL, b Detroit, Mich, May 28, 42; m 64; c 3. TOXICOLOGY, ENVIRONMENTAL HEALTH. *Educ:* Univ Mich, BS, 64; Wayne State Univ, MS, 66, PhD(pharmacol, physiol), 68. *Prof Exp:* Indust hygienist, Indust Bur Indust Hyg, Detroit, Mich, 64; fel toxicol, Sch Pub Health, Univ Mich, 68-69; prof assoc, Nat Acad Sci, 69-72; toxicologist, Environ Protection Agency, 72-73; INDUST TOXICOLOGIST, FORD MOTOR CO, 73- *Concurrent Pos:* Comnr, Toxic Substances Control Comn, State Mich, 79- *Mem:* AAAS; Am Indust Hyg Asn; Am Bd Indust Hyg; Soc Toxicol; Am Bd Toxicol. *Res:* Pharmacology; industrial hygiene; physiology; biochemistry; toxicology. *Mailing Add:* Ford Motor Co PLTW-900 Dearborn MI 48121

ROSLYCKY, EUGENE BOHDAN, b Tovste, Ukraine, May 8, 27; Can citizen; wid; c 2. MICROBIOLOGY. *Educ:* Univ Man, BSc, 53, MSc, 55; Univ Wis-Madison, PhD(microbiol), 60. *Prof Exp:* Asst microbiol, Univ Man, 53-55; asst bact, Univ Wis-Madison, 55-59; RES SCIENTIST MICROBIOL, RES CTR, AGR CAN, 59- *Concurrent Pos:* Hon lectr, Fac Med & Dent, Univ Western Ont, 65-, lectr, dept russ studies, 71-74; chmn, Int Subcomt Agrobacterium & Rhizobium, Int Asn Microbiol Socs, 74-83. *Mem:* Can Soc Microbiol; Am Soc Microbiol; Soc Gen Microbiol; Ukrainian Free Acad Sci; Shevchenko Sci Soc. *Res:* Characterization of bacteriophages; bacteriocins; serological reactions; herbicides in relation to microbes; rhizobia, agrobacteria & azotobacter mutations; herbicide residues. *Mailing Add:* 195 Tarbart Terr London ON N6H 3B3 Can

ROSMAN, HOWARD, b New York, NY, July 17, 29; m 52; c 2. PHYSICAL CHEMISTRY. *Educ:* State Univ NY, BA, 51, MA, 52; Columbia Univ, PhD(chem), 58. *Prof Exp:* Asst chem, Columbia Univ, 52-56; from instr to assoc prof chem, 56-72, chmn dept, 67-75 & 77-80, PROF CHEM, HOFSTRA UNIV, 72- *Concurrent Pos:* Lectr gen studies sch, Columbia Univ, 57-60. *Mem:* Am Chem Soc; AAAS; Sigma Xi. *Res:* Photochemistry of luminescent species. *Mailing Add:* 193 Dogwood Rd Valley Stream NY 11580

ROSNER, ANTHONY LEOPOLD, b Greensboro, NC, Nov 13, 43; m 66. ENZYMOLOGY, ENDOCRINOLOGY. *Educ:* Haverford Col, BS, 66; Harvard Univ, PhD(med sci), 72. *Prof Exp:* Teaching fel biochem, Harvard Univ, 69-72; fel molecular biol, Nat Inst Neurol Dis & Stroke, NIH, 72-73, staff fel, 73-74; res fel endocrinol, Med Sch, Tufts Univ, 74-75; res fel, 75-77, res assoc, 75-81, instr path, 81-83, tech dir, Clin Chem Lab, Beth Israel Hosp, Boston, 81-83; instr path, Harvard Med Sch, 81-83; tech dir, New England Path Serv, 83-86; DEPT ADMIN, DEPT CHEM, BRANDEIS UNIV, 86- *Concurrent Pos:* Fel, Nat Inst Gen Med Sci, 72; res assoc biol chem, Harvard Med Sch, 78-81; dir, Estrogen Receptor Assay Lab & tech dir, Clinical Chem Lab, Beth Israel Hosp, Boston, 81-83; consult, New England Path Serv, 86-; bd dirs, New England Chap Am Chem Soc, 90- *Mem:* AAAS; Am Chem Soc; Clin Ligand Assay Soc; Am Soc Cell Biol; Soc Res Admin. *Res:* Kinetics and control mechanisms attendant upon the binding of steroid hormones to receptor proteins in target tissues, comprised of either normal or cancerous cells. *Mailing Add:* 1443 Beacon St Apt 201 Brookline MA 02146-4707

ROSNER, DANIEL E(DWIN), b New York, NY, Oct 30, 33; m 58; c 2. CHEMICAL ENGINEERING, MECHANICAL ENGINEERING. *Educ:* City Col New York, BME, 55; Princeton Univ, MA, 57, PhD(aero eng), 61. *Hon Degrees:* MA, Yale Univ, 74. *Prof Exp:* Head interface kinetics & transport group, AeroChem Res Labs, Inc, 58-69; assoc prof, 69-74, PROF CHEM ENG & APPL SCI, YALE UNIV, 74- *Concurrent Pos:* Vis prof mech eng, Polytech Inst Brooklyn, 64-67; consult, Gen Elec Corp, Exxon & AeroChem Res Labs; fel, Guggenheim, jet propulsion, Barcelow Univ, 55-57, Gen Elec Co, 57; consult, Gen Elec Corp, SCM chemicals, Babcock & Wilcox, AVCO-Lycoming-Textron, Vortec & Shell; chmn, chem eng dept, Yale Univ, 84-87. *Mem:* Am Soc Eng Educ; Am Inst Chem Engrs; Am Inst Aeronaut & Astronaut; Sigma Xi; Combustion Inst; Am Chem Soc. *Res:* Energy and mass transfer in chemically reacting flow systems; experimental and theoretical studies in heterogeneous chemical kinetics; combustion; aerosol deposition, nucleation, growth, dispersion and coagulation; transport phenomena in materials processing; transport processes in chemically reacting flow systems; aeronautical engineering. *Mailing Add:* Dept Chem Eng Yale Univ PO Box 2159 Yale Sta New Haven CT 06520

ROSNER, FRED, b Berlin, Germany, Oct 3, 35; m 59; c 4. HEMATOLOGY, MEDICAL ETHICS. *Educ:* Yeshiva Univ, BA, 55; Albert Einstein Col Med, MD, 59; Am Col Physicians, FACP(internal med), 70. *Prof Exp:* Asst dir, div hemat, Maimonides Med Ctr, 65-70; dir, div hemat, 70-78, DIR, DEPT MED, QUEENS HOSP CTR, LONG ISLAND MED CTR, 78- *Concurrent Pos:* Prof med, State Univ NY, Stony Brook, 78-, consult, Univ Hosp, 81-; lectr, State Univ NY, Brooklyn, 72-; vis mem, Albert Einstein Col Med, Bronx NY, 75-; consult, Med Res Coun Can, 81- *Honors & Awards:* Revel Mem Award, Yeshia Univ, 71. *Mem:* Am Soc Hemat; Am Med Assoc; Am Soc Clin Oncol; Am Assoc Hist Med. *Res:* Internal medicine and hematology; medical education and medical ethics; medical history and Jewish medicine. *Mailing Add:* 82-68 164 St Jamaica NY 11432

ROSNER, JONATHAN LINCOLN, b New York, NY, July 23, 41; m 65; c 2. PHYSICS. *Educ:* Swarthmore Col, AB, 62; Princeton Univ, MA, 63, PhD(physics), 65. *Prof Exp:* Res asst prof physics, Univ Wash, 65-67; vis lectr, Tel-Aviv Univ, 67-69; from asst prof to prof, Univ Minn, Minneapolis, 69-82; PROF PHYSICS, UNIV CHICAGO, 82- *Concurrent Pos:* Consult, Argonne Nat Lab, 71-74, Fermi Nat Accelerator Lab, 75-78 & Brookhaven Nat Lab, 79-82; transl, Am Inst Physics, 71-73 & 78-; Alfred P Sloan res fel, Univ Minn, Minneapolis, Europ Orgn Nuclear Res, Geneva & Stanford Linear Accelerator Ctr, 71-73; mem, Inst Advan Study, 72, 76-77, 88-89; vis res assoc, Calif Inst Technol, 70, 72 & 75; mem, High Energy Physics adv panel, 75, 87-91, subpanel on new facil, 77 & subpanel on long range planning, 81; mem bd trustees, Aspen Ctr Physics, 78-; mem natural sci & eng grant selection comt, High Energy Physics, Can, 80-82; mem exec comt, Div Particles & Fields, Am Phys Soc, 84-87; prog comt, Cornell Electron Storage Ring, 86-88. *Mem:* Fel Am Phys Soc. *Res:* Theoretical and elementary particle physics. *Mailing Add:* Enrico Fermi Inst Univ Chicago 5640 S Ellis Ave Chicago IL 60637

ROSNER, JUDAH LEON, b New York, NY, Oct 18, 39; div; c 1. MOLECULAR BIOLOGY. *Educ:* Columbia Col, BA, 60; Yale Univ, MS, 64, PhD(biol), 67. *Prof Exp:* Biologist, 67, scientist, 67-70, staff fel, 70-74, RES BIOLOGIST, NIH, 74- *Mem:* Am Soc Microbiol. *Res:* Molecular biology and genetics of bacteria, bacterial viruses and transposable genetic elements; antibiotic-resistance mechanisms; outer membrane protein regulation, aspirin effects. *Mailing Add:* NIH Bldg 2 Rm 210 Bethesda MD 20892

ROSNER, MARSHA R, b Springfield, Mass, Nov 8, 50; m; c 2. CELL GROWTH. *Educ:* Mass Inst Technol, PhD(biochem), 78. *Prof Exp:* asst prof toxicol, Mass Inst Technol, 82-87; ASSOC PROF, BEN MAY INST PHARMACOL & PHYSIOL SCI, UNIV CHICAGO, 87- *Mem:* Am Soc Biol Chemists; Am Soc Cellular Biol; Am Cancer Soc; Am Soc Microbiol. *Mailing Add:* 4950 S Greenwood Chicago IL 60615

ROSNER, ROBERT, b Garmisch-Partenkirchen, WGer, June 26, 47; US citizen; m 71; c 2. PLASMA ASTROPHYSICS, SOLAR PHYSICS. *Educ:* Brandeis Univ, BA, 69; Harvard Univ, PhD(physics), 75. *Prof Exp:* From instr to assoc prof astron, Harvard Univ, 77-86; PROF ASTRON, UNIV CHICAGO, 87- *Concurrent Pos:* Woodrow Wilson Found fel, 69; res fel solar x-ray astron, Harvard Col Observ, 76-78. *Mem:* Fel Am Phys Soc; Int Astron Union; Am Astron Soc; Soc Indust & Appl Math. *Res:* Solar and stellar x-ray astronomy; plasma heating, transport processes, diagnostics; astrophysical magnetic field generation and diffusion; fluid dynamics. *Mailing Add:* Enrico Fermi Inst Univ Chicago 5640 S Ellis Ave Chicago IL 60637

ROSNER, SHELDON DAVID, b Toronto, Ont, Feb 19, 41; m 67; c 2. PHYSICS. *Educ:* Univ Toronto, BSc, 62; Harvard Univ, AM, 63, PhD(physics), 68. *Prof Exp:* Nat Res Coun Can fel, Clarendon Lab, Oxford Univ, 69-71; ASST PROF PHYSICS & NAT RES COUN CAN GRANT, UNIV WESTERN ONT, 71- *Concurrent Pos:* Vis fel, Univ Colo, 78-79. *Mem:* Am Phys Soc; Can Asn Physicists. *Res:* Atomic physics; optical pumping; atomic and molecular beams; precise resonance measurements of atomic and molecular structure of fundamental interest. *Mailing Add:* Dept Physics Univ Western Ont London ON N6A 3K7 Can

ROSNER, WILLIAM, b Brooklyn, NY, Jan 7, 33. ENDOCRINOLOGY. *Educ:* Univ Wis, BA, 54; Albert Einstein Col Med, MD, 62. *Prof Exp:* Instr med, 67-69, assoc, 69-70, asst clin prof med, 70-72, asst prof, 72-73, assoc prof, 73-82, PROF MED, COL PHYSICIANS & SURGEONS, COLUMBIA UNIV, 82- *Concurrent Pos:* Asst attend physician, Roosevelt Hosp, 67-79, assoc attend physician, 69-72, attend physician & dir, Div Endocrinol, 72- *Mem:* Fel Am Col Physicians; Am Fedn Clin Res; Am Soc Biol Chemists; Am Soc Clin Invest; Endocrine Soc. *Res:* Steroid hormones, including the steroids secreted by the adrenal cortex, the ovaries, and the testis, their transport in plasma and the specific plasma proteins to which they are bound. *Mailing Add:* Dept Med Columbia Univ 360 W 168th St New York NY 10032

ROSOFF, BETTY, b New York, NY, May 28, 20; m 42; c 1. ENDOCRINOLOGY. *Educ:* Hunter Col, BA, 42, MA, 60; City Univ New York, PhD(biol), 66. *Prof Exp:* Res chemist, Nat Aniline Div, Allied Chem Corp, NY, 43-48; res assoc cancer, Montefiore Hosp, Bronx, 52-62; instr biol, Bronx Community Col, 64-65; lectr, Hunter Col, 65-67; assoc prof, Paterson State Col, 67-68; from asst prof to prof, 74-86, EMER PROF BIOL, STERN COL, YESHIVA UNIV, 86- *Concurrent Pos:* Lectr physiol, Hunter Col, 61-64, adj prof, 68-75; mem working cadre, Nat Prostatic Cancer Proj, 74-79; vis investr, Am Mus Natural Hist, 81-82. *Mem:* AAAS; NY Acad Sci; Am Physiol Soc; Sigma Xi. *Res:* Zinc and trace metal metabolism; zinc and prostatic pathology; gonadotrophic control of prostate; thymus gland in development; radioactive isotope decontamination. *Mailing Add:* 280 Ninth Ave Apt 17H New York NY 10001

ROSOFF, MORTON, b Brooklyn, NY, Feb 19, 22. PHYSICAL CHEMISTRY, SURFACE CHEMISTRY. *Educ:* Brooklyn Col, BA, 42, MA, 49; Duke Univ, PhD(chem), 55. *Prof Exp:* Res assoc & lectr colloid chem, Columbia Univ, 52-54; adj asst prof biophys chem, Sloan Kettering Inst Cancer Res, 54-58; sr scientist, IBM Watson Labs, NY, 60-64; sr scientist surface chem, Columbia Univ, 64-68, assoc prof phys chem & chmn dept, Col Pharmaceut Sci, 68-76; group leader phys pharm, USV Pharmaceut Corp, 76-84; PROF, PHYS PHARMACY, ARNOLD & MARIE SCHWARTZ COL PHARMACEUT SCI, LONG ISLAND UNIV, 84- *Concurrent Pos:* Adj prof, City Univ New York, 56-75; sr scientist, Lever Bros Inc, NJ, 59-60; consult, Vet Admin Hosp, East Orange, NJ, 64, Dr Madaus Inc, WGer, 65-68, Am Stand & Testing, NY, 66-69 & Carter Wallace Inc, NJ, 69-74, Pennelt Inc, Pa, 85-86. *Mem:* Am Chem Soc; AAAS. *Res:* Physical chemistry of biological macromolecules; surface chemistry of biological macromolecules, microemulsions; metastable colloidal systems; physical pharmacy. *Mailing Add:* 15 Wellesley Ave Yonkers NY 10705-3841

ROSOLOWSKI, JOSEPH HENRY, b Fall River, Mass, May 18, 30; m 52; c 2. PHYSICAL PROPERTIES OF CERAMICS. *Educ:* Rensselaer Polytech Inst, BS, 52, PhD(solid state physics), 61. *Prof Exp:* Assoc physicist, Res & Develop Ctr, Vitro Corp Am, 52-54; PHYSICIST, CORP RES & DEVELOP CTR, GEN ELEC CO, 60- *Concurrent Pos:* AEC exchange scientist, Poland, 62-63. *Honors & Awards:* Ross Coffin Purdy Award, Am Ceramic Soc, 76. *Mem:* Am Phys Soc; Am Ceramic Soc; Sigma Xi. *Res:* Diffusion in solids; physical properties of ceramics; radiation effects in metals; technical management. *Mailing Add:* 1435 Myron St Schenectady NY 12309

ROSOMOFF, HUBERT LAWRENCE, b Philadelphia, Pa, Apr 11, 27; m 50; c 2. MEDICINE. *Educ:* Univ Pa, AB, 48; Hahnemann Med Col, MD, 52; Columbia Univ, DMedSci, 60. *Prof Exp:* Intern, Hahnemann Hosp, Pa, 52-53; asst resident surgeon, Presby Hosp, New York, 53-54, asst resident neurosurgeon, 54-55; asst resident neurosurgeon, Neurol Inst, New York, 57-58, resident neurosurgeon, 58-59; from asst prof to assoc prof neurol surg, Sch Med, Univ Pittsburgh, 59-66; prof & chmn dept, Albert Einstein Col Med & chief hosp, 66-71; PROF NEUROL SURG & CHMN DEPT, SCH MED, UNIV MIAMI, 71-, MED DIR, COMPREHESIVE PAIN CTR, 78- *Concurrent Pos:* Neurosurgeon, Presby Hosp & chief neurosurg, Vet Admin Hosp, Pittsburgh, 59-66; chief neurol surg, Bronx Munic Hosp Ctr & chief, Montefiore Hosp & Med Ctr, 66-71; chief neurosurg sect, Vet Admin Hosp, Miami, 71-; chmn dept neurol surg, Jackson Mem Hosp, Miami, 71-; chief, pain & rehab servs, SShore Med Hosp & Med Ctr, 85- *Honors & Awards:* Jeremiah Fix Award, 82; Am Acad Neur Surg Award, 56. *Mem:* AMA; Am Col Surg; Cong Neurol Surg; Soc Neurol Surg; NY Acad Sci; Am Acad Neurosurg; Am Asn Neurosurg; Am Pain Soc. *Res:* Neurological surgery; physiology; hypothermia; transplantation; pain. *Mailing Add:* Dept Neurol Surg Univ Miami Sch Med 1501 NW Ninth Ave Miami FL 33136

ROSOWSKY, ANDRE, b Lille, France, Mar 3, 36; US citizen; m 62; c 3. ORGANIC CHEMISTRY, MEDICINAL CHEMISTRY. *Educ:* Univ Calif, Berkeley, BS, 57; Univ Rochester, PhD(org chem), 61. *Prof Exp:* NSF res fel, Harvard Univ, 61-62; res assoc path, Children's Cancer Res Found & Children's Hosp Med Ctr, Dana-Farber Cancer Inst, 62-74; assoc biol chem, 64-78, prin assoc pharmacol, 78-87, ASSOC PROF, BIOL CHEM & MOLECULAR PHARMACOL, HARVARD MED SCH, 87- *Concurrent Pos:* Adj assoc prof med chem, Northeastern Univ, 76-82, adj prof, 82- *Mem:* Am Chem Soc; Am Asn Cancer Res. *Res:* Design and synthesis of new biologically active compounds, especially nitrogen heterocycles; medicinal chemistry. *Mailing Add:* Div Cancer Pharmacol Dana-Farber Cancer Inst 44 Binney St Boston MA 02115

ROSS, ALAN, b Hamilton, Ohio, Aug 25, 26; m 50; c 3. BIOSTATISTICS. *Educ:* Brown Univ, AB, 50; Iowa State Univ, MS, 52, PhD(statist), 60. *Prof Exp:* Res assoc biostatist, Sch Pub Health, Univ Pittsburgh, 54-56; from asst prof to assoc prof behav sci, Sch Med, Univ Ky, 56-64; assoc prof, 64-67, chmn dept, 67-81, PROF BIOSTATIST, SCH HYG & PUB HEALTH, JOHNS HOPKINS UNIV, 67- *Concurrent Pos:* Prin investr, Med Comput Planning grant, HEW, 62- 63; chief biostatist consult, Off Aviation Med, Fed Aviation Admin, 66-70; statist adv, Int Collab Study Med Care, WHO, 66-73; prog dir, Biostatist Training grant, Nat Inst Gen Med Sci, 66-76 & Pub Health Statist Training grant, HEW, 69-74; chief statistician, Afghan Demog Study, USAID proj, 71-76; statist adv, Precursors of Learning Disability, Kennedy Inst, Johns Hopkins, 76-87; statist adv, Chagasic Infection & Myocardiopathy, WHO, Venezuela, 77-; vis prof, Univ Adelaide, Australia, 82, Univ Nairobi, Kenya, 87 & Univ Padua, Italy, 90. *Mem:* Fel Am Statist Asn; Biomet Soc; Int Asn Surv Statisticians. *Res:* Sampling theory and methods. *Mailing Add:* Dept Biostatist Johns Hopkins Univ Baltimore MD 21205

ROSS, ALBERTA B, b Moores Hill, Ind, July 26, 28; m 56; c 4. CHEMISTRY, INFORMATION SCIENCE. *Educ:* Purdue Univ, BS, 48; Wash Univ, BS, 51; Univ Md, PhD(chem), 57. *Prof Exp:* Tech librn, Monsanto Chem Co, 48-53; res assoc, Univ Mich, 57-58; supvr, 64-71, consult, 71-72, SUPVR, RADIATION CHEM DATA CTR, UNIV NOTRE DAME, 72- *Mem:* Am Chem Soc; Inter-Am Photochem Soc; Europ Photochem Asn; Sigma Xi. *Res:* Radiation chemistry; chemical kinetics; data compilation; chemical literature. *Mailing Add:* Radiation Chem Data Ctr Radiation Lab Univ Notre Dame Notre Dame IN 46556-0768

ROSS, ALEX R, physical geology, geomorphology; deceased, see previous edition for last biography

ROSS, ALEXANDER, b St Louis, Mo, Feb 7, 20; m 47; c 1. ORGANIC CHEMISTRY. *Educ:* Wayne Univ, BS, 42; Univ Mich, MS, 50, PhD(org chem), 53. *Prof Exp:* Control chemist, Swift & Co, 42; instr chem, Univ Mich, 51; res chemist, Ethyl Corp, 52-55, asst res supvr, 55; sr res chemist, Olin-Mathieson Chem Corp, Conn, 56-57, sect chief, 57-59; head org res, Metal & Thermit Corp, 59-61, Europ tech mgr, M&T Chem Inc, Switz, 61-63, mgr org chem res & develop, NJ, 63-64, mgr chem res & develop, 64-65, dir res, 65-68, tech dir, 68-74; dir res & develop, Spencer Kellogg Div, Textron Inc, 74-76, vpres res & develop, 77-86; RETIRED. *Concurrent Pos:* Consult & Lectr, 86- *Mem:* AAAS; Am Chem Soc; Soc Plastics Engrs; Asn Res Dirs; NY Acad Sci. *Res:* Mechanism of antiknock action; organic phosphorus and agricultural chemistry; stereochemistry of carbocyclic systems; organohydrazine compounds; organometallics; stabilizers for polymers; organotin biocides; urethane chemicals; metal treatment; ceramics; polymers; catalysts; resins for coatings, inks, adhesives; urethanes, polyesters. *Mailing Add:* 400 N Cherry St Falls Church VA 22046

ROSS, ALONZO HARVEY, b Louisville, Ky, Sept 25, 50; m 74. GROWTH FACTOR RECEPTORS, DIFFERENTIATION. *Educ:* Cornell Univ, BA, 72; Stanford Univ, PhD(chem), 77. *Prof Exp:* Res assoc, Wistar Inst, 81-84, asst prof, 85-88; SR SCIENTIST, WORCESTER FOUND EXP BIOL, 88- *Concurrent Pos:* Asst prof, Dept Neurol, Univ Pa Sch Med, 86-88; assoc prof, Dept Biochem, Univ Mass Med Col, 89- *Honors & Awards:* Schweisguth Prize, Int Soc Pediat Oncol, 89. *Mem:* Am Asn Cancer Res; Am Asn Immunologists; Am Chem Soc; Am Soc Neurochem; Int Soc Differentiation; Soc Neurosci. *Res:* Developed the first anti-phosphotyrosine antibodies; study the structure of the nerve growth factor receptor and differentiation of neural tumor cells. *Mailing Add:* Worcester Found 222 Maple Ave Shrewsbury MA 01545

ROSS, ALTA CATHARINE, b Santa Monica, Calif, Jan 19, 47; m 69. VITAMIN METABOLISM, ATHEROSCLEROSIS. *Educ:* Univ Calif, Davis, BS, 70; Cornell Univ, MNS, 72, PhD(biochem), 76. *Prof Exp:* Staff & res assoc med, Columbia Univ, 76-78; from asst prof to assoc prof, 78-88, PROF BIOCHEM, MED COL PA, 88-, DIR DIV NUTRIT. *Concurrent Pos:* Coun fel, Am Heart Asn, 81- *Honors & Awards:* Mead Johnson Award, Am Inst Nutrit, 86. *Mem:* AAAS; Sigma Xi; Am Inst Nutrit; Am Heart Asn; Am Soc Cell Biol. *Res:* Vitamin A transport and metabolism; lipoprotein metabolism; hepatic retinoid metabolism; nutrition and immune function. *Mailing Add:* 3300 Henry Ave Philadelphia PA 19129

ROSS, ARTHUR LEONARD, b New York, NY, Mar 9, 24; m 48; c 2. DESIGN REVIEW & FAILURE PREVENTION. *Educ:* NY Univ, BAeE, 48, MAeE, 50, EngScD(aeronaut eng), 54. *Prof Exp:* Consult engr struct anal methods, Aircraft Nuclear Propulsion Dept, Gen Elec Co, 54-61, mgr S5G struct eval, Knolls Atomic Power Lab, 61-66, staff engr struct mech, Re-entry Systs Div, 66-91; CONSULT ENG, 91- *Concurrent Pos:* Chmn, Reliablity, Stress Anal & Failure Prev Comt, Am Soc Mech Engrs, 78-82, Philadelphia Sect, 82-83, Design Eng Div, 85-86; mem, Structures Tech Comn, Am Inst Aeronaut & Astronaut. *Mem:* Fel Am Soc Mech Engrs; Am Acad Mech; Sigma Xi; Am Inst Aeronaut & Astronaut. *Res:* Research and development of mechanical stress analysis, shock and vibration, hypervelocity impact, thermal stress, pressure vessels and composite materials; consultation and design reviews for failure prevention and mechanical design support. *Mailing Add:* Consult Eng 122 Maple Ave Bala Cynwyd PA 19004

ROSS, BERNARD, b Montreal, Que, Nov 10, 34; US citizen; m 68; c 2. MECHANICAL & STRUCTURAL ENGINEERING. *Educ:* Cornell Univ, BME, 57; Stanford Univ, MSc, 59, PhD(aeronaut eng), 65; Nat Sch Advan Aeronaut Studies, France, dipl aeronaut eng, 60; Univ Edinburgh, dipl eng mech, 61. *Prof Exp:* Assoc engr, Marquardt Aircraft Co, 57-58; res asst aeronaut eng, Standord Univ, 60-64, res assoc, 64-65; eng physicist, Stanford Res Inst, 65-70; sr mem staff & exec, 70-78, pres, 78-82, CHMN BD, FAILURE ANALYSIS ASSOCS, 78- *Concurrent Pos:* Consult, Failure Analysis Assocs, 68-70; vis lectr, Univ Santa Clara, 70-78; distinguished lect speaker, Am Inst Aeronaut & Astronaut, 90-91. *Mem:* AAAS; Am Soc Mech Engrs; Am Inst Aeronaut & Astronaut; Sigma Xi; Nat Soc Prof Engrs; Soc Automotive Engrs. *Res:* Experimental and theoretical research in the buckling of thin shell structures and the penetration of solids by projectiles; structural and mechanical analysis of failures; fracture mechanics; design analysis. *Mailing Add:* PO Box 3015 Menlo Park CA 94025

ROSS, BERND, solid state physics, solid state electronics; deceased, see previous edition for last biography

ROSS, BRADLEY ALFRED, b Broolyn, NY, June 5, 52. COMPUTER SIMULATION & MODELING. *Educ:* Univ Pa, BS, 74, MS, 76, PhD(chem eng), 79. *Prof Exp:* SR ENGR, MONSANTO CO, 79- *Mem:* Am Inst Chem Engrs; Sigma Xi. *Res:* Computer programs to aid in design, maintenance and operation of chemical plants, with emphasis on steady-state simulation. *Mailing Add:* 705 General Scott Rd King of Prussia PA 19406

ROSS, BRUCE BRIAN, b Bryn Mawr, Pa, June 30, 44; m 68; c 2. DYNAMIC METEOROLOGY. *Educ:* Brown Univ, ScB, 66; Princeton Univ, MA, 68, PhD(aerospace eng), 71. *Prof Exp:* PHYSICIST METEOROL, GEOPHYS FLUID DYNAMICS LAB, NOAA PRINCETON UNIV, 71- *Concurrent Pos:* NSF fel, 66-70. *Mem:* Am Meteorol Soc; Sigma Xi; Am Geophys Union. *Res:* Mesoscale meteorology; numerical modeling of generation and maintenance of severe storm systems. *Mailing Add:* Geophys Fluid Dynamics Lab Princeton Univ PO Box 308 Princeton NJ 08542

ROSS, CHARLES ALEXANDER, b Urbana, Ill, Apr 16, 33; m 59. GEOLOGY, PALEOBIOLOGY. *Educ:* Univ Colo, BA, 54; Yale Univ, MS, 58, PhD(geol), 59. *Prof Exp:* Res assoc, Yale Univ, 59-60; from asst geologist to assoc geologist, Ill State Geol Surv, 60-64; from asst prof to prof geol, Western Wash Univ, 64-82, chmn dept, 77-82; staff geologist to sr staff geologist, Stratigraphic Sci, Gulf Oil Explor Prod Co, 82-83, dir, Stratigraphic Sci, 83-85; SR BIOSTRATIGRAPHER, CHEVRON USA, 85- *Concurrent Pos:* NSF res grants late Paleozoic fusulinaceans, 64-68; consult petrol explor, Geol Surv Can, 75; mem, Int Subcomn Permian Stratig, 75-88. *Mem:* Fel AAAS; fel Geol Soc Am; hon mem Soc Econ Paleont & Mineral (secy-treas, 82-84); Am Asn Petrol Geol; Soc Study Evolution; Cushman Found Foraminiferal Res (pres, 81-82, 90-91). *Res:* Biostratigraphy and paleontology, including foraminiferal faunas; paleoclimatology; paleoecology; recent large calcareous Foraminifera; paleobiogeography. *Mailing Add:* Chevron USA PO Box 1635 Houston TX 77251

ROSS, CHARLES BURTON, b Rochester, NY, July 23, 34; m 62; c 2. COMPUTER SCIENCES. *Educ:* Villanova Univ, BS, 57; Purdue Univ, Lafayette, MS, 63, PhD(physics), 69. *Prof Exp:* AEC fel, Los Alamos Sci Lab, 69-71; asst prof, 71-80, ASSOC PROF COMPUT SCI, WRIGHT STATE UNIV, 80- *Mem:* Asn Comput Mach. *Res:* Digital design. *Mailing Add:* Dept Comput Sci Wright State Univ Dayton OH 45435

ROSS, CHARLES W(ARREN), electrical engineering, for more information see previous edition

ROSS, CLAY CAMPBELL, JR, b Lexington, Ky, June 17, 36; m 64. MATHEMATICS. *Educ:* Univ Ky, BS, 59; Univ NC, MA, 61, PhD(math), 64. *Prof Exp:* From asst prof to assoc prof math, Emory Univ, 67-73; assoc prof, 73-81, dir acad comput, 77-81, PROF MATH & CONSULT, FAC COMPUT, UNIV SOUTH, 81- *Concurrent Pos:* Vis prof math, Univ Mo, Rolla, 79-80. *Mem:* Math Asn Am; Soc Indust & Appl Math; Asn Comput Mach; Sigma Xi. *Res:* Singular differential equations and systems. *Mailing Add:* SPO Box 1220 Sewanee TN 37375

ROSS, CLEON WALTER, b Driggs, Idaho, May 27, 34; m 56; c 3. PLANT PHYSIOLOGY, PLANT BIOCHEMISTRY. *Educ:* Brigham Young Univ, BS, 56; Utah State Univ, MS, 59, PhD(plant physiol), 61. *Prof Exp:* From asst prof to assoc prof, 60-70, PROF BOT & PLANT PATH, COLO STATE UNIV, 70- *Mem:* Am Soc Plant Physiol. *Res:* Mechanisms of hormone-induced plant growth. *Mailing Add:* Dept Biol Colo State Univ Ft Collins CO 80523

ROSS, DAVID A, b New York, NY, Aug 8, 36. GEOLOGICAL OCEANOGRAPHY. *Educ:* City Col New York, BS, 58; Univ Kans, MS, 60; Univ Calif, San Diego, PhD(oceanog), 65. *Prof Exp:* Res asst geol, Univ Kans, 58-60; res oceanogr, Scripps Inst Oceanog, Calif, 60-65; from asst scientist to assoc scientist, 67-78, sr scientist & dir, Marine Policy & Ocean Mgt Prog, 78-85, Chmn, Dept Geol & Geophys, Woods Hole Oceanog Inst, 85-90; CONSULT, 90- *Concurrent Pos:* Instr, Mass Inst Technol, 71-78 & Fletcher Sch Law & Diplomacy, Tufts Univ, 71-78; mem exec bd, Law of the Sea Inst, 74-; Ocean Affairs Adv Comn, Dept of State, 75-, Sea Grant Coordr, 77-; mem, Ocean Policy Comt, Nat Acad Sci, 77-82, Ocean Studies Bd, 85- *Mem:* AAAS; Geol Soc Am; Am Geophys Union. *Res:* Distribution and movement of sediments on the continental shelf and deep sea; Black Sea, Mediterranean Sea, Red Sea and Persian Gulf; geology and geophysics of marginal seas; marine affairs. *Mailing Add:* Woods Hole Oceanog Inst Falmouth MA 02543

ROSS, DAVID SAMUEL, b Los Angeles, Calif, Dec 5, 37; m 66; c 1. PHYSICAL ORGANIC CHEMISTRY. *Educ:* Univ Calif, Los Angeles BS, 59; Univ Wash, PhD(org chem), 64. *Prof Exp:* PHYS ORG CHEMIST, SRI INT, 64- *Mem:* Am Chem Soc. *Res:* Chemical reaction kinetics; thermochemistry and mechanism gas and solution phases; acid-base catalysis; coal dissolution and liquefaction; chemistry related to synthetic fuels; fundamental studies in aromatic nitration and oxidation. *Mailing Add:* SRI Int PS269 Menlo Park CA 94025

ROSS, DAVID STANLEY, b DuBois, Pa, Apr 16, 47; div; c 2. AGRICULTURAL ENGINEERING. *Educ:* Pa State Univ, BS, 69, MS, 71, PhD(agr eng), 73. *Prof Exp:* Res asst agr eng, Pa State Univ, 69-71, NSF trainee, 71-72; asst prof, 73-78, ASSOC PROF AGR ENG, UNIV MD, 78- *Concurrent Pos:* Exten agr engr, Univ Md, 73-; 1977 & 1978 Yearbk Agr Comt, USDA, 76-78. *Honors & Awards:* Blue Ribbon Awards, Am Soc Agr Engrs, 77, 78, 79, 80 & 83. *Mem:* Am Soc Agr Engrs; Irrigation Asn; Coun Agr Sci & Technol. *Res:* Extension education programs in nursery and greenhouse structures, environment and equipment; equipment for turf, fruit and vegetable producers; trickle irrigation; energy conservation; post harvest cooling and handling of fruits and vegetables. *Mailing Add:* Dept Agr Eng Univ Md College Park MD 20742-5711

ROSS, DAVID WARD, b Detroit, Mich, Aug 11, 37; m 59; c 2. THEORETICAL PHYSICS. *Educ:* Univ Mich, BS, 59; Harvard Univ, AM, 60, PhD(physics), 64. *Prof Exp:* Res assoc physics, Univ Ill, Urbana, 64-66; univ fel, 66-67, asst prof physics, 66-70, res scientist, 66-74, ASST DIR THEORET PROG, FUSION RES CTR, UNIV TEX, AUSTIN, 74- *Concurrent Pos:* Mem, Inst Advan Study, Princeton Univ, 70-71; asst dir, Inst Fusion Studies, Univ Tex, Austin, 80-83. *Mem:* AAAS; fel Am Phys Soc. *Res:* Linear and nonlinear stability theory, radio frequency heating, and thermal transport of fusion plasmas. *Mailing Add:* Fusion Res Ctr Univ Tex Austin TX 78712

ROSS, DENNIS KENT, b Hebron, Nebr, May 4, 42; m 66; c 1. THEORETICAL PHYSICS. *Educ:* Calif Inst Technol, BS, 64; Stanford Univ, PhD(physics), 68. *Prof Exp:* From instr to assoc prof, 68-79, PROF PHYSICS, IOWA STATE UNIV, 79- *Mem:* Am Phys Soc; Int Astron Union. *Res:* Radar reflection test of general relativity; absorption in high energy photo production; large angle multiple scattering; magnetic monopoles; scalar-tensor theory of gravitation; charge quantization; Weyl geometry and quantum electrodynamics; gauge supersymmetry; fiber bundles and supergravity; holonomy groups and spontaneous symmetry breaking. *Mailing Add:* Dept Physics Iowa State Univ Ames IA 50011

ROSS, DONALD CLARENCE, b Buffalo, NY, Mar 15, 24; m 48; c 3. GEOLOGY. *Educ:* Univ Iowa, BA, 48, MS, 49; Univ Calif, Los Angeles, PhD(geol), 52. *Prof Exp:* Asst, Univ Iowa, 48-49; geologist, State Geol Surv, Iowa, 49; asst, Univ Calif, Los Angeles, 49-52; geologist, US Geol Surv, 52-89; RETIRED. *Res:* Igneous and metamorphic petrology; granitic rocks of the White and Inyo Mountains, California; granitic basement of the San Andreas fault region, Coast and Tranverse Ranges; tectonic framework, Southern Sierra Nevada, California. *Mailing Add:* 301 Barclay Ct Palo Alto CA 94306

ROSS, DONALD JOSEPH, b Brooklyn, NY, Mar 26, 28; m 51; c 3. NUTRITIONAL BIOCHEMISTRY. *Educ:* Fordham Univ, BS, 49, PhD(physiol), 56; Boston Col, MS, 50. *Prof Exp:* Asst biol, Boston Col, 49-50; from instr to prof, 50-64, chmn dept, 60-70, 77-80, Decamp prof, 79-85, PROF BIOL, FAIRFIELD UNIV, 85- *Concurrent Pos:* Dir biochem lab, St Vincent's Hosp, Bridgeport, Conn, 57-59, consult, 59- *Mem:* AAAS; NY Acad Sci; fel Am Inst Chem; Am Inst Biol; Sigma Xi. *Res:* Biochemistry of insect metamorphosis; invertebrate physiology; cardiac enzymology; enzyme induction of cardiac and vascular atherosclerotic lesions. *Mailing Add:* 473 Wormwood Rd Fairfield CT 06430-4556

ROSS, DONALD K(ENNETH), b St Louis, Mo, Apr 15, 25; m 51; c 1. ELECTRICAL & INDUSTRIAL ENGINEERING. *Educ:* Univ Minn, BSEE, 46; Mass Inst Technol, MSEE, 48; Wash Univ, DSc(indust eng), 60. *Prof Exp:* Asst, Mass Inst Technol, 46-48; PRES, ROSS & BARUZZINI, INC, 62- *Concurrent Pos:* Prof assoc, Bldg Res Adv Bd, Nat Acad Sci, 70- *Honors & Awards:* Outstanding Achievement Award, Inst Elec & Electronics Engrs Indust Appln Soc, 89. *Mem:* Inst Elec & Electronics Engrs; fel Am Consult Engrs Coun; Sigma Xi. *Res:* Effects of illumination on task performance; interaction of lighting and HVAC (heating, ventilating and air conditioning) systems. *Mailing Add:* Nine Crosswinds St Louis MO 63132

ROSS, DONALD LEWIS, b Paris, Tex, Aug 11, 36. ORGANIC CHEMISTRY. *Educ:* Univ Tex, BA, 58, PhD(chem), 63; Golden Gate Univ, MBA, 79. *Prof Exp:* Chemist, 62-70, GROUP MGR, SRI INT, 70- *Mem:* Am Chem Soc. *Res:* Synthesis of propellant ingredients; chemistry of organic fluorine and nitro compounds; synthesis of metabolite antagonists; process development. *Mailing Add:* SRI Int Menlo Park CA 94025

ROSS, DONALD MORRIS, b Kenosha, Wis, Aug 22, 23; m 46; c 6. INDUSTRIAL HYGIENE. *Educ:* Univ Tex, BS, 43; Univ Pittsburgh, MPH, 53, ScD(indust hyg), 56. *Prof Exp:* Tech supvr, Tenn Eastman Co, 43-46; res assoc physics, Carbide & Carbon Chem Co Div, Union Carbide & Carbon Corp, 46-48, indust hygienist, 48-52; res asso, Univ Pittsburgh, 56-58; indust hygienist, 58-62, chief health protection br, US AEC & US Energy Res & Develop Admin, 62-77, chief, 77-81, dir occup safety & health br, div occup safety & health, 81-85, DIR, DIV OCCUP SAFETY & HEALTH, US DEPT ENERGY, 85- *Concurrent Pos:* Fel indust hyg, AEC, Pa, 52-53, res fel, 53-56. *Mem:* Health Physics Soc; Am Indust Hyg Asn; Am Conf Govt Indust Hygienists. *Res:* Radiation protection; human engineering; control of industrial environment. *Mailing Add:* US Dept Energy EV-133 Washington DC 20545

ROSS, DONALD MURRAY, zoology; deceased, see previous edition for last biography

ROSS, DORIS LAUNE, b Thorndale, Tex, Apr 20, 26. BIOCHEMISTRY. *Educ:* Tex Woman's Univ, BS, 47; Baylor Univ, MS, 58; Univ Tex, PhD(biochem), 67. *Prof Exp:* Med technologist, 47-54, sect head chem, Clin Labs, 54-56, chief med technologist & teaching supvr, 56-64, clin biochemist, Hermann Hosp & prof clin lab sci, Med Sch, Univ Tex Health Sci Ctr, 67-77, PROF PATH & LAB MED & ASSOC DEAN SCH ALLIED HEALTH SCI, MED SCH, UNIV TEX, 77- *Concurrent Pos:* Chief med technologist, Care Medico, Algeria, 62. *Mem:* Am Soc Med Technol; Am Asn Clin Chemists; fel Am Acad Clin Biochem; fel Am Soc Allied Health Professions. *Res:* Clinical chemistry; clinical laboratory methodology; protein structure; immunoglobulins. *Mailing Add:* Dept Clin Lab Sci Univ Tex Health Sci Ctr PO Box 20708 Houston TX 77225

ROSS, DOUGLAS TAYLOR, b Canton, China, Dec 21, 29; US citizen; m 51; c 3. STRUCTURED ANALYSIS & DESIGN, COMPUTER-AIDED DESIGN. *Educ:* Oberlin Col, AB, 51; Mass Inst Technol, SM, 54. *Prof Exp:* Head computer applications group, Mass Inst Technol Elec Syst Lab, 52-69; pres, 69-75, chmn, 75-89, EMER CHMN, SOFTECH, INC 89- *Concurrent Pos:* Lectr, Dept Elec Eng, Mass Inst Technol, 60-69, Dept Elec Eng & Computer Sci, 83-; mem, working group 23 prog methods, Int Fedn Info Processing, working group algol NATO Software Eng Conf Garmisch & Rome, 68 & 69, various comts, Asn Comp Mach, syst lang & discrete mfg; SofTech prin investr, Air Force AFCAM & ICAM projs, 74-77; gen chmn, Specif Reliable Software Conf, Inst Elec & Electronic Engrs, 79, Inst Computer & Sci Technol eval panel, Nat Res Coun, 81-83; chmn bd, SofTech Microsysts, Inc, 79-81; trustee & dir, Charles Babbage Inst, 84-; dir, Cognition, Inc, 85-89. *Honors & Awards:* Joseph Marie Jaquard Mem Award, Numerical Control Soc, 75; Distinguished Contrib Award, Soc Mfg Engrs, 80. *Mem:* Asn Comput Mach; AAAS; Sigma Xi. *Res:* Created and directed development APT system for automatic programming of numerically controlled machine tools, now an international standard; created first software engineering language and software production tools. *Mailing Add:* 460 Totten Pond Rd Waltham MA 02154

ROSS, EDWARD WILLIAM, JR, b Jackson Heights, NY, July 3, 25; m 55; c 5. MATHEMATICS, APPLIED MATHEMATICS. *Educ:* Webb Inst Naval Archit, BSc, 45; Brown Univ, ScM, 49, PhD(appl math), 54. *Prof Exp:* Fel, Exp Towing Tank, Stevens Inst Technol, 46-47; res engr, 51-52; mathematician, US Army Res Agency, Watertown Arsenal, 55-68; STAFF MATHEMATICIAN, NATICK RES & DEVELOP COMMAND, US ARMY, 68-; RES PROF, DEPT MATH SCI, WORCESTER POLYTECH INST, 85- *Concurrent Pos:* Secy Army res & study fel, 60-61. *Res:* Plasticity; elasticity; thin shell theory; differential equations; asymptotic theory; mechanical vibrations; biological statistics. *Mailing Add:* Dept Math Sci Worcester Polytech Inst Worcester MA 01609

ROSS, ELDON WAYNE, b Elana, WVa, Oct 7, 34; m 59; c 2. FOREST PATHOLOGY. *Educ:* WVa Univ, BS, 57, MS, 58; Syracuse Univ, PhD(forest path), 64. *Prof Exp:* Instr forest bot, State Univ NY Col Forestry, Syracuse Univ, 63-64; plant pathologist, Forestry Sci Lab, Ga, 64-71, forest pathologist, Div Forest Insect & Disease Res, 71-73, asst dir forest insect & disease res, 73-74, asst dir continuing res, Northeast Forest Exp Sta, 74-79, dir, Southeastern Forest Exp Sta, Forest Serv, USDA, 79-85, ASSOC DEP CHIEF RES, USDA FOREST SERV, 85- *Mem:* Soc Am Foresters. *Res:* Dieback diseases of forest trees; Fomes annosus root rot; research administration. *Mailing Add:* Assoc Dep Chief Res USDA Forest Serv PO Box 96090 Washington DC 20090-6090

ROSS, ELLIOTT M, b Stockton, Calif, Jan 16, 49; m 73; c 1. BIOLOGICAL SIGNAL TRANSDUCTION, RECEPTORS. *Educ:* Univ Calif, Davis, BSc, 70; Cornell Univ, PhD(biochem), 75. *Prof Exp:* Postdoctoral fel pharmacol, Univ Va, 75-78, asst prof biochem & pharmacol, 78-81; assoc prof, 81, PROF PHARMACOL, UNIV TEX SOUTHWESTERN MED CTR, 85- *Concurrent Pos:* Mem, Pharmacol Sci Rev Comt, Nat Inst Gen Med Sci, 89-92. *Mem:* Am Chem Soc; Am Soc Biochem & Molecular Biol; Am Peptide Soc; Am Soc Pharmacol & Exp Therapeut. *Res:* Regulation of membrane-bound enzymes and receptors; GTP-binding regulatory proteins, G proteins, and G protein- coupled receptors; receptor-mimetic peptides. *Mailing Add:* Dept Pharmacol Univ Tex Southwestern Med Ctr 5323 Harry Hines Blvd Dallas TX 75235-9041

ROSS, ERNEST, b New York, NY, Dec 23, 20; m 49; c 4. POULTRY NUTRITION. *Educ:* Ohio State Univ, PhD(poultry nutrit), 55. *Prof Exp:* Asst poultry nutrit, Ohio State Univ, 51-55; asst poultry scientist, 57-60, assoc poultry scientist, 60-65, POULTRY SCIENTIST, UNIV HAWAII, 65- *Mem:* Poultry Sci Asn; World Poultry Sci Asn. *Res:* Physiological aspects of nutrition; tropical feedstuffs; environmental housing; management. *Mailing Add:* Dept Animal Sci Univ Hawaii 1800 East-West Rd Honolulu HI 96822

ROSS, FRED MICHAEL, b New York, NY, Aug 26, 21; m 54; c 3. INDUSTRIAL DIAMOND PROCESSING, DIAMOND RECLAMATION. *Educ:* Mich Technol Univ, BS, 43. *Prof Exp:* Sr gas analyst, Pure Oil Co, 43-44; electronics technician's mate radar, US Navy, 44-45; chem engr, Multiplate Glass Corp, 45-51; chief chemist, Diamond Dust Co, Inc, 52-80, pres & chief exec officer, 54-80; CHMN BD DIRS, ROBONARD, INC, 80- *Concurrent Pos:* Dir, Indust Diamond Asn Am, 77-78. *Honors & Awards:* Silver Medal, Bd Control, Mich Technol Univ, 78. *Mem:* Fel Am Inst Chemists. *Res:* Industrial diamond process for the manufacture of Ovate diamonds for use in petroleum bits & geological core drills; processes for the reclamation of diamonds from industrial diamond-bearing waste materials. *Mailing Add:* 10325 Crosswind Rd Boca Raton FL 33498

ROSS, FREDERICK KEITH, b Red Oak, Iowa, Nov 11, 42; m 64. STRUCTURAL CHEMISTRY, CRYSTALLOGRAPHY. *Educ:* Western Wash State Col, AB, 64; Univ Ill, Urbana, MS, 66, PhD(chem), 69. *Prof Exp:* Res assoc chem, Brookhaven Nat Lab, 69-71, State Univ NY Buffalo, 71-72; asst prof, Va Polytech Inst & State Univ, 72-79; SR RES SCIENTIST, UNIV MO, 79- *Concurrent Pos:* Adj assoc prof chem, Univ Mo-Columbia. *Mem:* Am Chem Soc; The Chem Soc; Am Crystallog Asn. *Res:* X-ray, neutron and gamma-ray diffraction; electronic distributions in solids by diffraction methods; diffraction physics of materials; low temperature x-ray and neutron diffraction. *Mailing Add:* Res Reactor Univ Mo Columbia MO 65211

ROSS, GERALD FRED, b New York, NY, Dec 14, 30; m 53; c 3. ELECTRONICS. *Educ:* City Col New York, BEE, 52; Polytech Univ, MEE, 55, PhD(electronics), 63. *Prof Exp:* Res asst microwave receivers, Univ Mich, 52-53; sr staff engr, W L Maxson Corp, NY, 54-58; res sect head phased array radar, Sperry Gyroscope Co, NY, 58-65, mem res staff microwaves, Sperry Rand Res Ctr, 65-68, mgr sensory & EM systs dept, Sperry Res Ctr, Sudbury, 68-81; PRES, ANRO ENG, INC, SARASOTA, FLA, 81- *Concurrent Pos:* Bd fels, Polytech Univ. *Mem:* Fel Inst Elec & Electronics Engrs; Sigma Xi. *Res:* Transient behavior studies related to microwave devices and high resolution radar systems; subnanosecond pulse technology and impulse radar. *Mailing Add:* 455 Longboat Club Rd Longboat Key FL 34228

ROSS, GILBERT STUART, b New York, NY, Nov 4, 30; m 56; c 3. NEUROLOGY. *Educ:* Franklin & Marshall Col, AB, 51; State Univ NY, MD, 55. *Prof Exp:* From instr to asst prof neurol, Univ Minn, 61-64; assoc prof, 64-66, PROF NEUROL & CHMN DEPT, STATE UNIV NY UPSTATE MED CTR, 66- *Concurrent Pos:* Consult, Vet Admin Hosps, Hot Springs, SDak, 61-63 & Syracuse, NY, 64-; Nat Inst Neurol Dis & Stroke res & training grants, 64- *Mem:* AAAS; Am Acad Neurol; Animal Behav Soc. *Res:* Electroencephalography; pain perception in subhuman forms by the use of operant conditioning techniques. *Mailing Add:* 750 E Adams St Syracuse NY 13210

ROSS, GORDON, b London, Eng, Sept 24, 30; m 54; c 2. PHYSIOLOGY, INTERNAL MEDICINE. *Educ:* Univ London, BSc, 51, MB, BS, 54. *Prof Exp:* Demonstr pharmacol, St Bartholomew's Hosp, London, 51-55; house physician & surgeon, St Mary Abbot's Hosp, 55-56; med registr, Mayday Hosp, 58-60 & King's Col Hosp, 60-63; from asst to assoc res pharmacologist, 63-66, from asst prof to assoc prof physiol & med, 66-72, PROF PHYSIOL & MED, SCH MED, UNIV CALIF, LOS ANGELES, 72- *Mem:* Am Fedn Clin Res; Am Physiol Soc; Sigma Xi. *Res:* Coronary circulation; adrenergic mechanisms. *Mailing Add:* Dept Physiol Sch Med Univ Calif Los Angeles CA 90024

ROSS, GRIFF TERRY, SR, endocrine physiology, medicine; deceased, see previous edition for last biography

ROSS, HARLEY HARRIS, b Chicago, Ill, Apr 18, 35; m 59, 82; c 4. RADIO ANALYTICAL CHEMISTRY, CHEMICAL INSTRUMENTATION. *Educ:* Univ Ill, Urbana, BS, 57; Wayne State Univ, MS, 58, PhD(radiochem), 60. *Prof Exp:* Scientist, Spec Training Div, Oak Ridge Inst Nuclear Studies, 60-63, chemist, Oak Ridge Nat Lab, 63-66, asst group leader nuclear & radiochem, 66-72, group leader anal instrumentation, 72-85, prof, Univ Tenn, 85-90; HEAD SPEC RADIOCHEM RES, OAK RIDGE NAT LAB, 85- *Honors & Awards:* IR-100 Award, 67. *Mem:* Sigma Xi. *Res:* Analytical applications of radioisotopes; liquid scintillator technology; organic luminescent properties; analytical instrumentation; radioisotope energy conversion; atomic and molecular spectroscopy; spectroscopic instrumentation. *Mailing Add:* Analysis Chem Div Oak Ridge Nat Lab Box X Oak Ridge TN 37830

ROSS, HOWARD PERSING, b Stockbridge, Mass, Oct 26, 35; m 58; c 3. GEOPHYSICS, ECONOMIC GEOLOGY. *Educ:* Univ NH, BA, 57; Pa State Univ, MS, 63, PhD(geophys), 65. *Prof Exp:* Computer, United Geophys Corp, Calif, 58-60; res gen phys scientist, Air Force Cambridge Res Lab, Mass, 65-67; sr res geophysicist, Explor Serv Div, Kennecott Explor, Inc, 67-77; SECT HEAD, APPL GEOPHYS, EARTH SCI LAB, UNIV UTAH RES INST, 77-; GEOPHYS CONSULT, 80- *Concurrent Pos:* Investr, Apollo Appln Prog, NASA, 66-67, mem peer rev panels, nuclear waste isolation, 79-87. *Mem:* Am Geophys Union; Soc Explor Geophys; Europ Asn Explor Geophys; Am Asn Petrol Geologists; Geothermal Resource Coun. *Res:* Potential field methods; research geophysics; mathematical models of geologic and geophysical processes; remote sensing; porphyry copper geology; geoelectric studies; mineral exploration; geothermal research; nuclear waste isolations studies. *Mailing Add:* 7089 Pinecone St Salt Lake City UT 84121

ROSS, HUGH COURTNEY, b Turlock, Calif, Dec 31, 23; m 50, 54, 84; c 3. ELECTRICAL ENGINEERING, HIGH VOLTAGE CONTROL. *Educ:* Stanford Univ, BSEE, 50. *Prof Exp:* Instr, San Benito High Sch & Jr Col, 50-51; chief engr, Vacuum Power Switches, Jennings Radio Mfg Corp, San Jose, Calif, 51-62; CHIEF ENGR HV CONSULT, SARATOGA, CALIF, 64-; PRES & GEN MGR, ROSS ENG CORP, CAMPBELL, 64- *Concurrent Pos:* Chmn, Santa Clara Valley Subsection Inst Elec & Electronics Engrs, 60-61. *Mem:* Fel Inst Elec & Electronics Engrs; Am Vacuum Soc; Am Soc Metals. *Res:* High voltage devices; major development in high power vacuum relays; vacuum switches; vacuum breakers; high voltage votmeters; digital and analog; high voltage control and measurement; author of many articles for Institute and Electric and Electronic Engineers transactions and other technical journals; measurement and safety. *Mailing Add:* Ross Eng Corp 540 Westchester Dr Campbell CA 95008

ROSS, IAN KENNETH, b London, Eng, May 22, 30; m 56; c 2. MYCOLOGY, CELL BIOLOGY. *Educ:* George Washington Univ, BS, 52, MS, 53; McGill Univ, PhD(bot), 57. *Prof Exp:* Lectr bot, McGill Univ, 57; res assoc mycol, Univ Wis, 57-58; from instr to asst prof bot, Yale Univ, 58-64; from asst prof to assoc prof, 64-73, vchmn, dept 86-87, PROF BIOL, UNIV CALIF, SANTA BARBARA, 73- *Concurrent Pos:* Sect ed, Can J Microbiol, 88-90; BP venture res fel, 88-91. *Mem:* AAAS; Mycol Soc Am; Am Soc Cell Biol; Am Soc Microbiol; Soc Indust Microbiol. *Res:* Biochemistry and molecular biology of differentiation and development of filamentous fungi; transduction of light reception into morphogenetic and genetic changes. *Mailing Add:* Dept Biol Sci Univ Calif Santa Barbara CA 93111

ROSS, IAN M(UNRO), b Southport, Eng, Aug 15, 27; nat US; m 55; c 3. ELECTRICAL ENGINEERING. *Educ:* Cambridge Univ, BA, 48, MA & PhD(elec eng), 52. *Prof Exp:* Mem tech staff, AT&T Bell Labs, 52-59, dir, Semiconductor Lab, Murray Hill, NJ, 59-62, dir Semiconductor Device & Electron Tube Lab, Allentown, Pa, 62-64, managing dir, Bellcomm, Inc, 64-68, pres, 68-71, exec dir, Network Planning Div, Bell Labs, 71-73, vpres, Network Planning & Customer Serv, 73-76, exec vpres, 76-79, pres, 79-91, PRES EMER, AT&T BELL LABS, 91- *Concurrent Pos:* Mem, bd dirs, Sandia Nat Lab, BF Goodrich Co & Thomas & Betts Corp. *Honors & Awards:* Morris N Liebermann Mem Prize Award, Inst Radio Engrs, 63; Indust Res Inst Medal, 87; Inst Elec & Electronics Engrs Founders Medal, 88- *Mem:* Nat Acad Sci; Nat Acad Eng; fel Inst Elec & Electronics Engrs; fel Am Acad Arts & Sci. *Res:* Systems engineering and development. *Mailing Add:* AT&T Bell Labs Murray Hill NJ 07974

ROSS, IRA JOSEPH, b Live Oak, Fla, Apr 8, 33; m 55; c 3. AGRICULTURAL ENGINEERING. *Educ:* Univ Fla, BAgE, 55; Purdue Univ, MS, 57, PhD, 60. *Prof Exp:* Asst, Purdue Univ, 55-56, instr & asst in agr eng, 56-59; asst prof agr eng & asst agr engr, Univ Fla, 59-65, assoc prof & assoc agr engr, 65-67; assoc prof, 67-70, PROF AGR ENG, UNIV KY, 70- *Res:* Electric power and processing. *Mailing Add:* Dept Agr Eng Univ Ky Lexington KY 40506

ROSS, IRVINE E, b Needham, Mass, Sept 28, 08. ELECTRICAL ENGINEERING. *Educ:* Mass Inst Technol, SB, 30 & SM, 31. *Prof Exp:* Eng mgr, Gen Elec, 48-68; RETIRED. *Mem:* Fel Inst Elec & Electronics Engrs. *Mailing Add:* 4115 Spanish Trail Ft Wayne IN 46815

ROSS, JAMES NEIL, JR, b Akron, Ohio, Dec 18, 40; m 64; c 3. CARDIOVASCULAR PHYSIOLOGY, VETERINARY CARDIOLOGY. *Educ:* Ohio State Univ, DVM, 65, MSc, 67; Baylor Col Med, PhD(physiol), 72; Am Col Vet Internal Med, dipl, 73 & 74; Am Col Vet Emergency & Critical Care, dipl, 90. *Prof Exp:* Res asst, dept vet physiol & pharm, Ohio State Univ, 61-65, res assoc vet cardiol, 65-67, lectr clin cardiol, 65-67; from instr to asst prof, Dept Surg & Physiol, Baylor Col Med, 67-74, asst dir vivarium, 67-69, asst chief exp surg, Taub Labs Mech Circulatory Support, 70-74; assoc prof physiol, Med Col Ohio, 74-81; PROF & CHMN, DEPT MED, SCH VET MED, TUFTS UNIV, 81- *Concurrent Pos:* Consult, Vet Admin Hosp, Houston, 70-73; mem, Coun Clin Cardiol, Am Heart Asn; mem adv bd vet specialties & coun educ, Am Vet Med Asn, 89- *Mem:* Acad Vet Cardiol (pres, 73-75); Am Soc Artificial Internal Organs; Am Vet Med Asn; Am Col Vet Internal Med; Am Physiol Soc. *Res:* Hemodynamics; circulatory assist-replacement devices and techniques; biomaterials for blood interfacing; cardiovascular models; congenital heart disease; veterinary cardiology; cardiovascular surgery. *Mailing Add:* Dept Med Sch Vet Med Tufts Univ 200 Westboro Rd N Grafton MA 01536

ROSS, JAMES WILLIAM, b Ft Lewis, Wash, June 23, 28; m 70; c 3. ANALYTICAL CHEMISTRY, PHYSICAL CHEMISTRY. *Educ:* Univ Calif, Berkeley, BA, 51; Univ Wis-Madison, PhD(chem), 57. *Prof Exp:* Chemist, Tidewater Assoc Oil Co, 51-53; asst prof chem, Mass Inst Technol, 57-62; VPRES & DIR RES, ORION RES INC, 62- *Concurrent Pos:* Dir bd, Orion Res Inc, 62-; A D Williams distinguished vis scholar, Med Col Va, Va Commonwealth Univ, 72-73. *Mem:* Am Chem Soc; Am Electrochem Soc; AAAS. *Res:* Ion selective electrodes; theory, modes of fabrication and application to chemical and medical analysis. *Mailing Add:* 61 Gatehouse Rd Chestnut Hill MA 02167-1320

ROSS, JEFFREY, b St Louis, Mo, Feb 25, 43; m 68; c 2. MOLECULAR BIOLOGY. *Educ:* Princeton Univ, BA, 65; Wash Univ, MD, 69. *Prof Exp:* Intern med, Cedars-Sinai Med Ctr, Los Angeles, Calif, 69-70; res assoc, NIH, 70-74; asst prof, 74-78, ASSOC PROF, MCARDLE LAB CANCER RES, UNIV WIS-MADISON, 78- *Res:* Mechanism of synthesis of messenger RNA in eukaryotes with emphasis on the characterization of a precursor of globin messenger RNA; role of heme in development of authentic erythroid cells and of cultured erythroleukemic, Friend cells; mRNA synthesis and stability in the thalassemias. *Mailing Add:* Oncol-625 McArdle Res Univ Wis Med Sch 450 N Randall Ave Madison WI 53706

ROSS, JOHN, b Vienna, Austria, Oct 2, 26; nat US; m 50; c 2. PHYSICAL CHEMISTRY. *Educ:* Queens Col, NY, BS, 48; Mass Inst Technol, PhD, 51. *Hon Degrees:* Weizmann Inst Sci, 84; Queens Col, SUNY, 87; Univ Bordeaux, France, 87. *Prof Exp:* Res assoc phys chem, Mass Inst Technol, 50-52; res fel, Yale Univ, 52-53; from asst prof to prof chem, Brown Univ, 53-66; prof & chmn dept, Mass Inst Technol, 66-71, Frederick George Keyes prof chem, 71-80; PROF CHEM, STANFORD UNIV, 80-, CHMN DEPT, 83-, CAMILLE & HENRY DREYFUS PROF, 85- *Concurrent Pos:* Fel, NSF, 52-53, Guggenheim Found, 59-60 & Sloan Found, 60-64; vis Van der Waals prof, Univ Amsterdam, 66; mem bd govs, Weizmann Inst Sci, 71-; mem, NSF Adv Panel, 74-; chmn fac, Mass Inst Technol, 75-77. *Mem:* Nat Acad Sci; Am Chem Soc; fel Am Phys Soc; fel Am Acad Arts & Sci. *Res:* Chemical instabilities; oscillatory reactions; thermodynamics of systems far from equilibrium; efficiency of thermal, chemical, and biological engines. *Mailing Add:* Dept Chem Stanford Univ Stanford CA 94305

ROSS, JOHN, JR, b New York, NY, Dec 1, 28; m 58; c 3. MEDICINE, CARDIOLOGY. *Educ:* Dartmouth Col, AB, 51; Cornell Univ, MD, 55. *Prof Exp:* Clin assoc & sr investr, Nat Heart Inst, 56-60; asst resident, Columbia-Presby Med Ctr, 60-61 & Med Col, Cornell Univ, 61-62; attend physician & chief sect cardiovasc diag, Cardiol Br, Nat Heart Inst, 62-68; PROF MED & DIR CARDIOVASC DIV, SCH MED, UNIV CALIF, SAN DIEGO, 68- *Concurrent Pos:* Consult, Nat Conf Cardiovasc Dis, 63; mem cardiol adv comt, Nat Heart & Lung Inst, 74- *Mem:* Am Physiol Soc; fel Am Col Cardiol (vpres, 72-73); Am Soc Clin Invest; fel Am Col Physicians; Asn Am Physicians. *Res:* Cardiovascular research; cardiac catheterization techniques; mechanics of cardiac contraction; physiology of coronary circulation. *Mailing Add:* Dept Med Univ Calif San Diego M-013 Box 109 La Jolla CA 92093

ROSS, JOHN B(YE), b St Louis, Mo, June 14, 39; m 72; c 2. ELECTRICAL ENGINEERING. *Educ:* Pa State Univ, BS, 61, MS, 63, PhD(elec eng), 67. *Prof Exp:* Sr physicist, 67-80, RES ASSOC, EASTMAN KODAK CO, 80- *Mem:* Inst Elec & Electronics Engrs. *Res:* Applications of instrumentation and computers to analysis of the quality of film emulsion systems. *Mailing Add:* Res Engr B59 K P Eastman Kodak Co 343 State St Rochester NY 14650

ROSS, JOHN BRANDON ALEXANDER, b Suffern, NY, Feb 18, 47. BIOCHEMISTRY. *Educ:* Antioch Col, BA, 70; Univ Wash, PhD(biochem), 76. *Prof Exp:* Res assoc, Dept Chem, Univ Wash, 76-78; assoc res scientist, dept biol, Johns Hopkins Univ, 78-80; sr res fel, Dept Lab Med, Sch Med, Univ Wash, 80-82; asst prof, dept biochem, 82-86, ASSOC PROF, MT SINAI SCH MED, 87- *Mem:* Sigma Xi; Am Chem Soc; Am Soc Photobiol; Biophys Soc; Optical Soc Am. *Res:* Time resolved luminescence of biomolecules; structure of polypetides and proteins in solution; lipid, protein, and steroid interactions; luminescence assay in medicine. *Mailing Add:* Dept Biochem Mt Sinai Sch Med One Gustave L Levy Pl New York NY 10029

ROSS, JOHN EDWARD, b Swissvale, Pa, Jan 1, 29; m 60; c 4. INDUSTRIAL HYGIENE, HEALTH PHYSICS. *Educ:* Univ Pittsburgh, BS, 49, MS, 59; Am Bd Health Physics, cert, 64; Bd Cert Safety Prof, cert, 72. *Prof Exp:* Indust hygienist, Bettis Atomic Power Lab, Westinghouse Elec Corp, 53-61; sr health & safety engr, Nuclear Mat & Equip Corp, 61-62; mgr reactor serv, Controls for Radiation, Inc, 62-65, gen mgr occup health, 65-67, gen mgr, Teledyne Isotopes, 67-89; DEP DIV MGR, PLUMBROOK DIV, SVERDRUP TECHNOL, 89- *Mem:* Am Indust Hyg Asn; Health Physics Soc. *Res:* Sample collection, analysis, evaluation and corrective action for industrial health hazards; heat and noise stress; occupational health program development; nuclear facility decommissioning. *Mailing Add:* 308 Shawnee Pl Huron OH 44839

ROSS, JOHN FRANKLIN, chemistry, for more information see previous edition

ROSS, JOHN PAUL, b New Rochelle, NY, Apr 29, 27; m 51; c 4. PLANT PATHOLOGY. *Educ:* Univ Vt, BS, 52; Cornell Univ, PhD(plant path), 56. *Prof Exp:* Asst plant path, Cornell Univ, 52-56; from asst prof to assoc prof, NC State Univ, 67-67, prof, 67-; plant pathologist, Sci & Educ Admin-Agr Res, USDA, 56-; RETIRED. *Mem:* Am Phytopath Soc; Crop Sci Soc Am. *Res:* Diseases of soybeans caused by nematodes, viruses, fungi and bacteria. *Mailing Add:* 2008 Nakoma Pl NC State Univ Raleigh NC 27607

ROSS, JOHN STONER, atomic spectroscopy, for more information see previous edition

ROSS, JOSEPH C, b Tompkinsville, Ky, June 16, 27; m 52; c 5. INTERNAL MEDICINE. *Educ:* Univ Ky, BS, 50; Vanderbilt Univ, MD, 54; Am Bd Internal Med, dipl & cert pulmonary dis. *Prof Exp:* Intern med, Vanderbilt Univ Hosp, 54-55; resident, Duke Univ Hosp, 55-57, USPHS fel, 57-58; from instr to prof, Sch Med, Ind Univ, 58-70; prof med & chmn dept, Med Univ SC, 70-80; PROF MED, VANDERBILT UNIV, 81-, ASSOC VCHANCELLOR HEALTH AFFAIRS, 82- *Concurrent Pos:* Mem, President's Nat Adv Panel on Heart Dis, 72, Vet Admin Merit Rev Bd Respiration, 72-76, Cardiovasc Study Sect, Nat Heart Inst, 66-70, Comt Respiratory Physiol, Nat Acad Sci 66-67, Prog Proj Comt, NIH, 71-75; mem, Pulmonary Subspecialty Bd, 72-78, Am Bd Internal Med, 72-81, Adv Coun, Nat Heart, Lung & Blood Inst, NIH, 81-85. *Mem:* Asn Am Physicians; Am Soc Clin Invest; Am Physiol Soc; Am Col Physicians (pres, 78); Am Col Cardiol; AMA; Am Col Chest Physicians. *Res:* Pulmonary physiology, diffusion, function tests and capillary bed; emphysema. *Mailing Add:* D-3300 Med Ctr N Vanderbilt Univ Nashville TN 37232

ROSS, JOSEPH HANSBRO, b Houston, Tex, May 24, 25; m 56; c 4. ORGANIC CHEMISTRY. *Educ:* Rice Inst, BS, 46; Univ Tex, MA, 48; Univ Md, PhD(chem), 57. *Prof Exp:* Res assoc & fel chem, Univ Mich, 57-58; res chemist, Stamford Labs, Am Cyanamid Co, 58-63; asst prof, 63-67, from asst chmn to chmn dept, 67-73, ASSOC PROF CHEM, IND UNIV, SOUTH BEND, 67- *Mem:* Am Chem Soc; AAAS; Royal Soc Chem; Sigma Xi. *Res:* Reaction mechanisms; amine oxidations and condensations; nitrogen heterocycles; chromatography and solvent interactions; dyes. *Mailing Add:* Dept Chem Ind Univ South Bend IN 46634-7111

ROSS, JUNE ROSA PITT, b Taree, NSW, Australia, 31; m 59. MARINE SCIENCES, PALEOBIOGEOGRAPHY. *Educ:* Univ Sydney, BSc, 54, PhD(geol), 59, DSc, 74. *Prof Exp:* Res fel, Yale Univ, 57-60; res assoc, Univ Ill, 60-65; res assoc, Western Wash Univ, 65-67, assoc prof, 67-70, chair, 89-90, PROF BIOL, WESTERN WASH UNIV, 70- *Concurrent Pos:* NSF res grants, 60-71; pres, Western Wash Univ Fac Sen, 84-85; conf host, Int Bryozool Asn, 86. *Mem:* Soc Study Evolution; Marine Biol Asn UK; Int Bryozool Asn; Australian Marine Sci Asn; Electron Microscope Soc Am; Paleont Soc (treas, 87-); Paleont Asn; Am Soc Zoologists. *Res:* Ectoprocta; evolution biogeography and ecology; evolution of marine invertebrates and marine communities. *Mailing Add:* Dept Biol Western Wash Univ Bellingham WA 98225-9060

ROSS, KEITH ALAN, b Sturgis, SDak, Dec 19, 52; m 87. THIN FILMS, MICROELECTRONICS. *Educ:* SDak Sch Mines & Technol, BS, 75, MS, 77, PhD(elec eng), 83. *Prof Exp:* Assoc eng, Int Bus Mach, 77-79; instr, 82-83, asst prof, 83-90, ASSOC PROF TEACHING, SDAK SCH MINES & TECHNOL, 90- *Res:* Deposition and characterization of thin semiconducting films. *Mailing Add:* SDak Sch Mines & Technol Elec Eng Dept 501 E St Joseph St Rapid City SD 57701-3995

ROSS, KENNETH ALLEN, b Chicago, Ill, Jan 21, 36; m 86; c 2. MATHEMATICS. *Educ:* Univ Utah, BS, 56; Univ Wash, Seattle, MS, 58, PhD(math), 60. *Prof Exp:* Res instr math, Univ Wash, 60-61; asst prof, Univ Rochester, 61-64; assoc prof, 64-70, PROF MATH, UNIV ORE, 70- *Concurrent Pos:* Vis asst prof, Yale Univ, 64-65; Alfred P Sloan Found fel, 67-70; vis prof, Univ BC, 80-81. *Mem:* Am Math Soc; Math Asn Am (secy, 84-90). *Res:* Abstract harmonic analysis over topological groups. *Mailing Add:* Dept Math Univ Ore Eugene OR 97403-1222

ROSS, LAWRENCE JAMES, b Brooklyn, NY, Sept 4, 29; m 59; c 3. PHYSICAL ORGANIC CHEMISTRY. *Educ:* Fordham Univ, BS, 50; Rutgers Univ, PhD(chem), 61. *Prof Exp:* Chemist, Nat Starch & Chem Corp, 50-55, supvr control methods lab, 55-57; from res chemist to sr res chemist, Am Cyanamid Agr Ctr, 60-75, group leader, 75-78, mgr formulations, 78-83, sr formulations coordr, 83-87; RETIRED. *Mem:* Am Chem Soc; fel Am Inst Chem. *Res:* Preparation and characterization of starch derivatives; mechanism of organic reactions; preparation of sulfonamides; preparation of brighteners of the styryl benzoxazole type; organic intermediates; process development; agricultural chemicals. *Mailing Add:* 1789 Woodfield Rd Martinsville NJ 08836

ROSS, LEONARD LESTER, b New York, NY, Sept 11, 27; m 51; c 2. ANATOMY, NEUROBIOLOGY. *Educ:* NY Univ, AB, 46, MS, 49, PhD(anat), 54. *Prof Exp:* Res assoc, NY Univ, 49-52; asst prof anat, Med Col Ala, 52-57; from assoc prof to prof, Med Col, Cornell Univ, 57-73; prof anat & chmn dept, 73-90, EXEC VPRES, ANNENBERG DEAN & CHIEF ACAD OFFICER, MED COL PA, 90- *Concurrent Pos:* NIH fel, Cambridge Univ, 67-68. *Honors & Awards:* Linback Award; Founders Award. *Mem:* Electron Micros Soc Am; Histochem Soc; Am Asn Anatomists; Am Soc Cell Biol; Soc Neurosci; Col Physicians. *Res:* Nervous system; electron microscopy; histochemistry; biogenic amines; psychoneuroimmunology. *Mailing Add:* Off Dean Med Col Pa Philadelphia PA 19129

ROSS, LOUIS, b Akron, Ohio, Mar 24, 12; m 42; c 1. MATHEMATICS, STATISTICS. *Educ:* Univ Akron, AB, 34, BS, 35, MAEd, 39; Western Reserve Univ, PhD, 55. *Prof Exp:* Teacher pub sch, Ohio, 35-41; assoc prof, 46-77, EMER PROF MATH, UNIV AKRON, 77- *Concurrent Pos:* Dir Ohio-Ky-Tenn region, US Metric Asn, Inc, 72- *Mem:* Math Asn Am; Am Statist Asn; fel Am Soc Qual Control; Sigma Xi. *Res:* Quality control and reliability; impact of metrication on education, business, and industry in the US. *Mailing Add:* 2400 Burnham Rd Akron OH 44333

ROSS, LYNNE FISCHER, b New Orleans, La, Sept 14, 44. ENZYMATIC MODIFICATION OF FOOD PROTEINS. *Educ:* La State Univ, Baton Rouge, 66; Univ Tenn Med Ctr, Memphis, MS, 70; La State Univ Med Ctr, New Orleans, PhD(clin biochem), 80. *Prof Exp:* Med technologist, Clin Chem Lab, Oschner Found Hosp, New Orleans, 66-68; supvr, Clin Chem Lab, City Memphis Hosp, 68-70; asst prof clin chem, Dept Med Technol CAHP, Temple Univ, Philadelphia, 70-75; lab dir, Laser Res Found, New Orleans, 81-83; RES CHEMIST, AGR RES SERV SOUTHERN REGIONAL RES CTR, USDA, 83- *Mem:* Am Soc Advan Sci; Am Chem Soc; Am Soc Microbiol; Am Oil Chemists Soc; Am Soc Plant Physiologists; Plant Growth Regulatory Soc Am. *Res:* The isolation of enzymes and characterization of enzymes as well as other protein-ligand interactions in severak physiological systems both in plants and mammals. *Mailing Add:* PO Box 6130 Destin FL 32541

ROSS, MALCOLM, b Washington, DC, Aug 22, 29; m 56; c 2. MINERALOGY & PETROLOGY, CRYSTALLOGRAPHY & GEOCHEMISTRY. *Educ:* Utah State Univ, BS, 51; Univ Md, MS, 59; Harvard Univ, PhD(geol), 62. *Prof Exp:* PHYS CHEMIST GEOCHEM, US GEOL SURV, 54-59 & 61- *Concurrent Pos:* Distinguished regional lectr, US Geol Surv, 83-84. *Honors & Awards:* Distinguished Serv Award, US Dept Interior, 86; Distinguished Pub Serv Award, Mineral Soc Am, 90. *Mem:* AAAS; fel Mineral Soc Am (treas, 76-80, pres, 91); fel Geol Soc Am; Clay Minerals Soc; Am Geophys Union; Mineral Asn Can. *Res:* Mineralogy, petrology and crystallography of rock-forming silicates; subsolidus phase changes in silicates; pyroxene and amphibole geothermometry; relationships between human health and exposure to mineral particulates; mineralogy and petrology of alkalic rocks; material effects of acid rain. *Mailing Add:* US Geol Surv Nat Ctr 959 Reston VA 22092

ROSS, MARC HANSEN, b Baltimore, Md, Dec 24, 28; m 49; c 3. THEORETICAL PHYSICS. *Educ:* Queens Col, NY, BS, 48; Univ Wis, PhD, 52. *Prof Exp:* NSF fel, Cornell Univ, 52-53; assoc physicist, Brookhaven Nat Lab, 53-55; from asst prof to prof physics, Ind Univ, 55-63; dir residential col, 74-77, PROF PHYSICS, UNIV MICH, ANN ARBOR, 63-; SR SCIENTIST, ARGONNE NAT LAB, 84- *Concurrent Pos:* NSF fel, Univ Rome, 60-61. *Mem:* Am Phys Soc. *Res:* Physics of energy use; environmental physics. *Mailing Add:* Dept Physics Univ Mich Ann Arbor MI 48109

ROSS, MARTIN RUSSELL, b New York, NY, Aug 23, 22; m 46; c 3. VIROLOGY, IMMUNOLOGY. *Educ:* Univ Conn, BA, 50; Univ Mich, MS, 52; George Washington Univ, PhD(virol), 57. *Prof Exp:* Res virologist, US Army Biol Labs, 52-56; asst prof virol, WVa Univ, 56-58; res microbiologist, 58-59, CHIEF VIROLOGIST, CONN STATE DEPT HEALTH, 59- *Honors & Awards:* J Howard Brown Award, Am Soc Microbiol, 56. *Mem:* Am Soc Microbiol. *Res:* Antigenic structure of psittacosis group of agents; development of fluorescent antibody techniques; epidemiology and laboratory diagnosis of viral diseases. *Mailing Add:* PO Box 795 Glastonbury CT 06033

ROSS, MARVIN, b Brooklyn, NY, June 27, 31; m 58; c 2. HIGH PRESSURE PHYSICS. *Educ:* Brooklyn Col, BS, 55; Pa State Univ, PhD(chem), 60. *Prof Exp:* Res assoc, Univ Calif, Berkeley, 60-63; PHYS CHEMIST, LAWRENCE LIVERMORE NAT LAB 63-, DIV LEADER, CONDENSED MATTER PHYSICS, 87- *Mem:* Am Phys Soc; Sigma Xi; Am Geophys Union. *Res:* Theoretical studies of condensed media at high pressures and temperatures; applications of shock waves to high pressure and high temperature research. *Mailing Add:* Lawrence Livermore Lab L-299 Univ Calif Livermore CA 94550

ROSS, MARVIN FRANKLIN, b Sacramento, Calif, Sept 27, 51. CLIFFORD ALGEBRAS & LIE ALBEBRAS. *Educ:* Univ Calif, San Cruz, BA, 72; Univ Calif, Davis, MA, 74, PhD(physics), 80. *Prof Exp:* Lectr physics, Univ Calif, Davis, 79-84; ASST PROF PHYSICS, UNIV PAC, 84- *Mem:* Am Phys Soc; Sigma Xi. *Res:* Clifford algebras and relativistic quantum mechanics and elementary particles; cluster-bethe-lattice calculation of the density of states of amorphous silicon. *Mailing Add:* 1237 42nd Ave Sacramento CA 95822

ROSS, MARY HARVEY, b Albany, NY, Apr 1, 25; m 47; c 3. ENTOMOLOGY. *Educ:* Cornell Univ, BA, 45, MA, 47, PhD(paleont), 50. *Prof Exp:* Biologist, Oak Ridge Nat Lab, 51; from instr to assoc prof, 59-80, PROF ENTOM, VA POLYTECH INST & STATE UNIV, 80- *Concurrent Pos:* NSF grants, 71-73 & 80-83; Naval Facil Eng Command grant, 74; Off Naval Res grant, 77-86; adj assoc prof, NC State Univ, 77-79; mem orgn comt, Genetics Sect, XV Int Cong Entom; counr, Am Genetics Asn, 77-80. *Mem:* Am Genetic Asn (secy, 80-82, pres, 84); Entom Soc Am; Genetics Soc Am; Sigma Xi; Genetics Soc Can; AAAS. *Res:* Genetics, cytogenetics and behavior of the cockroach. *Mailing Add:* Dept Entom Va Polytech Inst & State Univ Blacksburg VA 24061

ROSS, MERRILL ARTHUR, JR, b Montrose, Colo, June 2, 35; m 55, 73; c 5. WEED SCIENCE. *Educ:* Colo State Univ, BS, 57, MS, 59, PhD(plant physiol), 65. *Prof Exp:* Instr bot & plant path & jr plant physiologist, Colo State Univ, 59-65; asst prof bot, 65-69, assoc prof plant physiol, 69-75, ASSOC PROF BOT & PLANT PATH, PURDUE UNIV, WEST LAFAYETTE, 75-, EXTEN WEED SPECIALIST, 65- *Mem:* Weed Sci Soc Am. *Res:* Weed control; agronomy; botany; control of Johnson grass, Canada thistle and other perennial weeds in Indiana crop production systems. *Mailing Add:* Dept Bot & Plant Path Purdue Univ West Lafayette IN 47907

ROSS, MICHAEL H, b Jamaica, NY, Oct 23, 30; m 58; c 3. ANATOMY, CELL BIOLOGY. *Educ:* Franklin & Marshall Col, BS, 51; NY Univ, MS, 59, PhD(biol), 60. *Prof Exp:* From instr to assoc prof anat, Sch Med, NY Univ, 60-71; prof path & dir, 71-77, CHMN DEPT ANAT, SCH MED, UNIV FLA, 77- *Concurrent Pos:* Chmn, Fla State Anat Bd, 71-; assoc ed, Am J Anat, 70-73 & Anat Record, 73- *Mem:* Am Asn Anatomists; Am Soc Cell Biol; Pan-Am Asn Anat; Soc Study Reprod; hon mem Bolivian Soc Anat. *Res:* Cell biology; tissue development; testicular function and developmental changes. *Mailing Add:* Dept Anat Med Col Univ Fla J Hillis Miller Health Ctr Box J-235 Gainesville FL 32610

ROSS, MICHAEL RALPH, b Middletown, Ohio, Apr 19, 47; m 70; c 2. AQUATIC ECOLOGY, FISH BIOLOGY. *Educ:* Miami Univ, BS, 69; Ohio State Univ, MS, 71, PhD(zool), 75. *Prof Exp:* Asst prof, 75-80, ASSOC PROF FISHERIES BIOL, UNIV MASS, 80- *Concurrent Pos:* Vis asst prof, Itasca Biol Sta, 79-81. *Mem:* Sigma Xi; Am Soc Ichthyologists & Herpetologists; Ecol Soc; AAAS; Am Fisheries Soc. *Res:* Behavior, mating systems, life history and population dynamics of fishes. *Mailing Add:* Dept Forestry & Wildlife Holdsworth Hall Univ Mass Amherst MA 01003

ROSS, MONTE, b Chicago, Ill, May 26, 32; m 57; c 3. COMMUNICATIONS. *Educ:* Univ Ill, BSEE, 53; Northwestern Univ, MSEE, 62. *Prof Exp:* Elec engr, Chance-Vought Corp, 53-54 & Corps of Engrs, 54-56; electronic engr, Motorola, Inc, 56-57; sr engr, Hallicrafters Co, 57-59, group mgr electronics res & develop, 59-61, from assoc dir res to dir res, 61-66; mgr laser technol, 66-74, prog mgr, 74-81, dir, Laser Commun Systs, 81-87, PRES, LASER DATA TECHNOL, MCDONNELL DOUGLAS ASTRONAUT, CO, 87- *Concurrent Pos:* Adv ed, Laser Focus Mag, 66-70; tech ed, Laser Appln Ser, Acad Press, 70-; reviewer of grants, Nat Sci Found, 75-; affil prof, Wash Univ. *Honors & Awards:* Technol Award, Am Inst Aeronaut & Astronaut; McDonnell Douglas Fel Award, 85. *Mem:* Fel Inst Elec & Electronics Engrs; Am Inst Aeronaut & Astronaut. *Res:* Laser communications systems; fiber optic systems; space communications; laser research and development, especially communications and guidance; diode laser systems. *Mailing Add:* 19 Beaver Dr St Louis MO 63141

ROSS, MYRON JAY, b Winthrop, Mass, June 10, 42; m 68; c 2. ELECTRICAL ENGINEERING. *Educ:* Northeastern Univ, BS, 64, MS, 66, PhD(elec eng), 70. *Prof Exp:* Staff engr, Proj Apollo, Draper Lab, Mass Inst Technol, 63-66; test dir test & eval, Mat Testing Directorate, Aberdeen Proving Ground, Md, 70-72; advan res & develop engr secure voice systs, GTE Sylvania, Needham, Mass, 72-74; eng specialist commun systs, CNR, Inc, Newton Mass, 74; ENG DEPT MGR, TELECOMMUN SYSTS, GTE GOVT SYSTS, 74- *Concurrent Pos:* NASA trainee fel, 66-69; lectr elec eng, Northeastern Univ, 68-70; NDEA fel, 70; assoc prof & lectr mgt sci, George Washington Univ, 71-72; lectr comput sci, Boston Univ, 78; lectr digital switching, Univ Calif, Los Angeles & Univ Md, 78- *Honors & Awards:* NASA Commendation, Apollo Prog; Centennial Medal, Inst Elec & Electronics Engrs, 84. *Mem:* Inst Elec & Electronics Engrs; Sigma Xi. *Res:* Telecommunications switching systems; data communication systems; integrated digital voice and data switching; secure voice systems; military communications networks. *Mailing Add:* 43 Condor Rd Sharon MA 02067

ROSS, NORTON MORRIS, clinical pharmacology; deceased, see previous edition for last biography

ROSS, PETER A, mechanical engineering, for more information see previous edition

ROSS, PHILIP, b Newton, Mass, Nov 2, 26; m 52; c 2. ECOLOGY. *Educ:* Brown Univ, AB, 49; Univ Mass, MS, 51; Harvard Univ, PhD(plant ecol & geog), 58, MPH, 68. *Prof Exp:* Fulbright res scholar trop agr, Imperial Col, 57-58; botanist, Mil Geol Br, US Geol Surv, DC, 58-62; asst chief training grants sect, Nat Inst Dent Res, 62-63, chief res grants sect, 63-65, asst head spec int progs sect, Off Int Res, NIH, 65-68; pres, Int Sugar Res Found, Inc, 68-70; exec secy, Comt Effects of Herbicides in Vietnam, 70-74, exec secy, Bd Agr & Renewable Resources, Nat Acad Sci-Nat Res Coun, 74-86; SCI ADV, OFF FED ACTIV, EPA, 87- *Concurrent Pos:* Prof lectr, Am Univ, 59-65. *Mem:* Ecol Soc Am; Int Soc Trop Ecol; Explorers Club. *Res:* Plant ecology and geography; vegetation mapping; microclimate; human ecology; science administration; agriculture. *Mailing Add:* 9108 Seven Locks Rd Bethesda MD 20817

ROSS, PHILIP NORMAN, JR, b Washington, DC, Oct 31, 43; div; c 1. CHEMICAL ENGINEERING, ELECTROCHEMISTRY. *Educ:* Yale Univ, BS, 65; Univ Del, MS, 69; Yale Univ, PhD(eng, appl sci), 72. *Prof Exp:* Proj engr chem eng, Explor Develop Div, Procter & Gamble Corp, 66-69; sr res assoc electrochem, Pratt & Whitney Aircraft & Power Systs Div, United Technol Corp, 72-76, sr scientist, United Technol Res Ctr, 76-78; PRIN INVESTR ELECTROCHEM, LAWRENCE BERKELEY LAB, UNIV CALIF, BERKELEY, 78- *Concurrent Pos:* Detailee to US Dept Energy, Off Basic Energy Sci, Div Mat Sci, 87-88. *Mem:* Am Chem Soc; Am Vacuum Soc; Electrochem Soc. *Res:* Electrochemical energy conversion and storage systems; electrodeposition of metals and semiconductors. *Mailing Add:* Mat & Chem Sci Div Lawrence Berkeley Lab Berkeley CA 94720

ROSS, REUBEN JAMES, JR, b New York, NY, July 1, 18; m 42; c 4. PALEONTOLOGY, STRATIGRAPHY SEDIMENTATION. *Educ:* Princeton Univ, AB, 40; Yale Univ, MS, 46, PhD(geol), 48. *Prof Exp:* Field asst, Newfoundland Geol Surv, 38; asst prof geol, Wesleyan Univ, 48-52; geologist, Paleont & Stratig Br, US Geol Surv, 52-80; ADJ PROF GEOL, COLO SCH MINES, 80- *Concurrent Pos:* Chmn subcomn Ordovician Stratigraphy, Int Union Geol Sci, 76-82. *Mem:* Paleont Soc; Geol Soc Am; Am Asn Petrol Geologists; Brit Palaeontograph Soc; Soc Econ Paleont & Mineral. *Res:* Invertebrate paleontology and Ordovician stratigraphy of Basin Ranges. *Mailing Add:* 5255 Ridge Trail Littleton CO 80123

ROSS, RICHARD FRANCIS, b Washington, Iowa, Apr 30, 35; m 57; c 2. VETERINARY MICROBIOLOGY. *Educ:* Iowa State Univ, DVM, 59, MS, 60, PhD(vet microbiol), 65; Am Col Vet Microbiol, dipl, 67. *Prof Exp:* Res assoc vet microbiol, Iowa State Univ, 59-61; operating mgr, Vet Labs, Inc, Iowa, 61-62; from instr to asst prof vet path, 62-65, assoc prof, 66-72, prof-in-chg, 85-90, PROF VET MICROBIOL, IOWA STATE UNIV, 72-, ASSOC DIR & ASSOC DEAN, 90- *Concurrent Pos:* NIH fel, Rocky Mt Lab, Nat Inst Allergy & Infectious Dis, Mont, 65-66; vis prof, Tierarztl Hochschule, Hanover & sr fel Humboldt Found, WGer, 75-76; distinguished prof, Iowa State Univ, 82; vchair bd gov, Am Col Vet Microbiologists, 74-75; chair, Int Orgn Mycoplasmology, 90-92. *Honors & Awards:* Howard Dunne Mem Lectr, Am Asn Swine Pract, 84, Howard Dunne Mem Award, 88. *Mem:* Am Vet Med Asn; Conf Res Workers Animal Dis; Am Soc Microbiol; Int Orgn Mycoplasmology; Am Col Vet Microbiologists (secy-treas, 77-83); AAAS. *Res:* Investigations on pneumonia in swine due to mycoplasma hypneumoniae and haemophilus pleuropneumoniae; agalactia in swine caused by coliform mastitis; arthritis in swine. *Mailing Add:* Vet Med Res Inst Iowa State Univ Ames IA 50011

ROSS, RICHARD HENRY, b Centre Hall, Pa, Sept 19, 16; m 43; c 3. DAIRY SCIENCE. *Educ:* Pa State Univ, BS, 38, PhD(dairy), 47; WVa Univ, MS, 40. *Prof Exp:* Asst dairy, WVa Univ, 38-40 & Pa State Univ, 40-42; from assoc prof to prof dairy husb, Univ Idaho, 47-53, head dept dairy sci, 60-70, prof dairy sci, 53-78; RETIRED. *Concurrent Pos:* Exten dairyman, Univ Idaho, 74-78. *Mem:* Am Soc Animal Sci; Am Dairy Sci Asn. *Res:* Dairy cattle nutrition and management; pasture. *Mailing Add:* 1464 Alpowa Ave Moscow ID 83843-2402

ROSS, RICHARD HENRY, JR, b Bellefonte, Pa, June 5, 46; m 80; c 2. ENVIRONMENTAL DATA FOR PESTICIDE REGISTRATION, PESTICIDE RISK ASSESSMENTS. *Educ:* Univ Idaho, Moscow, BS, 68; Mich State Univ, East Lansing, MS, 70, PhD(entom & insect biochem), 72. *Prof Exp:* Postdoctoral insect aging, Argonne Nat Lab, Ill, 72-73; field res rep, Agr Div, Ciba-Geigy Corp, 73-76, sr field res rep, 76-81, sr field res rep II, 81-85, sr environ specialist II, 85-88, MGR ENVIRON & CONTRACT STUDIES, AGR DIV, CIBA-GEIGY CORP, 88- *Mem:* Entom Soc Am; Soc Environ Toxicol & Chem; Am Soc Testing & Mat. *Res:* Environmental and crop tolerance data to support registrations agricultural chemicals; author of various publications. *Mailing Add:* 5424 Bunch Rd Summerfield NC 27358

ROSS, RICHARD STARR, b Richmond, Ind, Jan 18, 24; m 50; c 3. CARDIOLOGY. *Educ:* Harvard Med Sch, MD, 47. *Prof Exp:* From instr to assoc prof, 54-65, dean med fac & vpres med, 75-90, PROF MED, SCH MED, JOHNS HOPKINS UNIV, 65-, EMER DEAN, 90- *Concurrent Pos:* Consult, Nat Heart & Lung Inst, 61-78, training grant comt, 71 & adv coun, 74; chmn cardiovasc study sect, NIH, 66; mem bd dirs, Waverly Press, Merck & Co, Johns Hopkins Hosp, Francis Scott Key Med Ctr, Equitablr Life Assurance Soc of Us. *Mem:* Inst Med-Nat Acad Sci; Am Soc Clin Invest; Asn Am Physicians; Am Physiol Soc; master Am Col Physicians; Am Fedn Clin Res. *Res:* Cardiology utilizing physiological and radiological techniques. *Mailing Add:* Johns Hopkins Univ Sch Med 1830 E Monument St Baltimore MD 21205

ROSS, ROBERT ANDERSON, b Auchinleck, Scotland, Sept 6, 31; m 58; c 3. SURFACE CHEMISTRY, APPLIED CHEMISTRY. *Educ:* Univ Glasgow, BS, 54, PhD(phys chem), 58; Univ Strathyclyde, ARCST, 54, DSc, 78. *Prof Exp:* Res engr, Marconi's Res Labs, Chelmsford, Eng, 57-59; asst res mgr, Joseph Crosfields Ltd, Warrington, 59-60; lectr inorg chem, Univ Strathyclyde, 60-64; head, Dept Chem, Col Technol, Univ Belfast, 64-69; dean sci, Lakehead Univ, 70-75, prof chem, 69-81; dir corp res, Domtar Inc, Montreal, 81-90; ADJ PROF CHEM, QUEENS UNIV, KINGSTON, 90- *Concurrent Pos:* Grants, Nat Res Coun Can, Dept Energy, Imp Oil Ltd, Ont Dept Environ & Int Nickel Co Ltd. *Mem:* Royal Soc Chem; Can Inst Chem; Can Soc Chem Eng. *Res:* Heterogeneous catalysis and adsorption; technological and combustion chemistry; air and water pollution studies. *Mailing Add:* RR 1 King Pitt Rd Box 45 Kingston ON K7L 4V1 Can

ROSS, ROBERT EDGAR, b Trenton, NJ, Aug 25, 48; m 79; c 2. FOOD SCIENCE. *Educ:* Rutgers Univ, BS, 70; Univ Mass, MS, 72, PhD(food sci), 73. *Prof Exp:* Res asst prod develop, PepsiCo, Inc, 68; food technologist, Prod Develop, Hunt-Wesson Foods, Inc, 73-74, sr food technologist, 74-75, group leader, 75-77; mgr corp new prod, Int Nabisco Brands, Inc, 78-79, asst dir, 79-81, assoc dir, 81-82, group dir, Corp Food Res, 81-82, group dir, Baked Goods Res & Develop, 82-84, sr dir, Biscuit Res & Develop, Nabisco Brands,

85-86, dir tech serv, 87-89; vpres, Res & Develop Qual Assurance, Estee Corp, 89-90; DIR, RES & DEVELOP, PEPPERIDGE FARM, 90- *Concurrent Pos:* Vpres, Gum & Confectionery Res & Develop, Warner-Lambert Co, 86-87; George H Cook scholar, Rutgers. *Mem:* Inst Food Technologists; Sigma Xi. *Res:* New product development and brand maintenance; technical evaluation of new business opportunities, joint ventures and technology licensing; operations support; quality assurance. *Mailing Add:* 102 Coventry Lane Trumbull CT 06611

ROSS, ROBERT EDWARD, b Rochester, NY, Nov 19, 37; m 59. ORGANIC CHEMISTRY. *Educ:* St John Fisher Col, BS, 61; Princeton Univ, MA, 63, PhD(org chem), 66. *Prof Exp:* RES ASSOC, EASTMAN KODAK CO, 66- *Mem:* Am Chem Soc; AAAS; Royal Soc Chem. *Res:* Organic synthesis of heterocyclic compounds, photographic couplers and redox materials; evaluation of photographically useful fragments in a film format; precipitation; sensitization of photo-sensitive silver halide crystals for practical imaging systems. *Mailing Add:* 2587 Browncroft Blvd Rochester NY 14625

ROSS, ROBERT GORDON, b Oxford, NS, July 14, 22; m 52; c 2. PHYTOPATHOLOGY. *Educ:* McGill Univ, BSc, 49, MSc, 54; Univ Western Ont, PhD(bot), 56. *Prof Exp:* Res officer & plant pathologist, Agr Can, Kentville, NS, 49-76, asst dir res sta, 76-85; RETIRED. *Mem:* Am Phytopath Soc; Can Phytopath Soc (pres, 75-76). *Res:* Tree fruit diseases. *Mailing Add:* Group Box 579 RR 2 Wolfville NS B0P 1X0 Can

ROSS, ROBERT TALMAN, b San Diego, Calif, Oct 10, 40; m 62; c 3. BIOPHYSICAL CHEMISTRY. *Educ:* Calif Inst Technol, BS, 62; Univ Calif, Berkeley, PhD(chem), 66. *Prof Exp:* Lectr chem, Univ Calif, Berkeley, 66-67; asst prof, Am Univ Beirut, 67-70; assoc prof biophys, 70-75, assoc prof, 75-81, PROF BIOCHEM, OHIO STATE UNIV, 81- *Concurrent Pos:* Chemist, Lawrence Berkeley Lab, Univ Calif, 66-67; dir, Beirut Study Ctr, Univ Calif, 69-70 & Biophys Prog, Ohio State Univ, 81-; vis scholar, Univ Calif, Berkeley, 83-84. *Mem:* Biophys Soc; Am Chem Soc. *Res:* Thermodynamics, statistical mechanics and kinetics of biological systems; photosynthesis; fluorescence spectroscopy; multilinear data analysis. *Mailing Add:* Dept Biochem Ohio State Univ 484 W 12th Ave Columbus OH 43210

ROSS, RODERICK ALEXANDER, b Ottawa, Ont, June 27, 26. APPLIED MATHEMATICS. *Educ:* Univ Toronto, PhD(math), 58. *Prof Exp:* From lectr appl math to assoc prof math, 57-85, PROF MATH, UNIV TORONTO, 85- *Res:* Wave propagation; differential equations. *Mailing Add:* Dept Math Univ Toronto Toronto ON M5S 1A1 Can

ROSS, RONALD RICKARD, b Minneapolis, Minn, July 11, 31; m 52; c 3. PARTICLE ASTROPHYSICS. *Educ:* Univ Calif, Berkeley, BA, 56, PhD(physics), 61. *Prof Exp:* Res physicist, Lawrence Radiation Lab & lectr physics, 61-63, asst prof, 63-67, assoc prof, 67-72, FAC SR SCIENTIST, LAWRENCE BERKELEY LAB, 63-, PROF PHYSICS, UNIV CALIF BERKELEY, 72- *Concurrent Pos:* Vis prof, Inst High Energy Physics, Univ Heidelberg, 66-67. *Honors & Awards:* IBM Fel, 57. *Mem:* AAAS; Am Phys Soc; Am Asn Physics Teachers. *Res:* Electron-positron interactions; fundamental interactions of particles; search for dark matter. *Mailing Add:* Dept Physics Univ Calif Berkeley CA 94720

ROSS, RUSSELL, b St Augustine, Fla, May 25, 29; m 56; c 2. EXPERIMENTAL PATHOLOGY, CELL BIOLOGY. *Educ:* Cornell Univ, AB, 51; Columbia Univ, DDS, 55; Univ Wash, PhD(exp path), 62. *Hon Degrees:* DSc, Med Col Pa, 87. *Prof Exp:* Intern, Presby Hosp, New York, 55-56; staff mem in chg oral surg sect, USPHS Hosp, Seattle, 56-58; from asst prof to assoc prof, 62-69, assoc dean sci affairs, 71-78, PROF PATH, SCH MED, UNIV WASH, 69-, ADJ PROF BIOCHEM, 78-, CHMN, PATH, 82- *Concurrent Pos:* NIH career develop res award, 62-67; Guggenheim fel, 66-67; vis fel, Clare Hall, Cambridge Univ, 66-68; mem cell biol study sect B, NIH, 66-69, mem, Molecular Biol Study Sect, 69-71, chmn, 74-76; mem adv coun, Nat Heart, Lung & Hl Blood Inst, 78-81; res comt, Am Heart Asn, 83; mem, Nat Res Adv Bd & Cleveland Clin Fedn; Int Recognition Award, Heart Res Found, 86; dir, Ctr Vascular Biol, Univ Wash Sch Med, 90. *Honors & Awards:* Gordon Wilson Medal, Am Clin & Climatol Asn, 81; Nat Res Achievement Award, Am Heart Asn, 90. *Mem:* Inst Med-Nat Acad Sci; Histochem Soc; Electron Micros Soc Am; Am Soc Exp Pathologists; Am Asn Pathologists; hon mem Harvey Soc; AAAS; Am Soc Cell Biol. *Res:* Wound healing; connective tissue; inflammation and atherosclerosis; autoradiography and biochemical analyses; cell and molecular biology of growth factors. *Mailing Add:* Dept Path SM-30 Univ Wash Sch Med Seattle WA 98195

ROSS, SAM JONES, JR, b Tompkinsville, Ky, July 4, 31; m 56; c 3. SOIL MORPHOLOGY. *Educ:* Western Ky State Col, BS, 58; Univ Ky, MS, 61; Purdue Univ, PhD(soil), 73. *Prof Exp:* Asst soils, Univ Ky, 59-61; soil scientist, Va Polytech Inst, 61 & Soil Surv Lab, 62-68; instr soils, Purdue Univ, 68-73; soil scientist, Soil Surv Invest Lab, 73-75; SOIL SCIENTIST, NAT SOIL SURV LAB, 75- *Mem:* Am Soc Agron; Soil Sci Soc Am. *Res:* Field hydraulic conductivity work and soil physical work in laboratory. *Mailing Add:* 3100 N 35th St No 6 Lincoln NE 68504

ROSS, SHEPLEY LITTLEFIELD, b Sanford, Maine, Nov 5, 27; m 54; c 4. MATHEMATICS. *Educ:* Univ Boston, AB, 49, AM, 50, PhD(math), 53. *Prof Exp:* Lectr math, Univ Boston, 50-53; instr, Northeastern Univ, 53-54 & Univ Boston, 54-55; from instr to assoc prof, 55-70, PROF MATH, UNIV NH, 70- *Mem:* Am Math Soc; Math Asn Am. *Res:* Ordinary differential equations. *Mailing Add:* Dept Math Univ NH Cate Rd Barrington NH 03825

ROSS, SIDNEY, b Philadelphia, Pa, May 12, 26; m 48; c 3. PHYSICS, RESEARCH ADMINISTRATION. *Educ:* Pa State Univ, BS, 48; Univ Pa, MS, 53; Temple Univ, PhD(physics), 61. *Prof Exp:* Physicist interior ballistics res, Pitman-Dunn Labs, US Dept Army, 48-50, weapon systs res, 51-52, chief interior ballistics br, 53-54, ballistics div, 55-56, tech asst dir labs group, 57, asst dir physics res lab, 58-66, dir appl sci lab, 66-68, tech dir, Frankford Arsenal, 68-77; STAFF TECH ADV, GOVT SYSTS DIV ENG, RCA CORP, 77- *Mem:* Am Phys Soc; Sigma Xi; Am Chem Soc. *Res:* Plasma physics; magnetohydrodynamics; hyperballistic propulsion and impact systems; advanced weapon systems; solid state physics; lasers; defense sciences and engineering. *Mailing Add:* 6811 Kindred St Philadelphia PA 19149

ROSS, SIDNEY DAVID, b Lynn, Mass, Jan 31, 18; m 42; c 1. CHEMISTRY. *Educ:* Harvard Univ, AB, 39, PhD(org chem), 44; Boston Univ, MA, 40. *Prof Exp:* Asst chem, Boston Univ, 39-40; asst chem, Harvard Univ, 40-41, res assoc, Naval Ord Res Coun, 41-45, Comt Med Res, 45 & Pittsburgh Plate Glass Co fel, 46; dir org res, 46-56, res assoc, 56-71, dir corp res & develop, 71-85, Sprague fel, Sprague Elec Co, 85-89, CONSULT, 90- *Mem:* Am Chem Soc; Electrochem Soc; The Chem Soc. *Res:* Mechanism of organic reactions; polymerization; kinetics; dielectrics; synthetic organic chemistry. *Mailing Add:* MRA Labs Inc North Adams MA 01247

ROSS, STANLEY ELIJAH, b Paterson, NJ, Nov 27, 22; m 61; c 1. ORGANIC POLYMER CHEMISTRY, TEXTILES. *Educ:* Rutgers Univ, BS, 43; NY Univ, MS, 49. *Prof Exp:* Res chemist, Cent Res Lab, Celanese Corp Am, 47-52 & Allied Chem Corp, 52-59; head polymer dept, J P Stevens & Co, Inc, 60-65, head process develop dept, 65-67, res assoc, Explor Tech, 67-75, prod develop mgr, 71-74, mgr nonwoven tech, 75-79, mgr fiber technol, Res & Develop Div, 79-84; mgr fiber finishes, Emery Div, Quantum Chem Corp, 84, MGR FIBER FINISHES, ORG PRODS DIV, HENKEL CORP, 84- *Mem:* Am Chem Soc; Fiber Soc (vpres, 74, pres, 75); NY Acad Sci. *Res:* Polymers, fibers and textiles and their characterizations; mechanical behavior; thermoanalysis; extrusion and orientation of thermoplastics; crystallinity behavior; carbonization and pyrolysis of cellulose; development of nonwoven textile structures; cotton dust and its measurement, frictional meas, fiber finish development. *Mailing Add:* 201 Bloomfield Lane Greer SC 29650-3808

ROSS, STEPHEN T, b Chicago, Ill, Mar 21, 31; m 52; c 2. MEDICINAL CHEMISTRY. *Educ:* Univ Ill, BS, 53; Rutgers Univ, MS, 60, PhD(org chem), 61. *Prof Exp:* Chemist, Johnson & Johnson, 55-59; assoc sr investr, 61-74, SR INVESTR, SMITH KLINE & FRENCH LABS, 74- *Mem:* Am Chem Soc; AAAS; NY Acad Sci. *Res:* Structure and activity relationships; synthetic heterocyclic chemistry; organo-analytical chemistry; organophosphorus chemistry; infrared and nuclear magnetic resonance spectroscopy; mass spectrometry; high-performance liquid and gas chromatography. *Mailing Add:* 718 Old State Rd Berwyn PA 19312-1441

ROSS, STEPHEN THOMAS, b Hollywood, Calif, Nov 9, 44; m 69; c 1. ICHTHYOLOGY, FISH BIOLOGY. *Educ:* Univ Calif, Los Angeles, BA, 67, Calif State Univ, Fullerton, MA, 70; Univ SFla, PhD(biol), 74. *Prof Exp:* Teaching asst biol, Calif State Univ, Fullerton, 67-69; teaching asst, Univ SFla, 70-71, instr, 72; from asst prof to assoc prof, 74-83, PROF BIOL, UNIV SOUTHERN MISS, 84- *Concurrent Pos:* Vis res prof, Univ Okla Biol Sta, 81-82; vis scholar, Dept Wildlife Fisheries Biol, Univ Calif, Davis, 91- *Mem:* Am Soc Ichthyologists & Herpetologists; Am Fisheries Soc; Soc Study Evolution; Soc Systematic Zool; Ecol Soc Am. *Res:* Ecological and evolutionary relationships of fishes. *Mailing Add:* Dept Biol SS Box 5018 Univ Southern Miss Hattiesburg MS 39406-5018

ROSS, STEWART HAMILTON, geology, for more information see previous edition

ROSS, STUART THOM, b Green Bay, Wis, July 6, 23; m 49; c 2. MECHANICAL METALLURGY, MATERIALS SCIENCE. *Educ:* Purdue Univ, BSMetE, 47, MSMetE, 49, PhD(metall), 50. *Prof Exp:* Asst chief metallurgist, Harrison Radiator Div, Gen Motors Corp, 50-52; head metall res, Eng Div, Chrysler Corp, 52-59; chief mat engr, Aeronutronic Div, Ford Motor Corp, 59-61; vpres eng & res, Brooks & Perkins, Inc, 61-65; dir eng & develop, Wolverine Tube Div, Universal Oil Prod, 65-69; vpres & tech dir, Crucible Specialty Metals Div, Colt Industs, 69-74 & Int Mill Serv, IU Int, 74-76; V PRES & TECH DIR, ITT MEYER INDUSTS, 76- *Mem:* Sigma Xi; fel Am Soc Metals; Am Inst Metall Engrs; Am Nuclear Soc; Brit Iron & Steel Inst. *Res:* Hot isostatic compaction; prealloyed powders; shapes/cavities; mechanical metallurgy of steel; fracture toughness; failure analysis and causation. *Mailing Add:* 2247 Hallquist Rd Red Wing MN 55066

ROSS, SYDNEY, b Glasgow, Scotland, July 6, 15. COLLOID CHEMISTRY. *Educ:* McGill Univ, BSc, 36; Univ Ill, PhD(chem), 40. *Prof Exp:* Instr chem, Monmouth Col, 40-41; res assoc & fel chem, Stanford Univ, 41-45; assoc prof phys chem, Univ Ala, 45-46; sr chemist, Oak Ridge Nat Lab, Tenn, 46-48; assoc prof, 48-51, PROF COLLOID SCI, RENSSELAER POLYTECH INST, 51- *Concurrent Pos:* Mem adv bd, Gordon Res Conf, 54; vis prof, Univ Strathclyde, Scotland, 75-78. *Mem:* Am Chem Soc; Royal Soc Chem. *Res:* Foams and antifoams; electrokinetic phenomena; dispersions and emulsions; adsorption of gases by solids. *Mailing Add:* Dept Chem Rensselaer Polytech Inst Troy NY 12181

ROSS, THEODORE WILLIAM, b Oak Park, Ill, May 9, 35; m 62; c 2. GEOLOGY. *Educ:* Ind Univ, BS, 57, MA, 62; Wash State Univ, PhD(geol), 69. *Prof Exp:* Instr, 66-67, chmn dept, 67-70, ASST PROF GEOL, LAWRENCE UNIV, 70-, CHMN DEPT, 74- *Mem:* Soc Econ Geol; Nat Asn Geol Teachers; Geol Soc Am. *Res:* Economic geology of metals; determination of pre-Pleistocene bedrock topography of Wisconsin. *Mailing Add:* Dept Geol Lawrence Univ Box 599 Appleton WI 54912

ROSS, THOMAS EDWARD, b Bud, WVa, Apr 25, 42; m 64; c 2. WATER PROBLEMS. *Educ:* Marshall Univ, BA, 68, MS, 69; Univ Tenn, PhD(geog), 77. *Prof Exp:* PROF & CHAIR GEOG, PEMBROKE STATE UNIV, 69- *Concurrent Pos:* Consult, Off Econ Develop, Pembroke State Univ, 87-92; bd dirs, Am Indian Specialty Group, Asn Am Geographers, 87-91. *Mem:* Asn Am Geogr. *Res:* Environmental issues; groundwater pollution and water supplies in southeastern United States; environmental impacts upon economic development. *Mailing Add:* 132 Bee Gee Rd Lumberton NC 28358

ROSS, TIMOTHY JACK, b Spokane, Wash, Feb 17, 49; m 81; c 4. STRUCTURAL ENGINEERING, COMPUTATIONAL SOFTWARE. *Educ:* Wash State Univ, BS, 71; Rice Univ, MS, 73; Stanford Univ, PhD(struct eng), 83. *Prof Exp:* Vulnerability engr, Defense Intel Agency, 73-78; res engr, Air Force Weapons Lab, 78-86; ASSOC PROF STRUCT, UNIV NMEX, 87- *Concurrent Pos:* Adv, Res Assoc Prog, Nat Res Coun, 84-86; dir & pres, IntelliSys Corp, 86-88. *Mem:* Am Soc Civil Engrs; Int Fuzzy Systs Asn. *Res:* Exploratory and experimental studies of microfracture in brittle materials; fuzzy logic and expert systems; dynamics of space structures; probability theory; risk assessment. *Mailing Add:* Dept Civil Eng Univ NMex Albuquerque NM 87131

ROSS, WILLIAM D(ANIEL), b Elmira, NY, Nov 22, 17; m 61; c 1. LIGHT SCATTERING. *Educ:* Columbia Univ, BA, 38, BS, 39. *Prof Exp:* Res chemist, E I Du Pont de Nemours & Co, Inc, 39-82, res fel, 68-82; RETIRED. *Honors & Awards:* Roon Awards, Fedn Socs Paint Technol, 69 & 71, Bruning Award, 78. *Mem:* AAAS; Am Chem Soc; Optical Soc Am; Am Inst Chem Engrs; NY Acad Sci. *Res:* Titanium pigments; optics of pigments; chemistry of titanium; surface chemistry; thermodynamics; heat transfer; light scattering. *Mailing Add:* 36 Ridgewood Circle Wilmington DE 19809

ROSS, WILLIAM DONALD, b Hamilton, Ont, Sept 13, 13; nat US; m 39, 79; c 1. PSYCHOANALYSIS. *Educ:* Univ Man, BSc(med) & MD, 38; McGill Univ, dipl, 49; FRCP(C); Am Bd Psychiat & Neurol, dipl, 49; Chicago Inst Psychoanal, cert, 58, Am Psychoanal Asn, 77. *Prof Exp:* Demonstr histol, Univ Man, 33-34, demonstr physiol, 35-38, demonstr clin path & med, 38-39; demonstr neurol, McGill Univ, 40-43, demonstr psychiat, 43-48; asst prof, Univ Cincinnati, 48-53; prof, Univ BC, 53; assoc prof, 54-60, assoc prof environ health, 61-76, prof, 76-80, EMER PROF PSYCHIAT & ENVIRON HEALTH, COL MED, UNIV CINCINNATI, 80- *Concurrent Pos:* Attend psychiatrist & clinician, Outpatient Dept, Univ Hosp Cincinnati, dir & psychiat consult-liaison Serv; consult, Vet Admin Hosp, Cincinnati. *Mem:* AAAS; fel Soc Personality Assessment; Am Sociol Asn; Am Psychiat Asn; Am Psychoanal Asn; Sigma Xi. *Res:* Industrial mental health; psychosomatic medicine; psychopharmacology; psychoanalysis. *Mailing Add:* 854 Rue De La Paix B-1 Cincinnati OH 45220

ROSS, WILLIAM J(OHN), b Auckland, NZ, June 11, 30; m 59; c 1. ELECTRICAL ENGINEERING. *Educ:* Univ Auckland, BSc, 51, MSc, 53, PhD(physics), 55. *Prof Exp:* Res asst, Ionosphere Res Lab, 55-56, from asst prof to assoc prof elec eng, 56-63, PROF ELEC ENG, PA STATE UNIV, UNIVERSITY PARK, 63-, HEAD DEPT, 71- *Mem:* Am Geophys Union; Inst Elec & Electronics Engrs; Am Soc Eng Educ. *Res:* Ionospheric radio propagation; ionospheric theory. *Mailing Add:* Dept Elec Eng Pa State Univ 121 University Park PA 16802

ROSS, WILLIAM MAX, b Farmington, Ill, Mar 20, 25; m 52; c 5. PLANT BREEDING, CROP PRODUCTION. *Educ:* Univ Ill, BS, 48, MS, 49, PhD(agron), 52. *Prof Exp:* Res asst agron, Univ Ill, 49-51, Kans State Univ, 51-52; res agronomist & res geneticist, Ft Hays Exp Sta, Kans, 52-69, res geneticist & prof, USDA, Dept Agron, Univ Nebr, Lincoln, 69-85; RETIRED. *Concurrent Pos:* Assoc ed, Crop Sci, 80-82. *Mem:* Fel Am Soc Agron; Crop Sci Soc Am. *Res:* Production, breeding and genetics of sorghum; population improvement for yield, grain quality and insect resistance. *Mailing Add:* 1631 Karlee Dr Lincoln NE 68527

ROSS, WILLIAM MICHAEL, resource management; deceased, see previous edition for last biography

ROSSA, ROBERT FRANK, b Kankakee, Ill, Aug 17, 42; m 69; c 1. ALGEBRA. *Educ:* Univ Okla, BA, 63, MA, 66, PhD(math), 71. *Prof Exp:* Asst math, Univ Okla, 63-69; from asst prof to assoc prof, 69-84, PROF COMPUT SCI & MATH, ARK STATE UNIV, 84- *Mem:* Am Math Soc; Asn Comput Mach; Inst Elec & Electronics Engrs; Math Asn Am. *Res:* Ring theory; radicals. *Mailing Add:* Box 151 State University AR 72467

ROSSALL, RICHARD EDWARD, b Lytham St Annes, Eng, Jan 11, 26; Can citizen; m 52; c 2. INTERNAL MEDICINE, CARDIOLOGY. *Educ:* Univ Leeds, BSc, 47, MB, ChB, 50, MD, 58; MRCP, 52; FRCP(C), 58; FRCP, 73. *Prof Exp:* House physician internal med, Gen Infirmary, Leeds, Eng, 50, Postgrad Med Sch, Univ London, 50-51 & Brompton Hosp, London, 51; asst chest physician, Bromley & Farnboro, Kent, 51-52; registr, Hull & East Riding Hosp Bd, 54, Gen Infirmary, Leeds, 54-55, sr registr, 55-57; from instr to prof, 57-71, from asst dean to assoc dean, 69-78, internal med, Univ Alta, DIR, DIV CARDIOL, UNIV HOSP, 69- *Concurrent Pos:* Consult cardiologist, Edmonton Gen, Royal Alexandria & Misericordia Hosps, 57- & Dept Vet Affairs, 59-; fel coun clin cardiol, Am Heart Asn. *Mem:* Fel Am Col Cardiol; Can Soc Clin Invest; fel Am Col Physicians; Can Cardiovasc Soc (pres, 80-82). *Res:* Pulmonary lymphatic and blood circulation in health and disease. *Mailing Add:* 2C2 W C MacKenzie Health Sci Ctr Univ Alta Edmonton AB T6G 2R7 Can

ROSSAN, RICHARD NORMAN, b Chicago, Ill, May 15, 28; m 66; c 1. PARASITOLOGY. *Educ:* Univ Ill, BS, 49, MS, 51; Rutgers Univ, PhD(zool), 59. *Prof Exp:* Mem staff, Naval Med Res Inst, 52-56; res training assoc, Howard Univ, 59-60; res assoc, Inst Med Res, Christ Hosp, 60-63; asst prof biol & res parasitologist, Nat Ctr for Primate Biol, Univ Calif, Davis, 63-69; mem prof staff, Gorgas Mem Lab, 69-90, actg chief, Div Parasitol, 87-90, actg dir, 89-90; RETIRED. *Concurrent Pos:* WHO traveling fel, 65. *Mem:* AAAS; Am Soc Parasitol; Am Soc Trop Med & Hyg; Sigma Xi. *Res:* Biology, chemotherapy and immunity of avian and primate malarias; serum changes and immunity associated with visceral leishmaniasis; evaluation of experimental antimalarial drugs in monkey models. *Mailing Add:* PSC Box 2209 APO Miami FL 34002

ROSSANO, AUGUST THOMAS, b New York, NY, Feb 1, 16; m 44; c 8. ENVIRONMENTAL ENGINEERING. *Educ:* Mass Inst Technol, BS, 38; Harvard Univ, SM, 41, ScD(air sanit), 54; Am Bd Indust Hyg, dipl, 58, Am Acad Environ Engr, dipl. *Prof Exp:* Asst civil eng, Univ Ill, 39-40; sanit engr, USPHS, 41-62; vis prof environ health eng, Calif Inst Technol, 60-63; prof, 63-81, EMER PROF CIVIL ENG, AIR RESOURCE PROG, UNIV WASH, 81- *Concurrent Pos:* Consult, WHO, World Bank, 24 foreign govts, US govt & various pvt corps, 60-; mem int sect, Air Pollution Control Asn, Pacific, NW, 82. *Honors & Awards:* Spec Serv Award, USPHS, 58. *Mem:* Am Pub Health Asn; Am Indust Hyg Asn; Air Pollution Control Asn; Am Acad Environ Engrs; Sigma Xi. *Res:* Industrial hygiene; air pollution; radiological health. *Mailing Add:* Dept Civil Eng FX-10 Univ Wash Seattle WA 98195

ROSSANT, JANET, b Chatham, UK, July 13, 50; m 76; c 2. MAMMALIAN DEVELOPMENT. *Educ:* Oxford Univ, BA, 72; Cambridge Univ, PhD(embryol), 76. *Prof Exp:* Vis prof, Dept Immunol, Univ Alta, 77; from asst prof to assoc prof biol, Brock Univ, 77-85; ass prof, 85-88, PROF DEPT MED GENETICS, UNIV TORONTO, 88-; SR SCIENTIST, MT SINAI HOSP, 85- *Concurrent Pos:* Asst prof, Dept Path, McMaster Univ, 81-85; E W R Steacie Fel, 83. *Mem:* Brit Soc Develop Biol; Can Soc Cell Biol; Am Soc Cell Biol. *Res:* Early mammalian development, using micromanipulative and molecular genetic approaches to studying early differentiation; teratocarcinoma differentation. *Mailing Add:* Mt Sinai Hosp Res Inst 600 University Ave Toronto ON M5G 1X5 Can

ROSSBACHER, LISA ANN, b Fredericksburg, Va, Oct 10, 52; m 79. PLANETARY GEOLOGY, SCIENCE WRITING. *Educ:* Dickinson Col, BS, 75; State Univ NY Binghamton, MA, 78; Princeton Univ, MA, 79, PhD(geol), 83. *Prof Exp:* Geologist archeol dig, Dickinson Col & Hartwick Col, 74-75; instr geol, Dickinson Col, 76-77; from asst prof to assoc prof, 84-91, PROF, CALIF STATE POLYTECH UNIV, 91-, ASSOC VPRES ACAD AFFAIRS, 87-; RES ASSOC, WHITTIER COL, 79- *Concurrent Pos:* Consult, Repub Geothermal Inc, 79-81; vis researcher, Univ Uppsala, Sweden, 84. *Mem:* AAAS; Asn Women Geoscientists; Asn Earth Sci Ed; Geol Soc Am; Sigma Xi. *Res:* Geomorphology of planetary surfaces, especially solutional landforms (karst and thermokarst) on Earth and Mars; writing on careers in geology and space exploration. *Mailing Add:* Acad Affairs Calif State Polytech Univ Pomona CA 91768

ROSSBY, HANS THOMAS, b Boston, Mass, June 8, 37; m 62; c 2. PHYSICAL OCEANOGRAPHY. *Educ:* Royal Inst Technol, Sweden, BS, 62; Mass Inst Technol, PhD(oceanog), 66. *Prof Exp:* Res asst mech physics, Royal Inst Technol, Sweden, 60-62; res asst geophys, Mass Inst Technol, 62-66, res assoc oceanog, 66-68; from asst prof to assoc prof geophys, 68-75; PROF OCEANOG, UNIV RI, 75- *Honors & Awards:* Bigelow Gold Medal, Woods Hole Oceanog Isnt. *Mem:* Sigma Xi; Am Meteorol Soc; fel Am Geophys Union; Oceanog Soc. *Res:* Structure and variability of ocean currents; Gulf stream; North Atlantic current; oceanic microstructure. *Mailing Add:* Dept Oceanog Univ RI Kingston RI 02881

ROSSE, CORNELIUS, b Csorna, Hungary, Jan 13, 38. ANATOMY, IMMUNOLOGY. *Educ:* Bristol Univ, MB, ChB, 64, MD, 74. *Prof Exp:* House surgeon, United Hosps, Bristol Univ, 64-65, demonstr anat, Univ, 65-67; from asst prof to assoc prof biol struct, 67-75, PROF BIOL STRUCT, SCH MED, UNIV WASH, 75-, CHMN BIOL STRUCT, 81- *Mem:* Exp Hemat Soc; Am Asn Anatomists; Anat Soc Gt Brit & Ireland; Sigma Xi. *Res:* Experimental hematology; problems relating to hemopoietic stem cell and to the role of the bone marrow in producing potentially immunocompetent cells. *Mailing Add:* Dept Biol Struct SM 20 Univ Wash Seattle WA 98195

ROSSE, WENDELL FRANKLYN, b Sidney, Nebr, June 5, 33; m 59; c 4. MEDICINE, IMMUNOLOGY. *Educ:* Univ Omaha, AB, 53; Univ Nebr, MS, 56; Univ Chicago, MD, 58. *Prof Exp:* NIH clin assoc, Nat Cancer Inst, 60-63; vis res fel, Med Sch, Univ London, 63-64; sr investr med, Nat Cancer Inst, 64-66; from asst prof to assoc prof, 66-72, assoc prof immunol, 70-74, chief immuno-hemat, 71-76, PROF MED, MED CTR, DUKE UNIV, 72-, PROF IMMUNOL, 74-, DIR BLOOD BANK, 71-, CHIEF HEMAT-MED ONCOL, 76- *Mem:* Am Soc Hemat; Am Asn Physics; Am Soc Clin Invest; Am Fedn Clin Res. *Res:* Immune hemolytic anemia and thrombocytopenia; complement-dependent mechanisms of hemolysis. *Mailing Add:* Box 3934 Duke Univ Med Ctr Durham NC 27710

ROSSEN, JOEL N(ORMAN), b Detroit, Mich, June 22, 27; m 52; c 3. PROPULSION SYSTEMS DEVELOPMENT, PROGRAM MANAGEMENT. *Educ:* Mass Inst Technol, BS, 48, MS, 49. *Prof Exp:* Res engr adhesives, Div Indust Coop, Mass Inst Technol, 49-50; chem engr, Tracerlab, Inc, 50-51; chem engr, Atlantic Res Corp, 51-56, proj engr, 56-57, head, Rocket Ballistics Group, 57-58, asst dir, Solid Propellant Div, 58-60, dir, 60-62; asst mgr, Solid Rocket Br, United Tech Ctr, United Aircraft Corp, 62-64, tech asst, Opers Dept, 64, asst chief engr, Solid Rockets, 64-65, asst prod develop mgr, 65-66, mgr advan technol br, 66-67, mgr ICM progs, 67-71, mgr FMB propulsion progs, 71-73, mgr ballistic missile progs, 73-78, mgr solid rocket progs, Chem Systs Div, United Technologies Corp, 78-84; TECH & MGT CONSULT, J N ROSSEN, INC, 84- *Mem:* Am Chem Soc; Am Inst Aeronaut & Astronaut; Am Inst Chem Engrs. *Res:* Solid and hybrid rocket propulsion technology development and design; systems development; solid rocket propulsion systems management. *Mailing Add:* 26763 Palo Hills Dr Los Altos Hills CA 94022-1927

ROSSEN, ROGER DOWNEY, b Cleveland, Ohio, June 4, 35; m 61; c 3. IMMUNOLOGY, INTERNAL MEDICINE. *Educ:* Yale Univ, BA, 57; Western Reserve Univ, MD, 61. *Prof Exp:* From intern to asst resident internal med, Columbia-Presby Hosp, N Y, 61-63, sr resident internal med, 66-67; from clin assoc to clin investr, Lab Clin Invest, Nat Inst Allergy & Infectious Dis, 63-66; from instr to assoc prof, 67-73, PROF MICROBIOL, IMMUNOL & MED, BAYLOR COL MED, 73-; CHIEF CLIN IMMUNOL, VET ADMIN HOSP, 72- *Concurrent Pos:* Nat Inst Allergy & Infectious Dis spec res fel, Baylor Col Med, 67-68; mem adv comt transplantation & immunol, Nat Inst Allergy & Infectious Dis, 70-74; grants, Vet Admin, Dept Health & Human Servs & NIH. *Mem:* Am Asn Immunologists; Am Soc Microbiol; Soc Exp Biol & Med; Am Soc Pathologists. *Res:* Regulation of antibody formation, especially the regulatory influences of the monocyte in B cell differentiation; investigation of monocyte abnormalities in AIDS and related diseases. *Mailing Add:* Dept Microbiol & Immunol Baylor Col Med Houston TX 77030

ROSSER, JOHN BARKLEY, mathematics, computer sciences; deceased, see previous edition for last biography

ROSSETTI, LOUIS MICHAEL, b Chicago, Ill, Aug 20, 48; m 72; c 1. SPEECH PATHOLOGY, DEVELOPMENTAL PSYCHOLOGY. *Educ:* Northern Mich Univ, BS, 70, MA, 72; Southern Ill Univ, PhD(speech path), 78. *Prof Exp:* Chief speech pathologist, Children's Develop Ctr, 72-74; res asst, Southern Ill Univ, 74-76; from asst prof to assoc prof speech path, Northwest Mo State Univ, 84-87; CHMN COMMUN DISORDERS, UNIV WIS, OSHKOSH, 87- *Concurrent Pos:* Lectr & consult, Kirksville Col Osteop Med, 77-; diag consult, Kirksville Diag Ctr, 77- & Head Start Prog Mo, 78; assoc ed, Mo Speech & Hearing Asn J, 78-; res consult, Knoxville Vet Admin Hosp, 78- *Mem:* Am Speech & Hearing Asn. *Res:* Clinical application of distinctive feature theory; infant screening, especially high risk infants. *Mailing Add:* Dept Speech Univ Wis 800 Algoma Blvd Oshkosh WI 54901

ROSSETTOS, JOHN N(ICHOLAS), b Nisyros, Greece, Mar 11, 32; US citizen; m 63; c 2. SOLID MECHANICS, APPLIED MATHEMATICS. *Educ:* Mass Inst Technol, BS & MS, 56; Harvard Univ, MA, 60, PhD(solid mech), 64. *Prof Exp:* Staff engr, Mass Inst Technol, 54-56; res engr, United Aircraft Res Ctr, 56-57; sr engr, Allied Res, Inc & Am Sci & Eng, Inc, 57-59; res scientist appl mech, Langley Res Ctr, NASA, 64-66; sr staff scientist, Mech & Comput, Avco Corp, 66-69; assoc prof mech & appl math, 69-77, PROF MECH ENG, NORTHEASTERN UNIV, 77- *Concurrent Pos:* Lectr, Univ Va, 64-66; adj prof, Boston Univ, 67-69; vis assoc prof, Mass Inst Technol, 70; NASA res grant, Northeastern Univ, 71-; consult mech, Textron Defense Systs, 81-; hon res assoc struct mech, Harvard Univ, 79-80; vis scientist, Army Mat Technol Lab, 86-87. *Mem:* Am Acad Mech; assoc fel Am Inst Aeronaut & Astronaut; Am Soc Mech Engrs. *Res:* Mechanics of damage in composites; composite joints; application of finite element method and computers in engineering science; vibration and buckling of structural components. *Mailing Add:* Dept Mech Eng Northeastern Univ Boston MA 02115

ROSSI, BRUNO B, b Venice, Italy, Apr 13, 05; nat US; m 38; c 3. ASTROPHYSICS. *Educ:* Bologna Univ, PhD(physics), 27. *Hon Degrees:* PhD, Univ Palermo, 64; Dr, Univ Durham, Eng, 74 & Univ Chicago, 77. *Prof Exp:* Asst physics, Univ Florence, 28-32; prof, Univ Padua, 32-38; res assoc, Victoria Univ, 38-39; res assoc, Univ Chicago, 39-40; assoc prof, Cornell Univ, 40-43; mem staff, Los Alamos Proj, Univ Calif, 43-46; prof physics, Mass Inst Technol, 46-66, inst prof, 66-70, emer inst prof, 70-85; RETIRED. *Concurrent Pos:* Hon prof, La Paz; consult, AEC & NASA; mem space sci bd, Nat Acad Sci. *Honors & Awards:* Gold Medal, Italian Phys Soc, 78; Int Feltrinelli Award, Nat Acad Lincei, 71; Cresson Medal, Franklin Inst, 74; Rumford Prize Award, Am Acad Arts & Sci, 76; Nat Metal Sci Award, 85; Wolff Award in Physics, 87. *Mem:* Nat Acad Sci; Am Philos Soc; Am Phys Soc; Nat Acad Lincei; Royal Astron Soc Eng. *Res:* Cosmic rays; space research. *Mailing Add:* Ctr Space Res Mass Inst Technol 37-667 Cambridge MA 02139

ROSSI, EDWARD P, b Cleveland, Ohio, Jan 28, 34; div; c 1. ORAL PATHOLOGY, DENTISTRY. *Educ:* Kent State Univ, BA, 57; Western Reserve Univ, DDS, 62, MS, 65. *Prof Exp:* Intern, Crile Vet Admin Hosp, 62-63; from instr to asst prof, 64-71, ASSOC PROF ORAL PATH & CHMN DEPT, SCH DENT & ASST PROF, SCH MED, CASE WESTERN RESERVE UNIV, 71- *Concurrent Pos:* Am Cancer Soc clin fel, 65-66, advan clin fel, 66-69; consult, Cleveland Vet Admin Hosp; mem fel awards comt, Am Fund Dent Educ. *Mem:* Am Dent Asn; Am Acad Oral Path; Am Asn Cancer Educ; Int Acad Path. *Res:* Teaching oral pathology; oral pathology diagnostic laboratory; programmed instruction in dental education. *Mailing Add:* Sch Dent 2123 Abington Rd Case Western Reserve Univ 2040 Adelbert Rd Cleveland OH 44106

ROSSI, ENNIO CLAUDIO, b Madison, Wis, Apr 3, 31; m 57; c 2. MEDICINE. *Educ:* Univ Wis, BA, 51, MD, 54; Am Bd Internal Med, dipl, 64. *Prof Exp:* Intern, Ohio State Univ, 54-55; Fulbright scholar, Univ Rome, 55-56; resident med, Univ Wis, 58-61, res fel hemat, 61-63; from instr to asst prof, Marquette Univ, 63-66; assoc prof, 66-72, PROF MED, SCH MED, NORTHWESTERN UNIV, CHICAGO, 72- *Mem:* Am Fedn Clin Res; Am Soc Hemat. *Res:* Hematology; platelet metabolism and function; abnormal hemoglobins. *Mailing Add:* Dept Med Sch Med Northwestern Univ 303 E Chicago Chicago IL 60611

ROSSI, GEORGE VICTOR, b Ardmore, Pa, Mar 8, 29; m 56; c 1. PHARMACOLOGY. *Educ:* Philadelphia Col Pharm & Sci, BSc, 51, MSc, 52; Purdue Univ, Lafayette, PhD(pharmacol), 54. *Prof Exp:* Res assoc pharmacol, Nat Drug Co, 54-55; assoc prof, 55-61, dir, Dept Biol Sci, 65-81, PROF PHARMACOL, PHILADELPHIA COL PHARM & SCI, 61-, ASSOC DEAN RES & GRAD STUDIES, 77-, DIR, DEPT PHARMACOL & TOXICOL, 81- *Concurrent Pos:* Lindback Found Award, 81. *Mem:* AAAS; Am Soc Pharmacol & Exp Therapeut; Am Pharmaceut Asn; fel Acad Pharmaceut Sci; NY Acad Sci. *Res:* Cardiovascular pharmacology; antihypertensive, antiarrhythmic and antianginal drug mechanisms; histamine receptor mechanisms; gastric secretory mechanisms. *Mailing Add:* Dept Pharmacol Philadelphia Col Pharm & Sci 43rd St Kingsessing Mall Philadelphia PA 19104

ROSSI, HARALD HERMAN, b Vienna, Austria, Sept 3, 17; US citizen; m 46; c 3. RADIATION BIOPHYSICS. *Educ:* Johns Hopkins Univ, PhD(physics), 42. *Prof Exp:* Instr physics, Johns Hopkins Univ, 40-45; res scientist radiol physics, 46-60, from asst prof to prof, 49-87, PROF RADIOL, COL PHYSICIANS & SURGEONS, COLUMBIA UNIV, 87- *Concurrent Pos:* Res physicist, Nat Bur Stand, 45; physicist & radiation protection officer, Presby Hosp, 54-60; mem, Nat Coun Radiation Protection, 54-, bd dirs, 71-76; chmn, Health Comnrs, Tech Adv Comn Radiation, NY, 58-83; dir Dept Energy contract radiation physics, biophysics & radiation biol, 60-87; mem, Int Comn Radiation Units, 59-, chmn liaison comt, 75-77; chmn radiation study sect, NIH, 62-67; consult, Off Radiation Control, NY, 63-83; consult, Defense Atomic Support Agency, Armed Forces Radiol Res Inst, 65-70; consult, Brookhaven Nat Lab, 65-80; consult & chmn adv comn radiation aspects SST, Fed Aviation Agency, 68-74; chmn rev comn biol & med res, Argonne Nat Lab, 71-76; consult med use nuclear powered pacemakers, AEC, 71-75; mem, Comt Biol Effects of Iionizing Radiations, Nat Res Coun, 76-82; chmn, Int Comt Radiol Protection/Int Comt Radiol Units joint task group on radiation protection quantities, 80-86. *Honors & Awards:* Shonka Award, Shonka Mem Found, 72; 8th Lauriston Taylor Lectr, Nat Comt Radiol Protection, 84; L H Gray Medal, Int Comt Radiol Units, 85; Edith Quimby Mem Lect, 82; Wright Langham Lect, 83; Landauer Mem Lect, 88. *Mem:* Radiation Res Soc (pres, 74); Am Radium Soc; Radiol Soc NAm; Am Col Radiol; Sigma Xi. *Res:* Radiation dosimetry; radiobiology; radiation protection; biophysics. *Mailing Add:* 105 Larchdale Ave Upper Nyack NY 10960

ROSSI, HUGO, b 1935. MATHEMATICS. *Educ:* City Col, NY, BS, 56; Mass Inst Technol, MS, 57 & Phd(math), 60. *Prof Exp:* Teaching asst, Mass Inst Technol, 57-58; asst prof, Princeton Univ, 60-63; assoc prof, Brandeis Univ, 63-66, prof, 66; assoc ed, Transactions of Am Math Soc, 73-78; DEAN, COL SCI, UNIV UTAH, 87- *Concurrent Pos:* Ed, Pac J Math, 80-85 & Univ Lect Series, Am Math Soc, 87-; mem, comt sci policy, Am Math Soc, 83-85; vis, Inst Advan Study, Princeton, 83-84; mem, comt status prof, 85-87; Gardner fel, 83; asst prof, Univ Calif, Berkeley, 60. *Res:* Complex analysis. *Mailing Add:* Univ Utah JTB 302 Math Dept Salt Lake City UT 84112

ROSSI, JOHN JOSEPH, b Washington, DC, July 8, 46; m 69; c 4. MOLECULAR GENETICS. *Educ:* Univ NH, BA, 69; Univ Conn, MS, 71, PhD(genetics), 76. *Prof Exp:* Res assoc, Div Biol & Med, Brown Univ, 76-80; RES SCIENTIST & DIR, DEPT MOLECULAR GENETICS, BECKMAN RES INST, CITY OF HOPE, 80- *Concurrent Pos:* Adj prof, Dept Biochem, Univ Med Sch-Loma Linda, Calif. *Mem:* Am Soc Microbiol; AAAS. *Res:* Mechanisms of gene regulation in prokaryotic and simple eukaryotic systems; messenger RNA splicing; RNA processing; ribozyme inhibition of acquired immune deficiency syndrome viral function. *Mailing Add:* Dept Molecular Genetics City of Hope Res Inst 1450 E Duarte Rd Duarte CA 91010

ROSSI, MIRIAM, b Asti, Italy, Mar 8, 52; US citizen. CRYSTALLOGRAPHY. *Educ:* Hunter Col, BA, 74; Johns Hopkins Univ, MA, 76, PhD(chem), 79. *Prof Exp:* Res asst, Fox Chase Cancer Ctr, 78-82; asst prof, 82-89, ASSOC PROF CHEM, VASSAR COL, 89- *Mem:* Am Chem Soc; Am Crystallog Asn. *Res:* Study of anti-carcinogenic molecules using x-ray crystallographic techniques. *Mailing Add:* Vassar Col Box 484 Poughkeepsie NY 12601

ROSSI, NICHOLAS PETER, b Philadelphia, Pa, July 17, 27; m 57; c 4. CARDIOVASCULAR SURGERY. *Educ:* Univ Pa, 51; Hahnemann Med Col, MD, 55. *Prof Exp:* Instr surg, Univ Iowa, 60-61, assoc, 61-62; instr, Univ Ky, 63-64; from asst prof to assoc prof, 64-72, PROF SURG, UNIV IOWA, 72- *Concurrent Pos:* Fel cardiovasc surg, Univ Ky, 63-64; consult, Oakdale State Sanatorium, 64- *Mem:* AAAS; Am Col Cardiol; AMA; Am Col Surg; Am Col Chest Physicians. *Mailing Add:* Dept Surg Univ Iowa Iowa City IA 52242

ROSSI, ROBERT DANIEL, b Philadelphia, Pa, Nov 4, 50; m 77. ORGANIC CHEMISTRY. *Educ:* Philadelphia Col Pharm & Sci, BS, 72; Temple Univ, PhD(chem), 78. *Prof Exp:* Sr chemist, Borg-Warner Chem, 77-80; proj supvr, 80-83, res assoc, 84-88, TECH DIR, NAT STARCH & CHEM CORP, 88- *Mem:* Am Chem Soc; Soc Advan Mat Process Eng; Int Soc Hybrid Microelectronics; Inst Elec & Electronics Engrs. *Res:* Preparation of unique monomers and their respective homo- and copolymers; chemistry of organophosphorus compounds; conducting polymers; high temperature polymers, polymers for electronic applications; polyimide chemistry. *Mailing Add:* Nat Starch & Chem Corp Ten Finderne Ave Bridgewater NJ 08807

ROSSIER, ALAIN B, b Lausanne, Switz, Nov 29, 30; m 58. PHYSICAL MEDICINE, REHABILITATION. *Educ:* Univ Lausanne, BS, 50, Med Sch, Fed Dipl, 57, MD, 58. *Prof Exp:* Asst chief, Univ Hosp Geneva & chief paraplegic ctr, Beau-Sejour Hosp, Univ Hosp Geneva, 64-73; chief, Spinal Cord Injury Serv, Vet Admin Med Ctr & prof spinal cord rehab, Harvard Med Sch, 73-84; chief, Paraplegic Ctr, Univ Orthop Clin Balgrist, Zurich, Switz, 86-89; PROF PARAPLEGIOLOGY, MED UNIV ZURICH, SWITZ, 86- *Concurrent Pos:* Fel, Spinal Cord Injury Serv, Vet Admin Hosp, Long Beach, Calif & Swiss Acad Méd Sci, 62-63; consult neurosurg & orthopaedic surg, Children's Hosp Med Ctr, 73-, Spinal Cord Injury, Braintree Hosp & Mass Rehab Hosp; assoc staff mem, New England Med Ctr Hosp, Boston; consult, Paraplegic Ctr, Univ Orthop Clin Balgrist, Zurich, Switz, 86- *Mem:* Int Med Soc Paraplegia; Am Asn Neurol Surg; Am Urol Asn; Am Cong Rehab Med; Int Continence Soc. *Res:* Spinal cord regeneration, heterotopic bone ossification; urodynamic problems in neurogenic bladder; deep venous thrombosis; spinal fusion; wheelchair bioengineering. *Mailing Add:* 32 Quai Gustave Ador Geneva 1207 Switzerland

ROSSIGNOL, PHILIPPE ALBERT, b Sherbrooke, Que, Jan 15, 50; m 82. MEDICAL ENTOMOLOGY. *Educ:* Univ Ottawa, BSc, 71; Univ Toronto, MSc, 75, PhD(parasitol), 78. *Prof Exp:* Res fel, 78-81, RES ASSOC MED ENTOM, DEPT TROP PUB HEALTH, SCH PUB HEALTH, HARVARD UNIV, 81- *Mem:* AAAS; Am Soc Trop Med & Hyg; Royal Soc Trop Med & Hyg. *Res:* Physiology of salivatim, oogenesis and parasite transmission of mosquitoes. *Mailing Add:* Dept Entom Ore State Univ Corvallis OR 97331

ROSSIN, A DAVID, b Cleveland, Ohio, May 5, 31; m 66; c 2. NUCLEAR ENGINEERING. *Educ:* Cornell Univ, BS, 54; Mass Inst Technol, MS, 55; Northwestern Univ, MBA, 63; Case Western Reserve Univ, PhD(metall), 66. *Prof Exp:* Sr scientist metall, Argonne Nat Lab, 55-72; syst nuclear res engr, Commonwealth Edison Co, 72-78, dir res, 78-81; dir, Nuclear Safety Anal Ctr, Elec Power Res Inst, 81-86; asst sec nuclear energy, US Dept Energy, 86-87; PRES, ROSSIN & ASSOC, LOS ALTOS HILLS, CALIF, 87- *Concurrent Pos:* Fel, Adlai Stevenson Inst, 67-68. *Mem:* Am Nuclear Soc; Am Soc Testing & Mat; AAAS. *Res:* Reliability of nuclear reactor piping and pressure vessels; radiation embrittlement of reactor materials; nuclear reactor safety systems; nuclear waste disposal and spent fuel management; nuclear nonproliferation policy. *Mailing Add:* 24129 Hillview Dr Los Altos Hills CA 94024

ROSSIN, P(ETER) C(HARLES), b New York, NY, Sept 29, 23; m 46; c 2. PHYSICAL METALLURGY. *Educ:* Lehigh Univ, BS, 48; Yale Univ, MS, 50. *Prof Exp:* Res metallurgist, Remington Arms Co, 48-51; res assoc, Gen Elec Co, 51-55; from gen mgr refractomet div to works mgr, Titusville Plant, Universal-Cyclops Steel Corp, 55-64; vpres opers, Fansteel Metall Corp, 64-65; asst vpres prod, Crucible Steel Co Am, 65-67, CHMN, CHIEF EXEC OFFICER & PRES, DYNAMET INC, 67- *Mem:* Am Soc Metals. *Res:* Metallurgical processes associated with reactive and refractory metals. *Mailing Add:* Dynamet Inc 195 Museum Rd Washington PA 15301

ROSSING, THOMAS D(EAN), b Madison, SDak, Mar 27, 29; m 52; c 5. PHYSICS, ACOUSTICS. *Educ:* Luther Col, BA, 50; Iowa State Univ, MS, 52, PhD(physics), 54. *Prof Exp:* Physicist, Univac Div, Sperry Rand Corp, 54-57; prof physics, St Olaf Col, 57-71; PROF PHYSICS, NORTHERN ILL UNIV, 71- *Mem:* Am Phys Soc; Am Asn Physics Teachers; fel Acoust Soc Am; Sigma Xi. *Res:* Magnetic materials and devices; ferromagnetic resonance; surface effects in fusion reactors; ultrasonic dispersion in gases; musical acoustics. *Mailing Add:* Dept Physics Northern Ill Univ DeKalb IL 60115

ROSSINGTON, DAVID RALPH, b London, Eng, July 13, 32; US citizen; m 55; c 4. PHYSICAL CHEMISTRY. *Educ:* Bristol Univ, BSc, 53, PhD(chem), 56. *Prof Exp:* Res fel phys chem, State Univ NY Col Ceramics, Alfred Univ, 56-58; tech officer paints div, Imp Chem Industs, Eng, 58-60; from asst prof to assoc prof, 60-69, head, Div Ceramic Eng & Sci, 76-79 & 82-84, PROF PHYS CHEM, STATE UNIV NY COL CERAMICS, ALFRED UNIV, 69-, DEAN, SCH ENG, 84- *Concurrent Pos:* Fulbright traveling fels, 56 & 58. *Mem:* Am Chem Soc; fel Am Ceramic Soc; Nat Inst Ceramic Engrs; Am Soc Eng Educ. *Res:* Surface chemistry; catalysis; adsorption phenomena; chemistry of cement hydration; nuclear waste encapsulation. *Mailing Add:* Sch Eng Alfred Univ Alfred NY 14802

ROSSINI, FREDERICK ANTHONY, b Wash, DC, Sept 20, 39. TECHNOLOGY & SOCIAL FORECASTING, INTERDISCIPLINARY RESEARCH PROCESSES. *Educ:* Spring Hill Col, BS, 62; Univ Calif, Berkeley, PhD(physics), 68. *Prof Exp:* Actg asst prof physics, Univ Calif, Berkeley, 68, NIMH postdoctoral fel philos, 69-71; Nat Acad Sci res assoc, NASA-Ames Res Ctr, 71-72; from asst prof to prof technol & sci policy, Ga Inst Technol, 72-89, dir, Technol Policy & Assessment Ctr, 81-89, assoc dir, Off Interdisciplinary Progs, 83-85, dir, 85-89; VPROVOST & PROF INFO SYSTS & SYSTS ENG, GEORGE MASON UNIV, 89- *Concurrent Pos:* Assoc ed, Technol Forecastng & Social Change, 80-90; ed-in-chief, Impact Assessment Bull, 81-84; actg dir, Software Eng Res Ctr, Ga Inst Technol, 87-89. *Mem:* Int Asn Impact Assessment (treas, 81-88); Sigma Xi; sr mem Inst Elec & Electronics Engrs. *Mailing Add:* Provost's Off George Mason Univ Fairfax VA 22030-4444

ROSSINI, FREDERICK DOMINIC, physical chemistry; deceased, see previous edition for last biography

ROSSIO, JEFFREY L, b Cleveland, Ohio, May 22, 47; m 72; c 2. IMMUNOLOGY, MICROBIOLOGY. *Educ:* Univ Mich, BS, 69; Ohio State Univ, MS, 71, PhD(microbiol), 73. *Prof Exp:* Instr biochem, Univ Tex Med Br, 76-78; asst prof immunol, Wright State Univ, 78-81; SCIENTIST, NAT CANCER INST, 81- *Mem:* Sigma Xi; Reticuloendothelial Soc; Am Soc Microbiol. *Res:* Thymic hormones; immune regulation; immunotherapy. *Mailing Add:* Biol Resp Mod Prog Bldg 567 Nat Cancer Inst Fcrf PO Box B Frederick MD 21701

ROSSITER, BRYANT WILLIAM, b Ogden, Utah, Mar 10, 31; m 51; c 8. ORGANIC CHEMISTRY. *Educ:* Univ Utah, BA, 54, PhD(org chem), 57. *Prof Exp:* Res assoc, Eastman Kodak Co, 57-69, assoc head, 69-70, dir chem div, Res Labs, 70-84, dir sci & technol develop, Res Labs, 84-86; pres, Viratek Inc, Costa Mesa, Calif, 86-88; Sr Vpres, ICN Pharmaceut, 88-90. *Concurrent Pos:* Mem, US Nat Comt, Int Union Pure & Appl Chem, 73-81, finance comt, 75-79, chmn, 77-80; mem bd trustees, Eastman Dent Ctr, 74-, chmn, 82-85 & Eyring Res Inst, 78-79; chmn, Chemrawn comt, 74-86, Int Conf Chem & World Food Supplies, Manila, Philippines, 82; mem, US Nat Acad Adv Comt, Int Coun Sci Unions, 81-; mem, res adv comt, Agency Int Develop, 82-, chmn, 89-; sr ed, John Wiley & Sons; ed, Phys Methods Chem; mem & fel lectr award, Am Inst Chemists, 88; chmn bd, Nuclear Acid Res Inst, 87-89. *Honors & Awards:* Will Judy Lect Award, Juniata Col, 78. *Mem:* Am Chem Soc; Sigma Xi; fel Am Inst Chem; NY Acad Sci. *Res:* Chemistry of photographic processes; mechanisms of organic reactions; research management; photochemical processes, chemical instrumentation; drug research and development. *Mailing Add:* 25662 Dillon Rd Laguna Hills CA 92653

ROSSITTO, CONRAD, b Siracusa, Italy, Sept 4, 26; m 53; c 3. POLYMER & ORGANIC CHEMISTRY. *Educ:* Univ Palermo, Dr(chem), 51. *Prof Exp:* Res chemist, BB Chem Co, 54-59; sr res chemist, USM Chem Co, 60-68, mgr polymer synthesis group, 68-70, sr tech adv, 70-73, sr res chemist, USM Corp, 73-78, group leader, Bostic Div, 78-85, lab mgr, 85-88, DIR TECH DEVELOP, EMHART CORP, 88- *Mem:* Am Chem Soc; Adhesion Soc; Soc Mfg Engrs; Tech Asn Pulp & Paper Industr. *Res:* Synthesis of polymers to be used in adhesives and coatings. *Mailing Add:* 363 N Main St Andover MA 01810-2610

ROSSKY, PETER JACOB, b Philadelphia, Pa, Apr 15, 50; div; c 1. LIQUID STATE THEORY. *Educ:* Cornell Univ, BA, 71; Harvard Univ, MA, 72, PhD(chem physics), 78. *Prof Exp:* Fel, State Univ NY, Stony Brook, 77-79; from asst prof to prof chem, 79-90, WATT CENTENNIAL PROF, UNIV TEX, AUSTIN, 90- *Concurrent Pos:* Alfred P Sloan Found fel, 82; Res Career Develop Award, NIH, 83; Camille & Henry Dreyfus Teacher-Scholar grant, 84. *Honors & Awards:* Presidential Young Investr Award, NSF, 84. *Mem:* Am Chem Soc; Am Phys Soc. *Res:* Theoretical chemistry; statistical mechanics of solution structure and dynamics; aqueous systems; solvent effects in biochemical systems; quantum effects in liquids. *Mailing Add:* Dept Chem Univ Tex Austin TX 78712-1167

ROSSMAN, AMY YARNELL, b Spokane, Wash, Sept 20, 46; m 88; c 1. MYCOLOGY, SYSTEMATICS. *Educ:* Grinnell Col, BA, 68; Ore State Univ, PhD(bot), 75. *Prof Exp:* Teaching fel mycol, Cornell Univ, 78-79; res assoc bot, New York Bot Garden, Bronx, NY, 79-80; mycologist, Animal & Plant Health Inspection Serv, 80-83; MYCOLOGIST, AGR RES SERV, USDA, 83- *Mem:* AAAS; Mycol Soc Am (treas, 83-86); Brit Mycol Soc; Am Phytopath Soc. *Res:* Systematics of Ascomycetes and other micro fungi, with emphasis on plant pathogenic and tropical species. *Mailing Add:* B011A Rm 304 USDA/Agr Res Sta Beltsville MD 20705

ROSSMAN, DOUGLAS ATHON, b Waukesha, Wis, July 4, 36; m 57, 90; c 2. SYSTEMATIC HERPETOLOGY. *Educ:* Southern Ill Univ, BA, 58; Univ Fla, PhD(zool), 61. *Prof Exp:* Instr zool, Univ NC, 61-63; from asst prof to assoc prof zool, 63-67, assoc prof zool & physiol & assoc cur, Mus Zool, 67-76, ADJ PROF ZOOL & PHYSIOL, LA STATE UNIV, 76-, CUR REPTILES, MUS NATURAL SCI, 76- *Concurrent Pos:* Res grants, Am Philos Soc, 63, NSF, 65-67; dir, Mus Nat Sci, 84-86. *Mem:* Herpetologists League; Am Soc Ichthyol & Herpet; Soc Study Amphibians & Reptiles. *Res:* Taxonomy and evolution of colubrid snake subfamily Natricinae (garter snakes); taxonomy of other colubrid snakes, especially Neotropical; snake osteology. *Mailing Add:* Mus Natural Sci La State Univ Baton Rouge LA 70803

ROSSMAN, ELMER CHRIS, plant breeding; deceased, see previous edition for last biography

ROSSMAN, GEORGE ROBERT, b LaCrosse, Wis, Aug 3, 44; m. MINERALOGY, INORGANIC CHEMISTRY. *Educ:* Wis State Univ, BS, 66; Calif Inst Technol, PhD(chem), 71. *Prof Exp:* From instr to assoc prof, 71-83, PROF MINERAL, CALIF INST TECHNOL, 83- *Mem:* Mineral Soc Am. *Res:* Physical and chemical properties of minerals and related synthetic materials; color; spectroscopy; radiation effects; role of trace water. *Mailing Add:* 170-25 Calif Inst Technol Pasadena CA 91125

ROSSMAN, ISADORE, b Elizabeth, NJ, Mar 29, 13; m 43; c 1. GERIATRICS. *Educ:* Univ Wis, BA, 33; Univ Chicago, PhD(anat), 37, MD, 42. *Prof Exp:* From assoc prof to prof, 70-84, EMER PROF COMMUNITY HEALTH, ALBERT EINSTEIN COL MED, 84-; MED DIR DEPT HOME CARE & EXTENDED SERV, MONTEFIORE HOSP, BRONX, NY, 54- *Concurrent Pos:* Ed, Clin Geriat, 71, 79 & 86. *Honors & Awards:* Coggins Award, 79; Jos Freeman Award, Geront Soc Am, 81, Leavitt Award, 87. *Res:* Mortality studies; patterns of medical care. *Mailing Add:* Montefiore Hosp & Med Ctr 111 E 210th St Bronx NY 10467

ROSSMAN, TOBY GALE, b Weehawken, NJ, June 3, 42; m 90. GENETIC TOXICOLOGY. *Educ:* Washington Sq Col, AB, 64; NY Univ, PhD(microbiol), 68. *Prof Exp:* Assoc res scientist, 71-73, from asst prof to assoc prof, 73-85, PROF ENVIRON MED, SCH MED, NY UNIV, 85- *Concurrent Pos:* NIH trainee, Med Ctr, NY Univ, 69-71. *Mem:* AAAS; Am Soc Microbiol; Asn Women Sci; Environ Mutagen Soc. *Res:* DNA repair and mutagenesis; genetic toxicology; environmental carcinogenesis. *Mailing Add:* Dept Environ Med NY Univ Med Ctr New York NY 10016

ROSSMANN, MICHAEL G, b Frankfurt, Germany, July 30, 30; m 54; c 3. CRYSTALLOGRAPHY. *Educ:* Univ London, BSc, 50, hons, 51, MSc, 53; Univ Glasgow, PhD(chem), 56. *Hon Degrees:* PhD, Univ Uppsala, 83, Univ Strasbourg, 84, Vrije Univ Brussel, 90. *Prof Exp:* Asst lectr physics, Univ Strathclyde, 52-56; Fulbright traveling fel chem, Univ Minn, Minneapolis, 56-58; res worker, MRC Lab Molecular Biol, Eng, 58-64; from assoc prof to prof biol, 64-67, prof biol sci, 67-78, HANLEY DISTINGUISHED PROF BIOL SCI, PURDUE UNIV, 78-, PROF BIOCHEM, 75- *Honors & Awards:* Keilin Lectr, Biochem, Brit Biochem Soc, 83; Fankuchen Award, Am Crystallog As, 86; Gairdner Found Int Award, 87; Merck Award, Am Soc Biochem & Molecular Biol, 89; Computerworld Smithsonian Award in Med, 90; Louisa Gross Horwitz Prize, Columbia Univ, 90. *Mem:* Nat Acad Sci; Am Chem Soc; Am Soc Biochem & Molecular Biol; Biophys Soc; Brit Inst Physics; Am Crystallog Asn; fel Am Acad Arts & Sci. *Res:* Determination of the structure of biological macromolecules, particularly proteins and viruses, by means of x-ray crystallography. *Mailing Add:* Dept Biol Sci Purdue Univ West Lafayette IN 47907

ROSSMILLER, JOHN DAVID, b Elkhorn, Wis, Mar 12, 35; m 57. BIOCHEMISTRY. *Educ:* Univ Wis, BS, 56, MS, 62; PhD(biochem), 65. *Prof Exp:* chmn dept, Wright State Univ, 80-88, asst prof biol sci, 65-71, interim dean, Col Sci & Eng, 84-86, ASSOC PROF BIOL SCI, WRIGHT STATE UNIV, 71- *Mem:* AAAS. *Res:* Biochemical physiological mechanisms of regulation. *Mailing Add:* Dept Biol Sci Wright State Univ Dayton OH 45431

ROSSMOORE, HAROLD W, b New York, NY, June 15, 25; m 46; c 4. BACTERIOLOGY. *Educ:* Univ Mich, BS, 49, MS, 51, PhD(bact), 55. *Prof Exp:* From instr to assoc prof, 54-68, PROF BIOL, WAYNE STATE UNIV, 68- *Concurrent Pos:* Fulbright & NSF fels, 64-65; consult. *Mem:* AAAS; Soc Indust Microbiol; Am Soc Microbiol; Soc Invert Path; fel Royal Soc Health. *Res:* Environmental microbiology, especially detection and control of deterioration water and oil systems; mechanisms of infection and resistance, especially in insects. *Mailing Add:* Dept Biol 210 Sci Hall Wayne State Univ 425 Life Sci 5950 Cass Ave Detroit MI 48202

ROSSNAGEL, BRIAN GORDON, b Gladstone, Man, May 19, 52. PLANT BREEDING, AGRONOMY. *Educ:* Univ Man, BSAgr, 73, PhD(plant breeding, agron), 78. *Prof Exp:* FEED GRAIN BREEDER BARLEY & OAT, CROP DEVELOP CTR, UNIV SASK, 77- *Mem:* Agr Inst Can; Can Soc Agron; Am Soc Agron; Crop Sci Soc Am. *Res:* Breeding and development of genotypes and agronomic practices for the optimum production of barley and oat in Saskatchewan; specializing in the development of hulless barley culture. *Mailing Add:* Crop Develop Ctr Univ Sask Saskatoon SK S7N 0W0 Can

ROSSNER, LAWRENCE FRANKLIN, b St Louis, Mo, Dec 17, 38; m 60; c 3. ASTROPHYSICS. *Educ:* Univ Chicago, SB, 60, SM, 61, PhD(astrophys & astron), 66. *Prof Exp:* Res assoc astron, Columbia Univ, 66-68; asst prof physics, Brown Univ, 68-75; DIR COMPUT CTR, EPA ENVIRON RES LAB, NARRAGANSETT, RI, 75- *Mem:* Am Astron Soc. *Res:* Application of fluid dynamics to problems of astrophysical interest, particularly galactic structure. *Mailing Add:* Environ Protection Agency Environ Res Lab 27 Tarzwell Dr Narragansett RI 02882

ROSSO, PEDRO, b Genoa, Italy, Aug 27, 41; Chilean citizen; m 67; c 3. PERINATAL BIOLOGY. *Educ:* Cath Univ Chile, BS, 62; Univ Chile, MD, 66. *Prof Exp:* Intern & resident pediat, Univ Chile, 66-69; fel growth & develop, Dept Pediat, Col Med, Cornell Univ, 70-72; adj asst prof nutrit, 72-73, ASST PROF PEDIAT, INST HUMAN NUTRIT, COLUMBIA UNIV, 73- *Concurrent Pos:* Consult, Regional Grad Appl Nutrit Course, Seameo Proj, Djakarta, Indonesia, 72 & Nutrit Prog, Univ PR, 75; NIH career develop award, 75. *Mem:* Soc Pediat Res; Am Inst Nutrit; Am Soc Clin Nutrit; Harvey Soc. *Res:* Control of prenatal growth and the influence of maternal nutrition on fetal growth and development. *Mailing Add:* Dept Pediat Sch Med Cath Univ Casilla 114-D Santiago Chile

ROSSOL, FREDERICK CARL, b New York, NY, Feb 6, 33; m 54; c 1. MAGNETISM. *Educ:* Univ Calif, Berkeley, BSEE, 59; NY Univ, MEE, 61; Harvard Univ, MA, 63, PhD(appl physics), 66. *Prof Exp:* Mem tech staff, Bell Labs, Inc, 59-90; RETIRED. *Mem:* Inst Elec & Electronics Engrs; Am Phys Soc; Sigma Xi. *Res:* Magnetic properties of materials. *Mailing Add:* 4106 Spring Brook Dr Edison NJ 08820-4229

ROSSOMANDO, EDWARD FREDERICK, b New York, NY, Feb 26, 39; m 65; c 2. ENZYMOLOGY, DIAGNOSTICS. *Educ:* Univ Pa Sch Dental Med, DDS, 64; Rockefeller Univ, PhD(biomed sci), 69. *Prof Exp:* Res assoc, Nat Inst Dental Res, NIH, Bethesda, Md, 68-70; special fel, Brandeis Univ, Waltham, Mass, 70-72; from asst prof to assoc prof, 72-83, PROF BIOCHEM, DEPT BIOSTRUCT & FUNCTION, UNIV CONN HEALTH CTR, 83-; DIR, INST ORAL BIOL, 83- *Concurrent Pos:* Vis scientist, Dept Biochem & Biophysics, Univ Pa, 78-79, Dept Med Biochem, Rockefeller Univ, 86, Lab Biochem, Nat Inst Med Res, London, Eng, 86. *Mem:* Am Soc Biochem & Molecular Biol; Am Chem Soc; Am Soc Cell Biol; Am Soc Dental Res. *Res:* Biochemical changes in body fluids associated with diseases; development of immunological methods; studies on cytokines; application to periodental diseases; enzymatic control of morphogenesis; development of high performance liquid chromatography methods for enzymatic studies; analysis of semerotic units in biomolecules; application to dictyostelium discoidem. *Mailing Add:* Dept Biostruct & Function Univ Conn Health Ctr Farmington CT 06030

ROSSON, H(AROLD) F(RANK), b San Antonio, Tex, Apr 4, 29; m 51; c 3. CHEMICAL ENGINEERING. *Educ:* Rice Inst Technol, BS, 49, PhD, 58. *Prof Exp:* From asst prof to assoc prof chem eng, 57-66, chmn dept chem & petrol eng, 64-70, PROF CHEM ENG, UNIV KANS, 66- *Mem:* Am Chem Soc; Am Inst Chem Engrs. *Res:* Rate processes. *Mailing Add:* Dept Chem & Petrol Eng Univ Kans Lawrence KS 66045

ROSSON, REINHARDT ARTHUR, b Santa Monica, Calif, Oct 5, 49; m 77. BIOGEOCHEMISTRY, MARINE MICROBIAL ECOLOGY. *Educ:* Univ Calif, Los Angeles, AB, 71, PhD(microbiol), 78. *Prof Exp:* Res biologist, Scripps Inst Oceanog, 78-81; asst prof marine microbiol, Dept Marine Studies, Marine Sci Inst, Univ Tex, Austin, 81-; AT BIO-TECH RESOURCES. *Mem:* AAAS; Am Geophys Union; Am Soc Microbiol. *Res:* Laboratory studies on the physiology (biochemistry, mechanisms, and regulation), of bacterial manganese oxidation and combined area field studies in local bays, lagoons, nearshore Gulf of Mexico, and the deep sea on tracemetal cycling catalyzed by bacteria. *Mailing Add:* Bio-Tech Resources 1035 S Seventh St Manitowoc WI 54220

ROSSOW, PETER WILLIAM, b Los Angeles, Calif, Dec 10, 48. CELL CYCLE REGULATION, HORMONE ACTION. *Educ:* Mass Inst Technol, SB, 71; Harvard Univ, PhD(microbiol & molecular genetics), 76. *Prof Exp:* Fel pharmacol, Med Sch, Harvard Univ, 75-79; res assoc, Sidney Farber Cancer Inst, 79-80; assoc staff scientist, Jackson Lab, 80-84; res assoc prof, Univ Paris, France, 84-86; staff scientist, Inst Med Res, San Jose, Calif, 86-87; PRES, BOFFIN RESOURCES, 88- *Concurrent Pos:* Coop prof zool, Univ Maine, Orono, 81- *Mem:* Am Soc Cell Biol; AAAS; NY Acad Sci. *Res:* Hormonal regulation of normal cell growth and the earliest events in the acquisition of the malignant phenotype in model cell culture systems. *Mailing Add:* 224 Hollister Ave Rutherford NJ 07070

ROSSOW, VERNON J, b Danbury, Iowa, July 22, 26; m 48; c 4. FLUID DYNAMICS. *Educ:* Iowa State Col, BS, 47; Univ Mich, MS, 49; Swiss Fed Inst Technol, DrTechSci, 57. *Prof Exp:* Res scientist, 49-70, STAFF SCIENTIST, AMES RES CTR, NASA, MOFFETT FIELD, 70- *Concurrent Pos:* Mem res adv comt fluid mech, NASA, 63-69. *Honors & Awards:* NRC Fel, 51-52; Super Performance Award, NASA, 67. *Mem:* Am Helicopter Soc; Am Geophys Union; assoc fel Am Inst Aeronaut & Astronaut; Am Meteorol Soc. *Res:* Aerodynamics; author of 60 research papers in fluid mechanics, one patent. *Mailing Add:* 549 Arboleda Dr Los Altos CA 94024

ROST, ERNEST STEPHAN, b Breslau, Ger, June 3, 34; US citizen; m 63; c 2. NUCLEAR PHYSICS. *Educ:* Princeton Univ, AB, 56; Univ Pittsburgh, PhD(physics), 61. *Prof Exp:* From instr to asst prof physics, Princeton Univ, 61-66; assoc prof, 66-70, PROF PHYSICS, UNIV COLO, BOULDER, 70- *Concurrent Pos:* Vis staff mem, Los Alamos Sci Lab, 67- *Mem:* Am Phys Soc. *Res:* Theoretical nuclear physics; nuclear structure and reactions. *Mailing Add:* Dept Physics Dp/6t F1017 Univ Colo Box 390 Boulder CO 80309

ROST, THOMAS LOWELL, b St Paul, Minn, Dec 28, 41; m 63; c 3. PLANT PHYSIOLOGY, PLANT CYTOLOGY. *Educ:* St John's Univ, BS, 63; Mankato State Univ, MA, 65; Iowa State Univ, PhD(bot), 71. *Prof Exp:* From asst prof to prof, 72-88, fac asst chancellor, 81-82, PROF BOT, UNIV CALIF, DAVIS, 83- *Concurrent Pos:* Vis fel, Australia Nat Univ, Canberra, 79-80; co ed, Mech & Control Cell Div, 77; consult, Food & Agr Org, Univ Uruguay, 79; vis prof, Univ Wroclaw, Wroclaw, Poland, 88, Fac Sci, Univ Uruguay, 89. *Mem:* Sigma Xi; Bot Soc Am; Am Inst Biol Sci; Soc Exp Biol; Soc Develop Biol. *Res:* Tissue culture studies; the cell cycle in root meristems and the effects of stress factors on structure and cell cycle progression; in site localization of RNA's involved in differentiation events in root meristems. *Mailing Add:* Dept Bot Univ Calif Davis CA 95616-8537

ROST, WILLIAM JOSEPH, b Fargo, NDak, Dec 8, 26; m 51; c 3. ORGANIC CHEMISTRY, PHARMACOLOGY. *Educ:* Univ Minn, BS, 48, PhD(pharmaceut chem), 52. *Prof Exp:* From asst prof to prof pharmaceut chem, Univ Mo-Kansas City, 52-89; RETIRED. *Mem:* Am Chem Soc; Am Pharmaceut Asn. *Res:* Synthesis of chemicals that have possible medicinal activity. *Mailing Add:* 709 W 115th Terr 5100 Rockhill Rd Kansas City MO 64114

ROSTAMIAN, ROUBEN, b Tehran, Iran, Oct 27, 49. APPLIED MATHEMATICS. *Educ:* Arya-Mehr Univ, Tehran, BS, 72; Brown Univ, PhD(appl math), 78. *Prof Exp:* Asst prof, Purdue Univ, 77-81; asst prof math, Pa State Univ, 81-83, assoc prof, 83-85; assoc prof, 85-87, PROF MATH, UNIV MD, 87- *Res:* Qualitative study of solutions of degenerate parabolic equations; asymptotic behavior of solutions in large time; linear and nonlinear elasticity, existence of solutions for problems with internal constraints. *Mailing Add:* Dept Math Univ MD Baltimore MD 21228

ROSTENBACH, ROYAL E(DWIN), b Buffalo, Iowa, Sept 20, 12; m 40. CHEMISTRY, ENGINEERING. *Educ:* St Ambrose Col, BS, 35; Univ Iowa, MS, 37, PhD(chem, eng), 39. *Prof Exp:* Asst chem, Catholic Univ, 35-36; sanit eng, Univ Iowa, 37-39; consult engr, Clark, Stewart & Wood Co, 39-40; chem & pub health engr, USPHS, 40-42; engr, Chem Div, US War Prod Bd, 42-43; chief, Copolymer Br, Off Rubber Reserve, Reconstruct Finance Corp, 43-53; sr engr & specialist, Hanford Atomic Prod Oper, Gen Elec Co, 53-59; sr proj engr, Bendix Aviation Corp, 59-60; res dir, Mast Develop Co, 60-61; prog dir, Div Eng, NSF, 62-88; CONSULT, 89- *Mem:* Am Chem Soc; fel Am Inst Chem; Am Inst Chem Engrs; Water Pollution Control Fedn; Am Nuclear Soc; Health Physics Soc. *Res:* Chemical, environmental, energy and nuclear engineering. *Mailing Add:* 6111 Wiscasset Rd Bethesda MD 20816-2119

ROSTOKER, GORDON, b Toronto, Ont, July 15, 40; m 66; c 3. SPACE PHYSICS. *Educ:* Univ Toronto, BSc, 62, MA, 63; Univ BC, PhD(geophys), 66. *Prof Exp:* Nat Res Coun Can fel space physics, Royal Inst Technol, Sweden, 66-68; from asst prof physics to assoc prof, 68-79, PROF PHYSICS, UNIV ALTA, 79-, DIR, INST EARTH & PLANETARY PHYSICS, 85- *Concurrent Pos:* Int Union Geod & Geophys appointee, Steering Comt, Int Magnetospheric Study, 73-79; mem grant selection comt space & astron, Nat Res Coun Can, 73-76, chmn, 75-76, mem assoc comt space res, 75-80; Nat Res Coun Can travel fel, 74; ed, Can J Phys, 80-86; McCalla prof, Univ Alberta, 83-84; chmn steering comt, Scostep, Solar-Terr Energy Prog, 88-; mem, Nat Sci Eng Res Coun Can Phys & Astron Comt, 88- & Grant Selection Comt Sci Publ, 88- *Honors & Awards:* E W R Steacie Prize, 79. *Mem:* Am Geophys Union; Can Asn Physicist. *Res:* Study of the solar-terrestrial interaction, with emphasis on the investigation of magnetospheric substorms and magnetosphere-ionosphere coupling using magnetometer arrays and satellite data. *Mailing Add:* Inst Earth & Planetary Physics Univ Alta Edmonton AB T6G 2J1 Can

ROSTOKER, NORMAN, b Toronto, Ont, Aug 16, 25; m 48; c 2. PHYSICS. *Educ:* Univ Toronto, BASc, 46, MA, 47; Carnegie Inst Technol, DSc(physics of solids), 50. *Prof Exp:* Res physicist, Carnegie Inst Technol, 48-53, Armour Res Found, 53-56 & Gen Atomic Div, Gen Dynamics Corp, 56-62; prof physics, Univ Calif, San Diego, 62-65; res physicist, Gen Atomic Div, Gen Dynamics Corp, 65-67; IBM prof eng, Cornell Univ, 67-73, chmn Dept Appl Physics, 67-70; PROF PHYSICS, UNIV CALIF, IRVINE, 73- *Mem:* Fel Am Phys Soc. *Res:* Band theory; design of fission reactors; plasma physics; controlled thermonuclear research; intense electron beams. *Mailing Add:* Dept Physics Univ Calif Irvine CA 92717

ROSTOKER, WILLIAM, b Hamilton, Ont, June 21, 24; nat US; m 49; c 4. METALLURGY, BIOMATERIALS. *Educ:* Univ Toronto, BASc, 45, MASc, 46; Lehigh Univ, PhD(metall eng), 48. *Prof Exp:* Res fel indust metall, Univ Birmingham, 48-49, lectr, 49-50; asst prof, Ill Inst Technol, 50-51, from res metallurgist to sr metall adv, Metals Div, IIT Res Inst, 51-65; actg dean grad col, 68-70, PROF METALL & BIOENG, UNIV ILL, CHICAGO CIRCLE, 65- *Concurrent Pos:* Consult, Frankford Arsenal, 58-73, mem refractory metals panel, Mat Adv Bd, 59-63, chmn metal-working processes comt, 63-69; chmn rev comts eng metall, Idaho Facility, Argonne Nat Lab, 65-68; mem nat mat adv bd, Nat Acad Sci, 69-71; mem & consult, Army Sci Adv Panel, 71-76; prof orthopedic surg, Rush Med Sch, Chicago, 72-; res assoc, Field Mus, 85- *Honors & Awards:* Kappa Delta Award, Am Acad Orthop Surg, 70; Pub Serv Group Award, NASA, 81. *Mem:* Fel Am Soc Metals. *Res:* Fracture processes; powder metallurgy; biomedical engineering; brittle fracture; refractory metals; physical metallurgy; metal processing; skeletal reconstruction; alloy development; ancient metallurgical processes. *Mailing Add:* Dept Biol-Eng Univ Ill Chicago IL 60680

ROSWELL, DAVID FREDERICK, b Evansville, Ind, Dec 5, 42; m 65; c 2. CHEMISTRY. *Educ:* Johns Hopkins Univ, AB, 64, PhD(chem), 68; Loyola Col, MBA, 85. *Prof Exp:* From asst prof to assoc prof, 68-74, PROF CHEM, LOYOLA COL, MD, 74-, DEAN, COL ARTS & SCI, 80- *Concurrent Pos:* Cottrell Res Corp grant, Loyola Col, 69-70; res scientist, Johns Hopkins Univ, 71- *Honors & Awards:* Md Chemist Award, 86. *Mem:* Am Chem Soc; Am Inst Chem. *Res:* Chemiluminescence; photochemistry; biochemistry. *Mailing Add:* Dean Arts & Sci Col Arts & Sci Loyola Col Baltimore MD 21210

ROSZEL, JEFFIE FISHER, b Amarilllo, Tex, Apr 5, 26; m 49; c 1. CYTOPATHOLOGY, VETERINARY PATHOLOGY. *Educ:* Univ Pa, VMD, 63; Okla State Univ, PhD(comp path), 75. *Prof Exp:* NIH fel, Hahnemann Med Col & Hosp, 63-65; from instr to asst prof path, Sch Vet Med, Univ Pa, 65-70; vis prof, 71-72, from asst prof to assoc prof, 72-78, PROF PATH, COL VET MED, OKLA STATE UNIV, 78- *Concurrent Pos:* Res assoc, Hahnemann Med Col & Hosp, 66-68; lectr path, Philadelphia Col Pharm & Sci, 68-70; mem vet med rev comt, Bur Health Manpower Educ, NIH, 72 & co-prin investr, Registry Canine & Feline Neoplasms, 72-78. *Mem:* Am Soc Cytol; Vet Cancer Soc. *Res:* Comparative cytopathology of neoplasms. *Mailing Add:* Dept Vet Path Okla State Univ Stillwater OK 74078

ROSZKOWSKI, ADOLPH PETER, b Chicago, Ill, July 27, 28; m 51; c 4. PHARMACOLOGY. *Educ:* DePaul Univ, BS, 51, MS, 54; Loyola Univ, PhD(pharmacol), 56. *Prof Exp:* Mem staff, Dept Pharmacol, G D Searle & Co, 52-53; asst dir biol res, McNeil Labs, Inc, Div Johnson & Johnson, 57-66; head pharmacol, Kendall Res Ctr, Ill, 66-67; dir dept pharmacol & asst dir inst clin med, 67-77, vpres, 81-87, DIR, INST EXP PHARMACOL SYNTEX CORP, 77- *Concurrent Pos:* McNeil fel, Univ Pa, 56-57; instr, Sch Med, Temple Univ, 57-64; lectr, Sch Med, Stanford Univ, 75-79. *Honors & Awards:* Johnson Medal Award, 65. *Mem:* AAAS; Am Soc Pharmacol & Exp Therapeut; Am Soc Clin Pharmacol & Therapeut; Soc Toxicol. *Res:* Psychopharmacology; autonomic pharmacology; gastrointestinal pharmacology. *Mailing Add:* 15060 Sobey Rd Saratoga CA 95070

ROSZMAN, LARRY JOE, b Marion, Ohio, June 9, 44; m 66. ATOMIC PHYSICS, PLASMA PHYSICS. *Educ:* Bowling Green State Univ, BS, 66; Univ Fla, PhD(physics), 71. *Prof Exp:* Teaching asst physics, Univ Fla, 66-70, res asst physics, 70-71; PHYSICIST, NAT BUR STANDARDS, 71- *Mem:* Am Phys Soc; AAAS. *Res:* Calculation of electron-atom/ion collision cross sections and rates for important processes in thermonuclear plasmas; calculation of plasma line broadening effects; analysis of plasma-ion interaction; theoretical physics; thermal physics. *Mailing Add:* Nat Bur Standards Bldg 221 Rm A267 Gaithersburg MD 20899

ROTA, GIAN-CARLO, b Italy, Apr 27, 32; nat US; div. MATHEMATICS. *Educ:* Princeton Univ, BA, 53; Yale Univ, MA, 54, PhD(math), 56. *Hon Degrees:* DSc, Univ Strasbourg, France, 84; Univ L'Aquild, Italy, 90. *Prof Exp:* Fel, Courant Inst Math Sci, NY Univ, 56-57; Benjamin Pierce instr math, Harvard Univ, 57-59; from asst prof to assoc prof, Mass Inst Technol, 59-65; prof, Rockefeller Univ, 65-67; prof math, 67-74, PROF APPL MATH & PHILOS, MASS INST TECHNOL, 74- *Concurrent Pos:* Sloan fel, 62-64; mem comt math adv, Off Naval Res, 63-67; Hedrick lectr, Am Math Asn, 67; ed-in-chief, Advan in Math, 68-; vis prof, Univ Colo, 69-80; Andre Aisenstadt vis prof, Univ Montreal, 71; Taft lectr, Univ Cincinnati, 71; fel, Los Alamos Sci Lab, 71-; Hardy lectr, London Math Soc, 73; consult, Los Alamos Nat Lab, 63-71, Rand Corp, 64-71; ed, Studies Appl Math, 70-, J Math Anal & Applns; chmn, Math Sect, AAAS, 88. *Honors & Awards:* Steele Prize, Am Math Soc, 89. *Mem:* Nat Acad Sci; Am Math Soc; fel Academia Argentina de Ciencias; Soc Indust & Appl Math (vpres, 75); fel Inst Math Statist; fel AAAS; fel Am Acad Arts & Sci. *Res:* Combinatorial theory; probability; phenomenology. *Mailing Add:* 1105 Massachusetts Ave Apt 8F Cambridge MA 02138-5217

ROTAR, PETER P, b Omaha, Nebr, June 9, 29; m 56; c 4. CROP BREEDING, CYTOGENETICS. *Educ:* Washington State Univ, BSc, 54, MSc, 57; Univ Nebr, PhD(agron), 60. *Prof Exp:* Asst prof biol, Salve Regina Col, 60-62; asst prof agron & asst agronomist, 62-72, chmm, Dept Agron & Soil Sci, 76-85, PROF AGRON & AGRONOMIST, UNIV HAWAII, 72- *Concurrent Pos:* Vis prof, Univ Fla, Gainesville, 83; consult & agr expert, UN, 84 & 85. *Mem:* Crop Sci Soc Am; Am Soc Agron; Bot Soc Am; Sigma Xi. *Res:* Cytology; cytogenetics of tropical legumes and grasses used for forage and pasture; taxonomy of grasses; crop management. *Mailing Add:* Dept Agron & Soil Sci Univ Hawaii Honolulu HI 96822

ROTARIU, GEORGE JULIAN, b Los Angeles, Calif, Aug 24, 17; m 48; c 3. PHYSICAL CHEMISTY, INSTRUMENTATION. *Educ:* Univ Chicago, BS, 39, MS, 40; Univ Ill, PhD(chem), 50. *Prof Exp:* Asst chem & pharmacol, Univ Chicago, 41-45; res chemist, Lever Bros Co, 45-46; instr phys sci, Univ Chicago, 46-48; asst phys chem, Univ Ill, 48-50; res assoc, Univ Calif, 50-52; asst prof phys chem, Loyola Univ, Ill, 52-55; dir inland testing labs & nuclear eng, Cook Elec Co, 55-57; dir, Nuclear Tech & Phys Chem, Booz-Allen Appl Res Inc, Ill, 57-62; chief, Anal & Appl Br, Div Isotopes Develop, US Atomic Energy Comn, Off Environ, Safety & Health, Dept Energy, 62-64, Chief Systs Eng Sect, Radiation & Thermal Appl Br, 64-66, prog mgr, Process Radiation Staff, 66-72, Radiation Applications Br, Div Applications Technol, 72-73, prog mgr instrumentation, Div Biomed & Environ Res, US Energy Res & Develop Admin, 73-76, environ progs, Div Technol Assessments, 76-82, environ protection scientist, Off Nuclear Safety, Off Environ Safety & Health, Dept Energy, 82-84; RETIRED. *Concurrent Pos:* US Rep Int Atomic Energy Agency Conf, Warsaw, Poland & panel, Cracow, Poland, 65 & Munich, 69; Comn Europ Communities Conf, Brussels, Belg, 71. *Mem:* Am Chem Soc; Am Nuclear Soc; Sigma Xi. *Res:* Nuclear radiation testing; nuclear radiation effects on materials; high altitude environments; environmental testing; lubricants; films and coatings; radioisotope applications in industry and; environmental assessments for magnetohydrodynamics, enhanced gas recovery, enhanced oil recovery and oil shale technologies; health physics appraisal of Department of Energy laboratories, and regulatory aspects; radiobioassay criteria; evaluation of nuclear public information/safety activities of Department of Energy facilities and of selected nuclear industry. *Mailing Add:* 4609 Woodfield Rd Bethesda MD 20814

ROTEM, CHAVA EVE, b Jan 15, 28; wid; c 2. CARDIOLOGY, EXPERIMENTAL MEDICINE. *Educ:* Univ Lausanne, cert bact, 50, MD, 52; MRCP(Edin), 58, MRCP(L), 58, FRCP(C), 68. *Prof Exp:* House officer pediat, Kantonsspital, Zurich, Switz, 52-53; sr house surgeon, Leicester Chest Unit, UK, 54-55, med registr, Leicester Isolation Hosp & Chest Unit, 55-59, sr med registr cardiol, Cardiac Invest Ctr, 59-60; consult physician, Scottish Mission Hosp, Nazareth, 60-61; cardiologist, NIH Surv Ischeamic & Hypertensive Heart Dis, Rambam Govt Hosp, Haifa, Israel, 62-63; res fel cardiol, Stanford Univ, 63-64; cardiologist, St Paul's Hosp, Vancouver, BC, 64-65; cardiologist, Shaughnessy Hosp, Vancouver, BC, 65-68, head, Div Cardiol, 78-82, In Chg Intensive Coronary Care Univ & Cardiac Catheterization Lab, 68-89. *Concurrent Pos:* Physician, Rothschild Hadassa Munic Hosp, Haifa, 62-63. *Mem:* Can Cardiovasc Soc; Soc Microcirc; Israel Soc Cardiol; NY Acad Sci; fel Am Col Cardiol; NAm Soc Pacing & Electrophysiol. *Res:* Investigative cardiology; hydrodynamics of the great blood vessels, especially clinical applications; clinical investigation of new cardiac drugs. *Mailing Add:* Dept Cardiol Shaughnessy Site Univ Hosp 4500 Oak St Vancouver BC V6H 3N1 Can

ROTENBERG, A DANIEL, b Toronto, Ont, July 21, 34; m 62; c 4. MEDICAL PHYSICS. *Educ:* Univ Toronto, PhD(med biophys), 62. *Prof Exp:* Biophysicist, Montreal Gen Hosp, 63-66, physicist, 66-69; lectr therapeut radiol, Sch Med, McGill Univ, 69-71; head dept biomed physics & physicist, Jewish Gen Hosp, 69-77; asst prof diag radiol, Sch Med, McGill Univ, 70-77; dir res & develop, Coinamatic Inc, Montreal, 77-89; VPRES TECHNOL & DEVELOP, TEAMWORK ENTERPRISES, 91- *Concurrent Pos:* Consult, Jewish Gen Hosp, 66-69, Queen Elizabeth Hosp, 68-, St Mary's Hosp, 68-, Reedy Mem Hosp, 69- & Etobicoke Gen Hosp, 72-; assoc scientist, Royal Victoria Hosp, 70- & Northwestern Gen Hosp, 74-; vpres, Beique, Rotenberg & Radford Inc, 66-80; dir, BRRCM Inc, 80- *Mem:* Soc Nuclear Med; Can Asn Physicists (secy-treas, 66-69); Asn Advan Med Instrumentation. *Res:* Hospital physics and instrumentation; microprocessor developments; laser photocoagulation; diagnostic radiology. *Mailing Add:* 54 Elmridge Dr Toronto ON M6B 1A4 Can

ROTENBERG, DON HARRIS, b Portland, Ore, Mar 31, 34; m 58; c 2. POLYMER CHEMISTRY, RESEARCH ADMINISTRATION. *Educ:* Univ Ore, BA, 55; Harvard Univ, AM, 56; Cornell Univ, PhD(org chem), 60. *Prof Exp:* Res chemist, Enjay Chem Intermediate Div, 60-67, sr res chemist, Enjay Polymer Lab, Esso Res & Eng Co, 67-71; mgr polymer sci & eng, Am Optical Co, 71-75, dir polymer res & develop, 75-78, dir mat sci & process lab, 78-80, vpres res & develop, 85, vpres res & develop & gen mgr, Tech Div, Precision Prods, 85-88; TECH DIR, COBURN OPTICAL INDUSTS, 88- *Mem:* Am Chem Soc; AAAS; Sigma Xi; Soc Plastics Engrs. *Res:* Optical plastics, optical and abrasion resistant coatings, photochromic materials; polyethylene-propylene, butyl and chlorobutyl elastomers; engineering plastics; polyurethanes; chemical and polymer synthesis, analysis, characterization and physical properties; product and process development; contact, ophthalmic, and precision optical lenses and processes. *Mailing Add:* 4507 E 108th St S Tulsa OK 74137-6850

ROTENBERG, KEITH SAUL, b San Francisco, Calif, Oct 10, 50; m 76; c 1. PHARMACOKINETICS, BIOPHARMACEUTICS. *Educ:* Univ Calif, Berkeley, BA, 72; Univ Md, PhD (pharmacokinetics & biopharmaceut), 77. *Prof Exp:* Reviewer, 77-80, TECH SUPVR BIOAVAILABILITY & BIOEQUIVALENCY, FOOD & DRUG ADMIN, 80- *Mem:* Am Pharmaceut Asn; Acad Am Pharmaceut Asn; AAAS; NY Acad Sci; Am Chem Soc. *Res:* Pharmacokinetics of nicotine in animals and man; pharmacokinetics and bioavailability of drugs in man; in vivo-in vitro correlations; formulation factors which affect a drugs bioavailability and in vitro dissolution. *Mailing Add:* Lorex Pharmaceut 4930 Oakton St Skokie IL 60077-2900

ROTENBERG, MANUEL, b Toronto, Ont, Mar 12, 30; nat US; m 52; c 2. ATOMIC PHYSICS, BIOPHYSICS. *Educ:* Mass Inst Technol, SB, 52, PhD(physics), 55. *Prof Exp:* Staff mem, Los Alamos Sci Lab, 55-57; instr physics, Princeton Univ, 57-58, staff mem, Proj Matterhorn, 58; asst prof physics, Univ Chicago & Inst Comput Res, 50-61; asst prof physics, 61-65, asst dir inst radiation physics & aerodyn, 65-67, assoc prof, 67-70, assoc dean grad studies & res, 71-75, dean grad studies & res, 75-83, PROF APPL PHYSICS, UNIV CALIF, SAN DIEGO, 70-, CHAIR, DEPT ELEC & COMPUTER ENG, 88- *Concurrent Pos:* Co-ed, Methods Computational Physics, 63-77; founding co-ed, J Computational Physics, 64. *Mem:* Fel Am Phys Soc; Sigma Xi; AAAS. *Res:* Atomic theory; scattering theory; numerical techniques; population theory; cell kinetics. *Mailing Add:* Dept Elec & Comput Eng Univ Calif San Diego La Jolla CA 92093-0407

ROTENBERRY, JOHN THOMAS, b Roanoke, Va. AVIAN ECOLOGY, POPULATION & COMMUNITY ECOLOGY. *Educ:* Univ Tex, Austin, BA, 69; Ore State Univ, MS, 74, PhD(ecol), 78. *Prof Exp:* Res assoc, Dept Biol, Univ NMex, 78-80; assoc prof, dept Biol Sci, Bowling Green State Univ, 80-90; DIR, NATURAL RESERVE SYST, UNIV CALIF-RIVERSIDE, 90- *Mem:* Ecol Soc Am; Am Ornithologists Union; Am Soc Naturalists; Sigma Xi; Cooper Ornith Soc; Asn Field Ornithologists. *Res:* Animal-environment relationships, particularly their influence on population dynamics and community structure. *Mailing Add:* Dept Biol Univ Calif Riverside CA 92521

ROTERMUND, ALBERT J, JR, b St Louis, Mo, June 20, 40; m 68; c 3. CELL BIOLOGY, CELL PHYSIOLOGY. *Educ:* St Louis Univ, BS, 62, MS, 66; State Univ NY Buffalo, PhD(biol), 69. *Prof Exp:* NIH fel, Brookhaven Nat Lab, 68-70; asst prof, 70-76, ASSOC PROF BIOL, LOYOLA UNIV, CHICAGO, 77- *Mem:* Am Soc Zool, Sigma Xi. *Res:* Cellular physiology and biochemistry; hibernation. *Mailing Add:* 407 E Burr Oak Dr Arlington Heights IL 60004-2125

ROTH, ALLAN CHARLES, b St Joseph, Mo, Dec 23, 46; m 68; c 2. PHYSIOLOGY, BIOMEDICAL ENGINEERING. *Educ:* Iowa State Univ, BS, 69, MS, 74, PhD(biomed eng), 77. *Prof Exp:* Design engr, Rockwell Int, Calif, 69-72; res asst, Biomed Eng Prog, Iowa State Univ, 72-74, teaching asst, Dept Zool, 74-77; res assoc, Dept Physiol & Biophys, Univ Wash, 77-80; ASST PROF, DIV BIOMED ENG, UNIV VA, 80- *Mem:* Sigma Xi; Biomed Eng Soc Am. *Res:* Fluid mechanics of arterial stenosis; comparative circulatory hemodynamics; gas transport in the tissue; circulatory transport and instrumentation. *Mailing Add:* Dept Biomed Eng Univ Va Med Ctr Box 377 Charlottesville VA 22908

ROTH, ARIEL A, b Geneva, Switz, July 16, 27; nat US; m 52; c 2. BIOLOGICAL OCEANOGRAPHY, PALEONTOLOGY. *Educ:* Pac Union Col, BA, 48; Univ Mich, MS, 49, PhD(zool), 55. *Prof Exp:* From instr to assoc prof biol, Pac Union Col, 50-57; res assoc, Loma Linda Univ, 57-58; from assoc prof to prof biol & chmn dept, Andrews Univ, 58-63; chmn, Dept Biol, 63-73, PROF BIOL, LOMA LINDA UNIV, 63-, DIR, GEOSCI RES INST, 80- *Concurrent Pos:* Mem, Geosci Res Inst, 73-; ed, Origins. *Mem:* Sigma Xi; Geol Soc Am; Soc Econ Paleontologists & Mineralogists; Am Asn Petrol Geologists; Int Asn Sedimentologists. *Res:* Factors affecting rate of coral growth; parasitology; schistosomiasis; invertebrate zoology; coral reef development; identification of coral reefs; history and philosophy of science. *Mailing Add:* Geosci Res Inst Loma Linda Univ Loma Linda CA 92354

ROTH, ARTHUR JASON, b Brooklyn, NY, Oct 8, 49; m 72; c 3. MATHEMATICAL STATISTICS. *Educ:* Cornell Univ, BA, 71; Univ Minn, PhD(statist), 75. *Prof Exp:* Asst prof statist, Carnegie-Mellon Univ, 75-77; asst prof math, Syracuse Univ, 77-80; mem staff, Ciba-Geigy Corp, 80-88; MEM STAFF, G D SEARLE CORP, 88- *Mem:* Inst Math Statist; Am Statist Asn. *Res:* Carcinogenecity testing, categorical data analysis, survival analysis, robust test for trend; group testing, particularly Bayesian and sequential aspects; sequential analysis; nonparametric statistics; ranking and selection problems. *Mailing Add:* G D Searle Corp 4901 Searle Pkwy Skokie IL 60077

ROTH, BARBARA, b Milwaukee, Wis, June 9, 16. ORGANIC CHEMISTRY. *Educ:* Beloit Col, BS, 37; Northwestern Univ, MS, 39, PhD(org chem), 41. *Prof Exp:* Lab asst, Northwestern Univ, 38-41; res chemist, Calco Chem Div, Am Cyanamid Co, 41-51; group leader keratin res, Toni Div, Gillette Co, 51-55; sr res chemist, Wellcome Res Labs, NY, 55-71, group leader, Burroughs Wellcome Co Inc, Res Triangle Park Inc, 71-86; ADJ PROF, CHEM DEPT, UNIV NC, CHAPEL HILL, 87- *Concurrent Pos:* Instr, Lake Forest Col, 40-41; adj prof, sch pharm, Univ NC, 71-86. *Mem:* Fel AAAS; Am Chem Soc; NY Acad Sci. *Res:* Pyrimidine and medicinal chemistry; synthetic organic chemistry. *Mailing Add:* Seven Lone Pine Rd Chapel Hill NC 27514

ROTH, BEN G, b Bloomington, Ill, Oct 5, 42; m 83; c 4. MATHEMATICS. *Educ:* Occidental Col, AB, 64; Dartmouth Col, AM, 66, PhD(math), 69. *Prof Exp:* From asst prof to assoc prof, 69-79, PROF MATH, UNIV WYO, 79- *Mem:* Am Math Soc; Math Asn Am. *Res:* Spaces of continuous and differentiable functions; spaces of distributions; rigidity. *Mailing Add:* Dept Math Univ Wyo Laramie WY 82071

ROTH, BENJAMIN, b New York, NY, Feb 6, 09; m 41; c 1. ELEMENTARY PARTICLE PHYSICS. *Educ:* City Col New York, BS, 29; Columbia Univ, MA, 31; Polytech Inst Brooklyn, BEE, 47; Cornell Univ, PhD(physics), 51. *Prof Exp:* Instr physics, Univ Conn, 51-56; from asst prof to assoc prof, Okla State Univ, 57-61; from assoc prof to prof, 62-76, EMER PROF PHYSICS, BROOKLYN COL, 76- *Mem:* Am Phys Soc. *Res:* Elementary particle physics. *Mailing Add:* Dept Physics Brooklyn Col Brooklyn NY 11210

ROTH, BERNARD, b New York, NY, May 28, 33; m 55; c 2. MECHANICAL ENGINEERING. *Educ:* City Col New York, BS, 56; Columbia Univ, MS, 58, PhD(mech eng), 62. *Prof Exp:* Lectr mech eng, City Col New York, 56-58; from asst prof to assoc prof, 62-71, PROF MECH ENG, STANFORD UNIV, 71- *Concurrent Pos:* Consult, Atlantic Design Co, 57, Columbia Univ, 62-64, Int Bus Mach Corp, 64-67, Univ Neger, Israel, 73 & Atomic Energy Comn, France, 76; prin investr, NSF Grants, 63-; vis prof, Technol Univ Delft, 68-69, Kanpur, India, 77, Shanghai, China, 79 & Bangalore, India, 84. *Honors & Awards:* Melville Medal, Am Soc Mech Engrs, 67, Machine Design Award, 84; Joseph F Engleberger Award, 86. *Mem:* Am Soc Mech Engrs; Int Fedn Theory Mach & Mechanisms (pres, 81-84). *Res:* Robotics; interpersonal relations; kinematics; numerical methods, especially computer aided design; machine design, especially analytical techniques. *Mailing Add:* Dept Mech Eng Stanford Univ Stanford CA 94305

ROTH, CHARLES, b Huncovce, Czech, Dec 25, 39; Can citizen; m 63; c 2. APPLIED MATHEMATICS, THEORETICAL PHYSICS. *Educ:* McGill Univ, BSc, 61, MSc, 62; Hebrew Univ, Israel, PhD(theoret physics), 65. *Prof Exp:* Lectr math & physics, Hebrew Univ, Israel, 63-65; from asst prof to assoc prof, 65-83, PROF MATH, MCGILL UNIV, 83- *Mem:* Can Math Cong. *Res:* Applications of group representations and irreducible tensorial sets to atomic and nuclear spectroscopy. *Mailing Add:* Dept Math & Statist McGill Univ Sherbrooke St W Montreal PQ H3A 2M5 Can

ROTH, CHARLES BARRON, b Columbia, Mo, Apr 27, 42; m 64; c 2. SOIL CHEMISTRY, MINERALOGY. *Educ:* Univ Mo, BS, 63, MS, 65; Univ Wis, PhD(soil sci), 69. *Prof Exp:* Res assoc, 68-70, from asst prof to assoc prof, 70-85, PROF SOIL CHEM & MINERAL, PURDUE UNIV, 85- *Concurrent Pos:* Vis assoc prof, Tex A&M Univ, 78-79; vis prof & consult, Univ Nebr, Lincoln, 90. *Mem:* AAAS; Am Soc Agron; Int Asn Study Clays; Soil Sci Soc Am; Int Soc Soil Sci; Mineral Soc Am; Clay Mineral Soc. *Res:* Physical and chemical effects of iron, aluminum and silica sesquioxides on the physicochemical properties of soils and clays; redox reactions of iron in clay mineral structures; prediction of soil erosion from reclaimed minelands; infrared analysis of water and organic compound interaction with soils and clays. *Mailing Add:* Dept Agron Purdue Univ West Lafayette IN 47907

ROTH, DANIEL, b New York, NY, Oct 27, 20; m 50; c 2. MOLECULAR BIOLOGY, CYTOLOGY. *Educ:* Columbia Univ, AB, 40; New York Med Col, MD, 43. *Prof Exp:* Asst pathologist, United Hosp, 51-56; asst prof path, Med Ctr, NY Univ, 56-59; dir lab, Bergen Pines County Hosp, 59-62 & St Barnabas Hosp, Bronx, 62-67; assoc prof path, Med Ctr & Univ Hosp, NY Univ, 67-73, assoc clin prof environ med, 73-80; pathologist, Goodwin Inst Cancer Res, 74-84; RETIRED. *Mem:* Am Asn Path & Bact; Am Soc Exp Path; Am Soc Photobiol; NY Acad Sci; Environ Mutagen Soc. *Res:* Studies into conformational changes in cellular DNA produced by chemical and physical carcinogenic agents; utilizing a fluorescent probe of DNA secondary structure developed in our laboratory. *Mailing Add:* 29 Hickory Hill Lane Tappan NY 10983

ROTH, DONALD ALFRED, b Slinger, Wis, July 31, 18; m 51; c 4. MEDICINE. *Educ:* Univ Wis, PhD(chem), 44; Marquette Univ, MD, 52; Am Bd Internal Med, dipl, 63; Am Bd Nephrol, dipl, 76. *Prof Exp:* Asst chem, Univ Wis, 41-44, instr, 46-48; chief renal sect, Wood Vet Admin Hosp, 58-86; from instr to assoc prof med, 61-86, CLIN PROF MED, MED COL WIS, 86- *Mem:* AMA; Am Diabetes Asn; Am Fedn Clin Res; Am Heart Asn; fel Am Col Physicians; Am Soc Nephrology; Int Soc Nephrology. *Res:* Renal disease; hemodialysis; hypertension, including retinal photography and ultrastructure of kidney. *Mailing Add:* 1620 Revere Dr Brookfield WI 53045

ROTH, ELDON SHERWOOD, b St Paul, Minn, Nov 7, 29; m 56. GEOMORPHOLOGY. *Educ:* Univ Calif, Los Angeles, BA, 53; Univ Southern Calif, MS, 59, PhD(higher educ), 69. *Prof Exp:* Instr geol & chem, Barstow Col, 60-69; assoc prof physics & dir environ sci, 69-, CTR EXCELLANCE EDUC, NORTHERN ARIZ UNIV. *Mem:* Geol Soc Am; Nat Asn Geol Teachers. *Res:* Local and regional geomorphology; delineation of geomorphic regions by quantitative means; adaptation of computer methods to geomorphic research; energy resources and utilization. *Mailing Add:* Ctr Excellance Educ Northern Ariz Univ Box 5774 Flagstaff AZ 86011

ROTH, FRANK J, JR, microbiology, mycology, for more information see previous edition

ROTH, FRIEDA, microbiology, virology, for more information see previous edition

ROTH, GEORGE STANLEY, b Honolulu, Hawaii, Aug 5, 46; m 72; c 2. GERONTOLOGY, ENDOCRINOLOGY. *Educ:* Villanova Univ, BS, 68; Temple Univ, PhD(microbiol), 71. *Prof Exp:* Asst microbiol, Sch Med, Temple Univ, 68-71; fel,biochem, Fels Res Inst, 71-72; staff fel, Geront Res Ctr, 72-76, RES BIOCHEMIST GERONT, NAT INST AGING, 76-, CHIEF, MOLECULAR PHYSIOL & GENETICS SECT, 84- *Concurrent Pos:* Res consult, George Washington Univ, 77-81; co-ed, Chem Rubber Co Press, 77-; ed, Neurobiol Aging, 80-85; exchange scientist, Nat Acad Sci, 77-80; Alpha Omega Alpha prof, Univ PR, 86; Sigma Chi Scholar in residence, Miami Univ, 89. *Honors & Awards:* Ann Res Award, Am Aging Asn, 81; Sandoz Prize for Geront Res, 89; Third Age Award, Int Asn Geront, 89. *Mem:* Fel Geront Soc. *Res:* Effect of aging on hormone action; molecular mechanisms of aging. *Mailing Add:* Geront Res Ctr Francis Scott Key Med Ctr Baltimore MD 21224

ROTH, GERALD J, b Winona, Minn, Apr 26, 41; m; c 5. MEDICINE. *Educ:* Harvard Col, AB, 63; Harvard Med Sch, MD, 67; Am Bd Internal Med, dipl, 72, dipl hemat, 82. *Prof Exp:* Intern med, Univ Utah Hosp, 67-68, clin fel hemat, 70-71; resident med, Univ Wash Hosp, 68-70; res investr, US Army Med Res Lab, Ft Knox, Ky, 71-74; res fel hemat, Sch Med, Wash Univ, 74-76; from asst prof to assoc prof med, Sch Med, Univ Conn, 76-84; assoc prof, 84-89, PROF MED, UNIV WASH, SEATTLE, 89-; CHIEF, HEMAT SECT, SEATTLE VET ADMIN CTR, 84- *Concurrent Pos:* Staff physician, Univ Conn Health Ctr, Farmington, 76-84; estab investr, Am Heart Asn, 81-86; mem ad hoc hematol study sect, NIH, 87, 88, 89 & 90, subcomt hemostasis, Am Soc Hemat, 84-87, Vascular Biol Burart Subcomt, Study Sect, Am Heart Asn, 90-93, subcomt on thrombosis, mem & chmn, 90-93, Vet Admin Med Res Serv, Am Soc Hemat, 91-94. *Mem:* Am Fedn Clin Res; Am Soc Hemat; Am Soc Biol Chemists; Am Heart Asn; Am Soc Clin Invest; AAAS; Int Soc Thrombosis & Hemostasis. *Res:* Internal medicine. *Mailing Add:* Dept Med Univ Wash Seattle Med Ctr 111 1650 S Columbia Way Seattle WA 98108

ROTH, HAROLD, b Wilkes-Barre, Pa, Jan 26, 31; m 52; c 3. SOLID STATE PHYSICS. *Educ:* Mass Inst Technol, BS, 52; Univ Pa, MS, 54, PhD(physics), 59. *Prof Exp:* Asst physics, Univ Pa, 53-54; res assoc, Gen Atomic Div, Gen Dynamics Corp, Calif, 56-59, staff scientist, 59; staff scientist, Raytheon Res Div, Mass, 59-63, prin scientist, 63-65; chief adv res br, Electronics Res Ctr, NASA, 65-68, chief electronic components lab, 68-70; dir res & advan develop, 70-73; dir res & eng, Electronics Div, Allen-Bradley Co, 73-81; DIR, SOLID STATE SCI DIRECTORATE, ROME AIR DEVELOP CTR, HANSCOM AFB, MASS, 81- *Concurrent Pos:* Mem adv subcomt electrophys, NASA, 68-70; chmn, task force automotive solid state electronics, Soc Automotive Engrs & indust adv comt, Milwaukee Sch Eng. *Mem:* Am Phys Soc; Sigma Xi; Inst Elec & Electronics Engrs; Am Inst Physics. *Res:* Semiconductors; galvanomagnetic effects; radiation damage; electrical properties of junctions; passive electronic components; photonics. *Mailing Add:* Solid State Sci Directorate Rome Lab Hanscom AFB MA 01731

ROTH, HAROLD PHILMORE, b Cleveland, Ohio, Aug 2, 15; m 52; c 2. GASTROENTEROLOGY, EPIDEMIOLOGY. *Educ:* Western Reserve Univ, BA, 36, MD, 39; Harvard Univ, MS, 67. *Prof Exp:* Intern, Cincinnati Gen Hosp, 39-40; house officer, Fifth Med Serv, Boston City Hosp, 40-42; asst resident med, Barnes Hosp, St Louis, Mo, 42-43; from clin instr to sr clin instr med, Case Western Reserve Univ, 49-55, from asst prof to assoc prof, 55-74, assoc prof community health, 71-74; assoc dir digestive dis & nutrit, 74-83, dir, 83-85, DIR, EPIDEMIOL & DATA SYSTS PROG, NAT INST DIABETES, DIGESTIVE & KIDNEY DIS, 85- *Concurrent Pos:* Asst med, Wash Univ, 42-43; asst physician, Out-Patient Dept, Univ Hosps, Cleveland,

49-69, assoc physician, Dept Med, 69-74; chief gastroenterol sect, Vet Admin Hosp, 47-74, dir gastroenterol training prog, Univ Hosps & Vet Admin Hosp, 63-74; USPHS spec fel, Sch Pub Health, Harvard Univ, 66-67; mem, Digestive Dis Comt, USPHS, 74-; mem, Nat Comn Digestive Dis, 76-78 & Nat Digestive Dis Adv Bd, 80- *Honors & Awards:* Spec Recognition Award, Am Gastroenterol Asn, 84. *Mem:* Am Gastroenterol Asn; fel Am Col Physicians; Am Asn Study Liver Dis; Soc Clin Trials (pres, 78-80); Ctr Soc Clin Res. *Res:* Composition of bile and formation of gallstones; patient care, factors influencing patients' cooperation with medical regimens; clinical epidemiology of digestive diseases; esophageal spasm; peptic ulcer; gallstones; improving the medical record in the light of new technologies including the computer work with the committees of the Institute of Medicine and presentations to and through the Medical Records Institute. *Mailing Add:* Fed Bldg Rm 106 NIH Bethesda MD 20892

ROTH, HEINZ DIETER, b Rheinhausen, Ger, Oct 25, 36; m 64; c 2. PHYSICAL ORGANIC CHEMISTRY. *Educ:* Univ Karlsruhe, BS, 58; Univ Cologne, MS, 62, Dr rer nat, 65. *Prof Exp:* Fel org chem, Yale Univ, 65-67; mem tech staff org chem, Bell Labs, 67-88; PROF CHEM, RUTGERS UNIV, 88- *Mem:* Am Chem Soc; Ger Chem Soc. *Res:* Reactivity and structure of carbenes, radicals and radical ions; photochemistry; chemically induced magnetic polarization; nuclear magnetic resonance; electron paramagnetic resonance. *Mailing Add:* Dept Chem Wright Rieman Labs Rutgers Univ PO Box 939 Piscataway NJ 08855-0939

ROTH, HOWARD, b New York, NY, Oct 11, 25; m 43; c 2. FOOD CHEMISTRY. *Educ:* City Col New York, BS, 53. *Prof Exp:* Sr proj mgr, DCA Industs Inc, 53-75, assoc dir res & develop, Cent Res Lab, 75-77, dir Corp Res Ctr, 77-83; CONSULT, 83- *Mem:* Am Chem Soc; Am Oil Chem Soc; Int Microwave Power Inst (treas); Am Asn Cereal Chem. *Res:* Chemistry and physics of fats, oils and cereals; confectionary and bakery products; microwave applications; food product and machine development. *Mailing Add:* 206 Devoe Ave Yonkers NY 10705

ROTH, IVAN LAMBERT, b Nixon, Tex, Feb 21, 28; m 51; c 2. MICROBIOLOGY. *Educ:* Tex Lutheran Col, BA, 50; Univ Tex, MA, 56; Baylor Univ, PhD(microbiol), 63. *Prof Exp:* Anal chemist, Texaco Inc, Tex, 51-54; instr biol, Univ Houston, 58-62; asst prof microbiol, Med Ctr, Univ Ala, 62-66; assoc prof, 66-75; PROF MICROBIOL, UNIV GA, 75- *Mem:* AAAS; Am Soc Microbiol; Electron Micros Soc Am. *Res:* Ultrastructure of animal tissue infected with bacteria; bacterial virulence, avirulence and pathogenesis in animals; ultrastructure of microbial cells; ultrastructure of bacterial capsules and slime; scanning electron microscopy of slime molds (myxomycetes). *Mailing Add:* Dept Microbiol Univ Ga 821 Bio Sci Athens GA 30602

ROTH, J(OHN) REECE, b Washington, Pa, Sept 19, 37; m 72; c 2. INDUSTRIAL PLASMA ENGINEERING, FUSION ENERGY. *Educ:* Mass Inst Technol, SB, 59; Cornell Univ, PhD(eng physics), 63. *Prof Exp:* Aerospace res scientist, Phys Sci Div, Lewis Res Ctr, NASA, 63-78; vis prof elec eng, 78-82, res prof, dept physics, 82-83, PROF ELEC ENG, UNIV TENN, KNOXVILLE, 83- *Concurrent Pos:* Assoc ed, Trans Plasma Sci, Inst Elec & Electronics Engrs, 73-88; prin investr, Off Naval Res, Air Force Off Sci Res, Army Res Off & Tenn Valley Authority contracts, 80-; mem, Nat Acad Sci-Nat Res Coun Comt Advan Fusion Power, 87- 88. *Mem:* AAAS; Am Nuclear Soc; fel Inst Elec & Electronics Engrs; Am Phys Soc; assoc fel Am Inst Aeronaut & Astronaut; Am Soc Eng Educ; Inst Elec & Electronics Engrs Nuclear & Plasma Sci Soc; Archaeol Inst Am; Sigma Xi. *Res:* High temperature plasma science related to controlled fusion; plasma cloaking of military radar targets; plasma heating and confinement; effects of electric fields on toroidal plasmas; public policy issues in fusion energy; industrial plasma engineering; space applications of fusion energy. *Mailing Add:* Dept Elec & Computer Eng Univ Tenn 409 Ferris Hall Knoxville TN 37996-2100

ROTH, JACK A, b LaPorte, Ind, Jan 29, 45; c 2. MOLECULAR BIOLOGY OF CANCER, ONCOGENES. *Educ:* Cornell Univ, Ithaca, BA, 67; Johns Hopkins Univ, MD, 71; Am Bd Surg, cert, 81; Am Bd Thoracic Surg, cert, 83. *Prof Exp:* PROF & CHMN, DEPT THORACIC SURG, UNIV TEX M D ANDERSON CANCER CTR, 86- *Concurrent Pos:* Bud S Johnson chair & prof tumor biol, Univ Tex M D Anderson Cancer Ctr; NIH res grant, 87; mem, Esophageal Cancer Strategy Group, Nat Cancer Inst, NIH, 89 & prog comt, Am Soc Clin Oncol, 89-90; chmn, Task Force Acad Surg Oncol Specialties, Soc Surg Oncol, 90; Spec Achievement Award, US Dept Health & Human Serv, 81. *Mem:* Am Asn Cancer Res; Am Asn Immunologists; Am Asn Thoracic Surg; Am Surg Asn; NY Acad Sci. *Res:* Identifying the molecular events critical to the genesis and progression of thoracic cancers; developing technology for modifying expression of cancer genes for therapeutic application. *Mailing Add:* M D Anderson Cancer Ctr Univ Tex 1515 Holcombe Blvd Box 109 Houston TX 77030

ROTH, JAMES A, b Grinnell, Iowa, Mar 1, 51; m; c 3. VETERINARY MICROBIOLOGY. *Educ:* Iowa State Univ, DVM, 75, MS, 79, PhD(vet microbiol), 81. *Prof Exp:* Vet, Belle Plaine, Iowa, 75-77; from instr to assoc prof vet microbiol & prev med, 77-86, chmn, Interdept Prog Immunobiol, 84-87, PROF VET MICROBIOL & PREV MED, COL VET MED, IOWA STATE UNIV, 86-, PROF-IN-CHG, CTR IMMUNITY ENHANCEMENT DOMESTIC ANIMALS, 87- *Concurrent Pos:* Prin investr, Nat Cancer Inst, NIH, 89-84; interim chmn, Dept Vet Microbiol & Prev Med, Iowa State Univ, 90-91 & Dept Microbiol, Immunol & Prev Med, 91-; mem bd gov, Am Col Vet Microbiologists, 90-93; prog dir, Nat Inst Allergy & Infectious Dis, NIH, 90-95; proj dir, Agr Res Serv, USDA, 90-91; chmn, Sci Adv Bd, Biotechnol Res & Develop Corp, 91-; mem bd gov, Am Col Vet Microbiologists, 90-91. *Honors & Awards:* Beecham Award, 86. *Mem:* Am Vet Med Asn; Am Soc Microbiol; Am Asn Immunologists; Am Asn Vet Immunologists (pres, 89); Am Soc Animal Sci; Soc Leukocyte Biol; Am Col Vet Microbiologists. *Res:* Veterinary immunology; author of numerous technical publications; awarded one US patent. *Mailing Add:* Dept Vet Microbiol & Prev Med Iowa State Univ Ames IA 50011

ROTH, JAMES FRANK, b Rahway, NJ, Dec 7, 25; m 50; c 3. PHYSICAL CHEMISTRY. *Educ:* WVa Univ, BA, 47; Univ Md, PhD(chem), 52. *Prof Exp:* Sr res chemist, Franklin Inst, 51-54; res chemist, Lehigh Paints & Chem Inc, 54-56 & cent res lab, Gen Aniline & Film Corp, 56-59; mgr chem br, Franklin Inst, 59-60; mgr catalysis res, Cent Res Dept, Monsanto Co, 60-69, mgr catalysis res, 69-73, dir catalysis res, 73-76, dir process sci, Corp Res Labs, 76-80; corp chief scientist, Air Prod & Chemicals, Inc, 80-90; RETIRED. *Honors & Awards:* E V Murphree Award in Indust & Eng Chem, Am Chem Soc, 76 & Award Indust Chem, 91; R J Kokes Award, Johns Hopkins Univ, 77; Perkin Medal, Soc Chem Indust, 88; Chem Pioneer Award, Am Inst Chem, 86. *Mem:* Nat Acad Eng; Catalysis Soc NAm; Am Chem Soc. *Res:* Heterogeneous and homogeneous catalysis. *Mailing Add:* 4436 Calle Serena Sarasota FL 34238

ROTH, JAMES LUTHER AUMONT, b Milwaukee, Wis, Mar 8, 17; m 38, 83; c 3. GASTROENTEROLOGY. *Educ:* Carthage Col, BA, 38, Univ Ill, MA, 39; Northwestern Univ, MD, 44, PhD(physiol), 45; Am Bd Internal Med, dipl, 54; Am Bd Gastroenterol, dipl, 55. *Hon Degrees:* DSc, Carthage Col, 57; MSc, Univ Pa, 71. *Prof Exp:* Instr physiol, Med Sch Northwestern Univ, 42-44; intern med, Mass Gen Hosp, 44-45; resident physician, Grad Hosp, Philadelphia, 45-46; resident physician, Univ Hosp, 48-49, from instr to assoc gastroenterol, Div Grad Med, 50-54, asst prof physiol, 53-66, from asst prof to prof 54-68, prof clin gastroenterol, 59-68, chief, Gastroenterol Serv & dir, Inst Gastroenterol, Presby-Univ PA Med Ctr, 65-86, prof clin med, 68-86, EMER PROF, SCH MED, UNIV PA, 86- *Concurrent Pos:* Res fel med, Grad Hosp, Philadelphia, 49-50; dir, Gastrointestinal Res Lab, Grad Hosp, 50-66, chief, Gastrointestinal Clin, 58-66; spec consult, Off Surgeon Gen, US Army, 48-49; mem adv comt rev, US Pharmacopeia XV, 50-60, 70-; exchange prof, Med Sch, Pontif Univ Javeriana, 60, prof extraordinary, 60-; dir grad div gastroenterol, Univ Pa, 61-69; consult, subcomt digestive syst, AMA, 62-64; mem drug efficacy panel, Food & Drug Admin, 66-68; consult, US Navy, Bethesda & Philadelphia Navy Hosps, 67-; mem bd dirs & chmn prog comt, Digestive Dis Found, 69-72; assoc ed, Bockus' Gastroenterol, ed, 4th ed; sr consult, Presby-Univ PA Med Ctr, 86-; James I A Roth vis prof, Presby Med Ctr, 91. *Honors & Awards:* Bronze Medal, AMA, Sigma Xi Prize, Joseph Capps Award, Inst Med, Chicago, 44; Order of Christopher Columbus, Govt Dominican Repub, 69; Clin Achievement Award, Am Col Gastroenterologists, 89. *Mem:* AMA; Am Gastroenterol Asn; fel Am Col Physicians; Pan-Am Med Asn; Bockus Int Soc Gastroenterol (secy-gen, 58-67, vpres, 67-71, pres elect, 71-73, pres, 73-75); Am Col Gastroenterol (vpres, 81-82); hon mem Gastroenterol Socs of Venezuela, Colombia, Dominican Republic, Florida. *Res:* Intermediary metabolism of phenylalanine; caffeine potentiation of gastric secretion; caffeine gastric analysis; cold environment metabolic balances; penetration and hemorrhage in peptic ulcer; pancreatitis; hepatic coma; hazards of anti-cholinergic drugs; salicylate erosion and ulceration; ulcerative colitis. *Mailing Add:* Inst Gastroenterol Suite W-390 Presby Med Ctr Philadelphia Philadelphia PA 19104

ROTH, JAN JEAN, environmental physiology, paleontology, for more information see previous edition

ROTH, JAY SANFORD, b New York, NY, June 10, 19; c 4. BIOCHEMISTRY. *Educ:* City Col New York, BS, 40; Cornell Univ, MS, 41; Purdue Univ, PhD(org chem), 44. *Prof Exp:* Asst, Purdue Univ, 41-44; asst prof chem, Univ Idaho, 44-47 & Rutgers Univ, 47-50; from asst prof to assoc prof biochem, Hahnemann Med Col, 50-60; PROF BIOCHEM, UNIV CONN, 60- *Concurrent Pos:* Brit-Am Cancer Res fel, Strangeways Res Lab, Cambridge, England, 53-54; assoc, Marine Biol Lab, Woods Hole; Nat Cancer Inst career fel, 62-; assoc ed, Cancer Res, 71-76. *Mem:* AAAS; Am Soc Biol Chem; Am Chem Soc; Am Asn Cancer Res. *Res:* Nucleic acids and nucleases in relation to cell division; control mechanisms in cancer; deoxynucleotide metabolism in growth; virus tumor biochemistry. *Mailing Add:* Dept Molecular & Cell Biol Univ Conn U-125 Storrs CT 06268

ROTH, JEROME A, b Springfield, Ill, Aug 13, 40; m 64; c 2. ORGANIC CHEMISTRY. *Educ:* Loyola Univ, Ill, BS, 62; Ill Inst Technol, PhD(org chem), 66. *Prof Exp:* Assoc catalysis chem, Northwestern Univ, 66-67, instr org chem, 67-68; from asst prof to assoc prof, 68-78, PROF CHEM, NORTHERN MICH UNIV, 78- *Concurrent Pos:* Am Chem Soc res grant, 68-71; vis prof, Univ Cincinnati, 78-79. *Mem:* Am Chem Soc; Org Reactions Catalysis Soc. *Res:* Organic chemical synthesis; mechanisms; transition-metal organic chemical compounds; heterogeneous and homogeneous catalysis. *Mailing Add:* Dept Chem Northern Mich Univ Marquette MI 49855

ROTH, JEROME ALLAN, b New York, NY, Aug 20, 43; m 85; c 4. BIOCHEMISTRY, NEUROPHARMACOLOGY. *Educ:* State Univ NY Col New Paltz, BS, 65; Cornell Univ, MNS, 67, PhD(biochem), 71. *Prof Exp:* Res assoc biochem, Vanderbilt Univ, 71-72; res assoc to asst prof anesthesiol, Sch Med, Yale Univ, 72-76; from asst prof to assoc prof, 76-85, PROF, DEPT PHARMACOL & THERAPEUT, STATE UNIV NY, BUFFALO, 85- *Mem:* Int Soc Neurochem; Am Soc Pharmacol & Exp Therapeut; Am Soc Neurochem; Soc Neurosci. *Res:* Structure and properties of monomine oxidase, catechol-o-methyltransferase and phenolsulfotransferase; effect of drugs on enzymes involved in biosynthesis and degradation of biogenic amines; post-translational modification of proteins. *Mailing Add:* Dept Pharmacol & Therapeut State Univ NY Buffalo NY 14214

ROTH, JESSE, b New York, NY, Aug 5, 34; m; c 3. MEDICAL SCIENCES GENERAL. *Educ:* Columbia Univ, BA, 55; Albert Einstein Col Med, MD, 59. *Hon Degrees:* Dr, Univ Uppsala, 80. *Prof Exp:* From intern to asst resident, Barnes Hosp, Wash Univ, 59-61; Am Diabetes Asn res fel, Radioisotope Serv, Vet Hosp, Bronx, NY, 61-63; clin assoc, 63-65, sr investr, 65-66, chief diabetes sect, Clin Endocrinol Br, 66-74, chief diabetes br, Nat Inst Arthritis, Metab & Digestive Dis, 74-83, DIR, DIV INTRAMURAL RES, NAT INST DIABETES & DIGESTIVE & KIDNEY DIS, NIH, 83- *Honors & Awards:* Eli Lilly Award, Am Diabetes Asn, 74; Ernst Oppenheimer Mem Award, Endocrine Soc, 74; Spec Achievement Award, US Dept Health, Educ & Welfare, David Rumbough Mem Award, Juvenile

Diabetes Found, 77; Regents' lectr, Univ Calif, 77; G Burroughs Mider lectr, NIH, 78; Diaz Cristobal Prize, Int Diabetes Fedn, 79; Gairdner Found Annual Award, 80; A Cressy Morrison Award, NY Acad Sci, 80; Joslin Medal, New Eng Diabetes Asn, 81; Hazen Prize, Mt Sinai Sch Med, City Univ, 79; Banting Medal, Am Diabetes Asn, 82; Koch Award, Endocrine Soc, 85. *Mem:* Endocrine Soc; Am Diabetes Asn; Am Fedn Clin Res; Am Soc Clin Invest (pres, 78-80). *Res:* Diabetes; clinical research. *Mailing Add:* Nat Inst Diabetes & Digestive & Kidney Dis Bldg 10 Rm 9N222 Bethesda MD 20892

ROTH, JOHN AUSTIN, b Louisville, Ky, May 14, 34; m 59; c 2. CHEMICAL ENGINEERING, ENVIRONMENTAL ENGINEERING. *Educ:* Univ Louisville, BChE, 56, MChE, 57, PhD(chem eng), 61. *Prof Exp:* Teaching asst chem eng, Univ Louisville, 56-59; from asst prof to prof chem eng, 62-74, from asst dean to assoc dean Sch Eng, 68-72, chmn chem fluid & thermal sci div, 71-75, dir, Ctr Environ Qual Mgt, 74-80, PROF CHEM & ENVIRON ENG, VANDERBILT UNIV, 71- *Concurrent Pos:* Year-in-indust partic, 76-68, Savannah River Lab, E I du Pont de Nemours & Co, Inc, Del, 64; comnr, Ky Bur Environ Protection, 77-78; chmn bd & vpres, Chem & Environ Serv, Inc, 86- *Mem:* Water Pollution Control Fedn; Am Inst Chem Engrs; Am Chem Soc; Nat Soc Prof Engrs. *Res:* Water and waste water treatment by ozonation; carbon adsorption; chemical engineering kinetics; mass transfer processes; hazardous materials; mixing processes. *Mailing Add:* Ctr Environ Qual Mgt Vanderbilt Univ Nashville TN 37235

ROTH, JOHN L, JR, b Trenton, NJ, July 5, 49; m 73; c 3. ANGIOSPERM EVOLUTION, PALYNOLOGY. *Educ:* Brigham Young Univ, BS, 73, MS, 75; Ind Univ, MA, 79, PhD(bot), 81. *Prof Exp:* Vis lectr bot, Ind Univ, 80; res assoc bot, Univ Mass, Amherst, 81-87; BILINGUAL SCI TEACHER, HOLYOKE HIGH SCH, 87- *Mem:* Bot Soc Am; Am Soc Plant Taxonomists; Int Org Paleobot; AAAS. *Res:* Angiosperm evolution, systematics, paleobotany and palynology; foliar epidermal anatomy of Magnoliales, Laurales and various Jurassic-Eocene plants; palynology of primitive monocotyledohs living and fossil; paleoecology, foliar physiognomy, depositional selection of plant remains. *Mailing Add:* 100 High Point Dr Amherst MA 01002

ROTH, JOHN PAUL, b Detroit, Mich, Dec 16, 22; div; c 2. COMPUTER SCIENCE, MATHEMATICS. *Educ:* Univ Detroit, BME, 46; Univ Mich, MS, 48, PhD(math), 54. *Prof Exp:* Instr math, Wayne State Univ, 46-47; res assoc appl math, Univ Mich, 47-53; Pierce instr math, Univ Calif, 53-55; staff mathematician, Inst Adv Study, 55-56; mem res staff, Thomas J Watson Res Ctr, IBM Corp, 56-90; RETIRED. *Concurrent Pos:* Res engr, Continental Aviation & Eng Corp, 46-47; consult, Shell Develop Co, 54-55; adj prof, City Univ New York, 81-82. *Honors & Awards:* NASA Award. *Mem:* Am Math Soc; Soc Indust & Appl Math; fel Inst Elec & Electronics Engrs. *Res:* Mathematical computer design; combinatorial topology, especially conceptual and practical use in the solution of physical problems. *Mailing Add:* 59 Touchstone Way Millwood NY 10546

ROTH, JOHN R, b Winona, Minn, Mar 14, 39; m 61; c 2. SALMONELLA BACTERIA. *Educ:* Harvard Univ, BA, 61; Johns Hopkins Univ, PhD, 65. *Prof Exp:* postdoct, NIH, 65-67; from asst prof to prof, Univ Calif, Berkeley, 67-75; PROF MICROBIOL, UNIV UTAH, 76-, CHMN, DEPT BIOL, 89- *Concurrent Pos:* Guggenheim fel, Col Spring Harbor Lab, 74-75; mem, Genetics Study Sect, NIH, 72-76, Microbiol Physiol & Genetics Study Sect, 83-87; coordr grad prog molecular biol, Univ Utah, 75-88. *Honors & Awards:* Rosenblatt Prize, 90. *Mem:* Nat Acad Sci; Genetics Soc Am; Am Soc Microbiol. *Res:* Genetic analysis of the Salmonella typhimurium bacterium; histidine-purine biosynthesis; chromosomal rearrangements; anaerobic metabolism. *Mailing Add:* Dept Biol 201 Biol Bldg Univ Utah Salt Lake City UT 84112

ROTH, JONATHAN NICHOLAS, b Albany, Ore, Mar 2, 38; m 59; c 3. PLANT PATHOLOGY, MARINE BIOLOGY. *Educ:* Goshen Col, AB, 59; Ore State Univ, PhD(plant path), 62. *Prof Exp:* From asst prof to assoc prof, 62-70, PROF BIOL, GOSHEN COL, 70-, CHMN, DEPT BIOL, 80- *Concurrent Pos:* NIH fel, Inst Marine Sci, Miami, Fla, 64-65. *Mem:* Am Inst Biol Sci. *Res:* Phytopathology, mycology; marine algae; invertebrate zoology; development of temperature-independent substitute for agar-agar as a microbiological medium gelling agent. *Mailing Add:* Dept Nurs/Baccalaureate Goshen Col 1700 S Main St Goshen IN 46526

ROTH, LAURA MAURER, b Flushing, NY, Oct 11, 30; m 52; c 2. SOLID STATE PHYSICS. *Educ:* Swarthmore Col, BA, 52; Radcliffe Col, MA, 53, PhD(physics), 57. *Prof Exp:* Staff physicist, Lincoln Lab, Mass Inst Technol, 56-62; prof physics, Tufts Univ, 62-67; physicist, Res & Develop Ctr, Gen Elec Co, 67-72; Abby Rockefeller Mauze vis prof physics, Mass Inst Technol, 72-73; res prof sci & math, 73-77, PROF PHYSICS, STATE UNIV NY, ALBANY, 77- *Concurrent Pos:* Consult, Lincoln Lab, Mass Inst Technol, 62-67. *Mem:* Fel Am Phys Soc; Sigma Xi. *Res:* Band structure of solids; Bloch electrons in magnetic fields; magnetooptics; magnetism; liquid and amorphous metals. *Mailing Add:* 1270 Ruffner Rd Schenectady NY 12309

ROTH, LAWRENCE MAX, b McAlester, Okla, June 25, 36; m 65; c 2. PATHOLOGY, ELECTRON MICROSCOPY. *Educ:* Vanderbilt Univ, BA, 57; Harvard Med Sch, MD, 60; Am Bd Path, dipl & cert anat path, 66, cert clin path, 68, cert dermatopath, 74. *Prof Exp:* Intern, Univ Ill Res & Educ Hosps, Chicago, 60-61; resident anat path, Barnes Hosp, St Louis, Mo, 61-63, resident surg path, 63-64; resident clin path, Univ Calif Med Ctr, San Francisco, 67-68; from asst prof to assoc prof path, Sch Med, Tulane Univ, 68-71; assoc prof, 71-75, PROF PATH & DIR SURG PATH DIV, SCH MED, IND UNIV, INDIANAPOLIS, 75- *Concurrent Pos:* Nat Inst Gen Med Sci sr res trainee, Hormone Lab, Karolinska Inst, Stockholm, Sweden, 64-65; asst path, Sch Med, Wash Univ, 61-64; series ed, Contemporary Issues Surg Pathol. *Mem:* Am Asn Path; Am Soc Clin Path; Int Acad Path; Int Soc Gynec Pathologists. *Res:* Gynecological and endocrine pathology; steroid chemistry; electron microscopy; ovarian tumors. *Mailing Add:* Dept Path Ind Univ Med Ctr Indianapolis IN 46202-5280

ROTH, LAWRENCE O(RVAL), b Hillsboro, Wis, June 7, 28; m 54; c 2. AGRICULTURAL ENGINEERING. *Educ:* Univ Wis, BS, 49 & 51; Okla State Univ, MS, 56, PhD(eng), 65. *Prof Exp:* From instr to assoc prof, 51-72, PROF AGR ENG, OKLA STATE UNIV, 72- *Concurrent Pos:* Ford Found residency eng pract fel, 67-68. *Mem:* Am Soc Agr Eng; Am Soc Eng Educ; Weed Sci Soc Am; Nat Soc Prof Engrs. *Res:* Farm power and machinery; machine design and development for drift control of pesticides and mechanization of horticultural crop production. *Mailing Add:* Dept Agr Eng Okla State Univ Stillwater OK 74074

ROTH, LEWIS FRANKLIN, b Poplar, Mont, Apr 12, 14; m 45; c 2. FOREST PATHOLOGY, MYCOLOGY. *Educ:* Miami Univ, BA, 36; Univ Wis, PhD(plant path), 40. *Prof Exp:* Asst plant path, Univ Wis, 36-38; from instr to assoc prof, 40-57, prof, 57-79, EMER PROF BOT & PLANT PATH, ORE STATE UNIV, 79-, CONSULT. *Concurrent Pos:* Sci aide, Forest Prod Lab, US Forest Serv, 40, collab, 58- *Mem:* Fel Am Phytopath Soc; Mycol Soc Am; Soc Am Foresters. *Res:* Epidemiology; life history and control of forest diseases; root diseases; dwarf mistletoe; pine needle blight; aquatic fungi. *Mailing Add:* 4798 Becker Circle Albany OR 97321

ROTH, LINWOOD EVANS, b Ft Wayne, Ind, Mar 8, 29; m 49; c 2. CELL BIOLOGY, ELECTRON MICROSCOPY. *Educ:* Univ Ind, AB, 50; Northwestern Univ, MS, 55; Univ Chicago, PhD(zool), 57. *Prof Exp:* Electron microscopist, Med Sch, Univ Ind, 50-52; sr res technician, Div Biol & Med Res, Argonne Nat Lab, 52- 54, asst scientist, 54-60; from assoc prof to prof biochem & biophys, Iowa State Univ, 60-67, asst dean grad col, 62-67; prof biol & dir, div biol, Kans State Univ, 67-76; vchancellor grad studies & res, 76-84, PROF CELL BIOL, UNIV TENN, KNOXVILLE, 76- *Concurrent Pos:* Ed, Europ J Cell Biol, 74-; dir, Hungarian Exchange Prog, 84-; adv bd, Int Review Cytol. *Mem:* AAAS; Am Soc Cell Biol; Sigma Xi; Am Sci Affil; Electron Micros Soc Am. *Res:* Cell biology; analytical electron microscopy; nitrogen fixation in legumes; aluminum toxicity. *Mailing Add:* 507 Cherokee Blvd Knoxville TN 37919

ROTH, MARIE M, b Boston, Mass, Apr 30, 26; m 51; c 4. ORGANIC & GENERAL CHEMISTRY. *Educ:* Mt Holyoke Col, BA, 45, MA, 47; Univ Wis, PhD(org chem), 52. *Prof Exp:* Res librn, Pittsburgh Plate Glass Co, 51-52; abstractor, Chem Abstr, 52-59; lectr, Univ Wis Ctr Syst, Waukesha County, 71 & 73, lectr gen chem, Washington County, 72-79, 86; lectr, Marquette Univ, 81-82, res assoc, 82-84; RETIRED. *Concurrent Pos:* Adj asst prof, Univ Wis-Milwaukee, 81. *Mem:* Am Chem Soc. *Res:* Heterocyclic compounds and synthesis of antimalarials; synthesis of steroid intermediates; trace solubility studies. *Mailing Add:* 1620 Revere Dr Brookfield WI 53045-2315

ROTH, MARK A, b Elgin, Ill, Aug 12, 57; m 80; c 3. OBJECT-ORIENTED DATABASES. *Educ:* Ill Inst Technol, BS, 78; Air Force Inst Technol, MS, 79; Univ Tex Austin, PhD(computer sci), 86. *Prof Exp:* Simulation systs developer, Computer Sci Div, Data Automation Directorate, Headquarters Air Univ, 80-83; asst prof, 83-86, ASSOC PROF COMPUTER SCI, DEPT ELEC & COMPUTER ENG, SCH ENG, AIR FORCE INST TECHNOL, 86- *Mem:* Asn Comput Mach; Inst Elec & Electronics Engrs Computer Soc; Air Force Asn; Mil Opers Res Soc. *Res:* Database management systems; applications for non-traditional databases; computer-simulated wargames; integration of programming languages, software engineering, computer graphics and data base systems. *Mailing Add:* Sch Eng Air Force Inst Technol Wright-Patterson AFB OH 45433

ROTH, MICHAEL WILLIAM, b Davenport, Iowa, June 30, 52; m 73. APPLIED PHYSICS, NEURAL NETWORKS. *Educ:* MacMurray Col, BA, 71; Univ Ill, Urbana, MS, 72, PhD(physics), 75. *Prof Exp:* Res assoc, Fermi Nat Accelerator Lab, 75-77; sr physicist, 77-88, PRIN STAFF PHYSICIST, APPL PHYSICS LAB, JOHNS HOPKINS UNIV, 88- *Concurrent Pos:* Proj mgr, Appl Physics Lab, Johns Hopkins Univ, 80-83, sect supvr & proj dir, 81-83, lead eng, 83-86, mem, Adv Tech Subcomt Prog Rev Bd, 86-88, prin investr, 89-; Stuart S Janney fel, Johns Hopkins Univ, 87; tech expert, Technol Initiative Game, Naval War Col, 88; mem, Adv Bd Fels & Profs, Johns Hopkins Univ, 89-90, chair, 90-91; assoc ed, Trans Neural Networks, Inst Elec & Electronics Engrs, 89-90, ed-in-chief, 91-; mem, Neural Networks Coun Exec Comt, Inst Elec & Electronics Engrs, 91-; co-chair, Conf on Appln Artificial Neural Networks, Soc Photo-Optical Instrumentation Eng, 91. *Mem:* Am Asn Artificial Intelligence; Inst Elec & Electronic Engrs; Sigma Xi; Int Neural Network Soc; Soc Photo-Optical Instrumentation Eng. *Res:* Neural networks; systems analysis. *Mailing Add:* 5022 Hayload Ct Columbia MD 21044

ROTH, NILES, b New York, NY, Sept 27, 25; m 52; c 3. PHYSIOLOGICAL OPTICS, OPTOMETRY. *Educ:* Univ Calif, Berkeley, BS, 55, MOpt, 56, PhD(physiol optics), 61. *Prof Exp:* Asst res biophysicist, Univ Calif, Los Angeles, 61-69; assoc prof, 69-76, PROF PHYSIOL OPTICS, COL OPTOM, PAC UNIV, 76- *Concurrent Pos:* Res grants, USPHS, 61-65, Am Cancer Soc, 66-67; consult, Long Beach Vet Admin Hosp, 61-69. *Mem:* AAAS; Am Optom Asn; Optical Soc Am; fel Am Acad Optom; Sigma Xi. *Res:* Factors affecting resting ocular refractive state and pupil size; psychophysical and photometric aspects of vision testing. *Mailing Add:* 2113 15th Ave Forest Grove OR 97116

ROTH, NORMAN GILBERT, b Chicago, Ill, Dec 11, 24; m 50; c 6. TECHNICAL MANAGEMENT. *Educ:* Univ Chicago, BS, 47; Univ Ill, MS, 49, PhD(bact), 51. *Prof Exp:* Bacteriologist, Ft Detrick, Md, 51-57; sr res bacteriologist, Whirlpool Corp, 57-60, dir life support, 60-73, waste mgt, 73-77, res & eng spec projs, 77-88, res & eng tech support, 88-91; CONSULT, 91- *Res:* Aerospace life support, Gemini, Apollo, Skylab, Shuttle, Space Station; food, waste, water, sanitation management systems; microbial deterioration; appliance sanitation; psychrophilic bacteria; bacterial spores. *Mailing Add:* 1801 Briarcliff St Joseph MI 49085

ROTH, PAUL FREDERICK, b Pittsburgh, Pa, Feb 9, 32. COMPUTER SIMULATION LANGUAGES. *Educ:* Univ Pittsburgh, BS, 53; Univ Pa, MS, 65. *Prof Exp:* Electronics engr, Bell Aircraft Corp, 53-54; develop engr, Goodyear Aircraft Corp, 55-57; mem sr staff engr, Bendix Corp, 57-60; advan systs engr, Gen Elec Co, 60-66; sr tech staff, Burroughs Corp, 66-72; comput scientist, Nat Bur Standards, US Dept Com, 72-79, chief, Div Comput, 80-82; oper res analyst, US Dept Energy, 79-80; ASSOC PROF COMPUT SCI, VA POLYTECH INST & STATE UNIV, 82- *Concurrent Pos:* Adj lectr comput sci, Villanova Univ, 65-68; adj prof, Univ Md, 75-77 & Va Polytech Inst & State Univ, 75-80; consult, Nat Acad Sci, 81-82. *Mem:* Asn Comput Mach. *Res:* Applications of simulation and modeling to network and computer performance evaluation; computer models for teaching computer science systems topics. *Mailing Add:* Dept Comput Sci & Eng Univ SFla Tampa FL 33620

ROTH, PETER HANS, b Zurich, Switz, June 25, 42; m 88; c 3. MICROPALEONTOLOGY, MARINE GEOLOGY. *Educ:* Swiss Fed Inst Technol, Zurich, dipl, 65, PhD(geol), 71. *Prof Exp:* Asst res geologist, Scripps Inst Oceanog, 71-75; assoc prof, 75-82, PROF GEOL & GEOPHYS, UNIV UTAH, 82- *Concurrent Pos:* Mem, Nat Acad Sci-Acad Sci USSR Exchange, 87. *Honors & Awards:* Kern Prize, Swiss Fed Inst Technol, 70. *Mem:* Am Geophys Union; AAAS; Geol Soc Am; Swiss Geol Soc; Am Asn Petrol Geologists. *Res:* Biostratigraphy, paleoecology and preservation of calcareous nannofossil; early diagenesis of deep-sea carbonates; paleoceanography. *Mailing Add:* Dept Geol & Geophys Univ Utah Salt Lake City UT 84112

ROTH, RAYMOND EDWARD, b Rochester, NY, Oct 29, 18; m 42; c 5. STATISTICS. *Educ:* St Bonaventure Univ, BS, 40, MS, 42; Rochester Univ, PhD(statist), 63. *Prof Exp:* Consult, Gen Elec Co, 43-46; res assoc, Univ Notre Dame, 46-47; asst prof physics, Univ Dayton, 47-49; physicist, Wright-Patterson Air Force Base, 49-50; res assoc, Atomic Energy Proj, Med Sch, Rochester Univ, 53-57; assoc prof math & head dept, St Bonaventure Univ, 58-63, prof & chmn dept, 63-66; prof statist & dir, Comput Ctr, State Univ NY Col Geneseo, 66-68; Archibald Granville Bush prof math, 68-84, EMER PROF, ROLLINS COL, 84- *Concurrent Pos:* Vis staff, med sch & math dept, Univ Okla, 62-63; consult, Civil Aeromed Res Inst, Fed Aviation Agency, Oklahoma City, 62-63 & Med Div, Oak Ridge Inst Nuclear Studies, 64-65. *Mem:* Sigma Xi. *Res:* Flash burn effects; quantal response in a Latin square design of experiment; models. *Mailing Add:* PO Box 940535 Maitland FL 32751-0535

ROTH, RENÉ ROMAIN, b Timisoara, Romania, Feb 24, 28; Can citizen; m 61; c 3. ANIMAL PHYSIOLOGY, HISTORY & PHILOSOPHY OF BIOLOGY. *Educ:* Univ Cluj, MSc, 50; Univ Alta, PhD(comp endocrinol), 69. *Prof Exp:* Lectr bact, Inst Vet Med, Bucharest, Romania, 50-52; lectr zool & parasitol, Inst Vet Med, Arad, 52-56; res fel endocrinol, Univ Timisoara Hosp, 56-60 & Sch Med, Hebrew Univ, Israel, 60-62; asst prof, 66-78, ASSOC PROF ZOOL, UNIV WESTERN ONT, 78- *Concurrent Pos:* Hon lectr, His Med & Sci, Univ Western Ont, 85-89; mem exec comt, Can Soc Theoret Biol, 85- *Mem:* Hist Sci Soc; Am Inst Biol Sci; NY Acad Sci; Can Soc Zool; Can Soc Endocrinol & Metab; Can Soc Theoret Biol. *Res:* Influence of light and temperature on gonad activity and nutrition in the red-back vole; effect of protein intake levels on reproduction and dietary self-selection in rats; theoretical biology; history of biology; comparative physiology of growth and nutrition. *Mailing Add:* Dept Zool Univ Western Ont London ON N6A 5B7 Can

ROTH, RICHARD FRANCIS, b St Louis, Mo, Jan 18, 38; m 63; c 3. PHYSICS. *Educ:* Rockhurst Col, BS, 59; Princeton Univ, MA, 61, PhD(physics), 64. *Prof Exp:* Fel physics, Princeton Univ, 63-64, res assoc, 64-65, instr, 65-66; staff physicist, Comn Col Physics, Univ Mich, 66-69; ASSOC PROF PHYSICS & ASTRON, EASTERN MICH UNIV, 69- *Mem:* Am Phys Soc; Am Asn Physics Teachers. *Res:* Experimental elementary particle physics; use of computers. *Mailing Add:* Dept Physics & Astron Eastern Mich Univ Ypsilanti MI 48197

ROTH, RICHARD LEWIS, b New York, NY, Feb 24, 36. MATHEMATICS. *Educ:* Harvard Univ, BA, 58; Univ Calif, Berkeley, MA, 60, PhD(math), 63. *Prof Exp:* Asst prof, 63-68, assoc prof, 68-86, PROF MATH, UNIV COLO, BOULDER, 86- *Concurrent Pos:* Regional specialist, NSF, Cent Am, 65-66; Fulbright lectr, Colombia, 69. *Mem:* Am Math Soc; Math Asn Am. *Res:* Representations of finite groups; algebra; color symmetry. *Mailing Add:* Dept Math Univ Colo Campus Box 426 Boulder CO 80309-0426

ROTH, ROBERT ANDREW, JR, b McKeesport, Pa, Aug 15, 46; div; c 2. BIOCHEMICAL PHARMACOLOGY, ENVIRONMENTAL MEDICINE. *Educ:* Duke Univ, BA, 68; Johns Hopkins Univ, PhD(biochem toxicol), 75; Am Bd Toxicol, dipl. *Prof Exp:* Toxicol test specialist, US Army Environ Hyg Agency, 69-71; res fel pulmonary pharmacol, Dept Anesthesiol, Yale Univ, 75-77; from asst prof to assoc prof, 82-87, PROF PHARMACOL & TOXICOL, MICH STATE UNIV, EAST LANSING, 87- *Concurrent Pos:* Toxicology Study Section, NIH; Assoc ed, Toxicol Appl Pharmacol. *Mem:* Am Soc Pharmacol Exp Therapeut; Soc Toxicol. *Res:* Removal and metabolism of drugs and hormones by lung; pulmonary toxicology; effect of carbon monoxide and of hypoxic hypoxia on hepatic drug metabolism; role of platelets and neutrophils in toxic responses; hepatic toxicology. *Mailing Add:* Dept Pharmacol Mich State Univ East Lansing MI 48824

ROTH, ROBERT EARL, b Wauseon, Ohio, Mar 30, 37; m 59; c 2. SCIENCE EDUCATION, NATURAL RESOURCES. *Educ:* Ohio State Univ, BS, 59 & 61, MS, 60; Univ Wis, PhD(environ educ), 69. *Prof Exp:* Teacher & conserv educ supvr, Ethical Cult Schs, 61-63; teacher & naturalist, Edwin Gould Found for Children, 63-65; instr outdoor teacher educ, Northern Ill Univ, 65-67; res asst environ educ, Wis Res & Develop Ctr for Cognitive Learning, 67-69; from asst prof to assoc prof, Ohio State Univ, 69-78, chmn dept, 72-85, sr coordr, int affairs, 85-89, PROF NATURAL RESOURCES, ENVIRON MGT EDUC, SCH NAT RES, OHIO STATE UNIV, 79-, ASST DIR/SCH SECY, 89- *Concurrent Pos:* Res assoc, Educ Resources Info Ctr Environ Educ, 70-74; from assoc prof to prof, Ohio Agr Res & Develop Ctr & Fac Sci & Math Educ; vis scholar, Midwest Univ Consortium Int Affairs, Univ Develop Proj, Indonesia, 88; prin investr, US AID Natural Resource Mgt Proj, Environ Educ, Dominican Repub, 81-88. *Honors & Awards:* Publ Prize, J Environ Educ, 73. *Mem:* NAm Asn Environ Educ (pres elect, 75-76, pres, 76-77); Nat Sci Teachers Asn; Asn Interpretive Naturalists; Conserv Educ Asn; AAAS. *Res:* Concept development and attitude formation in environmental management; curriculum development; program modeling and interpretive skill development and evaluation; information analysis in environmental education; program and international environmental education development. *Mailing Add:* 570 Morning St Worthington OH 43085

ROTH, ROBERT EARL, b Springfield, Ill, Mar 3, 25; m 48; c 5. RADIOLOGY. *Educ:* Univ Ill, BS, 47, MD, 49. *Prof Exp:* Intern, St Louis County Hosp, Clayton, Mo, 49-50; resident, US Naval Hosp, San Diego, Calif, US Vet Admin Hosp, Nashville, Tenn & Vanderbilt Univ Hosp, 50-54, asst chief radiol, US Vet Admin Hosp, 54-55; from asst prof to assoc prof, 55-59, prof radiol & chmn dept, 59-69, PROF RADIATION ONCOL & CHMN DEPT, MED COL, UNIV ALA, BIRMINGHAM, 69- *Concurrent Pos:* Actg chief radiol, Vet Admin Hosp, Birmingham, Ala; consult, Vet Admin Hosps, Birmingham & Tuskegee, Ala. *Mem:* AAAS; fel Am Col Radiol; Soc Nuclear Med; Radiol Soc NAm; NY Acad Sci. *Res:* Radiation therapy. *Mailing Add:* 1528 N 26th St Birmingham AL 35234

ROTH, ROBERT HENRY, JR, b Hackensack, NJ, Sept 18, 39; m 63; c 2. NEUROPHARMACOLOGY. *Educ:* Univ Conn, BS, 61; Yale Univ, PhD(pharmacol), 65. *Prof Exp:* From instr to asst prof, 66-69, assoc prof pharmacol, 69-77, PROF PSYCHIAT & PHARMACOL, YALE UNIV, 77- *Concurrent Pos:* Nat Inst Gen Med Sci fel physiol, Karolinska Inst, Sweden, 65-66; USPHS res grants, 66-72. *Mem:* Am Soc Pharmacol & Exp Therapeut. *Res:* Neuropharmacology and neurochemistry, especially related to central nervous system depressants and sleep; monoamines and chemical transmission in the nervous system; endogenous factors in control of neurohumors in the nervous system. *Mailing Add:* Dept Pharmacol Yale Univ Psych 333 Cedar St New Haven CT 06510

ROTH, ROBERT MARK, b Brooklyn, NY, Apr 9, 43; m 64; c 3. BIOLOGY, GENETICS. *Educ:* Brooklyn Col, BS, 63; Brandeis Univ, PhD(biol), 67. *Prof Exp:* Fel biol, Univ Wis-Madison, 68; from asst prof to assoc prof, 68-76, PROF BIOL, ILL INST TECHNOL, 76-, CHMN DEPT, 78- *Concurrent Pos:* Res assoc, Univ Calif, Berkeley, 70; USPHS res career develop award, 72; vis lectr biol chem, Harvard Med Sch, 73-74; vis scientist, Free Univ Berlin, 84, 86, Argonne Nat Lab, 84-87. *Mem:* Genetics Soc Am; Am Soc Photobiol; Am Soc Microbiol. *Res:* Biochemical genetics of yeast; photodynamic action and photosensitivity in microorganisms. *Mailing Add:* Dept Biol Ill Inst Technol Chicago IL 60616

ROTH, ROBERT S, b Chicago, Ill, Aug 21, 26; m 54; c 3. GEOLOGY. *Educ:* Coe Col, BA, 47; Univ Ill, MS, 50, PhD(geol), 51. *Prof Exp:* Res assoc, Eng Exp Sta, Univ Ill, 51; geologist, 51-56, solid state physicist, 57-61, res chemist, 62-68, supvr chemist, 69-81, RES CHEMIST, NAT BUR STANDARDS, 81- *Mem:* Geol Soc Am; Mineral Soc Am; Am Crystallog Asn; Mineral Soc Gt Brit & Ireland; Am Ceramic Soc. *Res:* X-ray crystallography and phase equilibria of ceramic materials. *Mailing Add:* Nat Inst Standards & Technol Ceramics Div Gaithersburg MD 20899

ROTH, ROBERT STEELE, b Philadelphia, Pa, July 3, 30; m 66; c 1. APPLIED MECHANICS. *Educ:* Kenyon Col, AB, 53; Carnegie-Mellon Univ, MS, 54; Harvard Univ, PhD(appl math), 62. *Prof Exp:* Engr math, Aberdeen Proving Ground, 54-56; scientist & group leader mech, Systs Div Avco Corp, 62-74; TECH STAFF APPL MATH, CHARLES STARK DRAPER LAB, INC, CAMBRIDGE, 74- *Concurrent Pos:* Assoc ed, Math Biosci, 72-76. *Mem:* Sigma Xi; Am Acad Mech. *Res:* Structural mechanics; dynamic buckling of thin shells; plastic buckling of thin shells; numerical analysis; nonlinear differential equations; system identification; segmental differential approximation; bioengineering analysis; author of 4 books. *Mailing Add:* 192 Commonwealth Ave Boston MA 02116-2752

ROTH, RODNEY J, b Brockway, Pa, Mar 13, 27; m 54; c 2. MATHEMATICS. *Educ:* Pa State Univ, BA, 51; Univ Iowa, MFA, 53; Duke Univ, PhD(math), 62. *Prof Exp:* Asst prof math, Univ Ky, 61-63 & Univ SFla, 63-66; assoc prof & chmn dept, Upsala Col, 66-70; assoc prof, 71-77, PROF MATH, RAMAPO COL, NJ, 77- *Mem:* Am Math Soc; Math Asn Am; Sigma Xi. *Res:* Algebra; applications of mathematics in behavioral and social science. *Mailing Add:* 32 Clinton Ave Montclair NJ 07042

ROTH, ROLAND RAY, b Stuttgart, Ark, Jan 9, 43; m 64; c 3. ECOLOGY, VERTEBRATE BIOLOGY. *Educ:* Univ Ark, Fayetteville, BS, 66; Univ Ill, Urbana, MS, 67, PhD(zool), 71. *Prof Exp:* Res assoc zool, Univ Ill, Urbana, 71; asst prof, 71-77, ASSOC PROF UNIV DEL, 77- *Mem:* Am Inst Biol Sci; Ecol Soc Am; Am Ornith Union; Cooper Ornith Soc; Soc Conserv Biol. *Res:* Habitat quality, selection and use, especially of birds; breeding ecology and population ecology of wood thrush; urban wildlife; pesticides and birds. *Mailing Add:* Dept Entom & Appl Ecol Univ Del Newark DE 19717-1303

ROTH, RONALD JOHN, b New York, NY, Feb 1, 47. ORGANIC CHEMISTRY. *Educ:* City Col NY, BS, 67; Columbia Univ, PhD(chem), 72. *Prof Exp:* Res assoc chem, Univ Chicago, 72-73; instr chem, Brown Univ, 74-75; asst prof, 75-80, ASSOC PROF CHEM, GEORGE MASON UNIV, 81- *Mem:* Am Chem Soc; Sigma Xi. *Res:* Synthesis of small strained hydrocarbons and their metal catalyzed transformations. *Mailing Add:* George Mason Univ Dept Chem Fairfax VA 22030

ROTH, ROY WILLIAM, b Collingswood, NJ, May 27, 29; m 55; c 3. POLYMER CHEMISTRY. *Educ:* Mass Inst Technol, BS, 50; Univ Mich, MS, 51; Mass Inst Technol, PhD(chem), 55. *Prof Exp:* Res chemist, Stamford Labs, Am Cyanamid Co, Conn, 55-60, group leader, 60-64, mgr prod res sect, 64-67, mgr prod develop, Davis & Geck Dept, Lederle Labs, 68-75; dir prod develop, Kendall Co, Colgate-Palmolive, 76-78; res assoc, Pall Corp, 78-79; chief fiber & fabric tech, Natick Labs, US Army, 79-85; HQ STAFF, US ARMY RES OFF, 86- *Mem:* Am Chem Soc; Soc Plastics Eng. *Res:* Polymer chemistry; medical specialties; product development; non-woven fabrics; specialty filters; sterile-disposable products; chemical protective garments, ballistic protection. *Mailing Add:* US Army Res Off Research Triangle Park NC 27709

ROTH, SANFORD IRWIN, b McAlester, Okla, Oct 14, 32; m 61; c 4. PATHOLOGY, ELECTRON MICROSCOPY. *Educ:* Harvard Univ, MD, 56. *Prof Exp:* From intern to asst resident, Mass Gen Hosp, 56-58, actg asst resident, 58-60, asst, 62-64, asst pathologist, 64-70, assoc pathologist, 70-75; prof path & chmn dept, Col Med, Univ Ark, 75-80; PROF, DEPT PATH, MED SCH, NORTHWESTERN UNIV, 80-; ATTEND PATHOLOGIST, NORTHWESTERN MEM HOSP, CHICAGO, ILL, 80- *Concurrent Pos:* USPHS res trainee, Mass Gen Hosp, 58-60, teaching fel, Harvard Med Sch, 58-60; instr, Sch Med, Tufts Univ, 58-60; asst, Harvard Med Sch, 62-63, instr, 63-64, assoc, 64-67, from asst prof to assoc prof path, 67-75; Am Cancer Soc fac res assoc, 67-72; attend pathologist, Northwestern Univ Med Ctr, 80-; fac, res assoc, Am Cancer Soc, US & Can, 67-72; chief lab serv, Vet Admin Lakeside Med Ctr, 80-85. *Mem:* Electron Micros Soc Am; Int Acad Pathologists; Am Soc Cell Biol; Soc Invest Dermat; Am Asn Pathologists; AAAS; Am Soc Bone & Mineral Res. *Res:* Experimental pathology; pathology of the parathyroid glands; dermatopathology; surgical pathology. *Mailing Add:* Dept Path Northwestern Univ Col Med 303 E Chicago Ave Chicago IL 60611

ROTH, SHELDON H, b Toronto, Ont, June 19, 43; m 68; c 2. NEUROPHARMACOLOGY, DEVELOPMENTAL NEUROTOXICOLOGY. *Educ:* Univ Toronto, BSc, 66, MSc, 69 & PhD (med & pharmacol), 71. *Prof Exp:* Grad fel, Prov of Ont, 68-71; res fel, Can Med Res Coun, 71-73; lectr pharmacol, Oxford Univ, Eng, 72-73; asst prof, 73-77, actg head, 77-78, assoc prof, 77-82, PROF PHARMACOL & THERAPEUT, UNIV CALGARY, 82-, ASSOC PROF ANESTHESIA, 81- *Concurrent Pos:* Med Staff, Foothills Hosp, 76-; hon res fel, Univ London, 81; vis lectr, Harvard Med Sch, Boston, 82, vis prof, 86; vis prof, Can Med Res Coun, 83; div head toxical, Univ Calgary, 87- *Mem:* Pharmacol Soc Can; Am Asn Advan Sci; Soc Neurosci; Int Brain Res Orgn; Anaesthesia Res Soc. *Res:* Mechanisms of action of anaesthetic agents; toxicology of environmental chemicals and drugs on developing nervous system. *Mailing Add:* Dept Pharmacol-Therapeut Univ Calgary 3330 Hospital Dr NW Calgary AB T2N 4N1 Can

ROTH, SHIRLEY H, chemistry; deceased, see previous edition for last biography

ROTH, STEPHEN, b New York, NY, Sept 3, 42; m 81; c 3. MEMBRANE BIOCHEMISTRY, CARBOHYDRATE BIOCHEMISTRY. *Educ:* Johns Hopkins Univ, AB, 64; Case Western Reserve Univ, PhD(embryol), 68. *Hon Degrees:* MA, Univ Pa, 80. *Prof Exp:* Fel biochem, Johns Hopkins Univ, 68-70, from asst prof to assoc prof develop biol, 70-80; chmn, Dept Biol, 82-88, PROF DEVELOP BIOL, UNIV PA, 80- *Concurrent Pos:* Mem, Cell Biol Study Sect, NIH, 74-78; chmn sci adv bd, Neose Pharmaceut Co, Inc; Res Career Develop Award, NIH, 77-81; res award, Tokyo Med Soc, 87. *Mem:* Am Soc Develop Biol; AAAS; Int Soc Develop Biol; Am Soc Cell Biologists. *Res:* Membrane biochemistry as it controls morphogenesis in vertebrate embryos; glycosyltcans fecases. *Mailing Add:* Dept Biol Univ Pa Philadelphia PA 19104

ROTH, THOMAS ALLAN, b Cudahy, Wis, Dec 12, 37; m 67. METALLURGICAL ENGINEERING, MATERIALS SCIENCE. *Educ:* Univ Wis, BS, 60, MS, 61, PhD(metall eng), 67. *Prof Exp:* Teaching asst metall eng, Univ Wis, 61-62, instr, 64-65; from asst prof to assoc prof indust eng, 65-76, ASSOC PROF CHEM ENG, KANS STATE UNIV, 76- *Concurrent Pos:* Prin investr, Kans State Univ Res Coord Coun grants, 66-68; proj dir, Nat Sci Found Instr Sci Equip grant, 68-70; acad year res grant, Kans State Univ, 68-70; Kans State Univ Bur Gen Res grant, 70; consult, 74-; prin investr, US Air Force Off Sci Res grant, 75-77; bk reviewer J Electrochem Soc, 77-80; fac intern, Am Soc Testing & Mat, 81; US Air Force res assoc & Southereastern Ctr Elec Eng Educ fac fel, Wright-Patterson AFB, Ohio, 81. *Mem:* Am Soc Testing & Mat; Am Soc Metals; Metall Soc. *Res:* Physical metallurgy; influence of adsorbed gases on the surface properties of metals; hydrogen embrittlement of iron and steel; ionic thermoconductivity; interfacial free energy. *Mailing Add:* Dept Chem Eng Durland Hall Kans State Univ Manhattan KS 66506

ROTH, THOMAS FREDERIC, b Detroit, Mich, Feb 28, 32; m 63. CELL BIOLOGY, DEVELOPMENTAL BIOLOGY. *Educ:* Tufts Univ, BS, 54; Harvard Univ, MA, 59, PhD(biol), 64. *Prof Exp:* USPHS fel biol, Harvard Univ, 64; fel, Univ Calif, San Diego, 64-66, asst res biologist, 66-67, asst prof biol, 67-72; assoc prof, 72-77, PROF BIOL, UNIV MD BALTIMORE COUNTY, 77- *Mem:* Soc Develop Biol; Am Soc Chem & Molecular Biol; Am Soc Cell Biol. *Res:* Molecular biology; biochemistry and ultra structure of receptor mediated protein transport; regulation of coated vesicles. *Mailing Add:* Dept Biol Sci Univ Md Baltimore County Baltimore MD 21228

ROTH, WALTER, b New York, NY, Dec 4, 22; m 47; c 2. PHYSICAL CHEMISTRY. *Educ:* City Col New York, BS, 44; NY Univ, MS, 47; Rensselaer Polytech Inst, PhD(chem), 54. *Prof Exp:* Chemist, Kellex Corp, NJ, 44-45; chemist, Carbide & Carbon Chem Co, Tenn, 45-46; phys chemist, US Bur Mines, Pa, 48-53; phys chemist, Res Lab, Gen Elec Co, NY, 54-59; sr physicist, Armour Res Found, Ill Inst Technol, 59-63; mgr gaseous electronics res, Xerox Corp, 63-68; mgr, San Diego Opers, KMS Technol Ctr, 68-70; vpres & tech dir, Diag Instruments, Inc, 70-73; INDEPENDENT CONSULT, 73- *Concurrent Pos:* Mem, Sci Adv Coun, Rensselaer Polytech Inst, 64-66; arbitrator, Am Arbitration Asn, 71- *Mem:* Am Chem Soc; Am Phys Soc; Sigma Xi. *Res:* Chemical kinetics and mechanisms of light emission; relaxation processes behind shock waves in gases; combustion kinetics and spectroscopy; processes in gas discharges; photochemistry. *Mailing Add:* 8241 El Paseo Grande La Jolla CA 92037

ROTH, WALTER JOHN, b Ann Arbor, Mich, July 20, 39; m 59; c 2. OPERATIONS RESEARCH. *Educ:* Univ Mich, Ann Arbor, BS, 61; Univ NMex, MS, 63, PhD(math), 71. *Prof Exp:* Engr, Bendix Systs Div, Bendix Corp, 59-61; staff mem, Sandia Corp, 61-65; ASST PROF MATH, UNIV NC, CHARLOTTE, 68- *Mem:* Am Math Soc; Oper Res Soc Am. *Res:* Partial differential equations. *Mailing Add:* Dept Math Univ NC University Sta Charlotte NC 28223

ROTH, WILFRED, b New York, NY, June 24, 22; m 44; c 4. ELECTRONICS. *Educ:* Columbia Univ, BS, 43; Mass Inst Technol, PhD(physics), 48. *Prof Exp:* Mem staff, Radiation Lab, Mass Inst Technol, 43-45, assoc, Res Lab Electronics, 46-47; chief physicist, Rieber Res Lab, NY, 48; develop group leader, Harvey Radio Labs, Mass, 48-49; sect head, Res Div, Raytheon Mfg Co, 49-50; co-dir, Rich-Roth Labs, 50-55; dir, Roth Lab Phys Res, 55-64; chmn dept, 64-80, PROF ELEC ENG, UNIV VT, 64- *Concurrent Pos:* Partner, Rich Roth Labs, 50-55; treas & dir res, Ultra Viscoson Corp, 52-53; chmn bd dirs, Roth Lab Phys Res, 55-68. *Mem:* AAAS; fel Acoust Soc Am; fel Inst Elec & Electronics Engrs; Sigma Xi. *Res:* Transducers; biomedical engineering; ultrasonic engineering; system dynamics. *Mailing Add:* Univ Vt Burlington VT 05401

ROTHAUGE, CHARLES HARRY, electrical engineering; deceased, see previous edition for last biography

ROTHAUS, OSCAR SEYMOUR, b Baltimore, Md, Oct 21, 27; m 53; c 1. MATHEMATICS. *Educ:* Princeton Univ, AB, 48, PhD(math), 58. *Prof Exp:* Mathematician, Dept Defense, 53-60; mathematician, Inst Defense Anal, 60-65, dep dir, 63-65; vis prof math, Yale Univ, 65-66; chmn dept, 73-76, PROF MATH, CORNELL UNIV, 66- *Mem:* Am Math Soc. *Res:* Lie groups; geometry; several complex variables. *Mailing Add:* Dept Math Cornell Univ White Hall Ithaca NY 14853-7901

ROTHBART, HERBERT LAWRENCE, b Feb 5, 37; US citizen; m 61; c 1. PHYSICAL & ANALYTICAL CHEMISTRY, RESEARCH ADMINISTRATION. *Educ:* Brooklyn Col, BS, 58; Rutgers Univ, PhD(chem), 63. *Prof Exp:* From instr to asst prof chem, Rutgers Univ, 62-66; head separation & compos invests, 66-76, chief phys chem lab, 76-80, dir eastern regional res lab, 80-83, DIR N ATLANTIC AREA, AGR RES SERV, USDA, 84- *Concurrent Pos:* Consult. *Honors & Awards:* Presidential Ran Vi Award, 89. *Mem:* Am Chem Soc; Inst Food Technol; AAAS. *Res:* Study of the fundamentals of separation processes including equilibrium and transport phenomena; spectroscopy; electron microscopy; computer applications; development of mathematical representations to describe systems and predict efficient separations; physical properties of food components. *Mailing Add:* 411 Norfolk Rd Flourtown PA 19031

ROTHBERG, GERALD MORRIS, b NJ, May 14, 31; m 54; c 2. SOLID STATE PHYSICS. *Educ:* Mass Inst Technol, BS, 52; Columbia Univ, PhD(physics), 59. *Prof Exp:* Adams res fel, Univ Leiden, 58-59; asst prof physics, Rutgers Univ, 59-66; assoc prof, 66-70, head metallurgy dept & dir, Cryogenics Ctr, 74-77, prof physics, 70-80, PROF MAT & METALL ENG, STEVENS INST TECHNOL, 80- *Concurrent Pos:* Fulbright lectr & res, Univ Barcelona, 64-65, Fulbright sr lectr, 72-73. *Honors & Awards:* Jess H Davis Mem Res Award, Stevens Inst Technol, 75. *Mem:* AAAS; Am Phys Soc; Sigma Xi; Am Soc Metals. *Res:* Mossbauer effect; high pressures; cryogenics. *Mailing Add:* Dept Mat & Metall Eng Stevens Inst Technol Hoboken NJ 07030

ROTHBERG, JOSEPH ELI, b Philadelphia, Pa, May 15, 35; m 58. PHYSICS. *Educ:* Univ Pa, BA, 56; Columbia Univ, MA, 58, PhD(physics), 63. *Prof Exp:* Res assoc physics, Yale Univ, 63-64, from instr to asst prof, 64-69; assoc prof, 69-74, PROF PHYSICS, UNIV WASH, 74- *Mem:* Fel Am Phys Soc. *Res:* Elementary particle physics; experimental physics. *Mailing Add:* Dept Physics Univ Wash Seattle WA 98195

ROTHBERG, LEWIS JOSIAH, b New York, NY, Sept 22, 56. QUANTUM CONFINED SEMICONDUCTORS, SURFACE SCIENCE. *Educ:* Univ Rochester, BS, 77; Harvard Univ, PhD(physics), 84. *Prof Exp:* MEM TECH STAFF, AT&T BELL LABS, 83- *Mem:* Am Phys Soc; Optical Soc Am; Electrochem Soc. *Res:* Transient optical spectroscopy of processes in condensed matter including carrier dynamics in reduced dimensionality semiconductors, protein dynamics and surface adsorbate relaxation. *Mailing Add:* AT&T Bell Labs 600 Mountain Ave Murray Hill NJ 07974-2070

ROTHBERG, SIMON, b New York, NY, Mar 7, 21; m 46; c 3. BIOCHEMISTRY OF THE SKIN. *Educ:* Columbia Univ, BS, 48; Georgetown Univ, MS, 52, PhD(biochem), 56. *Prof Exp:* Phys chemist, Nat Bur Stand, 48-53; biochemist, Nat Heart Inst, 53-56; biochemist, Nat Cancer Inst, 57-70; res prof, 70-81, EMER PROF, MED COL VA, VA COMMONWEALTH UNIV, 81- *Concurrent Pos:* NIH res grants, 71-74, 74-78; vis scientist, Cambridge, 64-65; Med Col Va support grants, 78-80. *Mem:* AAAS; Am Chem Soc; Am Soc Biol Chem; Brit Biochem Soc; Sigma Xi; Am Soc Cell Biol. *Res:* Mechanisms of enzymatic reactions; decarboxylation; oxygenases; structure of normal and abnormal keratin; enzymes of epidermis; biochemical regulation of epidermal proliferation and keratinization in normal and pathological skin; role of chalone; DNA catabolism; cyto skeletal proteins of the epidermis. *Mailing Add:* Med Col Va Va Commonwealth Univ 2717 Kenmore Rd Richmond VA 23225

ROTHBLAT, GEORGE H, b Willimantic, Conn, Oct 6, 35; m 57; c 2. BIOCHEMISTRY. *Educ:* Univ Conn, BA, 57; Univ Pa, PhD(microbiol), 61. *Prof Exp:* Assoc, Wistar Inst, 61-66, assoc mem, 66-71, mem, 71-76; PROF BIOCHEM, MED COL PA, 76- *Concurrent Pos:* Asst prof, Sch Med, Univ Pa, 66-71, assoc prof, 71-76; estab investr, Am Heart Asn, 70-75, fel coun arteriosclerosis. *Mem:* AAAS; Tissue Cult Asn; Am Soc Biol Chemists; Sigma Xi. *Res:* Cellular lipid metabolism; cholesterol metabolism in tissue culture cells. *Mailing Add:* Dept Physiol & Biochem Med Col Pa 3300 Henry Ave Philadelphia PA 19129

ROTHCHILD, IRVING, b New York, NY, Dec 2, 13; m 35, 58; c 1. REPRODUCTIVE ENDOCRINOLOGY. *Educ:* Univ Wis, BA, 35, MA, 36, PhD, 39; Ohio State Univ, MD, 54. *Prof Exp:* Asst dir chem, Michael Reese Hosp, Chicago, Ill, 41-43; physiologist, USDA, Md, 43-48; asst prof physiol, Sch Med, Univ Md, 48-49; asst prof physiol & obstet & gynec, Ohio State Univ, 49-53; assoc prof obstet & gynec, 55-66, prof, 66-82, EMER PROF REPRODUCTIVE BIOL, SCH MED, CASE WESTERN RESERVE UNIV, 82- *Concurrent Pos:* Boerhaave prof, Univ Leiden, 77-78. *Mem:* Fel AAAS; Am Soc Zool; Am Physiol Soc; Endocrine Soc; Soc Study Reproduction; Soc Study Fertility. *Res:* Physiology of reproduction. *Mailing Add:* Dept Ostet/Gynec Case Western Reserve Univ 2441 Kenilworth Rd Cleveland Heights OH 44118

ROTHCHILD, ROBERT, b New York, NY, May 12, 46. ORGANIC CHEMISTRY, INSTRUMENTATION. *Educ:* City Col New York, BS, 67; Columbia Univ, MA, 68, MPhil, 74, PhD(org chem), 75. *Prof Exp:* Instr org chem, Schwartz Col Pharm & Health Sci, Long Island Univ, 68-72; res asst org chem & instrumentation, Columbia Univ, 72-73; adj asst prof adv pharmaceut synthesis, Schwartz Col Pharm & Health Sci, Long Island Univ, 75-76; lectr org chem, Tex A&M Univ, 76-77; adj asst prof, org chem, Brandeis Univ, 77-78; ASST PROF ORG CHEM, JOHN JAY COL CRIMINAL JUSTICE, CITY UNIV NEW YORK, 78- *Res:* Organic mass spectrometry; organic mechanism and synthesis; nuclear magnetic resonance; chromatographic techniques. *Mailing Add:* Dept Sci John Jay Col Criminal Just N 445 W 59th St New York NY 10019

ROTHE, CARL FREDERICK, b Lima, Ohio, Feb 6, 29; m 52; c 2. PHYSIOLOGY, BIOMEDICAL ENGINEERING. *Educ:* Ohio State Univ, BSc, 51, MSc, 52, PhD(physiol), 55. *Prof Exp:* Sr asst scientist, USPHS, 55-58; from instr to assoc prof, 58-70, PROF PHYSIOL, IND UNIV SCH MED, IND UNIV-PURDUE UNIV, INDIANAPOLIS, 70- *Concurrent Pos:* Estab investr, Am Heart Asn, 63-69; indust consult biomed instrumentation, 67-; consult, NIH, 71-75, mem cardiovasc renal study sect, Nat Heart & Lung Inst. *Mem:* Fel Am Physiol Soc; Microcirculatory Soc; Biomed Eng Soc; AAAS; Am Asn Univ Prof; fel Am Heart Asn. *Res:* Cardiovascular physiology; instrumentation for physiological research; computer simulation of physiological systems; bioengineering; control of venous pressure and capacitance vessels. *Mailing Add:* Dept Physiol & Biophys Ind Univ Sch Med 635 Barnhill Dr Indianapolis IN 46202-5120

ROTHE, ERHARD WILLIAM, b Breslau, Ger, Apr 15, 31; US citizen; m 59; c 2. CHEMICAL PHYSICS. *Educ:* Univ Mich, BS, 52, MS, 54, PhD(chem), 59. *Prof Exp:* Staff scientist physics, Gen Dynamics Convair, 59-69; PROF ENG, CHEM ENG, WAYNE STATE UNIV, 69- *Concurrent Pos:* Lectr, San Diego State Col, 65-66; consult, Phys Dynamics, Inc, 75-; adj prof chem, Wayne State Univ, 75-; summer guest researcher, Max Plank Inst, 79-; Gershenson Distinguished Prof, Wayne State Univ, 86-88. *Honors & Awards:* Max Planch Res Prize. *Mem:* AAAS; Am Chem Soc; fel Am Phys Soc; Sigma Xi. *Res:* Physics and chemistry of atomic and molecular collisions; laser-driven chemistry; laser interaction with surfaces; laser combustion-diagnostics. *Mailing Add:* Dept Chem Eng Wayne State Univ Detroit MI 48202

ROTHE, KAROLYN REGINA, b Fayetteville, NC, Jan 28, 47. LIMNOLOGY, ECOLOGY. *Educ:* Univ Fla, BS, 67, PhD(zool), 70. *Prof Exp:* ASST PROF PHYS SCI, CALIF STATE UNIV, CHICO, 71- *Mem:* Ecol Soc Am; Am Soc Limnol & Oceanog. *Res:* Limnology, especially of Littoral Zone; water resources management. *Mailing Add:* Dept Geol & Phys Sci Calif State Univ Chico CA 95929

ROTHEIM, MINNA B, b New York, NY, Dec 27, 33. MICROBIAL GENETICS. *Educ:* Queens Col NY, BS, 54; Amherst Col, MA, 56; Univ Rochester, PhD(biol), 61. *Prof Exp:* Nat Cancer Inst res fel microbial genetics, Univ Rochester, 61-62, res assoc, 62-68, asst prof, 64-68; res assoc, Rockefeller Univ, 68-70; ASSOC PROF MICROBIAL GENETICS, STATE UNIV NY HEALTH SCI CTR, SYRACUSE, 70- *Concurrent Pos:* Nat Inst Allergy & Infectious Dis fel, 61-62. *Mem:* Genetics Soc Am; Am Soc Microbiol; NY Acad Sci; Sigma Xi. *Res:* Molecular genetics of an N group plasmid; molecular basis of plasmid host range. *Mailing Add:* Dept Microbiol & Immunol State Univ NY Health Sci Ctr 750 E Adams St Syracuse NY 13210

ROTHENBACHER, HANSJAKOB, b Blaubeuren, WGer, Jan 21, 28, US citizen; c 3. VETERINARY PATHOLOGY, VETERINARY MICROBIOLOGY. *Educ:* Univ Munich, dipl vet med, 52, DMV, 53; Mich State Univ, MS, 55, PhD(path), 62; Am Col Vet Pathologists, dipl, 62. *Prof Exp:* Asst prof vet sci, Univ Ark, 55-58; res vet path, Mich State Univ, 58-61, pathologist in chg necropsy, 61-63; from assoc prof to prof, 63-88, EMER PROF VET SCI, PA STATE UNIV, UNIVERSITY PARK, 88-, PATH CONSULT, 88- *Mem:* Am Asn Avian Path; Wildlife Dis Asn; Int Acad Path; Am Vet Med Asn; Poultry Sci Asn; Animal Sci Asn. *Res:* Ecologic pathology; diseases of fish; pathology of viral, bacterial and nutritional diseases of animals; endocrine pathology. *Mailing Add:* Dept Vet Sci Pa State Univ University Park PA 16802

ROTHENBERG, ALAN S, b Harvey, Ill, Jan 30, 51. ORGANIC CHEMISTRY. *Educ:* Univ Utah, BS, 72; Pa State Univ, PhD(org chem), 77. *Prof Exp:* Res chemist monomer synthesis, 77-80, RES GROUP LEADER PROD & PROCESS DEVELOP, STAMFORD RES LABS, AM CYANAMID CO, 81- *Mem:* Am Chem Soc. *Res:* Organic synthesis; natural products chemistry; iso- quinoline alkaloids; monomer synthesis; cationic monomers; water treating chemicals. *Mailing Add:* 185 Range Rd Wilton CT 06897-3920

ROTHENBERG, ALBERT, b New York, NY, June 2, 30; m 70; c 3. PSYCHIATRY. *Educ:* Harvard Univ, AB, 52; Tufts Univ, MD, 56; Am Bd Psychiat & Neurol, Dipl Psychiat, 65. *Prof Exp:* Intern, Pa Hosp, 56-57; resident, Dept Psychiat, Sch Med, Yale Univ, 57-60, instr psychiat, 60-61 & 63-64, from asst prof to assoc prof, 64-74; prof psychiat, Sch Med, Univ Conn & chief psychiat clin serv, Univ Health Ctr, 75-79; DIR RES, AUSTEN RIGGS CTR INC, 79-; CLIN PROF PSYCHIAT, HARVARD MED SCH, 86- *Concurrent Pos:* Chief resident, Yale Psychiat Inst, 60-61, asst med dir, 63-64; attend psychiatrist, PR Inst Psychiat, 61-63, chief neuropsychiat, Rodriguez US Army Hosp, 61-63; asst attend psychiatrist, Yale New Haven Hosp, 63-68, attend psychiatrist, 68-84; USPHS res career develop awards, Dept Psychiat, Yale Univ, 64-69 & 69-74; attend psychiatrist, West Haven Vet Admin Hosp, 68-84; vis prof, Pa State Univ, 71, adj prof, 71-79; clin prof psychiat, Yale Univ Sch Med, 74-84; Guggenheim Found fel, 74-75; attend physician, Dempsey Hosp, Farmington, Conn, 75-79; lectr, Sch Med, Harvard Univ, 82-86; fel, Ctr Advan Study Behav Sci, 86-87. *Mem:* Fel Am Psychiat Asn; Am Soc Aesthetics; fel Am Col Psychoanalysts; AAAS; Sigma Xi. *Res:* Psychological and psychiatric basis of the creative process in pure and applied science, literature, visual and graphic arts, and music; psychotherapy; schizophrenia. *Mailing Add:* Austen Riggs Ctr Inc Stockbridge MA 01262

ROTHENBERG, HERBERT CARL, b Brooklyn, NY, Mar 9, 19; m 45; c 3. RESEARCH ADMINISTRATION. *Educ:* Univ Minn, AB, 38; Pa State Col, PhD(physics), 49. *Prof Exp:* Geophysicist & mathematician, Standard Oil Co, Venezuela, 40-42; res engr, Div Phys War Res, Duke Univ, 42-45; res assoc, Acoust Lab, Pa State Univ, 45-49; fel, Mellon Inst, 49-51; physicist & mgr, Electronic Devices Lab, Gen Elec Co, 51-66; phys scientist, Cent Intel Agency, 66-85; CONSULT, 85- *Concurrent Pos:* Mem res proj, Off Sci Res & Develop. *Mem:* Am Phys Soc; Sigma Xi; AAAS. *Res:* Geophysics; acoustics; electronics; microwaves; solid state. *Mailing Add:* 918 Leigh Mill Rd Great Falls VA 22066

ROTHENBERG, LAWRENCE NEIL, b Philadelphia, Pa, July 30, 40; m 71; c 2. MEDICAL PHYSICS. *Educ:* Univ Pa, BA, 62; Univ Wis-Madison, MS, 64, PhD(nuclear physics), 70; Am Bd Radiol, cert radiol physics, 76. *Prof Exp:* Teaching asst physics, Univ Wis-Madison, 62-63, from res asst to res assoc nuclear physics, 66-70; from instr to asst prof, 71-79, ASSOC PROF PHYSICS IN RADIOL, MED COL, CORNELL UNIV, 79-; ASSOC ATTEND PHYSICIST MED PHYSICS, MEM HOSP, MEM SLOAN-KETTERING CANCER CTR, 78-, ASSOC CLIN MEM, 85- *Concurrent Pos:* Am Cancer Soc fel med physics, Mem Hosp, 70; asst physicist, Mem Hosp, Sloan-Kettering Cancer Ctr, 70-73, asst attend physicist, 73-78; asst attend physicist, NY Hosp, 71-; assoc ed, Med Physics; chmn bd dirs, Am Col Med Physics, 87; mem, Nat Coun on Radiation Protection & Measurements, 87- *Mem:* NY Acad Sci; Am Col Radiol; Health Physics Soc; Am Phys Soc; Am Asn Physicists in Med; Am Col Med Physics; Radiol Soc N Am. *Res:* Diagnostic x-ray physics; development of x-ray test methods for United States Public Health Service; computed tomography, mammography, phantums for abdominal radiography, and magnetic resonance imaging. *Mailing Add:* Dept Med Physics Mem Hosp Mem-Sloan Kettering Cancer Ctr 1275 York Ave New York NY 10021

ROTHENBERG, MORTIMER ABRAHAM, b New York, NY, June 3, 20; m 44; c 4. NEUROCHEMISTRY, ENVIRONMENTAL SCIENCES. *Educ:* Univ Louisville, BA, 41; NY Univ, MS, 42; Columbia Univ, PhD(biochem), 49. *Prof Exp:* Res assoc neurol & biochem, Columbia Univ, 47-49; res assoc & asst prof, Inst Radiobiol & Biophys, Univ Chicago, 49-51; chief biochemist, Chem Corps Proving Ground, US Dept Army, 51-53, chief chem div, 53-55, dir res, 55-57, sci dir, 57-63, sci dir, Dugway Proving Ground, 63-68, sci dir, Deseret Test Ctr, Ft Douglas, 68-73, sci dir, Dugway Proving Ground, 73-81; CONSULT, NAT DEFENSE & ENVIRON SCI, 81- *Concurrent Pos:* Mem corp, Woods Hole Marine Biol Lab, Mass. *Mem:* Fel AAAS; emer mem Am Chem Soc; fel Am Inst Chemists; NY Acad Sci. *Res:* Physiology and biochemistry of nerve conduction and transmission using inhibitors, drugs pesticides; enzymology and protein isolation and purification; transport and dilution of gases and aerosols in the atmosphere; atmospheric dynamics. *Mailing Add:* 2233 East 3980 S Salt Lake City UT 84124

ROTHENBERG, RONALD ISAAC, b New York, NY; m 77. MATHEMATICS, OPERATIONS RESEARCH. *Educ:* City Col New York, BSChE, 58; Northwester Univ, Evanston, MS, 60; Univ Calif, Davis, PhD(eng), 64. *Prof Exp:* Lectr, 60-62 & 64-66, asst prof, 66-80, ASSOC PROF MATH, QUEEN'S COL, NY, 80- *Concurrent Pos:* Vchair Metrol, NY chap Math Asn Am, 87-89. *Mem:* Math Asn Am; Sigma Xi. *Res:* Ordinary and partial differential equations; probability and statistics; engineering mathematics; mathematics for optimization; applied mathematics; author of four books; computing relating to mathematics. *Mailing Add:* Dept Math Queens Col Flushing NY 11367

ROTHENBERG, SHELDON PHILIP, b New York, NY, May 28, 29; m 56; c 2. HEMATOLOGY, INTERNAL MEDICINE. *Educ:* NY Univ, AB, 50; Chicago Med Sch, MD, 55. *Prof Exp:* From instr to prof med, New York Med Col, 61-80, chief hemat/oncol sect, 69-80; PROF MED, STATE UNIV NY DOWNSTATE, 80- *Mem:* Am Soc Clin Oncol; Am Soc Hemat; Am Soc Clin Invest; Am Fedn Clin Res; Asn Cancer Res; Asn Am Prof. *Res:* Study of metabolism and adsorption of vitamin B-12 and folic acid and the interrelationship of these cofactors in enzymatic reactions in health and disease. *Mailing Add:* Dept Med State Univ NY Downstate Med Ctr 450 Clarkson Ave Brooklyn NY 11203

ROTHENBERG, STEPHEN, b New York, NY, Feb 3, 41; div; c 1. COMPUTER SYSTEMS. *Educ:* Carnegie Inst Technol, BS, 62; Univ Wash, PhD(phys chem), 66. *Prof Exp:* Res assoc, Princeton Univ, 66-68; mgr chem applns, Univ Comput Co, 68-71; mgr tech serv, Info Systs Design, 71-73, vpres comput servs, 73-76; PRES, ROTHENBERG COMPUT SYST, INC, 76- *Concurrent Pos:* Nat Ctr Atmospheric Res fel, 68. *Mem:* Int Word Processing Asn. *Res:* Computer calculation of molecular electronic structure; chemical information systems; programming languages; computer-based office systems; computer systems and applications. *Mailing Add:* 4154 Interdale Way Palo Alto CA 94306

ROTHENBUHLER, WALTER CHRISTOPHER, b Monroe Co, Ohio, May 4, 20; m 44; c 4. ZOOLOGY. *Educ:* Iowa State Col, BS, 50, MS, 52, PhD(genetics, zool), 54. *Prof Exp:* Res assoc genetics, Iowa State Univ, 50-54, from asst prof to prof apicult, 54-62; prof, 62-85, EMER PROF APICULT, ETHOLOGY & BEHAV GENETICS, OHIO STATE UNIV, 85- *Concurrent Pos:* Mem comt on African honey bee, Nat Res Coun. *Mem:* Entom Soc Am; Int Bee Res Asn; Sigma Xi. *Res:* Biology of honey bees, particularly genetics, gynandromorphism, behavior and disease resistance; animal behavior; behavior genetics of honey bees and other animals. *Mailing Add:* 6226 Alrojo St Worthington OH 43085

ROTHENBURY, RAYMAND ALBERT, b London, Eng, May 14, 37; m 63; c 2. ACADEMIC ADMINISTRATION. *Educ:* Univ Exeter, BSc, 58; Univ London, PhD(chem), 61. *Prof Exp:* Fel inorg chem with Prof R J Gillespie, McMaster Univ, 61-63; res chemist, Dow Chem Can Ltd, 63-66; lectr chem, 66-68, chmn technol, 68-75, dir technol & appl arts, 75-81, DEAN TECHNOL & APPL SCI, LAMBTON COL APPL ARTS & TECHNOL, 81- *Mem:* Fel Chem Inst Can; Prof Engr Province Ont. *Mailing Add:* Lambton Col Appl Arts & Technol Box 969 Sarnia ON N7T 7K4 Can

ROTHER, ANA, b Brandenburg/Havel, Ger, July 14, 31. PHARMACOGNOSY. *Educ:* Nat Univ Colombia, BS, 51; Univ Munich, Dr rer nat(pharm), 58. *Prof Exp:* Teaching asst pharm chem, Nat Univ Colombia, 51-53; chemist-analyst, Beneficencia, Bogota, 53-54 & Squibb & Sons, Calif, 58-59; fel, 59-61, res asst, 61-64, from instr to asst prof, 64-70, RES ASSOC PHARMACOG, SCH PHARM, UNIV CONN, 70- *Mem:* Am Chem Soc; Am Soc Pharmacog; Am Soc Plant Physiol; AAAS; Tissue Cult Asn. *Res:* Isolation, structure and biosynthesis of secondary plant constituents; alkaloids; plant tissue culture. *Mailing Add:* Sch Pharm Univ Conn U-92 Storrs CT 06268

ROTHERHAM, JEAN, b Pompton Lakes, NJ, Jan 4, 22. BIOCHEMISTRY, HISTONE CHEMISTRY. *Educ:* Montclair State Teachers Col, BA, 42; Univ NC, MS, 52, PhD, 54. *Prof Exp:* Fel, Nat Cancer Inst, 54 & Nat Heart & Lung Inst, 55, biochemist, Nat Cancer Inst, 56-81; CONSULT, 81- *Mem:* Sigma Xi. *Res:* Nucleic acids; intermediary metabolism of normal and tumor tissue; proteins. *Mailing Add:* 1307 Noyes Dr Silver Spring MD 20910-2720

ROTHERMEL, JOSEPH JACKSON, b Reading, Pa, Nov 4, 18; m 43; c 5. CHEMISTRY, GLASS TECHNOLOGY. *Educ:* Franklin & Marshall Col, BS, 40; Univ Pittsburgh, PhD(chem), 48. *Prof Exp:* Asst chem, Univ Pittsburgh, 40-44; res chemist, Manhattan proj, Sch Med & Dent, Univ Rochester, 44-46; asst chem, Univ Pittsburgh, 46-48; res chemist, Res Labs, Corning Glass Works, 48-58, mgr chem serv, 58-66, mgr tech anal, 66-68, sr eng assoc, Mfg & Eng Div, 68-83; RETIRED. *Honors & Awards:* Eugene C Sullivan Award, Corning Sect, Am Chem Soc, 73. *Mem:* Am Chem Soc; Am Ceramic Soc; Soc Glass Technol. *Res:* Non-silicate glass compositions; glass melting; borate and phosphate glasses; x-ray absorbing transparent barriers. *Mailing Add:* 540 Powder House Rd Corning NY 14830

ROTHFELD, LEONARD B(ENJAMIN), b New York, NY, Feb 20, 33; m 67; c 1. COAL TECHNOLOGY, ENGINEERING & CONSTRUCTION. *Educ:* Cornell Univ, BChE, 55; Univ Wis, MS, 56, PhD(chem eng), 61. *Prof Exp:* Res engr, Emeryville Res Ctr, Shell Develop Co, 61-62, group leader eng res, Houston Res Lab, Shell Oil Co, 62-67, sr engr, NY, 67-70, spec assignment, Shell Int Petrol Maatschapij, The Hague, Holland, 70-71, group leader, Martinez Refinery, Shell Oil Co, Calif, 71-72, spec assignment, Bellaire Res Ctr, Shell Develop Co, 72-73; dir process eng, Synthetic Crude & Minerals Div, Atlantic Richfield Co, Los Angeles, 73-74, mgr process eng, Synthetic Crude & Minerals Div, ARCO Coal Co, 74-81, mgr gen eng, 81-86; CONSULT, 86- *Mem:* Am Inst Chem Engrs; Soc Mining Engrs-Inst Mining Engrs. *Res:* Synthetic fuels; shale oil; coal processing; coal mine facilities. *Mailing Add:* 7215 W Eighth Pl Denver CO 80202

ROTHFIELD, LAWRENCE I, b New York, NY, Dec 30, 27; m 53; c 4. MICROBIOLOGY, BIOCHEMISTRY. *Educ:* Cornell Univ, AB, 47; NY Univ, MD, 51. *Prof Exp:* From intern to asst resident internal med, Bellevue Hosp, New York, 51-53; asst resident, Presby Hosp, New York, 55-56; asst prof med, NY Univ, 62-64; from asst prof to assoc prof molecular biol, Albert Einstein Col Med, 64-68; chmn dept, 68-80, PROF MICROBIOL, SCH MED, UNIV CONN, 68- *Concurrent Pos:* Consult molecular biol sect, NIH, 70-74; mem microbiol & immunol adv comt, President's Biomed Res Panel, 75; chmn, Microbial Physiol Div, Am Soc Microbiol, 75; consult biochem, Biophys Rev Comt, NSF, 79-83. *Mem:* Am Soc Biol Chem; Am Soc Microbiol. *Res:* Membrane molecular biology. *Mailing Add:* Dept Microbiol Univ Conn Health Ctr Farmington CT 06032

ROTHFIELD, NAOMI FOX, b New York, NY, Apr 5, 29; m 53; c 4. MEDICINE. *Educ:* Bard Col, BA, 50; NY Univ, MD, 55. *Prof Exp:* Clin asst med, Sch Med, Univ Frankfurt, 54-55; intern, Lenox Hill Hosp, New York, 55-56; asst in med, Sch Med, NY Univ, 59-61, from instr to asst prof med, 61-68; assoc prof, 68-73, PROF MED, SCH MED, UNIV CONN, 73-, HEAD ARTHRITIS SECT, 71- *Concurrent Pos:* Fel, Sch Med, NY Univ, 56-59, Arthritis Found clin scholar, 64-68; consult, Hartford Hosp, 71- *Mem:* Am Soc Clin Invest; Am Rheumatism Asn; Am Asn Immunol; Am Fedn Clin Res; Asn Am Physicians. *Res:* Connective tissue diseases and their pathogenesis; antinuclear and anti-DNA antibodies. *Mailing Add:* Dept Med Univ Conn Sch Med Farmington CT 06107

ROTHFUS, JOHN ARDEN, b Des Moines, Iowa, Dec 25, 32; m 59; c 2. RESEARCH ADMINISTRATION, BIOCHEMISTRY. *Educ:* Drake Univ, BA, 55; Univ Ill, PhD(chem), 60. *Prof Exp:* Asst biochem, Univ Ill, 55-59; USPHS fel, Col Med, Univ Utah, 59-61, instr, 61-63; asst prof, Sch Med, Univ Calif, Los Angeles, 63-65; prin res chemist, USDA, 65-70, invest head, 70-73, res leader, 73-90, LEAD SCIENTIST, NAT CTR AGR UTIL RES, AGR RES SERV, USDA, 90- *Mem:* AAAS; Am Chem Soc; NY Acad Sci; Am Soc Plant Physiologists; Am Oil Chemists Soc. *Res:* Protein chemistry; structure and function of conjugated macromolecules; lipid chemistry; plant cell and tissue culture; plant biochemistry. *Mailing Add:* Northern Regional Res Ctr 1815 N University Peoria IL 61604

ROTHFUS, ROBERT R(ANDLE), b Rochester, NY, May 13, 19; m 42; c 3. CHEMICAL ENGINEERING. *Educ:* Univ Rochester, BS, 41; Carnegie Inst Technol, MS, 42, DSc(chem eng), 48. *Prof Exp:* Chem engr, Eastman Kodak Co, NY, 42-46; from instr to assoc prof chem eng, 47-59, head dept, 71-78, PROF CHEM ENG, CARNEGIE-MELLON UNIV, 59- *Mem:* Am Chem Soc; Am Soc Mech Engrs; Am Inst Chem Engrs. *Res:* Flow of fluids in conduits; heat transfer by convection; fine particle technology; process dynamics and control. *Mailing Add:* Dept Chem Eng Carnegie-Mellon Univ 5000 Forbes Ave Pittsburgh PA 15213-3876

ROTHKOPF, MICHAEL H, b New York, NY, May 20, 39; m 60; c 2. OPERATIONS RESEARCH. *Educ:* Pomona Col, BA, 60; Mass Inst Technol, MS, 62, PhD(opers res), 64. *Prof Exp:* Mathematician, Shell Develop Co, Shell Oil Co, 64-67, res supvr, 67-71, head planning methods & models div, Shell Int Petrol Co, 71-73; scientist, Xerox Palo Alto Res Ctr, 73-82; prog leader & energy anal, Lawrence Berkeley Lab, 82-88; RUTGERS CTR OPER RES, 88- *Concurrent Pos:* Instr, Univ Calif Exten, 66-78; asst prof math, Calif State Univ, Hayward, 67-69, lectr, Sch Bus & Econ, 74, 76-78; lectr, Dept Med Info Sci, Univ Calif, San Francisco, 74-75; consult, assoc prof, Dept Eng, Econ Syst, Stanford Univ, 80-; lectr, Grad Sch Bus, Univ Santa Clara, 80. *Mem:* Opers Res Soc Am; Inst Mgt Sci; Am Econ Asn; Int Asn Energy Econ. *Res:* Mathematical modeling useful for improving decisions; management science generally; energy economics; microeconimics. *Mailing Add:* Rutgers Ctr Oper Res Rutgers Univ New Brunswick NJ 08903

ROTHLEDER, STEPHEN DAVID, b New York, NY, Mar 7, 37; m 60; c 2. SYSTEM DESIGN & SYSTEM SCIENCE, UNDERWATER ACOUSTICS. *Educ:* NY Univ, BS, 57; Mass Inst Technol, SM, 59, PhD(plasma physics), 62. *Prof Exp:* Res assoc plasma physics lab, Princeton Univ, 62-66; DISTINGUISHED MEM TECH STAFF, OCEAN SYSTS RES DEPT, BELL LABS, INC, 66- *Mem:* Am Phys Soc. *Res:* Plasma physics; reentry physics; ocean physics; nuclear engineering. *Mailing Add:* Three Trouville Dr Parsippany NJ 07054

ROTHLISBERGER, HAZEL MARIE, b Elgin, Iowa, May 10, 11. MATHEMATICS. *Educ:* Iowa State Teachers Col, BA, 38; Univ Wis, MA, 50. *Prof Exp:* Actg prin & teacher high sch, Owasa, Iowa, 40-43; prof & head dept, 43-76, EMER PROF MATH, UNIV DUBUQUE, 76- *Mem:* Math Asn Am; Nat Coun Teachers Math. *Mailing Add:* Elgin IA 52141

ROTHMAN, ALAN BERNARD, b Pittsburgh, Pa, July 5, 27; m 63; c 1. PHYSICAL CHEMISTRY, NUCLEAR ENGINEERING. *Educ:* Univ Pittsburgh, BS, 49; Carnegie Inst Technol, MS, 52, PhD(chem), 54. *Prof Exp:* Sr scientist, Bettis Atomic Power Lab, Westinghouse Elec Co, 53-54; res chemist, Glass Div Res Lab, Pittsburgh Plate Glass Corp, 54-57; assoc chemist, Argonne Nat Lab, 57-60; sr scientist, Astronuclear Lab, Westinghouse Elec Corp, 60-61, mgr flight safety sect, 61-62, adv engr, 62-64, adv scientist, Bettis Atomic Power Lab, 64-65; assoc scientist, Space Div, Chrysler Corp, 65-68; CHEMIST, ARGONNE NAT LAB, ARGONNE, 68- *Mem:* Am Nuclear Soc. *Res:* Fields of reactor safety experiments; materials research and development; reactor safety and safeguards analyses; nuclear waste management. *Mailing Add:* 301 Lake Hinsdale Dr Apt 403 Willowbrook IL 60514

ROTHMAN, ALAN MICHAEL, b Philadelphia, Pa, July 7, 43; m 72; c 2. ANALYTICAL CHROMATOGRAPHY, RADIOTRACER CHEMISTRY. *Educ:* Purdue Univ, BS, 65; Pa State Univ, PhD(org chem), 69. *Prof Exp:* Instr radiation safety, Defense Atomic Support Agency Nuclear Weapons Sch, 69-71; SR CHEMIST, ROHM AND HAAS RES LABS, 73- *Mem:* Am Chem Soc; Health Physics Soc; Asn Off Anal Chemists. *Res:* Utilization of radiotracers and chromatography to further research goals of all company research groups, including major fields of plastics, agricultural products, health products, and analysis. *Mailing Add:* 519 Pine Tree Rd Jenkintown PA 19046

ROTHMAN, ALBERT J(OEL), b Brooklyn, NY, Jan 16, 24; m 47; c 3. HIGH TEMPERATURE CERAMICS & INSULATION, CHEMISTRY GENERAL. *Educ:* Columbia Univ, BS, 44; Polytech Univ NY, MChE, 50; Univ Calif, Berkeley, PhD(chem eng), 54. *Prof Exp:* Chem engr, Stamford Res Labs, Am Cyanamid Co, 44-48; process design engr, Colgate-Palmolive Co, 48-50; res chem engr, Shell Oil Co, 53-56; proj mgr & res assoc corrosion of metals, Eng Res Labs, Columbia Univ, 56-58; chemist ceramic res & develop, 58-59, group leader, 59-61, assoc div leader, Inorg Mat, 61-72, dep proj leader, Oil Shale, 72-81, proj leader, nuclear waste isolation, 81-84, sr res chem engr, Lawrence Livermore Lab, Univ Calif, 84-86; CONSULT, 87- *Mem:* Am Chem Soc. *Res:* High temperature material for lasers; nuclear waste isolation; research and development shale oil pyrolysis and recovery from oil shale; heat transfer and thermal conductivity of gases; refrigerant substitutions for freon. *Mailing Add:* 503 Yorkshire Dr Lawrence Livermore Lab Livermore CA 94550

ROTHMAN, ALVIN HARVEY, b Brooklyn, NY, Feb 25, 30; m 54; c 3. BIOLOGY. *Educ:* Univ Calif, Los Angeles, BA, 52, MA, 54; Johns Hopkins Univ, ScD(pathobiol), 58. *Prof Exp:* USPHS fel, Johns Hopkins Univ, 59-60; res assoc biol, Rice Univ, 60-64; asst prof, 64-73, PROF BIOL, CALIF

STATE UNIV, FULLERTON, 73- *Mem:* Soc Protozool; Am Soc Parasitol; Soc Syst Zool; Wildlife Dis Asn; Sigma Xi. *Res:* Parasite physiology; functional-ultrastructural relationships of helminths. *Mailing Add:* Dept Biol Calif State Col Fullerton CA 92634

ROTHMAN, ARTHUR I, b Montreal, Que, Apr 8, 38; m 60; c 3. MEDICAL EDUCATION. *Educ:* McGill Univ, BSc, 59; Univ Maine, MS, 65; State Univ NY Buffalo, EdD(sci educ), 68. *Prof Exp:* Res assoc sci educ, Grad Sch Educ, Harvard Univ, 67-68; res assoc med educ, Educ Res Unit, 68-69, DIR & PROF MED EDUC, DIV STUDIES MED EDUC, FAC MED, UNIV TORONTO, 69-, PROF HIGHER EDUC GROUP, ONT INST STUDIES EDUC, 76- *Concurrent Pos:* Hon lectr med educ, Dept Internal Med, Univ Toronto, 69- consult, Can Asn Can Coun fel, 75-76; vis scholar, Univ Teaching Methods Univ, Inst Educ, Univ London, 75-76. *Mem:* Am Educ Res Asn; Can Soc Study Higher Educ; Asn Study Med Educ. *Res:* Examination of programs directed at the development of faculty as teachers; relationships between student characteristics, learning and career selection in medicine; methods of curriculum evaluation in medicine; programs of patient education. *Mailing Add:* Dept Higher Educ 252 Bloor St W Toronto ON M5S 1V6 Can

ROTHMAN, FRANCOISE, b Paris, France, 23; US citizen. MEDICINE. *Educ:* Paris Med Sch, MD, 51. *Prof Exp:* Asst med dir, Med Tribune, 65-73; asst dir clin res, Sandoz Labs, 74-75; area med dir, Abbott Int, 76-77; assoc dir clin res, Merrell Int, 78; assoc med dir, Ayerst Labs, 79-89; RETIRED. *Mem:* Am Asn Study Headache; Drug Info Asn; AMA; NY Acad Sci. *Mailing Add:* 1441 Third Ave Apt 22A New York NY 10028-1977

ROTHMAN, FRANK GEORGE, b Budapest, Hungary, Feb 2, 30; US citizen; m 53; c 4. EDUCATIONAL ADMINISTRATION, BIOCHEMISTRY & GENETICS. *Educ:* Univ Chicago, AB, 48, MS, 51; Harvard Univ, PhD(chem), 55. *Prof Exp:* Res assoc chem, Univ Wis, 57; res assoc biol, Mass Inst Technol, 57-61; from asst prof to assoc prof, 61-70, dean biol, 84-90, PROF BIOL, BROWN UNIV, 70- PROVOST, 90- *Concurrent Pos:* NSF fel, 56-58; Am Cancer Soc fel, 67-68. *Mem:* Genetics Soc Am; Gerontol Soc Am; AAAS. *Res:* Biochemical genetics; biology of aging. *Mailing Add:* Provost Off Brown Univ Box 1862 Providence RI 02912

ROTHMAN, HERBERT B, b New York, NY, May 27, 24. TRANSPORTATION ENGINEERING. *Educ:* Rensselaer Polytech Inst, BSCE, 44. *Prof Exp:* Engr railroad design, Gibbs & Hill, 45; struct & hwy engr, Parsons, Brinckenhoff, Quade & Douglas, 45-46; designer to partner bridge, tunnel, transp & spec struct eng, Ammann & Whitney, 46-78; PRIN & DIR TRANSP & BRIDGES, WEIDLINGER ASSOCS, 78- *Concurrent Pos:* Mem, Res Comt, Inst Bridge Tunnel & Turnpike Asn; dep mayor & hwy comnr, Village Laurel Hollow, NY. *Honors & Awards:* Roebling Award, Am Soc Civil Engrs, Rowland Medal; Arthur J Boase 25th Annual Lectr, Univ Colo, Boulder, 90. *Mem:* Nat Acad Eng; Sigma Xi; Am Soc Civil Engrs; Am Concrete Inst; Am Road & Transp Builders Asn; Am Inst Steel Construct; Inst Bridge Integrity & Safety; Struct Stability Res Coun; Int Bridge & Tunnel Turnpike Asn. *Res:* Author of various publications. *Mailing Add:* Weidlinger Assoc 333 Seventh Ave New York NY 10001

ROTHMAN, HOWARD BARRY, b New York, NY, July 17, 38; m 72; c 3. SPEECH & HEARING SCIENCE. *Educ:* City Col New York, BA, 61; NY Univ, MA, 64; Stanford Univ, PhD(speech & hearing sci), 71. *Prof Exp:* Clinician aphasia, Inst Phys Med & Rehab, 63-64; res instr, Commun Sci Lab, 69-70, from asst prof to assoc prof 71-88, PROF, DEPT COMMUN PROCESSES & DIS, INST ADVAN STUDY COMMUN PROCESSES, UNIV FLA, 88- *Concurrent Pos:* Consult, Speech Transmission Lab, Royal Inst Technol, Sweden, 72, Defense Res Bd, Defense & Civil Inst Environ Med, Ont, 72, Swed Develop Corp & Swed Nat Defense Works on Underwater Commun, 73, Harbor Br Found/Smithsonian Inst, 73, US Dist Court, Western Dist Tenn, 75 & Alachua County Sheriff's Dept, 75. *Mem:* Acoust Soc Am; Am Asn Phonetic Sci; Am Speech & Hearing Asn; NY Acad Sci; Sigma Xi. *Res:* Acoustic and perceptual aspects of speaker identification; underwater speech communication; acoustic and electromyographic aspects of atypical speech behavior; tape decoding and authentication; perceptual and acoustic analysis of the singing voice. *Mailing Add:* Inst Advan Study Commun & Dept Commun Processes & Dis ASB 63 Univ Fla Gainesville FL 32611

ROTHMAN, JAMES EDWARD, b Haverhill, Mass, Nov 3, 50; m 71; c 1. BIOCHEMISTRY. *Educ:* Yale Univ, BA, 71; Harvard Univ, PhD(biochem), 76. *Prof Exp:* Fel, Dept Biol, Mass Inst Technol, 76-78; asst prof, 78-81, ASSOC PROF, DEPT BIOCHEM, STANFORD UNIV, 81- *Mem:* Am Soc Biol Chemists. *Mailing Add:* Dept Biol Princeton Univ Princeton NJ 08544

ROTHMAN, LAURENCE SIDNEY, b New York, NY, Jan 20, 40. ATOMIC PHYSICS, ATMOSPHERIC CHEMISTRY. *Educ:* Mass Inst Technol, BS, 61; Boston Univ, MA, 64, PhD(physics), 71. *Prof Exp:* Res physicist infrared physics, Block Assoc Inc, Cambridge, Mass, 61-63; ATOMIC & MOLECULAR PHYSICIST ATMOSPHERIC TRANSMISSION, OPTICAL PHYSICS DIV, AIR FORCE GEOPHYSICS LAB, 68- *Concurrent Pos:* Invited prof, Univ Paris, 85, 87, 90. *Mem:* Sigma Xi; fel Optical Soc Am; Soc Photo Optical Instrum Eng. *Res:* Development of theoretical and analytical methods for the study of molecular physics and infrared absorption; study of mechanisms of attenuation and emission of infrared radiation in model atmospheres. *Mailing Add:* Optical Physics Div Hanscom AFB Bedford MA 01731

ROTHMAN, MILTON A, b Philadelphia, Pa, Nov 30, 19; m 50; c 2. PLASMA PHYSICS. *Educ:* Ore State Col, BS, 44; Univ Pa, MS, 48, PhD(physics), 52. *Prof Exp:* Physicist, Bartol Res Found, Franklin Inst, Pa, 52-59 & Plasma Physics Lab, Princeton Univ, 59-68; prof physics, Trenton State Col, 68-79; sr res scientist, Franklin Res Ctr, 79-85; RETIRED. *Mem:* Am Phys Soc; AAAS. *Res:* Nuclear physics; plasma physics; heating of magnetically confined plasma by ion cyclotron waves; experimental basis of the fundamental laws of physics. *Mailing Add:* 2020 Chancellor St Philadelphia PA 19103

ROTHMAN, NEAL JULES, b Philadelphia, Pa, Nov 20, 28; m 55; c 3. MATHEMATICAL ANALYSIS. *Educ:* Univ Del, BS, 51; Tulane Univ, MS, 54; La State Univ, PhD, 58. *Prof Exp:* Sr comput analyst, Burroughs Corp, 54-56; instr & asst math, La State Univ, 56-58; from instr to asst prof, Univ Rochester, 58-62; from asst prof to prof math, Univ Ill, Urbana, 62-82; prog dir modern anal, NSF Washington, 79-82; chmn dept, 82-86, PROF, MATH DEPT, IND UNIV PURDUE UNIV-INDIANAPOLIS, 82- *Concurrent Pos:* Vis prof, Israel Inst Technol, Haifa, Israel, 71-72, 76-77, Tel-Aviv Univ, Tel Aviv, Israel, 89. *Mem:* Am Math Soc; Math Asn Am. *Res:* Topological algebra. *Mailing Add:* Math Sci Ind Univ-Purdue Univ 1125 E 38th St Indianapolis IN 46205

ROTHMAN, PAUL GEORGE, b Detroit, Mich, Apr 16, 23; m 51; c 5. CEREAL PATHOLOGY. *Educ:* Mich State Univ, BS, 50, MS, 52; Univ Ill, PhD, 55. *Prof Exp:* Asst farm crops, Mich State Univ, 50-52; asst agron, Univ Ill, 52-55; agronomist, Agr Res Serv, USDA, 55-67, res pathologist, 67-; RETIRED. *Mem:* Am Soc Agron; Am Phytopath Soc. *Res:* Plant breeding and pathology. *Mailing Add:* 1455 Chelmsford St Paul MN 55108

ROTHMAN, RICHARD HARRISON, b Philadelphia, Pa, Dec 2, 36; m 60; c 2. ORTHOPEDIC SURGERY, ANATOMY. *Educ:* Univ Pa, BA, 58, MD, 62; Jefferson Med Col, PhD(anat), 65. *Prof Exp:* NIH fel, Jefferson Med Col, 63-65, dir orthop res lab, 65-71; asst prof, 71-73, ASSOC PROF ORTHOP SURG, SCH MED, UNIV PA, 73- *Mem:* Am Asn Anat. *Res:* Blood flow in tendon and bone. *Mailing Add:* 800 Spruce St Philadelphia PA 19107

ROTHMAN, SAM, b Brooklyn, NY, Feb 1, 20; m 43; c 2. PHYSICAL CHEMISTRY, RESEARCH ADMINISTRATION. *Educ:* Long Island Univ, BS, 43; Am Univ, MA, 54, PhD, 59. *Prof Exp:* Chemist, Nat Inst Standards & Technol, 46-55; dep res coordr, Off Naval Res, 55-61, asst res & eng off, Bur Naval Weapons, 61-63, dep dir explor develop, Hq, Naval Mat Command, 63-70, dep dir prog mgt off, 70-74; res prof, George Washington Univ, 74-75, chmn dept, 75-84, prof, 75-79, EMER PROF ENG ADMIN, GEORGE WASHINGTON UNIV, 90- *Concurrent Pos:* Prof lectr, Am Univ, 59-67, adj prof, 68-74. *Mem:* AAAS; Am Chem Soc; Sigma Xi. *Res:* Polymers in solution; adsorption of macromolecules; research and development management. *Mailing Add:* Dept Eng Mgt George Washington Univ Washington DC 20052

ROTHMAN, SARA WEINSTEIN, b Winthrop, Mass, July 29, 29; div; c 2. BACTERIAL TOXINS, PATHOGENESIS. *Educ:* Simmons Col, BS, 65; Boston Univ, AM, 67, PhD(microbiol), 70. *Prof Exp:* Am Cancer Soc fel, Dept Biochem & Pharmacol, Sch Med, Tufts Univ, 70-73; res assoc, Dept Path, Boston Univ, 73-74, res asst prof, 74-75, res asst prof, Dept Microbiol, Sch Med, Boston Univ, 75-78; res chemist, dept biol chem, 78-89, RES CHEMIST, DEPT MOLECULAR PATH, WALTER REED ARMY INST RES, 89- *Concurrent Pos:* Res assoc, Mallory Inst Path, 76-78; mem spec sci staff, Boston City Hosp, 76-78; vis prof, Ore Health Sci Univ, 86-87; secy, Army Sci & Eng fel, 86-87; Found microbiol lectr, Am Soc Microbiol, 87-88, Am Acad Microbiol, 90- *Mem:* Am Soc Microbiol; Sigma Xi; Asn Women in Sci; Int Soc Toxicol; The Protein Soc. *Res:* Purification of bacterial toxins and isolation of receptors; biological, physical and genetic characterization of bacterial toxins; molecular pathogenesis of enteric disease; development of cell culture and immunoassays; anaerobic bacteriology; animal models of enteric disease; diagnosis of antibiotic-associated colitis. *Mailing Add:* Div Path Walter Reed Army Inst Res Washington DC 20307-5100

ROTHMAN, STEPHEN SUTTON, b New York, NY, July 10, 35; m 57; c 2. CELL PHYSIOLOGY, CELLULAR BIOPHYSICS. *Educ:* Univ Pa, AB, 56, DDS, 61, PhD(physiol), 64. *Prof Exp:* Instr physiol, Univ Pa, 64-65; instr, Harvard Med Sch, 65-66, assoc, 66-67, from asst prof to assoc prof, 67-71; PROF PHYSIOL, UNIV CALIF, SAN FRANCISCO, 71-; SR FAC SCIENTIST, LAWRENCE BERKELEY LAB, 87- *Concurrent Pos:* Mem, joint prog biophysics & med physics, Univ Calif, San Francisco, 72-83; exec comt, joint prog biophysics, 81-83, exec comt, Group in Biophysics, 83, joint prog bioeng, Univ Calif, San Franciso & Univ Calif, Berkeley, 87; fac mem, Sch Eng, Univ Calif, Berkeley, 88-; mem Ctr X-ray Optics, Lawrence Berkeley Lab, 87- *Mem:* AAAS; Am Physiol Soc; NY Acad Sci; Biophys Soc; Sigma Xi; Soc Gen Physiologists; Am Soc Zool. *Res:* Soft x-ray microscopy and holography for high resolution imaging and chemical mapping of cellular substructure in its natural state. Properties of protein transport. *Mailing Add:* Dept Physiol Univ Calif San Francisco CA 94143

ROTHMAN, STEVEN J, b Giessen, Ger, Dec 18, 27; US citizen; m 51; c 2. PHYSICAL METALLURGY. *Educ:* Univ Chicago, PhB, 47; Stanford Univ, BS, 51, MS, 53, PhD(metall eng), 55. *Prof Exp:* Asst metallurgist, 54-60, assoc metallurgist, 60-77, METALLURGIST, ARGONNE NAT LAB, 77- *Concurrent Pos:* NSF fel, 62-63; Fulbright lectr, 73; ed, J Appl Physics, 90- *Mem:* Am Inst Mining, Metall & Petrol Engrs; Am Phys Soc. *Res:* Diffusion in solids. *Mailing Add:* Argonne Nat Lab Bldg 212-C 216 Argonne IL 60439

ROTHMAN-DENES, LUCIA B, b Buenos Aires, Arg, Feb 17, 43; US citizen; m 68; c 2. BIOCHEMISTRY, MOLECULAR BIOLOGY. *Educ:* Univ Buenos Aires, Lic, 64, PhD(biochem), 67. *Prof Exp:* Fel biochem, Res Inst Biochem, Buenos Aires, 67; fel, Nat Inst Arthritis, Metab & Digestive Dis, 67-69, vis fel genetics, 69-70; fel molecular biol, dept biophys, 70-72, res assoc, 72-74, from asst prof to prof molecular biol, dept biophys & theoret biol, 74-84, PROF, DEPT MOLECULAR GENETICS & CELL BIOL & DEPT BIOCHEM & MOLECULAR BIOL, UNIV CHICAGO, 84- *Concurrent Pos:* Nat Res Coun fel, Buenos Aires, 67; NIH spec res fel, 70-72 & res career develop award, 75-80; mem, microbiol genetics & physiol study sect, NIH, 80-84, genertic basis dis study sect, 85-89; chmn, Bacteriophage Div, Am Soc Microbiol, 85, div group 1 rep, 90-92; counr, Am Soc Virol, 88-90; mem, Damon Ronyon, Walter Winchell cancer fund sci adv bd, 89-; biochem panel, NSF, 90- *Mem:* Am Soc Biol Chemists; Am Soc Microbiol; AAAS; Am Soc Virol. *Res:* Biochemistry and regulation of transcription and DNA replication in bacteriophage infected bacteria; regulation of rat and fetoprotein expressions. *Mailing Add:* Dept Molecular Genetics & Cell Biol Univ Chicago 920 E 58th St Chicago IL 60637

ROTHMEIER, JEFFREY, medicine, computer science, for more information see previous edition

ROTHROCK, GEORGE MOORE, b Columbus, Ohio, May 5, 19; m 46, 80; c 3. POLYMER CHEMISTRY, PATENT LIAISON. *Educ:* De Pauw Univ, AB, 41; Purdue Univ, PhD(org chem), 45. *Prof Exp:* Polymer chemist, E I Du Pont De Nemours & Co, Inc, 45-60, sr patent chemist, 60-82; RETIRED. *Mem:* Am Chem Soc. *Res:* Chlorination and fluorination of hydrocarbons; polymer chemistry; polymer solvents; synthetic fibers; non-woven textiles. *Mailing Add:* 29 Commodore Point Rd Lake Wylie SC 29710

ROTHROCK, JOHN WILLIAM, b Bangor, Pa, Jan 6, 20; m 46; c 2. BIOCHEMISTRY. *Educ:* Pa State Col, BS, 41; Univ Ill, PhD(biochem), 49. *Prof Exp:* Res chemist, Merck Sharp & Dohme Res Labs, Merck & Co, 49-86; RETIRED. *Mem:* Am Chem Soc. *Res:* Antibiotics; growth factors; steroids; enzymes; drugs to lower blood pressure (Enalapril) and cholesterol (Lovastatin). *Mailing Add:* 51 Cardinal Dr Watchung NJ 07060-6108

ROTHROCK, LARRY R, b Feb 6, 40; m 60; c 2. LASER PHYSICS, CRYSTAL GROWTH. *Educ:* Oakland City Col, BS, 63. *Prof Exp:* Res physicist, 66-68, res assoc, 68-72, mgr, 72-75, mgr res, 75-80, mgr new technol, 80-85, TECHNOL MGR, UNION CARBIDE CORP, 85- *Mem:* Optical Soc Am; Am Asn Crystal Growth. *Res:* Single crystal mat; laser crystals and sapphire; electro-optics. *Mailing Add:* 13808 Belvedere Dr Poway CA 92064

ROTHROCK, PAUL E, b Newark, NJ, Oct 17, 48; m 73; c 1. BOTANY, PHYTOPATHOLOGY. *Educ:* Rutgers Univ, BA, 70; Pa State Univ, MS, 73, PhD(bot), 76. *Prof Exp:* Prof biol, Montreat-Anderson Col, 76-81; from asst prof to assoc prof, 81-88, PROF BIOL, TAYLOR UNIV, 88- *Mem:* Nat Asn Biol Teachers; Am Sci Affil; Am Soc Plant Taxonomists. *Res:* Systematics of the genus Carex (Cyperaceae). *Mailing Add:* Dept Biol Taylor Univ Upland IN 46989

ROTHROCK, THOMAS STEPHENSON, b Springdale, Ark, Sept 7, 28; m 54; c 2. ORGANIC CHEMISTRY. *Educ:* Univ Ark, BA, 50, MS, 56, PhD(org chem), 58. *Prof Exp:* Res chemist, Tenn Eastman Corp, 53-54; sr res chemist, Celanese Corp Am, Tex, 57-63; res chemist, Monsanto Co, St Louis, 63-86; RETIRED. *Mem:* Am Chem Soc. *Res:* Carbon-14 tracer studies. *Mailing Add:* 602 E Claymont Dr Ballwin MO 63011

ROTHSCHILD, BRIAN JAMES, b Newark, NJ, Aug 14, 34; m 62; c 2. FISH BIOLOGY, AQUATIC ECOLOGY. *Educ:* Rutgers Univ, BS, 57; Univ Maine, MS, 59; Cornell Univ, PhD(vert zool), 62. *Prof Exp:* Asst fishery biol, NJ Div Fish & Game, 54-57, zool, Univ Maine, 57-59 & vert zool, Cornell Univ, 59-62; chief skipjack tuna ecol prog, Honolulu Biol Lab, Bur Com Fisheries, 62-68; from assoc prof to prof, Fishery Res Inst & Ctr Quant Sci, Univ Wash, 68-71; dep dir, Northwest Fisheries Ctr, Nat Marine Fisheries Serv, 71-72, dir Atmospheric Admin, 72-75, dir, Extended Jurisdiction Prog Staff, 75-76, dir, Off Policy & Long-Range Planning, 76-77; sr policy adv, Off Admin, Nat Oceanic & Atmospheric Admin, 77-80; PROF, CTR ENVIRON & ESTUARINE STUDIES, CHESAPEAKE BIOL LAB, UNIV MD, 80- *Concurrent Pos:* Asst, Inst Fishery Res, Univ NC, 59; affil grad fac, Univ Hawaii, 64-, lectr, 66-67; vis fel biomet, Cornell Univ, 65-66. *Mem:* Fel AAAS; Am Fisheries Soc. *Res:* Ecology of fishes; marine ecology; biology and population dynamics; resource policy. *Mailing Add:* Chesapeake Biol Lab Univ Md Box 38 Solomons MD 20688

ROTHSCHILD, BRUCE LEE, b Los Angeles, Calif, Aug 26, 41. MATHEMATICS. *Educ:* Calif Inst Technol, BS, 63; Yale Univ, PhD(math), 67. *Prof Exp:* Instr math, Mass Inst Technol, 67-69; asst prof, 69-73, assoc prof, 73-77, PROF MATH, UNIV CALIF, LOS ANGELES, 77- *Concurrent Pos:* US Air Force grant, Mass Inst Technol, 67-69; consult, Bell Tel Labs, 68-71 & Network Anal Corp, 69; NSF grant, Univ Calif, Los Angeles, 69-; ed, J Combinatorial Theory, 70-; Sloan Found fel, 73-75. *Mem:* AAAS; Am Math Soc; Math Asn Am; Soc Indust & Appl Math; Sigma Xi. *Res:* Combinatorial theory; finite geometries; asymptotic enumeration; Ramsey theory; graph theory; network flows. *Mailing Add:* Dept Math Univ Calif Los Angeles CA 90024

ROTHSCHILD, DAVID (SEYMOUR), b New York, NY, Nov 28, 41; m 74. ELECTRICAL ENGINEERING, SYSTEMS SCIENCE. *Educ:* City Col New York, BEE, 63; NY Univ, MSEE, 66, PhD(elec eng), 71. *Prof Exp:* Asst engr, Sperry Gyroscope Corp, 63; electronics engr, Grumman Aerospace Corp, 63-66, res scientist control systs res, 66-75; cash mgt systs consult, Chase Manhattan Bank, 75-81; DECISION SUPPORT SYSTS ANALYST, LEHMAN BROS, KUHN LOEB, 81- *Mem:* Inst Elec & Electronics Engrs. *Res:* Linear optimal control theory; sensitivity analysis; nonlinear stability; estimation theory; research on the design of flight control systems which meet specified goals despite existing disturbances. *Mailing Add:* Mabon Nugent & Co 161 Berty Plaza New York NY 10006

ROTHSCHILD, GILBERT ROBERT, b Chicago, Ill, Dec 21, 15. ELECTRICAL ENGINEERING. *Educ:* Armour Inst Technol, BS, 36. *Prof Exp:* Jr engr, Repub Flow Meters Co, 36-37; test engr, Electromotive Div, Gen Motors Corp, 38-42; welding engr, Goodyear Aircraft Corp, 42-46; res engr welding res, Air Reduction Co, Inc, 46-52, sect head, 52-56, asst mgr, 56-57, asst dir, 57-62, asst mgr process & equip develop dept, Air Reduction Sales Co, 62-63, mgr eng & develop dept, Airco Indust Gases Div, Air Reduction Co, Inc, 63-66, gen mgr advan prod develop dept, Airco Welding Prod Div, Airco, Inc, 66-67, tech consult, 67-68, chief welding engr, 68-71, sr staff engr, Cent Res Labs, 71-80; RETIRED. *Concurrent Pos:* Consult. *Mem:* Am Welding Soc; Am Soc Metals. *Mailing Add:* 198 Hillside Ave Berkeley Heights NJ 07922

ROTHSCHILD, HENRY, b Horstein, Germany, June 5, 32; US citizen. MOLECULAR BIOLOGY, MEDICINE. *Educ:* Cornell Univ, BA, 54; Univ Chicago, MD, 58; Johns Hopkins Univ, PhD(biol), 68. *Prof Exp:* Intern, Univ Chicago, 58-59; asst resident, Univ Hosp, Baltimore, 59-61; asst physician, Johns Hopkins Univ, 63-66; instr med, Mass Gen Hosp, 70-71; assoc prof med, 71-75, assoc prof anat, 72-75, ASSOC GRAD FAC, MED CTR, LA STATE UNIV, NEW ORLEANS, 73-, PROF MED & ANAT, 75- *Concurrent Pos:* Fel, Johns Hopkins Univ, 67-68; USPHS spec fel, Mass Gen Hosp, 68-70; King Trust Award, 70-71; vis physician, Charity Hosp, New Orleans, 72-; dir, La Ethnogenetic Dis Screening Prog; consult, La Sickle Cell Educ & Screening Prog; Wellcome res travel grant, 79; vis prof, Fac Med, Univ Autonoma Nuevo Leon, Mex, 81; med dir, New Orleans Home & Rehab Ctr & St Margaret's Daughters Home. *Mem:* Am Fedn Clin Res; Am Col Physicians; Am Asn Cancer Res; Am Soc Human Genetics; Am Soc Exp Path. *Res:* Molecular medicine; oncogenic virology; genetics; gerontology. *Mailing Add:* Dept Med La State Univ Med Ctr New Orleans LA 70112

ROTHSCHILD, KENNETH J, b New York, NY, Feb 9, 48; m 74; c 2. BIOPHYSICS. *Educ:* Rensselaer Polytech Inst, BS, 69; Mass Inst Technol, PhD(physics), 73. *Prof Exp:* Res assoc biophys, Harvard-Mass Inst Technol Prog Health Sci & Technol, 73-75; asst prof physiol, Sch Med & asst prof physics, 76-81, & assoc prof physiol & assoc prof physics, 82-85, PROF PHYSICS & DIR CELLULAR BIOPHYS, BOSTON UNIV, 86- *Concurrent Pos:* NIH fel, Nat Eye Inst, 74, coprin investr, 75; NIH fel, 72-75; established investr, Am Heart Asn; vis scientist, Neth Found Basic Sci. *Honors & Awards:* Whitaker Found Award. *Mem:* Biophys Soc; Sigma Xi; AAAS. *Res:* Molecular mechanisms of membrane transport; conformational analysis of photoreceptor membranes using raman spectroscopy, fourier transform IR spectroscopy; purple membrane; nonequilibrium thermodynamics. *Mailing Add:* Dept Physics Boston Univ 590 Commonwealth Ave Boston MA 02215

ROTHSCHILD, MARCUS ADOLPHUS, b New York, NY, June 2, 24; m 65; c 2. INTERNAL MEDICINE. *Educ:* Yale Univ, BS, 45; NY Univ, MD, 49; Am Bd Nuclear Med, dipl, 75. *Prof Exp:* Intern, Beth Israel Hosp, New York, 49-50; resident med, Mt Sinai Hosp, New York, 50-51; resident, Beth Israel Hosp, Boston, Mass, 51-52; chief med resident, Beth Israel Hosp, New York, 52-53; from clin instr to assoc prof med, 55-71, PROF MED, SCH MED, NY UNIV, 71- *Concurrent Pos:* Dazian Found fel, Radioisotope Serv, Vet Admin Hosp, NY, 53-55; sect chief gen med & chief radioisotope serv, Vet Admin Hosp, NY, 55-; lectr physiol, Hunter Col, 54-55; USPHS grants, 56-78; adj prof, Rockefeller Univ, 77-; spec asst dir med res, Vet Admin Alcohol Res, 78-; ed, Alcoholism Clin Exp Res. *Mem:* Fel Am Col Physicians; Am Soc Clin Invest; Am Asn Study Liver Diseases (secy-treas, 76-81, pres, 85-); Soc Nuclear Med; Asn Am Physicians. *Res:* Protein metabolism. *Mailing Add:* 21770 Cypress Dr Boca Raton FL 33433

ROTHSCHILD, RICHARD EISEMAN, b St Louis, Mo, Nov 27, 43; m 68; c 1. X-RAY & GAMMA RAY ASTRONOMY. *Educ:* Washington Univ, AB, 65; Univ Ariz, MS, 68, PhD(physics), 71. *Prof Exp:* Res assoc physics, Univ Ariz, 71-73; Nat Acad Sci/Nat Res Coun fel astrophys, Goddard Space Flight Ctr, NASA, 73-75, astrophysicist, 75-77; RES PHYSICIST, UNIV CALIF, SAN DIEGO, 77- *Concurrent Pos:* Co-investr, Cosmic X-ray Exp High Energy Astron Observ, 76-86; prin investr, High Energy X-Ray Timing Exp on X-Ray Timing Explorer, 83- *Honors & Awards:* NASA Group Achievement Award, Goddard Space Flight Ctr, 78 & 79. *Mem:* Am Phys Soc; Am Astron Soc. *Res:* Astronomical objects emitting x-rays and gamma rays; data analysis from detector systems. *Mailing Add:* CASS C-011 Univ Calif San Diego La Jolla CA 92093

ROTHSCHILD, WALTER GUSTAV, b Berlin, Ger, Aug 18, 24; nat US. RELAXATION PHENOMENA, LIQUID CRYSTALS. *Educ:* Tech Univ Berlin, BS, 49; Max Planck Inst, 50; Columbia Univ, MA, 58, PhD, 61. *Prof Exp:* Res scientist, Electronized Chem Corp, NY, 50-52, Tracerlab, Inc, Mass, 52-54, Brookhaven Nat Lab, 54-57 & Columbia Univ, 57-61; STAFF SCIENTIST, FORD MOTOR CO, 61- *Concurrent Pos:* Guest prof, Dept Physics, Univ Buenos Aires, 69; Fulbright fel, 76-77, NATO res fel, 79-84, 88-90; guest prof, Univ Bordeaux I, France, 76-77, 86 & Univ Vienna, Austria, 77; vis prof, Univ Paris VI, France, 79, 80, 82 & 84; fac, Wayne State Univ, 90- *Honors & Awards:* Donald Julius Groen Prize, Inst Mech Engrs, London, Eng, 86. *Mem:* Am Phys Soc; Am Chem Soc. *Res:* Molecular interactions, dynamics and structure; liquid and amorphous-state phenomena; catalysis; infrared and Raman relaxation phenomena and spectroscopy. *Mailing Add:* 1454 Brookfield Dr Ann Arbor MI 48103-6085

ROTHSTEIN, ASER, b Vancouver, BC, Apr 29, 18; nat US; m 40; c 3. BIOPHYSICS, CELL PHYSIOLOGY. *Educ:* Univ BC, BA, 38; Univ Rochester, PhD(biol), 43. *Hon Degrees:* DSc, Univ Rochester, 83. *Prof Exp:* Asst biol, Univ Rochester, 40-42, asst physiol, Off Sci Res & Develop Contract, 43, asst, Manhattan Proj, 44-45, assoc, 46-47, from instr to assoc prof pharmacol, 46-72, prof radiation biol, 59-72, co-chmn dept, 65-72, chief physiol sect, 48-60, assoc dir Atomic Energy Proj, 60-65, co-dir, 65-72; dir res inst, Hosp Sick Children, 72-86; prof med biophys, 73-86, univ prof, 80-86, prof, 80-86, EMER PROF, UNIV TORONTO, 86- *Concurrent Pos:* NSF sr fel, 59-60; vis prof, Dept Radiation Biol & Biophys, Med Sch, Univ Rochester & Swiss Fed Univ, Weizman Inst. *Honors & Awards:* Gardner Award, 86. *Mem:* Am Physiol Soc; Biophys Soc; Int Union Physiol Sci; Soc Gen Physiologists; Royal Soc Can. *Res:* Structure and function of cell membranes, especially role of membrane proteins; role of membrane in cell growth and differentiation. *Mailing Add:* Res Inst Hosp Sick Children Toronto ON M5G 1X8 Can

ROTHSTEIN, EDWIN C(ARL), b Brooklyn, NY, Nov 17, 33; m 58; c 2. TOXICOLOGY, ENVIRONMENTAL SCIENCE. *Educ:* Brooklyn Col, PA, 54; Mass Inst Technol, MS, 56; Polytech Inst Brooklyn, PhD(chem eng), 64. *Prof Exp:* Res technologist, Cent Res Labs, Socony Mobil Oil Co, 56-59; res engr, Res Labs, Keuffel & Esser Co, 59-65; asst tech develop mgr, Indust Chem Div, Geigy Chem Corp, 65-67; chief chemist, Sinclair & Valentine Div, Martin-Marietta Corp, 67-69, dir res & develop, 69-70; tech dir, Equitable Bag Co, Inc, Long Island City, 70-78; tech develop mgr, Sun Chem Corp, 78-82; PRES, LEBERCO TESTING INC, 82- *Honors & Awards:* Roon Award, Fedn Paint Socs, 64. *Mem:* AAAS; Am Chem Soc; Soc Plastics Engrs; NY Acad Sci; Am Soc Testing & Mat; Asn Off Anal Chemists; Soc Cosmetic Chemists. *Res:* Printing inks and coatings; pigment dispersion rheology; photographic processes; plastics stabilization and testing; toxicology; analytical chemistry. *Mailing Add:* Leberco Testing Inc 123 Hawthorne St Roselle Park NJ 07204-0206

ROTHSTEIN, HOWARD, b Brooklyn, NY, Aug 25, 35; m 56; c 3. CELL PHYSIOLOGY. *Educ:* Johns Hopkins Univ, BA, 56; Univ Pa, PhD(gen physiol), 60. *Prof Exp:* Asst instr zool, Univ Pa, 57-60; fel ophthal, Col Physicians & Surgeons, Columbia Univ, 60-62; from asst prof to prof zool, Univ Vt, 62-77; assoc prof ophthal, Kresge Eye Inst, Wayne State Univ Sch Med, Detroit, 77-81; prof & chmn Dept Biol Sci, 81-86, PROF, FORDHAM UNIV, BRONX, NY, 86- *Mem:* Am Soc Zool; AAAS; Soc Gen Physiol. *Res:* Cell division; hormonal control of growth; vitro culture and wound healing of ocular tissues. *Mailing Add:* Dept Biol Sci Fordham Univ Bronx NY 10458

ROTHSTEIN, JEROME, b New York, NY, Dec 14, 18; m 41; c 3. PLASMA PHYSICS, SOLID STATE PHYSICS. *Educ:* City Col New York, BS, 38; Jewish Theol Sem Am, BHL, 39; Columbia Univ, AM, 40. *Prof Exp:* Res assoc, Columbia Univ, 41-42; physicist, Evans Signal Corps Lab, US Dept Army, 42-57; sr sci exec, Edgerton, Germeshausen & Grier, Inc, 57-61; vpres & dir res, Maser Optics, 61-62; sr staff scientist, Lab For Electronics, Inc, Mass, 62-67; prof Comput & Info Sci & Biophys, 67-89, EMER PROF, OHIO STATE UNIV, 89- *Concurrent Pos:* Tutor, City Col New York, 41-42; distinguished lectr, Inst Elec & Electronics Engrs, 81-82. *Mem:* AAAS; Am Phys Soc; Biophys Soc; Inst Elec & Electronics Engrs; Sigma Xi. *Res:* Physical electronics; vacuum techniques; plasma and arc spot; solid state and radiation physics; lasers; photography; statistical mechanics; methodology, biophysics; neural modeling; information and systems theory; computer and information science; formal languages; automata theory; pattern recognition; artificial intelligence; thermodynamics and evolution. *Mailing Add:* Dept Comput & Info Sci Ohio State Univ Columbus OH 43210

ROTHSTEIN, LEWIS ROBERT, b New York, NY, Oct 4, 20; m 46; c 2. CHEMISTRY. *Educ:* Queens Col NY, BS, 42; Ill Inst Technol, MS, 44, PhD(chem), 49. *Prof Exp:* Tech dir, Mason & Hanger-Silas Mason Co, 48-60; dir spec projs, Amcel Propulsion, Inc, 60-65; dir chem & munitions dept, Northrop-Carolina Inc, 65-71; spec asst, Naval Weapons Sta, Yorktown, Va, 71-74, asst tech dir, 74-88; RETIRED. *Mem:* Am Chem Soc. *Res:* Physical organic chemistry. *Mailing Add:* 124 Selden Rd Newport News VA 23606

ROTHSTEIN, MORTON, b Vancouver, BC, Sept 8, 22; nat US; m 47; c 3. BIOCHEMISTRY. *Educ:* Univ BC, BA, 46; Univ Ill, MS, 47, PhD(org chem), 49. *Prof Exp:* Res assoc tracer chem, Sch Med, Univ Rochester, 49-54; from asst res biochemist to assoc res biochemist, Sch Med, Univ Calif, San Francisco, 54-60; assoc res scientist, Res Inst, Kaiser Found, 60-65; chmn dept, 69-70, prof, 65-89, EMER PROF BIOL SCI, STATE UNIV NY, BUFFALO, 89- *Concurrent Pos:* Lectr, Sch Med, Univ Calif, San Francisco, 58-65; Nat Inst Child Health & Human Develop res grant, 72-77; Nat Inst Aging res grant, 77-88. *Honors & Awards:* Gold Medal for Res, Am Aging Asn, 88. *Mem:* AAAS; Am Soc Biol Chem; fel Geront Soc. *Res:* Biochemistry of aging. *Mailing Add:* Dept Biol Sci State Univ NY Buffalo NY 14260

ROTHSTEIN, ROBERT, b Philadelphia, Pa, Sept 22, 25; m 46; c 2. MEDICAL ADMINISTRATION, AGRICULTURE. *Educ:* Brooklyn Col, BA, 47, MA, 53. *Prof Exp:* Head labs, Brooklyn Women's Hosp, NY, 47-57; dir labs, Key Clin Labs Inc, 57-60; lectr biol, Brooklyn Col, 60-61; teacher, Farmingdale High Sch, 61-64; from asst prof to assoc prof, State Univ NY Agr & Tech Col, 64-68, prof biol & coordr, Med Lab Technol, 68-71, chmn dept, 71-75, chmn div human serv, 75-78, PROF MED LAB TECHNOL, STATE UNIV NY AGR & TECH COL FARMINGDALE, 71-, DEAN SCH AGR & HEALTH SCI, 78- *Concurrent Pos:* Lectr, Sch Gen Studies, Brooklyn Col, 53-73; vis prof, Univ RI, 64-77. *Mem:* AAAS; Am Inst Biol Sci; Nat Asn Biol Teachers; Am Asn Higher Educ; Am Soc Allied Health Professions. *Res:* Histochemical identification of phospholipid in situ in frog adrenal cortices under varying physiological conditions utilizing carbowax techniques; behavior of Japanese quail, especially nesting activity in captivity; utilization of classroom interaction analysis as an instrument to determine effective teaching. *Mailing Add:* Dept Med Technol State Univ NY Agr & Tech Col Farmingdale NY 11735

ROTHSTEIN, RODNEY JOEL, b Seattle, Wash, Nov 4, 47; m 69; c 2. MOLECULAR GENETICS. *Educ:* Univ Ill, Chicago, BS, 69; Univ Chicago, PhD(genetics), 75. *Prof Exp:* Fel molecular genetics, Dept Radiation Biol & Biophys, Sch Med & Dent, Univ Rochester, 75-77; fel biochem, Sect Biochem, Cell & Molecular Biol, Cornell Univ, 77-79; asst prof microbiol, NJ Med Sch, Newark, 79-; ASSOC PROF, DEPT GENETICS & DEVELOP, COL PHYSICIANS & SURGEONS, COLUMBIA UNIV. *Concurrent Pos:* Assoc ed, Molecular & Cellular Biol, Current Genetics, Genetica. *Mem:* Genetics Soc Am. *Res:* Investigation of the mechanisms of genetic recombination and gene rearrangements in eukaryotic genomes; effects of changes in DNA topology on recombination. *Mailing Add:* Dept Genetics & Develop Col Physicians & Surgeons Columbia Univ 701 W 168th St Rm 1602 New York NY 10032

ROTHSTEIN, SAMUEL, materials science, corrosion; deceased, see previous edition for last biography

ROTHWARF, ALLEN, b Philadelphia, Pa, Oct 1, 35; m 57; c 3. THEORETICAL SOLID STATE PHYSICS. *Educ:* Temple Univ, AB, 57; Univ Pa, MS, 60, PhD(physics), 64. *Prof Exp:* Instr physics, Rutgers Univ, 60-62; mem tech staff, RCA Labs, 64-72; fel Sch Metall & Mat Sci, Univ Pa, 72-73; sr scientist, Mgr, Inst Energy Conversion, Univ Del, 73-79; assoc prof, Drexel Univ, Philadelphia, Pa, 79-83, prof elec eng, Drexel Univ, Philadelphia, PA, 83-91; CONSULT, 91- *Mem:* Am Phys Soc; Am Vacuum Soc; fel Inst Elec & Electronics Engrs. *Res:* Theory of superconductivity; transport properties of fermion systems; energy losses in solids; theory of amorphous solids; photovoltaics; localized wave functions in metals; modeling of new solar cell structures and materials; amorphous silicon hydride; metal-insulator-semiconductor solar cells; thin film transistors. *Mailing Add:* Dept Elec & Comput Eng Drexel Univ 32nd & Chesnut Sts Philadelphia PA 19104

ROTHWARF, FREDERICK, b Philadelphia, Pa, Apr 23, 30; m 51; c 4. SOLID STATE PHYSICS, MATERIALS SCIENCE. *Educ:* Temple Univ, AB, 51, AM, 53, PhD(physics), 60. *Prof Exp:* Physicist antisubmarine warfare, US Naval Air Develop Ctr, Pa, 51-52; asst, Temple Univ, 52-56; physicist, Frankford Arsenal, 56-71; PHYSICIST SOLID STATE PHYSICS, ELECTRONICS TECHNOL & DEVICES LAB, US ARMY, FT MONMOUTH, 71- *Concurrent Pos:* Physicist thin film res, Burroughs Corp, Philadelphia, 52-53; health physicist, Radiobiol & X-ray Dept, Temple Univ Hosp, 54-55; instr physics, Ogontz Ctr, Pa State Univ, 55-56; consult, Dept Otolaryngol, Presby Hosp, Philadelphia, 62-70; Secy of Army fel, Univ Paris-Orsay, France, 65-66; consult superconductivity, Naval Ships Res & Develop Ctr, Annapolis, Md, 68-72; pres, Terra Systs, Inc, Toms River, 70-; consult magnetics, Harry Diamond Labs, Washington, DC, 72-74. *Honors & Awards:* Sci Achievement Award, Secy of Army, 74. *Mem:* Am Phys Soc; Inst Elec & Electronics Engrs; Electrochem Soc; Int Asn Hydrogen Energy. *Res:* Semiconductor technology of silicon and gallium arsenide; rare earth-cobalt and amorphous magnetic materials; magnetic circuit designs; superconductivity; metal physics; metal hydrides; cryogenics; electronics; health physics. *Mailing Add:* 11722 Indian Ridge Reston VA 22091

ROTHWELL, FREDERICK MIRVAN, b Arena, Wis, May 7, 23; m 45; c 3. MYCOLOGY, PHYSIOLOGICAL ECOLOGY. *Educ:* Eastern Ky Univ, BS, 49; Univ Ky, MS, 51; Purdue Univ, PhD(mycol), 55. *Prof Exp:* USPHS fel, Purdue Univ, 55-56; res microbiologist, Buckman Labs, Inc, Tenn, 56-57, tech rep, 59-61; assoc prof animal dis, Agr Exp Sta, Miss State Univ, 57-59; prof life sci, Ind State Univ, Terre Haute, 61-74; consult, 74-75; microbiologist, US Forest Serv, 75-87; RETIRED. *Concurrent Pos:* Res grants, Com Solvents Corp, 64-66, Tenn Valley Auth, 65-68 & USDA, 67-68; adj prof, Ind Univ Sch Med, Terre Haute Ctr Med Educ & Ind State Univ, 72-75. *Res:* Microbiology associated with revegetation of surface-mined lands; mycorrhizal associates and asymbiotic and symbiotic nitrogen-fixing bacterial species. *Mailing Add:* 504 Center Berea KY 40403-1739

ROTHWELL, NORMAN VINCENT, b Passaic, NJ, Sept 30, 24. CYTOGENETICS. *Educ:* Rutgers Univ, BS, 49; Univ Ind, PhD(bot), 54. *Prof Exp:* From lectr to instr bot, Univ Ind, 53-54; assoc prof biol, Southeastern Mo State Col, 54-55; lab instr, Cancer Inst, Univ Miami, Fla, 55-56; prof biol, Long Island Univ, 56-; RETIRED. *Concurrent Pos:* Fulbright lectr, Univ Ceylon, 65-66 & 68-69. *Res:* Cytogenetic studies on Claytonia virginica; cellular differentiation in the grass root tip of epidermis. *Mailing Add:* Dept Biol Long Island Univ Brooklyn NY 11201

ROTHWELL, PAUL L, b Norwood, Mass, Apr 2, 38; m 65; c 2. SPACE SCIENCE. *Educ:* Harvard Univ, BA, 60; Northeastern Univ, MS, 62, PhD(physics), 67, MBA, 78. *Prof Exp:* Res fel elementary particles, Northeastern Univ, 62-67, RES PHYSICIST, AIR FORCE GEOPHYSICS LABS, 67- *Mem:* Am Geophys Union; Am Phys Soc; Sigma Xi. *Res:* Validity of quantum electrodynamics and the decay modes of the boson resonances; trapped radiation in the Van Allen belts; solar particles; substorm dynamics; magnetosphere-ionosphere coupling. *Mailing Add:* 15 George St Littleton MA 01460

ROTHWELL, WILLIAM STANLEY, b Wabasha, Minn, May 3, 24; m 46; c 4. OPTICAL OBSERVABLES, X-RAY DIFFRACTION ANALYSIS. *Educ:* US Naval Acad, BS, 45; Univ Wis, MS, 48, PhD(physics), 54. *Prof Exp:* Instr influence mines, US Navy Explosive Ord Disposal Sch, 46-47; instr physics, Racine Exten Ctr, Univ Wis, 49-52; res physicist, Corning Glass Works, 54-57; res group leader, Res Labs, Allis-Chalmers Mfg Co, 57-62; staff scientist, 62-81, sr staff scientist, Lockheed Palo Alto Res Lab, 81-86; INDEPENDENT CONSULT & AUTHOR, 86- *Res:* Solid state physics; small angle x-ray scattering; materials science; aerospace sciences; educational books and software. *Mailing Add:* 343 S Gordon Way Los Altos CA 94022

ROTI ROTI, JOSEPH LEE, b Newport, RI, Oct 12, 43; div. RADIATION BIOPHYSICS, THEORETICAL BIOLOGY. *Educ:* Mich Technol Univ, BS, 65; Univ Rochester, PhD(biophys), 72. *Prof Exp:* Res instr, 73-76, asst prof, 76-79, ASSOC PROF RADIOL, UNIV UTAH, 79- *Concurrent Pos:* Res collabr, Brookhaven Nat Lab, 73-75. *Mem:* Cell Kinetics Soc; Radiation Res Soc. *Res:* Simulation of cell kinetics in vitro and in vivo using matrix algebra; studies of the effects of x-irradiation and hyperthermia on DNA, chromosomal proteins and cell progression through the cell cycle. *Mailing Add:* 7050 Cornell Ave University City MD 63130

ROTKIN, ISADORE DAVID, b Chicago, Ill, June 26, 21; m 42; c 3. EPIDEMIOLOGY, CANCER. *Educ:* Univ Chicago, BS, 46; Univ Calif, Berkeley, MS, 49, PhD(genetics), 54. *Prof Exp:* Researcher, Cancer Res Lab, Univ Calif, Berkeley, 50-54; consult sci & indust, 55-59; dir cancer res, Res Inst, Kaiser Found, 59-68; dep chief oper studies, Nat Cancer Control Prog, USPHS, 68; staff consult, Calif Regional Med Progs, 68-70; from assoc prof to prof, 73-86, EMER PROF PREV MED, UNIV ILL COL MED, 86- *Concurrent Pos:* Consult bd med, Nat Acad Sci, Vet Admin Hosp, WSide, Chicago, Zellerbach-Saroni Tumor Inst, Mt Zion Hosp, San Francisco, unit epidemiol & biostatist, WHO, France, inst behav res, Tex Christian Univ, med dept, Pac Tel Co, San Francisco, Dept Chronic Dis, State of Conn & Stanford Res Inst; lectr epidemiol & community health, Sch Med, Univ Calif, San Diego, 70-72; consult, Am Cancer Soc, Alameda & San Francisco Counties, Calif, Oak Forest Hosp, Oak Forest, Ill, Ill Cancer Coun, Comprehensive Cancer Ctr, Chicago, Head & Neck Cancer Network, Rush Med Cancer Ctr, Chicago & Northwestern Univ Med Sch, Chicago, Ill Inst Technol, Chicago & Pan Am Health Orgn, WHO, Washington, DC; chmn, Task Force Epidemiol & Statist, Ill Comprehensive Cancer Coun, Chicago, 73-; chmn, Panel Anal Epidemiol, XI Int Cancer Congress, Italy, 74; chmn, Legis & Tumor Registry Panels, Ill Cancer Coun, Chicago, 74-; mem, Comt Comprehensive Cancer Ctr Patient Data Syst, Nat Cancer Inst, NIH, 74-; vis prof, Univ Calif, Los Angeles, 80-81. *Mem:* Am Asn Cancer Res; Soc Epidemiol Res; Am Soc Human Genetics; Am Inst Biol Sci; Soc Study Social Biol. *Res:* Cancer and chronic disease epidemiology; etiology and epidemiology of cervical and prostatic cancers; benign prostatic hypertrophy; case control studies; multidimensional risk studies; disease prevention and control; medical systems; community health; sexual data and counselling. *Mailing Add:* Dept Prev Med & Community Health Univ Ill Col Med Chicago IL 60690

ROTMAN, BORIS, b Buenos Aires, Arg, Dec 4, 24; nat US; wid; c 1. MOLECULAR BIOLOGY, IMMUNOLOGY. *Educ:* Valparaiso Tech Univ, Chile, MS, 48; Univ Ill, PhD(bact), 52. *Prof Exp:* Res assoc, Univ Ill, 52-53; Enzyme Inst, Madison, Wis, 53-56; vis prof, Sch Med, Univ Chile, 56-59; res dir, Radioisotope Serv, Vet Admin Hosp, Albany, NY, 59-61; res assoc genetics, Sch Med, Stanford Univ, 61; head biochem sect, Syntex Inst Molecular Biol, Calif, 61-66; PROF MED SCI, BROWN UNIV, 66- *Concurrent Pos:* Am Soc Microbiol pres fel, 58; fel, Harvard Univ, 58. *Mem:* Am Soc Biol Chem; Am Soc Microbiol; Am Asn Cancer Res; Fed Am Scientists; AAAS. *Res:* Chemosensitivity testing; chemotherapy cancer research; short term micro-organ culture to assess chemosensitivity of tumors from individual cancer patients. *Mailing Add:* Div Biol & Med Sci Brown Univ Providence RI 02912

ROTMAN, HAROLD H, EXPERIMENTAL BIOLOGY. *Prof Exp:* PROF MED & CHIEF MED SERV, DUKE UNIV, 85- *Mailing Add:* Dept Med Vet Admin Med Ctr Riceville & Tunnel Rds Ashville NC 28805

ROTMAN, JOSEPH JONAH, b Chicago, Ill, May 26, 34; m 78; c 2. MATHEMATICS. *Educ:* Univ Chicago, AB, 54, MS, 56, PhD(math), 59. *Prof Exp:* Res assoc, 59-61, from asst prof to assoc prof, 61-68, PROF MATH, UNIV ILL, URBANA, 68- *Concurrent Pos:* Vis prof, Queen Mary Col, Univ London, 65-66 & Hebrew Univ, Jerusalem, 70, 78; ed, Proc, Am Math Soc, 70-73; Lady Davis prof, Israel Inst Technol, Haifa, 77-78, Tel Aviv Univ, 82; vis prof, Bar-ilan Univ Ramat Gan, Israel, 84 & Queen Mary Col, Univ London, 85-86; ann vis, SAfrican Math Soc, 85, Univ Oxford, 90. *Mem:* Math Asn Am; Am Math Soc. *Res:* Algebra. *Mailing Add:* 803 Brighton Urbana IL 61801

ROTMAN, MARVIN Z, b Philadephia, Pa, Sept 3, 33; c 3. MEDICINE. *Educ:* Ursinus Col, BS, 54; Jefferson Med Col, MD, 58. *Prof Exp:* Assoc prof clin radiol, NY Med Col, 71-75, dir radiation oncol, 75-79; PROF & CHMN RADIATION ONCOL, STATE UNIV NY, BROOKLYN, 79- *Concurrent Pos:* Dir radiation oncol, Univ Hosp, Long Island Col Hosp & Kings County Hosp, 79-; consult, Clin Cancer Prog Proj Grant, NIH, 76-85; ed, Int J Radiation Oncol, Biol & Physics, 79-; examnr, Oral Specialty Bds, Am Bd Radiol, 80-85 & 87; pres, SCAROP, 84-86; vpres, Radiol Soc NAm, 87-88. *Mem:* Fel Am Col Radiol; Soc Chmn Acad Radiother Prog (pres, 77-78). *Mailing Add:* 450 Clarkson Ave Box 1211 Brooklyn NY 11203

ROTMAN, WALTER, b St Louis, Mo, Aug 24, 22; m 54; c 2. ELECTRICAL ENGINEERING. *Educ:* Mass Inst Technol, BS, 47, MS, 48. *Prof Exp:* Asst elec eng, Res Lab Electronics, Mass Inst Technol, 47-48; electronic scientist, Air Force Cambridge Res Labs, 48-76; electronic scientist, Rome Air Development Ctr, 76-80; ELECTRONIC STAFF ENGR, LINCOLN LAB, MASS INST TECHNOL, 80- *Concurrent Pos:* Mem comn VI, Int Sci Radio Union. *Mem:* Sigma Xi; fel Inst Elec & Electronics Engrs. *Res:* Development of microwave antennas for radar and communications; millimeter wave satellite communication antennas. *Mailing Add:* 17 Gerald Rd Brighton MA 02135

ROTT, NICHOLAS, b Budapest, Hungary, Oct 6, 17; nat US; m 44; c 2. AERODYNAMICS, ACOUSTICS. *Educ:* Swiss Fed Inst Technol, Zurich, MME, 40, Dr Sc Tech, 43. *Hon Degrees:* Dr, Fed Inst Technol, Lausanne, 85. *Prof Exp:* Res assoc, Swiss Fed Inst Technol, 43-47, privat-docent, 47-51; from assoc prof to prof aeronaut eng, Cornell Univ, 51-59; prof eng, Univ Calif, Los Angeles, 59-67; prof eng, Swiss Fed Inst Technol, 67-83; AFFIL, AERO/ASTRO DEPT, STANFORD UNIV, 83- *Mem:* Am Phys Soc; Am Inst Aeronaut & Astronaut; Acoust Soc Am. *Res:* High speed aerodynamics; boundary layers; rotating flow and vortices; acoustics, particularly thermal effects and non-linear effects; non-linear dynamics. *Mailing Add:* 1865 Bryant St Palo Alto CA 94301

ROTTENBERG, HAGAI, b Haifa, Israel, Aug 21, 36; US citizen. BIOENERGETICS, BIOMEMBRANES. *Educ:* Hebrew Univ, MSc, 63; Harvard Univ, PhD(biophys), 68. *Prof Exp:* Asst prof biochem, Brooklyn Col, 67-69; res assoc biochem, Weizmann Inst, 69-72; assoc prof biochem, Tel Aviv Univ, 72-77 & Univ Pa, 75-76; staff scientist biochem, Bell Labs, 77-78; PROF BIOCHEM, HAHNEMANN UNIV, 78- *Mem:* Biophys Soc; Am Soc Biol Chemists; NY Acad Sci; AAAS. *Res:* Biological membrane transport in cell metabolism and function in animal cells; effect of alcohol and other drugs on biological membrane transport. *Mailing Add:* Hahnemann Univ Broad & Vine Philadelphia PA 19102

ROTTER, JEROME ISRAEL, b Los Angeles, Calif, Feb 24, 49; m 70; c 3. MEDICAL GENETICS, INTERNAL MEDICINE. *Educ:* Univ Calif, Los Angeles, BS, 70, MD, 73. *Prof Exp:* Intern, Harbor Gen Hosp, Torrance, Calif, 73-74; med resident, Wadsworth Vet Admin Hosp, 74-75; fel med genetics, Los Angeles County Harbor Med Ctr, Univ Calif, Los Angeles, 75-77, sr res fel, 77-78, asst res pediatrician med genetics, 78-79, from asst prof to assoc prof, 79-87, PROF MED & PEDIAT, SCH MED, UNIV CALIF, LOS ANGELES, 87-, DIR, DIV MED GENETICS, CEDARS-SINAI MED CTR, 86- *Concurrent Pos:* Res fel, Ctr Ulcer Res & Educ, 76-77, investr, 77-80, key investr, 80-89; res fel, Nat Inst Arthritis, Metab & Digestive Dis, USPHS, 77-79, clin investr, 79-82; dir, Genetic Epidemiol Care, UCLA Ctr Study Inflammatory Bowel Dis, Harbor-UCLA, 85-; chmn, Med Genetics Sect, Med Knowledge Self-Assessment Prog VIII, Am Coll Physicians, 86-89; dir, Genetic Epidemiology Care, UCLA prog proj Molecular Biol Arteriosclerosis, 88-; Cedars-Sinai Bd of Govs Endowed Chair, Med Genetics. *Honors & Awards:* Richard Weitzman Mem Res Award for Outstanding Young Investr, 83. *Mem:* Am Soc Human Genetics; fel Am Col Physicians; Am Fedn Clin Res; Am Gastroenterol Asn; Am Diabetes Asn; Am Soc Clin Invest. *Res:* Genetics of the gastrointestinal disorders and of common diseases, with special interest in diabetes. *Mailing Add:* Div Med Genetics Cedars-Sinai Med Ctr 8700 W Beverly Blvd Los Angeles CA 90048

ROTTINK, BRUCE ALLAN, b Minneapolis, Minn, Feb 15, 47; m 71. TREE PHYSIOLOGY, PLANT ECOLOGY. *Educ:* Univ Minn, St Paul, BS, 69; Mich State Univ, PhD (forestry), 74. *Prof Exp:* Res biologist, Dow Corning Corp, 73-75; RES FORESTER, CROWN ZELLERBACH CORP, 75- *Mem:* Am Soc Plant Physiologists; Soc Am Foresters; Sigma Xi. *Res:* Plant water relations; reproductive biology of conifers; physiological differences between genotypes of trees; biological nitrogen fixation. *Mailing Add:* 14 Touchstone Lake Oswego OR 97034

ROTTMAN, FRITZ M, b Muskegon, Mich, Mar 29, 37; m 59; c 3. BIOCHEMISTRY. *Educ:* Calvin Col, BA, 59; Univ Mich, PhD(biochem), 63. *Prof Exp:* From asst prof to assoc prof, 66-74, Mich State Univ, prof biochem, 74-; AT DEPT MOLECULAR & MICROBIOL, SCH MED, CASE WESTERN RESERVE UNIV. *Concurrent Pos:* NIH fel, 63-64; Am Cancer Soc fel, NIH, 64-66; vis prof biochem, Univ BC, 74-75. *Mem:* AAAS; Am Soc Biol Chem; Am Chem Soc. *Res:* RNA chemistry; protein biosynthesis; RNA processing and the presence and role of trace nucleotides in RNA molecules. *Mailing Add:* Dept Molecular Bio & Microbiol Sch Med Case Western Reserve Univ Cleveland OH 44106

ROTTMAN, GARY JAMES, b Denver, Colo, Sept 21, 44; m 78; c 2. ASTROPHYSICS, ATMOSPHERIC SCIENCE. *Educ:* Rockhurst Col, BA, 66; Johns Hopkins Univ, MS, 69, PhD (physics), 72. *Prof Exp:* Res asst physics, Johns Hopkins Univ, 69-72, teaching asst, 72; RES ASSOC PHYSICS, LAB ATMOSPHERIC & SPACE PHYSICS, UNIV COLO, 72-, LECTR, DEPT PHYSICS & ASTROPHYS, 74- *Concurrent Pos:* Scientist, NASA Sounding Rocket Exp, Johns Hopkins Univ & Colo Univ, 69-; co-investr, NASA Orbiting Solar Observ-8, 75-; proj scientist, Solar Mesosphere Explorer Satellite Prog, Lab Atmospheric & Space Physics, NASA, 75-, prin investr, Solar Extreme Ultraviolet Rocket Prog, 77-; consult, Univ Space Res Asn, 77. *Honors & Awards:* Group Achievement Award, Nat Aeronaut & Space Admin, 77. *Mem:* Am Astronaut Soc; Am Geophys Union. *Res:* Study of the solar atmosphere, in particular the measurement of persistant flows and oscillatory motions in the chromosphere and corona, these observations are made using ultraviolet spectrophotometry from sounding rockets and satellite platforms. *Mailing Add:* Lab Atmospheric & Space Physics Univ Colo Campus Box 392 Boulder CO 80309

ROTTMANN, WARREN LEONARD, b New York, NY, Dec 23, 43; m 65; c 1. CELL BIOLOGY. *Educ:* State Univ NY Binghamton, BA, 65; Col William & Mary, MA, 67; Univ Ore, PhD(biol), 71. *Prof Exp:* NIH fel cell biol, Johns Hopkins Univ, 71-73, assoc res fel, 73-74; asst prof zool, Univ Minn, Minneapolis, 74-76; asst prof genetics & cell biol, Univ Minn, St Paul, 76-; AT DEPT BIOSCI, 3M CO. *Res:* Role of cell surface and plasma membrane in growth control and differentiation; biochemical and cellular mechanisms of intercellular adhesion and cellular invasiveness. *Mailing Add:* Dept Biosci 3M Ctr 270-35-05 3M Co St Paul MN 55144

ROTTY, RALPH M(CGEE), b St Louis, Mo, Aug 1, 23; m 44; c 4. ENGINEERING THERMODYNAMICS, METEOROLOGY. *Educ:* Univ Iowa, BS, 47; Calif Inst Technol, MS, 48 & 49; Mich State Univ, PhD(mech eng), 53. *Prof Exp:* From instr to assoc prof mech eng, Mich State Univ, 49-58; prof & head dept, Tulane Univ, 58-66; prof eng & dean, Old Dominion Univ, 66-72; CHIEF SCIENTIST, INST ENERGY ANALYSIS, OAK RIDGE ASSOC UNIVS, 74- *Concurrent Pos:* Nat Res Coun resident res assoc, Nat Oceanic & Atmospheric Admin, 72-73. *Mem:* Fel Am Soc Mech Engrs; AAAS; Am Meteorol Soc. *Res:* Irreversible and atmospheric thermodynamics; energy policy analysis, especially relation of energy use to global and regional climate; atmospheric consequences of anthropogenic activities, fossil fuels and atmospheric CO2, energy requirements in the developing world. *Mailing Add:* Mech Eng Dept Univ New Orleans Lake Front New Orleans LA 70148

ROTZ, CHRISTOPHER ALAN, b Van Nuys, Calif, Feb 29, 48; m 73; c 3. POLYMER ENGINEERING, MANUFACTURING PROCESSING. *Educ:* Mass Inst Technol, BS, 73, SM, 76, PhD(mech eng), 78. *Prof Exp:* Asst prof mech eng, Univ Tex, Austin, 78-85; ASSOC PROF MECH ENG, BRIGHAM YOUNG UNIV, PROVO, UTAH, 85- *Mem:* Assoc mem Am Soc Mech Engrs; Soc Plastics Engrs. *Res:* Characterization and prediction of viscoelastic properties; injection molding of polymer blends; engineering design with polymers and composites; modeling and control of manufacturing processes; finite element analysis. *Mailing Add:* 1648 S 270 West Orem Provo UT 84058

ROUBAL, RONALD KEITH, b Omaha, Nebr, Mar 22, 35; wid; c 3. INORGANIC CHEMISTRY. *Educ:* Creighton Univ, BS, 57, MS, 59; Univ Iowa, PhD(chem), 65. *Prof Exp:* Assoc prof, 64-76, PROF CHEM, UNIV WIS-SUPERIOR, 76-, CHMN, DIV SCI & MATH, 81- *Mem:* Am Chem Soc; Sigma Xi. *Res:* Synthesis and studies of new borazine compounds; preparation of coordination compounds; natural waters chemistry. *Mailing Add:* Chem Prog Area Univ Wis Superior WI 54880-2898

ROUBAL, WILLIAM THEODORE, b Eugene, Ore, Dec 20, 30; m 53; c 4. BIOORGANIC CHEMISTRY. *Educ:* Ore State Univ, BA, 54, MS, 59; Univ Calif, Davis, PhD(biochem), 64. *Prof Exp:* RES CHEMIST, ENVIRON CONSERV DIV, NAT MARINE FISHERIES SERV, 60- *Concurrent Pos:* Affil prof, Col Fisheries, Univ Wash, 74-; instr chem technol, Seattle Cent Community Col, 78- *Honors & Awards:* A E McGee Award; Am Oil Chemist Soc, 64. *Mem:* Sigma Xi; Am Chem Soc. *Res:* Analysis of xenobiotic free radicals in the environment and biota of Puget Sound; studies on uptake, depuration and metabolism of xenobiotics in aquatic organisms. *Mailing Add:* Environ Conserv Div Nat Marine Fisheries Serv 2725 Montlake Blvd E Seattle WA 98112

ROUBENOFF, RONENN, US citizen. RHEUMATOLOGY. *Educ:* Northwestern Univ, BS, 81, MD, 83; Johns Hopkins Univ, MHS, 90. *Prof Exp:* Intern-resident, Osler Med Serv, Johns Hopkins Hosp, 83-86, asst chief serv, 86-87; fel epidemiol, Johns Hopkins Sch Pub Health Hyg, 88-90; RES ASSOC NUTRIT, TUFTS UNIV SCH MED, 90- *Concurrent Pos:* Instr med, Johns Hopkins Univ Sch Med, 87-88, fel rheumatology, 87-90. *Mem:* Fel Am Col Physicians; Am Inst Nutrit; Am Col Rheumatology; AAAS; Am Soc Parenteral & Enteral Nutrit. *Res:* Effect of chronic autoimmune diseases, chiefly rheumatoid arthritis, on body composition and nutritional metabolism, via interleukin-l and tumor necrosis factor production. *Mailing Add:* USDA Human Nutrit Res Ctr Tufts Univ 711 Washington St Boston MA 02111

ROUBICEK, RUDOLF V, b Prosnitz, Austria, Feb 11, 18; Can citizen. DESIGN OF BIOREACTORS, PHOTOSYNTHESIS. *Educ:* Tech Univ, Prague, Dipl Ing, 46; Tech Univ Czech, Dr Technol(chem eng), 48. *Prof Exp:* Dir res & develop, Spofa-United Pharmaceut Industs, Prague, Czech, 52-68; head fermentation group, Res Ctr, Can Packers Inc, 68-82; DIR BIOCHEM ENG, CTR BIOCHEM ENG RES, NMEX, 82-; EMER PROF BIOCHEM ENG, INDUST WASTE TREATMENT & TRANSP OPERS, DEPT CHEM ENG, NMEX STATE UNIV, 82-, LUKE SHIRES PROF, 85- *Concurrent Pos:* Pres, Bio/Film Technol Inc. *Mem:* Soc Indust Microbiol; Am Inst Chem Engrs. *Res:* Biosynthesis and recovery processes of primary and secondary metabolites; photosynthesis; fermented food and single cell proteins; design of bioreactors and photobioreactors; transport operations in biological systems; biochemical engineering fermentation processes. *Mailing Add:* NMex State Univ Box 3805 Las Cruces NM 88003

ROUBIK, DAVID WARD, b Schenectady, NY, Oct 3, 51; m 74; c 3. PALYNOLOGY, POLLINATION. *Educ:* Ore State Univ, BS, 75; Univ Kans, PhD(entom), 79. *Prof Exp:* RES ENTOMOLOGIST, SMITHSONIAN TROP RES INST, 79- *Mem:* Ecol Soc Am. *Res:* Ecological impact of Africanized honeybees in the Americas; general biology of bees; pollination ecology. *Mailing Add:* Smithsonian Trop Res Inst APO Miami FL 34002-0011

ROUF, MOHAMMED ABDUR, b Dacca, Bangladesh, May 2, 33; m 65. BACTERIOLOGY. *Educ:* Univ Dacca, BS, 54, MS, 55; Univ Calif, Davis, MA, 59; Wash State Univ, PhD(bact, biochem), 63. *Prof Exp:* Lectr bot, Govt Col Sylhet, Bangladesh, 55-57; asst bacteriologist, Pullman Div Indust Res, Wash State Univ, 63-64; from asst prof to assoc prof, 64-68, PROF BACT, UNIV WIS-OSHKOSH, 68-, CHMN, DEPT BIOL, 70- *Concurrent Pos:* Received grants from various state and fed agencies; NSF fac fel, 76-78. *Honors & Awards:* Duncan Res Award, Univ Wis-Oshkosh, 85. *Mem:* AAAS; Am Soc Microbiol; Sigma Xi; fel Am Acad Microbiol; fel Am Inst Chem. *Res:* Microbial physiology; degradation of uric acid by bacteria; iron and manganese oxidizing bacteria; methanogens. *Mailing Add:* Dept Biol & Microbiol Univ Wis Oshkosh WI 54901

ROUFA, DONALD JAY, b St Louis, Mo, Apr 8, 43; m 67; c 2. BIOCHEMISTRY, GENETICS. *Educ:* Amherst Col, AB, 65; Johns Hopkins Univ, PhD(biol), 70. *Prof Exp:* Chemist, Nat Inst Child Health & Human Develop, 69-70; res scientist biochem, NY State Dept Health, Albany, 70-71; from instr to asst prof biochem & med, Baylor Col Med, 71-75; assoc prof biol, 75-81, PROF BIOL, KANS STATE UNIV, 81- *Mem:* NY Acad Sci; Am Soc Microbiol; Am Soc Biol Chemists. *Res:* Biochemical mechanisms involved in protein synthesis; molecular mechanisms participating in genetic processes; animal cell somatic genetics. *Mailing Add:* Div Biol Kans State Univ Manhattan KS 66506

ROUFFA, ALBERT STANLEY, b Boston, Mass, Mar 30, 19; m 43; c 3. BOTANY. *Educ:* Mass State Col, BS, 41; Rutgers Univ, MS, 47, PhD(bot), 49. *Prof Exp:* Asst bot, Rutgers Univ, 41-42 & gen biol, 46-49; from asst prof to prof, 49-83, chmn dept, 61-63, exec secy dept, 64-68, EMER PROF BIOL SCI, UNIV ILL, CHICAGO, 83- *Concurrent Pos:* Dir James Woodworth Prairie Preserve, 69- *Mem:* Am Asn Biol Sci. *Res:* prairie preserve monitoring and management. *Mailing Add:* 4426 Florence Ave Downers Grove IL 60515

ROUGH, GAYLORD EARL, b Cochranton, Pa, Nov 17, 24; m 49; c 3. ENVIRONMENTAL SCIENCES. *Educ:* Univ Pittsburgh, BS, 50, MS, 52, PhD(animal ecol), 61. *Prof Exp:* Instr biol, Alfred Univ, 52-56; asst, Univ Pittsburgh, 56-57, spec lectr, 57-58; from asst prof to chmn dept, Alfred Univ, 69-74, dir environ studies prog, 78-81, from prof to emer prof biol, 66-87; RETIRED. *Mem:* AAAS; Ecol Soc Am; Am Soc Ichthyologists & Herpetologists; Sigma Xi; Am Fisheries Soc; Int Asn Great Lakes Res. *Res:* Radioecology; physiological ecology; respiratory metabolism and thyroid activity in small mammals; animal ecology, especially behavior; limnology, aquatic productivity; population dynamics and meristic characters of freshwater fishes. *Mailing Add:* Dept Biol Alfred Univ Alfred NY 14802

ROUGHGARDEN, JONATHAN DAVID, b Paterson, NJ, Mar 13, 46; m 69; c 1. BIOMATHEMATICS. *Educ:* Univ Rochester, AB & BS, 68; Harvard Univ, PhD(biol), 71. *Prof Exp:* Asst prof biol, Univ Mass, Boston, 70-72; from asst prof to assoc prof, 72-81, PROF BIOL, STANFORD UNIV, 81- *Honors & Awards:* Guggenheim Fel. *Res:* Theoretical population biology; population ecology and genetics; mathematical theory of density-dependent natural selection, niche width, community structure, faunal buildup, population spatial structure and species borders; field work on anolis lizards; intertidal marine community ecology. *Mailing Add:* Dept Geophys Stanford Univ Herrin Labs 413 Stanford CA 94305

ROUGHLEY, PETER JAMES, b Doncaster, Eng, July 22, 47; m 77; c 2. CARTILAGE PROTEOGLYCAN BIOCHEMISTRY. *Educ:* Nottingham Univ, BSc, 69, PhD(chem), 72. *Prof Exp:* Fel, Charing Cross Hosp, Eng, 72-74, Strangeways Res Lab, Eng, 74-77; RES SCIENTIST, SHRINERS HOSP, CAN, 77- *Concurrent Pos:* From asst prof to assoc prof, McGill Univ, 77-91, prof, 91- *Mem:* Biochem Soc Eng; Orthopaedic Res Soc; Royal Soc Chem Eng. *Res:* Structure and function of the proteoglycans isolated from human articular cartilage; proteinases and proteinase inhibitors present in cartilage; synthesis and degradation of human articular cartilage; genetic disorders of cartilage proteoglycans. *Mailing Add:* Genetics Unit Shriners Hosp 1529 Cedar Ave Montreal PQ H3G 1A6 Can

ROUGVIE, MALCOLM ARNOLD, b Newton, Mass, Feb 4, 28; m 59. BIOPHYSICS. *Educ:* Mass Inst Technol, SB & SM, 51, PhD(biophys), 54. *Prof Exp:* Instr biophys, Mass Inst Technol, 54-55; res assoc, Iowa State Univ, 55-56, asst prof physics, 57-60, from asst prof to assoc prof biophys, 60-91; RETIRED. *Mem:* Biophys Soc; AAAS. *Res:* Structure of proteins by optical and physicochemical methods. *Mailing Add:* 2233 McKinley Ct Ames IA 50010

ROUKES, MICHAEL L, b Redwood City, Calif, Oct 9, 53. QUANTUM TRANSPORT IN NANOSTRUCTURES, ULTRA-LOW-TEMPERATURE PHYSICS. *Educ:* Univ Calif, Santa Cruz, BA(chem) & BA(physics), 78; Cornell Univ, PhD(physics), 85. *Prof Exp:* Collab researcher, T J Watson Res Ctr, IBM, 83-85; MEM TECH STAFF & PRIN INVESTR LOW TEMPERATURE PHYSICS & QUANTUM STRUCT RES, BELL COMMUN RES, 85- *Mem:* Am Phys Soc. *Res:* Electronic and thermal transport physics of nonstructures: theory, low temperature experimentation, techniques for microfabrication; ultra low noise instrumentation design. *Mailing Add:* Bellcore Rm 3X-285 331 Newman Springs Rd Red Bank NJ 07701-7040

ROULEAU, WILFRED T(HOMAS), b Quincy, Mass, May 3, 29; m 54; c 2. FLUID MECHANICS, ENERGY CONVERSION. *Educ:* Carnegie Inst Technol, BS, 51, MS, 52, PhD(mech eng), 54. *Prof Exp:* From asst prof to assoc prof mech eng, 54-65, PROF MECH ENG, CARNEGIE MELLON UNIV, 65- *Concurrent Pos:* Consult, Gulf Res & Develop Labs, Bituminous Coal Res, Inc, Westinghouse Res Labs, Whirlpool Corp, PPG Industs, Mine Safety Appliances Co, Pittsburgh Corning Corp, Colt Industs, Inc, Celanese Fibers Corp, Mat Eng & Testing Co, Air Technologies, Inc. *Mem:* Am Soc Mech Engrs; Sigma Xi. *Res:* Hydrodynamic stability; wave propagation in viscous fluids; viscous flows; biological flows; hydrodynamic lubrication; porous bearings; numerical analysis of fluid flow; jet mixing; magnetohydrodynamics; flow and erosion in turbomachinery; energetics; power. *Mailing Add:* Dept Mech Eng Carnegie Mellon Univ Schenley Park Pittsburgh PA 15213-3890

ROULIER, JOHN ARTHUR, b Cohoes, NY, May 3, 41; m 62; c 4. MATHEMATICS. *Educ:* Siena Col, NY, BS, 63; Syracuse Univ, MS, 66, PhD(math), 68. *Prof Exp:* Technician, Gen Elec Co, 63-64; asst prof math, Mich State Univ, 68; NSF res assoc math, Rensselaer Polytech Inst, 69; asst prof, Union Col, NY, 69-73; from asst prof math to assoc prof math, NC State Univ, 73-79, prof, 79-80; prof math, 80-84, interim head, 89-90, PROF COMPUTER SCI & ENG, DEPT COMPUTER SCI & ENG, UNIV CONN, 84- *Mem:* Math Asn Am; Am Math Soc; Soc Indust & Appl Math. *Res:* Approximation theory; approximation by polynomials satisfying linear restrictions; weighted approximation; Chebyshev rational approximation on infinite line segments; shape preserving spline interpolation; geometric modeling; computational geometry. *Mailing Add:* Dept Computer Sci & Eng Univ Conn Storrs CT 06269-3155

ROULSTON, DAVID J, b London, Eng, Nov 3, 36; div; c 3. SEMICONDUCTORS. *Educ:* Queen's Univ, Belfast, BSc, 57; Univ London, PhD(elec eng) & DIC, 62. *Prof Exp:* Sci officer, Civil Serv, Portland, Eng, 57-58; engr, CSF Dept, RPC, France, 62-67; assoc prof, 67-72, PROF ELEC ENG, UNIV WATERLOO, 72- *Concurrent Pos:* Vis assoc prof, Univ Waterloo, 66-67; consult engr, Thomson-CSF, France, 73- & Res & Develop Labs in US, Can, Japan, UK & India, UN Indust Develop Orgn, 77-; mem elec eng grants comt, Nat Res Coun Can, 75-78; vis fel, Wolfson Col, Oxford, UK, 88-89. *Mem:* Fel Inst Elec Engrs UK; sr mem Inst Elec & Electronics Engrs. *Res:* Semiconductor devices and circuits; device characterization and modeling using computer aided techniques, microwave bipolar and field effect transistors, integrated circuit modelling; photo diodes and solar cells. *Mailing Add:* Elec & Comp Eng Dept Univ Waterloo Waterloo ON N2L 3G1 Can

ROULSTON, THOMAS MERVYN, b Armagh, Northern Ireland, July 23, 20; m 47; c 3. OBSTETRICS & GYNECOLOGY. *Educ:* Queen's Univ Belfast, MB, BCh & BAO, 43; FRCOG, 64; FACOG, 64; FRCS(C), 69; FSOGC, 86. *Prof Exp:* Chmn dept, 64-78, prof, 64-89, EMER PROF OBSTET & GYNEC, UNIV MAN, 89- *Concurrent Pos:* Obstetrician & gynecologist, Health Sci Ctr, Winnipeg, 56-87; consult, Misericordia Gen, Grace, St Boniface Gen & Children's Hosps, Winnipeg, 64-87; pres, Family Planning Fedn Can, 70-72; vpres, Western Hemisphere Region, Int Planned Parenthood Fedn, 80-87; hon obstetrician & gynecologist, Health Sci Ctr Winnipeg, 87- *Honors & Awards:* Queen Elizabeth II Silver Jubilee Medal; Alan Guttmacher Mem Medal. *Mem:* Fel Am Col Obstet & Gynec; Soc Obstet & Gynec Can (pres, 72); Can Med Asn. *Res:* Fertility regulation; population problems; genital cancer; medical education; ultrasonics in obstetrics and gynecology. *Mailing Add:* Dept Obstet & Gynec Univ Man 59 Emily St Winnipeg MB R3E 1Y9 Can

ROUND, G(EORGE) F(REDERICK), b Worcestershire, Eng, Jan 31, 32; m 56; c 5. CHEMICAL & MECHANICAL ENGINEERING. *Educ:* Univ Birmingham, BSc, 54, PhD(chem eng), 57, DSc(chem eng), 74. *Prof Exp:* Asst res officer, Res Coun Alta, 57-61, assoc res officer, 61-67, sr res officer, 67-68; assoc prof, 68-71, PROF MECH ENG, MCMASTER UNIV, 71- *Concurrent Pos:* Sr vis, Dept Appl Math & Theoret Physics, Univ Cambridge, 73-74; sr consult, Kuwait Inst Sci Res 78-80; prof chem eng, Kuwait Univ, 78-80; assoc ed, J Pipelines, 85- *Honors & Awards:* Henry R Worthington Award, 77; Distinguished Lectr Award, IFPS, 89. *Mem:* Chem Inst Can; Royal Soc Arts. *Res:* Fluid dynamics; thermodynamics; author of 75 papers and three books on fluid mechanics. *Mailing Add:* Fac Eng McMaster Univ Hamilton ON L8S 4L7 Can

ROUNDS, BURTON WARD, b Milan, NH, May 6, 24; m 46; c 3. FISH & WILDLIFE MANAGEMENT, NATURAL RESOURCE CONSERVATION. *Educ:* Colo A&M Col, BS, 48. *Prof Exp:* Wildlife res biologist effects fed projs wildlife, US Fish & Wildlife Serv, 48-58; supvry wildlife res biologist determination causes wetland losses, Bur Sport Fisheries & Wildlife, US Dept Interior, 58-61, supvry wildlife biologist wetlands protection, 61-68, wetlands prog coordr, 68-72, regional supvr, Wildlife Serv, 72; area mgr fish & wildlife admin, US Fish & Wildlife Serv, 72-79; FISH & WILDLIFE CONSERV CONSULT, 79- *Concurrent Pos:* Vchmn fish & wildlife resources div, Soil Conserv Soc Am, 72, chmn, 73; fac mem gyroscope, US Fish & Wildlife Serv, 75 & 76, mem pathfinder, 78; mem adv coun to dean sch forestry, Univ Mont, 77-79. *Mem:* Wildlife Soc; Am Soc Mammalogists; Soil Conserv Soc Am; Trumpeter Swan Soc; Nat Wildlife Refuge Asn. *Res:* Ecology of prairie wetlands; ecology of wild canids. *Mailing Add:* HC54 Box 7C Columbus MT 59019

ROUNDS, DONALD EDWIN, b Maywood, Ill, Jan 17, 26; m 51; c 2. IMMUNOLOGY. *Educ:* Occidental Col, BA, 51; Univ Calif, Los Angeles, PhD(zool), 58. *Prof Exp:* Lab technician, AEC Proj, 51-54; res assoc exp embryol, Univ Calif, Los Angeles, 54-56; investr tissue cult, Med Br, Univ Tex, 58-59; investr, Tissue Cult Lab, Pasadena Found Med Res, 59-60, assoc dir, Div Cell Biol, 60-64, res dir, 64-90; CHIEF SCIENTIST, AIR MUNITIONS DEVELOP LAB, 90- *Concurrent Pos:* Adj asst prof anat, Univ Southern Calif, 62-70, adj assoc prof, 70-83; consult, air & indust hyg lab, Dept Pub Health, Calif, 64-66; biomed consult, Electro-Optical Systs, Inc, 65-68; asst clin prof path, Sch Med, Loma Linda Univ, 65-74, clin prof, 74-; mem bd dirs, Laser Inst Am, 73- *Mem:* Fel AAAS; Tissue Cult Asn; Am Soc Cell Biol; Soc Gen Physiol; Am Sci Film Asn. *Res:* Development of immunological test systems for tumor markers; tissue culture; cell physiology. *Mailing Add:* 1261 Sonoma Dr Altadena CA 91001

ROUNDS, FRED G, b Colfax, Wash, June 4, 25; m 59; c 2. CHEMICAL ENGINEERING. *Educ:* State Col Wash, BS, 49. *Prof Exp:* From jr res engr to sr res engr, 49-75, SR STAFF RES ENGR, RES LABS, GEN MOTORS CORP, 75- *Honors & Awards:* Alfred E Hunt Award, Am Soc Lubrication Engrs, 64. *Mem:* Am Chem Soc; fel Am Soc Lubrication Engrs. *Res:* Automotive engine exhaust gas emissions; diesel odor; engine combustion; additive mechanisms; lubricant coking; lubricant composition effects on rolling contact fatigue, friction and wear; diesel soot effects on wear. *Mailing Add:* 554 Bridge Park Troy MI 48098-1804

ROUNDS, RICHARD CLIFFORD, b Rockford Ill, Feb 16, 43; m 65; c 2. BIOGEOGRAPHY, RESOURCE GEOGRAPHY. *Educ:* Ill State Univ, BSc, 65, MSc, 67; Univ Colo, Boulder, PhD(geog), 71. *Prof Exp:* Teaching assoc phys geog, Ill State Univ, 66-67; from asst prof to prof geog, 70-89, DIR, RURAL DEVELOP, BRANDON UNIV, 89- *Honors & Awards:* Res Award, N Am Bluebird Soc, 85; Conserv Award, Ducks Unlimited, 87. *Mem:* Can Wildlife Soc; Asn Am Geographers. *Res:* Wildlife resources; tourism and recreation. *Mailing Add:* Rural Develop Inst Brandon Univ 270 18th St Brandon MB R7A 6A9 Can

ROUNSAVILLE, BRUCE J, b Ancon, CZ, May 21, 49; US citizen. DRUG ABUSE, PSYCHIATRIC EPIDEMIOLOGY. *Educ:* Yale Univ, BA, 70; Sch Med, Univ Md, MD, 74; Am Bd Psychiat & Neurol, dipl, 79. *Prof Exp:* Asst prof psychiat, 78-84, dir res, Drug Dependence Unit, 78-84, ASSOC PROF PSYCHIAT & DIR RES, SUBSTANCE TREATMENT UNIT, SCH MED, YALE UNIV, 84-; PROF PSYCHIAT, SCH MED, UNIV CONN, 84- *Honors & Awards:* Anna Monika Award, Anna Monika Found, 86; Res Scientist Award, Nat Inst Drug Abuse, 88. *Mem:* Asn Clin Psychosocial Res; Am Psychiat Asn; Soc Psychother Res; Am Psychopathol Asn. *Res:* Clinical and epidemiological research on the diagnosis and treatment of substance use disorders and effective disorders. *Mailing Add:* Res Off Yale Univ 27 Sylvan Ave New Haven CT 06519

ROUNSLEY, ROBERT R(ICHARD), b Detroit, Mich, Jan 11, 31; m 53; c 5. DRYING, SIMULATION. *Educ:* Mich Technol Univ, BS, 52, MS, 54; Iowa State Univ, PhD(chem eng), 57. *Prof Exp:* Res asst, Argonne Nat Lab, 53-54; instr, Iowa State Univ, 54-57; sr res engr, 57-79, RES FEL, MEAD CORP, 79- *Concurrent Pos:* Prof math & eng, Chillicothe Br, Ohio Univ, 63-68. *Mem:* Soc Comput Simulation; Am Inst Chem Engrs; Am Chem Soc; Tech Asn Pulp & Paper Indust. *Res:* Application of computer control in process industries; process simulation; statistics; finishing of paper; drying of paper and coatings. *Mailing Add:* Cent Res Labs Mead Corp Chillicothe OH 45601

ROUNTREE, JANET, b Chicago, Ill, Aug 14, 37. SPECTROSCOPY, ELECTRO-OPTICAL INSTRUMENTS. *Educ:* Cornell Univ, AB, 58; Univ Chicago, PhD(astron, astrophys), 67. *Prof Exp:* Res assoc astron, Yerkes Observ, Univ Chicago, 67-68; sci officer astron, Leiden Observ, Univ Leiden, Netherlands, 68-70; astronr adjoint, Meudon Observ, Observ Paris, France, 70-71; vis fel astrophys, Joint Inst Lab Astrophys, Univ Colo, 71-72; lectr astron & res astronr, Univ Denver, 72-77, dir observ opers, 74-77; sr res assoc, NASA Goddard Space Flight Ctr, 77-79; PHYS SCIENTIST, US DEPT AIR FORCE, 79- *Concurrent Pos:* Eng lang ed & translr, Astron & Astrophys, 69-72; translr, D Reidel Co, Dordrecht, Netherlands, 69-72 & Joint Publ Res Serv, 73-79. *Mem:* Int Astron Union; Am Astron Soc; Royal Astron Soc; Netherlands Astron Soc. *Res:* Spectral classification of early type stars; short-period variable stars; local galactic structure; stellar associations and clusters; infrared astronomy. *Mailing Add:* 6001 Wynnwood Rd Bethesda MD 20816

ROUQUETTE, FRANCIS MARION, JR, b Aransas Pass, Tex, Dec 25, 42; m 64; c 4. ANIMAL HUSBANDRY, RANGE SCIENCE & MANAGEMENT. *Educ:* Tex A&I Univ, BS, 65; Tex Tech Univ, MS, 67; Tex A&M Univ, PhD(soil & plant sci), 70. *Prof Exp:* From asst prof to assoc prof, 70-83, PROF FORAGE PHYSIOL, AGR RES & EXTEN CTR, TEX A&M UNIV, 83- *Honors & Awards:* Cert Merit, Am Forage & Grassland Coun. *Mem:* Am Soc Agron; Sigma Xi; Am Soc Range Sci; Am Soc Animal Sci; Am Forage & Grassland Coun. *Res:* Investigation of forage animal systems under various levels of grazing intensity; animal science and nutrition. *Mailing Add:* Agr Res & Exten Ctr Drawer E Tex A&M Univ Overton TX 75684

ROURKE, ARTHUR W, b Boston, Mass, Oct 8, 42; m 65; c 2. CELL PHYSIOLOGY. *Educ:* Lafayette Col, AB, 64; Univ Conn, PhD(cell biol), 70. *Prof Exp:* Fel cell physiol, Univ Conn, 70-72; asst prof biol, 72-77, ASSOC PROF ZOOL & BIOCHEM, UNIV IDAHO, 77- *Concurrent Pos:* Muscular Dystrophy Asn Am grant, 73; Heart Asn grant, 75. *Mem:* Biophys Soc; Sigma Xi; AAAS; Am Soc Cell Biologists. *Res:* Cellular turnover in eukaryotes. *Mailing Add:* Dept Life Sci Univ Idaho Moscow ID 83843

ROUS, STEPHEN N, b New York, NY, Nov 1, 31; m 66; c 2. UROLOGY. *Educ:* Amherst Col, AB, 52; New York Med Col, MD, 56; Univ Minn, MS, 63. *Prof Exp:* Fel, Mayo Grad Sch Med, Univ Minn, 60-63; assoc prof urol, New York Med Col, 68-72, asst dean, 68-70, assoc dean, 70-72; prof surg & chief div urol, Col Human Med, Mich State Univ, 72-75; prof & chmn dept urol, Med Univ SC, 75-88; urologist-in-chief, Med Univ & Charleston County Hosps, SC, 75-88; adj prof, 88-91, PROF SURG, MED SCH, DARTMOUTH COL, 91-; STAFF UROLOGIST, DARTMOUTH-HITCHCOCK MED CTR, 91- *Concurrent Pos:* Chief urol, Metrop Hosp Ctr, New York, 68-72; consult, Vet Admins, Roper & St Francis Hosps, Charleston, 75-88; adj prof urol, Med Univ SC, 88-; ed dir, Med Books Div, W W Norton & Co, Pub, 88-; chief urol, Vet Admin Med Ctr, White River Junction, Vt. *Mem:* Am Urol Asn; Am Col Surg; Soc Univ Urol; Pan-Pac Surg Asn; Int Soc Urol; hon Ger Urol Asn; Sigma Xi. *Mailing Add:* 32 Sanborn Rd Hampton Falls NH 03844

ROUSE, BARRY TYRRELL, b Jan 9, 42; Brit & Can citizen; m 65; c 2. IMMUNOLOGY, VIROLOGY. *Educ:* Bristol Univ, BVSc, 65; Univ Guelph, MSc, 67, PhD(immunol), 70. *Prof Exp:* Houseman vet med, Bristol Univ, 65-66; fel immunol, Walter & Eliza Hall, Inst Med Sci, Melbourne, Australia, 70-72; asst prof, Univ Sask, 72-73, assoc prof, 73-77; assoc prof, 77-78, PROF IMMUNOL, UNIV TENN, KNOXVILLE, 78- *Concurrent Pos:* Med Res Coun Can fel, 70-72, mem study sect immunol & transplantation, Med Res Coun Can, 74- 77; ad hoc mem, Study Sect Exp Virol, NIH, 78; mem grant rev panel, Morris Animal Found, 78-80; grants, NIH, Morris Animal Found, Nat Hog Producers. *Mem:* Royal Col Vet Surgeons; Am Asn Immunologists; Infectious Dis Soc Am; Reticuloendothelial Soc; Can Soc Immunol. *Res:* Mechanisms of recovery from herpesvirus infections; treatment methods for canine allergy; immunological diseases of domestic animals. *Mailing Add:* Dept Microbiol Univ Tenn Knoxville TN 37996-0845

ROUSE, CARL ALBERT, b Youngstown, Ohio, July 14, 26; m 55; c 3. THEORETICAL ASTROPHYSICS, THEORETICAL NUCLEAR PHYSICS. *Educ:* Case Western Reserve Univ, BS, 51; Calif Inst Technol, MS, 53, PhD(physics), 56. *Prof Exp:* Sr res engr, NAm Aviation Inc, 56-57; theoret physicist, Lawrence Radiation Lab, Univ Calif, 57-65, res assoc theoret physics, Space Sci Lab, Univ Calif, Berkeley, 65-68; staff physicist, Gulf Radiation Technol, 68-74; staff scientist, Ga Technol Inst, 74-87; staff physicist, Mission Res, 87-89; CONSULT, 89- *Concurrent Pos:* Instr, Exten Div, Univ Calif, Los Angeles, 57; NSF res assoc, E O Hulburt Ctr Space Res, US Naval Res Lab, 65-68; instr, Exten Div, Univ Calif, San Diego, 75; prin investr, NSF grant, 81-83. *Mem:* Fel Am Phys Soc; Am Astron Soc; Int Astron Union. *Res:* Ionization equilibrium equations of state for monatomic matter; exploding wire phenomena; numerical solutions to the Schrodinger equations; solar and stellar interiors; radiation transport; solar oscillations; original theoretical studies of new shielding materials for fission and fusion nuclear reactors. *Mailing Add:* 627 15th St Del Mar CA 92014

ROUSE, DAVID B, b Lafayette, Ind, July 15, 49; m 71; c 4. AQUACULTURE, MARICULTURE. *Educ:* Auburn Univ, BS, 71, MS, 73; Tex A&M Univ, PhD(fisheries), 81. *Prof Exp:* Aquatic biologist, Water Imp Comt, State of Ala, 73-78; res assoc, Tex A&M Univ, 79-81; adj instr biol & aquatic biol, 77-78, ASST PROF CRUSTACEAN AQUACULT, AUBURN UNIV, 81- *Concurrent Pos:* Coop scientist, Univ Agr Sci, India, 84-89; aquacult consult, Traverse Group Inc, Mich, 84-; shrimp specialist, USAID, Honduras, Panama, Ecuador, Madagascar, 81- *Mem:* Am Fisheries Soc; World Aquacult Soc. *Res:* Aquaculture development; fish and shrimp culture; culture of freshwater and saltwater fish and shrimp with applications to the United States and overseas. *Mailing Add:* Dept Fisheries Auburn Univ Auburn AL 36830

ROUSE, GEORGE ELVERTON, b Chugwater, Wyo, May 4, 34; m 61; c 3. MINERALS EXPLORATION. *Educ:* Colo Sch Mines, Geol Engr, 61, DSc, 68. *Prof Exp:* Mining geologist, Anglo Am Corp SAfrica, Ltd, 61-64; proj mgr, Mineral Explor Earth Sci Inc, 68-71, vpres & sr proj mgr, 71-81; spec proj mgr, Minerals, Inc, 83-86; vpres, Benton Resources Ltd, 87-89; PRES, BIGHORN EXPLOR, 81-; PRES, ROUSE, BANER & ASSOC, INC, 89- *Mem:* AAAS; Asn Explor Geochemists; Asn Inst Mining Engrs. *Res:* Global tectonics; gas geochemistry for mineral exploration; geochemistry of metalliferous shales; hydrometallurgy. *Mailing Add:* 9254 Fern Way Blue Mountain Estates Golden CO 80403

ROUSE, GLENN EVERETT, b Hamilton, Ont, Aug 1, 28; m 52; c 1. PALYNOLOGY, PALEOBOTANY. *Educ:* McMaster Univ, BA, 51, MSc, 52, PhD(palynology, paleobot), 56. *Prof Exp:* From instr to assoc prof, 57-68, PROF PALYNOLOGY & PALEOBOT, UNIV BC, 68- *Concurrent Pos:* Palynological consult, Oil, Coal & Mineral Explor. *Mem:* Am Asn Stratig Palynologists; Geol Asn Can. *Res:* Palynology of Mesozoic and Tertiary strata of Arctic and Western Canada. *Mailing Add:* 3529 6270 University Blvd Dept Bot Univ BC Vancouver BC V6T 2B1 Can

ROUSE, HUNTER, b Toledo, Ohio, Mar 29, 06; m 32; c 3. FLUID MECHANICS, HYDRAULICS. *Educ:* Mass Inst Technol, SB, 29, SM, 32; Dr Ing, Karlsruhe, 32; Dr Sci, Univ Paris, 59. *Hon Degrees:* Dr Ing, Karlsruhe, 75. *Prof Exp:* Asst hydraul, Mass Inst Technol, 31-33; instr civil eng, Columbia Univ, 33-35; asst prof fluid mech, Calif Inst Technol, 35-39; prof, 39-72, assoc dir, Inst Hydraul Res, 42-44, dir, 44-66, dean, Col Eng, 66-72, Carver Prof, 72-74, EMER DEAN ENG, UNIV IOWA, 72-, EMER CARVER PROF HYDRAUL, 74- *Concurrent Pos:* Assoc hydraul engr, Soil Conserv Serv, USDA, 36-39; consult, US Off Naval Res, 48-66 & Waterways

Exp Sta, US Corps Engrs, 49-68; Fulbright exchange prof, Univ Grenoble, 52-53; NSF sr fel, Univs Goettingen, Rome, Cambridge & Paris, 58-59; sr scholar, Australian-Am Educ Found, 73; vis prof, Col State Univ, 75-86. *Honors & Awards:* Norman Medal, Am Soc Civil Engrs, 38, Von Karman Medal, 63 Hist & Hertiage Award,80; Westinghouse Award, Am Soc Eng Educ, 48 & Bendix Award, 58; John Fritz Award, 91. *Mem:* Nat Acad Eng; hon mem Am Soc Civil Engrs; Am Soc Eng Educ; fel Am Acad Arts & Sci; hon mem Am Soc Mech Engrs; hon mem Venezuela Soc Hydraul Engrs; hon mem Int Asn Hydraul Res. *Res:* Engineering hydraulics; history of hydraulics; engineering education; human ecology. *Mailing Add:* 10814 Mimosa Dr Sun City AZ 85373

ROUSE, JOHN WILSON, JR, b Kansas City, Mo, Dec 7, 37; m 56; c 1. ELECTRICAL ENGINEERING. *Educ:* Purdue Univ, BS, 59; Univ Kans, MS, 65, PhD(elec eng), 68. *Prof Exp:* Engr, Bendix Corp, 59-64; res coordr, Ctr Res, Inc, Univ Kans, 64-68; from asst prof & actg dir to prof elec eng & dir, Remote Sensing Ctr, Tex A&M Univ, 68-78; distinguished prof elec eng, chmn dept & dir, Bioeng Prog, Univ Mo-Columbia, 78-81; DEAN ENG, UNIV TEX, ARLINGTON, 81- *Concurrent Pos:* Mem comn F, Int Union Radio Sci, 69- *Mem:* Inst Elec & Electronics Engrs. *Res:* Electromagnetic and acoustic scattering; radar systems; remote sensor systems; geoscience applications. *Mailing Add:* Southern Res Inst 2000 Ninth Ave S PO Box 55305 Birmingham AL 35255

ROUSE, LAWRENCE JAMES, JR, b New Orleans, La, Oct 26, 42; m 74; c 2. COASTAL OCEANOGRAPHY, REMOTE SENSING. *Educ:* Loyola Univ, New Orleans, BS, 64; La State Univ, Baton Rouge, PhD(physics), 72. *Prof Exp:* Asst prof physics, dept physics, LA State Univ, 72-73, asst prof, Coastal Studies Inst, 73-79, asst prof phys oceanog, dept marine sci, 78-79, assoc prof phys oceanog, dept marine sci, Coastal Studies Inst, 79-85, chmn dept marine sci, 83-84, ASSOC PROF PHYS OCEANOG, DEPT GEOL & GEOPHYS, COASTAL STUDIES INST, LA STATE UNIV, 85- *Mem:* Sigma Xi; Am Geophys Union; Am Meteorol Soc; Inst Elec & Electronics Engrs; The Oceanog Soc. *Res:* Coastal circulation processes and air-sea interactions on continental shelves; remote sensing of these processes; sediment transport processes. *Mailing Add:* Coastal Studies Inst La State Univ Baton Rouge LA 70803

ROUSE, ROBERT ARTHUR, b St Louis, Mo, Sept 4, 43; m 66; c 3. THEORETICAL CHEMISTRY. *Educ:* Wash Univ, AB, 65; Northwestern Univ, PhD(chem), 68. *Prof Exp:* Res assoc chem, Harvard Univ, 68-69; asst prof chem, Univ Mo-St Louis, 69-76; ASSOC PROF COMPUTER SCI, WASH UNIV, 77- *Mem:* Am Chem Soc; Am Phys Soc; Soc Comput Mach; Sigma Xi. *Res:* Molecular quantum mechanics; interacting molecular systems; reactions in theoretical organic chemistry. *Mailing Add:* Wash Univ CSDP Box 1141 St Louis MO 63130

ROUSE, ROBERT S, b Northampton, Mass, Sept 2, 30; m 51; c 4. ORGANIC CHEMISTRY. *Educ:* Yale Univ, BS, 51, MS, 53, PhD(chem), 57. *Prof Exp:* Lab asst, Yale Univ, 51-53 & 55, asst instruction, 56; asst prof chem, Lehigh Univ, 56-62; group leader, Plastics Div, Allied Chem Corp, NJ, 62-66, tech supvr, 66-67; assoc dean fac, 68-73, chmn dept, 67-73, dean fac, 73-80, vpres acad affairs, 73-81, provost, 80-81, PROF CHEM, MONMOUTH COL, 67- *Concurrent Pos:* Consult, Allied Chem Corp, 68-70; Sabbatical leave, 81-82. *Mem:* Am Chem Soc; fel NY Acad Sci; Sigma Xi. *Res:* Organic reactions and synthesis; environment; energy; science education. *Mailing Add:* 407 Cynthia Ct Princeton NJ 08540

ROUSE, ROY DENNIS, b Andersonville, Ga, Sept 20, 20; m 46; c 2. SOILS AND SOIL SCIENCE. *Educ:* Univ Ga, BSA, 42, MSA, 47; Purdue Univ, PhD(soil chem), 49. *Prof Exp:* Assoc soil chemist, 49-56, prof soils, 56-66, assoc dir, Agr Exp Sta & asst dean, Sch Agr, 66-72, dir & dean, 72-81, EMER DIR, AGR EXP STA & EMER DEAN, SCH AGR, AUBURN UNIV, 82- *Concurrent Pos:* Chmn Exp Sta Comt on Org and Policy, 77; mem Food Adv Comt, Off Tech Assessment, 78. *Mem:* Fel Soil Sci Soc Am; fel Am Soc Agron. *Res:* Soil chemistry of potassium; potassium nutrition of agronomic plants; resource allocation and research administration. *Mailing Add:* 827 Salmon St Auburn AL 36830

ROUSE, THOMAS C, b Milwaukee, Wis, Sept 24, 34; m 71; c 2. PARASITOLOGY, PHYSIOLOGY. *Educ:* Univ Wis-Milwaukee, BS, 59; Univ Wis-Madison, MS, 63, PhD(zool), 67. *Prof Exp:* Asst prof, Wis State Univ, Platteville, 65-67; from asst prof to assoc prof, 67-85, PROF BIOL, UNIV WIS-EAU CLAIRE, 85- *Mem:* Sigma Xi. *Res:* Effect of exercise on the normal and diseased cardiovascular system. *Mailing Add:* Dept Biol Univ Wis Eau Claire WI 54701

ROUSE, WILLIAM BRADFORD, b Fall River, Mass, Jan 20, 47; m 68. HUMAN-SYSTEM INTERACTION, INFORMATION SYSTEMS. *Educ:* Univ RI, BS, 69; Mass Inst Technol, SM, 70, PhD(eng), 72. *Prof Exp:* Asst prof syst eng, Tufts Univ, 73; prof syst eng, Univ Ill, 74-81; prof syst eng, Georgia Inst Tech, 81-88; CHIEF EXEC OFFICER, SEARCH TECHNOL, INC, 80- *Concurrent Pos:* Consult, Electronic Assoc Inc, 73, Xerox & Barber-Colman Inc, 77; assoc ed, Transactions, Inst Elec & Electronics Engrs, 76-78; vis prof, Delft Univ Technol, 79-80. *Honors & Awards:* Schuck Award, Am Automatic Control Coun, 79; Norbert Wiener Award, Syst Soc, 86; Centennial Medal, Inst Elec & Electronics Engrs, 84. *Mem:* Nat Acad Engrs; fel Inst Elec & Electronics Engrs; Systs, Man & Cybernet Soc; Human Factors Soc; Inst Indust Engrs. *Res:* Human decision making; human-computer interactions; design of information systems; author of many articles. *Mailing Add:* 4725 Peachtree Corners Circle Suite 200 Norcross GA 30092

ROUSEK, EDWIN J, b Burwell, Nebr, Sept 8, 17; m 45; c 2. AGRICULTURE. *Educ:* Univ Nebr, BSc, 40; Cornell Univ, MSc, 43. *Prof Exp:* Chmn dept animal sci, 48-63, PROF ANIMAL SCI, CALIF STATE UNIV, FRESNO, 63- *Concurrent Pos:* Chmn univ senate, Calif State Univ, Fresno, 69-71. *Mem:* Am Meat Sci Asn. *Res:* Animal nutrition and meats. *Mailing Add:* 1175 W San Madele Fresno CA 93711

ROUSELL, DON HERBERT, b Winnipeg, Man, Sept 4, 31; m 57; c 2. STRUCTURAL GEOLOGY. *Educ:* Univ Man, BSc, 52, Univ BC, MSc, 56; Univ Man, PhD(geol), 65. *Prof Exp:* Geologist, Tidewater Oil Co, 52-53 & Chevron Oil Co Venezuela, 56-59; from asst prof to assoc prof, 63-83, chmn dept 77-78 & 87-90, PROF GEOL, LAURENTIAN UNIV, 83- *Concurrent Pos:* Nat Res Coun Can grants, 68-69, 71-73 & 83-84; grants, Ont Dept Univ Affairs, 68-69, indust, 74-75 & 79-80, Ont Geol Surv, 80-81, Energy Mines & Resources, 86 & 89, Can Geol Found, 87 & Ministry Northern Develop & Mines, 89 & 90; fel, Ctr Tectonophys, Tex A&M Univ, 69-70. *Mem:* Fel Geol Soc Am; fel Geol Asn Can. *Res:* Geology of Northwestern Venezuela; Precambrian geology of Northern Manitoba; geology of the Sudbury Basin; structure of the Greville Front. *Mailing Add:* Dept Geol Laurentian Univ Sudbury ON P3E 2C6 Can

ROUSH, ALLAN HERBERT, b Hardin Mont, Feb 24, 18; m 44; c 2. BIOCHEMISTRY. *Educ:* Mont State Col, BS, 40; Univ Wash, PhD(biochem), 51. *Prof Exp:* From asst prof to assoc prof, 51-62, prof, 62-82, EMER PROF BIOCHEM, ILL INST TECHNOL, 82- *Mem:* Am Chem Soc; Am Soc Biol Chemists; Mycol Soc Am. *Res:* Enzymology; nucleic acid metabolism; active transport. *Mailing Add:* 1615 S Black No 72 Bozeman MT 59715

ROUSH, FRED WILLIAM, b Brooklyn, NY, May 7, 47. MATHEMATICS. *Educ:* Univ NC, AB, 66; Princeton Univ, PhD(math), 72. *Prof Exp:* Asst prof math, Univ Ga, 70-74; ASST PROF MATH, ALA STATE UNIV, 76- *Mem:* Am Math Soc. *Res:* Boolean matrix theory; combinatorics. *Mailing Add:* 615 Lynwood Dr Montgomery AL 36111

ROUSH, MARVIN LEROY, b Topeka, Kans, Dec 26, 34; m 55; c 3. RISK ASSESSMENT. *Educ:* Ottawa Univ, BSc, 56; Univ Md, PhD(nuclear physics), 64. *Prof Exp:* Asst prof physics, Baker Univ, 59-61 & Tex A&M Univ, 65-66; from asst prof to assoc prof, 66-80, PROF NUCLEAR ENG, UNIV MD, COLLEGE PARK, 80- *Mem:* AAAS; Am Phys Soc; Am Asn Physics Teachers; Am Nuclear Soc. *Res:* Risk assessment of nuclear power plants; neutron, gamma-ray and charged particle spectroscopy; trace-element analysis by x-ray fluorescence spectroscopy. *Mailing Add:* Dept Chem & Nuclear Eng Univ Md College Park MD 20742

ROUSH, WILLIAM BURDETTE, b Sheridan, Wyo, Apr 26, 45; m 73; c 2. ANIMAL HUSBANDRY. *Educ:* Brigham Young Univ, BS, 72, MS, 75; Ore State Univ, PhD(poultry sci), 79- *Prof Exp:* asst prof, 79-88, ASSOC PROF POULTRY SCI, PA STATE UNIV, 88- *Mem:* Poultry Sci Asn; Opers Res Soc Am; Am Inst Nutrit. *Res:* Management systems analysis for the biological and economical optimization of poultry production. *Mailing Add:* 204 Animal Indust Bldg Dept Poultry Sci Pa State Univ University Park PA 16802

ROUSLIN, WILLIAM, b Providence, RI, Nov 10, 38; m 70. BIOCHEMISTRY, BIOENERGETICS. *Educ:* Brown Univ, AB, 60; Univ Conn, PhD(biochem), 68. *Prof Exp:* NIH res fel, Cornell Univ, 68-70; asst prof biol, Douglass Col, Rutgers Univ, New Brunswick, 70-77; asst prof, 77-85, ASSOC PROF PHARMACOL & CELL BIOPHYS, COL MED, UNIV CINCINNATI, 85- *Concurrent Pos:* Res grants, NIH, 84-87, 87-91 & 91-96. *Mem:* AAAS; Am Asn Pathologists; Am Soc Cell & Develop Biol; Am Physiol Soc; Am Soc Biochem & Molecular Biol; Biophys Soc; Cardiac Muscle Soc; Int Soc Heart Res. *Res:* Mitochondrial ATPase in cardiac muscle. *Mailing Add:* Dept Pharmacol & Cell Biophys Univ Cincinnati Col Med 231 Bethesda Ave Cincinnati OH 45267-0575

ROUSOU, J A, SURGERY. *Prof Exp:* BAY STATE MED CTR. *Res:* Surgery. *Mailing Add:* Bay State Med Ctr 759 Chestnut St Springfield MA 01107

ROUSSEAU, CECIL CLYDE, b Philadelphia, Pa, Jan 13, 38; m 65; c 2. MATHEMATICS, PHYSICS. *Educ:* Lamar Univ, BS, 60; Tex A&M Univ, MS, 62, PhD (physics), 68. *Prof Exp:* Asst prof physics, Baylor Univ, 68-70; asst prof math, Memphis State Univ, 70-75; Carnegie fel, Univ Aberdeen, 75-76; from asst prof to assoc prof, 76-81, PROF MATH, MEMPHIS STATE UNIV, 81- *Concurrent Pos:* Collab ed, Probs & Solutions, Soc Indust & Appl Math, 73-; vis prof, Univ Waterloo, 87-88. *Mem:* Math Asn Am; Soc Indust & Appl Math; Am Math Soc. *Res:* Graph theory; analysis; mathematical physics; mathematical statistics. *Mailing Add:* Dept Math Sci Memphis State Univ Memphis TN 38152

ROUSSEAU, DENIS LAWRENCE, b Franklin, NH, Nov 18, 40; m 63. BIOLOGICAL PHYSICS, PHYSICAL CHEMISTRY. *Educ:* Bowdoin Col, BA, 62; Princeton Univ, PhD(phys chem), 67. *Prof Exp:* Res fel, Univ Southern Calif, 67-69; DISTINGUISHED MEM STAFF, BIOPHYS RES, BELL LABS, INC, 69- *Mem:* Biophys Soc; fel Am Phys Soc; AAAS; fel Optical Soc Am; Sigma Xi. *Res:* Raman scattering from biological materials; heme proteins; molecular physics. *Mailing Add:* Bell Labs Inc Murray Hill NJ 07974

ROUSSEAU, RONALD WILLIAM, b Sept 28, 43; US citizen; m 63; c 3. SEPARATION PROCESSES, PHASE EQUILIBRIA. *Educ:* La State Univ, BS, 66, MS, 68, PhD(chem eng), 69. *Prof Exp:* Instr chem eng, La State Univ, 67; res engr, Westvaco Corp, 69; from asst prof to prof chem eng, NC State Univ, 69-86; DIR, CHEM ENG, GA INST TECHNOL, 87- *Concurrent Pos:* Vis prof, Princeton Univ, 82-83. *Honors & Awards:* Outstanding Chem Eng, Am Inst Eng, 86. *Mem:* Am Inst Chem Engrs; Am Chem Soc. *Res:* Nucleation, growth and crystal size distributions; separation process technology; conditioning gases produced from coal by absorption and stripping; phase equilibria, crystallization and precipitation. *Mailing Add:* Sch Chem Eng Ga Inst Technol Atlanta GA 30332-0100

ROUSSEAU, VIATEUR, b Baie des Sables, Que, July 5, 14; nat US; m 58; c 3. ORGANIC CHEMISTRY. *Educ:* Am Int Col, BS, 39; NY Univ, PhD(org chem), 48. *Prof Exp:* Analytical chemist, Chapman Valve Mfg Co, Mass, 40-42; asst chem, NY Univ, 42-47; instr, Col Mt St Vincent, 47-50, asst prof, 50-56; from assoc prof to prof, 56-82, chmn sci div, 66-68, EMER PROF CHEM, IONA COL, 82- *Concurrent Pos:* Adj prof, Fordham Univ, 58-73. *Mem:* Am Chem Soc. *Res:* Chemistry of indazoles; synthesis; structural relationships; absorption spectra. *Mailing Add:* Dept Chem Iona Col New Rochelle NY 10801

ROUSSEL, JOHN S, b Hester, La, Nov 23, 21; m 46; c 5. ENTOMOLOGY, AGRONOMY. *Educ:* La State Univ, BS, 42, MS, 48; Tex A&M Univ, PhD(entom), 50. *Prof Exp:* From asst prof to assoc prof, 49-57, PROF ENTOM, LA STATE UNIV, BATON ROUGE, 57-, COORDR COTTON RES & ASST TO DIR AGR EXP STA, 61- *Concurrent Pos:* NSF grant, 57-60. *Mem:* AAAS; Entom Soc Am; Am Inst Biol Sci. *Res:* Cotton production practices and insects. *Mailing Add:* 5032 Perkins Rd Baton Rouge LA 70808

ROUSSEL, JOSEPH DONALD, b Paulina, La, Apr 28, 29; m 58; c 2. REPRODUCTIVE PHYSIOLOGY. *Educ:* La State Univ, BS, 58, MS, 60, PhD(reprod physiol), 63. *Prof Exp:* Res asst, La State Univ, 59-63; res assoc, Univ Ark, 63-65 & Delta Regional Primate Res Ctr, 66-72; PROF REPROD PHYSIOL, LA STATE UNIV, BATON ROUGE, 72- *Concurrent Pos:* Res grant, 63-66. *Mem:* Am Fertil Asn; Am Dairy Sci Asn; Int Fertil Asn; Brit Soc Study Fertil; Am Soc Animal Sci. *Res:* Environmental reproductive physiology; semen metabolism; enzyme activity in spermatogenesis and physiology factors influencing reproduction; reproduction in primates; nutritional influence in reproduction; ovulation; semen preservation and artificial insemination; embryo transfer. *Mailing Add:* Dept Dairy Sci La State Univ Baton Rouge LA 70803

ROUSSIN, ROBERT WARREN, b Columbia, Mo, Jan 27, 39; m 62. NUCLEAR ENGINEERING, INFORMATION SCIENCE. *Educ:* Univ Mo-Rolla, BS, 62; Univ Ill, MS, 64, PhD(nuclear eng), 69. *Prof Exp:* Co-op student mech eng, McDonnell Aircraft Corp, 57-62; engr mech & nuclear eng, Allis Chalmers, 62; DIR, RADIATION SHIELD INFO CTR, OAK RIDGE NAT LAB, MARTIN MARIETTA ENERGY SYSTS, 68- *Concurrent Pos:* Chmn shielding subcomt, Cross Sect Eval Working Group, 75- *Mem:* Am Nuclear Soc. *Res:* Promote the exchange of information and computing technology in the field of radiation transport. *Mailing Add:* Oakridge Nat Lab Radiation Shielding Info Ctr PO Box 2008 Oak Ridge TN 37831-6362

ROUSSOS, CONSTANTINE, b Kingston, NY, Sept 22, 47; m 69; c 2. COMPUTER SCIENCE. *Educ:* Old Dominion Univ, BA, 65; Col William & Mary, MS, 74; Univ Va, PhD(comput sci), 79. *Prof Exp:* Instr math, Tidewater Community Col, 74-75; instr comput sci & math, Wash & Lee Univ, 77-78; grad asst, Univ Va, 75-79; vpres, software systs, Three Ridges Corp, 79-81; dir, comput serv, 81-85, CHMN, COMPUT SCI, LYNCHBURG COL, 85- *Concurrent Pos:* Vis researcher, Naval Surface Warfare Ctr, 88; vis prof computer sci, Univ Limerick, Ireland, 89. *Mem:* Asn Comput Mach; Digital Equip User's Soc. *Res:* Data structures and network theory; development of database software; database user interfaces; algorithms for AI applications. *Mailing Add:* Rte 1 Box 186 Arrington VA 22922

ROUTBORT, JULES LAZAR, b San Francisco, Calif, May 15, 37; m 66; c 2. MATERIALS SCIENCE. *Educ:* Univ Calif, Berkeley, BS, 60; Cornell Univ, PhD(eng physics), 65. *Prof Exp:* Sci Res Coun fel physics, Cavendish Lab, Cambridge Univ, 64-66; AEC fel, Rensselaer Polytech Inst, 66-68; PHYSICIST, MAT SCI DIV, ARGONNE NAT LAB, 68- *Concurrent Pos:* Humboldt fel, Res Inst Transurane, Karlsruhe, 73-74 & Univ Hamburg, Ger, 82; Euratom fel, Res Inst Transurane, 81-82; adj prof mat sci, NC State Univ, Raleigh & mat eng, IIT, Chicago; proj mgr, Off Basic Energy Sci, Dept Energy, Washington, DC, 87-88; assoc ed, Appl Physics Lett. *Mem:* Fel Am Ceramic Soc; Am Phys Soc; Material Res Soc. *Res:* Mechanical properties of ceramics; diffusion and elastic properties of ceramics. *Mailing Add:* Mat Sci Div Argonne Nat Lab 9700 S Cass Ave Argonne IL 60439-4838

ROUTH, DONALD KENT, b Oklahoma City, Okla, Mar 30, 37; m 60; c 2. PEDIATRIC PSYCHOLOGY, CLINICAL CHILD PSYCHOLOGY. *Educ:* Univ Okla, BA, 62; Univ Pittsburgh, MS, 65, PhD(psychol), 67. *Prof Exp:* Asst prof psychol, Univ Iowa, 67-70; assoc prof psychol, Bowling Green State Univ, Ohio, 70-71 & Univ NC, Chapel Hill, 71-77; prof psychol, Univ Iowa, 77-85, PROF PSYCHOL, UNIV MIAMI, FLA, 85- *Concurrent Pos:* Ed, J Pediat Psychol, 76-82, J Clin Child Psychol, 87-91; chmn, Behav Med Study Sect, NIH, 83-85; pres, Div Child, Youth & Family Serv, Am Psychol Asn, 84, Div Ment Retardation, 87-88. *Honors & Awards:* Distinguished Contrib Award, Soc of Pediat Psychol, 81. *Mem:* Soc Pediat Psychol (pres, 73- 74); Am Psychol Asn. *Res:* Conditioning of infant vocalizations; development of activity level in children; effects of mother presence on children's response to medical stress; phonemic awareness, reading and spelling in children. *Mailing Add:* Dept Psychol Univ Miami PO Box 248185 Coral Gables FL 33124

ROUTH, JOSEPH ISAAC, b Logansport, Ind, May 8, 10; m 37, 76; c 2. CLINICAL BIOCHEMISTRY, PATHOLOGY. *Educ:* Purdue Univ, BSChE, 33, MS, 34; Univ Mich, PhD(biochem), 37. *Prof Exp:* From instr to prof, 37-78, dir, clin biochem lab, 52-64, dir spec clin chem lab, 70-78, EMER PROF BIOCHEM, UNIV IOWA, 78- *Concurrent Pos:* Consult clin chem, Vet Admin Hosp, Iowa City, Iowa, 52- *Mem:* Am Chem Soc; Am Soc Biol Chemists; Am Asn Clin Chem (pres, 57-58); Am Bd Clin Chem (pres, 59-73); fel Am Inst Chem. *Res:* Purification and properties of trypsin inhibitor, levodopa metabolites in Parkinson's disease, enzyme inhibitors; methodology, metabolism and protein binding of analgesic drugs; effects of drugs on clinical chemistry parameters. *Mailing Add:* Dept Biochem 4-450 BSB Univ Iowa Iowa City IA 52242-1193

ROUTIEN, JOHN BRODERICK, b Mt Vernon, Ind, Jan 23, 13; m 44, 67. MICROBIOLOGY. *Educ:* DePauw Univ, AB, 34; Northwestern Univ, AM, 36; Mich State Col, PhD(mycol), 40. *Prof Exp:* Instr bot, Univ Mo, 39-42; mycologist, Chemotherapeut Res Labs, Pfizer Inc, 46-73, res adv, Cent Res, 73-77; RETIRED. *Concurrent Pos:* Mem, Antimicrobial Agents & Chemother, 74-77. *Honors & Awards:* Com Solvents Award, 50. *Mem:* Bot Soc Am; Mycol Soc Am; Am Soc Microbiol; Soc Indust Microbiol; NY Acad Sci. *Res:* Isolation of microorganisms; culture collection; taxonomy of microorganisms. *Mailing Add:* 318 Grassy Hill Rd Lyme CT 06371

ROUTLEDGE, RICHARD DONOVAN, b Toronto, Ont, Aug 15, 48; m 70; c 2. STATISTICAL METHODOLOGY. *Educ:* Queen's Univ, Kingston, Ont, BSc, 70; Univ Alta, Edmonton, MSc, 72; Dalhousie Univ, Halifax, PhD(biol), 75. *Prof Exp:* Killam fel math, Univ Alta, 75-77, asst prof, 75-80; asst prof math, 80-83, ASSOC PROF MATH & STATIST, SIMON FRASER UNIV, 83- *Mem:* Biomet Soc; Statist Soc Can. *Res:* Development of statistical methodology and use of mathematical optimization to study problems in population ecology and resource management. *Mailing Add:* Dept Math & Stat Simon Fraser Univ Burnaby BC V5A 1S6 Can

ROUTLEY, DOUGLAS GEORGE, b BC, Apr 26, 29; m 58; c 2. HORTICULTURE. *Educ:* Univ BC, BSA, 52; Pa State Univ, MS, 55, PhD(agr, biol chem), 57. *Prof Exp:* Asst prof biochem, 57-64, assoc prof biochem & plant sci, 64-70, PROF PLANT SCI, UNIV NH, 70- *Concurrent Pos:* NIH spec fel, 66-67. *Mem:* Am Soc Hort Sci. *Res:* Plant growth regulators; greenhouse crops. *Mailing Add:* Dept Plant Sci & Biochem Univ NH Durham NH 03824

ROUTLY, PAUL MCRAE, b Chester, Pa, Jan 4, 26; m 51; c 2. ASTROPHYSICS. *Educ:* McGill Univ, BSc, 47, MSc, 48; Princeton Univ, AM, 50, PhD(astrophys), 51. *Prof Exp:* Nat Res Coun Can fel, 51-53; res fel, Calif Inst Technol, 53-54; chmn dept & dir observ, 54-63, from asst prof to prof astron, Pomona Col, 54-63; exec off, Am Astron Soc, 62-68; dir, Div Astrometry & Astrophys, US Naval Observ, 68-77, head explor develop staff, 77-85; RETIRED. *Concurrent Pos:* Vpres, Comn 38, Int Astron Union; 38; vis prof, Rutgers In-Serv Inst, 66; pres, Comn 38, Int Astron Union, 73- *Mem:* Am Astron Soc; Int Astron Union. *Res:* Astrometry; molecular spectroscopy; absolute and relative atomic transition probabilities; stellar parallaxes. *Mailing Add:* 9401 Kentsdale Dr Potomac MD 20854

ROUTSON, RONALD C, b Chewelah, Wash, Dec 12, 33; m 58; c 1. SOIL PHYSICAL CHEMISTRY & MINERALOGY. *Educ:* Wash State Univ, BS, 58, PhD(soil chem), 70. *Prof Exp:* Sr res scientist & prog mgr, Battelle Pac Northwest Labs, 65-77; staff scientist soil chem, Rockwell Hanford Co, 77-90; CONSULT, 91- *Mem:* Am Soc Agron; Soil Sci Soc Am; Clay Minerals Soc. *Res:* Modeling the movement of radionuclides through soil systems; disposal and fate of wastes in soil; moisture movement in soil systems; plant uptake of pollutants. *Mailing Add:* Rte 1 Box 1351 Benton City WA 99320

ROUTTENBERG, ARYEH, b Reading, Pa, Dec 1, 39; c 2. NEUROSCIENCE. *Educ:* McGill Univ, BA, 61; Northwestern Univ, MA, 63; Univ Mich, PhD(neurosci & behav), 65. *Prof Exp:* PROF PSYCHOL & NEUROBIOL/PHYSIOL & DIR UNDERGRAD NEUROSCI PROG, COL ARTS & SCI & GRAD NEUROSCI PROG, PSYCHOL DEPT, NORTHWESTERN UNIV, 73- *Concurrent Pos:* Mem res adv bd, NIMH, 76-80. *Mem:* Neurosci Soc; Am Physiol Soc; Am Soc Neurochem; Am Asn Anatomists; Int Soc Neurochem. *Res:* Brain Chemistry; memory and learning. *Mailing Add:* Dept Psychol & Physiol Northwestern Univ 633 Clark St Evanston IL 60208

ROUVRAY, DENNIS HENRY, b Rochford, Essex, UK, Aug 22, 38; m 62; c 3. TOPOLOGICAL STRUCTURE ANALYSIS. *Educ:* Imperial Col, BSc, 61, ARCS, 61, DIC, 64, PhD(chem), 64. *Prof Exp:* Fel chem, Dalhousie Univ, Halifax, NS, 64-66; fel chem, Univ Liverpool, 66-67; lectr, Univ Witwatersrand, S Africa, 67-75; res fel, Max Planck Inst, Muelheim, Ger, 75-78; vis prof, Yarmouk Univ, Jordan, 78-79; dir res publ, Diebold Europe Comput Consult, London, 79-84; RES SCIENTIST CHEM, UNIV GA, ATHENS, 84- *Concurrent Pos:* Vis prof, Univ Oxford, UK, 70-71 & 78; ed, J Math Chem, 87-; consult, Du Pont Chem Co, Waynesboro, Va, 88-, Cornell Univ, NY Hosp, 83-87, Glaxo, Research Triangle Park, NC, 90-; co-prin investr, US Off Naval Res, Univ Ga, 84-88. *Mem:* Int Soc Math Chem (secy, 86-). *Res:* The characterization of molecular structure and the prediction of physical, chemical and biological properties of materials using mathematical methods, especially topological and graph-theoretical analysis. *Mailing Add:* Dept Chem Univ Ga Athens GA 30602

ROUX, KENNETH H, b Philadelphia, Pa, May 12, 48; m 70; c 1. IMMUNOGENETICS, IMMUNOREGULATION. *Educ:* Del Valley Col, BS, 70; Tulane Univ, MS, 72, PhD(immunol), 74. *Prof Exp:* Fel immunol, Univ Ill Med Ctr, 75-78; from asst prof to assoc prof, 78-88, PROF IMMUNOL, DEPT BIOL SCI, FLA STATE UNIV, 89- *Concurrent Pos:* Prin investr, dept biol sci, Fla State Univ, 79- *Mem:* Fedn Am Soc Exp Biol; Am Asn Immunologists; AAAS. *Res:* Genetics and regulation of immunoglobulin molecules; electronmicroscopy of immunoglobulin molecules and immune complexes, molecular genetics of antibody diversity; monoclonal antibody (hybridoma) production; electron microscopy. *Mailing Add:* Dept Biol Sci Fla State Univ Tallahassee FL 32306

ROUX, STANLEY JOSEPH, b Houston, Tex, Feb 9, 42; m 75; c 2. PHOTOBIOLOGY, MEMBRANE BIOLOGY. *Educ:* Spring Hill Col, BS, 66; Loyola Univ, New Orleans, 68; Yale Univ, PhD(biol), 71. *Prof Exp:* Fel, Yale Univ, 71-73; asst prof biol, Univ Pittsburgh, 73-78; from asst prof to assoc prof, 78-86, PROF BOT, UNIV TEX, AUSTIN, 86-, CHMN, BOT DEPT, 86- *Mem:* AAAS; Am Soc Plant Physiologists; Am Chem Soc. *Res:* Identifying cellular mechanisms for the control of plant growth and development by light and for the control of plant tropisms by light and gravity. *Mailing Add:* Dept Bot Univ Tex Austin TX 78712

ROUZE, STANLEY RUPLE, metal physics, electron microscopy, for more information see previous edition

ROVAINEN, CARL (MARX), b Virginia, Minn, Mar 13, 39; m 66; c 3. NEUROPHYSIOLOGY, ENDOTHELIAL CELL BIOLOGY. *Educ:* Calif Inst Technol, BS, 62; Harvard Univ, PhD(physiol), 67. *Prof Exp:* NIH training fel biochem, 67-68, from instr to asst prof physiol & biophys, 68-73, assoc prof, 73-79, PROF CELL BIOL & PHYSIOL, SCH MED, WASH UNIV, 79- *Concurrent Pos:* Nat Inst Neurol Dis & Stroke res grant, 73- *Mem:* Soc Neurosci; Soc Gen Physiol; Am Physiol Soc. *Res:* Physiological and anatomical organization of lamprey brain and spinal cord; brain angiogenesis. *Mailing Add:* Dept Cell Biol & Physiol 660 S Euclid Ave Wash Univ Sch Med St Louis MO 63110

ROVELLI, CARLO, b Verona, Italy, May 3, 56. QUANTUM GRAVITY, GRAVITATION. *Educ:* Univ Bologna, Italy, Laurea, 81; Univ Padova, Italy, PhD(physics), 84. *Prof Exp:* INFN postdoctoral, Univ Roma, Italy, 87-88; postdoctoral fel, Yale Univ, 87; contractual prof mech, Univ Dell'aquila, Italy, 89; RESEARCHER, UNIV TRENTO, ITALY, 90-; PROF PHYSICS, UNIV PITTSBURGH, PA, 90- *Concurrent Pos:* Vis scientist, SISSA, Triestre, Italy, 89-90; prin investr, NSF Res Proj, Non Perturbative Quantum Gravity, 90-92. *Mem:* Sigma Xi. *Res:* Construction of a quantistic theory of the gravitational field; general relativity; definition of the concept of time. *Mailing Add:* Physics Dept 100 Allen Hall Univ Pittsburgh Pittsburgh PA 15260

ROVELSTAD, GORDON HENRY, b Elgin, Ill, May 19, 21; m 45; c 3. DENTISTRY. *Educ:* Northwestern Univ, DDS, 44, MSD, 48, PhD(dent), 60; Am Bd Pedodont, dipl. *Hon Degrees:* DSc, Georgetown Univ, 69. *Prof Exp:* Instr dent, Northwestern Univ, 46-49, asst prof pedodont & consult, Cleft Palate Inst, 49-53; res officer, Dent Res Lab, Dent Corps, USN, 54-58, head res & sci div, Dent Sch, 60-65, dir dent res, Dent Res Facil, Training Ctr, 65-67, officer-in-charge & sci dir, Naval Dent Res Inst, Ill, 67-69, prog mgr dent res, Dent Div, Bur Med & Surg, Washington, DC, 69-74; prof pediat dent & assoc prof physiol & biophysics, Sch Dent, Univ Miss Med Ctr, Jackson, 74-81, actg chmn, Dept Pediat Dent & assoc dean educ prog, 77-81; EXEC DIR, AM COL DENTISTS, 81- *Concurrent Pos:* Chief dent staff, Children's Mem Hosp, Chicago, 49-53; consult, Herrick House Rheumatic Fever Inst, 51-53. *Mem:* AAAS; Am Soc Dent Children; Am Dent Asn; Am Acad Pedodontics; NY Acad Sci; Sigma Xi. *Mailing Add:* Am Col Dentists 7315 Wisconsin Ave Bethesda MD 20814

ROVELSTAD, RANDOLPH ANDREW, b Elgin, Ill, Mar 11, 20; m 45; c 5. MEDICINE. *Educ:* St Olaf Col, BA, 40; Northwestern Univ, MD, 44; Univ Minn, PhD(med), 54. *Prof Exp:* From instr to assoc prof med, Mayo Med Sch, 53-85; RETIRED. *Concurrent Pos:* Consult, Mayo Clin, 52-85. *Mem:* Am Gastroenterol Asn. *Res:* Gastroenterology; gastric secretion; gastric and duodenal pH; enterocutaneous potentials; Composition of ascitic fluid. *Mailing Add:* 200 First St SW Rochester MN 55901

ROVERA, GIOVANNI, b Cocconato, Italy, Sept, 23, 40; m 79; c 1. HEMATOLOGY. *Educ:* Univ Torino, Italy, MD, 64. *Prof Exp:* Fel biochem, Fels Inst, 68-70; resident path, Temple Univ, 70-73, asst prof, 73-75; assoc prof, 75-78, PROF CANCER RES, WISTAR INST, 79- *Concurrent Pos:* Scholar, Leukemia Soc Am, 74-79; mem, Molecular Biol Grad Group, Univ Pa, 77-; assoc ed, J Cellular Physiol, 78- *Res:* Proliferation and differentiation of leukemia cells; expression of globin genes; mechanisms of tumor pomotion. *Mailing Add:* Wistar Inst 3601 Spruce St Philadelphia PA 19104

ROVETTO, MICHAEL JULIEN, b Challis, Idaho, Mar 20, 43; m 67, 87; c 2. PHYSIOLOGY, BIOCHEMISTRY. *Educ:* Utah State Univ, BS, 65; Univ Idaho, MS, 68; Univ Va, PhD(physiol), 70. *Prof Exp:* Res assoc physiol, Hershey Med Ctr, Pa State Univ, 71-73, asst prof, 73-74; from asst prof to assoc prof physiol, Jefferson Med Col, 74-80; ASSOC PROF PHYSIOL, SCH MED, UNIV MO, 80- *Concurrent Pos:* NIH fel, Hershey Med Ctr, Pa State Univ, 70-71. *Mem:* Am Physiol Soc; Biophys Soc; Cardiac Muscle Soc; Int Soc Heart Res. *Res:* Myocardial energy metabolism; regulation of cardiovascular function; adenine nucleotide metabolism. *Mailing Add:* Physiol Dept Univ Mo Sch Med Columbia MO 65251

ROVICK, ALLEN ASHER, b Chicago, Ill, Feb 11, 28; m 49; c 4. PHYSIOLOGY, SCIENCE EDUCATION. *Educ:* Roosevelt Univ, BS, 51; Univ Ill, MS, 54, PhD(physiol), 58. *Prof Exp:* Instr physiol, Stritch Sch Med, Loyola Univ Chicago, 58-59, assoc, 59, from asst prof to assoc prof, 66-67; Univ Ill Proj, Thailand, 67-68; assoc prof physiol, Univ Ill Med Sch, 69-70; exec secy cardiovasc study sect, Div Res Grants, NIH, 70-71, chief cardiac dis br, Nat Heart & Lung Inst, 71-72; assoc prof biomed eng, 75-80, PROF PHYSIOL, RUSH MED SCH, RUSH UNIV, 80- *Mem:* AAAS; Am Physiol Soc; Am Heart Asn; Sigma Xi. *Res:* Metabolic control of local blood circulation; influence of hemodynamics on tissue water partition; local control of blood flow; effect of arterial pulse on blood flow; computer based instruction. *Mailing Add:* Dept Physiol Rush-Presby-St Lukes Med Ctr Chicago IL 60612

ROVIT, RICHARD LEE, b Boston, Mass, Apr 3, 24; m 53; c 3. NEUROSURGERY. *Educ:* Jefferson Med Col, MD, 50; McGill Univ, MSc, 61; Am Bd Neurol Surg, dipl, 62. *Prof Exp:* Asst neurosurgeon, Montreal Neurol Inst, 60-61; assoc prof & surgeon, Jefferson Med Col & Hosp, 61-66, head div neurosurg, 61-65; assoc prof neurosurg, 66-71, PROF CLIN NEUROSURG, SCH MED, NY UNIV, 71-; DIR, DEPT NEUROSURG, ST VINCENT'S HOSP, 66- *Concurrent Pos:* Consult, US Vet Admin Hosps, Philadelphia, 62-64 & Coatesville, 62-66; chief sect neurosurg, Philadelphia Gen Hosp, New York, Columbus Hosp & St Vincent's Hosp; attend neurosurgeon, New York Univ Hosp, 66- *Mem:* AAAS; Am Acad Neurol; Am Epilepsy Soc; Asn Res Nerv & Ment Dis; Am Asn Neurol Surg; Am Col Surgeons; Soc Neurol Surgeons. *Res:* Surgery of epilepsy and neuroendocrinology. *Mailing Add:* St Vincents Med Ctr Dept NS 153 W 11th St New York NY 10011

ROVNER, DAVID RICHARD, b Philadelphia, Pa, Sept 20, 30. ENDOCRINOLOGY, METABOLISM. *Educ:* Temple Univ, AB, 51, MD, 55. *Prof Exp:* Intern, San Francisco Hosp, Univ Calif Serv, 55-56; resident med, Med Ctr, Univ Mich, 56-57; internist radiobiol, USAF Sch Avaition Med, 57-59; resident med, Med Ctr, Univ Mich, 59-60, from fel to prof endocrinol & metab, 60-71; vchmn dept med, 71-80, PROF ENDOCRINOL & METAB, COL HUMAN MED, MICH STATE UNIV, 71-, CHIEF, DIV ENDROCRINOL & METAB, 80- *Concurrent Pos:* Distinguished faculty Award, Mich State Univ, 84. *Mem:* Endocrine Soc; Am Fedn Clin Res; fel Am Col Physicians; Inst Elec & Electronics Engrs; Cent Soc Clin Res. *Res:* Decision analysis as applied to clinical medicine; endocrine hypertension; data base usage. *Mailing Add:* Med Dept B-234 Life Sci I Mich State Univ East Lansing MI 48824-1317

ROVNER, JEROME SYLVAN, b Baltimore, Md, July 15, 40; m 62; c 2. ARACHNOLOGY, ANIMAL BEHAVIOR. *Educ:* Univ Md, BS, 62, PhD(zool), 66. *Prof Exp:* NSF fel zool, Gutenberg Univ, Mainz, 66-67; asst prof, 67-71, assoc prof, 71-77, PROF ZOOL, OHIO UNIV, 77- *Concurrent Pos:* Res grant, NSF, 72 & 74; guest researcher, Goethe Univ, Frankfurt, 80. *Mem:* AAAS; Animal Behav Soc; Am Arachnological Soc (pres, 85-87); Brit Arachnological Soc. *Res:* Predatory behavior in spiders; acoustic and chemical communication in spiders; flooding survival in spiders. *Mailing Add:* Dept Zool Ohio Univ Athens OH 45701-2979

ROVNYAK, GEORGE CHARLES, b Ford City, Pa, Jan 31, 41; m 63; c 3. MEDICINAL CHEMISTRY, ORGANIC CHEMISTRY. *Educ:* St Vincent Col, AB, 62; Univ Pittsburgh, BS, 65, PhD(chem), 70. *Prof Exp:* Chemist, Neville Chem Co, 63-66; res investr, 70-80, sr res investr med chem, 80-88, RES FEL, E R SQUIBB & SONS, INC, 80- *Mem:* Am Chem Soc; NY Acad Sci; Sigma Xi. *Res:* Synthesis and structure-activity relationship of biologically active organic compounds; reaction mechanisms; application of small ring compounds to chemical synthesis; heterocyclic chemistry. *Mailing Add:* Bristol-Myers Squibb PO Box 4000 Princeton NJ 08543

ROVNYAK, JAMES L, b Ford City, Pa, Jan 9, 39; m 63; c 2. MATHEMATICS. *Educ:* Lafayette Col, AB, 60; Yale Univ, MA, 62, PhD(math), 63. *Prof Exp:* Asst prof math, Purdue Univ, 63-67; assoc prof, 67-73, PROF MATH, UNIV VA, 73- *Concurrent Pos:* NSF fel, Inst Advan Study, 66-67; Alexander von Humboldt Award, US Sr Scientist, Fed Repub Ger, 79. *Mem:* Am Math Soc. *Res:* Hilbert space; complex analysis. *Mailing Add:* Dept Math & Math Astron Univ Va Bldg 202 Charlottesville VA 22903-3199

ROW, CLARK, b Washington, DC, July 24, 34; m 72. FOREST ECONOMICS. *Educ:* Yale Univ, BS, 56; Duke Univ, MF, 58; Tulane Univ La, PhD(econ), 73. *Prof Exp:* Res forester, Southern Forest Exp Sta, US Forest Serv, 58-62, proj leader forest prod & mkt res, 62-65; chief, Forest Prods Demand & Price Anal Br, 65-67, chief, Forest Econ Br, 68-75, LEADER, EVAL METHODS RES GROUP, US FOREST SERV, 75- *Concurrent Pos:* Spec lectr, Duke Univ, 81. *Mem:* Am Econ Asn; Soc Am Foresters. *Res:* Economics of forest, range, watershed and outdoor recreation wilderness management; pest control economics; demand for forest products; economics of forest products industries. *Mailing Add:* 5503 Boxhill Lane Baltimore MD 21210-2001

ROW, THOMAS HENRY, b Blacksburg, Va, Feb 9, 35; m 75; c 4. NUCLEAR ENGINEERING. *Educ:* Roanoke Col, BS, 57; Va Polytech Inst & State Univ, MS, 59. *Prof Exp:* Instr math, Roanoke Col, 57; res staff, 59-67, nat prog coordr, Reactor Containment Spray Syst Prog, Atomic Energy Comn, 67-71, dir, Environ Statements Proj, 71-75, head, Environ Impact Sect, Energy Div, 75-81, DIR, NUCLEAR & CHEM WASTE PROG, OAK RIDGE NAT LAB, 81- *Concurrent Pos:* Consult, Adv Comt Reactor Safeguards, US Atomic Energy Comn, 68-71; mem Standard Comt, Am Nat Standards Inst, 71-72. *Mem:* Am Nuclear Soc. *Res:* Nuclear waste. *Mailing Add:* 412 Virginia Rd PO Box X Oak Ridge TN 37830

ROWAN, DIGHTON FRANCIS, b Amsterdam, NY, Dec 31, 14. MEDICAL VIROLOGY. *Educ:* San Jose State Col, BA, 48; Stanford Univ, MA, 53, PhD(bact & exp path), 54. *Prof Exp:* Instr microbiol virol, Stanford Univ Sch Med, 53-56; instr epidemiol, Sch Trop Med & Pub Health, Tulane Univ, 56-57; asst prof microbiol & virol, Sch Med, Univ Vt, 57-59; prin virologist develop & res, NJ State Dept Health, 59-61; dir, virol lab, Mont State Bd Health, 61-63; prof dir virol dept, Col Dent, Baylor Univ & Med Ctr, 63-73; PROF DIR MICROBIOL VIROL, ADV MICROBIOL INFECTIOUS DIS DIV, SCH MED, SOUTHERN ILL UNIV, 73- *Honors & Awards:* Sect Award, Am Pub Health Asn, 61; Outstanding Serv in Educ Recognition Award, Am Soc Clin Pathologists, 68. *Mem:* Fel Am Acad Microbiol; fel Am Pub Health Asn; Am Soc Microbiol; AAAS; Sigma Xi. *Res:* The role of the herpes viruses in vulvitis; birth defects and sudden infant death; role of coxsackieviruses in adult myocarditis and pericarditis; respiratory disease and encephalitis surveillance. *Mailing Add:* 2313 Westchester Blvd Springfield IL 62704

ROWAN, NANCY GORDON, b Plainfield, NJ, Apr 6, 46; wid; c 2. INORGANIC CHEMISTRY. *Educ:* Mt Holyoke Col, AB, 68; Boston Univ, PhD(inorg chem), 74. *Prof Exp:* Vis asst prof chem, Carnegie-Mellon Univ, 73-74; from asst prof to assoc prof chem, Am Univ, 78-; ASST PROF CHEM, UNIV SOUTHERN MAINE. *Mem:* Am Chem Soc. *Res:* Kinetics of transition metal reactions; models for metal binding sites in proteins. *Mailing Add:* Chem Dept Univ Southern Maine 96 Fallmouth Portland ME 04103

ROWAND, WILL H, b Mar 20, 08. ENGINEERING. *Educ:* Cornell Univ, ME, 29. *Prof Exp:* Mem staff, Babcock & Wilcox, 29-48, chem engr, 48-53, vpres eng, 53-61, vpres mkt, 61-66, vpres nuclear power, 66-72; RETIRED. *Honors & Awards:* Newcomen Medal, 54. *Mem:* Nat Acad Eng; fel Am Soc Mech Engrs. *Mailing Add:* 10802 S Singletree Trail Dewey AZ 86327

ROWE, ALLEN MCGHEE, JR, b Columbus, Ohio, May 15, 32. THERMODYNAMICS. *Educ:* Ohio State Univ, BPetrolEng & MS, 56; Univ Tex, PhD(petrol eng), 64. *Prof Exp:* Jr engr, Texaco, Inc, 56-57, res engr, 57-58; res engr, Esso Prod Res Co, 58-61 & Tex Petrol Res Comt, 61-64; asst prof petrol eng, Okla State Univ, 64-71, assoc prof mech eng, 71-76; sr res engr, Atlantic Richfield Co, 76-85; CONSULT, MICROSIM INT INC, 87- *Concurrent Pos:* Consult, Continental Oil Co, Okla, 64, Intercomp, Tex, 71 & Marathon Oil Co, 74; NASA fel, 67 & 68; lectr, Stanford Univ, 74. *Mem:* Inst Mining, Metall & Petrol Engrs. *Res:* Calculation of equilibrium compositions of hydrocarbon mixtures; desalination research. *Mailing Add:* 2510 Parkhaven Plano TX 75075

ROWE, ANNE PRINE, b Detroit, Mich, Feb 1, 27; m 50; c 2. PHYSICAL CHEMISTRY. *Educ:* Univ Mich, BS, 50, MS, 68, PhD(mat eng), 73. *Prof Exp:* Asst res chem, Chem Dept, Univ Mich, 61-68, res assoc, Lab Metall & Mat Eng, 68-71, sr res assoc dent mat, 71-75; res mat engr, NASA Lewis Res Ctr, 76-78; assoc prof mech eng, Fla Inst Technol, 79-83; assoc prof, Purdue Univ, Calumet, 83-87; ASSOC PROF CHEM, LAROCHE COL, 87- *Concurrent Pos:* Mat engr, NASA Kennedy Space Ctr, 80- *Mem:* Am Soc Metals; Electron Micros Soc Am; Soc Women Engrs. *Res:* Electon metallography of a cobalt-chromium-nickel-tantalum alloy; dental amalgams; corrosion and erosion of superalloys for application in a pressurized fluidized bed coal combustor and turbine combined cycle system; corrosion testing of candidate paint systems for resistance to marine environment corrosion of ground support systems. *Mailing Add:* 416 Forest Highlands Dr Pittsburgh PA 15238

ROWE, ARTHUR W(ILSON), b Newark, NJ, Sept 14, 31; m 57; c 2. BIOCHEMISTRY, CRYOBIOLOGY. *Educ:* Duke Univ, AB, 53; Rutgers Univ, PhD, 60. *Prof Exp:* Res chemist, Linde Div, Union Carbide Corp, 60-64; investr & dir crybiol, NY Blood Ctr, 64-88; assoc prof, 69-83, PROF SCH MED, NY UNIV, 83-; ADJ PROF, STATE UNIV NY, BINGHAMTON, 88- *Concurrent Pos:* Consult, Nat Cancer Inst, NIH, 62-66; consult, Lab Exp Med & Surg in Primates, Sch Med, NY Univ, 67-; ed-in-chief, J Cryobiol, 73-; deleg, Am Blood Comn, 76-82; vis prof, Univ Damascus, 77, Helmholtz Inst Aachen, Fed Repub Ger, 84, Jadaupur Univ, 90. *Mem:* Fel AAAS; Am Chem Soc; Soc Cryobiol (treas, 69-72, vpres, 73-76); fel Am Inst Chemists; Transplantation Soc; NY Acad Sci; Sigma Xi. *Res:* Low temperature preservation of bone marrow, stem cells, blood, leukocytes, platelets and tissues and embryos; cryobiology; cellular metabolism and isotopic techniques; immunohematology; cryogenic freezing of tissues. *Mailing Add:* Dept Forensic Med NY Univ Sch Med 550 First Ave New York NY 10016

ROWE, BRIAN H, b London, Eng, May 6, 31; c 3. AIRCRAFT ENGINE DEVELOPMENT. *Educ:* Durham Univ, BS, 55. *Prof Exp:* Var eng pos, Gen Elec Co, 57-68, gen mgr, CF6 Proj Dept, 68-72, vpres & gen mgr, Com Engine Proj Div, 72-74, vpres & gen mgr, Airline Prog Div, 74-76, vpres & gen mgr, Aircraft Eng Div, 76-79, SR VPRES, AIRCRAFT ENGINES, GEN ELEC CO, CINCINNATI, OHIO, 79- *Mem:* Nat Acad Eng; fel Royal Aeronaut Soc. *Mailing Add:* Gen Elec Co Mail Drop J101 One Neumann Way Cincinnati OH 45215-6301

ROWE, CARLETON NORWOOD, b Halifax, Pa, Apr 1, 28; m 56; c 3. TRIBOLOGY, LUBRICANTS & LUBRICATION. *Educ:* Juniata Col, BS, 51; Pa State Univ, MS, 53, PhD(phys chem), 55. *Prof Exp:* Res chemist, Tex Co, 55-60; sr res chemist, 60-65, res assoc, 65-86, SR RES ASSOC, MOBIL RES & DEVELOP CORP, PRINCETON, NJ, 86- *Mem:* Am Chem Soc; fel Am Soc Lubrication Engrs (pres, 85-86); Soc Automotive Engrs. *Res:* Lubrication; friction and wear; additive chemistry; contact fatigue; base stock properties; synthetic lubricants. *Mailing Add:* Mobil Res & Develop Corp Box 1028 Princeton NJ 08543-1028

ROWE, CHARLES DAVID, b Winchester, Ind, Dec 6, 39; m 61; c 1. POLYMER CHEMISTRY. *Educ:* Purdue Univ, BS, 61, MS, 64; Univ Mich, PhD(chem), 69. *Prof Exp:* Group leader plastics, Rohm & Haas Co, 68-72; mgr, polymer prod adhesives, Daubert Chem Co, 72-80, dir, res & develop, 80-82; mgr new prods, 82-86, VPRES RES & DEVELOP, SWIFT ADHESIVES, REICHHOLD CHEM INC, 86- *Mem:* Am Chem Soc. *Mailing Add:* Swift Adhesives Reichhold Chems Inc 3100 Woodcreek Dr Downers Grove IL 60515-5400

ROWE, DAVID JOHN, b Totnes, Eng, Feb 4, 36; m 58; c 2. MATHEMATICAL PHYSICS. *Educ:* Cambridge Univ, BA, 59; Oxford Univ, BA, 59, MA & DPhil(nuclear physics), 62. *Prof Exp:* Ford Found fel, Niels Bohr Inst, Copenhagen, Denmark, 62-63; UK Atomic Energy Authority res fel, Atomic Energy Res Estab, Harwell, Eng, 63-66; res assoc, Univ Rochester, 66-68; assoc prof, 68-74, PROF PHYSICS, UNIV TORONTO, 74-, ASSOC DEAN, SCH GRAD STUDIES, 85- *Concurrent Pos:* Vis prof Univ Sao Paulo, 71-72; assoc chmn, dept physics, Univ Toronto; Alfred P Sloan fel, 70; chmn, Theoret Physics Div, Can Asn Physists, 70, 71; Isaac Walton Killam fel, 90-92. *Honors & Awards:* Rutherford Memorial Medal & Prize, Royal Soc Can, 84. *Mem:* Can Asn Physicists; Am Physicists Soc; fel Royal Soc Can. *Res:* Theory of nuclear structure and reactions; collective motion; group theory. *Mailing Add:* Dept Physics Univ Toronto Toronto ON M5S 1A7 Can

ROWE, EDWARD C, b Oakland, Calif, Dec 23, 33; m 55; c 3. NEUROPHYSIOLOGY. *Educ:* Wesleyan Univ, BA, 55; Univ Mich, MS, 57, PhD(zool), 64. *Prof Exp:* Assoc prof, 61-73, PROF BIOL, EMPORIA STATE UNIV, 73- *Concurrent Pos:* NIH res grant physiol ganglia, 66-71. *Mem:* AAAS; Am Soc Zool. *Res:* Comparative neurophysiology. *Mailing Add:* Dept Biol Emporia State Univ 1200 Coml St Emporia KS 66801

ROWE, EDWARD JOHN, b Racine, Wis, Dec 28, 10; m 43. PHARMACY. *Educ:* Univ Wis, BS, 37, PhD(pharm), 41. *Prof Exp:* From asst prof to assoc prof, Col Pharm, Butler Univ, 41-45, prof pharm & head dept, 45-82; RETIRED. *Mem:* Acad Pharmaceut Sci. *Res:* Pharmaceutical chemistry and dispensing. *Mailing Add:* 5332 Brendonridge Rd Indianapolis IN 46226

ROWE, ELIZABETH SNOW, b Seattle, Wash, Dec 28, 43; m 66; c 2. PHYSICAL BIOCHEMISTRY. *Educ:* Duke Univ, BA, 66, PhD(biochem), 71. *Prof Exp:* Fel calorimetry, Med Sch, Johns Hopkins Univ, 71-72; res assoc elec birefringence, Georgetown Univ, 72-75; fel membrane phys chem, Chem Dept, Johns Hopkins Univ, 76-77; asst prof, 77-78, adj asst prof, 80-84, RES ASSOC PROF BIOCHEM, MED SCH, UNIV KANS, 84-; ASSOC RES CAREER SCIENTIST, VET ADMIN, 88- *Concurrent Pos:* Dir, Molecular Mech Alcoholism Lab, Kans City Vet Admin Med Ctr, 78- *Mem:* Sigma Xi; Am Soc Biol Chemists; Biophys Soc; Res Soc Alcoholism. *Res:* Physical properties of membrane components; effect of anesthetics and alcohol on membranes; thermodynamics of protein structure and function. *Mailing Add:* Phys Biochem Res Lab Vet Admin Med Ctr 4801 Linwood Blvd Kansas City MO 64128

ROWE, ENGLEBERT L, b Highland Park, Mich, Dec 18, 25; m 49; c 5. PHYSICAL PHARMACY. *Educ:* Wayne State Univ, BS, 50, MS, 58. *Prof Exp:* Chemist, City of Dearborn, 49-53 & Minn Mining Mfg Co, 53-58; res scientist res & develop, Drug Delivery, Upjohn Co, 58-; RETIRED. *Mem:* Am Chem Soc; Am Asn Pharmaceut Scientists; Sigma Xi. *Res:* Coagulation of colloids; kinetics of drug degradation; surface tension measurement and correlation with drug activity; sustained action dosage forms; particle size and surface area of emulsions and solids. *Mailing Add:* 3743 Cedaridge Kalamazoo MI 49008

ROWE, GEORGE G, b Vulcan, Alta, May 17, 21; nat US; m 47; c 3. INTERNAL MEDICINE. *Educ:* Univ Wis, BA, 43, MD, 45; Am Bd Internal Med, dipl, 55; Am Bd Cardiovasc Dis, dipl, 69. *Prof Exp:* Instr anat, Sch Med, Wash Univ, 48-50; resident, Med Sch, Univ Wis, 50-52, Am Heart Asn fel, 52-54, res assoc med, 54-55; voluntary res assoc, Hammersmith Hosp, London, 56-57; from asst prof to assoc prof, 57-64, PROF MED, MED SCH, UNIV WIS, MADISON, 64- *Concurrent Pos:* Markle scholar, 55-60. *Mem:* Am Physiol Soc; Am Soc Pharmacol & Exp Therapeut; fel Am Col Physicians; Am Fedn Clin Res; Am Soc Clin Invest. *Res:* Hemodynamics of the systemic and coronary circulations; congenital and acquired heart disease. *Mailing Add:* Dept Med Med Sch Univ Wis 600 N Highland Ave Madison WI 53792

ROWE, H(ARRISON) E(DWARD), b Chicago, Ill, Jan 29, 27; m 51; c 4. ELECTRICAL ENGINEERING. *Educ:* Mass Inst Technol, BS, 48, MS, 50, ScD(elec eng), 52. *Hon Degrees:* ME, Stevens Inst Technol, 88. *Prof Exp:* Mem tech staff, Radio Res Lab, Bell Labs, Holmdel, NJ, 52-84; ANSON WOOD BURCHARD PROF ELEC ENG, STEVENS INST TECHNOL, HOBOKEN, NJ, 84- *Concurrent Pos:* Mem comn 6, Int Union Radio Sci; vis lectr, Univ Calif, Berkeley, 63 & Imp Col, Univ London, 68 & 81; res asst, Mass Inst Technol, 48-52; mem, Defense Sci Bd Task Force, 72-74. *Honors & Awards:* Microwave Prize, Inst Elec & Electronics Engrs, 72 & David Sarnoff Award, 77. *Mem:* Fel Inst Elec & Electronics Engrs; Sigma Xi. *Res:* Communications systems; wave guides; optical communication systems; noise and antenna theory; optimizing radio-astronomical observations of incoherent fields with an antenna; terrain mapping by a radiometer; author of 44 publications. *Mailing Add:* Stevens Inst Technol Castle Pt Hoboken NJ 07730

ROWE, IRVING, physics, electronics; deceased, see previous edition for last biography

ROWE, JAMES LINCOLN, b Chicago, Ill, Nov 14, 17; m 48; c 3. ORGANIC CHEMISTRY. *Educ:* Princeton Univ, BA, 39; Univ Chicago, PhD(org chem), 46; Ind Univ, DJ, 68. *Prof Exp:* Chemist, Nat Defense Res Comt, Univ Chicago, 42 & Off Sci Res & Develop, 44-46; res chemist, 46-54, patent agent, 54-68; patent attorney, Eli Lilly & Co, 68-85. *Concurrent Pos:* Mem coun, Woodard, Emhardt, Naughton, Moriarty & McNett, 90- *Mem:* AAAS; Am Chem Soc. *Res:* Mechanisms of organic reactions; free radical reactions; synthesis of synthetic drugs; war gases; nitrogen mustards; decomposition of di-acetyl peroxide in alcohols. *Mailing Add:* 111 Monument Circle Suite 3700 Indianapolis IN 46204

ROWE, JAY ELWOOD, b Tacoma, Wash, Jan 10, 47; m 65; c 3. INDUSTRIAL ORGANIC CHEMISTRY. *Educ:* Bucknell Univ, BS, 68; Lehigh Univ, PhD(org chem), 73. *Prof Exp:* Instr chem, Muhlenberg Col, 72-73, asst prof, 73-74; RES ASSOC, CROMPTON & KNOWLES CORP, 74- *Mem:* Am Chem Soc; Am Asn Textile Chemists & Colorists; Sigma Xi. *Res:* Chemistry and theory of acid dyes. *Mailing Add:* Amity Gardens Four Welsh Ct Douglasville PA 19518

ROWE, JOHN EDWARD, b Jacksonville, Fla, Sept 25, 41; m 65; c 3. EXPERIMENTAL SOLID STATE PHYSICS. *Educ:* Emory Univ, BS, 63; Brown Univ, PhD(physics), 71. *Prof Exp:* Mem tech staff, 69-80, RES HEAD SURFACE PHYSICS, BELL LABS, 80- *Concurrent Pos:* Mem, Synchrotron Users Exec Comt, Brookhaven Nat Lab; affil prof physics, Univ Fla. *Mem:* Fel Am Phys Soc; Am Vacuum Soc. *Res:* Electron spectroscopy on surfaces and bulk solids using photoemission, electron energy loss and Auger spectroscopies; low energy electron diffraction; studies of chemisorption and of film growth; synchrotron radiation; surface states on semiconductors; chemisorption electronic and structural properties; surface vibrational modes using surface enhanced Raman spectroscopy and high resolution electron energy loss spectroscopy. *Mailing Add:* Bell Labs Rm 1C-323 600 Mountain Ave Murray Hill NJ 07974

ROWE, JOHN JAMES, b Washington, DC, Aug 2, 44; m 68; c 1. MICROBIAL BIOCHEMISTRY, NITRATE REDUCTION. *Educ:* Colo State Univ, BS, 67; Ariz State Univ, MS, 71; Univ Kans, PhD(microbiol), 75. *Prof Exp:* Fel microbiol, Dept Biol, Univ Ga, 75-77; asst prof, 77-83, ASSOC PROF MICROBIOL, DEPT BIOL, UNIV DAYTON, 83- *Mem:* Sigma Xi; Am Soc Microbiol. *Res:* Physiological and genetic studies of bacteria in the genus Pseudomonas; denitrification; inorganic nitrogen metabolism; Pseudomonas aeruginosa vaccine. *Mailing Add:* Dept Biol Univ Dayton 300 College Park Dayton OH 45469

ROWE, JOHN MICHAEL, b Oakville, Ont, Apr 9, 39. SOLID STATE PHYSICS. *Educ:* Queen's Univ, Ont, BSc, 62; McMaster Univ, PhD(solid state physics), 66. *Prof Exp:* Fel physics, Argonne Nat Lab, 66-67, asst staff physicist, 67-72, assoc physicist, 72-73; RES PHYSICIST, NAT BUR STANDARDS, 73- *Mem:* AAAS; Am Phys Soc. *Res:* Study of lattice and liquid dynamics by slow neutron scattering. *Mailing Add:* Nat Inst Standards & Technol A-104-235 Gaithersburg MD 20899

ROWE, JOHN STANLEY, b Hardisty, Alta, June 11, 18; m 54; c 2. PLANT ECOLOGY. *Educ:* Univ Alta, BSc, 41; Univ Nebr, MSc, 48; Univ Man, PhD(ecol), 56. *Prof Exp:* Forest ecologist, Can Dept Forestry, 48-67; PROF PLANT ECOL, UNIV SASK, 67- *Mem:* Can Inst Forestry; Can Bot Asn; Ecol Soc Am. *Res:* Ecology of northern Canada; boreal forest, tundra and peatlands. *Mailing Add:* Dept Plant Ecol Univ Sask Saskatoon SK S7N 0W0 Can

ROWE, JOHN W, MEDICINE. *Educ:* Canisius Col, BS; Univ Rochester, MD. *Prof Exp:* Prof med & dir Div Aging, Harvard Med Sch; PRES, MT SINAI MED CTR, MT SINAI HOSP & SCH MED, 88-, PROF MED, GERIAT & ADULT DEVELOP, 88- *Concurrent Pos:* Mem bd gov, Am Bd Internal Med; clin assoc, NIH, res & clin fel, Mass Gen Hosp & res fel med, Harvard Med Sch, 72-75; sr physician, Brigham & Women's Hosp & Beth Israel Hosp; dir, Geriat Res Educ Clin Ctr, Vet Admin, Boston. *Mem:* Inst Med-Nat Acad Sci. *Res:* Geriatrics; internal medicine. *Mailing Add:* Mt Sinai Med Ctr One Gustave L Levy Pl New York NY 10029

ROWE, JOHN WESTEL, b New York, NY, Sept 3, 24; m 49; c 3. WOOD CHEMISTRY. *Educ:* Mass Inst Technol, BS, 48; Univ Colo, MS, 52; Swiss Fed Inst Technol, Zurich, ScD(org chem), 56. *Prof Exp:* Proj leader, Forest Prod Lab, USDA, 57-84; RETIRED. *Concurrent Pos:* Mem, Nat Acad Sci Corrim Comt, 74-75; chmn, Wis Sect, Am Chem Soc, 68-69 & alt counr, 76-78. *Honors & Awards:* Wood Salutes Award, 75. *Mem:* Fel AAAS; Soc Econ Bot; Phytochem Soc NAm; fel Am Inst Chemists; fel Int Acad Wood Sci; Sigma Xi; Am Chem Soc. *Res:* Natural products, especially extractives of wood and bark; higher terpenoids and steroids; improved chemical utilization of wood. *Mailing Add:* 1001 Tumalo Trail Madison WI 53711-3024

ROWE, JOSEPH E(VERETT), b Highland Park, Mich, June 4, 27; m 50; c 2. ELECTRICAL & COMPUTER ENGINEERING. *Educ:* Univ Mich, BSE(elec eng) & BSE(math), 51, MSE, 52, PhD(elec eng), 55. *Prof Exp:* Asst, Electron Tube Lab, Eng Res Inst, Univ Mich, Ann Arbor, 51-53, lectr, 52-55, res assoc, 53-55, dir electron physics lab, 58-68, from asst prof to prof elec eng, 55-74, chmn dept elec & comput eng, 68-74; vprovost & dean eng, Case Inst Technol, 74-76, provost, 76-78; vpres & gen mgr technol, Harris Controls Div, Harris Corp, 78-82; exec vpres res & defense systs, Gould Inc, 82-83, vchmn & chief tech officer, 83-86; VPRES & CHIEF SCIENTIST, PPG INDUST INC, 87- *Concurrent Pos:* Chmn, Coalition Adv Indust Technol, 85-86; mem, Army Sci Bd, 85- *Honors & Awards:* Curtis W McGraw Res Award, Am Soc Eng Educ, 64. *Mem:* Nat Acad Eng; fel Inst Elec & Electronics Engrs; Am Phys Soc; Am Soc Eng Educ; fel AAAS. *Res:* Microwave circuits; traveling-wave tubes; crossed-field devices; electromagnetic field theory; noise; computers; lasers; solid state devices; plasmas; integrated circuits. *Mailing Add:* 416 Forest Highlands Dr Pittsburgh PA 15238

ROWE, KENNETH EUGENE, b Canon City, Colo, Feb 8, 34; m 70; c 4. EXPERIMENTAL STATISTICS, BIOMETRY. *Educ:* Colo State Univ, BS, 57; NC State Univ, MS, 60; Iowa State Univ, PhD(animal breeding), 66. *Prof Exp:* Geneticist, Regional Swine Breeding Lab, USDA, Ames, Iowa, 61-64; from asst prof to assoc prof exp statist, 64-70, assoc prof, 70-80, PROF STATIST, ORE STATE UNIV, 80- *Concurrent Pos:* NSF fac develop fel, NC State Univ, 69-70; sr statist adv, Special Studies Staff, IERL, US Environ Protection Agency, NC, 78-79. *Mem:* Biomet Soc; Am Statist Asn; Sigma Xi. *Res:* Applications of statistics, particularly biological problems and quantitative genetics; statistical computation. *Mailing Add:* Dept Statist Ore State Univ Corvallis OR 97331

ROWE, LAWRENCE A, b Boston, Mass, Apr 11, 48. HUMAN-COMPUTER INTERFACES, COMPUTER-INTEGRATED MANUFACTURING. *Educ:* Univ Calif, Irvine, BS, 70, PhD(info & computer sci), 76. *Prof Exp:* PROF COMPUTER SCI, UNIV CALIF, BERKELEY, 76- *Concurrent Pos:* Prin investr, NSF, 78-; founder & dir, Ingres Corp, 80-90; consult, Siemens Corp Res, 87- & Harris Corp, 88-; chmn, Software Systs Awards Comt, Asn Comput Mach, 90. *Mem:* Asn Comput Mach; Inst Elec & Electronics Engrs. *Res:* Design and implementation of computer systems (software and hardware) to solve challenging problems; worked on application development systems, programming languages, databases and computer-integrated manufacturing. *Mailing Add:* Dept Elec Eng & Computer Sci Univ Calif Berkeley CA 94720

ROWE, LEONARD C, corrosion, for more information see previous edition

ROWE, MARK J, b Oakland, Calif, July 16, 43; m 66; c 5. BIOCHEMISTRY. *Educ:* Brigham Young Univ, BS, 68, PhD(biochem), 72. *Prof Exp:* Fel molecular biol, Stanford Univ, 72-73; asst prof, 73-78, ASSOC PROF BIOCHEM, EASTERN VA MED SCH, 78- *Concurrent Pos:* NIH fel, Dept Biol, Stanford Univ, 72-73. *Mem:* Am Chem Soc; AAAS; Sigma Xi. *Res:* Nucleic acids, protein biosynthesis; mitochondrial membrane biogenesis, protein synthesis and genetics and biochemical genetics; ovarian molecular endocrinology. *Mailing Add:* Dept Biochem Eastern Va Med Sch Box 1980 Norfolk VA 23501

ROWE, MARVIN W, b Amarillo, Tex, July 6, 37; c 4. NUCLEAR GEOCHEMISTRY, ARCHAEOLOGICAL CHEMISTRY. *Educ:* NMex Inst Mining & Technol, BS, 59; Univ Ark, PhD(chem), 66. *Prof Exp:* Res asst radioactiv, Los Alamos Sci Lab, 60-63; Miller res fel physics, Univ Calif, Berkeley, 66-68; asst prof chem, Univ Wash, 68-69; from asst prof to assoc prof, 75-87, PROF CHEM, TEX A&M UNIV, 87- *Concurrent Pos:* Vis prof, Univ Antwerp, Belgium, 85 & 90, Max Planck Inst, Mainz, WGer, 86-88, Los Alamos Nat Lab, 86-88. *Mem:* Am Chem Soc; Am Soc Archaeology; fel Meteoritical Soc; Int Asn Geochem & Cosmochem. *Res:* Radioactivity, magnetism and noble gas mass spectrometry in meteorites; chronology of early solar system; dating pictographs. *Mailing Add:* Dept Chem Tex A&M Univ College Station TX 77843-3255

ROWE, MARY BUDD, b Jersey City, NJ, Mar 24, 25. SCIENCE EDUCATION. *Educ:* NJ State Col Montclair, BA, 47; Univ Calif, Berkeley, MA, 55; Stanford Univ, PhD(sci educ), 64. *Prof Exp:* Teacher, Am High Sch, Munich Ger, 55-58; consult, Colo State Dept Educ, 59-61; lectr sci educ, Stanford Univ, 64-65; assoc prof sci, Teachers Col, Columbia Univ, 65-72; assoc prof, 72-73, PROF SCI EDUC, INST DEVELOP HUMAN RESOURCES, UNIV FLA, 73- *Concurrent Pos:* Fel, Univ Fla, 70-71; prog dir, Res Sci Educ, NSF, 78-80. *Honors & Awards:* Outstanding Res Award in Sci Educ, Nat Asn Res Sci Teaching, 74; Robert Carlton Award, Nat Sci Teachers Asn. *Mem:* Fel AAAS; Nat Asn Res Sci Teaching; Nat Sci Teachers Asn; Am Educ Res Asn; Sigma Xi. *Res:* Influence of pausing phenomena on quality of enquiry; rewards as a factor in inquiry; relation of science to fate control orientations. *Mailing Add:* Inst Develop Human Resources Univ Fla 421 NW 32nd St Gainesville FL 32607

ROWE, NATHANIEL H, b Hibbing, Minn, May 26, 31; div; c 4. ORAL PATHOLOGY, ANTIVIRAL CHEMOTHERAPY. *Educ:* Univ Minn, BS, DDS, 55, MSD, 58; Am Bd Oral Path, dipl. *Prof Exp:* Instr, Univ Minn, 58-59; from asst prof to assoc prof gen & oral path & chmn dept, Sch Dent, Wash Univ, 59-69; assoc prof path, 69-76, PROF DENT & ORAL PATH, SCH DENT, UNIV MICH, ANN ARBOR, 68-, PROF PATH, SCH MED, 76-, ASSOC DIR, DENT RES INST, 70-, SR RES SCIENTIST, 77- *Concurrent Pos:* Res fel oral path, Univ Minn, 55-58; consult, Vet Admin, 64- & Ellis Fischel State Cancer Hosp, 66-; assoc res scientist, Cancer Res Ctr, Columbia, Mo, 67-71; mem sci adv bd, Cancer Res Ctr, Columbia Mo, 75-79; civilian prof consult, Off Surgeon, Fifth US Army, 67-83; mem prof adv coun cancer, Mich Asn Regional Med Progs, 69-73, comt cancer control, Mich Dent Asn, 71- & coun dent educ, Am Dent Asn, 71-77, coun on hosp & inst dent, 77-81, coun dental therapeut, 84- *Honors & Awards:* Tiffany Div Nat Award, Am Cancer Soc, 79. *Mem:* AAAS; Am Cancer Soc; Am Dent Asn; fel Am Acad Oral Path (pres, 77-78); Am Asn Cancer Res; fel Int Col Dent. *Res:* Effect of environmental variables upon oral cancer induction; etiology and pathogenesis of dental caries; Herpes Simplex virus, antiviral chemotherapy. *Mailing Add:* Univ Mich 5223 N University Ave Dent Res Bldg Ann Arbor MI 48109-1078

ROWE, PAUL E, b Marlboro, Mass, Nov 16, 27; m 53; c 3. PHYSICAL ORGANIC CHEMISTRY. *Educ:* Mass Inst Technol, SB, 48; Boston Univ, PhD(chem), 59. *Prof Exp:* Proj chemist, Nat Northern Div, Am Potash & Chem Co, 53-59; head, Org Develop Dept, Emerson & Cuming, Inc, 69-74, chief chemist, 74-78, dir res & develop, 78-80, partner, Cuming Corp, 80-81; CONSULT, 81- *Mem:* Am Chem Soc. *Res:* Research and development in the explosive and propellant fields and in the plastics and ceramic fields with emphasis on electronic and microwave materials. *Mailing Add:* 71 West Way Mashpee MA 02649

ROWE, RANDALL CHARLES, b Baltimore, Md, Sept 26, 45; m 67; c 2. VEGETABLE PATHOLOGY, POTATO PATHOLOGY. *Educ:* Mich State Univ, BS, 67; Ore State Univ, PhD(plant path), 72. *Prof Exp:* Res assoc peanut path, NC State Univ, 72-74; asst prof, 74-79, ASSOC PROF VEG PATH, OHIO AGR RES & DEVELOP CTR & OHIO STATE UNIV, 79- *Mem:* Am Phytopath Soc; Potato Asn Am. *Res:* Biology, ecology and control of potato and vegetable diseases with emphasis on soil-borne fungi, verticillium, fusarium, rhizoctonia, fungal-nematode interactions, and fungicide evaluation. *Mailing Add:* Plant Path Ohio State Univ 201 Kottman Hall Columbus OH 43210

ROWE, RAYMOND GRANT, b Seattle, Wash, Oct 24, 41; m 71; c 3. METALLURGY. *Educ:* Wash State Univ, BS, 65; Univ Ill, MS, 68, PhD(metall eng), 75. *Prof Exp:* Scientist metall, Battelle-Northwest Labs, 65-67; fels, Jones & Laughlin Steel Co, 67-68 & Am Vacuum Soc, 70-72; adv eng metall, Westinghouse-Hanford Co, 74-76; staff metallurgist, 76-78, METALLURGIST, RES & DEVELOP CTR, GEN ELEC CORP, 78- *Mem:* Am Soc Metals; Metall Soc-Am Inst Mining & Mat. *Res:* Titanium alloys; intermetallic compounds; mechanical properties; rapid solidification. *Mailing Add:* Res & Develop Ctr Gen Elec Corp Schenectady NY 12301

ROWE, RICHARD J(AY), b Lackawanna, NY, June 12, 30; m 53; c 4. AGRICULTURAL ENGINEERING. *Educ:* Cornell Univ, BS, 52, PhD, 69; Iowa State Univ, BS, 57, MS, 59. *Prof Exp:* Agr engr, Agr Res Serv, USDA, 57-59; PROF AGR ENG, UNIV MAINE, ORONO, 59- *Mem:* Am Soc Agr Engrs; Am Soc Eng Educ; Sigma Xi. *Res:* Power and mechanization of agricultural operations; systems modeling and simulation. *Mailing Add:* Dept Bio-Resource Eng Univ Maine Orono ME 04469

ROWE, ROBERT S(EAMAN), b Wilmington, Del, Jan 31, 20; m 42; c 2. CIVIL ENGINEERING. *Educ:* Univ Del, BCE, 42; Columbia Univ, MS, 49; Yale Univ, MEng, 50, DEng, 51. *Prof Exp:* Design engr, Triumph Explosives, Inc, 38-42; from instr to assoc prof civil eng, Princeton Univ, 46-56, dir rivers & harbors sect, 51-56; J A Jones prof eng & chmn dept civil eng, Duke Univ, 56-60; dean eng, Vanderbilt Univ, 60-70, prof civil, environ & water resources eng, 70-; RETIRED. *Concurrent Pos:* Vis prof, Univ Del, 48-49; asst to dean, NY Univ, 51-52; indust educator, E I du Pont de Nemours & Co, Inc, 55-56; consult scientist, Land Locomotion Res Lab, Detroit Arsenal; consult & mem adv panel, Off Ord Res, US Army. *Mem:* Am Soc Civil Engrs; Am Soc Eng Educ; Soc Am Mil Engrs; Nat Soc Prof Engrs; Am Concrete Inst. *Res:* Structural engineering; land mobility. *Mailing Add:* Dept Civil & Environ Eng Vanderbilt Univ Nashville TN 37240

ROWE, RONALD KERRY, b 1951; m 73; c 2. SOIL MECHANICS, ROCK MECHANICS. *Educ:* Univ Sydney, BSc, 73, BE, 75, PhD(geotech eng), 79. *Prof Exp:* Cadet engr, Australian Dept Construct, 71-74, engr, 75-78; from asst prof to assoc prof, 78-86, PROF CIVIL ENG, UNIV WESTERN ONT, 86- *Honors & Awards:* Collingwood Prize, Am Soc Civil Engrs, 85; Can Geotech Soc Prize, 84 & 87, honorable mention, 90. *Mem:* Inst Engrs Australia; Asn Prof Engrs Ont; Int Soc Soil Mech & Found Eng; Can Geotech Soc; Eng Inst Can; Int Soc Rock Mech; Int Geotextile Soc; Can Tunnelling Asn; NAm Geosynthetics Soc. *Res:* Geotechnical & hydrogeologic engineering, with particular emphasis on design and analysis of landfills and contaminant migration through soil and rock; soil and rock-structure interaction problems including tunnelling, shallow foundations, embankments and geosynthesis. *Mailing Add:* Fac Eng Sci Univ Western Ont London ON N6A 5B9 Can

ROWE, THOMAS DUDLEY, b Missoula, Mont, June 25, 10; m 34; c 1. PHARMACY. *Educ:* Univ Mont, BS, 32, MS, 33; Univ Wis, PhD(pharm), 41. *Prof Exp:* Instr pharm, Univ Nebr, 34-35; from instr to asst prof, Med Col Va, 35-40, assoc prof & actg chmn dept, 40-43, asst dean & chmn dept, 43-45; prof & chmn dept, Rutgers Univ, 45-51; dean, 51-75, prof pharm, 51-75, EMER DEAN, COL PHARM, UNIV MICH, ANN ARBOR, 75- *Concurrent Pos:* From asst dean to dean col pharm, Rutgers Univ, 45-51; consult, Off Surgeon Gen, US Dept Army, 58- *Mem:* Am Asn Cols Pharm (pres, 57-58); Am Pharmaceut Asn (1st vpres, 52). *Res:* Plant chemistry; pharmacology of medicinal plants; phytochemical and pharmacological investigation of fresh Aloe leaves; pharmacy education; socio-economic problems in pharmacy; ethics. *Mailing Add:* Col Pharm Univ Mich Ann Arbor MI 48109

ROWE, VERALD KEITH, b Warren, Ill, Oct 5, 14; m 37; c 2. TOXICOLOGY, INDUSTRIAL HYGIENE. *Educ:* Cornell Col, AB, 36; Univ Iowa, MS, 38. *Hon Degrees:* ScD, Cornell Col, 71. *Prof Exp:* Biochemist, Dow Chem Co, 37-44, proj leader, Dow Chem USA, 44-49, toxicologist, 49-52, tech expert, 52-54, lab div leader, 54-64, asst dir, 64-70, dir toxicol & indust hyg, Chem Biol Res, 70-74, res scientist, 73-78, dir toxicol affairs & health & environ sci, 74-79; res fel, 78-79, CONSULT INDUST TOXICOL, 79- *Concurrent Pos:* Mem comt toxicol, Nat Acad Sci-Nat Res Coun, 64-72; mem hazardous mat adv comt, Environ Protection Agency, 75-76. *Honors & Awards:* Cummings Award, Am Indust Hyg Agency, 79; Borden Award, Am Indust Hyg Asn, 84. *Mem:* Am Soc Pharmacol & Exp Therapeut; Soc Toxicol (pres, 66-67); Am Chem Soc; Am Indust Hyg Asn; Am Acad Indust Hyg (pres, 72-73). *Res:* Determination of physiological effects of chemicals on animals and man; metabolism of chemicals. *Mailing Add:* 9605 Sandstone Dr Dow Chem USA 1803 Bldg Sun City AZ 85351

ROWE, VERNON DODDS, b Washington, DC, July 11, 44; m 66; c 2. DEVELOPMENTAL NEUROBIOLOGY. *Educ:* Duke Univ, BS, 65, MD, 69. *Prof Exp:* Resident neurol, Johns Hopkins Hosp, 71-72 & 75-77; res assoc develop neurobiol, Nat Inst Child Health & Human Develop, 72-75; ASST PROF NEUROL, UNIV KANS MED CTR, 77- *Concurrent Pos:* Staff neurologist, Kansas City Vet Admin Hosp, 77-; Basil O'Connor Starter Grant, Nat Found, 78-80. *Mem:* Sigma Xi. *Res:* Tissue culture of sympathetic and pineal tissue; developmental neurotoxicology. *Mailing Add:* 4320 Wornall Rd Suite 52-II Kansas City KS 64111

ROWE, WILLIAM BRUCE, b Canon City, Colo, Nov 12, 35; m 78; c 3. BIOCHEMISTRY. *Educ:* Colo State Univ, BS, 57; Univ Rochester, MS, 59, PhD(biochem), 67. *Prof Exp:* Instr, Med Col, Cornell Univ, 69-72, asst prof biochem, 72-78; assoc dir, 78-84, dir biochem, Baxter-Travenol Labs, 84-87; DIR, BASIC RES, CLINTEC NUTRIT CO, 87- *Concurrent Pos:* Fel, Sch Med, Tufts Univ, 67 & Med Col, Cornell Univ, 67-69. *Mem:* AAAS; Am Chem Soc; Sigma Xi; NY Acad Sci; Am Soc Parenteral & Enteral Nutrit. *Res:* Amino acid and protein metabolism; enzymology and control mechanisms of intermediary metabolism; clinical nutrition; biotechnology. *Mailing Add:* Clintec Mgt Serv Three Parkway N Suite 200 Deerfield IL 60015

ROWE, WILLIAM DAVID, b Orange, NJ, Jan 7, 30; m 87; c 7. RISK ANALYSIS, MANAGEMENT SCIENCE. *Educ:* Wesleyan Univ, BS, 52; Univ Pittsburgh, MS, 52; Univ Buffalo, MBA, 61; Am Univ, PhD(bus admin/opers res), 73. *Prof Exp:* Jr engr, Westinghouse Elec Corp, 52-53, asst engr, Anal Dept, 53-55, assoc engr, 55-56, engr, 56-57, supv engr, Systs Control Dept, 57-61; eng specialist, Advan Systs Lab, Sylvania Electronic Systs, Mass, 61-62, mgr, Digital Systs Dept & Minuteman syst task mgr, 62-63, tech dir, Minuteman WS-133B Prog, 63, mem info processing staff, 63-64, sr eng specialist, 63-66, dir advan technol, 66-68; head, Spec Studies Subdept, Mitre Corp, Wash Oper, 68-69, dept head & assoc, 69-72; dep asst adminr radiation progs, US Environ Protection Agency, 72-80; prof decisions & rick anal, Ctr Technol Admin, Am Univ, Washington, DC, & dir, Inst Risk Anal, 80-87; PRES, ROWE RES ENG ASSOCS, INC, 84- *Concurrent Pos:* Head, Environ Systs Dept, Mitre Corp, 69-72; adj prof, Am Univ, 72-80. *Honors & Awards:* Elizur Wright Award, Risk & Ins Asn. *Mem:* AAAS; Systs, Man & Cybernet Soc; Soc Risk Anal; sr mem Inst Elec & Electronics Engrs; Am Risk & Ins Asn; Sigma Xi; Soc Sci Explor; Am Fedn Info Processing Soc (treas, 66-67). *Res:* Risk analysis and management; environmental science and radiation protection; risk analysis philosophy and application. *Mailing Add:* 309 N Alfred St Alexandria VA 22314

ROWELL, ALBERT JOHN, b Ely, Eng, July 19, 29; m 54; c 4. INVERTEBRATE PALEONTOLOGY. *Educ:* Univ Leeds, BSc, 50, PhD(geol), 53. *Prof Exp:* From asst lectr to lectr geol, Univ Nottingham, 55-64, reader, 64-67; PROF GEOL, UNIV KANS, 67- *Concurrent Pos:* Vis prof, Univ Kans, 64-65. *Mem:* Geol Soc Am; Am Paleont Soc; Soc Syst Zool; Geol Soc London; Brit Paleont Asn. *Res:* Application of numerical methods in paleontology; Cambrian stratigraphy and biogeography; Paleozoic inarticulate brachiopods. *Mailing Add:* Dept Geol Univ Kans Rm 120 Lindley Hall Lawrence KS 66045

ROWELL, CHARLES FREDERICK, b Lowville, NY, May 29, 35; m 55; c 2. PHYSICAL ORGANIC CHEMISTRY. *Educ:* Syracuse Univ, BS, 56; Iowa State Univ, MS, 59; Ore State Univ, PhD(org chem), 64. *Prof Exp:* From asst prof to assoc prof, US Naval Postgrad Sch, 62-79, assoc chmn dept, 81-84, chmn dept, 84-88, PROF CHEM & CONSULT FORENSIC, US NAVAL ACAD, 79- *Concurrent Pos:* Vis scientist, Chicago Br, Off Naval Res, 74-75; vis prof, US Naval Acad, 75-76. *Mem:* AAAS; Am Chem Soc; Royal Soc Chem. *Res:* Photochemistry; cyclopropane reactions; water pollution; nitramine chemistry; liquid propellants; azulene derivatives. *Mailing Add:* Dept Chem US Naval Acad Annapolis MD 21402

ROWELL, CHESTER MORRISON, JR, b Burnet, Tex, Dec 2, 25. TAXONOMY. *Educ:* Univ Tex, BS, 47; Agr & Mech Col, Tex, MS, 49; Okla State Univ, PhD, 67. *Prof Exp:* Asst herbarium, Univ Mich, 53-54; asst prof biol, Agr & Mech Col, Tex, 49-57, assoc prof, Tex Tech Univ, 57-70; PROF BIOL & HEAD DEPT, ANGELO STATE UNIV, 70- *Mem:* Sigma Xi. *Res:* Seedplants of Texas and Mexico. *Mailing Add:* PO Box 817 Marfa TX 79843

ROWELL, JOHN BARTLETT, b Pawtucket, RI, Nov 26, 18; m 44; c 3. CEREAL RUST PATHOLOGY. *Educ:* RI State Col, BS, 41; Univ Minn, PhD(plant path), 49. *Prof Exp:* Jr res asst plant path, Exp Sta, RI State Col, 40; asst, Univ Minn, St Paul, 41-42 & 46-47; asst, Exp Sta, RI State Col, 47-48, asst res prof, 48-49; res assoc, Univ Minn, St Paul, 49-55, assoc prof, 55-67; plant physiologist, Crops Res Div, USDA, 55-69, res plant pathologist & leader cereal rust lab, Sci & Educ Admin-Agr Res, 69-80; prof plant path, Univ Minn, St Paul, 67-81; RETIRED. *Concurrent Pos:* Consult, Ford Found, 71. *Mem:* Fel AAAS; fel Am Phytopath Soc; Bot Soc Am; Indian Phytopath Soc. *Res:* Control of cereal rusts. *Mailing Add:* 1963 Eustis St St Paul MN 55113

ROWELL, JOHN MARTIN, b Linslade, Eng, June 27, 35; m 59; c 3. SUPERCONDUCTIVITY, SUPERCONDUCTING ELECTRONICS. *Educ:* Oxford Univ, Eng, BSc, 57, MA & DPhil(physics), 61. *Prof Exp:* Mem staff, Bell Labs, 61-69, dept head, 69-81, dir, 81-83; asst vpres, Bellcore, 83-89; chief tech officer, 89-91, PRES, CONDUCTUS INC, 91- *Concurrent Pos:* Vis prof, Stanford Univ, 75. *Honors & Awards:* Fritz London Mem Low Temperature Physics Prize, 78. *Mem:* Fel Am Phys Soc. *Res:* Superconductivity and tunneling; tunneling spectroscopy. *Mailing Add:* Conductus Inc 969 W Maude Ave Sunnyvale CA 94086

ROWELL, LORING B, b Lynn, Mass, Jan 27, 30; m 56; c 2. PHYSIOLOGY. *Educ:* Springfield Col, BS, 53; Univ Minn, PhD(physiol), 62. *Prof Exp:* NIH fel cardiovasc physiol & cardiol, 62-63; from res instr to res asst prof med, 63-68, assoc prof physiol, biophys & med, 70-72, PROF PHYSIOL & BIOPHYS, SCH MED, UNIV WASH, 72- *Concurrent Pos:* Estab investr, Am Heart Asn, 66-71; vis prof, Univ Copenhagen, 78-79; vis prof, Ege Univ, 79; vis prof, Univ Antwerp, 79; Ida Bean vis prof, Univ Iowa, 80; assoc ed, J Appl Physiol; adj prof med, Univ Wash, 72- *Honors & Awards:* Joseph B Wolfe Mem Lectr, Am Col Sports Med, 80; Citation Award, Am Col Sports Med, 83; James M Schwinghammer Mem Lectr, Mich State Univ, 84. *Mem:* Am Physiol Soc; Sigma Xi; Am Heart Asn. *Res:* Human cardiovascular function; total and regional blood flow; temperature regulation; skeletal muscle and hepatic metabolism; hemodynamics-pressure regulation. *Mailing Add:* Dept Physiol & Biophys Univ Wash Sch Med SJ-40 Seattle WA 98195

ROWELL, NEAL POPE, b Mobile, Ala, Jan 11, 26; m 47; c 5. PHYSICS. *Educ:* Univ Ala, BS, 49, MS, 50; Univ Fla, PhD(physics), 54. *Prof Exp:* Instr physics, Univ Ala, 50-51; physicist, Mine Countermeasures Sta, USN, 51-52; instr physics, Univ Fla, 53-54; physicist, Courtaulds, Inc, 54-68; dir dev eng, 68-72, prof eng, 72-74, PROF PHYSICS, UNIV SALA, 74- *Mem:* Am Phys Soc; Am Asn Physics Teachers. *Res:* Physical properties of viscose rayon; electron diffraction study of alloys produced by simultaneous vacuum evaporation of aluminum and copper. *Mailing Add:* 354 McDonald Ave Mobile AL 36604

ROWELL, PETER PUTNAM, b St Petersburg, Fla, July 24, 46; m 72; c 2. NEUROCHEMISTRY, TOXICOLOGY. *Educ:* Stetson Univ, BS, 68; Univ Fla, PhD(pharmacol), 75. *Prof Exp:* Res assoc pharmacol, Vanderbilt Univ, 75-77; asst prof, 77-83, ASSOC PROF PHARMACOL, UNIV LOUISVILLE, 83- *Mem:* Am Soc Pharmacol & Exp Therapeut; Soc Neurosci; Soc Exp Biol & Med; Int Asn Biomed Geront; Am Inst Biol Sci; AAAS. *Res:* Neuropharmacology of nicotine; modulation of neurotransmitter release in the brain; autonomic pharmacology. *Mailing Add:* Dept Pharmacol & Toxicol Univ Louisville Sch Med Louisville KY 40292

ROWELL, ROBERT LEE, b Quincy, Mass, July 29, 32; m 56; c 4. PHYSICAL CHEMISTRY, COLLOID SCIENCE. *Educ:* Mass State Col Bridgewater, BS, 54; Boston Col, MS, 56; Ind Univ, PhD(phys chem), 60. *Prof Exp:* From instr to assoc prof, 60-78, dir, Res Comput Ctr, 61-64, actg head, 83-86, PROF CHEM, UNIV MASS, AMHERST, 78- *Concurrent Pos:* Int Bus Mach Corp fel, Comput Ctr, Mass Inst Technol, 63-64; res assoc, Clarkson Col, 66-70; vis prof, Univ Bristol, Eng, 73-74, Unilever vis prof, 80; assoc ed, Langmuir, 86- *Mem:* Am Chem Soc; Sigma Xi; AAAS; Golden Key Nat Hon Soc. *Res:* Laser light scattering by molecules, macromolecules and colloidal particles; characterization of particle shape and particle size distribution; stability, structure and rheology of coal slurry fuels. *Mailing Add:* Dept Chem Univ Mass Amherst MA 01002

ROWEN, BURT, b New York, NY, Mar 30, 21; m 42; c 3. AEROSPACE MEDICINE. *Educ:* Lafayette Col, BA, 42; NY Univ, MD, 45; Am Bd Prev Med, dipl, 65. *Prof Exp:* With Med Corps, USAF, 46, instr, USAF Sch Aviation Med, 49-51, asst air attache, Stockholm, Sweden, 52-55, med dir, X-15 Proj, USAF Flight Test Ctr, 56-62, med dir, Dynasoar Proj, Aeronaut Syst Div, 62-64, dep comdr, Aerospace Med Res Labs, 64-65, flight surgeon, Vietnam, 65-66, Hq, 12th Air Force, Waco, Tex, 66-68 & Hq, 17th Air Force, Ramstein, Ger, 68-69, dep surgeon, Hq, USAF Europ, Wiesbaden, Ger, 69-72, comdr, USAF Sch Health Care Sci, Sheppard AFB, 72-74, mem Phys Eval Bd, Air Force Military Manpower Personnel Ctr, Randolph AFB, 74-81,

CHIEF MED STANDARDS DIV, MANPOWER & PERSONNEL CTR, USAF, RANDOLPH AFB, 81- *Mem:* Fel Aerospace Med Asn; Am Col Prev Med; Int Acad Aerospace Med. *Res:* Development of life support systems, including air to ground telemetry of life support parameters and real time ground readout; health care training and management of health care resources; medical manpower & support of tactical air operations; adjudication of medical conditions in relation to the United States Air Force Disability System; medical standards management of United States Air Force active and reserve forces. *Mailing Add:* Air Force Manpower & Personnel Ctr Hq AFMPC/SGM Randolph AFB TX 78150-6001

ROWEN, WILLIAM H(OWARD), electrical & nuclear engineering; deceased, see previous edition for last biography

ROWIN, GERALD L, synthetic organic chemistry; deceased, see previous edition for last biography

ROWLAND, ALEX THOMAS, b Kingston, NY, Feb 25, 31; m 53; c 2. STEROID CHEMISTRY. *Educ:* Gettysburg Col, AB, 53; Brown Univ, PhD(chem), 58. *Prof Exp:* From asst prof to prof, 58-82, chmn dept, 68-82, OCKERSHAUSEN PROF CHEM, GETTYSBURG COL, 82- *Mem:* Am Chem Soc; AAAS. *Res:* Chemistry of highly substituted cyclohexanones. *Mailing Add:* Dept Chem Gettysburg Col Gettysburg PA 17325

ROWLAND, DAVID LAWRENCE, b Philadelphia, Pa, Sept 30, 50; c 1. PSYCHOENDOCRINOLOGY, ETHOLOGY. *Educ:* Southern Ill Univ, BA, 72; Univ Chicago, MA, 75, PhD(biopsychol), 77. *Prof Exp:* Asst prof psychol & dir, Psychol Animal Lab, Millikin Univ, 76-80; res fel, State Univ NY, Stony Brook, 80-81; assoc prof psychol, 81-89, chmn dept, 83-88, PROF PSYCHOL, VALPARAISO UNIV, 90- *Concurrent Pos:* Proj dir, NSF grants, 79-81; res fel, Stanford Univ, 84-85; vis scientist, Erasmus Univ, Rotterdam. *Mem:* Soc Comp Psychol; Am Asn Univ Prof. *Res:* Neural and hormonal basis of reproductive behavior. *Mailing Add:* Dept Psychol Valparaiso Univ Valparaiso IN 46383

ROWLAND, E(LBERT) S(ANDS), metallurgy, for more information see previous edition

ROWLAND, F SHERWOOD, b Delaware, Ohio, June 28, 27; m 52; c 2. ATMOSPHERIC KINETICS. *Educ:* Ohio Wesleyan Univ, AB, 48; Univ Chicago, MS, 51, PhD, 52. *Hon Degrees:* Various from US & Can univs, 89-91. *Prof Exp:* Instr chem, Princeton Univ, 52-56; from asst prof to assoc prof, 56-63, prof chem, Univ Kans, 63-64; prof chem & dept chmn, 64-70, Daniel G Aldridge Jr, chair prof, 85-89, PROF CHEM, UNIV CALIF, IRVINE, 64-, DONALD BREN PROF, 89- *Concurrent Pos:* Mem, Fachbeirat, Max Planck Inst Nuclear Phys & Chem, 82-; mem, Ozone Comm, Int Asn Meteorol & Atmospheric Physics, 80-; mem, Comm Atmospheric Chem & Global Pollution, 79-; mem, Bd Environ Studies & Toxicol, US Acad Sci, 86-; mem, US Nat Comt, Sci Comt Probs Environ, 86-; mem, Ozone Trends Panel, NASA, 86-; mem, Comt Atmospheric Chem, US Nat Acad Sci, 87-; Zucker fel, Yale Univ, 91. *Honors & Awards:* Tolman Medal, Am Chem Soc, 76; Gordon Billard Award in Environ Sci, NY Acad Sci, 77; Humboldt Sr Scientist Award, WGer, 81; Cert Commendation, US Fed Aviation Admin, 82; Tyler World Prize in Environ; Charles A Dana Award Pioneering Achievement in Health, 87; McGregor Lectr, Colgate Univ, 75; Philips Lectr, Haverford Col, 75; Snider Lectr, Univ Toronto, 79; Whitehead Lectr in Chem, Univ Ga, 86; Gustavson Lectr Chem, Univ Chicago, 88; Rachel Carson Award, Soc Environ Toxicol & Chem, 88; King Lectr, Kans State Univ, 91. *Mem:* Fel AAAS; fel Am Geophys Union; Am Acad Arts & Sci; Nat Acad Sci. *Res:* Co-discover with Dr Mario J Molena of theory that chlorofluorocarbon gases deplete the ozone layer of the stratosphere; research in chemical kinetics; latitudinal variation of trace tropospheric gases; global studies of trace gas concentrations; oxidative capacity of the earths atmosphere. *Mailing Add:* Dept Chem Univ Calif Irvine CA 92717

ROWLAND, IVAN W, b Pocatello, Idaho, Apr 1, 10; m 58; c 1. PHARMACY, MICROBIOLOGY. *Educ:* Idaho State Col, BS, 32; Univ Colo, MS, 47; Univ Wash, PhD(pharm), 54. *Prof Exp:* Dean col pharm, Idaho State Col, 54-56; dean sch pharm, 56-81, EMER DEAN, UNIV PAC, 81- *Mem:* Am Pharmaceut Asn. *Res:* Antiseptics; germicides; antibiotics; pharmacognosy; pharmaceutical chemistry; pharmacy administration. *Mailing Add:* 3355 Fairway Dr Stockton CA 95204

ROWLAND, JAMES RICHARD, b Muldrow, Okla, Jan 24, 40; m 63; c 2. ELECTRICAL ENGINEERING. *Educ:* Okla State Univ, BS, 62; Purdue Univ, Lafayette, MS, 64, PhD(elec eng), 66. *Prof Exp:* Instr elec eng, Purdue Univ, Lafayette, 64-66; from asst prof to assoc prof, Ga Inst Technol, 66-71; from assoc prof to prof elec eng, Okla State Univ, 71-85; chmn, Elec & Comput Eng, 85-89, PROF ELEC & COMPUT ENG, UNIV KANS, 85- *Concurrent Pos:* Consult, Lockheed-Ga Co, 66-71, US Army Missile Command, 69-79 & Sandia Nat Lab, 79. *Honors & Awards:* Centennial Medal, Inst Elec & Electronics Engrs, 84, Educ Soc Achievement Award, 86, Frontiers Educ Conf Award, 88. *Mem:* Inst Elec & Electronics Engrs; Am Soc Eng Educ; Nat Soc Prof Engrs. *Res:* Stochastic modeling; optimal estimation and control. *Mailing Add:* Elec & Comput Eng Univ Kans Lawrence KS 66045

ROWLAND, JOHN H, b Bellefonte, Pa, Feb 20, 34; m 56; c 4. MATHEMATICS, COMPUTER SCIENCE. *Educ:* Pa State Univ, BS, 56, PhD(math), 66; Univ Wash, MS, 58. *Prof Exp:* Instr math, Univ Nev, 58-61; staff mathematician, HRB Singer, Inc, 61-66; assoc prof math & comput sci, 66-77, head dept comput sci, 73-79, PROF MATH & COMPUT SCI, UNIV WYO, 77- *Concurrent Pos:* NSF fac residency in comput sci, Santa Monica, Calif, 70-71; vis prof appl math, Brown Univ, 78-79; contractor, IBM Tech Educ, Austin, Tex, 85-86; vis prof info & comput sci, Ga Inst Technol, 86-87. *Mem:* Inst Elec & Electronics Engrs Comput Soc; Soc Indust & Appl Math; Asn Comput Mach. *Res:* Theory of program testing; numerical analysis; approximation theory. *Mailing Add:* Dept Comput Sci Univ Wyo Laramie WY 82071

ROWLAND, LENTON O, JR, b Mobile, Ala, Sept 29, 43. POULTRY NUTRITION. *Educ:* Univ Fla, BSA, 65, MS, 67, PhD(animal nutrit), 72. *Prof Exp:* Poultry serviceman layers & pullets, Cent Soya Inc, 65-66; staff mem res & develop animal nutrit, Dow Chem Co, USA, 72-73; mem tech serv, Dow Chem Co, Latin Am, 73-74; ASST PROF POULTRY NUTRIT, TEX A&M UNIV, 75- *Mem:* Poultry Sci Asn; Am Soc Animal Sci; World Poultry Sci Asn. *Res:* Basic hen and broiler nutrition with particular emphasis on new feed ingredients, energy and amino acid requirements of poultry. *Mailing Add:* 15137 Post Oak Bend Dr College Station TX 77845

ROWLAND, LEWIS PHILLIP, b Brooklyn, NY, Aug 3, 25; m 52; c 3. NEUROLOGY. *Educ:* Yale Univ, BS, 45, MD, 48. *Hon Degrees:* Dr, Univ Aix-marseilles, 86. *Prof Exp:* Res asst neuroanat, Columbia Univ, 48; asst med, Yale Univ, 49-50; asst resident, Columbia Univ, 50-52, asst neurol, Col Physicians & Surgeons, 53; clin instr, Georgetown Univ, 53-54; instr, Col Physicians & Surgeons, Columbia Univ, 54-56, assoc, 56-57, from asst prof to prof, 56-67; prof neurol & chmn dept, Univ Pa, 67-73; PROF NEUROL & CHMN DEPT, COL PHYSICIANS & SURGEONS, COLUMBIA UNIV, 73-, DIR NEUROL SERV, PRESBY HOSP, 73- *Concurrent Pos:* Clin investr, Nat Inst Neurol Dis & Blindness, 53-54; asst neurologist, Montefiore Hosp, New York, 54-57; neurologist, Presby Hosp, 54-67; vis fel, Med Res Coun Labs, London, Eng, 56; co-dir, Neurol Clin Res Ctr, Columbia Univ, 61-67; mem med adv bd, Myasthenia Gravis Found, 63-, pres, 71-73; sci adv bd, Muscular Dystrophy Asns Am, 69-86 & med adv bd, 86-; med adv bd, Multiple Sclerosis Soc, 69-89, hon bd, 89-, Res Progs Adv Comt, 80-82 & 84-88; mem neurol res training comt B, Nat Inst Neurol Dis & Stroke, 71-73, Nat Adv Coun, 85-89, bd sci counc, Nat Inst Neurol Commun Dis Stroke, 78-82, chmn, 81-82; ed-in-chief, Neurol, 77-87; attend/consult, Harlem Hosp, 73-; co-dir, H Houston Merritt Clin Res Ctr Muscular Dystrophy & Related Dis, Columbia-Presby Med Ctr, 74-; secy-treas, Asn Univ Profs Neurol, 69-73, trustee, 73-77 & pres, 77-78; med adv bd, Comt Combat Huntington's Dis, 74-84; consult, Josiah Macy, Jr Found, 80-82 & Klingenstein Found, 81; Steven W Swank vis prof, Univ Oregon, 77, Robert B Aird vis prof neurol, Univ Calif, San Francisco, 79. *Honors & Awards:* Numerous Lect Awards, US & Can Insts. *Mem:* AAAS; Asn Res Nerv & Ment Dis (pres, 69-70, vpres, 79-80); hon mem Am Neurol Asn (pres-elect, 79, pres, 80-81); fel Am Acad Neurol (pres-elect, 87-89, pres, 89-90); Soc Neurosci; Am Soc Human Genetics; Sigma Xi; Muscular Dystrophy Asn; Nat Multiple Sclerosis Soc; AMA. *Res:* Neuromuscular disease. *Mailing Add:* Neurol Inst 710 W 168th St New York NY 10032

ROWLAND, NEIL EDWARD, b London, UK, Feb 20, 47; m 79; c 3. FEEDING BEHAVIOR, THIRST & SODIUM APPETITE. *Educ:* Univ Col London, BSc, 68, MSc, 72; Sussex Univ, MSc, 71; London Univ, PhD(exp psychol), 74. *Prof Exp:* Res assoc, psychobiol, Univ Pittsburgh, 74-81; assoc prof, 81-86, PROF PSYCHOL, UNIV FLA, 86- *Mem:* Am Physiol Soc; Am Inst Nutrit; Am Psychol Asn; Am Diabetes Asn; Soc Neurosci; Soc Study Ingestive Behav. *Res:* Physiological and neural basis of food and fluid intake and selection in rodents. *Mailing Add:* Dept Psychol Univ Fla Gainesville FL 32611

ROWLAND, NEIL WILSON, b Singapore, July 5, 19; US citizen; m 43; c 3. BOTANY, ECOLOGY. *Educ:* Union Col, Nebr, BA, 47; Univ Nebr, MA, 52, PhD, 61. *Prof Exp:* Head dept, 52-67, acad dean, 67-77, PROF BIOL, UNION COL, NEBR, 52- *Mem:* Sigma Xi. *Res:* Plant physiology and ecology. *Mailing Add:* 5300 Locust St Lincoln NE 68516

ROWLAND, RICHARD LLOYD, b Delaware, Ohio, May 31, 29; m 64. PHYSICAL CHEMISTRY. *Educ:* Ohio Wesleyan Univ, BA, 51; Univ Chicago, PhD(chem), 60. *Prof Exp:* Res chemist, Stand Oil Co, Ohio, 61-62, sr res chemist, 62-64; asst prof, Robert Col, Istanbul, 64-67, assoc prof, 67-71; assoc prof, Univ Calif, Irvine, 71-72; MEM FAC CHEM, COSUMNES RIVER COL, 72- *Mem:* Am Chem Soc. *Res:* Solid state diffusion; electrochemistry. *Mailing Add:* 8756 Leo Virgo Ct Elk Grove CA 95624-1796

ROWLAND, ROBERT EDMUND, b St Charles, Ill, Jan 10, 23; m 44; c 3. RADIATION BIOPHYSICS. *Educ:* Cornell Col, BS, 47; Univ Ill, MS, 49; Univ Rochester, PhD, 64; Univ Chicago, MBA, 75. *Prof Exp:* Assoc physicist, Argonne Nat Lab, 50-62; sr tech assoc, Univ Rochester, 62-64; assoc dir, Radiol Physics Div, Argonne Nat Lab, 64-67, sr biophysicist, 66-83, dir, Radiol & Environ Res Div, 67-81, interim assoc lab dir, biomed & environ res, 81-83; RETIRED. *Concurrent Pos:* Mem, Nat Coun Radiation Protection & Measurements, 71-83. *Mem:* Radiation Res Soc; Health Physics Soc. *Res:* Radiation biology; epidemiology and toxicology of internally deposited radium. *Mailing Add:* 700 W Fabyan-8C Batavia IL 60510

ROWLAND, SATTLEY CLARK, b San Jose, Calif, May 15, 38; m 61; c 2. MATERIALS SCIENCE, PHYSICS. *Educ:* Pac Union Col, BA, 60; Univ Utah, PhD(mat sci), 66. *Prof Exp:* Assoc prof, 66-74, PROF PHYSICS, ANDREWS UNIV, 74- *Concurrent Pos:* NSF grant, Andrews Univ, 69-72; vis prof mat sci, Stanford Univ, 80 & 82; vis scientist, Argonne Nat Lab, 89- *Mem:* AAAS; Am Asn Physics Teachers; Am Phys Soc. *Res:* X-ray diffraction studies of the structure of amorphous and crystalline semiconductors; lattice parameter measurements; high pressure studies of fatigue in metals. *Mailing Add:* 8486 S Hillcrest Dr Berrien Springs MI 49103

ROWLAND, STANLEY PAUL, b LaCrosse, Wis, Feb 25, 16; m 43; c 4. CELLULOSE STRUCTURE & CHEMISTRY, POLYMER. *Educ:* Univ Minn, BChem, 38; Univ Ill, PhD(org chem), 43. *Prof Exp:* Res chemist, Rohm and Haas Co, 43-56; res supvr, US Indust Chem Co, 56-59, asst mgr org chem, 59-61, mgr explor polymer res, 61-63; res leader crosslink struct, Southern Regional Res Ctr, 63-74, res leader nat polymer struct res, 74-84; RETIRED. *Concurrent Pos:* Instr, Tulane Univ, 65-68. *Honors & Awards:* Anselme Payen Award, Cellulose Div, Am Chem Soc, 81; Olney Medal, Am Asn Textile Chemists & Colorists, 85. *Mem:* Am Chem Soc; Fiber Soc. *Res:* Chemical modification of cellulose; structural characterization of chemically modified and crosslinked celluloses; chemical reactivity of cellulose and synthetic polymers; synthesis of intermediates, resins and polymers; plasticizers; plastics; fabrication; characterization and testing of polymers and plastics; dehydration of hydrobenzoins. *Mailing Add:* 16343 Old Hwy 99 SE Tenino WA 98589

ROWLAND, THEODORE JUSTIN, b Cleveland, Ohio, May 15, 27; m 52, 68; c 3. ELECTRONIC STRUCTURE. *Educ:* Western Reserve Univ, BS, 48; Harvard Univ, MA, 49, PhD(appl physics), 54. *Prof Exp:* Res physicist, Res Labs, Union Carbide Metals Co Div, Union Carbide Corp, 54-61; PROF PHYS METALL, UNIV ILL, URBANA, 61- *Concurrent Pos:* Fac res participant, Bell Labs, Murray Hill, NJ, 73, Argonne Nat Lab, 78-84, Ames Lab, Univ Iowa, 81-82. *Mem:* Fel Am Phys Soc; Sigma Xi; Mat Res Soc; Minerals, Metals & Mat Sci; AAAS; Am Asn Univ Prof. *Res:* Radiospectroscopy; nuclear magnetic resonance in metals; physics of metals, especially solid solutions; nuclear relaxation in polymers. *Mailing Add:* Dept Mat Sci & Eng Univ Ill 1304 W Green St Urbana IL 61801

ROWLAND, VERNON, b Clevelend Heights, Ohio, Aug 11, 22; m 49; c 2. PSYCHIATRY. *Educ:* Harvard Med Sch, MD, 46; Am Bd Psychiat & Neurol, dipl. *Prof Exp:* From intern to asst resident med, Univ Hosps, Cleveland, Ohio, 46-48, resident neuropsychiat, 52-53; resident, Cleveland Vet Admin Hosp, 50-52; assoc prof, 60-71, PROF PSYCHIAT, MED SCH, CASE WESTERN UNIV, 71- *Concurrent Pos:* NIMH & univ res fels psychiat, Western Reserve Univ, 53-55; NIMH career develop award, 55-60; assoc physician, Univ Hosps, 54- *Mem:* AAAS; Pavlovian Soc NAm; Am Psychiat Asn; Sigma Xi. *Res:* Conditioned electrographic response in the brain. *Mailing Add:* Univ Hosps Hanna Pavilion Cleveland OH 44106

ROWLAND, WALTER FRANCIS, b Decatur, Ill, Nov, 9, 31; c 3. CIVIL ENGINEERING, WATER RESOURCES. *Educ:* Univ Ill, BS, 54, MS, 57; Stanford Univ, PhD(civil eng), 67. *Prof Exp:* Instr civil eng, Univ Ill, 56-59; actg instr, Stanford Univ, 59-61; actg asst prof, Univ Calif, Berkeley, 63-64; asst prof, Colo State Univ, 64-67; assoc prof, 67-72, PROF CIVIL ENG, CALIF STATE UNIV, FRESNO, 72- *Concurrent Pos:* Dir atmospheric water resources res, Fresno State Col Found, 71-73. *Mem:* Am Soc Civil Engrs. *Res:* Surface and atmospheric water resources research and development. *Mailing Add:* Dept Civil Eng Calif State Univ 6241 N Maple Ave Fresno CA 93740

ROWLAND, WILLIAM JOSEPH, b Brooklyn, NY, Dec 15, 43; m 71; c 2. ETHOLOGY, BEHAVIORAL ECOLOGY. *Educ:* Adelphi Univ, BA, 65; State Univ NY Stony Brook, PhD(biol), 70. *Prof Exp:* Sci co-worker ethol, Zool Lab, Rijksuniversiteit te Groningen, Haren, Neth, 70-71; asst prof, 71-77, ASSOC PROF ZOOL, IND UNIV, BLOOMINGTON, 77- *Concurrent Pos:* Guest co-worker ethol, Zool Lab, Univ Leiden, Neth, 84. *Mem:* Sigma Xi; Animal Behav Soc; AAAS; Int Asn Fish Ethologists. *Res:* Causation, evolution and function of behavior in fishes and other lower vertebrates; social behavior and behavioral ecology of fishes. *Mailing Add:* Dept Biol Ind Univ Bloomington IN 47401

ROWLANDS, DAVID T, JR, b Wilkes-Barre, Pa, Mar 22, 30; m 58; c 2. PATHOLOGY, IMMUNOLOGY. *Educ:* Univ Pa, MD, 55. *Prof Exp:* Asst prof path, Univ Colo, 62-64; asst prof biochem, Rockefeller Univ, 64-66; assoc prof path, Duke Univ, 66-70; prof path, Sch Med, Univ Pa, 70-78, chmn dept, 73-78; STAFF MEM, DEPT PATH, COL MED, UNIV SFLA, 79- *Mem:* Am Asn Path & Bact; Fedn Am Socs Exp Biol; Int Acad Path; Am Soc Clin Path. *Res:* Developmental immunology; transplantation immunity. *Mailing Add:* Dept Path Col Med Univ SFla 12901 N 30th St Box 11 Tampa FL 33612

ROWLANDS, JOHN ALAN, b Altrincham, Eng, May 4, 45; m 72; c 2. SOLID STATE PHYSICS, MEDICAL PHYSICS. *Educ:* Leeds Univ, BSc, 66, PhD(physics), 71. *Prof Exp:* Killam fel physics, Univ Alta, 71-73, res assoc physics, 71-77; vis asst prof, dept physics, Mich State Univ, 77-79; ASST PROF, DEPT RADIOL, UNIV TORONTO, 79- *Mem:* Can Asn Physicists; Am Asn Physicists Med; Soc Electro Optical Eng. *Res:* Physics of diagnostic radiology. *Mailing Add:* Med Physics Reichmann Res Bldg Sunnybrook Health Sci Ctr 2075 Bayview Ave Toronto ON M4N 3M5 Can

ROWLANDS, R(ICHARD) O(WEN), b Llangefni, Wales, Apr 26, 14; nat US; m 39; c 2. MATHEMATICS, ELECTRICAL ENGINEERING. *Educ:* Univ Wales, BS, 36, MS, 50. *Prof Exp:* Head filter design group, Gen Elec Co, Eng, 37-48; sr lectr studio sect, Eng Training Dept, Brit Broadcasting Co, 48-57; assoc prof elec eng, 57-58, assoc prof eng res, 58-64, chmn dept eng acoust, 67-70, prof eng res, 64-79, EMER PROF ENG RES, PA STATE UNIV, 79- *Concurrent Pos:* Liaison scientist, US Off Naval Res, Eng, 70-71. *Mem:* Acoust Soc Am; Inst Elec & Electronics Engrs; Brit Inst Elec Engrs; Sigma Xi. *Res:* Circuit theory; communications; information theory as applied to signal detection; electroacoustics; acoustic telemetry. *Mailing Add:* Appl Res Lab PO Box 30 State College PA 16801

ROWLANDS, ROBERT EDWARD, b Trail, BC, July 7, 36; m 59; c 2. STRESS ANALYSIS, FAILURE-STRENGTH. *Educ:* Univ BC, BASc, 59; Univ Ill, Urbana, MS, 64, PhD(mech), 67. *Prof Exp:* From res engr to sr res engr, IIT Res Inst, 67-74; from asst prof to assoc prof, 74-80, PROF MECH, UNIV WIS-MADISON, 74- *Concurrent Pos:* Consult various orgns, 67- *Honors & Awards:* Hetenyi Award, Soc Exp Stress Anal, 70 & 76. *Mem:* Am Soc Testing & Mat; Am Acad Mech; Am Soc Mech Engrs; Soc Exp Stress Anal; NAm Photonics Asn. *Res:* Experimental stress analysis; photomechanics; fatigue; fracture; composite and advanced materials; numerical processing of experimental data; energy storage; materials at cryogenic environments; wood and paper engineering. *Mailing Add:* Dept Eng Mech Univ Wis-Madison Madison WI 53706

ROWLANDS, STANLEY, medical biophysics, nuclear medicine, for more information see previous edition

ROWLETT, ROGER SCOTT, b Chickasha, Okla, Jan 1, 55; m. ENZYME KINETICS, COMPUTATIONAL BIOCHEMISTRY. *Educ:* Univ Ala, BS, 76, PhD(chem), 81. *Prof Exp:* Fel biochem, Col Med, Univ Fla, 81-82; asst prof, 82-88, ASSOC PROF CHEM, COLGATE UNIV, 88- *Mem:* Am Chem Soc; NY Acad Sci. *Res:* Mechanistic studies of carbonic anhydrase; molecular modeling. *Mailing Add:* Dept Chem Colgate Univ Hamilton NY 13346

ROWLETT, RUSSELL JOHNSTON, JR, b Richmond, Va, Sept 19, 20; m 43; c 2. INFORMATION SCIENCE. *Educ:* Univ Va, BS, 41, MS, 43, PhD(org chem), 45. *Prof Exp:* Asst chem, Univ Va, 40-42, assoc, 42-46; chemist, E I du Pont de Nemours & Co, Del, 46; asst ed, Chem Abstr, Ohio State Univ, 47-48, assoc ed, 49-52; patent coordr, Va-Carolina Chem Corp, 52-55, asst dir res, 55-57, asst dir res & develop, 57-60, dir, 60; asst dir, Va Inst Sci Res, 60-67; ed, Chem Absr Serv, 67-79, dir publ & serv, 79-82; RETIRED. *Concurrent Pos:* Mem comt chem info, Nat Res Coun, 68-73; chmn, Gordon Conf Sci Info Prob Res, 74. *Honors & Awards:* Presidential Res Citation, 45; Miles Conrad Award, Nat Fedn Abstracting & Indexing Serv, 80; Herman Skolnik Award, Am Chem Soc, 83 & Distinguished Serv Award, 90. *Mem:* Fel AAAS; Am Chem Soc; Sigma Xi; hon fel Nat Fedn Abstr & Indexing Serv (secy, 74-76, pres 77-78). *Res:* Synthetic organic chemistry; antimalarial drugs; organic chemical nomenclature; patents; chemistry of phosphorus compounds; scientific information storage and retrieval. *Mailing Add:* Covenant Towers 5001 Little River Rd Myrtle Beach SC 29577

ROWLETT, RUSSELL JOHNSTON, III, b Charlottesville, Va, June 26, 45; m 67; c 1. TOPOLOGY. *Educ:* Univ Va, BA, 67, PhD(math), 70. *Prof Exp:* Instr math, Princeton Univ, 70-74; asst prof, 74-78, ASSOC PROF MATH, UNIV TENN, 78-, DIR, LIBERAL ARTS COOP PROG, 79- *Mem:* Am Math Soc; Math Asn Am; Am Soc Eng Educ. *Res:* Classification of smooth actions of compact lie groups on compact manifolds; related questions in the theory of compact transformation groups. *Mailing Add:* Two Brightleaf Ct Durham NC 27701

ROWLEY, DAVID ALTON, b Rochester, NY, July 21, 40; m 62; c 2. PHYSICAL INORGANIC CHEMISTRY. *Educ:* State Univ NY Albany, BS, 63, MS, 64; Univ Ill, Urbana, PhD(inorg chem), 68. *Prof Exp:* From asst prof to assoc prof, 68-81, PROF CHEM, GEORGE WASHINGTON UNIV, 81-, ASST DEAN, GRAD SCH ARTS & SCI, 74- *Mem:* Am Chem Soc. *Res:* Inorganic electronic absorption spectroscopy; reaction mechanism and reactions of coordinated ligands. *Mailing Add:* Dept Chem George Washington Univ Washington DC 20006

ROWLEY, DONALD ADAMS, b Owatonna, Minn, Feb 4, 23; m 48; c 4. EXPERIMENTAL PATHOLOGY, IMMUNOLOGY. *Educ:* Univ Chicago, BS, 45, MS, MD, 50. *Prof Exp:* Sr asst surgeon, Nat Inst Allergy & Infectious Dis, 51-54; from instr to asst prof, 54-69, PROF PATH & PEDIAT, UNIV CHICAGO, 69-, DIR, LA RABIDA CHILDREN'S HOSP & RES CTR, LA RABIDA-UNIV CHICAGO INST, 77- *Concurrent Pos:* USPHS sr res fel, 59-69; vis scientist, Sir William Dunn Sch Path, Oxford Univ, 61-62 & 70-71; dir res, La Rabida Children's Hosp & Res Ctr, La Rabida-Univ Chicago Inst, 74-77. *Mem:* Am Soc Exp Path; Am Asn Path & Bact; Am Asn Immunol. *Res:* Immunologic networks and regulation of the immune responses, specific enhancement and suppression by antigen and antibody. *Mailing Add:* Dept Pathol-Pediat Univ Chicago 950 E 59th St Chicago IL 60637

ROWLEY, DURWOOD B, b Walton, NY, Aug 11, 29; m 50; c 3. FOOD MICROBIOLOGY. *Educ:* Hartwick Col, BS, 51; Syracuse Univ, MS, 53, PhD(microbiol), 62. *Prof Exp:* Asst prof biol, Hartwick Col, 56-59; asst microbiol, Syracuse Univ, 59-62; asst prof, Univ Mass, 62-63; res microbiologist, US Army Natick Labs, 63-70, chief microbiol div, Food Lab, 70-74, head food microbiol group, 74-82, CHIEF BIOL SCI DIV, SCI & ADVAN TECHNOL LAB, US ARMY NATICK RES & DEVELOP CTR, 82- *Concurrent Pos:* Mem comt nitrate & alternative curing agents in food, Nat Acad Sci; mem, Comn Microbiol Criteria for Foods, Food & Drug Admin, USDA, Nat Marine Fisheries Serv & US Army Natick Res & Develop Ctr. *Mem:* Sigma Xi; Am Soc Microbiol; Inst Food Technol. *Res:* Bacterial spores; radiation microbiology; microbiological safety of mass feeding systems; rapid recovery and estimation of injured and uninjured food-borne bacteria; thermal resistance of food-borne pathogenic and spoilage microorganisms. *Mailing Add:* 144 Spring St Millis MA 02054

ROWLEY, GEORGE RICHARD, b Rahway, NJ, Aug 21, 23. BIOCHEMISTRY, PHYSIOLOGY. *Educ:* Upsala Col, AB, 49; Rutgers Univ, PhD(biochem), 55. *Prof Exp:* Chem qual control chemist, E R Squibb & Sons, 49-51; USPHS grant psychiat, Coatesville Vet Admin Hosp & Med Sch, Univ Pa, 55-56; assoc res specialist agr chem, Rutgers Univ, 56-57; instr physiol, Col Physicians & Surgeons, Columbia Univ, 57-60; sr res chemist, Colgate-Palmolive Res Lab, 60-64; asst prof, 64-68, ASSOC PROF BIOCHEM & ACTG CHMN DEPT, SCH DENT, FAIRLEIGH DICKINSON UNIV, 68- *Mem:* AAAS; Int Asn Dent Res; Am Chem Soc; NY Acad Sci. *Res:* Blood coagulation and platelet aggregation during hyperlipemia and effects of lipolytic activity; metabolism of vitamin E and vitamin C. *Mailing Add:* 14 Thrumont Rd West Caldwell NJ 07006

ROWLEY, GERALD L(EROY), biochemistry, organic chemistry, for more information see previous edition

ROWLEY, JANET D, b New York, NY, Apr 5, 25; m 48; c 4. CYTOGENETICS. *Educ:* Univ Chicago, BS, 46, MD, 48. *Hon Degrees:* DSc, Univ Ariz, 89; Univ Pa, 89; Knox Col, 91. *Prof Exp:* Intern med, Marine Hosp, 50-51; res fel, Cook County Hosp, 55-61; clin instr neurol, Med Sch, Univ Ill, 57-61; res assoc, 62-69, assoc prof, 69-77, PROF MED, SCH MED, UNIV CHICAGO, 77- *Concurrent Pos:* Julian D Levinson Res Found fel, 55-58; USPHS spec trainee, 61-62; vis scientist, Genetics Lab, Oxford Univ, 70-71; mem, Nat Cancer Adv Bd & bd dirs, Am Soc Human Genetics; mem MIT Corp vis comt, Dept Appl Biol Sci, Frederick Cancer Res Fac; mem bd,

Med Genetics, Am Soc Human Genetics; coun mem, Inst Med, 88. *Honors & Awards:* Esther Langer Award, Ann Langer Cancer Res Found, 83; Kuwait Cancer Prize, 84; A Cressy Morrison Award, NY Acad Sci, 85; Woodward Award, Mem Sloan-Kettering Cancer Ctr, 86; Antoinne Lacassagne Prize, Nat League Against Cancer, 87; Co-recipient King Fasial Int Prize in Med, 88; Dameshek Prize, Am Soc Hemat, 82; Karnofsky Prize, Am Soc Clin Oncol, 87; GHA Clowes Award, Am Asn Cancer Res, 89; Wm Proctor Prize, Sigma Xi, 89; Mary Harris Thompson Prize, 90. *Mem:* Nat Acad Sci; Am Soc Human Genetics; Am Soc Hemat; Inst Med; Am Asn Cancer Res. *Res:* Human chromosomes; chromosome abnormalities in pre-leukemia as well as in leukemia and lymphoma; quinacrine and Giemsa stains to identify chromosomes in maliqnant cells; molecular genetic analysis of chromosome translocations and deletions. *Mailing Add:* Section Hemat-Oncol Dept Med Univ Chicago 5841 S Maryland Box 420 Chicago IL 60637

ROWLEY, PETER DEWITT, b Providence, RI, Dec 6, 42; m 87; c 2. VOLCANOLOGY. *Educ:* Carleton Col, BA, 64; Univ Tex, Austin, PhD(geol), 68. *Prof Exp:* Temp instr geol, Kent State Univ, 68-69; asst prof, Carleton Col, 69-70; GEOLOGIST, US GEOL SURV, 70- *Concurrent Pos:* Leader, NSF-USGS geol expeds, Antarctica, 70-87. *Honors & Awards:* Antarctic Serv Medal, US Govt, 72; Meritorious Serv Award, Dept of Interior, 86. *Mem:* Soc Econ Geologists; fel Geol Soc Am; Am Geophys Union; Am Geol Inst; Rocky Mountain Asn Geologists. *Res:* Geology of Antarctic Peninsula, Antarctica; Antarctic mineral deposits; geology of Iron Springs mining district, Utah; structural geology and volcanic stratigraphy of High Plateaus of Utah; geology of eastern Uinta Mountains, Utah-Colorado; geology of Marysvale mining district, Utah; geology of Mount St Helen's volcano; geology of Caliente caldera complex, Nevada; Rowley Massif named geographic feature, Antarctica. *Mailing Add:* US Geol Surv Mail Stop 913 Box 25046 Fed Ctr Denver CO 80225-0046

ROWLEY, PETER TEMPLETON, b Greenville, Pa, Apr 29, 29; m 67; c 2. HUMAN GENETICS, INTERNAL MEDICINE. *Educ:* Harvard Col, AB, 51; Columbia Univ, MD, 55. *Prof Exp:* Intern med, NY Hosp-Cornell Med Ctr, 55-56; resident, Boston City Hosp, 58-60; asst prof med, Stanford Univ, 63-70; assoc prof, 70-75, PROF MED, PEDIAT, GENETICS & MICROBIOL, CHMN, DIV GENETICS, SCH MED, UNIV ROCHESTER, 75- *Concurrent Pos:* Hon res asst, Univ Col, Univ London, 60-61; physician & pediatrician, Strong Mem Hosp, 70- *Mem:* Am Soc Human Genetics; Am Fedn Clin Res; Am Col Physicians; Am Soc Hemat. *Res:* Genetics of human cancer; hemoglobinopathies; hematopoietic differentiation; evaluation of genetic screening counseling. *Mailing Add:* Div Genetics Univ Rochester Sch Med Rochester NY 14642

ROWLEY, RICHARD L, b Salt Lake City, UT, Sept 1, 51; m 72; c 6. THERMODYNAMICS, TRANSPORT PHENOMENA. *Educ:* Brigham Young Univ, BS, 74; Mich State Univ, PhD(phys chem), 78. *Prof Exp:* Res asst thermodynamics, Ctr Thermochem Studies, 74; ASST PROF CHEM ENG, RICE UNIV, 78- *Concurrent Pos:* Consult, Eng Dept, Texaco, Inc, 81. *Mem:* Am Chem Soc; Am Inst Chem Engrs; AAAS; Sigma Xi. *Res:* Measurement and prediction of thermophysical properties, particularly transport coefficients, in liquid mixtures; formulation of liquid mixture models and theories as related to thermodynamic properties. *Mailing Add:* 350 CB Brigham Young Univ Provo UT 84602

ROWLEY, RODNEY RAY, b Cedar City, Utah, Mar 25, 34; m 61; c 4. AUDIOLOGY, PSYCHOACOUSTICS. *Educ:* Brigham Young Univ, BS, 59; Univ Md, MA, 61; Univ Okla, PhD, 66. *Prof Exp:* Assoc prof speech, Univ Utah, 63-73; DIR AUDIOL, MED CTR, LOMA LINDA UNIV, 73- *Concurrent Pos:* Res audiologist, Vet Admin Hosp, Oklahoma City, 65-66; consult, Utah State Training Sch, 68-73, Mt Fuel Supply, 71-73 & Patton State Hosp, 74- *Mem:* Am Speech & Hearing Asn. *Res:* Loudness; noise and man; behavior modification and hearing loss. *Mailing Add:* 409 Arrowview Dr Redlands CA 92373

ROWLEY, WAYNE A, b Spring Glen, Utah, Aug 27, 33; m 57; c 3. ENTOMOLOGY. *Educ:* Utah State Univ, BS, 60, MS, 62; Wash State Univ, PhD(med entom), 65. *Prof Exp:* Res entomologist, US Army Biol Labs, Ft Detrick, Md, 65-67; asst prof, 67-71, assoc prof, 71-75, PROF ENTOM, IOWA STATE UNIV, 75- *Mem:* Sigma Xi; Entom Soc Am; Am Mosquito Control Asn. *Res:* Insect transmission of vertebrate pathogens; mosquito biology and flight; biology and taxonomy of bloodsucking midges. *Mailing Add:* Dept Entom Iowa State Univ Ames IA 50010

ROWND, ROBERT HARVEY, b Chicago, Ill, July 4, 37; m 59; c 3. BIOCHEMISTRY, MOLECULAR BIOLOGY. *Educ:* St Louis Univ, BS, 59; Harvard Univ, MA, 61, PhD(biophys), 63. *Prof Exp:* USPHS fel molecular biol, Med Res Coun Unit, Cambridge Univ, Eng, 63-65; Nat Acad Sci-Nat Res Coun res fel biochem, Pasteur Inst, Paris, France, 65-66; from asst prof to prof biochem & molecular biol, Univ Wis-Madison, 66-81, chmn, Molecular Biol Lab, 70-81; JOHN G SEARLE PROF & CHMN, DEPT MOLECULAR BIOL, MED & DENT SCH, NORTHWESTERN UNIV, 81- *Concurrent Pos:* Mem, NSF adv panel for develop biol, 68-71; USPHS res career develop award, NIH, 68-73; mem adv panel grad fel, NSF, 74-77, chmn, 76-77, NATO fel, 79; assoc ed, Plasmid, 77-87; mem, NIH Microbiol Biochem Study Sect, 78-; mem comt human health effects subtherapeut antibiotic use in animal feed, Nat Res Coun, 79-81; ed, J Bacteriol, 81-; mem People to People Prog del microbiologist to China, 83; vis prof, Saint Louis Univ, 84; vchmn Gordon Conf, Extrachromasomal Elements, 84, chmn, 86; assoc ed, J Biotechnol & Med Eng, 86-; sr tech adv recruitment consult, UN Develop Prog, China, 87; hon res prof, biotechnol res ctr, Chines Acad Agr Sci, Beijing, 87- *Mem:* Am Soc Microbiol; Am Acad Microbiol; NY Acad Sci. *Res:* Structure, function and replication of nucleic acids; cellular regulatory mechanisms; genetics and mechanism of drug resistance in bacteria; macromolecular chemistry and biology. *Mailing Add:* Dept Biochem Northwestern Univ Med & Dent Schs 303 E Chicago Ave Chicago IL 60611

ROWNTREE, ROBERT FREDRIC, b Columbus, Ohio, Feb 8, 30; m 56; c 2. TECHNOLOGY BASE MANAGEMENT, LABORATORY PLANNING. *Educ:* Miami Univ, BA, 52; Syracuse Univ, MPA, 53; Ohio State Univ, PhD(physics), 63. *Prof Exp:* Physicist, Wright Air Develop Ctr, Wright-Paterson AFB, Ohio, 53; asst physics, Ohio State Univ, 56-58, Univ Res Found, 58-59, res assoc, 61-63; physicist, Naval Weapons Ctr, China Lake, Calif, 63-64, assoc missions anal, 64-68, prog dir air strike warfare, 68-72, res & develop planning, Weapons Planning Group, 72-75, head regt group, Off Resource & Technol, 75-76, head, Off Plans & Progs, 76-78, technol base coordr, Lab Directorate, 78-89; RETIRED. *Honors & Awards:* Michelson Lab Award, Naval Weapons Ctr, 78. *Mem:* Optical Soc Am; Am Phys Soc; Sigma Xi. *Res:* Optical properties of solids; instrumentation for infrared spectroscopy; arms control research; military operations research; laboratory management; science policy; understanding complex, technology-influenced organizations and environments, elucidating the critical features in their futures, and communicating findings and suggestions to diverse constituancies. *Mailing Add:* 808 Murray Ave San Louis Obis CA 93405

ROWOTH, OLIN ARTHUR, b Trenton, Mo, May 6, 21; m 46; c 3. POULTRY NUTRITION, POULTRY HUSBANDRY. *Educ:* Univ Mo, BA, 43, MA, 48. *Prof Exp:* Vitamin chemist, Beacon Milling Co, Spencer Kellogg & Sons, Inc, 47-50, poultry nutritionist, 50-59, asst dir poultry & small animal res, 59-64, dir poultry res, Beacon Feeds, Textron Inc, 64-66, DIR RES & TECH SERV, BEACON MILLING CO, INC, 66-, VPRES SERV, 68- *Mem:* Poultry Sci Asn; Animal Nutrit Res Coun; World Poultry Sci Asn; Am Soc Animal Sci; Am Poultry Hist Soc. *Res:* Nutrition of chickens, ducks, turkeys, game birds and dogs; feeding systems; nutrient requirements of poultry. *Mailing Add:* 179 E Genesee St Auburn NY 13021-4227

ROWSELL, HARRY CECIL, b Toronto, Ont, May 29, 21; m 46; c 4. VETERINARY PATHOLOGY. *Educ:* Univ Toronto, DVM, 49, DVPH, 50; Univ Minn, PhD(vet path), 56; Am Col Lab Animal Med, dipl, 79. *Hon Degrees:* PhD, Univ Sask, 80; PhD, Univ Guelph, 87. *Prof Exp:* Asst res bact, Ont Vet Col, Univ Guelph, 50-51, asst prof path, 53-56, prof res, 56-57 & physiol sci, 58-65; prof vet path, Univ Sask, 65-68; EXEC DIR, CAN COUN ANIMAL CARE, 68- *Concurrent Pos:* Consult, Blood & Vasc Dis Res Unit, Dept Med, Univ Toronto, 58-64 & Med Sci Complex, 64-65, res assoc, 64-65; permanent secy, Can Coun Animal Care, 70-; hon prof, Peking Union Med Col, China. *Honors & Awards:* Medal, Acad Med Sci USSR; Humane Award, Can Vet Med Asn. *Mem:* Am Heart Asn; Am Soc Exp Path; Can Vet Med Asn; Can Physiol Soc; Can Fedn Biol Sci; Med Acad Sci USSR; hon mem Am Col Lab Animal Med; hon assoc Royal Col Vet Surgeons UK. *Res:* Comparative medicine; pathogenesis of bacterial and mycotic infections; comparative pathology and cardiology; hemostatic mechanism; coagulation defects in animals; experimental and naturally occuring atherosclerosis; laboratory animal science, care and management. *Mailing Add:* Exec Dir Can Coun Animal Care 1000-151 Slater St Ottawa ON K1P 5H3 Can

ROWTON, RICHARD LEE, b Springfield, Mo, May 29, 28; m 51; c 2. ORGANIC CHEMISTRY. *Educ:* Univ Mo-Rolla, BS, 50, MS, 52; Okla State Univ, PhD(org chem), 59. *Prof Exp:* Sr res chemist, Jefferson Chem Co, 59-77, SR TECH SERV REP, TEXACO CHEM CO, 77- *Mem:* Am Chem Soc. *Res:* Petrochemicals epoxide reaction mechanisms and isotope effects in chemical reactions; urethane chemistry, foams and elastomers; urethane foam seating. *Mailing Add:* 7607 Delafield Ln Austin TX 78752-1312

ROWZEE, E(DWIN) R(ALPH), b Washington, DC, May 17, 08; m 35; c 3. CHEMICAL ENGINEERING. *Educ:* Mass Inst Technol, MS, 31. *Hon Degrees:* DSc, Laval Univ, 55. *Prof Exp:* Chem engr, Goodyear Tire & Rubber Co, 31-35, in-charge synthetic rubber develop, 35-42; mgr, Co-Polymer Plant, Can Synthetic Rubber, Ltd, 42-44; dir res, Polymer Corp Ltd, 44-47, mgr prod eng & res, 47-51, mem bd dirs, 50, vpres & mgr, 51-57, pres & managing dir, 57-71, chmn bd dirs, Polymer Corp, 71-78; chmn bd dirs, Urban Transp Develop Corp, 73-83; RETIRED. *Concurrent Pos:* mem bd gov, Univ Windsor & Ont Res Found; mem, Sci Coun Can, 66-69. *Honors & Awards:* Purvis Mem Lectr, Soc Chem Indust, 47; R S Jane Mem Lectr, Chem Inst Can, 60; Found Lectr, Brit Inst Rubber Indust, 63. *Mem:* Am Chem Soc; Chem Inst Can (pres, 54-55). *Mailing Add:* 580 Woodrowe Ave Sarnia ON N7V 2W2 Can

ROXBY, ROBERT, b Abington, Pa, June 4, 40; m 65; c 2. BIOCHEMISTRY, MOLECULAR BIOLOGY. *Educ:* Gettysburg Col, BA, 62; Univ NC, Chapel Hill, MA, 65; Duke Univ, PhD(biochem), 70. *Prof Exp:* Fel biochem & biophys, Ore State Univ, 70-72; asst prof biochem, Temple Univ Sch Med, 72-75; asst prof biochem, 75-81, chmn dept, 84-89, ASSOC PROF BIOCHEM, UNIV MAINE, ORONO, 81- *Concurrent Pos:* Vis prof, State Univ, Gruningen, Neth, 81-82; vis scientist, Max Planck Inst Plant Breeding, 89-90. *Mem:* Sigma Xi; AAAS. *Res:* Developmental biology of plants. *Mailing Add:* Dept Biochem Univ Maine Hitchner Hall Orono ME 04473

ROXIN, EMILIO O, b Buenos Aires, Arg, Apr 6, 22; div; c 2. MATHEMATICS. *Educ:* Univ Buenos Aires, Engr, 47, Dr(math), 58. *Prof Exp:* Asst math & physics, Univ Buenos Aires, 48-56, prof math, 56-67; PROF MATH, UNIV RI, 67- *Concurrent Pos:* Researcher, AEC, Arg, 53-60 & Res Inst Advan Study, Md, 60-64; res assoc, Brown Univ, 64-65; vis prof, Univ Mich, 67. *Mem:* Am Math Soc; Arg Math Union; Math Asn Am; Soc Indust & Appl Math. *Res:* Ordinary differential equations, especially applied to control theory. *Mailing Add:* Dept Math Univ RI Kingston RI 02881

ROY, ARUN K, b Ganraganis, India, Dec 29, 38; US citizen; m 68; c 1. MOLECULAR BIOLOGY, ENDOCRINOLOGY. *Educ:* Univ Calcutta, BS, 58, MS, 60; Wayne State Univ, PhD(biochem), 65. *Prof Exp:* From asst prof to prof biol sci, Oakland Univ, 69-87; PROF OBSTET/GYNEC & CELLULAR & STRUCT BIOL, HEALTH SCI CTR, UNIV TEX, 88- *Concurrent Pos:* Res career award, NIH, 76, develop award, 89; mem, Cancer Prog Proj Rev Comt, NIH, 83-87 & Endocrinol Study Sect, 89-93. *Mem:* Fel AAAS. *Mailing Add:* Dept Obstet & Gynec Health Sci Ctr Univ Tex 7703 Floyd Curl Dr San Antonio TX 78284-7836

ROY, CLAUDE CHARLES, b Quebec City, Que, Oct 21, 28; m 62; c 3. PEDIATRICS, GASTROENTEROLOGY. *Educ:* Laval Univ, BA, 49, MD, 54, FRCPS(C), 59. *Prof Exp:* NIH res fel, Med Ctr, Univ Colo, Denver, 64-66, from asst prof to assoc prof pediat, 66-70; asst to dir, Dept Pediat, 77-80, dir pediat res ctr, 80-84, CHMN, STE JUSTINE HOSP, 84-; PROF PEDIAT, UNIV MONTREAL, 70- *Concurrent Pos:* Chmn med adv bd, Can Cystic Fibrosis Found; mem, med adv bd, Can Found Ileitis & Colitis & Can Liver Found; assoc ed, J Pediat Gastoenterol & Nutrit; chmn, clin investr study sect, Med Res Counc Can. *Honors & Awards:* Harry Shuachman Award, NAm Soc Pediat Gastroenterol, 87. *Mem:* Am Col Nutrit; Can Soc Clin Invest; Soc Pediat Res; Am Gastroenterol Asn; Am Pediat Soc; Soc Exp Biol & Med; Can Asn Gastroenterol. *Res:* Bile salt metabolism and function; bile salt in clinical and experimental disorders of the liver, pancreas and gastrointestinal tract; role of taurine in clinical nutrition, lipoprotein metabolism in malabsorptive disorders; free radicals in chronic liver disease and in disorders of nutrition. *Mailing Add:* Ste Justine Hosp 3175 Ste Catherine Rd Montreal PQ H1W 2C4 Can

ROY, DAVID C, b Baton Rouge, La, Nov 14, 37; m; c 3. STRATIGRAPHY, SEDIMENTATION. *Educ:* Iowa State Univ, BS, 61; Mass Inst Technol, PhD(geol), 70. *Prof Exp:* ASSOC PROF, GEOL, BOSTON COL, 70-, DEPT CHAIR, 89- *Concurrent Pos:* Geologist Consult, 70- *Mem:* Fel Geol Soc Am; Sigma Xi; Nat Asn Geol Teachers. *Res:* Stratigraphy in New England geology and experimental sedimentology. *Mailing Add:* Dept Geol & Geophys Boston Col Chestnut Hill MA 02167

ROY, DELLA M(ARTIN), b Merrill, Ore, Nov 3, 26; m 48; c 3. GEOCHEMISTRY, MATERIALS SCIENCE. *Educ:* Univ Ore, BS, 47; Pa State Univ, MS, 49, PhD(mineral), 52. *Prof Exp:* Asst mineral, 49-52, res assoc geochem, 52-59, sr res assoc, 59-69, sr res assoc, Mat Res Lab, 63-69, assoc prof mat sci, 69-75, PROF MAT SCI, MAT RES LAB, PA STATE UNIV, UNIVERSITY PARK, 75- *Concurrent Pos:* Mem comn A2EO6, Hwy Res Bd, Nat Acad Sci; ed, Cement & Concrete Res; ed-in-chief, Univ Cement & Concrete Res, 71-; chair, NMAB Nat Acad Sci Res Comt on Concrete, 80-83; adv, NMAB Comt on Concrete Durability, 86-87; Sir Frederick Lea mem lectr, 87; hon fel, Inst Concrete Technol, 87; coun, Mat Res Soc, 88; bd trustees, Am Ceramic Soc, 90. *Honors & Awards:* Slag Award, Am Concrete Inst, 89. *Mem:* Nat Acad Eng; Mat Res Soc; fel Mineral Soc Am; fel Am Concrete Soc; fel Am Ceramics Soc; Clay Minerals Soc; mem, Concrete Soc; Am Nuclear Soc; Am Soc Testing & Mat; Soc Women Engrs; fel AAAS. *Res:* Phase equilibria; materials synthesis; crystal chemistry and phase transitions; crystal growth; cement chemistry, hydration and microstructure; concrete durability; biomaterials; special glasses; radioactive waste management; geologic isolation; chemically bonded ceramics. *Mailing Add:* 217 Mat Res Lab Pa State Univ University Park PA 16802

ROY, DEV KUMAR, b Patna, India, July 15, 51; m 85. RECURSIVE FUNCTION THEORY. *Educ:* Indian Inst Technol, Kharagpur, BS, 71; Indian Inst Technol, Delhi, MS, 73; Univ Rochester, PhD(math), 80. *Prof Exp:* Lectr math, Univ Wis-Milwaukee, 79-81; asst prof, 81-86, ASSOC PROF MATH, FLA INT UNIV, 86-, CHMN, DEPT MATH, 87- *Mem:* Asn Symbolic Logic. *Res:* Investigations in recursive algebra and model theory; constructions in recursive ordered structures, hierarchies, and presentations; coding and priority arguments in the context of ordered structures. *Mailing Add:* Dept Math Fla Int Univ Tamiami Trail Miami FL 33199

ROY, DIPAK, b WBengal, India, Aug 4, 46; m 75; c 2. HAZARDOUS WASTE MANAGEMENT. *Educ:* Jadavpur Univ, India, BCE, 68; Indian Inst Technol, MTech, 71; Univ Ill, Urbana, PhD(civil & environ eng), 79. *Prof Exp:* Scientist Res Environ Eng, Nat Environ Eng Res Inst, 71-73; design engr civil eng, Catalytic, Inc, 73-74, Air Pollution Control, Johns & March, 74-75; asst prof, 79-85, ASSOC PROF ENVIRON ENG, LA STATE UNIV, 85- *Concurrent Pos:* Consult, NY Assoc, New Orleans, 80; Woodward-Clyde Consults, 88. *Honors & Awards:* Fulbright fel. *Mem:* Water Pollution Control Fedn; Am Water Works Asn; Am Soc Civil Engrs; Asn Environ Eng Prof; AAAS; Am Chem Soc. *Res:* Destruction of hazardous wastes by photolytic ozonation; removal and inactivation of viruses from water and wastewater; methane generation from waste and organic biomass; mathematical modelling of environmental engineering processes; biodegradation of hazardous waste. *Mailing Add:* Amoco Chem Co PO Box 3011 Naperville IL 60566

ROY, DONALD H, b Raleigh, NC, July 13, 36; m 65. NUCLEAR PHYSICS, REACTOR PHYSICS. *Educ:* NC State Univ, BS, 58, PhD(nuclear eng), 62; Mass Inst Technol, MS, 59. *Prof Exp:* Reactor physicist, 59-60, sr reactor physicist, 62-63, GROUP SUPVR THEORET PHYSICS & EXP PHYSICS ANALYSIS, BABCOCK & WILCOX CO, 63- *Res:* Advanced computational methods for resonance absorption calculations and prediction of high energy nuclear reaction cross sections and distribution matrices. *Mailing Add:* 1045 Greenway Ct Lynchburg VA 24503

ROY, GABRIEL D, b India, July 7, 39; US citizen; m 61; c 4. COMBUSTION & PROPULSION SCIENCE, PULSE POWER PHYSICS. *Educ:* Univ Kerala, BS, 60, MS, 65; Indian Inst Technol, cert heat transfer, 71; Univ Tenn, PhD(eng sci), 77. *Prof Exp:* Lectr mech eng, Col Eng, Trivandrum, 60-65, asst prof, 65-71; res asst aerospace eng, Univ Tenn, 71-76, sr res engr energy eng, Univ Tenn Space Inst, 76-78, group leader heat transfer, 78-81; prog mgr energy conversion, TRW Inc, 81-87; PROG MGR PROPULSION & PULSE POWER, OFF NAVAL RES, 87- *Concurrent Pos:* Staff adv, Col Eng, Univ Kerala, 65-71; secy, Trivandrum Arts Group, 69-71; asst prof, Tenn State Univ, 77-78, Univ Tenn, Nashville, 78-79 & Motlow State Col, 79-81; vis prof, Calif State Univ, 83-87; prog reviewer, Dept Energy, Strategic Defense Initiative Orgn, 87-; assoc ed, J Propulsion & Power, Am Inst Aeronaut & Astronaut, 89-, vchmn, Propellants & Combustion. *Mem:* Assoc fel Am Inst Aeronaut & Astronaut; Am Soc Mech Engrs; Sigma Xi. *Res:* Propulsion; pulse power; magnetohydrodynamics; underwater propulsion; fluid dynamics; recipient of two patents; author of numerous publications. *Mailing Add:* 9944 Great Oaks Way Fairfax VA 22030

ROY, GABRIEL L, b Otterburne, Man, Apr 27, 38; c 2. ANIMAL BREEDING, GENETICS. *Educ:* Univ Man, BSA, 64, MSc, 66; Univ Sask, PhD(animal breeding), 73. *Prof Exp:* Supt swine prod, Govt Nfld & Labrador, 66-69; RES SCIENTIST ANIMAL BREEDING, RES STA AGR CAN, 72- *Mem:* Am Soc Animal Sci; Can Soc Animal Sci. *Res:* Dairy protein yield, lifetime production, crossbreeding and mating scheme; dairy-beef crossbreeding; cow productivity; carcass fat distribution; dystocia. *Mailing Add:* Agr Can Res Sta PO Box 90 Lennoxville PQ J1M 1Z3 Can

ROY, GUY, b Que, Apr 10, 39; m 63; c 3. BIOPHYSICS, PHYSIOLOGY GENERAL. *Educ:* Univ Laval, BSc, 64; Univ Calif, PhD(biophys), 69. *Prof Exp:* From asst prof to assoc prof, 68-80, PROF BIOPHYS, UNIV MONTREAL, 80- *Concurrent Pos:* Vis prof, Dept Physiol, Kyoto Univ, Kyoto, Japan, 75-76, Univ Otago, New Zealand, 83-84. *Mem:* Biophys Soc. *Res:* Ionic currents across artificial and biological membranes; theory and experiments. *Mailing Add:* Dept Physics Univ Montreal Box 6128 Montreal PQ H3C 3J7 Can

ROY, HARRY, b Cincinnati, Ohio, Aug 29, 43; m; c 1. PLANT MOLECULAR BIOLOGY & BIOCHEMISTRY. *Educ:* Brown Univ, AB, 65, ScM, 66; Johns Hopkins Univ, PhD(cell biol), 70. *Prof Exp:* Fel plant biochem, Johns Hopkins Univ, 70-71; NIH fel plant physiol, Cornell Univ, 71-74; asst prof life sci, Polytech Inst NY, 74-76; from asst prof to assoc prof, 76-88, PROF BIOL, RENSSELAER POLYTECH INST, 89- *Concurrent Pos:* Mem, Capital Dist Plant Res Group; Nat Res Serv Awards fel, Harvard Univ, 83-84. *Mem:* Am Soc Plant Physiologists; Int Soc Plant Molecular Biol. *Res:* Synthesis and assembly of chloroplast proteins. *Mailing Add:* Dept Biol Rensselaer Polytech Inst Troy NY 12180-3590

ROY, JEAN-CLAUDE, b Quebec, Que, Jan 6, 27; m 56; c 2. RADIOCHEMISTRY, ANALYTICAL CHEMISTRY. *Educ:* Univ Laval, BA, 47, BSc, 51; Univ Notre Dame, PhD(chem), 54. *Prof Exp:* Vis chemist, Carnegie Inst Technol, 54-55; res officer radiochem, Atomic Energy Can, 55-62 & Int Bur Weights & Measures, France, 62-64; prof radiochem, 64-74, PROF CHEM, UNIV LAVAL, 74- *Mem:* Chem Inst Can. *Res:* Methodology and measurement of radioactivity in waters and sediments; x-ray fluorescence analysis. *Mailing Add:* Dept Chem Fac Sci Univ Laval Quebec PQ G1K 7P4 Can

ROY, M S, b Libourne, France, May 3, 46; Can citizen; m 74; c 1. OPHTHALMOLOGY. *Educ:* Univ Bordeaux, MD, 73; Univ London, DO, 76; FRCS(C); 80; FCOphth, 89; Am Bd Ophthal, cert, 83. *Prof Exp:* VIS SCIENTIST RETINAL DIS SECT, NAT EYE INST, 82- *Concurrent Pos:* Int Res Group Color Vision Deficiency, 88. *Mem:* Asn Res Vision; Int Soc Ocular Fluorophotom. *Res:* Pathogenesis and treatment of age related maculor degeneration; sickle cell retinopathy and diabetic retinopathy. *Mailing Add:* Dept Ophthal NJ Med Sch Univ Med & Dent 15 S Ninth St Newark NJ 07107

ROY, MARIE LESSARD, b Fall River, Mass, Mar 1, 44; m 69; c 2. OCCUPATIONAL MEDICINE, GENERAL PRACTICE. *Educ:* Albertus Magnus Col, BA, 65; Univ Conn, PhD(org chem), 69, MD, 79; FRCPS(C), 89. *Prof Exp:* Lectr org chem & chem eng, Univ Toronto, 69-73, lectr pharmacol, 73-77; internship, 79-80, resident internal med, 80-81, SR MED TOXICOLOGIST, ONT MINISTRY LABOR, 81- *Concurrent Pos:* Lectr, Univ Toronto, 84- *Mem:* Am Chem Soc; Sigma Xi; Soc Toxicol Can; Royal Soc Chem; Am Conf Govt Indust Hygienists; Ont Med Asn; NY Acad Sci; Med Asn Can; Am Col Occup Med. *Res:* Evaluation of health effects of hazardous substances in the workplace; health effects of toxic substances; supervised a long-range epidemiology study of health effects on hospital staff due to exposure to waste anaesthetic gases; toxicology of lead. *Mailing Add:* 89 Poyntz Ave Willowdale ON M2N 1J3 Can

ROY, PAUL-H(ENRI), b Quebec City, Que, Apr 19, 24; m 56. CHEMICAL ENGINEERING. *Educ:* Laval Univ, BASc, 48; Univ Mich, MSE, 49; Ill Inst Technol, PhD(chem eng), 55. *Prof Exp:* Proj engr, Res Inst, Univ Mich, 49-51; res engr, E I du Pont de Nemours & Co, Inc, 55-60; from asst prof to assoc prof chem eng, 60-64, vdean fac sci, 61-69, PROF CHEM ENG, LAVAL UNIV, 64- *Concurrent Pos:* Consult, E I du Pont de Nemours & Co, Inc, 64-69; mem, Hydro-Que Res Comt, 65-66 & Laval Admin Comn, 65-69; Laval rep, Can Res Mgt Asn, 67-72; rep sci adv planning br, Ministry of State Sci & Technol, Govt Can, 72- *Mem:* Am Inst Chem Engrs; Can Chem Eng Soc; Can Coun Prof Engrs; Sigma Xi. *Res:* Mixing of liquids as related to dynamics of chemical reactors; mechanism of liquid; liquid dispersion; waste water treatment; chemical process economics. *Mailing Add:* 380 Chemin St Louis No 503 Quebec PQ G1S 4M1 Can

ROY, PRADIP KUMAR, b Hazaribagh, Bihar, India, Feb 27, 43; m; c 3. THIN DIELECTRIC PROCESS, TECHNOLOGY DEVELOPMENT & CHARACTERIZATION. *Educ:* Indian Inst Technol, Kharagpur, India, BTech, 64; Univ Alta, Edmonton, Can, MSc, 67; Columbia Univ, New York, NY, PhD(mat sci & eng), 73. *Prof Exp:* Sr res fel alloys & intermetallics, Nat Metall Lab, 64-66; postdoctoral fel thin film superconductivity & dielectrics, T J Watson IBM Res Ctr, 73-75; sr scientist thin film solar photovoltaics, SES, Inc, Shell Subsid, 76-81; MEM TECH STAFF MICROELECTRONICS, RES & DEVELOP, AT&T BELL LABS, 82- *Mem:* Sr mem Inst Elec & Electronics Engrs; Mat Res Soc; Am Vacuum Soc; Am Phys Soc. *Res:* Thin oxide process technology for future generations of submicron IC's and CCD's; conceptualized, developed and implemented to manufacturing a multilayering process in semiconducting and metal thin films for their structural and electrical property control. *Mailing Add:* 2102 Riverbend Rd Allentown PA 18103

ROY, PRODYOT, b Calcutta, India, 35; US citizen; m 63; c 2. MATERIALS SCIENCE, PHYSICAL CHEMISTRY. *Educ:* Univ Calcutta, BS; Univ Calif, Berkeley, MS, PhD(mat sci). *Prof Exp:* Res engr surface chem, Univ Calif, Berkeley, 63-65, res engr chem, 67-68; res fel, Max Planck Inst Phys Chem, 65-67; engr chem, 68-80, mgr plant mat, 80-83, PRIN SCIENTIST, GEN

ELEC CO, SAN JOSE, CA. *Mem:* Am Inst Mining, Metall & Petrol Engrs; fel Am Soc Metals. *Res:* Thermodynamics of metals and alloys; surface chemistry and adsorption phenomenon; solid state electrochemistry; fast breeder coolant chemistry; mechanical properties of materials; corrosion and stress corrosion; terrestial & space power conversion; power storage technology e.g. fuel cells; primary & secondary batteries. *Mailing Add:* 14333 Saratoga Ave No 77 Saratoga CA 95070

ROY, RABINDRA (NATH), b July 31, 39; m 68. PHYSICAL CHEMISTRY, ANALYTICAL CHEMISTRY. *Educ:* Jadavpur Univ, India, BSc, 59, MSc, 61; La State Univ, Baton Rouge, PhD(chem), 66. *Prof Exp:* Indian Dept Health res scholar phys chem, Jadavpur Univ, India, 61-63; teaching asst chem, La State Univ, Baton Rouge, 65-66; from asst prof to assoc prof phys & anal chem, Drury Col, 66-71; petrol res fund res assoc phys chem, Univ Fla, 71-73; PROF CHEM & CHMN DEPT, DRURY COL, 73- *Mem:* Am Chem Soc; Royal Soc Chem; Am Soc Test & Mat. *Res:* Thermodynamics and analytical processes of electrolytes and nonelectrolytes in aqueous mixed and nonaqueous solvents from physicochemical measurements; buffer solutions; ion-selective electrode and mixed strong electrolytes. *Mailing Add:* Dept Chem Drury Col 900 N Benton Springfield MO 65802

ROY, RADHA RAMAN, b Calcutta, India, Feb 8, 21; US citizen; m 56; c 2. NUCLEAR PHYSICS. *Educ:* Univ Calcutta, BS, 40, MS, 42; Univ London, PhD(nuclear physics), 46. *Prof Exp:* AEC fel, Univ London, 43-49; in chg courses nuclear physics & dir nuclear lab, Free Univ Brussels, 49-57; from asst prof to prof nuclear physics, Pa State Univ, 58-63, dir nuclear physics lab, 58-63; PROF NUCLEAR PHYSICS, ARIZ STATE UNIV, 63- *Mem:* Fel Am Phys Soc. *Res:* Interactions of photons and leptons with matter; nuclear structure; fission; pair and triplet production; scattering of light particles; nuclear instrumentation. *Mailing Add:* Dept Physics Ariz State Univ Tempe AZ 85287

ROY, RAJARSHI, b Calcutta, India, June 4, 54; m 82. LASER PHYSICS, QUANTUM OPTICS. *Educ:* Delhi Univ, Bsc, 73, MSc, 75; Univ Rochester, MA, 77, PhD(physics), 81. *Prof Exp:* Teaching fel physics, Joint Inst Lab Astrophysics, 81-82; asst prof, 82-87, ASSOC PROF PHYSICS, SCH PHYSICS, GA INST TECHNOL, 87- *Concurrent Pos:* Vis scientist, Bell Labs, 87. *Mem:* Am Phys Soc; Optical Soc Am. *Res:* Laser physics; quantum optics; experimental tests of statistical theories and non-linear phenomena; quantum fluctuations and their effects on optical devices. *Mailing Add:* Sch Physics Ga Inst Technol Atlanta GA 30332

ROY, RAM BABU, b Patna, India, Jan 15, 33; US citizen;; c 3. ORGANIC CHEMISTRY, ANALYTICAL CHEMISTRY. *Educ:* Univ Bihar, India, BSc, 56; Patna Univ, MSc, 58; Univ Newcastle, Eng, PhD(org chem), 67. *Prof Exp:* Head dept chem, Magadh Univ, India, 60-68; vis prof, Univ Mich, Ann Arbor, 68-69 & Univ Wash, 69-71; sr res fel biochem sci, Princeton Univ, 71-72; res chemist food eng, Mass Inst Technol, 72-73; SR STAFF SCIENTIST, TECHNICON INSTRUMENT CORP, 73- *Mem:* Am Chem Soc; fel Am Inst Chemists; fel Inst Food Technol; Royal Inst Chem. *Res:* Application of autoanalyzer and near infrared in analytical works, especially development of methods for food, agriculture, pharmaceutical and water samples; use of NMR, ESR and MS for determining structures of organic compounds. *Mailing Add:* 36 Dwight Ave Hillcrest Spring Valley NY 10977

ROY, RAMAN K, b Digboi, India, Jan 14, 47; m 72; c 2. BIOCHEMISTRY. *Educ:* Calcutta Univ, BSc, 66, MSc, 68, PhD(biochem), 73. *Prof Exp:* Lectr biochem, Burdwan Univ Med Col, India, 72-73; staff fel, Boston Biomed Res Inst, 74-76, res assoc, 77-79, staff scientist, 80-83; SR SCIENTIST, ASTRA RES CTR INDIA, 84- *Concurrent Pos:* Vis scientist, Indian Inst Sci, Bangalore, 84-86. *Mem:* NY Acad Sci; Am Soc Biochem & Molecular Biol. *Res:* Virulence mechanisms of enteropathogens; bacterial cell wall biosynthesis. *Mailing Add:* Astra Res Ctr India 18th Cross Malleswaram Bangalore 560003 India

ROY, ROB, b Brooklyn, NY, Jan 2, 33; m 59; c 3. BIOENGINEERING & BIOMEDICAL ENGINEERING. *Educ:* Cooper Union, BSEE, 54; Columbia Univ, MSEE, 56; Rensselaer Polytech Inst, DEng Sc(elec eng), 62, Albany Med Col, MD, 76, Am Bd Anesthesiologists, dipl, 79. *Prof Exp:* Sr engr character recognition, Control Instrument Div, Burroughs Corp, 56-60; from instr to assoc prof elec eng, 60-68, prof syst eng, 68-79, PROF & CHMN BIOMED ENG, RENSSELAER POLYTECH INST, 79- *Concurrent Pos:* Consult, Raytheon Corp, 66-, US Air Force, 67- & Cornell Aeronaut Lab, 67-; NIH spec fel, 72-74; consult, Searle Corp, 79-, Kendall Corp, 81-, Am Edwards, 85. *Mem:* Inst Elec & Electronics Engrs; Am Soc Anesthesiologists. *Res:* Pattern recognition; digital signal processing; process identification; adaptive control systems; biomedical research; radar systems. *Mailing Add:* Ten Sevilla Dr Clifton Knolls Elnora NY 12065

ROY, ROBERT FRANCIS, b Boston, Mass, Aug 13, 30; m 60; c 4. GEOPHYSICS. *Educ:* Harvard Univ, AB, 52, MS, 60, PhD(geophys), 63. *Prof Exp:* Geologist, Anaconda Co, 52-57; res fel geophys, Harvard Univ, 63-66; sr res fel, Calif Inst Technol, 66-68; assoc prof, Univ Minn, Minneapolis, 68-71; prof geophys, Purdue Univ, Lafayette, 71-77; L A NELSON PROF GEOL SCI, UNIV TEX, EL PASO, 77- *Mem:* Am Geophys Union; Soc Econ Geol; Soc Explor Geophysicists. *Res:* Geothermal studies; economic geology; geochemistry; petrology. *Mailing Add:* Dept Geol Sci Univ Tex El Paso TX 79968

ROY, ROBERT MICHAEL MCGREGOR, b Nanaimo, BC, Can, June 10, 42; m 67; c 1. RADIATION BIOLOGY, CELL BIOLOGY. *Educ:* Univ Toronto, BSc, 63, MA, 65, PhD(zool), 68. *Prof Exp:* Chmn dept biol, 77-82, dean sci, 82-85, actg vice-rector acad, 84-85, ASSOC PROF BIOL, CONCORDIA UNIV, 70- *Res:* Radiation biology; physiological and biochemical effects of ionizing radiation on plants and animals; post radiation modification of damage and repair. *Mailing Add:* Dept Biol Rm 1260 Concordia Univ 1455 de Maisonneuve Blvd W Montreal PQ H3G 1M8 Can

ROY, RUSTUM, b Ranchi, India, July 3, 24; nat US; m 48; c 3. NEW MATERIALS, CRYSTAL CHEMISTRY. *Educ:* Patna Univ, India, BSc, 42, MSc, 44; Pa State Univ, PhD(ceramics), 48. *Hon Degrees:* DSc, Toyko Inst Technol, 87. *Prof Exp:* Fel, Pa State Univ, 48-49, res assoc, 50-51, from asst prof to assoc prof geochem, 51-57, prof solid state, 67-81, dir mat res lab, 62-85, dir sci technol & soc prog, 84-90, EVAN PUGH PROF SOLID STATE, PA STATE UNIV, 81- *Concurrent Pos:* Sr sci officer, Cent Glass & Ceramic Res Insr, Calcutta, India, 50; prof geochem, Pa State Univ, 57-, chmn Solid State Technol Prog, 60-67, dir Mat Res Lab, 62-85, chmn Sci, Technol & Soc Prog, 69-; sci policy fel, Brookings Inst, Washington DC, 82-83; ed, J Mat Educ, 79-; consult, Xerox Corp, 63-83, Carborundum Co, 62-, Bausch & Lomb, 62-, Lanxide Corp, 84- *Honors & Awards:* Mineral Soc Am Award, 57; Welch Lectr, London Univ, 74; Fairchild Lectr, Lehigh Univ, 76; Hibbert Lectr, London Univ, 79. *Mem:* Nat Acad Eng; foreign mem Royal Swed Acad Eng Sci; foreign fel Indian Nat Sci Acad; Am Ceramic Soc. *Res:* New materials preparation and characterization; crystal chemistry, synthesis, stability, phase equilibria and crystal growth in non-metallic systems; author of 5 books, over 500 scientific and 100 technology and society papers and chapters in books and magazines. *Mailing Add:* Mat Res Lab Pa State Univ University Park PA 16802

ROY, WILLIAM ARTHUR, b Elkins, WVa, July 24, 48. TERATOLOGY, BONE BIOLOGY. *Educ:* Fairmont State Col, BS, 70; WVa Univ, PhD(anat), 76. *Prof Exp:* Nat res serv award anat, Univ Va, 76-78; asst prof anat & embryol, Sch Dent, Marquette Univ, 78-86; Sch Pediatric Med, Barry Univ, Miami Shores, Fla, 86-87; STAFF MEM, DEPT ANAT, UNIV MISS MED CTR, 87- *Mem:* Teratology Soc; AAAS; Am Asn Anatomists. *Res:* Normal and abnormal craniofacial development during the prenatal and early postnatal periods; pathogenesis and treatment of craniofacial anomalies such as premature craniosynostosis. *Mailing Add:* Dept Phys Ther Univ Miss Med Ctr Jackson MS 39216-4505

ROY, WILLIAM R, b 1926. OBSTETRICS & GYNECOLOGY. *Prof Exp:* Mem staff, Stormont-Vail Hosp & St Francis Hosp, Topeka, Kans, 55-89; RETIRED. *Mem:* Inst Med-Nat Acad Sci. *Mailing Add:* 6137 SW 38th Terr Topeka KS 66610

ROYAL, GEORGE CALVIN, JR, b Williamston, SC, Aug 5, 21; m 69; c 6. MEDICAL MICROBIOLOGY. *Educ:* Tuskegee Inst, BS, 43; Univ Wis, MS, 47; Univ Pa, PhD, 57. *Prof Exp:* Instr bact, Tuskegee Inst, 47-48; asst immunol, Ohio Agr Exp Sta, 50-52; asst prof bact, Agr & Tech Col NC, 52-53, prof, 57-65; from asst prof to assoc prof, 66-82, PROF MICROBIOL, COL MED, HOWARD UNIV, 82- *Concurrent Pos:* Fel, Jefferson Med Col, 65-66; dir, Sr Res Proj, US AEC, 58-65; dir, Undergrad Res Partic Prog, NSF, 59-61; mem, Nat Adv Food Comt, Food & Drug Admin, 72- *Mem:* NY Acad Sci; Am Soc Microbiol; fel Am Acad Microbiol; Sigma Xi. *Res:* Virulence factors of candida albicans. *Mailing Add:* Dept Microbiol Howard Univ Col Med Washington DC 20059

ROYAL, HENRY DUVAL, b Norwich, Conn, May 14, 48. MEDICINE. *Educ:* Providence Col, BS, 70; St Louis Univ, MD, 74. *Prof Exp:* From instr to assoc prof radiol, Harvard Univ, 79-86; ASSOC PROF RADIOL, WASH UNIV, 87- *Concurrent Pos:* Radiologist, Beth Israel Hosp, 76-86 & Mallinckrodt Inst Radiol, 87- *Mem:* Soc Nuclear Med; Radiol Soc NAm; Health Physics Soc. *Res:* Effects of radiation; health care policy. *Mailing Add:* 510 S Kingshighway Blvd St Louis MO 63110

ROYALL, RICHARD MILES, b Elkin, NC, Aug 13, 39; m 59; c 2. MATHEMATICAL STATISTICS. *Educ:* NC State Univ, BS, 62; Stanford Univ, MS, 64, PhD(statist), 66. *Prof Exp:* From asst prof to assoc prof biostatist, 66-74, assoc prof math sci, 74-76, PROF BIOSTATIST, JOHNS HOPKINS UNIV, 76- *Mem:* Int Statist Inst; Biomet Soc; fel Am Statist Asn; Inst Math Statist; Am Public Health Asn. *Res:* Finite population sampling theory; foundations of statistical inference. *Mailing Add:* Dept Biostatist Johns Hopkins Univ 615 N Wolfe St Baltimore MD 21205

ROYALS, EDWIN EARL, b Climax, Ga, Jan 23, 19; m 42; c 4. ORGANIC CHEMISTRY. *Educ:* Emory Univ, AB, 40, MS, 41; Univ Wis, PhD(org chem), 44. *Prof Exp:* Instr chem, Ga Inst Technol, 44-46; from asst prof to assoc prof, Emory Univ, 46-62; res chemist, Heyden Newport Chem Corp, 62-65; prof chem, Pensacola Jr Col, 65-84, head dept, 67-84; RETIRED. *Concurrent Pos:* Consult citrus indust. *Mem:* Am Chem Soc. *Res:* Base catalyzed condensations; terpene and epoxide chemistry; reactions of olefinic linkage. *Mailing Add:* 3016 Swan Lane Pensacola FL 32504

ROY-BURMAN, PRADIP, b Comillah, India, Nov 12, 38; m 63; c 2. MOLECULAR BIOLOGY, MOLECULAR GENETICS. *Educ:* Univ Calcutta, BSc, 56, MSc, 58, PhD(biochem), 63. *Prof Exp:* Sr res chemist, Dept Bot, Univ Calcutta, 62-63; res assoc biochem, 63-66, asst prof, 67-70, from asst prof to assoc prof biochem & path, 70-78, chmn, grad comt exp path, 78-84, PROF BIOCHEM & PATH, MED SCH, UNIV SOUTHERN CALIF, 78-, CHMN, BIOMED RES SUPPORT GRANT COMT, 84-, VICE CHMN, DEPT PATH, 84- *Concurrent Pos:* Dernham sr fel oncol, Am Cancer Soc, 66-71, mem, spec grants comt, 74-78, prin investr, res grants & contracts, 68-70; prin investr, res grants & contracts, NIH, 70-, ad hoc mem NIH Path B study sect, 88; prin investr, res grants & contracts, priv founds, 83-; travel grants, Int Mgt, 76-79 & 88; vis scientist, Govt India, 86; mem, NIH Path B Study Sect, 90-94. *Mem:* Am Soc Biol Chem; Am Soc Microbiol; Int Asn Comp Res Leukemia & Related Dis; Am Soc Virol. *Res:* Molecular biology of retroviruses and endogenous retrovirus genes; viral and cellular oncogenes; pathogenetic mechanisms in leukemia; oncogene alleles and genetic susceptibility to leukemogenesis; recombinant DNA technology in molecular pathology; molecular genetics of lung cancer. *Mailing Add:* Dept Path Univ Southern Calif Med Sch Los Angeles CA 90033

ROYCE, BARRIE SAUNDERS HART, b Bishop's Stortford, Eng, Jan 10, 33; US citizen; m 64; c 2. SOLID STATE PHYSICS. *Educ:* Univ London, BSc, 54, PhD(physics), 57. *Prof Exp:* Res assoc physics, Carnegie Inst Technol, 57-60; res assoc, 60-61, asst prof, 61-66, assoc prof, 66-77, PROF SOLID STATE SCI, 77-, MASTER DEAN, MATHEY COL, PRINCETON UNIV, 87- *Concurrent Pos:* Vis prof, Univ Sao Paulo, 62 & 69 & Nat Polytech Inst, Mex, 67 & 78; Sci Res Coun sr fel, Clarenden Labs, Oxford Univ, Eng, 69-70; vis prof, Solid State Physics Lab, Orsay, France, 73 & 78. *Mem:* Am Phys Soc; Sigma Xi. *Res:* Radiation damage in ionic solids; color centers and radiation induced reactions; molecular solids; optical properties; physical properties of biological materials; surface properties of insulators; photoacoustic spectroscopy of solids; photothermal deflection spectroscopy; catalytic materials; electrochemical problems; atomic level structure of composite materials and their macroscopic performance; development of photoacoustic technique, and the equivalent "mirage" technique, for use in studies of photoelectrochemical reactions, surface groups, and adsorbates on high, specific surface area and cosmetic materials such as catalysts and polymetric fillers; the combination of optical methods with other physical probes to study high temperature ceramics as catalytic materials, photopolymerization and photo-degradation and photocorrosion, thermosetting polymers and phase seperation in modified thermosetting matrices. *Mailing Add:* Mat Lab Princeton Univ D416 Duffield Hall Princeton NJ 08540

ROYCE, GEORGE JAMES, b Petoskey, Mich, Sept 30, 38. ANATOMY. *Educ:* Mich State Univ, BA; Ohio State Univ, MA, 63, PhD(anat), 67. *Prof Exp:* NIH fel, Case Western Reserve Univ, 68-69; instr, Albany Med Col, 69-70, asst prof anat, 70-74; ASSOC PROF ANAT, UNIV WIS-MADISON, 74- *Mem:* Soc Neurosci. *Res:* Neuroanatomy. *Mailing Add:* Dept Anat Univ Wis-Madison 153 Bardeen Lab 1300 University Dr Madison WI 53706

ROYCE, PAUL C, b Minneapolis, Minn, July 2, 28; m 56; c 3. HEALTH CARE DELIVERY. *Educ:* Univ Minn, BA, 46, MD, 52; Case Western Reserve Univ, PhD(physiol), 59. *Prof Exp:* Asst prof med, Albert Einstein Col Med, 61-69; assoc prof, Hahnemann Med Univ, 73-81; dean & prof clin sci, Duluth Sch Med, Univ Minn, 81-87; SR VPRES, MONMOUTH MED CTR, LONG BRANCH, NJ, 87- *Concurrent Pos:* Dir med educ, Guthrie Med Ctr, Sagre, Pa, 70-81; clin prof med, Upstate Med Ctr, State Univ NY, 79-81. *Mem:* Am Physiol Soc; AAAS; AMA; Fedn Am Scientists; Physicians Social Responsibility. *Res:* Health care delivery. *Mailing Add:* Monmouth Med Ctr 300 Second Ave Long Branch NJ 07740

ROYCE, WILLIAM FRANCIS, b DeBruce, NY, Jan 5, 16; m 40; c 3. FISHERIES, RESOURCE MANAGEMENT. *Educ:* Cornell Univ, BS, 37, PhD(vert zool), 43. *Prof Exp:* From aquatic biologist to fishery res biologist, US Fish & Wildlife Serv, 42-58; prof fisheries, Univ Wash, 58-72, dir fisheries res inst, 58-67, assoc dean col fisheries, 67-72; assoc dir, Nat Marine Fisheries Serv, 72-76; CONSULT, FISHERIES SCI & DEVELOP, AQUATIC ENVIRON, 76- *Mem:* Fel AAAS; hon mem Am Fisheries Soc. *Res:* Population studies of fish; biological oceanography; life history of fishes; measurement and sampling techniques of fisheries; fishery policy; history of fishery management. *Mailing Add:* 10012 Lake Shore Blvd NE Seattle WA 98125-8159

ROYCHOUDHURI, CHANDRASEKHAR, b Barisal, India, Apr 7, 42; m 77. PHYSICS, OPTICS. *Educ:* Jadavpur Univ, BSc, 63, MSc, 65; Univ Rochester, PhD(optics), 73. *Prof Exp:* Sr lectr physics, Univ Kalyani, W Bengal, India, 65-68; scientist optics, Nat Inst Astrophys & Optics, Puebla, Mex, 74-78; SCIENTIST, TRW SYSTS, 78- *Mem:* Optical Soc Am; Soc Photo-optical Instrumentation Engrs; Inst Elec & Electronics Engrs. *Res:* Interferometry; holography; Fourier optics; spectroscopy; interference, diffraction and spectroscopy with ultra-short pulses; conceptual foundations of quantum mechanics; lasers; laser communication. *Mailing Add:* 100 Wooster Heights Rd Danbury CT 06810

ROYDEN, HALSEY LAWRENCE, b Phoenix, Ariz, Sept 26, 28; m 48; c 3. MATHEMATICAL ANALYSIS. *Educ:* Stanford Univ, BS, 48, MS, 49; Harvard Univ, PhD(math), 51. *Prof Exp:* From asst prof to assoc prof, 51-58, assoc dean, 62-65, actg dean, 68-69, dean, Sch Humanities & Sci, 73-81, PROF MATH, STANFORD UNIV, 58- *Concurrent Pos:* Ed, Pac J Math, 54-58 & 67-69; NSF fel, 58-59; vis prof, Mid East Tech Univ, Ankara, 66; mem math sci div, Nat Res Coun, 71-75 & adv comt res, NSF, 74-76; adv coun, NSF, 78-81. *Mem:* Am Math Soc; Math Asn Am. *Res:* Functions of a complex variable; conformal mapping; Riemann surfaces; differential geometry; complex manifolds. *Mailing Add:* Dept Math Stanford Univ Stanford CA 94305-2125

ROYER, DENNIS JACK, b Lock Haven, Pa, May 24, 41; c 2. CHEMICAL ENGINEERING. *Educ:* Pa State Univ, BS, 62, PhD(chem eng), 67; Carnegie Inst Technol, MS, 64. *Prof Exp:* dir, Continental Oil Co, 68-80, dir eng res, Chem Res Div, Conoco Inc, 80-85; STAFF MEM, DUPONT CO, 85- *Mem:* Am Inst Chem Engrs. *Res:* Petrochemical processes; production scale gas chromatography. *Mailing Add:* Du Pont Co 1007 Market St Wilmington DE 19898

ROYER, DONALD JACK, b Newton, Kans, May 7, 28. INORGANIC CHEMISTRY. *Educ:* Univ Kans, PhD(chem), 56. *Prof Exp:* From asst prof to assoc prof, 56-78, PROF CHEM, GA INST TECHNOL, 78- *Mem:* AAAS; Am Chem Soc. *Res:* Complex inorganic compounds and inorganic stereochemistry. *Mailing Add:* Dept Chem Ga Inst Technol Atlanta GA 30332

ROYER, GARFIELD PAUL, b Waynesboro, Pa, Dec 2, 42; m 66; c 3. BIOCHEMISTRY. *Educ:* Juniata Col, BS, 64; WVa Univ, PhD(biochem), 68. *Prof Exp:* NIH fel, Northwestern Univ, 68-70; from asst prof to prof biochem, Ohio State Univ, 70-83; MGR, BIOTECHNOL DIV, NAPERVILLE, ILL, 83- *Mem:* Am Soc Biol Chemists; Am Chem Soc. *Res:* Enzymology; immobilized enzymes and synthetic enzyme models. *Mailing Add:* Amoco Technol Co 305 E Shuman Blvd Naperville IL 60563-8467

ROYER, THOMAS CLARK, b Battle Creek, Mich, Jan 2, 41; m 68; c 2. PHYSICAL OCEANOGRAPHY. *Educ:* Albion Col, AB, 63; Tex A&M Univ, MS, 66, PhD(phys oceanog), 69. *Prof Exp:* Asst prof, 69-74, assoc prof, 74-81, PROF PHYS OCEANOG, UNIV ALASKA, FAIRBANKS, 81- *Concurrent Pos:* Assoc ed, J Geophys Res (oceans); consult, Sci Applns Inc, Int, Dobrocky-Seatech, NSF, Greenhoun & O'Mora, Exxon. *Mem:* Am Geophys Union; Am Meteorol Soc; Sigma Xi; AAAS; Oceanog Soc. *Res:* Ocean circulation, especially the Alaskan Gyre; measurement of currents, water masses and air-sea interactions; long period ocean waves including tsunamis and storm surges. *Mailing Add:* Inst Marine Sci Univ Alaska Fairbanks AK 99775-1080

ROYS, CHESTER CROSBY, b Milwaukee, Wis, Nov 19, 12. INVERTEBRATE PHYSIOLOGY. *Educ:* Univ Mich, BS, 34, MS, 35; Univ Iowa, PhD(zool), 50. *Prof Exp:* Asst zool, Mus Zool, Univ Mich, 35; asst entom, Chicago Natural Hist Mus, 36; independent collector insects, Mex, Cent Am & WIndies, 36-37; vpres, Osborn & Roys, Inc, 38-42; res assoc, 50-78, EMER RES ASSOC BIOL, TUFTS UNIV, 78- *Concurrent Pos:* Rockefeller Found res fel, Univ Sao Paulo, 54. *Mem:* AAAS; Entom Soc Am; Am Soc Zool; Am Inst Biol Sci. *Res:* Sensory physiology, particularly chemical senses of insects and marine invertebrates. *Mailing Add:* PO Box 171 Moline IL 61265-0171

ROYS, PAUL ALLEN, b Evanston, Ill, Apr 3, 26; m 52; c 4. PHYSICS. *Educ:* Ill Inst Technol, BS, 48, MS, 50, PhD(physics), 58. *Prof Exp:* Instr physics, Ill Inst Technol, 51-52; sr scientist, Atomic Power Div, Westinghouse Elec Corp, 52-57; from asst prof to assoc prof physics, Univ Wichita, 57-61; PROF PHYSICS & CHMN DEPT, CARROLL COL, WIS, 61- *Concurrent Pos:* Consult, Beech Aircraft Corp, 58-61 & Dynex Co, 62-65. *Mem:* Am Phys Soc; Am Asn Physics Teachers; Sigma Xi. *Res:* Low energy nuclear physics; mechanics; instrumentation. *Mailing Add:* Dept Physics Carroll Col Waukesha WI 53186

ROYSE, DANIEL JOSEPH, b Olney, Ill, Aug 31, 50; m 72; c 2. PLANT PATHOLOGY, MYCOLOGY. *Educ:* Eastern Ill Univ, BS, 72, MS, 74; Univ Ill, Urbana, PhD(plant path), 78. *Prof Exp:* Res asst plant path, Univ Ill, 74-78, teaching asst, 75; teaching asst air pollution, Nat Univ Bogota, Colombia, 75; from instr to asst prof 78-84, ASSOC PROF PLANT PATH, PA STATE UNIV, 84- *Concurrent Pos:* Int consult to mushroom indust. *Mem:* Am Phytopath Soc; Mycol Soc Am; Am Mushroom Inst. *Res:* Specialty mushrooms; biochemical systematics and selective breeding of Agaricus bisporus and Lentinus edodes; cultivation of Shiitake, Pleurotus and morels on synthetic substrate; edible mushroom cultivation. *Mailing Add:* 316 Buckhout Lab Pa State Univ University Park PA 16802

ROYSTER, JULIA DOSWELL, b Chicago, Ill, Apr 9, 51. AUDIOLOGY & SPEECH PATHOLOGY. *Educ:* Univ NC, Chapel Hill, BA, 72, MS, 75; NC State Univ, PhD, 81. *Prof Exp:* Speech pathologist & audiologist, Savannah Speech & Hearing Ctr, 75-76, Whitaker Rehab Ctr, Forsyth Mem Hosp, Winston-Salem, NC, 76-77 & Hearing & Speech Ctr, NC Mem Hosp, Chapel Hill, 77-78; instr, Dept Psychol, Meredith Col, Raleigh, NC, 79; intern, Gov Advocacy Coun for Persons with Diabetes, Raleigh, NC, 80; vis instr, Dept Psychol & Dept Speech Commun, NC State Univ, Raleigh, 81-82; vis instr, 86-88, ADJ ASST PROF, DIV SPEECH & HEARING SCI, UNIV NC, CHAPEL HILL, 89-; PRES, ENVIRON NOISE CONSULTS, INC, RALEIGH, NC, 77- *Concurrent Pos:* Mem, Tech Comt on Noise, Acoust Soc Am, 87-, prog comt, Nat Hearing Conserv Asn, 90-91; chair, Comt Regional Chap, Acoust Soc Am, 88-, Accredited Standards Comt Bioacoust, Am Nat Standards Inst, S3, 91; consult, numerous co. *Mem:* Fel Acoust Soc Am; Am Speech-Lang-Hearing Asn; Nat Hearing Conserv Asn. *Res:* Author of numerous publications. *Mailing Add:* 4706 Connell Dr Raleigh NC 27612

ROYSTER, L(ARRY) H(ERBERT), b Durham Co, NC, Sept 22, 36; m 57; c 2. ENGINEERING. *Educ:* NC State Univ, BS, 59, PhD(eng), 68. *Prof Exp:* Res engr, NC State Univ, 59-61; sr dynamics engr, N Am Aviation, Inc, 61-64; res engr, 64-67, from instr to prof eng, 67-76, PROF MECH ENG & AEROSPACE ENG, NC STATE UNIV, 76- *Concurrent Pos:* Consult to numerous industs. *Mem:* Fel Acoust Soc Am. *Res:* Vibrations, noise control, hearing conservation. *Mailing Add:* Dept Mech & Aerospace Eng NC State Univ Raleigh NC 27695-7910

ROYSTER, ROGER LEE, b Shelby, NC, Feb 10, 49; m 80. ANESTHESIOLOGY, CRITICAL CARE MEDICINE. *Educ:* Univ NC, Chapel Hill, AB, 71; Bowman Gray Sch Med, MD, 76. *Prof Exp:* Resident internal med, NC Baptist Hosp, 76-78, fel cardiol, 78-80, resident anesthesiol, 80-82; asst prof, 82-86, ASSOC PROF, DEPT ANESTHESIA, BOWMAN GRAY SCH MED, 86- *Concurrent Pos:* Oral bd examr, Am Soc Anesthesiol, 88- *Mem:* Am Heart Asn; Am Col Cardiol; Int Anesthesia Res Soc; Am Soc Anesthesiol. *Res:* Study of inotropic agents in the cardiac surgery paticut; measurements of myocardial blood flow during cardiac surgery; the cardiac electrophysiologic effects of anesthetic agents. *Mailing Add:* Dept Anesthesia Wake Forest Univ Med Ctr 300 S Hawthorne Winston Salem NC 27103

ROYSTER, WIMBERLY CALVIN, b Robards, Ky, Jan 12, 25; m 50; c 2. MATHEMATICS. *Educ:* Murray State Univ, BS, 46; Univ Ky, MA, 48, PhD(math), 52. *Prof Exp:* Asst math, Univ Ky, 46-48, instr, 48-52; asst prof, Auburn Univ, 52-56; from asst prof to assoc prof, Univ Ky, 56-62, chmn dept math, 63-69, dean col arts & sci, 69-72, dean grad sch, 72-88, vchancellor res, 83-88, vpres res, 88-90, PROF MATH, UNIV KY, 62- *Concurrent Pos:* Mem, Inst Advan Study, 62 & Nat Res Coun, 72-75; mem bd dirs, Oak Ridge Assoc Univs, 78-84; mem bd dirs, Coun Grad Schs, 80-85; mem, Grad Rec Exam Bd, 80-85; chair, bd trustees, South-East Consortium Int Develop, 82-83 & Southeastern Univs Res Asn, 90-92. *Mem:* AAAS; Am Math Soc; Math Asn Am. *Res:* Geometric function theory; univalent function theory; expansion in orthonormal functions; approximate methods in difference equations; summability. *Mailing Add:* 512 H Robotics Bldg Univ Ky Lexington KY 40506

ROYSTON, RICHARD JOHN, b Windlesham, Eng, June 18, 31; US citizen; div; c 2. COMPUTER SCIENCE. *Educ:* Oxford Univ, BA, 52, MA, 58; Univ London, BSc, 66. *Prof Exp:* Sci officer appl math, Atomic Energy Res Estab, 52-61; asst mathematician comput sci, Argonne Nat Lab, 62-65; vis scientist, Europ Orgn Nuclear Res, 65-66; assoc mathematician, Argonne Nat Lab, 66-69; sr consult, Scicon Ltd, 69-71; div dir appl math, Argonne Nat Lab, 71-85; VPRES RES, AOI SYSTS INC, 85- *Mem:* Asn Comput Mach; fel Brit Comput Soc; sr mem Inst Elec & Electronics Engrs. *Res:* Networking; pattern recognition; operating systems; vision systems. *Mailing Add:* 68 Baldwin St Apt 22 Charlestown MA 02129-1709

ROYT, PAULETTE ANNE, b Brooklyn, NY, June 14, 45; m 71. MICROBIOLOGY. *Educ:* Am Univ, BS, 67, MS, 71; Univ Md, PhD(microbiol), 74. *Prof Exp:* Fel metab, Nat Heart, Lung & Blood Inst, 75-77; asst prof, 77-83, ASSOC PROF, DEPT BIOL, GEORGE MASON UNIV, 83- *Mem:* Am Soc Microbiol. *Res:* Protein degradation in microorganisms; glucose transport in yeast; iron transport in pseudomonas aeruginose. *Mailing Add:* Dept Biol George Mason Univ 4400 University Blvd Fairfax VA 22030

ROYTBURD, VICTOR, b Moscow, USSR, Nov 28, 45; US citizen; c 2. PARTIAL DIFFERENTIAL EQUATIONS. *Educ:* Moscow State Univ, USSR, MS, 67; Univ Calif, Berkeley, PhD(appl math), 81. *Prof Exp:* Sr researcher, Krzhizhanovsky Energy Inst, Moscow, USSR, 70-79; asst prof, 81-87, ASSOC PROF MATH, RENSSELAER POLYTECH INST, 87- *Concurrent Pos:* Vis mem, Courant Inst Math Sci, 85-86; sr res mathematician, Princeton Univ, 88-89. *Mem:* Soc Indust & Appl Math. *Res:* Partial differential equations of combustion theory and continuum mechanics; numerical methods for these equations. *Mailing Add:* Dept Math Sci Rensselaer Polytech Inst Troy NY 12180-3590

ROZANSKI, GEORGE, b Bronx, NY, Apr 21, 12; m 51; c 3. GEOLOGY. *Educ:* City Col New York, BS, 36; Columbia Univ, AM, 38. *Prof Exp:* Inspector construct, Tenn Valley Authority, 41-42; geologist, Empresa Petrolera Fiscal Peru, 46-48 & US Geol Surv, 49-65; geologist, US Corps Engrs, 65-77; ENG GEOLOGIST & CONSULT, 77- *Mem:* Asn Eng Geologists. *Res:* Engineering geology. *Mailing Add:* 10503 Hutting Pl Silver Spring MD 20902-4953

ROZDILSKY, BOHDAN, b Wola Ceklynska, Ukraine, Nov 22, 16; m 59; c 3. NEUROPATHOLOGY. *Educ:* Univ Lvov, MD, 41; McGill Univ, MSc, 56; Univ Sask, PhD(neuropath), 58; FRCP(C). *Prof Exp:* From assoc prof to prof neuropath, 64-88, res assoc prof, 56-88, EMER PROF NEUROPATH, UNIV SASK, 88- *Concurrent Pos:* Mem staff, Univ Sask Hosp, 57-, neuropathologist, 62-; lectr, Univ Sask, 59-64; consult, Ment Hosps, Sask. *Mem:* Am Asn Neuropath; Am Acad Neurol; fel Am Soc Clin Path; Can Asn Neuropath; Int Acad Path. *Res:* Cerebrovascular permeability; kernicterus; mental subnormality. *Mailing Add:* 515 Lake Crescent Saskatoon SK S7H 3A3 Can

ROZE, ULDIS, b Riga, Latvia, Jan 3, 38; US citizen; m 66; c 1. ECOLOGY. *Educ:* Univ Chicago, BS, 59; Washington Univ, PhD(pharm), 64. *Prof Exp:* From instr to assoc prof, 64-90, PROF BIOL, QUEENS COL, NY, 91- *Mem:* AAAS; Ecol Soc Am; Am Soc Mammalogists; Wilderness Soc. *Res:* Natural history of the porcupine. *Mailing Add:* Dept Biol Queens Col Flushing NY 11367

ROZEBOOM, LLOYD EUGENE, b Orange City, Iowa, Oct 17, 08; m 39; c 2. MEDICAL ENTOMOLOGY. *Educ:* Iowa State Col, SB, 31; Johns Hopkins Univ, ScD(med entom), 34. *Hon Degrees:* DSc, Northwestern Col, 76. *Prof Exp:* Asst med entom, Sch Hyg & Pub Health, Johns Hopkins Univ, 31-34; med entomologist, Gorgas Mem Lab, Panama, 34-37; assoc prof med entom, Okla Agr & Mech Col, 37-39; assoc prof parasitol, 39-58, prof 58-77, EMER PROF MED ENTOM, SCH HYG & PUB HEALTH, JOHNS HOPKINS UNIV, 77- *Concurrent Pos:* Guggenheim fel, 54; exchange prof, Univ Philippines, 54-55. *Honors & Awards:* Ashford Award, 41; John M Belkin Mem Award, Am Mosquito Control Asn, 82. *Mem:* AAAS; fel Am Soc Trop Med & Hyg (pres, 74). *Res:* Taxonomy, genetics and biology of mosquitoes; transmission and epidemiology of arthropod-borne diseases. *Mailing Add:* 3196 Laverne Circle Hampstead MD 21074

ROZEE, KENNETH ROY, b Halifax, NS, Feb 7, 31. MICROBIOLOGY, VIROLOGY. *Educ:* Dalhousie Univ, BSc, 53, MSc, 55, PhD(microbiol), 58; Univ Toronto, dipl bact, 58. *Prof Exp:* Res asst biol, Dalhousie Univ, 55-59, asst prof, 59-62; assoc prof microbiol, Univ Toronto, 62-67; head dept, Dalhousie Univ, 74-84, prof microbiol, 68-; STAFF MEM, DEPT MICROBIOL, LCDC, TUNEY'S PASTURE. *Mailing Add:* Dept Microbiol LCDC Tuney's Pasture Ottawa NS K1A 0L2 Can

ROZELLE, RALPH B, b West Wyoming, Pa, July 9, 32; m 57; c 3. PHYSICAL CHEMISTRY. *Educ:* Wilkes Col, BS, 54; Alfred Univ, PhD(chem), 61. *Prof Exp:* Instr chem, Alfred Univ, 60-62; from asst prof to assoc prof, 62-66, chmn dept chem, 65-70, dir grad studies, 67-72, chmn div natural sci & math, 67-77, PROF CHEM, WILKES COL, 66-, DEAN HEALTH SCI, 74-, PROJ DIR, WILKES-HAHNAMANN COOP MED EVAL PROG FAMILY MED, 77- *Mem:* Am Chem Soc. *Res:* Fuel cells; heterogeneous catalysis; inorganic chemistry of water pollution. *Mailing Add:* Dept Chem Wilkes Col Wilkes-Barre PA 18766

ROZEMA, EDWARD RALPH, b Chicago, Ill, Nov 16, 45; m 74; c 2. APPROXIMATION THEORY, NUMERICAL ANALYSIS. *Educ:* Calvin Col, AB, 67; Purdue Univ, MS, 69, PhD(math), 72. *Prof Exp:* Vis asst prof math, Purdue Univ, Calumet Campus, 72-73; PROF MATH, UNIV TENN, CHATTANOOGA, 73- *Mem:* Math Asn Am; Soc Indust & Appl Math. *Res:* Numerical analysis; methods of teaching. *Mailing Add:* Dept Math Univ Tenn Chattanooga TN 37401

ROZEN, JEROME GEORGE, JR, b Evanston, Ill, Mar 19, 28; m 48; c 3. SYSTEMATIC ENTOMOLOGY. *Educ:* Univ Kans, BA, 50; Univ Calif, Berkeley, PhD, 55. *Prof Exp:* Asst instr biol, Univ Kans, 50, instr, 51; asst, Univ Calif, Berkeley, 51-55; entomologist, USDA, US Nat Mus, 56-58; asst prof entom, Ohio State Univ, 58-60; assoc cur hymenoptera, 60-65, chmn dept entom, 60-71, dep dir res, 72-87, CUR, AM MUS NATURAL HIST, 65- *Concurrent Pos:* Ed pub, Entom Soc Am, 59-60; pres, NY Entom Soc, 64-65. *Mem:* Fel AAAS; Entom Soc Am; Soc Syst Zool; Soc Study Evolution; Orgn Biol Field Stas (pres, 90). *Res:* Ethology, taxonomy and phylogeny of bees; taxonomy, ethology, phylogeny and morphology of Coleoptera larvae; evolution; insect morphology; zoogeography. *Mailing Add:* Am Mus Natural Hist Cent Park W at 79th St New York NY 10024

ROZENBERG, J(UDA) E(BER), b Bocicoul-Mare, Romania, Oct 12, 22; US citizen; m 55; c 4. STRUCTURAL ENGINEERING, CIVIL ENGINEERING. *Educ:* Berchem Tech Sch Com & Admin, Antwerp, dipl, 39; Univ Notre Dame, BSCE, 63, MSCE, 65, PhD(civil eng), 67. *Prof Exp:* Traffic mgr int freight, Reliable Shipping Co, New York, 55-61; asst prof civil eng, Univ Notre Dame, 67-68 & Christian Bros Col, 68-70; assoc prof, 70-79, PROF CIVIL ENG, TENN STATE UNIV, 80- *Concurrent Pos:* Consult, Wright-Patterson AFB Mat Lab, 78-79, Stone & Webster Engr Corp, Boston, Mass, 81-83. *Mem:* Am Soc Civil Engrs; Biblical Archeol Soc; Asn Archit & Eng Israel. *Res:* Systems analysis of structures; composites. *Mailing Add:* 400 N Wilson Blvd Nashville TN 37205

ROZENDAL, DAVID BERNARD, b Chamberlain, SDak, Feb 4, 37; m 57; c 2. CIVIL ENGINEERING. *Educ:* SDak Sch Mines & Technol, BS, 58; Univ Minn, MS, 60; Purdue Univ, PhD(struct eng), 74. *Prof Exp:* Asst prof, 60-65, ASSOC PROF CIVIL ENG, UNIV TEX, EL PASO, 69- *Mem:* Am Soc Eng Educ. *Res:* Stress analysis; structural engineering. *Mailing Add:* Dept Civil Eng Univ Tex El Paso TX 79968

ROZGONYI, GEORGE A, b Brooklyn, NY, Apr 24, 37; m 63; c 3. MATERIALS SCIENCE. *Educ:* Univ Notre Dame, BS, 58, MS, 60; Univ Ariz, PhD(field emission micros), 64. *Prof Exp:* Teaching fel eng sci, Univ Notre Dame, 58-60; instr mech eng, Univ Ariz, 60-61, res assoc, Field Emission Lab, 61-63; mem tech staff struct anal & thin film mat, Bell tel Labs, 63-82; PROF, NC STATE UNIV, 82- *Concurrent Pos:* Electronics ed, J Electrochem Soc, 77-; vis scientist, Max Plank Inst, Stuttgart, WGer, 79-80, Ctr Nat d'Etudes Sci, Microelectronics Lab, Grenoble, France, 81. *Honors & Awards:* Recipient Electronics Div Award, Electrochem Soc, 81. *Mem:* Am Phys Soc; Electrochem Soc; Am Asn Crystal Growth; Mat Res Soc. *Res:* Application of x-ray, electron diffraction and optical microscopy to the study of native and process-induced defects in films and substrates of metals and semiconductors; thin film growth; vacuum physics; crystal defects. *Mailing Add:* Dept Mat Sci & Eng NC State Univ Raleigh NC 27695-7916

ROZHIN, JURIJ, b Kharkiv, Ukraine, USSR, Mar 19, 31; US citizen; m 60; c 1. BIOCHEMISTRY. *Educ:* Wayne State Univ, BS, 55, PhD(biochem), 67. *Prof Exp:* Res assoc biochem, Child Res Ctr Mich, Detroit, 61-63; fel biochem, Med Sch, Wayne State Univ, 67-69, instr, 71-77, asst prof, Dept Biochem, 77-88; CONSULT, 88- *Concurrent Pos:* Res scientist, Mich Cancer Found, 69-81. *Mem:* Am Chem Soc; Sigma Xi; Am Asn Cancer Res. *Res:* Sulfation of steroids in normal and tumor tissue; promotion of carcinogens in the colon; hydrolases in the plasma membrane of tumor cells. *Mailing Add:* Dept Pharmacol Wayne State Univ Med Sch Detroit MI 48201

ROZIER, CAROLYN K, b Fulton, Mo, Aug 17, 44; div; c 1. ANATOMY. *Educ:* Univ Mo, BS, 66; Univ Okla, MS, 71, PhD(anat sci), 72. *Prof Exp:* Staff phys therapist, Gen Leonard Wood Army Hosp, Mo, 66-67; educ coordr & phys therapist, St Anthony Hosp, Oklahoma City, 67-69; asst anat sci, Med Ctr, Univ Okla, 70-72; asst prof anat, Sch Med, Tex Tech Univ, 72-73; DEAN, SCH PHYS THER, TEX WOMAN'S UNIV, 73-, PROVOST, INST HEALTH SCI, 80- *Concurrent Pos:* Res asst, Civil Aeromed Inst, Fed Aviation Admin, Dept Transp, Oklahoma City, 72; instr, Oscar Rose Jr Col, 72. *Mem:* Am Phys Ther Asn. *Res:* Functional and neuroanatomical problems and problems in the field of rehabilitation. *Mailing Add:* 3882 Regent Dallas TX 75229

ROZMIAREK, HARRY, b Pulaski, Wis, Mar 27, 39; m 62; c 4. LABORATORY ANIMAL MEDICINE, GNOTOBIOLOGY. *Educ:* Univ Minn, BS, 60, DVM, 62; Ohio State Univ, MS, 69, PhD(microbiol-immunol), 76. *Prof Exp:* Attend vet, Ft Meyer, Va, 62-64; pvt pract, 64-65; post vet, Ft Wadsworth, NY, 65 & Ft Hamilton, NY, 65-67; chief, vet med & surg, Med Res Lab, Edgewood Arsenal, Md, 69-72; chief, lab animal med, Seato Med Res Lab, Bangkok, Thailand, 72-74; dir, Animal Resources Div, US Army Med Res Inst Infectious Dis, Fredrick, Md, 76-83; DIR, UNIV LAB ANIMAL RESOURCES, OHIO STATE UNIV, 83- *Concurrent Pos:* Consult, Am Asn Accreditation Lab Animal Care, 71-, Anemia & Malnutrit Res Ctr, Chiang Mai Univ, 72-74, US Army Surgeon Gen, 79-83 & US AID, 81-; prog site visitor & consult, Div Res Resources, NIH, 72-; assoc prof, Col Med, Pa State Univ, 76-83; mem, comt care & use of animals, Nat Res Coun, 82-; prof, Ohio State Univ, 83-; vchmn, Am Ans Accreditation Lab Animal Care. *Honors & Awards:* Res Award, Am Asn Lab Animal Sci, 80. *Mem:* Am Asn Lab Animal Sci (pres, 83-84); Am Vet Med Asn; Am Soc Lab Animal Practr(secy-treas, 79-81); Am Col Lab Animal Med; Am Asn Accreditation Lab Animal Care. *Res:* Laboratory animal medicine; spontaneous, zoonotic and latent diseases in laboratory animals; comparative medicine studies using animals as models for human disease and illness; containment of hazardous agents and disease organisms; malaria. *Mailing Add:* 6491 Drexel Rd Philadelphia PA 19151

ROZSNYAI, BALAZS, b Szekesfehervar, Hungary, Nov 24, 29; US citizen. PHYSICS, MATHEMATICS. *Educ:* Eotvos Lorand Univ, Budapest, MS, 52; Univ Calif, Berkeley, PhD(physics), 60. *Prof Exp:* Physicist, Cent Res Inst Physics, Budapest, Hunagry, 52-56 & Int Bus Mach Corp, Calif, 60-64; SR PHYSICIST, LAWRENCE LIVERMORE LAB, UNIV CALIF, 64- *Concurrent Pos:* Asst prof, San Jose State Col, 61-63. *Mem:* Am Phys Soc.

Res: Quantum chemistry; electron-molecular scattering; scattering theory; nuclear physics, especially nuclear shell model; optics; thermodynamics; mathematical physics. *Mailing Add:* 1104 Ave de Las Palmas Livermore CA 94550

ROZZELL, THOMAS CLIFTON, b Gastonia, NC, Apr 5, 37; m 59; c 3. ENVIRONMENTAL BIOLOGY, RADIATION BIOLOGY. *Educ:* Fisk Univ, BS, 59; Univ Cincinnati, MS, 60; Univ Pittsburgh, ScD(environ radiation), 68. *Prof Exp:* Radiochemist, USPHS, 60-65; res assoc environ radiation, Univ Pittsburgh, 66-68, asst prof, 68-71; sci officer, Off Naval Res, 71-86; ASSOC DIR, POSTDOCTORAL PROG, NAT RES COUN, 86- *Concurrent Pos:* Consult, Bur Health, Manpower & Educ, USPHS, 69-71; spec asst to vchancellor health professions, Univ Pittsburgh, 70-71. *Mem:* Health Physics Soc; Int Microwave Power Inst; Bioelectromagnetics Soc; AAAS. *Res:* Radioecology; effect of environmental radiation on man; biological effects of electromagnetic energy, including microwaves, lasers and radio frequency fields. *Mailing Add:* 3422 Barkley Dr Fairfax VA 22031

ROZZI, TULLIO, b Civitanova, Italy, Sept 13, 41; Brit & Ital citizen; m 67; c 2. WAVEGUIDE FIELDS & DISCONTINUITIES, MILLIMETRIC CIRCUITS & ANTENNAS. *Educ:* Pisa Univ, Italy, dottore, 65; Leeds Univ, UK, PhD(elec eng), 68; Bath Univ, DSc, 87. *Prof Exp:* Res scientist, Philips Res Labs, Eindhoven, Holland, 68-78; prof elec eng, Univ Liverpool, UK, 78-81; prof & head electronics, Univ Bath, UK, 81-91; PROF ELEC ENG, UNIV ANCONA, ITALY, 86- *Concurrent Pos:* Vis prof, Dept Elec Eng, Univ Ill, Urbana, 75 & Sch Elec Eng, Bath Univ, UK, 92- *Honors & Awards:* Microwave Prize, Inst Elec & Electronics Engrs, 75. *Mem:* Fel Inst Elec & Electronics Engrs; fel Inst Elec & Electronics Engrs UK. *Res:* Microwave theory and techniques, in particular waveguide fields and discontinuities; integrated optics, planar waveguides; millimetric circuits and antennas. *Mailing Add:* Dept Electronics & Control Fac Eng Univ Ancona Via Brecce Bianche Ancona 60131 Italy

RUBANYI, GABOR MICHAEL, b Budapest, Hungary, Jan 7, 47; m 73; c 2. ANIMAL PHYSIOLOGY, MEDICINE. *Educ:* Semmelweis Med Univ, Hungary, MD, 71; Hungarian Nat Acad Sci, PhD, 80. *Prof Exp:* Vis prof physiol, dept obstet gynecol, Washington Univ, 75-76; assoc prof physiol, dept physiol, Semmelweis Med Univ, 71-82; vis prof physiol, Univ Cincinnati Col Med, 82-83; asst prof, dept physiol & biophys, Mayo Clin & Found, 83-86; head cardiovasc pharmacol, Schering-Plough Co, 86-87; dir pharmacol, Belex Labs Inc, 87-90; DIR, INST PHARMACOL, SCHERING AGR, 90- *Concurrent Pos:* Ed, Int J Cardiol & Am J Physiol, Heart & Circulation, 86-, Hypertension, 89-, J Vascular Med Biol, 90- *Mem:* Am Physiol Soc; fel Am Heart Asn; Int Union Pure & Applied Chem; European Soc Clin Invest; Int Soc Oxygen Transport & Tissue. *Res:* Role of trace metals in ischemic coronary heart diseases; role of endothelium and endothelium-derived vasoactive factors in the normal control and pathophysiology of the cardiovascular system. *Mailing Add:* Berlex Labs Inc 110 E Hanover Ave Cedar Knolls NJ 07927

RUBATZKY, VINCENT E, b New York, NY, Oct 24, 32; m 56; c 2. PLANT PHYSIOLOGY, HORTICULTURE. *Educ:* Cornell Univ, BS, 56; Va Polytech Inst, MS, 60; Rutgers Univ, PhD(plant physiol, biochem, hort), 64. *Prof Exp:* EXTEN VEG CROPS SPECIALIST, UNIV CALIF, DAVIS, 64-, AGRICULTURALIST, COOP EXTEN SERV, 73-, VPRES AGR SCI, VEG CROPS EXTEN, 73- *Mem:* AAAS; Am Soc Hort Sci. *Res:* Herbicidal selectivity; biochemistry of tomato internal browning related to tobacco mosaic virus; spacing relating to mechanization and use of growth regulating chemicals. *Mailing Add:* 3233 Chesapeake Bay Ave Davis CA 95616

RUBBERT, PAUL EDWARD, b Minneapolis, Minn, Feb 18, 37; m; c 3. AERODYNAMICS. *Educ:* Univ Minn, BS, 58, MS, 60; Mass Inst Technol, PhD(aerodynamics), 65. *Prof Exp:* Res scientist aerodynamics, 60-62 & 65-72, SUPVR AERODYNAMICS RES, BOEING CO, 72- *Concurrent Pos:* Assoc ed, J Am Inst Aeronaut & Astronaut, 75-78. *Honors & Awards:* Arch T Colwell Merit Award, Soc Automotive Engrs, 68. *Mem:* Fel Am Inst Aeronaut & Astronaut; Sigma Xi. *Res:* Computational fluid dynamics including hydrodynamics; subsonic, transonic and supersonic flows; viscous and inviscid phenomena; steady and unsteady flows. *Mailing Add:* 20131 S E 23rd Pl Issaquah WA 98027

RUBBIA, CARLO, b Gorizia, Italy, March 31, 34. PHYSICS. *Educ:* Rome Univ, PhD(physics). *Prof Exp:* Prof physics, Harvard Univ, 72-87; mem staff, Lab Particle Physics, Europ Org Nuclear Res, 87-89, GEN DIR, CERN EUROP LAB PARTICLE PHYSICS, 89- *Honors & Awards:* Nobel Prize, 84. *Mem:* Nat Acad Sci. *Mailing Add:* CERN Europ Lab Particle Physics CH 1211 Geneva 23 Switzerland

RUBEGA, ROBERT A, b Blackstone, Mass, Aug 2, 27; m 54; c 8. ACOUSTICS. *Educ:* Univ RI, BS, 51; Univ Rochester, MS, 61, PhD(elec eng), 66. *Prof Exp:* Physicist, US Navy Underwater Sound Lab, 51-56; staff mem res acoust, Gen Dynamics/Electronics, 56-63, mgr acoust labs, 63-65, mgr info sci lab, 65-68; dir marine tech serv, Marine Resources Inc, 68-71; supvr physicist, US Naval Underwater Systs Ctr, 71-87; RETIRED. *Concurrent Pos:* Asst, Univ Rochester, 63-65. *Honors & Awards:* Stromberg Carlson Award, 59. *Mem:* Acoust Soc Am; Inst Elec & Electronics Engrs. *Res:* Acoustics and techniques of signal processing. *Mailing Add:* 19 Woodland Dr E Groton CT 06340

RUBEL, EDWIN W, b Chicago, Ill, May 8, 42; m 63; c 1. NEUROBIOLOGY. *Educ:* Mich State Univ, BS, 64, MA, 67, PhD(psychol), 69. *Prof Exp:* Asst physiol psychol, Mich State Univ, 64-66, res asst neurobiol, 66-68; NIMH fel, Univ Calif, Irvine, 69-71; asst prof psychobiol, Yale Univ, 71-77; ASSOC PROF OTOLARYNGOL & PHYSIOL, UNIV VA MED CTR, 77- *Mem:* AAAS; Am Asn Anatomists; Psychonomic Soc; Sigma Xi. *Res:* Neuroembryology; behavior development; ethology. *Mailing Add:* Dept Otolaryngol Univ Va Med Hosp Box 430 Charlottesville VA 22908

RUBEL, LEE ALBERT, b New York, NY, Dec 1, 28; m 54; c 2. MATHEMATICS. *Educ:* City Col, BS, 50; Univ Wis, MS, 51, PhD(math), 54. *Prof Exp:* Instr math, Cornell Univ, 54-56; mem, Inst Advan Study, 56-58; asst prof, 58-65, mem ctr advan studies, 64-65, 73-74 & 89-90, PROF MATH, UNIV ILL, URBANA, 65- *Concurrent Pos:* NSF fel, 56-58; vis assoc prof, Columbia Univ, 60-62; NSF sr fel, Univ Paris, 65-66; vis prof, Princeton Univ, 67-68, mem inst advan study, 67-68; vis prof, Inst Math Sci, Madras, India, 68; ed proc, Am Math Soc, 72-73; vis prof, Flinders Univ SAustralia, 74; ed, Ill J Math, 73-81, Int J Math & Math Sci, 78, Transactions Ill State Acad Sci, 81 & Am Math Monthly, 87-92. *Mem:* Am Math Soc. *Res:* Complex variables; harmonic and functional analysis; number theory; logic and topology; differential equations. *Mailing Add:* Dept Math Univ Ill 1409 W Green St Urbana IL 61820

RUBEN, JOHN ALEX, b Los Angeles, Calif, Jan 5, 47; m 73. MORPHOLOGY. *Educ:* Humboldt State Col, BS, 68; Univ Calif, Berkeley, MA, 70. *Prof Exp:* ASST PROF ZOOL, ORE STATE UNIV, 75- *Mem:* Soc Vert Paleontologists; AAAS. *Res:* Functional morphology and physiology of reptiles. *Mailing Add:* Dept Zool Ore State Univ Corvallis OR 97331

RUBEN, LAURENS NORMAN, b New York, NY, May 14, 27; m 50; c 3. IMMUNOLOGY. *Educ:* Univ Mich, AB, 49, MS, 50; Columbia Univ, PhD(zool), 54. *Prof Exp:* Asst, Columbia Univ, 50-53; Nat Cancer Inst fel, Princeton Univ, 54-55; from instr to assoc prof, 55-67, chmn dept, 73-87, PROF BIOL, REED COL, 67- *Mem:* AAAS; Am Soc Zool; Soc Develop Biol; Int Soc Develop & Comp Immunol (pres elect, 86); Am Asn Immunologists. *Res:* The evolution and development of humoral immune responses; hapten-carrier immunization of amphibian model systems which represent evolutionary, developmental, anatomical and physiological immunologic progressions. *Mailing Add:* Dept Biol Reed Col Portland OR 97202

RUBEN, MORRIS P, b East Liverpool, Ohio, Aug 27, 19; m 43; c 1. PERIODONTOLOGY, ORAL BIOLOGY. *Educ:* Ohio State Univ, BSc, 40; Loyola Univ, La, DDS, 43; Univ Pa, cert periodont & Boston Univ, cert periodont oral med, 61; Am Bd Periodont, dipl, 66. *Prof Exp:* From asst prof to prof stomatol, 60-73, asst dean, Sch Grad Dent, 61-64, assoc dean grad studies, 69-77, PROF ORAL BIOL, CHMN DEPT & PROF PERIODONT, SCH GRAD DENT, BOSTON UNIV, 73- *Concurrent Pos:* Nat Inst Dent Res fel periodont, Grad Sch Med, Univ Pa, 59-60 & fel stomatol, Sch Med, Boston Univ, 60-61; consult pediat periodont, Kennedy Mem Hosp, Boston; attend dent surgeon, Beth Israel Hosp, Boston. *Mem:* Fel Am Col Dent; Am Acad Periodont; Am Dent Asn; fel Am Acad Dent Sci; fel Int Col Dent. *Mailing Add:* 80 Elinor Rd Newton MA 02161

RUBEN, REGINA LANSING, b Newark, NJ, Jan 2, 50; m 75; c 1. EXPERIMENTAL ONCOLOGY, TISSUE CULTURE. *Educ:* Univ Rochester, BA, 72; Ohio State Univ, MS, 74, PhD(anat), 75. *Prof Exp:* Fel anat, Col Med, Univ Ill, 75-78; sr instr, 78-79, ASST PROF ANAT, HAHNEMANN MED COL, 79- *Mem:* Tissue Culture Asn; Am Asn Anatomists; Sigma Xi; NY Acad Sci. *Res:* Invitro characterization of human tumor cells; assessment of differential invivo susceptibility of mammals in different physiological states to topically applied carcinogen; carcinogen susceptibility of cells cultured from mammals in different phases of their normal physiological cycles. *Mailing Add:* DuPont Co Med Prod Dept BMP 25 2110 PO Box 80026 Wilmington DE 19880-0026

RUBEN, ROBERT JOEL, b New York, NY, Aug 2, 33; m 56; c 4. OTOLARYNGOLOGY, CELL BIOLOGY. *Educ:* Princeton Univ, AB, 55; Johns Hopkins Univ, MD, 59. *Prof Exp:* Dir neurophysiol, Sch Med Johns Hopkins Univ, 58-64; res assoc exp embryol, NIH, 65-68; asst prof otolaryngol, Med Ctr, NY Univ, 66-70; PROF OTOLARYNGOL & CHMN DEPT, ALBERT EINSTEIN COL MED, 70- *Concurrent Pos:* Attend otolaryngol surg, Montefiore & Morrisania Hosps, 70-; attend & chmn, Dept Otolaryngol, Lincoln Hosp, 71-77; chmn dept, Bronx Munic Hosp Ctr, 71-; attend physician, Albert Einstein Col Med, 71- *Mem:* Acoust Soc Am; Am Asn Anat; Am Acad Ophthal & Otolaryngol; Am Otological Soc; Am Asn Hist Med; Sigma Xi. *Res:* Normal and diseased states of the inner ear, particularly physiological, genetic, cellular pathological, embryological and behavior aspects of genetic deafness. *Mailing Add:* 1025 Fifth Ave Apt 12C New York NY 10028

RUBENFELD, LESTER A, b New York, NY, Dec 30, 40; m 64. MATHEMATICAL PHYSICS. *Educ:* Polytech Inst Brooklyn, BS, 62; NY Univ, MS, 64, PhD(math), 66. *Prof Exp:* Fel, Courant Inst Math Sci, NY Univ, 66-67; asst prof, 67-72, ASSOC PROF MATH, RENSSELAER POLYTECH INST, 72- *Mem:* Am Math Soc; Soc Indust & Appl Math. *Res:* Wave propagation in elasticity and electromagnetic theory and asymptotic expansions arising from these fields. *Mailing Add:* Dept Math Rensselaer Polytech Inst Troy NY 12181

RUBENS, SIDNEY MICHEL, b Spokane, Wash, Mar 21, 10; m 44; c 1. PHYSICS. *Educ:* Univ Wash, BS, 34, PhD(physics), 39. *Prof Exp:* Instr physics, Univ Southern Calif, 39-40; res assoc, Univ Calif, Los Angeles, 40-41; from assoc physicist to physicist, US Naval Ord Lab, 41-46; sr physicist, Eng Res Assocs, Inc, 46-51, staff physicist, 51-52; dir physics, Univac Div, Sperry Rand Corp, 52-55, mgr physics dept, 55-60, mgr phys res, 60-61, dir res, 61-66, staff scientist, Defense Systs Div, 66-69, dir spec projs, Univac Fed Systs Div, 69-71, dir spec tech activities, Defense Systs Div, 71-75; CONSULT, 75- *Concurrent Pos:* Mem res & technol adv subcomt instrumentation & data processing, NASA, 67-69; hon fel, Univ Minn, 77- *Honors & Awards:* Meritorious Civilian Serv Award, USN, 45; Info Storage Award, Inst Elec & Electronics Engrs Magnetics Soc, 87. *Mem:* AAAS; Am Phys Soc; Optical Soc Am; Am Geophys Union; fel Inst Elec & Electronics Engrs. *Res:* Plasma physics; solid state; ferromagnetism; thin films; optics; magnetic measurements; digital computer components; information processing. *Mailing Add:* 1077 Sibley Mem Hwy Apt 506 St Paul MN 55118

RUBENSON, J(OSEPH) G(EORGE), b Newburgh, NY, Aug 7, 20; m 43; c 3. ELECTRONIC & SYSTEMS ENGINEERING. *Educ:* City Col New York, BS, 40; Polytech Inst Brooklyn, MEE, 46. *Prof Exp:* Proj engr, Nat Union Radio Corp, 41-46 & Polytech Res & Develop Corp, 46-51; sect head, Airborne Instruments Lab, Inc, 51-58; eng specialist, Sylvania Elec Prod, Inc, Gen Tel & Electronics Corp, 58-60; mgr systs div, Watkins-Johnson Co, 60-64; sr res engr, Stanford Res Inst, SRI Int, Calif, 64-70, dir syst eval dept, 70-71, dir, Washington, 71-78; prog mgr, Syst Planning Corp, San Diego, 78-84, dir, 84-89; RETIRED. *Concurrent Pos:* Instr, Polytech Inst Brooklyn, 46-50. *Mem:* Sigma Xi; Inst Elec & Electronics Engrs. *Res:* Electron tubes; industrial electronics; electronic countermeasures; microwave components; electronic reconnaisance; electronic systems; radar systems. *Mailing Add:* 3826 Liggett Dr San Diego CA 92106-2024

RUBENSTEIN, ABRAHAM DANIEL, b Lynn, Mass, Nov 19, 07; m 37; c 3. PUBLIC HEALTH. *Educ:* Harvard Univ, AB, 28, MPH, 40; Boston Univ, MD, 33; Am Bd Prev Med, dipl. *Prof Exp:* Intern & res med, Brockton Hosp, 33-35; house officer internal med, Boston City Hosp, 35-36; epidemiologist, Mass Dept Pub Health, 37-42, dist health officer, 42-47, dir, Div Hosp Surv & Construct, 47-50, dir, Div Hosps, 50-54, dir, Bur Hosp Facils & dep comnr, 54-69; HOSP CONSULT, 69- *Concurrent Pos:* From asst prof to assoc clin prof, Sch Pub Health, Harvard Univ, 48-60, Simmons Col, 50-59, Boston Col, 50-71 & Sch Med, Tufts Univ, 71-73; sr lectr food sci & nutrit, Mass Inst Technol, 71-74, vis prof, 68-70; pres & trustee, New Eng Sinai Hosp; consult, Jewish Mem Hosp & Beth Israel Hosp, Boston. *Mem:* AMA; Am Pub Health Asn; Am Hosp Asn; Asn Teachers Prev Med; fel Am Col Prev Med. *Res:* Epidemiology of Salmonellosis; medical care programs. *Mailing Add:* 164 Ward St Newton Center MA 02159

RUBENSTEIN, ALBERT HAROLD, b Philadelphia, Pa, Nov 11, 23; m 49; c 2. INDUSTRIAL ENGINEERING, TECHNOLOGY MANAGEMENT. *Educ:* Lehigh Univ, BS, 49; Columbia Univ, MS, 50, PhD(indust eng & mgt), 54. *Prof Exp:* Asst to pres, Perry Equip Corp, 40-43; res assoc indust eng, Columbia Univ, 50-53; asst prof indust mgt, Mass Inst Technol, 54-59; PROF INDUST ENG & MGT SCI, NORTHWESTERN UNIV, 59-, WALTER P MURPHY PROF, 86- *Concurrent Pos:* Ed, Transactions, Inst Elec & Electronics Engrs, 59-85; dir studies col res & develop, Inst Mgt Sci, 60; dir, Narragansett Capital Corp, 60-83; vis prof, Sch Bus, Univ Calif, Berkeley, 64; pres, Int Appl Sci & Technol Assocs, Inc, 77- *Mem:* Fel Soc Appl Anthrop; Inst Elec & Electronics Engrs; Inst Mgt Sci. *Res:* Organization, economics and management of research and development and innovation; field research on organizational behavior. *Mailing Add:* Dept Indust Eng & Mgt Sci Northwestern Univ Evanston IL 60201

RUBENSTEIN, ALBERT MARVIN, b New York, NY, May 9, 18; m 44; c 2. PHYSICS. *Educ:* Brooklyn Col, BA, 39; Univ Md, MS, 49. *Prof Exp:* Lab asst & specifications engr, Micamold Radio Corp, NY, 41-42; unit head submarine radar design, Bur Ships, US Dept Navy, 45-51, mem planning staff in charge of air defense & atomic warfare, Electronics Div, 51-54; head planning staff, 54-56, surveillance warfare syst coordr, Off Naval Res, 56-59; asst chief, Ballistic Missile Defense Br, Adv Res Projs Agency, US Dept Defense, 59-60, actg & asst dir, Ballistic Missile Defense Off, 60-61, asst dir, Ballistic Missile Defense Eng Off, 61-64; tech staff mem, Inst Defense Anal, 64-68; asst dir, Advan Sensor Off, Advan Res Proj Agency, 68-71, dep dir, 71-73; RES STAFF MEM SCI & TECHNOL DIV, INST DEFENSE ANALYSIS, 73- *Concurrent Pos:* Lectr exten div, Univ Md, 53 & Dept Agr grad sch, 54-60; lectr exten div, Univ Va, 58-59. *Honors & Awards:* Civilian Meritorious Awards, US Navy Bur Ships, 54 & Off Naval Res, 59. *Mem:* Am Phys Soc; NY Acad Sci. *Res:* Electronics; radar; air defense; ballistic missile defense systems and space technology; advanced sensors; optics; electro-optics solid state detectors; space systems; reconnaissance and systems analyses. *Mailing Add:* 2709 Navarre Dr Chevy Chase MD 20815

RUBENSTEIN, ARTHUR HAROLD, b Germiston, SAfrica, Dec 28, 37; m 62; c 2. MEDICINE, ENDOCRINOLOGY. *Educ:* Univ Witwatersrand, MBBCh(med), 60. *Prof Exp:* From asst prof to assoc prof, 68-74, assoc chmn dept, 74-81, PROF MED, UNIV CHICAGO, 74-, CHMN DEPT, 81- *Concurrent Pos:* Smith & Nephew fel, Postgrad Sch Med, Univ London, 65-66; Schweppe Award, Schweppe Found, 70-73; estab investr, Am Diabetes Asn, 75; mem adv coun, Nat Inst Health, Arthritis, Metab Diabetes & Digestive Dis, 77-80 & chmn, Nat Diabetes Adv Bd, 81- *Honors & Awards:* Lilly Award, Am Diabetes Asn, 73; Sci Award, Juv Diabetes Found, 77; Banting Medal, Am Diabetes Asn, 83, S Berson Prize, 85. *Mem:* Am Diabetes Asn; master Am Col Physicians; Asn Am Physicians; Endocrine Soc; Am Soc Clin Invest; Am Inst Med; Am Acad Arts & Sci. *Res:* Diabetes mellitus; insulin biosynthesis and secretion; proinsulin; c-peptide; obesity; insulin resistance; immunoassay of hormones and lipoprotein peptides; lipoprotein metabolism. *Mailing Add:* Dept Med Univ Chicago Box 348 Chicago IL 60637

RUBENSTEIN, EDWARD, b Cincinnati, Ohio, Dec 5, 24; m 54; c 3. INTERNAL MEDICINE. *Educ:* Univ Cincinnati, MD, 47. *Prof Exp:* Asst, Univ Cincinnati, 46-47; intern internal med, Cincinnati Gen Hosp, Ohio, 47-48, asst resident, 48-50; asst resident internal med, Ward Serv, Barnes Hosp, 52-53; clin instr, Univ Cincinnati, 53-54; lectr & clin prof med, 54-77, ASSOC DEAN POSTGRAD MED EDUC, SCH MED, STANFORD UNIV, 72-, PROF MED (CLIN), 77- *Concurrent Pos:* Fel metab, May Inst Med Res, Cincinnati, Ohio, 50; head, dept clin physiol, San Mateo County Gen Hosp, 57-70, chief med, 60-70; ed-in-chief, Sci Am Med. *Mem:* Inst Med-Nat Acad Sci; master Am Col Physicians; Am Clin & Climat Asn. *Res:* Synchrotron radiation; thromboembolism. *Mailing Add:* Dept Med Stanford Univ Sch Med Stanford CA 94305

RUBENSTEIN, HOWARD S, b Chicago, Ill, June 14, 31; m 68; c 4. MEDICAL SCIENCES, GENERAL MEDICINE. *Educ:* Carleton Col, BA, 53; Harvard Univ, MD, 57. *Prof Exp:* Intern & resident med, Los Angeles County Gen Hosp, Calif, 57-60; res fel exp surg & bact, Harvard Med Sch, 60-62, Harold C Ernst fel bact, 62-64; res assoc bact, Harvard Univ, 64-65, res assoc path, 65-67, clin instr med & chief allergy clin, Univ Health Serv, 67-89, physician, 66-89; MED CONSULT, DEPT SOCIAL SERV, STATE CALIF, 89- *Concurrent Pos:* Round the world corresp, Lancet, 88-89; mem, Physicians for Human Rights. *Mem:* Fel Am Acad Allergy; Int Physicians Prevention Nuclear War; Am Col Allergists. *Res:* Mechanism of lethal action of endotoxin; differentiating reticulum cell in antibody formation; influence of environmental temperature on resistance to endotoxin; protection antisera afford against endotoxic death; controlling behavior of patients with bronchial asthma and their overcontrolling mothers; ethical problems posed by bee stings; clinical allergy in China versus United States. *Mailing Add:* 2175 Euclid Ave El Cajon CA 92019-2664

RUBENSTEIN, IRWIN, b Kansas City, Mo, Sept 6, 31; m 56; c 3. MOLECULAR BIOLOGY. *Educ:* Calif Inst Technol, BS, 53; Univ Calif, Los Angeles, PhD(biophys), 60. *Prof Exp:* Fel, Johns Hopkins Univ, 60-63; from asst prof to assoc prof molecular biophys, Yale Univ, 63-70; prof genetics & cell biol, 70-88, PROF & HEAD PLANT BIOL, COL BIOL SCI, UNIV MINN, ST PAUL, 88- *Concurrent Pos:* NIH fel, Carnegie Inst Wash, 69-70; mem biol sci comt, World Book Encyclop. *Mem:* AAAS; Biophys Soc; Genetics Soc Am; Am Soc Microbiol; Am Soc Plant Physiologists; Int Soc Plant Molecular Biol; Sigma Xi. *Res:* Molecular biology of plants. *Mailing Add:* Dept Plant Biol Univ Minn St Paul MN 55108-1095

RUBENSTEIN, KENNETH E, organic chemistry, biochemistry, for more information see previous edition

RUBENTHALER, GORDON LAWRENCE, b Gothenburg, Nebr, Sept 30, 32; m 53; c 4. CEREAL CHEMISTRY. *Educ:* Kans State Univ, BS, 60, MS, 62. *Prof Exp:* Phys sci aid, 58-62, cereal technologist, 62-66, res cereal technician, 66-68, RES CEREAL TECHNICIAN IN CHG WHEAT RES, WHEAT QUAL LAB, AGR RES SERV, USDA, 68- *Mem:* Sigma Xi; Am Asn Cereal Chemists; Crop Sci Soc Am. *Res:* Improvement of milling, baking, protein and amino acid balance of experimental wheat varieties; efforts to improve nutritional value and utilization of wheat for human consumption in domestic and foreign market. *Mailing Add:* Wilson Hall Rm 7 Wash State Univ Pullman WA 99163-4004

RUBER, ERNEST, b Berlin, Ger, Aug 21, 34; US citizen; m 55; c 2. ECOLOGY. *Educ:* Brooklyn Col, BA, 59; Rutgers Univ, PhD(zool), 65. *Prof Exp:* Instr biol, Franklin & Marshall Col, 64-65; asst prof zool, Howard Univ, 65-68; assoc prof, 68-80, PROF BIOL, NORTHEASTERN UNIV, 80- *Mem:* AAAS; Am Soc Limnol & Oceanog; Ecol Soc Am; Sigma Xi; Am Mosquito Control Asn. *Res:* Coastal ecology; ecology and taxonomy of salt marsh Microcrustacea; effects of pesticides. *Mailing Add:* Dept Biol Northeastern Univ Boston MA 02115

RUBIN, ALAN, b Philadelphia, Pa, Nov 10, 23; m 47; c 3. MEDICINE. *Educ:* Univ Pa, MD, 47; Am Bd Obstet & Gynec, dipl, 56. *Prof Exp:* Intern, Univ Hosp, 47-48, resident obstet & gynec, 49-51, instr, 51-52, res assoc, 52-72, assoc, 72-79, CLIN ASSOC PROF OBSTET & GYNEC, SCH MED, UNIV PA, 80- *Concurrent Pos:* Res fel pharmacol, Sch Med, Univ Pa, 48-49; Nat Cancer Inst trainee, 51-52; res fel & chief resident, Gynecean Hosp Inst Gynec Res, Univ Pa, 51-52, res assoc, 53-; assoc, Albert Einstein Med Ctr, 55, asst dir, Div Gynec, 68-, actg chmn, Div Obstet & Gynec, 78-80; gynecologist, Grad Hosp Univ Pa, 73-; mem, Am Comn Maternal & Infant Health; dir gynec div, Wills Eye Hosp, 77-80; actg chmn dept gynec, Grad Hosp, 77-78, chmn, 79-; clin prof obstet & gynec, Med Sch, Temple Univ, 81-83. *Mem:* Am Geriat Soc; Am Fertil Soc; Am Soc Human Genetics; Am Col Surgeons; Am Col Obstet & Gynec. *Res:* Obstetrics and gynecology; infertility; cancer. *Mailing Add:* 1905 Spruce St Philadelphia PA 19103

RUBIN, ALAN, b Boston, Mass, Mar 4, 38; m 61; c 2. PHARMACOLOGY. *Educ:* Univ Mass, BS, 59; Univ Wis, PhD(pharmacol), 64. *Prof Exp:* Pharmacologist, Smith Kline & French Labs, 64-66; sr pharmacologist, Midwest Res Inst, 66-68; sr pharmacologist, 68-73, RES PHARMACOLOGIST, LILLY LABS CLIN RES, 73- *Concurrent Pos:* Instr, Dept Pharmacol, Sch Med, Ind Univ, Indianapolis, 68-74, asst prof & assoc mem grad fac, 74-78; adj prof pharmacol, Sch Med, WVa Univ, Morgantown, 86- *Mem:* Am Soc Pharmacol & Exp Therapeut; Am Soc Clin Pharmacol & Therapeut. *Res:* Drug metabolism and disposition; biochemical pharmacology. *Mailing Add:* Lilly Lab for Clin Res Eli Lilly & Co Wm Wishard Mem Hosp 1001 W 10th St Indianapolis IN 46202

RUBIN, ALAN A, b New York, NY, July 10, 26; m 53; c 3. PHARMACOLOGY. *Educ:* NY Univ, BA, 50, MS, 53, PhD(biol), 59. *Prof Exp:* Pharmacologist, Schering Corp, 54-64; dir pharmacol, Endo Labs, 64-71, vpres res, 71-74; dir res, 74-82, dir sci info & technol, 82-87, DIR LICENSING TECHNOL, DU PONT-MERCK PHARMACEUT, 87- *Concurrent Pos:* Ed, Search New Drugs, 72, New Drugs: Discovery & Develop, 78. *Mem:* AAAS; Am Soc Pharmacol & Exp Therapeut; Am Heart Asn; NY Acad Sci; Soc Exp Biol Med. *Res:* Characterization and evaluation of drugs with special emphasis on the cardiovascular system and central and autonomic nervous systems; assessment of biomedical technology. *Mailing Add:* Du Pont-Merck Pharmaceut Barley Mill Plaza Wilmington DE 19880-0025

RUBIN, ALAN BARRY, b Brooklyn, NY, Mar 24, 41; m 63; c 3. ORGANIC CHEMISTRY. *Educ:* Cornell Univ, BA, 62; Univ Rochester, PhD(org chem), 67. *Prof Exp:* Teaching asst org chem, Univ Rochester, 62-66; res chemist, Walter Reed Army Inst Res, Walter Reed Army Med Ctr, 67-69; sect chief, 77-84, BR CHIEF, US ENVIRON PROTECTION AGENCY, 84-; HEAD SPEC CHEM, BRADLEE MED LABS, 69-; CHIEF RES CHEMIST, ADAMS LABS, INC, 69- *Mem:* Am Chem Soc. *Res:* Free radical chemistry; synthesis of antimalarials; medicinal chemistry. *Mailing Add:* 3304 Mill Cross Ct Oakton VA 22124

RUBIN, ALAN J, b Yonkers, NY, Mar 20, 34; m 62; c 1. ENVIRONMENTAL CHEMISTRY, SANITARY ENGINEERING. *Educ:* Univ Miami, BSCE, 59; Univ NC, Chapel Hill, MSSE, 62, PhD(environ chem), 66. *Prof Exp:* Civil engr, Fed Aviation Agency, Tex, 59-60; asst prof environ chem, Univ Cincinnati, 65-68; assoc prof, Water Resources Ctr, 68-74, PROF CIVIL ENG, OHIO STATE UNIV, 74- *Concurrent Pos:* Harry D Pierce vis prof civil eng, Technion-Israel Inst Technol, Haifa, 84. *Mem:* Am Chem Soc; Am Water Works Asn; Water Pollution Control Fedn; Int Asn Water Pollution Res & Control. *Res:* Water chemistry, coagulation and colloid stability; foam separations and flotation; adsorption; aluminum(III) speciation and metal ion hydrolysis, disinfection and inactivation of Naegleria and Giardia cysts by chloramines, chlorine, ozone, chlorinated isocyanurates, iodine and chlorine dioxide, dewatering of inorganic sludges; physical-chemical water and waste treatment. *Mailing Add:* Water Resources Ctr Ohio State Univ 590 Woody Hayes Dr Columbus OH 43210

RUBIN, ALBERT LOUIS, b Memphis, Tenn, May 9, 27; m 53; c 1. INTERNAL MEDICINE. *Educ:* Cornell Univ, MD, 50; Am Bd Internal Med, dipl, 57. *Prof Exp:* Assoc prof med, 59-68, PROF BIOCHEM & SURG, MED COL, CORNELL UNIV, 69-, PROF MED, 76-; DIR, ROGOSIN LABS, NY HOSP CORNELL MED CTR, 63-, DIR ROGOSIN KIDNEY CTR, 71- *Concurrent Pos:* Estab investr, Am Heart Asn, 58-63; attend surgeon, NY Hosp, 69- *Res:* Heart and kidney diseases. *Mailing Add:* Dept Biochem/Surg Rogosin Lab Cornell Univ Med Col 525 E 68th St New York NY 10021

RUBIN, ALLEN GERSHON, b Lewiston, Maine, July 4, 30. SPACE PHYSICS. *Educ:* Boston Univ, PhD(physics), 57. *Prof Exp:* Physicist, Oak Ridge Nat Lab, 57-58 & Williamson Develop Co, 58-63; PHYSICIST, AIR FORCE CAMBRIDGE RES LABS, 63- *Mem:* Am Phys Soc; Am Geophys Union; Sigma Xi; Am Inst Aeronaut & Astronaut. *Res:* Nuclear spectroscopy and reactions; plasma acceleration; laboratory astrophysics; lasers; spacecraft changing; computer simulation codes; computer code validation; space environment. *Mailing Add:* Air Force Geophys Lab Hanscom AFB Bedford MA 01731

RUBIN, ARTHUR I(SRAEL), b New York, NY, Dec 3, 27; m 50; c 2. COMPUTER SCIENCES. *Educ:* City Col New York, BS, 49; Stevens Inst Technol, MS, 53. *Prof Exp:* Physicist, Picatinny Arsenal, Ord Corps, 50-55; appln engr, Electronic Assocs, Inc, NJ, 55-56, supvr, 57-59, dir comput ctr, 59-62; chief automatic comput, Martin Co, 62-67, mgr hybrid comput sci dept, Orlando Div, Martin Marietta Corp, 67-69; sr tech staff consult, Electronic Assocs, Inc, 69, dir comput ctr, 69-70, mgr anal eng dept, 71-77; dir planning, 78-80, DIR PROJ CONTROL, AUTODYNAMICS, INC, 80- *Concurrent Pos:* Chmn libr comt, Simulation Coun, 64-77; mem surv comt, Am Fedn Info Processing Socs, 66-69, secy, 68-69. *Mem:* Inst Elec & Electronics Engrs; sr mem Soc Comput Simulation; assoc fel Am Inst Aeronaut & Astronaut. *Res:* Analog and hybrid computer programming and design; thermal physics. *Mailing Add:* 917 Stuart Rd Princeton NJ 08540

RUBIN, BARNEY, b Kansas City, Mo, Jan 1, 24; m 48, 84; c 2. CHEMICAL ENGINEERING, TECHNICAL MANAGEMENT. *Educ:* Univ Calif, BS, 46, MS, 48, PhD(chem eng), 50. *Prof Exp:* Res chem engr, Radiation Lab, Univ Calif, 50-51 & Calif Res & Develop Co, 51-53; leader, Process & Mat Develop Div, 53-71, sr staff scientist, 71-81, asst assoc dir, 81, communs scientist, 82-83, sr staff scientist, Lawrence Livermore Nat Lab, Univ Calif, 86; RETIRED. *Concurrent Pos:* Consult, Arms Control & Disarmament, The Pentagon, 84-85. *Mem:* Am Chem Soc. *Res:* Computer networking assessment; energy technology research and development assessment. *Mailing Add:* 477 Woodbine Ln Danville CA 94526-2653

RUBIN, BENJAMIN ARNOLD, b New York, NY, Sept 27, 17; m 51. MICROBIOLOGY, EPIDEMIOLOGY. *Educ:* City Col, BS, 37; Va Polytech Inst, MS, 38; Yale Univ, PhD(microbiol), 47. *Prof Exp:* Asst bacteriologist, US War Dept, 40-44; res microbiologist, Schenley Res Inst, Ind, 44-45; microbiol chemist, Off Sci Res & Develop, Yale Univ, 45-47; assoc microbiologist, Brookhaven Nat Lab, 47-52; chief microbiologist, Syntex SAm, Mex, 52-54; asst prof pub health, Col Med, Baylor Univ, 54-60; mgr, Biol Prod Develop Dept, Wyeth Labs, Inc, Am Home Prod, 60-84; PROF MICROBIOL, PHILADELPHIA COL OSTEOP MED, 84- *Honors & Awards:* John Scott Award & Medal, 81. *Mem:* Am Soc Microbiol; Am Asn Immunol; Harvey Soc; Genetics Soc Am; Radiation Res Soc. *Res:* Immunogenetics; virology; chemical and viral carcinogenesis; bioengineering; immunosuppressants; vaccine development; epidemiology and public health; radiobiology; chemotherapy; injection devices; biomed devices. *Mailing Add:* Dept Microbiol & Immunol Philadelphia Col Osteop Med 4150 City Ave Philadelphia PA 19131

RUBIN, BERNARD, b New York, NY, Feb 15, 19; m 45; c 2. PHARMACOLOGY. *Educ:* Brooklyn Col, BA, 39; Yale Univ, PhD(pharmacol), 50. *Prof Exp:* Bact asst respiratory dis, Bur Labs, New York City Health Dept, 40-42, food & drug inspector, Bur Food & Drugs, 44-45; med lab technician path, Marine Hosp, USPHS, 43-44; res asst pharmacol & chemother, Nepera Chem Co, 45-48; sr res group leader pharmacol, squibb inst med res, 50-84; CONSULT LICENSING, BRISTOL-MYERS SQUIBB PHARMACEUT CO, 84- *Mem:* AAAS; Am Soc Pharmacol; Soc Exp Biol & Med; NY Acad Sci; Sigma Xi; Am Heart Asn. *Res:* Bioassay of natural products, as veratrum viride and Rauwolfia; chemotherapeutic agents (isoniazid, etc) and their pharmacological characteristics; anesthetics; analgetics; antihistaminics; biometrics; phenothiazine tranquilizers; central nervous system stimulants; antiserotonins; polypeptides; autonomic, vascular and gastrointestinal pharmacology; (teprotide and captopril) inhibitors of angiotensin-converting enzyme; worldwide surveys of drug candidates available for licensing; preparation of licensing candidate summary profile; special reports on drug candidates in a special category, e.g. antidepressants, hypoglycemics, etc. *Mailing Add:* Two Pin Oak Dr Lawrenceville NJ 08648

RUBIN, BRUCE JOEL, b Brooklyn, NY, Nov 24, 42; m 64; c 2. ELECTROPHOTOGRAPHY. *Educ:* City Univ New York, BEchE, 64; Polytech Univ Brooklyn, MchE, 65; Univ Rochester, MBA, 73; Rochester Inst Technol, MEE, 88. *Prof Exp:* Chemist, 65-73, sr chemist, 73-78, RES ASSOC, EASTMAN KODAK CO, 79- *Mem:* Soc Photog Scientists & Engrs; Am Chem Soc. *Res:* Theory of electrophotographic systems; testing techniques to accurately describe new electrophotographic processes. *Mailing Add:* Ten Chadwick Dr Rochester NY 14618

RUBIN, BYRON HERBERT, b Chicago, Ill, July 25, 43. BIOPHYSICAL CHEMISTRY. *Educ:* Reed Col, BA, 65; Duke Univ, PhD(chem), 71. *Prof Exp:* Fel, Dept Biochem, Duke Univ Med Ctr, 70-73; assoc, Inst Cancer Res, 73-77; asst prof chem, Emory Univ, 77-85; SR RES SCIENTIST, EASTMAN KODAK, 85- *Concurrent Pos:* Assoc prof, Dept Chem, Univ Rochester. *Mem:* Am Chem Soc; Am Crystallog Asn. *Res:* X-ray defraction. *Mailing Add:* Eastman Kodak Res Labs Rochester NY 14650

RUBIN, CHARLES STUART, b Scranton, Pa, Nov 24, 43; m 67. BIOCHEMISTRY. *Educ:* Univ Scranton, BS, 65; Cornell Univ, PhD(biochem), 71. *Prof Exp:* Asst prof neurosci & molecular biol, 72-77, ASSOC PROF MOLECULAR PHARMACOL, ALBERT EINSTEIN COL MED, 78- *Concurrent Pos:* Damon Runyon fel, Albert Einstein Col Med, 70-72; City of New York Health Res Coun career scientist award, 73; NIH res career develop award, 76. *Mem:* AAAS; Am Chem Soc; Am Soc Biol Chemists. *Res:* Enzymology; enzymes involved in the metabolism and action of cyclic AMP; structure and function of membrane-associated enzymes, especially in the central nervous system; biochemical genetics. *Mailing Add:* 168 W Pinebrook Dr New Rochelle NY 10804

RUBIN, CYRUS E, b Philadelphia, Pa, July 20, 21; m 47; c 2. MEDICINE, GASTROENTEROLOGY. *Educ:* Brooklyn Col, AB, 43; Harvard Med Sch, MD, 45. *Prof Exp:* Intern, Beth Israel Hosp, Boston, Mass, 45-46, resident radiol, 50-51; resident med, Cushing Vet Admin Hosp, Framingham, Mass, 48-50; from instr to assoc prof, 54-62, PROF MED, UNIV WASH, 62- *Concurrent Pos:* Dazian fel, Univ Chicago, 51-52, Runyon res fel, 52-53; Nat Cancer Inst career res award, 62-; consult, Univ, Children's Orthop, Vet Admin, King County & USPHS Hosps, Seattle, Wash. *Mem:* Am Soc Clin Invest; Am Soc Cell Biol; Am Fedn Clin Res; Am Gastroenterol Asn; Asn Am Physicians; Sigma Xi. *Res:* Structure and function of the human gastrointestinal tract. *Mailing Add:* RR 104-UH RG-24 Univ Wash Seattle WA 98195

RUBIN, DAVID CHARLES, b Brooklyn, NY, Feb 9, 43; m 68; c 2. HERPETOLOGY, ECOLOGY. *Educ:* Cornell Univ, BS, 63; Ind State Univ, Terre Haute, MA, 65, PhD(systs & ecol), 69. *Prof Exp:* Actg instr biol, Ind State Univ, 66-67; proj writer, interrelated math-sci proj, Broward County, Fla, 69-70; from asst prof to assoc prof, 70-84, PROF BIOL, CENT STATE UNIV, 84- *Concurrent Pos:* Consult interrelated math-sci proj, Broward County, Fla 71 & interdisciplinary environ educ proj, 72; mem Interuniv Comt Environ Qual, Ohio Bd Regents, 72-75; mem adv bd, Ohio Biol Surv, 75-; pres, Cent State Univ chap, Am Asn Univ Profs, 75-77 & 84-86; consult, MacMillian Publ Co, 79-80; consult, State NJ, Dept Educ, 84; chief negotiator, Cent State Univ chap, Am Asn Univ Professors, 88-, mem, nat coun, 88-91. *Mem:* Soc Study Amphibians & Reptiles. *Res:* Systematics of Plethodontid salamanders; amphibian and reptile distribution; amphibian and reptile food habits. *Mailing Add:* Dept Biol Cent State Univ Wilberforce OH 45384

RUBIN, DONALD BRUCE, b Washington, DC, Dec 22, 43; m 75; c 2. STATISTICS. *Educ:* Princeton Univ, AB, 65; Harvard Univ, MS, 66, PhD(statist), 70. *Prof Exp:* Lectr statist, Harvard Univ, 70-71; res statistician, Educ Testing Serv, 71-80, chmn statist group, 75-80; mem staff, Math Res Ctr, Univ Wis, 80-81; prof dept statist & educ, Univ Chicago, 81-84; PROF DEPT STATIST, HARVARD UNIV, 84-, CHMN DEPT, 85- *Concurrent Pos:* Vis lectr statist, Princeton Univ, 71-74; assoc ed, J Am Statist Asn, 75-79, coord & applns ed, 79-82, assoc ed, J Educ Statist, 76-; vis scholar, Univ Calif, Berkeley, 75; vis assoc prof, Univ Minn, 76 & Harvard Univ, 78; Guggenheim fel, 77-78. *Mem:* Fel Am Statist Asn; Biomet Soc; fel Inst Math Statist; Psychometric Soc; fel Int Statist Inst; fel AAAS. *Res:* Inference for causal effects in randomized and non-randomized studies; analysis of incomplete data; non response in sample surveys. *Mailing Add:* Dept Statist Harvard Univ Cambridge MA 02138

RUBIN, DONALD HOWARD, b New York, NY, Feb 3, 48; m 72; c 3. VIRAL PATHOGENESIS, MUCOSAL IMMUNE RESPONSE. *Educ:* State Univ NY, BA, 69; Cornell Univ Med Col, MD, 74. *Prof Exp:* Employee health physician, Boston Lying-In, 78-80; instr med, Harvard Med Sch, 79-80; asst prof, 80-88, ASSOC PROF, MED & MICROBIOL, UNIV PA, 88-; PHYSICIAN, INFECTIOUS DIS, VET AFFAIRS MED CTR, 85- *Concurrent Pos:* Res assoc, Vet Affairs Med Ctr, 86-90. *Mem:* Am Soc Clin Invest; fel Infectious Dis Soc; fel Am Col Physicians; Am Soc Microbiol; Am Soc Virol; Am Asn Appl Psychol. *Res:* Viral pathogenesis; infection of gastrointestinal mucosa and hepatocytes with Reovirus; cellular and molecular analysis of viral pathogenesis; mucosal immune response to viral infection. *Mailing Add:* Vet Affairs Med Ctr Res Med (ISIF) 4100 University Ave Philadelphia PA 19104

RUBIN, EDWARD S, b New York, NY, Sept 19, 41; c 2. MECHANICAL ENGINEERING, ENERGY. *Educ:* City Col New York, BE, 64; Stanford Univ, MS, 65, PhD(mech eng), 69. *Prof Exp:* NSF trainee & res assoc, High Temp Gas Dynamics Lab, Dept Mech Eng, Stanford Univ, 64-68; asst prof mech eng, 69-72, asst prof mech eng & pub affairs, 72-74, assoc prof mech eng & eng & pub policy, 74-79, PROF MECH ENG & ENG & PUB POLICY, CARNEGIE-MELLON UNIV, 79-, DIR, CTR ENERGY & ENVIRON STUDIES, 78- *Concurrent Pos:* Consult, 70-; mem energy systs task group, Gov Energy Coun, Harrisburg, Pa, 74-76; vis mech engr, Ctr Energy Policy Anal, Brookhaven Nat Lab, 75 & 77; mem air & water qual tech adv comt, Pa Dept Environ Resources, 77-; mem US steering comt-coal task force, Int Inst Appl Syst Anal, Laxenburg, Austria, 77-78; vis prof, Dept Physics (Energy Res Group), Cambridge Univ, Eng, 79-80; vis fel, Churchill Col, 79-80; mem comt advan fossil energy technol, Nat Acad Sci/Nat Acad Eng/Nat Res Coun, 83-84, panel on clean coal use, Energy Res Adv Bd, Dept Energy, 84-85, panel on energy resources & environ, Eng Res Bd, Nat Acad Eng/Nat Res Coun, 84-85, Nat Air Pollution Control Tech Adv Comt, Environ Protection Agency, 85-89, Work Group on Clean Coal Technol, Nat Coal Coun, 88. *Mem:* Am Soc Mech Engrs; Air Pollution Control Asn; AAAS. *Res:* Environmental impacts of coal conversion and utilization; modelling of energy and environmental systems; air quality management; technology assessment and public policy. *Mailing Add:* 128A Baker Hall Carnegie-Mellon Univ 5000 Forbes Ave Pittsburgh PA 15213

RUBIN, EMANUEL, b New York, NY, Dec 5, 28; m 55; c 4. PATHOLOGY. *Educ:* Villanova Univ, BS, 50; Harvard Univ, MD, 54. *Prof Exp:* Asst pathologist, Col Physicians & Surgeons, Columbia Univ, 61-64; assoc attend pathologist, Mt Sinai Hosp, 62-68, attend pathologist, 69-77, prof path, Mt Sinai Sch Med, 66-77, chmn dept, 72-76; PROF & CHMN, DEPT PATH & LAB MED, HAHNEMANN MED COL & HOSP, 77- *Concurrent Pos:* Consult, Path B Study Sect, NIH & Cardiomyopathies Study Sect, Nat Heart, Lung & Blood Inst, 81-86; ed, Lab Invest, 81- *Mem:* Int Acad Path; Am Soc Biol Chemists; Am Asn Path; Am Col Physicians; Am Asn Study Liver Dis. *Res:* Function and structure in human and experimental liver disease; effects of ethanol on the liver, heart and other organs. *Mailing Add:* Dept Path Thomas Jefferson Univ Jefferson Med Col 1020 Locust St Suite 271 Philadelphia PA 19107

RUBIN, G A, b Leipzig, Ger, Dec 30, 26; m 55; c 3. PHYSICS. *Educ:* Univ Heidelberg, BA, 46; Univ Saarbrucken, dipl, 53, PhD(physics), 55. *Prof Exp:* Instr physics, Univ Saarlandes, 53-56; res scientist, Roechling Steelworks, 56-57; proj group leader mat res, Union Carbide Corp, 57-60; assoc prof physics, Clarkson Tech, 60-65; sr physicist, IIT Res Inst, 65-67; PROF PHYSICS & CHMN DEPT, LAURENTIAN UNIV, 67- *Concurrent Pos:* Consult, Steinzeug A G, Ger, 62-64; NASA & Nat Res Coun Can grants, 65-; consult, IIT Res Inst, 67-72. *Mem:* Am Phys Soc; Am Ceramic Soc; NY Acad Sci; Can Phys Soc; Ger Phys Soc. *Res:* Acoustic emissions from particulate matter; surface and fine particle physics; charge transportion in solids; ultrasonic fragmentation of rock. *Mailing Add:* Dept Physics & Astron Laurentian Univ Ramsey Lake Rd Sudbury ON P3E 2C6 Can

RUBIN, GERALD M, b Mar 31, 50. BIOCHEMISTRY. *Educ:* Mass Inst Technol, BS, 71; Univ Cambridge, Eng, PhD(molecular biol), 74. *Prof Exp:* Helen Hay Whitney Found fel, Dept Biochem, Sch Med, Stanford Univ, 74-76; asst prof biol chem, Sidney Farber Cancer Inst, Harvard Med Sch, 77-80; staff mem, Dept Embryol, Carnegie Inst Washington, Baltimore, Md, 80-83; JOHN D MACARTHUR PROF GENETICS, DEPT MOLECULAR & CELL BIOL, UNIV CALIF, BERKELEY, 83-, HEAD, DIV GENETICS, 87-; INVESTR, HOWARD HUGHES MED INST, 87- *Concurrent Pos:* Tutor, Bd Tutors in Biomed Sci, Harvard Col, 77-80; assoc ed, Develop Biol, 82-85, Cell, 85-, J Neurosci, 86-, Neuron, 88-; mem bd sci counrs, Nat Inst Environ Health Sci, 84-87; adj prof, Dept Biochem & Biophys, Sch Med, Univ Calif, San Francisco, 87-; mem, Vis Comt, Dept Biochem & Molecular Biol, Harvard Univ, 88-91, Dept Biol, Mass Inst Technol, 88-, Univ Calif, Davis, 89; mem, Comt Drosophila Info Serv, Genetics Soc Am, 89-; ed, Develop, 91- *Honors & Awards:* Passano Found Young Scientist Award, 83; Newcomb Cleveland Prize, AAAS, 84; Eli Lilly Award in Biol Chem, Am Chem Soc, 85; US Steel Found Award in Molecular Biol, Nat Acad Sci, 85; Genetics Soc Am Medal, 86. *Mem:* Nat Acad Sci. *Res:* Molecular and cell biology. *Mailing Add:* Dept Molecular & Cell Biol Univ Calif 539 Life Sci Addn Bldg Berkeley CA 94720

RUBIN, HARRY, b New York, NY, June 23, 26; m 52; c 4. CELL BIOLOGY. *Educ:* Cornell Univ, DVM, 47. *Hon Degrees:* DSc, Guelph, 87. *Prof Exp:* Veterinarian, Foot & Mouth Dis Proj, Mex, 47-48; sr asst veterinarian, Virus Lab, USPHS, Ala, 48-52; res fel virol, Nat Found Infantile Paralysis, Biochem & Virus Lab, Calif, 52-53; res fel, Biol Div, Calif Inst Technol, 53-55, sr res fel, 55-58; assoc prof, 58-60, PROF VIROL, UNIV CALIF, BERKELEY, 60-, RES VIROLOGIST MOLECULAR BIOL, 73-, PROF MOLECULAR & CELL BIOL, 88- *Concurrent Pos:* Dyer lectr, NIH, 64; Harvey lectr, NY Acad Med, 66. *Honors & Awards:* Rosenthal Award, AAAS, 59; Eli Lilly Award, 61; Lasker Award, 64. *Mem:* Nat Acad Sci; Am Acad Arts & Sci. *Res:* Cell growth regulation and malignancy. *Mailing Add:* Dept Molecular & Cell Biol Stanley Donner ASU Univ Calif Berkeley CA 94720

RUBIN, HARVEY LOUIS, b San Diego, Calif, Apr 28, 14; m 41; c 1. VETERINARY MICROBIOLOGY, PATHOLOGY. *Educ:* Auburn Univ, DVM, 39; Univ Ky, MS, 40; Johns Hopkins Univ, MPH, 52; Am Col Vet Prev Med, dipl. *Prof Exp:* Sr technician, Univ Ky, 40; instr vet anat & histol, Agr & Mech Col, Tex, 40-42; instr, Vet Corps, US Army, 42-68; dir, Animal Dis Diag Lab, Live Oak, Fla, 68-77; CHIEF, BUR DIAG LABS, FLA DEPT AGR, 77- *Concurrent Pos:* Adj prof, Col Vet Med, Univ Fla, 77- *Honors & Awards:* EP Pope Award, Am Asn Vet Lab Diag, 88. *Mem:* Am Vet Med Asn; Am Asn Vet Lab Diagnosticians; Am Col Vet Toxicol; Sigma Xi; Am Leptospirosis Res Conf. *Res:* Salmonella in hogs; immunology; leptospirosis; zoonoses. *Mailing Add:* Animal Dis Diag Lab PO Box 420460-0460 Kissimmee FL 32742-0460

RUBIN, HERBERT, b New York, NY, Oct 9, 23; wid; c 2. PHYSICAL CHEMISTRY. *Educ:* NY Univ, BA, 44; Polytech Inst Brooklyn, MS, 49, PhD(phys chem), 53. *Prof Exp:* Res assoc radiation chem, Syracuse Univ, 53-54; sr chemist, Ozalid Div, Gen Aniline & Film Corp, 54-56; sr res engr, Schlumberger Well Surv Corp, 56-63; specialist phys chem, Repub Aviation Corp, 63-64; sr res chemist, Reaction Motors Div, Thiokol Chem Corp, 64-66; prin scientist, Inmont Corp, 66-75, res assoc, 75-87; RETIRED. *Mem:* AAAS; Am Chem Soc; Sigma Xi. *Res:* Wide-line nuclear magnetic resonance; electrochemistry; photoreproductive processes; photopolymerization and ultraviolet absorption of inks and coatings; ultraviolet lamp photometry; solvent-polymer solution behavior. *Mailing Add:* 35 Byron Pl Livingston NJ 07039

RUBIN, HERBERT, b New York, NY, June 11, 30; m 57; c 3. COMMUNICATIONS, SPEECH PATHOLOGY. *Educ:* Queens Col, NY, BA, 51; Brooklyn Col, MA, 55; Harvard Univ, PhD(psychol), 59. *Prof Exp:* Clin coord speech & audiol, Univ Pittsburgh, 59-65, assoc prof speech, 65-72, dir commun prog, 70-74, PROF SPEECH, UNIV PITTSBURGH, 72-, ASSOC DEAN GRAD FAC ARTS & SCI, 78-, PROF COMMUN. *Concurrent Pos:* Asst dean, Grad Fac Arts & Sci, Univ Pittsburgh, 73-78. *Mem:* Acoust Soc Am; fel Am Speech & Hearing Asn; Speech Commun Asn; Am Psychol Asn; Sigma Xi. *Res:* Psychoacoustics; psycholinguistics; language development; self-communication. *Mailing Add:* 4720 Bayard St Pittsburgh PA 15213

RUBIN, HERMAN, b Chicago, Ill, Oct 27, 26; m 52; c 2. STATISTICS, MATHEMATICS. *Educ:* Univ Chicago, BS, 44, MS, 45, PhD(math), 48. *Prof Exp:* Res asst, Cowles Comn Res Econ, Chicago, 44-46, res assoc, 46-47 & 48-49; asst prof statist, Stanford Univ, 49-55; assoc prof math, Univ Ore, 55-59; prof statist, Mich State Univ, 59-67; PROF STATIST, PURDUE UNIV, LAFAYETTE, 67- *Mem:* AAAS; Am Math Soc; Math Asn Am; fel Inst Math Statist. *Res:* Mathematical statistics; statistical decision theory; mathematical logic and set theory; stochastic processes. *Mailing Add:* Dept Statist Purdue Univ West Lafayette IN 47907

RUBIN, HOWARD ARNOLD, b Baltimore, Md, Jan 4, 40; m 65; c 2. PHYSICS. *Educ:* Mass Inst Technol, SB, 61; Univ Md, College Park, PhD(physics), 67. *Prof Exp:* Asst prof, 66-72, assoc prof, 72-88, PROF PHYSICS, ILL INST TECHNOL, 88- *Mem:* Am Phys Soc. *Res:* High energy physics experiment. *Mailing Add:* Dept Physics Ill Inst Technol 3300 S Fed St Chicago IL 60616

RUBIN, ISAAC D, b Krakow, Poland, Nov 8, 31; US citizen; m 57; c 1. POLYMER CHEMISTRY, ORGANIC CHEMISTRY. *Educ:* Brooklyn Col, BS, 54; Brooklyn Polytech Inst, MS, 57, PhD(polymer chem), 61. *Prof Exp:* Chemist, Acralite Co, Inc, 54-56, plant mgr, 56-57; develop chemist, Gen Elec Co, 60-62; sr chemist, Texaco Inc, 62-64, res chemist, 64-67, group leader, 67-78, technologist, 78-82, res mgr, 82-85, coordr, 85-87, sr res assoc, 87-88, fel, 88-89, RES FEL, TEXACO INC, 90- *Mem:* NY Acad Sci; Am Chem Soc. *Res:* Polymerization mechanisms; structure-property relationships in polymers; lubricants, fuels, energy and petrochemicals; technical writing; science and technology. *Mailing Add:* Texaco Inc PO Box 509 Beacon NY 12508

RUBIN, IZHAK, b Haifa, Israel, May 22, 42; m 65; c 3. ELECTRICAL ENGINEERING. *Educ:* Israel Inst Technol, BS, 64, MS, 68; Princeton Univ, PhD(elec eng), 70. *Prof Exp:* Electronics engr, Israel Aircraft Industs, 67-68; res asst elec eng, Princeton Univ, 69-70; PROF ENG & APPL SCI, UNIV CALIF, LOS ANGELES, 70-; PRES, IRI CORP, 80- *Concurrent Pos:* Off Naval Res grant, Univ Calif, Los Angeles, 72-, NSF res grant, 75-, Naval Res Lab grant, 78-; actg chief scientist, Xerox Telecoms Network, 79-80; consult, Comput Commun Networks, 70-, C3 Systems and Networks, 80- *Mem:* Fel Inst Elec & Electronics Engrs. *Res:* Data and computer communication networks; satellite communication networks; queueing systems; network flows; communication and information theory; stochastic processes; C3 systems and networks. *Mailing Add:* Elec Eng Dept Univ Calif Eng IV 58-115 Los Angeles CA 90024-1594

RUBIN, JACOB, b Wloclawek, Poland, Feb 1, 19; US citizen; m 43; c 3. SOIL PHYSICS, HYDROLOGY. *Educ:* Univ Calif, Berkeley, BS, 42, PhD(soil sci), 49. *Prof Exp:* Soil physicist, Agr Res Sta, Rehovoth, Israel, 50-61, chmn div irrig & soil physics, 56-58, dir inst soils & water, 59-61; adv solute transport, 74-76, SOIL PHYSICIST, WATER RESOURCES DIV, US GEOL SURV, 62- *Concurrent Pos:* Adj lectr, Israel Inst Technol, 52-55, adj sr lectr, 56-58, adj assoc prof, 59-61; external lectr, Fac Agr, Hebrew Univ, Israel, 52-61; leader, UN Proj Tech Assistance to Cyprus Govt, 57; consult, Govt Israel, 57-61 & Italian Ministry Agr, 60; consult prof, Stanford Univ, Calif, 75- *Honors & Awards:* O E Meinzer Award, Geol Soc Am, 77. *Mem:* AAAS; Soil Sci Soc Am; Am Geophys Union; Sigma Xi. *Res:* Movement of water and solutes in the unsaturated zone; flow of fluids and transport of reacting solutes in porous media; evapotranspiration; irrigation practices; soil physical properties in relation to plant growth. *Mailing Add:* 3247 Murray Way Palo Alto CA 94303

RUBIN, JEAN E, b New York, NY, Oct 29, 26; m 52; c 2. MATHEMATICS. *Educ:* Queens Col, NY, BS, 48; Columbia Univ, MA, 49; Stanford Univ, PhD(math), 55. *Prof Exp:* Instr math, Queens Col, NY, 49-51; instr, Stanford Univ, 53-55; lectr, Univ Ore, 55-59; asst prof, Mich State Univ, 61-68; assoc prof, 68-75, PROF MATH, PURDUE UNIV, 75- *Concurrent Pos:* Reviewer math reviews, 65-79; vis lectr, Math Asn Am, 76-86. *Mem:* Am Math Soc; Asn Symbolic Logic; Math Asn Am. *Res:* Set theory; axiom of choice; cardinal and ordinal numbers; models of set theory; textbook on undergraduate logic; two advanced monographs on set theory. *Mailing Add:* Dept Math Purdue Univ West Lafayette IN 47907

RUBIN, JOEL E(DWARD), b Cleveland, Ohio, Sept 5, 28; m 53; c 3. ELECTRICAL ENGINEERING. *Educ:* Case Inst Technol, BS, 49; Yale Univ, MFA, 51; Stanford Univ, PhD(theatre eng), 60. *Prof Exp:* VPRES ILLUMINATING ENG, KLIEGL BROS LIGHTING, LONG ISLAND CITY, 54- *Concurrent Pos:* Mem comt theatre eng, US Inst Theatre Technol, 60- *Honors & Awards:* Founder's Award, US Inst Theatre Technol, 72. *Mem:* US Inst Theatre Technol; Illum Eng Soc; Int Orgn Scenographers & Theatre Technicians (pres, 71-79); fel Am Theatre Asn. *Res:* Theatre technology and illuminating engineering. *Mailing Add:* 24 Edgewood Ave Hastings-on-Hudson NY 10706

RUBIN, KARL C, b Urbana, Ill, Jan 27, 56. ALGEBRAIC NUMBER THEORY, ARITHMETIC OF ELLIPTIC CURVES. *Educ:* Princeton Univ, AB, 76; Harvard Univ, MA, 77, PhD(math), 81. *Prof Exp:* Vis res fel, Princeton Univ, 81-82, instr math, 82-83; prof, Columbia Univ, 88-89; asst prof, 84-87, PROF MATH, OHIO STATE UNIV, 87- *Concurrent Pos:*

Alfred P Sloan fel, 85; NSF presidential young investr, 88; vis scholar, Harvard Univ, 90-91. *Mem:* Am Math Soc. *Res:* Arithmetic of elliptic curves; Iwasawa theory; special values of L-functions, especially questions related to the Birch and Swinnerton-Dyer conjecture. *Mailing Add:* Dept Math Ohio State Univ Columbus OH 43210

RUBIN, KENNETH, b New York, NY, Feb 28, 28; m 51; c 3. PHYSICS. *Educ:* NY Univ, BS, 50, MS, 52, PhD(physics), 59. *Prof Exp:* Asst panel electron tubes, Res Div, NY Univ, 52-53 & Radiation Lab, Columbia Univ, 53-54; instr physics, City Col New York, 54-55; from instr to asst prof, NY Univ, 55-64; assoc prof, 64-72, PROF PHYSICS, CITY COL NEW YORK, 72- *Mem:* Am Phys Soc. *Res:* Atomic beam resonance and atomic scattering experiments utilizing atomic beam techniques. *Mailing Add:* Dept Physics City Col New York Convent & 138th St New York NY 10031

RUBIN, LAWRENCE G, b Brooklyn, NY, Sept 17, 25; m 51; c 3. PHYSICS, LOW TEMPERATURE THERMOMETRY. *Educ:* Univ Chicago, BS, 49; Columbia Univ, MA, 50. *Prof Exp:* Staff mem solid state physics, Res Div, Raytheon Co, Mass, 50-58, mgr instrumentation group, 58-64; GROUP LEADER INSTRUMENTATION & OPERS, NAT MAGNET LAB, MASS INST TECHNOL, 64-, DIV HEAD, 77- *Concurrent Pos:* Mem adv panel, Nat Acad Sci, Nat Bur Standards, 77-82, 85-90; chmn & organizer, Instrument & Measurement Sci Group, Am Phys Soc, 85; chmn Physics Today, Buyers Giude Adv Comt, 87- *Mem:* Fel Am Phys Soc; fel Inst Elec & Electronics Engrs; Instrument Soc Am; Am Vacuum Soc. *Res:* Instrumentation in electronics, vacuum, cryogenics and visible and infrared spectrometry; temperature measurement and control; bulk properties of semiconductors; low level signal processing; high field magnetometry. *Mailing Add:* Nat Magnet Lab Bldg NW14 Mass Inst Technol 170 Albany St Cambridge MA 02139

RUBIN, LEON E, b Winthrop, Mass, Apr 24, 21; m 47; c 3. ANALYTICAL CHEMISTRY. *Educ:* Mass Inst Technol, BS, 42; Boston Univ, MS, 47; Columbia Univ, PhD(chem), 51. *Prof Exp:* Res assoc phys org chem, Columbia Univ, 51-52; scientist, 52-65, res group leader, 65-67, MGR ANALYTICAL CHEM, POLAROID CORP, 67- *Mem:* Am Chem Soc. *Res:* Spectrophotometry; titrimetry; chromatography; absorption spectroscopy; kinetics of reactions; photographic chemistry; polymer analysis. *Mailing Add:* 40 Lafayette Rd Newton MA 02162-1017

RUBIN, LEON JULIUS, b Poland, Nov 22, 13; Can citizen; m 71; c 2. FOOD SCIENCE, FOOD ENGINEERING. *Educ:* Univ Toronto, BASc, 38, MASc, 39, PhD(org chem), 45. *Prof Exp:* Lectr chem eng, Univ Toronto, 39-45; res chemist, Res Ctr, Can Packers Inc, 45-49, res dir, 49-78; prof, 79-85, EMER PROF FOOD ENG, DEPT CHEM & APPL CHEM, UNIV TORONTO, 85- *Concurrent Pos:* Mem, Can Comt Food, Expert Comt Meats & chmn, Indust/Govt Nitrites & Nitrosamines; working group res mgt, Int Union Food Sci & Technol; organizing comt, World Cong Food Sci & Technol, 91. *Honors & Awards:* Montreal Medal, Chem Inst Can, 70, Charles Honey Award, 75; William J Eva Award, Can Inst Food Sci & Technol, 78; John Labatt Award, Chem Inst Can, 80; Sci & Technol Award, Can Meat Coun, 91. *Mem:* Am Chem Soc; fel Inst Food Technologists; fel Chem Inst Can; fel Can Inst Food Sci & Technol; Am Oil Chemists Soc; Am Meat Sci Asn. *Res:* Food chemistry; food processes and products; food additives; lipid chemistry; oilseed processing, meat curing; chemistry of meat flavor. *Mailing Add:* Dept Chem Eng Univ Toronto Toronto ON M5S 1A4 Can

RUBIN, LEONARD ROY, b Brooklyn, NY, Nov 16, 39; m 60; c 2. TOPOLOGY. *Educ:* Tulane Univ, BS, 61; Univ Miami, MS, 63; Fla State Univ, PhD(math), 65. *Prof Exp:* From asst prof to assoc prof, 67-78, interim dept chmn, 83-85, PROF MATH, UNIV OKLA, 78- *Concurrent Pos:* Staff mem, Okla Univ Res Inst, 68; vis assoc prof, Univ Utah, 73-74. *Honors & Awards:* Fulbright lectr, Univ Zagreb, Yugoslavia, 85-86. *Mem:* Am Math Soc. *Res:* Geometric topology and dimension theory. *Mailing Add:* Dept Math Univ Okla 601 Elm Ave Rm 423 Norman OK 73019

RUBIN, LEONARD SIDNEY, b New York, NY, Aug 27, 22; m 50; c 3. PSYCHOPHYSIOLOGY, PSYCHOPHARMACOLOGY. *Educ:* NY Univ, PhD(neurosci), 51. *Prof Exp:* Asst instr biochem, NY Med Col, 44-45; tutor psychol, City Col New York, 45; instr, Wash Sq Col, NY Univ, 47-50, res assoc col eng, 50-53; chief psychol & human eng br, Med Labs, US Chem Corps, 53-57; head psychobiol, Eastern Pa Psychiat Inst, 57-80; Assoc prof psychait, Sch Med, Univ Pa, 65-81, PROF PHYSIOL & PHARM, PHILADELPHIA COL OSTEOP, 80- *Concurrent Pos:* Adj prof, Sch Med, Temple Univ, 70-74; foreign exchange scholar, Nat Acad Sci, Yugoslavia, 74. *Honors & Awards:* Soc Sci Res Coun Award, Inst Math, Stanford Univ, 55; Hon Award, Soc Biol Psychiat, 59. *Mem:* Fel Acad Psychosomat Med; Soc Biol Psychiat; fel Am Psychol Asn; Am Psychopathol Asn; Soc Psychophysiol Res; Am Physiol Soc. *Res:* Autonomic neurohumoral concomitants of schizophrenia; other behavior disorders; chronic alcoholism; drug abuse; migraine. *Mailing Add:* 706 Powder Mill Lane Wynnewood PA 19096

RUBIN, LOUIS, b New York, NY, Aug 19, 22; m 46; c 2. CHEMISTRY. *Educ:* NY Univ, AB, 43; Univ Mo, AM, 47, PhD(chem), 50. *Prof Exp:* Asst chem, Univ Mo, 47-49; chemist, Winthrop Labs, 51-52, dir, Process Develop Lab, 52-59; sr chemist, Rohr Aircraft Corp, 59-61, chief tech staff, 61-64; mem tech staff, 64-75, STAFF SCIENTIST, AEROSPACE CORP, 75- *Concurrent Pos:* Lectr, Univ Calif, Los Angeles & Exten Div, 63-64; pres, Southern Calif Sect, Soc Plastics Engrs, 76-77. *Mem:* Am Chem Soc; Am Inst Aeronaut & Astronaut; Soc Plastics Engrs; Am Ceramic Soc; Soc Advan Mat & Process Eng. *Res:* Organic chemistry; reinforced plastics; adhesives; polymer chemistry; ablative materials; hypervelocity erosion; metal-matrix composites; carbon-carbon composites. *Mailing Add:* 2502 E Willow St Signal Hill CA 90806

RUBIN, MARTIN ISRAEL, b New York, NY, Nov 2, 15; m 42; c 4. BIOCHEMISTRY. *Educ:* City Col New York, BS, 36; Columbia Univ, PhD(org chem), 42. *Hon Degrees:* DSc, Univ Louis Pasteur, France, 76. *Prof Exp:* Herenscheim fel, Mt Sinai Hosp, 39-40; res chemist, Wallace & Tiernan Prod, Inc, NJ, 41-46 & Schering Corp, 46-48; prof bioclin chem, Med Sch, 48-52, sch med dent, 58-81, PROF BIOCHEM, GRAD SCH, 52-, EMER PROF, GEORGETOWN UNIV, 81- *Concurrent Pos:* US Chem; consult, WHO, Pan Am Health Orgn, Food & Drug Admin & NIH Comn Clin Chem, Int Union Pure & Appl Chem; pres, Int Fedn Clin Chem. *Honors & Awards:* Smith Kline & French Award, Fisher Award, Roe Award & Gemlot Award, Am Asn Clin Chemists. *Mem:* AAAS; Am Chem Soc; Soc Exp Biol & Med; Am Asn Clin Chem; Am Rheumatism Asn. *Res:* Synthetic drugs; use of radioisotopes in biochemistry and medicine; steroid chemistry; clinical chemistry; synthesis of unsaturated lactones related to the cardiac aglycones; chelate compounds; mineral metabolism. *Mailing Add:* 3218 Pauline Dr Georgetown Univ Sch Med & Dent Washington DC 20015

RUBIN, MARYILN BERNICE, b St Peter, Ill. GENERAL PHYSIOLOGY, NURSING. *Educ:* Wash Univ, BSN, 59, MSN, 60; Southern Ill Univ, Carbondale, PhD(physiol), 71; Lutheran Hosp Sch Nursing, dipl nursing, 54. *Prof Exp:* Staff nurse, instr & supvr, Operating Rm nursing, Lutheran Hosp, St Louis, Mo, 54-57, instr, Med-Surg Nursing, Team Nursing, 60-61, curric head, Med-Surg Nursing, 65-67; instr, med-surg nursing, grad & undergrad, Wash Univ, St Louis, Mo, 61-65, asst dean, Undergrad Nursing Prog, 62-65, asst prof, Med-Surg Nursing Grad Prog, 68-69; teaching asst physiol, Southern Ill Univ, Carbondale, Ill, 67-68, res asst, 69-70; lectr, Med-Surg Nursing Undergrad Prog, Southern Ill Univ, Edwardsville, Ill, 69-70, asst prof, 70-71; chairperson, Sect Med-Surg Nursing, St Louis Univ, 71-77, assoc prof, Med-Surg Nursing, Grad Prog, 71-75, actg chairperson, 75, coordr, Grad Med-Surg Nursing, 71-85, PROF, SCH NURSING, ST LOUIS UNIV, ST LOUIS, MO, 75-, DIR RES, 85- *Concurrent Pos:* Consult, Ind State Univ, 84-87; appointee, Bd Sci Counselor, Nat Inst Occup Safety & Health, 84-87; prin investr, Bedrest Res, 90-91. *Mem:* Am Physiol Soc; AAAS. *Res:* Physiologic effects of bedrest; cutaneous blood perfusion at critical sites on the skin are examined for vasomotor change as well as changes in blood volume, flood flow and blood velocity. *Mailing Add:* 3525 Caroline St St Louis MO 63104

RUBIN, MAX, b New York, NY, Jan 6, 16; m 42; c 2. ANIMAL NUTRITION, BIOCHEMISTRY. *Educ:* Rutgers Univ, BS, 38; Univ Md, MS, 40, PhD(poultry nutrit), 42. *Prof Exp:* Chemist, Chem Warfare Serv, 42-43, physiologist, 43-44; biologist, US Bur Animal Indust, 44-46; nutritionist, Schenley Distillers Corp, 46-47; fel, 69-72, FAC RES ASSOC ANIMAL NUTRIT, UNIV MD, COLLEGE PARK, 72- *Concurrent Pos:* Fel animal nutrit, Univ Md, 42-43. *Mem:* AAAS; Poultry Sci Asn; World Poultry Sci Asn; Am Chem Soc. *Res:* Amino acid metabolism; buphthalmos as it is related to glycine in the diet; lysine and tryptophane from grains which have genetically improved amino acid composition. *Mailing Add:* 15107 Interlachem No 914 Silver Spring MD 20906-5612

RUBIN, MELVIN LYNNE, b San Francisco, Calif, May 10, 32; m 53; c 3. MEDICINE, OPHTHALMOLOGY. *Educ:* Univ Calif, Berkeley, BS, 53; Univ San Francisco, MD, 57; Univ Iowa, MS, 61; Am Bd Ophthal, cert. *Prof Exp:* Exec secy vision res training comn, Nat Inst Neurol Dis & Blindness, 61-63; from asst prof to assoc prof, 63-67, PROF OPHTHAL, COL MED, UNIV FLA, 67-, CHMN DEPT, 79- *Honors & Awards:* Statesman Award, Joint Common Allied Health Personnel Opthal, 87. *Mem:* Asn Res Vision & Ophthal (pres); Am Acad Ophthal (secy, pres 88); Am Ophthal Soc; Sigma Xi. *Res:* Visual physiology and visual optics; retinal anatomy; microscopic histology and photochemistry; effects of therapy for retinal detachment surgery on the microstructrue of the eye. *Mailing Add:* Dept Ophthal MSB J284 JHM HC Univ Fla Med Ctr Gainesville FL 32610

RUBIN, MEYER, b Chicago, Ill, Feb 17, 24; m 44; c 3. GEOLOGY, GEOMORPHOLOGY & GLACIOLOGY. *Educ:* Univ Chicago, BS, 47, MS, 49, PhD, 56. *Prof Exp:* Asst geol, Univ Chicago, 49-50; geologist, Mil Geol Br, 50-53, GEOLOGIST, C-14 LAB, ISOTOPE GEOL BR, US GEOL SURV, 53- *Concurrent Pos:* Instr, Ill Inst Technol, 49 & USDA Grad Sch, 51-52. *Mem:* Geol Soc Am; Am Geophys Union; Am Quaternary Asn; Fedn Am Sci. *Res:* Radiocarbon age determinations of Wisconsin glaciations; sea-level changes; dating volcanic eruptions; tandem accelerator mass spectrometry. *Mailing Add:* US Geol Surv Nat Ctr 971 Reston VA 22092

RUBIN, MILTON D(AVID), b Boston, Mass, July 4, 14; m 44; c 2. SYSTEMS ENGINEERING, COMPUTER GRAPHICS. *Educ:* Harvard Univ, BS, 35. *Prof Exp:* Proj engr, Raytheon Corp, Mass, 44-48; res dir comput, Philbrick Res, Inc, 48-49; proj leader radar & comput, Lab Electronics, Inc, Mass, 49-51; group leader radar, Lincoln Lab, Mass Inst Technol, 51-55; dept mgr anal, Sylvania Elec Prod, Inc, 55-57; systs engr, Lincoln Lab, Mass Inst Technol, 57-59 & Mitre Corp, Mass, 59-69; consult scientist, Raytheon Co, 69-75; pres, Energy Conserv Res Inst, 75-77; staff mem, Lincoln Lab, Mass Inst Technol, 77-85; sr systs engr, Teledyne Brown Eng, Lexington, Mass, 85-87; SYST ENGR, MITRE CORP, 87- *Mem:* Fel AAAS; Inst Elec & Electronics Engrs; Soc Gen Systs Res (secy-treas, 64-67, pres, 68-69); Inst Mgt Sci; Opers Res Soc Am. *Res:* Radar systems engineering; printed circuit techniques; computer and radar circuits; analysis of networks of radars for detection of aircraft, missiles and satellites; general systems research; air traffic control; organization theory and technology; communication systems; energy utilization efficiency; computer graphics. *Mailing Add:* 19 Dorr Rd Newton MA 02158

RUBIN, MITCHELL IRVING, pediatrics, nephrology; deceased, see previous edition for last biography

RUBIN, MORTON HAROLD, b Albany, NY, Mar 10, 38; div; c 2. THEORETICAL PHYSICS. *Educ:* Mass Inst Technol, SB, 59; Princeton Univ, PhD(physics), 64. *Prof Exp:* Jr scientist physics, Avco-Everett Res Corp, 60-61; fel physics, Univ Wis, 64-66, instr, 66-67; asst prof physics, Univ

Pa, 67-73; ASSOC PROF PHYSICS, UNIV MD BALTIMORE COUNTY, 73- *Concurrent Pos:* Vis asst prof physics, Pahlavi Univ, Shiraz, Iran, 70-71. *Mem:* Am Phys Soc; Am Asn Physics Teachers. *Res:* Critical phenomena; nucleation and condensation; statistical mechanics; non-equilibrium thermodynamics; quantum theory of scattering; quantum measurement theory. *Mailing Add:* Dept Physics Univ Md Baltimore County 5401 Wilkens Ave Catonsville MD 21228

RUBIN, MORTON JOSEPH, b Philadelphia, Pa, May 15, 17; m 40, 75; c 3. METEOROLOGY. *Educ:* Pa State Univ, BA, 42; Mass Inst Technol, MS, 52. *Prof Exp:* Meteorologist, Pan-Am Grace Airways, 42-49; res assoc meteorol, Mass Inst Technol, 49-52; res meteorologist, US Weather Bur, Environ Sci Serv Admin, 52-65, sr staff scientist meteorol, 65-67, dep chief plans & requirements div, 67-69, chief off spec studies, 67-69, chief res group, Nat Oceanic & Atmospheric Admin, 69-74; sci officer, World Meteorol Orgn, 74-81; vis scholar, Scott Polar Res Inst, 81-82; RETIRED. *Concurrent Pos:* Mem heat & water panel, Comn Polar Res, Nat Acad Sci, 63-65; chmn, Working Group Antarctic Meteorol, Sci Comt Antarctic Res, 64-82; pres, Int Comn Polar Meteorol, 64-72. *Honors & Awards:* Antarctic Serv Medal, 65; Am Meteorol Soc Spec Award, 65; Silver Medal, Dept Com, 69. *Mem:* Am Meteorol Soc; Am Geophys Union. *Res:* Polar meteorology; heat and water budget; southern hemisphere atmospheric circulation and climate. *Mailing Add:* 8013 Westover Rd Bethesda MD 20814

RUBIN, NATHAN, organic chemistry; deceased, see previous edition for last biography

RUBIN, RICHARD LEE, b Cleveland, Ohio, Sept 29, 46. HARMONIC ANALYSIS. *Educ:* Wash Univ, AB, 68, MA, 70, PhD(math), 74. *Prof Exp:* Asst prof math, Oakland Univ, 74-76; asst prof, 76-79, ASSOC PROF MATH, FLA INT UNIV, 79- *Concurrent Pos:* Vis prof, Nat Res Coun, Turin, Italy, 82-83; lectr, Fla State Univ Study Ctr, Florence, Italy, 85. *Mem:* Am Math Soc. *Res:* Harmonic analysis on Lie groups; behavior of multiplier operators and singular integrals; application of function space techniques to complex analysis. *Mailing Add:* Dept Math Fla Int Univ Miami FL 33199

RUBIN, RICHARD MARK, b Pensacola, Fla, July 26, 37; m 63; c 2. NUCLEAR ENGINEERING, PHYSICS. *Educ:* Univ Mich, BSE, 59; Univ Okla, MNE, 61; Kans State Univ, PhD(nuclear eng), 70. *Prof Exp:* Exp supvr Off Civil Defense contract, Nuclear Eng Shielding Facil, Kans State Univ, 67-69, instr nuclear eng, 69; asst prof, Miss State Univ, 70-77; PROJ PHYSICIST, RADIATION RES ASSOCS, INC, 77- *Concurrent Pos:* Adj mem grad fac, Tex Christian Univ, 77- *Mem:* Am Nuclear Soc; Am Soc Eng Educ; Sigma Xi. *Res:* Radiation shielding; radiation transport; neutron activation analysis; dosimetry. *Mailing Add:* 4309 Sarita Dr Ft Worth TX 76109

RUBIN, ROBERT HOWARD, b Philadelphia, Pa, Mar 26, 41. THEORETICAL ASTROPHYSICS, INFRARED ASTRONOMY. *Educ:* Case Inst Technol, BS, 63; Case Western Reserve Univ, PhD(astrophys), 67. *Prof Exp:* Res assoc, Nat Radio Astron Observ, 67-69; res assoc, Univ Ill, Urbana, 69-71, asst prof astron, 71-72; assoc prof physics, Calif State Univ, Fullerton, 72-81; sr nat res coun assoc, 81-83, ASSOC, AMES RES CTR, NASA, 83- *Concurrent Pos:* Prin investr, NSF grant, 70-72, NASA IRAS grant, 85-89 & NASA IUE grant; fac fel, NASA Ames, Santa Clara Univ, 79 & Stanford Univ, 80 & 81; vis prof & res astromr, Univ Calif, Los Angeles, 83-89. *Mem:* Am Astron Soc; Int Astron Union; Int Sci Radio Union; Sigma Xi. *Res:* Interstellar medium; theoretical computer modeling of gaseous nebulae with emphasis on predicting line intensities; extracting physical information such as elemental abundances and properties of exciting stars. *Mailing Add:* NASA Ames Res Ctr Mail Stop 245-6 Moffett Field CA 94035-1000

RUBIN, ROBERT JAY, b Boston, Mass, Mar 25, 32; m 83; c 3. ENVIRONMENTAL TOXICOLOGY, BIOCHEMISTRY. *Educ:* Univ Mass, BS, 53; Boston Univ, AM, 55, PhD(pharmacol), 60. *Prof Exp:* Asst prof, Univ Kans, 63-64; from asst prof to assoc, 64-73, Johns Hopkins Univ, 64-73, dir, div toxicol, 77-79, dir, biochem toxicol prog, 83-87, dep dir, Ctr Environ Health Sci, 87-89, PROF ENVIRON HEALTH SCI, SCH HYG & PUB HEALTH, JOHNS HOPKINS UNIV, 73- *Concurrent Pos:* USPHS fel pharmacol, Sch Med, Yale Univ, 60-63; Res Career Develop Award, Nat Inst Environ Health Sci, 69-74; vis prof, Hebrew Univ, Jerusalem, Israel, 76, Univ Milan Italy, 84, Bethune Univ, Changchun PRC, 86. *Mem:* AAAS; Soc Occup & Environ Health; Soc Toxicol; Am Soc Pharmacol & Exp Therapeut. *Res:* Biochemical toxicology; biochemical mechanisms of action of pharmacologic and toxic agents; drug metabolism and disposition; carcinogenesis. *Mailing Add:* Dept Environ Health Sci Johns Hopkins Sch Hyg & Pub Health Baltimore MD 21205

RUBIN, ROBERT JOSHUA, b New York, NY, Aug 17, 26; m 48; c 4. PHYSICAL CHEMISTRY. *Educ:* Cornell Univ, AB, 48, PhD(phys chem), 51. *Prof Exp:* Sr mem staff, Appl Physics Lab, Johns Hopkins Univ, 51-55; vis asst prof phys chem, Univ Ill, 55-57; physicist, Nat Bur Standards, 57-87; spec expert, 87-89, SPEC VOL, NIH, 89- *Concurrent Pos:* NSF sr fel, 63-64; vis prof, Kyoto Univ, Japan, 68; res physicist, NIH, 76-77; vis prof chem eng, Univ Calif, Berkeley, 81. *Mem:* Am Phys Soc; Am Chem Soc; Biophys Soc; Am Math Soc. *Res:* Chemical physics; statistical and quantum mechanics; mathematical physics; biophysics. *Mailing Add:* 3308 McKinley St NW Washington DC 20015

RUBIN, ROBERT TERRY, b Los Angeles, Calif, Aug 26, 36; m 62; c 3. PSYCHIATRY. *Educ:* Univ Calif, Los Angeles, AB, 58; Univ Calif, San Francisco, MD, 61; Univ Southern Calif, PhD, 77; Am Bd Psychiat & Neurol, dipl & cert psychiat, 68. *Prof Exp:* Intern, Philadelphia Gen Hosp, 61-62; resident psychiat, Sch Med, 62-65, asst prof, 65-71, vis prof, 72-74, adj prof, 74-77, PROF PSYCHIAT, SCH MED, UNIV CALIF, LOS ANGELES, 77-, MEM, BRAIN RES INST, 69- *Concurrent Pos:* Prof psychiat, Col Med, Pa State Univ, 71-74. *Honors & Awards:* Res Scientist Develop Award, Nat Inst Mental Health, 72, 77, 82; Fulbright-Hays Res Scholar Award, 83; UK Med Res Coun Sr Vis Scientist Award, 83. *Mem:* Fel Am Psychiat Asn; World Psychiat Asn; fel Am Col Psychiat; Int Soc Psychoneuroendocrinol; fel AAAS. *Res:* Biochemical and neuroendocrine correlates of stress and mental illness. *Mailing Add:* Dept Psychiat Harbor Univ Calif Los Angeles Torrance CA 90509

RUBIN, RONALD PHILIP, b Newark, NJ, Jan 4, 33; m 56; c 3. CELLULAR PHARMACOLOGY, SECRETORY MECHANISMS. *Educ:* Harvard Univ, AB, 54, AMT, 58; Yeshiva Univ, PhD(pharmacol), 63. *Prof Exp:* From instr to assoc prof pharmacol, Downstate Med Ctr, State Univ NY, 64-74; PROF PHARMACOL & HEAD DIV CELLULAR PHARMACOL, MED COL VA, 74- *Mem:* AAAS; Am Soc Pharmacol & Exp Therapeut; Endocrine Soc; Harvey Soc; NY Acad Sci. *Res:* Cellular and biochemical events that regulate the secretory process. *Mailing Add:* Dept Pharmacol State Univ NY Farber Hall Rm 102 Sch Med & Biomed Sci Buffalo NY 14214

RUBIN, SAMUEL H, b New York, NY, July 24, 16; m 43; c 2. INTERNAL MEDICINE. *Educ:* Brown Univ, AB, 38; Univ Chicago, MS, 55; St Louis Univ, MD, 43; Am Bd Internal Med, dipl. *Prof Exp:* Assoc prof, 60-65, dir, Cardiovasc Training Prog, 66-71, assoc dean, 71-72, vpres acad affairs, 75-77, dean col, 75-83, provost, 77-83, dir, Inst Med Ethics, 84-86, PROF INTERNAL MED, NY MED COL, 65-, PROVOST & EMER DEAN, 84- *Concurrent Pos:* Exec dean, NY Med Col, 62- *Mem:* Fel Am Col Physicians; Asn Am Med Cols; NY Acad Sci. *Res:* Clinical and internal medicine; medical education. *Mailing Add:* Scarborough Manor Scarborough NY 10510

RUBIN, SAUL H, b Cleveland, Ohio, Nov 24, 23; m 47; c 4. ENGINEERING. *Educ:* Case Inst Technol, BS, 44, MS, 47. *Prof Exp:* Res engr, Nat Adv Comt Aeronaut, Ohio, 44-45; asst, Case Inst Technol, 45-46; sci engr, Joy Mfg Co, 47-50; sci engr, Tyroler Metals, Inc, 50-64, pres, 64-88; RETIRED. *Mem:* Am Soc Mech Engrs; Sigma Xi; Nat Soc Prof Engrs; Org Spare Parts & Equipment. *Res:* Test and analysis of aircraft compressors and navy refrigeration coils; analytical design and analysis of axial-flow fans and compressors; design of equipment for salvage and reclamation. *Mailing Add:* 2445 Queenston Rd Cleveland Heights OH 44118

RUBIN, SAUL HOWARD, b New York, NY, Oct 15, 12; m 32; c 2. BIOCHEMISTRY. *Educ:* City Col, BS, 31; NY Univ, MS, 32, PhD(biochem), 39. *Prof Exp:* Biochemist, Littauer Res Fund, Harlem Hosp, New York, 32-34; res biochemist, Metab Lab, NY Univ, 36-41; dir, Nutrit Labs, Hoffmann-LaRoche, Inc, 41-49, coordr pharmaceut res, 49-53, dir, New Prods Div, 53-58, corp dir prod develop, 58-77; RETIRED. *Concurrent Pos:* Mem liaison & sci adv bd, Quartermaster Corps Food & Container Inst, US Army. *Mem:* Asn Res Dirs (past pres); Am Soc Clin Nutrit; Am Pharm Asn; Am Soc Biol Chemists; Am Inst Nutrit; NY Acad Med. *Res:* Acid base and lipid metabolism; vitamin assays and metabolism; drug formulation and metabolism. *Mailing Add:* 12401 Rock Garden Lane Miami FL 33156

RUBIN, SHELDON, b Chicago, Ill, July 19, 32; m 55; c 3. STRUCTURAL DYNAMICS, LIQUID ROCKET DYNAMICS. *Educ:* Calif Inst Technol, BS , 53, MS, 54, PhD(mech eng, physics), 56. *Prof Exp:* Res engr sound & vibration, Lockheed Aircraft Co, 56-58; head, Tech Eng Sect, Hughes Aircraft Co, 58-62; PRIN ENG SPECIALIST, ENG GROUP, AEROSPACE CORP, LOS ANGELES, 62- *Concurrent Pos:* Consult, 56-; lectr, Univ Calif, Los Angeles, 64-; chmn, Aerospace Shock & Vibration Comt, Soc Automotive Engrs, 62- *Honors & Awards:* Shuttle Flight Cert Award, NASA, 81. *Mem:* Assoc fel Am Inst Aeronaut & Astronaut; fel Soc Automotive Engrs. *Res:* Test and analysis of structural vibration; influence of propulsion system on vibration stability of liquid rocket vehicles; dynamic behavior at friction surfaces. *Mailing Add:* Aerospace Corp PO 92957 M4/899 Los Angeles CA 90009-2957

RUBIN, STANLEY G(ERALD), b Brooklyn, NY, May 11, 38; m 63; c 3. COMPUTATIONAL FLUID MECHANICS, VISCOUS FLOWS. *Educ:* Polytech Inst Brooklyn, BAE, 59; Cornell Univ, PhD(aerospace eng), 63. *Prof Exp:* Res scientist, Boeing Airplane Co, 63; NSF fel, Henri Poincare Inst, Univ Paris, 63-64; from asst prof to assoc prof eng & appl mech, Polytech Inst New York, 64-73, prof mech & aerospace eng, 73-79, assoc dir, Aerodyn Labs, 77-79; head dept, 79-89, PROF AEROSPACE & ENG MECH, UNIV CINCINNATI, 79-, DIR, NASA SPACE ENG RES CTR, 88- *Concurrent Pos:* Vis prof mech eng, Old Dominion Univ, 73-74; consult, Fluid Dynamics Tech Comt, Am Inst Aeronaut & Astronaut, 75-78, Aerospace Corp, 77-82, Allison Gas Turbine, 84-86, NASA aeronaut res technol sub comt, 86-89; vis scientist, Air Force Wright Aeronaut Labs, 87; mem adv comt, Inst Comput Methods in Propulsion, NASA Lewis Res Ctr, 87-; contractor, USAF Off Sci Res, 67- & NASA, 73-; chief ed, Int J Computers & Fluids. *Mem:* Am Soc Mech Engrs; Soc Indust & Appl Math; Am Soc Eng Educ; assoc fel Am Inst Aeronaut & Astronaut; Sigma Xi. *Res:* High-speed gasdynamics; three-dimensional viscous interactions; asymptotic expansions in viscous flow; numerical analysis of flow problems. *Mailing Add:* 10695 Deershadow Lane Cincinnati OH 45242

RUBIN, VERA COOPER, b Philadelphia, Pa, July 23, 28; m 48; c 4. ASTRONOMY. *Educ:* Vassar Col, BA, 48; Cornell Univ, MA, 51; Georgetown Univ, PhD(astron), 54. *Hon Degrees:* DSc, Creighton Univ, 78, Harvard Univ, 88 & Yale Univ, 90. *Prof Exp:* Instr math & physics, Montgomery County Jr Col, 54-55; res assoc astron, Georgetown Univ, 55-65, lectr, 59-62, asst prof, 62-65; STAFF MEM, DEPT TERRESTRIAL MAGNETISM, CARNEGIE INST, 65- *Concurrent Pos:* Assoc ed, Astron J, 72- & Astrophys J Letters, 77-; mem, space astron comt, Nat Acad Sci, 72-, US Nat Comt, Int Astron Union, 72-76, Bd of Dirs, Asn Univs Res Astron, Inc, 73-76 & Space Sci Bd, Nat Acad Sci, 74-76; mem, Smithsonian Coun, 79-84; distinguished vis astron, Cerro Tololo, 78; Chancellor's distinguished prof astron, Univ Calif, Berkeley, 81; mem coun, Am Women in Sci, 84-; president's distinguished visitor, Vassar Col, 87; Tinsley vis prof, Univ Tex, 88; mem bd physics & astron, Nat Acad Sci. *Mem:* Nat Acad Sci; Int Astron Union; Am Astron Soc. *Res:* External galaxies; galactic dynamics; spectroscopy. *Mailing Add:* Dept Terr Magnet Carnegie Inst 5241 Broad Branch Rd Washington DC 20015

RUBIN, WALTER, b Worcester, Mass, Aug 15, 33; m 58; c 5. GASTROENTEROLOGY, CELL BIOLOGY. *Educ:* Mass Inst Technol, BS, 55; Cornell Univ, MD, 59. *Prof Exp:* From instr to asst prof med, Med Col, Cornell Univ, 64-69, asst prof anat, 68-69; assoc prof med, 69-70, assoc prof anat, 69-75, CHIEF GASTROENTEROL & VCHMN DEPT MED, MED COL PA, 69-, PROF MED, 70-, PROF ANAT, 75- *Concurrent Pos:* Res fel med, Med Col, Cornell Univ, 62-64; Nat Inst Arthritis & Metab Dis res grant & Am Cancer Soc res grant, Med Col, Cornell Univ, 66-69 & Med Col Pa, 69-88; consult gastroenterol, Mem Hosp, New York, 69-70 & Vet Admin Hosp, Philadelphia, 69-; mem gastrointestinal syst res eval comt, US Vet Admin, 69-71, mem merit rev bd gastroenterol, 72-74; mem ed coun, Ital J Gastroenterol, 70-84; mem gastroenterol adv panel, Subcomt Scope of US Pharmacopeia Comt of Revision, 75-85; chmn, Gen Med A Study Sect, NIH, 82-85. *Mem:* Am Soc Cell Biol; Histochem Soc; fel Am Col Physicians; Am Gastroenterol Asn; Am Soc Clin Invest; Am Asn Study Liver Dis. *Res:* Gastrointestinal epithelia, particularly their fine structure, ontogenesis, differentiation and function in health and disease. *Mailing Add:* Dept Med Med Col Pa 3300 Henry Ave Philadelphia PA 19129

RUBINK, WILLIAM LOUIS, b Sterling, Colo, Apr 5, 47; m 75; c 3. AFRICANIZED HONEY BEE ECOLOGY, TRACHEAL MITE ECOLOGY. *Educ:* Utah State Univ, BS, 69, MS, 73; Colo State Univ, PhD(entom), 78. *Prof Exp:* Res asst, Utah State Univ, 71-73; teaching asst, Colo State Univ, 75-78, res assoc, 78-79; res assoc, Ohio State Univ, 79-82; RES ENTOMOLOGIST, AGR RES SERV, USDA, 83- *Mem:* AAAS; Entom Soc Am; Sigma Xi; Am Entom Soc. *Res:* Basic ecology of solitary wasps, including nest site selection; ecology of maize pests and parasitoids, including a tachinid fly, Bonnetia comta; population studies of the honey bee during the process of Africanization. *Mailing Add:* 2116 Nightingale McAllen TX 78504

RUBINO, ANDREW M, b Brooklyn, NY, Apr 30, 22; m 44; c 3. INORGANIC CHEMISTRY, PHYSICAL CHEMISTRY. *Educ:* St Johns's Col, NY, BS, 49. *Prof Exp:* Res chemist, Reheis Co Inc, NJ, 51-56, mgr tech serv, 56-60 & res & develop, 60-65; vpres & tech dir, Dragoco Inc, 65-66; dir res & develop, 66-73, vpres-tech dir, 73-78, CONSULT, REHEIS CHEM CO DIV, ARMOUR PHARMACEUT CO, 78- *Mem:* AAAS; Am Chem Soc; Soc Cosmetic Chem; NY Acad Sci. *Res:* Aluminum chemistry; chelates; cosmetics; pharmaceuticals and other fine chemicals. *Mailing Add:* Reheis Inc 235 Snyder Ave Berkeley Heights NJ 07922

RUBINOFF, IRA, b New York, NY, Dec 21, 38; m 78; c 3. ZOOLOGY. *Educ:* Queens Col, BS, 59; Harvard Univ, AM, 61, PhD(biol), 63. *Prof Exp:* Fel evolutionary biol, Harvard Univ, 64, asst cur ichthyol, Mus Comp Zool, 65; asst dir sci, 65-73, DIR, SMITHSONIAN TROP RES INST, 73- *Concurrent Pos:* Assoc ichthyol, Mus Comp Zool, Harvard Univ, 65-79; mem adv sci bd, Gorgas Mem Inst, 74-90; mem bd dirs, Charles Darwin Res Found, Int Sch Panama, 84-86, 90-; trustee, Rare Animal Relief Effort, 76-86; vis fel, Wolfson Col, Oxford Univ, 80-81; hon dir, Latin Am Inst Advan Studies; vis scholar, Mus Comp Zool, Harvard Univ, 87-88; mem bd dir, Nat Asn Conserv of NATVRG (Panama), 85- *Honors & Awards:* Order of Vasco Nunez De Balboa. *Mem:* Fel AAAS; Am Soc Nat; Am Soc Ichthyol & Herpet; Soc Study Evolution; fel Linnean Soc London; NY Acad Sci. *Res:* Biological implications of the construction of a sea level canal; zoogeography of the eastern tropical Pacific; strategies for preservation of world's tropical forests; diving physiology of sea snakes. *Mailing Add:* Smithsonian Trop Res Inst Box 2072 Balboa Panama

RUBINOFF, MORRIS, b Toronto, Ont, Aug 20, 17; nat US; m 41; c 3. ELECTRICAL ENGINEERING. *Educ:* Univ Toronto, BA, 41, MA, 42, PhD(external ballistics), 46. *Prof Exp:* Teach tech sch, Ont, 41; asst, Univ Toronto, 42-44; res fel & instr physics, Harvard Univ, 46-48; res engr, Inst Advan Study, 48-50; from asst prof to assoc prof, 50-63, PROF ELEC ENG, MOORE SCH ELEC ENG, UNIV PA, 63-; PRES, PA RES ASSOCS, 60- *Concurrent Pos:* Asst dir, Naval Ord Eng, 43-44; chief engr, Philco Corp, 57-59; trustee, Eng Index, 67-; mem res adv comt commun, instrumentation & data processing, NASA. *Honors & Awards:* Inst Elec & Electronics Engrs Award, 55. *Mem:* Asn Comput Mach; Inst Elec & Electronics Engrs. *Res:* Computer logical design; electronic circuit design; mathematical analysis; information retrieval. *Mailing Add:* Comp Info Sci 2636/D2 Univ Pa Philadelphia PA 19174

RUBINOFF, ROBERTA WOLFF, b New York, NY, Aug 26, 39. MARINE BIOLOGY, EVOLUTIONARY BIOLOGY. *Educ:* Queen's Col, BS, 59; Duke Univ, MEM, 81. *Prof Exp:* Biologist res, 66-75, marine sci coordr admin & res, Smithsonian Trop Res Inst, 75-80; asst dir, 80-84, actg dir, 84-85, DIR OFF FEL & GRANTS ADMIN, SMITHSONIAN INST, 85- *Mem:* AAAS; Am Inst Biol Sci. *Res:* Behavior and evolution of transisthmian species of intertidal organisms; taxonomic group speciality is tropical marine inshore fishes. *Mailing Add:* Smithsonian Inst 7300 L'Enfant Plaza Washington DC 20560

RUBINS, ROY SELWYN, b Manchester, Eng, Nov 11, 35; m 63. MAGNETIC RESONANCE. *Educ:* Oxford Univ, BA, 57, MA & DPhil(physics), 64. *Prof Exp:* Res physicist, Hebrew Univ Jerusalem, 61-63; vis physicist, Battelle Mem Inst, Geneva, 63; res assoc physics, Syracuse Univ, 64-66; asst res physicist & asst prof in residence physics, Univ Calif, Los Angeles, 66-68; lectr Calif State Col, Los Angeles, 69; from asst prof to assoc prof, 69-82, PROF PHYSICS, UNIV TEX, ARLINGTON, 82- *Concurrent Pos:* Michael & Anna Wix fel, Hebrew Univ Jerusalem, 61-62; adj prof, Mont State Univ, 84- *Mem:* Am Phys Soc; Am Asn Physics Teachers. *Res:* Experimental and theoretical treatment of electron paramagnetic resonance in solids; nuclear relaxation in solid hydrogen deuteride; magnetic studies of organometallic compounds; modulated microwave absorption and magnetization studies of conventional high temperature conductors. *Mailing Add:* Dept Physics Box 19059 Univ Tex Arlington TX 76019

RUBINSON, JUDITH FAYE, b Rose Hill, NC, Nov 14, 52; m 80; c 2. ELECTROCHEMISTRY. *Educ:* Univ NC, Chapel Hill, BS, 74; Univ Cincinnati, PhD(anal chem), 81. *Prof Exp:* Chemist, Velsicol Chem Corp, 74-75 & Merrell-Nat Labs, 76-77; postdoctoral assoc, Charles F Kettering Res Lab, 81-84; asst prof, Wright State Univ, 85-89 & Univ Dayton, 89-90; ASST PROF, COL MT ST JOSEPH, 90- *Mem:* Am Chem Soc; Coun Undergrad Res. *Res:* Electrochemistry of biological molecules. *Mailing Add:* 5701 Delhi Rd Cincinnati OH 45233-1670

RUBINSON, KALMAN, b New York, NY, Dec 24, 41; wid; c 2. NEUROANATOMY, RETINAL DEVELOPMENT. *Educ:* Columbia Univ, AB, 62; State Univ NY Downstate Med Ctr, PhD(anat), 68. *Prof Exp:* Vis asst prof anat, Sch Med, Univ PR, San Juan, 67-68; vis prof, Sch Med, Univ PR, San Juan, 69; asst prof, 69-74, assoc prof cell biol, 74-78, ASSOC PROF PHYSIOL & BIOPHYS, SCH MED, NY UNIV, 78- *Concurrent Pos:* Nat Inst Neurol Dis & Stroke fel, Lab Perinatal Physiol, PR, 68-69; Nat Inst Neurol Dis & Stroke res grants, Sch Med, NY Univ, 70-72 & Pub Health Res Inst NY, 74-76; assoc, Dept Neurobiol & Behav, Pub Health Res Inst, NY, 69-76. *Mem:* Soc Neurosci; Asn Res Vision & Ophthal. *Res:* Comparative neuroanatomy; development of sensory pathways and retina. *Mailing Add:* Dept Physiol & Biophys Sch Med NY Univ 550 First Ave New York NY 10016

RUBINSON, KENNETH A, b June 20, 44; m 81; c 2. PHYSICAL BIOCHEMISTRY, BIOPOLYMER KINETICS. *Educ:* Oberlin Col, BA, 66; Univ Mich, PhD(phys-inorg chem), 72. *Prof Exp:* Vis assoc prof & adj prof, Dept Chem, Univ Cincinnati, 79-82, adj asst prof, Sch Med, 85-91; FIVE OAKS RES INST, 82-; ADJ ASSOC PROF, WRIGHT STATE UNIV, 85-; ADJ ASSOC PROF, SCH MED, UNIV CINCINNATI, 91- *Concurrent Pos:* Res asst, Univ Chem Lab, Cambridge, UK, 73-78. *Honors & Awards:* Postdoctoral Fel, NIH, 73-75; English SRC Postdoctoral Fel, 75-76; Res Serv Award, NIH, 76-77; Postdoctoral Fel, NATO, 77-78; Hon Ramsey Mem Fel, 76-78. *Mem:* Royal Soc Chem; Am Phys Soc. *Res:* Magnetic resonance; chromatographies; chemical modification of living tissues; biological macromolecular kinetics. *Mailing Add:* Five Oaks Res Inst 354 Oakwood Park Dr Cincinnati OH 45238

RUBINSTEIN, ASHER A, b Kishinev, USSR, Sept 16, 47; m 80; c 1. FRACTURE MECHANICS. *Educ:* Leningrad Polytech Inst, Dipl, 72; Israel Inst Technol, MSc, 77; Brown Univ, PhD(eng), 81. *Prof Exp:* Design engr equip design, Lenigrad Turbine Blades Plant, 72-74; res engr mat testing, Israel Inst Metals, 74-75; asst prof mech eng, State Univ NY, Stony Brook, 81-87; ASSOC PROF DEPT MECH ENG, TULANE UNIV, 87- *Concurrent Pos:* Prin investr, NSF grant, 83-85 & NASA, 86- *Mem:* Soc Eng Sci; Am Soc Mech Engrs; Am Acad Mech. *Res:* Application of theoretical mechanics of materials to problems at interface of continuum mechanics and material science, especially the analysis of problems in the deformation and fracture of solids at both macro and micro scales. *Mailing Add:* Dept Mech Eng Tulane Univ New Orleans LA 70118-5674

RUBINSTEIN, CHARLES B(ENJAMIN), b New York, NY, Dec 25, 33; m 57; c 3. HUMAN FACTORS, SYSTEMS ENGINEERING. *Educ:* City Col New York, BEE, 59; NY Univ, MEE, 61. *Prof Exp:* Mem tech staff, Res Area, Bell Tel Labs, 59-77, head, Human Factors Dept, 77-82, Int Planning Dept, 82-90, HEAD, NETWORKING & HUMAN FACTORS DEPT, AT&T BELL LABS, 90- *Honors & Awards:* Leonard G Abraham Prize Paper Award, Inst Elec & Electronics Engrs Commun Socs, 73. *Mem:* Fel Optical Soc Am; Asn Res Vision & Ophthal. *Res:* Magneto, electro, geometrical and physical optics; optical properties of materials; crystallographic properties of rare earth crystals; holography; color perception and rendition in complex scenes; digital coding of color pictures; visual threshold; human factors; systems engineering. *Mailing Add:* AT&T 307 Middletown-Lincroft Rd Lincroft NJ 07738

RUBINSTEIN, EDUARDO HECTOR, b Buenos Aires, Arg, July 10, 31; c 2. PHYSIOLOGY. *Educ:* Univ Buenos Aires, BS, 48, MD, 58, PhD(med), 61. *Prof Exp:* Asst physiol, Sch Med, Univ Buenos Aires, 56-58, asst med, Med Res Inst, 58-61, head neurophysiol sect, 64-67, assoc researcher, 66-67; from asst prof physiol to assoc prof, 67-77, PROF-IN-RESIDENCE ANESTHESIOL & PHYSIOL, UNIV CALIF, LOS ANGELES, 77- *Concurrent Pos:* Res fel physiol, Sch Med, Yale Univ, 61-62; res fel physiol, Sch Med, Gotenburg Univ, 62-64; WHO med res grant, Med Res Inst, Univ Buenos Aires, 65-67; estab investr, Nat Res Coun, Arg, 64; asst researcher & lectr, Cardiovasc Res Inst, Univ Calif, San Francisco, 66; sr res investr, Los Angeles County Heart Asn, 67; prin investr, Am Heart Asn grant, Univ Calif, Los Angeles, 69-72. *Mem:* Am Physiol Soc; Soc Neurosci. *Res:* Central nervous control of autonomic functions, especially cardiovascular and gastrointestinal. *Mailing Add:* Dept Physiol 53170 Chs Univ Calif 405 Hilgard Ave Los Angeles CA 90024

RUBINSTEIN, HARRY, b Cologne, Germany, Dec 19, 30; nat US; m 54; c 3. ORGANIC CHEMISTRY. *Educ:* Brooklyn Col, BS, 53; Purdue Univ, MS, 56, PhD(org chem), 58. *Prof Exp:* Res chemist, Wyandotte Chem Corp, 58-59, Keystone Chemurgic Corp, 59-60 & Merck & Co, 60-63; asst prof chem, Springfield Col, 64-65; PROF CHEM, UNIV LOWELL, 65-, dean grad sch, 77-87. *Concurrent Pos:* Vis prof, Harvard Sch Pub Health, 87. *Mem:* Am Chem Soc; Sigma Xi. *Res:* Heterocyclic compounds; natural products; general organic syntheses; mass spectroscopy; capillary electrophoresis. *Mailing Add:* 15 Fairbanks Rd Chelmsford MA 01824

RUBINSTEIN, LAWRENCE VICTOR, b Washington, DC, Oct 10, 46; m 80; c 3. MEDICAL STATISTICIAN. *Educ:* Cornell Univ, BA, 69; Univ Md, MA, 74, PhD(math statist), 78. *Prof Exp:* Programmer, Univac, Apollo Proj, 69-73; teaching asst math, Univ Md, College Park, 73-78; math statistician, Nat Inst Neurologic Dis & Stroke, 83-85; statist fel, 78-83, MATH STATISTICIAN, NAT CANCER INST, 85- *Mem:* Am Statist Soc; Biometric Soc. *Res:* Statisical design and analysis of clinical studies in particular, sample size determination and trial monitoring strategies; in-vitro cell-line screens for anti-cancer agents. *Mailing Add:* 5504 Manorfield Rd Rockville MD 20853

RUBINSTEIN, LUCIEN JULES, neuropathology; deceased, see previous edition for last biography

RUBINSTEIN, LYDIA, b Buenos Aires, Arg, Jan 30, 36; m 59; c 2. ENDOCRINOLOGY, NEUROENDOCRINOLOGY. *Educ:* Univ Buenos Aires, MD, 59, PhD(physiol), 61. *Prof Exp:* Estab investr, Inst Biol & Exp Med, 61-76; asst prof, 76-81, ASSOC PROF, DEPT OBSTET & GYNEC, SCH MED, UNIV CALIF, 81- *Concurrent Pos:* Nat Coun Sci Res fel anat, Med Sch, Yale Univ, 61-62 & fel physiol, Gotenburg Univ, 62-64; Ford Found fel anat, Med Sch, Univ Calif, Los Angeles, 68-69. *Mem:* Endocrine Soc; Arg Biol Soc; NY Acad Sci. *Res:* Regulatory mechanisms of reproduction and growth processes. *Mailing Add:* 1100 Glendon Ave Suite 950 Los Angeles CA 90024

RUBINSTEIN, MARK, b Brooklyn, NY, Dec 26, 35; m 57; c 3. SOLID STATE PHYSICS. *Educ:* Univ Colo, BA, 57; Univ Calif, PhD(physics), 63. *Prof Exp:* Res assoc solid state physics, Atomic Energy Res Estab, Eng, 63; RES PHYSICIST, US NAVAL RES LAB, 64- *Mem:* Am Phys Soc. *Res:* Nuclear magnetic resonance in magnetically ordered systems. *Mailing Add:* 3302 Wessynton Way Alexandria VA 22309

RUBINSTEIN, MOSHE FAJWEL, b Miechow, Poland, Aug 13, 30; nat US; m 53; c 2. STRUCTURAL DYNAMICS. *Educ:* Univ Calif, Los Angeles, BS, 54, MS, 57, PhD, 61. *Prof Exp:* Designer, Murray Erick Assocs, 54-56; struct designer, Victor Gruen Assocs, 56-61; from asst prof to assoc prof eng, 61-69, coord prof & prog dir continuing educ, Mod Eng for Execs Prog, 65-70, chmn eng systs dept, 70-75, PROF ENG, SCH ENG & APPL SCI, UNIV CALIF, LOS ANGELES, 69- *Concurrent Pos:* Consult, Pac Power & Light Co, Ore, Northrop Corp, US Army, NASA Res Ctr, Langley, Tex Instruments Co, Hughes Space Syst Div, US Army Sci Adv Comn, Kaiser Aluminum & Chem Corp & IBM; Sussmann chair distinguished vis, Technion Israel Inst Technol, 67-68; Fulbright-Hays fel, Yugoslavia & Eng, 75-76; mem acad rev bd, IBM. *Honors & Awards:* Award for Excellence, Am Soc Eng Educ, 65. *Mem:* Am Soc Chem Engrs; Am Soc Eng Educ; Seismol Soc Am; Sigma Xi; NY Acad Sci. *Res:* Use of computers in structural systems, analysis and synthesis; problem solving and decision theory; author of seventy publications and six books which have been translated to foreign languages. *Mailing Add:* Dept of Eng Systs Univ of Calif Los Angeles CA 90024

RUBINSTEIN, ROY, b Darlington, Eng, Sept 12, 36; m 68; c 2. EXPERIMENTAL HIGH ENERGY PHYSICS. *Educ:* Cambridge Univ, BA, 58; Birmingham Univ, PhD(physics), 61. *Prof Exp:* Res assoc, Birmingham Univ, 61-62; res assoc & actg asst prof physics, Cornell Univ, 62-66; from assoc physicist to physicist, Brookhaven Nat Lab, 66-73; PHYSICIST, FERMI NAT ACCELERATOR LAB, 73- & ASST DIR. *Mem:* Am Phys Soc. *Res:* Elastic scattering and total cross sections of nucleons and mesons on nucleons in the 100 GeV and greater region. *Mailing Add:* Fermi Nat Accelerator Lab PO Box 500 Batavia IL 60510

RUBIO, RAFAEL, b Queretaro, Mex, Feb 15, 28; m 55, 79; c 5. CARDIOVASCULAR PHYSIOLOGY. *Educ:* Univ Mex, BS, 63; Univ Va, PhD(physiol), 68. *Prof Exp:* Res assoc physiol, Syntex, Mex, 52-55; res assoc, Inst Nac Cardiol, Mex, 55-63; res assoc, Case Western Reserve Univ, 64-66; from asst prof to assoc prof, 69-76, PROF PHYSIOL, UNIV VA, 76- *Concurrent Pos:* NIH fel, Univ Va, 68-69, Am Heart Asn fel, 74; peer rev comt, NIH, 79-82. *Honors & Awards:* Fogarty Fel, 83. *Mem:* Mex Physiol Sci Soc; Am Physiol Soc; Biophys Soc; Soc Gen Physiol. *Res:* Blood flow regulation; muscle contraction and metabolism. *Mailing Add:* Dept Physiol Univ Va Sch Med 441 Jordan Hall Charlottesville VA 22903

RUBIS, DAVID DANIEL, b Jackson, Minn, May 30, 24; m 59; c 3. PLANT BREEDING, GENETICS. *Educ:* Univ Minn, BS, 48; Iowa State Univ, MS, 50, PhD(crop breeding), 54. *Prof Exp:* Res asst, Corn Breeding Proj, Iowa State Univ, 48-52; agent-agronomist, Spec Crops Sect, Agr Res Serv, USDA, 52-56; from asst prof agron & asst agronomist to assoc prof & assoc agron, Univ Ariz, 56-64, prof plant sci & agronomist, 64-86; RETIRED. *Concurrent Pos:* Consult agron. *Mem:* Am Soc Agron; Crop Sci Soc Am; Am Genetic Asn; Genetics Soc Am. *Res:* Genetics and breeding of safflower, soybeans, guayule and plantago; development of new crops; pollination of entomphilous plants; genetics of water use in plants; development of new crops. *Mailing Add:* PO Box 42544 Tucson AZ 85733

RUBLEE, PARKE ALSTAN, b Buffalo, NY, June 25, 49; m 77; c 4. MARINE BIOLOGY, LIMNOLOGY. *Educ:* Dartmouth Col, BA, 71; NC State Univ, MSc, 74, PhD(zool), 78. *Prof Exp:* Res assoc, Rosenstiel Sch Marine & Atmospheric Sci, 79-80, Chesapeake Bay Ctr Environ Studies, Smithsonian Inst, 80-82; assoc prof biol, Whitman Col, 82-89; ASST PROF BIOL, UNIV NC, GREENSBORO, 90- *Mem:* Am Soc Limnol & Oceanog; Am Soc Microbiol; AAAS. *Res:* The role of heterotrophic microorganisms in aquatic systems. *Mailing Add:* Biol Dept Univ NC Greensboro Greensboro NC 27412

RUBLOFF, GARY W, b Peoria, Ill, June 13, 44; m 66; c 2. ELECTRONIC MATERIALS & PROCESSING, INTERFACES. *Educ:* Dartmouth Col, AB, 66; Univ Chicago, SM, 67, PhD(physics), 71. *Prof Exp:* Res assoc, Physics Dept, Brown Univ, 71-73; res staff mem, IBM Res Div, T J Watson Res Ctr, 73-84, tech asst to vpres, 84-85, mgr, Growth, Interfaces & Offline Processing, 85-87, mgr, Thin Film Mat & Characterization, MGR, GROWTH, ETCHING & INTEGRATED PROCESSING, IBM RES DIV, T J WATSON RES CTR, YORKTOWN HEIGHTS, NY, 89- *Concurrent Pos:* Prin investr, Off Naval Res, 75-; ed, Deposition & Growth: Limits for Microelectronics, AIP Conf, Repub of China, 87; chmn, Electronic Mat & Processing, Div Am Vacuum Soc, 87, secy, 90- *Mem:* Fel Am Phys Soc; Am Vacuum Soc; Mat Res Soc; Inst Elec & Electronics Engrs; Electron Device Soc. *Res:* Fundamental chemistry and physics of electronic materials growth, processing and interfaces; ultraclean and integrated processing; CUD & MOS structures; interfacial reactions; surface science; optical properties of solids and surfaces. *Mailing Add:* IBM Res PO Box 218 Yorktown Heights NY 10598

RUBNITZ, MYRON ETHAN, b Omaha, Nebr, Mar 2, 24; m 52; c 4. PATHOLOGY. *Educ:* Univ Nebr, BSc, 45, MD, 47. *Prof Exp:* Fel path, Med Sch, Northwestern Univ, 53-55, assoc path, 55-59, asst prof, 59-63; assoc prof, 63-69, PROF PATH, STRITCH SCH MED, LOYOLA UNIV, CHICAGO, 69- *Concurrent Pos:* Adj prof, Ill State Univ, 79-, Col St Francis, 88-; prof, Northern Ill Univ, 79-; clin prof Dent Sch, 70- *Mem:* Am Soc Clin Path; Col Am Path; Int Acad Path. *Res:* Surgical pathology. *Mailing Add:* Lab Serv Vet Admin Hosp Hines IL 60141

RUBOTTOM, GEORGE M, b London, Eng, Mar 19, 40; US citizen; m 67, 89; c 2. ORGANIC CHEMISTRY. *Educ:* Middlebury Col, AB, 62; Mass Inst Technol, PhD(org chem), 67. *Prof Exp:* Fel org chem, Calif Inst Technol, 66-68; instr, Bucknell Univ, 68-70; from asst prof to assoc prof, Univ PR, 70-74; assoc prof, 75-78, prof org chem, Univ Idaho, 78-85; PROG DIR CHEM, NSF, 85- *Concurrent Pos:* NIH fel, 67-68. *Mem:* Am Chem Soc. *Res:* Chemistry of organo-silicon compounds; chemistry of small ring compounds; chemistry of natural products. *Mailing Add:* Dept Chem NSF Washington DC 20550

RUBY, EDWARD GEORGE, b Rochester, NY, Aug 31, 49. MICROBIAL SYMBIOSES, MICROBIAL PHYSIOLOGY. *Educ:* Stetson Univ, BS, 71; Scripps Inst Oceanog, PhD(marine biol), 77. *Prof Exp:* Res fel, Harvard Univ, 77-79, Woods Hole Oceanog Inst, 79-81; res assoc, Univ Calif, Los Angeles, 81-82; ASSOC PROF BIOL SCI, UNIV SOUTHERN CALIF, 82- *Concurrent Pos:* Adj asst prof microbial physiol, Boston Univ, 78-79. *Mem:* Am Soc Microbiol; Am Soc Limnol & Oceanog; Am Chem Soc; NY Acad Sci. *Res:* Procaryotic cellular differentiation including in the dimorphic growth cycle of the bacterial genus Bdellovibrio; physiological and biochemical studies of their transformation from one cell form to another to understand the processes of development in bacteria. *Mailing Add:* 893 1/2 Ellis Ave Los Angeles CA 90034

RUBY, JOHN L, b Indianapolis, Ind, Mar 1, 12; m 39; c 1. FOREST GENETICS. *Educ:* Purdue Univ, BSF, 34; Mich State Univ, MS, 59, PhD(forest genetics), 64. *Prof Exp:* Instr forestry, 63-64, asst prof natural sci, 65-71, prof, 71-77, EMER PROF NATURAL SCI, MICH STATE UNIV, 77- *Mem:* AAAS; Soc Am Foresters; Sigma Xi. *Res:* Forest genetics; species variability studies through parental and provenance testing. *Mailing Add:* RR No 1 Box 272 Irons MI 49644

RUBY, JOHN ROBERT, b Elida, Ohio, May 26, 35; m 57; c 3. ANATOMY. *Educ:* Baldwin-Wallace Col, BS, 57; St Louis Univ, MS, 59; Univ Pittsburgh, PhD(anat), 63. *Prof Exp:* From instr to asst prof anat, Univ Cincinnati, 63-67; from asst prof to assoc prof, 67-77, PROF ANAT, LA STATE UNIV MED CTR, NEW ORLEANS, 77-, ASSOC DEAN FAC AFFAIRS, LA STATE UNIV SCH MED, 84- *Mem:* Am Asn Anat; Electron Micros Soc Am. *Res:* Development of female reproductive system; ultrastructure of salivary glands; pancreatic cancer. *Mailing Add:* Dept Anat La State Univ Med Ctr 1542 Tulane Ave New Orleans LA 70112

RUBY, LAWRENCE, b Detroit, Mich, July 25, 25; m 51; c 3. NUCLEAR PHYSICS. *Educ:* Univ Calif, Los Angeles, BA, 45, MA, 47, PhD(physics), 51. *Prof Exp:* Lectr, 60-61, from assoc prof to prof nuclear eng, Univ Calif, Berkeley, 61-87, physicist, Lawrence Berkeley Lab, 50-87; PROF NUCLEAR SCI & REACTOR DIR, REED COL, PORTLAND, ORE, 87- *Mem:* Am Phys Soc; Am Asn Physics Teachers; Am Nuclear Soc. *Res:* Nuclear spectroscopy; accelerators; reactor dynamics; nuclear fusion. *Mailing Add:* 663 Carrera Lane Lake Oswego OR 97034

RUBY, MICHAEL GORDON, b Muskogee, Okla, May 27, 40; m 68. AIR POLLUTION CONTROL, ENVIRONMENTAL ECONOMICS. *Educ:* Univ Okla, BS, 62; Univ Wash, MS, 65, MSE, 78, PhD(civil eng), 81. *Prof Exp:* Environ specialist, City Seattle, 72-76; prin, Environ Res Group, 76-81; asst prof civil eng, Univ Cincinnati, 81-84; PRES, ENVIROMETRICS, INC, 84- *Concurrent Pos:* Consult, WHO, AID, 79-; dir, Int Environ Eng Inst, 88- *Mem:* Air & Waste Mgt Asn; Am Acad Environ Engrs; Am Meteorol Asn; Sigma Xi. *Res:* Air pollution measurement and control technology, particularly with reference to combustion and odors; indoor air pollution measurement; benefit-cost analysis of environmental policies and projects. *Mailing Add:* 4128 Burke Ave N Seattle WA 98103-8320

RUBY, PHILIP RANDOLPH, b Aurora, Ill, May 23, 25; m 53; c 1. ORGANIC CHEMISTRY. *Educ:* Univ Ill, BA, 49; Univ Iowa, MS, 51, PhD(org chem), 53. *Prof Exp:* Res chemist, 53-60, GROUP LEADER, PIGMENTS DIV, AM CYANAMID CO, BOUND BROOK, 60- *Mem:* Am Chem Soc. *Res:* Aromatic organic chemistry; organic pigments. *Mailing Add:* 22 Circle Dr Millington NJ 07946

RUBY, RONALD HENRY, b San Francisco, Calif, Dec 1, 32; m 57; c 4. BIOPHYSICS. *Educ:* Univ Calif, Berkeley, AB, 54, PhD(physics), 62. *Prof Exp:* Actg asst prof physics, Univ Calif, Berkeley, 62-64; NSF fel biol, Mass Inst Technol, 64-65; from asst prof to assoc prof, 65-79, PROF PHYSICS, UNIV CALIF, SANTA CRUZ, 79- *Concurrent Pos:* Sloan Found fel, 66-68. *Mem:* Am Phys Soc. *Res:* Solid state physics; problems of physics in biological systems; energy conversion process in photosynthesis. *Mailing Add:* Nat Sci Div Univ Calif Santa Cruz CA 95068

RUBY, STANLEY, b Brooklyn, NY, Nov 26, 20; m 46; c 3. PHYSICAL CHEMISTRY, MATERIALS SCIENCE. *Educ:* City Col New York, BS, 41; Columbia Univ, PhD(mining, metall), 54. *Prof Exp:* Res assoc, Manhattan Proj Sam Labs, 43-46; res assoc, Brookhaven Nat Lab, 50-52; group leader ore concentration, Minerals Beneficiation Lab, Columbia Univ, 52-54; sr engr, Sylvania Elec Prod Inc, 54-56; sect chief reentry mat, Res & Adv Develop Div, Avco Corp, 56-62; sr scientist, Allied Res Assocs, 62-64; prog mgr, Advan Res Proj Agency, Arlington, 64-77; prog mgr, Dept Energy, Washington, DC, 78-87; RETIRED. *Concurrent Pos:* Res Assoc, Columbia Univ, 50-51. *Mem:* Am Chem Soc; Am Inst Mining, Metall & Petrol Eng; Sigma Xi; NY Acad Sci. *Res:* Surface chemistry; high temperature chemistry; electrochemistry; ore concentration; winning of metals; radiochemistry. *Mailing Add:* 10913 Kenilworth Ave Garrett Park MD 20896-0025

RUCH, RICHARD JULIUS, b Perryville, Mo, June 9, 32; m 54; c 4. PHYSICAL CHEMISTRY. *Educ:* Southeast Mo State Col, BS, 54; Iowa State Univ, MS, 56, PhD(chem), 59. *Prof Exp:* Asst prof chem, Univ SDak, 59-62; asst prof, Southern Ill Univ, 62-66; asst chmn dept, 72-86, ASSOC PROF CHEM, KENT STATE UNIV, 66- *Mem:* Fedn Soc Coatings Technol; Am Chem Soc. *Res:* Surface and colloid chemistry; the stability of colloidal dispersions is studied with the aid of dielectric measurements and with various interfacial techniques. *Mailing Add:* Dept Chem Kent State Univ Kent OH 44242-0002

RUCH, RODNEY R, b Springfield, Ill, Aug 18, 33; m 56; c 3. ANALYTICAL CHEMISTRY. *Educ:* Univ Ill, BS, 55; Southern Ill Univ, MA, 59; Cornell Univ, PhD(chem), 65. *Prof Exp:* Res asst chem, Gen Atomic Div, Gen Dynamics Corp, 61-63; res chemist, Nat Bur Standards, 65-66; assoc chemist, Ill State Geol Surv, 66-71, chemist, 71-73, head anal chem sect, 73-84, minerals eng sect, 84-89, ASST BR CHIEF, ILL STATE GEOL SURV, 89- *Concurrent Pos:* Prin res officer, Broken Hill Proprietary Co, Australia, 80-81. *Honors & Awards:* Gov Award, 71. *Mem:* Am Chem Soc; Am Soc Testing & Mat. *Res:* Coal analysis; radiochemical separations; activation analysis; radiochemistry; trace element analysis. *Mailing Add:* Ill State Geol Surv 615 E Peabody Champaign IL 61820

RUCHKIN, DANIEL S, b New Haven, Conn, June 29, 35; m 58, 85; c 1. BIOMEDICAL ENGINEERING, ELECTRICAL ENGINEERING. *Educ:* Yale Univ, BE, 56, MEng, 57, DEng, 60. *Prof Exp:* Instr elec eng, Yale Univ, 59-61; asst prof, Univ Rochester, 61-64; res assoc prof psychiat, NY Med Col, 64-70; assoc prof physiol & comput sci, 71-77, PROF PHYSIOL, SCH MED, UNIV MD, BALTIMORE CITY, 77- *Concurrent Pos:* J Javits neurosci investr, Nat Inst Neurol Dis & Stroke, 86-93. *Mem:* AAAS; Inst Elec & Electronics Engrs; Biomed Engr Soc; fel Am EEG Soc; Soc Psychophysiol Res. *Res:* Brain research, specifically analysis of evoked and event-related potentials; psychophysiology; signal analysis; development of computer software of evoked potential analysis. *Mailing Add:* Dept Physiol Univ Md Sch of Med Baltimore MD 21201

RUCHMAN, ISAAC, b New York, NY, July 2, 09; m 40; c 1. MICROBIOLOGY, IMMUNOLOGY. *Educ:* City Col New York, BSc, 37; Univ Cincinnati, MSc, 41, PhD(bact), 44; Am Bd Microbiol, dipl. *Prof Exp:* Tech asst, Rockefeller Inst, 30-39; asst, Children's Hosp Res Found, Cincinnati, Ohio, 39-44, res assoc, 46-52; from instr to asst prof bact, Univ Cincinnati, 44-55; head microbiol, Wm S Merrell Co, 55-63; prof, 63-74, EMER PROF MICROBIOL, UNIV KY, 74- *Concurrent Pos:* USPHS grant in aid; attend bacteriologist, Cincinnati Gen Hosp, 46-55; adj prof, Col Eng, Off Continuing Educ, Univ Ky, 75-; adj prof, Transylvania Univ, 76- *Mem:* Fel Am Acad Microbiol; Soc Exp Biol & Med; Am Asn Immunol; Am Soc Microbiol; NY Acad Sci. *Res:* Toxoplasmosis; tularemia; virology; immunity in virus infections; herpes simplex; neurotropic viruses; antimicrobial screening; upper respiratory infections; chemotherapy. *Mailing Add:* 365 Garden Rd Lexington KY 40502-2417

RUCHTI, RANDAL CHARLES, b Janesville, Wis, Oct 27, 46; m 70; c 2. HIGH ENERGY PHYSICS, ELEMENTARY PARTICLE PHYSICS. *Educ:* Univ Wis-Madison, BS, 68; Univ Ill, Urbana, MS, 70; Mich State Univ, PhD(physics), 73. *Prof Exp:* Res assoc physics, Northwestern Univ, 73-76, asst prof, 76-77; ASST PROF PHYSICS, UNIV NOTRE DAME, 77- *Mem:* Am Inst Physics; Sigma Xi; Am Phys Soc. *Res:* Strong interactions; charm particle production and charm decay processes; new quantum number production. *Mailing Add:* Dept Physics Univ Notre Dame Notre Dame IN 46556

RUCINSKA, EWA J, b Poland; US citizen; c 1. RESEARCH ADMINISTRATION, PHARMACOLOGY. *Educ:* Med Sch, Gdansk, Poland, MD, 60, PhD(pharmacol), 66. *Prof Exp:* Assoc prof, dept pharm, Med Sch, Gdansk, Poland, 67-73; assoc dir, Squibb Inst Med Res, 75-78; dir int, 78-84, dir cardiovasc, 84-85, DIR CARDIO-RENAL, MERCK SHARP & DOHME RES LABS, 85- *Concurrent Pos:* Dir Med Teaching Prog for Students, Gdansk, 63-73; coordr clin res POLFA-Pharm & Dept Pharm, Poland, 67-73; sr lectr, continuing educ prog for physicians & pharmacists, Poland, 70-73. *Mem:* AMA; Am Fedn Clin Res; Am Soc Pharmacol & Exp Therapeut; NY Acad Sci. *Res:* Evaluation of new drugs regarding efficacy and safety in a variety of cardiovascular syndromes; evaluation of the effects of various agents in the renin-angiotension-aldosterone system and the relationships with other humeral systems. *Mailing Add:* Merck Sharp & Dohme Res Labs BL3-1 West Point PA 19486

RUCKEBUSCH, GUY BERNARD, b Lille, France, June 28, 49; m 76; c 1. SYSTEM THEORY, DIGITAL SIGNAL PROCESSING. *Educ:* Ecole des Mines de Paris, Engr, 72; Univ Pierre et Marie Curie, Paris, PhD(probability), 75, Dr es Sci, 80. *Prof Exp:* Res assoc syst theory, Cent di Automatique de Ecole des Mines de Paris, 73-75, Inst Nat de la Recherche en Info et Automatique, 75-78; engr syst sci, 78-80, RES ENGR PETROL SCI, DOLL RES, SCHLUMBERGER, 80- *Concurrent Pos:* Prof probability, Ecole des Mines de Nancy, 75-78. *Res:* Stochastic system theory; develop new signal processing techniques to process schlumberger subsurface measurements. *Mailing Add:* Schlumberger Doll Res PO Box 307 Ridgefield CT 06877

RUCKENSTEIN, ELI, b Botosani, Romania, Aug 13, 25; US citizen; m 48; c 2. CHEMICAL ENGINEERING. *Educ:* Bucharest Polytech Inst, MS, 49, Dr Eng(chem eng), 66. *Prof Exp:* Prof chem eng, Bucharest Polytech Inst, 49-69; NSF sr scientist, Clarkson Col Technol, 69-70; prof, Univ Del, 70-73; fac prof eng, appl sci & chem eng, 73-81, DISTINGUISHED PROF, STATE UNIV NY, BUFFALO, 81- *Honors & Awards:* George Spacu Award, Romanian Acad Sci, 63; Alpha Chi Sigma Award, Am Inst Chem Engrs, 77; Sr Humboldt Award, Alexander Humboldt Found, 85; Kendall Award, Am Chem Soc, 86; Walker Award, Am Inst Chem Engrs, 88. *Mem:* Nat Acad Eng; Am Inst Chem Engrs; Am Chem Soc. *Res:* Transport phenomena in fluids and solids; supported metal catalysts; oxide catalysis; zeolite catalysis, enzymatic catalysis, heterogeneous kinetics; thermodynamics and kinetics of interfacial phenomena; micellization; microemulsions and colloids; transport phenomena in colloidal systems; deposition of cells on surfaces; separation processes; polymer composites; membrane for separation processes. *Mailing Add:* Dept Chem Eng 504 Furnas Hall State Univ NY N Campus Amherst NY 14260

RUCKER, JAMES BIVIN, b Emporia, Kans, Mar 15, 35; m 57; c 2. GEOLOGY, OCEANOGRAPHY. *Educ:* Univ Mo, BA, 60, MA, 61; La State Univ, PhD(geol), 66. *Prof Exp:* Oceanographer, US Naval Oceanog Off, 63-72; dir, Miss Marine Resources Coun, 72-74; ecologist, Nat Oceanic & Atmospheric Admin, 75-90; PROJ MGR, TGS, 90- *Concurrent Pos:* Prof lectr, George Washington Univ, 66-84, Johns Hopkins Univ, 84- *Mem:* AAAS; Geol Soc Am; Sigma Xi; Am Soc Limnol & Oceanog. *Res:* Recent marine sediments and benthic biota; environmental impact of deep ocean dumping, remote sensing of environment. *Mailing Add:* 43 Lakeside Carriere MS 39426

RUCKER, ROBERT BLAIN, b Oklahoma City, Okla, Mar 29, 41; m 67; c 2. NUTRITIONAL BIOCHEMISTRY. *Educ:* Oklahoma City Univ, BA, 63; Purdue Univ, MS, 66, PhD(biochem), 68. *Prof Exp:* Fel biochem, Univ Mo-Columbia, 68-70; from asst prof to assoc prof nutrit & biol chem, 70-78, chmn dept nutrit, 81-88, PROF NUTRIT & BIOL CHEM, UNIV CALIF, DAVIS, 78- *Concurrent Pos:* Chair, Gordon Conf Elastins, co-chair, Fedn Am Soc Exp Biol Conf Trace Mineral Metab. *Mem:* AAAS; Am Inst Nutrit; Am Chem Soc; Am Soc Biochem & Molecular Biol; Soc Exp Biol Med. *Res:* Connective tissue metabolism; selected functions of nutritionally essential trace elements and vitamins; biochemistry of elastin. *Mailing Add:* Dept Nutrit Univ Calif Davis CA 95616

RUCKERBAUER, GERDA MARGARETA, b Steyr, Austria, Dec 7, 26; Can citizen. VETERINARY IMMUNOLOGY. *Educ:* Univ Toronto, BA,51, MA, 55; Ont Vet Col, DUM, 61. *Prof Exp:* Tech asst bot, Univ Toranto, 51-52, tech off genetics, 52-53; res fel, Hosp for Sick Children, Toronto, 53-55; res off carcinogens, Ont Vet Col, 55-57; vet, Animal Dis Res Inst, Can Dept Agr, 61-78, res scientist serol, Food Prod & Inspection Div, 68-89; RETIRED. *Mem:* NY Acad Sci; Genetics Soc Can; Can Vet Med Asn; Can Soc Immunol; Can Micros Soc. *Res:* Methods for diagnosis of animal diseases using serological, elisa and fluorescence microscopical techniques. *Mailing Add:* 19 Barran Nepean ON K2T 1G5 Can

RUCKLE, WILLIAM HENRY, b Neptune, NJ, Oct 29, 36; m 60; c 1. MATHEMATICS. *Educ:* Lincoln Univ, Pa, AB, 60; Fla State Univ, MS, 62, PhD(math), 63. *Prof Exp:* Asst prof math, Lehigh Univ, 63-69; assoc prof, 69-74, PROF MATH SCI, CLEMSON UNIV, 74- *Concurrent Pos:* Vis prof math, Western Washington Univ, 78-79; Fulbright vis scholar, Trinity Col, Dublin; Fulbright fel, 83-84; William C Foster vis fel, US Arms Control & Disarmament Agency, 89-90. *Mem:* Am Math Soc; Math Asn Am; Soc Indust Appl Math; Irish Math Soc; Opers Res Soc Am. *Res:* Functional analysis, summability theory, game theory and risk analysis. *Mailing Add:* Dept Math Sci Clemson Univ Clemson SC 29631

RUCKLIDGE, JOHN CHRISTOPHER, b Halifax, Eng, Jan 15, 38; m 62; c 4. MINERALOGY, CRYSTALLOGRAPHY. *Educ:* Cambridge Univ, BS, 59; Univ Manchester, PhD(mineral), 62. *Prof Exp:* Res assoc, Univ Chicago, 62-64; res asst, Oxford Univ, 64-65; from lectr to assoc prof, 65-77, PROF MINERAL, UNIV TORONTO, 77-, ASSOC DIR, ISOTRACE LAB, 81- *Mem:* Mineral Soc Am; Geol Asn Can; Mineral Asn Can; Microbeam Anal Soc; Sigma Xi. *Res:* Electron probe and x-ray crystallographic studies on minerals; ultrasensitive mass spectrometry with tandem accelerators. *Mailing Add:* Dept Geol Univ Toronto St George Campus Toronto ON M5S 1A1 Can

RUCKNAGEL, DONALD LOUIS, b St Louis, Mo, May 30, 28; m 55; c 2. HUMAN GENETICS. *Educ:* Wash Univ, AB, 50, MD, 54; Univ Mich, PhD(human genetics), 64. *Prof Exp:* From intern to jr resident med, Duke Univ Hosp, NC, 54-56; investr human genetics, Nat Inst Dent Res, 57-59; from asst prof to assoc prof, 64-70, PROF HUMAN GENETICS, MED SCH, UNIV MICH, ANN ARBOR, 70-; PROF INTERNAL MED, SIMPSON MEM INST, 77- *Concurrent Pos:* Fel, Univ Pittsburgh, 56-57; USPHS sr res fel human genetics, Med Sch, Univ Mich, Ann Arbor, 59-62, res career develop award, 62-69; res assoc internal med, Simpson Mem Inst, 68-77. *Mem:* Am Soc Human Genetics; Am Soc Hemat; Eugenics Soc; Physicians Social Responsibility; Am Fedn Clin Res. *Res:* Internal medicine; hematology; human genetics, especially hemoglobinopathies. *Mailing Add:* Comprehensive Sickle Cell Ctr Childrens Hosp Med Ctr Bethesda & Elland Ave Cincinnati OH 45229

RUDAT, MARTIN AUGUST, b Burbank, Calif, June, 11, 52; m 75; c 3. MASS SPECTROMETRY. *Educ:* Harvey Mudd Col, BS, 74; Cornell Univ, MS, 76, PhD(anal chem), 78. *Prof Exp:* Res chemist, 78-85, GROUP MGR, E I DU PONT DE NEMOURS & CO, INC, WILMINGTON, DEL, 85- *Mem:* Am Chem Soc; Am Soc Mass Spectrometry; Sigma Xi. *Res:* Ionization techniques; mass spectral methods; mass spectral reactions and mechanisms; two dimensional mass spectrometry; linking of mass spectrometry with other techniques; secondary ion mass spectrometry. *Mailing Add:* Textile Fibers AS&T 302 Du Pont Exp Sta Wilmington DE 19898

RUDAVSKY, ALEXANDER BOHDAN, b Poland, Jan 17, 25; US citizen; m 55; c 1. CIVIL ENGINEERING. *Educ:* Univ Minn, BS, 53, MS, 55; Hanover Tech Univ, Dr Ing(civil eng), 66. *Prof Exp:* Civil engr, Justin & Courtney, Philadelphia, 56-57 & Iran, 57-58; prof eng, San Jose State Univ, 60-85; DIR & OWNER, HYDRO RES SCI, 64- *Concurrent Pos:* Vis assoc prof, Stanford Univ, 70. *Mem:* Am Soc Civil Engrs; Am Soc Mech Engrs. *Res:* Hydraulic research, through model studies of engineering problems related to hydraulic structures, rivers, ports and harbors, coastal protection, and sedimentation; development of instrumentation; library documentation system and service; consultative services. *Mailing Add:* Hydro Res Sci 3334 Victor Ct Santa Clara CA 95050

RUDAZ, SERGE, b Verdun, Que, Aug 19, 54; m 83; c 1. ELEMENTARY PARTICLE PHYSICS. *Educ:* Cornell Univ, MS, 79, PhD(physics), 79. *Prof Exp:* Res fel, European Orgn Nuclear Res, Geneva, 79-81; asst prof, 81-85, assoc prof, 85-87, PROF PHYSICS, UNIV MINN, 87- *Honors & Awards:* Pres Young Investr Award, 84; Herzberg Medal, Can Asn Physicists, 85. *Mem:* Inst Particle Physics Can; Am Phys Soc. *Res:* Unified field theories of elementary particle interactions; astroparticle physics and cosmology; models of high energy processes; relativistic many-body theory. *Mailing Add:* Sch Physics & Astron Univ Minn Minneapolis MN 55455

RUDBACH, JON ANTHONY, b Long Beach, Calif, Sept 23, 37; m 59; c 2. MICROBIOLOGY, IMMUNOLOGY. *Educ:* Univ Calif, Berkeley, BA, 59; Univ Mich, MS, 61, PhD(microbiol), 64, Lake Forest Grad Sch Mgt, MBA, 86. *Prof Exp:* Nat Inst Allergy & Infectious Dis fel biophys, Rocky Mountain Lab, USPHS, 64-66, res scientist, 66-67; lectr, Univ Mont, 67-70, assoc prof microbiol, 70-75, prof, 75-, dir, Stella Dincen Mem Res Inst, 79-85; head, Infectious Dis & Immunol Res Dept, Abbott Labs; VPRES RES & DEVELOP, RIBI IMMUNOCHEM RES LABS, 85- *Concurrent Pos:* Head microbiol lab, Abbott Lab, 77-79. *Mem:* AAAS; Soc Exp Biol & Med; Am Asn Immunol; Am Soc Microbiol. *Res:* Molecular biology of endotoxins from Gram-negative bacteria; relation of structure of endotoxins to biological activity; detoxification of endotoxins by human plasma; immunology of lipopolysaccharides. *Mailing Add:* Vpres Res & Develop Ribi Immunochem Res Inc PO Box 1409 Hamilton MT 59840

RUDD, D(ALE) F(REDERICK), b Minneapolis, Minn, Mar 2, 35; m 64; c 2. CHEMICAL ENGINEERING. *Educ:* Univ Minn, BS, 56, PhD(chem eng), 60. *Prof Exp:* Asst prof chem eng, Univ Mich, 60-61; fel, 61-62, from asst prof to prof, 62-68, SLICHTER PROF CHEM ENG, UNIV WIS-MADISON, 80- *Concurrent Pos:* J S Guggenheim fel, 70. *Honors & Awards:* Allan P Colburn Award, 71. *Mem:* Nat Acad Eng. *Res:* Process Engineering. *Mailing Add:* Dept Chem Eng Univ Wis 1415 Johnson Dr Madison WI 53706

RUDD, DEFOREST PORTER, b Boston, Mass, Aug 17, 23; m 50; c 3. INORGANIC CHEMISTRY. *Educ:* Harvard Univ, BA, 47; Univ Calif, PhD(phys chem), 51. *Prof Exp:* Res assoc, Northwestern Univ, 50-52; PROF CHEM, LINCOLN UNIV, PA, 52- *Concurrent Pos:* NSF sci fac fel, Cornell Univ, 59-60 & Stanford Univ, 66-67. *Mem:* AAAS; Am Chem Soc. *Res:* Solutions and critical phenomena; reactions of transition metal complexes. *Mailing Add:* Lincoln Univ RD 1 Lincoln University PA 19352

RUDD, MILLARD EUGENE, b Fargo, NDak, Sept 29, 27; m 53; c 3. ATOMIC PHYSICS. *Educ:* Concordia Col, BA, 50; Univ Buffalo, MA, 55; Univ Nebr, PhD(physics), 62. *Prof Exp:* From asst prof to prof physics, Concordia Col, 54-65; assoc prof physics, 65-68, actg chmn dept, 70-72, PROF PHYSICS, UNIV NEBR, LINCOLN, 68- *Concurrent Pos:* NSF fel, 60-61; mem comm atomic & molecular sci, Nat Acad Sci, 80; vchmn, Div Electron & Atomic Physics, Am Phys Soc, 79, chmn, 80. *Mem:* Am Asn Physics Teachers; fel Am Phys Soc. *Res:* Atomic collisions; ion-atom collisions; electron spectroscopy; history of science. *Mailing Add:* Behlen Lab Physics Univ Nebr Lincoln NE 68588-0111

RUDD, NOREEN L, b Vancouver, BC, Sept 3, 40; m 74. AREUPLOIDY. *Educ:* Univ British Columbia, MD, 65; Am Bd Pediat, dipl, 70; FRCP(C), 73; Can Col Med Geneticists, fel, 76. *Prof Exp:* From asst prof to assoc prof, pediat, Univ Toronto, 73-80; assoc prof, pediat, Univ Calgary, 80-84; assoc staff pediat, Alta Children's Hosp, 80-91; prof, pediat/obstet, Univ Calgary, 84-91; RETIRED. *Concurrent Pos:* Dir, Genetics Clin, Hosp for Sick Children, Toronto, 75-80; assoc staff, Obstet, Toronto Gen Hosp, 78-80; nat comts, Med Res Coun, 86-, Sci Coun Can, 87-; consult, Foothills Hosp, 83-91. *Mem:* Can Col Med Geneticists; Genetics Soc Can; Am Soc Human Genetics; Royal Col Physicians & Surgeons Can. *Res:* Research into the cellular basis of aneuploidy, a common chromosome error recognizable for recurrent abortions, Down Syndrome and some types of cancer. *Mailing Add:* Med Genetics 1820 Richmond Rd SW Alta Childrens Hosp Calgary AB T2T 5C7 Can

RUDD, ROBERT L, b Los Angeles, Calif, Sept 18, 21; m 47; c 2. ZOOLOGY. *Educ:* Univ Calif, PhD(zool), 53. *Prof Exp:* Asst, Univ Calif, 47-50, assoc zool, 51-52, asst specialist, 52-56, from asst prof to assoc prof, 56-86, EMER PROF ZOOL, UNIV CALIF, 86- *Mem:* Fel AAAS; Soc Study Evolution; Am Soc Mammal; Am Ornith Union; Ecol Soc Am; Cooper Ornith Soc; Asn Trop Biol. *Res:* Pesticides; pollution ecology; vertebrate evolution; population phenomena, chiefly mammals; comparative endocrine structures; tropical ecology; conservation. *Mailing Add:* Dept Zool Univ Calif Davis CA 95616-8755

RUDD, VELVA ELAINE, b Fargo, NDak, Sept 6, 10. BOTANY. *Educ:* NDak Agr Col, BS, 31, MS, 32; George Washington Univ, PhD(bot), 53. *Prof Exp:* Asst bot, NDak Agr Col, 31-32 & Univ Cincinnati, 32-33; instr, Sch Hort, Ambler, Pa, 33-34; supvr sci courses, State Dept Supervised Studies, NDak, 35-37; asst bot, Univ Cincinnati, 37-38; asst sci aide, Bur Plant Indust, USDA, 38-42, personnel classification investr, 42-43; agr prog officer, Food Supply Mission, Inst Inter-Am Affairs, Venezuela, 43-45; agr prog analyst, UNRRA, 45-48; from asst cur to cur, Div Phanerograms, Dept Bot, US Nat Mus, Smithsonian Inst, 48-73; SR RES FEL, DEPT BIOL, CALIF STATE UNIV, NORTHRIDGE, 73- *Concurrent Pos:* Res assoc, Div Phanerogams, Dept Bot, Smithsonian Inst, 73- *Mem:* AAAS; Am Soc Plant Taxonomists; Int Bur Plant Taxon; Soc Bot Mexico. *Res:* Systematic botany; plant geography. *Mailing Add:* PO Box 19 Reseda CA 91337

RUDD, WALTER GREYSON, b Teaneck, NJ, Dec 24, 43; m 71; c 1. COMPUTER SCIENCE. *Educ:* Rice Univ, BA, 66, PhD(phys chem), 69. *Prof Exp:* Res assoc phys chem, State Univ NY Albany, 69-70, biophys, 70-71; from asst prof to assoc prof, La State Univ, Baton Rouge, 71-81, prof & chem comput sci, 81-; DEPT COMPUT SCI, ORE STATE UNIV. *Mem:* AAAS; Asn Comput Mach. *Res:* Statistical mechanics and thermodynamics; theoretical biophysics; analog and hybrid computational techniques; computer simulation of biological systems; small computer system development. *Mailing Add:* Dept Comput Sci Ore State Univ Corvallis OR 97331

RUDDAT, MANFRED, b Insterburg, Ger, Aug 21, 32; m 62; c 2. PLANT MOLECULAR BIOLOGY. *Educ:* Univ Tubingen, Dr rer nat(bot), 60. *Prof Exp:* Sci asst bot, Univ Tubingen, 60-61; NSF res fel plant physiol, Calif Inst Technol, 61-64; asst prof bot, Univ Chicago, 64-68, from asst prof to assoc prof biol, 68-84, assoc prof molecular genetics, cell biol, ecol & evolution, 84-90, ASSOC DEAN STUDENTS, UNIV CHICAGO, 88-, ASSOC PROF ECOL, EVOLUTION & DEVELOP BIOL, 90- *Concurrent Pos:* Ed, Bot Gazette, 74- *Mem:* Am Bot Soc; AAAS; Am Soc Plant Physiol; Japanese Soc Plant Physiol; Int Plant Growth Substances Asn. *Res:* Developmental biology of plants; physiology and biochemistry of plant growth regulators; gene expression, host/pathogen interactions. *Mailing Add:* Dept Ecol & Evolution Barnes Lab Univ Chicago 5630 S Ingleside Chicago IL 60637

RUDDELL, ALANNA, b Ottawa, Ont, Feb 16, 56. MOLECULAR BIOLOGY. *Educ:* Queen's Univ, BSc, 77; Case Western Reserve Univ, PhD(develop genetics, anat), 83. *Prof Exp:* Postdoctoral, Dept Zool, Univ BC, 83-85, Dept Genetics, Fred Hutchinson Cancer Res Ctr, 85-90; ASST PROF, DEPT MICROBIOL & IMMUNOL, UNIV ROCHESTER, 90- *Mem:* AAAS; Am Soc Microbiol. *Res:* Analysis of role of transcription factors in retroviral oncogenesis, using avian leukosis virus induction of lymphoma as a model system. *Mailing Add:* Dept Microbiol & Immunol Univ Rochester Box 672 Rochester NY 14642

RUDDICK, JAMES JOHN, b Elmira, NY, June 11, 23. SEISMOLOGY. *Educ:* Woodstock Col, AB, 46, STL, 56; St Louis Univ, MS, 50, PhD(physics), 52. *Prof Exp:* Instr physics, St Peter's Col, 47-48; ASSOC PROF PHYSICS, CANISIUS COL, 57-, DIR, SEISMOL STA, 69- *Mem:* Am Phys Soc; Am Asn Physics Teachers. *Res:* Earthquakes of western New York. *Mailing Add:* Dept Physics Canisius Col Buffalo NY 14208-1098

RUDDICK, KEITH, b Haltwhistle, Eng, Dec 2, 39; m 88. PHYSICS. *Educ:* Univ Birmingham, BSc, 61, PhD(physics), 64. *Prof Exp:* Res assoc physics, Univ Mich, 64-66; from asst prof to assoc prof, 66-76, PROF PHYSICS, UNIV MINN, MINNEAPOLIS, 76- *Mem:* Inst Physics; Am Phys Soc. *Res:* Experimental high energy physics. *Mailing Add:* Sch Physics Univ Minn Minneapolis MN 55455

RUDDLE, FRANCIS HUGH, b West New York, NJ, Aug 19, 29; m 64; c 2. GENETICS, CELL BIOLOGY. *Educ:* Wayne State Univ, BA, 53, MS, 55; Univ Calif, Berkeley, PhD, 60. *Hon Degrees:* MA, Yale Univ, Lawrence Univ, 82, Weizmann Inst, Israel, 83. *Prof Exp:* From asst prof to assoc prof, Yale Univ, 61-72, chmn, Dept Biol, 77-83, Ross Granville Harrison prof biol, 83-88, PROF BIOL & HUMAN GENETICS, YALE UNIV, 72-, STERLING PROF BIOL, 88-, CHMN, DEPT BIOL, 88- *Concurrent Pos:* NIH fel biochem, Univ Glasgow, 60-61; mem, Cell Biol Study Sect, NIH, 65-70, chmn adv comt, Human Mutant Cell Bank, Nat Inst Gen Med Sci, 72-77, bd sci counselors, Nat Heart, Lung & Blood Inst, 80-84, chmn, Genome Study Sect, NIH, 91-; mem bd dirs, Am Soc Human Genetics, 71-75; mem adv comt, Biol Div, Oak Ridge Nat Lab, 77-80; mem, Panel Cell Biol, Nat Res coun, 85-86; ed-in-chief, J Exp Zool, 85-; mem coun, Human Genome Orgn, 88-; bd sci overseers, Jackson Lab, 89-91. *Honors & Awards:* Dyer lectr, NIH, 78; Conden lectr, Univ Ore, 81; Dickson Prize, Univ Pittsburgh, 81; Herman Beerman Award, Soc Invest Dermat, 82; Allan Award, Am Soc Human Genetics, 83; NY Acad Sci Award in Biol & Med Sci, 87; Katherine Berkan Judd Award, 88. *Mem:* Nat Acad Sci; Inst Med-Nat Acad Sci; Soc Develop Biol (pres, 71-72); Am Soc Cell Biol (pres-elect, 86, pres, 87); Am Soc Human Genetics (pres-elect, 84, pres, 85); fel AAAS; fel NY Acad Sci; Am Asn Cancer Res; Am Genetic Asn; Am Soc Cell Biol. *Res:* Somatic cell genetics and differentiation. *Mailing Add:* Dept Biol Yale Univ 260 Whitney Ave New Haven CT 06511

RUDDLE, NANCY HARTMAN, b St Louis, Mo, Apr 3, 40; m 64; c 2. IMMUNOLOGY. *Educ:* Mt Holyoke Col, AB, 62; Yale Univ, PhD(microbiol), 68. *Prof Exp:* Res assoc & lectr tissue typing, Surgery Dept, 68-71, fel & lectr microbiol, 71-74, asst prof, 75-80, assoc prof epidemiol, 80-90, PROF BIOL & IMMUNOL, YALE UNIV, 91- *Concurrent Pos:* Fel, Damon Runyon Mem Found, Yale Univ, 72, Am Cancer Soc, 73-74 & fac res award, 79-84; vis investr, Basel Inst Immunol, 83; mem review bd, NSF, ACS, 87-90, NIH, 90- *Mem:* Am Soc Microbiol; Sigma Xi; Am Asn Immunol. *Res:* Delayed hypersensitivity; tumor immunology; leukemia virus-lymphocyte interactions; lymphokines; lymphotoxin; tumor necrosis factor. *Mailing Add:* Dept Epidemiol & Pub Health Yale Univ Sch Med New Haven CT 06510

RUDDON, RAYMOND WALTER, JR, b Detroit, Mich, Dec 23, 36; m 61; c 3. CANCER BIOLOGY, PHARMACOLOGY. *Educ:* Univ Detroit, BS, 58; Univ Mich, PhD(pharmacol), 64, MD, 67. *Prof Exp:* From instr to prof pharmacol, Univ Mich, Ann Arbor, 64-76; dir biol marker prog, Nat Cancer Inst, Frederick, Md, 76-81; PROF & CHMN, DEPT PHARMACOL, UNIV MICH MED SCH, 81- *Concurrent Pos:* Teaching fel chem, Univ Detroit, 58-59, NIH predoctoral fel, 59-64; Am Cancer Soc scholar, 64-67. *Mem:* AAAS; Am Soc Pharmacol & Exp Therapeut; Am Asn Cancer Res; Am Soc Biol Chemists; Endocrine Soc. *Res:* Biological markers of neoplasia; differentiation of neoplastic cells; mechanism of action of anticancer agents. *Mailing Add:* Dept Pharmacol Univ Mich Med Sch M6322 Med Sci I Ann Arbor MI 48109-0626

RUDDY, ARLO WAYNE, b Humboldt, Nebr, Aug 28, 15; m 43, 86; c 4. MEDICINAL CHEMISTRY. *Educ:* Univ Nebr, BSc, 36, MSc, 38; Univ Md, PhD(med chem), 40. *Prof Exp:* Sharp & Dohme res assoc, Northwestern Univ, 40-41; res chemist, Sharples Chem, Inc, Pa Salt Mfg Co, Mich, 41-42; sr res chemist, Sterling-Winthrop Res Inst, 42-51; dir org chem res, Chilcott Labs, 51-52, dir chem develop, Warner-Chilcott Res Labs, 52-57, dir org chem, 57-61, dir chem develop, Warner-Lambert Res Inst, 61-78; RETIRED. *Mem:* Am Chem Soc; Am Pharmaceut Asn; Acad Pharmaceut Sci; fel NY Acad Sci. *Res:* Medicinal chemistry, including mercurials, arsenicals, pressor amines, radiopaques, local anesthetics and antispasmodics. *Mailing Add:* 11 Juniper Dr Morris Plains NJ 07950

RUDE, PAUL A, b Los Angeles, Calif, Nov 4, 30; m 51; c 6. ELECTRONICS ENGINEERING, COMPUTER SCIENCE. *Educ:* Univ Calif, Los Angeles, BS, 55; Univ Pittsburgh, MS, 57, PhD(elec eng), 62. *Prof Exp:* Sr engr, Westinghouse Elec Corp, 55-63; asst prof elec eng, 63-64, ASSOC PROF ELEC ENG & CHMN DEPT, LOYOLA UNIV, LOS ANGELES, 64- *Concurrent Pos:* Lectr, Univ Pittsburgh, 60-61; consult, Space-Gen Corp, 63-64, Pac Tel Co, 64-65 & Hughes Aircraft Co, 66. *Mem:* Am Soc Eng Educ; Inst Elec & Electronics Engrs. *Res:* Medical-optical systems; computer design. *Mailing Add:* Dept Elec Eng & Comput Sci Loyola Marymount Univ 7101 W 80th St Los Angeles CA 90045

RUDE, THEODORE ALFRED, b Chuquicamata, Chile, Mar 10, 25; US citizen; m 46; c 5. PATHOLOGY, MICROBIOLOGY. *Educ:* Univ Pa, VMD, 52; Am Col Vet Path, dipl. *Prof Exp:* Pvt pract, Wis, 52-59; diagnostician, Cent Animal Health Lab, Wis, 59-61, lab supvr, 61-63; staff vet, Norwich Pharmacal Co, NY, 63-64; staff vet, Salsbury Labs, 64-69, mgr path & toxicol, 69-71, biol devleop mgr, 71-72, gen mgr, Fromm Labs, 72-80, vpres & gen mgr, 80-82; asst to pres, Salsbury Labs, 83-86; VET CONSULT, VET CONSULT SERVS, 86- *Mem:* Am Vet Med Asn; Am Asn Avian Path. *Res:* Pathology associated with avian, canine and feline diseases; development of diagnostic reagents, vaccines and bacterins for use with canines and felines; toxicology of chemical compounds. *Mailing Add:* 1716 Laurel Ave PO Box 261 Hudson WI 54016

RUDEE, MERVYN LEA, b Palo Alto, Calif, Oct 4, 35; m 58; c 2. MATERIALS SCIENCE. *Educ:* Stanford Univ, BS, 58, MS, 62, PhD(mat sci), 65. *Prof Exp:* From asst prof to prof mat sci, Rice Univ, 64-74, master, Wiess Col, 69-74; prof mat sci & provost, Warren Col, 74-82, DEAN ENG, UNIV CALIF, SAN DIEGO, 82- *Concurrent Pos:* Guggenheim fel, Cavendish Lab, Cambridge Univ, 71-72; vis scientist, IBM Watson Res Ctr, Yorktown Heights, NY. *Mem:* Am Phys Soc; Electron Micros Soc Am; Mat Res Soc; Sigma Xi. *Res:* Defects in crystals, studies primarily by x-ray diffraction and electron microscopy; high temperature superconductors. *Mailing Add:* Univ Calif San Diego La Jolla CA 92093-0403

RUDEL, LAWRENCE L, b Salt Lake City, Utah, Sept 19, 41; m 68; c 3. PLASMA LIPOPROTEIN METABOLISM, CHOLESTEROL METABOLISM. *Educ:* Colo State Univ, BS, 63; Univ Ark Med Ctr, MS, 65, PhD(biochem), 69. *Prof Exp:* NIH fel, dept biochem, Univ Ark Med Ctr, 67-69; res fel, Banting & Best Dept Med Res, Can Heart Found, Univ Toronto, 70-71 & Cardiovasc Res Inst, Univ Calif, San Francisco, 72-73; from asst prof to assoc prof, Wake Forest Univ, 73-82, assoc biochem, 78-87, PROF COMP MED, BOWMAN GRAY SCH MED, WAKE FOREST UNIV, 82-, PROF BIOCHEM, 88- *Concurrent Pos:* Assoc ed, J Lipid Res, 80-90, Atherosclerosis, 89-, Arterio & Thromb, 90-; mem & chmn, Metab Study Sect, NIH, 86-90. *Honors & Awards:* Fel, Arteriosclerosis Coun AHA. *Mem:* AAAS; Am Asn Pathologists; Am Heart Asn; Am Soc Biol Chemists; Sigma Xi. *Res:* Evaluation of the role of the liver and intestine in cholesterol absorption and transport by lipoproteins of the circulation, including the blood and the lymph; experiments are done in nonhuman primate models of atherosclerosis with emphasis on nutritional effects. *Mailing Add:* Dept Comp Med Bowman Gray Sch Med Wake Forest Univ Winston-Salem NC 27103

RUDENBERG, FRANK HERMANN, b Berlin, Ger, Dec 4, 27; nat US; m; c 6. NEUROPHYSIOLOGY. *Educ:* Harvard Univ, SB, 49; Univ Chicago, SM, 51, PhD(physiol), 54. *Prof Exp:* Asst physiol, Univ Chicago, 52-53; instr & asst prof, Mich State Univ, 54-58; asst prof, 58-62, ASSOC PROF, INTEGRATED FUNCTIONAL LAB & PHYSIOL & BIOPHYS, UNIV TEX MED BR GALVESTON, 62- *Concurrent Pos:* Consult, Southwest Res Inst, San Antonio, 67-74; mem bd dir, Sci, Inc, 76- & Coast Alliance, 80-; assoc prof, Sch Allied Health, Univ Tex Med Br, 76-82; invited mem Bioeng Deleg Peoples Repub China, 87; guest lectr, Univ Houston, Clear Lake, 84-88, 91- *Mem:* Biophys Soc; Am Inst Biol Sci; Neuroelec Soc; Soc Neurosci; Coastal Soc; AAAS; NY Acad Sci; Am Asn Univ Prof. *Res:* Neurotoxicology; cerebral circulation, intracranial pressure and edema; neurophysiology of central nervous system trauma; medical instrumentation including microsensors; medical education. *Mailing Add:* Integrated Functional Lab Dept Physiol & Biophys Univ Tex Med Br Galveston TX 77550-2779

RUDENBERG, H(ERMANN) GUNTHER, b Berlin, Ger, Aug 9, 20; nat US; m 52; c 3. PHYSICS, ELECTRONICS. *Educ:* Harvard Univ, SB, 41, AM, 42, PhD, 52. *Prof Exp:* Asst electronics, Harvard Univ, 41-42; asst scientist, Los Alamos Sci Lab, 46; physicist res div, Raytheon Mfg Co, 48-52; dir res, Transitron Electronic Corp, 52-62; sr staff mem, Arthur D Little, Inc, 62-83; PRIN, RUDENBERG ASSOCS, 83- *Concurrent Pos:* Consult, Spencer-Kennedy Labs, 45-48 & 60-62. *Honors & Awards:* Centennial Medal, Inst Elec & Electronics Engrs. *Mem:* Inst Elec & Electronics Engrs. *Res:* Semiconductors; electron ballistics; microwaves; microelectronics; management consulting; technological forecasting; integrated circuits. *Mailing Add:* Rudenberg Assocs Three Lanthorn Lane Beverly MA 01915

RUDERMAN, IRVING WARREN, b New York, NY, Jan 7, 20; m 45; c 4. ELECTROOPTICS. *Educ:* Columbia Univ, PhD(chem), 49. *Prof Exp:* Res chemist, St Regis Paper Co, NJ, 40-46; teaching fel, Columbia Univ, 46-47, lectr, 47-48, res scientist, 48-54; consult, 54-56; founder & pres, Isomet Corp, 56-73; FOUNDER & PRES, INRAD, INC, 73- *Mem:* Am Chem Soc; Am Phys Soc; Inst Elec & Electronics Eng; fel NY Acad Sci. *Res:* Solid state physics; crystal growth; lasers; active optics. *Mailing Add:* INRAD Inc 181 Legrand Ave Northvale NJ 07647-2404

RUDERMAN, JOAN V, b Mt Vernon, NY. MOLECULAR BIOLOGY. *Educ:* Barnard Col, Columbia Univ, BA, 69; Mass Inst Technol, PhD(biol), 74. *Prof Exp:* Jane Coffin Childs fel, Mass Inst Technol, 74-76; ASSOC PROF BIOL, HARVARD MED SCH, 76- *Concurrent Pos:* Instr, Marine Biol Lab, 76-, corp mem, 79- *Mem:* Soc Develop Biol. *Res:* Molecular analysis of the levels at which differential gene expression during development is regulated. *Mailing Add:* Dept Anat & Cell Biol Harvard Med Sch Boston MA 02115

RUDERMAN, MALVIN AVRAM, b New York, NY, Mar 25, 27; m 53; c 3. ASTROPHYSICS. *Educ:* Columbia Univ, AB, 45; Calif Inst Technol, MS, 47, PhD, 51. *Prof Exp:* Physicist, Radiation Lab, Univ Calif, 51-52; NSF fel, Columbia Univ, 52-53; from asst prof to prof physics, Univ Calif, Berkeley, 53-64; prof, NY Univ, 64-69; PROF PHYSICS, COLUMBIA UNIV, 69-, CENTENNIAL PROF, 80- *Concurrent Pos:* Guggenheim fel, Inst Theoret Physics, Copenhagen, 57, Univ, Rome, 58 & Cambridge Univ, Int Astron, 79-80; vis sr res fel, Imperial Col, London, 67-68; vis prof, NY Univ, 60 & 64, Stanford Univ, 84-85; Stanley H Klosk vis lectr, NY Univ, 77; mem adv bd, Nat Inst Theoret Physics, 78-81; coun-at-large, Am Phys Soc, 81-, Cong Fel Comt, 81; mem, Pres Peer Rev Comt Acid Rain, 82; mem Acid Rain Rev Panel, Environ Protection Agency, 83; trustee, Aspen Ctr Physics, 86- *Honors & Awards:* Goodspeed-Richards Mem Lectr, Univ Pa, 70; Goldhaber Mem Lectr, Univ Tel Aviv, 72; Boris Pregel Award, NY Acad Sci, 74; Boris Jacobson Lectr, Univ Wash, 81. *Mem:* Nat Acad Sci; Am Phys Soc; Am Acad Arts & Sci; Am Astron Soc; NY Acad Sci. *Res:* Elementary particles; astrophysics. *Mailing Add:* Dept Physics Columbia Univ New York NY 10027

RUDERSHAUSEN, CHARLES GERALD, b Jersey City, NJ, May 26, 28; m 54; c 4. CHEMICAL ENGINEERING. *Educ:* Univ Va, BChE, 49; Univ Wis, MS, 50, PhD(chem eng), 52. *Prof Exp:* Sr engr, Eastern Lab, DuPont Chemicals, E I du Pont de Nemours & Co Inc, 52, res chem engr, 53, group leader, 54, sect head, 55-59, tech asst, 59-60, task coordr, 60-61, res sect head, 61-64, res mgr, Newburgh Res Lab, NY & Poromerics Res Lab, Tenn, 64-69, mgr tech progs, Eastern Lab, NJ, 69-72, licensing specialist, 72-76, res assoc, 76-85, SR PATENTS & CONTRACTS ASSOC, DUPONT CHEMICALS, E I DU PONT DE NEMOURS & CO, INC, WILMINGTON, 85- *Mem:* Sigma Xi. *Res:* Chemical processing; explosives; metallurgy; leather replacement materials; patents; licensing; metal winning; synfuels; catalysis. *Mailing Add:* 109 Taylor Lane Kennett Square PA 19348

RUDESILL, JAMES TURNER, b Rapid City, SDak, Nov 26, 23; m 48; c 3. ORGANIC CHEMISTRY. *Educ:* SDak Sch Mines & Tech, BS, 48; Iowa State Col, MS, 50; Purdue Univ, PhD(org chem), 57. *Prof Exp:* Res chemist, Cudahy Packing Co, 50-53; asst, Purdue Univ, 53-57; from asst prof to assoc prof, 57-67, PROF ORG CHEM, NDAK STATE UNIV, 67- *Mem:* Am Chem Soc; Royal Soc Chem. *Res:* Determination of configuration in geometrical isomers; additions and substitutions of free radicals; stereochemistry of ring opening processes. *Mailing Add:* Ladd Hall 255 NDak State Univ Fargo ND 58105-5516

RUDGE, WILLIAM EDWIN, b New Haven, Conn, June 14, 39; m 62; c 3. COMPUTATIONAL PHYSICS. *Educ:* Yale Univ, BS, 60; Mass Inst Technol, PhD(physics), 68. *Prof Exp:* Physicist, Prod Develop Lab, Components Div, 60-63, STAFF MEM, RES LAB, IBM CORP, 68- *Res:* Molecular dynamics calculations using special purpose parallel computer. *Mailing Add:* 1187 Washoe Dr IBM 650 Harry Rd San Jose CA 95120-6099

RUDGERS, ANTHONY JOSEPH, b Washington, DC, May 23, 38; m 61; c 3. PHYSICS. *Educ:* Univ Md, BS, 61; Cath Univ Am, MSE, 69, PhD(acoust), 77. *Prof Exp:* Res physicist, Acoust Div, Washington, DC, 61-76, RES PHYSICIST ACOUST, UNDERWATER SOUND REF DETACHMENT, NAVAL RES LAB, 76- *Concurrent Pos:* Chmn, Tech Comt Phys Acoust, Acoust Soc Am, 84-87. *Mem:* Fel Acoust Soc Am; Am Asn Physics Teachers; Sigma Xi. *Res:* Acoustics; materials research; radiation, diffraction and scattering theory; random signal theory; elasticity and structural mechanics; linear systems; continuum mechanics. *Mailing Add:* Underwater Sound Ref Detachment Naval Res Lab PO Box 568337 Orlando FL 32856-8337

RUDGERS, LAWRENCE ALTON, soil science, for more information see previous edition

RUDIN, ALFRED, b Edmonton, Alta, Feb 5, 24; m 49; c 3. POLYMER SCIENCE. *Educ:* Univ Alta, BSc, 49; Northwestern Univ, Evanston, PhD(org chem), 52. *Prof Exp:* Res chemist, Can Industs, Ltd, 52-60, group leader plastics, 60-64, mgr plastics lab, 64-67; assoc prof, 67-69, PROF CHEM, UNIV WATERLOO, 69-, PROF CHEM ENG, 73-, ADJ PROF CHEM, CHEM ENG & PHYSICS, 89- *Honors & Awards:* Protective Coatings Award, 83; Roon Found Award, 88; Polysar Award, 89. *Mem:* Am Chem Soc; Chem Inst Can; Soc Plastics Eng; Rheology Soc. *Res:* Polymer chemistry and engineering. *Mailing Add:* Dept Chem Univ Waterloo Waterloo ON N2L 3G1 Can

RUDIN, BERNARD D, b Los Angeles, CA, Nov 18, 27. TECHNICAL MANAGEMENT, COMPUTER SCIENCE. *Educ:* Calif Inst Technol, BS, 49; Univ Southern Calif, MS, 51; Stanford Univ, PhD(math), 65. *Prof Exp:* Mgr, Lockhead Missile & Space Corp, 56-65; MGR, APPLN TECHNOL CTR, IBM, 86- *Mem:* Am Math Soc; Math Asn Am. *Res:* Computational methods in various engineering and scientific disciplines; approximation theory. *Mailing Add:* Dept 41UA/276 IBM Kingston Lab Neighborhood Rd Kingston NY 12401

RUDIN, MARY ELLEN, b Hillsboro, Tex, Dec 7, 24; m 53; c 4. MATHEMATICS. *Educ:* Univ Tex, PhD(topology), 49. *Prof Exp:* Instr math, Duke Univ, 50-53; asst prof, Univ Rochester, 53-57; LECTR & PROF MATH, UNIV WIS-MADISON, 58- *Honors & Awards:* Hedrick Lectr, Math Asn Am, 79. *Mem:* Am Math Soc; Math Asn Am; Am Math Soc (vpres, 80-83); Asn Symbolic Logic. *Res:* Set theoretic topology, particularly the construction of counter examples. *Mailing Add:* Dept Math Univ Wis Madison WI 53706

RUDIN, WALTER, b Vienna, Austria, May 2, 21; m 53; c 4. MATHEMATICS. *Educ:* Duke Univ, BA & MA, 47, PhD(math), 49. *Prof Exp:* Instr math, Duke Univ, 49-50; Moore instr, Mass Inst Technol, 50-52; from asst prof to prof, Univ Rochester, 52-59; PROF MATH, UNIV WIS-MADISON, 59- *Mem:* Am Math Soc; Math Asn Am. *Res:* Mathematical analysis, especially abstract harmonic analysis, Fourier series and holomorphic functions of one and several variables. *Mailing Add:* Dept Math Univ Wis 805 Van Vleck 480 Lincoln Dr Madison WI 53706

RUDINGER, GEORGE, b Vienna, Austria, May 30, 11; US citizen; m 47, 72; c 4. FLUID MECHANICS, ENGINEERING PHYSICS. *Educ:* Vienna Tech Univ, Ingenieur, 35. *Prof Exp:* Res assoc med radiol, Vienna Gen Hosp, Austria, 35-38; physicist, Sydney Hosp, Australia, 39-46; prin physicist fluid mech, Cornell Aeronaut Lab, Buffalo, 46-70; prin scientist, Textron Bell Aerospace Co, Buffalo, 71-76; CONSULT, 76- *Concurrent Pos:* Proj engr, Royal Australian Air Force & Australian Ministry Munitions, 44-46; teacher physics, Sydney Tech Col, 45-46; physicist, Australian Glass Mgrs Pty, Ltd, Sydney, 45-46; adj prof, Dept Mech & Aero Eng, State Univ NY Buffalo, 75- *Honors & Awards:* Aerospace Pioneer Award, Am Inst Aeroaut & Astronaut, 74; Centennial Medallion, Am Soc Mech Engrs, 80; Fluids Eng Award, Am Soc Mech Engrs, 83. *Mem:* Fel Am Soc Mech Engrs; fel Am Phys Soc; fel Am Inst Aeronaut & Astronaut; Int Soc Biorheology; fel Brit Inst Physics; fel AAAS. *Res:* Gas particle flow, nonsteady duct flow and blood flow; x-ray photography. *Mailing Add:* 47 Presidents Walk Buffalo NY 14221-2426

RUDISILL, CARL SIDNEY, b Lincolnton, NC, Mar 21, 29; m 60; c 2. MECHANICAL ENGINEERING. *Educ:* NC State Univ, BME, 54, MS, 61, PhD(mech eng), 66. *Prof Exp:* Instr mech eng, NC State Univ, 60-65; assoc prof, 65-75, NASA res grants, 70-73, 75-76, 77-79, PROF MECH ENG, CLEMSON UNIV, 75- *Mem:* Am Inst Aeronaut & Astronaut; Am Soc Mech Engrs. *Res:* Optimization of aircraft structures. *Mailing Add:* Dept Mech Eng Clemson Univ 201 Sikes Hall Clemson SC 29631

RUDKIN, GEORGE THOMAS, genetics, cytology; deceased, see previous edition for last biography

RUDKO, ROBERT I, b New York, NY, Apr 24, 42; m 64; c 2. ELECTRICAL ENGINEERING. *Educ:* Cornell Univ, BEE, 63, MS, 65, PhD, 67. *Prof Exp:* Sr res scientist, 67-77, PRIN RES SCIENTIST, RES DIV, RAYTHEON CO, 77-; PRES, LASER ENG INC. *Mem:* Inst Elec & Electronics Engrs; Am Soc Lasers Med & Surg. *Res:* Laser radars; infrared optics; infrared gas lasers; surgical lasers. *Mailing Add:* Laser Eng Inc 113 Cedar St Suite S-2 Milford MA 01757

RUDMAN, ALBERT J, b New York, NY, Nov 14, 28; m 51; c 3. GEOPHYSICS. *Educ:* Ind Univ, BS, 52, MA, 54, PhD(geophys), 63. *Prof Exp:* Geophysicist, Carter Oil Co, 54-57 & Ind Geol Surv, 57-65; assoc prof geophysics, 65-77, PROF GEOL, IND UNIV, BLOOMINGTON, 77- *Mem:* Soc Explor Geophys; Am Geophys Union. *Res:* Solid earth geophysics, especially exploration. *Mailing Add:* Dept Geol Ind Univ Bloomington IN 47401

RUDMAN, DANIEL, b Pittsfield, Mass, Jan 25, 27; m 55; c 2. INTERNAL MEDICINE. *Educ:* Yale Univ, BS, 46, MD, 49; Am Bd Internal Med, dipl, 59. *Prof Exp:* Intern med, New Haven Hosp, 49-50; sr asst surgeon, Nat Cancer Inst, 51-53; resident, Jewish Hosp Brooklyn, 53-54; resident, Columbia Univ Res Serv, Goldwater Mem Hosp, 54-55; assoc med, 57-58, from asst prof to assoc prof, Col Physicians & Surgeons, Columbia Univ, 58-68; prof med, Emory Univ & dir, Clin Res Ctr, Emory Univ Hosp, 68-83; chief-of-staff geriat med, Vet Admin Med Ctr, N Chicago, 83-88; PROF MED, MED COL WIS, 88- *Concurrent Pos:* Res fel, Columbia Univ Res Serv, Goldwater Mem Hosp, 55-68. *Mem:* Harvey Soc; Am Fedn Clin Res; Am Soc Clin Invest; Endocrine Soc; Am Inst Nutrit; Am Asn Physic. *Res:* Cyclic nucleoride metabolism; amino acid metabolism in liver disease; biologically active peptides in animal tissues; cancer related proteins in body fluids; metabolic effects of human growth hormone; melanotropic peptides in the central nervous system; total intravenous hyperalimentation; endoneine, metabolic and nutritional abnormalities in the elderly. *Mailing Add:* Dept Med Med Col Wis Vet Admin Med Ctr 5000 W National Ave 18 Milwaukee WI 53295

RUDMAN, PETER S, b Passaic, NJ, Apr 30, 29; m 56; c 3. SOLID STATE PHYSICS. *Educ:* Mass Inst Technol, BSc, 51, MSc, 53, DSc, 55. *Prof Exp:* Res engr, Res Lab, Westinghouse Elec Corp, Corp, 55-56; lectr physics, Israel Inst Technol, 56-61; fel, Battelle Mem Inst, 61-66; assoc prof mat sci, Vanderbilt Univ, 67-68; prof physics, Technion-Israel Inst Technol, 68-80; MEM STAFF, INCO RES & DEVELOP INC, 80- *Mem:* Am Phys Soc; Am Soc Metals; Israel Crystallog Soc (secy, 60). *Res:* Metal physics; x-ray diffraction. *Mailing Add:* 321 Cupsaw Dr Ringwood NJ 07456

RUDMAN, REUBEN, b New York, NY, Jan 18, 37; m 58; c 5. X-RAY CRYSTALLOGRAPHY, STRUCTURAL CHEMISTRY. *Educ:* Yeshiva Univ, BA, 57; Polytech Inst Brooklyn, PhD(chem), 66. *Prof Exp:* Res assoc chem, Brookhaven Nat Lab, 66-67; from asst prof to assoc prof, 67-75, PROF CHEM, ADELPHI UNIV, 75- *Concurrent Pos:* Chmn comn on crystallog apparatus, Int Union Crystallog, 75-78; vis prof, Hebrew Univ, 73-74; pres, Asn Orthodox Jewish Scientists, 79-80; chmn orientational disorder crystals, Gordon Res Conf, 80; ed, Semi-annual J Assoc Orthodox Jewish Scientists, 85- *Mem:* Sigma Xi; AAAS; Am Crystallog Asn. *Res:* X-ray structure analysis; x-ray instrumentation and low-temperature apparatus; study of phase transitions and molecular complexes. *Mailing Add:* Dept Chem Adelphi Univ Garden City NY 11530

RUDMOSE, H WAYNE, b Cisco, Tex, Mar 16, 15; m 40; c 2. PHYSICS. *Educ:* Univ Tex, BA, 35, MA, 36; Harvard Univ, PhD(physics), 46. *Prof Exp:* Asst physics, Univ Tex, 32-35, tutor & chg electronic labs, 35-37; instr, Harvard Univ, 38-39, asst, 39-41, res assoc & group leader, Electro-Acoustic Lab, 41-43; assoc dir, 43-45; from asst prof to prof physics, Southern Methodist Univ, 46-63; group vpres, Sci & Systs Group, Tracor, Inc, 63-80; RETIRED. *Concurrent Pos:* Acoust consult, 46-80; mem comt on hearing & bio-acoustics, Nat Acad Sci, 53, chmn, 75-76. *Mem:* Acoust Soc Am; Am Acad Indust Hyg; Am Indust Hyg Asn. *Res:* Audio communication; hearing; architectural acoustics; noise measurement control. *Mailing Add:* 2802 Scenic Dr Austin TX 78703-1041

RUDNER, RIVKA, b Ramat-Gan, Israel, Apr 9, 35; m 56; c 3. GENETICS. *Educ:* NY Univ, BA, 57; Columbia Univ, MS, 58, PhD(zool), 61. *Prof Exp:* Res worker biochem, Col Physicians & Surgeons, Columbia Univ, 61-62, res assoc, 62-63, asst prof, 63-68; assoc prof, 68-74, PROF BIOL SCI, HUNTER COL, 74- *Mem:* Fedn Am Biochemists. *Res:* Molecular doning in B subtiles of L and H specific operons; sequence analyis and transcriptioner mapping of promotion regions; variation of nucleotide sequences among related Bacillus genomes in conserved and non conserved regions. *Mailing Add:* Dept Biol Sci Hunter Col 695 Park Ave New York NY 10021

RUDNEY, HARRY, b Toronto, Ont, Apr 14, 18; nat US; m 46; c 2. BIOCHEMISTRY. *Educ:* Univ Toronto, BA, 47, MA, 48; Western Reserve Univ, PhD(biochem), 52. *Prof Exp:* Res asst biochem, Univ Toronto, 46-48; sr instr, Sch Med, Western Reserve Univ, 51-53, from asst prof to prof, 53-67; prof biol chem & dir dept, 67-88, EMER PROF, MOLECULAR GENETICS, BIOCHEM & MICROBIOL, COL MED, UNIV CINCINNATI, 88- *Concurrent Pos:* Am Cancer Soc scholar, 55-57; NSF sr res fel, 57-58; USPHS res career develop award, 58-63 & res career award, 63-; vis prof, Case Inst Technol, 65-66; ed, Archiv Biochem & Biophys, 65- & J Biochem, 75-80; mem panel metab biol, NSF, 68-71; mem res career award comt, NIH, 69-71; mem biochem test comt, Nat Bd Med Examr, 74; pres, Dept Biochem, Assoc Med Sch, 81-82. *Honors & Awards:* G Rieveschel Jr Award, 77. *Mem:* Am Chem Soc; Am Soc Biol Chemists; Am Soc Microbiol; Brit Biochem Soc. *Res:* Biosynthesis of isoprenoid precursors of sterols. *Mailing Add:* Dept Molecular Genetics Biochem & Microbiol Univ Cincinnati Col Med Cincinnati OH 45267-0524

RUDNICK, ALBERT, b Manchester, NH, Apr 26, 22. VIROLOGY. *Educ:* Univ NH, BS, 42, MS, 44; Univ Calif, PhD(parasitol), 59. *Prof Exp:* Res asst, Hooper Found, Univ Calif, San Francisco, 47-50; entomologist, State Bur Vector Control, Calif, 52; med entomologist, US Chem Corps, Ft Detrick, Md, 54; res asst, Sch Pub Health, Univ Calif, 55; res assoc, Grad Sch Pub Health, Univ Pittsburgh, 56-60; asst res virologist, Hooper Found, Univ Calif, San Francisco, 60-65, assoc res virologist, 65-71; assoc prog dir, Univ Calif Int Ctr Med Res & Training, 72-74, prog dir, 74-80; prof virol, Dept Epidemiol & Internal Health, Hooper Found, Univ Calif, San Francisco, 74-82; RETIRED. *Concurrent Pos:* Rep, Comn Viral Infections, US Armed Forces Epidemiol Bd, Guam, 48, Tokyo, 49, Manila, 56, assoc mem, 66-72; consult, US Opers Mission, Bangkok, 58; consult to chief surgeon, 13th Air Force, Philippines, 58; actg sr virologist, Inst Med Res, Kuala Lumpur, 62-63; temporary adv, Western Pac Regional Off, WHO, Bangkok, 64; consult, 80- *Mem:* Am Soc Trop Med & Hyg; Wildlife Dis Asn; Am Mosquito Control Asn; Malaysian Soc Parasitol & Trop Med; Pac Sci Asn; Sigma Xi. *Res:* Ecology of dengue and other arboviruses of southeastern Asia. *Mailing Add:* 101 Lombard St Suite 503 W San Francisco CA 94111

RUDNICK, GARY, b Philadelphia, Pa, Sept 14, 46; m 81; c 2. BIOCHEMISTRY, PHARMACOLOGY. *Educ:* Antioch Col, BS, 68; Brandeis Univ, PhD(biochem), 74. *Prof Exp:* Fel biochem, Roche Inst Molecular Biol, 73-75; asst prof, 75-80, ASSOC PROF PHARMACOL, SCH MED, YALE UNIV, 80- *Concurrent Pos:* Estab investr, Am Heart Asn, 79-84. *Mem:* Am Soc Biol Chemists. *Res:* Membrane function; mechanisms of solute transport across biological membranes; uptake and storage of neurotransmitters. *Mailing Add:* Dept Pharmacol Yale Univ 333 Cedar St PO Box 3333 New Haven CT 06510-8066

RUDNICK, ISADORE, b New York, NY, May 8, 17; m 39; c 5. ACOUSTICS, LOW TEMPERATURE PHYSICS. *Educ:* Univ Calif, Los Angeles, BA, 38, MA, 40, PhD(physics), 44. *Prof Exp:* Res physicist, Brown Univ, 42; res physicist, Duke Univ, 42-45; from instr to asst prof physics, Pa State Univ, 45-48; from asst prof to assoc prof, Univ Calif, Los Angeles, 48-58, prof physics, 59-87; RETIRED. *Concurrent Pos:* Fulbright fel, Royal Inst Technol, Copenhagen, 57-58 & Israel Inst Technol, 65; vis prof, Univ Paris, 72, Israel Inst Technol, 73, Univ Tokyo, 77 & Univ Nanjing, China, 79; fac res lectr, Univ Calif, Los Angeles, 75-76; distinguished lectr, Acoust Soc Am, 80- *Honors & Awards:* Biennial Award, Acoust Soc Am, 48, Silver Medal, 75, Gold Medal, 82; Fritz London Award, Int Union Pure & Appl Physics, 81. *Mem:* Nat Acad Sci; fel Am Phys Soc; fel Acoust Soc Am (vpres, 62, pres, 69). *Res:* Ultrasonics; high intensity acoustics; cavitation; elastic wave damping in metals; low temperature physics; quantum liquids; non-linear physics. *Mailing Add:* Dept Physics Univ Calif Los Angeles CA 90024

RUDNICK, JOSEPH ALAN, b Durham, NC, Feb 1, 44; m 68; c 3. ACOUSTICS. *Educ:* Univ Calif, Berkeley, BA, 65; Univ Calif, San Diego, PhD(physics), 70. *Prof Exp:* Res assoc, Univ Wash, 69-72, Technion, Isreal Inst Technol, 72-74 & Case Western Reserve Univ, 74-78; asst prof physics, Tufts Univ, 78; prof, Univ Calif, Santa Cruz, 78-84; chmn, Dept Physics, 86-89, PROF PHYSICS, UNIV CALIF, LOS ANGELES, 84- *Concurrent Pos:* Consult, Jet Propulsion Lab, Calif Inst Technol, 85- *Mem:* Am Phys Soc; Mat Res Soc. *Res:* Phase transitions and critical phenomena; physics of disordered systems; structure and dynamics of deposited films; polymer physics; nonlinear dynamics and chaos. *Mailing Add:* Dept Physics Univ Calif 405 Hilgard Ave Los Angeles CA 90024-1547

RUDNICK, LAWRENCE, b Philadelphia, Pa, Mar 17, 49; m 70; c 2. RADIO ASTRONOMY, ASTROPHYSICS. *Educ:* Cornell Univ, BA, 70; Princeton Univ, MA, 72, PhD(physics), 74. *Prof Exp:* Res fel radio astron, Nat Radio Astron Observ, 74-76, asst scientist, 76-78, assoc scientist, 78; from asst prof to assoc prof, 79-86, PROF ASTRON, UNIV MINN, 86- *Mem:* Am Astron Soc; Int Astron Union. *Res:* Extragalactic radio astronomy; radio galaxies; supernova remnants; relativistic particle acceleration. *Mailing Add:* Sch Physics & Astron Univ Minn 116 Church St SE Minneapolis MN 55455

RUDNICK, MICHAEL DENNIS, b Huntington, NY, Aug 5, 45; c 2. ANATOMICAL SCIENCES, OTOLOGY. *Educ:* State Univ NY Buffalo, BA, 71, PhD(anat), 78. *Prof Exp:* Resident, Dept Pediat, Univ SFla, Tampa, 87-89; fel pediat med, Univ Colo, 89-91; DIR, WESTSIDE NEIGHBORHOOD HEALTH CTR, DENVER, 91- *Concurrent Pos:* Clin instr, State Univ NY, Buffalo, 78- *Res:* Anatomy, physiology and pathology of the inner ear; aminoglycoside antibiotic ototoxicity; cochlear implant and the function of kinocilia in the vestibular system. *Mailing Add:* 2288 E Eastman Ave Englewood CO 80110

RUDNICK, STANLEY J, b Chicago, Ill, Oct 10, 37; m 58; c 5. INSTRUMENTATION ENGINEERING. *Educ:* Northwestern Univ, BS, 59; Univ Ill, MS, 61; Univ Chicago, MBA, 78. *Prof Exp:* Engr electronics, Motorola, Inc, 59-60; engr instrumentation, 61-74, dir electronics div, 74-85, RES PROG MGR, ARGONNE NAT LAB, 85- *Concurrent Pos:* Mem, Nat Instrumentation Methods Comt, 64- *Mem:* Inst Elec & Electronics Engrs; Nuclear & Plasma Sci Soc; Eng Mgt Soc; Aerospace & Electronic Systs Soc. *Res:* Nuclear instrumentation and systems. *Mailing Add:* Argonne Nat Lab Bldg 207 9700 S Cass Ave Argonne IL 60439

RUDNYK, MARIAN E, b Long Island, NY, June 14, 60. PLANETARY GEOLOGY, HISTORICAL ASTRONOMY. *Educ:* Calif State Polytech Univ, BS, 83. *Prof Exp:* Planetary photogeologic consult, Path Sect, 83-85 & 86-88, astrom, Comet & Asteroid Res Prog, 85-86, MGR, PLANETARY IMAGE FACIL, JET PROPULSION LAB, NASA, 88- *Concurrent Pos:* Lectr astron; mem, Planetary Sci & Imaging Teams, NASA. *Res:* Astronomical asteroids; historical information and data interpretation of historical NASA and international planetary missions. *Mailing Add:* 732 W Hillcrest Blvd Monrovia CA 91016

RUDO, FRIEDA GALINDO, b New York, NY, Nov 13, 23; m 45; c 2. PHARMACOLOGY. *Educ:* Goucher Col, AB, 44; Univ Md, MS, 60, PhD(pharmacol), 63. *Hon Degrees:* DSc, Goucher Col, 76. *Prof Exp:* From instr exp surg to asst prof pharmacol, Univ Md, Sch Med, 60-68, from asst prof to assoc prof, Sch Dent, 68-75, prof, 70-90, EMER PROF PHARMACOL, SCH DENT, UNIV MD, BALTIMORE, 90- *Concurrent Pos:* Consult, Ohio Chem Co, 60, res grant, 64 - *Mem:* Fel Am Col Dentists; fel Explorer's Club. *Res:* Cardiovascular research using artificial heart; cardiac output studies on new nitrate compounds; pharmacology and toxicity of new anesthetic agents and analgesics. *Mailing Add:* Dept Pharmacol Univ Md Sch Dent Baltimore MD 21201

RUDOLF, LESLIE E, b Pelham, NY, Nov 12, 27; m 55; c 4. SURGERY. *Educ:* Union Col, NY, BS, 51; Cornell Univ, MD, 55; Am Bd Surg, dipl, 63. *Prof Exp:* Intern & to resident surg, Peter Bent Brigham Hosp, Boston, 55-59 & 60-61; asst, Harvard Med Sch, 59-60; instr, New York Hosp-Cornell Univ, 61-63; from asst prof to assoc prof, 63-72, PROF SURG, SCH MED, UNIV VA, 72-, VCHMN DEPT SURG, 76- *Concurrent Pos:* Mem, Nat Bd Med Examr; Markle scholar acad med, 66-71. *Mem:* Am Col Surg; Soc Surg Alimentary Tract; Am Soc Nephrology; Am Soc Artificial Internal Organs; Transplantation Soc. *Res:* Organ and tissue transplantation and preservation; long-term organ storage and histocompatibility typing; evaluation of current medical and surgical methods used in treatment of deep thrombophlebitis. *Mailing Add:* Dept Surg Univ Va Med Ctr Univ Va Sch Med Charlottesville VA 22904

RUDOLF, PAUL OTTO, b LaCrosse, Wis, Nov 4, 06; m 32; c 2. VARIATION IN PINUS & PICEA. *Educ:* Univ Minn, BS, 28; Cornell Univ, MF, 29. *Prof Exp:* Res asst forestry, Cornell Univ, 28-29; jr forester, US Forest Serv, Southern Forest Exp Sta, 29-30 & Lake States Forest Exp Sta, 30-35; asst silviculturist, Lake States Forest Exp Sta, 35-37, assoc silviculturist, 37-42, silviculturist, 42-49, forester, 49-59, res forester, 59-65, prin silviculturist, 65; prin silviculturist, N Cent Forest Exp Sta, 66-67, expert, 67-71; lectr appl silvicult & forest ecol, Univ Minn, 67-75, res assoc, 75-82; RETIRED. *Concurrent Pos:* Exec secy, Lake States Forest Tree Improv Comt, 55-66; US rep & chmn, Tree Seed Experts Meeting, Orgn Econ Coop & Develop, 63 & 65; mem, World Forestry Cong, Seattle, 60, Soc of Am Foresters; chmn, Div Silviculture, 61, actg chmn, 65, chmn Tree Seed Comt, 63-64. *Honors & Awards:* Super Serv Award, US Dept Agr, 65. *Mem:* Fel Soc Am Foresters; fel AAAS; Sigma Xi. *Res:* Reforestation; racial variation in forest trees; forest seeds; silviculture of northern trees; distribution of tree species; forest research history; forest tree improvement. *Mailing Add:* 7244 York Ave S Apt 423 Edina MN 55435-4417

RUDOLPH, ABRAHAM MORRIS, b Johannesburg, SAfrica, Feb 3, 24; nat US; m 49; c 3. PHYSIOLOGY. *Educ:* Univ Witwatersrand, MB, BCh, 46, MD, 51; Am Bd Pediat, dipl, 53; FRCP(E), 66, FRCP(L). *Prof Exp:* Res fel pediat, Harvard Med Sch, 51-53, res fel physiol, 53-54, Am Heart Asn res fel, 54-55, instr pediat, 55-57, assoc, 57-60; assoc prof, Albert Einstein Col Med, 60-63, prof pediat & assoc prof physiol, 63-66; PROF PEDIAT & PHYSIOL, UNIV CALIF, SAN FRANCISCO, 66-, PROF OBSTET & GYNEC, 74- *Concurrent Pos:* Fel, Children's Med Ctr, Boston, 51-53, from asst cardiologist in chg to assoc cardiologist in chg, Cardiopulmonary Lab, 55-66; Am Heart Asn estab investr, 56-; mem, Nat Adv Heart & Lung Coun. *Honors & Awards:* Mead Johnson Award; Borden Award; Arvo Yippo Award, Helsinki. *Mem:* Soc Clin Invest; Am Pediat Soc; Am Physiol Soc; Soc Pediat Res; Am Acad Pediat. *Res:* Cardiovascular physiology, particularly physiology of the fetus and newborn; physiology of congenital heart disease; pediatric cardiology. *Mailing Add:* Dept Pediat-Obstet-Gynec Univ Calif Med Ctr San Francisco CA 94143

RUDOLPH, ARNOLD JACK, b Johannesburg, SAfrica, Mar 28, 18; m 51; c 4. PEDIATRICS, NEONATAL-PERINATAL MEDICINE. *Educ:* Univ Witwatersrand, MB, BCh, 40. *Prof Exp:* Resident neurol & dermat, Univ Witwatersrand, 41; resident pediat, Transvaal Mem Hosp Children, 41, casualty & outpatient dept officer, 42-46, clin fel med, 47-49, clin asst pediat, 49-56; res fel pediat, Boston-Lying-In Hosp, Children's Med Ctr & Harvard Med Sch, 57-59; from asst prof to prof, 61-91, DISTINGUISHED FAC MEM, BAYLOR COL MED, 81- *Mem:* AMA; Brit Med Asn; Am Pediat Soc; hon fel Phillipines Pediat Soc; Am Acad Pediat. *Res:* Newborn physiology; newborn problems. *Mailing Add:* Dept Pediat Baylor Col Med One Baylor Plaza Texas Med Ctr Houston TX 77030-3498

RUDOLPH, EMANUEL DAVID, b Brooklyn, NY, Sept 9, 27; m 62. BOTANY, HISTORY OF BIOLOGY. *Educ:* NY Univ, AB, 50; Wash Univ, PhD(bot), 55. *Prof Exp:* Asst, Hunter Col, 50; docent, Brooklyn Children's Mus, 50-51; asst, cryptogamic herbarium, Mo Bot Garden, 51-55, asst librn, 54-55; from instr to asst prof bot, Wellesley Col, 55-61; from instr to assoc prof, Ohio State Univ, 61-69, res assoc, 61-69, dir, Inst Polar Studies, 69-73, dir environ biol prog, 72-78, actg chmn, 78-79, chmn, 79-87, prof bot, 69-89, EMER PROF PLANT BIOL, OHIO STATE UNIV, 90- *Concurrent Pos:* Spec asst, Washington Univ, 53-54; fel mycol, NSF, Univ Wis, 59; ed, Plant Sci Bull, Bot Soc Am, 80-85. *Honors & Awards:* Antarctic Medal, US Govt, 70. *Mem:* AAAS; Bot Soc Am; Am Bryol & Lichenological Soc (vpres, 72-73, pres, 74-75); fel Arctic Inst NAm; fel Linnean Soc London; Hist Sci Soc. *Res:* Lichenology; antarctic and arctic botany; history of botany and biology. *Mailing Add:* Dept Plant Biol Ohio State Univ 1735 Neil Ave Columbus OH 43210

RUDOLPH, FREDERICK BYRON, b St Joseph, Md, Oct 17, 44; m 71; c 2. ENZYMOLOGY, METABOLIC REGULATION. *Educ:* Univ Mo, Rolla, BS, 66; Iowa State Univ, PhD(biochem), 71. *Prof Exp:* NSF postdoctoral fel biochem, Univ Wis, 71-72; from asst prof to assoc prof, 72-85, PROF BIOCHEM, RICE UNIV, 85-, DIR, MABEE LAB BIOTECH & GENETIC ENG, 86- *Concurrent Pos:* Consult, World Book Encycl, 72-; assoc ed, Yearbk Cancer, 76-84; adj prof, Univ Tex Med Sch, Houston, 81-; mem, Biochem Study Sect, NIH, 83-87. *Mem:* Am Soc Biochem & Molecular Biol; Am Chem Soc; Am Soc Biotechnol; Coun Undergrad Res; Asn Biol Lab Educ. *Res:* Enzyme studies on nucleotide metabolism; structure and function of enzymes; role of dietary nucleotides in immune function; anaerobic fermentation; protein purification; antimetabolite action. *Mailing Add:* 4414 Silverwood Houston TX 77035

RUDOLPH, GUILFORD GEORGE, b Kiowa, Kans, Jan 2, 18; m 44; c 3. BIOCHEMISTRY. *Educ:* Univ Colo, BA, 40; Wayne State Univ, MS, 42; Univ Utah, PhD(biochem), 48. *Prof Exp:* Teaching fel biochem, Wayne State Univ, 40-42; instr, Univ Utah, 46-48; asst prof, Vanderbilt Univ, 49-57; assoc prof, Univ Md, 57-60; assoc prof, Vanderbilt Univ, 60-67; asst dean basic sci, 67-73, head dept, 67-85, PROF BIOCHEM, LA STATE UNIV MED CTR, SHREVEPORT, 67- *Concurrent Pos:* Am Cancer Soc res assoc, Univ Chicago, 48-49; prin scientist, Vet Admin Hosp, Nashville, Tenn, 49-57; mem grad fac, Med Ctr, La State Univ, 68-; consult, Vet Admin Hosp, 68- *Mem:* AAAS; Am Physiol Soc; Am Asn Clin Chem; Am Chem Soc. *Res:* Clinical chemistry. *Mailing Add:* 550 Dunmoreland Dr Shreveport LA 71106-6125

RUDOLPH, JEFFREY STEWART, b Chicago, Ill, Oct 30, 42; m 67; c 2. PHARMACY, CHEMISTRY. *Educ:* Univ Ill, Chicago, BS, 66; Purdue Univ, Lafayette, MS, 69, PhD(pharm), 70. *Prof Exp:* Sr res pharmacist, Ciba-Geigy Pharmaceut Corp, 70-72; sr scientist, McNeil Labs Div, Johnson & Johnson, 72-75, group leader pharm, Pilot Plant, 75-76; asst dir, 77-80, dir pharmaceut develop, Stuart Pharmaceut Div, ICI Am, Inc, 80-87; VPRES PHARMACEUT RES & DEVELOP, ICI PHARMACEUT, 88- *Mem:* Acad Pharmaceut Sci; Am Pharmaceut Asn; Am Asn Pharmaceut Scientist. *Res:* Optimization of drug delivery systems; development of new dosage forms with emphasis on optimum bioavailability; evaluation of pharmaceutical processing equipment. *Mailing Add:* 2201 Patwynn Ct Wynnwood Wilmington DE 19810

RUDOLPH, LEE, b Cleveland, Ohio, March 28, 48; c 2. KNOT THEORY, COMPLEX PLANE CURVES. *Educ:* Princeton Univ, AB, 69, Mass Inst Technol, PhD(math), 74. *Prof Exp:* Researcher, Proj Logo, Artificial Intelligence Lab, Mass Inst Technol, 74; instr math, Brown Univ, 74-77; asst prof math, Columbia Univ, 77-82; vis asst prof, Brandeis Univ, 83; vis asst prof, 86-87, asst prof, 87-90, ASSOC PROF, CLARK UNIV, 90- *Concurrent Pos:* Vis researcher, Univ Geneva, Switzerland, 82, vis prof math, 83-84; mem, Math Sci Res Inst, Calif, 84-85; vis lectr, Univ Md, 85; vis prof math, Univ Zaragoza, Spain, 86, Univ Nacional Autonoma de Mex, 86, Univ Paul Sabatier, Toulouse, France, 91. *Mem:* Am Math Soc. *Res:* Relationships between low-dimensional topology and several complex variables; knot theory of complex plane curves; theories of braided surfaces, fibered links, enhanced Milnor number, generalized Jones polynomials; homology of arithmetic groups. *Mailing Add:* Box 251 Adamsville RI 02801-0251

RUDOLPH, LUTHER DAY, b Cleveland, Ohio, Aug 10, 30; m 54; c 3. INFORMATION SCIENCE. *Educ:* Ohio State Univ, BS, 58; Univ Okla, MEE, 64; Syracuse Univ, PhD(systs & info sci), 68. *Prof Exp:* Engr, Gen Elec Co, 58-64; res engr, Res Corp, 64-68, from asst prof to assoc prof systs & info sci, 68-75, PROF SYSTS & INFO SCI, SCH COMPUT & INFO SCI, SYRACUSE UNIV, 75- *Mem:* Inst Elec & Electronics Engrs; Am Soc Psychical Res; Parapsychol Asn; Am Math Soc. *Res:* Theory and implementation of error-correcting codes; application of combinatorial mathematics to problems in communication and system science; application of information theory to extrasensory communication. *Mailing Add:* Dept Comput Info Sci Syracuse Univ 212 Sims Hall Syracuse NY 13210

RUDOLPH, PHILIP S, b Syracuse, NY, May 10, 12; m 42; c 2. PHYSICAL & RADIATION CHEMISTRY, MASS SPECTROMETRY. *Educ:* Syracuse Univ, AB, 33, PhD(chem), 51. *Prof Exp:* Instr, Syracuse Univ, 46; chemist, Oak Ridge Nat Lab, 51-73; RETIRED. *Mem:* Am Chem Soc; Radiation Res Soc; NY Acad Sci. *Res:* Mass spectrometric kinetic studies; chemical kinetics and reaction mechanisms; alpha particle induced radiolyses in the mass spectrometer. *Mailing Add:* 106 E Damascus Rd Oak Ridge TN 37830-4019

RUDOLPH, RAY RONALD, b Lock Haven, Pa, Feb 6, 27. MATHEMATICS. *Educ:* Johns Hopkins Univ, BE, 50. *Prof Exp:* Instr math, McCoy Col, John Hopkins Univ, 54-59, res asst oper res, Inst Coop Res, 50-53, res staff asst, 53-58, res assoc, 58-63, res scientist, 63-69; sr res analyst, Thor Div, Falcon Res & Develop Co, 69-81; SR RES ANALYST, KETRON, INC, 81- *Mem:* Am Ord Asn. *Res:* Aircraft vulnerability; weapons effectiveness; ballistics; operations research. *Mailing Add:* 219 Quaker Ridge Rd Timonium MD 21093

RUDOLPH, RAYMOND NEIL, b Lansing, Mich, May 19, 46; m 71; c 2. PHYSICAL CHEMISTRY. *Educ:* Univ NMex, BS, 68; Univ Colo, MS, 71, PhD(chem), 77. *Prof Exp:* Lab asst chem, Univ Colo, 68-71; sec teacher physics & math, Koidu Sec Sch, Sierra Leona, Africa, 71-74; asst prof, 77-83, ASSOC PROF CHEM, ADAMS STATE COL, 83- *Concurrent Pos:* Am Soc Eng Educ/NASA res fel, 79 & 80. *Mem:* Sigma Xi; Am Chem Soc. *Res:* Ultraviolet/visible spectroscopy; photophysics and photochemistry of small molecules. *Mailing Add:* 511 Brown Alamosa CA 81101

RUDOLPH, WILLIAM BROWN, b St Paul, Minn, Dec 14, 38; m 61; c 3. MATHEMATICS. *Educ:* Bethany Col, WVa, BA, 60; Purdue Univ, Lafayette, MS, 65, PhD(math educ), 69. *Prof Exp:* Teacher, Shaker Heights Bd Educ, 61-63; instr math, Menlo Col, 63-66; instr, Univ Santa Clara, 64-66; instr & res asst math educ, Purdue Univ, Lafayette, 66-69; from asst prof to assoc prof, 69-81, PROF MATH & EDUC, IOWA STATE UNIV, 81- *Concurrent Pos:* Consult, Iowa Dept Pub Instr, 71-; consult & res grant evaluator, North Cent Asn, ESEA Title III & State of Iowa, 71- *Mem:* Sch Sci & Math Asn; Nat Coun Teachers Math. *Res:* Information theory concepts as applied to language analysis; mathematics learning theory; computer assisted instruction; textbook author. *Mailing Add:* Dept Math Iowa State Univ Ames IA 50011

RUDVALIS, ARUNAS, b Bavaria, Ger, June 8, 45; US citizen. ALGEBRA. *Educ:* Harvey Mudd Col, BS, 65; Dartmouth Col, MA, 67, PhD(math), 69. *Prof Exp:* Eng assoc, Gen Atomic Div, Gen Dynamics, 65; vis asst prof math, Dartmouth Col, 69; res assoc, Mich State Univ, 69-70, asst prof, 70-72; asst prof, 72-74, ASSOC PROF MATH, UNIV MASS, 75- *Concurrent Pos:* NSF res grants, 73, 74 & 75. *Mem:* Am Math Soc. *Res:* Finite simple groups; representations of finite groups; finite geometries; coding theory. *Mailing Add:* Dept Math & Statist Univ Mass Amherst MA 01003

RUDY, BERNARDO, b Mexico City, Mex, March 21, 48; m. NEUROBIOLOGY, MEMBRANE PHYSIOLOGY. *Educ:* Nat Univ, Mex, MD, 71; Centro Invest Estud Avanzados, Mex, PhD(biochem), 72; Cambridge Univ, UK, PhD(physiol), 76. *Prof Exp:* Res assoc, Univ Pa, 76-78; ASST PROF PHYSIOL & MEMBRANE PHYSIOL, NY UNIV MED CTR, 79- *Mem:* Biophys Soc; Soc Neurosci. *Res:* Molecular understanding of brain function including studies of the structure of the molecules involved in excitation, as well as their metabolism and genetic control. *Mailing Add:* Dept Physiol & Biophys NY Univ Med Ctr 550 First Ave New York NY 10016

RUDY, CLIFFORD R, b Cleveland, Ohio, Jan 27, 43; m 64. NUCLEAR MEASUREMENTS. *Educ:* Univ Ariz, BS, 64; Univ Wash, PhD(chem), 70. *Prof Exp:* Postdoctoral res assoc, Kans State Univ, 70-72, Purdue Univ, 72-76; sr chemist, Monsanto-Mound, 77-83, prof mgr, 83-85, 85-88, SR RES SPECIALIST, EG&G MOUND, 88- *Mem:* Am Phys Soc; Am Chem Soc; Am Nuclear Soc; Inst Nuclear Mat Mgt. *Res:* Perform applied research in nuclear material safeguards measurements; establish measurement control systems for nuclear accountability. *Mailing Add:* PO Box 3000 Mound Miamisburg OH 45343-3000

RUDY, LESTER HOWARD, b Chicago, Ill, Mar 6, 18; m 50; c 1. PSYCHIATRY. *Educ:* Univ Ill, BS, 39; Univ Ill Col Med, MD, 41; Northwestern Univ, MSHA, 57. *Prof Exp:* Resident psychiat, Downey Vet Admin Hosp, Ill, 46-48, chief serv, 48-54; supt, Galesburg State Res Hosp, Ill, 54-58; dir, Ill Ment Health Insts, 72-75; dir, Ill State Psychiat Inst, 58-75; dir, Univ Ill Hosp, 81-82; PROF & HEAD, DEPT PSYCHIAT, COL MED, UNIV ILL, 75- *Concurrent Pos:* Prof, Dept Psychiat, Univ Ill, 59-; comnr, Joint Comn Accreditation Hosps, 67-76; chmn, NIMH Res Serv Comt, 72-73; sr consult, Vet Admin; exec dir, Am Bd Psychiat & Neurol, 72-86. *Mem:* Fel Am Col Psychiat; fel Am Psychiat Asn; Am Asn Social Psychiat. *Res:* Educational standards and evaluation. *Mailing Add:* 3200 Highland Ave Downers Grove IL 60515

RUDY, PAUL PASSMORE, JR, b Santa Rosa, Calif, Aug 29, 33; m 54; c 3. COMPARATIVE PHYSIOLOGY. *Educ:* Univ Calif, Davis, AB, 55, MA, 59, PhD(zool), 66. *Prof Exp:* Res asst pesticides, Univ Calif, Davis, 53-54, isopods, 60, lab technician, Marine Aquaria, 62-65; teacher pub sch, Calif, 56-62; fel salt & water balance in aquatic animals, Univ Birmingham, 66-67; fel, Univ Lancaster, 67-68; asst prof, 68-71, ASSOC PROF BIOL, UNIV ORE, 71-, DIR ORE INST MARINE BIOL, 69- *Concurrent Pos:* Partic, Int Indian Ocean Exped, Stanford Univ, 64; asst dir, Ore Inst Marine Biol, 68-69; NSF grant, 69-71. *Mem:* AAAS; Am Inst Biol Sci; Am Soc Zool; Sigma Xi. *Res:* Marine biology. *Mailing Add:* Ore Inst Marine Biol Charleston OR 97420

RUDY, RICHARD L, b Covington, Ohio, Aug 15, 21; m 52; c 5. VETERINARY SURGERY. *Educ:* Ohio State Univ, DVM, 43, MSc, 47. *Prof Exp:* From instr to assoc prof vet surg, 44-57, prof vet surg & radiol & chmn dept, 57-70, assoc dir vet clin, 61, prof vet clin sci, 71-84, EMER PROF, OHIO STATE UNIV, COL VET MED, 84- *Mem:* Am Asn Vet Clinicians; Am Col Vet Surg; Am Animal Hosp Asn; Am Vet Med Asn; Vet Orthop Soc. *Res:* Veterinary surgery, including orthopedics, thoracic and neurologic. *Mailing Add:* 3000 Oldham Rd Columbus OH 43221

RUDY, THOMAS PHILIP, b Chicago, Ill, Mar 14, 24; m 51. ORGANIC CHEMISTRY, SOLID ROCKET PROPELLANTS. *Educ:* Univ Chicago, MS, 50, PhD(chem), 52. *Prof Exp:* Chemist, Shell Develop Co, 52-56 & 58-62; asst prof chem, Univ Chicago, 56-58; head org chem, United Technol Chem Systs, 62-67, prin scientist, 67-85, chief scientist, 85-90; CONSULT, 90- *Mem:* Am Chem Soc. *Res:* Organic chemistry; lubricants; fuels; antioxidants; polymers; rocket propellant ingredients and combustion. *Mailing Add:* 21142 Sarahills Dr Saratoga CA 95070

RUDY, YORAM, b Tel-Aviv, Israel, Feb 12, 46. CARDIAC ELECTROPHYSIOLOGY, ELECTROCARDIOGRAPHY. *Educ:* Israel Inst Technol, BSc, 71, MSc, 73; Case Western Reserve Univ, PhD(biomed eng), 78. *Prof Exp:* Res assoc, Case Western Reserve Univ, 78-79, vis asst prof, 79-81, from asst prof to assoc prof, 81-89, PROF BIOMED ENG, CASE WESTERN RESERVE UNIV, 89- *Concurrent Pos:* Vis prof, biomed eng dept, Technion, Israel, 82-83; mem, Cardiovascular & Pulmonary Study Sect, NIH, 84-88. *Mem:* AAAS; Inst Elec & Electronics Engrs; sr mem Biomed Eng Soc; Cardiac Electrophysiologic Soc; Biophys Soc; Am Physiol Soc; Am Heart Asn; Sigma Xi. *Res:* Model studies of the electrical activity of the heart on the cellular and tissue level; multi-electrode mapping of heart and body surface potentials; forward and inverse problems in electrocardiography; models of arrhythmias and neural control of the heart. *Mailing Add:* Dept Biomed Eng Case Western Reserve Univ Cleveland OH 44106

RUDZIK, ALLAN D, b Mundare, Alta, Nov 30, 34; m 60; c 2. PHARMACOLOGY. *Educ:* Univ Alta, BSc, 56, MSc, 58; Univ Wis, PhD(pharmacol), 62. *Prof Exp:* Pharmacologist, Ayerst Labs, Que, 62-63; pharmacologist, Pitman-Moore Div, Dow Chem Co, 63-65, sr pharmacologist, 65-66; res scientist cent nerv syst, 66-72, sr scientist, 72-74, res head, 74-79, mgr cent nerv syst res, 79-81, GROUP MGR THERAPEUT, UPJOHN CO, 81- *Mem:* AAAS; Am Soc Pharmacol & Exp Therapeut; NY Acad Sci; Sigma Xi. *Res:* Autonomic pharmacology as applied to smooth muscle, cardiovascular system and the central nervous system. *Mailing Add:* 17 Black Birch Dr Randolph NJ 07869

RUDZINSKA, MARIA ANNA, b Dabrowa, Poland; nat US; m 30. ZOOLOGY, PROTOZOOLOGY. *Educ:* Jagiellonian Univ, MS & PhD(zool). *Prof Exp:* Res assoc cell physiol, NY Univ, 46-52, instr, 47-49; vis investr electron micros, 52-56, res assoc parasitol, 56-60, from asst prof to assoc prof, 60-75, EMER PROF ROCKEFELLER UNIV, 75- *Honors & Awards:* Alfred Jurzykowski Award for Outstanding Achievements in Biol & Sci, Alfred Jurzykowski Found, NY, 75. *Mem:* Harvey Soc; hon mem Soc Protozool; fel Geront Soc; fel NY Acad Sci; Sigma Xi. *Res:* Morphogenesis, aging and cell biology of free-living and parasitic protozoa; cell biology; parasitology. *Mailing Add:* Rockefeller Univ New York NY 10021

RUE, EDWARD EVANS, b Harrisburg, Pa, Oct 3, 24; m 44; c 3. EXPLORATION GEOLOGY, RESOURCE MANAGEMENT. *Educ:* Berea Col, AB, 48; Colo Sch Mines, MS, 49. *Prof Exp:* Geologist, Magnolia Petrol Co, 49-53; CONSULT GEOLOGIST, 53- *Honors & Awards:* Martin Van Couvering Mem Award, Am Inst Prof Geologists, 86. *Mem:* Fel Geol Soc Am; Am Asn Petrol Geol; Soc Petrol Eng; Am Inst Mining, Metall & Petrol Eng; hon mem Am Inst Prof Geologists (secy-treas, 66-67, pres, 79). *Res:* Petroleum geology and engineering; evaluations for industry and governmental agencies; industrial minerals exploration and programming for major producers; geological research coordinated with field work. *Mailing Add:* POD 647 Mt Vernon IL 62864

RUE, JAMES SANDVIK, b Sheyenne, NDak, Nov 19, 29; m 57; c 2. MATHEMATICS. *Educ:* Mayville State Col, BS, 51; Univ NDak, MS, 55; Iowa State Univ, PhD(math), 65. *Prof Exp:* From instr to asst prof math, Univ NDak, 55-60; mathematician, Boeing Airplane Co, 57-58; instr math, Iowa State Univ, 60-65; asst prof, Univ Wyo, 65-66 & Wash State Univ, 66-70; ASSOC PROF MATH, UNIV NDAK, 70- *Mem:* Am Math Soc; Math Asn Am. *Res:* Functional analysis. *Mailing Add:* PO Box 8135 Univ Sta Grand Forks ND 58202

RUE, ROLLAND R, b Marshfield, Wis, Apr 25, 35; m 58; c 2. PHYSICAL CHEMISTRY. *Educ:* Macalester Col, BA, 57; Iowa State Univ, PhD(phys chem), 62. *Prof Exp:* Asst prof, 62-70, ASSOC PROF CHEM, SDAK STATE UNIV, 70- *Mem:* AAAS; Am Chem Soc. *Res:* Interpretation of molecular wave functions; theoretical chemistry; thermodynamic properties of solutions; electrochemistry. *Mailing Add:* 2043 Elmwood Dr Brookings SD 57006-2736

RUEBNER, BORIS HENRY, b Dusseldorf, Ger, Aug 30, 23; US citizen; m 57; c 2. PATHOLOGY. *Educ:* Univ Edinburgh, MB, ChB, 46, MD, 56. *Prof Exp:* Asst prof path, Dalhousie Univ, 57-59; from asst prof to assoc prof, Johns Hopkins Univ, 59-68; PROF PATH, SCH MED, UNIV CALIF, DAVIS, 68- *Mem:* Col Am Path; Int Acad Path; Am Asn Pathologists; Am Asn Study Liver Dis; AMA. *Res:* Liver pathology; hepatic carcinogenesis. *Mailing Add:* Dept Path Univ Calif Sch Med Davis CA 95616

RUECKERT, ROLAND R, b Rhinelander, Wis, Nov 24, 31; m 59; c 1. VIROLOGY. *Educ:* Univ Wis, BS, 53, MS, 57, PhD(oncol), 60. *Prof Exp:* Mem res staff, McArdle Mem Lab Cancer Res, Univ Wis, 59-60; asst res virologist, Univ Calif, Berkeley, 62-64, lectr molecular biol, 64-65; from asst prof to assoc prof, 65-72, PROF BIOCHEM, UNIV WIS-MADISON, 72-, CHMN, INST MOLECULAR BIOL, 88- *Concurrent Pos:* Fel, Max Planck Res Inst Biochem, Munich, 60-61; Max Planck Res Inst Virol, Tubingen, 61-62. *Mem:* AAAS; Am Soc Microbiol; Am Soc Biol Chemists; Am Soc Virol. *Res:* Structure of animal viruses; mechanism of virus neutralization; structure of biological antivirals. *Mailing Add:* Inst Molecular Virol Univ Wis 1525 Linden Dr Madison WI 53706

RUEDENBERG, KLAUS, b Bielefeld, Germany, Aug 25, 20; m 48; c 4. THEORETICAL CHEMISTRY, THEORETICAL PHYSICS. *Educ:* Univ Fribourg, Lic rer nat, 44; Univ Zurich, PhD(theoret physics), 50. *Hon Degrees:* PhD Univ Basel, Switz, 75. *Prof Exp:* Asst, Univ Zurich, 46-47; res assoc physics, Univ Chicago, 51-55; from asst prof to assoc prof chem & physics, Iowa State Univ, 55-62; prof chem Johns Hopkins Univ, 62-64; PROF CHEM & PHYSICS, IOWA STATE UNIV, 64-, DISTINGUISHED PROF SCI & HUMANITIES, 78- *Concurrent Pos:* Guggenheim fel, 66-67; vis prof, Swiss Fed Inst Technol, 66-67, Wash State Univ, 70, Univ Calif, Santa Cruz, 73 & Univ Bonn, Germany, 74; adv ed, Int J Quantum Chem & Chem Phys Letts, 67-84; assoc ed, Theoretica Chimica Acta, 67-84, ed-in-chief, 84-; Fulbright Sr Scholar, Monash Univ & CSIRO Chem Physics Lab, Melbourne, Australia, 82; vis prof, Univ Kaiserslautern, Ger, 87. *Mem:* Fel AAAS; Am Chem Soc; Sigma Xi; fel Am Phys Soc; fel Am Inst Chem; Am Asn Univ Profs. *Res:* Atomic and molecular quantum mechanics; chemical binding and reactions; molecular structure and spectra; quantum chemistry; many-body quantum theory. *Mailing Add:* Dept Chem Iowa State Univ Ames IA 50011-0061

RUEDISILI, LON CHESTER, b Madison, Wis, Feb 7, 39; m 65; c 2. HYDROGEOLOGY, ENERGY RESOURCES. *Educ:* Univ Wis, Madison, BS, 61, MS, 65 & 71 PhD(geol), 68. *Prof Exp:* Geologist & geophysicist, Standard Oil Co Calif, 67; asst prof environ geol, Univ Wis, Parkside, 72-74; assoc prof, 74-79, PROF HYDROGEOL, UNIV TOLEDO, 79- *Concurrent Pos:* Fel water resources specialist, Environ Protection Agency, 71-72; consult hydrogeol, 71-; consult, US Aid Pres, Pakistan, 90-91. *Mem:* Fel Geol Soc Am; Am Water Resources Asn; AAAS; Am Inst Hydrol; Asn Groundwater Scientists & Engrs. *Res:* Applied water, land, and energy resources management; investigating geologic controls to water quality and quantity problems and geologic factors in engineering studies; solid and liquid waste management; environmental impact analysis; petroleum recovery; water law; environmental geology. *Mailing Add:* Dept Geol Univ Toledo 2801 W Bancroft Toledo OH 43606

RUEGAMER, WILLIAM RAYMOND, b Huntington, Ind, Dec 15, 22; m 46. BIOCHEMISTRY. *Educ:* Ind Univ, BS, 43; Univ Wis, MS, 44, PhD(biochem), 48. *Prof Exp:* Biochemist, Swift & Co, Ill, 48-49; instr biophys, Univ Colo, 49-51; asst chief radioisotope labs, Vet Admin Hosp, Denver, Colo, 51-54; from assoc prof to prof biochem, State Univ NY Upstate Med Ctr, 54-68, actg chmn dept, 67-68; prof biochem & chmn dept, Univ Nebr Med Ctr, Omaha, 68-85, assoc dean, sch allied health prof, 74-85; RETIRED. *Mem:* Am Soc Biol Chem; Am Inst Nutrit; Endocrine Soc; Soc Exp Biol & Med; Am Chem Soc. *Res:* Thyroid metabolism and atherosclerosis. *Mailing Add:* 24 Pine St Sugarmill Woods Homosassa FL 32646

RUEGER, LAUREN J(OHN), b Archbold, Ohio, Dec 30, 21; m 44; c 4. SPACECRAFT SYSTEM ENGINEERING. *Educ:* Ohio State Univ, BSc, 43, MSc, 47. *Prof Exp:* Mem staff, Radiation Lab, Mass Inst Technol, 43-45; asst physics, Ohio State Univ, 46-47; res engr, Battelle Mem Inst, 47-49; proj leader, Nat Bur Standards, 49-53; PRIN PROF STAFF MEM, APPL PHYSICS LAB, JOHNS HOPKINS UNIV, 53- *Concurrent Pos:* mem, US Study Group 7, Int Radio Consultive Comt; proceedings ed, Dept Defense Precise Time & Time Interval Appl & Planning Conf. *Mem:* Eng Physics Soc (secy, 41, pres, 42); Am Phys Soc; fel Inst Elec & Electronics Engrs. *Res:* Satellite system engineering and ground station instrumentation; microwave radar system and component design; shipboard satellite navigation equipment design; electronic system reliability engineering; analysis of nuclear radiation effects in electronic systems; hydrogen maser frequency standard design and applications; precision time frequency technology. *Mailing Add:* 1415 Glenallan Ave Silver Spring MD 20902

RUEGSEGGER, DONALD RAY, JR, b Detroit, Mich, May 29, 42; m; c 4. RADIOLOGICAL PHYSICS, MEDICAL IMAGING. *Educ:* Wheaton Col, BS, 64; Ariz State Univ, MS, 66, PhD(nuclear physics), 69. *Prof Exp:* RADIOL PHYSICIST, MIAMI VALLEY HOSP, DAYTON, OHIO, 69-, CHIEF, MED PHYSICS SECT, 82- *Concurrent Pos:* Consult, Vet Admin Hosp, Dayton, 70-77 & Wright Patterson AFB Med Ctr, 82-84; lectr, Sch X-ray Technol, Miami Valley Hosp, 70-76 & 79-85; from clin asst prof to clin assoc prof radiol, Sch Med, Wright State Univ, Dayton, 76- & dir group radiol physics, 77-; pres, Ohio River Valley Chpt, Am Asn Physicists in Med, 82-83 & co-chmn, 84-85. *Mem:* Am Asn Physicists Med; Am Col Med Physics; Am Col Radiol; Health Physics Soc; Am Phys Soc; AAAS. *Res:* Development of new techniques to treat tumors with both ionizing and non-ionizing radiation; clinical program of combining x-rays and microwaves (hyperthemia) for cancer therapy. *Mailing Add:* Med Physics Sect Miami Valley Hosp One Wyoming St Dayton OH 45409

RUEHLE, JOHN LEONARD, b Winter Haven, Fla, Feb 4, 31; m 54; c 4. PLANT PATHOLOGY. *Educ:* Univ Fla, BSA, 53, MS, 57; NC State Col, PhD(plant path), 61. *Prof Exp:* PLANT PATHOLOGIST, FORESTRY SCI LAB, SOUTHEASTERN FORESTRY EXP STA, US FOREST SERV, 61- *Mem:* Am Phytopath Soc; Soc Nematol. *Res:* Forest nematology, especially host-parasite relationships; Mycorrhiza. *Mailing Add:* Forestry Sci Lab Univ Ga Carlton St Athens GA 30601

RUEHLI, ALBERT EMIL, b Zurich, Switz, June 22, 37; US citizen. ELECTRICAL ENGINEERING. *Educ:* Zurich Tech Sch, Telecom Engr, 63; Univ Vt, PhD(elec eng), 72. *Prof Exp:* Res staff mem semiconductor circuits & devices, IBM T J Watson Res Ctr, 63-66, eng & math analyst, Develop Lab, 66-71, res staff mem design automation, 71-78, MGR & RES STAFF MEM COMPUT AIDED DESIGN, IBM T J WATSON RES CTR, 78- *Honors & Awards:* Outstanding Contrib Award, IBM, 75, Invention Achievement Awards, 74 & 76, Res Div Outstanding Contrib Award, 78; Guillemin-Cauer Award, Inst Elec & Electronics Engrs, 82- *Mem:* Fel Inst Elec & Electronics Engrs. *Res:* Electrical circuit theory; microwave theory; computer aided design. *Mailing Add:* IBM T J Watson Res Ctr PO Box 218 Yorktown Heights NY 10598

RUEL, MAURICE M J, b Quebec City, Que, Feb 19, 37; m 64; c 2. ENVIRONMENTAL SCIENCES. *Educ:* Laval Univ, BScA, 61, MScA, 65, DSc(chem eng), 68. *Prof Exp:* From asst prof to assoc prof chem eng, Univ Sherbrooke, 68-72; chief res & develop div, environ emergencies, Dept Environ, 73-74; asst dir, Northern Natural Resources & Environ Br, Northern Prog, Dept Indian Affairs & Northern Develop, 74-76, dir, Northern Renewable Resources & Environ Br, 76-77, dir gen, Northern Environ Br, 77-81; dir gen, environ protection br, Can Oil & Gas Lands Admin, 82-87; DIR GEN, ENERGY CONSERV BR, MINES & RESOURCES, 87- *Concurrent Pos:* Lectr, chem eng dept, Laval Univ, 67-68; consult, mining, pulp & paper industs, govt agencies, 68-73; mem bd, Mgt Res Ctr, Sherbrooke, Que, 70-73 & Que Comn Water in Agr, 71-73; mem bd dirs, Ctr Land Use Planning, Univ Sherbrooke, 70-73,; mem, Nat Surv & Mapping Comn, 75-81; chmn, environ studies revolving fund, Energy Mines & Resources, 83-87. *Honors & Awards:* Queen's Silver Jubilee Medal, Govt Can, 77. *Mem:* Fel Chem Inst Can; Chem Eng Soc Can. *Res:* Biological and physical environment; research and development in marine engineering; oil and gas exploration and development offshore and in the Canadian Arctic; used water treatment; energy efficiency and conservation. *Mailing Add:* Oil & Gas Br Mines & Resources 580 Booth St Rm 1890 Ottawa ON K1A 0E4 Can

RUELIUS, HANS WINFRIED, b Worms, Ger, Feb 18, 15; nat US; m 46; c 2. DRUG METABOLISM. *Educ:* Univ Geneva, DSc(chem), 42. *Prof Exp:* Res chemist synthesis pharmaceuts, Kast & Ehinger, Ger, 44-45; res assoc, Med Res Chem Dept, Max-Planck Inst, 46-51; sr res scientist, Wyeth Inst Med Res, Wyeth Labs, 51-66, mgr drug metab dept, res div, 66-78, assoc dir biol res, drug metab, 78-85; vis prof pharmacol, 85-87, CONSULT PHARMACOL, MED SCH JOHNS HOPKINS UNIV, 88- *Concurrent Pos:* Res assoc, Univ Pa, 51-53. *Mem:* AAAS; Am Chem Soc; Am Soc Pharmacol & Exp Therapeut; Int Soc Study Xenobiotics. *Res:* Isolation, characterization and structure proof of natural substances; drug metabolism; chemical carcinogens; biochemical mechanisms of drug toxicity. *Mailing Add:* 132 Sherburn Circle Weston MA 02193-1058

RUELKE, OTTO CHARLES, b Oshkosh, Wis, Feb 18, 23; m 59; c 3. AGRONOMY, PLANT PHYSIOLOGY. *Educ:* Univ Wis, BS, 50, MS, 52, PhD(agron), 55. *Prof Exp:* Instr high sch, 50-51; res asst agron, Univ Wis, 52-55; from asst prof to assoc prof, 55-69, PROF AGRON, UNIV FLA, 69- *Concurrent Pos:* Mem staff crop ecol, Forage & Pasture Sci, 55-; mem exec comt, Southern Pasture & Forage Crop Improv Conf, 63-65, chmn, 64; mem, Am Forage & Grassland Coun. *Mem:* Am Soc Agron; Crop Sci Soc Am; Sigma Xi. *Res:* Cold injury; plant growth regulation; microclimatology; forage management and quality evaluations. *Mailing Add:* 1125 SW Ninth Rd Gainesville FL 32601

RUENITZ, PETER CARMICHAEL, b Los Angeles, Calif, Nov 10, 43. MEDICINAL CHEMISTRY, ORGANIC CHEMISTRY. *Educ:* Univ Minn, BS, 66; Univ Kans, PhD(med chem), 74. *Prof Exp:* Res assoc med chem, Upjohn Co, 66-68; from asst prof to assoc prof, 74-88, PROF MED CHEM, SCH PHARM, UNIV GA, 88- *Mem:* Am Chem Soc; Fedn Am Soc Exp Biol. *Res:* Chemistry of bicyclic amines; nonsteroidal estrogen-antiestrogen metabolism and mechanism of action. *Mailing Add:* Sch Pharm Univ Ga Athens GA 30602

RUEPPEL, MELVIN LESLIE, b Rolla, Mo, Sept 18, 45; m 69; c 3. AGRICHEMICAL DISCOVERY & DEVELOPMENT, HUMAN RESOURCE MANAGEMENT. *Educ:* Univ Mo, BS, 66; Univ Calif, Berkeley, PhD(chem), 70. *Prof Exp:* NIH fel biochem, Cornell Univ, 70-71; sr res chemist metab, Monsanto Co, 71-75, group leader, 75-77, res mgr environ process, 77-80, res dir synthesis, 80-82, res dir process technol, 82-85, tech adv patent litigation, 82-86, dir plant protection, 85-86, dir herbicide technol, 86-89, dir global prod develop, 89-90, DIR TECHNOL, ROUNDUP DIV, MONSANTO CO, 91- *Concurrent Pos:* Chmn, United Way & Leadership Training Prog. *Honors & Awards:* Thomas & Hochwalt Award, 85; Mac Award, 88. *Mem:* Am Chem Soc; AAAS; Sigma Xi; Weed Sci Soc; Int Union Pure & Appl Chem. *Res:* Environmental fate and safety of pesticides; agricultural and pesticide chemistry; agrichemical synthesis bioevaluation and development; management, administration and strategy of research; breakthrough project technology; coaching performance. *Mailing Add:* Monsanto Co C35H 800 N Lindbergh Blvd St Louis MO 63167

RUESCH, JURGEN, b Naples, Italy, Nov 9, 09; nat; m 37; c 1. PSYCHIATRY, NEUROSCIENCES. *Educ:* Univ Zurich, MD, 35. *Prof Exp:* Asst neurol, neuroanat & neuropath, Med Sch, Univ Zurich, 36-38; asst psychiat, Univ Basel, 38-39; fel, Rockefeller Found, 39-41; asst, Mass Gen Hosp, 41; res fel neuropath, Harvard Med Sch, 41-43; lectr, 43-48, from assoc prof to prof, 48-77, EMER PROF PSYCHIAT, SCH MED, UNIV CALIF, SAN FRANCISCO, 77- *Concurrent Pos:* Res psychiatrist, Langley Porter Neuropsychiat Inst, 43-58, dir treatment res ctr, 58-64, dir sect social psychiat, 65-75. *Honors & Awards:* Hofheimer Award, Am Psychiat Asn, 51. *Mem:* AAAS; fel AMA; hon mem German Soc Psychiat & Neurol; Am Col Psychiat; fel Am Psychiat Asn; Asn Res Nerv & Ment Dis. *Res:* Social psychiatry; communication; human behavior. *Mailing Add:* 2543 Vallejo St San Francisco CA 94123

RUESINK, ALBERT WILLIAM, b Adrian, Mich, Apr 16, 40; m 63; c 2. PLANT PHYSIOLOGY. *Educ:* Univ Mich, BA, 62; Harvard Univ, MA, 65, PhD(biol), 66. *Prof Exp:* NSF fel bot, Inst Gen Bot, Swiss Fed Inst Technol, 66-67; asst prof, 67-72, assoc prof bot, prof, 72-80, PROF PLANT SCI, IND UNIV, BLOOMINGTON, 80- *Concurrent Pos:* Dir undergrad educ biol sci, Ind Univ, 72-74, 79-81, actg chmn dept biol, 81. *Mem:* AAAS; Am Soc Plant Physiol; Am Inst Biol Sci; Bot Soc Am. *Res:* Relationships between the plant cell plasma membrane and cell wall, especially as related to wall elongation. *Mailing Add:* Dept Biol Ind Univ Bloomington IN 47405

RUETMAN, SVEN HELMUTH, b Rakvere, Estonia, June 14, 27; nat US; m 57; c 2. ORGANIC CHEMISTRY. *Educ:* Millikin Univ, BS, 53; Univ Utah, PhD(chem), 57. *Prof Exp:* Res chemist, Shell Develop Co, 57-61; res chemist, Narmco Res & Develop Telecomput Corp, 61-63; res chemist, 63-70, res specialist, 70-77, RES LEADER, CENTRAL RES, DOW CHEM USA, 77- *Res:* Heterocyclic chemistry; polymer chemistry; process development; pesticides. *Mailing Add:* El Camino Corto Walnut Creek CA 94598-5409

RUEVE, CHARLES RICHARD, b Springfield, Ohio, May 25, 18. MATHEMATICS. *Educ:* St Joseph's Col, Ind, AB, 47; Univ Notre Dame, MS, 49, PhD(math), 63. *Prof Exp:* Chmn dept math & physics, 46-74, PROF MATH, ST JOSEPH'S COL, IND, 46-, CHMN DEPT, 77- *Mem:* AAAS; Math Asn Am; Am Math Soc. *Res:* Modern abstract algebra; number theory; real analysis; algebraic topology. *Mailing Add:* Dept Math St Joseph's Col Rensselaer IN 47978

RUF, ROBERT HENRY, JR, b Malden, Mass, Aug 30, 32; m 56. HORTICULTURE. *Educ:* Univ Mass, BS, 55; Cornell Univ, MS, 57, PhD(veg crops), 59. *Prof Exp:* From asst prof to assoc prof hort & from asst horticulturist to assoc horticulturist exp sta, Univ Nev, Reno, 59-75; PRES, GREENHOUSE GARDEN CTR, 74-; VPRES, GEOTHERMAL DEVELOP ASN, 78- *Res:* Greenhouse management; propagation; geothermal-agricultural research. *Mailing Add:* 4201 Palomino Circle Reno NV 89509

RUFENACH, CLIFFORD L, b Ronan, Mont, Nov 16, 36. IONOSPHERIC PHYSICS, PHYSICAL OCEANOGRAPHY. *Educ:* Mont State Univ, BS, 62, MS, 63; Univ Colo, PhD(elec eng), 71. *Prof Exp:* Res engr, Electronics Res Lab, Mont State Univ, 62-67; res physicist, Space Environ Lab, Boulder, Colo, 67-75; res physicist, Ocean Remote Sensing Lab, Miami, 75-76, staff mem, Wave Propagation Lab, Nat Oceanic & Atmospheric Admin, Boulder, 76-88; CONSULT, 88- *Mem:* Sr mem Inst Elec & Electronics Engrs; Int Sci Radio Union; Am Geophys Union. *Res:* Remote sensing of the atmosphere and ocean; space science. *Mailing Add:* 1102 Third Ave Longmont CO 80501

RUFF, ARTHUR WILLIAM, JR, b Newark, NJ, Aug 18, 30; c 3. METAL PHYSICS. *Educ:* Rice Univ, BS, 52; Univ Ariz, MS, 53; Univ Md, PhD(physics), 63. *Prof Exp:* Physicist, Shell Develop Co, 53-54; physicist, Aberdeen Proving Ground, 55-56; sect chief microstruct characterization, Nat Bur Standards, 63-76, chief Metal Sci & Standards Div, 76-79, group leader, Wear & Mech Properties, Nat Inst Standards & Technol, 80-87, PHYSICIST, NAT BUR STANDARDS, 57- *Honors & Awards:* Dept Com Spec Award, 63 & 71, Silver Medal Award, 71. *Mem:* Fel Am Soc Metals; fel Am Soc Testing & Mat. *Res:* Dislocation; defects in crystals; plastic deformation; electron microscopy; surface physics; chemical dissolution; metal physics; wear; tribology. *Mailing Add:* A215 Mat Nat Inst Standards & Technol Gaithersburg MD 20899

RUFF, GEORGE ANTONY, b Bay Shore, NY, May 10, 41; m 66; c 4. PHYSICS. *Educ:* Le Moyne Col, NY, BS, 62, Princeton Univ, MA, 64, PhD(physics), 66. *Prof Exp:* Res assoc physics, Cornell Univ, 66-68; from asst prof to assoc prof, 68-79, chmn physics dept, 76-85, chmn, Div Nat Sci & Math, 77-82, PROF PHYSICS, BATES COL, 79-, CHMN, DIV NAT SCI & MATH, 85-, CHARLES A DANA PROF PHYSICS, 85- *Concurrent Pos:* Vis assoc prof, Univ Ariz, 75-76; vis prof, Univ Freiburg, Ger, 82-83; vis res prof, Univ Va, 89-90. *Mem:* Am Phys Soc; Am Asn Physics Teachers. *Res:* Optical and atomic physics; quantum electronics. *Mailing Add:* Dept Physics Bates Col Lewiston ME 04240

RUFF, GEORGE ELSON, b Wilkes-Barre, Pa, Jan 12, 28; m 51; c 5. PSYCHIATRY. *Educ:* Haverford Col, AB, 48; Univ Pa, MD, 52. *Prof Exp:* Intern, Univ Mich, 52-53, resident psychiat, 53-56; vchmn, Sch Med, Univ Pa, 74-82, actg chmn, 82-84, assoc dean, Med Student Progs, 76-80, PROF PSYCHIAT, SCH MED, UNIV PA, 56- *Concurrent Pos:* USPHS fel, Inst Neurol Sci, Sch Med, Univ Pa, 56-57; USPHS career investr, 59-64; consult, US Air Force, 60-63, NASA, 62-63 & 88- & Vet Admin, 64-82. *Honors & Awards:* Longacre Award, Aerospace Med Asn, 59. *Mem:* Am Col Psychoanalysts; Am Psychiat Asn; Am Col Psychiat; Geront Soc Am; Am Geriat Soc. *Res:* Psychiatric and psychophysiologic studies of human stress; medical education; rehabilitation; geriatric psychiatry. *Mailing Add:* Univ Pa Hosp 3615 Chestnut St Philadelphia PA 19104-2683

RUFF, IRWIN S, b New York, NY, Oct 11, 32; m 59; c 2. METEOROLOGY. *Educ:* City Col New York, BS, 53; NY Univ, MS, 57. *Prof Exp:* Asst res scientist, Dept Meteorol, NY Univ, 55-59; res meteorologist, US Weather Bur, 59-65 & Nat Earth Satellite Serv, 65-85, RES METEOROLOGIST, SATELLITE RES LAB, DEPT RES, NAT OCEANOG & ATMOSPHERIC ADMIN, 85- *Mem:* Am Meteorol Soc; Am Geophys Union; Asn Orthodox Jewish Scientists. *Res:* Reflection properties of the earth and atmosphere in solar wavelengths; satellite determinations of terrestrial radiative properties and influence on climate; climatic change; scene identification from satellites. *Mailing Add:* 5200 Auth Rd Rm 711 Marlow Heights MD 20023

RUFF, JOHN K, b New York, NY, Feb 19, 32; m 54; c 3. INORGANIC CHEMISTRY. *Educ:* Haverford Col, BS, 54; Univ NC, PhD(chem), 59. *Prof Exp:* Asst phys chem, Univ NC, 54-56; chemist, Rohm & Haas Co, 57-68; ASSOC PROF CHEM, UNIV GA, 68- *Mem:* Am Chem Soc; Royal Soc Chem. *Res:* Organometallic chemistry of boron, aluminum and gallium; fluorine chemistry; nitrogen fluorides; hypofluorites; sulfur oxyfluoride derivatives. *Mailing Add:* Dept Chem Univ Ga Athens GA 30601-3040

RUFF, MICHAEL DAVID, b Newton, Kans, July 22, 41; m 63; c 2. PARASITOLOGY. *Educ:* Kans State Univ, BS, 64, MS, 66, PhD(parasitol), 68. *Prof Exp:* Instr biol, Marymount Col, 67-68; parasitologist, 406th Med Lab, Japan, 68-71; NIH res fel, Rice Univ, 71-72; from asst prof to assoc prof poultry sci, Univ Ga, 72-77; CHIEF, PROTOZOAN DIS LAB, ANIMAL PARASITOL INST, USDA, 77- *Mem:* Poultry Sci Asn; Am Soc Parasitol; Am Asn Vet Parasitol; Am Micros Soc; Am Asn Avian Pathologists. *Res:* Physiology and biochemistry of parasites; metabolism of larval trematodes; schistosomiasis; avian coccidia, host parasite interactions. *Mailing Add:* Agr Res Ctr LPSI Bldg 1040 Beltsville MD 20705

RUFF, ROBERT LAVERNE, b Laurel, Mont, Mar 4, 39; m 60; c 2. WILDLIFE ECOLOGY, WILDLIFE MANAGEMENT. *Educ:* Univ Mont, BS, 61, MS, 63; Utah State Univ, PhD(ecol), 71. *Prof Exp:* Furbearer biologist, Mont Fish & Game Dept, 61; NIH res asst parasitol bighorn sheep, Univ Mont, 61-62; res asst magpie predation on pheasant nests, Mont Coop Wildlife Res Unit, 62-63, res assoc ecol grizzly bears & elk, 64-66; NIH res asst behav ground squirrels, Utah State Univ, 66-70; from asst prof to assoc prof, 70-81, EXTENT WILDLIFE SPECIALIST, UNIV WIS-MADISON, 70-, WATER RESOURCES MGR FAC, 71-, PROF WILDLIFE ECOL, 81-, CHNM DEPT, 87- *Concurrent Pos:* Chmn, wildlife mgt on pvt lands comt, The Wildlife Soc, 84-87. *Mem:* Audubon Soc; Animal Behav Soc; Wildlife Soc; Bear Biol Asn. *Res:* Effects of social behavior on the dynamics of animal populations; human dimensions of wildlife management on private lands; environmental impact assessment; ecology of black bear populations; ecology of European brown bears in National Parks of Yugoslavia. *Mailing Add:* Dept Wildlife Ecol Univ Wis 226 Russell Labs Madison WI 53706

RUFFA, ANTHONY RICHARD, b Pittsburgh, Pa, Dec 12, 33; m 59; c 5. CHEMICAL PHYSICS, SOLID STATE PHYSICS. *Educ:* Carnegie Inst Technol, BS, 55, MS, 57; Catholic Univ, PhD(physics), 60. *Prof Exp:* Theoret solid state physicist, Nat Bur Standards, 60-66 & Naval Res Lab, 66-83; THEORET SOLID STATE PHYSICIST, ALGORITHMS RES INST, 83- *Concurrent Pos:* Lectr, Am Univ, 59-60 & Georgetown Univ, 63-64. *Mem:* Phys Soc; NY Acad Sci; Sigma Xi. *Res:* Thermal, optical and magnetic properties of crystals; theory of chemical bonding in crystals; quantum theory of atomic and molecular structure. *Mailing Add:* PO Box 4587 Silver Spring MD 20914

RUFFER, DAVID G, b Archbold, Ohio, Aug 25, 37; m 58; c 3. VERTEBRATE ZOOLOGY. *Educ:* Defiance Col, BS, 59; Bowling Green State Univ, MA, 60; Univ Okla, PhD(zool), 64. *Prof Exp:* From instr to assoc prof biol, Defiance Col, 64-73, dean, 69-73; PROVOST, provost, Elmira Col, 73-78; PRES, ALBRIGHT COL, 78- *Mem:* AAAS; Am Soc Mammal; Ecol Soc Am; Animal Behav Soc. *Res:* Ecology and behavior of Cricetid rodents; evolution of behavior in the grasshopper mice. *Mailing Add:* Albright Col Reading PA 19612

RUFFIN, SPAULDING MERRICK, b Emporia, Va, Apr 8, 23; m 58; c 3. BIOLOGICAL CHEMISTRY. *Educ:* Hampton Inst, BS, 43; Mich State Univ, MA, 53, PhD(animal nutrit), 56. *Prof Exp:* Teacher pub schs, NC, 46-52; PROF CHEM, SOUTHERN UNIV, BATON ROUGE, 56- *Mem:* Am Chem Soc. *Res:* Vitamins; amino acids; intermediary metabolism; biochemistry; organic chemistry. *Mailing Add:* Dept Chem Southern Univ A&M Col Baton Rouge LA 70813

RUFFINE, RICHARD S, b New York, NY, July 28, 28; m 60; c 3. PHYSICS. *Educ:* Queens Col, NY, BS, 50; Syracuse Univ, MS, 53; NY Univ, PhD(physics), 60. *Prof Exp:* Instr physics, Hunter Col, 54-55; instr, NY Univ, 58-59, res assoc, 59-60; staff scientist, GC Dewey Corp, 60-62; mem tech staff, RCA Labs, 62-67; prog mgr, Advan Res Proj Agency, US Dept Defense, 67-68; asst dir & chief reentry physics div, US Army Advan Ballistic Missile Defense Agency, 68-75; SPECIALIST TECHNOL & ANALYSIS OFF DIR, DEFENSE RES ENG, 75- *Concurrent Pos:* Adj asst prof, NY Univ, 60-61; consult, Inst Defense Anal, 66-67. *Honors & Awards:* Meritorious Civilian Serv, US Army, 75. *Mem:* AAAS; Am Phys Soc; Am Geophys Union; Am Inst Aeronaut & Astronaut. *Res:* Atomic physics; electromagnetic scattering, re-entry physics. *Mailing Add:* 4050 N 27th Rd Arlington VA 22207

RUFFNER, JAMES ALAN, b Akron, Ohio, Sept 6, 30; m 59; c 2. HISTORY OF SCIENCE, SCIENCE INFORMATION. *Educ:* Ohio State Univ, BSc, 51; Univ Mich, MS, 58; Ind Univ, MA, 63, PhD(hist sci), 66. *Prof Exp:* Physicist, Battelle Mem Inst, 51-52; instr earth sci, ETex State Col, 58-60; from asst prof to assoc prof natural sci, Monteith Col, 64-78, ACAD SERV OFFICER, SCI LIBR, WAYNE STATE UNIV, 78- *Mem:* Soc Hist Technol; Am Libr Asn; Am Meteorol Soc; Hist Sci Soc; Air Pollution Control Asn; Soc Social Study Sci. *Res:* Bibliometrics; communication networks in science. *Mailing Add:* Sci Libr Wayne State Univ Detroit MI 48202

RUFFOLO, JOHN JOSEPH, JR, b Chicago, Ill, Jan 4, 42. CELL BIOLOGY, PROTOZOOLOGY. *Educ:* Loyola Univ Chicago, BS, 66; Univ Iowa, MS, 69, PhD(zool), 72. *Prof Exp:* Fel zool, Univ Wis-Madison, 72-73, res assoc, 73-74; asst prof biol, Va Commonwealth Univ, 74-75, res assoc biophys, Med Col Va, 75-76, asst prof biophys, Med Col Va, 76-80; MEM STAFF, VET ADMIN MED CTR, 80- *Concurrent Pos:* HEW fel, 73-74; prin investr, Nat Eye Inst, 78-81; co-prin investr, US Army Med Res & Develop Command, 78-80; investr, HEW, 78-81, USPHS, 79-82 & Am Cancer Soc, 79-80. *Mem:* Am Micros Soc; Am Soc Cell Biol; Am Soc Zoologists; Biophys Soc; Soc Protozoologists. *Res:* Developmental cell biology of ciliate protozoa; endosymbiosis; intracellular calcification; ocular photopathology and photochemical lesions in mammals. *Mailing Add:* Dept Biol Univ Wis Whitewater WI 53190

RUFFOLO, ROBERT RICHARD, JR, b Yonkers, NY, Apr 14, 50. PHARMACOLOGY, NEUROBIOLOGY. *Educ:* Ohio State Univ, BS, 73, PhD(pharmacol), 76. *Prof Exp:* Res assoc, Am Found Pharmaceut Educ, 73-76; res assoc pharmacol, NIH, 77-78; sr phamacologist, Lilly Res Labs, 78-82, res scientist, 82-84; DIR, CARDIOVASC & RENAL PHARMACOL, SMITH KLINE & FRENCH LABS, 84- *Concurrent Pos:* Res assoc & fel pharmacol, Nat Inst Gen Med Sci, 77-78. *Mem:* Fedn Am Scientists; Am Soc Pharmacol & Exp Therapeut. *Res:* Pharmacology of vascular smooth muscle; adrenergic receptors; hypertension; congestive heart failure. *Mailing Add:* 709 Swedeland Rd PO Box 1539 King of Prussia PA 19406-2799

RUGGE, HENRY F, b South San Francisco, Calif, Oct 28, 36; m 67. MEDICAL TECHNOLOGY, PLASMA PHYSICS. *Educ:* Univ Calif, Berkeley, AB, 58, PhD(physics), 63. *Prof Exp:* Res asst plasma physics, Lawrence Radiation Lab, Univ Calif, 59-63; staff physicist, Physics Int Co, Calif, 63-69; staff physicist, Arkon Sci Labs, 70-72; vpres, Link Assocs, 72-73; vpres, Norse Systs, Inc, 73-75; vpres & gen mgr, Rasor Assocs Inc, 75-80; pres, Ultra Med Inc, 80-81; pres, Berliscan Inc, 81-82; exec vpres, 83-88, PRES & CHIEF EXEC OFFICER RASOR ASSOCS, INC, 89- *Res:* Plasma physics and gaseous electronics; atomic physics; medical and scientific instrumentation development; energy conversion, research and development. *Mailing Add:* Rasor Assocs Inc 253 Humboldt Ct Sunnyvale CA 94089

RUGGE, HUGO R, b San Francisco, Calif, Nov 7, 35; m 69; c 2. PHYSICS. *Educ:* Univ Calif, Berkeley, AB, 57, PhD(physics), 63. *Prof Exp:* Mem tech staff, 62-68, dept head, 68-79, prin dir, Lab Opers, 79-81, dir, Space Sci Lab, 81-89, VPRES, LAB OPERS, AEROSPACE CORP, 90- *Mem:* Int Astron Union; fel Am Phys Soc; Am Geophys Union; Am Astron Soc. *Res:* Space science; solar x-rays; upper atmosphere, infrared astronomy, high energy physics; satellite instrumentation. *Mailing Add:* Lab Opers M2-264 Aerospace Corp PO Box 92957 Los Angeles CA 90009

RUGGE, RAYMOND A(LBERT), electrical engineering, for more information see previous edition

RUGGERI, ZAVERIO MARCELLO, b Bergamo, Italy, Jan 7, 45; m 71. MEDICINE, MEDICINAL SCIENCES. *Educ:* Univ Milan, Italy, MD, 70. *Prof Exp:* Res fel, hemophilia, Hemophilia Ctr, Univ Milan, 70-72, asst prof, 72-80; assoc dir FVIII/vWF, hemophilia & thrombosis ctr, Policlinico Hosp Milan, 80-82; asst mem, dept immunol & basic & clin res, Scripps Clin & Res Found, 82-85, ASSOC MEM, DEPT BASIC & CLIN RES, DIV EXP HEMOSTASIS, SCRIPPS CLIN & RES FOUNDM 85- *Concurrent Pos:* Postdoctoral hemat, Univ Pavia, Italy, 73; vis investr, St Thomas Hosp, London, 74-75, St Bantholomeuis Hosp, London, 76, Res Inst Scripps Clin, 79-80. *Mem:* NY Acad Sci; Int Soc Thrombosis & Hemostasis; Am Heart Asn; World Fed Hemophilia; AAAS; Am Soc Hemat. *Res:* Platelet adhesive functions; platelet receptors for adhesive molecules; mechanisms of cell adhesion. *Mailing Add:* Basic & Clin Res Scripps Clin & Res Found 10666 N Torrey Pines Rd La Jolla CA 92037

RUGGERO, MARIO ALFREDO, b Resistencia, Argentina, Nov 7, 43; m 73. NEUROPHYSIOLOGY, AUDITORY PHYSIOLOGY. *Educ:* Cath Univ Am, BA, 65; Univ Chicago, PhD(physiol), 72. *Prof Exp:* Asst prof, 75-87, ASSOC PROF OTOLARYNGOL & NEUROSCI, UNIV MINN, 87- *Concurrent Pos:* NIH fel, neurophysiol, Univ Wis, 72-75; assoc ed, J Neurosci, 89-; mem, Commun Disorders Rev Comt, Nat Inst Deafness & other Commun Disorders, NIH, 90-94. *Mem:* AAAS; Soc Neurosci; Acoust Soc Am; Asn Res Otolaryngol; Am Asn Univ Prof. *Res:* Physiology of the ear and of the auditory nerve. *Mailing Add:* Dept Otolaryngol Med Sch Univ Minn 2630 University Ave SE Minneapolis MN 55414

RUGGIERI, MICHAEL RAYMOND, b Quantico, Va, Mar 2, 54; m 81; c 1. NEUROPHARMACOLOGY, INFECTIOUS DISEASE. *Educ:* Temple Univ, BA, 76; Univ Pa, PhD(pharmacol), 84. *Prof Exp:* Res assoc, 85-86, dir urol basic res, 87-90, RES ASST PROF SURG, UNIV PA, 86-; ASSOC PROF & DIR UROL RES, TEMPLE UNIV MED CTR, 90- *Concurrent Pos:* Prin investr, NIH, 85- *Mem:* AAAS; Am Soc Microbiol; Am Urol Assoc; Am Soc Pharmacol & Exp Therapeut; Soc Neurosci. *Res:* Pharmacological, biochemical and physiological investigation of the role of purinergic innervation in the function of the urinary bladder as a model for innervation of smooth muscle in general; pharmacology of bacterial adherence in urinary tract infection; biochemistry and pathophysiology of interstitial cystitis; smooth muscle; extitation contraction coupling. *Mailing Add:* Urol Res Labs 3400 N Broad St Philadelphia PA 19140

RUGGIERO, ALESSANDRO GABRIELE, b Rome, Italy, Apr 10, 40; m 65; c 2. PARTICLE PHYSICS, ACCELERATOR PHYSICS. *Educ:* Univ Rome, PhD(physics), 64. *Prof Exp:* Physicist, Nat Lab Frascati, Italy, 62-65, Europ Orgn Nuclear Res, Geneva, Switz, 66-69 & Nat Lab Frascati, Italy, 69-70; Physicist, Fermi Nat Accelerator Lab, Batavia, Ill, 70-84; sr physicist, Argonne Nat Lab, Argonne, Ill, 85-86; SR PHYSICIST & HEAD ACCELERATOR, PHYSICS DIV, BROOKHAVEN, UPTON, LI, NY, 87- *Mem:* Am Physics Soc. *Res:* Accelerator physics; design construction and operation of large accelerators, proton-proton and electron-positron storage and colliding devices for high energy physics experiment; design of proton-antiproton colliders; study of methods to collect anti-matter conceptual design of Tevatron I; new methods of acceleration; invention of the wakeatron; design of the relativistic heavy ion collidor (RHIC) at Brookhaven National Laboratory; feasibility studies of advanced hadron facilities; cooling techniques for heavy ions. *Mailing Add:* 33 Inlet View East Moriches NY 11940

RUGGIERO, DAVID A, b New York, NY, May 2, 49; m 77. CENTRAL NEUROTRANSMISSION. *Educ:* Queens Col, NY, BA, 72; Col Physicians & Surgeons, Columbia Univ, MA, 76, MPhil & PhD(anat), 77. *Prof Exp:* Instr neuroanat & gross anat, Col Physicians & Surgeons, Columbia Univ, 74-76; course dir neurosci, NY Col Podiatric Med, 76-77; fel neurol, Med Col, 77-79, instr, 80-81, asst prof, 81-88, ASSOC PROF NEUROL, MED COL, CORNELL UNIV, 88- *Concurrent Pos:* Reviewer, study sect, Drug Abuse, Biomed Res Review Comt, 87. *Honors & Awards:* Harriet Ames Award, NY Heart Asn, 79. *Mem:* Soc Neurosci; Sigma Xi; Am Asn Anatomists; Int Brain Res Orgn. *Res:* Central regulation of arterial blood pressure and cerebrovascular mechanisms; immunochemical identification of structures in brain which play a role in the tonic and reflex control of the circulation; central regulation of arterial blood pressure and cerebral blood flow. *Mailing Add:* Dept Neurol Col Med Cornell Univ New York NY 10021

RUGGLES, IVAN DALE, b Omaha, Nebr, Dec 7, 27; m 60; c 2. MATHEMATICS. *Educ:* Nebr Wesleyan Univ, AB, 49; Univ Wyo, MA, 51; Iowa State Col, PhD(math), 58. *Prof Exp:* Instr math, Iowa State Col, 57-58; from asst prof to assoc prof, San Jose State Col, 58-65; opers analyst, Stanford Res Inst, 65-66; SCI PROF SPECIALIST APPL MATH, LOCKHEED MISSILES & SPACE CO, 66- *Mem:* AAAS; Am Math Soc; Soc Indust & Appl Math; Math Asn Am; Asn Comput Mach. *Res:* Real variables; elementary number theory; numerical analysis. *Mailing Add:* 127 Belvue Dr Los Gatos CA 95030

RUGH, WILSON J(OHN), II, b Tarentum, Pa, Jan 16, 44; m 76; c 2. ELECTRICAL ENGINEERING. *Educ:* Pa State Univ, BS, 65; Northwestern Univ, MS, 67, PhD(elec eng), 69. *Prof Exp:* From asst prof to assoc prof, 69-79, PROF ELEC ENG, JOHNS HOPKINS UNIV, 79- *Concurrent Pos:* Vis prof, Princeton Univ, 84, Beijing Inst Tech, 82. *Mem:* Inst Elec & Electronics Engrs; Soc Indust & Appl Math. *Res:* Systems and control theory. *Mailing Add:* Dept Elec & Comput Eng Johns Hopkins Univ 34th & Charles Sts Baltimore MD 21218

RUGHEIMER, NORMAN MACGREGOR, b Charleston, SC, Feb 10, 30; m 58, 82; c 4. PHYSICS. *Educ:* Col Charleston, BS, 50; Univ NC, PhD(physics), 65. *Prof Exp:* Instr physics, The Citadel, 57-59; from asst prof to assoc prof, 64-80, PROF PHYSICS, MONT STATE UNIV, 80-, ASST DEAN, COL LETTERS & SCI, 69- *Concurrent Pos:* Adv on energy to comnr higher educ, State Mont, 75- *Mem:* Am Asn Physics Teachers. *Res:* Transmission and reflection coefficients and properties of thin superconducting films; holography. *Mailing Add:* Dept Math Mont State Univ Bozeman MT 59717

RUH, EDWIN, b Westfield, NJ, Apr 22, 24; m 52; c 2. CERAMICS, CHEMISTRY. *Educ:* Rutgers Univ, BSc, 49, MSc, 53, PhD(ceramics, chem), 54. *Prof Exp:* Res engr, Harbison-Walker Refractories Co, 54-57, asst dir res, Garber Res Ctr, 57-70, dir res, 70-73, dir advan technol, 73-74; vpres res, Vesuvius Crucible Co, 74-76; sr lectr & assoc head dept metall & mat sci, Carnegie-Mellon Univ, 76-84; PRES, RUH INT INC, 76-; RES PROF, DEPT CERAMICS, RUTGERS UNIV, 84- *Concurrent Pos:* Ed, Metall Trans, 78-83. *Honors & Awards:* Pace Award, Nat Inst Ceramic Engrs, 63. *Mem:* AAAS; fel Am Ceramic Soc (pres, 85-86); Nat Inst Ceramic Engrs; Am Inst Mining, Metall & Petrol Engrs; fel Inst Ceramics UK; Sigma Xi; Am Soc Testing Mat; Keramos-Prof Ceramic Eng Fraternity (pres, 70-72); Metall Soc; Am Soc Metals Int. *Res:* Refractories and refractory technology, manufacture, applications, failure analysis; ceramics and refractories. *Mailing Add:* 892 Old Hickory Rd Pittsburgh PA 15243

RUH, MARY FRANCES, b Chicago, Ill, July 18, 41; m 68; c 2. PHYSIOLOGY, ENDOCRINOLOGY. *Educ:* Marquette Univ, BS, 63, MS, 66, PhD(physiol), 69. *Prof Exp:* Instr pub health, Univ Mass, Amherst, 66-67; instr physiol, Univ Ill, Urbana, 69-71; from asst prof to assoc prof, 7þ-84, PROF PHYSIOL, SCH MED, ST LOUIS UNIV, 84- *Mem:* Am Soc Cell Biol; Am Physiol Soc; Endocrine Soc; Sigma Xi. *Res:* Steroid hormone action; mammalian reproductive physiology. *Mailing Add:* Dept Pharm Phys Sci St Louis Univ Sch Med St Louis MO 63104

RUH, ROBERT, b Plainfield, NJ, Aug 2, 30; m 52; c 3. CERAMICS, MATERIALS SCIENCE. *Educ:* Rutgers Univ, BS, 52, MS, 53, PhD(ceramics), 60. *Prof Exp:* Res ceramist, Aerospace Res Labs, Wright-Patterson AFB, Ohio, 58-65 & Chem Res Labs, Commonwealth Sci & Indust Res Orgn, Australia, 65-66; res ceramist, Aerospace Res Labs, 66-67, res ceramist, Processing & High Temperature Mat Br, Air Force Mat Lab, 67-70, SR PROJ ENGR, PROCESSING & HIGH TEMPERATURE MAT BR, AIR FORCE MAT LAB, WRIGHT-PATTERSON AFB, 70- *Concurrent Pos:* Ian Potter Found fel, 65-66. *Mem:* Fel Am Ceramic Soc; Nat Inst Ceramic Engrs; Ceramic Educ Coun; fel Am Inst Chem. *Res:* Development of improved ceramic materials through fabrication, characterization and property studies. *Mailing Add:* 4225 Murrel Dr Dayton OH 45429

RUHE, CARL HENRY WILLIAM, b Wilkinsburg, Pa, Dec 1, 15; m 43, 74; c 3. MEDICAL ADMINISTRATION. *Educ:* Univ Pittsburgh, BS, 37, MD, 40. *Prof Exp:* From instr to assoc prof physiol & pharmacol, Sch Med, Univ Pittsburgh, 37-60, from asst dean to assoc dean, 55-60; from asst secy to secy, coun med educ, 60-76, group vpres 76, sr vpres, AMA, 76-82; RETIRED. *Mem:* AAAS; Am Physiol Soc; Sigma Xi. *Res:* Human blood values; blood volume; hypothermia. *Mailing Add:* PO Box 31431 Rio Verde AZ 85263

RUHE, ROBERT VICTORY, b Chicago Heights, Ill, Nov 7, 18; m 43; c 3. GEOLOGY. *Educ:* Carleton Col, BA, 42; Iowa State Col, MS, 48; Univ Iowa, PhD(geol), 50. *Prof Exp:* From instr to asst prof geol, Iowa State Col, 46-51; geomorphologist, Soil Mission, Econ Coop Admin, Belgian Congo, 51-52; res geologist soil surv, Soil Conserv Serv, USDA, 53-70; PROF GEOL & DIR WATER RESOURCES RES CTR, IND UNIV, BLOOMINGTON, 70- *Concurrent Pos:* Geologist, Iowa Geol Surv, 47-51; Nat Res Coun fel, 50-51; vis prof, Cornell Univ, 58; prof, Iowa State Univ, 63-70; vis prof, Johns Hopkins Univ, 66-67; pres comn on paleopedol, Int Union Quaternary Res, 69-73; mem work group on exp basins, Nat Acad Sci, 71-75, mem panel on land burial radioactive wastes, 73- *Honors & Awards:* Kirk Bryan Award, Geol Soc Am, 74. *Mem:* Fel AAAS; fel Geol Soc Am; Soil Sci Soc Am; Int Asn Quaternary Res. *Res:* Geomorphology, quaternary geology; pedology; hydrogeology. *Mailing Add:* Dept Geol Ind Univ Bloomington IN 47401

RUHL, ROLAND LUTHER, b Chicago, Ill, Dec 13, 42; m 65; c 3. MECHANICAL DESIGN, DYNAMICS OF PHYSICAL SYSTEMS. *Educ:* Cornell Univ, BME, 65, MBA, 66, PhD(eng), 70. *Prof Exp:* PRES, RUHL & ASSOC CONSULT ENGRS, 73- *Concurrent Pos:* Adj prof, Univ Ill, Urbana-Champaign, 70- *Mem:* Am Soc Mech Engrs; Am Soc Safety Engrs; Soc Automotive Engrs; Soc Mfg Engrs. *Res:* Design automation and in particular simulation and dynamics of mechanical systems. *Mailing Add:* 1906 Fox Dr Suite G Champaign IL 61820

RUHLING, ROBERT OTTO, b Takoma Park, Md, Dec 3, 42; m 64; c 9. EXERCISE PHYSIOLOGY. *Educ:* Univ Md, BS, 64, MA, 66; Mich State Univ, PhD(phys educ), 70. *Prof Exp:* NIH trainee cardiovasc physiol, Inst Environ Stress, Univ Calif, Santa Barbara, 70-71, res physiologist, 71-72; dir, Human Performace Res Lab, 72-84, from asst prof to assoc prof exercise physiol, 72-87, chmn dept phys educ, 84-86, assoc dean, Col Health, Univ Utah, 86-87; CHMN, DEPT HEALTH, SPORT & LEISURE STUDIES, GEORGE MASON UNIV, 87- *Concurrent Pos:* Mem Utah Gov's Adv Coun on Phys Fitness; mem, sports med deleg Peoples' Repub China, 85; distinguished vis scholar, James Cook Univ, Townsville, Australia, 86. *Mem:* Fel Am Alliance Health, Phys Educ & Recreation; fel Am Col Sports Med; Sigma Xi; NY Acad Sci; fel Human Biol Coun. *Res:* Investigate the effects of exercise and the environment on the cardiovascular, respiratory, muscular, and nervous systems of the mammalian body. *Mailing Add:* Dept Health Sport & Leisure Studies George Mason Univ 4400 University Dr Fairfax VA 22030-4444

RUHMANN WENNHOLD, ANN GERTRUDE, b Brooklyn, NY, June 12, 32; m 67; c 2. SUBSTANCE ABUSE, ENDOCRINOLOGY. *Educ:* Seton Hill Col, BA, 54; State Univ NY Downstate Med Ctr, MD, 58. *Prof Exp:* Res asst radiobiol, 60-61, USPHS res fel, 61-63, from instr to assoc res prof anat, 65-78, res scientist med, 67-81, ASSOC RES PROF MED, UNIV UTAH, 78-, ASST PROF PSYCHIAT, 85-; STAFF PSYCHIATRIST, SALT LAKE VET ADMIN MED CTR, 84- *Concurrent Pos:* Chief resident, dept psychiat, Univ Utah, 84- *Mem:* Endocrine Soc; Am Fedn Clin Res. *Res:* Clinical in nature, involving malondial- dehyde in the urine of alcohol and drug abusers; use of B-blockers in post-traumatic stress disorder and bromocriptine treatment of cocaine cravings. *Mailing Add:* 2180 Belaire Dr Salt Lake City UT 84109

RUHNKE, LOTHAR HASSO, b Ger, Mar 2, 31; US citizen; m 59; c 2. ATMOSPHERIC PHYSICS. *Educ:* Tech Univ, Munich, MS, 55; Univ Hawaii, PhD(geosci), 69. *Prof Exp:* Res asst, Electrophys Inst, Munich, 54-57; res physicist, Meteorol Div, US Army Res & Develop Lab, Ft Monmouth, 57-61; res scientist, Appl Sci Div, Litton Indust Inc, 61-65; sci dir, Mauna Loa Observ, Nat Oceanic & Atmospheric Admin, Hawaii, 65-68, res physicist, Atmospheric Physics & Chem Lab, Environ Res Lab, 68-72; BR HEAD ATMOSPHERIC PHYSICS, NAVAL RES LAB, 72- *Concurrent Pos:* Mem several subcomns & working groups, Int Comn Atmospheric Elec, 63-, secy, 75- *Honors & Awards:* Outstanding Achievement Award, Nat Oceanic & Atmospheric Admin & NASA, 72. *Mem:* Am Geophys Union; Int Union Geophys & Geodesy; Sigma Xi. *Res:* Basic and applied research in atmospheric electricity, electrostatics, cloud physics, electrooptics propagation and marine meteorology; research management and administration in atmospheric physics. *Mailing Add:* 11208 Wedge Dr Reston VA 22090

RUIBAL, RODOLFO, b Cuba, Oct 27, 27; nat US; c 1. VERTEBRATE ZOOLOGY. *Educ:* Harvard Univ, AB, 50; Columbia Univ, MA, 52, PhD(zool), 55. *Prof Exp:* Instr biol, City Col NY, 52-54; from instr to assoc prof, 54-70, PROF ZOOL, UNIV CALIF, RIVERSIDE, 70- *Concurrent Pos:* Guggenheim fel, 70; res assoc, Smithsonian Inst; vis prof, Univ Chile, 68; ed, J Herpet, 80-; chmn, Nongame Adv Comt, Calif Dept Fish & Game. *Mem:* Soc Study Evolution; Am Soc Zool; Am Soc Ichthyol & Herpet. *Res:* Evolution and ecology. *Mailing Add:* Dept Biol Univ Calif Riverside CA 92521

RUINA, J(ACK) P(HILIP), b Rypin, Poland, Aug 19, 23; nat US; m 47; c 3. ELECTRICAL ENGINEERING. *Educ:* City Col, BS, 44; Polytech Inst Brooklyn, MEE, 49, DEE, 51. *Prof Exp:* From instr to assoc prof elec eng, Brown Univ, 50-54; res assoc prof, Control Systs Lab, Univ Ill, 54-59, res prof, Coord Sci Lab & prof elec eng, 59-63; vpres spec labs, 66-70, PROF ELEC ENG, MASS INST TECHNOL, 63- *Concurrent Pos:* Dep for res to Asst Secy Res & Eng, USAF, 59-60; asst dir defense res & eng, Off Secy Defense, 60-61, dir adv res projs agency, US Dept Defense, 61-63; pres, Inst Defense Anal, 64-66; mem gen adv comt, Arms Control & Disarmament Agency, 69-73; mem, Int Sci Radio Union; consult, var govt agencies; mem panel on telecommun, Nat Acad Eng; sr consult, Off Sci & Technol Policy, 77- *Honors & Awards:* Fleming Award, 62. *Mem:* Fel AAAS; fel Am Acad Arts & Sci; fel Inst Elec & Electronics Engrs; Inst Strategic Studies. *Res:* Statistical theory of noise; radar systems. *Mailing Add:* Dept Elec Eng Mass Inst Technol Cambridge MA 02139

RUIZ, CARL P, b Santa Barbara, Calif, Feb 1, 34; m 56; c 2. REACTOR CHEMISTRY, RADIOCHEMISTRY. *Educ:* Univ Calif, Santa Barbara, BA, 56, Univ Calif, Berkeley, PhD(chem), 61. *Prof Exp:* Chemist, GE Nuclear Energy, Vallecitos Nuclear Ctr, 61-69, tech specialist, 69-73, mgr radiol & process eng, 73-82, consult engr, 82-84, mgr process & radiation chem, 84-87, MGR CHEM & RADIOLOGICAL PROCESSES, GE NUCLEAR ENERGY, VALLECITOS NUCLEAR CTR, 87- *Mem:* Am Chem Soc; Am Nuclear Soc; fel Am Inst Chem; Sigma Xi; AAAS. *Res:* Development of irradiated nuclear fuel measurement methods; Methods and systems for the non-destructive measurement of uranium fuel enrichment; spent nuclear fuel, radioisotope and nuclear reactor decontamination chemical process studies; process for control of Co-60 in boiling water reactor recirculation piping; water radiolysis computer modeling applied to boiling water reactors; 10 patents. *Mailing Add:* 1901 Ocaso Camino Fremont CA 94539-5643

RUIZ-PETRICH, ELENA, b Mendoza, Arg, Nov 19, 33; wid. CARDIAC ELECTROPHYSIOLOGY. *Educ:* Univ Cuyo, MD, 59. *Hon Degrees:* DSc, Univ Cuyo, 76. *Prof Exp:* Sr instr physiol, Fac Med, Univ Cuyo, 65-68; from asst prof to assoc prof, 68-77, chmn, dept biophys, 78-84, PROF BIOPHYS, FAC MED, UNIV SHERBROOKE, 77-, CHMN, DEPT PHYSIOL BIOPHYS, 87- *Concurrent Pos:* Res fel heart physiol, Fac Med, Univ Cuyo, 59-63; Arg Nat Coun Sci & Technol Invest fel, 60-62, grant, 66-67, fel heart electrophysiol, Univ Southern Calif, 63-65; Med Res Coun Ottawa grants, 67-71; Que Heart Found grant, 69-73; fel, Can Heart Found, 69-70, res scholar, 71-75. *Mem:* Int Soc Heart Res; Can Physiol Soc; Am Physiol Soc. *Res:* Electrical activity of the heart and ionic distribution; anoxia and ischemia; atrioventricular conduction; membrane impedance and cytoplasmic resistivity in skeletal and cardiac muscle. *Mailing Add:* 1-1682 Simard Sherbrooke PQ J1J 3X1 Can

RUKAVINA, NORMAN ANDREW, b Ft William, Ont, June 20, 37; m 60; c 4. SEDIMENTOLOGY, LIMNOLOGY. *Educ:* Univ Toronto, BA, 59; Univ Western Ontario, MSc, 61; Univ Rochester, PhD(geol), 65, PhD(eng), 72. *Prof Exp:* NSF grant fel & lectr basic mineral, Univ Rochester, 65-66; RES SCIENTIST, CAN CENTRE INLAND WATERS, ENVIRON DEPT CAN, 66- *Concurrent Pos:* Ed, Proc Conf Great Lakes Res, 72-75; assoc ed, Geosci Can, 73-76 & J Great Lakes Res, 76-79. *Mem:* Sigma Xi; fel Geol Soc Am; Soc Econ Paleont & Mineral; Int Asn Gt Lakes Res; fel Geol Asn Can. *Res:* Nearshore sedimentology of the Great Lakes; lakeshore erosion; coastal geomorphology; acoustics of sediments; sedimentology of contaminated sediments. *Mailing Add:* Can Ctr Inland Waters PO Box 5050 Burlington ON L7R 4A6 Can

RUKHIN, ANDREW LEO, b Leningrad, USSR, Oct 1, 46; US citizen; m 73; c 2. STATISTICAL DECISION THEORY. *Educ:* Leningrad State Univ, MS, 67; Steklov Math Inst, PhD(math & statist), 70. *Prof Exp:* Res assoc, Steklov Math Inst, Acad Sci USSR, 70-74; assoc prof statist, Purdue Univ, 77-80, prof, 82-86; PROF MATH STATIST, UNIV MD BALTIMORE COUNTY, 89- *Concurrent Pos:* Lectr, Leningrad Technol Inst, 68; vis asst prof, Leningrad State Univ, 72; vis prof, Rome Univ, Italy, 77, Univ Mass, Amherst, 82. *Honors & Awards:* Sr Distinguished Scientist Award, Alexander von Humboldt Found, 90. *Mem:* fel Inst Math Statist; Am Statist Asn. *Res:* Description of universal statistical estimators; adaptive procedures for a finite parameter; characterizations of probability distributions. *Mailing Add:* Dept Math Univ Md Baltimore County Baltimore MD 21228

RUKNUDIN, ABDUL MAJEED, b India, Feb 16, 52; m 80; c 2. PHYSIOLOGY OF FERTILIZATION, ION CHANNELS IN MEMBRANES. *Educ:* Univ Madras, India, BS, 72, MS, 74, PhD(physiol), 82. *Prof Exp:* Asst prof, Univ Madras, India, 75-84; res assoc electron micros, Univ Paris VI, 85-86; vis researcher path, Univ Bristol, UK, 86-87; postdoctoral res assoc electron micros, 88-91, RES ASST PROF BIOPHYS, STATE UNIV NY, 91- *Mem:* Electron Micros Soc Am; Am Cell Biol; Soc Study Reproduction; Soc Study Fertil. *Res:* Physiology of reproduction in insects; ultrastructural aspects of mammalian fertilization; biophysical aspects of mechanotransducing ion channels in heart cells. *Mailing Add:* 320 Cary Hall Dept Biophys State Univ NY Med Sch Buffalo NY 14214

RULAND, NORMAN LEE, physical chemistry, for more information see previous edition

RULE, ALLYN H, b New York, NY, June 18, 34; m 70; c 4. IMMUNOCHEMISTRY, BIOCHEMISTRY. *Educ:* Cent Conn State Col, BSA, 56; Boston Univ, PhD(biol), 65. *Prof Exp:* NIH fel biochem, Brandeis Univ, 65-66; asst prof immunochem & biochem, Boston Col, 66-74; res fel dermat & res assoc med, Sch Med, Tufts Univ, 74-77; ASSOC PROF GRAD DEPT BIOL, BOSTON COL, 77-; ASST PROF OBSTET-GYNEC, TUFTS UNIV SCH MED, 80- *Concurrent Pos:* Aid to Cancer res grant, 67-71; NASA res grant, 70-73; asst res prof, Mt Sinai Med Sch, 70-74. *Mem:* Am Asn Immunologists; AAAS; Path Soc. *Res:* Antigenic purification, characterization and haptenic inhibitions of glycoproteins from blood group substances, lymphocytes, muscle, skin, tumors, carcinoembryonic antigens as well as studies in immune response, suppression and autoimmunity. *Mailing Add:* 43 Grove Hill Ave Newton MA 02160

RULF, BENJAMIN, b Jerusalem, Israel, Nov 6, 34; US citizen; m 68; c 4. ELECTRONICS ENGINEERING, PHYSICAL MATHEMATICS. *Educ:* Technion Israel Inst Technol, BSc, 58, MSc, 61; Polytech Univ, PhD(electroph), 65. *Prof Exp:* Res scientist, Courant Inst Math Sci, 65-67; sr lectr appl math, Tel Aviv Univ, 67-71, assoc prof, 73-79; assoc prof, Rensselaer Polytech Inst, 71-73; res officer, Nat Res Coun Can, 76-77; mem tech staff, MITRE Corp, 79-83; dir res, Radant Technols, 83- 85; mgr advan technol, Lockheed Electronics Co, 85-90; SR STAFF SCIENTIST, GRUMMAN CO, 90- *Concurrent Pos:* Adj prof, Princeton Univ, 87, NJ Inst Technol, 86, Tufts Univ, 83-85; adj assoc prof, Univ Mass, Amherst, 81-83; lectr, Tufts Univ, 78-81. *Mem:* Sr mem Inst Elec & Electronics Engrs. *Res:* Electromagnetic and acoustic wave propagation and diffraction theory; antenna theory and design; radar theory; engineering and science education; technical management. *Mailing Add:* Ten Evergreen Ct Westfield NJ 07090

RULFS, CHARLES LESLIE, b St Louis, Mo, Oct 21, 20; m 42; c 5. ANALYTICAL CHEMISTRY, INORGANIC CHEMISTRY. *Educ:* Univ Ill, BS, 42; Purdue Univ, PhD(chem), 49. *Prof Exp:* Res chemist, Res Labs, Linde Air Prods Co Div, Union Carbide Corp, 42-45; from asst prof to assoc prof, 49-61, PROF CHEM, UNIV MICH, ANN ARBOR, 61- *Concurrent Pos:* Consult, Los Alamos Sci Lab, 60- *Mem:* Am Chem Soc. *Res:* Chemistry of technetium and rhenium; polarographic theory and applications; unusual oxidation levels; microanalytical techniques. *Mailing Add:* 1297 Newport Rd Ann Arbor MI 48103-2313

RULIFFSON, WILLARD SLOAN, b Balaton, Minn, July 19, 18; m 41; c 2. BIOCHEMISTRY. *Educ:* Buena Vista Col, BS, 40; Univ Iowa, MS, 48, PhD(biochem), 53. *Prof Exp:* From asst prof chem to assoc prof biochem, 53-68, PROF BIOCHEM, KANS STATE UNIV, 68- *Mem:* AAAS; Am Chem Soc. *Res:* Radio iron transport; biochemical applications of mass spectrometry; monoamine oxidase enzymes and inhibitors. *Mailing Add:* Dept Biochem Kans State Univ Manhattan KS 66502

RULIFSON, JOHNS FREDERICK, b Bellefontaine, Ohio, Aug 20, 41; m 64; c 2. INFORMATION SCIENCE. *Educ:* Univ Wash, BS, 66; Stanford Univ, PhD(comput sci), 73. *Prof Exp:* Res mathematician comput sci, Stanford Res Inst, 66-73; mem res staff comput sci, Xerox Palo Alto Res Ctr, 73-80; eng mgr, Rohm Corp, 80-85; eng mgr artificial intel, Syntelligence, 85-87; ENG DIR, SOFTWARE PROD, SUN MICROSYSTS, 87- *Mem:* Asn Comput Mach; Inst Elec & Electronics Engrs. *Res:* Office automation, especially written communications in managerial and office environments and the design and evaluation of computer systems; human interface technologies; expert systems; window systems. *Mailing Add:* 3785 El Centro Palo Alto CA 94304

RULIFSON, ROGER ALLEN, b Manchester, Iowa, Nov 13, 51; m 81; c 2. ANADROMOUS FISHES, PENAEID SHRIMPS. *Educ:* Univ Dubuque, BS, 73; NC State Univ, MS, 75, PhD(marine sci eng), 80. *Prof Exp:* Res asst, NC Coop Fishery Res Unit, Dept Zool, NC State Univ, 73-75; leader fish distribution & vulnerability assessment task, Ecol Serv Group, Tex Instruments, Inc, 75-77; res asst, Dept Zool, NC State Univ, 77-80, res assoc, NC Coop Fishery Res Unit, Dept Zool, 80-81; asst prof fisheries, marine sci, Ctr Environ Sci, Unity Col, 81-83; ASSOC SCI INST COASTAL MARINE RES, ECAROLINA UNIV, 83-, ASSOC PROF, BIOL DEPT, ECAROLINA UNIV, 87- *Concurrent Pos:* Consult, US Fish & Wildlife Serv, Fishery Resources, Region 4, Atlanta, Ga, 80-82; vis scientist, Acadia Univ, Nova Scotia. *Mem:* Am Fisheries Soc; Estuarine Res Fedn; Sigma Xi. *Res:* Life history aspects of striped bass, alewife, blueback herring, and American shad; evolutionary ecology of man-created wetlands for mitigation purposes; tidal power effects on fish and fisheries; hydroelectric generation effects on fisheries. *Mailing Add:* Inst Coastal Marine Res ECarolina Univ Greenville NC 27858

RULON, RICHARD M, b Babylon, NY, Oct 29, 22; m 43; c 4. PHYSICAL CHEMISTRY, CERAMICS. *Educ:* Alfred Univ, BS, 43; Univ Pittsburgh, PhD(phys chem), 51. *Prof Exp:* Engr, Radio Corp Am, Pa, 43-45; teacher high sch, NY, 45-46; instr physics, Alfred Univ, 46-47; AEC fel binary metal alloys, Univ Pittsburgh, 50-51; from sr engr to engr-in-charge, Sylvania Elec Prod Inc, Gen Tel & Electronics Corp, Mass, 51-60; res dir glass to metal seals, Hermetite Corp, 60-62; from assoc prof to prof chem, Alfred Univ, 62-84; CONSULT, D G O'BRIEN NH, CORNING GLASS NY, 84- *Concurrent Pos:* Lectr & coordr, Northeastern Univ, 55-61; consult, Mass & Tetron, Inc, NY, 62-76 and others; mem comt nomenclature for electron-optical devices, Joint Comt Inst Elec Eng & Inst Radio Eng, 57-60. *Mem:* Am Ceramic Soc; Electrochem Soc; Nat Inst Ceramic Eng; Sigma Xi. *Res:* Electroluminescent and electron optical devices; dielectric ceramic materials; glass to metal seals for electronic and hermetic devices. *Mailing Add:* 3811 Ebury St Houston TX 77066

RULON, RUSSELL ROSS, b Apr 26, 36; m 63; c 3. PHYSIOLOGY. *Educ:* Luther Col (Iowa), BA, 58; Univ Iowa, MS, 60, PhD(physiol), 61. *Prof Exp:* Instr physiol, Col Med, Univ Iowa, 61-63; asst prof, 63-72, PROF BIOL, LUTHER COL, IOWA, 72- *Concurrent Pos:* USPHS trainee, Med Sch, Univ Va, 69-70; vis scientist cardiol res, Mayo Clin, 75-76; vis scientist, Inst Arctic Biol, Fairbanks, Alaska, 89. *Mem:* Am Physiol Soc. *Res:* Muscle physiology; physiological ecology; electrophysiology; comparative physiology of neurogenic and myogenic hearts; environmental physiology; effect of toxic materials on heart metabolism. *Mailing Add:* Dept Biol Luther Col Decorah IA 52101

RUMBACH, WILLIAM ERVIN, b Weston, WVa, Mar 15, 34; m 55; c 3. SCIENCE ADMINISTRATION. *Educ:* Glenville State Col, BS, 55; WVa Univ, MS, 58. *Prof Exp:* Instr genetics, WVa Univ, 57-60; instr biol, Cent Fla Jr Col, 60-67; dir div natural sci, 67-78, chmn & prof, Dept Sci, 78-87, PROF BIOL & BOTANY, CENT FLA COMMUNITY COL, 88- *Concurrent Pos:* Consult, Crystal River Marine Sci Res Sta, 66-68. *Mailing Add:* Dept Sci Cent Fla Community Col PO Box 1388 Ocala FL 32678

RUMBARGER, JOHN H, b Norfolk, Va, Dec 27, 25; m 51, 82; c 9. BALL & ROLLER BEARING TECHNOLOGY. *Educ:* Lehigh Univ, BSME, 49; Univ Pa, MSME, 63. *Prof Exp:* Prin engr, Franklin Inst Res Labs, 65-75, dir eng, 76-85, program mgr, Calspan, 86-88, exec engr, 89-90; CONSULT ENGR, JOHN H RUMBARGER, PE, INC, 91- *Honors & Awards:* Silver Snoopy Astronauts Personal Achievement Award, NASA, 90. *Mem:* Am Soc Mech Engrs. *Res:* Ball and roller bearing design, application, test and evaluation; tribology. *Mailing Add:* 123 Poplar Ave Wayne PA 19087

RUMBAUGH, MELVIN DALE, b Pella, Iowa, Sept 13, 29; m 53; c 4. CROP BREEDING. *Educ:* Cent Col (Iowa), BS, 51; Univ Nebr, MS, 53, PhD(agron), 58. *Prof Exp:* Asst prof agron, Colo State Univ, 58-59; from asst prof to prof, SDak State Univ, 59-76; RES GENETICIST, USDA-AGR RES SERV, LOGAN, UT. *Concurrent Pos:* Vis scientist, People's Repub China, 87, 90. *Mem:* Fel Crop Sci Soc Am; fel Am Soc Agron; Soc Range Mgt. *Res:* Biometrical genetics and breeding of plants, especially of Medicago sativa; utilization of legume species for improvement of range. *Mailing Add:* Agr Res Serv USDA Forage & Range Res Lab Utah State Univ Logan UT 84322-6300

RUMBLE, DOUGLAS, III, b Atlanta, Ga, June, 15, 42; m 67; c 2. METAMORPHIC PETROLOGY, STABLE ISOTOPE GEOCHEMISTRY. *Educ:* Columbia Col, NY, BA, 64; Harvard Univ, PhD(geol), 69. *Prof Exp:* Asst prof geol, Univ Calif, Los Angeles, 71-73; PETROLOGIST, GEOPHYS LAB, CARNEGIE INST WASH, 73- *Concurrent Pos:* Prog dir, Nat Sci Found, 85-87; coun, Mineral Soc Am, 86-89; pres, Geol Soc Washington, 88; ed, J Petrol, 89- *Mem:* Mineral Soc Am; Am Geophys Union. *Res:* Nature of fluid-rock interaction during metamorphism through chemical thermodynamics and stable isotope geochemistry. *Mailing Add:* Geophys Lab 5251 Broad Branch Rd NW Washington DC 20015-1305

RUMBLE, EDMUND TAYLOR, III, b Philadelphia, Pa, Oct 26, 42; m 76. NUCLEAR ENGINEERING. *Educ:* US Naval Acad, BS, 65; Univ Calif, Los Angeles, MS, 71, PhD(nuclear eng), 74. *Prof Exp:* Mem staff reactor safety, 74-77, DIV MGR NUCLEAR MAT, SCI APPLNS, INC, 77- *Mem:* Am Nuclear Soc. *Res:* Reactor safety and performance. *Mailing Add:* 580 Madison Palo Alto CA 94303

RUMBURG, CHARLES BUDDY, b Welch, WVa, Dec 12, 31; m 54; c 3. AGRONOMY. *Educ:* Colo State Univ, BS, 54; Rutgers Univ, MS, 56, PhD(farm crops), 58. *Prof Exp:* Res asst farm crops, Rutgers Univ, 54-58; res agronomist, Agr Res Serv, 58-70, supt, Colo Mountain Meadow Res Ctr, 70-77, AGRONOMIST, COOP STATE RES SERV, USDA, 77-, DEP ADMIN, 85- *Concurrent Pos:* Fel, Ore State Univ, 65-66; assoc prof agron, Colo State Univ, 70-77. *Mem:* Am Soc Agron. *Res:* Quantity and quality of forage from native meadows; increasing the efficiency of nitrogen fertilizer. *Mailing Add:* Coop State Res Serv USDA Washington DC 20250

RUMELY, JOHN HAMILTON, b New York, NY, Jan 14, 26; m 48; c 3. PLANT TAXONOMY, PLANT ECOLOGY. *Educ:* Oberlin Col, AB, 48; Wash State Univ, PhD(bot), 56. *Prof Exp:* Instr biol, Wash State Univ, 55-56; from asst prof to prof, 56-85, cur herbartium, 73-88, EMER PROF BOT, MONT STATE UNIV, 85- *Concurrent Pos:* Ed, Mont Acad Sci, 84-88. *Res:* Flora of Montana; vegetation constitution and succession; plant life histories. *Mailing Add:* Dept Biol Mont State Univ Bozeman MT 59717-0346

RUMER, RALPH R, JR, b Ocean City, NJ, June 22, 31; m 53; c 4. CIVIL ENGINEERING, HYDRAULIC ENGINEERING. *Educ:* Duke Univ, BS, 53; Rutgers Univ, MS, 59; Mass Inst Technol, ScD(civil eng), 62. *Prof Exp:* Engr, Luken Steel Co, Pa, 53-54; instr civil eng, Rutgers Univ, 56-59; engr, Soil Conserv Serv, USDA, NJ, 57-59, res asst hydraul, Mass Inst Technol, 61-62, asst prof civil eng, 62-63; assoc prof eng, 63-69, actg head dept civil eng, 66-67, chmn dept, 67-74, prof, 69-76, chmn dept, 84-87, PROF CIVIL ENG, STATE UNIV NY, BUFFALO, 78-; EXEC DIR, NY CTR HAZARDOUS WASTE MGMT, 87- *Concurrent Pos:* Ford Found fel, 62-63; NIH res grant, 65-69; sr res fel, Calif Inst Technol, 70-71; prin tech consult, Lake Erie Wastewater Mgt Study, Buffalo Dist, US Army Corps Engrs, 73-; prof civil eng & chmn dept, Univ Del, 76-78; dir great lake prog, 86- *Mem:* Am Soc Civil Engrs; Am Geophys Union; Nat Soc Prof Engrs; Int Asn Hydraulic Res; Int Asn Gt Lakes Res. *Res:* Water resources, flow through porous media; lake dynamics; hydraulic modelling; hydraulic processes related to water quality control; ice engineering. *Mailing Add:* 821 Eggert Rd Buffalo NY 14226

RUMFELDT, ROBERT CLARK, b Shawinigan Falls, Que, Nov 28, 36; m 59; c 5. PHOTOCHEMISTRY, INORGANIC CHEMISTRY. *Educ:* Loyola Univ, Can, BSc, 59; Univ Alta, PhD(radiation chem), 63. *Prof Exp:* Gen Elec Res Found fel, Univ Leeds, 63-65; asst prof, 65-68, ASSOC PROF CHEM, UNIV WINDSOR, 68- *Mem:* Chem Inst Can. *Res:* Spectroscopy and photochemistry of inorganic systems. *Mailing Add:* Dept Chem Univ Windsor Windsor ON N9B 3P4 Can

RUMMEL, ROBERT EDWIN, physical chemistry; deceased, see previous edition for last biography

RUMMEL, ROBERT WILAND, b Dakota, Ill, Aug 4, 15; m 39; c 5. AERONAUTICAL ENGINEERING. *Educ:* Curtiss Wright Tech Inst, Dipl aero eng, 34. *Prof Exp:* Stress analyst, Hughes Aircraft Co, Calif, 35 & Lockheed Aircraft Corp, 36; detail designer, Aero Eng Corp, 36 & Nat Aircraft Co, 37; chief engr, Rearwin Aircraft Co, Mo, Ken Royce Eng Co & Rearwin Aircraft & Engines, 37-43; chief engr, Trans World Airlines, 43-46, vpres eng, 56-59, vpres planning & res, 59-69 & tech develop, 69-77; PRES, ROBERT W RUMMEL ASSOCS, INC, 77- *Concurrent Pos:* Mem comn on aircraft operating probs, NASA, 63-69, mem res comn on aeronaut, 69-70; mem panel on Supersonic Transport environ effects, Commerce Tech adv bd, 71-72; consult, NASA, 72-; mem, Presidental Comn Space Shuttle Challenger Accident. *Honors & Awards:* Distinguished Pub Serv Medal, NASA, 79. *Mem:* Nat Acad Eng; fel Am Inst Aeronaut & Astronaut (treas, 71-72); fel Soc Automotive Engrs (vpres, 56). *Res:* Aeronautical research relating to aircraft transport design and operations. *Mailing Add:* Robert W Rummel Assocs Inc PO Box 7330 Mesa AZ 85206

RUMMENS, F H A, b Eindhoven, Netherlands, May 20, 33; m 60; c 3. MOLECULAR SPECTROSCOPY, PHYSICAL CHEMISTRY. *Educ:* Univ Leiden, Drs, 58; Oxford Univ, Brit Coun bursary, 58; Eindhoven Technol Univ, DSc, 63; PhD, Univ Sask, 72. *Prof Exp:* Sci co-worker, Eindhoven Technol Univ, 59-67; assoc prof, 67-72, PROF CHEM, UNIV REGINA, 72- *Concurrent Pos:* Niels Stensen fel, 63-64; Nat Res Coun Can fel, 64-65; vis prof chem, Univ Colo, 67; France-Can exchange fel, 74. *Mem:* Fel Chem Inst Can; Spectros Soc Can; Int Soc Magnetic Resonance. *Res:* Nuclear magnetic resonance, infrared and ultraviolet spectroscopy; structure of simple molecules; effects of medium on spectra; analytical spectroscopy. *Mailing Add:* Dept Chem Univ Regina Regina SK S4S 0A2 Can

RUMMERY, TERRANCE EDWARD, b Brockville, Ont, Nov 16, 37; m 67; c 2. SOLID STATE CHEMISTRY. *Educ:* Queen's Univ, Ont, BSc, 61, PhD(chem), 66. *Prof Exp:* Nat Res Coun Can overseas fel, Univ Col, Univ London, 66-67; assoc res scientist mat chem, Ont Res Found, 68-69; scientist phys chem, Airco Speer Res Labs, 69-71; assoc res officer phys chem, Atomic Energy Can, Ltd, 71-76, sr res officer & head, Res Chem Br, 76-79, dir, Waste Mgt Div, 79-84, gen mgr, Marketing & Sales, 86-87, gen mgr, Marine Propulsion Unit, 87-90, PRES, ATOMIC ENERGY CAN LTD, 90- *Mem:* Fel Chem Inst Can; Can Nuclear Soc. *Res:* Physical chemistry of power reactor coolant systems; basic science underlying nuclear waste management. *Mailing Add:* Atomic Energy Can Ltd Res 344 Slater St Ottawa ON K1A 0S4 Can

RUMPEL, MAX LEONARD, b WaKeeney, Kans, Mar 17, 36; m 61; c 2. INORGANIC CHEMISTRY, CHEMICAL EDUCATION. *Educ:* Ft Hays Kans State Col, AB, 57; Univ Kans, PhD(chem), 62. *Prof Exp:* From instr to assoc prof, 61-68, Chmn dept, 72-88 PROF CHEM, FT HAYS STATE UNIV, 68- *Concurrent Pos:* Mem, NSF Res Partic Prog Col Teachers, Univ Colo, 65-66 & Wash State Univ, 71. *Mem:* Am Chem Soc; Sigma Xi. *Res:* Unusually low oxidation states of metals; electrochemistry; computer programming for chemistry; instrumentation. *Mailing Add:* Dept Chem Ft Hays State Univ 600 Park St Hays KS 67601-4099

RUMPF, JOHN L, b Philadelphia, Pa, Feb 21, 21; m 44; c 2. CIVIL ENGINEERING. *Educ:* Drexel Inst Technol, BS, 43; Univ Pa, MS, 54; Lehigh Univ, PhD, 60. *Prof Exp:* From instr to assoc prof civil eng, Drexel Inst Technol, 47-56; res instr, Lehigh Univ, 56-60; prof, Drexel Univ, 60-64, head dept, 64-69; dean, Col Eng Technol, Temple Univ, 69-76, vpres acad affairs, 76-82, vpres, 82-84, EMER PROF CIVIL ENG, TEMPLE UNIV, 86- *Concurrent Pos:* Chmn, Res Coun Riveted & Bolted Struct Joints, 65-71; consult struct eng, 50- *Mem:* Am Soc Civil Engrs; Am Soc Eng Educ; Nat Soc Prof Engrs; Am Concrete Inst; Sigma Xi. *Res:* Behavior of steel structures and their component parts and connections. *Mailing Add:* Dept Civil-Construct Eng Temple Univ Col Eng Compt Sci & Archit 12th & Norris St Philadelphia PA 19122

RUMPF, R(OBERT) J(OHN), b Auburn, NY, July 31, 16; m 40. AERONAUTICAL ENGINEERING. *Educ:* Univ Notre Dame, BS, 39. *Prof Exp:* Design engr, Stinson Aircraft Co, 39-45 & Bendix Aviation Corp, 45-46; res engr, Univ Mich, 46-55; prin res engr, Ford Motor Co, 55-70; dir res & develop, Hamill Mfg Co Div, Firestone Tire & Rubber Co, 70-86; ENG CONSULT, TRW VEHICLE SAFETY SYSTS INC, 86- *Mem:* Soc Automotive Engrs. *Res:* Aircraft development; guided missile, air defense and highway control systems; automotive safety; automotive restraint systems. *Mailing Add:* 37 Fisher Rd Grosse Pointe MI 48230

RUMSEY, THERON S, b Whitley Co, Ind, Mar 11, 39. RUMINANT NUTRITION. *Educ:* Purdue Univ, BS, 61, MS, 63, PhD(animal sci), 65. *Prof Exp:* Res animal scientist, 65-86, RES LEADER, RUMINANT NUTRIT LAB, LIVESTOCK & POULTRY SCI INST, AGR RES SERV, USDA, 86- *Mem:* Am Soc Animal Sci; Am Inst Nutrit; Am Dairy Sci Asn; Am Registry Prof Animal Scientists. *Mailing Add:* Ruminant Nutrit Lab Livestock & Poultry Sci Inst Agr Res Serv USDA Bldg 200 Rm 124 BARC-E Beltsville MD 20705

RUMSEY, VICTOR HENRY, b Devizes, Eng, Nov 22, 19; m 42; c 3. APPLIED PHYSICS. *Educ:* Cambridge Univ, BA, 41, DSc, 73; Tohoku Univ, Japan, PhD, 62. *Prof Exp:* From asst to sr sci officer, Sci Civil Serv, Gt Brit, 41-45; mem staff theoret physics, Atomic Energy Estab, 45-48; from asst prof to assoc prof elec eng, Ohio State Univ, 48-54; prof, Univ Ill, 54-57 & Univ Calif, Berkeley, 57-69; prof appl physics, Univ Calif, San Diego, 69-88; RETIRED. *Concurrent Pos:* Head, Antenna Lab, Ohio State Univ, 48-54; Guggenheim fel, 65. *Honors & Awards:* Liebmann Prize, Inst Elec & Electronics Engrs, 62; George Sinclair Award, Ohio State Univ, 83. *Mem:* Nat Acad Eng; Inst Elec & Electronics Engrs; Am Astron Soc; Int Sci Radio Union. *Res:* Theory of electromagnetic waves; antennas; physics of atomic piles; wave propagation in a turbulent medium. *Mailing Add:* 2199 Bohemian Hwy Occidental CA 95465

RUMSEY, WILLIAM LEROY, b Philadelphia, Pa, Nov 5, 51; m 72; c 2. NUCLEAR MEDICINE, CELLULAR PHYSIOLOGY. *Educ:* Pa State Univ, BS, 73; Temple Univ, MEd, 80, PhD(physiol & biochem), 85. *Prof Exp:* Therapeut actg supvr psychol, Norristown State Hosp, 73-80; postdoctoral, 85-89, RES ASSOC BIOCHEM, SCH MED, UNIV PA, 89-; SR RES INVESTR RADIOPHARMACEUT, BRISTOL-MYERS SQUIBB PHARMACEUT RES INST, 89- *Concurrent Pos:* Co-prin investr, Dept Biochem & Biophys, Sch Med, Univ Pa, 89-; reviewer, Am J Physiol, 89- *Mem:* Am Physiol Soc; AAAS; Int Soc Oxygen Transport to Tissue. *Res:* Mechanisms responsible for oxygen sensing; cellular events that bring about a change in oxygen delivery to tissue. *Mailing Add:* Dept Radiopharmaceut Bristol-Myers Squibb Co New Brunswick NJ 08903

RUNCO, PAUL D, b Pittsburgh, Pa, Feb 22, 57. PHYSICS. *Educ:* Carnegie Mellon, BS, 81, MS, 82. *Prof Exp:* SCIENTIST, CARNEGIE MELLON, 84- *Res:* Design of testing apparatus and procedures for research and development projects in the areas of sensor technology and materials science; utilizing applied physics; computer science; electrical engineering. *Mailing Add:* 6339 Marchand St No 4 Pittsburgh PA 15206

RUND, HANNO, b Schwerin, Ger. MATHEMATICS, MATHEMATICAL PHYSICS. *Educ:* Univ Cape Town, BSc, 47, PhD(math), 50; Univ Freiburg, Habil, 52. *Hon Degrees:* DSc, Univ Natal, 82; Dr, Univ Waterloo, 84. *Prof Exp:* Lectr appl math, Univ Cape Town, 49-51; docent math, Univ Bonn, 52-54; prof appl math, Univ Toronto, 54-56; prof & head dept math, Univ Natal, 56-60, dean sci, 57-60; res prof math, Univ South Africa, 61-66; prof & head dept pure math, Univ Witwatersrand, 67-69; prof & head dept appl math, Univ Waterloo, 70; head dept, 71-78, PROF MATH, UNIV ARIZ, 71- *Concurrent Pos:* Exhib 1851 scholar, Oxford Univ, 52; mem bd gov, Univ Durban, 60-69; vis prof, Univ Waterloo, 64, adj prof, 70-; vis prof, Univ Toronto, 65, Univ Ariz, 67, Univ Witwatersrand, 80 & Univ Cape Town, 83. *Honors & Awards:* SAfrican Math Soc Award, 83. *Mem:* Am Math Soc; Can Math Soc; Ger Math Asn; SAfrican Math Soc; SAfrican Acad Arts & Sci. *Res:* Differential geometry; calculus of variations; theory of relativity. *Mailing Add:* Dept Math Univ Ariz Tucson AZ 85721

RUND, JOHN VALENTINE, b Champaign, Ill, Mar 9, 38; m 61. INORGANIC CHEMISTRY. *Educ:* Univ Ill, BS, 59; Cornell Univ, PhD(chem), 62. *Prof Exp:* Asst prof, 63-69, ASSOC PROF CHEM, UNIV ARIZ, 69- *Mem:* Am Chem Soc; Royal Soc Chem; Sigma Xi. *Res:* Reaction mechanisms of coordination complexes; photochemistry; metalloenzyme models. *Mailing Add:* Dept Chem Univ Ariz Tucson AZ 85721

RUNDEL, PHILIP WILSON, b Palo Alto, Calif, Aug 7, 43; m 80; c 3. PLANT ECOLOGY, ECOSYSTEM STUDIES. *Educ:* Pomona Col, BA, 65; Duke Univ, AM, 67, PhD(bot), 69. *Prof Exp:* Instr bot, Duke Univ, 68-69; asst prof pop & environ biol, 69-74, from assoc prof to prof ecol & evolutionary biol, Univ Calif, Irvine, 74-83; assoc dir, 83-89, PROF BIOL, LAB BIOMED ENVIRON SCI, UNIV CALIF, LOS ANGELES, 83- *Concurrent Pos:* Vis prof, Univ Chile, Santiago, 72, Lehrstuhl Botanik II Univ, WGer, 76 & Univ Hawaii, 83. *Mem:* AAAS; Ecol Soc Am; Am Bryol & Lichenological Soc; Bot Soc Am; Brit Ecol Soc; Asn Trop Biol. *Res:* Physiological plant ecology; arid ecosystems; tropical ecosystems. *Mailing Add:* Lab Biomed & Environ Sci Univ Calif Los Angeles CA 90024

RUNDEL, ROBERT DEAN, b Palo Alto, Calif, Jan 9, 40. PHYSICS. *Educ:* Dartmouth Col, BA, 61; Univ Wash, PhD(physics), 65. *Prof Exp:* Res assoc physics, Culham Lab, Eng, 65-68; sr res assoc space sci, Rice Univ, 68-69, instr physics, 69-70, asst prof, 70-77, adj prof space physics & astron, 77-80; assoc prof, 80-85, PROF PHYSICS, MISS STATE UNIV, 85- *Concurrent Pos:* Staff scientist, Johnson Space Ctr, NASA, 74-80. *Mem:* Am Asn Physics Teachers; Am Soc Photobiol; Sigma Xi. *Res:* Atomic collision physics. *Mailing Add:* 1100 Friar Tuck Starkville MS 39759

RUNDELL, CLARK ACE, b Verndale, Minn, Sept 1, 38; m 61; c 2. CLINICAL CHEMISTRY. *Educ:* St Cloud State Col, BS, 61; Univ NDak, MS, 63, PhD(phys chem), 66; Am Bd Clin Chem, dipl, 76. *Prof Exp:* Dept Defense res fel, Purdue Univ, Lafayette, 65-66; res chemist, W R Grace & Co, Clarksville, 66-73; res assoc, Univ Md Hosp, 73-75; CLIN CHEMIST, MAINE MED CTR, 75-, DIR CHEM, 80- *Mem:* Am Chem Soc; Am Asn Clin Chemists; Clinical Legend Assay Soc. *Res:* Enzymology catalysis; kinetics; data processing. *Mailing Add:* Maine Med Ctr Portland ME 04102

RUNDELL, HAROLD LEE, b Hurley, SDak, Dec 1, 22; m 47; c 1. ZOOLOGY. *Educ:* SDak State Col, BS, 52; Univ Iowa, PhD(zool), 57. *Prof Exp:* Assoc prof biol, Parsons Col, 57-59; assoc prof, 59-71, head dept, 59-81, PROF BIOL, MORNINGSIDE COL, 71- *Res:* Fine structure of animal parasites, chiefly tapeworms. *Mailing Add:* 14 Eastview Dr Morningside Col Sioux City IA 51106

RUNDELL, MARY KATHLEEN, b Cleveland, Ohio, Nov 19, 46; m 76. MICROBIOLOGY, VIROLOGY. *Educ:* Univ Rochester, BA, 68; Case Western Reserve Univ, PhD(microbiol), 73. *Prof Exp:* Fel biochem, Univ Calif, Berkeley, 74; fel pharmacol, Case Western Reserve Univ, 74-75; fel microbiol, State Univ NY Stony Brook, 75-76; asst prof, 76-82, ASSOC PROF MICROBIOL, MED CTR, NORTHWESTERN UNIV, 82- *Concurrent Pos:* Nat Cancer Inst fel, 75-76 & res grant, 77-; Am Cancer Soc res grant, 76-77; NSF res grant, 80-82. *Mem:* Am Soc Microbiol; AAAS. *Res:* Molecular virology and viral genetics; viral transformation. *Mailing Add:* Dept Microbiol-Immunol Northwestern Univ Med Sch 303 E Chicago Ave Chicago IL 60611

RUNDLES, RALPH WAYNE, b Urbana, Ill, Sept 10, 11; m 36; c 4. INTERNAL MEDICINE. *Educ:* DePauw Univ, AB, 33; Cornell Univ, PhD(anat), 37; Duke Univ, MD, 40. *Prof Exp:* Asst & instr anat, Med Col, Cornell Univ, 33-37; intern, Univ Hosp, Univ Mich, 40-41, asst resident internal med, Med Sch, 41-42, resident, 42-43, instr, 43-45; assoc, 45-57, PROF MED, SCH MED, DUKE UNIV, 57-, DIR HEMATOL, 77- *Concurrent Pos:* Consult, NIH, Am Cancer Soc, Am Bur Med Advan China & Burroughs-Wellcome Co. *Mem:* Am Soc Hemat; Am Soc Clin Invest; AMA; Asn Am Physicians; Int Soc Hemat. *Res:* Hematology; cancer chemotherapy. *Mailing Add:* Duke Univ Sch Med Box 3096 Durham NC 27706

RUNDO, JOHN, b London, Eng, Dec 27, 25; US citizen; m 53; c 3. RADIOLOGICAL PHYSICS, HUMAN RADIOBIOLOGY. *Educ:* Univ London, BSc, 49, PhD(radiation biophys), 58. *Hon Degrees:* DSc, Univ London, 80. *Prof Exp:* Sci officer, Atomic Energy Res Estab, Harwell, UK, 49-51; sr sci officer, Finsen Lab, Copenhagen, Denmark, 52-54; prin sci officer, Atomic Energy Res Estab, Harwell, UK, 55-69; assoc scientist, Argonne Nat Lab, 69-74, head, Ctr Human Radiobiol, 80-83, interim assoc lab dir, 83, dep head, 83-84, actg head, Human Radiobiol Sect, 84-85, prog mgr, Int Prog, 85- 90, SR BIOPHYSICIST, ARGONNE NAT LAB, 74-, HEAD, ENVIRON HEALTH SECT, 90- *Concurrent Pos:* Task group mem, Int Comn Radiol Units & Measurements, 65-70; consult, 81- *Mem:* AAAS; Health Physics Soc. *Res:* Metabolism, dosimetry, and late biological effects of natural and artificial radioactivity in the human body. *Mailing Add:* Argonne Nat Lab 9700 S Cass Ave Argonne IL 60439

RUNECKLES, VICTOR CHARLES, b London, Eng, Sept 2, 30; m 53; c 2. PLANT PHYSIOLOGY. *Educ:* Univ London, BSc, 52, PhD(plant physiol), 55, Imp Col, dipl, 55. *Prof Exp:* Nat Res Coun Can fel plant physiol, Queen's Univ, Ont, 55-57; plant biochemist, Imp Tobacco Co Can Ltd, 57-63, res coordr, 63-65, asst mgr res, 65-66, mgr res & prod design, 66-69; head dept, 69-88, PROF PLANT SCI, UNIV BC, 69- *Concurrent Pos:* Lectr, Sir George Williams Univ. *Mem:* Am Soc Plant Physiol; Phytochem Soc NAm (secy, Plant Phenolics Group NAm, 61-66, vpres, 66, pres, 67, ed-in-chief, 69-); Am Phytopath Soc; Air Pollution Control Asn; AAAS; Nuffield Foud Fel. *Res:* Chemistry and metabolism of secondary plant products; pyrolysis of natural products; effects of air pollution on vegetation, cold hardiness; free radicals in plants. *Mailing Add:* Dept Plant Sci Univ BC 2075 Wesbrook Pl Vancouver BC V6T 1W5 Can

RUNEY, GERALD LUTHER, b Charleston, SC, Feb 16, 38; m 69. PHYSIOLOGY. *Educ:* Col Charleston, BS, 60; Univ SC, MS, 63, PhD(physiol), 67. *Prof Exp:* Assoc prof biol, 67-80, PROF BIOL, THE CITADEL, 80- *Mem:* Sigma Xi. *Res:* Endocrine control of lipid metabolism. *Mailing Add:* Dept Biol The Citadel Charleston SC 29409

RUNG, DONALD CHARLES, JR, b Rome, NY, Sept 12, 32; m 56; c 6. MATHEMATICAL ANALYSIS. *Educ:* Niagara Univ, BA, 54; Univ Notre Dame, MS, 57, PhD(math), 61. *Prof Exp:* From asst prof to assoc prof, 61-72, PROF MATH, PA STATE UNIV, UNIVERSITY PARK, 72-, HEAD DEPT, 75- *Concurrent Pos:* Sr Fulbright lectr, Nat Tsing Hua Univ, Taiwan, 67-68; vis sr res scientist, Carleton Univ, Ottawa, 74-75. *Mem:* Math Asn Am; Am Math Soc. *Res:* Complex function theory, especially cluster set theory. *Mailing Add:* Dept Math Pa State Univ 203 McAllister Bldg University Park PA 16802

RUNGE, EDWARD C A, b St Peter, Ill, Aug 4, 33; m 56; c 2. AGRONOMY, SOILS & SOIL SCIENCE. *Educ:* Univ Ill, BS, 55, MS, 57; Iowa State Univ, PhD(agron-soil), 63. *Prof Exp:* Instr & res assoc, Iowa State Univ, 59-63, asst prof, 63; from asst prof to prof, Univ Ill, 63-73; prof & chmn dept agron, Univ Mo, 73-80; PROF SOIL & CROP SCI & HEAD DEPT, TEX A&M UNIV, 80- *Concurrent Pos:* Mem bd dirs, Coun Agr, Sci & Technol, 82-85, Tropsoils, Soil Mgt Entity, 83-, Found Agr Res, 85- & Miss Chem Corp, 86-; mem adv coun, Potash & Phosphate Inst, 83-88; chmn, Tex State Seed & Plant Bd, 83-, Agron Sci Found, 89- *Mem:* Fel Soil Sci Soc Am (pres, 85); fel AAAS; fel Am Soc Agron (pres, 89); Soil Conserv Soc Am. *Res:* Soil genesis and classification; soil chemical relationships; soil water-climate-crop yield modeling; dynamic nature of soil development processes; interdependence of soil moisture-rainfall-temperature on corn and soybean yields. *Mailing Add:* Dept Soil & Crop Sci Tex A&M Univ College Station TX 77843-2474

RUNGE, RICHARD JOHN, b Buffalo, NY, Sept 20, 21; m 57; c 3. WELL LOGGING THEORY & PHYSICAL BASIS, DOWNHOLE EXPLORATION. *Educ:* Univ Chicago, BS, 44; Univ NMex, MS, 49, PhD(physics), 52. *Prof Exp:* Asst prof physics, Univ NMex, 49-52 & Tulsa Univ, 53-55; res asst, Stanolind Amoco Tulsa Labs, 55-56; res assoc, Chevron Oil Field Res Co, 56-86; RES CONSULT, PETROSOFT CO, 86- *Mem:* Soc Prof Well Log Analysts. *Res:* Exploration geophysics and well logging with emphasis on computer applications; nmr and epr logging theory; long range electric logging. *Mailing Add:* 2106 E Sycamore Anaheim CA 92806

RUNGE, THOMAS MARSCHALL, b Mason, Tex, Jan 24, 24; m 47; c 3. CARDIOVASCULAR DISEASES. *Educ:* Univ Tex Med Br Galveston, MD, 47; Am Bd Internal Med, dipl, 55; Am Bd Cardiovasc Dis, dipl, 67. *Prof Exp:* Intern med, Milwaukee County Gen Hosp, Wis, 47-48; resident, Hosp Univ Pa, 48-51; pvt pract, 53-68; PROF BIOMED ENG, UNIV TEX, AUSTIN, 68- *Concurrent Pos:* Fel, Hosp Univ Pa, 48-51; fel, cardiol, St Luke's Hosp, Tex Med Ctr, 65-66; med dir, Noninvasive Cardiol, Brackenridge Hosp. *Mem:* AMA; Am Heart Asn; fel Am Col Physicians; fel Am Col Cardiol; fel

Am Col Chest Physicians. *Res:* Cardiac devices; pulsatile flow cardiopulmonary bypass pumps; contrasting pharmacodynamic action of polar and nonpolar cardiac glycosides as delineated by noninvasive techniques; pulsatile flow hemodialysis. *Mailing Add:* 2630 Exposition 117 Austin TX 78703

RUNION, HOWELL IRWIN, b Ann Arbor, Mich, Oct 26, 33; m 59; c 2. ELECTROPHYSIOLOGY. *Educ:* Col of the Pac, BA, 56; Univ Ore, MS, 63; Univ Glasgow, PhD(electrophysiol), 68; Stanford Univ Med Sch, PA, 80. *Prof Exp:* Chmn, Lincoln Unified Sch Dist, Calif, 58-65; res asst electrophysiol, Univ Glasgow, 65-68; res specialist, Univ Calif, Berkeley, 68-69; assoc prof, 69-80, PROF ELECTROPHYSIOL, SCH PHARM, UNIV PAC, 80- *Concurrent Pos:* Sci adv, Esten Corp, 60-65, Aquatic Res Inst, Port Stockton, 63-65, & Etec Corp, 72-75 & Elec Hazards, Underwriters Med-Dent Inst Bd, 75-; mem bd dirs, Alcoholism Coun Calif, 77-87 & San Joaquin County Alcoholism Adv Bd, 78-80. *Mem:* Am Inst Biol Sci; Brit Soc Exp Biol; Int Asn Elec Inspectors; Am Acad Physician Assts; Sigma Xi. *Res:* Pathological mechanisms involved in atrophy and neuropathy secondary to electrical injury and electrical burns. *Mailing Add:* 6324 Plymouth Rd Stockton CA 95207

RUNK, BENJAMIN FRANKLIN DEWEES, b Germantown, Pa, Apr 10, 06. BOTANY. *Educ:* Univ Va, BS, 29, MS, 30, PhD(biol), 39. *Prof Exp:* Instr sociol, Univ Wis, 30-33; econ analyst, Fed Emergency Relief Admin, 33-34; from asst prof to prof biol, 34-76, registr, 56-59, dean univ, 59-68, EMER PROF BIOL, UNIV VA, 76- *Concurrent Pos:* Instr, Marine Biol Lab, Woods Hole, Mass, 37-41. *Mem:* Fel AAAS; Bot Soc Am; Phycol Soc Am; Sigma Xi. *Res:* Morphology and taxonomy of the marine algae of the east coast of the United States. *Mailing Add:* The Union Rte 1 Box 283 Troy VA 22974

RUNKE, SIDNEY MORRIS, b Greenwood, SDak, Dec 23, 11; m 42; c 2. METALLURGY. *Educ:* Univ Ariz, BS, 35, MS, 36, EMet, 56. *Prof Exp:* Mill foreman, Cia Huanchaca de Bolivia, 36-39 & Coconino Copper Co, Ariz, 40; educ analyst, US Dept Educ, Washington, DC, 42; metall engr, US Bur Mines, Mo, 42-45 & Ark, 45-47, metallurgist, SDak, 47-56; mill supt, Rare Metals Corp Am, Ariz, 57-60, chief metallurgist, Utah, 60-62 & El Paso Natural Gas Co, 62-77; RETIRED. *Mem:* Am Inst Mining, Metall & Petrol Engrs. *Res:* Ore benefication, geology and chemistry. *Mailing Add:* 632 Londonderry Rd El Paso TX 79907

RUNKEL, RICHARD A, b La Crosse, Wis, Aug 21, 32; m 57; c 6. INDUSTRIAL PHARMACY. *Educ:* Univ Wis, BS, 58, PhD(pharm), 67. *Prof Exp:* Lectr, Univ Wis, 66; staff researcher, Syntex Res, 67-75, head dept drug metab, 75-89; RETIRED. *Mem:* Am Pharmaceut Asn; Pharmaceut Soc Japan. *Res:* Drug availability, disposition, metabolism, absorption and excretion. *Mailing Add:* 741 Garland Dr Palo Alto CA 94304

RUNKLE, JAMES READE, b Grove City, Pa, July 3, 51; m 73; c 4. FOREST TREE DYNAMICS, LANDSCAPE ECOLOGY. *Educ:* Ohio Wesleyan Univ, BA, 73; Cornell Univ, PhD(ecol), 79. *Prof Exp:* Vis asst prof biol & ecol, Univ Ill, Chicago, 78-79; asst prof, 79-85, ASSOC PROF BIOL, WRIGHT STATE UNIV, 85- *Concurrent Pos:* Vis sci, Res Inst, Christchurch, NZ, 88-89. *Mem:* AAAS; Am Inst Biol Sci; Am Soc Naturalists; Brit Ecol Soc; Ecol Soc Am; Int Asn Vegetation Sci; Natural Areas Asn. *Res:* Distribution and role of small scale disturbances (gaps) in forest regeneration; effect of human landscape patterning on forest composition. *Mailing Add:* Dept Biol Sci Wright State Univ Dayton OH 45435

RUNKLES, JACK RALPH, b San Angelo, Tex, Sept 4, 22; m 45; c 4. SOIL PHYSICS. *Educ:* Agr & Mech Col Tex, BS, 50, MS, 52; Iowa State Col, PhD(soil physics), 56. *Prof Exp:* Res assoc, Iowa State Col, 53-55; from asst prof to assoc prof soil physics, SDak State Col, 55-64; PROF SOIL SCI, TEX A&M UNIV, 64-, DIR WATER RESOURCES INST, 74- *Mem:* Soil Sci Soc Am; Am Soc Agron; Am Geophys Union. *Res:* Agronomy; soils; mathematics; physics; physical chemistry; hydrology. *Mailing Add:* Water Res Inst Tex A&M Univ College Station TX 77843

RUNNALLS, NELVA EARLINE GROSS, research administration, education administration; deceased, see previous edition for last biography

RUNNALLS, O(LIVER) JOHN C(LYVE), b Barrie Island, Ont, June 26, 24; m 47; c 2. METALLURGY, CERAMICS. *Educ:* Univ Toronto, BASc, 48, MASc, 49, PhD(extractive metall), 51. *Prof Exp:* Res officer, Atomic Energy Can Ltd, 51-56, head fuel develop br, 56-59, rep to Nat Defence Col, 59-60, head res metall br, 61-67, asst dir chem & mat div, Chalk River Nuclear Labs, 67-69, chief liaison off, Europe, 69-71; sr adv uranium & nuclear energy, Dept Energy, Mines & Res, Can, 71-79; PROF ENERGY STUDIES, UNIV TORONTO, CAN, 79-, CHMN, CTR NUCLEAR ENGR, 83- *Concurrent Pos:* mem, Int Adv Comt, Nuexco Int Corp. *Honors & Awards:* Queen's Silver Jubilee Medal, 77; BTA Bell Commemorative Medal, Can Mining Jour, 79; Jan F McRae Award, Can Nuclear Asn, 80. *Mem:* Can Inst Mining & Metall; Can Nuclear Asn; Can Nuclear Soc; fel Royal Soc Can. *Res:* Energy systems with emphasis on uranium resources and nuclear energy development. *Mailing Add:* Dept Metall & Mat Univ Toronto Toronto ON M5S 1A1 Can

RUNNELLS, DONALD DEMAR, b Eureka, Utah, Dec 30, 36; m 58; c 2. GEOCHEMISTRY, HYDROGEOLOGY. *Educ:* Univ Utah, BS, 58; Harvard Univ, MA, 60, PhD(geol), 64. *Prof Exp:* Geochemist, Shell Develop Co, 63-67; asst prof geol, Univ Calif, Santa Barbara, 67-69; assoc prof, 69-75, PROF, 75-, CHAIR GEOL SCI, UNIV COLO, BOULDER, 90- *Concurrent Pos:* Nat Sci Found grad fel, 58-62; regional ed, J Explor Geochem, 71-75; consult, numerous pvt co & govt agencies Los Alamos Sci Lab, 75-76 & Argonne dat Lab, 80-84; assoc ed, J Chem Geol,80-88. *Mem:* Fel Geol Soc Am; Asn Explor Geochem; Soc Econ Paleont & Mineral; Geochem Soc; Asn Ground Water Scientists & Engrs; Asn Explor Geochemists (pres, 90-91). *Res:* Geochemistry of natural waters; low-temperature geochemistry; water pollution; geochemical exploration; geochemistry of trace substances. *Mailing Add:* Dept Geol Sci Univ Colo Boulder CO 80309-0250

RUNNELS, JOHN HUGH, b Mize, Miss, Mar 30, 35; m 60; c 3. ANALYTICAL CHEMISTRY. *Educ:* Univ Denver, BS, 63; Colo State Univ, PhD(chem), 68. *Prof Exp:* Chemist, W P Fuller & Co, 58-60 & Marathon Oil Co, 60-64; sect supvr, Anal Br, 68-91, TECH ENVIRON SPECIALIST, PHILLIPS PETROL CO, 91- *Mem:* Am Chem Soc. *Res:* Analytical methods for determining trace components in natural and synthetic materials. *Mailing Add:* Phillips Petrol Co Rm 117-CPL Bartlesville OK 74004

RUNNELS, LYNN KELLI, b Perry, Okla, June 9, 38; m 58; c 5. CHEMICAL PHYSICS, MATHEMATICAL PHYSICS. *Educ:* Rice Univ, BA, 60; Yale Univ, MS, 61, PhD(chem), 63. *Prof Exp:* From asst prof to assoc prof, 63-72, PROF CHEM, LA STATE UNIV, BATON ROUGE, 72- *Concurrent Pos:* Sloan Found fel, 66-70. *Mem:* Am Phys Soc; Am Chem Soc. *Res:* Statistical mechanics of liquids, dense gases and surface-adsorbed phases; theory of phase transitions; theory of liquid crystals; properties of ice. *Mailing Add:* Dept Chem La State Univ 307 Choppin Hall Baton Rouge LA 70803

RUNNELS, ROBERT CLAYTON, b Houston, Tex, Oct 19, 35; m 61; c 3. METEOROLOGY. *Educ:* Univ Houston, BS, 60; Tex A&M Univ, MS, 62, PhD(meteorol), 68. *Prof Exp:* Space scientist planetary atmospheres, Johnson Space Ctr, NASA, Houston, 65-66; from instr to asst prof, 66-85, ASSOC PROF METEOROL, TEX A&M UNIV, 85- *Concurrent Pos:* Vis scientist, Johnson Space Ctr, NASA, 83 & 84. *Mem:* Sigma Xi; AAAS; Am Geophys Union; Am Meteorol Soc. *Res:* Physical meteorology; science education; rainfall distribution as detected by radar and rain gauges; improvement of learning experience for undergraduate students, such as self-paced teaching materials. *Mailing Add:* Dept Meteorol Tex A&M Univ College Station TX 77843

RUNNER, MEREDITH NOFTZGER, b Schenectady, NY, Jan 7, 14; m 41; c 6. LIMB MORPHOGENESIS. *Educ:* Ind Univ, AB, 37, PhD(zool), 42. *Prof Exp:* Instr zool, Univ Conn, 42-46; Finney-Howell fel, Jackson Mem Lab, 46-48, res assoc, 48-57, staff scientist, 57-62; chmn dept, 62-63, dir inst develop biol, 66-71, PROF BIOL, UNIV COLO, BOULDER, 62- *Concurrent Pos:* Res biologist, Roswell Park Mem Inst, 55-56; prog dir develop biol, NSF, 59-62, mem panel develop biol, 62-64; mem study sect cell biol, NIH, 64-68, human embryol & develop, 69-73; mem, Brest Cancer Task Force, 75-78. *Mem:* AAAS; Genetics Soc Am; Am Soc Zool; Am Asn Anat; Teratol Soc (pres, 67); Soc Develop Biol (pres, 68-69). *Res:* Transplantation of embryonic primordia in mammals; transplantation and explantation of the mouse ovum; physiology of reproduction and hormonal balance in the mouse; mechanism of action of teratogenic agents; genetics of development; limb morphogenesis; staging system for mouse embryo morphogenesis; retinoic acid interaction with gene morpho-regulatory networks. *Mailing Add:* Dept Molecular Cellular & Develop Biol Univ Colo Box 347 Boulder CO 80309

RUNQUIST, OLAF A, b Lohrville, Iowa, Apr 11, 31; m 51; c 4. ORGANIC CHEMISTRY, CHEMICAL EDUCATION. *Educ:* Iowa State Univ, BS, 52; Univ Minn, PhD(chem), 56. *Prof Exp:* From instr to asst prof org chem, Col St Thomas, 55-57; asst prof, 57-73, prof org chem, 73-77, PROF CHEM, HAMLINE UNIV, 77- *Concurrent Pos:* Consult, Minn Mining-Rayette, Inc, 56-57; instr, Exten Div, Univ Minn, 58-59. *Mem:* Am Chem Soc. *Res:* Chemistry of glycosylamines; base strengths of amines. *Mailing Add:* 1536 Hewitt Ave St Paul MN 55104-1205

RUNSER, RICHARD HENRY, physiology, medicine, for more information see previous edition

RUNSTADLER, PETER WILLIAM, JR, b San Francisco, Calif, Jan 19, 34. FLUID DYNAMICS. *Educ:* Stanford Univ, BA, 55, MS, 56, PhD, 61. *Prof Exp:* Teaching asst mech eng, Stanford Univ, 58, res assoc, 60-61; vpres res & develop, 68-75, VPRES & TECH DIR, CREARE PROD INC, 75- *Concurrent Pos:* Adj prof, Thayer Sch Eng, Dartmouth Col, 68- *Mem:* Am Soc Mech Engrs; Sigma Xi. *Mailing Add:* Four Freeman Rd Hanover NH 03755

RUNYAN, JOHN WILLIAM, JR, b Memphis, Tenn, Jan 23, 24; m 49; c 3. INTERNAL MEDICINE. *Educ:* Washington & Lee Univ, AB, 44; Johns Hopkins Univ, MD, 47; Am Bd Internal Med, dipl, 55. *Prof Exp:* Intern, Johns Hopkins Hosp, 47-48; resident internal med, Albany Hosp, 48-49, chief resident, 49-50; res assoc & fel metab dis, Thorndike Mem Lab, Harvard Med Sch, 50-53; from instr to asst prof, Albany Med Col, 53-60; from assoc prof to prof med, 60-73, chief sect endocrinol, 64-73, PROF COMMUNITY MED & CHMN DEPT, COL MED, UNIV TENN, MEMPHIS, 73-, DIR, DIV HEALTH CARE SCI, 72-, PROG DIR GERONT, 80- *Concurrent Pos:* Clin dir, Albany Med Ctr Group Clin, 59-60; consult, Memphis Vet Admin Hosp, 62- *Honors & Awards:* John D Rockefeller III Award Pub Serv Health, 77; Rosenthal Award, Am Col Physicians, 80; Upjohn Award, Am Diabetes Asn, 81. *Mem:* Am Diabetes Asn; fel Am Col Physicians; Am Fedn Clin Res. *Res:* Endocrinology and metabolism. *Mailing Add:* 951 Court Ave Col Med Univ Tenn Memphis TN 38163

RUNYAN, THORA J, b Lemont Furnace, Pa, Sept 11, 31; m 54; c 2. MEDICAL SCIENCES. *Educ:* Univ Idaho, BS, 61; Harvard Univ, DSc(nutrit), 68. *Prof Exp:* Res assoc nutrit, Sch Pub Health, Harvard Univ, 62-63; from asst prof to assoc prof Nutrit, Iowa State Univ, 68-90; NUTRENDS CONSULT, 86- *Concurrent Pos:* Consult & expert witness, health fraud litigation, States of Iowa & Ill. *Mem:* Am Inst Nutrit; Am Col Nutrit; Nat Coun Against Health Fraud; AAAS. *Res:* Age, sex and exercise in nutrient metabolism; biological factors in food section. *Mailing Add:* Nutrends Consult 1201 Orchard Dr Ames IA 50010

RUNYAN, WILLIAM SCOTTIE, b Merrill, Wis, June 24, 31; m 54; c 2. NUTRITION, CELL BIOLOGY. *Educ:* Univ Idaho, BS, 60, MS, 62; Harvard Univ, DSc(nutrit), 68. *Prof Exp:* Res asst tissue cult, Harvard Univ, 61-63; asst prof food & nutrit, 68-74, Nutrit Found future leader grant, 70-71, ASSOC PROF FOOD & NUTRIT, IOWA STATE UNIV, 74- *Concurrent Pos:* Mem, Am Heart Asn. *Mem:* Tissue Cult Asn; NY Acad Sci. *Res:* Relationships between nutritional status and physical activity; diet and coronary disease. *Mailing Add:* Dept Food & Nutrit Iowa State Univ 107 Mackay Ames IA 50011

RUOF, CLARENCE HERMAN, b Hummelstown, Pa, Sept 6, 19; m 45; c 2. CHEMISTRY. *Educ:* Gettysburg Col, BA, 41; Haverford Col, MS, 42; Pa State Univ, PhD(org chem), 48. *Prof Exp:* Asst chem, Haverford Col, 41-42 & Pa State Univ, 42-48; mem staff, Coal Res Lab, Carnegie Inst Technol, 48-54; sr fel, Mellon Inst, 54-60; staff scientist, 60-69, prin staff engr & supvr fuels & lubricants, Ford Motor Co, 69-88; RETIRED. *Mem:* Soc Automotive Engrs; Am Soc Testing & Mat; Am Chem Soc; Sigma Xi. *Res:* Steroids; high octane gasoline components; aromatic acids from coal; structure of coals; plastics; automotive lubricants. *Mailing Add:* 147 Claremont Dearborn MI 48124

RUOFF, ARTHUR LOUIS, b Ft Wayne, Ind, Sept 17, 30; m 54; c 5. PHYSICAL CHEMISTRY. *Educ:* Purdue Univ, BS, 52; Univ Utah, PhD(phys chem), 55. *Prof Exp:* From asst prof to assoc prof eng mat, 55-65, PROF MAT SCI & APPL PHYSICS, CORNELL UNIV, 65-, DIR & CHAIRED PROF, 77- *Concurrent Pos:* NSF sci fac fel & vis assoc prof, Univ Ill, 61-62. *Mem:* Fel Am Phys Soc. *Res:* High pressures; nature of imperfections in solids; transport phenomena in solids; deformation properties of solids; audio-tutorial techniques of instruction. *Mailing Add:* Dept Mat Sci & Eng Cornell Univ Ithaca NY 14853

RUOFF, WILLIAM (DAVID), b Reading, Pa, Apr 4, 40; m 60; c 3. ORGANIC CHEMISTRY. *Educ:* Albright Col, BS, 62; Univ Del, MS, 65, PhD(chem), 67. *Prof Exp:* From asst prof to assoc prof, 66-74, PROF CHEM, FAIRMONT STATE COL, 74-, CHMN DIV SCI, 68- *Mem:* Am Chem Soc. *Res:* Physical organic and synthetic organic chemistry. *Mailing Add:* Fairmont State Col Fairmont WV 26554-2489

RUOTSALA, ALBERT P, b Morse Twp, Minn, Sept 16, 26; m 50; c 3. MINERALOGY. *Educ:* Univ Minn, BA, 52, MS, 55; Univ Ill, PhD, 62. *Prof Exp:* Jr geologist, Bear Creek Mining Co, 52-54; geologist, C&NW Rwy, 55-56; instr geol, Tex Tech Col, 56-57; asst prof, Tex Western Col, 57-60 & Northern Ill Univ, 62-64; from asst prof to assoc prof, 64-74, PROF MINERAL, MICH TECHNOL UNIV, 74- *Mem:* AAAS; Geochem Soc; Am Geophys Union; Mineral Soc Am; Geol Soc Am. *Res:* Clay mineralogy; igneous and metamorphic petrology; economic geology; mineralogy and chemistry of soils. *Mailing Add:* N 1130 River Rd Menominee MI 49858

RUPAAL, AJIT S, b Sangrur, Panjab, India, June 25, 33; m 63; c 2. NUCLEAR PHYSICS. *Educ:* Panjab Univ, India, BSc, 54, MSc, 55; Univ BC, PhD(nuclear physics), 63. *Prof Exp:* Lectr, Ramgarhia Col, Panjab, 55-56; sr res scholar physics, Panjab Univ, 56-57; from asst prof to assoc prof physics, 64-73, chmn, 80-89, PROF PHYSICS, WESTERN WASH UNIV, 74- *Concurrent Pos:* Nat Res Coun Can fel, Chalk River Nuclear Labs, 63-64. *Mem:* Am Asn Physics Teachers; Can Asn Physicists. *Res:* Neutron gamma angular correlations and gamma branching ratios in C; interpretations of Fresnel's equations; neutron time-of-flight spectrometer; positron transmission in thin foils; negative work function of positrons in metals; electrical breakdown of ceramic insulators; low energy positive muon beam development; solid oxide fuel cell. *Mailing Add:* Dept Physics & Astron Western Wash Univ Bellingham WA 98225

RUPERT, CLAUD STANLEY, b Porterville, Calif, Feb 24, 19; m 54; c 2. BIOPHYSICS. *Educ:* Calif Inst Technol, BS, 41; Johns Hopkins Univ, PhD(physics), 51. *Prof Exp:* Eng trainee & sr detail draftsman, Lockheed Aircraft Corp, 41-42; jr instr physics, Johns Hopkins Univ, 46-50, res asst to Prof Strong, 50-52, Am Cancer Soc fel biophys, 52-54, asst prof, 54-57, res assoc, Sch Hyg & Pub Health, 57-58, from asst prof to assoc prof, 58-65; prof biol, Southwest Ctr Advan Studies, 65-69; dean natural sci & math, 75-80, Lloyd Viel Berkner prof, 81-89, PROF BIOL, UNIV TEX, DALLAS, 69-, ASSOC TO VPRES ACAD AFFAIRS, 88- *Concurrent Pos:* USPHS sr res fel, 58-65; lab guest, Inst Microbiol, Copenhagen, Denmark, 61-62; mem impact stratospheric change comt, Nat Res Coun, 75-79. *Honors & Awards:* Finsen Medal, Int Comn Photobiol, 64. *Mem:* AAAS; Biophys Soc; Am Soc Photobiol. *Res:* Infection and transformation of cells by nucleic acids; cell biology; photobiology; photoenzymology; radiation biology. *Mailing Add:* Prog Biol PO Box 830688 Richardson TX 75083-0688

RUPERT, EARLENE ATCHISON, b McCalla, Ala, July 22, 21; m 51; c 3. PLANT GENETICS, TAXONOMY. *Educ:* Huntingdon Col, AB, 41; Univ Ala, Tuscaloosa, MA, 43; Univ Va, PhD(biol), 46. *Prof Exp:* Nat Res Coun fel, Harvard Univ, 46-47; Rockefeller Found fel, Mex Agr Prog, Mexico City, 47-48; from asst prof to assoc prof bot, Univ NC, Chapel Hill, 48-51; res assoc & lectr agron, Univ Calif, Davis, 70-74; assoc prof, 74-78, PROF AGRON & SOILS, CLEMSON UNIV, 78- *Mem:* Am Inst Biol Sci; Bot Soc Am; Genetics Soc Am; Genetics Soc Can; Torrey Bot Club. *Res:* Interspecific hybridization among Leguminosae plant cytogenetics and cytotaxonomy; tissue culture systems for forage legumes. *Mailing Add:* Dept Agron & Soils Clemson Univ Clemson SC 29634

RUPERT, GERALD BRUCE, b Akron, Ohio, Aug 23, 30; m 54; c 2. GEOPHYSICS. *Educ:* Ind Univ, BS, 56, MA, 58; Univ Mo-Rolla, PhD(geophys), 64. *Prof Exp:* Sr seismic computer, Texaco, Inc, 57-60; from asst instr to instr mining engr, Univ Mo-Rolla, 60-64, asst prof geophys, 64-66; res geophysicist, Western Geophys Co, 66-67, mgr, Milano Digital Ctr, Italy, 67-69, sr seismic analyst, 69; assoc prof rock mech res & mining eng, 69-74, SR INVESTR ROCK MECH & EXPLOR RES CTR, 74-, PROF GEOPHYS & CHMN DEPT GEOL & GEOPHYS, UNIV MO-ROLLA, 76- *Mem:* Soc Explor Geophys; Am Geophys Union; Seismol Soc Am; Earthquake Eng Res Inst. *Res:* Exploration geophysics; digital filtering; rock mechanics, particularly wave propagation; earthquake mechanisms; viscoelasticity. *Mailing Add:* Dept Geol & Geophysics Univ Mo Box 249 Rolla MO 65401

RUPERT, JOHN PAUL, b Delphos, Ohio, Oct 14, 46; m 68; c 2. POLYMER SCIENCE. *Educ:* Heidelberg Col, BS, 68; Akron Univ, PhD(polymer sci), 75. *Prof Exp:* Tech supvr eng, Goodyear Tire & Rubber Co, 68-72; fel chem, Inst Polymer Sci, Akron Univ, 75; chemist polymer physics, Union Carbide Corp, 75-76, sr chemist polyurethane raw mat facia develop, 76-78; sr res chemist new appln urethane technol, BASF Wyandotte, 78-; AT MONSANTO INDUS CHEM CO. *Mem:* Am Chem Soc. *Res:* Effect of polymerization variables on anionic polymerizations, polymer characterization and structure-property and rheology-property relationships of engineering thermoplastics; nuclear magnetic resonance of polymers. *Mailing Add:* 312 Baird Ave Wadsworth OH 44281-2223

RUPERT, JOSEPH PAUL, physical chemistry, for more information see previous edition

RUPF, JOHN ALBERT, JR, b Wichita, Kans, Apr 8, 39; m 60; c 3. ELECTRICAL ENGINEERING. *Educ:* Univ Kans, BSEE, 61; Mass Inst Technol, MSEE & EE, 64; Purdue Univ, Lafayette, PhD(elec eng), 69. *Prof Exp:* Instr elec eng, Purdue Univ, Lafayette, 64-69; asst prof, 69-72, ASSOC PROF ELEC ENG & RES ASSOC, BUR CHILD RES, UNIV KANS, 72- *Mem:* Acoust Soc Am; Inst Elec & Electronics Engrs; Inst Noise Control Eng. *Res:* Speech perception, analysis and synthesis; human factors engineering; digital signal processing; noise control. *Mailing Add:* 4921 White Settlement Rd No 34 Ft Worth TX 76114

RUPICH, MARTIN WALTER, b Youngstown, Ohio, Feb 3, 52; m; c 2. ELECTRONIC MATERIAL SYNTHESIS, OPTICAL & ELECTROCHEMICAL SENSORS. *Educ:* John Carroll Univ, Cleveland, Ohio, BS, 74; Northeastern Univ, PhD(inorg chem), 80. *Prof Exp:* Chemist, New Eng Nuclear Corp, 75-76; SR SCIENTIST, EIC LABS INC, 80- *Mem:* Am Chem Soc; AAAS; Mat Res Soc; Electrochem Soc. *Res:* Synthesis and characterization of advanced materials for optical, electronic and catalytic applications, specifically high temperature oxide superconductors, non-linear optical materials, supported catalyst, battery cathodes; development of sensors for atmospheric gases. *Mailing Add:* EIC Labs Inc 111 Downey St Norwood MA 02062

RUPLEY, JOHN ALLEN, b Brooklyn, NY, July 15, 33; m 60; c 1. PROTEIN CHEMISTRY, ENZYME CHEMISTRY. *Educ:* Princeton Univ, AB, 54; Univ Wash, PhD(biochem), 59. *Prof Exp:* NIH fel, Cornell Univ, 59-61; from asst prof to prof chem, 61-78, PROF BIOCHEM, UNIV ARIZ, 78- *Concurrent Pos:* Consult, Fel Panel, NIH, 68-70 & Biochem Study Sect, 71-75; consult, W R Grace & Co, 68-; NIH spec fel, Oxford Univ, 70; mem biochem carcinogen panel, Am Cancer Soc, 72-76; mem pub affairs comt, Fedn Am Soc Exp Biol, 74-77; hon assoc mem, Inst Josef Stefan, Yugoslavia, 74-; co-chmn proteins, Gordon Conf, 83. *Mem:* Am Chem Soc; Am Soc Biol Chemists; AAAS; Am Inst Chemists; NY Acad Sci; Sigma Xi. *Res:* Correlation of protein structure and properties; mechanism of action of enzymes; thermochemistry. *Mailing Add:* Dept Biochem 30 E Calle Belleza Tucson AZ 85716

RUPP, FRANK ADOLPH, microbiology, for more information see previous edition

RUPP, JOHN JAY, b Archbold, Ohio, Sept 28, 40; m 63; c 2. INORGANIC CHEMISTRY, CHEMICAL EDUCATION COMPUTER ASSISTED INSTRUCTION. *Educ:* Ohio Univ, BS, 62; Northwestern Univ, PhD(chem), 67. *Prof Exp:* from assft prof to assoc prof, 66-87, PROF CHEM, ST LAWRENCE UNIV, 87- *Concurrent Pos:* NSF instrnl equip prog grant, 69-71; vis prof, Univ Hawaii, Manoa, 84-85. *Mem:* Am Chem Soc. *Res:* Preparative and physical inorganic chemistry, especially unusual Lewis acid-base addition compounds, organometallics and x-ray crystallography. *Mailing Add:* 52 Spears St Canton NY 13617

RUPP, RALPH RUSSELL, b Saginaw, Mich, Apr 12, 29; m 55; c 2. AUDIOLOGY, SPEECH PATHOLOGY. *Educ:* Univ Mich, BA, 51, MA, 52; Wayne State Univ, PhD(audiol, speech path), 64. *Prof Exp:* Speech & hearing consult, Detroit Pub Schs, Mich, 52-59; exec dir, Detroit Hearing Ctr, 59-62; assoc audiol, Henry Ford Hosp, 62-65; PROF SPEC EDUC & AUDIOL, SPEECH & HEARING SCI, SCH EDUC, UNIV MICH, ANN ARBOR, 65-; COORDR AUDIOL, EASTERN MICH UNIV, 85- *Concurrent Pos:* Consult audiol, C S Mott Children's Health Ctr, Mich, 66-, Ann Arbor Vet Admin Hosp, 67-, St Joseph Mercy Hosp, Ann Arbor, 67 & Dept Hearing Speech & Lang, Kenny-Mich Rehab Found, Pontiac Gen Hosp. *Honors & Awards:* Distinguished Serv Award and Honors Asn, Michigan Speech- Lang-Hearing Asn. *Mem:* Fel Am Speech & Hearing Asn. *Res:* Effect of excessively loud rock 'n roll music on the human hearing mechanisms; improvement of hearing efficiency of the elderly; audiological assessment techniques; language ability of hearing-impaired children; speech audiometry; auditory processing and listening skills in children. *Mailing Add:* 1004 W Cross Ypsilanti MI 48197

RUPP, W DEAN, b Archbold, Ohio, Aug 24, 38; m 62; c 1. MOLECULAR GENETICS, GENETIC ENGINEERING. *Educ:* Oberlin Col, AB, 60; Yale Univ, PhD(pharmacol), 65. *Prof Exp:* Res assoc radiobiol, 65-69, asst prof, 69-74, assoc prof, 74-81, PROF THERAPEUT RADIOL, MOLECULAR BIOPHYS & BIOCHEM, SCH MED, YALE UNIV, 81- *Concurrent Pos:* Res assoc biol chem, Harvard Univ, 67-68. *Mem:* Biophys Soc; Radiation Res Soc; Am Soc Microbiol. *Res:* Cloning of genes and characterization of enzymes involved in DNA repair; expression of proteins from cloned genes. *Mailing Add:* Dept Radiol Yale Univ 333 Cedar St New Haven CT 06511

RUPP, WALTER H(OWARD), b Pittsburgh, Pa, Dec 22, 09; m 37; c 3. CHEMICAL ENGINEERING. *Educ:* Univ Pittsburgh, BS, 30. *Prof Exp:* Jr engr, Standard Oil Co, NJ, 30-36, group head, Esso Res & Eng Co, 36-44, asst div head, 44-49, supv engr, 49-54, tech adv, 54-56, staff engr, 56-62, head tech info ctr, 62-68; PRES, HYLO CO, 68- *Concurrent Pos:* Independent consult, 68- *Mem:* Am Inst Chem Engrs. *Res:* Petroleum refining; design engineering; air pollution; advanced information systems for engineers. *Mailing Add:* 359 Dogwood Way Mountainside NJ 07092

RUPPEL, EARL GEORGE, b Milwaukee, Wis, Nov 10, 32; m 58; c 3. PLANT PATHOLOGY, VIROLOGY. *Educ:* Univ Wis-Milwaukee, BS, 58; Univ Wis-Madison, PhD(plant path), 62. *Prof Exp:* Wis Alumni Res Found-Am Cancer Soc res grant, 62-63; plant pathologist, Tropic & Subarctic Res Br, Crops Res Div, Sci & Educ Admin-Agr Res, 63-65, PLANT PATHOLOGIST, AGR RES SERV, USDA, 65-; MEM AFFIL FAC, PLANT PATHOL & WEED SCI DEPT, COLO STATE UNIV, 70- *Concurrent Pos:* Mem grad fac, Colo State Univ, 71- *Honors & Awards:* Meritorious Serv Award, Am Soc Sugar Beet Technologists, 87. *Mem:* Am Phytopath Soc; Mycol Soc Am; Am Soc Sugar Beet Technol; Int Soc Plant Path. *Res:* Epidemiology of sugarbeet diseases; physiological and biochemical nature of disease resistance; physiology and properties of pathogens; breeding for disease resistance in sugarbeet. *Mailing Add:* USDA-Agr Res Serv Crops Res Lab 1701 Center Ave Ft Collins CO 80526

RUPPEL, EDWARD THOMPSON, b Ft Morgan, Colo, Oct 26, 25; m 56; c 4. GEOLOGY. *Educ:* Univ Mont, BA, 48; Univ Wyo, MA, 50; Yale Univ, PhD(geol), 58. *Prof Exp:* Geologist, US Geol Surv, 48-86, DIR & STATE GEOLOGIST, MONT BUR MINES & GEOL, 86- *Mem:* Fel Geol Soc Am; fel Soc Econ Geologists. *Res:* Structural geology; economic geology; geomorphology. *Mailing Add:* Mont Bur Mines & Geol Butte MT 59701

RUPPEL, ROBERT FRANK, b Detroit, Mich, June 2, 25; m 48; c 2. ENTOMOLOGY. *Educ:* Mich State Col, BS, 48; Ohio State Univ, MS, 50, PhD(entom), 52. *Prof Exp:* Res asst, Mich State Col, 48; from asst entomologist to entomologist, Rockefeller Found, Columbia Univ, 52-62; res biologist, Niagara Chem Div, FMC Corp, 62; assoc prof, 63-68, liaison agr proj, 66-67, PROF ENTOM, MICH STATE UNIV, 68- *Concurrent Pos:* Tech dir, Entom Prog, Dept Agr Res, Govt Colombia, 58-62; res assoc, Univ Calif, 60. *Mem:* AAAS; Entom Soc Am; Am Inst Biol Sci. *Res:* Economic entomology; insect ecology; chemical control; taxonomy. *Mailing Add:* Dept Entomol Mich State Univ East Lansing MI 48824

RUPPEL, THOMAS CONRAD, b Pittsburgh, Pa, June 30, 30; m 57; c 4. PHYSICAL CHEMISTRY. *Educ:* Duquesne Univ, BS, 52; Univ Pittsburgh, BS, 59, MS, 68. *Prof Exp:* Res asst, Heat Insulation fel, Mellon Inst, 52-53 & phys chem dept, 53-57, res assoc, Food Packaging fel, 57-58 & Protective Coatings fel, 58-60; chemist, Res Ctr, Koppers Co, Inc, 60-63; sr res asst water pollution, Grad Sch Pub Health, Univ Pittsburgh, 63-65; res chemist, 65-76, CHEM ENGR, DEPT ENERGY, 76- *Concurrent Pos:* Chmn, Pittsburgh Sect, Am Chem Soc, 88. *Mem:* Air & Waste Mgt Asn; Am Chem Soc; Am Inst Chem Engrs. *Res:* Thermal diffusion; physical adsorption at elevated pressure; chemistry of gaseous non-disruptive electrical discharges; chemical reaction engineering; thermodynamic coal conversion process calculations; on-line engineering database development; environmental engineering and regulations. *Mailing Add:* US Dept Energy PO Box 10940 Pittsburgh PA 15236

RUPPERT, DAVID, STOCHASTIC APPROXIMATION, REGRESSION MODELING. *Educ:* Cornell Univ, BA, 70; Univ Vet, MA, 73; Mich State Univ, PhD(statist), 77. *Prof Exp:* from asst prof to assoc prof statist, Univ NC, 83-87; PROF, OPERS RES, CORNELL UNIV, 87- *Honors & Awards:* Wilcoxon Prize, Am Soc Quality Control, 86. *Mem:* Inst Math Statist; Am Statist Asn; Math Asn Am; Opers Res Soc Am; Sigma Xi. *Res:* Rècursive estimation and stochastic approximation, robustness and regression; biological modeling and natural research management. *Mailing Add:* 154 Ellis Hollow Creek Rd Ithaca NY 14850

RUPPRECHT, KEVIN ROBERT, b Trenton, NJ, Apr 14, 55; m 79; c 1. MOLECULAR GENETICS, GENETIC ENGINEERING. *Educ:* Cornell Univ, BS, 76; Univ Notre Dame, MS, 80, PhD(microbiol), 81. *Prof Exp:* RES ASSOC, UNIV CHICAGO, 81- *Mem:* Am Soc Microbiol; AAAS. *Res:* Discerning the role of the E coli K12 gene ion in capsule production and cell division using recombinant DNA techniques. *Mailing Add:* Abbott Labs Dept 9TV Bldg AP20 Abbott Park IL 60064

RUSAY, RONALD JOSEPH, b New Brunswick, NJ, Dec 21, 45; m 67; c 2. ORGANIC CHEMISTRY, BIOLOGICAL CHEMISTRY. *Educ:* Univ NH, BA, 67, MS, 69; Ore State Univ, PhD(org chem, oceanog), 76. *Prof Exp:* Teacher chem, Bridgton Acad, 71-73; Am Chem Soc-Petrol Res Fund res fel, Ore State Univ, 74-76; res chemist synthesis, 76-80, bus analyst, 81-83, MGR PROD DEVELOP, PAC BASIN, STAUFFER CHEM CO, 84- *Mem:* AAAS; Am Chem Soc; Sigma Xi; NY Acad Sci. *Res:* Development of organic compounds of potential agricultural pharmaceutical importance. *Mailing Add:* 1030 Leland Dr LaFayette CA 94549

RUSCELLO, DENNIS MICHAEL, b Washington, Pa, Sept 2, 47; m 68; c 3. SPEECH PATHOLOGY. *Educ:* Calif State Col, BS, 69; WVa State Univ, MS, 72; Univ Ariz, PhD(speech path), 77. *Prof Exp:* Speech clinician speech path, Allegheny Intermediate 3, Pittsburgh, Pa, 69-74; res assoc, Univ Ariz, 74-77; asst prof, 77-80, assoc prof speech path, 80-85, PROF SPEECH PATH, WVA UNIV, 85- *Concurrent Pos:* WVa Univ Found, Inc grant, 77-78; ed consult, Cleft Palate J, 81- *Mem:* Fel Am Speech & Hearing Asn; Am Cleft Palate Asn. *Res:* Investigation of speech problems exhibited by young children and persons with cranio-facial anomalies. *Mailing Add:* Dept Speech Path & Audiol WVa Univ Morgantown WV 26506

RUSCH, DONALD HAROLD, b Appleton, Wis, Dec 22, 38; m 65; c 1. WILDLIFE ECOLOGY, VERTEBRATE BIOLOGY. *Educ:* Univ Wis-Madison, BS, 62, PhD(wildlife ecol, zool), 70; Utah State Univ, MS, 65. *Prof Exp:* Res specialist birds, Man Dept Natural Resources, 71-72, actg chief wildlife res, 72-73; asst leader, 73-74, LEADER, WIS COOP WILDLIFE RES UNIT, US FISH & WILDLIFE SERV, 74- *Concurrent Pos:* Asst prof, Dept Wildlife Ecol, Univ Wis, 73-79, assoc prof, 79-83, prof, 83- *Mem:* Wildlife Soc; Ecol Soc; Sigma Xi. *Res:* Population ecology; vertebrate predation; waterfowl migration; grouse ecology; population indices; analysis of wildlife habitat. *Mailing Add:* Wis Coop Wildlife Res Unit Univ Wis Madison WI 53706

RUSCH, WILBERT H, SR, b Chicago, Ill, Feb 19, 13; m 37; c 5. BIOLOGY, GEOLOGY. *Educ:* Ill Inst Technol, BS, 34; Univ Mich, MS, 52; Eastern Mich Univ, SpS, 69. *Hon Degrees:* LLD, Concordia Sem, Mo, 75. *Prof Exp:* Instr music & math, Concordia Teachers Col, Nebr, 32-33; instr physics & math, Concordia Col, Ind, 37-46, assoc prof biol, 46-57; assoc prof, Concordia Teachers Col, Nebr, 57-60, prof biol & geol, 60-63; div chmn, 63-73 & 74-75, actg pres, 73-74, acad dean, 75-77, prof biol & geol, 63-80, EMER PROF BIOL & GEOL, CONCORDIA LUTHERAN COL, MICH, 80- *Concurrent Pos:* Kellogg Found fel, 73. *Mem:* Nat Asn Biol Teachers; Nat Asn Geol Teachers; Nat Sci Teachers Asn; fel Creation Res Soc (pres). *Res:* Trees of Great Lake States. *Mailing Add:* 2717 Cranbrook Rd Ann Arbor MI 48104

RUSCHAK, KENNETH JOHN, b Homestead, Pa, March 3, 49; m 78; c 2. CHEMICAL ENGINEERING. *Educ:* Carnegie-Mellon Univ, BS, 71; Univ Minn, PhD(chem eng), 74. *Prof Exp:* RES SCIENTIST, EASTMAN KODAK CO, 74- *Honors & Awards:* C E K Mees Award for Sci, Eastman Kodak Co, 83. *Mem:* Am Inst Chem Engrs. *Res:* Capillary hydrodynamics, perturbation methods, and numerical simulation of fluid flow with applications to coating technology. *Mailing Add:* 236 Wimbledon Rd Rochester NY 14617

RUSCHER, PAUL H, b Mt Vernon, NY, June 8, 55; m 75. SYNOPTIC & MESOSCALE METEOROLOGY, BOUNDARY LAYER STUDIES. *Educ:* State Univ NY, Oneonta, BS, 76; Ore State Univ, MS, 81, PhD(atmospheric sci), 88. *Prof Exp:* Vis asst prof meteorol, Tex A&M Univ, 83-84, Fla State Univ, 84; instr atmospheric sci, Creighton Univ, 85-86; res assoc & instr atmospheric sci, Ore State Univ, 86-88; ASST PROF METEOROL, FLA STATE UNIV, 88- *Concurrent Pos:* Prin investr res grants, Fla State Univ, 88-; mem, Comt Boundary Layers & Turbulence, Am Meteorol Soc, 90-, Comt Undergrad Scholarships & Awards, 91-; proj dir, Tallahassee Area Rain Gauge Proj, 91- *Mem:* Am Meteorol Soc; Am Geophys Union; Can Meteorol & Oceanog Soc; Nat Weather Asn. *Res:* Synoptic studies; mesoscale meteorology especially of coastal winds; boundary layer studies, observational and modelling aspects; weather education; physical and applied climatology; severe storms; turbulence; diffusion and air pollution problems. *Mailing Add:* Dept Meteorol B-161 Fla State Univ Tallahassee FL 32306

RUSCHMEYER, ORLANDO R, b Stewart, Minn, Feb 27, 25; m 51; c 2. PUBLIC HEALTH BIOLOGY, AQUATIC MICROBIOLOGY. *Educ:* Univ Minn, Minneapolis, BA, 51, MS, 56, PhD(environ health), 65. *Prof Exp:* Asst microbiol, 52-56, instr pub health biol, 59-65, ASST PROF PUB HEALTH BIOL, UNIV MINN, MINNEAPOLIS, 66- *Concurrent Pos:* Consult biologist, Int Joint Comn Boundary Waters of US & Can, 61-62. *Mem:* AAAS; Am Soc Limnol & Oceanog; Am Soc Microbiol; Int Asn Theoret & Appl Limnol; Sigma Xi. *Res:* Transformations of organic compounds by soil microflora; limnology of western Lake Superior; water pollution biology; environmental microbiology. *Mailing Add:* 1798 Carl St St Paul MN 55113

RUSH, BENJAMIN FRANKLIN, JR, b Honolulu, Hawaii, Jan 14, 24; m 48; c 2. SURGERY. *Educ:* Univ Calif, AB, 44; Yale Univ, MD, 48. *Prof Exp:* Res asst, Sloan-Kettering Div, Med Col, Cornell Univ, 54-57; instr surg, Sch Med, Johns Hopkins Univ, 57-59, asst prof, 59-62; from assoc prof to prof, Col Med, Univ Ky, 62-69; prof, 69-71, JOHNSON & JOHNSON PROF SURG, COL MED & DENT NJ, NEWARK, 71-, CHMN DEPT, 69- *Concurrent Pos:* Resident, Mem Ctr Cancer & Allied Dis, 53-57; asst chief surg, Baltimore City Hosps, 59-62; consult, Nat Cancer Plan, Nat Cancer Inst. *Mem:* Am Fedn Clin Res; fel Am Col Surgeons; Am Asn Cancer Educ (pres); Am Surg Asn; Sigma Xi; Soc Surg Oncol. *Res:* Fluid balance and renal physiology in relation to surgery; surgical oncology. *Mailing Add:* 100 Bergen St Newark NJ 07103

RUSH, CECIL ARCHER, b Dillwyn, Va, Apr 14, 17; m 57; c 1. MICROCHEMISTRY. *Educ:* Col William & Mary, BS, 38. *Prof Exp:* Anal chemist, Edgewood Arsenal, Chem Res Labs, 40-44, microchemist, 44-50, supvr microanal lab, 50-60, chief microchem lab, Chem Res Labs, 60-80; CONSULT MICROCHEM, ANALYTICAL CHEM, CHEM & PHYS MICROSCOPY, 81- *Mem:* Am Microchem Soc; Am Chem Soc; Sigma Xi; Am Crystallog Asn. *Res:* Analytical chemistry; crystallography; chemical microscopy; chemical warfare agents. *Mailing Add:* 1410 Northgate Rd Baltimore MD 21218-1549

RUSH, CHARLES KENNETH, b Toronto, Ont, Jan 15, 21; m 46; c 4. MECHANICAL ENGINEERING. *Educ:* Queen's Univ, Ont, BSc, 44; McGill Univ, Dipl, 62; Carleton Univ, Can, MEng, 63. *Prof Exp:* Res engr, Nat Res Coun Can, 44-63; from assoc prof to prof, 63-86, EMER PROF MECH ENG, QUEEN'S UNIV, ONT, 86- *Mem:* Am Soc Eng Educ; Am Soc Heating Refrigerating Air-Conditioning Engrs; Int Solar Energy Soc; Can Soc Mech Eng. *Res:* Aircraft icing; energy utilization; solar energy. *Mailing Add:* Dept Mech Eng Queen's Univ Kingston ON K7L 3N6 Can

RUSH, CHARLES MERLE, b Philadelphia, Pa, Oct 10, 42; m 64; c 2. IONOSPHERIC PHYSICS. *Educ:* Temple Univ, BA, 64; Univ Calif, Los Angeles, PhD(meteorol), 67. *Prof Exp:* Staff scientist, Space Systs Div, Avco Corp, Mass, 67-69; res physicist, Air Force Cambridge Res Labs, 69-77; MEM STAFF, NAT OCEANIC & ATMOSPHERIC ADMIN, 77- *Mem:* Am Meteorol Soc; Am Geophys Union; Sigma Xi. *Res:* Dynamics and structure of the earth's ionized atmosphere. *Mailing Add:* 5106 Forsythe Pl Boulder CO 80303

RUSH, DAVID EUGENE, b Carthage, Mo, Mar 5, 43; m 66; c 2. MATHEMATICS. *Educ:* Southwest Mo State Col, BSEd, 65; Western Wash State Col, MS, 68; La State Univ, Baton Rouge, PhD(math), 71. *Prof Exp:* High sch teacher, Mo, 65-67; from asst prof to assoc prof, 71-82, PROF MATH, UNIV CALIF, RIVERSIDE, 82- *Concurrent Pos:* Dept Chair, Univ Calif, Riverside, 85-88. *Mem:* Am Math Soc; Math Asn Am. *Res:* Commutative algebra. *Mailing Add:* Dept Math Univ Calif Riverside CA 92521

RUSH, FRANK E(DWARD), JR, b Washington, Pa, July 25, 21; m 44; c 5. CHEMICAL ENGINEERING. *Educ:* Washington & Jefferson Col, BA, 43; Mass Inst Technol, BS, 44; Univ Del, MChE, 53. *Prof Exp:* Res engr, Eng Res Lab, Exp Sta, E I du Pont de Nemours & Co, Inc, 46-53, res proj engr, 53-55, res proj supvr, 55-56, res supvr, 56-61, res assoc, 61-63, chem eng consult, Eng Serv Div, 63-68, sr consult, Eng Dept, 68-76, prin consult, Eng Dept, 76-86; RETIRED. *Concurrent Pos:* Adj prof, Univ Del, 71-81. *Mem:* Am Chem Soc; fel Am Inst Chem Engrs. *Res:* Diffusional operations; mass transfer. *Mailing Add:* Eight Briar Lane Newark DE 19711-3102

RUSH, JAMES E, b Warrensburg, Mo, July 18, 35; m 58; c 4. LIBRARY & INFORMATION SCIENCE, SYSTEM DESIGN. *Educ:* Cent Mo State Univ, BS, 57; Univ Mo, PhD(org chem), 62. *Prof Exp:* Asst ed, Org Index Ed Dept, Chem Abstr Serv, Am Chem Soc, 62-65, asst head, Chem Info Procedures Dept, 65-68; from asst prof to assoc prof comput & info sci, Ohio State Univ, 68-73; dir res & develop, OCLC, Inc, 73-80; pres, James E Rush Assoc, Inc, 80-88; EXEC DIR, PALINET, 88- *Concurrent Pos:* Ed, Chem Lit, 70-73; adj prof comput & info sci, Ohio State Univ, 73-80; adj prof libr & info sci, Univ Ill, 80-83. *Mem:* Sigma Xi; Am Chem Soc; Am Soc Info Sci. *Res:* Organoboron compounds; organometallic and coordination chemistry; stereochemistry; information storage and retrieval; library automation; telecommunication and networks. *Mailing Add:* 673 Old Eagle Sch Rd Stafford PA 19087

RUSH, JOHN EDWIN, JR, b Birmingham, Ala, Aug 11, 37; m 63, 82; c 5. THEORETICAL PHYSICS. *Educ:* Birmingham-Southern Col, BS, 59; Vanderbilt Univ, PhD(physics), 65. *Prof Exp:* From instr to asst prof physics, Univ of the South, 64-67; from asst prof to assoc prof physics, Univ Ala, Huntsville, 67-83, chmn dept, 69-72, dean sch grad studies & res, 72-76; RES SCIENTIST, KAMAN SCI CORP, HUNTSVILLE, ALA, 83- *Mem:* Am Phys Soc; Sigma Xi. *Res:* Nuclear weapons effects. *Mailing Add:* 12034 Chicamauga Trail Huntsville AL 35803

RUSH, JOHN JOSEPH, b Brooklyn, NY, Apr 20, 36; m 61. SOLID STATE PHYSICS, PHYSICAL CHEMISTRY. *Educ:* St Francis Col, NY, BS, 57; Columbia Univ, MA, 58, PhD(phys chem), 62. *Prof Exp:* Asst, Columbia Univ, 57-62; phys chemist, 62-71, chief neutron-solid state physics sect, Nat Measurements Lab, 71-81, LEADER, NEUTRON-CONDENSED MATTER SCI GROUP, NAT BUR STANDARDS, 81- *Concurrent Pos:* Guest scientist, Solid State Sci Div, Argonne Nat Lab, 62-65. *Honors & Awards:* Gold & Silver Medals, Commerce Dept. *Mem:* AAAS; Am Phys Soc; fel Nat Inst Standards & Technol. *Res:* Study of molecular solids, catalysts and hydrogen in metals by thermal neutron scattering and other spectroscopic techniques; vibrations and rotations in condensed systems; structure; phase transitions. *Mailing Add:* Div 566 0 MALS Sci & Eng Lab Nat Inst Standards & Technol Gaithersburg MD 20899

RUSH, KENT RODNEY, b Quakertown, Pa, Sept 5, 38; m 59; c 3. ORGANIC CHEMISTRY. *Educ:* Franklin & Marshall Col, BS, 60; Univ Minn, PhD(org chem), 63. *Prof Exp:* Res chemist, Distillation Prod Indust Div, Eastman Kodak Co, 63-65, res assoc res labs, 65-78, prod develop mgr, 78-90; CONSULT, 90- *Mem:* Am Chem Soc. *Res:* Chemistry of nitrogen heterocycles, polyenes; carotenoids; photographic chemistry. *Mailing Add:* Eastman Kodak Co 343 State St Rochester NY 14650

RUSH, RICHARD MARION, b Bristol, Va, Dec 5, 28; m 55; c 2. PHYSICAL CHEMISTRY. *Educ:* Princeton Univ, AB, 49; Univ Va, MS, 52, PhD(chem), 54. *Prof Exp:* Asst chem, Mass Inst Technol, 53-54; asst prof, Haverford Col, 54-56; CHEMIST, OAK RIDGE NAT LAB, 56- *Mem:* Fel AAAS; Am Chem Soc; Am Nuclear Soc; fel Am Inst Chemists; Sigma Xi. *Res:* Physical chemistry; inorganic solution chemistry; thermodynamics of electrolyte solutions; environmental impact assessment. *Mailing Add:* Oak Ridge Nat Lab PO Box 2008 Oak Ridge TN 37831-6291

RUSH, RICHARD WILLIAM, b Austin, Minn, July 14, 21; div; c 4. GEOLOGY. *Educ:* Univ Iowa, BA, 45; Columbia Univ, MA, 48, PhD (geol), 54. *Prof Exp:* Consult, River Prod Corp, 61-64; assoc prof, Northern Ariz Univ, 63-69; CONSULT GEOL, 69- *Mem:* Am Inst Prof Geol; AAAS; Sigma Xi; Geol Soc Am; Am Geophys Union; Am Asn Petrol Geologists. *Res:* Siluvian stratigraphy; geomorphology; conceptual design of new surface mining techniques; regional tectonics; suspect or exotic terranes; computer techniques to aid fundamental geologic research in regional tectonics; expert systems. *Mailing Add:* 337 W Pasadena No 16 Phoenix AZ 85013

RUSH, STANLEY, b New York, NY, June 17, 20; m 52; c 1. ELECTRICAL ENGINEERING. *Educ:* Brooklyn Col, BA, 42; Syracuse Univ, MEE, 58, PhD(elec eng), 62. *Prof Exp:* Design engr, RCA Victor Div, Radio Corp Am, 46-47; supvry electronic scientist, Rome Air Develop Ctr, US Air Force, 47-57; instr elec eng, Syracuse Univ, 57-62; assoc prof, 62-67, PROF ELEC ENG, UNIV VT, 67- *Mem:* Inst Elec & Electronics Engrs; NY Acad Sci. *Res:* Current flow in body from heart and brain generators; interpretation of electrocardiogram on physical basis; electrophysiology of the heart; tissue impedance at gross and cellular level; electromagnetic field theory. *Mailing Add:* Dept Elec Eng Univ Vt Agr Col 85 S Prospect Burlington VT 05401

RUSHFORTH, CRAIG KNEWEL, b Ogden, Utah, Sept 4, 37; m 58; c 5. ELECTRICAL ENGINEERING. *Educ:* Stanford Univ, BS, 58, MS, 60, PhD(elec eng), 62. *Prof Exp:* Res asst radio propagation, Stanford Univ, 58-59; res engr, Stanford Res Inst, 63; asst prof elec eng, Utah State Univ, 62-66; assoc prof, Mont State Univ, 66-72, prof, 72-73; PROF ELEC ENG, UNIV UTAH, 73- *Concurrent Pos:* Lectr, Stanford Univ, 63; consult panel synthetic aperture optics, Nat Acad Sci, 67; mem staff, Int Defense Analyses, 67-68. *Mem:* Inst Elec & Electronics Engrs. *Res:* Communication theory; digital signal processing. *Mailing Add:* Dept Elec Eng Univ Utah Salt Lake City UT 84112

RUSHFORTH, NORMAN B, b Blackpool, Eng, Dec 27, 32; m 56; c 2. ANIMAL BEHAVIOR, EPIDEMIOLOGY. *Educ:* Univ Birmingham, BSc, 54; Cornell Univ, MS, 58, PhD(statist), 61. *Prof Exp:* From instr to assoc prof biol, 61-72, asst prof biostatist, 63-72, ASSOC PROF BIOSTATIST & PROF BIOL, CASE WESTERN RESERVE UNIV, 72-, CHMN DEPT, 71- *Honors & Awards:* Brit Johnson Found Fel, 56; Woodrow Wilson Fel, 59. *Res:* Application of quantitative methods in biological research; animal behavior; epidemiological studies of violent behavior. *Mailing Add:* Dept Biol Case Western Reserve Univ 2040 Adelbert Rd Cleveland OH 44106

RUSHFORTH, SAMUEL ROBERTS, b Salt Lake City, Utah, Nov 24, 45; m 64; c 3. ENVIRONMENTAL SCIENCES. *Educ:* Weber State Col, BS, 66; Brigham Young Univ, MS, 68, PHD(bot), 70. *Prof Exp:* From asst prof to assoc prof, 70-80, PROF BOT, BRIGHAM YOUNG UNIV, 80- *Mem:* Bot Soc Am; Int Phycol Soc; Sigma Xi; Phycol Soc Am. *Res:* Taxonomic and ecological investigation of the algae of western America; effects of air quality on aquatic communities. *Mailing Add:* Dept Bot Brigham Young Univ Provo UT 84602

RUSHING, ALLEN JOSEPH, b Charlottesville, Va, Oct 23, 44; m 73; c 3. CONTROL SYSTEMS. *Educ:* Univ Denver, BSEE, 66; Univ Mo-Rolla, MSEE, 70, PhD(elec eng), 73. *Prof Exp:* Elec engr instruments, Monsanto Co, 67-70; RES ASSOC, EASTMAN KODAK CO, 73- *Concurrent Pos:* Mem adj fac, Rochester Inst Technol, 75-87. *Mem:* Sr mem Inst Elec & Electronics Engrs. *Res:* Applications of control theory to electrophotography. *Mailing Add:* Eastman Kodak Res Labs Bldg 951 Kodak Park Rochester NY 14650

RUSHING, FRANK C, b Nordheim, Tex, July 11, 06; m 34; c 3. ELECTRICAL ENGINEERING. *Educ:* Univ Tex, BS, 28; Univ Pittsburgh, MS, 30. *Prof Exp:* Eng consult, Westinghouse Corp, 28-90; RETIRED. *Mem:* Fel Am Soc Mech Engrs; fel Inst Elec & Electronics Engrs; Sigma Xi. *Mailing Add:* 6346 Belleview Dr Columbia MD 21046

RUSHING, THOMAS BENNY, b Marshville, NC, Oct 30, 41; m 62; c 2. GEOMETRIC TOPOLOGY. *Educ:* Wake Forest Univ, BS, 64, MA, 65; Univ Ga, PhD(math), 68. *Prof Exp:* Asst prof math, Univ Ga, 68-69; from asst prof to assoc prof, 69-77, PROF MATH, UNIV UTAH, 77-, CHMN DEPT, 85- *Concurrent Pos:* Vis prof, Univ Fla, 74 & 81; Nat Acad Sci exchange scientist, Univ Zagreb, Yugoslavia, 75-76 & 83; vis fel, Warwick Math Inst, Eng, 76; David P Gardner fac fel, Univ Utah, 77; visitor, Inst Math, Nat Univ, Mexico, 82, Banach Math Ctr, Warsaw, Poland, 84 & Math Inst, Univ Heidelberg, Germany, 85; mem, Inst Advan Study, Princeton, 82-83 & 88-89. *Mem:* Am Math Soc; Math Asn Am. *Res:* Topology of manifolds; embedding problems; piecewise linear topology; shape, fibrations and bundle theories, dynamical systems. *Mailing Add:* Dept Math Univ Utah Salt Lake City UT 84112

RUSHMER, ROBERT FRAZER, b Ogden, Utah, Nov 30, 14; m 42; c 3. BIOENGINEERING. *Educ:* Univ Chicago, BS, 35; Rush Med Col, MD, 39. *Hon Degrees:* PhD(Hon), Univ Linkoping, Sweden, 77. *Prof Exp:* Intern, St Luke's Hosp, San Francisco, 39-40; fel pediat, Mayo Found, Univ Minn, 40-42, asst, 41-42; assoc prof aviation med, Sch Med, Univ Southern Calif, 46-47; from asst prof to prof physiol, Sch Med, Univ Wash, 47-68, prof bioeng, 68-86, dir, Ctr Advan Studies Biomed Sci, 78-84, EMER PROF BIOENG, UNIV WASH, 86- *Honors & Awards:* Ida B Gould Award, AAAS, 62; Modern Med Award, 62. *Mem:* Inst Med-Nat Acad Sci; Am Physiol Soc; Am Heart Asn; AAAS; fel Am Col Cardiol; Biomed Eng Soc. *Mailing Add:* 7050 56th Ave NE Seattle WA 98115

RUSHTON, BRIAN MANDEL, b Sale, Eng, Nov 16, 33; m 58; c 3. CHEMISTRY. *Educ:* Univ Salford, ARIC, 57; Univ Minn, MS, 59; Univ Leicester, PhD(chem), 63. *Prof Exp:* Sr res chemist, Petrolite Corp, 63-65, group leader, 65-66; sect mgr, Asland Chem Co, 66-69; corp res mgr, Hooker Chem Corp, 69-72, dir polymer & plastics res & develop, 72-74, vpres res & develop, 74-75; pres, Celanese Res Corp, 75-80, corp vpres-technol, 80-81; VPRES RES & DEVELOP, AIR PROD & CHEM, INC, 81- *Concurrent Pos:* Mem, Nat Mats Adv Bd, Nat Res Coun, Nat Acad Sci/Nat Acad Eng, 80-84; mem gov bd, Coun Chem Res, 81-87; mem, Life Sci Vis Comt, Lehigh Univ, 83-86; chairperson, Vis Comt, Ctr Surface & Coatings, 86-89; mem, Adv Comt Pa, Ben Franklin Prog, Northeast Tier, 86-88; mem bd trustees, Textile Res Inst, 86-89; mem bd dir, Indust Res Inst, 90- & Mich Molecular Inst, 91- *Mem:* Coun Chem Res (treas, 81-87); Textile Res Inst; Am Chem Soc; Soc Chem Indust; Indust Res Inst (pres-elect, 90-). *Res:* Polymer chemistry; general chemistry. *Mailing Add:* Air Prod & Chem Inc 7201 Hamilton Blvd Allentown PA 18195-1501

RUSHTON, PRISCILLA STRICKLAND, b Clarksdale, Miss, Dec 19, 42; m 66; c 1. RADIOBIOLOGY. *Educ:* Southwestern at Memphis, BA, 63; Emory Univ, MS, 64, PhD(biol), 67. *Prof Exp:* Asst prof, 67-71, ASSOC PROF GENETICS, MEMPHIS STATE UNIV, 71- *Mem:* AAAS; Am Inst Biol Sci; Am Genetic Asn; Sigma Xi. *Res:* Modification and repair of x-irradiation induced chromosome and chromatid aberrations in plants. *Mailing Add:* 95 Hollyoke Memphis TN 38117

RUSINKO, FRANK, JR, b Nanticoke, Pa, Oct 12, 30; m 57; c 2. FUEL TECHNOLOGY, CARBON & GRAPHITE. *Educ:* Pa State Univ, BS, 52, MS, 54, PhD(fuel technol, phys chem), 58. *Prof Exp:* Res assoc, Pa State Univ, 58-59; scientist, Speer Carbon Co, Air Reduction Co, 59-61, mgr develop, Carbon Prod Div, 61-62; mgr develop, IGE Div, Airco, Inc, 62-67, dir develop & technol serv, 67-70, vpres & tech dir, Airco Speer Carbon-Graphite Div, 70-76; vchmn, Powder Tech, 85-86, pres, Electrotools, Inc, UTI Inc, 76-89, PRES & CHMN, EDIMAX TRANSOR, 82-, PRES, INTECH EDM ELECTROTOOLS, 89- *Concurrent Pos:* Bd mem, C-Cor Electronics, State Col, Pa. *Mem:* Am Chem Soc; Am Inst Mining, Metall & Petrol Engrs; Am Soc Testing & Mat; Am Mgt Asn. *Res:* Carbon and graphite; gas-solid reactions; gas adsorption; coal carbonization; irradiation of coal and graphite; nuclear graphite; catalysis of heterogeneous carbon reactions; fuel cells; surface chemistry; research and development management; high temperature materials technology; electrical discharge machine technology, including electrode materials, accessories, dielectrics and dielectric oil filtration. *Mailing Add:* Intech EDM Electrotools 2001 W 16th Ct Broadview IL 60153

RUSKAI, MARY BETH, b Cleveland, Ohio, Feb 26, 44. MATHEMATICAL PHYSICS, OPERATOR THEORY. *Educ:* Notre Dame Col, Ohio, BS, 65; Univ Wis-Madison, MA, 69, PhD(chem), 69. *Prof Exp:* Battelle fel math physics, Theoret Physics Inst, Univ Geneva, 69-71; res assoc math, Mass Inst Technol, 71-72; res assoc theoret physics, Univ Alta, 72-73; asst prof math, Univ Ore, 73-76; from asst prof to assoc prof, 77-86, PROF MATH, UNIV LOWELL, 86- *Concurrent Pos:* Consult, Bell Labs, NJ, 72, 83 & 88-89; vis asst prof, Rockefeller Univ, 80-81; guest prof, Univ Vienna, Austria, 81; sci scholar, Mary Ingraham Bunting Inst, 83-85; vis assoc math, Calif Inst Technol, 84; vis mem, Courant Inst Math Sci, NY Univ, 88-89; vis scientist, Inst Theoret Atomic & Molecular Physics, Harvard-Smithsonian Ctr Astrophys, 90. *Mem:* Am Math Soc; Math Asn Am; Am Phys Soc; Int Asn Math Physics; Asn Women in Math; Sigma Xi; AAAS. *Res:* Operator theory; statistical mechanic; multi-particle Coulomb system. *Mailing Add:* 35 A Pine St Arlington MA 02174

RUSKIN, ARNOLD M(ILTON), b Bay City, Mich, Jan 4, 37. PROJECT MANAGEMENT, SYSTEM ENGINEERING. *Educ:* Univ Mich, BSE(chem eng) & BSE(mat eng), 58, MSE, 59, PhD(eng mat), 62; Claremont Grad Sch, MBE, 70. *Prof Exp:* Instr mat eng, Univ Mich, 61-62; lectr appl physics, Rugby Col Eng Technol, Eng, 62-63; from asst prof to prof eng, Harvey Mudd Col, 63-73; eng mgr, Everett/Charles, Inc, 73-74; vpres & prog mgr, Claremont Eng Co, 74-78; network syst engr, Jet Propulsion Lab, Calif Inst Technol, 78-80, mgr, Network Strategy Develop, 80-86, dep mgr, Syst Eng Resource Ctr, 86-90, training mgr, Systs, Software & Opers Resource Ctr, 87-90; PARTNER, CLAREMONT CONSULT GROUP, 79- *Concurrent Pos:* Res engr, E I du Pont de Nemours & Co, Inc, 58; dir joint col & indust libr study, Harvey Mudd Col, 63-65, Union Oil Co fel eng, 66-73, asst dir, Freshman Div Fac, 71-72, dir, 72-73; fac fel, Pac Coast Banking Sch, Univ Wash, Seattle, 67; visitor, Prog Indust Metall & Mgt Techniques, Univ Aston, Birmingham, Eng, 69; assoc instnl res, Claremont Univ Ctr, 65-67, from assoc prof to prof econ & bus, Grad Sch, 70-73; continuing educ specialist, Univ Calif, Los Angeles, 74-84, lectr, 75-77, adj prof eng, 77-84, dir eng exec prog, 78-84; Zambelli fel, syst eng & proj mgt, Royal Melbourne Inst Technol, Australia, 90. *Mem:* Proj Mgt Inst; Am Soc Engr Mgt; assoc fel Am Inst Aeronaut & Astronaut; Am Inst Chem Engrs; Metall Soc; Am Soc Metals. *Res:* Engineering management; project management; system engineering methodology; strategic and tactical planning. *Mailing Add:* 4525 Castle Lane La Canada CA 91011-1436

RUSKIN, ASA PAUL, rehabilitation medicine, neurology; deceased, see previous edition for last biography

RUSKIN, RICHARD A, b New Rochelle, NY, Oct 1, 24; m 46, 66; c 1. OBSTETRICS & GYNECOLOGY. *Educ:* Duke Univ, BA, 40, MD, 44; Am Bd Obstet & Gynec, dipl, 54. *Prof Exp:* Resident obstet & gynec, New York Lying-In-Hosp, 47-52; from instr to assoc prof, 52-72, PROF OBSTET & GYNEC, MED COL, CORNELL UNIV, 72- *Concurrent Pos:* Resident, Kings County Hosp, New York, 46-47; resident, New York Polyclin Med Sch & Hosp, 47-52, clin prof, 65-; attend obstetrician & gynecologist, New York Lying-In-Hosp, 65-, attend, 72-, clin prof, 72, prof, 72-; consult, Workmans Compensation Bd; attend obstet & gynec, Roosevelt Hosp, 76-; prof & attend, New York Polyclin Med Sch & Hosp, 65-76. *Mem:* Fel Am Col Obstet & Gynec; fel Am Col Surg; Geriat Soc Am. *Res:* Oxytocin and vasopressin in clinical obstetrics. *Mailing Add:* 850 Park Ave New York NY 10021

RUSKIN, ROBERT EDWARD, b Sioux Falls, SDak, Oct 30, 16; m 42; c 2. ATMOSPHERIC PHYSICS. *Educ:* Kans State Col, AB, 40. *Prof Exp:* Physicist, Uranium Isotope Separation Proj, Naval Res Lab, 42-47, head instrument sect, Atmospheric Physics Br, 47-71, actg head, 71-72, asst head, 72-79. *Concurrent Pos:* Vis prof, Colo State Univ, 70; consult atmospheric physicist, 82- *Honors & Awards:* Meritorious Civilian Serv Award, US Navy, 45, Superior Accomplishment Award, 65. *Mem:* Fel AAAS; Am Phys Soc; Sigma Xi; Am Meteorol Soc; fel Instrument Soc Am. *Res:* Cloud physics; physics of interactions of the atmosphere; aircraft instrumentation; air pollution; marine fog and haze interactions with electrooptical systems; marine salt aerosol in relation to gas turbine ship engines. *Mailing Add:* 1406 Ruffner Rd Alexandria VA 22302

RUSLING, JAMES FRANCIS, b Philadelphia, Pa, Nov 4, 46. BIOELECTROCHEMISTRY & ELECTROCATALYSIS. *Educ:* Drexel Univ, BSc, 69; Clarkson Col Technol, PhD(anal chem), 79. *Prof Exp:* Spectroscopist anal chem, Sadtler Res Co, 72-73; anal chemist, Wyeth Labs, 73-76; asst prof, 79-85, assoc prof, 85-89, PROF ANALYTICAL CHEM, UNIV CONN, 90- *Concurrent Pos:* Mem, Inst Mat Sci & Environ Res Inst, Univ Conn, 81- *Mem:* Am Chem Soc; Am Asn Univ Professors; Electrochem Soc; Soc Electroanal Chem. *Res:* Development of electrocatalytic systems in organized media for redox transformations of organic compounds; applications of such systems to destroying pollutants and modeling of biological redox events; development of computerized methods for data interpretation. *Mailing Add:* Univ Conn Box U 60 Storrs CT 06268

RUSOFF, IRVING ISADORE, b Newark, NJ, Jan 29, 15; m 41; c 2. NUTRITION. *Educ:* Univ Fla, BS, 37, MS, 39; Univ Minn, PhD(physiol chem), 43. *Prof Exp:* Asst, Nutrit Lab, Exp Sta, Univ Fla, 37-39, asst dairy res, 39-40; fels, Nat Found Infantile Paralysis, Univ Minn, 43-44, Off Sci Res & Develop, 44-45 & US Naval Ord, 45-46; head nutrit lab, Standard Brands, Inc, 46-47; head biochem sect, Res Ctr, Gen Foods Corp, 47-57, head nutrit sect, 57-61, group leader fats & oils, 61-62; mgr res & res serv, DCA Food Industs, 62-63; mgr nutrit & biochem & coordr spec proj res, Beech-Nut Life Savers, Inc, 63-66; head biol sci & asst to res dir, Nat Biscuit Co, 66-67, mgr develop res, 67-68; dir basic studies, Nabisco Brands Inc, 68-70, dir sci, 71-76, sr scientist, 76-85, dir nutrit sci, 78-85; CONSULT, 85- *Concurrent Pos:* Assoc ed, Am Oil Chemists Soc J, 60-62; mem, ad hoc comt food sci abstracts, Food & Nutrit Bd, Nat Res Coun, 79-80, mem chmn indust, Laison Panel, 79-83; chmn & liason, Nat Inventors Hall of Fame, Inst Food Technologists, 80-; chmn, Food & Nutrit Conf, Gordon Res Conf, 81. *Honors & Awards:* Charles N Frey Award, Am Asn Cereal Chemists, 84. *Mem:* Am Chem Soc; Am Inst Nutrit; fel Inst Food Technologists; NY Acad Sci; Am Pub Health Asn; Am Asn Cereal Chemists. *Res:* Human and animal nutrition; vitamin and mineral metabolism; trace elements and nutrients; biomedicine. *Mailing Add:* 65 Central Blvd Brick NJ 08724-2451

RUSOFF, LOUIS LEON, b Newark, NJ, Dec 23, 10; m 45; c 2. ANIMAL NUTRITION, NUTRITION. *Educ:* Rutgers Univ, BS, 31; Pa State Col, MS, 32; Univ Minn, PhD(agr biochem, nutrit), 40. *Prof Exp:* Asst, Nutrit Lab, Exp Sta, Univ Fla, 32-35, lab asst & instr animal nutrit, 35-37, asst & asst prof, 37-42; assoc dairy nutritionist, Exp Sta, La State Univ, Baton Rouge, 42-50, assoc prof dairy nutrit, 48-50, prof dairy nutrit & dairy nutritionist, 50-81; RETIRED. *Honors & Awards:* Charles E Coates Award, Am Chem Soc, 59; Borden Award, Am Dairy Sci Asn, 65; Hon Scroll, Am Inst Chem, 82. *Mem:* AAAS; Am Chem Soc; Am Soc Animal Sci; Am Dairy Sci Asn; Am Inst Nutrit. *Res:* Vitamin assays; minerals and alkaloids; biochemistry of blood; trace elements in animal nutrition; milk analysis; feeding and digestion trials; molasses; urea feeding; antibiotics in food and silage preservation; fluoridation of milk; nutritive value of aquatic plants as foodstuffs for animals and man. *Mailing Add:* 1704 Myrtledale Ave Baton Rouge LA 70808

RUSS, CHARLES ROGER, b New London, Wis, July 2, 37; m 61; c 3. INORGANIC CHEMISTRY. *Educ:* Marquette Univ, BS, 59, MS, 61; Univ Pa, PhD(chem), 65. *Prof Exp:* Asst chem, Marquette Univ, 59-61 & Univ Pa, 61-65; asst prof, 65-71, ASSOC PROF CHEM, UNIV MAINE, ORONO, 71- *Mem:* Am Chem Soc. *Res:* Synthesis and properties of silicon or germanium compounds. *Mailing Add:* Dept Chem Univ Maine 333 Aubery Hall Orono ME 04473

RUSS, DAVID PERRY, b Wilmington, Del, May 7, 45. TECTONIC GEOMORPHOLOGY, QUATERNARY GEOLOGY. *Educ:* Pa State Univ, BS, 67, PhD(geol), 75; WVa Univ, MS, 69. *Prof Exp:* Geologist, Waterways Exp Sta, US Army Corps Engrs, 70-72; GEOLOGIST, US GEOL SURV, 75- *Mem:* Geol Soc Am. *Res:* Geological and geophysical investigations of earthquake hazards; seismotectonics of the New Madrid seismic zone. *Mailing Add:* US Geol Surv 911 Nat Ctr 12202 Sunrise Velley Dr Reston VA 22092

RUSS, GERALD A, b Washington, DC, Oct 17, 36; m 65; c 3. NUCLEAR MEDICINE. *Educ:* Univ Md, BS, 64; Georgetown Univ, PhD(biochem), 74. *Prof Exp:* Res asst physiol, Georgetown Univ, 62-64; res technician cytogenetics, Radiol Health, USPHS, 64; res assoc biochem, Georgetown Univ, 64-72; res assoc biophys, 73-76, assoc biophys, Mem Sloan-Kettering Cancer Ctr, 76-82; ASST PROF RADIOL, UNIV ROCHESTER MED CTR, 82- *Concurrent Pos:* Instr biophys, Sloan-Kettering Div, Grad Sch Med Sci, Cornell Univ, 74-77, asst prof biophys, 77-82, asst prof pharmacol exp ther, 80-82. *Mem:* Am Chem Soc; Soc Nuclear Med; Asn Univ Radiologist; NY Acad Sci. *Res:* Distribution and kinetics of radiolabeled compounds and the physiological interpretations of these; emphasis on short-lived, gamma-emitting compounds. *Mailing Add:* Dept Nuclear Med Ctr Univ Rochester Rochester NY 14642

RUSS, GUSTON PRICE, III, b Mobile, Ala, Apr 5, 46; m 71; c 2. MASS SPECTROMETRY, ISOTOPE GEOCHEMISTRY. *Educ:* Univ of the South, BA, 68; Calif Inst Technol, PhD(chem), 74. *Prof Exp:* Fel, Univ Calif, San Diego, 74-77; asst prof chem, Univ Hawaii, 77-81; CHEMIST, LAWRENCE LIVERMORE NAT LAB, 81- *Mem:* Am Geophys Union. *Res:* Nuclear geochemistry; lunar regolith studies; inductively coupled plasma mass spectrometry; chronology; isotope ratio mass spectrometry. *Mailing Add:* Lawrence Livermore Nat Lab MS L-310 PO Box 808 Livermore CA 94550

RUSS, JAMES STEWART, b Canton, Ohio, Aug 22, 40; m 63, 81; c 3. HIGH ENERGY PHYSICS. *Educ:* Ind Univ, BS, 62; Princeton Univ, MA, 64, PhD(physics), 66. *Prof Exp:* Instr physics, Princeton Univ, 66-67; from asst prof to assoc prof, 67-80, PROF PHYSICS, CARNEGIE-MELLON UNIV, 80- *Concurrent Pos:* Vis scientist, Cern, 85-86. *Mem:* Am Phys Soc; Sigma Xi; Am Asn Physics Teachers. *Res:* Elementary particle physics; experiments using electronic detectors; computer-oriented data-handling systems; computer simulation of experimental data. *Mailing Add:* Dept Physics Carnegie-Mellon Univ Pittsburgh PA 15213-3890

RUSSEK, ARNOLD, b New York, NY, July 13, 26; m 56; c 2. PHYSICS. *Educ:* City Col New York, BS, 47; NY Univ, MS, 48, PhD(physics), 53. *Prof Exp:* Instr physics, Univ Buffalo, 53-55; from asst prof to assoc prof, 55-65, PROF PHYSICS, UNIV CONN, 65- *Mem:* Fel Am Phys Soc. *Res:* Nuclear structure; atomic structure and atomic collisions; electromagnetic diffraction theory. *Mailing Add:* Dept Physics Univ Conn Storrs CT 06268

RUSSEK-COHEN, ESTELLE, b Brooklyn, NY, July 23, 51; m; c 1. MULTIVARIATE METHODS, BIOASSAY METHODS. *Educ:* State Univ NY, Stony Brook, BS, 72; Univ Wash, Seattle, PhD(biomath), 79. *Prof Exp:* Programmer, NY Life Ins, 72-73; res asst biostat, Univ Wash, 73-75, teaching asst, 75-76, instr, 77-78; asst prof, 78-83, ASSOC PROF BIOSTATIST, UNIV MD, 83- *Concurrent Pos:* Consult, Univ Md Sea Grant Prog & Joint Consult Lab, USDA, Univ Md, 81- *Mem:* Am Statist Asn; Biomet Soc; Classification Soc. *Res:* Biostatistics. *Mailing Add:* Dept Animal Sci Univ Md College Park MD 20742

RUSSEL, DARRELL ARDEN, b McPherson, Kans, May 26, 21; m 49; c 9. AGRONOMY. *Educ:* Kans State Univ, BS, 43; Univ Ill, MS, 47, PhD(soil fertil, anal chem), 55. *Prof Exp:* Spec asst soil fertil, Univ Ill, 46-47, asst soils, 51-55; instr soils, Iowa State Univ, 47-49, dist exten dir, Agr Exten Serv,

49-51; asst prof soil chem, N La Hill Farm Exp Sta, La State Univ, 55-60; agriculturist, Div Agr Develop, 60-78, asst adminr, Int Fertilizer Prog, 78-81, HEAD, EDUC & COMMUN SERV STAFF, DIV AGR DEVELOP, NAT FERTILIZER DEVELOP CTR, TENN VALLEY AUTHORITY, 81- *Concurrent Pos:* AID consult, Morocco, 66, Bolivia, 69, Paraguay, 70, Indonesia, 71, Ghana, 72 & 74, Cent Treaty Orgn, 74, UN Indus Develop Orgn, 76 & Int Fertilizer Develop Ctr, 81. *Mem:* Fel Soil Sci Soc Am; fel Am Soc Agron; Crop Sci Soc Am; Am Chem Soc; Coun Agr Sci & Technol. *Res:* Analytical procedures for soil testing; effect of fertilizers on soil fertility, crop growth and crop quality; fertilizer education. *Mailing Add:* 501 Cleveland Ave Florence AL 35630

RUSSEL, MARJORIE ELLEN, b New York, NY, July 16, 44; m 81; c 1. MOLECULAR GENETICS. *Educ:* Oberlin Col, BA, 66; Univ Wis, MS, 68; Univ Colo, PhD(molecular biol), 77. *Prof Exp:* Res asst molecular biol, Dept Genetics, Univ Wash, 68-70 & Dept Molecular Biol, Univ Geneva, 71-73; fel genetics, 77-87, ASST PROF, ROCKEFELLER UNIV, 87- *Concurrent Pos:* Damon Runyon-Walter Winchell Cancer Fund fel, 77-78. *Mem:* Am Soc Microbiol. *Res:* Control of gene expression in prokaryotes; mechanisms of protein transport into and across bacterial membranes. *Mailing Add:* Rockefeller Univ Dept Genetics 1230 York Ave New York NY 10021

RUSSEL, WILLIAM BAILEY, b Corpus Christi, Tex, Nov 17, 45; m 72; c 2. CHEMICAL ENGINEERING. *Educ:* Rice Univ, BA & MChE, 69; Stanford Univ, PhD(chem eng), 73. *Prof Exp:* NATO fel appl math, dept appl math & theoret physics, Cambridge Univ, 73-74; from asst prof to assoc prof, 74-83, PROF CHEM ENG, PRINCETON UNIV, 83- *Concurrent Pos:* Olaf A Hougen prof, Univ Wis, 84. *Mem:* Am Inst Chem Engrs; Soc Rheology; Am Chem Soc; Mat Res Soc. *Res:* Fluid mechanics; colloidal suspensions; polymer solutions. *Mailing Add:* Dept Chem Eng Princeton Univ Princeton NJ 08544

RUSSELL, ALAN JAMES, b Salford, Eng, Aug 8, 62; m 87; c 1. APPLIED ENZYMOLOGY, PROTEIN ENGINEERING. *Educ:* Univ March Inst Technol Brit, BS, 84, Imp Col, PLD, 87, DIC, 87. *Prof Exp:* NATO fel chem, Mass Inst Technol, 87-89; ASST PROF CHEM ENG, UNIV PITTSBURGH, 89- *Concurrent Pos:* NSF presidential young investr, 90; Fulton C Noss fel, Univ Pittsburgh, 90. *Mem:* Am Chem Soc; Am Inst Chem Engrs; Biochem Soc. *Res:* Study of enzymes in extreme environments: using protein engineering to design enzymes in organic solvents and at high temperatures and pressures. *Mailing Add:* Dept Chem Eng (Biotech) Univ Pittsburgh Pittsburgh PA 15261

RUSSELL, ALLAN MELVIN, b Newark, NJ, Feb 2, 30; m 51; c 5. PHYSICS. *Educ:* Brown Univ, ScB, 51, ScM, 53; Syracuse Univ, PhD(physics), 57. *Prof Exp:* Res assoc physics, Syracuse Univ, 57-58; asst prof, Univ Calif, 58-64; assoc prof, Wesleyan Univ, 64-67; assoc prof, 67-70, assoc provost, 67-68, provost & dean fac, 68-72, PROF PHYSICS, HOBART & WILLIAM SMITH COLS, 70- *Mem:* Am Phys Soc; Am Asn Physics Teachers; Sigma Xi. *Res:* Low energy electron diffraction; field emission; measurement; molecular beams; paleomagnetism; epistemology; philosophy of science. *Mailing Add:* 69 Snell Rd Geneva NY 14456

RUSSELL, ALLEN STEVENSON, b Bedford, Pa, May 27, 15; m 41. CHEMISTRY, ENGINEERING. *Educ:* Pa State Univ, BS, 36, MS, 37, PhD(phys chem), 41. *Prof Exp:* Chemist, Bell Tel Labs, 37; asst, Pa State Univ, 37-40; chemist, Aluminum Co Am, 40-44, asst chief phys chem div, 44-53, chief, 53-55, chief process metall div, 55-69, asst dir res, 69-74, assoc dir, 74, vpres, 74-78, vpres sci & technol, 78-81, vpres & chief scientist, Alcoa Labs, 81-82; RETIRED. *Concurrent Pos:* Adj prof, Univ Pittsburgh. *Honors & Awards:* Karl J Bayer Medallist, 81; Gold Medal, Am Soc Metals, 82; Chem Pioneer Award, 83; James Douglas Gold Medal, 87. *Mem:* Nat Acad Eng; Am Chem Soc; fel Am Soc Metals; Sigma Xi; fel Am Inst Mining, Metall & Petrol Engrs. *Res:* Process metallurgy of aluminum. *Mailing Add:* Nine N Caliboque Cay Rd Hilton Head Island SC 29938

RUSSELL, ANTHONY PATRICK, b London, Eng, Sept 10, 47; m 70; c 3. HERPETOLOGY, MAMMALOGY. *Educ:* Univ Exeter, BSc, 69; Univ London, PhD(zool), 72. *Prof Exp:* Lectr zool, Univ Botswana, Lesotho & Swaziland, 73; asst prof, Univ Calgary, 73-80, assoc prof biol, 80-87, asst dean, 85-89, ASSOC DEAN FAC SCI, UNIV CALGARY, 89-, PROF BIOL, 87- *Concurrent Pos:* Res assoc, Dept Herpetol, Royal Ont Mus, 80-, Royal Tyrrell Mus Paleont, Drumheller, Alta, 87- *Mem:* Am Soc Zoologists; Soc Syst Zool; Linnean Soc London; Zool Soc London; Can Soc Zoologists; Soc Study Evolution; Am Soc Ichthyol & Herpetol; Herpetol League; Paleontol Soc; Soc Preserv Nat Hist Collections. *Res:* Investigation of the functional biology and evolutionary morphology of reptiles and mammals, particularly locomotor and feeding mechanisms. *Mailing Add:* Dept Biol Sci Univ Calgary Calgary AB T2N 1N4 Can

RUSSELL, B DON, b Denison, Tex, May 25, 48; m 73; c 4. ELECTRIC POWER SYSTEMS, REAL-TIME COMPUTER CONTROL. *Educ:* Tex A&M Univ, BS, 70, ME, 71; Univ Okla, PhD(elec eng), 75. *Prof Exp:* Instr elec eng, Tex A&M Univ, 70-71; instr physics, Abilene Christian Univ, 71-73; instr elec eng, Univ Okla, 73-75; res engr power systs, Okla Gas & Elec, & Elec Power Res Inst, 75; from asst prof to prof, Tex A&M Univ, 76-88; RES ENGR, TEX ENG EXP STA, 76-, PROF ELEC ENG, 88- *Concurrent Pos:* Consult numerous indust orgns, 73-90; prin investr, Tex A&M Univ Res Found, 76-90; pres, Micon Eng, Inc, 77-90; assoc dir, Inst for Ventures & New Technol, 83-85; assoc ed, Elec Power Systs Res, 85-90. *Mem:* Fel Inst Elec & Electronics Engrs; sr mem Power Eng Soc; Instrument Soc Am. *Res:* Applications of advanced computer technology to solution of power systems problems; automation, control and protection of power systems; development of fault detection systems for power lines to improve efficiency and safety of power system operation. *Mailing Add:* Dept Elec Eng Tex A&M Univ College Station TX 77846

RUSSELL, CATHERINE MARIE, b Tuckahoe, NY, Nov 20, 10. MEDICAL MICROBIOLOGY, MEDICAL PARASITOLOGY. *Educ:* Col of Mt St Vincent, BS, 32; Columbia Univ, MA, 48; Univ Va, PhD(biol), 51; Am Bd Med Microbiol, dipl. *Prof Exp:* Technician, Res & Diag Labs, State Dept Health, NY, 39-41; technician, New York Med Col, 41-42, instr bact & parasitol, 42-48; from instr to assoc prof microbiol, 48-77, from assoc prof to prof clin path, 63-77, EMER PROF PATH, MED SCH, UNIV VA, 77- *Mem:* Am Soc Trop Med & Hyg; Am Soc Parasitol; Am Soc Microbiol; Sigma Xi. *Res:* Trematode infections; development of the egg of Plagitura salamandra; leptospirosis; pathogenesis and identification of the organism; virus-cell relationship. *Mailing Add:* 511 N First St No 505 Charlottesville VA 22901

RUSSELL, CHARLES ADDISON, b Danielson, Conn, Jan 12, 21; m 47; c 3. POLYMER CHEMISTRY, ANALYTICAL CHEMISTRY. *Educ:* Yale Univ, BSc, 42, MSc, 44, PhD(org chem), 49. *Prof Exp:* Assoc prof phys chem, Bucknell Univ, 48-52; chemist, Nat Lead Co, 52-59; mem tech staff, 59-66, supvr atmospheric effects, microchem & contact group, 66-83, DIST RES MGR, BELL COMMUN RES, 84- *Mem:* AAAS; Am Chem Soc. *Res:* Effects of contamination on telephone equipment; interaction of various materials with environment and each other; behavior of telephone contacts. *Mailing Add:* 106 Kenley Way Sun City Center FL 33573

RUSSELL, CHARLES BRADLEY, b Evanston, Ill, Apr 8, 40; m 68. MATHEMATICS, STATISTICS. *Educ:* Univ of the South, BA, 62; Fla State Univ, MS, 63, PhD(statist), 68. *Prof Exp:* Asst prof, 67-72, ASSOC PROF MATH SCI, CLEMSON UNIV, 72- *Mem:* Inst Math Statist; Am Statist Asn; Opers Res Soc Am; Inst Mgt Sci. *Res:* Probability theory; stochastic processes; management science. *Mailing Add:* Dept Math Sci Clemson Univ 201 Sikes Hall Clemson SC 29634

RUSSELL, CHARLES CLAYTON, b Key West, Fla, Oct 9, 37; m 58; c 2. NEMATOLOGY, ENTOMOLOGY. *Educ:* Univ Fla, BSA, 60, MSA, 62, PhD(nematol), 67. *Prof Exp:* NEMATOLOGIST, OKLA STATE UNIV, 67-, PROF, DEPT PLANT PATH, 80- *Mem:* Am Phytopath Soc; Soc Nematol; Sigma Xi. *Res:* Plant parasitic; free-living nematodes. *Mailing Add:* Dept Plant Path Okla State Univ Stillwater OK 74074

RUSSELL, CHARLOTTE SANANES, b Brooklyn, NY, Jan 4, 27; m 47; c 2. BIOCHEMISTRY, ORGANIC CHEMISTRY. *Educ:* Brooklyn Col, AB, 46; Columbia Univ, AM, 47, PhD(org chem), 51. *Prof Exp:* Lectr org chem, Brooklyn Col, 49; res worker biochem, Col Physicians & Surgeons, Columbia Univ, 51-54, res assoc, 54; from instr to assoc prof, 54-72, PROF CHEM, CITY COL NEW YORK, 72- *Concurrent Pos:* Vis asst prof, Col Physicians & Surgeons, Columbia Univ, 63- 64; prin investr, res grants; reviewer, prof journals & granting agencies. *Mem:* Sigma Xi; AAAS; Am Chem Soc; Am Soc Biol Chemists; Royal Soc Chem. *Res:* Biochemistry of heme compounds; enzymology of heme biosynthesis; invertebrate lectins; lipid hemagglutinins; solid state reactions in organic chemistry. *Mailing Add:* Dept Chem City Col New York New York NY 10031

RUSSELL, CHRISTOPHER THOMAS, b London, Eng, May 9, 43; m 66; c 2. GEOPHYSICS, MAGNETOPHERIC PHYSICS. *Educ:* Univ Toronto, BSc, 64; Univ Calif, Los Angeles, PhD(space physics), 68. *Prof Exp:* Res geophysicist, 68-81, PROF GEOPHYS, INST GEOPHYS & PLANETARY PHYSICS, UNIV CALIF, LOS ANGELES, 82- *Concurrent Pos:* Prin investr, Int Sun Earth Explorer, 72-87; mem, USNC Int Union Radio Sci, 75-81; prin investr, Pioneer Venus Orbiter, 75-; assoc ed, J Geophys Res, 76-78, Geophys Res Letter, 79-81, ed, Solar Wind Three, 74, Auroral Process, 79, IMS Source Book, 82, Space Sci Revs, 83-88, Planetary & Space Sci, 84-88, Solar Wind Interactions, 86; chair, NAS/Space Sci Bd Comt on Data Mgmt & Comput, 85-88; chair, Int Sci Comn D Cospar, 82-86; pres, Solar Terrestrial Relationships Sect, Am Geophys Union, 88-90; Interdisciplinary scientist, Galileo mission, 77- *Honors & Awards:* Macelwane Award; Harold Jeffreys lectr, Royal Astron Soc, 87. *Mem:* Fel Am Geophys Union; Int Union Radio Sci; fel AAAS; Am Astron Soc; Space Sci Bd; Europ Geophys Soc. *Res:* Magnetospheric physics; solar-terrestrial relationships; interplanetary physics; lunar physics; planetary physics. *Mailing Add:* Inst Geophys & Planetary Physics Univ Calif Los Angeles CA 90024-1567

RUSSELL, DALE A, b San Francisco, Calif, Dec 27, 37; m 64; c 3. VERTEBRATE PALEONTOLOGY. *Educ:* Univ Ore, BA, 58; Univ Calif, Berkeley, MA, 60; Columbia Univ, PhD(geol), 64. *Prof Exp:* NSF fel, Yale Univ, 64-65; cur fossil vert, 65-77, CHIEF PALEOBIOL DIV, NAT MUS NATURAL SCI, NAT MUS CAN, 77- *Mem:* Soc Vert Paleont. *Res:* Mesozoic, particularly Cretaceous, reptiles; Cretaceous-Tertiary boundary problems. *Mailing Add:* Nat Mus Natural Sci Metcalfe & McLeod Sts Ottawa ON K1A 0M8 Can

RUSSELL, DAVID A, b St John, NB, April 25, 35; US citizen; m 57; c 3. AERONAUTICAL & ASTRONAUTICAL ENGINEERING, ENGINEERING PHYSICS. *Educ:* Univ Southern Calif, BEng, 56; Calif Inst Technol, MSc, 57, PhD(aeronauts & physics), 61. *Prof Exp:* Sr scientist, Jet Propulsion Lab, Calif Inst Technol, 61-67; res assoc prof, 67-70, assoc prof, 70-74, PROF AERONAUT & ASTRONAUT, UNIV WASH, 74-, CHAIR DEPT, 77- *Concurrent Pos:* Mem, Plasma Dynamics Tech Comt, Am Inst Aeronauts & Astronauts, 70-72 & 86-; exec coun, Fluid Dynamics Div, Am Phys Soc, 77; mem Kirkland Planning Comn, Wash, 71-79, chair, 77-79, chair, Kirkland Land Use Policy Plan Comn, 77-79, mem City Coun, 84-92; prin investr, NASA, 69-72 & 84-86, Dept Defense, 69-71 & 73-81, indust, 85-86; consult, pvt firms & govt labs, 67- *Mem:* Fel Am Phys Soc; assoc fel Am Inst Aeronauts & Astron. *Res:* Fluid mechanics and gas physics with applications to aerodynamics, shock processes and gas lasers. *Mailing Add:* 4507 105th Ave NE Kirkland WA 98033

RUSSELL, DAVID L, b Orlando, Fla, May 1, 39; m 60; c 2. MATHEMATICS. *Educ:* Andrews Univ, BA, 60; Univ Minn, PhD(math), 64. *Prof Exp:* Asst prof, 64-69, assoc prof math & comput sci, 69-77, PROF L&S, MATH, UNIV WIS-MADISON, 77- *Concurrent Pos:* Res consult, Honeywell, Inc, 63- *Mem:* Am Math Soc; Soc Indust & Appl Math. *Res:* Control theory of ordinary and partial differential equations; asymptotic theory of ordinary differential equations. *Mailing Add:* 7400 28th Ave Kenosha WI 53706

RUSSELL, DENNIS C, b Southampton, Eng, Sept 4, 27; m 51; c 2. MATHEMATICAL ANALYSIS & APPROXIMATION THEORY. *Educ:* Univ Sheffield, BSc, 48; Univ London, MSc, 52, PhD(math anal), 58, DSc, 72. *Prof Exp:* Asst lectr math, Northampton Col Advan Technol, 48-52; demonstr, Univ Col London, 53-55; asst lectr, Keele Univ, 55-57, lectr, 57-60; assoc prof, Mt Allison Univ, 60-62; chmn dept, 62-69, prof, 62-89, EMER PROF MATH, YORK UNIV, 89- *Concurrent Pos:* Nat Res Coun Can sr res fel, 68; Can Coun leave fel, 69 & 76-77; hon res fel, Birkbeck Col, London, 68-69; German Acad Exchange Serv res fel, 71; Nuffield Found res travel award, 73 & Nat Sci Eng Res Coun Can travel grants, 76, 80 & 83, collaborative res grant, 82, res operating grants, 69-91; consult grad prog, Univ Calgary, 78; vis prof, Tel-Aviv Univ, 81. *Mem:* Am Math Soc; Math Asn Am; Can Math Soc; London Math Soc; fel Inst Math & Applns. *Res:* Mathematical analysis; matrix transformations on sequence spaces, and summability of sequences, series and integrals; approximation theory; inequalities. *Mailing Add:* Dept Math York Univ Toronto ON M3J 1P3 Can

RUSSELL, DIANE HADDOCK, animal physiology, pharmacology; deceased, see previous edition for last biography

RUSSELL, DONALD GLENN, b Kansas City, Mo, Nov 24, 31; m 53; c 2. PETROLEUM ENGINEERING. *Educ:* Sam Houston State Univ, BS, 53; Univ Okla, MS, 55. *Prof Exp:* Var eng assignments, Tex, La & NY, Shell Oil, 56-72, gen mgr, info & comput serv, Houston, 72-76, vpres corp planning, 77-78, vpres, Int Explor & Prod, 78-80, vpres prod, Houston, 80-87; CHMN & CHIEF EXEC OFFICER, SONAT EXPLOR CO, 88- *Concurrent Pos:* Mem bd dirs, Am Petrol Inst. *Honors & Awards:* John Franklin Carll Award, Soc Petrol Engrs, Am Inst Mining Metall & Petrol Engrs, 80, Cedric K Ferguson Medal & DeGolyer Medal. *Mem:* Nat Acad Eng; Nat Gas Supply Asn; Am Petrol Asn; Soc Petrol Engrs; Am Inst Mining Metall & Petrol Engrs (vpres, 73, pres, 74). *Mailing Add:* Sonat Exploration Co PO Box 1513 Houston TX 77251-1513

RUSSELL, DONALD HAYES, psychiatry, for more information see previous edition

RUSSELL, DOUGLAS STEWART, b Georgetown, Ont, June 16, 16; m 45; c 3. ANALYTICAL CHEMISTRY, TRACE INORGANIC ANALYSIS. *Educ:* Univ Toronto, BA, 40, MA, 41. *Prof Exp:* Asst chem, Univ Toronto, 40-42; supvr anal chem, Welland Chem Works, Ltd, 42-44; asst res officer radiochem, Can Atomic Energy Proj, 44-46; asst res officer, Nat Res Coun Can, 46-51, sr res officer, 52-70, head anal sect, Div Chem, 52-81, prin res officer, 70-81; RETIRED. *Honors & Awards:* Fisher Sci Lectr Award, Chem Inst Can, 79. *Mem:* Am Chem Soc; Chem Inst Can; hon mem Spectros Soc Can (vpres, 68-69, pres, 69-70); Int Colloquial Spectros (secy, 67). *Res:* Development of methods for the determination of trace impurities in highly purified metals, ultrapure acids, reagents and semiconductor materials; using optical emission spectrometry, spark source mass spectrometry and stable isotope dilution. *Mailing Add:* 44 Tower Rd Nepean ON K2G 2E7 Can

RUSSELL, ELIZABETH SHULL, b Ann Arbor, Mich, May 1, 13; m 36; c 4. MAMMALIAN GENETICS. *Educ:* Univ Mich, AB, 33; Columbia Univ, MA, 34; Univ Chicago, PhD(zool), 37. *Hon Degrees:* DSc, Univ Maine, Farmington, 75, Colby Col, 84, Med Col Ohio & Bowdoin Col. *Prof Exp:* Asst zool, Univ Chicago, 35-37; independent investr, Jackson Lab, 39-40, res assoc, 46-57, sr staff scientist, 57-82, emer sr scientist, 82-88; RETIRED. *Concurrent Pos:* Nourse fel, Am Asn Univ Women, 39-40; Finney-Howell fel, 47; Guggenheim fel, 58-59. *Mem:* Nat Acad Sci; Am Acad Arts & Sci; Genetics Soc Am; Am Soc Naturalists; Soc Develop Biol. *Res:* Mammalian physiological genetics; action of deleterious genes; mouse anemias and hemoglobins; coat color; muscular dystrophy; genetic effects on aging. *Mailing Add:* Jackson Lab 600 Main St Bar Harbor ME 04609

RUSSELL, ERNEST EVERETT, b Jackson, Miss, Apr 16, 23; m 49; c 3. GEOLOGY. *Educ:* Miss State Univ, BS, 49, MS, 55; Univ Tenn, PhD(geol), 65. *Prof Exp:* Instr geol, Univ Tenn, 54-55; from asst prof to assoc prof, 55-68, prof, 68-85, EMER PROF GEOL, MISS STATE UNIV, 85- *Concurrent Pos:* Consult geologist, Tenn Div Geol, 58-70, WAE, US Geol Surv, 61-68, Tenn Valley Authority, 75-83. *Mem:* Fel Geol Soc Am; Paleont Soc; Am Inst Prof Geologists. *Res:* Upper Cretaceous Litho and Biostratigraphy of the eastern Gulf Coastal Plain; Lithofacies relations in Mesozoic and Cenozoic sediments. *Mailing Add:* Dept Geol & Geog Miss State Univ Mississippi State MS 39762

RUSSELL, FINDLAY EWING, b San Francisco, Calif, Sept 1, 19; m 50; c 5. PHYSIOLOGY, TOXINOLOGY. *Educ:* Walla Walla Col, BA, 41; Loma Linda Univ, MD, 50; Univ Santa Barbara, PhD, 74. *Hon Degrees:* LLD, Univ Santa Barbara, 88. *Prof Exp:* Intern, White Mem Hosp, Los Angeles, 50-51; res fel biol, Calif Inst Technol, 51-53; physiologist, Inst Med Res, Huntington Mem Hosp, Pasadena, 53-55; res prof neurosurg, Loma Linda Univ, 55-66; prof neurol, physiol & biol, Univ Southern Calif, 66-80; RES PROF HEALTH SCI, UNIV ARIZ, 80- *Concurrent Pos:* Dir lab neurol res, Los Angeles County Hosp, 55-66; vis physiologist, Lab Marine Biol Asn, Eng, 58; vis prof, Cambridge Univ, 62-63 & 70-71, Ain Shams Univ Cairo, 63 & Univ Ljubljana, 75-76; ed, Toxicon, 62-68; G Griffith Scholar, Am Col Physicians, 86. *Honors & Awards:* F Redi Medal, 66; J Stefan Award, 78; C H Thienes Award, 85. *Mem:* Fel Am Col Physicians; fel NY Acad Sci; Int Soc Toxinology (pres, 62-66); fel Royal Soc Trop Med & Hyg; Am Physiol Soc. *Res:* Venoms. *Mailing Add:* Dept Pharmacol & Toxicol Col Pharm Univ Ariz Tucson AZ 85721

RUSSELL, FREDERICK A(RTHUR), b New York, NY, Apr 18, 15; m 47; c 2. ELECTRICAL ENGINEERING. *Educ:* Newark Col Eng, BS, 35, EE, 39; Stevens Inst Technol, MS, 41; Columbia Univ, ScD(eng), 53. *Prof Exp:* From instr to asst prof elec eng, Newark Col Eng, 37-44; sr proj engr, Div War Res, Columbia Univ, 44-45; exec assoc, NJ Inst Technol, 45-56, from asst prof to assoc prof elec eng, 45-53, chmn dept, 56-75, prof, 53-80, asst to pres & dir planning, 75-80, adj prof, 80-85; RETIRED. *Concurrent Pos:* Consult, Franklin Inst, 45-46 & Bell Tel Labs, Inc, 57-70. *Mem:* Am Soc Eng Educ; fel Inst Elec & Electronics Engrs; Sigma Xi. *Res:* Digital and analog computer simulation; control systems. *Mailing Add:* 101 Elkwood Ave New Providence NJ 07974

RUSSELL, GEORGE A, b Bertrand, Mo, July 12, 21; m 44; c 4. PHYSICS. *Educ:* Mass Inst Technol, BS, 47; Univ Ill, MS, 52, PhD(physics), 55. *Prof Exp:* Physicist, Bur Aeronaut, US Navy, 55-59 & Antisubmarine Warfare Develop Squadron, 59-60; assoc prof physics, Southern Ill Univ, 60-62; assoc prof, 62-65, prof physics & assoc dir mat res lab, 65-74, head dept physics, 68-74, assoc dean grad col, 70-72, assoc vchancellor res & develop, 72-74, DEAN GRAD COL, UNIV ILL, URBANA, 72-, VCHANCELLOR RES, 74- *Concurrent Pos:* Consult, Off Naval Res, 60; chancellor, Univ Mo-Kansas City, 77- *Mem:* Am Phys Soc; Am Asn Physics Teachers; Sigma Xi. *Res:* Color centers in ionic crystals; infrared radiation; antisubmarine warfare; utilization of educated manpower; university administration. *Mailing Add:* Chancellor's Off Univ Mo 5100 Rockhill Rd Kansas City MO 64110

RUSSELL, GEORGE ALBERT, b Chicago Heights, Ill, Aug 29, 36; m 57; c 3. AUTOMATIC CONTROL SYSTEMS. *Educ:* Mass Inst Technol, BS, 58; Ariz State Univ, MS, 61; Univ Conn, PhD(elec eng), 67. *Prof Exp:* Develop engr, AiRes Mfg Co Div, Garrett Corp, 58-61, eng specialist, 64-67; asst prof mech eng, 67-70, ASSOC PROF MECH ENG, UNIV MASS, AMHERST, 70- *Concurrent Pos:* Consult, AiRes Mfg Co Div, Garrett Corp, Springfield Wire Inc. *Mem:* Acoust Soc Am; Am Soc Mech Engrs. *Res:* Gas turbine hydromechanical controls. *Mailing Add:* Dept Mech & Aerospace Eng Univ Mass Amherst MA 01003

RUSSELL, GEORGE K(EITH), b Bronxville, NY, Dec 13, 37; m 70. PLANT PHYSIOLOGY, BIOCHEMISTRY. *Educ:* Princeton Univ, AB, 59; Harvard Univ, PhD(biol), 63. *Prof Exp:* NSF fel biochem, Cornell Univ, 63-64; NSF fel biol, Brandeis Univ, 64-65; asst prof, Princeton Univ, 65-67; from asst prof to assoc prof, 67-77, PROF BIOL, ADELPHI UNIV, 77- *Mem:* Am Soc Plant Physiologists; AAAS; Sigma Xi. *Res:* Genetic and biochemical control of photosynthesis and chloroplast development. *Mailing Add:* Dept Biol Adelphi Univ Garden City NY 11530

RUSSELL, GERALD FREDERICK, b Edmonton, Alta, Feb 29, 44. FOOD SAFETY, COMPUTERS IN FOOD SCIENCE & NUTRITION. *Educ:* Univ Alta, BSc, 64; Univ Calif, PhD(agr chem), 68. *Prof Exp:* ASSOC PROF FOOD SCI, UNIV CALIF, DAVIS, 68- *Concurrent Pos:* Vis assoc prof, Eppley Inst Res in Cancer, Med Col, Univ Nebr, 75-76. *Mem:* Am Chem Soc; Inst Food Technologists; Am Soc Mass Spectrometry; AAAS; Sigma Xi. *Res:* Chemistry of food volatiles; chemistry of Maillard reaction in foods; structure and mechanisms of flavor compounds; agents and mechanisms of toxic and carcinogenic substances in foods; computers in food science and nutrition. *Mailing Add:* Dept Food Sci & Technol Univ Calif Davis CA 95616

RUSSELL, GLEN ALLAN, b Rensselaer Co, NY, Aug 23, 25; m 53; c 2. ORGANIC CHEMISTRY. *Educ:* Rensselaer Polytech Inst, BChE, 47, MS, 48; Purdue Univ, PhD(chem), 51. *Prof Exp:* Res assoc, Res Lab, Gen Elec Co, 51-58; assoc prof, 58-62, prof chem, 62-72, DISTINGUISHED PROF, IOWA STATE UNIV, 72- *Concurrent Pos:* Sloan Found fel, 59-62; vis prof, Univ Wyo, 67, Univ Grenoble, France, 85; Guggenheim fel, Nuclear Res Ctr, Grenoble, France, 72; fel Japan Soc Prom Sci, 83. *Honors & Awards:* Am Chem Soc Award, 65 & 72, James Flack Norris Award, 83; Fulbright-Hays Lectr, Univ Wurzburg, 66; Reilly Lectr, Univ Notre Dame, 66. *Mem:* AAAS; Am Chem Soc; The Chem Soc. *Res:* Physical organic chemistry; reactions of free radicals and atoms; chlorination; oxidation; application of electron spin resonance spectroscopy to organic molecules. *Mailing Add:* Dept Chem Iowa State Univ Ames IA 50010

RUSSELL, GLENN C, b Taber, Alta, June 6, 21; m 47; c 5. SOIL CHEMISTRY. *Educ:* Brigham Young Univ, BS, 43; Univ Mass, MS, 49; Purdue Univ, PhD(soil chem), 52. *Prof Exp:* Instr, Univ Mass, 47-49; fel, Purdue Univ, 49-51; res officer, Agr Res Sta, Lethbridge, 51-54, head soils sect, 54-66, dir exp farm, PEI, 66-70, dir res sta, Ont, 70-75, DIR RES STA, SUMMERLAND, BC, CAN DEPT AGR, 75- *Mem:* Soil Sci Soc Am; Am Soc Agron; Can Soc Soil Sci; Agr Inst Can; Can Soc Agron. *Res:* Effect of fertilizers and soil amendments on chemical changes in the soil and on crop production. *Mailing Add:* 1448 Dartmouth St Penticton BC V2A 4B6 Can

RUSSELL, GLENN VINTON, neuroanatomy; deceased, see previous edition for last biography

RUSSELL, GRANT E(DWIN), b Asheville, NC, June 5, 16; m 42; c 6. CHEMICAL ENGINEERING. *Educ:* Wash Univ, BSChE, 38, MSChE, 50. *Prof Exp:* Chemist, Reardon Co, 38-40; anal chemist, Monsanto Co, 40-41, tech asst, 42-45, chem engr, 45-48, supv engr, 48-52, asst eng supt, 52-57, eng mgr, 57-65, sr eng specialist, 65-68, sr res specialist, 68-74; sr chem eng, 75-80, CONSULT PROCESS ECON, SRI INT, 80- *Mem:* Am Inst Chem Engrs. *Res:* Economic evaluation of research projects; preliminary process design; design and economic evaluation of chemical process. *Mailing Add:* 1033 Havre Ct Sunnyvale CA 94087

RUSSELL, HENRY FRANKLIN, b Glenolden, Pa, July 18, 40; m 68; c 2. ORGANIC CHEMISTRY, HETEROCYCLIC CHEMISTRY. *Educ:* Univ Del, BS, 63, MS, 65; Univ Va, PhD(org chem), 73. *Prof Exp:* Instr chem, US Naval Acad, 66-68; process & develop chemist, Am Cyanamid Co, 73-79; PROF CHEM, JOHNSON C SMITH UNIV, 79- *Concurrent Pos:* MBRS prin investr, 82- *Mem:* Am Chem Soc. *Res:* Synthesis of melatonin analogs, substituted indoles and precursors; anti-mitotic activity. *Mailing Add:* 3739 Severn Ave Charlotte NC 28210-6213

RUSSELL, HENRY GEORGE, b Eng, June 12, 41. STRUCTURAL ENGINEERING. *Educ:* Univ Sheffield, BEng, 62, PhD(struct eng), 65. *Prof Exp:* Jr res fel struct eng, Bldg Res Sta Eng, 65-68; struct engr, 68-74, mgr, 74-79, dir, Portland Cement Asn, 79-86; exec dir, 87-88, PRES, CONSTRUCT TECHNOL LABS, 89- *Honors & Awards:* Martin P Korn Award, Prestressed Concrete Inst, 80; Delmar L Bloem Award, Am Concrete Inst, 86. *Mem:* Fel Am Concrete Inst; Prestressed Concrete Inst; Transp Res Bd. *Res:* Shrinkage compensating concretes; time dependent behavior of reinforced concrete columns and posttensioned concrete bridges; high strength concrete; bridge failures. *Mailing Add:* Construct Technol Labs Inc 5420 Old Orchard Rd Skokie IL 60077-1030

RUSSELL, JAMES, b Leeds, Eng, Feb 14, 28; m 51; c 2. POLYMER CHEMISTRY, PHYSICAL CHEMISTRY. *Educ:* Univ London, BSc, 50, PhD(polymer chem), 53. *Prof Exp:* Fel, Cornell Univ, 53-55; res chemist, Am Viscose Corp, 55-61; res assoc, St Regis Paper Co, NY, 61-67, mgr res & develop, 67-71, dir corp prod develop, 71-75; vpres res & develop, Sylvachem Corp, 75-86; CONSULT, 86- *Mem:* Am Chem Soc; Tech Asn Pulp & Paper Indust. *Res:* Wood chemistry and its application to pulping and byproducts; polymer chemistry and its application to packaging. *Mailing Add:* 2938 Jenks Ave Panama City FL 32405

RUSSELL, JAMES A(LVIN), JR, b Lawrenceville, Va, Dec 25, 17; m 43; c 2. ELECTRICAL ENGINEERING. *Educ:* Oberlin Col, AB, 40; Bradley Univ, BS, 41, MS, 50; Univ Md, EdD(indust ed), 67. *Hon Degrees:* LLD, St Paul's Col, Va, 84. *Prof Exp:* Instr elec technol, US Naval Training Sch, 42-45 & St Paul's Col, 45-50; assoc prof electronics eng, Hampton Inst, 50-58, div dir technol, 63-68, prof electronics, 67-71; pres, St Paul's Col, Va, 71-81; prof indust technol & chmn div professional studies, WVA State Col, 82-86, acting pres, 86-87, exec asst pres, 87-88; RETIRED. *Concurrent Pos:* Res engr, Thomas J Watson Res Ctr, IBM Corp, 68-69; chmn dept electronics technol, Hampton Inst, 58-68, chmn dept eng & dir eng & technol div, 68-71. *Mem:* Am Soc Eng Educ; Inst Elec & Electronics Engrs; Tech Educ Asn. *Res:* Industrial education; investigation into changes in critical thinking and achievement in electronics as the result of exposure of subjects to specific techniques of critical thinking. *Mailing Add:* 811 Grandview Dr Dunbar WV 25064

RUSSELL, JAMES CHRISTOPHER, b Montreal, Que, Oct 24, 38; m 61; c 1. ATHEROSCLEROSIS, CLINICAL CHEMISTRY. *Educ:* Dalhousie Univ, BSc, 58; Univ Sask, MSc, 59, PhD(radiation chem), 62. *Prof Exp:* Gen Elec Co fel phys chem, Univ Leeds, 62-64; univ fel chem, 64-67, asst prof biochem in surg & dir biochem lab, Surg-Med Res Inst, 67-72, assoc prof, 72-81, PROF SURG, DIV EXP SURG, UNIV ALTA, 81- *Concurrent Pos:* Consult med staff sci & res, Univ Alta Hosp, 69- *Mem:* Can Soc Clin Chemists; Can Pysiol Soc; Can Soc Clin Invest; Can Athero Soc; Can Acad Clin Biochem; Nat Acad Clin Biochem. *Res:* Clinical chemistry in both humans and animals; animal models of cardiovascular disease; biochemistry of exercise; diabetes and obesity in rodents; maintain colony of JRC; LA-corpulent rats. *Mailing Add:* Dept Surg Univ Alta 275 Heritage Med Res Ctr Edmonton AB T6G 2S2 Can

RUSSELL, JAMES E(DWARD), b Rapid City, SDak, May 20, 40; m 63; c 2. ROCK MECHANICS, MINING ENGINEERING. *Educ:* SDak Sch Mines & Technol, BS, 63, MS, 64; Northwestern Univ, Evanston, PhD(theoret & appl mech), 66. *Prof Exp:* Sr res engr, Southwest Res Inst, 66-67; from asst prof to assoc prof civil eng, SDak Sch Mines & Technol, 67-71, from assoc prof to prof civil & mining eng, 71-76; proj mgr rock mech, Off Waste Isolation, Union Carbide Corp, 77-78; PROF PETROL ENG & GEOPHYS, TEX A&M UNIV, 78- *Concurrent Pos:* Vpres struct & mech systs, RE/SPEC, Inc, 69-76; Brokett prof, 82-83, Halliburton prof, 85-86. *Mem:* Am Soc Civil Engrs; Am Inst Mining, Metall & Petrol Engrs; Am Acad Mech; Am Geophys Union. *Res:* Rock mechanics; nuclear waste disposal and underground storage; in situ gasification of coal-subsidence. *Mailing Add:* Dept Petrol Eng Tex A&M Univ College Station TX 77843

RUSSELL, JAMES EDWARD, b Ft Wayne, Ind, Sept 27, 31; m 77. PHYSICS. *Educ:* Yale Univ, BS, 53, MS, 54, PhD(physics), 58. *Prof Exp:* Fel, Univ Va, 57-58; res assoc physics, Ind Univ, 58-60; res physicist, Carnegie-Mellon Univ, 60-62; vis asst prof, Univ Padua, 62-63; sr res officer, Univ Oxford, 63-65; assoc prof, 65-74, PROF PHYSICS, UNIV CINCINNATI, 74- *Concurrent Pos:* Vis scientist, Consum Educ Resource Network, 74; consult, Lawrence Berkeley Labs, 75. *Res:* Exotic atoms; chemical physics. *Mailing Add:* Dept Physics Univ Cincinnati Cincinnati OH 45221

RUSSELL, JAMES MADISON, III, b Newport News, Va, June 12, 40; m 60; c 3. ATMOSPHERIC SCIENCE, ELECTRICAL ENGINEERING. *Educ:* Va Polytech Inst & State Univ, BSEE, 62; Univ Va, MEE, 66; Univ Mich, PhD(aeronomy), 70. *Prof Exp:* Aerospace engr, NASA Langley Res Ctr, 62-68; res sci, Univ Mich, 68-70; res scientist, 70-75, head, chem & dynamics br, 75-84, HEAD, THEORET STUDIES BR, NASA LANGLEY RES CTR, 84- *Concurrent Pos:* Lectr physics, Christopher Newport Col, 71-73 & 84, remote sensing, George Washington Univ, 73-; co-leader, Limb Infrared Monitoring Stratosphere Sci Team, 75-83; prin invest, Halogen Occultation Exp Sci Team, 75-; co-investr, Atmospheric Trace Molecule Exp Sci Team, 78-88; mem, Nat Acad Sci Commun Solar & Space Physics, 84-86, Nat Acad Sci Panel Middle Atmospheric Prog, 85-87, prin investr, Spectroscopy Atmosphere, Emission Sci Team, 89- *Honors & Awards:* Sci Achievement Medal, NASA, 82. *Mem:* Am Meteorol Soc; Sigma Xi. *Res:* Satellite remote sensing of the atmosphere; atmospheric physics, including dynamics, radiation and transport; integrity of the ozone layer; tropospheric studies and stratosphere-troposphere exchange. *Mailing Add:* NASA Langley Res Ctr Mail Stop 401B Hampton VA 23665

RUSSELL, JAMES N(ELSON), JR, b Hereford, Tex, Dec 26, 07; m 39; c 3. CHEMICAL ENGINEERING. *Educ:* Univ Okla, BS, 31. *Prof Exp:* Technician, Oil Refining Lab, Pure Oil Co, Okla, 26-28, anal chemist & chief technician, 32-35, chemist in charge phys dept, Control Lab, Ill, 35-36, refining design engr, 36-42; inspector & pipeline engr, Panhandle Eastern Pipeline Co, Mo, 31-32; process & proj engr, Stone & Webster Eng Corp, Mass, 42-56; supv process engr, Bechtel Corp, Calif, 56-59; sr chem engr, C F Braun & Co, 59-75; consult engr, Fluor, 75-78 & Jacobs Eng Group, 78-80; refinery mgr & consult Huntway Refining Co & planning dir, Edgington Oil Co, 80-81; consult, Kinetic Technol Int, 80-82; RETIRED. *Concurrent Pos:* Consult to Pres, Advan Extraction Technologies Inc, Houston, 86-87; consult & tech adv, Tsai Int Co, San Marino, Calif, 88-89. *Mem:* Am Inst Chem Engrs. *Res:* Design, construction and initial operation of refining units; gasoline and chemical plants. *Mailing Add:* 1602 N Robinson Rd Texarkana TX 75501

RUSSELL, JAMES T, b Nagercoil, India, Sept 26, 44; US citizen; m 70; c 1. MEMBRANE TRANSPORT PROCESS, SIGNAL TRANSDUCTION. *Educ:* Univ Madras, BVSc, 66; Post Grad Inst Med Educ & Res, MSc, 71; Copenhagen Univ, Lic Med, 74. *Prof Exp:* Adj lectr physiol, Copenhagen Univ, Denmark, 74-76; fel biochem, St Louis Univ, 76-77; sr staff fel neurobiol, 78-83, RES CHEMIST, NIH, 84- *Concurrent Pos:* Vet asst surgeon, Madras Animal Husb Serv, India, 66-67; vet officer, Amul Dairy, India, 67-68; vis scientist, Med Physiol Inst, Copenhagen Univ, Denmark, 80. *Mem:* Soc Neurosci; Scand Soc Physiologists. *Res:* Presynaptic mechanisms of neurotransmitter secretion & its modulation; molecular description of neunonal excitability and presynaptic plasticity; functional organization of the nerve terminal; neuronal cell biology; protein chemistry; muscle biochemisty. *Mailing Add:* NIH Bldg 36 Rm B-316 Bethesda MD 20892

RUSSELL, JAMES TORRANCE, b Bremerton, Wash, Feb 23, 31; m 53; c 3. APPLIED PHYSICS, ELECTRONICS. *Educ:* Reed Col, BA, 53. *Prof Exp:* Physicist, Hanford Atomic Prod Oper, Gen Elec Co, 53-60, sr physicist, 60-65; res assoc exp physics, Pac Northwest Labs, Battelle Mem Inst, 65-66, sr res assoc appl physics, 66-80; vpres & tech dir, Digital Rec Corp, 80-85; CONSULT PHYSICS, RUSSELL ASSOC INC, 85- *Mem:* AAAS; Am Phys Soc; Inst Elec & Electronics Engrs; Soc Photo-Optical Instrumentation Engrs; Optical Soc Am. *Res:* Physics of instrumentation; sensor development; digital computer methods; high resolution optical systems; laser devices; development of techniques for experimental physics; 35 United States patents. *Mailing Add:* Russell Assocs 15305 SE 48th Dr Bellevue WA 98006

RUSSELL, JOEL W, b Elkhart, Ind, May 18, 39; m 61; c 5. PHYSICAL CHEMISTRY. *Educ:* Northwestern Univ, BA, 61; Univ Calif, Berkeley, PhD(chem), 65. *Prof Exp:* Res fel, Univ Minn, 65-66; from asst prof to assoc prof, 66-82, interim dean grad study, 84-85, PROF CHEM, OAKLAND UNIV, 82-, INTERIM DEAN SCH HEALTH SCI, 83- *Concurrent Pos:* Vis mem staff, Australian Nat Univ, 72-73; res fel, Southampton Univ, 80-81; vis scholar, Univ Mich, 88-89. *Mem:* Am Chem Soc; Sigma Xi; Am Asn Univ Prof. *Res:* Molecular structure and dynamics; infrared and raman spectroscopy; studies of means to enhance teaching and learning for chemistry. *Mailing Add:* Dept Chem Oakland Univ Rochester MI 48309-4401

RUSSELL, JOHN ALBERT, b Ludington, Mich, Mar 23, 13; m 36; c 2. ASTRONOMY. *Educ:* Univ Calif, AB, 35, AM, 37, PhD(astron), 43. *Prof Exp:* Guide, Griffith Observ, 35; asst, Univ Calif, 36-39; instr astron, Pasadena City Col, 39-41, Exten Div, Univ Calif, Los Angeles, 41, Santa Ana Army Air Base, 42 & Pasadena City Col, 46; head dept astron, 46-69, from asst prof to prof, 46-78, chmn div phys sci & math, 59-62, assoc dean natural sci & math, Col Lett, Arts & Sci, 63-68, emer prof, 78-83, DISTINGUISHED EMER PROF ASTRON, UNIV SOUTHERN CALIF, 83- *Concurrent Pos:* Fac res lectr, Univ Southern Calif, 57-; mem comn interplanetary dust, Int Astron Union. *Honors & Awards:* Univ Assocs Award, 60. *Mem:* AAAS; Am Astron Soc; Meteoritical Soc (secy, 49-58, pres, 58-62). *Res:* Meteor spectroscopy and statistics. *Mailing Add:* 5654 Coliseum St Los Angeles CA 90016

RUSSELL, JOHN ALVIN, b San Antonio, Tex, Aug 15, 34; m 58; c 2. TRIBOLOGY. *Educ:* Univ Tex, BS, 57; St Mary's Univ, MS, 64. *Prof Exp:* Assoc engr, Convair Astronaut, San Diego, 57; proj officer nuclear weapons, US Air Force, 57-60; mgr special proj fuels & lubricants, Alcor Inc, 60-71; MGR SYNTHETIC FUELS DEVELOP, SOUTHWEST RES INST, 71- *Concurrent Pos:* Sr res engr tribology, Southwest Res Inst, 60-66; mem fuels & lubricants comt, Coord Res Coun, 67-; mem tech adv panel, Alternative Fuels, US Dept Energy, 76- *Mem:* Soc Automotive Engrs (pres, 72-73); Am Soc Mech Engrs (vpres, 65-66); Am Soc Lubrication Engrs; Sigma Xi. *Res:* Synthetic fuels and lubricants performance in aviation and automotive power plants; cryogenic tribology; hydrocarbon fuels utilization projection analysis. *Mailing Add:* 7215 Brookside Lane San Antonio TX 78209

RUSSELL, JOHN BLAIR, b Rochester, NY, Dec 13, 29; m 55; c 1. INORGANIC CHEMISTRY. *Educ:* Oberlin Col, AB, 51; Cornell Univ, PhD(chem), 56. *Prof Exp:* Instr & res assoc chem, Cornell Univ, 55-56; from asst prof to assoc prof, 56-65, PROF CHEM, HUMBOLDT STATE UNIV, 65- *Mem:* Am Chem Soc; Sigma Xi. *Res:* Solutions of metals in liquid ammonia; electrochemistry in aqueous and non-aqueous solvents; complex ions; author, general chemistry texts. *Mailing Add:* Dept Chem Humboldt State Univ Arcata CA 95521

RUSSELL, JOHN GEORGE, b Manila, Philippines, Dec 19, 41; US citizen; m 66. ORGANIC CHEMISTRY. *Educ:* Purdue Univ, BS, 63; Univ Minn, PhD(org chem), 68. *Prof Exp:* Res assoc chem, Ohio State Univ, 68; asst prof, 69-74, ASSOC PROF CHEM, CALIF STATE UNIV, SACRAMENTO, 74- *Mem:* Am Chem Soc. *Res:* Nuclear magnetic resonance spectroscopy; conformational analysis; synthesis, rate and equilibria studies of organometallic compounds. *Mailing Add:* Dept Chem Calif State Univ Sacramento CA 95819-4698

RUSSELL, JOHN HENRY, b Roswell, NMex, Aug 15, 19; m 42; c 4. ENGINEERING. *Educ:* Univ Tex, ChemE, 42. *Prof Exp:* Jr engr, Holston Ord Works, Tenn Eastman Corp, 42-44; sr engr, Manhattan Proj, 44-45; sect leader nuclear weapons res & develop, Manhattan Proj, 45-46, asst group leader, Los Alamos Sci Lab, 46-60, group & assoc div leader, nuclear reactor develop, 60-70, STAFF MEM WEAPON DEVELOP, LOS ALAMOS SCI LAB, UNIV CALIF, 70- *Mem:* Am Chem Soc; Am Inst Chem Engrs; Am Nuclear Soc. *Res:* Ultra high temperature reactor experiment; all aspects of high temperature gas-cooled reactors; classified application of high explosives to nuclear weapons development. *Mailing Add:* 1199A 41st St Los Alamos NM 87544-1912

RUSSELL, JOHN LYNN, JR, b Woodsborough, Tex, Aug 5, 30; m 51; c 3. NUCLEAR PHYSICS. *Educ:* Univ Tex, BS, 51; Rice Univ, MA, 54, PhD(physics), 56. *Prof Exp:* Theoret physicist, Vallecitos Atomic Lab, Gen Elec Co, 56-60, mgr exp reactor physics, 60-62; mgr spec projs, Gen Atomic Co, San Diego, Calif, 62-78; prof, dept nuclear eng, Ga Inst Technol, 78-83, dir, Frank H Neely Nuclear Res Ctr, 79-83; PRES & CHMN BD, THERAGENICS CORP, 83- *Mem:* Am Nuclear Soc; Int Asn Hydrogen Energy; Int Solar Energy Soc; Am Phys Soc; Int Soc Neutron Capture Ther. *Res:* Pulsed neutron measurements; fast neutron spectra in bulk media; fast reactors; control of nuclear reactors; solar energy; hydrogen production; charged particle scattering; radiopharmaceuticals. *Mailing Add:* 4165 Big Creek Overlook Alpharetta GA 30201

RUSSELL, JOHN MCCANDLESS, b Drumright, Okla; m 62; c 2. PHYSIOLOGY. *Educ:* Univ NMex, BS, 66; Univ Utah, PhD(pharmacol), 71. *Prof Exp:* NIH fel, Dept Physiol & Biophys, Sch Med, Washington Univ, 72-73; from asst prof to assoc prof, Univ Tex Med Br, 74-85, prof & actg chmn, 85-86; CONSULT, 86- *Concurrent Pos:* Mem, Physiol Study Sect, NIH, 88-92, Cellular Cardiovasc Physiol & Pharmacol Study Sect, Am Heart Asn, 88-89. *Mem:* Biophys Soc; Soc Gen Physiologists; AAAS; Red Cell Club; Am Physiol Soc. *Res:* Mechanisms involved in the membrane translocation of ions, especially the chloride anion and its role in intracellular pH and volume regulation in biological tissue. *Mailing Add:* Dept Physiol & Biophys Univ Tex Med Br Galveston TX 77550

RUSSELL, JOSEPH L, chemical engineering, for more information see previous edition

RUSSELL, JOSEPH LOUIS, b Vicksburg, Miss, June 18, 36; m 62; c 1. FLUORINE CHEMISTRY. *Educ:* Alcorn State Univ, BS, 60; Marquette Univ, MS, 71, PhD(inorg chem), 74. *Prof Exp:* Teacher math, Natchez Pub Schs, 60-64; control chemist, Liquid Glaze Chem Co, 65; scientist chem, Battelle-Northwest Lab, 66-69; teaching asst chem, Marquette Univ, 69-74; asst prof, 74-77, ASSOC PROF CHEM, ALCORN STATE UNIV, 77- *Concurrent Pos:* Lab instr chem, Ala State Col, 65. *Mem:* Am Chem Soc. *Res:* Reactions in anhydrons liquid hydrogen fluoride; transition metal bonding to plant harmones; photochemical catalyzed decomposition of volatile Freons. *Mailing Add:* Alcorn State Univ PO Box 232 Lorman MS 39096-9402

RUSSELL, KENNETH CALVIN, b Greeley, Colo, Feb 4, 36; m 63; c 2. METALLURGY, MATERIALS SCIENCE. *Educ:* Colo Sch Mines, MetE, 59; Carnegie Inst Technol, PhD(metall eng), 64. *Prof Exp:* Asst engr, Westinghouse Elec Corp, 59-61; NSF fel, Oslo, 63-64; from asst prof to assoc prof metall, 64-78, PROF METALL & NUCLEAR ENG, MASS INST TECHNOL, 78- *Concurrent Pos:* Ford Found fel eng, 64-66. *Mem:* Am Inst Mining, Metall & Petrol Engrs; Am Phys Soc; fel Am Soc Metals; Am Nuclear Soc; Am Soc Testing & Mat; fel Am Soc Metals. *Res:* Phase transformations; lattice defects in solids; radiation damage. *Mailing Add:* 21 Taft Ave Lexington MA 02173

RUSSELL, KENNETH EDWIN, b Barnwell, Eng, Dec 9, 24; m 55; c 3. POLYMER CHEMISTRY. *Educ:* Cambridge Univ, BA, 45, PhD(chem), 48. *Prof Exp:* Vis asst prof chem, Pa State Col, 48-50; asst lectr, Manchester Univ, 50-52; fel, Princeton Univ, 52-54; lectr, 54-56, from asst prof to prof, 56-90, EMER PROF CHEM, QUEEN'S UNIV, ONT, 90- *Mem:* Am Chem Soc; fel Chem Inst Can; Royal Soc Chem. *Res:* hydrogen abstraction reactions in solution; studies of ethylene copolymers; phase structure of polyethylene; polymer grafting. *Mailing Add:* Dept Chem Queen's Univ Kingston ON K7L 3N6 Can

RUSSELL, KENNETH HOMER, b Portland, Ore, June 7, 33; m 54; c 3. PHYSICAL CHEMISTRY. *Educ:* Portland State Col, BS, 58; Washington State Univ, PhD(phys chem), 64. *Prof Exp:* Instr chem, Portland State Col, 58-59; from asst prof to assoc prof, 63-72, PROF CHEM, CALIF STATE UNIV, FRESNO, 72- *Concurrent Pos:* Lectr solar energy; partner, Solar Design & Mfg Co. *Mem:* AAAS; Am Chem Soc; Royal Soc Chem; Int Solar Energy Soc. *Res:* Molecular structure and infrared spectroscopy of inorganic compounds in the solid state. *Mailing Add:* Dept Chem Calif State Univ 6241 N Maple Ave Fresno CA 93740

RUSSELL, LEONARD NELSON, b Coldwater, Mich, Jan 15, 22; m 47; c 2. NUCLEAR PHYSICS. *Educ:* Kalamazoo Col, AB, 47; Ohio State Univ, PhD(physics), 52. *Prof Exp:* Res physicist, Mound Lab, 47-54, from asst prof to prof physics & math, Ohio Wesleyan Univ, 54-85; RETIRED. *Res:* Proton capture studies with van de Graaff generator; beta ray spectroscopy. *Mailing Add:* 45 Westgate Dr Delaware OH 43015

RUSSELL, LEWIS KEITH, b E Liverpool, Ohio, Nov 15, 31; m 57; c 8. PHYSICS, ELECTRONICS ENGINEERING. *Educ:* Univ Calif, Berkeley, BA, 56, MA, 65. *Prof Exp:* Physicist solid-state, Int Bus Mach Corp, Poughkeepsie, 56-58; nuclear physicist, Lawrence Livermore Labs, 58-62; sr engr transistor devices, Raytheon Co, Mountain View, Calif, 62-67; dept mgr circuit res, Signetics Corp, 67-; MGR CIRCUIT DESIGN, GEN ELEC. *Concurrent Pos:* Lectr, Stanford Univ, 79- *Mem:* Sr mem Inst Elec & Electronics Engrs. *Res:* Advanced electronic circuits and systems; computer aided design and simulation; solid state electronic devices; device physics; computer architecture; artificial intelligence. *Mailing Add:* Circuit Design Dept Gen Elec Mircoelectronics Ctr One Micron Dr Research Triangle Park NC 27709

RUSSELL, LIANE BRAUCH, b Vienna, Austria, Aug 27, 23; nat US; m 47; c 2. MUTAGENESIS. *Educ:* Hunter Col, AB, 45; Univ Chicago, PhD(zool), 49. *Prof Exp:* Asst, Jackson Mem Lab, 45, 46; asst, dept zool, Univ Chicago, 46-47; biologist, 48-75, corp res fel, 83-88, HEAD, MAMMALIAN GENETICS & DEVELOP SECT, OAK RIDGE NAT LAB, 75-, SR RES FEL, 88- *Concurrent Pos:* Sci adv to US deleg, First Atoms-for-Peace Conf, 55; mem comt energy & the environ, Nat Acad Sci, 75-77, biological effects of ionizing radiations, 77-80; mem sci comt 1 task group, Nat Coun for Radiation Protection & measurements, 75-77; assoc ed, Mutation Res, 76-, Environ Mutagenesis, 80-83; mem comt, Int Comn for Protection against Mutagens & Carcinogens, 77-83; mem, Int Comt Standardized Genetic Nomenclature for Mice, 77-90; mem, Gene-Toxicol Prog Coord Comt, Environ Protection Agency, 79-, mem sci adv bd, Rev Panel on Mutagenicity Guidelines, 85-86; mem, Bd Toxicol & Environ Health Hazards, Nat Acad Sci, 81-; mem sci adv panel, Off Technol Assessment, 85; mem bd, Environ Studies & Toxicol, Nat Acad Sci, 86-90; fel, Environ Health Inst, 87-; chmn, Mammalian Mutagenesis Group, Int Agency Res Cancer, 79; mem, Tech Support Group, Environ Protection Agency Risk Assessment, Reproductive & Teratogenic Effects, 80-81. *Honors & Awards:* Int Roentgen Medal, 73. *Mem:* Nat Acad Sci; fel AAAS; Genetics Soc Am Soc; Environ Mutagen Soc (pres, 84). *Res:* Induced and spontaneous genetic changes in mammalian germ cells; genetic analysis of the mouse genome; functional mapping with deletion complexes; genetic activity of the mammalian X chromosome; developmental genetics; teratogenesis. *Mailing Add:* Biol Div Oak Ridge Nat Lab PO Box 2009 Oak Ridge TN 37831-8077

RUSSELL, LORIS SHANO, b Brooklyn, NY, Apr 21, 04; m 38. PALEONTOLOGY. *Educ:* Univ Alta, BSc, 27, LLD, 58; Princeton Univ, AM, 29, PhD(paleont), 30. *Hon Degrees:* LLD, Univ Alta, 58. *Prof Exp:* Field officer, Geol Div, Res Coun, Alta, 28-29; asst paleontologist, Geol Surv Can, 30-37; from asst dir to dir, Royal Ont Mus, 37-50; chief zoologist, Nat Mus Can, 50-56, dir, 56-63; chief biologist, 63-71, EMER CUR, ROYAL ONT MUS, 71- *Concurrent Pos:* From asst prof to prof paleont, Univ Toronto, 37-70, emer prof, 70- *Honors & Awards:* Willet G Miller Medal, Royal Soc Can, 59; Can Jubilee Medal, 78; Elkanah Billings Medal, Geol Soc Can, 84. *Mem:* Geol Soc Am; Paleont Soc; Soc Vert Paleont; Can Mus Asn; Royal Soc Can. *Res:* Vertebrate paleontology; fossil vertebrates and mollusks of western North America; Cretaceous and Tertiary stratigraphy of western North America. *Mailing Add:* 55 Erskine Toronto ON M4P 1Y7 Can

RUSSELL, MARVIN W, b Poole, Ky, Aug 26, 27; m 48; c 4. PHYSICS, APPLIED MATHEMATICS. *Educ:* Western Ky Univ, BS, 50; Univ Fla, MS, 52, PhD(physics), 54. *Prof Exp:* Physicist, Gen Elec Co, 54-61; sr res scientist physics, Kaman Nuclear, 61-62; head dept physics, 62-65, dean, 65-80, PROF PHYSICS, OGDEN COL SCI & TECHNOL, WESTERN KY UNIV, 64- *Concurrent Pos:* Vis lectr, Ky Wesleyan Col, 56-60. *Mem:* Am Phys Soc; Inst Elec & Electronics Engrs; Am Asn Physics Teachers. *Res:* Electron and heat transfer physics; mass spectrometry; applied mathematics; ion mobility and recombination phenomena; mathematical models; atmospheric physics; science education. *Mailing Add:* Ogden Col Sci & Technol Western Ky Univ Bowling Green KY 42101

RUSSELL, MICHAEL W(ILLIAM), b Epsom, Eng, July 12, 44. MUCOSAL IMMUNOLOGY, ORAL MICROBIOLOGY. *Educ:* Univ Cambridge, Eng, BA, 66, MA, 70; Univ Reading, Eng, PhD(microbiol), 73. *Prof Exp:* Researcher, Nat Inst Res Dairying, Univ Reading, UK, 68-72; res fel, dept oral immunol, Med & Dent Schs, Guy's Hosp, UK, 72-79; vis investr, Inst Dent Res, 79-81, res asst prof, 82-86, RES ASSOC PROF, DEPT MICROBIOL, UNIV ALA, BIRMINGHAM, 86- *Concurrent Pos:* Investr, Inst Dent Res, Univ Ala, Birmingham, 82-86, Res Ctr Oral Biol, 87-; vis assoc prof, Dept Oral Biol, Royal Dent Col, Aarhus, Denmark, 87-88; assoc scientist, Ctr AIDS Res, Univ Ala Birmingham, 88- *Mem:* Brit Soc Immunol; Int Asn Dent Res; Am Soc Microbiol; Am Asn Immunologists; Soc Mucosal Immunol. *Res:* Induction and biological functions of immunoglobulin A antibodies; immunity to dental caries; protein antigens of Streptococcus mutans. *Mailing Add:* Dept Microbiol Univ Ala Birmingham Univ Sta Birmingham AL 35294

RUSSELL, MORLEY EGERTON, b Los Angeles, Calif, June , 29; m 63; c 2. PHYSICAL CHEMISTRY. *Educ:* Col Wooster, AB, 51; Mass Inst Technol, SM, 53; Univ Mich, PhD(chem), 58. *Prof Exp:* Fel phys chem, Nobel Inst Chem, Sweden, 58-59; asst prof, Mich State Univ, 59-64; ASSOC PROF PHYS CHEM, NORTHERN ILL UNIV, 65- *Mem:* Am Soc Mass Spectrometry; Am Chem Soc. *Res:* Kinetics of homogeneous reactions; mass spectrometry. *Mailing Add:* Dept Chem Northern Ill Univ DeKalb IL 60115

RUSSELL, NANCY JEANNE, b Virginia, Minn, June 12, 38; m 65. BIOLOGY, PHARMACOLOGY. *Educ:* Univ Minn, BA, 60, MA, 67, PhD(zool), 72. *Prof Exp:* Res assoc develop biol, Reed Col, 67-68; res asst cutaneous biol, Ore Regional Primate Ctr, 68-69; res assoc neuropharmacol, 72-78, res assoc & fel neuropharmacol-morphol, 73-75, res instr pharmacol, 75-78, RES ASST PROF PHARMACOL, SCH MED, UNIV ORE, 78- *Concurrent Pos:* Pharmaceut Mfrs Asn Found fel pharmacol morphol, 73; fel pharmacol, Sch Med, Univ Ore Health Sci Ctr, 73-75. *Mem:* Int Fedn Electron Microscopy Soc; Am Soc Zoologists. *Res:* Neuropharmacology neuroanatomy and neuropathology as related to drug action on sensory system, cochlear and vestibular system, and myenteric plexus of gastro intestinal tract. *Mailing Add:* Dept Pharmacol Ore Health Sci Univ 3181 SW Sam Jackson Park Rd Portland OR 97201-3798

RUSSELL, PAUL E(DGAR), b Roswell, NMex, Oct 10, 24; m 43; c 3. ELECTRICAL ENGINEERING. *Educ:* NMex State Univ, BS, 46 & 47; Univ Wis, MS, 50, PhD(elec eng), 51. *Prof Exp:* From instr to asst prof elec eng, Univ Wis, 47-52; sr dynamics engr, Gen Dynamics/Convair, 52-54; prof elec eng, Univ Ariz, 54-63, head dept, 58-63, dir appl res lab, 56-58, 60-63; prof elec eng & dean col eng, Kans State Univ, 63-67; prof eng, Ariz State Univ, 67-90, dir, Sch Construct & Tech, 88-90; CONSULT ENGR, 90- *Concurrent Pos:* Consult, NSF, Dynamic Sci, Westinghouse, Motorola,

AiResearch, US Army Electronic Proving Grounds & numerous univs, UN. *Mem:* Am Soc Eng Educ; fel Inst Elec & Electronics Engrs; Nat Soc Prof Engrs; Sigma Xi. *Res:* Control systems analysis and design; computers; photovoltaic power systems. *Mailing Add:* 5902 E Caballo Lane Scottsdale AZ 85253

RUSSELL, PAUL SNOWDEN, b Chicago, Ill, Jan 22, 25; m 52; c 4. SURGERY, IMMUNOLOGY. *Educ:* Univ Chicago, PhB, 44, BS, 45, MD, 47; Am Bd Surg, dipl, 57; Am Bd Thoracic Surg, dipl, 60. *Hon Degrees:* MA, Harvard Univ, 62. *Prof Exp:* Surg intern, Mass Gen Hosp, 48-49, asst surg resident, 49-51 & 53-55; teaching fel surg, Harvard Med Sch, 56, instr, 57-59; clin assoc surg & tutor med sci, 59-60; assoc prof surg, Col Physicians & Surgeons, Columbia Univ, 60-62; JOHN HOMANS PROF SURG, HARVARD MED SCH, 62- *Concurrent Pos:* USPHS res fel, 54-55; resident, Mass Gen Hosp, 56, asst surg, 57-60; assoc attend surgeon, Presby Hosp & assoc vis surgeon, Francis Delafield Hosp, 60-62; USPHS career develop award, 60-62; chief gen surg serv, Mass Gen Hosp, 62-69, vis surgeon, 69-; mem comts trauma & tissue transplantation, Div Med Sci, Nat Acad Sci-Nat Res Coun, 63; mem allergy & immunol study sect, Div Res Grants, NIH, 63-65, chmn, 65; secy, Dept Surg, Harvard Med Sch, 65-70. *Mem:* Fel Am Col Surg; Am Asn Thoracic Surg; Soc Clin Surg; fel Royal Soc Med; Transplantation Soc (pres, 70-72); Sigma Xi. *Mailing Add:* Dept Surg Mass Gen Hosp Boston MA 02114

RUSSELL, PAUL TELFORD, b San Francisco, Calif, June 5, 35; m 59; c 3. BIOCHEMISTRY. *Educ:* Univ Calif, Berkeley, BA, 59; Univ Ore, MS, 61, PhD(biochem), 63. *Prof Exp:* Trainee, Steroid Training Prog, Clark Univ & Worcester Found Exp Biol, 63-64; fel chem, Univ Miss, 64-66; asst prof obstet & gynec, Ohio State Univ, 66-68; asst prof obstet, Gynec & biol chem, 68-76, ASSOC PROF RES OBSTET & GYNEC, MED SCH, UNIV CINCINNATI, 76-, ASSOC PROF PEDIAT, 78- *Res:* Pathways and mechanisms of biosynthesis and metabolism of steroidal compounds; metabolic pathways of prostaglandins. *Mailing Add:* Reproduction & Infertility Christ Hosp 2139 Auburn Ave Cincinnati OH 45219

RUSSELL, PERCY J, b New York, NY, May 29, 26; m 55; c 3. BIOCHEMISTRY. *Educ:* City Col New York, BS, 50; Brooklyn Col, MA, 55; Western Reserve Univ, PhD(biochem), 59. *Prof Exp:* Res fel bact, Harvard Univ, 59-61; from asst prof to assoc prof biochem, Univ Kans, 61-70; ASSOC PROF BIOCHEM, UNIV CALIF, SAN DIEGO, 70- *Res:* Immunochemistry; enzyme mechanisms; phospholipid metabolism. *Mailing Add:* Dept Biol 2130 Bonner Hall Univ Calif San Diego M-001 La Jolla CA 92093

RUSSELL, PETER BYROM, b Manchester, Eng, Oct 24, 18; nat US; m 45; c 2. ORGANIC CHEMISTRY. *Educ:* Univ Manchester, BSc, 40, MSc, 41, PhD, 45, DSc, 54. *Prof Exp:* Sr res chemist, Burroughs Wellcome & Co, Inc, 47-56; dir res, John Wyeth & Brother Ltd, Eng, 56-57; mgr org develop lab, 57-59, DIR RES, WYETH LABS DIV, AM HOME PROD CORP, 59- *Concurrent Pos:* Chmn ad hoc study group med chem, Walter Reed Army Inst Res, 73-75. *Mem:* Am Chem Soc. *Res:* Isolation and synthesis of natural products; synthesis of chemotherapeutically active compounds; medicinal chemistry; pharmaceutical chemistry. *Mailing Add:* 1708 Sherwook Circle Villanova PA 19085-1910

RUSSELL, PETER JAMES, b Kent, Eng, Oct 30, 47; div; c 2. MOLECULAR BIOLOGY, MICROBIOLOGY. *Educ:* Univ Sussex, BSc, 68; Cornell Univ, PhD(genetics), 72. *Prof Exp:* From asst prof to assoc prof, 72-84, PROF BIOL & GENETICS, REED COL, 84- *Concurrent Pos:* Vis assoc prof, dept physiol chem, Univ Wis, 80-81; ed-in-chief, Fungal Genetics Newslett, 80-; mem bd, Ore Res & Technol Develop Corp, 86-89; vis prof, Dept Molecular, Cellular & Develop Biol, Univ Colo, Boulder, 89-91. *Mem:* Am Soc Microbiol. *Res:* Regulation of ribosomal DNA copy number and expression in eukaryotes; biochemistry of morphogenesis of Candida albicans. *Mailing Add:* Dept Biol Reed Col Portland OR 97202-8199

RUSSELL, PHILIP BOYD, b Buffalo, NY, Mar 15, 44; div. REMOTE SENSING. *Educ:* Wesleyan Univ, BA, 65; Stanford Univ, MS, 67, 90, PhD(physics), 71. *Prof Exp:* Fel, Nat Ctr Atmospheric Res, 72-72; physicist, 72-76, sr physicist, SRI Int, 76-82; chief, Atmosphere Exp Br, 82-89, actg chief & actg dep chief, Earth Syst Sci Div, 88-89, CHIEF, ATMOSPHERIC CHEM & DYNAMICS BR, NASA AMES RES CTR, 89- *Concurrent Pos:* Prin investr grants, NSF, NASA & ARO NASA & Army Res Off, 74-82; mem, Army Basic Res Adv Comn, Nat Res Coun, 79-81, Solar Terrestrial Observ Sci Study Group, NASA, 79-80 & Atmospheric Lidar Working Group, 77-79; chmn, Comm Laser Atmosphere Studies, Am Meteorol Soc, 79-82, mem, Comn Radiation Energy, 79-81 & Comm Atmospheric Environ, Am Inst Aeronaut & Astronaut, 84-; mem, Interagency Task Group Airborne Geosci; guest ed, 3 J Geophys Res issues on satellite validation & stratosphere-troposphere exchange. *Mem:* Am Meteorol Soc; Optical Soc Am; Am Geophys Union; AAAS; Sigma Xi; Am Inst Aeronaut & Astronaut. *Res:* Atmospheric science; remote sensing and radiative transfer; radiative and climatic effects of aerosols and gases; aircraft and stellite measurements; laser and acoustic radar; data validation and error analysis; atmospheric multi-sensor experiment design. *Mailing Add:* 960 Terrace Dr Los Altos CA 94024-5939

RUSSELL, PHILIP KING, b Syracuse, NY, Jan 26, 32; m 55; c 3. INFECTIOUS DISEASES, VIROLOGY. *Educ:* Johns Hopkins Univ, AB, 54; Univ Rochester, MD, 58; Am Bd Internal Med, dipl. *Prof Exp:* Med Corps, US Army, 58-, intern med, NC Mem Hosp, 58-59, lab officer, Walter Reed Army Inst Res, 59-61, resident med, Univ Hosp, Baltimore, 61-63, clin investr infectious dis, Pakistan Med Res Ctr, 63-64, res officer, Walter Reed Army Inst Res, 64-65, mem, Seato Med Res Lab, 65-68, chief dept virus dis, 68-71, dep dir, 76-79, dir & commandant, 79-83, DIR DIV COMMUN DIS & IMMUNOL, WALTER REED ARMY INST RES, US ARMY, 71-; COMMANDER, FITZSIMMONS ARMY MED CTR, 83- *Concurrent Pos:* Mem bd dir, Gorgas Mem Inst; sci adv comt denque & yellow fever, Pan Am Health Orgn; tech adv comt, Denque haemorrhnqic Fever, WHO, Infectious Dis Comt & Task Force Virology, Nat Inst Allergy & Infectious Dis, NIH; chmn steering comt, Scientific Working Group on Immunology of Malaria, WHO, 80-; sci adv group experts, WHO, 85- *Honors & Awards:* Gorgas Medal; Smadel Mem Medal; Paul Siple Medal. *Mem:* Am Soc Trop Med & Hyg; Am Soc Microbiol; Am Asn Immunol; Am Epidemiol Soc; Infectious Dis Soc Am. *Res:* Virus diseases; immunology; pathogenesis of virus diseases; epidemiology of arboviruses. *Mailing Add:* US Army Res-Dev Command Ft Detrick Frederick MD 21701

RUSSELL, PHILLIP K, geological research, for more information see previous edition

RUSSELL, RAYMOND ALVIN, b Buffalo, NY, Jan 16, 17; m 43; c 2. PHYSIOLOGY. *Educ:* Hamilton Col, AB, 38; Univ Rochester, MS, 41, PhD(physiol), 51. *Prof Exp:* Asst agr biochem, Univ Minn, 41-42; asst invest chem, NY Agr Exp Sta, 42-44; from instr to assoc prof, La State Univ Med Ctr, New Orleans, 51-80; RETIRED. *Concurrent Pos:* Consult, Southern Baptist Hosp, New Orleans, La, 59-76. *Mem:* Am Physiol Soc; AAAS. *Res:* Secretion of the small intestine; myocardial metabolism; circulatory shock. *Mailing Add:* Box 507 Kiln MS 39556

RUSSELL, RICHARD DANA, b Pomona, Calif, Nov 15, 06; m 33; c 2. SEDIMENTOLOGY, OCEANOGRAPHY. *Educ:* Pomona Col, BA, 27; Univ Calif, PhD(geol), 32. *Prof Exp:* From instr to assoc prof geol, La State Univ, 31-42; res assoc, Div War Res, Univ Calif, 42-44, oceanogr, 44-45, publ mgr, 45-46; tech ed & head publ dept, US Navy Electronics Lab, 46-47, sr staff geophysicist, 47-50, sr consult geophys & head sci planning bd, 50-55; mgr geol res, Marathon Oil Co, 55-63, assoc dir explor res, 63-70, consult, 70-71; RETIRED. *Concurrent Pos:* At Bikini Sci Reserve, 47; lectr, Calif Inst Technol, 48 & Scripps Inst, Univ Calif, 49-54; mem comt sedimentation, Nat Res Coun, 37-46; hon comt, Ctr Oceanog Res & Study; vpres, Am Geol Inst, 72, pres, 73; mem, Nat Comm Geol, 73-77. *Mem:* Fel AAAS; fel Geol Soc Am; hon mem, Soc Paleontologists & Mineralologists (vpres, 43, pres, 48); Am Inst Prof Geologists (pres, 69); hon mem Rocky Mountain Asn Geologists; hon mem Am Asn Petrol Geologists. *Res:* Petrology of recent sediments; oceanography; stratigraphy; field and marine geology; sedimentary petrology and petrography; military oceanography; lake evaporation; petroleum exploration; research administration. *Mailing Add:* 6597 Meadowridge Dr Santa Rosa CA 95409

RUSSELL, RICHARD DONCASTER, b Toronto, Ont, Feb 27, 29; m 51; c 3. GEOPHYSICS, INSTRUMENTATION. *Educ:* Univ Toronto, BA, 51, MA, 52, PhD(physics/geophys), 54. *Prof Exp:* Lectr physics, Univ Toronto, 54-56, asst prof, 56-58; assoc prof, Univ BC, 58-62; prof, Univ Toronto, 62-63; head dept, 68-79, assoc vpres, acad, 83-86, PROF GEOPHYS, UNIV BC, 63- *Mem:* Am Geophys Union; fel Royal Soc Can; Geol Asn Can; Can Soc Explor Geophysicists. *Res:* Mass spectrometry; seismology; electronics; histroy of the earth; geochronology; physics of geophysical instruments and geochemistry. *Mailing Add:* 5450 University Blvd Apt 107 Vancouver BC V6T 1K4 Can

RUSSELL, RICHARD LAWSON, b Bar Harbor, Maine, Nov 24, 40; m 62, 77; c 3. DEVELOPMENTAL NEUROGENETICS. *Educ:* Harvard Col, BA, 62; Calif Inst Technol, PhD(genetics), 67. *Prof Exp:* Asst prof genetics, Cornell Univ, 66-67; mem staff, Lab Molecular Biol, Med Res Coun, Cambridge Univ, 67-70; asst prof biol, Calif Inst Technol, 70-76; assoc prof, 76-84, PROF BIOL SCI, UNIV PITTSBURGH, 84- *Concurrent Pos:* NSF fel, 67-68; NATO fel, 68-69; Am Heart Asn fel, 69-70; mem, NIH Molecular Cytol Study Sect, 80-84. *Mem:* AAAS; Fedn Am Scientists; Genetics Soc Am. *Res:* Genetic analysis of structure and function in simple nervous systems. *Mailing Add:* Dept Biol Sci Univ Pittsburgh Pittsburgh PA 15260

RUSSELL, RICHARD OLNEY, JR, b Birmingham, Ala, July 9, 32; m 63; c 4. CARDIOLOGY. *Educ:* Vanderbilt Univ, BA, 53, MD, 56. *Prof Exp:* From instr to prof med, 64-81, CLIN PROF, UNIV ALA, BIRMINGHAM, 81-; CARDIOLOGIST, CARDIA VASCULAR ASN, 81- *Concurrent Pos:* Co-dir myocardial infarction res unit, Univ Ala, Birmingham, 70-75. *Mem:* Am Fedn Clin Res. *Res:* Ischemic heart disease; acute myocardial infarction; hemodynamics of acute and chronic ischemic heart disease. *Mailing Add:* Cardia Vascular Asn 880 Montclair Rd Suite 170 Birmingham AL 35213

RUSSELL, ROBERT JOHN, b Ballymena, Northern Ireland, Jan 21, 38; US citizen; m 71; c 2. LABORATORY ANIMAL MEDICINE. *Educ:* Univ Ill, BS, 60, DVM, 62; Tex A&M Univ, MS, 69. *Prof Exp:* Clin pract vet med, Roseland Animal Hosp, Chicago, 62; base vet, USAF, Norton AFB, Calif, 62-64 & Royal Air Force, Wethersfield, Eng, 64-67; resident lab animal med, Tex A&M Univ & Brooks AFB, 67-69; dir, Vet Serv, Cam Rahn Bay 20th Hosp, USAF, Vietnam, 69-70, dir, Lab Animal Med, Armed Forces Inst Path, Washington, 70-76, Naval Med Res Inst, Bethesda, Md, 76-80 & Prog Resources, Inc, Frederick, Md, 82-85; DIR, ANIMAL MED LAB, HARLAN SPRAGUE DAWLEY, INC, FREDERICK, 84- *Concurrent Pos:* Consult, NIH, Bethesda, 72-82, Bioqual, Inc, Rockville, Md, 78-; adj fac, NVa Community Col, Herndon, Va, 76- *Mem:* Am Vet Med Asn; Am Asn Lab Animal Sci. *Res:* Laboratory animal management; infectious disease; animal reproduction; hypobaric medicine; author of various publications. *Mailing Add:* Harlan Sprague Dawley Inc 7121 English Muffin Way Frederick MD 21701

RUSSELL, ROBERT JULIAN, JR, mammalogy, for more information see previous edition

RUSSELL, ROBERT LEE, b Independence, Mo, June 27, 27; m 50; c 2. PHYSIOLOGY, PHARMACOLOGY. *Educ:* Univ Mo, PhD(physiol, pharmacol), 54. *Prof Exp:* Asst physiol & pharmacol, 50-53, from asst instr to assoc prof pharmacol, 53-66, PROF PHARMACOL, UNIV MO-COLUMBIA, 66- *Mem:* Fel AAAS; Soc Exp Biol & Med; Am Soc Pharmacol & Exp Therapeut. *Res:* Drugs affecting lipid metabolism. *Mailing Add:* M572 Med Ctr Univ Mo Columbia MO 65212

RUSSELL, ROBERT M, b Boston, Mass, Apr 9, 41; c 2. GASTROENTEROLOGY. *Educ:* Harvard Univ, BA, 63; Columbia Univ, MD, 67. *Prof Exp:* Vis asst prof med, Shiraz Univ, Shiraz, Iran, 74-75; from asst prof to assoc prof, Sch Med, Univ Md, 75-81; res & develop assoc, Vet Admin Med Ctr, Baltimore, 75-79; staff physician gastroenterol, Univ Md Hosp & Vet Admin Hosp, 75-81; assoc prof med, Sch Med, Tufts Univ, 81-86, adj assoc prof nutrit, 82-86; DIR HUMAN STUDIES, HUMAN NUTRIT RES CTR-TUFTS UNIV, USDA, 81-, PROF MED & NUTRIT, 88- *Concurrent Pos:* Consult, Food & Drug Admin, 74-79 & Nat Iranian Cancer Found, 78; clin investr, Vet Admin Career Develop Prog, Baltimore, 79-81; mem, Comt Nutrit Res, Am Soc Clin Nutrit, 80-84; staff physician, New Eng Med Ctr, Boston, 81; mem bd dirs, Am Col Nutrit, 81-; actg dir, Human Nutrit Res Ctr-Tufts Univ, USDA, 83-85 & 85-86, assoc dir, 87-; mem, Nat Dairy Coun Sci Adv Bd, 87-91, chmn, 90-91; nutrit consult, Am Bd Internal Med, 90- *Honors & Awards:* Res Award, Chicago Soc Gastroenterol, 74. *Mem:* Am Gastroenterol Asn; fel Am Col Physicians; Am Inst Nutrit; Am Fedn Clin Res; Am Soc Clin Nutrit; Gastroenterol Res Group. *Res:* Gastrointestinal absorptive function of micronutrients; effect of aging on gastrointestinal function; folic acid and retnoid metabolism. *Mailing Add:* Human Nutrit Res Ctr Tufts Univ-USDA 711 Washington St Boston MA 02111

RUSSELL, ROBERT RAYMOND, b Beach, Wash, July 3, 20; m 43; c 3. CHEMISTRY. *Educ:* Graceland Col, AB, 43, MA, 46; Univ Kans, PhD(chem), 49. *Prof Exp:* prof chem, Univ Mo-Rolla, 48-; RETIRED. *Mem:* Am Chem Soc. *Res:* Organic chemical reaction mechanisms. *Mailing Add:* Box 766 Rolla MO 65401

RUSSELL, ROSS F, b Auburn, Nebr, May 7, 19; m 44; c 4. CHEMICAL ENGINEERING, POLYMER CHEMISTRY. *Educ:* Peru State Col, BA, 41; Iowa State Univ, PhD(chem eng), 50. *Prof Exp:* Engr, E I du Pont de Nemours & Co, Inc, 50-57, process supvr, 57-67, sr res engr, 67-82; VPRES, MESERAN CO, 83- *Mem:* Am Chem Soc; Am Inst Chem Engrs. *Res:* Plant installations of solvent extraction, distillation and polymerization equipment; computerized analytical instruments. *Mailing Add:* 701 Highview Dr Chattanooga TN 37415

RUSSELL, RUTH LOIS, b San Diego, Calif, July 22, 28; m 54; c 1. MICROBIOLOGY. *Educ:* Univ Calif, Los Angeles, BA, 52, PhD(microbiol), 63. *Prof Exp:* Asst bact, Univ Calif, Los Angeles, 52-55; instr biol, Occidental Col, 56-57; med bacteriologist, San Fernando Vet Admin Hosp, Calif, 59-61; res microbiologist, Olive View Hosp, 61-62; med bacteriologist, San Fernando Vet Admin Hosp, Calif, 62-63; from asst prof to assoc prof microbiol, 63-72, prof, 72-73, actg dean, Grad Studies, 74-77, DIR ENVIRON STUDIES, CALIF STATE UNIV, LONG BEACH, 77- *Concurrent Pos:* Consult microbiologist, var clin & biomed labs. *Mem:* AAAS; Am Soc Microbiol; Am Pub Health Asn; NY Acad Sci; Am Soc Qual Control. *Res:* Pub health microbiology; medical bacteriology; environmental microbiology. *Mailing Add:* Dept Microbiol Calif State Univ 1250 Bellflower Blvd Long Beach CA 90840

RUSSELL, SCOTT D, b Milwaukee, Wis, Dec 8, 52; c 1. CYTOLOGY. *Prof Exp:* Asst prof, 81-87, DIR S R NOBLE ELEC MICROS LAB, 84-, ASSOC PROF STRUCTURAL BOT, 87- *Concurrent Pos:* prin investr, 2 NSF grants, 2 USDA grants, 1 Dept Educ grant, 82-; vis scientist, Univ Melbourne, Australia, 84-; vis prof, Ecole Normal Superieurede, Lyon, France, 89. *Honors & Awards:* Jeanette Siron Pelton Award, Bot Soc Am, 88. *Mem:* Bot Soc Am; AAAS; Int Soc Plant Morpholog; Elec Microsc Soc Am; Microbeam Anal Soc. *Res:* Structural basis of double fertilization and male cytoplasmic transmission in angiosperms; reproductive plant cytology and ultrastructure. *Mailing Add:* Dept Bot & Microbiol Univ Okla Norman OK 73019

RUSSELL, SHARON MAY, b Klamath Falls, Ore, May 14, 44; m 75. GROWTH HORMONES. *Educ:* Stanford Univ, BA, 66, PhD(physiol), 71. *Prof Exp:* Res asst, Dept Physiol, Stanford Univ, Palo Alto, Calif, 66-68; Postdoctoral res fel, Dept Physiol-Anat, Univ Calif, Berkeley, 71-73, asst res physiologist, 73-81, assoc res physiologist, 81-89, ASSOC RES PHYSIOLOGIST, DEPT INTEGRATIVE BIOL, UNIV CALIF, BERKELEY, 89- *Concurrent Pos:* Lectr, Dept Physiol-Anat, Univ Calif, Berkeley, 74-89 & Dept Integrative Biol, 91-; vis asst prof, Mills Col, Oakland, Calif, 84-85 & 87. *Mem:* Endocrine Soc; Am Soc Zoologists; Coalition Animals & Animal Res. *Res:* Hormonal control of growth and development in mammals. Specifically, growth-promoting mechanisms of growth hormone and prolactin; interactions among hormones of the anterior pituitary gland and pancreas and liver-derived growth factors; roles of peptide growth factors in mammalian embryonic and fetal development; theoretical analyses of the evolution of prolactin and growth hormone molecules and their receptors. *Mailing Add:* Dept Integrative Biol LSA 281 Univ Calif Berkeley CA 94720

RUSSELL, STEPHEN MIMS, b Hot Springs, Ark, Sept 16, 31. ORNITHOLOGY. *Educ:* Va Polytech Inst, BS, 53; La State Univ, PhD(zool), 62. *Prof Exp:* From instr to asst prof biol, La State Univ, 58-64; asst prof, 64-70, ASSOC PROF ZOOL, UNIV ARIZ, 70-, CUR BIRDS, 64- *Mem:* fel Am Ornith Union; Cooper Ornith Soc; Wilson Ornith Soc; Sigma Xi. *Res:* Biology of desert and neotropical birds, especially behavior, ecology, distribution. *Mailing Add:* Dept of Ecol & Evolutionary Biol Univ of Ariz Tucson AZ 85721

RUSSELL, STUART JONATHAN, b Portsmouth, UK, Feb 25, 62. MACHINE LEARNING, LIMITED RATIONALITY. *Educ:* Oxford Univ, BA Hons, 82; Stanford Univ, PhD(computer sci), 86. *Prof Exp:* ASST PROF COMPUTER SCI, UNIV CALIF, BERKELEY, 86- *Concurrent Pos:* Consult, MCC, 86-89; vis prof, Oxford Univ, 90; mem, fel panel, Nat Res Coun, 90; NSF presidential young investr, 90. *Mem:* Am Asn Artificial Intel; Asn Comput Mach. *Res:* Artificial intelligence; machine learning, knowledge-based induction; limited rationality-decision-theoretic-metareasoning; analogical reasoning; real time decision making. *Mailing Add:* Computer Sci Div Univ Calif Berkeley CA 94720

RUSSELL, T(HOMAS) L(EE), b Pomona, Calif, Dec 15, 30; m 55; c 3. MECHANICAL ENGINEERING. *Educ:* Calif Inst Technol, BS, 52, MS, 53, PhD, 58. *Prof Exp:* Lectr metall, Calif Inst Technol, 56-58; res engr, Chevron Res Co Div, Standard Oil Co Calif, 58-60, group supvr, 60-63, supvr res eng, 63-67, staff financial analyst, 67-71, sr adv Mid East, 71-77, pres, Iran Chevron Oil Co, 73-77, MGR ANALYSIS DIV, STANDARD OIL CO CALIF, 77- *Mem:* Am Inst Mining, Metall & Petrol Engrs; Soc Petrol Engrs. *Res:* Metallurgy; oceanography; applied mechanics; reservoir and petroleum engineering; economic analysis. *Mailing Add:* 1683 Caminito Alvivado La Jolla CA 92037

RUSSELL, T W FRASER, b Moose Jaw, Sask, Aug 5, 34; m 56; c 3. CHEMICAL ENGINEERING. *Educ:* Univ Alta, BSc, 56, MSc, 58; Univ Del, PhD(chem eng), 64. *Prof Exp:* Res chem engr, Res Coun Alta, 56-58; design engr, Union Carbide Can, 58-61; prof chem eng, 64-81, ALLAN P COLBURN PROF CHEM ENG, UNIV DEL, 81-, DIR, INST ENERGY CONVERSION, 79-, CHMN, DEPT CHEM ENG, 86- *Concurrent Pos:* Consult, E I du Pont de Nemours & other chem & electronics firms, 61- *Honors & Awards:* Leo Friend Award, Am Chem Soc, 82; 3M Chem Eng Div Award, Am Soc Eng Educ, 84; Award in Chem Eng Pract, 87, Thomas H Chilton Award, AIChE, 88. *Mem:* Nat Acad Eng; Am Chem Soc; Am Soc Eng Educ; Am Inst Chem Engrs. *Res:* Gas-liquid system design; economics of the chemical process industries; development of a continuous process for solar cell manufacture; semiconductor reaction and reactor engineering analysis and photovoltaic unit operations; multiphase fluid mechanics. *Mailing Add:* Inst Energy Conversion Univ Del Newark DE 19716-0001

RUSSELL, TERRENCE R, b Chicago, ILL, Dec 22, 40. SCIENCE ADMINISTRATION. *Educ:* Southern Ill Univ, BA, 67, MS, 69, PhD(soc sci), 79. *Prof Exp:* Res assoc, Nat Surv Community Corrections Progs Ctr Study Crime, Delinquency & Corrections, 71-72; instr social & co-founder, co-dir Urban Studies Prog, Ill Wesleyan Univ, 72-77; asst prof social & dir res internship prog, Ill State Univ, 77-80; asst prof soc, Northern Ky Univ, 80-84; sr res assoc, Work Force Analysis, 84-85, prof relations prog mgr, 85-88, mgr, Am Chem Soc, 88-91; EXEC DIR, ASSN INSTITUTIONAL RES, 91- *Concurrent Pos:* Mem, Comt Measuring Nat Needs for Scientists to yr 2000, chair, Social Sci Sect, 87; adv comt Div Sci Resources Studies, NSF, 87-89; comn Prof in Sci & Technol, Prof Soc Ethis Goup, AAAS; vis assoc prof, Georgetown Univ, 87-, chmn, Engrs & Scientists' Joint Comt on Pensions, Am Asn Eng Soc, 91; grant, Occup Mobility Studies, NSF, 88-89. *Mem:* Nat Acad Sci; Acad Mgt; Am Chem Soc; Am Inst Chemists; Am Soc Asn Exec; Am Sociol Asn; Asn Inst Res; Soc Social Study Sci. *Res:* Development, implementation and marketing of programs serving the professional and information needs of American Chemical Society members; studies on issues concerning the scientific and technical labor force and ethics in science, business and the professions. *Mailing Add:* Assn Institutional Res 314 Stone Bldg Florida State Univ Talahassee FL 32306

RUSSELL, THOMAS EDWARD, b Tucson, Ariz, May 8, 42; m 65; c 1. PLANT PATHOLOGY. *Educ:* Univ Ariz, BS, 65, MS, 67; Tex A&M Univ, PhD(plant path), 70. *Prof Exp:* Agr specialist pesticide res, Buckman Labs, Tenn, 70-71; asst plant pathologist, 71-76, ASSOC RES SCIENTIST PLANT PATH, UNIV ARIZ, 76- *Mem:* Am Phytopath Soc. *Res:* Diseases of vegetable crops, turf and cotton. *Mailing Add:* 6244 E Berneil Paradise Valley AZ 85253

RUSSELL, THOMAS J, JR, b Dec 1, 31. BIOCHEMISTRY. *Educ:* Rutgers Univ, AB, 57, PhD(biochem), 61. *Prof Exp:* Res assoc biochem, Bur Biol Res & instr, Rutgers Univ, 61-62; founder & pres, Bio/Dynamics, Inc, 62-82; PRES, LIFE SCI DIV, IMS INT, 75- *Mem:* AAAS; NY Acad Sci; Am Chem Soc. *Res:* Metabolism; toxicology; application of computer technology to biological research. *Mailing Add:* Bio/Dynamics Inc PO Box 43 East Millstone NJ 08873

RUSSELL, THOMAS PAUL, b Boston, Mass, Nov 18, 52; m 78; c 2. POLYMER PHYSICS. *Educ:* Boston State Col, BS, 74; Univ Mass, Amherst, MA, 77, PhD(polymer sci), 79. *Prof Exp:* Res fel polymer physics, Univ Mainz, W Germany, 79-81; RES STAFF MEM POLYMER PHYSICS, IBM RES LABS, 81- *Concurrent Pos:* Mem, Nat Steering Comt for Advan Neutron Source, Prog Adv Comt, Los Alamos Nat Lab; chmn, Div Polymer Physics Prog Comt, Am Phys Soc. *Honors & Awards:* A Doolittle Award, Am Chem Soc, 85. *Mem:* Fel Am Phys Soc; Am Chem Soc; Am Crystallog Asn; Sigma Xi; Mat Res Soc. *Res:* The physics of high polymers in the solid state; characteristics of high temperature polymers for use in the microelectronics industry; compatibility, phase separation kinetics and crystallization of polymer mixtures and their interdiffusion by use of small angle x-ray, light and neutron scattering; time resolved scattering techniques using synchrohon radiation; x-ray and neutron reflectivity on thin polymer films. *Mailing Add:* Almaden Res Ctr K93-802 IBM Res Labs 650 Harry Rd San Jose CA 95120-6099

RUSSELL, THOMAS SOLON, b Bracey, Va, June 27, 22; m 49; c 4. STATISTICS. *Educ:* Wake Forest Col, BS, 44; Va Polytech Inst, MS, 53, PhD(statist), 56. *Prof Exp:* Instr math, Capitol Radio Engrs Inst, 46-47; instr, Va Polytech Inst, 47-51, asst statistician, Agr Exp Sta, 53-56; from asst statistician to assoc statistician, 56-63, PROF AGR & STATISTICIAN, AGR EXP STA, WASH STATE UNIV, 63- *Mem:* Biomet Soc; Am Statist Asn; Inst Math Statist; Sigma Xi. *Res:* Mathematical statistics. *Mailing Add:* SW 910 Fountain Pullman WA 99163

RUSSELL, THOMAS WEBB, b Greenville, Tex, Dec 17, 40; m 64. HETEROGENEOUS CATALYSIS, ORGANIC REDUCTIONS. *Educ:* Tex Col Arts & Indust, BSc, 62; Univ Colo, MSc, 64, PhD(chem), 66. *Prof Exp:* Vis asst prof chem, La State Univ, 66-67; asst prof chem, Eastern N Mex Univ, 67-72, assoc prof, 72-79, prof, 79-81; sr res chemist, El Paso Petrochem Co, 81-83; PRES, LONE STAR VINEYARD CO, 84- *Concurrent Pos:* Am Chem Soc Petrol Res Fund fel, NSF fel. *Mem:* AAAS; Am Chem Soc. *Res:* Heterogeneous hydrogenation and methanation catalysts; organic synthesis; groundwater quality and control. *Mailing Add:* HC64 Box 7 Gardendale TX 79758

RUSSELL, VIRGINIA ANN, b Oneonta, NY, Aug 11, 25. ANALYTICAL CHEMISTRY, INORGANIC CHEMISTRY. *Educ:* Westminster Col, BS, 47; Syracuse Univ, MS, 49, PhD(chem), 53. *Prof Exp:* Fel boron chem, Syracuse Univ, 52-55; chemist, Gen Elec Co, 55-85. *Mem:* Am Chem Soc; NAm Thermal Anal Soc; Sigma Xi. *Res:* Instrumental analysis; auger, x-ray diffraction and thermal. *Mailing Add:* PO Box 4976 Syracuse NY 13221-4976

RUSSELL, WILBERT AMBRICK, b Lenore, Man, Aug 3, 22; nat US; m 43; c 3. PLANT BREEDING, GENETICS. *Educ:* Univ Man, BSA, 42; Univ Minn, MS, 47, PhD(plant breeding, genetics), 52. *Prof Exp:* Agr res officer, Exp Sta, Can Dept Agr, Morden, 47-52; from asst prof to assoc prof, 52-62, PROF AGRON, COL AGR, IOWA STATE UNIV, 62- *Honors & Awards:* Crop Sci Res Award, Crop Sci Soc Am, 81; Res Award,, Nat Coun Com Plant Breeders, 79. *Mem:* Fel Am Soc Agron; Crop Sci Soc Am. *Res:* Breeding and genetics of field corn. *Mailing Add:* Dept Agron Iowa State Univ Ames IA 50011

RUSSELL, WILLIAM CHARLES, b New Brunswick, NJ, Nov 11, 55; m 86; c 2. THIN FILM SCIENCE, CHEMICAL VAPOR DEPOSITION PROCESS MODELING. *Educ:* Rutgers Univ, BS, 77; Univ Ky, MS, 86, PhD(mat sci), 88. *Prof Exp:* Ceramic engr, Corhart Refractories Co, 77-78; SR PROJ ENGR, GTE VALENITE CORP, 88- *Mem:* Mat Res Soc. *Res:* Thin refractory ceramic coatings deposition by chemical vapor deposition and related methods on cemented carbides for the metal cutting industry; thin film analytical techniques; thermodynamics. *Mailing Add:* GTE Valenite Corp 1711 Thunderbird Troy MI 48084

RUSSELL, WILLIAM LAWSON, b Newhaven, Eng, Aug 19, 10; nat US; m 36, 47; c 6. GENETIC RISK FROM RADIATION & CHEMICALS. *Educ:* Oxford Univ, BA, 32; Univ Chicago, PhD, 37. *Prof Exp:* Asst zool, Univ Chicago, 34-36; res assoc, Jackson Mem Lab, 37-47; prin geneticist, 47-77, CONSULT GENETICS, OAK RIDGE NAT LAB, 77- *Concurrent Pos:* Adv, US deleg to UN Sci Comt Effects Atomic Radiation, 62-86; mem, various other int and nat comts on genetic risks of radiation and chemicals. *Honors & Awards:* Enrico Fermi Award, 77; Roentgen Medal, 73; Distinguished Achievement Award of Health, Physics Soc, 76; Environ Mutagen Soc Award, 89. *Mem:* Nat Acad Sci; Genetics Soc Am (pres, 65); Environ Mutagen Soc. *Res:* Genetics of the mouse; genetic effects of radiation and chemicals. *Mailing Add:* Biol Div Oak Ridge Nat Lab PO Box 2009 Oak Ridge TN 37831-8077

RUSSELL, WILLIAM T(RELOAR), b Medford, Ore, Dec 21, 20; m 46; c 3. MECHANICAL & ELECTRICAL ENGINEERING. *Educ:* Univ Wash, BS, 42; Calif Inst Technol, MS, 47, PhD(eng), 50. *Prof Exp:* Asst gen mgr, Space Vehicles Div, 54-71, vpres & gen mgr, Defense Systs Div, 71-73, VPRES PROD ASSURANCE, TRW DEFENSE & SPACE SYSTS GROUP, REDONDO BEACH, 73- *Mem:* Am Inst Aeronaut & Astronaut; sr mem Inst Elec & Electronics Engrs. *Res:* Systems engineering; spacecraft development; aerospace general management; reliability and quality assurance. *Mailing Add:* 13065 Mindanao Way No 10 Marina del Rey CA 90291

RUSSEY, WILLIAM EDWARD, b Kalamazoo, Mich, Apr 5, 39; div; c 2. ORGANIC CHEMISTRY. *Educ:* Kalamazoo Col, AB, 61; Harvard Univ, AM, 64, PhD(chem), 67. *Prof Exp:* From asst prof to assoc prof, 66-75, chmn dept, 68-75, PROF CHEM, JUNIATA COL, 75- *Concurrent Pos:* NSF fac sci fel, Max Planck Inst Coal Res, Muelheim/Ruhr, WGer, 75-76, 80 & 81; Fulbright vis prof, Fachhochschule Muenster & Univ Marburg, W Germany, 82-83. *Mem:* Am Chem Soc; Royal Soc Chem. *Res:* Steroid biogenesis; olefin cyclization reactions; molecular rearrangements; fossil fuel chemistry. *Mailing Add:* Dept Chem Juniata Col Huntingdon PA 16652

RUSSFIELD, AGNES BURT, b Portland, Ore, Jan 9, 17; m 54. PATHOLOGY. *Educ:* Reed Col, BA, 35; Univ Calif, MA, 37; Univ Chicago, PhD(zool), 43; Cornell Univ, MD, 49. *Prof Exp:* Asst zool, Univ Calif, Los Angeles, 35-36; asst, Univ Chicago, 37-39 & 40-41, asst, Off Sci Res & Develop Proj, 42-43; intern, Mass Gen Hosp, 49-50, asst resident, 50-51, fel path, 51-54, asst, 54-57, asst pathologist, 57-58; assoc, Bio-Res Inst Inc, 59-61; res assoc, Children's Cancer Res Found, 61-67; resident, Mallory Inst Path, 67-68, asst pathologist, 68-71; res assoc, Bio-Res Inst, Inc, 72-73; staff pathologist, St Vincent Hosp, 74-75; SCIENTIST, MASON RES INST, 75- *Concurrent Pos:* From instr to assoc clin prof, Harvard Med Sch, 52-68; coop scientist, Worcester Found Exp Biol, 65-67 & 68-; assoc prof, Med Sch, Tufts Univ, 68-71 & Med Sch, Univ Mass, 75-78. *Mem:* Am Asn Cancer Res; Endocrine Soc; Am Asn Path; Am Soc Cytol; Soc Toxicol Path. *Res:* Endocrine changes in cancer; carcinogenesis. *Mailing Add:* 65 Commons Dr Apt 608 Shrewsbury MA 01545

RUSSI, GARY DEAN, b Canton, Ohio, Apr 23, 46; m 67; c 2. PHARMACOLOGY, TOXICOLOGY. *Educ:* Southwestern Okla State Univ, BS, 69; Kans Univ, PhD(pharmacol, toxicol), 72. *Prof Exp:* Asst pharmacol, Col Pharm, Kans Univ, 69-72; asst prof, 72-75, assoc prof, 75-80, PROF PHARMACOL, COL PHARM, DRAKE UNIV, 80- *Concurrent Pos:* Dir clin externship & dir comput-assisted instr, Col Pharm, Drake Univ, 75-, coordinator prog develop, 79- *Mem:* Am Asn Col Pharm; Sigma Xi. *Res:* Interaction of drugs and autonomic neurotransmitter release; mechanisms of drug-drug interactions; drug utilization review. *Mailing Add:* 1546 NW 90th St Des Moines IA 50322

RUSSI, SIMON, b Rowne, Poland, Jan 5, 11; nat US; m 37; c 2. ANATOMIC PATHOLOGY. *Educ:* Royal Univ Modena, Italy, MD, 35. *Prof Exp:* Asst path, Wash Univ, 42-46; from instr to asst prof clin path, Med Col Va, 46-50, from asst prof to assoc prof path, 50-68, clin prof path & clin path, 68-81; CLIN PROF PATH, GEORGE WASHINGTON UNIV MED CTR, 81- *Concurrent Pos:* Chief path sect, Vet Admin Hosp, Richmond, Va, 46-58; consult, US Army Hosp, Ft Lee, 58-81; dir labs, Petersburg Gen Hosp, 58-78 & John Randolph Hosp, Hopewell, Va, 78-81. *Mem:* Am Soc Clin Path; emer mem Am Asn Path & Bact; emer mem AMA; emer mem Col Am Pathologists; emer mem Int Acad Path. *Res:* Adrenal cortical adenomas; pulmonary hemosiderosis; intestinal lipodystrophy; chemistry of arteriosclerosis. *Mailing Add:* 101 Queen St Alexandria VA 22314

RUSSIN, NICHOLAS CHARLES, b Butler, Pa, Feb 6, 22; m 47; c 3. PHYSICAL CHEMISTRY. *Educ:* Washington & Jefferson Col, AB, 43; Carnegie Inst Technol, MS, 49, DSc, 50. *Prof Exp:* develop assoc, Tenn Eastman Co, 50-85; RETIRED. *Mem:* Am Chem Soc; fel Am Inst Chemists; Am Asn Textile Technol. *Res:* Thermodynamic properties of solutions; man-made fibers. *Mailing Add:* 312 McTeer Dr Kingsport TN 37663-2013

RUSSO, ANTHONY R, b Trenton, NJ, May 14, 37; m 61; c 3. REEF ECOLOGIST. *Educ:* US Naval Acad, BS, 58; Univ Wis, MS, 66; Univ Hawaii, MS, 78; Fla Inst Technol, PhD(oceanog), 86. *Prof Exp:* Nuclear engr, Gulf Gen Atomic Lab, 66-69; researcher, Univ Calif, San Diego, 68-70; ASST PROF OCEANOG, LEEWARD COL, UNIV HAWAII, 70- *Concurrent Pos:* Consult, Univ Hawaii, 72-; vis prof, Oxford Univ, Eng, 76-77, Univ Naples, 80 & Univ Thessaloniki, Greece, 89-90. *Mem:* Am Soc Naturalists; Nature Conservancy. *Res:* Structure and dynamics of coral reef invertebrate fauna; factors which regulate epifaunal invertebrate populations; effects of pollution stress on coral reef communities. *Mailing Add:* 1088 Bishop St Apt 2007 Honolulu HI 96813

RUSSO, DENNIS CHARLES, b Cleveland, Ohio, Feb 11, 50; m 85; c 1. BEHAVIORAL MEDICINE, PEDIATRIC PSYCHOLOGY. *Educ:* Univ Calif, Santa Barbara, BA, 72, PhD(educ psychol), 75. *Prof Exp:* Asst prof psychiat & pediat, Sch Med, Johns Hopkins Univ, 77-79; asst prof psychiat, 79-86, ASSOC PROF PSYCHIAT, HARVARD MED SCH, 86-; dir behav med, 79-89, SR ASSOC PSYCHOL, THE CHILDREN'S HOSP, BOSTON, 82- *Concurrent Pos:* Dir training & clin servs, behav pscyhol, John F Kennedy Inst, Baltimore, 75-79; distinguished vis prof, US Air Force Wilword Hall Med Ctr, 81-88. *Mem:* Fel Am Psychol Asn; fel Soc Behav Med; Asn Advan Behav Ther (secy-treas, 83-86, pres, 87-88); dipl, Am Bd Behav Psychol. *Res:* Psychological sequelae of chronic illness in children with particular reference to the relationship between physiology and behavior; development of psychological treatment for chronic illness management. *Mailing Add:* New Medico Assocs 100 Federal St 29th floor Boston MA 02110

RUSSO, EDWIN PRICE, b New Orleans, La, June 4, 38; m 61; c 4. MECHANICAL & ELECTRICAL ENGINEERING. *Educ:* Tulane Univ, BS, 60, MS, 62; La State Univ, PhD(mech eng), 68. *Prof Exp:* Elec engr, Chevron Oil Co, 62-63; res engr, Boeing Co, 63-66; instr eng mech, La State Univ, Baton Rouge, 66-68; assoc prof eng sci, La State Univ, New Orleans, 68-77; PROF MECH ENG, UNIV NEW ORLEANS, 77- *Mem:* Am Soc Mech Engrs. *Res:* stress analysis; fluid mechanics; heat transfer; oceanography. *Mailing Add:* Dept Mech Eng Univ New Orleans New Orleans LA 70148

RUSSO, EMANUEL JOSEPH, b Philadelphia, Pa, Jan 23, 34; m 65. PHARMACEUTICS. *Educ:* Philadelphia Col Pharm, BSc, 55; Temple Univ, MSc, 57; Univ Wis, PhD(pharm), 60. *Prof Exp:* Chemist, Phys Chem Dept, Schering Corp, NJ, 60-63; group leader pharmaceut develop, 63-78, MGR ORAL & TOPICAL PROD, WYETH LABS, 78- *Mem:* Am Pharmaceut Asn; Acad Pharmaceut Sci. *Res:* Study of physical and chemical properties of drugs, before and after their inclusion into a finished dosage form in order to formulate the most stable and efficacious product. *Mailing Add:* Wyeth Labs PO Box 8299 Philadelphia PA 19101

RUSSO, JOHN A, JR, b New York, NY, June 24, 33; m 59; c 7. METEOROLOGY. *Educ:* City Col New York, BS, 55; Univ Conn, MBA, 68. *Prof Exp:* Weather forecaster, US Weather Bur, 55-56; res asst statist & meteorol, Univ Ariz, 57-60; res assoc radar meteorol, Travelers Res Corp, Conn, 60-63, assoc scientist, 63-66, res scientist, 66-68, dir opers res, 68-71; MGR RES SUPPORT, OPERS RES DEPT, HARTFORD INS GROUP, 71- *Mem:* Am Meteorol Soc. *Res:* Meteorological statistics; design and development of information systems; weather sensitivity analysis; business administration. *Mailing Add:* 311 Cedarwood Lane Newington CT 06111

RUSSO, JOSE, b Mendoza, Argentina, Mar 24, 42; m 69; c 1. EXPERIMENTAL PATHOLOGY, PATHOLOGY. *Educ:* A Alvarez Col, Argentina, BA & BS, 59; Sch Med, Univ Cuyo, Argentina, Physician, 67, Med Dr, 68; Bd Cert Path, 82. *Prof Exp:* Instr path, Inst Gen Exp Path, Sch Med, Mendoza, Argentina, 61-66, instr histol & embryol, Inst Histol & Embryol, 66-69, chief instr embryol, 69-71; res fel, Inst Molecular Cell Evolution, Univ Miami, Fla, 71-73; sr res scientist, Mich Cancer Found, 73-74, chief, Exp Path Lab, 74-79, chmn, dept path, 79-91; CHMN, DEPT PATH, FOX CHASE CANCER CTR, 91- *Concurrent Pos:* Fel, Nat Coun Res, Argentina, 67-69 & 69-71; assoc mem, Mich Cancer Found, 70-; assoc clin prof path, Wayne State Univ, 79-; grants, Am Cancer Soc & Nat Cancer Inst. *Mem:* Foreign mem Soc Study Reproduction; Am Soc Cell Biol; Soc Exp Biol & Med; Electron Micros Soc Am; Am Asn Cancer Res; Col Am Path; Am Soc Clin Pathologists; Am Cancer Soc, Nat Cancer Inst; Am Col Pathologists, Int Acad Pathologists. *Res:* Study of the etiology and pathogenesis of human breast cancer, and pathogenesis of the same disease in experimental models; chemical carcinogenesis; oncogenes; cell transformation. *Mailing Add:* Dept Path Fox Chase Cancer Ctr 7701 Burholme Ave Philadelphia PA 19111

RUSSO, JOSEPH MARTIN, b Middletown, Conn, Feb 6, 49. AGRICULTURAL METEOROLOGY, SYSTEMS SCIENCE. *Educ:* St Louis Univ, BS, 71; McGill Univ, MSc, 74; Cornell Univ, PhD(agrometeorol), 78. *Prof Exp:* Res assoc, Agron Dept, Cornell Univ, 78-79, Dept Entom, NY State Agr Exp Sta, 79-81; ASST PROF, DEPT HORT, PA STATE UNIV, 81- *Concurrent Pos:* Res consult, 78-79; vis fel, Agron Dept, Cornell Univ, 79. *Mem:* Am Meteorol Soc; Am Soc Agron; Soil Sci Soc Am; Am Soc Hort Sci; Int Soc Biometeorol. *Res:* Theory and experimental designs for agricultural production systems. *Mailing Add:* Dept Hort Pa State Univ University Park PA 16802

RUSSO, MICHAEL EUGENE, b St Louis, Mo, Aug 5, 39. CHEMISTRY. *Educ:* Wash Univ, AB, 61, PhD(chem), 70. *Prof Exp:* Investr chem res & develop, Mallinckrodt Chem Works, 67-74, RES MGR, MALLINCKRODT, INC, 74- *Concurrent Pos:* Vpres & dir, Parkside Develop Corp; dir, Montecello Corp. *Res:* Physical and inorganic chemistry; molecular spectroscopy. *Mailing Add:* Ten Kingsbury Pl St Louis MO 63112

RUSSO, RALPH P, b New York, NY, Dec, 29, 52; m 76; c 1. MATHEMATICAL STATISTICS. *Educ:* State Univ NY, Binghamton, MA & PhD(math), 80. *Prof Exp:* Asst prof Statist, State Univ NY, Buffalo, 80-; AT DEPT STATIST & MATH, UNIV IOWA. *Mem:* Am Statist Asn; Inst Math Statist. *Mailing Add:* Dept Statist Univ Iowa Iowa City IA 52242

RUSSO, RAYMOND JOSEPH, b St Louis, Mo, May 30, 44; m 70; c 2. ENTOMOLOGY, ECOLOGY. *Educ:* Southeast Mo State Univ, BS, 66; Northeast Mo State Univ, MA, 71; Univ Notre Dame, PhD(entom), 76. *Prof Exp:* ASSOC PROF ECOL, IND UNIV-PURDUE UNIV, INDIANAPOLIS, 76- *Concurrent Pos:* Consult, Ind State Bd Health, 78. *Mem:* Entom Soc Am; Soc Neurosci; Sigma Xi. *Res:* Insect neurobiology; computer applications to biology; medical entomology. *Mailing Add:* Dept Biol Ind Univ-Purdue Univ 1125 E 38th St Indianapolis IN 46205

RUSSO, RICHARD F, b Somerville, Mass, Apr 22, 27; m 56; c 7. OPERATIONS RESEARCH. *Educ:* Mass Maritime Acad, BS, 47; Boston Col, AB, 51, AM, 53. *Prof Exp:* Develop physicist, Am Optical Co, 52-55; staff mem, Lincoln Lab, Mass Inst Technol, 55-57; sr engr, Shipbldg Div, Bethlehem Steel Co, 57-64; mathematician, Dewey & Almy Chem Div, W R Grace & Co, 64-66; mgr corp mgt sci, Itek Corp, 66-69, sr analyst innovative software, 70-71; sr systs analyst, New Eng Life Inst Co, Mass, 71-73; systs planning specialist, United Illum Co, 73-75; pvt pract, 75-77; sr bus analyst, Honeywell Info Systs, 77-79, tech support adminr, 79-83; PROJ MGR, BLUE CROSS MASS, 83- *Mem:* Opers Res Soc Am. *Res:* Application of advanced mathematical techniques and computer technology to the solution of business and industrial problems. *Mailing Add:* 170 Crosby St Arlington MA 02174

RUSSO, ROY LAWRENCE, b Kelayres, Pa, Nov 6, 35; m 59; c 4. ELECTRICAL ENGINEERING. *Educ:* Pa State Univ, BS, 57, MS, 59, PhD(elec eng), 64. *Prof Exp:* From instr to asst prof elec eng, Pa State Univ, 59-65; res staff mem, IBM Corp, 65-68, 83-85, mgr design automation, 68-78, sr engr, 78-81, mgr design automation strategy, 81-82, MGR DESIGN AUTOMATION LAB, IBM CORP, 85- *Concurrent Pos:* Consult prof elec eng, Stanford Univ, 82-83. *Honors & Awards:* Leonard A Doggett Award, 66; Int Bus Mach Corp Outstanding Contrib Award, 68, Invention Achievement Award, 78; Centennial Medal, Inst Elec & Electronic Engrs, 84, Distinguished Serv Award, Computer Soc, 85, 90. *Mem:* Fel Inst Elec & Electronics Engrs Computer Soc (pres, 86-87). *Res:* Logic design of computers; computer reliability; design automation. *Mailing Add:* 1793 Blossom Ct Yorktown Heights NY 10598

RUSSO, SALVATORE FRANKLIN, b Hartford, Conn, Feb 6, 38; m 67; c 2. PHYSICAL BIOCHEMISTRY. *Educ:* Wesleyan Univ, BA, 60; Northwestern Univ, PhD(phys chem), 64. *Prof Exp:* Instr, Northwestern Univ, 63-64; res assoc phys biochem, Univ Wash, 64-67; from asst prof to assoc prof, 68-83, PROF CHEM, WESTERN WASH UNIV, 83- *Concurrent Pos:* NIH fel, 65-67; adjoint prof, Univ Colo, 77-78, vis prof, 84-85; vis fac, Wash State Univ, 80; vis scientist, COBE Labs, Inc, 81. *Mem:* Am Chem Soc; Sigma Xi; Am Asn Univ Prof; Protein Soc. *Res:* Conformational changes in proteins; fluorescent probes of protein structure; hydrogen bonds between model peptide groups in solution; blood plasma exchange; biochemistry of blood pressure regulation; site specific mutagenesis in cro protein. *Mailing Add:* Dept Chem Western Wash Univ Bellingham WA 98225

RUSSO, THOMAS JOSEPH, b Brooklyn, NY, Feb 10, 36; m 88; c 4. ORGANIC CHEMISTRY. *Educ:* Polytech Inst Brooklyn, BSc, 57; Pa State Univ, PhD(org chem), 65. *Prof Exp:* Instr chem, Bucknell Univ, 62 & Juniata Col, 62-64; from instr to asst prof, 64-71, ASSOC PROF CHEM, PA STATE UNIV, ALTOONA, 71- *Mem:* Am Chem Soc. *Res:* Physical organic chemistry. *Mailing Add:* Dept Chem Pa State Univ Altoona PA 16603

RUSSOCK, HOWARD ISRAEL, b Philadelphia, Pa, Dec 22, 47. ETHOLOGY. *Educ:* Western Md Col, BA, 69; Pa State Univ, MS, 71; WVa Univ, PhD(biol), 75. *Prof Exp:* Eli Lilly Endowment Inc internship, Purdue Univ, 75-76; asst prof, 76-81, assoc prof, 81-87, PROF BIOL, WESTERN CONN STATE UNIV, 88- *Concurrent Pos:* Investr, NSF fac fel awards, Rockefeller Univ, 79-81; vis fac fel, Yale Univ, 84-85; Conn St Univ-Am Asn Univ Professors res grants, 86, 87-88, 88-89, 90-91. *Mem:* AAAS; Am Inst Biol Sci; Am Soc Zoologists; Animal Behav Soc; Sigma Xi. *Res:* Developmental and social behavior; effects and importance of early experience in birds and cichlid fish. *Mailing Add:* Dept Biol Sci Western Conn State Univ Danbury CT 06810

RUSSU, IRINA MARIA, b Bucharest, Romania, Aug 15, 48. BIOPHYSICS. *Educ:* Univ Bucharest, MS, 70; Univ Pittsburgh, PhD(biophys), 79. *Prof Exp:* Res assoc, Univ Pittsburgh, 79-80; res asst prof, Carnegie-Mellon Univ, 80-86; ASST PROF, WESLEYAN UNIV, 86- *Mem:* Am Biophys Soc. *Res:* Structure-function relationships in proteins and nucleic acids by means of nuclear magnetic resonance spectroscopy. *Mailing Add:* Dept Molecular Biol & Biochem Wesleyan Univ Middletown CT 06457

RUST, CHARLES CHAPIN, b Medford, Wis, June 13, 35; m 56; c 3. ZOOLOGY. *Educ:* Wis State Univ, BS, 60; Univ Wis-Madison, MS, 61, PhD(zool), 64. *Prof Exp:* From asst prof to prof zool, Univ Wis Ctr Syst, 64-90; RETIRED. *Concurrent Pos:* Vis scientist, Regional Primate Ctr, 68-69. *Honors & Awards:* Am Soc Mammalogists Annual Award, 64. *Mem:* AAAS; Am Soc Zoologists; Am Soc Mammalogists. *Res:* Endocrinology of mammalian reproductive and pelage cycles. *Mailing Add:* 3501 Hackdarth Rd Janeville WI 53545

RUST, DAVID MAURICE, b Denver, Colo, Dec 9, 39; m 63; c 2. ASTROPHYSICS. *Educ:* Brown Univ, ScB, 62; Univ Colo, PhD(astrophys), 66. *Prof Exp:* Res fel solar physics, Nat Ctr Atmospheric Res, 63-66; Carnegie fel astrophys, Mt Wilson & Palomar Observ, 66-68; astrophysicist, Sacramento Peak Observ, Air Force Cambridge Res Labs, 68-74; sr staff scientist, Am Sci & Eng, Inc, 74-83; PHYSICIST, PRIN PROF STAFF, APPL PHYS LAB, JOHNS HOPKINS UNIV, 83- *Concurrent Pos:* Consult, Lockheed Calif, 66-70; vis assoc prof, Univ Md, 71; chmn, Working Group Solar Maximum Year, Int Astron Union, 73; vis astronr, Observ of Paris, 78; consult, NASA, 78-; assoc ed, Geophys Res Lett, 90- *Mem:* Am Geophys Union; Am Astron Soc; Int Astron Union. *Res:* Solar physics and magnetic fields; astronomical instrumentation; solar flares and x-ray emission; satellite-borne telescopes. *Mailing Add:* Applied Phys Lab Johns Hopkins Univ Johns Hopkins Rd Laurel MD 20723-6099

RUST, JAMES HAROLD, b Peoria, Ill, Sept 19, 36. NUCLEAR ENGINEERING. *Educ:* Purdue Univ, BS, 58, PhD(nuclear eng), 65; Mass Inst Technol, SM, 60. *Prof Exp:* Asst prof nuclear eng, Univ Va, 64-67; assoc prof, 67-76, PROF NUCLEAR ENG, GA INST TECHNOL, 76- *Mem:* Am Soc Eng Educ; Am Nuclear Soc; Am Soc Mech Engrs; Nat Soc Prof Engrs. *Res:* Heat transfer and fluid mechanics pertinent to nuclear reactors, especially liquid metals and two phase systems. *Mailing Add:* 340 Garden Lane NW Atlanta GA 30309

RUST, JOSEPH WILLIAM, b Butler, Ky, Oct 1, 25. ANIMAL HUSBANDRY, ANIMAL NUTRITION. *Educ:* Univ Ky, BS, 53, MS, 57; Iowa State Univ, PhD(animal nutrit), 63. *Prof Exp:* Instr dairying, Univ Ky, 53-59; res assoc dairy sci, Iowa State Univ, 59-63; asst prof dairy husb, 63-64, from asst prof to assoc prof animal husb, 64-80, PROF ANIMAL HUSB & SUPT, NCENT EXP STA, UNIV MINN, 80- *Mem:* Am Dairy Sci Asn; Am Soc Animal Sci. *Mailing Add:* N Cent Exp Sta Univ Minn 1861 Hwy 169 E Grand Rapids MN 55744

RUST, LAWRENCE WAYNE, JR, b St Paul, Minn, Mar 25, 37; m 64; c 3. MATHEMATICS, STATISTICS. *Educ:* Univ Minn, BS, 59, MS, 61, PhD(aeronaut eng), 64. *Prof Exp:* Sr engr, Ventura Div, Northrop Corp, 64-65 & Appl Sci Div, Litton Industs, 65-66; res engr, NStar Res & Develop Inst, 66-76; pres, Appl Anal Serv, 76-83; DP MGR, FILMTEC CORP, 84- *Res:* Mathematical modeling; computer simulation; statistical data analysis; heat and mass transfer; fluid mechanics. *Mailing Add:* 1826 N Alameda St St Paul MN 55113

RUST, MICHAEL KEITH, b Akron, Ohio, Aug 26, 48; m 70; c 2. ENTOMOLOGY. *Educ:* Hiram Col, BA, 70; Univ Kans, MA, 73, PhD(entom), 75. *Prof Exp:* PROF ENTOM, UNIV CALIF, RIVERSIDE, 75- *Concurrent Pos:* Pres, Pac Br, Entom Soc Am, 85-86. *Mem:* Entom Soc Am; Animal Behav Soc; Sigma Xi. *Res:* Urban entomology, especially biology and control of cockroaches, fleas and termites; chemical factors influencing termite feeding. *Mailing Add:* Dept Entom Univ Calif Riverside CA 92521

RUST, PHILIP FREDERICK, b Oakland, Calif, 1947; m. BIOSTATISTICS, STATISTICS. *Educ:* Calif Inst Technol, BS, 69; Univ Calif, Berkeley, MA, 71, PhD(biostatist), 76. *Prof Exp:* Lieutenant biostatist, Ctr Dis Control, USPHS, 71-73; asst prof statist, Univ Mo-Columbia, 76-79; asst prof, 79-83, ASSOC PROF BIOMETRY, MED UNIV SC, 83- *Concurrent Pos:* Consult, Med Ctr, Univ Mo, 76-78, Med Univ SC, 79-; Environ Protection Agency; Rippel Found grant, 82-84. *Mem:* Am Statist Asn; Biomet Soc; Sigma Xi; Soc Epidemiol Res. *Res:* Applied stochastic processes; biometry; epidemiology; statistical distributions. *Mailing Add:* Dept Biometry Med Univ SC Charleston SC 29425-2503

RUST, RICHARD HENRY, b Bunker Hill, Ill, Oct 12, 21; m 42; c 5. SOIL SCIENCE. *Educ:* Univ Ill, BS, 47, MS, 50, PhD(agron), 55. *Prof Exp:* Asst prof soil physics, Univ Ill, 55-56; asst prof, 56-71, PROF SOILS, UNIV MINN, ST PAUL, 71- *Concurrent Pos:* Consult to indust & govt. *Mem:* Fel Am Soc Agron; Soil Conserv Soc Am. *Res:* Soil genesis, classification and physical chemistry; properties related to clay mineralogy of soils; soil productivity evaluation. *Mailing Add:* 1922 Autumn St St Paul MN 55113

RUST, RICHARD W, b Logan, Utah, Oct 11, 42. INSECT ECOLOGY. *Educ:* Utah State Univ, BS, 65, MS, 67; Univ Calif, PhD(entom), 72. *Prof Exp:* Res asst pollination, Wild Bee Pollination Lab, USDA, 65-67; field biologist parasites, Field Mus Nat Hist, Chicago, 69; teaching asst zool, Univ Calif, 72; fac mem entom, Univ Del, 73-78; FAC MEM BIOL, UNIV NEV, 78- *Mem:* Entom Soc Am; Am Entom Soc (rec secy, 74-); Soc Study Evolution; Soc Syst Zoologists; Ecol Soc Am. *Res:* Pollination ecology. *Mailing Add:* Dept Biol Univ Nev Reno NV 89557

RUST, STEVEN RONALD, b US citizen. EXTENSION, FEEDLOT MANAGEMENT. *Educ:* Univ Wis-River Falls, BS, 77; Okla State Univ, MS, 78, PhD(animal nutrit), 83. *Prof Exp:* Asst prof animal nutrit, Mont State Univ, 82-84; ASSOC PROF ANIMAL NUTRIT, MICH STATE UNIV, 85- *Mem:* Am Soc Animal Sci; Animal Nutrit Res Coun; Am Registry Prof Animal Scientists. *Res:* Ruminant nutrition; efficient utilization of high grain diets in beef cattle; forage utilization. *Mailing Add:* 113 Anthony Hall Mich State Univ East Lansing MI 48824

RUST, VELMA IRENE, b Edmonton, Alta, May 22, 14; m 55. MATHEMATICAL STATISTICS, ECONOMETRICS. *Educ:* Univ Alta, BSc, 34, MEd, 44, BEd, 47; Univ Ill, PhD(math, math educ), 59. *Prof Exp:* Teacher math & sci, Alta, 36-44; admin secy fac educ, Univ Alta, 44-52, from lectr to asst prof math educ & dir student teaching prog, 52-56; researcher personnel planning, Royal Can Air Force Hq, Can Dept Nat Defence, Ont, 60-62, chief staff training, Inspection Serv, 62-65; statistician, Origin & Destination Statist, Govt Can, 65-67; sr policy analyst, Can Dept Health & Welfare, 67-79; RETIRED. *Concurrent Pos:* Asst, Univ Ill, 56-59; sessional lectr, Carleton Univ, 59-62. *Res:* Social security; aviation; education. *Mailing Add:* 811 Adams Ave Ottawa ON K1G 2Y1 Can

RUST, WALTER DAVID, b Randolph AFB, Tex, Oct 8, 44; m 66; c 2. ATMOSPHERIC PHYSICS. *Educ:* Southwestern Univ, BS, 66; NMex Inst Mining & Technol, MS, 69, PhD(physics), 73. *Prof Exp:* US Nat Res Coun res assoc, 73-75, ATMOSPHERIC PHYSICIST, US DEPT COM, NAT OCEANIC & ATMOSPHERIC ADMIN, 75- *Mem:* Am Geophys Union; Am Meteorol Soc; Royal Meteorol Soc Eng; Sigma Xi. *Res:* Cloud electrification; lightning suppression by chaff dispersal within thunderstorms; distribution of electric fields; use of atmospheric electric measurements to assess possibility of triggered lightning by rockets launched near thunderstorms; remote detection of corona; severe storm electricity. *Mailing Add:* Nat Severe Storm Lab 1313 Halley Dr Norman OK 73069

RUSTAD, DOUGLAS SCOTT, b Juneau, Alaska, Sept 25, 40; m 67. INORGANIC CHEMISTRY, PHYSICAL CHEMISTRY. *Educ:* Univ Wash, BS, 62, MS, 64; Univ Calif, Berkeley, PhD(chem), 67. *Prof Exp:* Lectr inorg chem, Univ West Indies, 67-69; from asst prof to assoc prof, 69-79, PROF INORG CHEM, SONOMA STATE UNIV, 79- *Mem:* Am Chem Soc. *Res:* High temperature spectroscopic and thermodynamic studies of transition metal halides. *Mailing Add:* Dept Chem Sonoma State Univ Rohnert Park CA 94928

RUSTAGI, JAGDISH S, b Sikri, India, Aug 13, 23; m 49; c 3. BIOSTATISTICS. *Educ:* Univ Delhi, BA, 44, MA, 46; Stanford Univ, PhD(statist), 56. *Prof Exp:* Lectr math, Hindu Col, Univ Delhi, 46-52; asst prof, Carnegie Inst Technol, 55-57; asst prof statist, Mich State Univ, 57-58; reader, Aligarh Muslim Univ, 58-60; assoc prof math, Univ Cincinnati, 60-63; from assoc prof to prof math, Ohio State Univ, 63-70, prof statist, 70-88, chmn dept, 79-88, EMER PROF & CHMN STATIST, OHIO STATE UNIV, 88- *Concurrent Pos:* NIH res grants, 62-68; consult, Toxic Hazards Unit, Wright-Patterson AFB, Ohio, 64-65; Air Force Off Sci Res grants, 67-; vis scholar, Stanford Univ, 71-72 & 83 & Off Naval Res, 78-; adj prof prev med, Col Med, Ohio State Univ, 67-75; vis prof, Univ Philippines, 88-89; vis scientist, IBM Corp, San Jose, Calif, 90-91. *Mem:* Fel Inst Math Statist; Biomet Soc; fel Am Statist Asn; fel Indian Soc Med Statist; Indian Soc Agr Statist; Int Statist Inst. *Res:* Mathematical statistics; medical statistics; operations research; optimization techniques in statistics. *Mailing Add:* Dept Statist Ohio State Univ Columbus OH 43210-1247

RUSTAGI, KRISHNA PRASAD, b Khurja, India, Jan 1, 32; m 56; c 2. FOREST BIOMETRICS, LAND USE PLANNING. *Educ:* Agra Univ, India, BSc, 51, MSc, 53; Yale Univ, PhD(forestry), 73. *Prof Exp:* Asst conservator, Forest Dept, India, 55-61, dep conservator, 61-64; lectr, Forest Res Inst & Col, 64-69; asst prof, 73-79, ASSOC PROF FORESTRY, UNIV WASH, 79- *Mem:* Soc Am Foresters. *Res:* Biometric investigations in growth and yield of forest stands and trees; operations research application in land use; forest management planning. *Mailing Add:* Dept Forest Resources Univ Wash Seattle WA 98195

RUSTED, IAN EDWIN L H, b Nfld, July 12, 21; m 49; c 2. ENDOCRINOLOGY, HEALTH SCIENCES EDUCATION. *Educ:* Univ Toronto, BA, 43; Dalhousie Univ, MD, CM, 48; McGill Univ, MSc, 49; FRCP(C), 53. *Hon Degrees:* LLd, Dalhousie Univ; LLd, Mt Allison Univ; DSL, Univ Toronto. *Prof Exp:* Med consult, Dept Health, Nfld, 52-67; actg dir postgrad & continuing med educ, Mem Univ Nfld, 66-67, coodr med sch planning, 66-67, dean med, 67-74, vpres health sci, 74-79, pro vice chancellor & vpres, Health Sci & Prof Sch, 81-88, prof med, 67-89, EMER DEAN, MEM UNIV NFLD, 89- *Concurrent Pos:* Physician & dir med educ, St John's Gen Hosp, 53-67, chmn dept med, 67-68; mem coun, Royal Col Physicians & Surgeons Can, 61-70, vpres, 68-70; vis prof, Univ Toronto & Laval Univ, 74-75; vchmn, Am Col Physicians, Bd Regents, 85-86; chmn, Int Med Activ subcomt, 85-88; officer, Order of Can, 85; mem, Nat Coun Bioethics Human Res, 88- *Mem:* Col Family Physicians Can; Asn Can Med Col (vpres, 73-74); fel Am Col Physicians. *Res:* Thyroid disorders; hypertension, especially epidemiology. *Mailing Add:* Health Sci Ctr Mem Univ St John's NF A1B 3V6 Can

RUSTGI, MOTI LAL, b Delhi, India, Sept 29, 29; m 52; c 2. NUCLEAR & MEDICAL PHYSICS, ATOMIC & SOLID STATE PHYSICS. *Educ:* Univ Delhi, India, BSc, 49, MSc, 51; La State Univ, PhD, 57. *Prof Exp:* Instr physics, La State Univ, 57; res assoc, Yale Univ, 57-60, asst prof, 64-66; fel, Nat Res Coun Can, 60-61; reader, Banaras Hindu Univ, 61-63; asst prof, Univ Southern Calif, 63-64; assoc prof, 66-68, PROF PHYSICS, STATE UNIV NY, BUFFALO, 68- *Concurrent Pos:* Res Assoc, Harvard Univ, 60-61; vis prof, State Univ NY Stony Brook, 73; vis scientist, Oak Ridge Nat Lab, 80-; prin investr, NASA Res Grants, 84- *Mem:* Fel Am Phys Soc; Am Inst Physics; Health Physics Soc Am. *Res:* Electromagnetic interactions with nuclei; nuclear structure; energy losses of high energy particles in matter; change particle interaction with tissues; quantum wells. *Mailing Add:* Dept Physics State Univ NY Buffalo NY 14260

RUSTGI, OM PRAKASH, b Delhi, India, Aug 1, 31; US citizen; m 63; c 2. SPECTROSCOPY, SOLID STATE PHYSICS. *Educ:* Univ Delhi, BS, 52, MS, 54; Univ Southern Calif, PhD(physics), 60. *Prof Exp:* Head optics div, Proj Celescope, Smithsonian Astrophys Observ, 60-62; physicist ultraviolet radiation, Northrop Corp, 63-67; mem prof staff space res, TRW Systs Inc, 67-71; vis assoc prof physics, Univ Ill, 71-73; from asst prof to assoc prof, 73-85, PROF PHYSICS, STATE UNIV NY COL BUFFALO, 85- *Concurrent Pos:* Consult, Smithsonian Astrophys Observ, 63-64 & Lawrence Livermore Lab, 74-76; prog mgr, Northrop Corp, 65-67; vis assoc prof physics, Univ Nebr, 81-82. *Mem:* Am Phys Soc; Optical Soc Am; Sigma Xi; Am Asn Physics Teachers. *Res:* Optical properties of thin films; photoionization in gases and light elements; lasers and holography; x-rays and crystal structure; laser plasma interaction studies. *Mailing Add:* 362 Sunrise Blvd Williamsville NY 14221

RUSTIGIAN, ROBERT, b Boston, Mass, July 26, 15; m 56; c 2. MEDICAL MICROBIOLOGY. *Educ:* Univ Mass, BS, 38; Brown Univ, MS, 40, PhD, 43. *Prof Exp:* Nat Res Coun fel bact & immunol, Harvard Med Sch, 46-48, instr, 48-49, assoc instr, 49; asst prof microbiol, Univ Chicago, 49-55; asst prof, Sch Med, Tufts Univ, 55-61, assoc prof, 61-67; chief virologist, Vet Admin Hosp, 67-86; assoc prof microbiol & moelcular genetics, Harvard Med Sch, 75-86; RETIRED. *Mem:* Am Soc Microbiol; Am Asn Immunol. *Res:* Biological and biochemical studies of virus-host interactions in chronic viral infections of neural and non-neural cells; virology and immunology. *Mailing Add:* Gilmore Rd North Easton MA 02356

RUSTIONI, ALDO, b Porto Ceresio, Italy, July 22, 41; m 69; c 2. NEUROBIOLOGY, ANATOMY. *Educ:* Univ Parma, MD, 65. *Prof Exp:* Asst prof neuroanat, Erasmus Univ, Rotterdam, Holland, 68-72, sr asst prof, 72-73; ASSOC PROF ANAT & PHYSIOL, UNIV NC, CHAPEL HILL, 73- *Concurrent Pos:* Vis investr, Nat Inst Health & Med Res, Paris, 72-73; Europ Training Prog Brain & Behav fel, 72-73; grants, Dutch Orgn Fundamental Res Med, 72-73, USPHS, 75-81 & 80-85 & Nat Found March Dimes, 78-80. *Mem:* Europ Neurosci Asn; Am Asn Anatomists; Soc Neurosci; AAAS; Am Acad Neurobiol. *Res:* Neuroanatomy; neurocytology; neurohistochemistry; electron microscopy; neurophysiology; somatosensory system. *Mailing Add:* Dept Anat & Physiol Univ NC Chapel Hill NC 27514

RUSTON, HENRY, b Lodz, Poland, July 23, 29; US citizen; m 59; c 3. SOFTWARE & ELECTRICAL ENGINEERING. *Educ:* Univ Mich, BSE(math) & BSE(elec eng), 52, PhD(elec eng), 60; Columbia Univ, MS, 55. *Prof Exp:* Intermediate test engr, Wright Aeronaut Div, Curtiss-Wright Corp, NJ, 55; elec engr, Reeves Instrument Co, NY, 55-56; res asst, Eng Res Inst, Univ Mich, 56-58, res assoc, 58-60, assoc res engr, 60; asst prof elec eng, Moore Sch Elec Eng, Univ Pa, 60-64; assoc prof, 64-86, PROF, POLYTECH UNIV, 86- *Concurrent Pos:* Lectr, Univ Mich, 60; indust consult, 61- *Mem:* Sr mem Inst Elec & Electronics Engrs; Asn Comput Mach. *Res:* Circuit theory; computer simulation; software engineering; programming languages. *Mailing Add:* Dept Comp Sci Polytech Univ 333 Jay St Brooklyn NY 11201

RUSY, BEN F, b Sturgeon Bay, Wis, July 12, 27; m 57; c 4. PHARMACOLOGY, ANESTHESIOLOGY. *Educ:* Univ Wis, BS, 52, MD, 56; Temple Univ, MS, 59; Am Bd Anesthesiol, dipl, 62. *Prof Exp:* Asst instr anesthesiol, Med Sch, Temple Univ, 59-60, instr anesthesiol & pharmacol, 60-62, from asst prof to prof anesthesiol & pharmacol, 62-77; PROF, DEPT ANESTHESIOL, UNIV HOSP, UNIV WIS-MADISON, 77- *Concurrent Pos:* NIH fel, 59-61; NIH res grant, 62-, 75-78, 82-85; mem coun basic sci & coun circulation, Am Heart Asn, 65-; USPHS spec res fel, Heart & Lung Inst, 71-72; hon res fel, Dept Physiol, Univ Col, Univ London, 71-72. *Mem:* AAAS; Am Soc Anesthesiol; Am Soc Pharmacol & Exp Therapeut. *Res:* Effect of anesthetics and antihypertensive drugs on cardiac function. *Mailing Add:* Dept Anesthesiol Univ Hosp B6/387 Madison WI 53792

RUTAN, ELBERT L, b June, 1943. RESEARCH ADMINISTRATION. *Educ:* Calif Polytech Univ, BS, 65, DSc, 87; Daniel Webster Col, Dr, 87; Lewis Univ, Dr, 88; Delft Univ Technol, Dr, 90. *Prof Exp:* Flight test proj engr, Air Force Flight Test Ctr, Edwards AFB, 65-72; dir, Bede Test Ctr, Bede Aircraft, Newton, Kans, 72-74; PRES, RUTAN AIRCRAFT FACTORY, INC, 74-; PRES & CHIEF EXEC OFFICER, SCALED COMPOSITES, INC, 82- *Honors & Awards:* Air Medal, 70; Stan Dzik Design Contrib Trophy, 72; Omni Aviation Safety Trophy, 73; Outstanding New Design, Exp Aircraft Asn, 75, 76 & 78; Aircraft Design Cert of Merit, Am Inst Aeronaut & Astronaut, 86; Presidential Citizen's Medal, 86; Award for Unique & Useful Plastic Prod, Soc Plastics Engrs, 87; W Randolph Lovelace Award, Soc NASA Flight Surgeons, 87; Distinguished Serv Award, Acad Model Aeronaut, 87; J H Doolittle Award, Soc Exp Test Pilots, 87; Collier Trophy, Nat Aeronaut Asn & Nat Aviation Club, 87; Meritorious Serv Award, Nat Bus Aircraft Asn, 87; Spirit St Louis Medal, Am Soc Mech Engrs, 87; Brit Gold Medal Aeronauts Royal Aeronaut Soc, 87; Outstanding Eng Achievement Award, Nat Soc Prof Engrs, 88. *Mem:* Nat Acad Eng; Exp Aircraft Asn; Soc Flight Test Engrs; Acad Model Aeronaut; Soc Exp Test Pilots; Aircraft Owners & Pilots Asn. *Res:* Granted 3 patents. *Mailing Add:* Scaled Composites Inc Hangar 78 Airport Mojave CA 93501

RUTENBERG, AARON CHARLES, b Chicago, Ill, July 28, 23; m 64; c 2. PHYSICAL CHEMISTRY. *Educ:* Univ Chicago, PhD(chem), 50. *Prof Exp:* Jr inspector, US War Dept, 42; res assoc, Univ Chicago, 50-51; chemist, Oak Ridge Nat Lab, 51-71; DEVELOP CHEMIST, MARTIN MARIETTA ENERGY SYSTS, Y-12 PLANT, 72- *Mem:* Am Chem Soc. *Res:* Nuclear magnetic resonance spectroscopy. *Mailing Add:* 101 Monticello Rd Oak Ridge TN 37830

RUTENBERG, MORTON WOLF, b Philadelphia, Pa, Jan 10, 21; m 42; c 1. STARCH CHEMISTRY, CARBOHYDRATE CHEMISTRY. *Educ:* Univ Pa, BS, 42, MS, 47, PhD(org chem), 49. *Prof Exp:* Jr chemist, Eastern Regional Res Lab, USDA, 42-43; res group supvr, Nat Starch Prod, Inc, 49-58, res sect leader, 58-75, assoc dir, 75-80, dir, starch res, 80-84, vpres natural polymer res, 84-86, div vpres res & develop, Food Prod & Indust Starch Divisions, 87-90, CONSULT, STARCH CHEM & TECHNOL, NAT STARCH & CHEM CORP, 90- *Honors & Awards:* Melville L Wolfrom Award, Div Carbohydrate Chem, Am Chem Soc, 87. *Mem:* AAAS; Am Chem Soc; fel Am Inst Chem; Royal Soc Chem; Am Asn Cereal Chemists; Sigma Xi. *Res:* Synthetic organic, carbohydrate and polysaccharide chemistry; starch chemistry and technology; polysaccharides. *Mailing Add:* 587 Rockview Ave North Plainfield NJ 07063-1850

RUTFORD, ROBERT HOXIE, b Duluth, Minn, Jan 26, 33; m 54; c 3. GEOLOGY. *Educ:* Univ Minn, BA, 54, MA, 63, PhD(geol), 69. *Prof Exp:* Res assoc geol, Univ Minn, 62-67; from asst prof to assoc prof, Univ SDak, 67-72, actg chmn dept geol, 68-69, chmn, 69-71, chmn,dept geol & physics, 71-72; co-dir, Ross Ice Shelf Proj, Univ Nebr-Lincoln, 72-73, dir, 73-75; head, Off Polar Progs, NSF, 75-77; interim chancellor, Univ Nebr-Lincoln, 80-81, vchancellor res & grad studies, 77-82; PRES, UNIV TEX, DALLAS, 82-

Concurrent Pos: NSF res grant, 68-69; mem,panel geol & geophys, comt polar res, Nat Acad Sci, 68-73, mem, Ross Ice Shelf Proj steering group, 72-73, antarctic adv panel, Deep Earth Sampling Proj, Joint Oceanog Inst; chmn, Interagency Arctic Res Coord Comt, 75-77; mem, comt int rels, Nat Res Coun, Nat Acad Sci, 77-81 & Polar Res Bd, 80-; mem, Antartic Sect, Ocean Affairs Adv Comt, Dept State, 78- *Honors & Awards:* Antarctic Serv Medal, US Secy Defense, 68. *Mem:* Fel Geol Soc Am; Arctic Inst NAm; Sigma Xi; Nat Coun Univ Res Admin. *Res:* Antarctic geology; geomorphology; glacial geology in eastern South Dakota. *Mailing Add:* 6809 Briar Cove Dr Dallas TX 75240

RUTGER, JOHN NEIL, b Noble, Ill, Mar 3, 34; m 58; c 2. GENETICS, PLANT BREEDING. *Educ:* Univ Ill, BS, 60; Univ Calif, Davis, MS, 62, PhD(genetics), 64. *Prof Exp:* Asst prof plant breeding, Cornell Univ, 64-70, assoc prof, 70; res geneticist, Dept Agron & Range Sci, Univ Calif, Davis, 70-88, ASSOC AREA DIR, MIDSOUTH AREA, AGR RES SERV, USDA, 89- *Mem:* AAAS; fel Am Soc Agron; fel Crop Sci Soc Am. *Res:* Rice genetics and breeding; inheritance of semi-dwarfism, male sterility and other hybrid rice mechanisms; cold tolerance. *Mailing Add:* USDA Agr Res Serv PO Box 225 Stoneyville MS 38776

RUTGERS, JAY G, b Holland, Mich, Feb 16, 24; m 55; c 1. ANALYTICAL CHEMISTRY. *Educ:* Hope Col, BA, 49; Northwestern Univ, PhD(anal chem), 55. *Prof Exp:* Res assoc, Merck Sharp & Dohme Res Labs, 54-61; RES SCIENTIST, WYETH LABS, INC, 61- *Mem:* Am Chem Soc. *Res:* Analytical method development for pharmaceuticals; purity testing of new drugs. *Mailing Add:* 232 Upper Valley North Wales PA 19454-2445

RUTH, BYRON E, b Chicago, Ill, Mar 25, 31; m 60; c 2. CIVIL ENGINEERING, TRANSPORTATION. *Educ:* Mont State Univ, BSCE, 55; Purdue Univ, MSCE, 59; WVa Univ, PhD(civil eng), 67. *Prof Exp:* Asst dir res & develop, Symons Mfg Co, 60-61; from instr to asst prof civil eng, WVa Univ, 61-70; assoc prof, 70-77, PROF CIVIL ENG, UNIV FLA, 77- *Concurrent Pos:* Mem, Hwy Res Bd, Nat Acad Sci-Nat Res Coun. *Mem:* Am Soc Civil Engrs; Am Soc Testing & Mat; Am Soc Photogram & Remote Sensing; Asn Asphalt Paving Technol (pres, 86). *Res:* Bituminous materials; concrete and aggregate materials; soil exploration and testing; remote sensing applications to terrain analysis and site selection; flexible and rigid pavement response; nondestructive testing and evaluation of pavements; low temperature behavior of asphalt and mixtures for pavement analysis; asphalt age hardening prediction and effects. *Mailing Add:* Dept Civil Eng Univ Fla Gainesville FL 32611

RUTH, JAMES ALLAN, b Wichita, Kans, Dec 24, 46. PHARMACOLOGY, MEDICINAL CHEMISTRY. *Educ:* Univ Kans, BS, 68; Northwestern Univ, Evanston, PhD(org chem), 74. *Prof Exp:* Res assoc med chem, Univ Kans, 74-76, res assoc pharmacol, 76-78; ASST PROF MED CHEM & PHARMACOL, SCH PHARM, UNIV COLO, 78- *Concurrent Pos:* NIH fels med chem, 74-76; Am Heart Asn fels pharmacol, 76-78. *Mem:* Am Chem Soc; Royal Soc Chem. *Res:* Neurochemistry; neuropharmacology. *Mailing Add:* Dept Pharmacol Box 297 Univ Colo Boulder CO 80309-0297

RUTH, JOHN MOORE, b Pittsboro, NC, May 26, 24. PHYSICAL CHEMISTRY. *Educ:* Univ NC, BS, 44, PhD(phys chem), 59. *Prof Exp:* Chemist, Dockery Labs, 45-46; jr chemist, Oak Ridge Nat Lab, 47-48, assoc chemist, 48-50, 52; staff mem, Los Alamos Sci Lab, 54-55; staff mem dacron res lab, E I du Pont de Nemours & Co, Inc, 57-59; sr chemist, Res Dept, Liggett & Myers Tobacco Co, 59-66; RES CHEMIST, AGR ENVIRON QUAL INST, AGR RES SERV, USDA, 66- *Concurrent Pos:* Staff mem, Los Alamos Sci Lab, 53, 55 & E I du Pont de Nemours & Co, 56. *Mem:* Fel AAAS; Am Chem Soc; fel Am Inst Chem; Am Soc Mass Spectrometry. *Res:* Mass spectrometry; molecular spectroscopy; identification of organic compounds; determination of molecular structures. *Mailing Add:* PO Box 450 College Park MD 20740-0450

RUTH, ROYAL FRANCIS, b Des Moines, Iowa, Oct 3, 25; m 50; c 1. EMBRYOLOGY, IMMUNOLOGY. *Educ:* Grinnell Col, AB, 49; Univ Wis, MS, 53, PhD(zool), 54. *Prof Exp:* Fel embryol, Ind Univ, 54-56; staff mem, Carnegie Inst Technol, 56-61; from asst prof to assoc prof, 61-71, PROF ZOOL, UNIV ALTA, 71- *Concurrent Pos:* Vis prof zool, Univ Wis, 61 & 62; mem, Animal Welfare Comt, Alta, 61-72; mem, Grants Comt Cell Biol & Genetics, Nat Res Coun Can, 69-72, chmn, 71-72; prof biol & chmn dept, Washington & Lee Univ, 75-76; mem, Comt Post-Doctoral Fel, Natural Sci & Eng Res Coun Can, 79-82, chmn, 81-82; mem, Sci Comt, Alta Heritage Found Med Res, 80-81, chmn, Capital Grants Comt, 81-82. *Mem:* Radiation Soc. *Res:* Immediate lethal effects of ionizing radiation; clinical radiation-hypotension and early reactor-accident deaths. *Mailing Add:* Dept Zool Univ Alta Edmonton AB T6G 2M7 Can

RUTHERFORD, CHARLES, b Trenton, Mo, May 17, 39; m 63; c 1. CELL BIOLOGY, BIOCHEMISTRY. *Educ:* William Jewell Col, BA, 61; Col William & Mary, MA, 63; Univ Miami, PhD(cell & molecular biol), 68. *Prof Exp:* Fel biochem, Med Sch, Washington Univ, 68-69; fel biochem develop, Retina Found, 69-71; from asst prof to assoc prof, 72-83, PROF BIOL, VA POLYTECH INST & STATE UNIV, 83- *Concurrent Pos:* Fel, Boston Biomed Res Inst, 69-72. *Mem:* Am Soc Biochem & Molecular Biol. *Res:* Molecular biology of development. *Mailing Add:* Dept Biol Va Polytech Inst & State Univ Blacksburg VA 24061

RUTHERFORD, JAMES CHARLES, b Oakland, Calif, Aug 27, 46. INVERTEBRATE ECOLOGY. *Educ:* Calif State Col, Hayward, BS, 68; Univ Calif, Berkeley, MA, 71, PhD(zool), 75. *Prof Exp:* Asst prof zool, Ore State Univ, 75-76; asst prof biol, Hilo Col, Univ Hawaii, 76-80; asst prof biol, Hawaii Prep Acad, 85-86; math-sci teacher, Honokaa High Sch, 87-88; SCI DEPT HEAD, PARKER SCH, KAMUELA, 88- *Mem:* Ecol Soc Am; Sigma Xi; AAAS. *Res:* Quantitative invertebrate natural history with emphasis on ecological and evolutionary theory especially as it relates to marine intertidal and benthic invertebrates, echinoderms in particular. *Mailing Add:* PO Box 725 Kamuela HI 96743

RUTHERFORD, JOHN GARVEY, b Baltimore, Md, Sept 25, 42; m 65; c 2. NEUROANATOMY. *Educ:* Cornell Univ, AB, 64; Syracuse Univ, MS, 68; State Univ NY Upstate Med Ctr, PhD(anat), 72. *Prof Exp:* Asst prof, 70-73, ASSOC PROF ANAT, FAC MED, DALHOUSIE UNIV, 73- *Mem:* NY Acad Sci; Am Asn Anatomists; Soc Neurosci. *Res:* Ultrastructure and connections of mammalian brain stem nuclei. *Mailing Add:* Dept Anat Dalhousie Univ Halifax NS B3H 4H7 Can

RUTHERFORD, JOHN L(OFTUS), b Philadelphia, Pa, Mar 6, 24; m 47; c 2. METALLURGY, MATERIALS SCIENCE. *Educ:* Univ Pa, BA, 52, MS, 61, PhD(metall), 63; Rutgers Univ, MA, 81. *Prof Exp:* Proj engr, Sharples Corp, Pa, 45-52; res physicist, Franklin Inst, 52-60; sr staff scientist, Aerospace Group, Gen Precision, Inc, 63-69; res mgr, 69-80, DIR, MAT & PROCESSES LAB, KEARFOTT DIV, SINGER CO, 80- *Mem:* Am Soc Metals; Am Inst Mining, Metall & Petrol Engrs; Am Inst Aeronaut & Astronaut; Sigma Xi. *Res:* Micro-mechanical properties of metals, polymers, and fiber-reinforced composites; relationships between atomic structure and properties; deformation and mechanisms in materials; surface topology in friction and wear. *Mailing Add:* 161 Pershing Ave Ridgewood NJ 07450

RUTHERFORD, KENNETH GERALD, b Lindsay, Ont, May 10, 24; m 50; c 2. ORGANIC CHEMISTRY. *Educ:* Univ Western Ont, BA, 49; Wayne State Univ, PhD(chem), 54. *Prof Exp:* Res assoc org chem, Ohio State Univ, 54-56; asst prof chem, Univ Tulsa, 56-58; PROF CHEM, UNIV WINDSOR, 58- *Concurrent Pos:* Consult, Howick Chem Co, 70- *Mem:* Chem Inst Can; Am Chem Soc. *Res:* Synthetic organic chemistry; optically active trityl systems; infrared attenuation-laser designators; low temperature pyrolitic elimination reactions; potential routes to silicon to carbon compounds. *Mailing Add:* Dept Chem Univ Windsor 735 Lounsbrough Windsor ON N6G 1G1 Can

RUTHERFORD, MALCOLM JOHN, b Durham, Ont, July 6, 39; m 62; c 2. PETROLOGY. *Educ:* Univ Sask, BScEng, 61, MSc, 63; Johns Hopkins Univ, PhD(geol), 68. *Prof Exp:* Fel, Univ Calif, Los Angeles, 68-69, asst prof geol, 69-70; asst prof, 70-75, ASSOC PROF GEOL, BROWN UNIV, 75- *Mem:* AAAS; Geol Soc Am; Mineral Soc Am; Am Geophys Union; Mineral Soc Can. *Res:* Chemistry variations and origin of igneous and metamorphic rocks in the earth's crust and the lunar crust through a combination of analytical studies and laboratory synthesis; origin and processes involved in formation of economic metal; sulfide deposits in association with igneous rocks. *Mailing Add:* Dept Geol Sci Brown Univ Brown Sta Providence RI 02912

RUTHERFORD, PAUL HARDING, b Shipley, Eng, Jan 22, 38; m 59; c 2. PLASMA PHYSICS. *Educ:* Cambridge Univ, BA, 59, PhD(plasma physics), 63. *Prof Exp:* Res assoc plasma physics, Princeton Univ, 62-63; res assoc, Culham Lab, UK Atomic Energy Auth, 63-65; mem res staff, 65-68, res physicist, 68-72, head theoret div, 72-79, LECTR PLASMA PHYSICS, PRINCETON UNIV, 68-, SR RES PHYSICIST, PLASMA PHYSICS LAB, 72-, ASSOC DIR, 80- *Honors & Awards:* E O Lawrence Mem Award, US Dept Energy, 83. *Mem:* Fel Am Phys Soc. *Res:* Theoretical plasma physics. *Mailing Add:* Plasma Physics Lab Princeton Univ PO Box 451 Princeton NJ 08543

RUTHERFORD, WILLIAM M(ORGAN), b Lake Charles, La, Oct 5, 29. CHEMICAL ENGINEERING. *Educ:* Univ Ill, BS, 51, MS, 52, PhD(chem eng), 54. *Prof Exp:* Chemist, Explor & Prod Res Lab, Shell Develop Co, Tex, 54-62; res specialist, 62-66, sr res specialist, 66-69, Monsanto fel, 69-79, SR MONSANTO FEL, MOUND LAB, MONSANTO RES CORP, 79- *Mem:* AAAS; Am Chem Soc; Am Inst Chem Engrs; Sigma Xi; Am Nuclear Soc. *Res:* Isotope separation; thermal diffusion; equilibrium and transport properties of fluids; properties of fluids at high pressure. *Mailing Add:* 410 Marylhurst Dr Dayton OH 45459

RUTHVEN, DOUGLAS M, b England, Oct 9, 38; m 68; c 1. CHEMICAL ENGINEERING, PHYSICAL CHEMISTRY. *Educ:* Univ Cambridge, BA, 60, MA, 63, PhD(chem eng), 66. *Prof Exp:* Design engr, Power Gas Corp, Eng, 61-63; from asst prof to assoc prof, 66-74, PROF CHEM ENG, UNIV NB, 74- *Mem:* Can Soc Chem Engrs; Am Inst Chem Engrs. *Res:* Sorption and diffusion in molecular sieve zeolites; adsorption separation processes. *Mailing Add:* Dept Chem Eng Univ NB Fredericton NB E3B 5A3 Can

RUTISHAUSER, URS STEPHEN, b Altadena, Calif, Feb 27, 46; m 74. BIOCHEMISTRY, EMBRYOLOGY. *Educ:* Brown Univ, ScB, 67; Rockefeller Univ, PhD(biochem), 73. *Prof Exp:* Jane Coffin Childs fel, Weizmann Inst Sci, Israel, 73-74; asst prof biochem, 74-79, ASSOC PROF, ROCKEFELLER UNIV, 79-; AT DEPT GENETICS, CASE WESTERN RESERVE UNIV. *Honors & Awards:* McKnight Scholar. *Mem:* Soc Neurosci; Am Asn Cell Biol. *Res:* Immunology; developmental and cell biology; immunoglobulin structure; cell-cell interactions; neurobiology. *Mailing Add:* 2119 Abington Rd Cleveland OH 44106

RUTKIN, PHILIP, b New York, NY, Sept 17, 33; m 57; c 4. ORGANIC CHEMISTRY. *Educ:* City Col New York, BS, 55; NY Univ, PhD(org chem), 60. *Prof Exp:* Res chemist, Faberge, 58-60, dir res, 60-66, vpres res, 66-70; VPRES, ESTEE LAUDER, 75- *Concurrent Pos:* Eve sessions instr, Farleigh Dickinson Univ, 59-, NY Univ, 61 & City Col New York, 62-; gen mgr, Chemspray Div, ATI; vpres, Revlon-Int. *Mem:* Am Chem Soc; Soc Cosmetic Chemists; AAAS; Acad Sci. *Res:* Cosmetic chemistry. *Mailing Add:* Six Henhawk Rd Kings Point NY 11024-2107

RUTLAND, LEON W, b Commerce, Tex, Aug 24, 19; m 44; c 2. MATHEMATICS. *Educ:* Univ Colo, PhD, 54. *Prof Exp:* From asst prof to assoc prof appl math, Univ Colo, 54-64; chmn dept, 64-70, PROF MATH, VA POLYTECH INST & STATE UNIV, 64- *Mem:* Am Math Soc; Am Soc Eng Educ; Soc Indust & Appl Math; Math Asn Am. *Res:* Applied mathematics. *Mailing Add:* 1391 Locust Ave Blacksburg VA 24060

RUTLEDGE, CARL THOMAS, b Fayetteville, Ark, Sept 17, 44; m 67; c 2. MICROCOMPUTER INTERFACING. *Educ:* Univ Ark, BS, 66, MS, 69, PhD(physics), 71. *Prof Exp:* prof physics, Southern Ark Univ, 70-81; assoc prof, 81-87, PROF PHYSICS, E CENT UNIV, 87- *Mem:* Sigma Xi; Am Asn Physics Teachers. *Res:* X-ray diffraction studies of liquids; interferometry and physical optica; improvements in teaching basic astronomy and physics; microcomputer software and interfacing. *Mailing Add:* Dept Physics E Cent Univ Ada OK 74820-6899

RUTLEDGE, CHARLES O, b Topeka, Kans, Oct 1, 37; m 61; c 4. PHARMACOLOGY. *Educ:* Univ Kans, BS, 59, MS, 61; Harvard Univ, PhD(pharmacol), 66. *Prof Exp:* NATO fel pharmacol, Gothenburg Univ, 66-67; from asst prof to assoc prof, Med Ctr, Univ Colo, Denver, 67-75; prof pharmacol & toxicol & chmn dept, Sch Pharm, Univ Kans, 75-87; PROF PHARMACOL, SCH PHARM, PURDUE UNIV, 87-, DEAN SCH PHARM, NURSING & HEALTH SCI, 87- *Mem:* AAAS; Am Soc Pharmacol & Exp Therapeut (secy-treas, 91-92); Am Soc Neurochem; Soc Neurosci; Am Asn Cols Pharm; Am Asn Pharmaceut Scientists. *Res:* Autonomic, behavioral and biochemical pharmacology; interactions of drugs with the synthesis, storage, uptake, release and metabolism of neurotransmitters in the central nervous system to produce alterations in behavior. *Mailing Add:* Sch Pharm Purdue Univ Robert E Heine Bldg Lafayette IN 47907

RUTLEDGE, DELBERT LEROY, b Mooreland, Okla, July 20, 25; m 47; c 2. THERMAL PHYSICS. *Educ:* Univ NMex, BS, 46; Okla State Univ, MS, 48, EdD, 58. *Prof Exp:* Asst prof physics, Cent State Col, Okla, 47-57; from asst prof to prof, 57-88, EMER PROF PHYSICS, OKLA STATE UNIV, 88- *Concurrent Pos:* Assoc prog dir, NSF Div Undergrad Educ, 65-66. *Mem:* Am Asn Physics Teachers; Am Phys Soc; Am Soc Eng Educ. *Mailing Add:* 1924 McElroy Stillwater OK 74074

RUTLEDGE, DOROTHY STALLWORTH, b Tuscaloosa, Ala, May 3, 30; m 49; c 2. MATHEMATICS, COMPUTER SCIENCE. *Educ:* Birmingham-Southern Col, BA, 51; Emory Univ, MS, 60, PhD(math), 66. *Prof Exp:* Asst prof math, Agnes Scott Col, 66-69; asst prof, 69-72, ASSOC PROF MATH, GA STATE UNIV, 72- *Mem:* Am Math Soc; Math Asn Am. *Res:* Analysis; functional analysis. *Mailing Add:* Dept Math Computer Sci Ga State Univ Atlanta GA 30303

RUTLEDGE, FELIX N, b Anniston, Ala, Nov 20, 17; m 50; c 1. GYNECOLOGIC ONCOLOGY. *Educ:* Univ Ala, BS, 39; Johns Hopkins Univ, MD, 43; Am Bd Obstet & Gynec, dipl, 51. *Prof Exp:* PROF GYNEC, UNIV TEX M D ANDERSON HOSP & TUMOR INST, 54- *Mem:* Am Asn Obstet & Gynec; Am Gynec Soc; Am Radium Soc; Soc Pelvic Surg; Soc Gynec Oncol. *Res:* Diagnosis and treatment of female pelvic malignancies. *Mailing Add:* 6723 Bertner Houston TX 77030

RUTLEDGE, GENE PRESTON, b Spartanburg, SC, Dec 3, 25; m 50; c 4. PHYSICAL CHEMISTRY, RESEARCH ADMINISTRATION. *Educ:* Wofford Col, BS, 46; Univ Tenn, MS, 48. *Hon Degrees:* DSc, Wofford Col, 71. *Prof Exp:* Res chemist & proj engr, Lab & Eng Div, Union Carbide Nuclear Co, Tenn, 48-54; supvry engr, Develop Eng Div, Goodyear Atomic Corp, Ohio, 54-56, sr scientist, S5W & Pressurized Water Reactor Projs, Westinghouse Elec Corp, 56-59, supvry scientist, Naval Reactors Facility, Bettis Atomic Power Lab, 60-67; exec dir, Idaho Nuclear Energy Comn, 67-76; energy scientist, Off Gov, State of Alaska, 76-78; RES DIR & OWNER, PAC POLAR RIMS, 78- *Concurrent Pos:* Mem bd dirs, Western Interstate Nuclear Compact, 69; mem, Gov Task Force Radioactive Waste Mgt, 71; consult, high technol firms throughout USA. *Mem:* Am Nuclear Soc; Am Chem Soc; Am Inst Chem Eng. *Res:* Energy development including nuclear, geothermal, solar, geosolar, hydrogen and wind and environmental; forestry, agriculture, mining and research using nuclear methods; author and co-author three books. *Mailing Add:* 6930 Oakwood Dr Anchorage AK 99507

RUTLEDGE, HARLEY DEAN, b Omaha, Nebr, Jan 10, 26; m 54; c 5. SOLID STATE PHYSICS. *Educ:* Tarkio Col, AB, 50; Univ Mo, MS, 56, PhD(photoelec emission), 66. *Prof Exp:* Assoc prof physics, Cent Methodist Col, 57-60; instr, Univ Mo, 60-61; assoc prof, 63-67, PROF PHYSICS, SOUTHEAST MO STATE UNIV, 67-, HEAD DEPT, 64- *Mem:* Am Asn Physics Teachers. *Res:* Experimental study of photoelectric emission from strontium oxide sprayed cathodes; theoretical analysis of photoelectric emission from semiconductors. *Mailing Add:* Dept Physics Southeast Mo State Univ Cape Girardeau MO 63701

RUTLEDGE, JACKIE JOE, b Woodward, Okla, Dec 20, 41; m 66; c 2. ANIMAL BREEDING. *Educ:* Okla State Univ, BS, 68; NC State Univ, MS, 71, PhD(animal breeding), 73. *Prof Exp:* Res asst animal breeding, NC State Univ, 68-72; res assoc animal breeding, Univ Wis, 72-74; asst prof animal breeding, Univ Vt, 74-75; asst prof, 75-77, assoc prof, 77-80, prof animal breeding, 80-87, PROF & CHMN, UNIV WIS, 88- *Mem:* Am Soc Animal Sci; Genetics Soc Am; Sigma Xi; Am Dairy Sci Asn; Am Genetic Asn; Biomet Soc. *Res:* Dynamics of genetic variances and covariances among and within populations; multivariate statistical theory applications to genetic problems; realized heritability of and correlated predictors of fecundity in economic and laboratory animals. *Mailing Add:* Dept Meat & Animal Sci Univ Wis Animal Sci Bldg 438 Madison WI 53706

RUTLEDGE, JAMES LUTHER, b Woodward, Okla, Oct 1, 37; m 63; c 2. SOLID STATE PHYSICS, ELECTRONICS. *Educ:* Okla State Univ, BS, 63, MS, 66, PhD(physics), 68. *Prof Exp:* Sr physicist, Motorola Semiconductor Prod, Inc, 67-74, proj leader res & develop, Fairchild Semiconductor, 74, mgr, Adv Prod Res & Develop Labs, 74-84, vpres tech staff, 84-91; DIR UNIV & NAT LABS PROGS, SEMATECH, 91- *Mem:* Inst Elec & Electronics Engrs; Am Phys Soc. *Res:* Solid state surface physics; electrical properties of semiconductors; physics of semiconductor devices. *Mailing Add:* 8367 S Forest Tempe AZ 85284

RUTLEDGE, JOSEPH DELA, b Selma, Ala, Aug 9, 28; m 54; c 4. MATHEMATICS, COMPUTER SCIENCES. *Educ:* Swarthmore Col, BA, 50; Cornell Univ, PhD(math), 59. *Prof Exp:* Sr systs engr, Remington Rand Univac Div, Sperry Rand Corp, 50-53; asst bact, Univ Hosp, Univ Pa, 53-55; asst math, Cornell Univ, 55-58; STAFF MEM, RES DIV, T J WATSON RES CTR, IBM CORP, 58- *Concurrent Pos:* Vis lectr, Wesleyan Univ, 65-67; lectr, Univ Grenoble, 67-68; adj prof, NY Univ, 69-70; IBM vis prof, Spelman Col, 71-72. *Mem:* Asn Symbolic Logic; Asn Comput Mach; Math Asn Am; Sigma Xi. *Res:* Theory of automata and computation; computer design and applications, especially man-machine communication and problem specification; programming and programming language. *Mailing Add:* 11 Sycamore Terr RD 2 Mahopac NY 10541

RUTLEDGE, LESTER T, b Big Sandy, Mont, June 12, 24; m 49. PHYSIOLOGY. *Educ:* Univ Utah, MA, 52, PhD(physiol, psychol), 53. *Prof Exp:* Res assoc & res instr physiol, Univ Utah, 53-56; from res assoc to assoc prof, 56-67, chmn, Neurosci Prog, 72-84, PROF PHYSIOL, MED SCH, UNIV MICH, ANN ARBOR, 67- *Concurrent Pos:* NIH sr res fel, 58-62, career develop award, 63-68. *Mem:* AAAS; Soc Neurosci; Am Physiol Soc. *Res:* Integrative processes of central nervous systems; electrophysiology of cortical and subcortical relationships; neural basis of learning; spinal and supraspinal reflexes; association cortex; epilepsy; neurological teaching; alcohol effects on cortex. *Mailing Add:* Dept Physiol-Med Sci Bldg Univ Mich Med Sch Ann Arbor MI 48109

RUTLEDGE, LEWIS JAMES, b McComb, Miss, Apr 24, 24; m 47; c 3. OTOLARYNGOLOGY, PEDIATRICS. *Educ:* Tulane Univ, MD, 47; Am Bd Pediat, dipl, 55; Am Bd Otolaryngol, dipl, 61. *Prof Exp:* Intern, Charity Hosp, New Orleans, La, 47-48, resident pediat, 48-50, resident otolaryngol, 55-57; instr pediat, Sch Med, Tulane Univ, 53-55; NIH spec trainee otolaryngol, Tulane Univ & Univ Chicago, 57-58; asst prof, 59-62, ASSOC PROF OTOLARYNGOL, SCH MED, TULANE UNIV, 62-, ASST PROF PEDIAT, 59- *Concurrent Pos:* Instr rhinoplasty, Columbia Univ at Mt Sinai Hosp, New York, 59- *Res:* Histopathology of human and animal temporal bones. *Mailing Add:* 7330 Sycamore Ln New Orleans LA 70118

RUTLEDGE, ROBERT B, b St Louis, Mo, Dec 9, 35; m 60; c 4. MATHEMATICS, ELECTRICAL ENGINEERING. *Educ:* St Louis Univ, BS, 58, MS, 59, PhD(math), 62. *Prof Exp:* From asst prof to assoc prof math & appl sci, 62-69, assoc prof eng, 69-73, PROF MATH, SOUTHERN ILL UNIV, 73- *Concurrent Pos:* Consult electromagnetic sensor lab, Emerson Elec Co, 62- *Mem:* Math Asn Am; Soc Indust & Appl Math. *Res:* Analysis of advanced radar processing techniques using stochastic models; digital simulation techniques. *Mailing Add:* Div Eng Southern Ill Univ Edwardsville IL 62026

RUTLEDGE, ROBERT L, b Pocahontas, Miss, June 23, 30; m 58; c 5. PHYSICAL CHEMISTRY. *Educ:* Miss State Univ, BS, 52; Univ Ill, MS, 53, PhD(phys chem), 58. *Prof Exp:* Sr res chemist, Socony Mobil Oil Co, Tex, 58-60; res chemist, 60-70, res specialist, 70-78, SR RES SPECIALIST, MINN MINING & MFG CO, 78- *Mem:* Soc Photog Sci & Eng. *Res:* Image evaluation of photographic media; novel imaging systems. *Mailing Add:* 2220 Powers Ave St Paul MN 55119

RUTLEDGE, THOMAS FRANKLIN, b Cordova, Tenn, June 1, 21; m 43; c 3. ORGANIC CHEMISTRY, CATALYSIS. *Educ:* Univ Ark, BS, 43; Univ Del, PhD(org chem), 50. *Prof Exp:* Chemist, Res Labs, Socony-Vacuum Oil Co, 43-44, 47-48; sr res chemist, Monsanto Chem Co, 50-51; sect head org div, Res Labs, Air Reduction, Inc, 51-57; mgr org & catalysis res, Corp Res Dept, ICI-Am, 57-81; RETIRED. *Concurrent Pos:* Vpres, Develop Solar Elec Corp, 84. *Mem:* Am Chem Soc. *Res:* Mechanisms of chemical oxidations via chromic acid; preparation of synthetic lubricants of hydrocarbon type; catalyst preparation; acetylene and hydrocarbon chemistry; urea chemistry; polyurethanes; polymers; catalytic oxidation; polymerization; electrochemistry. *Mailing Add:* 124 Uxbridge Cherry Hill NJ 08034

RUTLEDGE, WYMAN COE, b Abrahamsville, Pa, Dec 15, 24; m 45; c 3. APPLIED PHYSICS, INSTRUMENTS. *Educ:* Hiram Col, AB, 44; Univ Mich, MS, 48, PhD(physics), 52. *Prof Exp:* Lab instr, Hiram Col, 43-44; res asst, Oceanog Inst, Woods Hole, 46-47; res assoc, Univ Mich, 47-48, res asst, 48-50; jr physicist, Argonne Nat Lab, 50-52; res physicist, Philip Labs, Inc, 52-56; sr res physicist, 56, sr res fel, Cent Res Labs, 56-68, inst syst consult, 68-86, PRIN SCIENTIST, MEAD CORP, 86- *Concurrent Pos:* Part-time instr, Ohio Univ, 58-68; mem, Simulation Coun, Inc; mem measurement technol comt, Am Inst Paper Res, process systs & controls comt, Tech Asn Pulp & Paper Indust; mem particulate subcomt, Inter Soc Air Sampling Comt, 72-84; mem, Environ Res Tech Comt, Instrument Soc Am; Adv Bd Trustees, Ohio Univ, Chillicothe, vpres, 80-82, pres, 82-84, emer, 85-; mem, Bd Trustees, Hiram Col, 80-86; mem, Advan Sensor Comt, Dept Energy, 82- *Honors & Awards:* Phoenix Post Doct Fel. *Mem:* Am Phys Soc; Optical Soc Am; fel Instrument Soc Am; Measurements & Data Soc; Sigma Xi; fel Tech Asn Pulp Paper Indust. *Res:* Underwater research; upper atmosphere; nuclear spectroscopy; thermionic emission; instrumentation and automation; computer systems application; basic phenomena; air and water quality instrumentation; sensors. *Mailing Add:* 704 Ashley Dr Chillicothe OH 45602

RUTMAN, ROBERT JESSE, b Kingston, NY, June 23, 19; wid; c 6. BIOCHEMISTRY, MOLECULAR BIOLOGY. *Educ:* Pa State Univ, BS, 40; Univ Calif, PhD(biochem), 50. *Hon Degrees:* MS, Univ Pa, 72. *Prof Exp:* Asst prof biochem, Jefferson Med Col, 50-53; res assoc biol, 54-56, chem, 57-64, assoc prof, 64-69, chmn lab biochem, Dept Animal Biol, 71-73, & 78-79, prof, 69-87, EMER PROF MOLECULAR BIOL, SCH VET MED, UNIV PA, 87- *Concurrent Pos:* Prof biochem & chmn dept, Philadelphia Col Osteopath, 59-62; consult, Hartford Found, Presby Hosp, Philadelphia, 63-65; vis prof, Univ Ibadan, Nigeria, 73-74, external examnr, 81; legal consult, Environ Carcinogenesis. *Mem:* AAAS; Am Chem Soc; Am Soc Biol Chem; Am Asn Cancer Res; Am Asn Vet Educr; Vet Oncol Soc. *Res:* Cancer chemotherapy; mechanism of action of chemotherapeutic drugs; nucleic acid metabolism;

biochemical thermodynamics; nucleic acid metabolism in nucleus and mitochondria with particular reference to effect of clinically important anti-cancer drugs; biological response modifiers, liposomes/transport. *Mailing Add:* Dept Animal Biol Sch Vet Med Univ Pa Philadelphia PA 19104-6046

RUTNER, EMILE, b Budapest, Hungary, Apr 28, 21; nat US. CHEMICAL PHYSICS. *Educ:* Carnegie Inst Technol, BS, 43; Cornell Univ, PhD(chem), 51. *Prof Exp:* Chemist, Control Lab, Koppers Co, 43; anal develop, Publicker Industs, 43-45; asst, Cornell Univ, 47-50; phys chemist, Wright Air Develop Ctr, US Dept Air Force, 51-56; physicist, Lewis Res Ctr, NASA, 56-60; chemist, 60-75, physicist, US Air Force Mat Lab, Wright Patterson AFB, 75-83; RETIRED. *Concurrent Pos:* Res assoc, Forrestal Res Ctr, Princeton Univ, 51-52. *Mem:* Am Chem Soc; Am Phys Soc. *Res:* Spectra; thermophysics; solid state; radiation effects; kinetics; high energy laser effects. *Mailing Add:* 34 Columbia Ave Takoma Park MD 20912

RUTOWSKI, RONALD LEE, b Van Nuys, Calif, May 16, 49; m 74. ANIMAL BEHAVIOR. *Educ:* Univ Calif, Santa Cruz, BA, 71; Cornell Univ, PhD(behav), 76. *Prof Exp:* Asst prof, 76-80, ASSOC PROF ZOOL, ARIZ STATE UNIV, 80- *Concurrent Pos:* Prin investr, NSF grants, 78-80, 80-82, 83-85 & 86-88; vis scientist, Archbold Biol Sta, Lake Placid, Fla, 81, Long Marine Lab, Univ Calif, Santa Cruz, 82, Dept Zool, James Cook Univ, Townsville, Queensland, Australia, 89; lectr, Univ Stockholm, 84. *Mem:* Animal Behav Soc; AAAS; Am Soc Naturalists; Lepidopterists Soc. *Res:* Animal communication, especially in invertebrates; reproductive behavior. *Mailing Add:* Dept Zool Ariz State Univ Tempe AZ 85287

RUTSCHKY, CHARLES WILLIAM, b Pottstown, Pa, May 2, 23; m 45; c 4. ENTOMOLOGY. *Educ:* Pa State Univ, BS, 43; Cornell Univ, PhD(entom), 49. *Prof Exp:* Investr fruit insects, NY Agr Exp Sta, 47-49; from asst prof to assoc prof entom, Pa State Univ, 49-62; prof, Univ Hawaii, 62-64; prof, 65-86, EMER PROF ENTOM, PA STATE UNIV, 86- *Mem:* Entom Soc Am. *Mailing Add:* Dept Entom Pa State Univ University Park PA 16802

RUTSTEIN, MARTIN S, b Boston, Mass, Apr 1, 40; m 62; c 2. MINERALOGY, GEOCHEMISTRY. *Educ:* Boston Univ, BA, 61, MA, 62; Brown Univ, PhD(mineral), 69. *Prof Exp:* Teaching asst geol, Boston Univ, 60-62; chief geol & soils sect, US Army Eng Sch, 62-64; res asst geol-mineral, Brown Univ, 64-68; asst prof mineral, Juniata Col, 68-70; asst prof, 70-72, chmn dept, 72-74, actg assoc vpres acad affairs, 74-75, ASSOC PROF GEOL SCI, STATE UNIV NY COL NEW PALTZ, 72-, CHMN DEPT, 76- *Concurrent Pos:* Pa Res Found fel, Juniata Col, 69-70; NY Res Found & Geol Soc Am Penrose Fund fels, State Univ NY Col New Paltz, 71-72, NSF fel, 72-73. *Mem:* Mineral Soc Am. *Res:* High temperature-pressure chain silicate phase relations; crystal chemistry of sulphides; environmental geology and human health; petrology. *Mailing Add:* Dept Geol Sci State Univ NY Col New Paltz NY 12561

RUTTAN, VERNON W, b Alden, Mich, Aug 16, 24. AGRICULTURAL ECONOMICS. *Educ:* Yale Univ, BA, 48; Univ Chicago, MA, 50, PhD, 52. *Hon Degrees:* LLD, Rutgers Univ, 78; Doktor der Agrarwissenschaften ehrenhalber, Christian Albrechts Universitat, Kiel, Ger, 86; DAgr, Purdue Univ, 91. *Prof Exp:* Economist, Div Regional Studies, Tenn Valley Authority, 51-53, Off Gen Mgr, 54-55; from asst prof to prof, Dept Agr Econ, Purdue Univ, 55-63; agr economist, Rockefeller Found, Int Rice Res Inst, Philippines, 63-65; trustee, Agr Develop Coun, Inc, 67-73, pres, 73-77; prof & head, Dept Agr Econ, Univ Minn, 65-70, prof & dir, Econ Develop Ctr, 70-73, prof, Dept Agr & Appl Econ & Dept Econ, Hubert H Humphrey Inst Pub Affairs, 78-86, REGENTS PROF, UNIV MINN, 86- *Concurrent Pos:* Assoc agr economist, Giannini Found Agr Econ, Univ Calif, Berkeley, 58-59; mem, Comt New Orientations Agr Econ Res, Am Agr Econ Asn, 59-63, Comt Prof Problems Int Res, 67-69, chmn, 68-69; US coun mem, Int Asn Agr Economists, 69-72; mem, Tech Adv Comt, Consult Group Int Agr Res, 73-77, bd dirs, Int Serv Nat Agr Res, 79-86, Res Adv Comt, USAID, 67-75 & 83-86 & var agr comts, Nat Res Coun, Social Sci Res Coun, Nat Planning Asn, 71-; Alexander von Humboldt award, 84. *Honors & Awards:* B Y Morrison Mem Lectr, USDA, 83, Distinguished Serv Award, 86; Nat Award Agr Excellence, Nat Agri-Mkt Asn, 89. *Mem:* Nat Acad Sci; fel Am Acad Arts & Sci; fel Am Agr Econ Asn (vpres, 67-68, pres, 71-72); Am Econ Asn; fel AAAS; Int Asn Agr Economists. *Res:* Agricultural economics; author and co-author of nine books and numerous technical publications. *Mailing Add:* Dept Agr & Appl Econ Univ Minn 1994 Buford Ave St Paul MN 55108

RUTTENBERG, HERBERT DAVID, b Philadelphia, Pa, June 14, 30; m 55; c 4. PEDIATRICS, CARDIOLOGY. *Educ:* Univ Calif, Los Angeles, BA, 52, MD, 56. *Prof Exp:* Fel pediat, Med Sch, Univ Minn, 57-58 & 60-61, USPHS fel pediat cardiol, 61-63; NIH trainee cardiovasc res technol, Wash Univ, 63-64; asst prof pediat cardiol, Sch Med, Univ Calif, Los Angeles, 64-69; assoc prof pediat cardiol & chmn div, 69-78, prof pediat & chmn, div pediat cardiol, 78-84, PROF PEDIAT & PATH, SCH MED, UNIV UTAH, 84- *Concurrent Pos:* Res assoc cardiovasc path, Charles T Miller Hosp, St Paul, Minn, 62-63; regional vpres, Am Heart Asn, 87-88. *Mem:* Fel Am Acad Pediat; fel Am Col Cardiol; fel NY Acad Sci; Am Fedn Clin Res; fel Am Pediat Soc. *Res:* Neurohumoral control of cardiac function; adreneric receptors; cardiovascular function in clinical and experimental complete heart block; exercise physiology and preventive cardiology. *Mailing Add:* Div Pediat Cardiol Univ Utah Col Med 100 N Medical Dr Salt Lake City UT 84113

RUTTENBERG, STANLEY, b St Paul, Minn, Mar 12, 26; m 55; c 2. ATMOSPHERIC PHYSICS, METEOROLOGY. *Educ:* Mass Inst Technol, BS, 46; Univ Calif, Los Angeles, MA, 51. *Prof Exp:* Res asst geophys, Inst Geophys, Univ Calif, Los Angeles, 49-55; prog officer, US Nat Comt, Int Geophys Yr, Nat Acad Sci, 55-56, head prog officer, 57-60; secy, Panel World Magnetic Surv, Spec Comt Int Yrs Quiet Sun, Geophys Res Bd, 61-62, exec secy, US Nat Comt, 62-64; asst to dir, 64-73, mem sr mgt, Nat Ctr Atmospheric Res, 73-79, dir off sci prog develop, UN Corp for Atmospheric Res, 79-84, dir, 84-86, SR MGT, UN CORP FOR ATMOSPHERIC RES PROJ, 86- *Concurrent Pos:* Consult, US Nat Comt, Spec Comn Int Yrs Quiet Sun, Nat Acad Sci, 64-68; mem Geophys Film Comt, US Nat Acad Sci, 84-86; chmn, ICSU Panel on World Data Ctrs, 88- *Mem:* Int Asn Meteorol & Atmospheric Physics (secy gen, 75-87); AAAS; Am Geophys Union. *Res:* Atmospheric physics; space-based observational techniques; geophysical data storage and access. *Mailing Add:* UN Corp Atmospheric Res PO Box 3000 Boulder CO 80307

RUTTER, EDGAR A, JR, b Newark, NJ, Feb 3, 37; m 65. MATHEMATICS. *Educ:* Marietta Col, BA, 59; Iowa State Univ, PhD(math), 65. *Prof Exp:* Asst prof math, NMex State Univ, 65-66; asst prof math, Univ Kans, 66-69, assoc prof, 69-76, prof, 76-77; PROF MATH & CHMN DEPT, WRIGHT STATE UNIV, 77- *Mem:* Am Math Soc. *Res:* Ring theory. *Mailing Add:* Dept Math Wright State Univ Colonel Glenn Hwy Dayton OH 45435

RUTTER, HENRY ALOUIS, JR, b Richmond, Va, Mar 28, 22; m 55; c 1. TOXICOLOGY, BIOCHEMISTRY. *Educ:* Va Polytech Inst, BS, 43; Univ Richmond, MS, 47; Georgetown Univ, PhD(biochem), 52. *Prof Exp:* Chemist, Standard Oil Co, La, 43; chemist, US Naval Ord Lab, 43-45; asst, Univ Tenn, 48-49; instr chem, Am Univ, 49-50; assoc prof org chem, Carson-Newman Col, 50-51; instr phys chem, Georgetown Univ, 51-52; biochemist & mem staff, Biochem Res Found, Franklin Inst, 52-55; chemist, Walter Reed Army Inst Res, 56-59; chief res chemist, Dept Surg, Baltimore City Hosps, 59-60; res coordr chem pharmacol, 60-61, sr chemist, 61-63, sci dir chem dept, 63-64, res coordr pharmacol dept, 64-66, PROJ MGR TOXICOL DEPT, HAZLETON LABS, INC, 66- *Mem:* Am Chem Soc; Soc Cryobiol; Soc Toxicol; Am Inst Chemists. *Res:* Drugs and industrial chemicals; medical application of plastics; environmental sciences. *Mailing Add:* 2110 Mc Kay St Falls Church VA 22043-1522

RUTTER, MICHAEL L, PSYCHIATRY. *Hon Degrees:* Dr, Univ Leiden, 85, Cath Univ, 90; DSc, Univ Birmingham, 90, Univ Chicago, 91; MD, Univ Edinburgh, 90. *Prof Exp:* Res fel, Dept Pediat, Albert Einstein Col Med, NY, 61-62; social psychiat unit, Med Res Coun, 62-65; PROF & HEAD, DEPT CHILD & ADOLESCENT PSYCHIAT, INST PSYCHIAT, UNIV LONDON, 73- *Mem:* Foreign assoc mem Inst Med; hon fel Brit Psychol Soc; hon fel Am Acad Pediat; hon mem Am Acad Child Psychiat; fel Royal Soc; foreign hon mem Am Acad Arts & Sci; foreign assoc mem Nat Acad Educ. *Res:* Stress resistance in children; developmental links between childhood and adult life; schools as social institutions; interviewing skills; neuropsychiatry; infantile autism and psychiatric epidemiology; author of 28 books and over 220 papers. *Mailing Add:* Inst Psychiat Univ London 16 De Crespigny Park Denmark Hill London SE5 8AF England

RUTTER, NATHANIEL WESTLUND, b Omaha, Nebr, Nov 22, 32; Can citizen; m 61; c 2. QUATERNARY GEOLOGY. *Educ:* Tufts Univ, BS, 55; Univ Alaska, MS, 62; Univ Alta, PhD(geol), 66. *Prof Exp:* Explor geologist, Venezuelan Atlantic Ref Co, 55-58; res scientist, Geol Surv, Can, 65-74, head urban projs, 73-74; environ adv, Nat Energy Bd, 74-75; assoc prof, 75-77, PROF & CHMN GEOL, UNIV ALTA, 77- *Concurrent Pos:* Instr, Dept Archaeol, Univ Calgary, 67-73; consult, R S Peabody Found Archaeol, 70; pres, Westlund Consults Ltd; ed bd, Quaternary Rev; mem subcomt, NAm Stratig; pres, Int Quaternary Asn Cong, Ottawa, 87; assoc ed, Geosci Can, & Arctic; pres Int Union, Quaternary Res, 87-91; mem Can Nat Comt, Int Geol correction prog, 87-91; mem adv bd, Alta Geol Surv; mem, ICSU-IGBP Sci Steering Comt Global Changes of the Past; mem, Task Force on Global Change, Int Union Geol Sci. *Mem:* Fel Geol Soc Am; fel Geol Asn Can; Soc Econ Paleontologists & Mineralogists; fel Arctic Inst NAm; Sigma Xi. *Res:* Quaternary sedimentary and stratigraphy of the Rocky Mountains, Canada; paleosol investigations of glaciated and unglaciated surfaces; terrain analysis and land classification; northern Canada; amino acid dating methods; sea level changes, Argentina; less statigraphy, China; quaternary sedimentation, Southwest Africa. *Mailing Add:* Dept Geol Univ Alta Edmonton AB T6G 2G7 Can

RUTTER, WILLIAM J, b Malad City, Idaho, Aug 28, 28; m 71; c 2. MOLECULAR BIOLOGY. *Educ:* Harvard Univ, BA, 49; Univ Utah, MS, 50; Univ Ill, PhD(biochem), 52. *Prof Exp:* USPHS fel, Inst Enzyme Res, Univ Wis, 52-54; USPHS fel biochem, Nobel Inst, Sweden, 54-55; from asst prof to prof biochem, Univ Ill, Urbana, 55-65; prof biochem & genetics, Univ Wash, 65-69; chmn dept biochem & biophys, 69-82, dir, Hormone Res Inst, 83-89, HERTZ STEIN PROF BIOCHEM, DEPT BIOCHEM & BIOPHYS, UNIV CALIF, SAN FRANCISCO, 69- *Concurrent Pos:* Consult, Abbott Labs, 60-75, Eli Lilly Co, 77-80 & Merck & Co, 77-81; vis scientist, Guggenheim Mem fel, Stanford Univ, 62-63; mem sci adv bd, Hagedorn Res Lab, Copenhagen, 80-86 & German Ctr Molecular Biol, Univ Heidelberg, 83-87; mem & chmn basic sci adv comn, Nat Cystic Fibrosis Res, 69-74, mem, Pres Adv Coun, 74-75; mem develop biol panel, NSF, 71-73; pres, Pac Slope Biochem Conf, 71-73; mem adv bd, Revista Pan-Am Asn Biochem Soc, 71-; mem biomed adv comt, Los Alamos Sci Lab, 72-75 & biol div adv comt, Oak Ridge Nat Lab, 76-80; mem bd sci counrs, Nat Inst Environ Health Sci, NIH, 76-81; mem bd dirs, Keystone Life Sci Study Ctr, 83-, Hana Biologics, Berkeley, Calif, 80-83, chmn bd dirs & founder, Chiron Corp, Emeryville, Calif, 81- & Meridian Instruments, Ann Arbor, Mich, 82-88; Kroc vis prof, Joslin Diabetes Ctr, Harvard Univ, Cambridge, Mass & Univ Tex Health Sci Ctr, Dallas, 86. *Honors & Awards:* Pfizer Award, Am Chem Soc, 67; J J Berzelius Award, Karolinska Inst, 83. *Mem:* Nat Acad Sci; Am Soc Cell Biol; Am Soc Biol Chemists (treas, 70-76); Am Soc Develop Biol (pres-elect, 74-75, pres, 75-76); Am Chem Soc; Am Acad Arts & Sci. *Res:* Control of gene expression; mechanisms of cytodifferentiation; regulation of cell proliferation; mechanism of enzyme action; macromolecular variation and evolution. *Mailing Add:* Hormone Res Inst Univ Calif San Francisco CA 94143-0534

RUTTIMANN, URS E, b Switz, 9138. MEDICAL IMAGE PROCESSING. *Educ:* HTL, Burgdorf, Switz, BS, 61; Swiss Fed Inst Technol, MS, 66; Univ Pa, PhD(biomed eng), 72. *Prof Exp:* Asst prof, Dept Clin Eng, 73-79, ASSOC PROF CHILD HEALTH & DEVELOP, GEORGE WASHINGTON UNIV, 80-; SR SCIENTIST, DIAG SYSTS BR, NAT INST DENT RES,

NIH, 86- *Mem:* Biomed Eng Soc; Inst Elec & Electronics Engrs Eng Med & Biol Soc; Inst Elec & Electronics Engrs Computer Soc; Inst Elec & Electronics Engrs Acoust Speech & Signal Processing Soc; Soc Photo-Optical Instrumentation; Int Asn Dent Res. *Mailing Add:* Lab Clin Studies NIAAA NIH Bldg 10-1N104 9000 Rockville Pike Bethesda MD 20892

RUTZ, LENARD O(TTO), b Franklin, Wis, Jan 27, 24; m 53; c 2. CHEMICAL ENGINEERING. *Educ:* Marquette Univ, BChE, 45; Univ Wis, BS, 52, MS, 53; Univ Iowa, PhD(chem eng), 58. *Prof Exp:* Engr, Allis-Chalmers Mfg Co, 45-47, asst engr mining & mineral dressing div, 47-50; res chem engr, E I du Pont de Nemours & Co, 53-54; instr chem eng, Univ Iowa, 56-58, asst prof, 58-61; res engr, Missile & Space Systs Div, Douglas Aircraft Co, Inc, 61, supvr adv space tech, 62-63, chief eng res sect adv biotech, 63-65, sr staff specialist, Astropower Lab, Newport Beach, 65-77; chem engr, Procon Int Inc, 77-83; CONSULT CHEM ENGR, SAN MARINO, 84- *Concurrent Pos:* Res fel, Eudora Hull Spalding Lab, Calif Inst Technol, 61-62; consult, Douglas Aircraft Co, Inc, 61-62; adj prof chem eng, Calif Polymer Univ, 78-, Univ Southern Calif, 84- *Mem:* Am Inst Aeronaut & Astronaut; Am Chem Soc; Am Inst Chem Eng; Am Soc Eng; Nat Soc Prof Engrs. *Res:* Applied thermodynamics; batteries; bioscience and biotechnology; desalination; fuel cells; heat transfer; membrane separation processes; nuclear engineering; physical and flow porperties of porous media; sorption processes; transport phenomena; ultra-high vacua; petroleum refining processing; synfuels; alternate energy systems. *Mailing Add:* 2075 Del Mar Ave San Marino CA 91108-2809

RUTZ, RICHARD FREDERICK, b Alton, Ill, Feb 9, 19; m 45; c 3. SEMICONDUCTOR DEVICES & MATERIALS. *Educ:* Shurtleff Col, BA, 41; State Univ Iowa, MS, 46. *Prof Exp:* Staff mem electronics, Sandia Corp, 48-51; mgr semiconductor devices, T J Watson Res Ctr, IBM Corp, 51-80, mem staff, 80-87; RETIRED. *Mem:* Am Inst Physics; fel Inst Elec & Electronics Engrs. *Res:* Experimental semiconductor device technology; method of fabrication of crystalline shapes; high speed switching transistors; 20 publications and 50 US patents on transistors, lasers and tunnel diode devices and materials. *Mailing Add:* Rte 1 Box 70 Cold Spring NY 10516

RUUD, CLAYTON OLAF, b Glasgow, Mont, July 31, 34; div; c 2. NONDESTRUCTIVE MATERIALS CHARACTERIZATION. *Educ:* Wash State Univ, BS, 57; San Jose State Univ, MS, 67; Denver Univ, PhD(mat sci), 70. *Prof Exp:* Asst remelt met, Kaiser Aluminum & Chem Corp, 57-58; develop eng, Boeing Aircraft Co, 58-61; res engr, Lockheed Missiles & Space Corp, 62-64, FMC Corp, 64-67; sr res scientist, Denver Univ, 70-79; PROF INDUST ENG, DIR, NDT&E PROG & ASST DIR, MAT RES LAB, PA STATE UNIV, 79- *Concurrent Pos:* Ed, Advances X-Ray Anal, 70-80, X-Ray Spectros, 77-; dir, Denver X-ray conf, 70-79; chmn, Particulates Subcomt, Safe Drinking Water Comt, Nat Acad Sci, dir, 75-77; chmn, Educ Subcomt, Joint Comt Powder Diffraction Standards, 80-85; pres, Denver X-ray Instruments, Inc. *Honors & Awards:* IR 100 Award, Int Ctr Diffraction Data Pattern Contribution. *Mem:* Am Soc Metals; Am Soc Metall Engrs; Soc Exp Stress Anal; Am Soc Testing Mat; Soc Mfg Engrs. *Res:* Materials characterization using x-ray diffraction; ultrasonic velocity and attenuation, eddy current and magentic techniques; cause and effect of residual stresses in metallic components, especially measurement methods and instrumentation; nondestructive materials characterization. *Mailing Add:* 159 Mat Res Lab Pa State Univ University Park PA 16802

RUUS, E(UGEN), b Parnu, Estonia, Aug 19, 17; Can citizen; m 45; c 3. HYDRAULICS. *Educ:* Univ Technol, Tallinn, Estonia, Dipl civil eng, 41; Karlsruhe Tech Univ, DrEng(civil eng), 57. *Prof Exp:* Sr asst & lectr civil eng, Univ Technol, Tallinn, Estonia, 42-43; designer, Dept War, Finland, 44; A B Skanska Cementgjuteriet, Sweden, 45-50 & Wagner & Oliver Consult Engrs, Ont, 50-51; hydraul & sr design engr, B C Eng Co, 51-56; lectr civil eng, 57-58, from asst prof to prof, 58-84, HON PROF CIVIL ENG, UNIV BC, 84- *Res:* Water power development; hydraulic transients and turbine governing; pumping and pumped discharge lines. *Mailing Add:* Dept Civil Eng Univ BC Vancouver BC V6T 1W5 Can

RUVALCABA, ROGELIO H A, b Tepic, Mex, Apr 16, 34; US citizen; m 61; c 3. PEDIATRIC ENDOCRINOLOGY. *Educ:* Univ Guadalajara, MD, 57. *Prof Exp:* Rotating intern med & surg, Hotel Dieu Hosp, New Orleans, 59, resident pediat, 60; resident, Children's Mem Hosp, Omaha, Nebr, 61; resident, Creighton Mem Hosp, 62; trainee pediat endocrinol & metab dis, 62-64, chief trainee, 64-66, from instr to assoc prof pediat endocrinol, 65-77, PROF PEDIAT ENDOCRINOL, UNIV WASH, 77-, ACTG HEAD, DIV PEDIAT ENDOCRINOL, 77- *Concurrent Pos:* Dir endocrinol clin, Rainier Sch, 66-; mem human res rev bd, Dept Social & Health Serv, 75-78; dir pediat endocrine clin, Mary Bridge Children's Hosp, Tacoma, 75- *Mem:* Endocrine Soc; Lawson Wilkins Pediat Endocrine Soc; Western Soc Pediat Res; Am Acad Pediat. *Res:* Metabolic and endocrine disorder. *Mailing Add:* 1280 SW 301st St Federal Way WA 98003

RUVALDS, JOHN, b Jelgava, Latvia, Mar 26, 40; US citizen; div. PHYSICS. *Educ:* Grinnell Col, BA, 62; Univ Ore, MA, 65, PhD(physics), 67. *Prof Exp:* Res assoc physics, James Franck Inst, Univ Chicago, 67-69; from asst prof to assoc prof, 69-76, PROF PHYSICS, UNIV VA, 76- *Mem:* Am Phys Soc; NY Acad Sci. *Res:* Theoretical solid state physics; superconductors. *Mailing Add:* Dept Physics Jesse W Beams Lab 206 Univ Va Charlottesville VA 22903

RUWART, MARY JEAN, b Detroit, Mich, Oct 16, 49. GASTROENTEROLOGY, DRUG DELIVERY. *Educ:* Mich State Univ, BS, 70, PhD(biophys), 74. *Prof Exp:* NIH trainee biophys, Mich State Univ, 70-73; res assoc surg, Med Sch, St Louis Univ, 74-75, from instr to asst prof, 75-76; res scientist gastroenterol, 76-84, sr res scientist diabetes & gastroenterol res, 84-87, SR RES SCIENTIST, UPJOHN CO, 87- *Honors & Awards:* Fred Kagen Lead Finding Award, 86. *Mem:* Am Gastroenterol Asn; Am Fedn Clin Res; AAAS; Am Asn Study Liver Dis; Am Asn Pharmaceut Sci. *Res:* Intestinal motility; cholesterol metabolism; gallstone disease; organ preservation and function; carbohydrate and lipid metabolism; integration of basic and applied sciences; nutrition; prostaglandins; liver disease; peptide absorption. *Mailing Add:* Drug Delivery Syst Res Upjohn Co Kalamazoo MI 49001

RUWE, WILLIAM DAVID, b Lafayette, Ind, Feb 18, 53; m 90. THERMOREGULATION, FEVER & ANTIPYRESIS. *Educ:* Wabash Col, AB, 75; Purdue Univ, MS, 77, PhD(neurobiol), 80. *Prof Exp:* Res asst physiol psychol, Purdue Univ, W Lafayette, Ind, 75-78; res asst, Univ NC, 78-80, fel, 80-81; fel, Univ Calgary, Alta, Can, 80-85; res assoc, Ft Detrick, Frederick, Md, 85-86; ASST PROF PHYSIOL, UNIV ARK MED SCH, 86- *Concurrent Pos:* Prin investr, Nat Inst Neurol & Commun Disorders & Stroke, NIH, 88- & Univ Ark Med Sch, 88- *Mem:* Am Physiol Soc; Soc Neurosci; AAAS; Soc Exp Biol & Med; Sigma Xi; Int Brain Res Orgn. *Res:* Thermoregulatory processes and their possible mediation by neuroactive substances of the hypothalamus; the febrile state and associated acute response; mechanisms of antipyresis, both endogenous and exogenous; addictive processes, especially withdrawal; the alterations occurring in some of these processes that are altered during aging. *Mailing Add:* Dept Physiol & Biophys Slot 505 Univ Ark Med Sch 4301 W Markham St Little Rock AR 72205-7199

RUYLE, WILLIAM VANCE, b Malcolm, Nebr, Feb 20, 20; m 47; c 6. ORGANIC CHEMISTRY, PHARMACEUTICAL CHEMISTRY. *Educ:* Univ Nebr, AB, 42, MS, 43; Univ Ill, PhD(org chem), 49. *Prof Exp:* Res assoc synthetic med chem, Merck & Co, Inc, 49-61, res fel, 61-77, sr res fel, Merck Sharp & Dohme Res Labs, 77-85; RETIRED. *Mem:* Am Chem Soc. *Res:* Research and development in synthetic organic and medicinal chemistry. *Mailing Add:* 7134 Dudley St Lincoln NE 68505

RUZE, JOHN, b New York, NY, May 24, 16; m 56; c 4. ELECTRONICS ENGINEERING. *Educ:* City Col New York, BS, 38; Columbia Univ, MS, 40; Mass Inst Technol, ScD(electronic eng), 52. *Prof Exp:* Sect head antenna design, Signal Corps Radar Labs, 40-46; asst lab head, Air Force Cambridge Res Labs, 46-52; dir, Gabriel Lab, 52-54; pres, Radiation Eng Lab, 54-62; Sr Staff Mem, Lincoln Labs, Mass Inst Technol, 62-; RETIRED. *Concurrent Pos:* Ed, Inst Elec & Electronics Engrs Trans Antennas & Propagation, 67-69. *Mem:* Fel Inst Elec & Electronics Engrs; Int Union Radio Sci. *Res:* Microwave optics; large antenna systems; electromagnetic theory. *Mailing Add:* Concord Greene Apt No 4-6 Concord MA 01742

RUZICKA, FRANCIS FREDERICK, JR, b Baltimore, Md, June 30, 17; m 41; c 6. RADIOLOGY. *Educ:* Col Holy Cross, AB, 39; Johns Hopkins Univ, MD, 43. *Prof Exp:* Roentgenologist, Cancer Detection Ctr, Univ Minn, 48-49, instr radiol, Univ Hosps, 49-50; chmn dept radiol, St Vincent's Hosp & Med Ctr, New York, 50-73; assoc chmn dept, 73-76, chmn dept, 76-81, PROF RADIOL, MED CTR, UNIV WIS-MADISON, 73- *Concurrent Pos:* Consult radiol, St Elizabeth's Hosp, Elizabeth, NJ, Overlook Hosp, Summit, NJ & Columbus Hosp, New York; assoc clin prof radiol, Sch Med, NY Univ, 50-53, clin prof, 54-73; consult comt Vet Admin health resources, Nat Acad Sci, 75-76. *Mem:* Radiol Soc NAm; AMA; fel Am Col Radiol; NY Acad Sci; Am Roentgen Ray Soc. *Res:* Roentgen aspects of the portal venous system; xeromammagraphy; vascular roentgenology; double contrast techniques in gastrointestinal radiology including the use of drugs; digital video angiography with special reference to renal artery surgery; motility studies of alimentary tract using radiologic techniques. *Mailing Add:* Univ Wis Clin Ctr NIH Bethesda MD 20205

RYALL, ALAN S, JR, b San Mateo, Calif, July 6, 31; m 57; c 2. SEISMOLOGY. *Educ:* Univ Calif, Berkeley, AB, 56, MA, 59, PhD(geohpys), 62. *Prof Exp:* Res seismologist, Univ Calif, Berkeley, 60-61; geophysicist, US Geol Surv, 62-63; prof seismol, seismologist & dir, Seismol Lab, Mackay Sch Mines, Univ Nev, Reno, 64-76; prog mgr, Defense Adv Res Proj Agency, Washington, DC, 76-78; PROF SEISMOL, SEMISMOLOGIST & DIR, SEISMOL LAB, MACKAY SCH MINES, UNIV NEV, RENO, 78- *Concurrent Pos:* Geophysicist, Defense Adv Res Proj Agency, Washington, DC, 76-78; del, Comprehensive Test Ban Treaty Negotiations, Geneva, 78. *Mem:* AAAS; fel Am Geophys Union; Geol Soc Am; Seismol Soc Am (vpres, 80-81, pres, 81-82); Earthquake Eng Res Inst. *Res:* Earthquake seismology; seismic regionalization; earthquake mechanisms; crust-mantle structure; test ban treaty verification. *Mailing Add:* Dept Geol & Geog Univ Nev Reno NV 89557

RYALS, GEORGE LYNWOOD, JR, b Erwin, NC, Nov 14, 41; m 67; c 1. INVERTEBRATE ECOLOGY. *Educ:* Elon Col, AB, 66; Appalachian State Univ, MA, 67; Clemson Univ, PhD(zool), 75. *Prof Exp:* Asst prof biol, Lees-McRae Col, 67-69 & Southeastern Col, 69-70; assoc prof biol & chmn dept, Elon Col, 73; DIR, GENETIC DESIGN. *Mem:* Am Inst Biol Sci; AAAS; Am Soc Zoologists; NAm Benthological Soc. *Res:* Structure and function of benthic communities; seasonal regulation in Chironomidae and Odonata. *Mailing Add:* Genetic Design 7017 Albert Pick Rd Greensboro NC 27409

RYAN, ALLAN JAMES, b Brooklyn, NY, Dec 9, 15; m 42; c 3. MEDICAL EDITING. *Educ:* Yale Univ, BA, 36; Columbia Univ, MD, 40; Am Bd Surg, dipl, 47. *Hon Degrees:* Doctor Sports Sci, US Sports Acad, 83. *Prof Exp:* From assoc prof to prof rehab med & phys educ, Univ Wis-Madison, 65-76, athletic teams physician, 65-76; ed-in-chief, Postgrad Med, 76-79, The Physician & Sports Med, 73-86, Fitness in Bus, 87-89; RETIRED. *Concurrent Pos:* Pvt pract, gen surg, Meriden, Conn, 46-65. *Mem:* Hon fel Brit Asn Sport & Med; fel Am Col Sports Med. *Res:* Sports medicine; physical education. *Mailing Add:* 4510 W 77th St Edina MN 55435

RYAN, ANNE WEBSTER, b Lowell, Mass, Apr 17, 27; wid; c 2. ORGANIC CHEMISTRY, INFORMATION SCIENCE. *Educ:* Simmons Col, BS, 49. *Prof Exp:* Asst ed, 49-51 & 52-64, group leader, 64-65, asst dept head, 65-67, MGR, CHEM ABSTRACTS SERV, 67- *Concurrent Pos:* Tech librn res & develop, Lever Bros Co, 51-52. *Mem:* Am Chem Soc. *Res:* Chemical information, storage and retrieval. *Mailing Add:* 1785 F Northwest Ct Columbus OH 43210

RYAN, BILL CHATTEN, b Long Beach, Calif, Oct 2, 28; m 50; c 4. METEOROLOGY. *Educ:* Univ Nev, Reno, BS, 50; Tex A&M Univ, MS, 64; Univ Calif, Riverside, PhD(geog), 74. *Prof Exp:* Officer electronics, US Air Force, 51-58, officer meteorol, 58-65; res meteorologist, Meteorol Res Inc,

65-67; RES METEOROLOGIST, USDA FOREST SERV, 67- *Concurrent Pos:* Mem, Sci Comt Riverside Air Pollution Control Dist, 69- *Honors & Awards:* Outstanding Meteorologist Award, Riverside-San Bernardino Chap, Am Meteorol Soc, 75. *Mem:* Am Meteorol Soc. *Res:* Development of a mathematical model to diagnose and predict wind in remote areas of mountainous terrain. *Mailing Add:* 2291 Quartz Pl Riverside CA 92507

RYAN, CARL RAY, b Gateway, Ark, Mar 3, 38; m 64; c 2. SATELLITE COMMUNICATION SYSTEMS FOR MILITARY APPLICATIONS. *Educ:* Univ Ark, BSEE, 62; Iowa State Univ, MS, 63; Univ Mo, Rolla, PhD, 69. *Prof Exp:* Prof elec, Mich Technol Univ, 77-79; mem tech staff, 63-77 & 79-89, vpres, 89-90, DIR, STRATEGIC ELECTRONICS DIV, MOTOROLA INC, 90- *Concurrent Pos:* Dan Nobel fel, Motorola Inc, 75; adj prof, Ariz State Univ, 81-89; gen chmn, Phoenix Conf Computers & Commun, 88; mem, Eng Acceleration Comn, 88- *Mem:* Fel Inst Elec & Electronics Engrs; Am Defense Preparedness Asn. *Res:* Very high data rate communication equipment; channel models, computer simulation, coders, modems and circuits; computer aided design of communication system hardware. *Mailing Add:* Strategic Electronics Div Motorola Inc Rte 2 Box 170 Cassville MO 65625

RYAN, CECIL BENJAMIN, b Runge, Tex, Aug 4, 16; m 41; c 2. POULTRY SCIENCE. *Educ:* Tex Col Arts & Indust, BS, 38; Tex A&M Univ, MS, 47, PhD, 62. *Prof Exp:* Teacher pub sch, 38-42; from asst prof to prof, 47-81, EMER PROF POULTRY SCI, TEX A&M UNIV, 81- *Concurrent Pos:* Piper Prof, 66. *Mem:* Fel AAAS; Poultry Sci Asn (secy-treas, 54-77, first vpres, 78-79, pres, 79-80); Genetics Soc Am; World Poultry Sci Asn. *Res:* Environmental physiology and management. *Mailing Add:* Dept Poultry Sci Tex A&M Univ College Station TX 77843

RYAN, CHARLES EDWARD, JR, b Crestline, Ohio, Mar 30, 38; m 62; c 5. ELECTRICAL ENGINEERING. *Educ:* Case Inst Technol, BSc, 60; Ohio State Univ, MSc, 61, PhD(elec eng), 68. *Prof Exp:* Eng asst, North Elec Co, 56-60; res asst, Electrosci Lab, Ohio State Univ, 60-62, res assoc, 62-68, asst supvr, 68-71; chief, Electromagnetic Effectiveness Div, 80-88, SR RES ENGR, ENG EXP STA, GA INST TECHNOL, 71-, PRIN RES ENGR, 76- *Mem:* Inst Elec & Electronics Engrs; Sigma Xi. *Res:* Electromagnetic theory applied to antennas and scattering; radar cross section studies; computer analysis. *Mailing Add:* 2499 Kingsland Dr Doraville GA 30360

RYAN, CHARLES F, b Terre Haute, Ind, Dec 30, 41; m 62; c 1. CLINICAL PHARMACOLOGY, DRUG DEVELOPMENT. *Educ:* Purdue Univ, BS, 63, MS, 66, PhD(pharmacol), 69. *Prof Exp:* Asst prof pharmacol, Univ Wisc, 68-72; assoc prof & chmn pharmacol, Univ Nebr, 72-79; prof & dep dean pharm, Wayne State Univ, 79-82; dir clin res, Harris Labs, 82-89; DIR CLIN DEVELOP, PHARMACO, 89- *Concurrent Pos:* Adj prof, Univ Nebr, 82-89. *Mem:* Acad Pharmaceut Sci; Am Asn Pharmaceut Scientists; Am Soc Clin Pharmacol & Therapeut; Am Soc Pharmacol & Exp Therapeut; Am Pharmaceut Asn; Drug Info Asn. *Res:* Clinical research and drug development. *Mailing Add:* Pharmaco 4009 Banister Lane Austin TX 78704

RYAN, CLARENCE AUGUSTINE, JR, b Butte, Mont, Sept 29, 31; m 54; c 4. CHEMISTRY. *Educ:* Carroll Col, Mont, BA, 53; Mont State Univ, MS, 56, PhD(chem), 59. *Prof Exp:* Fel, Ore State Univ, 59-61; fel, Western Regional Lab, USDA, 61-63, chemist, 63-64; asst agr chemist, Wash State Univ, 64-68, assoc prof biochem & assoc agr chemist, 68-72, chmn, Dept Agr Chem, 78-80, PROF BIOCHEM & FEL, INST BIOL CHEM, WASH STATE UNIV, 72- *Concurrent Pos:* USPHS Career Develop Award, 64-74; prog mgr competitive grants, Biol Stress, USDA, 83-84; mem bd dirs, Int Soc Plant Molecular Biol, 88-91; consult, Kimin Ind, Des Moines, Iowa. *Honors & Awards:* Merck Award, 59. *Mem:* Nat Acad Sci; Am Soc Biol Chemists; Am Chem Soc; Am Soc Plant Phys; Int Soc Plant Molecular Biol; Int Soc Chem Ecol; AAAS. *Res:* Protein chemistry; plant proteolytic enzymes, naturally occurring proteinase inhibitors; plant molecular biology; signal transduction in plants. *Mailing Add:* Inst Biol Chem Wash State Univ Pullman WA 99164-6340

RYAN, CLARENCE E, JR, b Moline, Ill; m 41. ORGANIC COATINGS FOR METAL. *Educ:* St Ambrose Univ, BS, 43; John Marshall & Blackstone Law, LLB, 56. *Prof Exp:* Navy comn, naval sci, Notre Dame, 44; res fel physics & chem, Augustana Col, Univ Ala, Univ Ill & Mich State Univ, 46; Instr inorg chem, Univ Ill, Moline, 46-47; apprentice pharmacist, Rexall Drugs, 47-48; chemist analyst & spectrogr, Aluminum Co Am, 48, org coatings, aluminum seal div, 48-50; tech dir, metal cont div, Ball Metal & Coatings Co, 50-74; TECH MGR, BENNETT INDUSTS, STEEL CONTAINERS, 78- *Concurrent Pos:* Res asst, Off Naval Res, Naval Reserve, 50-62; consult, coatings, Weyerhauser Corp, 75-76, Par Enterprises, 77-78; chief developer, Bennett Volatile Emission Plant Control, Ill Environ Protection Agency, 82-85; chief researcher, container exemptions, Dept Transp, 83-86, tech adv, develop new containers, 85-86; mil spec adv, container coating technol, US Army, 84-85; mem, Hazardous Mat Adv Coun. *Mem:* Am Chem Soc; fel Am Inst Chemists; Am Soc Testing & Mat; Am Defense Preparedness Asn. *Res:* Exterior functional and lining materials for steel, aluminum and tinplate containers; interpretation of proper containers for hazardous materials or food products; appropriate coatings for various products. *Mailing Add:* PO Box 247 Oak Forest IL 60452

RYAN, DALE SCOTT, b Pasadena, Calif, Sept 7, 47; m 70; c 1. FOOD SCIENCE. *Educ:* Occidental Col, AB, 69; Univ Calif, Davis, MS, 71, PhD(biochem), 74. *Prof Exp:* Chemist, Agr Res Serv, USDA, 69-70; ASST PROF FOOD CHEM, UNIV WIS-MADISON, 74- *Mem:* Inst Food Technologists. *Res:* Characterization of the biochemical determinants of nutritional value; isolation and characterization of the main proteins of economically important plant protein resources; protein structure and function. *Mailing Add:* 814 Harvey Rd Madison WI 53704

RYAN, DAVE, Can citizen. COMMUNICATION ELECTRONICS, REMOTE SENSING. *Educ:* St Francis Xavier Univ, BSc Hon, 79; Carleton Univ, BSc Hon, 83, MEng, 84. *Prof Exp:* Qual assurance, Res & Develop Group, Epitek Electronics, Kanata, Ont, 82-83; teaching asst electronics, Dept Elec Eng, Carleton Univ, 83-84; res assoc isotope mass spectrometry, Dept Geol, Univ Ottawa, 84-85; vpres stable isotopes, Spectrion Anal Trace Elements, 84-86; systs engr electron microscopes, Cambridge Instruments Ltd, PLC, 85-89; PRES MICRO COMPUTERS, COMSOL CORP, 89-; GEN MGR ELECTRONICS, DAVE RYAN CONSULT ENG, 89- *Concurrent Pos:* Consult, Canmet, Energy Mines & Resources, Govt Can, 89- *Mem:* Am Phys Soc; Can Asn Physicists; hon mem Can Med Asn; hon mem Radio Soc Gt Brit; Can Amateur Radio Fedn; Am Radio Relay League. *Res:* Engineering analysis and design for electronic systems in scientific instruments and communication electronics applications; prototype and testing of electronic structures; reliability engineering; production design implementation; mass spectrometry techniques; electron microscopes; image analysis; x-ray optics; remote sensing techniques; radio frequency device modeling and field propagation. *Mailing Add:* Comsol Corp 5010 Merivale Depot Ottawa ON K2C 3H3 Can

RYAN, DAVID GEORGE, b Quebec, Que, Dec 31, 37. PARTICLE PHYSICS. *Educ:* Queen's Univ, Ont, BSc, 59, MSc, 61; Univ Birmingham, PhD(physics), 65. *Prof Exp:* Res fel physics, Univ Birmingham, 65-66; res assoc, Cornell Univ, 66-67; from asst prof to assoc prof, 67-81, PROF PHYSICS, MCGILL UNIV, 81- *Mem:* Am Phys Soc. *Res:* Interactions and decays of elementary particles. *Mailing Add:* Dept Physics McGill Univ 3600 University Montreal PQ H3A 2T8 Can

RYAN, DONALD EDWIN, b San Diego, Calif, July 1, 35. MATHEMATICS, ASTRONOMY. *Educ:* Univ Tex, BA, 57, MA, 61, PhD(math), 64. *Prof Exp:* Aero-engr, Pensacola Naval Air Sta, 55; test engr, Convair Astronaut Div, Gen Dynamics Corp, 57-59; instr math, Univ Tex, 59-63, 63-64; asst prof math, Eastern Mich Univ, 64-65; asst prof, Bowling Green State Univ, 65-68; assoc prof, 68-74, PROF MATH, NORTHWESTERN STATE UNIV, 74- *Concurrent Pos:* NSF consult, India, 67; chmn gifted & talented adv comt, Northwestern State Univ. *Mem:* Am Math Soc; Math Asn Am. *Res:* Functions of a complex variable which are ecart fini and their topological structures. *Mailing Add:* Dept Math Northwestern State Univ Natchitoches LA 71497

RYAN, DONALD F, b Syracuse, NY, July 24, 30; m 59; c 4. PHYSICS. *Educ:* LeMoyne Col, NY, BS, 57; Cath Univ Am, MS, 60, PhD(physics), 63. *Prof Exp:* Res fel physics, Cath Univ Am, 63-64, from res asst prof to res assoc prof, 64-66; from asst prof to assoc prof, 66-72, chmn dept physics & earth sci, 70-72, PROF PHYSICS, STATE UNIV NY COL PLATTSBURGH, 72- *Mem:* Am Phys Soc. *Res:* Cosmic ray and elementary particle research. *Mailing Add:* Four Valcocr Blvd Plattsburgh NY 12901

RYAN, DOUGLAS EARL, b Can, Jan 21, 22; m 45; c 3. INORGANIC CHEMISTRY, ANALYTICAL CHEMISTRY. *Educ:* Univ NB, BSc, 44; Univ Toronto, MA, 46; Imp Col, London, dipl & PhD(chem), 51; Univ London, DSc, 65. *Prof Exp:* Asst prof chem, Univ NB, 46-48; from asst prof to assoc prof, 51-63, chmn, dept chem, 69-73, MCLEOD PROF, DALHOUSIE UNIV, 63-, DIR, TRACE ANALYSIS RES CTR, 71-, DIR, SLOWPOKE FACIL, 76-, EMER PROF CHEM, 87- *Concurrent Pos:* Nat Res Coun Can traveling fel, 59-60; exec dir, Ctr Anal Res & Develop, Univ Colombo, Sri Lanka, 80- *Honors & Awards:* Fisher Sci Lect Award, Chem Inst Can, 72. *Mem:* Fel Chem Inst Can. *Res:* Metal chelates; molecular spectroscopy; neutron activation and trace analysis in general. *Mailing Add:* Dept Chem Dalhousie Univ Halifax NS B3H 4J1 Can

RYAN, EDWARD MCNEILL, b St Louis, Mo, May 24, 20; m 76; c 3. GEOLOGY. *Educ:* Miami Univ, BA, 41; Univ Mo-Columbia, MA, 43. *Prof Exp:* Aerial phototopographer, US Army Corp Engrs, 43-46; PROF GEOL, STEPHENS COL, 46- *Mem:* Nat Asn Geol Teachers. *Res:* Geologic travel and color photography in fifty states, most provinces of Canada, Virgin Islands and East Africa. *Mailing Add:* 505 S Garth Columbia MO 65203

RYAN, FREDERICK MERK, b Pittsburgh, Pa, Jan 20, 32; m 52; c 2. PHYSICS. *Educ:* Carnegie-Mellon Univ, BS, 54, MS, 56, PhD(physics), 59. *Prof Exp:* Res physicist, Westinghouse Elec Corp, 59-61, sr physicist, 61-66, fel scientist, 66-72, adv scientist, 72-85, consult scientist, 85-90, LEADER INSTRUMENT DEVELOP GROUP, WESTINGHOUSE ELEC CORP, 90- *Concurrent Pos:* Rosemount anal instr, 90- *Mem:* Am Phys Soc; Electrochem Soc. *Res:* Experimental solid state physics; luminescence; optical physics; optical analytical instrumentation. *Mailing Add:* PO Box 406 New Alexander PA 15670

RYAN, GEORGE FRISBIE, b Yakima, Wash, July 28, 21; m 47; c 1. HORTICULTURE. *Educ:* State Col Wash, BS, 47; Univ Calif, Los Angeles, PhD(hort sci), 53. *Prof Exp:* Res asst hort, Univ Calif, Los Angeles, 48-52, instr & jr horticulturist, 52-54, asst prof & asst horticulturist, 54-60, assoc specialist, 60-61; asst prof & asst horticulturist, Citrus Exp Sta, Univ Fla, 61-67; assoc horticulturist, 67-79, HORTICULTURIST, WESTERN WASH RES & EXTEN CTR, WASH STATE UNIV, 79- *Mem:* Am Soc Hort Sci; Weed Sci Soc Am; Int Plant Propagation Soc; Sigma Xi. *Res:* Chemical regulation of plant growth and flowering; nutrition of ornamental plants; chemical weed control. *Mailing Add:* 1877 Skyline Dr Tacoma WA 98406

RYAN, JACK A, b Pittsburgh, Pa, Nov 3, 29; m 56; c 2. GEOPHYSICS. *Educ:* Rice Univ, BS, 51; Pa State Univ, PhD(geophys), 59. *Prof Exp:* Res scientist, Douglas Aircraft Co, 59-61, chief lunar & planetary sci sect, 61-63, chief environ sci br, McDonnell Douglas Astronaut Co, Huntington Beach, 69-76, chief lunar & planetary sci br, 73-76; lectr earth sci, 75, team leader, Viking Mars Meteorol Sci, 77-79, prof geol sci & chair dept, Calif State Univ, Fullerton, 76-89. *Concurrent Pos:* Consult, US Army Corps Engrs, 64-69 & McDonnell Douglas Astronaut Co, 78-; mem resources & environ subgroup, Working Group on Extraterrestrial Resources, NASA, 65-71. *Honors &*

Awards: Newcombe-Cleveland Award, AAAS, 77. *Mem:* AAAS; Am Geophys Union; Asn Eng Geologists. *Res:* Planetary physics, Viking 1976 Mars Landers Meteorology Team; atmospheric science, especially Mars atmospheric dynamics; solid earth geophysics. *Mailing Add:* 10182 La Sierra Pl Santa Ana CA 92705

RYAN, JACK LEWIS, b Dallas, Ore, May 14, 33; m 64; c 1. INORGANIC CHEMISTRY. *Educ:* Ore State Univ, BS, 53, MS, 56. *Prof Exp:* Chemist, Hanford Lab, Gen Elec Co, 55-60, sr scientist, 60-65; sr res scientist, 65-69, RES ASSOC, PAC NORTHWEST LABS, BATTELLE MEM INST, 69- *Concurrent Pos:* Consult, Lawrence Berkeley Lab, Univ Calif, 74; lectr, Joint Ctr Grad Study, Richland, Wash, 74- *Mem:* AAAS; Am Chem Soc. *Res:* Inorganic and physical chemistry of actinide and lanthanide elements; coordination chemistry; absorption spectroscopy; ion exchange; solvent extraction; non-aqueous solutions; electrochemistry, chemical processing of actinide elements. *Mailing Add:* 1326 Broadview Dr West Richland WA 99352

RYAN, JAMES ANTHONY, b Cairo, Ill, Mar 16, 43; m 63; c 3. SOIL CHEMISTRY, SOIL BIOCHEMISTRY. *Educ:* Murray State Univ, BS, 66; Univ Ky, MS, 68, PhD(soils), 71. *Prof Exp:* Fel soil sci, Univ Wis-Madison, 71-74; soil scientist, Munic Environ Res Lab, 74-77, MEM STAFF, ULTIMATE DISPOSAL RES PROG, ENVIRON PROTECTION AGENCY, 77- *Mem:* Am Soc Agron; Soil Sci Soc Am; Water Pollution Control Fedn; AAAS; Sigma Xi. *Res:* Nitrogen transformations in soils; sewage sludge disposal on agricultural lands; transformation of heavy metals in soils in relation to their phytotoxic effects. *Mailing Add:* Risk Reduction Eng Lab Environ Protection Agency Cincinnati OH 45268

RYAN, JAMES M, b Milwaukee, Wis, Feb 20, 32; m 55, 89; c 8. CHEMICAL ENGINEERING. *Educ:* Univ Mich, BS & MS, 55; Mass Inst Technol, ScD(chem eng), 58. *Prof Exp:* Technologist, Shell Chem Co, 58-63, res engr, 63-65; proj leader gas chromatography, Abcor, Inc, 65-66, prog mgr, 66-70, mgr eng, 70-71, mgr commercial plants eng, 71-73, vpres, Abcor Japan, 73-75, dir technol, Abcor, Inc, 75-78; mgr process & prod develop, Helix Process Systs Inc, 78-81; mgr res & develop, 81-88, dir process design & evaluation, 88-91, ENG FEL, KOCH PROCESS SYSTS, INC, 91- *Mem:* Am Chem Soc; Am Inst Chem Engrs; Soc Petrol Engrs. *Res:* Chemical process development; membrane process and equipment design, engineering, plants; separation processes; gas chromatography; large scale liquid chromatography; reaction kinetics; chemical manufacturing economics; process simulation; mass transfer; cryogenic gas field separations; large scale simulation. *Mailing Add:* Seven Swallow Lane Wichita KS 67230-6619

RYAN, JAMES MICHAEL, b Chicago, Ill, Oct 9, 47; m 79. GAMMA RAY ASTRONOMY, SOLAR PHYSICS. *Educ:* Univ Calif Riverside, BA, 70, PhD(physics), 78; Univ Calif San Diego, MS, 73. *Prof Exp:* Res scientist, 78-84, RES PROF PHYSICS, UNIV NH, 84- *Concurrent Pos:* Prin investr, Univ NH, 88- *Mem:* Am Geophys Union; Am Astron Soc; Sigma Xi. *Res:* Comptel instrument on gamma ray observatory; gamma ray astronomy; solar flare physics (gamma and x-ray measurements); cosmic ray production and acceleration theory; atmospheric gamma ray and neutron measurements. *Mailing Add:* Space Sci Ctr Univ NH Durham NH 03824

RYAN, JAMES PATRICK, b Philadelphia, Pa, Jan 10, 47; m 68; c 2. GASTROINTESTINAL PHYSIOLOGY. *Educ:* Villanova Univ, BS, 68, MS, 70; Hahnemann Med Col, PhD(physiol), 74. *Prof Exp:* Fel, Sch Med, Univ Pa, 74-75; asst prof, 75-80, ASSOC PROF, SCH MED, TEMPLE UNIV, 80- *Concurrent Pos:* Lectr, Gwynedd Mercy Col, 76- *Mem:* NY Acad Sci; Sigma Xi. *Res:* Neural and hormonal control of gastrointestinal smooth muscle motility, including how motor patterns are affected during pregnancy. *Mailing Add:* Dept Physiol Sch Med Med Sch Temple Univ Philadelphia PA 19140

RYAN, JAMES WALTER, b Amarillo, Tex, June 8, 35; div; c 3. MEDICINE, BIOCHEMISTRY. *Educ:* Dartmouth Col, AB, 57; Cornell Univ, MD, 61; Oxford Univ, DPhil(biochem), 67. *Prof Exp:* Intern med, Montreal Gen Hosp, Can, 61-62, resident, 62-63; res assoc, NIH, 63-65; hon med officer to Regius prof med, Oxford Univ, 65-67; asst prof biochem, Rockefeller Univ, 67-68; assoc prof, 68-79, PROF MED, SCH MED, UNIV MIAMI, 79- *Concurrent Pos:* USPHS fel, Oxford Univ, 65-67; USPHS spec fel, Rockefeller Univ, 67-68, career develop award, 68; sr scientist, Papanicolaou Cancer Res Inst, 72-77; vis prof, Clin Res Inst Montreal, 74; mem coun cardiopulmonary dis & med adv bd, Coun High Blood Pressure, Am Heart Asn; investr, Howard Hughes Med Inst, 68-71; Pfizer travelling fel, Univ Montreal, 74; vis fac, Mayo Clin, 74. *Honors & Awards:* Louis & Artur Luciano Award, McGill Univ, 85-85. *Mem:* Am Heart Asn; Am Chem Soc; Am Soc Biol Chemists; Brit Biochem Soc; AAAS; Biochem Soc; NY Acad Sci; Sigma Xi; Europ Microcirulation Soc. *Res:* Action and metabolism of the vasoactive polypeptides; bradykinin and angiotensin, and their relation to diseases of high blood pressure. *Mailing Add:* Dept Med Univ Miami Sch Med Miami FL 33101

RYAN, JOHN DONALD, b Norristown, Pa, July 9, 21; m 43; c 3. GEOLOGY. *Educ:* Lehigh Univ, BA, 43, MS, 48; Johns Hopkins Univ, PhD(geol), 52. *Prof Exp:* Asst, Lehigh Univ, 46-48; asst, Johns Hopkins Univ, 49-50; tech expert, State Dept Geol, Md, 50-52; from instr to prof, 52-84, chmn dept, 61-76, EMER PROF GEOL, LEHIGH UNIV, 84- *Concurrent Pos:* Coop geologist, State Topol & Geol Surv, Pa, 47-48, 53; geologist, US Geol Surv, 48-; Fulbright lectr, Cent Univ Ecuador, 71. *Mem:* Geol Soc Am; Am Asn Geol Teachers; Soc Econ Paleont & Mineral; Geol Soc Finland. *Res:* Recent sediments in Chesapeake Bay; studies in the stratigraphic control of uranium deposits; environmental geology; Cretaceous and Tertiary sedimentation in the Wyoming Rockies. *Mailing Add:* 305 Walnut St Hellertown PA 18055

RYAN, JOHN F, b Boston, Mass, May 16, 35; m 59; c 3. ANESTHESIOLOGY. *Educ:* Boston Col, AB, 57; Columbia Univ, MD, 61. *Prof Exp:* Asst prof anesthesiol, Col Physicians & Surgeons, Columbia Univ & Columbia-Presby Med Ctr, 68-69; ASST PROF ANESTHESIOL, HARVARD MED SCH & MASS GEN HOSP, 69- *Res:* Pediatric anesthesia; biochemistry of the myoneural junction; hypotensive anesthesia and malignant hyperpyrexia. *Mailing Add:* Mass Gen Hosp Fruit St Boston MA 02114

RYAN, JOHN PETER, b St Paul, Minn, July 28, 21; m 44; c 5. PHYSICAL CHEMISTRY. *Educ:* Col St Thomas, BS, 43; Univ Minn, PhD(phys chem), 52. *Prof Exp:* Instr chem, Col St Thomas, 46-51; res chemist radiochem, 51-61, mgr nuclear prod dept, 61-73, MGR STATIC CONTROL SYSTS DEPT, 3M CO, 73- *Mem:* Am Nuclear Soc; Am Chem Soc. *Res:* Development of commercial and medical products containing radioactive isotopes. *Mailing Add:* 360 Edith Dr West St Paul MN 55118-3008

RYAN, JOHN WILLIAM, b La Crosse, Wis, Sept 8, 26; m 52; c 3. ORGANIC CHEMISTRY. *Educ:* Loras Col, BS, 48; Univ Iowa, MS, 51; Univ Ky, PhD(org chem), 57. *Prof Exp:* Proj leader org chem res, 57-65, supvr, 65-68, personnel coordr res, develop & eng, 68-69, mgr tech serv & develop, resins & chem bus, 69-73, MGR RES & DEVELOP SILICONE FLUIDS, DOW CORNING CORP, 73- *Mem:* Sigma Xi. *Res:* Organosilicone and organometallic compounds. *Mailing Add:* 514 Linwood Midland MI 48640

RYAN, JON MICHAEL, b Ottumwa, Iowa, Nov 14, 43; m 65; c 1. CELL BIOLOGY. *Educ:* William Penn Col, BA, 65; Univ Nebr, Lincoln, MS, 67; Iowa State Univ, PhD(cell biol), 70. *Prof Exp:* NIH fel, 70-71, res asst, 72-76, res assoc cell biol, Wistar Inst Anat & Biol, 76; res scientist, Res Inst, Ill Inst Technol, 76-78, sr scientist, 78-81; cell biologist, 81-83, cell biol group leader, 84-89, ASSOC RES FEL, ABBOTT LABS, 89- *Mem:* AAAS; Tissue Cult Asn; Sigma Xi. *Res:* Control of cellular proliferation in tissue culture; aging; large scale cell culture; virus production; plasminogen activators. *Mailing Add:* 266 Woodstock Clarendon Hills IL 60014

RYAN, JOSEPH DENNIS, organic chemistry, for more information see previous edition

RYAN, JULIAN GILBERT, b Metamora, Ill, Oct 6, 13; m 36; c 1. PETROLEUM CHEMISTRY. *Educ:* Univ Ill, BS, 35. *Prof Exp:* Res chemist, Shell Oil Co, 35-42, supvr fuels & lubricants develop, 46-63, tech adv to res dir, 63-70, staff res engr, 70-78; RETIRED. *Concurrent Pos:* Group leader, Coord Res Coun, 53-66, mem diesel div, 66-76; chmn, St Louis Sect, Soc Auto Engrs, 64-65. *Mem:* AAAS; Am Chem Soc; Soc Automotive Eng. *Res:* Corrosive and abrasive wear of engines; gasoline oxidation stability; tetraethyl lead antagonism by sulfur compounds; antiknock performance and abnormal combustion phenomena; aviation and automotive fuels and lubricants; automotive engines exhaust emission; patents and publications in fuels and lubricants. *Mailing Add:* 664 Halloran Ave Wood River IL 62095

RYAN, KENNETH JOHN, b New York, NY, Aug 26, 26; m 48; c 3. ENDOCRINOLOGY, BIOCHEMISTRY. *Educ:* Harvard Med Sch, MD, 52; Am Bd Obstet & Gynec, dipl, 64. *Prof Exp:* Intern med, Mass Gen Hosp, Boston, 52-53, resident, 53-54, fel biochem, 54-55, Am Cancer Soc fel, 54-56; asst resident, Columbia-Presby Med Ctr, 56-57; resident, Boston Lying-In-Hosp & Free Hosp Women, 57-60; dir, Fearing Lab, Free Hosp Women, 60-61; Arthur H Bill prof obstet & gynec & chmn dept, Sch Med, Case Western Reserve Univ, 61-70, chmn dept reprod biol, 68-70, coordr biol sci, 69-70; prof reprod biol & chmn, Dept Obstet & Gynec, Univ Calif, San Diego, 70-72, chief-of-staff, Boston Hosp Women, 73-80; KATE MACY LADD PROF OBSTET & GYNEC & CHMN DEPT, HARVARD MED SCH, 73-, DIR LAB HUMAN REPROD & REPROD BIOL, 74-; CHMN DEPT OBSTET & GYNEC, BRIGHAM & WOMEN'S HOSP, 80- *Concurrent Pos:* Fel med, Harvard Med Sch, 53-54 & 55-56, teaching fel obstet & gynec, 57-60, instr, 60-61; dir, Fearing Res Lab, 60-70; dir dept obstet & gynec, Univ Hosps Cleveland, 61-70; mem nat adv coun, USPHS; mem, Pres Comt Ment Retardation, 68-72; chmn, Nat Comn Protection Human Subjects, 74-78; Henry J Kaiser Sr fel, 82; chmn, ethnics comt, Am Col Obstet, & Gynec, 84-88. *Honors & Awards:* Ernst Oppenheimer Award, Endocrine Soc, 64; Weinstein Award United Cerebral Palsy, 71. *Mem:* Inst Med-Nat Acad Sci; Am Soc Biol Chemists; Endocrine Soc; fel Am Col Obstet & Gynec; Soc Gynec Invest; Henry J Kaiser Sr Fel, 82. *Mailing Add:* Brigham & Women's Hosp 75 Francis St Boston MA 02115

RYAN, MICHAEL PATRICK, JR, physics, for more information see previous edition

RYAN, MICHAEL T, b Tipperary, Ireland, Sept 29, 25; Irish & Can citizen; m 55; c 3. BIOCHEMISTRY, CHEMISTRY. *Educ:* Univ Col, Dublin, BSc, 46, MSc, 47; McGill Univ, PhD(biochem), 55. *Prof Exp:* Asst chem, Univ Col, Galway, 47-49; from lectr to assoc prof, Fac Med, 49-71, PROF BIOCHEM, FAC MED & SCI, UNIV OTTAWA, 71- *Mem:* Can Biochem Soc. *Res:* Interaction of steroids and proteins. *Mailing Add:* 2380 Georgina Dr Ottawa ON K2B 7M7 Can

RYAN, NORMAN W(ALLACE), b Casper, Wyo, Feb 9, 19; m 42; c 4. CHEMICAL ENGINEERING. *Educ:* Cornell Univ, BChem, 41, ChemE, 42; Mass Inst Technol, ScD, 49. *Prof Exp:* Chem engr, Standard Oil Co, Ind, 42-46; assoc prof chem eng, 48-57, PROF CHEM ENG, UNIV UTAH, 57- *Concurrent Pos:* Consult. *Mem:* Am Chem Soc; Am Inst Chem Engrs; Sigma Xi. *Res:* Combustion; fluid dynamics. *Mailing Add:* Dept Chem Eng Univ Utah Salt Lake City UT 84112

RYAN, PATRICK WALTER, b Chicago, Ill, Nov 14, 33; m 56; c 4. ORGANIC POLYMER CHEMISTRY. *Educ:* Loyola Univ Chicago, BS, 55; Purdue Univ, PhD(chem), 59. *Prof Exp:* From res chemist to sr res chemist, Sinclair Res Inc, 59-65, group leader, 65-68, div dir, 68-69; res mgr, Arco

Chem Co, 69-84; mgr, Technol Transfer, Atlantic Richfield Co, 84-86; TECH CONSULT, 86- *Mem:* Am Chem Soc; Sigma Xi. *Res:* Syntheses of low molecular weight organic polymers and their applied chemistry; chemicals for tertiary oil recovery and basic petrochemicals. *Mailing Add:* 3567 Via La Primavera Thousand Oaks CA 91360

RYAN, PETER MICHAEL, b Quincy, Mass, Aug 6, 43; m 68; c 2. MATHEMATICS. *Educ:* Calif Inst Technol, BS, 65; Dartmouth Col, AM, 67, PhD(math), 70. *Prof Exp:* Asst prof math, Gustavus Adolphus Col, 69-76; asst prof math, Moorhead State Univ, 76-78; ASSOC PROF MATH & COMPUT SCI, JACKSONVILLE UNIV, 78- *Mem:* AAAS; Am Math Soc; Math Asn Am; Soc Indust Appl Math. *Res:* Frames; complete Brouwerian lattices. *Mailing Add:* Jacksonville Univ 2800 University Blvd N Jacksonville FL 32211

RYAN, RICHARD ALEXANDER, b Detroit, Mich, Feb 27, 25; m 47; c 3. VERTEBRATE ZOOLOGY. *Educ:* Cornell Univ, AB, 48, MS, 49, PhD(zool), 51. *Prof Exp:* From instr to assoc prof, 52-64, chmn dept, 66-69, 74-75, PROF BIOL, HOBART & WILLIAM SMITH COLS, 64- *Mem:* AAAS; Am Soc Ichthyol & Herpet; Am Soc Mammal; Wildlife Soc. *Res:* Life history of vertebrates. *Mailing Add:* 4532 Lakeview Rd Dundee NY 14837

RYAN, RICHARD PATRICK, b Decatur, Ill, May 4, 38; m 60; c 3. PATENT AGENT. *Educ:* Millikin, BA, 60; Univ Ky, PhD(org chem), 68. *Prof Exp:* Chemist, Neisler Labs, Inc, Union Carbide Corp, 60-63; sr scientist, 67-71, group leader, 71-75, sr clin res assoc, Mead Johnson Res Ctr, 75-80, patent coord, 80-82, patent agent, 82-86, SR PATENT AGENT, BRISTOL MYERS CO, 86- *Concurrent Pos:* Lectr, Univ Evansville, 69-73. *Mem:* Fel Am Geriat Soc; Am Chem Soc; Sigma Xi; Am Heart Asn; AAAS. *Res:* Preparation and prosecution of patent applications covering pharmaceutical and nutritional inventions; synthesis of biologically active heterocyclic compounds; correlation of chemical structure with biological activity; drug therapy of cardiovascular, central nervous system, respiratory, neoplastic and nutritional diseases. *Mailing Add:* 15 Ash Ct Middletown CT 06457

RYAN, ROBERT DEAN, b Upland, Calif, Mar 3, 33; m 61; c 2. MATHEMATICS. *Educ:* Calif Inst Technol, BS, 54, PhD(math), 60; Harvard Univ, MPA, 71. *Prof Exp:* Res assoc math, Calif Inst Technol, 60-61; asst prof, US Army Math Res Ctr, Wis, 61-63, asst prof, Univ, 63-65; mathematician, Math Br, Washington, DC, 65-69, prog dir oper res, 69-71, spec asst res, 71-86, DIR, SPEC PROG OFF, OFF NAVAL RES, 86- *Mem:* Am Math Soc; Math Asn Am. *Res:* Harmonic analysis and measure theory. *Mailing Add:* Off Naval Res European Off 223 Old Marylaborne Rd London NW1 5TH England

RYAN, ROBERT F, b Hoquiam, Wash, June 23, 22; div. SURGERY. *Prof Exp:* Resident surg, Emergency Hosp, Washington, DC, 50-51; from instr to asst prof, Tulane Univ, 56-63, assoc prof, Tulane Univ, plastic surg, 63-67, prof surg, 67-87, CHIEF SECT PLASTIC SURG, 69-, EMER PROF, 87- *Honors & Awards:* Hoekton Gold Medal, AMA, 59. *Mem:* AMA; Am Surg Asn; Am Asn Plastic Surgeons; Am Soc Plastic & Reconstructive Surg. *Res:* Plastic surgery; tissue transplantation; wound healing; cancer. *Mailing Add:* Windward Apt 301 16777 Perdidokey Dr Pensacola FL 32407

RYAN, ROBERT J, b Cincinnati, Ohio, July 18, 27; m 54; c 6. BIOCHEMISTRY, ENDOCRINOLOGY. *Educ:* Univ Cincinnati, MD, 52. *Prof Exp:* Intern, Henry Ford Hosp, Detroit, 52-53; res fel, Univ Ill Col Med, 53-54, resident med, Res & Educ Hosp, 54-57; Am Col Physicians res fel endocrinol, New Eng Ctr Hosp, 57-58; from instr to assoc prof, Univ Ill Col Med, 58-67; from assoc prof to prof med, 67-79, chmn dept molecular med, Mayo Clin, 73-79, prof cell biol, Mayo Grad Sch Med, Univ Minn, 79-85; prof, 85-90, EMER PROF BIOCHEM, MAYO MED SCH, 90- *Concurrent Pos:* Mem staff, Mayo Clin, 67-90, chmn dept endocrine res, 70-79; mem reproductive biol study sect, NIH; mem pop res comn, Nat Inst Child Health & Human Develop; mem med adv bd, Nat Pituitary Agency. *Honors & Awards:* Robert H Williams Award, Endocrine Soc, 84. *Mem:* Endocrine Soc (vpres, 77-78); Am Soc Biol Chemists; Soc Study Reprod (pres, 87-88); Soc Exp Biol & Med; Am Soc Clin Invest. *Res:* Structure and function of human gonadotropic hormones. *Mailing Add:* Dept Biochem & Molecular Biol Mayo Med Sch Rochester MN 55905

RYAN, ROBERT PAT, b Pensacola, Fla, May 1, 25; m 50; c 2. ULTRASOUND. *Educ:* Rice Univ, BS, 45; Brown Univ, ScM, 59, PhD(physics), 63. *Prof Exp:* Physicist, US Navy Mine Defense Lab, Fla, 47-58; asst prof physics, Univ Ky, 63-65; Nat Acad Sci-Nat Res Coun res assoc solid state physics, Res Dept, Naval Ord Lab, Md, 65-66; physicist, NASA Electronics Res Ctr, 66-70; physicist, mech eng div, 70-82, STAFF ENGR, TRANSP SYSTS CTR, DEPT TRANSP, 82- *Mem:* Acoust Soc Am; Inst Elec & Electronics Engrs; Am Soc Nondestructive Test; Sigma Xi. *Res:* Underwater sound noise measurement; analysis; guided mode and finite-amplitude propagation; ultrasonic relaxations in glasses; second-order optical effects; piezo and ferroelectricity; ultrasonic techniques for nondestructive testing; ultrasonic imaging; signal processing; pattern recognition. *Mailing Add:* 110 Quincy Ave Braintree MA 02184

RYAN, ROBERT REYNOLDS, b Klamath Falls, Ore, July 29, 36. ACTINIDE CHEMISTRY, ORGANIC EXPLOSIVES. *Educ:* Portland State Univ, BS, 61; Ore State Univ, PhD(chem), 65. *Prof Exp:* Fel x-ray diffraction, Swiss Fed Inst Technol, Zurich, 65-66; fel vibrational spec, 66-67, staff mem, 67-80, DEP GROUP LEADER, LOS ALAMOS NAT LAB, 80- *Mem:* Am Chem Soc; Am Crystallog Asn. *Res:* Trasition metal chemistry; actinide chemistry; continuous phase changes; small molecule activation; organic explosives; vibrational spectroscopy; x-ray diffraction; gas phase electron diffraction. *Mailing Add:* 391 Navajo Los Alamos NM 87544-2621

RYAN, ROGER BAKER, b Port Chester, NY, May 5, 32; m 55; c 4. INSECT ECOLOGY & BIOLOGICAL CONTROL. *Educ:* State Univ NY Col Forestry, Syracuse, BS, 53; Ore State Univ, MS, 59, PhD(entom, plant path), 61. *Prof Exp:* Entomologist, Forestry & Range Sci Lab, US Forest Serv, 61-65, res entomologist, 65-89; PROF, DEPT FOREST SCI, ORE STATE UNIV, 89- *Mem:* Entom Soc Am; Entom Soc Can; Int Orgn Biol Control. *Res:* Biological control, physiology and behavior of insects; biological control of forest insect pests using introduced parasitoids and predators (acquisition, propagation, release, evalation and population dynamics). *Mailing Add:* Dept Forest Sci Ore State Univ Corvallis OR 97331-5705

RYAN, SIMEON P, b New York, NY, May 30, 22. BIOLOGY. *Educ:* St Francis Col, BS, 51; St Louis Univ, MS, 53, PhD(biol), 57. *Prof Exp:* From instr to assoc prof biol, St Francis Col, NY, 57-70, dir pre-med & pre-dent training, 58-66, actg head dept biol, 66-70; asst prof, 70-74, ASSOC PROF BIOL, NASSAU COMMUNITY COL, 74- *Mem:* AAAS; NY Acad Sci. *Res:* Cytological nutritional effects; cytopathology. *Mailing Add:* Dept Biol Nassau Community Col Stewart Ave Garden City NY 11530

RYAN, STEWART RICHARD, b Schenectady, NY, Jan 26, 42; m 66; c 3. ATOMIC & MOLECULAR PHYSICS. *Educ:* Univ Notre Dame, BS, 64; Univ Mich, MS, 65, PhD(physics), 71. *Prof Exp:* Res staff physicist, Yale Univ, 71-73, instr, 73-74; res assoc, Univ Ariz, 74-76, staff physicist, 76-77; asst prof, 77-82, ASSOC PROF PHYSICS, UNIV OKLA, 82- *Mem:* Am Phys Soc; Am Asn Physics Teachers; Am Asn Eng Educ; Sigma Xi; Am Solar Energy Soc. *Res:* Molecular dissociation processes; atom-molecule collisions; electron-molecule collisions; low temperature properties of helium 3 & helium 4 mixtures; applied physics; instrumentation; energy conservation. *Mailing Add:* Dept Physics & Astron Univ Okla Norman OK 73019

RYAN, THOMAS ARTHUR, JR, b Ithaca, NY, June 12, 40; m 66. STATISTICS. *Educ:* Wesleyan Univ, BA, 62; Cornell Univ, PhD(math), 68. *Prof Exp:* Instr math statist, Columbia Univ, 67-68, asst prof, 68-69; asst prof, 69-75, ASSOC PROF STATIST, PA STATE UNIV, UNIVERSITY PARK, 75- *Mem:* Inst Math Statist; fel Am Statist Asn; Asn Comput Mach; Inst Elec & Electronics Engrs. *Res:* Statistical computing. *Mailing Add:* Dept Statist Pa State Univ University Park PA 16802

RYAN, THOMAS JOHN, b New York, NY, June 12, 43; m 70; c 3. BIOCHEMISTRY, ORGANIC CHEMISTRY. *Educ:* Manhattan Col, BS, 65; State Univ NY, Stony Brook, PhD(org chem), 70. *Prof Exp:* Res assoc org chem, Johns Hopkins Univ, 69-71; assoc, Rensselaer Polytech Inst, 71-72; fel biochem, 72-74, RES SCIENTIST BIOCHEM, DIV LABS & RES, DEPT HEALTH, ALBANY, NY, 74- *Mem:* Am Chem Soc; AAAS. *Res;* Fibrinolysis; coagulation; enzyme mechanism and structure. *Mailing Add:* Dept Health Empire State Plaza Albany NY 12201

RYAN, THOMAS WILTON, b San Mateo, Calif, March 20, 46; m 69; c 2. SYSTEMS DESIGN, SYSTEMS SCIENCE. *Educ:* Univ Santa Clara, BS, 68; Univ Ariz, MS, 71, PhD(elec eng), 81. *Prof Exp:* Teacher math & physics, Green Fields Sch, 71-76; res asst, Univ Ariz, 76-79; STAFF SCIENTIST, SCI APPLN, INC, 79- *Mem:* Inst Elec & Electronics Engrs; Soc Photo-Optical Instrumentation Engrs. *Res:* Applications of signal processing in radar signal analysis; image processing. *Mailing Add:* Sci Appln Int Corp 5151 E Broadway Suite 1100 Tucson AZ 85711

RYAN, UNA SCULLY, b Kuala Lumpur, Malaysia, Dec 18, 41; Brit citizen; m 73; c 2. CELL BIOLOGY, PULMONARY DISEASES. *Educ:* Bristol Univ, BSc, 63; Cambridge Univ, PhD(cell biol), 68. *Prof Exp:* From instr to assoc prof, 67-80, PROF MED, SCH MED, UNIV MIAMI, 80- *Concurrent Pos:* Vis investr, Lab Cardiovasc Res, Howard Hughes Med Inst, Miami, 67-71, dir, Lab Ultrastruct Studies, 70-71; adj asst prof biol, Univ Miami, 68-; sr scientist, Papanicolaou Cancer Res Inst, 72-77; estab investr, Am Heart Asn, 72-77, mem basic sci coun, 77-, mem, Microcirculation Coun, 78-; mem pulmonary dis adv comt, Nat Heart, Lung & Blood Inst, 73-76, mem res rev comt A, 77-81, chmn, 80-81. *Honors & Awards:* Louis & Artur Lucian Award, 85. *Mem:* Royal Entom Soc London; Am Soc Cell Biol; Europ Soc Endocrinol; Tissue Cult Asn; NY Acad Sci. *Res:* Application of advanced techniques of electron microscopy and cell culture to studies of the endocrine or non-ventilatory functions of the lung, with particular emphasis on correlations of fine structure of endothelial cells with specific metabolic activities. *Mailing Add:* Dept Med Sch Med Univ Miami PO Box 016960 Miami FL 33101

RYAN, VICTOR ALBERT, b Laramie, Wyo, Aug 11, 20; m 50; c 4. NUCLEAR CHEMISTRY, ACADEMIC & INDUSTRIAL SAFETY. *Educ:* Univ Wyo, BS, 42; Univ Minn, PhD(phys chem), 51. *Prof Exp:* Chemist, E I du Pont de Nemours & Co, Tenn, 51-53; chemist, Dow Chem Co, Colo, 53-58; from asst prof to prof chem, 58-84, prof chem eng, Univ Wyo, 80-, safety dir, 74-; RETIRED. *Mem:* Am Chem Soc; Am Nuclear Soc; Am Asn Univ Profs; Sigma Xi. *Res:* Photochemistry; flash photolysis; industrial chemistry; lanthanide and actinide chemistry; prompt and ordinary activation analysis. *Mailing Add:* PO Box 2052 Laramie WY 82070

RYAN, WAYNE L, b Corning, Iowa, June 14, 27; m 48; c 5. BIOCHEMISTRY, IMMUNOLOGY. *Educ:* Creighton Univ, BS, 49, MS, 51; Univ Mo, PhD(biochem), 53. *Prof Exp:* Instr microbiol & asst prof biochem, Sch Med, Creighton Univ, 53-61, assoc prof biochem, 61-64; from assoc prof to prof biochem, 64-87, res assoc prof obstet & gynec, 64-67, asst dean res, 73-79, RES PROF OBSTET & GYNEC, UNIV NEBR CTR, OMAHA, 67-; PRES, STRECK LABS, 87- *Mem:* Am Chem Soc; Am Asn Cancer Res; Soc Exp Biol & Med. *Res:* Cell fusion; diagnostic control methods. *Mailing Add:* 42nd/Dewey Col Med Univ Nebr Med Ctr Omaha NE 68105

RYAN, WILLIAM B F, b Troy, NY, Sept 1, 39; m 62; c 2. OCEANOGRAPHY, GEOLOGY. *Educ:* Williams Col, BA, 61; Columbia Univ, PhD(geol), 71. *Prof Exp:* Res asst oceanog, Woods Hole Oceanog Inst, 61-62; res asst geol, 62-74, SR RES ASSOC GEOL, LAMONT-DOHERTY GEOL OBSERV, COLUMBIA UNIV, 74- *Mem:* Am Geophys Union; Sigma Xi. *Res:* Marine geology and geophysics. *Mailing Add:* 12 Clinton Ave South Nyack NY 10960

RYAN, WILLIAM GEORGE, b Berkeley, Calif, Jan 7, 51; div; c 1. PSYCHOPHARMACOLOGY. *Educ:* Ind Univ, AB, 74, Ind Univ Sch Med, MD, 78. *Prof Exp:* Asst prof, 82-87, ASSOC PROF PSYCHIAT, SCH MED, UNIV ALA, 87- *Mem:* Am Psychiat Asn; Soc Biol Psychiat. *Res:* Psychopharmacology and the neuroendorine (particularly thyroid) aspects of schizophrenia and the major mood disorders. *Mailing Add:* 218 Smolian Clin Univ Ala Birmingham AL 35294

RYANT, CHARLES J(OSEPH), JR, b Chicago, Ill, Apr 1, 20; m 77. ENGINEERING, ENVIRONMENTAL ENGINEERING. *Educ:* Armour Inst Technol, BS, 40; Ill Inst Technol, MS, 41, PhD(chem eng), 47. *Prof Exp:* Instr chem eng, Ill Inst Technol, 40-41, instr chem, 46-47; chem engr, Sinclair Refining Corp, Ind, 41 & Wurster & Sanger, Inc, 41-43; sr proj engr, Standard Oil Co, Ind, 43-59; consult, 59-68; tech consult, Joint Ill House & Senate Air Pollution Study Comt, 68-69; vpres & dir eng, Sparkleen Systs, Inc, 69-70; exec dir, Midwest Legis Coun Environ, Univ Ill, Chicago Circle, 70-80; CONSULT, C J RYANT, JR & ASSOC, 80- *Concurrent Pos:* Civilian with AEC, 44; instr, Ill Inst Technol, 53-; chmn, Calumet City Environ Control Comn, 74-81; secy, Cleveland Twp Planning Comn, 81-87. *Mem:* Am Chem Soc; Am Inst Chem Engrs. *Res:* Filtration and sedimentation; heat transfer; conditioning of air; heat transfer in a double exchanger; petroleum refinery furnaces. *Mailing Add:* 504 Ryant Rd Maple City MI 49664-9721

RYASON, PORTER RAYMOND, b Bridgeport, Nebr, Jan 18, 29; div; c 4. PHYSICAL CHEMISTRY. *Educ:* Reed Col, BA, 50; Harvard Univ, MA, 52, PhD(chem), 54. *Prof Exp:* Sr res assoc, Chevron Res Co, Stand Oil Co Calif, 53-73; mem tech staff, Jet Propulsion Lab, Calif Inst Technol, 73-78; RES SCIENTIST, CHEVRON RES CO, CHEVRON CORP, 78- *Concurrent Pos:* Mem, Coop Air Pollution Eng Proj Group, 58-68, 68-; mem, Nat Air Pollution Control Admin Adv Comt Chem & Physics, Dept Health, Educ & Welfare, 68-70; mem, Task Force Hydrocarbon Reactivities, Am Petrol Inst, 69-70, Task Force Aerometric Data Anal, 70- *Mem:* AAAS; Am Chem Soc; Am Phys Soc. *Res:* Molecular spectroscopy; gas phase chemical kinetics; fast reactions; combustion and flame; infrared spectra of adsorbed species; air pollution; aerochemistry; solar photochemical conversion; tribology. *Mailing Add:* 60 Madrone Rd Fairfax CA 94930-2120

RYBA, EARLE RICHARD, b Elyria, Ohio, June 27, 34; m 54; c 4. X-RAY CRYSTALLOGRAPHY. *Educ:* Mass Inst Technol, BS, 56; Iowa State Univ, PhD(phys metall), 60. *Prof Exp:* Asst prof, 60-65, ASSOC PROF METALL, PA STATE UNIV, UNIVERSITY PARK, 65- *Mem:* Am Crystallog Asn; Am Soc Metals Int; Am Inst Metall Engrs; Mat Res Soc; Int Ctr Diffraction Data. *Res:* Crystal structures, properties and theory of bonding in intermetallic compounds; quasicrystals; characterization of materials by x-ray diffraction techniques. *Mailing Add:* Rm 304 Steidle Bldg Pa State Univ University Park PA 16802

RYBACK, RALPH SIMON, b Detroit, Mich, Oct 17, 40; m 76; c 3. CLINICAL LABORATORY CHEMISTRIES. *Educ:* Wayne State Univ, BA, 63, MD, 66. *Prof Exp:* Res fel, 67-68, teaching fel psychiat, 70-72, instr, 72-73, asst prof, Harvard Med Sch, 73-77; vis scientist, 76-77, med officer, lab clin studies, Nat Inst Alcohol Abuse & Alcoholism, 77-84; ASSOC CLIN PROF, UNIFORM SERV UNIV HEALTH SCI, 81- *Concurrent Pos:* Dir, Alcohol & Drug Abuse Serv, McLean Hosp, Div Mass Gen Hosp, 72-76; assoc clin prof, Med Sch, George Washington Univ, 79-80. *Mem:* Am Psychiat Asn. *Res:* Interrelationships of commonly order clinical chemistries for the diagnosis of alcoholism and related illness including nonalcoholic liver diseases. *Mailing Add:* 11607 Springride Rd Potomac MD 20854

RYCHECK, MARK RULE, b Racine, Wis, Dec 30, 37; m 62; c 3. INORGANIC CHEMISTRY, TECHNICAL INFORMATION SCIENCE. *Educ:* St Francis Col, Pa, BS, 59; Univ Cincinnati, PhD(chem), 67. *Prof Exp:* Res asst spectros, Mellon Inst Sci, 61-62; Ohio State Univ Res Found fel, Ohio State Univ, 66-68; from res chemist to sr res chemist, Phillips Petrol Co, 68-77, sr patent develop chemist, 78-84, sr info scientist, Petrol Res Ctr, 85-89; SR INFO SCIENTIST, UPJOHN CO, 89- *Mem:* Am Chem Soc; Am Soc Info Sci. *Res:* Coordination compounds; catalysis. *Mailing Add:* Corp Tech Libr Upjohn Co 7284-267-23 Kalamazoo MI 49001

RYCHECK, RUSSELL RULE, b Racine, Wis, June 11, 32; m 62; c 2. MEDICINE, EPIDEMIOLOGY. *Educ:* Univ Pittsburgh, MD, 57, MPH, 59, DrPH, 64; Am Bd Prev Med, dipl, 68. *Prof Exp:* Teaching fel, Children's Hosp, Pittsburgh, 59-61, res fel pediat, 61-62; res fel epidemiol, 62-64, asst prof, 64-67, ASSOC PROF EPIDEMIOL, GRAD SCH PUB HEALTH, UNIV PITTSBURGH, 67-, ASST PROF PEDIAT & MED, 72- *Concurrent Pos:* Consult, Allegheny Co Health Dept, Pa, 64- *Mem:* Am Col Prev Med; Am Pub Health Asn; Soc Epidemiol Res. *Res:* Epidemiology of infectious diseases in hospitals and civilian communities; maternal and child health. *Mailing Add:* Grad Sch Pub Health A455 Univ Pittsburgh 4200 Fifth Ave Pittsburgh PA 15213

RYCKMAN, DEVERE WELLINGTON, b South Boardman, Mich, May 27, 24; m; c 3. ENVIRONMENTAL ENGINEERING, BIOCHEMISTRY. *Educ:* Rensselaer Polytech Inst, BS, 44; Mich State Univ, MS, 49; Mass Inst Technol, ScD, 56; Environ Eng Intersoc, dipl, 78. *Prof Exp:* From instr to asst prof civil & sanit eng, Mich State Univ, 46-53; res asst sanit eng, Mass Inst Technol, 53-55; from assoc prof environ & sanit eng to prof & dir dept, Wash Univ, 56-72; consult, Ryckman, Edgerley, Tomlinson & Assocs, 72-75, CONSULT & PRES, RYCKMAN'S EMERGENCY ACTION & CONSULT TEAM, 75- *Concurrent Pos:* USPHS res & prog grants & consult, 56-; mem eng coun prof develop, Sanit Eng Accreditation Comt, 62; Mo Gov's Sci Adv Comt, 62-; vis lectr, Univ Hawaii, 63 & Vanderbilt Univ, 64; US Air Force Sch Aerospace Med, 64; Mfg Chem Asn res grant, 65-68; mem Nat Sci Traineeship Rev Comt, 65; consult Nat Energy Res & Develop Admin, 77- *Honors & Awards:* Resources Div Award, Am Water Works Asn, 62 & George Warren Fuller Award, 65; Grand Conceptor Award, Am Consult Engrs Coun, 69. *Mem:* Am Water Works Asn; Am Pub Health Asn; Am Soc Civil Engrs; Am Soc Eng Educ; Dipl Am Acad Environ Engrs. *Res:* Industrial water; wastewater; solid waste problems; significance of chemical structure in biodegradation of pesticides, synthetic detergents and other organic chemicals; environmental and energy engineering, hazardous substances and environmental crises engineering research. *Mailing Add:* PO Box 27310 St Louis MO 63141

RYCKMAN, RAYMOND EDWARD, b Shullsburg, Wis, June 19, 17; m 43; c 3. MEDICAL ENTOMOLOGY, PARASITOLOGY. *Educ:* Univ Calif, BS, 50, MS, 57, PhD, 60. *Prof Exp:* Asst prof entom & head dept, Sch Trop & Prev Med, 50-59, from asst prof to assoc prof, 59-72, chmn dept, 80-87, PROF MICROBIOL, LOMA LINDA UNIV, 72- *Mem:* Entom Soc Am; Am Soc Trop Med & Hyg; Soc Vector Ecologists. *Res:* Biosystematics; ecology of blood feeding insects and their vertebrate hosts; taxonomy, ecology and vector potential of insects of medical and veterinary importance; world bibliography and literature to the parasitic Hemiptera. *Mailing Add:* Dept Microbiol Sch Med Loma Linda Univ Loma Linda CA 92350

RYDEN, FRED WARD, b Boulder, Colo, Dec 20, 19; m 49; c 1. MICROBIOLOGY. *Educ:* Univ Colo, BA, 47; Vanderbilt Univ, MS, 50, PhD(virol), 56, MD, 60. *Prof Exp:* Asst parasitic dis, 47-48, instr, 49-52, microbiol, 52-56, asst prof, 56-60, path, 60-71, dir clin bact & serol lab, Univ Hosp, 60-67, ASST CLIN PROF PATH, SCH MED, VANDERBILT UNIV, 71- *Concurrent Pos:* Dir serol lab, Vanderbilt Univ Hosp, 50-55; consult, Vet Admin Hosp, 60-67; pathologist & dir lab, Nashville Mem Hosp, 67- *Mem:* AAAS; Am Med Asn; fel Am Soc Clin Path; fel Col Am Path. *Res:* Virology; mode of infection and replication of mammalian viruses. *Mailing Add:* Dept Path Nashville Mem Hosp 612 W Due West Ave Madison TN 37115

RYDER, D(AVID) F(RANK), b Seekonk, Mass, Aug 22, 19; m 57; c 1. CHEMICAL ENGINEERING. *Educ:* Tufts Univ, BS, 41. *Prof Exp:* Res engr, E I Du Pont de Nemours & Co, Inc, 41-55, sr res engr, 55-75, develop assoc, 75-80; RETIRED. *Mem:* Am Chem Soc. *Res:* Synthetic and textile fibers; industrial fibers. *Mailing Add:* 101 Watford Rd Wilmington DE 19808

RYDER, EDWARD JONAS, b New York, NY, Oct 6, 29; m 62; c 3. GENETICS, PLANT BREEDING. *Educ:* Cornell Univ, BS, 51; Univ Calif, PhD(genetics), 54. *Prof Exp:* Location Leader, 82-89, GENETICIST, PAC WEST AREA, AGR RES SERV, USDA, 57-, RES LEADER, VEG PROD RES, US AGR RES STA, 72- *Concurrent Pos:* Instr, Monterey Peninsula Col, 58-64; mem hort adv bd, AVI Publ Co, 76-85. *Mem:* AAAS; Crop Sci Soc; fel Am Soc Hort Sci. *Res:* Quantitative genetics; evolution; breeding new cultivars of lettuce with improvements in disease, insect and stress resistance, horticultural quality and uniformity; gene identification and linkage studies in lettuce; genetic-physiological basis for resistance. *Mailing Add:* USDA Agr Res Serv 1636 E Alisal St Salinas CA 93905

RYDER, JOHN DOUGLASS, b Columbus, Ohio, May 8, 07; m 33; c 2. ELECTRONICS. *Educ:* Ohio State Univ, BEE, 28, MS, 29; Iowa State Univ, PhD(elec eng), 44; Tri-State Col, DEng, 63. *Prof Exp:* Electronic engr, Gen Elec Co, NY, 29-31; in chg electronic res, Bailey Meter Co, Ohio, 31-41; asst prof elec eng, Iowa State Univ, 41-44, prof, 44-49; prof elec eng & head dept, Univ Ill, 49-54; dean eng, 54-68, prof 68-72, EMER PROF ELEC ENG, MICH STATE UNIV, 72- *Concurrent Pos:* Asst dir, Eng Exp Sta, Iowa State Col, 47-49; ed, J, Inst Elec & Electronics Engrs, 58-59, 63-64. *Honors & Awards:* Medal Honor, Nat Electronics Conf, 70; Haraden Pratt Award, Inst Elec & Electronics Engrs, 79. *Mem:* Am Soc Eng Educ; Inst Elec & Electronics Engrs (pres, Inst Radio Eng, 55); Ralph Batcher Award, Radio Club Am, 87. *Res:* Vacuum tube and electric circuits; electronic applications. *Mailing Add:* 1839 SE 12th Ave Ocala FL 32671

RYDER, KENNETH WILLIAM, JR, b Mobile, Ala, May 1, 45; m 71; c 3. CLINICAL CHEMISTRY, LABORATORY MANAGEMENT. *Educ:* Knox Col, BA, 67; Ind Univ, PhD(biochem), 72; Univ Ill, MD, 75. *Prof Exp:* From asst prof to assoc prof, 78-88, PROF PATH, SCH MED, IND UNIV, 88-, ASSOC CHMN DEPT, 87- *Concurrent Pos:* Dir labs, Wishard Hosp, Indianapolis, 86- *Honors & Awards:* Joseph Kleiner Award, Am Soc Med Technol, 87. *Mem:* AMA; Am Soc Clin Pathologists; Am Asn Clin Chem; Col Am Pathologists; Clin Lab Mgt Asn. *Res:* Improvement of clinical laboratory tests by reducing interferences and enhancing specificity. *Mailing Add:* Dept Path Wishard Hosp 1001 W Tenth St Indianapolis IN 46202

RYDER, OLIVER A, b Alexandria, Va, Dec 27, 46; m 70; c 2. CONSERVATION BIOLOGY, ENDANGERED SPECIES. *Educ:* Univ Calif, Riverside, BA, 68, San Diego, PhD(biol), 75. *Prof Exp:* Fel geneticist, Sch Med, Univ Calif, 75-77; geneticist, 78-86, KLEBERG GENETICS CHAIR, SAN DIEGO ZOO, 86- *Concurrent Pos:* Species coordr, Asian Wild Horse Species Survival Plan, 82-; mem, Wildlife Conserv Comn, Am Asn Zool Parks & Aquaria, 84-90, Conserv Comn, Am Soc Mammalogists, 85-92; adj assoc prof biol, Univ Calif, San Diego, 87-; coun mem, Am Genetics Asn, 90-; secretariat, Przewalski's Horse Global Mgt Plan Working Group, 90- *Mem:* Am Genetics Asn; Am Soc Mammalogists; Am Asn Zool Parks & Aquaria; Soc Study Evolution; Soc Conserv Biol. *Res:* Genetics in support of conservation programs for endangered species in captivity and in the wild; species conservation; molecular and chromosomal evolution; conservation education in primary and secondary schools, colleges and for general public. *Mailing Add:* CRES Box 551 San Diego CA 92112

RYDER, RICHARD ARMITAGE, b Windsor, Ont, Feb 25, 31; m 56; c 2. FISHERIES ECOLOGY, LIMNOLOGY. *Educ:* Univ Mich, BSc, 53, MSc, 54. *Prof Exp:* Res asst, US Fish & Wildlife Serv, 53-54; dist biologist fisheries, 54-58, biologist-in-chg inventory, 58-61, res scientist, 62-71, COORDR FISHERIES, ONT DEPT LANDS & FORESTS, 61- *Concurrent Pos:* Res scientist, Ont Ministry Natural Resources, 71-91; consult to various int pub & pvt orgn, 71-88; comt mem, Man & Biosphere, 73-76 & Int Joint Comn, 73-90; bd tech experts, Great Lakes Fishery Comn, 82-87. *Mem:* Am Soc Fishery Res Biologists; Am Fisheries Soc (pres, 80); Int Asn Theoret & Appl Limnol; Am Soc Limnol & Oceanog; Int Asn Great Lakes Res; Can Conf Fish Res (pres, 88). *Res:* Methods of determining levels of fish production from global waters; defining concepts necessary for a basic understanding of fish communities in fresh water. *Mailing Add:* Ministry Natural Resources PO Box 2089 Thunder Bay ON P7B 5E7 Can

RYDER, RICHARD DANIEL, b Providence, RI, Oct 1, 44; m 81; c 1. PHYSICS. *Educ:* Univ RI, BS, 66; Brown Univ, ScM, 73, PhD(physics), 74. *Prof Exp:* STAFF MEM PHYSICS, LOS ALAMOS NAT LAB, UNIV CALIF, 77- *Mem:* Am Phys Soc; Inst Elec & Electronics Engrs; Plasma Sci Soc; Sigma Xi. *Res:* Accelerator physics; particle beam optics; underwater acoustics. *Mailing Add:* 390 Richard Ct Los Alamos NM 87544

RYDER, ROBERT J, b Olean, NY, Apr 19, 31; m 54; c 5. CERAMICS. *Educ:* Alfred Univ, BS, 53; Pa State Univ, MS, 55, PhD(ceramics), 59. *Prof Exp:* Res scientist, Brockway Glass Co, Inc, 59-62, asst dir res develop, 62-69, dir res & develop, 69-78, vpres res & develop, 78-88; STAFF, OWENS ILL CO, TOLEDO, 88- *Concurrent Pos:* Chmn air & water qual comt, Glass Container Mfrs Inst, 68-75. *Mem:* Fel Am Ceramic Soc; Soc Glass Technol; Am Chem Soc. *Res:* Physical chemistry of glass melting process; colored glasses; physical properties of glasses; reactivity of glass surfaces. *Mailing Add:* Owens Brockway Glass One Seagate 25L-GC Toledo OH 43666-0001

RYDER, ROBERT THOMAS, b Bowling Green, Ohio, Sept 23, 41; m 68; c 2. STRATIGRAPHY, PETROLEUM GEOLOGY. *Educ:* Mich State Univ, BS, 63; Pa State Univ, PhD(geol), 68. *Prof Exp:* Geologist, Shell Develop Co, Tex, 69-72, Shell Oil Co, 72-74; GEOLOGIST, US GEOL SURV, 74- *Concurrent Pos:* Mem, Am Sedimentary Basins Deleg, Peoples Repub China, 85. *Mem:* Geol Soc Am; Am Asn Petrol Geologists; Soc Econ Paleontologists & Mineralogists. *Res:* Seismic detection of stratigraphic traps; sedimentation and tectonics; regional stratigraphy; nonmarine depositional environments. *Mailing Add:* US Geol Surv MS 955 Nat Ctr Reston VA 22092

RYDER, RONALD ARCH, b Kansas City, Kans, Feb 3, 28; m 55; c 2. WILDLIFE MANAGEMENT. *Educ:* Colo Agr & Mech Col, BS, 49, MS, 51; Utah State Univ, PhD(wildlife mgt), 58. *Prof Exp:* Wildlife technician, Dept Game & Fish, Colo, 48-49, 54-55; instr, Wartburg Col, 57-58; from asst prof wildlife mgt to prof wildlife biol, 58-85, chmn dept biol majors, 69-72, EMER PROF WILDLIFE BIOL, COL FORESTRY & NATURAL RESOURCES, COLO STATE UNIV, 85- *Concurrent Pos:* NSF fel, Univ BC, 65-66; vis prof, Mem Univ Nfld, 72-73; vis prof, Univ Otago, New Zealand, 81. *Mem:* Wildlife Soc; Am Soc Mammal; Cooper Ornith Soc; Wilson Ornith Soc; Am Ornith Union. *Res:* Waterfowl ecology and management; bird banding; distribution of Colorado birds; nongame wildlife management. *Mailing Add:* Dept Fishery & Wildlife Biol Colo State Univ Ft Collins CO 80523

RYDGREN, A ERIC, astronomy, for more information see previous edition

RYDZ, JOHN S, b Milwaukee, Wis, May 7, 25; m 46; c 2. PHYSICS, ELECTRONICS. *Educ:* Mass Inst Technol, BS, 52; Univ Pa, MS, 56. *Prof Exp:* Proj engr, Radio Corp Am, 52-56, eng group leader, Appl Physics Group, 56-59, mgr new bus develop, 59-61; exec vpres & dir, Nuclear Corp Am, 61-63; mgr new prod develop, 63-66; vpres res & develop, Diebold, Inc, Ohio, 66-71; vpres & tech dir, N Atlantic Consumer Prod Group, NY, 71-74, VPRES ENG, SEWING PROD GROUP, SINGER CO, 74-; VPRES TECHNOL, EMHART CORP. *Mem:* Inst Elec & Electronics Engrs; Optical Soc Am. *Res:* Spectrophotometry and colorimetry; masers; color electrofax; molecular resonance; zener diodes; nuclear instrumentation; electrostatic printing; closed circuit television techniques; information search and retrieval; consumer products. *Mailing Add:* 29 Ariel Way Avon CT 06001

RYE, DANNY MICHAEL, b Glendale, Calif, Feb 21, 46; m 67; c 4. GEOCHEMISTRY. *Educ:* Occidental Col, AB, 67; Univ Minn, PhD(geol), 72. *Prof Exp:* Fel geol, Purdue Univ, 72; res staff geologist, Yale Univ, 72-74, instr, 74-75, asst prof, 75-80, assoc prof, 80-90, PROF GEOL, YALE UNIV, 90- *Concurrent Pos:* Ed, Am J Sci, 86- *Honors & Awards:* Lindgren Award, Soc Econ Geologists, 76. *Mem:* Soc Econ Geologists; Geochem Soc; Geol Soc Am. *Res:* Isotopic composition of carbon, hydrogen, oxygen and sulfur to obtain information about the volatile phases present during hydrothermal ore formation, or during the formation of metamorphic rocks. *Mailing Add:* Dept Geol & Geophys Yale Univ Box 6666 New Haven CT 06511

RYE, ROBERT O, b Los Angeles, Calif, Nov 29, 38; m 64; c 3. GEOCHEMISTRY. *Educ:* Occidental Col, AB, 60; Princeton Univ, PhD(geol), 65. *Prof Exp:* GEOLOGIST, US GEOL SURV, 64- *Mem:* Geol Soc Am; Geochem Soc. *Res:* Stable isotope studies of ore deposits; sulfur isotope studies. *Mailing Add:* 11863 W 27th Dr Lakewood CO 80215

RYEBURN, DAVID, b Cincinnati, Ohio, June 19, 35; m 57; c 3. GENERAL TOPOLOGY. *Educ:* Kenyon Col, AB, 54; Ohio State Univ, PhD(math), 62. *Prof Exp:* Instr math, Kenyon Col, 56-57; asst, Ohio State Univ, 58-61, instr, 61-62; asst prof, Kenyon Col, 62-66; ASST PROF MATH, SIMON FRASER UNIV, 66- *Mem:* AAAS; Am Math Soc; Math Asn Am; Can Math Soc. *Mailing Add:* PO Box 387 Lynden WA 98264-0387

RYEL, LAWRENCE ATWELL, b Farmington, Mich, Feb 22, 30; m 52; c 2. BIOSTATISTICS, WILDLIFE RESEARCH. *Educ:* Mich State Univ, BS, 51, MS, 53, PhD(zool), 71. *Prof Exp:* Game biologist, Mich Dept Conserv, 53-61, biometrician, 61-64, biomet supvr, Res & Develop Div, 64-72, chief div, 72-73, chief, Off Surv & Statis Serv, 73-78, chief, Surv & Statist Serv, Environ Serv Div, 78-80, HEAD, SURV & STATIST SERV WILDLIFE DIV, MICH DEPT NATURAL RESOURCES, 80- *Concurrent Pos:* Vis prof wildlife sci, Utah State Univ, 77. *Mem:* Biomet Soc; Wildlife Soc; Am Soc Mammal; Wilson Ornith Soc; Sigma Xi. *Res:* Wildlife population dynamics; sample survey design; consultation in design and analysis of fisheries and wildlife research studies. *Mailing Add:* 882 N 300 E Logan UT 84321

RYERSON, GEORGE DOUGLAS, b East Orange, NJ, Apr 29, 34. ORGANIC CHEMISTRY. *Educ:* Lehigh Univ, BS, 55; Mass Inst Technol, PhD(org chem), 60. *Prof Exp:* Chemist, Esso Res & Eng Co, 60-67; sr assoc ed, 67-75, SR ED, CHEM ABSTR SERV, 75- *Mem:* AAAS; Am Chem Soc; Am Soc Info Sci; Sigma Xi. *Res:* Macromolecular chemistry. *Mailing Add:* 2407 Ravenel Dr Columbus OH 43209-3308

RYERSON, JOSEPH L, b Goshen, NY, Oct 20, 18; m 41; c 3. ELECTRICAL ENGINEERING. *Educ:* Clarkson Col Technol, BEE, 41; Syracuse Univ, MEE, 54 & PhD(elec eng), 67. *Prof Exp:* Tech dir, NATO Thinktank, SHAPE Tech Ctr, Haag, Neth, 64-68,; tech dir electronics, 51-64, commun, 69-73, sr scientist electronics, 73-76, tech dir surveillance, Rome Air Develop Ctr, 76-78; PROF ENGR, JOSEPH L RYERSON, 78- *Concurrent Pos:* Mem commun syst eng comt, 69-88 & Alexander Graham Bell Medal comt, 83-87; consult, Energy Independence Inc, 74- *Mem:* Fel Inst Elec & Electronics Engrs; fel AAAS; Nat Soc Prof Engrs; Am Asn Physics Teachers. *Res:* Theories for aircraft automatic landing and missile launching based on combined inertial and radio navigation coordinate information and development of early methods of satellite communications. *Mailing Add:* RD 2 Box 4 Holland Patent NY 13354-9603

RYFF, JOHN V, b Jersey City, NJ, Oct 18, 32; m 58. MATHEMATICAL ANALYSIS. *Educ:* Syracuse Univ, AB, 57; Stanford Univ, PhD(math), 62. *Prof Exp:* Benjamin Peirce instr math, Harvard Univ, 62-64; Off Naval Res assoc, Univ Wash, 64-65, asst prof, 65-67; Inst Advan Study, 67-68 & Inst Defense Anal, 68; prog dir math sci sect, NSF, 69-72; prof math & head dept, Univ Conn, 72-79; PROG DIR, MATH SCI SECT, NSF, 79- *Mem:* Soc Indust Appl Math; Math Asn Am. *Res:* Real and complex function theory; functional analysis. *Mailing Add:* Div Math Sci NSF Washington DC 20550

RYGE, GUNNAR, dentistry, physics; deceased, see previous edition for last biography

RYKBOST, KENNETH ALBERT, b Marion, NY, June 5, 41; m; c 2. POTATO VARIETY DEVELOPMENT. *Educ:* Cornell Univ, BS, 63, MS, 66; Ore State Univ, PhD(soil sci), 73. *Prof Exp:* Res assoc, Long Island Res Sta, Cornell Univ, 73-76; crop scientist, McCain Foods Ltd, NB, Can, 76-87; SUPT, KLAMATH EXP STA, ORE STATE UNIV, 87- *Concurrent Pos:* Mem, Western Regional Potato Variety Develop Comn, 87-91, chmn, 90. *Mem:* Potato Asn Am; Am Soc Agron. *Res:* Cultural management of potatoes with emphasis on fertility, population density, control of diseases and pests, and seed handling; new cultivars. *Mailing Add:* Klamath Exp Sta Ore State Univ 6941 Washburn Way Klamath Falls OR 97603

RYKER, LEE CHESTER, b Indianapolis, Ind, July 1, 40. BIOACOUSTICS, FOREST ENTOMOLOGY. *Educ:* Franklin Col, Ind, BA, 63; Univ Mich, MS, 65; Univ Ore, MS, 71; Ore State Univ, PhD(entom), 75. *Prof Exp:* Res assoc bioacoust, Entom Dept, Ore State Univ, 75-82; res assoc forest entom, Simon Fraser Univ, 83-84. *Mem:* Animal Behav Soc; Coleopterists Soc. *Res:* Chemoacoustic communication research on species of bark beetles destructive to economically important western coniferous trees; ecology of birds-insects; song behavior of birds. *Mailing Add:* 478 Willow Rd Ashland OR 97520

RYLANDER, HENRY GRADY, JR, b Pearsall, Tex, Aug 23, 21; m 43; c 4. MECHANICAL ENGINEERING. *Educ:* Univ Tex, BS, 43, MS, 52; Ga Inst Technol, PhD(mech eng), 65. *Prof Exp:* Design engr, Steam Div, Aviation Gas Turbine Div, Westinghouse Elec Corp, 43-47; from asst prof to assoc prof, 47-68, res scientist, 50, JOE J KING PROF MECH ENG, UNIV TEX, AUSTIN, 68- *Concurrent Pos:* Design engr, Fargo Eng Co, 49-50; eng consult, Mobil Oil Corp, 56-70, Tracor, Inc, 60-72 & CMI Corp, 70- *Mem:* Fel Am Soc Mech Engrs; Am Soc Lubrication Engrs. *Res:* Machine design lubrication and bearing performance including the effects of solids in bearing lubrication. *Mailing Add:* Dept Mech Eng Univ Tex Austin TX 78712

RYLANDER, MICHAEL KENT, b Hillsboro, Tex, Dec 25, 35; div; c 2. VERTEBRATE BEHAVIOR, VERTEBRATE NEUROANATOMY. *Educ:* North Tex State Univ, BA, 56, MS, 62; Tulane Univ, PhD(biol), 65. *Prof Exp:* From asst prof to assoc prof, 65-75, PROF BIOL, TEX TECH UNIV, 75-, CUR BIRDS, MUS, 67-, ADJ PROF, DEPT ANAT, SCH MED, 75- *Res:* Comparative neuroanatomy; behavior of vertebrates, chiefly birds and mammals. *Mailing Add:* Dept Biol Sci Tex Tech Univ Lubbock TX 79409

RYMAL, KENNETH STUART, b Winnepeg, Man, Sept 13, 22; m 49; c 5. FOOD SCIENCE. *Educ:* Mass Inst Technol, BS, 49; Univ Fla, MS, 66; Univ Ga, PhD(food sci), 73. *Prof Exp:* Chemist fish prods, Assoc Fish By-Prods Inc, 49-51; res & develop food mfg, John E Cain Co, Inc, 51-52; owner-mgr, Rymal's Restaurant, 52-64; res assoc, Univ Fla, 64-66; from asst prof to prof hort, Auburn Univ, 66-90; RETIRED. *Mem:* Inst Food Technologists; Am Soc Hort Sci; Asn Anal Chem; Asn Off Anal Chemists. *Res:* Chemistry of horticultural crops; composition, flavor and nutritive content; chemical nature of insect resistance in horticultural crops. *Mailing Add:* 900 Neal Rd Auburn AL 36830

RYMER, WILLIAM ZEV, b Melbourne, Australia, June 3, 39; m 77; c 3. NEUROPHYSIOLOGY, NEUROLOGY. *Educ:* Melbourne Univ, Australia, MBBS, 62; Monash Univ, Australia, PhD(neurophysiol), 73. *Prof Exp:* Resident internal med, Dept Med, Monash Univ, Australia, 63-67 & grad scholar neurophysiol, 67-72; Fogarty fel, Lab Neural Control, Nat Inst Neurol & Commun Dis & Stroke, NIH, 72-74; res assoc, Med Sch, Johns Hopkins Univ, 75-76; asst prof neurosurg & physiol, Med Sch, State Univ NY, 77-78; asst prof physiol & nuerol, 78-81, assoc prof rehab med & assoc prof biomed eng, 83-87, ASSOC PROF PHYSIOL & NEUROL, MED SCH, NORTHWESTERN UNIV, 81-, PROF PHYSIOL REHAB MED, 87-; JOHN G SEARLE PROF REHAB & DIR RES, REHAB INST CHICAGO, 89- *Concurrent Pos:* Instr, Cold Spring Harbor Lab, 80-81; mem Muscle Skeletal and orthopedics study section, NIH, 83-87; consult spasticity res, Warner-Lambert Drug, Co. *Mem:* Soc Neurosci; AAAS. *Res:* Neural control of movement using animal models, normal and neurologically impaired human subjects; interneuronal circuity of the spinal cord; neurophysiological basis of spasticity. *Mailing Add:* Dept Res Rehab Inst Chicago Rm 1406 Ward 345 E Superior Ave Chicago IL 60611

RYMON, LARRY MARING, b Portland, Pa, Nov 16, 34; m 62; c 2. ANIMAL ECOLOGY, CONSERVATION. *Educ:* East Stroudsburg Univ, BS, 58, MEd, 64; Ore State Univ, PhD(biol), 69. *Prof Exp:* Test dept expediter, Electronics, Electro-Mech Res, Fla, 58; assoc prof, 68-71, PROF BIOL, E STROUDSBURG UNIV, 71-, CUR, NATURAL HIST MUS, 71- *Concurrent Pos:* Danforth Assoc, 70; coordr, Environ Studies Inst, 74- *Mem:* Wildlife Soc; Am Inst Biol Sci; Sigma Xi. *Res:* Osprey reintroduction; river otter reintroduction and management. *Mailing Add:* Dept Biol E Stroudsburg Univ East Stroudsburg PA 18301

RYNASIEWICZ, JOSEPH, b Pawtucket, RI, May 22, 17; m 47; c 2. ANALYTICAL CHEMISTRY. *Educ:* Univ RI, BS, 41, MS, 44. *Prof Exp:* Asst agr chem, Exp Sta, Univ RI, 41-45; asst org & anal chem, Northwestern Univ, 45-46; asst anal chem, Knolls Atomic Power Lab, Gen Elec Co, 46-52, res assoc, 52-57, mgr, 57-62, consult anal chem, 62, corrosion engr, 62-64, supvr chem anal, Lamp Metals & Components Dept, 64-71, mgr anal chem, refratory metals prod dept, 71-83; RETIRED. *Mem:* Am Chem Soc; Am Soc Test & Mat. *Res:* Soil chemistry; analytical chemistry of nuclear reactor materials; micro separations of uranium and transuranium elements; high temperature thermal analysis; analytical chemistry of tungsten, molybdenum; corrosion of zircaloy and reactor fuels. *Mailing Add:* 5245 E Farnhurst Rd Lyndhurst OH 44124

RYNBRANDT, DONALD JAY, b Jamestown, Mich, Apr 8, 40; m 67; c 1. TOXICOLOGY, PATHOLOGY. *Educ:* Hope Col, BA, 62; Mich State Univ, PhD(biochem), 67. *Prof Exp:* Assoc dir chem, Drug Anal Sects, St Luke's Hosp, 68-90; SCI DIR, FORENSIC TOXICOL ASSOCS, 87- *Concurrent Pos:* Instr, Case Western Reserve Univ, 68-70, adj sr instr, 70-75, adj asst prof, 75-; Ohio Thoracic grant, 72-73. *Mem:* Am Asn Clin Chemists; Clin Ligand Assay Soc; Clin Lab Mgt Asn. *Res:* Development of biochemical mechanisms for pulmonary diseases; isolation of lung proteolytic enzymes and proteolytic enzyme inhibitors; development of clinical chemistry tests; development of forensic toxicol tests. *Mailing Add:* Forensic Toxicol Assocs 11311 Shaker Blvd Cleveland OH 44104

RYND, JAMES ARTHUR, b Chicago, Ill, Nov 8, 42; m 68; c 2. BIOORGANIC CHEMISTRY, PROTEIN STRUCTURE. *Educ:* Univ Ill, BS, 66; Univ Calif, Riverside, PhD(chem), 71. *Prof Exp:* PROF CHEM, BIOLA UNIV, 70- *Concurrent Pos:* Fel, Univ Calif, Riverside, 72 & 73; res assoc, Col Med, Univ Calif, Irvine, 77-78, Calif State Univ, Fullerton, 80. *Mem:* Am Chem Soc; AAAS; Am Sci Affil. *Res:* X-ray crystal structure of organic compound; enzyme model systems involving flavins and prostaglandins. *Mailing Add:* Dept Chem Biola Univ 13800 Biola Ave LaMirada CA 90639-0002

RYNN, NATHAN, b New York, NY, Dec 2, 23; div; c 3. FUNDAMENTAL PLASMA PHYSICS EXPERIMENTS, CONTROLLED FUSION. *Educ:* City Col New York, BEE, 44; Univ Ill, MS, 47; Stanford Univ, PhD(elec eng), 56. *Prof Exp:* Res asst, Univ Ill, 46-47; res engr, RCA Labs, 47-52; res asst, Stanford Univ, 52-56, res assoc, 58; mem tech staff, Ramo Wooldridge Corp, 56-57; supvr, Huggins Labs, 57-58; res staff physicist, Princeton Univ, 58-66; prof elec eng & physics, 66-69, PROF PHYSICS, UNIV CALIF, IRVINE, 69- *Concurrent Pos:* Consult, Curtiss-Wright Corp, 64; vis lectr, Univ Calif, Berkeley, 65-66; consult, Maxwell Labs, 72-73; consult, Lawrence Livermore Labs, 75-80; vis scientist, Fontenay aux Roses Nuclear Res Ctr, France & Ecole Polytech, Paris, 75; consult Hughes Aircraft, Malibu, Calif, 77-78; Fulbright sr res fel, Ecole Polytech, Paliseau, France, 78; consult, TRW, Inc, 80-; vis prof, Ecole Polytechnique, Lausanne, Switz, 86-; prin investr, NSF grants. *Mem:* Fel Am Phys Soc; fel Inst Elec & Electronics Engrs; Am Geophys Union; Sigma Xi; AAAS. *Res:* Experimental plasma physics in the Q-machine; wave interaction in plasmas; plasma turbulence; controlled fusion; plasma diagnostics; physics of the magnetosphere; plasma applications; stochasticity and chaos in plasma. *Mailing Add:* Dept Physics Univ Calif Irvine CA 92717

RYNTZ, ROSE A, b Detroit, Mich, July 23, 57. PAINTS, COATINGS. *Educ:* Wayne State Univ, BS, 79; Univ Detroit, PhD(chem), 83. *Prof Exp:* Res chemist, Dow Chem Co, 83-85 & Ford Motor Co, 85-86; sr res chemist, E I Du Pont de Nemours, 86-88; sr proj chemist, Dow Corning Corp, 88-89; TECH DIR, AKZO COATINGS, INC, 89- *Concurrent Pos:* Adj prof, Univ Detroit, 86-; tech chair, Fedn Socs Coatings Technol-Detroit Soc, 87-, prof develop comt chair, 91-; chair, Younger Chemists Comt, Am Chem Soc, 88-90; prof Develop Comt, Fedn Soc Coatings Technol, 91-; lectr, Univ Wis, 91- *Mem:* Am Chem Soc; Soc Automotive Engrs. *Res:* Innovative coating development in areas of waterborne primers and basecoats; new crosslinking technology clearcoats; adhesion to plastics; high solids dispersion technology. *Mailing Add:* 1845 Maxwell Troy MI 48084

RYON, ALLEN DALE, b Republic, Ohio, Mar 18, 20; m 42; c 3. CHEMICAL ENGINEERING, INORGANIC CHEMISTRY. *Educ:* Heidelberg Col, BS, 41. *Prof Exp:* Chemist, Basic Refractories Inc, Ohio, 41-42, Basic Magnesium Inc, Nev, 42-44 & Tenn Eastman Corp, Tenn, 44-48; chem engr, Oak Ridge Nat Lab, 48-62, asst sect chief nuclear fuel processing, 62-83; RETIRED. *Concurrent Pos:* Consult, 83- *Mem:* Am Chem Soc. *Res:* Separation and purification by solvent extraction and chromatography. *Mailing Add:* 125 Goucher Circle Oak Ridge TN 37830-4898

RYPKA, EUGENE WESTON, b Owatonna, Minn, May 6, 25; m 67; c 2. MICROBIOLOGY. *Educ:* Stanford Univ, BA, 50, PhD(med microbiol), 58. *Prof Exp:* Asst prof biol, Univ NMex, 57-62; assoc bacteriologist, Leonard Wood Mem Leprosy Res Lab, Johns Hopkins Univ, 62-63; sr scientist, 63-70, HEAD SECT MICROBIOL, LOVELACE MED CTR, 70- *Concurrent Pos:* Vis lectr, Univ NMex, 67-68 & 70-71; adj prof biol, Univ NMex, 72- *Mem:* AAAS; Am Soc Microbiol; Am Soc Cybernet; Inst Elec & Electronics Engrs; Sigma Xi; Int Soc Systs Sci. *Res:* Bacterial physiology; immunology; medical microbiology; systems and cybernetics; genetics definition and comprehension; NP-complete problems. *Mailing Add:* Sect Microbiol 5400 Gibson Blvd SE Albuquerque NM 87108

RYSCHKEWITSCH, GEORGE EUGENE, b Frankfurt, Ger, July 8, 29; nat US; m 50; c 3. INORGANIC CHEMISTRY. *Educ:* Univ Dayton, BS, 52; Ohio State Univ, PhD(chem), 55. *Prof Exp:* Fel, Ohio State Univ, 56; from asst prof to assoc prof, 56-65, PROF CHEM, UNIV FLA, 65- *Mem:* Am Chem Soc. *Res:* Chemistry of Group III elements; boron hydrides; molecular addition compounds; inorganic reaction mechanisms. *Mailing Add:* Dept Chem Univ Fla Gainesville FL 32601

RYSER, FRED A, JR, b Albion, Mich, Feb 29, 20; m 45; c 4. ZOOLOGY. *Educ:* Univ Wis, PhD(zool), 52. *Prof Exp:* Res assoc, Univ Wis, 52-53; from asst prof to assoc prof biol, 53-70, PROF BIOL, UNIV NEV, RENO, 70- *Mem:* Am Soc Mammal; Cooper Ornith Soc; Am Ornith Union; Sigma Xi. *Res:* Temperature regulation and metabolism of mammals; avian and mammalian ecology. *Mailing Add:* Box 8156 University Sta Reno NV 89507

RYSER, HUGUES JEAN-PAUL, b La Chaux-de-Fonds, Switz, June 11, 26; m 61; c 3. CELL BIOLOGY, PHARMACOLOGY. *Educ:* Univ Berne, MD, 53, DrMed(pharmacol), 55. *Prof Exp:* Asst biochem, Med Sch, Univ Berne, 52-55; Swiss Nat Found Sci Res fel med, Univ Med Hosp, Univ Lausanne, 55-56, asst, 56-58; res fel, Harvard Med Sch, 58-60, from instr to asst prof pharmacol, 60-69; assoc prof cell biol & pharmacol, Med Sch, Univ Md, Baltimore, 69-70, prof, 70-72; PROF PATH & PHARMACOL, BOSTON UNIV, 72-, PROF PUB ENVIRON HEALTH, SCH PUB HEALTH, 80-, PROF BIOCHEM, SCH MED, 81- *Concurrent Pos:* Clin & res fel med, Mass Gen Hosp, Boston, 58-64; prin investr, Nat Inst Gen Med Sci grant, 61-67; Lederle med fac award, 64-67; consult, George Washington Univ, 67-69; career develop award, Nat Cancer Inst, 68-69, grant, 68-, mem adv group biol & immunol segment carcinogenesis prog, 72-75; mem adv comt inst grants, Am Cancer Soc, 76-81; mem med res comt, Mass Div, Am Cancer Soc, 79-, mem bd dirs, 83- *Mem:* Histochem Soc; Am Asn Cancer Res; Am Soc Cell Biol; Soc Gen Physiol; Am Soc Pharmacol & Exp Therapeut; Am Asn Path. *Res:* Enzymatic studies on milligram amounts of diseased human liver; penetration and fate of macromolecules into mammalian cells in culture; interaction of basic polymers with cell membranes; molecular mechanisms in chemical carcinogenesis; macromolecular carriers for cancer chemotherapy and drug targeting. *Mailing Add:* 503 Annursnac Hill Rd Concord MA 01742

RYSTEPHANICK, RAYMOND GARY, b Erickson, Man, Aug 24, 40; m 63. THEORETICAL PHYSICS. *Educ:* Univ Man, BSc, 62, MSc, 63; Univ BC, PhD(physics), 65. *Prof Exp:* Fel, McMaster Univ, 65-67; asst prof, 67-71, assoc prof, 71-81, PROF PHYSICS, UNIV REGINA, 81- *Mem:* Am Phys Soc; Can Asn Physicists; AAAS. *Res:* Calculation of soft x-ray emission spectra in metals; gravitational effects in metals. *Mailing Add:* Dept Physics & Astron Univ Sask Regina SK S7N 0W0 Can

RYTAND, DAVID A, b San Francisco, Calif, Nov 4, 09; m 37; c 3. CARDIOLOGY. *Educ:* Stanford Univ, AB, 29, MD, 33. *Prof Exp:* From instr to prof med, 36-75, exec, 54-60, Arthur L Bloomfield prof med, 58-75, EMER PROF MED, SCH MED, STANFORD UNIV, 75- *Concurrent Pos:* Vis prof physiol, Dartmouth Med Sch, 63-64; chief cardiol, Santa Clara Valley Med Ctr, 75-82. *Mem:* Asn Am Physicians; Am Soc Clin Investrs; Soc Exp Biol & Med; Am Soc Pharmacol & Exp Therapeut; Am Fed Clin Res. *Res:* Studies on the mechanism of atrial flutter. *Mailing Add:* Stanford Univ Med Ctr Rm C 248 Stanford CA 94305-5233

RYTTING, JOSEPH HOWARD, b Rexburg, Idaho, June 12, 42; m 65; c 12. PHYSICAL CHEMISTRY, PHARMACEUTICS. *Educ:* Brigham Young Univ, BA, 66, PhD(phys chem), 69. *Prof Exp:* From asst prof to assoc prof, 69-80, PROF PHARM, UNIV KANS, 80- *Concurrent Pos:* Prof, Upjohn Co, 78; assoc ed, Int J Pharmaceut, 84- *Honors & Awards:* Tensiochimica Int Prize Surfactant Chem, Italian Oil Chemist's Soc, 74. *Mem:* Am Chem Soc; Sigma Xi; Calorimetry Conf; fel Am Asn Pharm Scientists; Controlled Release Soc; Am Asn Cols Pharm. *Res:* Application of solution thermodynamics to drug design and delivery; physical chemistry of biologically active agents; effects of pressure and temperature on biologicals and pharmaceutical products; stability; rectal, intraoral and intestinal drug absorption; insulin absorption. *Mailing Add:* Pharmaceut Chem Dept Univ Kans Lawrence KS 66045

RYU, JISOO VINSKY, b Hamhung City, Korea, Mar 11, 41; US citizen; m 71; c 3. EXPLORATION GEOPHYSICS. *Educ:* Univ Utah, BS, 65; Univ Minn, MS, 67; Univ Calif, PhD(eng geosci), 71. *Prof Exp:* Asst res prof geophys, Dept Geol & Geophys Sci, Univ Utah, 71-73; res geophysicist, Space Sci Lab, Univ Calif, Berkeley, 73; res specialist geophys, Exxon Prod Res Co, 73-77; consult, Saratoga, Calif, 77-78; geophysicist, Chevron Oil Field Res Co, 78-82, Chevron USA Co, 82-84, GEOPHYSICIST, CHEVRON OIL FIELD RES CO, 84- *Concurrent Pos:* Repub Korea Army,

61-62; Jane Lewis fel, Univ Calif, Berkeley, 70. *Mem:* Soc Explor Geophys; Europ Asn Explor Geophys; Am Geophys Union. *Res:* Seismic signal processing; wide aperature seismology; geophysical exploration of hydrocarbon, groundwater and mineral deposits. *Mailing Add:* 657 W Valencia Mesa Dr Fullerton CA 92635

RYUGO, KAY, b Sacramento, Calif, Apr 10, 20; m 55; c 5. POMOLOGY, PLANT PHYSIOLOGY. *Educ:* Univ Calif, BS, 49, MS, 50, PhD, 54. *Prof Exp:* From asst pomologist to assoc pomologist, 55-69, POMOLOGIST, UNIV CALIF, DAVIS, 69-, LECTR, 76- *Concurrent Pos:* NATO fel, Univ Bologna, 72. *Mem:* Int Soc Hort Sci; Am Soc Hort Sci; Bot Soc Am; Scand Soc Plant Physiol; Am Soc Plant Physiol. *Res:* Physiology and biochemistry; native growth regulators in fruits and trees. *Mailing Add:* Dept Promol Univ Calif Davis CA 95616

RYZLAK, MARIA TERESA, b Augustow, Poland, Feb 27, 38; US citizen; m 59; c 2. ENZYMOLOGY, RECEPTOR PROTEINS. *Educ:* Univ Manchester, Eng, BSc, 66; NJ Inst Technol, Newark, MS, 70; Univ Med & Dent NJ, PhD(biochem), 86 & Rutgers Univ, New Brunswick, PhD(biochem), 86. *Prof Exp:* Res chemist org synthesis, Troy Chem Co, Newark, NJ, 64-68; teaching fel, Chem Lab, NJ Inst Technol, Newark, 68-69; res asst, Endocrinol Lab, Worcester Found Exp Biol, 69-71, sr res asst molecular biol, 71-74; res assoc biochem, Med Sch, Temple Univ, 74-77; chemist, US Naval Air Develop Ctr, 77-78; res assoc biochem, McNeil Pharmaceut, 78-81; grad res asst, 82-86, RES ASSOC BIOCHEM, RUTGERS UNIV, 89- *Concurrent Pos:* Busch fel, Waksman Inst, Rutgers Univ, 86-89; postdoctoral fel, Charles & Johanna Busch Mem Found, 86-89. *Mem:* Sigma Xi; Soc Exp Biol & Med; Res Soc Alcoholism. *Res:* Purification and characterization of aldehyde dehydrogenases from the human liver and prostate; study of the metabolism of acetaldehyde and biogenic aldehydes arising from the oxidation of biogenic anines, putrescine, spermine and spermidine; study of oxysterols, specific receptor proteins, regulation of HMG-CoA-reductase and carcinogenesis by oxysterols of the human and rat prostates; mechanism of carcinogenesis of the human liver and prostate due to ethanol metabolism. *Mailing Add:* Waksman Inst Rutgers State Univ NJ PO Box 759 Piscataway NJ 08855-0759

RZAD, STEFAN JACEK, b Warsaw, Poland, Mar 15, 38; m 63; c 2. PHYSICAL CHEMISTRY. *Educ:* Univ Louvain, MS, 60, PhD(phys chem), 64. *Prof Exp:* Fel radiation lab, Univ Notre Dame, 64-66 & radiation labs, Mellon Inst, Carnegie-Mellon Univ, 67-74; PHYS CHEMIST, CORP RES & DEVELOP CTR, GEN ELEC CO, 74- *Mem:* Am Chem Soc; Inst Elec & Electronics Engrs; Sigma Xi. *Res:* Reactions of radicals in electron irradiated liquids; gas-phase vacuum ultraviolet photochemistry; ionic processes; conduction and mechanisms of electrical breakdown in liquids, solids and gases; high voltage phenomena; plasma enhanced chemical vapor deposition of organic and inorganic materials. *Mailing Add:* Corp Res & Develop Ctr Gen Elec Co Schenectady NY 12301

RZESZOTARSKI, WACLAW JANUSZ, b Warsaw, Poland, Apr 14, 36; US citizen. RADIOPHARMACEUTICALS. *Educ:* Gdansk Inst Technol. MS, 59; Polish Acad Sci, PhD(chem), 64. *Prof Exp:* Res assoc med chem, Univ Calif, Santa Barbara, 66-68 & Wash Hosp Ctr, 68-75; from asst prof to prof med chem, George Washington Univ, 75-85; DIR MED CHEM, NOVA PHARMACEUT CORP, 85- *Mem:* Am Chem Soc; AAAS; Nuclear Med Soc; Polish Inst Arts & Sci; Pilsudski Inst Am. *Res:* Medicinal chemistry; drug design, synthesis and development; pharmaceutical effect on central nervous system. *Mailing Add:* Radiopharm Chem Ross 662 George Washington Univ Med Ctr Washington DC 20037

S

SAACKE, RICHARD GEORGE, b Newark, NJ, Oct 31, 31; m 54; c 5. REPRODUCTIVE PHYSIOLOGY, CYTOLOGY. *Educ:* Rutgers Univ, BS, 53; Pa State Univ, MS, 55, PhD(dairy sci), 62. *Prof Exp:* Dairy exten specialist, Univ Md, 57-58; from instr to asst prof dairy physiol, Pa State Univ, 58-65; asst prof, 65-68, PROF DAIRY SCI, VA POLYTECH INST & STATE UNIV, 68- *Concurrent Pos:* Nat Inst Child Health & Human Develop res grant, 65-69. *Mem:* AAAS; Am Dairy Sci Asn; Am Soc Animal Sci. *Res:* Cytology and physiology of bovine spermatozoa; ultrastructural study of the bovine mammary gland with emphasis on milk synthesis and secretion. *Mailing Add:* Dept Dairy Sci Va Polytech Inst & State Univ Blacksburg VA 24061

SAADA, ADEL SELIM, b Heliopolis, Egypt, Oct 24, 34; US citizen; m 60; c 2. SOIL MECHANICS. *Educ:* Ecole Centrale de Paris, France, Engr, 58; Univ Grenoble, MS, 59; Princeton Univ, PhD(soil mech), 61. *Prof Exp:* Res assoc civil eng, Princeton Univ, 61-62; from asst prof to assoc prof, 62-88, FRANK H NEFF PROF CIVIL ENG, CASE WESTERN UNIV, 88-CHMN DEPT, 78- *Concurrent Pos:* NSF res grant soil mech, 65-, US-France res grant civil eng & Army Res Off grant fracture mech, 83-; soils & foundations consult; Air Force Off Sci Res grant, 88. *Honors & Awards:* Richard J Carrol Lectr, Johns Hopkins Univ, 90. *Mem:* Am Soc Testing & Mat; fel Am Soc Civil Engrs; Int Soc Soil Mech. *Res:* Mechanical behavior of soils under different stress systems; fracture mechanics applied to soils. *Mailing Add:* Dept Civil Eng Case Western Reserv Univ Univ Circle Cleveland OH 44106

SAALFELD, FRED ERIC, b Joplin, Mo, Apr 9, 35; m 58; c 1. PHYSICAL CHEMISTRY, INORGANIC CHEMISTRY. *Educ:* Southeast Mo State Col, BS, 57; Iowa State Univ, MS, 59, PhD(chem), 61. *Prof Exp:* Fel, Iowa State Univ, 61-62; sect head mass spectrometry, 62-74, head, Phys Chem Br, 74-76, supt, Chem Div, 76-82, dir res prog, 82-87, DIR, OFF NAVAL RES, 87- *Concurrent Pos:* Chief scientist, Off Naval Res, London Br Off, 80. *Mem:* Fel AAAS; Am Chem Soc; Am Soc Mass Spectrometry (secy, 70-74); fel Chem Soc London; Mass Spectros Soc Japan. *Res:* Application of mass spectrometry to chemical problems; investigation of ion-molecule reaction and the kinetics of chemical reactions occuring in flames and chemical laser; analysis of exceedingly complex mixtures. *Mailing Add:* Off Naval Res Code 10 800 N Quincy St Arlington VA 22217-5000

SAARI, DONALD GENE, b Ironwood, Mich, Mar 9, 40; m 66; c 2. MATHEMATICAL ECONOMICS, CELESTIAL MECHANICS. *Educ:* Mich Technol Univ, BS, 62; Purdue Univ, MS, 64, PhD(math), 67. *Hon Degrees:* PhD(appl math & econ), Purdue Univ, 88. *Prof Exp:* Res staff astronr, Yale Univ, 67-68; from asst prof to assoc prof, 68-74, PROF MATH, NORTHWESTERN UNIV, 74- *Concurrent Pos:* Consult, Nat Bur Standards, 79-86; ed, Soc Indust & Appl Math, 81-86; Guggenheim fel, 88. *Honors & Awards:* Lester Ford Award, Math Asn Am, 85. *Mem:* Am Math Soc; Math Asn Am; Soc Indust & Appl Math; Am Astron Soc; Economet Soc. *Res:* Qualitative behavior of the N-body problem of celestial mechanics; collisions and singularities; behavior of expanding gravitational systems; dynamical systems; mathematical economics; decision analysis. *Mailing Add:* Dept Math Northwestern Univ Evanston IL 60208

SAARI, EUGENE E, b Grand Rapids, Minn, July 17, 36; m 60; c 4. PHYTOPATHOLOGY. *Educ:* Univ Minn, BS, 59, MS, 62, PhD(plant path), 66. *Prof Exp:* Instr plant path, Okla State Univ, 62-65; res asst, Univ Minn, 65-66; res assoc, Mich State Univ, 66-67; plant pathologist, India, 67-73, regional wheat pathologist, Int Maize & Wheat Improv Ctr-Lebanon, 73-76, regional plant pathologist, Egypt, 76-80, SE Asia regional wheat rep, Thailand, 80-84, plant pathologist, Mex, 84-86, WHEAT SPECIALIST, TURKEY, 86- *Mem:* Am Phytopath Soc; Am Soc Agron; Crop Sci Soc Am; Sigma Xi. *Res:* Wheat diseases, breeding and production. *Mailing Add:* Cimmyt Apdo Postal 6-641 Deleg Cuauhtemoc Mexico DF 06600 Mexico

SAARI, JACK THEODORE, b Virginia, Minn, Jan 1, 43; m 74; c 2. CARDIOVASCULAR PHYSIOLOGY. *Educ:* Univ Minn, Minneapolis, BChE, 65, PhD(physiol), 70. *Prof Exp:* Instr physiol, Univ Minn, Minneapolis, 70-71; res assoc biophys, Univ Calgary, 71-75; asst prof physiol, Sch Dent, Marquette Univ, 75-78; from asst prof to assoc prof physiol, Sch Med, Univ NDak, 78-87; RES PHYSIOLOGIST, USDA AGR RES SERV HUMAN NUTRIT RES CTR, GRAND FORKS, ND, 87- *Concurrent Pos:* William H Davies Mem Res fel, Div Med Biophys, Univ Calgary, 72-74, Med Res Coun Can prof asst, 74-75. *Mem:* Sigma Xi; Am Physiol Soc; Am Heart Asn. *Res:* Microvascular permeability and vasoactivity; calcium kinetics in cardiac muscle; trace mineral metabolism and cardiovascular function. *Mailing Add:* USDA Agr Res Serv Human Nutrit Res Ctr PO Box 7166 University Sta Grand Forks ND 58202

SAARI, WALFRED SPENCER, b Lonsdale, RI, Feb 6, 32; m 53; c 4. ORGANIC CHEMISTRY. *Educ:* Brown Univ, ScB, 53; Mass Inst Technol, PhD(org chem), 57. *Prof Exp:* Res chemist, Shell Develop Co, Calif, 57-59; sr res chemist, 59-75, SR INVESTR, MERCK SHARP & DOHME RES LABS, 75- *Mem:* Am Chem Soc. *Res:* Medicinal chemistry. *Mailing Add:* Merck Sharp & Dohme Res Labs West Point PA 19486

SAARLAS, MAIDO, b Tartu, Estonia, Feb 5, 30; US citizen; m 56; c 2. AERONAUTICAL & MECHANICAL ENGINEERING. *Educ:* Univ Ill, Urbana, BS, 53, MS, 54 & 57; Univ Cincinnati, PhD(mech eng), 66. *Prof Exp:* Res engr, Univ Chicago, 54-56 & Douglas Aircraft Co, 56; sr engr, Autonetics Div, NAm Rockwell Corp, 58-59 & Aeronutronic Div, Ford Motor Co, 59-61; res assoc, Univ Cincinnati, 61-65 & Kinetics Corp, 65-67; sr engr, Gen Elec Co, 67-69; PROF AERONAUT ENG, US NAVAL ACAD, 80- *Concurrent Pos:* Lectr, Univ Calif, Los Angeles, 58-61; consult, Douglas Aircraft Co, 65, Gen Elec Co, 65-67, Trident Eng, 69- & Cadcom, Inc, 70- *Mem:* Am Inst Aeronaut & Astronaut; Am Soc Mech Engrs. *Res:* Fluid mechanics and heat transfer, flight mechanics and dynamics; turbo-machinery. *Mailing Add:* 225 Mill Harbor Dr Arnold MD 21012

SAATY, THOMAS L, b Mosul, Iraq, July 18, 26; US citizen; m 48; c 5. MATHEMATICS. *Educ:* Catholic Univ, MS, 49; Yale Univ, MA, 50, PhD, 53. *Prof Exp:* Mathematician, Melpar, Inc, 53-54; mathematician & sci analyst, Mass Inst Technol, 54-57; mathematician, US Dept Navy, 57-58, sci liaison officer, Off Naval Res, Eng, 58-59, dir adv planning, 59-61, head math br, 61-63, mathematician, US Arms Control & Disarmament Agency, Dept State, 63-69; prof statist & oper res, 69-79, UNIV PROF, UNIV PITTSBURGH, 79- *Concurrent Pos:* Lectr, Am Univ & USDA Grad Sch, 54-; prof lectr, Catholic Univ, George Washington Univ & Exten Div, Univ Calif, Los Angeles; Ford Found lectr, Nat Planning Inst, Cairo, 59 & 64; mathematician, US Dept Air Force & Nat Bur Standards, 51-52; consult var corps, US govt agencies & depts & foreign govts; exec dir, Conf Bd Math Sci, AAAS, 65-67. *Honors & Awards:* Lester R Ford Award, Math Asn Am, 73; Inst Mgt Sci Award, 77. *Mem:* Fel AAAS; Am Math Soc; Math Soc Am; Opers Res Soc Am; Royal Span Acad. *Res:* Optimization; nonlinear processes; graph theory; queueing theory and stochastic processes; operations research, especially in underdeveloped countries; mathematical methods and military uses; models of arms reduction; systems; planning; decision making; conflict resolution. *Mailing Add:* Mervis Hall 322 Univ Pittsburgh 4200 Fifth Ave Pittsburgh PA 15260

SAAVEDRA, JUAN M, b Buenos Aires, Arg, Oct 5, 41; m 70. PHARMACOLOGY, PSYCHIATRY. *Educ:* Univ Buenos Aires, MD, 65. *Prof Exp:* Vis fel, NIH, 71-73, vis scientist pharmacol, 73-79, med officer, 79-89, CHIEF, SECT PHARMACOL, NIMH, 89- *Concurrent Pos:* Assoc prof clin psychiat, Uniformed Serv Univ Health Sci. *Honors & Awards:* Adminr Award Meritorious Achievement, Dept Health Human Serv, Alcohol Drug Abuse & Mental Health Admin, 87. *Mem:* Int Soc Hypertension; Int Soc Neurochem; Soc Neurosci; Int Brain Res Orgn. *Res:* Psychopharmacology; biological psychiatry; neural regulation of blood pressure; central control of automatic functions; stress and central nervous system. *Mailing Add:* NIH Rm 2D-45 Bldg 10 9000 Rockville Pike Bethesda MD 20892

SABA, GEORGE PETER, II, b Wilkes-Barre, Pa, Sept 30, 40; m 67; c 4. RADIOLOGY, BIOPHYSICS. *Educ:* Pa State Univ, BS, 62, MS, 64; State Univ NY Buffalo, MD, 68. *Prof Exp:* Asst physics, Pa State Univ, 62-64; intern med, Buffalo Gen Hosp, 68-69; clin assoc, Nat Heart Inst, 69-71; ASST PROF RADIOL, JOHNS HOPKINS HOSP, 74- *Concurrent Pos:* Consult physician radiol, Union Mem Hosp, 73-, Havre de Grace Hosp, 74- & Md Gen Hosp, 74- *Res:* Clinical medicine; pancreatic studies and gastrointestinal radiology. *Mailing Add:* Dept Radiol Johns Hopkins Univ 720 Rutland Ave Baltimore MD 21205

SABA, SHOICHI, b Tokyo, Japan, Feb 28, 19; m; c 3. ELECTRICAL ENGINEERING, SCIENCE POLICY. *Educ:* Univ Tokyo, BSc, 41. *Prof Exp:* Mem, Toshiba Corp, 42-68, chief engr, Heavy Apparatus Div, 68-70, dir & gen mgr, 70-80, pres & chief exec officer, 80-86, chmn bd & exec officer, 86-87, ADV TO BD, TOSHIBA CORP, 87- *Concurrent Pos:* Dir, Japan Atomic Power Co, 81, Tokyo Bay Hilton Co Ltd, 85, Mutsu Ogawara Develop Inc, 86; vchmn, bd trustees, Int Christian Univ, 88; pres, Global Infrastruct Fund Res Found, 89; chmn, Japanese Panel, Japan-US Conf Cult & Educ Interchange, 91. *Mem:* Foreign assoc Nat Acad Eng; fel Inst Elec & Electronics Engrs. *Res:* Author or co-author of 4 books & 10 technical papers; holder of 2 patents. *Mailing Add:* Toshiba Corp 1-1 Shibaura 1-Chome Minato-Ku Tokyo 105 Japan

SABA, THOMAS MARON, b Wilkes Barre, Pa, Mar 8, 41; m 63; c 3. MEDICAL PHYSIOLOGY, BIOPHYSICS. *Educ:* Wilkes Col, BA, 63; Univ Tenn, PhD(physiol, biophys), 67. *Prof Exp:* Lab asst biol, Wilkes Col, 61-62; instr physiol & biophys, Med Units, Univ Tenn, 67-68; from asst prof to assoc prof physiol, Univ Ill Col Med, 68-73; PROF PHYSIOL & CHMN DEPT, ALBANY MED COL, 73- *Concurrent Pos:* Consult physiologist, Vet Admin Hosp, Hines, Ill, 70-73; clinical physiologist, Albany Med Ctr, 73-; adj prof biomed eng, Rensselaer Polytech Inst. *Mem:* AAAS; Reticuloendothelial Soc; Soc Exp Biol & Med; Am Asn Study Liver Dis; Am Physiol Soc; Shock Soc; Surg Infection Soc. *Res:* Cardiovascular and metabolic aspects of the liver; physiology and physiopathology of the reticuloendothelial system; physiological mechanisms of host-defense; plasma fibronection and phagocytores; lung vascular injury; pathophysiology of traumatic shock; lung and peripheral vascular permeability. *Mailing Add:* Dept Physiol & Cell Biol Albany Med Col Albany NY 12208

SABA, WILLIAM GEORGE, b Wilkes-Barre, Pa, Aug 15, 32; m 60; c 4. PHYSICAL CHEMISTRY. *Educ:* Wilkes Col, BS, 54; Univ Pittsburgh, PhD(phys chem), 61. *Prof Exp:* Asst chem, Univ Pittsburgh, 54-55, phys chem, 55-61; phys chemist, Heat Div, Nat Bur Stand, 61-69, PATENT EXAM, PATENT OFF, DEPT COM, 69- *Mem:* Am Chem Soc; Am Phys Soc; AAAS. *Res:* Thermodynamic properties of solid solutions; low-temperature calorimetry; solid state electronics; semiconductor design and processing. *Mailing Add:* 2623 Kinderbrook Lane Bowie MD 20715

SABACKY, M JEROME, b Cedar Rapids, Iowa, June 22, 39; m 67; c 2. ORGANIC CHEMISTRY. *Educ:* Coe Col, BA, 61; Univ Ill, MS, 63, PhD(chem), 66. *Prof Exp:* Sr res chemist, Org Chem Div, Monsanto Co, 66-75, from res specialist to sr res specialist, Nutrit Chem Div, 75-85, MONSANTO FEL, ANIMAL SCI DIV, MONSANTO INDUST CHEM CO, 85- *Honors & Awards:* Thomas & Hochwalt Award, 81. *Mem:* Am Chem Soc. *Res:* Magnetic resonance studies of ortho-substituted derivatives of triphenylmethane; homogeneous catalysis; catalytic asymmetric synthesis employing transition metal complexes. *Mailing Add:* 324 Holloway Ballwin MO 63011

SABADELL, ALBERTO JOSE, b Barcelona, Spain, Oct 31, 29; m 56; c 3. CHEMICAL ENGINEERING. *Educ:* Univ Buenos Aires, Lic chem, 54; Princeton Univ, MSE, 63. *Prof Exp:* Chemist, Invests Inst Sci & Tech Armed Forces, 55-56, chief div explosives and propellants, 57-60; prof theory explosives, Eng Sch Army, Argentinian Army, 60-61; res asst rocket propulsion, Forrestal Res Ctr, Princeton Univ, 62-63; assoc prof combustion, Sch Eng, Univ Buenos Aires, 64-65; res engr, Princeton Chem Res, 66 & Aerochem Res Labs Inc, Sybron Corp, 67-75; tech staff, Metrek Div, Mitre Corp, 76-82; phys scientist, bur sci & technol, 83-87, DEP DIR OFF ENERGY, BUR SCI & TECHNOL, USAID, 88- *Concurrent Pos:* NASA fel, Princeton Univ, 61-63; hon mem space tech comt, Arg Space Nat Comn, 64-65; panel mem, Nat Res Coun, 82- *Mem:* Am Chem Soc. *Res:* Energy. *Mailing Add:* 1210 Meadow Green Lane McLean VA 22102

SABATH, LEON DAVID, b Savannah, Ga, July 24, 30; div; c 3. MEDICINE, MICROBIOLOGY & INFECTIOUS DISEASES. *Educ:* Harvard Univ, AB, 52; Harvard Med Sch, MD, 56. *Prof Exp:* Intern med, Peter Bent Brigham Hosp, Boston, 56-57; jr resident, Bellevue Hosp, New York, 59-60; res fel, Harvard Med Sch, 60-62; sr resident, Peter Bent Brigham Hosp, 62-63; spec fel, Oxford Univ, 63-65; assoc med, Harvard Med Sch, 65-68, from asst prof to assoc prof, 68-74; head sect infectious dis, 74-83, PROF MED, UNIV MINN, MINNEAPOLIS, 74-; STAFF PHYSICIAN, UNIV MINN HOSPS, 74- *Concurrent Pos:* Spec fel, Nat Inst Allergy & Infectious Dis, 63-67, Res Career Develop Award, 69-74; attend physician, Vet Admin Hosp, West Roxbury, 68-74; assoc physician, Boston City Hosp, 69-74. *Mem:* Am Fedn Clin Res; Soc Gen Microbiol; Am Soc Microbiol; Soc Clin Pharmacol & Therapeut; fel Am Col Physicians; Am Soc Clin & Infectious Dis; fel Infectious Dis Soc Am. *Res:* Antibiotics; infectious diseases; bacterial resistance to antibiotics; antibiotic assays; clinical pharmacology; penicillins, penicillinases; bacterial cell walls. *Mailing Add:* Univ Minn Hosp Minneapolis MN 55455

SABATINI, DAVID DOMINGO, b Bolivar, Arg, May 10, 31; m 60; c 2. CELL BIOLOGY, BIOCHEMISTRY. *Educ:* Nat Univ Litoral, MD, 54; Rockefeller Univ, PhD(biochem), 66. *Prof Exp:* Instr, lectr & assoc prof histol, Inst Gen Anat & Embryol & dir admis, Med Sch, Univ Buenos Aires, Arg, 57-60; Rockefeller Found fel, Med Sch, Yale Univ, 61 & Rockefeller Inst, 61-62; res assoc, Cell Biol Lab, Rockefeller Univ, New York, NY, 61-63, from asst prof to assoc prof cell biol, 66-72; prof, 72-74, FREDERICK L EHRMAN PROF, DEPT CELL BIOL, MED CTR, NY UNIV, 75-, CHMN, 72-, DIR, MD-PHD PROG, SCH MED, 87- *Concurrent Pos:* Fel, Nat Acad Med, Arg, 56; UNESCO fel, Biophys Inst, Rio de Janeiro, 57; Pfizer travelling fel, 72; mem, Molecular Biol Study Sect, NIH, 73-77, chmn, 76-77; coun mem, Am Soc Cell Biol, 74-77; ed, J Cellular Biochem, 80-84, Molecular & Cellular Biol, 80-82, Biol Cell, 86- & Current Opinions Cell Biol, 90-; bd dirs, Pub Health Res Inst, 80-88; mem, Bd Basic Biol, Nat Res Coun, 86- *Honors & Awards:* Wendell Griffith Mem Lectr, St Louis Univ, Mo, 77; Mary Peterman Mem Lectr, Mem-Sloan Kettering Inst, New York, NY, 77; 25th Robert J Terry Lectr, Wash Univ, 78; Samuel Roberts Noble Res Recognition Award, 80; E B Wilson Award, Am Soc Cell Biol, 86; Seventh Annual Kenneth F Naidorf Mem Lectr, Columbia Univ, 89. *Mem:* Nat Acad Sci; Am Soc Cell Biol (pres, 78-79); Am Soc Biol Chemists; Tissue Cult Asn; Asn Anat Chmn; fel NY Acad Sci; Am Soc Microbiol; Harvey Soc (vpres, 85-86, pres, 86-87); fel Am Acad Arts & Sci. *Res:* Electron microscopy; membrane and organelle biogenesis and protein synthesis in free and membrane bound ribosomes; structure of endoplasmic reticulum membrane; mechanisms of cellular aging; author of various publications. *Mailing Add:* Dept Cell Biol Sch Med NY Univ New York NY 10016

SABBADINI, EDRIS RINALDO, b Anzio, Italy, Apr 1, 30; Can citizen; m 60; c 2. IMMUNOBIOLOGY. *Educ:* Univ Pavia, MD, 54; McGill Univ, PhD(exp surg), 67. *Prof Exp:* Asst surg, Univ Pavia, 58-63; lectr exp surg, McGill Univ, 68-69; asst prof, 69-72, assoc prof, 72-80, PROF IMMUNOL, UNIV MAN, 80- *Mem:* Can Soc Immunol; Am Asn Immunologists; Transplantation Soc; NY Acad Sci. *Res:* Transplantation and tumor immunology; regulation of cell-mediated immunity. *Mailing Add:* Dept Immunol Univ Man Winnipeg MB R3T 2N2 Can

SABBAGH, HAROLD A(BRAHAM), b W Lafayette, Ind, Jan 9, 37; m 66; c 3. ELECTRICAL ENGINEERING, PHYSICS. *Educ:* Purdue Univ, BSEE & MSEE, 58, PhD(elec eng), 64. *Prof Exp:* Instr elec sci, US Naval Acad, 59-61; asst prof elec eng, Rose-Hulman Inst Technol, 64-67, assoc prof, 67-70, prof elec eng & physics, 70-72; electronics engr, Naval Weapons Support Ctr, 72-80; PRES, SABBAGH ASSOC, INC, 80- *Concurrent Pos:* Consult, Crane Naval Ammunition Dept, Ind. *Mem:* Inst Elec & Electronics Engrs; Sigma Xi. *Res:* Electromagnetic waves; electroacoustic waves; nondestructive evaluation, computational electromagnetics. *Mailing Add:* 2609 Spicewood Lane Bloomington IN 47401

SABBAGHIAN, MEHDY, b Tehran, Iran, Nov 22, 35; m 63; c 1. ENGINEERING SCIENCE. *Educ:* Abadan Inst Technol, Iran, BSc, 58; Case Inst Technol, MSc, 63; Univ Okla, PhD(eng sci), 64. *Prof Exp:* Proj engr, Iranian Oil Ref Co, 58-60 & Viking Air Prod, 61-62; from asst prof to assoc prof, 64-74, PROF ENG SCI, LA STATE UNIV, BATON ROUGE, 74- *Concurrent Pos:* Consult, Hydro Vac Inc, 70-71, Gamma Indust, 72- & Nuclear Systs Inc, 73. *Mem:* Am Soc Mech Engrs; Am Soc Eng Educ. *Res:* Viscoelastic materials and their physical behaviors; thermoviscoelasticity with time dependent properties. *Mailing Add:* Dept Mech Eng 2523 Ceba Bldg La State Univ Baton Rouge LA 70803

SABBAN, ESTHER LOUISE, b Detroit, Mich, July 21, 48. NEUROCHEMISTRY. *Educ:* Hebrew Univ, BSc, 70, MSc, 72; New York Univ, PhD (biochem), 77. *Prof Exp:* Asst res scientist cell biol, Med Ctr, New York Univ, 77-80, res asst prof cell biol & psychol, 80-83; asst prof, 83-86, ASSOC PROF BIOCHEM & MOLECULAR BIOL, NY MED COL, 86- *Honors & Awards:* NIH Career Develop Award. *Mem:* NY Acad Sci; Am Soc Biol Chemists; Am Soc Cell Biol; Soc Neurosci; AAAS. *Res:* Regulation of biosynthesis of dopamine beta-hydroxylase and tyrosine hydroxylase. *Mailing Add:* Dept Biochem & Molecular Biol NY Med Col Valhalla NY 10595

SABEL, CLARA ANN, b Louisville, Ky, Aug 19, 32. MATHEMATICS. *Educ:* Spalding Col, BA, 54; Xavier Univ, Ohio, MA, 61; Syracuse Univ, PhD(math), 70. *Prof Exp:* Teacher, LaSalette Acad, 57-61 & Nazareth Col & Acad, 61-64; asst prof math, Spalding Col, 69-77, chmn dept, 72-77; PROF, DEPT MATH, NAZARETH COL, KY, 77- *Mem:* Math Asn Am; Am Math Soc. *Res:* Abstract algebra; ring theory; group theory. *Mailing Add:* 9300 Shelbyville Rd Suite 501 Louisville KY 40222

SABELLI, HECTOR C, b Buenos Aires, Arg, July 25, 37; nat US; m 60; c 2. PSYCHIATRY, NEUROPHARMACOLOGY. *Educ:* Univ Buenos Aires, MD, 59, DrMed, 61. *Prof Exp:* Res fels, Arg Soc Advan Sci, 59-60 & Arg Coun Res, 60-61; asst prof pharmacol, Chicago Med Sch, 62-64; career investr, Arg Coun Res, 64-66; vis prof pharmacol, Chicago Med Sch, 66-67, actg chmn dept, 70, chmn dept, 71-75; asst prof psychiat, 79-84, ASSOC PROF PSYCHIAT & PROF PHARMACOL, RUSH UNIV, 84- *Concurrent Pos:* Prof & chmn, Inst Pharmacol, Nat Univ Litoral, 65-66; psychiatrist, Rush-Presby, St Lukes Hosp, dir, Psychobiol Lab, 79- *Honors & Awards:* Soc Biol Psychiat Award, 63; Sci Res Award, Interstate Postgrad Med Sch Asn NAm, 70; Clin Res Award, Am Acad Clin Psychiatrists, 84. *Mem:* Soc Biol Psychiat; Am Soc Pharmacol & Exp Therapeut; Soc Neurosci. *Res:* Biogenic amines psychiatric disorders and drug therapy; psychodynamics and pharmacotherapy of depression. *Mailing Add:* Dept Psychiat Rush Univ 1753 W Harrison Chicago IL 60612

SABELLI, NORA HOJVAT, b Buenos Aires, Argentina, Dec 22, 36; m 60; c 2. THEORETICAL CHEMISTRY, COMPUTER SCIENCE. *Educ:* Univ Buenos Aires, MS, 58, PhD(chem), 64. *Prof Exp:* Res assoc chem, Univ Chicago, 61-63; instr phys chem, Univ Buenos Aires, 64-65; asst prof, Nat Univ Litoral, 65-66; chemist, Univ Chicago, 67-69; instr comput sci, 69-75, ASSOC PROF COMPUT SCI & CHEM, UNIV ILL CHICAGO CIRCLE, 75-; SR RES SCIENTIST, NCSA, 89- *Concurrent Pos:* Career investr, Argentine Nat Res Coun, 64-66; vis resident assoc, Argonne Nat Labs, 74-84. *Mem:* Am Chem Soc; AAAS. *Res:* Theoretical organic chemistry; molecular orbital and semiempirical methods; computational chemistry; ab initio methods; potential curves. *Mailing Add:* NCSA Univ Ill 605 E Springfield Champaign IL 61820

SABER, AARON JAAN, b London, Eng, Aug 20, 46; Can citizen; m 83; c 2. THERMOFLUIDS. *Educ:* Eng Univ Toronto, BASc, 69; Princeton Univ, MA, 71, PhD(aerospace), 74. *Prof Exp:* Mem res staff, Guggenheim Labs, Princeton Univ, 74-75; PROF ENG THERMODYNAMICS, CONCORDIA UNIV, 75- *Concurrent Pos:* Consult, Govt Can, 76-; pres, Lignasco Resources Ltd, 80-; consult, Pulp & Paper Indust. *Mem:* Am Inst Aeronaut & Astronaut; Can Aeronaut & Space Inst. *Res:* Aerospace propulsion; fluid structures; electromagnetic engineering. *Mailing Add:* 4827 Grand Blvd Montreal PQ H3X 3S1 Can

SABERSKY, ROLF H(EINRICH), b Berlin, Ger, Oct 20, 20; nat US; m 46; c 2. HEAT TRANSFER, FLUID MECHANICS. *Educ:* Calif Inst Technol, BS, 42, MS, 43, PhD, 49. *Prof Exp:* Develop engr, Aerojet-Gen Corp, Gen Tire & Rubber Co, 43-46; from instr to prof, 49-88, PROF EMER MECH ENG, CALIF INST TECHNOL, 88- *Concurrent Pos:* Consult, Aerojet-Gen Corp, 49-71, var aerospace & high tech co, 71- *Honors & Awards:* Heat Transfer Mem Award, Am Soc Mech Engrs, 77. *Mem:* Fel Am Soc Mech Engrs. *Res:* Heat transfer to granular materials and complex fluids. *Mailing Add:* 1060 Fallen Leaf Rd Arcadia CA 91006-1903

SABES, WILLIAM RUBEN, b St Paul, Minn, Jan 18, 31; m 51; c 3. ORAL PATHOLOGY. *Educ:* Univ Minn, BS & DDS, 59, MSD, 61; Am Bd Oral Path, dipl, 69. *Prof Exp:* Fel, Univ Minn, 59-61; asst prof oral histopath, Sch Dent, Temple Univ, 61-63; asst prof histol & path & chmn sect, Sch Dent, Univ Detroit, 63-65, from asst prof to assoc prof histopath & diag, 65-68, chmn dept, 66-68; assoc prof, Col Dent, Univ Ky, 68-71, prof oral path & chmn dept, 71-78; PROF PATH DEPT, SCH DENT, UNIV DETROIT, 78- *Concurrent Pos:* Consult, Vet Admin Hosps, Philadelphia, 61-63, Dearborn, Mich, 65-68, Lexington, Ky, 73-78, Allen Park, Mich, 78- & Mich Dent Asn Comn Cancer Control, Hosp & Inst Dent Servs, 85-86; vis prof, Sch Dent, Univ Calif, Los Angeles, 76; clin assoc prof, Dept Path Sch Med, Wayne State Univ, 79-84; coun mem, Am Acad Oral Path, 80-83; consult, Am Cancer Soc, Mich Div, Prof Educ Comm, 85-86; dent chmn, Am Asn Cancer Educ, 77-78, exec coun mem, 90-91; sec path sect, Am Asn Dent Schs, 72-73, chmn elect path sect, 73-74, chmn path sect, 74-75. *Honors & Awards:* Award, Am Acad Dent Med, 59. *Mem:* Am Dent Asn; Am Acad Oral Path; Am Asn Cancer Educ; AAAS; Am Asn Univ Professors; Am Asn Dent Schs; fel Am Col Dentists. *Res:* Experimental carcinogenesis. *Mailing Add:* Dept Path Sch Dent Univ Detroit 2985 E Jefferson Detroit MI 48207-4282

SABESIN, SEYMOUR MARSHALL, b Riga, Latvia, Nov 27, 32; US citizen; m 57; c 3. INTERNAL MEDICINE, GASTROENTEROLOGY. *Educ:* City Col New York, BS, 54; NY Univ, MD, 58. *Prof Exp:* Clin fel exp path, Lab Path, Nat Cancer Inst, 59-61; resident internal med, New York Hosp-Cornell Med Ctr, 62-63; instr med, Harvard Med Sch, 63-65, assoc, 65-69; assoc prof med & path & dir electron micros lab, Jefferson Med Col, 69-73; prof med & dir div gastroenterol, Col Med, Univ Tenn, Memphis, 73-85; DYRENFORTH PROF MED & CHMN DIV GASTROENTEROL, RUSH-PRESBY-ST LUKE'S MED CTR, CHICAGO, 85- *Concurrent Pos:* Nat Inst Arthritis & Metab Dis fel, Mass Gen Hosp & Harvard Med Sch, 62-64, clin & res fel gastroenterol, 63-65; assoc investr, Metab Res Ctr, Mass Gen Hosp, 65-69; consult, Vet Admin Hosp, 69-; NIH & Am Heart Asn grants, Jefferson Med Col, 72-; mem, Gen Med A Study Sect, NIH, 78-; fel Coun Arteriosclerosis, Am Heart Asn; assoc ed, Lipids; counr, cent soc clin res, 84. *Honors & Awards:* Rorer Award, Am Col Gastroenterol, 70 & 71. *Mem:* Am Soc Cell Biol; Am Gastroenterol Asn; fel Am Col Physicians; Am Fed Clin Res; Am Heart Asn; Am Asn Study Liver Dis; fel Am Col Gastroenterol; fel Am Heart Asn Coun Arteriosclerosis. *Res:* Biochemical pathology of the liver; mechanisms of lipid transport in intestine and liver; lipoprotein metabolism; experimental liver injury. *Mailing Add:* Rush-Presby-St Luke's Med Ctr 1725 W Harrison Suite 256 Chicago IL 60612

SABET, TAWFIK YOUNIS, b Egypt, Nov 24, 26; nat US; m 53; c 1. IMMUNOBIOLOGY. *Educ:* Cairo Univ, BSc, 48, MS, 52, PhD(microbiol), 55. *Prof Exp:* Asst bacteriologist, Cairo Univ, 48-51, instr microbiol, 55-56; res assoc, Univ Ill, 57-64; NIH spec fel, 65 & 66; assoc prof, 67-72, PROF HISTOL, COL DENT, UNIV ILL MED CTR, 72- *Concurrent Pos:* Guest investr, US Naval Res Unit 3, 55-56; res assoc, Presby-St Luke's Hosp, Chicago, 57-64. *Mem:* AAAS; Am Soc Microbiol; Am Asn Immunol; Reticuloendothelial Soc; Sigma Xi. *Res:* Macrophage activation; macrophage phagocytosis; neutrophil function. *Mailing Add:* Col Dent Univ Ill-Chicago PO Box 6998 Chicago IL 60680

SABEY, BURNS ROY, b Magrath, Alta, May 17, 28; m 48; c 6. SOILS. *Educ:* Brigham Young Univ, BSc, 53; Iowa State Col, MS, 54, PhD, 58. *Prof Exp:* Instr soils, Iowa State Col, 54-58; from asst prof to prof soil microbiol, Univ Ill, Urbana, 58-69; PROF SOIL SCI, COLO STATE UNIV, 69- *Concurrent Pos:* NSF fac fel, 67-68. *Mem:* Fel Am Soc Agron; fel Soil Sci Soc Am. *Res:* Soil microbes and their influence on plant nutrient transformations in the soil; nitrification; denitrification; ammonification; organic waste recycling on land; mine land reclamation; revegetation of oil shale retorted. *Mailing Add:* Dept Agron Colo State Univ Ft Collins CO 80523

SABHARWAL, CHAMAN LAL, b Ludhiana, India, Aug 15, 37; m 68; c 2. MATHEMATICS, COMPUTER SCIENCE. *Educ:* Panjab Univ, India, BA, 59, MA, 61; Univ Ill, Urbana, MS, 66, PhD(math), 67. *Prof Exp:* Lectr math, D A V Col, Hoshiarpur, 61-63; teaching asst, Univ Ill, Urbana, 63-67; from asst prof to prof math, St Louis Univ, 67-83; software engr, McDonnell Douglas McAir, 83-85; PROF, UNIV MO, ROLLA, 85- *Concurrent Pos:* NSF res grant, McDonnell Douglas Lab, 79; A I grants, McDonnell Douglas Res Lab. *Mem:* Am Math Soc; Math Asn Am; Asn Comput Mach; Am Asn Artificial Intel; Inst Elec & Electronic Engrs. *Res:* Mathematical physics; functional analysis; software engineering; algorithm development; robotics; graphics; artificial intelligence. *Mailing Add:* 5892 Chrisbrook Dr St Louis MO 63128

SABHARWAL, KULBIR, b Punjab, India, Jan 5, 43; m 76; c 2. FOOD SCIENCE, TECHNOLOGY. *Educ:* Punjab Agr Univ, India, BSc, 64, MSc, 66; Ohio State Univ, MS, 69, PhD(food sci nutrit), 72. *Prof Exp:* Res asst, Ohio State Univ, 67-69, res assoc, 69-72; dir res & develop prod develop, An Amfac Co, Wapakoneta, Ohio, 72-86; food consult, 86-87; dir, Tech Serv, Gilardis Frozen Foods & Bakery, 87-90; DIR, TECH SERV, NU-TEK FOODS, 90- *Mem:* Inst Food Technologists; Am Dairy Sci Asn; Am Cult Dairy Prod Inst; Am Chem Soc; Am Oil Chemists Soc. *Res:* Development of various cheese substitutes in addition to process cheese and cheese products; functional food ingredients of dairy and non-dairy source; improvement of pizza crust, sauce and other toppings; setting up of new cheese operation. *Mailing Add:* 3366 Muirfield Pl Lima OH 45805

SABHARWAL, PRITAM SINGH, b Jehlum, Punjab, India, Apr 22, 37; US citizen; m 65; c 2. DEVELOPMENTAL BIOLOGY. *Educ:* Univ Delhi, BSc, 57, MSc, 59, PhD(bot), 63. *Prof Exp:* Res asst bot, Univ Delhi, 59-63, asst prof, 63-64; NSF fel & res assoc, Univ Pittsburgh, 64-65 & Ind Univ, Bloomington, 65-66; asst prof, 66-71, ASSOC PROF BOT, UNIV KY, 71- *Concurrent Pos:* USDA contract, 69-72. *Mem:* AAAS; Tissue Cult Asn; Int Soc Plant Morphol; Bot Soc Am. *Res:* Control of differentiation in plants. *Mailing Add:* 604 Lakeshore Dr Lexington KY 40502

SABHARWAL, RANJIT SINGH, b Dhudial, Pakistan, Dec 11, 25; nat US; m 48; c 3. PURE MATHEMATICS. *Educ:* Sikh Nat Col, Lahore, BA, 44; Punjab Univ, India, MA, 48; Univ Calif, Berkeley, MA, 62; Wash State Univ, PhD(math), 66. *Prof Exp:* Lectr math, Khalsa Col, Bombay, 51-58; teaching asst, Univ Calif, 58-62; instr, Portland State Col, 62-63; instr, Wash State Univ, 63-66; asst prof, Kans State Univ, 66-68; assoc prof, 68-74, PROF MATH, CALIF STATE UNIV, HAYWARD, 74- *Concurrent Pos:* Trustee-secy, Sikh Found, USA. *Mem:* Am Math Soc; Math Asn Am; Sigma Xi; AAAS. *Res:* Non-Desarguesian planes. *Mailing Add:* Dept Math Calif State Univ Hayward CA 94542

SABIA, RAFFAELE, b Procida, Italy, Aug 27, 33; US citizen; m 68; c 4. MATERIALS SCIENCE. *Educ:* St Francis Col, BS, 56; Polytech Inst Brooklyn, PhD(polymer chem), 60. *Prof Exp:* Chemist, Polymer Chem Div, W R Grace & Co, 59-60, sect head polymers, 60-63; mem tech staff, 63-67, supvr Organic Mat Eng, AT&T Bell Tel Labs, NJ, 67-88; SR STAFF ENGR, AT&T NASSAU METALS, NY, 88- *Concurrent Pos:* AT&T fel. *Mem:* Soc Rheol; Soc Plastics Eng. *Res:* Mechanical and physical properties of polymers and materials in general; materials for wire and cable applications; materials reclamation. *Mailing Add:* AT&T Nassau Metals One Nassau Place Staten Island NY 10307

SABIDUSSI, GERT OTTO, b Graz, Austria, Oct 28, 29; m 72; c 1. MATHEMATICS, ALGEBRA. *Educ:* Univ Vienna, PhD, 52. *Prof Exp:* Mem, Inst Adv Study, NJ, 53-55; instr, Univ Minn, 55-56; res instr math, Tulane Univ, 56-57, asst prof, 57-60; from assoc prof to prof, McMaster Univ, 60-69; dir math res ctr, 71-72, PROF MATH, UNIV MONTREAL, 69- *Concurrent Pos:* Fulbright grant, 53-54. *Mem:* Can Math Soc; Sigma Xi. *Res:* Graph theory; combinatorics; automata. *Mailing Add:* Dept Math CP6128 Univ Montreal Montreal PQ H3C 3J7 Can

SABIN, ALBERT B(RUCE), b Russia, Aug 26, 06; nat US; m 35, 72; c 2. INFECTIOUS DISEASES, VIROLOGY. *Educ:* NY Univ, BSc, 28, MD, 31. *Hon Degrees:* Numerous from US & foreign univs, 59-75. *Prof Exp:* Res assoc bact, Sch Med, NY Univ, 26-31; house physician, Bellevue Hosp, 32 & 33; Nat Res Coun fel, Lister Inst London, Eng, 34; asst, Rockefeller Inst, 35-37, assoc, 37-39; assoc prof pediat, Col Med, Univ Cincinnati, 39-46, prof res pediat, 46-60, distinguished serv prof, 60-70, emer distinguished serv prof, 71-; distinguished res prof biomed, Med Univ SC, 74-82, emer distinguished res prof, 82; sr expert consult, Fogarty Int Ctr, NIH, 82-86; RETIRED. *Concurrent Pos:* Chief div infectious dis, Children's Hosp Res Found, Cincinnati, 39-69; consult to US Army serving on comn virus & rickettsial dis, Armed Forces Epidemiol Bd & spec mission, Middle East, Italy, Panama, Japan, Korea, China & Ger, 41-62; consult, USPHS, 47-70; mem, Armed Forces Epidemiol Bd, 63-69; mem nat adv coun, Nat Inst Allergy & Infectious Dis, 65-70; mem bd gov, Weizmann Inst Sci & Hebrew Univ, Israel, 65-; trustee, NY Univ, 66-70; Ohio State Regents prof, 68-69; pres, Weizmann Inst Sci, 70-72; mem bd gov, Israel Inst Technol, 70-77 & Tel Aviv Univ, 71-; Fogarty scholar, NIH, 73; mem adv comt med res, Pan-Am Health Orgn, 73-77; expert consult, Nat Cancer Inst, 74; mem, US Army Med Res Develop Adv Panel, 74-80; consult to asst secy health, Dept Health, Educ & Welfare, 75-77. *Honors & Awards:* Mangia d'Oro Medal, City of Siena, Italy, 68; Order of the Sacred Treasure of Japan, 68; Walter Reed Medal, Am Soc Trop Med & Hyg, 69; Gold Medal, Royal Soc Health, 69; Decoration of the Aztec Eagle, Sash First Class, Govt Mex, 70; Ordem do Cruzeiro do Sul, Govt Brazil, 70; US Nat Medal of Sci, 70; Howland Award, Am Pediat Soc, 74; Presidential Medal of Freedom, 86. *Mem:* Nat Acad Sci; fel Am Acad Arts & Sci; fel Am Soc Trop Med & Hyg; Infectious Dis Soc Am (pres, 68-69); fel Royal Soc Health; Acad Med Sci USSR. *Res:* Pneumococcus infection; pleuropneumonia group; experimental arthritis; toxoplasmosis; dengue; sandfly fever; neurotropic viruses; poliomyelitis; live, oral polio vaccine; role of viruses in human cancer; aerosolized measles vaccine. *Mailing Add:* 3101 New Mexico NW Apt 1001 Washington DC 20016-5902

SABIN, JOHN ROGERS, b Springfield, Mass, Apr 29, 40; m 88; c 4. QUANTUM CHEMISTRY, THEORETICAL CHEMISTRY. *Educ:* Williams Col, BA, 62; Univ NH, PhD(radiation chem), 66. *Prof Exp:* NIH fel, Quantum Chem Group, Univ Uppsala, 66-67; fel chem, Northwestern Univ, 67-68; asst prof, Univ Mo-Columbia, 68-71; from assoc prof to prof physics, 77-80, PROF PHYSICS & CHEM, UNIV FLA, 80- *Concurrent Pos:* Assoc ed, Int J Quantum Chem, 73-; consult, Phys Sci Dir, Micom-Redstone Arsenal, 72-77; vis prof, Odense Univ, Denmark, 80-; ed, Adv Quantum Chem, 85- *Mem:* Am Chem Soc; fel Am Phys Soc. *Res:* Energy deposition characteristics of swift, massive charged particles in materials; determination of microscopic and macroscopic properties of materials; calculational quantum mechanics of thin metal films, polymers and small molecules. *Mailing Add:* Dept Physics Univ Fla Gainesville FL 32611

SABIN, THOMAS DANIEL, b Webster, Mass, Apr 28, 36; m 58; c 3. NEUROLOGY. *Educ:* Tufts Univ, BS, 58, MD, 62. *Prof Exp:* Resident neurol, Boston City Hosp, 64-66; assoc chief rehab, USPHS Hosp, Carville, La, 67-70; assoc dir, 70-75, DIR NEUROL UNIT, BOSTON CITY HOSP, 75-; PROF NEUROL & PSYCHIAT, BOSTON UNIV, 81- *Concurrent Pos:* Asst prof neurol, Tufts Univ, 70-; assoc prof, Sch Med, Boston Univ, 75-; lectr neurol, Harvard Med Sch, 75-80; consult ed, J Phys Ther, 75-80. *Mem:* AAAS; Am Acad Neurol; Soc Clin Neurologists; Int Leprosy Asn. *Res:* Clinical problems in peripheral nerve disorders; application of computerized tomography of the brain to behavioral disorders and dementia; treatment of end stage parkinsonism; pathogenesis of peripheral neuropathies. *Mailing Add:* Neurol Unit Boston City Hosp 818 Harrison Ave Boston MA 02118

SABINA, LESLIE ROBERT, b Fort Erie, Ont, Nov 28, 28; m 55; c 2. VIROLOGY, MICROBIOLOGY. *Educ:* Cornell Univ, AB, 52; Univ Nebr, MS, 56, PhD(microbiol), 60. *Prof Exp:* Bacteriologist, Ont Dept Health, 52-53; bacteriologist, Mt Sinai Hosp, Toronto, 53-54; asst animal path, Univ Nebr, 56-59, instr vet sci, 59-60; sr res asst virol, Connaught Med Res Labs, Toronto, 60-62, res assoc, 62-63; Upjohn Co, 64-65; asst prof, 65-71, PROF VIROL, UNIV WINDSOR, 71- *Concurrent Pos:* Registered, Nat Registry Microbiol. *Mem:* Am Soc Microbiol; Can Soc Microbiol; Can Col Microbiol. *Res:* Host-virus interactions; viral chemotherapeutics; tissue culture cell nutrition. *Mailing Add:* Dept Biol Univ Windsor Windsor ON N9B 3P4 Can

SABINS, FLOYD F, b Houston, Tex, Jan 5, 31; m 54; c 2. GEOLOGY. *Educ:* Univ Tex, BS, 52; Yale Univ, PhD(geol), 55. *Prof Exp:* Sr res geologist, Chevron Res Co, 55-67, sr res assoc, 67- 88, SR RES SCIENTIST, CHEVRON OIL FIELD RES CO, 88- *Concurrent Pos:* Asst prof, Calif State Col Fullerton, 65-66; adj prof, Univ Southern Calif, 66-; regents prof, Univ Calif, Los Angeles, 75- *Honors & Awards:* Pecora Award, 83; Alan Gordon Award, 81. *Mem:* Fel Geol Soc Am; Am Soc Photogram; Am Asn Petrol Geologists. *Res:* Remote sensing; sedimentary petrology; stratigraphy; structural geology. *Mailing Add:* Chevron Oil Field Res Co PO Box 446 La Habra CA 90631

SABISKY, EDWARD STEPHEN, b Middleport, Pa, Sept 11, 32; m 55; c 2. SOLID STATE PHYSICS. *Educ:* Pa State Univ, BS, 56; Univ Southern Calif, MS, 59; Univ Pa, PhD(physics), 65. *Prof Exp:* Mem tech staff, Hughes Aircraft Co, 56-59; mem tech staff, RCA Res Labs, 59-80; mem staff, Solar Energy Res Inst, 80-88, sr scientist & prog mgr, 88; VPRES, IDM, 89- *Mem:* Am Phys Soc; Inst Elec & Electronics Engrs. *Res:* Spin-phonon interaction; liquid helium films; tunable phonon spectrometer; dispersion in sound velocity in liquid helium; masers; double resonance employing optical-microwave techniques; spin memory; circular dichroism in solids; paramagnetic resonance of ions in solids; atomic hydrogen maser. *Mailing Add:* 11 Carnation Pl Trenton NJ 08648

SABISTON, CHARLES BARKER, JR, b Wake Forest, NC, July 22, 33; m 59; c 2. MICROBIOLOGY, DENTISTRY. *Educ:* Wake Forest Univ, BS, 53; Univ NC, DDS, 57; Va Commonwealth Univ, PhD(microbiol), 68; Univ Iowa, cert periodont, 75. *Prof Exp:* Pvt pract dent, NC, 60-64; Nat Inst Dent Res fel & Dent Res Training Prog fel, Va Commonwealth Univ, 64-67, assoc prof periodont, Med Col Va, 67-72; from assoc prof to prof periodont, 72-87, PROF FAMILY DENT, COL DENT, UNIV IOWA, 87- *Concurrent Pos:* Nat Inst Dent res grants, Va Commonwealth Univ, 71-72 & Col Dent, Univ Iowa, 72-78; consult, coun therapeut, Am Dent Asn, 71-89; consult, J Am Dent Asn, 78-89. *Mem:* Am Soc Microbiol; Am Dent Asn; Int Asn Dent Res; Am Asn Dent Res; Am Acad Periodont. *Res:* Microbial factors in periodontal disease etiology; non-sporing anaerobic bacteria; clinical dental microbiology; dentinal hypersensitivity. *Mailing Add:* Dept Family Dent Univ Iowa Col Dent Iowa City IA 52242

SABISTON, DAVID COSTON, JR, b Jacksonville, NC, Oct 4, 24; m 55; c 3. GENERAL SURGERY, CARDIOTHORACIC SURGERY. *Educ:* Univ NC, BS, 43; Johns Hopkins Univ, MD, 47. *Prof Exp:* Intern surg, Johns Hopkins Hosp, 47-48, from asst to assoc prof, Univ, 55-59; PROF, DUKE UNIV, 64-, JAMES B DUKE PROF SURG, MED CTR, 71-, CHMN DEPT, 64- *Concurrent Pos:* Asst surg, Johns Hopkins Univ, 48-49, Cushing fel, 49-50; NIH res career award, 62-64; investr, Howard Hughes Med Inst, 55-61; consult, NIH & Womack Army Hosp; chmn, Surg Study sect, NIH, 70-72; ed, Annals Surg, co-ed, Surg Chest; chmn, Am Bd Surg, 71-72; chmn, Accreditation Coun Grad Med Educ, 85-86. *Honors & Awards:* Sci Councils' Distinguished Achievement Award, Am Heart Asn, 83; Michael E DeBakey Award for Outstanding Achievement, 84; Col Medalist, Am Col Chest Physicians, 87. *Mem:* Inst Med-Nat Acad Sci; Am Asn Thoracic Surgeons (pres, 84-85); Am Col Surg (pres, 85-86); Soc Univ Surg (pres, 68-69); Am Surg Asn (pres, 77-78); Soc Surg Chairman (pres, 74-76). *Res:* General and cardiovascular surgery. *Mailing Add:* Dept Surg Duke Univ Med Ctr Durham NC 27710

SABLATASH, MIKE, b Bienfait, Sask, Sept 30, 35; m 61; c 3. ELECTRICAL ENGINEERING, APPLIED MATHEMATICS. *Educ:* Univ Man, BScEng, 57, MSc, 64; Univ Wis-Madison, PhD(elec eng), 68. *Prof Exp:* Commun engr, Sask Power Corp, 57; mem common sci staff, Res & Develop Labs, Northern Elec Co, 61-65; asst prof elec eng, Univ Toronto, 68-72; statistician V, Energy Bd Can, 72-76; RES SCIENTIST, COMPUT COMMUN IMAGE PROCESSING & INFO TECHNOL, DEPT COMMUN & COMMUN RES CENT, 76- *Concurrent Pos:* Lab demonstr, Univ Man, 57-60, res asst, 58-59; lectr, Univ Ottawa, 64-65; teaching asst, Univ Wis, 65-68; consult, Consociates Ltd, 68-72; Nat Res Coun grant, Univ Toronto, 68-72; sci info technol adv, Commun Res Ctr, Dept Commun, Govt Can, 80- *Mem:* Sigma Xi; Inst Elec & Electronics Engrs. *Res:* Communication signals, networks and systems and their optimal design; statistical communication and information theory; mathematical prof programming; functional analysis; computer communications in information systems and technology. *Mailing Add:* 23 A Vertona St Nepean ON K2G 4G6 Can

SABLE, EDWARD GEORGE, b Rockford, Ill, Dec 12, 24; m 54; c 2. STRUCTURAL GEOLOGY, STRATIGRAPHY-SEDIMENTATION. *Educ:* Univ Minn, BA, 48; Univ Mich, MS, 59, PhD, 65. *Prof Exp:* Geologist petrol explor, 48-56, geologist mineral explor, 57-81, GEOLOGIST FRAMEWORK MAPPING, US GEOL SURV, 81- *Concurrent Pos:* Instr, Univ Ky, Elizabethtown, 65-66. *Honors & Awards:* Case Mem Award, Univ Mich, 59; Meritorious Serv Award, Dept Interior, 87. *Mem:* AAAS; Geol Soc Am; Arctic Inst NAm. *Res:* Mesozoic and Paleozoic stratigraphy and structural and economic geology of Arctic Alaska; tectonics of arctic regions; petroleum exploration of Alaska; granite emplacement; regional Mississippian stratigraphy of Eastern Interior Basin, United States; coordination of Devonian stratigraphic data; petroleum source rock studies; oil and gas appraisal; precambrian of southern Arabian shield; stratigraphy and structure of southern Utah; coal, north slope of Alaska. *Mailing Add:* US Geol Surv Box 25046 Denver Fed Ctr Lakewood CO 80225

SABLE, HENRY ZODOC, b Toronto, Ont, Apr 1, 18; nat US; m 42; c 2. BIOCHEMISTRY, ORGANIC CHEMISTRY. *Educ:* Univ Toronto, BA, 39, MD, 43; Univ Ill, MS, 47; Wash Univ, PhD(biochem), 50. *Prof Exp:* Instr biochem, Sch Med, Tufts Univ, 50-51, asst prof, 51-53; from asst prof to assoc prof biochem, 53-66, prof chem, 67, co-dir dept biochem, 67-75, actg dir, 66-67 & 75-78, PROF BIOCHEM & CHEM, SCH MED, CASE WESTERN RESERVE UNIV, 66- *Concurrent Pos:* Markle scholar med sci, Sch Med, Case Western Reserve Univ, 56-61; vis prof & NSF fel, Univ Geneva, 59-60. *Mem:* AAAS; Am Chem Soc; Am Soc Biol Chemists; NY Acad Sci. *Res:* Coenzyme mechanisms; cyclitol analogues of glycerolipids; magnetic resonance spectroscopy; chemistry of cyclitols; conformational analysis; thiamin-polyphosphate requiring enzymes; spectroscopy. *Mailing Add:* 2614 Dysart Rd University Heights OH 44118

SABLIK, MARTIN J, b Brooklyn, NY, Oct 21, 39; m 65; c 4. MAGNETIC PROPERTIES OF MATERIALS, COMPUTER MODELING IN APPLIED PHYSICS. *Educ:* Cornell Univ, BA, 60; Univ Ky, MS, 65; Fordham Univ, PhD(theoret solid state physics), 72. *Prof Exp:* Jr engr, Martin Co, Orlando, Fla, 62-63; instr physics, Univ Ky, Lexington, 63-65; res assoc, Fairleigh Dickinson Univ, Teaneck, NJ, 65-67, from instr to assoc prof, 67-80; sr res scientist, 80-87, STAFF SCIENTIST, SOUTHWEST RES INST, SAN ANTONIO, TEX, 87- *Concurrent Pos:* Mem adv comt, Conf Properties & Applications of Magnetic Mat, 90- *Honors & Awards:* Imagineer Award, Mind Sci Found, 89. *Mem:* Am Phys Soc; Am Soc Nondestructive Testing; Inst Elec & Electronics Engrs; Am Asn Physics Teachers; Am Geophys Union. *Res:* Magnetism-modeling effects of stress on hysteresis of magnetic properties for use in detection of residual stress; space science-computer simulation of electrostatic analyzers; geophysics-electrical resistance tomographic imaging; superconductivity-melt processing of ceramic superconductors; nondestructive evaluation-modeling for electromagnetic characterization of defects; acoustics-statistical energy analysis of structure-borne sound; condensed matter theory-rare earth physics. *Mailing Add:* Southwest Res Inst PO Drawer 28510 San Antonio TX 78228-0510

SABNIS, ANANT GOVIND, b Chandgad, India, Feb 22, 44; m 73; c 1. ELECTRONICS, SOLID STATE PHYSICS. *Educ:* Univ Bombay, BE, 65; SDak Sch Mines & Technol, MS, 71, PhD(elec eng), 74. *Prof Exp:* Elec engr elec mach, Cropton-Greaves Ltd, Bombay, 65-66; lectr elec eng, Shri Bhagubhai Mafatlal Polytech, Bombay, 66-69; vis asst prof elec eng, Univ Pittsburgh, 74-75, asst prof, 75-80; MEM STAFF, BELL LABS, 80- *Mem:* Inst Elec & Electronics Engrs; Sigma Xi. *Res:* Solid state device physics; thin-films; integrated circuits design and modeling; material characterization. *Mailing Add:* 2E 245 Bell Labs 555 Union Blvd Allentown PA 18103

SABNIS, GAJANAN MAHADEO, b Belgaum, India, June 11, 41; nat US; m 69. STRUCTURAL ENGINEERING. *Educ:* Univ Bombay, BE, 61; Indian Inst Technol, Bombay, MTech, 63; Cornell Univ, PhD(struct eng), 67. *Prof Exp:* Res assoc struct eng, Cornell Univ, 67, fel, Univ Pa, 68; res engr, Am Cement Corp, Calif, 68-69; eng supvr, Bechtel Power Corp, 70-73, sr engr, 73-74; assoc prof, 74-80, PROF CIVIL ENG, HOWARD UNIV, 80- *Concurrent Pos:* Consult engr, Bombay, 63-64; res, McGill Univ; pres, KC Eng, PC. *Mem:* Fel Am Concrete Inst; Am Soc Civil Engrs; Am Soc Eng Educ; Soc Exp Stress Anal; Inst Engrs India. *Res:* Structural models, shear strength of concrete slabs; deflection of structures; properties of concretes; nuclear power plants; structural failure investigations; ferrocement. *Mailing Add:* 13721 Townline Rd Silver Spring MD 20906

SABNIS, SUMAN T, b Rajkot, India, Nov 27, 35; US citizen; m 60; c 2. CHEMICAL ENGINEERING, STATISTICS. *Educ:* Christ Church Col, India, BSc, 54; Harcourt Butler Tech Inst, BS, 57; Lehigh Univ, MS, 60, PhD(chem eng), 67; Rutgers Univ, MS, 64. *Prof Exp:* Chem engr, Union Carbide Plastics Co, NJ, 60-63; sr chem engr, Monsanto Co, 66-68, group leader kinetics of polymer systs, 68-74; mgr mat res & develop, Kerite Co, Harvey Hubbell Inc, 74-76, mgr process eng, 76-81; VPRES ENG, LARIBEE WIRE MFG CO INC, 81- *Mem:* Am Inst Chem Engrs; Am Chem Soc; Inst Elec & Electronics Engrs; Soc Plastics Engrs; Sigma Xi. *Res:* Mixing studies in polymeric systems, kinetics of polymer systems, statistics and process development in area of polymers; compounding of rubber and plastic compounds; vulcanization techniques for insulated wire and cable. *Mailing Add:* 843 Garden Rd Orange CT 06477

SABO, JULIUS JAY, b Cleveland, Ohio, May 27, 21; m 48; c 2. ENVIRONMENTAL ENGINEERING, POLLUTION CONTROL. *Educ:* Fenn Col, BME, 48; Univ Colo, MPA, 81. *Prof Exp:* Chem engr, Repub Steel Corp, 48-52; consult engr, Reserve Mining Co, 52-55; develop engr, Gen Elec Co, 55-58; chief monitoring, Reactor Testing Sta, US Pub Health Serv, Idaho, 58-60, chief nuclear anal, 60-64, asst chief res grants, Div Radiol Health, 65-67, chief res grants prog & exec secy radiol health study sect, 67-69; dir off grants admin, Environ Control Admin, 69-71; grants info off, Environ Protection Agency, 71-75; mem fac, Colo Tech Col, 78-79; CONSULT ENGR HAZARDOUS WASTE CONTROL, 80- *Concurrent Pos:* Tech chmn, Non Conventional Energy Resources Comt & Conf, 75-77; mgr dir,

Analysis Non Conventional Energy Resources. *Mem:* Am Acad Environ Engrs; fel Royal Soc Health. *Res:* Environmental aspects of developing non-conventional energy resources. *Mailing Add:* 6935 Blackhawk Pl Colorado Springs CO 80919

SABOL, GEORGE PAUL, b Clairton, Pa, Oct 17, 39; m 63; c 4. NUCLEAR MATERIALS. *Educ:* Pa State Univ, BS, 61; Carnegie Inst Technol, MS, 64, PhD(mat sci), 67. *Prof Exp:* Engr superalloy res & develop, Colwell Res Ctr, TRW Inc, 65-67; sr engr phys metall, 67-72, fel engr, 72-76, adv engr, 76, MGR CORE MAT DEVELOP NUCLEAR MAT, 76-, MGR MAT APPLICATIONS, WESTINGHOUSE RES & DEVELOP CTR, 87- *Mem:* Am Soc Metals; Am Inst Mining & Metall Engrs; Am Nuclear Soc. *Res:* Processing, corrosion response and mechanical behavior of zirconium-based alloys; performance of nuclear fuel in light water reactors; physical metallurgy of nickel-based superalloys. *Mailing Add:* 4350 Northern Pike Monroeville PA 15146

SABOL, STEVEN LAYNE, b Phoenix, Ariz, Sept 21, 44; div; c 1. BIOCHEMISTRY. *Educ:* Yale Col, BS, 66; NY Univ, MD & PhD(biochem), 73. *Prof Exp:* Intern med, Duke Univ Med Ctr, 73-74, res assoc, 74-77; sr staff fel, 77-79, MED OFFICER RES, LAB BIOCHEM GENETICS, NAT HEART LUNG & BLOOD INST, NIH, 79- *Mem:* Am Soc Biochem & Molecular Biol; Soc Neurosci; Am Soc Neurochem. *Res:* Protein and peptide biosynthesis; regulation of gene expression in the nervous system. *Mailing Add:* NIH Bldg 36 Rm 1C-06 Bethesda MD 20892

SABOUNGI, MARIE-LOUISE JEAN, b Tripoli, Lebanon, Jan 1, 48; m 76, 89; c 2. PHYSICAL CHEMISTRY, PHYSICS. *Educ:* Univ Aix, Marseille, France, PhD(thermodyn), 73. *Prof Exp:* CHEMIST, CHEM TECH DIV, ARGONNE NAT LAB, 73- *Mem:* Electrochem Soc; AAAS. *Res:* Molten salt chemistry; alloy thermodynamics; statistical mechanics; electrochemistry. *Mailing Add:* Argonne Nat Lab 9700 S Cass Ave Argonne IL 60439

SABOURIN, THOMAS DONALD, b Bay City, Mich, May 31, 51; m 74. PRODUCT REGISTRATION. *Educ:* Univ Mich, BA, 73; Calif State Univ, MA, 77; La State Univ, PhD(physiol), 81. *Prof Exp:* Anal chemist, Sel Rex Div Occidental Petrol, 73-75; teaching asst physiol, Calif State Univ, Hayward, 75-76; teaching asst biol, physiol & marine, La State Univ, 78-80; Nat Inst Environ Health Sci fel, Environ Health Sci Ctr, Ore State Univ, 81-82; res scientist, Battelle Columbus Opers, 82-85, prin res scientist, 85-88, mgr, Agr Chem Projs, 88-89, VPRES, CHEM REGIST, BATTELLE COLUMBUS OPERS, 89- *Concurrent Pos:* Prin investr, Sigma Xi grant, 80-81, Lerner Fund Marine Res grant, 80-82 & alternatives toxicity testing, 82-88; consult, Browning & Ferris Industs, 80-81; adj asst prof zool, Ohio State Univ, 86- *Mem:* Sigma Xi; AAAS; Am Soc Testing & Mat; Soc Toxicol; Soc Environ Toxicol & Chem. *Res:* Development, validation and implementation of test systems with nonmammalian organisms for use in screening and monitoring potential adverse effects of environmental agents to humans; environmental physiology. *Mailing Add:* Battelle Columbus Opers 505 King Ave Columbus OH 43201-2693

SABOURN, ROBERT JOSEPH EDMOND, b Sturgeon Falls, Ont, July 6, 26; m 53; c 2. GEOLOGY. *Educ:* Univ Ottawa, BSc, 47; Laval Univ, BAppSc, 51, MSc, 52, DSc(geol), 55. *Prof Exp:* From asst prof to assoc prof, 55-67, chmn dept geol & mineral, 65-71, PROF GEOL, LAVAL UNIV, 67-; CONSULT, 88- *Concurrent Pos:* Mem, AID Proj, Senegal, WAfrica, 63. *Mem:* Geol Soc Am; fel Geol Asn Can; Can Inst Mining & Metall. *Res:* Engineering, field and areal geology; geomorphology. *Mailing Add:* 1914 Bourbonniere Sillery PQ G1S 1N4 Can

SABROSKY, CURTIS WILLIAMS, b Sturgis, Mich, Apr 3, 10; wid; c 1. ENTOMOLOGY, TAXONOMY. *Educ:* Kalamazoo Col, AB, 31; Kans State Univ, MS, 33. *Hon Degrees:* ScD, Kalamazoo Col, 66. *Prof Exp:* From instr to asst prof entom, Mich State Univ, 36-45; entomologist, Entom & Plant Quarantine, 46-53 & Entom Res Br, 53-67, dir syst entom lab, 67-73, res entomologist, 73-80, COOP SCIENTIST, USDA, 80- *Concurrent Pos:* Mem, Int Comn Zool Nomenclature, 63-85, pres, 77-83; mem permanent comt, Int Cong Entom, 60-80; pres & chmn orgn comt, XV Int Cong Entom, 73-76. *Honors & Awards:* Distinguished Serv Award, Kans State Univ, 65; Superior Serv Award, USDA, 62, Distinguished Serv Award, 80. *Mem:* AAAS; hon mem Entom Soc Am (pres, 69); Soc Syst Zool (pres, 62); hon fel Entom Soc Can; hon foreign mem Entom Soc USSR; hon mem Int Cong Entom; hon pres Entom Soc Washington. *Res:* Taxonomy of higher flies; problems of zoological nomenclature. *Mailing Add:* 205 Medford Leas Medford NJ 08055-2236

SABRY, ISMAIL, b Alexandria, Egypt, Aug 23, 52. NEUROENDOCRINOLOGY, IN SITU LIGAND BINDING FOR RECEPTOR DETERMINATION. *Educ:* Alexandria Univ, BSc, 73, MSc, 78, PhD(exp zool), 83. *Prof Exp:* Teaching asst, 73-82, asst prof, 83-88, ASSOC PROF ZOOL & PHYSIOL, FAC SCI, ALEXANDRIA UNIV, EGYPT, 88- *Concurrent Pos:* USAID postdoctoral, Univ Tex Health Sci Ctr, 87; vis scientist, Dept Physiol, Inst Endocrinol, Gunma Univ, Maebashi, Japan, 89-91, lectr, Neuroendocrinol Workshop, 90. *Mem:* Am Soc Zoologists. *Res:* Physiological aspects of the neuropeptide somatortatin in the pineal and harderian glands; regulation of pituitary growth hormone release by hypothalmic somatortatin and growth hormone releasing factor with reference on somatortatin receptors under several physiological conditions. *Mailing Add:* Zool Dept Fac Sci Alexandria Univ Alexandria Egypt

SABRY, ZAKARIA I, b Tanta, Egypt, Aug 16, 32; Can citizen; m 56; c 2. NUTRITION, BIOCHEMISTRY. *Educ:* Univ Ain Shams, Cairo, BSc, 52; Univ Mass, MSc, 54; Pa State Univ, PhD(biochem), 57. *Prof Exp:* From asst prof to assoc prof food technol & nutrit, Am Univ Beirut, 57-64, head dept, 61-64; from assoc prof to prof nutrit, Univ Toronto, 64-72; pres, Nutrit Res Consults, Inc, 74-79; dir food policy & nutrit div, Food & Agr Orgn, UN, 79-84; PROF PUB HEALTH, UNIV CALIF, BERKELEY, 84- *Concurrent Pos:* Nat Res Coun Can fel, 61-62; nat coordr, Nutrit Can, Health Protect Br, Can Dept Health & Welfare, 69-74; prof appl human nutrit, Univ Guelph, 76-79. *Mem:* Am Inst Nutrit; AAAS; Soc Nutrit Educ; Sigma Xi. *Res:* Relationships of diet and disease; development of risk factors of coronary heart disease and cancer; nutrition survey; assessment of nutritional status in man. *Mailing Add:* 421 Warren Hall Univ Calif Berkeley CA 94720

SABSHIN, MELVIN, b New York, NY, Oct 28, 25; m 55; c 1. PSYCHIATRY. *Educ:* Univ Fla, BS, 44; Tulane Univ, MD, 48. *Prof Exp:* Resident psychiat, Tulane Univ, 49-52; res psychiatrist, Psychosom & Psychiat Inst, Michael Reese Hosp, 53-55, asst dir, 55-57, assoc dir, 57-61; prof psychiat & head dept, Col Med, Univ Ill, 61-74; MED DIR, AM PSYCHIAT ASN, 74- *Concurrent Pos:* Fel, Ctr Advan Study Behav Sci, 67-68; assoc ed, Am J Psychiat, 71-74; actg dean, Abraham Lincoln Sch Med, Univ Ill, 73-74. *Honors & Awards:* Distinguished Serv Award, Am Psychiat Asn, 86, Admin Psychiat Award, 88. *Mem:* AAAS; Am Col Psychiat (pres, 73-74); Am Psychiat Asn; AMA; Am Psychosom Soc; Group Advan Psychiat. *Res:* Social psychiatry; empirical studies of adaptive behavior. *Mailing Add:* Am Psychiat Asn 1400 K St NW Washington DC 20005

SABY, JOHN SANFORD, b Ithaca, NY, Mar 21, 21; m 45; c 4. PHYSICS, TECHNICAL MANAGEMENT. *Educ:* Gettysburg Col, AB, 42; Pa State Univ, MS, 44, PhD(physics), 47. *Hon Degrees:* ScD, Gettysburg Col, 69. *Prof Exp:* Lab instr physics, Gettysburg Col, 40-42; asst, Pa State Univ, 42-47; instr, Cornell Univ, 47-50; res physicist, Electronics Lab, Gen Elec Co, 51-52, supvr semiconductor components develop, 52-55, mgr, Semiconductor & Solid State, 55-56, Lamp Res Lab, 56-71 Lamp Phenomena Res Lab, 71-82; CONSULT, GEN ELEC CO, 82- *Honors & Awards:* Centennial Award, Inst Elec & Electronics Engrs, 87. *Mem:* Am Phys Soc; fel Inst Elec & Electronics Engrs; Sigma Xi. *Res:* X-ray liquid diffraction; atmospheric ultrasonics; wave mechanics of interacting particles; x-ray solid state; p-n-p transistor; p-n junction studies; power transistors; luminescence; electroluminescence; gas discharge physics; electron emission. *Mailing Add:* Eight Tamarac Terr Hendersonville NC 28739

SACCO, ANTHONY G, b Utica, NY, Nov 2, 44; m 67; c 2. IMMUNO-REPRODUCTION. *Educ:* Univ Rochester, BA, 66; Univ Tenn, MS, 68, PhD(zool), 71. *Prof Exp:* Res assoc, Inst Cancer Res, 72-74; from asst prof to assoc prof, 74-85, PROF, DEPTS OBSTET, GYNEC, IMMUNOL & MICROBIOL, SCH MED, WAYNE STATE UNIV, 85- *Concurrent Pos:* Non-clin assoc, Dept Path, Hutzel Hosp, Wayne State Univ, 75- & co-dir, in vitro fertil prog, 83- *Mem:* Soc Study Reprod; Sigma Xi; Am Fertil Soc. *Res:* Immuno-reproduction: antigenic properties and possible reproductive roles of antigens present in female and male reproductive tissues, tract secretions, and ova and sperm; fertility control: immunocontraception; mammalian reproductive biology: sperm-egg/ova interaction and recognition during the fertilization process. *Mailing Add:* 1682 Hallmark Dr Troy MI 48098-4349

SACCO, LOUIS JOSEPH, JR, b Chicago, Ill, Mar 24, 24; m 51; c 5. ORGANIC CHEMISTRY. *Educ:* DePaul Univ, ScB, 48, ScM, 50. *Prof Exp:* Res chemist, Baxter Lab, Inc, 50-56 & Nalco Chem Co, 56-57; res & develop chemist, Alkydol Lab, Inc, 57-58; process res chemist & sr res asst, 58-70, tech counr, 70-72, process res chemist & sr res asst, 73-75, supvr, Process Improv Lab, 75-80, SUPVR SAFETY, TRAINING & WASTE MGT, G D SEARLE & CO, 81- *Concurrent Pos:* Lectr, Chicago City Jr Col, 65-70 & DePaul Univ, 69-71 & 78-81. *Mem:* Am Chem Soc. *Res:* Amino acids; carbohydrate chemistry; fatty acids; steroids; prostaglandins. *Mailing Add:* 3348 N Panama Ave Chicago IL 60634

SACCOMAN, FRANK (MICHAEL), b Hibbing, Minn, July 31, 31; c 4. ANATOMY, ZOOLOGY. *Educ:* Bemidji State Col, BS, 58; Univ Minn, PhD(anat), 64. *Prof Exp:* Instr anat, Univ Mich, 64-66; from asst prof to prof biol, 66-80, chmn dept, 66-80, DEAN, DIV SCI & MATH, BEMIDJI STATE UNIV, 80- *Concurrent Pos:* NSF fel, Univ Minn, 70. *Res:* Radioautographic studies of developing extraembryonic membranes in the mouse; cellular migration in developing extraembryonic membranes with the use of radioautography; DNA synthesis in freshwater algae. *Mailing Add:* Dean Sci & Math Bemidji State Univ Bemidji MN 56601

SACCOMAN, JOHN JOSEPH, b Paterson, NJ, Sept 10, 39; m 63; c 2. MATHEMATICAL ANALYSIS. *Educ:* Seton Hall Univ, BS, 60; NY Univ, MS, 62, PhD(math educ), 74. *Prof Exp:* From instr to asst prof, 61-81, ASSOC PROF MATH, SETON HALL UNIV, 81- *Mem:* Am Math Soc; Math Asn Am. *Res:* Development of a set of normability conditions for topological vector spaces using a generalized Hahn-Banach theorem; applications of non-standard analysis to functional analysis; historical aspects of the Hahn Banach theorem; historical aspects of Krein-Milman theorem; historical aspects of functional analysis and topological vector spaces, eg, origin of weak convergence and the Banach-Alaogu theorem; precursors of the Krein-Rutman theorem; origin and ramifications of the Banach-Saks property. *Mailing Add:* Dept Math & Comput Sci Seton Hall Univ South Orange NJ 07079

SACHAN, DILEEP SINGH, b Makhauli, India, Dec 18, 38; US citizen; m 68; c 3. NUTRITIONAL BIOCHEMISTRY, MICROBIOLOGY. *Educ:* M P Vet Col, India, BVSc, 61, MVSc, 63; Univ Ill, Urbana, MS, 66, PhD(nutrit), 68. *Prof Exp:* Lectr obstet & gynec, M P Vet Col, India, 63-64; res asst nutrit, Univ Ill, Urbana, 64-69; res assoc pharmacol & microbiol, Case Western Reserve Univ, 69-71; asst prof pharmacol, Meharry Med Col, 71-76; res chemist, Vet Admin Hosp, 76-78; fel nutrit & gastroenterol, Vanderbilt Univ Med Ctr, 78-79; assoc prof, 79-87, PROF NUTRIT & BIOCHEM, DEPT NUTRIT & FOOD SCI, UNIV TENN, KNOXVILLE, 87- *Concurrent Pos:* NIH fel, Case Western Reserve Univ, 69-71; consult clin path, Vet Admin Hosp, 78-80; prin investr res grants, NSF, NIH, USDA & AID. *Mem:* Am Inst Nutrit; NY Acad Sci; Am Col Nutrit; AAAS; Biochem Pharmacol; Am Soc Clin Nutrit. *Res:* Nutrient-nutrient and nutrient-drug interactions; lipid metabolism; carnitine nutriture; bioavailability of nutrients; nutritional status. *Mailing Add:* Dept Nutrit & Food Sci Col Human Ecol Univ Tenn Knoxville TN 37996-1900

SACHDEV, GOVERDHAN PAL, b Lahore, India, July 17, 41; m 68; c 2. BIOCHEMISTRY, BIO-ORGANIC CHEMISTRY. *Educ:* Univ Delhi, BSc, 61, MSc, 63, PhD(chem), 67. *Prof Exp:* Lectr chem, Ramjas Col, Univ Delhi, 63-64; res fel, Univ Delhi, 64-67, sr res fel, 67-68; res assoc biochem, Yale Univ, 68-75; asst mem, Okla Med Res Found, 76-84; ASSOC RES PROF, COL PHARM, UNIV OKLA HSC, 84- *Honors & Awards:* Eason Award, Eason Oil Co-Okla Geol Soc, 78. *Mem:* Am Chem Soc; AAAS; Sigma Xi; Soc Complex Carbohydrates; Am Asn Col Pharm. *Res:* Structure and function of biological membrane-bound enzymes; role of glycoproteins and proteins in obstructive pulmonary diseases. *Mailing Add:* Col Pharm Univ Okla PO Box 26901 Oklahoma City OK 73190

SACHDEV, SHAM L, b Hoshiarpur, India, Dec 21, 37; m 67; c 2. CHEMISTRY. *Educ:* Panjab Univ, India, BS, 59, MS, 60; La State Univ, PhD(chem), 66; Am Bd Indust Hyg, cert; Int Hazard Control Cert Bd, Hazard Control Mgr. *Prof Exp:* Vis res assoc chem, La State Univ, 65-66, vis asst prof, 66-70; anal specialist, 70-74, mgr methods & develop, Kem-Tech Labs, Inc, 74-78; MEM STAFF, ENV HEALTH SERV, 78- *Concurrent Pos:* USPHS fel, La State Univ, 65-67, Air Pollution Control Admin fel, 68-69, NSF fel, 69-70. *Mem:* AAAS; Am Chem Soc; Air Pollution Control Asn. *Res:* Determination and significance of trace elements in environmental samples such as air, water, food and others; study of trace elements in the environment. *Mailing Add:* 334 Woodstone Ct Baton Rouge LA 70808

SACHDEV, SUBIR, b New Delhi, India, Dec 2, 61; m 85; c 1. PHYSICS. *Educ:* Mass Inst Technol, Cambridge, SB, 82; Harvard Univ, MS, 84, PhD(physics), 85. *Prof Exp:* Postdoctoral fel, AT&T Bell Labs, Murray Hill, 85-87; asst prof, 87-89, ASSOC PROF PHYSICS, YALE UNIV, 89- *Mem:* Am Phys Soc. *Res:* Theory of strongly correlated electronic systems with applications to superconductors, semiconductors and magnetism. *Mailing Add:* 54 Sloane Physics Lab Yale Univ PO Box 6666 New Haven CT 06511

SACHDEV, SURESH, b New Delhi, India, Sept 17, 54; m 80; c 2. SEMICONDUCTOR PROCESS DEVELOPMENT, CHEMICAL VAPOR DEPOSITION & THIN FILMS. *Educ:* Inst Technol, Banaras Hindu Univ, Varanasi, India, BTech Hon, 76; Univ Ill, MS, 80; Univ Calif, Berkeley, cert, 89. *Prof Exp:* Prod develop engr super refractories, Carborundum Universal, Madras, India, 76-77; res & teaching asst elec ceramics & glass, Univ Ill, Urbana, 78-80; dir applications & develop, Genus Inc, Mountain View, Calif, 84-85; prog mgr/sr engr thin films & engr, CVD technol develop, Intel Corp, Livermore, Calif, 80-84, prog mgr technol develop, Santa Clara, Calif, 85-90, mfg mgr programmable logic oper, Folsom, Calif, 90-91, MFG INTEGRATION MGR, TECHNOL & DEVELOP, INTEL CORP, SANTA CLARA, CALIF, 91- *Mem:* Inst Elec & Electronics Engrs; Am Ceramic Soc. *Mailing Add:* 3609 Maidu Pl Davis CA 95616

SACHDEVA, BALDEV KRISHAN, b India, Oct 15, 39; m 78. APPLIED MATHEMATICS. *Educ:* Univ Delhi, India, BSc Hons, 59, MA, 61; Pa State Univ, PhD(math), 73. *Prof Exp:* Lectr math, Univ Delhi, 64-69; res assoc math & elec eng, Carleton Univ, Ottawa & Nat Res Coun, 73-74; lectr math, Univ Wis-Milwaukee, 75-77; vis prof math, Panjab Univ, Chandigarh, India, 77-79; dir, acad affairs, 82-83, chmn dept, 84-89, PROF MATH, UNIV NEW HAVEN, CONN, 79- *Mem:* Am Math Soc; Soc Indust & Appl Math; Math Asn Am. *Res:* Scattering of acoustic and electromagnetic waves; general applied mathematics; numerical analysis. *Mailing Add:* Dept Math Univ New Haven 300 Orange Ave West Haven CT 06516

SACHER, ALEX, b Brooklyn, NY, July 24, 22; m 43; c 4. BIOENGINEERING & BIOMEDICAL ENGINEERING. *Educ:* City Col NY, BS, 43; Polytech Univ, MS, 46, PhD(polymer chem), 48. *Prof Exp:* Asst, Polytech Inst Brooklyn, 45-46 & 47-48; assoc tech dir, Maybunn Chem Co, NY, 48-49; fel synthetic rubber, Mellon Inst, 49-51; asst res mgr, Irvington Varnish & Insulator Div, Minn Mining & Mfg Co, 51-55; tech dir, Stand Insulation Co, 55-58; vpres commercial develop, Hudson Pulp & Paper Corp, NY, 58-59; pres, Dimensional Pigments, Inc, 60-62; PRES, UNIVERSAL PETROCHEM, INC, WHIPPANY, 63- *Concurrent Pos:* Adj prof, NY Inst Technol, 59. *Honors & Awards:* Spec Award, Soc Plastics Engrs, 59. *Mem:* Fel AAAS; fel Am Inst Chemists; Am Soc Testing & Mat; fel NY Acad Sci; Soc Plastics Engrs; Am Chem Soc; Am Soc Biomech. *Res:* Emulsion polymerization for application in floor polish and aqueous based Inks Industries; polymer chemistry; specialty chemicals floor maintenance; biomechanics and reconstruction of slip, trip, stumble, stick and fall accidents. *Mailing Add:* 92 Van Ness Ct Maplewood NJ 07040

SACHER, EDWARD, b New York, NY, June 3, 34; m 62; c 3. PHYSICAL CHEMISTRY. *Educ:* City Col New York, BS, 56; Pa State Univ, PhD(phys chem), 60. *Prof Exp:* Teaching asst chem, Pa State Univ, 56-57; fel, Ohio State Univ, 60-61 & Ottawa Univ, Ont, 61-63; res chemist, E I du Pont de Nemours & Co, Inc, 63-68; staff chemist, IBM Corp, 68-70, adv chemist, Mat Lab, Systs Prod Div, 70-80, Gen Technol Div, 80-82; RES PROF, DEPT ENG & PHYS, ECOLE POLYTECHNIQUE, 82- *Mem:* Am Phys Soc; Royal Soc Chem; Sigma Xi; Inst Elec & Electronics Engrs. *Res:* Chemical kinetics; AC and DC dielectric properties of polymeric solids; structure and motions of polymeric solid surfaces. *Mailing Add:* Eng Physics Ecole Polytech Montreal PQ H3C 3A7 Can

SACHER, ROBERT FRANCIS, b Chicago, Ill, July 23, 47; m 73. PLANT PHYSIOLOGY. *Educ:* Univ Ill, BS, 75, MS, 77; Wash State Univ, PhD(hort), 80. *Prof Exp:* Assoc, 80-81, RES ASSOC, BOYCE THOMPSON INST PLANT RES, CORNELL UNIV, 81- *Mem:* Am Soc Plant Physiol; Am Soc Hort Sci; AAAS. *Res:* Plant stress responses and partitioning the contributions of individual systems to overall tolerance; using genetically defined breeding lines with known differences in salt tolerance as model populations for comparative studies. *Mailing Add:* Beatrice/Huntwesson 1645 W Valencia Dr Fullerton CA 92633-3899

SACHLEBEN, RICHARD ALAN, b Madison, Ind, Jan 9, 56; m 90. CHEMICAL SEPARATIONS, SOLVENT EXTRACTION. *Educ:* Ga Inst Technol, BSci, 79; W M Rice Univ, PhD(org chem), 85. *Prof Exp:* Student res assoc, US Environ Protection Agency, 76-79; vis asst prof, Colo Sch Mines, 84-86, Univ Colo, Boulder, 85; postdoctoral fel, 86-87, RES STAFF, OAK RIDGE NAT LAB, US DEPT ENERGY, 87- *Mem:* Am Chem Soc; AAAS. *Res:* Design and synthesis of metal specific ligands; chemical and structural principles of separations by solvent extraction; incorporation of metals into nucleic acids and DNA. *Mailing Add:* MS-6119/Bldg 4500 PO Box 2008 Oak Ridge TN 37831-6119

SACHS, ALLAN MAXWELL, b New York, NY, July 13, 21; m 49; c 3. PHYSICS. *Educ:* Harvard Univ, BA, 42, MA, 47, PhD(physics), 50. *Prof Exp:* Instr, 49-50, assoc, 50-51, from asst prof to assoc prof, 51-60, chmn dept, 67-71, PROF PHYSICS, COLUMBIA UNIV, 60- *Res:* High energy particle physics; experimental intermediate energy particle physics. *Mailing Add:* Pupin Lab Columbia Univ New York NY 10027

SACHS, BENJAMIN DAVID, b Madrid, Spain, Mar 4, 36; US citizen; m 65; c 1. BIOLOGICAL PSYCHOLOGY. *Educ:* City Col New York, BA, 57, MSEd, 61; Univ Calif, Berkeley, PhD(comp psychol), 66. *Prof Exp:* From asst prof to assoc prof, 68-76, PROF PSYCHOL, UNIV CONN, 76- *Concurrent Pos:* Nat Inst Child Health & Human Develop fel, Rutgers Univ, Newark, 66-68; mem adv bd, Current Contents/Life Sci, 70-; consult ed, J Comp & Physiol Psychol, 79-81, J Comp Psychol, 88; prin investr, NICHD res grants, 69- *Mem:* Fel Am Psychol Soc; Animal Behav Soc; Int Soc Psychoneuroendocrinol; Soc Neurosci; Int Acad Sex Res. *Res:* Neuroendocrine aspects of reproductive behavior; sexually dimorphic behavior patterns. *Mailing Add:* Dept Psychol U-20 Univ Conn 406 Babbidge Rd Storrs CT 06269-1020

SACHS, DAVID, b Chicago, Ill, Aug 18, 33. MATHEMATICS. *Educ:* Ill Inst Technol, BS, 55, MS, 57, PhD(math), 60. *Prof Exp:* Instr math, Ill Inst Technol, 59-60; from instr to asst prof, Univ Ill, Urbana, 60-66; assoc prof, 66-71, PROF MATH, WRIGHT STATE UNIV, 71- *Mem:* Am Math Soc; Math Asn Am; Sigma Xi. *Res:* Lattice theory; exchange geometries; foundations of geometry. *Mailing Add:* Dept Math Wright State Univ Dayton OH 45431

SACHS, DAVID HOWARD, b New York, NY, Jan 10, 42; m 69; c 4. TRANSPLANTATION, IMMUNOGENETICS. *Educ:* Harvard Col, AB, 63, MD, 68; Univ Paris, DES, 64. *Prof Exp:* Intern surg, Mass Gen Hosp, 68-69, res fel, 69-70; res assoc biochem, lab chem biol, Nat Inst Arthritis & Metab Dis, NIH, 70-72; sr investr, Immunol Br, Nat Cancer Inst, NIH, 72-74, sect chief, Transplant Biol Sect, 74-82, BR CHIEF, IMMUNOL BR, NAT CANCER INST, NIH, 82- *Concurrent Pos:* Teaching asst org chem, Harvard Univ, 62-65; vis prof, dept cell res, Wallenberg Lab, Univ Uppsala, Sweden, 84-85. *Mem:* Transplantation Soc; Am Asn Immunologists; Am Soc Biochem & Molecular Biol. *Res:* Investigations in basic immunology with particular emphasis on studies of the major histocompatibility complex and its role in transplantation in animal models and in man. *Mailing Add:* Nat Cancer Inst NIH Bldg 10 Rm 4313 Bethesda MD 20892

SACHS, DONALD CHARLES, physics, for more information see previous edition

SACHS, FREDERICK, b New York, NY, Jan 8, 41; m 64; c 2. BIOPHYSICS. *Educ:* Univ Rochester, BA, 62; State Univ NY Upstate Med Ctr, PhD(physiol), 72. *Prof Exp:* Assoc engr electromagnetic compatibility, Douglas Aircraft Co, 62-64; jr researcher biophys, dept biochem & biophys, Univ Hawaii, 69-71; staff fel, Biophys Lab, Nat Inst Neurol Dis & Stroke, 71-75; asst prof pharmacol, 75-80, asst prof biophys, 80-81, assoc prof biolphys, 81-88, PROF BIOPHYS, STATE UNIV NY BUFFALO, 88- *Mem:* Biophys Soc; AAAS; Soc Gen Physiol. *Res:* Mechanisms of ion transport in cells, excitability; mechanical transduction; instrumentation design. *Mailing Add:* Dept Biophys Sci State Univ NY 105 Parker Hall 3435 Main St Buffalo NY 14214

SACHS, FREDERICK LEE, b Brooklyn, NY, Feb 28, 38; m 60; c 3. INTERNAL MEDICINE. *Educ:* Princeton Univ, AB, 59; Columbia Univ, MD, 63. *Prof Exp:* From intern to chief resident med, Yale-New Haven Hosp, 63-69; instr, 69-70, asst prof med, 70-77, ASSOC CLIN PROF MED, SCH MED, YALE UNIV, 77- *Concurrent Pos:* Winchester fel chest dis, Sch Med, Yale Univ, 69-70; consult, Vet Admin Hosp, West Haven, 69- *Honors & Awards:* Upjohn Award, 68. *Mem:* AAAS; Am Thoracic Soc; Sigma Xi. *Res:* Alveolar macrophages. *Mailing Add:* 42 Indian Trail Woodbridge CT 06525

SACHS, GEORGE, b Vienna, Austria, Aug 26, 36; US citizen; m 63; c 4. BIOCHEMISTRY. *Educ:* Univ Edinburgh, BSc, 57, MB, ChB, 60,. *Hon Degrees:* DSc, Univ Edinburgh, 80. *Prof Exp:* Instr biochem, Albert Einstein Col Med, 61-62; from asst prof to assoc prof med & physiol, 63-70, assoc prof physiol & biophys, 77-80, PROF MED, SCH MED, UNIV ALA, BIRMINGHAM , 70-, PROF PHYSIOL & BIOPHYS & DIR, MEMBRANE BIOL UNIT, 80- *Concurrent Pos:* Fel biochem, Albert Einstein Col Med, 61; res fel, Columbia Univ, 62-63. *Mem:* Biophys Soc; Brit Biochem Soc; NY Acad Sci; Soc Exp Biol & Med; Am Physiol Soc. *Res:* Physiology and biochemistry of transport. *Mailing Add:* Membrane Biol Lab Dept Med Bldg 113 Rm 324 Univ Calif Cure Wadsworth Vet Admin Med Ctr Los Angeles CA 90073

SACHS, HARVEY MAURICE, b Atlanta, Ga, Dec 10, 44; m 67; c 1. GEOLOGY, ENERGY. *Educ:* Rice Univ, AB, 67; Brown Univ, PhD(geol), 73. *Prof Exp:* Fel oceanog, Ore State Univ, 72-74; asst prof, Case Western Res Univ, 74-76; asst prof geol, Princeton Univ, 77-82, consult, 82-84; mem tech staff, AT&T Bell Labs, 84-87; ASST COMNR ENERGY, STATE NJ, 87-; CONSULT, 90- *Concurrent Pos:* Prin investr various grants, NSF & Pa Power & Light Co, 81-82. *Mem:* AAAS; Am Geophys Union; Geol Soc Am; Sigma Xi. *Res:* Engineering and evaluation of energy efficiency modifications for dwellings; side effects of conservation; geology of radon distribution; energy efficiency and policy; geology, applied, economic and engineering; oceanography; stratigraphy-sedimentation. *Mailing Add:* 20 Wynnewood Dr Cranbury NJ 08512

SACHS, HERBERT K(ONRAD), b Chemnitz, Ger, Mar 4, 19; nat US; wid; c 3. ENGINEERING MECHANICS. *Educ:* Tech Col Zurich, Switz, dipl, 41; Wayne State Univ, MS, 56; Brunswick Tech Inst, DrEng(mech), 63. *Prof Exp:* Designer, St Louis Car Co, Mo, 48-49 & Am Car & Foundry Div, ACF Industs, Inc, 49; design analyst, Int Harvester Co, Ind, 50-53; head, vehicle dynamics, Truck & Coach Div, Gen Motors Corp, Mich, 53-58; from assoc prof to prof, 58-84, EMER PROF MECH ENG, WAYNE STATE UNIV, 84- *Concurrent Pos:* Consult, Atomic Power Develop Assocs, Mich, Dana Corp & Rockwell-Standard Corp; chmn & ed proc, Int Conf Vehicle Mech Mich, 68, hon chmn Paris, 71; ed, J Vehicle Systs Dynamics, 71-; res engr, Dept Transp, Nat Hwy Traffic Safety Admin, 71-72; fac fel eng, Dept Transp, Transp Systs Ctr, Mass, 78-79. *Mem:* Indust Math Soc (vpres); Am Soc Mech Engrs; Am Asn Univ Professors; Int Asn Vehicle Syst Dynamics (pres, 77-81); Nat Forensic Ctr. *Res:* Nonlinear and vehicle mechanics; general dynamics; vehicle dynamics; theory of controls applied to active suspension design; adaptive controls; stability theory; dynamics and vibration analysis; forensic engineering. *Mailing Add:* Dept Mech Engr 655 Merrick Wayne State Univ 5950 Cass Ave Detroit MI 48202

SACHS, HOWARD GEORGE, b New York, NY, Dec 12, 43; m 68; c 2. DEVELOPMENTAL BIOLOGY, ANATOMY. *Educ:* Worcester Polytech Inst, BS, 65; Clark Univ, PhD(biol), 71. *Prof Exp:* Fel develop biol, Carnegie Inst, 70-72; from asst prof to assoc prof & asst dean, Grad Col, Univ Ill, 72-80; ASSOC PROVOST & ASSOC PROF ANAT, OHIO STATE UNIV, 80- *Mem:* Am Asn Anatomists; NY Acad Sci; AAAS; Sigma Xi; Am Heart Asn. *Res:* Cardiac biology and pathology. *Mailing Add:* Res & Grad Studies Pa State Univ Middletown PA 17057

SACHS, JOHN RICHARD, b Brooklyn, NY, July 29, 34; m 59; c 3. HEMATOLOGY, PHYSIOLOGY. *Educ:* Manhattan Col, BS, 56; Columbia Univ, MD, 60. *Prof Exp:* Res hematologist, Walter Reed Army Inst Res, 66-69; asst prof physiol, Sch Med, Yale Univ, 69-72, assoc prof, 72-75; assoc prof, 75-77, PROF MED, SCH MED, STATE UNIV NY, STONY BROOK, 77- *Mem:* Am Physiol Soc; Biophys Soc; Soc Gen Physiologists. *Res:* Cation transport, membrane physiology. *Mailing Add:* Dept Med State Univ NY Stony Brook NY 11794

SACHS, LESTER MARVIN, b Chicago, Ill, May 16, 27; m 57; c 1. THEORETICAL PHYSICS, INFORMATION SCIENCE. *Educ:* Ill Inst Technol, BS, 50, MS, 54, PhD(physics), 61. *Prof Exp:* Instr physics, Univ Ill, 55-58; student resident assoc, Argonne Nat Lab, 59-60; asst prof physics, Wayne State Univ, 60-64; scientist, Res Inst Advan Studies, 65-69; pres, Comput Prog Assocs, Inc, Md, 69-71; independent consult, 71-72; comput systs analyst, Bur Labor Statist, 72-75; syst develop specialist, 75-80, TECH ADV, SOCIAL SECURITY ADMIN, 80- *Mem:* Asn Comput Mach; Am Phys Soc; Sigma Xi. *Res:* Atomic and molecular structure; solid state theory; microcomputers. *Mailing Add:* 8823 Stonehaven Rd Randallstown MD 21133-4223

SACHS, MARTIN WILLIAM, b New Haven, Conn, Sept 30, 37; m 68. COMPUTER SCIENCE. *Educ:* Harvard Univ, AB, 59; Yale Univ, MS, 60, PhD(physics), 64. *Prof Exp:* Dept guest nuclear physics, Weizmann Inst, 64, inst fel, 64, NATO fel, 65, res asst, 66; res assoc, Nuclear Struct Lab, Yale Univ, 67-72, sr res assoc & lectr, 72-76; RES STAFF MEM, DEPT COMPUT SCI, IBM T J WATSON RES CTR, 76- *Concurrent Pos:* Mem panel on-line comput in nuclear res, Nat Res Coun, 68-70. *Mem:* Asn Comput Mach; Am Phys Soc; Sigma Xi; sr mem Inst Elec & Electronics Engrs. *Res:* Computer systems and communications. *Mailing Add:* 28 Warnock Dr Westport CT 06880

SACHS, MARVIN LEONARD, b Allentown, Pa, Aug 31, 26; m 74; c 1. INTERNAL MEDICINE, VASCULAR DISEASES. *Educ:* Yale Univ, BA, 46; Harvard Univ, MD, 50; Am Bd Internal Med, dipl, 58. *Hon Degrees:* MA, Univ Pa, 71. *Prof Exp:* Intern surg, obstet & pediat, Allentown Gen Hosp, Pa, 50; intern med, Univ Hosps Cleveland, 51-52, resident, 52-53; instr, 55-61, physician-in-chg, Univ Pa Div Med Clin, Philadelphia Gen Hosp, 60-78, assoc, 61-69, ASST PROF MED, SCH MED, UNIV PA, 69-, CHIEF, MED VASCULAR SERV & DIR, MED VASCULAR LAB, HOSP UNIV PA. *Concurrent Pos:* Nat Heart Inst fel cardiovasc dis, Hosp Univ Pa, 55-57; mem cardiovasc sect, Dept Med, Univ Pa, 57-; consult, Food & Drug Admin, Dept Health, Educ & Welfare, 64; consult, Archit Res Unit, Univ City Sci Ctr, Philadelphia, 65-; lectr, Wharton Sch, 66-; mem res in nursing in patient care rev comt, USPHS, 67-70; attend physician, Vet Admin Hosp, 67-; mem coun arteriosclerosis, coun circulation & mem adv bd, Am Heart Asn. *Mem:* Fel Am Col Physicians; AMA; Am Fedn Clin Res; Am Thyroid Asn; Sigma Xi; Soc Vascular Med & Biol. *Res:* Peripheral vascular diseases; patient care research and education; organization and design of health care facilities. *Mailing Add:* Hosp Univ Pa 3400 Spruce St Philadelphia PA 19104

SACHS, MENDEL, b Portland, Ore, Apr 13, 27; m 52; c 4. THEORETICAL PHYSICS. *Educ:* Univ Calif, Los Angeles, AB, 49, MA, 50, PhD(physics), 54. *Prof Exp:* Theoret physicist, Radiation Lab, Univ Calif, 54-56; res scientist, Lockheed Missiles & Space Co, 56-61; res prof, McGill Univ, 61-62; assoc prof physics, Boston Univ, 62-66; PROF PHYSICS, STATE UNIV NY, BUFFALO, 66- *Concurrent Pos:* Asst prof, San Jose State Col, 57-61. *Res:* Relativity; field theory; quantum electrodynamics; elementary particles; philosophy of science; physical applications of group theory; astrophysics and cosmology. *Mailing Add:* Dept Physics State Univ NY Buffalo Amherst NY 14260

SACHS, MURRAY B, HEARING SCIENCES. *Educ:* Mass Inst Technol, BS, 62, MS, 64, PhD(elec eng & auditory physiol), 66. *Prof Exp:* Postdoctoral fel, Univ Cambridge, Eng, 68-69; from asst prof to assoc prof biomed eng, 70-80, DIR, CTR HEARING SCI, JOHNS HOPKINS UNIV, 86-, MASSEY PROF & DIR, DEPT BIOMED ENG, 91- *Concurrent Pos:* Mem, Comn Commun & Control, Int Union Pure & Appl Biophysics, 75-80; mem, Commun Dis Panel & Basic Sci Task Force, Nat Inst Neurol & Commun Dis & Stroke, NIH, 77-78, chmn, Commun Dis Rev Comt, 77-79, ad hoc adv comt, Commun Dis Prog, 79-86, sci prog adv comt, 84-86; prof biomed eng, Johns Hopkins Univ, 80-, prof neurosci, 81-, prof otolaryngol-head & neck surg, 83-; mem, Sensory Physiol & Perception Panel, NSF, 82-85; Jacob Javitz neurosci investr, 85; mem, Comt Hearing & Bioacoust, Nat Acad Sci, 85-88; mem ad hoc prog adv comt, Cochlear Implant Prog, Mass Inst Technol, 86- *Mem:* Inst Med; Sigma Xi. *Res:* Neural mechanisms of auditory perception; mathematical models for signal processing in the nervous system. *Mailing Add:* Sch Med Johns Hopkins Univ 720 Rutland Ave Baltimore MD 21205

SACHS, RAINER KURT, b Frankfurt, Ger, June 13, 32; US citizen; div; c 4. THEORETICAL PHYSICS. *Educ:* Mass Inst Technol, BSc, 53; Syracuse Univ, PhD(physics), 58. *Prof Exp:* Fel physics, Univ Hamburg, 59-60; fel, Univ London, 60-61; asst prof, Stevens Inst Technol, 62-63; from assoc prof to prof, Univ Tex, 63-69; PROF PHYSICS & MATH, UNIV CALIF, BERKELEY, 69- *Mem:* Am Phys Soc; Am Astron Soc. *Res:* General relativity; cosmology. *Mailing Add:* Dept Math Univ Calif 2120 Oxford St Berkeley CA 94720

SACHS, ROBERT GREEN, b Hagerstown, Md, May 4, 16; m 41, 50, 68; c 8. HIGH ENERGY PHYSICS, PARTICLES & FIELDS. *Educ:* Johns Hopkins Univ, PhD(physics), 39. *Hon Degrees:* DSc, Purdue Univ, 67, Univ Ill, 77, Elmhurst Col, 87. *Prof Exp:* Res fel theoret physics, George Washington Univ, 39-41 & Univ Calif, 41; instr physics, Purdue Univ, 41-43; chief air blast sect, Ballistic Res Lab, Aberdeen Proving Ground, 43-46; dir theoret physics div, Argonne Nat Lab, 46-47; from assoc prof to prof physics, Univ Wis, 47-64; assoc lab dir high energy physics, Argonne Nat Lab, 64-68, dir Nat Lab, 73-79; dir, Enrico Fermi Inst, 68-73; prof, 64-86, EMER PROF PHYSICS, UNIV CHICAGO, 86- *Concurrent Pos:* Consult, Argonne Nat Lab, 46-52, 60-63; consult, Ballistic Res Lab, Aberdeen Proving Ground, 46-59; Higgins vis prof, Princeton Univ, 55-56; consult, Lawrence Radiation Lab, Univ Calif, 55-59; mem adv panel physics, NSF, 58-61; Guggenheim fel & vis prof, Ecole Normale Superieure, Univ Paris, 59-60; Guggenheim fel, Europ Orgn Nuclear Res, 59-60; mem sci policy comt, Stanford Linear Accelerator Ctr, 66-70; high energy physics adv panel, US AEC, 67-69; mem physics surv comt & chmn elem particle physics panel, Physics Surv Comt, Nat Acad Sci, 69-72, chmn physics sect, 77-80 & chmn, Class I Math & Phys Sci, 81-83. *Mem:* Nat Acad Sci; AAAS; Am Phys Soc; fel Am Acad Arts & Sci (vpres, 79-83); Am Inst Physics. *Res:* High energy physics; fundamental particles; nuclear theory; solid state; terminal ballistics; nuclear power reactors; physics of time reversal; author or co-author of over 90 publications and 5 books. *Mailing Add:* 5490 S Shore Dr Chicago IL 60615

SACHS, ROY M, b New York, NY, Apr 1, 30; m 53; c 5. PLANT PHYSIOLOGY. *Educ:* Mass Inst Technol, BS, 51; Calif Inst Technol, PhD(plant physiol), 55. *Prof Exp:* Fulbright fel, Univ Parma, 55-56; jr res botanist, Univ Calif, Los Angeles, 56-58, asst plant physiologist, 58-61; from asst plant physiologist to assoc plant physiologist, 58-70, PLANT PHYSIOLOGIST & PROF ENVIRON HORT, UNIV CALIF, DAVIS, 70- *Mem:* Am Soc Plant Physiol; Am Soc Hort Sci; Agron Soc Am. *Res:* Vegetative growth; flowering; growth substances. *Mailing Add:* Dept Environ Hort Univ Calif Davis CA 95616

SACHS, THOMAS DUDLEY, b St Louis, Mo, Jan 29, 25; m 61, 88; c 4. PHYSICAL ACOUSTICS, BIOPHYSICS. *Educ:* Univ Calif, Berkeley, BA, 51; Innsbruck Univ, PhD(physics, math), 60. *Prof Exp:* Pub sch teacher, Calif, 52-53 & high sch & jr col teacher, 53-55; res investr chem, Western Reserve Univ, 60-62; asst prof, 62-77, ASSOC PROF PHYSICS, UNIV VT, 77- *Concurrent Pos:* Consult, Jr High Sch Sci Exp Prog, 62-65, Ladd Res, IBM Corp & Varian Corp; pres, Electronic Educator, Inc, 69-79; mem, Vt Regional Cancer Ctr, 80-; mem, Cell Biol Group, Univ Vt, 84-; res dir, Wellen Assocs, Inc, 88- *Mem:* AAAS; Am Phys Soc; Acoust Soc Am; Inst Elec & Electronics Engrs. *Res:* Perturbed acoustic propagation parameter measurements in liquids, plasmas and biological tissues with applications to liquid structure, cavitation, solar dynamics and mechanics; thermo-acoustic tissue characterization, non-invasive biopsy, perfusion and non-invasive temperature measurement by thermo-acoustic sensing technique (TAST). *Mailing Add:* 88 Lamore Rd Essex Junction VT 05452

SACHSE, WOLFGANG H, b Berlin-Charlottenburg, Ger, Mar 22, 42; US citizen; m 70; c 3. PHYSICAL ACOUSTICS. *Educ:* Pa State Univ, University Park, BS, 63; Johns Hopkins Univ, MSE, 66, PhD(mech), 70. *Prof Exp:* Ger acad exchange fel, Inst Metall, Aachen, 69-70; from asst prof to assoc prof, 70-83, PROF APPL MECH, CORNELL UNIV, 83- *Concurrent Pos:* Consult; Ger acad exchange fel, Nat Bur Standards, 77-78. *Mem:* Acoust Soc Am; Inst Acoust; Inst Elec & Electronics Engrs; Sigma Xi; Exp Mech Soc. *Res:* Mechanics of materials; wave propagation in solids; ultrasonics; acoustic emission; non-destructive testing of materials; transducers. *Mailing Add:* Dept Theoret & Appl Mech Thurston Hall Cornell Univ Ithaca NY 14853

SACHTLEBEN, CLYDE CLINTON, b Lincoln, Nebr, May 4, 36; m 58; c 2. PHYSICS, SCIENCE EDUCATION. *Educ:* Nebr Wesleyan Univ, BA, 57; Univ Nebr, MA, 60; Univ Iowa, PhD(sci educ), 67. *Prof Exp:* From instr to assoc prof, 60-68, PROF PHYSICS, HASTINGS COL, 68-, CHMN DEPT, 62- *Concurrent Pos:* US AEC lectr, Oak Ridge Radioisotope-Mobile Lab, 65-, res assoc, Oak Ridge Assoc Univs, 72; vis prof physics, Univ Nebr, 65-81, Univ Colo, 90-91; NSF fel, Univ Iowa, 66-67. *Mem:* Am Asn Physics Teachers. *Res:* Atomic and nuclear physics. *Mailing Add:* 623 N Shore Dr Hastings NE 68901

SACHTLER, WOLFGANG MAX HUGO, b Delitzsch, Ger, Nov 8, 24; m 53; c 3. CATALYSIS. *Educ:* Univ Technol Braunschweig, MS, 49, PhD(chem), 52. *Prof Exp:* Res chemist, Kon-Shell Lab, Amsterdam, 52-60, Shell Res Ctr, Emeryville, LA, 60-61, sect leader, Kon-Shell Lab, Amsterdam, 61- 72, dir, 73-83; PROF CHEM, NORTHWESTERN UNIV, 83- *Concurrent Pos:* Prof chem, Univ Leiden, 63-84. *Honors & Awards:* Numerous invited lects throughout the world; R L Burwell Award, NAm Catalysis Soc, 85; E V Murphree Award, Am Chem Soc, 87. *Mem:* Royal Acad Sci Neth; Am Chem Soc; Neth Chem Soc; Sigma Xi. *Res:* Heterogeneous catalysis; surface science; thermodynamics; electron emission. *Mailing Add:* Dept Chem Northwestern Univ 2145 Sheridan Rd Evanston IL 60208-3120

SACK, E(DGAR) A(LBERT), JR, b Pittsburgh, Pa, Jan 31, 30; m 52; c 2. ELECTRICAL ENGINEERING, SOLID STATE ELECTRONICS. *Educ:* Carnegie Inst Technol, BS, 51, MS, 52, PhD(elec eng), 54. *Prof Exp:* Res engr, Carnegie Inst Technol, 51-54; res engr, Res Labs, Westinghouse Elec Corp, 54-56, proj leader, TV Sect, 56-57, sect mgr, Dielec Devices, 57-60, dept mgr, Electronics, 60-61, dept mgr, Solid State Devices, 61-62, mgr eng, Molecular Electronics Div, 62-65, mgr tech opers, 65-66, asst gen mgr, 66-67, gen mgr, Integrated Circuits Div, 67-69; vpres & gen mgr, Integrated Circuits Div, Gen Instrument Corp, 69-70, vpres opers, Microelectronics, 70-71, vpres comput prod, 71-73, vpres & gen mgr, Microelectronics, 73-76, sr vpres, 76-84; CHMN & CHIEF EXEC OFFICER, ZILOG CORP, 84- *Concurrent Pos:* Dir, Regional Indust Tech Educ Coun, 80-82; fel, Polytechnic Inst, 81; dir, Semiconductor Indust Asn, 80-82. *Mem:* Fel Inst Elec & Electronics Engrs. *Res:* Ferroelectrics; electroluminescence; nonlinear circuits; integrated circuit design and processing; technical and general management; MOS memory device; solid state devices. *Mailing Add:* 21412 Sarahills Ct Saratoga CA 95070

SACK, FRED DAVID, b New York, NY, May 22, 47. PLANT CELL BIOLOGY, PLANT STRUCTURE & FUNCTION. *Educ:* Antioch Univ, BA, 69; Cornell Univ, PhD(plant biol), 82. *Prof Exp:* Res fel, Boyce Thompson Inst Plant Sci, Cornell, 81-84; ASSOC PROF PLANT ANAT, DEPT BOT, OHIO STATE UNIV, 84- *Mem:* Am Soc Plant Physiologists; Am Soc Gravitational & Space Biol; Am Soc Cell Biol; AAAS; Bot Soc Am. *Res:* Stomatal structure and function; plant graviperception. *Mailing Add:* Dept Plant Biol Ohio State Univ 1735 Neil Ave Columbus OH 43210-1293

SACK, GEORGE H(ENRY), JR, b Baltimore, Md, Apr 17, 43. MEDICAL GENETICS, GENE ORGANIZATION. *Educ:* Johns Hopkins Univ, BA, 65, MD, 68, PhD(molecular biol & microbiol), 75. *Prof Exp:* Intern med, Johns Hopkins Hosp, 68-69, asst resident, 69-70; fel microbiol, Johns Hopkins Univ, 70-73; major med, US Army Med Corps, 73-75; fel med genetics, Johns Hopkins Hosp, 75-76; asst prof med pediat, 76-84, asst prof physiol chem, 80-84, ASSOC PROF MED, PEDIAT & BIOL CHEM, JOHNS HOPKINS UNIV, 84- *Concurrent Pos:* Biochemist, J F Kennedy Inst, 81- *Mem:* Am Soc Human Genetics; AAAS. *Res:* Molecular biology: human gene structure, organization, expression, polymorphisms; clinical applications of genetic principles and technology. *Mailing Add:* Div Med Genetics Blalock 1008 Johns Hopkins Hosp Baltimore MD 21205

SACK, RICHARD BRADLEY, b Le Sueur, Minn, Oct 25, 35; m 55; c 4. MEDICINE, MICROBIOLOGY. *Educ:* Lewis & Clark Col, BS, 56; Univ Ore, MS & MD, 60; Johns Hopkins Univ, ScD(pathobiol), 68. *Prof Exp:* Internship, Univ Wash, Seattle, 60-61, int med residency, 61-62 & 64-65; John Hopkins Hosp fel med, Calcutta, 62-64; assoc prof, Sch Med, Univ Ore, 70-72; fel pathiol, Sch Pub Health, 65-68, from instr to assoc prof, Sch Med, 66-79, PROF MED, JOHNS HOPKINS UNIV, 79-, HEAD, DIV GEOG MED, 77- *Concurrent Pos:* Consult, WHO. *Mem:* Infectious Dis Soc Am; Am Soc Microbiol; Am Fedn Clin Res; Am Soc Clin Invest. *Res:* Cholera; diarrheal diseases; bacterial enterotoxins. *Mailing Add:* Dept Int Health Johns Hopkins Univ 615 N Wolfe St Baltimore MD 21205

SACK, ROBERT A, b Mar 14, 44; US citizen; m 78; c 1. BIOCHEMISTRY, VISUAL SCIENCES. *Educ:* NY Med Col, PhD(biochem), 72. *Prof Exp:* ASSOC PROF BASIC SCI, STATE UNIV NY, 72- *Concurrent Pos:* Guest res assoc, Brookhaven Nat Lab, 71- *Res:* Regeneration of visual pigments; cryobiochemistry; ocular microbiology and immunology. *Mailing Add:* Dept Biol Sci Col Optom 100 E 24th St New York NY 10010

SACK, RONALD LESLIE, b Minneapolis, Minn, Mar 29, 35; m 58; c 2. SOLID MECHANICS, STRUCTURAL ENGINEERING. *Educ:* Univ Minn, BS, 57, MSCE, 58, PhD(civil eng), 64. *Prof Exp:* Asst prof civil eng, Clemson Univ, 64-65; res engr, Boeing Co, 65-70; assoc prof, 70-74, PROF CIVIL ENG, UNIV IDAHO, 74- *Concurrent Pos:* Vis lectr, Dept Mech Eng, Seattle Univ, 67-68 & Dept Civil Eng, Univ Wash, 68-70; NASA-Am Soc Eng Educ summer fel, Stanford Univ & Moffet Field, 71; Royal Norwegian Coun Sci & Indust Res fel, Trondheim, Norway, 76-77. *Mem:* Am Soc Civil Engrs; Am Soc Eng Educ; Sigma Xi. *Res:* Application of approximate numerical methods to structural engineering problems; investigation of structural stability problems; seismic design; wind and snow loading on structures. *Mailing Add:* Sch Civil Eng 202 W Boyd St Rm 334 Norman OK 73019-0631

SACK, WOLFGANG OTTO, b Leipzig, Ger, Mar 17, 28; US citizen; m 55; c 2. ANATOMY, VETERINARY MEDICINE. *Educ:* Univ Toronto, DVM, 57; Univ Edinburgh, PhD(vet anat), 62; Univ Munich, Dr med vet, 72. *Prof Exp:* Asst prof vet anat, Ont Vet Col, Toronto, 58-60, assoc prof, 62-64; assoc prof, 64-73, prof, 73-77, MEM FAC VET ANAT, CORNELL UNIV, 64- *Concurrent Pos:* Mem, Int & Am Comts Vet Anat Nomenclature, 64-; guest prof, Univ Munich, 71-72; chair, Int Comt Vet Embryol Nomenclature, 87- *Mem:* Am Asn Vet Anat (pres, 81-82); Am Asn Anat; Am Vet Med Asn; World Asn Vet Anat (gen secy, 83-); Royal Col Vet Surg. *Res:* Developmental anatomy of domesticated animals; gross anatomy of the horse. *Mailing Add:* Dept Vet Anat Cornell Univ Ithaca NY 14853

SACKEIM, HAROLD A, b Hackensack, NJ, July 13, 51; m 77; c 1. CLINICAL PSYCHOLOGY, NEUROPSYCHOLOGY. *Educ:* Columbia Col, BA, 72; Oxford Univ, BA & MA, 74; Univ Pa, PhD(psychol), 77. *Prof Exp:* Asst prof psychol, Columbia Univ, 77-79; asst prof, NY Univ, 79-81; dep chief biol psychiat, NY State Psychiat Inst, 81-91; assoc prof psychol, NY Univ, 81-87; PROF PSYCHIAT, COLUMBIA UNIV, 91-; CHIEF, BIOL PSYCHIAT, NY STATE PSYCHIAT INST, 91- *Concurrent Pos:* Lectr psychiat, Col Physicians & Surgeons, Columbia, 80-; consult ed, Imagination, Cognition & Personality, 80-; assoc ed, J Social & Clin Psychol, 81, Convulsive Ther. *Honors & Awards:* Estab Investr Award, Nat Alliance Res Schizophrenia & Depressions, 89; Merit Award, NIMH, 90. *Mem:* Am Psychol Asn; AAAS; Int Neuropsychol Soc; Soc Biol Psychiat; Am Psychopathological Asn; Asn Res Nerv & Ment Dis; Am Col Neuropsychopharmacol. *Res:* The role of functional brain asymmetry in the regulation of emotion; psychobiology and treatment of affective disorders; dissociation and consciousness. *Mailing Add:* Dept Biol Psychiat NY State Psychiat Inst 722 W 168th St New York NY 10032

SACKETT, W(ILLIAM) T(ECUMSEH), JR, b Xenia, Ohio, Jan 28, 21; m 46; c 4. ELECTRICAL ENGINEERING. *Educ:* Johns Hopkins Univ, BEE, 41, DrEng(elec eng), 50. *Prof Exp:* Engr, Duquesne Light Co, 41-42; elec engr, US Naval Ord Lab, 42-50; asst chief elec eng div, Battelle Inst, 50-55; mgr res, Kuhlman Elec Co, 55-59; dir res, 60-77, dir, Syst & Res Ctr, 77-80, vpres, Corp Technol Ctr, 80- 84, vpres, corp res, Honeywell Inc, 84-86; assoc dean, Inst Technol, Univ Minn, 86-89; PRES, XOX CORP, 89- *Concurrent Pos:* Instr, Johns Hopkins Univ, 47-48. *Mem:* AAAS; Inst Elec & Electronics Engrs. *Res:* Contact resistance; instrumentation; insulation; research administration; aerospace sciences. *Mailing Add:* 1349 Pikelake Dr St Paul MN 55112

SACKETT, WILLIAM MALCOLM, b St Louis, Mo, Nov 14, 30; m 56; c 2. GEOCHEMISTRY, CHEMISTRY. *Educ:* Wash Univ, BA, 53, PhD(chem), 58. *Prof Exp:* Chemist, Carbide & Carbon Chem Corp, Ky, 53-54 & Mallinckrodt Chem Co, Mo, 54; vis asst res chemist, Scripps Inst Oceanog, Univ Calif, San Diego, 58-59; sr res engr, Pan Am Petrol Corp, 59-61, tech group supvr, 61-62; asst prof geol, Columbia Univ, 62-64; assoc prof chem, Univ Tulsa, 65-68; from assoc prof to prof oceanog, Tex A&M Univ, 68-79; prof marine sci & chmn dept, 79-83, GRAD RES PROF, UNIV SFLA, 83- *Concurrent Pos:* Von Humboldt fel, 76-77; Am Geophys Union field fel, 86-87. *Mem:* Am Geophys Union; Geochem Soc; Sigma Xi. *Res:* Isotope geochemistry of carbon, uranium-thorium series of radioactive elements; organic geochem marine pollutions. *Mailing Add:* Dept Marine Sci Univ SFla 140 7th Ave S St Petersburg FL 33701

SACKMAN, GEORGE LAWRENCE, b Baxley, Ga, Mar 15, 33; m 63; c 2. ELECTRICAL ENGINEERING. *Educ:* Univ Fla, BME, 54, BEE, 57, MSE, 59; Stanford Univ, PhD(elec eng), 64. *Prof Exp:* Res engr, Electron Tube Div, Litton Industs, Inc, 64-65; from assoc prof to prof elec eng, Naval Postgrad Sch, 65-84; PROF ELEC ENG, WATSON SCH, 84- *Concurrent Pos:* Consult, Lansmont Corp, 74- *Mem:* Inst Elec & Electronics Engrs; Acoust Soc Am; Sigma Xi; Res Soc Am. *Res:* Underwater acoustics; ultrasonic image systems; acoustic signal processing. *Mailing Add:* Dept Elec Eng State Univ NY PO Box 6000 Binghamton NY 13902-6000

SACKMAN, JEROME L(EO), b Rockaway Beach, NY, June 16, 29; m 51; c 2. ENGINEERING. *Educ:* Cooper Union, BCE, 51; Columbia Univ, MS, 55, ScD, 59. *Prof Exp:* Civil engr, Eng Res & Develop Labs, US Dept Army, Ft Belvoir, Va, 51-52; asst thermal inelasticity, Inst Flight Struct, Columbia Univ, 56-57, from instr to asst prof appl mech, 57-60; from asst prof to assoc prof, 60-66, PROF CIVIL ENG, UNIV CALIF, BERKELEY, 66-, VCHMN DIV STRUCT ENG & STRUCT MECH, 67- *Concurrent Pos:* Sci consult, Paul Weidlinger, Consult eng, NY, 59-, Math Sci, Wash, 67-, Lockheed Propulsion Co, Calif, 67- & Physics Int Co, 70-; mem at large, US Nat Comt Theoret & Appl Mech, 75-78. *Mem:* AAAS; Am Soc Civil Engrs; Am Soc Mech Engrs; Soc Eng Sci; Soc Exp Stress Anal; Sigma Xi. *Res:* Mechanics of solids; stress and stability analysis of deformable solids; wave propagation in deformable solids. *Mailing Add:* Dept Civil Eng Univ Calif Berkeley CA 94720

SACKMANN, I JULIANA, b Schoenau, EGer, Feb 8, 42; US citizen; m 73; c 2. ASTROPHYSICS. *Educ:* Univ Toronto, BA, 63, MA, 65, & PhD(astrophys), 68. *Prof Exp:* Nat Res Coun Can fel astrophys, Univ Observ, Gottingen, WGer, 68-71; Alexander von Humboldt fel, Max Planck Inst Physics & Astrophys, Munich, WGer, 69-71; res fel, 71-74, res assoc, Jet Propulsion Lab & vis assoc, 74-76, sr res fel astrophys, 76-81, FAC ASSOC, CALIF INST TECHNOL, 81- *Mem:* Am Astron Soc; Can Astron Soc. *Res:* Violent helium shell flashes in stars and their consequences on new element nucleosynthesis explaining carbon stars; new convective breakthroughs in the interior; observable surface variabilities on short timescales; FG Sagittae; mass loss and the sun. *Mailing Add:* Kellogg Radiation Lab 106-38 Calif Inst Technol Pasadena CA 91125

SACKNER, MARVIN ARTHUR, b Philadelphia, Pa, Feb 16, 32; m 56; c 3. PULMONARY PHYSIOLOGY. *Educ:* Temple Univ, BS, 63; Jefferson Med Col, MD, 57; Am Bd Internal Med, dipl, 65; Am Bd Pulmonary Dis, dipl, 69. *Prof Exp:* Intern med, Philadelphia Gen Hosp, 57-58, resident, 58-61; Am Col Physicians res fel physiol, Grad Sch Med, Univ Pa, 61-64, instr, 63-64; chief div pulmonary dis, 64-78, DIR MED SERV, MT SINAI MED CTR, 74-; PROF MED, UNIV MIAMI, 73- *Concurrent Pos:* Pa Heart Asn fel cardiol, 58-59; Am Col Physicians Brower traveling fel, 66; mem pulmonary dis adv comt, Nat Heart & Lung Inst; mem, Am Bd Pulmonary Dis, 74; mem, Am Bd Internal Med, chmn, Subspecialty Bd Pulmonary Dis, 78-79. *Mem:* Am Fedn Clin Res; fel Am Col Physicians; Am Thoracic Soc (pres, 80-81); Am Physiol Soc. *Res:* Pulmonary circulation; mucociliary clearance; mechanics of breathing, non-invasive cardiopulmonary monitoring. *Mailing Add:* Dept Med Univ Miami Sch Med Miami Beach FL 33101

SACKS, CLIFFORD EUGENE, b Carlisle, Pa, Oct 18, 53; m 77; c 3. ORGANIC CHEMISTRY & MANAGEMENT. *Educ:* Purdue Univ, BS, 75; Calif Inst Technol, PhD(chem), 80. *Prof Exp:* res chemist, 79-85, ASSOC DIR PROCESS RES & DEVELOP, UPJOHN CO, 85- *Concurrent Pos:* Assoc dir, Chem Res Prep. *Mem:* Am Chem Soc. *Res:* Process research and development including steroids, heterocycles and insecticidal agents. *Mailing Add:* Upjohn Co 1500-91-2 Portage MI 49081

SACKS, DAVID B, b Cape Town, SAfrica, Mar 20, 50; US citizen. PATHOLOGY. *Educ:* Univ Cape Town, MBChB, 76. *Prof Exp:* Instr path & med, Sch Med, Wash Univ, 88-89; ASST PROF PATH, HARVARD MED SCH, 89- *Concurrent Pos:* Young investr award with distinction, Acad Clin Lab Physicians & Scientists, 87; asst pathologist, Barnes Hosp, 88-89; med dir clin chem, Brigham Womens Hosp, 89- *Mem:* Am Soc Biochem & Molecular biol; fel Am Col Physicians; fel Am Col Pathologists; Am Diabetes Asn; Am Fedn Clin Res; Am Asn Clin Chem. *Res:* Mechanism of insulin action, including intracellular signal transduction; calmodulin function. *Mailing Add:* Dept Clin Path-Clin Lab Bldg Brigham & Women's Hosp 75 Francis St Boston MA 02115

SACKS, GERALD ENOCH, b Brooklyn, NY, Mar 22, 33; m 55, 83; c 4. COMPUTER PROGRAMMING. *Educ:* Cornell Univ, BEE & MEE, 58, PhD(math), 61. *Hon Degrees:* MA, Harvard Univ, 73. *Prof Exp:* NSF fel math, Inst Advan Study, 61-62; from asst prof to assoc prof, Cornell Univ, 62-67; PROF MATH, MASS INST TECHNOL, 67-; PROF MATH, HARVARD UNIV, 72- *Concurrent Pos:* Guggenheim fel, 66-67; Inst Advan Study fel, 73-74. *Mem:* Am Math Soc; Asn Symbolic Logic. *Res:* Mathematical logic; recursive function theory; set theory. *Mailing Add:* Dept Math Harvard Univ One Oxford St Cambridge MA 02138

SACKS, JEROME, b New York, NY, May 8, 31. MATHEMATICS. *Educ:* Cornell Univ, BA, 52, PhD, 56. *Prof Exp:* Instr math, Calif Inst Technol, 56-57; asst prof math statist, Columbia Univ, 57-60; asst prof math, Cornell Univ, 60-61; prof statist, Rutgers Univ, 79-81; assoc prof, 61-66, PROF MATH, NORTHWESTERN UNIV, 66- *Mem:* Am Statist Asn; Inst Math Statist. *Res:* Statistics; calibration; regression analysis; time series; robustness. *Mailing Add:* Statist 101 Ill Hall Univ Ill 725 S Wright Champaign IL 61820

SACKS, JONATHAN, b Worcester, S Africa, Oct 26, 43; m 69; c 1. GEOMETRY, TOPOLOGY. *Educ:* Univ Cape Town, BSc, 68, MSc, 70; Univ Calif, Berkeley, PhD(math), 75. *Prof Exp:* Vis lectr math, Univ Ill Urbana-Champaign, 75-77; LECTR MATH, UNIV CHICAGO, 77- *Mem:* Am Math Soc. *Res:* Eigenvalues of the Laplacian on Riemannian manifolds; application of the study of the topology of function spaces to variational problems in Riemannian geometry. *Mailing Add:* 63 Clark St Newton MA 02159

SACKS, LAWRENCE EDGAR, b Los Angeles, Calif, Mar 9, 20; m 63; c 2. MICROBIOLOGY. *Educ:* Univ Calif, Los Angeles, AB, 41; Univ Wash, MS, 43; Univ Calif, PhD(microbiol), 48. *Prof Exp:* Asst bacteriologist, Comt Lignin & Cellulose Res, 43-44; bacteriologist, 48-62, prin chemist, 62-72, MICROBIOLOGIST, WESTERN REGIONAL RES CTR, SCI & EDUC ADMIN-AGR RES, USDA, 72- *Mem:* AAAS; Am Soc Microbiol; Am Chem Soc. *Res:* Action of antibiotics and surface active agents on bacteria; bacterial denitrification; Arthrobacter; spores; Clostridium perfringens; mutagen screening. *Mailing Add:* USDA Western Reg Res Ctr 800 Buchanan St Albany CA 94710

SACKS, LAWRENCE J, b Newark, NJ, June 12, 28; m 55; c 3. INORGANIC CHEMISTRY. *Educ:* Drew Univ, AB, 52; Pa State Univ, MS, 58; Univ Ill, PhD(inorg chem), 64. *Prof Exp:* Instr chem, Pa State, 53-55; res chemist, Monsanto Chem Co, 55-56; asst prof chem, Reed Col, 60-63 & Rose Polytech, 64-65; assoc prof, State Univ NY Col Buffalo, 65-68; prof, Hampton Inst, 68-70; PROF CHEM, CHRISTOPHER NEWPORT COL, 70- *Concurrent Pos:* Assoc dir, Lab Chem Evolution, Dept Chem, Univ Md, College Park, 85-86. *Mem:* AAAS; Am Chem Soc. *Res:* Thermodynamics; theoretical calculations of bond energies, dipole movements and steric effects. *Mailing Add:* Dept Chem Christopher Newport Col Newport News VA 23606

SACKS, MARTIN, b Brooklyn, NY, Aug 30, 24; m 50; c 2. INVERTEBRATE ZOOLOGY. *Educ:* City Col New York, BS, 49; Univ Ill, MS, 50, PhD, 53. *Prof Exp:* Lectr, 53-59, from instr to assoc prof, 59-69, head dept, 69-72, PROF BIOL, CITY COL NEW YORK, 69- *Mem:* AAAS; Am Soc Zoologists; Soc Syst Zool; Am Micros Soc; Sigma Xi. *Res:* Taxonomy, ecology and embryology of marine and aquatic gastrotricha. *Mailing Add:* 27 Cranford Pl Teaneck NJ 07666

SACKS, MARTIN EDWARD, b Bronx, NY, Nov 22, 43; m 67; c 1. CHEMICAL ENGINEERING, COAL SCIENCE. *Educ:* Cooper Union, BChE, 65; Univ Mich, MSE, 66; Stevens Inst Technol, PhD(chem eng), 72. *Prof Exp:* Res engr coal conversion, FMC Corp, 66-70; sr res engr, Cogas Develop Co, 72-78, process design supvr coal gasification, 78-80, eng mgr, 80-81; prin process engr & sect mgr, FMC Corp, 81-89; ENG CONSULT, 89- *Mem:* Am Inst Chem Engrs; Am Chem Soc. *Res:* Coal conversion, especially pyrolysis and gasification. *Mailing Add:* FMC Corp PO Box 8 Princeton NJ 08543

SACKS, WILLIAM, b Toronto, Ont, Jan 30, 26; m 52; c 4. POLYMER CHEMISTRY, POLYMER ENGINEERING. *Educ:* Univ Toronto, BASc, 48, MASc, 49; McGill Univ, PhD(chem), 54. *Prof Exp:* Res officer, Nat Res Coun Can, Ottawa, 49-52, fel, McGill Univ, 52-54; res engr & res sect mgr, Visking Div, Union Carbide Corp, 54-61, group leader, Plastics Div, 61-65; res assoc & res mgr, Gen Chem & Fabricated Prod Div, Allied Corp, 66-73, tech dir, Films Dept, 73-79, dir, New Bus Develop, 79-87; EXEC DIR, PLASTICS INST AM, 88- *Concurrent Pos:* Mem, Nat Res Coun Can, 52, 53. *Mem:* Am Chem Soc; Chem Inst Can; Soc Plastics Engrs; Sigma Xi; Com Develop Asn. *Res:* Physical chemistry of polymers; properties and applications of polymer films; processing behavior of synthetic polymers. *Mailing Add:* 686 Long Hill Rd Gillette NJ 07933

SACKS, WILLIAM, b Philadelphia, Pa, Feb 17, 24; m 54; c 3. BIOCHEMISTRY. *Educ:* Pa State Univ, BS, 47, MS, 48, PhD, 51. *Prof Exp:* Dir chem lab, Southern Div, Einstein Med Ctr, Philadelphia, 51-54, res investr, 54-58; prin res scientist, Kline Inst Psychiat Res, 58-90; RETIRED. *Concurrent Pos:* Res assoc prof psychiat, Sch Med, NY Univ, 79- *Mem:* AAAS; Am Chem Soc; Sigma Xi; Soc Neurosci; Int Soc Neurochem; Am Soc Neurochem; Am Soc Biol Chemists. *Res:* Cerebral metabolism in vivo in mental disease. *Mailing Add:* One Mary Ann Lane New City NY 10956

SACKSTEDER, RICHARD CARL, b Muncie, Ind, Feb 11, 28; m 52; c 2. MATHEMATICS. *Educ:* Univ Chicago, BS, 48, PhB, 46; Johns Hopkins Univ, PhD, 60. *Prof Exp:* Res mathematician, Ballistics Res Lab, US Dept Army, Aberdeen Proving Ground, Md, 54-59; vis mem, Inst Math Sci, NY Univ, 60-62; asst prof math, Barnard Col & Columbia Univ, 62-65; assoc prof, 65-67, PROF MATH, CITY UNIV NEW YORK, 67- *Concurrent Pos:* Jr instr, Johns Hopkins Univ, 57-59; lectr, Goucher Col, 58. *Mem:* Am Math Soc; Catgut Acoust Soc. *Res:* Differential geometry; fluid dynamics. *Mailing Add:* Dept Math City Univ New York Grad Ctr 33 W 42nd St New York NY 10036

SACKSTON, WALDEMAR E, b Manitoba, Jan 4, 18; m 41; c 2. PLANT PATHOLOGY. *Educ:* Univ Manitoba, BSA, 38; McGill Univ, MSc, 40; Univ Minn, PhD(plant path), 49. *Prof Exp:* Asst, Macdonald Col, McGill Univ, 38-40; agr asst, Dom Lab Plant Path, Can Dept Agr, 41-46; from asst plant pathologist to plant pathologist, 46-58, sr plant pathologist & head plant path sect, 58-60, chmn dept, 60-69, prof, 60-83, EMER PROF PLANT PATH, MACDONALD COL, MCGILL UNIV, 83- *Concurrent Pos:* Specialist & consult, Point IV Prog, Chile, 54 & Uruguay, 56-57; at Res Ctr on Oilseed Crops, Cordoba, Spain, 72-77. *Honors & Awards:* Pustovoit Award, Int Sunflower Asn, 82; D L Bailey Award, Can Phytopath Soc, 83. *Mem:* fel Am Phytopath Soc; Indian Phytopath Soc; Int Sunflower Asn (pres, 78-80); fel Can Phytopath Soc (vpres, 58-60, pres, 60-61); Sigma Xi. *Res:* Diseases of oilseed crops; soilborne and seedborne diseases; epidemiology. *Mailing Add:* Dept Plant Sci Macdonald Col McGill Univ Ste Anne de Bellevue PQ W9X 1C0 Can

SACKTOR, BERTRAM, biochemistry, physiology; deceased, see previous edition for last biography

SADAGOPAN, VARADACHARI, b Uppiliappan Koil, Madras, India; US citizen; m 62. MATERIALS SCIENCE. *Educ:* Univ Madras, BSc, 53; Annamalai Univ, Madras, MA, 55; Indian Inst Sci, DIISc, 58; Mass Inst Technol, SM, 60, MetE, 61. *Hon Degrees:* ScD, Mass Inst Technol, 64. *Prof Exp:* Asst econ affairs off, Hq, UN, 62, spec asst to secy-gen & chief liaison off for dels from Asia & Far East, UN Europ Off, Geneva, Switz, 62-63; res assoc mat sci, Mass Inst Technol, 64-67; prin res scientist, Avco Corp, Mass, 67-68; res staff mem phys sci, T J Watson Res Ctr, 68-72, mgr univ rel, 72-76, mgr tech rel, Off Res & Develop Coord, IBM Europe, 76-79, mgr tech rel, 79-88, DIV MGR UNIV REL, IBM RES DIV, IBM CORP, 88- *Concurrent Pos:* Consult, Asian Bank, Manila, 71- *Mem:* Am Phys Soc; Inst Elec & Electronics Engrs. *Res:* Electronic materials. *Mailing Add:* IBM Res Div PO Box 218 Yorktown Heights NY 10598

SADANA, AJIT, b Rawalpindi, Pakistan, Feb 14, 47; US citizen; m 73; c 2. ENZYMES, MATHEMATICAL MODELING. *Educ:* Indian Inst Technol, BTech, 69; Univ Del, MCE, 72, PhD(chem eng), 75. *Prof Exp:* Sr sci officer, Nat Chem Lab, Coun Sci & Indust Res, India, 75-80; vis assoc prof chem eng, Auburn Univ, 80-81; assoc prof, 81-89, PROF CHEM ENG, UNIV MISS, 90- *Concurrent Pos:* Consult, First Chem Corp, 87; vis res scientist, EI du Pont de Nemours & Co, 88; sr fel, Naval Res Lab, Wash, DC, 90. *Mem:* Am Inst Chem Engrs. *Res:* The structure-function relationships of enzymes; mathematical modeling of enzyme inactivations to obtain physical insights into enzymes; protein/enzyme inactivation during bioseparation. *Mailing Add:* Chem Eng Dept Univ Miss University MS 38677-9740

SADANA, YOGINDER NATH, b Peshawar, India, May 15, 31; m 62; c 2. INORGANIC CHEMISTRY. *Educ:* Univ Agra, BSc, 51, MSc, 53; Univ BC, PhD(chem), 63. *Prof Exp:* Sr res fel electrodeposition of metals & alloys, Nat Metall Lab, India, 54-58; res asst, Res Inst Precious Metals & Metall Chem, Ger, 58-59; instr chem, Wash State Univ, 62-63; res engr, Cominco Ltd, Can, 64-66; from asst prof to assoc prof, 66-77, PROF CHEM, LAURENTIAN UNIV, 77-; hon prof, 89, Nanchang Inst Aero-Technol, China. *Concurrent Pos:* Nat Res Coun Can res grants, 66-; fels, Chem Inst Can & Inst Metal Finishing; hon prof, Nanchang Inst Aero-Technol, China, 89. *Mem:* Am Chem Soc; Chem Inst Can. *Res:* Electrodeposition of metals and alloys; surface finishing techniques; electrodeposition of gold and gold alloys and their x-ray structures. *Mailing Add:* Dept Chem Laurentian Univ Ramsey Lake Rd Sudbury ON P3E 2C6 Can

SADANAGA, KIYOSHI, b Onomea, Hawaii, Feb 6, 20; m 66; c 2. GENETICS. *Educ:* Univ Hawaii, BS, 42; Iowa State Univ, MS, 51, PhD(genetics), 55. *Prof Exp:* Asst genetics, Sugar Cane Exp Sta, Hawaii Sugar Planters Asn, 52-53; geneticist, Oat Proj, 56-76, GENETICIST SOYBEAN RES, USDA, 76-; PROF GENETICS, IOWA STATE UNIV, 59- *Concurrent Pos:* Fulbright scholar, Kyoto Univ, 60-61. *Res:* Genetics and cytogenetics of soybeans. *Mailing Add:* 1307 Sequoia Pl Davis CA 95616

SADAVA, DAVID ERIC, b Ottawa, Ont, Mar 14, 46; m 72; c 1. CELL BIOLOGY. *Educ:* Carleton Univ, BSc, 67; Univ Calif, San Diego, PhD(cell biol), 71. *Prof Exp:* Researcher entomol, Can Dept Agr, 65-66; adv sci policy, Sci Secretariat, Can, 67; teaching asst biol, Univ Calif, San Diego, 67-71; researcher marine biol, Scripps Inst Oceanog, 72; from asst prof to assoc prof, 72-83, PROF BIOL, CLAREMONT COLS, 83-, CHMN JOINT SCI DEPT, 80- *Concurrent Pos:* Woodrow Wilson Found fel, 67; vis prof, Univ Colo, 77-, Univ Calif, 86. *Mem:* AAAS; Soc Pediat Res; Asn Politics Life Sci. *Res:* Biochemical genetics of thoroughbred horses; isozymes in human development; plant developmental biochemistry; pharmacology. *Mailing Add:* Joint Sci Dept Claremont Cols Claremont CA 91711

SADEE, WOLFGANG, b Bad Harzburg, WGer, Mar 25, 42; m 74. DRUG METABOLISM, ANALYTICAL BIOCHEMISTRY. *Educ:* Free Univ Berlin, Dr rer nat, 68. *Prof Exp:* NATO fel, Univ Calif, San Francisco, 69; res assoc fel clin pharmacol, Free Univ Berlin, 70-71; asst prof pharm & med, Univ Southern Calif, 71-74; from asst prof to assoc prof, 74-81, PROF PHARM & PHARMACEUT CHEM, UNIV CALIF, SAN FRANCISCO, 81- *Concurrent Pos:* Ed in chief, Pharmaceut Res. *Mem:* Am Chem Soc; AAAS; Am Asn Pharmaceut Scientists. *Res:* Opiates, neurotransmitter receptors, neuroblastoma. *Mailing Add:* Sch Pharm Univ Calif San Francisco CA 94143

SADEH, WILLY ZEEV, b Galatz, Romania, Oct 13, 32; nat US; m 56; c 2. ENGINEERING. *Educ:* Israel Inst Technol, BSc, 58, MSc, 64; Brown Univ, PhD(eng), 68. *Prof Exp:* Res engr, Gen Aero M Dassault, Paris, 58-59, Nat Sci Res Ctr, 59-60 Desalination Plants, Israel, 60-62 & Negev Inst Arid Zone Res, 62; instr mech eng, Israel Inst Technol, 62-64; fel eng, Brown Univ, 64-65, res asst, 65-68; from asst prof to assoc prof, 68-76, PROF CIVIL ENG, COLO STATE UNIV, 76- *Honors & Awards:* Cert appreciation, Technol Utilization Prog, NASA, 71. *Mem:* AAAS; Am Soc Mech Engrs; Am Asn Univ Professors; assoc fel Am Inst Aeronaut & Astronaut; Am Soc Eng Educ; Sigma Xi. *Res:* Fluid mechanics; turbulent flow; instrumentation; structural aerodynamics; atmospheric turbulence; air pollution; turbomachinery; boundary-layer flow. *Mailing Add:* Dept Civil Eng Colo State Univ Ft Collins CA 80523

SADIK, FARID, b Taibeh, Palestine, July 3, 34; US citizen; m 59; c 4. PHARMACEUTICS. *Educ:* Univ Ga, BSPharm, 58; Univ Miss, PhD(pharmaceut), 68. *Prof Exp:* Pharmacist, 58-65; instr pharmaceut, Univ Miss, 65-68, asst prof, 68; asst prof, Northeast La Univ, 68-71; assoc prof, Univ Miss, 71-73; assoc prof, 73-76, PROF PHARMACEUT, ASSOC DEAN & DIR GRAD STUDIES, COL PHARM, UNIV SC, 76- *Concurrent Pos:* Fulbright scholar, 87-88. *Mem:* Am Pharmaceut Asn. *Res:* Clinical pharmacy; effect of particle size of drugs on absorption. *Mailing Add:* Col Pharm Univ SC Columbia SC 29208

SADIK, SIDKI, plant physiology, for more information see previous edition

SADJADI, FIROOZ AHMADI, b Tehran; US citizen. SIGNAL & IMAGE PROCESSING, AUTOMATIC OBJECT RECOGNITION. *Educ:* Purdue Univ, BS, 72, MS, 74; Univ Southern Calif, EE, 76; Univ Tenn, PhD(elec eng). *Prof Exp:* Res asst image processing, Image Processing Inst, Univ Southern Calif, 74-77; res asst image & signal processing, Image & Pattern Anal Lab, Univ Tenn, 77-83; PRIN RES SCIENTIST SIGNAL & IMAGE PROCESSING, HONEYWELL SYSTS & RES CTR, 83- *Concurrent Pos:* Consult image processing, Oak Ridge Nat Lab; prin investr, Defense Advan Res Projs Agency, Army & Air Force, 84-; lectr, George Washington Univ, Univ Md, Univ Calif, Los Angeles, Int Soc Optical Eng, Gov Res Centers, 88-; guest ed, Optical Eng J, 91. *Mem:* Sigma Xi; sr mem Inst Elec & Electronics Engrs; Int Soc Optical Eng. *Res:* Modeling, algorithm design, sensor fusion, performance evaluation for the development of adaptive milimerwave radar, infrared, radar & sonar signal processing systems for a variety of applications: automatic object recognition, enhanced vision for autonomous landing, remote sensing, seafloor mapping, etc. *Mailing Add:* 3400 Highcrest Rd NE Minneapolis MN 55418

SADLER, ARTHUR GRAHAM, b Pontefract, Eng, Oct 20, 25; Can citizen; m 57; c 4. CHEMISTRY, CERAMICS. *Educ:* Univ Leeds, BSc, 51, PhD(ceramics), 56. *Prof Exp:* Mem, Electronic Ceramics, Dept Mines & Tech Surv, Govt Can, 57-59, sr sci staff, 59-62; mem sci staff ferrites, Northern Elec Co Ltd, Ont, 62-63, dept chief magnetic mat, 63-67, mgr phys sci, 67-69; mgr electronic mat & processes, 69-72, MGR STA APPARATUS BR LAB, BELL-NORTHERN RES LTD, 72- *Mem:* Can Ceramic Soc (pres, 69-70); fel Brit Inst Ceramics; fel Chem Inst Can. *Res:* Chemistry of high temperature reactions in inorganic materials; chemistry and physics of electronic ceramics; research management. *Mailing Add:* 480 Cloverdale Rd Rockcliffe ON K1M 0Y6 Can

SADLER, CHARLES ROBINSON, JR, b Richmond, Va, June 17, 50; m; c 1. LOW BACK PAIN, COMPUTERS-IN-MEDICINE. *Educ:* Rice Univ, Houston, Tex, BSEE, 72, MEE, 73; Baylor Col Med, MD, 76. *Prof Exp:* ASST CLIN PROF ORTHOP SURG, UNIV SOUTHERN CALIF, 90- *Concurrent Pos:* Consult orthop surg, Rancho Los Amigos Hosp, 84-; pres, Charles Sadler MD PC, 86-; vpres, US Sect Int Col Surgeons, 90- *Res:* Orthopedic surgery. *Mailing Add:* 8447 Wilshire Blvd No 424 Beverly Hills CA 90211

SADLER, G(ERALD) W(ESLEY), b Kindersley, Sask, Sept 19, 25; m 55. MECHANICAL ENGINEERING. *Educ:* Univ Sask, BSc, 47; Univ Ill, MS, 51. *Prof Exp:* Lectr & engr for supt bldgs off, 47-50, lectr thermodyn, 51-52, asst prof, 52-55, supt bldgs, 55-59, assoc prof, 60-77, PROF MECH ENG, UNIV ALTA, 77- *Mem:* Am Soc Heating, Refrig & Air-Conditioning Engrs; Eng Inst Can; Can Soc Mech Engrs; Solar Energy Soc. *Res:* Thermodynamic principles in relation to mechanical engineering; heat transfer, power production and fluid flow; heating, air conditioning and solar energy. *Mailing Add:* Dept Mech Eng Univ Alta Edmonton AB T6G 2M7 Can

SADLER, GEORGE D, b Long Beach, Miss, Mar 17, 52; m 75; c 4. POLYMER GAS PERMEATION, OXIDATION OF FOODS. *Educ:* Fla State Univ, BS, 74; Brigham Young Univ, MS, 80; Purdue Univ, PhD(food chem), 84. *Prof Exp:* Res & develop technician, Energy Systs Div, Olin Industs, 74-76, opers control technician, 76-77; res fel, Purdue Univ, 84-86; asst prof, 86-90, ASSOC PROF FOOD CHEM, UNIV FLA, 91- *Mem:* Inst Food Technologists; Am Chem Soc. *Res:* Gas permeation in polymers used in food packaging; quality consequences of polymer gas permeation on food quality. *Mailing Add:* 700 Experiment Station Rd Lake Alfred FL 33850

SADLER, J(ASPER) EVAN, b Huntington, WVa, Nov 9, 51; m 81; c 2. HEMATOLOGY. *Educ:* Princeton Univ, AB, 73; Duke Univ, PhD(biochem), 78, MD, 79. *Prof Exp:* Intern internal med, Duke Univ Med Ctr, 79-80, resident, 80-81; fel hemat, Univ Wash, Seattle, 81-84; asst prof, 84-89, ASSOC PROF MED, SCH MED, WASH UNIV, 89-, ASST PROF BIOCHEM & MOLECULAR BIOPHYS, 85-; ASSOC INVESTR, HOWARD HUGHES MED INST, 84- *Concurrent Pos:* Mem Coun Thrombosis, Am Heart Asn. *Mem:* Am Soc Clin Invest; Am Heart Asn; Int Soc Thrombosis & Hemostasis; Am Fedn Clin Res; Am Soc Hemat; Am Soc Biochem & Molecular Biol. *Res:* Regulation and structure-function relationships of hemostatic proteins. *Mailing Add:* Dept Med Wash Univ Sch Med 660 S Euclid Box 8045 St Louis MO 63110

SADLER, JAMES C, b Silver Point, Tenn, Feb 9, 20; m 41; c 2. METEOROLOGY. *Educ:* Tenn Polytech Inst, BS, 41; Univ Calif, Los Angeles, MA, 47. *Prof Exp:* Mil dir, Sacramento Peak Observ, US Air Force, 51-53, res meteorologist, Air Force Sch Aviation Med, 53-55; dir trop meteorol course, Air Weather Serv, Univ Hawaii, 55-59; chief satellite meteorol br, Air Force Cambridge Res Labs, 59-62, chief satellite utilization, Int Indian Ocean Exped, 62-65; from assoc prof to prof meteorol, 65-87, EMER PROF, UNIV HAWAII, 87- *Concurrent Pos:* Consult, US Air Force, 67- & US Navy, 68- *Honors & Awards:* Banner Miller Award, Am Meteorol Soc. *Mem:* AAAS; fel Am Meteorol Soc; Am Geophys Union. *Res:* Kinematic description of the general circulation of the tropics and its relation to satellite determined cloud climatology and life history of tropical cyclones. *Mailing Add:* Dept Meteorol Hig 340 Univ Hawaii-Manoa 2500 Campus Rd Honolulu HI 96822

SADLER, MICHAEL ERVIN, b Tex, May 18, 48. PHYSICS, ELEMENTARY PARTICLE PHYSICS. *Educ:* Tex Tech Univ, BA, 71; Ind Univ, MS, 74, PhD(physics), 77. *Prof Exp:* Res physicist, Univ Calif Los Angeles, 77-80; ASST PROF PHYSICS, ABILENE UNIV, 88- *Mem:* Am Phys Soc; Am Asn Phys Teachers; Coun Undergrad Res. *Mailing Add:* ACU Station Box 7646 Abilene TX 79699

SADLER, MONROE SCHARFF, b Natchez, Miss, Oct 2, 20; m 43; c 2. PHYSICAL CHEMISTRY. *Educ:* Mass Inst Technol, SB, 42; Carnegie Inst Technol, MS, 48, DSc(chem), 49. *Prof Exp:* Res chemist, E I du Pont de Nemours & Co, Inc, 49-53, res supvr, 53-59, lab dir, 59-63, dir mat res, 63-66, asst dir res & develop, 66-68, asst dir, Develop Dept, 68-70, dir develop dept, 70-75, asst dir, Cent Res & Develop Dept, 75-80; RETIRED. *Mem:* Am Chem Soc; Am Phys Soc. *Res:* Physics of high pressure; ultrasonics; nuclear magnetic resonance; ferromagnetic materials. *Mailing Add:* 8510 Sandy Oak Lane Sarasota FL 34238

SADLER, STANLEY GENE, b Spring Lake, Utah, Mar 6, 38; m 63; c 5. MECHANICAL & AEROSPACE SCIENCES. *Educ:* Univ Utah, BS, 62; Univ Rochester, MS, 64, PhD(mech & aerospace sci), 68. *Prof Exp:* Lab asst, High Velocity Impact Lab, Univ Utah, 62-63; res engr, Rochester Appl Sci Assocs, 67-71, group head aerodyn & hydrodyn res, 71-72; sr proj engr, Homelite Textron, 72-78; dynamic staff specialist, 78-80, chief rotor dynamics, 81-83, group engr aeromechanics, 83-89, STAFF ENGR, AEROMECHANICS, BELL HELICOPTER TEXTRON, 89- *Mem:* Am Soc Mech Engrs; Am Helicopter Soc; Am Inst Aeronaut & Astronaut. *Res:* Statics and dynamics of elastic systems; fluid dynamics and stability; helicopter rotor dynamics and noise; shell analysis; rotating system vibration and stability. *Mailing Add:* 1002 Curtis Ct Arlington TX 76012-5327

SADLER, THOMAS WILLIAM, b Portsmouth, Ohio, Feb 25, 49; m 76; c 2. TERATOLOGY, DEVELOPMENTAL BIOLOGY. *Educ:* Wake Forest Univ, BS, 71; Univ Va, PhD(anat), 76. *Prof Exp:* Asst prof anat, Univ Va, 76-79; assoc prof anat, Col Med, Univ Cincinnati, 79-82; assoc prof anat, 82-88, PROF CELL BIOL & ANAT, SCH MED, UNIV NC. *Concurrent Pos:* Vis prof, Downing Col, Cambridge Univ, Eng, 76. *Mem:* Teratol Soc; Am Asn Anatomists; AAAS. *Res:* Investigation of normal and abnormal events during embryogenesis; development of techniques for maintaining mammalian embryos in culture during organogenesis. *Mailing Add:* Dept Cell Biol & Anat Sch Med Univ NC 108 Taylor Bldg CB No 7090 Chapel Hill NC 27599

SADLER, WILLIAM OTHO, b Chunky, Miss, July 23, 03. LIMNOLOGY. *Educ:* Miss Col, BA, 29, LittD, 69; Cornell Univ, PhD(limnol), 32. *Prof Exp:* Asst prof zool, Miss Col, 29-30; investr, US Bur Fisheries, 30-32; prof zool, 32-72, EMER PROF ZOOL, MISS COL, 72-, CHMN DEPT BIOL SCI, 61- *Mem:* AAAS; Am Fisheries Soc. *Res:* Aquiculture; limnology; vertebrate taxonomy. *Mailing Add:* PO Box 125 Clinton MS 39056

SADOCK, BENJAMIN, b New York, NY, Dec 22, 33; m 63; c 2. MEDICINE, PSYCHIATRY. *Educ:* Union Col, NY, AB, 55; New York Med Col, MD, 59; Am Bd Psychiat, dipl, 66. *Prof Exp:* Instr psychiat, Univ Tex Southwestern Med Sch Dallas, 64-65; from instr to assoc prof psychiat, NY Med Col, 65-75, prof, 75-80, co-dir, Sexual Ther Ctr, 72-80, dir continuing educ psychiat, 75-80; PROF PSYCHIAT & DIR STUDENT HEALTH PSYCHIAT, NY UNIV, 80-, DIR, UNDERGRAD EDUC, 86- *Concurrent Pos:* Consult, Wichita Falls State Hosp, Tex, 64-65; chief psychiat consult, Student Health Serv, New York Med Col, 66-, dir div group process, 68-; assoc examr, Am Bd Psychiat, 67-; clin asst prof, Sch Med, NY Univ, 69-73; vchmn, Dept Psychiat, NY Univ, 84-, co-dir, grad med educ, 86- *Mem:* Am Psychiat Asn; Am Pub Health Asn; Am Orthopsychiat Asn; NY Acad Med; fel Am Col Physicians. *Mailing Add:* Four E 89th St New York NY 10028

SADOCK, VIRGINIA A, b Bulgaria, Nov 25, 38; US citizen; m 63; c 2. PSYCHIATRY. *Educ:* Bennington Col, AB, 60; New York Med Col, MD, 70. *Prof Exp:* asst clin prof psychiat, New York Med Col, 74-80; ASSOC CLIN PROF PSYCHIAT, NY UNIV MED CTR, 80- *Concurrent Pos:* Dir postgrad prog human sexuality, Dept Psychiat, New York Univ Med Ctr, 80. *Mem:* AMA; Am Psychiat Asn; Am Med Womens Asn; Am Asn Sex Educrs & Counrs. *Mailing Add:* Four E 89th St New York NY 10128

SADOFF, AHREN J, b Ithaca, NY, May 26, 36. ELEMENTARY PARTICLE PHYSICS, PHYSICS. *Educ:* Mass Inst Technol, BS, 58; Cornell Univ, PhD(physics), 64. *Prof Exp:* Res assoc Lab Nuclear Studies, Cornell Univ, 64-65; PROF PHYSICS, ITHACA COL, 88- *Concurrent Pos:* Vis scientist, Lawrence Livermore Radiation Lab, Berkeley, Calif, 80; vis prof, Cornell Univ, 86-88. *Mem:* Am Phys Soc; Sigma Xi. *Mailing Add:* Dept Physics Ithaca Col Ithaca NY 14850

SADOFF, HAROLD LLOYD, b Minneapolis, Minn, Sept 17, 24; m 46; c 4. MICROBIOLOGY, BIOCHEMISTRY. *Educ:* Univ Minn, BChEng, 47; Univ Ill, MS, 52, PhD(microbiol, biochem), 55. *Prof Exp:* Res assoc chem eng, Univ Ill, 54-55; from asst prof to assoc prof, 55-65, PROF MICROBIOL, MICH STATE UNIV, 65-, PROF PUB HEALTH, 77- *Concurrent Pos:* USPHS fel, Univ Wash, 61-62 & Stanford Univ, 70-71; mem microbiol chem study sect, NIH, 71-73. *Mem:* AAAS; Am Soc Microbiol; Am Soc Biol Chemists; NY Acad Sci; Soc Appl Bact; Sigma Xi. *Res:* The biochemistry and molecular biology of cell differentiation of microorganisms. *Mailing Add:* Dept Microbiol & Pub Health Mich State Univ East Lansing MI 48823

SADOSKY, THOMAS LEE, b Ft Thomas, Ky, Sept 23, 39; m 64; c 1. INDUSTRIAL ENGINEERING, HUMAN FACTORS. *Educ:* Univ Ohio, BS, 62, MS, 64; Univ Mich, PhD(indust eng), 68. *Prof Exp:* Actg instr eng graphics, Univ Ohio, 62-64; instr, Univ Mich, 64-65; assoc prof indust eng, Ga Inst Technol, 68-80. *Res:* Human performance; operations research. *Mailing Add:* Dept Ind Sys Eng Ga Inst Tech 225 North Ave Atlanta GA 30332

SADOULET, BERNARD, b Nice, France, Apr 23, 44. ASTROPHYSICS. *Educ:* Serie A & Math Elementaires, Lyon, Baccalaureat, 60-61, Univ Paris, License, 65; Ancien Eleve l'Ecole Polytech, Paris, dipl, 63-65; Univ Orsay, France, dipl, 65-66, PhD(phys sci), 71. *Prof Exp:* Fel, Lawrence Berkeley Lab, Univ Calif, Berkeley, 73-76, physicist, 76-81, sr physicist, 81-84; fel & staff, Europ Orgn Nuclear Res, 66-73; sabbatical, 84-85, PROF PHYSICS, UNIV CALIF, BERKELEY, 85-, DIR, CTR PARTICLE ASTROPHYS, NSF SCI & TECHNOL CTR, 88- *Res:* Search for dark matter particles with ionization and cryogenic detectors and development of gas scintillation drift chambers for x-ray astrophysics. *Mailing Add:* Ctr Particle Astrophys Univ Calif Berkeley 301 LeConte Hall Berkeley CA 94720

SADOWAY, DONALD ROBERT, b Toronto, Can, Mar 7, 50; m 73; c 3. MOLTEN SALT CHEMISTRY. *Educ:* Univ Toronto, BASc, 72, MASc, 73, PhD(chem metall), 77. *Prof Exp:* NATO fel, 77-78, asst prof, 78-82, ASSOC PROF MAT ENG, MASS INST TECHNOL, 82- *Mem:* Minerals, Metals, & Mat Soc; Electrochem Soc; Can Inst Mining & Metall; Int Soc Electrochem; AAAS. *Res:* Electrochemical processing of materials in molten salts and in cryogenic media. *Mailing Add:* 77 Massachusetts Ave Rm 8-109 Cambridge MA 02139

SADOWSKI, CHESTER M, b Toronto, Ont, Aug 10, 36; m 79; c 2. PHYSICAL CHEMISTRY. *Educ:* Univ Toronto, BA, 57, PhD(chem kinetics), 61. *Prof Exp:* Res asst high temperature chem kinetics, Cornell Univ, 61-62; defence serv sci officer, Can Armament Res & Develop Estab, 62-67; asst prof natural sci, York Univ, 67-70, chair, natural sci, 77-80 & sci studies, 86-89, ASSOC PROF CHEM, YORK UNIV, 70- *Concurrent Pos:* Guest worker, Nat Oceanic & Atmospheric Admin, 73-74; sabbaticant, Univ Cambridge, UK, 81-82. *Mem:* Fel Chem Inst Can; Inter-Am Photochem Soc. *Res:* Laser assisted kinetic studies of free radical reactions; laser photochemistry of small molecules; environmental analysis of polynuclear aromatic hydrocarbons. *Mailing Add:* Dept Chem York Univ 4700 Keele St North York ON M3J 1P3 Can

SADOWSKI, IVAN J, b Sask, Can, Oct 31, 60; m 83; c 1. TRANSCRIPTION, SIGNAL TRANSDUCTION. *Educ:* Univ Sask, BSc, 82; Univ Man, MSc, 84; Univ Toronto, PhD(med biophys), 87. *Prof Exp:* Postdoctoral fel molecular biol, Harvard Univ, 87-90; ASST PROF BIOCHEM, DEPT BIOCHEM, UNIV BC, 90- *Mem:* Am Soc Microbiol; Am Soc Genetics; AAAS. *Res:* Regulation to transcription by signal transduction pathways. *Mailing Add:* Dept Biochem Univ BC Vancouver BC V6T 1W5 Can

SADOWSKY, JOHN, b Worcester, Mass, Aug 27, 49; m 72; c 3. SIGNAL & IMAGE PROCESSING, DESIGN & ANALYSIS OF ALGORITHMS. *Educ:* Johns Hopkins Univ, AB, 71; Univ Md, College Park, MA, 73, PhD(math), 80. *Prof Exp:* Mathematician, Soc Sec Admin, 73-77 & US Bur Census, 77-79; software engr, Hadron, Inc, 79-81; prin scientist & sect mgr, Syst Eng & Develop Corp, 81-89; SR MATHEMATICIAN & ASST GROUP SUPVR, M S EISENHOWER RES CTR, APPL PHYSICS LAB, JOHNS HOPKINS UNIV, 89- *Concurrent Pos:* Adj grad fac computer sci, Whiting Sch Eng, Johns Hopkins Univ, 81-; mem, Exec Comt, Capitol Area Chap, Asn Comput Mach, 90- *Mem:* Am Math Soc; Inst Elec & Electronics Engrs; Sigma Xi; Asn Comput Mach Comput Soc. *Res:* Signal and image processing; diophantine approximation; applied number theory; computational complexity and algorithm analysis; signal analysis, representations, and decomposition; computational architectures. *Mailing Add:* 9006 Scotch Pine Ct Columbia MD 21045

SADTLER, PHILIP, b Flourtown, Pa, July 19, 09; m 40, 64; c 3. ANALYTICAL CHEMISTRY. *Educ:* Lehigh Univ, BS, 34. *Prof Exp:* Pres, Sadtler Res Labs Inc, 34-69, Sandia, Inc, 70-90; PRES, SANDA CORP, 70- *Concurrent Pos:* Consult fluorine damage, infra-red & thermometric titration. *Mem:* Franklin Inst; Am Chem Soc; Am Soc Testing & Mat; Soc Appl Spectros; Am Mgt Asn. *Res:* Molecular studies with infrared spectrophotometry, ultraviolet, nuclear magnetic resonance; analyses, diesel engines; polymers; air pollution; explosion; synthesis of new materials; computerization of analytical instruments; digitization of spectral data. *Mailing Add:* 3555 W School House Lane Philadelphia PA 19144

SADUN, ALBERTO CARLO, b Atlanta, Ga, Apr 28, 55. EXTRAGALACTIC JETS, FLUX VARIABILITY FROM ACTIVE GALACTIC NUCLEI. *Educ:* Mass Inst Technol, BS, 77, PhD(physics), 84. *Prof Exp:* ASSOC PROF ASTRON, AGNES SCOTT COL, 84-; DIR ASTROPHYS BRADLEY OBSERV, 84- *Concurrent Pos:* Adj prof, Ga State Univ, 86-; res affil, ANSA-Caltech Jet Propulsion Lab. *Mem:* Am Astron Soc; Astron Soc Pac; fel Royal Astron Soc; Int Astron Union; Sigma Xi; NY Acad Sci. *Res:* High-energy extragalactic astrophysics, particularly observations and some theoretical modeling; dynamics and morphology of optical and radio jets; flux variations of active galactic nuclei. *Mailing Add:* 206 E Davis St Decatur GA 30030

SADURSKI, EDWARD ALAN, b Detroit, Mich, June 12, 49; m 71; c 2. INORGANIC CHEMISTRY, ORGANOMETALLIC CHEMISTRY. *Educ:* Oakland Univ, BS, 71; Wayne State Univ, PhD(inorg chem), 78. *Prof Exp:* Instr chem, Mich Christian Col, 70-72; vis asst prof inorg chem, Miami Univ, 78-80; ASSOC PROF CHEM, OHIO NORTHERN UNIV, 80- *Mem:* Am Chem Soc; Sigma Xi. *Res:* Main group organometallics; nuclear magnetic resonance spectroscopy and single crystal x-ray crystallography. *Mailing Add:* Dept Chem Ohio Northern Univ Ada OH 45810

SAE, ANDY S W, b Hong Kong, Jan 5, 41; c 1. BIOCHEMISTRY. *Educ:* Kans State Univ, BS, 64, MS, 66, PhD(biochem), 69. *Prof Exp:* Assoc prof, 69-80, chmn dept sci, 77-87, PROF CHEM, EASTERN NMEX UNIV, 80-, ACTG GRAD ASST DEAN, 88- *Concurrent Pos:* Tour speaker, Am Chem Soc. *Mem:* Am Chem Soc. *Res:* Chemical education; immobilized enzymes; enzyme isolation purification; peroxidases. *Mailing Add:* Dept Phys Sci Eastern NMex Univ PO Box 2266 Portales NM 88130

SAEGEBARTH, KLAUS ARTHUR, b Berlin, Ger, Jan 5, 29; nat US; m 53; c 3. ORGANIC CHEMISTRY, AGRICULTURAL CHEMISTRY. *Educ:* Univ Calif, BS, 53; Univ Wash, PhD(chem), 57. *Prof Exp:* Chemist, Radiation Lab, Univ Calif, 50; res chemist, Elastomer Chem Dept, 57-65, res & develop supvr, Co, 65-67, res div head, 67-69, lab dir, Fabrics & Finishes Dept, Exp Sta, 69-70, res & develop mgr, Marshall Lab, 71-72, asst nat mgr indust finishes, E I Du Pont de Nemours & Co, Inc, 72-73, nat mgr trade finishes, 73-74, asst dir, Finishes Div, 74-78, dir res & develop, Fabrics & Finishes Dept, 78-80, dir, Agrichem Res & Develop Div, 80-88, dir, 88-90, VPRES FIBERS RES & DEVELOP, E I DU PONT DE NEMOURS & CO INC, 90- *Concurrent Pos:* Trustee, Textile Res Inst, Princeton. *Mem:* Am Chem Soc; Sigma Xi; Soc Chem Indust. *Res:* Oxidation mechanisms; organometallics; catalysis; elastomers; finishes; agricultural chemicals. *Mailing Add:* 604 Haverhill Rd Sharpley Wilmington DE 19803

SAEGER, VICTOR WILLIAM, b Kansas City, Mo, May 17, 33; m 62; c 2. ENVIRONMENTAL CHEMISTRY, ANALYTICAL CHEMISTRY. *Educ:* Univ Mo-Kansas City, BA, 53; Iowa State Univ, PhD(phys chem), 60. *Prof Exp:* RES SPECIALIST, MONSANTO CO, ST LOUIS, 60- *Mem:* Am Chem Soc; Am Soc Microbiol; Sigma Xi. *Res:* Environmental fate testing and assessment. *Mailing Add:* 870 N Dickson Kirkwood MO 63122

SAEKS, RICHARD E, b Chicago, Ill, Nov 30, 41. ELECTRICAL ENGINEERING, APPLIED MATHEMATICS. *Educ:* Northwestern Univ, Evanston, BS, 64; Colo State Univ, MS, 65; Cornell Univ, PhD(elec eng), 67. *Prof Exp:* From asst prof to assoc prof elec eng, Univ Notre Dame, 67-73; from assoc prof to prof elec eng & math, Tex Tech Univ, 73-79, Paul Whitfield prof eng, math & comput sci, 79-; AT DEPT ELEC ENG & COMPUT SCI, ARIZ STATE UNIV. *Concurrent Pos:* NASA fel, Marshall Space Flight Ctr, 69-70; consult, Res Triangle Inst, 78-; ed-at-large, Marcell Dekker Inc, 78-; distinguished fac res award, Tex Tech Univ, 78- *Mem:* Am Soc Eng Educ; fel Inst Elec & Electronics Engrs; Soc Indust & Appl Math; Am Math Soc. *Res:* Fault analysis; large-scale systems; mathematical system theory. *Mailing Add:* Ill Inst Technol Armour Col Eng Chicago IL 60616

SAELENS, DAVID ARTHUR, b Camden, NJ, June 10, 43; m 76. AUTONOMIC NERVOUS SYSTEM, DRUG METABOLISM. *Educ:* Albany Col Pharm, BS, 69; Med Univ SC, PhD(pharm), 74. *Prof Exp:* Fel pharm, Med Univ SC, 74-76, instr, 76-77; asst prof pharm, Univ Houston, 77-78; asst prof pharm, Eastern Va Med Sch, 78-; AT COL PHARM, DRAKE UNIV. *Concurrent Pos:* Adj asst prof, Old Dominion Univ, 78- *Mem:* Sigma Xi. *Res:* Factors important in the local control of neurotransmitter release, specifically the role of presynaptic receptors and calcium metabolism, in normal and diseased states and the impact of drug therapy. *Mailing Add:* 111 S Dawes Ave Kingston PA 18704

SAEMAN, W(ALTER) C(ARL), b Norlina, NC, Apr 15, 14; m 40; c 1. CHEMICAL ENGINEERING, GLASS MAKING. *Educ:* Ga Inst Technol, BS, 40. *Prof Exp:* From jr to assoc chem engr, Chem Eng Div, Wilson Dam, Tenn Valley Authority, 40-46, proj leader, 47-52; sr chem engr, Oak Ridge Nat Lab, 46-47; engr, Olin Indust, Inc, 52-58, res proj mgr, Olin Corp, New Haven, 58-70, asst res dir, 70-77, prin engr, 77-81; RETIRED. *Mem:* Am Chem Soc; Am Inst Chem Eng; NY Acad Sci. *Res:* Drying; crystallization; ammonium nitrate; process control; nuclear energy development design of experiments; mathematics; natural philosophy. *Mailing Add:* Carolina Meadows Villa 123 Whippoorwill Lane Chapel Hill NC 27514

SAEMANN, JESSE C(HARLES), JR, b Adell, Wis, Sept 16, 21; m 58; c 2. ENGINEERING MECHANICS. *Educ:* Univ Wis, BS, 43, MS, 50, PhD(eng mech), 55. *Prof Exp:* From instr to prof, 46-83, EMER PROF ENG MECH, UNIV WIS-MADISON, 83- *Mem:* Sigma Xi. *Res:* Shrinkage cracking of concrete block; variation of concrete masonry; soil mechanics. *Mailing Add:* 4421 Waite Lane Madison WI 53711

SAENGER, EUGENE L, b Cincinnati, Ohio, Mar 5, 17; m 41; c 2. RADIOLOGY, NUCLEAR MEDICINE. *Educ:* Harvard Univ, AB, 38; Univ Cincinnati, MD, 42; Am Bd Radiol, dipl, 46, cert nuclear med, 72. *Prof Exp:* From asst to assoc prof, Univ Cincinnati, 43-62, prof radiol, 62-87, dir radioisotope lab, 50-87, EMER PROF RADIOL & EMER DIR,

RADIOISOTOPE LAB, COL MED, UNIV CINCINNATI, 87- *Concurrent Pos:* Radiation therapist, Children's Hosp, Cincinnati, 46-47; pvt pract, 46-62; mem adv comt, Biol Effects Ionizing Radiation, 64-72; mem subcomt sealed gamma sources, Nat Comt Radiation Protection; Am Roentgen Ray Soc rep, Nat Coun Radiation Protection & Measurements, mem bd dirs, chmn sci comt brachyther, 68-; Aubrey Hampton lectr, Mass Gen Hosp, 71; prog dir radiol sci, Nat Inst Gen Med Sci; consult to dir, Bur Radiol Health & to Surgeon Gen, US Air Force. *Honors & Awards:* Twenty Eighth George Charles de Hevesy Nuclear Pioneer Award, 87. *Mem:* Am Roentgen Ray Soc; fel Am Col Radiol; Health Physics Soc; Soc Nuclear Med; Am Radium Soc; Sigma Xi. *Res:* Radiobiology; cancer; radiological sciences; public health. *Mailing Add:* 9160 Given Rd Cincinnati OH 45243

SAENZ, ALBERT WILLIAM, b Medellíin, Colombia, Aug 27, 23; nat US; m 57. THEORETICAL PHYSICS. *Educ:* Univ Mich, BS, 44, MA, 45, PhD(physics), 49. *Prof Exp:* Physicist, Naval Res Lab, 50-51; physicist, Inst Appl Math, Ind Univ, 51-52; physicist, Naval Res Lab, 52-64, head anal & theory br, Radiation Div, 64-66, head theory br, Nuclear Sci Div, 66-74, head theory br, Nuclear Tech Div, 74-76, HEAD THEORY CONSULT STAFF, CONDENSED MATTER & RADIATION SCI DIV, NAVAL RES LAB, 76-; RES PROF PHYSICS DEPT, CATH UNIV, WASHINGTON, DC, 81- *Concurrent Pos:* Vis scientist, Mass Inst Technol, 57 & Oak Ridge Nat Lab, 58; vis fel, Johns Hopkins Univ, 64 & Princeton Univ, 76-77; lectr, Univ Md grad prog, Nat Res Lab, 50-51, 54-55 & 63-64; lectr, Cath Univ Am, 55-56 & Univ Md, 62. *Honors & Awards:* Pure Sci Award,Naval Res Lab, 69. *Mem:* Fel Am Phys Soc; NY Acad Sci; Sigma Xi. *Res:* Special and general relativity; symmetry and degeneracy in quantum mechanics; equilibrium and nonequilibrium statistical mechanics; spin-wave theory and inelastic magnetic scattering of neutrons; quantum scattering theory. *Mailing Add:* Code 6603S Nuclear Tech Div Naval Res Lab Washington DC 20375

SAENZ, REYNALDO V, b San Juan, Tex, Sept 29, 40; m 66. PHARMACEUTICAL CHEMISTRY. *Educ:* Univ Tex, BS, 62, MS, 65, PhD(pharmaceut chem), 67. *Prof Exp:* Assoc prof pharmaceut chem, Sch Pharm, Northeast La Univ, 66-89; ASST DEAN, UNIV NMEX, 89- *Mem:* Am Pharmaceut Asn; Am Chem Soc; Am Asn Col Pharm. *Res:* Synthesis of agents acting on the central nervous system or peripheral nervous system based on known compounds already established as active. *Mailing Add:* Col Pharm Univ NMex Albuquerque NM 87131

SAETHER, OLE ANTON, b Kristiansand, Norway, Dec 9, 36; m 66; c 3. LIMNOLOGY, ENTOMOLOGY. *Educ:* Univ Oslo, Cand Mag, 60, Cand Real, 63. *Prof Exp:* Sci asst limnol, Univ Oslo, 61-63, lectr hydrobiol, 63-69, Nansen Fund fel, 62-69; res scientist, Freshwater Inst, Fisheries Res Bd Can, 69-77; PROF SYST ZOOL & HEAD DEPT, UNIV BERGEN, 77- *Concurrent Pos:* Vis scientist, Fisheries Res Bd Can, 67-68; consult, Indust Bio-Test Lab, Inc, 71-; adj prof entom, Univ Manitoba, 73-77; adj res scientist, Freshwater Inst, Dept Environ, Man, Can, 78-; ed, Fauna Morveg Ser B, 79-85. *Mem:* Int Asn Theoret & Appl Limnol; Am Soc Limnol & Oceanog; Entom Soc Can; Norweg Entom Soc; NAm Benthol Soc. *Res:* Taxonomy, morphology and ecology of Chironomidae, Chaoboridae and Hydracarina; benthic fauna; zooplankton; phytoplankton; systematic zoology. *Mailing Add:* Dept Syst Zool Univ Bergen North 5007 Norway

SAETTLER, ALFRED WILLIAM, plant pathology; deceased, see previous edition for last biography

SAEVA, FRANKLIN DONALD, b Rochester, NY, Nov 28, 38; m 63; c 3. CHEMISTRY. *Educ:* Bucknell Univ, BS, 60; State Univ NY Buffalo, PhD(org chem), 68. *Prof Exp:* Sr scientist, Xerox Corp, 68-79; SR SCIENTIST, EASTMAN KODAK CO, 79- *Mem:* Am Chem Soc. *Res:* Organic reaction mechanisms and photochemistry. circular dichroism studies; liquid crystals; organic synthesis; electrochemical behavior of organics. *Mailing Add:* 1219 Gerrads Cross Webster NY 14580

SAEZ, JUAN CARLOS, b Osorno, Chile, Feb 2, 56; m 78; c 2. CELL BIOLOGY, PHOSPHORYLATION. *Educ:* Univ Concepcion, Chile, MS, 79; Albert Einstein Col Med, MS, 85, PhD(neurosci), 86. *Prof Exp:* Instr physiol, Univ Concepcion, Chile, 80-83; postdoctoral neurosci, 86-87, instr, 87-89, ASST PROF NEUROSCI, DEPT NEUROSCI, ALBERT EINSTEIN COL MED, 89-, INVESTR, LIVER CTR, 90- *Concurrent Pos:* Prin investr, NIH, 90- *Mem:* Am Soc Cell Biol; Biophys Soc. *Res:* Regulation and function of gop junction in excitable and non-excitable cells. *Mailing Add:* Dept Neurosci Albert Einstein Col Med 1300 Morris Park Ave Bronx NY 10461

SAFAI, BIJAN, DERMATOLOGY. *Prof Exp:* MEM STAFF, MEM SLOAN KETTERING CANCER CTR. *Mailing Add:* Mem Sloan Kettering Cancer Ctr 1275 York Ave New York NY 10021

SAFANIE, ALVIN H, b Washington, DC, Dec 27, 24; m 50; c 4. ANATOMY, HISTOLOGY. *Educ:* Cornell Univ, DVM, 47; Mich State Univ, MS, 50; Univ Ill, PhD(anat), 62. *Prof Exp:* Instr anat, Col Vet Med, Mich State Univ, 47-51, asst prof, 51-52; from instr to prof anat, 52-77, PROF VET ANAT & HISTOL, COL VET MED, UNIV ILL, URBANA, 77- *Mem:* Am Vet Med Asn; Am Asn Vet Anatomists; Am Soc Zoologists. *Res:* Vagus nerve of pig; embryology of domestic animals. *Mailing Add:* 361 Vet Med Univ Ill Champaign IL 61820

SAFAR, PETER, b Vienna, Austria, Apr 12, 24; nat US; m 50; c 2. ANESTHESIOLOGY, CRITICAL CARE MEDICINE. *Educ:* Univ Vienna, MD, 48; Am Bd Anesthesiol, dipl. *Hon Degrees:* Dr, Univ Mainz, 72. *Prof Exp:* Resident path & surg, Univ Vienna, 48-49; fel surg, Yale Univ, 49-50; resident anesthesiol, Univ Pa, 50-52, chief dept, Nat Cancer Inst, Lima, Peru, 52-53; asst prof, Sch Med, Johns Hopkins Univ, 54-61; prof anethesiol/critical care med & chmn dept, Sch Med, 61-78, DISTINGUISHED PROF RESUSCITATION MED & FOUND DIR, RESUSCITATION RES CTR, UNIV PITTSBURGH, 78- *Concurrent Pos:* Clin assoc prof, Sch Med, Univ Md, 55-61; chief, Baltimore City Hosps, Md, 55-61; res contractor, US Army Res & Develop Div, Off Surgeon Gen, 57-69; vis scientist, Cardiovasc Res Inst, Univ Calif, San Francisco, 69-70; mem comts resuscitation, emergency & critical care, Nat Res Coun, Am Heart Asn & other nat orgns; mem, Interagency White House Comt Emergency Med Serv; Wattie prof, NZ, 81; F S Cheeves prof, Pittsburgh. *Mem:* Am Physiol Soc; Am Soc Anesthesiol; Soc Critical Care Med (pres, 73); Sigma Xi. *Res:* Resuscitation; neurosciences; author or coauthor of over 500 publications. *Mailing Add:* Resuscitation Res Ctr Univ Health Ctr 3434 Fifth Ave 2nd Floor Pittsburgh PA 15260

SAFDARI, YAHYA BHAI, b Amravati, India, July 25, 30; US citizen; m 62; c 4. ENGINEERING. *Educ:* Aligarh Muslim Univ, India, BSME, 52, BSEE, 53; Univ Wash, MSME, 59; NMex State Univ, DSc(mech eng), 64. *Prof Exp:* Asst engr, Delhi Cloth Mills, India, 54-56; res asst mech engr, Univ Wash, Seattle, 57-58; design engr, Refrig Co, Wash, 58-61; instr mech eng, NMex State Univ, 61-64; assoc prof, 64-69, PROF, DEPT MECH ENG, BRADLEY UNIV, 69-; PRES, SAF ENERGY CONSULTS, INC, 79- *Concurrent Pos:* Pres, Sun Systs Inc, 76-79; vis prof & planner, dept mech eng, Makkah Univ, Saudi Arabia, 83-85. *Mem:* Am Soc Heating, Refrig & Air-Conditioning Engrs; Am Soc Mech Engrs; Int Solar Energy Soc; Int Solar Energy Soc; Solar Energy Indust Asn; Am Solar Energy Soc; Sigma Xi. *Res:* Heat transfer in transparent media; convection heat transfer in nuclear reactors; solar energy; energy management; photovollair remote application; solar grain drying. *Mailing Add:* Dept Mech Eng Bradley Univ Peoria IL 61606

SAFDY, MAX ERROL, b Brooklyn, NY, Nov 9, 41; m 68; c 1. ORGANIC CHEMISTRY, MEDICINAL CHEMISTRY. *Educ:* Polytech Inst Brooklyn, BS, 63; Univ NC, PhD(org chem), 69. *Prof Exp:* Res scientist, Corp Res Div, Miles Labs, Inc, 70-76, sr res scientist, Ames Div, 77-83, supvr, diag res & develop, 84-87, mfg develop diag, 87-88, MGR, PILOT PLANT OPERS, DIAG DIV, MILES INC, 88- *Mem:* Am Chem Soc; AAAS; Sigma Xi. *Res:* Medical diagnostics, cardiovascular agents, biogenic amines, amino acids and peptides; microencapsulation; medical instrumentation. *Mailing Add:* Miles Inc 3400 Middlebury St Elkhart IN 46516

SAFE, STEPHEN HARVEY, b Belleville, Ont, May 14, 40; m 62; c 2. CHEMISTRY, TOXICOLOGY. *Educ:* Queen's Univ, Ont, BSc, 62, MSc, 63; Oxford Univ, DPhil(chem), 65. *Prof Exp:* Sci Res Coun res asst chem, Oxford Univ, 66-67; NIH res assoc biochem, Harvard Univ, 67-68; assoc res officer microbiol-chem, Nat Res Coun Can, 68-73; prof biochem, Univ Guelph, 73-81; PROF, TEX A&M UNIV, 81-, DISTINGUISHED PROF, 85- *Honors & Awards:* Safety Health Environ Chem Award, Royal Soc Chem, Distinguished Achievement Award Res, 87-88. *Mem:* FASEB; Am Chem Soc; Soc Toxicol; Am Asn Cancer Res; Chem Indust Coun; Chem Soc. *Res:* Biochemistry and toxicology of pollutants. *Mailing Add:* 1207 Charles Ct College Station TX 77840

SAFER, BRIAN, b Brooklyn, NY, Dec 3, 42; m 69. MOLECULAR BIOLOGY, PHARMACOLOGY. *Educ:* Columbia Univ, BA, 64; Baylor Col Med, MS, 67, MD, 69; Univ Pa, PhD(molecular biol), 72. *Prof Exp:* USPHS fel, Univ Pa, 69-71, Pa Plan fel, 71-73; sr staff fel, Nat Heart & Lung Inst, 73-79, CHIEF SECT PROTEIN BIOSYNTHESIS, NAT HEART, LUNG, & BLOOD INST, NIH, 79- *Honors & Awards:* Louis N Katz Prize, Am Heart Asn, 73. *Mem:* Am Chem Soc; Sigma Xi; Am Heart Asn; Am Soc Biol Chemists. *Res:* Regulation of protein synthesis; initiation factor characterization; eucaryotic transcription factors. *Mailing Add:* Nat Heart Lung & Blood Inst NIH Bldg 10 Rm 7D03 Bethesda MD 20892

SAFERSTEIN, LOWELL G, b Newark, NJ, July 25, 40; m 76; c 1. POLYMER CHEMISTRY, ORGANIC CHEMISTRY. *Educ:* Rutgers Univ, BS, 62, MS, 65, PhD(org chem), 67. *Prof Exp:* Res chemist, Celanese Res Co, 67-76; SR RES CHEMIST, ETHICON INC, 76- *Mem:* Am Chem Soc; Am Inst Chemists; Sigma Xi. *Res:* Synthesis of high performance polymers; synthesis of biopolymers. *Mailing Add:* Three Timber Rd Edison NJ 08820

SAFERSTEIN, RICHARD, b Brooklyn, NY, July 17, 41; m 75; c 2. FORENSIC SCIENCE. *Educ:* City Col New York, BS, 63, MA, 66, PhD(org chem), 70. *Prof Exp:* Chemist, US Treasury Dept, NY, 64-69; chemist, Shell Chem Co, 69-70; CHIEF CHEMIST, FORENSIC SCI BUR, NJ STATE POLICE, 70- *Concurrent Pos:* Instr, Trenton State Col & Ocean County Col, 72- *Mem:* Am Chem Soc; Am Acad Forensic Sci; Forensic Sci Soc; Int Asn Forensic Sci; Can Soc Forensic Scientists. *Res:* Application of chemical ionization mass spectroscopy to forensic science; forensic characterization of polymers by pyrolysis-gas chromatography. *Mailing Add:* NJ State Police Lab PO Box 7068 West Trenton NJ 08625

SAFF, EDWARD BARRY, b New York, NY, Jan 2, 44; m 66; c 3. MATHEMATICAL ANALYSIS. *Educ:* Ga Inst Technol, BS, 64; Univ Md, College Park, PhD(math), 68. *Prof Exp:* Fulbright grant, Imp Col, Univ London, 68-69; dir, Ctr Math Serv, 78-83, from assoc prof to prof, 69-86, UNIV DISTINGUISHED PROF MATH, UNIV SFLA, 86-, GRAD RES PROF, 86- *Concurrent Pos:* NSF grant, Univ South Fla, 69-72 & 80-; Air Force res grant, 73-79; Guggenheim fel, Oxford Univ, 78; assoc dir, Ctr for Excellence in Math, Sci Comput & Technol, 83-; ed-in-chief, Construct Approximation J, 83-; dir, Inst Constructive Math, 85-; Fulbright fel, 82; hon prof math, Zhejiang Normal Univ, China. *Honors & Awards:* Sigma Xi Outstanding Researcher Award, Univ South Fla, 84. *Mem:* Am Math Soc; Math Asn Am; Sigma Xi. *Res:* Approximation in complex domain; approximate solutions of differential equations; Pade approximants; geometry of polynomials. *Mailing Add:* 11738 Lipsey Rd Tampa FL 33618-3620

SAFFER, ALFRED, b New York, NY, Dec 3, 18; m 42, 85; c 2. PHYSICAL CHEMISTRY. *Educ:* NY Univ, AB, 39, MS, 41, PhD(phys chem), 43. *Prof Exp:* Asst chem, NY Univ, 40-43; res assoc, Princeton Univ, 43-45; sr res chemist, Firestone Tire & Rubber Co, 45-48; sr vpres mfg, Halcon Int Inc, NY, 48-70; pres, Oxirane Int, NJ, 71-78; vchmn, Halcon Int Inc, 78-81;

RETIRED. *Concurrent Pos:* Pres, Catalyst Develop Corp, NJ, 57-70. *Honors & Awards:* Chem Pioneer Award, Am Inst Chemists, 82. *Mem:* Nat Acad Eng; Am Chem Soc; Am Inst Chem Eng; Soc Chem Indust; Am Inst Chem; Sigma Xi. *Res:* Oxidation of hydrocarbons in gas and liquid phases; processes for manufacture of petrochemicals; manufacture of catalysis; mechanisms of organic reactions. *Mailing Add:* 16629 Ironwood Dr Delray Beach FL 33445

SAFFER, CHARLES MARTIN, JR, b Salem, Mass, Dec 15, 14; m 42; c 1. INDUSTRIAL CHEMISTRY. *Educ:* Mass Inst Technol, SB, 36, SM, 37, PhD(chem), 38. *Hon Degrees:* MA, Christ Church, Oxford Univ, 73. *Prof Exp:* Moore traveling fel, Oxford Univ, 39-40; res asst, Harvard Univ, 40-41; res chemist, Aerojet Eng Corp, Calif, 46-48; scientist, Bur Aeronaut, US Dept Navy, 48-51; vpres, Microcard Corp, 51-54; dir res, Nat Fireworks Ord Corp, 53-57; asst mgr nat northern div, Am Potash & Chem Corp, 57-58, head propellant chem res, 58-60; asst dir res planning, Thiokol Chem Corp, 60-65, tech consult, 65-66; tech dir, Sonneborn Div, Witco Chem Corp, 66-70, Activated Carbon Div, 70-75, Inorganic Specialties Div, 75-82; RETIRED. *Concurrent Pos:* Officer, US Navy, 41-46. *Mem:* Am Chem Soc. *Res:* Organo-silicon synthesis; synthetic estrogens; sodium triphenylmethyl; smokeless propellants; explosives. *Mailing Add:* Two Rockwood Rd Levittown PA 19056

SAFFER, HENRY WALKER, b New York, NY, Apr 4, 35; m 63; c 3. PHYSICAL CHEMISTRY, TEXTILE CHEMISTRY. *Educ:* NC State Col, BS, 56, MS, 58; Princeton Univ, MA, 60, PhD(phys chem), 63. *Prof Exp:* Res chemist, 62-66, sr res chemist, 66-72, res supvr, 72-75, mkt develop supvr, 75-76, strategist, Textile Fibers Dept, 76-77, prog mgr, 78-79, 80-82, div mgr, bus serv mgr, 83-85, SITE MGR, E I DU PONT DE NEMOURS & CO, INC, 85-, ADMIN MGR, CENT RES & DEVELOP DEPT. *Res:* Protein chemistry; nuclear magnetic resonance spectroscopy of fibers; textile warp sizing; engineering fibers for special end uses; nonwoven products and processes; business diversification and development; laboratory administration. *Mailing Add:* DuPont Experimental Sta E I du Pont de Nemours & Co Inc PO Box 80328 Wilmington DE 19880-0328

SAFFERMAN, ROBERT S, b Bronx, NY, Dec 19, 32; m 58; c 3. MICROBIOLOGY. *Educ:* Brooklyn Col, BS, 55; Rutgers Univ, PhD(microbiol), 60. *Prof Exp:* Res fel microbiol, Inst Microbiol, Rutgers Univ, 55-59; microbiologist, USPHS, 59-64 & US Dept Interior, 64-70; microbiologist, 70-74, chief virol sect, 74-88, CHIEF VIROL BR, ENVIRON MONITORING SYSTS LAB, US ENVIRON PROTECTION AGENCY, 88- *Honors & Awards:* Gans Medal, Soc Water Treatment & Examination, UK, 70. *Mem:* AAAS; Am Soc Microbiologists; Phycol Soc Am; Sigma Xi; fel Am Acad Microbiol. *Res:* Survival and persistence of viruses in water sources; phycoviruses; virus monitoring methodology; medical aspects of phycology. *Mailing Add:* US Environ Protection Agency Nat Environ Res Ctr Cincinnati OH 45268

SAFFIOTTI, UMBERTO, b Milan, Italy, Jan 22, 28; US citizen; m 58; c 2. ONCOLOGY, CYTOLOGY. *Educ:* Univ Milan, MD, 51, dipl occup med, 57. *Prof Exp:* Fel, Inst Path Anat, Univ Milan, 51-52; res asst oncol, Chicago Med Sch, 52-54, res assoc, 54-55; chief pathologist, Inst Occup Med & asst occup med, Univ Milan, 56-60; fel, Inst Gen Path, 57-60; from asst prof to prof oncol, Chicago Med Sch, 60-68; assoc sci dir carcinogenesis, Etiology Area, Nat Cancer Inst, 68-72, assoc dir carcinogenesis, Div Cancer Cause & Prev, 72-76, chief, Exp Path Br, 74-78, CHIEF LAB EXP PATH, DIV CANCER ETIOLOGY, NAT CANCER INST, 78-, ACTG HEAD, REGISTRY EXP CANCERS, 88- *Concurrent Pos:* Mem comt young scientists in cancer res, Int Union Against Cancer, 56-58, mem comt cancer prev, 59-66 & mem panel on carcinogenicity, 63-66; NIH career develop award, 64-68; mem path B study sect, NIH, 64-68; partic panel carcinogenesis, Secy's Comn Pesticides & Environ Health, Dept Health, Educ & Welfare, 69, chmn ad hoc comt eval low levels environ carcinogenesis, Surg Gen, 69-70 & mem comt coord toxicol & related prog, 73-76; mem working groups, Eval Carcinogenic Risk of Chem to Man, Int Agency Res Cancer, 70-88; mem adv comt to scholars-in-residence prog, Fogarty Int Ctr, NIH, 73-77; mem bd dirs, Rachel Carson Trust for Living Environ, Inc, 76-79; chmn carcinogenesis contract prog mgt group, Div Cancer Cause & Prev, Nat Cancer Inst, 68-76, mem occup cancer task force & chmn comt carcinogenesis, 78-80; mem, Cancer Prev Task Force, 79-; partic, Work Group on Assessment, Interagency Regulatory Liaison Group, 79-80 & Work Group on Regulation of Carcinogens, US Regulatory Coun, 79-80. *Honors & Awards:* Superior Serv Honor Award, HEW, 71; Pub Interest Sci Award, Environ Defense Fund, 77; Pub Health Serv Spec Recognition Award, HEW, 80. *Mem:* AAAS; Am Asn Cancer Res; Am Asn Pathologists; fel NY Acad Sci; Sigma Xi; Int Comn Occup Health; Soc Occup & Environ Health; Soc Toxicol. *Res:* Experimental pathology of chemical carcinogenesis, especially respiratory; cell culture carcinogenesis models; combined effects of carcinogens; identification and evaluation criteria for carcinogens; occupational and environmental carcinogenesis; pneumoconioses. *Mailing Add:* 5114 Wissioming Rd Bethesda MD 20816

SAFFIR, ARTHUR JOEL, b Chicago, Ill, May 11, 41; m 63; c 2. DENTAL SCIENCE, NUTRITION. *Educ:* Mass Inst Technol, PhD(nutrit), 70; Tufts Univ, DMD, 64. *Prof Exp:* Res assoc, Mass Inst Technol, 64-70; dir res & develop, Mat Analysis, 70; pvt consult, 70-73; dir oral health res, 73-81, PRES BIOSYST RES, COOPER LABS, 81- *Mem:* Int Asn Dent Res. *Res:* Dental research; statistical analysis; electron optics; nutrition clinical research. *Mailing Add:* 2057 Summit Dr SW Lake Oswego OR 97034

SAFFMAN, PHILIP GEOFFREY, b Leeds, Eng, Mar 19, 31; m 54; c 3. APPLIED MATHEMATICS, FLUID MECHANICS. *Educ:* Cambridge Univ, BA, 53, PhD(appl math), 56. *Prof Exp:* Lectr appl math, Cambridge Univ, 58-60; reader, King's Col, Univ London, 60-64; prof fluid mech, 64-69, PROF APPL MATH, CALIF INST TECHNOL, 69- *Concurrent Pos:* Res fel, Trinity Col, Cambridge Univ, 55-59; vis prof, Mass Inst Technol, 70-71. *Mem:* Fel Am Acad Arts & Sci; fel Royal Soc London. *Res:* Turbulence; viscous flow; wave interactions; vortex motion. *Mailing Add:* 399 Ninita Pkwy Pasadena CA 91106

SAFFO, MARY BETH, b Inglewood, Calif, Apr 8, 48; m 78; c 1. INVERTEBRATE ZOOLOGY, MARINE BIOLOGY. *Educ:* Univ Calif, Santa Cruz, BA, 69; Stanford Univ, PhD(biol), 77. *Prof Exp:* Miller res fel, dept bot, Univ Calif, Berkeley, 76-78; asst prof biol, Swarthmore Col, 78-85; ASSOC RES MARINE BIOLOGIST, INST MARINE SCI, UNIV CALIF, SANTA CRUZ, 85-, LECTR, 88- *Concurrent Pos:* Independent investr, Marine Biol Lab, 79-80; Am Assoc Univ Women, postdoctoral fel, Univ Calif Berkeley, 81-82 & Univ Wash, 83-84; prog officer, Div Invertebrate Zool, Am Soc Zoologists, 85-87, comt ensure equal opportunity, 88- *Mem:* Sigma Xi; Am Soc Zoologists; Soc Study Evolution; Mycol Soc Am; Soc Evolutionary Protistology; fel AAAS. *Res:* Biology symbiosis especially microbial-invertebrate symbioses; nitrogen excretion; physiological ecology,; tunicate biology; protist biology. *Mailing Add:* Inst Marine Sci Univ Calif 272 Appl Sci Santa Cruz CA 95064

SAFFORD, LAWRENCE OLIVER, b Bremen, Maine, Dec 27, 38; div; c 3. FORESTRY, ECOLOGY. *Educ:* Univ Maine, BS, 61, PhD(plant sci), 68; Yale Univ, MFor, 62. *Prof Exp:* Res forester, Maine, 62-68, 69-70, RES FORESTER, NORTHEASTERN FOREST EXP STA, US FOREST SERV, USDA, NH, 70- *Concurrent Pos:* WVa Pulp & Paper Co fel, Yale Univ, 68-69. *Mem:* Soc Am Foresters; AAAS; Am Soc Agron; Sigma Xi. *Res:* Soil-tree relationships; soil moisture and nutrient requirements of forest trees. *Mailing Add:* Forestry Sci Lab PO Box 640 Durham NH 03824-0640

SAFFORD, RICHARD WHILEY, b New York, NY, Sept 1, 24; m 83; c 2. SYSTEMS SCIENCE, OPERATIONS RESEARCH. *Educ:* Union Col, NY, BS, 45; Univ Mich, MS, 46; Mass Inst Technol, PhD(physics), 53. *Prof Exp:* Jr scientist, Brookhaven Nat Lab, 47-48; mem staff, Lab Nuclear Sci, Mass Inst Technol, 48-53; res engr & head adv bomber studies, Boeing Co, 53-56; from sr staff physicist to mgr airborne ltd war systs, Hughes Aircraft Co, 56-62; advan projs mgr solar physics, Space Systs Div & chief space & electronics planning, Repub Aviation Corp, 62-65; MEM TECH STAFF, MITRE CORP, C3 DIV, INTELLIGENCE & ELECTRONIC WARFARE SYSTS, BEDFORD, 65- *Concurrent Pos:* Consult & lectr, business admin. *Mem:* Sigma Xi. *Res:* Describing analyzing, evaluating and modelling military command, control and communications systems; computer modelling of air warfare, electronic combat and intelligence systems. *Mailing Add:* 12 Meriam St Lexington MA 02173

SAFFRAN, JUDITH, b Montreal, Que, Nov 5, 23; m 47; c 4. BIOCHEMISTRY. *Educ:* McGill Univ, BSc, 44, PhD(biochem), 48. *Prof Exp:* Melville Trust fel biochem, Univ Edinburgh, 58-59; biochemist, Jewish Gen Hosp, Montreal, 55-58; biochemist, Jewish Gen Hosp, Montreal, 61-69; sr res fel, Med Res Inst, Toledo Hosp, 69-74; asst prof, Med Col Ohio, 74-76, assoc prof obstet, gynec & biochem, 76-79, clin chemist, 79-85, TOXICOLOGIST, MED COL OHIO, 85- *Concurrent Pos:* Adj asst prof, Med Col Ohio, 70-74. *Mem:* Can Biochem Soc; Am Soc Biol Chemists; Am Asn Clin Chem. *Res:* Steroid hormones; mechanism of action; reproductive physiology. *Mailing Add:* 2331 Hempstead Rd Toledo OH 43606

SAFFRAN, MURRAY, b Montreal, Que, Oct 30, 24; US citizen; m 47; c 4. DRUG DELIVERY. *Educ:* McGill Univ, BSc, 45, MSc, 46, PhD(biochem), 49. *Prof Exp:* Lectr psychiat, McGill Univ, 48-52; Life Ins Med Res Fund fel, Copenhagen Univ, 52-53; asst prof psychiat, McGill Univ, 53-58; Founds Fund Res Psychiat fel, Univ Edinburgh, 58-59; assoc prof biochem & psychiat, McGill Univ, 59-65, bldg dir, McIntyre Med Sci Bldg, 63-65, prof biochem, 65-69; chmn dept, 69-80, PROF BIOCHEM, MED COL OHIO, 69- *Concurrent Pos:* Mem endocrinol study sect, NIH, 64-68, mem neurol A study sect, 77-81; mem biochem test comt, Nat Bd Med Examrs, 65-69, fel, 79; vis lectr, Ctr Pop Studies, Sch Pub Health, Harvard Univ, 67-68; mem, Int Brain Res Orgn; chmn biochem test comt, Nat Bd Podiatry, 78-; consult, pharmaceut indust; vis prof, Ben Gurion Univ, 81, Armenian Acad Sci, 88; bk rev ed, Trends in Endocrinol & Metab, 89- *Honors & Awards:* Ayerst Award, Endocrine Soc, 67. *Mem:* AAAS; Endocrine Soc; Am Soc Biol Chemists; Am Diabetes Asn. *Res:* Neuroendocrinology; peptide hormones; intestinal absorption of peptides; oral adimistration of peptide drugs. *Mailing Add:* Dept Biochem & Molecular Biol Med Col Ohio PO Box 10008 Toledo OH 43699-0008

SAFFREN, MELVIN MICHAEL, b Brooklyn, NY, Sept 13, 29; div; c 3. MATHEMATICAL PHYSICS, FLUID DYNAMICS. *Educ:* City Col New York, BS, 51; Mass Inst Technol, PhD(physics), 59. *Prof Exp:* Asst physics, Brookhaven Nat Lab, 52; asst physics, Mass Inst Technol, 52-59; physicist, Res Lab, Gen Elec Co, 59-62; mem tech staff, Jet Propulsion Lab, 62-67, supvr theoret physics group, 67-71, space sci div rep to off res & advan develop, 71-76, STAFF SCIENTIST, JET PROPULSION LAB, CALIF INST TECHNOL, 76- *Concurrent Pos:* Consult & mem vis fac, Univ Southern Calif, 66-68; mem physics subcomt, Bluebook Update Task, NASA, 70-71, mem physics & chem in space working group, 72-, proj scientist, Drop Dynamics Module Proj, 75-; chmn steering comt, Int Colloquium on Drops & Bubbles, 74-, co-ed proceedings, 76; exec secy, Int Symp Relativity Exp Space, 77. *Mem:* Am Phys Soc; Inst Elec & Electronics Engrs; Math Asn Am; Am Math Soc. *Res:* Theory and computation of energy bands in solids; superconductivity; many body problem; interaction of radiation with matter; low temperature physics; superfluidity; physics and chemistry experiments in earth-orbiting laboratories; tunneling; dynamics of liquid drops and bubbles; theory of theta functions; theory of Hilbert transforms. *Mailing Add:* 100 Lockwood Lane Apt 419 Scotts Valley CA 95066

SAFIR, SIDNEY ROBERT, b Trenton, NJ, June 17, 16; m 42; c 2. MEDICINAL CHEMISTRY. *Educ:* Univ Mich, BS, 37, MS, 38, PhD(org chem), 40. *Prof Exp:* Fuller fel, Univ Mich, 40-41; org chemist, Am Cyanamid Co, 41-46 & Schenley Distillers Co, 46-47; org chemist, Lederle Labs, Am Cyanamid Co, 47-82; RETIRED. *Mem:* Am Chem Soc; Sigma Xi. *Res:* Synthesis of novel antipsychotic, anxiolytic and analgetic agents. *Mailing Add:* 7775 Beltane Dr San Jose CA 95135

SAFKO, JOHN LOREN, b San Diego, Calif, Oct 29, 38; m 64; c 5. PHYSICS. *Educ:* Case Inst Technol, BS, 60; Univ NC, PhD(physics), 65. *Prof Exp:* From asst prof to assoc prof, 64-78, PROF PHYSICS & ASTRON, UNIV SC, 79- *Concurrent Pos:* NSF grants, mem, NSF panels. *Mem:* Am Phys Soc; Am Asn Physics Teachers; Am Astron Soc; Int Astron Union; Int Soc Gen Relativity & Gravitation; AAAS. *Res:* General relativity and gravitation theory; teaching methods; astrophysical investigations related to relativity; astronomy education. *Mailing Add:* Dept Physics & Astron Univ SC Columbia SC 29208

SAFLEY, LAWSON MCKINNEY, JR, b Fayetteville, Tenn, Jan 13, 50; m 73; c 3. SANITARY & ENVIRONMENTAL ENGINEERING. *Educ:* Univ Tenn, BS, 72; Cornell Univ, MS, 74, PhD(agr eng), 77. *Prof Exp:* Res support specialist, Cornell Univ, 76-77; asst prof agr eng, Univ Tenn, 77-81; assoc prof, 81-87, PROF BIO & AGR ENG, NC STATE UNIV, 87- *Mem:* Am Soc Agr Engrs; Sigma Xi; Am Soc Agr Consults. *Res:* Manurial nutrient loss during storage; land application of manure; systems analysis of animal manure systems; anaerobic digestion systems. *Mailing Add:* Bio & Agr Eng Dept NC State Univ Box 7625 Raleigh NC 27695-7625

SAFONOV, MICHAEL G, b Pasadena, Calif, Nov 1, 48; m 68, 85; c 2. CONTROL THEORY, SIGNAL PROCESSING. *Educ:* Mass Inst Technol, BS & MS, 71, PhD(elec eng), 77. *Prof Exp:* Res & teaching asst, Mass Inst Technol, 75-77; from asst prof to assoc prof, 77-88, PROFF ELEC ENG, UNIV SOUTHERN CALIF, 88-, ASSOC DEPT CHAIR, 89- *Concurrent Pos:* Consult, Anal Sci Corp, Systs Control, Honeywell, Northrup, TRW, Lear-Siegler, Lear Astronics & United Technologies; vis prof, Cambridge Univ, 83-84, Imperial Col, 87, Caltech, 90-91. *Mem:* Fel Inst Elec & Electronics Engrs; Sigma Xi. *Res:* Control and feedback theory; aircraft flight control; hierarchical decomposition methods; multivariable control synthesis; stability theory. *Mailing Add:* Dept Elec Eng Systs Univ Southern Calif Los Angeles CA 90089-2563

SAFRAN, SAMUEL A, b Brooklyn, NY, Nov 22, 51; m 75; c 3. COLLOID PHYSICS. *Educ:* Yeshiva Univ, BA, 73; Mass Inst Technol, PhD(physics), 78. *Prof Exp:* Postdoctoral fel, mem tech staff, Bell Labs, 78-80; sr staff physics, Exxon Res & Eng, 80-90; actg dept head, Polymer Dept, 90, PROF, WEIZMANN INST SCI, 90- *Concurrent Pos:* Coordr, Inst Theoret Physics prog self-assembling syst, 89. *Mem:* Fel Am Phys Soc; Am Chem Soc. *Res:* Theoretical condensed matter physics; complex fluid physics; structure and phase behavior of microemulsion surfactants, polymers, colloids; structure of interfaces in both fluids and solids. *Mailing Add:* Polymer Res Dept Weizmann Inst Sci Rehovot 76100 Israel

SAFRANYIK, LASZLO, b Besenyszog, Hungary, Feb 13, 38; Can citizen; m 66; c 2. FOREST ENTOMOLOGY. *Educ:* Univ BC, BSF, 61, MF, 63, PhD(pop dynamics), 69. *Prof Exp:* Res officer forest entom, Can Dept Forestry, 64-69, RES SCIENTIST FOREST ENTOM, CAN FORESTRY SERV, 69- *Mem:* Entom Soc Can; Can Inst Forestry. *Mailing Add:* 141 Durrance Dr Victoria BC V8X 4M6 Can

SAFRON, SANFORD ALAN, b Chicago, Ill, July 24, 41; m 80; c 2. PHYSICAL CHEMISTRY, SURFACE CHEMISTRY & PHYSICS. *Educ:* Univ Calif, Berkeley, BS, 63; Harvard Univ, MA, 65, PhD(chem), 69. *Prof Exp:* Guest researcher, Physics Inst, Univ Bonn, 69-70; from asst prof to assoc prof, 70-91, PROF CHEM, FLA STATE UNIV, 91- *Concurrent Pos:* Res Corp-Cottrell grant, Fla State Univ, 71, Petrol Res Fund-Am Chem Soc grant, 71-74; Res Corp-Cottrell grant, 76, NSF-URP grant, 79 & DOE grant, 85 & 88; vis prof, Max Planck Inst, Gottingen. *Mem:* AAAS; Am Phys Soc; Am Chem Soc; Am Vacuum Soc. *Res:* Dynamics of chemical reactions; dynamics of crystal surfaces; helium atom-surface scattering. *Mailing Add:* Dept Chem Fla State Univ Tallahassee FL 32306

SAGAL, MATTHEW WARREN, b Brooklyn, NY, Nov 23, 36; m 59; c 3. PHYSICAL CHEMISTRY. *Educ:* Cornell Univ, BChE, 58; Mass Inst Technol, PhD(phys chem), 61. *Prof Exp:* Mem tech staff chem, Bell Tel Labs, 61-66; res leader, Western Elec Co, 66-67, asst dir mat & chem processes, 67-69, dir mat & chem process, 69-76, mgr prod planning, 76-79, dir eng, Allentown Works, 79-83; dir, Int Strategic Planning & Bus Develop, AT&T Tech Systs, 83-88, VPRES BUS DEVELOP AT&T MICROELECTRONICS, 88- *Mem:* Electrochem Soc; sr mem Inst Elec & Electronics Engrs. *Res:* Electronic materials; manufacturing processes; environmental analysis; plastics; ceramics. *Mailing Add:* AT&T Microelectronics Two Oak Way Berkeley Heights NJ 07922

SAGALYN, PAUL LEON, b New York, NY, Mar 21, 21; div; c 2. SOLID STATE PHYSICS, MATERIALS SCIENCE. *Educ:* Harvard Univ, BS, 42; Mass Inst Technol, PhD(physics), 52. *Prof Exp:* Staff mem, Radiation Lab, Mass Inst Technol, 43-45, res assoc, Dept Physics, 52-56; res physicist, 56-85, SR SCIENTIST, MAT TECHNOL LAB, US ARMY, 85- *Mem:* Am Phys Soc; Mat Res Soc. *Res:* Use of nuclear magnetic resonance as a tool for studying the electronic structure of solids; ion implantation as a tool for modifying the surface related properties of materials. *Mailing Add:* US Army Mat Technol Lab Watertown MA 02172-0001

SAGALYN, RITA C, b Lowell, Mass, Nov 24, 24; m 52; c 2. SPACE PHYSICS, IONOSPHERIC PHYSICS. *Educ:* Univ Mich, Ann Arbor, BS, 48; Radcliffe Col, MS, 50. *Prof Exp:* Res physicist, Air Force Cambridge Res Labs, 48-58 & Aeronomy & Ionospheric Physics Labs, 58-69, br chief, space physics, Ionospheric Physics Lab, 69-75, br chief Elec Processes Bd, 75-81, DIR SPACE PHYSICS DIV, GEOPHYS DIRECTORATE, USAF CAMBRIDGE RES LABS, 81- *Honors & Awards:* Guenther Loeser Award, Air Force Cambridge Res Labs, 58; Patricia Kayes Glass Award, USAF, 66. *Mem:* Sigma Xi; Am Geophys Union. *Res:* Upper atmospheric and space research; study experimentally and theoretically of the influence of soft particle fluxes, solar ultraviolet, terrestrial electric and magnetic fields, plasma motions and instabilities on spatial distribution and temporal behavior of environmental plasma; develop Solaterrestrial physics. *Mailing Add:* 555 Annursmac Hill Rd Concord MA 01742

SAGAN, CARL, b Brooklyn, NY, Nov 9, 34. PLANETARY SCIENCES. *Educ:* Univ Chicago, AB, 54, BS, 55, MS, 56, PhD(astron, astrophys), 60. *Hon Degrees:* Various from US & foreign univs, cols & insts, 75-80. *Prof Exp:* Miller res fel astron, Inst Basic Res Sci, Univ Calif, Berkeley, 60-62; asst prof, Harvard Univ, 62-68; astrophysicist, Smithsonian Astrophys Observ, 62-68; assoc prof astron, Ctr Radiophysics & Space Res, 68-70, prof, 70-77, assoc dir, Ctr Radiophys & Space Res, 72-81, DAVID DUNCAN PROF ASTRON & SPACE SCI, CORNELL UNIV, 77-, DIR, LAB PLANETARY STUDIES, 68- *Concurrent Pos:* Vis asst prof, Sch Med, Stanford Univ, 62-63; Alfred P Sloan Found res fel, 63-67; vchmn working group moon & planets, comt space res, Int Coun Sci Unions, 68-75; lectr, Astronaut Training Prog, NASA, 68-72; chmn US deleg, Joint Nat Acad Sci-Soviet Acad Sci Conf Commun Extraterrestrial Intel, Armenia, 71; vis assoc, Calif Inst Technol, 71-72 & 76-77, consult, Jet Propulsion Lab; mem bd dirs, Coun Advan Sci Writing, 72-75; lectr var US & foreign univs, cols & insts, 59-; mem adv coun, Smithsonian Inst, 75-; mem exobiology comt, Planetary Atmospheres Study Group, Nat Acad Sci, mem steering comt exobiol study, Space Sci Bd & Panel Origins Life & Astron Surv Comt & comt sci & pub policy; mem organizing comt, Comn 16 Phys Study Planets, Int Astron Union; pres, Planetology Sect, Am Geophys Union, 80-82; NSF & Am Astron Soc vis prof; experimenter, Mariner II Venus probe; Mariner IX Mars orbiter, Viking Mars lander; Voyager, Outer Solar Syst Probes; pres, Carl Sagan Prods, Inc, 77- *Honors & Awards:* A Calvert Smith Prize, 64; Apollo Achievement Award, NASA, 69, Except Sci Achievement Medal, 72; Int Astronaut Prize, Galabert Found, Paris, 73; John W Campbell Mem Award, 74; Klumpke-Roberts Prize Popularization Astron, Astron Soc Pac, 74; Golden Plate Award, Am Acad Achievement, 75; Joseph Priestley Award, 75; Newcomb Cleveland Prize, 77; Rittenhouse Medal, Franklin Inst-Rittenhouse Astron Soc, 80; Pulitzer Prize, 78; George Foster Peabody Award, 81; Glenn Seabory Award, Am Platform Asn, 81; Ralph Coats Roe Medal, Am Soc Mech Engrs, 81. *Mem:* Fel AAAS; fel Am Acad Arts & Sci; Am Phys Soc; fel Am Geophys Union (pres, Planetology Sect, 80-82); fel Am Inst Aeronaut & Astronaut; Sigma Xi; Planetary Soc (pres, 74-). *Res:* Physics of planetary atmospheres; planetary surface conditions; production of organic molecules in astronomical environments; origin of life; extraterrestrial biology; space vehicle exploration of the solar system. *Mailing Add:* Lab Planetary Studies Cornell Univ Space Sci Bldg Ithaca NY 14853

SAGAN, HANS, b Vienna, Austria, Feb 15, 28; nat US; m 54; c 1. MATHEMATICS. *Educ:* Univ Vienna, PhD, 50. *Prof Exp:* Asst prof math, Vienna Tech Univ, 50-54 & Mont State Univ, 54-57; assoc prof, Univ Idaho, 57-61, prof & head dept, 61-63; PROF MATH, NC STATE UNIV, 63- *Concurrent Pos:* Math Asn Am lectr, 63-73, 77-; vis prof, Munich Tech Univ, 64; vis prof, Univ Vienna, 72. *Honors & Awards:* Poteat Award, NC Acad Sci, 66. *Mem:* Math Asn Am. *Res:* Eigenvalue problems; calculus of variations and optimal control theory; functional analysis; partial difference equations; space-filling curves. *Mailing Add:* Dept Math NC State Univ Raleigh NC 27695-8205

SAGAN, LEON FRANCIS, b Chicopee Falls, Mass, May 23, 41. MATHEMATICS. *Educ:* Towson State Col, BS, 62; Col William & Mary, MA, 64; Univ Md, College Park, PhD(math educ), 71. *Prof Exp:* PROF MATH, ANNE ARUNDEL COMMUNITY COL, 64- *Mem:* Math Asn Am; Nat Coun Teachers Math; Am Math Asn. *Res:* Remedial math; college algebra and trigonometry; calculus. *Mailing Add:* Dept Math Anne Arundel Community Col 101 College Pkwy Arnold MD 21012

SAGAN, LEONARD A, b San Francisco, Calif, Feb 18, 28; m 54; c 3. INTERNAL MEDICINE, ENVIRONMENTAL MEDICINE. *Educ:* Stanford Univ, AB, 50; Univ Chicago, MD, 55; Harvard Univ, MPH, 65; Am Bd Internal Med, dipl, 64. *Prof Exp:* Intern, Univ Calif Hosp, 55-56, resident internal med, 56-61; physician, Atomic Bomb Casualty Comn, Japan, 61-64; physician, US AEC, Washington, DC, 65-68; physician, 68-78, assoc dir dept environ med, Palo Alto Med Clin, 71-78; SR SCIENTIST, ELEC POWER RES INST, 78- *Mem:* Fel Am Col Physicians; Nat Coun Radiation Protection. *Res:* Late effects of radiation; non-ionizing radiation. *Mailing Add:* 3412 Hillview Ave Palo Alto CA 94303

SAGAR, WILLIAM CLAYTON, b Columbus, Ohio, Oct 17, 29; m 53; c 3. SYNTHETIC ORGANIC CHEMISTRY. *Educ:* Capital Univ, BS, 51; Ohio State Univ, MSc, 54, PhD(chem), 58. *Prof Exp:* Res chemist, Ethyl Corp, 58-61; from asst prof to assoc prof chem, 61-70, PROF CHEM, CENTRE COL KY, 70-, CHMN, DIV SCI & MATH, 72- *Mem:* Am Chem Soc. *Res:* Synthesis of insecticides; aldol condensations. *Mailing Add:* Dept Chem Centre Col Danville KY 40422-1394

SAGARAL, ERASMO G, b Camiguin, Philippines, June 2, 36; m 60; c 5. PLANT GROWTH REGULATORS PGRS, MANAGEMENT OF SOYBEANS & SMALL GRAINS BREEDING. *Educ:* Cent Mindanao Univ, BSA, 59; Univ Philippines, MS, 68; Va Polytech Inst & State Univ, PhD(plant physiol), 78. *Prof Exp:* Farm mach teacher agr, Bohol Agr Col, 59-62; voc agr teacher farm mech, Cent Mindanao Univ, 62-65; asst prof agron, Xavier Univ, 69-73, head dept, 69-75 & 78-79, assoc prof agron, 73-75 & 78-79; sr scientist, Dole Philippines, Inc, 79-85; res specialist physiol, 85-87, SUPT RES & ADMIN, EASTERN VA AGR EXP STA, VA POLYTECH INST & STATE UNIV, 87- *Concurrent Pos:* Prof lectr, Southeast Asia Rural Social Leadership Inst, 78-79; consult, Coconut Fedn Philippines, 78-79. *Mem:* Am Soc Agron; Plant Growth Regulator Soc Am. *Res:* Weed control in rice (rainfed and flooded conditions); developed research programs on nutrition, weed control and cultural management practices for pineapples and bananas; evaluated the performance of plant growth regulators in soybeans, wheat and barley; coordinating research on soybeans and small grains breeding program at Eastern Virginia Agricultural Experiment Station. *Mailing Add:* PO Box 338 Warsaw VA 22572

SAGAWA, KIICHI, physiology, biomedical engineering; deceased, see previous edition for last biography

SAGAWA, YONEO, b Keeau, Hawaii, Oct 11, 26; wid; c 2. CYTOGENETICS. *Educ:* Washington Univ, AB, 50, MS, 52; Univ Conn, PhD(cytogenetics), 56. *Prof Exp:* Res assoc biol, Brookhaven Nat Lab, 55-57; from asst prof to assoc prof bot, Univ Fla, 57-64; PROF HORT, UNIV HAWAII, 64-, DIR, HAROLD L LYON ARBORETUM, 67- *Concurrent Pos:* Dir undergrad sci educ prog & undergrad res participation & independent study, NSF, Univ Fla, 64; res assoc, Univ Calif, Berkeley, 70; ed, Hawaii Orchid J, 72-; fel, Agr Univ, Neth, 79-80. *Honors & Awards:* Am Orchid Soc Inc. *Mem:* Bot Soc Am; Int Soc Hort Sci; AAAS; Am Soc Hort Sci; Int Asn Plant Tissue Culture; Sigma Xi. *Res:* Cytogenetics of cultivated plants, especially subtropical plants; morphogenesis; tissue culture; tissue culture for micropropagation, germplasm storage and disease elimination. *Mailing Add:* Univ Hawaii H L Lyon Arboretum 3860 Manoa Rd Honolulu HI 96822

SAGE, ANDREW PATRICK, b Charleston, SC, Aug 27, 33; m 62; c 3. DECISION SUPPORT SYSTEMS. *Educ:* The Citadel, BS, 55; Mass Inst Technol, MS, 56; Purdue Univ, PhD(elec eng), 60. *Hon Degrees:* DEngr, Univ Waterloo, Can, 87. *Prof Exp:* Instr elec eng, Purdue Univ, 56-60; assoc prof, Univ Ariz, 60-63; tech staff mem, Aerospace Corp, Calif, 63-64; prof elec eng, Univ Fla, 64-67, prof nuclear eng, 66-67; prof & dir info & control sci ctr, Inst Technol, Southern Methodist Univ, 67-74, head dept elec eng, 72-74; chmn, Dept Chem Eng, Univ Va, 74-75, assoc dean, 74-80, Lawrence R Quarles Prof Eng & Appl Sci & Chmn, Dept Eng Sci & Systs, 77-84; assoc vpres acad affairs, 84-85, FIRST AM BANK PROF INFO TECHNOL, GEORGE MASON UNIV, 84-, DEAN, SCH INFO TECHNOL & ENG, 85- *Concurrent Pos:* Consult var corp & insts, 57-; ed, Trans on Systs, Man & Cybernetics, Inst Elec & Electronics Engrs; ed, Automatica, 80-; co-ed-in-chief, Chief Info & Decision Technologies, 82- *Honors & Awards:* Barry Carlton Award, Inst Elec & Electronics Engrs, 70, Norbert Wiener Award, 81 & Centennial Medal, 84; Frederick Emmonds Terman Award, Am Soc Eng Educ, 70; Outstanding Serv Award, Int Fedn Automotive Control, 90. *Mem:* Fel AAAS; Inst Mgt Sci; fel Inst Elec & Electronics Engrs; Am Soc Eng Educ; Am Inst Decision Sci. *Res:* Software systems engineering, expert systems; decision support systems; systems engineering; education; optimization and estimation theory; information technology and management. *Mailing Add:* Sch Info Technol & Eng George Mason Univ Fairfax VA 22030

SAGE, GLORIA W, b Brooklyn, NY, Mar 7, 36; m 58; c 1. PHYSICAL CHEMISTRY, ANALYTICAL CHEMISTRY. *Educ:* Cornell Univ, AB, 57; Radcliffe Col, AM, 58; Harvard Univ, PhD(phys chem), 63. *Prof Exp:* Jr chemist, Res & Adv Develop Div, Avco Corp, 57-58; res assoc chem, Univ Ore, 61-63; instr, 63-66, res assoc, 66-67; res assoc, Syracuse Univ, 67-70; res assoc biochem, State Univ NY Upstate Med Ctr, 70-72, asst prof med technol, 72-76, res assoc pediat, 76-77; res assoc chem, Tel Aviv Univ, 77-78; SR SCIENTIST, SYRACUSE RES CORP, 80- *Concurrent Pos:* Consult, 78-80. *Mem:* AAAS; Am Chem Soc. *Res:* Phosphorescence; magnetic circular dichroism; ultraviolet and fluorescence spectroscopy of proteins; conformation of proteins; molecular spectroscopy; clinical chemistry method development and lab management evaluation; environmental fate of chemicals; data base development; writing assessments of the environmental fate of chemicals and exposure for data bases and technical support documents; project accounting system development for toxic chemical testing. *Mailing Add:* Syracuse Res Corp Merrill Lane Syracuse NY 13210

SAGE, HARVEY J, b New York, NY, Jan 5, 33; m 68; c 1. IMMUNOCHEMISTRY, BIOCHEMISTRY. *Educ:* Polytech Inst Brooklyn, BS, 54; Yale Univ, PhD(chem), 58. *Prof Exp:* Res assoc hemat, Sch Med, Yale Univ, 58-60; res assoc biochem, St Luke's Hosp, Cleveland, 60-62; res assoc, Brandeis Univ, 62-64; asst prof biochem, 64-71, ASST PROF PATH, DUKE UNIV, 64-, ASSOC PROF BIOCHEM, 71-, ASSOC PROF IMMUNOL, 74- *Mem:* AAAS; Am Soc Biochem; Am Asn Immunologists. *Res:* Specificity of antigen-antibody reactions; in vitro lymphocyte culture and isolation of lymphocyte surface membrane proteins; protein structure; use of synthetic polypeptides as models for protein structure and as immunogens. *Mailing Add:* Dept Biochem Duke Univ 3711 Med Ctr Durham NC 27710

SAGE, HELENE E, b Philadelphia, Pa, Oct 6, 46; m 85. VASCULAR BIOLOGY, CONNECTIVE TISSUE PROTEINS. *Educ:* Mt Holyoke Col, AB, 69; Univ Utah, PhD(biol sci), 77. *Prof Exp:* Res asst prof, dept biochem, 80-82, asst prof, dept biol struct, 82-85, ASSOC PROF, DEPT BIOL STRUCT, UNIV WASH, 85- *Concurrent Pos:* Investr, Am Heart Asn, 81-86; vis assoc prof, Univ Med & Dent NJ, 85 & Nat Ctr Sci Res-LGME, Strasbourg, France, 85-86. *Mem:* Am Soc Cell Biol; Am Chem Soc; Am Soc Biochem & Molecular Biol. *Res:* Protein synthesis by cells in vitro; protein structure, especially elastin and collagen; relationship of cell behavior to extracellular matrix; control of gene expression and cellular phenotypic modulation. *Mailing Add:* Dept Biol Struct Univ Wash SM-20 Seattle WA 98195

SAGE, JAY PETER, b Pittsburgh, Pa, Nov 8, 43; m 71; c 2. SUPERCONDUCTING ELECTRONICS, ARTIFICIAL NEURAL NETWORKS. *Educ:* Harvard Univ, BA, 64, MA, 65, PhD(physics), 69. *Prof Exp:* Sr res scientist, Res Div, Raytheon Co, Waltham, Mass, 68-81; MEM TECH STAFF, LINCOLN LAB, MASS INST TECHNOL, 81- *Concurrent Pos:* Raytheon exchange scientist, Toshiba Res & Develop Ctr, Japan, 73-74. *Mem:* Inst Elec & Electronics Engrs. *Res:* Superconducting circuits, charge-coupled devices and metal oxide semiconductor integrated circuitry for signal processing. *Mailing Add:* Lincoln Lab Mass Inst Technol PO Box 73 Lexington MA 02173-9108

SAGE, JOSEPH D, b Leonardo, NJ, July 14, 31; m 51; c 7. SOIL MECHANICS, ENGINEERING GEOLOGY. *Educ:* Rutgers Univ, BS, 53, MS, 58; Clark Univ, PhD(geog), 74. *Prof Exp:* Pres & mem bd, Geotechnics Inc, 59-65; from instr to assoc prof, 57-77, PROF CIVIL ENG, WORCESTER POLYTECH INST, 77- *Concurrent Pos:* Prin investr, Dept Transp & NSF; partner, Sage & Dandrea,77- *Mem:* Sigma Xi; Am Soc Civil Engrs; Portuguese Soc Geotechnol. *Res:* Rock mechanics; frost action in particulate systems; mathematical synthesis of climatological time series. *Mailing Add:* Dept Civil Eng Worcester Polytech Inst Worcester MA 01609

SAGE, MARTIN, b Torquay, Eng, Dec 6, 35; m 65; c 2. ZOOLOGY, PHYSIOLOGY. *Educ:* Univ Nottingham, BSc, 57, PhD(zool), 60. *Prof Exp:* Demonstr zool, Univ Nottingham, 59-60; from asst lectr to lectr zool, Univ Leicester, 60-66, lectr physiol, 66-69; assoc prof zool, Univ Tex, Austin, 69-74, assoc prof marine studies, 73-74, res scientist, Marine Sci Inst, Port Aransas, 69-74; assoc prof, 74-77, chmn dept, 75-81, PROF BIOL, UNIV MO, ST LOUIS, 77- *Concurrent Pos:* Tutor, Univ Nottingham, 60; resident tutor, Univ Leicester, 60-65; Wellcome fund travel grant & asst zoologist, Cancer Res Genetics Lab & Bodega Marine Lab, dept zool, Univ Calif, Berkeley, 68-69; reader, Marine Biol Lab, Woods Hole, 81-82; assoc dean, Col Arts & Sci, Univ Mo, St Louis, 83-85 & 86- *Mem:* Res Defense Soc; Am Soc Zool. *Res:* Comparative endocrinology and physiology; evolution of vertebrate endocrine and neuroendocrine control systems; evolution of biological activity of hormones; endocrine control of osmoregulation; hormones and behavior; biological rhythms. *Mailing Add:* Dept Biol Univ Mo St Louis MO 63121-4499

SAGE, MARTIN LEE, b New York, NY, Mar 4, 35; m 58; c 1. CHEMICAL PHYSICS. *Educ:* Cornell Univ, AB, 55; Harvard Univ, MA, 48, PhD(chem physics), 59. *Prof Exp:* Fel physics, Brandeis Univ, 59-61; asst prof chem & theoret sci, Univ Ore, 61-67; assoc prof, 67-85, PROF CHEM, SYRACUSE UNIV, 85-, DIR TECHNOL & PUB AFFAIRS PROG, 88- *Concurrent Pos:* Vis assoc prof chem, Tel Aviv Univ, Israel, 77-78; vis, Dept of Theoret Chem, Univ Oxford, Eng, 85-86. *Mem:* AAAS; Am Phys Soc; Am Chem Soc; Am Asn Univ Prof; Nat Asn Sci Technol & Soc. *Res:* Quantum chemistry; intramolecular dynamics; multiphoton photochemistry; computer algebra. *Mailing Add:* Dept Chem Syracuse Univ Syracuse NY 13244

SAGE, NATHANIEL MCLEAN, JR, b Boston, Mass, Feb 4, 18; m 55, 72; c 5. RESEARCH ADMINISTRATION. *Educ:* Mass Inst Technol, SB, 41, SM, 51, PhD, 53. *Prof Exp:* Teacher high sch, Conn, 46-49; asst to dir admis, Mass Inst Technol, 49-50; from instr to asst prof geol, Amherst Col, 51-55; from asst prof to assoc prof, Univ NH, 55-60, chmn dept, 57-60; assoc dir sponsored res, Mass Inst Technol, 60-68; coordr res, Univ RI, 68-83, EMER COORDR RES, UNIV RI, 83- *Mem:* Fel Geol Soc Am; Am Asn Petrol Geologists. *Res:* Invertebrate paleontology; carboniferous of Nova Scotia and Pennsylvania anthracite region. *Mailing Add:* 957 Saugatucket Rd PO Box 3726 Peace Dale RI 02883-0394

SAGE, ORRIN GRANT, JR, b Los Angeles, Calif, May 31, 46; m 70. ENVIRONMENTAL GEOLOGY. *Educ:* Univ Calif, BA, 69, MA, 71, PhD(geol), 73. *Prof Exp:* Environ scientist environ mgt, Multran Am Corp & Henningson, Durham & Richardson, 72-75; LECTR ENVIRON STUDIES, UNIV CALIF, SANTA BARBARA, 73- *Concurrent Pos:* NSF fel, 70-73; consult environ & agr, Environ Corp & Sage Assocs, 75-; extension lectr wilderness survival, Univ Calif, 75- *Mem:* Geol Soc Am; Wilderness Soc. *Res:* Tectonic evolution of western California; environmental assessment and land use planning; environmental effects of California agriculture. *Mailing Add:* 1396 Danielson Rd Santa Barbara CA 93108

SAGER, CLIFFORD J, b New York, NY, Sept 28, 16; m; c 4. PSYCHIATRY. *Educ:* Pa State Col, BS, 37; NY Univ, MD, 41; Am Bd Psychiat & Neurol, dipl, 48; NY Med Col, cert psychoanal, 49. *Prof Exp:* Consult psychiat, Family Welfare Orgn, Allentown, Pa, 46-47; assoc dean & dir therapeut serv, Postgrad Ctr Ment Health, 48-60; dir clin serv, NY Med Col, 60-63, chief family treatment & study unit, 64-70, prof psychiat & dir partial hosp prog, 66-70; clin prof psychiat, Mt Sinai Sch Med, City Univ New York, 70-80, attend psychiat, Mt Sinai Hosp, 74-80; CLIN PROF PSYCHIAT, NY HOSP/CORNELL MED CTR, 80-, ATTEND PSYCHIAT, 80- *Concurrent Pos:* Asst adj psychiatrist, Beth Israel Hosp, 48-50; vis psychiatrist, Metrop Hosp, 60-70; attend psychiatrist, Flower & Fifth Ave Hosp, 60-70; chief family treatment & study unit, Beth Israel Med Ctr, 70-74, assoc dir family & group ther, 71-73, chief behav sci serv prog, Ctr & Hosp, 71-73; dir psychiat, Gouverneur Hosp, 70-73; psychiat dir, Jewish Family Serv, New York, 73-; ed, J Sex & Marital Ther, 74-; dir family psychiat, Jewish Bd Family & Children's Serv, New York, 78-, dir, Sex Therapy Clin & Remarried Consult Ser; psychiatric counr, Corp Health Prog, Jewish Bd for Family Childrens Serv. *Honors & Awards:* Ann Award for Distinguished Prof Contrib to Family Therapy, 83, Am Asn Marriage & Family Therapists, 83. *Mem:* Fel Am Psychiat Asn; fel Am Med Asn; fel Am Acad Psychoanal; fel Am Orthopsychiat Asn; fel Am Group Psychother Asn; charter fel Am Family Therapy Asn; fel Am Asn Marriage & Family Therapists; Soc Med Psychiat; Soc Sex Therapy & Res. *Res:* The marital couple and the development of suitable methods of bringing psychiatric treatment to those segments of the population previously not reached by effective psychological and social forms of treatment; new methods of treating the sexual dysfunctions; family process; marital interaction; typography of marriages; problems of remarriage. *Mailing Add:* 65 E 76th St New York NY 10021

SAGER, EARL VINCENT, b Buffalo, NY, Sept 24, 45; m 79. ENGINEERING PHYSICS. *Educ:* State Univ NY, Buffalo, BA, 67, MA, 69; Univ Md, PhD(physics), 79. *Prof Exp:* RES ANALYST, SYST PLANNING CORP, 78- *Mem:* Am Phys Soc. *Res:* Radar digital signal processing. *Mailing Add:* 6730 White Post Rd Centreville VA 22020

SAGER, JOHN CLUTTON, b New Castle, Pa, Mar 15, 42; m 64; c 2. ENVIRONMENTAL CONTROL. *Educ:* Pa State Univ, BS, 64, MS, 70, PhD(agr eng), 73. *Prof Exp:* AGR RES ENG, RADIATION BIOL LAB, SMITHSONIAN INST, 73- *Concurrent Pos:* Instr pilot, Ag Rotors Inc, 69; res asst, Agr Eng Dept, Pa State Univ, 68-69 & 70-73, instr, 70; adj asst prof, Agr Eng Dept, Univ Md, 84- *Mem:* Am Soc Agr Eng; Am Soc Photobiol; AAAS. *Res:* Direct measurement of environmental parameters and execution of a research program on environmental effects on plants with emphasis on growth and productivity; design new or modify equipment to provide control of environmental factors critical to the research program with emphasis on electromagnetic radiation (light). *Mailing Add:* NASA KSC MD Res L Kennedy Space Center FL 32899

SAGER, RAY STUART, b Cuero, Tex, Feb 24, 42; m 62; c 2. INORGANIC CHEMISTRY, PHYSICAL CHEMISTRY. *Educ:* Tex Lutheran Col, BS, 64; Tex Christian Univ, PhD(chem), 68. *Prof Exp:* Asst prof chem, Concordia Col, Moorhead, Minn, 68-69; from asst prof to assoc prof, Capital Univ, 69-74; ASSOC PROF CHEM, PAN AM UNIV, 75- *Mem:* Am Chem Soc. *Res:* Structure and properties of copper II complexes of schiff bases containing amino acids. *Mailing Add:* Dept Sci Victoria Col 2200 E Red River Victoria TX 77901

SAGER, RONALD E, b Adrian, Mich, Aug 8, 47; m 69; c 2. CRYOGENICS, MAGNETIC MEASUREMENTS. *Educ:* Mich State Univ, BS, 69; Univ Calif, San Diego, MS, 74, PhD(physics), 77. *Prof Exp:* Res physicist, Physical Dynamics, Calif, 77-79 & S H E Corp, Calif, 79-82; sr res physicist, Quantum Design, Inc, Calif, 82-87; VPRES, QUANTUM MAGNETICS, INC, CALIF, 87- *Concurrent Pos:* Prin investr, Dept Energy, 79-82, Dept Defense, 85-91 & Dept Transp, 87-91; guest instr, Mgt Technol Sem, Univ Minn, 91. *Mem:* Am Phys Soc. *Res:* Cryogenic instrumentation development for commercial applications; superconducting quantum interference devices to make extremely sensitive magnetic measurements in field from 1 millioersted to 70,000 oersted and over temperatures from 1.7 kelvin to 800 kelvin. *Mailing Add:* Quantum Magnetics Inc 11578 Sorrento Valley Rd No 30 San Diego CA 92121

SAGER, RUTH, b Chicago, Ill, Feb 7, 18; m 73. GENETICS. *Educ:* Univ Chicago, BS, 38; Rutgers Univ, MS, 44; Columbia Univ, PhD(genetics), 48. *Prof Exp:* Merck fel, Nat Res Coun, Rockefeller Univ, 49-51, staff mem, 51-55; res assoc zool, Columbia Univ, 55-60, sr res scientist, 60-66; prof biol, Hunter Col, City Univ New York, 66-75; PROF CELLULAR GENETICS, HARVARD MED SCH & CHIEF DIV GENETICS, DANA FARBER CANCER INST, 75- *Concurrent Pos:* Guggenheim fel, Imp Cancer Res Fund Lab, London, 72-73; nonresident fel, Edinburgh Univ; mem, Bd Sci Counr, Nat Inst Arthritis, Diabetes, & Digestive & Kidney Dis, 79-81; mem, Genetics Study Sect, NIH, 81-83; mem, Presidential Young Investr Award Panel, NSF, 84. *Honors & Awards:* Gilbert Morgan Smith Medal, Nat Acad Sci, 88; Schneider Mem Lectr, Univ Tex, 90. *Mem:* Nat Acad Sci; Am Soc Cell Biol; Genetics Soc Am; Am Acad Arts & Sci; Sigma Xi; Am Soc Biol Chemists; Am Asn Cancer Res; Am Soc Human Genetics. *Res:* Molecular genetics; organelle genetics and biogenesis; mammalian cell genetics; genetic mechanisms of carcinogenesis; tumor suppressor genes; breast cancer. *Mailing Add:* Sidney Farber Cancer Inst 44 Binney St Boston MA 02115

SAGER, THOMAS WILLIAM, US citizen. STATISTICS. *Educ:* Univ Iowa, BA, 68, MS, 71, PhD(statist), 73. *Prof Exp:* Asst prof statist, Stanford Univ, 73-78; vis asst prof math & bus, 78-79, asst prof, 79-82, ASSOC PROF STATIST, UNIV TEX, AUSTIN, 82- *Mem:* Inst Math Statist; Am Statist Asn; Sigma Xi. *Res:* Spatial patterns; density estimation; isotonic regression; environmental statistics; computational statistics; sampling. *Mailing Add:* 2301 Doral Dr Austin TX 78746

SAGER, WILLIAM FREDERICK, b Ill, Jan 22, 18; m 41; c 3. ORGANIC CHEMISTRY. *Educ:* George Washington Univ, BS, 39, MA, 41; Harvard Univ, PhD(chem), 48. *Prof Exp:* Chemist, Tex Co, 41-45; from asst prof to prof chem, George Washington Univ, 48-64; prof chem, Univ Ill, Chicago, 65-86, head dept, 65-80; RETIRED. *Concurrent Pos:* Guggenheim fel, Oxford Univ, 54-55; consult, Bur Weapons, US Dept Navy, Army Chem Ctr, NIH, W Grace Co & Houdry Process Co. *Mem:* Am Chem Soc. *Res:* Mechanisms of organic reactions; chemistry of high explosives. *Mailing Add:* 1552 John Anderson Dr Ormond Beach FL 32176

SAGERMAN, ROBERT H, b Brooklyn, NY, Jan 23, 30; m 54; c 4. MEDICINE, RADIOLOGY. *Educ:* NY Univ, BA, 51, MD, 55; Am Bd Radiol, dipl, 61. *Prof Exp:* Clin instr radiol, Med Sch, Tulane Univ, 56-57; instr, Sch Med, Stanford Univ, 61-64; asst prof, Columbia-Presby Med Ctr, 64-68; PROF RADIOL & DIR RADIOTHER DIV, STATE UNIV NY UPSTATE MED CTR, 68- *Mem:* AAAS; Am Soc Therapeut Radiol; Radiol Soc NAm; Am Radium Soc; Radiation Res Soc. *Res:* Therapeutic radiology; radiation biology. *Mailing Add:* Dept Radiol State Univ NY Upstate Med Ctr 155 Elizabeth Blackwell St Syracuse NY 13210

SAGERS, RICHARD DOUGLAS, b Tooele, Utah, Dec 19, 28; m 50; c 6. BIOCHEMISTRY. *Educ:* Brigham Young Univ, BS, 54, MS, 55; Univ Ill, PhD(bact), 58. *Prof Exp:* From asst prof to assoc prof, 58-64, assoc dean, Col Biol & Agr, 80-86, PROF MICROBIOL, BRIGHAM YOUNG UNIV, 64- *Honors & Awards:* NIH career development award, 63-68. *Mem:* Am Soc Microbiologists; Am Soc Biol Chemists. *Res:* Metabolic pathways, energy relationships and biosynthetic mechanisms in anaerobic microorganisms; metabolism of natural products. *Mailing Add:* Dept Microbiol Brigham Young Univ Provo UT 84602

SAGERT, NORMAN HENRY, b Midland, Ont, Mar 31, 36; m 59; c 3. PHYSICAL & RADIATION CHEMISTRY, SURFACE SCIENCE. *Educ:* Queen's Univ, BSc, 59, MSc, 60; Ottawa Univ, PhD(phys chem), 63. *Prof Exp:* Fel phys chem, Cambridge Univ, 63-64; asst res officer, 64-68, assoc res officer, 68-76, sr res officer, 76-90, MGR, RES CHEM, ATOMIC ENERGY CAN LTD, 90- *Concurrent Pos:* Ed, Electrochem, Solution Chem & Thermochemistry, Can J Chem, 84-89; counr, Chem Inst Can, 87-90. *Mem:* Fel Chem Inst Can. *Res:* Surface and colloid science; radiation chemistry of hydrocarbons and iodine containing solutions; high temperature mass spectrometry. *Mailing Add:* Whiteshell Nuclear Res Estab Pinawa MB R0E 1L0 Can

SAGGIOMO, ANDREW JOSEPH, b Philadelphia, Pa, Mar 20, 31; m 53; c 3. MEDICINAL CHEMISTRY, QUALITY ASSURANCE. *Educ:* La Salle Col, BA, 52; Temple Univ, MA, 54. *Prof Exp:* Chemist, Philadelphia Qm Depot, 52; asst, Duquesne Univ, 52-53; res fel, Res Inst, Temple Univ, 53-56; res assoc, Germantown Labs, Inc, 56-61, proj dir, 61-69, financial mgr, 69-72, vpres & treas, 72-80; admin mgr, Franklin Res Ctr, 80-85, mgr, qual assurance, 82-85, QUAL ASSURANCE SPECIALIST, DEFENSE PERSONNEL SUPPORT CTR, 86- *Mem:* Am Chem Soc. *Res:* Organic fluorine chemistry; dyes; polymers; organometallics; medicinals; anticancer and anti-inflammatory agents; psychotropic drugs; antimalarials; polychlorinated biphenyl disposal methods. *Mailing Add:* 1817 Schley St Philadelphia PA 19145

SAGI, CHARLES J(OSEPH), b Phillipsburg, NJ, Mar 10, 35; m 59; c 2. MECHANICAL ENGINEERING. *Educ:* Lehigh Univ, BS, 56; Stanford Univ, MS, 61, PhD(mech eng), 65. *Prof Exp:* Develop engr, Ingersoll-Rand Co, 56-59; asst prof mech eng, Stanford Univ, 64-65; assoc sr res engr, Gen Motors Corp, 65-67; sr res engr, Creare, Inc, 67-68; sr res engr, Spec Progs Dept, 68-77, sr res engr, 77-80, staff res engr, Fluid Dynamics Res Dept, 80-85, STAFF RES ENGR, FLUID MECH DEPT, GEN MOTORS RES LABS, 85- *Mem:* Am Soc Mech Engrs. *Res:* Fluid mechanics of internal flow; aerodynamics of ground vehicles. *Mailing Add:* 5607 Thorny Ash Rochester MI 48063

SAGIK, BERNARD PHILLIP, b New York, NY, May 8, 25. VIROLOGY. *Educ:* City Col New York, BS, 47; Univ Ill, MS, 48, PhD, 52; Am Bd Med Microbiol, dipl pub health & virol. *Prof Exp:* Asst bact, Univ Ill, 48-52; Nat Found Infantile Paralysis fel & instr biophys, Sch Med, Univ Colo, 52-54; sect head virol, Upjohn Co, 54-62; dir viral chemother, Ciba Pharmaceut Co, 62-66; from assoc prof to prof microbiol, Univ Tex, Austin, 66-73; prof microbiol, Univ Tex Health Sci Ctr, San Antonio, 73-80, prof life sci, 73-80, dean, Col Sci & Math, 73-80; prof biol sci & vpres acad affairs, 80-88, PROF BIOSCI & BIOTECH, DREXEL UNIV, 88- *Concurrent Pos:* Vis scholar, Univ Ill, 60-61; lectr, City Univ New York, 63-66 & Drew Univ, 66; adj prof environ health eng, Univ Tex, Austin, 73-80. *Mem:* AAAS; Am Soc Microbiologists; Am Acad Microbiol. *Res:* Virus-host cell interactions; pathogenesis of virus infections; arbovirus genetics; viruses, sewage and terrestrial waste disposal. *Mailing Add:* Drexel Univ 32nd & Chestnut St Philadelphia PA 19104

SAGLE, ARTHUR A, b Honolulu, Hawaii; m 60; c 1. DIFFERENTIAL GEOMETRY, SYSTEMS THEORY. *Educ:* Univ Wash, BS, 56, MS, 57; Univ Calif, Los Angeles, PhD(math), 60. *Prof Exp:* Instr math, Univ Chicago, 60-62; asst prof, Syracuse Univ, 62-64; ONR fel & res instr, Univ Calif, Los Angeles, 64-65; res fel, Yale Univ, 65-66; prof, Univ Minn, 66-72; PROF MATH, UNIV HAWAII, 72- *Concurrent Pos:* NSF grants, 60-72; invit lectr, Am Math Soc, 64; vis prof, Univ Tex, 71; mem, Hadronic Mech Conf, Como, Italy, 84; ed, J Algebras Groups Geometries, 85; mem, Int Cong Math, 86; lect, Math Theory Network Systs, 87; vis scholar, Univ Utah, 88. *Honors & Awards:* Oberwolfach lectr, Ger Govt, 68. *Mem:* Sigma Xi; Am Math Soc; London Math Soc. *Res:* Investigation of the interdependency of differential geometry; systems; lie groups and non-associative algebras; stability and bifurcations of quadratic systems. *Mailing Add:* 20 Manu Pl Hilo HI 96720

SAH, CHIH-HAN, b Peiping, China, Aug 16, 34; US citizen; m 66; c 3. MATHEMATICS. *Educ:* Univ Ill, BS, 54, MS, 56; Princeton Univ, PhD(math), 59. *Prof Exp:* Instr math, Princeton Univ, 59-60; Benjamin Peirce instr, Harvard Univ, 60-63; from asst prof to prof, Univ Pa, 63-70; PROF MATH, STATE UNIV NY, STONY BROOK, 70- *Concurrent Pos:* Vis lectr, Harvard Univ, 67-68; vis prof, Univ Calif, Berkeley, 69-70 & 76-77. *Mem:* Am Math Soc. *Res:* Finite groups; algebraic number theory; rings; cohomology of groups. *Mailing Add:* Dept Math State Univ NY Stony Brook NY 11794

SAH, CHIH-TANG, b Beiling, China, Nov 10, 32; nat US; m 59; c 2. ENGINEERING PHYSICS, ELECTRICAL ENGINEERING. *Educ:* Univ Ill, BS(eng physics) & BS(elec eng), 53; Stanford Univ, MS, 54, PhD(elec eng), 56. *Hon Degrees:* Dr, Univ Leuven, Belg, 75. *Prof Exp:* Res asst, Electronics Lab, Stanford Univ, 54-56, res assoc, 56-57; mem sr staff, Semiconductor Lab, Shockley Transistor Corp, 56-59; sr mem tech staff, Fairchild Semiconductor Corp, 59-61; prof elec eng & physics, Univ Ill, Urbana, 63-88; PITTMAN EMINENT SCHOLAR CHAIR, UNIV FLA, GAINESVILLE, 88-, GRAD RES PROF CHAIR, 88- *Concurrent Pos:* Mgr & head physics dept, Fairchild Semiconductor Res Lab, 61-65; US Nat Acad Sci Committeeman, 75-78; life fel, Franklin Inst of Philadelphia. *Honors & Awards:* Browder J Thompson Prize, Inst Radio Eng, Inst Elec & Electronic Engrs, 62; Franklin Inst Award Develop Stable MOS Transistors, 75; J J Ebers Award, Electron Device Soc, 81; Achievement Award High Technol, Asian Am Mfg Asn, 84; Jack Morton Award, Inst Elec & Electronics Engrs, 88. *Mem:* Nat Acad Eng; fel Am Phys Soc; fel Inst Elec & Electronics Engrs. *Res:* Solid state and semiconductor electronics and physics. *Mailing Add:* Dept Elec Eng Univ Fla Gainesville FL 32611

SAHA, ANIL, b Calcutta, India, Mar 1, 30; US citizen. IMMUNOLOGY. *Educ:* Presidency Col, BS, 49; Univ Calcutta, MS, 52, PhD(appl chem), 61. *Prof Exp:* Res asst heme-proteins, Indian Coun Med Res, 53-56; res assoc, Med Col, Cornell Univ, 56-57; res fels, Calif Inst Technol, 60-64; asst prof med & allergy, McGill Univ, 65-69; asst prof, 69-70, ASSOC PROF MICROBIOL, COL MED & DENT NJ, 70-, ASST DEAN, GRAD SCH BIOMED SCI, 73- *Mem:* Am Chem Soc; Am Soc Biol Chemists; Am Asn Immunol; Am Soc Microbiol; Transplantation Soc; Sigma Xi. *Res:* Heme-proteins; immunoglobulins; cellular mediators. *Mailing Add:* Microbiol Dept NJ Coll Med Newark NJ 07103

SAHA, BIJAY S, b India. MAGNETIC CERAMICS, XEROGRAPHY. *Educ:* Banaras Hindu Univ, BS, 72, MS, 74; Indian Inst Technol, MTech, 76; State Univ NY, Stony Brook, PhD(mat sci), 85. *Prof Exp:* Inspection engr, Metall, Indian Oil Corp, 77-80; SCIENTIST, MAT SCI, EASTMAN KODAK CO, 85- *Mem:* Fel Inst Elec & Electronics Engrs; Am Phys Soc; Am Ceramic Soc; Am Soc Metals Int; Metall Soc. *Res:* Magnetic materials; xerography, metallurgical and ceramic materials design for electronics applications; surface science; coatings for specifics surface properties, materials characterization techniques; image science. *Mailing Add:* Eastman Kodak Res Labs 66 Eastman Ave Bldg 82 Rochester NY 14650-2129

SAHA, GOPAL BANDHU, b Chittagong, Bangladesh, Apr 30, 38; US citizen; m 65; c 2. NUCLEAR CHEMISTRY, RADIOPHARMACY. *Educ:* Dacca Univ, Bangladesh, BSc, 59, MSc, 60; McGill Univ, PhD(chem), 65; Am Bd Radiol, cert med nuclear physics, 75, Am Bd Nuclear Med Sci, 79. *Prof Exp:* Teaching fel biochem, Dacca Univ, 60-61; asst prof chem, Purdue Univ, 65-66; teaching asst chem, McGill Univ, 61-64, res assoc chem, 66-69, asst prof diag radiol, 70-75; from assoc prof to prof radiol & nuclear med, Univ Ark Med Sci, 76-82, dir radiopharmaceut prog & assoc prof, Col Health Related Profs, 76-82, prof pharm, Col Pharm, Univ NMex, 82-84; staff nuclear chemist, 84-88, DIR, NUCLEAR CHEM & PHARM, DEPT NUCLEAR MED, CLEVELAND CLIN FOUND, 88- *Concurrent Pos:* Radiopharmacist nuclear med, Royal Victoria Hosp, Montreal, 70-75. *Mem:* Soc Nuclear Med; Sigma Xi; Am Chem Soc; AAAS; Am Asn Physicists Med; Radiol Soc NAm. *Res:* Radiochemistry; reactor and cyclotron production of radionuclides and their chemical processing; surface chemistry; separation chemistry of different elements and compounds; activation analysis; preparation of radiopharmceuticals and study of their in vivo distribution; dosimetry of various radionuclides; quality control of radiopharmaceuticals; high performance liquid chromatography. *Mailing Add:* Dept Nuclear Med Cleveland Clin Found 9500 Euclid Ave Cleveland OH 44195

SAHA, JADU GOPAL, b Bengal, India, Dec 1, 31; Can citizen; m 57; c 1. PESTICIDE CHEMISTRY. *Educ:* Univ Calcutta, BSc, 53, MSc, 56; Univ Sask, PhD(org chem), 62. *Prof Exp:* Sr sci asst, Cent Fuel Res Inst, Dhanbad, India, 56-59; res fel org chem, Radiation Lab, Univ Notre Dame, 62-63; res fel, Univ Sask, 63-64; res officer, 65-66, res scientist, 67-75, DIR, CHEM & BIOL RES INST, CAN DEPT AGR, 75-; DIR GEN, HEALTH & WELFARE CAN. *Mem:* AAAS; Am Chem Soc; Chem Inst Can; Royal Soc Chem. *Res:* Utilization of coal tar and mechanism of aromatic substitution reactions; persistence, translocation, photodecomposition and metabolism of pesticides. *Mailing Add:* 11 Weatherwood Crescent Nepean ON K2E 7C5 Can

SAHA, PAMELA S, b Washington, DC, July 10, 51; m 72; c 2. MEDICINE. *Educ:* Stanford Univ, BS, 72; La State Univ, MD, 89. *Prof Exp:* ASST RES GEOL, YALE UNIV, 74-, RES ASST BIOCHEM, 78- *Concurrent Pos:* Resident pgy I&II, Vet Hosp, Shreveport, 89-91; resident physician, La State Univ Med Ctr, Shreveport, 89-91; resident doctor, Martin Luther King/Drew Med Ctr, 91- *Mem:* Am Med Asn; Am Psychiat Asn. *Res:* Published papers in national and international journals on bioethics, ethical issues in bioengineering, ethical issues related to animal research, ethical issues related to aids treatment. *Mailing Add:* 1373 S Center St Redlando CA 92373

SAHA, SUBRATA, b Kushita, India, Nov 2, 42; US citizen; m 72; c 2. BIOMECHANICS, BIOMATERIALS. *Educ:* Calcutta Univ, BE, 63; Tenn Tech, MS, 69; Stanford Univ, PhD(appl mech), 73. *Prof Exp:* Teaching & res asst, Stanford Univ, 71-73; res assoc, Yale Univ, 73-74, asst prof eng & appl sci, 74-79; ASSOC PROF BIOMED ENG, LA TECH, 79-; PROF & COORDR BIOENG, 79-, ASSOC PROF PHYSIOL & BIOPHYS, LA STATE UNIV MED CTR, SHREVEPORT, 80- *Concurrent Pos:* Mem Sch Grad Studies, La State Univ Med Ctr, 80-, Biomech Comt, Am Soc Civil Engrs, 81-, Long-Range Planning Comt, Shreveport Sigma Xi, 87-88; Grad fac, La State Univ Med Ctr, 87-; chmn steering comt, Southern Biomed Eng Conf, 81-83 & 84; mem-at-large, health care tech policy comt, Inst Elec & Electronics Engrs. *Honors & Awards:* Fulbright Award, 82; W C Hall Res Award, 87. *Mem:* Am Soc Mech Engrs; Alliance Eng Med Biol; Soc Biomat; Am Soc Civil Engrs. *Res:* Author of over 300 publications in national and international journals and conference proceedings in bioengineering, biomechanics and biomaterials. *Mailing Add:* Dept Orthop Surg La State Univ Med Ctr PO Box 33932 Shreveport LA 71130

SAHASRABUDDHE, CHINTAMAN GOPAL, b Shirpur, India, July 18, 36; m 60; c 2. PHYSICS, BIOCHEMISTRY. *Educ:* Agra Univ, India, BSc, 57; Vikram Univ, MSc, 59, MSc, 65; Ore State Univ, MS, 72, PhD(biochem-biophys), 74. *Prof Exp:* Lectr sci, MGGHSS, Mandleshwar, 59-63; lectr physics, Birla Inst Technol & Sci, 65-69; res assoc, MD Anderson Hosp & Tumor Inst, Univ Tex, 77-78, asst biochemist, 78-91; SR SCIENTIST, TANOX BIOSYSTS, 91- *Mem:* Biophys Soc; Am Soc Cell Biol; AAAS. *Res:* Structure-function relationship of chromatin, a genetically active complex of DNA, proteins and RNA in eukaryotes. *Mailing Add:* 4926 Willow Bend Houston TX 77035

SAHASRABUDHE, MADHU R, b Apr 1, 25; Can citizen; m 50; c 2. FOOD SCIENCE & TECHNOLOGY, NUTRITION. *Educ:* Agra Univ, BSC, 44; Banaras Hindu Univ, MSc, 46; Univ Bombay, PhD(biochem & nutrit), 52. *Prof Exp:* Res assoc food sci, Univ Ill, 54-57; tech officer, Kraft Foods Ltd, 57-58; head food additives, Nat Health & Welfare, Can, 58-68; mgr res & develop, Salada Foods Ltd, Salada Kelloggs, 69-72; actg dir, 73-74, sr res scientist food sci, Food Res Inst, 74-83, res coordr food, 83-84, actg dir, 84-85, ASST DIR FOOD RES, AGR CAN, 85- *Mem:* Am Oil Chemists' Soc; Can Inst Food Sci & Technol; Chem Inst Can; Inst Food Technol; Asn Food Scientist & Technologists India. *Res:* Plant lipids; chemistry; nutrition/processing; analytical methods; toxic compounds; safety food additives. *Mailing Add:* 15 Cramer Dr Nepean Ottawa ON K2H 5X2 Can

SAHATJIAN, RONALD ALEXANDER, b Cambridge, Mass, Oct 1, 42; m 66. ORGANOMETALLIC CHEMISTRY, POLYMER SCIENCE. *Educ:* Tufts Univ, BS, 64; Univ Mass, MS, 68, PhD(chem), 69. *Prof Exp:* Res chemist, Film Dept, E I du Pont de Nemours & Co, Inc, 69-71; scientist & supvr positive evaluation group, 71-75, res group leader, 75-80, res lab mgr, Polaroid Corp, Cambridge, 80-84 & Chem Fabrics, 84-86; VPRES RES, MEDI-TECH, 87- *Mem:* AAAS; Soc Photog Scientist & Engr; Am Chem Soc. *Res:* Organometallic carbonium ions; polymeric Schiff bases; dye diffusion processes in photography; non-silver imaging systems; organometallic polymers; medical polymers. *Mailing Add:* 29 Saddle Club Rd Lexington MA 02173

SAHBARI, JAVAD JABBARI, b Marand, Iran, Nov 13, 44; m 66; c 3. MICRO-ELECTRONICS CHEMICALS RESEARCH & DEVELOPMENT, PHOTOLITHOGRAPHY & OPTICAL SPECTROSCOPY. *Educ:* Univ Tehran, Iran, BS, 71; Univ Shiraz, Iran, MS, 75; Univ Calif, Davis, PhD(chem), 83. *Prof Exp:* Lectr gen/phys chem, Tabriz Univ, Iran, 75-77; assoc chem/gen, phys & anal, Univ Calif, Davis, 78-85; sr appln consult, Molecular Design Ltd, 85-86; sr chemist anal, Carter Anal Labs Inc, 86-87; res & develop mgr, EMT, Inc (Brent Chem Int), 87-89; CHIEF EXEC OFFICER & PRES, SILICON VALLEY CHEM LABS, INC, 89- *Concurrent Pos:* Res assoc, NSF, Univ Calif, Davis, 80-84. *Mem:* Am Chem Soc. *Res:* Author of several publications; X-ray crystallographic structures of acetylacetonate complexes and optical detection on magnetic resonance spectroscopy; granted several patents. *Mailing Add:* Silicon Valley Chem Labs Inc 3446 De La Cruz Blvd Santa Clara CA 95054

SAHINEN, WINSTON MARTIN, b Butte, Mont, Aug 4, 31; m 57; c 4. MINING ENGINEERING, GEOLOGICAL ENGINEERING. *Educ:* Mont Col Mineral Sci & Technol, BS, 53. *Prof Exp:* Res engr, Zonolite Co, 57-60; mine supt, Werdenhoff Mining Co, 60-61; sr mining engr, Pac Power & Light Co, 61-73; mining & geol engr, John T Boyd Co, 73-76, vpres, 76-80; PRES, SAHINEN MINING & GEOL SERV CO, 80- *Mem:* Am Inst Mining, Metall & Petrol Engrs. *Mailing Add:* 510 Bayou Knoll Dr Houston TX 77079

SAHLI, BRENDA PAYNE, b Richmond, Va, Sept 28, 42; m 67; c 3. OCCUPATIONAL HEALTH, TOXICOLOGY. *Educ:* Richmond Prof Inst, BS, 64; Med Col Va, MS, 67; Va Commonwealth Univ, PhD(pharmaceut chem), 74. *Prof Exp:* Res asst anal res, Am Tobacco Co, 64-65; chemist, Firestone Synthetic Fibers & Textiles Co, 67-69; teaching asst, Health Sci Ctr, Va Commonwealth Univ, 70-73; res chemist polymer res, Textile Fibers Dept, E I du Pont de Nemours & Co, 74-77; toxicologist, Va Dept Health, 77-82, vol compliance dir occup health, 82-83, dir toxic substance info, 83-84; CONSULT OCCUP HEALTH & ENVIRON TOXICOL, 84- *Concurrent Pos:* Adj prof, Acad Ctr, Va Commonwealth Univ, 75-77 & Falls Church Regional Ctr, Univ Va, 84-87; mem comt D-22, Am Soc Testing & Mat & chmn task force E-34; Gubernatorial appointee, Va Pesticide Control Bd, 89-91. *Mem:* Am Col Toxicol; Sigma Xi; Am Conf Govt Indust Hygienists; Soc Occup & Environ Health; Am Pub Health Asn. *Res:* Development of analytical test procedures for raw materials and fibers; bovine albumin tryptic hydrolyzate; isothermal compressibility of organic liquids; analytical ultracentrifugation; flame retardants; coatings for spun-bonded products; health hazard evaluation of substances with respect to conditions and circumstances of use. *Mailing Add:* 2900 Wicklow Lane Richmond VA 23236-1338

SAHLI, MUHAMMAD S, b Haifa, Palestine, June 8, 35; m 67; c 3. ORGANIC CHEMISTRY, POLYMER CHEMISTRY. *Educ:* Am Univ Beirut, BSc, 60; Univ SC, PhD(org chem), 66. *Prof Exp:* Instr chem, Am Univ Beirut, 60-61; res chemist, Film Res & Develop Lab, E I du Pont de Nemours & Co, Inc, 66-75; sr res scientist monomer technol, Fibers & Plastics Co, Allied Corp, 75-83; pres, Copy Van, Inc, 83-90; PRES, HARMON/ COMMONWEALTH CORP, 90- *Concurrent Pos:* Adj prof, Va Commonwealth Univ, 66-; mem, Sharp Electronics Corp Adv Coun, 84-89. *Mem:* Sigma Xi. *Res:* Organic synthesis; elucidation of structure of alkaloids; mechanism of pyrolysis of sulfoxides; emulsion polymerization and properties of dispersion coatings; formulation of coatings and characterization of polymers; industrial toxicology; Bechman rearrangement byproducts and mechanisms. *Mailing Add:* 2900 Wicklow Lane Richmond VA 23236-1338

SAHNEY, VINOD K, b Amritsar, India, Nov 16, 42; US citizen; m 70; c 2. OPERATIONS MANAGEMENT & PLANNING. *Educ:* Ranchi Univ, BSc, 63; Purdue Univ, MSME, 65; Univ Wis-Madison, PhD(indust eng), 70. *Prof Exp:* From asst prof to assoc prof indust eng, Wayne State Univ, 70-77; assoc prof health policy & mgt, Harvard Univ, 77-79; PROF INDUST ENG, WAYNE STATE UNIV, 79-; VPRES, CORP PLANNING & MARKETING, HENRY FORD HEALTH CARE CORP, 84- *Concurrent Pos:* Vis lectr, Exec Prog Health Policy & Mgt, Harvard Univ, 79-81; mem, Health Care Technol Study Sect, Dept Health & Human Serv, 80-; consult, Nat Ctr Health Serv Res, 80-; adminr, Henry Ford Hosp, Detroit, 81-; pres, Fairlane Health Serv Corp; bd dirs, Fairlane Health Serv Corp & Health Alliance Plan, Fla. *Mem:* Opers Res Soc; fel Inst Indust Engrs; fel Hosp Mgt Systs Soc. *Res:* Strategic planning and operations management in health services delivery organizations; developing better methods of planning and management control, including operations planning, staffing, and scheduling and managing the introduction of technology. *Mailing Add:* 4727 Burnley Dr Bloomfield Hills MI 48013

SAHNI, SARTAJ KUMAR, b Poona, India, July 22, 49; m 75; c 2. ALGORITHMS, DESIGN AUTOMATION. *Educ:* Indian Inst Technol, BTech, 70; Cornell Univ, MS, 72, PhD(comput sci), 73. *Prof Exp:* PROF COMPUT SCI, UNIV MINN, MINNEAPOLIS, 81- *Mem:* Asn Comput Mach; fel Inst Elec & Electronics Engrs; Soc Indust & Appl Math. *Res:* Design and analysis of computer algorithms; parallel computing; design automation of electronic circuits. *Mailing Add:* Univ Minn 136 Lind Hall Minneapolis MN 55455

SAHNI, VIRAHT, b Lahore, India, Dec 31, 44; c 1. MANY-BODY THEORY. *Educ:* Indian Inst Technol, India, BTech, 65; Polytech Inst Brooklyn, MS, 68, PhD(physics), 72. *Prof Exp:* Polytech fel elec engr, Polytech Inst Brooklyn, 65-68; instr, Pratt Inst, 68-70; sr res asst physics, Polytech Inst Brooklyn, 70-72; from instr to assoc prof, 72-81, PROF PHYSICS, BROOKLYN COL, 82- *Concurrent Pos:* Instr & fel, Brooklyn Col, 72-74; City Univ New York Res Found fac res grants, 73-74 & 75-87; vis res physicist, Inst Theoret Physics, Univ Calif, 83. *Mem:* Am Phys Soc; Sigma Xi. *Res:* Many-body theory of the inhomogeneous electron gas in atoms, molecules and metallic surfaces. *Mailing Add:* Dept Physics Brooklyn Col Brooklyn NY 11210

SAHU, SAURA CHANDRA, b Cuttack, India, June 29, 44; US citizen; m 66; c 3. PROTEOGLYCANS, LIPIDS. *Educ:* Uktal Univ, India, BS, 64; Columbia Univ, MS, 67; Univ Pittsburgh, PhD(chem), 71. *Prof Exp:* Res assoc biophys, Mich State Univ, 71-72 & C F Kettering Res Lab, 72-74; asst res prof biochem, Duke Univ, 74-79; res biochemist, US Consumer Prod Safety Comn, 79-88; RES CHEMIST, USDA, 88- *Concurrent Pos:* Fel, Indian Atomic Energy Comn, 64-66. *Honors & Awards:* Fel, Indian Atomic Energy Comn, 64-66. *Mem:* Am Soc Biol Chemists; Am Soc Pharmacol & Exp Therapeut; Soc Toxicol; Soc Complex Carbohydrates; NY Acad Sci; Oxygen Soc. *Res:* Lung biochemistry; inhalation toxicology; structure, function and metabolic activity of lungs; cancer biochemistry and free radical biochemistry. *Mailing Add:* 13321 Kurtz Rd 200 C St SW Woodbridge VA 22193

SAHYOUN, NAJI ELIAS, b Brummana, Lebanon, Nov 27, 49; m 88. MOLECULAR NEUROBIOLOGY, CELLULAR NEUROBIOLOGY. *Educ:* Am Univ Beirut, BSc, 69, MSc, 72, MD, 74. *Prof Exp:* Fel, Johns Hopkins Sch Med, 74, 75, JHH Dept Med Div Clin Phamacol, 75-77; asst prof, dept Biochem, Am Univ Beirut, 78-79; sr scientist, 77-85, PRIN SCIENTIST, MOLECULAR BIOL DEPT, BURROUGHS WELLCOME CO, 85-, SECT HEAD, DIV CELL BIOL, 88- *Concurrent Pos:* Vis prof, Am Univ Beirut Sch Med, 79-83. *Mem:* Am Soc Biochem & Molecular Biol; Soc Neurosci; NY Acad Sci; AAAS. *Res:* Identification and characterization of molecular switches regulating neuronal growth, differentiation synapse formation, communication and plasticity. *Mailing Add:* Div Cell Biol Burroughs Wellcome Co 3030 Cornwallis Rd Research Triangle Park NC 27709

SAHYUN, MELVILLE RICHARD VALDE, b Santa Barbara, Calif, Feb 11, 40; m 66; c 2. PHOTOGRAPHIC CHEMISTRY. *Educ:* Univ Calif, Santa Barbara, AB, 59; Univ Calif, Los Angeles, PhD(chem), 63. *Prof Exp:* Sr asst scientist, Nat Cancer Inst, 62-65; NIH res fel chem, Calif Inst Technol, 65-66; res specialist, Imaging Res Lab, 66-74, sr res specialist, Systs Res Lab, 74-81, STAFF SCIENTIST, CORP RES LAB, 3M CO, 81- *Mem:* Am Chem Soc; fel Soc Photog Sci & Eng; Chem Soc London. *Res:* Photographic science; photography; reaction kinetics; organic photochemistry; solid state science; computer modelling. *Mailing Add:* Corp Res Lab 3M Co Box 33221 St Paul MN 55133-3221

SAIBEL, EDWARD, mathematics, mechanics; deceased, see previous edition for last biography

SAID, RUSHDI, b Cairo, Egypt, May 12, 20; US citizen; m 53; c 2. GENERAL EARTH SCIENCE, GEOLOGY. *Educ:* Cairo Univ, BSc, 41 & MSc, 44; Harvard Univ, PhD(geol), 50. *Hon Degrees:* Dr, Tech Univ, Berlin, 86. *Prof Exp:* Lectr geol, Cairo Univ, 51-57, asst prof, 57-64; prof geol, Alexandria Univ, 64-68; pres, Mining Orgn, Egypt, 68-70, Geol Surv Egypt, 70-78; CONSULT, 78- *Concurrent Pos:* Mem bd, Int Geol Correlation Prog, 73-75; vpres, Arab Mining Co, Amman, Jordan, 75-78; sr res scientist, Inst Earth & Man, Southern Methodist Univ, Dallas, 78-; assoc ed, J African Earth Sci, 83-; fel, Inst Advan Studies, Berlin, 88- *Honors & Awards:* Medal Sci & Arts, Egyptian Govt, 62; Nachtigal Medal, Geog Soc Berlin, 86. *Mem:* Egyptian Acad Sci; hon fel Geol Am Soc; fel Royal Geog Soc; hon mem, Geol Soc Africa; emer mem, Am Asn Petrol Geol; !nst Egypt. *Res:* Author of a large number of publications dealing primarily with different aspects of geology of Egypt and evaluation of River Nile. *Mailing Add:* 3801 Mill Creek Dr Annandale VA 22003

SAID, SAMI I, b Cairo, Egypt, Mar 25, 28. PULMONARY DISEASE, PEPTIDES. *Educ:* Univ Cairo, MB, BCh, 51. *Prof Exp:* Intern, Univ Hosp, Univ Cairo, 51-52, resident internal med, 53; asst resident, Bellevue & Univ Hosps, Postgrad Med Sch, NY Univ, 53-55, instr med, Sch Med, 55, NY Heart Asn res fel, Bellevue Hosp, 55-57; asst physician & fel, Sch Med, Johns Hopkins Univ & Johns Hopkins Hosp, 57-58; from asst prof to prof, Med Col Va, 58-71; prof internal med & pharmacol, Univ Tex Health Sci Ctr, 71-81; chief, Pulmonary Dis Sect, Dallas Vet Admin Med Ctr, 71-81; prof med, chief, pulmonary dis & critical care sect, Univ Okla Health Sci Ctr, 81-87; PROF & ASSOC HEAD RES, DEPT MED, UNIV ILL COL MED & VET ADMIN WESTSIDE MED CTR, 87-, MED INVESTR, VET ADMIN, 88- *Concurrent Pos:* Fulbright res fel, Naval Med Res Unit 3, 52-53; Fulbright traveling fel, 53; Nat Heart Inst res career develop award, 62-71; vis scientist, Karolinska Inst, Sweden, 68-70; dir, Pulmonary Specialized Ctr Res, Dallas, 71-81; mem, merit rev bd respiration, Vet Admin, 72-75; mem, pulmonary-allergy clin immunol adv comt, Bur Drugs, Food & Drug Admin, 72-76 & pulmonary dis adv comt, Vet Admin, 79-81; exchange scientist, Vet Admin-INSERM, France, 73; mem, rev panel nat res & demonstration ctrs, Nat Heart & Lung Inst, NIH, 74; assoc ed, Peptides, 80- *Honors & Awards:* William S Middleton Award, Vet Admin, 81; Smith-Kline Award, Can Cong Clin Chem, 83. *Mem:* Am Soc Clin Invest; Asn Am Physicians; Soc Neurosci; Am Gastroenterol Asn; Endocrine Soc; hon mem Royal Belgian Soc Gastroenterol; Sigma Xi; Am Psysiol Soc; Am Soc Pharmacol Exp Ther. *Res:* Mediators of pulmonary responses in health and disease; biology and biochemistry of vasoactive intestinal peptide and related peptides. *Mailing Add:* Dept Med Col Med MC 789 Univ Ill 1940 W Taylor St Chicago IL 60612

SAIDAK, WALTER JOHN, b Ottawa, Ont, May 10, 30; m 56; c 2. WEED SCIENCE. *Educ:* Ont Agr Col, BSA, 53; Cornell Univ, MS, 55, PhD(veg crops), 58. *Prof Exp:* Res officer, Plant Res Inst, Can Dept Agr, 58-62, res scientist, Res Sta, 62-73, res coordr weeds, Res Br, Cent Exp Farm, 73-89; RETIRED. *Mem:* Weed Sci Soc Am; Agr Inst Can. *Res:* Weed control in field and horticultural crops; translocation of herbicides. *Mailing Add:* 50 Kilmory Cr Nepean Ottawa ON K2E 6N1 Can

SAIDE, JUDITH DANA, b Worcester, Mass, Feb 21, 44. MUSCLE STRUCTURE & CHEMISTRY. *Educ:* Vassar Col, AB, 65; Boston Univ, PhD(physiol), 72. *Prof Exp:* Res fel, Dept Med, Mass Gen Hosp, 72-75; asst biochem, 75-77, asst prof, 77-82, ASSOC PROF, DEPT PHYSIOL, SCH MED, BOSTON UNIV, 82- *Concurrent Pos:* Res fel biol chem, Harvard Med Sch, 72-75, instr, Dept Physiol, 75-77; establ investr, Am Heart Asn, 77. *Mem:* AAAS; Am Heart Asn. *Res:* Identification, characterization and assembly of proteins of the z-band of striated muscle. *Mailing Add:* L-713 Dept Physiol Sch Med Boston Univ 80 E Concord St Boston MA 02118

SAIDEL, GERALD MAXWELL, b New Haven, Conn, May 27, 38; m 69; c 2. RESPIRATORY SYSTEM. *Educ:* Rensselaer Polytech Inst, BChE, 60; Johns Hopkins Univ, PhD(chem eng), 65. *Prof Exp:* From asst prof to assoc prof, 67-81, PROF BIOMED ENG, CASE WESTERN RESERVE UNIV, 81-, CHMN, 87- *Concurrent Pos:* Res engr, Vet Admin Med Ctr, Cleveland, 70-85; sect ed, Annals of Biomed Eng, 79-85; fel, City Col, NY, 85-86; vis prof, Techion-Israel Inst Tech, 85-86. *Mem:* Am Inst Chem Eng; Biomed Eng Soc (pres, 87-88); Am Asn Univ Professors; Inst Elec & Electronics Engrs Engr Biol Med Soc. *Res:* Transport processes in biomedical systems; modeling and computer simulation; parameter estimation of dynamic systems; optimal experiment design. *Mailing Add:* Dept Biomed Eng Case Western Reserve Univ Wickenden Bldg Cleveland OH 44106

SAIDEL, LEO JAMES, b Lanark, Ill, Aug 22, 16; m 43; c 3. BIOCHEMISTRY. *Educ:* Univ Chicago, BS, 38; Georgetown Univ, MS, 41, PhD(biochem), 46. *Prof Exp:* Lab aide, Food & Drug Admin, USDA, 38 & Bur Dairy Indust, 38-40, jr chemist, 40-42; res assoc, Col Physicians & Surgeons, Columbia Univ, 42-46; chemist, G Barr & Co, Ill, 46-47; assoc, Univ Chicago, 47; from instr to prof biochem, Chicago Med Sch, 47-82, chmn dept, 75-76; RETIRED. *Mem:* Am Chem Soc; Am Soc Biol Chemists. *Res:* Composition, structure and properties of peptides and proteins; ultraviolet absorption spectra of proteins and related materials. *Mailing Add:* PO Box 1205 Lyons CO 80540

SAIDUDDIN, SYED, b Kakinada, India, Dec 7, 38; m 67; c 1. ENDOCRINOLOGY, REPRODUCTIVE PHYSIOLOGY. *Educ:* Sri Venkateswara Univ, India, BVSc, 59; Indian Vet Res Inst, NDAG, 62; Univ Nev, Reno, MS, 64; Univ Wis-Madison, PhD(endocrinol), 68. *Prof Exp:* Proj assoc, Univ Wis, 67-69; NIH res fel, Med Sch, Tufts Univ, 69-71; asst prof, 71-76, assoc prof, 76-79, PROF VET PHYSIOL, COL VET MED, OHIO STATE UNIV, 79- *Mem:* Soc Study Reproduction; fel Am Col Vet Pharmacol & Therapeut; Endocrine Soc; Am Physiol Soc. *Res:* Endocrine control of ovarian follicular growth; mechanism of action of estrogen and progesterone. *Mailing Add:* Dept Vet Physiol Ohio State Univ Col Vet Med Columbus OH 43210

SAIED, FAISAL, b Karachi, Pakistan, July 14, 51; US citizen; m 82; c 2. NUMERICAL ANALYSIS, PARALLEL COMPUTATION. *Educ:* Trinity Col, Cambridge, Eng, BA Hons, 73; Gottingen Univ, Ger, dipl math, 77; Yale Univ, PhD(computer sci), 90. *Prof Exp:* ASST PROF COMPUTER SCI, UNIV ILL, URBANA-CHAMPAIGN, 89- *Mem:* Soc Indust & Appl Math. *Res:* Numerical analysis; scientific computation; partial differential equations; parallel algorithm; ocean acoustics. *Mailing Add:* 2206 Fletcher Urbana IL 61801

SAIER, MILTON H, JR, b Palo Alto, Calif, July 30, 41; m 61; c 3. MOLECULAR TRANSPORT, CELL REGULATION. *Educ:* Univ Calif, Berkeley, BS, 63, PhD(biochem), 68. *Prof Exp:* From asst prof to assoc prof, 72-82, PROF BIOL, UNIV CALIF, SAN DIEGO, 82- *Mem:* Am Soc Microbiol; Am Soc Cell Biol; Am Soc Biol Chemists. *Res:* Mechanism and regulation of sugar transport in bacteria; mechanism and regulation of salt transport in kidney cells. *Mailing Add:* Dept Biol C-016 Univ Calif San Diego Box 109 La Jolla CA 92093

SAIF, LINDA JEAN, b Columbus, Ohio, June 29, 47; m 70; c 1. MICROBIOLOGY, IMMUNOLOGY. *Educ:* Col Wooster, BA, 69; Ohio State Univ, MS, 71, PhD(microbiol), 76. *Prof Exp:* Res asst, Dept Microbiol, Case Western Reserve Univ, 69-70; instr microbiol, Ohio Agr Res & Dev Ctr, 72-74, res assoc, 75-76, res assoc fel, 77-78, asst prof, dept vet sci, 79-85, assoc prof, Food Animal Health Res Prog, 85-90, PROF, OHIO AGR RES & DEVELOP CTR, 90- *Concurrent Pos:* Vis prof, Col Vet Med, Univ Guelph, Ont, Can; consult, FAO & Interamerican Develop Bank. *Mem:* Am Soc Microbiol; Conf Res Workers Animal Dis; Am Soc Virol; Am Asn Vet Immunol. *Res:* Basic mechanisms of the immune response of swine and cattle; mechanisms of protection against enteric viral infections; identification and purification of bovine and porcine enteric viruses and immunoglobulins. *Mailing Add:* Food Animal Health Res Prog Ohio Agr Res & Develop Ctr/ Ohio State Univ Wooster OH 44691

SAIF, YEHIA M(OHAMED), b Minia, Egypt, Dec 23, 34; m 70; c 1. VETERINARY IMMUNOLOGY. *Educ:* Cairo Univ, DVM, 58; Ohio State Univ, MSc, 64, PhD(vet med), 67; Am Col Vet Microbiol, dipl. *Prof Exp:* Teaching asst vet med, Cairo Univ, 59-62; res asst vet sci, 65-67, fel, 67-68, from asst prof to assoc prof, 68-77, PROF VET PREV MED, OHIO AGR RES & DEVELOP CTR, OHIO STATE UNIV, 77- *Honors & Awards:* Upjohn Res Achievement Award, Am Asn Avian Pathologists; Res Award, Nat Turkey Fedn; Beecham Award for Res Excellence. *Mem:* AAAS; Am Soc Microbiol; Poultry Sci Asn; Am Vet Med Asn; NY Acad Sci. *Res:* Immune response of poultry; poultry diseases. *Mailing Add:* Food Animal Health Res Prog Ohio Agr Res & Develop Ctr Ohio State Univ Wooster OH 44691

SAIFER, MARK GARY PIERCE, b Philadelphia, Pa, Sept 16, 38; m 61; c 2. SUPEROXIDE DISMUTASE, ORGOTEIN. *Educ:* Univ Pa, AB, 60; Univ Calif, Berkeley, PhD(biophys), 67. *Prof Exp:* Actg asst prof zool, Univ Calif, Berkeley, 66; sr cancer res scientist, Roswell Park Mem Inst, 68-70; lab dir enzym, Diag Data, Inc, 70-78; VPRES & TECH DIR, DDI PHARMACEUT, INC, 78- *Concurrent Pos:* Am Cancer Soc fel, Dept Bacteriol & Immunol, Univ Calif, Berkeley, 67-68; fel, Int Lab Genetics & Biophys, Naples, 67; Damon Runyon Mem Fund grant, 68-70; res develop award, Health Res Inc, Buffalo, 68-70. *Mem:* AAAS; NY Acad Sci; Parenteral Drug Asn. *Res:* Regulation of synthesis, compartmentalization and secretion of proteins;

biological and medical effects of superoxide dismutase; immunology and pharmacology of proteins; suppression of immunogenicity by chemical modification of proteins; long-acting derivatives of proteins. *Mailing Add:* 518 Logue Ave Mountain View CA 94043

SAIFF, EDWARD IRA, b New Brunswick, NJ, Oct 11, 42; m 67; c 2. ZOOLOGY, ANATOMY. *Educ:* Rutgers Univ, BA, 64, PhD(zool), 73; State Univ NY Buffalo, MA, 68. *Prof Exp:* Instr zool, Rutgers Univ, 70-71, lectr, 71-72; asst prof, 72-75, assoc prof, 75-80, PROF BIOL, RAMAPO COL, NJ, 80- *Concurrent Pos:* Dir, Sch Theoret & Anal Sci, Ramapo Col, 84- *Mem:* Am Ornith Union; Am Soc Zoologists; Soc Syst Zoology; Linnean Soc London; Sigma Xi; AAAS; NY Acad Sci; Soc Neurosci. *Res:* Avian anatomy, particularly of the middle ear region; anatomical correlates of hearing in birds; evolutionary theory; CNS receptor sites; neurohistology. *Mailing Add:* Theoret & Appl Sci Ramapo Col 505 Ramapo Valley Rd Mahwah NJ 07430

SAIGAL, SUNIL, b Karnal, India, July 13, 57; m. COMPUTATIONAL MECHANICS, NUMERICAL METHODS. *Educ:* Punjab Eng Col, Chandigarh, India, BS, 78; Indian Inst Sci, Bangalore, India, MS, 80; Purdue Univ, PhD(aerospace eng), 85. *Prof Exp:* Postdoctoral fel aerospace eng, Purdue Univ, 85-86; asst prof mech eng, Worcester Polytech Inst, 86-89; asst prof, 89-91, ASSOC PROF CIVIL ENG, CARNEGIE MELLON UNIV, 91- *Concurrent Pos:* NASA-CASE res fel, Lewis Res Ctr, Ohio, 89; NSF presidential young investr, 90; Ladd res award, 90. *Mem:* Am Soc Civil Engrs; Am Soc Mech Engrs; Am Inst Aeronaut & Astronaut. *Res:* Integration of finite elements and boundary elements for unbounded domains; inverse problems in engineering and shape optimization with boundary elements; computational bounds for equivalent material properties of woven composites. *Mailing Add:* Dept Civil Eng Carnegie Mellon Univ Pittsburgh PA 15213

SAIGER, GEORGE LEWIS, epidemiology, medical statistics, for more information see previous edition

SAIGO, ROY HIROFUMI, b Sacramento, Calif, Aug 6, 40; m 67. PLANT ANATOMY, PLANT PATHOLOGY. *Educ:* Univ Calif, Davis, BA, 62; Ore State Univ, PhD(plant anat), 69. *Prof Exp:* Sci & agr dean, 84-90, PROF BIOL, PROVOST & VPRES ACAD & STUDENT AFFAIRS, SOUTHEASTERN LA UNIV, 90-; ASSOC PROF BIOL, UNIV WIS-EAU CLAIRE, 67- *Concurrent Pos:* Fac res grants, Univ Wis-Eau Claire, 68-69 & 71-72, teacher improv assignment, 70, asst to dean arts & sci, 80-; acad affairs intern, Univ Wis Syst, 75-76; sci & agr dean, Univ Northern Iowa, 84-90. *Honors & Awards:* Charles E Bessey Award, Bot Soc Am. *Mem:* AAAS; Bot Soc Am; Am Inst Biol Sci. *Res:* Effect of insects on the bark of coniferous trees; ultrastructural investigation of the phloem of lower vascular plants and protein body development in oats. *Mailing Add:* Southeastern La Box 768 University Sta Hammond LA 70402

SAI-HALASZ, GEORGE ANTHONY, b Budapest, Hungary, Dec 7, 43; m 70; c 1. MICROELECTRONICS. *Educ:* Eotovos Roland Sci Univ, Budapest, dipl, 66; Case Western Reserve Univ, PhD(physics), 72. *Prof Exp:* Fel, Univ Pa, 72-74; res staff mem, 74-84, RES MGR, IBM WATSON RES LAB, 84- *Mem:* Am Phys Soc; fel Inst Elec & Electronics Engrs. *Res:* Low temperature physics; transport in solid helium, nonequilibrium phenomena in superconductivity; semiconductor superlattices; band structure, optical and transport properties; physics of semiconductor devices; computer systems. *Mailing Add:* IBM T J Watson Res Ctr PO Box 218 Yorktown Heights NY 10598

SAILA, SAUL BERNHARD, b Providence, RI, May 23, 24; m 49; c 3. FISH BIOLOGY. *Educ:* Univ RI, BS, 49; Cornell Univ, MS, 50, PhD(fishery biol), 52. *Prof Exp:* Res assoc zool, Ind Univ, 52-54; fishery biologist, Div Fish & Game, RI Dept Agr & Conserv, 54-56; from asst prof marine biol to assoc prof oceanog, 56-67, coordr comput lab, 59-76, dir marine exp sta, 66-76, PROF OCEANOG, UNIV RI, 67- *Honors & Awards:* Am Fisheries Soc Award, 59. *Mem:* AAAS; Am Fisheries Soc; Am Soc Limnol & Oceanog; Inst Fishery Res Biologists; Int Asn Theoret & Appl Limnol. *Res:* Fish population dynamics. *Mailing Add:* Grad Sch Oceanog/Zool Univ RI Kingston RI 02881

SAILOR, SAMUEL, civil engineering, photogrammetry; deceased, see previous edition for last biography

SAILOR, VANCE LEWIS, b Springfield, Mo, June 28, 20; m 43; c 3. NUCLEAR SCIENCE. *Educ:* DePauw Univ, AB, 43; Yale Univ, MS, 47, PhD(physics), 49. *Prof Exp:* From assoc physicist to physicist, Brookhaven Nat Lab, Upton, 49-67, sr physicist, 67-85; CONSULT, NUCLEAR POWER PLANT SAFETY, 85- *Concurrent Pos:* Dir systs anal proj, Int Energy Agency. *Mem:* Am Phys Soc; Am Nuclear Soc; Am Asn Physics Teachers; Sigma Xi. *Res:* Neutron and reactor physics; charged particle reactions; low temperature physics; nuclear energy; environmental effects of energy production and usage. *Mailing Add:* 100 Durkee Lane East Patchogue NY 11772

SAIN, MICHAEL K(ENT), b St Louis, Mo, Mar 22, 37; m 63; c 5. CONTROL SYSTEMS. *Educ:* St Louis Univ, BS, 59, MS, 62; Univ Ill, PhD(elec eng), 65. *Prof Exp:* From asst prof to prof, 65-82, FRANK M FREIMANN PROF ELEC ENG, UNIV NOTRE DAME, 82- *Concurrent Pos:* Vis scientist, Univ Toronto, 72-73; res grants, NASA, 75-85, Airforce Off Sci Res, 76-78, NSF, 66-71, 73-77, 81-85 & 91-92, Off Naval Res, 79-81, Army Res Off, 88-89, Clark Components Int, 90-; mem rev panel, NSF, 76, 79 & 84; ed, Transactions on Automatic Control, Inst Elec & Electronics Engrs, 79-83; distinguished vis prof, Ohio State Univ, 87. *Honors & Awards:* Centennial Medal, Inst Elec & Electronics Engrs, 84; distinguished mem Control Syst Soc, Inst Elec & Electronics Engrs, 83. *Mem:* Fel Inst Elec & Electronics Engrs; Soc Indust & Appl Math; Am Soc Eng Educ. *Res:* Nonlinear multivariable control systems; hysteretic circuits, systems and control; structural control for earthquake hazard mitigation; gas turbine engine control; pressure modelling in internal combustion engines; global zeros and feedback properties; hybrid models and autonomous control; algebraic system theory and applications. *Mailing Add:* Dept Elec Eng Univ Notre Dame Notre Dame IN 46556

SAINI, GIRDHARI LAL, b Hariana, India, Aug 2, 31; m 49; c 3. MATHEMATICS. *Educ:* Panjab Univ, India, BS, 55, MA, 57; Indian Inst Technol, Kharagpur, PhD(math), 61. *Prof Exp:* Assoc lectr math, Indian Inst Technol, Kharagpur, 60, lectr, 60-62; vis asst prof, Math Res Ctr, Univ Wis-Madison, 62-64; asst prof, Indian Inst Technol, New Delhi, 64-67; assoc prof, 67-76, PROF MATH, UNIV SASK, 76- *Concurrent Pos:* Res mem, US Army Res Ctr, Univ Wis-Madison, 62-64. *Mem:* Am Math Soc; Can Math Soc. *Res:* Relativistic fluid mechanics. *Mailing Add:* Dept Math Univ Sask Saskatoon SK S7N 0W0 Can

SAINI, RAVINDER KUMAR, b Hoshiarpur, India, Jan 28, 46; US citizen; m 71; c 2. CLINICAL CARDIOVASCULAR PHARMACOLOGY, CLINICAL CARDIOLOGY. *Educ:* Col Vet Med, Hissar, India, DVM, 68; Postgrad Med Res Inst, Chandigarh, India, MS, 71; Univ Naples, Italy, PhD(pharmacol), 73. *Prof Exp:* Res assoc pharmacol, Sch Med, Univ Pa, 73-74; res fel pharmacol, Univ Wis, Madison, 74-76; res assoc pharmacol, Sch Med, Univ Miami, Fla, 76-78; res pharmacologist, Merrell Res Ctr, Cincinnati, Ohio, 78-79; SR RES INVESTR PHARMACOL, SQUIBB INST MED RES, 79-; ASSOC DIR CARDIOVASC CLIN, BRISTOL-MYERS SQUIBB CO. *Concurrent Pos:* Res grant investr, Tobacco Inst, 73-74, Am Lung Asn, 74-76 & NIH Cardiovasc Training, 76-78. *Mem:* Am Soc Pharmacol & Exp Therapeut; fel Am Col Angiol; Int Soc Heart Res; fel Am Col Cardiol. *Res:* Planning, development and management of cardiovascular clinical programs for arrhythmias, hypertension and lipid-lowering agents. *Mailing Add:* Dept Cardiovasc Clin Res Bristol Myers Squibb Co Rte 206 & Provinceline Rd Princeton NJ 08543

SAINSBURY, ROBERT STEPHEN, b Halifax, NS, Apr 16, 43; m 81; c 2. NEUROPSYCHOLOGY. *Educ:* Mt Allison Univ, BA, 63; Dalhousie Univ, MA, 65; McMaster Univ, PhD(psychol), 69. *Prof Exp:* From asst prof to assoc prof, 69-80, PROF PSYCHOL, UNIV CALGARY, 80- *Mem:* Can Psychol Asn. *Res:* The effects of brain lesions on species typical behavior in small mammals. *Mailing Add:* Dept Psychol Univ Calgary 2500 University Dr Calgary AB T2N 1N4 Can

ST AMAND, PIERRE, b Tacoma, Wash, Feb 4, 20; m 45; c 4. GEOPHYSICS. *Educ:* Univ Alaska, BS, 48; Calif Inst Technol, MS, 51, PhD(geophys, geol), 53. *Prof Exp:* Magnetic observer, Carnegie Inst, Alaska, 41-42, mem geophys inst, 46-48; physicist, 50-61, head, Earth & Planetary Sci Div & Spec Projs Off, 61-81, sr exec serv, Off Tech Dir, 81-88; ST-AMAND SCI SERV, 80- *Concurrent Pos:* Asst, Seismol Lab, Calif Inst Technol, 52-54; Fulbright scholar, France, 54-55; Int Coop Admin prof sch geol, Chile, 58-61; consult, UN Chilean & Argentine Govts, 60, Mex, Can & States of Calif, SDak, NDak, Ore & Wash; consult, Orgn Am States, 65-72; adj prof atmospheric sci, Univ Ndak, geol, McKay Sch Mines, Univ Nev, Reno. *Honors & Awards:* Distinguished Civilian Serv Medal, US Navy, 67 & Meritorious Serv Medal; Spec Award, Philippine Air Force; L T E Thompson Award, Naval Weapon Ctr, 74; Distinguished Pub Serv Award, 76; Thunderbird Award, Weather Modification Asn; Dipl de Honor, Soc Geol de Chile, 65. *Mem:* Fel AAAS; Seismol Soc Am; fel Geol Soc Am; Am Geophys Union; Weather Modification Asn; Earthquake Eng Res Inst. *Res:* Auroral height measurement; atmospheric refraction; terrestrial magnetism; ionosphere; light of night sky; seismology; earthquakes; structural geology; electronics and instrumentation; circum pacific tectonics; weather modification; oceanography; deep sea research; Arctic research. *Mailing Add:* 1748 Las Flores Ridgecrest CA 93555

ST AMAND, WILBROD, b Old Town, Maine, May 5, 27; m 50. CYTOGENETICS. *Educ:* Univ Maine, BA, 48; Univ Tenn, MS, 49, PhD(zool, entom), 54. *Prof Exp:* Asst zool, Univ Tenn, 48-49, instr, Exten Serv, 50; res assoc radiation biol, Oak Ridge Nat Lab, 54-55, biologist, 55-58; from assoc prof to prof, 58-88, EMER PROF BIOL, UNIV MISS, 88- *Mem:* AAAS; Am Micros Soc; Am Soc Zoologists; Genetics Soc Am; Am Inst Biol Sci. *Res:* Radiation cytology; radiosensitivity of the stages of mitosis; mouse genetics. *Mailing Add:* Rte 4 Box 226 Oxford MS 38655

ST ANGELO, ALLEN JOSEPH, b New Orleans, La, April 11, 32; m 58; c 3. LIPIDOXIDATION, ENZYMOLOGY. *Educ:* Southeastern La Univ, BS, 57; Tulane Univ, MS, 65, PhD(biochem), 68. *Prof Exp:* Res chemist, 58-79, actg res leader, 79-82, RES CHEMIST, SOUTHERN REGIONAL RES CTR, USDA, 82- *Concurrent Pos:* Vis adj prof, Food Sci Dept, Va Polytech Inst & State Univ, Blacksburg, Va. *Honors & Awards:* Invention Award, USDA, 89. *Mem:* Am Chem Soc; Am Oil Chemists Soc; Inst Food Technol; Sigma Xi. *Res:* Develop methodology for assessing overall quality of meat, poultry and fish; determine mechanism for formation of warmed-over flavor in meat and design procedure for its prevention. *Mailing Add:* Southern Regional Res Ctr USDA PO Box 19687 New Orleans LA 70179

SAINT-ARNAUD, RAYMOND, b Shawinigan, Que, Sept 23, 35; m 62; c 2. ELECTRICAL ENGINEERING. *Educ:* Laval Univ, BA, 55, BScAppl, 61, Dipl Adm, 72; Univ Strathclyde, PhD(elec eng), 66. *Prof Exp:* Engr, Hydro-Quebec, Montreal, 61-62; res asst eng, Univ Strathclyde, 62-65; asst prof elec eng, 65-70, ASSOC PROF ELEC ENG, LAVAL UNIV, 70- *Concurrent Pos:* Res assoc, Dept Exp Med, Laval Univ, 71- *Mem:* Inst Elec & Electronics Engrs. *Res:* High voltage engineering; ionization phenomena; gas lasers; electrostatics; bioelectricity; biometeorology. *Mailing Add:* Head Genie Elec Laval Univ St Foy PQ C1K 7P4 Can

ST ARNAUD, ROLAND JOSEPH, Can citizen. SOILS. *Educ:* Univ Sask, BSA, 48, MSc, 50; Mich State Univ, PhD(soils), 61. *Prof Exp:* Res officer, Sask Soil Surv, Can Dept Agr, 50-56; from asst prof to assoc prof, 56-70, PROF SOIL SCI, UNIV SASK, 70- *Concurrent Pos:* Sci ed, Can Soc Soil Sci, 63-66. *Mem:* Fel Can Soc Soil Sci (pres, 74-75); Am Soc Agron; Agr Inst Can. *Res:* Soil classification; mineralogical studies and micropedology. *Mailing Add:* 1302 Preston Ave Saskatoon SK S7H 2V4 Can

ST CLAIR, ANNE KING, b Bluefield, WVa, May 31, 47; m 71; c 1. BIOCHEMISTRY. *Educ:* Queens Col, BA, 69; Va Polytech & State Univ, MS, 72. *Prof Exp:* Res assoc, Nat Aeronaut & Space Admin, 72-77; polymer res chemist, 77-81, sr scientist, 81-86, MGR, ADVAN AIRCRAFT PROG, MAT DIV, LANGLEY RES CTR, NASA, 86- *Concurrent Pos:* Mem, Speakers Bur, Soc Advan Mat Process Engrs, 80-; lectr, State Univ NY, 81 & 82, sci prog chmn & lectr, course on high temperature polymers, 84-, Ehime Univ, Tokyo, Japan, 87; chmn, tech sessions, Am Chem Soc, Soc Advan Mat Process Engrs & Gordon Res Conf; chmn, Potentially Hazardous Mat Comt, NASA-Langley, 83-87; elected vchmn, Gordon Res Conf Films & Coatings, 89. *Honors & Awards:* IR-100 Award, 81. *Mem:* Am Chem Soc; Soc Advan Mat Process Engrs; Asn Women Sci. *Res:* Synthesis, characterization and development of high-performance aerospace materials for applications as structural adhesives, advanced composites, films and fibers; author or coauthor of 35 publications; granted 16 patents. *Mailing Add:* Langley Res Ctr NASA Mail Stop 227 Hampton VA 23665-5225

ST CLAIR, MARY BETH GENTER, b Watertown, NY, Mar 2, 62; m 87. PATHOLOGY, NEUROTOXICOLOGY. *Educ:* St Lawrence Univ, BS, 84; Duke Univ, PhD(path, toxicol), 88. *Prof Exp:* Res fel, 88-90, ASST PROF TOXICOL, NC STATE UNIV, 90- *Mem:* Soc Toxicol. *Res:* Pathology. *Mailing Add:* NC State Univ Box 7633 Raleigh NC 27695-7633

ST CLAIR, RICHARD WILLIAM, b Sioux Falls, SDak, Oct 10, 40; m 62; c 2. BIOCHEMISTRY, PATHOBIOLOGY. *Educ:* Colo State Univ, BS, 62, PhD(physiol), 65. *Prof Exp:* PROF PATH & PHYSIOL, BOWMAN GRAY SCH MED, WAKE FOREST UNIV, 65- *Concurrent Pos:* NIH fel aging, Bowman Gray Sch Med, 65-67; fel coun arteriosclerosis, Am Heart Asn, estab investr, 70-75, chmn, 90-92; Fogerty int fel, 85-86. *Mem:* Am Soc Exp Path; Tissue Culture Asn; Soc Exp Biol Med; Sigma Xi; AAAS. *Res:* Atherosclerosis research; arterial metabolism; lipoprotein metabolism in nonhuman primates; cellular lipoprotein metabolism. *Mailing Add:* Dept Path Bowman Gray Sch Med Wake Forest Univ Winston-Salem NC 27103

ST CLAIR, TERRY LEE, b Roanoke, Va, June 18, 43; m 71; c 1. POLYMER CHEMISTRY. *Educ:* Roanoke Col, BS, 65; Va Polytech Inst & State Univ, PhD(org chem), 73. *Prof Exp:* Chemist quality control, E I du Pont de Nemours & Co, Inc Orlon, 65-67;; solid propellants engr, Hercules Inc-Radford Army Ammo Plant, 67-68; chemist adhesives, 72-75, aerospace technologist polymers, 75-80, res chemist polymers, 80-84, HEAD-POLYMERIC MAT BR, LANGLEY RES CTR, NASA, 84- *Concurrent Pos:* Chmn, Gordon Conf-Adhesion, 86. *Honors & Awards:* Sci Achievement Medal, NASA; IR-100, 79 & 81. *Mem:* Am Chem Soc; Sigma Xi; Soc Aerospace Mat & Process Engrs; Adhesion Soc (pres, 90-92). *Res:* Preparation and development of adhesives and composite matrix resins for aerospace applications. *Mailing Add:* 17 Roberts Landing Poquoson VA 23662

SAINTE-MARIE, GUY, b Montreal, Que, May 22, 28; m 55; c 3. LYMPHOLOGY, HEMATOLOGY. *Educ:* Col Ste-Marie, BA, 50; Univ Montreal, MD, 55; McGill Univ, PhD(histol), 62. *Prof Exp:* From asst prof to assoc prof anat, Univ Western Ont, 61-66; assoc prof, 66-69, chmn dept, 77-85, PROF ANAT, UNIV MONTREAL, 69- *Concurrent Pos:* Nat Cancer Inst Can fel, 58-60; res fel bact & immunol, Harvard Univ, 60-61; Med Res Coun Can res assoc, 66. *Mem:* Am Soc Anat. *Res:* Histology and physiology of lymph nodes. *Mailing Add:* 3112 Brighton Ave Montreal PQ H3S 1T9 Can

SAINT-JACQUES, ROBERT G, b Que, Can, Dec 15, 41; m 66; c 3. FUSION ENERGY MATERIALS, PLASMA SURFACE INTERACTION. *Educ:* Univ Montreal, BScA 67; Univ BC, MScA, 69; Univ Laval, DSc, 71. *Prof Exp:* Asst, Ctr Nuclear Studies, Saclay, 71-72; from asst prof to assoc prof, 72-82, PROF MAT, NAT INST SCI RES, UNIV QUE, 82- *Concurrent Pos:* Res assoc, Electronic Optics Lab, Toulouse, 78 Atomic Energy Res Estab, Harwell, 85-86. *Mem:* Can Micros Soc. *Res:* Materials for fusion reactors; effect of thermal shock on titanium carbide sprayed coatings; transmission electron microscopy of GaAs; hydride formation in H implanted Si. *Mailing Add:* INRS-Energie Univ Que CP 1020 Varennes PQ J3X 1S2 Can

ST JEAN, JOSEPH, JR, b Tacoma, Wash, July 24, 23; m 71. MICROPALEONTOLOGY, INVERTEBRATE PALEONTOLOGY. *Educ:* Col Puget Sound, BS, 49; Ind Univ, AM, 53, PhD(geol), 56. *Prof Exp:* Instr geol, Kans State Col, 51-52; from instr to asst prof, Trinity Col, Conn, 55-57; from asst prof to assoc prof, Univ NC, Chapel Hill, 57-66, prof geol, 66-90; RETIRED. *Concurrent Pos:* Partic, Nat Acad Sci-USSR Acad Sci Exchange Prog, 65. *Mem:* Paleont Res Inst; Paleont Soc; Soc Econ Paleont & Mineral; Int Paleont Union; Paleont Asn London. *Res:* Stromatoporoidea; systematics, evolution and paleobiology of Paleozoic Stromatoporiodea (Porifera). *Mailing Add:* Dept Geol CB 3315 Univ NC Chapel Hill NC 27599-3315

ST JOHN, BILL, b Wink, Tex, Jul 27, 32. PETROLEUM EXPLORATION. *Educ:* Univ Tex, Austin, BS, 53, MA, 60, PhD(geol), 65. *Prof Exp:* Sr geologist, Esso Explor/Esso Prod Res, 65-73; chief frontier geologist, 73-74, vpres, LVO Int, 74-78; exec vpres, Agri Petro, 78-80, pres, 80-81; PRES, PRIMARY FUSES INC, 81- *Mem:* Am Asn Petrol Geologists; Am Geophys Union; Geol Soc Am. *Res:* Sedimentary basins of the world, giant oil & gas fields, petroleum potential of antarctica. *Mailing Add:* 22523 Wildwood Grove Houston TX 77225

ST JOHN, DANIEL SHELTON, b San Diego, Calif, June 25, 23; m 43; c 4. PHYSICAL CHEMISTRY. *Educ:* Univ Calif, BS, 43; Univ Wis, PhD(chem), 49. *Prof Exp:* Jr technologists, Shell Oil Co, 43-44; engr, Los Alamos Sci Lab, Univ Calif, 44-47; chemist, E I du Pont de Nemours & Co, 49-58, res mgr, 58-64, mem develop dept, 64-66; pres, Holotron Corp, 66-69; res mgr, E I Dupont de Nemours & Co, Inc, 70-74, res fel, Polymer Intermediates Dept, 74-80, res fel, Petrochem Dept, 80-85; RETIRED. *Mem:* Fel Am Nuclear Soc; Am Chem Soc; AAAS. *Res:* Theoretical reactor physics; chemical engineering; physical optics; heterogeneous catalysis; chemical reaction kinetics; computer programming. *Mailing Add:* 532 Ashland Ridge Rd Hockessin DE 19707-9662

ST JOHN, DOUGLAS FRANCIS, b Toledo, Ohio, May 4, 38; m 60; c 5. MATERIALS SCIENCE, MECHANICS. *Educ:* Univ Toledo, BSME, 60, MSME, 65, MBA, 71; Mich State Univ, PhD(mech), 69. *Prof Exp:* Exp engr, Pratt & Whitney Aircraft, 60; assoc mech engr, 64-65, mat scientist, 65-69; PRES, OI-NEG TV PRODS INC, 88- *Mem:* Am Phys Soc; Am Inst Physics; Sigma Xi; Tech Asn Graphic Arts. *Res:* Glass melting, refining and homogenizing process analysis and design; solid-to-solid adhesion; motivational research methodology. *Mailing Add:* 707 E Jenkins Ave OI-NEG TV Prod Inc Columbus OH 43232-0497

ST JOHN, FRAZE LEE, b Lebanon, Ohio, May 23, 39; m 62; c 2. ZOOLOGY. *Educ:* Miami Univ, BS, 61; Ind Univ, Bloomington, MA, 63; Ohio State Univ, PhD(zool), 70. *Prof Exp:* Instr zool, Miami Univ, 63-65; naturalist, biol, Wahkeena State Mem, Ohio, 68; res specialist, 69-70, asst prof, 70-79, ASSOC PROF ZOOL, OHIO STATE UNIV, 79- *Mem:* Sigma Xi. *Res:* Invertebrate ecology. *Mailing Add:* Dept Zool Ohio State Univ Newark OH 43055

ST JOHN, HAROLD, b Pittsburgh, Pa, July 25, 92; m 22; c 4. BOTANY. *Educ:* Harvard Univ, AB, 14, AM, 15, PhD(biol), 17. *Prof Exp:* Asst bot, Gray Herbarium, Harvard Univ, 13-17 & 19-20, Radcliffe Col, 13-15; asst prof, Wash State Univ, 20-23, assoc prof & cur herbarium, 23-29; prof, 29-58, EMER PROF BOT, UNIV HAWAII, 58-; ACTG CUR BOT, BISHOP MUS, 66- *Concurrent Pos:* With Geol Surv Can, 15, 17; botanist, Bishop Mus, 29-65; vis prof, Yale Univ, 39-40; botanist, Foreign Econ Admin, Colombia, 43-44; assoc dir, Manoa Arboretum, 53-58; Whitney vis prof, Chatham Col, 58-59; vis prof, Saigon, 59-61. *Mem:* AAAS; Bot Soc Am; Am Soc Plant Taxon; Ger Dendrol Soc. *Res:* Taxonomy and phytogeography of vascular plants; ferns and flowering plants; nomenclature of plants; weeds of pineapple fields of the Hawaiian Islands; flora of southeast Washington and adjacent Idaho; revision of genus Pandanus; list of flowering plants of the Hawaiian Islands; nonugraph of Cyrtandra. *Mailing Add:* 2365 Hoomaha Way Honolulu HI 96822

ST JOHN, JUDITH BROOK, b Memphis, Tenn, Aug 15, 40; m 67; c 2. BIOCHEMISTRY, PLANT PHYSIOLOGY. *Educ:* Millsaps Col, BS, 62; Univ Fla, PhD(bot), 66. *Prof Exp:* Res fel plant physiol, Univ Fla, 66-67; PLANT PHYSIOLOGIST, RES LEADER WEED SCI LAB, AGR RES, USDA, 67- *Mem:* Am Soc Plant Physiologists; Am Inst Biol Sci; Weed Sci Soc Am; Am Chem Soc. *Res:* Mechanisms of herbicide action; plant lipid biochemistry. *Mailing Add:* Agr Res Ctr Rm 425 NAL 10301 Baltimore Ave BARC-W Beltsville MD 20705

ST JOHN, PETER ALAN, b Ashtabula, Ohio, May 11, 41; m 67; c 2. ANALYTICAL CHEMISTRY. *Educ:* Univ Fla, BS, 63, PhD(anal chem), 67. *Prof Exp:* Sr res chemist, instrument develop, AMINCO, 67-70, proj mgr, New Prod Develop, 71-79, prog mgr, 79-81, asst vpres eng, Instrument Div, Baxter Travenol Labs, Inc, 81-82; vpres eng, Am Res Prod Inc, 82-83; PRES, ST JOHN ASSOCS, INC, 84- *Mem:* Am Chem Soc; Sigma Xi. *Res:* Atomic and molecular spectroscopy; chromatography; biochemistry; microbiology. *Mailing Add:* 3306 Sellman Rd Adelphi MD 20783

ST JOHN, PHILIP ALAN, b Lexington, Mass, Feb 13, 24; m 51; c 2. CELL PHYSIOLOGY, VERTEBRATE REGENERATION. *Educ:* Univ NH, BS, 49, MS, 51; Harvard Univ, PhD, 56. *Prof Exp:* From instr to asst prof biol, Brandeis Univ, 56-67; prof biol, Butler Univ, 67-; RETIRED. *Mem:* AAAS. *Res:* Invertebrate cell culture; physiology of regeneration of Turbellaria and vertebrates parasite chemotherapy. *Mailing Add:* 1557 Brewster Rd Indianapolis IN 46260

ST JOHN, RALPH C, b Ft Kent, Maine, Aug 29, 42; m 69; c 2. STATISTICS. *Educ:* Univ Maine, Orono, BS, 64; Univ Mass, MS, 68; Univ Wis, PhD(statist), 73. *Prof Exp:* Mathematician statist, IBM, 64-66; from asst prof to assoc prof, 73-83, dir, statist consult ctr, 77-84, PROF STATIST, BOWLING GREEN STATE UNIV, 83- *Mem:* Am Statist Asn; Am Soc Qual Control. *Res:* Regression; design for regression; experiments with mixtures. *Mailing Add:* 610 Rosewood Dr Bowling Green OH 43402-1471

ST JOHN, ROBERT MAHARD, b Westmoreland, Kans, Mar 20, 27; m 49; c 2. EXPERIMENTAL ATOMIC PHYSICS. *Educ:* Kans State Univ, BS, 50, MS, 51; Univ Wis, PhD(physics), 54. *Prof Exp:* From asst prof to prof, 54-90, EMER PROF PHYSICS, UNIV OKLA, 90- *Mem:* Fel Am Phys Soc; Am Asn Physics Teachers. *Res:* Gaseous electronics; atomic and electronic collisions. *Mailing Add:* Dept Physics & Astron Univ Okla Norman OK 73019

ST JOHN, WALTER MCCOY, b Providence, RI, Apr 23, 44; m 70; c 2. NEUROPHYSIOLOGY. *Educ:* Brown Univ, AB, 66; Univ NC, Chapel Hill, PhD(physiol), 70. *Prof Exp:* From instr to asst prof physiol, Med Ctr, Univ Ark, Little Rock, 70-74; sr fel & staff assoc, Col Physicians & Surgeons, Columbia Univ, 74-75, res assoc pharmacol, 75-76; Parker B Francis Found fel & res assoc physiol, 76-77, from asst prof to assoc prof, 77-83, PROF PHYSIOL, DARTMOUTH MED SCH, 83- *Concurrent Pos:* Vis scientist, Fac St Jerome, Marseille, France, 80-81. *Mem:* Am Physiol Soc. *Res:* Neural control of respiration. *Mailing Add:* Dept Physiol Dartmouth Med Sch Hanover NH 03756

ST LAWRENCE, PATRICIA, b New York, NY, July 22, 22. GENETICS. *Educ:* Bryn Mawr Col, BA, 44; Columbia Univ, PhD(zool), 52. *Prof Exp:* USPHS fel, Yale Univ, 52-54, res asst microbiol, 55-57; res biologist, Stanford Univ, 57-59; asst prof, 59-65, ASSOC PROF GENETICS, UNIV CALIF, BERKELEY, 65-, ASSOC GENETICIST, AGR EXP STA, 69- *Mem:* AAAS; Genetics Soc Am; Am Soc Naturalists. *Res:* Genetics and cytogenetics of Neurospora. *Mailing Add:* Dept Genetics 345 Milford Hall Univ Calif Berkeley CA 94720

ST LOUIS, ROBERT VINCENT, b Los Angeles, Calif, Dec 15, 32; m 60; c 1. PHYSICAL CHEMISTRY. *Educ:* Univ Calif, Los Angeles, BS, 54; Univ Minn, PhD(phys chem), 62. *Prof Exp:* Res assoc far-infrared spectros, Johns Hopkins Univ, 62-63; res chemist, US Borax Res Corp, 63-66; res assoc far-infrared spectros, Univ Southern Calif, 66-68; PROF CHEM, UNIV WIS-EAU CLAIRE, 68- *Concurrent Pos:* Res fel, Ctr Interdisciplinary Res, Univ Bielefeld, WGer, 85-86. *Mem:* Am Chem Soc; Am Phys Soc. *Res:* Chemical infrared spectroscopy; colloid chemistry; solvent effects. *Mailing Add:* Dept Chem Univ Wis Eau Claire WI 54701

ST MARY, DONALD FRANK, b Lake Charles, La, July 22, 40; m 63; c 3. NUMERICAL SOLUTIONS OF PARTIAL DIFFERENTIAL EQUATIONS, SCIENTIFIC COMPUTATION. *Educ:* McNeese State Col, BS, 62; Univ Kans, MA, 64; Univ Nebr, Lincoln, PhD(math), 68. *Prof Exp:* Instr math, Univ Nebr, Lincoln, 66-67 & Iowa State Univ, 67-68; from asst prof to assoc prof, 68-83, PROF MATH, UNIV MASS, AMHERST, 83- *Concurrent Pos:* Vis fac, Univ Okla, 75-76; vis res fel, Yale Univ, 83-90; summer fac res assoc, Navy/ASEE, 83, 84 & 87; consult, Naval Underwater Systs Ctr, 83-87; mem, Comt Equal Opportunity Sci & Eng, NSF, 86- *Mem:* Am Math Soc; Math Asn Am; Soc Indust & Appl Math; Am Acoust Soc; Nat Asn Math; Asn Women Math. *Res:* Numerical solutions of partial differential equations; current applications being parabolic and elliptic problems in underwater acoustics. *Mailing Add:* Dept Math Univ Mass Amherst MA 01003

ST MAURICE, JEAN-PIERRE, b Valleyfields, Que, Mar 25, 49; m 72; c 4. IONOSPHERIC PHYSICS. *Educ:* Col Valleyfield, Que, BA, 68; Univ Montreal, BSc, 71; Yale Univ, PhD(geophys), 75. *Prof Exp:* Res asst geophys, Dept Geol & Geophys, Yale Univ, 71-74; scholar ionospheric physics, atmospheric & oceanic sci, Univ Mich, Ann Arbor, 74-76; from res asst prof to res prof physics, Utah State Univ, 77-87; assoc prof physics, 87-90, PROF PHYSICS, UNIV W ONT, 90- *Concurrent Pos:* Vis scientist, Max Plank Inst fur Aeromie, WGer, 82 & ISAS, Tokyo, Japan, 87; SERC vis fel, Eng, 89. *Mem:* Am Geophys Union; Can Asn Physics. *Res:* Theory and measurement of non-equilibrium ion velocity distributions in the ionosphere; transport properties of the ionosphere; neutral winds near auroral regions; anomalous ionospheric heating; ionospheric irregularities and waves. *Mailing Add:* Dept Physics Univ W Ont London ON N6A 3K7 Can

ST OMER, VINCENT VICTOR, b Castries, BWI, Nov 16, 34; m 62; c 4. NEUROSCIENCE, VETERINARY PHARMACOLOGY. *Educ:* Univ Guelph, DVM, 62, PhD(pharmacol), 69; Univ Man, MSc, 65. *Prof Exp:* Res assoc bur child res, Univ Kans, 68-71; asst prof vet pharmacol, Col Vet Med, Kans State Univ, 72-74; assoc prof vet anat-physiol, Col Vet Med & asst prof pharmacol, 74-84, PROF VET BIOMED SCI & ASSOC PROF PHARMACOL, SCH MED, UNIV MO, COLUMBIA, 84- *Concurrent Pos:* Adj prof, Univ Kans, 70-73. *Mem:* Sigma Xi; Am Soc Vet Physiologists & Pharmacologists; Soc Neurosci; fel Am Acad Vet Pharmacol & Therapeut; NY Acad Sci; Behav Teratology Soc. *Res:* Adverse drug interaction; behavioral and development toxicology and neurotoxicology. *Mailing Add:* Dept Pharmacol Univ Mo Columbia Med Sch Columbia MO 65212

ST ONGE, G H, ENGINEERING. *Prof Exp:* PRIN, ST ONGE ASSOC INC. *Mailing Add:* St Onge Assoc Inc 60 Olcott Ave Bernardsville NJ 07924

ST PIERRE, GEORGE R(OLAND), b Cambridge, Mass, June 2, 30; m 57, 76; c 4. METALLURGY. *Educ:* Mass Inst Technol, SB, 51, ScD(metall), 54. *Prof Exp:* Sr res metallurgist, Inland Steel Co, 54-56; proj off, USAF, 56-57; from asst prof to assoc prof, 57-64, assoc dean, 64-66, PROF METALL ENG, OHIO STATE UNIV, 64-, CHMN DEPT, 84- *Concurrent Pos:* Consult, Off Technol Assessment, US Environ Protection Agency, LTV, Sohio & others. *Honors & Awards:* Bradley Stoughton Award, Am Soc Metals, 61, Gold Medal, 87; Mining Indust Educ Award, Am Inst Mining Engrs, 87. *Mem:* Fel Am Soc Metals; fel Am Inst Mining, Metall & Petrol Engrs; Sigma Xi; Am Composite Soc; Mat Res Soc. *Res:* Chemical and process metallurgy; high-temperature materials. *Mailing Add:* Dept Metall 141 Fontana Lab 116 W 19th Ave Columbus OH 43210

ST PIERRE, JEAN CLAUDE, b St Jean, Que, Aug 21, 42; m 63; c 2. AGRONOMY, PLANT PHYSIOLOGY. *Educ:* Laval Univ, BScA, 64, MSc, 67; Cornell Univ, PhD(agron), 70. *Prof Exp:* Res scientist forage crops, 70-81, prog anal res mgt, 81-, DIR GEN EXP FARM RES CTR, AGR CAN, OTTAWA. *Mem:* Am Soc Agron; Can Soc Agron; Crop Sci Soc Am. *Res:* Forage crop production and quality; management and fertilization of grasses; plant physiology applied to breeding. *Mailing Add:* Three Des Genevriers Hull PQ J9A 2P2 Can

ST PIERRE, LEON EDWARD, b Edmonton, Alta, Sept 1, 24; m 49; c 7. POLYMER CHEMISTRY. *Educ:* Univ Alta, BSc, 51; Notre Dame Univ, PhD, 55. *Prof Exp:* Res chemist, Res Lab, Gen Elec Co, 54-65; PROF POLYMER CHEM, McGILL UNIV, 65-, CHMN DEPT, 72- *Honors & Awards:* Notre Dame Univ Centennial of Sci Award, 65. *Mem:* Chem Inst Can; Soc Plastics Eng; Am Chem Soc. *Res:* Polymers from epoxides; chemo-rheology and radiation of polymers; surface chemistry on films generated in ultra high vacuum. *Mailing Add:* Dept Chem McGill Univ Sherbrooke St W PO Box 6070 Sta A Montreal PQ H3A 2M5 Can

ST PIERRE, P(HILIPPE) D(OUGLAS) S, b Liverpool, Eng, Sept 10, 25; nat US; m 48; c 2. POWDER TECHNOLOGY, ABRASIVES. *Educ:* Royal Sch Mines, BS, 45; Univ London, PhD, 52. *Prof Exp:* Metallurgist, Mines Br, Govt Can, 48-55; metallurgist, Gen Elec Res Lab, 55-67, mgr diamond eng, Gen Elec Co, Mich, 67-68, mgr eng, Specialty Mat Dept, 68-88, CONSULT GEN ELEC SUPERABRASIVES, 88- *Concurrent Pos:* Nuffield traveling fel, 47. *Mem:* Fel Am Ceramic Soc. *Res:* Special ceramics; materials processing; education. *Mailing Add:* Comtek 235 Medick Way Worthington OH 43085

ST PIERRE, RONALD LESLIE, b Dayton, Ohio, Feb 2, 38; m 61; c 2. HISTOLOGY, IMMUNOLOGY. *Educ:* Ohio Univ, BS, 61; Ohio State Univ, MSc, 62, PhD(anat), 65. *Prof Exp:* From instr to assoc prof, 65-72, assoc dir, Cancer Ctr, 74-78, prof anat & chmn dept, Ohio State Univ, 72-81; ASSOC VPRES HEALTH SERV & ACAD AFFAIRS, 81- *Concurrent Pos:* Lederle med fac award, 68-71; vis res assoc, Med Ctr, Duke Univ, 66-67. *Mem:* AAAS; Transplantation Soc; Am Asn Anat; Am Asn Immunologists; Reticuloendothelial Soc. *Res:* Role of lymphoid organs, especially bursa of Fabricius and thymus, in immunity; histocompatibility testing for organ transplantation; clinical and developmental immunology; cancer immunology. *Mailing Add:* Off Health Sci Ohio State Univ 370 W Ninth Ave Columbus OH 43210

ST PIERRE, THOMAS, chemistry; deceased, see previous edition for last biography

SAITO, TAKUMA, b Fukui, Japan, July 14, 32; m 62; c 2. CYTOLOGY. *Educ:* Kyoto Univ, BA, 52, MS, 54, PhD(biol), 66; Kansai Med Sch, MD, 69. *Prof Exp:* Res assoc anat, Kansai Med Sch, 57-64, from instr to asst prof, 64-67 & 69-71; res fel path, M D Anderson Hosp, 67-69; sect chief morphol, Inst Develop Res, 71-76; PROF ANAT, JICHI MED SCH, 76- *Honors & Awards:* Seto Prize, Japanese Soc Electron Micros, 90. *Mem:* Am Soc Cell Biol; Histochem Soc; Japanese Asn Anatomists; Japan Soc Histochem & Cytochem; Japanese Soc Develop Biol; Soc Develop Biol; Japan Soc Cell Biol. *Res:* Enzyme histochemistry of retina. *Mailing Add:* 4-26-18 Ekiminami-machi Oyama Tochigi 323 Japan

SAITO, THEODORE T, b Poston, Ariz, Sept 9, 42; m 68; c 2. OPTICS. *Educ:* Pa State Univ, PhD(physics), 70. *Prof Exp:* USAF, 70-, proj officer optical tech, Air Force Weapons Lab, 70-71, group leader optical eval facil, 71-73, optical coating, 73-74, leader, Energy Res & Develop Admin, Dept Defense, Lawrence Livermore Lab, 74-77, dir, Mfg Tech Transfer Prog, Air Force Mat Lab, Dept Energy, 77-79, tech mgr, Laser Hardening, Wright Aeronaut Lab, 79-80, dir Aerospace Mech, 80-84, Comdr, 84-87, dep prog leader, Precision Eng Prog, 87-88, DEP DEPT HEAD, DEFENSE SCI, USAF, LAWRENCE LIVERMORE NAT LAB, 88-, GROUP LEADER OPTICAL SCI & ENG. *Honors & Awards:* Tech Achievement Award, Photo Optical Instrumentation Engrs; Indust Res 100 Award. *Mem:* Fel Photo Optical Instrumentation Engrs (secy); fel Optical Soc Am; Am Soc Precision Eng. *Res:* Developing and commercializing diamond turning of optics; laser damage of optical materials; manufacturing technology and technology transfer; optical metrology; lasers; optical fabrication; spectroscopy. *Mailing Add:* Lawrence Livermore Nat Lab L-332 PO Box 808 Livermore CA 94550

SAJBEN, MIKLOS, b Bekescsaba, Hungary, July 31, 31; US citizen; m 63; c 3. FLUID MECHANICS. *Educ:* Budapest Tech Univ, Mech Engr, 53; Univ Pa, MS, 61; Mass Inst Technol, ScD(magnetohydrodyn), 64. *Prof Exp:* Develop engr, Westinghouse Elec Co, Pa, 57-61; res asst magnetohydrodyn, Mass Inst Technol, 61-64; asst prof aeronaut, Calif Inst Technol, 64-70; prin scientist, 70-83, McDonnell Douglas fel, 83-90, PROG DIR, MCDONNELL DOUGLAS RES LABS, 90- *Concurrent Pos:* Consult, Shock Hydrodyn, Inc, 66 & McDonnell Douglas Astronaut Co, 66-70; assoc ed, Am Inst Aeronaut & Astronaut J, 86-88; affil prof mech eng, Wash Univ, St Louis, 87-; mem adv comt, Bur Eng Res, Univ Tex, Austin, 88-89. *Mem:* Assoc fel Am Inst Aeronaut & Astronaut. *Res:* Internal and unsteady flows; aeronautical and astronautical engineering. *Mailing Add:* 441 Conway Meadows Dr Chesterfield MO 63017-9622

SAK, JOSEPH, b Zlin, Czech, Nov 20, 39; m 61; c 1. STATISTICAL MECHANICS. *Educ:* Charles Univ, Prague, MS, 61; Inst Solid State Physics, Prague, PhD(physics), 68. *Prof Exp:* Res assoc physics, Univ Chicago, 69-71; instr, Cornell Univ, 71-73; from asst prof to assoc prof, 73-85, PROF PHYSICS, RUTGERS UNIV, 85- *Mem:* Am Phys Soc. *Res:* Critical phenomena; renormalization group; kinetic theory; many body theory. *Mailing Add:* Dept Physics & Astron Rutgers Univ New Brunswick NJ 08903

SAKAGAWA, GARY TOSHIO, b Honolulu, Hawaii; c 2. FISHERIES. *Educ:* Univ Hawaii, BA, 63; Univ Mich, MSc, 67; Univ Wash, PhD(fisheries), 72. *Prof Exp:* Res assoc, Fisheries Res Inst, Univ Wash, 67-72; fishery biologist, 72-79, chief, Oceanic Fisheries Resources Div, 79-86, CHIEF, PELOGIC FISHERIES RESOURCES DIV, SOUTHWEST FISHERIES CTR, NAT MARINE FISHERIES SERV, 86- *Concurrent Pos:* Sci adv US deleg, Int Comn for Conserv Atlantic Tunas, 72 & 75-85 & Inter-Am Trop Tuna Comn, 73 & 79-86; mem, Billfish Mgt Plan Develop Team, Pac Fishery Mgt Coun, 78-79, convener, subcomt skipjack tuna, Int Comn for Conserv Atlantic Tunas, 78-84; dir, Am Int Fishery Res Biologists, 79-81; res assoc, Scripps Int Oceanog, 80-; consult stock assessment of tuna, Indian Ocean, Pac Tuna Prog, food & Agr Orgn,86- *Mem:* Am Fisheries Soc (pres, 81-82); Am Inst Fishery Res Biologists (dir, 79-81); Sigma Xi. *Res:* Stock assessment and fishery evaluation of tunas and billfishes; stock assessment and management of marine mammals. *Mailing Add:* 5871 Soledad Rd La Jolla CA 92037

SAKAGUCHI, DIANNE KOSTER, b Rockville Ctr, NY, Feb 27, 46; m 74. SYSTEM ENGINEERING. *Educ:* Hofstra Univ, BA, 66; Adelphi Univ, MS, 68, PhD(numerical anal), 72. *Prof Exp:* Mem tech staff, 72-80, proj engr, 80-86, MGR SYST REQUIREMENTS, AEROSPACE CORP, 86- *Res:* Intelligent systems; space flight requirements; system design. *Mailing Add:* Aerospace Corp 2350 E El Segundo Blvd El Segundo CA 90245

SAKAI, ANN K, b Boston, Mass, Jan 15, 51. EVOLUTIONARY ECOLOGY, PLANT ECOLOGY. *Educ:* Oberlin Col, AB, 72; Univ Mich, MS, 73, PhD(bot), 78. *Prof Exp:* Asst prof biol, Oakland Univ, 78-; RES SCIENTIST, BIOL STA, UNIV MICH. *Mem:* Soc Study Evolution; Ecol Soc Am. *Res:* Ecological and evolutionary relationships of sex expression in plants; biology of woody plants; dynamics of forest succession. *Mailing Add:* Univ Mich Biol Sta Pellston MI 49769

SAKAI, SHOICHIRO, b Japan, Jan 2, 28; m 58; c 2. MATHEMATICS. *Educ:* Univ Tohuku, Japan, BS, 53, PhD(math), 61. *Prof Exp:* Asst math, Univ Tohuku, Japan, 53-60; asst prof, Wasedu Univ, Japan, 60-64; vis lectr, Yale Univ, 62-64; assoc prof, 64-69, PROF MATH, UNIV PA, 69- *Concurrent Pos:* Guggenheim fel, 70 & 71; invited speaker, Int Cong Math, Helsinki, 78. *Mem:* Am Math Soc; NY Acad Sci; AAAS; Japan Math Soc. *Res:* Functional analysis. *Mailing Add:* 205 Green Hts 5-1-6 Odawara Sendai Japan

SAKAI, TED TETSUO, b Newell, Calif, April 12, 45. MEDICINAL CHEMISTRY. *Educ:* Univ Calif, Berkeley, BS, 67; Univ Calif, Santa Barbara, PhD(bioorg chem), 71. *Prof Exp:* Fel biochem & microbiol, Univ Colo Med Ctr, 71-76; asst prof biochem, 79-85, ASSOC SCIENTIST, COMPREHENSIVE CANCER CTR, UNIV ALA, 76-, RES ASSOC & PROF BIOCHEM, 85- *Mem:* Am Chem Soc; AAAS. *Res:* Mechanism of action of chemotherapeutic agents; drug-nucleic acid interactions; polyamines. *Mailing Add:* Comprehensive Cancer Ctr Univ Ala Univ Sta Rm CHSB B-31 Birmingham AL 35294

SAKAI, WILLIAM SHIGERU, b Cody, Wyo, Sept 9, 42; m 69. BOTANY, ELECTRON MICROSCOPY. *Educ:* Univ Mich, Ann Arbor, BS, 66; Univ Hawaii, PhD(bot), 70. *Prof Exp:* Teaching asst bot, Univ Hawaii, Manoa, 69-70, asst prof, 70-71, asst soil scientist agron & soil sci, 71-74; vpres prod & res, The Flower Cart, Inc, 74-76; from asst prof to assoc prof, 76-83, actg dean Col Agr, 82-83, PROF HORT, UNIV HAWAII, HILO, 83- *Mem:* AAAS; Bot Soc Am; Am Soc Plant Physiol; Sigma Xi; Am Soc Hort Sci; Int Plant Growth Regulator Soc. *Res:* Cell biology; plant anatomy and physiology; horticulture. *Mailing Add:* Dept Hort Univ Hawaii 1400 Kapiolani St Hilo HI 96720-4091

SAKAKINI, JOSEPH, JR, b Norfolk, Va, Oct 18, 32; m 58; c 5. MATERNAL-FETAL MEDICINE, CLINICAL RESEARCH. *Educ:* Va Mil Inst, BA, 55; Med Col Va, MD, 59. *Prof Exp:* Dir residency obstet-gynec, Madigan Army Hosp, 71-74; fel maternal-fetal med, William Beaumont Army Hosp, 74-76; chief obstst-gynec, Madigan Army Hosp, 77-81; dir, Tacoma Gen Hosp, 81-83; CHMN OBSTET-GYNEC MATERNAL-FETAL MED, TEX TECH MED SCH, EL PASO, 83- *Concurrent Pos:* Consult, Surgeon Gen US Army, 80-83; mem, Nurse Pract Exam Bd, Nurse Asn, Am Col Obstet & Gynec, 82-86, vchmn, AFD, 82-85; vis prof, Med Sch, Tex A&M Univ, 83; mem, Am Bd Obstet & Gynec, 84- *Mem:* Am Col Obstetricians & Gynecologists; Soc Perinatal Obstetricians; Nurse Asn-Am Col Obstet & Gynec; Am Bd Obstet & Gynec. *Res:* Pain medication to ease post partum and post operative discomfort; viral infections in pregnancy. *Mailing Add:* Dept Obstet & Gynec Tex Tech Univ Sch Med 4800 Alberta Ave El Paso TX 79905

SAKAKURA, ARTHUR YOSHIKAZU, b San Francisco, Calif, May 24, 28; m 56. THEORETICAL PHYSICS. *Educ:* Mass Inst Technol, BS, 49, MS, 50; Univ Colo, PhD(physics), 60. *Prof Exp:* Physicist, US Geol Surv, Colo, 50-58; physicist, Res Ctr, IBM Corp, 60-63; physicist, Joint Inst Lab Astrophys, Univ Colo, 63-65, physicist, Dept Physics, 65-66; physicist, 66-68, ASSOC PROF PHYSICS, COLO SCH MINES, 68- *Mem:* Am Phys Soc. *Res:* Statistical mechanics; quantum mechanical many-body problems. *Mailing Add:* Dept Physics Colo Sch Mines Golden CO 80401

SAKALOWSKY, PETER PAUL, JR, b Worcester, Mass, July 29, 42; m 69; c 3. PHYSICAL GEOGRAPHY. *Educ:* Worcester State Col, BSEd, 64; Clark Univ, MA, 66; Ind State Univ, PhD(geog), 72. *Prof Exp:* Instr geog, Bloomsburg State Col, 66; instr, Briarcliff Col, 66-68; assoc prof, 70-80, PROF GEOG, SOUTHERN CONN STATE COL, 80- *Res:* Geomorphology, climatology, coastal processes and beach morphology; historical coastline changes; interrelationship of coastal processes and man, microclimates; regional geography of Anglo-America; meandering streams; general physical geography. *Mailing Add:* Dept Geog Southern Conn State Col 501 Crescent New Haven CT 06515

SAKAMI, WARWICK, biochemistry; deceased, see previous edition for last biography

SAKAMOTO, CLARENCE M, b Lahaina, Hawaii, Nov 1, 31; m 64; c 2. METEOROLOGY, AGRICULTURE. *Educ:* Univ Hawaii, BS, 53; Pa State Univ, BS, 55; Rutgers Univ, MS, 62; Iowa State Univ, PhD(agr climat), 65. *Prof Exp:* Res asst, Rutgers Univ, 60-62 & Iowa State Univ, 62-65; adv agr meteorologist, Rutgers Univ Weather Bur, 65-67; res climatologist, Environ Data Serv, Reno, 67-68; state climatologist, Nat Weather Serv, Nev, 68-73; meteorologist, Environ Study Serv Ctr, Nat Oceanic & Atmospheric Admin, Auburn, Ala, 73-75; meteorologist, Nat Oceanic & Atmospheric Admin, 75-77, supv meteorologist, Ctr Climate & Environ Assessment, Environ Data Serv, 77-78, chief Models Br, Assessment & Info Serv Ctr, Columbia, Mo, 78-88; SR AGROMETEOROLOGIST, FOREIGN AGR ORGAN, 90- *Concurrent Pos:* Adj asst prof meteorol, Rutgers Univ, 65-67; adj res assoc, Desert Res Inst, Reno, 67-68; adj assoc prof, Univ Nev, Reno, 68-73; adj assoc prof, Auburn Univ, 73-75; res assoc, Univ Mo, 77-80, prof, 80-88; WMO Sr Agrometeorologist, 88-90. *Honors & Awards:* Unit Citation, Nat Oceanic & Atmospheric Admin, 77; Gold Medal, US Dept Com, 78. *Mem:* Fel Am Meteorol Soc. *Res:* Crop response and environment; climatological analysis; impact of climatic change on food production. *Mailing Add:* c/o Foreign Agr Organ Harare PO Box 3730 Harare Zimbabwe

SAKANO, THEODORE K, b Portland, Ore, Sept 24, 38; m 87. PHYSICAL CHEMISTRY. *Educ:* Ore State Col, BS, 60; Univ Wis, PhD(phys chem), 66. *Prof Exp:* From instr to assoc prof, 65-75, prof Chem, 75-87, dept chmn Rose-Hulman Inst Technol 80-83; PROF SCI DEPT, ROCKLAND COMMUNITY COL. *Mem:* Am Chem Soc. *Mailing Add:* Sci Dept Rockland Community Col Suffern NY 10901

SAKHARE, VISHWA M, b Belgaum, India, Aug 28, 32; m 56; c 3. MATHEMATICS. *Educ:* Karnatak Univ, India, BS, 51; Cambridge Univ, BA & MA, 54; Univ Tenn, Knoxville, PhD(math), 73. *Prof Exp:* Teaching asst math, Univ Idaho, 59-62, instr, 62-65; assoc prof, 65-76, PROF MATH, ETENN STATE UNIV, 76- *Mem:* Math Asn Am; Am Math Soc. *Res:* Applied mathematics. *Mailing Add:* Dept Math ETenn State Univ Johnson City TN 37614

SAKHNOVSKY, ALEXANDER ALEXANDROVITCH, b Asheville, NC, July 14, 26; m; c 2. PHYSICAL CHEMISTRY. *Educ:* Univ NC, AB, 49, MA, 50. *Prof Exp:* Jr chemist, Erwin Chem Lab, 51-52; res instr, Univ Miami, 52-56, res asst prof, Indust Chem Res Lab & Housing Res Lab, 56-68; PRES, CONSTRUCT RES LAB, 68- *Concurrent Pos:* Asst mgr paint proving grounds, Sun Tests, 50-55. *Honors & Awards:* Charles Martin Hall Award, Architectural Aluminum Mfr Asn, 74. *Mem:* Am Chem Soc; Am Soc Testing & Mat; fel Am Inst Chem. *Res:* Refrigeration chemistry; dehydration of refrigeration systems; desiccants; oil-refrigerant reactions; physical testing of building components; building water leakage. *Mailing Add:* Construct Res Lab 7600 NW 79th Ave Miami FL 33166

SAKITA, BUNJI, b Toyama-ken, Japan, June 6, 30; c 2. THEORETICAL HIGH ENERGY PHYSICS. *Educ:* Kanazawa Univ, Japan, BS, 53; Nagoya Univ, MS, 56; Univ Rochester, PhD(physics), 59. *Prof Exp:* Res assoc physics, Univ Wis, 59-62, asst prof, 62-64; assoc physicist, High Energy Physics Div, Argonne Nat Lab, 64-66; prof physics, Univ Wis-Madison, 66-70; DISTINGUISHED PROF PHYSICS, CITY COL NEW YORK, 70- *Concurrent Pos:* Guggenheim fel, 70; fel, Japan Soc Prom Sci, 75, 80 & 87. *Honors & Awards:* Nishina Prize, 74. *Mem:* Fel Am Phys Soc. *Res:* Elementary particles; symmetries of hadrons; weak interactions; field theories. *Mailing Add:* Dept Physics City Col New York Convent Ave & 138th St New York NY 10031

SAKITT, BARBARA, b New York, NY. VISION. *Educ:* Columbia Univ, PhD(physics), 65. *Prof Exp:* Assoc prof psychol, Stanford Univ, 76-79; PRIN RES SCIENTIST, MASS INST TECHNOL, 79- *Concurrent Pos:* NIH career develop award, 75- *Mem:* Optical Soc Am; Asn Res Vision & Ophthal; Am Psychol Asn; Psychonomic Soc; Soc Neurosci. *Res:* Vision and visual-motor coordination; visually triggered movements. *Mailing Add:* Dept Educ Psychol Fordham Univ Lincoln Ctr New York NY 10023

SAKITT, MARK, b Brooklyn, NY, Apr 7, 38; m 63. ELEMENTARY PARTICLE PHYSICS. *Educ:* Polytech Inst Brooklyn, BEE, 58; Univ Md, PhD(physics), 65. *Prof Exp:* Res assoc physics, 64-66, from asst physicist to physicist, 66-80, SR PHYSICIST, BROOKHAVEN NAT LAB, 80-, ASST DIR PLANNING & POLICY, 90- *Concurrent Pos:* Mem Adv Bd Arms Control, Disarmamant & Peace Studies, Res Ctr, State Univ NY, Stony Brook & adj lectr defense policy; adj lectr defense policy, State Univ NY, Stony Brook; sci fel, Ctr Int Security & Arms Control, Stamford Univ, 86-87; Carnegie sci fel, 86-87. *Mem:* Fel Am Phys Soc; NY Acad Sci; Sigma Xi. *Res:* High energy experimental physics; defense policy. *Mailing Add:* Dept Physics Brookhaven Nat Lab Upton NY 11973

SAKMAR, ISMAIL AYDIN, b Istanbul, Turkey, Sept 29, 25; div. ELEMENTARY PARTICLE PHYSICS & THEORETICAL PHYSICS, NUMBER THEORY. *Educ:* Istanbul Univ, MS, 51; Univ Calif, Berkeley, PhD(physics), 63. *Prof Exp:* Lectr physics, Univ Calif, Santa Barbara, 63-64; asst prof, Univ Miami, 64-67; from assoc prof to prof Appl Math, 67-86, NAT RES COUN CAN GRANTS, UNIV WESTERN ONT, 68- *Concurrent Pos:* Vis scientist, Int Ctr Theoret Physics, Italy, 65, 69, 70, 78 & 83, Europ Orgn Nuclear Res, Switz, 72 & Lab de Phys Theorique, Univ de Nice, France, 79; Near Eastern Col fel, Rockefeller Found. *Mem:* Am Phys Soc; Ital Phys Soc; Math Asn Am. *Res:* Elementary particles; S-matrix theory; Regge pole hypothesis; model independent calculations in particle physics; application of variational techniques with infinite variables and inequality constraints; number theoretical problems. *Mailing Add:* c/o Oryum Hess St 45 Munich 40 8000 Germany

SAKO, KUMAO, b Sebastopol, Calif, Oct 31, 24; m 66. MEDICINE, SURGERY. *Educ:* Univ Ill, BS, 50, MD, 52; Am Bd Surg, dipl, 59. *Prof Exp:* Intern, Cook County Hosp, Chicago, 52-53; resident surg, Augustana Hosp, 53-55; resident, 55-57, sr cancer res surgeon, 57-58, assoc cancer res surgeon, S8-61, ASSOC CHIEF HEAD & NECK SURGEON, ROSWELL PARK MEM INST, 61- *Concurrent Pos:* Asst, Sch Med, Univ Buffalo, 59-60, clin instr, 60-62, clin assoc, 62-65, res assoc, 65-75, res asst clin prof, 75- *Mem:* Fel Am Col Surg; Soc Head & Neck Surgeons; Am Soc Clin Oncol; Soc Surg Oncol; Am Asn Cancer Res. *Res:* Cancer research; head, neck and general surgery. *Mailing Add:* Dept Exp Path Roswell Park Mem Inst 666 Elm St Buffalo NY 14263

SAKO, YOSHIO, b Forestville, Calif, Jan 25, 18; m 54; c 3. SURGERY. *Educ:* Univ Minn, MD, 47, PhD(surg), 51. *Prof Exp:* From istr to assoc prof, 52-66, PROF SURG, MED SCH, UNIV MINN, MINNEAPOLIS, 66- *Concurrent Pos:* Nat Heart Inst trainee, 51-52; assoc chief surg & chief cardiovasc surg sect, Vet Admin Hosp, Minneapolis, 55- *Mem:* AMA; fel Am Col Surg. *Res:* Cardiovascular surgery. *Mailing Add:* Vet Admin Hosp Minneapolis MN 55417

SAKODA, WILLIAM JOHN, b Brooklyn, NY, Feb 25, 51; m 77. THEORETICAL COMPUTER SCIENCE. *Educ:* Harvard Col, AB, 72; Univ Calif, PhD(comput sci), 78. *Prof Exp:* Res asst, Univ Calif, 74-78, instr, 78; asst prof comput sci, Columbia Univ, 78-80; ASST PROF COMPUT SCI, PA STATE UNIV, 80- *Mem:* Asn Comput Mach; AAAS; Sigma Xi. *Res:* Parallel algorithms for picture analysis; artificial intelligence. *Mailing Add:* State Univ NY Stony Brook Lab Office Bldg Stony Brook NY 11794-4400

SAKS, NORMAN MARTIN, b New York, NY, May 31, 29. CELL PHYSIOLOGY, PHYSIOLOGICAL ECOLOGY. *Educ:* Brooklyn Col, BS, 51, MA, 56; NY Univ, PhD(biol), 64. *Prof Exp:* From instr to assoc prof, 64-84, PROF BIOL, CITY COL NEW YORK, 84- *Concurrent Pos:* Res assoc, NY Univ, 64-66; vis res prof, Univ Paris, 85. *Mem:* Phycol Soc Am. *Res:* Effects of abiotic factors including light irradiance, salinity and temperature on the growth rates of marine algae; cell metabolism, food web metabolism and dynamics. *Mailing Add:* Dept Biol City Col New York Convent Ave at 138th St New York NY 10031

SAKSENA, VISHNU P, b Shahjahanpur, India, July 15, 34; m 62. ZOOLOGY, AQUATIC BIOLOGY. *Educ:* Banaras Hindu Univ, BSc, 52, MSc, 54; Univ Okla, PhD(zool), 63. *Prof Exp:* Asst prof biol, Janta Vidyalaya Col, 54-55; res asst animal genetics, Indian Vet Res Inst, 55-57; from asst prof to assoc prof biol, Youngstown State Univ, 63-68; assoc prof, 68-76, chmn dept, 80-85, PROF BIOL, MUSKINGUM COL, 76- *Mem:* Am Inst Fishery Res Biologists; Am Fisheries Soc; Am Inst Biol Sci; World Maricult Soc; Indian Acad Life Sci. *Res:* Physiology of air breathing fishes; ecology of marine fish larvae; effect of pollution on fishes; fish reproduction and development. *Mailing Add:* 183 Highland Dr New Concord OH 43762

SAKSHAUG, EUGENE C, b Mandan, NDak, Oct 18, 23. LIGHTING PROTECTION. *Educ:* NC Univ, BEE, 52. *Prof Exp:* CONSULT, RAY CHEM, 86-, POWER TECH, 85- *Mem:* Nat Acad Eng; fel Inst Elec & Electronics Engrs; Sigma Xi. *Mailing Add:* PO Box 1531 Lanesboro MA 01237

SAKURA, JOHN DAVID, b Seattle, Wash, Mar 28, 36. BIOCHEMISTRY. *Educ:* Wheaton Col, BS, 58; Univ Ariz, PhD(biochem), 70. *Prof Exp:* Res asst, Dept Chem, Univ Ore, 61-64; RES ASSOC BIOCHEM, DEPT BIOL CHEM, HARVARD MED SCH, 71- *Concurrent Pos:* Nat Res Coun fel, Dept Biochem, Brandeis Univ, 70-71; asst biochemist, Ralph Lowell Labs, Harvard Med Sch, 71- *Mem:* Am Chem Soc; Am Soc Neurochem; Sigma Xi. *Res:* Structure and function of proteins; lipid-protein interactions; isolation and characterization of proteolipids; proteolipids in membrane transport. *Mailing Add:* 50 Hawthorne Ave Arlington MA 02174

SAKURAI, TOSHIO, b Osaka, Japan, Jan 17, 45; m 70; c 2. SURFACE SCIENCE. *Educ:* Univ Tokyo, BS, 67, MS, 69; Pa State Univ, PhD(physics), 74. *Prof Exp:* Mem tech staff, Bell Tel Labs, Murray Hill, 74-76; asst prof physics, Pa State Univ, 77-80; assoc prof, Inst Solid State Physics, Univ Tokyo, 81-89; PROF, INST MAT RES, TOHOKU UNIV, 89- *Concurrent Pos:* Mem bd dirs, Rev Sci Instrumentation, 86-89. *Mem:* Am Phys Soc; Am Vacuum Soc; Phys Soc Japan. *Res:* Development of an atom-probe field ion microscope and its applications to surface physics; experimental study of surface electronic structures of semiconductors by ultraviolet photoemission and ion neutralization spectroscopies and scanning tunneling microscopy. *Mailing Add:* Dept Mat Sci & Eng Pa State Univ 208 Steidle University Park PA 16802

SAKURAI, YOSHIFUMI, b Tokyo, Japan, Mar 30, 21; m 49; c 2. MAGNETIC MEMORY, CONTROL DEVICE. *Educ:* Osaka Univ, BS, 43, MS, 45, Dr Eng, 58. *Prof Exp:* Lectr elec eng, Fac Eng, Osaka Univ, 46-48, asst prof, 48-59, prof nuclear eng, 59-65, prof control eng, Fac Eng Sci, 65-84; PROF ELEC ENG, FAC ENG, SETSUNAN UNIV, 84- *Mem:* Fel Inst Elec & Electronics Engrs. *Res:* Magnetic materials, especially rare-earth transition metal alloy thin films used in large capacity magneto-optical memory. *Mailing Add:* Setsunan Univ Ikeda Nakamachi 17-8 Neyagawa Osaka 572 Japan

SALADIN, JURG X, b Solothurn, Switz, July 25, 29; m 63; c 2. NUCLEAR PHYSICS. *Educ:* Swiss Fed Inst Technol, dipl, 54, PhD(nuclear physics), 59. *Prof Exp:* Res asst physics, Swiss Fed Inst Technol, 54-59; res assoc, Univ Wis, 59-61; from asst prof to assoc prof, 61-69, PROF PHYSICS, UNIV PITTSBURGH, 69-, DIR NUCLEAR PHYSICS LAB, 80- *Concurrent Pos:* Vis prof, Univ Basel, Switz, 69-70, & Inst Atomic Physics, Univ Bucharest, Romania, 73; fel, Oak Ridge Assoc Univs, 86; chmn, Exec Comt User Group, 80-81, & Mem Exec Comt, NSCL, 86- *Mem:* Fel Am Phys Soc; Swiss Phys Soc; Sigma Xi; Europ Phys Soc. *Res:* Experimental nuclear physics; nuclear reactions and structure; electromagnetic properties of nuclei; nuclear shapes; collective properties of nuclei; coulomb excitation; physics of high spin states. *Mailing Add:* Dept Physics Univ Pittsburgh Pittsburgh PA 15213

SALADIN, KENNETH S, b Kalamazoo, Mich, May 6, 49; m 79; c 2. ETHOLOGY PARASITISM, INVERTEBRATE SENSORY PHYSIOLOGY. *Educ:* Mich State Univ, BS, 71; Fla State Univ, PhD(parasitol), 79. *Prof Exp:* From asst prof to assoc prof, 77-89, PROF BIOL, GA COL, 89- *Concurrent Pos:* Ed assoc, Humanist Mag, Amherst, NY, 79-82, sci columnist, 80-83; mem, bd dirs, Nat Ctr Sci Educ, 82-87. *Honors & Awards:* Elon E Byrd Award, Southeast Soc Parasitologists, 78. *Mem:* Am Soc Zoologists; Animal Behav Soc; AAAS; Nat Ctr Sci Educ (treas, 82-87); Nat Asn Adv Health Professions. *Res:* Sensory capacities and host-finding behavior of parasitic invertebrates, with emphasis on chemical and photosensory ecology of Digenea, Acari, Siphonaptera and Hymenoptera. *Mailing Add:* Dept Biol & Environ Sci Ga Col Milledgeville GA 31061

SALAFIA, W(ILLIAM) RONALD, b Baltimore, Md, Dec 28, 38; div; c 2. PSYCHOBIOLOGY, HUMAN FACTORS ENGINEERING. *Educ:* Loyola Col, BS, 60; Fordham Univ, MA, 63, PhD(exp psychol), 67. *Prof Exp:* Lectr psychol, Hunter Col City Univ New York, 63-65; from instr to assoc prof, 65-74, chmn, Psychol Dept, 75-77, Decamp prof health sci, 84-87, PROF PSYCHOBIOL, FAIRFIELD UNIV, 74- *Concurrent Pos:* Consult, Naval Underwater Syst Ctr, New London, Conn, 87- *Mem:* Psychonomic Soc; Soc Neurosci; AAAS; NY Acad Sci; Am Psychol Soc; Human Factors Soc. *Res:* Neural mechanisms of learning and memory; classical (Pavlovian) conditioning; human factors of sonar displays. *Mailing Add:* Dept Psychol Fairfield Univ Fairfield CT 06430-7524

SALAFSKY, BERNARD P, b Chicago, Ill, Dec 27, 35; m 61; c 3. PHARMACOLOGY. *Educ:* Philadelphia Col Pharm & Sci, BS, 58; Univ Wash, MS, 61, PhD(pharmacol-toxicol), 62. *Prof Exp:* Instr pharmacol, Univ Wash, 62-64; from asst prof to assoc prof, Univ Ill Col Med, 64-70; adj assoc prof, Med Sch, Univ Pa, 70-72; WHO consult, SEARO-Indo 001, 73-75; consult, Biomed Health Consult Inc, H K, 76; prof, Dept Biomed Sci, 77-82, DIR, COL MED, UNIV ILL, ROCKFORD, 82- *Concurrent Pos:* Nat Inst Neurol Dis & Stroke res grant, 65-71; Muscular Dystrophy Asn Am spec fel, 72; spec fel, Med Sch, Univ Bristol, 72-74; Fulbright lectr, Malaysia, 68; vis prof, Med Sch, Pahlavi Univ, Iran, 70-72; lectr toxicol, Univ Calif, Berkeley, 78-80; consult indust, 79-81. *Mem:* AAAS; Soc Trop Dis Hyg; Am Soc Pharmacol & Exp Therapeut; Royal Soc Trop Med & Hyg. *Res:* Biology of skin penentrating parasistes; tropical disease pharmacology. *Mailing Add:* Dept Off Col Med Univ Ill 1601 Parkview Rockford IL 61101

SALAH, JOSEPH E, b Jerusalem, Feb 27, 44; US citizen; m 65; c 1. INCOHERENT SCATTER RADAR SYSTEMS, RADIO ASTRONOMY TECHNIQUES. *Educ:* Univ Ill, Urbana, BSc, 65 & MSc, 66; Mass Inst Technol, PhD(meteorol), 72. *Prof Exp:* Group leader, Lincoln Lab, 66-83, DIR, HAYSTACK OBSERV, MASS INST TECHNOL, 83- *Concurrent Pos:* Mem, comt solar terrestrial res, Nat Astron Soc & Nat Res Coun, 85-88, adv comt astron sci, 85-88, Cedar Sci Steering Coun, Nat Sci Found, 87-; prin res scientist & sr lectr, Earth, Atmospheric & Planetary Sci Dept, Mass Inst Technol, 83- *Mem:* Am Geophys Union; Am Astron Soc; Am Meterol Soc; Union Radio Sci Int. *Res:* Structure and dynamics of the earth's inosphere and thermosphere; development of techniques and applications for high-power radar studies of the atmosphere; development of observational techniques for radio astronomy. *Mailing Add:* Haystack Observ Mass Inst Technol Route 40 Westford MA 01886

SALAHUB, DENNIS RUSSELL, b Castor, Alta, 1946; m 70. THEORETICAL CHEMISTRY, THEORETICAL SOLID STATE PHYSICS. *Educ:* Univ Alta, BSc, 67; Univ Montreal, PhD(chem), 70. *Prof Exp:* Fel chem, Univ Sussex, 70-72; res assoc, Univ Waterloo, 72-74 & Johns Hopkins Univ, 74; chemist, Res & Develop, Gen Elec Co, 75-76; from asst prof to prof, 76-90, MCCONNELL PROF CHEM, UNIV MONTREAL, 90- *Honors & Awards:* Noranda Award, Can Chem Soc, 87; Int Union Pure & Appl Chem Award, 84; Killam fel, 90. *Mem:* Am Chem Soc; fel Chem Inst Can; Can Asn Physicists. *Res:* Quantum theoretical studies of the electronic structure and properties of molecules, clusters and solids. *Mailing Add:* Dept Chem Univ Montreal Montreal PQ H3C 3J7 Can

SALAM, ABDUS, b Jhang, Pakistan, Jan 29, 26. ELEMENTARY PARTICLES. *Educ:* Panjab Univ, MA, 46; Cambridge, BA, 49; Cavendish Lab, Cambridge, PhD(theoret physics), 52. *Prof Exp:* Prof, Govt Col & Panjab Univ, 51-54; lectr, Cambridge Univ, 54-56; PROF THEORET PHYSICS, LONDON UNIV, 57-; FOUNDER & DIR, INT CTR THEORET PHYSICS, UNESCO, 64- *Honors & Awards:* Nobel Prize for physics, Nobel Found, 79; Hopkins Prize, Cambridge Univ, 58 & Adams Prize, 58; Hughes Medal, Royal Soc London, 64; Einstein Medal, UNESCO, Paris, 79; Gold Medal, Czech Acad Sci, 81. *Mem:* Foreign assoc Nat Acad Sci; Swed Acad Sci; Int Union Pure & Appl Physics (vpres, 72-78); AAAS; Third World Acad Sci; USSR Acad Sci; foreign mem Am Acad Arts & Sci. *Mailing Add:* Int Ctr Theoret Physics PO Box 586 Miramare Strada Costiera 11 Trieste 34100 Italy

SALAM, FATHI M A, NEURAL PROCESSORS, ADAPTIVE CONTROL SYSTEMS. *Educ:* Univ Calif, Berkeley, BS, 76, MS, 79, MA, 83, PhD(elec eng), 83. *Prof Exp:* Vis asst prof control systs, Univ Calif, Berkeley, 83; asst prof dynamical systs, Drexel Univ, 83-85; asst prof systs & circuit, 85-87, ASSOC PROF NONLINEAR SYSTS, MICH STATE UNIV, 87- *Concurrent Pos:* Prin investr, NSF, 84-90; assoc ed, Inst Elec & Electronics Engrs, Inc, 85-87, guest ed, 87-88; finance chmn, Inst Elec & Electronics Engrs & Systs, 88- *Mem:* Inst Elec & Electronics Engrs; Int Soc Optical Eng. *Res:* Nonlinear phenomena in circuits and systems; analysis and design of neural networks, adaptive systems, control and compliance of robot manipulators; stability of power systems. *Mailing Add:* Dept Elec Eng Mich State Univ East Lansing MI 48824

SALAMA, CLEMENT ANDRE TEWFIK, b Heliopolis, Egypt, Sept 27, 38; Can citizen; m 74; c 1. ELECTRICAL ENGINEERING. *Educ:* Univ BC, BAS, 61, MAS, 63, PhD(elec eng), 66. *Prof Exp:* Mem sci staff, Bell-Northern Res Labs, Ottawa, 66-67; from asst prof to assoc prof, 67-77, PROF SOLID STATE ELECTRONICS, UNIV TORONTO, 77-, J M HAM CHAIR MICROELECTRONICS, 87- *Concurrent Pos:* Natural Sci & Eng Res Coun grant, 67-, Defence Res Bd Can grant, 67-75; consult, Elec Eng Consociates, Ltd, 69-, dir, 69-76; vis prof, Cath Univ Leuven, Belgium, 75-76; chmn, Natural Sci & Eng Coun, Nat Microelectronics Fac Comt, 83-84, Inst Elec & Electronics Engrs, Toronto Sect, 85-87; mem bd dirs & chmn bd, Can Microelectronics Corp, 84-87; assoc ed, Inst Elec & Electronics Engrs Trans on Circuits & Systs, 87-89; Tech Prog Comt, Int Electron Devices Meeting, 87-88, Int Symp on Circuits & Systs, 88-, Can Conf on VLSI, 84-87, chmn, 85; res fel, Info Technol, ITAC/Nat Sci & Eng Res Coun Can; prin investr, Info Technol Res Ctr, 87-92. *Mem:* Fel Inst Elec & Electronics Engrs; Electrochem Soc. *Res:* Solid state electronics; integrated circuits; electronic circuits. *Mailing Add:* Dept Elec Eng Univ Toronto Toronto ON M5S 1A4 Can

SALAMA, GUY, b Cairo, Egypt, Apr 23, 47; US citizen. VOLTAGE-SENSITIVE DYES. *Educ:* City Col New York, BS, 68; Univ Pa, MS, 71, PhD(biophysics), 77. *Prof Exp:* Teaching asst physics, Univ Pa, 70-71, res fel physiol, 73-77, res fel biochem & biophysics, 77-80; asst prof, 80-87, ASSOC PROF PHYSIOL, SCH MED, UNIV PITTSBURGH, 87- *Concurrent Pos:* Lectr physics, Spring Garden Col, Philadelphia, Pa, 69-70; investr, Marine Biol Lab, Woods Hole, Mass, 80-81. *Honors & Awards:* Res Car Div Award, NIH. *Mem:* Biophys Soc; Soc Gen Physiologists; Am Heart Asn. *Res:* Development and application of voltage-sensitive dyes to measure transmembrane electrical potential in heart muscle and in the sarcoplasmic reticulum; optical recordings of cardiac action potentials, optical maps of action potential activation and repolarization in perfused hearts, cardiac arrhythmics; excitation-contraction coupling in mammalian skeletal muscle. *Mailing Add:* Dept Physiol Sch Med Univ Pittsburgh Pittsburgh PA 15261

SALAMA, KAMEL, b Bahgoura, Egypt, Apr 1, 32; m 71; c 2. MATERIALS ENGINEERING, PHYSICAL METALLURGY. *Educ:* Cairo Univ, BSc, 51, MSc, 55, PhD(physics), 59. *Prof Exp:* Res asst physics, Cairo Univ, 51-60, lectr, 60-65; res consult, Ford Sci Lab, 66-68; fel mat sci, Rice Univ, 68-71, sr res scientist, Mat Sci Dept, 71-73; PROF, UNIV HOUSTON, 73- *Concurrent Pos:* Partic, Int Seminar Res & Educ Physics, Uppsala Univ, Sweden, 62-63, fel, 62-64; distinguished vis scientist, NASA Langley Res Ctr, Hampton, Va, 83- *Honors & Awards:* Jacob Wallenberg Found Award, Swed Acad Eng, Stockholm, 82. *Mem:* Am Soc Metals; Am Inst Mining, Metall & Petrol Engrs; Am Phys Soc; Soc Metallurgical Engrs; Am Soc Nondestructive Testing. *Res:* Elastic and mechanical properties of solids; work hardening in environmental effects; mechanical behavior of metals and alloys; ultrasonic nondestructive characterization of materials properties and residual stresses. *Mailing Add:* Dept Mech Eng Univ Houston Houston TX 77004

SALAMŌ, GREGORY JOSEPH, b Brooklyn, NY, Sept 19, 44; m 87. QUANTUM OPTICS. *Educ:* Brooklyn Col, BS, 66; Purdue Univ, MS, 68; City Univ New York, PhD(physics), 74. *Prof Exp:* Resident visitor physics, Bell Labs-Murray Hill, 71-73; res assoc, Rochester Univ, 73-75; from asst prof to assoc prof, 75-85, PROF PHYSICS, UNIV ARK, 85- *Mem:* Sigma Xi; Am Phys Soc. *Res:* Design and use of lasers and related optical systems for coherent optical experiments and spectroscopic experiments with some useful application in mind. *Mailing Add:* Dept Physics Univ Ark Fayetteville AR 72701

SALAMON, IVAN ISTVAN, b Budapest, Hungary, Sept 10, 18; nat US; m 50, 68; c 1. BIOCHEMISTRY, ORGANIC CHEMISTRY. *Educ:* Swiss Fed Inst Technol, ChemEng, 41; Univ Basel, PhD(chem), 49. *Prof Exp:* Lab instr & asst, Univ Basel, 42-50; USPHS res fel, Med Col, Cornell Univ, 50-52; res fel biochem, Sloan-Kettering Inst, 52-54, from asst biochem to assoc biochem, 54-63; asst mem div endocrinol, Res Labs, Albert Einstein Med Ctr, Philadelphia, 63-68; BIOCHEMIST, HEKTOEN INST MED RES, COOK COUNTY HOSP, 68- *Mem:* Am Chem Soc; Swiss Chem Soc; Nat Acad Clin Biochem; Am Asn Clin Chemists. *Res:* Steroid chemistry and biochemistry; phospholipids. *Mailing Add:* Hektoen Inst Med Res Cook Co Hosp 627 S Wood St Chicago IL 60612-9985

SALAMON, KENNETH J, US citizen. TOXICOLOGY, ECOLOGY. *Educ:* Fordham Col, BS, 67, MS, 75, PhD(physiol ecol), 79. *Prof Exp:* Mem staff, Consolidated Edison Co, NY, 76-77, Inst Environ Med, NY Univ, 79-80; PROJ TOXICOLOGIST & ECOLOGIST, RAY J WESTON, INC, 80- *Concurrent Pos:* Mem, Advan Study Inst, NATO. *Mem:* Am Fisheries soc; Am Soc Limnol & Oceanog; AAAS; Sigma Xi. *Res:* Aquatic toxicology; fish physiology; bioassay and ecological investigations; estuarine biology; power plant impact studies; marine and freshwater phytoplankton and zooplankton physiology and ecology. *Mailing Add:* Life Systs Dept Roy F Weston Inc West Chester PA 19380

SALAMON, MYRON B, b Pittsburgh, Pa, June 4, 39; m 60; c 2. PHASE TRANSITIONS & CRITICAL PHENOMENA. *Educ:* Carnegie Inst Technol, BS, 61; Univ Calif, Berkeley, PhD(physics), 65. *Prof Exp:* NSF fel, 65-66; from asst prof to assoc prof, 66-74, PROF PHYSICS, UNIV ILL, URBANA, 74-, DIR, NSF-MAT RES LAB, 83- *Concurrent Pos:* Vis scientist, Inst Solid State Physics, Univ Tokyo, 71 & Tech Univ Munich, 74-75; Alfred P Sloan Found res fel, 71-72, Humboldt Found fel, 74; Inst Laue Langevin & Nat Ctr Sci Res, Grenoble, 81-82. *Mem:* Am Phys Soc; AAAS; Mat Res Soc. *Res:* Experimental studies of phase transitions in magnets, superconductors and modulated structures. *Mailing Add:* Dept Physics Univ Ill 1110 W Green St Urbana IL 61801

SALAMON, RICHARD JOSEPH, b Palmer, Mass, Apr 27, 32; m 60; c 1. SCIENCE EDUCATION. *Educ:* Col Holy Cross, BS, 53; Am Int Col, MA, 60; Univ Conn, Prof Dipl Educ, 64, PhD(sci educ), 68. *Prof Exp:* Instr, high sch, Mass, 59-65; PROF SCI & SCI EDUC, CENT CONN STATE UNIV, 66- *Mem:* Asn Educ Teachers Sci; Nat Sci Teachers Asn; Nat Asn Res Sci Teaching; Sch Sci & Math Asn. *Res:* Elementary, middle school, high school and college science instruction. *Mailing Add:* Dept Physics & Earth Sci Cent Conn State Univ New Britain CT 06050

SALAMONE, JOSEPH C, b Brooklyn, NY, Dec 27, 39; m 78; c 3. ORGANIC CHEMISTRY, POLYMER CHEMISTRY. *Educ:* Hofstra Univ, BSc, 61; Polytech Inst Brooklyn, PhD(org chem), 67. *Prof Exp:* Res Liverpool, 66-67; res assoc, Univ Mich, Ann Arbor, 67-70, secy, Macromolecular Res Ctr, 68-70; from asst prof to prof, Univ Lowell, 70-89, chmn dept, 75-78, actg dean, Col Pure & Appl Sci, 78-81, dean, Col Pure & Appl Sci, 81-84, chmn, Coun Deans, 81-83, EMER PROF CHEM, UNIV LOWELL, 89- *Concurrent Pos:* Secy-treas, Pac Polymer Fedn, 87-90, counr, 90-92; treas, Polymer Chem Div, Am Chem Soc, 74-78, chmn, 82. *Mem:* Am Chem Soc. *Res:* Syntheses of new monomers and polymers; copolymerization of ionic monomers; polyampholytes and ampholytic ionomers; solution properties of polymers. *Mailing Add:* Dept Chem Univ Lowell Lowell MA 01854

SALAMUN, PETER JOSEPH, b La Crosse, Wis, June 12, 19; m 46; c 8. BOTANY. *Educ:* Wis State Teacher's Col, BS, 41; Univ Wis-Madison, MS, 47, PhD(bot), 50. *Prof Exp:* Asst, Univ Wis, 45-48; from instr to prof, 48-84, chmn dept, 60-64, EMER PROF BOT, UNIV WIS-MILWAUKEE, 84- *Concurrent Pos:* Dir, Herbarium. *Mem:* AAAS; Am Inst Biol Sci; Bot Soc Am; Am Soc Plant Taxonomists; Am Meteorol Soc; Ecological Soc Am; AAAS. *Res:* Floristic and monographic work in plant taxonomy. *Mailing Add:* Dept Biol Sci Univ Wis Milwaukee WI 53201

SALANAVE, LEON EDWARD, b San Francisco, Calif, Nov 19, 17; m 49; c 2. OPTICS, SCIENCE EDUCATION. *Educ:* Univ Calif, AB, 40, MA, 47. *Prof Exp:* Assoc astron, Univ Calif, 42-47; instr astron & math, Sacramento Col, 47-52; lectr & consult astron, Morrison Planetarium, 49-53; assoc cur, Calif Acad Sci, 54-56; res assoc site surv, Nat Astron Observ, Asn Univs for Res Astron, 56-58; assoc res engr, Appl Res Lab, Univ Ariz, 58-60; res assoc optics, Inst Atmospheric Physics, 61-71; exec officer & ed jour, Astron Soc Pac, 71-74; prin investr, Lightning Atlas Proj, Off Naval Res, 75-80; instr, City Col San Francisco, 79-86; ASSOC ASTRON, CALIF ACAD SCI, 86- *Res:* Atmospheric and astronomical optics; optical spectrum of lightning; science writing and education. *Mailing Add:* Dept Astron Calif Acad Sci San Francisco CA 94118-4599

SALAND, LINDA C, b New York, NY, Oct 24, 42; m 64; c 2. ANATOMY, CYTOLOGY. *Educ:* City Col New York, BS, 63, PhD(biol), 68; Columbia Univ, MA, 65. *Prof Exp:* Res assoc anat, Col Physicians & Surgeons, Columbia Univ, 68-69; sr res assoc, 71-78, from asst prof to assoc prof, 78-89, PROF ANAT, SCH MED, UNIV NMEX, 89- *Concurrent Pos:* Assoc ed, Anat Record; ad hoc reviewer endocrinol & neuroendocrinol. *Mem:* Am Asn Anat; Am Soc Cell Biol; Am Soc Zool; Soc Neurosci. *Res:* Pituitary cytology; electron microscopic autoradiography; hypothalamic cytology; pancreatic islet cell ultrastructure; opiate peptides; neuroendocrine function; neural-immune interactions. *Mailing Add:* Dept Anat Univ NMex Sch Med Albuquerque NM 87131-5211

SALANECK, WILLIAM R, b Pottstown, Pa, Aug 19, 41; m 67; c 2. SOLID STATE PHYSICS. *Educ:* Albright Col, BS, 63; Univ Pa, MS, 64, PhD(physics), 68. *Prof Exp:* Scientist, Xerox Corp, Rochester, 68-72, mgr mat sci areas, 72-73, mgr photo & insulator physics area, 74, SR SCIENTIST MOLECULAR & ORG MAT AREA, XEROX WEBSTER RES CTR, 74-, PROF, SURFACE PHYSICS & CHEM, UNIV LINKÖPING, SWEDEN, 83- *Concurrent Pos:* NSF grants, NATO Advan Study Inst, 66 & Army res grant, 67; adj assoc prof physics, Univ Pa, 80- *Mem:* Am Phys Soc; European Phys Soc; Swedish Phys Soc; Europ Mat Res Soc. *Res:* Electron physics; photoelectron spectroscopy; electronic structure of molecular solids; electronic structure of conducting polymers; focused upon insulating or electrically conducting organic polymers, their surfaces, as well as polymer-metal interfaces, and model molecular systems for polymer-metal interfaces. *Mailing Add:* Inst Physics & Measurement Technol Univ Linkoping 58183 Linkoping Sweden

SALANITRE, ERNEST, b New York, NY, Feb 3, 15; m 45; c 2. MEDICINE, ANESTHESIOLOGY. *Educ:* City Col New York, BS, 36; Univ Rome, MD, 42. *Prof Exp:* Instr, Columbia Univ, 53-54, assoc, 55-58, from asst prof to prof, 59-86, EMER PROF ANESTHESIOL, COL PHYSICIANS & SURGEONS, COLUMBIA UNIV, 86- *Concurrent Pos:* From asst attend to assoc attend, Columbia-Presby Med Ctr, 53-71, attend, 71-86, emer consult, 90- *Honors & Awards:* Robert M Smith Award in Pediat Anesthesiol, Am Acad Pediat, 90. *Mem:* Am Soc Anesthesiol; AMA; Am Acad Pediat. *Res:* Pediatric anesthesiology; uptake and elimination of inhalational anesthetic agents in man. *Mailing Add:* 5933 Fieldston Rd New York NY 10471

SALANS, LESTER BARRY, b Chicago Heights, Ill, Jan 25, 36; m 58; c 2. MEDICINE, METABOLISM. *Educ:* Univ Mich, Ann Arbor, BA, 57; Univ Ill, Chicago, MD, 61. *Prof Exp:* Intern med, Stanford Med Sch, 61-62, resident, 62-64; asst prof, Rockefeller Univ, 67-68; from asst prof to assoc prof med, Dartmouth Med Sch, 71-78; chief, Lab Cellular Metab & Obesity & assoc dir diabetes, endocrinol & metab dis, 78-81, ACTG DIR, NAT INST ARTHRITIS, DIABETES, DIGESTIVE & KIDNEY DIS, NIH, 81-; AT SCH MED MT SINIA & MED CTR; VPRES, SANDOZ RES INST. *Concurrent Pos:* Fel, Stanford Med Sch, 63-64, USPHS fel, 64-65; USPHS spec fel, Rockefeller Univ, 65-67, Nat Inst Arthritis & Metab Dis grant, 68-77 & res career develop award, 72-77; adj prof med, Med Sch, Dartmouth Col, 78- *Mem:* Am Soc Clin Nutrit; Am Fedn Clin Res; Endocrine Soc; Am Soc Clin Invest; Am Diabetes Asn. *Res:* Intermediary carbohydrate/lipid metabolism related to diabetes mellitus, obesity and atherosclerosis. *Mailing Add:* NIH-NIAMDD Bldg 31 Rm 9A47 9000 Rockville Pike Bethesda MD 20814

SALANT, ABNER SAMUEL, b Cincinnati, Ohio, Mar 18, 30; m 52; c 3. FOOD TECHNOLOGY. *Educ:* NY Univ, BA, 50; Rutgers Univ, PhD(food technol), 53. *Prof Exp:* Res asst, Rutgers Univ, 51-53; assoc technologist, Gen Foods Corp, 53-56; proj leader, Tenco Div, Coca-Cola Co, 56-62; proj mgr org chem div, Monsanto Co, Inc, 62-68, vpres, Monsanto Flavor Essence Inc, 68-75; dir, food eng dir, 76-88, DIR, SCI & ADV TECH DIR, US ARMY NATICK RES DEVELOP ENG CTR, 88- *Mem:* AAAS; Am Chem Soc; Inst Food Technologists; Am Inst Chemists; fel NY Acad Sci; Soc Flavor Chemists. *Res:* Food chemistry; beverage and dessert products; dehydration; evaporation; extraction; classification; blending; commercial development; flavor and fragrance products; low-calorie foods and beverages; nutrition. *Mailing Add:* 20 Bowen Circle Sudbury MA 01776

SALANT, RICHARD FRANK, b New York, NY, Sept 4, 41; m 62; c 2. FLUID SEALING TECHNOLOGY. *Educ:* Mass Inst Technol, BS & MS, 63, DSc(mech eng), 67. *Prof Exp:* Asst prof mech eng, Univ Calif, Berkeley, 66-68; from asst prof to assoc prof mech eng, Mass Inst Technol, 68-72; mgr fluid mech res, 72-79, mgr fluid mech & heat transfer, 79-80, mgr fluid mech, heat transfer & appl physics, 80-83, head, dept fluid mech & head transfer, Borg-Warner Corp, 83-87; PROF MECH ENG, GA INST TECHNOL, 87- *Concurrent Pos:* Consult, Veriflo Corp, 67-68, United Aircraft Res Labs, 69-71, BW/IP Int, 87- & CR Industs, 89-; mem educ coun, Mass Inst Technol, 81-87; assoc ed, J Fluids Eng, 85-88. *Mem:* Fel Am Soc Mech Engrs; Am Soc Lubrication Engrs. *Res:* Fluid sealing technology; tribology; fluid mechanics; two-phase flow. *Mailing Add:* Sch Mech Eng Ga Inst Technol Atlanta GA 30332

SALARI, HASSAN, b Iran, Aug 27, 53; Can citizen; m 82; c 2. ENZYME BIOCHEMISTRY, INFLAMMATORY DISEASES. *Educ:* Tehran Univ, BSc, 76; Southampton Univ, Eng, PhD, 80. *Prof Exp:* Post doc immunol, McGill Univ, 82, biochem, Laval Univ, 85; res assoc, 85-87, ASST PROF MED & ASSOC PROF PHARMACOL, UNIV BC, 87- *Mem:* Can Soc Biochem; Can Soc Pharmacol. *Res:* Cellular biochemistry related to lung and heart diseases. *Mailing Add:* Vancouver Gen Hosp Res Inst 2660 Oak St Vancouver BC V6H 3Z6 Can

SALAS, PEDRO JOSE I, b Buenos Aires, Arg, Nov 29, 54; m 82; c 2. CELL BIOLOGY OF EPITHELIA, IMMUNOCYTOCHEMISTRY. *Educ:* Univ Buenos Aires, MD, 76, Master, 80, PhD(biophys), 81. *Prof Exp:* From instr to head instr cell biol, Univ Buenos Aires, 72-82; postdoctoral, State Univ NY Downstate, 82-84; res assoc, Cornell Univ Med Col, 84-88; STAFF RES MEM, INST INVEST BIOCHEM FUNDACIÓN CAMPOMAR, BUENOS AIRES, 88-; ASST PROF CELL BIOL, UNIV BUENOS AIRES, 90- *Concurrent Pos:* Res fel, Nat Res Coun, 78-82. *Mem:* Am Soc Cell Biol. *Res:* Plasma membrane polarization in epithelial cells; role of the cytoskeleton; early polarization; vacuolar apical compartment; selective anchoring of plasma membrane proteins to critical cytoskeleton. *Mailing Add:* Inst Invest Biochem Found Campomar Ave Patricias Argentinas 435 Buenos Aires 1405 Argentina

SALAS-QUINTANA, SALVADOR, b San Sebastián, PR, Sept 3, 55; US citizen; m 76; c 2. BIOTECHNOLOGY. *Educ:* Univ PR, BSA, 78, MS, 82; Rutgers Univ, PhD(hort), 88. *Prof Exp:* Postdoctoral res assoc & instr, Rutgers, State Univ NJ, 88-89; title III dir, 90, DEAN & DIR, LA MONTA13A REGIONAL COL, UNIV PR, 90-, ASST PROF HORT & DIR AGR TECHNOL DEPT, 90- *Concurrent Pos:* Vis prof, Mayaguez Campus, Univ PR, 90-, comt chmn grad students, 90- *Mem:* AAAS; Am Soc Plant Physiol; Am Soc Hort Sci; Plant Growth Regulator Soc Am. *Res:* Plant physiology; use of plant growth regulators in plant flowering and fruit ripening; effects of growth retardants and stimulants; genetic improvement of plants. *Mailing Add:* La Montaña Regional Col Univ PR Call Box 2500 Utuado PR 00761

SALATI, OCTAVIO M(ARIO), b Philadelphia, Pa, Dec 12, 14; m 51; c 3. ELECTRICAL ENGINEERING. *Educ:* Univ Pa, BS, 36, MS, 39, PhD(elec eng), 63. *Prof Exp:* Trainee, Radio Corp Am, 37-38, develop engr, 39; develop engr, C G Com Ltd, 39-42; sr engr, Hazeltine Electronics Corp, 42-48; from asst prof to assoc prof, 63-75, PROF ELEC ENG, UNIV PA, 75-, PROJ DIR, INST COOP RES, 48-, DIR, TV SYST, 70- *Concurrent Pos:* Vis prof, Pahlavi Univ, Iran, 69 & People's Repub China, 81; consult, USAF & Naval Med Res Inst, 77-; in charge design, construct & oper, Eng Schs Instrnl TV Syst, Univ Pa, dir eng schs, grad TV syst & dir continuing eng educ. *Mem:* AAAS; fel Inst Elec & Electronics Engrs; Sigma Xi. *Res:* Radio interference in communications and radar systems; microwave tubes; biological effects of microwave radiation; studies of out of band performance of antennas and radio propagation studies; BNC electrical connector and microwave absorber; shielding systems. *Mailing Add:* 570 Rosemary Circle Media PA 19063

SALAZAR, HERNANDO, b Ibague, Colombia, Nov 21, 31; m 56; c 4. PATHOLOGY, PUBLIC HEALTH. *Educ:* Col San Simon, Colombia, BS, 50; Nat Univ Colombia, MD, 58; Univ Pittsburgh, MPH, 71. *Prof Exp:* Instr morphol, Univ Valle, 59-61, asst prof, 63-66; from instr to asst prof, 66-70, ASSOC PROF PATH, UNIV PITTSBURGH, 70- *Concurrent Pos:* Rockefeller Found res fel anat, Wash Univ, 61-63, travel grant, 65; Health Res & Serv Found Pittsburgh grant, 68; NIH res grants, 70-72; Am Cancer Soc grant, 74-76; vis scientist, Sir William Dunn Sch Path, Oxford Univ, 73-74; pres, Latinamer Path Found, 81-82. *Mem:* Am Asn Path; Int Soc Gynec Path (pres, 84-86); Int Acad Path; Endocrine Soc; Am Soc Cell Biol. *Res:* Medical education; cancer; endocrinologic and reproductive pathology. *Mailing Add:* Dept Path Med-Chek Labs 3263 Nottingham Dr Pittsburgh PA 15325

SALBER, EVA JULIET, public health, community medicine; deceased, see previous edition for last biography

SALCE, LUDWIG, b New York, NY, Feb 3, 34; m 67; c 1. ORGANIC CHEMISTRY. *Educ:* Fordham Univ, BS, 55, PhD(chem), 67; NY Univ, MS, 61. *Prof Exp:* Chemist, Petrotex Div, Food Mach Chem Corp, 55-56, Burroughs Wellcome & Co, 56-59 & Col Physicians & Surgeons, Columbia Univ, 59-62; sr chemist, Merck & Co, Inc, 66-70; res chemist, Cybertek & Co, 71-72; RES CHEMIST, EVANS CHEMETICS, INC, CONN, 72-; RES DIR, ZOTOS INT INC, DARIEN, CONN, 85- *Mem:* Am Chem Soc; Soc Cosmetic Chemists. *Res:* Purines; pyrimidines; glycosides; amino acids; peptides; aromatic hydrocarbons; steroids; heterocyclics; synthesis of biologically active compounds; synthesis of sulfur compounds; cosmetics and hair research. *Mailing Add:* 32 Brynwood Lane Greenwich CT 06831

SALCH, RICHARD K, b Union City, NJ, Sept 10, 40; m 69; c 1. PLANT PATHOLOGY. *Educ:* Cent Col, Iowa, BA, 61; Rutgers Univ, MS, 62, PhD(plant biol), 69. *Prof Exp:* Lab technician, Boyce Thompson Inst Plant Res, NY, 63-66; from asst prof to assoc prof, 69-73, PROF BIOL, EAST STROUDSBURG STATE COL, 73- *Mem:* Sigma Xi; Am Inst Biol Sci. *Res:* Biological methods of plant disease control pollution problems and how they effect agriculture; control of fungus diseases of plants; physiology of fungus. *Mailing Add:* RD 3 Box 3578 Saylorsburg PA 18353

SALCMAN, MICHAEL, b Pilsen, Czechoslovakia, Nov 4, 46; US citizen; m 69; c 2. NEUROLOGICAL SURGERY, NEURO-ONCOLOGY. *Educ:* Boston Univ, BA & MD, 69. *Prof Exp:* Intern surg, Univ Hosp, Boston Univ Med Ctr, 69-70; res assoc neurophysiol, Lab Neural Control, Nat Inst Neurol Dis & Stroke, 70-72; resident neurosurg, Neurol Inst NY, Columbia Univ, 72-76; from asst prof to assoc prof neurosurg, 76-84, CHIEF NEURO-ONCOL, SCH MED, UNIV MD, 78-, PROF & HEAD NEUROSURG, 84- *Concurrent Pos:* Co-prin investr cats' visual cortex, Nat Eye Inst, NIH, 72-78 & hyperthermal radiotherapy, Am Cancer Soc, 78-80; assoc ed, Neurosurg, 82- *Mem:* Cong Neurol Surgeons; Am Asn Neurol Surgeons; Asn Advan Med Instrumentation; fel Am Col Surgeons; NY Acad Sci. *Res:* Biology and treatment of brain tumors using microsurgery, laser, microwave hyperthermia, reversal of blood-brain barrier and chemotherapy; chronic microelectrode technology for single unit neurophysiology and neural prosthesis development. *Mailing Add:* Div Neurol Surg Univ Md Med 655 W Baltimore St Baltimore MD 21201

SALCUDEAN, MARTHA EVA, b Cluj, Romania, Feb 26, 34; m 55; c 1. HEAT TRANSFER, FLUID FLOW. *Educ:* Univ Cluj, BEng, 56, MEng, 62; Inst Polytech, Brasov, Romania, PhD(mech), 69. *Prof Exp:* Engr, Armatura, Cluj, Romania, 56-60; design engr, Nat Ctr Indust Automatics, Bucharest, 60-63; sr res officer, Nat Res Ctr Metall, Bucharest, 63-75; lectr, Univ Ottawa, 76-77, from assoc prof to prof, 77-85; PROF & HEAD, UNIV BC, 85- *Concurrent Pos:* Res assoc, McGill Univ, 76-77; Lectr, Univ Ottawa, 76-77. *Mem:* Am Soc Mech Engrs. *Res:* Computational heat transfer and fluid dynamics; convective heat transfer in laminar and turbulent flows; heat transfer in buoyancy affected recirculatory flows; two phase flow and heat transfer in vertical and horizontal channels; void formation in two phase flow at low pressures; film cooling of turbine blades; flows in recovery boilers. *Mailing Add:* Dept Mech Eng Univ BC 2324 Main Mall Vancouver BC V6T 1W5 Can

SALDANHA, LEILA GENEVIEVE, b Bangalore, India, Jan 11, 55. DIETETICS. *Educ:* Univ Bombay, India, BSc, 74; Kans State Univ, Manhattan, MS, 83, PhD(food & nutrit), 85. *Prof Exp:* Prod supvr, Mafco Ltd, Bombay, India, 76-80; ASST PROF NUTRIT, DEPT ANIMAL SCI, FOOD & NUTRIT, SOUTHERN ILL UNIV, 85- *Concurrent Pos:* Clin dietician, Holy Family Hosp, New Delhi, 75; Fel, Gen Foods Fund, 83-85; Mae Baird Mem Scholar, 83-85; exten asst, Exten Serv, Kans State Univ, 84, asst instr dietetics, Dept Housing, 84. *Mem:* Am Dietetic Asn; Sigma Xi; Inst Food Technologists; Am Inst Nutrit. *Res:* Mechanism by which dietary calcium alters blood pressure; health habits and prevalence of anemia among residents in southern Illinois. *Mailing Add:* PO Box 1145 Battle Creek MI 49016

SALDARINI, RONALD JOHN, b Paterson, NJ, Nov 6, 39; m 62; c 3. PHYSIOLOGY, BIOCHEMISTRY. *Educ:* Drew Univ, BA, 61; Univ Kans, PhD(biochem, physiol), 67. *Prof Exp:* Sr res scientist, Metab Dis Ther Res Sect, 69-74, group leader respiratory & skin dis, 74, HEAD, DEPT BIOL, METAB DIS SECT, LEDERLE LABS, 74- *Concurrent Pos:* NIMH training fel biochem & physiol, Brain Res Inst, Sch Med, Univ Calif, Los Angeles, 67-68, NIH fel, 68-69. *Mem:* AAAS; Soc Study Reproduction; Brit Soc Study Fertil; Int Soc Fertil. *Res:* Estrogen and progestin receptors in uterus and corpus luteum; immunological regulation of immediate hypersensitivity disease states; cyclic nucleotide involvement in proliferative skin disease; sperm metabolism in monkeys. *Mailing Add:* Lederle Lab Dept of Biol One Cyanamid Plaza Wayne NJ 07470

SALDICK, JEROME, b Brooklyn, NY, Mar 24, 21; m 51; c 2. PHYSICAL CHEMISTRY. *Educ:* Brooklyn Col, BA, 40; Columbia Univ, MA, 41, PhD(phys chem, kinetics), 48. *Prof Exp:* Chemist, Kellex Corp, 48-51; assoc chemist, Brookhaven Nat Lab, 51-54; tech engr, Aircraft Nuclear Propulsion Dept, Gen Elec Co, 54-56; assoc, Astra, Inc, 56-58; pres, Gen Radionuclear Co, 58-59; resident scientist, Indust Reactor Labs, Am Mach & Foundry Co, 59-63; group leader, Princeton Chem Res Inc, 64-66; sr res chemist, FMC Corp, 66-73, res assoc, 73-83; RETIRED. *Mem:* Am Chem Soc. *Res:* Radiation chemistry; heterogeneous catalysis; applied microbiology. *Mailing Add:* 24 Randall Rd Princeton NJ 08540

SALE, PETER FRANCIS, b Jan 12, 41; Can citizen; m 71; c 1. BEHAVIORAL ECOLOGY, TROPICAL ECOLOGY. *Educ:* Univ Toronto, BSc, 63, MA, 64; Univ Hawaii, PhD(zool), 68. *Prof Exp:* Lectr, Univ Sydney,68-74, sr lectr, 75-81, assoc prof biol sci 82-87; asst dir, Inst Marine Ecol, 84-87; PROF & CHMN, DEPT ZOOL, UNIV NH, 88-; DIR, CTR MARINE BIOL, 90- *Concurrent Pos:* Adv Comt Crown-of-Thorns res grant, Australia, 72-73; Australian Res Grants Comt grant, 74-87; mem bd, Heron Island Res Sta, 74-85; mem consult bd, Lizard Island Res Sta, Australia, 75-87; Australian Marine Sci & Technol grant, 80-88, mem adv comt, 84-86; dir, Sydney Univ One Tree Island Field Sta, 74-87; mem, Crown-of-Thorns Starfish adv res comt, 86-88, Australasia Region Comt, CIES; Fulbright Awards Prog, 90-; prin investr, NSF, 91-94, Nat Oceanic & Atmospheric Admin, 91-94. *Honors & Awards:* Stoye Award, Am Soc Ichthyologists & Herpetologists, 68. *Mem:* Am Soc Naturalists; Soc Animal Behav; Sigma Xi; Ecol Soc Am; Am Soc Ichthyologists & Herpetologists. *Res:* Ecology of coral reef communities, especially fish and their high diversity; role of recruitment in mediating structure of communities; demography of coral reef fish; larval and juvenile ecology of temperate coastal fishes. *Mailing Add:* Dept Zool Univ NH Durham NH 03824

SALEEB, FOUAD ZAKI, b Toukh, Egypt, Sept 7, 34; m 66; c 3. SURFACE CHEMISTRY, PHYSICAL CHEMISTRY. *Educ:* Univ Alexandria, BS, 56; Univ London, PhD(phys chem) & DIC, 63. *Prof Exp:* Egyptian Govt fel, Imp Col, Univ London, 63-64; res scientist surface chem, Nat Res Ctr, Cairo, Egypt, 64-69; res assoc, Mass Inst Technol, 69-72; sr chemist, 72-80, PRINCIPAL SCIENTIST, TECH CTR, GEN FOODS CORP, 80- *Mem:* Am Chem Soc. *Res:* Surface chemistry of emulsions, foams and suspensions; electrokinetic properties of carbons, minerals and hydroxyapatites; monomolecular films and mass transfer. *Mailing Add:* Technical Ctr T22-1 Gen Foods Corp Tarrytown NY 10625

SALEEBY, JASON BRIAN, b Los Angeles, Calif, Oct 24, 48; m 78; c 1. GEOLOGY. *Educ:* Univ Calif, Santa Barbara, PhD(geol), 75. *Prof Exp:* Asst prof geol, Univ Calif, Berkeley, 75-78; ASSOC PROF GEOL, CALIF INST TECHNOL, 78- *Concurrent Pos:* Mem, US Geol Surv, 75- *Mem:* Geol Soc Am; Am Geophys Union. *Res:* Tectonic and paleogeographic development of western North America; processes of accretion of ocean floor and island areas to continental edges by use of geochronology; field structure and petrology. *Mailing Add:* Div Geol & Planetary Sci 170-25 Calif Inst Technol 1201 E Calif Blvd Pasadena CA 91125

SALEH, ADEL ABDEL MONEIM, b Alexandria, Egypt, July 8, 42; m 70. ELECTRICAL ENGINEERING. *Educ:* Univ Alexandria, BSc, 63; Mass Inst Technol, SM, 67, PhD(elec eng), 70. *Prof Exp:* Instr elec eng, Univ Alexandria, 63-65; MEM TECH STAFF ELEC ENG, BELL TEL LABS, 70- *Concurrent Pos:* Consult engr, Sylvania Elec Prod, Inc, Mass, 68-70. *Mem:* Sr mem Inst Elec & Electronics Engrs. *Res:* Microwave and millimeter-wave circuits and communication systems research, including power amplifiers nonlinearities and efficiency, power combining networks, mixers and frequency converters, automated microwave measurements, and quasi-optical components. *Mailing Add:* AT&T Bell Labs Hoh-R139 Crawford Hill Lab Holmdel NJ 07733

SALEH, BAHAA E A, b Egypt, Sept 30, 44; US citizen. OPTICAL PROCESSING, IMAGE PROCESSING. *Educ:* Cairo Univ, BS, 66; Johns Hopkins Univ, PhD(elec eng), 71. *Prof Exp:* Asst prof elec eng, Univ Santa Catarina, Brazil, 71-74; res assoc, Max Planck Inst, Gottingen, Germany, 74-77; from asst prof to assoc prof, 77-81, PROF ELEC ENG, DEPT ELEC & COMPUT ENG, UNIV WIS-MADISON, 81-, CHMN, 90- *Concurrent Pos:* Topical ed, J Optical Soc Am; Guggenheim fel, 84-85. *Mem:* Fel Optical Soc Am; sr mem Inst Elec & Electronics Engrs. *Res:* Image processing, optical information processing, optical communication, statistical optics and vision. *Mailing Add:* Dept Elec & Comput Eng Univ Wis 415 Johnson Dr Madison WI 53706

SALEH, FARIDA YOUSRY, b Cairo, Egypt, June 17, 39; m 59; c 2. ENVIRONMENTAL CHEMISTRY, ANALYTICAL CHEMISTRY. *Educ:* Ain-Shams Univ, Cairo, Egypt, BS, 59; Alexandria Univ, Egypt, MS 67; Univ Tex, Dallas, PhD(environ chem), 76. *Prof Exp:* Anal chemist, Alexandria Labs, Alexandria, Egypt, 62-68; res chemist, Res Ctr, Dallas Water Reclamation, 68-75; res scientist, Ctr Environ Studies, Univ Tex, Dallas, 75-77; res fel, Dept Chem, Tex A&M Univ, Col Sta, Tex, 77-78; res scientist, II-Res, NTex State Univ, Denton, 78-83, res scientist III-Res, Inst Appl Sci, 83-85; ASSOC PROF, DIV ENVIRON SCI, UNIV NTEX, 85- *Concurrent Pos:* Asst prof, Dept Chem, NTex State Univ, Denton,80-83, assoc prof, 83- *Mem:* Am Chem Soc; Sigma Xi; Int Humic Substances Soc; Asn Women Sci. *Res:* Environmental chemistry; transport and fate of chemicals in the environment; analytical chemistry of pollutants with emphasis on utilization of advanced spectroscopic techniques to identify and measure pollutants. *Mailing Add:* Inst Appl Sci PO Box 13078 NT Sta NTex State Univ Denton TX 76203-3078

SALEH, WASFY SELEMAN, b Egypt, Jan 23, 32; Can citizen; m 64; c 3. SURGERY, UROLOGY. *Educ:* Cairo Univ, MB, Bch, 57; McGill Univ, PhD(exp surg), 70. *Prof Exp:* Asst prof surg, Ottawa Univ, 70-73; ASST PROF UROL, UNIV SHERBROOKE, 73- *Mem:* Am Col Surgeons; Int Transplantation Soc; Royal Col Surgeons & Physicians Can; Am Urol Asn. *Res:* Studies in graft versus host reactions. *Mailing Add:* 3029 Carling Ave Ottawa ON K2B 8E8 Can

SALEHI, HABIB, b Iran, Jan 29, 35; m 62; c 2. MATHEMATICAL ANALYSIS. *Educ:* Univ Tehran, BA, 58; Ind Univ, MA, 62, PhD(math), 65. *Prof Exp:* Instr math, Univ Tehran, 58-60; from asst prof to assoc prof, 65-74, PROF MATH & STATIST, MICH STATE UNIV, 74- *Concurrent Pos:* Nat Inst Gen Med Sci grant, 66-; NSF grants, Mich State Univ, 67-71. *Mem:* Am Math Soc; Inst Math Statist. *Res:* Prediction theory of stochastic processes as initiated by N Wiener and A N Kolmogorov; powerful tools and techniques used in mathematical analysis to solve various problems in prediction and communication theory. *Mailing Add:* Dept Math 425 Wells Hall Mich State Univ East Lansing MI 48824

SALEM, HARRY, b Windsor, Ont, Mar 21, 29; US citizen; m 57; c 1. PHARMACOLOGY, TOXICOLOGY. *Educ:* Univ Western Ont, BA, 50; Univ Mich, BSc, 53; Univ Toronto, MA, 55, PhD(pharmacol), 58. *Prof Exp:* Res asst, Univ Toronto, 58-59; pharmacologist, Air-Shields Inc, 59-62; sr pharmacologist, Smith Kline & French Labs, 62-65; pres, Whittaker Toxigenics, Inc, Decatur, Ill, 80-84; CHIEF TOXICOL DIV, RES DEVELOP & ENG CTR, MD, 84-; dir phamacol & toxicol, Cooper Labs, Cedar Knolls, NJ, 72-77; pres & chief toxicologist, Cannon Labs, Inc, 77-80; pres, Whittaker Toxigenics, INC, Decatur, Ill, 80-; CHIEF TOXICOL DIV, RES, DEVELOPENG CTR, MD, 84- *Concurrent Pos:* From instr to asst prof, Sch Med, Univ Pa, 60-75, assoc prof, 75-; adj prof environ health, Sch Pharm, Temple Univ, 76; chmn, tech comt, Inhalation Specialty Sect, Soc Toxicol; ed-in-chief, J Appl Technol; councilor & chmn, tech comt, Specialty Sect on Inhalation, Soc Toxicol. *Honors & Awards:* Cert Appreciation, Soc Toxicol, 90. *Mem:* AAAS; Am Chem Soc; Am Col Clin Pharmacol; charter mem Am Col Toxicol; Am COnf Gout Indust Hyg Inc; Am Soc Clin Pharmacol & Therapeut; Am Soc Pharmacol & Exp Therapeut; chem Corps Asn Inc; fel NY Acad Sci; Soc Comp Ophthal (vpres); Sigma Xi. *Res:* Respiratory, cardiovascular, ocular and general pharmacology, physiology and toxicology. *Mailing Add:* Toxicol Div SMCCR-RST Res, Develop & Eng Ctr Edgewood Area Aberdeen Proving Ground MD 21010-5423

SALEM, NORMAN, JR, b Feb 14, 50; m. LIPID CHEMISTRY, POLYUNSATURATED FATTY ACID METABOLISM. *Educ:* Univ Rochester, PhD(neurobiol), 78. *Prof Exp:* CHIEF SECT ANAL CHEM, NAT INST ALCOHOL ABUSE & ALCOHOLISM, 83- *Concurrent Pos:* Adj prof, Dept Physiol & Biophys, Georgetown Univ. *Mem:* Am Soc Neurochem; Am Soc Molecular Biol & Biochem; Soc Exp Biol & Med. *Res:* Over 70 publications on lipid biochemistry; specialist in omega-3 fatty acid area. *Mailing Add:* NIH Bldg 10 Rm 3C 102 Lab Clin Studies Bethesda MD 20892

SALEM, SEMAAN IBRAHIM, b Bterram, Lebanon, Apr 4, 27; US citizen; c 3. PHYSICS. *Educ:* Am Univ, Cairo, BSc, 55; Univ Tex, PhD(physics), 59. *Prof Exp:* Teacher physics & math, Tripoly Col & Tripoly Boys Sch, 55-56; consult radiation, NAm Aviation, Inc, 58-59; asst prof physics, Univ Tex, Arlington, 59-61; chmn dept physics, 79-88, from asst prof to assoc prof, 61-68, PROF PHYSICS, CALIF STATE UNIV, LONG BEACH, 68- *Concurrent Pos:* Gen Elec Co educ grant, 64; Res Corp res grants, 64-66; res physicist, Lawrence Livermore Nat Lab, 67-68; vis scientist, Calif Inst Technol, 76. *Honors & Awards:* Lebanon Govt Nat Award for the Best Article of the Year, 63. *Mem:* Am Phys Soc; Nat Asn Physics Teachers; Arab Phys Soc. *Res:* Measurement of x-ray line width; interaction of charged particles with metals; x-ray spectra; channeling; transition probabilities; sets of energy levels in the rare earth and transition elements. *Mailing Add:* Dept Physics Calif State Univ Long Beach CA 90840-3901

SALEMME, ROBERT MICHAEL, b Boston, Mass, June 17, 43; c 1. CHEMICAL & SYSTEMS ENGINEERING. *Educ:* Tufts Univ, BS, 64, MS, 67, Case Western Reserve Univ, 70. *Prof Exp:* Staff engr, 69-75, tech adminr, 75-77, MGR ENERGY CONVERSION SYSTS, GEN ELEC CORP RES & DEVELOP, 77- *Mem:* Am Inst Chem Engrs. *Res:* Energy conversion systems analysis and development of advanced energy conversion technologies; development and analysis of synthetic fuel processes. *Mailing Add:* 1156 Stratford Schenectady NY 12308

SALERNI, ORESTE LEROY, b Bolivar, Pa, June 11, 34. ORGANIC CHEMISTRY, MEDICINAL CHEMISTRY. *Educ:* Duquesne Univ, BS, 57, MS, 59; Univ Ill, PhD(pharmaceut chem), 63. *Prof Exp:* Assoc chemist, Midwest Res Inst, 62-65, sr chemist, 65-69; assoc prof, 69-77, PROF MED CHEM, COL PHARM, BUTLER UNIV, 77- *Concurrent Pos:* Res grant, 66-67. *Mem:* Am Chem Soc; Royal Soc Chem. *Res:* Synthesis of organic compounds with potential biological activity. *Mailing Add:* Col of Pharm Butler Univ 4600 Sunset Ave Indianapolis IN 46208

SALERNO, ALPHONSE, b Newark, NJ, Apr 3, 23; m 66; c 2. SURGERY, BIOLOGY. *Educ:* Seton Hall, BS, 44; Philadelphia Col Osteopath Med, DO, 48; Guadalajara Mex, cert(gen surg), 67 & abdominal surg, 86. *Prof Exp:* Chmn bd cert surg & pres Acad Surg, 76; CHIEF SURG, LIVINGSTON COMMUNITY HOSP, 76- & CHIEF STAFF, 86- *Mem:* NY Acad Sci. *Mailing Add:* 613 Park Ave East Orange NJ 07017

SALERNO, JOHN CHARLES, b Troy, NY, May 23, 49; m 74; c 1. BIOENERGETICS, SPECTROSCOPY. *Educ:* Mass Inst Technol, BSc, 72; Univ Pa, PhD(biophys), 77. *Prof Exp:* Res assoc biophys, Univ Pa, 77-78; NIH fel biochem, Duke Univ, 78-80; asst prof, 80-86, ASSOC PROF BIOL, RENSSELAER POLYTECH INST, 86- *Concurrent Pos:* NIH fel, 78-80; adj asst prof, dept biol, State Univ NY, Albany, 83- *Mem:* Biophys Soc. *Res:* Structure and function of electron transfer complexes using spectroscopic methods and computer modeling; thermodynamics of electron and proton transfer in energy conserving systems; membrane bound and multicomponent enzymes-electron paramagnetic resonance. *Mailing Add:* Biol Dept Rensselaer Polytech Inst Troy NY 12180-3590

SALERNO, LOUIS JOSEPH, b San Mateo, Calif, Mar 5, 49. CRYOGENICS. *Educ:* Univ Calif, Santa Barbara, BA, 74; San Jose State Univ, BS, 79; Stanford Univ, MS, 82. *Prof Exp:* Res scientist, Space Proj Div, 79-89, sci instrument engr, 89-90, ATMOSPHERIC ENTRY PROJ ENGR, AMES RES CTR, NASA, 90- *Concurrent Pos:* Lectr, Dept Mech Eng, San Jose State Univ, 80- *Res:* Thermal conductance of pressed metallic contacts at liquid helium temperatures; instruments for atmospheric entry vehicles; liquid helium transfer in space, especially fluid management, metering. *Mailing Add:* NASA Ames Res Ctr Mail Stop 240-8 Moffett Field CA 94035

SALERNO, RONALD ANTHONY, b Philadelphia, Pa, Dec 13, 42; m 64; c 4. BIOLOGICAL QUALITY CONTROL, VIROLOGY. *Educ:* St Vincent Col, BA, 64; Villanova Univ, MS, 67; Univ Md, College Park, MS, 70, PhD(zool), 71. *Prof Exp:* Teaching asst biol, Villanova Univ, 64-66; teaching asst zool, Univ Md, College Park, 66-69; sr technologist, Microbiol Assocs, Inc, 69-70, asst scientist, 70-71, asst investr, 71-73; sr virologist, Merck Sharp & Dohme Res Labs, 73-81; SR PROJ DEVELOP BIOLOGIST, BIOL QUAL CONTROL TECH SERV, MERCK SHARP & DOHME, 81- *Concurrent Pos:* Mem fac, Dept Biol, Grad Sch, Villanova Univ, 81- *Mem:* AAAS; NY Acad Sci. *Res:* Biology of the type C RNA tumor and herpes simplex viruses; viral vaccine development; biological assay development. *Mailing Add:* Merck Sharp & Dohme Res Labs Ten Sentry Pkwy Blue Bell PA 19422

SALES, BRIAN CRAIG, b Durham NC, Dec 19, 47; m 74; c 2. HIGH-TEMPERATURE SUPERCONDUCTIVITY, PHOSPHATE GLASSES. *Educ:* Carnegie-Mellon Univ, BS, 69; Univ Calif San Diego, PhD(physics), 74. *Prof Exp:* Fel, Univ Cologne, WGer 74-76; res physicist, Univ Calif, San Diego, 76-81; staff scientist, 81-87, CO-GROUP LEADER, SOLID STATE DIV, OAK RIDGE NAT LAB, TENN, 87- *Honors & Awards:* Mat Sci Award, US Dept Energy, 84. *Mem:* Am Phys Soc; Mat Res Soc; Am Ceramic Soc; Am Asn Crystal Growth. *Res:* Fundamental structure properties of phosphate glasses & crystals; synthesis of cuprate-based high temperature superconductors. *Mailing Add:* Oak Ridge Nat Lab Solid St Div Bldg 2000 PO Box X Oak Ridge TN 37830

SALES, JOHN KEITH, b Syracuse, NY, Jan 4, 34; m 57; c 5. STRUCTURAL GEOLOGY. *Educ:* Syracuse Univ, BS, 56; Univ Nev, Reno, PhD(geol), 66. *Prof Exp:* Explor geologist, Mobil Oil Corp, Wyo, 66-68; asst prof earth sci, State Univ NY Col Oneonta, 68-73, assoc prof, 73-80. *Concurrent Pos:* Lectr, Mobil Oil Corp Explor Sch, 67-, struct consult, 68-, grant, 70-72. *Mem:* Geol Soc Am; Am Asn Petrol Geologists. *Res:* Regional tectonics, plate tectonic scale modeling and glacier dynamics. *Mailing Add:* Mobil Explor PO Box 510 Stavanger 4001 Norway

SALETAN, LEONARD TIMOTHY, b New York, NY, Jan 5, 15; m 37; c 3. CHEMISTRY. *Educ:* Univ Wis, BA, 36, MA, 38. *Prof Exp:* Asst chief chemist & supvr lab, Schwarz Labs, NY, 38-45; brewing technologist, Army Exchange Serv, Europe, 45-46; sect head, Standard Brands, Inc, NY, 47-48; chemist & lab supvr, Wallerstein Co, Staten Island, 48-58; chief brewing lab, 58-66, asst dir res, 66-70, dir brewing lab, 70-72; dir develop hop extracts, Kalamazoo Spice Extraction Co, Mich, 72-75; consult, 75-80; RETIRED. *Mem:* Am Chem Soc; Am Soc Brewing Chem; Inst Food Technol; fel Am Inst Chem. *Res:* Brewing; enzymes; food quality control and analysis. *Mailing Add:* 25 Bay Ave Sea Cliff NY 11579-1039

SALEUDDIN, ABU S, b Faridpur, Bangladesh, Jan 14, 37; m 66; c 1. INVERTEBRATE NEUROENDOCRINOLOGY. *Educ:* Univ Dacca, BSc, 55, MSc, 57; Univ Reading, PhD(marine ecol), 63. *Prof Exp:* Lectr invert zool, Univ Dacca, 58-60 & 63-64; Nat Res Coun Can fel invert physiol, Univ Alta, 64-66; instr cell physiol, Duke Univ, 66-67; from asst prof to assoc prof, 67-78, PROF CELL BIOL, YORK UNIV, 78- *Mem:* Am Soc Zoologists; Can Soc Zoologists. *Res:* Physiology of biological calcification with special interest in shell regeneration in molluscs; neurosecretion in invertebrates. *Mailing Add:* Dept Biol York Univ 4700 Keele St Downsview ON M3J 1P6 Can

SALGADO, ERNESTO D, b Spain, Nov 16, 23; m 52; c 4. PATHOLOGY. *Educ:* Univ Zaragosa, AB, 40; Univ Madrid, MD, 47; Univ Montreal, PhD, 55. *Prof Exp:* Resident, Univ Madrid Hosp, 47-50; asst, Univ Montreal, 50-52, res assoc, 52-54; sr biologist, Nepera Chem Co, NY, 54-55, dir biol res, 55-57; endocrinologist, Pfizer Therapeut Inst, NJ, 57-58; assoc prof, 61-68, PROF PATH, COL MED & DENT NJ, 68- *Concurrent Pos:* Fel path, Col Med & Dent NJ, 58-61; consult, Pfizer Therapeut Inst, 58-59. *Mem:* AAAS; Am Soc Exp Path; Am Physiol Soc; fel NY Acad Sci. *Res:* Experimental and clinical pathology. *Mailing Add:* Dept Path Univ Med Dent NJ Med Sch 185 S Orange Ave Newark NJ 07103

SALGANICOFF, LEON, b Buenos Aires, Arg, Sept 11, 24; US citizen; m 54; c 2. PHARMACOLOGY, BIOCHEMISTRY. *Educ:* Univ Buenos Aires, MS, 47, DSc, 55. *Prof Exp:* Chief clin path, Military Hosp, Buenos Aires, 55-59; chief neurochem, Med Sch, Univ Buenos Aires, 59-64; res fel biochem, Dept Biophys Chem, Univ Pa, 65-68; assoc prof, 68-76, PROF PHARMACOL, DEPT PHARMACOL, MED SCH, TEMPLE UNIV, 76- *Concurrent Pos:* Mem, Nat Res Coun, 60-64; sect head pharmacol, Thrombosis Res Ctr, Med Sch, Temple Univ, 72-84; vis prof, Dept Gen Path, State Univ Rome, Italy, 80-; mem Coun Thrombosis, Am Heart Asn. *Mem:* Am Heart Asn; Sigma Xi. *Res:* Second messenger control of platelet function. *Mailing Add:* Dept Pharmacol Med Sch Temple Univ Philadelphia PA 19140

SALHANICK, HILTON AARON, b Fall River, Mass, Sept 17, 24; m 55; c 2. OBSTETRICS & GYNECOLOGY. *Educ:* Harvard Univ, AB, 47, MA, 49, PhD(biol), 50; Univ Utah, MD, 56. *Prof Exp:* Asst endocrinol, Harvard Univ, 47-50; res assoc obstet & gynec, Col Med, Univ Utah, 52-56; intern, St Louis Maternity Hosp, 56-57; from resident to chief resident, Col Med, Univ Nebr, 57-59, from asst prof to assoc prof, 57-62, asst prof biochem, 57-62; obstetrician-gynecologist-in-chief, Beth Israel Hosp, Boston, 62-65; head, dept pop sci, 71-73, PROF OBSTET & GYNEC, SCH MED, HARVARD UNIV, 62-, FREDERICK LEE HISAW PROF REPRODUCTIVE PHYSIOL, 70- *Concurrent Pos:* Mem, Endocrinol Study Sect, NIH, 61-65, chmn, 65-66; mem, Ctr Pop Studies, Sch Pub Health, Harvard Univ, 65- *Honors & Awards:* Distinguished Serv Award, Endocrine Soc, 89. *Mem:* AAAS; Am Chem Soc; fel Am Col Obstet & Gynec; Endocrine Soc; Soc Gynec Invest. *Res:* Endocrinology of reproductive system; conception control. *Mailing Add:* Sch Pub Health Harvard Univ Boston MA 02115

SALHANY, JAMES MITCHELL, b Detroit, Mich, Mar 27, 47; m 84. BIOPHYSICS. *Educ:* Univ Fla, BS, 72; Univ Chicago, PhD(biophys), 74. *Prof Exp:* Fel biophys, Bell Tel Labs, 74-75; res asst prof biochem, 75-79, assoc prof, dept internal med, 79-89, PROF DEPTS INTERNAL MED & BIOCHEM, UNIV NEBR MED CTR, 89- *Concurrent Pos:* Established investr, Am Heart Asn, 80-85. *Mem:* Am Chem Soc; Biophys Soc. *Res:* Membrane biophysics, ion transport and protein associations; Band 3 protein. *Mailing Add:* Va Med Ctr Res Serv 4101 Woolworth Ave Omaha NE 68105

SALIHI, JALAL T(AWFIQ), b Sulaymania, Iraq, Dec 6, 25; m 57; c 2. ELECTRICAL ENGINEERING. *Educ:* Univ Leeds, BSc, 48; Univ Calif, Berkeley, MS, 54, PhD(elec eng), 58. *Prof Exp:* Staff engr, Lenkurt Elec Co, Calif, 57-59; asst prof elec eng, Univ Baghdad, 59-61, head dept, 61-63; supvry res engr, Gen Motors Defense Res Labs, Calif, 63-71, supvry res engr elec propulsion dept, Res Labs, Gen Motors Tech Ctr, Mich, 71; chief hybrid & elec systs br, Environ Protection Agency, 71-72, asst dir, Div Advan Automotive Power Systs Develop, 72-; DIR RES, OTIS ELEVATOR CO, FARMINGTON, CONN; VPRES RES & DEV, US ELEVATOR CO. *Mem:* Inst Elec & Electronics Engrs; Soc Automotive Engrs. *Res:* Magnetic amplifiers and other types of nonlinear magnetic circuits; servomechanisms and control; power conversion; electric propulsion. *Mailing Add:* US Elvator Co 10728 US Elevator Rd Spring Valley CA 92078

SALIK, JULIAN OSWALD, b Czech, Sept 17, 09; nat US; m 39. RADIOLOGY. *Educ:* Jagiellonian Univ, MD, 36; Cambridge Univ, DMRE, 41. *Prof Exp:* Clin clerk radiol, Holzknecht Inst, Vienna, Austria, 37, Panel Hosp, Cracow, Poland, 38 & Middlesex Hosp, London, Eng, 39; resident, Columbia-Presby Med Ctr, New York, 42-44; dir dept radiol, Halloran Vet Admin Hosp, Staten Island, NY, 47-49; from asst prof to assoc prof, 56-82, EMER ASSOC PROF RADIOL, SCH MED, JOHNS HOPKINS UNIV, 82-; EMER RADIOLOGIST-IN-CHIEF, SINAI HOSP, 82- *Concurrent Pos:* Asst resident, Col Physicians & Surgeons, Columbia Univ, 42-44; radiologist, Levindale Home for Aged & Chronic Dis Hosp, Baltimore, Md, 51-82 & Hopkins Hosp, 56-82, emer radiologist, 82-; radiologist-in-chief, Sinai Hosp, 49-82. *Mem:* Emer mem Radiol Soc NAm; emer mem AMA; emer fel Am Col Radiol; emer fel NY Acad Med; emer mem Brit Inst Radiol. *Res:* Diagnostic radiology of gastrointestinal tract, pancreas and lungs; vascular anatomy and pathology; genito-urinary tract. *Mailing Add:* 4000 N Charles St Apt 1502 Baltimore MD 21218

SALIN, MARVIN LEONARD, b Brooklyn, NY, July 14, 46; c 2. BIOCHEMISTRY. *Educ:* Brooklyn Col, City Univ New York, BS, 67; Fla State Univ, MS, 69, PhD(biol sci), 72. *Prof Exp:* Assoc biochem, Univ Ga, 73-74; NIH fel, Duke Univ, 74-76, assoc biochem, 76-78; asst prof, 78-80, ASSOC PROF, MISS STATE UNIV, 80-. PROF, 87- *Concurrent Pos:* Humboldt Found fel, Max Planck Inst Biochem, Munich, WGer, 86; Fulbright fel, Univ Malta, 90. *Honors & Awards:* Fel, Juan March Found, Madrid, Spain. *Mem:* Am Soc Photobiol; Am Soc Plant Physiologists; Sigma Xi; Am Soc Biol Chemists. *Res:* Plant biochemistry, free radicals of oxygen, bioenergetics metabolism, oxidases, oxygenases and peroxidases; protein biochemistry. *Mailing Add:* Dept Biochem Miss State Univ Mississippi State MS 39762

SALINAS, DAVID, b New York, NY, Feb 25, 32. ENGINEERING, APPLIED MATHEMATICS. *Educ:* Univ Calif, Los Angeles, BS, 59, MS, 62, PhD(eng), 68. *Prof Exp:* Sr res engr, Space & Info Div, NAm Aviation, Inc, 62-65; scholar eng mech, Univ Calif, Los Angeles, 68-70; ASSOC PROF MECH ENG, US NAVAL POSTGRAD SCH, 70- *Mem:* Am Soc Civil Engrs; Am Soc Mech Engrs; Am Acad Mech; Sigma Xi. *Res:* Optimization of structures; micromechanics of composite materials; finite element methods; inelastic behavior of structures; mechanics of composite materials; optimization of structures; finite element analysis of nonlinear field problems. *Mailing Add:* Mech Engr Code 69ZC US Naval Postgrad Sch Monterey CA 93940

SALINAS, FERNANDO A, b Santiago, Chile, Oct 30, 39; Can citizen; m 64; c 2. ONCOLOGY, IMMUNOPATHOLOGY. *Educ:* Univ Chile, BSc, 57, DVM, 63. *Prof Exp:* Instr biol, Sch Vet Med, Univ Chile, Santiago, 50-63, asst prof morphol, Sch Med, 63-67, asst prof exp med, 67-70; asst prof med, Sch Med, Univ Southern Calif, Los Angeles, 73-76; assoc prof, 76-83, PROF PATH, FAC MED, UNIV BC, 84-; SR ONCOLOGIST, DEPT ADVAN THERAPEUT, CANCER CONTROL AGENCY BC, VANCOUVER, CAN, 76- *Concurrent Pos:* Mem, biohazards comt, Cancer Control Agency of BC, 85- *Mem:* Int Soc Exp Hemat; Am Asn Cancer Res; Am Asn Immunologists; Am Soc Clin Oncol; Chilean Soc Biol; Int Soc Interferon Res; AAAS. *Res:* Pathogenic and prognostic role of circulating immune complexes in cancer patients; immunoregulatory implications of oncofetal antigens in human cancer; new approaches to cancer treatment by use of biological response modifiers, emphasis in monoclonal antibodies and interferon. *Mailing Add:* Cancer Control Agency BC 600 W Tenth Ave Vancouver BC V5Z 4E6 Can

SALINGAROS, NIKOS ANGELOS, b Perth, Australia, Jan 1, 52; m 86. FUSION REACTORS, SUPERCONDUCTING MAGNETS. *Educ:* Univ Miami, BSc, 71; State Univ Stony Brook, MA, 74, PhD(physics), 78. *Prof Exp:* Asst prof physics, Univ Mass, Boston, 79-80, Univ Crete, Greece, 80-81 & math, Univ Iowa, Iowa City, 81-83; asst prof, 83-86, ASSOC PROF MATH, UNIV TEX, SAN ANTONIO, 86- *Concurrent Pos:* Vis assoc prof physics, Univ Rochester, 88-89; vis scholar, Univ Tex, Austin, 89-91, mem, Ctr Fusion Eng, 89-91. *Mem:* Sr mem Inst Elec & Electronics Engrs; Am Phys Soc; Int Asn Math Physics. *Res:* Developed the Clifford Algebras into a practical tool for field theory; thermonuclear fusion reactors; author of 1 publication. *Mailing Add:* Div Math Univ Tex San Antonio TX 78285-0664

SALINGER, GERHARD LUDWIG, b Berlin, Ger, Aug 25, 34; US citizen; m 58; c 3. LOW TEMPERATURE PHYSICS. *Educ:* Yale Univ, BS, 56; Univ Ill, MS, 58, PhD(physics), 62. *Prof Exp:* Asst physics, Univ Ill, 56-61; vis res prof, Univ Sao Paulo, 61-64; from asst prof to assoc prof, 64-75, PROF PHYSICS, RENSSELAER POLYTECH INST, 75-, CHMN DEPT, 78- *Concurrent Pos:* Vis assoc prof, Iowa State Univ, 74-75. *Mem:* Am Phys Soc; Am Asn Physics Teachers. *Res:* Low temperature thermal properties of amorphous materials. *Mailing Add:* Dept Physics Rensselaer Polytech Inst Troy NY 12181

SALINGER, RUDOLF MICHAEL, b Berlin, Ger, July 24, 36; US citizen; m 61; c 2. CHEMISTRY, CHEMICAL ENGINEERING. *Educ:* Cooper Union, BChE, 58; Univ Wis, MS, 60; Univ Cincinnati, PhD(org chem), 63. *Prof Exp:* Fel, Stanford Univ, 63-64; res chemist, Dow Corning Corp, 64-68, group leader, Res Eng Sect, 68-75, sr res group leader, 75-77, sect mgr anal serv, 77-80, corp qual assurance mgr, 80-82, sect mgr metall silicon, 82-85, process res mgr, adv ceramics prog, 85-90, GLOBAL ANALYSIS TECHNOL COORDR, DOW CORNING CORP, 90- *Mem:* Am Inst Chem Engr; Am Chem Soc; Sigma Xi; Am Soc Qual Control. *Res:* Organosilicon chemistry. *Mailing Add:* Dow Corning Corp PO Box 0994 Midland MI 48686-0994

SALIS, ANDREW E, b Boston, Mass, Oct 10, 15; m 41; c 3. ELECTRICAL ENGINEERING. *Educ:* Auburn Univ, BSc, 39, MSc, 40, Prof degree, 48; Tex A&M Univ, PhD(elec eng), 51. *Prof Exp:* From instr to assoc prof elec eng, Tex A&M Univ, 40-51; sr group engr, Gen Dynamics Corp, 51-59; chmn dept elec eng, 59-70, PROF ELEC ENG, UNIV TEX, ARLINGTON, 59-, DEAN ENG, 70- *Concurrent Pos:* Consult, LTV Corp & Gen Dynamics-San Diego Dallas Power & Light Co, 59-71. *Mem:* Inst Elec & Electronics Engrs; Am Soc Eng Educ. *Res:* High frequencies and natural electrical phenomena. *Mailing Add:* Dept Elec Eng Univ Tex Arlington TX 76010

SALISBURY, FRANK BOYER, b Provo, Utah, Aug 3, 26; m 49; c 7. PLANT PHYSIOLOGY. *Educ:* Univ Utah, BS, 51, MA, 52; Calif Inst Technol, PhD(plant physiol, geochem), 55. *Prof Exp:* Asst prof bot, Pomona Col, 54-55; asst prof, Colo State Univ, 55-61, prof, 61-66; head dept plant sci, 66-70, PROF PLANT PHYSIOL, UTAH STATE UNIV, 66-, PROF BOT, 68- *Concurrent Pos:* AEC fel, 51-53, McCallum fel, 53-54 & Lady Davis fel, Hebrew Univ, Jerusalem, 83; NSF sr fel, Univ Tubingen, Ger & Univ Innsbruck, Austria, 62-63; plant physiologist, US AEC, 73-74; vis prof, Univ Innsbruck, Austria & Hebrew Univ, Jerusalem, 83; mem aerospace med adv comt, NASA; chmn, Controlled Ecol Life Support Syst Working Group, NASA; ed-in-chief, Am & Pac Rim Countries, J Plant Physiol, 89- *Honors & Awards:* Cert Merit, Bot Soc Am, 82. *Mem:* AAAS; Am Soc Plant Physiologists; Ecol Soc Am; Am Inst Biol Sci; Bot Soc Am; Sigma Xi; Am Soc Gravitational & Space Biol. *Res:* Physiology of flowering; space biology; physiological ecology; plant responses to gravity; maximum yield of wheat in controlled environments. *Mailing Add:* Dept Soils & Biometeorol Utah State Univ Logan UT 84322-4820

SALISBURY, GLENN WADE, b Sheffield, Ohio, June 2, 10; m 32; c 2. ANIMAL REPRODUCTION. *Educ:* Ohio State Univ, BS, 31; Cornell Univ, PhD(animal husb), 34. *Prof Exp:* Asst animal husb, Cornell Univ, 31-34, instr, 34-36, asst prof & asst animal husbandman, Exp Sta, 36-40, assoc prof & assoc animal husbandman, 40-44, prof, 44-47; prof dairy sci & head dept, 47-69, dir agr exp sta, 69-78, EMER PROF ANIMAL SCI, UNIV ILL, URBANA, 78- *Concurrent Pos:* Cornell fel, Iowa State Col, 41; mem bd consults, Milk Mkt

Bd Eng & Wales, 48-52; dir, Int Dairy Show, Chicago, 53-70; Fulbright lectr, State Agr Univ, Wageningen, 55-56; mem agr subpanel, President's Sci Adv Comt, 61-63; consult, US Off Sci & Technol, 62- *Honors & Awards:* Borden Award, Am Dairy Sci Asn, 45; Award of Merit & Knight, Order of Merit, Ital Govt, 64; Morrison Award, Am Soc Animal Sci, 64, Paul A Funk Award, 71; co-winner, Wolfe Prize in Agr, Israeli Knesset, 81. *Mem:* Nat Acad Sci; fel AAAS; Am Dairy Sci Asn; Am Genetic Asn; fel Am Soc Animal Sci. *Res:* Artificial insemination; reproductive physiology and genetics of dairy cattle. *Mailing Add:* 2110 Race St Urbana IL 61801

SALISBURY, JEFFREY L, b Sept 21, 50; US citizen. CELL BIOLOGY. *Educ:* Ind State Univ, BS, 73; Rutgers Univ, MS, 75; Ohio State Univ, PhD(bot), 78. *Prof Exp:* Postdoctoral fel, Albert Einstein Col Med, 78-82, asst prof anat, 82-84; asst prof develop genetics & anat, Case Western Reserve Univ, 84-88, assoc prof, Neurosci Ctr, 88-89; ASSOC PROF BIOCHEM & MOLECULAR BIOL, MAYO CLIN/FOUND, 89- *Mem:* AAAS; Phycol Soc Am; Soc Protozoologists; Am Soc Cell Biol. *Res:* Research of centrin, a calcium-binding phosphoprotein associated with centrosomes and mitotic spindle poles; centrin is involved with processes that affect the dynamic behavior of the cell's major microtubule organizing centers during the cell cycle. *Mailing Add:* Mayo Clin-Found 200 First St SW Rochester MN 55905

SALISBURY, JOHN WILLIAM, JR, b Palm Beach, Fla, Feb 6, 33; m 57; c 2. REMOTE SENSING. *Educ:* Amherst Col, BA, 55; Yale Univ, MS, 57, PhD(geol), 59. *Prof Exp:* Res scientist, Air Force Cambridge Res Labs, 59-61, chief, Lunar Planetary Res Br, 61-70 & Spectros Studies Br, 70-76; chief, Geothermal Energy Br, 76-81, div dir, Dept Energy, 80-81; chief, Earth Resources Observation Syst Off, US Geol Surv, 81-83, res geologist, Geophys Br, 83-89; RES PROF, DEPT EARTH & PLANETARY SCI, JOHNS HOPKINS UNIV, 89- *Concurrent Pos:* Vis prof, Purdue Univ, 61-64; guest lectr, Hayden Planetarium, Am Mus, 64-75. *Honors & Awards:* Outstanding Contribution to Mil Sci Award, US Air Force, 61; Gunter Loeser Award, Air Force Cambridge Res Labs, 69 & Sci Achievement Award, 74; Meritorious Serv Award, Dept Interior, 90. *Mem:* AAAS; Am Geophys Union; fel Geol Soc Am. *Res:* Terrestrial and extraterrestrial geological remote sensing; geology of moon and planets; nature and extent of geothermal resources. *Mailing Add:* 5529 Coltsfoot Ct Columbia MD 21045

SALISBURY, MATTHEW HAROLD, b Far Rockaway, NY, Mar 17, 43; m 67; c 1. GEOPHYSICS. *Educ:* Mass Inst Technol, BS, 68; Univ Wash, MS, 71, PhD(geol sci), 74. *Prof Exp:* Technician geophys, Woods Hole Oceanog Inst, 64, res asst, 65; res asst seismol, Dept Geol & Geophys, Mass Inst Technol, 66-67; sr observer satellite geodesy, Smithsonian Astrophys Observ, 68-70; asst prof geophys, State Univ NY, Binghamton, 74-76; asst res geol, 76-79, assoc chief scientist, Deep Sea Drilling Proj, Scripps Inst Oceanog, 79-; Dalhousie Univ, NS, 84-88. *Concurrent Pos:* Partic scientist, Leg 34 Deep Sea Drilling Proj, Scripps Inst Oceanog, 73-74; asst proj officer, Deep Sea Drilling Proj, NSF, 74-75. *Mem:* Am Geophys Union. *Res:* Determination of the petrology of the lower crust through comparisons of seismic velocity structure determined by refraction and logging with laboratory-determined physical properties of geologic materials at high confining pressures and temperatures. *Mailing Add:* 55 Peregrine Crescent Bedford NS B4A 3B9 Can

SALISBURY, NEIL ELLIOT, b New Orleans, La, Oct 27, 28; m 80; c 4. FLUVIAL GEOMORPHOLOGY, NATURAL RESOURCES. *Educ:* Univ Minn, BA, 52, PhD(geog), 57. *Prof Exp:* Prof geog, Univ Iowa, 55-79; PROF GEOG, UNIV OKLA, 79- *Mem:* Asn Am Geographers; Geol Soc Am; Am Geophys Union; Am Water Resources Asn; Nat Coun Geog Educ. *Res:* Fluvial geomorphology, slope development and mass-wasting processes; natural resources, agricultural land quality and water resources. *Mailing Add:* Dept Geog Univ Okla Norman OK 73019

SALISBURY, STANLEY R, b Milwaukee, Wis, Oct 2, 32; m 56; c 6. NUCLEAR & ATOMIC PHYSICS, ARTIFICIAL INTELLIGENCE. *Educ:* Marquette Univ, BS, 55; Univ Wis, MS, 56, PhD(nuclear physics), 61. *Prof Exp:* Res fel nuclear physics, Univ Wis, 61-62; res scientist, Lockheed Palo Alto Res Lab, 62-68, from staff scientist to sr staff scientist, 69-81, consult scientist, 81-86, SR SCIENTIST, LOCKHEED PALO ALTO RES LAB, 86- *Mem:* Am Phys Soc. *Res:* X-ray phenomenology; Van de Graaff accelerators; neutron induced reactions; neutron cross sections; level parameters; aurora phenomenon; charged particle x-ray and neutron flux measurements from satellites; measurement of x-rays for environmental effects experiments; program management; systems design and fabrication; artificial intelligence; expert systems. *Mailing Add:* Lockheed Res Labs Dept 52-11 3251 Hanover St Bldg 203 Palo Alto CA 94306

SALIVAR, CHARLES JOSEPH, b New York, NY, Feb 15, 23; m 45, 75; c 4. ORGANIC CHEMISTRY. *Educ:* Queens Col, NY, BS, 43. *Prof Exp:* Res chemist, Chas Pfitzer & Co, Inc, 45-55, pharmaceut res supvr, 55-62; dir pharm res & develop, Mallinckrodt Chem Works, St Louis, 62-69, dir opers, 69-73; vpres & tech dir, KV Pharmaceut Co, 73-75; dir develop & qual assurance, Emko Co, 75-76, vpres tech, 76-77, mem bd dirs, 76-78, VPRES MFG & TECH OPERS, EMKO CO, 78- *Concurrent Pos:* Dir opers, Schering-Plough Corp, St Louis, 78-86. *Mem:* Am Pharmaceut Asn; Am Chem Soc. *Res:* Antibiotics; vitamins; medicinals; pharmaceutical dosage forms. *Mailing Add:* 3462 Gill Ave St Louis MO 63122

SALK, DARRELL JOHN, b Ann Arbor, Mich, Mar 30, 47; m 76; c 2. MEDICAL GENETICS, MEDICAL CYTOGENETICS. *Educ:* Stanford Univ, BA, 69; Johns Hopkins Univ, MD, 74. *Prof Exp:* Intern pediat, Children's Orthop Hosp Med Ctr, Seattle, Wash, 74-75; resident, Univ Wash, 75-78, fel med genetics, 78-80; asst prof, Univ Wuerzburg, WGer, 80-81; asst prof, 81-86, assoc prof path & pediat, Univ Wash, 86-87; dir cytogenetics lab, Children's Orthop Hosp Med Ctr, 83-87; assoc med dir, 87-90, VPRES MED & REGULATORY AFFAIRS, NEORX CORP, 90- *Concurrent Pos:* Co-dir Cytogenetics Lab, Univ Hosp, Seattle, 81-83. *Mem:* Am Soc Human Genetics; AAAS. *Res:* monoclonal antibodies for cancer detection and therapy. *Mailing Add:* NeoRx Corp 410 W Harrison St Seattle WA 98119

SALK, JONAS EDWARD, b New York, NY, Oct 28, 14; m 39, 70; c 3. MEDICINE, IMMUNOLOGY. *Educ:* City Col New York, BS, 34; NY Univ, MD, 39. *Hon Degrees:* Numerous from US & foreign univs. *Prof Exp:* Intern, Mt Sinai Hosp, NY, 40-42; res assoc, Univ Mich, 44-46, asst prof, 46-47; from assoc res prof bact to res prof bact, Univ Pittsburgh, 47-55, Commonwealth Prof prev med, 55-57, Commonwealth Prof exp med, 57-63, head virus res lab, 47-63; dir, 63-75, res fel, 63-84, FOUNDING DIR, SALK INST BIOL STUDIES, 75-, DISTINGUISHED PROF INT HEALTH SCI, 84- *Concurrent Pos:* Fel bact, Col Med, NY Univ, 39-40; Nat Res Coun fel med sci, Sch Pub Health, Univ Mich, 42-43, res fel epidemiol, 43-44; consult, US Secy War, 44-47 & US Secy Army, 47-54; mem comn influenza, Army Epidemiol Bd, 44-54; adj prof, Univ Calif, San Diego, 70-; mem, expert adv panel virus dis, WHO, 51- *Honors & Awards:* Presidential Citation, 55; Congressional Gold Medal, 55; Criss Award, Mutual of Omaha Ins Co, 55; Albert Lasker Award, 56; Albert Galliton Award, NY Univ, 57; Howard Taylor Ricketts Award, Univ Chicago, 57; Gold Medal Award, Nat Inst Soc Sci, 59; Robert Koch Medal, 63; Truman Commendation Award, 66; Mellon Inst Award, 69; Presidential Medal of Freedom, 77. *Mem:* Inst Med-Nat Acad Sci; AAAS; Am Asn Immunol; Am Col Prev Med; Sigma Xi; hon mem Am Acad Pediat. *Res:* Immunization and immunological properties of the influenza virus; immunological problems of poliomyelitis; experimental medicine; mechanisms of delayed hypersensitivity; studies on control of HIV/AIDS by immunological means; author of over 150 publications and 4 books. *Mailing Add:* Salk Inst Biol Studies PO Box 85800 San Diego CA 92186

SALK, MARTHA SCHEER, b Detroit, Mich, Apr 16, 45; m 67; c 2. TERRESTRIAL PLANT ECOLOGY, PHYCOLOGY. *Educ:* Albion Col, BA, 67; Univ Iowa, MS, 69; Univ Louisville, PhD(bot & ecol), 75. *Prof Exp:* Terrestrial ecologist, Gilbert/Commonwealth, Reading, Pa, 74-75; RES STAFF, OAK RIDGE NAT LAB, 75- *Mem:* Ecol Soc Am. *Res:* Environmental laws and regulations. *Mailing Add:* Environ Sci Div Bldg 1505 MS 6036 Oak Ridge Nat Lab PO Box 2008 Oak Ridge TN 37831-6036

SALK, SUNG-HO SUCK, b Seoul, Korea, Apr 14, 39; m 68; c 2. PHYSICS, CHEMISTRY. *Educ:* Midwestern Univ, BS, 66; Univ Houston, MS, 68; Univ Tex, Austin, PhD(physics), 72. *Prof Exp:* Res fel, Univ Tex, 72-74, res assoc chem, 74-77; from asst prof to assoc prof, Univ Mo, 77-87; PROF PHYSICS, POHANG INST SCI & TECHNOL, S KOREA, 88- *Mem:* Am Phys Soc. *Res:* Condensed matter physics related superconductivity, surface physics and semi-conductivity; nucleation; collision theory; atomic and molecular physics. *Mailing Add:* Dept Physics Pohang Inst Sci & Technol PO Box 125 Pohang Kyungbuk 680 Republic of Korea

SALKIN, DAVID, b Ukraine, Russia, Aug 8, 06; nat US; m 34; c 1. MEDICINE. *Educ:* Univ Toronto, MD, 29. *Prof Exp:* Intern, St Mary's Hosp, Detroit, Mich, 29-30; pathologist, Mich State Sanitarium, 33-34; demonstr med, Sch Med, WVa Univ, 35-38, from instr to asst prof, 38-48; chief prof serv, San Fernando Vet Admin Hosp, 48-61, chief of staff, 61-71, hosp dir, 67-71; dir res & educ, 71-75, med dir, 75-77, dir med res, 77-83, MED RES, LA VINA HOSP, 83- *Concurrent Pos:* Fel path, H Kiefer Hosp, 33; med dir, Hopemont Sanitarium, 34-41, supt, 41-48; assoc clin prof med, Univ Calif, Los Angeles, 51-61; clin prof, Loma Linda Univ, 60- & Univ Southern Calif, 64-71 (emer). *Mem:* Fel Am Soc Clin Pharmacol & Therapeut; fel Am Thoracic Soc; fel AMA; fel Am Col Chest Physicians; fel Am Col Physicians. *Res:* Pneumoperitoneum; intestinal tuberculosis; pulmonary cavities; physiology of pneumothorax; chemotherapy of tuberculosis; bronchiectasis and bronchitis; coccidiodomycosis; BCG treatment of cancer; various mycobacteria (atypical tuberculosis). *Mailing Add:* Huntington Med Res Inst 660 S Fair Oaks Ave Pasadena CA 91105

SALKIN, IRA FRED, b Chicago, Ill, Dec 21, 41; m 64; c 1. MEDICAL MYCOLOGY. *Educ:* Northwestern Univ, Evanston, BA, 63, MS, 64; Univ Calif, Berkeley, PhD(bot), 69. *Prof Exp:* Lectr, biol, Univ Calif, Santa Barbara, 69-70; res scientist III, 70-77, res scientist IV, 77-84, RES SCIENTIST V, NY STATE DEPT HEALTH, 84- *Concurrent Pos:* Mem adj fac, Dept Biol, Russell Sage Col, 77-88 & Union Col, NY, 78-84; dept biomed sci, Sch Pub Health, State Univ NY, 87- *Mem:* Med Mycol Soc Am; Int Soc Human & Animal Mycol; Mycol Soc Am; Brit Mycol Soc; Sigma Xi. *Res:* Development and improvement of diagnostic procedures; taxonomy of zoopathogenic fungi and studies of the physiologic factors associated with pathogenicity in the fungi. *Mailing Add:* Ctr Labs & Res NY State Dept Health Empire State Plaza Albany NY 12201

SALKIND, ALVIN J, b New York, NY, June 12, 27; m 65; c 2. ELECTROCHEMICAL ENGINEERING, BIOENGINEERING. *Educ:* Polytech Inst New York, BChE, 49, MChE, 52, DChE, 58. *Prof Exp:* Engr energy res, US Elec Mfg Co, 52-54; sr scientist energy res, Sonotone Corp, 54-56; res assoc, Polytech Inst New York, 56-58; sr scientist energy conversion, ESB Inc, 58-63, head lab electrochem, 63-68, mgr electromed prod, 68-71, vpres technol, 71-79, pres, 77-79; PROF & CHIEF BIOENG SECT, DEPT SURG, UNIV MED & DENT, ROBERT WOOD JOHNSON MED SCH, 70- *Concurrent Pos:* Adj prof chem eng, Polytech Inst New York, 60-70; consult, Dept Space Sci, Univ Mo, 68-70; consult, Hahnemann Med Sch, 69-71, Nat Res Coun & Dept Energy, 79-; vis prof, Case Western Reserve Univ, 81-82; vis scientist, Yugoslavian Acad Sci, 82; prof chem & biochem eng, Rutgers Univ, 85-; assoc dean & dir, Bur Eng Res, 89- *Mem:* Fel AAAS; fel Am Col Cardiol; Asn Advan Med Instrumentation; Am Inst Chem Engrs; Electrochem Soc. *Res:* Energy storage devices; batteries; electromedical devices. *Mailing Add:* 51 Adams Dr Princeton NJ 08540

SALKIND, MICHAEL JAY, b New York, NY, Oct 1, 38; m 90; c 4. RESEARCH MANAGEMENT, STRUCTURES & MATERIALS. *Educ:* Rensselaer Polytech Inst, BMetE, 59, PhD(mat eng), 62. *Prof Exp:* Asst to chief res, US Army Watervliet Arsenal, 62-64; chief, advan metall, United Technol Res Ctr, United Technol Corp, 64-68, chief, structures & mat, Sikorsky Aircraft, 68-75; dir prod develop, syst div, Avco, 75-76; mgr structures, NASA HQ, 76-80; dir aerospace sci, Air Force Off Sci Res, 80-89; PRES, OHIO AEROSPACE INST, 90- *Concurrent Pos:* Lectr eng mat,

Trinity Col Conn, 67-68, Univ Md, 81-84 & Johns Hopkins Univ, 85-89; chmn Conn Dept Environ Protection Tech Adv Group, 72-75. *Honors & Awards:* Von Karman Award, TRE Corp, 74. *Mem:* Am Soc Mech Engrs; Am Inst Aeronaut & Astronaut; Am Inst Mining, Metall & Petrol Engrs; Am Soc Metals; Am Soc Testing & Mat; Sigma Xi. *Res:* Fatigue in composites; fatigue and fracture behavior of titanium. *Mailing Add:* Ohio Aerospace Instit 2001 Aerospace Pkwy Brook Park OH 44142

SALKOFF, LAWRENCE BENJAMIN, b Brooklyn, NY, Mar 3, 44; m 68; c 1. NEUROGENETICS, NEUROBIOLOGY. *Educ:* Univ Calif, Los Angeles, BA, 67; Univ Calif, Berkeley, PhD(genetics), 79. *Prof Exp:* Teaching assoc genetics, Univ Calif, Berkeley, 76-79; fel, Dept Biol, Yale Univ, 79-83; ASSOC PROF ANAT-NEUROBIOL & GENETICS, WASH UNIV, ST LOUIS, 84- *Honors & Awards:* John Belling Prize, 80. *Mem:* Genetics Soc Am; Soc Neurosci; AAAS. *Res:* Neurogenetics; identification and characterization of genes that affect the function and development of the nervous system. *Mailing Add:* Dept Anat-Neurobiol Sch Med Wash Univ 600 S Euclid Ave St Louis MO 63110

SALL, THEODORE, b Paterson, NJ, Feb 22, 27; m 53; c 3. MICROBIOLOGY. *Educ:* Univ Louisville, AB, 49, MS, 50; Univ Pa, PhD(microbiol), 55. *Prof Exp:* Biochemist, Vet Admin Hosp, Philadelphia, 55-56; microbiologist & res assoc, Sch Med, Univ Pa, 56-61; staff scientist, RCA Space Ctr, 61-64; chief bact, Pepper Lab, Univ Pa Hosp, 64-68; from asst prof to assoc prof microbiol, NY Med Col, 68-72; PROF LIFE SCI, RAMAPO COL, NJ, 72- *Concurrent Pos:* Consult, Vet Admin Hosp, Philadelphia; chief microbiol serv, Metrop Hosp, New York, 68-71, dir, 71-72; Ethicon Corp, 76-78; res adv, Biorecovery Technol, Inc, 82-85; Biometallics Inc, 87- *Mem:* AAAS; Am Chem Soc; Am Soc Microbiol; fel NY Acad Sci; dipl Am Acad Microbiol. *Res:* Diagnostic bacteriology; microchemical analysis of bacteria; bacterial morphology; electron microscopy; fermentation chemistry; sterilization techniques. *Mailing Add:* 94 Woodcrest Dr Woodcliff Lake NJ 07675

SALLAVANTI, ROBERT ARMANDO, b Scranton, Pa, July 26, 42; m 64; c 3. PHYSICAL CHEMISTRY. *Educ:* Wilkes Col, BS, 63; Univ Pa, PhD(chem), 66. *Prof Exp:* Advan Res Projs Agency fel, Univ Pa, 66-67; USPHS fel, Yale Univ, 67-69; ASSOC PROF CHEM, UNIV SCRANTON, 69- *Concurrent Pos:* Asst prof, Quinnipiac Col, 68-69. *Mem:* Am Chem Soc; fel Am Inst Chemists; Sigma Xi. *Res:* Molecular orbital theory of organic and biological molecules; intermolecular potentials of the rare gases; transport properties of systems involving critical phenomena. *Mailing Add:* 104 Miles St Dalton PA 18414

SALLAY, STEPHEN, organic chemistry; deceased, see previous edition for last biography

SALLEE, G THOMAS, b Ontario, Ore, Feb 21, 40; m 66; c 4. MATHEMATICS. *Educ:* Calif Inst Technol, BS, 62; Univ Calif, Berkeley, MA, 64; Univ Wash, PhD(math), 66. *Prof Exp:* From asst prof to assoc prof, 66-75, PROF MATH, UNIV CALIF, DAVIS, 75- *Mem:* Math Asn Am; Am Math Soc. *Res:* Geometry, especially geometry of convex sets, sets of constant width and polytopes. *Mailing Add:* Dept Math Univ Calif Davis CA 95616

SALLEE, VERNEY LEE, b Amarillo, Tex, June 26, 42; m 64; c 2. RECEPTORS, GLAUCOMA. *Educ:* Hardin-Simmons Univ, BA, 64; Univ NMex, PhD(med sci, physiol), 70. *Prof Exp:* Fel physiol, Southwestern Med Sch, Univ Tex Health Sci Ctr, 70-72, asst prof, 72-78; asst prof, Tex Col Osteop Med, 78-81, assoc prof physiol, 81-84; SR SCIENTIST, DEPT PHARMCOL, ALCON LABS, FT WORTH, 84- *Mem:* Am Physiol Soc; Asn Res Vision & Ophthal; Int Soc Eye Res. *Res:* Glaucoma pharmacology; ocular physiology; aqueous humor dynamics. *Mailing Add:* Dept Pharmacol Alcon Labs Inc 6201 South Freeway Ft Worth TX 76134

SALLER, CHARLES FREDERICK, b Wheeling, WVa, Jan 8, 50. NEUROPHARMACOLOGY, NEUROCHEMISTRY. *Educ:* Georgetown Univ, BS, 71; Univ Pittsburgh, MS, 78, PhD(biol sci & psychobiol), 79. *Prof Exp:* Pharmacol res assoc, Nat Inst Mental Health, 79-81; res pharmacologist, Stuart Pharmaceuts, 82-85, sr res pharmacologist, ICI Pharma, ICI Am Inc, 86-91; PRES, ANALTICAL BIOL SERV INC, 90- *Concurrent Pos:* Pres, Del area chap Neurosci, 89. *Mem:* NY Acad Sci; Sigma Xi; Soc Neurosci; Am Soc Pharmacol & Exp Therapeut; Int Brain Res Orgn. *Res:* Explored the biochemical, physiological and behavioral responses to manipulations of brain monoamine neurotransmitter systems; particular emphasis on the effects of antipsychotic drugs. *Mailing Add:* Analytical Biol Serv 104 Warwick Dr Wilmington DE 19803

SALLET, DIRSE WILKIS, b Washington, DC, Aug 10, 36; m 63; c 4. MECHANICAL ENGINEERING, FLUID DYNAMICS. *Educ:* George Washington Univ, BME, 61; Univ Kans, MSME, 63; Univ Stuttgart, Dr Ing, 66. *Prof Exp:* Instr mech eng, Univ Kans, 61-63; sci assoc fluid dynamics, Inst Aero- & Gas Dynamics, Univ Stuttgart, 63-66; res mech engr, US Naval Ord Lab, 66-67; from asst prof to assoc prof, 67-76, PROF MECH ENG, UNIV MD, COLLEGE PARK, 76- *Concurrent Pos:* Consult, US Naval Ord Lab, 67-74 & Gillette Co Res Inst, 71-78; vis scientist, Max Planck Inst & AVA, Gottingen, Ger, 73-74; eng consult, 75-; vis scientist, Ctr Nuclear Res, Kernforschungszentrum Karlsruhe, Inst Reactor Components, Ger, 80-81; vis prof, Tech Univ, Munich, Ger, 87 & 89. *Honors & Awards:* Alexander von Humboldt Prize, 86. *Mem:* AAAS; Am Soc Mech Engrs; Am Phys Soc. *Res:* Thermodynamics and heat transfer; flow induced vibrations; vortex motions; two-phase flow; gas dynamics; hydrodynamics; author or coauthor of various publications. *Mailing Add:* Dept Mech Eng Univ Md College Park MD 20742

SALLEY, JOHN JONES, b Richmond, Va, Oct 29, 26; m 50; c 3. ORAL PATHOLOGY. *Educ:* Med Col Va, DDS, 51; Univ Rochester, PhD(path), 54. *Hon Degrees:* DSc, Boston Univ, 75. *Prof Exp:* Instr histol, Eastman Sch Dent Hygiene, 52-54; instr path diag & therapeut, Med Col Va, 54-55, from asst prof to chmn & prof path, 55-63; assoc vpres, 74-80, vpres res & grad affairs, Va Commonwealth Univ, 80-85; actg pres & vpres, Va Ctr Innovative Technol, 85-87; prof path & dean, Sch Dent, 63-74, EMER DEAN, UNIV MD, BALTIMORE, 87-, CONSULT, 90-; EMER PROF ORAL PATH, MED COL VA, 91- *Concurrent Pos:* Consult, Off Chief Med Examr, Commonwealth of Va, 56-90, Vet Admin Hosp, 59, NIH, 62-66 & USPHS, 63-74; mem adv comt regional med prog in Md, 66-74; consult, WHO, 68-; hon prof, Univ Peru Cayetano Heredia, 70; pres, Am Asn Dent Sch, 71-72; dent educ rev comt, Bur Health Manpower Educ, 72-76; sr prog consult, Robert Wood Johnson Found, 79-85; pres, Conf Southern Grad Schs, 83-84; prof oral path, Med Col Va, 87-90. *Honors & Awards:* Award, Int Asn Dent Res, 53; Presidential Award, Am Asn Hosp Dentists, 84. *Mem:* Fel AAAS; Am Dent Asn; Am Acad Oral Path; Int Asn Dent Res; Nat Coun Univ Res Adminr; Sigma Xi; hon mem Am Acad Oral Med. *Res:* Dental school administration; experimental carcinogenesis in oral tissues, including predisposing factors to oral cancer; etiological factors in periodontal diseases; hospital dentistry. *Mailing Add:* PO Box 838 Urbana VA 23175

SALLEY, JOHN JONES, JR, b Rochester, NY, June 2, 54; m 78; c 2. PHARMACEUTICAL CHEMISTRY. *Educ:* Randolph-Macon Col, BS, 76; Med Col Va, PhD(med chem), 80. *Prof Exp:* Fel org chem, Univ Ala, 80-81; res scientist, 81-84, group leader, 84-85, SECT HEAD, NORWICH-EATON PHARMACEUT, DIV PROCTER & GAMBLE, 85- *Mem:* Am Chem Soc. *Res:* New ethical pharmaceutical products, specifically in the area of cardiovascular and anti-inflammatory agents; development of novel organic chemical methodology and techniques. *Mailing Add:* Norwich-Eaton Pharmaceut PO Box 191 Norwich NY 13815

SALLMAN, BENNETT, b New York, NY, Dec 10, 17; m 43; c 3. MICROBIOLOGY. *Educ:* NY Univ, BS, 37; Univ Mich, MS, 39; Ohio State Univ, PhD(bact), 48; Am Bd Microbiol, dipl. *Prof Exp:* Asst prof bact, Hahnemann Med Col & Hosp, Ill, 48-51; chief bacteriologist, Commun Dis Ctr, USPHS, Ga, 51-53; assoc prof bact, 53-58, chmn dept, 61-80, PROF MICROBIOL, SCH MED, UNIV MIAMI, 58- *Mem:* Fel Am Acad Microbiol; fel Geront Soc. *Res:* Biochemical mechanisms of infectious diseases; aging at the cellular level; heart tissue metabolism; metabolic and ecologic interactions of microorganisms in the marine environment. *Mailing Add:* Univ Miami Med Sch PO Box 520875 Miami FL 33152

SALLOS, JOSEPH, b Budapest, Hungary, Aug 19, 31; Can citizen; m 53; c 2. ELECTRONICS. *Educ:* Budapest Tech Univ, Dip Ing, 55. *Prof Exp:* Engr, Radio Budapest, 49-56 & BC Tel Co, 57-59; RES ELECTRONIC ENGR, UNIV BC, 60-, SUPVR, ELECTRONIC INSTRUMENTATION DIV, 80- *Mem:* Can Coun Prof Engrs. *Res:* Short, middle and long wave broadcasting transmitters; direct distance dialing telephone systems; electron spin resonance spectroscopy; scientific electronic instrumentation; nuclear magnetic resonance spectroscopy. *Mailing Add:* Dept Chem Univ BC 2036 Main Mall Vancouver BC V6T 1Y6 Can

SALMASSY, OMAR K, b McConnelsville, Ohio, Sept 22, 25; m 54; c 3. MECHANICAL ENGINEERING. *Educ:* Purdue Univ, BSME, 49, MSE, 51. *Prof Exp:* Asst dynamic strain anal, Purdue Univ, 49-50; mech engr, Repub Steel Corp, 50-51; prin mech engr, Mat Res, Battelle Mem Inst, 51-56; sr scientist, Avco Corp, 56-58, group leader, 58-59, from chief mat res, Res & Develop Div to sr consult scientist, Res & Advan Develop Div, 59-61 & 64-67; proj officer, Apollo Spacecraft Struct, Off Manned Space Flight, NASA Hq, Washington, DC, 61-64; br chief, Composites Mat Res & Develop, 67-73, mat specialist, 73-81, SR SCIENTIST, COMPOSITES, MCDONNELL DOUGLAS ASTRONAUT CO, 81- *Res:* High-modulus fiber reinforced composites; carbon/carbon, graphite and ceramic composites; cryogenic insulations; thermal protection systems; re-entry, antiballistic missile interceptor and space vehicle materials technology; ablation, vulnerability and hardening, inelasticity, dust erosion and fracture phenomenology; liquified natural gas insulation system development. *Mailing Add:* McDonnell Douglas Astronaut Co 5301 Bolsa Ave Huntington Beach CA 92647

SALMI, ERNEST WILLIAM, nuclear physics; deceased, see previous edition for last biography

SALMOIRAGHI, GIAN CARLO, b Gorla Minore, Italy, Sept 19, 24; nat US; m 70; c 1. PHYSIOLOGY. *Educ:* Univ Rome, MD, 48; McGill Univ, PhD(physiol), 59. *Prof Exp:* From med officer to sr med officer, Int Refugee Orgn, Italy, 49-52; res fel, Cleveland Clin, Ohio, 52-55, res assoc, 55-56; lectr physiol, McGill Univ, 56-58; from neurophysiologist to chief clin neuropharmacol res ctr, NIMH, 59-67, dir div spec ment health res, 67-73; assoc comnr res, NY State Dept Ment Hyg, 73-77; assoc dir res, Nat Inst Alcohol Abuse & Alcoholism, 77-84; ASST VPRES, RES AFFAIRS, HAHNEMANN UNIV, 84-, PROF & CHMN, DEPT PHYSIOL & BIOPHYS, 86- *Concurrent Pos:* Clin prof psychiat, Med Sch, George Washington Univ, 66-73; mem, Int Brain Res Orgn. *Mem:* AAAS; Am Col Neuropsychopharmacol; Am Physiol Soc; Am Soc Pharmacol & Exp Therapeut; Soc Biol Psychiat; Soc Neurosci. *Res:* Neurophysiology and neuropharmacology of mammalian central neurons. *Mailing Add:* Hahnemann Univ Broad & Vine Philadelphia PA 19102-1192

SALMON, CHARLES G(ERALD), b Detroit, Mich, Oct 28, 30; m 53; c 3. STRUCTURAL & CIVIL ENGINEERING. *Educ:* Univ Mich, BS, 52, MS, 54; Univ Wis, PhD(struct eng), 61. *Prof Exp:* From instr to assoc prof, 56-67, PROF CIVIL ENG, UNIV WIS-MADISON, 67- *Honors & Awards:* Delmar L Bloehm Award & Joe W Kelly Award, Am Concrete Inst; Western Elec Award, Am Soc Eng Educ. *Mem:* Fel Am Soc Civil Engrs; Int Asn Bridge & Structural Eng; fel Am Concrete Inst; Am Soc Eng Educ; Nat Soc Prof Engrs; Am Asn Univ Profs; Am Soc Testing & Mat; Am Welding Soc. *Res:* Stability and stresses in edge loaded triangular plates; design methods and behavior of reinforced and prestressed concrete; design methods and behavior of steel structures. *Mailing Add:* 1415 Johnson Dr Rm 2246 Univ Wis Madison WI 53706

SALMON, EDWARD DICKINSON, b Montclair, NJ, Mar 1, 44; m 67; c 1. CELL BIOLOGY. *Educ:* Brown Univ, BS, 67; Univ Pa, PhD(biomed eng), 73. *Prof Exp:* Staff scientist cell motility, Marine Biol Lab, Woods Hole, 73-77; ASST PROF ZOOL, UNIV NC, 77- *Concurrent Pos:* Corp mem, Bermuda Biol Sta Res & Marine Biol Lab, Woods Hole. *Mem:* Am Soc Cell Biol; Sigma Xi. *Res:* Molecular mechanisms of cell motility especially the mitotic mechanisms of motility and the physiological effects of deep-sea cold temperatures and high hydrostatic pressures on cellular processes. *Mailing Add:* 411 Brandywine Rd Chapel Hill NC 27514

SALMON, ELI J, b Jerusalem, Israel, Dec 15, 28; US citizen; m 56; c 2. ENVIRONMENTAL SCIENCES. *Educ:* Utah State Univ, BSc, 51; Univ Mich, PhD(environ sci), 64. *Prof Exp:* Dir, Soreg Nuclear Res Ctr, 56-69; sr scientist, World Health Orgn, 69-74, Nat Acad Sci, 74-77; sr res assoc, Resources for the Future, 77-78; sr scientist, Med Div, Nat Acad Sci, 78-81; DIR HEALTH, SAFETY & SOCIOECON, FLUOR ENG & CONSTRUCTORS INC, 82- *Concurrent Pos:* Sr lectr, Tech Israel Inst Technol, 64-69; vis prof, Univ Tel Aviv, 64-69; lectr mgt, Univ Md, 78-82; pres, Environ, Health, Energy, Resources Corp, 78-82; vis sr lectr environ sci, Univ Calif, Irvine, 84- *Mem:* Am Health Physics Soc; Am Indust Hyg Asn; Am Pub Health Asn; Am Chem Soc; NY Acad Sci. *Res:* Evaluation of the federal programs and funded research in health and biologic effects of radiation; evaluation of the health, safety, environmental and socioeconomic implication of various energy systems and technologies; evaluation of health and environmental policies and regulations; risk assessment of various industries and technologies. *Mailing Add:* 2262 Albares Mission Viejo CA 92691

SALMON, JAMES HENRY, b Centerville, Pa, Feb 25, 32; m 67; c 2. NEUROSURGERY. *Educ:* Pa State Univ, BS, 53; Hahnemann Med Col, MD, 57; Am Bd Neurol Surg, dipl, 68. *Prof Exp:* Resident neurosurg, Yale Univ, 61-65, instr neurol surg, Sch Med, 64-65; from instr to assoc prof neurosurg, Sch Med, Univ Cincinnati, 66-72; prof neurosurg & chmn div, Sch Med, Southern Ill Univ, 72-77; PVT PRACT NEUROSURG, 77- *Concurrent Pos:* Knight fel neuropath, Sch Med, Yale Univ, 62-63; NIH fel, Univ London, 65; chief neurosurg, Cincinnati Vet Admin Hosp, 66-72; asst dir neurosurg, Children's Hosp, Cincinnati; attend neurosurgeon, Cincinnati Gen Hosp & Christian R Holmes Hosp, Cincinnati; consult neurosurg, Bur Serv Crippled Children, Hamilton County Neuromuscular Diag Clin & Shriners Burns Inst. *Mem:* Cong Neurol Surg; Asn Acad Surg; Int Soc Pediat Neurosurg; fel Am Col Surg; Am Asn Neurol Surg. *Res:* Hydrocephalus, adult and childhood; cerebral blood flow, cortical function; electron microscopy, cerebral capillaries in traumatic encephalopathy; neonatal meningitis, treatment, ultrastructure of brain. *Mailing Add:* 220 Anderson Dr Erie PA 16509-3205

SALMON, MICHAEL, b New York, NY, Apr 8, 38. ANIMAL BEHAVIOR. *Educ:* Earlham Col, BS, 59; Univ Md, MS, 62, PhD(zool), 64. *Prof Exp:* Asst animal behav, Univ Md, 59-64; NIH fel, Univ Hawaii, 64-65; asst prof biol, DePaul Univ, 65-67; from asst prof to assoc prof zool, 67-77, PROF ECOL, ETHOLOGY & EVOLUTION, UNIV ILL, URBANA-CHAMPAIGN, 77- *Mem:* AAAS; Animal Behav Soc; Crustacean Soc; Am Soc Zoologists. *Res:* Marine bioacoustics of crustaceans and fishes; sensory mechanisms; orientation and navigation by sea turtle; mating systems and sexual selection within the crustacea. *Mailing Add:* Dept Ecol & Evolution Univ Ill Urbana Campus 505 S Goodwin Ave Urbana IL 61801

SALMON, OLIVER NORTON, b Syracuse, NY, Mar 24, 17; m 45; c 4. PHYSICAL CHEMISTRY. *Educ:* Cornell Univ, AB, 40, PhD(phys chem), 46. *Prof Exp:* Lab asst animal nutrit, Cornell Univ, 40-41; chemist, Corning Glass Works, NY, 41-43; res asst, Off Sci Res & Develop, Cornell Univ, 43-45, res assoc, Off Res & Inventions, US Dept Navy, 46-47; phys chemist, Knolls Atomic Power Lab, Gen Elec Co, 47-56 & Electronics Lab, 56-60; res specialist, 3M Co, 60-62, supvr, 62-64, mgr mat physics res, 64-72, mgr basic & pioneering res, 72-73, sr res specialist, 73-81, sr res specialist, 3M Indust & Consumer Sector Res Lab, 81-82; RETIRED. *Mem:* Am Chem Soc; Am Ceramic Soc; Int Solar Energy Soc. *Res:* Solar energy; solid state physics and chemistry; chemical thermodynamics; liquid metals; hydrogen isotopes. *Mailing Add:* Rte 7 Box 175 AA Bemidji MN 56601

SALMON, PETER ALEXANDER, b Victoria, BC, Aug 5, 29; US citizen; m 53; c 3. SURGERY. *Educ:* Univ Wash, BS, 51, MD, 55; Univ Minn, Minneapolis, MS & PhD(surg), 61; Am Bd Surg, dipl, 64; FACS, 68; FRCS(C), 71. *Prof Exp:* Asst prof surg, Med Sch, Univ Minn, Minneapolis, 62-66; assoc prof, 66-72, PROF SURG, UNIV ALTA, 72- *Concurrent Pos:* Hartford Found grant, Mt Sinai Hosp, Minneapolis, 63-67, dir surg educ & res, 63-66; USPHS res grant, 64-65; Med Res Coun Can grant, Univ Alta, 67-90; dir surg educ & res, Mt Sinai Hosp, Minneapolis, 63-66. *Mem:* AAAS; Soc Exp Biol & Med; fel Am Col Surg; fel Royal Col Surg; Soc Univ Surgeons; Assoc Acad Surg. *Res:* Gastrointestinal physiology, secretion, motility; gastrointestinal transplantation, especially intestine, pancreas, liver; surgical treatment of obesity. *Mailing Add:* 2D4-27 Walter Mackenzie HSC Univ Alta Edmonton AB T6G 2B7 Can

SALMON, RAYMOND EDWARD, b Vancouver, BC, Apr 14, 31; m; c 3. POULTRY NUTRITION. *Educ:* Univ BC, BSA, 54, MSA, 57; Univ Sask, PhD, 72. *Prof Exp:* Nutritionist, Buckerfield's Ltd, 58-67; res scientist, Can Agr Res Br, 67-91. *Mem:* Poultry Sci Asn; Worlds Poultry Sci Asn. *Res:* Turkey nutrition research; utilization of fats and oils; rapeseed meal research; author of over 50 science publications. *Mailing Add:* 1458 Ashley Dr Swift Current SK S9H 1N6 Can

SALMON, VINCENT, b Kingston, BWI, Jan 21, 12; nat US; wid; c 2. ACOUSTICS. *Educ:* Temple Univ, AB, 34, AM, 36; Mass Inst Technol, PhD(physics), 38. *Prof Exp:* Physicist, Jensen Radio Mfg Co, 37, physicist in charge res & develop, Jensen Mfg Co, 38-49; mgr, sonics sect, SRI Int, 49-65, mgr, sonic prog, 65-70, staff scientist physics, Sensory Sci Res Ctr, 70-76, CONSULT PROF STANFORD UNIV, 77- *Concurrent Pos:* Ed, Audio Eng Soc J, 54-55; consult, acoust, 46-, Nat Acad Sci, 65- & Nat Acad Eng, 73-74; pres, Nat Coun Acoust Consults, 69-71; mem, Inst Noise Control Eng & pres, 74; vpres, Indust Health Inc, 72-77. *Honors & Awards:* Biennial Award, Acoust Soc Am, 46 & Silver Medal, 84. *Mem:* Fel Acoust Soc Am (pres, 70-71); fel Audio Eng Soc; Inst Noise Control Eng. *Res:* Theory of acoustic radiators; electroacoustics; underwater sound; industrial acoustics; audio engineering; sound recording; noise control; architectural acoustics; nondestructive sonic testing. *Mailing Add:* 765 Hobart St Menlo Park CA 94025

SALMOND, WILLIAM GLOVER, b West Wemyss, Scotland, May 29, 41; US citizen; m 68. SYNTHETIC ORGANIC CHEMISTRY, STEROID CHEMISTRY. *Educ:* St Andrews Univ, BSc, 63, PhD(chem), 66. *Prof Exp:* Scientist chem, Sandoz-Wander Inc, 70-72; sr res scientist chem, 72-77, res mgr, 77-82, dir, chem res & develop, 82-84, vpres, 84-90, VPRES CHEM OPERS, UPJOHN CO, 90- *Mem:* Am Chem Soc. *Res:* Commercially viable synthesis of complex natural products; application of organometallic chemistry to organic synthesis. *Mailing Add:* Fine Chem Div Upjohn Co Kalamazoo MI 49001

SALMONS, JOHN ROBERT, b Climax Springs, Mo, Sept 12, 32; m 57; c 2. CIVIL ENGINEERING. *Educ:* Univ Mo, BSCE, 60; Univ Ariz, MSCE, 65, PhD(civil eng), 66. *Prof Exp:* Detailer, Bridge Div, Mo State Hwy Dept, 57-59; asst civil eng, Univ Ariz, 61-64; assoc prof, 65-80, PROF CIVIL ENG, UNIV MO-COLUMBIA, 80- *Concurrent Pos:* Res engr, Prestressed Div, United Mat Inc, Ariz; instr, Off Civil Defense, 63-; Am Soc Eng Educ-Ford Found engr in residency with Wilson Concrete Co, Nebr, 69-70; mem bd dirs & vpres, Wilson Concrete Co, Nebr, 73- *Mem:* Am Soc Civil Engrs; Am Concrete Inst; Prestressed Concrete Inst. *Res:* Evaluation of precast-prestressed composite u-beam bridge slabs; continuity of precast-prestressed bridge members; large panel prefabricated concrete building construction; design of precast-prestressed concrete structures. *Mailing Add:* 3234 S Oak Ave Springfield MO 65804

SALO, ERNEST OLAVI, b Butte, Mont, Dec 31, 19; m 79; c 2. FISHERIES, OCEANOGRAPHY. *Educ:* Univ Wash, BS, 47, PhD(fisheries), 55. *Prof Exp:* Res biologist, Minter Creek Biol Sta, State Dept Fisheries, Wash, 50-54, asst supvr salmon hatcheries, 54-55; from asst prof to prof fisheries, Humboldt State Col, 55-65; from assoc prof to prof, 65-85, EMER PROF FISHERIES, UNIV WASH, 85- *Concurrent Pos:* Fels, Ger, Finland & Eng, 64, Chile, 69; consult, pvt indust, 65-85, Govt Chile, 69-83 & AEC-Nat Res Coun, 72-77. *Mem:* Am Fisheries Soc; Am Inst Fishery Res Biologists (past secy). *Res:* Estuarine ecology; aquaculture; ecology of salmon. *Mailing Add:* Fisheries Res Inst Univ Wash Seattle WA 98195

SALO, WILMAR LAWRENCE, b Nichols Twp, Minn, Aug 22, 37; m 60; c 3. BIOCHEMISTRY. *Educ:* Univ Minn, BS, 59, PhD(biochem), 67; Univ Wis, MS, 62. *Prof Exp:* Staff fel, NIH, 67-69; asst prof, Univ, 69-74, ASSOC PROF BIOCHEM, SCH MED, UNIV MINN, DULUTH, 74- *Mem:* Am Soc Biol Chemists; AAAS. *Res:* Biochemistry of complex carbohydrates; enzymology. *Mailing Add:* 4303 W Trischer Rd Duluth MN 55803

SALOMAN, EDWARD BARRY, b New York, NY, May 30, 40; m 68. LASER SPECTROSCOPY. *Educ:* Columbia Univ, AB, 61, MA, 62, PhD(atomic physics), 65. *Prof Exp:* Res physicist, Columbia Univ, 65-66; asst prof physics, Brown Univ, 66-72; PHYSICIST FAR ULTRAVIOLET PHYSICS & PHOTON PHYSICS, ELECTRON & OPTICAL PHYSICS DIV, NAT INST STANDARDS & TECHNOL, 72- *Mem:* AAAS; Am Phys Soc; Optical Soc Am. *Res:* Atomic spectroscopy; radiometry; resonance physics; optical resonance studies of stable and radioactive atoms; relaxation of optically oriented atoms; far ultraviolet physics; effect of external fields on autoionizing states of atoms and molecules, soft x-ray cross section data; atomic physics with synchroton radiation; relativistic calculations of atomic energy levels and transition probabilities; resonance ionization spectroscopy. *Mailing Add:* Photon Physics Group Nat Inst Standards & Technol Gaithersburg MD 20899

SALOMON, LOTHAR L, b Buedingen, Ger, Nov 8, 21; nat US; c 3. BIOCHEMISTRY. *Educ:* Columbia Univ, BS, 49, MA, 50, PhD(chem), 52. *Prof Exp:* Asst chem, Columbia Univ, 49-52; instr biochem & nutrit, Univ Tex Med Br, 52-53, from asst prof to assoc prof, 53-64; chief biol div, Dugway Proving Ground, 64-72; dep dir test opers, Deseret Test Ctr, 72-78; dep dir mat test directorate, Dugway Proving Ground, 78-81, sci dir, 81-89; RETIRED. *Concurrent Pos:* Consult isotopically labelled carbohydrates, 58-62. *Mem:* Fel AAAS; assoc AMA; Am Chem Soc; Soc Exp Biol & Med; fel Am Inst Chem; Sigma Xi. *Res:* Biosynthesis and metabolism of ascorbic acid and carbohydrates; membrane transport of carbohydrates; aerobiology; research and development administration. *Mailing Add:* 9710 Grand Oak Dr Austin TX 78750-3803

SALOMON, MARK, b Brooklyn, NY, June 2, 35; m 59; c 3. PHYSICAL CHEMISTRY. *Educ:* Hunter Col, BA, 57; Brooklyn Col, MA, 61; Univ Ottawa, PhD(chem), 64. *Prof Exp:* Chemist, Leesona-Moos Labs, 58-61; NSF res assoc chem, Princeton Univ, 64-65; asst prof, Rutgers Univ, 65-67; chemist, NASA, 67-72; CHEMIST, US ARMY ELECTRONIC COMMAND, FT MONMOUTH, 72- *Concurrent Pos:* Res scientist physics, Boston Col, 72-; titular mem comn V8, Int Union Pure & Applied Chem. *Mem:* Electrochem Soc; Int Union Pure & Appl Chem. *Res:* Chemical kinetics; biochemical kinetics; electrochemistry; thermodynamics. *Mailing Add:* US Army Electronic Command Mail Code SLCET-PR Ft Monmouth NJ 07703-5000

SALOMON, ROBERT EPHRIAM, b Brooklyn, NY, June 8, 33; m 61; c 2. PHYSICAL CHEMISTRY. *Educ:* Brooklyn Col, BA, 54; Univ Ore, PhD(phys chem), 60. *Prof Exp:* Lectr chem, Brooklyn Col, 57; Sloan res fel, 60; from asst prof to assoc prof phys chem, 61-67, chmn dept chem, 68-74, PROF PHYS CHEM, TEMPLE UNIV, 67- *Concurrent Pos:* Consult, Frankford Arsenal, Pa, 61, Gen Elec Co, 66 & Nuclear Regulatory Comn, 79- *Mem:* Am Chem Soc; Am Asn Univ Prof; Sigma Xi; Electrochem Soc. *Res:*

Spectroscopic and electrical properties of solids; conversion of ocean wave energy and solar heat into electricity using electrochemical hydrogen concentration cells; inorganic superconductors. *Mailing Add:* Dept Physical Chem Temple Univ Philadelphia PA 19122

SALOMONE, RAMON ANGELO, organic chemistry, for more information see previous edition

SALOMONSON, VINCENT VICTOR, b Longmont, Colo, July 19, 37; m 63; c 5. METEOROLOGY, HYDROLOGY. *Educ:* Colo State Univ, BS, 59, PhD(atmospheric sci), 68; Univ Utah, BS, 60; Cornell Univ, MS, 64. *Prof Exp:* Res hydrologist, Lab Meteorol & Earth Sci, 68-74, head, Hydrospheric Sci Br, Lab Atmospheric Sci, 74-80, chief, Lab Terrestial Physics, 80-88, dep dir, Space and Earth Sci Directorate, 88-90, DIR, EARTH SCI DIRECTORATE, GODDARD SPACE FLIGHT CTR, NASA, 90- *Concurrent Pos:* Mem working group remote sensing in hydrol, US Int Hydrol Decade-Nat Acad Sci, 71-74; US co-chmn remote sensing subcomt, Int Field Year of Great Lakes, 72-74; proj scientist, Landsat 4 & 5, 77-88; sci team leader, Earth Observing Syst (EOS) Modis, 88- *Honors & Awards:* Cert for Outstanding Performance, Goddard Space Flight Ctr, NASA, 74, 75, 76 & 77, NASA, Except Sci Achievement Medals, 76, 83; Distinguished Achievement Award, Inst Elec & Electronics Engrs Geo Sci & Remote Sensing Soc, 86; William T Pecora Award, 87. *Mem:* Am Meterol Soc; Am Geophys Union; Am Soc Photogrammetry & Remote Sensing (vpres, 89-90, pres elect, 90-91, pres, 91-92). *Res:* Author of over 110 publications on remote sensing applications and studies in hydrology, atmospheric science and earth resource management. *Mailing Add:* Earth Sci Directorate Code 900 NASA Goddard Space Flight Ctr Greenbelt MD 20771

SALOT, STUART EDWIN, b Los Angeles, Calif, Oct 23, 37; m 68; c 2. INORGANIC CHEMISTRY. *Educ:* Univ Calif, Berkeley, BA, 60; Univ Southern Calif, PhD(chem), 69; Am Bd Indust Hyg, cert, 78. *Prof Exp:* Asst prof chem, San Fernando Valley State Col, 68-69; asst prof, Calif State Polytech Col, 69-72; dir tech serv, Daylin Corp, Los Angeles, 72-75; PRES, CTL ENVIRON SERV, 75- *Mem:* AAAS; Am Chem Soc; Am Crystallog Asn; Am Bd Indust Hyg; Am Indust Hyg Asn. *Res:* Industrial environment pollution abatement studies; synthetic inorganic chemistry; reactions in non-aqueous solvent systems; non-stoichiometric compounds. *Mailing Add:* 24404 S Vermont Ave No 307 Harbor City CA 90710

SALOTTO, ANTHONY W, b Yonkers, NY, Aug 28, 36; m 63; c 3. PHYSICAL CHEMISTRY. *Educ:* Mass Inst Technol, BS, 58, MS, 59; NY Univ, PhD(chem), 69. *Prof Exp:* Res engr, Calif Res Corp, 59-61; instr chem, Sacramento City Col, 62-63; asst prof, Dutchess Community Col, 63-66; assoc prof, 69-77, asst chmn dept, 75-77, PROF CHEM, PACE UNIV, WESTCHESTER CAMPUS, 77-, CHMN DEPT CHEM, 77- *Mem:* AAAS; Am Chem Soc. *Res:* Analytical chemistry; environmental chemistry. *Mailing Add:* Dept Chem Pace Univ 861 Bedford Ave Pleasantville NY 10570-2799

SALOVEY, RONALD, b New York, July 11, 32; m 54; c 3. PHYSICAL CHEMISTRY, POLYMER CHEMISTRY. *Educ:* Brooklyn Col, BS, 54; Harvard Univ, AM, 58, PhD(phys chem), 59. *Prof Exp:* Res chemist, Interchem Corp, 54-55; mem tech staff, Bell Tel Labs, Inc, 58-70; res supvr, Hooker Res Ctr, Occidental Petrol Corp, Niagara Falls, 70-73, mgr res, 73-75; actg chmn chem eng, 87-89, PROF CHEM ENG & MAT SCI, UNIV SOUTHERN CALIF, 75-, CHMN CHEM ENG, 89- *Concurrent Pos:* Dir, Los Angeles Rubber Group Inc Found, 75-; adv bd, Polymer Eng & Sci & J Appl Polymer Sci. *Mem:* Am Chem Soc; Am Phys Soc; Soc Plastics Eng; Am Inst Chem Eng. *Res:* Physical chemistry of polymers. *Mailing Add:* 6641 Monero Dr Rancho Palos Verdes CA 90274

SALPETER, EDWIN ERNEST, b Vienna, Austria, Dec 3, 24; nat US; m 50; c 2. ASTROPHYSICS. *Educ:* Univ Sydney, MSc, 45; Univ Birmingham, PhD, 48. *Hon Degrees:* DSc, Univ Chicago, 69 & Case Western Reserve Univ, 70. *Prof Exp:* Res fel sci & indust res, Univ Birmingham, 48-49; res assoc, 49-53, from assoc prof to prof physics & astrophys, 53-72, J G WHITE DISTINGUISHED PROF PHYS SCI, CORNELL UNIV, 72- *Concurrent Pos:* Vis prof, Australian Nat Univ, 53-54; mem, Int Sci Radio Union; mem, US Nat Sci Bd, 78-84. *Honors & Awards:* Award, Carnegie Inst, 59; Gold Medal, Royal Astronomical Soc, 63; J R Oppenheimer Mem Prize, 74; H N Russell lectr, 74. *Mem:* Nat Acad Sci; Akad Leopoldina; Am Philos Soc; Am Astron Soc (vpres, 71-73); Int Astron Union. *Res:* Quantum theory of atoms; quantum electrodynamics; nuclear theory; energy production stars; theoretical astrophysics. *Mailing Add:* 308 Newman Lab Nuclear Studies Cornell Univ Ithaca NY 14853

SALPETER, MIRIAM MIRL, b Riga, Latvia, Apr 8, 29; US citizen; m 50; c 2. NEUROBIOLOGY, CYTOLOGY. *Educ:* Hunter Col, AB, 50; Cornell Univ, AM, 51, PhD(psychobiol), 53. *Prof Exp:* NIH fel biol, 57-60, res assoc, 60-64, sr res assoc, 64-66, assoc prof neurobiol, 66-73, PROF NEUROBIOL & BEHAVIOR, CORNELL UNIV, 73- *Concurrent Pos:* NIH career develop award, 62- *Mem:* AAAS; Electron Micros Soc Am; Am Soc Cell Biol; Soc Neurosci. *Res:* Cell biology; electron microscopy; regeneration; neurocytology. neurocytology; molecular organization of neuromuscular junctions; neurotropic phenomena. *Mailing Add:* 116 Westbourne Ln Ithaca NY 14850

SALSBURY, JASON MELVIN, b Richmond, Va, June 12, 20; m 46; c 2. INDUSTRIAL ORGANIC CHEMISTRY. *Educ:* Univ Richmond, BS, 40; Univ Va, MS, 43, PhD(org chem), 45. *Prof Exp:* Lab asst, Univ Richmond, 38-40; asst, Nat Defense Res Comt, Univ Va, 41-44 & Off Sci Res & Develop, 44-45; chemist, Am Cyanamid Co, 46-53, group leader, 54, mgr textile resin res, 54-57, mgr tech dept, Santa Rosa Plant, 57-61, dir fibers res, Fibers Div, Stamford Labs, 61-63, dir res, 63-64, dir res & commercial develop, 64-65, dir res, 65-66, tech dir, 66-67, vpres res & develop, Formica Corp, 67-72, dir, Chem Res Div, 72-81; pres, Saljas Mgt & Consult, Inc, 81-; DIR, INDEPENDENT OPPORTUNITIES CTR, GA INST TECH, 82- *Concurrent Pos:* Mem, Indust Res Inst. *Mem:* Am Chem Soc. *Res:* Polymer chemistry; synthetic organic chemistry; analysis of organic compounds; B-alkyl aminoalkyl esters of alkozybenzoic acids for local anesthetics; synthetic fibers; textile chemicals; laminates; panels; paper chemistry; polymers; resins; catalysts. *Mailing Add:* 20110 Boca W Dr Suite 258 Boca Raton FL 33434-5201

SALSBURY, ROBERT LAWRENCE, b Vancouver, BC, July 4, 16; nat US; m 45. RUMINANT NUTRITION, POULTRY NUTRITION. *Educ:* Univ BC, BA & BSA, 42; Mich State Univ, PhD(animal nutrit), 55. *Prof Exp:* Jr chemist, Can Dept Pub Works, 42-45; control chemist, E R Squibb & Sons, 45-46; res assoc, NJ Agr Exp Sta, 46-47; bacteriologist, State Dept Health, Mich, 47-49, biochemist, 49-50; agent bur dairy indust, USDA, 52-54; asst prof agr chem, Mich State Univ, 55-61; assoc prof animal sci & agr biochem, 61-73, PROF ANIMAL SCI & AGR BIOCHEM, UNIV DEL, 73- *Mem:* Am Chem Soc; Am Dairy Sci Asn; Am Soc Animal Sci; Am Soc Microbiol; Poultry Sci Asn; Sigma Xi. *Res:* Mineral balance of poultry diets; interactions amoung dietary ingredients; physiological effects of ionophores in poultry. *Mailing Add:* 105 Cheltenham Rd Newark DE 19711

SALSER, WINSTON ALBERT, b Wichita, Kans, May 5, 39; m 63; c 3. MOLECULAR BIOLOGY. *Educ:* Univ Chicago, BS, 63; Mass Inst Technol, PhD(molecular biol), 66. *Prof Exp:* Helen Hay Whitney fel, Inst Molecular Biol, Geneva, Switz, 65-67 & Inst Biophys & Biochem, Paris, France, 67-68; from asst prof to assoc prof, 68-75, PROF MOLECULAR BIOL, UNIV CALIF, LOS ANGELES, 75- *Mem:* AAAS. *Res:* Chromosome structure and gene expression in mammalian genomes; insertion of selected mammalian genes into bacterial plasmids; nucleotide sequence analysis of hemoglobin mRNAs and mammalian satellite DNAs. *Mailing Add:* Dept Biol 2203 Life Sci Univ Calif 405 Hillgard Ave Los Angeles CA 90024

SALSIG, WILLIAM WINTER, JR, b Medford, Ore, Mar 16, 19; m 44; c 3. MECHANICAL ENGINEERING. *Educ:* Univ Calif, BS, 43. *Prof Exp:* Design engr, Lawrence Berkeley Lab, Univ Calif, 43-79, bevatron mech eng, 54-63, 200 Billion Electron Volt Accelerator Study, 63-67, in charge, Electron Ring Accelerator Mech Design, 67-72, in charge, Bevalac Construct, 73-75, adminr, Accelerator & Fusion Res Div, 73-79, consult, 79-84; RETIRED. *Concurrent Pos:* Engr, Oak Ridge Nat Lab, 44-45 & Oper Greenhouse-Eniwetok, 50-51. *Mem:* Am Soc Mech Engrs. *Res:* Design of equipment for physical research; cyclotron; synchrotron; calutrons. *Mailing Add:* Eight Anson Way Kensington CA 94707

SALSTEIN, DAVID A, b New York, NY. ATMOSPHERIC DYNAMICS. *Educ:* Mass Inst Technol, SB, 72, SM, 73, PhD(meteorol), 76. *Prof Exp:* Fel, Nat Ctr Atmospheric Res, 76-77; sr staff scientist meteorol, Environ Res & Technol, Inc, 78-82; SR STAFF SCIENTIST METEOROL, ATMOSPHERIC & ENVIRON RES, INC, 82- *Mem:* Am Meteorol Soc; Am Geophys Union. *Res:* General circulation of the atmosphere. *Mailing Add:* Atmospheric & Environ Res Inc 840 Memorial Dr Cambridge MA 02139

SALSTROM, JOHN STUART, b Greensboro, NC, Jan 25, 45; m 67; c 2. MOLECULAR BIOLOGY, BIOCHEMISTRY. *Educ:* Miami Univ, BA, 67, Univ Wis, MS, 70, PhD(molecular biol), 77. *Prof Exp:* Res scientist, Harvard Univ, 77-; AT MOLECULAR GENETICS, INC. *Concurrent Pos:* NIH fel, 77- *Res:* Molecular biology of transcriptional controls over gene expression in E-coli and its phages. *Mailing Add:* 1250 W Minnehaha Pkwy Minneapolis MN 55419

SALT, DALE L(AMBOURNE), b Salt Lake City, Utah, July 1, 24; m 50; c 3. CHEMICAL ENGINEERING. *Educ:* Univ Utah, BS, 48, MS, 49; Univ Del, PhD(chem eng), 59. *Prof Exp:* Chem engr, Utah Oil Refining Co, 49-51; res fel chem eng, Univ Del, 51-54; from asst prof to assoc prof, 54-70, PROF CHEM ENG, UNIV UTAH, 70- *Honors & Awards:* A E Marshall Award, Am Inst Chem Engrs, 48. *Mem:* Am Soc Eng Educ; Am Inst Chem Engrs; Combustion Inst. *Res:* Gaseous combustion; rheology; accelerated particle dynamics; interphase transfer processes. *Mailing Add:* Dept Chem Eng Univ Utah Salt Lake City UT 84112

SALT, GEORGE WILLIAM, b Spokane, Wash, Oct 9, 19; m 42; c 3. ANIMAL ECOLOGY. *Educ:* Univ Calif, Los Angeles, BA, 42; Univ Calif, MA, 48, PhD(zool), 51. *Prof Exp:* Asst zool, Univ Calif, Berkeley, 46-49; assoc, Univ Calif, Davis, 49-50, lectr, 50-51, instr, 51-53, from asst prof to prof, 53-90, EMER PROF ZOOL, UNIV CALIF, DAVIS, 90- *Concurrent Pos:* NSF sr fel, 59-60; Rockefeller Found affiliate & vis prof, Univ Valle, Colombia, 71-72; ed, Am Naturalist, 79-84. *Mem:* Brit Ecol Soc; Am Soc Limnol Oceanog; Cooper Ornith Soc; Am Ornith Union; Am Naturalist Soc; Ecol Soc Am. *Res:* Faunal analysis and community structure; feeding in Protozoa; predator-prey interactions. *Mailing Add:* Dept Zool Univ Calif Davis CA 95616

SALT, WALTER RAYMOND, b Eng, Oct 12, 05; m 33; c 1. HISTOLOGY. *Educ:* Univ Alta, BSc, 40, MSc, 48. *Prof Exp:* Prof zool, Mt Royal Col, 45-49; from asst to prof, 49-71, sessional lectr, 71-80, EMER PROF ANAT, UNIV ALTA, 71- *Res:* Cytology; striated muscle; avian flight. *Mailing Add:* 533 Fargo Pl Victoria BC V9C 2L9 Can

SALTER, LEWIS SPENCER, b Norman, Okla, Feb 4, 26; m 50; c 4. THEORETICAL PHYSICS. *Educ:* Univ Okla, BS, 49; Oxford Univ, BA, 51, DPhil, 56. *Prof Exp:* Instr math, Europ Div, Univ Md, 52-53; from asst prof to prof physics, Wabash Col, 53-68; acad dean, Knox Col, Ill, 68-69, prof physics, 68-78, dean col & vpres acad affairs, 69-78, exec vpres col, 75-78; PROF PHYSICS & PRES COL, WABASH COL, 78- *Concurrent Pos:* Vis assoc prof, Bandung Tech Inst, 58-60; vis res prof, Nat Res Coun Can, 63-64. *Mem:* Am Phys Soc. *Res:* Solid state and low temperature physics. *Mailing Add:* Off of Pres Wabash Col 301 W Wabash Crawfordsville IN 47933

SALTER, ROBERT BRUCE, b Stratford, Ont, Dec 15, 24; m 48; c 5. ORTHOPEDIC SURGERY. *Educ:* Univ Toronto, MD, 47, MS, 60; FRCPS(C), 55; FRACS, 77; FRCS(I), 78. *Prof Exp:* Clin teacher orthop, Univ Toronto, 55-57; orthop surgeon, Hosp Sick Children, 55-57, chief orthop surg, 57-66; asst prof, 62-66, PROF SURG, UNIV TORONTO, 66-; SURGEON-IN-CHIEF, HOSP SICK CHILDREN, 66- *Concurrent Pos:* R S McLaughlin traveling fel, London Hosp, Eng, 54-55; consult, Ont Soc Crippled Children, 55-; mem, Med Res Coun Can, 67-69. *Honors & Awards:* Medal Surg, Royal Col Physicians & Surg Can, 60; Centennial Medal, Govt Can, 67; Gairdner Int Award Med Sci, 69; hon fel, Royal Col Physicians & Surg Glasgow, 70, Royal Col Surgeons Edinburgh, 73, Col Surgeons SAfrica, 73 & Royal Col Surgeons Eng; Sir Arthur Sims Commonwealth Traveling Prof, 73; Nicolas Andry Award, Asn Bone & Joint Surgeons, 74; Charles Mickle Award, Univ Toronto; Officer of Order of Can, 77. *Mem:* Fel Am Col Surg; Am Orthop Asn; Int Soc Orthop Surg & Traumatol; Can Orthop Res Soc; Royal Col Physicians & Surgeons Can (vpres, 70-, pres, 78-80). *Res:* Articular cartilage degeneration; avascular necrosis of epiphyses; epiphyseal injuries, congenital dysplasia and dislocation of the hip; Legg Perthes disease; experimental arthritis. *Mailing Add:* 79 Rosedale Heights Dr Toronto ON M4T 1C4 Can

SALTER, ROBERT MUNKHENK, JR, b Morgantown, WVa, Apr 24, 20; m 77; c 3. ELEMENTARY PARTICLE PHYSICS, APPLIED PHYSICS. *Educ:* Ohio State Univ, BME, 41; Univ Calif, Los Angeles, MA, 58, PhD(nuclear physics), 65. *Prof Exp:* Res engr metall & adv eng, Gen Motors Res Lab, 41-42; Lt propulsion, US Navy Aeronaut Exp Sta, Philadelphia & Exp Engines Sect, Bur Aeronaut, Washington, DC, 42-46; res engr, Aerophysics Lab, NAm Aviation, Inc, 46-48; dir proj feedback space systs res, Rand Corp, 48-54; dept mgr satellite br, US Air Force Satellite Prog Develop, Lockheed Missiles & Space Co, 54-58; PHYS SCIENTIST APPL PHYSICS, RAND CORP, 68- *Concurrent Pos:* Pres & gen mgr, Sigma Corp, 59-60 & Quantatron, 60-62; consult, Lockheed Missiles & Space Co, 59-65 & RCA Labs, 62-65; chmn bd, Telic Corp, 68-71; mem, Nuclear Propulsion Comt, Am Inst Aeronaut & Astronaut, 70-72 & Ad Hoc Comt Early Warning Physics, Adv Res Proj Agency, 72-76; pres, Xerad Inc, 57- & Spectravision Inc, 71- *Honors & Awards:* Space Pioneer Medal, DOD, 85. *Mem:* Sigma Xi. *Res:* Experimental and theoretical determination of neutron-neutron forces in nucleus through decay of di-neutron formed from pion absorption in deuteron; application of advanced physics in conceptualization of new devices in optics, electronics and aero-space systems. *Mailing Add:* 1514 Sorrento Dr Pacific Palisades CA 90272

SALTHE, STANLEY NORMAN, b Oct 16, 30; US citizen; m 59; c 2. EVOLUTIONARY SYSTEMS. *Educ:* Columbia Univ, BS, 59, MA, 60, PhD(zool), 63. *Prof Exp:* Am Cancer Soc fel, Brandeis Univ, 63-65; from asst prof to assoc prof, 65-72, PROF BIOL, BROOKLYN COL, 72- *Concurrent Pos:* NSF res grants, 66-71; City Univ New York res grants, 71-72 & 72-73. *Mem:* NY Acad Sci; Am Soc Naturalists; Int Soc Systs Sci. *Res:* Application of hierarchy theory to evolutionary process; infodynamics and development theory; macroevolution. *Mailing Add:* Dept Biol Brooklyn Col Brooklyn NY 11210

SALTIEL, ALAN ROBERT, b New Brunswick, NJ, Nov 29, 53; m 81; c 3. BIOCHEMISTRY. *Educ:* Duke Univ, AB, 75; Univ NC, PhD(biochem), 80. *Prof Exp:* Res scientist, Burroughs-Wellcome Co, 80-84; asst prof, Rockefeller Univ, 84-90; ASSOC PROF, UNIV MICH, 90-; DIR SIGNAL TRANSDUCTION, PARKE-DAVIS/WARNER-LAMBERT CO, 90- *Concurrent Pos:* Irma T Hirschl Scholar, Hirschl/Caulier Trust, 86. *Honors & Awards:* John Jacob Abel Award, Am Soc Pharmacol & Exp Therapeut, 90. *Mem:* Am Soc Biochem & Molecular Biol; Am Soc Pharmacol & Exp Therapeut; Endocrine Soc; Harvey Soc. *Res:* Investigated the molecular events involved in the initial phase of receptor activation by the hormones insulin, epidermal growth factor and nerve growth factor, to understand the key mechanisms responsible for the regulation of cellular metabolism and growth; molecular characterization of receptors that are coupled to guanyl nucleotide binding G protein; growth factor receptors and protein phosphorylation; hormonal control of carbohydrate and lipid metabolism. *Mailing Add:* Parke-Davis Warner-Lambert Co 2800 Plymouth Rd Ann Arbor MI 48105-2430

SALTIEL, JACK, b Salonica, Greece, Feb 14, 38; US citizen; m 65. PHOTOCHEMISTRY. *Educ:* Rice Univ, BA, 60; Calif Inst Technol, PhD(chem), 64. *Prof Exp:* NSF fel, Univ Calif, Berkeley, 63-64; from asst prof to assoc prof, 65-75, PROF CHEM, FLA STATE UNIV, 75- *Concurrent Pos:* Consult, Eli Lilly & Co, 65-67; Alfred P Sloan fel, 71-73. *Honors & Awards:* R A Welch Found Lectr, 90. *Mem:* AAAS; Am Chem Soc; Royal Soc Chem; Sigma Xi; Inter-Am Photochem Soc; Am Soc Photobiol. *Res:* Photochemistry of organic molecules. *Mailing Add:* Dept Chem B-164 Fla State Univ Tallahassee FL 32306-3006

SALTMAN, DAVID J, b New York, NY, Mar 23, 51; m 72; c 3. FIELDS, DIVISION ALGEBRAS. *Educ:* Univ Chicago, BA & MS, 72; Yale Univ, PhD(math), 76. *Prof Exp:* Dickson instr math, Univ Chicago, 76-78; asst prof, Yale Univ, 78-82; assoc prof, 82-87, Joe B & Louis Cook prof math, 87-89, KERR PROF MATH, UNIV TEX, AUSTIN, 89- *Mem:* Am Math Soc; Asn Women Math. *Res:* Division algebras, fields, Braver group and related areas of algebra and algebraic geometry. *Mailing Add:* Math Dept Univ Tex Austin TX 78712

SALTMAN, PAUL DAVID, b Los Angeles, Calif, Apr 11, 28; m 49; c 2. BIOCHEMISTRY, NUTRITION. *Educ:* Calif Inst Technol, BS, 49, PhD(biochem), 53. *Prof Exp:* From instr to prof biochem, Univ Southern Calif, 53-67; provost, Revelle Col, 67-72, vchancellor acad affairs, 72-80, PROF BIOL, UNIV CALIF, SAN DIEGO, 67- *Concurrent Pos:* NIH sr fel, 60-; sr Fulbright scholar, Perth, Australia, 81; Lady Davis prof, Jerusalem, Israel, 87. *Mem:* Am Chem Soc; Am Soc Plant Physiol; Am Soc Biol Chemists; Am Inst Nutrit. *Res:* Biological transport mechanisms; trace metal metabolism; photosynthesis; metabolism of higher plants; plant growth hormones; communication of science through films, television and radio. *Mailing Add:* Dept Biol Q-022 Univ Calif San Diego La Jolla CA 92093

SALTMAN, ROY G, b New York, NY, July 15, 32; m 59; c 3. COMPUTER APPLICATIONS, DATA ADMINISTRATION. *Educ:* Rensselaer Polytech Inst, Troy, NY, BEE, 53; Mass Inst Technol, MSEE, 55; Columbia Univ, EE, 62; Am Univ, MPA, 76. *Prof Exp:* Res engr, Sperry Gyroscope Co, 55-64; adv syst analyst, IBM Corp, 64-69; COMPUT SCIENTIST, INST COMPUT SCI & TECHNOL, NAT BUR STANDARDS, 69- *Concurrent Pos:* Intergovt Affairs fel, State Minn/US Civil Serv Comn, 75; exec secy, Comt Automation Opportunities Serv Sector, Fed Coun Sci & Technol, 72-75; com sci fel, US Dept Com/US House Reps, 77-78; dep mem, US Bd on Geog Names, 79-87; John & Mary R Markle Found grant, 86. *Honors & Awards:* E U Condon Award, Nat Bureau Standards, 78. *Mem:* Inst Elec & Electronics Engrs. *Res:* Productivity in federal computer use; policy implications of information systems; computer security in election administration. *Mailing Add:* Nat Inst Standards & Technol Tech Bldg B154 Gaithersburg MD 20899

SALTMAN, WILLIAM MOSE, b Perth Amboy, NJ, Nov 19, 17; m 43; c 3. RUBBER CHEMISTRY, KINETICS. *Educ:* Univ Mich, BS(chem eng) & BS(eng math), 38, MS, 39; Univ Chicago, PhD(phys chem), 49. *Prof Exp:* Testing chemist, State Hwy Dept, NJ, 40-42; res assoc, Calif Inst Technol, 45; sr chemist, Shell Chem Co, Colo, 49-54; sr chemist, 55-64, sect head budene & ethylene-propylene rubbers, 64-75, mgr stereo rubbers dept, 75-77, mgr specialty polymers dept, Res Div, 77-80, mgr, Polymer Serv, Goodyear Tire & Rubber Co, 80-82; CONSULT, RUBBER & PLASTICS, 82- *Concurrent Pos:* Vis prof, Ohio State Univ, 67. *Mem:* Am Chem Soc. *Res:* Stereospecific catalysts; rubber technology; polymerization; kinetics; polymer physical properties. *Mailing Add:* 12973 Candela Fl San Diego CA 92130-1857

SALTON, GERARD, b Nuremberg, Ger, May 8, 27; nat US; m 50; c 2. APPLIED MATHEMATICS. *Educ:* Brooklyn Col, BA, 50, MA, 52; Harvard Univ, PhD(appl math), 58. *Prof Exp:* From instr to asst prof appl math, Harvard Univ, 58-65; assoc prof comput sci, 66-67, chmn dept, 71-77, PROF COMPUT SCI, CORNELL UNIV, 67- *Concurrent Pos:* Guggenheim fel, 63; consult, Sylvania Elec Prod, Inc & Arthur D Little, Inc; ed-in-chief, Commun J, Asn Comput Mach. *Honors & Awards:* Spec Interest Group Info Retrieval Award, Asn Comput Mach, 83; Award Merit, Am Soc Info Sci, 89. *Mem:* Asn Comput Mach; Inst Elec & Electronics Engrs; Am Soc Info Sci; Asn Comput Ling. *Res:* Electronic data processing; business applications of computers; mathematical linguistics; theory of information retrieval. *Mailing Add:* Dept Comput Sci Cornell Univ Upson Hall Ithaca NY 14850

SALTON, MILTON ROBERT JAMES, b NSW, Australia, Apr 29, 21; m 51; c 2. MICROBIOLOGY. *Educ:* Univ Sydney, BSc, 45; Cambridge Univ, PhD(biochem), 51; ScD, Cambridge Univ, 67. *Hon Degrees:* Dr Med, Univ Liege, 67. *Prof Exp:* Res officer microbiol, Commonwealth Sci & Indust Res Orgn, Australia, 45-48; res fel biochem, Cambridge Univ, 48-54, Beit Mem res fel, 50-52, demonstr, 57-61; prof, Univ New South Wales, 62-64; PROF MICROBIOL & CHMN DEPT, SCH MED, NY UNIV, 64- *Concurrent Pos:* Merck Int fel, Univ Calif, 52-53; reader, Manchester Univ, 57-61; lectr, Off Naval Res, 60. *Honors & Awards:* Ciba Lectr, Rutgers Univ, 60. *Mem:* Am Soc Microbiol; Am Soc Biol Chem; Harvey Soc; Royal Soc Med; Brit Biochem Soc; fel Royal Soc London; hon mem Brit Soc Antimicrobial Chemother. *Res:* Chemistry and biochemistry of microbial cell surfaces. *Mailing Add:* Dept Microbiol NY Univ Sch Med New York NY 10016

SALTONSTALL, CLARENCE WILLIAM, JR, b El Centro, Calif, Jan 26, 25; m 46; c 3. SYNTHETIC MEMBRANE TECHNOLOGY, POLYMER CHEMISTRY. *Educ:* Pomona Col, BA, 48; Columbia Univ, PhD(org chem), 57. *Prof Exp:* Res chemist, Photo Prod Dept, E I du Pont de Nemours & Co, Inc, 53-58; res chemist, Chem Div, Aerojet-Gen Corp, 58-62, asst sr chemist, Solid Rocket Res Div, 62-63, sr chemist, 63-64, sr chemist, Chem & Struct Prod Div, 64-67, mgr res & develop, Phys Processes Dept, Environ Systs Div, 67-70, mgr desalination res, Water Purification Systs Oper, Envirogenics Co, 70-75; dir res & develop, Envirogenics Systs Co, 75-78; CONSULT MEMBRANE TECHNOL, 78- *Mem:* Am Chem Soc; Sigma Xi; AAAS. *Res:* Synthesis of nomomers, polyelectrolytes, cellulose derivatives, elastomers and novel heterocyclic polymers; solid propellants; membranes for water desalination and the mechanism of reverse osmosis; manufacture of membranes and membrane systems. *Mailing Add:* 1634 Alaska St West Covina CA 91791

SALTSBURG, HOWARD MORTIMER, b New York, NY, Sept 12, 28; m 51; c 2. PHYSICAL CHEMISTRY. *Educ:* City Col New York, 50; Boston Univ, MA, 51, PhD(phys chem), 55. *Prof Exp:* Chemist, Geophys Res Directorate Air Force Cambridge Res Labs, 51-54; Henry & Camille Dreyfus Found fel chem, Univ Rochester, 54-55; chemist, Knolls Atomic Power Lab, Gen Elec Co, 55-57, missiles & space vehicles dept, 57-59; res & develop staff mem, Gen Atomic Div, Gen Dynamics Corp, 59-69; PROF MAT SCI, CHEM ENG & CHEM, UNIV ROCHESTER, 69- *Mem:* Am Phys Soc; Am Chem Soc; Sigma Xi. *Res:* Nucleation; surface thermodynamics; evaporation; adsorption on oxides; electrical conductivity of oxides; molecular beam scattering from surfaces. *Mailing Add:* Dept Chem Eng Gavett Hall Univ Rochester Rochester NY 14627

SALTVEIT, MIKAL ENDRE, JR, b Minneapolis, Minn, Nov 11, 44; m 78; c 1. STRESS PHYSIOLOGY, POSTHARVEST PHYSIOLOGY. *Educ:* Univ Minn, BA, 67, MS, 72; Mich State Univ, PhD(hort & bot), 77. *Prof Exp:* Mgr, Biol Labs Antartica, Antartic Res Prog, Nat Sci Found, 68-69 & 70-71; res botanist, NASA Space Prog, Agr Res Serv, US Dept Agr, 72-73; res asst plant physiol, Dept Hort, Mich State Univ, 73-77, res assoc ethylene physiol, Plant Res Lab, 77-78; asst prpf postharvest physiol, hort dept, NC State Univ, 78-83; asst, 83-89, ASSOC, VEG CROPS DEPT, UNIV CALIF, DAVIS, 89- *Honors & Awards:* Outstanding Publ Award, Am Soc Hort Sci, 88, 89. *Mem:* Am Soc Hort Sci; Am Soc Plant Physiol; AAAS; Sigma Xi. *Res:* Physiological response of plants to abiotic stresses: chilling injury, physical wounding, altered gaseous atmospheres; postharvest physiology of fruit and vegetable crops (apples, carrots, lettuce, tomatoes). *Mailing Add:* Mann Lab/Veg Crops Dept Univ Calif Davis CA 95616

SALTZ, DANIEL, b Chicago, Ill, July 25, 32; div; c 4. MATHEMATICS. *Educ:* Univ Chicago, BA, 52, BS, 53; Northwestern Univ, MS, 55, PhD(math), 58. *Prof Exp:* Instr math, Northwestern Univ, 58-59; assoc prof, 59-70, PROF MATH, SAN DIEGO STATE UNIV, 70- *Mem:* Am Math Soc; Am Math Asn. *Res:* Fourier analysis; differential equations; generalized function theory. *Mailing Add:* San Diego State Univ San Diego CA 92182

SALTZ, JOEL HASKIN, b Champaign, Ill, June 4, 56; m 79; c 2. COMPILERS, HIGH PERFORMANCE COMPUTING. *Educ:* Univ Mich, Ann Arbor, BS & MA, 78; Duke Univ, MD, 86, PhD(computer sci), 86. *Prof Exp:* Scientist ocean acoust modeling, Sci Applns Inc, 78-79; asst prof computer sci, Yale Univ, 86-89; staff scientist, 85-86, LEAD COMPUTER SCIENTIST, INST COMPUTER APPLNS SCI & ENG, LANGLEY RES CTR, NASA, 89- *Concurrent Pos:* Res scientist, Yale Univ, 89-; adj asst prof, William & Mary Col, 90-; ed, Appl Numerical Math, 90- *Mem:* Inst Elec & Electronics Engrs; Asn Comput Mach; AMA. *Res:* Development of tools and compilers capable of mapping sparse, irregular, adaptive and geometrically complex problems onto a variety of multiprocessor architectures; multiprocessor solution methods for sparse matrix problems; adaptive and unstructured mesh page-directory entry problems; molecular dynamics simulations. *Mailing Add:* Langley Res Ctr NASA MS 132 C Hampton VA 23665

SALTZBERG, BERNARD, b Chicago, Ill, Apr 21, 19; m 42; c 5. BIOMATHEMATICS, BIOENGINEERING. *Educ:* Ill Inst Technol, BS, 52, MSEE, 53; Marquette Univ, PhD(biomed & elec eng), 72. *Prof Exp:* Instr eng math, Ill Inst Technol, 53-56; mem sr sci staff, Space Technol Lab, Thompson Ramo-Wooldridge, Inc, 56-60; sr scientist res staff, Bissett-Berman Corp, 60-65; assoc prof psychiat & neurol & dir div med comput sci, Sch Med, Tulane Univ, 65-67, prof psychiat & neurol & dir biomath & neural sci res, 67-75; PROF PSYCHIAT & NEUROL, TEX RES INST MENT SCI, TEX MED CTR HOUSTON, 76- *Concurrent Pos:* Schleider scholar, Tulane Univ, 63, Nat Inst Neurol Dis & Stroke res grant, Sch Med, 71-74; supvry res engr, Res Dept, Am Mach & Foundry Co, 53-56; consult appl math & period anal, Baylor Univ, 56-; instr appl math, Univ Southern Calif & Univ Calif, Los Angeles, 58-61; vis lectr, Tulane Univ, 63-65; mem adv bd, Inst Comprehensive Med, 64; mem adv comt, Comput & Biomath Sci Study Sect, Div Res Grants, NIH, 70-74; consult ed, Soc Psychophysiol Res, 68; prof psychiat, Univ Tex Med Sch Houston; prof, Grad Sch Biomed Sci, Univ Tex; adj prof biomed eng, Rice Univ; adj prof, Univ Houston; head biomath, Epilepsy Res Ctr, Baylor Med Sch & Methodist Hosp; tech adv bd, Micro Focus Corp. *Mem:* Am EEG Soc; Neuroelec Soc (vpres, 67); Soc Neurosci; Soc Biol Psychiat; Soc Psychophysiol Res; Am Epilepsy Soc. *Res:* Electroencephalographic signal analysis; brain research; pattern recognition; time series analysis. *Mailing Add:* Univ Tex Ment Sci Inst 1300 Moursund Houston TX 77030

SALTZBERG, BURTON R, b New York, New York, June 20, 33; m 59; c 3. COMMUNICATION THEORY. *Educ:* NY Univ, BS, 54; Univ Wisconsin, MS, 55; New York Univ, ScD(elect engr), 64. *Prof Exp:* MEM TECH STAFF, AT&T BELL LABS, 57- *Mem:* Fel Inst Elec & Electronics Engrs; Sigma Xi. *Res:* Data communications. *Mailing Add:* AT&T Rm 3B201 200 Laurel Ave Middletown NJ 07748

SALTZBERG, THEODORE, b Chicago, Ill, Mar 9, 27; m 53; c 3. ELECTRICAL ENGINEERING. *Educ:* Ill Inst Technol, BS, 50, MS, 52, PhD, 63. *Prof Exp:* Develop engr commun systs, Armour Res Found, Ill Inst Technol, 52-54; proj engr control systs, Cook Elec Co, 54-56; group leader digital electronics, Motorola, Inc, 56-59, asst sect head, 59-60, asst chief engr, 60-63, chief engr, 63-66, prod mgr signaling prod, 66-69, MGR INDUST PROD ENG, COMMUN DIV, MOTOROLA, INC, 69-; SR VPRES & DIR RES, NEW BUS, MOTOROLA ELEC CO INC. *Mem:* Inst Elec & Electronics Engrs. *Res:* Digital communications, particularly radio systems. *Mailing Add:* Motorola Elec Co Inc 1301 Algonquin Rd Schaumberg IL 60671

SALTZER, CHARLES, b Cleveland, Ohio, Feb 3, 18; m 40; c 1. MATHEMATICS. *Educ:* Western Reserve Univ, BA, 41; Univ Nebr, MA, 42; Brown Univ, MSc, 45, PhD(math), 49. *Prof Exp:* Instr math, Brown Univ, 44-48; from instr to assoc prof, Case Western Reserve Univ, 48-60; prof appl math, Univ Cincinnati, 60-62; PROF MATH, OHIO STATE UNIV, 62- *Concurrent Pos:* Fulbright award, 50-51; consult, Electronics Lab, Gen Elec Co, 56-58, Thompson-Ramo-Wooldridge, Inc, 58-61 & Bell Tel, 61-63. *Mem:* Am Math Soc; Math Asn Am; Sigma Xi; Inst Elec & Electronics Engrs. *Res:* Numerical analysis; partial difference equations; conformal mapping; computer theory; network theory; theory of distributions; control and communication theory; automata theory. *Mailing Add:* Dept Math Ohio State Univ 231 W 18th Ave Columbus OH 43210

SALTZER, JEROME H(OWARD), b Nampa, Idaho, Oct 9, 39; m 61; c 3. COMPUTER SYSTEMS. *Educ:* Mass Inst Technol, SB, 61, SM, 63, ScD, 66. *Prof Exp:* From instr to assoc prof elec eng, 63-76, PROF COMPUT SCI, MASS INST TECHNOL, 76- *Concurrent Pos:* Tech dir, Mass Inst Technol, Proj Athena, 84-89. *Mem:* AAAS; Asn Comput Mach; fel Inst Elec & Electronics Engrs. *Res:* Design of computer systems for enterprise support; data communication networks; information protection and privacy; impact of computer systems on society. *Mailing Add:* Mass Inst Technol 545 Technology Square Cambridge MA 02139

SALTZMAN, BARRY, b New York, NY, Feb 26, 31; m 62; c 2. METEOROLOGY. *Educ:* City Col New York, BS, 52; Mass Inst Technol, SM, 54, PhD(meteorol), 57. *Hon Degrees:* MA, Yale Univ, 68. *Prof Exp:* Res asst meteorol, Mass Inst Technol, 52-57, res staff, 57-61; sr res scientist, Travelers Res Ctr, Inc, 61-66, res fel, 66-68; PROF GEOPHYS, YALE UNIV, 68-, CHMN DEPT, 88- *Concurrent Pos:* Assoc ed, J Geophys Res, 71-74; ed, Advances Geophys, 77-; chmn, Comt Atmospheric Sci & Biometeorol, Yale Univ, 74- *Mem:* Fel AAAS; fel Am Meteorol Soc; Am Geophys Union; Sigma Xi; hon foreign mem Acad Sci Lisbon. *Res:* Geophysical fluid dynamics; theory of climate and the atmospheric general circulation. *Mailing Add:* Dept Geol & Geophys Yale Univ New Haven CT 06511

SALTZMAN, BERNARD EDWIN, b New York, NY, June 24, 18; m 49; c 3. INDUSTRIAL HYGIENE, AIR POLLUTION. *Educ:* City Col New York, BChE, 39; Univ Mich, MS, 40; Univ Cincinnati, PhD(chem eng), 58; Am Bd Indust Hyg, cert. *Prof Exp:* Chem engr, Joseph E Seagram & Sons, Inc, 40-41; jr pub health engr, USPHS, 41-43, from asst engr to sr asst engr, 43-49, from sanit engr to sr sanit engr, 49-61, sanit engr dir & dep chief, Chem Res & Develop Sect, 61-67; res prof environ health, 67-71, prof, 71-86, EMER PROF ENVIRON HEALTH, KETTERING LAB, COL MED, UNIV CINCINNATI, 86- *Concurrent Pos:* Consult, Occup Safety & Health Admin, Nat Inst Occup Safety & Health, Natl Res Coun & WHO. *Honors & Awards:* Wiley Award, Asn Off Anal Chemists, 78. *Mem:* Am Chem Soc; Am Indust Hyg Asn; Am Conf Govt Indust Hygienists; Air Pollution Control Asn; Asn Off Anal Chem; Am Soc Testing & Mat. *Res:* Air pollution chemistry; administration of research; industrial hygiene. *Mailing Add:* Kettering Lab Univ Cincinnati Col Med Cincinnati OH 45267-0056

SALTZMAN, HERBERT A, b Philadelphia, Pa, Nov 27, 28; m 54; c 3. MEDICINE. *Educ:* Jefferson Med Col, MD, 52; Am Bd Internal Med, dipl, 60; Am Bd Pulmonary Dis, dipl, 71. *Prof Exp:* Chief pulmonary dis, Vet Admin Hosp, Durham, NC, 58-63; asst dir hyperbaric unit & asst prof med, 63-64, assoc prof, 65-69, dir, F G Hall Lab Environ Res, 64-77, PROF MED, MED CTR, DUKE UNIV, 69-, CO-DIR, E G HALL LAB ENVIRON RES. *Concurrent Pos:* Mem comt underwater physiol & med, Nat Res Coun, 72-75. *Mem:* Am Heart Asn; Am Physiol Soc; Am Soc Clin Invest; Undersea Med Soc; Am Thoracic Soc. *Res:* Environmental research; respiratory physiology and chest diseases. *Mailing Add:* Dept Med Med Ctr Duke Univ Box 3823 Durham NC 27710

SALTZMAN, MARTIN D, b Brooklyn, NY, Mar 20, 41; m 73; c 1. ORGANIC CHEMISTRY. *Educ:* Brooklyn Col, BS, 61, MA, 64; Univ NH, PhD, 68. *Prof Exp:* Res assoc, Brandeis Univ, 68-69; asst prof chem, 69-74, assoc prof found sci & spec lectr chem, 74-81, PROF NATURAL SCI & SPEC LECTR CHEM, PROVIDENCE COL, 81- *Res:* History of chemistry. *Mailing Add:* Dept Chem Providence Col River Ave & Eaton St Providence RI 02918

SALTZMAN, MAX, b Brooklyn, NY, Apr 17, 17; m 41, 52; c 2. COLOR TECHNOLOGY. *Educ:* City Col New York, BS, 36. *Prof Exp:* Inspector, NY Inspection Div, Chem Warfare Serv, 41-46; res chemist, B F Goodrich Chem Co, 46-52; vpres, Phipps Prod Corp, 52-55; develop supvr, Harmon Colors, 55-61, tech asst to vpres, Nat Aniline Div, 61-66, sr scientist, Indust Chem Div, 66-69, mgr color technol specialty, Chem Div, Allied Chem Corp, 69-73; RES SPECIALIST, INST PHYSICS & PLANETARY PHYSICS, UNIV CALIF, LOS ANGELES, 73- *Concurrent Pos:* Adj prof, Rensselaer Polytech, 66-84; chmn tech comt, Dry Color Mfrs Asn, 61-70. *Honors & Awards:* Bruning Award, Fedn Soc Coatings Technol; Hon Mem, Int Soc Color Coun, 85, MacBeth Award, 86. *Mem:* Am Chem Soc; Optical Soc Am; Brit Soc Dyers & Colourists; Fedn Socs Paint Technol; Am Asn Textile Chem & Colorists. *Res:* Color measurement; spectrophotometry; color technology; archaeological chemistry; analytical dyes in ancient textiles. *Mailing Add:* 16428 Sloan Dr Los Angeles CA 90049

SALU, YEHUDA, b Tel-Aviv, Israel, Feb 17, 41; m 64; c 2. MEDICAL PHYSICS. *Educ:* Hebrew Univ, MSc, 64; Tel-Aviv Univ, PhD(physics), 73. *Prof Exp:* Asst scientist, Univ Iowa, 73-78, assoc scientist, 78-80; ass prof, 80-82, ASSOC PROF PHYSICS, HOWARD UNIV, 82- *Concurrent Pos:* Prin investr, NIH, 78- *Res:* Bioelectricity of the human heart, its modeling, measurements and interpretation; models and simulations of nerve networks. *Mailing Add:* Dept Physics & Astron Howard Univ 2400 Sixth St NW Washington DC 20059

SALUJA, JAGDISH KUMAR, b Jhelum, Pakistan, Jan 14, 34; US citizen; m 67; c 2. ROBOTICS, NEW COAL TECHNOLOGIES. *Educ:* Univ Bombay, India, BSc, 55, Univ Mich, Ann Arbor, BSE, 57, BSE, 58, MSE, 59; Univ Fla, Gainesville, PhD(nuclear eng), 66. *Prof Exp:* Jr engr, Cornell Dublier, Los Angeles, Calif, 57-58; control engr, Argonne Nat Lab, 59-61; sr nuclear engr, Westinghouse Elec, Pittsburgh, Pa, 67-77; PRES, VIKING SYSTS INT, PITTSBURGH, PA, 78- *Honors & Awards:* Small Bus Innovation Res Award, Off Energy Res, US Dept Energy, 83 & 84. *Mem:* Am Nuclear Soc; AAAS; Soc Mfg Engrs; Am Chem Soc. *Res:* Robot systems for work in hazardous environments; coal conversion and biomass conversion technologies; aging of nuclear plant components; manufacture of electronics calibration equipment. *Mailing Add:* 105 Wynnwood Dr Fox Chapel PA 15215

SALUJA, PREET PAL SINGH, b Kanpur, India; Can citizen; m 74; c 2. HAZARDOUS & RADIOACTIVE WASTE, MANAGEMENT & INSTRUMENTATION. *Educ:* Banaras Hindu Univ, India, BSc Hons, 64, MSc, Hons, 66; Univ Pa, PhD(electrochem), 71. *Prof Exp:* NIH fel, Cornell Univ, 71-73; Robert A Welch fel, Univ Houston, 75-76; sr res assoc, Univ Alta, Can, 76-80; assoc res officer, Nat Res Coun, Can, 80-81; assoc res officer, 81-82, RES OFFICER, HAZARDOUS & RADIOACTIVE WASTE MGT, WHITESHELL NUCLEAR RES ESTAB, ATOMIC ENERGY CAN LTD, 83- *Concurrent Pos:* Lectr chem, Banaras Hindu Univ, India, 66, vis fac, 73-75; exec, Chem Inst Can & Can Nat Comn Steam, 87-89. *Mem:* Am Chem Soc; Chem Inst Can. *Res:* Gas phase ion chemistry and important mass-spectrometry applications; in situ chemical kinetic at high pressures; instrumentation and innovation; waste management both hazardous and nuclear; development of a particle mass-spectrometer. *Mailing Add:* 3112-197 Victor Lewis Dr Winnipeg MB R3P 2A4 Can

SALUNKHE, DATTAJEERAO K, b Kolhapur, India, Nov 7, 25; m 55; c 2. FOOD TECHNOLOGY. *Educ:* Univ Poona, BSc, 49; Mich State Univ, MS, 51, PhD(food technol), 53. *Prof Exp:* From asst prof to prof hort 54-67, prof nutrit & food sci, Utah State Univ, 67-75; vchancellor & pres, Marathwada Agr Univ, India, 75-76, Mahatma Phule Agr Univ, India, 80-86; prof, 76-80, PROF EMER, NUTRIT & FOOD SCI, UTAH STATE UNIV, 86- *Mem:* Fel Inst Food Technologists. *Res:* Post-harvest physiology, pathology and microscopy of fruits and vegetables; horticultural processing; radiation effects on horticultural plants and plant products; food toxicology; mycotoxins and naturally occurring toxicants in plant foods; food and nutrition. *Mailing Add:* Dept Nutrit & Food Sci Utah State Univ Logan UT 84322

SALUTSKY, MURRELL LEON, b Goodman, Miss, July 16, 23; m 66; c 3. WATER CHEMISTRY. *Educ:* Univ Ky, BS, 44; Mich State Univ, PhD(chem), 50. *Prof Exp:* Asst, Mich State Univ, 46-49; res chemist, Mound Lab, Monsanto Chem Co, 50-52, sr res chemist, 52-55, sr res chemist, Inorg Chem Div, 55-57; supvr chem res, Res Div, 57-65, dir res, Dearborn Chem Div, 65-69, vpres res, 69-71, exec vpres, Dearborn Chem Div, 71-76, group vpres & chief tech officer, 76-82, VPRES TECH OPERS, DEARBORN CHEM CO, SUBSID W R GRACE & CO, 82- *Mem:* AAAS; Am Chem Soc; Am Inst Chem; Marine Technol Soc; Int Oceanog Found. *Res:* Inorganic chemical separations; rare earths; radium and rare radioactive elements; phosphorus and phosphates; agricultural chemicals; precipitation from homogenous solution; by-products from the sea; desalination; pretreatment of saline waters; industrial water treatment chemicals; waste water and pollution control chemicals and services. *Mailing Add:* 1950 Berkeley Rd Highland Park IL 60035-2725

SALVADOR, RICHARD ANTHONY, b Albany, NY, May 19, 27; m 66; c 2. PHARMACOLOGY. *Educ:* St Bernadine of Siena Col, BS, 51; Boston Univ, AM, 53; George Washington Univ, PhD(pharmacol), 56. *Prof Exp:* Res instr pharmacol, Sch Med, Wash Univ, 58-60; sr pharmacologist, Wellcome Res Labs, 60-69; group chief, Hoffman-La Roche Inc, 70-75, asst dir, Dept Pharmacol, 75-79, dir exp therapeuts, 79-83, asst vpres & dir exp therapeuts, 83-85, VPRES & DIR PRECLIN DEVELOP, HOFFMAN-LA ROCHE INC, 85- *Concurrent Pos:* Nat Inst Neurol Dis & Blindness fel, 57. *Mem:* AAAS; Am Soc Pharmacol & Exp Therapeut; Am Chem Soc; NY Acad Sci; Fedn Am Socs Exp Biol; Sigma Xi. *Res:* Lipid metabolism; autonomic drugs; hypolipemic drugs; feasability of altering collagen metabolism with drugs. *Mailing Add:* Hoffmann-La Roche Inc 340 Kingsland St Nutley NJ 07110

SALVADOR, ROMANO LEONARD, b Montreal, Que, Dec 12, 28; m 54; c 3. MEDICINAL CHEMISTRY. *Educ:* St Mary's Col, AB, 50; Univ Montreal, BSc, 54, MSc, 56; Purdue Univ, PhD(pharmaceut chem), 60. *Prof Exp:* From asst prof to assoc prof, 59-70, PROF PHARMACEUT CHEM & VDEAN FAC PHARM, UNIV MONTREAL, 70- *Mem:* Am Chem Soc; Can Soc Chemother; Chem Inst Can. *Res:* Cholinergic-anticholinergic drugs; cholinesterase regenerators; analgesics; central nervous system drugs; acetylenic drugs. *Mailing Add:* 417 Des Prairies Blvd Laval PQ H7N 2W7 Can

SALVADORI, ANTONIO, b Cesena, Italy, Apr 1, 41; m 64; c 3. COMPUTER SCIENCE, SOLID STATE PHYSICS. *Educ:* Nat Univ Ireland, BSc, 62, MSc, 63; McMaster Univ, PhD(physics), 68. *Prof Exp:* Lectr math & physics, Univ Col, Dublin, 62-63; asst prof comput & info sci, 67-73, ASSOC PROF COMPUT & INFO SCI, UNIV GUELPH, 67- *Concurrent Pos:* Vis lectr, Trinity Col, Dublin, 68-69, Wollongong Univ, Australia, 80, 82. *Mem:* Asn Comput Mach; Brit Comput Soc; Ital Asn Automatic Calculus. *Res:* Program profiles; biological simulation techniques; ecological modelling; computer assisted instruction; microcomputer applications. *Mailing Add:* Dept Comput & Info Sci Univ Guelph Guelph ON N1G 2W7 Can

SALVADORI, M(ARIO) G(IORGIO), b Rome, Italy, Mar 19, 07; nat US; m 75; c 1. ENGINEERING, PHYSICS. *Educ:* Univ Rome, DCE, 30, DrMath, 33, Libero Docente, 37. *Hon Degrees:* DrSc, Columbia Univ, 78. *Prof Exp:* Secy, Civil Eng Div, Nat Italian Res Coun, 34-38; lectr civil eng, Columbia Univ, 40-41, from instr to prof, 41-59, chmn, Archit Tech Div, 65-73, prof, Sch Archit, 59-75, James Renwick prof, 60-68, EMER PROF ARCHIT & JAMES RENWICK EMER PROF CIVIL ENG, COLUMBIA UNIV, 68-; CHMN BD, WEIDLINGER ASSOCS, 83- *Concurrent Pos:* Consult, Calculus Applns Inst, Italy, 34-38; asst prof, Univ Rome, 37-38; fel, Int Inst Cult Rels, 38; lectr, Princeton Univ, 55-60; partner, Paul Weidlinger, Consult Eng, 56-83; hon prof, Univ Minas Gerais; spec lectr, Columbia Univ, 68-90; hon chmn bd, Salvador Educ Ctr Built Environ, 65-90. *Mem:* Nat Acad Eng; hon mem Am Soc Civil Engrs; fel Am Soc Mech Engrs; fel NY Acad Sci; hon mem Am Inst Architects. *Res:* Theory of structures; applied mathematics and mechanics. *Mailing Add:* Two Beekman Pl New York NY 10022

SALVAGGIO, JOHN EDMOND, b New Orleans, La, May 19, 33; m 58; c 5. INTERNAL MEDICINE, IMMUNOLOGY. *Educ:* Loyola Univ, BS, 54; La State Univ, MD, 57; Am Bd Internal Med & Am Bd Allergy, dipl. *Prof Exp:* From instr to assoc prof med, La State Univ, 63-72; chmn Dept Med, 83-88, HENDERSON PROF MED, TULANE MED SCH, NEW ORLEANS, 75-; PROF MED, SCH MED, LA STATE UNIV, 72-; DIR IMMUNOL & CLIN ALLERGY UNIT, CHARITY HOSP, 64-, VCHANCELLOR RES, 88. *Concurrent Pos:* NIH res fel immunol & allergy, Dept Med, Mass Gen Hosp & Harvard Med Sch, 61-63; NIH res spec fel, Sch Med, Univ Colo, 72; dir USPHS Training Prog Clin Immunol & Allergy, Tulane Med Sch, 64- *Mem:* AAAS; Am Fedn Clin Res; Am Soc Clin Invest; Am Thoracic Soc; Am Asn Immunol; Asn Am Physicians. *Res:* Immediate and delayed hypersensitivity. *Mailing Add:* Dept Med 321 1700 Perd Tulane Sch Med 1430 Tulane Ave New Orleans LA 70112

SALVENDY, GAVRIEL, b Budapest, Hungary, Sept 30, 38; m 66; c 2. HUMAN-COMPUTER INTERACTION, COMPUTERIZED MANUFACTURING. *Educ:* Brunel Univ, UK, dipl, 64; Univ Birmingham, UK, dipl, 65, MSc, 66 & PhD(human factors), 68. *Prof Exp:* Asst prof, State Univ NY, Buffalo, 68-71; from assoc prof to prof, 71-84, NEC PROF INDUST ENG, PURDUE UNIV, 84-; PRES & CHIEF SCIENTIST, ERGOTECH INC, 85- *Concurrent Pos:* Fulbright distinguished prof, Tel Aviv Univ, Israel, 79-81; mem, Nat Res Coun, 85-86; founding ed, Int J Human-Computer Interaction, 88- & Int J Human Factors Mfg, 89-; chmn, Int Comn on Human Aspect in Computer, 86- & Int Conf on Human-Computer Interaction, 84- *Mem:* Nat Acad Eng; Am Psychol Asn; fel Human Factors Soc; fel Ergonomics Soc; Asn Comput Mach; Am Inst Indust Eng. *Res:* Human aspects in computing; human-computer interactive tasks and cognitive engineering in the design and use of expert systems. *Mailing Add:* Sch Indust Eng Grissom Hall Purdue Univ West Lafayette IN 47907

SALVESEN, ROBERT H, b Staten Island, NY, Jan 31, 24; m 48; c 3. ORGANIC CHEMISTRY, POLYMER CHEMISTRY. *Educ:* Wagner Col, BS, 48; Univ Buffalo, MA, 51; Polytech Inst Brooklyn, PhD(org chem), 58. *Prof Exp:* Proj leader petrol specialties, Tech Serv Labs, Socony Mobil Oil Co, NY, 52-59; res assoc, Esso Agr Chem Lab, Exxon Res & Eng Co, 59-71, Govt Res Lab, 71-79, Prod Res Div, 79-82; CONSULT, MGT OILS, SOLVENTS & HAZARDOUS WASTES, 83- *Mem:* Am Chem Soc. *Res:* Research and development of agricultural products; wax emulsions and coatings; specialty petroleum products and refinery by-products; new product development activities in polymers; design of oily waste treatment and disposal equipment; hazardous waste mgt; petroleum technologies. *Mailing Add:* Four Palermo Dr Tinton Falls NJ 07724

SALVETER, SHARON CAROLINE, b Pasadena, Calif, June 9, 49; m 84; c 1. INTELLIGENT SYSTEMS, DATABASE SYSTEMS. *Educ:* Univ Calif, San Diego, BS, 69; Univ Ore, MS, 71; Univ Wis, PhD(comput sci), 78. *Prof Exp:* Instr comput sci, Univ Ore, 71-73; asst prof comput sci, State Univ NY, Stony Brook, 78-82; asst prof, 82-85, ASSOC PROF COMPUT SCI, BOSTON UNIV, 85- *Concurrent Pos:* Prin investr, NSF, 79-81 & 83-, co prin investr, 81-83; consult, Bell Labs, 80, co-prin investr, 87-; sabbatical fel comput sci, Int Bus Mach, 81. *Mem:* Sigma Xi; Asn Comput Mach; Asn Comput Linguistics; Inst Elec & Electronics Engrs; Cognitive Sci Soc. *Res:* Investigation of computer learning mechanisms; applying artificial intelligence techniques in support of natural language front ends to databases. *Mailing Add:* Comput Sci Dept Boston Univ 111 Cummington St Boston MA 02215

SALVI, RICHARD J, b Chisholm, Minn, June 30, 46. PHYSIOLOGICAL PSYCHOLOGY. *Educ:* NDak State Univ, BS, 68; Syracuse Univ, PhD(psychol), 75. *Prof Exp:* Asst prof otolaryngol, Upstate Med Ctr, State Univ NY, 75-80; MEM FAC, UNIV TEX, DALLAS, 80- *Mem:* AAAS; Acoust Soc Am; Neurosci Asn Res Otolaryngol. *Res:* Auditory physiology; auditory psychophysics. *Mailing Add:* Commun Disorders 105 Park Hall State Univ NY Buffalo NY 14260

SALVIN, SAMUEL BERNARD, b Boston, Mass, July 10, 15; c 3. IMMUNOLOGY, MYCOLOGY. *Educ:* Harvard Univ, AB, 35, EdM, 37, AM, 38, PhD(biol), 41. *Prof Exp:* Asst, Radcliffe Col & Harvard Univ, 37-41; instr, Harvard Univ, 41-43; immunologist & mycologist, Rocky Mountain Lab, Nat Inst Allergy & Infectious Dis, 46-64; head immunol, Res Div, Ciba Pharmaceut Co, 65-67; PROF MICROBIOL & IMMUNOL, SCH MED, UNIV PITTSBURGH, 67- *Mem:* Bot Soc Am; Am Soc Microbiol; Am Asn Immunol; NY Acad Sci; Am Acad Microbiol. *Res:* Hypersensitivity and cellular immunity; antibody formation; immunological tolerance; autoimmune disease; immunology of pathogenic fungi; immunoregulation; lymphokines. *Mailing Add:* Dept Microbiol Scaife Hall Rm E1240 Univ Pittsburgh Sch Med Pittsburgh PA 15261

SALVO, JOSEPH J, b Lawrence, Mass, May 4, 58; m 87. MOLECULAR BIOLOGY. *Educ:* Harvard Univ, BA, 80; Yale Univ, MPhil, 82, PhD(molecular biophys & biochem), 87. *Prof Exp:* MOLECULAR BIOLOGIST, BIOL SCI LAB, GEN ELEC CORP RES & DEVELOP CTR, SCHENECTADY, NY, 88- *Concurrent Pos:* Adj asst prof, Dept Microbiol & Immunol, Albany Med Col, 89-, State Univ NY, Albany, 90- *Mem:* Am Chem Soc. *Res:* Genetic regulation of secondary metabolism and sporulation in the filamentous fungus Aspergillus parasiticus; elucidating the enzymatic pathways which produce specific para-hydroxylated aromatic compounds; kinship of early human ancestors from pre-Columbian South America and the prevalence of endemic infectious diseases and environmental pollutants; author of several publications. *Mailing Add:* 1155 Avon Rd Schenectady NY 12308

SALWEN, HAROLD, b New York, NY, Jan 30, 28; m 50; c 6. THEORETICAL PHYSICS, FLUID DYNAMICS. *Educ:* Mass Inst Technol, SB, 49; Columbia Univ, PhD(physics), 56. *Prof Exp:* Asst physics, Columbia Univ, 50-53; asst, Watson Lab, Int Bus Mach Corp, 53-55; res assoc statist mech, Syracuse Univ, 55-57; res fel solid state physics, Div Eng & Appl Physics, Harvard Univ, 57-59; asst prof, 59-65, assoc prof, 65-81, PROF PHYSICS, STEVENS INST TECHNOL, 81- *Concurrent Pos:* Vis scientist math dept, Imp Col Sci & Technol, London, 80 & Nat Maritime Inst, Teddington, Eng, 82; adj prof oceanog, Old Dominion Univ, 81-; vis prof math, Rensselaer Polytechnic Inst, 82. *Mem:* AAAS; Am Phys Soc; Sigma Xi. *Res:* Magnetic resonance; molecular structure; statistical mechanics of irreversible processes; kinetic theory; solid state theory; hydrodynamics; quantum-mechanical many-body problem. *Mailing Add:* 703 Riverview Ave Teaneck NJ 07666

SALWEN, MARTIN J, b Brooklyn, NY, Sept 21, 31; m 79; c 5. CLINICAL PATHOLOGY, LABORATORY ADMINISTRATION. *Educ:* City Col NY, BS, 53; State Univ NY, MD, 57. *Prof Exp:* Chief lab, USAF Hosp, Tachikawa, Japan, 64-66; dir path, Monmouth Med Ctr, 67-78; DIR PATH, KINGS COUNTY HOSP CTR, 79-; DIR CLIN PATH, UNIV HOSP BROOKLYN, 86- *Concurrent Pos:* Attend path, Yale-New Haven Hosp, 61-67; instr, Yale Univ Sch Med, 61-64, asst prof, 66-67, asst clin prof, 67-71; prof path, Hahnemann Med Col, 71-79; clin prof, State Univ NY Sci Ctr Brooklyn, 79-; counr, Am Soc Clin Pathologists, 88- *Mem:* Fel Am Soc Clin Pathologists; fel Col Am Pathologists; Am Asn Clin Chem; Am Asn Blood Bank; Asn Clin Scientists; NY Acad Sci. *Res:* Laboratory computers. *Mailing Add:* 934 Albemarle Rd Brooklyn NY 11218-2708

SALWIN, ARTHUR ELLIOTT, b Chicago, Ill, Feb 18, 48; m 77; c 2. SOFTWARE ENGINEERING, CHEMICAL PHYSICS. *Educ:* Univ Md, BS, 70; Princeton Univ, PhD(phys chem), 75. *Prof Exp:* Sr analyst comput sci, Xonics Inc, 75-76; sr staff scientist, Appl Physics Lab, Johns Hopkins Univ, 76-78; mem res staff, Riverside Res Inst, 78-80; mem tech staff, 81-83, GROUP LEADER, MITRE CORP, 83- *Concurrent Pos:* Inst Elect & Electronics Engrs comt on software eng; canvassee Ada 9x, comput rev, Asn Comput Mach; lectr phys chem, Georgetown Univ Grad Sch, 80-82;

computer prog instr, Fairfax County Adult Educ, 80- *Mem:* Asn Comput Mach. *Res:* Development of prototype air traffic control; studies in state-of-the-art software engineering and Ada practices; applications of computer technology to scientific problems. *Mailing Add:* Ten Sunnymeade Ct Potomac MD 20854

SALWIN, HAROLD, b Kansas City, Mo, Nov 24, 15; m 43; c 2. FOOD TECHNOLOGY. *Educ:* Univ Chicago, BS, 41. *Prof Exp:* Chemist, Tenn Valley Authority, 42-43, Explosives Res Lab, US Bur Mines, 43-45 & US Customs Lab, 45-48; res chemist, Armed Forces, Qm Food & Container Inst, 48-58, head food biochem lab, 58-61, actg chief chem br, 61; res chemist, Div Food Chem, Food & Drug Admin, 61-64, head decomposition & preservation sect, 64-71, chief protein & cereal prod br, 71-79; RETIRED. *Honors & Awards:* Outstanding Employee Award, Dept Army, 60; Rohland A Isker Award, 62. *Mem:* Am Chem Soc; Inst Food Technol; fel Asn Official Anal Chemists. *Res:* Food technology; chemistry of food deterioration; food dehydration; analytical methods. *Mailing Add:* 706 Kerwin Rd Silver Spring MD 20901-4621

SALYER, DARNELL, analytical chemistry; deceased, see previous edition for last biography

SALYERS, ABIGAIL ANN, MOLECULAR BIOLOGY OF ANAEROBES. *Educ:* George Washington Univ, PhD(physics), 69. *Prof Exp:* ASSOC PROF MICROBIOL, UNIV ILL, 78- *Res:* Polysaccharide catabolism. *Mailing Add:* Dept Microbiol Univ Ill 407 S Goodwin Urbana IL 61801

SALZANO, FRANCIS J(OHN), b Brooklyn, NY, Mar 23, 33; m 58; c 3. CHEMICAL ENGINEERING, MATERIAL SCIENCE. *Educ:* City Univ New York, BChE, 55. *Prof Exp:* Aeronaut res scientist, Lewis Flight Propulsion Lab, Cleveland, 55-57; SR CHEM ENGR, BROOKHAVEN NAT LAB, 56- *Mem:* AAAS. *Res:* Fused salt chemistry; surface adsorption; inorganic carbon chemistry; solid state electrolytes; chemistry of the alkali-metals, especially sodium; atmospheric chemistry; synthetic clean fuels; industrial energy conservation; batteries, fuel cells and advanced materials; strategies for long range research; research/technology transfer to industry. *Mailing Add:* 144 Avery Ave Patchogue NY 11772

SALZANO, FRANCISCO MAURO, b Cachoeira do Sul, Brazil, July 27, 28; m 52; c 2. HUMAN GENETICS. *Educ:* Fed Univ Rio Grande do Sul, Brazil, BS, 50, Univ Sao Paulo, PhD (genetics), 55. *Prof Exp:* From instr to assoc prof, Fed Univ Rio Grand do Sul, 52-81, researcher, Inst Natural Sci, 52-62, head, Genetics Sect, 63-68, dir inst, 68-71, PROF GENETICS, INST BIOSCI, FED UNIV RIO GRANDE DU SUL, 81- *Concurrent Pos:* Rockefeller Found fel, Univ Mich, 56-57; mem bd dirs, Latin Am Asn Genetics, 72-76 & 83-85; pres, Latin Am Asn Human Biol, 90-92. *Honors & Awards:* Medal, Brazilian Asn Advan Sci, 73. *Mem:* Am Soc Human Genetics; Am Asn Phys Anthrop; Int Union Anthrop & Ethnol Sci (vpres, 78-83 & 83-88); Genetic Soc Chile; Brazilian Acad Sci; hon mem Royal Anthrop Inst Gt Brit Ireland. *Res:* Blood groups, serum proteins, hemoglobin and enzyme types; characteristics of anthropological interest; medical genetics; DNA polymorphisms. *Mailing Add:* Dept Genetics Univ Fed Rio Grande du Sul Caixa Postal 15053 Porto Alegre 91501 Brazil

SALZARULO, LEONARD MICHAEL, b Montclair, NJ, Oct 11, 27; m 71; c 1. PHYSICS, CHEMICAL ENGINEERING. *Educ:* Newark Col Eng, BS, 51, MS, 53; Polytech Inst Brooklyn, PhD(chem eng), 66. *Prof Exp:* Head qual control lab, Sun Chem Corp, NJ, 51-53; instr chem, Newark Col Eng, 53-55, chem eng, 55-56; proj engr, S B Penick & Co, NJ, 56-59; asst prof appl mech, Newark Col Eng, 59-60; res assoc chem eng, Polytech Inst Brooklyn, 60-61; asst prof appl mech, 61-63, from asst prof to assoc prof physics, 63-67, asst chmn dept physics, 63-68, assoc chmn, 68-74, chmn, 74-79, PROF PHYSICS, NJ INST TECHNOL, 67- *Concurrent Pos:* Consult, Bendix Corp, NJ, 65-66 & Deluxe-Reading Corp, NJ, 65-66. *Mem:* NY Acad Sci; Am Asn Physics Teachers. *Res:* Transport phenomena in ion exchange membranes; electrochemistry, especially ionic transport phenomena; biophysics; physics apparatus; educational physics. *Mailing Add:* Dept Physics NJ Inst Technol 323 High St Newark NJ 07102

SALZBERG, BERNARD, b New York, NY, July 22, 07; m 41. ELECTRONICS. *Educ:* Polytech Univ, EE, 29, MEE, 33, DEE, 41. *Prof Exp:* Engr res & develop, RCA Commun, Inc, 29-31 & RCA Mfg Co, 31-41; assoc supt, Radio Div, Naval Res Lab, 41-52, assoc supt & consult, Electronics Div, 52-56; chief scientist, AIL Div, Cutler-Hammer, Inc, 56-72; CONSULT, 72- *Honors & Awards:* Mod Pioneer Award, Nat Asn Mfrs, 40; Diamond Award, Inst Radio Engrs, 55; Meritorious Civilian Award, US Navy, 56. *Mem:* Am Phys Soc; fel Inst Elec & Electronics Engrs. *Res:* Semiconductors; microwaves; electrophysics. *Mailing Add:* 15 Ridge Rock Lane East Norwich NY 11732

SALZBERG, BETTY, b Denver, Colo, Jan 19, 44; m 78; c 1. DATA BASE. *Educ:* Univ Calif, Los Angeles, BA, 64; Univ Mich, MA, 66, PhD(math), 71. *Prof Exp:* From asst prof to assoc prof math, 71-82, assoc prof comput sci, 82-, PROF, NORTHWESTERN UNIV, 90- *Mem:* Am Math Soc; Asn Comput Mach. *Res:* Finite groups of Lie type; search structures for data bases. *Mailing Add:* Col Comput Sci Northeastern Univ Boston MA 02115

SALZBERG, BRIAN MATTHEW, b New York, NY, Sept 4, 42. NEUROBIOLOGY, BIOPHYSICS. *Educ:* Yale Univ, BS, 63; Harvard Univ, AM, 65, PhD(physics), 71. *Prof Exp:* Res asst high energy physics, Harvard Univ, 65-71; fel neurobiol, Sch Med, Yale Univ, 71-74, res assoc physiol, 74-75; from asst prof to assoc prof, 75-82, PROF PHYSIOL, SCH MED, UNIV PA, 82- *Concurrent Pos:* Investr neurobiol, Marine Biol Lab, Woods Hole, 72-, mem corp, 74-, trustee, 80-84, 88-91; mem, Inst Neurol Sci, Univ Pa, 76-; guest fel, Royal Soc, 91; mem coun, Soc Gen Physiologists. *Honors & Awards:* Marine Biol Lab Prize, 81; Arturo Rosenblueth Prof, Mex City, 87. *Mem:* Biophys Soc, Exec Bd; Soc Neurosci; fel Am Phys Soc; Soc Gen Physiologists; fel AAAS; fel Japan Soc Prom Sci. *Res:* Development of molecular probes of membrane potential and their application to neurophysiology; optical recording of neuronal activity; excitation-secretion coupling; light scattering. *Mailing Add:* Dept Physiol B-400 Richards Bldg Philadelphia PA 19104-6085

SALZBERG, DAVID AARON, b Kansas City, Mo, May 5, 20; m 44; c 4. BIOCHEMISTRY. *Educ:* Univ Chicago, BS, 40; Univ Calif, MS, 48; Stanford Univ, PhD(biochem), 50. *Prof Exp:* Chemist, Ala Ord Works, 42-43 & US Engr Dist, Hawaii, 43-45; USPHS res asst, Stanford Univ, 49-50; biochemist, Palo Alto Med Res Found, 52-62, head, Basic Cancer Res Div, 58-62; asst res prof biochem, Univ San Francisco, 62-64; dir res, Arequipa Found, 63-67; pres, Tahoe Col, 67-68; FEL NEUROBIOL, MED SCH, STANFORD UNIV, 69- *Concurrent Pos:* Am Cancer Soc fel, 50-52 & scholar cancer res, 53-58; mem adv bd, Miramonte Found Ment Health & Great Books Found. *Mem:* Am Asn Cancer Res; Am Chem Soc. *Res:* Diabetes and hormone relationships in cancer; mutagenesis; genetic changes in cancer induction; azo-dye hepatocarcinogenesis; maternal and foster nursing in carcinogenesis; nutritional evaluation of biochemical intermediates; physiology of stress and emotion; brain hormones; nervous system regulation of growth. *Mailing Add:* 815 N Humboldt Ave No 303 San Mateo CA 94401

SALZBERG, HUGH WILLIAM, b New York, NY, June 27, 21; m 52. PHYSICAL CHEMISTRY. *Educ:* City Col New York, BS, 42; NY Univ, MS, 47, PhD(phys chem), 50. *Prof Exp:* Res chemist, US Naval Res Lab, 50-53; res chemist, Columbia Univ, 53-54; from instr to assoc prof, 54-70, PROF CHEM, CITY COL NEW YORK, 70- *Mem:* AAAS; Am Chem Soc; Electrochem Soc; Sigma Xi. *Res:* Electrodeposition and preparative electrochemistry; kinetics. *Mailing Add:* Dept Chem City Col New York New York NY 10031

SALZBRENNER, RICHARD JOHN, b Douglas, Ariz, July 25, 48; m 76. MECHANICAL PROPERTIES, ALLOY DEVELOPMENT. *Educ:* Univ Notre Dame, BS, 70; Univ Denver, PhD(mat sci), 73. *Prof Exp:* Res assoc, Mass Inst Technol, 73-78; MEM TECH STAFF, WESTERN ELEC, SANDIA NAT LAB, 78- *Mem:* Am Soc Metals; Am Inst Metall Engrs. *Res:* Mechanical property measurement; structure-property relationships in ferrous and non-ferrous alloys; low alloy steel development; martensitic transformations; corrosion fatigue; formability; fracture toughness; internal friction. *Mailing Add:* 1400 Caballero Dr SE Albuquerque NM 87123

SALZENSTEIN, MARVIN A(BRAHAM), b Chicago, Ill, May 12, 29; m 58; c 2. SAFETY ENGINEERING, MACHINE DESIGN. *Educ:* Ill Inst Technol, BS, 51. *Prof Exp:* Asst res mech engr, Armour Res Found, Ill Inst Technol, 49-51; sales engr, Sci Instruments, W H Kessel & Co, 53-57; assoc engr, Walter C McCrone Assocs, Inc, 57-61; PRES, POLYTECH, INC, 61- *Concurrent Pos:* Res engr, White Sands Proving Grounds. *Mem:* Am Soc Mech Engrs; Am Soc Safety Engrs; Am Gas Asn; Nat Fire Protection Asn; Am Nat Standards Inst. *Res:* Industrial zoning; combustion; safety. *Mailing Add:* 3740 W Morse Ave Chicago IL 60645

SALZER, JOHN M(ICHAEL), b Vienna, Austria, Sept 12, 17; nat US; m 44; c 4. ELECTRICAL ENGINEERING. *Educ:* Case Inst Technol, BS, 47, MS, 48; Mass Inst Technol, ScD(elec eng), 51. *Prof Exp:* Instr elec eng, Case Inst Technol, 47-48; res assoc, Digital Comput Lab, Mass Inst Technol, 48-51; mem tech staff, Res & Develop Labs, Hughes Aircraft Co, 51-54; dir systs, Res Labs, Magnavox Co, 54-59; dir intellectronics labs, Thompson Ramo Wooldridge, Inc, 59-63; vpres tech & planning, Librascope Group, Gen Precision, Inc, 63-68; pres, Salzer Technol Enterprises, 68-72; prin, Darling & Alsobrook, 72-75; SR VPRES, DARLING, PATERSON & SALTZER, 75-; PRES, SALZER TECHNOL ENTERPRISES, INC. *Concurrent Pos:* Lectr, Univ Calif, Los Angeles, 52-62. *Mem:* Int Soc Hybrid Microelectronics; Semiconductor Equip & Mat Inst; Sigma Xi; fel Inst Elec & Electronics Engrs; fel Inst Advan Eng; Semiconductor Equip & Mat Inst. *Res:* Digital computers and control systems; sampled data systems; information systems; electronic components. *Mailing Add:* 909 Berkeley St Santa Monica CA 90403

SALZMAN, EDWIN WILLIAM, b St Louis, Mo, Dec 11, 28; m 54; c 3. SURGERY. *Educ:* Wash Univ, AB, 50, MD, 53; Harvard Univ, MA, 69. *Prof Exp:* Asst in surg, Mass Gen Hosp, 61-65, asst surgeon, 65-66; instr, 61-65, assoc prof, 69-71, PROF SURG, HARVARD MED SCH, 72-; SURGEON & ASSOC DIR SURG SERV, BETH ISRAEL HOSP, BOSTON, 66-; SR RES ASSOC, MASS INST TECHNOL, 67- *Concurrent Pos:* NIH fel, Radcliffe Infirmary, Oxford Univ, 59; Am Cancer Soc clin fel, Mass Gen Hosp, Boston, 60-61, Med Found, Inc fel, 62-65; Markle scholar acad med, 68; assoc, Univ Seminar Biomat, Columbia Univ, 67-, chmn, 68; consult, Am Nat Res Cross, 68- & thrombosis adv comt, NIH, 70-; mem steering comt, Harvard Univ-Mass Inst Technol Prog Health Sci & Technol, 70-; dep ed, New Eng J Med. *Mem:* Am Physiol Soc; Soc Univ Surg; Soc Vascular Surg; Am Surg Asn; Am Soc Clin Invest. *Res:* Hemostasis and thrombosis; biochemistry of blood platelets; surgical physiology. *Mailing Add:* Dept Surg Harvard Med Sch Beth Israel Hosp 330 Brookline Ave Boston MA 02215

SALZMAN, GARY CLYDE, b Palo Alto, Calif, May 25, 42; m 65. INTELLIGENT SYSTEMS. *Educ:* Univ Calif, Berkeley, AB, 65; Univ Ore, MS, 68, PhD(nuclear physics), 72. *Prof Exp:* Peace Corps teacher physics & math, Ghana, 65-67; mem staff & presidential intern fel physics, 72-73, staff mem, Biophys & Instrumentation Group, 73-86, BIOPHYS SECT LEADER, BIOCHEM BIOPHYS GROUP, LIFE SCI DIV, LOS ALAMOS NAT LAB, 87- *Mem:* AAAS; Biophys Soc; Am Phys Soc; Optical Soc Am; Am Asn Artificial Intel; Soc Anal Cytol. *Res:* Light scattering from biological cells, automated cytology, instrumentation and techniques for cell identification and cancer diagnosis, knowledge-based systems, artificial neural networks. *Mailing Add:* 108 Sierra Vista Dr Los Alamos NM 87544

SALZMAN, GEORGE, b Newark, NJ, Sept 8, 25; m 48; c 2. THEORETICAL PHYSICS. *Educ:* Brooklyn Col, BS, 49; Univ Ill, PhD(physics), 53. *Prof Exp:* Instr physics, Univ Ill, 53-55; res assoc, Univ Rochester, 55-56, asst prof, 55-58; from asst prof to assoc prof, Univ Colo, 58-65; PROF PHYSICS, UNIV MASS, BOSTON, 65- *Concurrent Pos:* Ford Found fel, Theory Group, Europ Orgn Nuclear Res, Switz, 58-59; Fulbright res scholar, Frascati Nat Lab, Italy, 61-62; NSF res grant, 62-; vis assoc prof, Northeastern Univ, 64-65. *Mem:* Am Phys Soc; AAAS. *Res:* Development of a science for humane survival; theory of elementary particles; high energy scattering theory; electromagnetic structure of nucleons; theory of relativity. *Mailing Add:* Dept Physics Univ Mass Boston MA 02125

SALZMAN, LEON, b New York, NY, July 10, 15; m 50; c 4. PSYCHIATRY. *Educ:* City Col New York, BS, 35; Royal Col Physicians & Surgeons, MD, 40. *Prof Exp:* Prof clin psychiat, Med Sch, Georgetown Univ, 45-67; prof psychiat, Med Sch, Tulane Univ, La, 67-69; clin prof psychiat, Albert Einstein Col Med, 69-75; PROF CLIN PSYCHIAT, MED SCH, GEORGETOWN UNIV, 75- *Concurrent Pos:* Dep dir, Bronx State Hosp, 70-75; pvt pract psychiat & psychoanal, 75- *Mem:* Am Acad Psychoanal (pres); Am Psychiat Asn; Am Psychoanal Asn. *Res:* Sex behavior; obsessive and compulsive states. *Mailing Add:* 6625 Braeburn Pkwy Bethesda MD 20817

SALZMAN, NORMAN POST, b New York, NY, Aug 14, 26; m 54; c 3. VIROLOGY. *Educ:* City Col New York, BS, 48; Univ Mich, MS, 49; Univ Ill, PhD(biochem), 53. *Prof Exp:* Res asst, Squibb Inst Med Res, 49-50; biochemist, Nat Heart Inst, 53-55; mem, Virol Study Sect, 71-73, BIOCHEMIST, NAT INST ALLERGY & INFECTIOUS DIS, 55-, CHIEF LAB BIOL VIRUSES, 68- *Concurrent Pos:* Ed, J Virol, 67-75; vis prof, Univ Geneva, 73-74; prof lectr, Georgetown Univ. *Mem:* Am Soc Cell Biol; Am Soc Biol Chem; Am Asn Immunol; Am Soc Microbiol. *Res:* Virus replication; viral oncology. *Mailing Add:* Dept Microbiol Bldg 3900 Rm LM12 Preclin Georgetown Univ Sch Med Reservoir Rd NW Washington DC 20007

SALZMAN, STEVEN KERRY, b New York, NY, Feb 19, 52; m 74; c 2. PHARMACOLOGY, NEUROPHYSIOLOGY. *Educ:* Univ Fla, BS, 74; Univ Conn, PhD(neuropharmacol), 79. *Prof Exp:* ASSOC SCIENTIST & DIR MED SCI, ALFRED I DU PONT INST; ASSOC PROF, THOMAS JEFFERSON UNIV. *Mem:* Soc Neurosci; Neurotrauma Soc; Am Soc Pharmacol Exp Therapeut. *Res:* Characterization, detection and prevention of acute responses to spinal trauma and ischemia; control of surgical stress. *Mailing Add:* Alfred I du Pont Inst Nemours Found Res Dept PO Box 269 Wilmington DE 19899

SALZMAN, WILLIAM RONALD, b Cutbank, Mont, Feb 27, 36; c 2. THEORETICAL PHYSICAL CHEMISTRY. *Educ:* Univ Calif, Los Angeles, BS, 59, MS, 64, PhD(chem), 67. *Prof Exp:* Res metallurgist, Southern Res Inst, Ala, 61-62; from asst prof to assoc prof, 67-79, from actg head dept to head dept, 77-83, PROF CHEM, UNIV ARIZ, 79- *Concurrent Pos:* Am Chem Soc-Petrol Res Fund grant, Univ Ariz, 68-70. *Mem:* AAAS; Am Phys Soc; Sigma Xi; Am Asn Univ Professors; Am Chem Soc. *Res:* Application of quantum theory to problems of chemical interest; semiclassical and quantum radiation theory. *Mailing Add:* Dept Chem Univ Ariz Tucson AZ 85721

SAM, JOSEPH, b Gary, Ind, Aug 15, 23; m 45; c 3. PHARMACEUTICAL CHEMISTRY. *Educ:* Univ SC, BS, 48; Univ Kans, PhD(pharmaceut chem), 51. *Prof Exp:* Instr org chem, Univ SC, 47-48; asst pharmaceut chem, Univ Kans, 48-49; sr res chemist, McNeil Labs, Inc, 51-54, Bristol Labs, 55-57 & E I du Pont de Nemours & Co, 57-59; assoc prof pharmaceut chem, 59-61, chmn dept, 63-69, PROF PHARMACEUT CHEM, UNIV MISS, 61-, DEAN GRAD SCH & DIR UNIV RES, 68-, ASSOC VCHANCELLOR RES, 81- *Concurrent Pos:* Fulbright lectr, Cairo Univ, 65-66; mem pharm rev comt, Bur Health Manpower Educ, Dept Health, Educ & Welfare, 67-71; mem exec comt, Coun Res Policy & Grad Educ, Nat Asn State Univs & Land Grant Cols, 69-71 & 74-76. *Honors & Awards:* Found Res Award Pharmaceut & Med Chem, Am Pharmaceut Asn, 68. *Mem:* Am Chem Soc; Am Pharmaceut Asn; fel Acad Pharmaceut Sci; Am Asn Cols Pharm. *Res:* Medicinal chemistry. *Mailing Add:* Rte 3 Box 141G Coldwater MS 38618

SAMAAN, NAGUIB A, b Girga, Egypt, Apr 2, 25; m 61; c 5. ENDOCRINOLOGY. *Educ:* Univ Alexandria, BA, 46, MB & ChB, 51, DM, 53; Univ London, PhD, 64; FRCP. *Prof Exp:* Intern med, Univ Alexandria, 52-54, resident, 54-56; postgrad training, Chest Inst, Brompton Hosp, London, Eng, 56 & gen med, Postgrad Med Sch London, 56; clin asst endocrinol, Dept Endocrinol & Therapeut, Royal Infirmary, 57; resident med, North Cambridge Hosp Eng, 58; clin asst prof, Postgrad Med Sch London, 60-64; res assoc endocrinol & asst physician, Western Reserve Univ, 64-66; asst prof endocrinol, Univ Iowa, 66-69; assoc prof med & physiol & assoc internist chief, 69-73, PROF MED & PHYSIOL & INTERNIST, SECT ENDOCRINOL, UNIV TEX M D ANDERSON HOSP & TUMOR INST HOUSTON, 73- & PROF MED, UNIV TEX MED SCH, 76- *Concurrent Pos:* Brit Med Res Coun sr res fel, Postgrad Med Sch London, 60-64; physician, Vet Admin Hosp, Iowa City, Iowa, 66-69; mem staff, Univ Tex Grad Sch Biomed Sci Houston, 69; consult, Hermann Gen Hosp, 70. *Mem:* Fedn Am Socs Exp Biol; Am Physiol Soc; Endocrine Soc; Am Diabetes Asn; Am Fedn Clin Res. *Res:* Diagnosis, management and investigation of diabetes and endocrine disorders; investigations of normal and abnormal pregnancy; metabolic and endocrine changes associated with tumors. *Mailing Add:* Univ Tex Sys Cancer Ctr M D Anderson Hosp Tumor Inst 1515 Holcombe Houston TX 77030

SAMAGH, BAKHSHISH SINGH, b Abohar, India, Oct 1, 38; Can citizen; m 66; c 2. VETERINARY MEDICINE, SEROLOGY & IMMUNOLOGY. *Educ:* Panjab Univ, BVSc & AH, 60, MSc, 63; Univ Guelph, MSc, 68, PhD(serol), 72, DVM, 75. *Prof Exp:* Dist vet, Panjab Govt, India, 60-61; lectr microbiol, various univs, India, 63-66; fel, Univ Guelph, 66-75; vet scientist, 75-78, HEAD DIAG SEROL, ANIMAL DIS RES INST, AGR CAN, 78- *Mem:* Can Soc Immunol; Can Vet Med Asn; Am Asn Vet Lab Diagnosticians; Am Asn Vet Immunologists. *Res:* Research and development on immune response and sero-diagnosis of bacterial, viral and protozoan diseases of animals. *Mailing Add:* Three Arbuckle Crescent Nepean ON K2H 8P9 Can

SAMANEN, JAMES MARTIN, b Detroit, Mich, June 17, 47. PEPTIDE CHEMISTRY, SYNTHETIC ORGANIC CHEMISTRY. *Educ:* Kalamazoo Col, BA, 69; Univ Mich, PhD(org chem), 75. *Prof Exp:* Res assoc, Mass Inst Technol, 75-77; RES CHEMIST, BIOPRODS DEPT, BECKMAN INSTRUMENTS INC, 77- *Mem:* Am Chem Soc. *Res:* Synthesis and chemistry of peptides, alkaloids, heterocycles, natural products. *Mailing Add:* Smith Kline & French L412 PO Box 1539 King of Prussia PA 19406-0939

SAMARA, GEORGE ALBERT, b Lenanon, Dec 5, 36; US citizen. PHYSICS, CHEMICAL ENGINEERING. *Educ:* Univ Okla, BS, 58; Univ Ill, Urbana, MS, 60, PhD(chem eng, physics), 62. *Prof Exp:* Staff mem phys res, Sandia Labs, 62-63, div supvr, High Pressure Physics Div, 67-71, dept mgr, condensed matter & device physics, 79-83, dept mgr, Physics of Solids Res, 71-89, dept mgr, Condensed Matter & Surface Sci Res, 83-89, DEPT MGR, CONDENSED MATTER RES, SANDIA LABS, 89- *Concurrent Pos:* US Army Signal Corps, 63-65. *Honors & Awards:* Ipatieff Prize, Nat Acad Eng, 74. *Mem:* AAAS; Am Phys Soc; Mat Res Soc; Am Inst Chem Engrs. *Res:* Effects of high pressure and temperature on the physical properties of solids especially ferroelectric, ferromagnetic and semiconductor properties. *Mailing Add:* Dept 1150 Sandia Labs Albuquerque NM 87185

SAMAROO, WINSTON R, electrical engineering, solid state physics; deceased, see previous edition for last biography

SAMBORSKI, DANIEL JAMES, b Hamton, Sask, Aug 9, 21; m 51; c 2. PLANT PATHOLOGY, PLANT PHYSIOLOGY. *Educ:* Univ Sask, BSA, 49, MSc, 51; McGill Univ, PhD(plant path), 55. *Prof Exp:* Res assoc plant physiol, Univ Sask, 53-56; RES SCIENTIST PLANT PATH, AGR CAN RES STA, 56- *Concurrent Pos:* Assoc ed jour, Am Phytopath Soc, 76-79. *Mem:* Fel Royal Soc Can; Can Phytopath Soc; Can Soc Plant Physiologists; Am Phytopath Soc; Sigma Xi. *Res:* Genetics and biochemistry of host-parasite interactions. *Mailing Add:* Agr Can Res Sta 195 Dafoe Rd Winnipeg MB R3T 2M9 Can

SAMEJIMA, FUMIKO, b Tokyo, Japan, Dec 25, 30. MATHEMATICAL PSYCHOLOGY & STATISTICS, PSYCHOMETRICS. *Educ:* Keio Univ, Tokyo, BA, 53, MA, 56, PhD(psychol), 65. *Prof Exp:* Res psychologist, Educ Testing Serv, 66-67; res fel, Psychometric Labs, Univ NC, 67-68; asst prof psychol, Univ NB, Can, 68-70; assoc prof, Bowling Green State Univ, 70-73; PROF PSYCHOL, UNIV TENN, KNOXVILLE, 73- *Concurrent Pos:* Nat Res Coun Can grant, 69-70; Off Naval Res grant, 77-; consult ed, Appl Psychol Measurement, 75-; assoc ed, Educ Statist, 78-81; bd trustees, Psychometric Soc, 90. *Mem:* Sigma Xi; Psychometric Soc; Am Statist Soc; Am Educ Res Asn; Am Asn Univ Professors. *Res:* Mathematical statistics and psychometrics; mathematical model buildings in many applied areas. *Mailing Add:* Univ Tenn Dept Psychol 310B Austin Peay Bldg Knoxville TN 37996-0900

SAMELSON, HANS, b Strassburg, Ger, Mar 3, 16; nat US; m 40, 56; c 3. MATHEMATICS. *Educ:* Swiss Fed Inst Technol, DSc(math), 40. *Prof Exp:* Mem, Inst Advan Study, 41-42, 52-54 & 60-61; instr math, Univ Wyo, 42-43; asst prof, Syracuse Univ, 43-46; from asst prof to prof, Univ Mich, 46-60; PROF MATH, STANFORD UNIV, 60- *Mem:* Am Math Soc; Math Asn Am. *Res:* Topology of group manifolds; differential geometry. *Mailing Add:* Stanford Univ Stanford CA 94305

SAMELSON, LAWRENCE ELLIOT, b Chicago, Ill, Apr 18, 51; m 80; c 2. BIOCHEMICAL PATHWAYS. *Educ:* Univ Rochester, BA, 72; Yale Univ, MA, 77. *Prof Exp:* Resident med, Univ Chicago Hosp & Clins, 77-80; fel, Lab Immunol, Nat Inst Allergy & Infectious Dis, NIH, 80-83, expert, 83-85, sr staff fel immunol, Cell Biol & Metab Br, Nat Inst Child Health & Develop, 85-88, SR INVESTR IMMUNOL, CELL BIOL & METAB BR, NAT INST CHILD HEALTH & HUMAN DEVELOP, NIH, 88- *Mem:* Am Soc Clin Invest; Am Asn Immunologists; AAAS; Am Fedn Clin Res. *Res:* How the T cell receptor is coupled to intracellular signalling and identify the components of these biochemical pathways. *Mailing Add:* Cell Biol & Metab Br Nat Inst Child Health & Human Develop Bldg 18T Rm 101 Bethesda MD 20892

SAMES, GEORGE L, zoology, for more information see previous edition

SAMES, RICHARD WILLIAM, b Louisville, Ky, Apr 13, 28; m 55; c 5. BACTERIOLOGY. *Educ:* Ind Univ, AB, 51, MA, 54, PhD, 56. *Prof Exp:* Assoc prof biol & chmn dept, Bellarmine Col, Ky, 56-66, prof biol & dir sci develop, 66-68; dean col, Benedictine Col, 68-72; dean natural sci, 72-75, asst vpres acad affairs, 75-78, actg vpres acad affairs, 78-79, PROF BIOL, SANGAMON STATE UNIV, 79- *Concurrent Pos:* Dir, Instructional Sci Equip Prog, NSF, 65-66, consult, 66-, consult, Col Sci Improv Prog, 71-74; consult, Developing Insts Prog, US Off Educ, 68-74; gov task force on sci & technol, 79-80. *Mem:* Am Soc Microbiol; AAAS. *Res:* Bacterial viruses for anaerobic bacteria; science and public policy. *Mailing Add:* Dept Biol Sangamon State Univ Springfield IL 62794

SAMET, PHILIP, b New York, NY, Jan 30, 22; m 47; c 3. INTERNAL MEDICINE, CARDIOLOGY. *Educ:* NY Univ, BA, 42, MD, 47. *Prof Exp:* Intern, Mt Sinai Hosp, New York, 47-48; resident internal med, Bronx Vet Hosp, 48-51; from instr to assoc prof, 55-70, PROF MED, SCH MED, UNIV MIAMI, 70- *Concurrent Pos:* Res fel, Cardiopulmonary Lab, Bellevue Hosp, 51-53; chief, Div Cardiol, Mt Sinai Hosp, Miami Beach, Fla, 55- *Mem:* Am Thoracic Soc; Am Physiol Soc; Am Heart Asn; AMA; Am Col Cardiol. *Res:* Cardiac and pulmonary physiology. *Mailing Add:* Dept Med Univ Miami Sch Med Miami Beach FL 33101

SAMFIELD, MAX, b Memphis, Tenn, Apr 20, 18; m 44; c 4. CHEMICAL ENGINEERING. *Educ:* Rice Inst, BS, 40; Univ Tex, MS, 41, PhD(chem eng), 45. *Prof Exp:* Chem engr, Tex, 45-47; unit engr, Servel, Inc, Ind, 47-52; supvr eng res & develop, Liggett & Myers Inc, 52-58, asst to dir res, 58-62, sr asst dir res, 62-73; prof officer, US Environ Protection Agency, 73-80; CONSULT, 77- *Mem:* AAAS; Sigma Xi. *Res:* Production of acetylene from gaseous hydrocarbons; carbon black; absorption refrigeration; physico-chemical properties of tobacco technology; environmental research; pollution control. *Mailing Add:* 915 W Knox St Durham NC 27701

SAMI, SEDAT, b Istanbul, Turkey, Oct 23, 28; m 58; c 2. FLUID MECHANICS, HYDRAULICS. *Educ:* Tech Univ Istanbul, MSCE, 51; Univ Iowa, MS, 57, PhD(fluid mech), 66. *Prof Exp:* Design engr, Chase T Main, Inc, Turkey, 51-53, asst chief engr, 54-56; chief design engr, Eti Yapi Ltd, Turkey, 58-60, tech dir, 60-62; asst prof civil eng, Middle East Tech Univ, Ankara, 62-63; from asst prof to assoc prof, 66-72, actg chmn, Dept Eng Mech & Mat, 78-79, PROF FLUID MECH & HYDRAUL, SOUTHERN ILL UNIV, 72- *Mem:* Fel Am Soc Civil Engrs; Int Asn Hydraul Res; Sigma Xi. *Res:* Turbulence, fluctuating velocities and pressures; turbulent flows; electro-osmotic dewatering of ultrafine coal; aircraft refueling systems. *Mailing Add:* Dept Civil Eng Southern Ill Univ Carbondale IL 62901-6603

SAMIOS, NICHOLAS PETER, b New York, NY, Mar 15, 32; m 58; c 3. PHYSICS. *Educ:* Columbia Univ, AB, 53, PhD(physics), 57. *Prof Exp:* Instr physics, Columbia Univ, 56-59; from asst physicist to physicist, 59-68, chmn dept physics, 75-81, dep dir high energy & nuclear physics, 81-82, SR PHYSICIST, BROOKHAVEN NAT LAB, 62-, DIR, 82- *Honors & Awards:* E O Lawrence Mem Award, 80; Phys & Math Sci Award, NY Acad Sci, 80. *Mem:* Nat Acad Sci; fel Am Phys Soc; fel Am Acad Arts & Sci; AAAS. *Res:* High energy particle and nuclear physics. *Mailing Add:* Dir Off 460 Brookhaven Nat Lab Upton NY 11973-5000

SAMIR, URI, b Tel-Aviv, Israel, Sept 14, 30; m 67. SPACE PHYSICS. *Educ:* Hebrew Univ, Israel, MSc, 60; Univ London, PhD(physics), 67. *Prof Exp:* Teaching asst physics, Israel Inst Technol, 58-60; researcher, Israeli Defence Syst, 60-62; res assoc ionospheric physics, Univ Col, Univ London, 62-67; sci consult Gemini 10 & 11 spacecraft, Electro-Optical Systs, Inc, Calif, 67-68; assoc res physicist, 68-69, RES PHYSICIST, SPACE PHYSICS RES LAB, UNIV MICH, ANN ARBOR, 69- *Concurrent Pos:* Prof, dept geophys & planetary sci, Tel-Aviv Univ, Isreal, 74-85; mem, Sci Adv Bd Plasma Physics Exp on Future Space-Shuttles & Space Sta, 72-; mem steering & working group, Atmospheric-Magnetospheric & Plasmas in Space, 73-; chmn, Plasma Interaction Sect, NASA-Aircraft Multispectral Photog Syst Sci Definition Working Group, 75-76 & Israeli Nat Comt Res & Technol, 85-; mem, Subsatellite Sci Definition Team, 77-78 & Tetler Shuttle Working Group, 84- *Mem:* Am Phys Soc; Am Geophys Union; fel Brit Interplanetary Soc. *Res:* Flows of space-plasmas over bodies; ionospheres of the earth and planets; physics of cosmic rays; laboratory simulation of space physics and rarefied plasma physics phenomena and processes; space plasma expansion into a vacuum; laboratory simulation and measurements from space platforms; application of plasma expansion properties to the wake structure behind artificial satellites and behind non-magnetized planets and moons; interactions between large spacecraft and the terrestrial ionosphere. *Mailing Add:* Space Physics Res Lab Univ Mich 2455 Hayward St Ann Arbor MI 48109-2143

SAMIS, HARVEY VOORHEES, JR, b Easton, Md, July 14, 31; m 56; c 4. BIOCHEMISTRY, PHYSIOLOGY. *Educ:* Wash Col, BS, 56; Brown Univ, PhD(biochem physiol), 63. *Prof Exp:* Instr biol, Washington Col, 55-56; res scientist, Masonic Med Res Lab, 63-75, dir exp geront prog, Utica, NY, 68-75; sr scientist, Med Res Serv, Vet Admin Med Ctr, Bay Pines, Fla, 75-87, coordr res & develop, 77-87; CONSULT, 87- *Concurrent Pos:* Vis lectr biol, Syracuse Univ, 73-75; vis assoc prof & adj prof chem, Univ SFla, Tampa, 77- *Mem:* AAAS; Soc Gen Physiol; fel Geront Soc; Soc Develop Biol; fel Am Inst Chemists; Inter-Am Soc Chemother. *Res:* Chelation chemistry of biomolecules; alcoholism effect of aging on properties of biological macromolecules and their functions; effects of age on the temporal organization of biological systems; molecular genetics and neoplasia; nucleic acid metabolism. *Mailing Add:* 5301 14th Ave S Gulf Port FL 33707

SAMITZ, M H, b Philadelphia, Pa, Dec 18, 09; m 45; c 2. DERMATOLOGY. *Educ:* Temple Univ, MD, 33; Univ Pa, MSc, 45. *Prof Exp:* From instr to asst prof dermat, Grad Sch Med, 40-53, from asst prof to assoc prof, Sch Med, 49-67, prof dermat & dir grad dermat, 67-75, EMER PROF DERMAT, SCH MED, UNIV PA, 75- *Concurrent Pos:* Med dir, Skin & Cancer Hosp, Philadelphia, 53-54; prin investr, USPHS res grants, 58-75; consult, US Naval Hosp, 65-; consult & vis prof, Pa Col Podiat Med, 65-75, emer prof dermat, 75-; vis prof, Hahnemann Med Col, 67-, Univ Dar es Salaam, 75-; chief dept dermat, Grad Hosp, Univ Pa, 67-; mem comt, Div Educ, Nat Prog Dermat, 71-; mem comt on nickel, Nat Res Coun; consult, Food & Drug Admin, Dermat Adv Comt, 77-78; US-Poland health scientist exchange fel, 80; mem bd dirs, Found Int Dermatol Educ, 75-, pres, 79- *Mem:* AAAS; Am Acad Dermat; Am Col Physicians; Am Col Allergists; Soc Invest Dermat. *Res:* Industrial dermatology, particularly effects of chromium salts and nickel on the skin; clinical investigations in various aspects of clinical dermatology. *Mailing Add:* 1715 Pine St Philadelphia PA 19103

SAMLOFF, I MICHAEL, b Rochester, NY, Jan 24, 32; m 54; c 2. MEDICINE, GASTROENTEROLOGY. *Educ:* State Univ NY, MD, 56. *Prof Exp:* Instr med, Sch Med & Dent, Univ Rochester, 62-64, sr instr med & psychiat, 64-65, asst prof, 65-68; assoc prof, 68-72, PROF MED, SCH MED, UNIV CALIF, LOS ANGELES, 72-; ASSOC CHIEF STAFF RES, VET ADMIN MED CTR, 80- *Concurrent Pos:* USPHS fel med, Strong Mem Hosp, 58-59, fel psychiat, 61-62 & trainee med, 61-63; Am Cancer Soc advan clin fel, 63-66; chief, Gastroenterol Div, Harbor Hosp, Univ Calif, Los Angeles Med Ctr, 68-80. *Mem:* Am Fedn Clin Res; Am Psychosom Soc; Am Gastroenterol Asn; Am Soc Clin Invest; fel Am Col Physicians. *Res:* Gastritis; ulcer. *Mailing Add:* Vet Admin Med Ctr 16111 Plummer St Sepulveda CA 91343

SAMMAK, EMIL GEORGE, b Brooklyn, NY, Apr 3, 27; div; c 3. EMULSION POLYMERIZATION, RHEOLOGY. *Educ:* Polytech Inst Brooklyn, BS, 49, PhD(polymer sci), 58. *Prof Exp:* Sr res chemist, Chem Div, Int Latex & Chem Corp, 55-60, mgr basic polymer res, 60-68; mgr basic res, Standard Brands Chem Industs, Inc, 68-76; MGR POLYMER & ANALUTICAL RES, EMULSION POLYMERS DIV, REICHHOLD CHEM, INC, 76- *Concurrent Pos:* Adj prof, Wesley Col, 82, Del State Col, 88- *Mem:* AAAS; Am Chem Soc. *Res:* Synthesis, characterization and mechanical properties of butadiene latex, polyelectrolytes, polyacrylates, polyurethanes, allyl and formaldehyde resins; ionic polymerization; grafting olefin copolymers; starch utilization; rheology, computer applications, pollution control, analytical research. *Mailing Add:* English Village No G 3 Dover DE 19901

SAMMAK, PAUL J, b Dover, Del, Feb 8, 56; m; c 2. SIGNAL TRANSDUCTION, CYTOSKELETON & CELL MOTILITY. *Educ:* Hampshire Col, BA, 78; Univ Wis-Madison, MS, 80, PhD(biophys), 88. *Prof Exp:* Res asst, Physics Dept, Amherst Col, 78, Chem Dept, Univ Wis, 79-80, Lab Molecular Biol, Univ Wis, 81-88; res pharmacologist, Dept Pharmacol, Univ Calif, San Diego, 89-90; res physiologist, NIH Cancer Res Lab, 89, RES PHYSIOLOGIST, DEPT MOLECULAR & CELLULAR BIOL, UNIV CALIF, BERKELEY, 91- *Concurrent Pos:* Teaching & lab asst, Physics Dept, Univ Wis, 78-80; lectr & reader, Physiol Dept, Univ Calif, Berkeley, 89; NIH fel, 91. *Mem:* Am Soc Cell Biol; AAAS; Sigma Xi. *Res:* Cell biology, motility and regulation; molecular dynamics of biological polymers; wound healing. *Mailing Add:* Dept Molecular & Cell Biol Univ Calif 235 Life Sci Annex Berkeley CA 94720

SAMMELWITZ, PAUL H, b Buffalo, NY, Mar 13, 33; m 62; c 3. REPRODUCTIVE PHYSIOLOGY. *Educ:* Cornell Univ, BS, 55; Univ Ill, MS, 57, PhD(reprod physiol), 59. *Prof Exp:* Asst prof, 59-68, ASSOC PROF AVIAN & MAMMALIAN PHYSIOL & GENETICS, UNIV DEL, 68- *Concurrent Pos:* AAAS; Am Soc Animal Sci; Poultry Sci Asn; Soc Study Reproduction. *Res:* Mammalian and avian reproductive physiology; avian heat stress physiology; endocrine factors influenced by moderate dietary vitamin A deficiency; genetic resistance to Marek's disease; computer assisted instruction; multimedia lecture tools. *Mailing Add:* Dept Animal Sci & Agr Biochem Univ Del Newark DE 19717-1303

SAMMET, JEAN E, b New York, NY, Mar 23, 28. SOFTWARE HISTORY, PROGRAMMING LANGUAGES. *Educ:* Mt Holyoke Col, BA, 48; Univ Ill, MA, 49. *Hon Degrees:* DSc, Mt Holyoke Col, 78. *Prof Exp:* Teaching asst math, Univ Ill, 48-51; dividend technician, Metrop Life Ins Co, 51-52; teaching asst math, Barnard Col, Columbia Univ, 52-53; engr, Sperry Gyroscope Co, 53-58; sect head, Mobidic Programming, Sylvania Elec Prod, 58-59, staff consult prog res, 59-61; Boston adv prog mgr, IBM Corp, 61-65, prog lang tech mgr, 65-68, prog technol planning mgr, 68-74, prog lang technol, 74-79, div software technol mgr, 79-83, prog lang technol mgr, IBM Corp, 83-88; CONSULT, PROG LANG, 89- *Concurrent Pos:* Lectr, Adelphi Col, 56-58, Northeastern Univ, 67, Univ Calif, Los Angeles, 67-72 & Mt Holyoke Col, 74; ed-in-chief, Comput Rev & ACM Guide Comput Lit, 79-87. *Mem:* Nat Acad Eng; Asn Comput Mach (vpres, 72-74, pres, 74-76); Math Asn Am. *Res:* High level programming languages; use of computers for non-numerical mathematics; formula manipulation systems; programming systems; language measurement; practical uses of artificial intelligence; use of natural language on a computer; history of software; history of programming languages. *Mailing Add:* PO Box 30038 Bethesda MD 20824-0038

SAMMONS, DAVID JAMES, b Columbus, Ohio, Sept 2, 46; m 70; c 2. AGRONOMY, PLANT BREEDING. *Educ:* Tufts Univ, BS, 68; Harvard Univ, AM, 72; Univ Ill, PhD(agron), 78. *Prof Exp:* Vol, Peace Corps, Philippines, 68-70; asst biol, Harvard Univ, 70-72; assoc dir natural hist, Norwalk Mus & Zoo, Conn, 72; teacher-naturalist, Nat Audubon Soc, 72-75; res asst, Univ Ill, 75-78; from asst prof to assoc prof agron, 78-90, ASSOC DEAN UNDERGRAD STUDIES, UNIV MD, 89-, PROF AGRON, 90- *Concurrent Pos:* Mem fac adv comt, Univ Ill, 75-76; curric consult, Sinclair Community Col, Dayton, Ohio, 73-74; mem environ qual bd, Dept Urban Develop, New Towns Prog, 73-74; mem curric develop, Govt Philippines, 68-70; fac adv & partic, Study Abroad Prog, Univ Md, Asia, 79, Caribbean, 81 & PR, 83; fac mem, Grad Sch, USDA, Washington, DC, 80-83; vis prof, Berkeley Col, Yale Univ, 81; consult, Off Int Coop & Develop, The Gambia, USDA, 84, Proj Sci Technol Coop, AID, 84-85; vis lectr, Egerton Univ, Njoro, Kenya, 86-87; mem bd dirs, Consortium Int Crop Protection, 86-89, chmn, Eastern Wheat Region, 87-90; mem, Nat Wheat Improv Comt, 87-; coordr, Grad Studies Agron, Univ Md, 87-; mem, Nat Barley Improv Comt, 88-91; distinguished scholar/teacher, Univ Md, 88-89. *Honors & Awards:* Fulbright Award, Kenya, 86-87. *Mem:* Crop Sci Soc Am; Am Soc Agron; Asn Asian Studies-Middle Atlantic Region. *Res:* Applied breeding research designed to improve barley and wheat cultivars for producers in Maryland and the Mid-Atlantic region; small grain production, breeding and physiology; factors affecting quality in wheat. *Mailing Add:* Dept Agron Univ Md College Park MD 20742

SAMMONS, JAMES HARRIS, b Montgomery, Ala, March 13, 27; c 4. FAMILY PRACTICE, MEDICINE. *Educ:* Washington & Lee Univ, BS; St Louis Univ, MD. *Hon Degrees:* LHD, Tex Univ Sci Ctr. *Prof Exp:* Dep med examr, Harris County, Tex, 62-74; clin asst prof family med, Dept Community Med, Baylor Col Med, 72-75, clin assoc prof family pract, Family Pract Ctr, 73-74, vis assoc prof family pract, 74-; RETIRED. *Mem:* Inst Med-Nat Acad Sci; Am Pub Health Asn; AMA; Am Acad Gen Pract; Am Asn Med Soc Execs. *Mailing Add:* 161 E Chicago No 49E Chicago IL 60611

SAMN, SHERWOOD, b Los Angeles, Calif, Apr 20, 41; m 68; c 2. MATHEMATICS. *Educ:* Univ Calif, Berkeley, BA, 63, PhD(math), 68. *Prof Exp:* Asst prof math, Ind Univ-Purdue Univ, Indianapolis, 68-74; MATHEMATICIAN, BROOKS AFB, USAF, SAN ANTONIO, TEX, 74- *Mem:* Am Math Soc; Soc Indust & Appl Math. *Res:* Differential equations; cardiovascular system modeling; operations research. *Mailing Add:* USAF SAM-NGSA Brooks AFB San Antonio TX 78235

SAMOILOV, SERGEY MICHAEL, b Baku, USSR, Dec 17, 25; US citizen; m 87; c 1. POLYMER CHEMISTRY, CATALYSIS. *Educ:* Moscow Inst Fine Chem Technol, MS, 49; USSR Acad Sci, PhD(chem), 58. *Prof Exp:* Engr catalysis, USSR Nat Res Inst Artificial Fuel, 49-51, USSR Petrochem Plant No 16, 51-55; res assoc, USSR Acad Sci, 58-61; sr res assoc polymers, USSR Nat Res Inst Petrochem, 62-76; res assoc, Columbia Univ, 77-78; res assoc polymers, Celanese Res Co, 78-81; sr res chemist polymers, Allied-Signal Corp, 81-90; RETIRED. *Concurrent Pos:* Abstractor, Chem Abstracts J, Inst Sci Info, USSR Acad Sci, 56-76. *Mem:* Am Chem Soc. *Res:* Synthesis and investigation of polyolefins; radical pressure copolymerization of lower olefins; polymer emulsions; metalorganic polymers; structural and relaxational properties of polymers; Ziegler catalysis. *Mailing Add:* 155 Acropolis Dr Apt 5 Athens GA 30605

SAMOLLOW, PAUL B, b San Francisco, Calif, Mar 20, 48; m 68; c 2. ECOLOGICAL GENETICS, EVOLUTIONARY BIOLOGY. *Educ:* Univ Calif, BA, 71; Ore State Univ, PhD(zool), 78. *Prof Exp:* Lectr genetics & evolution, Humboldt State Univ, 78; asst prof, Univ Mont, 79; fel, Hawaii Inst Marine Biol, 79-81; asst scientist biochem & genetics, Southwest Found Biomed Res, San Antonio, Texas, 81-87; ASST PROF EVOLUTION & GENETICS, LEHIGH UNIV, 87- *Concurrent Pos:* Adj asst scientist biochem & genetics, Southwest Found Biomed Res, San Antonio, Texas, 87. *Mem:* Genetics Soc Am; Soc Study Evolution; AAAS; Sigma Xi; Am Soc Naturalists. *Res:* Structure and genetic dynamics of populations in varying environments; evolution of developmental homeostasis; biochemical genetics of New World marsupials; linkage and functional relationships among genes in or near the major histocompatibility complex in the house mouse. *Mailing Add:* Dept Biol William Hall No 31 Lehigh Univ Bethlehem PA 18015

SAMOLS, DAVID R, b Washington, DC, Aug 31, 45; m 76; c 2. RECOMBINANT DNA. *Educ:* Earlham Col, BS, 67; Univ Chicago, PhD(biol), 76. *Prof Exp:* Fel, Roche Inst Molecular Biol, 76-79; res assoc, Inst Molecular Biol, Univ Ore, 79; asst prof, 80-88, ASSOC PROF BIOCHEM, MED SCH, CASE WESTERN RESERVE UNIV, 88- *Mem:* Am Soc Cell Biol; Am Soc Biochem & Molecular Biol. *Res:* Signal and mechanism by which tissue injury and infection induce the liver to secrete c-reactive protein; types of genetic rearrangements and alterations caused by chemical carcinogenesis. *Mailing Add:* Dept Biochem Sch Med Case Western Reserve Univ Cleveland OH 44106

SAMORAJSKI, THADDEUS, b Shelburne, Mass, Oct 29, 23; m 52; c 2. ANATOMY. *Educ:* Univ Mich, BS, 48; Univ Chicago, PhD(anat), 56. *Prof Exp:* Asst & instr, Univ Chicago, 55-56; instr anat, Ohio State Univ, 56-60; dir lab neurochem, Cleveland Psychiat Inst, 60-74; MEM STAFF, TEX RES INST MENT SCI, 74- *Concurrent Pos:* Adj prof, Dept Biol, Tex Women's Univ, 78- & Dept Neurobiol & Anat, Sch Med, Univ Tex, 80- *Honors & Awards:* Cralow Medal, Polish Acad Sci. *Mem:* Soc Neurosci; Geront Soc. *Res:* Neurochemistry, particularly in relation to neurobiology of aging; environmental modification of life span; radiation neuropathology; myelin formation and degeneration; catecholamine metabolism. *Mailing Add:* Dept Neurobiol Univ Tex Health Sci Ctr PO Box 20036 Houston TX 77225

SAMPLE, HOWARD H, b Dallas, Tex, Sept 20, 38; m 58; c 3. SOLID STATE PHYSICS. *Educ:* Iowa State Univ, BS, 60, PhD(physics), 66. *Prof Exp:* NATO fel sci, Clarendon Lab, Oxford Univ, 66-67; asst prof, 67-73, ASSOC PROF PHYSICS, TUFTS UNIV, 73- *Concurrent Pos:* Vis scientist, Francis Bitter Nat Magnet Lab, 68- *Res:* Low temperature physics, superconductivity, thermal properties of disordered solids, low temperature thermometry in high magnetic fields. *Mailing Add:* Dept Physics Tufts Univ Medford MA 02155

SAMPLE, JAMES HALVERSON, b Cicero, Ill, Feb 27, 14; m 40; c 3. ORGANIC POLYMER CHEMISTRY. *Educ:* Elmhurst Col, BS, 35; Univ Ill, MS, 36, PhD(org chem), 39. *Prof Exp:* Asst chem, Univ Ill, 36-39; prof, Ind Cent Col, 39-42; prof, Franklin Col, 42-44; res chemist, Sherwin Williams Co, 44-47, chem res supvr, 47-58, asst dir resin res dept, 58-65, dir, 66-73, dir polymer & mat res-coating, 74-79; CONSULT, 80- *Mem:* Am Chem Soc. *Res:* Resins for surface coatings; alkyds. *Mailing Add:* 1206 E 165th Pl South Holland IL 60473

SAMPLE, JOHN THOMAS, b Kerrobert, Sask, May 4, 27; m 53; c 5. NUCLEAR PHYSICS. *Educ:* Univ BC, BA, 48, MA, 50 & PhD, 55. *Prof Exp:* Sci officer, Defence Res Bd, Can, 55-58; dir res, Secretariat BC, 81-88; from asst prof to assoc prof, 58-66, chmn dept physics, 67-76, PROF NUCLEAR PHYSICS, UNIV ALTA, 66- *Concurrent Pos:* Vis scientist, Brookhaven Nat Lab, 65-66; bd mem, Pac Isotopes & Pharmaceut Ltd, 83-89; dir, TRIUMF, Univ BC, 76-81; gen mgr, Ebco Technologies, 88- *Mem:* Am Inst Physics; Am Phys Soc; Can Asn Physicists. *Res:* Reactions of low and intermediate energy nuclear physics; design of nuclear medicine equipment. *Mailing Add:* TRIUMF 4004 Wesbrook Mall Vancouver BC V6T 1W5 Can

SAMPLE, PAUL E(DWARD), b Chicago, Ill, Nov 24, 28; m 53; c 5. CHEMICAL ENGINEERING. *Educ:* Ill Inst Technol, BS, 51; Univ WVa, MS, 55, PhD(chem eng), 57. *Prof Exp:* Asst, Exp Sta, Univ WVa, 54, 55-57; res engr, Film Dept, 57-59, group leader, Chem Develop, 59-60, supvr, Mfg Div, 60-62, tech rep, Mkt Div, 62-66, res supvr, Res & Develop Div, 66-69, group mgr, 69-72, TECH CONSULT, RES & DEVELOP DIV, E I DU PONT DE NEMOURS & CO, INC, 72- *Mem:* Am Chem Soc; Am Soc Metals; Am Inst Chem Engrs; Sigma Xi. *Res:* Process automation and instrumentation; packaging systems and packaging materials. *Mailing Add:* 308 Walden Rd Wilmington DE 19803

SAMPLE, STEVEN BROWNING, b St Louis, Mo, Nov 29, 40; m 61; c 2. ELECTROHYDRODYNAMICS. *Educ:* Univ Ill, Urbana, BS, 62, MS, 63, PhD(elec eng), 65. *Prof Exp:* Scientist, Melpar, Inc, 65-66; asst prof elec eng, Purdue Univ, 66-70, assoc prof, 70-73; prof elec eng & exec vpres, Univ Nebr, 74-82, dean, Grad Col, 77-82; PRES & PROF ELEC ENG, STATE UNIV NY, BUFFALO, 82- *Concurrent Pos:* Dep dir, Ill Bd Higher Educ, 71-74; mem bd dirs, Design & Mfr Corp, Connersville, Ill, 77- & Moog Inc, 82-; mem ednl activ bd, Inst Elec & Electronics Engrs, 82-84; mem exec comt, Nat Asn State Univs & Land-Grant Colls, 85-, chmn coun pres's, 85-86, chmn ednl & telecommun comt, 85-86. *Mem:* Sigma Xi; Inst Elec & Electronics Engrs. *Res:* Electrohydrodynamic instability of liquid drops in electric fields; harmonic electrical spraying of liquids from capillaries; solid-state digital control systems for appliances. *Mailing Add:* 889 Lebrun Rd Amherst NY 14226

SAMPLE, THOMAS EARL, JR, b Magnolia Park, Tex, Dec 17, 24; m 53, 62; c 3. SURFACE CHEMISTRY, RHEOLOGY. *Educ:* Rice Inst, BA, 48; Univ Tex, Austin, MA, 54, PhD(chem), 59. *Prof Exp:* Chemist, Magnet Cove Barium Corp, Houston, Tex, 48-50; US Air Force res fel chem, Univ Tex, Austin, 51-53; res chemist, Plastics Div, Monsanto Co, Texas City, Tex, 54-56; res group supvr, Magnet Cove Barium Corp, 57-61; res proj leader, Champion Chem, Inc, Houston, Tex, 61-63; Robert A Welch fel chem, Univ Tex, Austin, 63-66; sr res chemist, Prod Res Labs, Texaco Inc, Bellaire, Tex, 66-71; tech adv, Oilfield Prod Div, Dresser Industs Inc, Houston, Tex, 71-78, sr scientist, Magcobar Group, 78-83; tech adv, Hematech Ltd, 85-86, TECH DIR, SERV-TECH, INC, HOUSTON, TEX, 88- *Concurrent Pos:* Indust chem consult, 83- *Mem:* Am Inst Chemists; Royal Soc Chem; Am Chem Soc; Am Phys Soc; German Soc Chem; NY Acad Sci. *Res:* Chemical modification of mineral surfaces; oilwell and geothermal drilling fluids; tribology; rheology of suspensions; organic silicon chemistry; particle size analysis. *Mailing Add:* 7400 Bellerive Dr Suite 101 Houston TX 77063-6135

SAMPLES, WILLIAM R(EAD), b Whipple, WVa, Oct 17, 31; m 53; c 4. SANITARY ENGINEERING, ENVIRONMENTAL ENGINEERING. *Educ:* Univ WVa, BS, 53; Harvard Univ, MS, 55, PhD(eng), 59. *Prof Exp:* Asst prof civil eng, Calif Inst Technol, 59-65; fel water resources, Mellon Inst, 65-68, sr fel & head water resources, 68-71; coordr, Wheeling-Pittsburgh Steel Corp, 71-78, mgr environ control, 78-85, eng & environ control, 85-90, DIR ENVIRON CONTROL, WHEELING-PITTSBURGH STEEL CORP, 90- *Concurrent Pos:* Nat Air Pollution Control Techniques Adv Comt, Environ Protection Agency. *Mem:* Am Soc Civil Engrs; Water Pollution Control Fedn; Am Inst Chem Engrs; Air Pollution Control Asn; Am Iron & Steel Inst. *Res:* Industrial waste water control; air pollution; industrial hygiene; water quality; water and sewage treatment. *Mailing Add:* 2293 Weston Dr Pittsburgh PA 15241

SAMPLEY, MARILYN YVONNE, b Ala. NUTRITION. *Educ:* Auburn Univ, BS, 57; Univ Ala, MS, 61; Tex Woman's Univ, PhD(nutrit, biochem & foods), 69. *Prof Exp:* Instr nutrit & chief dietician, Sacred Heart Dominican Col, St Joseph Hosp, Houston, 57-58; teacher, K J Clark Jr High Sch, Mobile Ala, 59-60; res asst, Univ Ala, 60-61; instr nutrit & chief dietician, Sacred Heart Dominican Col, St Joseph Hosp, Houston, 62; dir food serv, Dickinson Sch Dist, Dickenson, Tex, 63-67; res asst, Tex Woman's Univ, 67-69; ASSOC PROF & CHMN, DEPT HOME ECON, TEX A&I UNIV, 72- *Mem:* Am Pub Health Asn; AAAS; Soc Nutrit Educ; Nat Coun Admin Home Econ; Nat Educ Asn. *Res:* Eating and food buying habits. *Mailing Add:* Dept Home Econ Tex A&I Univ Kingsville TX 78363

SAMPSON, CALVIN COOLIDGE, b Cambridge, Md, Feb 1, 28; m 53; c 2. MEDICINE, PATHOLOGY. *Educ:* Hampton Inst, BS, 47; Meharry Med Col, MD, 51. *Prof Exp:* From asst prof to assoc prof, 58-69, PROF PATH, COL MED, HOWARD UNIV, 69- *Concurrent Pos:* Asst ed, J Nat Med Asn, 65-77, ed, 78- *Mem:* Nat Med Asn; Int Acad Path; fel Col Am Path. *Mailing Add:* 1614 Varnum Pl NE Washington DC 20017

SAMPSON, CHARLES BERLIN, b Iowa Falls, Iowa, Dec 15, 39; m 65; c 3. STATISTICS, INFORMATION SCIENCE. *Educ:* Univ Iowa, BS, 61, MS, 63; Iowa State Univ, PhD(statist), 68. *Prof Exp:* Sr statistician, 68-73, res scientist, 73-74, head, Statist & Math Serv, 74-81, head, Sci Info Serv, 81-87, mgr, med info serv & statist, 87-88, DIR, STATIST & MATH SCI, ELI LILLY & CO, 89- *Concurrent Pos:* Mem bd dir, Am Statist Asn, 82-83, 88-89, coun, Biometrics Soc, 84-86; vis lectr, Comt Pres of Statist Soc, 84-; chmn nominations comt, Am Statist Asn, 89, mem comt nat & int statist standards, 91-93. *Mem:* Fel Am Statist Asn; Biometric Soc. *Res:* Application of statistical and mathematical models to biological, medical, and chemical research; design of experiments; pharmaceutical quality control. *Mailing Add:* Eli Lilly & Co Lilly Corp Ctr-2233 Indianapolis IN 46285

SAMPSON, DAVID ASHMORE, METABOLISM, METHODOLOGY. *Educ:* Colo State Univ, PhD(nutrit biochem), 82. *Prof Exp:* RES NUTRIT SCIENTIST, WESTERN HUMAN NUTRIT RES CTR, AGR RES SERV, USDA, SAN FRANCISCO, 84- *Mailing Add:* Western Human Nutrit Res Ctr Agr Res Serv USDA PO Box 29997 San Francisco CA 94129

SAMPSON, DEXTER REID, b New Glasgow, NS, Can, Sept 9, 30; m 58; c 2. GENETICS, PLANT BREEDING. *Educ:* Acadia Univ, BSc, 51; Harvard Univ, AM, 54, PhD, 56. *Prof Exp:* Res officer hort crops, 55-66, RES SCIENTIST CEREAL CROPS, CAN DEPT AGR, 66- *Mem:* Genetics Soc Can; Can Bot Asn; Sigma Xi; Can Soc Agron; Agr Inst Can. *Res:* Genetics of self-incompatability in angiosperms; genetics of Brassica and oats; breeding soft white pastry winter wheat and hard red winter wheat for milling and baking quality, high yield, winter survival and disease resistance. *Mailing Add:* Plant Res Centre Can Dept Agr Cent Exp Farm Ottawa ON K1A 0C6 Can

SAMPSON, DOUGLAS HOWARD, b Devils Lake, NDak, May 19, 25; m 56; c 4. ATOMIC PHYSICS. *Educ:* Concordia Col, Moorhead, Minn, BA, 51; Yale Univ, MA, 53, PhD(theoret physics), 56. *Prof Exp:* Staff mem, Theoret Div, Los Alamos Sci Lab, NMex, 56-61; theoret physicist, Space Sci Lab, Gen Elec Co, Pa, 61-64, group leader atomic & radiation physics, 64-65; assoc prof, 65-70, PROF ASTROPHYS, PA STATE UNIV, 70- *Concurrent Pos:* Consult, Lawrence Livermore Nat Lab & Los Alamos Nat Lab. *Mem:* Am Phys Soc; Am Astron Soc; Int Astron Union. *Res:* Atomic physics of very highly charged ions; theoretical astrophysics; statistical mechanics and kinetic theory; radiative transport. *Mailing Add:* Dept Astron & Astrophys Pa State Univ University Park PA 16802

SAMPSON, HENRY T, b Jackson, Miss, Apr 22, 34; m 61; c 2. NUCLEAR ENGINEERING. *Educ:* Purdue Univ, BS, 56; Univ Calif, Los Angeles, MS, 61; Univ Ill, PhD(nuclear eng), 67. *Prof Exp:* Res engr, US Naval Weapons Ctr, 56-62; mem tech staff, 67-81, DIR PLANNING & OPERS, SPACE TEXT PROG, AEROSPACE CORP, 81- *Mem:* AAAS; Am Nuclear Soc; Am Inst Aeronaut & Astronaut. *Res:* Research and development of rocket propellants and plastic bonded explosives; direct conversion of nuclear energy to electrical energy; analysis of space electrical power systems. *Mailing Add:* Aerospace Corp PO Box 92957 Mail Stop M-5-120 Los Angeles CA 90009-2957

SAMPSON, HERSCHEL WAYNE, b Greenville, Tex, June 28, 44; m 65; c 2. CELL CALCIUM, MINERALIZATION. *Educ:* Arlington State Col, BS, 67; Baylor Univ, PhD(anat), 70. *Prof Exp:* Asst prof anat, Sch Med, Creighton Univ, 70-72; assoc prof, Baylor Col Dent, 72-77 & Col Med & Dent, Oral Roberts Univ, 77-78; ASSOC PROF ANAT, MED SCH, TEX A&M UNIV, 79- *Mem:* Am Asn Anat; Am Soc Bone & Mineral Res; Am Physiol Soc; Am Soc Cell Biol; Electron Microscope Soc Am; Am Asn Clin Anatomists. *Res:* Calcium transfer and homeostasis at the cell level; bone and joint disease mechanisms; mineralization. *Mailing Add:* Dept Human Anat Tex A&M Univ Med Sch College Station TX 77843-1114

SAMPSON, JOHN LAURENCE, b Lynn, Mass, Dec 14, 29; m 52; c 3. PHYSICS. *Educ:* Mass Inst Technol, BS, 51; Tufts Univ, MS, 54, PhD, 62. *Prof Exp:* Physicist, Air Force Cambridge Res Labs, 51 & 55-59; instr physics, Tufts Univ, 54-55, asst, 59-61; physicist, Arthur D Little, Inc, 61-62; physicist, Air Force Rome Air Develop Ctr, Bedford, 62-90; PHYSICIST, NORTHEAST PHOTOSCI, 90- *Mem:* Sigma Xi. *Res:* Holography and fiber optics. *Mailing Add:* Eight Bedford St Lexington MA 02173

SAMPSON, JOSEPH HAROLD, b Spokane, Wash, Sept 14, 25. MATHEMATICS. *Educ:* Princeton Univ, MA, 49, PhD(math), 51. *Prof Exp:* Res grant, Off Naval Res, 51-52; C L E Moore instr math, Mass Inst Technol, 52; asst prof, Johns Hopkins Univ, 52-64; prof assoc, Univ Strasbourg, 64-65 & 68-69, Univ Grenoble, 74-75; chmn dept, 70-80, PROF MATH, JOHNS HOPKINS UNIV, 65- *Concurrent Pos:* Prof assoc, Univ Strasbourg, 68-69; ed, Am J Math, 78- *Mem:* Am Math Soc; Math Soc France; Italian Math Union; Sigma Xi. *Res:* Algebraic geometry; geometry of manifolds; geometric applications of partial differential equations, especially as connected with the Laplace operator; gave the first general description of harmonic mappings; number-theoretic applications of algebraic geometry; global analysis. *Mailing Add:* Dept Math Johns Hopkins Univ Baltimore MD 21218

SAMPSON, PAUL, b Keighley, W Yorkshire, Eng, Jan 4, 59; UK citizen. FLUORINATED UNNATURAL PRODUCTS. *Educ:* Univ Birmingham, Eng, BSc, 80, PhD(org chem), 83. *Prof Exp:* Res assoc org chem, Univ Iowa, 83-85; ASST PROF ORG CHEM, KENT STATE UNIV, 85- *Mem:* Am Chem Soc; Royal Soc Chem. *Res:* Synthesis of natural and fluorine-containing "unnatural" products; organofluorine chemistry and development of new synthetic methodology. *Mailing Add:* 37 Heights Ave No 1 Northfield OH 44067-1377

SAMPSON, PHYLLIS MARIE, b New York, NY, Sept 13, 28; m 74. BIOCHEMISTRY. *Educ:* Hunter Col, BA, 50; Columbia Univ, PhD(biochem), 72. *Prof Exp:* Res assoc chem, Yeshiva Univ, 72-73; res assoc biochem, Columbia Univ, 73-76; res assoc med, 76-84, RES ASST PROF, UNIV PA, 84- *Mem:* Sigma Xi; Soc Complex Carbohydrates; AAAS; NY Acad Sci. *Res:* Proteoglycan and glycosaminoglycan distribution in normal and pathological lung tissue and production by lung cells in tissue and organ culture; biochemistry and chemistry of carbohydrates. *Mailing Add:* 235 S Third St Philadelphia PA 19106

SAMPSON, RONALD N, b Pittsburgh, Pa, Sept 16, 30; m 53; c 4. CHEMICAL ENGINEERING. *Educ:* Carnegie-Mellon Univ, BS, 52 & 57. *Prof Exp:* Engr, Mat Eng Dept, Westinghouse Elec Corp, 52-57, supvry engr, Chem Appln Sect, 57-62, mgr insulation, 62-80, mgr, Chem Sci Div, Res Labs, 80-88, tech dir, 88-89; RETIRED. *Mem:* Am Chem Soc; Soc Plastics Engrs; AAAS; Inst Elec & Electronics Engrs. *Res:* Research and development of polymers and plastics in areas of electrical insulation, laminates, molding materials, adhesives and films. *Mailing Add:* Res & Develop 4250 Bulltown Rd Murrysville PA 15668

SAMPSON, SANFORD ROBERT, b Los Angeles, Calif, Feb 27, 37; m 59; c 2. PHYSIOLOGY, PHARMACOLOGY. *Educ:* Univ Calif, Berkeley, BA, 59; Univ Utah, PhD(pharmacol), 64. *Prof Exp:* Lectr, 68-69, asst prof pharmacol, 69-71, asst prof physiol, 71-74, ASSOC PROF PHYSIOL, MED CTR, UNIV CALIF, SAN FRANCISCO, 74-; DEPT LIFE SCI, BAR-ILAN UNIV, ISRAEL. *Concurrent Pos:* Fel pharmacol, Albert Einstein Col Med, 64-66; res fel, Cardiovasc Res Inst, Med Ctr, Univ Calif, San Francisco, 66-69, Nat Heart Inst spec fel, 69-71; Macy fac scholar, 78-79; vis scientist, Weizmann Inst Sci, 78-79. *Mem:* Am Physiol Soc; Soc Neurosci; Am Soc Pharmacol & Exp Therapeut; Int Soc Develop Neurosci. *Res:* Membrane channels and electro-genic pumps in excitable membranes. *Mailing Add:* Dept Life Sci Bar-Ilan Univ Ramat-Gan 52900 Israel

SAMPSON, WILLIAM B, b Toronto, Ont, Aug 31, 34; m 55; c 2. PHYSICS. *Educ:* Univ Toronto, BA, 58, MA, 59, PhD(physics), 62. *Prof Exp:* PHYSICIST, ACCELERATION DEPT, BROOKHAVEN NAT LAB, 62- *Mem:* Am Phys Soc. *Res:* Superconductivity and applications to high energy physics. *Mailing Add:* Accelerator Dept Brookhaven Nat Lab Upton NY 11973

SAMPUGNA, JOSEPH, b Sept 27, 31; US citizen; m 57; c 2. BIOCHEMISTRY, NUTRITION. *Educ:* Univ Conn, BA, 59, MA, 62, PhD(biochem), 68. *Prof Exp:* Res asst biochem, Univ Conn, 62-68; asst prof, 68-72, ASSOC PROF BIOCHEM, UNIV MD, COLLEGE PARK, 72- *Concurrent Pos:* Vpres, Chem Asn Md, 73-80, treas, 80- *Mem:* Am Chem Soc; Am Oil Chem Soc; Am Inst Nutrit. *Res:* Lipid biochemistry; membrane structure and function; metabolism of dietary lipids. *Mailing Add:* Dept Chem & Biochem Univ Md College Park MD 20742

SAMS, BRUCE JONES, JR, b Savannah, Ga, Jan 24, 28; m; c 3. HEALTH ECONOMICS. *Educ:* Ga Inst Technol, BS, 51; Harvard Univ, MD, 55. *Prof Exp:* Intern, NC Mem Hosp, Chapel Hill, 55-56; asst resident, Univ Calif, San Fransisco, 56-57; res fel hemat, Mass Gen Hosp, Boston, 57-59; dir, Med Residency Prog, Kaiser Permanente Med Ctr, San Francisco, 64-71, chief med, 65-71, chief staff educ, 66-69, physician-in-chief, 71-75; exec dir-elect, Permanente Med Group, Oakland, Calif, 75-76, exec dir, 76-91; RETIRED. *Concurrent Pos:* Clin instr med, Univ Calif, San Francisco, 61-62, assoc clin prof, 63-; physician, Kaiser Permanente Med Ctr, 62-; mem, Clin Pract Subcomt, Am Col Physicians; mem bd, Group Health Asn Am. *Mem:* Inst Med-Nat Acad Sci; fel Am Col Physicians; AMA; Am Col Physician Execs. *Mailing Add:* 1950 Franklin St Oakland CA 94612

SAMS, BURNETT HENRY, III, b Seattle, Wash, Apr 30, 31; m 56; c 2. DATABASES, MULTIPROCESSING. *Educ:* Univ Wash, BS, 51; Univ Ill, MS, 53, PhD(math), 58. *Prof Exp:* Res assoc, Control Systs Lab, Univ Ill, 57-58 & Comput Ctr, Mass Inst Technol, 58-59; proj leader, Astro-Electronics Prods Div, Radio Corp Am, 59-61, mgr, Prog Sci Sect, Data Systs Ctr, 62-64, systs res lab, RCA Labs, 64-76, head comput aided mfg, Solid State Technol Ctr, 76-81; MGR INTERGRATED SYSTS, NAT BROADCASTING CO, 81- *Concurrent Pos:* Instr, Dartmouth Col, 58-59; prof, Dept Elec Eng & Comput Sci, Stevens Inst Technol, 72-82. *Honors & Awards:* David Sarnof Achievement Award Sci. *Mem:* AAAS; Asn Comput Mach; Math Asn Am; Am Math Soc; Inst Elec & Electronics Engrs; Soc Motion Picture TV Engrs. *Res:* Computer system architecture; distributed systems, multiprocessing; information storage and retrieval; data communications; process control; software engineering; television control systems; high definition TV; spectrum utilization. *Mailing Add:* 513 Prospect Ave Princeton NJ 08540

SAMS, CARL EARNEST, b Knoxville, Tenn, Dec 8, 51; m 71; c 3. STRESS PHYSIOLOGY HORTICULTURAL CROPS. *Educ:* Univ Tenn, BS, 74, MS, 76; Mich State Univ, PhD(hort), 80. *Prof Exp:* Res plant physiologist, USDA, 80-83; asst prof, 83-85, ASSOC PROF, UNIV TENN, 85- *Mem:* Am Soc Hort Sci; Int Soc Hort Sci; Am Soc Plant Physiologists. *Res:* Stress physiology of horticultural crops: mineral nutrition, water stress, temperature stress and the relationship between stress related disorders and crop productivity and senescence; postharvest physiology. *Mailing Add:* Dept Plant & Soil Sci Univ Tenn PO Box 1071 Knoxville TN 37901-1071

SAMS, EMMETT SPRINKLE, b Burnsville, NC, July 17, 20; m 46; c 2. MATHEMATICS EDUCATION. *Educ:* Western Carolina Univ, BS, 41; George Peabody Col, MA, 49, NC State Univ, 58; Cornell Univ, 60; Univ Kans, 64. *Prof Exp:* Teacher math, Yancey County Bd Educ, 41-45; teacher, Madison County Bd Educ, 45-47; from instr to assoc prof, 47-57, PROF MATH, MARS HILL COL, 57- *Honors & Awards:* Robert S Gibbs Distinguished Teacher Award, Mars Hill Col, 80, W W Rankin Award Excellence in Math Educ, 83. *Mem:* Math Asn Am; Nat Coun Teachers Math; NC Council of Teachers of Math. *Res:* Serial correlation. *Mailing Add:* Dept Math Mars Hill Col Mars Hill NC 28754

SAMS, JOHN ROBERT, JR, b Kinston, NC, Feb 16, 36; m 63; c 2. CHEMICAL PHYSICS, ORGANOMETALLIC CHEMISTRY. *Educ:* Amherst Col, BA, 58; Univ Wash, PhD(phys chem), 62. *Prof Exp:* NATO fel, Imp Col, Univ London, 62-63; from asst prof to assoc prof, 63-72, PROF CHEM, UNIV BC, 72- *Mem:* NY Acad Sci. *Res:* Moessbauer spectroscopy; theoretical chemistry; magnetochemistry. *Mailing Add:* Dept Chem Univ BC 2075 Westbrook Mall Vancouver BC V6T 1Z2 Can

SAMS, LEWIS CALHOUN, JR, b Dallas, Tex, Sept 13, 28; m 52; c 2. INORGANIC CHEMISTRY. *Educ:* Midwestern Univ, BS, 50; Tex A&M Univ, MS, 54, PhD(inorg chem), 61. *Prof Exp:* Microanalyst, Ft Worth Gen Depot, US Army, 54-56; chemist, Celanese Chem Corp, 56-57; instr chem, ETex State Univ, 57-59; asst prof chem, Trinity Col, Tex, 61-63; from asst prof to assoc prof inorg chem, 63-81, PROF CHEM, TEX WOMAN'S UNIV, 81- *Mem:* Am Chem Soc. *Res:* Microwave spectroscopy; inorganic fluorine synthesis. *Mailing Add:* Dept Chem Tex Woman's Univ Box 23973 Denton TX 76204

SAMS, RICHARD ALVIN, b Lebanon, Ohio, Aug 28, 46. ANALYTICAL CHEMISTRY. *Educ:* Ohio State Univ, BS, 69, PhD(pharm), 75. *Prof Exp:* Sr scientist anal res, Pharmaceut Div, Ciba-Geigy Corp, 74-75; ASST PROF PHARMACOL & VET CLIN SCI, COL VET MED, OHIO STATE UNIV, 76- *Mem:* Am Chem Soc. *Res:* Investigation of high-pressure liquid chromatography and gas liquid chromatography separation mechanisms; investigation of comparative pharmacokinetics in various animal species. *Mailing Add:* 3416 Polley Rd Columbus OH 43221-2133

SAMS, WILEY MITCHELL, JR, b Ann Arbor, Mich, Apr 15, 33; m 59; c 3. DERMATOLOGY. *Educ:* Univ Mich, BS, 55; Emory Univ, MD, 59; Am Bd Dermat, dipl. *Prof Exp:* Intern, Emory Univ Hosp, 59-60; asst resident & resident dermat, Duke Univ Hosp, 60-62, assoc, Med Ctr, 63-64; asst clin prof, Med Ctr, Univ Calif, San Francisco, 65-66; from asst prof to assoc prof, Mayo Grad Sch Med, Univ Minn, 66-72; prof dermat & head div, Univ Colo Med Ctr, Denver, 72-76; prof, Univ NC, Chapel Hill, 76-80; PROF DERMAT & CHMN DEPT, UNIV ALA, 81- *Concurrent Pos:* Nat Cancer Inst fel, 62-64. *Mem:* Am Acad Dermat; Soc Invest Dermat; Am Fedn Clin Res. *Res:* Immunology of skin diseases. *Mailing Add:* Dept Dermat Univ Ala University Sta Birmingham AL 35294

SAMSON, CHARLES HAROLD, b Portsmouth, Ohio, July 12, 24; m 47; c 2. SYSTEMS ENGINEERING, STRUCTURAL ENGINEERING. *Educ:* Univ Notre Dame, BS, 47, MS, 48; Univ Mo, PhD(struct eng), 53. *Prof Exp:* Asst to field rep, Loebl, Schlossman & Bennett, Ill, 48-49; struct engr, Gen Dynamics-Convair, Tex, 51-52, sr struct engr, 52-53; asst prof civil eng, Univ Notre Dame, 53-56; proj aerodyn engr, Gen Dynamics-Ft Worth, 56-58, proj struct engr, 58-60; head, Civil Eng Dept, 64-79, actg pres, 80-81, vpres planning, 81-82, PROF AEROSPACE & CIVIL ENG, TEX A&M UNIV, 60- *Concurrent Pos:* Lectr, Southern Methodist Univ, 52-53 & 56-60. *Mem:*

Am Soc Civil Engrs; Am Soc Eng Educ; Nat Soc Prof Engrs (pres, 87-88); Am Pub Works Asn; Soc Gen Systs Res. *Res:* Systems engineering; structural mechanics; engineering education; systems planning. *Mailing Add:* 2704 Camelot Dr Bryan TX 77802

SAMSON, FRED BURTON, b West Lafayette, Ind, Dec 10, 40; m 66. WILDLIFE ECOLOGY. *Educ:* Ind Univ, Bloomington, BS, 62, MA, 66; Utah State Univ, PhD(biol), 74. *Prof Exp:* Wildlife biologist, Bur Sport Fisheries & Wildlife, US Dept Interior, 68-69, res biologist, 69-70; asst prof wildlife ecol, Pa State Univ, University Park, 74-76; biologist & asst unit leader, 76-81, biologist & unit leader, Mo Coop Wildlife Res Unit, US Fish & Wildlife Serv, 81-; AT COOP WILDLIFE RES UNIT, COLO STATE UNIV. *Mem:* Am Ornithologists Union; Cooper Ornith Soc; Wilson Ornith Soc; Wildlife Soc; Am Soc Mammalogists. *Res:* Avian and mammalian population ecology; endangered species. *Mailing Add:* 21628 Juneo AK 99802-1628

SAMSON, FREDERICK EUGENE, JR, b Medford, Mass, Aug 16, 18; m 45; c 3. PHARMACOLOGY. *Educ:* Univ Chicago, PhD(physiol), 52. *Prof Exp:* From asst prof to prof physiol, Med Ctr, Univ Kans, 52-73, actg chmn dept biochem & physiol, 61-62, chmn dept physiol & cell biol, 62-73, dir, R L Smith Res Ctr, 73-89, EMER PROF, MED CTR, UNIV KANS, 89- *Concurrent Pos:* Staff scientist, Neurosci Res Prog, Mass Inst Technol, 65-82. *Mem:* Am Physiol Soc; NY Acad Sci; Soc Neurosci; Am Soc Neurochem; Am Soc Cell Biol; AAAS; Am Chem Soc. *Res:* Neurochemistry; brain metabolism; microtubular systems; axoplasmic transport; brain regional functional mapping; experimental studies on neurological systems involved in seizures, toxicity and anesthesia; role of the brain extracellular compartment and brain cell microenvironment in central nervous system functions, seizures and toxins; role of oxygen free radicals in brain damage. *Mailing Add:* R L Smith Res Ctr Univ Kans Med Ctr Kansas City KS 66103

SAMSON, JAMES ALEXANDER ROSS, b Scotland, Sept 9, 28; nat US; m 54; c 2. ATOMIC PHYSICS. *Educ:* Univ Glasgow, BSc, 52, DSc, 70; Univ Southern Calif, MS, 55, PhD(physics), 58. *Prof Exp:* Asst physics, Univ Southern Calif, 53-58, res assoc, 58-60; res physicist, Univ Mich & Harvard Univ, 60-61, GCA Corp, Mass, 61-70; prof, 70-81, REGENTS PROF PHYSICS, UNIV NEBR-LINCOLN, 81- *Concurrent Pos:* Assoc ed, J of the Optical Soc Am, 70-81; mem adv screening comt physics, Coun for Int Exchange of Scholars, 78-81; vis prof, Univ Southampton, Eng, 72, Bonn Univ, W Ger, 76, Daresbury Synchrotron Lab, Eng, 76-77, Phys Res Lab, Ahmedabad, India, 77, Univ Hawaii, Honolulu, 80 & Australian Nat Univ, Canberra, 82; mem, Comt Line Spectra Elements-Atomic Spectros, Nat Res Coun, 81-84; chmn, Int Prog Comt, 6th Int Conf Vacuum Ultraviolet Radiation Physics, 80, mem, 83; mem, Comt Appln Physics, Am Phys Soc, 81-85; mem, X-ray & Ultraviolet Tech Comt, Optical Soc Am & Mees Medal Comt, 83, chmn, 85. *Mem:* AAAS; fel Am Phys Soc; fel Optical Soc Am; Sigma Xi. *Res:* Vacuum ultraviolet spectroscopy; atomic and molecular physics; photoelectron spectroscopy. *Mailing Add:* Dept Physics Univ Nebr Lincoln NE 68588-0111

SAMSON, STEN, b Stockholm, Sweden, Mar 25, 16; m 48; c 2. CHEMISTRY. *Educ:* Univ Stockholm, Fil Kand, 53, Fil Lic, 56, Fil Dr, 68. *Prof Exp:* Res fel chem, Univ Stockholm, 48-53; res fel, 53-61, sr res fel, 61-73, res assoc, 73-80, sr res assoc chem, 80-86, EMER SR RES ASSOC CHEM, CALIF INST TECHNOL, 86- *Concurrent Pos:* Consult, Comn Crystal Data, Int Union Crystal, 67-73, Syst Anal Instruments, 69-71 & Advan Res & Applications Corp, 80-; mem, US Panel Joint US Brazil Study Group Grad Training & Res Brazil, 75-76. *Mem:* Am Crystallog Asn. *Res:* Crystal structures of very complex intermetallic compounds; crystallographic tranformations associated with changes in physical properties, paralelectric and ferroelectric, conductors and insulator transitions especially in one-dimensional conductors; structures of quasi-crystalline substances. *Mailing Add:* Beckman Inst Calif Inst Technol MC 139-74 Pasadena CA 91125

SAMSON, WILLIS KENDRICK, b Syracuse, NY, May 15, 47. NEUROENDOCRINOLOGY, PEPTIDE NEUROCHEMISTRY. *Educ:* Duke Univ, AB, 68; Univ Tex Health Sci Ctr Dallas, PhD(physiol), 79. *Prof Exp:* ASST PROF PHYSIOL, UNIV TEX HEALTH SCI CTR DALLAS, 81- *Mem:* Endocrine Soc. *Res:* Neuroendocrinology; brain and gut peptides; control of anterior pituitary function. *Mailing Add:* Dept Physiol Univ Tex Health Sci Ctr 5323 Harry Hines Blvd Dallas TX 75235

SAMTER, MAX, b Berlin, Ger, Mar 3, 08; nat US; m 47; c 1. CLINICAL MEDICINE, ALLERGY & CLINICAL IMMUNOLOGY. *Educ:* Univ Berlin, MD, 33; Univ Ill, MS, 47; Am Bd Internal Med & Am Bd Allergy & Immunol, dipl, 49 & 74. *Prof Exp:* Asst dispensary physician, Sch Med, Johns Hopkins Univ, 37-38; asst biochem, 46, from instr to prof med, 46-80, head sect allergy & clin immunol, 47-75, assoc dean clin affairs, 74-75, chief of staff, Univ Hosp, 74-75, EMER PROF MED, ABRAHAM LINCOLN SCH MED, UNIV ILL MED CTR, 80-; SR CONSULT, MAX SAMTER INST ALLERGY & CLIN IMMUNOL, GRANT HOSP, CHICAGO, 84- *Concurrent Pos:* Consult, Chicago West Side, Hines Vet Admin & West Suburban Hosps; dir, Max Samter Inst Allergy & Clin Immunol, Grant Hosp, Chicago, 75-84. *Honors & Awards:* Outstanding Clinician Award, Am Acad Allergy & Immunol, 90. *Mem:* Am Med Asn; fel Am Col Physicians; Am Acad Allergy & Immunol (treas, 54, pres, 58); Int Asn Allergol & Clin Immunol; Interasma; Sigma Xi. *Res:* Function of eosinophils; mechanism of drug reactions; pathogenesis of bronchial asthma. *Mailing Add:* 645 Sheridan Rd Evanston IL 60202-2533

SAMUEL, ALBERT, b Tanjore, India, Feb 27, 37; US citizen; m 69; c 2. MOLECULAR BIOLOGY. *Educ:* Univ Madras, India, BA, 59, MSc, 61; Oberlin Col, Ohio, MA, 65; Mich State Univ, PhD(entomol), 71. *Prof Exp:* Demonstr zool, Am Col, Madurai, India, 61-63; fel entom, Mich State Univ, 71-73; ASSOC PROF BIOL & CHMN DEPT SCI & MATH, ST PAUL'S COL, VA, 73- *Concurrent Pos:* Activities coordr, 16 Insts Health Sci Consortium & Health Serv Consortium, St Paul's Col, 73-, dir, Biomed Res Activities, 73-; NIH fac fel, 77-79; res biologist, Lawrence Livermore Lab. *Mem:* Tissue Culture Asn; AAAS; Cell Kinetic Soc; Soc Analytical Cytol. *Res:* Comparative study of chemically induced aging and natural aging in cells in vitro; kinetics of tumor cells and its application in chemotherapy; aging and non-aging mammalian cells are used; changes at the molecular level studied and compared; correlation studies of the expression of cell surface antigens and specific cell cycle phases of in vitro T cell line. *Mailing Add:* Dept Sci & Math Frederick Community Col 7932 Opposumtown Pike Fredrick MD 21701

SAMUEL, ARYEH HERMANN, b Hildesheim, Ger, Feb 19, 24; US citizen; Wid; c 1. OPERATIONS RESEARCH. *Educ:* Univ Ill, BS, 43; Northwestern Univ, MS, 46; Univ Notre Dame, PhD(chem), 53. *Prof Exp:* Scientist, Broadview Res, 56-60, Stanford Res Inst, 60-65; res leader phys chem, Gen Precision, 65-67; sr scientist, Stanford Res Inst, 67-72; criminalist, County Santa Clara, Calif, 72-74; sr scientist, Vector Res Inc, 74-77; prin res scientist, Battelle Mem Inst, 77-87; CONSULT, 88- *Honors & Awards:* Lanchester Prize, Oper Res Soc Am, 62. *Mem:* Oper Res Soc Am. *Res:* Operations research-public systems, especially military and postal; remote sensing; chemical effects of radiations. *Mailing Add:* 10861 Bucknell Dr Wheaton MD 20902

SAMUEL, CHARLES EDWARD, b Portland, Ore, Nov 28, 45; m 68; c 2. VIROLOGY, INTERFERON. *Educ:* Mont State Univ, BS, 68; Univ Calif, Berkeley, PhD(biochem), 72. *Prof Exp:* Damon Runyon Scholar, Duke Univ Med Ctr, 72-74; from asst prof to assoc prof biol, 74-83, DIR, PROG MOLECULAR BIOL & BIOCHEM, UNIV CALIF, SANTA BARBARA, 88- *Concurrent Pos:* Fel, Damon Runyon-Walter Winchell Cancer Fund, 72-74; prin investr, Nat Inst Allergy & Infectious Dis 75-, Am Cancer Soc, 75-85; Res Career Develop Award, NIH, 79-84; assoc ed, Virol, 80-, J Interferon Res, 80-, J Virol, 83- & J Biol Chem, 89-; consult, NIH, 80-; Vis prof, Univ Zurich, 86-87; Merit Award, NIH, 89- *Mem:* Am Soc Biol Chemists; Am Soc Microbiol; Am Soc Virol; Int Soc Interferon Res. *Res:* Biochemistry of animal virus-cell interactions; mechanism of interferon action; molecular biology of reoviruses; translational control mechanisms. *Mailing Add:* Prog Biochem Molecular Biol Dept Biol Sci Univ Calif Santa Barbara CA 93106

SAMUEL, DAVID EVAN, b Johnstown, Pa, July 28, 40; m 66. WILDLIFE BIOLOGY, ORNITHOLOGY. *Educ:* Juniata Col, BS, 62; Pa State Univ, MS, 64; Univ WVa, PhD(zool), 69. *Prof Exp:* Instr biol, Bethany Col, 64-66; instr zool, 68-69, ASSOC PROF WILDLIFE BIOL, WVA UNIV, 69-, ASSOC WILDLIFE BIOLOGIST, 76- *Concurrent Pos:* USDA grant. *Mem:* Am Ornith Union; Wildlife Soc; Wilson Ornith Soc; Nat Audubon Soc. *Res:* Behavior. *Mailing Add:* Dept Wildlife Mgt WVa Univ Morgantown WV 26506

SAMUEL, EDMUND WILLIAM, b Canton, Ohio, Sept 17, 24; m 63. DEVELOPMENTAL BIOLOGY. *Educ:* Case Western Reserve Univ, BSEE, 45, MS, 49; Princeton Univ, MS, 59, PhD(biol), 60. *Prof Exp:* Res investr theoret physics, Sperry Gyroscope Corp, 49-50; res asst med physics, Mass Gen Hosp, Boston, 52-53; from asst prof to assoc prof biol, 60-71, PROF BIOL, ANTIOCH COL, 72- *Concurrent Pos:* NSF instrumentation grant, 62-64, undergrad res partic grant, 65-66. *Res:* Biophysics; theoretical and molecular biology; history and philosophy of science; East Asian science such as Japanese medicine; bioethics. *Mailing Add:* Dept Biol Southern Conn St Col 501 Crescent St New Haven CT 06515

SAMUEL, JAY MORRIS, b Stuttgart, WGer, Jan 24, 46. WELDING METALLURGY, WELDING PROCESSES. *Educ:* Rensselaer Polytech Inst, BS, 67, PhD(mat eng), 79. *Prof Exp:* Adj prof mech tech, Hudson Valley Community Col, 77-79; ASST PROF MECH ENG, UNIV WIS, MADISON, 79- *Honors & Awards:* Clyde Sanders Award, Foundry Educ Found & Am Colloid Co, 82. *Mem:* Am Welding Soc; Am Soc Metals. *Res:* Solidification mechanics; physical metallurgy of weldments; control of welding processes; structure; properties of engineering materials. *Mailing Add:* Dept Mech Eng Univ Wis 1513 University Ave Madison WI 53706

SAMUEL, MARK AARON, b Montreal, Que, Jan 26, 44; div; c 2. THEORETICAL HIGH ENERGY PHYSICS. *Educ:* McGill Univ, BSc, 64, MSc, 66; Univ Rochester, PhD(physics), 69. *Prof Exp:* Asst prof, 69-75, assoc prof, 75-81, PROF PHYSICS, OKLA STATE UNIV, 81- *Concurrent Pos:* Consult, NSF Educ Res Grant, Okla State Univ, 72-75; vis scientist, Stanford Linear Accelerator Ctr, 73 & 75; res grant, US Energy Res & Develop Admin, Dept Energy, 76-; vis scientist, Niels Bohr Inst, Copenhagen, Denmark, 77; vis scientist, Aspen Ctr Physics, 81, 85, 86, 87 & 89. *Mem:* Am Phys Soc; Am Asn Physics Teachers; Can Asn Physicists. *Res:* Field theory; particle physics; atomic physics; tests of quantum electrodynamics and quantum chromodynamics; applied mathematical techniques. *Mailing Add:* Dept Physics Okla State Univ Stillwater OK 74078

SAMUEL, WILLIAM MORRIS, b Windber, Pa, July 28, 40; m 70; c 2. PARASITOLOGY. *Educ:* Juniata Col, BSc, 62; Pa State Univ, MSc, 65; Univ Wis, PhD(vet sci, zool), 69. *Prof Exp:* Fel parasitol, 69-71, asst prof zool, 71-75, assoc prof, 75-78, PROF ZOOL, UNIV ALTA, 81- *Concurrent Pos:* Assoc ed, Can J Zool, 84-; asst ed, J Wildlife Dis, 85- *Mem:* Am Soc Parasitol; Wildlife Dis Asn; Wildlife Soc; Can Soc Zoologists; Am Inst Biol Sci. *Res:* Epizootiology of wildlife parasites; importance for host populations; emphasis on big game. *Mailing Add:* Dept Zool Univ Alta Edmonton AB T6G 2M7 Can

SAMUEL-CAHN, ESTER, Oslo, Norway, May 16, 33; Israeli citizen; m 70; c 4. DECISION THEORY, SEQUENTIAL ANALYSIS. *Educ:* Hebrew Univ, BA, 58; Columbia Univ, NY, MA, 59, PhD(statist), 61. *Prof Exp:* PROF STATIST, HEBREW UNIV, 62- *Concurrent Pos:* Vis prof, Columbia Univ, 80 & Rutgers Univ, 81. *Mem:* Inst Math Statist; fel Am Statist Asn; Int Statist Inst; Israel Statist Asn; Israel Soc Oper Res. *Mailing Add:* Dept Statist Hebrew Univ Jerusalem 91905 Israel

SAMUELS, ARTHUR SEYMOUR, b New York, NY, July 24, 25; c 4. PSYCHOANALYTIC MEDICINE, PSYCHOSOMATIC MEDICINE. *Educ:* Cornell Univ, BA, 44, MA, 49; Tulane Univ, MD, 53. *Prof Exp:* Dir, New Orleans Mental Health Clinic, 56-58; teacher group psychotherap, New Orleans Ctr Psychotherap, 76-77; ASSOC PROF CLIN PSYCHIAT, LA STATE UNIV MED SCH, 79-; DIR, STRESS TREATMENT CTR, NEW ORLEANS, 87- *Concurrent Pos:* Pvt pract pyschiat, 58-; dir, Biofeedback Ctr New Orleans, 76- *Mem:* Fel Am Acad Psychoanal; Fel Am Psychiat Asn. *Res:* The production of essential hypertension through automatic conditioning techniques; evaluation of various phenothiazine drugs in psychosis; the role of the composition of group in the efficacy of group therapy; using a group approach for reducing inter-racial prejudice; the combined use of hypnosis and biofeedback in the treatment of stress related illnesses; geriatric medicine. *Mailing Add:* 4510 St Charles Ave New Orleans LA 70115

SAMUELS, MARTIN E(LMER), b Dayton, Ohio, Apr 24, 18; m 43; c 2. CHEMICAL ENGINEERING. *Educ:* Univ Dayton, BChE, 39. *Prof Exp:* Chemist, Dayton Tire & Rubber Co, 39-43; chemist, Copolymer Rubber & Chem Corp, 43-45, develop supvr, 45-47, develop mgr, 57-61, prod qual mgr, 61-67, mgr tech serv, 67-81, asst to vpres mkt, 81-82; RETIRED. *Concurrent Pos:* Lectr. *Mem:* Am Chem Soc. *Res:* Development, evaluation, quality control, end uses and utilization of synthetic latexes and elastomers. *Mailing Add:* 8021 Owen St Baton Rouge LA 70809

SAMUELS, MYRA LEE, b Chicago, Ill, Mar 23, 40; m 67; c 2. STATISTICS. *Educ:* Swarthmore Col, BA, 61; Univ Calif, Berkeley, PhD(statist), 69. *Prof Exp:* Mathematician, US Naval Res Lab, 61-63; res asst statist & biostatist, Univ Calif, Berkeley, 66-68; instr statist, Purdue Univ, Lafayette, 68-69, asst prof, 70-71, vis asst prof, 72-79, vis lectr statist, 79-87, VIS INSTR & ASST HEAD STATIST CONSULT, PURDUE UNIV, LAFAYETTE, 87- *Concurrent Pos:* Consult statist, Bur Drugs, US Food & Drug Admin, 80-87. *Mem:* Am Statist Asn; Biomet Soc. *Res:* Biostatistics; applied statistics. *Mailing Add:* Dept Statist Purdue Univ West Lafayette IN 47907

SAMUELS, ROBERT, b Philadelphia, Pa, June 12, 18; m 48; c 3. BIOLOGY. *Educ:* Univ Pa, AB, 38, MA, 40; Univ Calif, PhD(zool), 52. *Prof Exp:* Jr entomologist, USPHS, 41-43; teaching asst zool, Univ Calif, 46-49, assoc, 49-52; instr biol, Calif State Polytech Col, 53; from instr to asst prof microbiol, Sch Med, Univ Colo, 53-63, vis prof, 63; prof, Merharry Med Col, 63-67; prof biol, Purdue Univ, Indianapolis, 67-70; prof biol, Ind Univ-Purdue Univ, Indianapolis, 70-78, chmn dept, 72-76; prof biol & chmn dept, 79-83, prof, 83-88, EMER PROF, ETENN STATE UNIV, 88- *Concurrent Pos:* Sect rep, Purdue Univ, 69-70; lectr, Univ Calif, 50-52; lectr, Sch Med, Univ Colo; consult, Indian Health Surv, Wyo State Bd Health, USPHS, Wetherill Mesa Archaeol Proj, Nat Park Serv & Nat Geog Surv. *Mem:* Soc Protozool (asst treas, 58-60, treas, 60-66, pres, 72-73); Am Soc Microbiol; Am Micros Soc; Soc Exp Biol & Med; Am Soc Parasitol; Sigma Xi. *Res:* Protozoology; cytology; nutrition; morphogenesis. *Mailing Add:* Dept Biol Sci Box 23590A ETenn State Univ Johnson City TN 37614

SAMUELS, ROBERT BIRELEY, b Palo Alto, Calif, Feb 27, 40, m 60; c 3. FOOD CHEMISTRY. *Educ:* Calif State Polytech Col, BS, 62; Univ Ill, PhD(food sci), 65. *Prof Exp:* USPHS trainee fel, 65-66; res chemist, Beckmam Instruments, Inc, Palo Alto, 66-67, sr res chemist, 67, group supvr chromatography res, 67-68, group supvr chromotography res & appln, 68-71, prog coordr bioprod, 71-72, biochem prog mgr, 72-73, mgr bioprod, Spino Div, 73-; mem staff, Smith Kline Beckman, Philadelphia, 73-87; PRES, ALLERGAN HUMPHREY, 87- *Mem:* AAAS; Am Chem Soc; Inst Food Technologists; Sigma Xi. *Res:* Lipid chemistry; lipoprotein structure; peptide synthesis and purification; instrumentation. *Mailing Add:* Allergan Humphrey 3081 Teagarden St San Leandro CA 94577

SAMUELS, ROBERT JOEL, b Brooklyn, NY, Jan 8, 31; m 86; c 2. PHYSICAL CHEMISTRY, POLYMER PHYSICS. *Educ:* Brooklyn Col, BS, 52; Stevens Inst Technol, MS, 55; Univ Akron, PhD(polymer chem), 61. *Prof Exp:* Res chemist, Picatinny Arsenal, 52-55, Res Ctr, Goodyear Tire & Rubber Co, 57-59 & Inst Rubber Res, Akron, 59-60; res chemist, Res Ctr, Hercules Inc, 60-67, sr res chemist, 67-70, res scientist, 70-79; PROF, GA INST TECHNOL, 79- *Concurrent Pos:* Adj prof, Dept Chem Eng, Univ Del, 72-79; affil prof, Dept Chem Eng, Univ Wash, 78-80. *Honors & Awards:* Am Chem Soc Award, 71; Soc Plastics Engrs Award, 83. *Mem:* Am Chem Soc; Am Phys Soc; Soc Plastics Engrs; Polymer Processing Soc. *Res:* Polymer morphology and mechanics; small and wide angle x-ray diffraction; infrared spectroscopy; birefringence and refractometry; small-angle light scattering, sonic and mechanical properties of polymers; chemical stress relaxation of polymers; polymer chromatography; physical chemistry of stress relaxation of polmers; polymer chromatography; physical chemistry of dilute solutions; polyolefins, polimides, polyesters and others. *Mailing Add:* Sch Chem Eng Ga Inst Technol Atlanta GA 30332-0100

SAMUELS, ROBERT LYNN, b Goldfield, Iowa, Oct 20, 30; m 52; c 3. ELECTRICAL ENGINEERING. *Educ:* Iowa State Univ, BS, 59, MS, 60, PhD, 63. *Prof Exp:* Adv res engr, Sylvania Electronic Systs-West, 63; asst prof elec eng, 63-70, ASSOC PROF ELEC ENG, IOWA STATE UNIV, 70- *Res:* Thin ferromagnetic film materials. *Mailing Add:* Catawba Valley Community Col Rte 3 Box 283 Hickory NC 28602

SAMUELS, STANLEY, b New York, NY, Oct 13, 29; m 51; c 3. NEUROCHEMISTRY. *Educ:* Syracuse Univ, AB, 51, MS, 54, PhD(biochem), 58. *Prof Exp:* Asst instr zool, Syracuse Univ, 52-54 & 55-57; asst biochem, Col Med, Univ Ill, 54-55; instr ophthalmic res, Col Med, Western Reserve Univ, 57-59; instr, Albert Einstein Col Med, 61-63; asst prof, 64-69, ASSOC PROF EXP NEUROL, SCH MED, NY UNIV, 69- *Concurrent Pos:* Nat Inst Neurol Dis & Blindness fel neurol, Albert Einstein Col Med, 59-61; Nat Multiple Sclerosis Soc fel, 61-63; consult, NY Eye & Ear Infirmary, 61-63. *Mem:* AAAS; Harvey Soc; Asn Res Nerv & Ment Dis; Int Soc Neurochem; Am Soc Neurochem. *Res:* Amino acid transport; thin-layer and high performance liquid chromatography; inborn metabolic errors; brain biochemistry. *Mailing Add:* Dept Neurol NY Univ Med Ctr New York NY 10016

SAMUELSON, BENGT INGEMAR, b Halmstad, May 21, 34; m 58; c 3. MEDICINE, PHYSIOLOGICAL CHEMISTRY. *Hon Degrees:* Dr, Univ Chicago, 78, Univ Ill, 83. *Prof Exp:* Res fel, Harvard Univ, 61-62; asst prof, Karolinska Inst, 61-66, chair, Dept Physiol Chem, 73-83, dean, Fac Med, 78-83, PROF MED & PHYSIOL CHEM, KAROLINSKA INST, 72-, PRES FAC MED, 83- *Honors & Awards:* Nobel Prize in Physiol of Med, 82. *Mem:* Foreign assoc Nat Acad Sci; hon mem Asn Am Physicians; AAAS; Am Soc Biol Chemists. *Mailing Add:* Karolinska Inst Solnavagen 1 Stockholm 10401 Sweden

SAMUELSON, CHARLES R, b Crookston, Minn, Feb 10, 27; m 56; c 3. BOTANY, ZOOLOGY. *Educ:* Moorhead State Univ, BS, 52; Univ Northern Colo, Greeley, MA, 56; NDak State Univ, Fargo, PhD(entom), 76. *Prof Exp:* Teacher biol & chem, Dawson, Minn pub sch, 52-54 & Thief River Falls, 55-65; PROF BIOL, NORTHLAND COMMUNITY COL, THIEF RIVER FALLS, MN, 65- *Concurrent Pos:* Survey entomologist, Dept Agr, Minn, 58-81; consult entomologist, 81- *Mem:* Entom Soc Am; Nat Asn Biol Teachers; Am Registry Prof Entomologists; Nat Educ Asn; Am Fedn Teachers. *Res:* Life cycle and associated environmental factors of the sunflower midge, Contarinia Schulzi Gagne in northwest Minnesota and northeast North Dakota; diptera: chironomidae in municipal sewage lagoons, (chironomid midges) biology and control. *Mailing Add:* Northland Community Col Hwy 1-E Thief River Falls MN 56701

SAMUELSON, DON ARTHUR, b Boston, Mass, Aug 30, 48; m 77; c 2. ZINC, GLAUCOMA. *Educ:* Boston Univ, BA, 71; Univ Fla, PhD(mycol), 77, MS, 82. *Prof Exp:* Asst prof, 82-88, ASSOC PROF VET OPHTHAL, COL VET MED, UNIV FLA, 88- *Mem:* Asn Res Vision & Ophthal; Int Soc Eye Res. *Res:* Aqueous humor dynamics in vertebrate eyes; spontaneous glaucoma in the dog and monkey; zinc nutrition and age related macular degeneration. *Mailing Add:* Box J-126 HSC Univ Fla Gainesville FL 32610

SAMUELSON, DONALD JAMES, b Warren, Pa, May 15, 40; m 65; c 2. MATHEMATICS. *Educ:* Cornell Col, BA, 62; Univ Calif Berkeley, MA, 65; Univ Calif, Santa Barbara, PhD(math), 69. *Prof Exp:* Asst prof math, Cornell Col, 65-67, Univ Hawaii, 69-71 & Pa State Univ, McKeesport, 71-75; asst prof, 75-77, ASSOC PROF MATH, PA STATE UNIV, UNIVERSITY PARK, 77- *Mem:* Am Math Soc. *Res:* Universal algebra. *Mailing Add:* 2273 Fairhill Lane San Jose CA 95125

SAMUELSON, H VAUGHN, b Uniontown, Pa, Dec 1, 38; m 67; c 3. FIBER ENGINEERING, STATIC CHARGE CONTROL. *Educ:* Pa State Univ, BS, 60; Western Reserve Univ, MS, 63, PhD(physics), 65. *Prof Exp:* Res chemist, 65-67, sr res chemist, 67-79, RES ASSOC, PIONEERING RES LAB, DUPONT CO, 79- *Mem:* NY Acad Sci; Sigma Xi. *Res:* Research and development on fibers and fiber systems: regulate static charging, aesthetics, membranes and bioreactors, spinneret technology, fiber engineering and spinning dynamics, conductive polymers, polymer blends and multicomponent fibers; recipient of 10 patents. *Mailing Add:* 69 Bullock Rd Chadds Ford PA 19317

SAMULON, HENRY A, b Graudenz, Ger, Dec 26, 15; nat US; m 43; c 2. ELECTRONICS. *Educ:* Swiss Fed Inst Technol, MS, 39. *Prof Exp:* Res engr & instr, Acoust Lab, Swiss Fed Inst Technol, 43-44, assoc, Inst Commun Technol, 44-47; mem staff, Electronics Lab, Gen Elec Co, 47-51, mgr, Eng Anal Subsect, 51-55; gen mgr, Electronic Systs Div, TRW Systs, 55-71, vpres, 64-71, vpres electronics equip, TRW Electronics, 71-74; vpres & mgr, Electronics Div, Xerox Corp, 74-81; PRES, H A SAMULON CONSULT, 81- *Mem:* Fel Inst Elec & Electronics Engrs. *Res:* Circuitry; color television; space and missile guidance and communication systems; microelectronics; general management. *Mailing Add:* 575 Muskingum Ave Pacific Palisades CA 90272

SAMULSKI, EDWARD THADDEUS, b Augusta, Ga, May 23, 43; m 76; c 1. PHYSICAL CHEMISTRY, POLYMER CHEMISTRY. *Educ:* Clemson Univ, BS, 65; Princeton Univ, PhD(chem), 69. *Prof Exp:* NIH fel, State Univ Groningen, 69-70; univ fel, Univ Tex, Austin, 71-72; from asst prof to prof chem, Univ Conn, 72-88; PROF CHEM, UNIV NC, 88- *Mem:* AAAS; Am Chem Soc; Sigma Xi. *Res:* Liquid crystals; biological macromolecules and synthetic polymers; application of magnetic resonance techniques to study molecular dynamics of polymer solutions and liquid crystal phases. *Mailing Add:* Dept Chem Univ NC Chapel Hill NC 27599-3290

SAMWORTH, ELEANOR A, b Wilmington, Del, May 10, 36. PHYSICAL CHEMISTRY. *Educ:* Wilson Col, AB, 58; Johns Hopkins Univ, MA, 60, PhD(phys chem), 63. *Prof Exp:* Nat Cancer Inst fel phys chem, Harvard Univ, 63-64; from asst prof to assoc prof, 64-81, PROF CHEM, SKIDMORE COL 81-, DEPT CHAIR, 85- *Mem:* Am Phys Soc; Sigma Xi; Am Chem Soc. *Res:* Molecular structure; vibrational spectroscopy; inorganic synthesis; computer applications to chemical education. *Mailing Add:* Dept Chem & Physics Skidmore Col Saratoga Springs NY 12866

SANABOR, LOUIS JOHN, b Cleveland, Ohio, Dec 2, 20; m 45; c 1. CHEMICAL ENGINEERING. *Educ:* Case Inst Technol, BS, 42; Carnegie Inst Technol, MS, 54. *Prof Exp:* Asst shift foreman, Butadiene Div, 43-45, from pilot plant group leader to sr chem engr, Res Dept, 45-58, staff asst to mgr explor res sect, 58-65, pilot plant group leader, Cost Anal & Design Group, 65-81, SR CHEM ENGR, ENG EVAL GROUP, RES DEPT, KOPPERS CO, INC, 70- *Mem:* Am Inst Chem Engrs; Am Chem Soc. *Mailing Add:* 311 Parkridge Dr Pittsburgh PA 15235

SANADI, D RAO, b India, July 8, 20; nat US; m 50; c 2. BIOCHEMISTRY. *Educ:* Univ Calif, PhD(biochem), 49. *Prof Exp:* Fel, Nat Cancer Inst, 49-52, res assoc, 52-53; asst prof biochem, Univ Wis, 53-55; asst prof, Univ Calif, 55-58; chief sect comp biochem, NIH, 58-66; exec dir, 69-71 & 75-77, DIR, DEPT CELL PHYSIOL, BOSTON BIOMED RES INST, 66- *Concurrent Pos:* Estab investr, Am Heart Asn, 54-58; chmn, Gordon Res Conf Energy Coupling Mechanisms, 69, 72 & 74; chmn, Gordon Res Conf Biol of Aging,

74; mem adult develop & aging res & training comt, Nat Inst Child Health & Human Develop, 70-73; mem adv panel metab biol, NSF, 71-74; assoc prof, Dept Biol Chem, Harvard Med Sch, 75-; ed, J Bioenergetics Biomembrane, 75- *Mem:* AAAS; Am Chem Soc; Am Soc Biol Chem; fel Geront Soc; Biophys Soc. *Res:* Intermediary metabolism; bioenergetics; enzymology; isotopes; aging. *Mailing Add:* Boston Biomed Res Inst 20 Stanford St Boston MA 02114

SAN ANTONIO, JAMES PATRICK, b New York, NY, June 4, 25; m 51; c 3. HORTICULTURE, PHYSIOLOGY. *Educ:* Univ Chicago, SB, 48, PhD(physiol), 51. *Prof Exp:* Res assoc mycol physiol, Univ Chicago, 51-53; plant physiologist, Sect Cotton & Other Fiber Crops & Dis, Agr Res Serv, 53-55, Veg & Ornamentals Res Br, Crops Res Div, 55-72, horticulturist, Plant Genetics & Germplasm Inst, Sci & Educ Admin-Agr Res, USDA, 72-88; RETIRED. *Mem:* Am Soc Plant Physiol; Bot Soc Am; Am Soc Hort Sci; Mycological Soc Am. *Res:* Cultivation of edible fungi. *Mailing Add:* 12411 Salem Lane Bowie MD 20715

SANATHANAN, C(HATHILINGATH) K, b Kerala, India, Feb 17, 36; m 63, 88. COMPUTER CONTROL, CONTROL SYSTEM DESIGN. *Educ:* Univ Madras, BS, 59; Case Western Reserve Univ, MS, 62, PhD(eng), 64. *Prof Exp:* Asst nuclear engr, Argonne Nat Lab, Ill, 63-68; assoc prof info eng, 68-71, PROF ELEC ENG, UNIV ILL, CHICAGO, 71- *Mem:* Am Nuclear Soc; Inst Elec & Electronics Engrs. *Res:* Automation control theory; control of aircrafts chemical processes and power plants; simulation of large scale systems; control and simulation; urban mass transportation. *Mailing Add:* Dept Elec Eng Box 4348 Univ Ill Chicago IL 60680

SANATHANAN, LALITHA P, b Sandakan, NBorneo, Jan 21, 43; nat US; m 63; c 1. APPLIED STATISTICS, MATHEMATICAL STATISTICS. *Educ:* Univ Mysore, BSc, 60, MSc, 62; Univ Chicago, PhD(statist), 69. *Prof Exp:* Res asst, Nat Opinion Res Ctr, 67-68; asst prof, 68-72, ASSOC PROF STATIST, UNIV ILL, CHICAGO CIRCLE, 72- *Concurrent Pos:* Consult, Presby St Lukes Hosp, 69-70; vis mathematician, Argonne Nat Lab, 78-79. *Mem:* Am Statist Asn. *Res:* Multinomial analysis; regression; multivariate statistics; decision theory; optimal allocation of resources. *Mailing Add:* 10 S-261 Argonne Ridge Rd Hinsdale IL 60521

SANAZARO, PAUL JOSEPH, b Sanger, Calif, Sept 27, 22; m 74. MEDICINE. *Educ:* Univ Calif, AB, 44, MD, 46; Am Bd Internal Med, dipl, recert, 77. *Prof Exp:* From asst prof to assoc prof med, Sch Med, Univ Calif, San Francisco, 53-62; dir div educ, Asn Am Med Cols, Ill, 62-68; dir, Nat Ctr Health Serv, Res & Develop, 68-72; assoc dep adminr develop, Health Serv & Ment Health Admin, US Dept Health, Educ & Welfare, 72-73; dir pvt initiative, Prof Standards Rev Orgn, 73-77, dir, pvt initiative qual assurance, 78-82; CLIN PROF MED, SCH MED, UNIV CALIF, SAN FRANCISCO, 75- *Concurrent Pos:* Clin assoc prof, Univ Ill Col Med, 62-67, clin prof, 67-68; consult, NIH & Dept Health & Human Serv; pvt consult, health serv res & develop. *Mem:* Fel Am Col Physicians. *Res:* Health services research and development, especially quality of medical care. *Mailing Add:* 1126 Grizzly Peak Blvd Berkeley CA 94708

SANBERG, PAUL RONALD, b Coral Gables, Fla, Jan 4, 55. ANIMAL MODELS OF NEURODEGENERATIVE DISORDERS, BRAIN-BEHAVIOR RELATIONS. *Educ:* York Univ, BSc, 76; Univ BC, MSc, 79; Australian Nat Univ, PhD (behav biol), 81; GDipl, Curtin Univ, 85. *Prof Exp:* res neurosci, Johns Hopkins Med Sch, 81-83; asst prof psychol, Ohio Univ, Athens, 83-86; assoc prof, 86-89, PROF PSYCHIAT, UNIV CINCINNATI MED SCH, 89-; SCI DIR, CELLULAR TRANSPLANTS, INC, 90- *Concurrent Pos:* Vis scholar, dept neurosci, Univ Calif, San Diego, 80; prin investr grants, NIH, Huntingtons Dis Found, Hereditary Dis Found, Tourette Syndrome Asn, Pratt Found, 82-; mem sci adv bd, Tourette Syndrome Asn, 90-, Am Health Assistance Found, 90; prof psychiat, Brown Univ, 90- *Honors & Awards:* Young Investr Award, Am Col Neuropsychopharmacol. *Mem:* Soc Neurosci; Am Psychol Asn; AAAS; Psychonomic Soc; Sigma Xi; NY Acad Sci; Am Col Neuropsychopharmacol. *Res:* Understanding how the brain controls movement and behavior; develop animal models of various neuropsychiatric disorders, in order to test various experimental treatments; studying the ability of brain tissue transplants into animal models of Parkinsons, Huntingtons and Alzheimers diseases. *Mailing Add:* Cellular Transplants Inc Four Richmond Sq Providence RI 02906

SANBORN, ALBERT FRANCIS, b Calif, June 21, 13; m 36, 71; c 2. GEOLOGY. *Educ:* Fresno State Col, AB, 48; Stanford Univ, MS, 50, PhD(geol), 52. *Prof Exp:* Div stratigrapher, Western Opers, Standard Oil Co, Calif, 55-63, sr geologist, Western Div, Calif Oil Co, 63-66, sr explor geologist, Western Div, Chevron Oil Co, Denver, 66-78; GEOL CONSULT, 78- *Mem:* Soc Independent Prof Earth Scientists; Am Asn Petrol Geologists; Soc Independent Prof Earth Scientists. *Res:* Petroleum geology; stratigraphy; mineralogy-petrology. *Mailing Add:* 460 S Marion Pkwy No 1552B Denver CO 80209

SANBORN, CHARLES E(VAN), b Mankato, Minn, July 11, 19; m 41, 80; c 4. CHEMICAL ENGINEERING, MATHEMATICS. *Educ:* Univ Minn, BChE, 41, PhD(chem eng), 49. *Prof Exp:* Engr, Shell Develop Co, Calif, 49-72, Shell Oil Co, 72-73, staff res engr, 73-82; RETIRED. *Mem:* Am Inst Chem Engrs. *Res:* Process development; industrial chemicals. *Mailing Add:* 205 Castle Hill Ranch Rd Walnut Creek CA 94595

SANBORN, I B, b Pepperell, Mass, Apr 29, 32; m 53; c 6. AUTOMATIC CONTROL SYSTEMS. *Educ:* Rensselaer Polytech Inst, BChE, 54; Inst Paper Chem, MCh, 56, PhD, 61. *Prof Exp:* Proj engr, Consol Papers, Inc, 61-65, process engr, 65-67, mgr process control, 67-68, mgr process develop & control, 68-76, ASSOC DIR PROD DEVELOP, CONSOL PAPERS, INC, 76- *Mem:* Tech Asn Pulp & Paper Indust; Instrument Soc Am. *Res:* Application of automatic control principles to the control of the paper making process. *Mailing Add:* Consol Papers Inc PO Box 50 Wisconsin Rapids WI 54494

SANBORN, MARK ROBERT, b Mason City, Iowa, Mar 17, 46; m 68; c 2. VIROLOGY, IMMUNOLOGY. *Educ:* Univ Northern Iowa, BA, 68, MA, 71; Iowa State Univ, PhD(bacteriol), 76. *Prof Exp:* ASST PROF VIROL & IMMUNOL, OKLA STATE UNIV, 76- *Mem:* Am Soc Microbiol; Sigma Xi; NY Acad Sci; AAAS. *Res:* Early events in paramyxovirus infection; induction of neurological autoimmune disease; immunological assay for gene expression. *Mailing Add:* Microbiol Dept LSE 313 Okla State Univ Stillwater OK 74074

SANBORN, RUSSELL HOBART, b Laconia, NH, Mar 31, 30; m 90. PHYSICAL CHEMISTRY. *Educ:* Wesleyan Univ, BA, 52; Univ Calif, PhD(chem), 56. *Prof Exp:* chemist, Lawrence Livermore Lab, Univ Calif, 56-85; CONSULT, 86- *Mem:* Sigma Xi. *Res:* Infrared spectroscopy; molecular structure and spectra; numerical analysis; data processing; computer simulation; gas chromatography and mass spectrometry with analytical chemistry. *Mailing Add:* 19100 Crest Avenue #17 Castro Valley CA 94546

SANBORNE, PAUL MICHAEL, b London, Ont, Can, Jan 18, 50. SYSTEMATICS, EVOLUTION. *Educ:* Carleton Univ, Ottawa, Ont, BSc, 77; Lakehead Univ, Thunder Bay, Ont, MSc, 79; McMaster Uiv, Hamilton, Ont, PhD(entom), 82. *Prof Exp:* Nat Sci & Eng Res Coun fel, syst entom, Carleton Univ, 83-85, E B Eastburn fel, syst entom, 85-86; ASST PROF ENTOM, MACDONALD COL, MCGILL UNIV & CUR, LYMAN MUS, 86- *Concurrent Pos:* Vis cur, Biosyst Res Inst, Agr Can, Ottawa, Ont, 85. *Mem:* Entom Soc Can; Coleopterists Soc. *Res:* Systematics, evolution and biogeography of Ichneumonidae (hymenoptera); behavior and ecology of Diptera, Coleoptera and Mecoptera. *Mailing Add:* Dept Entomol MacDonald Col 21111 Lakeshore Rd Ste Anne de Bellevue PQ H9X 1C0 Can

SANCAR, AZIZ, DNA REPAIR, CHEMICAL CARCINOGENESIS. *Educ:* Istanbul Med Sch, MD, 69; Univ Tex, Dallas, PhD(molecular biol), 77. *Prof Exp:* ASSOC PROF BIOCHEM, SCH MED, UNIV NC, CHAPEL HILL, 82- *Res:* Enzyme mechanisms. *Mailing Add:* Dept Biochem Sch Med Univ NC Chapel Hill NC 27514

SANCAR, GWENDOLYN BOLES, b Waco, Tex, Sept 10, 49; m 78. DNA REPAIR, GENE REGULATION. *Educ:* Baylor Univ, BS, 72; Univ Tex-Dallas, MS, 74, PhD(molecular biol), 77. *Prof Exp:* Fel hemat, State Univ NY Downstate, Brooklyn, 77-80; fel radiobiol, Yale Univ Sch Med, 80-82, res asst prof biochem, 82-86; ASST PROF BIOCHEM, UNIV NC, CHAPEL HILL, 87- *Mem:* Am Soc Microbiol; Am Soc Photobiol; Am Soc Biochem & Molecular Biol. *Res:* Molecular mechanism of repair of ultraviolet induced DNA damage and the regulation of genes whose products are involved in repair. *Mailing Add:* Dept Biochem CB 7260 Univ NC Chapel Hill NC 27599-7260

SANCES, ANTHONY, JR, b Chicago, Ill, July 13, 32; m 65; c 3. BIOMEDICAL ENGINEERING. *Educ:* Am Inst Technol, BSEE, 53; DePaul Univ, MS, 59; Northwestern Univ, PhD(biomed eng), 63. *Prof Exp:* Res engr, Mech Res Dept, Am Mach & Foundry Co, 53-59; mgr, Advan Res Dept, Sunbeam Corp, 59-60; asst prof, Am Inst Technol, 60-61; consult numerous firms, 61-64; PROF ELEC ENG & NEUROSURG & DIR BIOMED ENG, MARQUETTE UNIV, 64-; PROF BIOMED ENG & CHMN PROG, MED COL WIS, 64- *Concurrent Pos:* Walter Murphy fel, 62-63; NIH fel, 63-64; staff, Wood Vet Admin Hosp, 64-, Milwaukee County Gen Hosp, 64- & Deaconess Hosp, Milwaukee, 72- *Mem:* Biophys Soc; sr mem Inst Elec & Electronics Engrs; Instrument Soc Am; Neuroelec Soc (pres); Alliance Eng Med & Biol (past pres); Sigma Xi. *Res:* Nervous system research related to biomedical engineering and biomechanics. *Mailing Add:* Neurosurg 3700 W Wisconsin County Gen Hosp Milwaukee WI 53226

SANCETTA, CONSTANCE ANTONINA, b Richmond, Va, Apr 17, 49; div. MARINE GEOLOGY, MICROPALEONTOLOGY. *Educ:* Brown Univ, BA, 71, MS, 73; Ore State Univ, PhD(oceanog), 76. *Prof Exp:* Res assoc geol, Stanford Univ, 76-78; assoc res scientist, 79-83, res scientist, Columbia Univ, 84-87, SR RES SCIENTIST, LAMONT-DOHERTY GEOL OBSERV, COLUMBIA UNIV, 88- *Concurrent Pos:* NSF Ocean Sci Exec Adv Comt, 81-86; assoc ed, Marine Micropaleont, 83-; mem, Arctic Marine Sci Comt, Nat Res Coun, 84-86; mem, Educ & Human Resouce Comt, Am Geophys Union, 81-88, Planning Comt, 84-88; mem comt, Geol Soc Am, 85-87; mem, Cent & E Pac Regional Ocean Drilling Prog, 86-89; co-chmn, geol sect, NY Acad Sci, 83-85; NSF Ocean Sci Adv Panel, 89; secy, Ocean Sci sect, Am Geophys Union, 88-90; counr, Am Quaternary Asn, 88-, Oceanograph Soc, 89- *Mem:* Fel Geol Soc Am; Am Geophys Union; fel AAAS; Oceanography Soc; AM Q; Int D. *Res:* Cenozoic paleoceanography; marine diatoms, biostratigraphy and paleoecology; depositional processes and generation of fossil assemblages. *Mailing Add:* Lamont-Doherty Geol Observ Palisades NY 10964

SANCHEZ, ALBERT, b Solomonsville, Ariz, Feb 10, 36; m 62; c 4. NUTRITION, BIOCHEMISTRY. *Educ:* Loma Linda Univ, BA, 59, MS, 62; Univ Calif, Los Angeles, DrPH, 68. *Prof Exp:* Biochemist, Int Nutrit Res Found, 61-65; from asst prof to prof nutrit, Loma Linda Univ, 68-74; prof biochem & nutrit, Sch Med, Montemorelos Univ, Mex, 74-80; PROF NUTRIT, LOMA LINDA UNIV, 80- *Concurrent Pos:* Pres, Int Nutrit Res Found, Inc. *Mem:* AAAS; Am Dietetic Asn; Latin Am Nutrit Soc; Am Inst Nutrit. *Res:* Protein and amino acid nutrition; role of nutritional factors as they relate to public health problems. *Mailing Add:* Dept Nutrit Sch Health Loma Linda Univ Loma Linda CA 92350

SANCHEZ, DAVID A, b San Francisco, Calif, Jan 13, 33; m 58; c 2. MATHEMATICS. *Educ:* Univ NMex, BS, 55; Univ Mich, MA, 60, PhD(math), 64. *Prof Exp:* Res asst, Radar Lab, Inst Sci & Technol, Univ Mich, 59-63; instr math, Univ Chicago, 63-65; from asst prof to prof math, Univ Calif, Los Angeles, 65-77; PROF MATH, UNIV NMEX, 77- *Concurrent Pos:* Vis lectr, Univ Manchester, 65-66. *Mem:* Am Math Soc; Math Asn Am; Soc Indust & Appl Math. *Res:* Direct methods in the calculus of variations; nonlinear ordinary differential equations. *Mailing Add:* Math & Phys Sci NSF 1800 G St NW Rm 512 Washington DC 20550

SANCHEZ, ISAAC CORNELIUS, b San Antonio, Tex, Aug 11, 41; m 76; c 2. POLYMER PHYSICS. *Educ:* St Mary's Univ, Tex, BS, 63; Univ Del, PhD(phys chem), 69. *Prof Exp:* Nat Res Coun-Nat Acad Sci assoc, Nat Bur Stand, Washington, DC, 69-71; assoc scientist polymer sci, Xerox Corp, 71-72; asst prof polymer sci & eng, Univ Mass, Amherst, 72-77; RES CHEMIST, INST MAT SCI & ENG, NAT BUR STANDARDS, 77- *Concurrent Pos:* Adj prof, Polymer Sci & Eng, Univ Mass, 77- *Mem:* Am Chem Soc; fel Am Phys Soc. *Res:* Application of statistical mechanics to problems in polymer science. *Mailing Add:* Chem Eng Dept Univ Tex Austin TX 78712-1104

SANCHEZ, JOSE, b Suffern, NY, June 30, 36; m 62; c 2. ORGANIC CHEMISTRY. *Educ:* Univ Rochester, BS, 58; Brown Univ, PhD(org chem), 66. *Prof Exp:* Res chemist, F & F Dept, E I du Pont de Nemours, 65-67; GROUP LEADER, PEROXIDES, LUCIDOL DIV PENNWALT CORP, 67- *Mem:* Am Chem Soc. *Res:* Synthesis, evaluation, process development and production troubleshooting in the areas of organic peroxides and organic specialty compounds. *Mailing Add:* 1624 Huth Rd Grand Island NY 14072-1799

SANCHEZ, PEDRO ANTONIO, b Havana, Cuba, Oct 7, 40; US citizen; m 65, 90; c 3. AGRONOMY, SOIL FERTILITY. *Educ:* Cornell Univ, BS, 62, MS, 64, PhD(soil sci), 68. *Prof Exp:* Asst soil sci, Philippine Prog, Cornell Univ, 65-68; asst prof soil sci & co-leader, Nat Rice Res Prog Mission to Peru, NC State Univ, 68-71, assoc prof, 71-77; coordr, Trop Pastures Prog, Cent Int Agr Trop, 77-79; PROF SOIL SCI & LEADER TROP SOILS PROG, NC STATE UNIV, 79- *Concurrent Pos:* Chief NC mission to Peru, Lima, 82-83; adj prof tropical conserv, Duke Univ, 90; dir, Ctr World Environ & Sustainable Develop, Duke Univ, NC State Univ & Univ NC-Chapel Hill, Raleigh, 91. *Mem:* Fel Am Soc Agron; fel Soil Sci Soc Am; Int Soc Soil Sci; Brazilian Soc Soil Sci; Colombian Soc Soil Sci. *Res:* Fertility and management of tropical soils; fertility and management of rice soils and tropical pastures; low input systems; agro forestry; tropical soil and climate change; tropical conservation and sustainable development. *Mailing Add:* Dept of Soil Sci NC State Univ Raleigh NC 27695-7619

SANCHEZ, ROBERT A, b Colombia, SAm, Feb 4, 38; US citizen; m 60; c 4. BIO-ORGANIC CHEMISTRY, ORGANIC CHEMISTRY. *Educ:* Pomona Col, BA, 58; Kans State Univ, PhD(org chem), 62. *Prof Exp:* Fel org chem, Univ Colo, 62-64, asst prof, 63-64; asst prof, Haverford Col, 64-65; sr res assoc, Salk Inst Biol Studies, 65-72; dir res, Terra-Marine Bioresearch, Inc, 72-74; dir bio-org res & develop, Calbiochem-Behring Corp, 74-80, SR RES SCIENTIST, BEHRING DIAGNOSTICS, 80- *Mem:* AAAS; NY Acad Sci; Am Chem Soc; Intra-sci Res Found; Sigma Xi. *Res:* Fundamental and developmental research; pharmaceuticals and biochemicals; pre-biological chemistry and origins of life; bio-organic chemistry and natural products; cancer chemotherapy. *Mailing Add:* 2601 Jacaranda Ave Carlsbad CA 92008

SANCHINI, DOMINICK J, aerospace engineering; deceased, see previous edition for last biography

SANCIER, KENNETH MARTIN, b New York, NY, June 21, 20. PHYSICAL CHEMISTRY. *Educ:* Polytech Inst Brooklyn, BS, 42; Johns Hopkins Univ, MA, 47, PhD(chem), 49. *Prof Exp:* Phys chemist gas chem, Linde Air Prod Co, Union Carbide & Carbon Corp, 42-46; lab instr, Johns Hopkins Univ, 46-49; phys chemist low temperature chem, Brookhaven Nat Lab, 49-53; PHYS CHEMIST, SRI INT, 54- *Concurrent Pos:* Vis instr chem, Conn Col, 48; fel, Standard Oil Ind, 48-49; NSF vis scientist, Univ Tokyo, 66-67. *Mem:* Am Chem Soc; Sigma Xi; AAAS. *Res:* Photochemistry related to solar energy conversion; energy transfer processes in solution and between gases and solids; hydrogen bonding; absorption spectroscopy, particularly at low temperature; heterogenous catalysis; electron spin resonance of gaseous atoms, gas-solid interactions, and biochemical systems. *Mailing Add:* SRI Int 333 Ravenswood Ave Menlo Park CA 94025

SANCILIO, LAWRENCE F, b Brooklyn, NY, Dec 13, 32; m 60; c 5. PHARMACOLOGY. *Educ:* St John's Univ, NY, BS, 54; Georgetown Univ, PhD(pharmacol), 60. *Prof Exp:* Teaching asst pharmacol, Georgetow Univ, 58-59; res pharmacologist, Miles Labs, Inc, Ind, 60-68; assoc res pharmacologist, 68-71, group mgr pharmacol res sect, 71-81, DIR PHARMACOL, A H ROBINS, INC, 81- *Mem:* Soc Exp Biol & Med; Am Soc Pharmacol & Exp Therapeut (chmn, 85); Inflammation Res Asn (pres, 78-80). *Res:* Analgesia; mechanism of inflammation and the evaluation of systemic nonsteroidal and topical steroidal anti-inflammatory agents; immediate hypersensitivity. *Mailing Add:* A H Robins Inc 1211 Sherwood Ave PO Box 26609 Richmond VA 23220

SAN CLEMENTE, CHARLES LEONARD, b Milford, Mass, May 27, 14; m 44; c 2. MICROBIOLOGY. *Educ:* Univ Mass, BS, 37; Mich State Univ, MS, 40, PhD(bact), 42. *Prof Exp:* Asst biochem, Mich State Univ, 37-39, bact, 39-42; res immunologist, Sch Med, Western Reserve Univ, 42-43; assoc prof chem & bact, Mich Tech Univ, 46-51; assoc prof, 51-61, prof microbiol, 61-, EMER PROF BIOL, MICH STATE UNIV, 82- *Mem:* AAAS; Am Chem Soc; Am Soc Microbiol; Soc Indust Microbiol; Am Inst Biol Sci; NY Acad Sci. *Res:* Complement; brucellosis; antibiotics; pesticides; enzymology; characterization of Staphylococci; environmental studies. *Mailing Add:* 4730 Ottawa Dr Okemos MI 48864

SANCTUARY, BRYAN CLIFFORD, b Yorkshire, Eng, Jan 25, 45; Can citizen; m 67; c 3. QUANTUM MECHANICS, STATISTICAL MECHANICS. *Educ:* Univ BC, BSc, 67, PhD(chem theory), 71. *Prof Exp:* Resident theorecian molecular physics, Kamerlingh Onnes Lab, Holland, 72-76; asst prof chem, Univ Wis, Madison, 76; from asst prof to assoc prof, 76-86, PROF CHEM, MCGILL UNIV, MONTREAL, 86- *Concurrent Pos:* Vis prof chem, Univ York, UK, 81-82; vis prof chem, Res Sch Chem, ANU, Australia, 89-90; hon prof, E China Inst Technol. *Mem:* Am Inst Physics; Can Inst Chem. *Res:* Theoretical investigation on systems containing angular momentum; development of the multirole theory of NMR; quantum dynamics and statistical mechanics applied to non-spherical molecules in all phases. *Mailing Add:* Dept Chem McGill Univ 801 Sherbrooke St W Montreal PQ H3A 2K6 Can

SAND, LEONARD B, mineralogy; deceased, see previous edition for last biography

SAND, RALPH E, b Stamford, Conn, May 16, 21; c 2. PHYSICAL ORGANIC CHEMISTRY. *Educ:* Univ Conn, BS, 42; Polytech Inst Brooklyn, MS, 48, PhD(org chem), 50. *Prof Exp:* Proj leader, Lab Advan Res, Remington Rand, Inc, 50-53; group leader, Res Ctr, Gen Foods, Inc, 53-61; sr chemist, Pratt & Whitney Div, United Aircraft Corp, 61-66; mgr pioneering res, Nat Biscuit Co, 66-70; dir tech develop, CPC Int, 70-71; sr chemist anal methods develop, Anderson Clayton Res Ctr, 71-88; RETIRED. *Mem:* Am Chem Soc; Sigma Xi; Inst Food Technologists; Am Asn Cereal Chemists. *Res:* Enzyme models; organic, photographic, lithographic, food, natural polymers, carbohydrates and proteins. *Mailing Add:* 182 Moonlight Dr Plano TX 75094

SANDAGE, ALLAN REX, b Iowa City, Iowa, June 18, 26; m 59; c 2. ASTRONOMY. *Educ:* Univ Ill, AB, 48, Dsc, 67; Calif Inst Technol, PhD(astron), 53. *Hon Degrees:* DSc, Yale Univ, 66, Univ Chicago & Univ Ill, 67; LLD, Univ Southern Calif, 71; ScD, Miami Univ, 74 & Graceland Col, 85. *Prof Exp:* Homewood prof physics, Johns Hopkins Univ, 87-89; asst astronr, 52-56, ASTRONR, CARNEGIE OBSERV, 56-; SR RES SCIENTIST, SPACE TELESCOPE SCI INST, 87- *Concurrent Pos:* Rouse Ball lectr, Cambridge Univ, 57; lectr, Harvard Univ, 57 & Haverford Col, 58 & 66; consult, NSF, 62-63; mem vis comt, Nat Radio Astron Observ, 63-64, chmn, 65; mem comt astron facil, Nat Acad Sci, 64-65 & comt sci & pub policy, 65; fels, Australian Nat Univ, 68-69 & 72; Fulbright-Hays scholar, Australia, 72-73; res scholar, Univ Basel, 85 & Univ Calif, San Diego, 85-86; adj astronr, Univ Hawaii, 86. *Honors & Awards:* Warner Prize, Am Astron Soc, 58; Eddington Medal, Royal Astron Soc, 63, Gold Medal, 70; Pope Pius XI Gold Medal, Pontifical Acad Sci, 66; Nat Medal of Sci, 71; Cresson Medal, Franklin Inst, 73; Russell Prize, Am Astron Soc, 73; Bruce Gold Medal, Astron Soc Pac, 75; Crafoord Prize, Swed Acad Sci, 91; Adion Medal, 91; Tomalla Prize, Swiss Phys Soc, 91. *Mem:* AAAS; Am Astron Soc; Royal Astron Soc; Int Astron Union. *Res:* Stellar evolution; photoelectric photometry; observational cosmology; galaxies; stellar kinematics; quasi-stellar radio sources; extragalactic distance scale; age and evolution of the universe; galactic structure; curvature of space. *Mailing Add:* 8319 Josard Rd San Gabriel CA 91775

SANDALL, ORVILLE CECIL, b Cupar, Sask, July 4, 39; m 61; c 4. CHEMICAL ENGINEERING. *Educ:* Univ Alta, BS, 61, MS, 63; Univ Calif, Berkeley, PhD(chem eng), 66. *Prof Exp:* PROF CHEM ENG, UNIV CALIF, SANTA BARBARA, 66- *Concurrent Pos:* Consult. *Mem:* Am Inst Chem Engrs; Am Chem Soc; Sigma Xi; Am Soc Eng Educ. *Res:* Gas absorption, heat and mass transfer in turbulent flow; separation processes. *Mailing Add:* Dept Chem & Nuclear Eng Univ Calif Santa Barbara CA 93106

SANDBERG, ANN LINNEA, b Denver, Colo. MICROBIOLOGY. *Educ:* Mont State Univ, BS, 60; Univ Chicago, PhD(pharmacol), 64. *Prof Exp:* USPHS training grant, Med Sch, Tufts Univ, 65-67; assoc immunol, Pub Health Res Inst, New York, 68-72; res biologist, 72-76, CHIEF PATHOGENIC MECHANISMS SECT, LAB MICROBIAL ECOL, NAT INST DENT RES, 76- *Concurrent Pos:* Res assoc prof, Med Sch, NY Univ, 71-72. *Mem:* AAAS; Am Asn Immunol. *Res:* Alternate pathway of complement activation; definition of system and its biological consequences; host defense systems; molecular mechanisms of bacterial attachment. *Mailing Add:* Lab Microbial Ecol Bldg 30 Rm 303 NIH Bethesda MD 20892

SANDBERG, AVERY ABA, b Poland, Jan 29, 21; nat US; m 43; c 4. INTERNAL MEDICINE. *Educ:* Wayne Univ, BS, 44, MD, 46; Am Bd Internal Med, dipl, 48. *Prof Exp:* Intern, Receiving Hosp, Detroit, Mich, 46-47; resident cardiol, Mt Sinai Hosp, New York, 49-50; resident med, Vet Admin Hosp, Salt Lake City, Utah, 50-51; resident instr med, Col Med, Univ Utah, 52-53, instr med, 53-54; assoc chief med, Roswell Park Mem Inst, 54-57, chief med, 57-88; DIR CANCER CTR, SOUTHWEST BIOMED RES INST, SCOTTSDALE, ARIZ, 88- *Concurrent Pos:* Dazian Found Med Res fel, 50; univ fel med & NIH res fel, Col Med, Univ Utah, 51-52, Dazian Found Med Res fel, 52-53, Am Cancer Soc scholar, 53-56; asst prof, Med Sch, Univ Buffalo, 54-, asst res prof, Grad Sch, 56-57, res prof, 57-; res prof, Canisius Col, 69- & Niagara Univ, 69-; consult, Med Found Buffalo, NY. *Mem:* Am Soc Clin Invest; Endocrine Soc; Am Soc Hemat; Am Asn Cancer Res; Am Fedn Clin Res. *Res:* Metabolism of steroids; metabolic changes in leukemia; cytogenetics of human leukemia and cancer. *Mailing Add:* Cancer Ctr 6401 E Thomas Rd Scottsdale AZ 85251

SANDBERG, CARL LORENS, b Aniwa, Wis, July 24, 22; m 51; c 4. POLYMER CHEMISTRY. *Educ:* Univ Wis, BS, 48, MS, 49. *Prof Exp:* Chemist, Minn Mining & Mfg Co, 50-54, res chemist, 54-67, res specialist, 67-76, sr res specialist, 3M Co, 76-86; RETIRED. *Mem:* Am Chem Soc. *Res:* Synthesis and characterization of organic and fluoro-organic polymers. *Mailing Add:* 1736 Rowe Pl St Paul MN 55106

SANDBERG, EUGENE CARL, b Ashtabula, Ohio, Jan 4, 24; m 53; c 2. OBSTETRICS & GYNECOLOGY. *Educ:* Univ Calif, AB, 45, MD, 48; Am Bd Obstet & Gynec, dipl, 59. *Prof Exp:* Intern, San Francisco City & County Hosp, 48-49; intern, Vanderbilt Univ Hosp, 49-50, asst resident, 50-51 & 53-54, resident, 54-55; from instr to asst prof, 55-64, ASSOC PROF OBSTET & GYNEC, SCH MED, STANFORD UNIV, 64- *Concurrent Pos:* Trainee steroid biochem, Worcester Found Exp Biol, 61-62; Macy Found fel, Col Physicians & Surgeons, Columbia Univ, 62-63. *Mem:* Am Col Obstet & Gynec; Soc Gynec Invest; Endocrine Soc. *Mailing Add:* Obstet & Gynec Med Ctr A352 Stanford Univ Sch Med 300 Pasteur Dr Stanford CA 94305

SANDBERG, I(RWIN) W(ALTER), b New York, NY, Jan 23, 34; m 58; c 1. NONLINEAR SYSTEMS. *Educ:* Polytech Inst Brooklyn, BEE, 55, MEE, 56, DEE, 58. *Prof Exp:* Mem tech staff, Bell Labs, 56 & 58-67, head, Systs Theory Res Dept, 67-72, mem staff, Math Sci Res Ctr, 72-86; HOLDER COCKRELL FAMILY REGENTS CHAIR, NO 1 ELEC & COMPUT ENG, UNIV TEX, AUSTIN, 86- *Concurrent Pos:* Vchmn, Inst Elec &

Electronic Engrs Group on Circuit Theory; guest ed, Inst Elec & Electronic Engrs Trans on Circuit Theory, Spec Issue on Active & Digital Networks; Distinguished lectr, Inst Elec & Electronic Engrs; adv, Inst Elec & Commun Engrs of Japan & Am Men & Women Sci. *Honors & Awards:* Centennial Medal, Inst Elec & Electronic Engrs; Tech Achievement Award, Circuits & Systs Soc. *Mem:* Nat Acad Eng; fel Inst Elec & Electronic Engrs; AAAS; Sigma Xi; Soc Indust & Appl Math. *Res:* Nonlinear analysis; network theory; theory of feedback systems; communication systems; differential equations; integral equations; functional analysis; numerical analysis; neural networks; published more than 140 papers and holds nine United States patents. *Mailing Add:* Eng Sci Bldg Univ Tex Austin TX 78712-1084

SANDBERG, PHILIP A, b Cincinnati, Ohio, Mar 29, 37; c 2. GEOLOGY, PALEONTOLOGY. *Educ:* La State Univ, BS, 60, MS, 61; Univ Stockholm, Fil Lic, 64, Fil Dr(geol, paleont), 65. *Prof Exp:* From asst prof to assoc prof, 65-75, actg dept head, 77-78, PROF GEOL, UNIV ILL, URBANA, 75- *Concurrent Pos:* Vis assoc prof, Univ Minn, 70; NATO fel, Brit Mus Natural Hist, London, 72-73; vis assoc, Calif Inst Technol, 75; vis lectr, Philipps Univ, Marburg, WGer, 82; vis prof, Univ Leiden, Netherlands, 86; assoc, Ctr Advan Study, Univ Ill, 87. *Mem:* Int Paleont Union; Soc Econ Paleont & Mineral (vpres, 80-81); Paleont Soc; Int Asn Sedimentol. *Res:* Ultrastructure and chemistry of fossil and modern invertebrate skeletons; carbonate diagenesis; scanning electro microscopy; Cenozoic microfossils; modern marine and estuarine microfaunas; molecular paleontology and geoimmunology. *Mailing Add:* Dept Geol Univ Ill 143 NHB 1301 W Green Urbana IL 61801

SANDBERG, ROBERT GUSTAVE, b Minneapolis, Minn, Mar 20, 39; m 59; c 3. CLINICAL BIOCHEMISTRY. *Educ:* Hamline Univ, BS, 61; Ohio State Univ, MS, 63. *Prof Exp:* Group leader polymers, Cargill Inc, 63-66; sr chemist, Hoerner-Waldorf Inc, 66-69; clin chemist, St John's Hosp, 69-71; mkt planning mgr automatic clin anal, 75-80, PROD SUPVR CLIN CHEM, E I DU PONT DE NEMOURS & CO, INC, 71-, TECH MGR MKT, 80- *Mem:* Am Asn Clin Chem; Biomed Mkt Asn (pres, 79); Am Chem Soc. *Res:* Standards and control materials for clinical chemistry; development and marketing; automated clinical chemistry analyzers; preparation of stable enzyme solutions for verification of standard methods; evaluation of automated clinical chemistry and immunoassay methods. *Mailing Add:* 110 Hobson Dr Hockessin DE 19707-2105

SANDBORN, VIRGIL A, b Conway Springs, Kans, Apr 30, 28; m 55; c 2. AERONAUTICAL ENGINEERING. *Educ:* Univ Kans, BS, 50; Univ Mich, MS, 54. *Prof Exp:* Aeronaut res scientist, Nat Adv Comt Aeronaut, 51-58; aerospace res scientist, NASA, 58-62; consult scientist, Avco Corp, 62-63; prof aerodyn & fluid mech, 62-77, prof civil eng, 77-88, EMER PROF, COLO STATE UNIV, 88- *Concurrent Pos:* Vis prof, Purdue Univ, 66- *Mem:* Am Inst Aeronaut & Astronaut. *Res:* Turbulent air flow; turbulent boundary layer flow; separation of boundary layers from surfaces. *Mailing Add:* Dept of Civil Eng Colo State Univ Ft Collins CO 80523

SANDE, RONALD DEAN, b Twin Falls, Idaho, July 3, 42; m 64; c 3. VETERINARY RADIOLOGY. *Educ:* Wash State Univ, DVM, 66, MS, 71, PhD(vet sci), 75. *Prof Exp:* Res vet radionuclides, Univ Utah, 66-67; resident vet med & surg, 68-69, instr, 69-70, from asst prof vet radiol to assoc prof, 71-77, assoc prof, 77-79, PROF VET CLIN MED & SURG, WASH STATE UNIV, 77- *Mem:* Am Col Vet Radiologists; Am Vet Radiol Soc; Am Vet Med Asn. *Res:* Inherited metabolic bone disorders in animals; animal models. *Mailing Add:* Dept Vet Med Wash State Univ Pullman WA 99163

SANDEFUR, KERMIT LORAIN, b Arkansas City, Kans, Sept 1, 25; m 46; c 2. OPTICS. *Educ:* Univ Kans, BS, 50 & 51. *Prof Exp:* Assoc physicist, Midwest Res Inst, 51-56; sr res scientist, Hycon Mfg Co, Calif, 56-69; res & develop eng, Lockheed-Calif Co, 69-87; CONSULT, 87- *Res:* Photographic reconnaissance and infrared systems; supersonic flow; ballistics; optical systems analysis. *Mailing Add:* 3635 Landfair Rd Pasadena CA 91107

SANDEL, BILL ROY, b Brady, Tex, Nov 19, 45; m 80; c 2. SPACE PHYSICS. *Educ:* Rice Univ, BA, 68, MS, 71, PhD(space sci), 72. *Prof Exp:* Sr assoc-in-res, Kitt Peak Nat Observ, 73-78; res assoc planetary atmospheres, Lunar & Planetary Lab, Univ Ariz, 78-79; res scientist, Earth & Space Sci Inst, Univ Southern Calif, 79-83; ASSOC RES SCIENTIST, LUNAR & PLANETARY LAB, UNIV ARIZ, 84- *Concurrent Pos:* Co-investr Voyager ultraviolet spectrometer exp, Galileo ultraviolet spectrometer exp. *Honors & Awards:* Exceptional Scientific Achievement Award, NASA, 81. *Mem:* Am Geophys Union; Am Astron Soc. *Res:* Atmospheric and space physics of the outer planets; ultraviolet spectroscopy; development of ultraviolet imaging detectors. *Mailing Add:* Lunar & Planetary Lab Gould Simpson Bldg Rm 901 Univ Ariz Tucson AZ 85721

SANDEL, VERNON RALPH, b Marquette, Mich, June 4, 33; m 59; c 2. PHYSICAL ORGANIC CHEMISTRY. *Educ:* Mich Tech Univ, BS, 55; Northwestern Univ, PhD(org photochem), 62. *Prof Exp:* Chemist, Ethyl Corp, 55-57; spec proj chemist, Dow Chem Co, Mich, 61-62; res chemist, Eastern Res Lab, Mass, 63-67; assoc prof chem, Mich Technol Univ, 67-84; RETIRED. *Mem:* Am Chem Soc; Sigma Xi. *Res:* Nuclear magnetic resonance of carbanions; carbanion chemistry; organic reaction mechanisms. *Mailing Add:* PO Box 467 Dollar Bay MI 49922

SANDELL, LIONEL SAMUEL, b Montreal, Que, May 29, 45; m 71; c 2. WATER GEL EXPLOSIVES. *Educ:* McGill Univ, BSc, 66, PhD(phys chem), 70. *Prof Exp:* From postdoctoral assoc to res assoc, Col Environ Sci & Forestry, State Univ NY, 70-73; res chemist colloid chem, Pigments Dept, 73-78, sr chemist water gel explosives, Petrochem Dept, 78-82, res assoc, Agr Prod Dept, 82-90, SR RES ASSOC, AGR PROD DEPT, DU PONT CO, 90- *Mem:* Am Chem Soc. *Res:* Development chemistry to prevent gel breakdown in water gel explosives; developed novel opacifying pigments for paint, paper, and other items; agricultural formulations for sulfonylureas. *Mailing Add:* 2900 Bodine Dr Wilmington DE 19810

SANDER, C MAUREEN, b Lansing, Mich, Mar 2, 33. MEDICINE, PATHOLOGY. *Educ:* Mich State Univ, BS, 55; Univ Mich, MD, 58; Univ London, dipl clin path, 64; Am Bd Path, dipl, 66. *Prof Exp:* From instr to assoc prof, 65-76, PROF PATH, COL HUMAN MED, MICH STATE UNIV, 76- *Concurrent Pos:* Fulbright Fel, 63-64; Dir, Mich placental tissue Registry, 76- *Res:* pathology of the placenta; perinatal pathology. *Mailing Add:* Dept Path A-203 Clin Ctr Mich State Univ Lansing MI 48824

SANDER, DONALD HENRY, b Creston, Nebr, Apr 21, 33; m 53; c 2. AGRONOMY. *Educ:* Univ Nebr, BS, 54, MS, 58, PhD(agron, soils), 67. *Prof Exp:* Soil scientist, US Forest Serv, 58-64; asst prof agron, Exten, Kans State Univ, 64-67; assoc prof, 67-73, PROF AGRON, EXTEN, UNIV NEBR, LINCOLN, 73- *Honors & Awards:* Agron Achievement Award-Soils, Am Soc Agron, 85. *Mem:* Fel Am Soc Agron, 85; Soil Sci Soc Am. *Res:* Influence soil properties, especially nutrients and nutrient interactions on plant growth and composition. *Mailing Add:* Dept Agron Univ Nebr Lincoln NE 68583

SANDER, DUANE E, b Sioux Falls, SDak, Feb 14, 38; m 60; c 4. ELECTRICAL ENGINEERING. *Educ:* SDak Sch Mines & Technol, BS, 60; Iowa State Univ, MS, 62, PhD(elec eng), 64. *Prof Exp:* Instr elec eng, Iowa State Univ, 60-63, res asst, 63-64; intel analyst, US Army Foreign Sci & Technol Ctr, Washington, DC, 65-67; assoc prof elec eng, 67-75, PROF ELEC ENG, SDAK STATE UNIV, 75-, HEAD DEPT GEN ENG, 85- *Concurrent Pos:* Consult, Med Eng Serv Asn, 77- *Mem:* Inst Elec & Electronics Engrs; Int Soc Hybrid Microelectronics; Am Soc Hosp Eng; Nat Soc Prof Eng. *Res:* Bioengineering and data acquisition; clinical engineering. *Mailing Add:* Dean Eng Dak State Univ Box 2219 Brookings SD 57007

SANDER, EUGENE GEORGE, b Fargo, NDak, Sept 17, 35; m 79; c 2. ENZYMOLOGY. *Educ:* Univ Minn, BS, 57; Cornell Univ, MS, 59, PhD(biochem), 65. *Prof Exp:* NIH fel biochem, Brandeis Univ, 65-67; asst prof, Dept Biochem, Univ Fla, 67-70, assoc prof, 70-75, prof, 75-76; prof & chmn dept, WVa Univ, 76-80; PROF & HEAD, DEPT BIOCHEM & BIOPHYS, TEX A&M UNIV, 80- *Mem:* Am Chem Soc; Am Soc Biol Chemists; Sigma Xi; Biophys Soc; Am Inst Nutrit. *Res:* Mechanism of action of enzymes involved in dihydropyrimidine synthesis and degradation along with related organic model systems. *Mailing Add:* Forbes Bldg Rm 306 Tucson AZ 85721

SANDER, GARY EDWARD, b New Orleans, LA, Feb 14, 47; m 73; c 3. OPIOID PEPTIDES, CLINICAL PHARMACOLOGY. *Educ:* Tulane Univ, PhD(biochem), 71, MD, 74. *Prof Exp:* PROF MED CARDIOL, SCH MED, TULANE UNIV, 80- *Concurrent Pos:* Counr, Southern Sect Am Fedn Clin Res. *Mem:* Fel Am Col Physicians; Am Soc Hypertension; fel Am Col Cardiol; Am Soc Pharmacol & Exp Therapeut; Am Heart Asn. *Res:* Clinical pharmacology of hypertension, heart failure and ischemic heart disease; various aspects of cardiomyopathy; basic research including cardiovascular regulation by peptides and the effects of diabetes on the heart. *Mailing Add:* Sch Med Tulane Univ Tulane Ave New Orleans LA 70112

SANDER, IVAN LEE, b Cape Girardeau, Mo, Mar 13, 28; m 49; c 3. SILVICULTURE, FOREST MANAGEMENT. *Educ:* Univ Mo, BSF, 52, MS, 53. *Prof Exp:* Res forester, Cent States Forest Exp Sta, 53-65, Northeastern Forest Exp Sta, 66-70 & NCent Forest Exp Sta, 71-74, PROJ LEADER, NCENT FOREST EXP STA, USDA FOREST SERV, 75-; RES ASSOC, SCH FORESTRY, FISHERIES & WILDLIFE, UNIV MO, 71- *Mem:* Soc Am Foresters; Sigma Xi. *Res:* Determination of the critical biological, environmental and ecological factors that control hardwood reproduction establishment and growth with emphasis on oaks; development of systems to ensure harvested stands will be regenerated to oaks and other important hardwoods. *Mailing Add:* 1-26 Agr Bldg Univ Mo Columbia MO 65201

SANDER, LEONARD MICHAEL, b St Louis, Mo, Aug 17, 41; m 64; c 1. STATISTICAL PHYSICS. *Educ:* Washington Univ, BS, 63; Univ Calif, Berkeley, MA, 66, PhD, 68. *Prof Exp:* NSF fel physics, Univ Calif, San Diego, 68-69; from asst prof to assoc prof, 69-80, PROF PHYSICS, UNIV MICH, ANN ARBOR, 80- *Mem:* Fel Am Phys Soc; Mat Res Soc. *Res:* Theory of physical processes far from equilibrium; theoretical solid state physics. *Mailing Add:* Physics Dept Univ Mich Ann Arbor MI 48109-1120

SANDER, LINDA DIAN, b Harrisburg, Pa, Sept 2, 47; m 70; c 2. MEDICAL PHYSIOLOGY, GASTROINTESTINAL ENDOCRINOLOGY. *Educ:* Ariz State Univ, BS, 69; Univ Okla, PhD(physiol), 73. *Prof Exp:* Fel, dept physiol, Med Sch, Univ Tex, Houston, 73-76; asst prof, dept physiol, Med Ctr, La State Univ, New Orleans, 76-82; ASSOC PROF, DEPT PHYSIOL, MEHARRY MED COL, 82- *Mem:* Am Physiol Soc; Soc Exp Biol & Med. *Res:* Role of gastrointestinal hormones and brain-gut polypeptides on pituitary adrenal hormonal secretion and circadian rhythms; influence of stress on gastrointestinal function; gastrointestinal endocrinology. *Mailing Add:* Meharry Med Col 1005 D B Todd Jr Blvd Nashville TN 37208

SANDER, LOUIS FRANK, b Rockville Centre, NY, Aug 1, 33; m 56; c 3. METALLURGY, MATERIALS SCIENCE. *Educ:* Marquette Univ, BSME, 55, MS, 61; Pa State Univ, PhD(metall), 66. *Prof Exp:* Engr, Refining Div, Mobil Oil Co, 58-59; instr eng, Marquette Univ, 59-61; from asst prof to assoc prof, 61-72, PROF ENG, VILLANOVA UNIV, 72-, CHMN, DEPT MECH ENG, 70- *Concurrent Pos:* NSF grant, 63-65; consult, Pitman-Dunn Labs, US Army, Frankford Arsenal, 62, 68, E W Bliss Co, 63, Titanium Metals Corp, 67-68 & Naval Air Eng Ctr, 71- *Mem:* AAAS; Am Soc Metals; Am Inst Mining, Metall & Petrol Engrs; Am Soc Eng Educ; fel Am Inst Chem. *Res:* Solid lubricants; metal sulfide-metal oxide equilibria; high temperature plasticity of titanium alloys; heat capacity of solids; sulfidation of metals; fracture toughness; mechanical engineering; failure analysis. *Mailing Add:* Clin Engr Assoc Inc 153 Mayer Dr Pittsburgh PA 15237

SANDER, LOUIS W, b San Francisco, Calif, July 31, 18; m 53; c 3. PSYCHIATRY. *Educ:* Univ Calif, AB, 39, MD, 42; Boston Psychoanal Inst, grad, 61; Am Bd Psychiat & Neurol, dipl, 51. *Prof Exp:* Asst psychiat, Sch Med, Boston Univ, 47-51, instr, 51-57, asst prof, 54-58, res psychiatrist, Univ Med Ctr, 54-78, from asst res prof to assoc res prof, 58-68, prof psychiat, 68-78; PROF PSYCHIAT, SCH MED, UNIV COLO, 78- *Concurrent Pos:* Asst, Mass Mem Hosps, 48-55, asst vis physician, 55-59, assoc vis physician, 59-78; mem staff, James Jackson Putnam Children's Ctr, 50-53; jr vis physician, Boston City Hosp, 53-57, assoc vis physician, Pediat Serv, 57-78, USPHS grants, 54-78, career develop award, 63-68; consult, Bd Missions, Methodist Church, 54-60; res scientist award, Nat Inst Ment Health, 68-73 & 73-78; vis prof, Univ Calif Med Sch, 84. *Honors & Awards:* Grete Simpson Award, Univ Calif Med Sch. *Mem:* Am Psychiat Asn; Am Acad Child Psychiat; Soc Res Child Develop; Am Col Psychoanalysts; AAAS. *Res:* Early personality development, especially in relation to the influence of maternal personality on mother-child interactions; investigation of neonatal state regulation in the caretaking systems by non-intrusive bassinet monitoring; twenty-five year longitudinal study of personality development. *Mailing Add:* Dept Psych Univ Colo Med Ctr 4200 E Ninth Ave Container C249 Denver CO 80262

SANDER, NESTOR JOHN, geology, paleontology, for more information see previous edition

SANDER, WILLIAM AUGUST, III, b Charleston, SC, May 11, 42; m 67; c 2. COMMUNICATIONS, SIGNAL PROCESSING. *Educ:* Clemson Univ, BS, 64; Duke Univ, MS, 67, PhD(elec eng), 73. *Prof Exp:* Res asst, Duke Univ, 67-70; engr, Commun & Electronics Bd, US Army Airborne, 70-75; STAFF SCIENTIST RES MGT, US ARMY RES OFF, 75- *Concurrent Pos:* Instr, Fayetteville State Univ, 74-75. *Mem:* Inst Elec & Electronics Engrs; Sigma Xi. *Res:* Signal processing; communications; computer-aided design of integrated circuits. *Mailing Add:* US Army Res Off PO Box 12211 Research Triangle Park NC 27709-2211

SANDERFER, PAUL OTIS, b Union City, Tenn, Mar 1, 37; m 59; c 1. ORGANIC CHEMISTRY. *Educ:* Union Univ, BS, 59; Univ Fla, PhD(org chem), 65. *Prof Exp:* Asst prof chem, 65-71, ASSOC PROF CHEM & PHYSICS, WINTHROP COL, 71-, ACTG CHMN, DEPT CHEM & PHYSICS, 81- *Mem:* Am Chem Soc. *Res:* Kinetic studies of reaction mechanisms; organic synthesis. *Mailing Add:* Dept Chem & Physics Winthrop Col Rock Hill SC 29733

SANDERS, AARON PERRY, cell physiology, radiobiology, for more information see previous edition

SANDERS, BARBARA A, b New Orleans, La, Oct 20, 47; m 73; c 1. TECHNICAL MANAGEMENT. *Educ:* Southern Univ, BS, 69; Rutgers Univ, MS, 72. *Prof Exp:* Dept head, Composites Mat Characterization, Gen Motors, 79-80, mgr composites processing, 80-82, CAD/CAM Tooling Group, 82-84, dir artificial intel, 84-85, prog mgr, Truck & Bus Group, 89-90, DIR, ADVAN MFG ENG, GEN MOTORS, 91- *Concurrent Pos:* Mem & ed, Marine Composites, Nat Res Coun, 89-91. *Mem:* Soc Automotive Engrs; Soc Mfg Engrs. *Res:* Research and development management; assembly systems-automotive advanced manufacturing engineering. *Mailing Add:* 30300 Mound Rd MDA-23 Warren MI 48090-9040

SANDERS, BENJAMIN ELBERT, b Bowersville, Ga, Oct 19, 18; m 46; c 2. BIOCHEMISTRY. *Educ:* Wofford Col, BS, 39; Univ Ga, MS, 42; Purdue Univ, PhD(chem), 49; Am Bd Clin Chemists, dipl. *Prof Exp:* Asst chem, Univ Ga, 39-42; res biochemist, Dept Labs, Henry Ford Hosp, 49-51; res assoc immunochem, Merck, Sharp & Dohme, 51-53, protein chem, Merck Inst Therapeut Res, 53-58, dir, 58-61; from assoc prof to prof, 61-85, EMER PROF BIOCHEM, STATE UNIV NY, BUFFALO, 85- *Mem:* Fel AAAS; Am Soc Biol Chem; Am Chem Soc; NY Acad Sci. *Res:* Isolation and characterization plasma proteins; biochemistry schizophrenia; clinical chemistry; immunoglobulins. *Mailing Add:* Seven Lakes Box 964 West End NC 27376

SANDERS, BOBBY GENE, b Rhodell, WVa, Apr 16, 32; m 51; c 4. IMMUNOGENETICS. *Educ:* Concord Col, BS, 54; Pa State Univ, MEd, 58, PhD(genetics), 61. *Prof Exp:* Asst prof biol, Lafayette Col, 61-64; fel immunogenetics, Calif Inst Technol, 64-66, asst prof biol, 66-68; assoc prof, 68-74, PROF ZOOL, UNIV TEX, AUSTIN, 74-, ASSOC DEAN, COL NATURAL SCI, 85- *Mem:* Am Asn Immunologists; Am Genetic Asn. *Res:* Cell surface antigens in normal erythroid differentiation and retrovirus-induced erythro leukemia; immunochemical characterization of chicken major histocompatibility complex products. *Mailing Add:* Dept Zool Univ Tex Austin TX 78712

SANDERS, BOBBY LEE, b Ben Wheeler, Tex, Jan 12, 35; m 54; c 2. APPLIED MATHEMATICS. *Educ:* ETex State Univ, BS, 56; Fla State Univ, MS, 58, PhD(math), 62; Southern Methodist Univ, JD, 77. *Prof Exp:* Instr math, Fla State Univ, 57-58; from asst prof to prof math, Tex Christian Univ, 62-77; PARTNER, SANDERS & SANDERS, 77- *Mem:* Am Math Soc; Math Asn Am. *Res:* Structure theory of Banach spaces; functional analysis; categorical applications; applications of finite mathematical structures; quantitative and qualitative applications of mathematics to judicial problems and processes. *Mailing Add:* Sanders & Sanders PO Box 416 Canton TX 75103

SANDERS, BRENDA MARIE, b Island Falls, Maine, Mar 31, 51; m 83; c 1. TOXICOLOGY, MOLECULAR BIOLOGY. *Educ:* Wesleyan Univ, BA, 75; Univ Del, PhD(marine biol), 81. *Prof Exp:* Res assoc, Duke Univ Marine Lab, 80-83, res asst prof, 83-86; assoc prof physiol, 86-91, ASSOC DIR, MOLECULAR ECOL INST, CALIF STATE UNIV, LONG BEACH, 86-, PROF PHYSIOL, 91- *Concurrent Pos:* NAm ed, Ecotoxicol Ser, 91-; ed, Rev Aquatic Sci, 91-93; consult, JSA Inc, 91- *Mem:* AAAS; Am Soc Zoologists; Sigma Xi; Soc Environ Toxicol & Chem; Am Physiol Soc. *Res:* Molecular mechanisms by which organisms adapt to their environment; strss proteins as biological markers; aquatic toxicology; ecological assessments. *Mailing Add:* 3729 Myrtle Ave Long Beach CA 90807

SANDERS, CHARLES ADDISON, b Dallas, Tex, Feb 10, 32; m 56; c 4. CARDIOLOGY. *Educ:* Univ Tex Southwestern Med Sch Dallas, MD, 55. *Hon Degrees:* DSc, Suffolk Univ, 77, Mass Coll Pharm, 82. *Prof Exp:* Intern & asst resident med serv, Boston City Hosp, Mass, 55-57, chief resident, 57-58; clin & res fel cardiol, Mass Gen Hosp, 58-60, prog dir myocardial infarction res unit, 67-72, prog dir medlab systs, 69-73, assoc physician, 70-73, chief cardiac catheter unit, 62-72, gen dir, 72-81, physician, 73-81; exec vpres, Squibb Corp, 81-88, vchmn, 88-89; CHIEF EXEC OFFICER, GLAXO, INC, 89- *Concurrent Pos:* Instr, Harvard Med Sch, 64-66, assoc, 66-68, from asst prof to prof, 69-84. *Mem:* Inst Med-Nat Acad Sci; Am Fedn Clin Res; Am Heart Asn; Am Col Physicians; Asn Univ Cardiol; Am Soc Clin Invest. *Mailing Add:* Glaxo Inc Five Moore Dr Research Triangle Park NC 27709

SANDERS, CHARLES F(RANKLIN), JR, b Louisville, Ky, Dec 22, 31; m 56; c 3. HEAT TRANSFER, COMBUSTION. *Educ:* Univ Louisville, BChE, 54, MChE, 58; Univ Southern Calif, PhD(chem eng), 70. *Prof Exp:* Engr, Esso Res & Eng Co, 55-62; from asst prof to prof eng, 62-83, chmn dept thermal-fluid systs, 69-72, dean, Sch Eng & Comput Sci, 72-81, EMER PROF ENG, CALIF STATE UNIV, NORTHRIDGE, 83-; EXEC VPRES, ENERGEO, 89- *Concurrent Pos:* Consult, KVB Eng Co, 70-80; dir, Rusco Industs Corp, 80-83, pres & chief exec officer, 81-82; exec vpres, Energy Systs Assocs, 82-89; dir, Datametrics Corp, 82-86; dir, Uniforms Unlimited, 85- *Mem:* Am Inst Chem Engrs; Am Soc Eng Educ; Combustion Inst; Nat Soc Prof Engrs. *Res:* Combustion; pollution from combustion; radiative transfer through particulate clouds; combustion modelling; fluid-bed combustion; biomass combustion. *Mailing Add:* 5126 Alder Irvine CA 92715-2301

SANDERS, CHARLES IRVINE, b Baltimore, Md, Feb 18, 36; m 57; c 3. CHEMISTRY. *Educ:* Clemson Univ, BS, 56; Iowa State Univ, MS, 59, PhD(phys chem), 61. *Prof Exp:* Asst chem, Iowa State Univ, 56-61; res chemist, 61-68, res supvr, 68-70, sr res scientist, 70-77, MGR, ANALYSIS DEPT, ARMSTRONG WORLD INDUST INC, 77- & MGR, CHEM PROCESS DEVELOP, 87-, TSCA COORDR, 89-, CHEM HYG OFFICER, 90- *Concurrent Pos:* Resident indust scientist, Pa Tech Assistance Prog, Pa State Univ, 69; vis indust scientist prog, Indust Res Inst, 83 & 85. *Mem:* Am Chem Soc; Sigma Xi. *Res:* Physical chemistry of polymers; analytical chemistry; scanning electron microscopy; radiochemistry. *Mailing Add:* Res & Develop Ctr Armstrong World Indust Inc PO Box 3511 Lancaster PA 17604

SANDERS, CHARLES LEONARD, JR, b Chicago, Ill, Dec 27, 38; m 63; c 3. RADIOBIOLOGY. *Educ:* Col William & Mary, BS, 60; Tex A&M Univ, MS, 63; Univ Rochester, PhD(radiobiol), 66. *Prof Exp:* AEC fel, 66-68, res assoc, 68-78, STAFF SCIENTIST INHALATION TOXICOL, LIFE SCI DEPT, PAC NORTHWEST LABS, BATTELLE MEM INST, 78- *Concurrent Pos:* Affil asst prof, Univ Wash, 75-80, affil prof, 80-; mem, Nat Coun Radiation Protection & Measurements. *Res:* Inhalation toxicology and carcinogenesis of plutonium and other transuranic elements, and of beryllium, lead, mercury, cadmium, asbestos, silica and volcanic ash; general radiobiology and toxicology of energy effluents. *Mailing Add:* 1943 Davison Richland WA 99352

SANDERS, CHRISTINE CULP, b Tampa, Fla, Sept 3, 48; m 74. MEDICAL MICROBIOLOGY, INFECTIOUS DISEASES. *Educ:* Univ Fla, BS, 70, PhD(med microbiol), 73. *Prof Exp:* Technician microbiol, Shands Teaching Hosp & Clin, 69-70; chief technologist, Alachua Gen Hosp, 70; from asst prof to assoc prof, 73-85, PROF MICROBIOL, SCH MED, CREIGHTON UNIV, 85- *Concurrent Pos:* Res award, Sigma Xi, 74. *Mem:* Am Soc Microbiol; Am Fedn Clin Res; Infectious Dis Soc Am; NY Acad Sci. *Res:* Bacterial drug resistance; evaluation of new antimicrobial agents; cell wall deficient bacteria; antimicrobial antagonisms. *Mailing Add:* Dept Med Microbiol Sch Med Creighton Univ Omaha NE 68178

SANDERS, DARRYL PAUL, b Arch, NMex, Feb 11, 36; m 62; c 2. ENTOMOLOGY. *Educ:* Tex Tech Col, BS, 59; Purdue Univ, MS, 64, PhD(entom), 67. *Prof Exp:* Exten entomologist, Purdue Univ, 65-67; teaching & res entomologist, Tex A&M Univ, 67-70; teaching & exten entomologist, Purdue Univ, West Lafayette, 70-75; PROF ENTOM & CHMN DEPT, TEX TECH UNIV, 76- *Mem:* Entom Soc Am; Am Mosquito Control Asn; Am Registry Prof Entomologists. *Res:* Insects affecting man and livestock. *Mailing Add:* Dept Entom Univ Mo 2-4 Agr Bldg Columbia MO 65211

SANDERS, DOUGLAS CHARLES, b Lansing, Mich, May 21, 42; m 65. HORTICULTURE, PLANT PHYSIOLOGY. *Educ:* Mich State Univ, BS, 65; Univ Minn, MS, 69, PhD(hort), 70. *Prof Exp:* Res fel hort, Univ Minn, 69-70; from exten asst prof to exten assoc prof, 70-82, PROF HORT, NC STATE UNIV, 82- *Mem:* Am Soc Hort Sci; Am Soc Agron; Crop Sci Soc Am; Sigma Xi; Coun Agr Sci Technol. *Res:* Vegetable crop cultural practices; crop microclimate modification; influence of climate on plant physiology. *Mailing Add:* Box 7609 NC State Univ Raleigh NC 27695-7609

SANDERS, F KINGSLEY, MOLECULAR BIOLOGY. *Educ:* Oxford Univ, Eng, DPhil, 42. *Prof Exp:* Prof cell biol, Grad Sch Med Sci, Cornell Univ, 67-80; mem, Mem Sloan-Kettering Inst Cancer Res, 67-80; RETIRED. *Mailing Add:* Box 37 Canaan NY 12029

SANDERS, FRANK CLARENCE, JR, b Tiquisate, Guatemala, Dec 26, 40; US citizen; m 63; c 2. ATOMIC PHYSICS, MOLECULAR PHYSICS. *Educ:* Univ Tex, Austin, BS, 63, PhD(physics), 68. *Prof Exp:* Asst prof physics, Univ Tex, Austin, 68-69; asst prof, 69-77, ASSOC PROF PHYSICS, UNIV SOUTHERN ILL, CARBONDALE, 77- *Mem:* Am Phys Soc; Am Asn Physics Teachers. *Res:* Theoretical atomic and molecular physics; perturbation theory and its application to simple atomic and molecular systems. *Mailing Add:* Dept Physics Southern Ill Univ Neckers CO471 Carbondale IL 62901

SANDERS, FREDERICK, b Detroit, Mich, May 17, 23; m 46; c 3. METEOROLOGY. *Educ:* Amherst Col, BA, 44; Mass Inst Technol, ScD, 54. *Prof Exp:* Aviation forecaster, US Weather Bur, 47-49; asst, 49-52, from instr to prof, 52-84, EMER PROF METEOROL, MASS INST TECHNOL, 84- *Honors & Awards:* Special Award, Am Meteorol Soc, 91. *Mem:* Fel Am Meteorol Soc; Royal Meteorol Soc. *Res:* Synoptic meteorology; maritime cyclogenesis; mesoscale meterology; fronts and frontogenetical processes; cumulus convective systems; symmetric instability; trends in forecast kill. *Mailing Add:* Nine Flint St Marblehead MA 01945

SANDERS, GARY HILTON, b New York, NY, Aug 27, 46; m 73; c 2. PARTICLE PHYSICS. *Educ:* Columbia Univ, AB, 67; Mass Inst Technol, PhD(physics), 71. *Prof Exp:* Res asst high energy physics, Mass Inst Technol, 67-71; res assoc physics, Princeton Univ, 71-73, asst prof, 73-78; STAFF MEM, LOS ALAMOS NAT LAB, 78- *Concurrent Pos:* Vis scientist, Deutsches Elektronen Synchrotron, Hamburg, 68-71; NSF fel, 71-72; vis scientist, Brookhaven Nat Lab, 71-74, 84- & Fermi Nat Accelerator Lab, Ill, 75-79. *Mem:* Am Phys Soc; Sigma Xi. *Res:* Experimental studies of elementary particles including quantum electrodynamics, hadronic production of dimuons, new particles, rare decays of muons and kaons; particle beam physics; neutrino physics. *Mailing Add:* Los Alamos Nat Lab MS H846 Los Alamos NM 87545

SANDERS, HARVEY DAVID, b Winnipeg, Man, June 22, 25; m 59; c 4. PHARMACOLOGY, PHYSIOLOGY. *Educ:* Univ BC, BSP, 59, MSP, 61, MD, 72; Univ Man, PhD(pharmacol), 63; FRCP(C), 77. *Prof Exp:* Lectr pharmacog, 59-60, lectr chem, 60-61, from instr to asst prof pharmacol, 64-69, ASSOC PROF PHARMACOL & MED, UNIV BC, 69- *Concurrent Pos:* Ford fel, 63-64; McEachern fel, 64-66. *Mem:* Am Soc Pharmacol & Exp Therapeut; Am Fedn Clin Res; NY Acad Sci; Can Med Asn; Am Col Physicians. *Res:* Effects of centrally active drugs on responses of the cerebral cortex to electrical stimulation; intracellular and extracellular recordings; cardiovascular effects of local anesthetics; pharmacology of the human urinary bladder. *Mailing Add:* Dept Med Univ BC Vancouver Gen Hosp Vancouver BC V5Z 1M9 Can

SANDERS, HOWARD L(AWRENCE), b Newark, NJ, Mar 17, 21; m 49; c 2. ECOLOGY, DEEP SEA BIOLOGY. *Educ:* Univ BC, BA, 49; Univ RI, MS, 51; Yale Univ, PhD(zool), 55. *Prof Exp:* Res assoc, Woods Hole Oceanog Inst, 55-63, assoc scientist, 63-65, sr scientist, 65-86, EMER SR SCIENTIST, WOODS HOLE OCEANOG INST, 86- *Concurrent Pos:* Instr, Marine Biol Lab, Woods Hole, 60-68; mem, Environ Biol Panel, NSF, 66-68; adj prof biol sci, State Univ NY, Stony Brook, 69-75, res affil, Marine Sci Res Ctr, 69-80; assoc invert zool, Harvard Univ, 69-80; mem adv panel, Cent Am Sea Level Canal, Nat Acad Sci, 69-70; res assoc, Smithsonian Trop Res Inst, 70; corresp, Nat Mus Natural Hist, 75-; coun deleg sect G, biol sci, AAAS, 80-82; mem, Int Asn Biol Oceanog Working Group, High Diversity Marine Ecosysts, UNESCO, 86. *Mem:* Nat Acad Sci; fel AAAS; Soc Am Naturalists. *Res:* Ecology as applied to marine benthic communities; crustacean phylogeny; protobranch bivalves; deep-sea biology; oil pollution biology; author or co-author of about 64 scientific publications. *Mailing Add:* Woods Hole Oceanog Inst Woods Hole MA 02543

SANDERS, J(OHN) LYELL, JR, b Highland, Wis, Sept 11, 24; m 60; c 3. STRUCTURAL MECHANICS. *Educ:* Purdue Univ, BSc, 45; Mass Inst Technol, ScM, 50; Brown Univ, PhD(appl math), 54. *Hon Degrees:* MA, Harvard Univ, 60. *Prof Exp:* Res engr, Nat Adv Comt Aeronaut, 47-57; vis lectr struct mech, 57-58, lectr, 58-60, assoc prof, 60-64, GORDON MCKAY PROF STRUCT MECH, HARVARD UNIV, 64- *Concurrent Pos:* NSF sr fel, Delft Technol Univ, 67-68. *Mem:* Fel Am Soc Mech Engrs; fel Am Acad Arts & Sci; fel Am Acad Mech. *Res:* Theory of thin shells; theory of plasticity and fracture mechanics. *Mailing Add:* Pierce Hall Harvard Univ Cambridge MA 02138

SANDERS, JAMES GRADY, b Norfolk, Va, June 10, 51; m 72. TRACE METAL BIOGEOCHEMISTRY, PHYTOPLANKTON ECOLOGY. *Educ:* Duke Univ, BS, 73; Univ NC, Chapel Hill, MS, 75, PhD(marine sci), 78. *Prof Exp:* Fel, Woods Hole Oceanog Inst, 78-80; vis asst prof, Chesapeake Biol Lab, Ctr Environ & Estuarine studies, Univ Md, 80-81; from asst curator to assoc cur, 81-90, LAB DIR, BENEDICT ESTUARINE RES LAB, DIV ENVIRON RES, ACAD NAT SCI, 83-, CUR, 90- *Concurrent Pos:* Vis lectr, Dept Earth Sci, Bridgewater State Col, 80; adj assoc prof, Chesapeake Biol Lab, Ctr Environ & Estuarine studies, Univ Md, 82-89. *Mem:* AAAS; Am Soc Limnol & Oceanog; Estuarine & Coastal Sci Asn; Estuarine Res Fedn; Phycol Soc Am; Am Geophys Union; Oceanog Soc. *Res:* Impact of marine phytoplankton on trace metal transfer in food webs; biogeochemical cycling of metals and metalloids; effects of sublethal concentrations of toxic substances on the morphology growth, and community structure of marine phytoplankton and zooplankton. *Mailing Add:* Benedict Estuarine Lab Div Environ Res Acad Natural Sci Benedict MD 20612

SANDERS, JAMES VINCENT, b Twinsburg, Ohio, July 24, 32; m 55; c 4. UNDERWATER ACOUSTICS, FLUID DYNAMICS. *Educ:* Kent State Univ, BS, 54; Cornell Univ, PhD(physics), 61. *Prof Exp:* Asst prof, 61-68, ASSOC PROF PHYSICS, NAVAL POSTGRAD SCH, 68- *Res:* Fluid mechanics of non-Newtonian fluids; large amplitude standing acoustic waves; interaction of acoustic waves with fluid flow; acoustic properties of the ocean. *Mailing Add:* 27965 Berwick Dr Carmel CA 93923

SANDERS, JAY W, b Baltimore, Md, July 26, 24; m 50; c 3. AUDIOLOGY. *Educ:* Univ NC, AB, 50; Columbia Univ, MA, 51; Univ Mo, PhD(speech path), 57. *Prof Exp:* Instr speech, Univ Mo, 52-57; from asst prof to assoc prof, Trenton State Col, 57-62; from asst prof to assoc prof, 64-70, Prof Audiol, Vanderbilt Univ, 70-; RETIRED. *Concurrent Pos:* Nat Inst Neurol Dis & Blindness fel audiol, Northwestern Univ, 62-64; consult, St Francis Hosp, Hearing & Speech Ctr, Trenton, NJ, 61-62; asst prof, Vanderbilt Univ, 64-65, assoc prof, 65-70; mem staff, Bill Wilkerson Hearing & Speech Ctr, 64- *Mem:* Am Speech & Hearing Asn. *Res:* Audition; disorders of audition and diagnostic audiology. *Mailing Add:* 5518 Vanderbilt Old Hickory NJ 37138

SANDERS, JOHN CLAYTOR, b Roanoke, Va, Oct 29, 14; m 54; c 2. MECHANICAL ENGINEERING, AERONAUTICS. *Educ:* Va Polytech Inst, BS, 36, MS, 37. *Prof Exp:* Indust engr, Aluminum Co Am, 37-39; aeronaut res scientist, NASA, 39-58, asst chief, Wind Tunnel & Flight Div, Lewis Res Ctr, 58-77; RETIRED. *Concurrent Pos:* Mem subcomt internal flow, Nat Adv Comt Aeronaut, 47-48, subcomt power plant controls, NASA, 53-58 & subcomt struct dynamics, 63-64; consult indust automation & control; pres city coun, Strongsville, Ohio, 66-67. *Mem:* Fel Am Soc Mech Engrs. *Res:* Dynamics and control of aircraft and missile propulsion systems, including propeller engines, jet engines, chemical rockets and nuclear rockets. *Mailing Add:* 15305 Forest Park Dr Strongsville OH 44136

SANDERS, JOHN D, b Louisville, Ky, Aug 2, 38; m 67; c 1. ELECTRICAL ENGINEERING. *Educ:* Univ Louisville, BEE, 61; Carnegie-Mellon Univ, MS, 62, PhD(elec eng), 65. *Prof Exp:* Develop engr, Receiving Tube Dept, Gen Elec Co, Ky, 61; mem tech staff, Radio Corp Am Labs, NJ, 62; instr elec eng, Carnegie-Mellon Univ, 62-64; Proj officer, US Cent Intel Agency, 64-68; VPRES, WACHTEL & CO, INC, 68- *Concurrent Pos:* Asst prof lectr, George Washington Univ, 67-68; financial adv, Indust Training Corp, Radiation Systs, Inc, Tork, Inc, Fla Glass Indust, Temporaries, Inc & Data Measurement Corp. *Mem:* Inst Elec & Electronics Engrs; Financial Analysts Fedn; Nat Security Traders Asn. *Res:* Aiding management of small technical companies, primarily in communications and electronics. *Mailing Add:* 412 Farragut St Washington DC 20011

SANDERS, JOHN ESSINGTON, b Des Moines, Iowa, May 5, 26; m 52; c 3. GEOLOGY. *Educ:* Ohio Wesleyan Univ, BA, 48; Yale Univ, PhD(geol), 53. *Hon Degrees:* DSc, Ohio Wesleyan, 68. *Prof Exp:* Nat Res Coun fel, geol surv, Smithsonian Inst, 52-53 & Brit Mus Natural Hist, Neth, 53-54; instr geol, Yale Univ, 54-56, asst prof, 56-64; sr res assoc, Hudson Labs, Barnard Col, Columbia Univ, 64-69, vis prof, 68-69, prof geol, 69-89; ADJ STAR PROF, HOFSTRA UNIV, 90- *Concurrent Pos:* NSF fac fel sci, Yale Univ & Mass Inst Technol, 62-63; assoc ed, J Sedimentary Petrol, 62-76; mem, bd dirs, Mutual Oil Am, Inc, 69-75; mem NY State Dept Environ Conserv Hudson River PCB Settlement Adv Comt, 76-, chmn, 77; mem Nat Res Coun Comt Assess PCB's in Environ, 78. *Mem:* Fel AAAS; fel Geol Soc Am; Soc Econ Paleontologists & Mineralogists; Am Asn Petrol Geologists; NY Acad Sci. *Res:* Sedimentology; stratigraphy; paleogeography; primary structures in sedimentary deposits; Mississippian of southern Appalachians; Mississippian Brachiopoda; Triassic-Jurassic of Connecticut, New York and New Jersey; nearshore marine sediments; geological applications of side-looking sonar; origin and occurrence of petroleum; geology of New York metropolitan area. *Mailing Add:* Dept Geol Hofstra Univ Hempstead NY 11550

SANDERS, JOHN P(AUL), SR, b Hope, Ark, July 4, 26; m 57; c 2. CHEMICAL & NUCLEAR ENGINEERING. *Educ:* Univ Ark, BSChE, 50, MS, 52; Ga Inst Technol, PhD(chem eng), 63. *Prof Exp:* Instr chem eng, Univ Ark, 50-52; engr, Oak Ridge Nat Lab, 52-55; assoc prof chem eng, Univ Ark, 57-65; res staff mem, 65-80, SR RES STAFF MEM, OAK RIDGE NAT LAB, 80- *Concurrent Pos:* Lectr, Oak Ridge Assoc Univ. *Mem:* Am Inst Chem Engrs; Sigma Xi. *Res:* Heat removal from nuclear reactor cores; transfer of heat through outer wall of annuli as a function of the system parameters; evaluation of the performance of the cores of nuclear reactors; computational procedures for systems analysis. *Mailing Add:* 116 Nebraska Ave Oak Ridge TN 37830-8141

SANDERS, JOHN STEPHEN, air pollution effects, soil-borne diseases, for more information see previous edition

SANDERS, KENTON M, b Oakland, Calif, June 16, 50. SMOOTH MUSCLE ELECTROPHYSIOLOGY. *Educ:* Univ Calif, Los Angeles, PhD(physiol), 76. *Prof Exp:* Assoc prof, 82-86, PROF PHYSIOL, SCH MED, UNIV NEV, 86- *Mem:* Biophys Soc; Am Physiol Soc; Am Gastroenterol Asn; Am Motility Soc. *Mailing Add:* Dept Physiol Univ Nev Sch Med Reno NV 89557

SANDERS, LOUIS LEE, b Little Rock, Ark, May 18, 29; m 58; c 4. INTERNAL MEDICINE, BIOCHEMISTRY. *Educ:* Univ Ark, BS, 51, MD, 55, MS, 61. *Prof Exp:* From instr to asst prof, 62-69, assoc prof, 69-80, PROF INTERNAL MED, SCH MED, UNIV ARK, LITTLE ROCK, 80-; CHIEF METAB SECT, LITTLE ROCK VET ADMIN HOSP, 69-, ASST CHIEF MED SERV, 75- *Concurrent Pos:* Asst dir clin res ctr, Univ Ark, Little Rock, 62-65; attend physician, Little Rock Vet Admin Hosp, 62-65, staff physician, 66-69. *Mem:* Fel Am Col Physicians; Am Rheumatism Asn; Am Fedn Clin Res. *Res:* Diabetes mellitus; insulin metabolism in adult-onset diabetes; effect of oral hypoglycemic agents on insulin metabolism. *Mailing Add:* Little Rock Vet Admin Hosp 4300 W Seventh Little Rock AR 72005

SANDERS, MARILYN MAGDANZ, b Norfolk, Nebr, Aug 12, 42; div. BIOCHEMISTRY, MOLECULAR BIOLOGY. *Educ:* Stanford Univ, BSc, 64; Univ Wash, PhD(biochem), 69. *Prof Exp:* From instr to res assoc biochem sci, Princeton Univ, 71-73; from instr to assoc prof, 73-87, PROF PHARMACOL, ROBERT WOOD JOHNSON MED SCH, UNIV MED & DENT NJ, 87- *Concurrent Pos:* Med Res Coun Can grant, Univ BC, 69-71. *Mem:* Am Chem Soc; Am Soc Cell Biol; Genetics Soc Am; Am Soc Biol Chemists. *Res:* Mechanism of induction of the heat shock response; gene regulation in heat shock in Drosophilia melanogaster; regulation of development and gene expression in eukaryotic organisms. *Mailing Add:* Pharm Dept Robert Wood Johnson Med Sch Univ Med & Dent NJ 675 Hoes Lane Piscataway NJ 08854

SANDERS, MARTIN E, b Appleton City, Mo, Feb 1, 54. IMMUNOLOGY. *Educ:* Univ Mo, AB, 75; Univ Chicago, MD, 79. *Prof Exp:* Med staff fel immunol, Nat Inst Allergy & Infectious Dis, NIH, 82-85, Nat Cancer Inst, 85-88; physician-scientist, Upjohn Co, 88-90; DIR RHEUMATOLOGY-IMMUNOL RES, CENTOCOR, INC, 90- *Concurrent Pos:* Fel rheumatology, Johns Hopkins Univ, 84-85. *Mem:* Am Fedn Clin Res; Am Col Rheumatology; Am Asn Immunologists; AAAS; Am Col Physicians; Clin Immunol Soc. *Res:* Pharmaceutical research; development with particular emphasis on therapies for autoimmune and inflammatory disease. *Mailing Add:* Centocor Inc 200 Great Valley Pkwy Malvern PA 19355

SANDERS, MARY ELIZABETH, b Kobe, Japan, Mar 1, 17; US citizen. GENETICS. *Educ:* Mt Holyoke Col, AB, 38; Cornell Univ, MS, 40; Smith Col, PhD(genetics), 47. *Prof Exp:* Asst, Dept Genetics, Carnegie Inst, 40-41; teacher sch, SDak, 41-42; asst bot, Conn Col, 42-43; asst genetics, Exp Sta, Smith Col, 43-46; instr bot, Mt Holyoke Col, 46-47; asst bot, Yale Univ, 47-48; instr & res assoc integrated lib studies & bot, Univ Wis, 48-54; instr biol sci, Northwestern Univ, 54-55; res assoc agron, SDak State Col, 55-62; res assoc, Arnold Arboretum, Harvard Univ, 62-65, Am Asn Univ Women fel, 64-65; from asst prof to prof bot, Mont Alto Campus, Pa State Univ, 65-77; RETIRED. *Res:* Genetics and embryo culture of Datura; origin of colchicine-induced diploid sorghum mutants with multiple changed characters, many mutants immediately true-breeding; investigations of tomato for responses to colchicine similar to those found in sorghum. *Mailing Add:* 216 McBath St State College PA 16801

SANDERS, OLIVER PAUL, b Caney, Okla, Dec 26, 24; m 45; c 2. MATHEMATICS. *Educ:* Southeastern State Col, BA, 47; Okla State Univ, MS, 49, PhD, 56. *Prof Exp:* Asst prof math, Arlington State Col, 49-51 & Southeastern State Col, 51-54; instr, Okla State Univ, 54-56; asst prof, Univ Ark, 56-57; assoc prof, La Polytech Univ, 57-59; prof & head dept, Hardin-Simmons Univ, 59-62; prof & head dept, 62-65, acad dean, 65-68, provost, 68-70, vpres acad affairs, 70-74, prof, 74-87, EMER PROF MATH, APPALACHIAN STATE UNIV, 87- *Mem:* Math Asn Am. *Res:* Partial differential equations. *Mailing Add:* 331 Grand Blvd Boone NC 28607

SANDERS, OTTYS E, b Hubbard, Tex, Apr 25, 03; m 30. HERPETOLOGY. *Educ:* Southern Methodist Univ, AB, 24. *Prof Exp:* Mgr, Southwestern Biol Supply Co, 27-80; RETIRED. *Concurrent Pos:* Res assoc, Strecker Mus, Baylor Univ, 66- *Mem:* Am Soc Ichthyol & Herpet; Soc Syst Zool; Herpetologists' League; Soc Study Amphibians & Reptiles. *Res:* Amphibia; Planaria; author of one book. *Mailing Add:* 5712 W Ledbetter Dr Dallas TX 75236

SANDERS, RAYMOND THOMAS, b Ogden, Utah, June 23, 23; m 47; c 1. PHYSIOLOGY. *Educ:* Univ Utah, BS, 49, MS, 50; Stanford Univ, PhD(biol), 56. *Prof Exp:* Nat Found Infantile Paralysis fel, Physiol Inst, Univ Uppsala, 56-58; from asst prof to prof physiol, Utah State Univ, 58-70, prof zool, 70-77, prof biol, 77-88; RETIRED. *Concurrent Pos:* Dir, Hons Prog, Utah State Univ, 74- *Mem:* Am Soc Zoologists; Soc Gen Physiol. *Res:* Permeability phenomena in cells and model systems; role of salts in metabolism; sensory physiology of invertebrates. *Mailing Add:* Dept Biol Utah State Univ Logan UT 84322

SANDERS, RICHARD MARK, data processing, for more information see previous edition

SANDERS, RICHARD PAT, b Chicago, Ill, Mar 18, 43; m 67. GEOLOGY. *Educ:* Northern Ill Univ, BS, 65, MS, 68; Univ Ill, PhD(geol), 71. *Prof Exp:* Asst prof geol, Univ Wis-Stevens Point, 71-74; ASST PROF GEOL, WGA COL, 74- *Mem:* Am Geophys Union; Geol Soc Am; Sigma Xi. *Res:* Igneous and metamorphic petrology; mode of implacement of igneous plutons and volcanics. *Mailing Add:* Dept Geol WGa Col Carrollton GA 30117

SANDERS, ROBERT B, b Augusta, Ga, Dec 9, 38; m 61; c 2. BIOLOGICAL CHEMISTRY. *Educ:* Paine Col, BS, 59; Univ Mich, MS, 61, PhD(biochem), 64. *Prof Exp:* Am Cancer Soc fel biochem, Univ Wis, 64-66; from asst prof to assoc prof, 66-86, PROF BIOCHEM, UNIV KANS, 86-, ASSOC DEAN GRAD SCH, 87-, ASSOC VCHANCELLOR, RES GRAD STUDIES PUB SERV, 89- *Concurrent Pos:* Battelle Mem Inst fel, 70 & 71; consult, Interex Res Corp, 72-80, NIH & NSF; vis assoc prof, Dept Pharmacol, Sch Med, Univ Tex, Houston, 74-75; NIH fel, 74-75; prog dir, Regulatory Biol Prog, NSF, Washington, DC, 78-79, assoc dir, Jr Sci Humanities Symp, Kan-Neb-Okla Region, 83- *Mem:* Sigma Xi; Am Soc Biol Chemists; Am Soc Pharmacol & Exp Therapeut. *Res:* Biochemistry of hormone action; cyclic nucleotides; biochemistry of reproduction. *Mailing Add:* Dept Biochem Univ Kans Lawrence KS 66045-2106

SANDERS, ROBERT CHARLES, b Anaconda, Mont, Dec 17, 42; m 70; c 1. NUCLEAR ENGINEERING, THERMAL HYDRAULICS. *Educ:* Ore State Univ, BS, 66. *Hon Degrees:* ScD, Mass Inst Techol, 70. *Prof Exp:* Nuclear engr, Div Naval Reactors, US Dept Energy, 70-75; asst prof, Univ Mo, 75-79; ENGR CONSULT, MPR ASSOCS, 79- *Concurrent Pos:* Consult, Consumers Power Co, 77. *Honors & Awards:* Gold Medal, Soc Am Military Engr. *Mem:* Am Nuclear Soc; Nat Soc Prof Engrs; Sigma Xi; NY Acad Sci. *Res:* Nuclear power plant thermal-hydraulics, safety, and reliability; waste heat utilization; neutron activation analysis; fluid systems design and analysis; design and analysis of heat transfer equipment. *Mailing Add:* 1050 Connecticut Ave NW Rm 400 Washington DC 20036

SANDERS, RONALD L, b Greenway, Ark, June 19, 47. BIOCHEMISTRY. *Educ:* Wash Univ, AB, 70; St Louis Univ, PhD(biochem), 75. *Prof Exp:* Res assoc, Dept Biochem, Univ Colo Med Ctr, 75-76; fel, W Alton Jones Cell Sci Ctr, 76-78; RES ASST PROF ANAT & CELL BIOL, SCH MED, TUFTS UNIV, 79-, RES ASST PROF BIOCHEM, 85- *Concurrent Pos:* Am Lung Asn grant, 77-79; lectr biochem, Med Sch Tufts Univ, 80-84. *Mem:* Am Chem Soc; Am Oil Chemists Soc; AAAS; Am Phys Soc; Am Soc Biol Chem. *Res:* Pulmonary biochemistry, phospholipid and surfactant metabolism, type II culture. *Mailing Add:* Burroughs Wellcome Co 3030 Cornwallis Rd Research Triangle Park NC 27709-4498

SANDERS, SAMUEL MARSHALL, JR, b Charleston, SC, July 26, 28; div; c 2. HEALTH PHYSICS, RADIOBIOLOGY. *Educ:* The Citadel, BS, 49; La State Univ, Baton Rouge, MS, 52; Am Bd Health Physics, dipl, 61; Med Col Ga, PhD(radiobiol), 74. *Prof Exp:* Sr supvr, Bio-Assay Lab, Savannah River Plant, E I Du Pont de Nemours & Co, Inc, 52-55, chemist, Health Physics Sect, 55-60 & 62-64, sr res chemist, Environ Sci Sect, Savannah River Lab, 64-89; CHEMIST, WESTINGHOUSE SAVANNAH RIVER CO, 89- *Concurrent Pos:* Partic traveling lect prog, Oak Ridge Asn Univs, 63-67; consult, Dept Radiol, Med Col Ga, 68-74. *Honors & Awards:* Elda E Anderson Mem Award, Health Physics Soc, 68. *Mem:* Health Physics Soc. *Res:* Metabolism of water with the fixing of hydrogen in non-labile positions in macromolecules of plants and animals; microbeam analysis of plutonium-bearing microaerosols from fuel reprocessing facilities. *Mailing Add:* 1220 Fernwood Ct Aiken SC 29803-5206

SANDERS, T H, JR, b Philadelphia, Pa, July 23, 43. PHYSICAL METALLURGY. *Educ:* Ga Inst Technol, BS, 66, MS, 69, PhD(metall), 74. *Prof Exp:* Res scientist, Aluminum Co Am, Alcoa Ctr, Pa, 74-79, Fracture & Fatigue Res Lab, Metall Prog, Ga Inst Technol, 79-81; assoc prof phys metall, Purdue Univ, 81-87, prof metall, 85-87; PROF MAT ENG, SCH MAT ENG, GA INST TECH, 87- *Mem:* Am Inst Mining, Metall & Petrol Engrs; Am Soc Metals; Am Soc Eng Educ. *Res:* Aluminum alloy development for aerospace applications; microstructural mechanisms of fracture and fatigue, solidification and microstructure, and primary processing to improve properties. *Mailing Add:* Sch Mat Engr Ga Inst Tech Atlanta GA 30332-0245

SANDERS, THEODORE MICHAEL, JR, b New York, NY, Sept 14, 27; wid; c 2. PHYSICS. *Educ:* Harvard Univ, AB, 48; Columbia Univ, MS, 51, PhD(physics), 54. *Prof Exp:* Asst physics, Columbia Univ, 49-51; res assoc, Stanford Univ, 53-55; from asst prof to prof, Univ Minn, 55-63; vis prof, 63-64, PROF PHYSICS, UNIV MICH, ANN ARBOR, 64- *Concurrent Pos:* Sloan fel, 58-62. *Mem:* Fel Am Phys Soc. *Res:* Physics of atoms, molecules and solids; radio frequency spectroscopy; low temperature physics. *Mailing Add:* H M Randall Physics Lab Univ Mich Ann Arbor MI 48109

SANDERS, TIMOTHY D, b Laramie, Wyo, Apr 1, 35; m 57; c 2. PHYSICS. *Educ:* Stanford Univ, BS, 57, MSc, 59, PhD(physics), 62. *Prof Exp:* Res assoc theoret physics, Washington Univ, 62-64; asst prof physics & math, 64-69, asst dean fac, 68-70, assoc prof, 69-77, PROF PHYSICS, OCCIDENTAL COL, 77- *Concurrent Pos:* NSF sci fac fel, Stanford Linear Accelerator Ctr, 71-72. *Mem:* AAAS; Am Phys Soc; Am Math Soc; Sigma Xi. *Res:* Elementary particles; hyper-nuclei and light nuclei; group theory and its applications to quantum theory. *Mailing Add:* 1912 Campus Rd Los Angeles CA 90041

SANDERS, W EUGENE, JR, b Frederick, Md, June 25, 34; m 56; c 4. MEDICINE, MICROBIOLOGY. *Educ:* Cornell Univ, AB, 56, MD, 60; Am Bd Internal Med, dipl, 68. *Prof Exp:* Intern med, Johns Hopkins Hosp, 60-61, resident, 61-62; epidemic intel serv officer, Commun Dis Ctr, USPHS, 62-64; chief resident med, Teaching Hosps, Col Med, Univ Fla, 64-65, from asst prof to assoc prof med & microbiol, 65-72; PROF MED & MED MICROBIOL & CHMN DEPT MED MICROBIOL, SCH MED, CREIGHTON UNIV, 72- *Concurrent Pos:* Am Soc Pharmacol & Exp Therapeut travel award, 67; NIH res career develop award, 68-; Markle scholar acad med, 68-; instr, Sch Med, Emory Univ, 62-64; ed, Am J Epidemiol, 84- *Mem:* AAAS; Am Pub Health Asn; Am Fedn Clin Res; Am Soc Microbiol. *Res:* Internal medicine; infectious diseases and epidemiology; bacterial interference; antimicrobial agents and chemotherapy; bacterial carrier states. *Mailing Add:* Creighton Micro Dept 2500 California St Omaha NE 68178

SANDERS, W(ILLIAM) THOMAS, b Owensboro, Ky, June 13, 33; m 54; c 2. SOLID MECHANICS, MECHANICAL ENGINEERING. *Educ:* Purdue Univ, BS, 54; NY Univ, MS, 57; Columbia Univ, ScD(mech eng), 62. *Prof Exp:* Engr, Dow Chem Co, 54; engr-trainee, Oak Ridge Nat Lab, 54-55; engr, Combustion Eng, Inc, 55-57 & Am Mach & Foundry Co, 57-60; asst prof mech eng, 62-66, ASSOC PROF MECH ENG, COLUMBIA UNIV, 66- *Concurrent Pos:* Consult, Am Mach & Foundry Co, Avion Corp & Columbia Broadcasting Syst Labs. *Mem:* Am Phys Soc; Am Soc Mech Engrs; Sigma Xi. *Res:* Solid state engineering; dislocations in crystals; solar energy applications. *Mailing Add:* Box 443 Bardonia NY 10954

SANDERS, W(ALLACE) W(OLFRED), JR, b Louisville, Ky, June 24, 33; m 56; c 2. CIVIL ENGINEERING. *Educ:* Univ Louisville, BCE, 55, MEng, 73; Univ Ill, MS, 57, PhD(civil eng), 60. *Prof Exp:* Asst civil eng, Univ Ill, 55-59, res assoc, 59-60, asst prof, 60-64; assoc prof, 64-70, asst dir, Engr Res Inst, 80-84, PROF CIVIL ENG, IOWA STATE UNIV, 70-, ASSOC DIR, ENG RES INST, 84-, ASSOC DEAN, COL ENG, 88- *Honors & Awards:* Adams Mem Award, Am Welding Soc, 70; R C Reese Res Prize, Am Soc Civil Engrs, 78. *Mem:* Am Soc Civil Engrs; Am Welding Soc; Am Rwy Eng Asn; Am Soc Eng Educ. *Res:* Fatigue of welded joints in structural metals; behavior of timber, steel and concrete bridges. *Mailing Add:* 104 Marston Hall Iowa State Univ Ames IA 50011

SANDERS, WALTER L, b Evansville, Ind, Aug 21, 37; m 60; c 1. SCIENCE TEACHING, COMPUTER PROGRAMMING. *Educ:* Univ Goettingen, DR, 70. *Prof Exp:* Res assoc astron, Yerkes Observ, Univ Chicago, 70-71; assoc, Kitt Peak Nat Observ, 71-72; from asst prof to assoc prof, 72-85, PROF ASTRON, NMEX STATE UNIV, 85- *Concurrent Pos:* Vis astronomer, Univ Munster, 80, Kitt Peak Nat Observ, Cerro Tolo Nat Observ, Mt Wilson & Paloman Observ, Europ Southern Observ, Chili & Los Campanos Observ, Lick Observ. *Mem:* Am Astron Soc; Royal Astron Soc; Int Astron Union; Astron Soc Pac; Astron Ges. *Res:* Galactic structure studies, including galactic cluster, spiral structure, and galactic distance scale; extragalactic distance scale. *Mailing Add:* Dept Astron NMex State Univ PO Box 30001 Dept 4500 Las Cruces NM 80003

SANDERS, WALTER MACDONALD, III, b Bluefield, WVa, Dec 5, 30; m 56; c 4. CIVIL & SANITARY ENGINEERING. *Educ:* Va Mil Inst, BS, 53; Johns Hopkins Univ, MS, 56, PhD(sanit eng), 64. *Prof Exp:* Eng asst, WVa Water Serv Co, 47-52; sanit engr, Greeley & Hansen Engrs, 53; res sanit engr, USAF MSc, Ft Detrich, Md, 53-55; asst, dept sanit eng, Johns Hopkins Univ, 55-56; sanit eng consult, USPHS, Div Int Health, 56-58, asst chief, Water Supply Sect, Dept Water Supply & Pollution Control, 58-60, res sanit engr, Southeast Region USPHS, Clemson Univ, 62-64, chief, Freshwater Ecosysts Br, US Environ Protection Agency/USPHS, 64-76, assoc dir water qual res, Athens Environ Res Lab USPHS, 76-85; SR CONSULT, MR CHASMAN

& ASSOCS PC, 87- *Concurrent Pos:* Adj prof, Div Interdisciplinary Studies, Clemson Univ, 62-75; res assoc & prof, Ecol Inst, Univ Ga, 67- & adj res assoc, Dept Microbiol, 72-; consult environ, 85- *Honors & Awards:* Gold Medal, US Environ Protection Agency, 85. *Mem:* Am Soc Civil Engrs; AAAS; Sigma Xi. *Res:* Aquatic ecosystem studies and use of controlled environmental chamber-stream ecosystem complex to develop predictive models for future water quality and stream conditions; environmental pollution; ecological engineering; sanitary engineering; water quality management; control of toxic chemicals; environmental exposure and risk assessments; environmental assessments; wetlands assessments; groundwater protection and assessment. *Mailing Add:* 195 Xavier Dr Athens GA 30606

SANDERS, WILLIAM ALBERT, b Lafayette, Ind, Apr 28, 33; m 56; c 4. THEORETICAL CHEMISTRY, CHEMICAL DYNAMICS. *Educ:* Purdue Univ, BSChE, 55, PhD(phys chem), 63; Georgetown Univ, MS, 61. *Prof Exp:* Instr chem, US Naval Acad, 57-60; NSF fel theoret chem, Univ Wis, 63-65, res assoc, 65; asst prof, 65-69, ASSOC PROF CHEM, CATH UNIV AM, 69-, CHMN DEPT, 81- *Concurrent Pos:* Consult, Phys Res Labs, Edgewood Arsenal, 68-70 & Naval Res Lab, 78- *Mem:* Sigma Xi; Am Chem Soc. *Res:* Gas phase kinetics; theory of molecular beam scattering; applications of perturbation theory; theory of inter- and intramolecular forces; chemical dynamics. *Mailing Add:* 16 Stone Cliff Dr Niantic CT 06357-1514

SANDERS, WILLIAM MACK, b West Point, Ark, June 12, 26; m 49; c 3. MATHEMATICS. *Educ:* Ark State Teachers Col, BS, 49; Univ Ark, MA, 52; Univ Ill, Urbana, PhD(math), 65. *Prof Exp:* Instr math, SMiss Univ, 52-55, assoc prof, 58-64; assoc prof, Lawrence Univ, 64-69; PROF MATH & HEAD DEPT, JAMES MADISON UNIV, 69- *Mem:* Math Asn Am; Am Math Soc. *Res:* Foundations of geometry, especially abstract models of geometries. *Mailing Add:* Dept Math James Madison Univ Harrisonburg VA 22801

SANDERS-BUSH, ELAINE, b Russellville, Ky, Apr 27, 40; m 67, 87; c 1. SEROTONIN. *Educ:* Western Ky Univ, BS, 62; Vanderbilt Univ, PhD(pharmacol), 67. *Prof Exp:* NIMH fel, 67-69, from instr to assoc prof, 68-80, PROF PHARMACOL, SCH MED, VANDERBILT UNIV, 80- *Concurrent Pos:* NIMH res scientist develop award, 74; adv ed, Psychopharmacology, 85-; counr, Serotonim Club, 86-90; ed bd, Neurochemistry Int, 88. *Mem:* Am Soc Pharmacol & Exp Therapeut; Am Col Neuropsychopharmocol; Neurosci Soc; NIMH, Ment Health Educ Rev Comt, 86-88, Psychopharmacology Res Rev Comt, 76-80; Serotonim Club (vpres, 90-). *Res:* Psychopharmacology; serotonim receptor subtypes; biogenic amines; second messengers; amphetamine derivatives. *Mailing Add:* Dept Pharmacol Vanderbilt Univ Sch Med Nashville TN 37232

SANDERS-LOEHR, JOANN, b New York, NY, Sept 2, 42; m 65. BIOCHEMISTRY. *Educ:* Cornell Univ, BS, 64, PhD(biochem), 69. *Prof Exp:* Fel biochem, Ore Health Sci Univ, 69-71; from asst prof to prof chem, Portland State Univ 71-84; PROF CHEM & BIOL SCI, ORE GRAD INST, 84- *Concurrent Pos:* NIH fel, 70-71; Cottrell res grant, 72-74; NIH grant, 75-; vis assoc chem, Calif Inst Technol, Pasadena, 78-79. *Mem:* Am Soc Microbiol; Am Chem Soc; Am Soc Biochem & Molecular Biol. *Res:* Role of metal ions in biological systems; metalloprotein structure and function. *Mailing Add:* Dept Chem & Biol Sci Ore Grad Inst Sci Technol Beaverton OR 97006-1999

SANDERSON, ARTHUR CLARK, b Providence, RI, Oct 23, 46; m 71; c 2. SIGNAL PROCESSING, ROBOTICS. *Educ:* Brown Univ, BS, 68; Carnegie-Mellon Univ, MS, 70, PhD(elec eng), 72. *Prof Exp:* Res engr, Res & Develop Lab, Westinghouse, 68-71; fel, Delft Univ Technol, 72-73; asst prof elec eng, Carnegie-Mellon Univ, 73-75; vis prof biomed eng, Univ Iberoamericana, Mex, 76-77; assoc prof, 77-81, prof elec eng & dir, Robotics Inst, Carnegie-Mellon Univ, 81-87; PROF & DEPT HEAD, ELEC, COMPUT & SYSTS ENG DEPT, RENSSELAER POLYTECH INST, 87- *Concurrent Pos:* Vis scientist, Inst Politech Nat, Mex, 76-77; adj prof, Med Sch, Univ Pittsburgh, 80- *Mem:* Fel Inst Elec & Electronics Engrs; AAAS; Sigma Xi. *Res:* Signal processing and pattern recognition applied to robotics and biomedicine; sensor based control of robots. *Mailing Add:* Rensselaer Polytech Inst Troy NY 12180-3590

SANDERSON, BENJAMIN S, b Buffalo, NY, Mar 18, 22; m 52; c 4. PHYSICAL CHEMISTRY. *Educ:* Hobart Col, BS, 42; Ohio State Univ, PhD(phys chem), 55. *Prof Exp:* Tech supvr, Holston Ord Works, Tenn Eastman Corp, 43-44; res chemist, 55-58, group leader, 58-68, sect mgr anal sect, 68-72, PROCESS SUPT, RES LAB, NL INDUSTS, INC, 72- *Concurrent Pos:* Lectr, Rutgers Univ, 62-71. *Mem:* Am Chem Soc; Am Statist Asn; fel Am Inst Chemists. *Res:* X-ray diffraction and spectroscopy; statistics; use of computers in research; infrared spectroscopy; quality control. *Mailing Add:* 32 Sherwood Circle Little Silver NJ 07739

SANDERSON, DONALD EUGENE, b Oskaloosa, Iowa, Feb 4, 26; m 49; c 3. GEOMETRIC, SET-THEORETIC. *Educ:* Cornell Col, BA, 49; Calif Inst Technol, MS, 51; Univ Wis, PhD(math), 53. *Prof Exp:* Asst instr math, Calif Inst Technol, 49-51; from instr to assoc prof, 53-64, PROF MATH, IOWA STATE UNIV, 64- *Concurrent Pos:* Vis assoc prof, Mich State Univ, 62-63. *Honors & Awards:* Allendoerfer Award, Math Asn Am, 80. *Mem:* Am Math Soc; Math Asn Am. *Res:* Topology of manifolds; general topology, infinite dimensional topology; set-theoretic topology. *Mailing Add:* Dept Math Iowa State Univ Ames IA 50011

SANDERSON, DOUGLAS G, bioanalytical chemistry, for more information see previous edition

SANDERSON, EDWIN S, b Mannville, Alta, Aug 19, 20; m 42; c 4. ORGANIC CHEMISTRY, POLYMER CHEMISTRY. *Educ:* Univ Alta, BSc, 50; McGill Univ, PhD(cellulose & wood chem), 53. *Prof Exp:* Res chemist, Visking Corp, 53-54, mgr cellulose & casing res, 54-57; plant mgr, Visking Div, Union Carbide Can, Ltd, 57-60, admin mgr, 60-62, dir res & develop plastic prod, 62-66; dir technol br, Dept Indust, Trade & Com, 66-81; RETIRED. *Mem:* Chem Inst Can; Am Chem Soc; Can Inst Food Technol; Sigma Xi. *Res:* Cellulose and high polymer chemistry; food packaging and preservation. *Mailing Add:* RR 3 Merrickville ON K0G 1N0 Can

SANDERSON, GARY WARNER, b Thermal, Calif, Dec 17, 34; m 53; c 4. FOOD CHEMISTRY. *Educ:* Univ Calif, Davis, BS, 56; Univ Nottingham, PhD(bot), 61. *Prof Exp:* Biochemist & head biochem div, Tea Res Inst, Ceylon, 62-66; mgr tea res, Thomas J Lipton, Inc, 66-71, asst dir tea res & develop, 71-75, dir beverage prod res, 76-78; vpres res, 78-88, VPRES TECHNOL, UNIVERSAL FOODS CORP, 89- *Concurrent Pos:* Adj prof, Col New Rochelle, 75; mem sci adv comt, Am Inst Baking, indust adv comt, Food Sci Dept, Univ Wis-Madison; chmn indust adv bd biol, Marguette Univ, Milwaukee, Wis. *Mem:* NY Acad Sci; fel Inst Food Technologists; Am Asn Cereal Chemists; Am Chem Soc; Brit Biochem Soc; World Aquacult Soc. *Res:* Plant biochemistry and physiology; food chemistry; tea chemistry and biochemistry; fermentation science and microbiology; biotechnology. *Mailing Add:* Universal Foods Corp 433 E Michigan St Milwaukee WI 53202

SANDERSON, GEORGE ALBERT, b New Haven, Conn, Aug 22, 26; m 61; c 3. GEOLOGY, PALEONTOLOGY. *Educ:* Trinity Col, Conn, BS, 49; Univ Wis, PhD(geol), 54. *Prof Exp:* Asst geol, Univ Wis, 50-54; geologist, Shell Develop Co, Tex, 54-55, jr stratigrapher, Shell Oil Co, 55, paleontologist, 55-64; sr res scientist, Pan Am Petrol Corp, 64-65, staff res scientist, 65-69, res group supvr, 69-77, spec res assoc, 77-83, sr res assoc, Amoco Prod Co, 83-89; CONSULT, 90- *Mem:* Paleont Soc; Soc Econ Paleontologists & Mineralogists; Paleont Asn. *Res:* Paleozoic micropaleontology; paleoecology and biostratigraphy, especially Fusulinidae and small Foraminifera; biometrics; detailed biostratigraphy, especially Pennsylvanian and Permian. *Mailing Add:* 2643 E 26th St Tulsa OK 74114

SANDERSON, GLEN CHARLES, b Wayne Co, Mo, Jan 21, 23; m 47; c 2. ZOOLOGY, PHYSIOLOGY. *Educ:* Univ Mo, BS, 47, MA, 49; Univ Ill, PhD, 61. *Prof Exp:* Game biologist, Iowa State Conserv Comn, 49-55; game biologist, Ill Dept Conserv & Ill Natural Hist Surv, 55-60, assoc wildlife specialist, Ill Natural Hist Surv, 60-63, wildlife specialist, 63-64, actg head, Sect Wildlife Res, 63-64, head Sect Wildlife Res, 64-90, EMER PRIN SCIENTIST, ILL NATURAL HIST SURV, 90- *Concurrent Pos:* Prof, Univ Ill, 65-; adj res prof, Southern Ill Univ, 64, adj prof, 64-84; ed, J Wildlife Mgt, 71-72. *Honors & Awards:* Oak Leaf Award, Nature Conservancy, 75. *Mem:* Am Soc Mammal; Wildlife Soc; AAAS; Am Inst Biol Sci. *Res:* Population dynamics of wild animals, especially furbearers; physiological factors of reproductive and survival rates; lead poisoning in waterfowl. *Mailing Add:* Ill Natural Hist Surv 607 E Peabody Champaign IL 61820

SANDERSON, HENRY PRESTON, b Midgell, PEI, Aug 28, 25; m 51; c 2. CHEMISTRY, ENVIRONMENTAL HEALTH. *Educ:* Dalhousie Univ, BSc, 49; Univ Minn, Minneapolis, MPH, 62, PhD(environ health), 69. *Prof Exp:* Plant chemist, Can Packers Ltd, 49; chemist sci serv, Can Dept Agr, NS, 49-51, chemist plant prod serv, Ont, 51-54; chemist, Int Joint Comn, 54-56; res scientist occup health div, Dept Nat Health & Welfare, 56-70; Res Coun Off environ secretariat, Div Biol, Nat Res Coun Can, 70-77; CHIEF, ATMOSPHERIC CHEM CRITERIA & STANDARDS, ATMOSPHERIC EVAL SERV, 77- *Mem:* Can Standards Asn; Am Indust Hyg Asn; Air Pollution Control Asn. *Res:* Compilation of scientific data related to cause and effects of pollutants on receptors; air pollution; atmospheric chemistry; acidification of precipitation; toxic chemicals; oxidants and photochemical reaction; effects of pollutants on vegetation. *Mailing Add:* 14 Revcoe Dr Willowdale ON M2M 2B8 Can

SANDERSON, JAMES GEORGE, b Somerset, NJ, Oct 10, 49. NUMERICAL ANALYSIS. *Educ:* Lafayette Col, BS, 71; Univ NMex, MA, 73, PhD(math), 76. *Prof Exp:* STAFF MEM, THEORET DESIGN DIV, LOS ALAMOS SCI LAB, 76- *Mem:* Soc Indust & Appl Math. *Res:* Numerical solutions to coupled nonlinear differential equations; eigenvalue problems; fully 3-D tomography; floating point correctness proofs; integral equations. *Mailing Add:* 388 Cheryl Ave Los Alamos NM 87544

SANDERSON, JUDSON, b Orrick, Mo, July 15, 21; m 54; c 3. MATHEMATICS. *Educ:* Univ Ill, BS, 47, MS, 48, PhD(math), 50. *Prof Exp:* Asst math, Univ Ill, 47-48; asst prof, Tulane Univ, 50-51; assoc prof, USAF Inst Technol, 51-56; assoc prof, 56-75, PROF MATH, UNIV REDLANDS, 75- *Concurrent Pos:* Lectr, Ohio State Grad Ctr, Wright-Patterson AFB, 53-56; consult, San Bernadino County Sch Syst, 63-66. *Mem:* Am Math Soc; Math Asn Am. *Res:* Real variable measure theory and foundations of mathematics. *Mailing Add:* 1408 Garden Redlands CA 92373

SANDERSON, KENNETH CHAPMAN, b Woodbury, NJ, Jan 9, 33; m 61; c 2. FLORICULTURE, WASTE UTILIZATION. *Educ:* Cornell Univ, BS, 55; Univ Md, MS, 58, PhD(hort), 65. *Prof Exp:* Teaching asst hort, Univ Md, 55-57; retail florist, C J Sanderson Florist, 58-60; greenhouse mgr, Univ Md, 60-65; asst prof floricult, La State Univ, 65-66; from asst prof to assoc prof, 66-76, PROF FLORICULT, AUBURN UNIV, 77- *Concurrent Pos:* Vis lectr, Calif Polytech State Univ, 76-77 & 84-85; assoc ed, Hortsci & J Am Soc Hort Sci, 77-81; chmn, Environ Pollution Comt, Am Soc Hort Sci, 73-75, Munic Agr & Indust Waste Comt, 80-84; chmn, Flor Work Group, Am Soc Hort Sci, 90- *Mem:* Sigma Xi; Plant Growth Regulator Soc Am; Am Soc Hort Sci; Prof Plant Growers Asn. *Res:* Greenhouse construction and management, florist crop production, pest control; floral design, retail flower shop management; plant propagation; environmental control, pollution; waste utilization; plant growth regulators. *Mailing Add:* 222 Green St Auburn AL 36830

SANDERSON, KENNETH EDWIN, b Holland, Man, Sept 14, 34; m 62; c 4. MICROBIAL GENETICS. *Educ:* Univ Man, BSA, 56, MSc, 57; Cornell Univ, PhD(genetics), 62. *Prof Exp:* Res assoc genetics, Cornell Univ, 61-62; AEC res assoc, Brookhaven Nat Lab, 62-64, vis biologist, 64-65; Wellcome Trust fel genetics, Lister Inst Prev Med, London, 65, microbial genetics res unit, Hammersmith Hosp, London, 65-66; from asst prof to assoc prof, 66-73, PROF BIOL, UNIV CALGARY, 73- *Concurrent Pos:* Nat Res Coun Can grants, 66-80; NSF grant, 68-76; Humboldt fel, Max Planck Inst, Freiburg, Ger, 72-73; vis scientist, Stanford Univ, 79-80, Univ Victoria, 87. *Mem:* Genetics Soc Am; Am Soc Microbiol; Genetics Soc Can; Can Soc Microbiol; Sigma Xi. *Res:* Mechanisms of parasexual recombination in fungi; genetic structure of the chromosome of Salmonella typhimurium; genetic basis of cell wall synthesis in Salmonella typhimurium; recombinant DNA methods. *Mailing Add:* Dept Biol Sci Univ Calgary Calgary AB T2N 1N4 Can

SANDERSON, MARIE ELIZABETH, b Chesley, Ont, Nov 16, 21; m 46; c 3. PHYSICAL GEOGRAPHY. *Educ:* Univ Toronto, BA, 44; Univ Md. MA, 46; Univ Mich, PhD(geog), 65. *Prof Exp:* Res scientist, Ont Res Found, 46-50; res assoc climat, C W Thornwaite Asn Lab Climat, NJ, 50-51; from asst prof to prof geol, Univ Windsor, 74-88, dir, Great Lakes Inst, 81-87; ADJ PROF, UNIV WATERLOO, 88- *Concurrent Pos:* Vis prof, Univ Hawaii, 79; dir, Water Network, 88- *Honors & Awards:* Award of Merit, Can Asn Geogrs. *Mem:* Can Asn Geogrs (pres, 80-81); Asn Am Geogrs; Am Water Resources Asn; Can Meteorol Soc; Am Geophys Union. *Res:* Climatology; water balance of the Great Lakes; hydrology. *Mailing Add:* Dept Geog Univ Waterloo Waterloo ON N2L 3G1 Can

SANDERSON, RICHARD BLODGETT, b Waltham, Mass, July 20, 35. PHYSICS. *Educ:* Mass Inst Technol, BS, 57; Syracuse Univ, PhD(physics), 63. *Prof Exp:* Res assoc physics, Ohio State Univ, 62-64, vis asst prof, 65, asst prof, 65-71; PHYSICIST, USAF WRIGHT-AERONAUT LAB, 72- *Mem:* Optical Soc Am. *Res:* Infrared spectroscopy; optical properties of materials; infrared sensors. *Mailing Add:* 2317 Pine Knott Dr Dayton OH 45431

SANDERSON, RICHARD JAMES, b Sydney, Australia, Aug 4, 33; m 58; c 1. BIOPHYSICS, CELL BIOLOGY. *Educ:* Univ Sydney, BE, 54; Univ Denver, MS, 71, PhD(appl physics), 74. *Prof Exp:* Aerodynamicist, English Elec Aviation, Eng, 55-61; sect chief, Stanley Aviation, Denver, 61-63; sr res scientist, Martin Marietta Corp, Denver, 63-72; STAFF INVESTR MICROBIOL, WEBB-WARING LUNG INST, SCH MED, UNIV COLO, 72- *Res:* Function of human peripheral cells in the immune response and their isolation into pure populations; red cell aging; biophysics of lung surfactant. *Mailing Add:* Dept Microbiol Univ Colo Med Ctr 4200 E Ninth Ave Denver CO 80220

SANDERSON, ROBERT THOMAS, b Bryson City, NC, Dec 25, 12; m 39, 78; c 3. INORGANIC CHEMISTRY. *Educ:* Yale Univ, BS, 34; Univ Chicago, PhD(chem), 39. *Prof Exp:* Chemist, Grasselli Chem Div, E I du Pont de Nemours & Co, 34-35; asst, Univ Chicago, 36-39; chief chemist, Western Geophys Co, 39-40; res chemist & proj leader, Tex Co, 40-49; assoc prof chem, Univ Fla, 49-50; prof inorg chem, Univ Iowa, 50-63; prof chem, 63-78, EMER PROF CHEM, ARIZ STATE UNIV, 78- *Honors & Awards:* CMA Award, 67. *Mem:* Fel AAAS; Am Chem Soc; Sigma Xi. *Res:* Aluminum borohydride; geochemical exploration for petroleum; synthetic fuels and lubricants; volatile hydrides and derivatives; organometallic chemistry; electronegativity; bond energies; interpretations of inorganic chemistry; theory of polar convalence. *Mailing Add:* 4725 Player Dr Ft Collins CO 80525

SANDFORD, MAXWELL TENBROOK, II, b Kansas City, Mo, Nov 7, 44; m 71; c 1. ASTROPHYSICS, MATHEMATICAL PHYSICS. *Educ:* Univ Kans, AB(math), AB(physics) & AB(astron), 66, MA, 68; Univ Ind, Bloomington, PhD(astrophys), 71. *Prof Exp:* STAFF MEM, LOS ALAMOS SCI LAB, 71- *Mem:* Sigma Xi; Royal Astron Soc; Am Astron Soc. *Res:* Radiative transfer; hydrodynamics of dusty objects; digital vidicon observations. *Mailing Add:* Los Alamos Sci Lab MS B231 Box 1663 Los Alamos NM 87545

SANDHAM, HERBERT JAMES, b Lethbridge, Alta, Sept 30, 32; m 55; c 4. ORAL BIOLOGY, MICROBIOLOGY. *Educ:* Univ Alta, DDS, 57; Univ Man, MSc, 63, PhD(oral biol), 67. *Prof Exp:* Asst prof med microbiol, Univ Man, 66-68; asst prof microbiol & dent, Univ Ala, Birmingham & investr microbiol, Inst Dent Res, 68-71; assoc prof prev dent, 71-77, PROF MICROBIOL, FAC DENT, UNIV TORONTO, 77- *Concurrent Pos:* Hon prof, Hubei Med Col, Wuhan, People's Repub China, 87. *Mem:* Int Asn Dent Res; Can Asn Dent Res (pres, 88-90). *Res:* Dental caries etiology and prevention; oral microbiology. *Mailing Add:* Fac Dent Univ Toronto Toronto ON M5G 1G6 Can

SANDHU, HARBHAJAN SINGH, b Sarih, India, May 1, 32; m 62; c 2. PHYSICS. *Educ:* Punjab Univ, India, BSc, 54, MSc, 55; Pa State Univ, PhD(physics), 61. *Prof Exp:* Asst prof physics, Col William & Mary, 61-62 & Univ Southern Calif, 62-64; asst prof, 64-68, PROF PHYSICS, CALIF STATE UNIV NORTHRIDGE, 68- *Mem:* Am Phys Soc. *Res:* Pair and multiplet production by 5-90 million electron volt x-rays; charged particle nuclear reaction cross sections. *Mailing Add:* Dept Physics & Astron Univ Victoria Victoria BC V8W 2Y2 Can

SANDHU, MOHAMMAD AKRAM, b Baddomahli, Pakistan, Mar 26, 36; m 62; c 3. ORGANIC POLYMER CHEMISTRY. *Educ:* Univ Punjab, Pakistan, BSc, 58, MSc, 61; Univ Strathclyde, PhD(chem), 67. *Prof Exp:* Res chemist, Pakistan Coun Sci & Indust Res, 61-64; res specialist, Univ Minn, Minneapolis, 67-70; sr res chemist, 70-77, SR STAFF, EASTMAN KODAK CO, 77- *Concurrent Pos:* Assoc, Royal Inst Chem, 66; Walter Reed Army Inst Res fel, Univ Minn, Minneapolis, 67-69, NIH fel, 69-70. *Honors & Awards:* Hamilton Barrett Res Prize, Univ Strathclyde, 65. *Mem:* Royal Soc Chem; Am Chem Soc; Sigma Xi. *Res:* Organic syntheses, polymer for photographic, electrographic and electronic imaging systems; aromatic, heterocyclic and natural products; organic chemistry of Ferrocene. *Mailing Add:* Winthrop Pharmaceut 90 Park Ave Fifth Floor New York NY 10016

SANDHU, RANBIR SINGH, b Lyallpur, Pakistan, Jan 19, 28; US citizen; m 57; c 3. CIVIL ENGINEERING, STRUCTURAL MECHANICS. *Educ:* Univ Punjab, Pakistan, BA, 46; E Panjab Univ, India, BSc, 49; Univ Sheffield, MEng, 62; Univ Calif, Berkeley, PhD(civil eng), 68. *Prof Exp:* Asst engr to dep dir designs, Irrig Dept, Govt of Punjab, India, 50-63; assoc prof civil eng, Punjab Eng Col, India, 63-65; sr engr, Harza Eng Co, Ill, 68-69; assoc prof, 69-73, PROF CIVIL ENG, OHIO STATE UNIV, 73- *Concurrent Pos:* Consult applications finite element method, var agencies, 66-; vis scientist, Norweg Geotech Inst, 77 & Univ Dayton, 79; guest prof, Univ Stuttgart, WGer, 80. *Mem:* Am Soc Civil Engrs; Am Acad Mech. *Res:* Soil and structural mechanics; mathematical, numerical and variational methods; approximate solution of boundary value problems; theoretical and applied mechanics; mechanics of continua; finite element methods; composite laminates. *Mailing Add:* Dept Civil Eng Ohio State Univ 2070 Neil Ave Columbus OH 43210

SANDHU, SHINGARA SINGH, b Pauhowind, India, Oct 10, 32; m 60; c 4. SOIL CHEMISTRY. *Educ:* Panjab Univ, BS, 52, MS, 54; Utah State Univ, PhD(soil chem), 70. *Prof Exp:* Asst prof chem, Punjab Agr Univ, India, 55-65; assoc prof, Alcorn Agr & Mech Col, 68-69; PROF CHEM, CLAFLIN COL, 70- *Concurrent Pos:* Res affil, Punjab Agr Univ, India, 72. *Mem:* AAAS; Am Chem Soc; Soil Sci Soc Am. *Res:* Pollutants in rural drinking water supplies. *Mailing Add:* Dept Chem Claflin Col College Ave NE Orangeburg SC 29115

SANDIFER, JAMES ROY, b Blakely, Ga, May 21, 45; m 68. ELECTROCHEMISTRY. *Educ:* Miss Col, BS, 67; Univ NC, Chapel Hill, PhD(anal chem), 73. *Prof Exp:* RES CHEMIST, EASTMAN KODAK CO, 73- *Mem:* Am Chem Soc. *Res:* Mass transport properties of membranes, their electrical characteristics and their use in the fabrication of ion selective electrodes. *Mailing Add:* 231 Pearson Lane Rochester NY 14612

SANDIFER, MYRON GUY, JR, b Lowrys, SC, Sept 4, 22; m; c 2. PSYCHIATRY. *Educ:* Davidson Col, BS, 43; Harvard Med Sch, MD, 47; Am Bd Psychiatry, dipl, 54; Am Bd Internal Med, dipl, 74. *Prof Exp:* Instr psychiat, Sch Med, Univ NC, 55-58, asst prof, 58-59, from clin asst prof to clin assoc prof, 59-65; clin prof, Columbia Univ, 65-66; assoc dean acad affairs, 69-75, PROF PSYCHIAT, UNIV KY, 66-, PROF FAMILY PRACT, 74- *Concurrent Pos:* Dir res, NC Dept Ment Health, 59-65. *Mem:* AMA; Am Psychiat Asn. *Res:* Psychiatric diagnosis. *Mailing Add:* Dept Psych Med Ctr Univ Ky Lexington KY 40506

SANDIFER, PAUL ALAN, b Cincinnati, Ohio, Jan 3, 47; m 66; c 4. MARINE ZOOLOGY. *Educ:* Col Charleston, BS, 68; Univ Va, PhD(marine sci), 72. *Prof Exp:* Asst marine scientist, 72-74, assoc marine scientist, 74-76, ASST DIR, DEPT WILDLIFE & MARINE RESOURCES, MARINE RESOURCES RES INST, SC, 76- *Mem:* Am Soc Zoologists, SE Estuarine Res Soc; World Maricult Soc (secy-treas, 75, pres, 79). *Res:* Culture, ecology and taxonomy of decapod crustacean larvae; mariculture of crustaceans; biology of commercially important crustaceans; ecosystem analysis; research management. *Mailing Add:* 691 Prentiss St James Island SC 29412-4521

SANDIFER, RONDA MARGARET, b Barnwell, SC, May 31, 54. ORGANIC CHEMISTRY. *Educ:* Newberry Col, BS, 76; Vanderbilt Univ, PhD(chem), 80. *Prof Exp:* NIH fel, Univ Utah, 80-82; asst prof, Dept Chem, Memphis State Univ, 82-84; asst prof, Chem Dept, Reed Col, 84-88. *Mem:* Am Chem Soc. *Res:* Mechanisms of the enzymatic synthesis of terpenes, in particular inhibition of squalene synthetase. *Mailing Add:* 1625 N Highland St Portland OR 97217

SANDIFER, SAMUEL HOPE, b Walterboro, SC, May 27, 16; c 3. MEDICINE. *Educ:* The Citadel, BS, 37; Med Univ SC, MD, 41. *Prof Exp:* Prof mil sci, Med Col Va, 47-49; asst clin prof med, Univ Ga, 55-56; assoc clin prof pediat cardiol, Sch Med, Univ Louisville, 64-65; prof, 68-81, EMER PROF PREV MED, MED UNIV SC, 81- *Mem:* AAAS; fel Am Col Cardiol; fel Am Col Prev Med; fel Am Col Physicians. *Res:* Preventive medicine; pesticide study. *Mailing Add:* Dept Family Med & Prev Med Div Med Univ SC 171 Ashley Ave Charleston SC 29425

SANDIN, THOMAS ROBERT, b Beloit, Wis, July 19, 39; m 60; c 5. SOLID STATE PHYSICS. *Educ:* Univ Santa Clara, BS, 60; Purdue Univ, Lafayette, MS, 62, PhD(physics), 68. *Prof Exp:* Instr physics, Purdue Univ, Indianapolis, 62-65; assoc prof, 68-77, PROF PHYSICS, NC A&T STATE UNIV, 77- *Concurrent Pos:* Lectr, Ind Univ, Kokomo Campus, 62-64. *Mem:* Am Asn Physics Teachers; Fedn Am Scientists. *Res:* Low temperature solid state; Mossbauer effect; author of textbooks on classical and modern physics. *Mailing Add:* Dept Physics NC A&T State Univ Greensboro NC 27411

SANDINE, WILLIAM EWALD, b Des Moines, Iowa, June 6, 28; m 55; c 1. BACTERIOLOGY, MICROBIOLOGY. *Educ:* Iowa State Univ, BS, 50; NC State Univ, MS, 55; Ore State Univ, PhD(bact), 58. *Prof Exp:* Instr bact, Ore State Col, 58-59; res assoc dairy biochem, Univ Ill, 59-60; from asst prof to assoc prof, 60-66, PROF MICROBIOL, ORE STATE UNIV, 66- *Honors & Awards:* Pfizer Paul Lewis Award, 64. *Mem:* Am Soc Microbiol; Am Dairy Sci Asn. *Res:* Lactic acid bacteria, especially bacteriophage, growth, taxonomy, metabolism and genetics; Staphylococci in food; microbiology of cheese; ecology of lactic acid bacteria including their role in human and animal nutrition. *Mailing Add:* Microbiol Dept Ore State Univ Corvallis OR 97331

SANDLER, HAROLD, b Cincinnati, Ohio, Nov 24, 29; m 61; c 2. MEDICINE, PHYSIOLOGY. *Educ:* Univ Cincinnati, BS, 51, MD, 55. *Prof Exp:* Intern med, Univ Chicago, 55-56; asst med, Univ Wash, 56-58, res fel cardiol, 58-61, from instr to asst prof med, 63-65; clin investr cardiol, Vet Admin Hosp, Seattle, Wash, 63-65; res med officer cardiovasc physiol, Biomed Res Div, 65-69, actg chief div, 69-72, CHIEF BIOMED RES DIV, NASA-AMES RES CTR, MOFFETT FIELD, 72- *Concurrent Pos:* Wash State Heart Asn fel, 58-59; NIH fel, 59-61; asst clin prof, Sch Med, Stanford Univ, 66-71, assoc clin prof, 71-78, clin prof med, 78- *Mem:* AMA; Am Fedn Clin Res; Aerospace Med Asn; Am Col Cardiol. *Res:* Internal medicine; cardiology; angiocardiography and cineangiocardiography; cardiovascular physiology; aerospace medicine; space bioscience; biophysics; bioengineering. *Mailing Add:* Biomed Res Div Cardiovasc Off NASA-Ames Res Ctr Moffett Field CA 94035

SANDLER, LAURENCE MARVIN, genetics; deceased, see previous edition for last biography

SANDLER, MELVIN, b Brooklyn, NY, July 1, 37; m 61; c 3. ELECTRICAL ENGINEERING, ENERGY CONVERSION. *Educ:* Polytech Inst Brooklyn, BEE, 58, MEE, 60, PhD(electrophysics), 65. *Prof Exp:* Instr elec eng, Polytech Inst Brooklyn, 59-60; res asst comput sci, Microwave Res Inst, 60-62, sr res assoc, Farmingdale Grad Ctr, 62-64; group leader, Airborne Instruments Lab, Div, Cutler-Hammer, 64-68; sr venture specialist, W R Grace & Co, 68-69; chmn admis comt, 71-77, assoc prof, 69-77, PROF ELEC

ENG, COOPER UNION, 77-, CHMN DEPT ELEC ENG, 75-77 & 86- *Mem:* Inst Elec & Electronics Engrs; Sigma Xi. *Res:* Communication theory; application of phase and injection lock techniques to signal processing and communication systems; power electronics. *Mailing Add:* Dept Elec Eng Cooper Union 51 Astor Pl New York NY 10003

SANDLER, RIVKA BLACK, b Warsaw, Poland, Feb 20, 18; US citizen; m 47; c 2. GERONTOLOGY. *Educ:* Hebrew Univ, MSc, 42, PhD(endocrinol), 50. *Prof Exp:* Instr, Hebrew Univ, 47-50; lectr pharmacol, Sch Med, Univ Ottawa, 50-52; from asst prof to assoc prof pharmacol, Univ Pittsburgh, 64-78, prof sci, Sch Health Related Professions, Interdisciplinary Progs, 78-89, prof, 83-89, EMER PROF GERONTOL & EPIDEMIOL, UNIV PITTSBURGH, 89- *Mem:* Fel Geront Soc Am. *Res:* Gerontolog; osteoporosis. *Mailing Add:* Dept Res Osteoporsis Muscle Strength Sch Health Res Professions Univ Pittsbrugh 114 Pa Hall Pittsburgh PA 15261

SANDLER, SAMUEL, b Lipivitz, USSR, Jan 1, 21; Can citizen; m 48; c 3. COMBUSTION ENGINEERING, ANALYTICAL CHEMISTRY. *Educ:* Univ Toronto, BASc, 44, MASc, 48. *Prof Exp:* Lab supvr, Defence Industs Ltd, 44-45; instr chem eng, Univ Toronto, 46-48; prin sci res officer, Defence Res Bd Can, 48-58; from asst prof to assoc prof chem eng, 58-69, PROF CHEM ENG, UNIV TORONTO, 69- *Concurrent Pos:* Instr, Can Voc Training Inst, 46-48; consult, Chem Eng Res Consult, Ltd, 62- *Mem:* Fel Chem Inst Can. *Res:* Kinetics and mechanisms of oxidation, decomposition, ignition and detonation of fuel vapors and gases; associated instrumental methods of chemical analysis; flame arrestor design. *Mailing Add:* Dept Chem Engr Univ Ont 35 St George St Toronto ON M5Z 1A4 Can

SANDLER, SHELDON SAMUEL, b Cleveland, Ohio, Dec 17, 32; m 58; c 1. BIOPHYSICS, APPLIED PHYSICS. *Educ:* Case Western Reserve Univ, BSEE, 54; Yale Univ, MEng, 55; Harvard Univ, MS, 58, PhD(appl physics), 62. *Prof Exp:* Res assoc, Horizons, Inc, 56-57; mem staff, Lincoln Lab, Mass Inst Technol, 58-59; sr engr, Electronic Commun, Inc, 59-60; from asst prof to assoc prof, 62-81, PROF ELEC & COMPUT ENG, NORTHEASTERN UNIV, 81- *Concurrent Pos:* Consult, Harvard Col Observ, 62-63; res fel appl physics, Harvard Univ, 63-; consult, Raytheon Corp, 63-64; guest prof, Swiss Fed Inst Technol, 64-65; consult, US Naval Res Lab, 65-69; vis scholar, Med Res Coun Lab Molecular Biol, Eng, 69-70; consult, Geosci Surv, 70-72, Block Eng, 73-74 & Am Sci & Eng, 74-75 & Geo-Ctr Inc; mem, Comt VI, Int Union Radio Sci; vis prof, Univ Zurich, 76-77, Robotics Res Ctr, Univ RI, 83-84; vis scholar, Harvard Univ, 90-91. *Mem:* Sigma Xi. *Res:* Pattern recognition; picture processing and reconstruction; electromagnetics; geophysical exploration; bioengineering; biomedical engineering. *Mailing Add:* Dept Elec & Comput Eng Northeastern Univ Boston MA 02115

SANDLER, STANLEY I, b New York, NY, June 10, 40; m 62; c 3. THERMODYNAMICS, CHEMICAL ENGINEERING. *Educ:* City Col New York, BChE, 62; Univ Minn, PhD(chem eng), 66. *Prof Exp:* NSF fel, Inst Molecular Physics, Univ Md, 66-67; from asst prof to prof & chmn dept, 82-86, H B DUPONT PROF CHEM ENG DEPT, UNIV DEL, 82- *Concurrent Pos:* Camille & Henry Dreyfus Fac Scholar, 71; consult, Mobil Res & Develop Corp, 77-, Chevron, 78-, Union Carbide Corp, 82- & Du Pont, 85-; vis prof, Imperial Col, London, 73-74, Tech Univ Berlin, 81 & 88-89, Univ Del Sur Argentina, 85, Univ Queensland, Australia, 89. *Honors & Awards:* Prof Prog Award, Am Inst Chem Eng, 84; 3M Lectureship Award, Chem Eng Div, Am Soc Eng Educ, 88. *Mem:* Am Phys Soc; Am Inst Chem Engrs; Am Chem Soc; Am Soc Eng Educ; AAAS; Sigma Xi. *Res:* Thermodynamic properties modelling and measurement; predict properties of fluids under extreme conditions; phase equilibrium prediction and measurement; computer-aided design. *Mailing Add:* Dept Chem Eng Univ Del Newark DE 19716

SANDLIN, BILLY JOE, b Hunt Co, Tex, Jan 10, 27; m 49; c 2. PHYSICS. *Educ:* ETex State Univ, BS, 48, MS, 49; Univ Tex, PhD, 60. *Prof Exp:* Instr physics, ETex State Teachers Col, 46-49; instr, LeTourneau Tech Inst, 49; instr physics, geol & math, Odessa Col, 49-53; asst prof physics, Tex Tech Col, 55-57; instr math, Univ Tex, 57-58; ASSOC PROF PHYSICS, TEX TECH UNIV, 59- *Mem:* Am Phys Soc; Am Asn Physics Teachers. *Res:* Measurements of high accuracy and precision involving electrical and electronic techniques; temperature and temperature-difference measurements; properties of solids at low temperatures; electronic circuit development. *Mailing Add:* Dept Physics Tex Tech Univ Lubbock TX 79409

SANDMANN, WILLIAM HENRY, b Yakima, Wash, Jan 15, 28; m 53; c 3. ASTRONOMY, EXPERIMENTAL PHYSICS. *Educ:* Reed Col, BA, 53; Univ Utah, PhD, 60. *Prof Exp:* Asst prof physics, Grinnell Col, 59-63; assoc prof, 63-72, PROF PHYSICS, HARVEY MUDD COL, 73- *Concurrent Pos:* Vis scholar, Univ Tex, Austin, 72-73 & 76-77; vis prof, Univ Capetown & Sutherland Observ, 76-77. *Mem:* Am Astron Soc. *Res:* Observational astronomy; astronomical instrumentation. *Mailing Add:* Dept Physics Harvey Mudd Col 12th & Columbia Claremont CA 91711

SANDMEIER, HENRY ARMIN, b Antwerp, Belg, Mar 17, 20; nat US; m 61; c 2. NUCLEAR PHYSICS. *Educ:* Swiss Fed Inst Technol, dipl, 49, DSc(elec eng), 54, PhD(physics), 59; Mass Inst Technol, SM, 52, EE, 54. *Prof Exp:* Mem res staff, Mass Inst Technol, 51-54; assoc physicist, Argonne Nat Lab, 56-61; liaison scientist, London Br, US Off Naval Res, London Embassy, UK, 61-63; PHYSICIST, NAT SECURITY PROG, LOS ALAMOS NAT LAB, 63- *Concurrent Pos:* Vis prof nuclear eng, Purdue Univ, 66; consult, US Naval Weapons Eval Facility, 66-, Army Res Off, 70-; vis prof, Univ Stuttgart, 68-69; consult, Defense Depts, Switz, Norway, Sweden & Ger, 68- *Mem:* Am Phys Soc; Am Nuclear Soc; Sigma Xi. *Res:* Reactor physics and engineering; engineering education; assay techniques of fissionable materials; vulnerability of nuclear weapons and nuclear weapons effects; international scientific liaison and consulting. *Mailing Add:* 809 Camino del Este Santa Fe NM 87501

SANDMEYER, ESTHER E, b Winterthur, Switz, Aug 9, 29; US citizen. PHARMACOLOGY, TOXICOLOGY. *Educ:* Winterthur Tech Univ, Switz, BSc, 51; Ohio State Univ, MSc, 60, PhD(biochem), 65. *Prof Exp:* Lab asst clin anal, Gen Hosp, Winterthur, 45-47; chemist, Nuffield Lab, Univ Birmingham, 52-53, Feldm hle, Rorschach, Switz, 53-55, Bell Tel Labs, Inc, NJ, 55-57 & Chem Abstr, Ohio, 58-60; asst prof chem, Friends Univ, 65 & biochem, Univ Nev, Reno, 65-70; trainee physiol & biophys, Hahnemann Med Col, 70-71 & toxicol, Sch Med, Univ Calif, San Francisco, 71-72; biochemist-toxicologist, Gulf Oil Corp, 72-75, toxicologist & dir Biochem Lab, 75. *Concurrent Pos:* Desert Res Inst grant, 66-67; indust consult, 75-; prog chmn, Int Cong Technol & Technol Transfer, 84. *Mem:* AAAS; Am Chem Soc; Soc Environ Health. *Res:* Biochemistry, chemical engineering; feasibility studies; literature surveys; training of toxicologists and toxicology managers; establishing organic analytical division for commercial laboratory. *Mailing Add:* 7305 Live Oak Dr Kelseyville CA 95451

SANDO, JULIANNE J, b Santa Maria, Calif, July 16, 52. CARCINOGENESIS & TUMOR PROMOTION. *Educ:* Ind Univ Pa, BS, 74; Univ Mich, Ann Arbor, PhD(pharmacol), 79. *Prof Exp:* Res fel immunol & carcinogenesis, NIH, Bethesda, Md, 79-81, staff fel, Nat Cancer Inst, 81; asst prof, 82-88, ASSOC PROF PHARMACOL, UNIV VA, CHARLOTTESVILLE, 88- *Concurrent Pos:* Prin investr, NIH grant, 82- *Mem:* Am Asn Cancer Res; Am Soc Pharmacol & Exp Therapeut; AAAS; Am Soc Biochem & Molecular Biol. *Res:* Understanding, at the cellular and biochemical level, the role of hormones, immunological mediators and tumor promoters in controlling cell proliferation and the expression of specific cell products; mechanisms by which phorbol ester tumor promoters activate protein kinase C and lead to transcriptional effects; molecular endocrinology. *Mailing Add:* Dept Pharmacol Univ Va Sch Med Box 448 Charlottesville VA 22908

SANDO, KENNETH MARTIN, b Oglivie, Minn, May 15, 41; m 66. THEORETICAL CHEMISTRY. *Educ:* Univ Minn, BCh, 61; Univ Wis, PhD(chem), 68. *Prof Exp:* Physicist, Smithsonian Astrophys Observ, 68-69; asst prof chem, 69-73, ASSOC PROF CHEM, UNIV IOWA, 73- *Concurrent Pos:* Guest worker, Nat Bur Standards, 75-76. *Mem:* Am Phys Soc; Am Chem Soc. *Res:* Quantum chemistry; atomic and molecular processes. *Mailing Add:* Dept Chem Univ Iowa Iowa City IA 52240

SANDO, WILLIAM JASPER, b Washington, DC, Apr 23, 27; m 85; c 2. STRATIGRAPHY, PALEONTOLOGY. *Educ:* Johns Hopkins Univ, BA, 50, MA, 51, PhD(geol), 53. *Prof Exp:* Fel Johns Hopkins Univ, 53-54; GEOLOGIST, US NAT MUS, US GEOL SURV, 54- *Mem:* Geol Soc Am; Paleont Soc; Am Asn Petrol Geologists. *Res:* Carboniferous stratigraphy, coral faunas, and carbonate systems. *Mailing Add:* Room E-325 US Nat Mus Natural Hist Washington DC 20560

SANDOK, PAUL LOUIS, b Rice Lake, Wis, Aug 18, 43; m 69; c 5. ZOOLOGY, IMMUNOLOGY. *Educ:* Univ Wis, Madison, BS, 68, MS, 71, PhD(bact), 74, Univ NC, Charlotte, BSEE, 87. *Prof Exp:* Proj asst, Med Sch, Univ Wis-Madison, 68-69, res asst, Dept Bact, 69-74; Hormel fel, Hormel Inst, Univ Minn, 74-77, res assoc, 77-78; asst prof microbiol, Univ NC, Charlotte, 78-85; PROCESS INSTRUMENT ENGR, C T MAIN, INC, 87- *Mem:* Sigma Xi; Am Soc Microbiol; Inst Elec & Electronics Engrs; AAAS. *Res:* Host-parasite interaction, at the cellular and molecular levels; portable cardiac monitoring devices; biochemistry; one US patent. *Mailing Add:* 1287 Starwood Ave Charlotte NC 28215

SANDOR, GEORGE N(ASON), b Budapest, Hungary, Feb 24, 12; US citizen; m 64; c 2. MECHANICAL ENGINEERING, DESIGN ENGINEERING. *Educ:* Univ Polytech, Budapest, Dipl Ing, 34; Columbia Univ, New York, DEngSc, 59. *Hon Degrees:* Dr, Technol Univ Budapest, 86. *Prof Exp:* Asst chief engr, Hungarian Rubber Co, Dunlop Ltd, 35-37, head mfg dept, 37-38; design engr, Babcock Printing Press Corp, Conn, 39-44; vpres & chief engr, Harry W Faeber Corp, NY, 44-50; chief engr, Graphic Arts Res Lab, Time Inc, Conn, 50-61 & Huck Design Co, NY, 61; assoc prof mech eng, Yale Univ, 61-66; Alcoa Found prof mech design, Rensselaer Polytech Inst, 66-75, chmn, Div Machines & Struct, Sch Eng, 67-74, dir, Ctr Eng Design, 74-75; res prof mech eng & dir, Mech Eng Design & Rotordynamics Labs, 76-89, EMER RES PROF, UNIV FLA, GAINESVILLE, 89- *Concurrent Pos:* Instr, Univ Conn, 41-44; lectr, Columbia Univ, 61-62; consult, Graphic Arts Res Lab, Time Inc, 61-63, McCall's Corp, 61-62 & Huck Design Co, 61-67; mem bd dirs, Huck Co, Inc, 67-71; consult, Xerox Corp, 71- & Instituto Politecnico Olivetti, Ivrea, Italy; consult engr, 71-; NSF, NASA & Army Res Off grants; mem, Graphic Arts Tech Found. *Honors & Awards:* Machine Design & Applied Mechanisms Awards, Am Soc Mech Engrs, 75. *Mem:* Fel Am Soc Mech Engrs; Am Soc Eng Educ; Nat Soc Prof Engrs; NY Acad Sci; Am Acad Mech; hon mem Int Fedn Theory Mach & Mechanisms. *Res:* Methodology and philosophy of engineering design; design and development of automatic machinery; printing, paper processing and allied machine design; kinematic and kineto-elastodynamic synthesis of planar and spatial mechanisms; hard automation for high productivity in manufacturing; computer-aided experimental design; design optimization; robotics. *Mailing Add:* 300A MEB Univ Fla Gainesville FL 32611

SANDOR, THOMAS, b Budapest, Hungary, Nov 3, 24; Can citizen; m 49; c 1. BIOCHEMISTRY, ENDOCRINOLOGY. *Educ:* Pazmany Peter Univ, Budapest, dipl chem, 48; Univ Toronto, PhD(path chem), 60. *Prof Exp:* Res biochemist, Clin Res Dept, Hotel-Dieu Hosp, 56-59; from res asst prof to res assoc prof, 61-70, RES PROF MED, UNIV MONTREAL, 70-; SR RES SCIENTIST, LAB ENDOCRINOL, HOSPITAL NOTRE DAME, 59- *Concurrent Pos:* Career investr, Med Res Coun Can, 62-; Nuffield Found traveling fel, 64; Sci Res Coun sr vis res fel, 66 & 79-80; Schering traveling fel, 66; Endocrine Soc traveling fel, 68; vis prof zool, Univ Sheffield, 70-71 & 79-80; hon vis prof, Dept Biol Chem, Fac Exact & Natural Sci, Univ Buenos Aires, 74; assoc mem exp med, McGill Univ, 69-; corresp ed, J Steroid Biochem, 70-79. *Mem:* Endocrine Soc; Can Biochem Soc; Can Soc Clin

Invest; Royal Soc Med; Brit Soc Endocrinol; Biochem Soc, Gt Brit; fel Royal Soc Can; Am Soc Zoologists; Europ Soc Camp Endocrinologists; Endocrine Soc. *Res:* Steroid biochemistry; comparative steroid endocrinology; mechanism of steroid hormone action; steroid hormone reception. *Mailing Add:* Hospital Notre Dame C P 1560 Sta C Montreal PQ H2L 4K8 Can

SANDOVAL, HOWARD KENNETH, b New York, NY, Aug 15, 31; m 51. MICROBIOLOGY. *Educ:* City Col New York, BS, 53; Columbia Univ, AM, 56; Cornell Univ, PhD(microbiol), 64. *Prof Exp:* Teacher high sch, NY, 56-58; res asst, Sloan-Kettering Inst Cancer Res, NY, 58-59; Sloan fel, 64-65; instr biol, Brooklyn Col, 65-67; microbiologist, Lederle Labs, NY, 67-69; asst prof, 69-72, ASSOC PROF BIOL, MIAMI-DADE COMMUNITY COL N, 72- *Mem:* Am Soc Microbiol. *Res:* Colicins; lysogeny; microbial genetics; electron microscopy of bacterial viruses. *Mailing Add:* Dept Biol Miami-Dade Comm Col N Campus 11380 NW 27 Ave Miami FL 33167

SANDOW, BRUCE ARNOLD, b Los Angeles, Calif, Jan 26, 45; m 71. REPRODUCTIVE BIOLOGY, HISTOLOGY. *Educ:* Univ Calif, Berkeley, BA, 67, MA, 70, PhD(endocrinol), 78. *Prof Exp:* Res fel, Ore Regional Primate Res Ctr, 76-81; ASST PROF ANAT & ASST PROF OBSTET GYNEC, EASTERN VA MED SCH, 81- *Mem:* Sigma Xi; Am Fertil Soc; Am Asn Anatomists. *Res:* In vitro fertilization; oocyte maturation; hormonal regulation of follicular function. *Mailing Add:* 8530 Culfor Crescent Norfolk VA 23503

SANDOZ, GEORGE, b Toledo, Ohio, Jan 24, 21; m 61; c 3. METALLURGICAL ENGINEERING. *Educ:* Wayne State Univ, BS, 43; Univ Mich, MS, 45; Univ Md, PhD(metall), 59. *Prof Exp:* Metallurgist, Chevrolet Motor Co, 43-44 & US Naval Res Lab, 46-72; mertallurgist, Chicago Br Off, Off Naval Res, 72-77, dir sci, 77-85; RETIRED. *Concurrent Pos:* Mem malleable iron comt, Welding Res Coun, 56-57. *Mem:* Am Soc Metals. *Res:* Reaction kinetics; mechanical properties; fracture and corrosion of steel and cast iron; protective coatings for refractory metals; mechanical properties of intermediate phases; electric and magnetic alloys; stress corrosion and hydrogen embrittlement. *Mailing Add:* 2030 Glencoe Wheaton IL 60187

SANDQUIST, GARY MARLIN, b Salt Lake City, Utah, Apr 19, 36; m 60; c 5. NUCLEAR & MECHANICAL ENGINEERING. *Educ:* Univ Utah, BSME, 60, PhD(mech eng), 64; Univ Calif, Berkeley, MS, 61. *Prof Exp:* Teaching asst mech eng, 59 & 61-62, instr, 62-63 & 64-65, from asst prof to assoc prof, 65-75, PROF MECH ENG, UNIV UTAH, 75-, DIR NUCLEAR ENG, 66-, RES ASSOC PROF SURG, 78- *Concurrent Pos:* Vis scientist, NSF fac fel, Mass Inst Technol, 69-70; sabbatical, Technion, Haifa, Israel & Ben Gurion Univ, Beer Shiva, Israel, 69-70; prin scientist, Rogers & Assocs Eng Corp, 80- *Honors & Awards:* Glen Murphy Award, Am Soc Eng Educ, 84. *Mem:* Fel Am Nuclear Soc; Am Soc Eng Educ; fel Am Soc Mech Engrs; Nat Soc Prof Engrs; Health Physics Soc. *Res:* Safety and environment aspects of nuclear energy; nuclear fusion; biomedical engineering; applied mathematics; system analysis. *Mailing Add:* Dept Mech Eng Univ Utah Salt Lake City UT 84112

SANDRA, ALEXANDER, CELL BIOLOGY, HORMONAL CONTROL OF DEVELOPMENT. *Educ:* Case Western Reserve Univ, PhD(anat), 76. *Prof Exp:* ASSOC PROF ANAT, UNIV IOWA, 83- *Res:* Hormonal cellular action. *Mailing Add:* Dept Anat Univ Iowa Iowa City IA 52242

SANDRAPATY, RAMACHANDRA RAO, b Eluru, India, Feb 15, 42; nat US; m 66; c 2. ENERGY & ENVIRONMENTAL ENGINEERING. *Educ:* Andhra Univ, Waltair, India, BE Hons, 63; Univ Roorkee, India, ME, 65; Univ SC, Columbia, MS, 71, PhD(mech eng), 74. *Prof Exp:* Proj engr, Prod Div, Cent Mech Eng Res Inst, Durgapur, India, 66-67; prod eng & bus partner, SGKR Mill-B, Elura, AP, India, 67-69; PROF MECH & INDUST ENG & CHMN DEPT, SC STATE COL, ORANGEBURG, 73- *Concurrent Pos:* Vis prof, IBM Corp, Res Triangle Park, NC, 74; adj staff mem, SC Energy Res Inst, 77-; engr consult, Appl Eng Co, Orangeburg, SC, 78; fac fel, Western Elec, Chicago, Ill, 79, US Dept Transp, Cambridge, Mass, 80 & NASA Lewis Res Ctr, Cleveland, Ohio, 82; vis fac scientist, Lawrence Berkeley Lab, Calif, 81. *Mem:* Am Soc Mech Engrs; Am Soc Eng Educ; Am Asn Univ Prof; Air & Waste Mgt Asn; Inst Indust Engrs. *Res:* Energy and thermal sciences; air pollution control; combustion research; environmental engineer. *Mailing Add:* SC State Col No 8164 Orangeburg SC 29117-8164

SANDRI, JOSEPH MARIO, b Chicago, Ill, Mar 9, 29; m 59; c 4. ORGANIC CHEMISTRY. *Educ:* Ill Inst Technol, BS, 52; Mich State Univ, PhD(chem), 56. *Prof Exp:* Res proj chemist, Am Oil Co, 56-61, sr proj chemist, 61-62; sr group leader, Nalco Chem Co, Ill, 62-70; dir res & develop, Ott Chem Co, 71-72, Story Chem Co, 72-80; AT CALLOWAY CHEM CO. *Mem:* AAAS; Am Chem Soc; Royal Soc Chem; Am Asn Textile Chemists & Colorists. *Res:* Isocyanates; phosgene and medicinal chemistry; amino acids; agricultural and paper chemicals; chemicals, latices and resins for textiles; surfactants; macrocyclic chemistry; fragrance chemicals; polymers; reaction mechanisms. *Mailing Add:* Calloway Chem Co 6601 Canal St PO Box 2335 Columbus GA 31993

SANDRIDGE, ROBERT LEE, b Junior, WVa, June 12, 32; m 53; c 5. ANALYTICAL CHEMISTRY, ORGANIC POLYMER CHEMISTRY. *Educ:* WLiberty State Col, BS, 54; WVa Univ, MS, 58, PhD(org chem), 69. *Prof Exp:* Chemist, NC State Univ, 54; group leader anal chem, Mobay Chem Corp, 58-73, mgr anal group, Process Res Dept, 73-78, mgr Anal & Environ Res, 78-88; CONSULT, 88- *Concurrent Pos:* Chmn, NAm Analytical Comt, Int Isocyanate Inst. *Mem:* Am Chem Soc; Am Soc Testing & Mat; Soc Plastics Indust; AAAS. *Res:* Isocyanates; urethanes; polycarbonates; amines; spectroscopy and spectrometry. *Mailing Add:* RD 1 Box 65 Proctor WV 26055

SANDRIK, JAMES LESLIE, b Chicago, Ill, July 7, 38; m 67; c 2. DENTAL MATERIALS. *Educ:* Northwestern Univ, PhB, 67, MS, 68, PhD(biol mat), 72. *Prof Exp:* From asst prof to assoc prof, 72-81, PROF DENT MAT & CHMN DEPT, SCH DENT, LOYOLA UNIV CHICAGO, 82- *Concurrent Pos:* NIH fel, Northwestern Univ, 72. *Mem:* Am Dent Asn; Am Soc Metals; Int Asn Dent Res; Sigma Xi. *Res:* Polymeric restorative and reconstructive materials; non-precious casting alloys. *Mailing Add:* 409 S Clifton Park Ridge IL 60068

SANDS, DAVID CHANDLER, b Los Angeles, Calif, Aug 30, 41; m 69; c 3. PLANT PATHOLOGY, BACTERIOLOGY. *Educ:* Pomona Col, AB, 63; Univ Calif, Berkeley, PhD(plant path), 69. *Prof Exp:* NSF fel soil microbiol, Div Soils, Commonwealth Sci & Indust Res Orgn, Australia, 69-70; asst plant pathologist, Conn Agr Exp Sta, 70-77; asst prof, 77-82, ASSOC PROF PLANT PATH, MONT STATE UNIV, 83- *Concurrent Pos:* Adj assoc prof microbiol, Mont State Univ, 81-; Environ Protection Agency Rev Panels, 86-88; proj leader, US Agency Int Develop-Int Ctr Agr Res Dry Areas, Middle East, N & W Africa & SAm. *Mem:* AAAS; Am Phytopath Soc; Am Soc Microbiol; Brit Soc Gen Microbiol; Am Inst Biol Sci; Arab soc Plant Protection. *Res:* Physiology, taxonomy and ecology of bacterial plant pathogens, especially Pseudomonas; ecology and general physiological differences between plant pathogenic pseudomonads and saprophytes; bacterial diseases of cereal crops; selection of high lysine lines of bacteria for food fermentation and high lysine lines of cereal crops; ice nucleating bacteria; biocontrol of weeds; expert systems in prediction of plant disease epidemics. *Mailing Add:* Dept Plant Path Mont State Univ Bozeman MT 59717

SANDS, DONALD EDGAR, b Leominster, Mass, Feb 25, 29; m 56; c 2. PHYSICAL CHEMISTRY. *Educ:* Worcester Polytech Inst, BS, 51; Cornell Univ, PhD, 55. *Prof Exp:* Res assoc, Cornell Univ, 55-56; crystallographer, Lawrence Radiation Lab, 56-62; from asst prof to assoc prof, Univ Ky, 62-65, dir gen chem, 74-75, assoc dean advan studies, Col Arts & Sci, 75-80, actg dean, 80-81, assoc vpres acad affairs, 81-84, vchancellor acad affairs, 84-89, PROF CHEM, UNIV KY, 68-; SECT HEAD, NETWORKING & TEACHER PREP, NSF, 89- *Mem:* Sigma Xi; AAAS; Am Chem Soc; Am Crystallog Asn; NY Acad Sci. *Res:* Crystallography; thermodynamics. *Mailing Add:* Dept Chem Univ Ky Lexington KY 40506-0055

SANDS, ELAINE S, b Brooklyn, NY, Jan 25, 40; m 64; c 2. SPEECH-LANGUAGE PATHOLOGY. *Educ:* Brooklyn Col, AB, 60; Univ Mich, MS, 61; NY Univ, PhD(speech path, audiol), 77. *Prof Exp:* Speech pathologist, Med Ctr, NY Univ, 61-70; asst prof, 70-80, ASSOC PROF SPEECH PATH & AUDIOL, ADELPHI UNIV, 80-, DEPT CHAIR, 89- *Concurrent Pos:* Consult, New York Vet Admin Hosp, 76- *Mem:* Acad Aphasia; NY Acad Sci; fel Am Speech, Hearing & Language Asn. *Res:* Neurological aspects of language; aphasia in adults; recovery from aphasia. *Mailing Add:* Dept Speech Arts & Commun Dis Adelphi Univ Garden City NY 11530

SANDS, GEORGE DEWEY, b Norfolk, Va, June 16, 19; m 42; c 3. PHYSICAL CHEMISTRY. *Educ:* Col William & Mary, BS, 39; Univ Richmond, MS, 41; Univ Ill, PhD(phys chem), 45. *Prof Exp:* Sr res chemist, Firestone Tire & Rubber Co, 45-48; assoc prof chem, Col William & Mary, 48-56; chief, Nuclear Br, Transp Res Command, US Dept Army, 56-59; dir sci requirements, Martin Co, 59-60; aerospace polymer chemist, 60-70, assoc proj scientist, Proj Viking, 70-76, CHIEF, SCI & TECH INFO PROGS DIV, LANGLEY RES CTR, NASA, 70-76. *Mem:* Am Chem Soc; Sigma Xi. *Res:* Physical chemistry of high polymers. *Mailing Add:* Four Fox Croft Rd Williamsburg VA 23185

SANDS, HOWARD, b New York, NY, Aug 20, 42; m 68; c 3. PHARMACOLOGY, BIOCHEMISTRY. *Educ:* Rutgers Univ, New Brunswick, BA, 64; Case Western Reserve Univ, PhD(pharmacol), 69. *Prof Exp:* NIH training grant renal dis, Northwestern Univ, 69-71; pharmacologist, Nat Jewish Hosp & Res Ctr, 71-81; group leader, pharmacol/toxicol, New England Nuclear, 81-83; group leader, Immunopharmaceuticals, 83-88; RES ASSOC, DUPONT MERCK PHARMACEUT, 89- *Concurrent Pos:* Pharmacologist, Vet Admin Res Hosp, 70-71; asst prof, Dept Oral Biol, Sch Dent, Univ Colo, 75-77. *Mem:* Am Asn Cancer Res. *Res:* Physiology of drug delivery; radio pharmaceutical research and related immunology. *Mailing Add:* Glenolden Labs DuPont Merck Pharmaceut Glenolden PA 19036

SANDS, JEFFREY ALAN, b Kingston, Pa, Jan 16, 48; m 73. BIOPHYSICS. *Educ:* Univ Del, BS, 69; Pa State Univ, MS, 71, PhD(biophys), 73. *Prof Exp:* From asst prof to assoc prof, 73-82, PROF BIOPHYSICS, LEHIGH UNIV, 82- *Mem:* Biophys Soc; Am Soc Microbiol. *Res:* Molecular virology, antiviral agents, virus-cell interactions, glycoprotein transport. *Mailing Add:* Dept Physics/Biol Lehigh Univ Bethlehem PA 18015

SANDS, MATTHEW, b Oxford, Mass, Oct 20, 19; c 3. PHYSICS. *Educ:* Clark Univ, BA, 40; Rice Inst, MA, 41; Mass Inst Technol, PhD(physics), 48. *Prof Exp:* Physicist, US Naval Ord Lab, 41-43 & Los Alamos Sci Lab, 43-46; res assoc, Mass Inst Technol, 46-48, asst prof physics, 48-50; sr res fel, Calif Inst Technol, 50-52, from assoc prof to prof, 52-63; prof & dep dir, Linear Accelerator Ctr, Stanford Univ, 63-69; vchancellor sci, 69-72, prof physics, 69-86, EMER PROF PHYSICS, UNIV CALIF, SANTA CRUZ, 86- *Concurrent Pos:* Fulbright scholar, Italy, 52-53; consult, Inst Defense Anal, 60-68, Off Sci & Technol, 61-66 & Arms Control & Disarmament Agency, 62-66; mem, Comn Col Physics, 60-66, chmn, 64-66; mem, Pugwash Conf Sci & World Affairs, 61-62; pres, Sands-Kidner Assoc, Inc, 86-; consult, Linear Accelerator Ctr, Stanford Univ, 86- *Mem:* Fel Am Phys Soc; Am Asn Physics Teachers; Fedn Am Sci. *Res:* Electronic instrumentation; cosmic rays; accelerators; high-energy physics; education; science and public affairs. *Mailing Add:* 160 Michael Lane Santa Cruz CA 95060

SANDS, RICHARD DAYTON, b Skaneateles, NY, Nov 18, 29; m 53; c 3. ORGANIC CHEMISTRY. *Educ:* Oberlin Col, AB, 51; Syracuse Univ, MS, 54, PhD(org chem), 59. *Prof Exp:* Asst instr chem, Syracuse Univ, 52-56; from asst prof to prof, 56-70, chmn div math & sci, 74-76, Ferro prof, 70-80, chmn div phys sci, 76-80, chmn dept chem, 80-85, PROF CHEM, ALFRED UNIV, 70- *Honors & Awards:* Scholes Lectureship, Alfred Sigma Xi Club. *Mem:* Am Chem Soc. *Res:* Ring size in the pinacol rearrangement of alicyclic glycols; synthesis and cleavage of aliphatic bicyclic compounds. *Mailing Add:* Dept Chem Alfred Univ Alfred NY 14802

SANDS, RICHARD HAMILTON, b San Diego, Calif, Sept 28, 29; m 51; c 4. BIOPHYSICS, ATOMIC PHYSICS. *Educ:* Univ Redlands, BS, 50; Washington Univ, PhD(physics), 54. *Prof Exp:* Res assoc & instr physics, Stanford Univ, 54-57; from asst prof to assoc prof, 57-65, chmn, Dept Physics, 77-82, PROF PHYSICS & RES BIOPHYSICIST, UNIV MICH, ANN ARBOR, 65- *Honors & Awards:* Cert of Literature Award, Philalethes Soc, 81. *Mem:* AAAS; Am Asn Physics Teachers; Am Phys Soc; Biophys Soc. *Res:* Magnetic and optical resonance fluorescence in atomic and solid state physics; biophysical applications of electron paramagnetic resonance spectrometry; Mossbauer spectroscopy and electron-nuclear double resonance spectrometry. *Mailing Add:* 11946 Weiman Dr Pinckney MI 48169-9013

SANDS, SEYMOUR, b New York, NY, Mar 16, 18. TEXTILE CHEMISTRY. *Educ:* City Col, BS, 39; NY Univ, MS, 48; Polytech Inst Brooklyn, PhD(chem), 53. *Prof Exp:* Org chemist, Fleischmann Distilling Co, 46-49; res chemist, 53-82, develop assoc, Textile Fibers Dept, E I du Pont de Nemours & Co, Inc, 68-82; RETIRED. *Mem:* Am Chem Soc; Sigma Xi. *Res:* Polymer chemistry pertaining to fiber technology. *Mailing Add:* 2300 Riddle Ave, No 304 Wilmington DE 19806-2151

SANDSON, JOHN IVAN, b Jeannette, Pa, Sept 20, 27; m 57; c 2. MEDICINE. *Educ:* Wash Univ, MD, 53; Am Bd Internal Med, dipl, 60. *Prof Exp:* From intern to asst resident med serv, Presby Hosp, New York, 53-56; fel rheumatol, Albert Einstein Col Med, Bronx, NY, 57-60, from assoc med to prof med, 60-74, from asst dean to assoc dean health servs, 69-74, hosp med dir, 69-74; PROF MED & DEAN, SCH MED, BOSTON UNIV, 74- *Concurrent Pos:* Nat Inst Arthritis & Metab Dis trainee, Albert Einstein Col Med, 57-59, Arthritis & Rheumatism Found fel, 57-60, chief resident, 57, asst vis physician, Bronx Munic Hosp Ctr, 57-66, vis physician, 66-, head, Arthritis Group, 68-72; attend physician, Bronx Vet Admin Hosp, 57-66; investr health res coun, City New York, 61-71; consult, study sect, NIH, 71-74, mem gen med A study sect & gen res support study sect, 85-89. *Honors & Awards:* Maimonides Award, Anti Defamation League, 86. *Mem:* AAAS; Am Soc Clin Invest; Harvey Soc; Am Rheumatism Asn; Am Asn Physicians; Am Asn Med Cols; AMA. *Res:* Chemical and immunological studies of the proteoglycans of synovial fluid and cartilage. *Mailing Add:* Sch Med Boston Univ 80 E Concord St Boston MA 02118

SANDSTEAD, HAROLD HILTON, b Omaha, Nebr, May 25, 32; m 58; c 3. INTERNAL MEDICINE, NUTRITION. *Educ:* Ohio Wesleyan Univ, BA, 54; Vanderbilt Univ, MD, 58; Am Bd Internal Med & Nutrit, dipl, 67. *Prof Exp:* From intern to asst resident med, Barnes Hosp, St Louis, Mo, 58-60; asst resident path, Vanderbilt Hosp, Nashville, Tenn, 60-61; first asst resident med, Vet Admin & Vanderbilt Hosps, 63-64; Hugh J Morgan resident, Vanderbilt Hosp, 64-65; from instr to asst prof med, Sch Med, Vanderbilt Univ, 65-67, asst prof biochem, 65-67, asst prof med & nutrit, 67-70, assoc prof nutrit, 70-71; adj prof biochem & clin internal med, Sch Med, Univ NDak, 71-84; dir, Grand Forks Human Nutrit Res Ctr, Agr Res Serv, USDA, 71-84; dir, Human Nutrit Res Ctr Aging & prof nutrit, Sch Nutrit, Tufts Univ, 84-85; PROF & CHMN, DEPT PREV MED & COMMUNITY HEALTH, MED BR, UNIV TEX, GALVESTON, 85- *Concurrent Pos:* Nutrit Found Future leader, 68-71. *Honors & Awards:* Hull Gold Medal Award, AMA, 70; Meade Johnson Award, Am Inst Nutrit, 72; W O Atwater Mem Lectr, USDA, 84; Ellen Swallow Richards Mem Lectr, Univ NC, 85; Sam & Mary E Roberts Nutrit Medal, 85. *Mem:* Fel Am Col Physicians; Am Inst Nutrit; Am Soc Clin Nutrit; Soc Exp Biol & Med; Cent Soc Clin Res; Neurol Sci Soc. *Res:* Clinical nutrition; zinc metabolism; essential and toxic trace elements; nutrition, brain development and function. *Mailing Add:* Prev Med & Community Health Sch Univ Tex Med Ctr Galveston TX 77550

SANDSTED, ROGER FRANCE, b Holdrege, Nebr, Aug 5, 18; m 49; c 3. VEGETABLE CROPS. *Educ:* Univ Nebr, BS, 48; Univ Minn, PhD, 54. *Prof Exp:* Asst hort, Univ Nebr, 48-50 & Univ Minn, 52-54; asst horticulturist, Parma Br Exp Sta, Idaho, 54-57; from asst prof to assoc prof, 57-77, PROF VEG CROPS, EXTEN, CORNELL UNIV, 77-, DEPT EXTEN LEADER, 77-; from asst prof to assoc prof, 57-77, prof veg crops, Exten, 77-83, EMER PROF, DEPT FRUIT & VEG SCI, 83- *Mem:* Am Soc Hort Sci. *Res:* Legume vegetables; dry beans; cultural problems. *Mailing Add:* 22 Dutcher Rd Freeville NY 13068

SANDSTROM, DONALD JAMES, b Chicago, Ill, July 26, 37; m 59; c 3. METALLURGY OF URANIUM ALLOYS, DEFORMATION PROCESSING OF REFRACTORY MATERIALS. *Educ:* Univ Ill, BS, 58; Univ NMex, MS, 68. *Prof Exp:* Group leader, Mat Technol Group, 75-81, assoc div leader, Mat Sci & Technol Div, 81-85, dep div leader, 85-89, DIV LEADER, MAT SCI & TECHNOL DIV, LOS ALAMOS NAT LAB, 89- *Mem:* Fel Am Soc Mat; Am Inst Mining Metall & Petrol Engrs. *Res:* Material science and technology; Dept Defense and Dept Energy activities associated with uranium. *Mailing Add:* 2279 Calle Cacique Santa Fe NM 87505

SANDSTROM, DONALD RICHARD, b Spokane, Wash, May 9, 40; m 63; c 1. PHYSICS. *Educ:* Wash State Univ, BS, 63, PhD(physics), 67. *Prof Exp:* NSF fel, Univ Bonn, 67; NSF & UK Atomic Energy Authority fel, Univ Liverpool, 67-69; from asst prof to assoc prof, 69-81, PROF PHYSICS & CHEM PHYSICS, WASH STATE UNIV, 81- *Concurrent Pos:* Guest worker, Nat Bur Standards, 74-75. *Mem:* Am Phys Soc; AAAS; Sigma Xi. *Res:* X-ray absorption spectroscopy; determination of local structure in condensed matter by analysis of the extended x-ray absorption fine structure. *Mailing Add:* 6130 E Mercer Way Mercer Island WA 98040

SANDSTROM, WAYNE MARK, b Seattle, Wash, Feb 17, 27; m 53; c 3. UNDERWATER ACOUSTICS, ORDNANCE. *Educ:* Univ Wash, BS, 48, PhD(physics), 53. *Prof Exp:* From assoc physicist to sr physicist, Appl Physics Lab, Univ Wash, 52-59, asst dir, 59-64, dep dir, 64-70; PRES, HENDERSON TECH CORP, 70- *Concurrent Pos:* Consult, US Naval Undersea Warfare Eng Sta, Wash, 70- *Honors & Awards:* David Bushnell Award, Am Defense Preparedness Asn, 90. *Mem:* Am Phys Soc. *Res:* Electromagnetics. *Mailing Add:* 6215 54th NE Seattle WA 98115

SANDULEAK, NICHOLAS, astronomy; deceased, see previous edition for last biography

SANDUS, OSCAR, b New York, NY, July 29, 24; m 46; c 3. PHYSICAL CHEMISTRY. *Educ:* Univ Ky, BS, 49; Univ Chicago, MS, 50; Ill Inst Technol, PhD(chem), 55. *Prof Exp:* Asst chemist, Argonne Nat Lab, 55-58; res assoc phys chem, Radiation Lab, Univ Mich, 58-60, assoc res phys chemist, 60-61; res phys chemist, Conductron Corp, 62-63; sr chemist, Chemotronics, Inc, 63-64; assoc res chemist, Infrared Physics Lab, Univ Mich, Ann Arbor, 64-67; res chemist, Energetics & Warheads Div, Army Armament Res Eng & Develop Ctr, US Army, 68-89; RETIRED. *Mem:* AAAS; Am Chem Soc; Sigma Xi. *Res:* Thermodynamics and properties of nonelectrolytic solutions; uranium fuel and feed materials process development; dielectric relaxation; electromagnetic materials; electroplating process and plastic foam process development; fundamental aspects of physics and chemistry relating to missile reentry; photochemistry; spectroscopy; explosives. *Mailing Add:* Nine Farmstead Dr Parsippany NJ 07054

SANDUSKY, HAROLD WILLIAM, b Baltimore, Md, Sept 12, 49; m 78. COMBUSTION, DETONATION PHYSICS. *Educ:* Georgia Tech, BAE, 71; Princeton Univ, MSE, 73, MA, 74, PhD(aerospace & mech sci), 76. *Prof Exp:* MECH ENGR, NAVAL SURFACE WARFARE CTR, 76- *Res:* Nitric oxide emissions from turbojet combustors and cigarette burning; deflagration and shock to detonation transition for explosives and propellants. *Mailing Add:* 9307 Scaggsville Rd Laurel MD 20723

SANDVIG, ROBERT L(EROY), b Lead, SDak, Sept 11, 23; m 57; c 3. CHEMICAL ENGINEERING. *Educ:* SDak Sch Mines & Technol, BS, 44; Univ Cincinnati, MS, 48; Univ Colo, PhD(chem eng), 53. *Prof Exp:* Mech engr, Nat Adv Comt Aeronaut, Va, 44; instr chem, SDak Sch Mines & Technol, 46-47; chem engr, Darling & Co, Ohio, 48-49; instr chem, 49-50, asst prof chem & chem eng, 51-55, from assoc prof to prof, 56-87, head dept, 73-87, EMER PROF CHEM ENG, SDAK SCH MINES & TECHNOL, 87- *Concurrent Pos:* Instr, Univ Colo, 52-53; consult, Rocky Flats Plant, Dow Chem Co, 69-75, Rocky Flats Plant, Rockwell Int, 75-87 & SD Forest Products, Inc. *Mem:* Am Chem Soc; Am Soc Eng Educ; Am Inst Chem Engrs; Sigma Xi. *Res:* Development of substitute for chloride road deicer and development of process for enhancing the color of softwoods throughout the entire cross section of the wood. *Mailing Add:* Dept Chem Eng SDak Sch Mines & Technol Rapid City SD 57701

SANDVIK, PETER OLAF, b Moose Lake, Minn, Sept 24, 27; m 53; c 7. GEOLOGY. *Educ:* Univ Alaska, BS, 50, BME, 51; Stanford Univ, MS, 61, PhD(geol), 64. *Prof Exp:* Mining engr-assayer, Alaska Territorial Dept Mines, 50-52 & 54-56; instr geol, Univ Alaska, 57-60; CHIEF GEOLOGIST & DIR MINERAL DEVELOP, INT MINERALS & CHEM CORP, 64- *Mem:* Am Inst Mining, Metall & Petrol Engrs; Geol Soc Am; Soc Econ Geologists; Can Inst Mining & Metall. *Res:* Trace and minor element content of sulphide ore minerals; application of economic geology in mineral exploration. *Mailing Add:* 2209 Swainwood Dr Glenview IL 60025

SANDWEISS, JACK, b Chicago, Ill, Aug 19, 30; m 56; c 3. PHYSICS. *Educ:* Univ Calif, BS, 52, PhD(physics), 56. *Prof Exp:* Physicist, Radiation Lab, Univ Calif, 56-57; instr, 57-59, from asst prof to assoc prof, 59-64, prof, 64-80, chmn dept, 77-80, DONNER PROF PHYSICS, YALE UNIV, 80- *Concurrent Pos:* Consult, Lab Marine Physics, Yale Univ, 57-60, Brookhaven Nat Lab, 61-, Argonne Nat Lab & Nat Accelerator Lab; chmn, high energy physics adv panel, US Dept Energy, 83- *Mem:* Nat Acad Sci; Am Phys Soc. *Res:* High energy physics; streamer chamber technique; physics of strange particles and charmed and beauty particles; counter and spark chamber techniques; high energy heavy ion reactions. *Mailing Add:* Sloane Physics Lab Yale Univ New Haven CT 06520

SANDWITH, COLIN JOHN, b Friday Harbor, Wash, Nov 9, 36; m 57; c 4. MECHANICAL ENGINEERING. *Educ:* Univ Wash, BSME, 61; Ore State Univ, PhD(mat sci), 67. *Prof Exp:* Draftsman, Com Airplane Div, Boeing Airplane Co, 57-58 & Duffy, Lawver & Kumpf Eng Consult Co, 59; mech engr, Hanford Atomic Prod Opers, Gen Elec Co, 61-62; mech engr, Albany Res Ctr, US Bur Mines, 63-66; asst prof mech engr, 66-74, RES ASSOC PROF MECH ENG, UNIV WASH, 74-, SR ENGR, MECH ENG, APPL PHYSICS LAB, 85- *Concurrent Pos:* NIH grant, Univ Wash, 67-69, NASA grant, 68-70; contracts, USN. *Honors & Awards:* Gold Medal Valor Award, US Dept Interior, 66; Ralf Teetor Award, Soc Automotive Engrs, 72. *Mem:* Am Soc Metals; Am Soc Mech Engrs; Nat Asn Corrosion Engrs; Am Soc Eng Educ; Soc Mfg Engrs. *Res:* Corrosion, failure analysis, design, materials, ceramic coatings; structural failures; biomedical instrumentation; fiberoptic cardiovascular catheters; marine and industrial corrosion; mechanical behavior of materials; manufacturing processes; video tape documentation and post-mortem analysis of Navy instruments to determine design wisdom; accident reconstruction and analysis. *Mailing Add:* 4030 NE 95 Seattle WA 98115

SANDZA, JOSEPH GERARD, b New York, NY, Feb 4, 17; m 42; c 5. BIOCHEMISTRY. *Educ:* Polytech Inst Brooklyn, BS, 37; Fordham Univ, MS, 40, PhD(biochem), 42. *Prof Exp:* Asst & instr, Fordham Univ, 37-42; Rockefeller Found res assoc, Northwestern Univ, 42; mem, Comt Med Res, Off Sci Res & Develop, 43-44; from res chemist to group leader, Lederle Labs, Am Cyanamid Corp, 44-46, head penicillin develop dept, 46-47; sr chemist,

Hoffmann La Roche, Inc, 48; asst mgr, Fermentation Dept, Stauffer Chem Co, 48-50, Biochem Sect, Eastern Res Div, 50-55, asst to dir, Eastern Res Lab, 55-62; vpres, Com-Dev Inc, 62-63; pres, Caribbean Tech Assocs, 63-83; PRES, CARIBTEC LABS, INC, 69- *Concurrent Pos:* Consult, 63-; prof chem & chmn div sci & technol, World Univ, 66- *Mem:* Am Chem Soc; Am Inst Chem Eng; Sigma Xi. *Res:* Nutrition; organic and fermentation chemistry; industrial biochemistry; bioengineering; research administration; process development; process and plant economics and feasibility; management; pollution control. *Mailing Add:* Caribtec Labs Inc GPO Box 362242 San Juan PR 00936-2242

SANES, JOSHUA RICHARD, b Buffalo, NY, Sept 5, 49; m 82; c 2. DEVELOPMENTAL NEUROBIOLOGY. *Educ:* Yale Univ, BA, 70; Harvard Univ, MA & PhD(neurobiol), 76. *Prof Exp:* Asst prof physiol, 80-85, assoc prof, 85-88, PROF ANAT, WASH UNIV, 89- *Concurrent Pos:* Counr, Soc Neurosci, 90- *Mem:* AAAS; Soc Neurosci. *Res:* Roles of cell surface molecules, extracellular matrix molecules, and cell lineage in synaptic specificity. *Mailing Add:* Dept Anat Med Ctr Wash Univ St Louis MO 63110

SAN FILIPPO, JOSEPH, JR, b Chicago, Ill, Feb 3, 44. ORGANIC CHEMISTRY, ORGANOMETALLIC CHEMISTRY. *Educ:* DePaul Univ, BS, 65; Mass Inst Technol, PhD(chem), 70. *Prof Exp:* NIH fel, Stanford Univ, 70-71; ASST PROF CHEM, RUTGERS UNIV, NEW BRUNSWICK, 71- *Mem:* Am Chem Soc. *Res:* Transition and main group metal hydrides and organometallic compounds; reactive organic intermediates. *Mailing Add:* 50 Harrison Ave Highland Park NJ 08904

SANFORD, ALLAN ROBERT, b Pasadena, Calif, Apr 25, 27; m 56; c 2. GEOPHYSICS. *Educ:* Pomona Col, BA, 49; Calif Inst Technol, MS, 54, PhD(geophys), 58. *Prof Exp:* PROF GEOPHYS, NMEX INST MINING & TECHNOL, 57-, COORDR GEOPHYS, 78- *Mem:* Fel AAAS; Soc Explor Geophysicists; Seismol Soc Am; Am Geophys Union; Sigma Xi. *Res:* Seismology and seismicity; crustal exploration; tectonophysics; gravity. *Mailing Add:* Dept Geosci NMex Inst Mining & Technol Socorro NM 87801

SANFORD, BARBARA ANN, b Beaumont, Tex, Aug 5, 41. MICROBIOLOGY, IMMUNOLOGY. *Educ:* Hardin-Simmons Univ, BA, 62; Baylor Univ, MS, 65, PhD(microbiol), 68. *Prof Exp:* From res asst to res assoc immunol, Med Ctr, Baylor Univ, 63-68; from instr to asst prof, 68-77, ASSOC PROF MICROBIOL, UNIV TEX MED SCH SAN ANTONIO, 77- *Mem:* Am Soc Microbiol; Sigma Xi. *Res:* Development of in vitro models of delayed hypersensitivity to chemical carcinogens in humans and experimental animal models. *Mailing Add:* Rte 1 Box 1488-A Boerne TX 78006-9642

SANFORD, BARBARA HENDRICK, b Brockton, Mass, Oct 17, 27. GENETICS, IMMUNOLOGY. *Educ:* Boston Univ, BS, 49; Brown Univ, MA, 60, PhD(biol), 63. *Prof Exp:* Health info specialist, Mass Dept Pub Health, 49-51; from cancer res scientist to sr cancer res scientist, Roswell Park Mem Inst, 61-63; res fel, Mass Gen Hosp & Harvard Med Sch, 63-65; fel, Harvard Med Sch, 65-66, from res assoc to prin res assoc path, 69-73; biologist, NIH, 73-75, prog dir for immunol, DCRRC, Nat Cancer Inst, 75-78; dir res, Sidney Farber Cancer Inst, 78-81; assoc prof path, Harvard Med Sch, 78-81; DIR, JACKSON LAB, 81- *Concurrent Pos:* USPHS fel, Mass Gen Hosp, 63-64 & grants, 67-, Am Cancer Soc res grant, 65-67; asst biologist, Mass Gen Hosp, 65-73; asst mem, Inst Health Sci, Brown Univ, 71-75; mem animal resources adv comt, NIH, 72- *Mem:* Am Genetic Asn; Genetics Soc Am; Am Asn Immunol; Transplantation Soc; Am Asn Cancer Res. *Res:* Immunogenetics; transplantation; cancer research; human genetics. *Mailing Add:* Box 455 Mt Desert ME 04660

SANFORD, EDWARD RICHARD, b Clifton, NJ, Feb 15, 28; m 57; c 1. PHYSICS. *Educ:* Iowa State Univ, BS, 49, MS, 50, PhD(physics), 59. *Prof Exp:* Sr scientist, Bettis Atomic Power Lab, Westinghouse Elec Corp, 53-61; assoc prof, 61-66, chmn dept, 78-90, PROF PHYSICS, OHIO UNIV, 66- *Mem:* Am Phys Soc; Am Nuclear Soc; Inst Elec & Electronics Engrs; Am Asn Physics Teachers. *Res:* Solid state physics; nuclear reactor physics and engineering. *Mailing Add:* Dept Physics & Astron Ohio Univ Athens OH 45701

SANFORD, JAMES R, b Zanesville, Ohio, Jan 29, 33; m 56; c 2. PHYSICS. *Educ:* Oberlin Col, AB, 55; Yale Univ, MS, 57, PhD(physics), 61. *Prof Exp:* Fel, Yale Univ, 61-62; physicist, Brookhaven Nat Lab, 62-69; head exp facil sect, Nat Accelerator Lab, 69-72, assoc dir, Fermi Nat Accelerator Lab, 72-76; assoc dir, Brookhaven Nat Lab, 76-81, sr physicist, 76-90; SR PHYSICIST & ASST DIR, SUPERCONDUCTING SUPER COLLIDER LAB, 89- *Concurrent Pos:* Mem high energy physics adv panel, Dept Energy; design team mem, Superconducting Super Collider, 83-89. *Mem:* Am Phys Soc. *Res:* High energy particle physics and the use of accelerators. *Mailing Add:* PO Box 281 Bellport NY 11713

SANFORD, JAY PHILIP, b Madison, Wis, May 27, 28; m 50; c 5. INFECTIOUS DISEASES. *Educ:* Univ Mich, MD, 52; Am Bd Internal Med, dipl, 62; recertified, 74, cert infectious dis, 77; Am Bd Microbiol, dipl & cert med microbiol, 64. *Prof Exp:* Res assoc med, Univ Mich, 50-52; med house officer, Peter Bent Brigham Hosp, 52-53, asst med, 53-54; sr asst resident med, Duke Univ Hosp, 56-57; from asst prof to prof internal med, Univ Tex Health Sci Ctr Dallas, 57-75; prof internal med & dean, 75-91, pres, 81-90, EMER DEAN, SCH MED, UNIFORMED SERV UNIV HEALTH SCI, 91- *Concurrent Pos:* Consult, Vet Admin Hosp, Dallas, 57-75, Vet Admin Ctr, Temple, 62-75, Wilford Hall USAF Hosp, Lackland AFB, 63-, Div Health Mobilization, USPHS, 64-72, Brooke Gen Hosp, Ft Sam Houston, 64- & Clin Ctr, NIH, 75-; mem bd trustees, Dallas Health & Sci Mus, 61-70; mem bact & mycol study sect, Div Res Grants, NIH, 62-66; district dir, Disaster Med Care, Tex District 1A, 63-75; mem bd dirs, Dallas County Chapter, Am Red Cross, 65-71, chmn disaster med nursing comt, 65-71; mem training grant comt, Nat Inst Allergy & Infectious Dis, 69-73, chmn, 71-73; mem, Am Bd Internal Med, 72-80, chmn, 79-80; mem, Accreditation Coun Grad Med Educ, 87-91, chmn, 90-91. *Honors & Awards:* Bristol Award, Infectious Dis Soc Am, 81; Laureate Award, Am Col Physicians, 87. *Mem:* Inst Med, Nat Acad Sci; Am Fedn Clin Res (pres, 68-69); Soc Exp Biol & Med; Am Soc Clin Invest; Asn Am Physicians; fel Am Acad Microbiol; AMA; Am Soc Microbiol; Soc Med Consult Armed Forces (pres, 76-77); Infectious Dis Soc Am (pres, 78-79); master Am Col Physician. *Res:* Bacteriology; immunology. *Mailing Add:* 4509 Edmondson Ave Dallas TX 75205

SANFORD, KARL JOHN, b New York, NY, Mar 10, 47; m 75; c 3. CLINICAL BIOCHEMISTRY, BIOTECHNOLOGY. *Educ:* Rutgers Univ, AB, 69; Univ Fla, PhD(chem), 72. *Prof Exp:* Res assoc biochem, Mass Inst Technol, 73-74; res chemist clin biochem, Eastman Kodak Co, 75-80, res assoc & lab head biochem, 80-85, mgr res & tech develop bio-prod, 85-88, mgr, Indust Biotech Ctr, 88-90; VPRES RES & DEVELOP, GENENCOR, 90- *Res:* Management responsibilities for directing research in industrial biotechnology center supporting several lines & business activities. *Mailing Add:* 206 Andiron Lane Rochester NY 14612-2245

SANFORD, KATHERINE KOONTZ, b Chicago, Ill, July 19, 15; m 71. GENERAL BIOLOGY. *Educ:* Wellesley Col, BA, 37; Brown Univ, MA, 39, PhD(zool), 42. *Hon Degrees:* DSc, Med Col Pa, Philadelphia, 74, Catholic Univ Am, Washington, DC, 88. *Prof Exp:* Asst biol, Brown Univ, 37-41; instr, Western Col, 41-42 & Allegheny Col, 42-43; asst dir sch nursing, Johns Hopkins Univ Hosp, 43-47; res biologist, 47-73, chief cell physiol & oncogenesis, Lab Biochem, 73-77, CHIEF IN VITRO CARCINOGENESIS SECT, CANCER ETIOLOGY DIV, LAB CELLULAR & MOLECULAR BIOL, NAT CANCER INST, 77- *Concurrent Pos:* With USPHS, 47; Ross Harrison fel, 54; mem cell cult collection adv comt, Am Type Cult Collection, 60- *Mem:* Tissue Cult Asn; Am Soc Cell Biol; Am Asn Cancer Res; Int Soc Cell Biol; Sigma Xi. *Res:* Physiological genetics of Cladocera; nutrition of tissue cells grown in culture; characteristics of malignant cells in vitro; carcinogenesis studies in vitro; DNA damage and repair. *Mailing Add:* In Vitro Carcinogenesis Sect Nat Cancer Inst Bethesda MD 20892

SANFORD, L G, b Parrish, Ala, Sept 16, 30; m 55; c 2. ENTOMOLOGY, VERTEBRATE ZOOLOGY. *Educ:* Florence State Col, BS, 57; Auburn Univ, MS, 63, PhD(entom), 66. *Prof Exp:* Biol aide fisheries dept, Tenn Valley Authority, 56-58; teacher high sch, Ala, 58-60; instr zool, Auburn Univ, 60-63, res asst entom, 63-65; assoc prof biol, 65-69, PROF BIOL, JACKSONVILLE STATE UNIV, 69- *Mem:* Am Inst Biol Sci. *Res:* Mammalogy and Siphonaptera. *Mailing Add:* Dept Biol Jacksonville State Univ Jacksonville AL 36265

SANFORD, MALCOLM THOMAS, b Miami Beach, Fla, Oct 23, 42; m 70. APICULTURE. *Educ:* Univ Tex, BA, 64; Thunderbird Grad Sch Int Mgt, Ariz, BFT, 67; Univ Ga, MA, 73, PhD, 77. *Prof Exp:* Res assoc geog, Univ Ga, 70-73, res asst apicult, 74-77, lectr, 77; asst prof apicult, Ohio State Univ, 78-81; ASSOC PROF & EXTEN APICULTURIST, UNIV FLA, 81- *Mem:* Am Beekeeping Fedn; Int Bee Res Asn; Am Asn Prof Apiculturists; Entom Soc Am. *Res:* Transmission and scanning electron microscopy to study abdominal glands of queen honey bee; honey bee management in tropical and temperate lands; beekeeping industry of Yucatan; modeling populations by computer simulation; financial analysis of apicultural operations by personal computer. *Mailing Add:* Entomol & Nematol Dept Bldg 970 Univ Fla Gainesville FL 32611-0312

SANFORD, PAUL EVERETT, b Milford, Kans, Jan 14, 17; m 42; c 3. POULTRY NUTRITION. *Educ:* Kans State Univ, BS, 41; Iowa State Univ, MS, 42, PhD(poultry nutrit), 49. *Prof Exp:* Student asst poultry husb, Kans State Univ, 37-41; grad asst, Iowa State Univ, 46-49; from assoc prof to emer prof poultry nutrit & nutritionist, Kans State Univ, 49-85; RETIRED. *Concurrent Pos:* Guest lectr, Univ PR, 57; lectr, US Feed Grain Coun, Tokyo, 63; mem, Animal Nutrit Res Coun, 73-; state coordr, Nat Asn Col & Teachers Agr, 76-82, dir, Midwest Region, 83-85. *Honors & Awards:* E Walter Morrison Award, Kans State Univ, 76; AVJ Tresler Outstanding Award, Nat Asn Col & Teachers Agr, 83. *Mem:* Fel AAAS; Poultry Sci Asn; hon mem Broiler Soc Japan; Sigma Xi (secy, 73-75, pres-elect, 75-76, pres, 76-77); Am Poultry Hist Soc (secy, 67-70). *Res:* Nutrition; growth and nutritional requirements of chicks; effects of feeding protein supplement; antibiotics; vitamin A; sorghum grain and wheat utilization; calcium and other mineral requirements for layers; feed additives for growth, feed efficiency and egg production. *Mailing Add:* 343 N 14th St Kans State Univ Manhattan KS 66502

SANFORD, RICHARD FREDERICK, b Bronxville, NY, July 29, 50. KINETICS, ORE DEPOSITS. *Educ:* Johns Hopkins Univ, BA & MA, 73; Harvard Univ, PhD(geol), 78. *Prof Exp:* GEOLOGIST, US GEOL SURV, 78- *Mem:* Geol Soc Am; Mineral Soc Am; Am Geophys Union. *Res:* Kinetics and mass transport in sedimentary, hydrothermal, and metamorphic ore deposits; mathematical and computer applications; igneous and metamorphic petrology; fluid inclusions; uranium ore formation; lunar impact phenomena; metamorphism of ultramafic and impure carbonate rocks. *Mailing Add:* Geol Survey Federal Ctr Mail Stop 905 Box 25046 Denver CO 80225

SANFORD, RICHARD SELDEN, b Mass; c 6. ELECTRICAL ENGINEERING, CONTROL SYSTEMS. *Educ:* Yale Univ, BEE, 49, MEE, 51; Worcester Polytech Inst, PhD(elec eng), 69. *Prof Exp:* Sr engr, Wayland Labs, Raytheon Mfg Co, 53-56; ASSOC PROF ELEC ENG, CLARKSON COL TECHNOL, 56- *Mem:* AAAS; Inst Elec & Electronics Engrs; Sigma Xi. *Res:* Electrical, mechanical and hydraulic control systems; acoustic networks, circuits and systems. *Mailing Add:* Dept Elec Eng & Comput Eng Clarkson Col Potsdam NY 13676

SANFORD, ROBERT ALOIS, b East St Louis, Ill, Mar 1, 22; m 46; c 4. ORGANIC CHEMISTRY. *Educ:* St Louis Univ, BS, 43; Purdue Univ, PhD(chem eng & chem), 49. *Prof Exp:* Asst org chem, Univ Pittsburgh, 43-44; res chemist, Manhattan Proj, Univ Rochester, 44-46; res chemist, Catalysis Res Div, Sinclair Res Labs, Inc, Ill, 49-52, res chemist, Petrochem Div, 52-53, group leader, 53-57, asst dir, Res Div, 57-60, dir, Explor Div, 60-66; dir, 66-

79, vpres res & develop, 79-83, VPRES SCI & TECHNOL, BROWN & WILLIAMSON TOBACCO CO, 83- *Res:* Chemical engineering; heterogeneous catalysis; petrochemicals; organic chemistry. *Mailing Add:* 7404 Shadwell Lane Prospect KY 40059

SANFORD, THOMAS BAYES, b Toledo, Ohio, Apr 22, 40; m 62; c 3. PHYSICAL OCEANOGRAPHY, MARINE GEOPHYSICS. *Educ:* Oberlin Col, AB, 62; Mass Inst Technol, PhD(oceanog), 67. *Prof Exp:* Physicist, NASA, 62-63; instr oceanog, Mass Inst Technol, 66-67; from asst scientist to assoc scientist, Woods Hole Oceanog Inst, 67-79; PRIN OCEANOGR, APPL PHYSICS LAB PROF, SCH OCEANOG, UNIV WASH, SEATTLE, 79- *Concurrent Pos:* Consult, Sippican Ocean Systs Inc, 79-85, Endeco, 85; dir res, Horizon Marine Inc, 82-90 & Transtrack Inc, 86-89; vis scientist, Inst fur Meereskunde Kiel FRG, 76. *Honors & Awards:* A F Bulgin Premium, Inst Electronic & Radio Engrs, UK, 72. *Mem:* AAAS; Am Geophys Union; Sigma Xi; Am Meteorol Soc; Oceanog Soc. *Res:* Motionally induced electromagnetic fields in the sea; marine magneto-tellurics; ocean circulation; internal waves; eddy motion; marine acoustics. *Mailing Add:* Appl Physics Lab Univ Wash 1013 NE 40th St Seattle WA 98105

SANFORD, WALLACE GORDON, b Pasadena, Calif, Aug 1, 23; m 57; c 3. PLANT PHYSIOLOGY. *Educ:* Pomona Col, BA, 47; Univ Md, MS, 49; Univ Calif, Los Angeles, PhD(plant sci), 52. *Prof Exp:* Asst gen bot & plant physiol, Univ Md, 47-49 & Univ Calif, Los Angeles, 49-52; assoc plant physiologist, Pineapple Res Inst, Hawaii, 52-54, plant physiologist, 54-65, head dept agron, 55-65, dir res, 65-67; PROF AGRON & SOIL SCI & CHMN DEPT, UNIV HAWAII, 67- *Mem:* AAAS; Bot Soc Am; Am Soc Plant Physiol. *Res:* Administration; plant nutrition. *Mailing Add:* 3694 Woodlawn Terrace Pl Honolulu HI 96822

SANGER, ALAN RODNEY, b Southampton, Eng, Apr 23, 43; Can citizen; m 80; c 1. CATALYSIS, ORGANOMETALLIC CHEMISTRY. *Educ:* Univ Sussex, BSc, 65, MSc, 66, DPhil(inorg chem), 69. *Prof Exp:* Teaching fel chem, Univ Alta, 69-72 & Simon Fraser Univ, 73; RES OFFICER CHEM, ALTA RES COUN, 73- *Mem:* Am Chem Soc; fel Chem Inst Can. *Res:* Preparation, characterization and evaluation of inorganic and organometallic materials of use as homogeneous or heterogeneous catalysts; c-1 chemistry. *Mailing Add:* Alta Res Coun PO Bag 1310 Devon AB T0C 1E0 Can

SANGER, FREDERICK, b Rendcomb, Eng, Aug 13, 18. MOLECULAR BIOLOGY. *Educ:* St John's Col. *Prof Exp:* Prof biochem, Cambridge Univ, 40-60; dir, Lab Molecular Biol, Med Res Coun, 60-83; RETIRED. *Honors & Awards:* Nobel Prize in Chem, 56, 80; Royal Medal, Royal Soc, 69; Wheland Award, 78; Gold Medal, Royal Soc Med, 83. *Mem:* Am Acad Arts Sci; Am Soc Biol Chemist. *Mailing Add:* Far Leys Fen Lane Swaffham Bulbeck Cambridge CB5 0NJ England

SANGER, GREGORY MARSHALL, b Spokane, Wash, Feb 2, 46; m 72; c 1. OPTICAL ENGINEERING, MATERIALS SCIENCE. *Educ:* Calif State Univ, BS, 68; Univ Ariz, MS, 71, PhD(optics), 76. *Prof Exp:* Res assoc, Optical Sci Ctr, Univ Ariz, 71-72, proj engr, 72-76; group leader, Lawrence Livermore Lab, 76-86; OPERS MGR, PERKIN-ELMER, 87-; DIR OPERS, CONTRAUES USA. *Mem:* Optical Soc Am; fel Soc Photo-Optical Instrumentation Engrs; Sigma Xi. *Res:* Physics, engineering, manufacturing and metrology of high precision optical surfaces as applied to advanced aerospace, astronomical and energy related applications. *Mailing Add:* 81 Bennett St Hudson MA 01749

SANGER, JEAN M, b New York, NY, June 13, 41; m 64; c 2. CYTOLOGY, ANATOMY. *Educ:* Marymount Col, BS, 63; Dartmouth Col, PhD(molecular biol), 68. *Prof Exp:* Res fel biol, Dartmouth Col, 68; res assoc cell biol, 74-80, sr res investr, 80-82, res asst prof, 82-88, RES ASSOC PROF ANAT, SCH MED, UNIV PA, 88- *Concurrent Pos:* Corp mem, Bermuda Biol Sta Res, 75-; vis scientist cell biol, Europ Molecular Biol Lab, 79-80; corp mem, Marine Biol Lab, 81- *Mem:* Am Soc Cell Biol. *Res:* Cell biology; cell motility; cell differentiation. *Mailing Add:* Dept Anat Univ Pa Sch Med Philadelphia PA 19104-6058

SANGER, JON EDWARD, b Minneapolis, Minn, May 31, 39. AQUATIC ECOLOGY. *Educ:* Univ Minn, Minneapolis, BS, 61, MS, 64, PhD(plant ecol), 68. *Prof Exp:* Asst prof bot, Univ Minn, 68-69; from asst prof to assoc prof 69-78, PROF BOT & MICROBIOL & CHMN DEPT, OHIO WESLEYAN UNIV, 78- *Concurrent Pos:* Prin co-investr, NSF Grant, Ohio Wesleyan Univ & Univ Minn, 70-72, Off Water Resources Res Grant, 73-77, 87- & US Geol Surv, 79-83. *Mem:* AAAS; Sigma Xi; Int Asn Pure & Appl Limnol; Ecol Soc Am; Am Soc Limnol & Oceanog. *Res:* Geochemical studies on lakes, with special emphasis on productivity indices; historical aspects of lake eutrophication; landscape erosion, soil weathering and climate change. *Mailing Add:* Dept Bot & Microbiol Ohio Wesleyan Univ 70 S Henry St Delaware OH 43015

SANGER, JOSEPH WILLIAM, b New York, NY, Feb 25, 41; m 64; c 2. CELL & MOLECULAR BIOLOGY, DEVELOPMENTAL BIOLOGY. *Educ:* Manhattan Col, BS, 62; Dartmouth Col, PhD(molecular biol), 68. *Hon Degrees:* MA, Univ Pa, 76. *Prof Exp:* Trainee physiol, Dartmouth Med Sch, 67-68; trainee cell differentiation, 68-71, assoc anat, 71-72, from asst prof to assoc prof, 72-85, PROF ANAT, SCH MED, UNIV PA, 85- *Concurrent Pos:* Pa Plan scholar, Sch Med, Univ Pa, 71-74; corp mem, Bermuda Biol Sta for Res trustee, 77-82, Marine Biol Lab trustee, 90-; vis scientist, Europ Molecular Biol Lab, 79-80; Alexander von Humboldt fel, 79-80; Pa Muscle Inst, 73-; chair, Univ Pa Cell Biol Grad Prog, 90-; ed staff, J Cell Motility and the Cytoskeleton, 86-; mem study sect, Biol Sci, NIH, 91- *Mem:* Am Soc Cell Biol; Am Asn Anatomists; fel AAAS. *Res:* Cell biology and differentiation; cytology; muscle structure and function; motility; embryology. *Mailing Add:* Dept Anat Sch Med Lab Cell Motility Studies Univ Pa Philadelphia PA 19104-6058

SANGER, WARREN GLENN, b Minden, Nebr, Oct 6, 45; m 69; c 1. MEDICAL GENETICS. *Educ:* Kearney State Col, BS, 67; Univ Nebr, Lincoln, MS, 69, PhD(genetics), 74. *Prof Exp:* Res assoc cytogenetics, 73-74, ASST PROF HUMAN GENETICS, UNIV NEBR MED CTR, OMAHA, 75-, PROF PEDIAT & PATH, 79- *Concurrent Pos:* Coordr, Nebr Tay-Sachs Screening Prog, Univ Nebr Med Ctr, Omaha, 73-, dir, Ctr Human Genetics, 75-79, lab dir, Genetic Semen Bank, 75-79, dir, Ctr Human Genetics & Cytogenetics & Semen Bank, 79-88. *Mem:* Sigma Xi; Am Soc Human Genetics; Am Genetics Asn; Tissue Cult Asn; Am Asn Tissue Banks. *Res:* Prenatal diagnosis of genetic defects and the cytogenetics of infertility; cancer cytogenetics. *Mailing Add:* Ctr Human Genetics Univ Nebr Med Ctr Omaha NE 68105

SANGIOVANNI-VINCENTELLI, ALBERTO LUIGI, b Milan, Italy, June 23, 47; c 1. COMPUTER-AIDED DESIGN, INTEGRATED CIRCUITS. *Educ:* Milan Polytech, Dr Eng, 71. *Prof Exp:* From asst prof to assoc prof elec eng & comput sci, Milan Polytech, 71-76; from asst prof to assoc prof, 76-83, PROF ELEC ENG & COMPUT SCI, UNIV CALIF, BERKELEY, 83- *Concurrent Pos:* Vchmn dept elec eng & comput sci, Univ Calif, Berkeley, 82-85; vis scientist, IBM, 80-81; consult, Honeywell, SGS-Thomson, 81-, IBM, JJ Watson Res Ctr, 80-86, Cadence Syst, 83-, Nynex, 87-; vis prof, Mass Inst Tech, 87; corp fel, Harris Co, 80-, Thinking Mach, 86-, Syropsys, 86-, Actel, 89-, Crosscheck, 89-, Greylock, 90-, Biocad, 90- *Honors & Awards:* Guillemin-Cauer Award, Inst Elec & Electronics Engrs, 83, Darlington Award, 88. *Mem:* Fel Inst Elec & Electronics Engrs; Asn Comput Mach. *Res:* Algorithms for computer-aided design of electronic systems including simulation, synthesis, layout, testing and formal verification; software systems for computer aided design and hardware platforms. *Mailing Add:* 200 Tunnel Rd Berkeley CA 94705

SANGREN, WARD CONRAD, b Kalamazoo, Mich, Apr 20, 23; m 44; c 3. APPLIED MATHEMATICS. *Educ:* Princeton Univ, AB, 43; Univ Mich, MA, 47, PhD(math), 50. *Prof Exp:* Asst prof math, Miami Univ, 49-51; sr mathematician, Oak Ridge Nat Lab, 51-55; chief math & comput res div, Curtiss-Wright Corp, 55-56, chief gen atomic div, Gen Dynamics Corp, 56-62; vpres, Comput Applns Inc, Calif, 62-70; coordr comput activities, Univ Calif, Berkeley, 70-76; DIR INFO SYST & ANALYSIS, SAN FRANCISCO STATE UNIV, 76- *Mem:* Am Math Soc; Soc Indust & Appl Math; Math Asn Am; Asn Comput Mach. *Res:* Boundary value and eigenvalue problems; high speed calculation of reactors; nonlinear differential equations. *Mailing Add:* 121 Bates Ct Orinda CA 94563

SANGREY, DWIGHT A, b Lancaster, Pa, May 24, 40; m 64; c 3. CIVIL & GEOTECHNICAL ENGINEERING. *Educ:* Lafayette Col, BS, 62; Univ Mass, MS, 64; Cornell Univ, PhD(civil eng), 68. *Prof Exp:* Engr, H L Griswold, Consult Engrs, 60-64; proj engr, Shell Oil Co, Tex, 64-65; asst prof civil eng, Queen's Univ, Ont, 67-70; assoc prof, 70-77, prof civil & environ eng, Cornell Univ, 77-; DEPT HEAD, DEPT CIVIL ENG, CARNEGIE-MELLON UNIV, 87- *Concurrent Pos:* Mem hwy res bd, Nat Res Coun. *Honors & Awards:* Res Award, Am Soc Testing & Mat, 69. *Mem:* Am Soc Civil Engrs; Am Soc Testing & Mat; Am Soc Eng Educ; Int Soc Soil Mech & Found Engrs. *Res:* Shear strength and stress-strain behavior of soils; repeated and dynamic loading of soils; organic soils; slope stability; offshore, marine and coastal engineering; engineering geology; solid waste management and disposal. *Mailing Add:* Dept Civil Eng Carnegie-Mellon Univ Pittsburgh PA 15213

SANGSTER, RAYMOND CHARLES, b Lyons, Kans, Mar 15, 28; m 55, 80; c 2. INORGANIC CHEMISTRY, METROLOGY & STANDARDS. *Educ:* Univ Chicago, PhB, 46, BS, 47; Mass Inst Technol, PhD(chem), 51. *Prof Exp:* Asst chem, Mass Inst Technol, 48-49, res assoc, 51-52; mem tech staff, Semiconductor Res & Develop Dept, Hughes Aircraft Co, 52-54; mem tech staff, Cent Res Lab, Tex Instruments, Inc, 54-57, dir, Mat Res Dept, 57-58, res assoc, 58-62, dir, Semiconductor Exp Lab, 62-65; dir res, Bayside Lab, Gen Tel & Electronics Labs, Inc, NY, 65-68; chief electromagnetics div, Inst Basic Standards, Nat Bur Standards, 69-74, prog mgr strategic planning, 74-78, sr scientist, 78-79; sr scientist, GE-Tempo, Santa Barbara, Calif, 79-80; guest scientist & staff mem, Gmelin Inst, Frankfurt, WGer, 80-85; sr scientist, Raytheon Serv Co, Hsinchu, Taiwan, 86-87; consult, Nat Acad Sci, Bangkok, Thailand, 88-89; CONSULT & TECH WRITER, 89- *Concurrent Pos:* Chmn, Gordon Conf Chem & Metall Semiconductors, 67; chmn, NASA Working Group Electronic Mat, 68-70; US chmn, Marine Commun & Electronics Panel, US-Japan Natural Resources Comn, 69-71. *Mem:* Am Chem Soc; Am Phys Soc; Inst Elec & Electronics Engrs; fel Am Inst Chemists; Sigma Xi. *Res:* Research management; semiconductor materials and devices; organic scintillators; science and technology for development. *Mailing Add:* Rte 2 Box 185 Templeton CA 93465

SANGSTER, WILLIAM M(CCOY), b Austin, Minn, Dec 9, 25; m 46; c 3. HYDRAULIC ENGINEERING, ENGINEERING MECHANICS. *Educ:* Univ Iowa, BS, 47, MS, 48, PhD, 64. *Prof Exp:* Asst instr, Univ Iowa, 48; from asst prof to prof civil eng, Univ Mo, Columbia, 48-67, assoc dean col eng & assoc dir eng exp sta, 64-67; prof civil eng & dir sch civil eng, 67-74, DEAN, COL ENG, GA INST TECHNOL, 74- *Mem:* Am Soc Civil Engrs (pres, 74-75); Am Soc Eng Educ; Eng Joint Coun; Engrs Coun Prof Develop; Sigma Xi. *Res:* Hydrodynamic stability of stratified flows; orbital mechanics; general hydraulics. *Mailing Add:* Dean Eng Ga Inst Tech Atlanta GA 30332

SANI, BRAHMA PORINCHU, b Trichur, India, Sept 13, 37; m 67; c 2. BIOCHEMISTRY, CANCER. *Educ:* Univ Kerala, BS, 60; Holkar Sci Col, MS, 62; Indian Inst Sci, Bangalore, PhD(biochem), 67. *Prof Exp:* Res scholar biochem, Indian Inst Sci, Bangalore, 62-65, Univ Grants Comn India jr res fel, 65-67, Coun Sci & Indust Res India sr res fel, 67-68; staff fel, Boston Biomed Res Inst, 68-71; res assoc, Inst Cancer Res, Philadelphia, 71-74; sr biochemist, 74-78, HEAD, PROTEIN BIOCHEM SECT, SOUTHERN RES INST, 79-, SR STAFF SCIENTIST, 88- *Concurrent Pos:* Res consult, Cornea Res Dept, Retina Found, Boston, 68-70; res grants, NIH, Coun Tobacco Res & WHO. *Mem:* AAAS; Am Asn Cancer Res; Am Soc Biol Chemists; Sigma

Xi. *Res:* Molecular mechanism of chemical carcinogenesis and anticarcinogenesis; anti-carcinogen-protein interactions; modern molecular biology and hybridoma techniques; characterization of a retinoic acid-binding protein, a retinol-binding protein and a selenium-binding protein which may be involved in the control of epithelial differentiation and anticarcinogenesis; discovered specific retinoid-binding proteins in several filarial parasites. *Mailing Add:* Southern Res Inst 2000 Ninth Ave S Birmingham AL 35255-5305

SANI, ROBERT L(E ROY), b Antioch, Calif, Apr 20, 35; m 66; c 3. CHEMICAL ENGINEERING. *Educ:* Univ Calif, Berkeley, BS, 58, MS, 60; Univ Minn, PhD(chem eng), 63. *Prof Exp:* Instr math, Rensselaer Polytech Inst, 63-64; from asst prof to assoc prof chem eng, Univ Ill, Urbana, 64-76; PROF CHEM ENG, UNIV COLO, BOULDER, 76- *Concurrent Pos:* Guggenheim fel, 70-71; consult, Atmospheric Sci Div, Lawrence Livermore Labs, 71- *Mem:* Am Inst Chem Engrs; Soc Appl & Indust Math. *Res:* Nonlinear aspects of the dynamics of physical systems exhibiting transport and transformation processes that are coupled at the macroscopic level; computational fluid dynamics. *Mailing Add:* Dept Chem Eng Campus Box 424 CHI-72 Univ Colo Boulder CO 80309

SANIEE, IRAJ, b Jan 21, 56; m 87. CONTROL OF EPIDEMIC PROCESSES, APPLIED OPTIMIZATION OF DISCRETE SYSTEMS. *Educ:* Cambridge Univ, Eng, BA, 79, MPhil, 81, MA, 83, PhD(opers res), 84. *Prof Exp:* Lectr applicable math, Churchill Co, Cambridge Univ, 83-84; MEM TECH STAFF SYSTS, BELL COMMUN RES, 85- *Concurrent Pos:* Adj prof, Fairleigh Dickinson Univ, 88-90. *Mem:* Soc Indust & Appl Math; Opers Res Soc Am. *Res:* Deterministic and stochastic optimization as applied to control or contain performance of static and dynamic systems; EG telecommunications networks, their associated economics and implications on regulatory issues. *Mailing Add:* MRE 2G-351 445 South St Morristown NJ 07960

SANK, DIANE, b New York, NY, Dec 22, 27; div; c 3. BIOLOGICAL ANTHROPOLOGY, GENETICS. *Educ:* Long Island Univ, BS, 49; Univ Ill, MS, 51; Columbia Univ, PhD(human variations), 63. *Prof Exp:* Sr res scientist, NY State Psychiat Inst, 52-67; chmn dept, 69-71 & 87-88, prof anthrop, Hunter Col, 67-88, PROF ANTHROP, CITY COL NEW YORK, CITY UNIV NEW YORK, 68-; CONSULT, HUMAN GENETICS, N S KLINE PSYCHIAT RES INST, 67- *Concurrent Pos:* Res assoc psychiat, Col Physicians & Surgeons, Columbia Univ, 53-67, lectr, 67-71, assoc, Univ Sem Genetics & Evol Man, 67-; prof doctoral prog, Grad Ctr, City Univ New York, 67-71. *Mem:* Fel Int Soc Twin Studies; Am Soc Human Genetics; Soc Study Human Biol; Am Soc Primatologists; Soc Anthrop Visual Commun; fel AAAS; fel Am Anthrop Asn; Nat & Int Conf Dermatologlyphics. *Res:* Hereditary and bio-social studies of mental illness, mental retardation, deafness, learning disability and normal human behavior; primate behavior. *Mailing Add:* Dept Anthrop City Col New York Convent Ave & W 138th St New York NY 10031

SANK, VICTOR J, b Washington, DC, Sept 17, 44; m 65; c 3. ENGINEERING PHYSICS. *Educ:* Polytech Inst Brooklyn, BS, 66, PhD(physics), 71. *Prof Exp:* Instr physics, Polytech Inst Brooklyn, 68-71; res asst, 70-71; asst prof, Quinnipiac Col, 71-75; asst prof, San Antonio Col, 75-79; res physicist neuroradiol & comput tomography, NIH, 79-82, res physicist, biomed eng & instrumentation br, 82-85; CONSULT, VJ SYSTS, 77- *Concurrent Pos:* Electronic engr, Schulz Controls, 74-75; res assoc, Dept Radiol, Univ Tex Health Sci Ctr, San Antonio, 75-78; consult, Satellite Commun, NASA, 85- *Mem:* Am Phys Soc; Inst Elec & Electronics Engrs; Soc Magnetic Resonance in Med. *Res:* Mathematics and computer interfacing for diagnostic medical imaging; computerized tomography; positron-emission tomography; nuclear magnetic resonance. *Mailing Add:* Five Bunker Ct Rockville MD 20854

SANKAR, D V SIVA, b Vizianagram, India, Apr 7, 27; nat US; m 59; c 3. BIOCHEMISTRY, CHEMICAL PATHOLOGY. *Educ:* Univ Madras, MSc, 49, PhD(biochem), 51; Am Bd Clin Chemists, dipl. *Prof Exp:* Fulbright fel, Mass Inst Technol, 53-55; NSF fel, Johns Hopkins Univ, 56; asst prof biochem, Adelphi Col, 56-58; sr res scientist & head biochem res lab, Children's Unit, Creedmoor State Hosp, 58-63, assoc res scientist, 63-69; chief lab res, Queen's Children's Hosp, 69-81; PRES, AM HEALTH SCI SYSTS CORP, 81- *Concurrent Pos:* Adj prof, Long Island Univ, 64-, Fordham Univ, 70-78 & St Johns Univ, 71-74, NY Col Osteop Med, 86, Schwartz Col Pharm; adj assoc prof psychiat, NY Univ Med Ctr, 76-; ed, J Med. *Honors & Awards:* Indian Chem Soc Gold Medal, 53; Dr Quinn Medal. *Mem:* AAAS; Am Chem Soc; Am Soc Microbiol; Am Soc Pharmacol & Exp Therapeut; Am Asn Clin Chem; Soc Biol Psychiat. *Res:* Molecular mechanisms of metabolism; drug actions and pathogenesis; biosynthetic mechanisms, vitamins, enzymes, structure-activity relations; neurochemical and psychobiological research on drug actions and on mental health; hypothesis that sleep is a detoxification mechanism for the metabolic oxidative burdens of the wakeful state; rapid eye movement state results from a hypoxic condition in the brain areas. *Mailing Add:* PO Box 966 Westbury NY 11590-0966

SANKAR, SESHADRI, b Udamalpet, India; Can citizen. COMPUTER AIDED DESIGN, VIBRATION CONTROL. *Educ:* Univ Madras, BEng, 70; Sir George Williams Univ, MEng, 71, DEng, 73. *Prof Exp:* Dynamics analyst, res & develop, United Aircraft Can Ltd, 73-74; sr proj engr, Naval Eng Test Estab, Ville LaSalle, 74-75; asst prof, 75-78, ASSOC PROF MECH ENG, CONCORDIA UNIV, 78- *Concurrent Pos:* Assoc ed, Simulation, Soc Comput Simulation, 79-; consult, Naval Eng Test Estab, 75-77, Recreational Vehicle Div, Bombardier Ltd, 79- & VIA Rail, Montreal, 81-; mem, Comn Educ, Can Rep, Int Fedn Theory Mach & Mech, 79- *Mem:* Can Coun Theory Mach & Mech; Am Soc Mech Engrs; Soc Comput Simulation; Int Soc Math & Comput Simulation; Corp Engrs Que. *Res:* Computer aided optimal design of mechanical and hydromechanical systems using digital, analog and hybrid computation; vibration control using passive, semi-active and active devices for off-road, on-road and rail vehicles; use of interactive graphics in computer aided design. *Mailing Add:* Dept Mech Eng H-929 Concordia Univ 1455 de Maisonneuve Blvd W Montreal PQ H3G 1M8 Can

SANKAR, SURYANARAYAN G, b Madras, India, July 1, 42; m 74; c 2. SOLID STATE CHEMISTRY. *Educ:* Andhra Univ, BSc, 62, MSc, 63; Poona Univ, PhD(solid state chem), 68. *Prof Exp:* Fel chem, Tex Tech Univ, 70; fel, Univ Pittsburgh, 70-73, from res asst prof to res assoc prof, 73-78; sr res chemist, Mat Res Ctr, Allied Corp, 78-82; sr res assoc, Mat Res Lab, Pa State Univ, 82-85; SR SCIENTIST & ADJ PROF, CARNEGIE MELLON UNIV, 86- *Mem:* Mat Res Soc; Am Chem Soc; fel Am Inst Chemists. *Res:* Study of structural and magnetic properties of permanent magnets; study of the influence of paramagnetic impurities in superconductors; heterogeneous catalysis; metal hydrides. *Mailing Add:* 1224 Ridgewood Dr Upper St Clair PA 15241

SANKAR, THIAGAS SRIRAM, b Udamalpet, India, Feb 18, 40; m 67; c 2. MECHANICAL ENGINEERING. *Educ:* Univ Madras, BE, 61; Indian Inst Sci, ME, 63; Univ Waterloo, PhD(solid mech), 67. *Prof Exp:* Sect officer, Neyveli Lignite Corp, Madras, 61-62; teaching asst & res fel civil eng, Univ Waterloo, 63-65, lectr, 65-67; res engr, United Aircraft Can, 67-68; from asst prof to assoc prof mech eng, 68-77, PROF MECH ENG & CHMN DEPT, CONCORDIA UNIV, 77-, MEM, BD GOV, 78- *Concurrent Pos:* Res fel solid mech, Nat Res Coun Can, 65-67; grants, Nat Res Coun Can, Concordia Univ, 68-, Defence Res Bd Can, 72-76 & Govt Que, 71-; treas, Can Coun Theory Mach & Mech, 71-; consult, eng & educ. *Mem:* Am Acad Mech; Can Soc Mech Engrs (vpres, 79-); Am Soc Mech Engrs; fel Eng Inst Can. *Res:* Stochastic dynamics of mechanical systems; surface mechanics; machine tool dynamics; vibrations and robotics. *Mailing Add:* Dept Mech Eng Concordia Univ 1455 Demaisonneuve Blvd W Montreal PQ H3G 1M8 Can

SANKOFF, DAVID, b Montreal, Quebec, Can, Dec 31, 42. MATHEMATICS. *Educ:* McGill Univ, PhD(math), 69. *Prof Exp:* RES, CTR RECH MATH, UNIV MONT, 69- *Honors & Awards:* Vincent Prize, Can Fr Asn Adv Sci,77; Can Inst Advan Res, Fel, 87- *Mem:* Soc Indust & Appl Math; Linguistics Soc Am. *Res:* Statistical procedures for the analysis of sociolinguistic data; algorithms for the study of macromolecular structure and evolution; mathematics applied to natural and social science and the humanities. *Mailing Add:* Ctr Rech Math Univ Montreal cp 6128 Montreal PQ H3C 3J7 Can

SANMANN, EVERETT EUGENE, b Geronimo, Okla, Nov 10, 37; m 65. OPTICAL PHYSICS. *Educ:* Univ Okla, BS, 59, MS, 62, PhD(physics), 74. *Prof Exp:* Res physicist, Brown Eng Co, 66-68; gen engr optics, Ballistic Missile Defense Advan Technol Ctr, 68-79; DIR SYSTS ANALYST, NICHOLS RES CTR, 79- *Mem:* Am Phys Soc; Sigma Xi; Optical Soc Am. *Res:* Remote sensing using infrared receivers in space environments; scattering of coherent radiation from dielectric materials with varying reflectivity and surface roughness; theoretical and experimental nature of backscatter statistics. *Mailing Add:* 1130 Point Pines Dr Colorado Springs CO 80919

SANN, KLAUS HEINRICH, b Driesen, Ger, Aug 6, 19; m 44; c 2. ELECTRONICS, RADAR SYSTEMS. *Educ:* Technische Universitat Berlin, Diplom-Ingenieur, 51. *Prof Exp:* Electronics engr, Telefunken GmbH, Berlin, 51-53, asst lab mgr, 53-56, lab mgr, 56-58; electronics engr, 59-69, chief, advan res br, Harry Diamond Labs, 69-; RETIRED. *Concurrent Pos:* Mem working group on Radar Minimum Performance, Radio Adv Comt, 67. *Honors & Awards:* Hinman Award, Harry Diamond Labs, 76. *Mem:* Sr mem Inst Elec & Electronics Engrs; affil Nat Soc Prof Engrs. *Res:* Advanced development of fuzzing systems and solid state microwave sources; conceives and coordinates research programs of substantial importance to installation. *Mailing Add:* 4910 Bangor Dr Kensington MD 20845

SANNELLA, JOSEPH L, b Boston, Mass, July 27, 33; m 59; c 3. MATERIALS SCIENCE ENGINEERING. *Educ:* Harvard Univ, AB, 55; Univ Mass, MS, 58; Purdue Univ, PhD(biochem), 63; Univ Del, MBA, 69. *Prof Exp:* Res chemist, Am Viscose Div, FMC Corp, Pa, 62-67; supvr chem res, 67-74, dir res, res develop & eng, 74-85, dir corp lab serv, 85-88, DIR LAB SERV, PACKAGING PRODS GROUP, BALL CORP, 88- *Mem:* Am Chem Soc; Soc Plastics Engrs; Nat Metal Decorators Asn. *Res:* Process chemistry; analytical chemistry; plastics; packaging; glass; coatings; materials. *Mailing Add:* Ball Corp Packaging Prod Group 1509 S Macedonia Ave Muncie IN 47302

SANNER, JOHN HARPER, b Anamosa, Iowa, Apr 29, 31; m 58; c 2. PROSTAGLANDINS. *Educ:* Univ Iowa, BS, 54, MS, 61, PhD(pharmacol), 64. *Prof Exp:* From res investr to res fel pharmacol, G D Searle & Co, 63-86; RETIRED. *Mem:* Soc Pharmacol & Exp Therapeut. *Res:* Pharmacological antagonists; pharmacology of vasoactive peptides and prostaglandins; pharmacological mechanisms of smooth muscle; anti-inflammatory mechanisms. *Mailing Add:* 959 Appletree Lane Deerfield IL 60015

SANNES, FELIX RUDOLPH, b Zaporozhe, Ukraine, Nov 20, 40; Can citizen; m 69; c 2. ELEMENTARY PARTICLE PHYSICS. *Educ:* Univ BC, BS, 63; McGill Univ, PhD(physics), 68. *Prof Exp:* Res asst physics, Atomic Energy Can, 63-64; fel, 69-71, asst prof, 71-76, assoc prof, 76-81, PROF PHYSICS, RUTGERS UNIV, 81- *Mem:* Am Phys Soc. *Res:* Studied energy dependence of proton-proton elastic and inelastic scattering and proton-antiproton annihilation. *Mailing Add:* Dept Physics/Astron Rutgers State Univ Frelinghuysen Rd Piscataway NJ 08854

SANNES, PHILIP LOREN, b Canton, Ohio, Mar 10, 48; m 73; c 1. ENVIRONMENTAL HEALTH, CELL BIOLOGY. *Educ:* Ohio State Univ, BA, 70, MSc, 73, PhD(anat), 75. *Prof Exp:* Instr path, Med Univ SC, 75-77, asst prof, 77-78; ASST PROF ENVIRON HEALTH SCI, JOHNS HOPKINS UNIV, 78-, ASST PROF CELL BIOL & ANAT, 79-; ASSOC PROF, DEPT ANAT, PHYSIOL SCI & RADIOL, NC STATE UNIV COL VET MED, 88- *Concurrent Pos:* NIH fel, Inst Gen Health Sci, 77-78; assoc sr investr, path, Smith Kline & French Lab, 88. *Mem:* Am Asn Anatomists; Histochem Soc; Soc Cell Biol; Electrophoresis Soc. *Res:* Cell biology: ultrastructure and cytochemistry; alteration of lung structure and function relationships by environmental agents. *Mailing Add:* Dept Anat Physiol Sci & Radiol NC State Univ Col Vet Med 4700 Hillsborough St Raleigh NC 27606

SANNUTI, PEDDAPULLAIAH, b Rajupalem, India, Apr 2, 41; m 65; c 1. CONTROL SYSTEMS, INFORMATION SCIENCE. *Educ:* Eng Col, Anantapur, India, BE, 63; Indian Inst Technol, Kharagpur, MTech, 65; Univ Ill, Urbana, PhD(elec eng), 68. *Prof Exp:* PROF ELEC ENG, RUTGERS UNIV, NEW BRUNSWICK, 68- *Mem:* Inst Elec & Electronics Engrs. *Res:* Singular perturbation method in the theory of optimal control; communication systems; filtering. *Mailing Add:* 64 Valley Forge Dr New Brunswick NJ 08816

SANNY, CHARLES GORDON, b Aug 25, 47; m 72; c 1. ALDEHYDE DEHYDROGENASES. *Educ:* Okla Baptist Univ, BS, 70; Univ Okla, PhD(biochem), 75. *Prof Exp:* Assoc prof biochem, Univ Osteop Med & Health Sci, Des Moines, IA, 81-85; assoc prof, 85-89, PROF BIOCHEM, COL OSTEOP MED, OKLA STATE UNIV, 89- *Mem:* Am Sci Affil; Am Soc Biochem & Molecular Biol; Res Soc Alcoholism; Sigma Xi. *Mailing Add:* Col Osteop Med Okla State Univ 1111 W 17th St Tulsa OK 74107

SANOFF, HENRY, b New York, Jan 16, 34; m 57; c 2. BEHAVIORAL DESIGN, PROGRAMMING & POST OCCUPANCY EVALUATION. *Educ:* Pratt Inst, BArchit, 57, MArchit, 62. *Prof Exp:* Asst prof archit, Univ Calif, Berkeley, 63-67; PROF ARCHIT, NC STATE UNIV, 67- *Concurrent Pos:* Founder & chmn, Environ Design Res Assoc, 68-72, bd mem, 72-75; distinguished univ prof, Univ Col, London, 81-82, Univ Melbourne, Australia, 87; vis prof, Oxford Univ, 82, Inst SAfrican Archit, 85 & 86; Chettle scholar, Univ Sydney, Australia, 90; distinguished Fulbright scholar, Seoul Nat Univ, Korea, 90-91. *Honors & Awards:* Award of Honor, Environ Design Res Asn, 77; Statute Victory, World Cult Prize for Lett, Arts & Sci, 85. *Mem:* Am Inst Archit. *Res:* Research and writing focus on citizen participation in design and planning decisions; developed techniques and strategies to effectively involve non-designers in the process of designing buildings and planning urban development. *Mailing Add:* Sch Design NC State Univ Raleigh NC 27695

SAN PIETRO, ANTHONY, b Apr 22, 22. PLANT BIOCHEMISTRY. *Educ:* NY Univ, BA, 42; Columbia Univ, PhD(biochem), 51. *Prof Exp:* Res assoc, Biochem Dept, Columbia Univ, 51-52; Nat Polio Found postdoctoral fel, Johns Hopkins Univ, 52-54, from asst prof to assoc prof biol & McCollum-Pratt Inst, 54-62; sr investr, Kettering Res Lab, 62-64, asst dir, 64-68; chmn & prof, Dept Plant Sci, 68-77, DISTINGUISHED PROF PLANT BIOCHEM, IND UNIV, 75-, PROF, DEPT BIOL, 77- *Concurrent Pos:* Prof chem, Antioch Col, 62-68; US sr scientist award, Humboldt Found, 77; US-China lectr exchange prog, Nat Acad Sci, 80; sci adv, Off Pres, Ind Univ, 80-, spec consult, Off Vchancellor Undergrad Educ, 90-; vis scientist, Inst Phys & Chem Res, Japan, 85; mem bd gov, Ben Gurion Univ, Israel, 85; mem, Bd Sci & Technol Int Develop, Nat Res Coun, Nat Acad Sci, 86-89; vchmn, Biotechnol Comt, Corp Sci & Technol, 87-88. *Mem:* Nat Acad Sci; Am Soc Biochem. *Mailing Add:* Dept Biol Ind Univ Bloomington IN 47405

SANSING, NORMAN GLENN, b Woodstock, Ala, Aug 17, 32; m 55; c 3. BIOCHEMISTRY, PLANT PHYSIOLOGY. *Educ:* Auburn Univ, BS, 54, MS, 59; Iowa State Univ, PhD(plant physiol), 62. *Prof Exp:* Res assoc nucleic acid enzym, Biol Div, Oak Ridge Nat Lab, 62-64; asst prof biochem & bot, 64-72, ASSOC PROF BIOCHEM, UNIV GA, 72- *Mem:* AAAS; Am Soc Plant Physiologists. *Res:* Nucleic acid enzymology; isolation and characterization of plant nucleases. *Mailing Add:* Dept Biochem Univ Ga Athens GA 30602

SANSLONE, WILLIAM ROBERT, b Vineland, NJ, Feb 16, 31; m 60; c 1. NUTRITIONAL BIOCHEMISTRY. *Educ:* Rutgers Univ, BS, 53, PhD(biochem), 61; Univ NH, MS, 55. *Prof Exp:* Asst, Univ NH, 53-55; asst, Rutgers Univ, 58-61; instr biochem, State Univ NY Downstate Med Ctr, 61-64, from asst prof to assoc prof, 64-71; sr proj scientist, NIH, 71-73, exec secy biochem study sect, Div Res Grants, 73-75, prog dir review, Div Extramural Activ, Nat Cancer Inst, 75-83, assoc dir sci prog opers, Div Lung Dis, Nat Heart, Lung & Blood Inst, 83-87, DIR, OFF PROG PLANNING & EVAL, NAT INST ARTHRITIS & MUSCULOSKELETAL & SKIN DIS, NIH, 87- *Concurrent Pos:* Vis assoc prof, Med Col Pa, 70-71. *Mem:* AAAS; Am Inst Nutrit; Soc Exp Biol & Med; Sigma Xi. *Res:* Muscle and nutritional biochemistry; health-science administration. *Mailing Add:* Bldg 316 Rm 4C11 Nat Inst Arthritis Musculoskelet & Skin Dis NIH Bethesda MD 20892

SANSOM, RICHARD E, b Dallas, Tex, Jan 1, 33. MATHEMATICS. *Educ:* Univ Tex, BS, 59. *Prof Exp:* GROUP DIR SYSTS & PROD QUAL, TRW, 63- *Mailing Add:* 1604 Harkness St Manhattan Beach CA 90266

SANSONE, ERIC BRANDFON, b New York, NY, Mar 26, 39; c 2. INDUSTRIAL HYGIENE, SAFETY. *Educ:* City Col New York, BChE, 60; Univ Mich, MPH, 62, PhD(indust health), 67. *Prof Exp:* From asst prof to assoc prof, Indust Hyg & Air Eng, Univ Pittsburgh, 67-74; MEM STAFF, FREDERICK CANCER RES DEVELOP CTR, NAT CANCER INST, 74-, DIR ENVIRON CONTROL & RES PROG, 79- *Mem:* AAAS; Am Indust Hyg Asn; NY Acad Sci; Brit Occup Hyg Soc; Soc Occup Environ Health; Am Chem Soc; Health Physics Soc. *Res:* Risk assessment and environmental monitoring. *Mailing Add:* Frederick Cancer Res Develop Ctr Nat Cancer Inst Bldg 426 PO Box B Frederick MD 21702-1201

SANSONE, FRANCES MARIE, b Birmingham, Ala, June 30, 31. HUMAN ANATOMY. *Educ:* Cath Univ Am, AB, 52; Marquette Univ, MS, 56; Univ Tex, PhD(anat), 65. *Prof Exp:* Teaching asst histol & neuroanat, Dent Br, Univ Tex, 57-60; instr, 64-68, asst prof histol & neuroanat, 68-76, clin asst prof, 76-81, ASSOC PROF ANAT SCI, SCH MED, STATE UNIV NY BUFFALO, 81- *Mem:* Am Asn Anatomists; Soc Neurosci; Am Soc Cell Biol; Sigma Xi; Electron Mic Soc Am. *Res:* environmental neurotoxins and the epidemiology of neural tumors. *Mailing Add:* Dept Anat Sci State Univ NY Sherman 310 A Buffalo NY 14214

SANSONE, FRANCIS JOSEPH, b Dayton, Ohio, July 25, 51. ORGANIC BIOGEOCHEMISTRY, CHEMICAL OCEANOGRAPHY. *Educ:* Rensselaer Polytech Inst, BS, 73; Univ NC, Chapel Hill, MS, 76, PhD(marine sci), 80. *Prof Exp:* Asst prof, 80-87, ASSOC PROF OCEANOG, UNIV HAWAII, MANOA, 87- *Mem:* Am Soc Limnol & Oceanog; Am Soc Microbiol; Am Geophys Union; Am Chem Soc. *Res:* Marine organic geochemistry; microbially-mediated anaerobic decomposition in sediments; tropical seawater chemistry; anaerobic diagenesis of marine carbonates. *Mailing Add:* Dept Oceanog Univ Hawaii 1000 Pope Rd Honolulu HI 96822

SANSONE, FRED J, b New York, NY, Apr 7, 34; m 57. MATHEMATICS. *Educ:* Univ Mich, BSE, 56, MSE, 59; Rutgers Univ, MS, 62, PhD(math), 64. *Prof Exp:* Mathematician, Melpar, Inc, 56-57; systs analyst, Bendix Corp, 59-60; asst prof math, Case Inst Technol, 63-65; asst prof, Ariz State Univ, 65-75, assoc prof math, 75-; RETIRED. *Mem:* Am Math Soc; Asn Symbolic Logic. *Res:* Recursive function theory; mathematical logic. *Mailing Add:* Dept Math Ariz State Univ Tempe AZ 85287

SANSONETTI, S JOHN, b Blairsville, Pa, Aug 18, 14; m 48; c 2. METALLURGY. *Educ:* Ind Univ, BS, 36; Carnegie Mellon Inst, BS, 41. *Prof Exp:* Dir metall res, Reynolds Metals Co, 45-80; RETIRED. *Concurrent Pos:* Instr, Am Inst Mgt, 58-78. *Mem:* Fel Am Soc Metals Int; Am Chem Soc; Am Phys Soc; Optical Soc Am; Am Inst Mech Engrs; Nat Corrosion Eng Soc. *Res:* Methods of applications of aluminum development of alloys. *Mailing Add:* 503 Ridge Top Rd Richmond VA 23229

SAN SOUCIE, ROBERT LOUIS, b Adams, Mass, Apr 30, 27; m 53; c 3. MATHEMATICS, RESEARCH ADMINISTRATION. *Educ:* Univ Mass, BA, 49; Univ Wis, MA, 50, PhD(math), 53. *Prof Exp:* Wis Alumni Res Found asst math, Univ Wis, 49-51, univ asst, 51-52; instr, Univ Ore, 53-55, asst prof, 55-57; head math sect, Sylvania Electronic Systs Div, Gen Tel & Electronics Corp, 57-59, mgr adv commun, 59-61, mgr systs projs, 61-62; vpres eng electronics & space div, Emerson Elec Co, 62-64, pres & gen mgr, 64-67, exec vpres, 67-71; pres, DLJ Capital Corp, Subsidiary of Donaldson, Lufkin & Jenrette, Inc, 71-75; MANAGING DIR, MILL RIVER VENTURES LTD, 75- *Concurrent Pos:* Trustee, St Louis Univ, 71-82. *Mem:* Am Math Soc; Am Mgt Asn; Sigma Xi; Phi Beta Kappa. *Res:* Right alternative rings; additive and multiplicative systems; error correcting codes and management information systems. *Mailing Add:* San Soucie 68 Dortmunder Dr Manalapan NJ 07726

SANTACANA-NUET, FRANCISCO, b Barcelona, Spain, Jan 16, 31; US citizen; div; c 4. ANALYTICAL CHEMISTRY. *Educ:* Barcelona Indust Col, BSc, 51; Purdue Univ, MS, 59. *Prof Exp:* Chemist, S A Rovira, Bachs & Macia, 50-53, chief anal sect, 53-56, consult dye chem, 59-60; res chemist, Org Chem Div, Am Cyanamid Co, 60-70, group leader anal res & develop, 70-71, group head, Chem Res Div, Bound Brook, NJ, 71-80, mgr anal serv, 80-88, DIR SCI SERV, CHEM RES DIV, AM CYANAMID CO, STAMFORD, CONN, 88- *Mem:* Am Chem Soc. *Res:* Chromatography; automation and instrumentation; spectrophotometry; identification of organic structures; nuclear magnetic resonance; mass spectrometry; management; microscopy. *Mailing Add:* 34 Spinning Wheel Rd Fairfield CT 06430

SANTAMARIA, VITO WILLIAM, b Cleveland, Ohio, Aug 19, 48; m 89; c 6. TECHNICAL MANAGEMENT. *Educ:* Cleveland State Univ, BS, 70. *Prof Exp:* Sr technician, Lamp Glass Div, Gen Elec, 69-70; sect head, Indust Coatings-Appliance, Glidden, 70-77; tech mgr, Indust Coatings, Standard T Chem, 77-80, Coatings Div, Whittaker Corp, Batavia, 80-81, Extrusion Coatings, Valspar Corp, 81-86; tech dir, Indust Coatings Div, Pratt & Lambert, 86-89; TECH DIR, JAMESTOWN PAINT & VARNISH CO, 90- *Concurrent Pos:* Mem, corrosion comt, Fedn Soc Coatings Technol, 87- *Mem:* Asn Finishing Processes; Fedn Soc Coatings Technol; Nat Paint & Coatings Asn; Chem Coaters Asn. *Res:* OEM finishes technology advancement of protective and ornamental products ecologically sound forming films through evaporation, oxidation, and thermal reactivity. *Mailing Add:* 321 E Butler St Mercer PA 16137

SANT'AMBROGIO, GIUSEPPE, b Milano, Italy, Nov 28, 31; m 58; c 3. RESPIRATORY PHYSIOLOGY, NEUROPHYSIOLOGY. *Educ:* Univ Milan, MD, 56. *Prof Exp:* From asst prof physiol to prof, Med Sch, Univ Milan, 57-76; assoc prof, 75-77, PROF, DEPT PHYSIOL, UNIV TEX MED BR, GALVESTON, 77- *Concurrent Pos:* Res fel, Dept Physiol, Univ Ky, 58-59 & 60-61, Univ Oxford, 63-64; vis prof, Dept Physiol, McGill Univ, 73. *Mem:* Italian Soc Physiol; Brit Physiol Soc; Europ Soc Respiratory Pathophysiol; Am Physiol Soc; Soc Exp Biol & Med. *Res:* Neural control of breathing in mammals. *Mailing Add:* Dept Physiol & Biophysics Univ Tex Med Br Galveston TX 77550-2781

SANTAMOUR, FRANK SHALVEY, JR, b Lowell, Mass, Mar 7, 32; m 52; c 1. PLANT GENETICS, BIOCHEMISTRY. *Educ:* Univ Mass, BS, 53; Yale Univ, MF, 54; Harvard Univ, AM, 57; Univ Minn, PhD(forestry, plant genetics), 60. *Prof Exp:* Geneticist northeast forest exp sta, US Forest Serv, 57-64; geneticist, Morris Arboretum, Univ Pa, 64-67; RES GENETICIST, AGR RES SERV, USDA, 67- *Concurrent Pos:* Am Philos Soc grants, 66, 71, 80; Holly Soc Am grants, 71, 85; Hort Res Inst grants, 76, 79; Int Soc Arboriculture grant, 82. *Honors & Awards:* Jackson Dawson Medal, Univ Ma, 84. *Mem:* Soc Am Foresters; Bot Soc Am; Am Hort Soc; Int Soc Arboriculture (pres, 84). *Res:* Genetics and breeding of shade trees and wood ornamentals for urban areas; biochemistry of incompatabilities and pest resistance; biochemical systematics; cytology. *Mailing Add:* US Nat Arboretum Washington DC 20251

SANTARE, MICHAEL HAROLD, b Brooklyn, NY, Nov 25, 59; m 81; c 2. ENGINEERING MECHANICS. *Educ:* Rensselaer Polytech Inst, BS, 81; Northwestern Univ, MS, 84, PhD(T&AM), 87. *Prof Exp:* Asst prof, 86-91, ASSOC PROF MECH ENG, UNIV DEL, 91- *Mem:* Am Acad Mech; Am Soc Mech Engrs; Sigma Xi. *Res:* Mechanical behavior of inhomogeneous materials including structural response and fracture; applications of non-classical continuum mechanics to complex material behavior; orthopaedic biomechanics. *Mailing Add:* Dept Mech Eng Univ Del Newark DE 19716

SANTE, DANIEL P(AUL), b Lackawanna, NY, Nov 16, 19; m 58. ELECTRONICS. *Educ:* Tri-State Col, BS, 41. *Prof Exp:* Jr engr, Colonial Radio Corp, 41-42; sr design engr, Sylvania Electric Prod Co, 43-48; assoc electronics engr, Cornell Aeronaut Lab, Inc, 48-52, res engr, 52-59; sect head, Sylvania Elec Prod Inc Div, Gen Tel & Electronics Corp, 59-63; asst prof elec eng, 64-72, ASSOC PROF ELEC ENG, ERIE COMMUNITY COL, 72- *Res:* Radio telemetry; communications systems; missile systems; establishment of reliability techniques in a circuit sense for electronics equipment. *Mailing Add:* 4530 Greenbriar Rd Buffalo NY 14221

SANTELMANN, PAUL WILLIAM, b Ann Arbor, Mich, Oct 18, 26; m 50; c 4. AGRONOMY, WEED SCIENCE. *Educ:* Univ Md, BS, 50; Mich State Col, MS, 52; Ohio State Univ, PhD(agron), 54. *Prof Exp:* Asst prof agron, Univ Md, 54-61, assoc prof, 61-62; from assoc prof to prof, 62-74, regents prof, 74-78, PROF AGRON & HEAD DEPT, OKLA STATE UNIV, 78- *Concurrent Pos:* Mem adv group pest mgt & res, President's Coun Environ Qual; mem herbicide study group, Environ Protection Agency; mem study probs pest control team, Nat Acad Sci; mem bd dirs, Coun Agr Sci & Technol, 75-78; ed, newslett, Weed Sci Soc Am; res award, Sigma Xi. *Mem:* Am Inst Biol Sci; AAAS; fel Weed Sci Soc Am (pres, 78); fel Am Soc Agron; Weed Sci Soc; Sigma Xi. *Res:* Crop and weed management and ecology; herbicide persistence and activity. *Mailing Add:* Dept Agron Okla State Univ Stillwater OK 74074

SANTEN, RICHARD J, b Cincinnati, Ohio, Apr 2, 39; c 3. ENDOCRINOLOGY. *Educ:* Holy Cross Col, AB, 61; Univ Mich, Ann Arbor, MD, 65; Am Bd Int Med, dipl, 70 & 76. *Prof Exp:* Instr med, Univ Wash Sch Med, 70-71; from asst prof to prof, 71-86, EVAN PUGH PROF MED, M S HERSHEY MED CTR, PA STATE UNIV, 86-, CHIEF DIV ENDOCRINOL, 79- *Concurrent Pos:* Vis prof, Univ Liege, Belgium, 78-79, Hosp Necker, Paris, 85-86; consult physician, Lebanon Vet Admin Hosp, Pa, 80- *Mem:* Endocrine Soc; fel Am Col Physicians; Am Soc Androl; Am Soc Clin Invest; Am Soc Clin Oncol. *Res:* Hormonal control of breast and prostate cancer; experimental therapy of breast and prostate cancer; control gonudotropin secretion. *Mailing Add:* Dept Med Div Endocrinol Pa State Univ Milton Hershey Med Ctr PO Box 850 Hershey PA 17033

SANTER, JAMES OWEN, b Benenden, Eng, May 3, 31; m 57; c 4. ORGANIC CHEMISTRY. *Educ:* Univ London, BSc, 55; Ill Inst Technol, PhD(chem), 61. *Prof Exp:* Res chemist, Shawinigan Resins Corp, 61-65; SR RES SPECIALIST, MONSANTO POLYMERS & RESINS CO, 65- *Mem:* Am Chem Soc. *Res:* Chemistry and technology of aminoplast resins in surface coatings. *Mailing Add:* 15 Pleasant Pl East Longmeadow MA 01028

SANTER, MELVIN, b Boston, Mass, Aug 23, 26; m 55; c 3. MICROBIOLOGY. *Educ:* St John's Univ, NY, BS, 49; Univ Mass, MS, 51; George Wash Univ, PhD(bact), 54. *Prof Exp:* Nat Found Infantile Paralysis fel, Yale Univ, 54-55, NIH fel, 55-56; from asst prof to assoc prof biol, 56-68, PROF BIOL, HAVERFORD COL, 68- *Concurrent Pos:* Lalor fac award, 58; NSF sr fel, 62-63; Weizmann Inst Sci fel, 69-70. *Mem:* AAAS; Am Soc Microbiol; Am Soc Biol Chemists. *Res:* Biochemistry of autotrophic bacteria; ribosome structure and RNA sequence work. *Mailing Add:* Dept Biol Haverford Col Haverford PA 19041

SANTERRE, ROBERT FRANK, b Hanover, NH, June 28, 40; m 58; c 4. CELL CULTURE, MICROBIOLOGY. *Educ:* Southern Conn State Col, BS, 65; Univ NH, MS, 67, PhD(zool), 70. *Prof Exp:* Fel biol, Mass Inst Technol, 70-73 & Univ Calif, San Diego, 73-75; RES SCIENTIST MOLECULAR BIOL, LILLY RES LABS, 75- *Concurrent Pos:* Muscular Dystrophy Res Fel, 72-73. *Honors & Awards:* Muscular Dystrophy Res Fel, 72-73. *Mem:* Sigma Xi; AAAS; Am Soc Microbiol. *Res:* Oxidative phosphorylation and lipid metabolism; biochemical characterization of non-muscle actins; in vitro studies of insulin biosynthesis; development of cloning vectors for gene analysis and expression in higher eukaryotic cells; cytokine gene regulation in transgenic animals; transgenics. *Mailing Add:* Biotechnol Res Div Lilly Res Labs Corp Ctr Indianapolis IN 46285

SANTI, DANIEL V, b Buffalo, NY, Feb 6, 42. ORGANIC CHEMISTRY, BIOCHEMISTRY. *Educ:* State Univ NY, Buffalo, BS, 63, PhD(med chem), 67. *Prof Exp:* Asst prof chem, Univ Calif, Santa Barbara, 66-70; assoc prof pharmaceut chem & biochem, 70-76, PROF PHARMACEUT CHEM & BIOCHEM, UNIV CALIF, SAN FRANCISCO, 76- *Mem:* Am Chem Soc. *Res:* Enzyme mechanisms; protein biosynthesis; design of enzyme inhibitors; model enzyme reactions; nucleic acids; heterocyclic chemistry. *Mailing Add:* Dept Biochem & Biophysics Univ Calif San Francisco CA 94143

SANTIAGO, JULIO VICTOR, b San German, PR, Jan 13, 42; m 63; c 4. DIABETES, PEDIATRIC ENDOCRINOLOGY. *Educ:* Manhattan Col, New York, BS, 63 & Univ PR, MD, 67; Univ Montevideo, Uraguay, PhD, 84. *Prof Exp:* DIR RES, DIABETES RES & TRAINING CTR, 77-, PROF PEDIAT, WASH UNIV MED SCH, 83- *Concurrent Pos:* Prin investr, Diabetes Cent & Complications Trial, 83- *Mem:* Am Soc Clin Invest; Soc Ped Res; Am Diabetes Asn. *Res:* Treatment of diabetes mellitus, longterm complications of diabetes, insulin induced hypoglycemia, pediatric endocrinology. *Mailing Add:* Four Forest Pkwy Manchester MO 63011

SANTIAGO, NOEMI, b San Juan, PR, Dec 28, 53; c 2. DRUG DELIVERY, ORAL VACCINATION. *Educ:* Univ PR, BS, 74, MS, 78, PhD(immunol & microbiol), 85. *Prof Exp:* Instr biol, Sacred Heart Univ, 78-80 & Univ PR, 80-82; asst prof immunol & parasitol, Cayey Sch Med, 86-88; DIR, CLIN TECHNOLOGIES ASSOCS, INC, 89- *Concurrent Pos:* Consult, Marc Prog, NIH, 83-84; vis prof, Univ PR, Cayey Campus, 87-88. *Mem:* Am Asn Immunologists. *Res:* Oral delivery system based on microencapsulation with a protein-like polymer for delivery of proteins and other drugs as well as for oral vaccination. *Mailing Add:* 536 Commerce St Hawthorne NY 10532

SANTIAGO-MELENDEZ, MIGUEL, b Corozal, PR, Sept 28, 30; m 54, 80; c 6. STRUCTURAL ENGINEERING. *Educ:* Univ PR, Mayaguez, BSCE, 54; Tex A&M Univ, MCE, 60, PhD(struct eng), 62. *Prof Exp:* Instr mech, Univ PR, Mayaguez, 54-59, prof struct eng, 62-69; exec dir, Commonwealth of PR, 69-73; chmn, Dept Civil Eng, 77-80, assoc dean eng, Univ PR, 83-86; PROF ENG, CARRIBEAN UNIV, BAYAMÓN. *Concurrent Pos:* Consult, PR Planning Bd, 63-69; vis scholar, Univ Calif, Berkeley, 76-77. *Mem:* Am Concrete Inst; Am Soc Civil Engrs; Am Soc Eng Educ; Earthquake Eng Res Inst. *Res:* Shear and diagonal tension in reinforced concrete members, especially beams. *Mailing Add:* Col Sta Box 5089 Mayaguez PR 00709

SANTIDRIAN, SANTIAGO, b Burgos, Spain, Nov 16, 50; m 79. NUTRITION, PROTEIN & CHOLESTEROL METABOLISM. *Educ:* Univ Navarra, Spain, BSc, 73, MB, 74, PhD(pharmaceut physiol), 76. *Prof Exp:* Postdoctoral fel nutrit biochem, Mass Inst Technol, 77-78, res assoc nutrit biochem physiol, 78-81; prof physiol, Univ Granada, Spain, 83-84; prof, Univ La Laguna, Spain, 84-87; asst prof biochem, Sch Med, 76-77, assoc prof physiol, 81-83, PROF PHYSIOL, UNIV NAVARRA, SPAIN, 87-, HEAD, DEPT PHYSIOL, 87- *Concurrent Pos:* Res award, Royal Acad Pharm, 88. *Mem:* Am Inst Nutrit; Am Soc Animal Sci; NY Acad Sci. *Res:* Effect of hormones and diets on protein metabolism; effects of legumes on protein metabolism, immune response and cholesterol metabolism; effect of antibiotics on intestinal absorption of sugars and amino acids. *Mailing Add:* Dept Physiol Univ Navarra Pamplona 31008 Spain

SANTILLI, ARTHUR A, b Everett, Mass, July 25, 29; m 64; c 1. ORGANIC CHEMISTRY. *Educ:* Boston Univ, AB, 51; Tufts Col, MS, 52; Univ Mass, PhD(chem), 58. *Prof Exp:* Asst chem, Univ Mass, 54-57; fel, Tufts Univ, 58-60; sr res scientist, 60-65, group leader, 65-86, res supvr & prin scientist, 86-88, RES FEL, WYETH-AYERST RES, 88- *Concurrent Pos:* Manuscript Reviewer, J Med Chem & Med Chem Res. *Mem:* Am Chem Soc; Int Union Pure & Appl Chem. *Res:* Synthesis of heterocyclic compounds of possible medicinal interest. *Mailing Add:* Chem Dept Wyeth-Ayerst Res Princeton NJ 08543-8000

SANTILLI, RUGGERO MARIA, theoretical physics, for more information see previous edition

SANTISTEBAN, GEORGE ANTHONY, b Mex, Apr 12, 18; nat US; m 42; c 3. NEUROENDOCRINOLOGY, CARDIOVASCULAR DISEASES. *Educ:* Univ Mont, BA, 45; Univ Utah, MA, 49, PhD(human anat), 51. *Prof Exp:* Asst zool, Univ Mont, 47-48; lectr anat, Univ Utah, 49-51, res instr anat & radiobiol, 51-53; asst prof anat, Med Col Va, 53-54; asst prof sch med, Univ Southern Calif, 54-59, asst prof physiol, 59-64; head dept biol, 64-68, assoc prof, 68-75, prof, 75-84, EMER PROF BIOL, SEATTLE UNIV, 88- *Concurrent Pos:* Sr biologist, Pac Northwest Res Found, 68-; USPHS spec fel, Gothenburg Univ, 66-67; affil investr, Fred Hutchinson Cancer Ctr, 73- *Mem:* AAAS; Am Asn Cancer Res; NY Acad Sci; Am Asn Anat. *Res:* Effects of early experience as a modifying influence upon the function of the hypothalamus; psychosocial stress and development of hypertension, cardiovascular disease and malignancy; interrelationship between stress imprinting and intraneuronal RNA; psychosocial stress; hyperglycemia and tumor growth. *Mailing Add:* 9116 227th SW Edmonds WA 98020

SANTNER, JOSEPH FRANK, b Chicago, Ill, Aug 19, 19; m 46; c 5. MATHEMATICAL STATISTICS, OPERATIONS RESEARCH. *Educ:* St Louis Univ, BS, 50, MS, 52. *Prof Exp:* Mathematician, McDonnell Aircraft Corp, 52-54; opers res analyst, NAm Aviation, Inc, 54-55; lectr, Xavier Univ, Ohio, 55-58, asst prof, 58-62; math statistician, Robert A Taft Sanit Eng Ctr, US Dept Health, Educ & Welfare, 62-65, head math sci, 65-70; head math sci, Environ Protection Agency, 70-82; RETIRED. *Mem:* Am Statist Asn. *Res:* Mathematical logic; non-parametric statistical methods; design and analysis of experiments; applied mathematics. *Mailing Add:* 2124 Glenside Ave Norwood OH 45212

SANTNER, THOMAS JOSEPH, b St Louis, Mo, Aug 29, 47; m 70; c 4. MATHEMATICAL STATISTICS. *Educ:* Univ Dayton, BS, 69; Purdue Univ, MS, 71, PhD(math statist), 73. *Prof Exp:* Fel, Nat Sci Found, 69-72; from asst prof to prof statist, Sch Opers Res & Indust Eng, Cornell Univ, 73-89; PROF STATIST, DEPT STATIST, OHIO STATE UNIV, 89- *Concurrent Pos:* Vis scientist & prin investr, NSF grant, 75-77, vis scientist & co-prin investr, 77-79; Vis assoc prof biostatist, Univ Wash, 81-82; vis scientist, biometry br, Nat Cancer Inst, 78-79. *Mem:* Inst Math Statist; fel Am Statist Asn; Biometric Soc. *Res:* Analysis of discrete data; selection and ranking theory applied statistics. *Mailing Add:* Dept Statist Ohio State Univ 1958 Neil Ave Columbus OH 43210

SANTO, GERALD S(UNAO), b Olaa, Hawaii, Dec 22, 44; m 68; c 3. PLANT NEMATOLOGY. *Educ:* Univ Hawaii, BS, 67, MS, 69; Univ Calif, Davis, PhD(plant path), 74. *Prof Exp:* NEMATOLOGIST, WASH STATE UNIV, 74- *Honors & Awards:* Ciba-Geigy Recognition Award, Am Soc Nematol. *Mem:* Soc Nematologists; Am Phytopath Soc; Coun Agr & Sci Technol; Orgn Trop Am Nematologists; Potato Asn Am; AAAS. *Res:* Study of biology, pathogenicity and control of plant-parasitic nematodes. *Mailing Add:* Dept Plant Path Wash State Univ Rte 2 Box 2953-A Prosser WA 99350-9687

SANTORA, NORMAN JULIAN, b Camden, NJ, Sept 17, 35; m 62; c 3. CHEMICAL INFORMATION, QUANTITATIVE DRUG DESIGN. *Educ:* Temple Univ, AB, 57, AM, 60, PhD(org chem), 65. *Prof Exp:* Asst chem, Temple Univ, 57-60 & 61-65; fel, Univ Pa, 65-68; res assoc med chem, Wm H Rorer, Inc, 68-81; chem info specialist, 81-83, SR INFO CHEMIST, SMITH KLINE & FRENCH CORP, 84- *Concurrent Pos:* Adj prof comput sci, Spring Garden Col, 84- *Mem:* Am Chem Soc; Am Pharmaceut Asn. *Res:* Pyrimidine chemistry; synthesis and characterization of analogs of sulfonamides and nucleosides; quantitative structure-activity relationship studies. *Mailing Add:* 1323 Partridge Rd Roslyn PA 19001

SANTORO, FERRUCIO FONTES, b Salvador, Bahia, Apr 15, 52; m 74; c 2. TOXOPLASMOSIS, TROPICAL DISEASES. *Educ:* Fed Univ Bahia, Brazil, Pharmacist, 73; Lille Univ, France, Pharmacy Doct, 77, PhD(med), 80. *Prof Exp:* Fel res, Oswaldo Cruz Inst, Salvador, Brazil, 73-74, res dir, 87-89; fel res, Pasteur Inst, Lille, France, 74-77, res asst, 77-80 & 83-85, res dir, 85-87; assoc res, NY Univ Med Ctr, New York, 80-83; RES DIR, FAC MED, GRENOBLE, FRANCE, 89- *Concurrent Pos:* Consult, WHO, 79-80, Acta Tropica, 79- *Mem:* Brazilian Soc Trop Med; Am Asn Immunologists; Soc France Parasitol. *Res:* Immune complexes in parasitic diseases; interactions between complement and parasites; vaccins against parasitic diseases; growth and differentiation of parasites. *Mailing Add:* CIBP Pasteur Inst 1 rue du Prof Albert Calmette Lille 59000 France

SANTORO, THOMAS, b Brooklyn, NY, Oct 22, 28; m 64; c 2. MICROBIOLOGY. *Educ:* Brooklyn Col, BS, 54; Univ Kans, MA, 57; Pa State Univ, PhD(microbiol), 61. *Prof Exp:* Res asst bact, Univ Kans, 54-57; soil microbiologist, Pa State Univ, 57-61, res assoc, 61-63; res asst, Brooklyn Bot Garden, 63-69; ASST PROF BIOL, STATE UNIV NY COL NEW PALTZ, 69- *Mem:* Am Soc Microbiol; Mycol Soc Am; Brit Soc Gen Microbiol. *Res:* Antibiotics by Mycorrhizal fungi; effect of pesticides and insecticides on the ecology of soil microorganisms; interactions between clay minerals and microbial cells. *Mailing Add:* Dept Microbiol State Univ NY Col New Paltz NY 12561

SANTOS, EUGENE (SY), b Manila, Philippines, Feb 15, 41; m 67; c 2. COMPUTER SCIENCE, MATHEMATICS. *Educ:* Mapua Inst Technol, BSME, 61; Univ Philippines, MSc, 63; Ohio State Univ, PhD(math), 65. *Prof Exp:* Instr math, Mapua Inst Technol, 62-63; teaching asst, Ohio State Univ, 63-65, asst prof, 65-68; assoc prof, 68-74, PROF MATH, YOUNGSTOWN STATE UNIV, 74- *Mem:* Asn Comput Mach. *Res:* Theory of automata, computability and formal languages; software systems, artificial intelligence. *Mailing Add:* Dept Computer Sci Brown Univ Box 1910 Providence RI 02912

SANTOS, GEORGE WESLEY, b Oak Park, Ill, Feb 3, 28; m 52; c 4. ONCOLOGY, IMMUNOLOGY. *Educ:* Mass Inst Technol, BS & MS, 51; Johns Hopkins Univ, MD, 55; Am Bd Internal Med, dipl, 62. *Prof Exp:* Intern med, Johns Hopkins Hosp, 55-56, asst resident, 58-60, fel, 60-62; from instr to assoc prof med, 62-68, PROF ONCOL & MED, SCH MED, JOHNS HOPKINS UNIV, 73-; PHYSICIAN, JOHNS HOPKINS HOSP, 62- *Concurrent Pos:* Leukemia Soc scholar, 61-66; asst dir med oncol unit, Baltimore City Hosps, 62-, asst physician-in-chief, 63-65; mem, Cancer Clin Investigative Rev Comt & Immunol-Epidemiol Spec Virus-Cancer Prog, NIH, 69-73 & Cell Biol-Immunol-Genetics Res Evaluations Comt, Vet Admin, 69-71; chmn, Bone Marrow Transplant Registry & mem, Int Comt Organ Transplant Registry, Am Col Surgeons-NIH, 69-73; mem bd dir, Leukemia Soc Am, 73-76 & Am Nat Bd, Trustees, 87. *Honors & Awards:* Bristol-Meyers Award, 88. *Mem:* Am Soc Hemat; Transplantation Soc; Am Asn Immunologists; Am Asn Cancer Res; Int Soc Exp Hemat. *Res:* Transplantation immunology. *Mailing Add:* Oncol Ctr Johns Hopkins Univ 600 N Broadway Baltimore MD 21205

SANTOS-BUCH, CHARLES A, b Santiago, Cuba, Mar 20, 32; US citizen; m 56; c 3. EXPERIMENTAL PATHOLOGY. *Educ:* Harvard Univ, BA, 53; Cornell Univ, MD, 57. *Prof Exp:* Asst path, Med Col, Cornell Univ, 58-61, instr neuropath, 61-62; from asst prof to prof path, Sch Med, Emory Univ, 62-68; assoc prof, 68-76, assoc dean, 70-74, PROF PATH, MED COL, CORNELL UNIV, 76-; DIR, PAPANICOLAOU CYTOL LAB, NEW YORK HOSP, 79- *Concurrent Pos:* USPHS res training fel, 59-62; Markle scholar acad med, 64. *Mem:* Am Soc Exp Path; Am Asn Path & Bact; NY Acad Sci; Pan-Am Med Asn; Am Soc Cytol. *Res:* Diseases of small arteries; high resolution enzyme histochemistry; immunology of hypersensitivity diseases; molecular and biochemical parasitology. *Mailing Add:* Dept Path Cornell Univ Med Col 1300 York Ave New York NY 10021

SANTOS-MARTINEZ, JESUS, b Vieques, PR, Mar 5, 24; m 55; c 4. PHYSIOLOGY, PHARMACOLOGY. *Educ:* Univ PR, BS, 46; Univ Ill, MS, 48; Purdue Univ, PhD(pharmacol), 54. *Prof Exp:* Asst instr pharm, Col Pharm, Univ PR, San Juan, 46-47, instr, 47-51, asst prof pharmacol, 51-55, from asst prof to prof physiol, Sch Med, 55-72, lectr pharmacol, Sch Pharm, 58, prof basic sci & chmn dept, Sch Dent, 72-76, prof pharmacol, Sch Med, 76-80; mem fac, dept physiol & pharmacol, 79-86, PROF & CHMN, DEPT PHARMACOL, SCH MED, UNIV CENT DEL CARIBE, 86-, DIR PROG GRAD STUDIES, 89- *Concurrent Pos:* Nat Inst Arthritis & Metab Dis fel, Med Ctr, Ind Univ, 66-67; vis prof, Univ Col WI, 60 & Sch Med, Univ Nicaragua, 62. *Mem:* AAAS; Am Physiol Soc; Am Soc Nephrology; Int Soc Nephrology; Soc Exp Biol Med. *Res:* Renal physiology; electrolyte distribution. *Mailing Add:* Dept Pharmacol Univ Caribe Sch Med Bayamón PR 00621-6032

SANTULLI, THOMAS V, b New York, NY, Mar 16, 15; m 43; c 2. SURGERY. *Educ:* Columbia Univ, BS, 35; Georgetown Univ, MD, 39; Am Bd Surg, dipl, 47, cert spec competence pediat surg, 75. *Prof Exp:* Assoc prof, 55-67, CHIEF PEDIAT SURG, COL PHYSICIANS & SURGEONS, COLUMBIA UNIV, 55-, PROF SURG, 67- *Concurrent Pos:* Consult, Monmouth Mem & Fitkin Mem Hosps, NJ, 50-, St Joseph's Hosp, Yonkers, NY, 53- & St Joseph's Hosp, Stamford, Conn, 59-; attend surgeon, Presby Hosp, New York, 60- *Mem:* Fel Am Col Surg; fel Am Acad Pediat; Am Surg Asn; Am Pediat Surg Asn (pres, 80-81); Brit Asn Pediat Surg. *Res:* Pediatric surgery. *Mailing Add:* Babies Hosp 3959 Broadway New York NY 10032

SANUI, HISASHI, b Orosi, Calif, Jan 7, 24; m 54; c 2. BIOPHYSICS, CELL PHYSIOLOGY. *Educ:* Univ Calif, AB, 51, PhD(biophys), 58. *Prof Exp:* Jr res physiologist, 57-59, lectr physiol, 60-61, asst res physiologist, 61-67, ASSOC RES PHYSIOLOGIST, UNIV CALIF, BERKELEY, 67- *Mem:* AAAS; Am Physiol Soc; Biophys Soc; NY Acad Sci; Sigma Xi. *Res:* Active ion transport by living cells; subcellular morphology and biochemistry; ion binding by biological materials; atomic absorption spectrophotometry; role of cell membrane and ions in cell transformation; role of inorganic actions in cell growth regulation; mechanisms of heavy metal action on cells. *Mailing Add:* 506 Albermerle St El Cerrito CA 94530

SANWAL, BISHNU DAT, TISSUE CULTURE, MOLECULAR BIOLOGY. *Educ:* Fed Inst Technol, Zurich, DSc(bot), 53. *Prof Exp:* PROF BIOCHEM, UNIV WESTERN ONT, 73- *Mailing Add:* Dept Biochem Univ Western Ont Med Sch London ON N6A 5C1 Can

SANYER, NECMI, b Konya, Turkey, Oct 5, 19; nat US; m 53; c 2. WOOD CHEMISTRY. *Educ:* Inst Agr, Ankara, Turkey, BS, 41; State Univ NY, MS, 50, PhD, 53. *Prof Exp:* Hibbert Mem fel, McGill Univ, 53-54; res chemist, Mead Corp, 54-59; supvry chemist, 59-73, SUPVR RES CHEM, FOREST PRODS LAB, US FOREST SERV, 73- *Mem:* Fel Am Inst Chemists; Am Chem Soc. *Res:* Lignin, cellulose and pulping chemistry. *Mailing Add:* 1442 Skyline Dr Madison WI 53705

SANZONE, GEORGE, b Brooklyn, NY, Jan 13, 34; m 56; c 3. CHEMICAL PHYSICS, CHEMICAL KINETICS. *Educ:* Univ Ill, Urbana, BS, 65, MS, 67, PhD(chem), 69. *Prof Exp:* Designer, Burton Rodgers, Inc, Ohio, 59-60; proj engr, Bendix Corp, 60-63; engr dept chem, Univ Ill, Urbana, 63-65; asst prof, 69-81, ASSOC PROF CHEM, VA POLYTECH INST & STATE UNIV, 81- *Mem:* Am Chem Soc; Am Phys Soc; Am Soc Mass Spectros. *Res:* High temperature, fast chemical reaction studies employing shock tubes with mass spectrometric and optical detection techniques; molecular beam fluorescence; reactions in high-velocity flows. *Mailing Add:* Dept Chem Va Polytech Inst & State Univ Davidson Hall Blacksburg VA 24060

SAPAKIE, SIDNEY FREIDIN, b Port Chester, NY, May 10, 45; m 72; c 2. CHEMICAL ENGINEERING, FOOD ENGINEERING. *Educ:* Univ Mich, BSChEng, 67; Univ Minn, MBA, 72. *Prof Exp:* Res engr, Gen Mills, Inc, 67-70; group leader, Betty Crocker Div, Gen Mills Inc, 70-73, asst prod mgr mkt, Protein Div, 73-74, develop leader, Spec Technol Activ, 74-77, dir res, Gorton Group, 77-78, dir, Subsid Res & Develop, 78-80, appl eng & new process develop, 80-84, res & develop, Big G Div, 84-90, VPRES, CEREAL RES & TECHNOL, GEN MILLS INC, 90- *Mem:* Am Inst Chem Engrs; Inst Food Technologists; Am Asn Cereal Chemists. *Res:* Food sterilization, especially thermal processing, aseptic processing and microwave processing; food extrusion, especially math modeling and development; engineering economics; cereal processing. *Mailing Add:* Gen Mills Inc 9000 Plymouth Ave N Minneapolis MN 55427

SAPEGA, A(UGUST) E(DWARD), b Bridgeport, Conn, Dec 10, 25; m 55; c 2. ELECTRICAL ENGINEERING. *Educ:* Columbia Univ, BS, 46, MS, 51; Worcester Polytech Inst, PhD, 72. *Prof Exp:* From instr to assoc prof, 51-67, chmn dept, 71-81, PROF ENG, TRINITY COL, CONN, 67- *Mem:* Am Soc Mech Engrs; Inst Elec & Electronics Engrs. *Res:* Electrical engineering circuits and devices; semiconductor physics and circuits; computer applications. *Mailing Add:* Dept Eng Trinity Col Hartford CT 06106

SAPER, CLIFFORD B, b Chicago, Ill, Feb 20, 52; m 73; c 3. NEUROBIOLOGY, NEUROLOGY. *Educ:* Univ Ill, BS & MS, 72; Wash Univ, PhD(neurobiol) & MD, 77. *Prof Exp:* Intern med, Jewish Hosp St Louis, 77-78; resident neurol, NY Hosp-Cornell Med Ctr, 78-81; asst prof neurol & anat, Sch Med, Wash Univ, 81-85; assoc prof, 85-88, PROF PHARM, PHYSIOL & NEUROL, 88-, CHMN NEURO BIOL, UNIV CHICAGO, 87- *Concurrent Pos:* Teacher-investr Develop Award, Nat Inst Neurol & Commun Dis & Stroke, 81; McKnight Found scholar, 83. *Honors & Awards:* Jacob Javits Neurosci Award, Nat Inst Neurol & Commun Dis & Stroke. *Mem:* Soc Neurosci; Am Acad Neurol; Am Physiol Soc; Am Asn Anatomists; AAAS; Am Neurol Asn. *Res:* Neuroanatomy and neurophysiology of central nervous system control of the cardiovascular system and its integration with ongoing behaviour and arousal. *Mailing Add:* Dept Pharmacol & Physiol Univ Chicago 947 E 58th St Chicago IL 60637

SAPER, MARK A, b New York, NY, Sept 29, 54; m 85; c 2. MACROMOLECULAR CRYSTALLOGRAPHY, PROTEIN STRUCTURE. *Educ:* Univ Conn, BS, 76; Rice Univ, PhD(biochem), 83. *Prof Exp:* Postdoctoral, Rice Univ, 83-84 & Weizmann Inst Sci, 84-86; res assoc, Howard Hughes Med Inst, Harvard Univ, 86-90; ASST PROF, UNIV MICH, 90- *Mem:* Am Crystallographic Asn. *Res:* Answering immunological questions using the techniques of structural biology; protein phosphatases and plant lectins with x-ray crystallography. *Mailing Add:* Biophys Res Div Univ Mich 2200 Bonisteel Ann Arbor MI 48109-2099

SAPERS, GERALD M, b Brookline, Mass, Jan 17, 35; m 60; c 2. FOOD SCIENCE. *Educ:* Mass Inst Technol, SB & SM, 59, PhD(food technol), 61. *Prof Exp:* Food scientist, Pioneering Res Div, US Army Natick Labs, 61-63; sr res assoc, Lever Bros Co, NJ, 63-64; unit leader food res, Corn Prod Food Technol Inst, Mass, 64-68; res chemist, 68-80, res leader, 80-85, RES FOOD TECHNOLOGIST, EASTERN REGIONAL RES CTR, AGR RES SERV, USDA, 85- *Mem:* AAAS; Inst Food Technologists; Am Chem Soc. *Res:* Flavor chemistry and stability of dehydrated potato products and other dehydrated foods; quality of fruit and vegetable products; fruit and vegetable processing; natural pigments; home canning safety. *Mailing Add:* Eastern Regional Res Ctr USDA 600 E Mermaid Lane Philadelphia PA 19118

SAPERSTEIN, ALVIN MARTIN, b Bronx, NY, June 3, 30; m 56; c 2. THEORETICAL PHYSICS, PEACE & CONFLICT STUDIES. *Educ:* NY Univ, BA, 51; Yale Univ, MS, 52, PhD(physics), 56. *Prof Exp:* Asst physics, Yale Univ, 52, res asst, 52-56; res physicist, Eng Res Inst, Univ Mich, 56-57; res assoc, Brown Univ, 57-59; asst prof physics, Univ Buffalo, 59-62; res assoc, Argonne Nat Lab, 62-63; assoc prof, Wayne State Univ, 63-68, dir, Prog Environ Studies, 78-80, chmn exec bd & dir res, Ctr Peace & Conflict Studies, 85- 87, PROF PHYSICS, WAYNE STATE UNIV, 68- *Concurrent Pos:* NSF res grant, 65-67 & 67-71; hon res assoc, Univ Col, Univ London, 69-70 & 76-77; prof sci & technol, Weekend Col, Wayne State Univ, 74-75; vis prof, Open Univ, Eng, 76-77, 84; vis res fel, Stockholm Int Peace Res Inst, 83, Int Inst Strategic Studies, London, 84; mem exec bd, Ctr Peace & Conflict Studies, Wayne State Univ; Fulbright res scholar, Peace Res Inst Oslo, Norway, 87. *Mem:* AAAS; fel Am Phys Soc; Fedn Am Sci; Am Asn Physics Teachers; Union Concerned Scientists. *Res:* Scattering of nucleons from nucleons and nuclei; general theory of nuclear reactions; elementary particle reactions; general scattering theory; interaction between science and society; chaos and models of war initiation and war fighting; mathematical models of military strategy and international relations. *Mailing Add:* Dept Physics Wayne State Univ Detroit MI 48202

SAPERSTEIN, DAVID DORN, b New York, NY, June 30, 46; m 76; c 1. PHYSICAL CHEMISTRY. *Educ:* Johns Hopkins Univ, BA, 67; NY Univ, PhD(chem), 73. *Prof Exp:* Sr res chemist, Merck Sharp & Dohme Res Labs, 73-77, res fel, 77-81; adv scientist, 81-84, tech support mgr, 84-85, advisory chemist, Gen Prod Div, 85-90, ADVAN DISK DEVELOP MGR, IBM INSTRUMENTS, IBM, 91- *Mem:* Am Chem Soc; AAAS; Soc Appl Spectros; Western Spectros Asn (Sec, 83-86, chmn, 87). *Res:* Molecular investigations using physical, spectroscopic and analytic chemical methods; vibrational spectroscopy of adsorbates on surfaces; structure and properties of amorphous carbon. *Mailing Add:* IBM 5600 Cottle Rd San Jose CA 95193

SAPERSTEIN, LEE W(ALDO), b New York, NY, July 14, 43; m 67; c 2. MINING ENGINEERING. *Educ:* Mont Sch Mines, BS, 64; Oxford Univ, DPhil(eng sci), 67. *Prof Exp:* From asst prof to assoc prof mining eng, Pa State Univ, 67-78, prof & sect chmn, 78-87; PROF & CHMN MINING & ENG, UNIV KY, 87- *Concurrent Pos:* Mem, comt mineral technol, Nat Res Coun, 76-78, comt on surface mining & reclamation, 78-79 & comt abandoned mine lands, 85-86; chmn, Eng Accreditation Comn, Accreditation Bd Eng & Technol, 89-90. *Mem:* Soc Mining Engrs. *Res:* Materials handling in mines and tunnels; rapid excavation; noise in underground mines; rock fragmentation; mined-land reclamation; surface mining for coal; quarries; pre-mining planning; mining training, subsidence, advances in mining sciences. *Mailing Add:* Univ Ky 230 Mining & Mineral Resources Bldg Lexington KY 40506-0107

SAPERSTEIN, SIDNEY, b Brooklyn, NY, Apr 2, 23; m 47; c 3. NUTRITION. *Educ:* Brooklyn Col, AB, 47; Univ Calif, Los Angeles, MA, 48; Univ Calif, PhD(microbiol), 53. *Prof Exp:* Asst bact, Univ Calif, 49-53; res assoc antileukemics, Col Dent, NY Univ, 53-54; bacteriologist, Borden Co, 54-56, supvr microbiol res & develop, Borden Spec Prod Co, 56-65, res assoc, 65-66, dir res pharmaceut div, Borden Foods Co, 67-71; asst dir nutrit sci, 71-73, prin scientist, Inst Agr Sci Nutrit, 73-78, clin trials mgr, Syntex Labs, Inc, 78-81, ASSOC DIR, SCI AFFAIRS, SYNTEX LABS, INC, 81- *Concurrent Pos:* Mem tech adv group, Comn Nutrit, Am Acad Pediat, 75-81. *Mem:* Am Chem Soc; Am Inst Nutrit; Am Soc Clin Nutrit; AAAS; Sigma Xi; Am Dietetic Asn. *Res:* Bacteriology; biochemistry; allergy; nutrition; pharmacology. *Mailing Add:* Syntex Labs Inc 3401 Hillview Ave Palo Alto CA 94304-1397

SAPHIER, DAVID, b Bristol, UK, July 7, 57. NEUROENDOCRINOLOGY, CENTRAL NERVOUS SYSTEM-IMMUNE SYSTEM INTERACTIONS. *Educ:* Univ Col N Wales, Bangor, BSc, 75; Inst Biol, UK, CBiol & MIBiol, 75; Univ Cambridge, UK, PhD(neuroendocrinol), 86. *Prof Exp:* Sr scientist neurol, Hadassah Univ Hosp, Jerusalem, Israel, 83-90; ASST PROF PHARMACOL, LA STATE UNIV MED CTR, SHREVEPORT, 90- *Concurrent Pos:* External lectr endocrinol, Hebrew Univ Med Sch, Jerusalem, Israel, 85-90; ed referee, numerous journals, 86-; consult, Teva Pharmaceut, Rehovot, Israel, 89-90. *Honors & Awards:* Curt P Richter Prize, Int Soc Psychoneuroendocrinol, 88. *Mem:* Int Soc Psychoneuroendocrinol; Europ Neurosci Asn; Soc Neurosci. *Res:* Neural regulation of adrenocortical secretion; effects of immunoactivation upon neural and neuroendocrine function; physiology of cotransmitter substances in relation to classical neurotransmission and neuroendocrine events. *Mailing Add:* Dept Pharmacol La State Univ Med Ctr Box 33932 Shreveport LA 71130-3932

SAPICO, FRANCISCO L, b Manila, Philippines, July 18, 40; US citizen; m 69; c 2. INFECTIOUS DISEASES, CLINICAL MICROBIOLOGY. *Educ:* Univ Philippines, MD, 65. *Prof Exp:* Rotating intern, Philippine Gen Hosp, 64-65 resident internal med, 65-67; resident internal med, State Univ NY Upstate Med Ctr, Syracuse, NY, 67-69; teaching fel infectious dis, Ctr Health Sci, Univ Calif, Los Angeles, 69-71, adj asst prof med, dept med, 72-77; res fel infectious dis, Wadsworth Vet Admin Ctr, 71-72, staff physician, dept med, 72-77; from asst prof to assoc prof, 77-90, PROF MED, DEPT MED, UNIV SOUTHERN CALIF, 90-; PHYSICIAN SPECIALIST, RANCHO LOS AMIGOS MED CTR, 77-, ASSOC CHIEF, INFECTIOUS DIS DIV, 77- *Concurrent Pos:* Asst prof med, dept med, Univ Calif, Los Angeles, 72-77; chmn, Pharm & Therapeut Comt, Ranchos Los Amigos Med Ctr, 85- *Mem:* Fel Am Col Physicians; fel Infectious Dis Soc Am; Am Soc Microbiol. *Res:* Anaerobic infections; soft tissue and bone infections; new antibiotics; methicillin-resistant staphylococci; infections in diabetics. *Mailing Add:* Rancho Los Amigos Med Ctr 7601 E Imperial Hwy Downey CA 90242

SAPINO, CHESTER, JR, b Troy, NY, Jan 28, 41; m 60; c 2. ORGANIC CHEMISTRY. *Educ:* Rensselaer Polytech Inst, BSc, 65, PhD(org chem), 69. *Prof Exp:* Res asst org chem, Sterling Winthrop Res Inst, 59; lab technician anal chem, B T Babbit & Co, Inc, 59-60; org chemist silicone prod develop ctr, Gen Elec Co, 65; assoc prof org chem, Hudson Valley Community Col, 68-69; SR DEVELOP CHEMIST, BRISTOL LABS INC, 69- *Mem:* Am Chem Soc. *Res:* Penicillin and cephalosporin chemistry; natural product chemistry; antibiotic research and development; spectroscopy; computer applications to chemical problems. *Mailing Add:* 6451 Pheasant Rd East Syracuse NY 13057

SAPIR, DANIEL GUSTAVE, b Brussels, Belgium, May 21, 35; US citizen; m 62; c 2. MEDICINE. *Educ:* Brown Univ, AB, 56; Johns Hopkins Univ, MD, 60. *Prof Exp:* Fel nephrol, Tufts New Eng Med Ctr, 64-66; instr, 66-68, asst prof, 68-73, ASSOC PROF, SCH MED, JOHNS HOPKINS UNIV, 73- *Concurrent Pos:* Consult var pvt & pub orgn, 67-; Irvine-Blum scholar, Sch Med, Johns Hopkins Univ, 77. *Mem:* Am Fedn Clin Res; Am Soc Nephrol; Sigma Xi. *Res:* Renal metabolism; nutrition. *Mailing Add:* 8024 Rider Ave Baltimore MD 21204

SAPIRIE, S(AMUEL) R(ALPH), b Indianapolis, Ind, June 25, 09; m 36; c 1. CIVIL ENGINEERING. *Educ:* Purdue Univ, BS, 30, CE, 42. *Hon Degrees:* DEng, Purdue Univ, 58. *Prof Exp:* Field engr, State Dept Conserv, Ind, 30-33; construct engr camps, Emergency Conserv Work, 33-35; regional engr & chief land develop div, USDA, 35-42, engr in charge design projs, Soil Conserv Serv, 45-46; civil engr, Milwaukee Dist, US War Dept, 42-43, chief construct br, Northwest Div, 43-44, asst dir opers, Manhattan Dist, 46; dir prod & eng div, Oak Ridge Opers, Atomic Energy Comn, 47-49, dep mgr, 49-51, mgr, 51-72; eng consult, 72-78; RETIRED. *Concurrent Pos:* Sr consult, Bechtel Corp, 73-77. *Honors & Awards:* Nat Civil Serv League Career Serv Award, US Atomic Energy Comn, 55; Distinguished Serv Award, Nat Res Coun, 57; Outstanding Serv Award, Soc Advan Mgt, 72. *Mem:* Am Nuclear Soc; Am Soc Civil Engrs. *Res:* Design of large dams; management responsibility for design, construction, process improvement and development and operation of facilities for production of uranium-235; processing of source materials as feed for production of fissionable and special materials; management of research, development and analytical control laboratories. *Mailing Add:* 100 Ogden Circle Oak Ridge TN 37830

SAPIRSTEIN, JONATHAN ROBERT, b Los Angeles, Calif, Mar 1, 51; m 72; c 2. ATOMIC & MOLECULAR PHYSICS. *Educ:* Stanford Univ, Calif, BS, 73, PhD(physics), 79. *Prof Exp:* Researcher physics, Univ Calif, Los Angeles, 79-80, adj asst prof, 80-82; res assoc, Cornell Univ, 82-84; asst prof, 84-87, ASSOC PROF PHYSICS, UNIV NOTRE DAME, 88- *Res:* Higher order quantum electrodynamics calculations in one electron atoms; many body perturbation theory calculations in heavy atomic systems. *Mailing Add:* Dept Physics Univ Notre Dame Notre Dame IN 46556

SAPOLSKY, ASHER ISADORE, b Naroditch, Ukraine, USSR, Oct 2, 09; US citizen; m 40; c 3. BIOCHEMISTRY. *Educ:* Univ Pa, BS, 33; Philadelphia Col Pharm & Sci, BS, 35; Univ Miami Med Sch, PhD(biochem), 69. *Prof Exp:* Res asst prof, Univ Miami Sch Med, 69-79; VIS SCIENTIST VOL RES, MIRIAM HOSP, 80- *Mem:* Am Soc Biochem & Molecular Biol; Am Col Rheumatology; Orthop Res Soc. *Res:* Cartilage proteases which may be involved in the pathogenesis of osteoarthritis; discovered and isolated from human cartilage the neutral, metal-dependent proteoglycanase and now investigating possible inhibitors. *Mailing Add:* Miriam Hosp 164 Summit Ave Providence RI 02906

SAPONARA, ARTHUR G, b Newark, NJ, Nov 27, 36; m 59; c 2. BIOCHEMISTRY, MOLECULAR BIOLOGY. *Educ:* Rutgers Univ, AB, 58; Univ Wis, MS, 61, PhD(biochem), 64. *Prof Exp:* BIOCHEMIST, LOS ALAMOS NAT LAB, UNIV CALIF, 64- *Mem:* AAAS; Am Chem Soc. *Res:* Amino acid activation in protein biosynthesis; ribonucleic acid biosynthesis and modification in mammalian cells. *Mailing Add:* 328 Venado Los Alamos NM 87544

SAPOROSCHENKO, MYKOLA, b Ukraine, May 19, 24; nat US; m 60; c 2. PHYSICS. *Educ:* Ursinus Col, BS, 52; Wash Univ, AM, 54, PhD(physics), 58. *Prof Exp:* Asst prof physics, Univ Ark, 58-59 & Ill Inst Technol, 59-60; res assoc, Wash Univ, 60-62, asst prof, 62-65; from asst prof to assoc prof, 65-82, PROF PHYSICS & ASTRON, SOUTHERN ILL UNIV, CARBONDALE, 82- *Mem:* Am Phys Soc. *Res:* Gaseous electronics; ion-molecule reactions; mass spectrometry; mro06ssbaner spectroscopy. *Mailing Add:* Dept Physics Neckers C0402 Southern Ill Univ Carbondale IL 62901

SAPORTA, SAMUEL, b Athens, Greece, Mar 30, 46; m 70; c 2. NEUROANATOMY, NEUROBIOLOGY. *Educ:* Univ Calif, Davis, BA, 67; Univ Southern Calif, PhD(physiol psychol), 73. *Prof Exp:* Instr anat, Univ Calif, Los Angeles, 76, res asst, 76-77; asst prof, 77-83, ASSOC PROF ANAT, UNIV SFLA, 83- *Concurrent Pos:* NIMH fel, 70-73, NIH fel, 74-76. *Mem:* Am Asn Anatomists; Soc Neurosci; Sigma Xi; AAAS; Int Asn Study Pain. *Res:* Neuroanatomical and physiological organization of the somatosensory system; functional interrelationships of neurons which produce coding and decoding of information. *Mailing Add:* Dept Anat Col Med Univ SFla Med Ctr Tampa FL 33612

SAPP, RICHARD CASSELL, b Kokomo, Ind, Sept 8, 28; m 57; c 2. LOW TEMPERATURE PHYSICS, MAGNETISM. *Educ:* Wilmington Col, BSc, 49; Ohio State Univ, PhD(physics), 55. *Prof Exp:* Res assoc physics, Ohio State Univ, 55; Welch Found fel, Rice Univ, 55-57; from asst prof to assoc prof, 57-67, PROF PHYSICS, UNIV KANS, 67- *Concurrent Pos:* Sloan Found fel, 62-64. *Mem:* Fel Am Phys Soc; Am Asn Physics Teachers. *Res:* Low temperature physics; magnetic relaxation; spin glasses; superconductivity. *Mailing Add:* Dept Physics Univ Kans Lawrence KS 66045

SAPP, WALTER J, b Shreveport, La, Feb 16, 34; m 62; c 3. CELL BIOLOGY. *Educ:* Wiley Col, BS, 61; Univ Wis-Madison, MS, 64, PhD(zool), 66. *Prof Exp:* From asst prof to assoc prof, Tuskegee Inst, 66-76, head dept, 68-76, res assoc, 73-76, prof biol & dean students, 76-88, ASST DEAN SPONSORED PROGS, OFF SPONSORED PROGS, VET MED, TUSKEGEE UNIV, 88- *Mem:* AAAS; Am Soc Cell Biol. *Res:* Ultrastructure of genetic systems; ultrastructure and cytogenetics of tumor systems. *Mailing Add:* Tuskegee Inst Tuskegee AL 36088

SAPPENFIELD, ROBERT W, b Bedford, Ind, Oct 2, 24; m 48; c 3. MEDICINE. *Educ:* Ind Univ, MD, 47. *Prof Exp:* Intern, Med Ctr, Ind Univ, 47-48 & 49, resident pediat, 49-51; resident, Chicago Contagious Dis Hosp, 48; resident, La Rabida Sanitarium Rheumatic Fever, 48-49; epidemiologist, State Bd Health, La, 51-52; from instr to assoc prof pub health & pediat, 53-63, prof prev med, pub health & pediat & head dept pub health & prev med, 63-72, assoc dean, 72-79, PROF PREV MED & PUB HEALTH, SCH MED, LA STATE UNIV MED CTR, NEW ORLEANS, 79- *Concurrent Pos:* Res fel, Children's Hosp, 52-53; proj dir, Collab Child Develop Prog, Charity Hosp New Orleans, 60-62. *Mem:* Am Col Prev Med; Asn Teachers Prev Med; Am Acad Pediat. *Res:* Preventive medicine; pediatrics; epidemiology, especially communicable disease and perinatal problems. *Mailing Add:* 1901 Perdido St New Orleans LA 70112

SAPPENFIELD, WILLIAM PAUL, b Lee's Summit, Mo, Apr 10, 23; m 56; c 4. AGRONOMY, PLANT BREEDING. *Educ:* Univ Mo, BS, 48, PhD(plant breeding), 52. *Prof Exp:* Instr agron, Univ Mo, 48-51; agronomist, NMex State, 51-54 & Univ Calif, Davis, USDA, 54-56; from assoc prof to prof agron, Univ Mo, Columbia, 56-88; RETIRED. *Concurrent Pos:* Mem, Nat Cotton Testing Comt, Crop Variety Regist Comt & Nat Comt Res Task Force Comt; chmn, Miss Delta Cotton Variety Testing Comt; rep comt cotton quality, State Agr Exp Sta, USDA; res grants, Cotton Inc. *Mem:* Am Soc Agron; Crop Sci Soc Am. *Res:* Cotton breeding, host plant resistance, and fiber technology; production systems. *Mailing Add:* 607 Holly Hill Dr Sikeston MO 63801

SAPRA, VAL T, b Beawar, India, Nov 15, 42; m 73. PLANT BREEDING, PLANT CYTOGENETICS. *Educ:* Kans State Univ, PhD(plant breeding, genetics), 72. *Prof Exp:* Res asst plant breeding, Govt of Rajastham, India, 65-68; res asst, Kans State Univ, 68-72; fel, 72-73, assoc prof plant breeding & plant cytogenetics, 73-80, PROF AGRON, ALA A&M UNIV, 80- *Concurrent Pos:* Consult, Somdiaa, Paris, France; agronomist, WCent Africa. *Mem:* Crop Sci Soc Am; Am Soc Agron; Can Soc Genetics; Can Soc Agron. *Res:* Triticale breeding and cytogenetics; development of new triticale strains through conventional and mutation breeding procedures. *Mailing Add:* Dept Plant Sci Ala A&M Univ PO Box 285 Normal AL 35762

SAPRU, HREDAY N, CIRCULATION, RESPIRATION. *Educ:* Columbia Univ, PhD(neuro-pharmacol), 74. *Prof Exp:* PROF PHARMACOL, UNIV MED & DENT, NJ MED SCH, NEWARK, 74- *Mailing Add:* Sect Neurosurg & Dept Pharmacol MSB H592 NJ Med Sch 185 S Orange Ave Newark NJ 07103

SAR, MADHABANANDA, b Palchakada, India, Dec 31, 33; m 56; c 3. VETERINARY MEDICINE, PHYSIOLOGY. *Educ:* Bihar Univ, BVSc & AH, 56; Mich State Univ, MS, 63, PhD(physiol), 68. *Prof Exp:* Vet asst surg, Dept Vet Serv & Animal Husb, India, 56-59; instr parasitol, Orissa Col Vet Sci & Animal Husb, 59-61; res assoc neuroendocrinol, Univ Chicago, 68-69, res assoc, 69-70, instr & res assoc pharmacol, 69; res assoc neuroendocrinol, Labs Reprod Biol, 69-77, res asst prof, 77-78, RES ASSOC PROF ANAT, SCH MED, UNIV NC, CHAPEL HILL, 78- *Mem:* AAAS; Endocrine Soc; Soc Study Reprod; Int Brain Res Orgn; Am Physiol Soc. *Res:* Neuroendocrinology; endocrinology; reproductive physiology; hormone localization in brain and peripheral target tissues by autoradiography and immunohistochemistry. *Mailing Add:* Dept Anat Div Health Affairs Univ NC 108 Taylor Hall Chapel Hill NC 27599

SARA, RAYMOND VINCENT, b Carbondale, Pa, Jan 24, 27; m 52; c 4. MATERIALS SCIENCE. *Educ:* Pa State Univ, BS, 50, MS, 52. *Prof Exp:* CORP FEL, UCAR CARBON CO, 52- *Mem:* Am Ceramic Soc. *Res:* Phase equilibria and thermal behavior of refractory materials, mechanical and thermal properties of metal/non-metal fiber and ceramic composite systems, oxidation and diffusion phenomenon, oxidation protective coatings, graphitization and intercalation; abradable seals; high performance ceramics. *Mailing Add:* Parma Tech Ctr Ucar Carbon Co Box 6116 Cleveland OH 44101

SARACENO, ANTHONY JOSEPH, b Reggio, Italy, June 20, 33; nat US; m 70; c 1. VENT STACK SAMPLING TECHNOLOGY, PROCESS CHEMISTRY. *Educ:* St Vincent Col, BS, 55; Univ Notre Dame, PhD(chem), 58. *Prof Exp:* Res assoc, Univ Notre Dame, 56-58; res chemist, Gulf Res & Develop Co, 58-61; sr chemist, Pennwalt Corp, 61-64, proj leader, 65-67; group leader, Goodyear Atomic Corp, 67-73, sect head, 73-82, supvr, Chem Dept, 82-84; supvr, Chem & Mat Tech Dept, 84-89, supvr, Chem Technol Dept, 89-90, HEAD, CHEM TECHNOL DEPT, MARTIN MARIETTA ENERGY SYSTS, INC, 90- *Mem:* AAAS; Am Chem Soc. *Res:* Coordination compounds; solid state inorganic chemistry; inorganic polymers; infrared spectroscopy; organometallic compounds; lubricants; coatings; process development; air-water pollution control; halogen chemistry; metal corrosion. *Mailing Add:* Martin Marietta Energy Systs Inc PO Box 628 Piketon OH 45661

SARACHEK, ALVIN, b Pittsburgh, Pa, July 29, 27; m 56, 76. MICROBIOLOGY. *Educ:* Univ Mo, Kansas City, BA, 48, MA, 50; Kans State Univ, PhD(microbiol genetics), 58. *Prof Exp:* Instr biol, Univ Mo, Kansas City, 50-51; res assoc biol res lab, Univ Southern Ill, 51-54; fel microbial biochem, Inst Microbiol, Rutgers Univ, 57-58; from asst prof to prof, 58-72, chmn dept biol, 61-74, DISTINGUISHED PROF NATURAL SCI, WICHITA STATE UNIV, 72- *Concurrent Pos:* Microbial geneticist, US AEC, 65-66, mem adv comt prog in food irradiation, 67-72; chmn panel student oriented prog, NSF, 71-75; chmn grants & awards comt, Am Cancer Soc, Kans Div, 70-76, mem exec comt, 70-78; prof assoc, Sci Educ Directorate, NSF, 77-78; mem, Dept Energy adv coun life sci prog, Argonne Nat Lab, 78-80 & NSF adv panel, 81-; NATO sci fel, 81-; consult, Div Biol Energy Res, US Dept Energy, 80-; chair, External Peer Oversight Comt; NATO postdoctoral fel prog, NSF, 87-90; vis scientist to minority insts, Am Soc Microbiol. *Mem:* Am Cancer Soc; Am Soc Microbiol; Genetics Soc Am; Int Soc Human & Animal Mycol; fel Am Acad Microbiol. *Res:* Genetics and physiology of microorganisms; radiobiology; radiation genetics and chemical mutagenesis in fungi. *Mailing Add:* Dept Biol Sci Wichita State Univ Wichita KS 67208

SARACHIK, EDWARD S, b New York, NY, Apr 22, 41; div. DYNAMIC METEOROLOGY, OCEANOGRAPHY. *Educ:* Queens Col, NY, BS, 60; Brandeis Univ, MS, 63, PhD(physics), 66. *Prof Exp:* Res assoc Linear Acceleration Ctr, Stanford Univ, 65-67; staff physicist, Electronics Res Ctr, NASA, 67-70; staff mathematician Transport Systs Ctr, Dept Transport, 70-71; NSF fel & res assoc meteorol, Mass Inst Technol, 71-73; res fel & lectr atmospheric physics, Harvard Univ, 73-78, res assoc & proj mgr, Ctr Earth & Planetary Physics, 78-83, sr res fel, Dynamical Meteorol & Oceanog, Harvard Univ, 83-85; oceanog, NOAA/PAC Marine Environ Lab, Seattle, 84-87; RES PROF, DEPT ATMOSPHERIC SCI, SCH OCEANOG, UNIV WASH, 88- *Concurrent Pos:* Mem, Climate Res Comt, Nat Acad Sci/Nat Res Coun, 85-88 & Panel Model Assimilated Data Sets, 89-90; assoc ed, J Atmospheric Sci, 81-90, J Phys Oceanog, 84- *Mem:* Am Phys Soc; Am Meteorol Soc; Sigma Xi. *Mailing Add:* Dept Atmospheric Sci AK-40 Univ Wash Seattle WA 98195

SARACHIK, MYRIAM PAULA, b Antwerp, Belgium, Aug 8, 33; US citizen; m 54; c 1. SOLID STATE PHYSICS. *Educ:* Barnard Col, Columbia Univ, BA, 54, Columbia Univ, MS, 57, PhD(physics), 60. *Prof Exp:* From res asst to res assoc exp solid state physics, IBM Watson Lab, Columbia Univ, 55-61; mem tech staff, Bell Tel Labs, NJ, 62-64; from asst prof to assoc prof, 65-71, PROF PHYSICS, CITY COL NEW YORK, 71- *Concurrent Pos:* Exec officer, PhD prog Physics, City Univ New York, 75-78. *Mem:* Fel Am Phys Soc; NY Acad Sci. *Res:* Properties of superconducting materials; magnetic and transport properties of alloys; disordered systems; semiconductors; the metal insulator transition. *Mailing Add:* Dept Physics City Col New York Convent Ave at 138th St New York NY 10031

SARACHIK, PHILIP E(UGENE), b New York, NY, Dec 3, 31; m 64; c 1. CONTROL SYSTEMS, COMMUNICATION NETWORKS. *Educ:* Columbia Univ, AB, 53, BS, 54, MS, 55, PhD(elec eng), 58. *Prof Exp:* Staff engr, Int Bus Mach Res Lab, NY, 58-60; from asst prof to assoc prof elec eng, Columbia Univ, 60-64; assoc prof, NY Univ, 64-67, prof elec eng, 67-73; PROF ELEC ENG, POLYTECH UNIV, 73- *Concurrent Pos:* Consult, Aerospace Group, Gen Precision, Inc, 63-68; vis prof, control, commun & comput systs dept, Tel Aviv Univ, 79-80. *Mem:* fel Inst Elec & Electronics Engrs; Soc Indust & Appl Math. *Res:* Applications of computers to real time control systems; problems in optimal and adaptive control systems; management and control of communication networks. *Mailing Add:* Dept Elec Eng Polytech Univ 333 Jay St Brooklyn NY 11201

SARACINO, DANIEL HARRISON, b Ellenville, NY, Mar 25, 47. EXISTENTIAL COMPLETENESS, HOMOGENEOUS STRUCTURES. *Educ:* Cornell Univ, AB, 68; Princeton Univ, MA, 70, PhD(math), 72. *Prof Exp:* Gibbs instr math, Yale Univ, 72-74; from asst prof to assoc prof, 74-84, PROF MATH, COLGATE UNIV, 84-, CHMN, 86- *Concurrent Pos:* Guest researcher, Math Inst, Heidelberg, Ger, 75; prin investr, NSF Res Grant, 76-78; vis asst prof, Wesleyan Univ, 77-78, vis assoc prof, 80-81, van Vleck vis assoc prof, 82-83, van Vleck vis prof, Wesleyan Univ, 90-; vis Inst Advan Study, Princeton, 86. *Mem:* Asn Symbolic Logic. *Res:* Model-theoretic algebra; existential completeness and homogeneity in various algebraic contexts. *Mailing Add:* Dept Math Colgate Univ Hamilton NY 13346-1398

SARADA, THYAGARAJA, b Madras, India, Apr 19, 29; US citizen. SPECTROSCOPY, POLYMERS. *Educ:* Annamalai Univ, Madras, BSc, 51; Am Univ, MS, 70, PhD(phys chem), 72. *Prof Exp:* Lectr & head, Dept Chem, St Mary's Col, Madras, 51-52, SPW Col, Tirupati, India, 52-63; prof, Am Univ, Wash, 63-67, head, Dept Chem, 52-67, fel phys chem, 73-74, res asst, 75-78, asst prof, 78-79; sr chemist, Celanese Res Co, 79-82; sr chemist, 82-84, MGR CHEM RES, PITNEY BOWES, 85- *Concurrent Pos:* Dreyfus & Camille Found fel, Am Univ, 73-74; res asst, ERDA, 75-76 & US Defense, 76-78; sci pool officer, CLRI, Madras, 74-75. *Mem:* Am Chem Soc; Tech Asn Paper Pulp Indust; Am Soc Test & Mat. *Res:* Optical properties of liquid crystals; physico-chemical, electrical properties and characterization of fuel cell electrolytes and electrodes; electrochemical corrosion, complex ion theory; membranes; fuel cells, batteries, lithium batteries; ink-paper interactions; adhesives; ink jet technology; piezoceramic materials; inks; flourescence; dyes. *Mailing Add:* Pitney Bowes 276 Main Ave (Location 19-00) Norwalk CT 06851

SARAKWASH, MICHAEL, b South River, NJ, Feb 20, 25; m 72. STATISTICS, RELIABILITY. *Educ:* Columbia Univ, BS, 50; Stevens Inst Technol, MS, 58, PE, 78. *Prof Exp:* Mathematician, Evans Signal Lab, 50-53; sr qual control engr propeller div, Curtiss-Wright Corp, 53-58; sr mathematician reaction motors div, Thiokol Chem Corp, Denville, 58-63, statist mathematician, 63-67; consult statist & exp design, MS Assocs, 67-74; sr reliability engr, Res-Cottrell, NJ, 74-77; consult statist, reliability, qual control & exp design, M S Assocs, 77-80; dir, statist & anal div, Mil Sealift Command, Motby, Bayonne, NJ, 80-87; CONSULT PROBABILITY THEORY, STATIST, & ASSOC FIELDS, MS ASSOCS, 87- *Mem:* Am Statist Asn; Am Soc Qual Control; NY Acad Sci. *Res:* Industrial quality control; marketing and market research; specialist in graphic statistics; reliability; industrial application of probability theory and statistics. *Mailing Add:* 377 Colfax Ave Clifton NJ 07013-1703

SARAN, CHITARANJAN, b Lucknow, India, Sept 22, 39; US citizen; m 69; c 1. SAFETY CONTROL, ERGONOMICS. *Educ:* Indian Inst Technol, BTech, 62, MTech, 63; NC State Univ, PhD(bioeng, agr eng), 67. *Prof Exp:* Asst, NC State Univ, 63-67; asst agr engr, Univ PR, 67-68, from instr to asst prof math, 68-71; USPHS Trainee, Med Ctr & Ctr Safety, NY Univ, 71-73, asst prof safety, Ctr Safety, 73-75; coordr, St Louis Degree Prog, 75-78 & 88-90, assoc prof, 75-80, PROF INDUST SAFETY & HYG, CENT MO STATE UNIV, WARRENSBURG, 80- *Concurrent Pos:* Vis prof/scholar, Sweden, 86 & 90, India, 86-87 & Raleigh, 89. *Mem:* Am Soc Agr Engrs; Inst Eng, India; Am Soc Safety Engrs; Human Factors Soc; Sigma Xi; Safety, Health & Environ Protection Int. *Res:* Ergonomics. *Mailing Add:* RFD 3 Box 291 Warrensburg MO 64093-9307

SARANTAKIS, DIMITRIOS, b Nafplion, Greece, May 20, 36; m 65; c 2. MEDICINAL CHEMISTRY. *Educ:* Nat Univ Athens, BSc, 59; Imp Col, dipl, 65, Univ London, PhD(org chem), 65. *Prof Exp:* Res fel org chem, Univ Leicester, 65-66; res assoc, Royal Res Estab, Greece, 66-67; res assoc, Univ Wash, 67-71; chief chemist, Fox Chem Co, 71; GROUP LEADER ORG CHEM, WYETH LABS INC, PHILADELPHIA, PA, 71- *Mem:* Am Chem Soc. *Res:* Synthesis of biologically active polypeptides; synthesis of biologically active organic compounds. *Mailing Add:* 262 Sentinel Ave Newtown PA 18940-1166

SARANTITES, DEMETRIOS GEORGE, b Athens, Greece, May 5, 33; m 65; c 1. NUCLEAR CHEMISTRY, NUCLEAR PHYSICS. *Educ:* Mass Inst Technol, PhD (nuclear and inorganic chem), 63. *Prof Exp:* Radiochemist, Cyclotron Lab, Mass Inst Technol, 60-63; res assoc nuclear chem, Washington Univ, 63-65, from asst prof to assoc prof, 65-74; vis prof, Res Inst Physics, Stockholm, 74-75; assoc prof, 75-76, PROF NUCLEAR CHEM, WASH UNIV, 76- *Mem:* Am Phys Soc; Sigma Xi. *Res:* Investigations of nuclear reaction mechanisms and of nuclear structure with emphasis on the electromagnetic and nuclear properties of the high spin states. *Mailing Add:* Dept Chem Wash Univ One Brookings Dr St Louis MO 63130

SARASON, DONALD ERIK, b Detroit, Mich, Jan 26, 33. MATHEMATICS. *Educ:* Univ Mich, BS, 55, AM, 57, PhD(math), 63. *Prof Exp:* Mem math, Inst Adv Study, 63-64; from asst prof to assoc prof, 64-70, PROF MATH, UNIV CALIF, BERKELEY, 70- *Concurrent Pos:* NSF fel, 63-64; Sloan fel, 69-71. *Mem:* Am Math Soc; Math Asn Am. *Res:* Functional analysis. *Mailing Add:* Dept Math Univ Calif 2120 Oxford St Berkeley CA 94720

SARASON, LEONARD, b Brooklyn, NY, May 29, 25; m 62. MATHEMATICS. *Educ:* Yale Univ, BS, 45, BMus, 48, MusM, 49; NY Univ, PhD(math), 61. *Prof Exp:* Res asst math, Courant Inst Math Sci, NY Univ, 60-63; actg asst prof, Stanford Univ, 63-65; from asst prof to assoc prof, 65-74, PROF MATH, UNIV WASH, 74- *Mem:* Am Math Soc. *Res:* Partial differential equations. *Mailing Add:* Dept Math GN-50 Univ Wash Seattle WA 98195

SARAVANAMUTTOO, HERBERT IAN H, b Monkton, Scotland, June 20, 33; Can citizen; m 60; c 3. MECHANICAL ENGINEERING. *Educ:* Glasgow Univ, BSc, 55; Bristol Univ, PhD(mech), 68. *Prof Exp:* Engr, Orenda Engines, 55-59; analyst, KCS Ltd, Toronto, 59; engr, Orenda Engines, 59-64; lectr mech eng, Bristol Univ, 64-70; assoc prof aerothermodyn, 70-75, prof mech & aeronaut eng & chmn dept, Carleton Univ, 75-80, 82-88; DIR, GAS TOPS LTD, OTTAWA, 80- *Concurrent Pos:* Consult, Rolls Royce Ltd, 66-70, Brit Aircraft Corp, 75 & Avionics Div, Smith's Indust, 75-76; vis res fel, Royal Naval Eng Col, 80-81. *Mem:* Fel Can Aeronaut & Space Inst (vpres, 78-79, pres 79-80); fel Brit Inst Mech Engrs; fel Am Soc Mech Engrs. *Res:* Dynamic response of gas turbines; improvement of part load performance; engine health monitoring. *Mailing Add:* Dept Mech & Aerospace Eng Carleton Univ Ottawa ON K1S 5B6 Can

SARAVIA, NANCY G, IMMUNOPARASITOLOGY. *Educ:* Vanderbilt Univ, PhD(microbiol & immunol), 78. *Prof Exp:* SCIENTIST & GEN COORDR, ICIDR, TULANE UNIV, 85- *Mailing Add:* Ctr Int Invest Med Apdo Aéreo 5390 Cali Colombia

SARAVIS, CALVIN, b Englewood, NJ, Feb 27, 30; m 54; c 4. IMMUNOLOGY. *Educ:* Syracuse Univ, AB, 51; WVa Univ, MS, 55; Rutgers Univ, PhD(zool), 58. *Prof Exp:* Asst physiol, serol & immunol, Rutgers Univ, 55-58; head antiserum prod & develop, Blood Grouping Lab, Mass, 58-59; dir immunochem lab, Blood Res Inst, Inc, 59-72; prin assoc surg & mem fac med, Harvard Med Sch, 71-84; res assoc, 74-83, ASSOC RES PROF PATH, SCH MED, BOSTON UNIV, 83-; ASSOC PROF SURG-BIOCHEM, HARVARD MED SCH, 84- *Concurrent Pos:* Assoc med, Peter Bent Brigham Hosp, 61-69; res assoc, Harvard Med Sch, 62-71; chief, Immunol Div, Harvard Surg Unit, Boston City Hosp, 66-68; sr res assoc & asst dir, Gastrointestinal Res Lab, Mallory Inst Path Found, 74-; mem spec sci staff, Boston City Hosp & Mallory Inst Path, 78-; sr res assoc, Cancer Res Inst, New Eng Deaconess Hosp, Boston, 79- *Mem:* Transplantation Soc; Am Asn Immunologists. *Res:* Transplantation immunology; cancer immunology; isolation and identification of antibodies and antigens; hepatitis; detection, isolation and characterization of human cancer markers. *Mailing Add:* Dept Surg Biochem Harvard Med Sch Lab Cancer Biol 50 Binney St Boston MA 02215

SARAZIN, CRAIG L, b Milwaukee, Wis, Aug 11, 50; m 71; c 2. THEORETICAL ASTROPHYSICS. *Educ:* Calif Inst Technol, BS, 72; Princeton Univ, MA, 74, PhD(physics), 75. *Prof Exp:* Millikan fel physics, Calif Inst Technol, 75; mem physics, Inst Advan Study, 75-77; from asst prof to assoc prof, 77-86, PROF ASTRON, UNIV VA, 86- *Concurrent Pos:* Vis fel, Inst Astron, Cambridge Univ, 76 & 87; vis scientist, Nat Radio Astron Observ, 77-81; vis asst prof, astron dept, Univ Calif, Berkeley, 79; vis mem, Inst Advan Study, 81-82; Joint Inst Lab Astrophys vis fel, Univ Colo & Nat Bur Standards, 85-86. *Honors & Awards:* Haren Fisher Physics Prize, 71. *Mem:* Am Astron Soc; Int Astron Union. *Res:* Interstellar medium; clusters of galaxies; x-ray emission; extragalactic astronomy. *Mailing Add:* Dept Astron Univ Va PO Box 3818 Charlottesville VA 22903-0818

SARBACH, DONALD VICTOR, b Lincoln, Nebr, Nov 3, 11; m 34; c 3. PRODUCT ENGINEERING, ELASTOMERS MARKETING. *Educ:* Univ Nebr, BSc, 34. *Prof Exp:* From mem tech staff to mgr new prod develop, B F Goodrich Co, Ohio, 37-55; dir res, Hewitt-Robins, Inc, Conn, 55-58; dir rubber tech, Goodrich-Gulf Chem, Inc, 58-69; prod mgr, BF Goodrich Chem Co, 69-73, elastomer specialist, 73-77; consult ed, Rubber & Plastics News, 77-81; RETIRED. *Mem:* AAAS; Am Chem Soc; fel Am Inst Chem. *Res:* Rubber; synthetic rubber and plastics technology; product development and applications; author of 15 published articles in the field; 31 US patents and 8 foreign patents in the field. *Mailing Add:* 242 River Rd Hinckley OH 44233-9628

SARBER, RAYMOND WILLIAM, b Hammond, Ind, Apr 15, 16; m 37; c 2. MEDICAL BACTERIOLOGY. *Educ:* Western Mich Univ, AB, 38; Univ Cincinnati, MS, 41. *Prof Exp:* Asst bact, Univ Mich, 40; asst prof col pharm, Univ Cincinnati, 40-42; sr res bacteriologist, Parke, Davis & Co, 42-59; exec secy, Am Acad Microbiol, 68-78; exec secy, 59-82, EMER EXEC SECY, AM SOC MICROBIOL, 82-, EXEC SECY, NAT PEDIAT INFECTIOUS DIS SEM, 81- *Concurrent Pos:* Consult, Clopay Corp, 41-43; ed, Am Soc Microbiol News, 59-72; registr, Nat Registry Microbiol, 68-74; secy, Am Bd Med Microbiol, 68-74; secy gen, XIII Int Cong Microbiol, 81. *Mem:* Am Soc Microbiol; Sigma Xi; AAAS. *Res:* Tuberculosis antigens; tuberculins; pertussis, biological and chemical tuberculosis antigens; germicide testing methods; tissue culture. *Mailing Add:* 2212 Great Falls St Falls Church VA 22046

SARCHET, BERNARD REGINALD, b Byesville, Ohio, June 13, 17; m 41; c 3. ENGINEERING MANAGEMENT. *Educ:* Ohio State Univ, BChE, 39; Univ Del, MChE, 41. *Prof Exp:* Asst, Univ Del, 39-41; operator, Eng & Construct Div, Koppers Co, Inc, Pa, 41-42, gen foreman, Butadiene Div, 42-45, supv chem engr, Res Div, 45-46, supvr, Eng & Construct Div, 46-47, mgr, Oil City Plant, 47-50, mgr, Kobuta Plant, 50-53, asst mgr sales, Chem Div, 54-56, mgr develop dept, 56-58, mgr prod develop, Plastics Div, 58-61, mgr panel dept, 61-64, dir com develop, 64-67; chmn dept eng mgt, 67-81, exec dir, External Affairs, 75-79, prof, 67-87, EMER PROF ENG MGT, UNIV MO, ROLLA, 87- *Mem:* Am Chem Soc; Am Soc Eng Educ; Am Inst Chem Engrs; fel Am Soc Eng Mgt. *Res:* Absorption distillation; robotics and impact on management; housing construction panels. *Mailing Add:* Eng Mgt Bldg Univ Mo Rolla MO 65401

SARCIONE, EDWARD JAMES, b Lawrence, Mass, Dec 3, 25; m 53; c 3. BIOCHEMISTRY. *Educ:* St Michael's Col, BS, 48; Univ Kans, PhD(biochem), 57. *Prof Exp:* Biochemist, Dept Pub Health, Mass, 49-50; asst instr biochem, Sch Med, Univ Kans, 50-55; assoc cancer res scientist, Roswell Park Mem Inst, 55-72, prin cancer res scientist & prof & chmn physiol prog, 72-91; RETIRED. *Mem:* AAAS; Am Soc Biol Chemists; Am Chem Soc; Am Asn Cancer Res; Soc Exp Biol & Med. *Res:* Biosynthesis of glycoproteins and fetal proteins; molecular diseases. *Mailing Add:* Physiol/Gratwick Basic Sci Roswell Park Mem Inst 666 Elm St Buffalo NY 14263

SARD, RICHARD, b Brooklyn, NY, Apr 19, 41; m 63; c 2. SURFACE CHEMISTRY, SCIENCE MANAGEMENT. *Educ:* Stevens Inst Technol, BE, 62, MS, 63, PhD(phys metall), 68. *Prof Exp:* Electron microscopist, Cent Res Labs, Air Reduction Corp, 63-64; mem tech staff, Bell Tel Labs, 68-73, supvr plated film properties & interface studies, 73-78, supvr mat develop & comput applns, 78-79; dir technol, Plating Systs Div, Occidental Petrol Corp, 79-83; vpres technol, OMI Int Corp, 83-88; EXEC VPRES INT, ETHONE-OMI INC, 89- *Mem:* AAAS; Electrochem Soc; Am Electroplaters Soc; Sigma Xi. *Res:* Structure and properties of electrodeposits and other coatings in relation to process conditions; electrodeposition; surface characterization; physical properties; device phenomena; process control sensors; microprocessor applications. *Mailing Add:* 1180 Willow Lane Birmingham MI 48009

SARD, ROBERT DANIEL, b New York, NY, Aug 23, 15; wid; c 3. EXPERIMENTAL HIGH ENERGY PHYSICS. *Educ:* Harvard Univ, SB, 35, AM, 40, PhD(physics), 42. *Prof Exp:* Asst, Kamerlingh Onnes Lab, Leiden, Neth, 36-38; instr physics, Harvard Univ, 38-42, res assoc radio res lab, 42-45; res assoc, Mass Inst Technol, 45-46; asst prof physics, Wash Univ, St Louis, 46-48, from assoc prof to prof, 48-61; prof, 61-85, EMER PROF PHYSICS, UNIV ILL, URBANA, 85- *Concurrent Pos:* Fulbright adv res scholar, Univ Manchester, 51-52 & Lawrence Radiation Lab, Univ Calif, 59-60; consult particle accelerator div, Argonne Nat Lab, 58-60; vis scientist, Europ Orgn Nuclear Res, Serpukhov, USSR & Geneva, Switz, 70-71. *Mem:* Fel Am Phys Soc. *Res:* Elementary particles; cosmic rays; experimental particle physics. *Mailing Add:* Dept Physics Univ Ill 1110 W Green St Urbana IL 61801

SARDELLA, DENNIS JOSEPH, b Lawrence, Mass, July 3, 41; m 66; c 4. ANALYTICAL CHEMISTRY, PHYSICAL CHEMISTRY. *Educ:* Boston Col, BS, 62; Ill Inst Technol, PhD(phys org chem), 67. *Prof Exp:* Nat Res Coun Can fel, Univ Western Ont, 66-67; from asst prof to assoc prof, 67-81, PROF CHEM, BOSTON COL, 81-, DIR, PRESIDENTIAL SCHOLARS PROG, 90- *Concurrent Pos:* Vis lectr biol chem, Sch Med, Harvard Univ, 73-75. *Mem:* Am Chem Soc. *Res:* Nuclear magnetic resonance spectroscopy; structural chemistry; theoretical organic chemistry. *Mailing Add:* Dept Chem Boston Col Chestnut Hill MA 02167

SARDESAI, VISHWANATH M, b Goa, India, Nov 17, 32; m 66. BIOCHEMISTRY, CLINICAL CHEMISTRY. *Educ:* Univ Bombay, BS, 54, MS, 57; Wayne State Univ, PhD(physiol chem), 62. *Prof Exp:* Instr chem, Univ Bombay, 54-57; res chemist, Zandu Pharmaceut Works Ltd, India, 57-59; teaching asst biochem, Wayne State Univ, 59-60; instr, Sch Med, Tulane Univ, 62-63; asst prof, 63-69, ASSOC PROF BIOCHEM, SCH MED, WAYNE STATE UNIV, 69- *Mem:* AAAS; Am Physiol Soc; Am Inst Nutrit; Am Chem Soc; NY Acad Sci; Sigma Xi; Am Asn Clin Chem; Shock Soc. *Res:* Porphyrin biosynthesis and metabolism; oxidative phosphorylation; alcohol metabolism; metabolism in shock; clinical methods; tryptophan metabolism. *Mailing Add:* Surg Sch Med Wayne State Univ Detroit MI 48201

SARDINAS, AUGUST A, b Bronx, NY, June 19, 22; m 44; c 3. MATHEMATICAL ANALYSIS. *Educ:* Brooklyn Col, BA, 43; Harvard Univ, MA, 47; Univ Pa, PhD(math), 62. *Prof Exp:* Res asst info theory, Univ Pa, 49-50; staff engr, Burroughs Corp, 50-63; ASSOC PROF MATH, VILLANOVA UNIV, 63- *Concurrent Pos:* Logical design consult, Burroughs Corp, 65- *Mem:* Math Asn Am. *Res:* Information theory; logical design; analysis. *Mailing Add:* 55 Eastwood Rd Berwyn PA 19312

SARDINAS, JOSEPH LOUIS, b Havana, Cuba, Aug 1, 19; US citizen; m 42; c 2. MICROBIOLOGY, BIOCHEMISTRY. *Educ:* Brooklyn Col, AB, 48, MA, 52; St John's Univ, NY, PhD(microbiol), 61. *Prof Exp:* Res asst, Merck Sharp & Dohme, NY, 48; microbiologist, Pfizer Inc, 48-83; RETIRED. *Mem:* Am Soc Microbiol; Am Chem Soc; Sigma Xi. *Res:* Industrial fermentations; microbiological transformations; isolation and purification of fermentation products. *Mailing Add:* Four Hillside Dr Bloomfield CT 06002

SARDISCO, JOHN BAPTIST, b Shreveport, La, July 27, 34; m 59; c 1. PHYSICAL CHEMISTRY. *Educ:* Spring Hill Col, BS, 56; La State Univ, MS, 58, PhD(phys chem), 60. *Prof Exp:* Sect supvr phys chem sect, Res Eng & Develop Dept, Pennzoil United, Inc, 61-75, DIV MGR INORGANIC RES, RES & DEVELOP DEPT, PENNZOIL CO, 75-, MGR ANALYSIS SECT, TECHNOL DIV, 85- *Concurrent Pos:* Consult, Anal Serv, Inc, 79- *Mem:* Am Chem Soc; Sigma Xi; Am Inst Chemists; Smithsonian Inst. *Res:* Corrosion research and control; inorganic process development; hydrometallurgical refining of metals; instrumental inorganic analysis; water pollution control and water purification; environmental analysis; analytical chemistry. *Mailing Add:* Pennzoil Prods Co PO Box 7569 The Woodlands TX 77387-7569

SAREM, AMIR M SAM, b Teheran, Iran, Sept 5, 30; US citizen; m 57; c 3. CHEMICAL & PETROLEUM ENGINEERING. *Educ:* Univ Tulsa, BSPE, 54, MSPE, 56; Univ Okla, PhD(eng sci), 64. *Prof Exp:* Jr engr, Sinclair Res Inc, Okla, 54-55, intermediate res engr, 55-59, res engr, 59-61; teaching asst petrol eng, Univ Okla, 61-64; res engr, 64-66, sr res engr, 66-77, res assoc, Union Res Ctr, Union Oil Co Calif, 77-83; SR RES ASSOC, 86-; SUPV RESERVOIR ENGR, 86- *Concurrent Pos:* Lectr, Sinclair Res Lab, Inc, 59-61 & 64; lectr exten sch, Univ Calif, Los Angeles, 65- *Mem:* Am Inst Mining, Metall & Petrol Engrs; Am Chem Soc; Am Inst Chem Engrs; Soc Rheol. *Res:* Petroleum reservoir engineering; P-V-T properties of hydrocarbon systems; fluid flow mechanics in porous media; viscous and surfactant water flood of oil fields; rheological properties of polymer solutions; pipeline flow drag reduction. *Mailing Add:* 18741 La Casita Ave Yorba Linda CA 92686

SARETT, HERBERT PAUL, b Brooklyn, NY, Feb 5, 16; m 48; c 3. NUTRITION, MEDICAL FOODS. *Educ:* Brooklyn Col, BA, 36; Cornell Univ, MS, 37; Duke Univ, PhD(biochem), 42; Am Bd Nutrit, dipl, 52. *Prof Exp:* Instr biochem, Sch Med, Duke Univ, 42-43; asst prof & res assoc, Ore State Col, 43-45; res assoc med div, Chem Warfare Serv, US Dept Army, Md, 45; asst prof biochem & med, Sch Med, Tulane Univ, 46-51, assoc prof, 51-52; dir nutrit res, Mead Johnson Res Ctr, 52-67, dir nutrit & biochem res, 58-62, vpres nutrit sci, 67-71, vpres nutrit sci resources, 71-81; CONSULT NUTRIT SCI, 81- *Concurrent Pos:* Mem tech adv group comt nutrit, Am Acad Pediat, 61-67, 69-81, chmn, 71-74; mem panel new foods, White House Conf Foods, Nutrit & Health, 69; mem food standards & fortification policy, Food & Nutrit Bd, Nat Acad Sci, 70-72; chmn nutrit sci comt, Infant Formula Coun, 71-81; indust adv to US deleg comt foods for spec dietary uses, Codex Alimentarius Comn, 71-81. *Mem:* Fel Am Inst Nutrit; Soc Exp Biol & Med; Am Soc Clin Nutrit; Am Chem Soc; Am Soc Biol Chem. *Res:* Foods for special dietary uses; protein evaluation; infant nutrition; milk substitute formulas; formulas for infants with metabolic disorders; formula diets; meal replacements; medium chain triglycerides; cholestyramine; nutrition regulations; vitamin, mineral and fluoride supplements. *Mailing Add:* 5413 Palm Aire Dr Sarasota FL 34243-3706

SARETT, LEWIS HASTINGS, b Champaign, Ill, Dec 22, 17; m 44, 69; c 4. RESEARCH ORGANIZATION, LIFE SCIENCES. *Educ:* Northwestern Univ, BS, 39; Princeton Univ, PhD(chem), 42. *Hon Degrees:* DSc, Northwestern Univ, 72, Bucknell Univ, 77. *Prof Exp:* Res chemist, Merck & Co, Inc, 42-48, asst dir dept org chem & biochem res, 48-52, dir dept med chem, 52-56, dir dept synthetic org chem, 56-62, exec dir fundamental res, 62-66, vpres basic res, 66-69, pres, Merck Sharp & Dohme Res Labs, 69-76, sr vpres sci & technol, Merck & Co, Inc, 75-82; DIR, START UP BIOTECHNOL COS, 82- *Concurrent Pos:* Consult med chem sect, USPHS, 64-67; mem res & eval comt, Nat Cystic Fibrosis Res Found, 64-69; trustee, Cold Spring Harbor Lab Quant Biol, 68-70; rep, Indust Res Inst, 68-, dir, 74-77; rep, Pharmaceut Mfrs Asn, 69-; mem vis comt div biol, Calif Inst Technol, 69-76, chmn, 72-76; mem, Dirs Indust Res, 70-, secy, 71-72, chmn, 72-73; mem indust adv comt, Univ Calif, San Diego, 71-; mem adv panel develop res develop & eng in developing countries, Nat Acad Sci, 71-72; mem, bd trustees, Med Ctr, Princeton Univ, 77-79; mem, Sci & Technol Panel, Reagan Transition Team, 80-81, Drugs for Rare Dis, Pharmaceut Mfrs Asn Comn, 81-; mem bd dirs, Immunex Corp, Resonex Corp, Vestar Corp & Hybertech Corp; indust consult life sci, 82- *Honors & Awards:* Julius W Sturmer Mem Lectr, 59; Sci Award, Bd Dirs, Merck & Co, Inc, 51; Baekeland Award, Am Chem Soc, 51, Award, 64; Medal, Synthetic Org Chem Mfrs Asn, 64; William Scheele Lectr, Royal Pharmaceut Inst, Stockholm, Sweden, 64; Chem Pioneer Award, Am Inst Chem, 72; Nat Medal Sci, 75; Perkin Medal Award, Soc Chem Indust, 76; Gold Medal, Am Inst Chemists, 81. *Mem:* Nat Acad Sci; Inst Med-Nat Acad Sci; Am Inst Chem. *Res:* Biomedical research organization. *Mailing Add:* 1488 Marshall Rd Viola ID 83872

SARFATTI, JACK, b New York, Sept 14, 39; div. THEORETICAL PHYSICS, SCIENCE EDUCATION & COMMUNICATION. *Educ:* Cornell, BA, 60; Univ Calif San Diego, MS, 67; Univ Calif Riverside, PhD(physics), 69. *Prof Exp:* Asst prof physics, San Diego State Univ, 67-71; dir physics, Esalen Inst, 75-77; INSTR MATH, USN PACE, 71-; INSTR PHYSICS, HEBREW ACAD SAN FRANCISCO, 90- *Concurrent Pos:* Vis prof math, Unicamp-Brasil, 85. *Res:* Quantum nonlocality; causality-violation; possibility of super-luminal and retro-active communication; cosmology; physics of mind-matter interaction. *Mailing Add:* PO Box 26548 San Francisco CA 94126

SARGE, THEODORE WILLIAM, b Taunton, Mass, Feb 4, 18; m 44; c 5. PHYSICAL ORGANIC CHEMISTRY. *Educ:* Col Holy Cross, BS, 40, MS, 41. *Prof Exp:* Chemist, Dow Chem Co, 41-44, chemist, Saran Develop Lab, 44-56, res admin asst petrochem & gen res, Saginaw Bay Res Dept, 56-67, res admin asst new prod, 67-69, res admin asst pilot plant & process develop & eng admin, 69-74, mem staff, Patent Dept, 74-83; RETIRED. *Mem:* AAAS; Am Chem Soc; Sigma Xi. *Res:* Vinyl polymerizations; polymer properties; polymeric film applications and properties, especially water vapor and gas transmissions; petrochemicals-hydrocarbons processing and extraction; research administration, especially research and development, patents, safety and training. *Mailing Add:* 3307 Kentwood Dr Midland MI 48640

SARGEANT, PETER BARRY, b Cedar Rapids, Iowa, Jan 18, 36; m 56; c 3. FIBER PRODUCTION, RESEARCH ADMINISTRATION. *Educ:* Iowa State Univ, BS, 58; Ohio State Univ, PhD, 62. *Prof Exp:* Res asst, Ohio State Univ, 58-59, Nat Sci Found fel, 60-62; res chemist cent res dept, Dacron Tech 62-66, sr res chemist, Textile Fibers Dept, Va, 66-69, res supvr, Dacron Res Lab, NC, 69-71, res supvr textile res lab, 71-73, sr supvr process, 73-75, sr supvr, Dacron Tech, 75-83, tech supvr, 83-86; TECH GROUP MGR, TEXTILE DIV, TEXTILE FIBERS DEPT, E I DUPONT DE NEMOURS & CO, KINSTON, NC, 86- *Res:* Polymer chemistry; fiber technology. *Mailing Add:* Textile Fibers Dept Dacron Res & Develop Ctr E I du Pont de Nemours & Co Inc Kinston NC 28501

SARGENT, ANNEILA ISABEL, b Kirkcaldy, Scotland; m 64; c 2. ASTRONOMY, ASTROPHYSICS. *Educ:* Univ Edinburgh, BSc, 63; Calif Inst Technol, MS, 67, PhD(astron), 77. *Prof Exp:* Sr res asst, 67-70 & 72-74, res fel astron, 77-81, mem prof staff, 81-88, sr res fel, 88-90, SR RES ASSOC, CALIF INST TECHNOL, 90- *Mem:* Am Astron Soc; Royal Astron Soc; Int Astron Union. *Res:* Millimeter-wave studies of regions of star formation; star formation in galaxies; proto-planetary disks. *Mailing Add:* Dept Physics Calif Inst Technol 320-47 Pasadena CA 91125

SARGENT, BERNICE WELDON, b Williamsburg, Ont, Sept 24, 06; m 40. NUCLEAR PHYSICS. *Educ:* Queen's Univ, Ont, BA, 26, MA, 27; Cambridge Univ, PhD(physics), 32. *Prof Exp:* Lectr physics, Queen's Univ, Ont, 30-36, asst prof, 36-43; res physicist, Atomic Energy Proj, Nat Res Coun Can, 43-49, prin res physicist, 49-51, asst dir, 51; prof physics & head dept, 51-67, R Samuel McLaughlin res prof, 54-72, EMER PROF PHYSICS, QUEEN'S UNIV, ONT, 72- *Concurrent Pos:* Mem, Nat Res Coun Can, 56-62. *Honors & Awards:* Order British Empire, 46; Coronation Medal, 53; Gold Medal, Can Asn Physicists, 59. *Mem:* Fel Am Phys Soc; fel Royal Soc Can; Can Asn Physicists (vpres, 54-55, pres, 55-56). *Res:* Radioactivity; neutron physics; nuclear reactors; photonuclear reactions; nuclear structure; history of physics. *Mailing Add:* Dept Physics Stirling Hall Queen's Univ Kingston ON K7L 3N6 Can

SARGENT, CHARLES, b Mitchell, Nebr, May 24, 13; m 38; c 1. CIVIL ENGINEERING. *Educ:* Univ Idaho, BS, 48, CE, 52; Stanford Univ, MS, 58. *Prof Exp:* Asst prof civil eng, Univ Idaho, 48-53; from asst prof to prof, 53-61, dean math phys sci & eng, 61-67, EMER DEAN MATH, PHYS SCI & ENG, UNIV ALASKA, COLLEGE, 70- *Concurrent Pos:* Consult engr, 48-; exec dir planning & opers, Univ Alaska, 67-70; prof construct mgt, NDak State Univ, 70-75. *Mem:* Am Soc Civil Engrs; Am Soc Eng Educ. *Res:* Economic problems in engineering construction, particularly concrete aggregates. *Mailing Add:* 2501 E Sherman Ave No 201 Coeur d'Alene ID 83814-5859

SARGENT, DAVID FISHER, b Victoria, BC, June 29, 45; m 67; c 1. X-RAY CRYSTALLOGRAPHY, MEMBRANE ELECTRICAL PROPERTIES. *Educ:* Univ BC, BSc, 66; Univ Western Ont, PhD(biophys), 71. *Prof Exp:* Fel biochem, Univ Sydney, 74-75; fel biophys, 71-74, RES ASST BIOPHYS, FED INST TECH, ZURICH, 75- *Concurrent Pos:* Med Res Coun Can fel, 71-74. *Mem:* Biophys Soc; Union Swiss Soc Exp Biol. *Res:* Structure/function of DNA/protein complexes (x-ray crystallography); lipid/peptide interactions (lipid bilayer membranes: electrical properties). *Mailing Add:* Inst Molecular Biol & Biophys ETH-Hoenggerberg Zurich 8093 Switzerland

SARGENT, FRANK DORRANCE, b Concord, NH, July 9, 35; m 59; c 2. QUANTITATIVE GENETICS. *Educ:* Univ NH, BS, 57; NC State Univ, MS, 60, PhD(animal sci), 65. *Prof Exp:* Res asst animal breeding, 57-63, instr dairy husb, 63-65, from asst prof exten dairy husb to prof exten dairy husb, 65-81, PROF ANIMAL SCI, NC STATE UNIV, 81- *Mem:* Am Dairy Sci Asn; Sigma Xi. *Res:* Dairy cattle breeding; dairy herd management; production record systems; reproductive physiology. *Mailing Add:* Dept Animal Sci NC State Univ 105 Polk Hall Raleigh NC 27695-7621

SARGENT, FREDERICK PETER, b Plymouth, UK, July 26, 40; m 67; c 2. NUCLEAR WASTE DISPOSAL. *Educ:* Univ Exeter, BS, 61; Univ Leeds, MS, 63, PhD(chem), 65. *Prof Exp:* Nat Res Coun Can fel, Univ Sask, 65-67; Sci Res Coun UK fel, Univ Leeds, 67-69; res officer chem, 69-77, mem plutonium chem group, 77-78, sect leader exp pathways anal radionuclide migration in geologic formations, 78-81, HEAD, GEOCHEM & APPL CHEM BR, ATOMIC ENERGY CAN LTD, 81- *Mem:* Chem Inst Can; Sigma Xi; Can Nuclear Soc. *Res:* Fundamental processes in radiation chemistry; electron spin resonance; leaching; nuclide sorption; ion exchange; geochemical and geophysical aspects of nuclear waste disposal; product and process development for immobilization of nuclear waste. *Mailing Add:* Atomic Energy Can Ltd Pinawa MB R0E 1L0 Can

SARGENT, GORDON ALFRED, b Winterton, Eng, Apr 8, 38; m 66; c 7. MATERIALS SCIENCE. *Educ:* Univ London, BSc, 60, DIC, 63, PhD(metall), 64; Royal Sch Mines, ARSM, 60. *Prof Exp:* Res fel, Mellon Inst, 63-67; from asst prof to assoc prof mat sci, Univ Ky, 67-77, prof metall eng & mat sci, 77-82, chmn dept, 82-85; DEAN ENG, UNIV DAYTON, 85- *Mem:* Fel Am Soc Metals; Am Inst Mining, Metall & Petrol Engrs. *Res:* Deformation and physical properties of materials under high pressures; properties of materials subjected to irradiation damage. *Mailing Add:* 3685 Olde Willow Dr Beavercreek OH 45431

SARGENT, HOWARD HARROP, III, b Hartford, Conn, June 12, 36; m 68; c 2. SOLAR-TERRESTRIAL PHYSICS. *Educ:* Univ Conn, BS, 63; Univ Colo, MS, 72. *Prof Exp:* Engr-in-charge ionospheric physics, Nat Bur Standards, S Pole Sta, Antarctica, 63-64; res asst, Stanford Univ, 65-66; instrumentation engr, Environ Sci Serv Admin, 68-71; instrumentation engr solar radio astron, 71-73, gen phys scientist solar physics, 73-77, SPACE SCIENTIST SOLAR PHYSICS, NAT OCEANIC & ATMOSPHERIC ADMIN, 77- *Honors & Awards:* NOAA Spec Achievement Award, Nat Oceanic & Atmospheric Admin, 75. *Mem:* Am Geophys Union; Sigma Xi. *Res:* Geophysical, including weather, effects of solar activity; intra-cycle changes in solar behavior; solar and geomagnetic activity prediction techniques; long-term solar variability; time series analysis. *Mailing Add:* Space Environ Lab 325 Broadway Boulder CO 80303

SARGENT, KENNETH ALBERT, b Ellsworth, Maine, Aug 16, 32; m 54; c 3. GEOLOGY. *Educ:* Bates Col, BS, 54; Univ Iowa, MS, 57, PhD(geol), 60. *Prof Exp:* Instr field geol, Univ Iowa, 60; petrol geologist, Texaco, Inc, 60-62; geologist Spec Proj Br, 62-72, supvr Tech Reports Unit, 72-75, geologist regional br, 75-84, chief, 84-86, ASSOC CHIEF OFFICER, REGIONAL BR, US GEOL SURV, 87- *Mem:* Fel Geol Soc Am. *Res:* Volcanic rock petrography and petrology; geology of south central Utah. *Mailing Add:* US Geol Surv APO New York NY 09697-7002

SARGENT, MALCOLM LEE, b Grayling, Mich, Sept 14, 37; c 3. BIOCHEMISTRY, GENETICS. *Educ:* Univ Mich, BS, 60; Stanford Univ, PhD(biol), 66. *Prof Exp:* NIH fel bot, Univ Mich, 66-68; asst prof bot, 68-73, ASSOC PROF GENETICS & DEVELOP, UNIV ILL, URBANA, 73- *Mem:* Am Bryolog & Lichenological Soc; Am Soc Plant Physiol; Genetics Soc Am. *Res:* Biochemical-genetics of circadian rhythms and development in Neurospora and other fungi; reproductive and developmental physiology of bryophytes. *Mailing Add:* Dept Genetics/Bot 515 Merrill Hall Univ Ill Urbana IL 61801

SARGENT, MURRAY, III, b New York, NY, Aug 18, 41; m 67; c 2. QUANTUM OPTICS, COMPUTER SCIENCE. *Educ:* Yale Univ, BS, 63, MS, 64, PhD(physics), 67. *Prof Exp:* Fel physics, Yale Univ, 67; mem tech staff, Bell Tel Labs, 67-69; asst prof optical sci, 69-72, assoc prof optical sci & comput sci, 72-74, assoc prof, 74-77, PROF OPTICAL SCI, UNIV ARIZ, 77-, PRES, SCROLL SYSTS, 81- *Concurrent Pos:* Humboldt sr scientist award, Fed Repub Ger, 75; guest prof, Univ Stuttgart, 75-76 & Max Planck Inst for Quantum Optics, 75-76 & 80-85. *Mem:* Am Phys Soc; fel Optical Soc Am. *Res:* Laser physics and applications; micro computer systems; technical word processing. *Mailing Add:* Dept Optical Sci Univ Ariz Tucson AZ 85721

SARGENT, ROBERT GEORGE, b Port Huron, Mich, June 14, 37; m 70; c 1. OPERATIONS RESEARCH, DISCRETE EVENT SIMULATION. *Educ:* Univ Mich, BSE, 59, MS, 63, PhD(indust eng), 66. *Prof Exp:* Electronics engr, Hughes Aircraft Co, 59-61; grad asst & lectr eng, Univ Mich, 62-66; from asst prof to assoc prof, 66-82, PROF INDUST ENG & OPERS RES, SYRACUSE UNIV, 82- *Concurrent Pos:* Vis assoc prof, Cornell Univ, 81-82; dept ed, Simulation Modeling & Statist Comput, 80-85; nat lectr, Asn Comput Mach, 85-89. *Mem:* Opers Res Soc Am; Inst Mgt Sci; Am Inst Indust Engrs; Assoc Comput Mach; Soc Comput Simulation. *Res:* Digital simulation; modelling and performance evaluation of computer systems; scheduling; production and inventory control; model validation. *Mailing Add:* Dept Indust Eng & Opers Res Syracuse Univ Syracuse NY 13244

SARGENT, ROGER GARY, b Sandborn, Ind, Mar 7, 39; m 62; c 2. PARASITOLOGY. *Educ:* Ind State Univ, BS, 62, MS, 64; Univ SC, PhD(parasitol), 71. *Prof Exp:* Off Econ Opportunity fel malnutrit & parasitism, 70-71, dir fac res develop & lectr biol, 71-74, asst dean, Col Health & Phys Educ, 74-81, ASSOC PROF BIOL, UNIV SC, 74-, ASSOC DEAN COL HEALTH, 81- *Concurrent Pos:* Jannsen Pharmaceut fel clin drug eval, 71-72, res grant, 75; mem, Nat Coun Univ Res Adminr. *Mem:* Am Soc Parasitol; Am Zool Soc; Am Asn Health, Phys Educ & Recreation; Sigma Xi. *Res:* Intestinal parasites Ascaris lumbricoides and Trichuris trichura with emphasis in drug regimens and ovicidal effects of various compounds; comparative drug studies establishing efficacy of current drugs of choice for Ascaris lumbricoides to investigational drugs. *Mailing Add:* 309 Waccamaw St Columbia SC 29205

SARGENT, ROGER N, b Stelton, NJ, June 3, 28; m 52; c 2. ANALYTICAL CHEMISTRY, PHYSICAL CHEMISTRY. *Educ:* Lafayette Col, BS, 51; Rutgers Univ, PhD(anal chem), 56. *Prof Exp:* Shift supvr styrene control lab, Koppers Co, 51-52; asst chem, Rutgers Univ, 52-56, res fel, 56-57; sr res chemist ion exchange div, Dow Chem Co, Mich, 57-68, group leader anal chem, 68-76, sr res specialist, Human Health Res & Develop Lab, 76-80; MEM FAC, COL HEALTH PHYS EDUC, UNIV SC, 80- *Mem:* Am Chem Soc; Sigma Xi. *Res:* Chromotographic separation and purification of organic compounds with ion exchange resins; ion-exchange chromatography and exclusion; salting-out chromatography. *Mailing Add:* Col Health Phys Educ Univ SC Columbia SC 29208

SARGENT, THEODORE DAVID, b Peabody, Mass, Oct 25, 36; m 67; c 2. ZOOLOGY. *Educ:* Univ Mass, BS, 58; Univ Wis, MS, 60, PhD(zool), 63. *Prof Exp:* Instr zool, 63-64, from asst prof to assoc prof, 64-75, PROF ZOOL, UNIV MASS, AMHERST, 75- *Concurrent Pos:* Ed, J. Lepidop Soc, 72-74. *Mem:* Animal Behav Soc; Soc Study Evolution; Lepidop Soc (pres, 79). *Res:* Cryptic moths, behavior, ecology, genetics; melanism in North American moths; bird behavior; moths of the genus Catocala. *Mailing Add:* Dept Zool Univ Mass Amherst MA 01003-0027

SARGENT, THORNTON WILLIAM, III, b St Louis, Mo, June 25, 28; m 52; c 2. NUCLEAR MEDICINE, BIOCHEMICAL PHARMACOLOGY. *Educ:* Reed Col, BA, 51; Univ Calif, Berkeley, PhD(biophys), 59. *Prof Exp:* Physicist, Michelson Lab, US Naval Ord Test Sta, Calif, 51-52; res biophysicist, 59-77, SR BIOPHYSICIST, LAWRENCE BERKELEY LAB, RES MED & RADIATION BIOPHYS DIV, UNIV CALIF, BERKELEY, 59- *Concurrent Pos:* Prin investr, Dept Energy, 59-, NIMH, 82- *Mem:* Soc Nuclear Med; Soc Biol Psychiat; Sigma Xi. *Res:* In vivo radionuclide research in human disease; chromium metabolism in diabetes; iron absorption in hematolgic disorders; bioamine, amino acid and methyl carbon metabolism in schizophrenia and manic depressive illness; positron emission tomography of metabolism and cerebral blood flow in psychosis and Alzheimers disease. *Mailing Add:* 1044 Siler Pl Berkeley CA 94705

SARGENT, WALLACE LESLIE WILLIAM, b Elsham, Eng, Feb 15, 35; m 64; c 2. ASTROPHYSICS. *Educ:* Univ Manchester, BSc, 56, MSc, 57, PhD(astron), 59. *Prof Exp:* Res fel astron, Calif Inst Technol, 59-62; sr res fel, Royal Greenwich Observ, Eng, 62-64; asst prof physics, Univ Calif, San Diego, 64-66; from asst prof to prof astron, 66-81, exec officer, 75-81, IRA S BOWEN PROF ASTRON, CALIF INST TECHNOL, 81- *Concurrent Pos:* Hon vis fel, Australian Nat Univ, 65 & 67; vis fel, Cambridge Univ, 68-72, 74-75, 79, 82, 87, Oxford Univ, 73, Univ Groningen, 78, Univ Florence, 81, Europ Southern Observ, 80, 83, 85, Astrophys Inst, Paris, 84; Alfred P Sloan Found fel, 68-70; mem staff, Owens Valley Radio Observ, 78-; fel Royal Soc, 81. *Honors & Awards:* Helen B Warner Prize, Am Astron Soc, 69; George Darwin lectr, RAS, 87; Heineman Prize, Am Astron Soc, 91. *Mem:* Am Astron Soc; fel Am Acad Arts & Sci; fel Royal Astron Soc; Int Astron Union. *Res:* Stellar and extragalactic spectroscopy; evolution of galaxies; clusters of galaxies; quasars; cosmology. *Mailing Add:* Dept Astron 105-24 Calif Inst Technol Pasadena CA 91125

SARGENT, WILLIAM QUIRK, b Bell, Calif, Jan 26, 45; m 69; c 3. FLUID METABOLISM, ELECTROLYTE METABOLISM. *Educ:* Johns Hopkins Univ, BA, 67; Univ Tenn, PhD(physiol), 73. *Prof Exp:* USPHS trainee, Col Basic Med Sci, Univ Tenn, 69-71, teaching fel, 71-73; fel, Alcohol & Drug Res Ctr, Tenn Psychol Hosp, 73-75; CENT REGION MGR, SCI & TECHNOL AFFAIRS, HOECHST-ROUSSEL PHARMACEUT, MEMPHIS. *Mem:* Am Physiol Soc. *Res:* Renal and gastrointestinal electrolyte metabolism during long term drug administrations (ie, ethanol). *Mailing Add:* Mgr Sci Affairs Hoechst Roussel Pharmaceut 1358 Rustic View Manchester MO 63011

SARGENTINI, NEIL JOSEPH, b San Francisco, Calif. MUTAGENESIS, DNA REPAIR. *Educ:* Calif State Univ, Fresno, BA, 69, MA, 73; Stanford Univ, PhD(med microbiol), 80. *Prof Exp:* Res assoc radiation oncol, Sch Med, Stanford Univ, 80-84, sr res assoc, 84-91; ASST PROF MICROBIOL & IMMUNOL, KIRKSVILLE COL OSTEOP MED, MO, 91- *Mem:* Am Soc Photobiol; Radiation Res Soc; Sigma Xi. *Res:* Spontaneous mutagenesis; mutagenesis, DNA repair and survival of gamma; UV-irradiated Escherichia coli cells. *Mailing Add:* Dept Microbiol & Immunol Kirksville Col Osteop Med Kirksville MO 63501

SARGES, REINHARD, b Siegen, Ger, July 25, 35; US citizen; m 80; c 3. ORGANIC CHEMISTRY. *Educ:* Univ Frankfurt, dipl, 60, PhD(peptide chem), 62. *Prof Exp:* Vis fel chem, NIH, 62-64, vis assoc, 64-65; res chemist, 65-75, proj leader, 75-81, prin investr, 81-85, RES ADV, PFIZER INC, 85- *Mem:* Am Chem Soc. *Res:* Peptide and medicinal chemistry; biochemistry; CNS agents; antidiabetic drugs; aldose reductase inhibitors. *Mailing Add:* Pfizer Inc Eastern Point Rd Groton CT 06340

SARHAN, FATHEY, b Cairo, Egypt, Mar 28, 45; Can citizen. PLANT PHYSIOLOGY. *Educ:* Cairo Univ, BSc, 65; Univ Montreal, MSc, 73, PhD(plant physiol), 77. *Prof Exp:* Agronomist, Ministry Agr, Cairo, Egypt, 66-70; PROF PLANT PHYSIOL & BIOCHEM, UNIV QUE, MONTREAL, 79- *Concurrent Pos:* Lab instr, Univ Montreal, 71-76; res asst, McGill Univ, Montreal, 72-73; lectr, Univ Que, Montreal, 76-78; vis prof, Inst Grad Studies & Res, Alexandria Univ, 84-85. *Mem:* Can Soc Plant Physiol; Am Soc Plant Physiol; NY Acad Sci; Egyptian Soc Genetics. *Res:* Protein and nucleic acid synthesis mechanisms relative to cold hardiness in wheat; molecular and genetic manipulations of plant genes. *Mailing Add:* Dept Biol Sci Univ Que CP8888 Succ A Montreal PQ H3C 3P8 Can

SARI, JAMES WILLIAM, b Buffalo, NY, Oct 13, 42; m 71. SPACE SCIENCE, PLASMA PHYSICS. *Educ:* Oberlin Col, BA, 64; Univ Md, PhD(physics), 72. *Prof Exp:* Physicist plasma physics, Cornell Aeronaut Lab, 72-77; SR PHYSICIST, APPL PHYSICS LAB, JOHNS HOPKINS UNIV, 77- *Mem:* Am Geophys Union. *Res:* Interplanetary magnetic fields; cosmic-ray propagation; geomagnetic micropulsations; ocean magnetic fields. *Mailing Add:* 6010 Windham Rd Laurel MD 20707

SARIASLANI, SIMA, MICROBIOLOGY, BIOCHEMISTRY. *Educ:* Univ Kent, UK, PhD(microbiol & biochem), 74. *Prof Exp:* PRIN INVESTR MICROBIOL & BIOCHEM, E I DU PONT DE NEMOURS, 84- *Mem:* Am Chem Soc; Am Soc Microbiol; Soc Indust Microbiol. *Res:* Biochemistr and molecular biology of microbial xenobiotic metabolism and biotransformation. *Mailing Add:* Cent Res & Develop Dept DuPont Exp Sta Wilmington DE 19880-0228

SARIC, WILLIAM SAMUEL, b Chicago, Ill, Sept 28, 40; m 62, 90; c 1. HYDRODYNAMIC STABILITY. *Educ:* Ill Inst Technol, BS, 63, PhD(mech), 68; Univ NMex, MS, 65. *Prof Exp:* Staff mem environ testing, Sandia Labs, 63-66; instr, Ill Inst Technol, 66-68; staff mem, Sandia Labs, 68-75; assoc prof, 75-79, PROF MECH ENG, VA POLYTECH INST & STATE UNIV, 79- *Concurrent Pos:* Adj prof, Univ NMex, 72 & Va Polytech Inst & State Univ, 74-75; mem, Nat Tech Comt Fluid Dynamics, Am Inst Aeronaut & Astronaut, 75-78; consult, Sandia Labs, 75-81, Ecodynamics Corp, 76, Ballistic Res Labs, 76 & Lockheed-Ga, 77-; invited guest & researcher, USSR Acad Sci, 76 & 81. *Mem:* Fel Am Inst Aeronaut & Astronaut; fel Am Phys Soc; fel Am Soc Mech Engrs; Am Soc Eng Educ. *Res:* Hydrodynamic stability; boundary-layer transition; nonlinear waves; nonlinear dynamics; transpiration cooling; boundary-layer flows; laminar flow control; electron-beam induced nuclear fusion; stability of stratified flows. *Mailing Add:* Dept Mech & Aerospace Eng Ariz State Univ Tempe AZ 85287-6106

SARICH, VINCENT M, b Chicago, Ill, Dec 13, 34; m 61; c 2. PHYSICAL ANTHROPOLOGY. *Educ:* Ill Inst Technol, BS, 55; Univ Calif, Berkeley, PhD(anthrop), 67. *Prof Exp:* Instr anthrop, Stanford Univ, 65; asst prof, 67-70, assoc prof, 70-81, PROF ANTHROP, UNIV CALIF, BERKELEY, 81- *Mem:* AAAS; Am Asn Phys Anthrop; Am Soc Mammal; Sigma Xi. *Res:* Construction of quantitative phylogenies by the use of comparative molecular data; evolutionary and selective bases of human variation. *Mailing Add:* Dept Anthrop Univ Calif Berkeley CA 94720

SARID, DROR, b Haifa, Israel, Dec 13, 38; US citizen; m 63; c 2. SCANNING TUNNELING MICROSCOPY, LIGHT SCATTERING. *Educ:* Hebrew Univ, Jerusalem, BSc, 66, MSc, 68, PhD(physics), 72. *Prof Exp:* Fel physics, Univ Calif, Santa Barbara, 72-74; scientist, Xerox Webster

Res Ctr, 74-78; sr lectr, Hebrew Univ, Jerusalem, 78-80; PROF PHYSICS, OPTICAL SCI CTR, UNIV ARIZ, 80- *Concurrent Pos:* Consult, EG&G, Santa Barbara, 74; vis scholar, Univ Calif, Santa Barbara, 75; consult, Xerox Webster Res Ctr, 78-81, US Army, 81-88. *Mem:* Optical Soc Am; Am Phys Soc. *Res:* Laser light scattering, nonlinear optics and scanning tunneling microscopy. *Mailing Add:* Optical Sci Ctr Univ Ariz Tucson AZ 85721

SARIDIS, GEORGE N, b Athens, Greece, Nov 17, 31; m. ELECTRICAL ENGINEERING. *Educ:* Athens Tech Univ, Dipl, 55; Purdue Univ, MSEE, 62, PhD(optimal control), 65. *Prof Exp:* Instr elec mech, Athens Tech Univ, 55-63; from instr to prof elec eng, Purdue Univ, West Lafayette, 63-81; PROF ELEC, COMPUT & SYSTS ENG & DIR, ROBOTICS & AUTOMATION LAB, RENSSELAER POLYTECH INST, 81- *Concurrent Pos:* Engr, Telecommun Orgn Greece, 55-56 & Pub Power Corp Greece, 57-63. *Honors & Awards:* Centennial Medal, Inst Elec & Electronics Engrs, 84. *Mem:* Fel Inst Elec & Electronics Engrs; NY Acad Sci; Am Soc Mech Engrs; Am Soc Eng Educ; Soc Mfg Engrs. *Res:* Optimal control theory and applications; adaptive and learning systems; self-organizing control systems; bioengineering systems; prosthetics and robotics; intelligent systems. *Mailing Add:* Dept Elec Comput & Systs Eng Rensselaer Polytech Inst Troy NY 12180-3590

SARIN, PREM S, US citizen; m 65; c 2. BIOCHEMISTRY, CELL BIOLOGY MEDICAL SCIENCES. *Educ:* Univ Delhi, BSc, 54, MSc, 56, PhD(chem), 59; Cambridge Univ, PhD(chem), 62. *Prof Exp:* Lectr chem, Univ Delhi, 57-60; res fel, Harvard Univ, 63-65; biochemist, Univ Toronto, 65-67; head sect nucleic acids, Merck Sharp & Dohme, NJ, 67-70; dir molecular biol, Litton Bionetics, Md, 71-72; vis scientist, Nat Cancer Inst, 72-74, dept chief, Lab Tumor Cell Biol, 74-91; AT GEORGE WASHINGTON UNIV. *Concurrent Pos:* Vis prof, Univ del Rosario, Columbia, 84; adj prof biochem, George Washington Univ Sch Med, 87. *Mem:* Am Soc Cell Biol; Am Chem Soc; Am Soc Biol Chem; Am Asn Cancer Res. *Res:* Cell biology of human normal and neoplastic cells, AIDS and human T-lymphotropic viruses related neurological diseases; oncogenic viruses; antiviral agents, AIDS and cancer chemotherapy; etiology of cancer, human leukemia and acquired immune deficiency syndrome; human t-lymphotropic retroviruses. *Mailing Add:* Lab Tumor Cell Biol Nat Cancer Inst Bethesda MD 20892

SARJEANT, PETER THOMSON, b Orillia, Ont, June 24, 29; m 56; c 2. PAPER CHEMISTRY, ENGINEERING. *Educ:* Queens Univ, BSc, 53, MSc, 56; Pa State Univ, PhD(mat sci), 67. *Prof Exp:* Prod supvr pharmaceut, Merck Sharpe & Dohme, 54-57; res engr wood prod, 57-60, prod develop supvr lignin prod, 60-63, res chemist electrocopy, 63-64, res chemist paper coatings, 67-69, group leader, 69-70, assoc res dir, 70-71, res dir, 71-78, RES GROUP MGR, WESTVACO CORP, 78- *Mem:* Tech Asn Pulp & Paper Indust; Am Chem Soc; AAAS. *Res:* Process control of papermaking and chemical recovery processing; wood pulping processes; refining; cleaning; fourdrinier optimization; converting equipment for paper and board; history of papermaking; wood by-product chemicals. *Mailing Add:* 2864 Ion Ave Sullivans Island SC 29482-8669

SARJEANT, WALTER JAMES, b Strathroy, Ont, Apr 7, 44; m 67; c 2. PHYSICS, ELECTRICAL ENGINEERING. *Educ:* Univ Western Ont, BSc, 66, MSc, 67, PhD(physics), 71. *Prof Exp:* Asst dir res & develop, Gen-Tec Inc, 71-73; scientist, Lumonics Res Ltd, 73-75; scientist, Nat Res Coun Can, 75-78; scientist, Los Alamos Sci Labs, 79-81; JAMES CLERK MAXWELL PROF DEPT ELEC ENG, STATE UNIV NY, BUFFALO, 81- *Concurrent Pos:* Indust fel, Gen-Tec Inc, 71-73; consult, Dept Nat Defence, 71-73, Atomic Energy Can Ltd, 75-; adj prof elec eng, Tex Tech Univ, 77-; adj prof physics, Univ Ill, 79-; scientific adv, DARDA, 81-; consult, Los Alamos Nat Lab, 81- & Defense Nuclear Agency, 85- *Mem:* Fel Inst Elec & Electronics Engrs; Asn Prof Engrs. *Res:* High repetition rate power conditioning systems; gas discharge laser physics and chemical kinetics; generation and measurement of picosecond electrical impulses; electrical insulation and breakdown processes. *Mailing Add:* Dept Elec Eng Bonner Hall Rm 312 State Univ NY Buffalo Buffalo NY 14260

SARJEANT, WILLIAM ANTONY SWITHIN, b Sheffield, Eng, July 15, 35; m 66; c 3. PALEONTOLOGY, HISTORY OF GEOLOGY. *Educ:* Univ Sheffield, BSc Hons, 56, PhD(geol), 59; Univ Nottingham, DSc(geol), 72. *Prof Exp:* Demonstr temp lectr geol, Univ Col N Staffordshire, Eng, 60-61; res fel, Univ Reading, 61-62; from asst lectr to lectr, Univ Nottingham, 63-72; assoc prof, 72-80, PROF GEOL SCI, UNIV SASK, 80- *Concurrent Pos:* Vis prof geol & geophys, Univ Okla, 68-69; consult numerous petrol industs, Eng, France, US & Brazil. *Honors & Awards:* Sue Tyler Friedman Medal, Geol Soc London, 90. *Mem:* Fel Geol Soc London; fel Linnean Soc London; Geol Soc France; Paleont Asn; Am Asn Stratigraphical Palynologists; fel Geol Soc Am; Explorers Club. *Res:* Dinoflagellate cysts and acritarchs of the Triassic to Quaternary, Great Britain, France, Germany, Greenland, Algeria, Iran, and Canada; fossil vertebrate footprints and trace-fossil classification, Canada, United States, British Isles and Brazil; history and bibliography of the geological sciences; history of science. *Mailing Add:* Univ Sask Dept Geol Sci Saskatoon SK S7N 0W0 Can

SARKANEN, KYOSTI VILHO, wood chemistry, pulp chemistry; deceased, see previous edition for last biography

SARKAR, BIBUDHENDRA, b Kushtia, India, Aug 2, 35; m 65; c 2. BIOCHEMISTRY, PHYSICAL CHEMISTRY. *Educ:* Banares Univ, India, BPharm, 56, MPharm, 57; Univ Southern Calif, PhD(biochem), 64. *Prof Exp:* Vis prof, Inst Phys Chem Biol, Univ Paris, 76-77 & Cambridge, 77; PROF, UNIV TORONTO, 78-; HEAD, BIOCHEM RES, HOSP SICK CHILDREN, 90- *Concurrent Pos:* Med Res Coun Can res scholar, 65-70. *Honors & Awards:* Nuffield Found Award. *Mem:* AAAS; Can Biochem Soc; Am Chem Soc; fel Chem Inst Can; Am Soc Biochem & Molecular Biol; Int Asn Bioinorg Sci; Protein Soc. *Res:* Chemical and physico chemical studies related to coordination compounds of metals with proteins, peptides, amino acids, sugars and nucleic acids conducted for understanding their biochemical role in physiological systems and pathological conditions; molecular design to mimic functional sites of biomolecules. *Mailing Add:* Res Inst Hosp for Sick Children 555 University Ave Toronto ON M5G 1X8 Can

SARKAR, FAZLUL HOQUE, b Naroshingapur, India, Jan 26, 52; US citizen; m 83; c 2. GENE EXPRESSION & REGULATION, VIROLOGY-HPV & HUMAN CANCER. *Educ:* Calcutta Univ, BS, 71; Aligarh Muslim Univ, MS, 74; Banaras Hindu Univ, PhD(biochem), 78. *Prof Exp:* Res assoc virol & molecular biol, Mem Sloan Kettering Cancer Ctr, 78-81, asst researcher interferon, 81-84; asst prof res gene expression steroid, Oakland Univ, 84-87; dir res mutagenesis, Oxford Biomed Res, 87-88; dir, tumor & molecular biol, Henry Ford Hosp, 88-90; ASSOC PROF PATH & DIR MOLECULAR BIOL, WAYNE STATE UNIV SCH MED, 90- *Concurrent Pos:* From adj asst prof to adj assoc prof, Oakland Univ, 88-93; sci dir res, Oxford Biomed Res, 88- *Mem:* AAAS; Am Soc Biochem & Molecular Biol; Am Asn Cancer Res. *Res:* Expression and regulation of cellular genes in human solid tumors; studies on the mechanism of the role of human papillomavirus in human cancer. *Mailing Add:* Dept Path Wayne State Univ Sch Med Detroit MI 48201

SARKAR, KAMALAKSHA, b Calcutta, India, Sept 27, 47; m 78; c 2. COMPUTER-AIDED DESIGN. *Educ:* Bengal Eng Col, Sibpur, India, BE, 68; Indian Inst Technol, Kanpur, MTech, 74; Univ Tenn, Knoxville, PhD(eng sci), 80. *Prof Exp:* Structural engr, Indian Space Res Orgn, Trivandrum, 69-74; develop engr, 80-84, sr res engr, 84-88, RES ASSOC, ALLIED SIGNAL INC, MORRISTOWN, NJ, 88- *Concurrent Pos:* Assoc ed, Appl Mech Rev, 84-86. *Mem:* Am Soc Mech Engrs. *Res:* Computer-aided design to develop materials and components using an interdisciplinary approach (mechanical, structural, metallurgical and materials); composite material; fracture mechanics; fluid dynamics; finite element analysis and computational fluid dynamics. *Mailing Add:* One Leigh Ct Randolph Township NJ 07869

SARKAR, NILIMA, b India, June 2, 35; US citizen; m 61; c 2. MOLECULAR BIOLOGY. *Educ:* Univ Calcutta, India, BS, 53, MS, 55; Northwestern Univ, PhD(biochem), 61. *Prof Exp:* Fel, Dept Biochem, Univ Chicago, 61-63; fel, Dept Microbiol, Tufts Univ, 63-64; res fel, 64-67, res assoc, 67-69, assoc, 69-75, LECTR, DEPT BIOL CHEM, HARVARD MED SCH, 75-; STAFF SCIENTIST, DEPT METABOLIC REGULATION, BOSTON BIOMED RES INST, 76- *Mem:* Am Soc Biol Chemists; Am Soc Cell Biol; Am Chem Soc; AAAS. *Res:* Mechanisms of replication of the chromosome, especially in relation to the involvement of RNA primers; poly(A) RNA in bacteria with respect to structure, function and biosynthesis, using the technique of cloning of DNA. *Mailing Add:* Anat Dept Tufts Med Sch Boston MA 02101

SARKAR, NITIS, b Gauhati, India, Dec 1, 38; m 70; c 2. WATER SOLUBLE POLYMERS, MINERAL ENGINEERING. *Educ:* Gauhati Univ, India, BS, 57; Univ Calcutta, MS, 60; Mass Inst Technol, ScD(mineral eng), 65. *Prof Exp:* Res trainee, Fuel Res Inst, India, 60-61; lectr, Barasat Govt Col, 61-62; res asst, Mass Inst Technol, 62-65; res chemist chem lab, Dow Chem Co, 65-68 & Betz Labs, Inc, 68; ASSOC SCIENTIST, FUNCTIONAL POLYMERS RES, DOW CHEM CO, MIDLAND, MI, 68- *Honors & Awards:* MASTL (Dow) Scientists' Award, 84. *Mem:* Am Chem Soc. *Res:* Flocculation; water treatment; characterization of water soluble polymers; rheology of polymer solutions; detergency; dispersion; adhesion; paints; latexes; thickeners; enhanced oil recovery, cellulosic polymers, surface and colloid chemistry flotation; use of polymer in food; suspension polymerization; engineering thermoplastics and blends. *Mailing Add:* 1604 Bldg Dow Chem USA Midland MI 48674

SARKAR, PRIYABRATA, plant cytogenetics, for more information see previous edition

SARKAR, SATYAPRIYA, b Khanjanagar, India, Mar 1, 34; m 61; c 2. MOLECULAR & CELLULAR BIOLOGY OF MUSCLE DEVELOPMENT. *Educ:* Univ Calcutta, BSc, 53, MSc, 56; Northwestern Univ, PhD(biochem), 61. *Prof Exp:* Sr staff scientist, Boston Biomed Res Inst, 72-86; mem grad prog, Cell & Develop Biol, Sch Med, Harvard Univ, 80-86; ASSOC PROF, DEPT ANAT & CELL BIOL, TUFTS UNIV MED SCH, 86- *Concurrent Pos:* Consult, NIH Nat Heart, Lung & Blood Inst, 81-, mem, Biomed Sci Study Sect, NIH, 87-, mem, res peer rev comt, Am Heart Asn Mass Affiliate, 87-90, prin investr, NIH grant, 69- *Mem:* Am Chem Soc; Am Soc for Biochem & Molecular Biol; Am Soc for Cell Biol; NY Acad Sci; AAAS. *Res:* Molecular mechanism of gene expression in muscle cell; regulation of human myogenesis in normal and pathological conditions. *Mailing Add:* Dept Anat & Cellular Biol Tufts Univ Med Sch 136 Harrison Ave Boston MA 02111

SARKAR, SIDDHARTHA, in vitro fertilization, animal reproduction, for more information see previous edition

SARKARIA, GURMUKH S, b 1925. ENGINEERING ADMINISTRATION. *Educ:* Punjab Univ, BS, 45; Polytech Inst, Brooklyn, MS, 47; Harvard Univ, MSE, 48. *Prof Exp:* Vpres, Morrison Knudsen Inc, 70-81, sr vpres, 81-86; CONSULT, 87- *Concurrent Pos:* Gen coordr, hydroelec proj bet Paraguay-Brazil, 75-81; consult to Chinese Govt for Three Gorges hydroelectric power plant. *Mem:* Nat Acad Eng; Am Soc Civil Engrs. *Mailing Add:* 2378 San Clemente Way Vista CA 92084

SARKES, LOUIS A(NTHONY), nuclear physics, gas energy; deceased, see previous edition for last biography

SARKO, ANATOLE, b Tallinn, Estonia, May 27, 30; m 55; c 1. POLYMER CHEMISTRY, POLYMER PHYSICS. *Educ:* Upsala Col, BS, 52; NY Univ, MS, 60; State Univ NY Col Forestry, Syracuse, PhD(phys chem), 66. *Prof Exp:* Mem biochem staff, Gen Foods Corp, NY, 52-60, sr chemist, 60-63; res assoc chem, 66-67, from asst prof to assoc prof, 67-76, PROF CHEM, STATE UNIV NY COL ENVIRON & FORESTRY, 76-, CHMN DEPT, 84- *Concurrent Pos:* Vis scientist, Univ Frieburg, Ger & Nat Ctr Sci Res-CERMAV, Grenoble, France, 78. *Mem:* AAAS; Am Chem Soc; Am Phys Soc. *Res:* Physical chemistry of polymers; conformation of cellulose and other polysaccharides in solid and solution states and their structure-function relationships; x-ray techniques of polymers; computer techniques in chemistry. *Mailing Add:* 212 Kittell Rd Fayetteville NY 13066

SARLES, F(REDERICK) WILLIAMS, b Cincinnati, Ohio, Sept 27, 31; m 60; c 2. ELECTRICAL ENGINEERING. *Educ:* Duke Univ, BSEE, 53; Mass Inst Technol, MS, 55, ScD(elec eng), 61. *Prof Exp:* Res asst comput components & systs group, Lincoln Lab, Mass Inst Technol, 53-61, mem staff div sponsored res, 61, mem staff space commun, 61-69, asst group leader, 69-71, group leader spacecraft technol group, 71-78, sr staff mem, 78-80; vpres, Tri Solar Corp, Bedford, Mass, 80-82; PRES & CONSULT ENGR, FWS ENG, LEXINGTON, MASS, 82- *Concurrent Pos:* Lectr, Northeastern Univ, 67-71. *Mem:* Sr mem Inst Elec & Electronics Engrs; AAAS; Sigma Xi. *Res:* Design of control and communication systems; telemetry systems; design of space experiments; photovoltare systems, industrial controls and communication. *Mailing Add:* 54 Ledgelawn Ave Lexington MA 02173

SARLES, LYNN REDMON, b Grand Forks, NDak, Jan 22, 30; m 51; c 4. INSTRUMENTATION DESIGN. *Educ:* Stanford Univ, PhD(physics), 57. *Prof Exp:* Sloan fel physics, Univ Calif, 57-58; res physicist, Varian Assocs, 58-62; mgr phys optics, Maser Optics-West, 62-63; mgr geophys res, Varian Assocs, 63-67; dir admin, Ore Grad Ctr, 67-68, vpres, 68-72; VPRES RES, SYSTS MGT ASSOCS, INC, 71- *Mem:* Fel AAAS. *Res:* Magnetic resonance; optical pumping; geomagnetism; lasers; instrumentation; technological forecasting. *Mailing Add:* 2855 SW 107th Ave Portland OR 97225

SARMA, ATUL C, b Mangaldoi, Assam, India, Feb 1, 39; US citizen; m 79; c 2. CERAMIC & REFRACTORY BINDING SYSTEMS. *Educ:* Cotton Col, India, BSc Hons, 61; Guwahati Univ, MSc, 63; Univ Minn, MS, 67; Univ Louisville, PhD(inorg chem), 71. *Prof Exp:* Lectr chem, Cotton Col, India, 63-65; res assoc environ chem, Univ Louisville, 71-73; RES ASSOC MAT SCI & DIR, RES & DEVELOP & PROD, WHIP MIX CORP, 73- *Concurrent Pos:* Instr, Univ Louisville, 71-73, adj assoc prof, 76-83; mem & secy, specif comt waxes, Am Nat Standards Inst, 81-, chmn, specif comt dent stones, plasters & casting refractories, 85- *Mem:* Am Chem Soc; Am Ceramic Soc; Int Asn Dent Res; Sigma Xi. *Res:* Solid state, environmental and ceramic chemistry; precision casting technology; reactions in solids; high-technology ceramics; waxes, gypsum products and dental materials. *Mailing Add:* Whip Mix Corp PO Box 17183 Louisville KY 40217

SARMA, DITTAKAVI S R, b Tenali, India, June 15, 36; m 72. BIOCHEMISTRY, EXPERIMENTAL PATHOLOGY. *Educ:* Andhra Univ, India, BSc, 54; Univ Nagpur, MSc, 57; Univ Madras, PhD(biochem), 62. *Prof Exp:* Coun Sci & Indust Res sr res fel, Dept Biochem, Indian Inst Sci, Bangalore, 62-65; res assoc exp path, Sch Med, Univ Pittsburgh, 65-70, asst res prof, 70-71; asst res prof chem carcinogenesis, Fels Res Inst, Med Sch, Temple Univ, 71-77; assoc prof, 77-86, PROF PATH, UNIV TORONTO, 86- *Mem:* Am Soc Exp Path; Am Asn Cancer Res; Environ Mutagen Soc; NY Acad Sci. *Res:* Chemical carcinogenesis; DNA repair. *Mailing Add:* Dept Path Med Sci Bldg Univ Toronto Toronto ON M5S 1A8 Can

SARMA, PADMAN S, b India, Dec 3, 31; US citizen; c 2. VIROLOGY, VETERINARY MEDICINE. *Educ:* Madras Vet Col, India, DVM, 53; Univ Minn, MS, 57, PhD(virol), 59. *Prof Exp:* Asst lectr animal husb & microbiol, Madras Vet Col, 53-55; res officer, Pasteur Inst, India, 59-61; asst prof microbiol, Univ Ky, 61-62; vis scientist, NIH, 62-64; proj dir viral carcinogenesis, Microbiol Assoc Inc, 64-68; res microbiologist & chief, Sec Ecol & Epizool, Viral Carcinogensis Br, 68-77, chief, Animal Virol & Field Studies Sect, Lab Cellular & Molecular Biol, 77-83; PROG DIR, RNA VIRUS STUDIES I, BIOL CARCINOGENESIS BR, NAT CANCER INST, NIH, 83- *Concurrent Pos:* Instr lectr, Med Virol Course, 83- *Mem:* AAAS; Am Vet Med Asn. *Res:* Elucidation of the prevalence, etiological role and control of retroviruses responsible for the causation of naturally-occurring cancers in a variety of avian and mammalian species, including man; more recently, scientific program administration of NIH extramural grants and contracts in this area; author of 93 scientific publications in leading scientific journals. *Mailing Add:* 14124 Rippling Brook Dr Silver Spring MD 20906

SARMA, RAGHUPATHY, b Udipi, India, Feb 18, 37; m 63; c 2. MOLECULAR BIOLOGY, BIOCHEMISTRY. *Educ:* Presidency Col, Madras, BSc, 57; Univ Madras, MSc, 58, PhD(physics), 63. *Prof Exp:* Fel & res assoc, Royal Inst, London, 63-66 & Oxford Univ, 66-68; vis assoc, NIH, 68-71; asst prof, 71-77, assoc prof, 78-80, PROF BIOCHEM, STATE UNIV NY, STONY BROOK, 80- *Mem:* Am Crystallog Asn. *Res:* Structure and function of biological macromolecules; x-ray crystallography; crystallographic computing. *Mailing Add:* Dept Biochem State Univ New York Stony Brook NY 11794

SARMA, RAMASWAMY HARIHARA, b Perumbavoor, India, May 10, 39; m 63. BIOCHEMISTRY. *Educ:* Univ Kerala, BSc, 59, MS, 61; Brown Univ, PhD(chem), 67. *Prof Exp:* Fel biochem, Brandeis Univ, 67-69; fel, Univ Calif, San Diego, 69-70; asst prof, 70-75, assoc prof, 75-77, PROF CHEM, STATE UNIV NY ALBANY, 77-, DIR, INST BIOMOLECULAR STEREODYNAMICS, 77-, DIR, CTR BIOL MACROMOLECULES, 81- *Mem:* AAAS; Am Inst Chem; NY Acad Sci; Am Soc Biol Chemists. *Res:* Nucleic acid structures and their complexes with anticancer agents. *Mailing Add:* Dept Chem State Univ NY 1400 Washington Ave Albany NY 12222

SARMIENTO, GUSTAVO SANCHEZ, b Córdoba, Arg, Jan 10, 47; m 74; c 5. COMPUTATIONAL MECHANICS, NUMERICAL HEAT TRANSFER. *Educ:* Nat Univ Cuyo, Arg, physicist, 73, Dr Nuclear Eng, 91. *Prof Exp:* Scholar neutron physics, Neutron Physics Div, Centro Atomico Bariloche, Arg, 74-75, sci res computational mech, Nat AEC, Arg, 76-81; head, Comp Mech Dept, Empresa Nuclear Arg Centrales Electricas, SAm, 81-87; CONSULT ENGR COMPUTER MODELING, ARG INST SIDERURGIA, 86-; CONSULT ENGR & DIR COMPUTATIONAL MECH, MECACOMP CONSULT ENGRS, 87- *Concurrent Pos:* Asst prof, Balseiro Inst, Nat Univ Cuyo, Arg, 74-81; reviewer, Appl Mech Rev, 78-83; acad adv, Int Ctr Mech Sci, Arg, 78-; lectr & vis prof, numerous eng enterprises & govt insts, 83-; vis prof computational mech, several univs, Arg, 84- *Honors & Awards:* First Nat Award Eng, Secy Cult, Arg, 86. *Mem:* Am Acad Mech; Int Soc Computational Methods Eng. *Res:* Computer modeling; finite element method; continuum mechanics; heat transfer; numerical methods; numerical analysis; co-author of over 70 papers in international journals with referee and in international scientific meetings. *Mailing Add:* Mecacomp Consult Engrs Fla 274 31 Buenos Aires Argentina

SARMIENTO, JORGE LOUIS, b Lima, Peru,Feb 7, 46; m; c 2. GEOPHYSICAL FLUID DYNAMICS. *Educ:* Swarthmore Col, BS, 48; Columbia Univ, MA, 74, MPh, 76, PhD(geol), 78. *Prof Exp:* Res assoc, 78-80, asst prof, 80-86; assoc prof, 86-91, PROF GEOL & GEOPHYS SCI DEPT, ATMOSPHERIC & OCEANIC SCI PROG, PRINCETON UNIV, 91- *Concurrent Pos:* Dir, Atmospheric & Oceanic Sci Prog, Princeton Univ, 80- *Mem:* Am Geophys Union; Am Soc Limnol & Oceanog; Sigma Xi; Am Meterol Soc. *Res:* Published numerous articles in various journals. *Mailing Add:* Atmospheric & Oceanic Sci Prog Princeton Univ PO Box CN710 Princeton NJ 08544-0710

SARMIENTO, RAFAEL APOLINAR, b Manila, Philippines, Jan 8, 37; m 66; c 2. CHROMATOGRAPHIC & SPECTROSCOPIC ANALYSIS. *Educ:* Wash Col, Md, BS, 61; Am Univ, Wash, DC, MS, 69, PhD(org-anal chem), 75. *Prof Exp:* Chemist, Entom Res Div, USDA, 61-66, res chemist, Agr Environ Qual Inst, 66-75; adv, Int Atomic Energy Agency, Vienna, 75-79, proj dir, 79-83; sect head, Chem Coord Unit, USDA, 85-86; anal chemist, Anal Chem Lab, Environ Protection Agency, 86-88, ASST TO DIR, FIELD MGT DIV, FED GRAIN INSPECTION SERV, ENVIRON PROTECTION AGENCY, 89-; RES CHEMIST, AGR ENVIRON QUAL INST, USDA, 83- *Concurrent Pos:* Lectr, Univ Md, 73-75. *Mem:* Am Chem Soc; Am Inst Chemists; Am Asn Univ Professors; Entom Soc Am. *Res:* Development of chromatographic and spectrometric analysis of agricultural and environmental contaminants and pest control agents. *Mailing Add:* 5426 30th St NW Washington DC 20015

SARNA, SUSHIL K, b Rawalpindi, India, Mar 2, 42; m; c 1. PHYSIOLOGY. *Educ:* Univ Alta, PhD(biomed eng), 71. *Prof Exp:* Asst prof, depts pharmacol & elec eng, Univ Alta, 71-74; assoc prof, depts surg & elec eng, McMaster Univ, 74-80; PROF SURG & PHYSIOL, MED COL WIS, 81-, CHIEF, DEPT SURG RES SECT, 84-, DIR, INTERDEPT RES UNIT DIGESTIVE SYST RES, 87- *Concurrent Pos:* Mem prog comt, Am Gastrointestinal Soc, 84 & 85, Am Motility Soc, 84; mem, Digestive Dis Core Ctr Site comts, NIH, 84 & 85; ad hoc mem, Physiol Study sect, NIH, 85; ed, J Gastrointestinal Motility. *Honors & Awards:* Career Scientist Award, Vet Admin, 84, Outstanding Performance Award, 86, 88 & 89. *Res:* Gastrointestinal smooth muscle motor function. *Mailing Add:* Dept Surg Med Col Wis 8700 W Wisconsin Ave Milwaukee WI 53226

SARNA-WOJCICKI, ANDREI M, b Gdynia, Poland, May 30, 37; US citizen; m 79; c 3. TEPHROCHRONOLOGY, NEOTECTONICS. *Educ:* Columbia Col, NY, BA, 59; Univ Calif, Berkeley, PhD(geol), 71. *Prof Exp:* GEOLOGIST, US GEOL SURV, 71- *Concurrent Pos:* Instr geol, Univ Calif, Berkeley, 71-73. *Mem:* Friends of the Pleistocene. *Res:* Correlation of volcanic ash layers (tephrochronology) in the western United States; assessment of volcanic and seismic hazards in the western United States; volcanic ash dispersal from Mount St Helens, Washington. *Mailing Add:* 708 Garland Dr Palo Alto CA 94303

SARNER, STANLEY FREDERICK, b New York, NY, Oct 15, 31; m 67; c 2. PHYSICAL CHEMISTRY. *Educ:* City Col New York, BSc, 52; Univ Cincinnati, MSc, 61; Univ Del, PhD, 70. *Prof Exp:* Chemist, Picatinny Arsenal, NJ, 52-54; res chemist flight propulsion labs, Gen Elec Co, Ohio, 56-61; sr chemist, Thiokol Chem Corp, Md, 61-66, staff chemist, 66-67; res chemist, F&M Sci Div, Hewlett-Packard Co, Pa, 67-68; res chemist exp sta, E I du Pont de Nemours & Co, Inc, 68-70; res chemist, Chem Data Systs, 70-71; RES CHEMIST, REACTION INSTRUMENTS, 71- *Concurrent Pos:* Sr res assoc & instr chem, Univ Del, 71-; Archmere Acad, 74- *Honors & Awards:* Bendix Aviation Corp Award, 57; Del Award, Am Chem Soc, 70. *Mem:* Am Chem Soc; Am Inst Aeronaut & Astronaut; Brit Interplanetary Soc. *Res:* High temperature reactions; thermodynamics and kinetics; rocket propellants; pyrolysis; gas chromatography; instrument research. *Mailing Add:* 2635 Sherwood Dr Wilmington DE 19808

SARNESKI, JOSEPH EDWARD, b East Orange, NJ, Oct 27, 44; m 75. ANALYTICAL CHEMISTRY, INORGANIC CHEMISTRY. *Educ:* Kings Col, BS, 66; Case Western Reserve Univ, PhD(chem), 71. *Prof Exp:* Res assoc, Univ NC, Chapel Hill, 72-75; vis asst prof anal chem, Duke Univ, 75-78; asst prof, 78-80, ASSOC PROF CHEM, FAIRFIELD UNIV, 80- *Mem:* Am Chem Soc. *Res:* Nuclear magnetic resonance; analytical and structural applications; chelate chemistry; conformation analysis; chromatography. *Mailing Add:* Dept Chem Fairfield Univ Fairfield CT 06430

SARNGADHARAN, MANGALASSERIL G, BIOCHEMISTRY, VIROLOGY. *Educ:* Univ Delhi, PhD(chem), 66. *Prof Exp:* DIR DEPT CELL BIOL, BIONETICS RESEARCH INC, 77- *Res:* Protein chemistry; immunochemistry. *Mailing Add:* Dept Cell Biol Bionetics Res Inc 5510 Nicholson Lane Kensington MD 20895-1078

SAROFF, HARRY ARTHUR, b New York, NY, Mar 8, 14; m 50; c 3. ORGANIC CHEMISTRY. *Educ:* Rensselaer Polytech Inst, BS, 36, MS, 37, PhD(org chem), 40. *Prof Exp:* Res exec chem, US Naval Med Res Inst, 47-50; from chemist to chief sect, 50-74, CHIEF SECT MACROMOLECULES, LAB BIOPHYS CHEM, NAT INST ARTHRITIS & METAB DIS, 74- *Mem:* AAAS; Am Soc Biol Chemists; Biophys Soc; Am Chem Soc. *Res:* Physical chemistry of proteins; protein modification reactions; binding of ions and small molecules to proteins; action of hemoglobin. *Mailing Add:* NIH Bldg 4 Bethesda MD 20892

SAROFIM, ADEL FARES, b Cairo, Egypt, Oct 21, 34; US citizen; m 67; c 1. CHEMICAL ENGINEERING. *Educ:* Oxford Univ, BA, 55; Mass Inst Technol, SM, 57, ScD(chem eng), 62. *Prof Exp:* From asst prof to assoc prof chem eng, 61-72, Joseph R Mares prof, 81-84, PROF CHEM ENG, MASS INST TECHNOL, 72- , LAMMOT DU PONT PROF, 89- *Concurrent Pos:* Vis prof, Scheffield Univ, Eng, 71, Univ Naples, Italy, 83; mem, Panel Hazardous Trace Substances, Off Sci & Technol, 72 & 74, Comt Health & Ecol Effect Increased Utilization, Dept HEW, 77 & Energy Eng Bd, Nat Acad Sci, 83-88; Chevron vis prof, Calif Inst Technol, 78; comt hazardous waste in lab, Nat Acad Sci, 81-84, Chem Eng Res Frontiers, 85-87; Hottel lectr, Combustion Inst, 86; Lacey lectr, Calif Inst Technol, 87. *Honors & Awards:* Kuwait Petrochem Eng Prize, 83; Sir Alfred Egerton Medal, Combustion Inst, 84. *Mem:* Am Chem Soc; Am Inst Chem Engrs; Combustion Inst. *Res:* Radiative heat transfer, combustion, fluidization, gas-solid reactions, aerosol formation. *Mailing Add:* Dept Chem Eng Mass Inst Technol Cambridge MA 02139

SARPHIE, THEODORE G, b Hattiesburg, Miss, July 18, 44; m 73; c 2. HISTOCHEMISTRY. *Educ:* Univ Southern Miss, BS, 66, MS, 68; Univ Miss, PhD(anat & path), 72. *Prof Exp:* Asst prof anat, Univ S Ala, 72-79; ASST PROF ANAT, HAHNEMANN MED COL, 79- *Mem:* Am Soc Cell Biol; Histochem Soc; Am Asn Anatmists. *Res:* Macromolecular response of endothelial cell surfaces to hemodynamic forces and pathological conditions; using transmission and scanning electron microscopy and histo-cytochemistry. *Mailing Add:* Dept Anat Med Ctr La State Univ 1901 Perdido St New Orleans LA 70112

SARPKAYA, TURGUT, b Turkey, May 7, 28; US citizen; m 62. MECHANICAL ENGINEERING, MATHEMATICS. *Educ:* Tech Univ Istanbul, BS & MS, 50; Univ Iowa, PhD(eng mech), 54. *Prof Exp:* Res engr, Hydrodyn Lab, Mass Inst Technol, 54-55; lectr hydrodyn, Univ Paris, 55-56; from asst prof to prof eng mech, Univ Nebr, Lincoln, 57-64, Fawick prof, 64-66, prof mech eng, 66-67; prof, 67-76, chmn dept, 67-71, DISTINGUISHED PROF MECH ENG, NAVAL POSTGRAD SCH, MONTEREY, CALIF, 76- *Concurrent Pos:* Res vis prof, Univ Manchester; aerodyn res inst, Univ Gottingen, 71-72; Sigma Xi res award, 71. *Honors & Awards:* L F Moody Award, Am Soc Mech Engrs, 67; Collingwood Prize, Am Soc Civil Engrs, 57; Fluids Eng Award, 90. *Mem:* Int Asn Hydraul Res; Am Soc Mech Engrs; Am Soc Civil Engrs; Am Inst Aeronaut & Astronaut; Sigma Xi. *Res:* Hydrodynamics; heat transfer; unsteady fluid motions; turbulence; biomedical engineering; vortex motion; stability of flows; fluidics; pulsating flows in rigid and elastic systems. *Mailing Add:* Dept Mech Eng Naval Postgrad Sch Monterey CA 93943

SARRAM, MEHDI, b Kerman, Iran, June 28, 42; m 68; c 2. NUCLEAR SYSTEMS ENGINEERING. *Educ:* Univ Mich, BS, 65, MS, 66; Univ Teheran, PhD, 70. *Prof Exp:* Asst prof nuclear eng, Univ Tehran, 67-73; dir nuclear safeguards, Atomic Energy Orgn, Iran, 73-81; consult nuclear safeguards, Int Atomic Energy Agency, Vienna, Austria, 81-82; MGR NUCLEAR ANALYSIS, UNITED ENGRS & CONSTRUCTORS INC, 82- *Concurrent Pos:* Nuclear reactor supv, Tehran Nuclear Ctr, 67-73; dir nuclear training, Atom Energy Orgn, Iran, 73-81; invited lectr, Sandia Lab, 82; invited prof, Univ Pa, 84. *Mem:* Am Nuclear Soc. *Res:* Nuclear safeguards and physical security. *Mailing Add:* United Engrs Co 30 S 17th St Philadelphia PA 19101

SARRAS, MICHAEL P, JR, CELL BIOLOGY, ANATOMY. *Educ:* La State Univ, PhD(anat & cell biol), 78. *Prof Exp:* ASST PROF ANAT, MED CTR, UNIV KANS, 82- *Res:* Developmental biology; biochemistry. *Mailing Add:* Dept Anat Div Cell Biol Univ Kans Med Ctr 39th & Rainbow Blvd Kansas City KS 66103

SARRIF, AWNI M, b Acre, Israel, Sept 2, 42; US citizen; m 71; c 2. PHARMACOLOGY, TOXICOLOGY. *Educ:* Am Univ Beirut, BS, 65; Univ Wis, MS, 69, PhD(pharmacol), 73. *Prof Exp:* Proj assoc oncol, Univ Wis-Madison, 72-74, res assoc, 74-76; asst prof pharmacol, Univ Ala, Birmingham, 76-81; AT HASKELL LAB TOXICOL & INDUST MED, E I DUPONT CO, 81- *Mem:* Am Asn Cancer; Environ Mutagen Soc; Am Soc Pharmacol & Exp Therapeut; Soc Toxicol; Am Chem Soc. *Res:* Chemical carcinogens and genetic toxicology. *Mailing Add:* Haskell Lab Toxicol & Indust Med E I Dupont Co PO Box 50 Newark DE 19714

SARTAIN, JERRY BURTON, b Walnut, Miss, May 29, 45; m 65; c 1. SOIL SCIENCE, STATISTICS. *Educ:* Miss State Univ, BS, 67, MS, 70; NC State Univ, PhD(soil sci), 74. *Prof Exp:* From asst prof to assoc prof, 74-86, PROF, DEPT SOIL SCI, UNIV FLA, 86- *Mem:* Am Soc Agron; Soil Sci Soc Am; Soil & Crop Sci Soc Fla. *Res:* Turf and ornamental soil fertility. *Mailing Add:* Dept Soil Sci 106 Newell Hall Univ Fla Gainesville FL 32611

SARTELL, JACK A(LBERT), b St Cloud, Minn, June 18, 24; m 51; c 1. METALLURGY. *Educ:* Univ Minn, BS, 49, MS, 51; Univ Wis, PhD(metall), 56. *Prof Exp:* Res engr, Res Lab, Aluminum Co Am, 51-52; instr, Univ Wis, 52-56; sr res scientist, 56-63, staff scientist, 63-64, head res sect, 64-67, mgr appl physics dept, 67-86; CONSULT, MATH & PROCESSES, HONEYWELL RES CTR, 87- *Mem:* Fel Am Soc Metals; Am Inst Mining, Metall & Petrol Engrs; Brit Inst Metals. *Res:* Oxidation of metals and alloys; magnetics, computer memories and process control. *Mailing Add:* Appl Physics Dept 10701 Lyndale Ave S Bloomington MN 55420

SARTIANO, GEORGE PHILIP, cancer chemotherapy, internal medicine, for more information see previous edition

SARTIN, AUSTIN ALBERT, b Texarkana, Tex, Apr 21, 36; m 61. CARBONATE PETROGRAPHY. *Educ:* Centenary Col, BS, 59; Univ Ark, MS, 66; Southern Methodist Univ, PhD(geol), 72. *Prof Exp:* Tex Eastern Corp, 60-64; instr geol, Univ Southwestern La, 66-67; from asst prof to prof geol, Stephen F Austin State Univ, 70-85; WILLIAM C WOOLF PROF & CHMN GEOL, CENTENARY COL, 85- *Concurrent Pos:* Adj prof, Stephen F Austin State Univ, 85-; mem, Acad Liaison Comt, Am Asn Petrol Geologists, 87-93, delegate 89-92. *Mem:* Fel Geol Soc Am; Sigma Xi; Soc Econ Paleontologists & Mineralogists (vpres, 76-86, pres, 86-90); Am Inst Prof Geologists. *Res:* Lithofacies and biofacies analysis of carbonate and terrigenous clastic depositional systems; significance of fossil algae in carbonate shelf sedimentation; analysis of selected oil producing areas of East Texas and Northwest Louisiana. *Mailing Add:* Dept Geol Centenary Col PO Box 41188 Shreveport LA 71134-1188

SARTOR, ALBIN FRANCIS, JR, b Bartlett, Tex, Oct 14, 19; m 50; c 1. PETROLEUM CHEMISTRY. *Educ:* Rice Inst, BA, 41. *Prof Exp:* Jr res chemist, Shell Oil Co, 41-44, from res chemist to sr res chemist, 44-56, supvr, 56-63, sr res chemist, 63-86; RETIRED. *Mem:* Am Chem Soc. *Res:* Process and catalyst development; hydrogenation; technical information. *Mailing Add:* Country Meadows Ct Pearland TX 77584

SARTOR, ANTHONY, b Englewood, NJ, Mar 28, 43; m 64; c 3. CHEMICAL ENGINEERING. *Educ:* Manhattan Col, BChE, 64; Univ Mich, Ann Arbor, MSE, 65, PhD(chem eng), 68. *Prof Exp:* Develop engr, Celanese Plastics Co, 68-70; sr water qual engr, Consol Edison of NY, Inc, 70-72; dir environ affairs, NY Power Pool, 72-77; pres, Sartor Assocs, 77-78; EXEC VPRES, PAULUS, SOKOLOWSKI & SARTOR INC, 78- *Mem:* Am Inst Chem Eng; Am Chem Soc. *Res:* Effects of power plant discharges on the environment. *Mailing Add:* Paulus Sokolowski & Sartor Inc 67A Mountain Blvd Ext Warren NJ 07059-0039

SARTORELLI, ALAN CLAYTON, b Chelsea, Mass, Dec 18, 31. BIOCHEMICAL PHARMACOLOGY. *Educ:* Northeastern Univ, BS, 53; Middlebury Col, MS, 55; Univ Wis, PhD(oncol), 58. *Hon Degrees:* MA, Yale Univ, 67. *Prof Exp:* Asst chem, Middlebury Col, 53-55; asst oncol, Univ Wis, 55-58; from res chemist to sr res chemist, Biomed Div, Samuel Roberts Noble Found, 58-61; from asst prof to assoc prof pharmacol, Sch Med, Yale Univ, 61-67, chmn dept, 77-84, dep dir, Yale Comprehensive Cancer Ctr, 82-84 & Cancer Prev Res Unit Conn, 89-91, head Develop Therapeut Prog, Cancer Ctr, 74-90, PROF PHARMACOL, SCH MED, YALE UNIV, 67-, DIR, YALE COMPREHENSIVE CANCER CTR, 84-, ALFRED GILMAN PROF PHARMACOL, 87- *Concurrent Pos:* Mem cancer clin invest rev comt, NIH, 68-72, consult psoriasis topical chemother planning comt, 72; consult biochem, Univ Tex M D Anderson Hosp & Tumor Inst, Houston, 70-75; mem adv comt, Cancer Res Ctr, Mallinckrodt Inst Radiol, Sch Med, Wash Univ, 71 & inst res grants comt, Am Cancer Soc, 71-75; mem bd dirs, Am Asn Cancer Res, 75-78 & 84-87; mgt consult, Off Dir, Div Cancer Treat, Nat Cancer Inst, 75-76, mem, Bd Sci Counr, 78-82; consult, Nat Inst Arthritis, Metab & Digestive Dis prog on develop of iron chelators for clin use, 75-76; regional ed, Am Continent, Biochem Pharmacol, 68-; assoc ed, Cancer Res, 71-78; exec ed, Pharmacol & Therapeut, 74-; mem exp therapeut study sect, USPHS, 73-77; Charles B Smith vis res prof, Mem Sloan Kettering Cancer Ctr, 79; mem, Nat Prog Comt, 13th Int Cancer Cong, 79-, Publ Comt, Am Asn Cancer Res, 79-82, chmn comt, 81-82; mem adv bd, Univ Iowa Cancer Ctr, 79-83, Drug & Vaccine Develop Corp, Ctr Pub Resources, 80-81, Clin Cancer Res Ctr, Brown Univ, 80-86 & Specialized Cancer Ctr, Mt Sinai Med Ctr, 81-; consult, Bristol-Myeres Squibb, 82-; mem, External Adv Comt, Duke Comprehensive Cancer Ctr, 83-, Univ Southern Calif, 83-, sci adv bd, Liposerne Co, 86-, bd dirs, Metastasis Res Soc, 84-90, bd adv, Grace Cancer Drug Ctr, Roswell Park Mem Inst, 86-89, award comt, Am Soc Pharmacol & Therapeut, 88, Forum Drug Develop & Regulation, Inst Med, 89-, bd vis, Moffitt Cancer Ctr, Univ SFla, 89-, consult group, Cancer Ctr Prog, Nat Cancer Inst, 89- & nat bd, Cosmetic, Toiletry & Fragrance Asn, 89-; chmn, Sci Adv Comt, Columbia Univ Comprehensive Cancer Ctr, 86-, Sci Adv Bd, Vira Chem, Inc, 86- & bd dirs, Asn Am Cancer Inst, 88; ed-in-chief, Cancer Commun, 89-; William N Creasy vis prof clin pharmacol, Wayne State Univ, 83 & Bowman Gray Sch Med, 86; Mayo Found vis prof oncol, Mayo Clin, 83; Wellcome vis prof basic sci, Univ Pittsburgh Sch Med, 90. *Honors & Awards:* Paul K Smith Lectr, George Washington Univ, 78; Rufus Cole Lectr, Rockefeller Univ, 80; Walter Hubert Lectr, Brit Asn Cancer Res, 85; Pfizer Lectr, Univ Conn Health Ctr, 85; Award in Exp Therapeut, Am Soc Pharmacol & Exp Therapeut, 86; Mike Hogg Award Lectr, Univ Tex Anderson Cancer Ctr, 89. *Mem:* Inst Med-Nat Acad Sci; Am Soc Cell Biol; Am Soc Pharmacol & Exp Therapeut; Am Soc Biol Chemists; fel NY Acad Sci; Am Asn Cancer Res (vpres, 85, pres, 86); Asn Am Cancer Insts (vpres, 86, pres, 87); fel AAAS. *Res:* Biochemistry; molecular pharmacology; action of growth-inhibitory agents; nucleotide and polynucleotide metabolism; mechanisms of cell death; cell membranes; oncology; mechanisms of cell differentiation; mechanisms of drug resistance. *Mailing Add:* Dept Pharmacol Yale Univ Sch Med New Haven CT 06510

SARTORI, LEO, b Milan, Italy, Dec 9, 29; nat US; m 61; c 2. THEORETICAL ASTROPHYSICS, DEFENSE POLICY. *Educ:* Mass Inst Technol, SB, 50. PhD(physics), 56. *Prof Exp:* Physicist, Brookhaven Nat Lab, 55-56; instr physics, Princeton Univ, 56-59; asst prof, Rutgers Univ, 59-63; mem res staff sci teaching ctr, Mass Inst Technol, 63-66, lectr, 66-68, assoc prof, 68-72; chmn dept, 72-78, PROF PHYSICS, UNIV NEBR-LINCOLN, 72-, PROF POLIT SCI, 83- *Concurrent Pos:* Consult, Lockheed Aircraft Corp, 58-63, Arms Control & Disarmament Agency, 78-81, Dept of Energy, 81-88 & 89; Fulbright fel, Univ Torino, 53; chmn, Forum Physics & Soc, 84-85; vis scholar, Stanford Ctr Int Security & Arms Control, 85, 86 & 88. *Mem:* Fel Am Phys Soc; Arms Control Asn; Am Asn Physics Teachers; Fedn Am Sci; Int Astron Union. *Res:* High-energy astrophysics; supernovas; radio and x-ray sources; defense policy and arms control; theory of synchroton radiation. *Mailing Add:* Behlen Lab Physics Univ Nebr Lincoln NE 68588-0111

SARTORIS, DAVID JOHN, b Chicago, Ill, Nov 25, 55; m. DIAGNOSTIC RADIOLOGY, MUSCULOSKELETAL IMAGING. *Educ:* Stanford Univ, BS, 76, Md, 80. *Prof Exp:* Asst prof, 85-87, ASSOC PROF RADIOL, UNIV CALIF SCH MED, SAN DIEGO, 87-, CHIEF MUSCULOSKELETAL IMAGING, MED CTR, 85- *Concurrent Pos:* Vis prof, Univ Calif, Davis, 85, Univ Calif, Irvine, Maricopa Med Ctr & St Joseph's Hosp, Phoenix, Ariz, 86,

Univ Pittsburgh Sch Med & Doris Palmer Arthritis Ctr, St Margaret Mem Hosp, Pittsburgh, 88; mem, ed adv bd, Chem Rubber Co Press Inc, 85-, Diag Imaging Mag, 87 & Thieme Med Publ Inc, 86-; mem, Res Grant Rev Bd, The Arthritis Soc, 86- & bd consult, Bd Trustees Res Initiative Comt, Radiol Soc NAm, 87-; mem, Comt Prof Self-Eval & Continuing Educ, Am Col Radiol, 85-88 & Prog Comt Subcomt Gen Diag, Radiol Soc NAm, 87-88. *Honors & Awards:* Wallace Graham Lectr, Univ Ottawa, 87. *Mem:* Am Col Radiol; Am Roentgen Ray Soc; Asn Univ Radiologists; Radiol Soc NAm; Int Skeletal Soc. *Res:* Osteoporosis; noninvasive bone densitometry; musculoskeletal magnetic resonance imaging; diagnostic imaging of the musculoskeletal system; imaging-pathologic correlation in musculoskeletal disease; sports medicine; musculoskeletal trauma; temporomandibular joint dysfunction. *Mailing Add:* Radiol Dept H-756 Univ Calif Med Ctr 225 Dickinson St San Diego CA 92103

SARTORIS, NELSON EDWARD, b Auburn, Ill, Aug 2, 41; m 64; c 2. ORGANIC CHEMISTRY. *Educ:* MacMurray Col, AB, 63; Northwestern Univ, PhD(org chem), 68. *Prof Exp:* From asst prof to assoc prof, 68-80, PROF CHEM, WITTENBERG UNIV, 80-, CHMN DEPT, 84- *Concurrent Pos:* Vis prof chem, Rice Univ, 81-82. *Mem:* Am Chem Soc; AAAS. *Res:* Base catalyzed reactions; addition of ketenes to olefins; stereochemistry. *Mailing Add:* Dept Chem Wittenberg Univ PO Box 720 Springfield OH 45501

SARTWELL, PHILIP EARL, b Salem, Mass, Sept 11, 08; m 36; c 2. EPIDEMIOLOGY. *Educ:* Boston Univ, MD, 32; Harvard Univ, MPH, 38. *Prof Exp:* From asst prof to prof, 47-73, EMER PROF EPIDEMIOL, SCH HYG & PUB HEALTH, JOHNS HOPKINS UNIV, 73- *Concurrent Pos:* Consult, WHO, NIH & USPHS; vis lectr, Harvard Sch Pub Health, 75-89; adj prof epidemiol & biostatist, Pub Health, Boston Univ, 83-89. *Honors & Awards:* John Snow Award, Am Pub Health Asn. *Mem:* Am Epidemiol Soc; hon mem Royal Soc Med. *Res:* Epidemiology of acute and chronic conditions, including adverse effects of oral contraceptives; ionizing radiation. *Mailing Add:* 38 Cloutman Lane Marblehead MA 01945

SARVER, EMORY WILLIAM, b Bluefield, WVa, Oct 1, 42; m 63; c 3. ANALYTICAL CHEMISTRY. *Educ:* WVa Univ, AB, 64; Marshall Univ, MS, 66; Lehigh Univ, PhD(chem), 69. *Prof Exp:* CHEMIST, PHYS RES DIV, CHEM SYST LAB, ABERDEEN PROVING GROUND, 71- *Mem:* Am Chem Soc; Sigma Xi. *Res:* Characterization of biological materials by pyrolysis-mass spectrometer; development of methods for analysis of these reactants in dilute aqueous and nonaqueous systems. *Mailing Add:* US Army Chem Res Develop & Eng Ctr SMECR-DDT Aberdeen Proving Ground MD 21010

SARVEY, JOHN MICHAEL, b North Tonawanda, NY, Dec 31, 46; m 80. NEUROPHARMACOLOGY, ELECTROPHYSIOLOGY. *Educ:* Williams Col, BA, 69; State Univ NY, Buffalo, PhD(pharmacol), 76. *Prof Exp:* Vis scientist, Max Planck Inst Brain Res, Ger, 76-79; asst prof, 79-85, ASSOC PROF PHARMACOL, UNIFORMED SERV UNIV HEALTH SCI, 85- *Mem:* Soc Neurosci. *Res:* Electrophysiological investigation of synaptic pharmacology neurotransmitter function, and neuronal plasticity in the rat hippocampus in situ and in thin hippocampal slices in vitro and in rat visual cortical slices; effects of anticholinesterases on hippocampol slices. *Mailing Add:* Dept Pharmacol Uniformed Serv Univ Health Sci 4301 Jones Bridge Rd Bethesda MD 20814

SARWAR, GHULAM, b DI Khan, Pakistan, Jan 8, 43; Can citizen; m 69; c 3. PROTEIN & AMINO ACID NUTRITION, SAFETY OF NUTRITIONAL ASSESSMENT OF NOVEL FOODS. *Educ:* Univ Peshawar, Pakistan, BSc, 64, MSc, 68; Univ Sask, MSc, 71, PhD(nutrit), 74. *Prof Exp:* Res assoc nutrit, Univ Sask, 75-77; sci evaluator regulatory affairs, 77-82, RES SCIENTIST NUTRIT, HEALTH & WELFARE CAN, 82- *Concurrent Pos:* Postdoctoral fel, Univ Alta, 73-75; mem, Codex Comt Veg Proteins, Can Deleg, 82-89; auxiliary prof, Laval Univ, 85-89, McGill Univ, 85-91; off analyst, Health Protection Br, Govt Can, 85-91; sci adv, Food & Agr Orgn, UN, 89-90; sci consult, UN Develop Prog, Pakistan, 90. *Mem:* Am Inst Nutrit; Can Soc Nutrit Sci. *Res:* Protein nutrition such as analysis of proteins and amino acids in foods and determination of their biological effects and requirements, with emphasis on infant formulas. *Mailing Add:* Nutrit Res Div Nat Health & Welfare Health Protection Br Sir Frederick Banting Res Ctr Tunney's Pasture Ottawa ON K1A 0L2 Can

SARWATE, DILIP VISHWANATH, b Nagpur, India, Dec 25, 45; c 2. ELECTRICAL ENGINEERING, COMPUTER SCIENCE. *Educ:* Univ Jabalpur, BSc, 65; Indian Inst Sci, Bangalore, India, BE, 68; Princeton Univ, PhD(elec eng), 73. *Prof Exp:* Res assoc, 73-74, asst prof, 74-80, assoc prof elec, 80-84, PROF, UNIV ILL, 84- *Concurrent Pos:* Sr investr, Joint Serv Electronics Prog, 73-90; fac investr, NSF grant, 73-76, co-prin investr, 76-80; co-prin investr, Army Res Off contract, 78-91, Technol Challenge grant, State Ill, 90-; co-chmn, 18th & 19th Ann Allerton Conf Commun, 80-81; consult ITT Corp, Aerospace & Optical Div Systs, 85-; treas, Inst Elec & Electronics Engrs Info Theory Group, 81-83; assoc ed, coding theory, Inst Elec Eng Electronics, info theory, 83-85; co-chmn, army res office, workshop for spread spectrum systs, 85-; Battelle Columbus Labs, short term analysis prog, 85-86. *Mem:* Fel Inst Elec & Electronics Engrs. *Res:* Multiple-access communications; spread-spectrum communication; coding theory; analysis of algorithms. *Mailing Add:* Coord Sci Lab Univ Ill 1101 W Springfield Ave Urbana IL 61801-3082

SARWER-FONER, GERALD, b Poland, Dec 6, 24; nat Can; m 50; c 5. PSYCHIATRY. *Educ:* Univ Montreal, BA, 45, MD, 51; Royal Col Physicians Can, cert psychiat, 55; Am Bd Psychiat & Neurol, dipl, 57; FRCP(C), 71, FRCPsychiat, 73. *Prof Exp:* Intern, Univ Montreal Hosps, 50-51, clin lectr psychiat, Fac Med, 53-55; sr asst resident, Western Reserve Univ Hosps, Cleveland, Ohio, 52-53; clin demonstr, Fac Med, McGill Univ, 54-55, demonstr, 55-58, lectr, 58-62, asst prof, 62-66, assoc prof psychiat, 66-71; dir dept psychiat & psychiatrist-in-chief, Queen Elizabeth Hosp, 66-71; dir, Dept Psychiat, Ottawa Gen Hosp, 71-88, chmn dept, 74-86, PROF PSYCHIAT, FAC MED, UNIV OTTAWA, 71-; DIR, LAFAYETTE CLIN, DETROIT, MI, 89; PROF PSYCHIAT, WAYNE STATE UNIV MED SCH, DETROIT, MI, 90- *Concurrent Pos:* Fel, Butler Hosp, Providence, RI, 51-52; asst resident, Queen Mary Vet Hosp, 53-54, chief resident, 54-55, dir psychiat res, 55-60; dir psychiat res, Jewish Gen Hosp, 55-65, assoc psychiatrist, 55-71; mem adv bd psychiat, Defence Res Bd Can, 58-62; vis prof, Fac Med, Laval Univ, 64-76; consult, Notre Dame Hosp, Montreal, 64-71; vis prof, Chicago Med Sch, 68-76; consult, Ottawa Civic Hosp, 71-, Royal Ottawa Hosp, 71-, Pierre Janet Ctr Hosp, 72- & Nat Defence Med Ctr, 74- *Honors & Awards:* Hassan Azima Mem Lectr, Soc Biol Psychiat, 63; Silver Apple Award, Am Acad Psychiat Law, 77; First Samuel Bellet Mem Lectr, Inst Law & Psychiat, Univ Pa, 78; Karl Stern Mem Lectr, Univ Ottawa Fac Med, 79; Simon Bolivar Lectr, Am Psychiat Asn, New Orleans, 81; Sigmund Freud Award, Am Asn Psychoanal Physicians, 82; Laughlin Award, Am Col Psychoanal; William Silverburg Award, Am Acad Psychoanal, 90; Knight Malta. *Mem:* Fel Royal Col Psychiatrists; fel Am Col Psychiat; fel Am Col Psychoanalysts (pres, 84-85); fel Int Col Psychom Med (secy, 79-81); Am Acad Psychiat & Law (pres, 75-77); Soc Biol Psychiat (pres, 83-84); Can Psychoanal Soc (pres, 77-81); Can Asn Prof Psychiat (pres, 76-77 & 82-86); Am Asn Social Psychiat (pres elect, 91). *Res:* Dynamics of psychiatric drug therapy; adaptive difficulties of immigrant groups; psychoanalytic psychotherapy of marital problems; object relationship classification of depressive illnesses; human territoriality and instinct theory; anal object relationships. *Mailing Add:* Psychiat Dept Wayne State Univ Med Sch 951 E Lafayette Ave Detroit MI 48207

SARWINSKI, RAYMOND EDMUND, b La Salle, Ill, Jan 11, 36; div; c 2. SUPERCONDUCTING INSTRUMENTS & MAGNETS. *Educ:* Univ Ill, Urbana, BS, 60, MS, 61, PhD(physics), 66. *Prof Exp:* Fel physics, Ohio State Univ, 66-67, asst prof, 67-72; mgr advan develop, SHE Corp, 72-82; CONSULT, CRYOGENIC DESIGNS, 82- *Concurrent Pos:* Consult, Gen Atomic, Appl Superconetics, Quantum Design, Jet Propulsion Lab, Hughes, Ball Aerospace, Aerojet Gen, GWR, Nat Radio Astron Observ. *Mem:* Am Phys Soc. *Res:* Cryogenic instruments and measurements; MRI superconducting magnets; squid instruments; special dewars and cryostats; JT refrigeration systems; thermometers and coil foil material. *Mailing Add:* 2655 Soderbloom Ave San Diego CA 92122

SARYAN, LEON ARAM, b Wilmington, Del, July 18, 48; m 81; c 2. BLOOD LEAD TESTING, INDUSTRIAL HYGIENE. *Educ:* Johns Hopkins Univ, BA, 70, PhD(biol chem), 75. *Prof Exp:* Lab tech chem, Allied Kid Co, 64-68; engr air pollution, DuPont Co Petrol Lab, 68-69; NSF fel cancer res, Roswell Park Mem Inst, 70; res fel chem, Univ Wis-Milwaukee, 75-81, adj asst prof chem, 81-82; TECH DIR TOXICOL, WEST ALLIS MEM HOSP, 82- *Concurrent Pos:* Consult & pres, Saryan Assocs, 87-; asst clin prof, Dept Path, Med Col Wis, 91- *Mem:* Am Chem Soc; Am Indust Hyg Asn; Armenian Engrs & Scientists Am. *Res:* Heavy metal toxicology including blood lead testing for industrial and environmental exposures; cancer research and trace element analysis and metabolism, asbestos, and environmental science. *Mailing Add:* Indust Toxicol Lab West Allis Mem Hosp 8901 W Lincoln Ave West Allis WI 53227

SAS, DARYL, tissue culture, antibody production, for more information see previous edition

SASAKI, CLARENCE TAKASHI, b Honolulu, Hawaii, Jan 24, 41; m 67; c 2. OTORHINOLARYNGOLOGY, NEUROPHYSIOLOGY. *Educ:* Pomona Col, BA, 62; Yale Univ, MD, 66. *Prof Exp:* Intern, Univ Calif, San Francisco, 66-67; resident surg, Dartmouth Med Ctr, 67-68 & Yale-New Haven Med Ctr, 70-73; from instr to assoc prof, 73-82, PROF SURG, YALE SCH MED, 82-, CHIEF, SECT OTOLARYNGOL, 81- *Concurrent Pos:* Attend surg, Yale-New Haven Hosp; attend surg, West Haven Vet Admin Hosp, 73-76, consult, 73-, Windham Community Hosp, 80- & Backas Hosp, 81-; mem, Commmun Sci Study Sect, NIH, 80-83; prin investr, NIH grant, Hearing Res Study Sect, 83-84, Sensory Dis & Lang Study Sect, 85-88. *Honors & Awards:* First Prize Clin Res, Am Acad Ophthalmol & Otolaryngol, 72; Edmund Prince Fowler Award, Triol Soc, 79. *Mem:* Sigma Xi; Am Col Surgeons; Am Soc Head & Neck Surg; Triol Soc; Am Laryngol Soc; Am Acad Otolaryngol; Soc Neurosci; Asn Res Otolaryngol; NY Acad Sci; Soc Neurovascular Surg; Soc Head & Neck Surgeons; Cartesian Soc; Collegium ORLAS; NAm Skull Base Soc; Am Bronchoesoph Asn. *Res:* Neurophysiology of the larynx; postnatal development as related to the sudden infant death syndrome; tinnitus, development of a neurophysiologic correlate. *Mailing Add:* Dept Surg Yale Sch Med PO Box 3333 New Haven CT 06510

SASAKI, GORDON HIROSHI, b Honolulu, Hawaii, July 27, 42; m 69; c 1. PLASTIC SURGERY, GENERAL SURGERY. *Educ:* Pomona Col, BA, 64; Yale Univ, MD, 68. *Prof Exp:* Intern, Health Sci Ctr, Univ Ore, 68-69, surg resident, 69-70, surg res fel, 72-74, gen surg, 74-77; chief res plastic surg, Med Sch, Yale Univ, 77-79; asst prof plastic surg, Southwestern Med Sch, Univ Tex, 79-82; asst prof, Surg & Plastic Surg, 82-85, CLIN ASSOC PROF, DEPT PLASTIC SURG, UNIV SOUTHERN CALIF, LOS ANGELES, 85- *Concurrent Pos:* Instr gen surg & res assoc, Med Sch, Yale Univ, 78-79; fel, Surg Hand, Univ Conn, 78-79; attend staff, Vet Admin Hosp & Children's Hosp, Dallas, Tex, 79-82; attend surg, Huntington Mem Hosp, 85- *Mem:* AMA; fel Am Col Surg; Asn Acad Surg; Am Asn Plastic Surg; Am Asn Hand Surg; Am Bd Surg; Am Bd Plastic Surg. *Res:* Steroid and peptide hormones in breast cancer and other target tissues; microcirculation in skin; prostaglandins and microcirculation; macrophage-myofitonblast and wound contraction. *Mailing Add:* 800 Fairmount Ave Suite 319 Pasadena CA 91105

SASAKI, HIDETADA, b Akita, Japan, July 28, 41; m 71; c 2. GERIATRIC MEDICINE, RESPIRATORY MEDICINE. *Educ:* Tohoku Univ, Dr, 66, PhD(med), 71. *Prof Exp:* PROF & CHMN, DEPT GERIAT MED, 87- *Honors & Awards:* Kumagai Award, Japanese Thoracic Soc, 83. *Mem:* Am Thoracic Soc; Am Physiol Soc; Am Col Chest Physicians. *Res:* Aspiration pneumonia; bronchial asthma; Alzheimer's disease; chronic obstructive pulmonary disease; fibrosing lung disease. *Mailing Add:* Dept Geriat Med Sch Med Tohoku Univ 1-1 Seiryo-machi Aobaku Sendai 980 Japan

SASAKI, YOSHI KAZU, b Akita, Japan, Jan 2, 27; m 54; c 4. DYNAMIC METEOROLOGY. *Educ:* Univ Tokyo, BS, 50, PhD, 55. *Prof Exp:* Res scientist meteorol res found, Agr & Mech Col, Tex, 56-60, prin investr, 58-60, mem fac, 59-60; res scientist, 60-62, adj assoc prof, 61-64, assoc prof, 64-67, PROF METEOROL, UNIV OKLA, 67-, GEORGE LYNN CROSS RES PROF, 74- *Concurrent Pos:* Res dir, Naval Environ Prediction Res Fac, 74-75; dir, Coop Inst Mesoscale Meteorol Studies, 80- *Honors & Awards:* Prize, Meteorol Soc Japan, 55. *Mem:* Fel Am Meteorol Soc; Meteorol Soc Japan. *Res:* Mesometeorology; numerical weather prediction; variational methods. *Mailing Add:* Dept Meteorol Univ Okla 660 Parrington Oval Norman OK 73019

SASAMORI, TAKASHI, b Tokyo, Japan, Feb 1, 30; m 58; c 3. METEOROLOGY. *Educ:* Tohoku Univ, Japan, BSc, 53, MSc, 55, DrSc(meteorol), 58. *Prof Exp:* Res assoc meteorol, Geophys Inst, Tohoku Univ, Japan, 58-60; staff scientist, Tech Inst, Japan Defense Agency, 60-64; res assoc dept astron & geophys, Univ Colo, 64-65; staff scientist, Japan Defense Agency, 65-66; staff scientist, Nat Ctr Atmospheric Res, 67-78; PROF METEOROL, UNIV ILL, URBANA-CHAMPAIGN, 78- *Concurrent Pos:* Fel, Univ Colo, 64-65. *Mem:* Am Geophys Union; Am Meteorol Soc; Meteorol Soc Japan. *Res:* Radiation transfer in planetary atmosphere; numerical modelling of planetary boundary layer. *Mailing Add:* Lab Atmospheric Res 5 127 Csl Univ Ill Urbana Campus 1101 W Springfield Urbana IL 61801

SASHIHARA, THOMAS F(UJIO), b Los Angeles, Calif, May 7, 29; m 57; c 3. CHEMICAL ENGINEERING. *Educ:* Ohio State Univ, BChE & MSc, 53, PhD(chem eng), 57. *Prof Exp:* Instr chem eng, Ohio State Univ, 55-56; chem engr, Polychem Dept, 56-63, sr res engr, Plastics Dept, 63-71, SUPT RES LAB, PLASTICS DEPT, E I DU PONT DE NEMOURS & CO, INC, 71- *Mem:* Am Chem Soc; Am Inst Chem Engrs. *Res:* Chemical processes. *Mailing Add:* Three Roan Ct Surrey Park Wilmington DE 19803

SASHIN, DONALD, b New York, NY, Dec 11, 37; m 67; c 2. RADIOLOGICAL PHYSICS, HEALTH PHYSICS. *Educ:* Mass Inst Technol, BS, 60; Carnegie Inst Technol, MS, 62; Carnegie-Mellon Univ, PhD(medium energy physics), 68. *Prof Exp:* Proj physicist, Carnegie Inst Technol, 62-67; from instr to assoc prof radiol & radiation health, 67-74, dir radiol imaging, 75-84, ASSOC PROF RADIOL, SCH MED, UNIV PITTSBURGH, 74- *Concurrent Pos:* Pa Lions Sight Conserv & Eye Res Found grants, 71; Nat Cancer Inst contract, 73-74; Am Cancer Inst grant, 77-79; Nat Heart, Lung & Blood Inst contracts, 77-79, 80-83, 83-88; Western Pa Advan Tech Ctr grant, 86-88. *Mem:* Sigma Xi; Am Phys Soc; Health Physics Soc; Am Asn Physicists in Med. *Res:* Development of an electronic radiographic imaging system to improve diagnostic radiography to reduce exposure, reduce procedure time and improving image quality; clinical evaluation of the system performance in neurosurgery, gastro-intestinal fluoroscopy, angiography, mammography, pelvimetry, intra-uterine fetal transfusions; intravenous angiography and chest radiography. *Mailing Add:* RC 405 Scaife Hall Univ Pittsburgh Pittsburgh PA 15261

SASHITAL, SANAT RAMANATH, b Bagalkot, India; US citizen. SOLID STATE PHYSICS, MATERIALS SCIENCE. *Educ:* Univ Bombay, BSc, 59, MSc, 61; Pa State Univ, PhD(solid state sci), 67. *Prof Exp:* Physicist, Res Labs, Zenith Radio Corp, 69-74; vis scholar, Dept Mat Sci, Northwestern Univ, 74-76; mem tech staff, Res Labs, Hughes Aircraft Co, 76-88; mgr, 88-90, SR SECT HEAD, MAT PROCESSING, TRW, 88- *Mem:* Am Asn Crystal Growth; Am Vacuum Soc. *Res:* Crystal growth; epitaxial growth; thin films, structure and properties; electro-optic materials; restricted access memory materials. *Mailing Add:* 13 Mallard Irvine CA 92714-3630

SASIELA, RICHARD, b Brooklyn, NY, June 1, 40; m 62; c 2. LASERS. *Educ:* Polytech Inst Brooklyn, BEE, 61, MS, 62, PhD(electrophys), 67. *Prof Exp:* Res engr, Microwave Assocs, Inc, 67-69; MEM STAFF, LINCOLN LAB, MASS INST TECHNOL, 69- *Mem:* Inst Elec & Electronics Engrs; Sigma Xi. *Res:* Analysis of electromagnetic wave propragation in turbulence and development of adaptive-optics systems. *Mailing Add:* Lincoln Lab PO Box 73 Lexington MA 02173

SASIN, RICHARD, b Warsaw, Poland, Nov 16, 22; nat US; m 57; c 1. ORGANIC CHEMISTRY. *Educ:* Drexel Inst, BS, 47; Temple Univ, MA, 49, PhD(chem), 54. *Prof Exp:* Asst chem, Temple Univ, 47-51; instr, Drexel Inst, 51-53, asst prof, 53-57, assoc prof, 57-60, prof, 60-68; PROF CHEM & DEAN DIV SCI & MATH, MILLERSVILLE STATE COL, 68- *Concurrent Pos:* Consult, Hardesty Industs, 55-57; Fatty Acid Producers Coun fel, Eastern Regional Res Lab, USDA, 57-58. *Mem:* AAAS; Am Chem Soc; Am Oil Chem Soc; Sigma Xi. *Res:* Nitrogen heterocyclics; organotin compounds; derivatives of long-chain fatty acids; sulfur compounds and phosphorus derivatives of fatty acids. *Mailing Add:* 1117 Amy Lane Lancaster PA 17601

SASKI, WITOLD, b Poland, Dec 4, 09; nat US; m 50. BIOPHARMACEUTICS. *Educ:* Batory Univ, Poland, MPharm, 33; Univ Bologna, DPharm, 46; Inst Optical Sci, London, Eng, dipl, 50; Univ Nebr, BSc, 54. *Prof Exp:* Practicing pharmacist, Poland, 33-36; pharmaceut inspector, Polish Ministry Soc Welfare, 37-39; chief pharmacist med clin, Eng, 47-48; sr pharmacist supplies div, Brit Ministry Health, 48-51; asst prof pharm, Mont State Univ, 51-52; from asst prof to prof, 52-75, PROF PHARM, COL PHARM, UNIV NEBR, MED CTR, 75- *Concurrent Pos:* Vis assoc prof sch pharm, Univ Calif, San Francisco, 59-60; Fulbright scholar, Univ Pisa, 67-68; Nat Acad Sci exchange scientist, Poland, 70. *Mem:* Fel Am Acad Pharmaceut Sci; Am Asn Cols Pharm. *Res:* Surface-active agents and drug absorption mechanisms; medicinal product formulation; experimental pharmaceutical technology; author of nearly 350 publications. *Mailing Add:* 2600 S 46th St Lincoln NE 68506-2520

SASLAW, LEONARD DAVID, b Brooklyn, NY, Aug 27, 27. BIOCHEMISTRY, TOXICOLOGY. *Educ:* City Col New York, BS, 49; George Washington Univ, MS, 54; Georgetown Univ, PhD(chem), 63. *Prof Exp:* Chemist, Nat Cancer Inst, NIH, 51-57; chemist, Div Biophys, Sloan-Kettering Inst, 57-58; chemist, Biochem Br, Armed Forces Inst Path, 58-64; dir div biochem pharmacol, Cancer Chemother Dept, Microbiol Assocs, Inc, 65-68; sr biochemist, Nat Drug Co, 68-69; chief, Lab Cellular Biochem, Albert Einstein Med Ctr, 69-70; clin lab dir, Med Diag Ctrs, Inc, 70-71; lab dir & res assoc, Renal Lab, New York Med Col, 71-73; mgr biochem invests, Bio Dynamics Inc, NJ, 73-74; prof assoc, Smithsonian Sci Info Exchange, 75-77; physiologist, div toxicol, bur foods, 78-83, PHYSIOLOGIST, CTR VET MED, FOOD & DRUG ADMIN, 83- *Concurrent Pos:* Consult, Burton Parsons, Inc, 77-78. *Mem:* Sigma Xi; Am Col Toxicol; Am Soc Pharmacol & Exp Therapeut Assay Soc; Am Asn Cancer Res; Am Inst Biol Sci. *Res:* Drug metabolism; analytical biochemistry; cancer research; oxidation of unsaturated fatty acids; clinical chemistry; toxicology. *Mailing Add:* Food & Drug Admin 7500 Standish Pl Rockville MD 20855

SASLOW, WAYNE MARK, b Philadelphia, Pa, Aug 11, 42; m 71; c 2. LOW TEMPERATURE PHYSICS, SOLID STATE SCIENCE. *Educ:* Univ Pa, BA, 64; Univ Calif, Berkeley, MA, 67; Univ Calif, Irvine, PhD(physics), 68. *Prof Exp:* Res assoc, Univ Pittsburgh, 69-71; asst prof, 71-77, assoc prof, 77-83, PROF PHYSICS, TEX A&M UNIV, 83- *Concurrent Pos:* Joliot-Curie fel & vis prof, Univ Paris, 80-81. *Mem:* Am Phys Soc. *Res:* Theory of liquid helium, solid state theory; surface physics; spin glasses; hydrodynamics of condensed matter systems; disordered magnetic systems. *Mailing Add:* Dept Physics Tex A&M Univ College Station TX 77843-4242

SASMAN, ROBERT T, b Plattsburgh, NY, July, 23, 23; m 51; c 3. HYDROLOGY & WATER RESOURCES. *Educ:* Univ Wis, Madison, BS, 47. *Prof Exp:* Soil analyst, Univ Wis, 47; soil scientist, US Soil Conserv Serv, 47-48, geologist, 48-51; asst hydrologist, Ill State Water Surv Div, 51-57, hydrologist, 57-87; HYDROLOGIST, 87- *Concurrent Pos:* Lectr numerous cols, univ & orgn, 57-88; dir, Am Water Works Asn, 75-78, Groundwater Protection Adv Workgroup, 87-91, chair, Groundwater Comt, 89-91. *Honors & Awards:* Fuller Award, Am Water Works Asn, 76. *Mem:* Hon mem Am Water Works Asn; Nat Water Well Asn. *Res:* Analysis of groundwater resources of northern Illinois including availability, quantity and quality of shallow and deep aquifers, groundwater withdrawals, water-level trends, and feasibility of artificial recharge. *Mailing Add:* 1217 Sunset Rd Wheaton IL 60187-6119

SASMOR, DANIEL JOSEPH, physical chemistry, materials science; deceased, see previous edition for last biography

SASNER, JOHN JOSEPH, JR, b Lawrence, Mass, June 15, 36; m 58; c 3. COMPARATIVE PHYSIOLOGY, INVERTEBRATE ZOOLOGY. *Educ:* Univ NH, BA, 57, MS, 59; Univ Calif, Los Angeles, PhD(comp physiol), 65. *Prof Exp:* Res assoc physiol & biophys, Univ Ill, 62-63; res & develop off, USAF Sch Aerospace Med, 63-65; asst prof zool, 65-69, assoc prof, 69-81, PROF ZOOL, UNIV NH, 81- *Res:* Effects of naturally occurring microorganism toxins on excitable membranes. *Mailing Add:* 310 Mendums Landing Barrington NH 03825

SASS, DANIEL B, b Rochester, NY, Mar 28, 19; m 59; c 2. PALEONTOLOGY, ENVIRONMENTAL SCIENCES. *Educ:* Univ Rochester, BA & MS, 51; Univ Cincinnati, PhD, 59. *Prof Exp:* From asst prof to prof geol, 52-82, chmn dept, 52-74, coordr environ studies prog, 74-80, EMER PROF GEOL, ALFRED UNIV, 82- *Concurrent Pos:* Scholes Sigma Xi lectr, 65; mus cur & consult, 82- *Mem:* Fel Geol Soc Am; Paleont Soc; Electron Micros Soc. *Res:* Devonian stratigraphy and paleontology; electron microscopy; ultrastructure of bivalve shells; biomedical research; kirlian photography. *Mailing Add:* 27 High St Alfred NY 14802

SASS, HEINZ, GENE REGULATION. *Educ:* Max Planck Inst, Tübingen, Germany, PhD(chromosome res), 78. *Prof Exp:* Res assoc, Harvard Univ, 84-89. *Res:* Molecular anatomy; chromosome structure and function. *Mailing Add:* Dept Cell Biol Worcester Found Exp Biol 222 Maple Ave Shrewsbury MA 01545

SASS, JOHN HARVEY, b Chatham, Ont, July 20, 37; m 61. GEOPHYSICS. *Educ:* Univ Western Ont, BS, 59, MS, 61; Australian Nat Univ, PhD(geophys), 65. *Prof Exp:* Chief, Earthquake Drilling Proj, 84-85, on-site sci mgr, Salton Sea Sci Drilling Proj, 85-86, GEOPHYSICIST, US GEOL SURV, 67-, CHIEF, GEOTHERMAL STUDIES PROJ, 84- *Concurrent Pos:* Nat Res Coun Can fel, Univ Western Ont, 64-66; vis fel, Australian Nat Univ, 71-72 & Stanford Univ, 72. *Mem:* AAAS; Am Geophys Union; Can Geophys Union. *Res:* Earth's heat and internal temperatures; continental scientific drilling. *Mailing Add:* US Geol Surv 2255 N Gemini Dr Flagstaff AZ 86001

SASS, LOUIS CARL, b Chicago, Ill, Dec 18, 10. EXPLORATION. *Educ:* Univ Chicago, BS, 32. *Prof Exp:* Explor geol, Gulf Oil Corp, 33-65; RETIRED. *Mem:* Geol Soc Am; Sigma Xi; Am Asn Prof Geologists; Soc Econ Paleontologist & Minerologists. *Mailing Add:* 945 Tenderfoot Hill Rd Colorado Springs CO 80906

SASS, NEIL LESLIE, b Baltimore, Md, Oct 24, 44; m 84; c 3. NUTRITIONAL BIOCHEMISTRY, TOXICOLOGY. *Educ:* Wake Forest Col, BS, 66; WVa Univ, MS, 69, PhD(biochem), 71; Johns Hopkins, MS, 84. *Prof Exp:* Res toxicologist, Chem-Toxicol, Biomed Lab, US Army Edgewood Arsenal, 71-73, res anal chemist explosives, Chem Lab, 73-74; chief lab serv, Med Res & Develop, William Beaumont Army Med Ctr, 74-77; toxicologist, Bur Foods, Food & Drug Admin, 77-82, SPEC ASST TO DIR, CTR FOOD SAFETY & APPL NUTRIT, FOOD & DRUG ADMIN, 82- *Concurrent Pos:* Fel, Appl Behav Sci/Orgn Develop, Johns Hopkins Univ. *Mem:* Sigma Xi; Soc Appl Spectroscopy; Am Chem Soc; Soc Armed Forces Med Lab Scientists; Am Col Toxicol; NY Acad Sci; Am Soc Training & Develop. *Res:* Biochemical mechanisms of the chemical pain response; mechanisms of idiopathic respiratory distress syndrome; reptilian venoms as presynaptic acetylcholine inhibitors; radioimmunoassay development for thyroid hormone precursors; thyroid hormone actions; biochemical toxicology; motivational aspects of human performance; psychosocial aspects of eating disorders. *Mailing Add:* 12900 Fork Rd Baldwin MD 21013-9345

SASS, RONALD L, b Davenport, Iowa, May 26, 32; m 52, 69; c 4. BIOPHYSICAL CHEMISTRY. *Educ:* Augustana Col, BA, 54; Univ Southern Calif, PhD, 57. *Prof Exp:* Res fel chem, Brookhaven Nat Lab, 57-58; from asst prof to assoc prof, 58-66, PROF CHEM & BIOL, RICE UNIV, 66- *Concurrent Pos:* Guggenheim fel, Cambridge Univ, 65; adj prof, Baylor Col Med, 70-; fel, Nat Res Coun, 88. *Honors & Awards:* Salgo-Noren Distinguished Prof Award, 66. *Mem:* Am Geophys Union. *Res:* Ecology; atmospheric science; biogeochemistry. *Mailing Add:* Dept Biol Rice Univ Houston TX 77251

SASS, STEPHEN L, b New York, NY, Mar 11, 40; m 66; c 2. MATERIALS SCIENCE. *Educ:* City Col New York, BChE, 61; Northwestern Univ, PhD(mat sci), 66. *Prof Exp:* Res asst mat sci, Northwestern Univ, 61-66; Fulbright scholar, Delft Univ, 66; from asst prof to assoc prof, 67-79, PROF MAT SCI, CORNELL UNIV, 79- *Concurrent Pos:* Max-Planck fel, 80-81; Kringel vis prof, Technion, 80-81. *Mem:* Am Inst Metall Engrs; Am Ceramic Soc; Electron Micros Soc Am; Am Soc Metals; fel Am Phys Soc; AAAS. *Res:* Phase transformations in solids; electron microscopy; electron diffraction; x-ray diffraction; diffraction from crystalline imperfections; internal interfaces. *Mailing Add:* Dept Mat Sci & Eng Cornell Univ Ithaca NY 14853

SASSA, SHIGERU, b Tokyo, Japan, Mar 3, 35; m 63; c 2. HEMATOLOGY, BIOCHEMISTRY. *Educ:* Univ Tokyo, MD, 61, DrMedSci, 66. *Prof Exp:* Res assoc med, Fac Med, Univ Tokyo, 68; res assoc, 68-71, asst prof hemat & biochem, 71-75, ASSOC PROF HEMAT & BIOCHEM & PHYSICIAN, ROCKEFELLER UNIV, 75- *Mem:* Am Soc Biol Chem. *Res:* Heme biosynthesis; the regulatory mechanism of enzyme induction. *Mailing Add:* Rockefeller Univ New York NY 10021

SASSAMAN, ANNE PHILLIPS, b LaGrange, Ga, Jan 7, 44; m 66, 83; c 2. BLOOD DISEASES, ENVIRONMENTAL HEALTH. *Educ:* Auburn Univ, BS, 65; Duke Univ, PhD(immunol), 70. *Prof Exp:* Res assoc, Dept Surg, Med Ctr, Duke Univ, 70-71, Dept Biochem, 71-74; chemist, Bur Biol, Food, Drug & Admin, 74-76; scientist-adminr, 76-79, chief, Blood Dis Br, Nat Heart, Lung & Blood Inst, 79-86; DIR, DIV EXTRAMURAL RES & TRAINING, NAT INST ENVIRON HEALTH SCI, 86- *Concurrent Pos:* Fel cardiol, Dept Med, Med Ctr, Duke Univ, 71-74. *Honors & Awards:* NIH Director's Award, 85. *Mem:* Int Soc Thrombosis & Hemostasis; Am Soc Hemat; Am Heart Asn; AAAS; Sigma Xi; Soc Occup & Environ Health. *Res:* Biochemistry of blood coagulation and fibrinolysis; blood diseases, including thrombosis and hemostasis and red cell disorders. *Mailing Add:* Nat Inst Environ Health Sci Box 12233 Res Triangle Park NC 27709

SASSAMAN, CLAY ALAN, b Washington, DC, Dec 27, 48. POPULATION BIOLOGY, COMPARATIVE PHYSIOLOGY. *Educ:* Col William & Mary, BS, 70; Stanford Univ, PhD(biol), 76. *Prof Exp:* Scholar, Woods Hole Oceanog Inst, 75-76; ASST PROF BIOL, UNIV CALIF, RIVERSIDE, 76- *Mem:* Genetics Soc Am; AAAS; Am Genetic Asn. *Res:* Population genetics; population ecology; comparative physiology of invertebrates and lower vertebrates. *Mailing Add:* Dept Biol 900 Univ Ave Univ Calif Riverside CA 92521

SASSCER, DONALD S(TUART), b Washington, DC, June 30, 29; m 58; c 4. OCEAN THERMAL ENERGY CONVERSION. *Educ:* Univ Utah, BS, 53; Iowa State Univ, MS, 59, PhD(nuclear eng), 64. *Prof Exp:* Instr eng mech, Iowa State Univ, 56-59, asst prof nuclear eng, 61-64; head, Dept Nuclear Eng, Univ PR, 64-74, prof, 74-80, chief scientist & head Div Ocean Thermal Energy, Ctr Energy & Environ Res, 79-83, PROF MECH ENG, UNIV PR, 80-, CHIEF SCIENTIST II & ASST DIR, CTR ENERGY & ENVIRON RES, 83- *Concurrent Pos:* Researcher nuclear eng, Ames Lab, US AEC, 61-64, chief scientist & head dir nuclear eng, PR Nuclear Ctr, 64-76; consult, Argonne Nat Lab, US Dept Energy, 79- *Mem:* Am Nuclear Soc; Am Soc Eng Educ; Health Physics Soc. *Res:* Alternative energy research; co-generation of electrical energy on a farm by using an anaerobic digestor to produce methane gas fuel; long-term, in-situ determination of the biofouling and corrosion of potential ocean thermal energy conversion heat exchanger materials. *Mailing Add:* Dept Mech Eng Nuclear Eng Univ PR Mayaguez PR 00709

SASSE, EDWARD ALEXANDER, b Amarillo, Tex, July 10, 38; m 60; c 2. CLINICAL CHEMISTRY, CLINICAL PATHOLOGY. *Educ:* Arlington State Col, BS, 63; Univ Tenn, Memphis, PhD(biochem), 68; Am Bd Clin Chem, dipl, 73. *Prof Exp:* From instr to asst prof clin path, Sch Med, Univ Ala, Birmingham, 68-70; asst prof, 70-74, ASSOC PROF PATH, MED COL WIS, 74- *Concurrent Pos:* Consult scientist, Univ Ala Hosps & Clins, 68-70; co-dir clin chem sect, Dept Path & Labs, Milwaukee County Gen Hosp, 70- *Mem:* AAAS; Am Chem Soc; Am Asn Clin Chem. *Res:* Clinical chemistry methodology; toxicology; endocrinology; physical and chemical properties of proteins; enzyme immunoassay. *Mailing Add:* Dept Path Med Col Wis 8701 Watertown Plank Rd Milwaukee WI 53226

SASSENRATH, ETHELDA NORBERG, b Dubuque, Iowa, Feb 21, 21; m 51; c 2. BEHAVIORAL PHYSIOLOGY, PSYCHOPHARMACOLOGY. *Educ:* Dubuque Univ, AB, 42; Iowa State Univ, PhD(chem), 49. *Prof Exp:* Asst res biochemist, Sch Med, Univ Calif, 49-59; res assoc psychopharmacol, Ind Univ, 59-64; res specialist, Nat Ctr Primate Biol, 64-68, lectr behav biol, Sch Med, 68-73, res behav biologist, Calif Primate Res Ctr, 73-76, ASSOC PROF BEHAV BIOL, SCH MED, UNIV CALIF, DAVIS, 76-, PROF PSYCHIAT, 80- *Mem:* AAAS; Soc Neurosci; Soc Exp Biol & Med; Sigma Xi. *Res:* Mechanisms of psychoactive drug action; endocrine correlates of behavior; psycho endocrine correlates of stress and aging. *Mailing Add:* 4234 Cowell Blvd Davis CA 95616

SASSER, JOSEPH NEAL, b Goldsboro, NC, May 19, 21; m 45; c 4. NEMATOLOGY. *Educ:* NC State Col, BS, 43, MS, 50; Univ Md, PhD, 53. *Prof Exp:* Asst nematologist, USDA, 51-53; from asst prof to prof, 53-84, EMER PROF PLANT PATH, NC STATE UNIV, 84- *Concurrent Pos:* Tech consult, Rockefeller Found, Chile, 63-64; ed, Am Phytopath Soc, 63-65; prin investr, Int Meloidogyne Proj, 75-84 & Crop Nematode Res & Control Proj, 84-89; hon mem, Orgn Trop Am Nematologists. *Honors & Awards:* OMax Gardner Award, 82; Adventurers in Agr Sci Award of Distinction, 79. *Mem:* Hon mem Soc Nematol (vpres, 61-63, pres, 63-64); Am Phytopath Soc. *Res:* Nematode diseases of plants. *Mailing Add:* Dept Plant Path NC State Univ Raleigh NC 27695-7616

SASSER, LYLE BLAINE, b Tremonton, Utah, Feb 20, 39; m 63; c 4. DEVELOPMENTAL TOXICOLOGY, NUTRITION. *Educ:* Univ Idaho, BS, 61; Colo State Univ, MS, 65, PhD(nutrit), 68. *Prof Exp:* Asst prof animal sci, Fresno State Col, 67; asst scientist, Univ Tenn-AEC Agr Res Lab, 68-72, from asst prof to assoc prof nutrit, Univ Tenn-Oak Ridge, 73-81; sr res scientist, 81-87, STAFF SCIENTIST, BATTELLE PAC NORTHWEST LABS, 87- *Mem:* Am Inst Nutrit; Soc Exp Biol & Med; Soc Toxicol. *Res:* Metabolism of trace elements in animal nutrition; radioisotope tracers; trace element absorption studies; toxicity of heavy metals; interaction of heavy metals and dietary nutrients; body composition; effects of radiation on domestic animals; developmental toxicology, cardiovascular toxicology. *Mailing Add:* Biol & Chem Dept Battelle Pac Northwest Labs Box 999 Richland WA 99352

SASSOON, HUMPHREY FREDERICK, b Barnes, Eng, Aug 18, 20; US citizen; m 56; c 3. TOXICOLOGY. *Educ:* Oxford Univ, BA & MA, 46, BA, 48; Bristol Univ, PhD(animal sci), 64. *Prof Exp:* Res assoc nutrit biochem, Univ Ill, 64-65; sr investr biochem, Okla Med Res Found & med sch, Univ Okla, 65-70; res specialist alcoholism, Tex Res Inst Ment Sci, 70-71; res assoc life sci res off, Fedn Am Socs Exp Biol, 71-73; sr scientist, Tracor Jitco Inc, 73-76; sr scientist, Envirocontrol Inc, 76-81; PROPRIETOR, JEFFERSON FRUIT FARM, 81- *Concurrent Pos:* Founder & dir, Wash Grove Singers, 75-80; consult, Merrill Math, Merrill Publ Co, 83-85. *Mem:* AAAS; Brit Nutrit Soc; Am Inst Nutrit. *Res:* Adaptation of enzyme systems to dietary carbohydrates and fats; color vision anomalies in school children and in families of alcoholics; nutritional toxicology; human cultural evolution. *Mailing Add:* Jefferson Farm Rte 1 Box 136A Clearbrook VA 22624

SASTRE, ANTONIO, b Los Angeles, Calif, June 14, 50. NEUROTRANSMITTER RECEPTORS. *Educ:* Cornell Univ, BA, 70, MS, 73, PhD(appl math), 74. *Prof Exp:* Lectr, Cornell Univ, 74-76; instr, pharmacol, Cornell Med Col, 76-77; ASST PROF PHYSIOL, SCH MED, JOHNS HOPKINS UNIV, 77-, ASST PROF NEUROSCI, 80- *Concurrent Pos:* Adj asst prof pharmacol, Cornell Med Col, 79-; mem, Basic Sci Coun, Am Heart Asn. *Mem:* Soc Neurosci. *Res:* Cholinergic and adrenergic receptors in the heart and vascular smooth muscle; synaptogenesis in and properties of neurons and cardiocytes in tissue-culture. *Mailing Add:* John Hopkins Univ 725 N Wolfe St John Hopkins Univ 720 Rutland Ave Baltimore MD 21205

SASTRI, SURI A, b Tanjore, India, Dec 26, 39; m 69. PHYSICAL METALLURGY, MATERIALS SCIENCE. *Educ:* Banaras Hindu Univ, BSc, 60; Univ London, PhD(eng) & DIC, 64. *Prof Exp:* Australian Inst Nuclear Sci & Eng fel mat physics, Australian Atomic Energy Comn Res Labs, Sydney, 64-66; AEC res assoc metall, Ames Inst Atomic Res, Iowa State Univ, 67-68; res & develop scientist, Microphys Div, Gillette Safety Razor Co, 68-69, chief res & develop scientist, 69-71, group mgr mat res, 71-76, dir, chem & mat res, Boston Res & Develop Labs, 76-83; PRES, SURMET CORP, 83- *Mem:* Electron Micros Soc Am; Am Soc Metals; Am Inst Mining, Metall & Petrol Engrs; Brit Iron & Steel Inst. *Res:* Application of electron microscopy to the understanding of the relationship between microstructure and mechanical properties of materials; phase transformations and strengthening mechanisms in materials. *Mailing Add:* Surmet Corp 33 B St Burlington MA 01803

SASTRY, BHAMIDIPATY VENKATA RAMA, b Andhra, India, Oct 21, 27; nat US; m 68; c 1. PHARMACOLOGY, MEDICINAL CHEMISTRY. *Educ:* Andhra Univ, India, BSc, 49, MSc, 50, DSc(med chem), 56; Emory Univ, MS, 59; Vanderbilt Univ, PhD(pharmacol), 62. *Prof Exp:* Demonstr pharmaceut chem, Andhra Univ, India, 51-52, lectr, 52-56; res asst pharmacol, Emory Univ, 56-59; res assoc, 59-60, from instr to assoc prof, 60-71, PROF PHARMACOL, VANDERBILT UNIV, 71- *Mem:* AAAS; Soc Exp Biol & Med; fel Am Inst Chem; Soc Toxicol; Am Soc Pharmacol & Exp Therapeut. *Res:* Physiology, pharmacology and toxicology of radionuclides and insecticides; synthesis and screening of psychotherapeutic agents; pharmacology and structure-activity drugs active on autonomic nervous system. *Mailing Add:* Dept Pharmacol Vanderbilt Univ Sch Med Nashville TN 37232

SASTRY, SHANKARA M L, b India, June 11, 46; m 74; c 3. METALLURGY. *Educ:* Bangalore Univ, BS, 65; Indian Inst Sci, BEng, 68, MEng, 70; Univ Toronto, PhD(metall & mat sci), 74. *Prof Exp:* Vis scientist metall, Air Force Mat Lab, Wright Patterson AFB, 74-76; from res scientist to sr scientist phys metall, 77-90, PROG DIR, METALS & COMPOSITES, MCDONNEL DOUGLAS RES LABS, 90- *Concurrent Pos:* Prin investr, var res contracts, 78- *Mem:* Metall Soc Am. *Res:* Rapid solidification processing of titanium and aluminum alloys; novel consolidation techniques (explosive and dynamic compaction) of rapidly solidified powders; advanced titanium fabrication techniques; laser processing of materials. *Mailing Add:* 1305 Sheperd Hollow Colencoe MO 63038

SASTRY, VANKAMAMIDI VRN, b Guntur, India, Sept 1, 40; m 68; c 2. PILE FOUNDATIONS, DYNAMICS OF PARTICLE. *Educ:* Osmania Univ, BE Hons, 60; Indian Inst Sci, ME, 62; Tech Univ NS, Can, PhD(civil eng), 77. *Prof Exp:* Asst engr, Heavy Eng Corp, India, 62-65; assoc prof civil eng, Osmania Univ, 65-82; res assoc, Tech Univ NS, 82-87; DIR, DIV ENG,

ST MARY'S UNIV, 87- *Concurrent Pos:* Consult, Nolan Davis & Assocs Ltd, Can, 82- *Mem:* Can Geotech Soc; Asn Prof Engrs. *Res:* Behavior of pile foundations driven in homogeneous and layered soil systems and subjected to inclined and eccentric loads by studying fully instrumented, rigid and flexible piles under such loadings. *Mailing Add:* Div Eng St Mary's Univ Halifax NS B3H 3C3 Can

SASYNIUK, BETTY IRENE, b Jan 19, 42; m; c 2. CARDIOVASCULAR PHARMACOLOGY. *Educ:* Univ Man, PhD(pharmacol), 68. *Prof Exp:* ASSOC PROF CARDIOVASC PHARMACOL, MCGILL UNIV, 78- *Concurrent Pos:* Med Res Coun fel, 68-71, scholar, 71. *Mem:* Pharmacol Soc Can; Am Soc Pharmacol & Exp Ther; Can Soc Clin Invest; Cardiac Electrophysiol Soc; Can Cardiovasc Soc. *Res:* Cardiac electro-physiology; antiarrhythmic drugs. *Mailing Add:* Dept Pharmacol McIntyre Med Sci Bldg Rm 1212 McGill Univ 3655 Drummond St Montreal PQ H3G 1Y6 Can

SATAS, DONATAS, b Lithuania, Apr 7, 29; US citizen; m 53; c 3. CHEMICAL ENGINEERING. *Educ:* Ill Inst Technol, BS, 53. *Prof Exp:* Res engr, Armour Res Found, 55-57; group leader, Kendall Co, 57-69; tech dir, Whitman Prod Ltd, 69-75; CONSULT, SATAS & ASSOCS, 75- *Mem:* Am Inst Chem Engrs; Soc Plastics Engrs; Asn Consult Chemists & Chem Engrs; Am Lithuanian Engrs & Architects Asn; fel Am Inst Chemists; Adhesion Soc; Tech Asn Pulp & Paper Indust. *Res:* Adhesives and coating technology; pressure sensitive adhesives; emulsion and solution polymerization; paper coating and saturation; coating and lamination equipment. *Mailing Add:* 99 Shenandoah Rd Warwick RI 02886

SATCHER, DAVID, b Anniston, Ala, Mar 2, 41; m; c 4. MEDICAL GENETICS, MEDICAL EDUCATION. *Educ:* Morehouse Col, BS, 63; Case Western Reserve Univ, MD & PhD(cytogenetics), 70; Am Bd Family Pract, cert, 76 & 82. *Hon Degrees:* DSc, Med Col Pa, 86. *Prof Exp:* Fac fel, Community Med, 72-74, prof & chmn, Dept Family Med, actg dean, Charles R Drew Postgrad Med Sch, Los Angeles, 75-79; prof & chmn, Dept Community Med, Sch Med, Morehouse Col, Atlanta, 79-82; PRES, MEHARRY MED COL, NASHVILLE, 82- *Concurrent Pos:* Assoc dir, King-Drew Sickle Cell Ctr, Los Angeles, Calif, 73-75, dir, 74-79; Robert Wood Johnson Clin scholar & sr family pract resident, Univ Calif, Los Angeles, 75-76, asst prof epidemiol, 74-76; prin investr grants, Dept Health & Human Serv, 85-90 & 88-91, Nat Cancer Inst, 88-91, NSF, 87-; pres, Asn Minority Health Professions Schs; mem bd dirs, Univ Pa Med Ctr, First Am Bank, Nashville. *Mem:* Inst Med-Nat Acad Sci; fel Am Acad Family Physicians; AAAS; Am Cancer Soc; Am Health Asn; AMA; Am Soc Human Genetics; Nat Med Asn. *Res:* Medical genetics, community and family medicine; author or co-author of over 35 publications. *Mailing Add:* Meharry Med Col 1005 D B Todd Blvd Nashville TN 37208

SATCHLER, GEORGE RAYMOND, b London, Eng, June 14, 26; m 48; c 2. THEORETICAL PHYSICS. *Educ:* Oxford Univ, BA & MA, 51, DPhil(physics), 55. *Hon Degrees:* DSc, Oxford Univ, 89. *Prof Exp:* Sr studentship, Clarendon Lab, Oxford Univ, 51-56, Imp Chem Industs res fel, 56-59; PHYSICIST, OAK RIDGE NAT LAB, 59-, CORP RES FEL, 76- *Concurrent Pos:* Res assoc, Univ Mich, 56-57. *Honors & Awards:* Tom W Bonner Prize, Am Phys Soc, 77. *Mem:* Fel Am Phys Soc. *Res:* Theory of nuclear structure and nuclear reactions. *Mailing Add:* Physics Div PO Box 2008 Oak Ridge TN 37831-6373

SATER, VERNON E(UGENE), b Rock Rapids, Iowa, Apr 10, 35; m 57; c 3. CHEMICAL ENGINEERING. *Educ:* Ill Inst Technol, BS, 57, MS, 59, PhD(chem eng), 63. *Prof Exp:* Instr chem eng, Ill Inst Technol, 62; from asst prof to assoc prof, 62-74, PROF CHEM ENG & ACTG CHMN, ARIZ STATE UNIV, 74- *Mem:* Am Chem Soc; Am Inst Chem Engrs. *Res:* Process control; process simulation; microcomputer. *Mailing Add:* Col Eng Sci Ariz State Univ Tempe AZ 85287

SATHE, SHARAD SOMNATH, b Bombay, India, Oct 10, 40; m 66; c 2. ORGANIC & MEDICINAL CHEMISTRY. *Educ:* Bombay Univ, BSc, 60; Banaras Hindu Univ, BPharm, 63; Ind Univ, PhD(org chem), 71. *Prof Exp:* Anal chemist, Hoffman-La Roche Co, India, 63-65; res asst chem, CIBA Res Ctr, India, 65-67; assoc, Res Triangle Inst, 71-73; res investr, 73-79, tech supvr, 79-81, GROUP LEADER RES & DEVELOP, MALLINCKRODT INC, 81- *Mem:* Am Chem Soc; Am Inst Chem; NY Acad Sci. *Res:* Research and development in organic chemistry related to drug products, drug intermediates and fine organic chemicals. *Mailing Add:* Mallinckrodt Inc 3600 N Second St St Louis MO 63147

SATHE, SHRIDHAR KRISHNA, b Pune, India, Oct 30, 50; m 84. AGRICULTURAL & FOOD CHEMISTRY, FOOD SCIENCE & TECHNOLOGY. *Educ:* Bombay Univ, BSc Hons, 71, BSc Tech Hons, 74, MSc Tech, 77; Utah State Univ, PhD(food sci), 82. *Prof Exp:* Assoc lectr food technol, Bombay Univ, 76-77; postdoctoral food sci & biochem, Univ Ariz, 81-85 & food sci, Purdue Univ, 86-88; asst prof, 88-91, ASSOC PROF FOOD SCI, FLA STATE UNIV, 91- *Concurrent Pos:* Invited speaker, Cambridge Univ, UK, 83 & Univ Irapuato, Mex, 85; mem, Ann Meeting Prog Comt, Inst Food Technologists, 90-93. *Mem:* Inst Food Technologists; Am Asn Cereal Chemists; Am Oil Chemists Soc; Am Dairy Sci Asn; Sigma Xi. *Res:* Chemical and biochemical aspects of foods and their relation to nutrition and functionality; biochemistry of legumes with special emphasis on proteins, carbohydrates, and antinutritional factors. *Mailing Add:* Dept Nutrit Food & Movement Sci Fla State Univ Tallahassee FL 32306-2033

SATHER, BRYANT THOMAS, b Wallace, Idaho, Feb 8, 35; m 63. PHYSIOLOGY, ECOLOGY. *Educ:* Univ Idaho, BS, 57; Univ Hawaii, PhD(zool), 65. *Prof Exp:* Asst physiologist, Pac Biomed Res Ctr, Univ Hawaii, 64-66; asst prof zool NC State Univ, 66-67; asst prof, 67-69, ASSOC PROF PHYSIOL, RUTGERS UNIV, NEWARK, 69- *Mem:* AAAS; Am Soc Zoologists; Am Physiol Soc. *Res:* Comparative and ecological physiology; osmoregulation and electrolyte balance; mineral metabolism; renal physiology. *Mailing Add:* Dept Physiol Rutgers Univ 195 University Ave Newark NJ 07102

SATHER, DUANE PAUL, b Minneapolis, Minn, Sept 19, 33; m 55; c 3. MATHEMATICS. *Educ:* Univ Minn, BPhys, 59, MS, 60, PhD(math), 63. *Prof Exp:* Instr math, Univ Minn, 63-64; res assoc, Univ Md, 64-65; asst prof, Cornell Univ, 65-68; assoc prof math, Math Res Ctr, US Army, Univ Wis-Madison, 68-70; assoc prof, 70-74, PROF MATH, UNIV COLO, BOULDER, 74- *Mem:* Am Math Soc; Soc Natural Philos. *Res:* Partial differential equations; applied mathematics. *Mailing Add:* Math Box 426 Univ Colo Boulder CO 80309

SATHER, GLENN A(RTHUR), b Franklin, Minn, Jan 18, 28; m 52; c 1. CHEMICAL ENGINEERING, SOLUTION THERMODYNAMICS. *Educ:* Univ Minn, BCE & BBA, 52, PhD(chem eng), 59. *Prof Exp:* From asst prof to prof, 59-91, EMER PROF CHEM ENG, UNIV WIS-MADISON, 91- *Concurrent Pos:* NSF sci fac fel, Imp Col, Univ London, 66-67; year-in-indust prof, E I du Pont de Nemours & Co, 73-74. *Mem:* Am Inst Chem Engrs; Am Chem Soc; Am Soc Eng Educ. *Res:* Cryogenics; thermodynamics. *Mailing Add:* Dept Chem Eng Univ Wis Madison WI 53706

SATHER, J HENRY, b Presho, SDak, July 12, 21; m 48; c 3. ZOOLOGY. *Educ:* Univ Nebr, BSc, 43, PhD(zool), 53; Univ Mo, AM, 48. *Prof Exp:* Sr biologist fur invests, Game, Forestation & Parks Comn, Nebr, 48-53, leader game res, 53-55; wetland ecologist, US Fish & Wildlife Serv, 80-88; prof biol & dean grad sch, 55-80, EMER DEAN GRAD SCH & PROF BIOL SCI, WESTERN ILL UNIV, 80- *Concurrent Pos:* Mem, Nat Wetlands Tech Coun, 76-; adv to proj leader, Nat Wetland Inventory, 76-88; mem environ adv bd to chief engrs, 79-83; wetland res adv, Bombay Natural Hist Soc, 80- *Honors & Awards:* Patriotic Civilian Serv Award, US CEngr, 83; Spec Recognition Serv Award, Wildlife Soc, 87. *Mem:* AAAS; Explorers Club; Ecol Soc Am; Sigma Xi; Am Inst Biol Sci. *Res:* Mammals, upland game birds and wetland ecosystems. *Mailing Add:* 103 Oakland Lane Macomb IL 61455

SATHER, NORMAN F(REDRICK), b Elmhurst, Ill, Sept 17, 36; m 57; c 3. BIOTECHNOLOGY, ENVIRONMENTAL CONTROL TECHNOLOGY. *Educ:* Univ Ill, BS, 58; Univ Minn, PhD(chem eng), 62. *Prof Exp:* Asst prof chem eng, Univ Wash, 62-68, assoc prof, 68-74, prof, 74; chem engr energy & environ systs, 74-79, assoc div dir, 79-83, dep div dir, 83-89, DIV DIR, ARGONNE NAT LAB, 89- *Concurrent Pos:* Fel, Univ Cambridge, Eng, 71-72; consult, Ocean Thermal Energy Conversion to Ministry, Int Trade & Indust, Japan, 79. *Mem:* Am Inst Chem Eng; AAAS; Am Soc Mech Engr. *Res:* Development of ocean thermal energy converison power systems components and designs; thermochemical and biological processes for conversion of biomass to fuels and chemicals, and environmental control technology for coal combustion and gasification. *Mailing Add:* Energy Systs Div Argonne Nat Lab Argonne IL 60439

SATHOFF, H JOHN, b Peoria, Ill, Sept 21, 31; m 54; c 2. GENERAL PHYSICS. *Educ:* Bradley Univ, BS, 53; Ohio State Univ, MS, 55, PhD(nuclear physics), 60. *Prof Exp:* Res assoc physics, Ohio State Univ, 61; asst prof physics & chem, 61-64, assoc prof physics, 64-69, chmn dept, 64-81, PROF PHYSICS, BRADLEY UNIV, 69- *Mem:* Am Asn Physics Teachers; Am Chem Soc; Sigma Xi. *Res:* Acoustics; physics; computing. *Mailing Add:* Dept Physics Bradley Univ Peoria IL 61625

SATINOFF, EVELYN, b Brooklyn, NY; div; c 2. PHYSIOLOGICAL PSYCHOLOGY, BEHAVIORAL NEUROSCIENCE. *Educ:* Brooklyn Col, NY, BS, 58; Univ Pa, Philadelphia, PhD(physiol psychol), 63. *Prof Exp:* NIH fel & res assoc psychol, Univ Pa, 63-73; PROF PSYCHOL, UNIV ILL, 73- *Concurrent Pos:* Vis investr, Inst Invests Cerebral, Mexico City, 67-68; sr res assoc, Ames Res Ctr, Moffett Field, Calif, 75-76; mem, Thermal Physiol Comn, Int Union Physiol Sci, 77-83, Exec Comn Div Six, 81-84 & Fels Comt Div Three, Six & Twenty-eight, Am Psychol Asn, 84-86 & Neurosci Steering Comn, Am Physiol Soc, 83-86; vis prof, Bar Ilan Univ, Ramat Gan, Israel & sch med, Tel Aviv Univ, 79, NY Hosp, Cornell Med Ctr, White Plains, NY, 87-88; mem biopsychol study sect, NIH, 79-83; J S Guggenheim Fel, 87-88. *Mem:* Am Psychol Asn; Am Physiol Soc; Soc Neurosci; Sleep Res Soc; Int Union Physiol Sci; Soc Exp Psychologists. *Res:* Neural and pharmacological substrates of motivated behavior; thermoregulation; sleep-wakefulness; circadian rhythms. *Mailing Add:* Dept Psychol Univ Ill 603 E Daniel St Champaign IL 61820

SATIR, BIRGIT H, b Copenhagen, Denmark, Mar 22, 34; m 62; c 2. MEMBRANE BIOLOGY, CELL BIOLOGY. *Educ:* Univ Copenhagen, Cand Phil, 55, Magistra (biochem), 61. *Prof Exp:* Res assoc, Fibiger Lab, Copenhagen, 61-62 & Univ Chicago, 62-66; asst res physiologist, Univ Calif, Berkeley, 67-74, assoc res physiologist, 74-76, adj assoc prof, 76-77; prof & dir, Anal Ultrastruct Ctr, 77-84, PROF, ALBERT EINSTEIN COL MED, NY, 77- *Concurrent Pos:* Res fel, Univ Geneva, Switz, 65-66; spec fel, Univ Tokyo, Japan, 72-73; ed-in-chief, Mod Cell Biol, 80-; dir biol, Electron Micros Soc, 82-84. *Mem:* Am Soc Cell Biol; Electron Micros Soc; Soc Protozool. *Res:* Regulation of signal transduction in stimulus-exocytosis-coupling using secretory mutants of the ciliated protozoa; ultrastructural and biochemical analyses. *Mailing Add:* Dept Anat & Struct Biol Albert Einstein Col Med 1300 Morris Park Ave Bronx NY 10461

SATIR, PETER, b New York, NY, July 28, 36; m 62; c 2. CELL BIOLOGY, CELL MOTILITY. *Educ:* Columbia Col, AB, 56; Rockefeller Inst, PhD, 61. *Prof Exp:* Instr biol & zool, Univ Chicago, 61-63, asst prof, 63-67; assoc prof anat, Univ Calif, Berkeley, 67-73, dir, Electron Micros Lab, 69-76, prof physiol-anat, 73-77; PROF ANAT & STRUCT BIOL & CHMN DEPT, ALBERT EINSTEIN COL MED, 77- *Concurrent Pos:* Mem Nat Bd Med Examrs, 85- *Mem:* AAAS; Am Soc Cell Biol; Soc Protozool; Am Asn Anatomists; Biophys Soc. *Res:* Cell biology, ciliary motility; cytoplasmic and membrane organization; signal transduction in control of cell movement. *Mailing Add:* Dept Anat Albert Einstein Col Med 1300 Morris Park Ave Bronx NY 10461

SATKIEWICZ, FRANK GEORGE, b Cambridge, Mass, Mar 6, 27; m 51; c 4. PHYSICAL CHEMISTRY. *Educ:* Northeastern Univ, BS, 47; Wesleyan Univ, MA, 49; Mass Inst Technol, PhD(phys chem), 58. *Prof Exp:* Radiochemist, Tracerlab, Inc, 49-52; head high sch math dept, 52-54; res assoc, Norton Co, 58-63; staff scientist, GCA Corp, 63-69, prin scientist, 69-73; SR STAFF, APPL PHYSICS LAB, JOHNS HOPKINS UNIV, 73- *Mem:* Am Chem Soc; Am Soc Mass Spectrometry. *Res:* Solid state chemistry and physics; sputter-ion source mass spectrometry of solids; ion-acoustics. *Mailing Add:* 3453 Nanmark Ct Ellicott City MD 21043

SATO, CLIFFORD SHINICHI, molecular pathology, biochemistry, for more information see previous edition

SATO, DAIHACHIRO, b Fujinomiya-Shi, Japan, June 1, 32; m 56; c 3. MATHEMATICS, MATHEMATICAL EDUCATION. *Educ:* Tokyo Univ Educ, BS, 55; Univ Calif, Los Angeles, MA, 57, PhD(math), 63. *Prof Exp:* Reader math, Univ Calif, Los Angeles, 57-58, from asst to assoc, 58-61; asst prof, San Fernando Valley State Col, 61; instr, Tokai Univ, Japan, 61-62; instr, 62-63, lectr, 63, from asst prof to assoc prof, 63-71, PROF MATH, UNIV REGINA, 71- *Concurrent Pos:* Vis asst prof & NSF fel, Univ Calif, Los Angeles, 64; Can Math Cong res fel, Queen's Univ, Ont, 65, Univ Alta, 66 & 71, Univ BC, 67, Univ Man, 68 & 69, Carlton Univ, 73, Univ Calgary, 74 & 75 & Res Inst Math Sci, Kyoto Univ, 76, 83. *Honors & Awards:* Lester R Ford Award, Math Asn Am, 77. *Mem:* Am Math Soc; Math Asn Am; Soc Indust & Appl Math; Can Math Cong; Math Soc Japan. *Res:* Integer valued entire functions; generalized interpolations by analytic functions; prime representing functions; function theory; number theory; p-adic analysis; transcendentality problems; mathematics education; computer sciences; operations research; translatable GCD and LCM identities of binomial and multinomial coefficients. *Mailing Add:* Dept Math & Statist Univ Regina Regina SK S4S 0A2 Can

SATO, GENTEI, b Sendai, Japan, Mar 15, 26. ANTENNAS, MICROWAVE TRANSMISSION DEVICES. *Educ:* Tohoku Univ, BS, 47, PhD(elec eng), 61. *Prof Exp:* Res asst elec eng, Tohoku Univ, 47-56, assoc prof, 56-57; dir, Yagi Antenna Co, Ltd, 57-64; PROF ELEC ENG, SOPHIA UNIV, 64- *Concurrent Pos:* Lectr, Shibaura Inst Technol, 65-68, vis prof, 87; vis prof, Tsinghua Univ, 90. *Mem:* Fel Inst Elec & Electronics Engrs; Electromagnetic Acad. *Res:* Phased array antennas; satellite communication antennas; direction finding antennas; broadcasting antennas; microwave devices such as directional couplers, phase shifters, attenuators. *Mailing Add:* 4-1-37 Kamikizaki Urawa 338 Japan

SATO, GORDON HISASHI, b Los Angeles, Calif, Dec 17, 27; m 52; c 6. BIOLOGY. *Educ:* Univ Southern Calif, BA, 51; Calif Inst Technol, PhD(biophys), 55. *Prof Exp:* Teaching asst microbiol, Calif Inst Technol, 53-55; jr res virologist, Univ Calif, 56; instr molecular genetics, Med Sch, Univ Colo, 56-58; from asst prof to prof, Dept Biochem, Brandeis Univ, Waltham, Mass, 58-69; prof, Biol Dept, Univ Calif, San Diego, 69-83; DIR, W ALTON JONES CELL SCI CTR, INC, LAKE PLACID, NY, 83- *Concurrent Pos:* Panel mem, Molecular Biol Study Sect, NIH, 68-73, Breast Cancer Task Force, Nat Cancer Inst, 70-74; adj fac, Med Sch, Univ Vt, Burlington, 85; adj prof, Dept Biochem, Albany Med Col Union Univ, NY, 85; distinguished res prof & dir, Lab Molecular Biol, Clarkson Univ, Potsdam, NY, 87. *Honors & Awards:* Edwin J Cohn Lectr, Harvard Univ; Rosenstiel Award, 82. *Mem:* Nat Acad Sci; AAAS; Am Asn Immunologists; Endocrine Soc; Asn Biol Chem; Sigma Xi; Tissue Cult Asn (pres, 84-86); Int Cell Res Orgn; Am Soc Biol Chemists; fel Am Acad Arts & Sci. *Res:* Animal cell culture; endocrinology; bacteriophage; author of various publications. *Mailing Add:* W Alton Jones Cell Sci Ctr Ten Old Barn Rd Lake Placid NY 12946-1099

SATO, HIROSHI, b Matsuzaka, Japan, Aug 31, 18; nat US; m 47; c 3. SOLID STATE PHYSICS. *Educ:* Hokkaido Univ, MSc, 41; Univ Tokyo, DSc, 51. *Prof Exp:* Res assoc physics, Hokkaido Univ, 42, asst prof, Inst Low Temperature Res, 42-43; res physicist, Inst Phys Chem Res, Tokyo, 43-45; prof metal physics, Res Inst Iron, Steel & Other Metals, Tohoku Univ, Japan, 45-57; prin res physicist, Sci Lab, Ford Motor Co, Mich, 56-74; Ross Distinguished prof eng, 84-89, PROF, SCH MAT ENG, PURDUE UNIV, WEST LAFAYETTE, 74-, EMER ROSS DISTINGUISHED PROF ENG, 89-; COLLABR, LOS ALAMOS NAT LAB, NMEX, 89- *Concurrent Pos:* Res physicist, Res Lab, Westinghouse Elec Corp, 54-56; Guggenheim Mem fel, 66-67; consult, Solid State Div, Oak Ridge Nat Lab, 78-80; vis prof, Univ Hannover, WGer. *Honors & Awards:* Prize, Japan Inst Metals, 51; Alexandar von Homboldt US sr scientist award, 80. *Mem:* Fel Am Phys Soc; Phys Soc Japan; Am Ceramic Soc; Metall Soc; NY Acad Sci; hon mem Japan Inst Metals. *Res:* Metal physics; magnetism; diffusion and ion transport phenomena; high temperature ceramic materials; super ionic conductors; composite sic materials; kinetics of phase transitions; crystal growth; transmission electron microscopy. *Mailing Add:* Sch Mat Eng Purdue Univ West Lafayette IN 47907

SATO, MAKIKO, b Nishinomiya, Hyogo, Japan, May 29, 47; m 69; c 1. PLANETARY ATMOSPHERES. *Educ:* Osaka Univ, BS, 70; Yeshiva Univ, MA, 72, PhD(physics), 78. *Prof Exp:* Res scientist, Columbia, Univ, 78; res assoc, State Univ NY, Stony Brook, 78-79; SCI ANALYST PLANETARY SCI, SIGMA DATA SERV CORP, NASA GODDARD INST SPACE STUDIES, 80- *Concurrent Pos:* Co-investr, Voyager Spacecraft Mission, 80- *Mem:* Am Astron Soc; Planetary Soc. *Res:* Determination of the chemical compositions, cloud-haze structure and temperature profiles of the atmospheres of the outer planets by analyzing visible and infrared spectra. *Mailing Add:* Goddard Inst Space Studies 2880 Broadway New York NY 10025

SATO, MASAHIKO, CELL MOTILITY, PROTEIN BIOPHYSICS. *Educ:* Dartmouth Col, PhD(cell biol), 83. *Prof Exp:* Res fel organ histol, Sch Med, Johns Hopkins Univ, 84-86; SR SCIENTIST, 86- *Mailing Add:* Dept Bone Biol Merck Sharp Dohme West Point PA 19486

SATO, MOTOAKI, b Tokyo, Japan, Oct 11, 29; div; c 3. GEOCHEMISTRY, EARTH SCIENCES. *Educ:* Univ Tokyo, BS, 53, MS, 55; Univ Minn, Minneapolis, PhD(geol), 59. *Prof Exp:* Res fel geophys, Harvard Univ, 58-61; assoc prof geol, Inst Thermal Springs Res, Okayama Univ, 61-63; res geologist, 63-65, PROJ LEADER, US GEOL SURV, 65- *Concurrent Pos:* Apollo 12-17 prin investr oxygen fugacity studies lunar basalts, NASA. *Mem:* Am Geophys Union; Geochem Soc; div Geochem Am Chem Soc. *Res:* Electrochemistry of minerals; redox evolution of rocks; origin of sulfide self-potentials; geochemistry of gas-forming elements; electrochemical sensors for volcanic gas monitoring; oxygen fugacities of planetary rocks and meteorites; thermochemistry of fossil fuels; earthquake prediction by gas monitoring. *Mailing Add:* Mail Stop 959 Nat Ctr US Geol Surv Reston VA 22092

SATO, PAUL HISASHI, ENZYME DEFICIENCY DISEASE, ENZYME ADMINISTRATION. *Educ:* New York Univ, PhD(pharmacol), 75. *Prof Exp:* ASSOC PROF PHARMACOL & TOXICOL, MICH STATE UNIV, 83- *Res:* Ascorbic acid biosynthesis. *Mailing Add:* Dept Pharmacol & Toxicol/8430 Life Sci Mich State Univ East Lansing MI 48824

SATOH, PAUL SHIGEMI, b Osaka, Japan, Nov 6, 36; US citizen; c 2. BIOCHEMISTRY. *Educ:* St Paul's Univ, Tokyo, BA, 59; Wayne State Univ, PhD(biochem), 64. *Prof Exp:* Res assoc immunochem, Wayne State Univ, 64-66; res staff tumor immunol, Aichi Cancer Ctr, Nagoya, Japan, 66-68; sr res assoc protein chem, Med Sch, Tufts Univ, 68-72; sr res scientist III immunol, 72-79, MGR RES & DEVELOP, UPJOHN DIAG, THE UPJOHN CO, 79- *Mem:* Sigma Xi; NY Acad Sci; Am Chem Soc. *Res:* Immunological diagnosis of human cancer; enzyme-immunoassay; immunology of mediaters; lymphocytemembrane; cell mediated immunology of cancer patients, and immunosuppressive drugs; radioimmunoassay; bioluminescence. *Mailing Add:* 1424 Surrey Rd Portage MI 49081

SATRAN, RICHARD, b New York, NY, Oct 3, 28; m 51; c 2. NEUROLOGY. *Educ:* Univ Louisville, BA, 49, MD, 56; NY Univ, MA, 51. *Prof Exp:* Instr neurol & EEG, Univ Rochester Med Ctr, 62-63, dir, EEG Lab, 62-70, sr instr neurol, 63-64, from asst prof to assoc prof, 64-75, actg chmn div, 66, actg chmn dept, 84-86, vchmn dept, 79- 86, PROF NEUROL, UNIV ROCHESTER MED CTR, 75-, ASSOC CHMN DEPT, 86- , ASSOC DEAN MED SCH ADMIS, 90- *Mem:* fel Am Col Physicians; fel Am Acad Neurol; Sigma Xi; hon mem Fr Soc Neurol; fel Am Heart Asn. *Res:* Electroencephalography; cerebrovascular disease; medical history. *Mailing Add:* Dept Neurol Univ Rochester Med Ctr Rochester NY 14642

SATTAR, SYED ABDUS, b Hyderabad, India, Mar 23, 38; m 70; c 2. MEDICAL & ENVIRONMENTAL VIROLOGY. *Educ:* Univ Karachi, BSc, 58, MSc, 60; Univ Toronto, dipl bact, 62, MA, 64; Univ Ottawa, PhD(microbiol), 67. *Prof Exp:* Asst lectr microbiol, Univ Karachi, 60-61; res fel, Univ Ottawa, 64-67; lectr, Univ Karachi, 68-70; assoc prof, 70-82, PROF MICROBIOL, UNIV OTTAWA, 70-82. *Mem:* Can Soc Microbiologists; Int Am Soc Microbiol; Am Soc Virol; Int Asn Water Pollution; Int Asn Aerobiol; Can Col Microbiologists. *Res:* Study of human pathogenic viruses in air, surface and the water environment; viral gastroenteritis; disinfection of viruses. *Mailing Add:* Dept Microbiol Fac Med Univ Ottawa Ottawa ON K1H 8M5 Can

SATTEN, ROBERT A, b Chicago, Ill, Aug 4, 22; m 46; c 2. SOLID STATE SPECTROSCOPY. *Educ:* Univ Chicago, BS, 44; Univ Calif, Los Angeles, MA, 47, PhD(physics), 51. *Prof Exp:* Instr physics, Univ Calif, Los Angeles, 51-52; asst prof, Mass Inst Technol, 52-53; from asst prof to assoc prof, 53-63, PROF PHYSICS, UNIV CALIF, LOS ANGELES, 63- *Concurrent Pos:* Consult, Argonne Nat Lab, 59-72, Hughes Res Lab, 59-67 & Lockheed Res Lab, 63-65; Fulbright res fel, France, 61-62 & Ger, 69-70; vchmn dept physics, Univ Calif, Los Angelos, 68-73 & 87-; vis Erskine fel, Univ Canterbury, 71. *Mem:* Fel Am Phys Soc; Am Asn Physics Teachers. *Res:* Rare earth and actinide spectra in solids, vibronic spectra in crystals; optical detection of spin-lattice relaxation. *Mailing Add:* 1358 Woodruff Ave Univ Calif Los Angeles Los Angeles CA 90024

SATTER, LARRY DEAN, b Madelia, Minn, July 30, 37; m 66; c 1. ANIMAL NUTRITION. *Educ:* SDak State Univ, BS, 60; Univ Wis, MS, 62, PhD(biochem, dairy sci), 64. *Prof Exp:* From asst prof to assoc prof dairy sci, Univ Wis-Madison, 64-73, prof, 73-81; mem staff, 81-87, DIR, US DAIRY FORAGE RES CTR, UNIV WIS, USDA, 87- *Honors & Awards:* Am Feed Mfrs Award, 77. *Mem:* Am Dairy Sci Asn; Am Soc Animal Sci; Am Inst Nutrit; Brit Nutrit Soc. *Res:* Digestive phenomena occurring in the rumen and quantitative aspects of the rumen fermentation. *Mailing Add:* Dairy Forage Res Ctr USDA Univ Wis 1925 W Linden Dr W Madison WI 53706

SATTER, RUTH, b New York, NY, Mar 8, 23; m 46; c 4. CHRONOBIOLOGY, PLANT PHYSIOLOGY. *Educ:* Barnard Col, Columbia Univ, AB, 44; Univ Conn, PhD(bot), 68. *Prof Exp:* Res fel biol, Yale Univ, 67-73, from res assoc to sr res assoc, 73-81; PROF-IN-RESIDENCE, UNIV CONN, 81- *Concurrent Pos:* Counr, Am Soc Photobiol, 74-77; vis assoc prof bot, Conn Col, 77; mem gov bd, Am Inst Biol Sci, 78-81, exec comt, 81-83; vis prof bot, Cornell Univ, 79; mem metab biol panel, NSF, 81, postdoctoral fel panel, 84, 85 & 87. *Mem:* AAAS; Am Soc Plant Physiologists; Am Inst Biol Sci; Int Soc Chronobiol; Am Soc Photobiol; Am Women Sci. *Res:* Time measurement in higher plants with emphasis on light-clock control of ion fluxes and mechanisms of phototransduction. *Mailing Add:* Molecular & Cell Biol U-42 Univ Conn Storrs CT 06268

SATTERFIELD, CHARLES N(ELSON), b Dexter, Mo, Sept 5, 21; m 46; c 2. CHEMICAL ENGINEERING. *Educ:* Harvard Univ, SB, 42; Mass Inst Technol, SM, 43, ScD(chem eng), 46. *Prof Exp:* Res engr, 43-45, from asst prof to assoc prof chem eng, 46-59, PROF CHEM ENG, MASS INST TECHNOL, 59- *Concurrent Pos:* Vis lectr, Harvard Univ, 48-57; consult, 48-; consult, US Res & Develop Bd, 52-53 & US Dir Defense Res & Eng, 53-60; mem comt chem kinetics, Nat Acad Sci-Nat Res Coun, 60-66; mem adv bd,

Indust & Eng Chem, 66-68 & comt air qual mgt & chmn ad hoc panel abatement nitrogen oxides emissions from stationary sources, Nat Acad Eng, 70-72, Nat Res Coun Panel, direct combustion coal, 75-77. *Honors & Awards:* Wilhelm Award, Am Inst Chem Engrs; Kelly Lectr, Purdue Univ, 71; Van Winkle Lectr, Univ Texas, 79; Plenary Lectr, Int Symposium Chem React Eng, 74. *Mem:* Am Chem Soc; Am Inst Chem Engrs; fel Am Acad Arts & Sci. *Res:* Applied chemical kinetics and heterogeneous catalysis; mass transfer in chemical reactors; author and co-author of six books; holds three patents. *Mailing Add:* 38 Tabor Hill Rd Lincoln MA 01773

SATTERLEE, JAMES DONALD, b Seattle, Wash, Feb 16, 48; c 1. BIOINORGANIC & PHYSICAL CHEMISTRY. *Educ:* Cent Wash Univ, BA, 70, MS, 71; Univ Calif, Davis, PhD(chem), 75. *Prof Exp:* Res fel chem biol, Dept Chem, Calif Inst Technol, 75-78; asst prof, Dept Chem, Northern Ill Univ, 78-81; asst prof, 81-84, ASSOC PROF, DEPT CHEM, UNIV NMEX, 84- *Concurrent Pos:* Alfred P Sloan Found fel, 83-86; NIH res career develop award, 86- *Mem:* Am Chem Soc; Biophys Soc; Protein Soc. *Res:* Chemistry; metal ions in biology; nuclear magnetic resonance spectroscopy in chemical and biochemical systems; electron transfer in biology. *Mailing Add:* Dept Chem Northern Ill Univ De Kalb IL 60115

SATTERLEE, LOWELL DUGGAN, b Duluth, Minn, July 30, 43; m 63; c 2. FOOD CHEMISTRY, BIOCHEMISTRY. *Educ:* SDak State Univ, BS, 65; Iowa State Univ, MS, 66, PhD(biochem), 68. *Prof Exp:* Asst prof food technol, Iowa State Univ, 68-69; from asst prof to assoc prof, 69-75, PROF FOOD SCI, UNIV NEBR, LINCOLN, 75-, HEAD FOOD SCI, 83- *Concurrent Pos:* Viobin Labs indust grant, 69-73; Nebr Agr Asn grants, 72-79; NSF grant, 74-79 & 79-82. *Mem:* AAAS; Inst Food Technologists; Am Chem Soc. *Res:* Isolation, characterization and utilization of human food proteins. *Mailing Add:* Dept Food Sci & Technol Pa State Univ Main Campus University Park PA 16802

SATTERLUND, DONALD ROBERT, b Polk Co, Wis, Apr 10, 28; m 55; c 3. FORESTRY. *Educ:* Univ Mich, BSF, 51, MF, 55, PhD(forestry), 60. *Prof Exp:* Asst forestry, Univ Mich, 53-58; instr forest influences, State Univ NY Col Forestry, Syracuse, 58-60, asst prof, 60-64; from asst prof to assoc prof, 64-71, PROF FORESTRY, WASH STATE UNIV, 71- *Mem:* Soc Am Foresters; Soil Conserv Soc Am; Am Geophys Union. *Res:* Watershed management; forest influences and ecology. *Mailing Add:* Dept Forestry & Range Mgt Wash State Univ Pullman WA 99164-6410

SATTERLY, GILBERT T(HOMPSON), b Detroit, Mich, Sept 27, 29; m 52; c 4. CIVIL ENGINEERING. *Educ:* Wayne State Univ, BS, 52, MS, 61; Northwestern Univ, PhD(transp eng), 65. *Prof Exp:* Detailer struct design, Giffels & Vallet, Inc, 53-54; asst civil engr, Bur Hwy & Expressways, City of Detroit, 54-56; struct engr, Stran-Steel Corp, 56-57; instr transp & struct eng, Wayne State Univ, 57-59; sr asst engr, Bur Hwy & Expressways, City of Detroit, 59-60; lectr transp & traffic eng, Northwestern Univ, 63-65, asst prof, 65-66; assoc prof, Univ Mich, Ann Arbor, 66-68; assoc prof, Wayne State Univ, 68-70; assoc prof, 70-76, PROF CIVIL ENG, PURDUE UNIV, 76- *Concurrent Pos:* Mem, Hwy Res Bd, Nat Acad Sci-Nat Res Coun. *Mem:* Am Soc Civil Engrs; Am Soc Eng Educ; Inst Traffic Engrs. *Res:* Transportation and traffic engineering. *Mailing Add:* Dept Civil Eng Purdue Univ West Lafayette IN 47906

SATTERTHWAITE, CAMERON B, b Salem, Ohio, July 26, 20; div; c 5. PHYSICAL CHEMISTRY, PHYSICS. *Educ:* Col Wooster, BA, 42; Univ Pittsburgh, PhD(phys chem), 51. *Prof Exp:* Res assoc cryogenics res found, Ohio State Univ, 44-45; res chemist, Mound Lab, 45-46, group leader, 46-47; res chemist, E I du Pont de Nemours & Co, 50-53; res physicist, Westinghouse Elec Co, 53-61; from assoc prof to prof, Dept Physics & Mat Res Lab, Univ Ill, Urbana, 61-79; chmn dept, 79-82, prof, 82-85, EMER PROF PHYSICS, VA COMMONWEALTH UNIV, 85-; EMER PROF PHYSICS, DEPT PHYSICS & MAT RES LAB, UNIV ILL, URBANA, 79- *Concurrent Pos:* On leave, Iowa State Univ, 70-71; prog dir, NSF, 75-76. *Mem:* AAAS; fel Am Phys Soc. *Res:* Low temperature properties of metals, particularly superconductors; properties of metal hydrides. *Mailing Add:* 3311 Kensington Ave Richmond VA 23284-2303

SATTERTHWAITE, FRANKLIN EVES, statistics; deceased, see previous edition for last biography

SATTERWHITE, RAMON S(TEWART), b Little Rock, Ark, Feb 9, 40; m 62; c 2. ELECTRICAL ENGINEERING. *Educ:* Univ Ark, BSEE, 62; Univ NMex, MS, 64; Ohio State Univ, PhD(elec eng), 69. *Prof Exp:* Staff mem, Sandia Corp, 62-66; from asst prof to assoc prof elec eng, Lamar Univ, 69-76; VPRES, SOUTHERN AVIONICS CO, 76- *Mem:* Inst Elec & Electronics Engrs; Sigma Xi. *Res:* Electromagnetic field theory. *Mailing Add:* 5210 Gail Dr Beaumont TX 77708

SATTIN, ALBERT, b Cleveland, Ohio, Oct 5, 31; m 62; c 2. NEUROBIOLOGY. *Educ:* Western Reserve Univ, BS, 53, MD, 57; Am Bd Psychiat & Neurol, dipl, 79. *Prof Exp:* Intern med, Barnes Hosp, Wash Univ Med Ctr, 57-58; resident psychiat, Univ Hosp, Western Reserve Univ, 58-62, teaching fel, 62-64; fel neurochem, dept biochem, Inst Psychiat, Univ London, 65-66; instr psychiat, Case Western Reserve Univ, 66-68, sr instr, 68-69, asst prof pharmacol, 69-70 & asst prof psychiat & pharmacol, 70-77; assoc prof psychiat, 77-84, ASSOC PROF PSYCHIAT & NEUROBIOL, SCH MED & GRAD SCH, IND UNIV, 84-; CHIEF, ANTIDEPRESSANT NEUROPHARMACOL LAB, DVA SEPULVEDA MED CTR, 91- *Concurrent Pos:* Res assoc, dept pharmacol, Case Western Reserve, 65-68, instr, 68-69, sr instr psychiat, 69-70; prin investr, NIMH, 68-74 & 77-82 & NSF, 75-77 & Vet Admin, 83-; staff physician, R L Roudebush Vet Admin Ctr, 77-91; vis assoc prof psychiat, Dept Psychiat Behav Sci, Sch Med, Univ Calif, Los Angeles, 91- *Mem:* AAAS; fel Am Psychiat Asn; Int Soc Neurochem; Soc Biol Psychiat; Soc Neurosci. *Res:* Role of thyrotopin releasing hormone in anti-depressant treatment; discovery of the adenosine receptor and its specific blockade by methylxanthines; mechanisms and functional implications of the specifically seizure-induced large and prolonged increases in thyrotropin releasing hormone in limbic and cortical regions of the brain. *Mailing Add:* 16111 Plummer St 116A Sepulveda CA 91343-2099

SATTINGER, DAVID H, b Ft Wayne, Ind, April 3, 40. DIFFERENTIAL EQUATIONS. *Educ:* Oberlin Col, BA, 62; Mass Inst Technol, PhD(math), 65. *Prof Exp:* PROF MATH, UNIV MINN, MINNEAPOLIS, 72- *Mem:* Am Math Soc. *Mailing Add:* Sch Math Univ Minn Minneapolis MN 55455

SATTINGER, IRVIN J(ACK), b Indianapolis, Ind, Nov 1, 12; m 37; c 2. ELECTRICAL ENGINEERING. *Educ:* Univ Mich, BSE, 33, MS, 35. *Prof Exp:* Jr engr, Cent Ohio Light & Power Co, 36-37; elec draftsman, Commonwealth & Southern Corp, Mich, 37-38 & Loup River Pub Power Dist, Nebr, 38-39; asst engr, Ind Serv Corp, 39-41; elec designer, Basic Magnesium, Inc, Nev, 41-43; design engr, Lear, Inc, Mich, 43-48; res engr, Willow Run Labs, Inst Sci & Technol, Univ Mich, Ann Arbor, 48-73; RES ENGR, ENVIRON RES INST MICH, ANN ARBOR, 73- *Mem:* Sr mem Inst Elec & Electronics Engrs. *Res:* Application of computers to scientific problems; electronic measurement and control systems, particularly for aerospace and ground vehicles; applications of airborne and spacecraft remote sensing systems; energy studies; infrared technology. *Mailing Add:* Environ Res Inst Mich PO Box 8618 Ann Arbor MI 48107

SATTIZAHN, JAMES EDWARD, JR, b Moline, Ill, June 26, 20; m 43; c 2. PHYSICAL CHEMISTRY. *Educ:* Lawrence Col, BA, 42; Univ NMex, PhD(chem), 57. *Prof Exp:* Chemist, E I du Pont de Nemours & Co, 42-46; mem staff, Los Alamos Nat Lab, 46-85; RETIRED. *Mem:* Am Chem Soc; Sigma Xi. *Res:* Fission product behavior in various matrices. *Mailing Add:* 1422 44th St Los Alamos NM 87544

SATTLER, ALLAN R, b Los Angeles, Calif, June 28, 32; m 59; c 1. ATOMIC PHYSICS, NUCLEAR PHYSICS. *Educ:* Univ Calif, Los Angeles, BA, 54; Pa State Univ, MS, 59, PhD(physics), 62. *Prof Exp:* Eng aide, Univ Calif, Los Angeles, 53-54; jr engr, Douglas Aircraft Co, Inc, 54; physicist atomic power equip dept, Gen Elec Co, 56-57; STAFF MEM, SANDIA CORP, 63- *Concurrent Pos:* Fulbright travel grant, 62-63; mem staff, Asse Nuclear Waste Repository, 77. *Mem:* Am Phys Soc. *Res:* Atomic particle energy; channelling; radiation effect; neutron cross section measurements; nuclear waste technology; earth sciences; fuel technology and petroleum engineering. *Mailing Add:* Sandia Corp Kirtland AFB PO Box 5800 Albuquerque NM 87185

SATTLER, CAROL ANN, b DuBois, Pa, Sept 23, 46; m 71; c 2. BIOLOGICAL STRUCTURE. *Educ:* Thiel Col, BA, 68; Univ Colo, PhD(biol), 74. *Prof Exp:* Res assoc path, Med Sch, 74-75, proj assoc oncol, McArdle Lab Cancer Res, 75-77, asst scientist, 77-82, ASSOC SCIENTIST, MCARDLE LAB CANCER RES, UNIV WIS-MADISON, 82- *Mem:* Am Soc Cell Biol; AAAS. *Res:* Ultrastructure of cilia and the oral cavity of Tetrahymena pyriformis; ultrastructure of cultured epithelial cells in rat hepatocytes and human mammary; Mitosis and gap junctions in cultured rat hepatocytes; ultrastructure of prosophila ommatidia. *Mailing Add:* McArdle Lab Cancer Res Univ Wis 1400 University Ave Madison WI 53706

SATTLER, FRANK A(NTON), b New England, NDak, July 5, 20; m 45; c 9. CHEMICAL ENGINEERING. *Educ:* Univ NDak, BS, 42. *Prof Exp:* Res engr, Res Labs, Westinghouse Elec Corp, 42-54, supv chemist, 54-73, mgr wire, 73-78, adv scientist, Westinghouse Res & Develop Ctr, 78-83; RETIRED. *Mem:* Am Chem Soc. *Res:* Development of electrical insulating materials. *Mailing Add:* 120 Jamison Lane Monroeville PA 15146

SATTLER, JOSEPH PETER, b New York, NY, Oct 19, 40; m 68; c 3. LASERS, RADAR. *Educ:* Iona Col, BS, 62; Georgetown Univ, MS, 66, PhD(physics), 69. *Prof Exp:* physicist, 66-88, CHIEF SCIENTIST, HARRY DIAMOND LABS, 88- *Concurrent Pos:* Army Dep (A) Technol & Req, 85-87. *Honors & Awards:* Res & Develop Achievement Award, US Army, 75; Hinman Tech Achievement Award, Harry Diamond Labs, 78. *Mem:* Am Phys Soc; Inst Elec & Electronics Engrs; Sigma Xi. *Res:* High resolution infrared and submillimeter wave spectroscopy; solid state physics; quantum electronics; electron paramagnetic resonance. *Mailing Add:* 1320 Woodside Pkwy Silver Spring MD 20910-1551

SATTLER, ROBERT E(DWARD), b St Louis, Mo, Mar 31, 25; m 50; c 2. CHEMICAL ENGINEERING, MATHEMATICS. *Educ:* Wash Univ, St Louis, BS, 49; Univ Mo, MS, 52. *Prof Exp:* Process engr, Lago Oil & Transport Co, Ltd, 49-51; process design engr, 52-57, planning & correlation engr, 57-61, theoret develop engr, 61-63, mgr process fundamentals sect, 63-66, rate processes sect, 66-69, sr engr kinetics & mass transfer sect, 69-73, res & develop engr, Hydrocarbon Processes Br, 73-77, RES & DEVELOP ENGR, COAL PROCESSES SECT, ALT ENERGY BR, RES & DEVELOP DEPT, PHILLIPS PETROL CO, 77- *Mem:* AAAS; Am Inst Chem Eng; Am Chem Soc. *Res:* Heat transfer; reaction kinetics; reactor design. *Mailing Add:* 1245 Grandview Bartlesville OK 74006

SATTLER, ROLF, b Goppingen, Ger, Mar 8, 36; div. PLANT MORPHOLOGY. *Educ:* Univ Munich, PhD(taxon), 61. *Prof Exp:* NATO fel, 62-64; from asst prof to assoc prof, 64-77, PROF BOT, MCGILL UNIV, 77- *Honors & Awards:* Lawson Medal, Can Bot Asn, 74. *Mem:* Bot Soc Am; Can Bot Asn; Int Soc Plant Morphol; fel, Linnean Soc London; Sigma Xi; Can Soc Study Hist & Philos Sci; fel Royal Soc Can; Int Soc Hist, Philos & Social Studies Biol. *Res:* Philosophy of biology: basic biological concepts (including complementarity) and their relevance to the human condition; process philosophy and process morphology; developmental plant morphology including its philosophical and theoretical foundations, methodology and dynamical aspects (dynamic morphology). *Mailing Add:* Dept Biol McGill Univ Montreal PQ H3A 1B1 Can

SATTSANGI, PREM DAS, b Ghazipur City, India, May 2, 39; m 68; c 2. ORGANIC CHEMISTRY. *Educ:* Univ Allahabad, India, BS, 58, MS, 60, PhD(chem), 64. *Prof Exp:* Res assoc chem, Univ Ill, Urbana, 65-68; pool officer, Pa State Univ, 68-70, asst prof chem, 70-73, res assoc, 73-77, ASST PROF CHEM, PA STATE UNIV, 77- *Mem:* Am Chem Soc; Royal Soc Chem; Sigma Xi. *Res:* Synthesis of fluorescent probes; fluorescent modification of polypeptides of physiological interest; synthesis of heterocyclic compounds of biological interest. *Mailing Add:* Pa State Univ Fayette Campus Rte 119 N Box 519 Uniontown PA 15401

SATTUR, THEODORE W, b Passaic, NJ, Oct 20, 20; m 41; c 3. ANALYTICAL CHEMISTRY. *Educ:* Rutgers Univ, BS, 42. *Prof Exp:* Chemist, Raritan Copper Works, 42-43; res chemist, Metal & Thermit Corp, 43-47; res chemist, Cent Res Labs, Am Smelting & Refining Co, NJ, 47-60, asst chief chemist, 60-73, sr res chemist, 73-76, res assoc, 76-82; CONSULT CHEM, 82- *Mem:* Am Chem Soc; Sigma Xi; Soc Appl Spectros; fel Am Inst Chemists. *Res:* Methods for trace determination of halogens, sulfur and arsenic; application of atomic absorption spectroscopy. *Mailing Add:* 21 Hudson Pkwy Whiting NJ 08759

SATURNO, ANTONY FIDELAS, b Rochester, NY, Apr 7, 31; m 56; c 4. CHEMICAL PHYSICS. *Educ:* Univ Rochester, BS, 54; Carnegie Inst Technol, MS, 57, PhD(chem), 59. *Prof Exp:* Instr chem, Carnegie Inst Technol, 58-59; from asst prof to assoc prof, Univ Tenn, 59-66; assoc prof, 66-71, PROF CHEM, STATE UNIV NY ALBANY, 71-, CHMN DEPT, 74- *Concurrent Pos:* Consult, Metal & Ceramics Div, Oak Ridge Nat Lab, 63-66. *Mem:* AAAS; Am Chem Soc; Am Phys Soc; Sigma Xi. *Res:* Quantum chemistry, especially the application of high speed computers to chemical problems concerning the electronic structure of small molecules and atoms. *Mailing Add:* Dept Chem State Univ NY Albany NY 12203

SATYA, AKELLA V S, b Madras, India, Nov 21, 39; m 64; c 2. MATERIALS SCIENCE, PHYSICAL METALLURGY. *Educ:* Indian Inst Technol, Kharagpur, BTech, 60, MTech, 62; Mich State Univ, PhD(mat sci), 69. *Prof Exp:* Assoc lectr phys metall, Indian Inst Technol, Kharagpur, 60-62; chief metallurgist, Midwest Mach Co Ind, Inc, Mich, 64-66, consult, 66-67; adv engr, 69-80, mgr, 80-82, SR ENGR, DIAG DEVELOP, IBM CORP, 82- *Concurrent Pos:* Hon lectr, Indian Inst Sci, Bangalore, 62; consult, Nat Aeronaut Labs, Bangalore, 62. *Honors & Awards:* Outstanding Innovation Award, IBM, 77. *Mem:* Am Phys Soc; Electrochem Soc; Indian Inst Metals (treas-secy, 60-62); Sigma Xi; Inst Elec & Electronics Engrs. *Res:* Microelectronic device design and processing; metal-SC contacts, ion implantation, thin films; low-temperature specific heats, electronic energy bands, metal and semiconductor physics; charge coupled devices; system reliability; semiconductor device diagnostics methodologies, yield modeling/forecasting and test structures, yield and reliability management; solid state physics; electronics engineering. *Mailing Add:* IBM Corp E Fishkill Facil Hopewell Junction NY 12533

SATYANARAYANA, MOTUPALLI, b Masulipatam, India, Feb 24, 28; m 55; c 3. ALGEBRA. *Educ:* Andhra Univ, BA, 47, MA, 49; Univ Wis, PhD(math), 66. *Prof Exp:* Asst prof math, Govt Cols, Andhra, India, 50-56; sr lectr, Sri Venkateswara Univ, 56-63; from asst prof to assoc prof, 66-71, PROF MATH, BOWLING GREEN STATE UNIV, 71- *Concurrent Pos:* Reviewer, Zentralbalatt F r Mathematik, 71- *Mem:* Am Math Soc; Calc Math Soc. *Res:* Semigroups; rings and ordered structures; topological algebra. *Mailing Add:* Dept Math State Univ Bowling Green OH 43402

SATYANARAYANAN, MAHADEV, b 1953; US citizen; m 90. DISTRIBUTED SYSTEMS, FILE SYSTEMS & DATABASES. *Educ:* Indian Inst Technol, Madras, BTech, 75, MTech, 77; Carnegie- Mellon Univ, PhD(computer sci), 83. *Prof Exp:* Syst designer, Info Technol Ctr, 83-86, asst prof, 86-89, ASSOC PROF COMPUTER SCI, SCH COMPUTER SCI, CARNEGIE-MELLON UNIV, 89- *Concurrent Pos:* NSF presidential young investr, 87. *Mem:* Asn Comput Mach; Inst Elec & Electronics Engrs Computer Soc; Sigma Xi; Usenix Asn. *Res:* Design, development and evaluation of distributed systems that provide shared access to information; distributed file systems and databases, network protocols, security, performance evaluation. *Mailing Add:* Sch Computer Sci Carnegie Mellon Univ Pittsburgh PA 15213-3890

SATYA-PRAKASH, K L, b Mysore City, India, Dec 10, 47; m 72; c 2. CANCER CYTOGENETICS. *Educ:* Univ Mysore, India, PhD(cell biol), 76. *Prof Exp:* ASST PROF PATH & DIR, CANCER CYTOGENETICS LAB, SCH MED, MED COL GA, 86- *Mem:* Am Soc Human Genetics; AAAS; NY Acad Sci; Genetic Toxicol Assoc. *Res:* Cancer cytogenetics-management of the cancer patient. *Mailing Add:* Dept Path Sch Med Med Col Ga 1120 15th St Augusta GA 30912-3605

SATZ, HELMUT T G, b Berlin, Germany, Apr 13, 36; m 66; c 2. ELEMENTARY PARTICLE THERMODYNAMICS. *Educ:* Mich State Univ, BSc, 56, MSc, 59; Univ Hamburg, Ger, Dr Rer Nat(physics), 63. *Prof Exp:* Physicist, Desy, Hamburg, Germany, 63-67; res fel, Univ Calif, Los Angeles, 67-68; res assoc, Cern, Geneva, Switz, 68-70; vis prof physics, Univ Helsinki, Finland, 70-71; PROF PHYSICS, UNIV BIELEFELD, GER, 71-; PHYSICIST, BROOKHAVEN NAT LAB, 85- *Concurrent Pos:* Ed-in-chief, J Physics C, Particles & Fields, 77- *Res:* States of matter and critical behavior in strong interaction physics; statistical aspects of elementary particle interactions. *Mailing Add:* Dept Physics Brookhaven Nat Lab Upton NY 11973

SATZ, RONALD WAYNE, b Seattle, Wash, May 24, 51. SOFTWARE PACKAGE DESIGN & PRODUCTION, ENGINE DESIGN. *Educ:* Rensselaer Polytech Inst, BSc, 74, MEng, 74; Columbia Pac Univ, PhD(systs eng), 91. *Prof Exp:* Design engr, Int Harvester Co, 73; res engr, Caterpillar Tractor Co, 75; advan proj engr, 3M Co, 77-78; prod res engr, Budd Co, 78-80; sr syst engr, Gen Elec Co, 80-82; chief prod engr, 76-77, PRES, TRANSPOWER CORP, 82- *Mem:* Am Soc Mech Engrs; Soc Automotive Engrs; Opers Res Soc Am; AAAS; Int Soc Unified Sci (secy, 71-91, pres, 91-). *Res:* Systems engineering; product engineering; project engineering; theoretical physics; computer simulation and design of complex machines and processes, including engines; linear and nonlinear optimization programs. *Mailing Add:* One Oak Dr Parkerford PA 19457

SAUBERLICH, HOWERDE EDWIN, b Ellington, Wis, Jan 23, 19; m 45; c 2. BIOCHEMISTRY, MICROBIOLOGY. *Educ:* Lawrence Univ, BA, 44; Univ Wis, MS, 46, PhD(biochem), 48. *Prof Exp:* Assoc animal nutritionist, Exp Sta, Auburn Univ, 48-50, prof & animal nutritionist, 50-58; assoc prof animal husb, Iowa State Univ, 59; chief chem div, Med Res & Nutrit Lab, Fitzsimons Gen Hosp, US Dept Army, Denver, 59-74; chief dept nutrit, Western Human Nutrit Res Ctr, USDA, Letterman Army Inst Res, Presidio of San Francisco, 74-83, 82-; PROF & DIR, DIV EXP NUTRIT, UNIV ALA, BIRMINGHAM. *Concurrent Pos:* Res fel, Univ Tenn, 51; prof, Univ Indonesia, 57-59; prof, Colo State Univ, 64-74; vis prof, Vanderbilt Univ, 70-71; Dept Army res & study fel award, 70; adj prof, Univ Calif, Berkeley. *Honors & Awards:* Johnson Award, 52 & Borden Award, 71, Am Inst Nutrit; Meritorious Civilian Serv Award, US Army, 64; McLester Award, 65; Diag Lab Award, Can Soc Clin Chemists, 83. *Mem:* Soc Exp Biol & Med; Am Soc Biol Chemists; Am Soc Animal Sci; Am Soc Microbiol; Am Soc Clin Nutrit; fel Am Inst Nutrit. *Res:* Protein, amino acid, vitamin and mineral metabolism in the human, rat, monkey and mouse; nutrition of microorganisms; nutritional assessment and surveillance; vitamin metabolism; human nutrition; mineral and lipid metabolism. *Mailing Add:* Dept Nutrit Sci Univ Ala Univ Sta Birmingham AL 35294

SAUCIER, ROGER THOMAS, b New Orleans, La, Aug 30, 35; m 57; c 2. PHYSICAL GEOGRAPHY, ARCHEOLOGICAL GEOLOGY. *Educ:* La State Univ, BA, 57, MA, 58, PhD, 68. *Prof Exp:* Res asst geol, Coastal Studies Inst, La State Univ, 59-61; geogr, Geol Br, Soils Div, 59-61, spec asst, 74-81, PHYS SCIENTIST, ENVIRON LAB, US ARMY ENGR WATERWAYS EXP STA, 81- *Concurrent Pos:* Consult earth scientist, 80- *Honors & Awards:* Roald Fryxell Medal Interdisciplinary Res, Soc Am Archeol, 85. *Mem:* Am Quaternary Asn; Soc Am Archaeol; fel Geol Soc Am; Asn Eng Geologists; Soc Prof Archeologists. *Res:* Applied research in geomorphology; alluvial and coastal morphology; sedimentology and areal geographic and geologic mapping as related to engineering design and construction activities; anthropology; quaternary and environmental geology; earth sciences. *Mailing Add:* Environ Lab US Army Engr Waterways Exp Sta 3909 Halls Ferry Rd Vicksburg MS 39180-6199

SAUCIER, WALTER JOSEPH, b Moncla, La, Oct 5, 21; m 43; c 7. METEOROLOGY. *Educ:* Univ Southwestern La, BA, 42; Univ Chicago, SM, 47, PhD(meteorol), 51. *Prof Exp:* Asst meteorol, Univ Chicago, 46-48, instr, 48-52; from asst prof to assoc prof, Tex A&M Univ, 52-60; prof, Univ Okla, 60-69; prof, 69-86, EMER PROF METEOROL, NC STATE UNIV, 86- *Concurrent Pos:* mem bd cert consult meteorologists, Am Meteorol Soc, 70-75, chmn, 71-75; consult, Nat Acad Sci-Nat Res Coun, 71-77; mem bd dirs, Triangle Univs Consortium on Air Pollution, 72-86. *Mem:* Fel AAAS; Sigma Xi; fel Am Meteorol Soc; Am Geophys Union. *Res:* Atmospheric circulations and weather analysis. *Mailing Add:* NC State Univ Box 8208 Raleigh NC 27650

SAUDEK, CHRISTOPHER D, b Bronxville, NY, Oct 8, 41; m 66; c 3. DIABETES, LIPID METABOLISM. *Educ:* Harvard Univ, BA, 63; Cornell Univ, MD, 67. *Prof Exp:* Intern med, Presby-St Luke's Hosp, 67-68, resident, 68-69; resident, Harvard Med Serv, Boston City Hosp, 69-70; fel metab, Thorndike Mem Lab, 70-72; instr med, Harvard Med Sch, 72-73; asst prof, Sch Med, Cornell Univ, 73-79, assoc clin prof, 79-81; ASSOC PROF MED, SCH MED, JOHNS HOPKINS UNIV, DIR, DIABETES CTR & PROG DIR, CLIN RES CTR, 81- *Concurrent Pos:* Dir, Metab Lab, Sch Med, Cornell Univ, 73-, dir, Clin Res Ctr, 74-; adj asst prof, Rockefeller Univ, 74-; mem, Coun Arteriosclerosis, Am Heart Asn; dir, Johns Hopkins, Clin Res Ctr. *Mem:* Am Fedn Clin Res; Am Heart Asn; NY Acad Sci. *Res:* Insulin delivery systems for diabetes. *Mailing Add:* Dept Med Johns Hopkins Univ 720 Rutland Ave Baltimore MD 21205

SAUDER, DANIEL NATHAN, b Hamilton, Ont, Apr 15, 49; c 3. DERMATOLOGY, INTERNAL MEDICINE. *Educ:* York Univ, Toronto, BA, 71, MA, 72; McMaster Univ, Hamilton, MD, 75; FRCP(C), 80. *Prof Exp:* Intern, Royal Victoria Hosp, PQ, 75-76; resident II internal med, McMaster Univ, 76-77; fel dermat, Cleveland Clin Found, Ohio, 77-79; from asst prof to prof dermat, McMaster Univ, Hamilton, 82-90; PROF & CHIEF DERMAT, UNIV TORONTO, 90- *Concurrent Pos:* Vis assoc dermat, Nat Cancer Inst, NIH, Bethesda, Md, 79-82. *Honors & Awards:* Lila Gruber Cancer Res Award, 84. *Mem:* Fedn Am Soc Exp Biol; Can Soc Immunologists; Can Soc Clin Invest; Am Soc Immunologists; Am Acad Dermat; Soc Investigative Dermat. *Res:* Purification to homogeneity of several other epidermal immunoregulatory cytokines, particularly B-cell growth and differentiation factor epidermal derived interleukin-3 and epidermal derived colony stimulating factors; evaluation of the basic biology of these homogeneous and cloned factors, and their potential significance in inflamatory and immunologic conditions that effect the skin. *Mailing Add:* Sunnybrook Health Sci Ctr 2075 Bayview Ave Toronto ON M4N 3M5 Can

SAUDER, WILLIAM CONRAD, b Wheeling, WVa, Jan 3, 34; m 55; c 2. PHYSICS, X-RAY SPECTROSCOPY. *Educ:* Va Mil Inst, BS, 55; Johns Hopkins Univ, PhD(physics), 63. *Prof Exp:* From instr to assoc prof, 55-68, chmn dept, 79-84, PROF PHYSICS, VA MIL INST, 68- *Concurrent Pos:* Consult, Nat Bur Standards, 65-81. *Mem:* Am Phys Soc; Am Asn Physics Teachers. *Res:* Atomic constants; x-ray and gamma ray spectroscopy; ultrasonic interferometry; acoustic interferometry. *Mailing Add:* Dept Physics & Astron Va Mil Inst Lexington VA 24450

SAUER, CHARLES WILLIAM, b Louisville, Ky, Oct 5, 19; m 61. ORGANIC CHEMISTRY, BIOCHEMISTRY CREATIVITY. *Educ:* Mass Inst Technol, SB, 41, PhD(organic chem), 49. *Prof Exp:* Sr chemist, Arthur D Little Inc, 49-62, bus mgr, Energy & Mat Div, 60-62; asst to mgr plans & liaison, Res & Develop, Missile & Space Systs Div, Douglas Aircraft Co, 62-65; dir res admin, Res Dept, Bell Aerosysts Co, 66, dir future systs res, 66-69; dir technol surv, Am Can Co, 69-72; prin scientist planning, Calspan Corp, 72-73; adj prof, Future Studies Acad, 77-90, ADS PROF, CREATIVE STUDIES ACAD, STATE UNIV NY COL, BUFFALO, 90- *Concurrent Pos:* Consult, 74- *Mem:* Fel AAAS; Am Chem Soc; Sigma Xi; World Future Soc. *Res:* Application of systems approach to analysis and synthesis of how and why man develops science and technology, how technology enters society and influences the future; strategic and longer range planning; research management; five US patents. *Mailing Add:* 4421 Chestnut Ridge Rd Amherst NY 14228-3238

SAUER, DAVID BRUCE, b Akron, Ohio, Sept 20, 39; m 61; c 5. PLANT PATHOLOGY. *Educ:* Kent State Univ, BA, 61; Univ Minn, MS, 64, PhD(plant path), 67. *Prof Exp:* RES PLANT PATHOLOGIST, US GRAIN MKT RES CTR, AGR RES SERV, USDA, 67- *Mem:* Am Phytopath Soc; Sigma Xi; Am Asn Cereal Chemists. *Res:* Ecology and control of microorganisms in stored grain, including grain quality surveys, grain drying, mycotoxins, moisture relations, odor detection, testing of grain preservatives. *Mailing Add:* US Grain Mkt Res Ctr 1515 College Ave Manhattan KS 66502

SAUER, DENNIS THEODORE, b Lamont, Wash, Oct 26, 44; m 69. INORGANIC CHEMISTRY. *Educ:* Whitworth Col, BS, 66; Cent Wash State Col, MS, 68; Univ Idaho, PhD(inorg chem), 72. *Prof Exp:* Asst chem, Cent Wash State Col, 66-68; res fel, Univ Idaho, 68-71; fac intern chem, Univ Utah, 71-72; staff scientist, 72-82, tech supt, 82-83, tech mgr lasers, 83-84, prog mgr, DS Propellants, 84-87, MGR, ROCKET TECHNOL, HERCULES, INC, 87- *Mem:* Am Chem Soc. *Res:* Fluorine chemistry; phosphorus and boron chemistry; laser systems. *Mailing Add:* 7854 Deer Creek Rd Salt Lake City UT 84121

SAUER, HARRY JOHN, JR, b St Joseph, Mo, Jan 27, 35; m 56; c 9. MECHANICAL & AEROSPACE ENGINEERING. *Educ:* Mo Sch Mines, BS, 56, MS, 58; Kans State Univ, PhD(heat transfer), 63. *Prof Exp:* From instr to asst prof mech eng, Mo Sch Mines, 57-60; instr, Kans State Univ, 60-62; assoc prof, 62-66, PROF MECH & AERO ENG, UNIV MO, ROLLA, 66-, DEAN GRAD STUDY, 84- *Concurrent Pos:* Sr eng & consult, Midwest Res Inst, 63-70. *Honors & Awards:* Hermann F Spoehrer Mem Award, Am Soc Heat, Refrig & Air-Conditioning Engrs, 79, E K Campbell Award, 83. *Mem:* Fel Am Soc Mech Engrs; fel Am Soc Heat, Refrig & Air-Conditioning Engrs; Soc Automotive Engrs; Am Soc Eng Educ; Nat Soc Prof Engrs; Soc Res Adminr. *Res:* Heat transfer; thermophysical properties; environmental control; photographic science. *Mailing Add:* Dept Mech & Aerospace Eng Univ Mo Rolla MO 65401-0249

SAUER, HELMUT WILHELM, b Kassel, WGer, Aug 12, 36; m 68; c 2. DEVELOPMENTAL BIOLOGY. *Educ:* Univ Marburg, WGer, Dr rer nat, 65. *Prof Exp:* Asst prof, Univ Heidelberg, 65-73; assoc prof, Univ Konstanz, 73-76; prof, Univ Wurzburg, 76-81; PROF BIOL, TEX A&M UNIV, 81- *Concurrent Pos:* Fel cell biol, McArdle Lab Cancer Res, Univ Wis, Madison, 67-69; hon prof, Univ Wurzburg, 85- *Mem:* Ger Soc Zool; Ger Soc Biol Chemists; Ger Soc Entwicklungsbiol; Am Soc Zoologists; Soc Develop Biol; Int Soc Develop Biologists; Am Soc Cell Biol; Int Cell Cycle Soc. *Res:* Control of cellular growth and differentiation, employing a simple eukaryotic model Physarum and analyzing the mechanism of genome expression. *Mailing Add:* Rte 4 Box 449A College Station TX 77840-9613

SAUER, HERBERT H, b Newark, NJ, Dec 9, 29; m 57; c 4. MAGNETOSPHERIC PHYSICS. *Educ:* Rutgers Univ, BSc, 53; Univ Iowa, PhD(physics), 62. *Prof Exp:* Vacuum tube engr, Fed Telecommun Labs, NJ, 53-54; physicist inst telecommun & aeronomy, Environ Sci Serv Admin, 63-70, PHYSICIST SPACE ENVIRON LAB, NAT OCEANIC & ATMOSPHERIC ADMIN, 70- *Concurrent Pos:* Vis lectr, Univ Colo, 65, 68, 72, 77; vis prof, Univ Calgary, 69. *Mem:* Am Phys Soc; Am Geophys Union; Sigma Xi. *Res:* Magnetospheric and cosmic ray physics. *Mailing Add:* Space Environ Lab Nat Oceanic & Atmospheric Admin 325 Broadway Boulder CO 80303

SAUER, JOHN A, b Oct 26, 12. PHYSICS. *Educ:* Rutgers Univ, BS, 34, MS, 36; Cambridge Univ, PhD(math physics), 42. *Prof Exp:* Instr math & mech, Union Jr Col, 34-38; instr, Rutgers Univ, 36-38; fel & sr fel, Mellon Inst Indust Res, Pittsburgh, 41-44; asst & dir res eng, Elastic Stop Nut Corp, 44-46; prof & chmn, Dept Eng Mech, 46-53, prof & chmn, Dept Physics, Pa State Univ, 53-63; prof & chmn, 63-83, EMER PROF, DEPT MECH & MAT SCI, RUTGERS UNIV, 83- *Concurrent Pos:* consult, Amerace-ESNA Corp, 51-84; vis lect, Am Inst Physics, 71-72; vis prof, Clarendon Lab, Oxford Univ, 52-53; Guggenheim fel, 59-60; Rutgers res fel, Dept Eng Sci, 69-70; vis prof, Dept Eng Sci, Oxford Univ, 77-78; invited prin lect, Ger Phys Soc, Wurtzburg, &$, conference high pressure, Kyotto, Japan, 74; vis prof, Dept Eng Sci, Oxford Univ, 83-85; chmn, prof II rev comt, 78-79. *Honors & Awards:* Linbach Award, 75; Rutger Univ Award, 88. *Mem:* NY Acad Sci; fel Am Phys Soc; fel AAAS; Am Chem Soc; Sigma Xi (vpres, 71-72, pres, 72-73). *Res:* Polymer physics; physical properties of polymers; effects of molecular structure, temperature, pressure, radiation and chemical environment on mechanical and relaxation behavior of polymers and relation of properties to structure; author of several books and articles. *Mailing Add:* 33 Patton Dr East Brunswick NJ 08816

SAUER, JOHN ROBERT, b Aberdeen, SDak, Aug 1, 36; m 62; c 3. ENTOMOLOGY, BIOCHEMISTRY. *Educ:* St John's Univ, Minn, BS, 59; NMex Highlands Univ, MS, 64; Tulane Univ, PhD(biol), 69. *Prof Exp:* From asst prof to assoc prof, 69-77, PROF INSECT PHYSIOL, OKLA STATE UNIV, 77- *Concurrent Pos:* Mem, Trop Med, Parasitol, NIH; NIH res grant. *Mem:* Sigma Xi; Entom Soc Am; Am Soc Zool; Am Soc Parasitol. *Res:* Insect physiology; tick physiology; role of salivary glands in tick feeding; control of tick salivary glands. *Mailing Add:* 1304 W Osage Stillwater OK 74075

SAUER, JON ROBERT, b Schenectady, NY, Nov 24, 40; m 72; c 2. HIGH ENERGY PHYSICS, ACCELERATOR PHYSICS. *Educ:* Stanford Univ, BS, 62; Tufts Univ, PhD(physics), 70. *Prof Exp:* Staff physicist accelerator physics, Stanford Linear Accelerator Ctr, 62-64, Cambridge Electron Accelerator, 69-70 & Fermilab, 70-77; sr res assoc high energy physics, Ind Univ, 77-78; asst physicist high energy physics, Argonne Nat Lab, 78-81; mem tech staff, Bell Labs, 81-83; prod mgr, Adv Systs, Denelcor Supercomput, 83-84; ON LOAN AS CHIEF SCIENTIST TO CTR OPTOELECTRONIC COMPUT SYSTS, COLO UNIV, BELL LABS, 84- *Concurrent Pos:* On loan as chief scientist, Ctr Optoelectronic Comput Systs, Colo Univ, Boulder, 84- *Mem:* Sigma Xi; Am Phys Soc. *Mailing Add:* 15005 E Grand Ave Aurora CO 80015

SAUER, JONATHAN DEININGER, b Ann Arbor, Mich, July 5, 18; m 46; c 1. BIOGEOGRAPHY. *Educ:* Univ Calif, AB, 39; Washington Univ, MA, 48, PhD(genetics), 50. *Prof Exp:* From instr to assoc prof bot, Univ Wis-Madison, 50-59, from assoc prof to prof bot & geog, 59-67; vis prof geog, La State Univ, 67; prof geol, 67-88, EMER PROF GEOG, UNIV CALIF, LOS ANGELES, 88- *Concurrent Pos:* Vis assoc cur, Herbarium, Univ Mich, 55-56; dir bot gardens & herbarium, Univ Calif, Los Angeles, 74-80. *Mem:* Brit Ecol Soc; Ecol Soc Am; Am Soc Plant Taxon; Org Trop Studies; Soc Econ Bot. *Res:* Recent plant migration and evolution; systematics of Amaranthus, Canavalia and Stenotaphrum; dynamics of seacoast and riverbank pioneer vegetation. *Mailing Add:* 659 Erskine Dr Pacific Palisades CA 90272

SAUER, KENNETH, b Cleveland, Ohio, June 19, 31; m 58; c 4. BIOPHYSICAL CHEMISTRY. *Educ:* Oberlin Col, AB, 53; Harvard Univ, MA, 54, PhD(chem), 58. *Prof Exp:* From instr to asst prof chem, Am Univ Beirut, 57-60; NIH res fel, 60-63, from asst prof to assoc prof, 63-72, MEM SR STAFF, LAB CHEM BIODYN, UNIV CALIF, BERKELEY, 62-, PROF CHEM, 72- *Concurrent Pos:* Guggenheim fel, 76-77; Alexander von Humboldt Award, 85 & 86. *Mem:* Fel AAAS; Am Chem Soc; Biophys Soc; Am Soc Plant Physiologists; Am Soc Photobiol; Sigma Xi. *Res:* Photosynthetic energy conversion; biological molecular structure; molecular spectroscopy; fluorescence liftimes; excitation transfer. *Mailing Add:* Dept Chem Univ Calif Berkeley CA 94720

SAUER, LEONARD A, b Schenectady, NY, Aug 20, 29; m 56; c 3. CELL BIOLOGY, BIOCHEMISTRY. *Educ:* Cornell Univ, BS, 56; Univ Rochester, MD, 60; Rockefeller Univ, PhD(cell biol), 66. *Prof Exp:* From instr to assoc prof med, Sch Med, Yale Univ, 67-73; RES PHYSICIAN, MARY IMOGENE BASSETT HOSP, 73- *Concurrent Pos:* USPHS spec fel, Univ Marburg, 66-67. *Mem:* Am Soc Biol Chemists; Am Soc Cell Biol; Endocrine Soc; Soc Exp Biol & Med; Am Asn Cancer Res. *Res:* Cell regulatory processes; mitochondrial physiology; adrenal steroidogenesis; tumor biology. *Mailing Add:* Bassett Res Inst Mary Imogene Bassett Hosp Cooperstown NY 13326

SAUER, MYRAN CHARLES, JR, b Pittsburgh, Pa, Nov 30, 33; m 59; c 3. RADIATION CHEMISTRY. *Educ:* Carnegie Inst Technol, BS, 55; Univ Wis, PhD(chem), 58. *Prof Exp:* Resident res assoc, 59-61, asst chemist, 61-63, ASSOC CHEMIST, ARGONNE NAT LAB, 63- *Mem:* Am Chem Soc; Radiation Res Soc. *Res:* Kinetics and mechanisms of reactions initiated by ionizing radiation and light. *Mailing Add:* Chem Div Argonne Nat Lab Argonne IL 60439

SAUER, PETER WILLIAM, b Winona, Minn, Sept 20, 46; m 69; c 2. ELECTRICAL ENGINEERING. *Educ:* Univ Mo, Rolla, BS, 69; Purdue Univ, MS, 74, PhD(elec eng), 77. *Prof Exp:* Design engr, US Air Force Tactical Air Command, 69-73; res asst, elec eng dept, Purdue Univ, 73-77; from asst prof to assoc prof, 77-85, PROF ELEC ENG, UNIV ILL, URBANA, 85- *Concurrent Pos:* Elec power consult, US Army Corps Eng Res Lab, 75-; prin investr, NSF grant, 78-; res dir, Ill Power Affil Prog, 78-; Grainger assoc, 82- *Mem:* Sigma Xi; Inst Elec & Electronics Engrs. *Res:* Electrical power system simulation and analysis; electric power system operation and planning methods; electric power system dynamics and control. *Mailing Add:* 337 Everitt Lab Univ Ill 1406 W Green St Urbana IL 61801

SAUER, RICHARD JOHN, b Walker, Minn, Nov 15, 39; m 62; c 4. ENTOMOLOGY. *Educ:* St John's Univ, Minn, BS, 62; Univ Mich, Ann Arbor, MS, 64; NDak State Univ, PhD(entom), 67. *Prof Exp:* Teaching asst zool, Univ Mich, Ann Arbor, 62-64; asst prof biol, St Cloud State Col, 67-68; asst prof biol & entom, Mich State Univ, 68-70, exten entom, 70-72, assoc prof exten entom & exten pesticide coordr, 72-76; prof entom & head dept, Kans State Univ, 76-80; dep vpres, 83-85, DIR, MINN AGR EXP STA, UNIV MINN, ST PAUL, 80-, VPRES, AGR, FORESTRY & HOME ECON, 85- *Concurrent Pos:* Entom consult, Coop State Res Serv, USDA, 74-75. *Mem:* Entom Soc Am. *Res:* Taxonomy and biology of spiders; clearance of minor use pesticides; pesticide usage and safety. *Mailing Add:* Pres 202 Morril Hall Univ Minn Minneapolis 100 Church St SE Minneapolis MN 55455

SAUERBRUNN, ROBERT DEWEY, b Jonesboro, Ill, Dec 27, 22; m 47; c 2. ANALYTICAL CHEMISTRY, POLYMER CHEMISTRY. *Educ:* Southern Ill Univ, BS, 47; Univ Minn, PhD(anal chem), 53. *Prof Exp:* Develop assoc res & develop, 53-62, sr res chemist polymer chem, 62-63, SUPVR RES & DEVELOP, E I DU PONT DE NEMOURS & CO, INC, 63- *Mem:* Am Chem Soc. *Res:* Electrochemical and spectrophotometric analyses; chemical kinetics and polymer chemistry. *Mailing Add:* 904 Robin Dr Seaford DE 19973

SAUERLAND, EBERHARDT KARL, b Ger, Dec 17, 33; US citizen; m 86; c 3. ANATOMY, PSYCHIATRY. *Educ:* Univ Kiel, MD, 60. *Prof Exp:* From intern to resident, St John Gen Hosp, NB, 60-62; res scientist aerospace med, Lockheed-Calif Co, Burbank, 62-64; asst prof anat, Sch Med, Univ Calif, Los Angeles, 64-70, assoc prof anat & oral med, 70-71; prof anat, Univ Tex Med Br Galveston, 71-80; dir, Clin Invest Fac, Wilford Hall, USAF Med Ctr,

Lackland AFB, Tex, 80-83, resident diag radiol, 80-81; resident psychiat, 84-87, chief resident psychiat, 86-87, fel acad psychiat, Loma Linda Univ Med Ctr, 87-88; ASSOC PROF PSYCHIAT & PROF ANAT, UNIV TEX HEALTH SCI CTR, SAN ANTONIO, 88- *Concurrent Pos:* Adj prof anat, Univ Tex Med Br Galveston, 80-85; vis prof anat, USUHS, 80. *Mem:* Am Asn Anat; Am Asn Psychiatrists. *Res:* Interaction of brain and reflex mechanisms; electromyography; psychodynamics. *Mailing Add:* Dept Anat Univ Tex Med Sch Galveston 301 University Blvd Galveston TX 77550

SAUERS, RICHARD FRANK, b Philadelphia, Pa, Apr 4, 39; m 66; c 4. ORGANIC CHEMISTRY. *Educ:* LaSalle Col, BA, 65; Univ Minn, PhD(org chem), 69. *Prof Exp:* Chemist, Smith Kline & French Labs, 63-65; sr res chemist, 69-80, RES SUPVR, E I DU PONT DE NEMOURS & CO, INC, 80- *Mem:* Am Chem Soc. *Res:* Biologically active materials. *Mailing Add:* 11 Polaris Dr Newark DE 19711

SAUERS, RONALD RAYMOND, b Pittsburgh, Pa, June 19, 32; div; c 2. ORGANIC CHEMISTRY. *Educ:* Pa State Univ, BS, 53; Univ Ill, PhD(chem), 56. *Prof Exp:* USPHS fel, Univ Ill, 56-57; from instr to assoc prof, 57-70, PROF CHEM, RUTGERS UNIV, NEW BRUNSWICK, 70- *Concurrent Pos:* Vis fel, Princeton Univ, 66-67; USPHS spec fel, Brandeis Univ, 72-73. *Mem:* Am Chem Soc. *Res:* Stereochemistry of organic reactions; polycyclic hydrocarbon systems; organic photochemistry. *Mailing Add:* Dept Chem Rutgers Univ New Brunswick NJ 08903

SAUL, FRANK PHILIP, b New York, NY, Oct 31, 30; m 64; c 2. BIOLOGICAL ANTHROPOLOGY, CONTINUING MEDICAL EDUCATION. *Educ:* Brooklyn Col, AB, 52; Harvard Univ, AM, 59, PhD, 72; Am Bd Forensic Anthrop, dipl, 78. *Prof Exp:* Field asst, archaeol exped to SDak, Univ Mus, Kans, 50; asst phys anthrop, Am Mus Natural Hist, 51-52; phys anthropologist, Aero Med Lab, Wright Air Develop Ctr, US Dept Air Force, 53-58 & Natick Qm Res & Eng Ctr, US Dept Army, 58-59; field study Hutterite morphol, Harvard Univ, 59, teaching fel anthrop, 59-62; instr, Pa State Univ, 62-67, asst prof anthrop & phys anthropologist, Eastern Pa Archaeol Projs, 67-69; asst prof, 69-71, asst dean res, 84-90, ASSOC PROF ANAT, MED COL OHIO, 72-, ASSOC DEAN, CONTINUING MED EDUC, 89- *Concurrent Pos:* Consult, forensic anthrop, 54-, human factors, 59- & anthrop, Lab Cent Nervous Syst Studies, NIH, 79-; res assoc, Boston Mus Sci, 60-62; phys anthropologist Maya area projs, Peabody Mus, Harvard Univ, 62-; biol anthropologist Maya area projs, Cambridge Univ, 70-; regional lectr, Sigma Xi, 71-76, 89; hon cur biomed anthrop, Toledo Mus Health & Natural Hist, 77; vis prof phys anthrop, Univ Cambridge, 78; guest cur, hist of dis in Mex & Cent Am, Nat Mus Health & Med, 87-; guest fac, forensic anthropology course, Am Regist Path & Armed Forces Inst Path, 88; mem, sci adv bd, Armed Forces Inst Path, 89- *Mem:* Fel Am Anthrop Asn; Am Asn Phys Anthrop; fel Royal Anthrop Inst; fel AAAS; fel Am Acad Forensic Sci; Paleopathology Asn. *Res:* Osteology; human factors; origin and evolution of the Maya; biomedical anthropology; paleopathology; forensic anthropology. *Mailing Add:* Continuing Med Educ Med Col Ohio PO Box 10008 Toledo OH 43699-0008

SAUL, GEORGE BRANDON, II, b Hartford, Conn, Aug 8, 28; m 53. GENETICS. *Educ:* Univ Pa, AB, 49, AM, 50, PhD(zool), 54. *Prof Exp:* Asst instr zool, Univ Pa, 50-52; from instr to assoc prof, Dartmouth Col, 54-67; chmn dept, 68-76, acad vpres, 76-79, PROF BIOL, MIDDLEBURY COL, 67- *Concurrent Pos:* NSF sci fac fel, Univ Zurich, 59-60; res fel biol, Calif Inst Technol, 64-65; vis scientist, Boyce Thompson Inst Plant Res, 72-73. *Mem:* Fel AAAS; Radiation Res Soc; Genetics Soc Am; Am Genetic Asn; Sigma Xi; NY Acad Sci. *Res:* Cytogenetics; biochemical genetics; embryological genetics of Mormoniella vitripennis; extranuclear genetics. *Mailing Add:* Dept Biol Middlebury Col Middlebury VT 05753

SAUL, JULIE MATHER, b Indianapolis, Ind, May 23, 41; m 64; c 2. FORENSIC ANTHROPOLOGY, PALEOPATHOLOGY. *Educ:* Pa State Univ, BA, 63, Universidad de Salamanca, Spain, 63. *Prof Exp:* Res technician agron, Pa State Univ, 60-61, res asst anthrop, 64-69; RES ASSOC BIOMED ANTHROPOLOGY, DEPT ANAT, MED COL OHIO, 69-; DIR, FORENSIC ANTHROPOLOGY LAB, LUCAS COUNTY CORONERS OFF, 90- *Concurrent Pos:* Co-leader, Nat Geog Soc Maya Res Projs, Mex, 79, Guatemala, 89; co-prin invest, NSF Maya Res Proj, 81-82, field investr, Ceren Res Proj, El Salvador, 89; consult forensic anthrop, Lucas Co, Ohio, Coroner, Monroe Co, Mich Chief Med Examr, 76-; guest fac, Am Registry Path & Armed Forces Inst Path, 88-; co-chmn ann meeting, Paleopath Asn, 82-83. *Mem:* Sigma Xi; AAAS; Am Anthrop Asn; Am Asn Phys Anthropologists; Paleopath Asn. *Res:* origin and evolution of the Maya; paleopathology; osteology; forensic anthropology. *Mailing Add:* Dir Forensic Anthropology Lab Lucas Co Coroner's Off 2025 Arlington Ave Toledo OH 43609

SAUL, LEON JOSEPH, b New York, NY, Apr 26, 01; m 34; c 3. PSYCHIATRY. *Educ:* Columbia Univ, AB, 21, MA, 23; Harvard Med Sch, MD, 28; Am Bd Psychiat & Neurol, dipl, 37. *Prof Exp:* Commonwealth fel, Boston Psychopath Hosp & Harvard Med Sch, 30-32; clin assoc, Inst Psychoanal, Chicago, 32-42; assoc prof psychiat, Sch Med, Temple Univ, 46-48; prof clin psychiat, 48-60, prof psychiat, 60-69, EMER PROF PSYCHIAT, SCH MED, UNIV PA, 69- *Concurrent Pos:* Assoc attend physician, Cook County Psychopath Hosp, 36-42; consult, Vet Admin Hosp, Pa, 46-54; staff & training analyst, Philadelphia Psychoanal Inst, 46-70, emer analyst, 70-; lectr, Bryn Mawr Col, 46-50; chief psychiat consult, Swarthmore Col, 48-71, emer psychiat consult, 71-; mem staff, Inst of Pa Hosp, 56-70, hon consult, 70- *Mem:* Am Physiol Soc; Am Psychosomatic Soc (pres, 48-49); fel Am Acad Psychoanal; fel Am Col Psychoanal; fel Am Col Psychiat. *Res:* Cerebral action potentials; emotional factors in essential hypertension asthma and urticaria; psychodynamics role of hostility in neuroses and social relations. *Mailing Add:* 275 Highland Ave Media PA 19063

SAUL, LOU ELLA RANKIN, b Lenox, Calif, July 28, 27; US citizen; m 49; c 2. EVOLUTION MOLLUSKS, PALEONTOLOGY. *Educ:* Univ Calif Los Angeles, BA, 49, MA, 59. *Prof Exp:* Mus scientist cur, Geol Dept, Univ Calif, Los Angeles, 51-85; COLLECTION MGR INVERT PALEONT NATURAL HIST MUS LOS ANGELES COUNTY, 85- *Concurrent Pos:* Lectr earth & space sci, Univ Calif Los Angeles, 82- *Mem:* Paleont Soc; Geol Soc; AAAS; Sigma Xi; Paleont Asn; Soc Study Evolution. *Res:* Study of mollusks, bivalves and gastropods of Cretaceous through early Tertiary age from the Pacific slope of North America. *Mailing Add:* Earth Sci Div Natural Hist Mus Los Angeles Co Los Angeles CA 90007

SAUL, ROBERT H, materials science, solid state physics, for more information see previous edition

SAUL, WILLIAM EDWARD, b New York, NY, May 15, 34; m. STRUCTURAL ENGINEERING, FOUNDATION ENGINEERING. *Educ:* Mich Technol Univ, BS, 55, MS, 61; Northwestern Univ, PhD(civil eng), 64. *Prof Exp:* Mech engr, Shell Oil Co, 55-59; teaching asst, Mich Technol Univ, 59-60, instr eng mech, 60-62; from asst prof to prof, Univ Wis, 64-84, chmn dept, 76-80; dean engr, Univ Idaho, 84-90; CHAIRPERSON & PROF, DEPT CIVIL & ENVIRON ENG, MICH STATE UNIV, 90- *Concurrent Pos:* Indust consult, 64-; vis prof, Inst Aircraft & Space Structs Eng, Univ Stuttgart, 70-71, Fulbright travel grant, Alexander von Humboldt Found & Univ Wis Alumni Res Found grant. *Mem:* Fel Am Soc Civil Engrs; Int Asn Bridge & Struct Engrs; Sigma Xi; Am Soc Eng Educ; Am Concrete Inst; Nat Soc Prof Engr. *Res:* Dynamic response of structures; computer methods in structural analysis; theory of structures; reinforced concrete structures; applications in the response of structures due to high intensity forces such as earthquake, blast or storm; pile foundations. *Mailing Add:* A 349 Engr Bldg East Lansing MI 48824

SAULL, VINCENT ALEXANDER, geophysics, for more information see previous edition

SAUNDERS, B DAVID, b Bryan, Tex, April 12, 44; m 68; c 3. COMPUTER ALGEBRA. *Educ:* Univ Wis, BA, 70, PhD(math), 75. *Prof Exp:* Prof math & comput sci, Rensselaer Polytech Inst, 75-84; COMPUT & INFO SCI, UNIV DEL, 85- *Mem:* Asn Comput Mach; Math Asn Am; AAAS. *Res:* Systems and algorithms for symbolic mathematical computation. *Mailing Add:* Dept Comput & Info Sci Univ Del Newark DE 19716

SAUNDERS, BURT A, b Rochester, NY, July 20, 49. ELECTRONIC IMAGING, COLOR REPRODUCTION. *Educ:* Rochester Inst Technol, BS, 72; State Univ NY-Geneseo, MA, 76. *Prof Exp:* PRIN IMAGING SCIENTIST, ROCHESTER INST TECHNOL RES CORP, 86- *Concurrent Pos:* Consult, 80- *Mem:* Tech Asn Graphic Arts; Soc Info Display. *Res:* Control and reproduction of color in electronic imaging. *Mailing Add:* RD 1 Box 109 Nunda NY 14517

SAUNDERS, DONALD FREDERICK, b Utica, NY, Nov 9, 24; m 50; c 3. EXPLORATION GEOLOGY, PETROLEUM. *Educ:* St Lawrence Univ, BS, 46; Univ Wis, PhD(chem), 50. *Prof Exp:* Proj assoc, Univ Wis, 50-53; sr engr, Tex Instruments, Inc, 53-57; chief res geochemist, Geophys Serv, Inc, 57-62; mem tech staff, Tex Instruments Inc, 62-67, mgr radiation sci, 67-70, mgr new prog develop, 70-73, sr geoscientist, 73-83; vpres, Petrominex, 83-84, pres, 84-87; MGR INTEGRATED EXPLOR, RECON EXPLORATION INC, 87- *Mem:* Asn Explor Geochemists; Am Asn Petrol Geologists; Asn Petrol Geochem Explorationists. *Res:* Thermoluminescence of rocks and minerals; geochemistry of the origin of uranium deposits; geochemical prospecting; nuclear arms control studies; remote sensing of natural resources; geological interpretation of satellite imagery; environmental studies. *Mailing Add:* 4057 Northaven Rd Dallas TX 75229

SAUNDERS, DONALD ROY, b Chicago, Ill, June 2, 40; m 79. TOXICOLOGY, ENVIRONMENTAL SCIENCES. *Educ:* Leland Stanford Jr Univ, BA, 63; Purdue Univ, MS, 70, PhD(pharmacol, toxicol), 73; Am Bd Toxicol, dipl, 81. *Prof Exp:* Sr toxicologist, Riker Labs Inc, 73-76; res toxicologist, Stauffer Chem Co, 76-77, sr toxicologist, 77-78, toxicol contract admin, 78-82, mgr toxicol planning & contracts, 82-84, dir, Toxicol Dept, 84-88; DIR ENVIRON HEALTH CTR, AGR DIV, CIBA-GEIGY CORP, 88- *Mem:* Soc Toxicol; AAAS. *Res:* Safety evaluation of agricultural chemicals. *Mailing Add:* 400 Farmington Ave Farmington CT 06032

SAUNDERS, EDWARD A, b Manilla, Iowa, Mar 30, 25; m 46; c 4. SOLID STATE ELECTRONICS, NUCLEAR SCIENCE. *Educ:* US Mil Acad, BSMSE, 46; Purdue Univ, MS, 51; Rensselaer Polytech Inst, PhD(nuclear sci), 65. *Prof Exp:* Instr electronics, 51-54, prof elec, 61-65, PROF PHYSICS & HEAD DEPT, US MIL ACAD, 65-; DEAN OF ACAD, TRIDENT TECH COL. *Mem:* Am Asn Physics Teachers; Am Soc Eng Educ. *Res:* Radiation damage on semiconductor materials. *Mailing Add:* Trident Tech Col Box 10367 North Charleston SC 29411

SAUNDERS, FRANK AUSTIN, b Suffolk, Va, Dec 4, 40; m 63, 86; c 3. REHABILITATION ENGINEERING, PSYCHOLOGY. *Educ:* Juilliard Sch, BS, 61; Ind Univ, Bloomington, PhD(psychol), 65. *Prof Exp:* Res psychologist, Langley Porter Neuropsychiat Inst, 66-68; RES ASSOC & SR SCIENTIST, SMITH-KETTLEWELL INST VISUAL SCI, 68- *Concurrent Pos:* Fel med psychol, Langley Porter Neuropsychiat Inst, San Francisco, 65-66; clin instr med psychol, Sch Med, Univ Calif, San Francisco, 68-; lectr psychol, San Francisco State Col, 68- *Honors & Awards:* Hektoen Award, Am Med Asn, 72 & 77. *Mem:* Am Psychol Asn; Acoust Soc Am; Inst Elec & Electronics Engrs; Biomed Eng Soc. *Res:* Development of electrotactile displays and sensory aids for deaf and blind persons. *Mailing Add:* 1555 Beach Park Blvd Foster City CA 94404

SAUNDERS, FRANK LINWOOD, b Moline, Ill, July 26, 26; m 49; c 2. PHYSICAL CHEMISTRY. *Educ:* Augustana Col, AB, 50; Case Western Reserve Univ, MS, 52, PhD(chem), 53. *Prof Exp:* Proj leader, 56-66, group leader, 66-75, sr res specialist, 75-80, res assoc, 80-86, ASSOC SCI, DOW CHEM CO, 86- *Mem:* Am Chem Soc; Sigma Xi. *Res:* Polymer and colloid chemistry; latexes; stereospecific polymers. *Mailing Add:* Cent Res Dow Chem Co 1712 Bldg Midland MI 48674

SAUNDERS, FRANK WENDELL, b Reidsville, NC, Sept 27, 22; m 51; c 3. MATHEMATICS. *Educ:* Univ NC, AB, 45, MA, 47. *Prof Exp:* Instr math, Univ NC, 47-49; prof, Coker Col, 49-61; dir grad studies in math, 70-76, actg chmn dept, 76-78, PROF MATH, ECAROLINA UNIV, 61- *Res:* Number theory. *Mailing Add:* 1713 Morningside Pl Greenville NC 27834

SAUNDERS, FRED MICHAEL, b Lawton, Okla; m; c 3. WATER & WASTEWATER TREATMENT, HAZARDOUS WASTE TREATMENT. *Educ:* Va Polytech Inst, BS, 67, MS, 69; Univ Ill, Urbana-Champaign, PhD(civil eng), 75. *Prof Exp:* Design engr, Wiley & Wilson Consults, Inc, 74; from asst prof to assoc prof, 74-89, PROF CIVIL ENG, GA INST TECHNOL, 89-, PROG COORDR ENVIRON ENG, 89- *Concurrent Pos:* Mem, Standard Methods Comt, Water Pollution Control Fedn, 75-82, Toxic Substances Comt, 77-82 & Prog Comt, chmn, Res Comt, 83-86; mem, Distiguished Lectr Comt, Asn Environ Eng Prof, 77-82, Pub Comt Abstract Rev Comt, Nat Environ Eng, 81-88 & US Mat Comt, Int Asn Water Pollution Res & Control & US Nat Comt, 88-; ed, Am Soc Civil Engrs Nat Conf Environ Eng, 81, US regional ed, Environ Technol Lett, 85-89; mem bd dirs, Asn Environ Eng Prof, 83-86; vchmn, Water Pollution Mgt Comt, Environ Eng Div, Am Soc Civil Engrs, 85, chmn, 86; mem organizing comt, Int Asn Water Pollution Res & Control 1992 Biennial Conf, 86-; mem, Task Force Wastewater Mgt, Atlanta Regional Comn, 88-91; assoc ed, J Environ Eng, Am Soc Civil Engrs, 88-89; ed, J Water Pollution Control Fedn, 89-; mem exec comt, Environ Eng Div, Am Soc Civil Engrs, 88-, secy, 88-91, vchair, 91-; liasion, US Nat Comt to Int Asn Water Pollution Res & Control, Am Soc Civil Engrs, 88-, secy-treas, 90-; mem publ comt, WPCF, 89-; dipl, Am Acad Environ Engrs. *Honors & Awards:* Excellence in Presentation Award, Am Electroplaters Soc, 80, Sam Wyman Mem Award, 84. *Mem:* Am Soc Civil Engrs; Int Asn Water Pollution Res & Control; Water Pollution Control Fedn; Am Chem Soc; Asn Environ Eng Prof (pres, 85-86); Am Water Works Asn; Am Acad Environ Engrs; Sigma Xi; Nat Water Well Asn. *Res:* Investigation of unit operations and processes used in treatment, reclamation and disposal of industrial and domestic waters, wastewaters and residues; in situ biological treatment of ground water; solid & liquid separations; hazardous waste management; residue treatment. *Mailing Add:* Environ Eng Ga Inst Technol Atlanta GA 30332-0512

SAUNDERS, GEORGE CHERDRON, b Flushing, NY, Jan 4, 40; m 67; c 3. IMMUNOLOGY, IMMUNOCHEMISTRY. *Educ:* Univ Pa, VMD, 64. *Prof Exp:* Fel immunopath, Univ Colo Med Ctr, Denver, 64-66, from instr to asst prof, 66-72; STAFF MEM DIAG IMMUNOL, LOS ALAMOS SCI LAB, UNIV CALIF, 73- *Concurrent Pos:* Adj asst prof path, Sch Med, Univ NMex, 73-77, adj assoc prof, 78- *Honors & Awards:* Inventors Awards, Los Alamos Nat Lab, 82, 87. *Mem:* Soc Analytical Cytol; Am Asn Path. *Res:* Design, development and implementation of rapid automated antibody screening tests; indirect enzyme labeled antibody concept as a major tool in this research; clinical applications of flow cytometry. *Mailing Add:* 306 1/2 Curtis Ave Gallup NM 87301

SAUNDERS, GRADY FRANKLIN, b Bakersfield, Calif, July 11, 38; m 59; c 1. MOLECULAR BIOLOGY. *Educ:* Ore State Univ, BS, 60, MS, 62; Univ Ill, Urbana, PhD(microbiol), 65. *Prof Exp:* USPHS fel, Inst Physicochem Biol, Univ Paris, 65-66; from asst prof to assoc prof, 66-78, PROF BIOCHEM, UNIV TEX SYST CANCER CTR, M D ANDERSON HOSP & TUMOR INST, 78- *Concurrent Pos:* US-USSR exchange scientist, Inst Molecular Biol, USSR Acad Sci, Moscow, 72. *Mem:* Am Chem Soc; Biophys Soc; Am Soc Biol Chemists; Am Soc Cell Biol. *Res:* Regulation of gene activity; chromosome anatomy. *Mailing Add:* Univ Tex Syst Cancer Ctr M D Anderson Hosp & Tumor Inst 6723 Bertner Ave Houston TX 77025

SAUNDERS, JACK PALMER, b London, Eng, Sept 11, 15; nat US; m 42; c 2. PHARMACOLOGY. *Educ:* City Col New York, BS, 36; Univ Md, MS, 49, PhD(biochem), 53. *Prof Exp:* Chemist, R H Macy & Co, NY, 39-41; pharmacologist, Pharmacol Br, Chem Corps Med Labs, US Army Chem Ctr, Md, 46-48, 50-56, dep chief, 54-56; biochemist, USPHS Nutrit Unit, State Dept Health, Md, 48-50; exec secy, Pharmacol Exp Therapeut Study Sect, NIH, 56-57, asst chief extramural progs, Nat Inst Allergy & Infectious Dis, 57, exec secy, Cancer Chemotherapy Study Sect, 57-59, exec secy, Metab Study Sect, 59, asst chief biol sci, Res Grants Rev Br, Div Res Grants, 59-60, chief, 61-64, assoc chief, Div Res Grants, 64-65, dep sci dir chemother, Nat Cancer Inst, 65-67, assoc dir extramural activities, 67-72, dir div cancer grants, 72-73, dir div cancer res resources & ctr, 73-74; prof & dean, 74-87, EMER PROF PHARMACOL & TOXICOL & DEAN, GRAD SCH BIOMED SCI, UNIV TEX MED BR, GALVESTON, 87- *Concurrent Pos:* Consult, Nat Cancer Inst, 74-; co-chmn, Int Symp Immunol Cancer, Univ Tex Med Br-Montpellier Univ, France, 80. *Mem:* AAAS; Am Soc Pharmacol & Exp Therapeut; Am Chem Soc; Soc Toxicol; fel Am Acad Forensic Sci. *Res:* Toxicology; pharmacology of pesticides; nutrition; mechanisms of atherosclerotic plaque formation. *Mailing Add:* 164 San Fernando Dr Galveston TX 77550

SAUNDERS, JAMES ALLEN, b Cleveland, Ohio, Oct 4, 49; m 87; c 2. PLANT BIOCHEMISTRY, PLANT PHYSIOLOGY. *Educ:* Univ SFla, BA, 71; Miami Univ Ohio, PhD(bot), 75. *Prof Exp:* Teaching assoc, Miami Univ Ohio, 71-75; res biochemist, Univ Calif, Davis, 75-77; RES BIOCHEMIST, BARC-WEST, USDA, 77-; PRES, NATIVE SEEDS, INC, 80- *Concurrent Pos:* Adj assoc prof entom, Univ Md, 83- *Mem:* Sigma Xi; Am Soc Plant Physiologists; Phytochem Soc NAm; AAAS. *Res:* Secondary natural products in plants and their biosynthetic enzyme complexes including flavonoids, cyanogenic glucosides, alkaloids and phenolics; health related problems from tobacco use; electrofusion and production of somatic hybrids in plants, pollen gene transfer. *Mailing Add:* Biotechnol Lab Bldg 9 Rm 5 Plant Sci Inst USDA Beltsville MD 20705

SAUNDERS, JAMES CHARLES, b Elizabeth, NJ, May 8, 41; m 67; c 2. PHYSIOLOGICAL PSYCHOLOGY, ANIMAL PHYSIOLOGY. *Educ:* Ohio Wesleyan Univ, BA, 63; Conn Col, MA, 65; Princeton Univ, PhD(psychol), 68. *Hon Degrees:* MA, Univ Pa, 80. *Prof Exp:* Asst prof psychol, Monash Univ, Australia, 69-72; res assoc, Cent Inst Deaf, St Louis, 72-73; from asst prof to assoc prof, 73-89, PROF OTORHINOLARYNGOL, SCH MED, UNIV PA, 89- *Concurrent Pos:* Fel, Auditory Labs, Princeton Univ, 68; res fel, Dept Physiol, Univ Western Australia, 70; med assoc, Philadelphia Gen Hosp, 75-77; res assoc, Philadelphia Vet Hosp, 77-78; actg dir, Inst Neurol Sci, Univ Pa, 80-83, assoc dir, 83-; fel, Am-Scand Found, 84-85 & Swed Med Res Coun, 84-85; guest researcher, Karolinska Inst, Stockholm, Sweden, 84-85; mem, Comt Hearing & Bio-Acoust, Nat Res Coun, 86-; mem, study sect, Nat Inst Neurol & Commun Dis, 87-, mem exec coun, Asn Res Otolaryngol, 88-; Award of Res Excellence, Nat Inst Neurol Commun Dis, 88. *Mem:* Acoust Soc Am; Asn Res Otolaryngol; AAAS; Neurosci Soc; NY Acad Sci; Am Acad Otolaryngol; Soc Gen Physiol. *Res:* Communicative science; communicative disorders; audition; auditory neurobiology; animal psychoacoustics; physiological acoustics; developmental neurobiology; auditory psychology; hearing sciences. *Mailing Add:* Dept Otorhinolaryngol Head Neck Surg 5 Silverstein OTO 3400 Spruce St Philadelphia PA 19104

SAUNDERS, JAMES HENRY, b Ames, Iowa, May 3, 23; m 46; c 4. POLYMER CHEMISTRY, ORGANIC CHEMISTRY. *Educ:* Univ Ky, BS, 44; Univ Ill, PhD(org chem), 46. *Prof Exp:* Spec asst rubber res, Univ Ill, 44-47, group leader, 46-47; chemist, Monsanto Chem Co, 47-50, group leader, 50-54; group leader, Mobay Chem Co, Pa, 54-55, asst dir res, 55-59, dir res, 59-67; mgr nylon res, Monsanto Textiles Co, 68-69, dir nylon & polyester res, 69-75, dir res, Tech Ctr, 76-78, dir polyester res & develop, 79-80, gen mgr technol, 80-82; gen mgr res & develop, Monsanto Fiber & Intermediates Co, 83-85; TECH CONSULT, 86- *Concurrent Pos:* Affil prof, Dept Eng & Policy, Wash Univ, St Louis, Mo, 87-89; adj prof, Univ WFla, Pensacola, 90- *Mem:* AAAS; Fel Am Inst Chemists; NY Acad Sci; Am Chem Soc; Fiber Soc; Sigma Xi; Soc Plastics Engrs. *Res:* Emulsion polymerization; synthesis of substituted styrenes and butadienes; biphenyl and phosgene chemistry; preparation, reactions and applications of isocyanates; polyesters; polycarbonates; polyethers; polyurethanes; polyamides; synthetic foams; elastomers; coatings; adhesives; thermoplastics; synthetic fibers; technical management. *Mailing Add:* 691 Tennyson Pl Pensacola FL 32503-3233

SAUNDERS, JAMES ROBERT, b Simcoe, Ont, Oct 6, 31; m 55; c 4. VETERINARY PATHOLOGY, VETERINARY MICROBIOLOGY. *Educ:* Ont Vet Col, Univ Guelph, DVM, 54; Univ Toronto, dipl vet pub health, 57; Univ Wis, PhD(vet sci, cell physiol), 61. *Prof Exp:* Vet practitioner, Sask, 54-56; res asst microbiol vet sci, Univ Wis, 57-60; vet pathologist, Sask Dept Agr, 60-61; asst prof vet path & microbiol, Purdue Univ, 61-65; assoc prof vet microbiol, 65-70, PROF VET MICROBIOL, UNIV SASK, 70-, CHMN DEPT, 74- *Concurrent Pos:* NIH gen res grant, 63-65; Nat Res Coun Can grant, 66-67. *Mem:* Am Asn Avian Path; Am Col Vet Path; Conf Res Workers Animal Dis; Am Vet Med Asn; Can Vet Med Asn. *Res:* Role of viruses in pneumonias of cattle; pathology of encephalitic diseases; pseudorabies in swine; toxoplasmosis and Marek's disease in chickens; etio-pathogenesis of clostridial infections in birds and animals; ocular diseases of animals; congenital disorders of central nervous system of swine. *Mailing Add:* 416 Lake Cres Saskatoon SK S7H 3A4 Can

SAUNDERS, JEFFREY JOHN, b Minneapolis, Minn, Dec 12, 43. PALEONTOLOGY. *Educ:* Univ Minn, Minneapolis, BA, 66; Univ Ariz, MS, 70, PhD(geosci), 75. *Prof Exp:* Res assoc vert paleont, Ill State Mus Soc, 75-78, CUR VERT PALEONT, ILL STATE MUS, 78- *Mem:* Soc Vert Paleont; Am Quaternary Asn; Sigma Xi. *Res:* Taphonomy of spring deposits and the paleoecology of fossil Proboscidea. *Mailing Add:* 12 S Carriage Hills Springfield IL 62707-9627

SAUNDERS, JOHN BERTRAND DE CUSANCE MORANT, history of medicine, human anatomy, for more information see previous edition

SAUNDERS, JOHN WARREN, JR, b Muskogee, Okla, Nov 12, 19; m 42; c 5. EMBRYOLOGY, TERATOLOGY. *Educ:* Univ Okla, BS, 40; Johns Hopkins Univ, PhD(embryol), 48. *Prof Exp:* Instr zool, Univ Chicago, 48-49; from asst prof to prof, Marquette Univ, 49-66, chmn dept, 57-65; prof anat, Univ Pa, 66-67; prof, 67-85, EMER PROF BIOL SCI, STATE UNIV NY, ALBANY, 85- *Concurrent Pos:* Consult develop biol prog, NSF, 62-66, div biol & med sci, 69-71, chmn, 71. *Mem:* Am Soc Zoologists (secy, 64-66); Soc Develop Biol (pres, 67-68); Am Asn Anatomists; fel AAAS. *Res:* Experimental morphogenesis; chick limb bud and feather tracts; cellular death in embryogenesis. *Mailing Add:* PO Box 381W Waquoit MA 02536

SAUNDERS, JOSEPH FRANCIS, b Mt Pleasant, Pa, Apr 2, 27; m 50; c 2. BIOCHEMISTRY. *Educ:* Duquesne Univ, BS, 50; Georgetown Univ, MS, 55, PhD(chem), 60. *Prof Exp:* Tissue technologist, Pittsburgh Hosp, Pa, 50-51; asst head med & dent br, Off Naval Res, US Dept Navy, Washington, DC, 52-60, head, 60-64; chief environ biol & biosatellite prog & scientist, Off Space Sci & Appln, Nasa Hq, 64-70, chief biol prog, Off Life Sci, Off Manned Space Flight, 71-73; PROG MGR US-USSR, US-CHINA & US-HUNGARY CANCER PROGS, NAT CANCER INST, 73-, DEP DIR INT AFFAIRS, 74- *Concurrent Pos:* Instr, Bus Training Col, 50-51; guest scientist, Naval Med Res Inst, Bethesda, Md, 58-60; mem adv panel biochem, Off Naval Res, 64-65; exec secy space biol subcomt, NASA, 66-70; mem US nat comt, Int Inst Refrig, 69-71; foreign affairs ed, Biosci Commun, 75-77; coord ed, J Soviet Oncol, 79-; prog mgr, Am Physiol Soc, 83-; exec dir, Am Asn Immunologists, 86- *Honors & Awards:* Arthur S Flemming Award, 62; NIH Dir Award, 79- *Mem:* Am Chem Soc; Am Physiol Soc; Fedn Am Socs Exp Biol. *Res:* Cholesterol metabolism in atherosclerosis and nerve tissue; space biology and medicine; radiation biology; molecular biology of cancer. *Mailing Add:* 8131 Greeley Blvd Springfield VA 22152

SAUNDERS, JOSEPH LLOYD, b Elk City, Okla, Oct 26, 35; m 63; c 3. ENTOMOLOGY, NEMATOLOGY. *Educ:* Colo State Univ, BS, 59; Univ Wis, MS, 60, PhD(entom), 63. *Prof Exp:* Proj assoc entom, Univ Wis, 64-66; asst entomologist, Wash State Univ, 66-69, assoc prof entom, 69-71; assoc prof entom & int agr develop, Cornell Univ, 71-76; entomologist, cropping systs prog, Trop Agron Ctr Res & Educ, 76-84, COORDR, REGIONAL IPM PROG, CENT AM, COSTA RICA, 84- *Mem:* Entom Soc Am. *Res:* Pest management. *Mailing Add:* CATIE Turrialba Costa Rica

SAUNDERS, KIM DAVID, b Chicago, Ill, Jan 21, 45; div; c 2. EXPLOSIVE WELDING. *Educ:* Rose Polytech Inst, BSc, 66; Mass Inst Technol, PhD(oceanog), 71. *Prof Exp:* Res assoc oceanog, Mass Inst Technol, 71-72; NATO fel, Inst Geophys, Univ Bergen, 72-73, Royal Norwegian Coun Sci & Indust Res fel, 73-74; asst environ scientist, Environ & Energy Systs Div, Argonne Nat Lab, 74-78; OCEANOGRAPHER, NAVAL OCEANOG & ATMOSPHERIC RES LAB, STENNIS SPACE CTR, MISS, 78- *Concurrent Pos:* Consult, underwater explosion damage prev & explosive welding, 80- *Mem:* AAAS; Am Geophys Union; Sigma Xi; Inst Elec & Electronics Engrs; Marine Technol Soc. *Res:* Near shore circulations; high frequency internal waves; oceanographic instrumentation; geophysical fluid dynamics; numerical analysis and data quality control; ocean simulation. *Mailing Add:* Naval Oceanog & Atmospheric Res Lab Stennis Space Center MS 39529-5004

SAUNDERS, LEON Z, b Winnipeg, Man, Dec 16, 19; nat US; m 65; c 1. VETERINARY PATHOLOGY, HISTORY OF PATHOLOGY. *Educ:* Univ Toronto, VS & DVM, 43; Iowa State Col, MS, 46; Cornell Univ, PhD(vet path), 51. *Hon Degrees:* Dr Med Vet, Vet Col Vienna, 68. *Prof Exp:* Instr vet path, Iowa State Col, 46-48; asst, Cornell Univ, 48-51; pathologist, Chem Corps Med Labs, US Army Chem Ctr, Md, 51-52; assoc vet, Brookhaven Nat Lab, 52-54, vet, 55-58; head path & toxicol sect, Smith Kline & French Labs, 58-68, dir path & toxicol, 68-80, vpres drug safety eval, 80-83, vpres & sr consult pathologist, 83-90, CONSULT TOXICOL PATHOLOGIST, SMITH KLINE BEECHAM PHARMACEUTICALS, 91- *Concurrent Pos:* Vis asst prof, Univ Pa, 58-62, vis assoc prof, 62-63, adj prof, 64-; vpres, World Fedn Vet Path, 59-67, pres, 67-71; ed, Pathologia Veterinaria, 63-67; managing ed, Vet Path, 68-89; mem, sci adv bd, Armed Forces Inst Path, Washington, DC, 85-91. *Honors & Awards:* Schofield Mem Medal, 73; Theodor Kitt Medal, Munich, 82. *Mem:* Am Vet Med Asn; Am Col Vet Path (vpres, 67-, pres, 68); Am Asn Hist Med; US-Can Acad Path; Am Asn Pathologists; distinguished mem Am Col Vet Path; Non mem Soc Tox Path. *Res:* Animal ophthalmic pathology; history of veterinary pathology. *Mailing Add:* Smith Kline Beecham Pharmaceuticals PO Box 1539 King of Prussia PA 19406

SAUNDERS, MARTIN, b Brooklyn, NY, Jan 10, 31; m 82; c 2. ORGANIC CHEMISTRY. *Educ:* City Col, BS, 52; Harvard Univ, PhD(org chem), 56. *Prof Exp:* From instr to assoc prof chem, 55-70, PROF CHEM, YALE UNIV, 70-, FEL, BRANFORD COL, 70- *Concurrent Pos:* Yale Univ jr fac fel sci, 62-63; Sloan fel, 65-69; spec award, von Humboldt Found, 77-78 & 85; Kharasch award, Univ Chicago, 82-83. *Mem:* Am Chem Soc; fel AAAS. *Res:* Applications of nuclear magnetic resonance spectroscopy to organic chemistry; study of stable carbonium ion solutions; molecular modeling. *Mailing Add:* Dept Chem Yale Univ New Haven CT 06520

SAUNDERS, MORTON JEFFERSON, b Norfolk, Va, Nov 19, 25; m 50; c 2. OPTICS. *Educ:* Univ Va, BS, 50, MS, 52; Univ Fla, PhD(physics), 56. *Prof Exp:* Lab asst physics, Univ Va, 50-51, res asst, 51-52; optical engr, Farrand Optical Co, NY, 52-53; res asst physics, USAF Contract, Univ Fla, 55-56 & US Army Ord Contract, 56; MEM TECH STAFF, BELL TEL LABS, 56- *Mem:* AAAS; Optical Soc Am; NY Acad Sci. *Res:* Phase contrast studies of turbulent media; visibility through the atmosphere; electromagnetic scattering from microscopic water droplets and dielectric cylinders; fiber optic structures, losses in fiber optic structures; fiber optic index of refraction measurement. *Mailing Add:* AT&T Bell Tel Labs Rm 1B35 2000 Northeast Expwy Norcross GA 30071

SAUNDERS, PETER REGINALD, b Wokingham, Eng, Aug 21, 28; m 57; c 4. PHYSICAL CHEMISTRY, PHYSICS. *Educ:* Univ London, BSc, 51. *Prof Exp:* Res officer, Brit Gelatin & Glue Res Asn, 51-55, sr res officer, 55-57; res assoc concentrated polymer solutions, Univ Wis, 57-58; res physicist, Chemstrand Res Ctr, Inc, 58-64, sr res physicist, 64-67; prin scientist, Brit Food Res Asn, 67-68; SR SCIENTIST & RES SUPVR, FIBERS DIV, TECH CTR, ALLIED CHEM CORP, 68- *Mem:* Am Chem Soc; Soc Rheology; Am Asn Textile Chemists & Colorists. *Res:* Molecular characterization of macromolecules; rheological properties of solutions and gels of macromolecules; physics of fiber formation; relation of fiber structure to mechanical properties; dye diffusion in fibers. *Mailing Add:* 6618 Philbrook Rd Richmond VA 23234

SAUNDERS, PRISCILLA PRINCE, b Monterey, Calif, Apr 21, 38; m 59; c 1. BIOCHEMICAL PHARMACOLOGY. *Educ:* Ore State Univ, BS, 60, MS, 61; Univ Ill, PhD(biochem), 65. *Prof Exp:* USPHS fel, Enzym Lab, Nat Ctr Sci Res, France, 65-66; res assoc, 66-75, asst prof, 75-80, ASSOC PROF BIOCHEM, UNIV TEX M D ANDERSON HOSP & TUMOR INST, 80- *Mem:* Am Asn Cancer Res; Am Soc Pharmacol & Exp Therapeut; Am Soc Cell Biol. *Res:* Metabolism and mechanism of action of antitumor agents in cultured mammalian cells. *Mailing Add:* Dept Med Oncol Univ Tex M D Anderson Cancer Ctr 1515 Holcombe Blvd Houston TX 77030

SAUNDERS, RICHARD HENRY, JR, gerontology, for more information see previous edition

SAUNDERS, RICHARD L DE C H, b Grahamstown, SAfrica, May 29, 08; m 36; c 1. NEUROANATOMY, RADIOLOGY. *Educ:* Univ Edinburgh, MB, ChB, 32; MD, 40. *Hon Degrees:* Dipl radiol, Univ Lisbon, 65. *Prof Exp:* Vis physician, Settlers Hosp, SAfrica, 32; house surgeon, Bradford Royal Infirmary, Eng, 33; lectr anat, Univ Edinburgh, 33-37; from asst prof to assoc prof anat, 38-48, prof path anat & dir med mus, 48-49, prof anat & head dept, 49-73, EMER PROF ANAT, DALHOUSIE UNIV, 74- *Concurrent Pos:* Res prof neuroanat, Radcliffe Infirmary, Oxford, Eng, 74-79. *Mem:* Fel Royal Micros Soc; Am Asn Anatomists; fel Royal Soc Edinburgh. *Res:* Microfocal radiography in experimental and clinical medicine, with special emphasis on cerebral microcirculation and neural structure. *Mailing Add:* West Jeddore Halifax County NS B0J 1P0 Can

SAUNDERS, RICHARD LEE, b Lynn, Mass, June 24, 28; m 55; c 3. PHYSIOLOGY. *Educ:* Univ Mass, BS, 51; Univ Toronto, MA, 53, PhD(zool), 60. *Prof Exp:* RES SCIENTIST, CAN DEPT FISHERIES & OCEANS, 60- *Mem:* Am Fisheries Soc; Can Soc Zoologists; World Aquacult Soc; Aquacult Asn Can. *Res:* Environmental physiology; fish respiration and metabolism; osmotic and ionic changes resulting from stress; endocrinological control of salmon smolting and growth; Atlantic salmon biology; salmonid genetics; salmonid aquaculture. *Mailing Add:* 309 Montague St Andrews NB E0G 2X0 Can

SAUNDERS, ROBERT M(ALLOUGH), b Winnipeg, Man, Sept 12, 15; nat US; m 43. ELECTRICAL ENGINEERING. *Educ:* Univ Minn, BEE, 38, MS, 42; Tokyo Inst Technol, DEng, 71. *Prof Exp:* Instr elec eng, Univ Minn, 42-44; from lectr to assoc prof, Univ Calif, Berkeley, 47-55, prof elec eng, 56-65, chmn dept, 59-63; asst to chancellor, 64-65, dean sch eng, 64-73, PROF ELEC ENG, UNIV CALIF, IRVINE, 65- *Concurrent Pos:* Vis assoc prof, Mass Inst Technol, 54-55; mem eng educ & accreditation comt, Eng Coun Prof Develop, 65-71, chmn, 69-70, mem bd dirs, 71-75; mem eng adv comt, NSF, 68-71; mem bd visitors, Army Transp Sch, Ft Eustis, Va, 70-73 & Secy Navy's Bd Educ & Training, 72-78; consult, Aerospace Corp, El Segundo, 71-, Gen Motors Corp, Apollo Support Dept, Gen Elec Co, Rohr Corp, Chula Vista & Hughes Aircraft Co, Fullerton; secy, Nat Comn Eng Films. *Honors & Awards:* Centennial Medal, Inst Elec & Electronics Engrs, 84, Itariden Pratt Award, 90. *Mem:* Am Soc Eng Educ; fel Inst Elec & Electronics Engrs; fel AAAS. *Res:* Electrical machinery theory; feedback control systems; applications of digital computers to electrical machine design; theory of electromechanical energy converters; system simulation and optimization. *Mailing Add:* Univ Calif Elec Eng Irvine CA 92717

SAUNDERS, ROBERT MONTGOMERY, food science; deceased, see previous edition for last biography

SAUNDERS, ROBERT NORMAN, b Fairbury, Ill, Sept 25, 38; m 81; c 3. ATHEROSCLEROSIS, THROMBOSIS. *Educ:* Purdue Univ, BS, 61, MS, 66, PhD(pharmacol), 68. *Prof Exp:* Res scientist, G D Searle Co, 68-77, group leader, 77-80; sect head, 80-84, dept head, 84-85, DIR, SANDOZ, INC, 85- *Concurrent Pos:* Res comt chmn, G D Searle Co, 71-80, proj coordr, 73-79. *Mem:* Am Soc Pharmacol & Exp Therapeut; Int Soc Thrombosis & Haemostasis; Soc Exp Biol & Med; Am Heart Asn. *Res:* Diabetes and intermediary metabolism cholesterol synthesis regulation; smooth muscle cell proliferation; growth factors especially PDGF; platelet function and activity modification; prostaglandin synthesis and pharmacological activity; platelet activating factor activity and inhibition. *Mailing Add:* Sandoz Corp Rte 10 East Hanover NJ 07936

SAUNDERS, RONALD STEPHEN, b Parsons, Kans, Oct 8, 40; m 65; c 3. GEOLOGY, PLANETOLOGY. *Educ:* Univ Wis-Madison, BS, 63; Brown Univ, MSc, 68, PhD(geol), 70. *Prof Exp:* Sr scientist, 69-74, mem tech staff, 74-86, SR RES SCIENTIST, JET PROPULSION LAB, 86- *Concurrent Pos:* Proj scientist, Magellan, 81- *Honors & Awards:* Except Serv Medal, NASA, 86. *Mem:* Sigma Xi; Am Geophys Union; Soc Econ Paleontologists & Mineralogists; AAAS; Geol Soc Am. *Res:* Planetary geology of the moon, Mars, and Venus. *Mailing Add:* 616 E Deodara Dr Altadena CA 91001

SAUNDERS, SAM CUNDIFF, b Richland, Ore, Feb 24, 31; m 54; c 3. MATHEMATICAL STATISTICS & PROBABILITY. *Educ:* Univ Ore, BS, 52; Univ Wash, PhD(math statist), 56. *Prof Exp:* Mathematician, Math Serv Unit, Boeing Airplane Co, 56-58 & Math Res Lab, 58-60; asst prof math, Math Res Ctr, Univ Wis, 60-61; staff mem, Math Res Lab, Boeing Sci Res Labs, Wash, 61-72; PROF PURE & APPL MATH, WASH STATE UNIV, 72- *Concurrent Pos:* Consult, nuclear regulatory comn. *Mem:* Am Math Soc; Soc Indust & Appl Math; Math Asn Am; Inst Math Statist; fel Am Statist Asn. *Res:* Non-parametric methods; reliability theory; statistical inference. *Mailing Add:* Dept Math Wash State Univ Pullman WA 99164-3113

SAUNDERS, SHELLEY RAE, b Toronto, Can, Feb 28, 50; m 71; c 2. EVOLUTIONARY THEORY. *Educ:* Univ Toronto, BA, 72, MA, 73, PhD(phys anthrop), 77. *Prof Exp:* Lectr & sr demonstr, McGill Univ, 76-79; asst prof, Univ Toronto, 79-81; asst prof, 81-84, ASSOC PROF ANTHROP, McMASTER UNIV, 85- *Concurrent Pos:* Ed, J Can Asn Phys Anthropologists, 78-81. *Mem:* Am Asn Phys Anthropologists; AAAS; Human Biol Coun; Can Asn Phys Anthropologists (secy treas, 83-). *Res:* Investigation of bone growth remodeling of human infracranial bone; morphological variation of human bone; evolutionary changes of past human populations. *Mailing Add:* Dept Anthrop McMaster Univ 1280 Main St W Hamilton ON L8S 4L8 Can

SAUNDERS, VIRGINIA FOX, b Roanoke, Va, May 31, 38; m 63; c 2. NEUROPSYCHOLOGY, NEUROSCIENCES. *Educ:* Univ Mich, BA, 60; Ind Univ, PhD(psychol & neurophysiol), 66. *Prof Exp:* Fel, Interdisciplinary Training Prog, Univ Calif Med Ctr, 65-67; from asst prof to assoc prof, 67-76, PROF PSYCHOL, SAN FRANCISCO STATE UNIV, 76- *Concurrent Pos:* Spec progs coordr, Kentfield Sch Dist, Calif, 74-76; mem, Res & Eval Comt, Redwood High Sch, Calif, 75-77. *Res:* Physiological and pharmacological factors in sensation, perception, learning and memory. *Mailing Add:* Dept Psychol San Francisco State Univ 1600 Holloway Ave San Francisco CA 94132

SAUNDERS, WILLIAM BRUCE, b Tuscaloosa, Ala, Nov 12, 42; m 64; c 1. INVERTEBRATE PALEONTOLOGY, GEOLOGY. *Educ:* Univ Ark, BSc, 66, MSc, 68; Univ Iowa, PhD(geol), 71. *Prof Exp:* Asst prof, 70-76, chmn geol, 80-81, ASSOC PROF GEOL, BRYN MAWR COL, 76- *Concurrent Pos:* Exchange scientist, Nat Acad Sci-USSR Acad Sci, 74-75; prin investr, NSF grants marine biol, paleobiol, 75-77, 77-79, 80-84 & RV Alpha Helix, Philippines, 79. *Mem:* Int Union Geol Sci; fel Geol Soc Am; Int Paleont Asn; Paleont Soc; fel Explorers Club. *Res:* Evolution and paleobiology of fossil cephalopods, cephalopod biostratigraphy and biology of cephalopods; particularly living nautilus. *Mailing Add:* Dept Geol Bryn Mawr Col Bryn Mawr PA 19010

SAUNDERS, WILLIAM H, b Omaha, Nebr, Jan 7, 20; c 4. OTOLARYNGOLOGY. *Educ:* Univ Omaha, AB, 39; Univ Iowa, MD, 43. *Prof Exp:* From asst prof to assoc prof, 54-60, actg chmn dept, 61-63, PROF OTOLARYNGOL, COL MED, OHIO STATE UNIV, 60-, CHMN DEPT, 63- *Concurrent Pos:* Dir, Am Bd Otolaryngol, 74. *Mem:* AMA; Am Acad Ophthal & Otolaryngol; Am Laryngol, Rhinol & Otol Soc; Am Laryngol Asn; Am Otol Soc. *Res:* Otology. *Mailing Add:* Dept Otolaryngol Ohio State Col Med 456 Clinic Dr Columbus OH 43210

SAUNDERS, WILLIAM HUNDLEY, JR, b Pulaski, Va, Jan 12, 26; wid; c 2. PHYSICAL ORGANIC CHEMISTRY. *Educ:* Col William & Mary, BS, 48; Northwestern Univ, PhD(chem), 52. *Prof Exp:* Res assoc chem, Mass Inst Technol, 51-53; from instr to assoc prof, 53-64, chmn dept, 66-70, PROF CHEM, UNIV ROCHESTER, 64- *Concurrent Pos:* Guggenheim fel & hon res assoc, Univ Col, Univ London, 60-61; Sloan Found fel, 61-64; NSF sr fel & guest researcher, Univ Gothenburg, 70-71; ed, Tech of Chem, 85- *Mem:* Am Chem Soc; Royal Soc Chem. *Res:* elimination reactions; isotope effects and isotopic tracers; the use of kinetic isotope effects and isotopic tracers to study the mechanisms of organic reactions, especially elimination and other proton transfer processes; the role of tunneling in proton transfers. *Mailing Add:* Dept Chem Univ Rochester Rochester NY 14627

SAUPE, ALFRED (OTTO), b Badenweiler, WGer, Feb 14, 25; m 63; c 3. PHYSICS, PHYSICAL CHEMISTRY. *Educ:* Univ Freiburg, MS, 55, Dr rer nat(physics), 58. *Prof Exp:* Ger Res Asn fel, Univ Freiburg, 58-61; res assoc physics, Electronics Inst, Freiburg, 61-62; sci asst, Univ Freiburg, 62-67, docent phys chem, 67-68; vis prof, 68-70, PROF PHYSICS, KENT STATE UNIV, 70- *Mem:* AAAS. *Res:* Physics of liquid crystals, particularly molecular theories and elastic and optical properties; nuclear magnetic resonance on oriented molecules. *Mailing Add:* Dept Physics 207 Lowry Hall Kent State Univ Kent OH 44242

SAURO, JOSEPH PIO, b New Rochelle, NY, Apr 4, 27; m 48; c 3. PHYSICS. *Educ:* Polytech Inst Brooklyn, BS, 55, MS, 58, PhD(physics), 66. *Prof Exp:* Instr physics, Polytech Inst Brooklyn, 56-65; assoc prof, 66-80, dean, Col Arts & Sci, 69-80, interim dean, Col Eng, 78-80, PROF PHYSICS, SOUTHEASTERN MASS UNIV, 80- *Mem:* Am Phys Soc; Sigma Xi. *Res:* Scattering of x-rays by thin films; dielectric properties of phosphors. *Mailing Add:* Col Arts & Sci Southeastern Mass Univ North Dartmouth MA 02747

SAUSE, H WILLIAM, b Baltimore, Md, Sept 29, 20; m 44; c 3. ORGANIC CHEMISTRY. *Educ:* Johns Hopkins Univ, AB, 48, MA, 50, PhD(chem), 53. *Prof Exp:* Jr instr chem, Johns Hopkins Univ, 48-52; res assoc, Northwestern Univ, 52-54; res chemist, 54-77, clin pharm assoc, Searle Labs, Chicago, 77-86; CONSULT, 86- *Mem:* AAAS; Am Chem Soc. *Res:* Natural products; pharmaceutical chemistry; heterocyclic chemistry; clinical pharmacy. *Mailing Add:* 1061 Springfield Ave Deerfield IL 60015-3030

SAUSEN, GEORGE NEIL, b St Paul, Minn, Aug 14, 27; m 51; c 5. ORGANIC CHEMISTRY. *Educ:* Col St Thomas, BS, 49; Univ Wis, PhD(chem), 53. *Prof Exp:* Res chemist, E I du Pont de Nemours & Co, Inc, 53-63, suprv, 64-67, sr res chemist, 68-71, patent liaison & suprv patents & tech info, 71-88, MGR, CORP INFO SCI, CENT RES & DEVELOP DEPT, E I DU PONT DE NEMOURS & CO, INC, 88- *Mem:* Am Chem Soc. *Res:* Angular methylation studies of steroid intermediates; cyanocarbons; fluorine chemistry; polymer intermediates. *Mailing Add:* Cent Res & Develop Dept E I du Pont de Nemours & Co Inc Barley Mill Plaza 14 Wilmington DE 19800-0014

SAUSVILLE, JOSEPH WINSTON, b Brooklyn, NY, Oct 17, 18; m 45; c 3. HIGH TEMPERATURE CHEMISTRY, ENGINEERING MANAGEMENT. *Educ:* Polytech Inst Brooklyn, BS, 41; Univ Iowa, PhD(phys chem), 48. *Prof Exp:* Chemist, Tenn Eastman Corp, 44-46 & Nuclear Energy Propulsion Aircraft Div, Fairchild Eng & Airplane Corp, 47-48; from asst prof to assoc prof phys chem, Univ Cincinnati, 48-56, chmn grad studies, 51-56; mgr nuclear mat, Res Div, Curtiss-Wright Corp, Pa, 56-61, eng head advan develop metall, Wright Aero Div, NJ, 62-67; mgr phosphor advan develop, 67-70, MGR PHOSPHOR DEVELOP, FLUORESCENT & VAPOR LAMP DIV, WESTINGHOUSE ELEC CORP, 70- *Mem:* Am Chem Soc; Am Inst Chemists; Illum Eng Soc; Am Soc Photobiol. *Res:* Fundamental properties of matter; high temperature materials fabrication and evaluation; effects of ionizing radiations. *Mailing Add:* 41 Goodviet Pl Glen Rock NJ 07452

SAUTE, ROBERT E, b West Warwick, RI, Aug 18, 29; m 57; c 3. PHARMACY, COSMETIC SCIENCES. *Educ:* RI Col Pharm, BS, 50; Purdue Univ, MS, 52, PhD(pharm), 53. *Prof Exp:* Tech asst to gen mgr, Lafayette Pharmacol Co, 55-56; res chemist, H K Wampole Co, 56-58; plant supt, Strong Cobb Arner Co, 58-60; res chemist, Avon Prod Inc, NY, 60-64, dir prod & process develop & dir res & develop admin, 64-68; res dir, Toiletries Div, Gillette Co, 68-71; group vpres, Dart Industs Inc, 71-74; OWNER, SAUTE CONSULTS, 74- *Honors & Awards:* Fel Soc Cosmetic Chemists. *Mem:* AAAS; Soc Invest Dermat; Am Pharmaceut Asn; Acad Pharmaceut Sci; NY Acad Sci; Soc Cosmetic Chemists; Sigma Xi. *Res:* Consulting in the cosmetic, drug, toiletries and fragrance industries. *Mailing Add:* 10236 Mossy Rock Circle Los Angeles CA 90077

SAUTER, FREDERICK JOSEPH, b Pittsburgh, Pa, Dec 12, 43; m 66; c 3. ORGANIC CHEMISTRY. *Educ:* Duquesne Univ, BS, 65; Mass Inst Technol, PhD(org chem), 69. *Prof Exp:* Sr res chemist, 69-76, RES ASSOC EASTMAN KODAK CO, 76- *Mem:* AAAS; Am Chem Soc. *Res:* Organic syntheses and reaction mechanisms as applied to conventional and unconventional imaging systems. *Mailing Add:* 203 Dohrcrest Dr Rochester NY 14612

SAUTHOFF, NED ROBERT, b Belleville, Ill, Apr 14, 49; m 75; c 1. PLASMA PHYSICS. *Educ:* Mass Inst Technol, SB & SM, 72; Princeton Univ, PhD(astrophys), 75. *Prof Exp:* Res physicist, Plasma Physics Lab, Princeton Univ, 75-84, head, Tokamak fusion test reactor exp, Comput Br, 80-84, dep head, Comput Div, 83-86, head comput div, 86-88, PRIN RES PHYSICIST, PRINCETON UNIV, 84-, HEAD, PRINCETON BETA EXP, 88-, HEAD, EXP PROJS DEPT, 90- *Mem:* Am Phys Soc; Sigma Xi; Inst Elec & Electronics Engrs. *Res:* Plasma physics in general; x-ray techniques; magnetohydrodynamics in tokamaks by x-ray imaging and tomography. *Mailing Add:* Plasma Physics Lab Princeton Univ PO Box 451 Princeton NJ 08543

SAUTTER, CHESTER A, b Scotia, Nebr, Nov 16, 33; m 59; c 4. EXPERIMENTAL ATOMIC PHYSICS, ENVIRONMENTAL PHYSICS. *Educ:* Nebr Wesleyan Univ, AB, 55; Univ Nebr, MA, 58, PhD(physics), 63. *Prof Exp:* Fulbright travel grant & guest physicist, Inst Physics, Aarhus, Denmark, 63-64; ASSOC PROF PHYSICS, CONCORDIA COL, MOORHEAD, MINN, 65- *Concurrent Pos:* Consult & vis prof, Wash State Univ, 74-75; consult, Lutheran Coun, USA & Lutheran World Ministries, 81-82. *Mem:* Am Asn Physics Teachers; Am Phys Soc; Int Solar Energy Soc; Hist Sci Soc; Sigma Xi. *Res:* Charge exchange of slow atoms and ions in gases; stopping power of ions and atoms in carbon and hydrocarbon films; ion channelling in single crystals and neutron activation analysis; residual environmental effects of Church Rock, New Mexico 1979 uranium mill tailings spill. *Mailing Add:* Dept Physics & Environ Studies Concordia Col Moorhead MN 56560

SAUTTER, JAY HOWARD, b Waynesburg, Ohio, Nov 11, 12; m 37; c 3. VETERINARY PATHOLOGY. *Educ:* Ohio State Univ, DVM, 44; Univ Minn, PhD(path), 48. *Prof Exp:* Instr vet med, 47-48, assoc prof vet path, 48-52, PROF VET PATH, UNIV MINN, ST PAUL, 52- *Concurrent Pos:* AID consult, Peru, 65; tech asst, Nebr Proj, Columbia Univ, 70-71; tech asst, Food & Agr Orgn, Dom Repub, 72; prof & head dept path & microbiol, Ahmadu Bella Univ, Nigeria, 74-76; tech asst, Kans AID Proj, Ahmadu Bella Univ, 74- *Mem:* Am Asn Avian Pathologists; Am Vet Med Asn; Am Col Vet Pathologists; Nigerian Vet Med Asn; Am Asn Pathologists. *Mailing Add:* 1550 Fulham Ave St Paul MN 55108

SAUVAGE, LESTER ROSAIRE, b Wapata, Wash Nov 15, 26; m 56; c 8. CARDIOVASCULAR SURGERY & RESEARCH. *Educ:* St Louis Univ Sch Med, MD, 48. *Hon Degrees:* ScD, Gonzaga Univ, 82. *Prof Exp:* FOUNDER & DIR, HOPE HEART INST, 59-; PVT PRACT CARDIOVASC PEDIAT SURG, 59- *Concurrent Pos:* Chmn dept surg, Providence Med Ctr, 65-74, dir surg educ, 68-73; dir cardiac surg, Childrens Orthop Hosp & Med Ctr, 65-74; clin prof surg, Univ Wash. *Honors & Awards:* Clemson Award, Soc Biomats, 82; Jefferson Award, Am Inst Pub Serv, 83. *Mem:* Am Col Surgeons; Int Cardiovasc Soc; Soc Vascular Surg; Neurovascular Soc NAm. *Res:* Synthetic blood vessel prostheses; vascular surgical techniques; prediction and prevention of thrombotic complications of atherosclerosis; endothelial cell function; vascular autografts; approximately 175 articles. *Mailing Add:* Jefferson Med Tower Suite 101 1600 E Jefferson St Seattle WA 98122

SAVAGE, ALBERT B, b Minneapolis, Minn, Dec 22, 12; wid; c 2. POLYMER CHEMISTRY, SYNTHETIC ORGANIC & NATURAL PRODUCTS CHEMISTRY. *Educ:* Univ Minn, BChE & BA, 35, MS, 37. *Prof Exp:* Sr res specialist, Cellulose Res, Designed Polymers Res Lab, Dow Chem Co, 37-76; RETIRED. *Mem:* Sigma Xi. *Res:* Etherification; cellulose; soluble polymers; plastics. *Mailing Add:* 122 Varner Ct Midland MI 48640-3508

SAVAGE, BLAIR DEWILLIS, b Mt Vernon, NY, June 7, 41; m 66; c 1. ASTRONOMY. *Educ:* Cornell Univ, BEngrPhysics, 64; Princeton Univ, PhD(astron), 67. *Prof Exp:* Res assoc astron, Princeton Univ, 67-68; from asst prof to assoc prof, 68-78, PROF ASTRON, UNIV WIS-MADISON, 78- *Concurrent Pos:* Vis fel, Joint Inst Lab Astrophys, Colo, 74-75. *Mem:* Am Astron Soc; Int Astron Union. *Res:* Ultraviolet space astronomy; interstellar matter, high resolution spectroscopy. *Mailing Add:* Washburn Observ Univ Wis Madison WI 53706

SAVAGE, CARL RICHARD, JR, b Bloomsburg, Pa, Dec 1, 42; m 65; c 2. BIOCHEMISTRY, ENDOCRINOLOGY. *Educ:* Gettysburg Col, BA, 64; State Univ NY Buffalo, PhD(biochem), 71. *Prof Exp:* Dir res biochem, Vet Admin Hosp, Albany, NY, 73-75; ASST PROF BIOCHEM, SCH MED, TEMPLE UNIV, 75- *Concurrent Pos:* Damon Runyon fel, 71-73; Dermat Found grant, Temple Univ, 75-76, Inst grant, 75-76, Am Cancer Soc Inst grant, 76-77 & NIH Diabetes grant, 78-81. *Mem:* Am Soc Biol Chemists; Tissue Cult Asn. *Res:* Mechanism of action of polypeptide hormones; metabololic regulation; control of growth and differentiation; primary cultures of liver parenchymal cells; tissue culture. *Mailing Add:* Dept Biochem One Health Sci Campus Temple Univ Sch Med Broad & Ontario Sts Philadelphia PA 19140

SAVAGE, CHARLES, b Berlin, Conn, Sept 25, 18; m 40; c 2. PSYCHIATRY. *Educ:* Yale Univ, BA, 39; Univ Chicago, MD, 44; Am Bd Psychiat & Neurol, dipl, 51. *Prof Exp:* Intern, Univ Chicago, 45; asst resident psychiat, Yale Univ, 46; resident psychiatrist, US Naval Hosp, Md, 47-48, chief psychiatrist, SC, 48-49; res psychiatrist, Naval Med Res Inst, Nat Naval Med Ctr, 49-52; actg chief adult psychiatrist, NIMH, 53-58; psychiatrist, Livermore Sanitarium, 58-60 & Stanford Vet Admin Hosp, 61-62; med dir, Int Found Advan Study, 62-64; psychiatrist, County of Santa Clara, 64-65; dir res, Spring Grove State Hosp, 65-68; from clin asst prof to assoc prof, Inst Psychiat & Human Behav,

Univ Md, 65-80, prof psychiat, 72-86; PSYCHIATRIST, DIV MENT HEALTH, DEPT HEALTH, VI GOVT, 86- *Concurrent Pos:* David C Wilson Soc lectr, Univ Va, 67; prin investr, Nat Inst Mental Health grant, Studies of Selected Narcotic Agonists & Antagonists, 72-75; chief, Psychiat Serv & Drug Treatment Ctr, Vet Admin Hosp, 72-82; fel, Ctr Advan Study Behav Sci, 57-58; emer asst prof psychiat, Johns Hopkins Univ, 82- *Mem:* Fel Am Psychiat Asn; Am Psychoanal Asn; Am EEG Soc; AAAS. *Mailing Add:* Coral Bay St John VI 00830

SAVAGE, CHARLES FRANCIS, b Maywood, Ill, Feb 24, 06; m 30, 74. AERONAUTICAL & ASTRONAUTICAL ENGINEERING. *Educ:* Ore State Univ, BSEE, 28. *Prof Exp:* Div engr, Gen Elec Co, 28-71; exec dir, Coun Eng Law, 71-77; staff consult eng law, Engrs Joint Coun, 77-80; RETIRED. *Concurrent Pos:* Chmn nuclear cong, Am Soc Mech Engrs, 54-55, conf bd indust, Nat Soc Prof Engrs, 56-59, eng sect, AAAS, 57-58; hon mem, Coun Eng & Sci Soc Exec, 70- *Honors & Awards:* Centennial Medal, Inst Elec & Electronics Engrs, 84. *Mem:* Inst Elec & Electronics Engrs; fel Am Soc Mech Engrs; Nat Soc Prof Engrs; Am Soc Eng Educ; fel AAAS; fel Am Inst Elec Engrs; hon mem Coun Eng & Sci Soc Execs; assoc fel Am Inst Aeronaut & Astronaut. *Res:* Electrical measurement, especially in aircraft engine, guidance and control. *Mailing Add:* 3609 Palmer Ct Clovis NM 88101

SAVAGE, DENNIS JEFFREY, b Warren, Ohio, Oct 13, 42; m 68; c 2. ORGANIC CHEMISTRY. *Educ:* Univ Colo, Boulder, BA, 65; Univ Ariz, PhD(chem), 71. *Prof Exp:* SR ANALYTICAL CHEMIST, EASTMAN KODAK CO, 70- *Res:* Organometallic chemistry; organic polymer synthesis. *Mailing Add:* Eastman Kodak Co 343 State St Rochester NY 14650

SAVAGE, DONALD ELVIN, b Floydada, Tex, May 28, 17; m 42; c 4. PALEONTOLOGY. *Educ:* WTex State Col, BS, 37; Univ Okla, MS, 39; Univ Calif, PhD(paleont), 49. *Prof Exp:* From asst prof to assoc prof paleont, 49-62, PROF PALEONT, UNIV CALIF, BERKELEY, 62-, CUR MUS, 49- *Mem:* Geol Soc Am; Soc Vert Paleont. *Res:* Late Cenozoic mammals; nonmarine stratigraphy of western United States. *Mailing Add:* 6104 Oliver Rd Paradise CA 95969

SAVAGE, DWAYNE CECIL, b Arco, Idaho, Aug 8, 34; m 57; c 2. MOLECULAR & MICROBIAL ECOLOGY. *Educ:* Univ Idaho, BS, 56; Univ Calif, Berkeley, MA, 61, PhD(bact), 65. *Prof Exp:* Guest investr & fel, Rockefeller Univ, 65-67; from asst prof to assoc prof, Univ Tex, Austin, 67-73; from assoc prof to prof microbiol, Univ Ill & Sch Basic Med Sci, Col Med, 73-88; PROF & HEAD MICROBIOL, UNIV TENN, 88- *Concurrent Pos:* Nat Inst Allergy & Infectious Dis res proj grant, 68-; vis assoc prof, Sch Med, Univ Colo, Denver, 70, consult, Div Gastroenterol, 70-73; Am Soc Microbiol Found Lectr, 72-73; res grant rev, NIH, 74-78, 77-78 & 79-85, Nat Health Med Res Coun, Australia, 79-, Med Res Coun, N Zealand, 80- & Am Inst Biol Sci, 85-; consult, Nat Res Coun-Nat Acad Sci, 79, NSF, 86; indust consult, 79-87; vis scientist, Australian Soc Microbiol, 82. *Mem:* Am Acad Microbiol; Am Soc Microbiol; NY Acad Sci; Sigma Xi; Asn Gnotobiotics; fel AAAS; Soc Gen Microbiol UK. *Res:* Molecular properties of anaerobic gastrointestinal microorganisms. *Mailing Add:* Dept Microbiol Univ Tenn M409 Walters Life Sci Bldg Knoxville TN 37996-0845

SAVAGE, E LYNN, b New York, NY. SEDIMENTOLOGY, MEDICAL GEOLOGY. *Educ:* Brooklyn Col, BS; NY Univ, MS; Rutgers Univ, PhD (sedimentol, stratig) 67. *Prof Exp:* Annual lectr, 66-67, from instr to assoc prof, 67-86, PROF GEOL, BROOKLYN COL, CITY UNIV NY, 87- *Concurrent Pos:* Adj assoc prof, Hunter Col, 78-; conf dir & chmn, Environ Health, Med Geol, Earth Day, 89. *Mem:* Fel Geol Soc Am; Soc Econ Paleontologists & Mineralogists; NY Acad Sci; Sigma Xi; Int Soc Sedimentologists. *Res:* Relevance of aluminum and acid rain to etiology of Alzheimer's disease; sedimentology of sandstones; application of geology to disease patterns; environmental health problems. *Mailing Add:* Dept Geol Brooklyn Col City Univ New York Brooklyn NY 11210

SAVAGE, EARL JOHN, b Uniontown, Pa, Feb 28, 31; m 61. PLANT PATHOLOGY, MYCOLOGY. *Educ:* Waynesburg Col, BS, 57; WVa Univ, MS, 60, PhD(plant path), 63. *Prof Exp:* Res asst plant path, WVa Univ, 57-62; instr biol & head dept, Lewis Col, 62-63; asst prof, Univ Notre Dame, 63-68; asst prof bot, 68-70, ASSOC PROF BOT, IND UNIV, SOUTH BEND, 71-, CHMN DEPT BIOL SCI, 69- *Concurrent Pos:* NSF res grant, 64-66; consult ball band div, US Rubber Co, Ind, 64-; consult, Bendix Corp, 64-65. *Mem:* AAAS; Am Inst Biol Sci; Mycol Soc Am; Sigma Xi. *Res:* Sexuality in genus Phytophthora which includes homothallism, heterothallism and inter and intraspecific matings; effects of various environmental factors upon the production and germination of oospores of Phytophthora species. *Mailing Add:* Dept Biol Sci Ind Univ Box 7111 1700 Mishawaka Ave South Bend IN 46634

SAVAGE, ELDON P, b Bedford, Iowa, Apr 4, 26; m 48; c 2. PUBLIC HEALTH. *Educ:* Univ Kans, BA, 50; Tulane Univ, MPH, 57; Univ Okla, PhD(prev med, pub health), 67. *Prof Exp:* Entomologist, Tech Br, Nat Commun Dis Ctr, Kans, 50-51, WVa, 51-52, Ga, 53-54, proj entomologist, Kans, 54-55, demonstration entomologist, Iowa, 56-57, proj dir environ control, 57-58, dir commun dis control demonstration, Pa, 58-64, asst chief state aids sect, 64-66, chief state serv pesticides prog, Ga, 67-70; dep dir, Inst Rural Environ Health, 70-84, PROF ENVIRON HEALTH & TOXICOL, COLO STATE UNIV, 70-, CHIEF CHEM EPIDEMIOL SECT, INST RURAL ENVIRON HEALTH, 70-, DIR, 84- *Concurrent Pos:* Prof & head, Dept Environ Health, dir, Environ Health Serv, Colo State Univ. *Mem:* Sigma Xi; Res Soc Am; Nat Environ Health Asn; Am Acad Sanitarians; NY Acad Sci. *Res:* Ecology of flies; environmental sanitation; human ecology; environmental toxicology; chemical pesticides; chemical pollutants in mother's milk; epidemiology; chronic effects of pesticide exposure. *Mailing Add:* Prof & Chair Dept Environ Health Colo State Univ Ft Collins CO 80523

SAVAGE, GEORGE ROLAND, b Ft Worth, Tex, Apr 2, 29; m 55; c 4. MICROBIOLOGY. *Educ:* NTex State Univ, BS, 49; Univ Tex, MA, 50; Nat Registry Microbiol, cert, 66. *Prof Exp:* Res technician, Samuel Roberts Noble Res Found, Okla, 50-51; bacteriologist & serologist, NMex Dept Pub Health, 51-52; chief clin lab, Carlsbad Mem Hosp, NMex, 52-55; asst dir admin, Carnegie Inst, Ohio, 55-58; microbiologist, Directorate Biol Opers, Dept Army, 59-62, chief bact & fungal develop sect, 62-64, chief viral & rickettsial labs, 64-67; sr microbiologist, Biol Sci Div, Midwest Res Inst, Mo, 67-73; ENVIRON SCIENTIST, BLACK & VEATCH ENGRS-ARCHITECTS, 73- *Mem:* Am Soc Microbiol; NY Acad Sci; Sigma Xi. *Res:* Environmental studies related to the establishment and operation of energy production facilities; physiological and genetic problems associated with large-scale growth of pathogenic bacteria, viruses and rickettsia. *Mailing Add:* Black & Veatch 1500 Meadow Lake Pkwy Kansas City MO 64114

SAVAGE, GODFREY H, b Niagara Falls, NY, June 13, 27; m 60; c 3. MECHANICAL ENGINEERING, OCEAN ENGINEERING. *Educ:* Princeton Univ, BSE, 50; Stanford Univ, MS, 51, PhD(earth sci), 70; Harvard Univ, MBA, 54. *Prof Exp:* Petrol engr, Standard Oil Co Calif, 51-52; staff mem, Arthur D Little, Inc, 54-55, assoc & proj leader, 55-58; asst to dean grad sch bus, Harvard Univ, 58-59; asst vpres overseas opers, Leesona Holt Ltd, Eng & Leesona Corp, RI, 59-60; staff engr, Nat Acad Sci, 60-61; res engr, Woods Hole Oceanog Inst, 63-65; PROF MECH ENG & DIR ENG DESIGN LAB, UNIV NH, 65- *Concurrent Pos:* Vis engr, Mass Inst Technol, 63-66; founding co-chmn, Maine-NH Bi-State Comn Oceanog, 66-68; chmn, NH State Port Authority & Tidelands Conserv Authority, 67-69; pres, Ocean Technol Explor Co, 73-; Fulbright fel & vis prof, Heriot Watt Univ, 75-76; chief eng adv, Ocean Margin Drill Prog, JOI Inc, 79-81. *Mem:* Marine Technol Soc; Inst Offshore Eng, Scotland. *Res:* Invention and development of structural and mechanical-electrical systems for ocean exploration and defense: in buoy systems, deep diving systems and vehicles; wave energy measurement and analysis. *Mailing Add:* Dept Mech Eng Univ NH Durham NH 03824

SAVAGE, HOWARD EDSON, cell biology, for more information see previous edition

SAVAGE, I RICHARD, b Detroit, Mich, Oct 26, 25; m 50; c 2. STATISTICS. *Educ:* Univ Chicago, BS, 44; Univ Mich, MS, 45; Columbia Univ, PhD(statist), 54. *Prof Exp:* Statistician, Nat Bur Standards, 51-54; actg asst prof statist, Stanford Univ, 54-57; from assoc prof to prof, Univ Minn, 57-63; prof, Fla State Univ, 63-74; PROF STATIST, YALE UNIV, 74- *Concurrent Pos:* Consult, Ctr Advan Study Behav Sci, 56, NSF sr fel, 70-71; vis assoc prof, Sch Bus, Harvard Univ, 60-61; vis prof, Yale Univ, 67-68; ed, Ann of Statist, 74-77. *Mem:* Fel Am Statist Asn (pres, 84); Am Math Soc; fel Inst Math Statist; Int Statist Inst. *Res:* Government statistics; non-parametric techniques; control theory. *Mailing Add:* Dept Statist Yale Univ PO Box 2179 New Haven CT 06520

SAVAGE, JANE RAMSDELL, b Boston, Mass, Sept 23, 25. NUTRITION, BIOCHEMISTRY. *Educ:* Simmons Col, BS, 46, MS, 49; Univ Wis, PhD(nutrit, biochem), 63. *Prof Exp:* Asst chem, Simmons Col, 46-49; asst prof nutrit & chem, Univ Tenn, 49-59; res asst nutrit, Univ Wis, 59-62; from asst prof to prof, 73-86, assoc dean, 84-86, EMER PROF NUTRIT, UNIV TENN, KNOXVILLE, 86- *Mem:* Am Chem Soc; Am Home Econ Asn; Soc Nutrit Educ; Asn Women Develop. *Res:* Tryptophan metabolism; niacin containing co-enzymes; amino acid imbalance. *Mailing Add:* 9131 Wesley Pl Dr Knoxville TN 37922

SAVAGE, JAY MATHERS, b Santa Monica, Calif, Aug 26, 28; m 81; c 2. HERPETOLOGY. *Educ:* Stanford Univ, AB, 50, MA, 54, PhD(biol sci), 55. *Prof Exp:* Asst gen biol, Stanford Univ, 50-53, asst comp vert anat & embryol, 54; asst prof zool, Pomona Col, 54-56; from instr to assoc prof biol, Univ Southern Calif, 56-64, vchmn, Nat Oceanog Lab Syst, 71-74, prof & assoc dir, Allan Hancock Found, 64-82; chmn dept, 82-87, PROF BIOL, UNIV MIAMI, 82- *Concurrent Pos:* Mem, Stanford field exped, Mex, 50; herpetologist, Sefton Found-Stanford Exped, Gulf of Calif & Mex, 52; actg asst cur herpet, Philadelphia Acad Natural Sci, 54; Guggenheim Found fel, 63-64; dir advan sci sem trop biol, Univ Costa Rica, 61-63, prof & tech adv to dept biol, 63-66; res assoc, Los Angeles County Mus; mem bd dirs, Orgn Trop Studies, Inc, 63-, exec secy, 63-65, pres, 74-80; chmn comt select biol problems in the humid tropics, Nat Sci/Nat Res Coun, 80-81; int dir, Sch Int Rel, 79-81; chair, dept biol, Univ Miami, 82-87; mem, Int Comn Zool Nomen, 82- *Mem:* AAAS; Soc Study Evolution; Am Soc Ichthyologists & Herpetologists (treas, 60-63, pres, 82); Soc Study Amphibians & Reptiles; Sigma Xi; Soc Syst Zool; Soc Europ Herp; fel Explorer's Club; Brit Herpetologists Soc; Asn Trop Biol. *Res:* Ecology and evolution of amphibians and reptiles; ecology of midwater fishes; herpetofauna of tropical America; biosystematics frog genus Eleutherodactylus; ecological dynamics and evolution in the tropics; biogeography; evolution of consciousness. *Mailing Add:* Biol Dept Univ Miami PO Box 249118 Coral Gables FL 33124

SAVAGE, JIMMIE EUEL, b Calico Rock, Ark, Feb 15, 20; m 48; c 2. POULTRY NUTRITION. *Educ:* Univ Ark, BSA, 43; Univ Mo, MA, 48, PhD(agr chem), 55. *Prof Exp:* Nutritionist, Farm Bur Mills, Ark, 50-54; from asst prof to assoc prof, Univ Mo, Columbia, 55-64, prof poultry nutrit, 64-90, chmn dept, 66-90, EMER PROF POULTRY NUTRIT, ANIMAL SCI DEPT, UNIV MO, COLUMBIA, 90- *Mem:* AAAS; Am Inst Nutrit; Poultry Sci Asn (pres, 73-74). *Res:* Amino acid; trace mineral and unrecognized vitamin requirements of poultry. *Mailing Add:* 116 Animal Sci Ctr Univ Mo Columbia MO 65211

SAVAGE, JOHN EDMUND, b Lynn, Mass, Sept 19, 39; m 66; c 4. ELECTRICAL ENGINEERING, COMPUTER SCIENCE. *Educ:* Mass Inst Technol, ScB & ScM, 62, PhD(elec eng), 65. *Prof Exp:* Res asst elec eng, Mass Inst Technol, 62-65; mem tech staff, Bell Tel Labs, NJ, 65-67; PROF COMPUT SCI & ENG, BROWN UNIV, 67-, CHMN, DEPT COMPUT SCI, 85- *Concurrent Pos:* Consult, Codex Corp, Mass, 67-68, Fed Systs Div,

IBM Corp, Md, 68-69, Jet Propulsion Lab, 69-74, Lincoln Lab, Mass Inst Technol, 72, Battelle Inst, 77-79 & Davis Hoxie, NY, 82-; Guggenheim fel & Fulbright-Hays fel, 73; assoc ed, Inst Elec & Electronics Engrs Transactions on Comput, 76-78; vis prof comput sci, Univ Paris, 80-81; NSF fel; mem bd dirs, Comput Res Asn, 90-93. *Mem:* Inst Elec & Electronics Engrs Comput Soc; Asn Comput Mach; Sigma Xi. *Res:* Applied theory of computation; algorithms and analysis for VLSI systems; computational complexity; information theory and coding; parallel computation. *Mailing Add:* Dept Comput Sci Brown Univ PO Box 1910 Providence RI 02912

SAVAGE, JOHN EDWARD, b Philadelphia, Pa, June 11, 07; wid; c 2. MEDICINE. *Educ:* Univ Md, BS, 28, MD, 32; Am Bd Obstet & Gynec, dipl, 40. *Prof Exp:* Asst clin prof obstet & gynec, Sch Med, Univ Md, 46-61; chief of staff, Greater Baltimore Med Ctr, 65-68, chief obstet, 65-73; RETIRED. *Concurrent Pos:* Lectr, Johns Hopkins Univ, 58-73, emer lectr, 73- *Mem:* Fel Am Gynec & Obstet Soc; fel Am Col Obstet & Gynec; fel Am Col Surgeons. *Mailing Add:* 1544 Burnstone Dr Stone Mountain GA 30088-3400

SAVAGE, MICHAEL, b Yonkers, NY, Sept 6, 41; m 66; c 4. MECHANICAL ENGINEERING. *Educ:* Manhattan Col, BME, 63; Purdue Univ, Lafayette, MSME, 65, PhD(mech eng), 69. *Prof Exp:* Consult engr, Res Div, United Shoe Mach Co, Mass, 65-66; asst prof, Wash State Univ, 69-70; asst prof eng, Case Western Reserve Univ, 70-76; chief engr, Erickson Tool Co, 76-77; assoc prof mech eng, Purdue Univ, Calumet Campus, 77-79; assoc prof, 79-86, PROF MECH ENGR, UNIV AKRON, 86- *Honors & Awards:* Tech Achievement Award, Cleveland Tech Soc Coun, 77. *Mem:* Am Soc Mech Engrs; Am Soc Eng Educ; Soc Exp Mech; Sigma Xi. *Res:* Kinematic analysis and synthesis of linkages and other machine components; applications of analysis to the design of machinery; dynamics of machinery; mechanical design; kinematics. *Mailing Add:* 2547 Celia Dr Stow OH 44224

SAVAGE, NEVIN WILLIAM, b Berwick, Pa, Feb 17, 28; m; c 2. MATHEMATICS. *Educ:* Pa State Univ, BS, 50, MA, 52; Univ Calif, Los Angeles, PhD(math), 56. *Prof Exp:* Asst prof math, Polytech Inst Brooklyn, 55-59; assoc prof, 59-66, PROF MATH, ARIZ STATE UNIV, 66-, CHMN DEPT, 69- *Concurrent Pos:* NSF fac fel, 64-65. *Mem:* Am Math Soc; Math Asn Am. *Res:* Riemann surface theory and functional analysis. *Mailing Add:* Dept Math Ariz State Univ Tempe AZ 85287

SAVAGE, NORMAN MICHAEL, b Dover, Eng, Aug 23, 36; m 64; c 3. PALEONTOLOGY, STRATIGRAPHY. *Educ:* Bristol Univ, BS, 59; Univ Sydney, PhD(geol), 68. *Prof Exp:* Teaching fel geol, Univ Sydney, 62-66; res fel, Univ Col Swansea, Wales, 66-68; lectr, Univ Natal, 68-71; assoc prof, 71-79, PROF GEOL & HEAD DEPT, UNIV ORE, 79- *Concurrent Pos:* Consult, US Geol Surv, 74-76; vis fel, Clare Hall, Cambridge, Eng, 85-86. *Mem:* Fel Geol Soc London; fel Geol Soc Am; Brit Palaeont Asn; Paleont Soc. *Res:* Lower Paleozoic Brachiopods and conondonts. *Mailing Add:* Dept Geol Univ Ore Eugene OR 97403

SAVAGE, PETER, b Gardner, Mass, May 13, 42. EPIDEMIOLOGY. *Educ:* Boston Col, BA, 64; Tufts Univ, MD, 68. *Prof Exp:* CHIEF, CLIN & GENETIC EPIDEMIOL BR, NAT HEART, LUNG & BLOOD INST, NIH, 86- *Mem:* Am Heart Asn; Am Diabetes Asn. *Mailing Add:* Clin & Genetic Epidemiol Br Nat Heart Lung & Blood Inst NIH Fed Bldg Rm 300B 7550 Wisconsin Ave Bethesda MD 20892

SAVAGE, ROBERT E, b Middlebury, Vt, Dec 8, 32; m 64; c 2. CELL BIOLOGY. *Educ:* Oberlin Col, BA, 54; Univ Wis, MS, 58, PhD(bot), 63. *Prof Exp:* Lectr biol, Queens Col, NY, 62-63, from instr to asst prof, 63-67; from asst prof to assoc prof, 71-76, chmn dept, 76-81, PROF BIOL, SWARTHMORE COL, 76-, ISAAC H CLOTHIER JR PROF BIOL, 83- *Concurrent Pos:* Vis researcher, Dept Med Cell Genetics, Karolinska Inst, Sweden, 70-71, 74, 86 & Inst Physiol Bot, Univ Uppsala, Sweden, 79, Bot Dept, Univ Mass, 90. *Mem:* AAAS; Am Soc Cell Biol; Tissue Cult Asn; Sigma Xi. *Res:* Chromosomal chemistry and structure; somatic cell hybridization; nucleolar proteins; auxin receptors. *Mailing Add:* Dept Biol Swarthmore Col 500 College Ave Swarthmore PA 19081-1397

SAVAGE, STEVEN PAUL, b Topeka, Kans, Apr 8, 50; m 71; c 3. PHYSICAL ANTHROPOLOGY. *Educ:* Univ Kans, BA, 71; Univ Colo, MA, 73, PhD(anthrop), 78. *Prof Exp:* Asst prof, 75-81, ASSOC PROF ANTHROP, EASTERN KY UNIV, 81-, DEPT CHMN, 87- *Concurrent Pos:* Proj dir, NSF grant, 78-81. *Mem:* Am Asn Phys Anthropologists; Soc Med Anthrop; Paleopath Asn. *Res:* Photobiology; growth; osteology. *Mailing Add:* Dept Anthrop Sociol & Social Work Eastern Ky Univ Richmond KY 40475

SAVAGE, STUART B, b Far Rockaway, NY, Oct 18, 32; m 68. APPLIED MECHANICS, FLUID MECHANICS. *Educ:* McGill Univ, BEng, 60, PhD(eng mech), 67; Calif Inst Technol, MSc, 61, AeroE, 62. *Prof Exp:* Prin aerodyn engr, Appl Res & Develop Lab, Repub Aviation Corp, NY, 62-64; lectr civil eng, 64-67, from asst prof to assoc prof, 67-77, PROF CIVIL ENG, MCGILL UNIV, 77- *Concurrent Pos:* Acad visitor, Imp Col, Univ London, 71-72 & Cambridge Univ, 77-80. *Mem:* Am Soc Civil Engrs; Am Soc Mech Engrs; Am Acad Mech; Int Asn Hydraul Res; Sigma Xi. *Res:* Storage and flow of bulk solids; pneumatic transport; incompressible fluid mechanics. *Mailing Add:* Dept Civil Eng McGill Univ 817 Sherbrook W Montreal PQ H3A 2K6 Can

SAVAGE, WILLIAM F(REDERICK), b Anchorage, Alaska, May 23, 23; m 49; c 2. NUCLEAR ENGINEERING, MECHANICAL ENGINEERING. *Educ:* Rensselaer Polytech Inst, BAeroE, 43; Purdue Univ, MAeroE, 49. *Prof Exp:* Aerodynamicist, Consol Aircraft Co, Tex, 44-46; instr mech & aeronaut eng, Univ Ky, 46-48, assoc prof, 49-52; chief engr, Kett Corp, 53-55; prin engr, Aircraft Nuclear Propulsion Dept, Gen Elec Co, 55-58, supvr preliminary design, 58, mgr applns tech anal, 59-60; dir nuclear prod area, Adv Progs, Martin Co, 61-64, mgr resources planning, 65-67; asst dir eng & develop, Off Saline Water, US Dept Interior, 67-74, chief advan systs eval, 74-81; dir utility coordr, US Dept Energy, 81-85, dep dir, Nuclear Plant Performance, 85-90; RETIRED. *Mem:* Am Nuclear Soc. *Res:* Advanced nuclear systems technology, power plant technology, desalting technology. *Mailing Add:* RR 1 Box 546 Jonesville VA 24263-9505

SAVAGE, WILLIAM RALPH, physics; deceased, see previous edition for last biography

SAVAGE, WILLIAM ZUGER, b Duluth, Minn, June 20, 42; m 68; c 2. ROCK MECHANICS, GEOMECHANICS. *Educ:* Lawrence Univ, BA, 65; Syracuse Univ, MS, 68; Tex A&M Univ, PhD(geol), 74. *Prof Exp:* Staff scientist, Systs, Sci & Software, La Jolla, Calif, 72-75; GEOLOGIST, US GEOL SURV, DENVER, COLO, 75- *Concurrent Pos:* Prin investr, Landslide Dynamics & Kinematics Proj, US Geol Surv, Denver, Colo, 83-; adj assoc prof, Dept Civil, Archit & Environ Eng, Sec & Third Summer Schs Hydrogeol Hazards Studies, Univ Perugia, Italy, 88-, adj vis prof, 89-90; supvry geologist, US Geol Surv, Denver, Colo, 88-90. *Mem:* Am Geophys Union; Int Soc Rock Mech; Int Asn Eng Geologists; Am Acad Mech. *Res:* Development and application of continuum mechanics concepts and methods to model deformation and flow of surficial geologic materials and the distribution of near-surface in-situ stresses associated with these processes. *Mailing Add:* US Geol Surv MS 966 Denver Fed Ctr Box 25046 Denver CO 80225

SAVAGEAU, MICHAEL ANTONIO, b Fargo, NDak, Dec 3, 40; m 67; c 3. BIOCHEMICAL GENETIC NETWORKS. *Educ:* Univ Minn, BS, 62; Univ Iowa, MS, 63; Stanford Univ, PhD(cell physiol, syst sci), 67. *Prof Exp:* Res asst, Univ Iowa, 62-63; res asst, Stanford Univ, 63-64, lectr, 69-70; from asst prof to assoc prof, Univ Mich, 70-78, actg chmn, 79-80, interum chmn, 82-85, PROF MICROBIOL, UNIV MICH, ANN ARBOR, 78-, DIR CELLULAR BIOTECHNOL LAB, 89- *Concurrent Pos:* NIH fel, Stanford Univ, 64-67, 68-69, Univ Calif, Los Angeles, 67-68; prin investr, NSF & NIH grants, Univ Mich, Ann Arbor, 71-; Guggenheim fel & Fulbright sr res fel, Max Planck Inst Biophys Chem, Gottingen, Ger, 76-77; consult, Upjohn, 80-81, Off Technol Assessments, 82-83, & Synergen, 85-86; NIH, Spec Study Sect Biochem Modelling, 81-82; vis scientist & fel, Commonwealth Sci & Indust Res Org, Div Comput Res, John Curtin Sch Med Res, Australian Nat Univ, Canberra, 83-86; consult, Synergen, 85-86. *Mem:* AAAS; Am Soc Microbiol; Biophys Soc; Inst Elec & Electronics Engrs; Soc Indust & Appl Math; Soc Math Biol. *Res:* Development of a general-purpose nonlinear system theory, including optimal strategies for nonlinear modeling and computer analysis of organizationally complex systems; application of such methods to understand function, design and evolution of integrated biological networks in terms of their underlying molecular determinants. *Mailing Add:* 900 Lincoln Ann Arbor MI 48104

SAVAIANO, DENNIS ALAN, b Pomona, Calif, Dec 28, 53; m 75; c 2. INTESTINAL METABOLISM OF NUTRIENTS. *Educ:* Claremont McKenna Col, BA, 75; Univ Calif, Davis, MS, 77, PhD(nutrit), 80. *Prof Exp:* Asst prof, 80-86, ASSOC PROF NUTRIT, UNIV MINN, 86- *Mem:* Am Inst Nutrit; Inst Food Technol. *Res:* Intestinal purine metabolism; lactose digestion and tolerance; intestinal carnitine metabolism. *Mailing Add:* Dept Food Sci & Nutrit Univ Minn 1334 Eckles Ave St Paul MN 55108

SAVAN, MILTON, b Manchester, NH, June 24, 20; m 55; c 3. VETERINARY MEDICINE, VIROLOGY. *Educ:* Univ NH, BS, 41; Ont Vet Col, DVM, 45; Univ Wis, MS, 49, PhD, 56. *Prof Exp:* Pvt pract, Farmington, Maine, 45-47; res vet, Europ Mission on Foot & Mouth Dis, USDA, Denmark, 51-54 & Plum Island Animal Dis Lab, 56-58; assoc prof virol, Ont Vet Col, Univ Guelph, 58-85; RETIRED. *Mem:* Sigma Xi; Wildlife Dis Asn; Can Wildlife Fedn. *Res:* Veterinary virology; virus diseases of fish; viruses of veterinary importance. *Mailing Add:* 471 Stevenson St N Guelph ON N1E 5C6 Can

SAVARA, BHIM SEN, dentistry, for more information see previous edition

SAVARD, FRANCIS GERALD KENNETH, biochemistry, endocrinology; deceased, see previous edition for last biography

SAVARD, JEAN YVES, b Quebec, Que, Jan 25, 35; m 58; c 2. SOLID STATE PHYSICS. *Educ:* Laval Univ, BASc, 57; Univ London, PhD(microwaves), 61. *Prof Exp:* Asst prof elec eng, 61-67, ASSOC PROF ELEC ENG, LAVAL UNIV, 67- *Mem:* Am Phys Soc. *Res:* Paramagnetic resonance in solids. *Mailing Add:* Dept Elec Eng Laval Univ University City Quebec PQ G1K 7P4 Can

SAVEDOFF, LYDIA GOODMAN, b New York, NY, Dec 23, 20. CHEMISTRY. *Educ:* Hunter Col, BA, 41; Columbia Univ, MA, 44, PhD(chem), 47. *Prof Exp:* Technician, Manhattan Proj, SAM Labs, Columbia Univ, 42-43, asst univ, 43-47; assoc supvr, Ohio State Univ, 47-49; res assoc, Sch Med, Univ Wash, 49-52; instr chem, Gonzaga Univ, 52-54, asst prof, 54-59; NSF fac fel, Univ Calif, Los Angeles, 59-60; from asst prof to assoc prof, 60-67, PROF CHEM, CALIF STATE UNIV, NORTHRIDGE, 67- *Mem:* AAAS; Am Chem Soc; Am Phys Soc; Sigma Xi. *Res:* Physical properties and conductance of electrolyte and polyelectrolyte solutions. *Mailing Add:* 3249 A San Amadeo Laguna Hills CA 92653

SAVEDOFF, MALCOLM PAUL, b New York, NY, July 4, 28; m 48; c 3. ASTROPHYSICS. *Educ:* Harvard Univ, AB, 48; Princeton Univ, MA, 50, PhD(astron), 51. *Prof Exp:* Nat Res Coun fel, Mt Wilson & Palomar Observs, Calif Inst Technol, 51-52; NSF fel, Leiden Observ, 52-53; res assoc & asst prof physics, 53-56, asst prof optics, 56-59, Sloan Found res fel, 56-60, from asst prof to assoc prof physics & astron, 57-64, PROF PHYSICS & ASTRON, UNIV ROCHESTER, 64- *Concurrent Pos:* Dir, C E Kenneth Mees Observ, 64; NSF fel, Univ Leiden, 64-65; Nat Res Coun sr assoc, Goddard Space Flight Ctr, NASA, 79-80. *Mem:* Sigma Xi; Am Phys Soc; Am Astron Soc; Int Astron Union. *Res:* Interstellar material; stellar interiors. *Mailing Add:* Ten Cranston Rd Pittsford NY 14534

SAVEREIDE, THOMAS J, b Rockford, Ill, Nov 18, 32; m 57; c 3. INDUSTRIAL ORGANIC CHEMISTRY. *Educ:* St Olaf Col, BA, 57; Northwestern Univ, PhD(org chem), 61. *Prof Exp:* Sr res chemist, St Paul, Minn, 61-64, supvr org res, 64-69, tech mgr chem div, 69-73, tech dir, Chem Resources Div, 73-80, TECH DIR, SUMITOMO DIV 3M CO, TOKYO, JAPAN, 81- *Mem:* Am Chem Soc. *Res:* Organic synthesis; polymer chemistry. *Mailing Add:* 211 Eastbank Ct N Hudson WI 54016-1084

SAVERY, CLYDE WILLIAM, b White Plains, NY, Jan 3, 35; m 58; c 2. MECHANICAL ENGINEERING. *Educ:* Univ Ill, Urbana, BS, 57; Univ Wash, MS, 60; Univ Wis-Madison, PhD(mech eng), 69. *Prof Exp:* Res & develop assoc, Gen Atomic Div, Gen Dynamics, 60-66; from asst prof to prof mech eng, Drexel Univ, 69-80; chair mech eng, 80-89, INTERIM VPROVOST GRAD STUDIES & RES, PORTLAND STATE UNIV, 89- *Concurrent Pos:* Consult, Choice Mag, 69- & Gilbert Assoc, Inc, 72-77; res grants, NSF, US Dept Energy, HUD & Elec Power Res Inst, 73-; Fulbright sr res fel, Univ Maribor, Yugoslavia, 88. *Honors & Awards:* Ralph Teetor Award, Soc Automotive Engrs, 71. *Mem:* Soc Automotive Engrs; Am Soc Mech Engrs; Instrument Soc Am; Sigma Xi. *Res:* Heat and mass transfer with applications to vaporization, combustion, air pollution, cooling towers; nuclear power reactor safety; control engineering. *Mailing Add:* Grad Studies Portland State Univ PO Box 751 Portland OR 97207

SAVERY, HARRY P, b Coffeyville, Kans, Jan 4, 20; m 50; c 4. PHYSIOLOGY, ENDOCRINOLOGY. *Educ:* Colo Agr & Mech Col, BS, 49; Univ Wyo, MS, 50; Tex A&M Univ, PhD(reproduction physiol), 54. *Prof Exp:* from assoc prof to prof biol, Cent Mo State Univ, 60-87, head dept, 71-87; RETIRED. *Mem:* AAAS; Am Soc Animal Sci; Am Soc Zoologists; Am Inst Biol Sci. *Res:* Cytological studies of normal and superovulated ova. *Mailing Add:* Rte 5 Warrensburg MO 64093

SAVIC, MICHAEL I, b Belgrade, Yugoslavia, Aug 4, 29; US citizen; m 61; c 1. AUTOMATIC SPEAKER RECOGNITION & SIGNAL RECOGNITION, DIGITAL SIGNAL & SPEECH PROCESSING. *Educ:* Univ Belgrade, dipl ing, 55, Dr Eng Sc, 65. *Prof Exp:* Res & develop engr indust electronics, Kretztech Zipf, Austria, 56; res engr vacuum tubes, Tungsram Vienna, Austria, 57-58; asst prof elec eng, Univ Belgrade, 59-67; researcher ultrasound, Yale Univ, New Haven, Conn, 67-78; prof elec eng, Western New Eng Col, Springfield, Mass, 68-82; assoc prof, 82-91, PROF ELEC ENG, RENSSELAER POLYTECH INST, TROY, NY, 91- *Concurrent Pos:* Prin investr, NSF grant, 75-76 & Computer Controlled Cryosurg, Zacarian Res Found, 78-80; chmn, Elec Eng Dept, Western New Eng Col, 77-82; vis prof elec eng, Rensselaer Polytech Inst, Troy, NY, 80, 81 & 82, prin investr, Signal Recognition, 84-, Speaker Verification, 85-, Speech Recognition, 88-, Pipeline Leak Detection, 90-, Voice Character Transformation, 89- & Lang Identification, 89- *Honors & Awards:* Cert Appreciation, Inst Elec & Electronics Engrs, 72. *Mem:* Sr mem Inst Elec & Electronics Engrs; Int Asn Sci & Technol Develop. *Res:* Digital signal processing algorithms; hardware and software; speaker verification; signal identification; speech recognition; language identification; pipeline leak detection; voice character transformation. *Mailing Add:* Elec Computer & Syst Eng Dept Rensselaer Polytech Inst Troy NY 12180-3590

SAVIC, STANLEY D, b Belgrade, Yugoslavia, Dec 30, 38; US citizen. PRODUCT SAFETY, ACCIDENT INVESTIGATIONS. *Educ:* Roosevelt Univ, BS, 62; Univ Ill, MS, 69. *Prof Exp:* Res scientist, Univ Chicago, 64-72, mgr, 72-78, DIR, ZENITH ELECTRONICS, 78- *Concurrent Pos:* Consult, Univ Chicago, 65-78 & pvt, 78- *Honors & Awards:* Distinguished Contrib Award, Electronics Industs Asn, 87. *Mem:* NY Acad Sci; sr mem Inst Elec & Electronics Engrs; Nat Fire Protection Asn; Am Soc Testing & Mat; Am Asn Physicists Med; Health Physics Soc. *Res:* Medical physics radiation treatment planning research using computer simulations; electronic product safety research and cataode ray tube x-radiation and phospitor research. *Mailing Add:* Zenith Electronics Co 1000 Milwaukee Ave Glenview IL 60025

SAVICKAS, DAVID FRANCIS, b Chicago, Ill, Nov 9, 40. THEORETICAL PHYSICS. *Educ:* St Mary's Col, BA, 62; Mich State Univ, MS, 64, PhD(physics), 66. *Prof Exp:* Asst prof, Bucknell Univ, 66-69; ASSOC PROF PHYSICS, WESTERN NEW ENG COL, 69- *Mem:* Am Phys Soc. *Res:* Radiation pressure on interstellar particles; relativity theory; astrophysical kinematics. *Mailing Add:* Dept Physics Western New Eng Col 1215 Wilbraham Rd Springfield MA 01119

SAVIDGE, JEFFREY LEE, b Omaha, Neb, Dec 27, 52; m 78; c 1. THERMODYNAMICS & MATERIAL PROPERTIES. *Educ:* Okla City Univ, BS, 77; Univ Okla, MS, 83, PhD(chem engr), 85. *Prof Exp:* Lab chemist, Okla Med Res Found, 78-80; postdoctoral, Univ Okla, 85-86, vis assoc prof chem engr, 86; res engr, 86-87, PROG MGR THERMODYN, GAS RES INST, 87- *Mem:* Am Inst Chem Engrs; Sigma Xi. *Res:* High accuracy thermodynamic properties of natural gas and related fluids. *Mailing Add:* 124 N Cady Dr Palatine IL 60067

SAVILE, DOUGLAS BARTON OSBORNE, b Dublin, Ireland, July 19, 09; m 39; c 2. BOTANY. *Educ:* McGill Univ, BSA, 33, MSc, 34; Univ Mich, PhD(bot), 39. *Hon Degrees:* DSc, McGill Univ, 78. *Prof Exp:* Agr asst, Agr Can, 36-39, agr scientist, 39-53, sr mycologist, 53-57, prin mycologist, 57-74, EMER RES ASSOC, BIOSYST RES INST, AGR CAN, OTTAWA, 74- *Honors & Awards:* Lawson Medal, Can Bot Asn, 76. *Mem:* AAAS; Mycol Soc Am; Am Soc Plant Taxon; fel Arctic Inst NAm; fel Royal Soc Can. *Res:* Mycology; taxonomy of parasitic fungi; biology of rusts; co-evolution of rusts and host plants; arctic biology; avian aerodynamics. *Mailing Add:* 357 Hinton S Ottawa ON K1Y 1A6 Can

SAVILLE, D(UDLEY) A(LBERT), b Lincoln, Nebr, Feb 25, 33; m 59; c 2. CHEMICAL ENGINEERING, FLUID MECHANICS. *Educ:* Univ Nebr, Lincoln, BS, 54, MS, 59; Univ Mich, Ann Arbor, PhD(chem eng), 66. *Prof Exp:* Engr chem eng, Union Carbide Corp, 54-55; res engr, Calif Res Corp, 59-61 & Shell Develop Co, 66-68; from asst prof to assoc prof, 68-71, PROF CHEM ENG, PRINCETON UNIV, 77- *Mem:* Am Inst Chem Engrs; Am Chem Soc; Am Phys Soc. *Res:* Electrokinetics and other colloidal phenomena; hydrodynamic stability; electrohydrodynamics; heat and mass transfer in particulate suspensions; crystal growth. *Mailing Add:* Dept Chem Eng Princeton Univ Princeton NJ 08540

SAVILLE, THORNDIKE, JR, b Baltimore, Md, Aug 1, 25; m 50; c 3. CIVIL & COASTAL ENGINEERING. *Educ:* Harvard Univ, AB, 47; Univ Calif, MS, 49. *Prof Exp:* Res asst, Univ Calif, 47-49; hydraul engr, Beach Erosion Bd & Coastal Eng Res Ctr, Corps Engrs, 49-81, asst chief res div, 53-64, chief gen proj br, 54-64, chief res div, 64-71, tech dir, 71-81, CONSULT, COASTAL ENG & COASTAL ENG RES APPLICATIONS, 81- *Concurrent Pos:* Mem coun wave res, Eng Found, 54-64; mem, Permanent Int Asn Navig Cong, 56-, secy, US Deleg, Sect II, London, 57 & Baltimore, Md, 61, liaison officer, Int Comn Force of Waves & chmn, Am Sect Subcomt, 64-72; US mem, Permanent Int Comn, 71-78; mem comt tidal hydraul, US Army Corps Engrs, 64-81; mem adv bd, Nat Oceanog Data Ctr, 64-72, chmn, 68; gen chmn, Specialty Conf Coastal Eng, Santa Barbara, 65 & Washington, DC, 71; liaison rep, Panel Coastal Eng & Inland Waters, Comt Earthquake Eng Res, Nat Acad Eng, 66-70; dir, Am Shore Beach Preserv Asn, 76-, vpres, 88-; adv ed, Coastal Engineering, 71-; mem, tech coun on res, Am Soc Civil Engrs, 83-88, chmn, 85-87, mem, coastal eng res coun, 64-; mem coun, AAAS, 71-77. *Honors & Awards:* Huber Res Award, Am Soc Civil Engrs, 63, Moffatt-Nichol Award, 79. *Mem:* Nat Acad Eng; fel AAAS; fel Am Soc Civil Engrs; Am Geophys Union; Int Asn Hydraul Res; hon mem Int Asn Navig Cong. *Res:* Basic laws governing wave and surge action on beaches and shore structures and application of these to engineering design; coastal erosion processes and littoral tranpost; hydraulic model studies. *Mailing Add:* 5601 Albia Rd Bethesda MD 20816-3304

SAVIN, SAMUEL MARVIN, b Boston, Mass, Aug 31, 40; m; c 2. GEOCHEMISTRY, GEOLOGY. *Educ:* Colgate Univ, BA, 61; Calif Inst Technol, PhD(geochem), 67. *Prof Exp:* From asst prof to assoc prof, 67-76, chmn dept, 77-82, PROF GEOL SCI, CASE WESTERN RESERVE UNIV, 76- *Concurrent Pos:* Assoc ed, Geuchimica et Cosmochimica Acta, 76-79; mem, Earth Sci Adv Panel, NSF, 78-81; ed, Marine Micropaleont, 79-87, Paleogeog Paleoclimatol Paleoecol, 87- *Mem:* Fel Geol Soc Am; Geochem Soc; fel AAAS; Am Geophys Union; Clay Minerals Soc; Am Asn Petrol Geologists. *Res:* Stable isotope geochemistry; low temperature geochemistry; shore erosion oceanography; stable isotopes in medicine; paleoceanography. *Mailing Add:* Dept Geol Sci Cleveland OH 44106

SAVINELLI, EMILIO A, b New York, NY, May 7, 30; c 8. CHEMISTRY. *Educ:* Manhattan Col, BCE, 50; Univ Fla, MSE, 51, PhD(chem), 55. *Prof Exp:* Engr, Du Pont Co, 56-60; res dir, 60-62, div mgr, 62-68, dir mkt, 64-68, vpres, 68-71, PRES, DREW CHEM CORP, 71- *Mem:* Nat Acad Sci; Am Chem Soc; Am Soc Mech Engrs. *Mailing Add:* 34 Molbrook Dr Witton CT 06897-4709

SAVIT, CARL HERTZ, b New York, NY, July 19, 22; m 46; c 3. GEOPHYSICS. *Educ:* Calif Inst Technol, BS, 42, MS, 43. *Prof Exp:* Chief mathematician, Western Geophys Co Am, Calif, 48-60, dir systs res, 60-65, vpres res & develop, Tex, 65-70; asst to President's Sci Adv & chmn, US Interagency Comt Atmospheric Sci, 70-71; sr vpres, Western Geophys Co Am, 71-86; CONSULT, 86- *Concurrent Pos:* Assoc prof, San Fernando Valley State Col, 59-60; mem panel on-site inspection unidentified seismic events, US Govt, 61, select panel initiatives transp, 71; mem, President's Panel Disposition Oil Leasing Santa Barbara Channel & Offshore Pollution, 69; mem comt seismol, Nat Acad Sci-Nat Res Coun, 71-75, chmn, 72-74, mem US nat comt tunneling technol, 72-76, chmn geophys data panel, Geophys Res Bd, 75-76, mem, Panel on Earthquake Prediction & Assembly Math & Phys Sci, 77-80; US deleg to USSR in explor geophys, 71; consult panel, President's Sci Adv, 71-74; mem nat adv comt, Univ Tex Marine Biomed Inst, 73-78; mem, US Coastal Zone Mgt Comt, 75-78; mem nat offshore opers indust adv comt, USCG, 75-77; mem sea-bottom surv panel, US-Japan Coop Prog Natural Resources, US Dept Com, 75-78; coastal zone mgt adv comt, US Dept Com, 75-78; energy resource adv bd, US Dept Energy, 78-81, chmn, Geothermal Standard Comt, 78-81; chmn vis comt, Inst Geophys, Univ Tex, 84-; distinguished lectr, Soc Expl Geophys, 84; OCS policy comt of OCS Adv Bd, US Dept Int, 85-89; marine facil panel, US-Japan Coop Prog in Nat Resources, 86-89; panel real time earthquake warning, Nat Res Coun, 87-90; computer systs tech adv comt, US Dept Com, 89-90, nat sea grant panel, 90-; vis prof geophys, Univ Utrecht, Neth, 90. *Honors & Awards:* Compass Award, Marine Technol Soc, 79; Kauffman Gold Medal, Soc Explor Geophysicists, 79; Litton Advan Technol Achievement Award, 80; Distinguished Achievement Award, Int Asn Geophys Contractors, 83. *Mem:* Hon mem Soc Explor Geophysicists (pres, 71-72); Europ Asn Explor Geophysicists; fel Geol Soc Am; Acoust Soc Am. *Res:* Signal detection; data processing systems; crustal studies; oceanography; seismic exploration. *Mailing Add:* 201 Vanderpool Lane Houston TX 77024

SAVIT, JOSEPH, b Chicago, Ill, Oct 23, 21; m 42; c 4. REPROGRAPHY, INFORMOGRAPHY. *Educ:* Univ Chicago, BS, 42. *Prof Exp:* Res assoc gas warfare, Univ Chicago, 42-46; pres & mgr, Travelers Hotel Co, Calif, 47-51; plant chemist, Reproduction Prod Co, 52-53; staff chemist, A B Dick & Co, 53-56; chief chemist, Huey Co, 56-57; mgr res & develop, Eugene Dietzgen Co, 58-63 & Colonial Carbon Co, 63-64; mgr, Microstatics Div, SCM Corp, 64-66; dir res & asst vpres, Apeco Corp, 66-73; PRES, SAVIT ENTERPRISES, INC, 73- *Mem:* Tech Asn Pulp & Paper Indust; Am Chem Soc; Soc Photog Sci & Eng; fel Am Inst Chem. *Res:* Reprography, especially electrography and electrophotography; office print out; non-impact computer print-out systems; paper and film coatings; offset printing systems; ink-jet printing systems. *Mailing Add:* 751 Vernon Ave Glencoe IL 60022

SAVIT, ROBERT STEVEN, b Chicago, Ill, Aug 21, 47. ECONOMICS & FINANCE. *Educ:* Univ Chicago, BA, 69; Stanford Univ, MS, 70, PhD(physics), 73. *Prof Exp:* Res assoc theoret physics, Stanford Linear Accelerator Ctr, 73 & Fermi Nat Accelerator Lab, 73-74; vis scientist, Europ Orgn Nuclear Res, 74-75; physicist, Fermi Nat Accelerator Lab, 75-78; assoc res scientist & lectr, 78-83, assoc prof, 83-89, PROF PHYSICS DEPT, UNIV MICH, 90- *Concurrent Pos:* NATO fel, NSF, 74-75; grant recipient, Am-Swiss Found Sci Exchange, 74-75; assoc ed, Nuclear Phys Field Theory & Statist Systs, 80-; res fel, Sloan Found, 81-85, Columbia Futures Ctr, 88-89; vis scientist, Inst Theoret Physics, Santa Barbara, 81-82; vis prof, Hebrew

Univ Jerusalem, Weizmann Inst Sci, 86-87. *Mem:* Am Phys Soc. *Res:* Field theory; critical phenomena; statistical mechanics; equilibrium and non-equilibrium growth; dynamical systems; chaos; finance and economics; time series analysis; applied mathematics and statistics. *Mailing Add:* Physics Dept Univ Mich Ann Arbor MI 48109

SAVITCH, WALTER JOHN, b Brooklyn, NY, Feb 21, 43. COMPUTER SCIENCE, MATHEMATICS. *Educ:* Univ NH, BS, 64; Univ Calif, Berkeley, MA & PhD(math), 69. *Prof Exp:* Assoc prof info & comput sci, 69-80, PROF ELEC ENG & COMPUT SCI, UNIV CALIF, SAN DIEGO, 80- *Concurrent Pos:* NSF res grant, 70-76. *Mem:* Am Math Soc; Asn Comput Mach; Asn Symbolic Logic; Soc Indust & Appl Math. *Res:* Theoretical computer science; complexity of algorithms; formal languages; automata theory; mathematical logic. *Mailing Add:* Dept Appl Physics & Info Sci Cse-C-014 Univ Calif San Diego La Jolla CA 92093

SAVITSKY, DANIEL, b New York, NY, Sept 26, 21; m 62; c 3. NAVAL ARCHITECTURE. *Educ:* City Col New York, BCE, 42; Stevens Inst Technol, MSc, 52; NY Univ, PhD, 71. *Prof Exp:* Struct engr, Edo Corp, 42-44; aeronaut res scientist, Nat Adv Comt Aeronaut, 44-47; DIR, DAVIDSON LAB, STEVENS INST TECHNOL, 47-, PROF OCEAN ENG, 67- *Concurrent Pos:* Ottens res award, Stevens Inst Technol, 68; consult, indust & USN. *Honors & Awards:* Adm Cochrane Award, Soc Naval Archit & Marine Eng, 67. *Mem:* Soc Naval Archit & Marine Eng; Am Soc Naval Eng; Sigma Xi. *Res:* Hydrodynamics; ocean science and engineering. *Mailing Add:* 597 Delcina Dr River Vale NJ 07675

SAVITSKY, GEORGE BORIS, b Harbin, China, Mar 10, 25; nat US; m 48; c 3. PHYSICAL CHEMISTRY. *Educ:* Aurora Univ, China, BS, 47; Univ Fla, PhD(chem), 59. *Prof Exp:* Res assoc chem, Princeton Univ, 59-61; asst prof, Univ Calif, Davis, 61-65; assoc prof, 65-71, PROF CHEM, CLEMSON UNIV, 71- *Mem:* AAAS; fel Am Inst Chem; Am Chem Soc; Sigma Xi. *Res:* Spectroscopy; nuclear magnetic resonance spectroscopy. *Mailing Add:* Dept Chem & Geol Clemson Univ Clemson SC 29631

SAVITSKY, HELEN, genetics, for more information see previous edition

SAVITZ, DAVID ALAN, b Hamilton, Ohio, Mar 14, 54; m 83; c 2. ENVIRONMENTAL EPIDEMIOLOGY, REPRODUCTIVE EPIDEMIOLOGY. *Educ:* Brandeis Univ, BA, 75; Ohio State Univ, MS, 78; Univ Pittsburgh, PhD(epidemiol), 82. *Prof Exp:* Researcher epidemiol, Battelle Mem Inst, 76-79; asst prof prev med, Sch Med, Univ Colo, 81-85; asst prof, 85-88, ASSOC PROF EPIDEMIOL, SCH PUB HEALTH, UNIV NC, 88- *Mem:* Soc Epidemiol Res (secy & treas, 87-91); Am Pub Health Asn; Int Epidemiol Asn. *Res:* Occupational and environmental exposures in relation to reproductive health outcomes and cancer; epidemiologic methods. *Mailing Add:* Dept Epidemiol CB No 7400 Univ NC McGavern-Greenberg Hall Chapel Hill NC 27599-7400

SAVITZ, JAN, b Sellersville, Pa, Apr 8, 41; m 68. LIMNOLOGY. *Educ:* Pa State Univ, BS, 63; Ind Univ, PhD(limnol), 67. *Prof Exp:* Asst prof biol, Rockford Col, 67-69; asst prof, 69-74, ASSOC PROF BIOL, LOYOLA UNIV CHICAGO, 74- *Honors & Awards:* Mary Ashby Cheek Award, 68. *Mem:* AAAS; Am Fisheries Soc; Am Inst Biol Sci; Am Soc Limnol & Oceanog; Ecol Soc Am. *Res:* Protein metabolism of fish; fish predation on benthic organisms. *Mailing Add:* Dept Biol Loyola Univ Lake Shore Campus 6525 N Sheridan Rd Chicago IL 60626

SAVITZ, MAXINE LAZARUS, b Baltimore, Md, Feb 13, 37; m 61; c 2. ORGANIC CHEMISTRY, ELECTROCHEMISTRY. *Educ:* Bryn Mawr Col, AB, 58; Mass Inst Technol, PhD(org chem), 61. *Prof Exp:* NSF fel, Univ Calif, Berkeley, 61-62; instr chem, Hunter Col, 62-63; res chemist, Elec Power Div, US Army Eng Res & Develop Lab, Ft Belvoir, 63-68; assoc prof chem, Fed City Col, 68-71, prof, 71-72; prof mgr, Res Appl to Nat Needs, NSF, 72-73; chief bldgs conserv policy res, Fed Energy Admin, 73-75; div dir bldgs & indust conserv, Energy Res & Develop Admin, 75-76, div dir bldgs & community systs, 76-79, dep asst secy conserv, Dept Energy, 79-83; pres, Lighting Res Inst, 83-85; asst vpres eng, 85-87, MANAGING DIR GARRETT CERAMIC COMPONENT DIV, ALLIED-SIGNAL AEROSPACE CORP, 87- *Concurrent Pos:* mem, Energy & Eng Bd, Nat Acad Eng, 86-92; mem, Off Technol Assessment, Demand Bd, 87-89, Nat Mat Adv Bd Nat Res Coun, 89-91, pres, US Advan Ceramic Asn, 91. *Mem:* AAAS; Am Ceramic Soc. *Res:* Free radical mechanisms; anodic hydrocarbon oxidation; fuel cells; more efficient use of energy in buildings; community systems; appliances; agriculture and industrial processes; transportation; batteries and other storage systems; new materials; advanced structural ceramic materials. *Mailing Add:* 10350 Wilshire Blvd Los Angeles CA 90024

SAVITZKY, ABRAHAM, b New York, NY, May 29, 19; m 42; c 2. ANALYTICAL CHEMISTRY, COMPUTER SCIENCE. *Educ:* State Univ NY, BA, 41; Columbia Univ, MA, 47, PhD(phys chem), 49. *Prof Exp:* Res assoc electron micros, Columbia Univ, 49-50; staff scientist, Perkin-Elmer Corp, 50-71, sr staff scientist corp comput facil, 71-79, prin invest, instrument group, 80-85; PRES, SILVERMINE RESOURCES INC, 85- *Concurrent Pos:* Mem, Nat Acad Sci-Nat Res Coun Eval Panel Atomic & Molecular Physics, Nat Bur Standards, 70-72; dir, Time Share Peripherals Corp, 70-77; Sci Apparatus Makers Asn rep, Am Nat Standards Comt X-3 Comput & Data Processing, 72- *Honors & Awards:* Williams-Wright Award, Coblentz Soc, 86. *Mem:* Am Chem Soc; Am Phys Soc; Optical Soc Am; Soc Appl Spectros; Asn Comput Mach. *Res:* Development of laboratory and process analytical instrumentation; infrared spectroscopy; computer aided experimentation and data reduction; time sharing systems and languages; computer plotting; user interfacing; personal computers. *Mailing Add:* Three Mail Coach Ct Wilton CT 06897

SAVITZKY, ALAN HOWARD, b Danbury, Conn, June 23, 50; m 72; c 2. HERPETOLOGY, EVOLUTIONARY BIOLOGY. *Educ:* Univ Colo, BA, 72; Univ Kans, MA, 74, PhD(biol), 79. *Prof Exp:* Fel, Nat Mus Natural Hist, Smithsonian Inst, 76-78; mem staff, Sect Ecol & Systemetics, Cornell Univ, 78-; AT BIOL SCI DEPT, OLD DOMINION UNIV, NORFOLK. *Mem:* Am Soc Ichthyologists & Herpetologists; Am Soc Zoologists; Herpetologists League; Soc Study Amphibians & Reptiles; Soc Syst Zool; Sigma Xi. *Res:* Relationship between phylogeny and adaptation, especially among snakes; morphological correlates of specialized feeding habits; parallel evolution of complex adaptations. *Mailing Add:* Biol Sci Dept Old Dominion Univ Norfolk VA 23529-0266

SAVKAR, SUDHIR DATTATRAYA, b Poona, India, Sept 27, 39; US citizen; m 64; c 2. FLUID MECHANICS, ACOUSTICS. *Educ:* Catholic Univ Am, BS, 61; Univ Mich, MS, 63, PhD(mech eng), 66. *Prof Exp:* Res engr, 66-78, mgr eng mech, 78-83, MGR FLUID SYSTS, GEN ELEC CORP RES & DEVELOP, 83- *Concurrent Pos:* Adj assoc prof, Nuclear Eng Dept, Rensselaer Polytech Inst, 77- *Mem:* Assoc fel Am Inst Aeronaut & Astronaut; Sigma Xi. *Res:* Unsteady flow and structural interaction; acoustics; combustion instability; electro-hydrodynamics; plasma physics. *Mailing Add:* 2344 Jade Lane Schenectady NY 12309

SAVOIE, RODRIGUE, b Carleton, Que, Oct 1, 36; m 60; c 3. PHYSICAL CHEMISTRY. *Educ:* Univ of the Sacred Heart, BA, 56; Laval Univ, BSc, 60, PhD(chem), 63. *Prof Exp:* From asst prof to assoc prof, 65-75, PROF CHEM, LAVAL UNIV, 75- *Concurrent Pos:* Vis prof, Univ Ore, 76-77. *Mem:* Chem Inst Can; Spectros Soc Can; French Can Asn Advan Sci. *Res:* Infrared and Raman spectroscopy; molecular and crystal structures. *Mailing Add:* Dept Chem Laval Univ Quebec PQ G1K 7P4 Can

SAVOL, ANDREJ MARTIN, b Slovakia, Feb 4, 40; US citizen; m 69; c 3. MACHINE VISION, IMAGE PROCESSING. *Educ:* Carnegie-Mellon Univ, BS, 67; Univ Pittsburgh, MS, 75, PhD(elec eng), 79. *Prof Exp:* Comput programmer, Westinghouse Indust Systs Div, 67-69; head, comput servs, Mellon Inst, 69-72; sr specialist engr, Boeing Aerospace Co, 77-90, ASSOC TECH FEL, BOEING COM AIRPLANE GROUP, 90- *Honors & Awards:* Outstanding Tech Achievement Award, Pac Northwest Sect, Am Inst Aeronaut & Astronaut, 78. *Mem:* Inst Elec & Electronics Engrs. *Res:* Applications of computerized pattern recognition; computer analysis of images in manufacturing and inspection; integration of machine vision subsystems into manufacturing and assembly cells. *Mailing Add:* 4710 Lakeridge Dr E Sumner WA 98390

SAVORY, JOHN, b Lancashire, Eng, Apr 4, 36; US citizen; m 77; c 2. CLINICAL CHEMISTRY, PATHOLOGY. *Educ:* Univ Durham, BSc, 58, PhD(chem), 61. *Prof Exp:* Res chemist, Chemstrand Res Ctr, NC, 63-64; dir clin chem & instr path, Univ Fla, 66-67, asst dir clin labs & asst prof path, 67-72; assoc prof med & dir clin chem, Univ NC, Chapel Hill, 72-77; med & dir clin chem, Univ NC, Chapel Hill, 72-77; MEM FAC, DEPT PATH CLIN LABS, MED CTR, UNIV VA, 77- *Concurrent Pos:* Res fel chem, Univ Fla, 61-63; sr fel biochem, Univ Wash, 64-66; dir exp & clin path grad prog, Col Med, Univ Fla, 66-72; consult, Vet Admin Hosp, Gainesville, Fla, 67-72. *Mem:* Am Asn Clin Chemists; Am Chem Soc; Asn Clin Sci; Royal Soc Chem; Sigma Xi. *Res:* Organic fluorine chemistry; biochemistry. *Mailing Add:* Dept Path Clin Labs Box 168 Med Ctr Univ Va Charlottesville VA 22908

SAVORY, LEONARD E(RWIN), b Denver, Colo, Jan 11, 20; m 47; c 3. CHEMICAL ENGINEERING. *Educ:* Univ Denver, BSChE, 42; Ill Inst Technol, MSChE, 44. *Prof Exp:* Asst, Manhattan Proj, SAM Labs, Columbia Univ, 44-45; res technologist, Tenn Eastman Corp, 45-46; asst prof chem eng, Univ Denver, 46-50; mem staff, Los Alamos Sci Lab, NMex, 50-53; res proj supvr, Res Dept, United Gas Corp, 53-60, asst dir res, 60-66, mgr admin div, Res, Eng & Develop Dept, Pennzoil United, Inc, 67-73, sr tech adv, Environ, Safety & Health Affairs Dept, Pennzoil Co, 73-81; RETIRED. *Mem:* AAAS; Am Chem Soc; Am Inst Chem Engrs. *Res:* Nuclear energy; natural gas technology; pollution control. *Mailing Add:* PO Box 1675 Estes Park CO 80517

SAVOS, MILTON GEORGE, b Nashua, NH, July 14, 27; m 62; c 1. ENTOMOLOGY. *Educ:* Am Int Col, BA, 52; Univ Mass, MS, 54; Ore State Univ, PhD, 58. *Prof Exp:* Res entomologist, Inst Zool, Nancy Univ, France, 58-59; exten PROF ENTOM, COL AGR & NATURAL RESOURCES, UNIV CONN, 72-, EXTEN ENTOMOLOGIST, 60- *Mem:* Entom Soc Am; AAAS; Sigma Xi. *Res:* Biology and taxonomy of Symphyla; bionomics of garden symphylid; taxonomy; economic entomology. *Mailing Add:* Col Agr & Natural Resources Box U-67 Univ Conn Storrs CT 06268

SAVRUN, ENDER, b Adana, Turkey, July 29, 53; m 79; c 1. ELECTRON MICROSCOPY & X-RAY MICROANALYSIS, FAILURE ANALYSIS. *Educ:* Istanbul Tech Univ, Turkey, BS, 76, MS, 78; Univ Wash, PhD(ceramic eng), 86. *Prof Exp:* Res scientist ceramics, Flow Industs, 85-87, Photon Sci, 87-88; res mgr, 88-89, VPRES RES CERAMICS, MONTEDISON, 89- *Concurrent Pos:* Consult, Charlton Industs, 84-85. *Mem:* Am Ceramic Soc; Am Soc Metals; Am Soc Mech Engrs. *Res:* Processing of monolithic ceramics & ceramic matrix composites; mechanical behavior of structural ceramics; ceramic faced composite armor systems; ceramic cutting tools. *Mailing Add:* 4231 S Fremont Ave Tucson AZ 85714

SAWAN, MAHMOUD EDWIN, b Damanhor, Egypt, July 28, 50; US citizen; m 76; c 1. SYSTEMS DESIGN & SYSTEMS SCIENCE, ELECTRICAL ENGINEERING. *Educ:* Univ Alexandria, Egypt, BS, 73, MS, 76; Univ Ill, Urbana, PhD(elec eng), 79. *Prof Exp:* Lectr automatic control, Univ Alexandria, 73-76; res asst, Univ Ill, Urbana, 76-79; asst prof, 79-85, ASSOC PROF ELEC ENG, WITCHITA STATE UNIV, 85- *Concurrent Pos:* Secy, Witchita Sect, Inst Elec & Electronics Engrs, 85-87, vchm, 87- *Mem:* Sr mem Inst Elec & Electronics Engrs. *Res:* Application of control theory: development of robust control design techniques for systems with slow and fast modes. *Mailing Add:* Elect Eng Dept Wichita State Univ Wichita KS 67208

SAWAN, SAMUEL PAUL, b Akron, Ohio, Apr 18, 50. POLYMER CHEMISTRY, BIOPOLYMERS. *Educ:* Univ Akron, BS, 72, PhD(polymer sci), 76. *Prof Exp:* Data analyst, Akron Regional Air Pollution Control Agency, 73-74; scholar, Dept Pharmaceut Chem, Univ Calif, San Francisco, 76-78; ASST PROF, DEPT CHEM, UNIV LOWELL, 78- *Concurrent Pos:* Prin investr, Am Chem Soc Petrol Res Fund, 78-80; co-investr, Xerox Corp grant, 80-82; lectr radiol, Harvard Med Sch, 80- *Mem:* Am Chem Soc; Sigma Xi; AAAS. *Res:* Synthesis and characterization of macromolecules that exhibit catalytic activity; biocompatible polymeric materials; nmr investigations of polymer conformations and enzyme active sites; new radiological imaging agents. *Mailing Add:* Dept Chem Univ Lowell One Univ Ave Lowell MA 01854

SAWARD, ERNEST WELTON, internal medicine; deceased, see previous edition for last biography

SAWARDEKER, JAWAHAR SAZRO, b Goa, India, Nov 22, 37; m 66; c 3. PHYSICAL PHARMACY, ANALYTICAL CHEMISTRY. *Educ:* Univ Bombay, BS, 57; Univ Iowa, MS, 61, PhD(pharm), 64. *Prof Exp:* Res chemist, USDA, Ill, 64-66; assoc scientist, Ortho Pharmaceut Corp, 67-69, group leader scientist, 69-73; group mgr, Whitehall Labs, 73-77, dir qual control, 77-81; DIR QUAL CONTROL & VPRES MFG, GLAXO INC, 81-, VPRES INT QUAL ASSURANCE, GLAXO HOLDINGS, 90- *Mem:* Am Pharmaceut Asn; Am Chem Soc; Am Soc Qual Control. *Res:* Gas chromatography application to pharmaceutical systems and carbohydrate chemistry; exploration of analytical techniques and development of procedures; preformulation research; bioavailability studies. *Mailing Add:* Glaxo Inc Five Moore Dr Research Triangle Park NC 27709

SAWATARI, TAKEO, b Okayama, Japan, Feb 7, 39; US citizen; m; c 3. OPTICS. *Educ:* Waseda Univ, Tokyo Japan, BS, 62, PhD(appl physics), 70. *Prof Exp:* Researcher, Canon Camera, 62-65; res assoc, Univ Tokyo, 65-66; physicist, Optics Technol Inc, 66-70; prin physicist, Bendix Res Lab, 70-77, mem tech staff, Bendix Advan Technol Ctr, 77-81; PRES, SENTEC CORP, 81- *Concurrent Pos:* Lectr, Univ Mich, Dearborn, 78, Oakland Univ, 81. *Mem:* Optical Soc Am. *Res:* Development of fiber optics and its applications; optics; signal processing; laser applications; author of over 50 technical papers and contributor to two fiber optic textbooks. *Mailing Add:* 6105 Gilbert Lake Rd Bloomfield Township MI 48301

SAWATZKY, ERICH, b Cholm, Poland, Apr 21, 34; US citizen; m 61; c 3. PHYSICS. *Educ:* Univ BC, BSc, 58, MSc, 60, PhD(physics), 62. *Prof Exp:* RES STAFF SCIENTIST, SAN JOSE RES LAB, IBM CORP, 62- *Mem:* Am Phys Soc; Inst Elec & Electronics Engrs. *Res:* Magnetism in solid state materials; nuclear magnetic resonance, preparation and characterization of physical properties of thin films; magnetic and magneto-optic properties of thin films. *Mailing Add:* 1077 Trevino Terr San Jose CA 95120

SAWCHUK, ALEXANDER ANDREW, b Washington, DC, Feb 20, 45; m 71; c 2. ELECTRICAL ENGINEERING, OPTICS. *Educ:* Mass Inst Technol, SB, 66; Stanford Univ, MS, 68, PhD(elec eng), 72. *Prof Exp:* Elec engr, Goddard Space Flight Ctr, NASA, Md, 66 & Commun Satellite Corp, Washington, DC, 67; from asst prof to assoc prof elec eng, 71-82, PROF ELEC ENG, UNIV SOUTHERN CALIF, 82-, DIR SIGNAL & IMAGE PROCESSING INST, 78-88 & 90- *Concurrent Pos:* Consult, TRW Defense & Space Systs Group, 77-; founder dir, Optivision Inc, 83. *Mem:* Inst Elec & Electronics Engrs; Optical Soc Am; Soc Photo-Optical Instrumentation Engrs; Soc Info Display. *Res:* Digital image processing; statistical optics; optical information processing; multidimensional signal processing and system theory. *Mailing Add:* Dept Elec Eng Univ Southern Calif Powell Hall 306 Los Angeles CA 90089-0272

SAWCHUK, RONALD JOHN, b Toronto, Ont, May 29, 40; m 67; c 3. PHARMACOKINETICS, MEDICAL RESEARCH. *Educ:* Univ Toronto, BScPhm, 63, MScPhm, 66; Univ Calif, San Francisco, PhD(pharmaceut), 72. *Prof Exp:* From asst prof to assoc prof, 72-83, PROF PHARMACEUT, COL PHARM, UNIV MINN, MINNEAPOLIS, 83-, DIR GRAD STUDIES PHARMACEUT, 83- *Concurrent Pos:* Actg chmn, Dept Pharmaceut, Univ Minn, 83-86, dir, Clin Pharmacokinetics Lab, 82-; mem comt rev, US Pharmacopeial Con. *Mem:* Fel AAAS; Am Pharmaceut Asn; Am Asn Cols Pharm; Acad Pharmaceut Sci; Controlled Release Soc; Am Asn Pharmaceut Sci; NY Acad Sci. *Res:* Pharmacokinetics and kinetic modeling; drug distribution, metabolism and excretion; drug-drug interactions; quantitative analysis of foreign compounds in biological fluids; absorption of drugs using in situ animal models; antiviral and anticonvulsant; drug pharmacokinetics. *Mailing Add:* Dept Pharmaceut Univ Minn Col Pharm 308 Harvard St SE Minneapolis MN 55455

SAWERS, JAMES RICHARD, JR, b Memphis, Tenn, Feb 4, 40; m 61; c 2. ENGINEERING PHYSICS, BIOENGINEERING & BIOMEDICAL ENGINEERING. *Educ:* Duke Univ, BS, 62, PhD(nuclear physics), 66. *Prof Exp:* Res physicist, Du Pont Instruments, Del, 66-71, sr scientist, 71-73; tech supvr, du Pont Biomed, 74-75, mgr microtomy prod, 76-77, mgr clinprod, 77-78, mgr instrument prod, E I du Pont de Nemours & Co Inc, 78-79, mgr instrument mfr, 79-83, mgr du Pont Learning Systs, 83-85; PRES, KNOWLEDGE TECHNOL, 85- *Mem:* Soc Magnetism & Magnetic Mat; Electron Micros Soc Am; Am Phys Soc; Sigma Xi; Soc Photog Scientists & Engrs; Inst Mgt Consults. *Res:* Neutron polarization; magnetism; photochemistry; electroluminescence; electron imaging; submicron particle size analysis; diamond knives; computer analysis; biomedical instrumentation; process and pollution control instrumentation; industrial and manufacturing engineering; general computer sciences; technical management; electrical engineering; engineering and general physics; electromagnetism. *Mailing Add:* Old Snug Hill PO Box 808 947 Sharpless Rd Hockessin DE 19707

SAWHILL, ROY BOND, b Tacoma, Wash, June 1, 22; m 43; c 3. TRANSPORTATION, CIVIL ENGINEERING. *Educ:* Univ Wash, BS, 50; Univ Calif, MEng, 52. *Prof Exp:* Jr hwy engr, State Hwy Dept, Wash, 50-51; jr res engr, Inst Transp & Traffic Eng, Univ Calif, 51-52; jr traffic engr, City Eng Dept, Seattle, 52, asst traffic engr, 52-53, assoc traffic engr, 53-56; from asst prof to prof civil eng, Univ Wash, 56-85; consult engr, 85-91; RETIRED. *Concurrent Pos:* Mem, Hwy Res Bd, Nat Acad Sci-Nat Res Coun. *Res:* Traffic engineering; operation characteristics, including fuel and travel time, of commercial vehicles and relation to highway design and economic determination of highway improvements. *Mailing Add:* 5454 Turnberry Pl SW Port Orchard WA 98366

SAWHNEY, VIPEN KUMAR, b India; c 1. PLANT DEVELOPMENT & PHYSIOLOGY. *Educ:* Univ Punjab, BSc, 65, MSc, 67; Univ Western Ont, PhD(plant sci), 72. *Prof Exp:* Post doc fel biol, Simon Fraser Univ, 72-74; from asst prof to assoc prof, 75-86, asst head biol dept, 87-90, PROF BIOL, UNIV SASK, 86- *Concurrent Pos:* Vis fel, Yale Univ, 82-83. *Mem:* Can Soc Plant Physiologists; Can Bot Asn; Bot Soc Am; Am Soc Plant Physiologists. *Res:* Growth regulation of vegetative and floral apecies of normal and mutant plants; physiological and biochemical studies on male sterility in plants. *Mailing Add:* Dept Biol Univ Sask Saskatoon SK S7N 0W0 Can

SAWICKA, BARBARA DANUTA, b Debica, Poland; m 64; c 1. PROPERTIES OF MATERIALS, SOLID STATE PHYSICS. *Educ:* Jagiellonian Univ, Cracow, Poland, MSc, 64, PhD(sciences), 70, Inst Nuclear Physics, Cracow, Habilitation(physics), 80. *Prof Exp:* From res asst to assoc prof, Inst Nuclear Physics, Cracow, Poland, 64-84; RES SCIENTIST, CHALK RIVER NUCLEAR LABS, ONT, CAN, 85- *Concurrent Pos:* Vis res scientist, Joint Inst Nuclear Physics, Dubna, USSR, 64-66, Johns Hopkins Univ, Baltimore, 80, Rijksuniversitet Groningen, Neth, 82-83, Max Planck Inst Plasma Physics, 84-85; vis res assoc, Tech Univ, Otaniemi, Finland, 71; vis asst prof, Inst Physics, Zurich Univ, Switz, 74-75; vis assoc prof, Universite Lyon, France, 81. *Honors & Awards:* Res Award, Atomic Energy Comn, Poland, 76-80. *Mem:* Polish Phys Soc; Am Phys Soc; Europ Phys Soc; Am Ceramic Soc. *Res:* Solid State physics problems and properties of materials studied by nuclear techniques; computed tomography; hyperfine interactions; particle-solid interactions; influence of overpetic beams on materials properties; advanced ceramics. *Mailing Add:* Chalk River Nuclear Labs Chalk River ON K0J 1J0 Can

SAWICKI, JOHN EDWARD, b Philadelphia, Pa, March 10, 44; m 71; c 2. EXPERIMENTALIST, REACTION KINETICS. *Educ:* Drexel Univ, BS, 67, MS, 68; Univ Va, PhD(chem eng), 72. *Prof Exp:* Sr res officer, Chem Eng Group, SAfrica Coun Sci & Indust Res, 72-74; res engr & group leader, Joseph Schlitz Brewing Co, 74-78; sr prin res engr, 78-81, technol mgr & group leader, 81-87, PROCESS TECHNOLOGIST, AIR PROD & CHEMICALS, INC, 87- *Mem:* Am Inst Chem Engrs. *Res:* Applied and basic research in aromatic nitgration processes and chemistry; unit operations and plant optimization; process development. *Mailing Add:* RD 2 Breinigsville PA 18031

SAWICKI, STANLEY GEORGE, b Oklahoma City, Okla, Feb 23, 42; m 69. VIROLOGY, TISSUE CULTURE. *Educ:* Georgetown Univ, BS, 64; Columbia Univ, MA & PhD(pathobiol), 74. *Prof Exp:* Fel molecular cell biol, Rockefeller Univ, 74-77; ASST PROF MICROBIOL, MED COL OHIO, 77- *Mem:* AAAS; Am Soc Microbiol; Sigma Xi. *Res:* RNA and protein synthesis in eukaryotic cells; control of the replication cycle in alphaviruses and adenoviruses. *Mailing Add:* 2940 Talmadge Rd Toledo OH 43606

SAWIN, CLARK TIMOTHY, b Boston, Mass, May 23, 34; m 82; c 3. THYROID DISEASE. *Educ:* Brandeis Univ, BA, 54; Tufts Univ, MD, 58. *Prof Exp:* CHIEF, ENDOCRINE-DIABETES SECT, BOSTON VET ADMIN MED CTR, 66-; PROF MED, TUFTS MED SCH, 81- *Concurrent Pos:* Prin investr res projs, Dept Vet Affairs, 66- *Honors & Awards:* Reynolds Award, Am Physiol Soc, 90. *Mem:* Endocrine Soc; Am Thyroid Asn; Am Asn Hist Med; Am Geront Soc; Am Diabetes Asn. *Res:* Endocrine changes with aging in man; history of endocrinology. *Mailing Add:* Endocrine-Diabetes Sect Vet Admin Med Ctr 150 S Huntington Ave Boston MA 02130

SAWIN, STEVEN P, b Mason City, Iowa, Oct 19, 44; m 66; c 4. POLYMER REACTOR DESIGN. *Educ:* Iowa State Univ, BS, 66; Univ Ill, MS, 68, PhD(chem eng), 71. *Prof Exp:* Sr engr res & develop, Polyolefins Div, 70-74, group leader, 74-81, assoc dir, 81-85, DIR POLYPROPYLENE RES & DEVELOP, POLYOLEFINS DIV, UNION CARBIDE CORP, 85- *Mem:* Am Inst Chem Engrs; Am Chem Soc. *Res:* The UNIPOL process technology for polyethylene extended to polypropylene; all types of polypropylene homopolymers, random and impact copolymers are produced; new products and advanced catalyst system being developed. *Mailing Add:* PO Box 670 Bound Brook NJ 08805

SAWINSKI, VINCENT JOHN, b Chicago, Ill, Mar 28, 25; m 52; c 2. PHYSICAL SCIENCES, ADMINISTRATION. *Educ:* Loyola Univ, Chicago, BS, 48, MA, 50, PhD(biochem), 62. *Prof Exp:* Asst prof biochem, Loyola Univ, 49-67; prof & chmn phys sci, Wilbur Wright Col, City Cols Chicago, 67-71; RETIRED. *Concurrent Pos:* Supvry res scientist, Vet Admin, 61-66. *Mem:* Sr mem Am Chem Soc; fel AAAS; sr mem Am Asn Univ Professors; Sigma Xi; sr mem Nat Sci Teachers Asn; fel Am Inst Chemist. *Res:* Biochemistry lab research and article abstracts; chemical research and education. *Mailing Add:* 1945 N 77th Ct Elmwood Park IL 60635-3623

SAWITSKY, ARTHUR, b Jersey City, NJ, Jan 31, 16; c 2. HEMATOLOGY, ONCOLOGY. *Educ:* NY Univ, BA, 36, MD, 40; Am Bd Internal Med, dipl, 52, cert hemat, 72. *Prof Exp:* Intern, Kings County Hosp, 40-42; fel hemat, Dept Therapeut, Col Med, NY Univ, 46-47; asst resident med, Goldwater Mem Hosp, NY Univ Div, 47-48; assoc med, NY Med Col, 50-54; chief div hemat, 55-84, EMER DIR CANCER PROGS, LONG ISLAND JEWISH MED CTR, 85-; PROF MED & CLIN PATH, STATE UNIV NY STONY BROOK, 71- *Concurrent Pos:* Asst vis physician, Goldwater Mem Hosp, NY

Univ Div, 48-52; assoc vis physician hemat & physician-in-chg blood bank, Queens Hosp Ctr, 48-64; assoc vis physician hemat, Jamaica Hosp, 49-54; asst vis physician hemat, Flushing Hosp, 50-55; attend hematologist, North Shore Hosp, Manhasset, NY, 54-67; attend physician-in-chg, Long Island Jewish Hosp-Queens Hosp Ctr Affil, 64-84; clin assoc prof, State Univ NY Downstate Med Ctr, 69-73; NIH grants, 61-84; fels, Pall Found, 64-72, Nat Leukemia Soc & United Leukemia Soc, 64-, Zelda Grossberg Found, 65-72 & Dennis Klar Mem Fund, 69-; res collabr, Brookhaven Nat Lab; sr immuno hematologist, Bur Labs, New York City Dept Health, 74-; consult hematologist, North Shore Hosp, Manhasset, Huntington Hosp, St Francis Hosp, Roslyn, Flushing Hosp & Peninsula Gen Hosp, Edgemere, NY. *Honors & Awards:* 00104077x. *Mem:* Am Soc Hemat; Soc Exp Biol & Med; Am Fedn Clin Res; Am Asn Cancer Res; Am Soc Clin Oncol; Sigma Xi. *Res:* Biology and classification of sub-populations of patients in the chronic leukemias; basic understanding of the defect in the syndrome of the sea-blue histiocyte. *Mailing Add:* Long Island Jewish 270-05 76th Ave New Hyde Park NY 11042

SAWOROTNOW, PARFENY PAVOLICH, b Ust-Medveditskaya, Russia, Feb 20, 24; nat US. HILBERT SPACES, BANACH ALGEBRAS. *Educ:* Harvard Univ, MA, 51, PhD, 54. *Prof Exp:* From instr to assoc prof, 54-67, PROF MATH, CATH UNIV AM, 67- *Concurrent Pos:* NSF grants, 67 & 70. *Mem:* Sigma Xi; Am Asn Univ Profs; Am Math Soc; Math Asn Am; NY Acad Sci. *Res:* Functional analysis; Hilbert spaces; Banach algebras; vector measures; probability. *Mailing Add:* Six Avon Pl Avondale MD 20782

SAWUTZ, DAVID G, b Montclair, NJ, July 14, 54. RECEPTOR PHARMACOLOGY. *Educ:* Univ Cincinnati, PhD(pharmacol), 84. *Prof Exp:* Res fel, Mass Gen Hosp, 84-87; SR RES INVESTR, DEPT ENZYMOL & RECEPTOR BIOCHEM, STERLING DRUG, INC, 87- *Mem:* Am Soc Pharmacol & Exp Therapeut; AAAS. *Mailing Add:* Sterling Drug Inc Nine Great Valley Pkwy Great Valley PA 19355

SAWYER, BALDWIN, b Naragansett Pier, RI, July 21, 22; m 47; c 4. PHYSICS. *Educ:* Yale Univ, BS, 43; Carnegie Inst Technol, DSc(physics), 52. *Prof Exp:* Jr metallurgist, Manhattan Proj, Univ Chicago, 43-46; instr & res assoc, Carnegie Inst Technol, 48-51; mem tech staff, Bell Tel Labs, Inc, 51-53, group supvr, 53-57; treas & chief engr, Sawyer Res Prods, Inc, 57-60, vpres eng, 60-64, exec vpres, 64-73, vchmn & dir technol, 84-86; RETIRED. *Concurrent Pos:* Chmn, quartz mats standards comt, Electronic Industs Asn, 85-, Int Electrotech Comn, 86-; mem, Nat Mats Adv Bd, Nat Acad Sci, 85; vchmn, Supermat Mfg Co, 88- *Mem:* AAAS; Am Phys Soc; Electrochem Soc; Inst Elec & Electronics Engrs; Sigma Xi. *Res:* Cultured quartz crystals; plasma spray coatings. *Mailing Add:* Berkshire Rd Box 96 Gates Mills OH 44040-0096

SAWYER, C GLENN, b New Bern, NC, Feb 27, 22; c 4. CARDIOLOGY. *Educ:* Bowman Gray Sch Med, MD, 44; Am Bd Internal Med, dipl, 52. *Prof Exp:* From intern to chief med resident, Peter Bent Brigham Hosp, Boston, Mass, 44-50; instr med, Harvard Med Sch, 50-51; from instr to assoc prof med, 51-63, chief cardiol, 63-81, PROF MED, BOWMAN GRAY SCH MED, 63- *Concurrent Pos:* Fel clin coun cardiol, Am Heart Asn. *Mem:* Am Heart Asn; AMA; fel Am Col Physicians; fel Am Col Cardiol; Asn Univ Cardiol. *Mailing Add:* Dept Med Bowman Gray Sch Med 300 Hawthorne Rd SW Winston-Salem NC 27103

SAWYER, CHARLES HENRY, b Ludlow, Vt, Jan 24, 15; m 41; c 1. NEUROENDOCRINOLOGY. *Educ:* Middlebury Col, AB, 37; Yale Univ, PhD(zool), 41. *Hon Degrees:* ScD, Middlebury Col, 75. *Prof Exp:* Instr anat, Stanford Univ, 41-44; assoc, Duke Univ, 44-45, from asst prof to prof, 45-51; prof, 51-85, EMER PROF ANAT, SCH MED, UNIV CALIF, LOS ANGELES, 85- *Concurrent Pos:* Consult, Vet Admin Hosp, Long Beach, Calif, 52-74; chmn dept anat, Univ Calif, Los Angeles, 55-63, fac res lectr, 66-67; commonwealth Found fel, 58-59; mem anat panel, Nat Bd Med Examr, 60-64, chmn, 64; mem-fel rev bd pharmacol & endocrinol, USPHS, 61-63, 68-70; mem neuroendocrine panel, Int Brain Res Orgn, 61-, mem cent coun, 64-67; mem neurol study sect, NIH, 63-67. *Honors & Awards:* Koch Res Medal, Endocrine Soc, 73. 73; Hartman Award, Soc Study Reprod, 77; Henry Gray Award, Am Asn Anatomists, 84. *Mem:* Nat Acad Sci; hon mem Hungarian Soc Endocrinol & Metab; Am Asn Anatomists (vpres, 68-70); Soc Study Reprod; fel Am Acad Arts & Sci; hon mem Japan Endocrine Soc. *Res:* Neuroendocrinology of reproduction; nervous control of pituitary secretion; function, distribution, ontogenesis and properties of cholinesterases; effects of hormones on brain function. *Mailing Add:* Dept Anat Med Sch Univ Calif Los Angeles CA 90024

SAWYER, CONSTANCE B, b Lewiston, Maine, June 3, 26; div; c 4. ASTROPHYSICS, OCEANOGRAPHY. *Educ:* Smith Col, AB, 47; Harvard Univ, AM, 48, PhD, 52. *Prof Exp:* Res asst, Sacramento Peak Observ, 53-55; mem res staff, High Altitude Observ, Univ Colo, 55-58; astronr, Space Environ Lab, 58-75, phys scientist, Atlantic Oceanog Meteorol Labs, 75-76, phys scientist, Pac Marine Environ Lab, Nat Oceanic & Atmospheric Admin, 76-79; staff scientist, High Altitude Observ, 79-82; SCIENTIST, RADIO PHYSICS INC, 83- *Concurrent Pos:* Satellite oper, NASA, 80; consult, D-Peek, 83; res assoc, Univ Colo, 84-85. *Mem:* Int Astron Union; Int Union Geod & Geophys; AAAS; Am Geophys Union; Am Astron Soc; Sigma Xi. *Res:* Solar physics and solar-terrestrial relations; ocean remote sensing; internal waves; planetary radio emission. *Mailing Add:* 850 20th St 705 Boulder CO 80302

SAWYER, DAVID ERICKSON, b Boston, Mass, Feb 6, 27; m; c 3. OPTICS. *Educ:* Clark Univ, BA, 53; Univ Ill, MS, 55; Worcester Polytech Inst, PhD, 76. *Prof Exp:* Res staff mem, Lincoln Lab, Mass Inst Technol, 55-59; develop staff mem, Int Bus Mach Corp, NY, 59-61; res staff mem, Sperry Rand Res Ctr, Mass, 61-66; sect head & sr scientist semiconductor device res, Electronics Res Ctr, NASA, 66-70; group leader, Electron Device Div, Nat Bur Standards, 70-75, sr staff mem, 75-79; sr res engr, Chevron Res Co, 79-85; optics res group leader & res & develop dept head, Systron Donner Co, 85-90; CONSULT, 90- *Honors & Awards:* Indust Res 100 Award, 76. *Mem:* Soc Photo-Optical Instrumentation Engrs; Inst Elec & Electronics Engrs. *Res:* Solid-state physics and devices; energy conversion; exploratory measurement techniques. *Mailing Add:* c/o D E Sawyer Consult 1021 Everette St El Cerrito CA 94530

SAWYER, DAVID W(ILLIAM), b Pittsburgh, Pa, Feb 1, 10; m 38; c 1. PETROLEUM, ORGANIC CHEMISTRY. *Educ:* Univ Pittsburgh, BS, 33. *Prof Exp:* Asst civil engr, City Planning Comn, Pittsburgh, Pa, 33-34; fel petrol, Mellon Inst, 34-35; res chemist, Gulf Res & Develop Co, 34-43; res engr, Alcoa Res Labs, Aluminum Co Am, 43-53, sr res engr, Lubricants Div, Alcoa Tech Ctr Pittsburgh, 53-75; CONSULT, 75- *Concurrent Pos:* Mem interim subcomt turbine-gear oils, US Dept Navy, 53- *Mem:* Am Chem Soc; fel Am Inst Chemists; Soc Triboligists & Lubrication Engrs; Sigma Xi. *Res:* Lubricants; lubrication; petroleum refining; corrosion. *Mailing Add:* 620 12th St Oakmont PA 15139

SAWYER, DONALD C, b New York, NY, Aug 23, 36; m 62; c 3. ANESTHESIOLOGY. *Educ:* Mich State Univ, BS, 59, DVM, 61, MS, 62; Colo State Univ, PhD(vet anesthesiol), 69; Am Col Vet Anesthesiologists, dipl. *Hon Degrees:* Hon Dipl, Am Bd Vet Practr. *Prof Exp:* Asst instr vet med, Mich State Univ, 61-62; pvt pract, 62-63; NIH spec fel, Colo State Univ & Univ Calif, 65-70; assoc prof & head anesthesia sect, 70-77, actg assoc dean, 74-75, PROF VET MED & ANESTHESIA, MICH STATE UNIV, 77-, COORDR, LIFE LONG EDUC, 86- *Concurrent Pos:* Support grant, 70-72, pharmacol res, 72-88; consult, NIH, 85. *Mem:* Am Vet Med Asn; Am Soc Vet Anesthesiol (pres, 72); Am Soc Anesthesiol; Am Col Vet Anesthesia (pres, 78); Am Bd Vet Practitioners (pres, 79-82). *Res:* Cardiovascular effects of anesthetics; metabolism of inhalation anesthetics; effects of analgesics for pain relief. *Mailing Add:* Col Vet Med A-132 E Fee Hall Mich East Lansing MI 48824-1316

SAWYER, DONALD TURNER, JR, b Pomona, Calif, Jan 10, 31; m 52; c 3. ANALYTICAL CHEMISTRY, BIOINORGANIC CHEMISTRY. *Educ:* Univ Calif, Los Angeles, BS, 53, PhD(chem), 56. *Prof Exp:* Guggenheim fel, Cambridge 75- 62-63; vis res fel, Merton Col, Oxford Univ, 70; chmn, Gordon Res Conf Anal Chem, 71; mem adv comt, Res Corp, 78-; fac res lectr, Univ Calif, Riverside, 79; sci adv, US Food & Drug Admin, Calif, 75-82; AT TEX A&M UNIV, 85- *Concurrent Pos:* Fac, Univ Calif, Riverside, 56-86. *Mem:* Fel AAAS. *Res:* Electroanalytical chemistry; physical chemical studies of metal chelates; chemical instrumentation; optical, nuclear magnetic resonance and electron spin resonance spectroscopy; model studies of metalloenzymes; nuclear magnetic resonance studies of coordination complexes; oxygen activation by metalloproteins and transition metal complexes; oxygen chemistry. *Mailing Add:* Dept Chem Tex A&M Univ College Station TX 77843-3255

SAWYER, FREDERICK GEORGE, b Brooklyn, NY, Mar 14, 18; m 50; c 2. CHEMICAL ENGINEERING. *Educ:* Polytech Inst Brooklyn, BChE, 39, MChE, 41, DChE, 43. *Prof Exp:* Res chem engr, Eastman Kodak Co, 41; instr chem eng, Polytech Inst Brooklyn, 42-43; chem engr, Am Cyanamid Co, Conn, 43-45; tech serv, NY, 45-46; assoc ed, Indust & Eng Chem & Chem & Eng News, Am Chem Soc, 46-50; adminr air pollution res, Stanford Res Inst, 50-51, asst to dir res, 51-53; dir info, Ralph M Parsons Co, 53-57; vpres, Jacobs Eng Co, 57-63; consult corp commun & mkt, 63-72; vpres, Reynolds Environ Group, 72-78; CONSULT, FREDERICK G SAWYER & ASSOCS, 78- *Concurrent Pos:* Lectr, Stanford Univ, 52-53, Univ Southern Calif, 67-69, Univ Calif, Los Angeles, 68, Saddleback Col, 77-, Calif State Univ, Pomona, 79- & Univ Calif, Irvine, 82-; pres, Vita-Cell Prod, 64; dir, Econ Int, Inc, 65-; mem, Advan Technol Consults Corp. *Mem:* AAAS; Am Inst Chem Engrs; Chem Mkt Res Asn; Asn Mgt Consults; Am Chem Soc. *Res:* Corporate and scientific communications; technical public relations; business development; lexical research; environmental planning and assessment; handmade paper. *Mailing Add:* 12922 Keith Pl Tustin CA 92680

SAWYER, FREDERICK MILES, b Brockton, Mass, Nov 30, 24; m 48. FOOD TECHNOLOGY, NUTRITION. *Educ:* Mass Inst Technol, SB, 48; Univ Calif, MS, 51, PhD(nutrit), 58. *Prof Exp:* Food technologist seafoods div, Gen Foods Corp, 48-49; sr lab technician, Univ Calif, 50-57; from asst prof to PROF RES FOOD SCI, UNIV MASS, AMHERST, 57- *Mem:* AAAS; Am Chem Soc; Inst Food Technologists. *Res:* Flavor chemistry; biochemistry of food spoilage; frozen foods; sensory analysis of foods. *Mailing Add:* Dept Food Sci & Nutrit Univ Mass Amherst MA 01002

SAWYER, GEORGE ALANSON, b Chicago, Ill, July 20, 22; m 47; c 2. PLASMA PHYSICS. *Educ:* Univ Mich, BSE, 44, MS, 48, PhD(physics), 50. *Prof Exp:* Mem staff, 50-72, group leader, 72-74, alt div leader, 74-79, ASSOC DIV LEADER, LOS ALAMOS NAT LAB, 79- *Concurrent Pos:* Vis scientist, Royal Inst Technol, Sweden, 59-60. *Mem:* Sigma Xi; Am Phys Soc. *Res:* Plasma physics; nuclear reactions; radioactivity; x-ray and visible spectroscopy; controlled thermonuclear research; particle accelerators. *Mailing Add:* 2519 35th St Los Alamos NM 87544

SAWYER, JAMES W, b Malone, NY, June 30, 33; m 66. MANAGEMENT SCIENCE, SYSTEMS ENGINEERING. *Educ:* Clarkson Col Technol, BChE, 54; Univ Pa, MS, 70, PhD(systs eng), 73. *Prof Exp:* Res chem eng, E I du Pont de Nemours & Co Inc, 54-68; res assoc, Resources for the Future, 73-75, sr res assoc, Qual Environ Prog, 75-76, fel & asst dir, 76-78; prin analyst, Congressional Budget Off, 78-81; SR SCIENTIST, ERCO, 81- *Concurrent Pos:* Sr energy specialist, Nat Transp Policy Studies Comn 77-78. *Mem:* Inst Elec & Electronics Engrs; Am Inst Chem Engrs; Inst Mgt Sci; Am Inst Chem Engrs; Am Econ Asn. *Res:* Environmental quality policy; water resources; secondary materials (scrap) studies; mathematical modeling of water, secondary materials, fossil fuel and synfuel studies with emphasis on environmental quality; policy aspects. *Mailing Add:* 430 Rockewell Rd Hampton VA 23669

SAWYER, JANE ORROCK, b Richmond, Va, June 15, 44; div; c 2. MATHEMATICS, STATISTICS. *Educ:* Va Polytech Inst & State Univ, BS, 66, MS, 68, PhD(math), 75. *Prof Exp:* From instr to assoc prof math, Mary Baldwin Col, 69-81; qual engr, 81-84, sr eng specialist, 84-89, MGR DATA PROCESSING, AM SAFETY RAZOR CO, 89- *Concurrent Pos:* Coordr & researcher, NASA, 75-76, prin investr, 76-77. *Mem:* Am Math Soc; Math Asn Am; Asn Women Math; Soc Mfg Engrs. *Res:* Rings of continuous functions, specifically pseudocompact topological spaces and pseudocompactifications. *Mailing Add:* HCR 33 Box 6 Churchville VA 24421

SAWYER, JOHN ORVEL, JR, b Chico, Calif, Nov 22, 39; m 60; c 2. PLANT ECOLOGY. *Educ:* Chico State Col, AB, 61; Purdue Univ, MS, 63, PhD(plant ecol), 66. *Prof Exp:* Ecologist, Wilson Nuttall Raimond Engrs, Inc, 64-66; from asst prof to assoc prof ecol, 66-75, PROF BOT, HUMBOLDT STATE UNIV, 75- *Mem:* Ecol Soc Am; Sigma Xi. *Res:* Vegetation of northern California. *Mailing Add:* Dept Biol Humboldt State Univ Arcata CA 95521

SAWYER, JOHN WESLEY, b Raleigh, NC, Nov 2, 17; m 39; c 1. OPERATIONS RESEARCH. *Educ:* Wake Forest Col, AB, 38, AM, 43; Univ Mo, AM, 48, PhD(math), 51. *Prof Exp:* Instr high sch math, NC, 38-46 & Mo, 46-50; from asst prof to assoc prof math, Univ Ga, 50-53; assoc prof, Univ Richmond, 53-56; assoc prof, Wake Forest Univ, 56-61, prof math, 61-88; RETIRED. *Concurrent Pos:* Consult, R J Reynolds Industs, 58-83. *Mem:* Math Asn Am; Opers Res Soc Am; fel AAAS. *Res:* Differential and distance geometry; operations research; computer science. *Mailing Add:* 116 Belle Vista Ct Winston-Salem NC 27106

SAWYER, PAUL THOMPSON, b Hanover, NH, Aug 7, 40. PLANT TAXONOMY. *Educ:* Univ Vt, BS, 65; Mont State Univ, MS, 67, PhD(bot), 70. *Prof Exp:* Asst prof, 69-76, ASSOC PROF BIOL, MONT COL MINERAL SCI & TECHNOL, 76- *Mem:* Bot Soc Am; Am Soc Plant Taxon; Int Asn Plant Taxon. *Res:* Systematic investigations of the genera Delphinium and Ledum using chromatographic procedures; investigations of environmental changes occurring when grassland-shrub communities influenced by mining activities are reforested. *Mailing Add:* Dept Biol Sci Mont Col Mineral Sci & Technol Butte MT 59701

SAWYER, PHILIP NICHOLAS, b Bangor, Maine, Oct 25, 25; m 53; c 4. THORACIC SURGERY. *Educ:* Univ Pa, MD, 49; Am Bd Surg, dipl, 58; Am Bd Thoracic Surg, dipl, 60. *Prof Exp:* Asst, Harrison Dept Surg Res, Univ Pa, 47-49, intern, Hosp Univ Pa, 49-50; staff mem, Naval Med Res Inst, 51-53; chief resident, St Luke's Hosp, New York, 56-57; from instr to assoc prof surg, 57-66, PROF SURG, STATE UNIV NY DOWNSTATE MED CTR, 66-, HEAD VASCULAR SURG SERV & BIOPHYS & ELECTROCHEM LABS, 64- *Concurrent Pos:* Fel surg, Univ Pa, 53-56, Nat Cancer Inst res fel biophys, Johnson Found Med Physics, 50; fel path, St Luke's Hosp, New York, 57; Markle scholar, 59-64; vis surgeon, Kings County Hosp & univ hosp, State Univ NY; assoc attend surgeon, St John's Episcopal Hosp & Methodist Hosp, Brooklyn; consult vascular surg, USPHS Hosp, Staten Island, NY, 66. *Honors & Awards:* Clemson Award Basic Res Biomat. *Mem:* AAAS; Am Chem Soc; Am Heart Asn; Am Physiol Soc; Am Soc Artificial Internal Organs. *Res:* Tissue electrical potential differences and metabolism; ionic movement across cellular membranes; etiology of vascular thrombosis; preservation of tissues; homotransplantation techniques; cardiovascular surgery. *Mailing Add:* 7600 Ridge Blvd Brooklyn NY 11209

SAWYER, RALPH STANLEY, b Gray, Maine, Jan 9, 21; m 51; c 2. PHYSICS, ENGINEERING. *Educ:* Tufts Univ, BS, 44. *Prof Exp:* Res & lab instr piezo elec, Tufts Univ, 46-47; res engr instrumentation, Nat Adv Comt Aeronaut Labs, Langley Field, Va, 47-48; electronic engr, Indust Instrument Div, Minneapolis-Honeywell Regulator Co, 50; electronics engr commun, Hastings Instrument Co, Hampton, Va, 50-51; supvry electronic control engr, US Naval Weapons Sta, Yorktown, Va, 51-53; head, Range Data & Homing Test Sect, US Naval Underwater Ordnance Sta, Newport, RI, 53-54; head Ballistic Instrumentation Br, & consult, Missile Br, Spec Projs Off, US Naval Bur Weapons Labs, Dahlgren, Va, 54-59; res engr & head, commun sect, Space Task Group, Nat Aeronaut & Space Admin, Langley Field, Va, 59-61; head, Elec Systs Br, Johnson Space Ctr, Houston, 61-62, chief, Instrumentation & Electronic Systs Div, 62-69, chief, Tracking & Commun Develop Div, 69-89; RETIRED. *Honors & Awards:* Exceptional Serv Medals, Nat Aeronaut & Space Admin, 69; Centennial Medal, Inst Elec & Electronics Engrs, 84; Outstanding Leadership Medal, Nat Aeronaut & Space Admin, 81. *Mem:* Fel Inst Elec & Electronics Engrs; Am Inst Aeronaut & Astronaut; Nat Telecommunications Conf (chmn, 72); Aerospace & Electronic Systs Soc. *Res:* Responsible for the research, development, and test of all tracking, television, telemetry and communications systems used on all current and past manned space flight programs sponsored by the United States. *Mailing Add:* Box 630 Friendswood TX 77546

SAWYER, RAYMOND FRANCIS, b Northfield, Minn, Aug 30, 32; m 56; c 2. THEORETICAL PHYSICS. *Educ:* Swarthmore Col, AB, 53; Harvard Univ, MA, 55, PhD, 58. *Prof Exp:* NSF fel physics, Europ Orgn Nuclear Res, Geneva, 58-59; Wis Alumni Res Found fel, Univ Wis, 59-60, from asst prof to prof, 60-65; PROF PHYSICS, UNIV CALIF, SANTA BARBARA, 65- *Mem:* Am Phys Soc. *Res:* High energy physics; fundamental particle theory. *Mailing Add:* Dept Physics Univ Calif Santa Barbara CA 93106

SAWYER, RICHARD LEANDER, physiology, for more information see previous edition

SAWYER, RICHARD TREVOR, b Washington, DC, Sept 14, 48; m 72. MONONUCLEAR PHAGOCYTE BIOLOGY, CELL KINETICS. *Educ:* Colo State Univ, BS, 71; Eastern Mich Univ, MS, 74; Mich State Univ, PhD(mycol), 78. *Prof Exp:* Fel microbiol, Dept Microbiol & Pub Health, Mich State Univ, 78; res assoc cell kinetics, Dept Path & Lab Med, Sch Med, ECarolina Univ, 78-80; res scientist IV immunol, Norwich-Eaton Pharmaceut, 80-82; ASST PROF MICROBIOL & IMMUNOL, SCH MED, MERCER UNIV, 82- *Concurrent Pos:* Sigma Xi res award, Mich State Univ, 75. *Mem:* Reticuloendothelial Soc; Sigma Xi. *Res:* Functional activities and cell kinetics of mononuclear phagocytes, in particular resident macrophages. *Mailing Add:* Mercer Univ Sch Med Macon GA 31207

SAWYER, ROBERT FENNELL, b Santa Barbara, Calif, May 19, 35; m 57; c 2. MECHANICAL ENGINEERING, COMBUSTION. *Educ:* Stanford Univ, BS, 57, MS, 58; Princeton Univ, MA, 63, PhD(aerospace & mech sci), 66. *Prof Exp:* Instr physics, Antelope Valley Col, 58-61; mem res staff, Princeton Univ, 65-66; from asst prof to assoc prof, 66-75, PROF MECH ENG, UNIV CALIF, BERKELEY, 75-, CLASS 1935 PROF ENERGY, 88- *Concurrent Pos:* Res engr, USAF Rocket Propulsion Lab, 58-61, chief liquid systs analysis, 61; consult to various co & govt agencies, 64-; mem, Calif Air Resources Bd, 75-76; chmn, Energy & Resources Group, Univ Calif, Berkeley, 84-88. *Mem:* Am Inst Aeronaut & Astronaut; Am Soc Mech Engrs; Am Soc Eng Educ; Soc Automotive Engrs; Combustion Inst (vpres, 88-). *Res:* Propulsion; combustion; air pollution; chemical kinetics; fire science. *Mailing Add:* Dept Mech Eng Univ Calif Berkeley CA 94720

SAWYER, ROGER HOLMES, b Portland, Maine, Sept 5, 42; m 61, 83; c 5. DEVELOPMENTAL BIOLOGY. *Educ:* Univ Maine, BA, 65; Univ Mass, PhD(zool), 70. *Prof Exp:* NIH fel, Univ Calif, Davis, 70-71, NSF res assoc genetics, 71-72, asst res anatomist develop biol, Calif Primate Res Ctr, 73-74; from asst prof to assoc prof, 75-83, PROF BIOL, UNIV SC, 83-, CHMN DEPT, 86- *Concurrent Pos:* Mem, NIH Study Sect Biol. *Mem:* Sigma Xi; Soc Develop Biol; Am Soc Zoologists; NY Acad Sci; Int Soc Develop Biologists; fel AAAS; Am Soc Cell Biol; Linnean Soc. *Res:* Control of biochemical and morphological differentiation by epithelial-mesenchymal interactions during skin organogenesis and the action of mutant genes and teratogens on these interactions. *Mailing Add:* Dept Biol Univ SC Columbia SC 29208

SAWYER, STANLEY ARTHUR, b Juneau, Alaska, Mar 19, 40. MATHEMATICAL STATISTICS. *Educ:* Calif Inst Technol, BS, 60, PhD(math), 64. *Hon Degrees:* AM, Brown Univ, 69. *Prof Exp:* Courant instr math, Courant Inst, NY Univ, 65-67; from asst prof to assoc prof, Brown Univ, 67-69; from asst prof to prof math, Yeshiva Univ, 69-77; prof math & statist, Purdue Univ, 78-84; PROF MATH, GENETICS & BIOSTATIST, WASH UNIV, 84- *Mem:* Am Math Soc; Genetics Soc Am; fel Inst Math Statist. *Res:* Probability; population genetics. *Mailing Add:* Dept Math Wash Univ St Louis MO 63130

SAWYER, WILBUR HENDERSON, b Brisbane, Australia, Mar 23, 21; US citizen; m 42, 82; c 4. ENDOCRINE PHARMACOLOGY, COMPARATIVE ENDOCRINOLOGY. *Educ:* Harvard Univ, AB, 42, MD, 45, PhD(biol), 50. *Prof Exp:* Instr biol, Harvard Univ, 50-53; asst prof physiol, Col Med, NY Univ, 53-57; from assoc prof to prof, 57-78, Gustavus A Pfeiffer prof pharmacol, 78-90, EMER PROF PHARMACOL, COL PHYSICIANS & SURGEONS, COLUMBIA UNIV, 91- *Concurrent Pos:* Lederle med fac award, 55-57; mem adv panel regulatory biol, NSF, 59-62; mem gen med B study sect, Div Res Grants, NIH, 70-74; sr res scholar, Australian-Am Educ Found, 74; vis res fel, Howard Florey Inst Exp Biol & Med, Univ Melbourne, Australia, 74. *Mem:* Endocrine Soc; Am Soc Pharmacol & Exp Therapeut; Soc Endocrinol; Am Soc Zool; Soc Gen Physiol; Am Physiol Soc. *Res:* Comparative endocrinology; renal physiology; pharmacology of neurohypophysial hormones. *Mailing Add:* 1490 Kings Lane Palo Alto CA 94303-2836

SAWYER, WILLIAM D, b Roodhouse, Ill, Dec 28, 29; m 51; c 2. MICROBIOLOGY, INTERNAL MEDICINE. *Educ:* Univ Ill, 47-50; Wash Univ, MD, 54. *Prof Exp:* Asst prof microbiol, Sch Med, Johns Hopkins Univ, 64-67; mem staff, Rockefeller Found, 67-73; prof microbiol & immunol & chmn dept, Sch Med, Ind Univ, Indianapolis, 73-80; prof, microbiol & immunol, dept med, Wright State Univ, Ohio & dean, Sch Med, 81-87; PRES, CHINA MED BD NY INC, 88- *Concurrent Pos:* Fel med, Sch Med, Wash Univ, 58-60; consult, Army Med Res & Develop Command, 65-67, Armed Forces Inst Path, 79- & Lobund Adv Bd, Notre Dame Univ, 79-86; vis prof & chmn, dept microbiol, Mahidol Univ, Thailand, 67-73; consult, WHO Immunol Res & Training Ctr, Singapore, 69-73; chmn, Consortium Affils Int Progs, AAAS; hon prof, Sun Yat Sen Univ Med Sci, Guangzhou, Peoples Rep China. *Mem:* Am Soc Microbiol; Am Col Physicians; Am Acad Microbiol; Soc Exp Biol & Med; Am Asn Pathologists; AAAS. *Res:* Bacterial infection, genetics and physiology. *Mailing Add:* China Med Bd 750 Third Ave New York NY 10017

SAWYERS, JOHN LAZELLE, b Centerville, Iowa, July 26, 25; m 57; c 3. SURGERY. *Educ:* Univ Rochester, BA, 46; Johns Hopkins Univ, MD, 49. *Prof Exp:* From asst prof to prof surg, 69-83, JOHN CLINTON FOSHEE DISTINGUISHED PROF SURG & CHMN DEPT, VANDERBILT UNIV, 83- *Concurrent Pos:* Dir, Sect Surg Scis, Vanderbilt Univ, 83- *Mem:* Soc Univ Surg; Am Asn Thoracic Surg; Am Col Surg; Am Surg Asn; Int Cardiovasc Soc. *Res:* Surgery of the alimentary tract. *Mailing Add:* Dept Surg Vanderbilt Univ Nashville TN 37232

SAWYERS, KENNETH NORMAN, b Chicago, Ill, May 19, 36; m 62; c 2. APPLIED MATHEMATICS. *Educ:* Ill Inst Technol, BS, 62; Brown Univ, PhD(appl math), 67. *Prof Exp:* Math physicist, Stanford Res Inst, 66-69; from asst prof to assoc prof, Ctr Appln Math, 69-82, PROF MECH ENGR & MECHS, LEHIGH UNIV, 83-, ASSOC DEAN & APPL SCI, 90- *Mem:* Am Acad Mechs. *Res:* Theoretical seismology; mechanics of fluids and solids; stability theory. *Mailing Add:* Packard Lab 19 Lehigh Univ Bethlehem PA 18015

SAX, KARL JOLIVETTE, b Ancon, CZ, Sept 7, 18; US citizen; m 42; c 2. CHEMISTRY. *Educ:* Harvard Univ, BS, 40; Yale Univ, MS, 53, PhD(org chem), 55. *Prof Exp:* Asst, Harvard Univ, 40; chemist, E I du Pont de Nemours & Co, NJ, 40-41, control chemist, Ind, 41-42, Okla, 42-43, res chemist, Ill, 43; res chemist, Clinton Lab, Tenn, 43-45 & Lederle Labs, Am Cyanamid Co, NY, 45-51; asst, Yale Univ, 51-53; chemist, Shell Develop Co, 54-60; chemist, Lederle Labs, Am Cyanamid Co, 60-82; RETIRED. *Mem:* Assoc Am Chem Soc. *Res:* Natural products; lipids of coelenterates; synthetic lubricants; isolation and identification of steroids; isolation and identification of antibiotics; process and preparations research. *Mailing Add:* 55 Wax Myrtle Ct Hilton Head Island SC 29926

SAX, MARTIN, b Wheeling, WVa. CRYSTALLOGRAPHY, MACROMOLECULAR & BIORELEVANT STRUCTURES. *Educ:* Univ Pittsburgh, BS, 41, PhD(phys chem), 61. *Prof Exp:* Res chemist, Trojan Powder Co, 41-44 & Glyco Prod Co, 51-59; asst biochemist, W Pa Hosp, 46-57; postdoctoral crystallog, Univ Pittsburgh, 61-63, asst res prof, 63-66; RES CHEMIST & DIR, BIOCRYSTALLOG LAB, VET AFFAIRS MED CTR, PITTSBURGH, 66-, ASSOC CHIEF STAFF RES & DEVELOP, 72- *Concurrent Pos:* Adj assoc prof crystallog, Univ Pittsburgh, 66-71, adj prof, 71- *Mem:* AAAS; Am Crystallog Asn; Am Chem Soc. *Res:* Three dimensional structures and the functions of biological macromolecules as ascertained by x-ray diffractions from single crystals. *Mailing Add:* PO Box 12055 Pittsburgh PA 15240

SAX, ROBERT LOUIS, b Wheeling, WVa, Apr 7, 28; m 58; c 2. GEOPHYSICS. *Educ:* Mass Inst Technol, BS, 52, PhD(geophys), 60. *Prof Exp:* Geologist-geophysicist, Standard Oil Co Calif, 52-56; mem tech staff, Ramo-Wooldridge Labs, 59-60 & Hughes Aircraft Corp, 60-62; sr geophysicist, United Earth Sci Div, Teledyne Indust, 62-68, SR SCIENTIST, TELEDYNE INC, GEOTECH CORP, 68- *Mem:* AAAS; Am Geophys Union; Soc Explor Geophys; Seismol Soc Am; Inst Elec & Electronics Engrs. *Res:* Interpretation of earth gravity measurements; design of systems for probing and analyzing planetary atmospheres; wave propagation and source mechanisms in a solid earth; theory of seismic noise. *Mailing Add:* 3707 Cameron Mills Rd Alexandria VA 22305

SAX, SYLVAN MAURICE, b Wheeling, WVa, Feb 8, 23; m 57; c 3. CLINICAL CHEMISTRY. *Educ:* Univ Pittsburgh, BS, 44, PhD(org chem), 53; Am Bd Clin Chem, Dipl, 60. *Prof Exp:* Group leader res, Glyco Prods Co, 53-55; clin biochemist, 55-85, CLIN CHEM CONSULT, WESTERN PA HOSP, 85- *Concurrent Pos:* Instr & adj asst prof, Sch Med, Univ Pittsburgh, 57- *Honors & Awards:* Fisher Award, Am Asn Clin Chem. *Mem:* Am Chem Soc; Am Asn Clin Chem. *Res:* Clinical chemistry methodology and control. *Mailing Add:* Western Pa Hosp 6490 Monitor St Pittsburgh PA 15217-2722

SAXE, HARRY CHARLES, b Long Island City, NY, Mar 18, 20; m 45; c 2. CIVIL & STRUCTURAL ENGINEERING. *Educ:* City Col New York, BCE, 42; Univ Fla, MSE, 49; Mass Inst Technol, ScD, 52. *Prof Exp:* Asst civil eng, Univ Fla, 48-49, instr, 49-50; res asst, Mass Inst Technol, 50-52; assoc prof, Ga Inst Technol, 52-56; engr, Praeger-Kavanagh & Assocs, NY, 56-57; assoc prof civil eng, Polytech Inst Brooklyn, 57 & Univ Cincinnati, 57-59; prof & head dept, Univ Notre Dame, 59-60 & 61-65, actg dean col eng, 60-61 & 66-67, chmn dept civil eng, 67-69; dean Speed Sci Sch, Univ Louisville, 69-80, prof 69-83 DEAN & EMER PROF CIVIL ENG, 83- *Concurrent Pos:* NSF sci fac fel & vis prof, Imp Col, Univ London, 65-66; pres, Univ Louisville Inst Indust Res, 69-74; mem, Ky Sci & Technol Adv Coun, 70-75; Ky Bd Registr for Prof Engrs & Land Surveyors, 72-80; mem, energy res bd, Ky Dept Energy, 76-80; Louis S Le Tellier distinguished vis prof, Dept Civil Eng, The Citadel, Mil Col SC, 83-85 & 86-; Fredrik Wachtmeister distinguished prof eng, Va Mil Inst, 83; vis prof civil, Univ Md, Col Park, 85-86. *Mem:* Fel Am Soc Civil Engrs; Am Soc Eng Educ; Nat Soc Prof Engrs; Soc Am Mil Engrs; Am Soc Testing & Mat; Sigma Xi. *Res:* Structural mechanics including computer applications. *Mailing Add:* Dept Civil Eng Norwich Univ Northfield VT 05663

SAXE, LEONARD, b New York, NY, June 12, 47; m 70; c 1. PSYCHOLOGY. *Educ:* Univ Pittsburgh, BS, 69, MS, 72, PhD(social psychol), 76. *Prof Exp:* assoc prof psychol, Boston Univ, 75; VIS PROF, BRANDEIS UNIV, 88- *Concurrent Pos:* Consult, US Cong, Off Technol Assessment, 80-88; Fulbright lectr, Univ Haifa, Israel, 81-82; dir, Ctr Appl Social Sci, Boston Univ, 85-88. *Mem:* Am Psychol Asn; Soc Psychol Study Social Sci. *Res:* Mental health policy; use of social research in the development of social policy; lie detectors, polygraph tests. *Mailing Add:* Heller Sch Brandeis Univ Waltham MA 02254

SAXE, RAYMOND FREDERICK, nuclear & electrical engineering, for more information see previous edition

SAXE, STANLEY RICHARD, b Chelsea, Mass, Feb 1, 32; m 58; c 3. GERIATRIC DENTISTRY, PERIODONTOLOGY. *Educ:* Boston Univ, AB, 53; Harvard Univ, DMD, 58; Univ Wash, MSD, 60. *Prof Exp:* Instr periodont, Sch Dent, Univ Wash, 60-62; from asst prof to assoc prof, 62-77, chmn dept, 66-69, PROF PERIODONT, COL DENT, UNIV KY, 77-, PROF GERIATRIC DENT ORAL HEALTH SCI, 87- *Concurrent Pos:* Consult, Vet Admin Hosp, Am Lake, Wash, 61-62, Lexington, Ky, 62- & USPHS Hosp, 62-76; vis prof, Hadassah Sch Dent Med, Hebrew Univ Jerusalem, 70 & Sanders-Brown Res Ctr on Aging, Lexington, Ky, 84. *Mem:* Am Dent Asn; Am Acad Periodont; Int Asn Dent Res; Am Soc Geriat Dent. *Res:* Oral health care strategies for the older adult; etiology, assessment and treatment of periodontal disease. *Mailing Add:* Dept Oral Health Sci Lexington KY 40536-0084

SAXENA, BRIJ B, b India, July 6, 30. BIOCHEMISTRY, ENDOCRINOLOGY. *Educ:* Agra Univ, BSc, 49; Lucknow Univ, MSc, 51, PhD(biochem), 54; Univ Munster, Dr rer nat(physiol), 57; Univ Wis, PhD(biochem, endocrinol), 61. *Prof Exp:* Lectr biol, Lucknow Univ, 54-55; res assoc endocrinol, Univ Wis, 57-62; from asst prof to assoc prof biochem, NJ Col Med, 62-66; assoc prof biochem, Dept Med, 66-72, assoc prof endocrinol, Dept Obstet & Gynec, 70-72, PROF BIOCHEM, DEPT MED & PROF ENDOCRINOL, DEPT OBSTET & GYNEC, MED COL, CORNELL UNIV, 72-, DIR, DIV REPRODUCTIVE ENDOCRINOL, DEPT OBSTET & GYNEC, 80- *Concurrent Pos:* Career sci award, Health Res Coun, New York, 69. *Mem:* AAAS; Endocrine Soc; NY Acad Sci; fel Royal Soc Med; Harvey Soc; Sigma Xi. *Res:* Physiology and biochemistry of pituitary hormones and gonadal receptors. *Mailing Add:* NY Hosp-Cornell Med Ctr 525 E 68th St New York NY 10021

SAXENA, NARENDRA K, b Agra, India, Oct 15, 36; US citizen; m 70; c 2. MARINE SURVEYS, GEODESY. *Educ:* Agra Univ, BSc, 55; Hannover Tech Univ, dipl Ing, 65; Graz Tech Univ, Dr tech (satellite geod), 72. *Prof Exp:* Proj-in-charge satellite geod, Space Sci & Technol Ctr, India, 69; res assoc geod, Ohio State Univ, 69-74; asst prof geod eng, Univ Ill, Urbana, 74-78; asst prof surv, 78-81, assoc prof, 81-86, PROF CIVIL ENG, UNIV HAWAII, 86- *Concurrent Pos:* Mem, Spec Study Group, Int Asn Geod, 73-80; ed-in-chief, Marine Geod J Int J Ocean Surv Mapping & Sensing, 77-; co-chmn, Pac Cong Marine Technol, 84, 86 & 88; adj res prof oceanog, Naval Postgrad Sch, Monterey, 84- *Honors & Awards:* Fel, Marine Technol Soc. *Mem:* Marine Technol Soc; Am Geophys Union; Inst Navig; Am Soc Civil Engrs. *Res:* Ocean survey and navigation; computational methods to solve large systems; engineering survey; space geodesy. *Mailing Add:* Dept Civil Eng Univ Hawaii Honolulu HI 96822

SAXENA, SATISH CHANDRA, b Lucknow, India, June 24, 34; US citizen; m 61; c 4. ENERGY ENGINEERING. *Educ:* Lucknow Univ, BSc, 51; MSc, 53; Calcutta Univ, PhD(physics), 56. *Prof Exp:* Res assoc, Univ Md, 56-58 & Columbia Univ, 58-59; res assoc physics, Yale Univ, 59; engr, Bhabha Atomic Res Ctr, India, 59-61; reader & head, Dept Physics, Rajasthan Univ, India, 61-66; assoc prof, Purdue Univ, 66-68; PROF CHEM ENG, UNIV ILL, 68- *Concurrent Pos:* Sr res assoc, Ames Res Ctr, NASA, 69 & 70; sr res assoc, Argonne Nat Lab, 77; consult, Purdue Univ, Argonne Nat Lab, E I du Pont de Nemours & Co Inc, Oak Ridge Nat Lab & Gen Elec Co, Morgantown Energy Technol Ctr, WVa, Inst Gas Technol, Chicago, Ortho Inc, Chicago & Maremac Corp, Pittsburgh Energy Technol Ctr. *Mem:* Am Inst Chem Engrs. *Res:* Transport properties of gases and gaseous mixtures; modeling of fluidized bed operations, coal combustion, coal gasification and indirect coal liquefaction; solid waste management. *Mailing Add:* Dept Chem Eng Univ Ill PO Box 4348 Chicago IL 60680

SAXENA, SUBHASH C, b Etawah, UP, India; US citizen; m 59; c 2. MATHEMATICS EDUCATION, CURRICULUM DEVELOPMENT & ADMINISTRATION. *Educ:* Univ Delhi, BA, 52, MA, 54, PhD(math), 58. *Prof Exp:* Instr math, Defence Acad, India, 58; from asst prof to assoc prof Atlanta Univ, 59-63; assoc prof Northern Ill Univ, 63-68, Univ Akron, 68-73; from assoc prof to prof math, 73-87, CHMN DEPT, DEPT MATH, UNIV SC-COASTAL, 87- *Concurrent Pos:* Post doctoral fel math, Univ Delhi, 58-59; exchange res prof, Univ SC, Columbia, 80; vis res scientist, Indian Inst Technol, Delhi, India, 81-82. *Mem:* Am Math Soc; Math Asn Am; Nat Coun Teachers Math. *Res:* Discrete math, graph theory and combinatorics; tournamenents and balanced incomplete block designs; math education. *Mailing Add:* 4407 Greenbay Trail Myrtle Beach SC 29577

SAXENA, SURENDRA K, b Bundi, India. GEOTECHNICAL ENGINEERING, CIVIL ENGINEERING. *Educ:* Aligarh Univ, BSc, 55; Duke Univ, MS, 65, PhD(geotech eng), 71. *Prof Exp:* Dist engr, state govt Rajasthan, India, 55-62; soils engr, Port Authority NY & NJ, 69-74; sr engr, Dames & Moore, Cranford, NJ, 74-76; assoc prof, 76-80, PROF CIVIL ENG & CHMN DEPT, ILL INST TECHNOL, 80- *Concurrent Pos:* Res assoc, Mass Inst Technol, 73-74; consult, Argonne Nat Lab, 78-, Nat Hwy Res Prog, Transp Res Bd, 77-; appointee, Gov Comn Sci & Technol, Gov of Ill, 85-89. *Honors & Awards:* Hwy Res Bd Award, 70. *Mem:* Am Soc Civil Engrs; Am Soc Testing & Mat; Earthquake Eng Res Inst; Sigma Xi; Int Soc Soil Mech & Found Eng. *Res:* Geotechnical engineering; earthquake engineering; soil-structure interaction; geo-energy; soil behavior and soil improvement. *Mailing Add:* Dept Civil Eng Ill Inst Technol Chicago IL 60616

SAXENA, UMESH, b Pilibhit, India; m 62; c 3. INDUSTRIAL ENGINEERING. *Educ:* Univ Roorkee, BS, 60; Univ Wis-Madison, MS, 65, PhD, 68. *Prof Exp:* Lectr mech eng, Indian Inst Technol, Delhi, 62-64; PROF INDUST ENG, UNIV WIS-MILWAUKEE, 68- *Mem:* Oper Res Soc Am; Am Inst Indust Engrs. *Res:* Health care systems and delivery; application of quantitative methods in analysis and control of industrial and service systems; optimal design of health care delivery systems. *Mailing Add:* Indust & Systs Eng Univ of Wis Col Appl Sci & Eng Milwaukee WI 53201

SAXENA, VINOD KUMAR, b Agra, India, May 23, 44; c 2. METEOROLOGY, CLOUD PHYSICS. *Educ:* Agra Univ, BS, 61, MS, 63; Univ Rajasthan, PhD(physics), 69. *Prof Exp:* Lectr physics, Agra Col, 63-64; asst prof pharmaceut sci, Univ Saugar, 67-68; Off Naval Res fel, Univ Mo, Rolla, 68-71; res assoc cloud physics, Univ Denver, 71-73, cloud physicist & lectr physics, Denver Res Inst, 73-77; res assoc prof meteorol, Univ Utah, 77-79; assoc prof meteorol, 79-88, PROF METEOROL, NC STATE UNIV, 88- *Concurrent Pos:* Fac mem invited sem, NC Ctr Advan Teaching, Western Carolina Univ; US grants, NSF, US Environ Protection Agency, NSF Dept Energy, US Bur Mines, NASA. *Honors & Awards:* Meterol Award, Univ Utah, 79. *Mem:* Am Meteorol Soc; Am Geophys Union; fel Royal Meteorol Soc, UK; Am Asn Areasol Res; Air Pollution Control Asn; European Asn Areasol Res. *Res:* Cloud and aerosol physics; air pollution meteorology; condensation and nucleation phenomena; cloud-aerosol interactions; instrumentation for simulating atmospheric environment and monitoring Aitken; cloud and ice-forming nuclei. *Mailing Add:* Dept Marine Earth & Atmosphere Sci Box 8208 NC State Univ Raleigh NC 27695-8208

SAXINGER, W(ILLIAM) CARL, b Chicago, Ill, Oct 4, 41; m 67; c 3. DIAGNOSTIC IMMUNOLOGY. *Educ:* Univ Ill, BS, 63, PhD(microbiol), 69. *Prof Exp:* Nat Acad Sci & Nat Res Coun res assoc exbiol & chem evolution, Ames Res Ctr, NASA, 69-71; res assoc chem evolution, dept chem, Univ Md, 71-72; staff fel tumor virol, 72-74, SR INVESTR HUMAN RETROVIROL, LAB TUMOR CELL BIOL, DIV CANCER TREAT, NAT CANCER INST, 74-; RES PROF, LAB CHEM EVOLUTION, DEPT CHEM, UNIV MD, 72- *Res:* Processes of infection and immunity to human retroviruses causing human leukemia and immunosuppression. *Mailing Add:* Nat Cancer Inst Bldg 37 Rm 6C21 Bethesda MD 20205

SAXON, DAVID STEPHEN, b St Paul, Minn, Feb 8, 20; m 40; c 6. THEORETICAL NUCLEAR PHYSICS. *Educ:* Mass Inst Technol, BS, 41, PhD(physics), 44. *Hon Degrees:* LHD, Hebrew Union Col, 76, Univ Judaism, 77; LLD, Univ SCalif, 78. *Prof Exp:* Mem staff, Radiation Lab, Mass Inst Technol, 42-46; assoc physicist, Philips Lab, Inc, NY, 46-47; physicist, Inst Numerical Anal, Nat Bur Stand, Calif, 50-53; from asst prof to prof physics, Univ Calif, Los Angeles, 47-83, chmn dept, 63-66, dean phys sci, 66-70, vchancellor, 68-74, exec vchancellor, 74-75, pres, 75-83; off chmn corp, 83-90, HON CHMN, MASS INST TECHNOL, CAMBRIDGE, 90- *Concurrent Pos:* Guggenheim Mem fel, 56-57 & 61-62; Fulbright lectr, 61-62; consult, Systs Corp Am, 58-63, Convair Div, Gen Dynamics Corp, 60-63 & E H Plessit Assoc, 61-63; emer pres & emer prof physics, Univ Calif, Los Angeles, 83-; mem bd dirs, Mass Ctrs Excellence Corp, Ford Tech Adv panel & Harvard Overseer's comt to visit Med Sch & Sch Dent Med. *Mem:* Fel Am Phys Soc; Am Asn Physics Teachers; Sigma Xi; fel Am Acad Arts & Sci; Am Philos Soc; fel AAAS; Am Inst Physics. *Res:* Electromagnetic theory; quantum theory; nuclear physics; author of 1 book and author or coauthor of 3 physics texts and many scientific articles in professional journals. *Mailing Add:* Hon Chmn Rm 9-235 Mass Inst Technol 77 Massachusetts Ave Cambridge MA 02139-4307

SAXON, JAMES GLENN, b Marlin, Tex, July 16, 41; m 68; c 2. ANATOMY. *Educ:* Baylor Univ, BS, 63, MS, 66; Tex A&M Univ, PhD(wildlife sci), 70. *Prof Exp:* Asst prof biol, Memphis State Univ, 69-70; asst prof & chmn dept, Erskine Col, 70-74; from asst prof to assoc prof, 74-82, PROF ANAT, SOUTHERN COL OPTOM, 83- *Mem:* AAAS. *Res:* Anatomy and reproductive physiology of amphibians and ocular reptiles. *Mailing Add:* Dept Optom Southern Col Optom 1245 Madison Ave Memphis TN 38104

SAXON, ROBERT, b Brooklyn, NY, June 3, 24; m 60; c 2. POLYMER CHEMISTRY. *Educ:* Northwestern Univ, PhD(chem), 52. *Prof Exp:* Res chemist, US Indust Chem Co, 43-49; asst chem, Northwestern Univ, 50-52; res assoc, Harris Res Labs, 52-56; res chemist, AM Cyanamid Co, 56-72, mgr res & develop, 72-86; CONSULT, 86- *Mem:* AAAS; Am Chem Soc. *Res:* Organic synthesis and mechanisms; urethane chemistry and technology; surface coatings; laminated plastics; elastomer synthesis and technology. *Mailing Add:* 199 Laurel Circle Princeton NJ 08540-2718

SAXTON, HARRY JAMES, b Bell, Calif, July 2, 39; m 61, 83; c 2. MATERIALS SCIENCE. *Educ:* Stanford Univ, MS, 62, PhD(mat sci), 69. *Prof Exp:* Systs analyst, Ctr Naval Anal, 69-71; mem tech staff, 71-74, div supvr, 74-76, dept mgr, 76-83, DIR, SANDIA NAT LABS, 83- *Concurrent Pos:* Consult, Lawrence Livermore Lab, 66-71. *Mem:* Am Soc Metals. *Res:* Relation of materials microstructure to macroscopic properties; development and production of integrated circuits; quartz oscillators; electromechanical devices; explosive devices; battery research and development. *Mailing Add:* Sandia Nat Labs Org 2100 Albuquerque NM 87185

SAXTON, KEITH E, b Crawford, Nebr, Dec 22, 37; m 57; c 2. AGRICULTURAL ENGINEERING. *Educ:* Univ Nebr, BS, 61; Univ Wis, MS, 65; Iowa State Univ, PhD, 72. *Prof Exp:* HYDRAUL ENGR, FED RES, SCI & EDUC ADMIN, USDA, 61- *Mem:* Am Soc Agr Engrs; Am Soc Civil Engrs. *Res:* Hydrologic research on small agricultural watersheds. *Mailing Add:* NW 1830 Deane Pl Pullman WA 99163

SAXTON, RONALD L(UTHER), chemical engineering, for more information see previous edition

SAXTON, WILLIAM REGINALD, b Montreal, Que, May 3, 28; m 53; c 2. PULP CHEMISTRY. *Educ:* McGill Univ, BSc Hons, 49. *Prof Exp:* Mgr rayon res, Indust Cellulose Res Ltd, 58-62, mgr res planning, 62-65; mgr new prod develop, Int Cellulose Res Ltd, 65-68; asst to vpres res & develop, Int Paper Co, 68-71; pres, Can-Pac-Forest Prod Res Ltd, 71-89; RETIRED. *Honors & Awards:* Environ Improv Award, Chem Inst Can. *Mem:* Can Pulp & Paper Asn; Can Res Mgt Asn; Tech Asn Pulp & Paper Indust. *Res:* Chemical and mechanical wood pulps for printing and industrial papers, paper board packaging, tissue products, conversion to rayon, cellophane and other cellulosic products. *Mailing Add:* CAN-PAC Forest Prod Res Ltd 179 Main St W Hawkesbury ON K6A 2H4 Can

SAYALA, CHHAYA, b Bangalore, India, Apr 16, 50; US citizen; m 78; c 1. OIL EXPLORATION CHEMICALS, NITRO COMPOUND EXPLOSIVES. *Educ:* Bangalore Univ, India, BSc, 70, MSc, 72; Indian Inst Technol, India, PhD(chem), 76. *Prof Exp:* Res fel, Univ Agr Sci, India, 77-78; res fel, Univ Nijmegen, Neth, 78-79; lectr, Diablo Valley Col, 84, Ohlone Col, 85, George Mason Univ, 86-88; PATENT EXAMR, US PATENT & TRADEMARK OFF, 89- *Concurrent Pos:* Res fel, Coun Sci & Indust Res, India, 76 & 77-78, Univ Nijmegen, Neth, 78-79; lectr, San Jose State Univ, 85, Northern Va Community Col, 88-89. *Res:* Synthetic organometallic chemistry: syntheses and characterization of thiourea, & dithiocarbamate compounds with selenium, tellurium, transition metals. *Mailing Add:* 1887 Cold Creek Ct Vienna VA 22182

SAYALA, DASHARATHAM (DASH), b Hyderabad, AP, India, Sept 12, 43; US citizen; m 78; c 1. MINERAL EXPLORATION, MINING GEOLOGY. *Educ:* Osmania Univ, India, BSc, 62, MSc, 64; Univ NMex, MS, 72; George Washington Univ, PhD(geochem), 79. *Prof Exp:* Teaching fel, Univ NMex, 66-68; proj mgr min explor, Uranium King Corp, 68-74; asst prof, Howard Univ, 75-77; lectr, George Washington Univ, 77-78; adv res geoscientist, Bendix Field Eng Corp, 79-83; sr geochemist, Woodward-Clyde Consult, 83-85; tech staff, Mitre Corp, 85-91; SR SCIENTIST, SCI APPLICATIONS INT CORP, 91- *Honors & Awards:* Cert Appreciation, US Dept Energy, 86. *Res:* Trace element geochemistry and isotope geochemistry; application in mineral exploration and environmental studies, including ground water hydrology; fate and transport of hazardous chemicals and radionuclides. *Mailing Add:* 1887 Cold Creek Ct Vienna VA 22182

SAYEED, MOHAMMED MAHMOOD, b Dec 8, 37; m 61; c 4. MEMBRANE TRANSPORT, CIRCULATORY SHOCK PHYSIOLOGY. *Educ:* Univ Miami Sch Med, PhD(physiol), 65. *Prof Exp:* from asst prof to assoc prof physiol surg, Wash Univ Sch Med, 68-76; PROF PHYSIOL, STRITCH SCH MED, LOYOLA UNIV, 76- *Concurrent Pos:* Assoc prof physiol surg, Wash Univ Sch Med, 74-76; dir div cellular physiol, Jewish Hosp, St Louis, 70-76; study sect on biomed sci, NIH, 86-90. *Mem:* Am Physiol Soc; Soc Exp Biol Med; Shock Soc. *Res:* Cellular mechanics of pathogenesis of sepsis/septic shock. *Mailing Add:* Dept Physiol Loyola Univ Stritch Sch Med 2160 S First Ave Maywood IL 60153

SAYEG, JOSEPH A, b Fresno, Calif, Sept 22, 25; m 56; c 2. BIOPHYSICS. *Educ:* Univ Calif, Berkeley, AB, 47, PhD(biophys), 54; Am Bd Health Physics, dipl, 66; Am Bd Radiol, dipl, 69. *Prof Exp:* Res asst biophys, Donner Lab Med Physics & Biophys, Univ Calif, Berkeley, 49-54; biophysicist, Los Alamos Sci Lab, 54-61; sci specialist, Santa Barbara Labs, Edgerton, Germeshausen & Grier Inc, 61-66; assoc prof health physics, Univ Pittsburgh, 66-67; ASSOC PROF RADIATION MED, COL MED, UNIV KY, 67-, CHMN DEPT HEALTH RADIATION SCI, COL ALLIED HEALTH PROF, 68- *Concurrent Pos:* Consult, Nat Comt Radiation Protection, 58-70, Armed Forces Radiobiol Res Inst, 63-64, Walter Reed Army Inst Res, 65-67 & Mercy Hosp, Pittsburgh, Pa, 66-67; lectr, Univ Calif, Santa Barbara, 62-65. *Mem:* Am Asn Physicists in Med; Soc Nuclear Med; Health Physics Soc; Radiation Res Soc. *Res:* Radiation dosimetry, including x-rays, neutrons and heavy charge particles; health physics; neutron dosimetry for personnel dose evaluation; radiobiology; effect of radiation on unicellular organisms. *Mailing Add:* 1948 Blairmore Rd Lexington KY 40502

SAYEGH, FAYEZ S, b Jordan, Feb 25, 27; US citizen; m 55; c 3. MICROSCOPIC ANATOMY. *Educ:* Manila Cent Univ, DMD, 61; Univ Rochester, MS, 66; Colo State Univ, PhD(anat), 72. *Prof Exp:* Instr microbiol, South Chicago Community Hosp, 56-58; chmn histol dept, Eastman Dent Ctr, 61-69; NIH fel anat, Colo State Univ, 69-72; PROF HISTOL & CHMN DEPT, DENT SCH, UNIV MO, 72-, PROF MED & ANAT, MED SCH, 75- *Concurrent Pos:* Consult, Kerr Mfg Co, 64-71 & Lee Pharmaceut, 72- *Mem:* Am Asn Anatomists; Am Dent Asn; Int Asn Dent Res. *Res:* Oral biology; skeletal tissue mineralization; tooth germ formation and calcification. *Mailing Add:* Dept Anat Cleveland Chiropractr Col 6401 Rockhill Rd Kansas City MO 64131

SAYEGH, JOSEPH FRIEH, b Rmeimeen, Jordan, Mar 5, 28; m 55; c 4. GAS CHROMATOGRAPHY, MASS-SPECTROMETRY. *Educ:* NY Univ, BA, 54, MS, 65, PhD(biol), 69. *Prof Exp:* Teacher sci & math, Al-Ahlyam Col, Jordan, 54-56, St George Col, 56-60; res scientist, Res Ctr, Rockland State Hosp, 60-69; sr res scientist, 70-78, SCIENTIST V, NATHAN KLINE INST, 78-85; ASST RES PROF, NY UNIV, 85- *Mem:* NY Acad Sci; Clin Ligand & Assay Soc; Am Soc Neurochem. *Res:* Development of analytical methodologies for biomolecules; hormones and psychoactive drugs; gas chromatography; interfaced gas chromatography-mass spectrometry; high pressure liquid chromatography; infrared ultra violet; fluorometry; radioimmunoassay and enzyme immunoassays; column liquid chromatography; protein synthesis and amino acids uptake by the brains. *Mailing Add:* Nathan Kline Inst Orangeburg Rd Orangeburg NY 10962

SAYEKI, HIDEMITSU, b Yokohama, Japan, July 2, 33; m 59; c 1. MATHEMATICS. *Educ:* Univ Tokyo, BSc, 56; Univ Warsaw, MA, 61, PhD, 65. *Prof Exp:* Nat Res Coun Can fel, Univ Montreal, 66-67, asst prof, 67-70, ASSOC PROF MATH, UNIV MONTREAL, 70- *Mem:* Asn Symbolic Logic; Math Soc Japan; Can Math Cong; Am Math Soc. *Res:* Mathematical logic; set theory. *Mailing Add:* Dept Math Univ Montreal Box 6128 Succursale A Montreal PQ H3C 3J7 Can

SAYER, JANE M, b Keene, NH, Mar 7, 42. BIO-ORGANIC CHEMISTRY. *Educ:* Middlebury Col, BA, 63; Yale Univ, PhD(biochem), 67. *Prof Exp:* NIH fel biochem, Brandeis Univ, 67-69, sr res assoc, 69-74; asst prof, Univ Vt, 74-79, res asst prof chem, 79-83; RES CHEMIST, NAT INST DIABETES, DIGESTIVE & KIDNEY DIS, 85- *Concurrent Pos:* Sabbatical leave, Nat Inst Diabetes, Digestive & Kidney Dis, NIH, 79-83, spec expert, 83-85. *Mem:* Sigma Xi; Am Chem Soc. *Res:* Mechanisms of catalysis in enzymatic and related non-enzymatic reactions of carbonyl, imino and acyl compounds and epoxides; organic oxidation-reduction reactions of biochemical interest, reaction mechanisms and structure-activity; relationships of ultimate carcinogens. *Mailing Add:* Nat Inst Diabetes Digestive & Kidney Dis NIH Bldg 8 Rm 1A11 LBC Bethesda MD 20892

SAYER, JOHN SAMUEL, b St Paul, Minn, July 27, 17; m 40; c 4. INFORMATION SCIENCE, SYSTEMS ENGINEERING. *Educ:* Univ Minn, BSME, 40. *Prof Exp:* Supvr plant develop, E I du Pont de Nemours & Co, 47-52, eng planning, 52-54, sect mgr plant eng, 54-56, mgt sci, 56-60; exec vpres, Documentation Inc, 60-62; vpres mgt sci, Auerbach Corp, 62-65; vpres, Info Dynamics Corp, 65-66; pvt consult info sci, Mass, 66-69; exec vpres, Leasco Systs & Res, Md, 69; pvt consult, 70-71; exec vpres, Leasco Info Prod, 72-80; pres, REMAC Info Corp, 80-84; PRES, DAKOTA MGT CORP, 85-, INDEPENDENT CONSULT. *Concurrent Pos:* Pres, John Sawyer Assocs. *Mem:* Am Mgt Asn; Am Soc Info Sci; Am Soc Mech Engrs. *Res:* Large scale information systems concepts and designs, users needs, value functions and use patterns, in management and in scientific and engineering fields. *Mailing Add:* 13209 Colton Lane Gaithersburg MD 20878

SAYER, MICHAEL, b Newport, Eng, Nov 6, 35; m 60; c 4. ENGINEERING PHYSICS, CERAMICS ENGINEERING. *Educ:* Univ Birmingham, BSc, 57; Univ Hull, PhD(physics), 61, PEng, 80. *Prof Exp:* Nat Res Coun Can fel, 60-62; dir res, Almax Industs Ltd, 87-88; from assoc prof to prof physics, Queen's Univ, Ont, 62-82, head dept, 75-82, assoc res dean, Fac Appl Sci, 84-87, PROF PHYSICS, QUEEN'S UNIV, ONT, 89-, PROF MAT & METALL ENG, 91- *Concurrent Pos:* Vis asst prof, Univ Trent, 65-66; vis fel physics, Univ Sheffield, 72-73; mem, Can Eng Accreditation Bd, 88-; ed, J Can Ceramic Soc, 88- *Mem:* Can Ceramic Soc; Am Ceramic Soc. *Res:* Superionic conductors; amorphous and vitreous semiconductors; dielectric; thin film devices; piezoelectric devices and materials; industrial instrumentation; materials science engineering. *Mailing Add:* Dept Physics Queen's Univ Kingston ON K7L 3N6 Can

SAYER, ROYCE ORLANDO, b Toccoa, Ga, Apr 29, 41; m 64; c 2. NUCLEAR PHYSICS. *Educ:* Furman Univ, BS, 62; Univ Tenn, PhD(physics), 68. *Prof Exp:* Physicist, US Army Nuclear Effects Lab, 68-70; asst prof physics, Furman Univ, 70-73; res assoc, Vanderbilt Univ, 73-74; comput appln specialist, 74-80, COMPUT CONSULT, OAK RIDGE NAT LAB, 80- *Concurrent Pos:* Oak Ridge Assoc Univs fac res partic, Oak Ridge Nat Lab, 71; consult, Union Carbide Corp, 72- *Mem:* Am Phys Soc; Am Chem Soc; Sigma Xi. *Res:* Coulomb excitation of nuclear levels; gamma ray angular correlations; accelerator ion optics; heavy-ion physics; MHD stability; graphical display systems. *Mailing Add:* FEDC Bldg Oak Ridge Nat Lab P O Box 2009t Lab Oak Ridge TN 37831

SAYERS, DALE EDWARD, b Seattle, Wash, Nov 29, 43; div; c 2. SOLID STATE PHYSICS. *Educ:* Univ Calif, Berkeley, BA, 66; Univ Wash, MS, 68, PhD(physics), 71. *Prof Exp:* Res engr physics, Boeing Aerospace Co, 72-73; sr res assoc physics, Univ Wash, 74-76; from asst prof to assoc prof, 76-84, PROF PHYSICS, NC STATE UNIV, 84- *Concurrent Pos:* Vis assoc prof, Univ Paris-Sud, 83-84; Case centennial scholar, Case Western Reserve Univ. *Honors & Awards:* Sidhu Award, Pittsburgh X-ray Diffraction Conf, 73; Bertram Eugene Warren Award, Am Crystallog Asn. *Mem:* AAAS; Am Phys Soc; Sigma Xi; Am Chem Soc. *Res:* Development of the extended x-ray absorption fine structure technique and its application to structural studies of amorphous materials, biological systems, catalysts and other systems. *Mailing Add:* Dept Physics NC State Univ PO Box 8202 Raleigh NC 27695

SAYERS, EARL ROGER, b Sterling, Ill, July 20, 36; m 58; c 3. GENETICS, PLANT BREEDING. *Educ:* Univ Ill, BS, 58; Cornell Univ, MS, 61, PhD(plant breeding), 64. *Prof Exp:* Asst prof genetics, 63-66, asst acad vpres, 71-76, dean acad develop, 72-76, assoc acad vpres, 76-80, ASSOC PROF GENETICS, UNIV ALA, 66-, ACAD VPRES, 80- *Concurrent Pos:* Dir arboretum, Univ Ala, 64-66, actg head dept biol, 66-68, asst dean col arts & sci, 68-70, spec asst off acad affairs, 70-71, dean spec progs, 71-72; fel, Am Coun on Educ, mem acad admin internship prog. *Mem:* Genetics Soc Am. *Res:* Self and cross incompatibility in cultivated alfalfa; genetics of the colonial green alga Volvox aureus. *Mailing Add:* Box 870100 Tuscaloosa AL 35487

SAYERS, GEORGE, b Glasgow, Scotland, June 10, 14; nat US; m 66; c 4. ENDOCRINOLOGY. *Educ:* Wayne Univ, BS, 34, MS, 41; Univ Mich, MS, 36; Yale Univ, PhD(biochem), 43. *Prof Exp:* Asst & instr biochem, Sch Med, Yale Univ, 43-45; from asst prof to assoc prof pharmacol, Sch Med, Univ Utah, 45-52; chmn dept, Sch Med, Case Western Reserve Univ, 52-76, prof physiol, 52-; ADJ PROF, DEPT PHYSIOL & ANAT, ENVIRON PHYSIOL LAB, UNIV CALIF, BERKELEY. *Honors & Awards:* Abel Award, Am Soc Pharmacol & Exp Therapeut, 47; Ciba Award, Endocrine Soc, 49. *Mem:* AAAS; Endocrine Soc; Soc Exp Biol & Med; Am Soc Pharmacol & Exp Therapeut; Am Physiol Soc. *Res:* Pituitary and adrenocortical physiology. *Mailing Add:* Physiol Dept 1441 Campus Dr Berkeley CA 94708

SAYETTA, THOMAS C, b Williamsport, Pa, Apr 12, 37; m 68; c 1. MAGNETIC RESONANCE. *Educ:* Univ SC, BS, 59, PhD(physics), 64. *Prof Exp:* Engr, Radio Corp Am, NJ, 59-60; ASSOC PROF PHYSICS, ECAROLINA UNIV, 64- *Concurrent Pos:* Asst ed, Int J Math & Math Sci, 79-83. *Mem:* Am Asn Physics Teachers; Sigma Xi. *Res:* Quantum chemistry and electron spin resonance. *Mailing Add:* Dept Physics ECarolina Univ Greenville NC 27834

SAYLE, WILLIAM, II, b Baytown, Tex, Sept 30, 41; c 1. ELECTRONICS ENGINEERING. *Educ:* Univ Tex, Austin, BSEE, 63, MSEE, 64; Univ Wash, PhD(elec eng), 70. *Prof Exp:* Res engr, Boeing Co, 65-67, sr engr, 70; from asst prof to assoc prof, 70-82, PROF ELEC ENG, GA INST TECHNOL, 82- *Concurrent Pos:* Consult prof, UNESCO, Venezuela, 73-74; consult, Hughes Aircraft, 78-80 , Hewlett-Packard, 81, Motorolo, 83, Lockheed, 85, 86. *Mem:* Sr mem Inst Elec & Electronics Engrs; Am Asn Univ Prof. *Res:* Solid-state power electronics; computer-aided electronic circuit design. *Mailing Add:* Sch Elec Eng Ga Inst Technol Atlanta GA 30332-0250

SAYLES, DAVID CYRIL, b Scollard, Alta, Mar 23, 17; nat US; m 51; c 2. EXPLOSIVE PROPELLANTS, COMPOSITE MATERIALS. *Educ:* Univ Alta, BSc, 39; Univ Chicago, MS, 41; Purdue Univ, PhD(chem), 46. *Prof Exp:* Asst, Univ Alta, 38-39, Univ Chicago, 40-41 & Purdue Univ, 43-44; instr gen eng chem, 45-46; prof org chem & head dept chem, Ferris Inst, 46-47; dep dir res, Lowe Bros Co, 47-52; chief chem prod & equip unit, Wright Air Develop Ctr, Wright-Patterson AFB, Ohio, 52-53, chief high explosives & propellants unit, 53-54, chief ammunition sect, Gun & Rocket Br, 54-56; from asst chief to chief, Gun & Ammunition Br, Munitions Develop Labs, Armament Ctr, Eglin AFB, Fla, 56-57, tech adv, Tech Planning Group, 57-58; chief, Res Plans Br, Ord Missile Command, US Dept Army, Redstone Arsenal, 58-60, actg dep chief, Res Plans Div, 61-63, group leader propellants & mat, Propulsion Lab, US Army Missile Command, 63-66, res phys scientist, Propulsion Technol & Mgt Ctr, 66-70, gen engr, Missile Develop Div, Advan Ballistic Missile Defense Agency, 70-75, gen engr, interceptor directorate, Ballistic Missile Defense Advan Technol Ctr, 75-82, GEN ENGR, US ARMY STRATEGIC DEFENSE COMMAND, 82- *Concurrent Pos:* Consult, C C Letroy & Co, 46-47, Harrow Enterprises, 46-49, Sayles Chem Consults, 46-49 & Glo-Rnz, Inc, 54-58; consult, Hopestone Co Inc, 46-49, Lockheely Distribrs, 50, Holland Chem Co, 56, Res & Develop Corp, 70-75, Saycore Int, 80-, Technol Recognition Corp, 87- *Honors & Awards:* Hon Scroll, Am Inst Chemists, 65, 67 & 68; Res Achievement Award, Chem Inst Can, 70. *Mem:* Am Chem Soc; Am Inst Chemists; Am Ord Asn; Sigma Xi. *Res:* Reactions of organometallic compounds; diethylstilbesterol analogs; lysine; alkyd resins; silicones; epichlorohydrin-bisphenol A resins; surface coatings; synthetic drying oils; lacquers; linoleum; laminates; high explosives; liquid propellants; solid and hybrid materials and propellants; ammunition development; missile, rocket and interceptor propulsion subsystems, materials and propulsion technology. *Mailing Add:* 9616 Dortmond Dr SE Huntsville AL 35803

SAYLES, EVERETT DUANE, b Hillsdale, Mich, July 27, 03; wid; c 4. ZOOLOGY. *Educ:* Kalamazoo Col, AB, 27; Kans State Col, MS, 28; Univ Chicago, PhD(zool), 42. *Hon Degrees:* LLD, Eastern Col, 75. *Prof Exp:* Instr biol, Va Jr Col, Minn, 30-45; prof, Thiel Col, 45-54; prof, 54-74, chmn, biol dept & sci div, 61-74, EMER PROF BIOL, EASTERN COL, 74- *Concurrent Pos:* Vis prof, Fla Mem Col, 61-62. *Mem:* Fel AAAS; Am Inst Biol Sci; emer mem Sigma Xi; emer mem Sci Res Soc NAm. *Res:* Biology of male in mammals; castration of male guinea pig; male guinea pig post natal sexual differentiation; identification and culturing of Archiannelid worms (Dinophilus) and other microscopic Annelid worms. *Mailing Add:* 2247 Sandrala Dr Sarasota FL 34231-4447

SAYLES, FREDERICK LIVERMORE, b New York, NY, May 1, 40. GEOCHEMISTRY, MARINE CHEMISTRY. *Educ:* Amherst Col, BA, 62; Univ Calif, Berkeley, MA, 66; Univ Manchester, PhD(geochem), 68. *Prof Exp:* Fel, 68-69, asst scientist, 69-73, ASSOC SCIENTIST MARINE CHEM, WOODS HOLE OCEANOG INST, 73- *Mem:* AAAS; Geochem Soc Am; Am Geophys Union. *Res:* Geochemical mass balances in the oceans; the cycling of elements in the oceans and the geochemistry of sediments. *Mailing Add:* Dept Chem Woods Hole Oceanog Inst Woods Hole MA 02543

SAYLOR, LEROY C, b Cedar Rapids, Iowa, July 17, 31; m 55; c 3. FORESTRY, GENETICS. *Educ:* Iowa State Univ, BS, 58; NC State Univ, MS, 60, PhD(genetics), 62. *Prof Exp:* Asst geneticist, 61-62, from asst prof to prof genetics & forestry, 62-69, asst dean, Sch Forest Resources, 69-74, ASSOC DEAN, SCH FOREST RESOURCES, 74-, PROF GENETICS & FORESTRY, 69-, ASST DIR, AGR RES SERV, NC STATE UNIV, 87- *Concurrent Pos:* NSF res grants, 61-66; McIntire-Stennis res grants, 64-72. *Honors & Awards:* NC State Col Chap Res Award, Sigma Xi, 67. *Mem:* Genetics Soc Am; Soc Am Foresters; AAAS; Am Forestry Asn; Sigma Xi. *Res:* Cytogenetics of forest tree species; speciation and introgression in forest tree species. *Mailing Add:* 809 Merwin Rd Raleigh NC 27606

SAYLOR, PAUL EDWARD, b Dallas, Tex, Mar 19, 39; m 64; c 2. NUMERICAL ANALYSIS. *Educ:* Stanford Univ, BS, 61; Univ Tex, MA, 63; Rice Univ, PhD(math), 68. *Prof Exp:* Asst prof, 67-73, ASSOC PROF COMPUTER SCI, UNIV ILL, URBANA, 73- *Concurrent Pos:* Vis prof, Inst Fluid Dynamics & Appl Math, Univ Md, College Park, 73-74; hydrologist, US Geol Surv, Reston, Va, 75- *Mem:* Am Math Soc; Soc Indust & Appl Math; Sigma Xi. *Res:* Solution of linear systems arising from partial differential equations. *Mailing Add:* Comput Sci 222 Digital Computer Lab Univ Ill 1304 W Springfield Urbana IL 61801

SAYRE, CLIFFORD L(EROY), JR, computer aided design, for more information see previous edition

SAYRE, CLIFFORD M(ORRILL), JR, b Springfield, Mass, Dec 17, 30; m 59; c 2. CHEMICAL ENGINEERING. *Educ:* Mass Inst Technol, SB, 52. *Prof Exp:* Develop engr, Tech Sect Polychem Dept, E I du Pont de Nemours & Co, Inc, 52-56, engr, Plants Tech Mgr Off, 56-57, asst tech supt, 57-58, res supvr, 58-59, res supvr, Indust & Biochem Dept, 59-63, sr res engr, Plastics Dept, 63-66, sr res supvr, 66, div supt process res, Victoria Plant, 66-69, tech supt, Pontchartrain Works, 69-72, mgr financial & bus anal, Polymer Intermediates, 72-73, intermediates studies mgr, 73-74, mat & distrib mgr, 74-76, planning mgr, Nylon Intermediates, 76-77, mgr, Div Transp & Distrib, 77-81, dir, Int Div, 82-83, Logistics, 83-88 & 89-90, Corp Studies, 88-89, VPRES, MAT, LOGISTICS & SERV, E I DU PONT DE NEMOURS & CO, INC, 90- *Concurrent Pos:* Mem, Marine Bd, Nat Res Coun; chmn, Sea Transp Comt, US Coun for Int Bus; dir, Nat Indust Transp League; chmn, Shippers Competitive Ocean Transp & Compressed Gas Asn; vis comt, Mass Inst Technol Ctr Transp Studies. *Mem:* Fel AAAS; fel Am Inst Chemists; Am Chem Soc; Am Inst Chem Engrs; Nat Freight Trans Asn; Coun Logistics Mgt. *Res:* Nylon and polymer intermediates; industrial chemicals; plastics and environmental control; business and financial analysis; petrochemicals; transportation and distribution of chemicals; maritime research; transportation policy, domestic and international. *Mailing Add:* Mat Logistics & Serv B-8235 E I du Pont de Nemours & Co Wilmington DE 19898

SAYRE, DAVID, b New York, NY, Mar 2, 24; m 47. X-RAY MICROSCOPY, X-RAY CRYSTALLOGRAPHY. *Educ:* Yale Univ, BS, 43; Ala Polytech Inst, MS, 49; Oxford Univ, DPhil(chem crystallog), 51. *Prof Exp:* Mem staff radiation lab, Mass Inst Technol, 43-46; consult, US Off Naval Res, 51; assoc biophys, Johnson Res Found, Univ Pa, 51-55; mathematician, IBM Corp, 55-60, dir prog, 60-62, mgr mach reasoning, 62-64, mgr exp prog, 64-69, res staff mem, 69-90; RETIRED. *Concurrent Pos:* Mem nat comt crystallog, Nat Res Coun, 57-59 & 81-86; vis fel, All Souls Col, Oxford Univ, 72-73; fac mem, Int Sch Crystallogr, Erice, 74, 76, 78, Prague, 75, Ottawa, 81; guest scientist physics, State Univ NY, Stony Brook, 78- *Honors & Awards:* Fankuchen Award, Am Crystallog Asn, 89. *Mem:* Am Crystallog Asn (treas, 53-55, pres, 81). *Res:* X-ray diffraction imaging with soft x-rays and non-periodic microspecimens; x-ray holography; x-ray microscopy with focusing optics; mathematical methods in x-ray crystallography. *Mailing Add:* Three Harbor Rd St James NY 11780

SAYRE, EDWARD VALE, b Des Moines, Iowa, Sept 8, 19; m 43. PHYSICAL CHEMISTRY. *Educ:* Iowa State Col, BS, 41; Columbia Univ, AM, 43, PhD(chem), 49. *Prof Exp:* Chemist, Manhattan Dist Proj, SAM Labs, Columbia Univ, 42-45 & Eastman Kodak Co, 49-52; sr chemist, Brookhaven Nat Lab, 52-84; RES PHYS SCIENTIST, SMITHSONIAN INST, 84- *Concurrent Pos:* Vis lectr, Stevens Inst, 55-63; consult fel, Conserv Ctr, Inst Fine Arts, NY Univ, 60-67, adj prof, 65-74; Guggenheim fel, 69; distinguished vis prof, Am Univ Cairo, 69-70; regents prof, Univ Calif, Irvine, 72; head res lab, Mus Fine Arts, Boston, 75-78, sr scientist, 78-; Alexander von Humboldt US sr scientist award, Berlin, 80. *Honors & Awards:* George Hevesy Medal, 84. *Mem:* Fel Int Inst Conserv Hist & Artistic Works; Am Chem Soc; fel Am Inst Conserv. *Res:* Technical study of fine art and archaeological materials; single crystal spectra, cryogenic measurements; catalyst exchange and surface chemistry studies. *Mailing Add:* 2106 Wilkinson Pl Alexandria VA 22306-2540

SAYRE, FRANCIS WARREN, b Larchwood, Iowa, Nov 20, 24; m 54; c 1. BIOCHEMISTRY. *Educ:* Univ Calif, AB, 49, PhD(biochem), 55; Col Pacific, MA, 51. *Prof Exp:* Chemist, Aluminum Co Am, 43-44; asst chem, Modesto Jr Col, 45-46 & Col Pacific, 50-51; chemist, Agr Lab, Shell Develop Co, 51; res scientist, Clayton Found Biochem Inst, Univ Tex, 51-52; asst biochem, Univ Calif, 52-55; res assoc, City of Hope Med Ctr, 55-58; assoc res scientist, 59-62, res scientist, Kaiser Found Res Inst, 62-68; PROF BIOCHEM, UNIV OF THE PAC, 68- *Concurrent Pos:* Mem biochem subsect, Am Chem Soc Exam Comt, 74- *Mem:* Am Chem Soc; Brit Biochem Soc; AAAS; NY Acad Sci; Sigma Xi. *Res:* Growth and metabolic regulation; specific recognition sites in proteins; cellular aging; carcinogenesis; specific growth factors; enzyme synthesis; nutrition; nutrient interrelationships. *Mailing Add:* Sch Pharm Univ Pac 3601 Pacific Ave Stockton CA 95211

SAYRE, GENEVA, b Guthrie, Iowa, June 12, 11. BOTANY. *Educ:* Grinnell Col, BA, 33; Univ Wyo, MA, 35; Univ Colo, PhD(bot), 38. *Prof Exp:* Asst biol, Univ Colo, 35-38, instr, 39-40; from instr to prof, 40-72, chmn dept biol, 46-69, EMER PROF BIOL, RUSSELL SAGE COL, 72- *Concurrent Pos:* Am Asn Univ Women fel, 49-50, NSF grants, 57-58, 63-66, 69-70; res assoc, Farlow Herbarium, Harvard Univ, 72- *Honors & Awards:* Hedwig Medal, 83. *Mem:* Am Bryol Soc (pres, 51-53); Int Asn Taxon. *Res:* Taxonomy of mosses; history of botanical publication. *Mailing Add:* South St Chesterfield MA 01012

SAYRE, RICHARD MARTIN, b Hillsboro, Ore, Mar 25, 28; m 62; c 1. PLANT PATHOLOGY. *Educ:* Ore State Univ, BS, 51, MS, 54; Univ Nebr, PhD(plant nematol), 58. *Prof Exp:* Nematologist, Harrow Res St, Can Dept Agr, 58-65; NEMATOLOGIST, USDA, 65- *Mem:* Am Phytopath Soc; Soc Nematol. *Res:* Biological control of plant parasitic nematodes. *Mailing Add:* Nematol Lab Biosci Bldg Agr Res Ctr-W Beltsville MD 20705

SAYRE, ROBERT NEWTON, b Cottonwood Falls, Kans, July 25, 32; m 54; c 4. AGRICULTURAL CHEMISTRY, HISTOLOGY. *Educ:* Kans State Univ, BS, 54; Univ Wis, MS, 61, PhD(biochem, meat sci), 62. *Prof Exp:* Asst county agr agent, Kans Agr Exten Serv, USDA, 56-58; res asst muscle biochem, Univ Wis, 58-62; food technologist, Am Meat Inst Found, Ill, 62-64; res chemist poultry meat invests, 64-70, freeze damage reduction in plant tissue, 70-71, res chemist food technol-potato invests, 72-79, RES CHEMIST, FOOD QUAL RES UNIT-GRAIN COMPOS, WESTERN REGIONAL RES CTR, USDA, 79- *Mem:* Inst Food Technol; Am Chem Soc; Am Asn Cereal Chemists. *Res:* Investigation of rice and rice by product processing and product development; processing of other cereal grains and vegetables. *Mailing Add:* Western Regional Res Ctr USDA 800 Buchanan St Albany CA 94710

SAYRE, WILLIAM WHITAKER, civil engineering, hydraulics; deceased, see previous edition for last biography

SAYRES, ALDEN R, b New York, NY, Mar 7, 32; m 72. NUCLEAR PHYSICS. *Educ:* Dartmouth Col, AB, 53; Columbia Univ, PhD(physics), 60. *Prof Exp:* Res assoc physics, Columbia Univ, 60-65; from asst prof to prof physics, Brooklyn Col, 65-90; RETIRED. *Mem:* NY Acad Sci; Am Phys Soc; Am Asn Physics Teachers; Sigma Xi. *Res:* Nuclear particle detection and nuclear spectroscopy. *Mailing Add:* 113 Crest Dr Summit NJ 07901-4115

SAZ, ARTHUR KENNETH, b New York, NY, Dec 2, 17; m 45; c 1. MICROBIOLOGY. *Educ:* City Col New York BS, 38; Univ Mo, MA, 39; Duke Univ, PhD(bact), 43; Am Bd Microbiol, dipl. *Prof Exp:* Asst pharmacol, Sch Med, Duke Univ, 40-42, instr bact, 42-43; instr, New York Med Col, 46-47; asst prof, Iowa State Col, 48-49; bacteriologist, Nat Inst Allergy & Infectious Dis, 48-57, chief med & physiol bact sect, 57-64; prof microbiol & chmn dept, Sch Med & Dent, 64-84, PROF MICROBIOL, GEORGETOWN UNIV, 84- *Concurrent Pos:* Fel, Rockefeller Inst, 47-49; endowed lectr, Mico-Immunol Ann. *Mem:* AAAS; Am Soc Microbiol; Am Acad Microbiol. *Res:* Microbial biochemistry; bacterial physiology and mode of action of antibiotics; penicillinase induction; resistance to penicillin in staphylococcus and gonococcus; bacteriophages of Bacillus cereus; microbial interactions in periodontitis and disease in general; biochemistry of the gonococcus. *Mailing Add:* Dept Microbiol Georgetown Univ Sch Med & Dent Washington DC 20007

SAZ, HOWARD JAY, b New York, NY, Sept 29, 23; m 46; c 3. BIOCHEMISTRY, PARASITE BIOCHEMISTRY. *Educ:* City Col, BS, 48; Western Reserve Univ, PhD(microbiol), 52. *Prof Exp:* Res assoc microbiol, Western Reserve Univ, 52-53,; Nat Found Infantile Paralysis fel, biochem & microbiol, Univ Sheffield, 53-54; res assoc pharmacol, La State Univ, 54-55, from asst prof to assoc prof, 55-60; assoc prof pathobiol, Sch Hyg & Pub Health, Johns Hopkins Univ, 60-69; PROF BIOL SCI, UNIV NOTRE DAME, 69- *Concurrent Pos:* Mem trop med & parasitol study sect, USPHS, 65-69, chmn, 69-70; mem Army Res & Develop Study Group, Parasitic Dis, 73-77; mem Nat Sci Found, Regulatory Biol Panel, 84-86. *Mem:* Am Soc Microbiol; Sigma Xi; Am Soc Biol Chem; Brit Biochem Soc; Am Soc Parasitol. *Res:* Bacterial and helminth biochemistry; comparative biochemistry employing isotopic tracers and enzyme purification techniques. *Mailing Add:* Dept Biol Sci Col Sci Univ Notre Dame Notre Dame IN 46556

SAZAMA, KATHLEEN, b Sutherland, Nebr, May 8, 41; m 62; c 2. BLOOD BANKING. *Educ:* Univ Nebr, Lincoln, BS, 62; Am Univ, Washington DC, MS, 69; Georgetown Univ, MD, 76; Cath Univ, JD, 90. *Prof Exp:* Pathologist, Suburban Hosp, Bethesda, Md, 79-84; chief path, Mem Med Ctr Mich, Ludington, 84-86; chief blood bank pract, Food & Drug Admin, 86-89; consult, Ober, Kales, Grimes & Shriver, Baltimore, Md, 89-90; ASSOC MED DIR, SACRAMENTO MED FOUND, 90- *Concurrent Pos:* Med dir, Metrop Wash Blood Banks, 82-84; clin asst prof, Uniformed Serv Univ Health Sci, 80-89; liaison, Nat Coun Health Lab Serv, 83-85; consult, CAMA, 84-85. *Mem:* AMA; Am Asn Blood Banks; Am Soc Clin Path; Col Am Pathologists; Am Soc Apheresis. *Res:* Write about laboratory reaction to emergencies, dealing with DRG payment system, fatalities and other errors in blood banks. *Mailing Add:* 1460 Klamath River Dr Rancho Cordova CA 95670

SBAR, MARC LEWIS, b Philadelphia, Pa, Oct 7, 44; m 68; c 3. SEISMOLOGY, TECTONICS. *Educ:* Lafayette Col, Easton, Pa, BS, 66; Columbia Univ, PhD(geophys), 72. *Prof Exp:* Res scientist, 72-73, res assoc, Lamont-Doherty Geol Observ, Columbia Univ, 73-77; asst prof seismol, Univ Ariz, 77-83; sr geophysicist, Sohio Petrol Co, 83-84, STAFF GEOPHYSICIST, STANDARD OIL PROD CO, 84- *Concurrent Pos:* Consult earthquake seismol. *Mem:* Am Geophys Union; Soc Explor Geophys. *Res:* Earthquake hazard evaluation; earthquake prediction; regional and global earth movements and causes (tectonics); measurement and interpretation of stress in the earth; application of compressional and shear wave propagation to detection of lithologic properties at depth in the earth. *Mailing Add:* 1500 Witte Rd No 94 Houston TX 77080

SBARRA, ANTHONY J, b Victor, NY, Sept 3, 22; m 56; c 3. BACTERIOLOGY. *Educ:* Siena Col, BS, 48; Ind Univ, BA, 51; Univ Ky, MS, 51; Univ Tenn, PhD(bact), 54; Am Bd Microbiol, dipl. *Prof Exp:* Assoc biologist, Oak Ridge Nat Lab, 53-56; res assoc, Harvard Med Sch, 58-60, instr, 60; asst prof bact, 59-65, assoc prof obstet & gynec, 65-73, PROF OBSTET & GYNEC, MED SCH, TUFTS UNIV, 73- *Concurrent Pos:* Res fel bact & immunol, Harvard Med Sch, 56-58; lectr, Univ Tenn, 54-55; assoc dir dept path & med res, St Margaret's Hosp, 59-74, dir dept, 74-; biochem sect ed, Res, 66-73, mem adv bd, 74-; ed, J Infection & Immunol, 73- & Proc Soc Exp Biol & Med, 74. *Mem:* Am Soc Microbiol; Am Soc Exp Path; Reticuloendothelial Soc; affil AMA; fel Infectious Dis Soc. *Res:* Biological and biochemical approach to the host-parasite relationships. *Mailing Add:* St Margaret's Hosp 90 Cushing Ave Boston MA 02125

SCACCIA, CARL, B Aug 23, 46; c 1. BIOENGINEERING, POLYMER CHEMISTRY. *Educ:* State Univ NY, Buffalo, BS, 68, PhD (eng), 73; Univ Rochester, MS, 70. *Prof Exp:* Supvr, process & prod develop chem eng, Union Carbide, 73-79; mgr technol, chem eng, Combustion Eng, 79-81; dir res & develop, 81-91, VPRES & GEN MGR, SPECIALITY POLYMERS & ADHESIVES DIV, ASHLAND CHEM, 91- *Concurrent Pos:* Adj prof, State Univ NY, Buffalo, 77-81, Ohio State Univ, 81-; bd mem adv comt, Cent Ohio Tech Col, 81-90. *Mem:* Am Ins Chem Eng; Am Chem Soc; Am Soc Mech Engrs; Soc Plastics Indust; Soc Advan Mat & Process Eng. *Res:* Research and commercial development in various chemicals fields; acquisition evaluations, technology feasibility analysis; operations research, strategic planning; first of a kind product process commercialization. *Mailing Add:* 8422 Tibbermore Ct Dublin OH 43017

SCADRON, MICHAEL DAVID, b Chicago, Ill, Feb 12, 38; m 60; c 2. THEORETICAL PHYSICS, HIGH ENERGY PHYSICS. *Educ:* Univ Mich, BS, 59; Univ Calif, Berkeley, PhD(physics), 64. *Prof Exp:* Fel physics, Lawrence Radiation Lab, 64-66; NSF fel, Imp Col, Univ London, 66-68; asst prof, Northwestern Univ, 68-70; vis prof, 70-71, PROF PHYSICS, UNIV ARIZ, 71- *Concurrent Pos:* Sr res fel, Imp Col, Univ London, 72 & 78, Univ Durham, 78; Fulbright Scholar to Pakistan, 79 & India, 85; Australian fel, NSF, Univ Tasmania & Melbourne, 79, 85, 86; vis scientist, Int Ctr for Theoret Physics, Trieste, 67, 72, 78, 83, 87, 89. *Mem:* Am Phys Soc; Fedn Am Sci. *Res:* elementary particle theory; strong, electromagnetic and weak interactions. *Mailing Add:* Dept Physics Univ Ariz Tucson AZ 85721

SCAFE, DONALD WILLIAM, b Highgate, Ont, Nov 30, 37; m 62. GEOLOGY, OCEANOGRAPHY. *Educ:* Univ Western Ont, BSc, 60; Univ Kans, MS, 63; Tex A&M Univ, PhD(oceanog), 68. *Prof Exp:* WITH RES OFF, ALTA RES COUN, 67- *Mem:* Clay Minerals Soc; Assoc Prof Engrs. *Res:* Bentonite; ceramic clays; volcanic ash; industrial minerals. *Mailing Add:* Alta Res Coun Box 8330 Sta F Edmonton AB T6H 5X2 Can

SCAGLIONE, PETER ROBERT, b Tampa, Fla, Oct 9, 25; m 56; c 4. PEDIATRICS & PEDIATRIC NEPHROLOGY. *Educ:* City Col New York, BS, 48; Columbia Univ, MD, 52. *Prof Exp:* Asst pediat, Columbia Univ, 54-57, instr, 57-61, assoc, 61-65; pediatrician-in-chief, Brooklyn-Cumberland Med Ctr, 65-74; prof clin pediat, Med Ctr, NU Univ, 74-82; DIR PEDIAT, ST VINCENT'S HOSP & MED CTR; PROF CLIN PEDIAT, NY MED COL, 82- *Concurrent Pos:* From clin asst prof to clin assoc prof pediat, State Univ NY Downstate Med Ctr, 65-74; mem coun kidney in cardiovasc dis, Am Heart Asn, 73- *Mem:* Am Soc Nephrology; Int Soc Nephrology; Am Soc Pediat Nephrology. *Res:* Pediatric renal and acid-base disorders. *Mailing Add:* St Vincent's Hosp & Med Ctr 130 W 12th St New York NY 10011

SCAIFE, CHARLES WALTER JOHN, b Williamsport, Pa, Jan 27, 38; m 64; c 2. INORGANIC CHEMISTRY, SCIENCE EDUCATION. *Educ:* Cornell Univ, BA, 59, PhD(inorg chem), 66. *Prof Exp:* NSF fel, York, Eng, 66-67; asst prof chem, Middlebury Col, 67-72; chmn & assoc prof, 72-78, PROF CHEM, UNION COL, NY, 82- *Concurrent Pos:* Vis prof, NMex State Univ, 78-79. *Mem:* Am Chem Soc. *Res:* New laboratory experiments in general and inorganic chemistry; crystal growth on NASA space shuttle. *Mailing Add:* Dept Chem Union Col Schenectady NY 12308

SCALA, ALFRED ANTHONY, b Brooklyn, NY, Apr 29, 36; m 57; c 3. PHOTOCHEMISTRY. *Educ:* Brooklyn Col, BS, 57, MA, 61; Polytech Inst Brooklyn, PhD(org chem), 65. *Prof Exp:* Res chemist, Pfister Chem Works, 57-61; res assoc, Nat Res Coun, Nat Bur Stand, 64-66; from asst prof to assoc prof, 66-75, head dept chem, 77-80, PROF ORG CHEM, WORCESTER POLYTECH INST, 75- *Mem:* Am Chem Soc. *Res:* Radiation chemistry; zeolite catalysis. *Mailing Add:* Dept Chem Worcester Polytech Inst Worcester MA 01609

SCALA, E(RALDUS), b Trieste, Italy, June 22, 22; nat US; m 48; c 4. METALLURGY, COMPOSITE MATERIALS ENGINEERING. *Educ:* City Col New York, BS, 43; Columbia Univ, MS, 48; Yale Univ, DEng(metall), 53. *Prof Exp:* Metall chemist, Ledoux & Co, 43; res assoc, AEC Proj, Columbia Univ, 47-48; res metallurgist, Chase Brass & Copper Co, 48-52, training dir, 52-53, head phys metall sect, 53-55; mgr, Mat Dept, Res & Adv Develop Div, Avco Corp, 55-61; prof metall & mat sci, Cornell Univ, 61-68; dir, US Army Mat & Mech Res Ctr, Mass, 68-70; prof mat sci & eng, Cornell Univ, 70-74; dir, Cortland Line Co, 74-81, PRES, CORTLAND CABLE CO, 81- *Concurrent Pos:* Guggenheim fel, Delft Technol Univ, 67-68; consult, Aerospace Corp, Man Labs & Air Force Ballistic Missile Reentry Systs, Battelle Mem Inst; mem adv bd mat, NASA, mem res adv comt mat; mem mat adv bd, Nat Acad Sci; consult eng, Scala & Co, 78- *Mem:* Am Inst Aeronaut & Astronaut; Am Soc Metals; Am Inst Mining, Metall & Petrol Engrs; Marine Technol Soc. *Res:* Physical metallurgy; high strength fibers and composites; cables and ropes; marine cable technology. *Mailing Add:* Cortland Cable Co PO Box 330 Cortland NY 13045

SCALA, JAMES, b Ramsey, NJ, Sept 16, 34; m 57; c 4. BIOCHEMISTRY & NUTRITION. *Educ:* Columbia Col, AB, 60; Cornell Univ, PhD(biochem), 64. *Prof Exp:* Biochemist, Miami Valley Labs, Procter & Gamble Co, 64-66; sr res scientist, Owens-Ill Glass Inc, 66-68, chief life sci, Tech Ctr, 68-69, dir fundamental res, 69-71; dir appl nutrit, Tech Res, Thomas J Lipton, Inc, 71-75; dir nutrit & health sci, Gen Foods Corp, Tarrytown, NY, 75-78; VPRES RES & DEVELOP, SHAKLEE CORP, 78- *Concurrent Pos:* Lectr & adj prof, Med Col, Ohio Univ; lect prof, Georgetown Univ Sch Med, 73-85; instr, Univ Calif, Berkeley. *Mem:* Am Chem Soc; Am Soc Cell Biol; Inst Food Technol; Am Inst Nutrit. *Res:* Selenium metabolism in microorganisms and mammalian systems; percutaneous absorption; control mechanisms in protein biosynthesis; intermediary metabolism, especially the interrelationship of carbohydrate and lipid metabolism; relationship between metabolism and exercise physiology with emphasis on stress; interrelationship between nutrition and dental health; nutritional supplementation. *Mailing Add:* 44 Los Arabis Circle Lafayette CA 94549

SCALA, JOHN RICHARD, b Rochester, NY, May 21, 58; m 86; c 1. NUMERICAL MODELING, FIELD METEOROLOGY. *Educ:* Univ Rochester, BS, 80; Univ Va, MS, 84, PhD(environ sci), 90. *Prof Exp:* POSTDOCTORAL FEL, GODDARD SPACE FLIGHT CTR, NASA, 90- *Mem:* Am Meteorol Soc; Am Geophys Union; Sigma Xi. *Res:* Field observation and study of convective clouds; convective-scale dynamics and transport structure; impact of convection on vertical distribution of trace gases. *Mailing Add:* Severe Storms Br Code 912 NASA Goddard Space Flight Ctr Greenbelt MD 20771

SCALA, LUCIANO CARLO, b Rome, Italy, July 24, 23; nat US; m 51; c 7. ORGANIC CHEMISTRY. *Educ:* Univ Bologna, DSc(chem), 48. *Prof Exp:* Instr chem, Univ Bologna, 47-49; res assoc, Mass Inst Technol, 50-51; res chemist, Conn Hard Rubber Co, 51-53; mgr res labs, 53-87, CONSULT SCIENTIST RES LABS, WESTINGHOUSE ELEC CORP, 87- *Concurrent Pos:* Fulbright fel, 50. *Mem:* Sr mem Am Chem Soc. *Res:* Organic synthesis; high temperature insulation; thermal degradation mechanisms; organic monolayers; photoresists; electrophoretic processes; liquid crystals; radiation chemistry; reverse osmotic membranes. *Mailing Add:* 3359 Fawnway Dr Murrysville PA 15668-1422

SCALA, ROBERT ANDREW, b Utica, NY, Nov 14, 31; m 57; c 4. ENVIRONMENTAL HEALTH. *Educ:* Hamilton Col, AB, 53; Univ Rochester, MS, 56, PhD(physiol), 58; Am Bd Toxicol, dipl, 80. *Prof Exp:* Res asst toxicol, Nat Acad Sci-Nat Res Coun, 58-60; sect supvr, Toxicol-Pharmacol Dept, Hazleton Labs, Inc, 60-61, asst chief, 62, dir lab opers, 62-65; toxicologist, Med Res Div, Esso Res & Eng Co, 65-74, dir toxicol, 74-81, SR SCI ADV, EXXON BIOMED SCI, INC, 81- *Concurrent Pos:* Adj prof environ med, NY Univ, 70, environ & community med, Robert Wood Johnson Med Sch, 82-; affil assoc prof pharmacol, Med Col Va, 78-; adj prof toxicol, Rutgers Univ, 82-; pres, Am Bd Toxicol, 84-85. *Mem:* Am Chem Soc; Soc Toxicol (pres, 76-77); Europ Soc Toxicol; fel Acad Toxicol Sci. *Res:* Relation of chemical structure to biological function or activity; toxicology of chemicals used in foods, drugs, pesticides, cosmetics, industry and military chemicals. *Mailing Add:* Exxon Biomed Sci Inc CN 2350 East Millstone NJ 08875-2350

SCALA, SINCLAIRE M(AXIMILIAN), b Charleston, SC, June 27, 29; m 51; c 3. AERONAUTICAL SCIENCE, SPACE SCIENCE. *Educ:* City Col New York, BME, 50; Univ Del, MME, 53; Princeton Univ, MA, 55, PhD, 57; Univ Pa, MBA, 78. *Prof Exp:* Design engr, Aviation Gas Turbine Div, Westinghouse Elec Corp, 51-53; res engr, Missile & Space Div, Space Sci Lab, Gen Elec Co, 56-58, consult res engr, 58-59, mgr high altitude aerodyn, 59-64, mgr theoret fluid physics, 64-68, mgr fluid physics projs, 68-69, mgr, Environ Sci Lab, 69-73, chief scientist, 73-74, sr consult scientist, 74-80, mgr advan weapons concepts, 80-82; vpres advan eng, Fairchild Repub Co, 82-83, vpres advan technol, 83-84, vpres res, 84-85, dir res & advan prod develop, 85-87; prog mgr, Aircraft Syst Div, 87-91, DIR, BUS PLANNING, GRUMMAN CORP, 91- *Concurrent Pos:* Consult, Princeton Univ, 56-58; mem, res & technol adv subcomt fluid mech, NASA, 65-70. *Mem:* Am Inst Aeronaut & Astronaut; NY Acad Sci; Sigma Xi; Asn Unmanned Vehicle Systs; World Future Soc. *Res:* Gas dynamics; aerodynamics; hypersonics; viscous flow; applied mathematics; multicomponent fluid processes; ablation; nonequilibrium flow; shock waves; transport phenomena; radiative energy transfer; systems analysis; solar energy; biomedical science; computer science; heat transfer; advanced aircraft, rocket motors, plasma propulsion, jet propulsion. *Mailing Add:* 16 Carman Lane St James NY 11780-1306

SCALAPINO, DOUGLAS J, b San Francisco, Calif, Dec 10, 33; m 55; c 5. PHYSICS. *Educ:* Yale Univ, BS, 55; Stanford Univ, PhD(physics), 61. *Prof Exp:* Res assoc physics, Wash Univ, 61-62; res assoc, Univ Pa, 62-64, from asst prof to prof, 64-69; PROF PHYSICS, UNIV CALIF, SANTA BARBARA, 69- *Concurrent Pos:* Sloan fel, 64-66, Guggenheim fel, 76-77; Consult, E I du Pont de Nemours & Co, Inc, 64-87, IBM, 88- *Mem:* Nat Acad Sci; fel Am Phys Soc. *Res:* Many-body problems; superconductivity; magnetism; surfaces; statistical mechanics; phase transitions. *Mailing Add:* Dept Physics Univ Calif Santa Barbara CA 93106

SCALES, JOHN ALAN, b Louisville, Ky, June 24, 57; m 79; c 1. COMPUTATIONAL PHYSICS, INVERSE THEORY. *Educ:* Univ Del, BS, 79; Univ Colo, PhD(physics), 84. *Prof Exp:* Consult, 85-86, res scientist, 86-89, SR RES SCIENTIST, AMOCO PROD CO, TULSA RES CTR, 89- *Mem:* Fel Royal Astron Soc; Am Phys Soc; Soc Indust & Appl Math; Am Geophys Union; Sigma Xi. *Res:* Geophysical inverse theory; statistical physics; global optimization methods; large scale linear algebra, especially sparse linear systems and eigenvalue problems; wave propagation in complex media. *Mailing Add:* Amoco Prod Co Res Ctr PO Box 3385 Tulsa OK 74102

SCALES, STANLEY R, b Summitville, NY, Apr 24, 23. HEAT TREATMENT OF METALS. *Educ:* Univ Mo, Rolla, BS, 50. *Prof Exp:* Chief metallurgist, Hughes Tool Co, 50-85; RETIRED. *Concurrent Pos:* Chmn, Houston Sect, Am Welding Soc, 70-71; dir, Houston Mat Conf, 87. *Mem:* Fel Am Soc Metals Int. *Res:* Heat treatment of alloy steels; hard surfacing as pertains to oil well drilling tools; author of numerous technical publications; awarded 11 US patents. *Mailing Add:* 410 Axilda Houston TX 77017

SCALES, WILLIAM WEBB, b Shreveport, La, Aug 20, 32; m 61; c 2. ADVANCED SIGNAL & DATA PROCESSING SYSTEMS, UNDERWATER ACOUSTICS. *Educ:* Columbia Univ, AB, 54; Rice Univ, MA, 56, PhD(physics), 58. *Prof Exp:* MEM TECH STAFF, AT&T BELL LABS, 58- *Mem:* AAAS; Acoust Soc Am; Sigma Xi. *Res:* Low temperature specific heats of crystals; underwater transmission of sound; underwater acoustic systems; avionics signal and data processing systems. *Mailing Add:* AT&T Bell Labs Whippany Rd Whippany NJ 07981

SCALET, CHARLES GEORGE, b Chicago, Ill, Sept 9, 42; m 66; c 3. FISH BIOLOGY, ICHTHYOLOGY. *Educ:* Southern Ill Univ, Carbondale, BA, 64, MA, 67; Univ Okla, PhD(zool), 71. *Prof Exp:* Asst prof zool, Cent State Univ, 71-72; instr, Iowa State Univ, 72-73; from asst prof to assoc prof, 73-82, PROF WILDLIFE & FISHERIES SCI, SDAK STATE UNIV, 82-, DEPT HEAD, 76- *Mem:* Am Fisheries Soc; Am Soc Ichthyologists & Herpetologists; Wildlife Soc. *Res:* Management of South Dakota waters for fish production; culture of fishes; description of the life histories and ranges of South Dakota fishes. *Mailing Add:* Dept Wildlife/Fisheries SDak State Univ Box 2206 Brookings SD 57007

SCALETTAR, RICHARD, b New York, NY, Dec 9, 21; m 58; c 2. THEORETICAL PHYSICS. *Educ:* City Col NY, BS, 41; Univ Wis, MA, 43; Cornell Univ, PhD(physics), 59. *Prof Exp:* Physicist, Metall Lab, Univ Chicago, 43-44 & Clinton Labs, Oak Ridge, Tenn, 44-46; asst prof physics, Univ Rochester, 49-51; physicist, Curtiss-Wright Corp, 52-53; asst prof physics, Univ Southern Calif, 53-59; sr theoret physicist, John Jay Hopkins Lab Pure & Appl Sci, Gen Atomic, 60-68; PROF PHYSICS & CHMN DEPT PHYSICS-ASTRON, CALIF STATE UNIV, LONG BEACH, 68- *Concurrent Pos:* Consult, Aerojet Gen Corp, 56; lectr, Edwards AFB, 57; consult, Atomics Int, 58-59. *Mem:* Am Phys Soc. *Res:* Electron and gamma ray transport; interaction of electromagnetic radiation with plasmas; statistical mechanics and transport theory; reactor kinetics; particle and field theory. *Mailing Add:* Lake Forest 24782 Winterwood Dr El Toro CA 92630

SCALETTI, JOSEPH VICTOR, b New London, Conn, July 22, 26; m 51; c 2. BACTERIOLOGY. *Educ:* Univ Conn, BA, 50, MS, 53; Cornell Univ, PhD(bact), 57. *Prof Exp:* Instr bact, Univ Conn, 52-53; asst, Cornell Univ, 53-56; bacteriologist, Am Cyanamid Co, 56-57; res assoc pub health, Univ Minn, St Paul, 57-58, asst prof animal husb, 58-64; assoc prof 64-70, PROF MICROBIOL, UNIV NMEX, 70-, CHMN DEPT, 76-, VPRES, 85- *Mem:* Am Soc Microbiol; Am Pub Health Asn; Am Acad Microbiol. *Res:* Nucleic acid metabolism as related to bacteriophage-bacterial systems; growth characteristics of psychorophilic microorganism. *Mailing Add:* Dept Microbiol Univ NMex Albuquerque NM 87131

SCALFAROTTO, ROBERT EMIL, b Alexandria, Egypt, June 4, 20; US citizen; m 46; c 1. PIGMENTS CHEMISTRY, PAPER CHEMISTRY. *Educ:* Univ Genoa, DSc(indust chem), 48. *Prof Exp:* Chemist, Lechner & Muratori Co, 46-50, tech dir pigments, 50-57; tech mgr gen chem, Mercantile Develop, Inc, 57-58; appl res chemist, Pigments Div, Am Cyanamid Co, 58-63; asst mgr pigments div, Ciba Chem & Dye Co, NJ, 63-67, promotion coordr, 67-70, mgr tech develop, Pigments Dept, Ciba-Geigy Corp, NY, 70-76; mgr int indust chems res & develop, Am Cyanamid Co, 76-78, mgr new prod testing, 78-80, mgr prod develop, 80-84; CONSULT, 84- *Concurrent Pos:* Consult, Shell Ital Chem Serv, 50-54. *Mem:* Am Chem Soc; NY Soc Coatings-Technol; Tech Asn Pulp & Paper Indust. *Res:* Paper chemicals; pigments chemistry. *Mailing Add:* Four Windmill Lane Ocean View DE 19970

SCALIA, FRANK, b Brooklyn, NY, Mar 18, 39; m 60; c 2. NEUROBIOLOGY. *Educ:* NY Univ, BA, 59; State Univ NY, PhD(anat), 64. *Prof Exp:* From instr to assoc prof, 63-77, PROF ANAT, STATE UNIV NY DOWNSTATE MED CTR, 77- *Mem:* AAAS; Am Asn Anat; Soc Neurosci. *Res:* Experimental neuroanatomy; neuroembryology; vision; olfaction. *Mailing Add:* Dept Anat/Cell Biol State Univ NY Downstate Med Ctr 450 Clarkson Ave Brooklyn NY 11203

SCALLAN, ANTHONY MICHAEL, b Blackpool, Eng, Apr 12, 36; m 61; c 5. PAPER CHEMISTRY. *Educ:* Univ Liverpool, BSc, 57, PhD(polymer chem), 63. *Prof Exp:* Chemist, Roan Antelope Copper Mines, Zambia, 57-59; PRIN SCIENTIST & HEAD FIBER CHEM SECT, PULP & PAPER RES INST CAN, 63- *Mem:* Can Pulp & Paper Asn; Tech Asn Pulp & Paper Indust. *Res:* Physical chemistry of pulping and papermaking with emphasis on aspects related to the porous structure of wood and paper. *Mailing Add:* Pulp & Paper Res Inst Can 570 St Johns Rd Pointe Claire PQ H9R 3J9 Can

SCALLEN, TERENCE, b Minneapolis, Minn, Jan 16, 35; m 57; c 4. BIOCHEMISTRY. *Educ:* Col St Thomas, BS, 57; Univ Minn, Minneapolis, MD, 61, PhD(biochem, org chem), 65. *Prof Exp:* From asst prof to assoc prof, 65-76, PROF BIOCHEM, SCH MED, UNIV NMEX, 76- *Concurrent Pos:* Am Col Cardiol young investr award, 69. *Mem:* Am Chem Soc; Am Soc Biol Chemists; Am Soc Cell Biol. *Res:* Mechanisms of sterol and lipid biosynthesis; application of physical techniques to problems of steroid structure; sterol carrier protein. *Mailing Add:* Dept Biochem Univ NMex Sch Med Albuquerque NM 87131

SCALLET, BARRETT LERNER, b St Louis, Mo, May 13, 16; m 43; c 3. FOOD CHEMISTRY, GRAIN PRODUCTS CHEMISTRY. *Educ:* Wash Univ, BS, 37, MS, 43, PhD(org chem), 46. *Prof Exp:* Control chemist, Anheuser-Busch, Inc, 37-38, res chemist, 38-46, res group leader, 46-47, res proj leader, 47-48, sect dir, 48-52, dir corn prod sect, Cent Res Dept, 52-55, assoc dir, 55-75, dir corn prod res, 75-80; PRES, CENT RES, INC, 80- *Mem:* AAAS; Am Chem Soc; Am Asn Cereal Chem; Tech Asn Pulp & Paper Indust; Am Ceramic Soc. *Res:* Physical study of proteins and starches; electrophoresis; ultracentrifugation; chemistry of zein; beer proteins; corn syrups and starches; industrial utilization of corn products; food products; consumer products; new corn genetic varieties; glass fracture. *Mailing Add:* Cent Res Inc 18 Carrswold Clayton MO 63105-2914

SCALORA, FRANK SALVATORE, b New York, NY, June 16, 27. MATHEMATICS. *Educ:* Harvard Univ, AB, 49; Univ Ill, AM, 51, PhD(math), 58. *Prof Exp:* Asst math, Univ Ill, 49-54; mathematician, Repub Aviation Corp, 54; mathematician, IBM Corp, 56-63, info planning mgr, World Trade, 63-67, data mgt mgr, 67-71, sr analyst, 71-74, sr analyst, Data Processing div, 74-79, corp staff, 79-82, sr analyst, Acad Info Systs, 82-87; PROF, FORDHAM UNIV, 87- *Concurrent Pos:* Adj prof, Polytech Inst Brooklyn, 60-61; adj asst prof, Courant Inst Math Sci, NY Univ, 61-63. *Mem:* Am Math Soc; Inst Math Statist. *Res:* Probability theory and stochastic processes; measure theory; statistics; operations research; bank asset and liability management. *Mailing Add:* 225 E 57th St Apt 10-S New York NY 10022

SCALZI, FRANCIS VINCENT, b Reading, Pa, Dec 4, 33; m 63; c 2. ORGANIC CHEMISTRY. *Educ:* Gettysburg Col, BA, 55; Univ Del, MS, 60, PhD(chem), 63. *Prof Exp:* Res chemist, Firestone Tire & Rubber Co, Pa, 56-58; lectr chem, Rutgers Univ, 63-64; PROF CHEM, HIRAM COL, 64- *Concurrent Pos:* NSF res participation col teachers grant, 64-66; vis assoc prof, Univ Wis-Milwaukee, 70-71; NSF spec proj grant, 70-72; prof, Univ Akron, 76-84, Case Western Reserve Univ, 85, 90 & Kent State Univ, 87. *Mem:* Am Chem Soc. *Res:* N-halamine chemistry; aromatic substitution; thermal degradation of organic compounds. *Mailing Add:* Dept Chem Hiram Col Hiram OH 44234

SCALZI, JOHN BAPTIST, b Milford, Mass, Nov 13, 15; m 40; c 2. STRUCTURAL ENGINEERING, CIVIL ENGINEERING. *Educ:* Worcester Polytech Inst, BS, 38; Mass Inst Technol, SM, 40, ScD, 51. *Prof Exp:* Field engr, Metcalf & Eddy, Mass, 39; struct engr, Curtiss-Wright Corp, NY, 40-45; engr, Eng Div, Nat Aniline Div, 45-46; prof struct eng, Case Inst Technol, 46-60; engr, Mkt Develop Div, US Steel Corp, 60-62, dir mkt tech serv, 62-71; engr res & technol, HUD, 71-73; PROG DIR, NSF, WASHINGTON, DC, 73- *Concurrent Pos:* Lectr, Exten Eng Sci & Mgt War Training Prog, Cornell Univ, 42-45 & Univ Buffalo, 45; prof lectr, Western Reserve Univ, 46-60; lectr, Carnegie-Mellon Univ, 65-71 & George Washington Univ, 72-83; struct consult, Cleveland eng firms. *Mem:* Am Soc Civil Engrs; Sigma Xi. *Res:* Earthquake engineering; general civil engineering; infrastructure. *Mailing Add:* 1805 Crystal Dr No 918 Arlington VA 22202-4407

SCAMEHORN, RICHARD GUY, b Elkhart, Ind, June 20, 42; m 64; c 2. ORGANIC REACTION MECHANISMS, CHEMICAL EDUCATION. *Educ:* Hanover Col, BA, 64; Northwestern Univ, PhD(org chem), 68. *Prof Exp:* From asst prof to assoc prof, 68-82, PROF CHEM, RIPON COL, 82-, CHMN DEPT, 85- *Concurrent Pos:* Res assoc, Univ Calif, Santa Cruz, 75-76; vis prof, Univ Wis, 84. *Mem:* Am Chem Soc; Sigma Xi. *Res:* Organic reaction mechanisms; carbanion rearrangement reactions; kinetics; radical chain substitution reactions. *Mailing Add:* 627 Sunset Circle Ripon WI 54971

SCANDALIOS, JOHN GEORGE, b Nysiros, Greece, Nov 1, 34; US citizen; m 61; c 3. GENETICS. *Educ:* Univ Va, BA, 57; Adelphi Univ, MS, 60; Univ Hawaii, PhD(genetics), 65. *Hon Degrees:* DSc, Aristotelian Univ Thessaloniki, Greece, 86. *Prof Exp:* Instr biol, Hunter Col, 59-60; res assoc bact genetics, Cold Spring Harbor Lab, 60-63; NIH res fel molecular genetics, Univ Hawaii, 65; from asst prof to assoc prof res genetics, AEC Plant Res Lab, Mich State Univ, 65-72; prof genetics & head dept biol, Univ SC, 72-75; prof genetics & head dept, 75-85, DISTINGUISHED UNIV PROF, NC STATE UNIV, 85- *Concurrent Pos:* Instr radiation biol & biol & genetics, Adelphi Univ, 60-62; vis prof, Univ Calif, Davis, 68 & Org Am States, Arg, 72; mem, NIH-Recombinant DNA Adv Comt, 80-85; ed, Develop Genetics & Advances Genetics. *Mem:* AAAS; Genetics Soc Am; Soc Develop Biol (treas, 79-81); Am Genetic Asn (pres, 81-82); Int Soc Differentiation; Am Soc Biochem & Molecular Biol; Am Soc Plant Physiol. *Res:* Developmental-molecular genetics of eukaryotes; genetics, structure and function of isozymes; genetic regulation; biotechnology; plant molecular biology. *Mailing Add:* Dept Genetics NC State Univ Raleigh NC 27695-7614

SCANDRETT, JOHN HARVEY, b Liberal, Kans, July 7, 33; m 54; c 3. PHYSICS. *Educ:* La State Univ, BS, 54; Univ Wis, MS, 56, PhD(physics), 63. *Prof Exp:* Lectr physics, Mich State Univ, 60-62; lectr, Ind Univ, 62-63, asst prof, 63-66; assoc prof, 66-81, PROF PHYSICS, WASH UNIV, 81- *Mem:* Am Phys Soc. *Res:* Medical physics; image processing. *Mailing Add:* 6829 Waterman St St Louis MO 63130

SCANDURA, JOSEPH M, b Bay Shore, NY, Apr 29, 32; m 60; c 4. INTELLIGENT SYSTEMS. *Educ:* Univ Mich, BA, 53, MA, 55; Syracuse Univ, PhD(math educ), 62. *Prof Exp:* Math teacher, NY, 53-55; asst prof math, State Univ NY Col Oswego, 55-56; instr, Syracuse Univ, 56-63, asst prof math & educ, State Univ NY Buffalo, 63-64; res prof math educ, Fla State Univ, 64-66; PROF & DIR, INTERDISCIPLINARY STUDIES IN STRUCT LEARNING & INSTRNL SCI, UNIV PA, 66- *Concurrent Pos:* Consult, Merge Res Inst; Pres, Intelligent Micro Systs. *Mem:* AAAS; fel Am Psychol Asn; Am Educ Res Asn. *Res:* Theory and research in structural learning, cognitive psychology, instructional systems design and intelligent computer-based instruction. *Mailing Add:* Education, A-55, C1 Univ Pa Philadelphia PA 19104

SCANES, COLIN G, b London, Eng, July 11, 47; m 76; c 3. NUTRITION. *Educ:* Univ Hull, UK, BS, 69; Univ Wales, PhD(zool), 72. *Hon Degrees:* DSc, Univ Hull, UK, 85. *Prof Exp:* Lectr animal physiol & nutrit, Univ Leeds, UK, 72-78; assoc prof physiol, 78-81, CHMN DEPT ANIMAL SCI, RUTGERS UNIV, 81-, PROF ANIMAL SCI, 82- *Concurrent Pos:* Res Award, Rutgers Univ,86. *Mem:* Am Physiol Soc; Poultry Sci Asn; Am Soc Zoologists; Endocrine Soc; Am Soc Animal Sci; Soc Exp Biol. *Res:* Hormonal control of growth metabolism and reproduction, particularly in the domestic fowl and other farm animals. *Mailing Add:* Dept Animal Sci Cook Col Box 231 Rutgers Univ New Brunswick NJ 08903

SCANIO, CHARLES JOHN VINCENT, b Ann Arbor, Mich, June 23, 40; m 65; c 2. ORGANIC CHEMISTRY. *Educ:* Univ Mich, BS, 62; Northwestern Univ, PhD(org chem), 66. *Prof Exp:* From instr to asst prof chem, Iowa State Univ, 66-72; staff chemist, Pfizer, Inc, 72-77; head process res, UpJohn, Inc, 77-84; corp dir, res & develop, Chem Design Corp, 84-86; EXEC VPRES, CHEMSULTANTS, INC, 86- *Mem:* Am Chem Soc; Brit Chem Soc. *Res:* Processes to manufacture fine chemicals and pharmaceutical intermediates. *Mailing Add:* 300 High St Winchendon MA 01475-0420

SCANLAN, J(ACK) A(DDISON), JR, b Chicago, Ill, Sept 25, 17; m 40; c 3. MECHANICAL ENGINEERING. *Educ:* Univ Tex, BS, 40, MS, 52; Northwestern Univ, PhD(eng), 57. *Prof Exp:* Asst proj engr, Wright Aeronaut Corp, 40-47; asst prof mech eng, Univ Tex, 47-52; vis lectr, Northwestern Univ, 52-54; assoc prof, Univ Tex, 54-67; PROF MECH ENG, MONT STATE UNIV, 67- *Concurrent Pos:* Consult, Union Carbide Nuclear Co Div, Union Carbide Corp; vis prof, Mont State Univ, 66-67. *Mem:* AAAS; Am Inst Aeronaut & Astronaut; Am Soc Eng Educ; Soc Automotive Engrs; Am Soc Mech Engrs. *Res:* Heat transfer; nuclear power; thermodynamics; alternative energy sources. *Mailing Add:* Dept Mech Eng Mont State Univ Roberts Hall Bozeman MT 59717

SCANLAN, MARY ELLEN, b New York, NY, Sept 20, 42. SCIENCE COMMUNICATIONS. *Educ:* Chestnut Hill Col, BS, 64; Univ RI, PhD(org chem), 70. *Prof Exp:* Sr assoc ed, dept org chem, Chem Abstr Serv, 70-82; asst mgr, jour dept, 82-90, MGR ED OFF, AM CHEM SOC, 91- *Mem:* Am Chem Soc. *Mailing Add:* Am Chem Soc PO Box 3330 Columbus OH 43210

SCANLAN, RICHARD ANTHONY, b Syracuse, NY, Dec 13, 37; m 59; c 5. FOOD SCIENCE. *Educ:* Cornell Univ, BS, 60, MS, 62; Ore State Univ, PhD, 67. *Prof Exp:* Res & develop coordr, US Army Natick Labs, Mass, 62-64; asst food sci, Ore State Univ, 64-67, from asst prof to assoc prof, 67-78, dept head, 85-89, PROF FOOD SCI, ORE STATE UNIV, 78- , DEAN RES, ADMIN BLDG, 89- *Concurrent Pos:* Prin investr, res grants. *Honors & Awards:* Sigma Xi Res Award, Ore State Univ, 83. *Mem:* AAAS; Am Chem Soc; Sigma Xi; Inst Food Technologists. *Res:* Food toxicology; chemistry of formation and inhibition of N-nitrosamines; development of analytical methodology for nitrosamines; biological effects of N-nitrosamines; food chemistry; flavor chemistry. *Mailing Add:* Admin Bldg Ore State Univ Corvallis OR 97331

SCANLAN, ROBERT HARRIS, b Chicago, Ill, Aug 15, 14; m 39; c 4. STRUCTURAL DYNAMICS. *Educ:* Univ Chicago, SB, 36, SM, 39; Mass Inst Technol, PhD(math), 43; Univ Paris, Dr es Sci(mech), 56. *Prof Exp:* Assoc prof aeronaut, Rensselaer Polytech Inst, 46-51; Nat Res Coun-Nat Adv Comt Aeronaut fel, France, 51-52; res fel aeronaut, Nat Sci Res Ctr, France, 52-55; res engr, Nat Off Aeronaut Studies & Res, France, 55-57; Schlumberger Corp, Tex, 58-60; prof mech, Case Inst Technol, 60-66; prof civil eng, Princeton Univ, 66-85; PROF CIVIL ENG, JOHNS HOPKINS UNIV, 85- *Concurrent Pos:* Sloan vis prof, Princeton Univ, 66-67; T R Higgins Nat Lectr, 76; vis prof, Rice Univ, Univ Calif, Berkeley, 78; consult, Pvt US & Japanese firms. *Honors & Awards:* State-of-the Art in Civil Eng Award, Am Soc Civil Engrs, 69, Wellington Prize, 86, Newmark Medal, 86. *Mem:* Nat Acad Engr; Am Inst Aeronaut & Astronaut; Am Soc Mech Engrs; hon mem Am Soc Civil Engrs; fel Am Acad Mech. *Res:* Acoustics; aeroelasticity; vibrations; applied mechanics; wind engineering; bridge aerodynamics. *Mailing Add:* Dept Civil Eng Johns Hopkins Univ Baltimore MD 21218

SCANLEY, CLYDE STEPHEN, b Milwaukee, Wis, June 16, 21; div; c 3. WATER SOLUBLE POLYMERS. *Educ:* Univ Wis, BS, 43, PhD(org chem), 49. *Prof Exp:* Chemist, Standard Oil Co, Ind, 49-53; group leader, Am Cyanamid Co, 53-73; mgr res, Drew Chem Corp, 74-83; PROPRIETOR, CHEM RES & DEVELOP CONSULT, JEITO RES, 83- *Mem:* Am Chem Soc; Tech Asn Pulp & Paper Indust. *Res:* Vinyl polymers, especially polyelectrolyte flocculants, polycrylamides, surfactants, organic synthesis and process development. *Mailing Add:* 330 Speedwell Ave Morristown NJ 07960

SCANLON, CHARLES HARRIS, b Austin, Tex, Oct 13, 37; m 65; c 5. MATHEMATICAL ANALYSIS. *Educ:* Univ Tex, BA, 61, MA, 63, PhD(math), 67. *Prof Exp:* Asst prof math, Univ Okla, 67-70; ASSOC PROF MATH, ARK STATE UNIV, 70- *Mem:* Am Math Soc; Math Asn Am. *Res:* Analysis in metric spaces; generalized Riemann and Stieltjes integration. *Mailing Add:* Dept Math/Computer Sci Ark State Univ PO Box 1990 State University AR 72467

SCANLON, JACK M, b Binghamton, NY, Jan 3, 42; m 63; c 4. COMPUTER SCIENCE, ELECTRICAL ENGINEERING. *Educ:* Univ Toronto, BASc, 64; Cornell Univ, MS, 65. *Prof Exp:* Mem staff elec eng, 65-68; supvr comput sci, Bell Labs, 68-74, dept head electronic switching syst design, 74-77, dir, Software & Syst Design Lab, 77-79, exec dir, Processor & Common Software Systs Div, 79-; vpres, prod develop, AT&T Info Syst; VPRES, PROCESSOR & SOFTWARE SYSTS, WESTERN ELECTRONIC CO; VPRES & GEN MGR, INT CELLULAR INFRASTRUCT, MOTOROLA, INC, ARLINGTON HEIGHTS, 90- *Concurrent Pos:* NSF fel, 64-65; comput sci & technol bd, Nat Res Coun-Nat Acad Sci. *Mem:* Inst Elec & Electronics Engrs. *Res:* Computer science; communications science; physics, mathematics. *Mailing Add:* 30 Ryderwood Rd North Barrington IL 60010

SCANLON, JOHN EARL, b New York, NY, Nov 29, 25; m 47; c 2. MEDICAL ENTOMOLOGY. *Educ:* Fordham Univ, BS, 50; Cornell Univ, MS, 55; Univ Md, PhD, 60. *Prof Exp:* Med entomologist, Far East Med Res Unit, US Army, Tokyo, 50-53, Med Field Serv Sch, San Antonio, 55-56, Walter Reed Army Inst, 56-58, SEATO, Bangkok, 60-64 & Walter Reed Army Inst, 64-69; prof med zool, 69-75, ASSOC DEAN, SCH PUB HEALTH, UNIV TEX, HOUSTON, 75- *Mem:* AAAS; Am Soc Trop Med & Hyg (secy-tres, 80-). *Res:* Epidemiology of malaria and arbovirus; taxonomy and ecology of mosquitoes. *Mailing Add:* 24810 Broken Trail San Antonio TX 78255

SCANLON, PATRICK FRANCIS, b Athlone, Ireland, Sept 16, 41; m 67; c 4. REPRODUCTIVE PHYSIOLOGY, WILDLIFE RESEARCH. *Educ:* Nat Univ Ireland, BAgrSci, 65, MAgrSci, 66, PhD(animal physiol), 70. *Prof Exp:* Res demonstr animal physiol, Fac Agr, Univ Col, Dublin, 65-66, res scholar, 66-69; res assoc appl physiol, Univ Guelph, 69-71; from asst prof to assoc prof wildlife physiol, 71-78, PROF, DEPT FISHERIES & WILDLIFE SCI, VA POLYTECH INST & STATE UNIV, 78- *Concurrent Pos:* Vis prof, Telemark, Norway. *Mem:* Am Soc Animal Sci; Wildlife Soc; Am Soc Mammal; Wildlife Dis Asn; Soc Environ Toxicol & Chem. *Res:* Reproductive physiology of wild animals; influences of environmental contaminants on wild animals; vertebrate pest control; control of reproduction in wild and domestic animals. *Mailing Add:* Dept Fisheries & Wildlife Sci Va Polytech Inst & State Univ Blacksburg VA 24061-0321

SCANNELL, JAMES PARNELL, b Oak Park, Ill, Jan 16, 31; m 56; c 4. BIOCHEMISTRY. *Educ:* Univ Ill, BA, 51; Univ Calif, Berkeley, PhD(biochem), 60. *Prof Exp:* Fel biochem, Univ Calif, San Francisco, 60; lectr chem, Southern Ill Univ, 60-61; sr chemist, Papst Res Biochem, Pabst Brewing Co, 61-63 & Burroughs Wellcome & Co, 64-65; sr chemist, 66-75, res fel, 75-81, SR RES FEL, HOFFMANN-LA ROCHE & CO, 81- *Mem:* Am Chem Soc. *Res:* Isolation and characterization of natural products; chemistry and metabolism of nucleosides, amino acids and antibiotics. *Mailing Add:* Four Canterbury Dr North Caldwell NJ 07006

SCANU, ANGELO M, b Bonnanaro, Italy, Dec 16, 24; m 58; c 2. MEDICINE, BIOCHEMISTRY. *Educ:* Univ Sassari, MD, 49. *Prof Exp:* From intern to resident, Med Sch, Univ Sassari, 49-52; asst prof internal med, Med Sch, Univ Naples, 53-55; res assoc, Cleveland Clin, 58, staff asst, 59-62; asst prof, 63-66, res assoc biochem, 65-66, assoc prof internal med & biochem, 66-70, PROF INTERNAL MED BIOCHEM & MOLECULAR BIOL, MED SCH, UNIV CHICAGO, 70-, DIR, LIPOPROTEIN STUDY UNIT, 73-, DIR, CTR MOLECULAR MED, 87- *Concurrent Pos:* Res fel, Med Sch, Univ Barcelona, 52 & Med Sch, Univ Lund, 53; Fulbright scholar & res fel, Res Div, Cleveland Clin, 55-57; Fulbright scholar, Univ Nice, 80. *Honors & Awards:* Res Career Develop Award, 65-75. *Mem:* Am Soc Biol Chemists; Am Chem Soc; Am Physiol Soc; Biophys Soc; Am Soc Clin Invest; Asn Am Physicians. *Res:* Structure and function of serum lipoproteins in normal and disease states; lipoprotein cell interactions; genetics of lipoprotein disorders. *Mailing Add:* 5841S Maryland Box 231 Chicago IL 60637

SCAPINO, ROBERT PETER, b Chicago, Ill, July 20, 36; m 58; c 3. ANATOMY, DENTISTRY. *Educ:* Univ Ill, BS, 59, DDS, 62, MS, 63, PhD(anat), 68. *Prof Exp:* From instr to assoc prof, 65-75, PROF ORAL ANAT, COL DENT, COL MED, UNIV ILL, 75- *Concurrent Pos:* Nat Inst Dent Res fel anat, 62-65. *Mem:* Fel AAAS; Sigma Xi; Am Dent Asn. *Res:* Biomechanics of feeding in carnivores; comparative and human anatomy; function and pathology of jaw joints. *Mailing Add:* 917 Foxworth Blvd Lombard IL 60148

SCARBOROUGH, CHARLES SPURGEON, b Goodman, Miss, May 20, 33; m 70. INVERTEBRATE ZOOLOGY, ACAROLOGY. *Educ:* Rust Col, BA, 55; Northwestern Univ, Evanston, MS, 58; Mich State Univ, PhD(zool), 69. *Prof Exp:* Instr bot & zool, Alcorn Agr & Mech Col, 57-59; asst zool, 59-63, from instr to assoc prof natural sci, 63-77, from asst dir resident instr to dir resident instr, 71-77, from asst dean to actg dean, 77-81, PROF NATURAL SCI, MICH STATE UNIV, 77-, DIR, LYMAN BRIGGS SCH, 81- *Mem:* AAAS. *Res:* Free-living mites associated with bracket fungi, their taxonomy and biology. *Mailing Add:* E-27 Holmes Hall Lyman Briggs Sch Mich State Univ East Lansing MI 48824-1107

SCARBOROUGH, ERNEST N, b Annapolis, Md, May 21, 22; m 44; c 4. AGRICULTURAL ENGINEERING, AUTOMOTIVE ENGINEERING. *Educ:* Iowa State Col, BS, 43, MS, 47. *Prof Exp:* Asst prof agr eng, NC State Col, 48-52; assoc prof, Tenn Polytech Inst, 52-53; sr test engr, Thompson Prod Co, 53-54, prod analyst, 54-55; from assoc prof to prof, 55-83, chmn dept, 68-81, EMER PROF AGR ENG, UNIV DEL, 83- *Honors & Awards:* Award, Christian R & Mary H Lindbach Found, 64. *Mem:* Am Soc Agr Engrs. *Res:* Farm machinery and power; crop processing; environmental control; soil and water conservation. *Mailing Add:* 1002 Lakeside Dr Newark DE 19711

SCARBOROUGH, GENE ALLEN, b Hugo, Colo, Oct 8, 40; m 66. BIOCHEMISTRY. *Educ:* Univ Ariz, BS, 63; Univ Calif, Los Angeles, PhD(biochem), 66. *Prof Exp:* From asst to assoc prof biochem, Sch Med, Univ Colo, 76-77; assoc prof pharmacol, 77-82, PROF PHARMACOL, UNIV NC, CHAPEL HILL, 82- *Concurrent Pos:* Whitney fel, Harvard Med Sch, 67-68. *Mem:* AAAS; Am Soc Biol Chemists; Am Soc Microbiol; Biophys Soc. *Res:* Phospholipid biosynthesis; structure and function of biomembranes. *Mailing Add:* Dept Pharmacol Univ NC Chapel Hill NC 27514

SCARDERA, MICHAEL, b Providence, RI, May 11, 35; m 62; c 4. INDUSTRIAL ORGANIC CHEMISTRY. *Educ:* Brown Univ, BS, 57; Univ Bridgeport, MBA, 63. *Prof Exp:* Res chemist fuels res, 57-63, sr res chemist, 64-75, SR RES ASSOC SURFACE ACTIVE AGENTS, OLIN CORP, 76- *Concurrent Pos:* Instr, Southern Conn State Col, 65-75. *Mem:* Sigma Xi; Am Oil Chem Soc. *Res:* Synthesis and application of surface active agents. *Mailing Add:* Olin Corp 350 Knotter Dr Cheshire CT 06410

SCARFE, COLIN DAVID, b Danbury, Eng, Dec 17, 40; Can citizen; m 67; c 2. ASTRONOMY. *Educ:* Univ BC, BSc, 60, MSc, 61; Cambridge Univ, PhD(astron), 65. *Prof Exp:* From asst prof to assoc prof, 65-81, PROF ASTRON, UNIV VICTORIA, BC, 81- *Concurrent Pos:* Mem comns, Int Astron Union, 26, 30, 42; sabbatical, Observs, Cambridge, Eng, 71-72 & Mt John Observ, NZ, 78-79, Dominion Astrophys Observ, Victoria, Can, 85-86; vis fel, Mt Stromlo Observ, Australia, 78-79; mem, Assoc Comt Astron, Nat Res Coun Can, 77-83, chmn, Optical Astron Subcomt, 80-83; chmn, organizing comt, Can Asn Physicists meeting, Can Astron Soc, 83. *Mem:* Am Astron Soc; Royal Astron Soc Can; fel Royal Astron Soc; Can Astron Soc; Astron Soc Pac. *Res:* Spectroscopy and photometry of binary and multiple stars and cepheids. *Mailing Add:* Dept Physics & Astron Univ Victoria Victoria BC V8W 3P6 Can

SCARFONE, LEONARD MICHAEL, b North Adams, Mass, Oct 5, 29; m 55; c 3. PHYSICS. *Educ:* Williams Col, BA, 53, MA, 55; Rensselaer Polytech Inst, PhD(physics), 60. *Prof Exp:* Instr physics, Rensselaer Polytech Inst, 60-61; fel, Fla State Univ, 61-62, asst prof, 62-63; from asst prof to assoc prof, 63-70, PROF PHYSICS, UNIV VT, 70- *Mem:* Am Phys Soc; Am Asn Physics Teachers. *Res:* Quantum field theory; quantum theory of scattering; mathematical physics; elementary particles; theoretical solid state. *Mailing Add:* Cook Phys Sci Univ of Vt Burlington VT 05401

SCARGLE, JEFFREY D, b Evanston, Ill, Nov 24, 41; div; c 2. ASTRONOMY. *Educ:* Pomona Col, BA, 63; Calif Inst Technol, PhD(astron), 68. *Prof Exp:* Fel astron, Univ Calif, Berkeley, 68; instr astron & jr astronr, Univ Calif, Santa Cruz, 68-69, asst prof astron & astrophys, 69-74; RES SCIENTIST, AMES RES CTR, NASA, 75- *Mem:* Am Astron Soc; Int Astron Union. *Res:* Plasma astrophysics; quasars; The Crab Nebula; radiative transfer; statistical analysis of random processes; infrared astronomy; time series analysis; planetary detection astrometry. *Mailing Add:* Theoret Studies MS-245-3 NASA Ames Res Ctr Moffett Field CA 94035

SCARINGE, RAYMOND PETER, b Albany, NY, July 31, 50. CHEMISTRY, X-RAY DIFFRACTION. *Educ:* State Univ NY, Plattsburgh, BA, 72; Univ NC, PhD(inorg chem), 76. *Prof Exp:* Fel chem, Northwestern Univ, 76-78; RES CHEMIST, EASTMAN KODAK CO, 78- *Mem:* Am Chem Soc; Am Crystallog Asn. *Res:* Structural, electrical and magnetic properties in the solid state. *Mailing Add:* 26 Moorland Rd Rochester NY 14612

SCARINGE, ROBERT P, US citizen. SOFTWARE SYSTEMS. *Educ:* Rensselaer Polytechnic Inst, BS, 74, MS, 75, PhD(mech eng), 78. *Prof Exp:* Res scientist, Gen Elec Res & Develop, 78-82; prof heat transfer, 82-86, RES PROF THERMODYN HEAT TRANSFER, FLA INST TECHNOL, 86-; PRES, MAINSTREAM ENG CORP, 86- *Concurrent Pos:* Adj prof thermodynamics heat transfer, Rensselaer Polytechnic Inst, 78-82; fel, Dupont, 76, Ford, 77. *Mem:* Am Soc Mech Engrs. *Res:* Thermal control; chemical research; software development; hardware development; working fluid research; computer simulation of thermal management systems. *Mailing Add:* Mainstream Eng Corp 200 Yellow Pl Rockledge FL 32955-5327

SCARL, DONALD B, b Easton, Pa, Sept 17, 35; m 79; c 1. PHYSICS. *Educ:* Lehigh Univ, BA, 57; Princeton Univ, PhD(physics), 62. *Prof Exp:* Res assoc physics, NY Univ, 62-63; instr, Cornell Univ, 63-65, instr elec eng, 65-66; from asst prof to assoc prof physics, Polytech Inst Brooklyn, 66-74; FROM ASSOC PROF TO PROF, POLYTECH UNIV, 74- *Concurrent Pos:* Consult, Hazeltine Corp, 82-84, Naval Res Lab, 84-89. *Mem:* Am Phys Soc; Optical Soc Am; AAAS. *Res:* Quantum optics, particularly photon correlations and temporal coherence, lasers. *Mailing Add:* Dept Physics Polytech Univ Farmingdale NY 11735

SCARPA, ANTONIO, b Padua, Italy, July 3, 42; US citizen. BIOENERGETICS, MEMBRANE TRANSPORT. *Educ:* Univ Padua, MD, 66, PhD(gen path), 70. *Hon Degrees:* MS, Univ Pa, 76. *Prof Exp:* Asst prof gen path, Univ Padua, 69-71; from asst prof to prof biochem & biophys, Sch Med, Univ Pa, 73-85, dir, Biomed Instrumentation Group, 82-85; PROF & CHMN DEPT PHYSIOL & BIOPHYS, SCH MED, CASE WESTERN RESERVE UNIV, 85- *Concurrent Pos:* Nat Res Coun Italy exchange fel biochem, Univ Bristol, 68; Dutch Orgn Advan Pure Res fel biochem, Univ Utrecht, 70; NATO fel, Johnson Found, Univ Pa, 71; estab investr, Am Heart Asn, 73-78; prog chmn, US Bioenergetics Group, Biophys Soc, 74-75 & 84-85; assoc ed, Biophys J; adv bd, Biophys Soc Coun, 79-83; consult, Phys Biochem Study Sect, NIH, 83-87; nat res comt, Am Heart Asn; mem, Review Comt A, Nat Heart, Lung, Blood Inst, NIH, 90- *Mem:* Am Phys Soc; Biophys Soc; Am Soc Biol Chemists; Soc Gen Physiologists; Am Soc Physiologists. *Res:* Structure and function of biological membranes; ion transport; regulation of contraction of heart muscle; mechanisms secretion; structure and function of isolated sarcoplasmic reticulum with x-ray crystallography and neutron diffraction; intracellular Ca2; mechanism by which catecholamines are accumulated in chromaffin granules; secretion coupling in isolated bovine parathyroid cells. *Mailing Add:* Physiol Dept Case Western Reserve Univ Sch Med 2219 Abington Rd Cleveland OH 44106

SCARPACE, PHILIP J, b Buffalo, NY, Jan 4, 48; c 2. AGING, ADRENERIGIC PHARMACOLOGY. *Educ:* Calif State Univ, San Jose, BS, 70; Univ Rochester, PhD(biophys), 74. *Prof Exp:* Asst prof geriat & pharmacol, Univ Calif, Los Angeles, 77-87; lab chief, Geriat Res Ctr, Vet Admin Med Ctr, Sepalveda, Calif, 77-87; ASSOC PROF PHARMACOL, UNIV FLA, 87-; RES DIR, GERIAT RES EDUC & CLIN CTR, VET ADMIN MED CTR, GAINESVILLE, 87- *Concurrent Pos:* Asst prof math, Calif State Univ, Northridge, 79-81; assoc dir, Ctr Res Oral Health & Aging, Univ Fla, 88- *Mem:* Am Soc Pharmacol & Exp Therapeut; AAAS; Geront Soc Am. *Res:* Loss of responsiveness to adrenerigic agents with age; defective receptor signal transduction in the membranes of heart cells. *Mailing Add:* Geriat Res Educ Clin Ctr VA Med Ctr Gainesville FL 32608

SCARPELLI, DANTE GIOVANNI, b Padua, Italy, Feb 5, 27; nat US; m 51; c 3. EXPERIMENTAL PATHOLOGY. *Educ:* Baldwin-Wallace Col, BS, 50; Ohio State Univ, MS, 53, MD, 54, PhD, 60. *Hon Degrees:* DSc, Baldwin-Wallace Col, 66. *Prof Exp:* From instr to prof path, Ohio State Univ, 58-66; dean fac & acad affairs, Univ Kans Med Ctr, Kansas City, 72-73, prof path & oncol & chmn dept, 66-76; prof path & chmn dept, Northwestern Univ Med Sch, Chicago, 76-; PRIN INVESTR, ARMED FORCES INST PATH, WALTER REED MED CTR, WASHINGTON, DC. *Honors & Awards:* Silver Medal, Am Soc Clin Path, 56. *Mem:* AAAS; Am Soc Clin Path; Am Asn Path; Histochem Soc; Soc Exp Biol & Med. *Res:* Ultrastructural cytochemistry; carcinogenesis; comparative pathology. *Mailing Add:* Dept Path Northwestern Univ Med Sch 303 E Chicago Ave Chicago IL 60611

SCARPELLI, EMILE MICHAEL, b New York, NY, July 24, 31; m 52; c 7. PEDIATRICS, CARDIOPULMONARY PHYSIOLOGY. *Educ:* Fordham Univ, BS, 51; Duke Univ, MD, 60, PhD(physiol), 62. *Prof Exp:* Instr pediat & physiol, Albert Einstein Col Med, 62-64, res asst prof pediat, 64-66, assoc prof, 68-72, asst prof physiol, 64-71, assoc prof, 71-, prof pediat, 73-, dir pediat pulmonary div, 64-; AT PULMONARY & CRITICAL CARE MED DIV, SCHNEIDER CHILDRENS'S HOSP, LONG ISLAND JEWISH-HILLSIDE MED CTR. *Concurrent Pos:* Res grants, New York Heart Asn, 63-76, John Polachek Found Med Res, 65-66 & NIH career develop award, 66-76; mem, Int Med Comt, Lourdes; Nat Heart & Lung Inst, NIH Prog-Proj, 73-78 & training grant, 75-80. *Mem:* AAAS; Am Physiol Soc; Soc Pediat Res; NY Acad Sci; Am Heart Asn. *Res:* Cardiovascular physiology and disease, including dynamics of cardiac arrhythmias and circulatory shunts; pulmonary physiology and disease, including airway dynamics and the physiology, chemistry and morphology of lung surfactant. *Mailing Add:* 128 Constitution Dr Orangeburg NY 10962-2730

SCARPINO, PASQUALE VALENTINE, b Utica, NY, Feb 13, 32; m 86; c 2. ENVIRONMENTAL ENGINEERING. *Educ:* Syracuse Univ, BA, 55; Rutgers Univ, MS, 58; PhD(microbiol), 61. *Prof Exp:* USPHS res asst microbiol, Rutgers Univ, 58-61; asst prof biol sci, Fairleigh Dickinson Univ, 61-63; from asst prof to assoc prof environ eng, 63-71, PROF ENVIRON ENG, UNIV CINCINNATI, 71-, PROF ENVIRON HEALTH, 86- *Concurrent Pos:* Water pollution control admin, USPHS res award, 65-68; NASA Space Inst grant, Univ Cincinnati, 67-69; Fed Water Pollution Control Admin res contract, 69-71; US Environ Protection Agency grants, 72-82 & 84-86; chmn, Water Subcomt, Environ Task Force, City of Cincinnati, 72-73, Citizen-Scientist Comt Drinking Water Qual & Water Comt Environ Adv Coun, 75-76, chmn, Environ Adv Coun, 80-82, chmn, "Right to Know" Ordinance Tech Adv Comt, 83-88; Dept Defense res grant, 83-84; chmn, Pub Interest Adv Comt, Ohio River Valley Water Sanit Comn, 84-85, Ohio comnr, 85-, chmn, 89-90; res grant, Nat Inst Environ Health Sci, 88- *Mem:* Am Water Works Asn; Int Asn Water Pollution Res & Control; Am Soc Microbiol; Sigma Xi; fel Am Soc Microbiol. *Res:* Environmental microbiology; halogen inactivation of viruses in water and waste water; microbial food production; microbial survival in water, landfills and landfill leachates; dissemination of microbes in sewage treatment plant aerosols; risk assessment; biodegradation; bioaerosols; outdoor and indoor air pollution. *Mailing Add:* Dept Civil & Environ Eng 720 Rhodes Hall Univ Cincinnati Cincinnati OH 45221-0071

SCARPONE, A JOHN, b Newark, NJ, May 10, 31; m 57; c 3. PHARMACY. *Educ:* Rutgers Univ, BS, 53, MS, 64, PhD(pharmaceut chem), 69. *Prof Exp:* Pharmacist in charge, Jennis Drugs, Union, NJ, 57-68; res pharmacist, 68-72, group leader res & develop, Lederle Labs, Div Am Cyanamid Co, 72-78; sect head pharmaceut technol, Squibb Inst Med Res, 78-85, mgr, Squibb Pharmaceut Prod Div, 85-88, DIR SOLID DOSAGE FORMS & PROCESS DEVELOP-ORAL DOSAGE FORMS, PHARMACEUT OPERS DIV, BRISTOL-MYERS SQUIBB, 88- & DEVELOP, 88- *Mem:* Am Pharmaceut Asn; Acad Pharmaceut Sci; Sigma Xi. *Res:* Development of new pharmaceutical products, specifically tablet dosage forms involving new drug delivery systems and tabletting techniques; air suspension tablet film coating and development of new all aqueous film coating systems; process equipment; development of solid dosage forms. *Mailing Add:* Process Develop Oral Dosage Forms Pharmaceut Opers Div PO Box 191 New Brunswick NJ 08903

SCARRATT, DAVID JOHNSON, b Liverpool, Eng, Dec 21, 35; m 62; c 2. MARINE BIOLOGY. *Educ:* Univ Wales, BSc, 58, PhD(marine zool), 61. *Prof Exp:* SCIENTIST, SCI BR, DEPT FISHERIES & OCEANS, HALIFAX FISHERIES LAB, 61- *Mem:* Aquacult Asn Can; World Aquacult Soc. *Res:* The ecology and behavior of larval, juvenile and adult stages of North American lobster; effects of pollutants; industrial developments and fishing practices on commercial fisheries; marine resource inventories and atlases; vertebrate and molluscan aquaculture research; uptake and elimination marine toxins by molluscs. *Mailing Add:* Dept Fisheries & Oceans Halifax Fisheries Lab Box 550 Halifax NS B3J 2S7 Can

SCATLIFF, JAMES HOWARD, b Chicago, Ill, July 9, 27; m 61; c 2. RADIOLOGY. *Educ:* Northwestern Univ, BS, 49, MD, 52. *Prof Exp:* Intern, Cook County Hosp, Chicago, 52-53; resident radiol, Michael Reese Hosp, Chicago, 53-56; from instr to assoc prof, Sch Med, Yale Univ, 57-66; PROF RADIOL & CHMN DEPT, SCH MED, UNIV NC, CHAPEL HILL, 66- *Concurrent Pos:* NIH fel neuroradiol, St George's Hosp, London, Eng, 62-63; consult radiologist, Watts Hosp, Durham, NC, 66- *Mem:* AMA; Asn Univ Radiol (pres, 71-72); fel Am Col Radiol; Radiol Soc NAm; Am Soc Neuroradiol; Sigma Xi. *Res:* Microvasculature, neuroradiology and cardiac radiology. *Mailing Add:* Dept Radiol Univ NC Sch Med Chapel Hill NC 27514

SCATTERDAY, JAMES WARE, b Westerville, Ohio, Dec 10, 35; m 59. GEOLOGY. *Educ:* Denison Univ, BS, 57; Ohio State Univ, PhD(geol), 63. *Prof Exp:* From asst prof to assoc prof geol, 63-70, chmn, dept geol sci, 69-77, PROF GEOL, COL ARTS & SCI, STATE UNIV NY GENESEO, 70- *Mem:* Soc Econ Paleontologists & Mineralogists; Geol Soc Am; Paleont Soc; Int Palaeont Asn; Sigma Xi; Paleont Res Inst. *Res:* Study of ecology and sediments of recent coral reefs; stratigraphic and conodont biostratigraphic study of the Mississippian System of Ohio; stratigraphic and paleontologic studies of the Silurian and Devonian Systems of New York. *Mailing Add:* Dept Geol Sci Col Arts & Sci State Univ NY Geneseo NY 14454

SCATTERGOOD, EDGAR MORRIS, b Philadelphia, Pa, Mar 2, 36; m 65; c 2. CHEMICAL ENGINEERING, BIOCHEMICAL ENGINEERING. *Educ:* Mass Inst Technol, BS, 58; Univ Wis-Madison, PhD(chem eng), 66. *Prof Exp:* Design engr, Standard Oil Co of Calif, 59-61; res asst transport properties ion-exchange membranes, Univ Wis, 62-66; res engr, Gen Mills, Inc, 66-69 & North Star Res & Develop Inst, 69-70; res fel, St Olaf Col, 70-72; chem engr, Calgon Havens Systs, 72-73; RES FEL, MERCK SHARP & DOHME RES LABS, 73- *Mem:* Am Chem Soc; Am Inst Chem Engrs. *Res:* Transport properties of ion-exchange membranes; food processing; membrane processes; artificial biological membranes; mass tissue culture, virology, bacterial fermentation and vaccines. *Mailing Add:* 1375 Steven Lane Lansdale PA 19446

SCATTERGOOD, LESLIE WAYNE, b Seattle, Wash, May 22, 14; m 40; c 4. FISH BIOLOGY. *Educ:* Univ Wash, BS, 36. *Prof Exp:* Aquatic biologist, US Fish & Wildlife Serv, 39-49, fisheries res biologist, 49-58, dir fisheries biol lab, 58-61, chief br of reports, 61-68, chief div publ, 68-70; chief ed br, Nat Oceanic & Atmospheric Admin, 70-71, chief, Sci & Tech Publ Div, 71-80; FISHERY CONSULT, 81- *Concurrent Pos:* Sci asst, Int Pac Salmon Comn, Can, 38; biologist, Wash State Dept Fisheries, 39; local coordr, Off Coord Fisheries, Washington, DC, 43-45; fishery off, Panama, 52-53; Fulbright scholar, Fiskeridirektoratets, Bergen, Norway, 53-54; mem res comt, Int Passamaquoddy Fisheries Bd, 56-59. *Mem:* AAAS; Am Fisheries Soc; Inst Fishery Res Biol. *Res:* Life history of marine animals; science information. *Mailing Add:* 2514 N 24th St Arlington VA 22207

SCATTERGOOD, RONALD O, b Philadelphia, Pa, June 27, 37; m 66; c 1. METALLURGY. *Educ:* Lehigh Univ, SB, 60; Mass Inst Technol, SM, 63, ScD(metall), 68. *Prof Exp:* Assoc metallurgist, Argonne Nat Lab, 68-81; PROF MAT ENG, NC STATE UNIV, 81- *Mem:* Am Inst Mining, Metall & Petrol Engrs; Am Soc Metals. *Res:* Deformation of metals; dislocation theory; radiation effects. *Mailing Add:* NC State Univ Box 7907 Mat Sci NC State Univ Raleigh NC 27695

SCATTERGOOD, THOMAS W, b Mt Holly, NJ. PLANETOLOGY, CHROMATOGRAPHY. *Educ:* Univ Del, BS, 68; State Univ NY, Stony Brook, MS, 72, PhD(chem), 75. *Prof Exp:* Res fel, 76, res assoc, 79-84, adj asst prof, 84-87, ADJ ASSOC PROF CHEM, STATE UNIV NY, STONY BROOK, 87 - *Concurrent Pos:* Assoc, Nat Res Coun, Nat Acad Sci, 77-79; vis prof, Ames Res Ctr, NASA, 79 -; prin investr, State Univ NY, Stony Brook, 84 - *Mem:* Am Geophys Union; AAAS; Am Astron Soc; Planetary Soc; Int Soc Study Origin Life, ISSOL. *Res:* Investigation of chemical and physical processes in planetary atmospheres that result in formation of aerosols and clouds; planning and selection of instruments to be carried aboard planetary probes for the study of atmospheres of outer planets in the solar system. *Mailing Add:* NASA Ames Res Ctr 239-4 Moffett Field CA 94035

SCAVIA, DONALD, b Schenectady, NY. LIMNOLOGY, MODELING. *Educ:* Rensselaer Polytech Inst, BS, 73, MS, 74; Univ Mich, PhD, 80. *Prof Exp:* Res assoc aquatic modeling, Freshwater Inst, Rensselaer Polytech Inst, 74-75; RES SCIENTIST NUTRIENTS & ECOSYST DYNAMICS, GREAT LAKES ENVIRON RES LAB, NAT OCEANIC & ATMOSPHERIC ADMIN, 75- *Concurrent Pos:* Adj asst prof, Div Biol Sci, Univ Mich, 81- *Mem:* Am Soc Limnol & Oceanog; Int Asn Great Lakes Res; AAAS; Int Soc Ecol Modelling; Soc Int Limnol. *Res:* Investigation of biological, chemical and physical controls of nutrient cycles and carbon flow in the aquatic environment with particular emphasis on the use of models in the analysis; experimental analysis of ecological aspects of phytoplankton and zooplankton interactions in the control of primary production and nutrient cycling. *Mailing Add:* Great Lakes Environ Res Lab 2300 Washtenaw Rd Ann Arbor MI 48104

SCAVUZZO, RUDOLPH J, JR, b Plainfield, NJ, Jan 21, 34; m 55; c 10. MECHANICAL ENGINEERING. *Educ:* Lehigh Univ, BSME, 55; Univ Pittsburgh, MSME, 59, PhD(mech eng), 62. *Prof Exp:* Sr engr, Bettis Atomic Power Lab, Westinghouse Elec Corp, 55-64; from asst prof to assoc prof mech eng, Univ Toledo, 64-70; assoc prof mech eng, Hartford Grad Ctr, Rensselaer Polytech Inst Conn, Inc, 70-73; prof mech eng & head dept, 73-83, PROF MECH ENG, UNIV AKRON, 83- *Concurrent Pos:* Adj instr, Univ Pittsburgh, 63-64. *Mem:* Am Soc Mech Engrs; Am Soc Eng Educ; Am Indust Arts Asn; Soc Mech. *Res:* Solid mechanics; mechanical vibrations; dynamic shock analysis. *Mailing Add:* 4366 Shaw Rd Akron OH 44313

SCHAAD, LAWRENCE JOSEPH, b Columbus, Ohio, Sept 23, 30; m 54. PHYSICAL CHEMISTRY. *Educ:* Harvard Univ, AB, 52; Mass Inst Technol, PhD, 57. *Prof Exp:* Res assoc & NIH fel, Math Inst, Oxford Univ, 56-58; res assoc chem, Univ Ind, 59-61; from asst prof to assoc prof, 61-72, PROF CHEM, VANDERBILT UNIV, 72- *Mem:* Am Chem Soc; Am Phys Soc. *Res:* Quantum chemistry. *Mailing Add:* Box 1575 Dept Chem Vanderbilt Univ Nashville TN 37235

SCHAAD, NORMAN W, b Myrtle Point, Ore, Nov 9, 40; div; c 2. BACTERIOLOGY, ECOLOGY. *Educ:* Univ Calif, Davis, BS, 64, MS, 66, PhD(plant path), 69. *Prof Exp:* Fel plant path, Univ Calif, Davis, 69-71; from asst prof to prof phytobacteria, taxon ecol & serol, Univ Ga, 71-82; prof dept plant sci, Univ Idaho, 82-88; DIR BIOTECHNOL, HARRIS MORAN SEED CO, 88- *Concurrent Pos:* NSF vis prof, Univ Brasilia, 77; mem, outstanding grant activ, Univ Ga, 79-80; adj prof, Hassen II, Morocco, 85- *Mem:* Am Soc Microbiol; Am Phytopath Soc; Int Seed Testing Asn; Int Soc Plant Path. *Res:* Serology of phytobacteria; ecology of phytobacteria; seed pathology. *Mailing Add:* Harris Moran Seed Co 100 Breen Rd San Juan Bantista CA 95045

SCHAAF, NORMAN GEORGE, b Buffalo, NY, May 29, 36; m 59; c 3. MAXILLOFACIAL PROSTHETICS, PROSTHODONTICS. *Educ:* Univ Buffalo, DDS, 60; Am Bd Prosthodontics, dipl, 71. *Prof Exp:* Resident, Sch Dent, 63-64, instr prosthodontics, 64-65, from asst prof to assoc prof, 65-74, PROF MAXILLOFACIAL PROSTHETICS, STATE UNIV NY BUFFALO, 74- DIR REGIONAL CTR MAXILLOFACIAL PROSTHETICS, 67-; CHIEF DENT SERV, ROSWELL PARK MEM INST, 68- *Concurrent Pos:* Prosthodontic consult, J Sutton Regan Cleft Palate Found, Children's Hosp, Buffalo, 63-, consult hosp, 65-; consult, Eastman Dent Ctr; univ assoc, Buffalo Gen Hosp, 73-; maxillofacial prosthodontic consult, Buffalo Vet Hosp, 74- *Mem:* Am Cleft Palate Asn; Am Acad Maxillofacial Prosthetics; Soc Head & Neck Surgeons; Am Dent Assoc. *Res:* Anatomic, functional and cosmetic reconstruction, by the use of non-living substitutes, of those regions of the head and neck that are missing or defective whether from congenital anomaly, injury or disease. *Mailing Add:* Dept Dent Maxillofacial Prosthet Roswell Park Cancer Inst 666 Elm St Buffalo NY 14263

SCHAAF, ROBERT LESTER, b Baptistown, NJ, June 17, 29. ORGANIC CHEMISTRY. *Educ:* Rutgers Univ, BS, 50; Univ Mich, MS, 52, PhD(pharmaceut chem), 55. *Prof Exp:* From res chemist to sr res chemist, 55-67, res assoc, 67-77, sr res staff mem, 77-82, QUAL ASSURANCE ENGR, BASF CORP, 82- *Mem:* Am Chem Soc. *Mailing Add:* BASF Corp Wyandotte MI 48192

SCHAAF, THOMAS KEN, b Louisville, Ky, July 17, 43; m 84; c 3. MEDICINAL CHEMISTRY. *Educ:* Kalamazoo Col, BA, 65; Stanford Univ, PhD(chem), 69. *Prof Exp:* Res fel, Harvard Univ, 69-70; res chemist, Pfizer Inc, 70-74, proj leader, 74-76, mgr, 76-81, asst dir med sci chem, 81- 83, dir Animal Health Med Chem, 84-89, DIR, US ANIMAL DISCOVERY RES, PFIZER INC, 90- *Mem:* Am Chem Soc. *Res:* Antibacterial, antiparasitic and immune modulating agents. *Mailing Add:* Pfizer Cent Res Eastern Point Rd Groton CT 06340

SCHAAF, WILLIAM EDWARD, b Martins Ferry, Ohio, Aug 9, 38; m 61; c 3. FISHERIES RESOURCE MODELING. *Educ:* Duke Univ, BS, 61; Univ NC, MS, 63; Univ Mich, PhD(fisheries), 72. *Prof Exp:* Biostatistician, NIH, 63-66; biometrician, Inst Fisheries Res, Mich Dept Natural Resources, 66-69; BIOMETRICIAN, NAT MARINE FISHERIES SERV, DEPT COM, 69- *Mem:* Am Statist Asn; Am Fisheries Soc. *Res:* Modeling the population dynamics of exploited marine fishes, emphasizing optimal management strategies; developing ecosystem models of multi-species fisheries; investigation impacts of environmental variability, and life history strategies on expected yields. *Mailing Add:* Sleepy Creek Marshallberg NC 28553

SCHAAL, BARBARA ANNA, b Berlin, Ger, Sept 17, 47; US citizen; m; c 2. POPULATION BIOLOGY. *Educ:* Univ Ill, Chicago, BS, 69; Yale Univ, MPhil, 71, PhD(pop biol), 74. *Prof Exp:* Asst prof, Univ Houston, 74-76; asst prof bot, Ohio State Univ, 76-80; PROF BIOL, WASH UNIV, ST LOUIS, 80- *Mem:* AAAS; Genetics Soc Am; Soc Study Educ; Brit Ecol Soc; Ecol Soc; Bot Soc Am. *Res:* Genetic structure of outbreeding plant population; plant demographic genetics; molecular evolution. *Mailing Add:* Dept Biol Wash Univ St Louis MO 63130

SCHAAP, A PAUL, b Scottsburg, Ind, June 4, 45. ORGANIC CHEMISTRY. *Educ:* Hope Col, AB, 67; Harvard Univ, PhD(org chem), 70. *Prof Exp:* Asst prof, 70-74, assoc prof, 74-79, PROF CHEM, WAYNE STATE UNIV, 79- *Concurrent Pos:* Alfred P Sloan Res Fel, 74-76. *Mem:* Am Chem Soc; Royal Soc Chem; Sigma Xi. *Res:* Photo-oxidation; chemistry of singlet oxygen; chemiluminescence; 1,2-dioxetanes. *Mailing Add:* PO Box 07339 Detroit MI 48207

SCHAAP, LUKE ANTHONY, b South Holland, Ill, Nov 11, 31; m 59; c 4. ORGANIC CHEMISTRY. *Educ:* Calvin Col, AB, 53; Northwestern Univ, PhD(org chem), 57. *Prof Exp:* Asst, Northwestern Univ, 53-54; res chemist, Am Oil Co, Whiting, Ind, 57-74, RES CHEMIST, RES CTR, AMOCO OIL CO, 74- *Concurrent Pos:* Instr, Trinity Christian Col, 64-65. *Mem:* Am Chem Soc; Am Sci Affiliation; Soc Tribologists & Lubrication Engrs; Soc Automotive Engrs. *Res:* Base catalyzed reactions of hydrocarbons; high pressure and petroleum chemistry; olefin hydration; chemistry of nitrogen fluorides; lubricant research and development; fuels research and development. *Mailing Add:* 463 E 163rd St South Holland IL 60473

SCHAAP, WARD BEECHER, b Holland, Mich, Sept 15, 23; m 44; c 3. INORGANIC CHEMISTRY, ANALYTICAL CHEMISTRY. *Educ:* Wheaton Col, Ill, BS, 44; Univ Ill, MS, 48, PhD(inorg chem), 50. *Prof Exp:* Anal chemist, Metall Lab, Univ Chicago, 44; res chemist, Manhattan Proj, Oak Ridge Nat Lab, 44-47; from instr to assoc prof chem, 50-63, assoc dean, Col Arts & Sci, 66-71, assoc dean, Res & Develop, 71-73, actg vchancellor admin & budgetary planning, 73-76, prof chem, 63-88, dean admin & budgetary planning, 76-88, EMER PROF CHEM, IND UNIV, BLOOMINGTON, 88- *Concurrent Pos:* Consult, Union Carbide Nuclear Corp, 56- & E I du Pont de Nemours & Co, 58-60; NSF fel, Univ Calif, 60-61. *Mem:* Am Chem Soc. *Res:* Metal chelate compounds; polarography; electrochemistry in nonaqueous solvents; kinetics of inorganic reactions. *Mailing Add:* 819 S Jordan Bloomington IN 47401

SCHABER, GERALD GENE, b Covington, Ky, May 29, 38; m 61; c 3. GEOLOGY, ASTROGEOLOGY. *Educ:* Univ Ky, BS; Univ Cincinnati, MS, 62, PhD(geol), 65. *Prof Exp:* GEOLOGIST LUNAR & PLANETARY EXPLOR RES, FLAGSTAFF FIELD CTR, US GEOL SURV, 65- *Honors & Awards:* Spec Commendation for Apollo Astronaut Training, Geol Soc Am, 73; Group Achievement Award for Apollo Traverse Planning, NASA, 71; Group Achievement Award, NASA, 82; Autometric Award, Am Soc Photogammetry, 82. *Mem:* Sigma Xi; Planetary Soc; Geol Soc Am; Am Geophys Union. *Res:* Lunar and planetary geologic mapping; terrestrial, lunar and planetary remote sensing; geochemistry; terrestrial and planetary radar research. *Mailing Add:* Flagstaff Field Ctr US Geol Surv 2255 N Gemini Drive Ave Flagstaff AZ 86001

SCHACH, STEPHEN RONALD, b Cape Town, S Africa, Dec 3, 47; US citizen; m 74; c 2. SOFTWARE ENGINEERING, SOFTWARE TESTING. *Educ:* Univ Cape Town, BSc, 66, MSc, 69, PhD(appl math), 73; Weizmann Inst Sci, MSc, 72. *Prof Exp:* Lectr appl math, Univ Cape Town, 72-75, from lectr to assoc prof comput sci, 76-83; ASSOC PROF COMPUTER SCI, VANDERBILT UNIV, 83- *Concurrent Pos:* Vis scientist, Weizmann Inst Sci, Israel, 78-79. *Mem:* Asn Comput Mach; Inst Elec & Electronics Engrs Computer Soc; Sigma Xi. *Res:* Software engineering; software testing. *Mailing Add:* Box 70 Sta B Vanderbilt Univ Nashville TN 37235

SCHACHER, GORDON EVERETT, b Portland, Ore, Aug 24, 35; m 79; c 6. ATMOSPHERIC SCIENCES, SOLID STATE PHYSICS. *Educ:* Reed Col, AB, 56; Rutgers Univ, PhD(physics), 61. *Prof Exp:* Instr physics, Rutgers Univ, 60-61; fel, Argonne Nat Lab, Ill, 61-64; asst prof, 64-74, assoc prof, 74-80, PROF PHYSICS, NAVAL POSTGRAD SCH, 80- *Mem:* Am Phys Soc; Am Asn Physics Teachers; Sigma Xi; Soc Photo-optical Instrumentation Engrs. *Res:* Marine atmospheric boundary layer processes; electromagnetic propagation in the atmosphere; transport and dispension. *Mailing Add:* Dept Physics Naval Postgrad Sch Monterey CA 93940

SCHACHER, JOHN FREDRICK, b Jamestown, NDak, July 17, 28; m 52; c 3. PARASITOLOGY, INFECTIOUS DISEASES. *Educ:* NDak State Univ, BS, 53; Tulane Univ, MS, 56, PhD(parasitol), 61. *Prof Exp:* Tulane Univ prof asst parasitol, Univ Malaya, Singapore, 56-58; from asst prof to assoc prof trop dis, Am Univ Beirut, 60-69; assoc prof, 69-72, prof infectious trop dis & epidemiol, Univ Calif, Los Angeles, 72-, asst dean Acad Affairs, 80-; PROF EMERITUS, UNIV CALIF, LOS ANGELES. *Concurrent Pos:* NIH res grant, Am Univ Beirut & Univ Calif, Los Angeles, 63-76; Interam fel trop med, La State Univ, 66; consult, La Fisheries & Wildlife Comn, 54-55; mem expert comt filariasis, WHO, 73-, mem sci tech comt onchocerciasis control, 74- *Mem:* Int Filariasis Asn; Am Soc Parasitol; Am Soc Trop Med & Hyg; Royal Soc Trop Med & Hyg; Malaysian Soc Parasitol Trop Med. *Res:* Parasitology, especially helminthology, filariasis; host-parasite relations and pathology, taxonomy and identification of parasites in tissue. *Mailing Add:* Dept Pub Health Univ Calif Los Angeles CA 90024

SCHACHMAN, HOWARD KAPNEK, b Philadelphia, Pa, Dec 5, 18; m 45; c 2. MOLECULAR BIOLOGY, BIOCHEMISTRY. *Educ:* Mass Inst Technol, BS, 39; Princeton Univ, PhD(phys chem). *Hon Degrees:* DSc, Northwestern Univ, 74; DMed, Univ Naples, 90. *Prof Exp:* Chem engr, Continental Distilling Corp, Pa, 39-40; tech asst, Rockefeller Inst, 41-44; from instr to assoc prof biochem, 48-59, chmn dept molecular biol & dir virus lab, 69-76, PROF BIOCHEM & MOLECULAR BIOL, UNIV CALIF, BERKELEY, 59- *Concurrent Pos:* Guggenheim fel, 57-58; Fogarty Int Sch, NIH, 77-78; mem, Bd Sci Counselors, NIADDKD, NIH, 83-87, Bd Sci Consults, Mem Sloan-Kettering Cancer Ctr, 88-, Sci Coun & Sci Adv Bd, Stazione Zoologica, Naples, Italy, 88- *Honors & Awards:* E H Sargent & Co Award, Am Chem Soc, 62; Warren Prize, Mass Gen Hosp, 65; Carter-Wallace Lectr, Princeton Univ, 76; Gardiner Mem Lectr, NMed State Univ, 76; Jesse W Beams Mem Lectr, Univ Va, Charlottesville, 78; Merck Award, Am Soc Biol Chemists, 86; Bernard Axelrod Lectr, Purdue Univ, Lafayette, 86; Behring Diag Lectr, Univ Calif, San Diego, 87; Alta Heritage Found Lectr, Univ Alta, 87; William Lloyd Evans Award Lectr, Ohio State Univ, 88. *Mem:* Nat Acad Sci; Sigma Xi; Am Chem Soc; Am Soc Biochem & Molecular Biol (pres, 87-88); Am Acad Arts & Sci; Fed Am Soc Exp Biol (pres, 88-89). *Res:* Physical chemistry of macromolecules of biological interest; structure, function and interactions of proteins, nucleic acids and viruses; development and application of the ultracentrifuge. *Mailing Add:* Dept Molecular & Cell Biol & Virus Lab Stanley Hall Univ Calif Berkeley CA 94720

SCHACHT, JOCHEN (HEINRICH), b Konigsberg, Ger, July 2, 39; m 67; c 2. NEUROCHEMISTRY. *Educ:* Univ Bonn, BS, 62; Univ Heidelberg, MS, 65, PhD(biochem), 68. *Prof Exp:* Asst res biochemist, 69-72, from asst prof to assoc prof, 73-84, PROF BIOL CHEM OTORHINOL, KRESGE HEARING RES INST, UNIV MICH, 84- , ASSOC DIR, 89- *Concurrent Pos:* Fogarty Int fel, 79-80; vis prof, Dept Otolaryngol, Karolinska Inst, Stockholm, Sweden, 79-80; Sen Jacob Javits Neurosci Investr Award, 84-91; chercheur entranger, INSERM, France, 86-87; mem, Hearing Res Study Sect, NIH/NINCDS, 86-89. *Honors & Awards:* Animal Welfare Award, Erna-Graff-Found, Berlin, Ger, 87. *Mem:* Ger Soc Biol Chem; Am Soc Biol Chemists; Soc Neurosci; Asn Res Otolaryngol; Int Soc Neurochem. *Res:* Biochemistry of hearing and deafness; metabolism and function of phosphoinositides. *Mailing Add:* Kresge Hearing Res Inst Univ Mich Ann Arbor MI 48109-0506

SCHACHT, LEE EASTMAN, b Detroit, Mich, Sept 13, 30. HUMAN GENETICS. *Educ:* Dartmouth Col, AB, 52; Univ NC, MA, 55, PhD(zool), 57. *Prof Exp:* Instr human genetics, Univ Mich, 57-60; SUPVR HUMAN GENETICS UNIT, STATE DEPT HEALTH, MINN, 60- *Concurrent Pos:* Lectr, Sch Pub Health, Univ Minn, 64-74, adj asst prof, 74-76, adj assoc prof, 76- *Mem:* Am Soc Human Genetics. *Res:* Genetics in public health; genetic counseling. *Mailing Add:* Minn Dept Health 717 Delaware St SE PO Box 9441 Minneapolis MN 55440

SCHACHTELE, CHARLES FRANCIS, b Kearny, NJ, May 3, 42. DENTAL RESEARCH, MICROBIOLOGY. *Educ:* Macalester Col, BA, 63; Univ Minn, Minneapolis, MS, 65, PhD(microbiol), 68. *Prof Exp:* Asst prof, 68-72, assoc prof, 72-78, assoc prof microbiol, 74-78, PROF DENT, SCH DENT, UNIV MINN, MINNEAPOLIS, 78- PROF MICROBIOL, 78- *Concurrent Pos:* USPHS res career develop award, 71. *Mem:* AAAS; Am Soc Microbiol; Int Asn Dent Res. *Res:* Dental caries and periodontal disease. *Mailing Add:* Dept Dent/18-228 Moos Tower Univ Minn Minneapolis MN 55455

SCHACHTER, DAVID, b New York, NY, Oct 29, 27; m 56. MEDICINE. *Educ:* NY Univ, BS, 46, MD, 49; Am Bd Internal Med, dipl, 55. *Prof Exp:* From asst prof to assoc prof med, 57-69, PROF PHYSIOL, COL PHYSICIANS & SURGEONS, COLUMBIA UNIV, 69- *Mem:* AAAS; Soc Exp Biol & Med; Am Soc Clin Invest; Am Gastroenterol Asn; Am Physiol Soc. *Res:* Calcium transport and metabolism; membrane function and organization; active transport mechanisms. *Mailing Add:* Col Physicians & Surgeons Columbia Univ 530 W 168th St New York NY 10032

SCHACHTER, E NEIL, b New York, NY, May 10, 43; m 69; c 2. INTERNAL MEDICINE. *Educ:* Columbia Col, AB, 64; NY Univ, MD, 68. *Prof Exp:* From asst prof to assoc prof, Yale Sch Med, 74-84; PROF MED, MT SINAI SCH MED, 84-, MED DIR RESPIRATORY CARE, 84- *Concurrent Pos:* E L Trudeau fel, Am Thoracic Soc, 75; Maurice Hexter chair pulmonary med, Mt Sinai Sch Med, 87, assoc dir, Pieler Div, 90- *Mem:* Am Physiol Soc; Am Fedn Clin Res; Am Thoracic Soc; fel Am Col Physicians; Nat Asn Med Dirs (pres-elect, 89-91); fel Am Col Chest Physicians. *Res:* Occupational and environmental lung disease using clinical studies as well as in vitro pharmacologic studies; pulmonary medicine; epidemiology. *Mailing Add:* Mt Sinai Med Ctr One Gustave L Levy Pl New York NY 10029

SCHACHTER, H, b Vienna, Austria, Feb 25, 33; Can citizen; m 58; c 2. BIOCHEMISTRY. *Educ:* Univ Toronto, BA, 55, MD, 59, PhD(biochem), 64. *Prof Exp:* From asst prof to assoc prof, 64-70, PROF BIOCHEM, UNIV TORONTO, 70- *Concurrent Pos:* Cystic Fibrosis Res Found fel, 66-68; head res div, Dept Biochem, Hosp Sick Children. *Mem:* Am Soc Biol Chem; Can Biochem Soc; UK Biochem Soc; Soc Complex Carbohydrates. *Res:* Glycoprotein biosynthesis and metabolism; differentiation. *Mailing Add:* 555 University Ave Toronto ON 5MG 1X8 Can

SCHACHTER, JOSEPH, b New York, NY, Aug 26, 25; m 49; c 3. PSYCHIATRY. *Educ:* Dartmouth Col, AB, 46; Harvard Univ, PhD, 55; NY Univ, MD, 52. *Prof Exp:* Res assoc psychiat, Col Physicians & Surgeons, Columbia Univ, 56-60, asst clin prof, Univ & Columbia Psychoanal Clin Training & Res, 60-68, dir postdoctoral res training prog, 65-68; dir res child psychiat, Sch Med, Univ Pittsburgh, 68-74, assoc prof, 68-76, res assoc prof psychiat, 76-86, assoc prof epidemiol, Grad Sch Pub Health, 79-86; RETIRED. *Concurrent Pos:* Found Fund Res Psychiat fel, Columbia Univ & NY State Psychiat Inst, 55-56; pres, Pittsburgh Psychoanal Ctr, 78-80. *Mem:* Am Psychosom Soc; Am Psychiat Asn; Am Psychoanal Asn; Soc Psychophysiol Res. *Res:* Hypertension; behavioral medicine; cardiovascular development; psychoanalytic group psychotherapy. *Mailing Add:* 5400 Darlington Rd Pittsburgh PA 15217

SCHACHTER, JULIUS, b New York, NY, June 1, 36; wid; c 2. PUBLIC HEALTH & EPIDEMIOLOGY. *Educ:* Columbia Col, BA, 57; Hunter Col, MA, 60; Univ Calif, PhD(bact), 65. *Prof Exp:* Grad res microbiol, 60-65, asst res microbiologist, 65-68, from asst prof to assoc prof epidemiol & med, 68-75, asst dir, G W Hooper Found, 72-76, actg dir, 77-81, PROF EPIDEMIOL, UNIV CALIF, SAN FRANCISCO, 75-, PROF LAB MED, 80- *Concurrent Pos:* Mem viral & rickettsial registry comt, Am Type Cult Collection; co-dir, Collab Centre Reference & Res on Trachoma & Other Chlamydial Infections, WHO, 72-77, dir, 77-; mem, expert panel trachea, WHO, 72-, expert panel venereal dis, 80-; bd govs, Am Acad Microbiol, 86-; ed, Sexually Transmitted Dis, 89- *Mem:* Infectious Dis Soc Am; Am Soc Microbiol; Soc Exp Biol & Med; Am Venereal Dis Asn; Am Acad Microbiol; Am Epidemiol Soc. *Res:* Chlamydial infections, including psittacosis, trachoma, inclusion conjunctivitis, lymphogranuloma venereum, venereal diseases and perinatal infection; microbiology; epidemiology; host-parasite relationships. *Mailing Add:* Dept Lab Med Univ Calif San Francisco Gen Hosp 1001 Potrero Ave San Francisco CA 94110

SCHACHTER, MELVILLE, b Sept 22, 20; Can citizen; m 44; c 3. PHYSIOLOGY. *Educ:* McGill Univ, BSc, 41, MSc, 42, MD, 46. *Prof Exp:* Asst prof physiol, Dalhousie Univ, 47-50; staff mem, Nat Inst Med Res, London, 50-53; reader, Univ Col, Univ London, 53-65; PROF PHYSIOL & HEAD DEPT, UNIV ALTA, 65- *Concurrent Pos:* Consult, Parke, Davis & Co, 60-65; vis prof, Stanford Univ, 71. *Mem:* Brit Physiol Soc; Brit Pharmacol Soc; Brit Soc Immunol; Can Physiol Soc; Can Pharmacol Soc. *Res:* Autopharmacology-mediators. *Mailing Add:* Dept Physiol Univ Alberta Edmonton AB T6G 2E2 Can

SCHACHTER, MICHAEL BEN, b Brooklyn, NY, Jan 15, 41; m 67, 82; c 5. MEDICAL NUTRITION, CHELATION THERAPY. *Educ:* Columbia Col, BA, 61, MD, 65. *Prof Exp:* Intern, Hosp Joint Dis & Med Ctr, 65-66; psychiatric resident, Downstate Med Ctr, King County Hosp, 66-69; co-dir, Psychiat Outpatient Dept, Keesler AFB, Biloxi, Miss, 69-71; dir, Psychiat Emergency & Admis Dept, Rockland County Community Health CTR, 71-72, psychiat outpatient clin, 72-74; founder, owner & dir, Mountainview Med Assocs, 74-90, FOUNDER, OWNER & DIR, MICHAEL B SCHACHTER MD, PC, 90- *Concurrent Pos:* Bd dirs, Am Bd Chelation Ther, 83-88. *Honors & Awards:* Carlos Lamar Pioneer Mem Award, Am Acad Med Preventics, 79. *Mem:* Am Col Adv Med (vpres, 85-87, pres elec, 87-89, pres, 89-91); Am Bd Chelation Therapy; Acad Orthomolecular Psychiat (vpres, 81-83); Am Psychiat Asn; Am Col Nutrit; Am Acad Environ Med. *Res:* Degenerative diseases are largely caused by environmental factors. Prevention & treatment should focus on eliminating toxic factors and instituting positive life style changes, including improved diet, nutritional supplements, exercise and stress management. *Mailing Add:* Mountain View Med Assoc PC Mountainview Ave Nyack NY 10960

SCHACHTER, ROZALIE, b Rumania, Oct 14, 46; US citizen; m 67; c 2. DEVICE PHYSICS & ENGINEERING, PRODUCT APPLICATIONS & MARKETING. *Educ:* Brooklyn Col, BS, 68; Yeshiva Univ, MS, 70; New York Univ, PhD(physics), 79. *Prof Exp:* APS fel, Stauffer Chem Co, 79-80, group leader, 80-86; tech dir, Am Cyanamid, 86-90; DIR BUS DEVELOP, GEN MICROWAVE, 90- *Mem:* Am Phys Soc; Inst Elec & Electronics Engrs. *Res:* Compound semiconductor (III-V) optoelectronic devices (lasers, detectors) for communication and instrumentation; research on MOCVD epitaxial growth, electro-optical characterization and device fabrication and test; application of safe chemical sources to semiconductor processes; application of microwave radar sensors. *Mailing Add:* 75-14 173rd St Flushing NY 11366

SCHACH VON WITTENAU, MANFRED, b Pennekow, Ger, June 19, 30; m 55; c 3. ORGANIC CHEMISTRY, DRUG METABOLISM. *Educ:* Univ Heidelberg, BS, 52, MS, 55, PhD(org chem), 57. *Prof Exp:* Fel, Mass Inst Technol, 57-58; res chemist, 58-64, group supvr drug metab, 64-67, mgr drug metab, 67-71, from asst dir to dir, Dept Drug Metab, 71-74, exec dir, Safety Eval & Drug Metab, 74-81, VPRES SAFETY EVAL, PFIZER, INC, 81- *Mem:* Am Chem Soc; Soc Ger Chem; Am Soc Pharmacol & Exp Therapeut; Soc Toxicol. *Res:* Medicinal chemistry; analytical chemistry; toxicology. *Mailing Add:* Pfizer Inc Cent Res Eastern Point Rd Groton CT 06340

SCHACK, CARL J, b St Louis, Mo, Dec 26, 36; m 59; c 2. INORGANIC CHEMISTRY. *Educ:* St Louis Univ, BSChem, 58; Polytech Inst Brooklyn, PhD(inorg chem), 64. *Prof Exp:* Sr res engr, 64-68, MEM TECH STAFF, ROCKETDYNE DIV, ROCKWELL INT, 68- *Mem:* Am Chem Soc; Royal Soc Chem. *Res:* Fluorine chemistry; synthesis and material characterization of inorganic oxidizers, fluorine compounds including fluorocarbons and boron-nitrogen species. *Mailing Add:* 20744 Tribune St Chatsworth CA 91311-1528

SCHACTER, BERNICE ZELDIN, b Philadelphia, Pa, June 20, 43; m 68; c 2. IMMUNOGENETICS, BIOCHEMISTRY. *Educ:* Bryn Mawr Col, AB, 65; Brandeis Univ, PhD(biol), 70. *Prof Exp:* Charles F Kettering fel photosynthetic membrane, Lawrence Berkeley Lab, 70-71, res chemist, 71; Fla Heart Asn fel, Sch Med, Univ Miami, 72, instr pharmacol of cell membranes, 72-73; fel tumor cell membrane immunol, Oncol Ctr, Johns Hopkins Univ, 73-74, instr med & oncol, Sch Med, 74-76; mem staff, Dept Immunopath, Cleveland Clin, 76-77; from asst prof to assoc prof exp path, Case Western Reserve Univ, 77-84; sr res scientist, 84-87, ASSOC DIR IMMUNOL DEPT, BRISTOL-MYERS, SQUIBB, 88- *Concurrent Pos:* Adj prof, Univ Conn, Wesleyan Univ. *Mem:* Am Asn Immunologists; Am Soc Human Genetics; Am Asn Clin Histocompatibility Testing. *Res:* Immunobiology and immunogenetics; transplantation; tumor cell biology. *Mailing Add:* Bristol-Myers Five Research Pkwy Wallingford CT 06492

SCHACTER, BRENT ALLAN, b Winnipeg, Man, June 1, 42; m 81; c 2. MEDICAL RESEARCH, HEMATOLOGY. *Educ:* Univ Man, BSc & MD, 65; FRCP(C), 71. *Prof Exp:* From asst prof to assoc prof, 72-87, PROF INTERNAL MED, UNIV MAN, 87- *Concurrent Pos:* Fel, Med Res Coun Can, 70-72, scholar, 75-80. *Mem:* Am Soc Clin Oncol; Am Fedn Clin Res; Can Soc Clin Invest; Am Soc Hemat; Am Asn Study Liver Dis. *Res:* Investigation of the mechanism of action of heme oxygenase and the nature of the regulation of heme catabolism by this enzyme. *Mailing Add:* Dept Internal Med Univ Man 100 Olivia St Winnipeg MB R3E 0V9 Can

SCHAD, GERHARD ADAM, b Brooklyn, NY, Apr 2, 28; m 53; c 2. PARASITOLOGY. *Educ:* Cornell Univ, BS, 50; McGill Univ, MSc, 52, PhD(parasitol), 55. *Hon Degrees:* MSc, Univ Pa, 75. *Prof Exp:* Scripps res fel, Biol Res Inst, San Diego Zool Soc, 53; parasitologist, USDA, 55-58; asst prof parasitol, Inst Parasitol, Macdonald Col, McGill Univ, 58-64; from asst prof to assoc prof pathobiol, Johns Hopkins Univ, 64-73; prin investr parasitol, 72-79, assoc prof, 73-77, head, Lab Parasitol, 74-83, chmn, Grad Group Paraistol, 78-84, PROF PATHOBIOL, SCH VET MED, UNIV PA, 77- *Concurrent Pos:* Prin investr parasitol, Johns Hopkins Univ Ctr Med Res & Training, Calcutta, 64-66 & 68-70, NIH, 79-81 & 85-, WHO, 79-81 & 84- & USDA, 81-85; mem, Nat Selection Comt, US Educ Found, India, 68-69. *Mem:* Fel AAAS; Am Asn Vet Parasitol; Am Soc Trop Med & Hyg; Royal Soc Trop Med & Hyg; Am Soc Parasitol (pres, 90). *Res:* Population ecology of parasitic helminths; systematics; evolution and biology of parasitic helminths; developmental biology of parasitic helminths. *Mailing Add:* Parasitol/213 Vet/H1 Sch Vet Med Univ Pa Philadelphia PA 19104

SCHAD, THEODORE M(ACNEEVE), b Baltimore, Md, Aug 25, 18; m 44; c 2. CIVIL ENGINEERING, HYDROLOGY. *Educ:* Johns Hopkins Univ, BE, 39. *Prof Exp:* Eng aide, US Army Corps Engrs, Md, 39-40, chief specifications sect, Wash, 42-45, engr, 45-46; jr engr, US Bur Reclamation, Colo, 40-42, engr, 46-49, chief coord plans sect, 50-51, asst chief prog coord div, 51-54; budget exam, US Bur Budget, 54-58; sr specialist eng & pub works, Legis Ref Serv, Libr Cong, 58-68, actg chief sci policy res div, 66-67, dep dir, 67-68; exec dir, US Nat Water Comn, 69-73; exec secy, Environ Studies Bd, 73-77, dep exec dir, Comn Nat Resources, Nat Res Coun, 77-83; exec dir nat ground water policy forum, 84-86, SR FEL, CONSERV FOUND, 86-, RONCO CONSULT CORP, 86- *Concurrent Pos:* Staff dir, Select Comt Nat Water Resources, US Senate, 59-61; consult, Comt Interior & Insular Affairs, 63, Comt Water Resources Res, Fed Coun Sci & Tech, 62-64, Comt Sci & Tech & Astron, House of Rep, 63-64, Off Saline Water, 64-67, A T Kearney, Inc, 79-80, Gambia River Basin Develop Orgn, 86- & Apogee Res, 88-90; mem US comt large dams, Int Comn Large Dams, 64-; mem permanent int comn, Permanent Int Asn Navig Cong, 63-70; vis fel, Woodrow Wilson Nat Fel Found, 73-81. *Honors & Awards:* Iben Award, Am Water Resource Asn, 78; Caulfield Award, 90- *Mem:* Fel Am Soc Civil Engrs; Am Geophys Union; Nat Speleol Soc; hon mem Am Water Works Asn; Am Acad Environ Engrs; Nat Acad Pub Admin. *Res:* Water resources policy; federal policies and programs; water resources research; environmental policy; ground water management. *Mailing Add:* 4138 26th Rd N Arlington VA 22207

SCHADE, HENRY A(DRIAN), b St Paul, Minn, Dec 3, 00; m 25; c 2. NAVAL ARCHITECTURE. *Educ:* US Naval Acad, BS, 23; Mass Inst Technol, SM, 28; Tech Univ Berlin, Dr Ing, 37. *Hon Degrees:* Dr Ing, Tech Univ Berlin, 72. *Prof Exp:* Chief tech mission, US Navy, Europe, 45, dir, Naval Res Lab, 46-49; prof naval archit, Univ Calif, Berkeley, 49-69, emer prof, 69-; RETIRED. *Honors & Awards:* David Taylor Gold Medal, Soc Naval Archit & Marine Engrs, 64; Gibbs Bros Medal, Nat Acad Sci, 71. *Mem:* Nat Acad Eng; Soc Naval Archit & Marine Engrs (vpres, 65). *Res:* Ship structure design. *Mailing Add:* 88 Norwood Ave Kensington CA 94707

SCHADLER, DANIEL LEO, b Dayton, Ky, Apr 5, 48. PLANT PATHOLOGY, BIOCHEMISTRY. *Educ:* Thomas More Col, AB, 70; Cornell Univ, MS, 72, PhD(plant path), 74. *Prof Exp:* Res assoc, Univ Wis, 74-75; from asst prof to assoc prof, 75-84, PROF BIOL, OGLETHORPE UNIV, 84-, CHMN, 83- *Mem:* Am Phytopath Soc; Int Soc Plant Path; Am Chem Soc; AAAS. *Res:* Physiology and biochemistry of plant disease and of plant pathogens, especially phytotoxins. *Mailing Add:* Dept Biol 4484 Peachtree Rd NE Atlanta GA 30319-2797

SCHADLER, HARVEY W(ALTER), b Cincinnati, Ohio, Jan 4, 31; m 54; c 3. METALLURGY. *Educ:* Cornell Univ, BMetE, 54; Purdue Univ, PhD, 57. *Prof Exp:* Metallurgist, Res Labs, 57-69, mgr surfaces & reactions br, 69-71, mgr phys metall br, 71-73, MGR, MAT RES CTR, GEN ELEC RES & DEVELOP CTR, 73- *Honors & Awards:* Geisler Award, Am Soc Metals. *Mem:* Nat Acad Eng; Am Inst Mining, Metall & Petrol Engrs; Sigma Xi; fel Am Soc Metals. *Res:* Physical metallurgy; brittle fracture; diffusion; crystall perfection; plastic deformation; superconductivity. *Mailing Add:* 1333 Lowell Rd Schenectady NY 12308

SCHADT, FRANK LEONARD, III, b Syracuse, NY, Feb 3, 47; m 88; c 1. PHYSICAL ORGANIC CHEMISTRY, PHOTOGRAPHIC CHEMISTRY. *Educ:* Le Moyne Col, BS, 68; Princeton Univ, MA, 71, PhD(chem), 74. *Prof Exp:* Res chemist, Photosysts & Electronic Prod Dept, 74-80, sr res chemist, 80-86, RES ASSOC, ELECTRONICS DEPT, E I DU PONT DE NEMOURS & CO, INC, 86- *Mem:* Am Chem Soc; Sigma Xi; Am Inst Chemists. *Res:* Mechanistic organic chemistry; photochemically induced polymerization and rearrangement; photographic chemistry; photopolymer resists for electronics; synthesis of specialized polymers and reagents; electrically conductive coatings; Agx imaging systems. *Mailing Add:* 2305 W 17th St Wilmington DE 19806-1330

SCHADT, JAMES C, CARDIOVASCULAR CONTROL, NEUROPHYSIOLOGY. *Educ:* Tex Tech Univ, PhD(physiol), 78. *Prof Exp:* ASST PROF & RES INVESTR NEUROPHYSIOL, DALTON RES CTR, 78- *Mailing Add:* Dalton Res Ctr Univ Mo Columbia MO 65211

SCHADT, RANDALL JAMES, b Highland Park, Ill, Apr 14, 60. SOLID STATE NUCLEAR MAGNETIC RESONANCE, NUCLEAR MAGNETIC RESONANCE SIMULATIONS. *Educ:* Kearney State Col, BS, 82; Univ Mo, MS, 84; NC State Univ, PhD(physics), 90. *Prof Exp:* Res asst physics, Physics Dept, Univ Mo, 83-85; teaching asst physics, Physics Dept, Univ Ill, 85-87; res asst polymer chem, Physics Dept, NC State Univ, 87-90; vis scientist polymer physics, Max Planck Inst Polymer Sci, 90-91; VIS SCIENTIST POLYMER PHYSICS, DUPONT CENT RES & DEVELOP, 91- *Mem:* Am Phys Soc; Am Chem Soc. *Res:* Modern solid state nuclear magnetic resonance methods are used as the principal tools to probe molecular dynamics/structural correlations at various length scales in polymers and their relationship to measured macroscopic degradation. *Mailing Add:* 415 W 35th Kearney NE 68847

SCHAECHTER, MOSELIO, b Milan, Italy, Apr 26, 28; US citizen; wid; c 2. MICROBIOLOGY, MOLECULAR BIOLOGY. *Educ:* Univ Kans, MA, 51; Univ Pa, PhD(microbiol), 54. *Prof Exp:* From instr to assoc prof microbiol, Univ Fla, 58-62; from assoc prof to prof, 62-87, CHMN DEPT, TUFTS UNIV, 70-, DISTINGUISHED PROF MICROBIOL, 87- *Concurrent Pos:* Am Cancer Soc grant, State Serum Inc, Copenhagen, Denmark, 56-58; mem bact & mycol study sect, NIH, 75-79, chmn, 78-79; mem, Microbiol Test Comt, Nat Bd Med Examr, 82-85; chmn, Asn Med Sch Microbiol & Immunol, pres, 84-85. *Mem:* Am Soc Microbiol (pres, 85-86); Soc Gen Microbiol; Sigma Xi. *Res:* The role of the cell membrane in bacterial DNA replication and segregation. *Mailing Add:* Dept Molecular Biol Tufts Univ Med Sch Boston MA 02111

SCHAEDLE, MICHAIL, b Tallinn, Estonia, Dec 27, 27; US citizen; m 66; c 1. PLANT PHYSIOLOGY. *Educ:* Univ BC, BSA, 57, MSA, 59; Univ Calif, Berkeley, PhD(plant physiol), 64. *Prof Exp:* Fel, Univ Calif, Berkeley, 64-65; from asst prof to assoc prof, 65-81, PROF BOT, STATE UNIV NY COL ENVIRON SCI & FORESTRY, 81- *Mem:* AAAS; Am Inst Biol Sci; Am Soc Plant Physiol; Sigma Xi. *Res:* Plant nutrition; ion transport and cell permeability; photosynthesis in tissues of perennial plants; calcium and aluminum problems. *Mailing Add:* State Univ NY Col Environ Sci & Forestry Syracuse NY 13210

SCHAEDLER, RUSSELL WILLIAM, b Hatfield, Pa, Dec 17, 27. MEDICINE, MICROBIOLOGY. *Educ:* Ursinus Col, BS, 49; Jefferson Med Col, MD, 53. *Prof Exp:* Intern hosp, Jefferson Med Col, 53-54; asst & asst physician med & microbiol, Rockefeller Univ, 54-57, asst prof & resident assoc physician, 57-61, assoc prof & physician, 61-68; PROF MICROBIOL & CHMN DEPT, JEFFERSON MED COL, 68- *Mem:* Harvey Soc; Am Soc Microbiol; Am Gastroenterol Asn; Infectious Dis Soc Am; Am Asn Immunol; Sigma Xi. *Res:* Influence of environmental factors and nutrition on host resistance; ecology of the flora of the digestive tract and diarrheal diseases. *Mailing Add:* Dept Microbiol Jefferson Med Col Philadelphia PA 19107

SCHAEFER, ALBERT RUSSELL, b Oklahoma City, Okla, Oct 13, 44; m 68; c 2. OPTICAL DETECTORS & CHARGE COUPLED DEVICES, RADIOMETRIC PHYSICS. *Educ:* Univ Okla, BS, 66, PhD(physics), 70. *Prof Exp:* Physicist, Optical Radiation Sect, Nat Bur Standards, 70-86; chief scientist, Western Res Corp, 86-87; SR SCIENTIST, SCI APPLICATIONS INT CORP, 87- *Concurrent Pos:* Instr, Montgomery Col, 74-77, adj prof, 78- *Honors & Awards:* Bronze Medal, Dept Com, 81. *Mem:* Am Phys Soc; Optical Soc Am; Soc Photo-Optical Instrumentation Engrs. *Res:* Transition probabilities; radiometric physics; electro-optic devices; lifetimes; radiometry; photometry; spectroscopy; silicon detectors; astrophysics; charge coupled devices (CCD's). *Mailing Add:* SAIC Elec Vision Syst Div 4161 Campus Point Dr San Diego CA 92121-1513

SCHAEFER, ARNOLD EDWARD, b Tripp, SDak, Dec 8, 17; m 42; c 3. NUTRITION. *Educ:* SDak State Col, BS, 39; Univ Wis, MS, 41, PhD(biochem), 47. *Prof Exp:* Asst, Exp Sta, SDak State Col, 39-40; asst biochem, Univ Wis, 40-41, asst animal nutrit, 46-47, instr, 47; assoc animal nutrit, Ala Polytech Inst, 47-51; head nutrit res dept, E R Squibb & Sons, 51-55; biochemist, Interdept Comt Nutrit for Nat Defense, NIH, 55-56, exec dir, 56-64, Off Int Res, 64-67, chief nutrit prog, Health Serv & Ment Health Admin, 67-71; consult nutrit surveillance, Pan Am Health Orgn, 71-73; DIR, SWANSON CTR NUTRIT, UNIV NEBR MED CTR, OMAHA, 73- *Concurrent Pos:* Mem interdept ad hoc adv group res & develop food for shelters, Off Civil & Defense Mobilization, UN; mem interdept group freedom from hunger, Food & Agr Orgn; mem US comt, Int Union Nutrit Sci; pres, Fedn Am Socs Exp Biol, 67. *Honors & Awards:* Conrad Elvehjem Award, Am Inst Nutrit, 70. *Mem:* AAAS; Am Cancer Soc; fel Am Pub Health Asn; Am Inst Nutrit (secy, 60-63, pres, 67); Animal Nutrit Res Coun; Sigma Xi; Am Bd Nutrit (pres, 80-81); Fed Am Soc Exp Biol & Med (pres, 68); Am Col Nutrit; hon mem, Am Dietetic Asn. *Res:* Biochemistry; nutritional requirements and metabolism in animals; nutritional appraisal of man. *Mailing Add:* 1060 Verdon Circle RR 1 Plattsmouth NE 68048

SCHAEFER, CARL FRANCIS, b Schenectady, NY, Mar 20, 41; m 75; c 2. CIRCULATORY SHOCK, ANESTHESIOLOGY. *Educ:* Univ Toronto, BA; Univ Rochester, PhD(physiol psychol), 72. *Prof Exp:* Asst prof psychol, Wilkes Col, 70-71; fel psychosom med, 71-74, asst prof res med, 74-76, asst prof res anesthesiol, 76-80, ASSOC PROF ANESTHESIOL, UNIV OKLA HEALTH SCI CTR, 80-, DIR ANESTHESIA RES LAB, 88- *Concurrent Pos:* NIMH res grant, 77-78, Vet Admin res grant, 87- *Mem:* Am Psychol Asn; Sigma Xi; NY Acad Sci; Am Soc Anesthesiol; Am Physiol Soc; Shock Soc. *Res:* Physiological mechanisms of circulatory shock; cardiovascular effects of anesthetic agents; oxidative metabolism, small intestinal pathology and circulatory shock; development of anesthetic and physiological monitoring methods for small animals such as rats. *Mailing Add:* Dept Anesthesiol PO Box 26901/RB 29 R Oklahoma City OK 73190-3000

SCHAEFER, CARL W, II, b New Haven, Conn, Sept 6, 34; div; c 2. ENTOMOLOGY. *Educ:* Oberlin Col, BA, 56; Univ Conn, PhD(entom), 64. *Prof Exp:* From instr to asst prof biol, Brooklyn Col, 63-66; from asst prof to assoc prof, 66-75, PROF BIOL, UNIV CONN, 76- *Concurrent Pos:* Grant, City Univ NY Grad Div, 64-66, Univ Conn Res Found, 66-67 & NSF, 67-69, 79-81; co-ed, Ann Entom Soc Am, 73- *Mem:* Entom Soc Am; Soc Syst Zool; Soc Study Evolution; Indian Entom Soc; Am Soc Naturalist. *Res:* Comparative morphology, biology and phylogeny of the terrestrial Heteroptera. *Mailing Add:* Dept Ecol Evolutionary Biol Univ Conn N-43 Storrs CT 06268

SCHAEFER, CHARLES HERBERT, b Albany, Calif, Sept 24, 35; m 57; c 2. ENTOMOLOGY, AGRICULTURAL CHEMISTRY. *Educ:* Univ Calif, Berkeley, BS, 58, PhD(entom), 62. *Prof Exp:* Insect physiologist, US Forest Serv, Md, 62-64 & Shell Develop Co, Calif, 64-67; DIR MOSQUITO CONTROL RES LAB, UNIV CALIF, 67- *Mem:* AAAS; Entom Soc Am; Entom Soc Can; Mosquito Control Asn Am. *Res:* New approaches to insect control; biochemical relationships between insects and their respective plant hosts; mechanisms of insecticide degradation by target insects and by the environment. *Mailing Add:* 7740 E Saginaw Way Fresno CA 93727

SCHAEFER, DALE WESLEY, b Willoughby, Ohio, May 17, 41; m 62; c 2. CHEMICAL PHYSICS. *Educ:* Wheaton Col, BS, 63; Mass Inst Technol, PhD(phys chem), 68. *Prof Exp:* Fel physics, Mass Inst Technol, 68-70 & T J Watson Res Ctr, IBM Corp, 70-72; mem tech staff, Sandia Nat Lab, 72-80, supvr, Corrosion Div, 80-82, supvr, Chem Physics Div, 82-88, MGR, ORG & ELEC MAT DEPT, PHYSICS DIV, SANDIA NAT LAB, 82- *Concurrent Pos:* Int Sci Orgn Comt, Statphys 16, 85-86; chmn, Symposium Fractal Aspects Mat, Mat Res Soc, 86; Los Alamos Neutron Scattering Ctr Adv Comt, 87-91; NSF Pre-Doctoral fel, 64. *Honors & Awards:* Chemist Award, Am Inst Chemists, 63; Outstanding Sustained Res Award, Dept Energy Basic Energy Sci, 86. *Mem:* Fel Am Phys Soc; Mat Res Soc; Am Chem Soc; Am Ceramic Soc; Sigma Xi; fel Am Inst Chemists. *Res:* Polymer physics light scattering; small angle x-ray scattering; colloid physics; structure of fluids; neutron scattering; ceramic and polymer materials science. *Mailing Add:* Dept 1810 Sandia Labs Albuquerque NM 87185

SCHAEFER, DANIEL M, b Kewaskum, Wis, Sept 12, 51; m 75; c 2. RUMEN BACTERIOLOGY, RUMINANT NUTRITION. *Educ:* Univ Wis-Madison, BS, 73, MS, 75; Univ Ill, PhD(nutrit sci), 79. *Prof Exp:* Asst prof animal sci, Purdue Univ, 79-81; from asst prof to assoc prof, 81-90, PROF ANIMAL SCI UNIV WIS-MADISON, 90- *Mem:* Am Soc Animal Sci; Am Dairy Sci Asn; Am Soc Microbiol. *Res:* The understanding and manipulation of ruminal microbes to obtain nutritional advantages for the host ruminant. *Mailing Add:* Meat & Animal Sci Dept Univ Wis Madison WI 53706

SCHAEFER, DONALD JOHN, b Sioux Falls, SDak, Dec 9, 32; m 55; c 2. COMPUTER SCIENCE, MATHEMATICS. *Educ:* San Jose State Col, AB, 57; Ohio State Univ, MA, 58, PhD(math), 63. *Prof Exp:* Res engr, Lockheed Aircraft Corp, 60-61, 62; instr math, Ohio State Univ, 63-64; assoc prof math, 64-77, dir, Res & Instr Computer Ctr, 69-87, PROF COMPUTER SCI & MATH, WRIGHT STATE UNIV, 77- *Mem:* Am Math Soc; Mat Asn Am; Asn Computer Mach. *Res:* Numerical analysis and computer applications. *Mailing Add:* Dept Computer Sci Wright State Univ Dayton OH 45435

SCHAEFER, ERNST J, b Bad Nauheim, Germany, Nov 18, 45. LIPOPROTEINS, ATHEROSCLEROSIS. *Educ:* Mt Sinai Sch Med, New York City, MD, 72. *Prof Exp:* ASSOC PROF MED & CHIEF, LIPID METAB LAB, TUFTS UNIV, 82- *Mem:* Am Heart Asn; Am Inst Nutrit; Am Fedn Clin Res. *Mailing Add:* Dept Med Tufts Univ Med Sch 711 Washington St Boston MA 02111

SCHAEFER, FRANCIS T, b Hamilton, Mont, Mar 11, 13; m 37; c 2. CIVIL ENGINEERING, HYDROLOGY. *Educ:* Univ Minn, BCE, 34. *Prof Exp:* Observer, US Coast & Geod Surv, 34-35; surveyman, Corps Engrs, New Orleans, La, 35-36; engr, Gulf Res & Develop Co, 36-37; hydraul engr, Surface Water Br, Ark, Tenn & Okla, US Geol Surv, 37-42, off engr, Nebr, asst dist engr, Ky, 49-54, dist engr, Wis, 54-60, br area chief, 60-64, asst regional hydrologist, Water Resources Div, 64-86; RETIRED. *Concurrent Pos:* Delaware River Master, US Geol Surv, 75-; consult, Govt Brazil, 73. *Mem:* Fel Am Soc Civil Engrs. *Res:* Science administration. *Mailing Add:* 6801 Lemon Rd McLean VA 22101

SCHAEFER, FRANK WILLIAM, III, b Dayton, Ohio, Sept 1, 42. ANIMAL PARASITOLOGY. *Educ:* Miami Univ, BA, 64; Univ Cincinnati, MS, 70, PhD(biol), 73. *Prof Exp:* Res assoc parasitol, Univ Notre Dame, 73-78; MICROBIOLOGIST, US ENVIRON PROTECTION AGENCY, 78- *Honors & Awards:* Bronze Medal, US Environ Protection Agency, 84. *Mem:* AAAS; Sigma Xi; Am Soc Parasitol; Am Soc Microbiol; Soc Protozoologists. *Res:* In vitro cultivation of helminth and protozoan parasites; physiological studies relating to these parasites. *Mailing Add:* Environ Monitoring Syst Lab 26 West M L King Dr Cincinnati OH 45268

SCHAEFER, FREDERIC CHARLES, b Syracuse, NY, Nov 28, 17; m 45; c 3. ORGANIC CHEMISTRY. *Educ:* Syracuse Univ, BS, 39; Univ Akron, MS, 40; Mass Inst Technol, PhD(org chem), 43. *Prof Exp:* Res chemist, Goodyear Tire & Rubber Co, Ohio, 40-41; res chemist, 43-56, RES ASSOC, AM CYANAMID CO, 56-, proj leader, 74- *Concurrent Pos:* Am Cyanamid Co fel, Inst Org Chem, Univ Munich, 62-63. *Mem:* Am Chem Soc. *Res:* Organic nitrogen chemistry; heterocycles; s-triazines; reactive polymerizable materials; synthetic resin intermediates; keratin chemistry; hair care science. *Mailing Add:* 506 Jadetree Ct W Columbia SC 29169-4962

SCHAEFER, FREDERICK VAIL, b Fort Dix, NJ, May 25, 49; m 69. DNA DIAGNOSTIC MEDICINE, DEVELOPMENTAL BIOLOGY. *Educ:* Univ Md, BS, 71; NC State Univ, PhD(biochem), 79. *Prof Exp:* Res technician, Electronucleonics Lab, 71-73; fel res, 79-81, res assoc, 81-84, CLIN ASST PROF PEDIAT, INST CANCER RES, 85-, DIR, MOLECULAR GENETICS, 85- *Concurrent Pos:* Res fel, NIH, 80-81. *Mem:* Sigma Xi; Am Soc Human Genetics; Am Chem Soc; AAAS. *Res:* Molecular basis of carcinogensis; develop testing protocols for cancer diagnosis; determine molecular changes in the growth hormone related genes. *Mailing Add:* 5429 S 72nd East Ave Tulsa OK 74145

SCHAEFER, GEORGE, b New York, NY, May 30, 13; m 44; c 2. OBSTETRICS & GYNECOLOGY. *Educ:* NY Univ, BS, 33; Cornell Univ, MD, 37; Am Bd Obstet & Gynec, dipl, 48. *Prof Exp:* Attend obstetrician & gynecologist, Sea View Hosp, 40-48; from asst prof to prof, 51-78, EMER PROF OBSTET & GYNEC, MED COL, CORNELL UNIV, 78- *Concurrent Pos:* Consult, Booth Mem Hosp, 63-; attend, NY Hosp, 62-78; dir, obstet & gynec residency training prog, Mercy Hosp & Med Ctr, San Diego, 80-86; clin prof reproductive med, San Diego, Calif, 83-86. *Mem:* Fel Am Col Surg; fel Am Col Obstet & Gynec; hon fel Span Gynec Soc. *Res:* Tuberculosis in obstetrics and gynecology; study of the expectant father. *Mailing Add:* 6144 Laport La Mesa CA 91942-4313

SCHAEFER, GERALD J, b Sheboygan, Wis, Aug 12, 44; m 68; c 1. BEHAVIORAL PHARMACOLOGY, REWARD SYSTEMS IN THE BRAIN. *Educ:* Univ Wis-Milwaukee, BA, 68; Univ Akron, Ohio, MA, 70; Vanderbilt Univ, PhD(neurosci), 74. *Prof Exp:* Postdoctoral fel pharmacol, 74-76, ASSOC PROF PSYCHIAT & ASST PROF PHARMACOL, EMORY UNIV, 82-; RES SCIENTIST NEUROPHARMACOL, DEPT HUMAN RESOURCES, STATE OF GA, 76- *Mem:* AAAS; Am Soc Pharmacol & Exp Therapeut; Sigma Xi; Soc Neurosci; Int Brain Res Orgn. *Res:* Behavioral and biochemical effects of psychoactive drugs in animals; new drugs to treat mental health problems; operant conditioning and intracranial self-stimulation techniques; biochemical indices including measurement of neurotransmitter levels in the brain. *Mailing Add:* Ga Ment Health Inst Rm 504-N Atlanta GA 30306

SCHAEFER, HENRY FREDERICK, III, b Grand Rapids, Mich, June 8, 44; m 66; c 5. THEORETICAL CHEMISTRY. *Educ:* Mass Inst Technol, SB, 66; Stanford Univ, PhD(chem), 69. *Prof Exp:* From asst prof to prof chem, Univ Calif, Berkeley, 69-87, staff mem, Nuclear Chem Div, Lawrence Berkeley Lab, 71-75, staff mem, Molecular & Mat Res Div, 75-87; Wilfred T Doherty prof chem & dir Inst Theoret Chem, Univ Tex, Austin, 79-80; GRAHAM PERDUE PROF CHEM & DIR CTR COMPUTATIONAL QUANTUM CHEM, UNIV GA, 87- *Concurrent Pos:* Consult, Eastman Kodak, 74, Lawrence Livermore Lab, 68-80 & Energy Conversion Devices, 82-86, Dow Chem Co, 87, IBM Corp, 87; joint study proj, Univ Calif-IBM Res Lab, 72-77; Alfred P Sloan Res Fel, 72-74; John Simon Guggenheim Fel, 76-77; chmn, Subdiv Theoret Chem, Div Phys Chem, Am Chem Soc, 82, mem exec comt, 84-87, vchair, 90; mem NSF adv comt chem, 89-92. *Honors & Awards:* Am Chem Soc Pure Chem Award, 79, Leo Hendrik Baekeland Award, 83; Ann Award, World Asn Theoret Org Chemists, 90; Albert Einstein Centennial Lectr, Nat Univ Mex, 79, Lester P Kuhn Lectr, Johns Hopkins Univ, 82, John Howard Appleton Lectr, Brown Univ, 85, J A Erskine Lectr, Univ Canterbury, Christchurch, New Zealand, 86, Edward Curtis Franklin Lectr, Univ Kans, 86, John Lee Pratt Lectr, Univ Va, 88, Louis Jacob Bircher Lectr, Vanderbilt Univ, 88, Harry Emmett Gunning Lectr, Univ Alta, 90, Guelph-Waterloo distinguished Lectr, Univ Guelph & Univ Waterloo, Ont, 91. *Mem:* Fel Am Phys Soc; fel Int Acad Quantum Molecular Sci. *Res:* Rigorous quantum mechanical studies of the electronic structure of atoms and molecules. *Mailing Add:* Ctr Computational Quantum Chem Univ Ga Chem Bldg Rm 404 Athens GA 30602

SCHAEFER, JACOB FRANKLIN, b San Francisco, Calif, Aug 13, 38; m 62; c 2. PHYSICAL CHEMISTRY. *Educ:* Carnegie Inst Technol, BS, 60; Univ Minn, PhD(phys chem), 64. *Prof Exp:* Res chemist phys chem, 64-74, SCI FEL, DEPT NUCLEAR MAGNETIC RESONANCE, MONSANTO CO, 74- *Mem:* Am Chem Soc. *Res:* Nuclear magnetic resonance spectroscopy of polymers. *Mailing Add:* Dept Chem Wash Univ St Louis MO 63130

SCHAEFER, JACOB W, b Paullina, Iowa, June 27, 19; m 41; c 3. COMMUNICATIONS, SYSTEMS ENGINEERING. *Educ:* Ohio State Univ, BME, 41. *Hon Degrees:* DSc, Ohio State Univ, 76. *Prof Exp:* Engr, Bell Tel Labs, 41-42, 46-50, supvr Nike Ajax & Hercules syst design, 50-57, dept head Nike Zeus res, 57-61, asst dir Nike Zeus proj, 61-63, dir Kwajalien Field Sta, 63-65, dir data commun lab, 65-68, exec dir data & PBX div, 68-70, exec dir customer switching servs, 70-81, exec dir, mil systs div, 81-84, CONSULT, MIL SYSTS DIV, AT&T BELL LABS, 84- *Concurrent Pos:* Mem subcomt, Air Force Sci Adv Bd, 61-63. *Mem:* Nat Acad Eng; Am Ord Asn; fel Inst Elec & Electronics Engrs; Sigma Xi. *Res:* Servomechanisms for missile control surfaces; missile guidance and air defense missile systems; design of data communication terminals. *Mailing Add:* 115 Century Lane Watchung NJ 07060

SCHAEFER, JOSEPH ALBERT, b Bellevue, Iowa, Dec 24, 40; m 65; c 2. SOLID STATE PHYSICS. *Educ:* Loras Col, BS, 62; Univ Toledo, MS, 64; Northwestern Univ, Evanston, PhD(physics), 72. *Prof Exp:* From instr to assoc prof, 64-80, PROF PHYSICS, LORAS COL, 80- *Concurrent Pos:* Res fel, Northwestern Univ, Evanston, 74; consult, John Deere Co, 80; NSF sci fac prof develop fel, Iowa State Univ, 81-82; assoc res scientist, Iowa Inst Hydraul Res, 85, 86, 87, 88, 89 & 90. *Mem:* Am Asn Physics Teachers; Am Soc Eng Educ. *Res:* Magnetoresistance of potassium; transport properties of metals; switching varistors; ice engineering. *Mailing Add:* Dept Physics Loras Col 1450 Alta Vista Dubuque IA 52004-0178

SCHAEFER, JOSEPH THOMAS, b Milwaukee, Wis, Oct 23, 43; m 68; c 3. METEOROLOGY, WIND ENGINEERING. *Educ:* St Louis Univ, BS, 65, PhD(meteorol), 73. *Prof Exp:* Meteorologist, Nat Weather Serv, St Louis, Mo, 66-69; res meteorologist, US Navy Weather Res Facil, 69-71; res meteorologist, Nat Severe Storms Lab, 71-76, chief tech develop unit, Nat Severe Storms Forecast Ctr, 76-83, CHIEF SCI SERV, NAT WEATHER SERV CENT REGION, NAT OCEANIC & ATMOSPHERIC ADMIN, 83- *Concurrent Pos:* Mem, US/Japan Panel Wind & Seismic Effects, 79-; instr, Univ Mo, Kansas City, 85-86. *Honors & Awards:* Bronze Medal, Dept Com. *Mem:* Fel Am Meteorol Soc; Am Geophys Union; Nat Weather Asn; Can Meteorol & Oceanog Soc. *Res:* Weather forecasting; tornado climatology; severe thunderstorm environment; severe thunderstorm dynamics and structure; atmospheric boundary layer structure; mesoscale numerical atmospheric modelling; cumulus dynamics. *Mailing Add:* 5040 N Flora Kansas City MO 64116

SCHAEFER, PAUL THEODORE, b Rochester, NY, Mar 7, 30; m 51. MATHEMATICS. *Educ:* Univ Rochester, AB, 51, MA, 56; Univ Pittsburgh, PhD(infinite series), 63. *Prof Exp:* Instr math, Rochester Inst Tech, 55-56; from asst prof to prof, State Univ NY Albany, 56-67; PROF MATH, STATE UNIV NY COL GENESEO, 67- *Concurrent Pos:* First vchmn seaway sect, Math Asn Am, 76-77, chmn, 77-78; vis prof, Calif State Univ, Los Angeles, 78-79. *Mem:* Am Math Soc; Math Asn Am. *Res:* Series; summability; real and complex analysis. *Mailing Add:* Dept Math State Univ Col Arts & Sci Geneseo NY 14454

SCHAEFER, PHILIP WILLIAM, b Baltimore, Md, Feb 16, 35; m 58; c 4. APPLIED MATHEMATICS. *Educ:* John Carroll Univ, BS, 56, MS, 57; Univ Md, PhD(math), 64. *Prof Exp:* Instr math, Loyola Col, Md, 59-60; asst prof, Univ SFla, 64-67; assoc prof, 67-77, PROF MATH, UNIV TENN, KNOXVILLE, 77- *Mem:* Am Math Soc; Soc Indust & Appl Math. *Res:* Elliptic partial differential equations; maximum principles and solution bounds. *Mailing Add:* Dept Math Univ Tenn Knoxville TN 37996-1300

SCHAEFER, ROBERT J, b New Rochelle, NY, June 9, 39; m 74; c 1. PHASE TRANSFORMATIONS. *Educ:* Harvard Univ, AB, 60, AM, 61, PhD(appl physics), 65. *Prof Exp:* Res physicist, US Naval Res Lab, 64-79; PHYSICIST, NAT INST STANDARDS & TECHNOL, 79- *Mem:* Metall Soc, Am Inst Mining Engrs; Am Soc Metals; Am Asn Crystal Growth. *Res:* Solidification of metals; kinetics of growth of alloy phases; crystal morphology development; effects of rapid solidification; hot isostatic pressing; phase equilibria. *Mailing Add:* Metall Div Nat Inst Standards & Technol Gaithersburg MD 20899

SCHAEFER, ROBERT WILLIAM, b Schenectady, NY, Dec 5, 27; m 54; c 4. ANALYTICAL CHEMISTRY. *Educ:* Siena Col, BS, 49; Union Col, NY, MS, 51. *Prof Exp:* Asst, 49-51, from asst prof to assoc prof, 52-81, chairperson, 79-85, PROF ANALytical CHEM, UNION COL, 81- *Concurrent Pos:* Develop chemist, Willsboro Mining Co, 50; asst, Univ Ky, 51; sr sanit chemist, State Dept Health, NY, 54-60; lectr, Bard Col, 55; consult, Power Technol Inc, Schenectady, NY, 80-84, Welch Chem Co, Schenectady, NY, 65-86. *Mem:* Am Chem Soc. *Res:* Instrumental analytical chemistry; sanitary chemistry; methods for trace metals. *Mailing Add:* Dept Chem Union Col Schenectady NY 12308

SCHAEFER, SETH CLARENCE, b Tripp, SDak, Feb 15, 23. METALLURGY, THERMODYNAMICS. *Educ:* SDak Sch Mines, BS, 47; Univ Mo, Rolla, MS, 63. *Prof Exp:* Jr metallurgist, Am Smelting & Refining Co, Nebr, 47-49, actg smelter supt, 50-55, asst lead refinery supt, 56, smelter supt, 57-59, asst lead refinery supt, NJ, 59-61; metallurgist, 63-66, proj leader chem processes, 67-69, METALLURGIST, ALBANY METALL RES CTR, US BUR MINES, 69- *Mem:* Sigma Xi; Am Inst Mining, Metall & Petrol Engrs; Am Soc Metals. *Res:* Metallurgical thermodynamics. *Mailing Add:* 1014 S Lawn Ridge Albany OR 97321

SCHAEFER, THEODORE PETER, b Gnadenthal, Man, July 22, 33; m 60; c 3. PHYSICAL CHEMISTRY. *Educ:* Univ Man, BSc, 54, MSc, 55; Oxford Univ, DPhil(chem), 58. *Hon Degrees:* DSc, Univ Winnipeg, 82. *Prof Exp:* From asst prof to prof, 58-82, UNIV DISTINGUISHED PROF CHEM, UNIV MAN, 82- *Concurrent Pos:* Nat Res Coun Can sr res fel, 64-65; mem, Chem Grants Comt, Nat Res Coun Can, 75-77; mem, Natural Sci & Eng Res Coun, Can, 80-84; mem, Publ Grants Comt, Natural Sci & Eng Res Coun, Can 87-89. *Honors & Awards:* Noranda Award, Chem Inst Can, 73; Herzberg Award, Spectroscopy Soc Can, 75. *Mem:* Fel Royal Soc Can; fel Chem Inst Can. *Res:* Nuclear magnetic resonance spectroscopy in chemistry. *Mailing Add:* Dept Chem Univ Man Winnipeg MB R3T 2N2 Can

SCHAEFER, WILBUR CARLS, b Beardstown, Ill, Sept 18, 25; m 49; c 2. AGRICULTURAL CHEMISTRY. *Educ:* Bradley Univ, BS, 49, MS, 50. *Prof Exp:* Res chemist, Agr Res Serv, USDA, 50-61, asst to dir, Northern Regional Res Ctr, 61-72, prog analyst, 72-87; CONSULT, 87- *Mem:* AAAS; Am Asn Cereal Chem (treas, 72-74); Am Chem Soc. *Res:* Utilization of cereal grains and oilseeds; starch and dextran structure; starch derivatives; wheat gluten properties; histochemistry of wheat endosperm; chemical research administration; carbohydrate and protein chemistry. *Mailing Add:* 2208 W Newman Pkwy Peoria IL 61604

SCHAEFER, WILLIAM PALZER, b Bisbee, Ariz, Jan 13, 31; m 54; c 2. SYNTHETIC INORGANIC & ORGANOMETALLIC CHEMISTRY. *Educ:* Stanford Univ, BS, 52; Univ Calif, Los Angeles, MS, 54, PhD(anal chem), 60. *Prof Exp:* From instr to asst prof chem, Calif Inst Technol, 60-66; asst prof, Univ Calif, Davis, 66-68; sr res fel chem, 68-77, registrar, 71-77, dir financial aid, 72-77, res assoc, 77-81, SR RES ASSOC, CALIF INST TECHNOL, 81-, DIR, X-RAY DIFFRACTION FACIL, 86- *Mem:* Am Chem Soc; Am Crystallog Asn; fel AAAS. *Res:* Structure and stability of complexes of the transition metals; x-ray crystallography; structural chemistry. *Mailing Add:* Beckman Inst 139-74 Calif Inst Technol Pasadena CA 91125

SCHAEFERS, GEORGE ALBERT, b Erie, Pa, Mar 19, 29; m 60; c 3. ENTOMOLOGY. *Educ:* Univ Calif, BS, 55, PhD, 58. *Prof Exp:* From asst prof to assoc prof, 58-74, PROF ENTOM, NY STATE COL AGR, CORNELL UNIV, 74-, CHMN DEPT, 83- *Concurrent Pos:* Vis scientist, Int Inst Trop Agr, 74-75 & Coop State Res Serv, USDA, 81-82. *Mem:* Entom Soc Am. *Res:* Economic entomology; small fruit insects; insect vectors plant diseases; Aphid biology. *Mailing Add:* Dept Entom Cornell Univ Agr Exp Sta Geneva NY 14456

SCHAEFFER, BOBB, b New Haven, Conn, Sept 27, 13; m 41; c 2. VERTEBRATE PALEONTOLOGY. *Educ:* Cornell Univ, BA, 36; Columbia Univ, MA, 37, PhD(zool), 41. *Prof Exp:* Demonstr histol & embryol, Jefferson Med Col, 41-42; asst cur vert paleont, Am Mus Natural Hist, 46-49, assoc cur fossil fishes, 49-55, chmn dept, 66-76; vis assoc prof zool, Columbia Univ, 55-57, adj prof, 57-59, prof, 59-78; EMER CUR VERT PALEONT, AM MUS NATURAL HIST, 76- *Concurrent Pos:* Cur vert paleont, Am Mus Natural Hist, 55-76. *Mem:* Soc Vert Paleont (actg secy, 52-53, pres, 53); Soc Syst Zool; Soc Study Evolution; fel Geol Soc Am; Sigma Xi. *Res:* Systematics; morphology and embryology of fishes; systematic theory. *Mailing Add:* 1400 East Ave Rochester NY 14610

SCHAEFFER, CHARLES DAVID, b Allentown, Pa, June 14, 48. INORGANIC CHEMISTRY. *Educ:* Franklin & Marshall Col, BA, 70; State Univ NY, Albany, PhD(chem), 74. *Prof Exp:* Fel inorg chem, Yale Univ, 74-76; asst prof, 76-81, ASSOC PROF CHEM, ELIZABETHTOWN COL, 81- *Concurrent Pos:* Res grant, Cottrell Col, 77-78; Petrol Res Fund, Am Chem Soc grant, 78-80 & 84-86; res grant, Cottrell Col, 82-84. *Mem:* Am Chem Soc; Royal Soc Chem, London. *Res:* Main group organometallic chemistry; nuclear magnetic resonance spectroscopy. *Mailing Add:* Chem Dept Elizabethtown Col Elizabethtown PA 17022

SCHAEFFER, DAVID GEORGE, b Cincinnati, Ohio, Oct 6, 42; m 70. MATHEMATICS. *Educ:* Univ Ill, Urbana, BS, 63; Mass Inst Technol, PhD(math), 68. *Prof Exp:* Instr math, Brandeis Univ, 68-70; asst prof, 70-75, ASSOC PROF MATH, MASS INST TECHNOL, 75- *Mem:* Am Math Soc. *Res:* Partial differential equations; approximations by finite differences; functional analysis. *Mailing Add:* Dept Math/135 C Physics Bldg Duke Univ Durham NC 27706

SCHAEFFER, DAVID JOSEPH, b Brooklyn, NY, Feb 7, 43; m 70; c 2. ENVIRONMENTAL TOXICOLOGY, RISK ASSESSMENT. *Educ:* Brooklyn Col, BS, 63; Northwestern Univ, MS, 65; City Univ New York, PhD(org chem), 69. *Prof Exp:* Res asst chem, Brooklyn Col, 60-63, lectr, 65-69; res assoc, State Univ NY Binghamton, 69-70; asst prof, Sangamon State Univ, 70-72; adv environ sci, Ill Environ Protection Agency, 72-85; ENVIRON TOXICOLOGIST, CORPS ENGRS, CONSTRUCT ENG RES LAB, 85- *Concurrent Pos:* Adj assoc prof toxicol, Southern Ill Univ, 79-; adj asst prof health serv, Sangamon State Univ, 80; Hill scholar in residence, Univ Minn, 85. *Mem:* Am Chem Soc; NY Acad Sci; Int Asn Gt Lakes Res; Am Statist Asn; Nat Speleol Soc. *Res:* Environmental chemistry of water pollutants; statistical properties, chance mechanisms and quality control of environmental data. *Mailing Add:* 3502 Roxford Champaign IL 61821-5248

SCHAEFFER, GENE THOMAS, b Reading, Pa, Jan 15, 32; m 57; c 3. METALLURGY. *Educ:* Albright Col, BS, 56; Syracuse Univ, MS, 61, PhD(solid state sci & technol), 65. *Prof Exp:* Process engr, Western Elec Co, Inc, 56-57; ADVAN DEVELOP ENGR, CHEM & METALL DIV, GTE SYLVANIA INC, 65- *Mem:* Am Inst Mining, Metall & Petrol Engrs; Am Soc Metals; Sigma Xi. *Res:* Heavy metals; mechanical properties of materials; properties and processing of incandescent lamp metals. *Mailing Add:* 16 Thomas St Towanda PA 18848

SCHAEFFER, HAROLD F(RANKLIN), b Philadelphia, Pa, Sept 21, 99; wid; c 1. CHEMICAL MICROSCOPY, PHYSICAL CHEMISTRY. *Educ:* Muhlenberg Col, BSc, 22; Univ NH, MSc, 26. *Prof Exp:* High sch teacher, Pa, 22-24; asst agr chem, Univ NH, 24-26; prof chem, Waynesburg Col, 26-42, head dept, 40-42; instr, Univ Mo, 42-43; prof, Col Our Lady of the Elms, 43-44; res assoc in chg sulfur-org res lab, Univ Ala, 44-48; asst prof chem, Valparaiso, 48-52; res scientist, US Army Ord, Rensselaer Polytech, 52-53; prof chem, Grove City Col, 53-55; prof & head dept, Col Emporia, 55-59; prof, 59-68, res prof chem, 68-76, EMER PROF CHEM, WESTMINSTER COL, 76-; CONSULT, SCHAEFFER RES ASSOC, 76- *Concurrent Pos:* Abstractor, Chem Abstracts, 48-76; Res Corp grants, 54, 57 & 60; vist scientist, Mo high schs, 65-67. *Mem:* AAAS; Am Chem Soc; Am Microchem Soc; Sigma Xi. *Res:* Direct sulfuration of organic compounds; microscopic methods in chemistry; chemical microscopy of platinum metals; quantitative microscopy; thermal microscopy; reactions of organic squarates and other heterocyclic amines. *Mailing Add:* 15 E Chestnut St Fulton MO 65251

SCHAEFFER, HOWARD JOHN, b Rochester, NY, Mar 14, 27; m 50; c 4. MEDICINAL CHEMISTRY. *Educ:* Univ Fla, PhD(pharmaceut chem), 55. *Prof Exp:* Sr scientist, Southern Res Inst, 55-57, head pharmaceut chem sect, 57-59; assoc prof med chem & actg chmn dept, State Univ NY Buffalo, 59-63, prof, 63-70, chmn dept, 65-70; dept head org chem, Wellcome Res Labs, Burroughs Wellcome Co, 70-74; mem fac, Med Col Va, Va Commonwealth Univ, 74-77; mem staff, 77-, VPRES RES, DEVELOP & MED, BURROUGHS WELLCOME CO, 86- *Honors & Awards:* Ebert Prize, Am Pharmaceut Asn; Bristol Award Chemotherapy. *Mem:* Am Chem Soc; Acad Pharmaceut Sci; Sigma Xi. *Res:* Enzyme inhibition; kinetics; stereochemistry; antiviral chemotherapy. *Mailing Add:* 123 Bruce Dr Cary NC 27511

SCHAEFFER, JAMES ROBERT, b Rochester, NY, Aug 2, 33; m 67; c 4. ORGANIC CHEMISTRY, MICROBIOLOGY. *Educ:* Univ Notre Dame, BS, 55; Univ Pa, PhD(org chem), 59. *Prof Exp:* Res chemist, Synthetic Chem Div, 60-62, from res chemist to sr res chemist, Chem Div, 62-69, RES ASSOC, CHEM DIV, RES LABS, EASTMAN KODAK CO, 69- *Concurrent Pos:* US Army Off Ord Res fel, Northwestern Univ, 59-60. *Mem:* Am Chem Soc. *Res:* Chelate, dye and microbiological chemistry; organic sulfur compounds; development laboratory studies in organic synthesis; application of fermentation techniques to the synthesis of organic compounds; preparation of biocatalysts; development of thin films for use in clinical analysis. *Mailing Add:* Eastman Kodak Co 343 State St Rochester NY 14650

SCHAEFFER, JOHN FREDERICK, EPITHELIAL TRANSPORT. *Educ:* Syracuse Univ, PhD(physiol), 70. *Prof Exp:* ASSOC PROF PHYSIOL & BIOPHYSICS, SCH MED, IND UNIV, 78- *Mailing Add:* Ind Univ Sch Med Evansville Ctr Med Educ PO Box 3287 Evansville IN 47732

SCHAEFFER, LEE ALLEN, b Allentown, Pa, June 20, 43; m 70. ORGANIC CHEMISTRY. *Educ:* Lehigh Univ, BS, 65, MS, 69, PhD(org chem), 72. *Prof Exp:* Chemist, Lubrizol Corp, Cleveland, Ohio, 65-67; RES ASSOC, CROMPTON & KNOWLES CORP, 73- *Mem:* Am Chem Soc; Am Asn Textile Chemists & Colorists; Sigma Xi. *Res:* Preparation of dyes and related chemicals. *Mailing Add:* 6335 Walker Rd Macungie PA 18062

SCHAEFFER, MORRIS, b Berdichev, Ukraine, Russia, Dec 31, 07; US citizen; m 43; c 4. MEDICINE, MICROBIOLOGY. *Educ:* Univ Ala, AB & MA, 30; NY Univ, PhD(microbiol), 35, MD, 44; Am Bd Prev Med, dipl, 50. *Prof Exp:* Asst instr, Univ Ala, 28-30; asst, Res Labs, Dept Health, New York, 30-31, bacteriologist, 35-36; res assoc, 36-41, mem, Pub Health Res Inst, 41-42; house officer, City Hosp, Boston, Mass, 44-45; asst prof microbiol & pediat, Sch Med, Western Reserve Univ, 46-49; med dir, Virus & Rickettsia Sect, Commun Dis Ctr, USPHS, 49-59; asst Comm Health, Pub Health Res Inst & dir, Bur Labs, Dept Health, New York, 59-71; consult, Pan-Am Health Orgn, 71-72; dir, Off Efficacy Rev, Bur Biologics, Food & Drug Admin, 72-82, dir, Off Sci Adv & Consults, Ctr Drugs & Biologics, 82-88; RETIRED. *Concurrent Pos:* Instr, Sch Med, NY Univ, 33-36, lectr, 36-37, adj prof, 59-; lectr, Sch Pub Health, Columbia Univ, 63-73; resident contagious div, City & Babies & Childrens Hosps, Cleveland, 45-48; dir contagious pavillion, City Hosp & librn, lectr & consult, Nursing Sch, 48-49; assoc prof, Sch Med, Emory Univ, 51-59; vis lectr, Sch Med, Tulane Univ, 52-59; vis prof, Univ Havana, 52, Univ Gothenberg, 55, Univ Hawaii, 59 & Univ Wash, 72; consult to Surgeon Gen, USAF, 59-64 & Surgeon Gen comt influenza res, USPHS, 59-63; mem expert panel on viruses, WHO, 60-79; assoc mem, Armed Forces Epidemiol Bd, Comn Respiratory Dis, 61-70; mem adv comt, Ctr Dis Control, USPHS, 62-71, bd sci counsr, Div Biol Standards, 63-67; panel on virus dis, US-Japan Comt Biomed Res, NIH, 65-69; vis prof microbiol, Univ Wash, 72- *Mem:* Am Col Prev Med; Am Soc Microbiol; Soc Exp Biol & Med; Am Pub Health Asn; Am Asn Immunol; Sigma Xi. *Res:* Virology; public health lab management; infectious diseases; vaccine development; public health and epidemiology. *Mailing Add:* 84870 S Willamette St Eugene OR 97405-9514

SCHAEFFER, RILEY, b Michigan City, Ind, July 3, 27; m 50; c 4. INORGANIC CHEMISTRY. *Educ:* Univ Chicago, BS, 46, PhD(chem), 49. *Prof Exp:* Res chemist, Univ Chicago, 49-52; from asst prof to assoc prof chem, Iowa State Univ, 52-58; from assoc prof to prof, Ind Univ, Bloomington, 58-75, chmn dept, 67-72; dean arts & sci, Univ Wyo, 76-77, prof chem, 76-81; PROF & CHMN, DEPT CHEM, UNIV NMEX, 81- *Concurrent Pos:* NSF sr fel, 61-62; Guggenheim fel, 65-66; bd mem, Petrol Res Found, 70-72; vis comt inorg mat, Nat Bur Standards, 73-79; mem adv comt, NSF, 77-83. *Mem:* Am Chem Soc; fel AAAS; Am Asn Univ Professors; Sigma Xi; hon fel Royal Soc Chem. *Res:* Inorganic and physical inorganic chemistry of hydrides; chemistry of covalent inorganic compounds; archaeology; metal-oxygen clusters; structural inorganic chemistry; x-ray diffraction. *Mailing Add:* 2065 Hillsdale Circle Boulder CO 80303

SCHAEFFER, ROBERT L, JR, b Allentown, Pa, Oct 31, 17. PLANT TAXONOMY. *Educ:* Haverford Col, BS, 40; Univ Pa, PhD(bot), 48. *Prof Exp:* Asst bot, Univ Pa, 42-44, instr, 46-47; from asst prof to assoc prof biol, Upsala Col, 48-52; from asst prof to prof biol, Muhlenberg Col, 52-83;

RETIRED. *Mem:* AAAS; Am Fern Soc; Bot Soc Am; Torrey Bot Club; Sigma Xi. *Res:* Floristics of eastern Pennsylvania and northern New Jersey; floristic, taxonomic and evolutionary botany. *Mailing Add:* 32 N Eighth St Allentown PA 18101

SCHAEFFER, WARREN IRA, b Newark, NJ, Aug 13, 38. BACTERIOLOGY, CELL BIOLOGY. *Educ:* Rutgers Univ, BS, 60, MS, 62, PhD(bact), 64. *Prof Exp:* Res assoc cell biol, Mass Inst Technol, 64-66; asst prof med microbiol, Univ Calif-Calif Col Med, 66-67; assoc prof, Univ Vt, 67-77, actg chmn, 77-79, chmn, 79-88, PROF MED MICROBIOL, COL MED, UNIV VT, 77- *Concurrent Pos:* Chair Terminology Comt, Tissue Cult Asn, Terminology Comt, Int Asn Cell Cult. *Mem:* AAAS; Am Soc Microbiol; Sigma Xi; Tissue Cult Asn (secy, 84-88). *Res:* Tissue culture; effects of biologically active agents in cell cultures; expression and maintenance of the carcinogenic state; in vitro aging of epithelial cells; mycoplasmology. *Mailing Add:* Dept Microbiol & Molecular Genetics Univ Vt Col Med & Col Agric & Life Sci Burlington VT 05405

SCHAEFFER, WILLIAM DWIGHT, b Reading, Pa, Dec 4, 21; m 46; c 3. COLLOID CHEMISTRY. *Educ:* Lehigh Univ, BS, 43, MS, 47, PhD(chem), 67. *Prof Exp:* Asst instr chem, Lehigh Univ, 46-47; chemist & group leader, Res & Develop Dept, Godfrey L Cabot, Inc, 47-55; assoc res dir, Nat Printing Ink Res Inst, Lehigh Univ, 55-69; res dir, Graphic Arts Tech Found, 69-89; EXEC DIR, ENVIRON CONSERV BD GRAPHIC COMMUN INDUSTS INC, 86- *Concurrent Pos:* Vchmn, Int Asn Res Insts Graphic Arts Indust, 71-77; exec dir, Environ Conserv Bd, Graphic Commun Industs Inc, 86- *Mem:* Am Chem Soc; Tech Asn Pulp & Paper Indust; NY Acad Sci; Tech Asn Graphic Arts (pres, 69-70); Inter-Soc Color Coun (pres, 81-82); Sigma Xi. *Res:* Physical colloid and surface chemistry; carbon blacks; printing and printing inks; dispersions of pigments; environmental science. *Mailing Add:* Graphic Arts Tech Found 4615 Forbes Ave Pittsburgh PA 15213

SCHAEFGEN, JOHN RAYMOND, b Wilmette, Ill, Apr 9, 18; m 45; c 9. POLYMER CHEMISTRY. *Educ:* Northwestern Univ, Ill, BS, 40; Ohio State Univ, PhD(phys org chem), 44. *Prof Exp:* Sr chemist, Res Lab, Goodyear Tire & Rubber Co, 44-51; jr res assoc, Textile Fibers Dept, Pioneering Res, E I DuPont de Nemours & Co, Inc, 51-58, res assoc, 58-77, res fel, 77-82; CONSULT, ELJAY ASSOC, INC, 83- *Mem:* Am Chem Soc. *Res:* Physical organic chemistry; mechanisms of organic reactions; synthesis and study of properties of high polymers; fiber technology. *Mailing Add:* 129 Cambridge Dr Wilmington DE 19803

SCHAEPPI, ULRICH HANS, neurotoxicology, neuropharmacology, for more information see previous edition

SCHAER, JONATHAN, b Bern, Switz, Oct 1, 29; m 63; c 2. MATHEMATICS. *Educ:* Univ Bern, dipl educ, 52 & 57, Dr Phil(theoret physics), 62. *Prof Exp:* Pub sch teacher, Bern, Switz, 55-56; asst physics, Univ Bern, 56-57; asst math, Swiss Fed Inst Technol, 57-58; asst physics, Univ Bern, 58-62; asst prof math, Univ Alta, 62-67; ASSOC PROF MATH, UNIV CALGARY, 67- *Mem:* Can Math Soc; Am Math Soc; Math Asn Am. *Res:* Geometry; convexity; combinatorics; theory of relativity. *Mailing Add:* Dept Math Univ Calgary 2500 University Dr NW Calgary AB T2N 1N4 Can

SCHAERF, HENRY MAXIMILIAN, b Rohatyn, Poland, Mar 17, 07; nat US. MEASURE THEORY, INTEGRATION THEORY. *Educ:* Univ Lwow, MA, 29; Univ Göttingen, actuary's cert, 31; Swiss Fed Inst Technol, Habilitated(math), 45. *Hon Degrees:* DSc, Swiss Fed Inst Technol, 43. *Prof Exp:* Head actuarial dept, Der Anker Ins Co, Poland, 31-37; chief actuary, Vita-Kotwica Life Ins Co, 37-39; lectr math, Polish mil internees, Switz, 41-44; privat docent, Swiss Fed Inst Technol, 45-48; instr, Mont State Col, 46-47; from asst prof to assoc prof, 47-75, EMER ASSOC PROF MATH, WASH UNIV, 75- *Concurrent Pos:* Ford Found fel, Inst Advan Study, NJ & Denmark, 53-54; mem, US Army Math Res Ctr, Univ Wis, 58 & 61, vis prof, Ctr, 62-64; assoc prof, McGill Univ, 64-72. *Mem:* Am Math Soc; Swiss Math Soc; Swiss Asn Actuaries; Polish Math Soc. *Res:* Actuarial theory; real variables; measure and integration theory; structure and cardinality of bases; topological properties of maps; invariant measures; mathematical foundations of actuarial theory. *Mailing Add:* 600 W Olympic Pl Apt 507 Seattle WA 98119

SCHAETTI, HENRY JOACHIM, petroleum exploration, for more information see previous edition

SCHAETZLE, WALTER J(ACOB), b Pittsburgh, Pa, Feb 17, 34; m 65; c 2. MECHANICAL ENGINEERING. *Educ:* Carnegie Inst Technol, BS, 57, MS, 58; Wash Univ, DSc(mech eng), 62. *Prof Exp:* Propulsion engr, McDonnell Aircraft Corp, 58-62; from asst prof to assoc prof, Univ Ala, Tuscaloosa, 62-66, prof mech eng, 66-; RETIRED. *Concurrent Pos:* Mem staff, Col Petrol, Saudi Arabia, 69-71. *Mem:* Am Soc Mech Engrs; Am Soc Heating, Refrig & Air Conditioning Engrs; Am Soc Eng Educ; Int Solar Energy Soc. *Res:* Rarefied gas flow; experiments in molecule surface interactions; thermal energy storage; community energy systems; man produced energy tornado correlations; heating and cooling systems; heat pump systems; home energy conservation; solar energy. *Mailing Add:* Nine Oak Bluff North Port AL 35476

SCHAFER, DAVID EDWARD, b Wichita, Kans, Mar 8, 31; m 49; c 2. MEDICAL RESEARCH. *Educ:* Friends Univ, AB, 48; Univ Minn, PhD, 59. *Prof Exp:* Asst English, Univ Minn, 48-51, asst physiol, 53-56, instr, 57-58; from instr to asst prof, Sch Med, NY Univ, 58-63; asst prof pathobiol, Johns Hopkins Univ Ctr Med Res & Training, Calcutta, India, 64-66; mem field staff, Rockefeller Found, 66-68; res physiologist, Exp Surg Lab, Vet Admin Hosp, Minneapolis, 68-73; RES PHYSIOLOGIST, VET ADMIN HOSP, WEST HAVEN, 73-; LECTR PHYSIOL, SCH MED, YALE UNIV, 73- *Concurrent Pos:* Fulbright lectr, Sci Col, Calcutta Univ, 63-64; vis prof physiol & actg head dept, Fac Med Sci, Bangkok, 66-68; asst prof physiol, Med Sch, Univ Minn, Minneapolis, 68-73. *Mem:* AAAS; Biophys Soc; Soc Math Biol; Sigma Xi. *Res:* Cell physiology; perfusion; membrane permeability; gastrointestinal physiology; cholera. *Mailing Add:* Med Serv Vet Admin Hosp West Haven CT 06516

SCHAFER, IRWIN ARNOLD, b Pittsburgh, Pa, Mar 22, 28; m 48; c 3. PEDIATRICS, GENETICS. *Educ:* Univ Pittsburgh, BS, 48, MD, 53; Am Bd Pediat, dipl, 59; Am Bd Med Genetics, dipl, 84. *Prof Exp:* Intern, Montefiore Hosp, Pittsburgh, 53-54; epidemic intel off & chief hepatitis invest unit, Commun Dis Ctr, USPHS, 54-56; jr asst resident, Children's Hosp Med Ctr, Boston, 56-58; from instr to asst prof pediat, Stanford Univ, 61-67; assoc prof, 67-74, PROF PEDIAT, SCH MED, CASE WESTERN RESERVE UNIV, 74- *Concurrent Pos:* Res fel med, Children's Hosp Med Ctr, Boston, 58-61; res fel prev med, Harvard Med Sch, 58-61; dir premature infant res ctr, USPHS, 63-64, dir birth defects study ctr, 64-69; pediatrician, Case Western Univ Hosp & Cleveland Metrop Gen Hosp; dir, Genetics Prog, Cleveland Metrop Gen Hosp. *Mem:* Am Pediat Soc; Am Soc Pediat Res; Am Soc Human Genetics; Am Soc Cell Biol; Tissue Cult Asn. *Res:* Metabolism of cells in culture; cell differentiation and biology; embryonic organogenesis; biochemical genetics. *Mailing Add:* Rm 346 Res Bldg Cleveland Metrop Gen Hosp Cleveland OH 44109

SCHAFER, JAMES A, b Rochester, NY, Dec 11, 39; m 61; c 2. MATHEMATICS. *Educ:* Univ Rochester, BA, 61; Univ Chicago, MS, 62, PhD(math), 65. *Prof Exp:* Asst prof math, Univ Mich, Ann Arbor, 65-70; ASSOC PROF MATH, UNIV MD, COLLEGE PARK, 70- *Concurrent Pos:* Vis lectr, Aarhus Univ, 69-71 & 78-79. *Mem:* Am Math Soc. *Res:* Algebraic topology; homological algebra. *Mailing Add:* Dept Math Univ Md College Park MD 20742

SCHAFER, JAMES ARTHUR, b Buffalo, NY, Oct 10, 41; m 64; c 2. PHYSIOLOGY, NEPHROLOGY. *Educ:* Univ Mich, BS, 63, PhD(physiol), 68. *Prof Exp:* Fel biochem, Gustav-Embden Ctr, WGer, 68-69; fel physiol, Duke Univ, 69-70; from asst prof to assoc prof physiol, 70-76, from asst prof to assoc prof med, 70-80, PROF PHYSIOL, UNIV ALA, 76-, PROF MED, 80-, SR SCIENTIST, NEPHROLOGY RES & TRAINING CTR, 77- *Concurrent Pos:* Ed bd, J Gen Physiol, 79-, ed, Am J Physiol, 83-89; Welcome vis prof, Dartmouth, 86; chmn res comt, Nat Kidney & UrolDis adv bd, 87-90. *Honors & Awards:* Estab Investr Award, Am Heart Asn, 71; Robert F Pitts Mem Award for Outstanding Res Kidney Physiol, Int Congress Physiol Sci, Australia, 83. *Mem:* Am Physiol Soc; Am Soc Nephrol (secy-treas, 89-); Biophys Soc; Am Fedn Clin Res; Am Heart Asn. *Res:* Membrane transport process as related to renal tubules and intracellular regulatory processes. *Mailing Add:* Dept Physiol & Biophys Univ Ala Birmingham Birmingham AL 35294

SCHAFER, JOHN FRANCIS, b Pullman, Wash, Feb 17, 21; m 47; c 3. PLANT PATHOLOGY. *Educ:* Wash State Univ, BS, 42; Univ Wis, PhD(plant path, agron), 50. *Prof Exp:* Agt, Bur Plant Indust, USDA, 39-42; asst plant path, Univ Wis, 46-49; from asst prof to prof plant path, Purdue Univ, 49-68; prof & head dept, Kans State Univ, 68-72; prof & chmn dept, Wash State Univ, 72-80; integrated pest mgt coordr, Sci & Educ Admin, 80-81, actg nat res prog leader plant path & nematol, Agr Res Serv, 81-82, dir, 82-87, COLLABR, CEREAL RUST LAB, AGR RES SERV, UNIV MINN, USDA, 87- *Concurrent Pos:* Vis res prof, Duquesne Univ, 65-66; adj prof, Plant Path Univ Minn, 82- *Mem:* Fel AAAS; fel Am Phytopath Soc (pres, 78-79); Am Soc Agron. *Res:* Diseases of cereal crops; cereal breeding; plant disease resistance. *Mailing Add:* Cereal Rust Lab Univ Minn St Paul MN 55108

SCHAFER, JOHN WILLIAM, JR, b Mt Pleasant, Mich, May 18, 37; m 67; c 3. SOIL SCIENCE, SOIL GEOGRAPHY. *Educ:* Mich State Univ, BS, 59, PhD(soil sci), 68; Kans State Univ, MS, 60. *Prof Exp:* Assoc prof, 68-80, PROF AGRON, IOWA STATE UNIV, 80- *Concurrent Pos:* Vis prof, Shenyang Agr Col, People's Repub China. *Honors & Awards:* Wilton Park Award; Agron Educ Award, Am Soc Agron. *Mem:* Am Soc Agron; Soil Sci Soc Am; fel Nat Asn Cols & Teachers Agr; Soil Conserv Soc Am. *Res:* Relationships of soils to landscapes and their influence on land use and food production; new methods for effective teaching of undergraduate soil science; use of indigenous, knowledge of soils in decision making. *Mailing Add:* Dept Agron Iowa State Univ Ames IA 50011

SCHAFER, LOTHAR, b Dusseldorf, WGer, May 5, 39; m 65; c 2. PHYSICAL CHEMISTRY, INORGANIC CHEMISTRY. *Educ:* Univ Munich, dipl, 62, PhD(inorg chem), 65. *Prof Exp:* NATO fel chem, Oslo, 65-67; res assoc, Ind Univ, Bloomington, 67-68; from asst prof to assoc prof phys chem, 68-75, PROF PHYS CHEM, UNIV ARK, FAYETTEVILLE, 75- *Concurrent Pos:* Teacher-scholar grant, Dreyfus Found, 71. *Honors & Awards:* IR-100 Award, 85. *Mem:* Am Chem Soc; Royal Soc Chem; Am Inst Physics. *Res:* Structural studies by abinitio calculations. electron diffraction and spectroscopy; theoretical investigations of biophysical phenomena. *Mailing Add:* Dept Chem Univ Ark Fayetteville AR 72701

SCHAFER, MARY LOUISE, b Shelburn, Ind, Nov 4, 15. FOOD CHEMISTRY. *Educ:* Purdue Univ, BS, 38, PhD(chem), 55; Columbia Univ, MS, 49. *Prof Exp:* Intern, Univ Hosp, Western Reserve Univ, 39, dietitian, 40-42; asst nutrit, Teachers Col, Columbia Univ, 47-51; asst chem, Purdue Univ, 51-54; chemist, USPHS, Ohio, 55-69; chemist, Cincinnati Res Labs, Food & Drug Admin, 69-84; RETIRED. *Mem:* Am Chem Soc; Am Dietetic Asn; Sigma Xi. *Res:* Detection and assay methods for organomercury compounds in the environment; gas chromatographic methods for microbiological assessment of canned foods. *Mailing Add:* 6522 Sherman St Cincinnati OH 45230

SCHAFER, RICHARD DONALD, b Buffalo, NY, Feb 25, 18; m 42; c 2. MATHEMATICS. *Educ:* Univ Buffalo, BA, 38, MA, 40; Univ Chicago, PhD(math), 42. *Prof Exp:* Instr math, Univ Mich, 45-46; mem, Inst Advan Study, 46-48; asst prof math, Univ Pa, 48-53; prof & head dept, Univ Conn, 53-59; dep head dept, 59-68, prof, 59-88, EMER PROF MATH, MASS INST TECHNOL, 88- *Concurrent Pos:* NSF sr fel, Inst Advan Study, 58-59; mem, Sci Manpower Comn, 59-63. *Mem:* Am Math Soc (assoc secy, 54-58); Math Asn Am. *Res:* Nonassociative and Lie algebras. *Mailing Add:* Dept Math Rm 2-263 Mass Inst Technol Cambridge MA 02139

SCHAFER, ROBERT LOUIS, b Burlington, Iowa, Aug 1, 37; m 59; c 1. AGRICULTURAL ENGINEERING, ENGINEERING MECHANICS. *Educ:* Iowa State Univ, BS, 59, MS, 61, PhD(agr eng, eng mech), 65. *Prof Exp:* Agr engr, Agr Res Serv, Ames, Iowa, 59-64, agr engr, Nat Tillage Mach Lab, 64-85, AGR ENG, NAT SOIL DYNAMICS LAB, AGR RES SERV, USDA, 85- *Concurrent Pos:* Res lectr, Auburn Univ, 70-81, adj prof, 81- *Mem:* Am Soc Agr Engrs; Int Soc Terrain Vehicle Systs; Inst Elec & Electronics Engrs Comput Soc. *Res:* Soil dynamics as a mechanics of the actions of tillage tools and traction devices in soil. *Mailing Add:* PO Box 57 Loachapoka AL 36865-0057

SCHAFER, ROLLIE R, b Denver, Colo, Feb 17, 42; m 64; c 2. NEUROSCIENCE. *Educ:* Univ Colo, Boulder, BA, 64, MA, 67, PhD(zool), 69. *Prof Exp:* Asst prof biol, Metrop State Col Colo, 68-69 & NMex Inst Mining & Technol, 69-73; asst prof biol sci, Div Biol Sci, Univ Mich, Ann Arbor, 73-76; asst grad dean, 78-82, assoc dean sci & technol, 86-87, PROF BIOL SCI, 84-, ASSOC VPRES RES & DEAN GRAD SCH, UNIV NTEX, 87- *Concurrent Pos:* NSF res grants, 72-73, 74-76, 81-83 & 85-88. *Mem:* AAAS; Sigma Xi; Am Soc Neurochem; Soc Neurosci; Asn Chemoreception Sci. *Res:* Sensory mechanisms; Olfactory reception/information processing. *Mailing Add:* Grad Sch Univ NTex Denton TX 76203-5446

SCHAFER, RONALD W, b Tecumseh, Nebr, Feb 17, 38; m 60; c 3. ELECTRICAL ENGINEERING. *Educ:* Univ Nebr, Lincoln, BSc, 61, MSc, 62; Mass Inst Technol, PhD(elec eng), 68. *Prof Exp:* Instr elec eng, Univ Nebr, Lincoln, 62-63, Mass Inst Technol, 64-68; mem tech staff, Bell Tel Labs, 68-74; REGENT'S PROF SCH ELEC ENG, GA INST TECHNOL, 74- *Honors & Awards:* Emanuel R Piore Award, Inst Elec & Electronics Engrs, 80, Centennial Medal, 84. *Mem:* Fel Inst Elec & Electronics Engrs; fel Acoust Soc Am; Acoust, Speech & Signal Processing Soc (pres, 77-79). *Res:* Digital signal processing for speech and image processing. *Mailing Add:* 1920 Mercedes Ct Atlanta GA 30345

SCHAFF, JOHN FRANKLIN, b Elgin, Ill, Mar 2, 31; m 54; c 4. SCIENCE EDUCATION, PLANT PHYSIOLOGY. *Educ:* Ill State Univ, BS, 53; Kans State Univ, MS, 55; Fla State Univ, EdD(sci educ), 68. *Prof Exp:* Teacher bot & gen sci, Proviso Twp High Sch, Maywood, Ill, 56-58, chem, Maine Twp High Sch E, Park Ridge, Ill, 58-61 & chem & biol, Deerfield High Sch, Ill, 61-65; nat teaching fel chem, NFla Jr Col, Madison, 65-67; asst prof sci educ, Syracuse Univ, 68-72; assoc prof, 72-75, assoc dean, Col Educ, 77-84, PROF SCI EDUC, UNIV TOLEDO, 75- *Concurrent Pos:* Vis prof, State Univ NY, Geneseo, 70, Univ Wyo, 71 & Owens Tech Col, Ohio, 72-74, Ohio State Univ, 87; dir, Grants Prog for Teachers, NSF, 71-86. *Mem:* Sigma Xi; Am Chem Soc; fel AAAS; Nat Asn Res in Sci Teaching; Asn Educ Teachers Sci (pres, 78-79). *Res:* Semimicro experiments for high school chemistry; middle and junior high school science curriculum development; computer applications to high school science teaching; brain physiology and learning science. *Mailing Add:* 7633 Gillcrest Sylvania OH 43560

SCHAFFEL, GERSON SAMUEL, b Braddock, Pa, Mar 17, 18; m 43; c 3. CHEMISTRY, RESEARCH ADMINISTRATION. *Educ:* Carnegie Inst Technol, BS, 39, MS, 40, DSc(org chem), 42. *Prof Exp:* Fel, Westinghouse Res Lab, 42-43, group leader, Plastics Sect, 43-46; instr chem, Carnegie Inst Technol, 45-46; dir plastics res & develop, Gen Tire & Rubber Co, 46-50, mgr mfg & develop, Chem Div, 50-54, dir res, Brea Chem, Inc, 54-56; dir res, Sci Design Co, Inc, 56-60, asst vpres, 60-63, vpres, New York, 63-73; dir technol mgmt, Badger Co, Inc, Cambridge, 73-83; consult, 83-88; RETIRED. *Mem:* Am Chem Soc; fel Am Inst Chemists; NY Acad Sci. *Res:* Kinetics and mechanisms of organic reactions; mechanism of formation of condensation polymers; polymerization of vinyl compounds; new polyesters; petrochemicals; research administration; process licensing. *Mailing Add:* 63 Country Club Lane Belmont MA 02178

SCHAFFER, ARNOLD MARTIN, b New York, NY, Sept 24, 42; m 63; c 3. PHYSICAL CHEMISTRY, SURFACE SCIENCE. *Educ:* Polytech Inst Brooklyn, BS, 63; Univ Wash, PhD(chem), 70. *Prof Exp:* Fel chem, Univ Houston, 71-74; SR RES CHEMIST, PHILLIPS PETROL CO, 74-, MGR CATALYTIC CRACKING, 81- *Concurrent Pos:* Instr, Univ Houston, 73-74; mgr Anal Serv, 88- *Mem:* Am Chem Soc; Am Phys Soc; Am Inst Chem Eng. *Res:* Properties of heterogeneous catalysts; development of spectroscopic techniques; molecular spectroscopy; molecular orbital calculations; development of new analytical techniques. *Mailing Add:* Phillips Petrol Co 248 A PL PRC Bartlesville OK 74004

SCHAFFER, BARBARA NOYES, b St Louis, Mo, June 15, 47; m 76; c 2. NEURO PEPTIDE CHEMISTRY, MOLECULAR BIOLOGY. *Educ:* Beloit Col, BS, 69; Wash Univ, PhD(biochem), 73. *Prof Exp:* Asst prof, Univ Chicago, 77-86; ASST PROF, DEPT PSYCHIAT, UNIV TEX, 86- *Mem:* Am Soc Biochem & Molecular Biol. *Res:* Structure and function of insect neuropeptides and their genes. *Mailing Add:* Dept Psychiat Univ Texas Health Sci Ctr 5323 Harry Hines Blvd Dallas TX 75235

SCHAFFER, ERWIN LAMBERT, b Milwaukee, Wis, Nov 26, 36; m 60; c 1. STRUCTURAL ENGINEERING, FOREST PRODUCTS & APPLICATIONS. *Educ:* Univ Wis, BS, 59, MS, 61, PhD(eng mech), 71. *Prof Exp:* Struct eng, concrete dam destruction/anal, Bur Reclamation, US Dept Interior, 61-63, res engr fire performance, 63-73, res proj leader, wood prod processing, 73-77; res proj leader, fire performance, Forest Prod Lab, Forest Serv, USDA, 77-81, wood construct specialist, State & Pvt Forestry, 81-84; vpres res & develop, consult & mgt, PFS Corp, 84-87; ASST DIR RES MGT, FOREST PROD LAB, FOREST SERV, USDA, 87- *Concurrent Pos:* Exec secy, Soc Wood Sci & Technol, 70-73; vis scientist, Nat Bur Standards, US Dept Com, 76-77; chmn & proc ed, Residential Fire & Wood Prod Sem, Forest Prod Res Soc, 79-80; tech adv, Tech Adv Comt, Nat Forest Prod Asn, 87-; vchmn, Comt D7, Am Soc Testing & Mat, 87-; chmn, Comt Fire Protection, Struct Div, Am Soc Civil Engrs, 89-; res team leader, Forest Prod Lab, Forest Serv, USDA. *Honors & Awards:* Presidential Design Award, Nat Endowment for the Arts, 84; Tech Transfer Award, Fed Lab Consortium, 85; L J Markwardt Award, Am Soc Testing & Mat, 86. *Mem:* Int Acad Wood Sci; Am Soc Civil Engrs; Soc Wood Sci & Technol (pres-elect, 76-77, pres, 77-78); Forest Prod Res Soc; Am Soc Testing & Mat. *Res:* Conducted and directed wood product and wood engineering research including structural fire performance; author of over 70 publications; editor, author, or contributor to several books in above fields. *Mailing Add:* One Gifford Pinchot Dr Madison WI 53705

SCHAFFER, ERWIN MICHAEL, b Dumont, Minn, July 9, 22; m 47; c 4. PERIODONTOLOGY. *Educ:* Univ Minn, DDS, 45, MSD, 51. *Prof Exp:* Clin prof & chmn div, 57-64, dean, 64-80, PROF PERIODONTICS, SCH DENT, UNIV MINN, MINNEAPOLIS, 64- *Honors & Awards:* William J Gies Award Periodont, 74. *Mem:* Am Dent Asn; fel Am Col Dent; Am Acad Periodont; Am Acad Oral Med; Int Asn Dent Res. *Res:* Bone regeneration in periodontal disease; cartilage or cementum and dentine grafts; etiology of periodontal disease; root curettage; circadian periodicity. *Mailing Add:* Periodont/17-172 Moos Univ Minn Sch Dent Minneapolis MN 55455

SCHAFFER, FREDERICK LELAND, b Kingsburg, Calif, July 5, 21; m 45; c 6. VIROLOGY, BIOCHEMISTRY. *Educ:* Univ Calif, Berkeley, BA, 43, PhD(biochem), 50. *Prof Exp:* Asst, Univ Calif, Berkeley, 47-50, from asst res biochemist to res biochemist, 54-85; RETIRED. *Concurrent Pos:* Res fel, Univ Calif, Berkeley, 50-53, lectr med microbiol, 64-77; mem, Int Comt Taxon Viruses, 81-90. *Mem:* Fel AAAS; Am Soc Microbiol; Tissue Cult Asn; Am Soc Virol. *Res:* Tissue culture; microanalysis; purification, properties and molecular biology of viruses. *Mailing Add:* 38 Hardie Dr Moraga CA 94556

SCHAFFER, HENRY ELKIN, b New York, NY, May 4, 38; m 64; c 2. GENETICS. *Educ:* Cornell Univ, BS, 59; NC State Univ, MS, 62, PhD(genetics), 64. *Prof Exp:* NIH fel gen med studies, Cornell Univ, 64-65; asst prof biol, Brandeis Univ, 65-66; from asst prof to assoc prof genetics, NC State Univ, 66-80, prof genetics, 74-, prof biomath, 80-; AT COMPUT CTR, NC STATE UNIV. *Mem:* AAAS; Biomet Soc; Genetics Soc Am; Soc Study Evolution; Am Soc Naturalists; Sigma Xi. *Res:* Population and mathematical genetics; biometrics; computing and bioinstrumentation. *Mailing Add:* Box 7109 Comput Ctr NC State Univ PO Box 5487 Raleigh NC 27695

SCHÄFFER, JUAN JORGE, b Vienna, Austria, Mar 10,30; m 59; c 1. MATHEMATICS. *Educ:* Univ Pa, MS, 51; Univ Repub Uruguay Ing Ind, 53, MS, 57; Swiss Fed Inst Technol, DrScTech, 56; Univ Zurich, DrPhil(math), 56. *Prof Exp:* Prof math & gen mech, Univ Repub Uruguay, 57-68; PROF MATH, CARNEGIE MELLON UNIV, PITTSBURGH, PA, 75-, ASSOC DEAN, 86- *Concurrent Pos:* Guggengeim fel, Univ Chicago & Res Inst Advan Study, 60; vis Carnegie prof, Carnegie Inst Technol, 64-65. *Mem:* Am Math Soc; Am Asn Univ Prof; Uruguayan Asn Adv Sci. *Res:* Differential equations; functional analysis; author and co-author of four books. *Mailing Add:* Dept Math Carnegie Mellon Univ Pittsburgh PA 15213

SCHAFFER, PRISCILLA ANN, b St Louis, Mo, Dec 28, 41. VIROLOGY. *Educ:* Hobart & William Smith Cols, BA, 64; Cornell Univ, PhD(microbiol), 69. *Prof Exp:* Asst prof virol, Dept Virol & Epidemiol, Baylor Col Med, 71-76; assoc prof, 76-81, PROF MICROBIOL, DEPT MICROBIOL & MOLECULAR GENETICS, HARVARD MED SCH, 81- *Concurrent Pos:* Fel, Baylor Col Med, 69-71; Found lectr, Am Soc Microbiol, 81-82. *Mem:* AAAS; Am Soc Microbiol; Am Soc Trop Med & Hyg; Brit Soc Gen Microbiol. *Res:* Genetics of DNA tumor viruses, especially herpesviruses. *Mailing Add:* Dana-Farber Cancer Inst Harvard Med Sch 44 Binney St Boston MA 02115

SCHAFFER, ROBERT, b New York, NY, Mar 8, 20; m 42; c 3. ANALYTICAL CHEMISTRY, BIO-ORGANIC CHEMISTRY. *Educ:* Brooklyn Col, AB, 43; Wash Univ, PhD(chem), 50. *Prof Exp:* Chemist, Int Hormones, Inc, NY, 41-44; chemist, Los Alamos Sci Lab, Univ Calif, 44-46; org chemist, Wash Univ, 46-50; org chemist, Nat Bur Standards, 50-66, chief org chem sect, 66-73, chief bio-org standards sect, 73-78, supvry res chemist, Anal Chem Ctr, 78-88, CONSULT, NAT INST STANDARDS & TECHNOL. *Concurrent Pos:* Lectr, Georgetown Univ, 59-61; fel, Cambridge Univ, 61-62; consult diag prod comt, Food & Drug Admin, 74-76; mem coun, Nat Ref Syst Clin Chem, 81- *Honors & Awards:* Outstanding Contrib to Standardization in Clin Chem Award, Am Asn Clin Chem, 79, Joseph A Roe Award, 81; Bennett Rosa Award, Nat Bur Standards, 85. *Mem:* AAAS; Am Chem Soc; Am Asn Clin Chem. *Res:* Reaction mechanisms; radiochemical applications; position labeled carbohydrates; organic chemical analysis and characterization; purity; chemical and clinical chemical standards. *Mailing Add:* Ctr Analytical Chem Nat Inst Standards & Technol Gaithersburg MD 20899

SCHAFFER, SHELDON ARTHUR, b Salt Lake City, Utah, June 12, 43; m 66; c 3. BIOLOGICAL CHEMISTRY. *Educ:* Univ Calif, Berkeley, BS, 65; Univ Ill, Urbana, PhD(chem), 70. *Prof Exp:* NIH res fel biol chem, Med Sch, Harvard Univ, 70-72, teaching fel biochem, 72-73; res biochemist immunol, Am Cyanamid Co, 73-74, group leader atherosclerosis res, 74-81, head, dept metab & endocrinol res, med res div, 81-85; DIR, ATHEROSCLEROSIS RES, CIBA-GEIGY CO, 85- *Mem:* Am Chem Soc; Am Heart Asn; Royal Soc Chem. *Res:* Lipids chemistry and metabolism; atherosclerosis; platelet structure and function; membrane biochemistry. *Mailing Add:* Cholestech Corp 3347 Investment Blvd Hayward CA 94545

SCHAFFER, STEPHEN WARD, b San Diego, Calif, Oct 15, 44; m 67; c 2. BIOCHEMISTRY. *Educ:* Buena Vista Col, BS, 66; Univ Minn, PhD(biochem), 70. *Prof Exp:* Vis prof biochem, Univ El Salvador, 70-71; fel, Johnson Found, Univ Pa, 71-73; from asst prof to assoc prof biochem, Lehigh Univ, 73-80; res assoc prof physiol, Hahnemann Med Col, 78-80; assoc prof, 80-88, PROF PHARMACOL, UNIV SALA, 88- *Concurrent Pos:* Pres, Southeast Pharmacol Soc. *Mem:* Am Chem Soc; Int Soc Heart Res; Am Soc Biol Chemists; Am Soc Pharmacol & Exp Therapeut; Am Col Clin Pharmacol. *Res:* Regulation of myocardial calcium, effects of taurine and sulfonylureas, and diabetic cardiomyopathy. *Mailing Add:* Dept Pharmacol Col Med Med Sci Bldg Univ SAla Mobile AL 36688

SCHAFFER, WILLIAM MORRIS, b Elizabeth, NJ, May 11, 45; m 70. ECOLOGY, EVOLUTIONARY BIOLOGY. *Educ:* Yale Univ, BS, 67; Princeton Univ, MS, 71, PhD(biol), 72. *Prof Exp:* Asst prof biol, Univ Utah, 72-75; asst prof, 75-77, assoc prof biol, 77-80, ASSOC PROF ECOL & EVOLUTIONARY BIOL, UNIV ARIZ, 80- *Mem:* AAAS; Sigma Xi; Soc Study Evolution; Ecol Soc Am; Am Soc Naturalists. *Res:* Evolutionary and ecological aspects of reproductive strategies; plant-pollinator interactions. *Mailing Add:* Dept Ecol & Evolutionary Biol Univ Ariz Tucson AZ 87521

SCHAFFNER, CARL PAUL, b Bayonne, NJ, Feb 13, 28. BIOCHEMISTRY. *Educ:* Columbia Univ, AB, 50; Univ Ill, PhD, 53. *Prof Exp:* Fel microbiol, Univ, 53-54, from instr to assoc prof, Inst, 54-72, PROF MICROBIOL, WAKSMAN INST MICROBIOL, RUTGERS UNIV, 72-, PROF BIOCHEM, 80- *Mem:* AAAS; Am Soc Microbiol; Am Soc Biol Chemists; Am Chem Soc; Royal Soc Chem; Sigma Xi; hon fel Phillipine Soc Microbiol, 75. *Res:* Antibiotic chemistry; chemotherapy mechanism of action; biosynthesis. *Mailing Add:* Walkman Inst Rutgers Univ PO Box 759 Piscataway NJ 08855-0759

SCHAFFNER, FENTON, b Chicago, Ill, Dec 8, 20; m 43, 78; c 4. MEDICINE, LIVER DISEASES. *Educ:* Univ Chicago, BS, 41, MD, 43; Northwestern Univ, MS, 49; Am Bd Internal Med, dipl, 52 & 77; Am Bd Gastroenterol, dipl, 59. *Prof Exp:* Asst path, Med Sch, Northwestern Univ, 48-53, instr med, 55-57; assoc med, Col Physicians & Surgeons, Columbia Univ, 58-61, assoc prof path, 61-66; from assoc prof to prof, 66-73, actg chmn dept, 72-74, PROF PATH, MT SINAI SCH MED, 66-, GEORGE BAEHR PROF MED, 73- *Concurrent Pos:* Chmn dept med, Woodlawn Hosp, Chicago, 50-57; from asst attend physician to attend physician, Mt Sinai Hosp, 58- *Mem:* Am Asn Path; Am Gastroenterol Asn; fel Am Col Physicians; Am Asn Study Liver Dis (secy, 58-72, vpres, 75-76, pres, 76-77); Int Asn Study Liver. *Res:* Liver disease, especially electron microscopy of liver, primary biliary cirrhosis. *Mailing Add:* Mt Sinai Sch Med Box 1101 New York NY 10029

SCHAFFNER, GERALD, b Chicago, Ill, May 14, 27; m 52; c 3. ELECTRICAL ENGINEERING. *Educ:* Purdue Univ, BS, 49, MS, 50; Northwestern Univ, PhD(electronics), 56. *Prof Exp:* Asst, Purdue Univ, 49-50; design engr, Thordarson Elec Co, 50-51; proj engr, Stewart-Warner Corp, 51-57 & Motorola, Inc, 57-66; opers mgr microwave semiconductor devices, 66-69, eng mgr microwaves & microelectronics, 69-90, TECH DIR ENG, TELEDYNE RYAN ELECTRONICS, 90- ENG MGR MICROWAVES & MICROELECTRONICS, TELEDYNE RYAN ELECTRON CO, 69- *Mem:* Inst Elec & Electronics Engrs. *Res:* Microwave applications of solid state devices; microwave radar systems; impatt diodes. *Mailing Add:* Teledyne Ryan Electron Co 8650 Balboa Ave San Diego CA 92123

SCHAFFNER, JOSEPH CLARENCE, b Paducah, Ky, Mar 20, 30; m 60; c 2. ENTOMOLOGY. *Educ:* Iowa Wesleyan Col, BS, 51; Iowa State Univ, MS, 53, PhD(entom), 64. *Prof Exp:* Instr entom, Ind Univ, 61 & Iowa State Univ, 61-62; asst prof, 63-68, assoc prof, 68-80, PROF ENTOM, TEX A&M UNIV, 80- *Mem:* Entom Soc Am; Soc Syst Zool; Royal Entom Soc London. *Res:* Systematic entomology; insect taxonomy. *Mailing Add:* Dept Entom Tex A&M Univ College Station TX 77843

SCHAFFNER, ROBERT M(ICHAEL), b Brooklyn, NY, Jan 30, 15; m 40; c 3. CHEMICAL ENGINEERING. *Educ:* Polytech Inst Brooklyn, BChE, 36; WVa Univ, MS, 39; Univ Pittsburgh, PhD(chem eng), 41. *Prof Exp:* Chem engr, A Hess & Co, NY, 36-37; asst chem eng, WVa Univ, 37-39; instr, Univ Pittsburgh, 39-41; chem engr, Res Dept, Standard Oil Co, Ind, 41-43; chem engr & head food dehydration, Miner Labs, 43-44; chem engr, Libby, McNeill & Libby, 45-48, asst to vpres prod, 49-53, asst gen supt, Eastern Div, 53-57, vpres res & qual standards, 57-70; ASSOC DIR TECHNOL, BUR FOODS, FOOD & DRUG ADMIN, 71- *Mem:* Am Chem Soc; Am Inst Chem Engrs; fel Inst Food Technol. *Res:* Drying; vacuum distillation; food dehydration; food canning; food freezing. *Mailing Add:* 6629 Van Winkle Dr Falls Church VA 22044

SCHAFFNER, WILLIAM ROBERT, b Winthrop, Mass, Feb 29, 36; m 59; c 2. AQUATIC ECOLOGY. *Educ:* Univ Hartford, BS, 64; Cornell Univ, MS, 66, PhD(aquatic ecol), 71. *Prof Exp:* RES ASSOC AQUATIC SCI, CORNELL UNIV, 71- *Concurrent Pos:* Lectr, Cornell Univ, 79- 84, 88- *Mem:* Am Soc Limnol & Oceanog; Ecol Soc Am; Sigma Xi; Int Asn Great Lakes Res; AAAS. *Res:* Plankton ecology. *Mailing Add:* Ecol & Systematics Corson Hall Cornell Univ Ithaca NY 14853

SCHAFFRATH, ROBERT EBEN, b Syracuse, NY, Feb 19, 22; m 61; c 3. ORGANIC CHEMISTRY, ANALYTICAL CHEMISTRY. *Educ:* Bates Col, BS, 44; Syracuse Univ, MS, 48, PhD(chem), 58. *Prof Exp:* Instr chem, New Eng Col, 48-51; instr org chem, Univ Mass, 51-54; vis lectr chem, State Univ NY Teachers Col, New Paltz, 58-59, assoc prof, 59-60; chmn dept, 60-65, assoc prof, 60-75, PROF CHEM, C W POST COL, LONG ISLAND UNIV, 75- *Concurrent Pos:* Adj prof, State Univ New York, Old Westbury, 82-85 & 90. *Mem:* AAAS; Am Chem Soc; NY Acad Sci; Sigma Xi. *Res:* Mannich reaction; hindered rotation in organic compounds; nitrogen-sulfur heterocycles; hydrazines; organic synthesis. *Mailing Add:* Dept Chem C W Post Col Greenvale NY 11548

SCHAFFT, HARRY ARTHUR, b New York, NY, May 21, 32; m 62; c 2. ELECTRONICS. *Educ:* NY Univ, BS, 54; Univ Md, MS, 58. *Prof Exp:* PHYSICIST, NAT INST STANDARDS & TECHNOL, 58- *Mem:* Inst Elec & Electronics Engrs; Am Phys Soc. *Res:* Materials, process, assembly, and device characterization and measurement for power transistors, integrated circuits, solar cells; technology transfer and information dissemination. *Mailing Add:* Nat Inst Standards & Technol Bldg 225 Rm B360 Gaithersburg MD 20899

SCHAIBERGER, GEORGE ELMER, b Saginaw, Mich, Oct 27, 28; m 50; c 2. ECOLOGY. *Educ:* Univ Fla, BS, 50, MS, 51; Univ Tex, PhD(microbiol), 55. *Prof Exp:* Asst, Univ Fla, 50-51 & Univ Tex, 51-52; instr microbiol, Univ Ark, 55-57; sr res scientist, Merck & Co, 57-61; from instr to asst prof, 62-69, assoc prof, 69-77, PROF MICROBIOL, SCH MED, UNIV MIAMI, 77- *Concurrent Pos:* Del, Int Cong Microbiol, 61; NASA & NIH grants, 66-67; res career develop award, USPHS, 67-72. *Honors & Awards:* Outstanding Res Award, Merck & Co, 61. *Mem:* Am Soc Microbiol; fel Geront Soc. *Res:* Microbial ecology; radiation biology; nucleic acid and protein synthesis; metabolic changes associated with aging; environmental effects on cells; environmental virology. *Mailing Add:* Dept Microbiol Univ Miami Univ Sta Coral Gables FL 33124

SCHAIBLE, ROBERT HILTON, b Horton, Kans, Apr 30, 31; m 73; c 2. COMPARATIVE MEDICAL GENETICS. *Educ:* Colo State Univ, BS, 53; Iowa State Univ, MS, 59, PhD(genetics, embryol), 63. *Prof Exp:* Res assoc, Hall Lab Mammalian Genetics, Univ Kans, 62-63; fel, Biol Dept, Yale Univ, 63-64; asst prof genetics, NC State Univ, 64-68; USPHS spec res fel, Dept Zool, Ind Univ, Bloomington, 68-70; asst prof biol, Ind Univ-Purdue Univ, Indianapolis, 70-73; asst prof, Dept Med Genetics, Ind Univ Sch Med, 73-89; MEM STAFF, DEPT ENVIRON MGT, STATE IND, 91- *Concurrent Pos:* Adj asst prof, Sch Vet Med, Purdue Univ, 77- *Mem:* AAAS; Am Genetic Asn; Int Pigment Cell Soc; Am Asn Univ Prof; Am Soc Human Genetics; Sigma Xi. *Res:* Comparative genetics of vertebrates; genetic control of morphogenesis; proliferation, migration, mutation and differentiation of pigment cells in the clonal development of the pigmentation of the integument. *Mailing Add:* 3640 Willsee Lane Plainfield IN 46168

SCHAICH, KAREN MARIE, b Hamilton, Ohio, Nov 21, 47. FOOD SCIENCE, BIOCHEMISTRY. *Educ:* Purdue Univ, BS, 69; Mass Inst Technol, ScD(food sci), 74. *Prof Exp:* Res assoc, Brookhaven Nat Lab, 74-76, asst scientist, 76-78, assoc scientist, 78-80, scientist, 80-86, GUEST SCIENTIST, MED DEPT, BROOKHAVEN NAT LAB, 86-; ASST PROF LIPID CHEM, DEPT FOOD SCI, RUTGERS UNIV, 88- *Mem:* Am Chem Soc; Inst Food Technologists; AAAS; NY Acad Sci; Sigma Xi; Am Oil Chemists Soc; Oxygen Soc. *Res:* Oxidizing lipids: especially free radical reactions, interaction with proteins and nucleic acids; oxygen radical species, production and roles in toxicity mechanisms; electron spin resonance studies; free radicals in biological systems; biochemistry of food deterioration. *Mailing Add:* Dept Food Sci Rutgers Univ PO Box 231 New Brunswick NJ 08903-0231

SCHAICH, WILLIAM LEE, b Springfield, Mass, Oct 15, 44; m 66; c 2. THEORETICAL SOLID STATE PHYSICS. *Educ:* Denison Univ, BS, 66; Cornell Univ, MS, 68, PhD(theoret physics), 70. *Prof Exp:* Fel physics, Air Force Off Sci Res, 70-71; res assoc, Univ Calif, San Diego, 71-73; from asst prof to assoc prof, 73-80, PROF PHYSICS, IND UNIV, BLOOMINGTON, 80- *Mem:* Am Phys Soc. *Res:* Fundamental problems in theory of electronic structure and processes. *Mailing Add:* Dept Physics Swain Hall Ind Univ Bloomington IN 47405

SCHAIRER, G(EORGE) S(WIFT), b Wilkinsburg, Pa, May 19, 13; m 35; c 4. AERONAUTICAL ENGINEERING. *Educ:* Swarthmore Col, BS, 34; Mass Inst Technol, MS, 35. *Hon Degrees:* DEng, Swarthmore Col, 58. 58. *Prof Exp:* Aeronaut engr, Bendix Prod Corp, 35-37 & Consol Aircraft Corp, 37-39; from aerodynamicist to dir res, Boeing Co, 39-59, vpres res & develop, 59-71, vpres, 71-78; consult, 78-89; RETIRED. *Concurrent Pos:* Mem sci adv group, USAF, 44-45; sci adv bd, 56-59; mem aerodyn comt, Power Plant Comt & chmn subcomt propellers for aircraft, Nat Adv Comt Aeronaut; mem steering comt, Adv Panel Aeronaut, US Dept Defense, 57-61; mem comt aircraft operating prog, NASA, 59-60; mem panel sci & tech manpower, President's Sci Adv Comt, 63-64; mem sci adv comt, Defense Intel Agency, 65-71; trustee, Univ Res Asn, 66-76; mem aeronaut & space eng bd, Nat Res Coun, 76-78. *Honors & Awards:* Reed Award, Am Inst Aeronaut & Astronaut, 49; Am Soc Mech Engrs Medal, 58; Daniel Guggenheim Medal; Museum of Flight Pathfinders Award, 85. *Mem:* Nat Acad Sci; Nat Acad Eng; Int Acad Astronaut; hon fel Am Inst Aeronaut & Astronaut; Am Physical Soc; Am Helicopter Soc; mem Soc Naval Architects & Marine Engrs. *Res:* Aerodynamic design of large aircraft. *Mailing Add:* 4242 Hunts Point Rd Bellevue WA 98004

SCHAIRER, ROBERT S(ORG), b Plum Twp, Pa, Sept 7, 15; m 47; c 6. AERONAUTICS. *Educ:* Swarthmore Col, BS, 36; Calif Inst Technol, MS, 37, PhD(aeronaut), 39. *Prof Exp:* Aerodynamicist & asst chief airborne vehicles sect, Proj Rand, Douglas Aircraft Co, 39-48; asst chief aircraft div, Rand Corp, 48-53, chief, 53-56; tech asst to corp dir develop planning, Lockheed Aircraft Corp, 56-60, asst dir, 60-61; chief scientist, Pac Missile Range, Dept Navy, Calif, 61-62; asst corp dir develop planning, Lockheed Aircraft Corp, 62-63, corp dir, 63-67, assoc dir corp plan, 76-81; RETIRED. *Mem:* Am Inst Aeronaut & Astronaut; Sigma Xi. *Res:* Unsymmetrical lift distribution on a stalled monoplane wing; stability and control in flight testing; airplane performance. *Mailing Add:* 11750 Chenault St Los Angeles CA 90049

SCHAKE, LOWELL MARTIN, b Marthasville, Mo, June 6, 39; m 59; c 2. ANIMAL SCIENCE. *Educ:* Univ Mo, BS, 60, MS, 62; Tex A&M Univ, PhD(animal nutrit), 67. *Prof Exp:* Asst prof animal sci, Tex A&M Univ, 65-72, area livestock specialist, 67-69, mem grad fac, 67-80, from assoc prof to prof, 72-84; PROF & HEAD, DEPT ANIMAL SCI, UNIV CONN, 84- *Concurrent Pos:* Prof consult, Feed Co, Oil Co & Feedlots; expert witness. *Mem:* Am Soc Animal Sci. *Res:* Feedlot management; cow body weight and compositional changes; beef cattle behavior and grain processing. *Mailing Add:* Dept Animal Sci Univ Conn 3636 Horsebarn Rd Ext Storrs CT 06269-4040

SCHALEGER, LARRY L, b Milwaukee, Wis, Nov 24, 34; m 58, 78; c 3. ORGANIC CHEMISTRY. *Educ:* Grinnell Col, BA, 57; Univ Minn, PhD(org chem), 61. *Prof Exp:* Res assoc phys chem, Cornell Univ, 61-62, vis asst prof, 62-63; asst prof, Univ Hawaii, 63-67, assoc prof org chem, 67-75; vis assoc prof chem, Calif State Univ, Long Beach, 75-76; NSF fac fel, 76-77, specialist wood chem, Forest Prod Lab, Univ Calif, 77-78, STAFF SCIENTIST, LAWRENCE BERKELEY LAB, UNIV CALIF, 78-; AT BROWN & CALDWELL, GLENDALE. *Concurrent Pos:* Res assoc, Univ Pittsburgh, 70-71. *Mem:* Am Chem Soc; AAAS; fel Am Inst Chem. *Res:* Development of methods for trace analysis of organic compound in environmental media. *Mailing Add:* 430 S Fuller Ave Apt 11H Los Angeles CA 90036

SCHALES, OTTO, chemistry; deceased, see previous edition for last biography

SCHALGE, ALVIN LAVERNE, b Akron, NY, Nov 23, 30; m 53; c 2. ANALYTICAL CHEMISTRY, OILFIELD CHEMISTRY. *Educ:* Eastern Mich Univ, AB, 54; Univ Ill, MS, 56; PhD(chem), 59. *Prof Exp:* Res chemist, Ohio Oil Co, 59-63, spectros supvr, Denver Res Ctr, Marathon Oil Co, 63-66, chem sect supvr, 66-72, advan res chemist, 72-76, SR RES CHEMIST, DENVER RES CTR, MARATHON OIL CO, 76- *Mem:* Am Chem Soc; Soc Appl Spectros; Sigma Xi. *Res:* Emission spectroscopy; x-ray diffraction and fluorescence analysis; infrared and ultraviolet spectrometric analysis; nuclear magnetic resonance spectroscopy; ion selective electrodes; chemical stimulation of oil wells. *Mailing Add:* Denver Res Ctr Marathon Oil Co PO Box 269 Littleton CO 80160

SCHALK, JAMES MAXIMILLIAN, b New York, NY, Dec 19, 32; m 61; c 4. ENTOMOLOGY. *Educ:* Univ Ga, BSA, 60; Cornell Univ, MS, 63; Univ Nebr, PhD(entom), 70. *Prof Exp:* Res asst entom, Cornell Univ, 61-63; res entomologist, Birds Eye Div, Gen Foods Corp, 63-65; res entomologist, Grain & Forage Div, USDA, 65-71, res entomologist, Plant Genetics & Germplasm Inst, Sci & Educ Admin-Agr Res, 71-78; res leader entomol, 78-85, RES ENTOMOLOGIST, US VEG BREEDING LAB, 85- *Concurrent Pos:* Res entomologist, USAID, 71 & Int Prog Div, USDA, 71-76. *Mem:* Entom Soc Am; Am Soc Hort Sci; Sigma Xi. *Res:* Developing vegetable germplasm with resistance to insects and mites and investigating the nature of arthropod resistance in these plants. *Mailing Add:* US Veg Breeding Lab 2875 Savannah Hwy Charleston SC 29414

SCHALK, MARSHALL, b Boston, Mass, Apr 25, 07; m 33; c 4. GEOLOGY. *Educ:* Harvard Univ, AB, 29, AM, 31, PhD(geol), 36. *Prof Exp:* From asst prof to prof, 41-72, EMER PROF GEOL, SMITH COL, 72- *Mem:* Fel Geol Soc Am; Nat Asn Geol Teachers; Am Geophys Union. *Res:* Beach sedimentation; arctic shoreline of Alaska. *Mailing Add:* Clark Sci Ctr Smith Col Northampton MA 01063-0010

SCHALK, TERRY LEROY, b Eldora, Iowa, Aug 3, 43; m 65; c 2. CHARMONIUM PHYSICS, COMPUTER ENVIRONMENT PHYSICS. *Educ:* Iowa State Univ, BS, 65, PhD(high energy physics), 69. *Prof Exp:* Res assoc, Univ Calif, Riverside, 69-72; Stanford Linear Accelerator Ctr, 72-75; RES PHYSICIST, UNIV CALIF, SANTA CRUZ, 75- *Concurrent Pos:* Consult, 78- *Mem:* Am Phys Soc; Asn Comput Mach. *Res:* Study of the charmonium systems & glueon spectroscopy; computer environment for physicists. *Mailing Add:* SLAC PO Box 4349 Sanford CA 94305

SCHALL, ELWYN DELAUREL, b Montpelier, Ohio, May 6, 18; m 42; c 4. ANALYTICAL CHEMISTRY. *Educ:* Ohio State Univ, BS, 40, Purdue Univ, MS, 42, PhD(biochem), 49. *Prof Exp:* From asst prof to prof, 49-83, EMER PROF BIOCHEM, PURDUE UNIV, WEST LAFAYETTE, 83- *Concurrent Pos:* State chemist & seed comnr, Ind, 65-82. *Mem:* Am Chem Soc; Asn Off Analytical Chemists. *Res:* Analytical methods. *Mailing Add:* 108 Jordan Ln West Lafayette IN 47906

SCHALL, JOSEPH JULIAN, b Philadelphia, Pa, June 18, 46; m 72. EVOLUTIONARY ECOLOGY, PARASITE-HOST ECOLOGY. *Educ:* Pa State Univ, BS, 68; Univ RI, MS, 72; Univ Tex, Austin, PhD(zool), 76. *Prof Exp:* NIH res serv award, Univ Calif, Berkeley, 77-80; asst prof, 80-87, ASSOC PROF ZOOL, UNIV VT, 87- *Mem:* AAAS; Soc Study Evolution; Ecol Soc Am; Soc Study Amphibians and Reptiles. *Res:* Interface between evolutionary theory and ecology, behavior and physiology; interspecific associations such as the parasite-host and plant-herbivore relationships; ecology of lizard malaria. *Mailing Add:* Dept Zool Univ Vt Burlington VT 05405

SCHALL, ROY FRANKLIN, JR, b Pittsburgh, Pa, June 4, 39. RADIOIMMUNOLOGY. *Educ:* Va Mil Inst, BS, 61; Carnegie Inst Technol, MS, 65; Carnegie-Mellon Univ, PhD(nuclear chem), 69. *Prof Exp:* Chief, health phys, Walter Reed Army Med Ctr, 68-71; mem staff radioimmunoassay chem, Mt Sinai Med Ctr, Miami Beach, 71-74; group leader radioimmunol, Corning Glass Works, Medfield, Mass, 74-76; sr scientist enzyme immunol, 76-81, TECH DIR, ORGANON DIAG, 81- *Mem:* Am Asn Clin Chemists; Am Chem Soc; Clin Radioassay Soc. *Res:* Development of new immunoassay systems using enzymes as labeling entitles. *Mailing Add:* 2208 E Linfield Glenrdora CA 91740

SCHALLA, CHARENCE AUGUST, b Cleveland, Ohio, May 25, 18; m 43; c 2. MAGNETOTHERMOELECTRIC REFRIGERATION HEATING VENTILATING & AIR CONDITIONING DESIGN. *Educ:* Case Western Reserve Univ, BS, 41. *Prof Exp:* Asst plant engr, Cleveland Diesel Engine Div, Gen Motors Corp, 41-43, physicist, 46-50; jr scientist, Univ Calif, Los Alamos, 44-46; res proj engr, Pesco Prod Div, Borg-Warner Corp, 50-59; res scientist, 59-75, DESIGN SPECIALIST, LOCKHEED MISSILES & SPACE CO, INC, 76- *Mem:* Am Soc Mech Engrs. *Res:* Energy conversion; combustion gas liquefaction; aerospace nuclear liquid metal and liquid hydrogen pumping; undergound nuclear testbeds; in-space materials outgrassing; seawater desalination; coal gasification; missiles amd spacecraft facilities; author over 30 papers and articles. *Mailing Add:* 3540 Louis Rd Palo Alto CA 94303-4405

SCHALLENBERG, ELMER EDWARD, b Rome, NY, Dec 30, 29; m 57; c 2. PETROLEUM CHEMISTRY, RESEARCH ADMINISTRATION. *Educ:* Cornell Univ, AB, 51; Univ Calif, PhD(chem), 54. *Prof Exp:* Mem staff lubricants res & develop, Texaco Inc, Beacon, NY, 54-80, sci planning, 80-81, sr coordr res, Texaco Serv Europ-London, 81-82, sr coordr res & asst to mgr dir, 82-88, GESCHAFTSFUEHRER, TEXACO SERV DEUTSCHLAND, 88- *Mem:* Am Chem Soc; Sigma Xi. *Res:* Lubricant additives and product development. *Mailing Add:* Texaco Serv Deutschland GmbH Baumwall Five Hamburg 11 2000 Germany

SCHALLER, CHARLES WILLIAM, b Holmen, Wis, June 8, 20; m 42; c 3. AGRONOMY. *Educ:* Univ Wis, BS, 41, MS, 43, PhD(agron), 46. *Prof Exp:* Asst agron, Univ Wis, 41-46; from jr agronomist to assoc agronomist, Exp Sta, 46-61, from instr to assoc prof agron, Univ, 46-61, PROF AGRON, UNIV CALIF, DAVIS, 61-, AGRONOMIST, EXP STA, 61- *Mem:* Am Phytopath Soc; Am Soc Agron; Genetics Soc Am. *Res:* Genetics of disease resistance in wheat and barley; production of disease resistant varieties of of barley. *Mailing Add:* Dept Agron Range Sci Univ Calif Davis CA 95616-8515

SCHALLER, DARYL RICHARD, b Milwaukee, Wis, Oct 21, 43; m 66; c 2. FOOD SCIENCE. *Educ:* Univ Wis, BS, 64, MS, 66, PhD(food sci), 69. *Prof Exp:* Res fel, Dept Food Sci, Univ BC, 69-72; group leader cereal chem, Kellogg Co, 72-77, dir res serv, 77-79, dir res, 79-81, vpres, 81-85, sr vpres, Sci & Technol, 86-90, SR VPRES, RES QUAL & NUTRIT, KELLOGG CO, 90- *Concurrent Pos:* Res bd visitors, Memphis State Univ & Western Mich Univ Eng; bd trustees, Mich Biotechnol Inst. *Mem:* Am Chem Soc; Am Asn Cereal Chemists. *Res:* Ultrastructural changes in meat; chemistry of plant polyphenols, pigments; chemistry and analysis of dietary fiber; food chemistry. *Mailing Add:* 19 Castle Dr Battle Creek MI 49015

SCHALLER, EDWARD JAMES, b Philadelphia, Pa, Mar 28, 39; m 68; c 1. COATINGS CHEMISTRY, COLLOID CHEMISTRY. *Educ:* Villanova Univ, BE, 61; Univ Pa, PhD(chem eng), 65. *Prof Exp:* Sr chemist, 65-73, proj leader, 73-80, SECT MGR TRADE SALES COATINGS, ROHM & HAAS CO, SPRING HOUSE, 80- *Honors & Awards:* Roon Award, 87. *Mem:* Am Chem Soc. *Res:* Development of binders, thickeners, dispersants and other additives for water based coatings (especially architectural), adhesives and sealants. *Mailing Add:* 640 Runnymede Ave Jenkintown PA 19046

SCHALLER, ROBIN EDWARD, b St Paul, Minn, July 10, 37; m 66; c 2. FLUID AND PARTICLE TECHNOLOGY. *Educ:* Univ Minn, BS, 59, MSAeroEngr, 63, PhD(environ health eng), 75. *Prof Exp:* Develop engr, Ord & Aerospace Div, Honeywell Inc, 63-65; prod engr, Res Ctr, 3M Co, St Paul, Minn, 65-66; MANAGING ENGR, RES & DEVELOP DIV, DONALDSON CO INC, 66- *Concurrent Pos:* Chmn, Am Inst Chem Engrs Classifier Equip Testing & Eval Procedure Comt, 76- *Mem:* Am Soc Mech Engrs; Am Inst Chem Engrs; Air Pollution Control Asn; Am Indust Hyg Asn. *Res:* Fluid and particle mechanics pertaining to the behavior, production, sampling and characterization of particles; recognition, evaluation and engineering control of environmental pollutants. *Mailing Add:* Schaller Serv Inc 1052 Pecan Ct Naperville IL 60540

SCHALLERT, WILLIAM FRANCIS, b Maplewood, Mo, Sept 3, 27; m 54; c 2. ENGINEERING. *Educ:* Washington Univ, St Louis, BS, 53; St Louis Univ, MS, 55, MBA, 60, PhD, 76. *Prof Exp:* Asst prof aeronaut eng & head elec prog, Parks Col Aeronaut Technol, St Louis, 53-61, asst prof basic eng, St Louis Univ, 61-64; assoc prof eng, Florissant Valley Community Col, 64-67, chmn div eng & eng technol, 64-78, prof eng, 67-82; dean acad affairs, 82-88, prof aero eng, 82-88, prof elec eng, 88, CHMN ELEC ENG DEPT, PARKS COL, ST LOUIS UNIV, 88- *Concurrent Pos:* Consult, Emerson Elec Mfg Co, 55-64, Electro-Core, Inc 69- & Sch Eng, Univ Mo, Columbia, 72-76; comnr higher educ, NCent Asn, 71-76; co-dir, NSF grant, 72-76; consult & examr, NCent Asn, 76- *Mem:* Inst Elec & Electronics Engrs; Am Soc Eng Educ; Nat Soc Prof Engrs; Am Soc Cert Eng Technicians. *Res:* Avionics and systems engineering; behavioral and educational systems; electrical power system; pulsed energy systems. *Mailing Add:* Parks Col St Louis Univ Cahokia IL 62206

SCHALLES, ROBERT R, b Durango, Colo, Mar 25, 35; m 56; c 4. ANIMAL BREEDING, POPULATION GENETICS. *Educ:* Colo State Univ, BS, 63; Va Polytech Inst, MS, 66, PhD(animal breeding), 67. *Prof Exp:* Mgr, Feedlot Serv, Inc, Colo, 61-62; asst prof animal husbandry & asst animal husbandman, 66-70, assoc prof, 70-80, PROF ANIMAL SCI & ASSOC, AGR EXP STA, KANS STATE UNIV, 80- *Mem:* Am Soc Animal Sci; Am Genetic Asn. *Res:* Genetic and environmental influences on growth and development of animals. *Mailing Add:* Dept Animal Sci & Indust Kans State Univ Manhattan KS 66506

SCHALLHORN, CHARLES H, b Saginaw, Mich, Mar 26, 44; m 73; c 1. PHOTOGRAPHIC CHEMISTRY. *Educ:* Univ Mich, Ann Arbor, BS, 66; Univ Calif, Berkeley, PhD(chem), 70. *Prof Exp:* Sr res chemist, Eastman Kodak, 70-79, res assoc, 79-84, dir photo technol, Kodak Japan, 84-88, PROG DIR, STRATEGIC INFO, EASTMAN KODAK, 88- *Concurrent Pos:* Lectr chem, Univ Rochester, 73-82; chmn comt IT9-3, Image Stability, Am Nat Standards Inst. *Mem:* Am Chem Soc; Soc Imaging Sci & Technol. *Res:* Photographic and imaging chemistry. *Mailing Add:* Strategic Info Eastman Kodak Co 343 State St Rochester NY 14650

SCHALLIOL, WILLIS LEE, b Elkhart, Ind, Dec 20, 19; m 42; c 3. MATERIALS SCIENCE, ELECTRICAL ENGINEERING. *Educ:* Purdue Univ, BS, 42; Stanford Univ, PhD(metall eng), 50. *Prof Exp:* Mat engr, Westinghouse Elec Corp, Calif, 48-50; actg instr x-ray tech, Stanford Univ, 49-50; supvr pile fuels, Hanford Works, Gen Elec Co, 50-53; mgr br, Nibco Inc, 53-54, dir eng, 55-59; dir res aerospace div, Bendix Corp, 59-63; proj engr, CTS Corp, 63-64, dir res, CTS Res, Inc, 64-69; mem staff, Dept Chem, 69-75, assoc prof mat eng, 76-80, assoc coordr coop eng educ, 76-80, mgr indust rels, Sch Elec Eng, 80-85; vis asst prof mech eng technol, Purdue Univ, West Lafayette, 85-86; RETIRED. *Concurrent Pos:* Dir critical needs prog,

Purdue Univ, 76-80. *Res:* Microtranspiration for protection of rocket nozzle throats; chromium-magnesia composites; high-power precision cermet resistor modules; thermoset polymers for mechanical applications; computer-aided cooperative engineering education administration; profile of graduates from the engineering co-op program. *Mailing Add:* 172 Pathway Lane West Lafayette IN 47906-2156

SCHALLY, ANDREW VICTOR, b Wilno, Poland, Nov 30, 26; US citizen; m 56, 76; c 2. ENDOCRINOLOGY. *Educ:* McGill Univ, BSc, 55, PhD(biochem), 57. *Hon Degrees:* Sixteen from various foreign & Can univs. *Prof Exp:* Asst protein chem, Nat Inst Med Res, Eng, 49-52; assoc endocrinol, Allan Mem Inst Psychiat, Can, 52-57; asst prof & res assoc protein chem & endocrinol, Col Med, Baylor Univ, 57-62; assoc prof med, 62-66, PROF MED, SCH MED, TULANE UNIV, 66-; CHIEF, ENDOCRINOL & POLYPEPTIDE LABS, VET ADMIN HOSP, 62-, SR MED INVESTR, 73- *Concurrent Pos:* NIH fel, 60-62; consult indust, 57- *Honors & Awards:* Nobel Prize in Med, 77; Award, Am Thyroid Asn, 69; Ayerst Squibb Endocrine Soc Award; Gardner Found Award, 74; Edward T Tyler Award, 75; Borden Award, 75; Albert Lasker Award, 75. *Mem:* Nat Acad Sci; Am Soc Biol Chem; Am Physiol Soc; AAAS; Endocrine Soc. *Res:* Chemistry and biology of protein and peptide hormones; control of release; determination of structure and synthesis of thyrotropin releasing hormone, luteinizing hormone and follicle-stimulating hormone-releasing hormone; pro-somato statin; endocrine dependent cancers. *Mailing Add:* Vet Admin Hosp 1601 Perdido St New Orleans LA 70146

SCHAMBERG, RICHARD, b Frankfurt, Germany, Dec 18, 20; nat US; m 47, 69; c 3. AERODYNAMICS, SYSTEMS ANALYSIS. *Educ:* Calif Inst Technol, BS, 43, MS, 44, PhD(aeronaut), 47. *Prof Exp:* Asst, Calif Inst Technol, 43-47, asst aerodyn, 46-47; aerodyn engr, Proj Rand, Douglas Aircraft Co, 47-48; assoc engr, Rand Corp, Calif, 48-50, tech asst to chief, Aircraft Div, 50-55, assoc head, Aero-astronaut Dept, 56-63, head, 63-68; mem sr tech staff-corp, Northrop Corp, 68-71, corp dir technol applns, 71-87; CONSULT, 87- *Concurrent Pos:* Consult tech adv panel on aeronaut, Off Asst Secy Defense Res & Eng, 57-63; mem advan technol panel, Defense Advan Res Proj Agency, 73-75. *Mem:* Opers Res Soc Am; Am Inst Aeronaut & Astronaut; Unmanned Vehicles Soc. *Res:* Aerodynamics of rarefied gases; research and development planning operations research. *Mailing Add:* 10630 Bradbury Rd Los Angeles CA 90064

SCHAMBERGER, ROBERT DEAN, b Rochester, NY, June 28, 48. B-QUARK SPECTROSCOPY, P-P INTERACTIONS. *Educ:* State Univ NY, BA, 70, PhD(physics), 76. *Prof Exp:* Postdoctoral asst, 77-85, SR SCIENTIST PHYSICS, STATE UNIV NY, 86- *Mem:* Am Phys Soc. *Res:* Quark and hadron interactions at high energies. *Mailing Add:* Seven Sycamore Dr Stony Brook NY 11790

SCHAMBRA, PHILIP ELLIS, b Saginaw, Mich, Nov 8, 34; m 67; c 3. BIOPHYSICS. *Educ:* Rice Univ, BA, 56; Yale Univ, PhD(biophys), 61. *Prof Exp:* Grants assoc, NIH, 67-68; budget examr, Off Mgt & Budget, Exec Off of the President, 68-71, staff mem, Coun Environ Qual, 71-74; assoc dir interagency progs, Nat Inst Environ Health Sci, NIH, 74-80; chief int coord & liaison br, Fogarty Int Ctr NIH, Bethesda Md, 81-84; sci attache & Int Health Rep, Dep Chief Sci Off, US Embassy, New Delhi,India, 84-88; DIR, FOGARTY INT CTR, 88- *Honors & Awards:* Superior Serv Award, USPHS, 89. *Mem:* AAAS; Am Soc Trop Med & Hyg. *Mailing Add:* Fogarty Int Ctr Bldg 31 Rm B1C 39 NIH Bethesda MD 20892

SCHAMP, HOMER WARD, JR, b St Marys, Ohio, June 23, 23; c 2. HIGH PRESSURE PHYSICS. *Educ:* Miami Univ, AB, 44; Univ Mich, PhD(physics), 52. *Prof Exp:* Physicist, Mound Lab, Monsanto Chem Co, 51-52; from assoc prof to prof, Inst Molecular Physics, 52-71, dir, Inst, 64-65, dean fac, 65-71, RES PROF EDUC, UNIV MD, BALTIMORE COUNTY, 71- *Mem:* Fel Am Phys Soc. *Res:* Self-diffusion and ionic conductivity in alkali halides; experiments in high pressure physics; thermodynamics. *Mailing Add:* 521 Overdale Rd Baltimore MD 21229

SCHANBACHER, FLOYD LEON, b Cherokee, Okla, Dec 19, 41; m 64; c 1. BIOCHEMISTRY, DIFFERENTIATION. *Educ:* Northwestern State Col, BS, 64; Okla State Univ, MS, 67, PhD(biochem), 70. *Prof Exp:* Res assoc biochem & entom, Okla State Univ, 70; from asst prof to assoc prof dairy sci, 70-87, PROF DAIRY SCI, OHIO AGR RES & DEVELOP CTR, OHIO STATE UNIV, 87- *Mem:* Am Dairy Sci Asn; Am Chem Soc; Am Soc Microbiol; AAAS. *Res:* Milk protein characterization, biosynthesis and function; control and initiation of mammary differentiation; immunobiology of mammary differentiation; immunotoxicology; biochemical responses to toxic agents. *Mailing Add:* Dept Dairy Sci 116 Plumb Hall Ohio State Univ Main Campus Columbus OH 43210

SCHANBERG, SAUL M, b Clinton, Mass, Mar 22, 33; m 55; c 2. NEUROPHARMACOLOGY. *Educ:* Clark Univ, BA, 54, MA, 56; Yale Univ, PhD(pharmacol), 61, MD, 64. *Prof Exp:* Teaching asst bot & zool, Clark Univ, 54-55 & physiol, 55-56; res assoc pharmacol, Yale Univ, 61-62, univ scholar, 61-64; intern pediat, Albert Einstein Col Med, 64-65; res assoc pharmacol, Lab Clin Sci, NIMH, 65-67; asst prof clin pharmacol, 67-69, asst prof neurol & chief sect neuropharmacol, 69-87, PROF PHARMACOL, SCH MED, DUKE UNIV, 69-, PROF BIOL PSYCHIAT & ASSOC DEAN MED SCH, 87- *Concurrent Pos:* Res consult psychoneuropharmacol, Inst Behav Res, Silver Spring, Md, 65-68; Neurosci Res Prog award, Cent Nerv Syst Intensive Study Unit, 66, NIMH res scientist award, 68. *Honors & Awards:* Rehme-Anna Monika Prize, Ger, 67. *Mem:* Am Fedn Clin Res; Am Soc Pharmacol & Exp Therapeut; Int Soc Biochem Pharmacol; Am Col Neuropsychopharmacol; Am Soc Neurochem. *Res:* Pharmacologic and toxic effects of drugs, hormones and environmental influences on the metabolism of biogenic amines in the central nervous system; the role biogenic amines play in controlling cell metabolism and in mediating the effects of drugs and hormones on brain metabolism and development in normal and disease states; assessment of multiple molecular and physiological parameters in developing organ systems as indices of altered function maturation; chemical mechanisms in the CNS mediating growth and maturation in the mammalian neonate as regulated by mother-infant interactions and early experiences, effects of endocrines, neuropeptides and drugs, and sympathetic and endocrine physiology as mediators of the stress response. *Mailing Add:* Dept Pharmacol Box 3813 Med Ctr Duke Univ Durham NC 27706

SCHANEFELT, ROBERT VON, b Abilene, Kans, Sept 21, 42; m 63; c 2. FOOD SCIENCE, CEREAL CHEMISTRY. *Educ:* Kans State Univ, BS, 66, MS, 67, PhD(food sci), 70. *Prof Exp:* Food technologist, A E Staley Mfg Co, 70-73, group leader, 73-76, dir food & agr prod res & develop, 76-85, dir food & indust prod, 85-88, VPRES, A E STALEY MFG CO, 88- *Mem:* Am Asn Cereal Chem; Inst Food Technologists. *Res:* Research and applications development of modified food starches and corn sweetness. *Mailing Add:* Six Allen Bend Pl Decatur IL 62521

SCHANFIELD, MOSES SAMUEL, b Minneapolis, Minn, Sept 7, 44; m 77; c 2. FORENSIC GENETICS, PATERNITY TESTING. *Educ:* Univ Minn, Minneapolis, BA, 66; Harvard Univ, MA, 69; Univ Mich, Ann Arbor, PhD(human genetics), 71. *Prof Exp:* Res fel genetics, Univ Calif, San Francisco, 71-72, NIH fel, 72-74, asst res geneticist, 74-75; dir, Transfusion Serv & Ref Lab, Milwaukee Blood Ctr, 75-78; asst sci dir, Am Red Cross Blood Serv, 79-83; lab dir, Genetic Testing Inst, 83-85; LAB DIR, ANALYTICAL GENETIC TESTING CTR INC, 85- *Concurrent Pos:* Asst clin prof, dept med, Med Col Wis, 76-78; consult immunohemat, Vet Admin Ctr, Wood, Wis, 76-78; adj assoc prof genetic prog, George Washington Univ, 79-83; adj assoc prof, dept pediat, Emory Univ Atlanta, Ga, 84-; dir, Am Soc Crime Lab. *Honors & Awards:* Gold Medal, 1st Latin Am Cong Hemother & Immunohemat. *Mem:* Am Soc Human Genetics; Am Asn Phys Anthropologists; Soc Study Human Biol; Am Asn Immunologists; Am Acad Forensic Sci; Am Soc Crime Lab. *Res:* Determination of the biological significance, or properties, of the genetic markers on antibodies and the evolutionary forces which act on them; applied research on non-isotopic DNA technology. *Mailing Add:* Analytical Genetic Test Ctr 7808 Cherry Creek Dr S No 201 Denver CO 80231

SCHANK, STANLEY COX, b Fallon, Nev, Oct 31, 32; m 54; c 5. CYTOGENETICS, PLANT BREEDING. *Educ:* Utah State Univ, BS, 54; Univ Calif, PhD(genetics), 61. *Prof Exp:* From asst prof to assoc prof, 61-72, PROF GENETICS, UNIV FLA, 72- *Concurrent Pos:* Consult, IRI Res Inst, Brazil, 66; vis res prof, Brazil-Fla Contract, 73-75; fac develop leave, Commonwealth Sci & Indust Res Orgn, Australia, 70 & 78; nat crop adv comt, Genetics Soc Can, 85-; consult, Holstein Asn, Thailand, 90. *Mem:* Am Soc Agron; Genetics Soc Am; Genetics Soc Can; Crop Sci Soc Am. *Res:* General genetics; forage grass cytogenetics and breeding, primarily Pennisetums, Brachiarias and Hemarthrias; biomass production of Napiergrass hybrids for methane. *Mailing Add:* 2199 McCarty Hall Univ Fla Gainesville FL 32601

SCHANKER, JACOB Z, b Brooklyn, NY; m 65; c 2. RADIO COMMUNICATIONS, METEOR BURST. *Educ:* City Col City Univ New York, BEE, 63, ME, 67. *Prof Exp:* Sr engr, Gen Dynamics-Electronics Div, 63-69; sr engr, Harris Corp, RF Commun Div, 69-71; vpres & dir eng Community Music Serv Inc, 71-76; chief engr, Sci Radio Systs Inc, 77-86, dir, prod planning & develop, 86-89; PRIN ENGR, METSCAN, INC, 89- *Concurrent Pos:* Res asst, Dept Elec Eng, City Col New York, 65-66; adj prof, Rochester Inst Technol, 69-79. *Honors & Awards:* Centennial Medal, Inst Elec & Electronics Engrs, 84. *Mem:* Inst Elec & Electronics Engrs; Appl Comput Electromagnetics Soc; Inst Radio & Elec Engr Australia. *Res:* The application of meteor burst propagation to data communications systems; packet data techniques in HF, VHF and meteor burst systems; broadband HF antenna design and broadband matching network synthesis. *Mailing Add:* 65 Crandon Way Rochester NY 14618

SCHANNE, OTTO F, b Stuttgart, Ger, Feb 21, 32; div; c 2. BIOPHYSICS, CARDIAC ELECTROPHYSIOLOGY. *Educ:* Univ Heidelberg, Dr med, 60; Univ Paris, Dr Etat, 79. *Prof Exp:* Res asst physiol, Univ Heidelberg, 58-60, instr, 60-64; assoc prof pharmacol, Univ Southern Calif, 64-65; asst prof biophys, Univ Montreal, 65-66; from asst prof to assoc prof, 66-71, chmn dept, 70-78, assoc dean res, 81-83 PROF, UNIV SHERBROOKE, 71- *Concurrent Pos:* Med Res Coun Can scholar, 66-70, mem comt physiol & pharmacol, 73-76, mem core comt heart res develop grants, 77-79; mem assoc comt biophys, Nat Res Coun Can, 68-69, mem nat comt biophys, 77-81; Edwards prof cardiol, 85- *Mem:* Biophys Soc; Am Physiol Soc; Can Physiol Soc; Inst Elec & Electronics Engrs. *Res:* Antiarrhythmic drugs and cellular cardiac electrophysiology; membrane properties of cultured cardiac cells; pacemaker mechanisms. *Mailing Add:* Dept Physiol & Biophysics Univ Sherbrooke Fac Med 3001 12th Ave N Sherbrooke PQ J1H 5N4 Can

SCHANO, EDWARD ARTHUR, b Buffalo, NY, Oct 8, 18; m 42, 67; c 15. POULTRY HUSBANDRY. *Educ:* Cornell Univ, BS, 51; Mich State Univ, MS, 58. *Prof Exp:* Prof poultry husb & exten specialist, Cornell Univ, 52-; RETIRED. *Mem:* Poultry Sci Asn; World Poultry Sci Asn. *Res:* Youth poultry science projects. *Mailing Add:* 513 Dryden Rd Ithaca NY 14853

SCHANTZ, EDWARD JOSEPH, b Hartford, Wis, Aug 27, 08; m 40; c 5. BIOCHEMISTRY. *Educ:* Univ Wis, BS, 31, PhD(biochem), 39; Iowa State Col, MS, 33. *Prof Exp:* Res asst, Wis Agr Exp Sta, 36-40; biochemist, Res Labs, Carnation Milk Co, 40-42; res chemist, US Army Biol Ctr, 46-72; prof, 72-80, EMER PROF BIOCHEM, FOOD RES INST, UNIV WIS-MADISON, 80- *Concurrent Pos:* Consult, USPHS, 54-72. *Mem:* Fel AAAS; Am Chem Soc; Am Soc Biol Chem; fel NY Acad Sci. *Res:* Isolation and characterization of toxins and poisons produced by microorganisms; diffusion of substances in gels and various biological systems; studies on poisons produced by certain dinoflagellates and other algae. *Mailing Add:* Food Res Inst Univ Wis 1925 Willow Dr Madison WI 53706

SCHANTZ, PETER MULLINEAUX, b Camden, NJ, Oct 22, 39; c 2. VETERINARY PUBLIC HEALTH, PARASITOLOGY. *Educ:* Univ Pa, AB, 61, VMD, 65; Univ Calif, Davis, PhD(comp path), 71. *Prof Exp:* Epidemiologist, Pan-Am Zoonosis Ctr, Pan-Am Health Orgn-WHO, Arg, 69-74; VET EPIDEMIOLOGIST, CTR DIS CONTROL, USPHS, 74- *Mem:* Am Vet Med Asn; Am Soc Parasitol; Am Soc Trop Med & Hyg; Sigma Xi. *Res:* Epidemiology, immunodiagnosis and pathology of parasitic zoonoses. *Mailing Add:* 2569 Circlewood Rd Atlanta GA 30345

SCHANUEL, STEPHEN HOEL, b St Louis, Mo, July 14, 33; m 58; c 2. MATHEMATICS. *Educ:* Princeton Univ, AB, 55; Univ Chicago, MS, 56; Columbia Univ, PhD(math), 63. *Prof Exp:* Instr math, Ill Inst Technol, 59-61; instr, Columbia Univ, 61-63; instr, Johns Hopkins Univ, 63-65; asst prof, Cornell Univ, 65-69; assoc prof, State Univ NY, Stony Brook, 69-72; ASSOC PROF MATH, STATE UNIV NY, BUFFALO, 72- *Concurrent Pos:* Mem, Inst Advan Study, 65-66. *Mem:* Am Math Soc. *Res:* Algebra and number theory, especially transcendental numbers. *Mailing Add:* Dept Math Diefendorf Hall Rm 324 State Univ NY 3435 Main St Buffalo NY 14214

SCHAPER, LAURENCE TEIS, b Beloit, Kans, Oct 6, 36; m 60; c 2. SOLID WASTE MANAGEMENT. *Educ:* Kans State Univ, BS, 59; Stanford Univ, MS, 62. *Prof Exp:* PARTNER, CIVIL ENG, BLACK & VEATCH, 62- *Mem:* Nat Soc Prof Engrs; fel Am Soc Civil Engrs; Am Water Works Asn; Water Pollution Control Asn; Am Pub Works Asn; Am Acad Environ Engrs. *Res:* Solid waste management. *Mailing Add:* 8421 Briar Prairie Village KS 66207

SCHAPERY, RICHARD ALLAN, b Duluth, Minn, Mar 3, 35; m 57; c 1. CONTINUUM MECHANICS, FRACTURE MECHANICS. *Educ:* Wayne State Univ, BS, 57; Calif Inst Technol, MS, 58, PhD(aeronaut), 62. *Prof Exp:* From asst prof to prof aeronaut & eng sci, Purdue Univ, 62-69; prof Aerospace Eng & Civic Eng, Tex A&M Univ, 69-80, alumni prof, 80-85, distinguished prof, 80-85, Eng Exp Sta Chair, 85-90, dir, Mech & Math Ctr, 71-90; PROF AEROSPACE ENG & ENG MECH, UNIV TEX, 90- *Concurrent Pos:* Indust consult, Struct Anal & Mat Characterization, 60-; mem, solid propellant struct integrity comt, 66-72, comts Nat Mat Adv Bd, 78-87; Cockrell Family Regents Chair, Eng, 90- *Mem:* Am Inst Aeronaut & Astronaut; Am Ceramic Soc. *Res:* Elasticity; viscoelasticity; fracture mechanics; thermodynamics; composite materials. *Mailing Add:* Dept Aerospace Eng & Eng Mech Univ Tex Austin TX 78712-1085

SCHAPIRO, HARRIETTE CHARLOTTE, b New York, NY, Feb 9, 35. BIOCHEMISTRY. *Educ:* Univ Miami, BS, 56, PhD(biochem), 62; Brandeis Univ, MA, 59. *Prof Exp:* Sr lab asst, Howard Hughes Med Inst, 59-60; res instr biochem, Univ Miami, 62-63; res fel, Scripps Clin & Res Found, 63-66; from asst prof to assoc prof, 66-77, PROF BIOL, SAN DIEGO STATE UNIV, 77- *Mem:* AAAS; Am Chem Soc. *Res:* Molecular interactions utilizing fluorescence polarization measurements; immunochemistry. *Mailing Add:* Dept Biol San Diego State Univ San Diego CA 92182

SCHAPPELL, FREDERICK GEORGE, b Pottsville, Pa, Dec 20, 38; m 62; c 2. ORGANIC CHEMISTRY, MARKETING. *Educ:* Franklin & Marshall Col, BS, 60; Northwestern Univ, PhD(chem), 64. *Prof Exp:* From res chemist to sr res chemist, 64-74, res supvr chem, 74-76, mgr acquisitions & planning org dept, 76-77, mgr corp mkt develop polypropylene new bus, 78-81, mgr com develop resins, 78-84, dir sales graphic arts resins, 84-86, INDUST DIR, HERCULES, INC, 86- *Mem:* Am Chem Soc. *Res:* Chemistry of reactive intermediates; free radicals-peroxide synthesis and reaction mechanism; nitrenes-synthesis and applications of nitrene precursors-azidoformates, sulfonylazides, azides; 1-3 dipoles-synthesis and application; toner resins and toner technology; printing ink vehicles and ink resin technology. *Mailing Add:* Res Ctr Hercules Inc Wilmington DE 19899

SCHAPPERT, GOTTFRIED T, b Mannheim, Ger, Sept 10, 34; US citizen; c 1. LASER PHYSICS, MATTER. *Educ:* Mass Inst Technol, BS, 56, MS, 58, PhD(physics), 61. *Prof Exp:* Fel, Max Planck Inst Astrophysics, 61-62; instr, Dept physics, Mass Inst Technol, 62-63; consult, Cambridge Res Labs, US Air Force, 63-65; res assoc, Brandeis Univ, 65-67; physicist, Electronics Res Ctr, NASA, 67-71; MEM STAFF & GROUP LEADER, LOS ALAMOS NAT LAB, 71- *Mem:* Am Phys Soc; AAAS; Sigma Xi. *Res:* Laser fusion physics; laser physics and engineering; laser matter interaction physics. *Mailing Add:* 145 San Juan St Los Alamos NM 87544

SCHAR, RAYMOND DEWITT, b Butler, Pa, Apr 9, 23; m 43; c 3. POULTRY SCIENCE. *Educ:* Pa State Univ, BS, 50. *Prof Exp:* Instr voc agr, Pub Schs, Pa, 45-49; sr poultry inspector, Pa Dept Agr, 50-59; poultry coordr, USDA, 59-71, proj leader random sample poultry testing, Animal Improv Progs Lab, Animal Physiol & Genetics Inst, Sci & Educ Admin-Agr Res, 60-79, sr poultry coordr, 71-83, poultry scientist, Animal & Plant Health Inspection Serv, 84-88; RETIRED. *Concurrent Pos:* Adv mem & secy, Nat Comt Random Sample Poultry Testing, 60-79. *Mem:* Poultry Sci Asn; World Poultry Sci Asn (secy-treas, 80-85). *Res:* Poultry production and diseases; study trends; measures to combat diseases through blood testing of breeder flocks. *Mailing Add:* Nat Poultry Improv Plan Animal & Plant Health Inspection Serv 11711 Roby Ave Beltsville MD 20705

SCHARBER, SAMUEL ROBERT, JR, b Winchester, Tenn, Mar 23, 33; m 54, 77; c 4. PHYSICAL CHEMISTRY. *Educ:* Univ Notre Dame, BS, 55; Harvard Univ, EdM, 63; Univ Tex, Austin, PhD(chem), 70. *Prof Exp:* High sch teacher, Ill, 56-59; consult & reseacher, Bud Toye Co, El Cajon, 71-80; PROF CHEM, SAN DIEGO COMMUNITY COLS, 60- *Concurrent Pos:* Consult & researcher, 80-; acad year inst fel & summer fel, NSF. *Mem:* Am Chem Soc; Optical Soc Am. *Res:* Solid waste recycle, chiefly plastics, metals and rubber; air pollution, Palladium/detwium system. *Mailing Add:* 10058 Pandora Dr La Mesa CA 91941

SCHARENBERG, ROLF PAUL, b Hamburg, Ger, Mar 11, 27; m 55; c 2. NUCLEAR PHYSICS. *Educ:* Univ Mich, Ann Arbor, BS, 49, MS, 50, PhD(physics), 55. *Prof Exp:* From instr to asst prof physics, Mass Inst Technol, 55-60; assoc prof, Case Inst Technol, 61-65; assoc prof, 65-70, PROF PHYSICS, PURDUE UNIV, WEST LAFAYETTE, 71- *Mem:* Fel Am Phys Soc. *Res:* Magnetic and electric structure of nuclei. *Mailing Add:* 144 E Navajo St West Lafayette IN 47907

SCHARER, JOHN EDWARD, b Monroe, Wis, Oct 11, 39; m 65; c 2. ELECTRICAL ENGINEERING, APPLIED PHYSICS. *Educ:* Univ Calif, Berkeley, BS, 61, MS, 63, PhD(elec eng), 66. *Prof Exp:* From asst prof to assoc prof, 66-78, assoc chmn grad studies, 80-82, PROF ELEC ENG, UNIV WIS-MADISON, 78- *Concurrent Pos:* NSF grant, Univ Wis-Madison, 67-, Dept Energy res contract, 67-; Fr Atomic Energy Comn vis scientist, Ctr Nuclear Studies, Fontenay-aux-Roses, France, 69-70. *Mem:* Am Phys Soc; Inst Elec & Electronics Engrs. *Res:* Plasma physics, particularly linear and nonlinear wave propagation and instabilities; microwave and RF heating of plasmas; lasers. *Mailing Add:* Dept Elec Eng/2414 Eng Bldg Univ Wis 1415 Johnson Dr Madison WI 53706

SCHARF, ARTHUR ALFRED, b Chicago, Ill, July 27, 27; m 56; c 4. BIOLOGY, BIOPHYSICS. *Educ:* Northwestern Univ, BS, 48, MS, 50, PhD(bot), 53. *Prof Exp:* Bacteriologist, NShore Sanit Dist, Ill, 48; res asst, Arctic Res Lab, Alaska, 52; instr biol & bot, Ill Teachers Col Chicago-North, 53-56; asst prof biol, Elmhurst Col, 56-57; from asst prof to assoc prof biol & bot, 57-65, chmn natural sci div, 62-63, chmn dept biol, 63-65, prof bot, 65-80, PROF BIOL, NORTHEASTERN ILL UNIV, 65- *Mem:* AAAS. *Res:* History of science. *Mailing Add:* Dept Biol Northeastern Ill Univ 5500 N St Louis Ave Chicago IL 60625

SCHARF, BERTRAM, b New York, NY, Mar 3, 31; m 65; c 2. PSYCHOACOUSTICS. *Educ:* City Col New York, BA, 53; Univ Paris, dipl, 55; Harvard Univ, PhD(exp psychol), 58. *Prof Exp:* PROF PSYCHOL, NORTHEASTERN UNIV, 58- *Concurrent Pos:* Res assoc, Tech Hochschule Stuttgart, 62; vis res assoc, Sensory Res Lab, Syracuse Univ, 66; vis scientist, Med Sch, Helsinki Univ, 71-72; chmn, Working Group, US Standards Inst, 71-78; assoc ed, J Acoust Soc Am, 77-80; vis prof, Univ Provence, Marseille, 78-79, 85-86, 88; vis scientist, Nat Ctr Sci Res, France, 82-83, & 90-91. *Mem:* Fel Acoust Soc Am; fel AAAS; Psychonomics Soc; Am Psychol Asn; Int Audiol Soc; hon mem Finnish Acoust Soc. *Res:* Effects of noise on people; sensory psychology; psychophysics; psychoacoustics; loudness; frequency analysis; sound localization; normal and pathological hearing. *Mailing Add:* 22 Chestnut Pl Brookline MA 02146

SCHARF, WALTER, b Vienna, Austria, July 19, 29; nat US. GEMOLOGY. *Educ:* City Col New York, BS, 52; Columbia Univ, MA, 54, PhD(chem), 57. *Prof Exp:* Asst chem, Columbia Univ, 52-53; lectr, City Col New York, 53-58, from instr to asst prof, 58-68; ASSOC PROF CHEM & CHMN DEPT NATURAL SCI, BARUCH COL, 68- *Concurrent Pos:* Proj leader, Evans Res & Develop Corp, 57-58; vis lectr, Columbia Univ, 65. *Mem:* Royal Soc Chem; Am Chem Soc; Sigma Xi; fel Gemol Asn Gt Brit. *Res:* Isolation, purification and mechanism of action of polyphenol oxidases and flavorese enzymes; effects of gamma radiation on proteins; biodegradation of ascorbic acid; contraceptive plant estrogens; crystal growth from melts and solutions. *Mailing Add:* Dept Natural Sci Box 291 Baruch Col 17 Lexington Ave New York NY 10010

SCHARFETTER, D(ONALD) L, b Pittsburgh, Pa, Feb 21, 34; m 55; c 3. ELECTRICAL ENGINEERING. *Educ:* Carnegie Inst Technol, BS, 60, MS, 61, PhD(elec eng), 62. *Prof Exp:* Mem tech staff, Bell Labs, 62-76; prof elec eng, Carnegie-Mellon Univ, 76-80; mem staff, Palo Alto Res Ctr, Xerox Corp, 78-83; at Fastek, San Carlos, Ca, 84-87; prof elec eng, Univ Calif, Berkeley, 84-87; MGR PROCESSER MODEL, INTEL CORP, 87- *Mem:* Inst Elec & Electronics Engrs. *Res:* Analysis of device characteristics; device physics as applied to semiconductors. *Mailing Add:* 2250 Mission College Blvd PO Box 58125 Santa Clara CA 95052

SCHARFF, MATTHEW DANIEL, b New York, NY, Aug 28, 32; m 54; c 3. CELL BIOLOGY, IMMUNOLOGY. *Educ:* Brown Univ, AB, 54, New York Univ, MD, 59. *Prof Exp:* Intern & resident, Boston City Hosp, Mass, 59-61; res assoc, Nat Inst Allergy & Infectious Dis, 61-63; assoc cell biol, 63-64, from asst prof to assoc prof, 64-71, chmn dept, 71-83, dir, Div Biol Sci, 74-82, PROF CELL BIOL, ALBERT EINSTEIN COL MED, 71-, DIR, CANCER CTR, 86- *Honors & Awards:* Harvey lectr, 74; Dyer lectr, NIH, 80. *Mem:* Nat Acad Sci; Harvey Soc; Am Asn Immunol; Am Soc Clin Invest; Am Acad Arts & Sci. *Res:* Immunobiology. *Mailing Add:* Dept Cell Biol Albert Einstein Col Med Bronx NY 10461

SCHARFF, THOMAS G, b Paterson, NJ, Mar 9, 23; m 46; c 5. PHARMACOLOGY. *Educ:* Trinity Col, Conn, BS, 48, MS, 51; Univ Rochester, PhD(pharmacol), 56. *Prof Exp:* Lab supvr, Bigelow-Sanford Carpet Co, 48-49; res assoc, US AEC Proj, Trinity Col, Conn, 51-52; from instr to prof pharmacol, Sch Med & Dent, Univ Louisville, 56-85; RETIRED. *Concurrent Pos:* Prin investr, Am Heart Asn, 63-68. *Mem:* AAAS; Am Soc Pharmacol & Exp Therapeut; Soc Exp Biol & Med; Soc Toxicol; Am Chem Soc. *Res:* Cellular pharmacology and biochemistry; cell metabolism and transport. *Mailing Add:* 3623 Windwar Way Louisville KY 40220

SCHARFSTEIN, LAWRENCE ROBERT, b New York, NY, July 21, 27; m 55; c 3. CHEMICAL METALLURGY. *Educ:* Pa State Univ, BS, 46; NY Univ, PhD(phys chem), 53. *Prof Exp:* Supvr phys chem res, Goodyear Atomic Corp, 53-55; lead engr, Bettis Plant, Westinghouse Elec Corp, 55-59; supvr corrosion res, Carpenter Steel Corp, 59-65, asst mgr chem res, 65-68, mgr chem technol res, Carpenter Technol Corp, 68-71, dir nuclear mat, 71-76; GROUP LEADER, MAT ENG, MOBIL RES & DEVELOP CORP, 76- *Mem:* Am Soc Metals; Am Soc Testing & Mat; Nat Asn Corrosion Engrs; Am Chem Soc; Am Nuclear Soc. *Res:* Corrosion; chemistry of metals and surfaces; electrochemistry; plating; nuclear chemistry; physical metallurgy of stainless steels; gas-metal reactions; heat treating. *Mailing Add:* 40 Clover Lane Princeton NJ 08540

SCHARN, HERMAN OTTO FRIEDRICH, b Germany, July 20, 11; US citizen; m 40; c 4. MATHEMATICS, PHYSICS. *Educ:* Univ Gottingen, MS, 48; Darmstadt Tech Univ, Dr rer nat(math, physics), 66. *Prof Exp:* Aeronaut engr, Aeronaut Res Inst, Ger, 40-45; high sch teacher, Hanover, 48-57; astronaut engr, Holloman AFB, USAF, 57-71 & Kirtland AFB, 71-73; RETIRED. *Res:* Theory of optimal trajectories in space navigation. *Mailing Add:* 8100 Connecticut St NE Albuquerque NM 87110

SCHARNHORST, KURT PETER, b Hamburg, Ger, Apr 19, 36; US citizen; m 62; c 2. ACOUSTICS. *Educ:* City Col New York, BS, 61; Univ Md, PhD(physics), 69. *Prof Exp:* Physicist superconductivity, 60-70, RES PHYSICIST SOLID STATE PHYSICS, NAVAL SURFACE WARFARE CTR, 70- *Mem:* Acoust Soc Am. *Res:* Electro-optics; superconductivity; wave propagation phenomena; solid state device physics; acoustics. *Mailing Add:* Naval Surface Warfare Ctr White Oak Silver Spring MD 20903-5000

SCHARPEN, LEROY HENRY, b Red Wing, Minn, Oct 15, 35; m 63; c 2. PHYSICAL CHEMISTRY, SURFACE SCIENCE. *Educ:* Harvard Univ, AB, 61; Stanford Univ, PhD(chem), 66. *Prof Exp:* Res scientist, McDonnell Douglas Corp, 66-68; appln chemist, Sci Instruments Div, Hewlett Packard Co, 68-73, mgr electron spectros, Chem Anal Appln Lab, 73-76; exec vpres, 76-85, PRES, SURFACE SCI LABS, INC, 85- *Mem:* AAAS; Am Chem Soc. *Res:* Application of surface analysis techniques to industrial research; materials and process problem solving. *Mailing Add:* 10145 McLaren Pl Cupertino CA 95014

SCHARPF, LEWIS GEORGE, JR, b Springfield, Mo, Sept 15, 40; m 65; c 2. FERMENTATION, FLAVOR CHEMISTRY. *Educ:* Southwest Mo State Univ, BS, 61; Iowa State Univ, MS, 63, PhD(biochem), 65. *Prof Exp:* Res chemist, Monsanto Co, 65-70, res specialist, Monsanto Indust Chem Co, 70-74, mgr res & develop, 74-77; dir develop & vpres res & develop, 77-80, vpres tech dir flavors,80-84, INT FLAVORS & FRAGRANCES, 80-84, VPRES & DIR FLAVOR RES, 84- *Mem:* Flavor & Extract Mfrs Asn; Am Chem Soc; Inst Food Technol; Sigma Xi. *Res:* Food and fermentation chemistry; food ingredient development; flavor chemistry. *Mailing Add:* 35 Lewis Point Rd Fair Haven NJ 07701

SCHARPF, ROBERT F, b St Louis, Mo, June 22, 31; m 57; c 2. PLANT PATHOLOGY. *Educ:* Univ Mo, BS, 54; Univ Calif, Berkeley, MS, 57, PhD(plant path), 63. *Prof Exp:* PLANT PATHOLOGIST, PAC SOUTHWEST FOREST & RANGE EXP STA, US FOREST SERV, 60- *Mem:* Am Phytopath Soc. *Res:* Forest diseases; epidemiology; hyperparasites. *Mailing Add:* Pac SW Forest & Range Exp Sta PO Box 245 Berkeley CA 94701

SCHARPF, WILLIAM GEORGE, b Baltimore, Md, Aug 24, 25; m 57; c 2. ORGANIC CHEMISTRY, PESTICIDE CHEMISTRY. *Educ:* Univ Md, BS, 50; Rider Col, MBA, 81. *Prof Exp:* Res chemist pesticides, US Indust Chem, 43-52; res chemist med, Johnson & Johnson, 53-56; plant chemist, Gen Elec Co, 57-58; sr process chemist boron fuels, Thiokol, 58-59; sr res chemist pesticides, FMC Corp, 59-78, res assoc, 78-86; abstracter, Inst Sci Info, 86-87; chemist, Huels Am, 87-89; RETIRED. *Honors & Awards:* Thomas A Edison Patent Award, Res & Develop Coun NJ, 79. *Mem:* Am Chem Soc; Org Reactions Catalysis Soc. *Res:* Structure biological activity correlations; process research related to organic pesticides, particularly carbamate and pyrethroid insecticides. *Mailing Add:* 804 Roelofs Rd Yardley PA 19067

SCHARRER, BERTA VOGEL, b Munich, Ger, Dec 1, 06; nat US; m 34. NEUROENDOCRINOLOGY. *Educ:* Univ Munich, PhD(biol), 30. *Hon Degrees:* DMed, Univ Giessen, WGer, 76; DSc, Smith Col, 80, Yeshiva Univ, 83, Mt Holyoke Col, 84, State Univ NY Old Westbury, 85; LLD, Univ Calgary, Can, 82, Harvard Univ, Northwestern Univ. *Prof Exp:* Asst, Res Inst Psychiat, Munich, 32-34; guest investr, Neurol Inst, Frankfurt, 34-37; guest investr, Dept Anat, Univ Chicago, 37-38 & Rockefeller Inst, 38-40; sr instr, Western Reserve Univ, 40-46; instr & asst prof res, Univ Colo, 46-54; prof, 55-78, EMER DISTINGUISHED PROF ANAT & NEUROSCI, ALBERT EINSTEIN COL MED, 78- *Concurrent Pos:* Fel, Western Reserve Univ, 40-46; Guggenheim fel, 47-48; USPHS fel, 48-50; NSF grant, 78-80. *Honors & Awards:* Kraepelin Gold Medal Award, 78; S C Koch Award, 80; Henry Gray Award, Am Asn Anat, 82; Schleiden Medal, German Acad Leopoldina, 83; Nat Medal Sci, NSF, 83. *Mem:* Nat Acad Sci; hon mem Am Soc Zoologists; Am Asn Anat (pres, 78-79); Am Acad Arts & Sci; hon mem Europ Soc Comp Endocrinol; Ger Acad Leopoldina; foreign mem Royal Neth Acad Arts & Sci; hon mem Int Soc Neuroendocrinol; hon mem Ger Anat Soc; hon mem Israeli Soc Anat Sci. *Res:* Comparative neuroendocrinology and neurosecretion; comparative endocrinology; ultrastructure; neuroimmunology. *Mailing Add:* Dept Anat Albert Einstein Col Med Bronx NY 10461

SCHARTON, TERRY DON, b York, Nebr, May 12, 39; m 64; c 3. ACOUSTICS, VIBRATIONS. *Educ:* Mass Inst Technol, BS, 62, MS, 64, ScD(mech eng), 66. *Prof Exp:* Regional mgr acoust, Bolt, Beranck & Newman, 66-77; PRIN SCIENTIST VIBRATIONS, ANCO ENGRS INC, 77- *Mem:* Acoust Soc Am; Am Inst Aeronaut & Astronaut; Am Inst Automotive Engrs. *Res:* Offshore tower vibration testing for nondestructive evaluation; tube bundle heat exchange vibration diagnosis. *Mailing Add:* 1102 Stanford Santa Monica CA 90403

SCHARVER, JEFFREY DOUGLAS, b Massillon, Ohio, Nov 3, 47; m 70; c 2. ORGANIC CHEMISTRY. *Educ:* Bowling Green State Univ, BS, 69; Duke Univ, PhD(chem), 75. *Prof Exp:* Res chemist, 74-81, SECT HEAD, CHEM DEVELOP LABS, BURROUGHS WELLCOME CO, USA, 81- *Mem:* Am Chem Soc. *Res:* The development of new organic processes and their application to pharmaceutical research. *Mailing Add:* Burroughs Wellcome Co 3030 Cornwallis Rd Research Triangle Park NC 27709

SCHATTEN, GERALD PHILLIP, b New York, NY, Nov 1, 49. CELL BIOLOGY, DEVELOPMENTAL BIOLOGY. *Educ:* Univ Calif, Berkeley, BS, 71, PhD(cell biol), 75. *Prof Exp:* Instr zool, Univ Calif, Berkeley, 75; fels reproduction, Rockefeller Found, NY, 76 & 77; guest researcher, Ger Cancer Res Ctr, Heidelberg, 76-77; from asst prof to prof biol sci, 77-86, dir, Electron Micros Lab, Inst Molecular Biophysics, Fla State Univ, 81-86, PROF, MOLECULAR BIOL ZOOL & DIR INTERGRATED MICROS RESOURCE BIOMED RES, UNIV WIS-MADISON. *Concurrent Pos:* Prin investr res grants, fertil, NSF, 78, 79 & 81, NIH, 79 & 81, Am Cancer Soc, 80, Environ Protection Agency, 81 & Fla State Univ Found; NIH career development award, 80. *Honors & Awards:* Micrograph Award, Exp Cell Res, Stockholm, 76; Boehringer Ingelheim Professorship, 84. *Mem:* AAAS; Am Soc Cell Biol; Soc Develop Biol; Soc Study Reproduction; Am Soc Zoologists. *Res:* Fertilization; cancer; egg activation; cell transformation; movement of sperm; pronuclear movements and fusion; intracellular calcium localization; mitosis; cytokinesis; mitotic apparatus; nuclear envelope; membranes; electron microscopy; video microscopy; ion localization. *Mailing Add:* Univ Wis-Madison 1117 W Johnston St Madison WI 53706

SCHATTEN, HEIDE, b Niederweidbach, WGer, Sept 24, 46; m 77. CELL BIOLOGY, CANCER. *Educ:* Ger Cancer Res Ctr, Inst Cell Res, dipl, 74, Dr rer nat, 77. *Prof Exp:* Fel cell biol, Univ Calif, Berkeley, 77; fac res assoc cell biol, Fla State Univ, 77-81, from asst res scientist to assoc res scientist cell biol, 81-86; SR SCIENTIST CELL BIOL, DEPT MOLECULAR BIOL & ZOOL, UNIV WIS-MADISON, 86- *Concurrent Pos:* Co prin investr, NSF grant, 79; sr investr res grants, NIH, 80- & Environ Protection Agency, 81-; travel grant, Am Soc Cell Biol, 81. *Mem:* Am Soc Cell Biol; Ger Soc Cell Biol. *Res:* Fertilization; pronuclear movements; mitotic apparatus; cell division; microtubules; protein chemistry; mitosis; synchronization; cell culture; transformation; activation; cancer drugs; reproduction; electron microscopy. *Mailing Add:* Zool Res Bldg Univ Wis 1117 W Johnson St Madison WI 53706

SCHATTEN, KENNETH HOWARD, b New York, NY, Feb 1, 44; m 70. SOLAR PHYSICS. *Educ:* Mass Inst Technol, SB, 64; Univ Calif, Berkeley, PhD(physics), 68. *Prof Exp:* Researcher space physics, Univ Calif, Berkeley, 68-69; Nat Acad Sci fel, Goddard Space Flight Ctr, NASA, 69-70, researcher, 70-72; SR LECTR PHYSICS, VICTORIA UNIV, WELLINGTON, 72- *Res:* Solar terrestrial relationships; sun, solar corona, interplanetary space, geophysics; planetary physics; electric and magnetic fields; plasma physics. *Mailing Add:* Dept Physics Victoria Univ Pvt Bag Wellington New Zealand

SCHATTENBURG, MARK LEE, b Colo, Apr 12, 56. X-RAY LITHOGRAPHY, MICRO-NANO STRUCTURE FABRICATION. *Educ:* Univ Hawaii, BS, 78; Mass Inst Technol, PhD(physics), 84. *Prof Exp:* Postdoctoral assoc, 84-85, sci res staff, 85-90, RES SCIENTIST, MASS INST TECHNOL CTR SPACE RES, 90- *Concurrent Pos:* Res staff, Mass Inst Technol Submicron Structures Lab, 84-; instrument scientist, NASA AXAF High Energy Transmission Gratings Spectrometer, 85- *Mem:* Am Vacuum Soc; Am Astron Soc; Int Soc Optical Eng. *Mailing Add:* Mass Inst Technol 37-421 70 Vassar St Cambridge MA 02139

SCHATTNER, ROBERT I, b New York, NY, June 4, 25; c 2. PHARMACEUTICAL CHEMISTRY, PESTICIDE CHEMISTRY. *Educ:* Univ Pa, DDS, 48; City Univ New York, BS, 49. *Prof Exp:* Dent surgeon, USPHS, 48-49; pvt pract, 49-59; RES DIR, R SCHATTNER CO, 63-, OFFICER & DIR, R SCHATTNER FOUND MED RES, 64-; OFFICER & RES DIR, SPORICIDIN CO, WASHINGTON, DC, 77- *Concurrent Pos:* Res dir, Chloraseptic Co, 52-63; consult, Norwich Pharmacal Co, 63-65. *Mem:* Am Soc Microbiol; Royal Soc Health; Int Dent Fedn; Am Dent Asn; Am Chem Soc; Dent Mfg Asn; Am Dent Trade Asn. *Res:* Chemical development of new sterilizing solution; aerosol disinfectant spray; antimicrobial additive and preservative; antiseptic and analgesic pharmaceutical preparation for skin, lips and vagina. *Mailing Add:* 7101 Pyle Rd Bethesda MD 20817

SCHATTSCHNEIDER, DORIS JEAN, b New York, NY, Oct 19, 39; m 62; c 1. DISCRETE GEOMETRY, CURRICULUM MATERIALS DEVELOPMENT. *Educ:* Univ Rochester, AB, 61; Yale Univ, MA, 63, PhD(math) 66. *Prof Exp:* Instr math, Northwestern Univ, 64-65; asst prof math, Univ Ill Chicago, 65-68; PROF MATH, MORAVIAN COL, 68- *Concurrent Pos:* Chair, Math Dept, Moravian Col, 71-74, 85-; gov, Math Asn Am, 81-89; ed, Math Mag, 81-85; sr assoc, Visual Geometry Proj, NSF, 86-91; prin investr, Humanities, Sci & Technol Proj, NEH, 88-90. *Honors & Awards:* Allendoerfer Award, Math Asn Am, 79, Cert Meritorious Serv, 91. *Mem:* Math Asn Am; Am Math Soc; Asn Women Math; Nat Coun Teachers Math. *Res:* Geometry, especially tiling and polyhedra; teaching of geometry; geometry and art; art of M C Escher. *Mailing Add:* Math Dept Moravian Col Bethlehem PA 18018

SCHATZ, EDWARD R(ALPH), b St Mary's, Pa, Nov 28, 21; m 48; c 2. ELECTRICAL ENGINEERING. *Educ:* Carnegie Inst Technol, BS, 42, MS, 43, DSc, 49. *Prof Exp:* Asst engr, Metall Lab, Univ Chicago, 44 & Los Alamos Sci Lab, Univ Calif, 44-46; instr elec eng, 46-49, asst prof elec eng & indust admin, 49-52, assoc prof elec eng, 53-61, head dept, 53-57, asst dean col eng & sci, 57-60, assoc dean, 60-61, dean res, 61-64, PROF ELEC ENG, CARNEGIE-MELLON UNIV, 61-, VPRES ACAD AFFAIRS, 64-, PROVOST, 73- *Mem:* Am Soc Eng Educ; sr mem Inst Elec & Electronics Engrs; Sigma Xi. *Res:* Skin effect in wires; atomic bomb; transients in tapered transmission lines; communications. *Mailing Add:* 410 Buckingham Rd Pittsburgh PA 15215

SCHATZ, GEORGE CHAPPELL, b Watertown, NY, Apr 14, 49; m 75; c 3. QUANTUM REACTIVE SCATTERING. *Educ:* Clarkson Univ, BS, 71; Calif Inst Technol, PhD(chem), 75. *Prof Exp:* Res assoc, Mass Inst Technol, 75-76; from asst prof to assoc prof, 76-82, PROF CHEM, NORTHWESTERN UNIV, 82- *Concurrent Pos:* Consult, Argonne Nat Lab, 78-82 & staff scientist appointee, 83-86, vis scientist, 86-; consult, Battelle Columbus Lab, 79-80 & Signal Res Ctr, 85-86; Sloan Res Fel, 80-82; vis fel,

Joint Inst Lab Astrophys, 88-89. *Honors & Awards:* Dreyfus Award, 81-86; Fresenius Award, 83. *Mem:* Am Chem Soc; Fel Am Phys Soc; Combustion Inst. *Res:* Theoretical chemistry; quantum reactive scattering; classical trajectory simulations; collision induced energy transfer; potential energy surfaces; combustion kinetics; state to state chemistry; surface enhanced spectroscopy; electrodynamics near rough metal surfaces. *Mailing Add:* Dept Chem Northwestern Univ Evanston IL 60208-3113

SCHATZ, IRWIN JACOB, b St Boniface, Man, Oct 16, 31; m 67; c 4. INTERNAL MEDICINE, CARDIOVASCULAR DISEASES. *Educ:* Univ Man, MD, 56. *Prof Exp:* Fel, Mayo Clin, 58-61; chief sect peripheral vascular dis, Henry Ford Hosp, Detroit, 61-68; chief sect cardiovasc dis, Sch Med, Wayne State Univ, 68-72; assoc prof cardiol, Med Ctr, Univ Mich, Ann Arbor, 72-75, assoc dir div cardiol, 72-75, prof med, 73-75; PROF MED & CHMN DEPT, JOHN A BURNS SCH MED, UNIV HAWAII, HONOLULU, 75- *Concurrent Pos:* Fel coun circulation & clin cardiol, Am Heart Asn; spec consult health manpower, Dept HEW, 68-69; past gov, Am Col Cardiol; past pres, Hawaii Health Asn. *Mem:* Fel Am Col Physicians; fel Am Col Cardiol; Asn Prof Med. *Res:* Ischemic heart disease; platelet morphology; atherosclerosis. *Mailing Add:* Dept Med Univ Hawaii Med Sch 1356 Lusitana St Honolulu HI 96813

SCHATZ, JOSEPH ARTHUR, b US, June 23, 24; m 48; c 4. MATHEMATICS. *Educ:* Va Polytech Inst, BS, 47; Brown Univ, PhD(math), 52. *Prof Exp:* Draftsman, E I du Pont de Nemours & Co, Inc, 41-42; inspector, Signal Corps, US Dept Army, 42-43, engr, Manhattan Proj, 46; ed asst, Math Rev, Am Math Soc, 48-52; instr math, Lehigh Univ, 52-55 & Univ Conn, 55-57; mem staff, Sandia Corp, 57-72; ASSOC PROF MATH & COMPUT SCI, UNIV HOUSTON, 72- *Mem:* Am Math Soc; Math Asn Am; Asn Comput Mach; Asn Symbolic Logic; Sigma Xi. *Res:* Applied mathematics; computer sciences; logic. *Mailing Add:* Dept Math Univ Houston Houston TX 77004

SCHATZ, PAUL NAMON, b Philadelphia, Pa, Oct 20, 28; m 54; c 3. PHYSICAL CHEMISTRY. *Educ:* Univ Pa, BS, 49; Brown Univ, PhD(chem), 52. *Prof Exp:* Jewett fel, Calif Inst Technol, 52-53; res assoc chem, Brown Univ, 53-54; from asst prof to assoc prof, 56-65, PROF CHEM, UNIV VA, 65- *Concurrent Pos:* NSF sr fel, Oxford Univ, 63-64, Guggenheim fel, 74-75; vis sr res fel, St Johns Col, Oxford, 88. *Mem:* Am Chem Soc; Am Phys Soc. *Res:* Molecular structure, especially spectroscopy and quantum mechanics. *Mailing Add:* Dept Chem Univ Va Chem Bldg 134 Charlottesville VA 22903

SCHATZKI, THOMAS FERDINANT, b Berlin, Ger, Oct 20, 27; nat US; m 52; c 2. CHEMICAL PHYSICS. *Educ:* Univ Mich, BS, 49; Mass Inst Technol, PhD(chem), 54. *Prof Exp:* Res assoc, Univ Wis, 54-55 & Univ Ill, 55-57; chemist, Shell Develop Co, Calif, 57-72; RES CHEMIST, WESTERN REGIONAL LAB, USDA, 72- *Mem:* Inst Elec & Electronics Engrs. *Res:* Image analysis; polymer physics. *Mailing Add:* USDA Western Regional Lab Albany CA 94710

SCHATZLEIN, FRANK CHARLES, b Oceanside, NY, July 6, 29; m 52; c 2. INVERTEBRATE PHYSIOLOGY. *Educ:* Colgate Univ, BA, 51; Ind Univ, PhD(zool), 62. *Prof Exp:* From asst prof to prof biol, 59-84, PROF PHYSIOL, CALIF STATE UNIV, LONG BEACH, 85- *Concurrent Pos:* USPHS trainee endocrinol, 62-63. *Mem:* AAAS; Am Soc Zoologists; Sigma Xi. *Res:* Hormonal regulation of metabolism in crustacean larvae and adults. *Mailing Add:* Dept Anat & Physiol Calif State Univ Long Beach CA 90840

SCHAUB, JAMES H(AMILTON), b Moundsville, WVa, Jan 27, 25; m 48. CIVIL ENGINEERING. *Educ:* Va Polytech Inst, BS, 48; Harvard Univ, SM, 49; Purdue Univ, PhD(civil eng), 60. *Prof Exp:* Soils engr, State Hwy Dept, Ore, 49-50 & 51-52; lab dir, Palmer & Baker, Inc, Ala, 52-55; asst prof civil eng, Va Polytech Inst, 55-58; instr & res engr, Purdue Univ, 58-60; prof & chmn dept, Univ WVa, 60-67, assoc dean eng, 67-69; prof, 69-84, chmn dept, 84-87, DISTINGUISHED SERV PROF CIVIL ENG DEPT, UNIV FLA, 84- *Concurrent Pos:* Fel, NSF, 75-76; conquest prof humanities, Va Mil Inst, 86; vis prof eng, Swarthmore Col, 88. *Honors & Awards:* William H Wisely Award, Am Soc Civil Engrs, 86. *Mem:* Fel Am Soc Civil Engrs; Am Soc Eng Educ; Nat Soc Prof Engrs; Am Pub Works Asn. *Res:* Soil mechanics; highway engineering; professional ethics. *Mailing Add:* Dept Civil Eng Univ Fla Gainesville FL 32611

SCHAUB, ROBERT GEORGE, b Belleuvue, Pa, Dec 16, 47; m 83; c 1. VASCULAR BIOLOGY, THROMBOSIS. *Educ:* Univ Nev, BS, 70; Wash State Univ, PhD(physiol), 73. *Prof Exp:* Fel thrombosis, Wash State Univ, 73-75 & SCOR Thrombosis Temple Univ, 75-77; from asst prof to assoc prof physiol, Univ Tenn, 77-82; sr res scientist, Upjohn Co, 82-86, sr scientist thrombosis, 86-90; HEAD RES PHARMACOL, GENETICS INST, 90- *Mem:* Am Physiol Soc; Soc Exp Biol & Med; Int Soc Thrombosis & Haemostasis; NY Acad Sci; Sigma Xi; Am Asn Pathologists. *Res:* Animal models of vascular injury; thrombosis and coagulation; pathophysiology of peripheral circulation; electron microscopy of blood vessels and blood formed elements. *Mailing Add:* Clin Res Genetics Inst 87 Cambridge Park Dr Cambridge MA 02140

SCHAUB, STEPHEN ALEXANDER, b Walla Walla, Wash, Sept 29, 40; m 65; c 2. VIROLOGY, PUBLIC HEALTH & EPIDEMIOLOGY. *Educ:* Wash State Univ, BS, 64; Univ Tex, Austin, MA, 70, PhD(microbiol), 72. *Prof Exp:* Scientist, USPHS, 64-66; res microbiologist, 72-73, SUPVRY MICROBIOLOGIST, US ARMY BIOMED RES & DEVELOP LAB, 74- *Concurrent Pos:* Consult, US Justice Dept. *Mem:* Am Soc Microbiol; Sigma Xi; AAAS; Am Water Works Asn. *Res:* Virus concentration and enumeration from water and wastewater; aerobiology of spray irrigation of wastewater; infiltration of viruses in land systems by wastewater application; mechanisms of microbial disinfection, point of use water purification; rapid toxicity testing. *Mailing Add:* 6708 Autumn View Ct Eldersburg MD 21784

SCHAUBERT, DANIEL HAROLD, b Galesburg, Ill, Feb 15, 47; m 68; c 1. ANTENNAS, MICROWAVE ENGINEERING. *Educ:* Univ Ill, BS, 69, MS, 70, PhD(elec eng), 74. *Prof Exp:* Sr res & develop eng, Harry Diamond Labs, 74-80; sr res & develop eng & prog mgr, US Bur Radiol Health, 80-82; assoc prof, 82-88, PROF ELEC ENG, UNIV MASS, AMHERST, 88- *Concurrent Pos:* Chmn, Inst Elec & Electronics Engrs Antenna & Propagation Soc, 79-81, newslett ed, 82-84, secy-treas, 84-89, assoc ed, 90-; mem, Natl Res Coun Comt US Army Basic Res, 85-88; vis researcher, Plessoy Res & Technol, 89-90. *Mem:* Fel Inst Elec & Electronics Engrs; Int Radio Sci Union. *Res:* Printed circuit antennas and phased array design and analysis; monolithic millimeter wave circuits; scattering and absorption of dielectric and metallic bodies; antenna and scattering characteristics for transient applications. *Mailing Add:* 149 Aubinwood Rd Amherst MA 01002

SCHAUBLE, J HERMAN, b Macomb, Ill, Jan 18, 32; m 58; c 2. ORGANIC CHEMISTRY. *Educ:* Western Ill Univ, BS, 54, MS, 56; Univ Ill, PhD(org chem), 64. *Prof Exp:* Instr chem, Springfield Jr Col, 56-59; res assoc, Mass Inst Technol, 63-65; assoc prof org chem, 65-80, PROF CHEM, VILLANOVA UNIV, 80- *Mem:* Am Chem Soc; Royal Soc Chem. *Res:* Synthesis and conformational analysis of four and five membered ring heterocycles; organosulfur and selenium chemistry; synthetic photochemistry in the crystal state and in solution; complex metal hydride reduction of carbon-carbon multiple bonds. *Mailing Add:* 13 Chetwynd Rd Paoli PA 19301

SCHAUER, JOHN JOSEPH, b Dayton, Ohio, Aug 5, 36; m 70; c 4. MECHANICAL ENGINEERING. *Educ:* Univ Dayton, BME, 58; Carnegie-Mellon Univ, MS, 59; Stanford Univ, PhD(mech eng), 64. *Prof Exp:* Assoc prin res engr, Technol, Inc, Ohio, 65-67; PROF MECH ENG, UNIV DAYTON, 68-, CHMN DEPT, 85- *Res:* Convective heat and mass transfer; acoustics. *Mailing Add:* 209 Telford Aven Dayton OH 45419

SCHAUER, RICHARD C, b Pittsburgh, Pa, July 21, 37; m 69; c 2. PHYSIOLOGY. *Educ:* Univ Pittsburgh, BS, 60; NC State Univ, MS, 68, PhD(physiol zool), 72. *Prof Exp:* From instr to asst prof physiol, Univ NC, Greensboro, 71-78; ASSOC PROF, GANNON UNIV, ERIE, 78- *Mem:* AAAS; Am Soc Zool; Sigma Xi. *Res:* Effects of estrogen on rat uterus and heart; molecular endocrinology. *Mailing Add:* 4139 Harvard Rd Erie PA 16509

SCHAUF, CHARLES LAWRENCE, b Chicago, Ill, Sept 16, 43. PHYSIOLOGY, BIOPHYSICS. *Educ:* Univ Chicago, SB, 65, PhD(physiol), 69. *Prof Exp:* Asst prof, 72-75, assoc prof, 75-78, PROF NEUROL SCI & PHYSIOL, RUSH UNIV, 78- *Concurrent Pos:* NIH fel, Univ Md, 70-72; NSF grant rev panel; Nat Mult Sclerosis Soc grant rev comt. *Mem:* AAAS; Biophys Soc; Soc Gen Physiol; Soc Neurosci. *Res:* Biophysics of excitable membranes; demyelinating diseases; neuropharmacology. *Mailing Add:* Dept Biol Purdue Univ 1125 E 38th St Indianapolis IN 46205

SCHAUF, VICTORIA, b New York, NY, Feb 17, 43; c 2. INFECTIOUS DISEASES, IMMUNOLOGY. *Educ:* Univ Chicago, BS, 65, MD, 69. *Prof Exp:* Prof pediat, Col Med, Univ Ill, 77-84; CHMN DEPT PEDIAT, NASSAU COUNTY MED CTR, 86- *Concurrent Pos:* Head, Sect Pediat Infectious Dis & Immunol, Univ Ill, 74-, dir, Chiang Mai-Ill Leprosy Res Proj, 79-84; Off Biol Res & Review, 84-86. *Mem:* Soc Pediat Res; Infectious Dis Soc Am; Cent Soc Clin Res. *Res:* Clinical and laboratory research in immunology, virology, and bacteriology; clinical trials with antimicrobial agents; leprosy, immunology and epidemiology. *Mailing Add:* Dept Pediat Nassau Co Med Ctr 2201 Hempstead Turnpike East Meadow NY 11554

SCHAUFELE, ROGER DONALD, b Woodbridge, NJ, Mar 30, 28; m 49; c 2. ENGINEERING DESIGN, AERODYNAMICS. *Educ:* Rensselaer Polytech Inst, BAE, 49; Calif Inst Technol, MS, 52. *Prof Exp:* Dir technol, 71-76, dir advan eng, 76-79, dir air conditioning design, 79-81, vpres eng, 81-87, VPRES & GEN MGR COM ADVAN PRODS, DOUGLAS AIRCRAFT CORP, 87- *Mem:* Fel Am Inst Aeronaut & Astronaut; Soc Aeronaut Eng; fel Inst Aeronaut Engrs; Nat Aeronaut Asn. *Res:* Engineering design; aerodynamics; commercial advanced design. *Mailing Add:* Douglas Aircraft Co 3855 Lakewood Blvd Long Beach CA 90846

SCHAUFELE, RONALD A, b Calgary, Alta, Oct 5, 30; m 53. MATHEMATICAL STATISTICS. *Educ:* Univ Alta, BEd, 56; Univ Wash, BSc, 57, MSc, 62, PhD(math), 63. *Prof Exp:* Asst prof statist, Stanford Univ, 63-64 & Columbia Univ, 64-66; asst prof, 66-68, ASSOC PROF MATH, YORK UNIV, 68- *Concurrent Pos:* Can Nat Res Coun grant, 68-69. *Mem:* Am Math Soc; Math Asn Am; Inst Math Statist; Can Math Soc; Can Statist Asn; Am Statist Asn. *Res:* Probability; stochastic processes. *Mailing Add:* Dept Math & Stat York Univ North York ON M3J 1P3 Can

SCHAUMANN, ROLF, b Nuremberg, Ger, July 29, 41; m 69. FILTER DESIGN, INTEGRATED CIRCUITS. *Educ:* Univ Stuttgart, DiplIng, 67; Univ Minn, Minneapolis, PhD(elec eng), 70. *Prof Exp:* prof elec eng, Univ Minn, Minneapolis, 70-88; PROF & CHAIR, ELEC ENG, PORTLAND STATE UNIV, 88- *Mem:* Inst Elec & Electronics Engrs. *Res:* Theory of active and passive networks; distributed networks; linear integrated circuits; digital filters. *Mailing Add:* Dept Elec Eng/101 PCAT Portland State Univ Portland OR 97208-0751

SCHAUMBERG, GENE DAVID, b Rochester, Minn, Oct 3, 39; m 88; c 2. ORGANOMETALLIC CHEMISTRY. *Educ:* Pac Lutheran Univ, BS, 61; Wash State Univ, PhD(chem), 65. *Prof Exp:* Teaching asst chem, Wash State Univ, 61-65; from asst prof to assoc prof, 65-72, chmn div natural sci, 69-78, PROF ORG CHEM, SONOMA STATE UNIV, 72- *Concurrent Pos:* Fulbright lectr, 71-72 & 80-81; Scientists & Engrs in Econ Develop & NSF grant, 75; vis prof, Univ Calif, Riverside, 79-80 & Univ Hawaii, 85-86; Indo Am fel, 81-82. *Mem:* Am Chem Soc; Soc Sigma Xi. *Res:* Synthesis of new organoboron compounds of possible biological interest; soil and environmental chemistry. *Mailing Add:* Dept Chem Sonoma State Univ Rohnert Park CA 94928

SCHAUMBERGER, NORMAN, b Brooklyn, NY, May 2 8, 29; m 54; c 2. MATHEMATICS. *Educ:* City Col New York, BS, 51, MA, 52; Brooklyn Col, MA, 58; Columbia Univ, EdD(math educ), 62. *Prof Exp:* Teacher, NY Pub Schs, 51-57; instr math, Cooper Union, 57; PROF MATH, BRONX COMMUNITY COL, 59- *Concurrent Pos:* Lectr, City Col New York, 55-63; instr, Teachers Col, Columbia Univ, 63; NSF acad year inst sec math teachers, Dominican Col, NY, 66-67. *Mem:* Math Asn Am; Am Math Soc. *Res:* Problems in number theory and complex variables. *Mailing Add:* Dept Math Bronx Community Col University Ave & W 181st St Bronx NY 10453

SCHAUMBURG, FRANK DAVID, b Watseka, Ill, Jan 15, 38; m 60; c 2. ENVIRONMENTAL ENGINEERING. *Educ:* Ariz State Univ, BSCE, 61; Purdue Univ, West Lafayette, MSCE, 64, PhD(sanit eng), 66. *Prof Exp:* Res engr, BC Res Coun, 66; asst prof environ eng, 67-69, assoc prof, 69-80, PROF CIVIL ENG, ORE STATE UNIV, 80-, HEAD DEPT, 72- *Concurrent Pos:* Consult, Govt Acct Off, 75- & UNESCO, 77-78. *Mem:* Am Soc Civil Engrs; Water Pollution Control Fedn; Am Soc Eng Educ. *Res:* Biological waste treatment; environmental tradeoffs and interactions. *Mailing Add:* Dept Civil Eng Ore State Univ Corvallis OR 97331

SCHAUMBURG, HERBERT HOWARD, b Houston, Tex, Nov 6, 32; m 66; c 2. NEUROLOGY, EXPERIMENTAL NEUROPATHOLOGY. *Educ:* Harvard Univ, AB, 56; Wash Univ, MD, 60. *Prof Exp:* Fel neurol, Albert Einstein Col Med, 67-69; instr path, Harvard Med Sch, 69-71; assoc prof, 72-76, PROF NEUROL, ALBERT EINSTEIN COL MED, 77-, VCHMN DEPT, 78- *Honors & Awards:* Moore Award, Am Asn Neuropath, 77. *Mem:* Am Soc Neurol; Am Neurol Asn; Am Asn Neuropathologists. *Res:* Experimental neuropathology of myelin disease and effects of toxic chemicals on nervous system. *Mailing Add:* Dept Neurol Albert Einstein Col Med 1300 Morris Park Ave Bronx NY 10461

SCHAWLOW, ARTHUR LEONARD, b Mt Vernon, NY, May 5, 21; m 51; c 3. LASERS. *Educ:* Univ Toronto, BA, 41, MA, 42, PhD, 49. *Hon Degrees:* DSc, State Univ Ghent, 68, Univ Bradford, 70, Univ Ala, 84-, Trinity Col, Dublin, 86; Univ Toronto, LLD, 70; DTech, Lund Univ, 88. *Prof Exp:* Demonstr physics, Univ Toronto, 41-44; physicist microwave develop, Res Enterprises, Ltd, 44-45; demonstr physics, Univ Toronto, 45-49; fel & res assoc, Columbia Univ, 49-51; res physicist, Bell Tel Labs, Inc, 51-61; chmn, dept physics, 66-70 & 73-74, prof physics, 61-78, J G JACKSON-C J WOOD PROF PHYSICS, STANFORD UNIV, 78- *Concurrent Pos:* Vis assoc prof, Columbia Univ, 60; Marconi int fel, 77. *Honors & Awards:* Nobel Prize Physics, 81; Nat Medal of Sci, 91; Thomas Young Medal & Prize, Brit Inst Physics, 63; Liebmann Mem Prize, Inst Elec & Electronics Engrs, 64; Frederick Ives Medal, Optical Soc Am, 76; Ballantine Medal, Franklin Inst, 62; Schawlow Medal, Laser Inst Am. *Mem:* Nat Acad Sci; fel AAAS; fel Am Phys Soc (pres, 81); hon mem & fel Optical Soc Am (pres, 75); fel Inst Elec & Electronics Engrs; fel Am Acad Arts & Sci; fel Am Philos Soc; hon mem Am Soc Laser Med & Surg. *Res:* Radio frequency, optical and microwave spectroscopy; lasers and quantum electronics. *Mailing Add:* Dept Physics Stanford Univ Stanford CA 94305

SCHAY, GEZA, b Budapest, Hungary, June 22, 34; US citizen. MATHEMATICAL PHYSICS. *Educ:* Eotvos Lorand Univ, Budapest, BA, 56; Princeton Univ, PhD(math physics), 61. *Prof Exp:* Instr physics, Tufts Univ, 58-59; staff mathematician, Int Bus Mach Corp, 60-63; from asst prof to assoc prof math, George Washington Univ, 63-66; assoc prof, 66-70, PROF MATH, UNIV MASS, BOSTON, 70- *Mem:* Am Math Soc; Inst Math Statist. *Res:* Stochastic processes; diffusion theory; relativistic mechanics. *Mailing Add:* 298 Waltham St Newton MA 02165

SCHAYER, RICHARD WILLIAM, b Sydney, Australia, Feb 3, 15; US citizen; m 39; c 2. BIOCHEMICAL PHARMACOLOGY. *Educ:* George Washington Univ, BS, 40; Columbia Univ, PhD(biochem), 49. *Hon Degrees:* MD, Univ Lund, 75. *Prof Exp:* Chemist, USPHS, 36-42; asst, Columbia Univ, 46-49; res assoc, Rheumatic Fever Res Inst, Med Sch, Northwestern Univ, 49-57; Merck Inst Therapeut Res, 57-64; USPHS spec fel, 64-65; PRIN RES SCIENTIST, ROCKLAND RES INST, 65- *Mem:* Am Soc Biol Chemists; Am Soc Pharmacol & Exp Therapeut; Am Physiol Soc; Soc Exp Biol & Med; Brit Pharmacol Soc. *Res:* Histamine; metabolism; physiological and pathological significance; mechanism of action of glucocorticoids. *Mailing Add:* Rockland Res Inst Orangeburg NY 10962

SCHEAFFER, RICHARD LEWIS, b Williamsport, Pa, July 13, 40; m 63; c 2. STATISTICS. *Educ:* Lycoming Col, AB, 62; Bucknell Univ, MA, 64; Fla State Univ, PhD(statist), 68. *Prof Exp:* From asst prof to assoc prof statist, 67-77, PROF STATIST & CHMN DEPT, UNIV FLA, 77- *Mem:* Fel Am Statist Asn; Inst Math Statist; Biomet Soc; Int Statist Inst. *Res:* Sampling theory; applied probability, especially in the areas of reliability and two and three dimensional sampling problems. *Mailing Add:* 907 NW 21st Terr Gainesville FL 32603

SCHEARER, LAIRD D, b Allentown, Pa, Nov 26, 31; m 50; c 4. ATOMIC PHYSICS. *Educ:* Muhlenberg Col, BS, 54; Lehigh Univ, MS, 58; Rice Univ, PhD(physics), 66. *Hon Degrees:* DSc, Muhlenberg Col, 80. *Prof Exp:* Instr physics, Lafayette Col, 55-59; sr scientist, Tex Instruments, Inc, 59-71; chmn, dept physics, 71-77, PROF PHYSICS, UNIV MO, ROLLA, 77-, CUR'S PROF PHYSICS, 83- *Concurrent Pos:* Vis fel, Joint Inst Lab Astrophys, 77-78; prog assoc, NSF, 79-80; Sr Fulbright award, 84. *Honors & Awards:* Humboldt Award, 89. *Mem:* Am Phys Soc; Sigma Xi. *Res:* Optical pumping; magnetometers; penning ionization; magnetic resonance; atomic beams; tunable lasers. *Mailing Add:* Dept Physics Univ Mo Rolla MO 65401

SCHEARER, SHERWOOD BRUCE, b Reading, Pa, Jan 27, 42; m 63; c 2. POPULATION STUDIES. *Educ:* Lafayette Col, AB, 63; Columbia Univ, PhD(biochem), 71. *Prof Exp:* From res assoc to staff scientist, Pop Resource Ctr, 71-74, asst dir, Biomed Div, 71-76, from assoc to sr assoc, Int Progs Div, Pop Coun, 76-81, pres, 81-87; EXEC DIR, SYNERGOS INST, 87- *Concurrent Pos:* Mem, bd dir, Western Hemisphere Region, Int Planned Parenthood Fedn, 77-81, Planned Parenthood of New York, Pop Resource Ctr, adv bd, Margaret Sanger Ctr; chmn, Western Hemisphere Reserve/Int Planned Fedn Info & Educ Panel, 79-81. *Mem:* AAAS. *Res:* Development, transfer, absorption, implementation and monitoring of contraceptive technology; design and evaluation of epidemiological studies of drugs and devices; social science research into attitudes and practices of populations; analysis and development of population policy. *Mailing Add:* 25 Rockland Ave Larchmont NY 10538-1322

SCHEARER, WILLIAM RICHARD, b Kutztown, Pa, July 19, 35; m 58; c 3. NATURAL PRODUCTS CHEMISTRY, CHEMISTRY OF NUTRITION. *Educ:* Ursinus Col, BS, 57; Princeton Univ, MA, 59, PhD(org chem), 63. *Prof Exp:* From asst prof to assoc prof chem, Hartwick Col, 61-65; sr chemist, Ciba Pharmaceut Co, 65-68; ASSOC PROF CHEM, DICKINSON COL, 68- *Concurrent Pos:* Vis prof, Lehigh Univ, 81-82. *Mem:* Am Chem Soc. *Res:* Natural products; chemical education; applied chemistry. *Mailing Add:* Dept Chem Dickinson Col Carlisle PA 17013-2896

SCHECHTER, ALAN NEIL, b New York, NY, June 28, 39; m 65; c 2. MEDICAL RESEARCH, PROTEIN CHEMISTRY. *Educ:* Cornell Univ, AB, 59; Columbia Univ, MD, 63. *Prof Exp:* Intern med, Bronx Munic Hosp Ctr, 63-64, asst resident, 64-65; res assoc, Nat Inst Diabetes, Digestive & Kidney Dis, 65-67, USPHS vis fel, 67-68, med officer, 68-72, CHIEF, SECT MACROMOLECULAR BIOL, LAB CHEM BIOL, NAT INST DIABETES, DIGESTIVE & KIDNEY DIS, 72-, CHIEF, LAB CHEM BIOL, 81- *Concurrent Pos:* Sr asst surgeon, USPHS, 65-67 & med dir, 83- *Mem:* AAAS; Am Fedn Clin Res; Am Soc Biol Chemists; Am Soc Clin Invest; Biophys Soc; Am Soc Human Genetics. *Res:* Structure-function relations in proteins; hemoglobin chemistry; molecular genetics; genetic disease. *Mailing Add:* Lab Chem Biol Nat Inst Diabetes Digestive & Kidney Dis Bethesda MD 20892

SCHECHTER, DANIEL, physics; deceased, see previous edition for last biography

SCHECHTER, GERALDINE POPPA, b New York, NY, Jan 16, 38; m 65; c 2. HEMATOLOGY. *Educ:* Vassar Col, AB, 59; Columbia Univ, MD, 63. *Prof Exp:* Intern & asst resident med, Columbia-Presby Hosp, 63-65; resident med & hemat, 65-67, res assoc hemat, 68-69, asst chief, 70-74, CHIEF HEMAT, VET ADMIN HOSP, WASHINGTON, DC, 74- *Concurrent Pos:* From asst prof to assoc prof, George Washington Univ, 70-81, prof med, 81- *Mem:* Am Fedn Clin Res; Am Soc Hemat; Am Soc Clin Oncol; Am Asn Immunol. *Res:* Lymphocyte biology; immunological response to blood transfusion; hematological malignancies. *Mailing Add:* Hemat Sect Vet Admin Med Ctr 50 Irving St NW Washington DC 20422

SCHECHTER, JOEL ERNEST, b Detroit, Mich, Apr 18, 39; div; c 2. CELL BIOLOGY, ELECTRON MICROSCOPY. *Educ:* Wayne State Univ, BA, 61; Johns Hopkins Univ, MA, 63; Univ Calif, Los Angeles, PhD(anat), 68. *Prof Exp:* Instr med art, Sch Med, Johns Hopkins Univ, 63-64; asst prof, 69-73, assoc prof histol, 73-80, ASSOC PROF ANAT, SCH MED, UNIV SOUTHERN CALIF, 73- *Concurrent Pos:* Univ Calif, Los Angeles ment health trainee & fel, Brain Res Inst, Los Angeles, 68-69; instr exten art prog, Univ Calif, Los Angeles, 69- *Mem:* AAAS; Am Soc Cell Biol; Tissue Cult Asn; Am Asn Anatomists. *Res:* Developmental cell biology; ultrastructural histochemical studies of human pituitary tumors. *Mailing Add:* Dept Anat & Cell Biol Univ Southern Calif 1333 San Pablo St Los Angeles CA 90033

SCHECHTER, JOSEPH M, b New York, NY, Sept 28, 38. PHYSICS. *Educ:* Cooper Union, BEE, 59; Univ Rochester, PhD(physics), 65. *Prof Exp:* Res assoc physics, Fermi Inst, Univ Chicago, 65-67; from asst prof to assoc prof, 67-74, PROF PHYSICS, SYRACUSE UNIV, 75- *Mem:* Am Phys Soc. *Res:* Theoretical elementary particle physics. *Mailing Add:* Dept Physics 315 Physics Bldg Syracuse Univ Syracuse NY 13244-1130

SCHECHTER, MARSHALL DAVID, b Sept 4, 21; US citizen; div; c 4. CHILD PSYCHIATRY. *Educ:* Univ Wis, BS, 42; Univ Cincinnati, MD, 44. *Hon Degrees:* MA, Univ Pa, 77. *Prof Exp:* Pvt pract, 49-64; clin instr psychiat, Sch Med, Univ Calif, Los Angeles, 53, asst clin prof, 57, assoc clin prof, 63-64; prof psychiat, vchmn & head div child psychiat, Health Sci Ctr, Univ Okla, 69-73, consult prof pediat, 64-73, prof biol psychol, 66-73; prof & dir div child & adolescent psychiat, State Univ NY Upstate Med Ctr, 73-76; PROF PSYCHIAT & DIR DIV CHILD & ADOLESCENT PSYCHIAT, SCH MED, UNIV PA, PHILADELPHIA, 76- *Concurrent Pos:* Consult, Vet Admin Hosp, Oklahoma City, 64-73, Children's Med Ctr, Tulsa, 64-73, Okla State Dept Pub Health, 64-73, Okla State Dept Ment Health, 64-73, Wilford Hall, Lackland AFB, 66-, Head Start, Off Econ Opportunity, Hutchings Psychiat Ctr, 73-, Vet Admin Hosp, NY, 74-76, Off Child Develop, 75, Nat Inst Ment Hyg, 75-76 & Vet Admin Hosp, Philadelphia, 76-; distinguished vis prof, Wilford Hall Lackland Air Force Base, 72- *Mem:* Am Psychiat Asn; Am Psychoanal Asn; Soc Res & Child Develop. *Res:* Adoption; adolescence; autism; learning disabilities. *Mailing Add:* Dept Child Psychol 1142 Morris Univ Pa Wynnewood PA 19096

SCHECHTER, MARTIN, b Philadelphia, Pa, Mar 10, 30; m 57; c 4. PARTIAL DIFFERENTIAL EQUATIONS. *Educ:* City Col New York, BS, 53; NY Univ, MS, 55, PhD(math), 57. *Prof Exp:* Assoc res scientist, NY Univ, 57-58, from instr to asst prof math, 58-61; vis assoc prof, Univ Chicago, 61-62; from assoc prof to prof, NY Univ, 62-66; prof math, 65-85, Yeshiva Univ, chmn dept, 66-69; PROF MATH, UNIV CALIF, 83- *Concurrent Pos:* NSF sr fel, 65-66; mem, Inst Advan Study, 65-66, mem, Asn, 74-; vis prof math, Hebrew Univ, 73 & Univ Mex, 79. *Mem:* Am Math Soc; Inst Adv Studies. *Res:* Partial differential equations; functional analysis; operator theory; quantum mechanics; scattering theory; spectral theory. *Mailing Add:* Dept Math Univ Calif Irvine CA 92717

SCHECHTER, MARTIN DAVID, b Brooklyn, NY, Feb 28, 45; m 68; c 1. PHARMACOLOGY. *Educ:* Brooklyn Col, BS, 65; State Univ NY Buffalo, PhD(pharmacol), 70. *Prof Exp:* Res assoc pharmacol, Med Col Va, 70-72; sr res fel, Univ Melbourne, 72-74; from asst prof to assoc prof pharmacol, Eastern Va Med Sch, 74-78; PROF & CHMN PHARMACOL, COL MED, NORTHEASTERN OHIO UNIV, 78- *Concurrent Pos:* Fel, NIMH, 68-70; fel, AMA Educ & Res Found, 70-72, sr res fel, 72-74. *Mem:* Sigma Xi; Soc Neurosci; Behav Pharmacol Soc; Am Soc Pharmacol & Exp Therapeut; Am Chem Soc. *Res:* Psychopharmacology with special interest in correlations between behavioral and biochemical events associated with centrally-active drugs; operant conditioning; stimulus properties of drugs; drug abuse; hyperactive and aggressive behavior; drug self-administration. *Mailing Add:* Dept Pharmacol Northeastern Ohio Univ Col Med Rootstown OH 44272

SCHECHTER, MILTON SEYMOUR, b Brooklyn, NY, Aug 9, 15; m 46; c 2. ORGANIC CHEMISTRY, AGRICULTURAL CHEMISTRY. *Educ:* Brooklyn Col, BS, 35. *Prof Exp:* Chemist, Insecticide Div, Bur Entom & Plant Quarantine, USDA, Northeastern Region, 37-53 & Pesticide Chem Res Br, Entom Res Div, 53-72, chief chem & biophys control lab, Agr Environ Qual Inst, Beltsville Agr Res Ctr, 72-75, consult, Chem & Biophys Control Lab, Beltsville, Agr Res Ctr, Sci & Educ Admin-Agr Res, 75-88; RETIRED. *Concurrent Pos:* Chmn pesticide monitoring subcomt, Fed Comt Pest Control, 66-67, mem fed working group pest mgt, Monitoring Panel; US mem, Collab Int Pesticide Anal Coun, 65-75. *Honors & Awards:* Harvey W Wiley Award, Asn Off Agr Chem, 62; Burdick & Jackson Int Award, Am Chem Soc, 80. *Mem:* Fel AAAS; Am Chem Soc; NY Acad Sci; Entom Soc Am. *Res:* Synthesis of organic insecticides; pyrethrin-type esters and allethrin; methods of analysis for traces of organic pesticides; disinsection of aircraft; insecticide formulation; photoperiodism; biological rhythms; diapause; effects of light on insects; analytical chemistry. *Mailing Add:* 10909 Hannes Ct Silver Spring MD 20901

SCHECHTER, MURRAY, b New York, NY, Dec 6, 35; m 59; c 2. MATHEMATICS. *Educ:* Brooklyn Col, BA, 57; NY Univ, PhD(math), 63. *Prof Exp:* Staff mathematician, Kollsman Instrument Corp, 62-63; asst prof, 63-68, assoc prof, 68-80, PROF MATH, LEHIGH UNIV, 80- *Mem:* Am Math Soc; Soc Indust & Appl Math. *Res:* Convexity and its applications to optimization problems. *Mailing Add:* Dept Math Lehigh Univ Bethlehem PA 18015

SCHECHTER, NISSON, b Detroit, Mich, May 11, 40. BIOCHEMISTRY. *Educ:* Western Mich Univ, BA, 63, MS, 67, PhD(biochem), 71. *Prof Exp:* Fel biochem, Col Med, Univ Cincinnati, 71-73 & Weizmann Inst Sci, 73-75; RES ASST BIOCHEM, STATE UNIV NY, STONY BROOK, 75- *Concurrent Pos:* Weizmann fel, Weizmann Inst Sci, 73-74, Ahron Katzir fel, 74-75. *Mem:* Am Chem Soc; AAAS; Royal Soc Chem. *Res:* Protein synthesis in brain and neural tissue. *Mailing Add:* Dept Psychiat State Univ NY Health Sci Ctr Stony Brook NY 11794

SCHECHTER, ROBERT SAMUEL, b Houston, Tex, Feb 26, 29; m 53; c 3. CHEMICAL ENGINEERING. *Educ:* Agr & Mech Col, Tex, BS, 50; Univ Minn, PhD(chem eng), 56. *Prof Exp:* From asst prof to assoc prof chem eng, Univ Tex, Austin, 56-63, chmn dept, 70-73, dir, Ctr Thermodyn & Statist Mech, 68-74, chmn petrol eng, 75-78, Ernest Cockrell, Jr prof chem & petrol eng, 75-81, Dula & Ernest Cockrell Sr chair, 81-84, Getty Oil Co centennial chair petrol eng, 84-89, PROF CHEM ENG, UNIV TEX, AUSTIN, 63-, W A MONTY MONCRIEF CENTENNIAL ENDOWED CHAIR PETROL ENG, 89- *Concurrent Pos:* Vis prof, Univ Edinburgh, Scotland, 65-66 & Univ Brussels, Belg, 69; sr res award, Am Soc Eng Educ, 91. *Honors & Awards:* Donald P Katz Lectr, Univ Mich, 79; Chevalier Order of Palmes Academiques, Prime Minister France, 80. *Mem:* Nat Acad Eng; Am Inst Chem Engrs; Soc Petrol Engrs; Am Inst Mining Engrs; Am Chem Soc; Sigma Xi. *Res:* Surface phenomena; applied mathematics; surfactants, adsorption, micelles, and microemulsions; oil well stimulation, oil recovery and mass transfer; author of six books and more than 190 technical publications. *Mailing Add:* Dept Chem Eng Univ Tex Austin TX 78712

SCHECHTMAN, BARRY H, b New York, NY, Jan 23, 43; m 63; c 2. SOLID STATE PHYSICS, ENGINEERING. *Educ:* Cooper Union, BEE, 63; Stanford Univ, MS, 64, PhD(elec eng), 69. *Prof Exp:* Mem tech staff & chief scientist, San Jose Res Lab, IBM Corp, 75-76, mgr org solids, 76-79, appl sci, 79-84, advan record technol, 85-90, MGR STORAGE MFG RES, SAN JOSE RES LAB, IBM CORP, 90- *Mem:* Am Phys Soc; Sigma Xi. *Res:* Electronic properties of organic solids; materials and device research for semiconductor fabrication, optical technologies, computer printing, display and data storage; electronic states of molecular solids; charge generation and transport in insulators. *Mailing Add:* 6617 Creek View Ct San Jose CA 95120

SCHECKLER, STEPHEN EDWARD, b Irvington, NJ, Mar 17, 44; m 68; c 3. MORPHOLOGY, PALEOBOTANY. *Educ:* Cornell Univ, BSc, 68, MSc, 70, PhD(bot), 73. *Prof Exp:* Asst prof, Univ Alta, 75-76; asst prof, 77-83, ASSOC PROF BOT, VA POLYTECH INST & STATE UNIV, 83- *Concurrent Pos:* Nat Res Coun Can fel, Dept Bot, Univ Alta, 73-75 & 76-77; prin investr, Nat Geog Soc, NSF, Va Ctr Coal & Energy & Sigma Xi grants. *Honors & Awards:* Dimond Fund Award, Bot Soc Am. *Mem:* Bot Soc Am; Can Bot Asn; Geol Soc Belgium; Geol Soc Am; Sigma Xi. *Res:* Structure and patterns of organization, phylogenies of biocharacters and paleoecology of early land plants, especially early ferns, lycopods, progymnosperms and gymnosperms. *Mailing Add:* Dept Biol Va Polytech Inst & State Univ Blacksburg VA 24061

SCHECTER, ARNOLD JOEL, b Chicago, Ill, Dec 1, 34; m 64; c 3. OCCUPATIONAL MEDICINE, ENVIRONMENTAL HEALTH. *Educ:* Univ Chicago, BA, 54, BS, 57; Howard Univ, MD, 62; Columbia Univ, MPH, 75. *Prof Exp:* Fel anat, Harvard Med Sch, 62-64, instr med, Harvard Med Sch & Mass Gen Hosp, 64-65; intern surg, Beth Israel Hosp, 66; med officer & aviation med officer, US Army, 67-69; physician med, West Pt Med Ctr, Ky, 69-70; exec dir health care delivery, Floyd County Comprehensive Health Serv Prog, Inc, Ky, 70-71; med dir drug & alcohol dependence, inpatient rehab prog, Region 8 Ment Health Ctr, State Univ NY, 71-72, asst prof psychiat, 72-75; assoc prof prev med, NJ Med Sch, Newark, 75-79; PROF PREV MED, UPSTATE MED CTR, CLIN CAMPUS, STATE UNIV NY, BINGHAMTON, 79- *Mem:* AAAS; Am Soc Cell Biol; Electron Micros Soc Am; Am Pub Health Asn; Am Occup Med Asn; fel Am Col Physicians; fel Am Col Prev Med; fel Am Col Physicians. *Res:* Biological markers of exposure to chemicals; measurement of human tissue levels of chlorinated dibenzo-p-dioxins and ultrastrctural subcellular response as a means of understanding mechanisms of action. *Mailing Add:* Dept Prev Med SUNY Health Sci Ctr Syracuse Clin Campus Box 1000 Binghamton NY 13902

SCHECTER, LARRY, b Montreal, Que, Nov 21, 20; US citizen; m 51; c 2. NUCLEAR PHYSICS. *Educ:* Univ Calif, AB, 48, MA, 51, PhD, 53. *Prof Exp:* Res physicist, Radiation Lab, Univ Calif, 52, Calif Res & Develop Co, Stand Oil Co, Calif, 53-54 & Radiation Lab, Univ Calif, 54-55; from asst prof to prof physics, Ore State Univ, 55-88, chmn dept, 71-77. *Mem:* Am Phys Soc; Am Asn Physics Teachers. *Res:* Particle scattering; intermediate energy nuclear physics. *Mailing Add:* Dept Physics Ore State Univ Corvallis OR 97331

SCHECTMAN, RICHARD MILTON, b Wilkes-Barre, Pa, Apr 9, 32; m 60; c 2. ATOMIC PHYSICS. *Educ:* Lehigh Univ, BS, 54; Pa State Univ, MS, 56; Cornell Univ, PhD(physics), 62. *Prof Exp:* Asst prof eng physics, 61-64, assoc prof physics & astron, 64-71, PROF PHYSICS & ENG PHYSICS, UNIV TOLEDO, 71- *Concurrent Pos:* Consult, Lawrence Radiation Lab, Univ Calif, 63-65; vis assoc prof, Univ Ariz, 68 & Hebrew Univ, 69; vis scientist, Argonne Nat Lab, 72 & 75, Weizmann Inst Sci, 82-83 & Université Libre de Bruxelles, 89. *Honors & Awards:* Clement O Miniger Outstanding Res Award, Sigma Xi, 84. *Mem:* Am Phys Soc. *Res:* Particle accelerators; atomic transition probability measurements; beam foil spectroscopy; high-lying Rydberg states. *Mailing Add:* Dept Physics & Astron Univ Toledo Toledo OH 43606

SCHEDL, HAROLD PAUL, b St Paul, Minn, Sept 17, 20; m 45; c 3. MEDICINE. *Educ:* Yale Univ, BS, 42, MS, 44, PhD(chem), 46; Univ Iowa, MD, 55. *Prof Exp:* Res chemist, J T Baker Chem Co, 45-46 & Calco Chem Co, 46-49; intern, Res & Educ Hosp, Univ Ill, 55-56; responsible investr, Nat Heart Inst, 56-59; from res asst prof to res assoc prof med, 59-67, PROF MED, COL MED, UNIV IOWA, 67- *Concurrent Pos:* Commonwealth Found overseas fel, Churchill Col, Cambridge, 65; mem clin res fel panel, NIH, 67-70; ed, J Lab Clin Med, 70-74; Macy fac scholar, Westminster Hosp Med Sch, London, 77-78. *Mem:* AAAS; Am Chem Soc; Am Physiol Soc; Am Gastroenterol Asn; Endocrine Soc. *Res:* Small intestinal transport mechanisms for hexose, amino acids and peptides and their regulation through gene expression in the gastrointestinal tract; gastroenterology; calcium metabolism and vitamin D, diabetes and the gastrointestinal tract. *Mailing Add:* Dept Internal Med Univ Iowa Hosp Rm C316 Iowa City IA 52242

SCHEEL, CARL ALFRED, b LaCrosse, Wis, May 4, 23; m 46; c 4. ENTOMOLOGY. *Educ:* Wis State Univ, LaCrosse, BS, 48; Univ Wis, MS, 49, PhD(entom), 56. *Prof Exp:* Instr biol, Wis State Univ, Platteville, 50; asst prof, Cent Mich Univ, 51-52 & Mt Union Col, 52-54; asst entom, Univ Wis, 54-56; from asst prof to assoc prof biol, Cent Mich Univ, 56-64, prof biol, 64-86; RETIRED. *Mem:* Entom Soc Am; Am Inst Biol Sci; Nat Asn Biol Teachers. *Res:* Nutritional physiology of insects. *Mailing Add:* 9711 E School Section Lake Dr Mecosta MI 49332

SCHEEL, KONRAD WOLFGANG, b Mougden, China, Dec 26, 32; US citizen; m 59; c 2. CARDIOVASCULAR PHYSIOLOGY. *Educ:* Tulane Univ, BS, 62; Univ Miss, PhD(physiol), 68. *Prof Exp:* Res instr physiol, Med Ctr, Univ Miss, 63-64, res assoc surg, 67-68, asst prof physiol & med, 68-72; asst prof, Univ Tenn, Memphis, 72-75, assoc prof physiol, biophys & med, Ctr Health Sci, 75-80; prof & chmn physiol, Kirksville Col Osteop Med, 81-; PROF PHYSIOL & MED, TEX COL OSTEOP MED. *Concurrent Pos:* Volkswagen grant, Univ Kiel, 70-71; Tenn Heart investr, Ctr Health Sci, Univ Tenn, Memphis, 72-74 & USPHS grant, 72-75; guest lectr cardiol, Univ Kiel, 70-71; consult, Vet Admin, 73-75 & NIH, 74. *Mem:* Am Physiol Soc; Am Heart Asn; Biomed Eng Soc. *Res:* Coronary and coronary collateral hemodynamics; computer simulations; mechanisms of coronary collateral formation. *Mailing Add:* Dept Physiol & Med Tex Col Osteop Med Camp Bowie Montgomery Ft Worth TX 76107

SCHEEL, NIVARD, b Baltimore, Md, Nov 8, 25. SCIENCE EDUCATION. *Educ:* Cath Univ Am, AB, 49, MS, 51, PhD(physics), 61. *Prof Exp:* Instr physics, Xaverian Col, Silver Spring, Md, 50-54 & 58-60, pres, 60-66; actg pres, Cath Univ Am, 68-69, vpres, 69-80; PROF PHYSICS, TRINITY COL, WASHINGTON, DC, 82- *Mem:* Sigma Xi; Am Asn Physics Teachers. *Mailing Add:* 914 Perry Pl NE Washington DC 20017

SCHEELE, GEORGE F(REDERICK), b Yonkers, NY, May 23, 35; m 70. CHEMICAL ENGINEERING, FLUID MECHANICS. *Educ:* Princeton Univ, BSE, 57; Univ Ill, MS, 59, PhD(chem eng), 62. *Prof Exp:* From asst prof to assoc prof, 62-86, PROF CHEM ENG, CORNELL UNIV, 87-, ASSOC DIR, SCH CHEM ENG, 82- *Concurrent Pos:* Year-in-Indust Prof, E I du Pont de Nemours & Co, Inc, 70-71, consult, 73-84; vis prof, Univ Calif, Berkeley, 77-78. *Mem:* Fel Am Inst Chem Engrs. *Res:* Fluid mechanics of immiscible liquid-liquid systems; drop coalescence. *Mailing Add:* Cornell Univ Sch Chem Eng Olin Hall Ithaca NY 14853-5201

SCHEELE, LEONARD ANDREW, b Ft Wayne, Ind, July 25, 07; m 81; c 3. PUBLIC HEALTH, MEDICINE. *Educ:* Univ Mich, AB, 31, Wayne State Univ, BS, 33, MD, 34; Am Bd Prev Med, dipl, 49. *Hon Degrees:* LLD, Georgetown Univ, 48; Jefferson Med Col, 52; DSc, Wayne State Univ, 50; Univ Mich, 51; Columbia Univ, 53; St Johns Univ, 57. *Prof Exp:* Intern, US Marine Hosp, Chicago, 33-34; asst quarantine officer, USPHS, Calif, 34-35 & Hawaii, 35-36, health officer, Queen Anne County, Md, 36-37, spec cancer fel, Mem Hosp, New York, 37-39, officer in chg nat cancer control prog, Nat Cancer Inst, 39-42, chief field casualty sect, Med Div, Off Civilian Defense, 42-43, asst chief, Nat Cancer Inst, 46-47, asst surgeon gen, NIH & dir, Nat Cancer Inst, 47-48, surgeon gen, USPHS, 48-56; pres, Warner-Chilcott Labs, Warner-Lambert Pharmaceut Co, 56-60, mem bd dirs, 57-62 & 63-68, sr vpres, 60-68, pres, Warner-Lambert Res Inst, 65-68; RETIRED. *Concurrent Pos:* Head nutrit mission, US Dept Army, Ger, 48; pres, Int Cong Trop Dis & Malaria, Wash, 48; chmn US deleg, World Health Assembly, Rome, 49, Geneva, 50-53, deleg, 54, pres assembly, 51. *Mem:* Fel Am Col Surg; Am Pub Health Asn; AMA; Asn Mil Surg US (pres, 54); fel Am Col Physicians. *Res:* Cancer therapy end results; epidemiology; public health administrative techniques; business management; pharmaceutical development, manufacturing and marketing. *Mailing Add:* 700 New Hampshire Ave NW Washington DC 20037

SCHEELE, ROBERT BLAIN, b New York, NY, Dec 23, 40; m 63; c 2. BIOPHYSICS, MOLECULAR BIOLOGY. *Educ:* Yale Univ, BS, 62; Univ Pittsburgh, PhD(biophys), 68. *Prof Exp:* NIH fel biophys, State Univ NY, Buffalo, 68-70; res assoc biophys, Univ Conn, 70-75; res assoc molecular biol, Univ Wis-Madison, 75-77, asst scientist molecular biol, Lab Molecular, 77-84. *Mem:* Biophys Soc. *Res:* Protein-protein interactions, self-assembly of protein polymers, virus self-assembly and molecular aspects of motility. *Mailing Add:* 1206 Frisch Rd Madison WI 53711

SCHEELINE, ALEXANDER, b Altoona, Pa, June 6, 52; m 84; c 1. EMISSION SPECTROSCOPY, ANALYTICAL PHYSICS. *Educ:* Mich State Univ, BS, 74; Univ Wis, Madison, PhD(chem), 78. *Prof Exp:* Nat Res Coun fel anal chem, Nat Bur Standards, 78-79; asst prof chem, Univ Iowa, 79-81; asst prof, 81-87, ASSOC PROF CHEM, UNIV ILL, 87- *Concurrent Pos:* Consult, Spectral Sci Inc, 82-84, 88- & Nat Bur Standards, 82-; asst ed, Spectrochimica Acta, Part B, 89-; prog officer, NSF, 90-91. *Honors & Awards:* W F Meggers Award, Soc Appl Spectros, 79. *Mem:* Am Chem Soc; Soc Appl Spectros; Optical Soc Am; Am Soc Testing & Mat. *Res:* Plasma physics technology for elemental analysis of solid materials; sparks and pinches are studied through spectroscopy and simulation; oscillating reactions for chemical analysis; spectroscopic diagnostics of chemical vapor deposition plasmas. *Mailing Add:* 79 Roger Adams Lab Box 48 1209 W California Ave Urbana IL 61801

SCHEER, ALFRED C(ARL), b Center, Nebr, Feb 1, 26; m 46; c 4. CIVIL ENGINEERING. *Educ:* Iowa State Univ, BS, 48, MS, 50. *Prof Exp:* Instr civil eng, Iowa State Univ, 48-50; from asst prof to assoc prof, SDak Sch Mines & Technol, 50-58; assoc prof, Mont State Univ, 58-72, prof civil eng, 72-; RETIRED. *Concurrent Pos:* Mem, Hwy Res Bd, Nat Acad Sci-Nat Res Coun. *Mem:* Am Soc Eng Educ; Am Soc Civil Engrs. *Res:* Soil mechanics; highway engineering. *Mailing Add:* 10001 Thomas Dr Bozeman MT 59715

SCHEER, BRADLEY TITUS, b Los Angeles, Calif, Dec 17, 14; m 36. PHYSIOLOGY. *Educ:* Calif Inst Technol, BS, 36; Univ Calif, PhD(comp physiol), 40. *Prof Exp:* Asst physiol, Scripps Inst, Univ Calif, 36-37, Med Sch, 37-38 & Inst, 38-40; instr zool, WVa Univ, 40-42; asst biochem, Col Physicians & Surgeons, Columbia Univ, 42-43; instr biol, Calif Inst Technol, 43-45; asst prof biochem, Univ Southern Calif, 45-48, lectr zool, 46-48; assoc prof, Univ Hawaii, 48-50; assoc prof biol, 50-53, head dept, 58-64, prof, 54-77, EMER PROF BIOL, UNIV ORE, 77- *Concurrent Pos:* Vis asst prof, Hopkins Marine Sta, Stanford, 47; vis prof, Univ Calif, 52; Fulbright fel, Italy, 53-54; Guggenheim fel, France, 57-58; instr, Westmont Col, 78-82, adj prof biol, 82-86. *Mem:* AAAS; Am Physiol Soc; Soc Gen Physiol; Am Soc Zoologists. *Res:* Comparative biochemistry of carotenoids; isolation of enzymes; blood proteins of invertebrates; ecology of marine fouling organisms; physiology of fertilization; hormones of crustaceans; ion transports; thermodynamics in biology; systems analysis of salt and water balance; science and religion. *Mailing Add:* 93 Quarterdeck Way Pacific Grove CA 93950-2146

SCHEER, DONALD JORDAN, b Louisville, Ky, July 25, 34; m 66; c 2. ELECTRICAL ENGINEERING. *Educ:* Univ Louisville, BEE, 57, MEE, 58, MEng, 72; Ohio State Univ, PhD(elec eng), 66. *Prof Exp:* From instr to assoc prof, 59-72, chmn dept elec eng, 78-87, PROF ELEC ENG, UNIV LOUISVILLE, 72- *Mem:* Inst ELec & Electronics Engrs; Am Astron Soc; Sigma Xi. *Res:* Antennas; microwave engineering; radio astronomy. *Mailing Add:* Dept Elec Eng Univ Louisville Louisville KY 40292

SCHEER, MILTON DAVID, b New York, Dec 22, 22; m 45; c 3. PHYSICAL CHEMISTRY. *Educ:* City Col New York, BS, 43; NY Univ, MS, 47, PhD, 51. *Prof Exp:* Chemist, US Bd Econ Warfare, Guatemala, 43-44; res asst, NY Univ, 47-50; phys chemist, US Naval Air Rocket Test Sta, 51-52, US Bur Mines, 52-55 & Gen Elec Co, 55-58; phys chemist, 58-68, chief photochem sect, Nat Bur Standards, 68-70, chief, Phys Chem Div, 70-77, dir, Ctr Thermodyn & Molecular Sci, 77-80, chem physicist, Chem Kinetics Div, 81-85; consult assoc, McNesby & Scheer Res Assoc, Inc, 85-88; RETIRED. *Concurrent Pos:* Vis prof, Chem Dept, Univ Md, 80-81; Sr Fulbright fel; vis prof, Univ Rome, Italy, 82-83. *Mem:* Fel AAAS; Am Chem Soc; Am Phys Soc. *Res:* Reaction kinetics and photochemistry; surface chemistry and physics; low temperature chemistry; high temperature thermodynamics; excited state chemistry. *Mailing Add:* 15100 Interlachen Dr No 512 Silver Springs MD 20906

SCHEERER, ANNE ELIZABETH, b Philadelphia, Pa, Aug 7, 24. MATHEMATICS. *Educ:* Univ Pa, BS, 46, MS, 47, PhD(math), 53. *Prof Exp:* Instr math, Temple Univ, 47-50 & Washington Univ, 53-55; assoc prof, Georgetown Univ, 55-66; asst dean, Col Eng, Boston Univ, 66-67; specialist higher educ, Md Coun Higher Educ, 67-69; dean summer sessions & lifelong educ, 69-82, DEAN, SUMMER SESSIONS & DIR, INST RES, CREIGHTON UNIV, 83- *Mem:* Am Math Soc; Math Asn Am; Sigma Xi. *Res:* Complex variable; probability; information theory. *Mailing Add:* Creighton Univ 2500 California St Omaha NE 68178

SCHEETZ, HOWARD A(NSEL), b Sturgis, Mich, Mar 20, 27; m 52; c 2. MATERIALS SCIENCE, ENGINEERING MECHANICS. *Educ:* Mich State Univ, BS, 50; Pa State Univ, MS, 59. *Prof Exp:* Process engr, pilot plant, Corning Glass Works, 50-52, sr engr, Res & Develop Lab, 52-56; res assoc shock & vibration, Pa State Univ, 57-58; res engr, Physics Dept, Cornell Aeronaut Lab, 58-60, Mat Dept, 60-62 & Physics Dept, Armstrong Cork Co, 62-69; res assoc, Packaging Prod Div, Kerr Glass Mfg Corp, 69-73; dir res, Sonobond Corp, 73-77; develop mgr, Tech Ctr, Polymer Corp, 77-89; CONSULT, 89- *Concurrent Pos:* Lectr grad fac, Pa State Univ, 66-67; corp rep, Am Ceramic Soc. *Mem:* AAAS; Inst Elec & Electronics Engrs; Acoust Soc Am. *Res:* Analytical modeling and computer-aided studies of composite physical properties of glass, ceramic, polymeric and mixed composite material systems; applications of ultrasonic techniques to materials research and glass technology. *Mailing Add:* Tech Ctr Polymer Corp PO Box 14325 Reading PA 19612-4235

SCHEFER, ROBERT WILFRED, b San Francisco, Calif, July 7, 46; m 74. MECHANICAL ENGINEERING. *Educ:* Univ Calif, Berkeley, BS, 68, MS, 70, PhD(mech eng), 76. *Prof Exp:* Res engr combustion, Lawrence Berkeley Lab, 76-81. *Concurrent Pos:* NSF energy fel, 76-77. *Mem:* Combustion Inst; Sigma Xi; Am Inst Aeronaut & Astronaut. *Res:* Combustion; combustion generated air pollution; chemistry (chemical kinetics); fire research; high temperature catalysis; coal utilization. *Mailing Add:* Sandia Nat Lab Div 8351 Livermore CA 94551

SCHEFF, BENSON H(OFFMAN), b New York, NY, May 16, 31; m 53; c 4. SOFTWARE ENGINEERING, SOFTWARE PRODUCTIVITY & RISK. *Educ:* Oberlin Col, BA, 51; Columbia Univ, MA, 52. *Prof Exp:* Training instr electronic data processing mach, Nat Security Agency, 52-55; res engr, Servomechanisms Lab, Mass Inst Technol, 55-59; head adv comput control systs implementation, Group Data Systs Eng, Radio Corp Am, Mass, 59-66; mgr comput anal & appln, Software Dept, Missile Systs Div, Bedford, 66-73, MGR DATA PROCESSING, EQUIP DIV, RAYTHEON CO, WAYLAND, MASS, 73- *Mem:* Asn Comput Mach; Am Mgt Asn. *Res:* Integrated software engineering methodology; software research and development. *Mailing Add:* Concord Rd Box 577 Lincoln MA 01773

SCHEFFER, JOHN R, b Missoula, Mont, Feb 28, 39; m 63; c 2. ORGANIC CHEMISTRY. *Educ:* Univ Chicago, BS, 62. Univ Wis, PhD(chem), 67. *Prof Exp:* ASSOC PROF CHEM, UNIV BC, 67- *Concurrent Pos:* Nat Res Coun Can grant, 67-; Res Corp grant, 68-; Petrol Res Fund grant, 70-73 & 75-77; Guggenheim fel, 73-74. *Mem:* Am Chem Soc; Royal Soc Chem; Chem Inst Can. *Res:* Organic photochemistry, including new sources and reactions of singlet oxygen. *Mailing Add:* Dept Chem Univ BC 2036 Main Mall Vancouver BC V6T 1Y6 Can

SCHEFFER, ROBERT PAUL, b Newton, NC, Jan 26, 20; m 51; c 2. PLANT PHYSIOLOGY. *Educ:* NC State Col, BS, 47, MS, 49; Univ Wis, PhD(plant path), 52. *Prof Exp:* Asst plant path, NC State Col, 47-49; asst, Univ Wis, 49-52, proj assoc, 52-53; from asst prof to assoc prof bot & plant path, 53-63, PROF BOT & PLANT PATH, MICH STATE UNIV, 63- *Concurrent Pos:* NIH fel & guest investr, Rockefeller Inst, 60-61; NSF consult, Panel Regulatory Biol, 65-68; distinguished prof, Mich State Univ, 85; assoc ed, Phytopath, 72-75; mem press bd, Am Phytopath Soc, 80-85. *Mem:* Fel Am Phytopath Soc; Am Soc Plant Physiol; fel Explorer's Club. *Res:* Physiology of disease development and and disease resistance; toxins in plant disease. *Mailing Add:* Dept Bot & Plant Path Mich State Univ East Lansing MI 48824

SCHEFFER, THEODORE COMSTOCK, b Manhattan, Kans, Feb 10, 04; m 27; c 2. FOREST PRODUCTS. *Educ:* Univ Wash, BS, 26, MS, 29; Univ Wis, PhD(forest path), 34. *Prof Exp:* Nat Res Coun fel, Johns Hopkins Univ, 34-35; pathologist, Forest Prod Lab, US Forest Serv, 35-65, in chg fungus & insect invests, 65-69; RES ASSOC, FOREST RES LAB, ORE STATE UNIV, 69- *Mem:* Forest Prod Res Soc; Soc Am Foresters. *Res:* Fundamental and applied control of fungus damage to wood and wood products; effects of fungi on physical-chemical properties; testing fungus resistance; preservation of wood with volatile fungicides; bioassay techniques for appraising quality of preservative treatment. *Mailing Add:* Dept Forest Prod Ore State Univ Corvallis OR 97331

SCHEFFER, VICTOR B, b Manhattan, Kans, Nov 27, 06; wid; c 3. MARINE MAMMALOGY. *Educ:* Univ Wash, Seattle, BS, 30, MS, 32, PhD(zool), 36. *Prof Exp:* Biologist, US Bur Biol Surv, 37-40, US Fish & Wildlife Serv, 40-56 & US Bur Com Fisheries, 56-69; lectr ecol, Univ Wash, Seattle, 66-72; SCI WRITER, 69- *Concurrent Pos:* Chmn, Marine Mammal Comn, 73-76. *Honors & Awards:* John Burroughs Medal, 70; John Wood Krutch Medal, 75. *Mem:* Hon mem Am Soc Mammalogists; Wildlife Soc; AAAS; hon mem Soc Marine Mammal; Am Inst Biol Sci. *Res:* Natural history of marine mammals; wildlife management; ethical treatment of animals; history of environmentalism. *Mailing Add:* 14806 SE 54th St Bellevue WA 98006

SCHEFFLAN, RALPH, chemical engineering, for more information see previous edition

SCHEFFLER, IMMO ERICH, b Dresden, Ger, Dec 17, 40; Can citizen; m 65; c 1. MOLECULAR BIOLOGY, SOMATIC CELL GENETICS. *Educ:* Univ Manitoba, BSc, 63, MSc, 64; Stanford Univ, PhD(biochem), 69. *Prof Exp:* Helen Hay Whitney fel, Harvard Med Sch, 68-70; fel, Pasteur Inst, Paris, 70-71; from asst prof to assoc prof, 71-81, PROF BIOL, UNIV CALIF, SAN DIEGO, 81- *Concurrent Pos:* NIH 71-74, 74-77 & 77-82, 82-87, 87-92 & 90-95; chmn biol, Univ Calif, San Diego, 90-92; Am Cancer Soc grant, 74-76 & 76-78; NSF grants, 78-81 & 81-84, mem adv panel genetic biol, 77-80; mem adv panel personnel res, Am Cancer Soc; Alexander von Humboldt Sr US scientist award, Ger, 84-85. *Mem:* AAAS; Am Soc Cell Biol; fel Am Soc Exp Biol. *Res:* Selection of mutants of mammalian cells grown in tissue culture; study of control of the cell cycle; mammalian cell genetics; biogenesis of mitochondrion; regulation of ornithine decarboxylase. *Mailing Add:* Dept Biol Univ Calif San Diego La Jolla CA 92093-0322

SCHEIB, RICHARD, JR, b New York, NY, May 24, 14; m 42; c 1. PHYSICS. *Educ:* Columbia Univ, AB, 36, MA, 39. *Prof Exp:* Engr & sr engr, Sperry Gyroscope Co, 41-49, eng dept head, 49-62, planning mgr, 62-64, prog mgr, 64-69; assoc prof basic sci, Acad Aeronaut, NY, 70-83, chmn dept, 73-83; RETIRED. *Mem:* Am Phys Soc; Sigma Xi. *Mailing Add:* 29 Crest Rd New Hyde Park NY 11040

SCHEIBE, MURRAY, b Bronx, NY, Feb 28, 32; m 55; c 3. PHYSICS. *Educ:* Brooklyn Col, BS, 53; Univ Md, College Park, PhD(physics), 59. *Prof Exp:* Staff scientist, Lockheed Palo Alto Res Lab, Lockheed Aircraft Co, 58-71; staff mem, Mission Res Corp, 71-87; PHYS RES CORP, 87- *Mem:* Am Phys Soc. *Res:* Atmospheric physics and chemistry; high temperature gas dynamics; optical radiation phenomena. *Mailing Add:* 1133 Palomino Rd Santa Barbara CA 93105

SCHEIBE, PAUL OTTO, b Marion, NDak, Apr 7, 34; m 54; c 3. ELECTRICAL ENGINEERING, SYSTEM ANALYSIS & DESIGN. *Educ:* Univ NDak, BS, 58, MS, 59; Stanford Univ, PhD(elec eng), 62. *Prof Exp:* Instr, NDak State Sch Sci, 53-56; from instr to asst prof elec eng, Univ NDak, 58-60; mem eng staff, Sylvania Electronic Defense Labs, 62-64; mem sr eng staff & dir technol, ESL, Inc, Calif, 64-70; vpres & tech dir, Adac Labs, 70-84; vpres & tech dir, Sterling Networks, 84-89; PRIN CONSULT, IXZAR, INC, 86- *Concurrent Pos:* Lectr, Univ Santa Clara, 62- *Mem:* Inst Elec & Electronics Engrs; Soc Nuclear Med; Asn Comput Mach. *Res:* Characterization and bounds on the performance of time-variant systems; automatic design and analysis of electrical networks; statistical parameter estimation; mathematical modeling of physiological systems. *Mailing Add:* Three Still Creek Rd Woodside CA 94062-9731

SCHEIBEL, ARNOLD BERNARD, b New York, NY, Jan 18, 23; m 50. NEUROPHYSIOLOGY, PSYCHOPHYSIOLOGY. *Educ:* Columbia Univ, BA, 44, MD, 46; Univ Ill, MS, 53; Am Bd Psychiat & Neurol, dipl, 52. *Prof Exp:* Intern, Mt Sinai Hosp, New York, 46-47; resident psychiat, Barnes Hosp, St Louis, 47-48; from asst prof to assoc prof psychiat & anat, Sch Med, Univ Tenn, 51-55; from asst prof to assoc prof, 55-68, PROF PSYCHIAT & ANAT, SCH MED, UNIV CALIF, LOS ANGELES, 68-; ACTG DIR, BRAIN RES INST. *Concurrent Pos:* Guggenheim fel, Inst Physiol, Italy, 53-54 & 60-61; consult physician, Brentwood Vet Admin Hosp, Calif & Sepulveda Vet Admin Hosp. *Mem:* AAAS; fel Am Psychiat Asn; Am Acad Neurol; Am Neurol Asn; fel Am Acad Arts & Sci; Norweg Acad Sci. *Res:* Distortions of perception and memory in psychoses; experimental study of structural patterns of neuropil in the central nervous system and its relation to functional activity; aging in human brain; neuropsychiatry; neuroscience. *Mailing Add:* Dept Anat & Psychiat 73 235 Chs Univ Calif 405 Hillgard Ave Los Angeles CA 90024

SCHEIBEL, EDWARD G(EORGE), b Ridgewood, NY, Jan 15, 17; m 45; c 4. CHEMICAL ENGINEERING. *Educ:* Cooper Union, BS, 37; Polytech Inst Brooklyn, MChE, 40, DChE, 43. *Prof Exp:* From asst chemist to plant supt, H C Bugbird Co, NJ, 37-42; chem engr, M W Kellogg Co, NY, 42-43; res assoc, Polytech Inst Brooklyn, 43-44; chem engr, Hydrocarbon Res, Inc, 44-45 & Hoffmann-LaRoche, Inc, 45-55; dir eng, York Process Equip Corp, 55-63; prof chem eng & head dept, Cooper Union, 63-70; mgr process technol, Gen Elec Co, Mt Vernon, Ind, 69-73; res engr, Suntech Inc, 74-81; mem staff, E G Scheibel, Inc, 81-84; RETIRED. *Concurrent Pos:* Instr & adj prof, Polytech Inst Brooklyn, 42-52; adj prof, Newark Col Eng, 53-63. *Mem:* Am Chem Soc; Am Inst Chem Eng. *Res:* Distillation, absorption and liquid extraction; liquid oxygen production; fractional liquid extraction; extractive distillation of petroleum; heavy and fine chemicals. *Mailing Add:* 410 Notson Terr Port Charlotte FL 33951

SCHEIBEL, LEONARD WILLIAM, b Hays, Kans, Jan 18, 38; m 76; c 2. BIOCHEMICAL PHARMACOLOGY, CLINICAL PHARMACOLOGY. *Educ:* Creighton Univ, Omaha, Nebr, BS, 60, MS, 62; Johns Hopkins Univ, PhD(biochem), 67; Univ Fla, Gainesville, MD, 73. *Prof Exp:* Captain res biochem, Walter Reed Army Inst Res, Washington, DC, 67-70; res assoc pharmacol, Univ Fla, Gainesville, 70-73; intern med, Gorgas Hosp, Balboa Heights, Canal Zone, Panama, 73-74, resident internal med, 74-77; asst prof trop med, Rockefeller Univ, NY, 77-81; from asst prof to assoc prof, 81-86, PROF PREV MED, SCH MED, UNIFORMED SERV UNIV HEALTH SCI, BETHESDA, MD, 86- *Concurrent Pos:* Guest prof, USN Bur Med & Surg Med Trop, Gorgas Mem Hosp, Panama, 73 & 77; vis asst prof, Div Infectious Dis, Dept Int Med, Cornell Med Sch, 77-81; mem, US Army Med Res & Develop Adv Comt, 82-85; mem, drug adv bd, Food & Drug Admin Anti-Infective Dis, 82-85; adj assoc prof, dept immunol infectious dis, Johns Hopkins Univ & dept microbiol, Sch Med, Univ Md, 82-86. *Mem:* Fel, Am Col Physicians; fel, Am Col Prev Med; Fedn Am Soc Exp Biol; Am Soc Pharmacol & Exp Therapeut; Am Soc Trop Med; Am Soc Parasitologists; Undersea Med Soc; Walter Reed Army Inst Res Asn; fel Infectious Dis Soc Am. *Res:* Tropical medicine emphasizing metabolism and chemotherapy of parasitic diseases; rational design of new pharmacologic agents, testing and employment in the field, working extensively throughout Central and South America; international health and tropical medicine. *Mailing Add:* Dept Prev Med Sch Med Uniformed Serv Univ Health Sci 4301 Jones Bridge Rd Bethesda MD 20814-4799

SCHEIBER, DAVID HITZ, b Cleveland, Ohio, Aug 16, 31; m 56; c 5. POLYMER CHEMISTRY. *Educ:* Univ Notre Dame, BS, 53, PhD, 56. *Prof Exp:* Res chemist, Electrochem Dept, E I du Pont de Nemours & Co, Inc, 56-61, staff scientist, 61-62, res assoc, 62-64, res supvr, 64-68, res assoc, Photosysts & Electronics Dept, 68-90; RETIRED. *Mem:* Am Chem Soc. *Res:* Electronic materials. *Mailing Add:* Ten Cardiff Rd Wilmington DE 19803

SCHEIBER, DONALD JOSEPH, b Ft Wayne, Ind, Jan 24, 32; m 54; c 8. UNDERWATER ACOUSTICS, MECHANICS. *Educ:* St Procopius Col, 53; Univ Notre Dame, 53-57, PhD(physics), 57. *Prof Exp:* Nat Res Coun res assoc, Nat Bur Standards, 57-58, physicist, 58-62; SR STAFF ENGR, MAGNAVOX CO, 62- *Mem:* Nat Security Indust Asn. *Res:* Underwater sound; directional sensors; dynamics of buoy systems; oceanographic sensors. *Mailing Add:* Magnavox Govt & Indust Group 1313 Production Rd MS 25-308 Ft Wayne IN 46808

SCHEIBNER, RUDOLPH A, b Escanaba, Mich, Oct 22, 26; m 61; c 3. ENTOMOLOGY. *Educ:* Mich State Univ, BS, 51, MS, 58, PhD(entom), 63. *Prof Exp:* Instr natural sci, Mich State Univ, 63-65; asst prof, 65-69, assoc prof, 69-77, EXTEN PROF ENTOM, UNIV KY, 78- *Mem:* Am Entom Soc; Sigma Xi. *Res:* Taxonomy and ecology of insects. *Mailing Add:* Dept Entom Univ Ky Lexington KY 40506

SCHEID, CHERYL RUSSELL, b Leonardtown, Md, 1948. CELL PHYSIOLOGY, SMOOTH MUSCLE PHYSIOLOGY. *Educ:* Boston Univ, PhD(biol), 76. *Prof Exp:* Asst prof med physiol, Tufts Univ Med Sch, Boston, Ma, 79-80; asst prof, 80-87, ASSOC PROF MED PHYSIOL, SCH MED, UNIV MASS, 87- *Mem:* Am Physiol Soc; AAAS. *Res:* Calcium handling in smooth muscles. *Mailing Add:* Dept Physiol Sch Med Univ Mass 55 Lake Ave Worcester MA 01605

SCHEID, FRANCIS, b Plymouth, Mass, Sept 24, 20; m 44; c 3. MATHEMATICS. *Educ:* Boston Univ, BS, 42, AM, 43; Mass Inst Technol, PhD(math), 48. *Prof Exp:* From instr to prof math, Boston Univ, 48-86, chmn dept, 56-69; RETIRED. *Concurrent Pos:* Fulbright lectr, Univ Rangoon, 61-62; TV lectr, 61-66. *Mem:* Oper Res Soc Am. *Res:* Numerical analysis; problems of measuring golfing ability and golf course difficulty and golf related problems. *Mailing Add:* 135 Elm St Kingston MA 02364

SCHEID, HAROLD E, b Whiting, Kans, May 12, 22; m 48; c 6. PET FOOD NUTRITION RESEARCH, PET FOOD PRODUCT DEVELOPMENT. *Educ:* Sterling Col, BS, 47; Kans State Univ, MS, 48. *Prof Exp:* Asst biochemist, Am Meat Inst Fedn, 49-54; chief chemist & nutritionist, Honeggers & Co, 54-59; proj leader, Quaker Oats Co, 60-75; dir res & develop, Theracon Inc, 75-, consult, 75-; RETIRED. *Concurrent Pos:* Mem, Gordon Res Conf, 63-70. *Mem:* Inst Food Technol; Am Chem Soc; Am Asn Feed Micros; Am Asn Clin Chem. *Res:* Biological function and methods of analysis of vitamin B12, B2 and B6; the role of vitamin B12 in the regeneration of liver tissue. *Mailing Add:* 3006 SE Arbor Dr Topeka KS 66605

SCHEID, STEPHAN ANDREAS, b Munich, Ger, May 22, 41. VIROLOGY. *Educ:* Univ Cologne, DrMed, 69. *Prof Exp:* Intern, Inst Physiol Chem, Univ Cologne, 67-68 & Univ Hosp, Cologne, 68-69; fel, Rockfeller Univ, 69-73, from asst prof to assoc prof virol, 73-83; PROF MICROBIOL & VIROL, INST MED MICROBIOL & VIROL, 83- *Concurrent Pos:* USPHS Int fel, 69-71; res fel, Ger Res Asn, 71-73. *Mem:* Am Soc Microbiol; Am Asn Immunologists; Soc Gen Microbiol; Am Soc Biol Chemists. *Res:* Animal virology; myxoviruses; paramyxoviruses; structure and replication of enveloped viruses. *Mailing Add:* Inst Med Microbiol & Virol Univ Dusseldorf Universitatsstr 1 Dusseldorf 4000 Germany

SCHEIDT, FRANCIS MATTHEW, b Streator, Ill, Mar 2, 22; m 52. ORGANIC CHEMISTRY. *Educ:* Univ Ill, BS, 50, MS, 54, PhD(org chem), 56. *Prof Exp:* Res asst chem, State Geol Surv, Ill, 50-54; chemist, Dow Chem Co, 56-66, sr res chemist, 66-72, res specialist, 72-77, sr res specialist, Britton Res Lab, 77-80, res assoc, 80-88; RETIRED. *Concurrent Pos:* Sect ed, Chem Abstr, 68-86. *Mem:* Fel Am Inst Chemists; Am Chem Soc; Sigma Xi; NY Acad Sci. *Res:* Synthetic organic chemistry; heterogeneous catalysis. *Mailing Add:* 1906 Norwood Dr Midland MI 48640

SCHEIDT, WALTER ROBERT, b Richmond Heights, Mo, Nov 13, 42; m 64; c 2. CHEMISTRY, X-RAY CRYSTALLOGRAPHY. *Educ:* Univ Mo-Columbia, BS, 64; Univ Mich, Ann Arbor, MS, 66, PhD(chem), 68. *Prof Exp:* Res fel, Cornell Univ, 68-70; assoc prof, 70-80, PROF CHEM, UNIV NOTRE DAME, 80- *Mem:* AAAS; Am Chem Soc; Am Crystallog Asn. *Res:* Structure and chemistry of metalloporphyrins; structure of inorganic complexes. *Mailing Add:* Dept Chem Univ Notre Dame Notre Dame IN 46556

SCHEIDY, SAMUEL F, veterinary medicine, for more information see previous edition

SCHEIE, CARL EDWARD, b Fosston, Minn, July 14, 38; m 59; c 3. ENGINEERING PHYSICS. *Educ:* Concordia Col, Moorhead, Minn, BA, 60; Univ NMex, MS, 62, PhD, 65. *Prof Exp:* Res assoc physics, Univ NMex, 62-65; res aide, Inst Paper Chem, 65-68; from asst prof to assoc prof physics, Concordia Col, Moorhead, Minn, 68-73; prod eval engr, MacGregor Div, Brunswick Corp, 73-79, staff scientist, 79-82; dir prod develop, 82-89, VPRES GOLF TECHNOL, WILSON SPORTING GOODS, 90- *Concurrent Pos:* Vis scientist, Argonne Nat Lab, 72-73. *Res:* Molecular motion in solids using nuclear magnetic resonance techniques; physics of gulf. *Mailing Add:* 1113 Dawes Libertyville IL 60048

SCHEIE, HAROLD GLENDON, b Brookings, SDak, Mar 24, 09; m 51; c 2. OPHTHALMOLOGY. *Educ:* Univ Minn, BS, 31, MB & MD, 35; Univ Pa, DSc(med ophthal), 40; Am Bd Ophthal, dipl, 40. *Prof Exp:* From intern to resident, Hosp Univ Pa, 35-40, from instr to prof, Grad Sch Med, 46-60, chmn dept ophthal, 60-75, William F Norris & George E Deschweinitz prof ophthal, Sch Med, Grad Sch Med, Univ Pa, 60-75, dir, Scheie Eye Inst, 72-75, EMER WILLIAM F NORRIS & GEORGE E DE SCHWEINITZ PROF OPHTHAL, UNIV PA & FOUNDING DIR, SCHEIE EYE INST, 75- *Concurrent Pos:* From instr to prof, Med Sch, Univ Pa, 40-60, prof & chmn dept, Div Grad Med, 64; ophthal consult to many hosps, labs & govt agencies, 44-; lectr, US Naval Hosp, Philadelphia; chief ophthal serv, Philadelphia Gen Hosp & Children's Hosp, 49-70; chief ophthal serv & consult, Vet Admin Hosp; numerous name lectureships, 52-74; mem adv coun reserve affairs, Surgeon Gen, US Army, 44; mem nat adv comt, Eye-Bank Sight Restoration, Inc & Nat Coun Combat Blindness; mem bd examr, Am Bd Ophthal, 59-66; mem med adv comt, Medic Alert Found Int, 66. *Honors & Awards:* Howe Award, AMA, 64; Distinguished Serv Award for Excellence in Ophthal, Am Soc Contemporary Ophthal, 74; Horatio Alger Award, Am Schs & Cols Asn, 74; Golden Plate Award, Am Acad Achievement, 75. *Mem:* AAAS; Am Acad Ophthal & Otolaryngol (vpres, 60-61); Am Asn Ophthal (3rd vpres, 70); fel Am Col Surg (vpres, 61-62); AMA. *Res:* Infantile glaucoma; anesthesia in ophthalmic surgery; ACTH and cortisone; arteriosclerosis; retrolental fibroplasia. *Mailing Add:* 1024 Keith Ave Berkeley CA 94708

SCHEIE, PAUL OLAF, b Minn, June 24, 33; m 63; c 2. PHYSICS, BIOPHYSICS. *Educ:* St Olaf Col, BA, 55; Univ NMex, MS, 57; Pa State Univ, PhD(biophys), 65. *Prof Exp:* Asst prof physics, Oklahoma City Univ, 58-63, chmn dept, 58-61; asst prof biophys, Pa State Univ, 66-73; assoc prof, 73-80, PROF PHYSICS & CHMN DEPT, TEX LUTHERAN COL, 80- *Concurrent Pos:* Vis prof, Med Fac, Univ Bergen, Norway, 80-81. *Mem:* AAAS; Am Asn Physics Teachers; Biophys Soc; Royal Micros Soc. *Res:* Properties of solid-liquid interfaces; physical properties of bacteria. *Mailing Add:* 207 Leonard Lane Seguin TX 78155

SCHEIG, ROBERT L, b Warren, Ohio, Mar 16, 31; m 52; c 1. INTERNAL MEDICINE, GASTROENTEROLOGY. *Educ:* Yale Univ, MD, 56. *Prof Exp:* From intern to sr asst resident med, Grace-New Haven Hosp, Conn, 56-61; from instr to assoc prof med, Sch Med, Yale Univ, 63-73, assoc dir liver study unit, 63-73, assoc dean regional affairs, 71-73; prof med & head, Div Gastroenterol, Sch Med, Univ Conn, Farmington, 73-81, actg chmn, Dept Med, 78-79; PROF MED, STATE UNIV NY, BUFFALO, 81-; HEAD, DEPT MED, BUFFALO GEN HOSP, 81- *Concurrent Pos:* Fel, Sch Med, Yale Univ, 61-62; res fel, Harvard Med Sch, 62-63; from assoc physician to attend physician, Yale-New Haven Med Ctr, 63-73, dir adult clin res ctr, 69-71; attend physician, West Haven Vet Admin Hosp, Conn, 66-73, John Dempsey Hosp, Conn, 73-81 & Buffalo Gen Hosp, Erie County Med Ctr, Buffalo & Buffalo Vet Admin Med Ctr, 81-; consult, Hosp of St Raphael, New Haven, 69-73 & Waterbury Hosp, 72-73; attend physician & chief med, Newington Vet Admin Med Ctr, 73-81. *Mem:* Fel Am Col Physicians; Am Gastroenterol Asn; Am Asn Study Liver Dis; Am Fedn Clin Res; Sigma Xi. *Res:* Toxic liver injury; effect of ethanol on lipid metabolism and biochemical and histopathological correlation with clinical disease. *Mailing Add:* Chief Dept Med Buffalo Gen Hosp 100 High St Buffalo NY 14203

SCHEIN, ARNOLD HAROLD, b New York, NY, June 10, 16; m 40; c 2. BIOCHEMISTRY, PHYSIOLOGY, MICROBIOLOGY. *Educ:* City Col, BS, 36; Univ Iowa, PhD(biochem), 43; Am Bd Clin Chem, dipl. *Prof Exp:* Instr biochem, NY Med Col, 43-46; dir res, Mouton Processors of Can, Ltd, 46-47; from asst prof to assoc prof biol chem, Col Med, Univ Vt, 47-69; chmn dept chem, Calif State Univ, San Jose, 69-73, prof chem, 69-84; RETIRED. *Concurrent Pos:* Exchange sr lectr, Dept Biochem, St Bartholomew's Hosp Col Med, Eng, 56-57; Commonwealth fel, Virus Res Labs, Cambridge, 63-64; vis prof chem, Dept Microbiol, Sch Med Hebrew Univ Jerusalem, 77. *Mem:* Fel Am Soc Clin Chem; Am Soc Biol Chem; Brit Biochem Soc; Am Chem Soc. *Res:* RNA and DNA structure and metabolism. *Mailing Add:* 22448 Salem Ave No 1 Cupertino CA 95040

SCHEIN, BORIS M, b Moscow, USSR, June 22, 38; US citizen; m 66; c 2. SEMIGROUPS OF TRANSFORMATIONS & RELATIONS. *Educ:* Saratov State Univ, USSR, BS, 59, MS, 60; Leningrad Pedagogical Inst, PhD(math), 62, DSc, 66. *Prof Exp:* Prof math, Saratov State Univ, 62-79; prof math, Tulane Univ, New Orleans, La, 80; DISTINGUISHED PROF MATH, UNIV ARK, 80- *Concurrent Pos:* Managing ed, Semigroup Forum, Springer Int, NY, 70-; ed, Algebra Universalis, Birkhäuser-Verlag, Basel, 74-, J Aequationes Mathematicae, 76-82, Simon Stevin, Univ Ghent, Belgium, 78-, Math Social Sci, 80-; vis prof, Technol Univ, Clausthal, WGer, 81; chmn, Transl Comt, Am Math Soc, 85-87; vis prof, Tel-Aviv Univ, Israel, 86, Polytech Univ Catalonia, Spain, 86, Univ Siena, Italy, 86, Univ Calgary, Can, 86 & Shimane Univ, Japan, 90; sr res fel, Univ St Andrews, Scotland, 84. *Mem:* Am Math Soc. *Res:* Relation algebra; systems of transformations (everywhere defined or partial, single- or multi-valued) closed under composition; other natural operations, semigroups, ordered sets, lattices, universal algebras, and algebraic automats. *Mailing Add:* Dept Math Sci Univ Ark SE-307 Fayetteville AR 72701

SCHEIN, JEROME DANIEL, b Minneapolis, Minn, May 27, 23; m 82; c 2. PSYCHOLOGY. *Educ:* Univ Minn, PhD(psychol), 58. *Prof Exp:* Instr psychol, Univ Wis, 58-59; asst prof psychol, Fla State Univ, 59-60; prof psychol, Gallaudet Col, 60-68; dean educ, Univ Cincinnati, 68-70; emer prof sensory rehab, NY Univ, 70-; PROF SENSORY REHAB, UNIV ALTA, 89- *Res:* Education and rehabilitation of persons with impaired vision, hearing and both deaf-blindness. *Mailing Add:* 1703 Andros Isle Apt J-2 Coconut Creek FL 33066

SCHEIN, LAWRENCE BRIAN, b Brooklyn, NY, Jan 31, 44; m 69; c 2. EXPERIMENTAL SOLID STATE PHYSICS. *Educ:* Pa State Univ, BS, 65; Columbia Univ, MA, 67; Univ Ill, PhD(physics), 70. *Prof Exp:* Mem tech staff solid state res, RCA Corp, 65-67; David Sarnoff res fel, 67-69; asst, Univ Ill, 69-70; mem tech staff, Xerox Res Labs, 70-79, mgr, Explor Making Area, 79-83, MEM STAFF & MGR, ELECTROPHOTOG PHYSICS, IBM RES DIV, SAN JOSE, 83- *Mem:* Am Phys Soc; Soc Photo Scientists & Engrs; Sigma Xi; fel Soc Imaging Sci & Technol. *Res:* Metal-semiconductor tunneling; physics of electrophotography; transport properties of molecular crystals and molecularly doped polymers; static electricity. *Mailing Add:* IBM Res Div K41-803 650 Harry Rd San Jose CA 95120

SCHEIN, MARTIN WARREN, b Brooklyn, NY, Dec 23, 25; m 61; c 3. ETHOLOGY, EDUCATIONAL ADMINISTRATION. *Educ:* Univ Iowa, AB, 49; Johns Hopkins Univ, ScD(vert ecol), 54. *Prof Exp:* Biol aide, USPHS, 47-48; animal climatologist, Exp Sta, La State Univ & USDA, 51-55; from asst prof to assoc prof animal behav, Pa State Univ, 55-65, prof zool, 65-68; clin prof behav med & psychiat, WVa Univ, 73-84, chmn, dept biol, 80-86, centennial prof biol, 68-88; CONSULT, NAT DEFENSE MED COL, SAITAMA, JAPAN, 88- *Concurrent Pos:* Comnr undergrad educ in biol sci, George Washington Univ, 62-64, vchmn comn, 64-65, dir, 65-68, vis prof biol, univ, 65-68; vis lectr, Univ Southern Ill, 64; vis prof biol, Banaras Hindu Univ, Varanasi, India, 82-83; Fulbright Fel, India, 82-83; teacher, Tokyo Metrop Kokusai High Sch, 90-91. *Mem:* AAAS; Am Soc Zoologists; Animal Behav Soc (secy, 56-62, pres-elect, 66, pres, 67); Nat Asn Biol Teachers; Sigma Xi. *Res:* Behavior of domestic animals; sexual, social and feeding behavior; education in biology. *Mailing Add:* 5800 Nicholson Lane Apt 1001 Rockville MD 20852

SCHEIN, PHILIP SAMUEL, b Asbury Park, NJ, May 10, 39; m 67; c 2. PHARMACOLOGY, ONCOLOGY. *Educ:* Rutgers Univ, AB, 61; State Univ NY Upstate Med Ctr, MD, 65; Am Bd Internal Med, dipl, 72, 73; FRCPS(G), 81. *Hon Degrees:* Dr, Nat Univ Rosario, Argentina, 80. *Prof Exp:* Intern med, Beth Israel Hosp, Boston, Mass, 65-66; res assoc pharmacol, Nat Cancer Inst, 66-68; asst resident med, Beth Israel Hosp, Boston, 68-69; res physician Radcliffe Infirmary, Oxford, 69-70; instr, Harvard Med Sch, 70-71; sr investr oncol, Nat Cancer Inst, 71-74, head clin pharmacol, 73-74; chief med oncol, Georgetown Univ Hosp & Lombardi Cancer Res Ctr, 74-; vpres World Wide Clin Res & Develop, Smith, Kline & French Labs, Philadelphia, Pa, 83-86, chief exec officer & pres US Biosci, 86-87; CHMN & CHIEF EXEC OFF, US BIOSCI INC, 87- *Concurrent Pos:* Chief resident med, Beth Israel Hosp, Boston, 70-71; clin asst prof, 71-74, assoc prof med & pharmacol, 74-77, prof med & pharmacol, Med Sch, Georgetown Univ, 77-; consult, Walter Reed Army Hosp, 71 & Clin Ctr, NIH, 74; chmn, Gastrointestinal Tumor Study Group, 74-, Oncol Comt Adv Comt, Food & Drug Admin, 78-81, Med Oncol Comt, Am Bd Int Med, 80-; prof med & pharm, Univ Pa, 83. *Mem:* Am Asn Cancer Res; Am Soc Clin Oncol; fel Am Col Physicians; Am Soc Hemat; fel Royal Soc Med; Sigma Xi; Am Soc Clin Invest; Am Assoc Physicians. *Res:* Laboratory and clinical investigations in cancer chemotherapy and endocrinology. *Mailing Add:* US Bioscience Inc One Tower Bridge Suite 400 West Conshohocken PA 19428

SCHEIN, RICHARD DAVID, b East St Louis, Ill, Nov 18, 27; m 55; c 3. PLANT PATHOLOGY, ECOLOGY. *Educ:* DePauw Univ, BA, 48; Univ Calif, PhD(plant path), 52. *Prof Exp:* Asst, Univ Calif, 48-52; asst plant pathologist, Ill Natural Hist Surv, 52-53; from asst prof to assoc prof plant path, Pa State Univ, 55-63, assoc prof bot, 63-66, asst dean col sci, 64, assoc dean, 65-71, prof bot, 66-76, dir off environ qual progs, 71-75, prof plant path, 76-85, EMER PROF, PA STATE UNIV, 85- *Concurrent Pos:* Sr res fel, Agr Univ Neth, 75-76. *Mem:* AAAS; Am Inst Biol Sci; Am Phytopath Soc. *Res:* Plant disease epidemiology; parasitic ecology; influences of physical environment on plant disease development; instrumentation for plant disease study; research and education; economic botany. *Mailing Add:* 526 W Nittany Ave State College PA 16801

SCHEINBERG, ELIYAHU, b Tel Aviv, Israel, Mar 30, 34. POPULATION GENETICS, APPLIED STATISTICS. *Educ:* Univ Calif, Davis, BS, 61, MS, 62; Purdue Univ, PhD(pop genetics, appl statist), 66. *Prof Exp:* Res assoc, NC State Univ, 65-66; res scientist, Animal Genetics Sect, Animal Res Inst, Cent Exp Farm, Univ Ottawa, 66-68; assoc prof, Univ Calgary, 68-74, prof biol, 74-83. *Concurrent Pos:* Vis prof, Tel Aviv Univ, 71; Nat Res Coun Can traveling fel, 71-72. *Mem:* AAAS; Genetics Soc Can; Am Inst Biol Sci. *Res:* Computer simulation of genetic systems and testing biometrical and population genetics models using Tribolium and Drosophila. *Mailing Add:* 40 Bowvillage Cir NW Calgary AB T3B 4X2 Can

SCHEINBERG, ISRAEL HERBERT, b New York, NY, Aug 16, 19; m 52, 57; c 3. MEDICINE. *Educ:* Harvard Univ, AB, 40, MD, 43; Am Bd Internal Med, dipl. *Prof Exp:* Intern & asst resident med, Peter Bent Brigham Hosp, 43-44; res assoc chem, Mass Inst Technol, 47-51; instr med, Harvard Med Sch, 51-53, assoc, 53; asst prof, Columbia Univ, 53-55; assoc prof, 55-57, PROF MED, ALBERT EINSTEIN COL MED, 58-, HEAD DIV GENETIC MED, 73- *Concurrent Pos:* Commonwealth Fund fel, 63-64; jr fel, Soc Fellows, Harvard Univ, 47-50; consult, WHO, 52, 67; prin res scientist, NY State Psychiat Inst, 53-55; vis physician, Bronx Munic Hosp Ctr, 55-; res collabr, Med Dept, Brookhaven Nat Lab, 58-; Miller lectr, Dartmouth Med Sch, 61; vis prof physics, Univ Calif, San Diego, 63-64; chmn subcomt copper, Comn Biol & Med Effects Atmospheric Pollutants, Nat Res Coun, 72-77; vis prof, Children's Hosp, Harvard Med Sch, 77- *Honors & Awards:* Asn Res Nerv & Ment Dis Award, 59. *Mem:* Am Soc Clin Invest (vpres, 64-65); Asn Am Physicians. *Res:* Protein chemistry; chemistry and genetics of copper metabolism. *Mailing Add:* Dept Med Ullmann Bldg Albert Einstein Col Med 1300 Morris Park Ave Bronx NY 10461

SCHEINBERG, PERITZ, b Miami, Fla, Dec 21, 20; m 42; c 3. NEUROLOGY. *Educ:* Emory Univ, AB, 41, MD, 44; Am Bd Internal Med, dipl, 51; Am Bd Psychiat & Neurol, dipl, 54. *Prof Exp:* Instr med neurol, Duke Univ, 49-50; res assoc, Med Res Univ, 50-51, res asst prof physiol, Sch Med, 51-55, assoc prof neurol & chief div, 55-59, PROF NEUROL, SCH MED, UNIV MIAMI, 59-, CHMN DEPT, 61- *Concurrent Pos:* Res fel med neurol, Am Col Physicians, Med Sch, Duke Univ, 48-49, Am Heart Asn res fel, 49-50; consult, Vet Admin Hosp; med adv bd, Nat Multiple Sclerosis Soc & Myasthenia Gravis Found; consult, Surgeon Gen US; examr, Am Bd Psychiat & Neurol; trustee, Asn Univ Prof Neurol; mem stroke coun, Am Heart Asn. *Mem:* Asn Res Nerv & Ment Dis; Am Neurol Asn (pres); fel Am Col Physicians; fel Am Acad Neurol; Am Fedn Clin Res; Asn Univ Prof Neurol (pres). *Res:* Medical neurology; blood flow and metabolism of the brain. *Mailing Add:* Dept Neurol Univ Miami Sch Med Miami FL 33124

SCHEINBERG, SAM LOUIS, b New York, NY, June 15, 22; m 45; c 3. GENETICS. *Educ:* Cornell Univ, BS, 49; Iowa State Univ, MS, Univ Wis, PhD(genetics, poultry husb & zool), 54. *Prof Exp:* Fel immunogenetics, NIH, Univ Wis, 54-56; res biologist immunogenetics & immunochem, Oak Ridge Nat Lab, 56-58; geneticist & group leader, USDA, 58-70; LEADER, PIONEERING RES LAB, 70- *Mem:* Genetics Soc Am. *Res:* Immunogenetics; cellular and serum antigens in birds and mammals; somatic variation in birds and man; immunochemistry. *Mailing Add:* Pioneering Res Lab 7409 Wellesley Dr College Park MD 20740

SCHEINDLIN, STANLEY, b Philadelphia, Pa, July 8, 26; m 54; c 3. PHARMACEUTICAL CHEMISTRY. *Educ:* Temple Univ, BS, 45; Philadelphia Col Pharm, MS, 47, DSc(pharmaceut chem), 55. *Prof Exp:* Res fel, Philadelphia Col Pharm, 47-48, asst pharm, 48-49, instr, 49-55; res assoc labs, Nat Drug Co, 55-64, dir pharmaceut res labs, 64-70 & pharmaceut res & develop, 70-71; independent pharmaceut consult, 71-72; dir res, 72-77, dir tech affairs, 78-84, DIR REGULATORY AFFAIRS, LEMMON PHARMACOL, CO, 85- *Concurrent Pos:* Lectr, Philadelphia Col Pharm, 68-

69 & Spring Garden Col, 79-80; assoc adj prof, Sch Pharm, Temple Univ, 90- *Mem:* Am Soc Pharmacog; Am Chem Soc; Am Pharmaceut Asn; Acad Pharmaceut Sci; Parenteral Drug Asn; Regulatory Affairs Prof Soc. *Res:* Plant constituents; interactions of vitamins; stability; compatibility and incompatibility of drugs; formulation of parenterals; cancer chemotherapy. *Mailing Add:* 3011 Nesper St Philadelphia PA 19152

SCHEINER, BERNARD JAMES, b Atlantic City, NJ, Mar 12, 38; m 59; c 2. METALLURGY, CHEMISTRY. *Educ:* Univ Nev, Reno, BS, 61, PhD(org chem), 69. *Prof Exp:* Proj leader & res chemist, Reno Metall Res Sta, 66-79, SUPV METALLURGIST, FINE PARTICLE TECHNOL GROUP, TUSCALOOSA RES CTR, US BUR MINES, 79- *Mem:* Am Inst Mining, Metall & Petrol Engrs. *Res:* Development of processes for the recovery of metals from low-grade, refractory, and sulfide ores by means of innovative hydrometallurgical techniques; development of dewatering techniques for mineral processing waste slurries. *Mailing Add:* US Bur Mines Univ Ala PO Box L Tuscaloosa AL 35486

SCHEINER, DONALD M, b New York, NY, Mar 12, 32; m 54; c 2. BIOCHEMISTRY, MICROBIOLOGY. *Educ:* Cornell Univ, BS, 53, MFS, 54, PhD(biochem), 60. *Prof Exp:* Res chemist, Rohm and Haas Co, 60-63; asst prof, 63-71, ASSOC PROF CHEM, RUTGERS UNIV, 71- *Mem:* AAAS; Am Chem Soc; NY Acad Sci. *Res:* Analytical biochemistry; microbial metabolism; plant pigments; immunology. *Mailing Add:* Dept Chem Rutgers Univ Camden NJ 08102

SCHEINER, PETER, b Brooklyn, NY, Mar 13, 35; m 60; c 4. ORGANIC CHEMISTRY. *Educ:* Cornell Univ, AB, 57; Univ Mich, MS, 60, PhD(chem), 61. *Prof Exp:* NIH fel, Mass Inst Technol, 61-62; asst prof chem, Carleton Col, 62-64; from res chemist to sr res chemist, Mobil Oil Corp, 64-69; from assoc prof to prof chem, York Col, NY, 69-76, chmn dept natural sci, 73-76; CONSULT, DEPT AIR RESOURCES, NEW YORK, 71- *Concurrent Pos:* NIH res grant, Carleton Col, 62-64, NY Col, 79-82 & York Col, NY, 79-85; Am Chem Soc-Petrol Res Fund res grant, York Col, NY, 71-73. *Mem:* Am Chem Soc; Sigma Xi. *Res:* Organic photochemistry; heterocycles; air pollution; atmospheric chemistry; nucleoside chemistry. *Mailing Add:* 150-14 Jamaica Ave Jamaica NY 11432

SCHEINER, STEVE, b New York, NY, Feb 27, 51. PHYSICAL CHEMISTRY, QUANTUM CHEMISTRY. *Educ:* City Col New York, BS, 72; Harvard Univ, AM, 74, PhD(chem physics), 76. *Prof Exp:* Weizmann Found fel chem, Ohio State Univ, 76-78; ASST PROF CHEM, SOUTHERN ILL UNIV, CARBONDALE, 78- *Concurrent Pos:* Prin investr, Res Corp grant, 79-, NIH grant, 81- *Mem:* Int Soc Quantum Biol. *Res:* Proton transfer; hydrogen bonding; protein structure; opiate activity; biomembranes; hydration. *Mailing Add:* Dept Chem & Biochem Southern Ill Univ Carbondale IL 62901

SCHEINOK, PERRY AARON, b The Hague, Neth, Sept 21, 31; nat US; m 53, 71; c 2. RESEARCH ADMINISTRATION, MATHEMATICS. *Educ:* City Col New York, BS, 57; Ind Univ, PhD(math, statist), 60. *Prof Exp:* Asst math, Ind Univ, 52-54 & 56-59; mathematician, Burroughs Corp, 56; vis res assoc, Brookhaven Nat Lab, 59; instr comput & math, Wayne State Univ, 59-60, asst prof math, 60-62; sr systs engr, Radio Corp Am, 62-64; res asst prof pharmacol, Hahnemann Med Col, 64-68, res assoc prof, 68-70, res prof physiol & biophys, 70-72, dir comput ctr, 64-72, div biomet & comput, 70-72; exec dir, Del Health Serv Authority, 72-74; proj dir, University City Sci Ctr, 74-77; sr dir med data control, CIBA-GEIGY Corp, 77-82; assoc prof dept math, NJ Inst Technol, 83-87, UNIV JUDAISM, 87- *Concurrent Pos:* Consult, Henry Ford Hosp, 61-62, Pa Hosp, 65-66; free-lane statist consult, 82-; prin investr, NIH grant biomed res, Comput Ctr, 65-72. *Mem:* Am Statist Asn; Math Asn Am. *Res:* Application of probabilistic models to medical diagnosis; computerization of clinical hospital functions; electrocardiogram; clinical labs; time series analysis; large scale biological data bases; health services research. *Mailing Add:* 10141-2 Valley Circle Blvd Chartsworth CA 91311

SCHEIRER, DANIEL CHARLES, b Lebanon, Pa, Dec 10, 46; m 67; c 2. BOTANY, ELECTRON MICROSCOPY. *Educ:* Wheaton Col, Ill, BS, 68; Pa State Univ, MS, 71, PhD(bot), 74. *Prof Exp:* Asst prof, 74-80, ASSOC PROF BIOL, NORTHEASTERN UNIV, 80-, DIR, ELECTRON MICROSCOPY CTR, 78- *Concurrent Pos:* Vis scholar, Harvard Univ, 81-82. *Mem:* AAAS; Am Bryol & Lichenol Soc; Bot Soc Am; Electron Microscopy Soc Am; Sigma Xi; Am Soc Cell Biol; Am Soc Plant Physiol. *Res:* Ultrastructure of conducting tissue in lower plants; anatomy and ultrastructure of polytrichaceae; rhizobium-legume symbiosis. *Mailing Add:* Dept Biol Northeastern Univ Boston MA 02115

SCHEIRER, JAMES E, b Harrisburg, Pa, Dec 3, 43. ELECTROCHEMISTRY. *Educ:* Ursinus Col, BS, 65; Univ Pa, PhD(phys chem), 71. *Prof Exp:* Postdoctoral fel chem, State Univ NY Buffalo, 70-72; chair, Chem Dept, 87-90, PROF CHEM, ALBRIGHT COL, 72- *Concurrent Pos:* Vis scientist, Victoria Univ, Wellington, New Zealand, 84-85. *Mem:* Am Chem Soc. *Res:* Affect of high pressure, temperature, and viscosity on ion conductance in both aqueous and non aqueous solutions. *Mailing Add:* Dept Chem Albright Col PO Box 15234 Reading PA 19612-5234

SCHEIRING, JOSEPH FRANK, b Puchbach, Austria, Apr 18, 45; m 69; c 1. ENTOMOLOGY, ECOLOGY. *Educ:* Kent State Univ, BS, 68, MA, 70; Univ Kans, PhD(entom), 75. *Prof Exp:* Res assoc, Dept Entom, Mich State Univ, 75-76; ASST PROF, DEPT BIOL, UNIV ALA, 76- *Concurrent Pos:* Prin investr, Geol Surv Ala, 78-79 & Res Grants Comt, Univ Ala, 77 & 78. *Mem:* Ecol Soc Am; Entom Soc Am; NAm Benthol Soc; Sigma Xi. *Res:* Ecology of aquatic and semi-aquatic insects; niche relations in insects; applications of multivariate statistical methods to biology. *Mailing Add:* Dept Biol Box 1927 Univ Ala Tuscaloosa AL 35487

SCHEITER, B JOSEPH PAUL, b Philadelphia, Pa, Aug 25, 35. INSTITUTIONAL RESEARCH, ENROLLMENT PROJECTION MODELS. *Educ:* La Salle Univ, Philadelphia, BA, 57, MA, 58; Cath Univ Am, Washington, DC, MS, 66; Univ Santo Tomas, Manila, Philippines, PhD(educ), 76. *Prof Exp:* Teacher math, physics, relig, D J O'Connell High Sch, Arlington, Va, 59-66; chmn & assoc prof physics, De La Salle Univ, Manila, Philippines, 66-76; coordr acad comput computer sci, La Salle Univ, Philadelphia, 77-78, dir instnl res computer sci, 78-91; DIR INSTNL RES COMPUTER SCI, UNIV ST LA SALLE, BACOLOD, 91- *Concurrent Pos:* Sci chmn, D J O'Connell High Sch, Arlington, Va, 64-66; dir, Nat Sci Develop Bd, Sci Teaching Proj, Philippines, 67-70 & Computer Assisted Instr Proj, De La Salle Univ, 74-76. *Mem:* Asn Inst Res. *Res:* Institutional research; higher education; enrollment projection; academic credentials; financial aid; retention; outcomes. *Mailing Add:* Univ St La Salle Bacolod 6100 Philippines

SCHEKEL, KURT ANTHONY, b Colorado Springs, Colo, Jan 3, 43; m 68; c 1. ORNAMENTAL HORTICULTURE, FLORICULTURE. *Educ:* Colo State Univ, BS, 65, PhD(floricult), 71; Univ Nebr, Lincoln, MS, 68. *Prof Exp:* Assoc prof ornamental hort, 71-80, ASSOC PROF DEPT HORT & LANDSCAPE ARCHIT, WASH STATE UNIV, 80- *Mem:* Am Soc Hort Sci. *Res:* Nutrition and growing temperatures of greenhouse crops. *Mailing Add:* Dept Hort & Landscape Archit Wash State Univ Pullman WA 99164

SCHEKMAN, RANDY W, b St Paul, Minn, Dec 30, 48; m 73; c 1. CELL BIOLOGY. *Educ:* Univ Calif, Los Angeles, BA, 70; Stanford Univ, PhD(biochem), 75. *Prof Exp:* Fel, Univ Calif, San Diego, 74-76; asst prof, 76-81, ASSOC PROF, UNIV CALIF, BERKELEY, 81- *Concurrent Pos:* Fel, Cystic Fibrosis Found, 74; sabbatical fel, John S Guggenheim Found, 82. *Mem:* Am Soc Microbiol; Am Soc Biol Chemists. *Res:* Molecular mechanism of secretion and membrane assembly in eucaryotic cells. *Mailing Add:* Dept Biochem Univ Calif 401 Barker Hall Berkeley CA 94720

SCHELAR, VIRGINIA MAE, b Kenosha, Wis, Nov 26, 24. CHEMICAL EDUCATION. *Educ:* Univ Wis, BS, 47, MS, 53; Harvard Univ, EdM, 62; Univ Wis, PhD, 69. *Prof Exp:* Instr chem, Univ Wis-Milwaukee, 47-51; info specialist, Abbott Labs, Ill, 53-56; instr phys sci, Wright Jr Col, 57-58; asst prof chem, Northern Ill Univ, 58-63; prof, St Petersburg Jr Col, 65-67; asst prof chem, Chicago State Col, 67-68; prof chem, Grossmont Col, 68-80; CONSULT, 80- *Concurrent Pos:* Mem adv panel eval proposals, NSF, Washington DC, dir summer prog, 61. *Mem:* Am Chem Soc; fel Am Inst Chem. *Res:* Analytical instrumentation computers; protein requirement in nutrition. *Mailing Add:* 5702 Baltimore Dr 282 LaMesa CA 91942

SCHELBERG, ARTHUR DANIEL, b New York, NY, Mar 8, 21; wid; c 5. PHYSICS. *Educ:* Princeton Univ, AB, 42; Ind Univ, MS, 48, PhD(physics), 51. *Prof Exp:* Asst, Princeton Univ, 42-43, Los Alamos Sci Lab, 43-46, Ind Univ, 46-50 & Radiation Lab, Univ Calif, 51-52; asst, Los Alamos Sci Lab, 52-82; STAFF MEM, EG&G INC, E MERCK LABS, LOS ALAMOS, 82- *Mem:* Sigma Xi. *Res:* Nuclear physics; resonance capture of neutrons in U-238 and Th-232; low level radiography using channel plate image intensifiers; neutron flux measurements on downhole events at the Nevada test site. *Mailing Add:* 470 Camino Encantado Los Alamos NM 87544

SCHELD, WILLIAM MICHAEL, b Middletown, Conn, Aug 15, 47; m 69; c 1. INFECTIOUS DISEASES. *Educ:* Cornell Univ, BS, 69; Cornell Univ Med Col, MD, 73. *Prof Exp:* Intern med, 73-74, resident, 74-76, fel, 76-79, from asst prof to assoc prof med, 79-88, PROF MED, UNIV VA, 88- *Concurrent Pos:* Ed, Europ J Clin Microbiol, 84, mem, ACP MKSAP II, Infectious Dis, 86-88. *Honors & Awards:* Young Clin Investr Award, Am Fed Clin Res, 86. *Mem:* NY Acad Sci; AAAS; Infectious Dis Soc Am; Asn Clin Path; Am Soc Microbiol; Am Fed Clin Res; Am Soc Clin Invest. *Res:* Basic pathogenesis; pathophysiology of the central nervous system infections. *Mailing Add:* Box 385 Univ Va Med Ctr Charlottesville VA 22908

SCHELDORF, JAY J(OHN), b Camden, NJ, Jan 22, 32; m 53; c 2. CHEMICAL ENGINEERING. *Educ:* Univ Ill, BS, 53; Kans State Univ, MS, 54; Univ Colo, PhD(chem eng), 58. *Prof Exp:* Res chemist, Chem Div, Corn Prod Refining Co, 54-55; instr chem eng, Univ Colo, 55-58, asst prof, 58-66; assoc prof, 66-74, PROF CHEM ENG & ENG SCI, UNIV IDAHO, 74- *Mem:* Am Inst Chem Engrs; Am Chem Soc; Am Soc Eng Educ. *Res:* Fluid dynamics and heat transfer; physical chemistry; thermodynamics. *Mailing Add:* Dept Chem Eng Univ Idaho Moscow ID 83843

SCHELL, ALLAN CARTER, b New Bedford, Mass, Apr 14, 34; m 57; c 2. ELECTROMAGNETISM. *Educ:* Mass Inst Technol, SB & SM, 56, ScD(elec eng), 61. *Prof Exp:* Res physicist, Microwave Physics Lab, Air Force Cambridge Res Labs, 56-76; dir, electromagnetics, Rome Air Develop Ctgr, 76-87; CHIEF SCIENTIST, AF SYSTS COMMAND, 87- *Concurrent Pos:* Guenter Loeser Mem lectr, 65; vis assoc prof, Mass Inst Technol, 74; ed press, Inst Elec & Electronics Engrs, 76-79, dir, 81-82; ed, Proc Inst Elec & Electronics Engrs, 90- *Honors & Awards:* J J Bolljahn Award, 66; Centennial Medal, Inst Elec & Electronics Engrs, 84; Meritorious Serv Award, USAF, 88. *Mem:* Fel Inst Elec & Electronics Engrs; Int Union Radio Sci. *Res:* Electromagnetic theory; antennas; angular resolution enhancement. *Mailing Add:* 8062 Croom Rd Upper Marlboro MD 20772-9748

SCHELL, ANNE MCCALL, b Waco, Tex, Apr 23, 42; c 2. PSYCHO-PHYSIOLOGY. *Educ:* Baylor Univ, BS, 63; Univ Southern Calif, MA, 68, PhD(psychol), 70. *Prof Exp:* Vis asst prof psychol, Univ Southern Calif, 70-71; asst prof, 71-78, PROF PSYCHOL, OCCIDENTAL COL, 78-; RES ASSOC, NAT CTR HYPERACTIVE CHILDREN, 80- *Concurrent Pos:* Statist consult, Pasadena Unified Sch Dist, 73-75; res consult, Gateways Hosp, 75-80. *Mem:* Am Psychol Asn; Soc Psychophysiol Res. *Res:* Study of physiological components and concommitants of cognitive and affective processes in humans; physiological aspects of psychopathology. *Mailing Add:* Occidental Col Dept Psychol 1600 Campus Rd Los Angeles CA 90041

SCHELL, FRED MARTIN, b Cincinnati, Ohio, Oct 6, 43; div; c 2. SYNTHETIC ORGANIC CHEMISTRY. *Educ:* Univ Cincinnati, BS, 66, MS, 68; Ind Univ, PhD(chem), 72. *Prof Exp:* ASSOC PROF CHEM, UNIV TENN, KNOXVILLE, 72- *Concurrent Pos:* Oak Ridge Nat Lab, 78- *Mem:* Am Chem Soc; AAAS. *Res:* Development of new synthetic reactions and their application to synthesis of natural products; development of expert systems for organic synthesis; neural net development and applications; identification of chemosensory agents. *Mailing Add:* Dept Chem Univ Tenn Knoxville TN 37996-1600

SCHELL, GEORGE W(ASHINGTON), b Easton, Pa, June 25, 21; m 46; c 2. CHEMICAL ENGINEERING. *Educ:* Lafayette Col, BS, 43. *Prof Exp:* Jr chem engr, Atlantic Refining Co, Pa, 44; asst chem, Univ Southern Calif, 44-45; jr chem engr, Atlantic Refining Co, Philadelphia, 45-46, asst chemist, 46-47, asst prod foreman, 47-57, foreman, 57-66, prod technologist, Atlantic Richfield Co, Pa, 66-71; supv chemist, Qual Control Lab, Pennzoil United, Inc, 71-73, chief chemist, Pennzoil Co, 73-77, supt oil movements, 77-79, mgr packaging, 79-84; RETIRED. *Mem:* Am Chem Soc. *Res:* Petroleum technology. *Mailing Add:* 301 E Carpenter Ave Myerstown PA 17067

SCHELL, JOSEPH FRANCIS, b Miamisburg, Ohio, Dec 24, 28; m 54; c 7. GEOMETRY. *Educ:* Univ Dayton, BS, 50; Ind Univ, MA, 52, PhD(math), 57. *Prof Exp:* Asst, Ind Univ, 50-54; instr math, Univ Dayton, 54-56; res mathematician, Wright-Patterson AFB, Ohio, 56-61; asst prof math, Fla State Univ, 61-64; assoc prof, 64-68, chmn dept, 68-80, PROF MATH, UNIV NC, CHARLOTTE, 68- *Concurrent Pos:* Resident dir, Eglin Grad Ctr, Eglin AFB, 63-64. *Mem:* Am Math Soc; Math Asn Am; Soc Indust & Appl Math. *Res:* Differential geometry; relativity theory; topology. *Mailing Add:* Dept Math-Comput Sci Univ NC Univ Sta Charlotte NC 28223

SCHELL, ROBERT RAY, b Perry, Iowa, Sept 12, 37; m 61; c 2. ELECTRICAL ENGINEERING. *Educ:* Univ Ariz, BS, 59, MS, 61, PhD(elec eng), 67. *Prof Exp:* Jr engr, Melabs, Calif, 61-62; instr elec eng, Univ Ariz, 62-67; assoc engr, Collins Radio Co, 67-72; assoc engr, 72-76, ENG ANALYST, ELECTROSPACE SYSTS, INC, 76- *Mem:* AAAS; Inst Elec & Electronics Engrs. *Res:* Electromagnetic theory; wave propagation; antenna theory diffraction; numerical methods. *Mailing Add:* Electrospace Systs Inc 1901 Piano Rd Box 1359 Richardson TX 75080

SCHELL, STEWART CLAUDE, b Reading, Pa, Feb 4, 12; m 41; c 2. PARASITOLOGY. *Educ:* Kans State Univ, BS, 39; NC State Col, MS, 41; Univ Ill, PhD(zool, parasitol), 50. *Prof Exp:* From asst prof to prof, 49-78, chmn dept, 74-78, EMER PROF ZOOL, UNIV IDAHO, 78- *Mem:* Am Soc Parasitologists. *Res:* Life histories, development and taxonomy of parasitic helminths especially the trematodes. *Mailing Add:* Dept Biol Sci Univ Idaho Moscow ID 83843

SCHELL, WILLIAM JOHN, b Buffalo, NY, Oct 19, 40; m 76; c 2. POLYMER CHEMISTRY. *Educ:* State Univ NY Col Forestry, Syracuse Univ, BS, 64, MS, 66; Univ Southern Calif, PhD(polymer chem), 69. *Prof Exp:* Mgr spec membrane prod, Envirogenics Systs Co, 69-78; partner, KS&W Consult, 78-79; pres, Spectrum Separations Inc, 79-86; mgr opers, Air Prod Separex Div, 86-90, BUS DIR, SEPARATIONS PROD DIV, HOECHST CELANESE CORP, 90- *Mem:* Am Chem Soc; Am Inst Chem Engrs; Creation Res Soc. *Res:* Membrane systems for fluid separations; pollution control processes; polymer characterization; relaxation behavior of polymers. *Mailing Add:* Separations Prod Div Hoechst Celanese Corp 2100 E Orangethorpe Ave Anaheim CA 92806

SCHELL, WILLIAM R, b Portland, Ore, Apr 17, 32; m 59; c 1. ENVIRONMENTAL, EARTH & MARINE SCIENCES. *Educ:* Ore State Univ, BS, 54; Univ Idaho, MS, 56; Univ Wash, PhD(inorg chem, nuclear chem), 63. *Prof Exp:* Res technician soils chem, Univ Idaho, 54-56; independent investr, US Naval Radiol Defense Lab, 56-59; res asst chem, Univ Wash, 60-63; sr radiochemist, Hazleton-Nuclear Sci Corp, Isotopes, Inc, Calif, 64-65, head div atmospheric & oceanog sci, 65-68; vis scientist, Radiochem Div, Lawrence Radiation Lab, Univ Calif, 68; with div res & labs, Int Atomic Energy Agency, 68-71; res assoc prof with lab radiation ecol, Col Fisheries, Univ Wash, 71-77, assoc prof fisheries, 78-79, prof, 79-; DEPT RADIATION HEALTH, GRAD SCH PUB HEALTH, UNIV PITTSBURGH. *Concurrent Pos:* Guest lectr, Colo State Col, 66; adj prof oceanog, Dept Oceanog & radiol scientist, Dept Environ Health, Univ Wash, 79-82; vis scientist, Gas & Particulate Sci Div, Ctr Anal Chem, Nat Bureau Standards, 81-82; fel, Fulbright Found, Spain, 88. *Mem:* AAAS; Am Chem Soc; Sigma Xi; Health Physics; NY Acad Sci. *Res:* Environmental radiochemistry; colloidal chemistry; fallout studies; radiotracer techniques; chemical and radiochemical instrumentation; gas technology; carbon-14 dating; geophysics; nuclear debris in meteorology; oceanography and air pollution; radiation ecology. *Mailing Add:* Dept Radiation Health Univ Pittsburgh Grad Sch Pub Health Pittsburgh PA 15261

SCHELLENBERG, KARL A, b Hillsboro, Kans, July 13, 31; m 55; c 4. BIOCHEMISTRY. *Educ:* Col William & Mary, BS, 53; Johns Hopkins Univ, MD, 57; Harvard Univ, PhD(biochem), 64. *Prof Exp:* Intern med, Grace-New Haven Community Hosp, Conn, 57-58; biochemist, NIH, 58-60; from asst prof to assoc prof physiol chem, Johns Hopkins Univ, 63-73; PROF BIOCHEM & CHMN DEPT, EASTERN VA MED SCH, 73- *Concurrent Pos:* Markle Found scholar med sci, 65-70. *Mem:* Am Chem Soc; Am Soc Biol Chem; NY Acad Sci. *Res:* Biochemical reaction mechanisms. *Mailing Add:* 1332 Lakeview Dr Virginia Beach VA 23455

SCHELLENBERG, PAUL JACOB, b Leamington, Ont, Can, Dec 31, 42; Can citizen; m 66; c 5. BLOCK DESIGNS, FINITE GEOMETRIES. *Educ:* Univ Waterloo, BSc, 65, MA, 66, PhD(math), 71. *Prof Exp:* Lectr math, Univ Waterloo, 66-67; systs programmer, Comput Ctr, Indian Inst Technol, Kanpur, 67-69; lectr, Univ Waterloo, 71, asst prof, 71-79, dept chmn, 82-88 ASSOC PROF COMBINATORICS & OPTIMIZATION, UNIV WATERLOO, 79- *Res:* Combinatorial mathematics and designs; room squares, balanced room squares, Latin squares, pairwise balanced designs, and decompositions of graphs. *Mailing Add:* Dept Combinatorics & Optimization Univ Waterloo Waterloo ON N2L 3G1 Can

SCHELLER, W(ILLIAM) A(LFRED), b Milwaukee, Wis, June 6, 29; m 51; c 2. CHEMICAL ENGINEERING. *Educ:* Northwestern Univ, BS, 51, PhD(chem eng), 55. *Prof Exp:* Res engr, Calif Res Corp, Standard Oil Co, Calif, 55-60, group supvr, 60-63; assoc prof, 63-69, chmn dept, 71-78, PROF CHEM ENG, UNIV NEBR, LINCOLN, 69- *Concurrent Pos:* Du Pont fac fel, 64; Univ Res Coun fac fel, 65; Off Water Resources res grant, Univ Nebr, Lincoln, 66-69; guest prof, Univ Erlangen, 69-70, Ger Res Asn grant, 70; consult, Northern Natural Gas Co & Brunswick Corp; consult, IRAS Develop Corp, NY, 76- *Mem:* Am Chem Soc; Am Inst Chem Engrs; Am Soc Eng Educ; Sigma Xi. *Res:* Phase equilibrium; direct energy conversion; reaction kinetics; thermodynamics; computer aided design; process economics; alcohol blended fuels, gasohol. *Mailing Add:* Dept Chem Eng Univ Nebr Lincoln NE 68588

SCHELLING, GERALD THOMAS, b Sterling, Ill, Mar 24, 41; m 63; c 2. NUTRITION, METABOLISM. *Educ:* Univ Ill, BS, 63, MS, 64, PhD(nutrit), 68. *Prof Exp:* Res assoc nutrit, Univ Ill, 68; nutritionist, Smith, Kline & French Labs, 68-70; prof nutrit, Univ Ky, 70-79; prof nutrit, Tex A&M Univ, 79-88; PROF & HEAD DEPT ANIMAL SCI, UNIV IDAHO, 88- *Mem:* Am Soc Animal Sci. *Res:* Ruminant and nonruminant nutrition, with emphasis on growth regulation and nitrogen metabolism. *Mailing Add:* Univ Idaho Dept Animal Sci Moscow ID 83843

SCHELLMAN, JOHN ANTHONY, b Philadelphia, Pa, Oct 24, 24; m 54; c 2. PHYSICAL CHEMISTRY. *Educ:* Temple Univ, AB, 48; Princeton Univ, MA, 49, PhD(phys chem), 51. *Hon Degrees:* Dr, Chalmers Univ, Sweden, 83; Univ Padua, Italy, 90. *Prof Exp:* USPHS fel, Univ Utah, 51, res assoc, 52; USPHS fel, Carlsberg Lab, Denmark, 53-55; Du Pont fel, Univ Minn, 55-56, asst prof chem, 56-58; assoc prof, 58-62, PROF CHEM, UNIV ORE, 62- *Concurrent Pos:* Sloan fel, 59-64; mem NIH study sect biophys & biophys chem, 62-67; sr fel, USPHS, Lab des Hautes Pressions, Bellevue, France, 63-64; Guggenheim fel, Weizmann Inst, 69-70; vis scientist, Lab Chem Physics, Bethesda, Md, 80. *Mem:* Nat Acad Sci; Am Chem Soc; Am Soc Biol Chemists; Am Acad Arts & Sci; fel Am Phys Soc; Biophys Soc. *Res:* Optical rotation and thermodynamics of biochemical molecules. *Mailing Add:* Inst Molecular Biol Univ Ore Eugene OR 97403

SCHELLY, ZOLTAN ANDREW, b Budapest, Hungary, Feb 15, 38; m 67; c 3. PHYSICAL CHEMISTRY, COLLOIDAL DYNAMICS. *Educ:* Vienna Tech Univ, BS, 62, DSc(phys chem), 67. *Prof Exp:* AEC fel, Lab Surface Studies, Univ Wis-Milwaukee, 68; Air Force Off Sci Res fel, Univ Utah, 69-70; asst prof phys chem, Univ Ga, 70-76; PROF PHYS CHEM, UNIV TEX, ARLINGTON, 77-, ACTG CHMN, 90- *Concurrent Pos:* Alexander von Humboldt lectr fel, Max-Planck Inst Biophys Chem, 74. *Honors & Awards:* Wilfred T Doherty Award, Am Chem Soc, 86. *Mem:* Am Chem Soc; Austrian Chem Soc; Am Phys Soc. *Res:* Reaction kinetics; dynamics of fast rate processes; relaxation spectrometry; diffusion; laser techniques; dynamics of reverse micelles and suspensions; chemical instabilities and bifurcations; deterministic chaos. *Mailing Add:* Dept Chem Univ Tex Arlington TX 76019-0065

SCHELP, RICHARD HERBERT, b Kansas City, Mo, Apr 21, 36; m 58; c 2. MATHEMATICS. *Educ:* Cent Mo Univ, BS, 59; Kans State Univ, MS, 61, PhD(math), 70. *Prof Exp:* Assoc math missile scientist, Appl Physics Lab, Johns Hopkins, 61-66; instr math, Kans State Univ, 66-70; asst prof, 70-74, assoc prof, 74-79, PROF MATH, MEMPHIS STATE UNIV, 79- *Concurrent Pos:* Managing ed, J Graph Theory, 81-83; vis res, Hungarian Acad Sci, Math Inst, 85, 90. *Mem:* Am Math Soc; Math Asn Am. *Res:* Graph theory and lattice theory; Ramsey theory and Hamiltonian graph theory. *Mailing Add:* Dept Math Sci Memphis State Univ Memphis TN 38152

SCHELPER, ROBERT LAWRENCE, b San Antonio, Tex, Oct 28, 48; m 84. MICROGLIA, NEUROIMMUNOLOGY. *Educ:* St Mary's Univ, San Antonio, Tex, BA, 71; Univ Tex, San Antonio, MD, 75, PhD(anat), 78. *Prof Exp:* Instr, dept anat, Univ Tex, San Antonio, 76-78; asst prof, 82-87, ASSOC PROF PATH, SCH MED, UNIV IOWA, 87-, DIR AUTOPSY SERV, 87- *Honors & Awards:* Weil Award, Am Asn Neuropathologists, 85. *Mem:* Am Asn Pathologists; Fed Am Socs Exp Biol; Am Asn Anatomists; Soc Neurosci; Reticuloendothelial Soc. *Res:* The origin, functions and pathologic reactions of microglia; studies of inflammatory reactions in nervous tissues injuries; lectin histochemistry for cell identification; blood brain barrier dysfunction and Alzheimer's disease. *Mailing Add:* Dept Path Sch Med Univ Iowa Iowa City IA 52242

SCHELSKE, CLAIRE L, b Fayetteville, Ark, Apr 1, 32; m 57; c 3. AQUATIC ECOLOGY, LIMNOLOGY. *Educ:* Kans State Teachers Col, AB, 55, MS, 56; Univ Mich, PhD(zool), 61. *Prof Exp:* Res assoc, Univ Ga Marine Inst, 60-62; fishery biologist, Radiobiol Lab, Bur Com Fisheries, 62-63, supvry fishery biologist, 63, chief estuarine ecol prog, 63-66; tech asst, Off Sci & Technol, Exec Off President, 66-67; from asst res limnologist to assoc res limnologist, Univ Mich, Ann Arbor, 67-71, asst prof radiol health, Sch Pub Health, 67-68, lectr, Biol Sta, 70, from asst dir to actg dir, Great Lakes Res Div, 70-76, res limnologist, Great Lakes Res Div, 71-72, assoc prof limnol, Dept Atmospheric & Ocean Sci & assoc prof, Natural Resources, Sch Natural Resources, Univ Mich, Ann Arbor, 76-87; CARL S SWISHER PROF WATER RESOURCES, DEPT FISHERIES & AGR, UNIV FLA, 87- *Concurrent Pos:* Adj asst prof, NC State Univ, 64-66; consult to Ill Atty Gen, US Dept Justice. *Mem:* Fel AAAS; Am Soc Limnol & Oceanog (secy, 76-85, vpres, 87-88, pres, 88-90); Ecol Soc Am; Int Asn Great Lakes Res; fel Am Inst Fishery Res Biologists. *Res:* Eutrophication limnology and paleolimnology of the Great Lakes, fresh-water ecosystem ecology; relationships among silica, nitrogen, phosphorus and phytoplankton production; nutrients and other factors limiting primary productivity; biogeochemistry of silica. *Mailing Add:* Dept Fisheries & Agr Univ Fla 7922 NW 71st St Gainesville FL 32601

SCHELTEMA, RUDOLF S, b Madison, Wis, May 27, 26; m 55; c 2. MARINE BIOLOGY, BIOGEOGRAPHY. *Educ:* George Washington Univ, BS, 51, MS, 54; Univ NC, PhD(zool), 60. *Prof Exp:* Marine biologist, Chesapeake Biol Lab, Md, 51-54; res assoc, Oyster Res Lab, Rutgers Univ, 59-60; res assoc marine biol, 60-63, from asst scientist to assoc scientist, 63-85, SR SCIENTIST, WOODS HOLE OCEANOG INST, 85- *Concurrent Pos:* Fac mem, Cape Cod Community Col, 61-62; sr Fulbright-Hays Scholar, James Cook Univ, NQueensland, Australia, 77-78; assoc ed, Proc Nat Shellfish Asn, 66-68; ed adv, Marine Ecol Prog Ser, 80-, Zool Scripta, 82-; Mellon Study Award, 80-81. *Mem:* Am Soc Zool; Soc Syst Zool; Am Soc Naturalists; Am Malacog Union; Systs Asn UK; Sigma Xi. *Res:* Invertebrate zoology; morphology, ecology and comparative physiology of the larvae of marine benthic invertebrates; biogeography and evolution; reproduction of deep-sea invertebrate benthos; life history and settlement of fouling organisms. *Mailing Add:* Woods Hole Oceanog Inst Woods Hole MA 02543

SCHELTGEN, ELMER, b Limerick, Sask, Feb 5, 30; m 53; c 2. BIOPHYSICS, MOLECULAR GENETICS. *Educ:* Univ BC, BA, 55; Ind Univ, Bloomington, AM, 65; Univ Tex, Houston, PhD(biomed sci), 68. *Prof Exp:* Lectr & instr physics, Univ BC, 54-60; NASA grant, Univ Houston, 68-69; USPHS fel, Univ Tex M D Anderson Hosp & Tumor Inst Houston, 69; asst prof cancer res, 69-71, res assoc bact, 71-77, RES ASSOC MICROBIOL, UNIV SASK, 78- *Mem:* Am Chem Soc; Am Inst Biol Sci. *Mailing Add:* 1606 Ave B N Saskatoon SK S7L 1H3 Can

SCHEMENAUER, ROBERT STUART, b Prince Albert, Sask, Nov 3, 46; m 70. CLOUD PHYSICS, ARID LANDS WATER. *Educ:* Univ Sask, BA, 67; Univ Toronto, MSc, 69, PhD(meteorol), 72. *Prof Exp:* Meteorologist, 67-69, RES SCIENTIST CLOUD PHYSICS, ATMOSPHERIC ENVIRON SERV, ENVIRON CAN, 72- *Concurrent Pos:* Consult, UNDP, World Meteorol Asn, CIDA & IDRC, 87-; assoc ed, J Appl Meteorol, 82, 89, J Climate & Appl Meteorol, 83-88 & J Atmos & Oceanic Technol, 89-; prin investr, Environ Can Chem High Elevation Fog Prog, IDRC & CIDA Fog Collection Projs, Chile & Peru, 87-90. *Mem:* Can Meteorol & Oceanog Soc; Am Meteorol Soc; Int Water Resources Asn. *Res:* Laboratory and airborne studies of the microphysical processes responsible for precipitation formation; weather modification; fog as an arid lands water resource; precipitation chemistry; acidic deposition to forests. *Mailing Add:* 92 Caines Ave Willowdale ON M2R 2L3 Can

SCHEMM, CHARLES EDWARD, b Baltimore, Md, Oct 30, 47; m 77. NUMERICAL MODELING, TURBULENCE. *Educ:* Loyola Col, Md, BS, 69; Princeton Univ, PhD(geophys fluid dynamics), 74. *Prof Exp:* Res assoc, Inst Phys Sci & Technol, Univ Md, 74-77; SR OCEANOGR, APPL PHYSICS LAB, JOHNS HOPKINS UNIV, 77- *Concurrent Pos:* Vis lectr, dept meteorol, Univ Md, 76-81. *Mem:* Am Meteorol Soc; Sigma Xi; AAAS; Am Phys Soc. *Res:* Ocean boundary layer modeling; planetary boundary layer modeling; experimental and numerical studies of submarine hydrodynamics. *Mailing Add:* 11314 Old Hopkins Rd Clarksville MD 21029

SCHEMMEL, RACHEL A, b Farley, Iowa, Nov 23, 29. NUTRITION. *Educ:* Clarke Col, BA, 51; State Univ Iowa, MS, 52; Mich State Univ, PhD(nutrit), 67. *Prof Exp:* Dietitian, Childrens Hosp Soc, Calif, 52-54; adminr, St Joseph's Hosp, Calif, 54-55; instr food & nutrit, 55-63, from asst prof to assoc prof, 68-76, PROF NUTRIT, MICH STATE UNIV, 77- *Concurrent Pos:* Res fel, Dunn Nutrit Lab, Cambridge, 68; Sabbatic endocrinol, UCLA, 78; Sabbatic diabetes, NIDDK, 88. *Honors & Awards:* Borden Award, 1986; Sr Res Award, Sigma Xi, 86. *Mem:* AAAS; Am Inst Nutrit; Soc Exp Biol & Med; Brit Nutrit Soc; Am Dietetic Asn; Am Home Econ Asn; Inst Food Tech; Soc Nutrit Educ. *Res:* Obesity and lipid and carbohydrate metabolism; hypertension; nutritional status of human subjects; dental caries; nutrition and exercise. *Mailing Add:* Dept Food Sci & Human Nutrit Mich State Univ East Lansing MI 48824

SCHEMNITZ, SANFORD DAVID, b Cleveland, Ohio, Mar 10, 30; m 58; c 3. WILDLIFE RESEARCH, WILDLIFE ECOLOGY. *Educ:* Univ Mich, BS, 52; Univ Fla, MS, 53; Okla State Univ, PhD(wildlife zool), 58. *Prof Exp:* Res game biologist, Bur Res & Planning, State Dept Conserv, Minn, 58-59; asst prof wildlife resources, Univ Maine, Orono, 60; asst prof wildlife mgt, Pa State Univ, 61; from asst prof to prof wildlife resources, Sch Forest Resources, Univ Maine, Orono, 63-75; head dept, 76-81, PROF WILDLIFE, DEPT FISHERY & WILDLIFE SCI, NMEX STATE UNIV, 76- *Concurrent Pos:* With sub group 108-wildlife habitat mgt, Int Union Forest Res Orgn; res partic, NSF, 62, 64 & 66; Fulbright prof ecol, Tribhuvan Univ, Nepal, 83-84; Fulbright prof wildlife mgt, Moi Univ, Kenya, 90. *Mem:* Ecol Soc Am; Wildlife Soc; Am Soc Mammalogists; Sigma Xi. *Res:* Wildlife conservation; ecology of birds and mammals; forest zoology and ecology; effects of off-road vehicles on environment. *Mailing Add:* Dept Fish & Wildlife Sci NMex State Univ Las Cruces NM 88003-0003

SCHEMPP, ELLORY, b Philadelphia, Pa. CHEMICAL PHYSICS, TECHNICAL MANAGEMENT. *Educ:* Tufts Univ, BS, 62; Brown Univ, PhD(physics), 68. *Prof Exp:* Fel physics, Brown Univ, 67-68; res physicist, Bell Tel Labs, NJ, 68-70; asst prof crystallog & res asst prof physics, Univ Pittsburgh, 70-77; vis prof physics, Univ Ill, Champaign-Urbana, 77-78; vis prof, Univ Geneva, Switz, 77-79; sr scientist, Lawrence Berkeley Lab, 80-83; sr scientist, Gen Elec Med Systs, 83-87; vpres opers, Auburn Int, 87-89; OWNER, HARVARD CONSULT GROUP, 89- *Concurrent Pos:* Course coordr, George Washington Univ, 83- *Mem:* Am Phys Soc; Sigma Xi; Instrument Soc Am; Soc Magnetic Resonance Med; Soc Magnetic Resonance Imaging. *Res:* Nuclear quadrupole resonance studies of chemical bonds; molecular and ionic field gradients in crystals; hydrogen bonding and lattice dynamics; nuclear magnetic resonance; applications of magnetic resonance imaging to ground water assessment, hazardous waste mitigation, and oil well drill cores. *Mailing Add:* 24 Boston Ave Medford MA 02155

SCHEMSKE, DOUGLAS WILLIAM, b Chicago, Ill, Sept 8, 48. POPULATION BIOLOGY. *Educ:* Univ Ill, BS, 70, PhD(ecol), 77. *Prof Exp:* Fel, Smithsonian Trop Res Inst, 77-78; asst prof evolution & ecol, Amherst Col, 78-79; ASST PROF EVOLUTION & ECOL, UNIV CHICAGO, 79- *Concurrent Pos:* Vis instr, Field Sta, Univ Minn, 79-80. *Mem:* AAAS; Am Soc Naturalists; Asn Trop Biol; Ecol Soc Am; Soc Study Evolution. *Res:* Evolutionary processes in plant populations, with particular emphasis on breeding systems, population structure, gene flow and the assessment of selection intensities. *Mailing Add:* Dept Biol Crb 208 Univ Chicago 940 E 57th St Chicago IL 60637

SCHENA, FRANCESCO PAOLO, b Foggia, Italy, Mar 24, 40; m 69; c 2. NEPHROLOGY, IMMUNOLOGY. *Educ:* Univ Bari, MD, 64. *Prof Exp:* Med asst internal med, Univ Bari, 70-71, asst prof internal med, 72-82, assoc prof med ther, 83-85, PROF NEPHROLOGY, UNIV BARI, 86-, CHMN, 89- *Concurrent Pos:* Fel, Dept Nephrology, Univ Louvain-Belg, 68-70; vis prof, Inst Path-Case Western Reserve Univ, Cleveland, 85 & Renal Unit-Guy's Hosp, London, UK, 86; prin investr, CNR-Bilateral Proj, Italy-USA, 85- & CNR- Biotechnol Proj, Rome-Italy, 89-; nephrology consult, IRCCS Sci Res Inst-Bari, 90-; chmn, Regional Comt Organ Transplant, 90- *Mem:* Europ Renal Asn; Int Soc Nephrology; Am Kidney Asn; Am Soc Nephrology; Am Asn Immunologists; NY Acad Sci. *Res:* Principal investigator in the immunological research of human glomerulonephritis; developed new hypothesis on the pathogenesis of IgA nephropathy as the permanence in the blood of circulating immune complexes, which are not solubilised for the presence of increased amount of polymeric IgA; high production of interleukin-2 by peripheral blood mononuclear cells evidences the presence of unknown antigen in the blood which is able to activate continuously lymphocyte T helper. *Mailing Add:* Via Delle Murge 59/A Bari 70124 Italy

SCHENCK, HARRY ALLEN, b San Diego, Calif, May 29, 38; m 59; c 4. ACOUSTICS. *Educ:* Pomona Col, BA, 59; Harvard Univ, SM, 60, PhD(appl physics), 64. *Prof Exp:* Lectr & res fel acoust, Harvard Univ, 64; res physicist, USN Electronics Lab, 64-69, surveillance systs prog mgr, Naval Undersea Ctr, 69-75, head undersea surveillance dept, 75-84, ASSOC UNDERSEA SURVEILLANCE, NAVAL OCEAN SYSTS CTR, 85- *Concurrent Pos:* Vis prof, US Naval Acad, 84-85; assoc ed, US Navy J Underwater Acoust, 86- *Mem:* Acoust Soc Am. *Res:* Electroacoustic transducers; acoustic radiation and scattering theory. *Mailing Add:* Ocean Surveillance Dept Code 701(S) Naval Ocean Systs Ctr San Diego CA 92152-5000

SCHENCK, HILBERT VAN NYDECK, JR, b Boston, Mass, Feb 12, 26; m 50; c 4. PHYSICS, MECHANICAL ENGINEERING. *Educ:* Williams Col, BA, 50; Stanford Univ, MS, 52. *Prof Exp:* Test engr, Pratt & Whitney Aircraft Div, United Aircraft Corp, Conn, 52-56, from asst prof to prof mech eng, Clarkson Col Technol, 56-67; prof mech eng & appl Mech & Ocean Eng, Univ RI, 7-83, Dir Scuba Safety Proj, 71-80; RETIRED. *Concurrent Pos:* NSF grant, 63-64, res grant, 65-; Food & Drug Admin res grant scuba safety, 69-71; US Coast Guard grant scuba safety, 71-72; Manned Undersea Sci & Technol grants, 72- *Mem:* Am Phys Soc. *Res:* Engineering heat transfer; statistics of experimentation; instrumentation; underwater photography and oceanographic optics; diving technology and safety; scuba tank corrosion. *Mailing Add:* 343 Delano Rd Marion MA 02738

SCHENCK, JAY RUFFNER, b Geneva, Ill, Jan 10, 15; m 48; c 2. BIOCHEMISTRY. *Educ:* Univ Ill, BS, 36, MS, 37; Cornell Univ, PhD(biochem), 41. *Prof Exp:* Sci asst, Soybean Res Lab, USDA, 36-37; asst biochem, Sch Med, George Washington Univ, 37-38 & Cornell Univ, 38-40; res biochemist, Abbott Labs, 41-83; RETIRED. *Mem:* Am Chem Soc; Am Soc Biol Chem; Sigma Xi. *Res:* Microbiological assay; intermediary metabolism; isolation and chemistry of antibiotics; immunochemistry. *Mailing Add:* 403 Hull St Waukegan IL 60085

SCHENCK, JOHN FREDERIC, b Decatur, Ind, June 7, 39; div; c 3. MEDICAL RESEARCH, SOLID STATE PHYSICS. *Educ:* Rensselaer Polytech Inst, BS, 61, PhD(solid state physics), 65; Albany Med Col, MD, 77. *Prof Exp:* Assoc prof elec eng, Syracuse Univ, 70-73; consult scientist, Gen Electric Electronics Lab, 65-70, MEM TECH STAFF, CORP RES & DEVELOP CTR, GEN ELEC CO, 73- *Mem:* Am Phys Soc; Inst Elec & Electronics Engrs; Sigma Xi. *Res:* Electrical and electronic technology applied to clinical medicine; nuclear magnetic resonance and solid state devices in medical diagnosis; electric potentials at biological interfaces; social implications of technology. *Mailing Add:* Gen Elec Corp Res & Develop Ctr PO Box eight Schenectady NY 12301

SCHENCK, NORMAN CARL, b Oak Park, Ill, July 8, 28; m 51; c 4. PLANT PATHOLOGY, VA MYCORRHIZAE. *Educ:* Univ Ill, BS, 51, PhD(plant path), 55. *Prof Exp:* From asst plant pathologist to assoc plant pathologist, 56-69, plant pathologist, Agr Res Ctr & prof plant path, 69-90, EMER PROF PLANT PATH, UNIV FLA, 91- *Concurrent Pos:* Mem, Nat Comt Microbiol Collections in Plant Sci. *Honors & Awards:* Fel Award, Am Phytopath Soc, 86. *Mem:* Am Phytopath Soc; Mycol Soc Am. *Res:* Soil-borne plant disease; endomycorrhizal fungi; VA mycorrhizal fungi collection. *Mailing Add:* PO Box 90190 Gainesville FL 32607-0190

SCHENGRUND, CARA-LYNNE, b New York, NY, Feb 18, 41; m 61; c 2. BIOCHEMISTRY. *Educ:* Upsala Col, BS, 62; Seton Hall Univ, MS, 65, PhD(chem), 66. *Prof Exp:* Part-time instr, Upsala Col, 67; res worker, Col Physicians & Surgeons, Columbia Univ, 67-68, res assoc biochem, 68-69; res assoc, Pa State Univ, 69-72, asst prof, 72-79, actg dept chmn, 86-87, ASSOC PROF BIOCHEM, HERSHEY MED CTR, PA STATE UNIV, 79- *Mem:* Am Soc Biochem & Molecular Biol; Am Chem Soc; Am Soc Neurochem. *Res:* Biological roles of gangliosides; neural cell differentiation; neurochemistry and the neurotoxins produced by Clostridium botulinum and clostridium tetane. *Mailing Add:* Milton S Hershey Med Ctr Pa State Univ Hershey PA 17033

SCHENK, ERIC A, CARDIOVASCULAR RESEARCH. *Educ:* Univ Wash, Seattle, MD, 59. *Prof Exp:* PROF PATH, UNIV WASH, SEATTLE, 73- *Mailing Add:* Dept Path Sch Med & Dent Univ Rochester Rochester NY 14642

SCHENK, H(AROLD) L(OUIS), JR, b Columbus, Ohio, Jan 27, 29; div; c 2. ELECTROMAGNETIC ANALYSIS, APPLIED MAGNETICS. *Educ:* Ohio State Univ, BSc, 51, MSc, 52. *Prof Exp:* Res physicist, Gen Motors Corp, 52-55; SR ENGR, WESTINGHOUSE SCI & TECHNOL CTR, 57- *Mem:* Am Phys Soc; Inst Elec & Electronics Engrs. *Res:* Magnetic phenomena and technology. *Mailing Add:* Westinghouse STC MS 401 4X9B 1310 Beulah Rd Churchill Borough Pittsburgh PA 15235

SCHENK, JOHN ALBRIGHT, b Stevens Point, Wis, Oct 22, 24; m 55; c 2. FOREST ENTOMOLOGY. *Educ:* Univ Mich, BS, 50; Univ Wis, MS, 56, PhD(entom), 61. *Prof Exp:* Relief model aid cartog, Relief Model Div, Army Map Serv, 51-53; forester, US Forest Serv, 53-54; res fel forest entom, Univ Wis, 54-59 & Wis Conserv Dept, 59-61; from asst prof to assoc prof, Univ Idaho, 61-66, prof forest entom, 71-83, asst dept head, 83; RETIRED. *Mem:* Entom Soc Am; Entom Soc Can; Smithsonian Inst; Soc Am Foresters; Entom Soc Brit. *Res:* Forest entomological research with emphasis on biology and ecology of forest pests and their control by silvicultural and biological methods; cone and seed insects; bark beetles. *Mailing Add:* N 5985 Altmnonte Dr Rathdrum ID 83858

SCHENK, PAUL EDWARD, b Stratford, Ont, Feb 26, 37; m 60; c 2. PETROLOGY, STRATIGRAPHY. *Educ:* Univ Western Ont, BSc, 59; Univ Wis, MS, 61, PhD(geol), 63. *Prof Exp:* From asst prof to assoc prof, Dalhousie Univ, 63-75, chmn dept, 81-83, prof geol, 75-85, CARNEGIE PROF GEOL, DALHOUSIE UNIV, 85- *Concurrent Pos:* Can leader, IGCP Proj Caledonian Orogeny, 74-85; Comnr, NAm Comn Stratig Nomenclature, 75-78. *Mem:* Geol Soc Am; Am Asn Petrol Geol; Soc Econ Paleont & Mineral; fel Geol Asn Can; Int Asn Sedimentologists. *Res:* Petrology and stratigraphy of evaporites; paleoecology and petrology of carbonate sediments; stratigraphy and sedimentology; paleocurrent study of deep-sea fans; sedimentology in Paleozoic Atlantic of southeastern Atlantic Canada and Northwestern Africa. *Mailing Add:* Dept Geol Dalhousie Univ Halifax NS B3H 4H6 Can

SCHENK, ROY URBAN, b Evansville, Ind, Nov 18, 29; div. BIOCHEMISTRY. *Educ:* Purdue Univ, BS, 51; Cornell Univ, MS, 53, PhD, 54. *Prof Exp:* Instr chem, Evansville Col, 54-55; chemist, Northern Regional Res Lab, Ill, 55-57; asst chemist, Univ Ga Exp Sta, 57-60; asst prof chem, Univ Ky, 60-62; mem staff org chem, Mat Lab, Wright-Patterson AFB, Ohio, 62-64; sr res chemist, Drackett Co, Ohio, 64-65; assoc prof pharmaceut chem, Univ Cincinnati, 65-67; res assoc, Univ Wis-Madison, 67-70; biochemist, Bjorksten Res Labs, 70-90; PRES, BIOENERGETICS, INC, 73-; EXEC DIR, GENDEN HARMONY NETWORK, 90- *Mem:* Am Chem Soc; Inst Food Technologists; Sigma Xi; Am Acad Forensic Sci. *Res:* Male perspective on gendor issues; nutritional aspects of health and disease; biochemistry; road de-icing alternatives; chemistry related issues for automotive intoxication defense. *Mailing Add:* Bioenergetics Inc PO Box 9141 Madison WI 53715-9141

SCHENK, WORTHINGTON G, JR, b Buffalo, NY, Feb 10, 22; m 46; c 7. SURGERY. *Educ:* Williams Col, BA, 42; Harvard Med Sch, MD, 45; Am Bd Surg, dipl, 54. *Prof Exp:* Assoc, 54-56, from asst prof to assoc prof, 56-66, actg chmn dept, 69-72, PROF SURG, SCH MED, STATE UNIV NY BUFFALO, 66-, CHMN DEPT, 72- *Concurrent Pos:* Buswell fel surg res, State Univ NY, Buffalo, 56-60; attend surgeon, E J Meyer Hosp, 54-66, from assoc dir to dir surg res labs & dir surg, 54-; mem surg study sect, NIH, 69-73. *Mem:* Soc Vascular Surg (treas, pres); Soc Surg Alimentary Tract; Soc Univ Surg; Soc Clin Surg (secy); Am Surg Asn. *Res:* Biophysics of surgical problems in hemodynamics. *Mailing Add:* Dept Surg B201 462 Grider St Aa State Univ NY 3435 Main St Buffalo NY 14215

SCHENKEIN, ISAAC, biological chemistry, for more information see previous edition

SCHENKEL, ROBERT H, b New York, NY, June 12, 44; m 71; c 2. PARASITOLOGY, IMMUNOBIOLOGY. *Educ:* Lafayette Col, BS, 65; Adelphi Univ, MS, 67; Univ Ill, PhD(zool), 71. *Prof Exp:* Res assoc immunoparasitol, Univ Ill, Urbana, 71-72; res scientist, Univ NMex, 72-75; sr res parpaitologist, Am Cyanamid Co, 75-77, group leader parasitol res, 77-80, group leader chemother & immunol, 80-82, mgr, animal indust discovery, 82-84, dir global animal prod develop, 85-, DIR, ANIMAL INDUST DISCOVERY & TECH ACQUISTIONS. *Mem:* Am Soc Parasitologists; AAAS. *Res:* Development of new drugs for use in treating parasitic infections; development of new host-parasite systems for testing drugs; monoclonal antibodies for antigen identification. *Mailing Add:* Am Cyanamid Co PO Box 400 Princeton NJ 08540

SCHENKEN, JERALD R, b Detroit, Mich, Oct 11, 33; m 59; c 3. PATHOLOGY. *Educ:* Tulane Univ, MD, 58. *Prof Exp:* PATHOLOGIST, DEPT PATH, NEBR METHODIST HOSP & CHILDREN'S MEM HOSP, 65- *Concurrent Pos:* Clin prof path, Univ Nebr Med Ctr, 75- & Creighton Univ, 78- *Mem:* AMA; Col Am Pathologist; Am Soc Clin Path; Am Col Physicians; Am Med Asn. *Mailing Add:* Nebr Methodist Hosp PO Box 14424 Omaha NE 68114

SCHENKENBERG, THOMAS, b St Louis, Mo, Nov 3, 43; m 81. NEUROPSYCHOLOGY. *Educ:* Rockhurst Col, AB, 65; Univ Utah, MA, 69, PhD(clin psychol), 70; Am Bd Prof Psychol, dipl & cert clin psychol, 82, dipl & cert clin neuropsychol, 85. *Prof Exp:* Asst prof, 73-80, ASSOC PROF, DEPT NEUROL, UNIV UTAH, 80-, RES ASSOC PROF, DEPT PSYCHOL, 73-; CHIEF PSYCHOL SERV, VET ADMIN HOSP, 70- *Concurrent Pos:* Nat Inst Aging grant, 77; adj assoc prof, dept psychol, Brigham Young Univ, 73; adj asst prof, Dept Psychiat, Univ Utah, 77- *Mem:* Am Psychol Asn; Nat Register Health Serv Providers Psychol. *Res:* Clinical neuropsychology; electrophysiology; medical psychology. *Mailing Add:* Dept Neurol Univ Utah Sch Med 50 N Medical Dr Salt Lake City UT 84132

SCHENKER, HENRY HANS, b Vienna, Austria, June 19, 26; nat US; m 55; c 3. TEXTILE CHEMISTRY. *Educ:* City Col, BS, 49; Rutgers Univ, PhD(chem), 53. *Prof Exp:* Asst, Rutgers Univ, 51-52; res chemist, 52-56, supvr, Analysis Lab, 56-61, sr res chemist, 61-91, RES ASSOC, E I DU PONT DE NEMOURS & CO, INC, 91- *Mem:* Am Chem Soc; NY Acad Sci; Soc Automotive Engr; Tech Asn Pulp & Paper Ind. *Res:* Ion-exchange; textile chemistry; polymer chemistry; friction products. *Mailing Add:* Fibers Dept PO Box 80701 Wilmington DE 19880-0701

SCHENKER, STEVEN, b Krakow, Poland, Oct 5, 29; US citizen; c 5. INTERNAL MEDICINE, GASTROENTEROLOGY. *Educ:* Cornell Univ, BA, 51, MD, 55. *Prof Exp:* From intern to sr resident, Harvard Med Serv, Boston City Hosp, 55-57; clin assoc gastroenterol, Nat Inst Allergy & Infectious Dis, 59-61; asst prof, Col Med, Univ Cincinnati, 63-64; from asst prof to assoc prof internal med, Univ Tex Southwestern Med Sch, 64-69; prof med biochem & head div gastroenterol, Sch Med, Vanderbilt Univ, 69-; PROF MED & PHARMACOL & CHIEF DEP, MED DIV GASTROENTEROL & NUTRIT, UNIV TEX HEALTH SCI CTR. *Concurrent Pos:* Fel gastroenterol, Col Med, Univ Cincinnati, 58-59; res fel med, Thorndike Mem Lab, Harvard Med Sch, 61-63; Markle scholar acad med, 63-68; USPHS res career develop award, 68; mem alcoholism & alcohol probs rev comt, NIMH, 67-71, chmn, 80-81, VA merit review comt, drugs & alcohol, 85-88, chmn, 87-88; ed, Hepatol, 85- *Honors & Awards:* Alcoholism Res Award, 87. *Mem:* Am Asn Study Liver Dis (pres, 80); Am Fedn Clin Res; Am Soc Clin Invest; Am Gastroenterol Asn; Am Acad Neurol. *Res:* Liver disease, especially bilirubin metabolism in maturation; metabolic encephalopathies, especially hepatic coma, drug metabolism in liver disease and thiamine deficiency; placental drugs and nutrient transport. *Mailing Add:* Div Gastroenterol & Nutrit Univ Tex Health Sci Ctr 7703 Floyd Curl Dr San Antonio TX 78284

SCHENKMAN, JOHN BORIS, b New York, NY, Feb 10, 36; m 60. BIOCHEMISTRY, PHARMACOLOGY. *Educ:* Brooklyn Col, BS, 60; State Univ NY, PhD(biochem), 64. *Prof Exp:* Phys biochemist, Johnson Res Found, Sch Med, Univ Pa, 64-66; NSF vis scientist, Osaka Univ, 67-68; from asst prof to assoc prof pharmacol, Sch Med, Yale Univ, 68-78; prof & head, 78-87, PROF PHARMACOL, SCH MED, UNIV CONN, FARMINGTON, 87- *Concurrent Pos:* NIH fel, Sch Med, Univ Pa, 66-67; res assoc, Inst Toxicol, Univ Tubingen, Ger, 68, pharmacol study sect, 74-78; mem Flex Comn, Nat Bd Med Examiners, 82-86, pharmacol rev comt, Nat Inst Gen Med Sci, NIH, 82-86; ed, Int Encycl Pharmacol Therapeut, 74-82; assoc ed, Biochem Pharmacol, 75-84. *Mem:* Am Soc Biol Chem; Am Soc Pharmacol & Exp Therapeut; Brit Biochem Soc; Int Soc Study Xenobiotics. *Res:* Biological oxidations; microsomal mixed function oxidations; hemoprotein oxidases; lipid peroxidation. *Mailing Add:* Dept Pharmacol Univ Conn Sch Med Farmington CT 06032

SCHENNUM, WAYNE EDWARD, b Elgin, Ill, Aug 23, 49. ECOLOGY, POPULATION BIOLOGY. *Educ:* Univ Ill, Chicago Circle, BS, 71, PhD(biol), 75. *Prof Exp:* Teaching asst, Univ Ill, Chicago Circle, 74-75, asst prof biol, 76; asst prof biol, Judson Col, 76-77; asst prof biol, Wheaton Col, 77-78; environ consult, The Nature Conservancy & Ill Natural Land Inst, 78-81, land steward, 81-83; community ecologist, Iowa Conserv Comn, 83-84; NATURAL RESOURCE MGR, MGENRY COUNTY CONSERV DIST, 85- *Concurrent Pos:* Vis asst prof, Judson Col, 76 & 78-79, Concordia Teacher's Col, 76, Wheaton Col, 79, Barat Col, 80-81, Governor's State Univ, 82-83 & Northeastern Ill Univ, 87. *Mem:* Ecol Soc Am; Natural Areas Asn; Soc Ecol Restoration & Mgmt. *Res:* Monitoring of long-term ecological changes in managed and restored natural communities of the Midwest; study of ecological relationships of insects in Midwestern ecosystems; natural areas management, management planning, and inventory. *Mailing Add:* 136 Wagner Dr Cary IL 60013

SCHENTER, ROBERT EARL, b St Louis, Mo, Jan 4, 37; c 4. NUCLEAR PHYSICS. *Educ:* Calif Inst Technol, BS, 58; Univ Colo, Boulder, PhD(physics), 63. *Prof Exp:* Res assoc nuclear physics, Case Inst Technol, 63-65; sr res scientist, Nuclear Physics & Fast Reactor Cross Sections, Battelle Mem Inst Pac Northwest Labs, 65-70; res assoc, 70-76, mgr nuclear anal, 76-80, FEL SCIENTIST, WESTINGHOUSE HANFORD CO, 80- *Concurrent Pos:* Lectr, Joint Ctr Grad Study, 66-67. *Mem:* Am Phys Soc. *Res:* Theoretical calculations of the nucleon-nucleus optical model potential in terms of the nucleon-nucleus interaction; calculation of neutron reaction cross sections for reactor analyses. *Mailing Add:* 2240 Davison Richland WA 99352

SCHENZ, ANNE FILER, b Sharon, Pa, Sept 16, 45; m 68. FLAVOR TECHNOLOGY, SENSORY EVALUATION. *Educ:* Westminster Col, Pa, BS, 67; Kent State Univ, PhD(phys chem), 74. *Prof Exp:* Teacher chem, Springfield Sch Dist, Akron, Ohio, 67-68; vis prof anal chem, King's Col, NY, 75-76; prin develop chemist laundry detergents, Lever Bros Co, NJ, 76-78; proj specialist, Gen Foods Corp, 78-81, group leader, texture group flavors & natural prods, tech appln, Culinova Meals Div, 81-87; SR GROUP LEADER, PROD RES & DEVELOP, DIV ABBOTT LABS, ROSS LABS, 87- *Concurrent Pos:* Vis prof phys chem, King's Col, NY, 79; adj asst prof, Food Sci, Ohio State Univ, 89- *Mem:* Am Chem Soc; Asn Chemoreception Sci; Inst Food Technologists. *Res:* Liquid crystals; surfactant and bleach chemistry; texture of liquid foods; surface rheology; artificial sweeteners and salt substitutes; physical preservation of foods; refrigerated meals (Culinova); flavor technology; sensory evaluation. *Mailing Add:* 485 Retreat Lane W Powell OH 43065-9768

SCHENZ, TIMOTHY WILLIAM, b Washington, DC, Jan 2, 46; m 68. PHYSICAL CHEMISTRY. *Educ:* Westminster Col, Pa, BS, 68; Kent State Univ, PhD(phys chem), 73. *Prof Exp:* Sr chemist, Gen Food Corp, 74-76, proj specialist, 76-79, res specialist phys chem, 79-87; Sr res scientist, 87-90, ASSOC RES FEL, ROSS LABS, 90- *Concurrent Pos:* Adj asst prof, Food Sci Dept, Ohio State Univ. *Mem:* Am Chem Soc; Sigma Xi; NAm Thermal Analysis Soc. *Res:* Physical adsorption from solution; interactions with proteins in disperse systems; instrumentation and automation; foams and emulsions; thermal analysis; image analysis. *Mailing Add:* 485 Retreat Lane W Powell OH 43065

SCHEPARTZ, ABNER IRWIN, b New York, NY, July 29, 22; m 49; c 2. BIOCHEMISTRY. *Educ:* Purdue Univ, BS, 43; Univ Pittsburgh, PhD(chem), 50. *Prof Exp:* Asst chem, Univ Pittsburgh, 43-44; res assoc, Manhattan proj, Univ Rochester, 44-46; Nat Heart Inst fel, Univ Wis, 50-51; res biochemist, Vet Hosp, Pittsburgh, Pa, 51-56; sr res fel, USDA, Pa, 56-60; res assoc in charge biochem & biophys, Merck Inst Therapeut Res, 60-62; res chemist, 62-65, RES LEADER LEAF RES, TOBACCO LAB, USDA, 65- *Concurrent Pos:* Asst chem, Univ Pittsburgh, 46-49, instr, 54-55; consult, Children's Hosp, Pittsburgh, 56. *Mem:* Am Chem Soc; Am Soc Biol Chem; AAAS. *Res:* Uranium toxicology; fat chemistry; isolation of natural products; allergens; electrophoresis of proteins; chemistry of tobacco smoke; biochemistry and biophysics of viruses; electron microscopy; enzymology; tobacco biochemistry. *Mailing Add:* 110 Sandstone Ct Athens GA 30605

SCHEPARTZ, BERNARD, b New York, NY, Nov 9, 18; m 44. HISTORY OF BIOCHEMISTRY. *Educ:* Ohio Wesleyan, BA, 41; Univ Mich, MS, 42; Univ Pa, PhD(biochem), 49. *Prof Exp:* Asst, Comt Med Res War Proj, Dept Surg Res, Univ Pa, 42-44, chemist, Nat Defense Res Comt War Proj, Towne Sci Sch, 44-46; from instr to assoc prof biochem, 48-65, prof, 65-80, EMER PROF BIOCHEM, JEFFERSON MED COL, 80- *Mem:* AAAS; Am Soc Biol Chem; Am Chem Soc; Sigma Xi. *Res:* Intermediary metabolism of amino acids; dimensional analysis; historical and biographical writing on chemistry and biochemistry. *Mailing Add:* 7607 Brous Ave Philadelphia PA 19152-3907

SCHEPARTZ, SAUL ALEXANDER, b Nutley, NJ, Mar 18, 29; m 56; c 3. BIOCHEMISTRY. *Educ:* Ind Univ, AB, 51; Univ Wis, MS, 53, PhD(biochem), 55. *Prof Exp:* Res assoc, Wistar Inst, Univ Pa, 55-57; biochemist, Microbiol Assocs, Inc, 57-58; biochemist, Sect on Screening, Cancer Chemother, Nat Serv Ctr, 58-61, head biochem sect, Drug Eval Br, 61-64, asst chief drug eval br, 64, asst chief cancer chemother, Nat Serv Ctr, 64-66, chief, 66-72, assoc sci dir, Drug Res & Develop Chemother, 72-73, assoc dir drug res & develop, 73-76, actg dep dir, 76-78, actg dir, 80-81, DEP DIR, DIV CANCER TREAT, NAT CANCER INST, 78- *Mem:* AAAS; Am Chem Soc; Am Soc Microbiol; NY Acad Sci; Am Asn Cancer Res. *Res:* Cancer chemotherapy; fermentation biochemistry; tissue culture; mode of action of antibiotics; microbial polysaccharides. *Mailing Add:* Cancer Inst Executive Plaza N Rockville MD 20892

SCHEPLER, KENNETH LEE, b Clinton, Iowa, Apr 1, 49; m 72; c 3. LASER PHYSICS, BIOPHYSICS. *Educ:* Mich State Univ, BS, 71; Univ Mich, MS, 73, PhD(physics), 75. *Prof Exp:* Res biophysicist, Laser Effects Br, Sch Aerospace Med, 75-79, nuclear physicist, McClellan AFB, 79-81, LASER PHYSICIST, WRIGHT LAB, WRIGHT-PATTERSON AFB, USAF, 81- *Mem:* Inst Elec & Electronics Engrs Lasers & Electro-Optical Soc; Am Phys Soc. *Res:* Optical spectroscopy of laser crystals; tunable lasing of solid state materials; excited state absorption of transition metal doped crystals; computer modeling of laser performance and laser damage; non linear frequency conversion; mechanisms of biological interactions with laser radiation, cataractogenesis. *Mailing Add:* WL/ELOS Wright-Patterson AFB OH 45433-6543

SCHEPPERS, GERALD J, b North Bend, Nebr, Apr 11, 33; m 60; c 5. ANALYTICAL CHEMISTRY. *Educ:* Nebr State Teachers Col, Wayne, BA, 62; Iowa State Univ, MS, 65, PhD(anal chem), 67. *Prof Exp:* Asst prof, 66-67, assoc prof, 67-77, PROF CHEM, UNIV WIS-PLATTEVILLE, 78-, CHMN DEPT, 80- *Mem:* Am Chem Soc. *Res:* Fluorescent indicators; complex formation in non-aqueous solvents. *Mailing Add:* Dept Chem Univ Wis Platteville WI 53818

SCHER, ALLEN MYRON, b Boston, Mass, Apr 17, 21; m 52; c 2. PHYSIOLOGY. *Educ:* Yale Univ, BA, 42, PhD, 50. *Prof Exp:* From instr to assoc prof, 50-62, PROF PHYSIOL, UNIV WASH, 62- *Concurrent Pos:* Mem comput study sect, NIH, 63-67 & cardiovasc A study sect, 67-71. *Mem:* AAAS; Am Physiol Soc; Am Heart Asn. *Res:* Cardiovascular control systems; cardiac electrophysiology. *Mailing Add:* Dept Physiol & Biophys SJ-40 Univ Wash Seattle WA 98195

SCHER, CHARLES D, b Newark, NJ, July 25, 39; m 64; c 2. GROWTH CONTROL, REGULATION. *Educ:* Brandeis Univ, BA, 61; Univ Pa, MD, 65. *Prof Exp:* Intern, Bronx Munic Hosp, 65-66, asst resident, 66-67; res assoc, Nat Cancer Inst, 67-71; asst resident, Children's Hosp Med Ctr, 71-72, fel hemat, 72-74; from asst prof to assoc prof, Med Sch, Harvard Univ, 74-82; PROF, CHILDREN'S HOSP PHILADELPHIA, MED SCH, UNIV PA, 82- *Concurrent Pos:* Spec fel, NIH, 72-74; scholar, Leukemia Soc Am, 77-82; mem staff, Sidney Farber Cancer Inst, 77-82. *Mem:* Tissue Cult Asn; AAAS; Am Soc Cell Biol; Am Soc Microbiol; Am Soc Clin Invest. *Res:* Control of cell replication by growth factors. *Mailing Add:* Childrens Hosp 34th St & Civic Ctr Blvd Philadelphia PA 19104

SCHER, HERBERT BENSON, b New York, NY, Dec 11, 37; m 59; c 2. PHYSICAL CHEMISTRY, COLLOID CHEMISTRY. *Educ:* Cornell Univ, BChEng, 60; Univ Minn, MS, 62, PhD(phys chem), 64. *Prof Exp:* Res chemist rheology, Chem Res & Develop Lab, US Army, Edgewood Arsenal, 64-66; sr res chemist, Res Lab, Eastman Kodak Co, 66-68; SR SCIENTIST, CONTROLLED RELEASE PESTICIDES & PESTICIDE DISPERSIONS, STAUFFER CHEM CO, 68- *Concurrent Pos:* Instr, Univ Calif, Berkeley, 71-77; adj assoc prof, dept chem, Univ San Francisco, 78-80, vis lectr, dept pharm, 81- *Mem:* Am Chem Soc; Controlled Release Soc. *Res:* Controlled release pesticides; microencapsulation; interfacial polymerization; coating technology; diffusion of organic molecules through polymers; pesticide formulations; emulsions and dispersions; rheology; organic molecule-clay interactions; kinetics of pesticide degradation; stabilization of pesticides. *Mailing Add:* 1028 Wickham Dr Moraga CA 94556

SCHER, MARYONDA E, b Oakland, Calif, Feb 26, 31; m 52; c 2. PSYCHIATRY. *Educ:* Univ Wash, BS, 50, MD, 54. *Prof Exp:* From clin asst to clin instr, 55-65, clin asst prof to clin assoc prof, 65-76, ASSOC PROF PSYCHIAT, UNIV WASH SCH MED, 76- *Concurrent Pos:* Staff psychiatrist, Vet Admin Hosp, Seattle, 59-80; active mem med staff, Harborview Hosp, Seattle, 65- *Mem:* Fel Am Psychiat Asn. *Res:* Medical education; women. *Mailing Add:* Harborview Med Ctr ZA 99 325 9th Ave Seattle WA 98104

SCHER, ROBERT SANDER, b Cincinnati, Ohio, May 24, 34; m 61; c 3. TECHNICAL MANAGEMENT. *Educ:* Mass Inst Technol, SB, 56, SM, 58, Mech Eng, 60, ScD(mech eng), 63. *Prof Exp:* Engr aerospace, Astro Electronics Div, RCA Corp, 63-65; dept mgr optical encoders, Sequential Info Syst, 65-70; tech dir, 70-77, vpres eng, 77-86, PRES, TELEDYNE GURLEY, 86- *Mem:* Am Soc Mech Engrs; Optical Soc Am. *Res:* Design and development of precision measuring instruments, particularly optical encoders. *Mailing Add:* Two Laurel Oak Lane Clifton Park NY 12065

SCHER, WILLIAM, b Cleveland, Ohio; m. HEMATOLOGY, CELL DIFFERENTIATION. *Educ:* Yale Univ, BS, 55, MS, 57; Univ Va, MD, 61. *Prof Exp:* Asst prof cell biol, 68-78, ASSOC PROF MED & MOLECULAR BIOL, MT SINAI SCH MED, NEW YORK, 86-; ASSOC PROF, MT SINAI GRAD SCH BIOL SCI, CITY UNIV NEW YORK, 86- *Concurrent Pos:* Reviewer, study sect differential agents in human malignancies, NIH, 86. *Mem:* Am Asn Cancer Res; Am Soc Cell Biol; Am Soc Microbiol; Soc Exp Biol & Med; Tissue Cult Asn; Sigma Xi. *Res:* Molecular basis of cell differentiation and its relationship to malignancy; model system for study of synthesis of a differentiation marker, hemoglobin, in an in vitro system: dimethyl sulfoxide-induced mouse erythroleukemia cells. *Mailing Add:* Dept Med Div Oncol Mt Sinai Med Ctr One Gustave L Levy Pl Box 1178 New York NY 10029

SCHERAGA, HAROLD ABRAHAM, b Brooklyn, NY, Oct 18, 21; m 43; c 3. BIOPHYSICAL CHEMISTRY. *Educ:* City Col New York, BS, 41; Duke Univ, AM, 42, PhD(chem), 46. *Hon Degrees:* ScD, Duke Univ, 61, Univ Rochester, 88. *Prof Exp:* Am Chem Soc fel, Harvard Med Sch, 46-47; from instr to prof, 45-58, chmn dept, 60-67, TODD PROF CHEM, CORNELL UNIV, 65- *Concurrent Pos:* Guggenheim fel & Fulbright scholar, Carlsberg Lab, Denmark, 56-57; vis lectr, Div Protein Chem, Wool Res Lab, Commonwealth Sci & Indust Res Orgn, Australia, 59; mem adv panel molecular biol, NSF, 60-62; co-ed, Molecular Biol, 61-86; Welch Found lectr, Univ Tex, 62; co-chmn, Gordon Res Conf Proteins, 63; mem biochem training comt, NIH, 63-65 & career develop comt, 67-71; Guggenheim fel & Fulbright scholar, Weizmann Inst Sci, 63, NIH spec fel, 70, mem bd gov, 70-; mem adv bd, Biopolymers, 63- & Biochem, 69-74, 85-; Harvey lectr, 68; Gallagher lectr, 68-69; mem-at-large coun, Gordon Res Conf, 69-71, Lemieux lectr, 73; mem tech adv panel, Xerox Corp, 69-71 & 74-79; Hill lectr, 76; vis prof, Japan Soc Promotion Sci, 77; distinguished invited lectr, Univ Calgary, 79; Fogarty scholar, NIH, 84, 86. *Honors & Awards:* Lilly Award, Am Chem Soc, 57, Nichols Medal, 74, Kendall Award, 78, & Pauling Medal, 85, Mobil Award, 90; Linderstrom-Lang Medal, Carlsberg Lab, 83; Kowalski Medal, Int Soc Thrombosis & Hemostasis, 83; Repligen Award, Chem Biol Processes, 90. *Mem:* Nat Acad Sci; AAAS; Am Acad Arts & Sci; Am Soc Biol Chem; hon mem NY Acad Sci. *Res:* Physical chemistry of proteins and other macromolecules; structure of water and dilute aqueous solutions; blood clotting. *Mailing Add:* Baker Lab Chem Cornell Univ Ithaca NY 14853-1301

SCHERB, FRANK, b Union City, NJ, Sept 17, 30; m 64; c 4. SPACE PHYSICS. *Educ:* Mass Inst Technol, SB, 53, PhD(physics), 58. *Prof Exp:* Res staff assoc space physics, Mass Inst Technol, 58-61, from asst prof to assoc prof physics, 61-65; assoc prof, 65-69, PROF PHYSICS, UNIV WIS-MADISON, 69- *Mem:* Fel Am Phys Soc; Am Geophys Union; Am Astron Soc. *Res:* Cosmic rays; physics of interplanetary medium, especially the solar wind. *Mailing Add:* Dept Physics 6203 Chamberlin Hall Univ Wis Sterling Hall Madison WI 53706

SCHERBA, GERALD MARRON, b Chicago, Ill, Feb 9, 27; m 51; c 3. ZOOLOGY. *Educ:* Univ Chicago, BS, 50, MS, 52, PhD(zool), 55. *Prof Exp:* From instr to assoc prof biol, Chico State Col, 55-62; prof & chmn natural sci div, 62-66, dean acad affairs, 66-68, vpres, 68-84, DIR, DESERT STUDIES CTR, CALIF STATE UNIV, SAN BERNARDINO, 85- *Concurrent Pos:* Res grants, NY Zool Soc, 55-56 & 59, Am Acad Arts & Sci, 57 & NSF, 62-64. *Res:* Animal ecology; animal behavior; biology of ants. *Mailing Add:* Dept Biol Calif State Col 5500 State College Pkwy San Bernardino CA 92407

SCHERBENSKE, M JAMES, b Jamestown, NDak, Jan 13, 37; m 59; c 4. PHYSIOLOGY, BIOCHEMISTRY. *Educ:* Jamestown Col, BS, 59; Univ SDak, MA, 64, PhD(physiol), 66. *Prof Exp:* Fel renal physiol, Med Ctr, Kans Univ, 66-68; health scientist adminr, Nat Heart Inst, 68-69, HEALTH SCIENTIST ADMINR KIDNEY & UROL, NAT INST DIABETES & DIGESTIVE & KIDNEY DIS, NIH, 69- *Concurrent Pos:* Consult, Coordr Coun Urol, Am Urol Asn, 71-79. *Mem:* Am Soc Nephrology; Soc Univ Urologists. *Res:* Renal physiology and transport. *Mailing Add:* Kidney Dis & Urol Prog & Digestive Dis NIH Bethesda MD 20205

SCHERBERG, NEAL HARVEY, b Minneapolis, Minn, Nov 10, 39; c 2. MOLECULAR BIOLOGY, BIOCHEMISTRY. *Educ:* Oberlin Col, AB, 61; Tufts Univ, PhD(biochem), 66. *Prof Exp:* From asst prof to assoc prof med, 71-80, res assoc, 76-80, TECH DIR, THYROID FUNCTION LAB, UNIV CHICAGO, 80- *Concurrent Pos:* Fel molecular biol, Univ Chicago, 66-71, Am Cancer Soc fel, 66-68; Sr Int Fogarty fel, 86. *Mem:* Am Soc Biochem; Int Isotope Soc. *Res:* Protein synthesis; detection of mutations in DNA. *Mailing Add:* Univ Chicago Thyroid Study Unit 950 E 59th St Chicago IL 60637

SCHERER, GEORGE ALLEN, b Kokomo, Ind, Apr 3, 07; m 29; c 3. CHEMISTRY. *Educ:* Earlham Col, BS, 27; Cornell Univ, MS, 28; Purdue Univ, PhD(chem), 33. *Prof Exp:* Asst chem, Cornell Univ, 27-28 & Purdue Univ, 28-33; prof, Pac Col, 33-34 & McKendree Col, 34-36; from instr to prof,

Earlham Col, 36-57; admin secy, Am Friends Bd Missions, 57-60; prof chem & head dept, Western Col, 60-72; adj prof chem, Ind Univ East, 72-77; chem technician, Earlham Col, 72-86; RETIRED. *Concurrent Pos:* Vis prof, Univ Col Women, Hyderabad, India, 65-66. *Mem:* Am Chem Soc. *Res:* Electrode potentials; free energy measurements. *Mailing Add:* 2030 Chester Blvd No 216 Richmond IN 47374-1215

SCHERER, GEORGE WALTER, b Teaneck, NJ, Apr 27, 49; m 71. MATERIALS SCIENCE, CERAMICS. *Educ:* Mass Inst Technol, SB, 72, SM, 72, PhD(mat sci), 74. *Prof Exp:* Sr ceramist, Corning Glass Works, 74-85. *Honors & Awards:* George W Morey Award, 85 & Ross Coffin Purdy Award, Am Ceramic Soc; Woldemar Weyl Award, Int Cong on Glass, 86; W H Zachariasen Award, J Non-Crystal Solids, 87; Fulrath Pac Award, 90. *Mem:* Am Ceramic Soc; NY Acad Sci; Mat Res Soc. *Res:* Kinetics of crystallization and glass formation; viscous sintering; thermal stress analysis; optical waveguide fabrication; sol-gel processing. *Mailing Add:* E I du Pont de Nemours & Co Cent Res & Dev Dept Expt Sta 356/384 Wilmington DE 19880-0356

SCHERER, HAROLD NICHOLAS, JR, b Plainfield, NJ, Apr 5, 29; m 52, 74; c 7. TRANSMISSION LINE ENGINEERING, SUBSTATION ENGINEERING. *Educ:* Yale Univ, BE, 51; Rutgers Univ, MBA, 55. *Prof Exp:* Var engr positions, Pub Serv Elec & Gas Co. 51-63; rect mgr, Am Elec Power Serv Corp, 65-68, asst chief, 68-69, chief, 69-73, vpres, 73-83, sr vpres, Elec Eng, 83-90; PRES, COMMONWEALTH ELEC CO, 90- *Concurrent Pos:* Dir & vchmn, Am Nat Standards Inst, 80-87; mem US-USSR Working Group High Voltage Power Transmission, 77-, US-Italy Working Group Ultra High Voltage Power Transmission, 78-89, engr review bd, Bonneville Power Admin, 84-; mem, Int Admin Coun & tech comts, Conf Int Des Grands Reseaux Elec a Haute Tension; pres, Inst Elec & Electronics Engrs Power Eng Soc, 90- *Honors & Awards:* William Habirshaw Award & Medal, Inst Elec & Electronics Engrs, 86. *Mem:* Nat Acad Eng; fel Inst Elec & Electronics Engrs (vpres). *Res:* Ultra high voltage power transmission. *Mailing Add:* Commonwealth Elec Co 2421 Cranberry Hwy Wareham MA 02571

SCHERER, JAMES R, b Kansas City, Mo, Dec 31, 31; div; c 6. PHYSICAL CHEMISTRY, VIBRATIONAL SPECTROSCOPY. *Educ:* St Mary's Col, Calif, BS, 53; Univ Minn, PhD(phys chem), 58. *Prof Exp:* Res chemist, Chem Physics Res Lab, Dow Chem Co, Mich, 58-63; res chemist, Western Regional Res Lab, Sci & Educ Admin-Agr Res, USDA, 63-87, UNIV CALIF, SAN FRANCISCO, 87- *Mem:* AAAS; Am Chem Soc; Am Phys Soc; Coblentz Soc (pres, 71-72); Am Optical Soc; Sigma Xi. *Res:* Molecular infrared and Raman spectroscopy; biophysics vibrational assignments; force constant calculations and application of normal coordinate calculations to group frequencies; laboratory data acquisition with digital computers. *Mailing Add:* 1309 Arch St Berkeley CA 94708

SCHERER, KIRBY VAUGHN, JR, b Evansville, Ind, Feb 7, 36; m 61; c 3. ORGANIC CHEMISTRY. *Educ:* Harvard Univ, AB, 58, AM, 59, PhD(chem), 63. *Prof Exp:* Asst prof chem, Univ Calif, Berkeley, 62-67; ASSOC PROF CHEM, UNIV SOUTHERN CALIF, 67- *Concurrent Pos:* Sr scientist, Jet Propulsion Lab, 75- *Mem:* AAAS; Am Chem Soc; Royal Soc Chem. *Res:* Organofluorine chemistry; synthesis and properties of strained ring systems; chlorocarbon derivatives; organic chemistry of nitrogen. *Mailing Add:* 482 Snuff Mill Lane Hockessin DE 19707-9643

SCHERER, PETER WILLIAM, b Palmerton, Pa, May 15, 42; m 72; c 1. RESPIRATORY PHYSIOLOGY, BIOFLUID MECHANICS. *Educ:* Haverford Col, BS, 64; Yale Univ, PhD(eng & appl sci), 71, MD, 73. *Prof Exp:* From asst prof to assoc prof bioeng, 76-90, PROF BIOENG, UNIV PA, 90- *Concurrent Pos:* Prof, Dept Anesthesia, Univ Pa Med Sch, 89- *Mem:* Sr mem Biomed Eng Soc; Am Physiol Soc. *Res:* Respiratory fluid mechanics; mass; heat transfer; gas exchange; aerosol transport; heating and humidification of air in the lung; interaction of the respiratory system with the environment. *Mailing Add:* Dept Bioeng Univ Pa 220 S 33rd St Philadelphia PA 19104-6392

SCHERER, ROBERT C, b Jersey Shore, Pa, Apr 26, 31; m 54; c 3. ANIMAL ECOLOGY. *Educ:* Haverford Col, BS, 53; Pa State Univ, MS, 63, PhD(zool), 65. *Prof Exp:* Assoc prof, Lock Haven State Col, 65-71, prof zool, 71-; RETIRED. *Mem:* Am Fisheries Soc; Ecol Soc Am. *Res:* Population dynamics as applied to fish populations. *Mailing Add:* Rd No 4 Box 201 Jersey Shore NJ 17740

SCHERER, RONALD CALLAWAY, b Akron, Ohio, Sept 11, 45; m 71; c 2. SPEECH & VOICE SCIENCE. *Educ:* Kent State Univ, BS, 68; Ind Univ, MA, 72; Univ Iowa, PhD(speech sci), 81. *Prof Exp:* Res scientist, Denver Ctr Performing Arts, 83-88; adj asst prof & consult, 83-88, ADJ ASSOC PROF & CONSULT, UNIV IOWA, 88-; ASST PROF ADJOINT SPEECH SCI, UNIV COLO, BOULDER, 84-, ASST CLIN PROF OTOLARYNGOL, SCH MED, 88-; SR SCIENTIST, DENVER CTR PERFORMING ARTS, 88- *Concurrent Pos:* Prin investr, grants from NIH, Voice Found, Duke Univ, 80-; rev consult, 12 jour & orgn, 82-; consult, 8 univ & book publ, 83-; lectr, Prof Voice: Use & Abuse, 84-; adj prof speech sci, Univ Denver, 84-86; auditor, Int Soc Phonetic Sci, 88-; exec & legis bd, Nat Ctr Voice & Speech, 90-; lectr voice & speech sci, Nat Theatre Conserv, 90- *Mem:* Am Speech Lang Hearing Asn; fel Int Soc Phonetic Sci; Int Arts Med Asn. *Res:* Acoustics, aerodynamics and biomechanics of the larynx and speech production in general; author of numerous scientific publications. *Mailing Add:* 1245 Champa St Denver CO 80204

SCHERFIG, JAN W, b Copenhagen, Denmark, Apr 24, 36. ENVIRONMENTAL & CHEMICAL ENGINEERING. *Educ:* Danish Tech Univ, MS, 59; Univ Calif, Berkeley, PhD(sanit eng), 68. *Prof Exp:* Res engr, Danish Defense Res Bd, 60-61; prod engr, Danish Mineral Oil Refinery, 61-63; teacher chem, Technol Inst, Copenhagen, 62-63; res engr, Eng Sci, Inc, Calif, 63-66; res asst, Univ Calif, Berkeley, 66-67; from asst prof to assoc prof, 67-77, chmn environ & resources eng, 70-77, PROF CIVIL & ENVIRON ENG, UNIV CALIF, IRVINE, 77- *Concurrent Pos:* Consult, City of Calexico, Calif, Irvine Ranch Water Dist, Lowry Eng-Sci, Santa Ana, Encibra, Rio de Janeiro & Lowry & Assocs, Santa Ana, 67, City of Laguna Beach, 71. *Mem:* Am Soc Civil Engrs; Am Inst Chem Engrs. *Res:* Eutrophication; marine waste disposal, planning and optimization of water and waste. *Mailing Add:* Dept Civil & Environ Eng Univ Calif Irvine CA 92717

SCHERGER, DALE ALBERT, b Toledo, Ohio, Aug 22, 49. ENVIRONMENTAL ENGINEERING, HYDROLOGY & WATER RESOURCES. *Educ:* Univ Mich, BSE, 71, MSE, 72. *Prof Exp:* Engr, 71-73, sr engr, 73-75, chief engr, 75-77, dir eng, 77-82, VPRES, ENVIRON CONTROL TECHNOL CORP, 82- *Mem:* Water Pollution Control Fedn; Am Water Resources Asn; Nat Prof Eng Soc. *Res:* Advanced waste treatment technology for industrial and municipal wastes; methods development and implementation for control, cleanup and disposal of hazardous substances; development of techniques for controlling urban non-point source runoff. *Mailing Add:* 2939 Briarcliff Ann Arbor MI 48105

SCHERK, PETER, b Berlin, Ger, Sept 2, 10; Can citizen; m 46; c 3. GEOMETRY. *Educ:* Univ Göttingen, PhD(geom), 35. *Prof Exp:* From instr to prof math, Univ Sask, 43-59; prof, 59-80, EMER PROF MATH, UNIV TORONTO, 80- *Concurrent Pos:* Ed, Can Math Cong Newsletter, 54-57; ed-in-chief, Can Math Bull, 58-61, managing ed, 61-62; ed-in-chief, Can J Math, 62-67. *Mem:* Fel Royal Soc Can; Am Math Soc; Can Math Cong. *Res:* Projects in the geometry of orders and in the foundations of geometry. *Mailing Add:* Dept Math Univ Toronto Toronto ON M5S 1A1 Can

SCHERLAG, BENJAMIN J, b Brooklyn, NY, Oct 31, 32; m 60; c 4. CARDIOVASCULAR PHYSIOLOGY. *Educ:* City Col New York, BS, 54; Brooklyn Col, MA, 61; State Univ NY, PhD(physiol), 63. *Prof Exp:* Asst physiol, State Univ NY Downstate Med Ctr, 56-63; res physiologist, Cardiopulmonary Lab, USPHS Hosp, NY, 65-68; res physiologist, Sect Cardiovasc Dis, Mt Sinai Hosp Greater Miami, 68-74; PROF MED & ADJ PROF PHYSIOL, UNIV OKLA HEALTH SCI CTR, 78-, CARDIOVASC PHYSIOLOGIST, VET ADMIN HOSP, OKLAHOMA CITY, 78- *Concurrent Pos:* NIH fel pharmacol, Col Physicians & Surgeons, Columbia Univ, 63-65; lectr, Brooklyn Col, 60-67; res assoc, Columbia Univ, 65-67; prof med, Med Sch, Univ Miami, 74-78; res physiologist, Vet Admin Hosp, Miami, 74-78; med investr, Vet Admin Med Ctr, 80-86; res career scientist, VA Med Ctr, 85. *Honors & Awards:* Pioneers in Cardiac Pacing & Electrophysiol, NAm Soc Pacing & Electrophysiol, 89. *Mem:* Am Physiol Soc; Am Fedn Clin Res; fel Am Col Cardiol; Am Heart Asn; NY Acad Sci; NAm Soc Pacing & Electrophysiol. *Res:* Cardiac electrophysiology; pharmacology; our major research interest is disordered rhythms of the heart; abnormal impulse formation and conduction due to cardiac ischemia and infarction. *Mailing Add:* 151F Med Ctr Vet Admin 921 NE 13th St Oklahoma City OK 73104

SCHERMER, EUGENE DEWAYNE, b Spokane, Wash, June 21, 34; m 58; c 2. ENVIRONMENTAL CHEMISTRY. *Educ:* Eastern Wash State Col, BA, 58; Ore State Univ, MS, 62; La State Univ, PhD(chem), 71. *Prof Exp:* Teacher high schs, Wash, 58-61; instr chem, 62-84, DEAN INSTR, GRAYS HARBOR COL, 84- *Concurrent Pos:* Investr, Wash State Dept Ecol, 74-76, co-investr with US CEngrs, 75, 79-80. *Mem:* Am Chem Soc. *Res:* The effects of woodwaste leachate on quality of ground and surface waters; effects of dredging on the Grays Harbor Estuary; water quality effects of ocean disposal of dredge spoils. *Mailing Add:* Grays Harbor Col Aberdeen WA 98520

SCHERMER, ROBERT IRA, b Brooklyn, NY, Sept 10, 34; m 58; c 4. NUCLEAR PHYSICS, CRYOGENICS. *Educ:* Cornell Univ, BEngPhys, 56; Mass Inst Technol, PhD(nuclear eng), 61. *Prof Exp:* Res assoc nuclear cryogenics, Brookhaven Nat Lab, 60-62, assoc physicist, 62-65, physicist, 65-70; chmn dept physics, Springfield Tech Community Col, 70-74; mem staff, Los Alamos Nat Lab, 74-80, asst group leader, 80-87, GROUP LEADER, LAWRENCE BERKELEY LAB, 89-, CHIEF SCIENTIST, MAGNET DIV, SSC, 90- *Mem:* Am Phys Soc. *Res:* Low temperature physics; magnetic measurements; design of superconducting magnets. *Mailing Add:* 5871 Harbord Dr Oakland CA 94611

SCHERMERHORN, JOHN W, b NJ, Sept 1, 20; m 45; c 4. PHARMACY, BIONUCLEONICS. *Educ:* Rutgers Univ, BS, 42; Univ Minn, PhD(pharmaceut chem), 49. *Prof Exp:* Assoc prof pharmaceut chem, George Washington Univ, 49-53; prof pharm & chmn dept, Mass Col Pharm, 53-66; prof, Col Pharm, Northeastern Univ, 66-71, dean div health sci, 69-71; PROF HEALTH CARE SCI & DEAN SCH ALLIED HEALTH SCI, HEALTH SCI CTR, UNIV TEX, 71-, ACTG CHMN, DEPT HEALTH CARE SCI, 74- *Concurrent Pos:* Consult, 53- *Mem:* AAAS; Am Pharmaceut Asn; Sigma Xi. *Res:* Pharmaceutical product development. *Mailing Add:* 3788 Townsend Dr Dallas TX 75229-3922

SCHERPEREEL, DONALD E, b South Bend, Ind, Dec 21, 37; m 60; c 3. MATERIALS SCIENCE, ENGINEERING MANAGEMENT. *Educ:* Univ Notre Dame, BS, 59, MS, 61, PhD(metall, mat sci), 64. *Prof Exp:* Instr metall, Univ Notre Dame, 60-62; asst prof metall & mat sci, Mich State Univ, 64-69; sr res mat scientist, Whirlpool Corp, 69-76, dir mech syst res, Res & Eng Ctr, 76-85, dir eng serv, 85-86, dir prod line eng, 86-87, DIR PROD ENG, WHIRLPOOL CORP, LA VERGNE DIV, 87- *Mem:* Am Soc Metals; Am Soc Mech Engrs; Sigma Xi. *Res:* X-ray diffraction; electron microscopy; research management; product simulation; automated design; structural analysis. *Mailing Add:* 1404 Arrowhead Dr Brentwood TN 37027-7478

SCHERR, ALLAN L, b Baltimore, Md, Nov 18, 40; m 80; c 2. APPLICATION PROGRAMMING STRUCTURES. *Educ:* Mass Inst Technol, BS & MS, 62, PhD (elec eng), 65. *Prof Exp:* Res asst proj MAC, Mass Inst Technol, 63-65; staff engr, Systs Archit Syst Develop Div, IBM, Poughkeepsie, NY, 65-66, mgr TSO design & performance, Syst Develop Div, 67-70, mgr MVS prog, 71-74, mgr advan systs prog design, 75-76, mgr distrib systs prog, Syst Develop Div, Kingston, 77-79, mem corp tech comt,

Armonk, NY, 80, dir commun prog, Systs Commun Div, Kingston, 80-81, dir commun systs, corp staff, Valhalla, NY, 82-83, dir advan systs, Systs Prods Div, Rochester, Minn, 84-85, dir integrated applns, Applns Systs Div, Milford, Conn, 86-88, vpres develop & integration, 88-89, DIR ARCHIT & DEVELOP, APPLN SOLUTIONS LINE BUS, IBM CORP, MILFORD, CONN, 90- *Concurrent Pos:* IBM fel, 84. *Honors & Awards:* Grace Murray Hopper Award, Asn Comput Mach, 75. *Mem:* Fel Inst Elec & Electronics Engrs. *Res:* Application software design, systems application architecture (SAA); distributed processing structures and the software development process; special work on managing technical projects so as to deliberately produce extraordinary, unprecedented results. *Mailing Add:* 53 Treadwell Lane Weston CT 06883

SCHERR, CHARLES W, b Philadelphia, Pa, Mar 19, 26; m 52, 70; c 2. PHYSICS. *Educ:* Univ Pa, BS, 49; Univ Chicago, MS, 51, PhD(chem physics), 54. *Prof Exp:* Res assoc physics, Univ Chicago, 54-56; from asst prof to assoc prof, 56-66, PROF PHYSICS, UNIV TEX, AUSTIN, 66- *Mem:* Am Phys Soc. *Res:* Quantum mechanical investigation of atomic and molecular structure. *Mailing Add:* Dept Physics Univ Tex Austin TX 78712

SCHERR, DAVID DELANO, b Columbia, Mo, Oct 15, 34; m 58, 77; c 3. ORTHOPEDIC SURGERY, MICROBIOLOGY. *Educ:* Univ Mo-Columbia, BA, 56, MD, 59; Univ Iowa, MS, 63, PhD(microbiol), 66; Am Bd Orthop Surg, dipl, 69. *Prof Exp:* Staff orthop surg, David Grant Med Ctr, Travis AFB, Calif, 67-69; from asst prof to assoc prof orthop surg & microbiol, Sch Med, Univ Mo, Columbia, 69-75; PVT PRACT, ORTHOP, 75- *Mem:* Am Soc Microbiol; Am Acad Orthop Surg; Orthop Res Soc; Asn Acad Surg; AMA; Am Rheumatism Asn; Int Col Surgeons. *Res:* Activity of antibiotics in clinical uses; role of autoimmune mechanisms in rheumatic diseases. *Mailing Add:* 1111 Madison St Jefferson City MO 65101

SCHERR, GEORGE HARRY, b New York, NY, Dec 30, 20; m 44; c 3. MICROBIOLOGY. *Educ:* Queens Col, NY, BS, 41; Univ Ky, MS, 49, PhD(bact), 51. *Prof Exp:* Bacteriologist, City Dept Health, New York, 41-42; chemist, Calco Chem Div, Am Cyanamid Co, 43-48; asst prof microbiol, Sch Med, Creighton Univ, 51-54; asst prof bact, Col Med, Univ Ill, 54-59; vpres & dir res, Consol Labs, Inc, Ill, 59-69; dir, Colab Labs, Inc, Ill, 69-71; pres, Mat & Technol Systs, Inc, 71-72; PRES, TECHNAM, INC, 72- *Concurrent Pos:* Community prof environ sci, Governors State Univ, 73- *Mem:* AAAS; Am Soc Microbiol; Soc Indust Microbiol; Genetics Soc Am; Mycol Soc Am. *Res:* Immunology and infectious disease; effect of carcinogens on microorganisms; effect of hormones on infectious diseases; microbial genetics. *Mailing Add:* 50 Monee Rd Park Forest IL 60466

SCHERR, HARVEY MURRAY, electronics engineering, mechanical engineering, for more information see previous edition

SCHERR, LAWRENCE, b New York, NY, Nov 6, 28; m 54; c 2. INTERNAL MEDICINE, CARDIOLOGY. *Educ:* Cornell Univ, AB, 50, MD, 57. *Prof Exp:* From asst prof to prof, 58-85, DAVID V GREENE DISTINGUISHED PROF MED, CORNELL UNIV, MED COL, 85-, ASSOC DEAN, 70- *Concurrent Pos:* NY Heart Asn fel, Med Col, Cornell Univ, 59-60; Am Heart Asn teaching scholar, 66-67; from intern to chief resident, Cornell Med Div, Bellevue Hosp & Mem Ctr, 57-61, co-dir cardiorenal lab & asst vis physician, 61-63, assoc vis physician, 63-65, vis physician, 66-67, dir cardiol & renal unit, 63-67; physician to outpatients, NY Hosp, 61-63, from asst attend to attend, 63-; attend, Manhattan Vet Admin Hosp, 64-69; asst attend, Mem Hosp, 66-69, consult, 69-; career scientist, Health Res Coun New York, 62-66; fel coun clin cardiol, Am Heart Asn; chmn, NY State Bd Med, 73-75; dir Dept Med, N Shore Univ Hosp, 67-; chmn, Res Reveiw Comt, Int Med, 80-82; chmn bd regents, Am Col Physicians, 85-86. *Mem:* Master Am Col Physicians (pres, 87-88); Am Fedn Clin Res; AMA; Am Bd Internal Med (secy-treas, 79-86). *Res:* Internal medicine, including cardiovascular and renal disease and fluid and electrolyte problems; medical education. *Mailing Add:* North Shore Univ Hosp 300 Community Dr Manhasset NY 11030

SCHERRER, JOSEPH HENRY, b Chicago, Ill, Sept 5, 31; m 60; c 3. POLYMER CHEMISTRY. *Educ:* DePaul Univ, BS, 53; Univ Kans, PhD(org chem), 57. *Prof Exp:* Res chemist, Spencer Chem Co, 57-64; RES CHEMIST, COOK PAINT & VARNISH CO, 64- *Mem:* Am Chem Soc; Royal Soc Chem. *Res:* Synthesis of organic nitrogen compounds. *Mailing Add:* 5726 Floyd Shawnee Mission KS 66202

SCHERRER, RENE, b Boulogne, France, June 15, 32. MICROBIOLOGY. *Educ:* Univ Lausanne, dipl med, 58; Univ Basel, Dr Med, 61. *Prof Exp:* Third asst, Inst Microbiol, Univ Basel, 59-60, second asst, 60-63; res assoc microbiol, Univ Mich, 63-65; from instr to asst prof, Mich State Univ, 65-72; CONTRIB SCIENTIST, WESTERN REGIONAL RES CTR, AGR RES SERV, USDA, BERKELEY, 74-84 & 87- *Concurrent Pos:* Supvr clin diag lab, Inst Microbiol, Univ Basel, 59-63; vis scientist, Univ Calif, Irvine, 75-76. *Mem:* AAAS; Am Soc Microbiol; Am Soc Cell Biol; Electron Micros Soc Am; Brit Soc Gen Microbiol. *Res:* Molecular and cell biology and biophysics of bacteria; chromosome structure; DNA replication; cell division and cell cycle; morphogenes; cell wall structure; endospore formation, dormancy and heat resistance; water properties of bacteria. *Mailing Add:* Western Regional Res Ctr USDA Berkeley CA 94710

SCHERRER, ROBERT ALLAN, b Sacramento, Calif, Nov 21, 32; m 54; c 4. ORGANIC CHEMISTRY. *Educ:* Univ Calif, BS, 54; Univ Ill, PhD(chem), 58. *Prof Exp:* From assoc res chemist to res chemist, Parke Davis & Co, 58-66; sr med chemist, Minn Mining & Mfg Co, 66-69; res specialist, 69-72, sr res specialist, Riker Labs Div, 72-86, DIV SCIENTIST, 3M PHARMACEUT DIV, 3M CO, 86- *Concurrent Pos:* Sr ed, J Med Chem, 86 & 87. *Mem:* Am Chem Soc. *Res:* Synthetic medicinal chemistry; antiarthritic agents; regression analysis; antiasthmatic agents; antioxidants. *Mailing Add:* 3M Pharmaceut 3M Co 3M Ctr 270-2S06 St Paul MN 55144-1000

SCHERTZ, CLETUS E, b El Paso, Ill, Apr 12, 30; m 58; c 5. AGRICULTURAL ENGINEERING. *Educ:* Univ Ill, Urbana, BS(agr sci) & BS(agr eng), 54; Iowa State Univ, PhD(agr eng & theoret & appl mech), 62. *Prof Exp:* Asst prof agr eng, Univ Calif, Davis, 62-67; assoc prof, 67-71, PROF AGR ENG, UNIV MINN, ST PAUL, 71- *Mem:* Am Soc Agr Engrs. *Res:* Machines for harvest of food and fiber crops. *Mailing Add:* 213 Agr Eng Univ Minn St Paul MN 55108

SCHERTZ, KEITH FRANCIS, b El Paso, Ill, Feb 25, 27; m 54; c 6. CYTOGENETICS. *Educ:* Univ Ill, BS, 49, MS, 50; Cornell Univ, PhD(plant breeding), 57. *Prof Exp:* Geneticist, Fed Exp Sta, Agr Res Serv, PR, 57-59, GENETICIST, DEPT SOIL & CROP SCI, TEX A&M UNIV, USDA, 60- *Mem:* Fel Am Soc Agron; fel Crop Sci Soc. *Res:* Genetics and cytogenetics of sorghum, apomixis, reproductive behavior and sterility systems. *Mailing Add:* Dept Soil & Crop Sci Tex A&M Univ College Station TX 77843

SCHERVISH, MARK JOHN, b Detroit, Mich, Oct 10, 53; m 79. FOUNDATIONS OF INFERENCE, STATISTICAL COMPUTING. *Educ:* Mich State Univ, BS, 74; Univ Mich, MS, 75; Univ Ill, PhD(statist), 79. *Prof Exp:* Asst prof, 79-84, ASSOC PROF STATIST, CARNEGIE-MELLON UNIV, 84- *Concurrent Pos:* Vis asst statistician, Statist Lab, Univ Calif, Berkeley, 79; researcher, OEIV, US Dept Energy, Washington, DC, 80; hon res fel, dept statist sci, Univ Col London, 85. *Mem:* Am Statist Asn; Royal Statist Soc; Inst Math Statist; AAAS; Asn Comput Mach; Inst Mgt Sci. *Res:* Comparison and evaluation of probability forecasters and the combination of expert opinions; programs for evaluating multivariate probabilities and theoretical work in discriminant analysis. *Mailing Add:* Dept Statist Carnegie Mellon Univ Pittsburgh PA 15213

SCHERY, STEPHEN DALE, b Rio de Janeiro, Brazil, July 1, 45; US citizen; m 86. NUCLEAR PHYSICS, NATURAL RADIOACTIVITY. *Educ:* Ohio State Univ, BS, 67; Univ Ark, MS, 70; Univ Colo, PhD(physics), 73. *Prof Exp:* Asst prof physics, Kenyon Col, 73-74, marine sci, Tex A&M Univ, Galveston, 74-79; from asst prof to assoc prof, 79-90, RES PHYSICIST, NMEX INST MINING & TECHNOL, 79-, PROF PHYSICS, 90- *Concurrent Pos:* Vis prof & consult, Cyclotron Lab, Mich State Univ, 75-78; vis sr scientist, Australian AEC, 86; vis scientist, Dept Geol & Geophys, Yale Univ, 87; vis scholar, Australian Nuclear Sci & Tech Orgn, 90. *Mem:* Am Phys Soc; Am Geophys Union; Health Physics Soc; Sigma Xi. *Res:* Experimental nuclear physics; natural radioactivity in Earth and atmospheric science applications; transport of radon and thoron. *Mailing Add:* Dept Physics NMex Inst Mining & Technol Socorro NM 87801

SCHERZ, JAMES PHILLIP, b Rice Lake, Wis, May 12, 37; m 62; c 1. CIVIL ENGINEERING. *Educ:* Univ Wis, BS, 59, MS, 61, PhD(civil eng), 67. *Prof Exp:* Instr civil eng, 65-66, res asst, 66-67, from asst prof to assoc prof, 67-77, PROF CIVIL & ENVIRON ENG, INST ENVIRON STUDIES, UNIV WIS-MADISON, 77- *Concurrent Pos:* Aerial monitoring systs consult, 71- *Res:* Remote sensing to include water quality analysis, especially with special photography; surveying of prehistoric calendon sites. *Mailing Add:* Dept Civil Eng 2205 Eng Bldg Univ Wis 1415 Johnson Dr Madison WI 53706

SCHETKY, L(AURENCE) M(CDONALD), b Baguio, Philippines, July 15, 22; m 68; c 1. METALLURGY, CHEMICAL ENGINEERING. *Educ:* Rensselaer Polytech Inst, BChE, 43, MMetE, 48, PhD(metall), 53. *Prof Exp:* Instr metall, Rensselaer Polytech Inst, 46-53; mem res staff, Mass Inst Technol, 53-56, dir mat res, Instrumentation Lab, 56-59; vpres & tech dir, Alloyd Electronics Corp, Mass, 59-63; tech dir metall, Int Copper Res Asn, Inc, 63-83; CHIEF SCIENTIST, MEMORY METALS, INC, STAMFORD, CONN, 83- *Concurrent Pos:* Lectr, Rensselaer Polytech Inst & Mass Inst Technol; US rep, Int Metall Cong, 53; World Exchange Lectr, Am Foundrymen's Soc, 67. *Mem:* Fel Am Soc Metals; Am Inst Mining, Metall & Petrol Engrs; fel Inst Metals UK. *Res:* Physical metallurgy; materials problems in instrumentation; metrology; vapor phase theory; welding and joining; copper research technology; shape memory alloy technology. *Mailing Add:* Memry Technol 83 Keeler Ave Norwalk CT 06854

SCHETTLER, PAUL DAVIS, JR, b Salt Lake City, Utah, Mar 31, 37; m 66; c 2. PHYSICAL CHEMISTRY. *Educ:* Univ Utah, BS, 58; Yale Univ, PhD(phys chem), 64. *Prof Exp:* Fel, Univ Utah, 63-66; teaching intern chem, Antioch Col, 66-67; assoc prof, 67-78, chmn dept, 75-85, PROF CHEM, JUNIATA COL, 76- *Concurrent Pos:* Consult, Columbia Gas Corp, 81-82. *Mem:* Am Chem Soc. *Res:* Natural gas production from microporous rocks; measurements of isotherms and degassing rates of microporous solids and calculations of their implications for production from natural gas; Devonian shale. *Mailing Add:* Dept Chem Juniata Col Huntingdon PA 16652

SCHETZ, JOSEPH A, b Orange, NJ, Oct 19, 36; m 59; c 4. AEROSPACE & OCEAN ENGINEERING. *Educ:* Webb Inst Naval Archit, BS, 58; Princeton Univ, MSE, 60, MA, 61, PhD(mech eng), 62. *Prof Exp:* Sr scientist, Gen Appl Sci Lab, NY, 61-64; assoc prof aerospace eng, Univ Md, Col Park, 64-69; W MARTIN JOHNSON PROF AEROSPACE & OCEAN ENG & CHMN DEPT, VA POLYTECH INST & STATE UNIV, 69- *Concurrent Pos:* Consult, Appl Physics Lab, Johns Hopkins Univ, 64- *Mem:* Fel Am Inst Aeronaut & Astronaut; fel Am Soc Mech Engrs; Soc Naval Architects & Marine Engrs. *Res:* Fluid dynamics; ocean engineering; combustion; wind energy. *Mailing Add:* Dept Aerospace & Ocean Eng Va Polytech Inst & State Univ Blacksburg VA 24061

SCHETZEN, MARTIN, b New York, NY, Feb 10, 28; m. NONLINEAR SYSTEM THEORY. *Educ:* NY Univ, BEE, 51; Mass Inst Technol, SM, 54, ScD(elec eng), 61. *Prof Exp:* Electronic scientist, Nat Bur Standards, 51-52; asst microwaves, Res Lab Electronics, Mass Inst Technol, 52-54; engr, Appl Physics Lab, Johns Hopkins Univ, 54-56; asst elec eng, Mass Inst Technol, 56-58, commun & nonlinear theory, Res Lab Electronics, 58-60, instr elec eng, 60-61, asst prof elec eng , Mass Inst Technol, 61-65, staff mem, Res Lab Electronics, 61-65; assoc prof, 65-69, PROF ELEC ENG, NORTHEASTERN UNIV, 69- *Concurrent Pos:* Consult, Atlantic Refining

Co, Tex, 61-66, Instrumentation Lab, Mass Inst Technol, 64-71 & Radio Corp Am, Mass, 68-72; vis prof elec eng, Univ Calif, Berkeley, 77-78; vis scientist, Dept Math, Weizmann Inst Sci, Rehovot, Israel, 82 & 84-85. *Honors & Awards:* Apollo Achievement Award, Apollo Certificate of Commendation. *Mem:* AAAS; Inst Elec & Electronics Engrs; Sigma Xi. *Res:* Nonlinear and communication theory; analysis and synthesis of nonlinear systems; determination of optimum nonlinear systems. *Mailing Add:* Dept Elec & Comput Eng Dana Res Bldg Northeastern Univ Boston MA 02115

SCHETZINA, JAN FREDERICK, b Moundsville, WVa, Nov 29, 40; m 68; c 1. PHYSICS. *Educ:* Gannon Col, BA, 63; Pa State Univ, MS, 65, PhD(physics), 69. *Prof Exp:* Res assoc solid state physics, Pa State Univ, 69-70; asst prof, 70-75, ASSOC PROF PHYSICS, NC STATE UNIV, 75- *Mem:* Am Phys Soc; Am Vacuum Soc. *Res:* Optical and electrical properties of semiconductors. *Mailing Add:* Dept Physics NC State Univ 408 Cox Hall Box 8202 Raleigh NC 27695

SCHEUCH, DON RALPH, b Seattle, Wash, Sept 12, 18; m 50; c 3. ELECTRICAL ENGINEERING. *Educ:* Univ Calif, BS, 43; Stanford Univ, MA, 46, PhD(elec eng), 49. *Prof Exp:* Res assoc, Radio Res Lab, Harvard Univ, 43-45; sr res engr, SRI Int, 49-51, supvr, 51-53, group head, 53-55, lab mgr, 55-59, asst dir eng, 59-60, exec dir electronics & radio sci, 60-68, vpres eng, 68-69, vpres & chmn, Off Res Opers, 77-80, sr vpres, 77-80; PVT CONSULT, 80- *Concurrent Pos:* Spec consult, USAF, 42-45. *Mem:* Sigma Xi; Inst Elec & Electronics Engrs. *Res:* Weapons systems evaluation; lasers; antennas; communications. *Mailing Add:* 430 Golden Oak Dr Portola Valley CA 94025

SCHEUCHENZUBER, H JOSEPH, b Lancaster, Pa, June 4, 44; m 68. BIOMECHANICS, PHYSICAL EDUCATION. *Educ:* West Chester State Col, BS, 68; Pa State Univ, MS, 70; Ind Univ, PhD(human performance), 74. *Prof Exp:* Phys dir, York YMCA, Pa, 68; instr scuba, Pa State Univ, 69-70; instr aquatics, York Col, Pa, 70-72; asst prof, 74-80, PROF BIOMECH, SPRINGFIELD COL, 80- *Concurrent Pos:* Spec consult acad appln comput sci, Springfield Col, 74- *Mem:* Am Asn Health Phys Educ & Recreation; Am Col Sports Med. *Res:* Biomechanical study of kinetic and kinematic factors present during normal human locomotive movements, and modification of similar abnormal motions based on that information. *Mailing Add:* Springfield Col Box 1726 Springfield MA 01109

SCHEUER, ERNEST MARTIN, b Germany, July 28, 30; US citizen; m 53, 72; c 2. STATISTICS, OPERATIONS RESEARCH. *Educ:* Reed Col, BA, 51; Univ Wash, MS, 54; Univ Calif, Los Angeles, PhD(math), 60. *Prof Exp:* Math statistician, Control Data Corp, Rand Corp, Space Technol Labs & US Naval Ord Test Sta, 51-70; assoc prof, 70-72, PROF MGT SCI, CALIF STATE UNIV, NORTHRIDGE, 72-, PROF MATH, 76- *Mem:* Fel Am Statist Asn; Math Asn Am; Inst Math Statist; Int Statist Inst. *Res:* Reliability, theory and applications; statistical distributions; generating random variables for simulations; testing goodness-of-fit; statistical and economic analysis of warranties. *Mailing Add:* Dept Mgt Sci Calif State Univ Northridge CA 91330

SCHEUER, JAMES, b New York, NY, Feb 21, 31. PHYSIOLOGY, BIOCHEMISTRY. *Educ:* Univ Rochester, BA, 55; Yale Univ, MD, 56. *Prof Exp:* Res assoc, Res Inst Muscle Dis, New York, 62-63; trainee metab & nutrit, Grad Sch Pub Health, Univ Pittsburgh, 63-64, from instr to assoc prof med, Sch Med, 64-72, assoc prof biochem, Fac Arts & Sci, 70-72; chief cardiol, Montefiore Hosp & Med Ctr, 72-87, interim chmn dept med, 87-90; assoc prof physiol, 72-78, vchmn med, 80-90, chmn med, 90, PROF MED, ALBERT EINSTEIN COL MED, 72-, PROF PHYSIOL, 78- *Concurrent Pos:* USPHS fel cardiol, Mt Sinai Hosp, New York, 59; USPHS fel myocardial metab, Cornell Univ, 62-63; USPHS fel biochem, Univ Pittsburgh, 64-65, career develop award myocardial metab, 68-; mem coun circulation, Rosie Sci Coun, Am Heart Asn; mem, Int Study Group Res Cardiac Metab. *Mem:* Am Soc Clin Invest; Am Physiol Soc; Soc Exp Biol & Med; Asn Am Physicians. *Res:* Correlation of biochemistry, metabolism and mechanical function of the heart, with emphasis on cardiac hypertrophy, the effects of physical condition aging, and the effects of diabetes. *Mailing Add:* Dept Med Albert Einstein Col Med 1300 Morris Park Ave Bronx NY 10461

SCHEUER, PAUL JOSEF, b Heilbronn, Ger, May 25, 15; nat US; m 50; c 4. ORGANIC CHEMISTRY. *Educ:* Northeastern Univ, BS, 43; Harvard Univ, MA, 47, PhD(chem), 50. *Prof Exp:* From asst prof to prof, 50-85, chmn dept, 59-62, EMER PROF CHEM, UNIV HAWAII, 85- *Concurrent Pos:* Barton lectr, Univ Okla, 67; vis prof, Univ Copenhagen, 77, 89; J F Toole lectr, Univ New Brunswick, 77; ed, Marine Natural Prod, 78-83, Bio Org Marine Chem, 87. *Mem:* AAAS; Am Chem Soc; Royal Soc Chem; Swiss Chem Soc. *Res:* Structure and biosynthesis of natural products; secondary metabolites of marine organisms; marine toxins; marine ecology. *Mailing Add:* Dept Chem Univ Hawaii at Manoa 2545 The Mall Honolulu HI 96822-2275

SCHEUING, RICHARD A(LBERT), b Lynbrook, NY, Aug 19, 27; m 50; c 3. AERONAUTICAL ENGINEERING. *Educ:* Mass Inst Technol, SB & SM, 48; NY Univ, PhD, 71. *Prof Exp:* Aerodyn res engr, 48-52, aerodyn res group leader, 52-56, head fluid mech sect, 56-70, dep dir, Res Dept, 61-77, DIR RES DEPT, GRUMMAN AEROSPACE CORP, 77- *Mem:* AAAS; assoc fel Am Inst Aeronaut & Astronaut. *Res:* Fluid dynamics; hypersonics; shock tunnels; magnetohydrodynamics. *Mailing Add:* 37 Quaker Path Cold Spring Harbor NY 11724

SCHEUPLEIN, ROBERT J, physical chemistry, biophysics, for more information see previous edition

SCHEUSNER, DALE LEE, b Watertown, SDak, Feb 10, 44; m 71; c 2. EDUCATION, FOOD MICROBIOLOGY. *Educ:* SDak State Univ, BS, 66; NC State Univ, MS, 68; Mich State Univ, PhD(food sci), 72. *Prof Exp:* Res microbiologist, S C Johnson & Son Inc, 72-83; HEAD SCI DEPT, CHRISTIAN LIFE SCH, 87- *Concurrent Pos:* Consult, 83- *Mem:* Am Soc Microbiol; Inst Food Technologists. *Res:* Methods development and application of environmental microbiology and microbial decontamination, especially as applied to health care and food processing facilities. *Mailing Add:* 4625 N Green Bay Rd Racine WI 53404

SCHEVE, BERNARD JOSEPH, b Cincinnati, Ohio, July 2, 45; m 69; c 2. PHOTOCHEMISTRY, RADIATION CHEMISTRY. *Educ:* Xaiver Univ, Ohio, BS, 67, MS, 68; Mich State Univ, PhD(photochem), 74. *Prof Exp:* Asst chemist drug chem, Merrell Nat Labs, Richardson Merrell Inc, 71-73; res chemist, 74-80, SR RES CHEMISTS, HERCULES INC, 80- *Mem:* Am Chem Soc; AAAS; NY Acad Sci. *Res:* Polymer modification; free radical chemistry; photo polymers; polysaccharides; polyole fins; terpenes. *Mailing Add:* 3863 Maywood Ct Cincinnati OH 45211-4424

SCHEVE, LARRY GERARD, b Palo Alto, Calif, Mar 1, 50; m. BIOCHEMISTRY. *Educ:* Seattle Pac Univ, BS, 72; Univ Calif, Riverside, PhD(biochem), 76. *Prof Exp:* Res asst biochem, Dept Surg, Vet Admin Hosp, Martinez, Calif, 77; lectr, 77-79, from asst prof to assoc prof, 79-86, PROF CHEM, DEPT CHEM, CALIF STATE UNIV, 87- *Mem:* Am Chem Soc; Sigma Xi; NY Acad Sci; AAAS. *Res:* Biochemistry of peroxidases (thyroid peroxidase and myeloperoxidase of the leukocyte); isolation and purification of membrane-bound proteins. *Mailing Add:* Dept Chem Calif State Univ Hayward CA 94542

SCHEVING, LAWRENCE EINAR, b Hensel, NDak, Oct 20, 20; m 49; c 4. ANATOMY, CHRONOBIOL. *Educ:* DePaul Univ, BS, 49, MS, 50; Loyola Univ Ill, PhD(anat), 57. *Prof Exp:* Asst embryol, DePaul Univ, 49-50; from instr to assoc prof biol sci & chmn dept, Lewis Col, 50-57; from instr to prof anat, Chicago Med Sch, 57-66; prof, Sch Med, La State Univ, New Orleans, 67-70; REBSAMEN PROF ANAT SCI, COL MED, UNIV ARK, LITTLE ROCK, 70- *Concurrent Pos:* Instr, Sch Nursing, Garfield Park Hosp, 49-50; vis prof, med Hochschule, Hanover, Ger & Univ Bergan Bergan, Norway,73; mem, Army Med Res & Develop Adv Comt, Wash, DC, 83- *Honors & Awards:* Alexander von Humboldt Sr Scientist Prize, 73. *Mem:* Int Soc Chronobiol (secy-tres, pres, 85-); Am Asn Anat. *Res:* Chronobiology; author of over 300 publications. *Mailing Add:* Dept Anat Univ Ark Col Med Little Rock AR 72205

SCHEWE, PHILLIP FRANK, b Evanston, Ill, July 7, 50; m 81; c 2. HIGH ENERGY PHYSICS. *Educ:* Univ Ill, BS & MS, 72; Mich State Univ, BA, 77, PhD(physics), 78. *Prof Exp:* Asst physicist, Brookhaven Nat Lab, 78-79; WITH AM INST PHYSICS, 79- *Concurrent Pos:* Ed, Physics News. *Mem:* Am Phys Soc. *Res:* Deep inelastic lepton scattering; development of superconducting magnets. *Mailing Add:* Am Inst Physics 335 E 45th St New York NY 10017

SCHEXNAYDER, MARY ANNE, b La, Nov 6, 48. ORGANIC CHEMISTRY, PHOTOCHEMISTRY. *Educ:* La State Univ, BS, 70; Rice Univ, PhD(chem), 74. *Prof Exp:* NIH trainee, Inst Lipid Res, Baylor Col Med, 75; res chemist, Hercules Inc, 75-84; asst to vpres res, 84-87, MGR RES ADMIN, MINONT, 89- *Mem:* Am Chem Soc. *Res:* Free radical chemistry; ziegler catalyst and polymerization. *Mailing Add:* 1712 Dahlia St Baton Rouge LA 70808-8825

SCHEY, HARRY MORITZ, b Chicago, Ill, Feb 20, 30. BIOSTATISTICS. *Educ:* Northwestern Univ, BS, 50; Harvard Univ, AM, 51; Univ Ill, PhD, 54. *Prof Exp:* Asst physics, Univ Ill, 52-54; sr physicist, Theoret Physics Div, Lawrence Livermore Lab, Univ Calif, 54-66; physicist, Educ Res Ctr, Mass Inst Technol, 66-73, co-dir, Proj, CALC, 73-75; fel, Dept Biostatist, Univ NC, Chapel Hill, 75-77; asst prof biostatist, Bowman Gray Sch Med, Wake Forest Univ, 78-84; ASSOC PROF MATH, ROCHESTER INST TECHNOL, 84- *Mem:* Am Statist Asn; Biomet Soc. *Res:* Kolmogorov-Smirnov goodness-of-fit tests; clinical studies; renal disease epidemiology; statistical methods in psychiatric diagnosis; obesity in children; leukemia clustering; blood pressure measurement techniques; statistical computing; geometric aspects of linear regression. *Mailing Add:* Rochester Inst Tech One Lomb Memorial Dr Rochester NY 14623

SCHEY, JOHN ANTHONY, b Sopron, Hungary, Dec 19, 22; US citizen; m 48; c 1. METAL DEFORMATION PROCESSES. *Educ:* Jozsef Nador Tech Univ, Hungary, dipl, 46; Acad Sci, Budapest, Hungary, PhD(metall), 53. *Hon Degrees:* Dr, Univ Stuttgart, Ger, 87, Univ Heavy Indust, Miskolc, Hungary, 89. *Prof Exp:* Supt metal works, Steel & Metal Works, Csepel, Budapest, 47-51; reader metals technol, Tech Univ, Miskolc, Hungary, 51-56; supvr fabrication, Res Lab, Brit Aluminum Co, 56-62; metall adv metal working, IIT Res Inst, Chicago, 62-68; prof metall eng, Univ Ill, Chicago Circle, 68-74; prof, 74-88, ADJ PROF MECH ENG, UNIV WATERLOO, ONT, 88- *Concurrent Pos:* Mem, Metalworking Processes Comt, Mat Adv Bd-Nat Acad Sci, 67-70; consult, 16 indust orgns, 68-; NAm ed, J Mech Working Technol, 77-87; course dir, Forging Indust Asn, Die Design Inst, 78-; assoc ed, J Lubrication Technol, Am Soc Mech Engrs, 81-87. *Honors & Awards:* W H A Robertson Medal, Inst Metals, London, Eng, 66; Gold Medal Award, Soc Mfg Engrs, 74; Dofasco Award, Can Inst Mining & Metall, 84. *Mem:* Nat Acad Eng; fel Am Soc Metals; fel Soc Mfg Engrs; Sigma Xi; Can Inst Mining & Metall. *Res:* Interactions between material properties and process conditions in metalworking processes, friction, lubrication and wear; development of new manufacturing processes; social implications of technology; tribology of metalworking. *Mailing Add:* Dept Mech Eng Univ Waterloo Waterloo ON N2L 3G1 Can

SCHIAFFINO, SILVIO STEPHEN, b Brooklyn, NY, Nov 1, 27; m 54; c 2. BIOCHEMISTRY, RESEARCH ADMINISTRATION. *Educ:* Georgetown Univ, BS, 46, MS, 48, PhD(biochem), 56. *Prof Exp:* Lab instr chem, Georgetown Univ, 46-48; biochemist, Div Nutrit, US Food & Drug Admin, Washington, DC, 48-50 & 54-60 & Chem Sect, Hazleton Labs, Inc, 60-61; scientist adminr, Nat Cancer Inst, 61-64, asst chief, Res Grants Rev Br, 64-69, chief, 69-72, assoc dir sci rev, 72-78, dep dir, 78-83, ACTG DIR, DIV RES GRANTS, NIH, 83- *Mem:* AAAS; Soc Res Adminrs. *Res:* Biochemistry and microbiology of nutritionally important substances, especially vitamins, amino acids and proteins; stability of vitamins; clinical chemistry; food additives. *Mailing Add:* Div Res Grants NIH Bethesda MD 20892

SCHIAGER, KEITH JEROME, b Hot Springs, SDak, Mar 29, 30; m 51; c 4. ENVIRONMENTAL HEALTH, HEALTH PHYSICS. *Educ:* Colo Agr & Mech Col, BS, 56; Univ Mich, MPH, 62, PhD(environ health), 64. *Prof Exp:* Health physicist, Argonne Nat Lab, 57-61; assoc prof radiation biol, Colo State Univ, 64-73; alt group leader environ studies, Los Alamos Sci Lab, 73-75; prof health physics, Univ Pittsburgh, 75-78; PRES, ALARA, INC, 78-; DIR, RADIOL HEALTH, UNIV UTAH, 82- *Concurrent Pos:* Pres, Am Acad Health Physics, 90. *Mem:* Am Nuclear Soc; Sigma Xi; Health Physics Soc (pres-elect, 91). *Res:* Radiological health; environmental radiation; inhalation exposure from radon progeny. *Mailing Add:* 3671 S Millbrook Terr Salt Lake City UT 84106

SCHIAVELLI, MELVYN DAVID, b Chicago, Ill, Aug 8, 42; m 66; c 2. PHYSICAL ORGANIC CHEMISTRY. *Educ:* DePaul Univ, BS, 64; Univ Calif, Berkeley, PhD(chem), 67. *Prof Exp:* Res assoc chem, Mich State Univ, 67-68; from asst prof to assoc prof, 68-80, chmn dept, 78-84, PROF CHEM, COL WILLIAM & MARY, 80-, DEAN FAC ARTS & SCI, 84- *Concurrent Pos:* Hon res fel, Univ Aberdeen, Scotland, 83- *Mem:* Am Chem Soc; Royal Soc Chem. *Res:* Secondary isotope effects; acid-catalysis; vinyl cations. *Mailing Add:* Dept Chem Col William & Mary Williamsburg VA 23185

SCHICK, JEROME DAVID, b Pontiac, Mich, Jan 23, 38; m 66; c 2. CHEMISTRY, PHYSICS. *Educ:* Wheaton Col, BS, 60; Wayne State Univ, MS, 65, PhD(phys chem), 68. *Prof Exp:* Res chemist, Henry Ford Hosp, Detroit, Mich, 60-63; proj chemist, Bendix Res Labs, 66-68; res chemist, Air Force Avionics Lab, Wright-Patterson AFB, 68-69; SR ENGR SEMICONDUCTOR DEVICES, IBM CORP, HOPEWELL JUNCTION, 69- *Mem:* Am Chem Soc; Electrochem Soc; Creation Res Soc; Microbeam Anal Soc. *Res:* Semiconductor devices and materials; electron spectroscopy; scanning electron microscopy; auger spectrometry; integrated circuits processing and failure studies; electron beam induced current; transistor and junction charaterization. *Mailing Add:* Kuchler Dr La Grangeville NY 12540

SCHICK, KENNETH LEONARD, b New York, NY, Feb 20, 30; m 57; c 3. SOLID STATE PHYSICS, BIOPHYSICS. *Educ:* Columbia Univ, BA, 51; Rutgers Univ, PhD(physics), 59. *Prof Exp:* Physicist, US Naval Air Missile Test Ctr, 51-52; from asst prof to assoc prof, 59-74, chmn dept, 71-77, PROF PHYSICS, UNION COL, NY, 74- *Concurrent Pos:* Vis res prof, State Univ Leiden, 65-66 & 72-73; NATO & NSF sr fel, 67; fel, Weizmann Inst, Israel, 79-80. *Mem:* Am Asn Physics Teachers. *Res:* Magnetic resonance; membrane structure in living systems; dynamic light scattering. *Mailing Add:* Dept Physics Union Col Schenectady NY 12308

SCHICK, LEE HENRY, b Philadelphia, Pa, Nov 23, 35; m 61; c 3. THEORETICAL NUCLEAR PHYSICS. *Educ:* Univ Pa, BS, 56; Univ Colo, MA, 58, PhD(theoret nuclear physics), 61. *Prof Exp:* Lectr math, Univ Birmingham, 61-63; res assoc, Univ Minn, 63-65; asst prof, Univ Southern Calif, 65-70; assoc prof, 70-74, PROF PHYSICS, UNIV WYO, 74- *Concurrent Pos:* Assoc dean, Col Arts & Sci, Univ Wyo, 83- *Mem:* Am Phys Soc. *Res:* Application of information-theory and scattering theory to geophysical and quantum mechanical problems. *Mailing Add:* Dept Physics & Astron Univ Wyo Laramie WY 82071

SCHICK, LLOYD ALAN, b Bluffton, Ohio, Mar 7, 45; m 75; c 2. BIOCHEMISTRY, CHEMISTRY. *Educ:* Ohio Northern Univ, BA, 66; Purdue Univ, MS, 68; Univ Notre Dame, PhD(biochem), 74. *Prof Exp:* Anal chemist, Ind State Chemist's Lab, 66-68; asst res scientist, Miles Labs, 68-71, assoc res scientist, 71-74, res scientist, 74-77, sr res scientist, 77-83, STAFF SCIENTIST, DIAG DIV MILES LABS, 83- *Concurrent Pos:* Res Award, Ohio Heart Asn, 65-66. *Mem:* Am Chem Soc; Am Asn Clin Chem. *Res:* Medical diagnostics; immunoassay; protein purification techniques; thyroid diagnostics; enzymology; hemolytic and fibrinolytic pathways. *Mailing Add:* Diag Div Miles Inc PO Box 70 Elkhart IN 46515

SCHICK, MARTIN J, b Prague, Czech, Oct 20, 18; nat US; div; c 2. COLLOID CHEMISTRY. *Educ:* Carnegie Inst Technol, BS, 42; Polytech Inst Brooklyn, PhD(chem), 48. *Prof Exp:* Res chemist, Shell Develop Co, 42-45 & 48-58; sr res assoc, Res & Develop Ctr, Lever Bros Co, 58-66; prin scientist, Cent Res Lab, Interchem Corp, Clifton, 66-69; res mgr, Surfactant & Org Chem Lab, Diamond Shamrock Corp, 69-78, sr scientist, process chem div, 78-83; CONSULT, 83- *Concurrent Pos:* Adj prof, Lehigh Univ, 84. *Mem:* Am Chem Soc; Fiber Soc; Am Oil Chemists Soc. *Res:* Surface and polymer chemistry; nonionic surfactants; surfactant synthesis; friction and lubrication of synthetic fibers. *Mailing Add:* 12 W 72nd St New York NY 10023-4163

SCHICK, MICHAEL, b Philadelphia, Pa, Mar 17, 39; m 81. STATISTICAL MECHANICS, SURFACE PHYSICS. *Educ:* Tufts Univ, BA & BS, 61; Stanford Univ, MS, 64 & PhD(physics), 67. *Prof Exp:* Postdoc fel, Case Western Reserve Univ, 67-69; from asst prof to assoc prof, 69-78, PROF PHYSICS, UNIV WASH, 78- *Concurrent Pos:* Vis prof, Lab Nat Technol, 77-78, Cen Saclay, 84-85 & 89, Nordita, 85 & Univ Oslo, 90. *Mem:* Fel Am Phys Soc. *Res:* Application of renormalization group to adsorbed systems; classification of order-disorder transitions in adsorbed systems; exact renormalization group; solution of two-dimensional ising model; theories of multilayer growth and wetting; theory of microemulsions. *Mailing Add:* Dept Physics FM-15 Univ Wash Seattle WA 98195

SCHICK, PAUL KENNETH, b Czech, Oct 12, 32; US citizen; m 62; c 2. HEMATOLOGY, ONCOLOGY. *Educ:* Boston Univ, MD, 61. *Prof Exp:* Intern & resident med, Kings County Hosp, Brooklyn, NY, 61-63; resident med, New York Med Col, 63-65; pvt pract, 65-69; hemat trainee, Montefiore Hosp, Bronx, 69-71; asst prof, 71-76, assoc prof med, 76-79, ASSOC PROF BIOCHEM, MED COL PA, 79-; PROF MED, SCH MED, TEMPLE UNIV, 82- *Concurrent Pos:* Assoc prof med, Sch Med, Temple Univ, 79-82. *Mem:* Int Soc Thrombosis & Hemostasis; fel Am Col Physicians; Am Soc Hemat; Sigma Xi; Am Fedn Clin Res; Am Soc Physiol. *Res:* Understanding the structure and function of platelet membranes in order to define the role of platelets in hemostasis and to develop anti-platelet drugs for the prevention of thrombosis; investigation of megakaryocyte maturation and biochemistry. *Mailing Add:* Cardeza Found Jefferson Med Col 1015 Walnut St Philadelphia PA 19107

SCHICKEDANTZ, PAUL DAVID, b Columbus, Ohio, Aug 20, 31; m 57; c 4. ORGANIC CHEMISTRY. *Educ:* Oberlin Col, AB, 53; Ohio State Univ, PhD(org chem), 59. *Prof Exp:* Res chemist, Am Cyanamid Co, 59-61; res chemist, Consumer Prod Div, Union Carbide Co, WVa, 61-63 & Chem & Plastics Div, 63-67; SR RES CHEMIST, LORILLARD TOBACCO CO, DIV LOEWS INC, 67- *Mem:* Am Chem Soc; Sigma Xi. *Res:* Synthesis and properties of bridgehead nitrogen compounds; stain repellant and wash and wear textile finishes; synthesis of S-triazines, insect repellents, and condensation polymers; liquid chromatographic analysis of polycyclic aromatic hydrocarbons in tobacco smoke; analysis of urinary drug metabolites; cigarette flavor chemistry; analytical biochemistry. *Mailing Add:* 2809 Watauga Dr Greensboro NC 27408

SCHIEBLER, GEROLD LUDWIG, b Hamburg, Pa, June 20, 28; m 54; c 6. MEDICINE, PEDIATRIC CARDIOLOGY. *Educ:* Franklin & Marshall Col, BS, 50; Harvard Med Sch, MD, 54. *Prof Exp:* Asst prof, 60-63, assoc prof, 63-66, PROF PEDIAT, COL MED, UNIV FLA, 66-, CHMN DEPT, 68- *Concurrent Pos:* Teaching fel med, Harvard Med Sch, 55-56; med fel pediat, Univ Minn Hosps, 56-57; med fel specialist, Med Ctr, Univ Minn, 57-59, med fel, Mayo Clin & Found, 59-60; Nat Heart Inst res fel, 59-60; mem study sect, Coun Rheumatic Fever & Congenital Heart Dis, Am Heart Asn, 61-64. *Mem:* Inst Med-Nat Acad Sci; Am Acad Pediat; Am Col Cardiol; Am Heart Asn; AAAS. *Res:* Heart disease in infants and children; cardiovascular physiology. *Mailing Add:* Dept Pediat Univ Fla Col Med J Hills Millers Health Ctr Gainesville FL 32610

SCHIEBOUT, JUDITH ANN, b Tampa, Fla, Oct 16, 46. VERTEBRATE PALEONTOLOGY, PALEOECOLOGY. *Educ:* Univ Tex, Austin, BA, 68, MA, 70, PhD(geol), 73. *Prof Exp:* Lectr, San Diego State Univ, 74-76; asst prof, 76-79, ADJ ASSOC PROF GEOL, LA STATE UNIV, 79-, DIR MUS GEOSCI, 79- *Concurrent Pos:* NSF, res grant, 83-88. *Mem:* Soc Vertebrate Paleont; Geol Soc Am; Sigma Xi; Paleont Soc; Soc Econ Paleontologists & Mineralogists; Asn Sci Mus Dirs. *Res:* Analysis of early Tertiary mammal distribution and paleogeography; relationship between vertebrate taphonomy and fluvial sedimentation. *Mailing Add:* Mus Geosci La State Univ Baton Rouge LA 70803

SCHIEFER, H BRUNO, veterinary pathology, toxicology, for more information see previous edition

SCHIEFERSTEIN, GEORGE JACOB, b Buffalo, NY, July 22, 42; m 73; c 1. PHARMACOLOGY & TOXICOLOGY. *Educ:* LeMoyne Col, BS, 64; Univ Buffalo, PhD(pharmacol), 70. *Prof Exp:* Pharmacologist, G D Searle, Inc, 68-71 & Wm H Rorer, Inc, 71-72; PHARMACOLOGIST, NAT CTR TOXICOL RES, FOOD & DRUG ADMIN, 72- *Mem:* Soc Toxicol. *Res:* Subchronic and chronic testing of chemical carcinogens in laboratory animals; benzidine 4-, diethylstibestrol & 1, 4 dithiane. *Mailing Add:* 8807 Manassas Circle Mabelvale AR 72103

SCHIEFERSTEIN, ROBERT HAROLD, b Klamath Falls, Ore, May 18, 31; m 50; c 5. PLANT PHYSIOLOGY, AGRONONOMY. *Educ:* Ore State Univ, BS, 54; Iowa State Univ, MS, 55, PhD(plant physiol), 57. *Prof Exp:* Asst plant physiol, Iowa State Univ, 54-56, res assoc, 56-57; tech rep agr chem, Chipman Chem Co, Inc, 57-61, tech mgr, 61-62; plant physiologist, Shell Develop Co, 62-65, supvr herbicides res, 65-67, chief plant physiologist, 67-68, dept head plant physiol, 68-70, supvr pesticide develop, 71, res & develop proj mgr agron prods, 71-72, prod rep, 72-73, tech support rep, 73-77, staff tech serv rep, Shell Chem Co, 77-85, field develop & tech serv rep & coordr, Field Res Resources, Shell Develop Co, 85-89; RETIRED. *Concurrent Pos:* Consult, 89- *Mem:* Am Soc Plant Physiol; fel Weed Sci Soc Am. *Res:* Herbicides; defoliants; plant growth regulators; plant surface wax and cuticle; research and development administration with agricultural chemicals. *Mailing Add:* PO Box 1291 Twain Harte CA 95383

SCHIENLE, JAN HOOPS, b Washington, DC, June 20, 45; c 3. ENVIRONMENTAL TOXICOLOGY, INDUSTRIAL TOXICOLOGY. *Educ:* Calif State Univ, Northridge, BS, 74, MS, 76. *Prof Exp:* BIOL SAFETY OFFICER & INDUST HYGIENIST, UNIV CALIF, SANTA BARBARA, 76-; ASST PROF ENVIRON OCCUP HEALTH, CALIF STATE UNIV, NORTHRIDGE, 81- *Concurrent Pos:* Environ sound consult, Santa Barbara Coun Bowl Orgn & Pvt developers, 76-; fac, Calif Spec Training Inst, 79-81; mem, Santa Barbara Coun Hazardous Mat task Force, 81-, Health Adv Comt, Santa Barbara Coun, 82- *Mem:* Am Indust Hygiene Asn; Am Conf Govt Hygienists; Nat Environ Health Asn. *Mailing Add:* Environ Health & Safety 18111 Nordhoff St North Ridge CA 91330

SCHIERMAN, LOUIS W, b Carlyle, Ill, Feb 16, 26; m 57; c 2. IMMUNOGENETICS. *Educ:* Univ Ill, BS, 51; Iowa State Univ, MS, 61, PhD(genetics, immunol), 62. *Prof Exp:* Res assoc immunogenetics, Iowa State Univ, 62-64, asst prof, 64-65; res assoc, Mt Sinai Hosp, New York, 65-66, asst prof, Mt Sinai Sch Med, 66-68; assoc prof path, New York Med Col, 68-77; PROF, UNIV GA, 77- *Mem:* AAAS; Genetics Soc Am; Am Asn Immunol. *Res:* Relationships of blood groups to histocompatibility; tumor immunology; genetic control of immune responses. *Mailing Add:* Dept Avian Med Univ Ga 953 College Station Rd Athens GA 30601

SCHIESSER, ROBERT H, b Niagara Falls, NY, Jan 12, 37; m 65; c 3. PHYSICAL CHEMISTRY, SURFACE CHEMISTRY. *Educ:* Clarkson Tech, BChE, 58, MChE, 60; Lehigh Univ, PhD(phys chem), 66. *Prof Exp:* Sr res chemist, Rohm & Haas Co, 65-68; sr res chemist, Betz Labs, Inc, 68-76; RES SCIENTIST, SCOTT PAPER CO, 76- *Mem:* Am Chem Soc; Tech Asn Pulp & Paper Indust. *Res:* Adsorption at solid-liquid interface; polymer solution properties; stability of lyophobic colloids; structure and properties of polyelectrolytes. *Mailing Add:* 1834 Mare Rd Warrington PA 18976

SCHIESSER, W(ILLIAM) E(DWARD), b Willow Grove, Pa, Jan 9, 34; m 58; c 2. CHEMICAL ENGINEERING. *Educ:* Lehigh Univ, BS, 55; Princeton Univ, MA, 58, PhD(chem eng), 60. *Prof Exp:* From asst prof to prof, 60-76, MCCANN PROF CHEM ENG, LEHIGH UNIV, 76-, MGR USER SERV, COMPUT CTR, 69- *Concurrent Pos:* Consult, Indust & Govt; mem, Am Automatic Control Coun. *Mem:* Am Inst Chem Engrs; Inst Elec & Electronics Engrs. *Res:* Applied mathematics; systems analysis. *Mailing Add:* Dept Chem Eng Bldg A Lehigh Univ Mountaintop Campus Bethlehem PA 18015

SCHIESSL, H(ENRY) W(ILLIAM), b Ingolstadt, Ger, Dec 1, 24; nat US. INORGANIC CHEMISTRY, CHEMICAL ENGINEERING. *Educ:* Cornell Univ, BChE, 50; Univ Heidelberg, DSc, 64. *Prof Exp:* Process design engr, 50-56, asst to dir res & develop, 56-61, res assoc, 64-70, RES MGR, OLIN CORP, 70- *Concurrent Pos:* Mem fac, Univ New Haven. *Mem:* Am Chem Soc; Nat Asn Corrosion Engrs; Sigma Xi. *Res:* Inorganic chemistry; heavy chemicals; statistical analysis of experimental data; thermochemistry; kinetics; hydrazine chemistry. *Mailing Add:* 79 Parsonage Hill Rd Northford CT 06472

SCHIEVE, WILLIAM, b Portland, Ore, Apr 28, 29; m 52; c 2. THEORETICAL PHYSICS, STATISTICAL MECHANICS. *Educ:* Reed Col, AB, 51; Lehigh Univ, MS, 57, PhD(physics), 59. *Prof Exp:* Asst physics, Lehigh Univ, 51-57; res fel, Bartol Res Found, 57-60; res physicist, US Naval Radiol Defense Lab, 61-67; actg dir statist mech & thermodyn, 69-77; ASSOC PROF PHYSICS, UNIV TEX, AUSTIN, 67- *Concurrent Pos:* Study fel with Prof Prigogine, Free Univ Brussels, 64-65. *Honors & Awards:* Silver Medal, US Naval Radiol Defense Lab, 64. *Mem:* Am Phys Soc. *Res:* Thermal conductivity of insulating crystals; statistical mechanics of phonons; fundamental theory of statistical mechanics; perturbation theory; reversibility; molecular dynamics. *Mailing Add:* Dept Physics Univ Tex Austin TX 78712

SCHIEWETZ, D(ON) B(OYD), b Dayton, Ohio, Nov 15, 27; m 56; c 1. TECHNOLOGY LICENSING & TRANSFER. *Educ:* Northwestern Univ, BS, 50; Univ Cincinnati, MS, 52, PhD(chem eng), 54. *Prof Exp:* Res engr, E I du Pont de Nemours & Co, Inc, 54-59, tech supvr, 59-63, area supvr, 63-65, plant supt, 66-70, tech supt, 70-73, mfg supt, 73-75, capacity mgr, 75-76, mfg & res mgr, 76-86, technol sales mgr, Polymer Prod Dept, 86-90, OPERS MGR, CHEM DEPT, E I DU PONT DE NEMOURS & CO, INC, 90- *Mem:* Am Inst Chem Engrs; Sigma Xi. *Res:* Polymer science and processing technology; chemical kinetics; process economics; venture analysis. *Mailing Add:* Five West Ct Beacon Hill Wilmington DE 19810

SCHIFERL, DAVID, US citizen. HIGH PRESSURE PHYSICS, SOLID STATE PHYSICS. *Educ:* Univ Chicago, BS, 66, MS, 69, PhD(physics), 75. *Prof Exp:* Res physicist, Dept Geophys Sci, Univ Chicago, 74-75, x-ray consult, 77; vis scientist, Max Planck Inst Solid Body Res, Stuttgart, WGer, 75-77; STAFF SCIENTIST, UNIV CALIF, LOS ALAMOS SCI LAB, 77- *Concurrent Pos:* NATO fel, Max Planck Inst Solid Body Res, Stuttgart, 75-76 & German Acad Exchange Serv, 76-77. *Res:* Experimental high pressure physics primarily with diamond-anvil cells; theoretical and experimental studies on crystal structure stability; materials, non-destructive testing. *Mailing Add:* Los Alamos Sci Lab PO Box 1663 Los Alamos NM 87545

SCHIFF, ANSHEL J, b Chicago, Ill, Sept 24, 36. EARTHQUAKE ENGINEERING. *Educ:* Purdue Univ, Lafayette, BSME, 58, MSESc, 61, PhD(eng sci), 67. *Prof Exp:* Res asst aeronaut, astronaut & eng sci, Purdue Univ, W Lafayette, 59-67, from asst prof to prof mech eng, 67-85; CONSULT, PROF CIVIL ENGR, STANFORD UNIV, STANFORD CA, 85- *Concurrent Pos:* Consult, principle of precision measurement industs; instr indust courses instrumentation & measurements, 74- *Mem:* Instrument Soc Am; Am Soc Eng Educ; Am Soc Mech Engrs; Am Soc Civil Engrs; Earthquake Eng Res Inst. *Res:* Design of instumentation for structural dynamics; experimental and analytical studies of dynamic systems; system identification; evaluation of the impact of natural disasters on community services. *Mailing Add:* Dept Civil Eng Stanford Univ Stanford CA 94305

SCHIFF, ERIC ALLAN, b Los Angeles, Calif, Aug 29, 50; m 73; c 2. SEMICONDUCTORS, PHOTOCONDUCTIVITY. *Educ:* Calif Inst Technol, BS, 71; Cornell Univ, PhD(physics), 79. *Prof Exp:* Res assoc, James Franck Inst, Univ Chicago, 78-81; asst prof, 81-87, ASSOC PROF, DEPT PHYSICS, SYRACUSE UNIV, 87- *Concurrent Pos:* Vis prof, Brown Univ, 88- 89. *Mem:* Am Phys Soc; Mat Res Soc; Sigma Xi. *Res:* Electron transport and photocarrier recombination in amorphous semiconductors, especially hydrogenated amorphous silicon; optical, photoconductive, and electron spin resonance characterization of semiconductors. *Mailing Add:* Dept Physics Syracuse Univ Syracuse NY 13244-1130

SCHIFF, GILBERT MARTIN, b Cincinnati, Ohio, Oct 21, 31; m 55; c 2. INFECTIOUS DISEASES, VIROLOGY. *Educ:* Univ Cincinnati, BS, 53, MD, 57. *Prof Exp:* Intern, Univ Hosp, Iowa City, 57-58, resident internal med, 58-59; med officer, Lab Br, Commun Dis Ctr, Ga, 59-61; head tissue cult invest unit, Sect Virol, Perinatal Res Br, Nat Inst Neurol Dis & Blindness, 61-64; dir clin virol lab, Univ Cincinnati, 64-78, asst prof med & microbiol, 64-67, assoc prof med, 67-71, asst prof microbiol, 67- 71, PROF MED, COL MED, UNIV CINCINNATI, 71-; PRES, JAMES N GAMBLE INST MED RES, 84- *Concurrent Pos:* Attend physician, Dept Med, Emory Univ, Atlanta, Ga, 59-61; prin investr, USPHS grant, 64-67 & Nat Found res grant, 65-67; consult, Comt Maternal Health, Ohio State Med Asn, 64-70, Hamilton County Neuromuscular Diag Clin, 66, 75, Contract Immunization Status in US, 75-77, Vet Admin Comt Viral Hepatitis Among Dent Personnel; Nat Inst Child Health & Human Develop career res develop award, 70-74; dir, Christ Hosp Inst Med Res, Cincinnati, Ohio, 74-83 & comt cancer prog, 79-, human res, 80-, infection control, 81- & chairperson, library comt, 74-; mem, NIH Study Sect, adv comt & rev comt; chairperson, Animal Care Comt, James N Gamble Inst Med Res, 74-; mem comt, Rubella Immunization, Ohio Dept Health, Rubella Control, Cincinnati Dept Health, Surgeon's Gen Adv Comt, Immunization Pract, 71-75, Subcomt Antimicrobial Agents, US Pharmacopeia, 77-80; mem, Univ Liaison Comt, Christ Hosp, 82-, res comt bd trustees, Children's Hosp Med Ctr, 85-, Hoxworth Community Adv Bd, Hoxworth Blood Ctr, 91-; comt, animal care, Univ Cincinnati Col Med, 67-77, human res, 75-78, continuing educ, 77-; chmn, search comt & dir radiother, Christ Hosp, 80-82. *Mem:* AAAS; Am Soc Microbiol; Sigma Xi; Am Fedn Clin Res (secy-treas, 67-70); Am Pub Health Asn; Sci Res Soc Am; Cent Soc Clin Res (secy-treas, 77-81, vpres, 83, pres, 84); Infectious Dis Soc Am; Am Soc Clin Invest; fel Am Col Physicians. *Res:* Clinical virology. *Mailing Add:* 2141 Auburn Ave Cincinnati OH 45219

SCHIFF, HAROLD IRVIN, b Kitchener, Ont, June 24, 23; m 48; c 2. PHYSICAL CHEMISTRY. *Educ:* Univ Toronto, BA, 45, MA, 46, PhD(chem), 48. *Prof Exp:* Asst chem, Univ Toronto, 45-48; fel, Nat Res Coun Can, 48-50; from asst prof to prof, McGill Univ, 50-65, dir upper atmosphere chem group, 59-65; prof & chmn dept chem & dir nat sci, 64-66, dean sci, 66-72, PROF CHEM, YORK UNIV, 72-, UNIV PROF, 80- *Concurrent Pos:* Nuffield fel, Cambridge Univ, 59-60; Eskine prof, Univ Canterbury, NZ, 73-74; Collisions; reporter, Working Group VII-Lab Data, Int Asn Geomagnetism & Aeronomy; mem comt stratospheric pollution, Govt Can; mem bd dirs, Scintrex, Ltd; chmn panel atmospheric chem & transport, US Acad Sci, 78-80; adv panel, Fed Aviation Admin High Altitude Pollution Prog; pres, Unisearch Assoc, Inc; titular mem, Atmospheric Chem Comm, Int Union Pure & Appl Chem. *Mem:* Am Meteorol Soc; fel Chem Inst Can; fel Royal Soc Can. *Res:* Mass spectrometry; chemical kinetics; atomic physics; upper and lower atmosphere; excitation processes; atmospheric measurements; acid deposition; photo oxidation. *Mailing Add:* Dept Chem York Univ 4700 Keele St Downsview ON M3J 1P6 Can

SCHIFF, HARRY, b Boryslaw, Poland, May 17, 22; Can citizen; m 52; c 1. THEORETICAL PHYSICS. *Educ:* McGill Univ, BSc, 49, MSc, 50, PhD(physics), 53. *Prof Exp:* Lectr, Univ Alta, 53-54, from asst prof to prof physics, 54-87, chmn dept, 64-67, dir, Theoret Phys Inst, 80-83; RETIRED. *Mem:* Am Phys Soc; Can Asn Physicists. *Res:* Unitary field theories, particularly modified Maxwell fields with application to elementary particle structure; solitons with virial constraints. *Mailing Add:* 13004 66th Ave Edmonton AB T6H 1Y7 Can

SCHIFF, JEROME A, b Brooklyn, NY, Feb 20, 31. PLANT PHYSIOLOGY. *Educ:* Brooklyn Col, BA, 52; Univ Pa, PhD(bot, biochem), 56. *Prof Exp:* Res assoc microbiol, 56-57, from instr to prof, 57-74, ABRAHAM & ETTA GOODMAN PROF BIOL, BRANDEIS UNIV, 74- *Concurrent Pos:* Carnegie fel, dept plant biol, 62-63; asst ed, Plant Physiol, 64-79; mem, develop biol grant rev panel, NSF, 65-68; dir, Exp Marine Bot Prog, Marine Biol Lab, Woods Hole, 74-79; mem, Grad Fel Prog Panel, NSF-Nat Res Coun, 74 & Biol Grant Rev Panel, US-Israel Binat Sci Found, 74-; consult grant applns, Dept Army, 74-77; chief co-ed, Plant Sci, 81-; consult metabolic biol, NSF, 82-86; chair, Biol Dept, Brandeis Univ, 72-75, dir, Inst Photobiol Cells & Organelles, 75-87; USPHS fel, 54-56; vis prof, Tel Aviv Univ & Hebrew Univ, 72 & Weizmann Inst, 77. *Mem:* Fel AAAS; Soc Develop Biol (secy, 64-66); Am Soc Cell Biol; Phycol Soc Am; Am Soc Plant Physiol; fel Am Acad Arts & Sci. *Res:* Sulfate metabolism; pigment formation in plants; physiology of algae and other Protista, particularly Euglena; development and inheritance of chloroplasts and other cell organelles. *Mailing Add:* Inst Photobiol Cells & Org Brandeis Univ Waltham MA 02254

SCHIFF, JOEL D, b New York, NY, July 17, 43; m 68; c 2. MUSCLE PHYSIOLOGY, NEUROPHYSIOLOGY. *Educ:* Columbia Univ, AB, 64, MA, 65, PhD(biophys), 72. *Prof Exp:* Res Assoc physiol, Naval Med Res Inst, 72-73; asst prof, 73-79, ASSOC PROF PHYSIOL, NEW YORK UNIV, 79- *Mem:* Sigma Xi; Am Physiol Soc; AAAS; Am Inst Biol Sci; NY Acad Sci. *Res:* Mechanisms that regulate contraction of smooth muscle organs in mammals, including the pharmacological-physiological regulation and intracellular mechanisms involved. *Mailing Add:* 42 Essex Rd Great Neck NY 11023

SCHIFF, LEON, b Riga, Latvia, May 1, 01; nat US; m 25; c 3. INTERNAL MEDICINE. *Educ:* Univ Cincinnati, BS, 22, MD, 24, MS, 27, PhD(med), 29. *Prof Exp:* Asst bact, 22-23, asst med, 25-26, from instr to assoc prof, 26-55, prof clin med, 55-58, prof med, 58-70, EMER PROF MED, COL MED, UNIV CINCINNATI, 70- *Concurrent Pos:* Prof med, Sch Med, Univ Miami, 70-72, clin prof, 73-; attend physician, Med Serv, Cincinnati Gen Hosp; consult, US Vet Hosp. *Honors & Awards:* Friedenwald Medal, Am Gastroenterol Asn, 73; Nat Comn Digestive Dis Award, 77; Leon Schiff Ann lectr, Univ Cincinnati, 81; Distinguished Serv Award, Am Asn Study Liver Dis, 81; Doris Faircloth Averbach Award, 85. *Mem:* Am Soc Clin Invest; Am Gastroenterol Asn; master Am Col Physicians; Am Asn Study Liver Dis; Int Asn Study Liver Dis; Asn Francaise pour l'etude de Foie. *Res:* Diseases of the digestive tract; liver disease and jaundice; clinical research. *Mailing Add:* Dept Med Sch Med Univ Miami Miami FL 33101

SCHIFF, LEONARD NORMAN, b New York, NY, Dec 7, 38; m 62; c 2. COMMUNICATIONS ENGINEERING. *Educ:* City Col New York, BEE, 60; NY Univ, MSc, 62; Polytech Inst Brooklyn, PhD(elec eng), 68. *Prof Exp:* Mem tech staff, Bell Tel Labs, Inc, 60-67; mem tech staff, RCA Labs, 67-78, head commun anal res, 78-83, dir commun res lab, 83-87; DIR COMMUN RES LAB, DAVID SARNOFF RES CTR, 87- *Mem:* Sr mem Inst Elec & Electronics Engrs; Sigma Xi. *Res:* Communications theory; systems science; satellite communications. *Mailing Add:* David Sarnoff Res Ctr Princeton NJ 08540

SCHIFF, PAUL L, JR, b Columbus, Ohio, Feb 3, 40; m 61; c 2. PHARMACOGNOSY, NATURAL PRODUCTS CHEMISTRY. *Educ:* Ohio State Univ, BSc, 62, MSc, PhD(pharmacog). *Prof Exp:* Asst prof pharmacog, Col Pharm, Butler Univ, 67-69, Dept Pharmacog, Sch Pharm, Univ Miss, 70; assoc prof, 70-74, PROF PHARMACOG, SCH PHARM, UNIV PITTSBURGH, 74-, CHMN DEPT, 70- *Mem:* Am Soc Pharmacog; Acad Pharmaceut Sci; Am Pharmaceut Asn; Am Chem Soc; Phytochem Soc

Am; Sigma Xi. *Res:* Isolation and identification of plant metabolites with potential pharmacological activity and in particular with the isolation and identification of benzylisoquinoline-derived alkaloids. *Mailing Add:* Dept Pharmacog 512 Salk Hall Sch Med Univ Pittsburgh Pittsburgh PA 15261

SCHIFF, SIDNEY, b Chicago, Ill, June 9, 29; m 54; c 2. ORGANIC CHEMISTRY. *Educ:* Ill Inst Technol, BS, 51; Ohio State Univ, MS, 54, PhD(chem), 58. *Prof Exp:* Res chemist, 58-77; supvr, Phillips Petrol Co, 77-86; AT CITGO PETROL CORP, 86- *Mem:* Soc Automotive Engrs; Am Chem Soc; Soc Tribologists & Lubrication Engrs. *Res:* Lubricating oil additives; fuels and lubricants; synthetic lubricants; catalytic cracking catalysts; metals passivation. *Mailing Add:* 2401 Cherokee Hills Circle Bartlesville OK 74006

SCHIFF, STEFAN OTTO, b Braunschweig, Ger, July 22, 30; US citizen; wid; c 2. RADIATION BIOLOGY. *Educ:* Roanoke Col, BS, 52; Univ Tenn, PhD(radiation biol), 64. *Prof Exp:* Instr zool, Univ Tenn, 63-64; asst prof, 64-71, assoc prof, 71-77, chmn, Dept Biol Sci, 77-87, PROF ZOOL, GEORGE WASHINGTON UNIV, 77-, CHMN GRAD PROG GENETICS, 71- *Mem:* Radiation Res Soc; Sigma Xi; AAAS. *Res:* Effects of microwave radiation on mammalian sensory structures. *Mailing Add:* Dept Biol George Washington Univ Washington DC 20052

SCHIFFER, JOHN PAUL, b Budapest, Hungary, Nov 22, 30; nat US; m 60; c 2. NUCLEAR PHYSICS. *Educ:* Oberlin Col, BA, 51; Yale Univ, MS, 52, PhD(physics), 54. *Prof Exp:* Asst physics, Yale Univ, 51-54; res assoc, Rice Inst, 54-56; from asst physicist to assoc physicist, 56-65, from assoc dir to dir, Physics Div, 64-82, SR PHYSICIST, ARGONNE NAT LAB, 65-, ASSOC DIR, PHYSICS DIV, 82-; PROF PHYSICS, UNIV CHICAGO, 69- *Concurrent Pos:* Guggenheim fel, 59-60; vis assoc prof, Princeton Univ, 64; vis prof, Univ Rochester, 67-68 & Tech Univ Munich, 73-74; ed, Comments on Nuclear & Particle Physics, 72-75 & Physics Letters, 78-; assoc ed, Revs Modern Physics, 72-77; sr US scientist award, Alexander von Humboldt Found, 73-74; mem, panel future nuclear sci, Nat Acad Sci-Nat Res Coun, 75-76; mem, Nuclear Sci Adv Comt, Dept Energy, NSF, 80-85, chmn, 83-85. *Honors & Awards:* Tom W Bonner Prize, Am Phys Soc, 76; Wilbur Cross Medal, Yale Univ, 85. *Mem:* Nat Acad Sci; fel Am Phys Soc; fel AAAS. *Res:* Experimental nuclear physics; nuclear reactions and structure; heavy ion reactions, pion reactions in nuclei. *Mailing Add:* Argonne Nat Lab 9700 S Cass Ave Argonne IL 60439

SCHIFFER, MARIANNE TSUK, b Budapest, Hungary, June 28, 35; US citizen; m 60; c 2. X-RAY CRYSTALLOGRAPHY. *Educ:* Petrik Lajos Chem Indust Tech Sch, Hungary, BS, 55; Smith Col, MA, 58; Columbia Univ, PhD(biochem), 65. *Prof Exp:* Res assoc biochem, Biol & Med Res Div, 65-67, asst biochemist, 68-74, biophysicist, 74-87, SR BIOPHYSICIST, BIOL, ENVIRON & MED RES DIV, AGRONNE NAT LAB, 87- *Concurrent Pos:* Vis scientist, Max Planck Inst Biochem, 73-74; lectr, Northwestern Univ, 82- *Mem:* Am Crystallog Asn; Am Asn Immunologists; NY Acad Sci. *Res:* Determination of protein structure by x-ray with special emphasis on the structures of immunoglobulins and the photosynthetic reaction center diffraction; correlation of amino acid sequence and conformation of protein molecules. *Mailing Add:* Biol Environ & Med Res Div Argonne Nat Lab Argonne IL 60439-4833

SCHIFFER, MENAHEM MAX, b Berlin, Ger, Sept 24, 11; nat US; m 37; c 1. MATHEMATICAL ANALYSIS. *Educ:* Hebrew Univ, Israel, MA, 34, PhD(math), 38. *Hon Degrees:* DS, Israel Inst Technol, 73. *Prof Exp:* Instr, Hebrew Univ, Israel, 34-38, sr asst, 38-43, lectr, 43-46, prof, 50-51; res lectr, Harvard Univ, 46-49; vis prof, Princeton Univ, 49-50; chmn dept, 53-60, prof, 51-76, EMER PROF MATH, STANFORD UNIV, 76- *Concurrent Pos:* Fulbright fel, 65-66. *Mem:* Nat Acad Sci; Am Math Soc; Am Acad Arts & Sci; foreign mem Finnish Acad Sci; Am Math Asn. *Res:* Theory of functions; partial differential equations; calculus of variations; applied mathematics. *Mailing Add:* 3748 Laguna Ave Palo Alto CA 94306

SCHIFFMACHER, E(DWARD) R(OBERT), b Lawrence, NY, July 8, 24; m 56; c 2. ELECTRICAL ENGINEERING. *Educ:* Union Col, NY, BS, 45; Cornell Univ, MS, 52. *Prof Exp:* Instr elec eng, Union Col, NY, 46-47; instr, Cornell Univ, 48-52, res assoc, 52-57; electronic scientist, Nat Bur Standards, 57-60, supvry electronic engr, 60-65; supvry electronic engr, Environ Sci Serv Admin, Nat Oceanic & Atmospheric Admin, 65-70, supvry electronic engr, 70-80, electronic engr, 80-82; CONSULT ENGR, 82- *Mem:* Am Astron Soc; Inst Elec & Electronics Engrs. *Res:* Ionospheric studies using radio astronomy techniques and several earth satellite experiments; solar radio astronomy; ground-based ionospheric sounding; geomagnetism. *Mailing Add:* 2155 Emerald Rd Boulder CO 80304

SCHIFFMAN, GERALD, IMMUNOLOGY, BACTERIAL VACCINES. *Educ:* New York Univ, PhD(biochem), 54. *Prof Exp:* PROF MICROBIOL & IMMUNOL, HEALTH SCI CTR, STATE UNIV NY, 70- *Mailing Add:* 1246 E 21st St Brooklyn NY 11210

SCHIFFMAN, LOUIS F, b Poland, July 15, 27; nat US; m 63; c 2. PHYSICAL CHEMISTRY. *Educ:* NY Univ, BChE, 48, MS, 52, PhD(phys chem), 55. *Prof Exp:* Res chem engr, Pa Grade Crude Oil Asn, 48-50; teaching fel chem, NY Univ, 50-54; res chemist, E I du Pont de Nemours & Co, Inc, 54-56 & Atlantic Refining Co, 56-59; res chemist, Amchem Prods Inc, 59-67, head corrosion group, 67-70; PRES & CONSULT, TECHNI RES ASSOCS, INC, WILLOW GROVE, 70- *Concurrent Pos:* Publ & ed, Patent Licensing Gazette, World Technol & Guide to Available Technols. *Mem:* Am Chem Soc; Am Inst Chem; NY Acad Sci; Licensing Exec Soc; Sigma Xi; Technol Transfer Soc. *Res:* Enhanced recovery, petroleum production; combustion; barrier separations; corrosion; surface treatment of metals; electroplating; industrial chemistry; radiochemistry. *Mailing Add:* 1837 Merritt Rd Abington PA 19001

SCHIFFMAN, ROBERT A, b New York, NY, July 4, 45; m; c 1. HIGH TEMPERATURE SCIENCE, MICROGRAVITY SCIENCE. *Educ:* Long Island Univ, BA, 68; NMex Inst Mining & Technol, MS, 73, PhD(chem metall), 78. *Prof Exp:* Metall consult, 69-79; res assoc, high temperature sci, Ames Labs, 79-81; res assoc fac, high temperature containerless sci, Yale Univ, 81-84; sr mat scientist, containerless mat sci, Midwest Res Inst, 84-87; dir res, Intersonics Inc, 87-90; PRES, RS RES INC, 90- *Concurrent Pos:* Pres, Robert Schiffman Res Inst, 82- *Mem:* Am Chem Soc; Am Soc Metals; Am Inst Metall & Mining Engrs; Metall Soc; Mat Res Soc. *Res:* High temperature containerless environments for materials research and processing; laser fluorescence spectroscopy; temperature measuring; high temperature thermodynamics and phase diagrams of multicomponent systems; microgravity experimental methods. *Mailing Add:* 1960 Saunders Rd Riverwoods IL 60015

SCHIFFMAN, ROBERT L, b New York, NY, Oct 27, 23; m 47; c 2. SOIL MECHANICS, COMPUTER SCIENCE. *Educ:* Cornell Univ, BCE, 47; Columbia Univ, MS, 51; Rensselaer Polytech Inst, PhD(soil mech), 60. *Prof Exp:* Asst civil engr, New York Dept Hosps, 47-49; res assoc, Columbia Univ, 49-54, instr civil eng, 54-55; asst prof, Lehigh Univ, 55-57 & Rensselaer Polytech Inst, 57-60, assoc prof, 60-63, prof, 63-66 & theoret soil mech, Univ Ill, Chicago Circle, 66-70; lectr civil eng & fac res assoc, Comput Ctr, 69-70, assoc dir res, 70-77, PROF CIVIL ENG, UNIV COLO, BOULDER, 70- *Concurrent Pos:* Prin investr, Res Projs, Am Iron & Steel Inst-Pa State Hwy Dept, 55-57, Off Naval Res, 58-66, land locomotion lab, Army Mat Command, 59-65, NSF, 61-, US Geol Surv, 61-64 & US Bur Pub Rds, 64-66; mem comt stress distrib in earth masses, Hwy Res Bd, Nat Acad Sci-Nat Res Coun, 56-63, dept soils, found & geol, comt composite pavements & res needs comt, 64-70, mem comt on mech of earth masses & layered systs, 64-, chmn, 64-70, mem comt theories of pavement design, 65-, comt design of composite pavements & struct overlays & task force on comt interaction on pavement design, 70-; vis assoc prof civil eng, Mass Inst Technol, 62-63, lectr, 63-65, vis prof, 65-66; ed, newsletter, Inst Soc Terrain-Vehicle Systs, 65-67; ed, Soil Mech & found div newslett, 66-68; lectr, Northwestern Univ, 67-68; indust consult; lectr. *Honors & Awards:* Hogentogler Award, Am Soc Testing & Mat, 60. *Mem:* Am Soc Civil Engrs; Int Soc Terrain-Vehicle Systs; Am Soc Cybernetics; Asn Comput Mach; Am Soc Testing & Mat. *Res:* Theoretical soil mechanics; theory of consolidation; elasticity; computer science. *Mailing Add:* Dept Civil Eng Ecot 525 Univ Colo Boulder CO 80309-0928

SCHIFFMAN, SANDRA, b Minneapolis, Minn, Feb 26, 37; m 57; c 2. BIOCHEMISTRY. *Educ:* Univ Calif, Berkeley, AB, 58; Univ Southern Calif, PhD(biochem, blood clotting), 61. *Prof Exp:* From instr to asst prof biochem, 61-75, assoc prof, 75-83, PROF MED & BIOCHEM, SCH MED, UNIV SOUTHERN CALIF, 83- *Concurrent Pos:* Estab investr, Am Heart Asn, 71-76, mem, Thrombosis Coun, Am Heart Asn. *Mem:* Am Soc Biol Chemists; Am Soc Hemat; Am Heart Asn; NY Acad Sci; AAAS; Sigma Xi. *Res:* Chemistry of proteins of blood coagulation. *Mailing Add:* 3920 Sapphire Dr Encino CA 91436

SCHIFFMAN, SUSAN S, b Chicago, Ill, Aug 24, 40; m 89; c 1. PSYCHOPHYSICS, CHEMORECEPTION. *Educ:* Syracuse Univ, BA, 65; Duke Univ, PhD(psychol), 70. *Prof Exp:* Fel aging & obesity, 70-72, from asst prof to assoc prof, 72-81, DIR, WEIGHT CONTROL UNIT, DEPT PSYCHIAT, DUKE MED CTR, 76-, PROF PSYCHIAT, 82- *Concurrent Pos:* Mem, Salt & Water Subgroup, Nat Hypertension Task Force, 76; Comt Sodium Restricted Diets, Food & Nutrit Bd, 74-79; vis scientist biochem, Oxford, 80-81. *Mem:* Sigma Xi; Asn Chemoreception Sci; Am Chem Soc; Europ Chemoreception Sci. *Res:* Physicochemical properties that relate to perception of taste and smell; changes of taste and smell with age, obesity and disease state. *Mailing Add:* Dept Psychol Duke Univ Durham NC 27706

SCHIFFMANN, ELLIOT, b Newark, NJ, Apr 23, 27; m 60. BIOCHEMISTRY, ORGANIC CHEMISTRY. *Educ:* Yale Univ, BA, 49; Columbia Univ, PhD(biochem), 55. *Prof Exp:* Scientist, Nat Heart Inst, 55-61; scientist, 61-64, biochemist, Nat Inst Dent Res, 64-; RES BIOCHEMIST, LAB PATH, NAT CANCER INST, NIH. *Concurrent Pos:* Lectr, Grad Biochem Lab, Georgetown Univ, 57-58. *Res:* Antimetabolites and cholesterol biosynthesis; mechanisms of calcification in non osseous tissue; chemotaxis in leucocytes; biosynthesis of connective tissue components; recognition processes in cells. *Mailing Add:* Lab Path Nat Cancer Inst NIH Bldg Ten Rm 2A33 Bethesda MD 20892

SCHIFFMANN, ROBERT F, b New York, NY, Feb 11, 35; m 56; c 3. FOOD SCIENCE, MICROWAVE ENGINEERING. *Educ:* Columbia Univ, BS, 55; Purdue Univ, MS, 59. *Prof Exp:* Anal develop chemist, DCA Food Industs, Inc, 59-60, res proj leader, 60-61; dir lab radiochem, Nucleonics Corp Am, Inc, 61-62, vpres & tech dir radiochem & health physics, 62-63; res scientist microwave processing & foods, DCA Food Industs, Inc, 63-68, sr proj mgr, 68-71; partner new prod res & develop consult, Bedrosian & Assocs, 71-78; PRES, R F SCHIFFMANN ASSOC, CONSULT MICROWAVE, 78- *Concurrent Pos:* Secy & mem exec directorate, Int Microwave Power Inst, 69-71, mem bd gov, 69-, pres, 73-81, chmn, 81-; assoc ed foods, J Microwave Power, 70-; vpres, Natural Pak Systs, 77-; fel, Int Microwave Power Indust, 83; pres, Innovative Opportunities Ltd, Inc, 84-; nat sci lectr, Inst Food Tech, 87-; fel, Int Microwave Inst, 84. *Honors & Awards:* Putman Award, Putman Publ Co, 72. *Mem:* Inst Food Technologists; Sigma Xi; Int Microwave Inst; Soc Plastics Indust. *Res:* New product and process research and development; microwave processing applications; food processing; extrusion technology; physical chemistry of food systems; fruit and vegetable storage; bakery and dairy production. *Mailing Add:* R F Schiffmann Assoc 149 W 88th St New York NY 10024

SCHIFFRIN, ERNESTO LUIS, b Buenos Aires, Arg, Aug 8, 46; Can citizen; m 71; c 2. HYPERTENSION, CARDIOVASCULAR MEDICINE. *Educ:* Univ Buenos Aires, MD, 69; McGill Univ, PhD(exp med), 80; FRCP(C), 82. *Hon Degrees:* FACP, 84. *Prof Exp:* Resident, internal med, Inst Med Investr, Buenos Aires, Argentina, 70-74, atten physician, 74-76; res fel, Clin Res Inst

Montreal, 76-80, sr investr, 80-83; asst prof, 81-86, ASSOC PROF MED, SCH MED, UNIV MONTREAL, 86-; LAB DIR, EXP HYPERTENSION, CLIN RES INST MONTREAL, 83- *Concurrent Pos:* Atten physician, Hotel-Dieu Hospital, Univ Montreal, 81-; mem, Sci Rev Comt, Can Heart Found, 84-87 & Coun High Blood Pressure Res, Am Heart Asn, 87-89; mem, sci rev comt, Med Res Coun Can, 89- *Honors & Awards:* Astra Young Investr Award, Can Hypertension Soc, 85. *Mem:* Can Hypertension Soc; Inter Am Hypertension Soc; Am Fedn Clin Res; Can Soc Clin Invest; Soc Exp Biol & Med; Int Hypertension Soc; Can Soc Internal Med; Am Hypertension Soc. *Res:* Investigation of mechanisms involved in experimental and clinical hypertension; role of vasoactive peptides in hypertension; regulation of aldosteron secretion. *Mailing Add:* 110 Pine Ave W Montreal PQ H2W 1R7 Can

SCHIFFRIN, MILTON JULIUS, b Rochester, NY, Mar 23, 14; m 42; c 2. DRUG DEVELOPMENT, ANALGESICS. *Educ:* Univ Rochester, BA, 37, MS, 39; McGill Univ, PhD(physiol), 41. *Prof Exp:* Instr physiol, Northwestern Univ, 41-42; dir clin res, Hoffman-LaRoche, 46-64, dir drug regulatory affairs, 64-79, asst vpres, 71-79; PRES, WHARRY RES ASN, INC, 79- *Concurrent Pos:* Porter fel, Am Physiol Soc, 41; lectr pharmacol, Sch Med, Univ Ill, 49-56, clin asst prof anesthesiol, 56-61; vis lectr, Rush Med Col, 70-79. *Mem:* Am Med Writer's Asn (pres, 73); Am Physiol Soc; Int Col Surgeons; Am Col Clin Pharmacol; Drug Info Asn; Pharmacol Soc Can. *Res:* Gastrointestinal physiology; pain; clinical research and regulation of new drugs. *Mailing Add:* Wharry Res Asn Inc 1001 Second Ave W Unit 401 Seattle WA 98119-3560

SCHIFREEN, RICHARD STEVEN, b Trenton, NJ, Mar 17, 52. CLINICAL CHEMISTRY, ANALYTICAL CHEMISTRY. *Educ:* Muhlenberg Col, BS, 74; Univ Ga, PhD(chem), 78. *Prof Exp:* Fel clin chem, Hartford Hosp, 78-80; MEM STAFF, E I DU PONT DE NEMOURS & CO, INC, 80- *Concurrent Pos:* Mem area comt clin chem, Nat Comt Clin Lab Standards. *Mem:* Am Chem Soc; Am Asn Clin Chem; Sigma Xi. *Res:* Immunoassay development; fibrinolysis; cancer diagnostics. *Mailing Add:* Life Tech Inc 8717 Grovement Circle PO Box 6009 Gaithersburg MD 20877-4117

SCHILB, THEODORE PAUL, b St Louis, Mo, May 13, 33; m 65; c 1. BIOPHYSICS, PHYSIOLOGY. *Educ:* Univ Louisville, BS, 55, MS, 57, PhD(biophys), 65. *Prof Exp:* Instr physiol, Univ Louisville, 65-67, instr biophys, 67-68; asst prof, Mt Sinai Sch Med, 68-74; asst prof physiol, La State Univ Sch Med, New Orleans, 74-; asst prof physiol, Downstate Med Ctr, State Univ NY; res coordr, Nephrol Div, Brookdale Med Ctr; asst dir spec projs, Nat Nephrol Found; RETIRED. *Concurrent Pos:* Career develop award, 69. *Mem:* Biophys Soc; Am Physiol Soc. *Res:* Active transport of inorganic ions across cell membranes; secretory physiology. *Mailing Add:* 1913 Trevilian Way Louisville KY 40205

SCHILD, ALBERT, b Hessdorf, Mar 3, 20; nat US; m 46; c 7. OPERATIONS RESEARCH. *Educ:* Univ Toronto, BA, 46; Univ Pa, MA, 48, PhD(math), 51. *Prof Exp:* Asst instr math, Univ Pa, 46-50; from instr to prof math, 50-90, chmn dept, 63-79, EMER PROF, TEMPLE UNIV, 90- *Concurrent Pos:* consult, 56-; lectr, 60- *Mem:* Am Math Soc; Math Asn Am. *Res:* Theory of functions; complex variables; operations research. *Mailing Add:* Dept Math Temple Univ Philadelphia PA 19122

SCHILD, RUDOLPH ERNEST, b Chicago, Ill, Jan 10, 40; m 82. ASTRONOMY, PHYSICS. *Educ:* Univ Chicago, BS, 62, MS, 63, PhD(astrophys), 66. *Prof Exp:* Res assoc, Calif Inst Technol, 66-69; ASTROPHYSICIST, SMITHSONIAN ASTROPHYS OBSERV, 69- *Concurrent Pos:* Res consult, Mass Inst Technol, 73-74; lectr astron, Harvard Univ, 73-83. *Mem:* Am Astron Soc; Int Astron Union. *Res:* Extragalactic energetic, especially quasars and x-ray sources; gravitational lenses. *Mailing Add:* Ctr Astrophys 60 Garden St Cambridge MA 02138

SCHILDCROUT, MICHAEL, b New York, NY, Feb 6, 43. PHYSICS, ELECTRICAL ENGINEERING. *Educ:* Hunter Col, BS, 66; Univ Pittsburgh, PhD(physics), 75; George Washington Univ, MS, 85. *Prof Exp:* engr, Naval Intel Support Ctr, 77-79; ELEC ENGR, NAVAL SECURITY GROUP, 79- *Mem:* Am Phys Soc; Inst Elec & Electronics Engrs. *Res:* Evaluating the performance of spread spectrum communication systems; their low probability of intercept; anti-jam capabilities. *Mailing Add:* Electronics Eng Naval Security Group Command 3801 Nebraska Ave NW Washington DC 20390

SCHILDCROUT, STEVEN MICHAEL, b Grand Rapids, Mich, July 18, 43; m 64; c 2. PHYSICAL INORGANIC CHEMISTRY. *Educ:* Univ Chicago, BS, 64; Northwestern Univ, PhD(chem), 68. *Prof Exp:* Res fel chem, Rice Univ, 68-69; from asst prof to assoc prof, 69-81, PROF CHEM, YOUNGSTOWN STATE UNIV, 81- *Mem:* Am Chem Soc; Am Soc Mass Spectrometry; Sigma Xi. *Res:* Mass spectrometry and chemistry of gaseous ions; chemical ionization mass spectrometry. *Mailing Add:* Dept Chem Youngstown State Univ Youngstown OH 44555-3663

SCHILDKRAUT, CARL LOUIS, b Brooklyn, NY, June 20, 37. BIOLOGICAL CHEMISTRY. *Educ:* Cornell Univ, AB, 58; Harvard Univ, AM, 59, PhD(chem), 61. *Prof Exp:* NSF fel, 61-63; asst prof, 64-70, assoc prof, 71-76, PROF CELL BIOL, ALBERT EINSTEIN COL MED, 76- *Concurrent Pos:* Kennedy scholar, 66-69; NIH career develop award, 69-74; mem molecular biol adv panel, NSF, 70-73. *Honors & Awards:* Hirschl Career Scientist Award, 75. *Mem:* AAAS; Am Chem Soc; Am Soc Cell Biol; Am Soc Biol Chem & Molecular Biol. *Res:* Physical chemistry and enzymology of DNA; analysis of the organization and replication of the DNA of mammalian cells; regulatory mechanisms in mammalian cells. *Mailing Add:* Dept Cell Biol Albert Einstein Col Med 1300 Morris Park Ave CH 416 Bronx NY 10461

SCHILDKRAUT, JOSEPH JACOB, b Brooklyn, NY, Jan 21, 34; m 66; c 2. PSYCHIATRY, NEUROPSYCHOPHARMACOLOGY. *Educ:* Harvard Col, AB, 55; Harvard Med Sch, MD, 59. *Prof Exp:* Intern med, Univ Calif Hosp, San Francisco, 59-60; teaching fel, Harvard Med Sch, 60-63; clin assoc, Lab Clin Sci, NIMH, 63-65, spec fel, 65-66, res psychiatrist, 66-67; asst prof, Harvard Med Sch, 67-70, assoc prof, 70-74; resident psychiat, 60-63, chief resident res unit, 61-63, SR PSYCHIATRIST & DIR NEUROPSYCHOPHARMACOL LAB, MASS MENT HEALTH CTR, 67-; PROF PSYCHIAT, HARVARD MED SCH, 74- *Concurrent Pos:* Prin investr numerous grants; consult comts & orgns. *Honors & Awards:* Anna-Monika Found Prize, Dortmund, Ger, 67; McCurdy-Rinkel Prize, Am Psychiat Asn, 69, Hofheimer Prize, 71; William C Menninger Mem Award, Am Col Physicians, 78. *Mem:* Am Psychiat Asn; Am Col Psychiatrists; World Psychiat Asn; Am Col Neuropsychopharmacol; Am Soc Pharmacol & Exp Therapeut. *Res:* Neuropsychopharmacology, biochemistry, and biology of psychiatric disorders, particularly the affective disorders (depressions and manias) and the schizophrenic disorders. *Mailing Add:* Mass Ment Health Ctr 74 Fenwood Rd Boston MA 02115

SCHILE, RICHARD DOUGLAS, b New Haven, Conn, Apr 3, 31; m 53; c 3. ENGINEERING MECHANICS, MATERIALS ENGINEERING. *Educ:* Rensselaer Polytech Inst, BAeroE, 53, MS, 57, PhD(mech), 67. *Prof Exp:* Res engr, Res Labs, United Aircraft, 57-59, group supvr, 59-62, sr mat scientist, 62-69; assoc prof eng, Dartmouth Col, 69-76; eng res dir, Ciba-Geigy Corp, 76-82; PROF MECH ENG, UNIV BRIDGEPORT, 83-, CHMN MECH ENG DEPT, 88- *Concurrent Pos:* Pres, Ardes Enterprises, 82- *Mem:* Am Soc Mech Engrs. *Res:* Engineering research and development of thermoplastic and thermosetting; polymers and additives for high performance composites. *Mailing Add:* 22 Bloomer Rd Ridgefield CT 06877

SCHILLACI, MARIO EDWARD, b Philadelphia, Pa, Feb 18, 40; m 62; c 3. RADIATION PHYSICS, PARTICLE PHYSICS. *Educ:* Drexel Univ, BS, 62; Brandeis Univ, MA, 64, PhD(physics), 68. *Prof Exp:* Fel, 67-69, STAFF SCIENTIST, MEDIUM-ENERGY PHYSICS DIV, LOS ALAMOS SCI LAB, UNIV CALIF, 70- *Concurrent Pos:* Prof physics, Univ NMex, Los Alamos. *Mem:* Am Phys Soc; Radiation Res Soc. *Res:* Medium energy particle physics; medical radiation physics. *Mailing Add:* Medium-Energy Physics Div MP-4 H846 Los Alamos Nat Lab Los Alamos NM 87545

SCHILLER, ALFRED GEORGE, b Irma, Wis, Dec 5, 18; m 44; c 2. VETERINARY MEDICINE. *Educ:* Mich State Univ, DVM, 43, MS, 56; Am Col Vet Surg, dipl, 66. *Prof Exp:* Practr small animal med, Minneapolis, Minn, 47-52; from instr to prof vet clin med, Small Animal Clin, Univ Ill, Urbana, 52-90, actg head dept, 76-78, actg assoc dean acd affairs, 78; RETIRED. *Mem:* Am Col Vet Surg (pres, 72, exec secy, 75-); Sigma Xi. *Res:* Veterinary surgery. *Mailing Add:* 405 Park Lane Dr Champaign IL 61820

SCHILLER, CAROL MASTERS, b St Augustine, Fla, Dec 31, 40; m 64; c 2. BIOCHEMICAL TOXICOLOGY. *Educ:* State Univ NY, Cortland, BSc, 62; Univ NC, Chapel Hill, MAT, 63, JD, 84; Univ Tex, Dallas, PhD(biochem), 70. *Prof Exp:* Instr chem & physics, Barlow High Sch, Redding, Conn & Jordan High Sch, Durham, NC, 63-65; fel med & tutor chem, Univ Toronto, 71-73; res assoc biochem, Univ NC, 73-75; sr staff fel, NIH, 75-78, res chemist, 79-85; CONSULT TOXICOL, 86- *Concurrent Pos:* Alt mem, Nutrit Coord Comt, NIH, 75-, coordr, Fed Women's Prog, 77-78, mem, Digestive Dis Coord Comt, 78-; adj asst prof biochem, Univ NC, 76-80, adj assoc prof, 80-, mem fac, Med Sch, 78-, mem grad fac toxicol, 80-; mem fac, W A Jones Cell Sci Ctr, 77-78, chmn, In Vitro Res & Human Values Comt, Tissue Cult Asn, 74-78; Cong fel, 85-86; expert witness, 86-; chair, Sect Toxicol, Regulatory Affairs & Legal Asst Ctr, 89-92. *Mem:* Am Chem Soc; AAAS; Asn Women Sci; Soc Toxicol; NY Acad Sci. *Res:* Effects of environmental chemicals/causation; premarket product development. *Mailing Add:* UCB Plaza Suite 220 3605 Glenwood Ave Raleigh NC 27612

SCHILLER, EVERETT L, medical parasitology, for more information see previous edition

SCHILLER, JOHN JOSEPH, b Philadelphia, Pa, Dec 10, 35; m 57; c 3. MATHEMATICS. *Educ:* La Salle Col, BA, 57; Temple Univ, MA, 60; Univ Pa, PhD(math), 66. *Prof Exp:* Physicist, US Naval Res & Develop Ctr, 57-59; from instr to asst prof, 59-71, ASSOC PROF MATH, TEMPLE UNIV, 71- & RES ASSOC PROF PHYSIOL & BIOPHYS, 78- *Concurrent Pos:* Mathematician, John D Kettelle Corp, 69. *Mem:* Am Math Soc. *Res:* Riemann surfaces; artificial intelligence. *Mailing Add:* 3630 Salina Rd Philadelphia PA 19154

SCHILLER, NEAL LEANDER, b Lowell, Mass, Nov 5, 49; m 73; c 3. MEDICAL MICROBIOLOGY, INFECTIOUS DISEASES. *Educ:* Boston Col, BS, 71; Univ Mass, Amherst, PhD(microbiol), 76. *Prof Exp:* Res fel, Div Infectious Dis, Cornell Med Ctr, New York Hosp, 76-78, clin lab tech trainee, Diag Microbiol Lab, 77-78; asst prof, 79-85, ASSOC PROF MED MICROBIOL, DIV BIOMED SCI, UNIV CALIF, RIVERSIDE, 85- *Concurrent Pos:* Sabbatical leave, Lab Clin Invest, Nat Inst Allergy & Infectious Dis, NIH, Bethesda, Md, 85-86; chmn, div gen med microbiol, Am Soc Microbiol, 86-87. *Mem:* Am Soc Microbiol; AAAS. *Res:* The interaction of pathogenic bacteria with host defense mechanisms; characterization of bacterial virulence factors; examination of host defenses including chemotaxis, serum killing, opsonization, complement activation, phagocytic uptake and intracellular killing; identifying ways to enhance the host defense mechanisms. *Mailing Add:* Div Biomed Sci Univ Calif 900 University Ave Riverside CA 92521-0121

SCHILLER, PETER WILHELM, b Frauenfeld, Switz, Feb 9, 42; Swiss & Can citizen. PEPTIDE CHEMISTRY, MOLECULAR PHARMACOLOGY. *Educ:* Swiss Fed Inst Technol, Zurich, dipl, 66, DSc, 71. *Prof Exp:* Res fel, dept biol, Johns Hopkins Univ, Baltimore, 71-72 & lab chem biol, Nat Inst Arthritis, Metab, & Digestive Dis, NIH, Bethesda, Md, 73-74; from asst prof to assoc prof, 75-85, PROF, DEPT MED, UNIV

MONTREAL, 85-; DIR, LAB CHEM BIOL & PEPTIDE RES, CLIN RES INST MONTREAL, 75- *Concurrent Pos:* Vis scholar, dept biochem, Univ Wash, Seattle, 69; vis prof, dept molecular biol, Swiss Fed Inst Technol, Zurich, 79; mem, Sci Rev Comt, Can Heart Found, 81-84, planning comt, Am Peptide Symposia, 81-87; consult, Inst Armand-Frappier, Laval, Que, Can, 84-; mem, Res Rev Comt, Nat Inst Drug Abuse, 86-; assoc mem, Dept Exp Med, McGill Univ, Montreal, 78-89, adj prof, 89- *Honors & Awards:* Kern Prize & Silver Medal, Swiss Fed Inst Technol, Zurich, 71; Max-Bergmann Medal for Achievements in Peptide Res, 87; Marcel-Piché Prize, 87. *Mem:* AAAS; Am Chem Soc; Am Soc Biochem & Molecular Biol; Can Biochem Soc; NY Acad Sci; Swiss Biochem Soc; Protein Soc; Am Peptide Soc; Am Soc Hypertension; fel Royal Soc Can. *Res:* Molecular pharmacology of peptide hormones and neurotransmitters; chemical synthesis; structure-activity relationships; conformational aspects of peptide-receptor interactions; characterization of receptors and peptide drug-development; chemistry and pharmacology of opioids. *Mailing Add:* Clin Res Inst Montreal 110 Pine Ave W Montreal PQ H2W 1R7 Can

SCHILLER, RALPH, b New York, NY, July 8, 26; m 50; c 3. THEORETICAL PHYSICS. *Educ:* Brooklyn Col, BA, 48; Syracuse Univ, MS, 50, PhD(physics), 52. *Prof Exp:* Asst physics, Syracuse Univ, 48-52; asst prof, Univ Sao Paulo, 52-54; from asst prof to prof physics, 54-90, head dept, 75-86, EMER PROF, STEVENS INST TECHNOL, 90- *Concurrent Pos:* Res assoc, Syracuse Univ, 60-61; vis prof, Weizmann Inst, 80. *Mem:* Am Phys Soc; Sigma Xi. *Res:* Optics. *Mailing Add:* Dept Physics Stevens Inst Technol Hoboken NJ 07030

SCHILLER, WILLIAM R, b Bennett, Colo, Jan 14, 37; m 60; c 2. SURGERY. *Educ:* Drury Col, BS, 58; Northwestern Univ, MD, 62. *Prof Exp:* Prof surg, Univ NMex, 78-83; dir, Trauma Ctr, St Joseph Hosp, Phoenix, Ariz, 83-89; DIR, BURN & TRAUMA CTR, MARICOPA MED CTR, PHOENIX, ARIZ, 89- *Concurrent Pos:* Bd dirs, Am Trauma Soc, 84-90; clin prof surg, Univ Ariz, 89-91. *Mem:* Am Trauma Soc; Am Col Surgeons; Am Asn Surg Trauma. *Res:* Investigation into the metabolic effects of injury; clinical aspects of burn and trauma. *Mailing Add:* Burn & Trauma Ctr Maricopa Med Ctr 2601 E Roosevelt Phoenix AZ 85008

SCHILLETTER, JULIAN CLAUDE, b Clemson, SC, Nov 1, 01. HORTICULTURE. *Educ:* Clemson Univ, BS, 22; Iowa State Univ, MS, 23, PhD, 30. *Prof Exp:* From instr to assoc prof, 22-45, prof hort & dir residence, 45-67, residence analyst, 67-72, EMER PROF HORT, IOWA STATE UNIV, 72- *Concurrent Pos:* Mem, Int Hort Cong, 30; horticulturist, Nat Res Proj, Works Progress Admin, 38. *Res:* Differentiation of flower bud in Dunlap strawberries; growth of Dunlap strawberries; general horticulture. *Mailing Add:* 111 Lynn Ames IA 50010

SCHILLING, CHARLES H(ENRY), b Louisville, Ky, June 3, 18; m 45; c 5. CIVIL ENGINEERING. *Educ:* US Mil Acad, BS, 41; Univ Calif, MS, 47; Rensselaer Polytech Inst, PhD(civil eng), 59. *Prof Exp:* US Army, 41-80, engr combat battalion, France & Ger, 44-45, engr aviation battalion, Ger, 47-50, instr mil art & eng, US Mil Acad, 51-52, assoc prof, 52-55, area engr, Eastern Ocean Dist, Corps Engrs, 55-56, prof mil art & eng, US Mil Acad, 56-69, head dept, 63-69, prof eng & head dept, US Mil Acad, 69-80; CIVIL ENGR, 81- *Concurrent Pos:* Vis prof, Univ Mich, 65 & Univ Stuttgart, 68-69. *Mem:* Am Soc Eng Educ; Soc Am Mil Engrs; Am Soc Civil Engrs; Nat Soc Prof Engrs. *Res:* Structural engineering; vibrations in suspension bridges; application of computers in engineering and education. *Mailing Add:* Four Trahern Terr Clarksville TN 37040

SCHILLING, CURTIS LOUIS, JR, b Goshen, NY, May 19, 40; m 61; c 2. POLYMER & ORGANOSILICON CHEMISTRY. *Educ:* Syracuse Univ, BS, 61, MS, 64; Univ Ariz, PhD(org chem), 67. *Prof Exp:* US Army Res Off grantee, Univ Iowa, 67-68; res chemist, 68-75, proj scientist, 76-79, res scientist, 80-84, SR RES CHEMIST, SPECIALTY CHEM DIV, UNION CARBIDE CORP, 84 - *Mem:* Am Chem Soc; Am Ceramic Soc. *Res:* Organosilicon chemistry; silicone surfactant stabilization of polyurethane foams; organosilicon routes to silicon carbide; organofunctional silanes. *Mailing Add:* Specialty Chem Div Union Carbide Corp PO Box 180 Sistersville WV 10591

SCHILLING, EDWARD EUGENE, b Los Angeles, Calif, Sept 23, 53. PLANT SYSTEMATICS, CHEMOTAXONOMY. *Educ:* Mich State Univ, BS, 74; Ind Univ, PhD(biol), 78. *Prof Exp:* Instr bot, Univ Tex, 78-79; ASST PROF BOT, UNIV TENN, 79- *Mem:* AAAS; Am Soc Plant Taxonomists; Bot Soc Am; Sigma Xi. *Res:* Systematics and evolution of weeds; cytoplasmic genetics; flavonoid chemistry. *Mailing Add:* Dept Bot Univ Tenn Knoxville TN 37996

SCHILLING, EDWARD GEORGE, b Lancaster, NY, Nov 9, 31; m 59; c 2. QUALITY CONTROL. *Educ:* Univ Buffalo, BA, 53, MBA, 54; Rutgers Univ, MS, 62, PhD(statist), 67. *Prof Exp:* Instr statist, Univ Buffalo, 57-59; engr, Radio Corp Am, 59-61; teaching asst statist, Rutgers Univ, 61-62; sr engr, Carborundum Corp, 62-64; instr statist, Rutgers Univ, 64-67; consult statistician, Lamp Bus Div, Gen Elec Co, 69-74, mgr statist & qual systs oper, 75-80, mgr lighting quality oper, Lighting Bus Group, 80-83; assoc prof, 67-69, PAUL A MILLER PROF & CHMN GRAD STATIST DEPT, ROCHESTER INST TECHNOL, 83- *Concurrent Pos:* Consult, 68-69, 83-; series ed, Marcel Dekker, Inc, NY, 84- *Honors & Awards:* Brumbaugh Award, Am Soc Qual Control, 74, 78, 79 & 81, Shewhart Medal, 83; Ellis R Ott Award, 84. *Mem:* Fel Am Statist Asn; fel Am Soc Qual Control; Inst Math Statist; Am Soc Testing & Mat; Am Econ Asn. *Res:* Mathematical statistics with applications in the physical and engineering sciences, quality control, business and economics. *Mailing Add:* Rochester Inst Technol Rochester NY 14623-0887

SCHILLING, GERD, b Hanover, Ger, Oct 6, 39; m 69; c 2. PLASMA HEATING, NEUTRAL BEAM TECHNOLOGY. *Educ:* Mass Inst Technol, BS, 61; Case Inst Technol, MS, 63, PhD(physics), 67. *Prof Exp:* Res assoc, Notre Dame Univ, 68-70; res physicist, Max Planck Inst Plasma Physics, WGer, 70-74 & Oak Ridge Nat Lab, Fusion Energy Div, 74-77; RES PHYSICIST, PLASMA PHYSICS LAB, PRINCETON UNIV, 77- *Mem:* Am Phys Soc. *Res:* Controlled thermonuclear fusion and plasma physics. *Mailing Add:* Plasma Physics Lab Princeton Univ PO Box 451 Princeton NJ 08544

SCHILLING, JESSE WILLIAM, crystallography, for more information see previous edition

SCHILLING, JOHN ALBERT, b Kansas City, Mo, Nov 5, 17; m 43; c 4. SURGERY. *Educ:* Dartmouth Col, AB, 37; Harvard Univ, MD, 41; Am Bd Surg, dipl, 48. *Prof Exp:* Resident, Roosevelt Hosp, Columbia Univ, 44; instr surg, Univ Rochester, 44-48, asst prof surg & surg anat, Sch Med & Dent, 48-56; prof surg & head dept, Med Ctr, Univ Okla, 56-74; PROF SURG, MED SCH, UNIV WASH, 74-, CHMN DEPT, 75- *Concurrent Pos:* Mem, Boyd-Bartlett Exped, Arctic, 41; mem adv bd, Am J Surg, 58-; mem surg study sect, Div Res Grants, NIH, 60-64, mem bd sci counr, Nat Cancer Inst, 66-71, chmn, 69-71, mem diag subcomt, Breast Cancer Task Force, 71-; mem, Am Bd Surg, Inc, 63-69, chmn, 68-69; mem comt metab in trauma, Surgeon Gen, US Army, 63-71, chmn, 67-71; consult, Div Hosp & Med Facil, Dept Health, Educ & Welfare, 66-; mem comt trauma, Div Med Sci, Nat Res Coun, 69-72; consult, Off Surgeon Gen, USAF. *Mem:* Am Soc Exp Path; Am Surg Asn; Soc Exp Biol & Med; Am Cancer Soc; Soc Univ Surg; Sigma Xi. *Res:* Intestinal obstruction; circulation of liver; transplantation of cancer; paper chromatography; peptic ulcer; visualization of biliary tract; wound healing; respiratory physiology; shock. *Mailing Add:* Dept Surg RF-25 Univ Wash Health Sci Ctr Seattle WA 98195

SCHILLING, JOHN H(AROLD), b Lincoln, Nebr, Sept 7, 27; m 53; c 2. GEOLOGY, MINING. *Educ:* Pa State Univ, BS, 51; NMex Inst Mining & Technol, MS, 52. *Prof Exp:* Geologist, NMex Bur Mines, 51-56 & 59-60 & Cerro de Pasco Corp, Peru, 56-58; mining geologist, Nev Bur Mines & Geol, 60-70, assoc dir, 70-73, dir & state geologist, 73-88; GEOL & MINERAL RESOURCES CONSULT, 88- *Concurrent Pos:* Mem adv bd, Ctr Water Res, 51-79; secy, Nev Oil & Gas Conserv Comn, 66-70, comnr, 70-76; ed, Isochron/West, 69-; chmn, Gov's Mapping Comt, 70-88 & State Bd Geog Names, 85-88; consult various companies & govt agencies; vpres, Nev Hist Press, 71-89; head, J S Enterprises, 89- *Mem:* Am Inst Mining, Metall & Petrol Engrs; Soc Econ Geol; Am Asn State Geologists. *Res:* Economic geology; metallogenics; molybdenum deposits; isotopic age of intrusive rocks. *Mailing Add:* 1301 Royal Dr Reno NV 89503

SCHILLING, PRENTISS EDWIN, experimental statistics; deceased, see previous edition for last biography

SCHILLING, ROBERT FREDERICK, b Adell, Wis, Jan 19, 19; m 46; c 5. MEDICINE. *Educ:* Univ Wis, BS, 40, MD, 43; Am Bd Nutrit, dipl; Am Bd Internal Med, dipl, 51. *Prof Exp:* Asst med, Harvard Med Sch, 49-51; from asst prof to assoc prof, 51-62, PROF MED, UNIV WIS-MADISON, 62- *Concurrent Pos:* Commonwealth Fund res fel, London Hosp, Eng, 59; consult, NIH, Food & Drug Admin & Vet Admin. *Mem:* Am Soc Clin Invest; Soc Exp Biol & Med; Am Fedn Clin Res; Asn Am Physicians. *Res:* Hematology; nutrition. *Mailing Add:* Dept Med Univ Wis Madison WI 53792

SCHILLING, WILLIAM FREDERICK, b Toledo, Ohio, Aug 21, 42; m 69; c 1. PHYSICAL METALLURGY. *Educ:* Case Inst Technol, BS, 64; Mass Inst Technol, ScD, 69. *Prof Exp:* Res engr phys metall, Alcoa Res Labs, 64-65; res asst, Mass Inst Technol, 65-69, staff mem, Div Sponsored Res, 69-71; sr res assoc, Lab Phys Sci, PR Mallory Co, 71-72; engr process develop, Mat & Processes Lab, 72-76, mgr mat develop, 76-79, mgr advan mat systs, Gas Turbine Div, 79-84, MGR MFG ENG & TECHNOL, TURBINE BUS OPER, GEN ELEC CO, 84- *Mem:* Am Soc Metals; Am Powder Metall Inst. *Res:* Materials and process development of high temperature materials for advanced industrial gas turbines, including superalloy development, directional solidification, hot corrosion resistant coatings, coating processes and hot isostatic pressing applications; plasma spray processing, especially vacuum plasma spraying deposition; design and development of flexible manufacturing systems, advanced metal removal technologies, laser-aided manufacturing. *Mailing Add:* 741 Richbourg Rd Greenville SC 29615

SCHILLINGER, EDWIN JOSEPH, b Chicago, Ill, July 14, 23; m 49; c 6. PHYSICS, SCIENCE EDUCATION. *Educ:* DePaul Univ, BS, 44; Univ Notre Dame, MS, 48, PhD(physics), 50. *Prof Exp:* From instr to prof, 50-88, chmn dept, 52-68 & 76-79, dean, Col Lib Arts & Sci, 66-70 & 80-81, EMER PROF PHYSICS, DEPAUL UNIV, 88- *Concurrent Pos:* Consult, NSF, 62-67 & Off Supt Pub Instr, State Ill, 69-72. *Mem:* AAAS; fel Am Phys Soc; Am Asn Physics Teachers. *Res:* Development of courses and curricula in physics and interdisciplinary science for general students, with emphasis on history, philosophy and methodology of science and its interaction with society and public policy. *Mailing Add:* 7724 Peterson Ave Chicago IL 60631

SCHILLINGER, JOHN ANDREW, JR, b Severn, Md, June 17, 38; m 59; c 2. AGRONOMY, CROP BREEDING. *Educ:* Univ Md, College Park, BS, 60, MS, 62; Mich State Univ, PhD(plant breeding), 65. *Prof Exp:* Assoc instr plant breeding, Mich State Univ, 63-65; res entomologist, Entom Res Div, USDA & Mich State Univ, 65-67; from asst prof to assoc prof plant breeding, Univ Md, College Park, 67-73; SOYBEAN & ALFALFA PROJ LEADER, ASGROW SEED CO, 73- *Mem:* Am Soc Agron; Entom Soc Am. *Res:* Development of improved varieties of alfalfa and soybeans with resistance to pests and with high yield potential. *Mailing Add:* 645 E Ridge Circle Kalamazoo MI 49009-9107

SCHILSON, ROBERT E(ARL), b Keokuk, Iowa, May 25, 27; m 52; c 1. CHEMICAL ENGINEERING. *Educ:* Univ Ill, BS, 50; Univ Minn, PhD(chem eng), 58. *Prof Exp:* Chem engr, Hanford Works, Gen Elec Co, Wash, 50-53; res engr, Marathon Oil Co, 58-61, adv res engr, 61-65, sr res engr, Denver Res Ctr, 65-73, mgr eng dept, 73-77, adv sr refining engr, La Refining Div, 77-82; lectr math, Univ Sierra Leone, WAfrica, US Peace Corps, 87-89; RETIRED. *Mem:* Am Inst Mining, Metall & Petrol Engrs; Am Inst Chem Eng. *Res:* Thermal methods of oil recovery; catalysis and chemical kinetics; refining and petrochemicals; coke and carbon technology. *Mailing Add:* 14 Anchorage Salem SC 29676

SCHILT, ALFRED AYARS, b Haigler, Nebr, Aug 30, 27; m 49; c 4. ANALYTICAL CHEMISTRY. *Educ:* Univ Colo, BA, 50, MA, 52; Univ Ill, PhD(chem), 56. *Prof Exp:* Anal chemist, Eastman Kodak Co, 51-53; asst, Univ Ill, 53-56; from instr to asst prof chem, Univ Mich, 56-62; from assoc prof to prof chem, Northern Ill Univ, 62-89; CHEM CONSULT, 89- *Concurrent Pos:* Vis res prof, Ind Univ, 70-71. *Mem:* Am Chem Soc. *Res:* Coordination compounds and their application in chemical analysis; analytical separations; spectroscopy; general analytical methods; perchloric acid and perchlorates. *Mailing Add:* 7704 NW 44th Pl Gainesville FL 32606

SCHIMA, FRANCIS JOSEPH, b Chicago, Ill, Apr 15, 35; m 67; c 2. NUCLEAR PHYSICS, RADIOACTIVITY METROLOGY. *Educ:* Ill Benedictine Col, BS, 57; Univ Notre Dame, PhD(physics), 64. *Prof Exp:* Res assoc, Ind Univ, 64-66; PHYSICIST, NAT BUR STANDARDS, 66- *Concurrent Pos:* Consult, Nat Coun Radiation Protection & Measurements, 75- *Mem:* Am Phys Soc; Sigma Xi. *Res:* Study nuclear structure through measurement of radioactive decay; prepare standards of radioactivity; low level radioactivity measurements for neutrino detectors. *Mailing Add:* Radiation Physics 245 C114 Nat Bur Standards Washington DC 20234

SCHIMEK, ROBERT ALFRED, b Beaver Falls, Pa, May 1, 26; m 50; c 2. MEDICINE. *Educ:* Franklin & Marshall Col, BS, 45; Johns Hopkins Univ, MD, 50. *Prof Exp:* Asst instr, Johns Hopkins Univ & house officer & resident, Hosp, 50-53; staff ophthalmologist, Henry Ford Hosp, Mich, 53-57; head dept ophthal, Ochsner Clin & Found Hosp, 57-77; assoc prof, 57-76, CLIN PROF OPHTHAL, SCH MED, TULANE UNIV, 76- *Concurrent Pos:* Clin prof, Sch Med, La State Univ, 78; mem vis staff, Eye, & Ear Inst, Charity, Touro & E Jefferson Hosp, 57- *Mem:* AMA; Asn Res Vision & Ophthal; Am Col Surgeons; Am Acad Ophthal. *Res:* Ophthalmology. *Mailing Add:* 4224 Houma Blvd No 110 Metairie LA 70006

SCHIMELPFENIG, CLARENCE WILLIAM, b Dallas, Tex, Apr 8, 30; m 56; c 3. ORGANIC CHEMISTRY, CHEMISTRY EDUCATION. *Educ:* NTex State Col, BS, 53, MS, 54; Univ Ill, PhD, 57. *Prof Exp:* Asst prof chem, George Wash Univ, 57-59 & NTex State Univ 59-62; res chemist, E I du Pont de Nemours & Co, Inc, 62-73; asst prof chem, State Univ NY Buffalo, 73-75 & Erskine Col, 75-76; from asst prof to assoc prof chem, Tex Wesleyan Col, 76-81; assoc prof chem, NTex State Univ, 81- 82; PROF CHEM, DALLAS BAPTIST UNIV, 82- *Concurrent Pos:* Robert A Welch Found grantee, 60-62, 77-81. *Mem:* Am Chem Soc; Royal Soc Chem; Sigma Xi. *Res:* Organic and polymer chemistry. *Mailing Add:* 2008 Silver Leaf Dr Pantego TX 76013

SCHIMERT, GEORGE, b Raemismuehle-Zell, Switz, Feb 19, 18; US citizen; m 56; c 8. CARDIOVASCULAR SURGERY. *Educ:* Univ Bonn, MD, 42; Pazmany Peter Univ, Budapest, MD, 43; Univ Minn, MSc, 60; Am Bd Surg, dipl, 61; Am Bd Thoracic Surg, dipl, 62. *Prof Exp:* Asst surg, Univ Md, 54-55; adv surg & chief thoracic surg, Med Col, Seoul Univ, 58-59; from instr to asst prof surg, Univ Minn, 60; from asst prof to prof surg, State Univ NY, Buffalo, 60-90; chief cardiovasc surg, Buffalo Gen Hosp, 65-90, surgeon, 68- 90; RETIRED. *Concurrent Pos:* Fel cardiovasc surg, Univ Md, 55-56; fel med, Hosps, Univ Minn, 56-58; attend thoracic surgeon, Vet Hosp, Buffalo, 61-; assoc surg, Buffalo Gen Hosp, 65-68; assoc attend cardiovasc surgeon, Children's Hosp, 67-; assoc prof, Sch Med, NY Univ, 69- *Mem:* AMA; fel Am Col Chest Physicians; fel Am Col Surg; fel Soc Thoracic Surg; fel Soc Vascular Surg. *Res:* Development of cardiopulmonary bypass equipment; design of prosthetic heart valves; multivalvular replacement; determination of myocardial sodium potassium ratios; correction of overwhelming heart failure by cardiac surgical procedures. *Mailing Add:* Dept Surg State Univ NY Health Sci Ctr 3435 Main St Buffalo NY 14214

SCHIMITSCHEK, ERHARD JOSEF, b Neutitschein, Czech, Dec 8, 31; m 56; c 2. PHYSICAL CHEMISTRY, PHYSICS. *Educ:* Univ Munich, Dr rer nat, 57. *Prof Exp:* Sr res engr, Convair Gen Dynamics Corp, 58-60, staff scientist, 60-62; PHYSICIST, NAVAL OCEAN SYSTS CTR, 62- *Mem:* Am Phys Soc. *Res:* Electrooptics; quantum electronics; liquid and gas discharge lasers. *Mailing Add:* 3930 Point Loma Ave San Diego CA 92106

SCHIMKE, ROBERT T, b Spokane, Wash, Oct 25, 32; div; c 4. BIOCHEMISTRY, MOLECULAR BIOLOGY. *Educ:* Stanford Univ, AB, 54, MD, 58. *Prof Exp:* Biochemist, NIH, 60-65, chief sect biochem regulation, 65-66; chmn dept pharmacol, 70-73, chmn dept biol, 78-82, PROF BIOL, STANFORD UNIV, 66- *Honors & Awards:* Charles Pfizer Award Enzyme Chem, Am Chem Soc, 69; Boris Pregal Award Res Biol, NY Acad Sci, 74; W C Rose Biochem Award, 83; A P Sloan Jr Prize, Gen Motors Cancer Res Found, 85. *Mem:* Nat Acad Sci; Inst Med Nat Acad Sci; Am Soc Biol Chemists; Am Acad Arts & Sci. *Res:* Mechanisms of actions of hormones in metabolic regulation and development; significance and control mechanisms of protein turnover in animals; gene amplification and resistance phenomena. *Mailing Add:* Dept Biol Sci Stanford Univ Stanford CA 94305

SCHIMMEL, ELIHU MYRON, b Bayonne, NJ, Dec 14, 29; m 55; c 2. MEDICINE. *Educ:* Univ Ill, Urbana, AB, 50; Yale Univ, MD, 54. *Prof Exp:* Instr med, Sch Med, Yale Univ, 60-64; PROF MED, SCH MED, BOSTON UNIV, 64-; LECTR, DEPT MED, TUFTS UNIV. *Concurrent Pos:* USPHS res fel, Mass Gen Hosp, Harvard Univ, 58-60; vis scientist, Dept Nutrit Biochem, Mass Inst Technol, 76-77, USDA Human Nutrit Res Ctr, Tufts, 86-87; contrib ed, Nutrit Reviews, 89- *Mem:* Am Fedn Clin Res. *Res:* Gastroenterology; nutrition. *Mailing Add:* 150 S Huntington Ave Boston MA 02130-4893

SCHIMMEL, HERBERT, b New York, NY, Sept 12, 09; m 34, 71; c 6. BIOMATHEMATICS, BIOPHYSICS. *Educ:* Univ Pa, BA, 30, MS, 32, PhD(physics), 36. *Prof Exp:* Engr-economist, US Nat Res Proj, Philadelphia, 36-41; staff dir & consult, US Cong, Washington, DC, 41-47; sr officer sci, technol & econ, UN, 48-52; independent consult, 53-63; assoc math & physics, 64-70, organizer & dir, Sci Comput Ctr, 67-71, assoc prof, 71-78, EMER PROF NEUROL, ALBERT EINSTEIN COL MED, 78-; CONSULT, 79- *Mem:* Inst Elec & Electronics Engrs. *Res:* Applications of math and physics to medicine and biology, especially stochastic processes to brain research and use of time series to estimate adverse effects of air pollution; relations of science to economic and social development. *Mailing Add:* 26 Usonia Rd Pleasantville NY 10570

SCHIMMEL, KARL FRANCIS, b Allentown, Pa, Mar 24, 36; m 61; c 4. POLYMER SYNTHESIS, ORGANIC SYNTHESIS. *Educ:* Muhlenberg Col, BS, 57; Duquesne Univ, PhD(org chem), 61. *Prof Exp:* Res chemist, E I du Pont de Nemours & Co, Inc, 62-64; SR RES ASSOC, PPG INDUSTS, 64- *Mem:* Am Chem Soc. *Res:* Coating and resins end uses. *Mailing Add:* PPG Industs Res Ctr Rosanna Dr PO Drawer 9 Allison Park PA 15101

SCHIMMEL, PAUL REINHARD, b Hartford, Conn, Aug 4, 40; m 61; c 2. BIOPHYSICS, BIOCHEMISTRY. *Educ:* Ohio Wesleyan Univ, AB, 62; Mass Inst Technol, PhD(phys biochem), 66. *Prof Exp:* Res assoc chem, Stanford Univ, 66-67; from asst prof to assoc prof, 67-76, PROF BIOL, MASS INST TECHNOL, 76- *Concurrent Pos:* Alfred P Sloan fel, 70-72; consult, NIH, 75-79; chmn, div biol chem, Am Chem Soc, 84-85. *Honors & Awards:* Pfizer Award, Am Chem Soc, 78. *Mem:* Am Soc Biochem & Molecular Biol; Am Chem Soc; fel AAAS; Am Acad Arts & Sci; Nat Acad Sci. *Res:* Gene, protein structure and function; aminoacyl acid synthetases; molecular recognition of transfer RNA; directed mutagenesis approach to structure-function relationships. *Mailing Add:* Dept Biol Mass Inst Technol Cambridge MA 02139

SCHIMMEL, WALTER PAUL, US citizen. AERONAUTICAL & ASTRONAUTICAL ENGINEERING. *Educ:* Purdue Univ, BS, 65; Univ Notre Dame, MS, 66, PhD(appl phys), 69. *Prof Exp:* Div supvr syst anal, Sandia Nat Labs, Albuquerque, NMex, 69-82; assoc dir electrothermal, Inst Res Hydro, Que, Montreal, 82-85; dept head, 85-88, PROF, EMBRY RIDDLE UNIV, DAYTONA BEACH, FLA, 88- *Concurrent Pos:* NDEA fel, Univ Notre Dame, 65-68. *Mem:* Fel Am Soc Mech Engrs; Accreditation Bd Eng Technol; Am Inst Aeronaut & Astronaut; Am Soc Eng Educ; Can Elec Asn. *Res:* Author of over 100 publications; thermophysical properties; aerothermodynamics and plasma applications; laser measurement techniques; holographic interparometry. *Mailing Add:* 549 Pelican Bay Dr Daytona Beach FL 32119

SCHIMMER, BERNARD PAUL, b Newark, NJ, June 14, 41; m 65; c 2. ENDOCRINOLOGY, MEDICAL RESEARCH. *Educ:* Rutgers Univ, BS, 62; Tufts Univ, PhD(pharmacol), 67. *Prof Exp:* Fel endocrinol & biochem, Brandeis Univ, 67-69; from asst prof to assoc prof, 69-80, PROF MED RES & PHARMACOL, UNIV TORONTO, 80- *Mem:* Am Soc Biol Chemists; Can Soc Biol Chemists; AAAS; Endocrine Soc. *Res:* Studies concerned with regulation of differentiated functions in mammalian somatic cell cultures, with principal emphasis on mechanism of ACTH action in adrenal cortex; hormone action studied through use of biochemistry and molecular genetics in cell culture systems. *Mailing Add:* Banting & Best Dept Med Res Univ Toronto 112 College St Toronto ON M5G 1L6 Can

SCHIMMERLING, WALTER, b Milan, Italy, Mar 10, 37; m 61; c 3. PHYSICS, BIOPHYSICS. *Educ:* Univ Buenos Aires, MS, 62; Rutgers Univ, PhD(radiation sci), 71. *Prof Exp:* Instr physics & Ger, Univ Buenos Aires, 59-62; res assoc physics, Atomic Energy Comn, Arg, 62-65; health physicist, Princeton Univ, 65-66, mem prof staff, Princeton-Pa Accelerator, 66-68, head, radiation measurements, 68-71, asst dir, 71-72; RES SR SCIENTIST, LAWRENCE BERKELEY LAB, UNIV CALIF, BERKELEY, 72- *Concurrent Pos:* Prin investr, NASA, 75- & Nat Cancer Inst, 78-; vis sci, Ctr Nuclear Studies, Saclay, France, 84-85 & 87; lectr, Univ Calif Exten, 78-80; vis sr scientist, NASA, 90-91. *Mem:* Am Phys Soc; Radiation Res Soc; Am Asn Physicists Med. *Res:* High energy heavy ions; nuclear physics and applications to cancer radiotherapy, space shielding and dosimetry. *Mailing Add:* Lawrence Berekely Lab Bldg 29 Rm 215C One Cyclotron Rd Berkeley CA 94720

SCHIMPF, DAVID JEFFREY, b Chicago, Ill, Oct 10, 48; m 71; c 2. PLANT ECOLOGY. *Educ:* Iowa State Univ, BS, 70; Utah State Univ, PhD(biol), 77. *Prof Exp:* Asst prof, 79-85, ASSOC PROF BIOL, UNIV MINN, DULUTH, 85-, DEPT HEAD, 89- *Mem:* AAAS; Am Inst Biol Sci; Ecol Soc Am. *Res:* Autecology, population ecology, and community ecology of vascular plants; vegetation of the Lake Superior region. *Mailing Add:* Dept Biol Univ Minn Duluth MN 55812-2496

SCHIN, KISSU, CELL BIOLOGY, DEVELOPMENTAL BIOLOGY. *Educ:* Univ Göttingen, Germany, PhD(biol), 62. *Prof Exp:* PROF BIOL, STATE UNIV NY, PLATTSBURGH, 70- *Mailing Add:* Dept Biol State Univ NY Plattsburgh Broad & Beekman Sts Plattsburgh NY 12901

SCHINAGLE, ERICH F, b Vienna, Austria, Sept 18, 32; m 57; c 3. PROJECT MANAGEMENT. *Educ:* Univ Vienna, MD, 61. *Prof Exp:* Intern, 61-62 & resident, Dept Internal Med, 62-65; med dir, 65-67, dir clin res, 67-69, med dir, 69-85, PROJ DIR, PHARMACIA INC, 85- *Mem:* AAAS; Am Heart Asn; Am Physiol soc. *Res:* Preclinical and clinical development of drugs and devices in allergy, auto immune disease, cardiology, dermatology, oncology and ophthalmology. *Mailing Add:* 20 Pickle Brook Rd Bernardsville NJ 07924

SCHINDLER, ALBERT ISADORE, b Pittsburgh, Pa, June 24, 27; m 51; c 3. PHYSICS. *Educ:* Carnegie Inst Technol, BS, 47, MS, 48, DSc(physics), 50. *Prof Exp:* Asst physics, Carnegie Inst Technol, 47-50, res physicist, 50-51; dir, Div Mat Res, NSF, 88-90; physicist, US Naval Res Lab, Purdue Univ, 51-60, head metal physics br, Metall Div, 60-75, assoc dir res for mat & gen sci, US Naval Res Lab, 75-85, Mat Res Lab, 85-88, DIR, MIDWEST SUPER CONDUCTIVITY CONSORTIUM, PURDUE UNIV, 90- *Honors & Awards:* Hulburt Award, 56; Nat Capital Award, 62; Naval Res Lab-Sci Res Soc Am Award, 65. *Mem:* Fel Am Phys Soc; Sigma Xi. *Res:* Solid state physics; physics of metals; ferromagnetism; low temperature properties; effects of alloying on physical properties. *Mailing Add:* Midwest Super Conductivity Consortium Purdue Univ West Lafayette IN 47907

SCHINDLER, CHARLES ALVIN, b Boston, Mass, Dec 27, 24; m 55; c 3. MICROBIOLOGY, BIOCHEMISTRY. *Educ:* Rensselaer Polytech Inst, BS, 50; Univ Tex, MA, 56, PhD(microbiol), 61. *Prof Exp:* Res & develop officer, Radiobiol Lab, Atomic Warfare Div, USAF, 52, asst prog dir, Armed Forces Spec Weapons Proj, 52-53, res scientist microbiol, Army Biol Warfare Lab, 56-58, Univ Tex, 61 & Armed Forces Inst Path, 62-68; asst prof microbiol, Univ Okla, 68-72; prof natural sci, Flagler Col, 72-73; sci teacher, Norman Pub Schs, 74-86; SCI SUPVR, OKLA CITY PUB SCHS, 89- *Concurrent Pos:* Charles E Lewis fel, 58. *Mem:* Am Soc Microbiol; Am Chem Soc; Brit Soc Gen Microbiol; NY Acad Sci; Sigma Xi. *Res:* Antibiotics and bacteriolytic enzymes, especially their production and mode of action on the bacterial cell; relationship of action of bacteriolytic enzymes to structure of microorganisms; recipient of United States and foreign patents. *Mailing Add:* 2000 Morgan Dr Norman OK 73069

SCHINDLER, DAVID WILLIAM, b Fargo, NDak, Aug 3, 40; m 64, 79; c 3. LIMNOLOGY, BIOGEOCHEMISTRY. *Educ:* NDak State Univ, BS, 62; Oxford Univ, DPhil(ecol), 66. *Hon Degrees:* DSc, NDak State Univ, 78. *Prof Exp:* Asst prof biol, Trent Univ, 66-68; dir, Exp Lakes Area Proj, 70-87, RES SCIENTIST, FRESHWATER INST, CAN DEPT FISHERIES & OCEANS, 68- *Concurrent Pos:* Res grants, Can Nat Res Coun, Ont Dept Univ Affairs, 66-68 & NSF, 67-68; adj prof zool, Univ Man, 72-, adj prof bot, 81-; vis sr res assoc, Lamont-Doherty Geol Observ, Columbia Univ, 76-; chmn, US Nat Acad Sci Comt Atmosphere & Biosphere, 79-81; chmn, Int Joint Comn Comt Ecol & Geochem; fel, Royal Soc Can, 83- *Honors & Awards:* Frank Rigler Mem Award, Soc Can Limnologists, 84; G E Hutchinson Medal, Am Soc Limnol & Oceanog, 85; Ken Doan Medal, Can Dept Fisheries & Oceans, 85. *Mem:* Am Soc Limnol & Oceanog (vpres, 81-82, pres, 82-83); Brit Ecol Soc; Int Asn Theoret & Appl Limnol; Am Geophys Union; Ecol Soc Am; Am Inst Biol Sci. *Res:* Ecosystems; biological and chemical ecology; biogeochemistry; experimental mainpulation of whole ecosystems. *Mailing Add:* Killan Mem Dept Zool Univ Alta Edmonton AB T6G 2E9 Can

SCHINDLER, GUENTER MARTIN, b Ebersdorf, Ger, Sept 15, 28; nat US; m 57; c 2. MATHEMATICS. *Educ:* Univ Gottingen, dipl, 53, Dr rer nat, 56. *Prof Exp:* Assoc math, Univ Gottingen, 53-56; res mathematician, Kernreactor, Karlsruhe, 56-57; sr mathematician, USAF Missile Develop Ctr, NMex, 57-58; sr scientist, Aerophys Develop Corp, Calif, 58-59; chief mathematician, Adv Tech Corp, 59-61; proj mgr, Gen Elec Co, 61-65; sr tech specialist, NAm Rockwell Corp, 65-68; prin scientist, Douglas Aircraft Co, Long Beach, Calif, 68-74; vis prof, Univ Southern Calif, 74-75; independent consult, 75-77; mem tech staff, Rockwell Int-nam Aircraft Opers, El Segundo, 77-89; RETIRED. *Concurrent Pos:* Lectr, Univ NMex, 58; assoc prof, Univ Calif, 58-59; consult, Avco-Crosley Corp, 59, Astro-Res Corp, 61-65 & US Navy Marine Eng Lab, 65. *Mem:* Am Math Soc; NY Acad Sci. *Res:* Pure and applied mathematics; theoretical physics. *Mailing Add:* 28026 Beechgate Dr Rancho Palos Verdes CA 90274

SCHINDLER, HANS, b Vienna, Austria, Sept 2, 11; nat US; m 39. PETROLEUM CHEMISTRY. *Educ:* Prague Ger Univ, DSc(org chem), 34. *Prof Exp:* Asst chief chemist, Julius Schindler Oil Works, Ger, 35-38; sr res chemist, Pure Oil Co, 38-46; sr res chemist, Witco Chem Corp, 46-52, mgr Petrolia Ref, 53-60,vpres, Sonneborn Div, 60-76; CONSULT, 76- *Concurrent Pos:* Mem, Tech Oil Mission, 45. *Mem:* NY Acad Sci; Am Inst Chem; Am Chem Soc. *Mailing Add:* One Wash Square Village New York NY 10012

SCHINDLER, JAMES EDWARD, b Fargo, NDak, Apr 20, 44; m 67; c 2. ZOOLOGY. *Educ:* NDak State Univ, BS, 66; Queen's Col, Oxford Univ, DPhil(zool), 69. *Prof Exp:* Asst prof Univ Ga, 69-76; ASSOC PROF ZOOL, CLEMSON UNIV, 76- *Mem:* Am Soc Limnol & Oceanog; Sigma Xi. *Res:* Aquatic ecology; limnology. *Mailing Add:* Dept Zool Long Hall Clemson Univ Clemson SC 29631

SCHINDLER, JOE PAUL, b Berlin, Ger, Apr 8, 27; US citizen; m 55; c 3. ELECTRONICS ENGINEERING, TECHNICAL MANAGEMENT. *Educ:* Polytech Inst, Brooklyn, BEE, 50, MEE, 58. *Prof Exp:* Eng develop dept head, Polarad Electronics Corp, 50-62, vpres mkt, 62-65; vpres mkt, Narda Microwave Corp, 65-79, Gen Microwave Corp, 87-89 & Bertan Assocs, 90-91; pres & chief exec officer, Rohde & Schwarz, Polarad Inc, 79-87; CONSULT, MICROWAVE & ELECTRONICS CO, 86- *Concurrent Pos:* Dir, N Hills Electronics, 81-91; dir, Safety First Systs, Ltd, 87-, secy & treas, 90- *Mem:* Sigma Xi; Am Mach Asn; Inst Elec & Electronics Engrs. *Res:* Electronics; microwave instrumentation and components. *Mailing Add:* 118 Old Mill Rd Great Neck NY 11023

SCHINDLER, JOEL MARVIN, b New York, NY, Oct 27, 50; m 88; c 1. DEVELOPMENTAL BIOLOGY, CELL BIOLOGY. *Educ:* Hebrew Univ Jerusalem, BSc, 73, MSc, 75; Univ Pittsburgh, PhD(biol), 78. *Prof Exp:* Teaching & res asst, develop biol, Hebrew Univ Jerusalem, 73-75; teaching fel biol, Univ Pittsburgh, 76-77, researcher, 77-78; fel develop biol, Roche Inst Molecular Biol, 78-81; asst prof, 81-85, assoc prof anat & cell biol, Col Med, Univ Cincinnati, 85-87; PROG OFF, GENETICS & TERATOLOGY BR NAT INST CHILD HEALTH DEVELP, NIH, 87-; ADJ PROF PEDIAT, MED SCH, GEORGETOWN UNIV, 89- *Concurrent Pos:* Mem, grad prog develop biol, 82-; vis fel, biol sci, Macquarie Univ, Sydney, Australia, 85; mem coord comt, Interagency Skin Dis, 87-; proj officer, Nat Transgenic Mouse Facil, 88- *Mem:* Sigma Xi; Soc Develop Biol; Am Asn Anatomists; AAAS; Am Soc Cell Biol; NY Acad Sci. *Res:* Molecular aspects of development and differentiation; cell-cell interaction; interaction between differentiating cells and their environment; changes in the expression of active gene sequences during differentiation. *Mailing Add:* Genetics & Teratology Br Nat Inst Child Health & Human Develop EPN Rm 643 Bethesda MD 20892

SCHINDLER, JOHN FREDERICK, b Chicago, Ill, Aug 23, 31; m 55; c 2. FRESH WATER BIOLOGY. *Educ:* Mich State Univ, BS, 53, MS, 54. *Prof Exp:* Res asst phycol, Mich State Univ, 53, 54, 59-60; test design aide & off biol warfare, US Army, Dugway Proving Ground, Utah, 56-57; chief foreman mining, Minnas Cerro Colo, Mex, 58; asst dir, Naval Arctic Res Lab, Univ Alaska-Off Naval Res, 60-71, dir, 71-73; chief scientist & environ eng, Alaskan Resource Sci Corp, 73-76; dir environ affairs, Husky Oil NPR Opers Inc, 76-83; CHIEF, ENVIRON ASSESSMENT SECT, ALASKA OCS, MINERALS MGT SERV, DEPT INTERIOR, 84- *Concurrent Pos:* Mem, NSF Chihuahua Biol Exped, 55; mem, Scott Polar Res; arctic consult & vpres, Pac Alaska Assoc Ltd, 73-, environ consult, Pipeline Coordr Off, State Alaska, 75-76. *Mem:* AAAS; fel Arctic Inst NAm; Explorers Club. *Res:* Arctic logistics and science support in the Arctic; Arctic oceanography; freshwater algae of the Flathead Basin, Montana; genus Staurastrum. *Mailing Add:* 2473 Captain Cook Dr Anchorage AK 99517-1254

SCHINDLER, MAX J, b Warnsdorf, Czech, June 21, 22; US citizen; m 55; c 3. COMPUTER SCIENCE, MICROWAVES. *Educ:* Vienna Tech Univ, Dipl Ing, 51, Dr Tech Sc, 53. *Prof Exp:* Asst sound recording & magnetics res, Vienna Tech Univ, 51-54; engr, Tungsram-Watt, Austria, 54-57; res scientist, Aeronaut Res Lab, Wright Air Develop Ctr, Ohio, 57-58; engr, Phys & Chem Lab, RCA Microwave, NJ, 58-61, lead engr, 62-67, mem tech staff, RCA Labs, David Sarnoff Res Ctr, 67-69, sr engr solid state & TWT subsysts eng, RCA Microwave, Harrison, 69-75; comput consult, 76; comput ed, Electronic Design, 76-79; PRES, PRIME TECHNOL INC, 80- *Concurrent Pos:* Eng physics consult, Nat Corp Sci, NY, 77-81; Microwave consult, Photovolt, NY, 78-82. *Honors & Awards:* Inst Elec & Electronics Engrs Award, 76 & Centennial Medal, 84; Jesse Neal Award, 86 & 88. *Mem:* Sr mem Inst Elec & Electronics Engrs; Asn Comput Mach. *Res:* Magnetic materials and measurements; magnetic focusing structures; high-efficiency traveling-wave tubes; microwave solid state amplifiers; electronic delay devices; computer-aided design; computer system design; software design methodologies; book author. *Mailing Add:* RD 3 Box 77 Rockaway Dr Boonton NJ 07005

SCHINDLER, STEPHEN MICHAEL, b New York, NY, Apr 9, 40; m 61; c 2. ASTROPHYSICS. *Educ:* Long Island Univ, BS, 68; Colo State Univ, PhD(physics), 74. *Prof Exp:* Sr scientist physics, Bettis Atomic Power Lab, 73-74; res assoc physics, Case Western Reserve Univ, 74-78; SR SCIENTIST & MEM PROF STAFF PHYSICS, CALIF INST TECHNOL, 79- *Mem:* Am Phys Soc. *Res:* High energy astrophysics, with emphasis in gamma-ray astronomy, cosmic ray origin theory, and solar particle production. *Mailing Add:* 220-47 Downs Lab Calif Inst Technol Pasadena CA 91125

SCHINDLER, SUSAN, b Brooklyn, NY, June 4, 42. MATHEMATICS. *Educ:* Mt Holyoke Col, AB, 63; Univ Wis, MA, 65, PhD(math), 69. *Prof Exp:* Asst prof math, Long Island Univ, 69-70; asst prof, 70-77; PROF MATH, BARUCH COL, 77- *Mem:* Am Math Soc; Math Asn Am; Soc Indust & Appl Math. *Res:* Decompositions of group representations. *Mailing Add:* Dept Math Baruch Col 17 Lexington Ave New York NY 10010

SCHINDLER, WILLIAM JOSEPH, b Cleveland, Ohio, Dec 11, 31; m 72; c 3. ENDOCRINOLOGY. *Educ:* Univ Calif, Los Angeles, BA, 55, PhD(anat), 59; Baylor Col Med, MD, 74. *Prof Exp:* Asst physiol, Sch Med, Univ Calif, Los Angeles, 55-56, asst anat, 56-57, interdisciplinary trainee neurol sci, 58-59; instr, 59-60, asst prof, 62-67, ASSOC PROF PHYSIOL, BAYLOR COL MED, 67- *Concurrent Pos:* NSF fel neuroendocrinol, Maudsley Hosp, London, 60-61, Found Fund for Res in Psychiat fel, 61-62; Nat Inst Arthritis & Metab Dis res career develop award, 66- *Mem:* AAAS; Endocrine Soc; Am Physiol Soc; Soc Exp Biol & Med; NY Acad Sci. *Res:* Neuroendocrinology; developmental endocrinology, especially pituitary-thyroid maturation and function; human infertility; growth hormone secretion and control. *Mailing Add:* 15 E Greenway Plaza No 20K Houston TX 77046

SCHINGOETHE, DAVID JOHN, b Aurora, Ill, Feb 15, 42; m 64; c 2. DAIRY NUTRITION. *Educ:* Univ Ill, Urbana, BS, 64, MS, 65; Mich State Univ, PhD(nutrit), 68. *Prof Exp:* From asst prof to assoc prof, 69-80, PROF DAIRY SCI & NUTRIT, SDAK STATE UNIV, 80- *Concurrent Pos:* Actg Dept Head, Dairy Sci & Nutrit, SDak State Univ, 86. *Honors & Awards:* Am Food Indust Asn Award, Am Dairy Sci Asn, 89. *Mem:* Am Dairy Sci Asn; Am Soc Animal Sci; Coun Agr Sci & Technol; Am Inst Nutrit. *Res:* Nutritional biochemistry of rumen metabolism; gastrointestinal digestion and absorption; milk synthesis; protein, energy, and vitamin E nutrition of dairy cattle; whey utilization; sunflower product utilization. *Mailing Add:* Dept Dairy Sci Box 2104 SDak State Univ Brookings SD 57007-0647

SCHINK, CHESTER ALBERT, b Portland, Ore, Feb 17, 20; m 47; c 2. ORGANIC CHEMISTRY. *Educ:* Reed Col, BA, 41; Ore State Col, MA, 43, PhD(org chem), 47. *Prof Exp:* Chemist, Exp Sta, Hercules Powder Co, Del, 43-44, Radford Ord Works, Va, 44-45; asst, Ore State Col, 45-47; res chemist, E I du Pont de Nemours & Co, Inc, 47-51; mgr, Krishell Labs, Inc, 51-56; mgr, Chem Support Lab, Tektronix, Inc, 56-70, sr chemist, 70-71, corp chemist, 71-85; CONSULT, 85- *Mem:* Am Indust Hyg Asn; Am Chem Soc; Am Electroplaters Soc; Am Soc Safety Eng. *Res:* Identification of constituents of natural products; biologically active compounds; synthetic resins and plastics primarily of the vinyl type; research chemicals, especially purines, pyrimidines and enzymes; air and water quality; water pollution control, industrial hygiene and chemical safety. *Mailing Add:* 3943 SE Cooper St Portland OR 97202

SCHINK, DAVID R, b Los Angeles, Calif, Aug 3, 31; m 51; c 4. CHEMICAL OCEANOGRAPHY. *Educ:* Pomona Col, BA, 52; Univ Calif, Los Angeles, MS, 53; Univ Calif, San Diego, PhD(oceanog), 62; Stanford Univ, MS, 58. *Prof Exp:* Res geochemist, Scripps Inst, Calif, 60-62; asst prof oceanog, Narragansett Marine Lab, Univ RI, 62-66; mgr air-ocean studies, Palo Alto Labs, Teledyne-Isotopes, 66-71; assoc prof, 72-76, assoc dean geosci, 84-88, PROF OCEANOG, TEX A&M UNIV, 76- *Concurrent Pos:* Admin judge, Atomic Safety & Licensing Bd Panel, Nuclear Regulatory Comn, 74-; assoc ed, J Geophys Res, 80-86, Progress Oceanog, 81- US Nat Report to Int Union Geod & Geophys (chem oceanog), 82-86; mem, expert group, methods, standards & intercalibration, Intergovt Oceanog Comn, UN, 82-87; mem, adv comt ocean sci, NSF, 84-88 & adv comt earth sci, 85-88. *Mem:* AAAS; Am Geophys Union; Am Chem Soc; Oceanog Soc. *Res:* Oceanic silicon budgets and behavior; air-ocean gas exchange; radon/radium in sea water; diagenesis of marine sediments; chemistry of warm-core rings; applications of accelerator mass spectrometry to oceanography. *Mailing Add:* Dept Oceanog Tex A&M Univ College Station TX 77843

SCHINK, F E, b Brooklyn, NY, May 4, 22. ELECTRICAL ENGINEERING. *Educ:* Polytech Inst, BEE, 52, MEE, 55. *Prof Exp:* asst chief, Port Authority NY & NJ, 76-84, chief elec eng, 84-88; RETIRED. *Honors & Awards:* Centennial Medal, Inst Elec & Electronics Engrs, 84. *Mem:* Power Eng Soc; fel Inst Elec & Electronics Engrs. *Mailing Add:* 14 Middlebury Lane Cranford NJ 07016

SCHINZINGER, ROLAND, b Osaka, Japan, Nov 22, 26; nat US; m 52; c 3. OPERATIONS RESEARCH. *Educ:* Univ Calif, BS, 53, MS, 54, PhD, 66. *Prof Exp:* Engr, Westinghouse Elec Corp, Pa, 54-58; from asst prof to assoc prof elec eng, Robert Col, Istanbul, 58-63; from asst prof to assoc prof, 66-80, assoc dean, 79-83 & 85-86, PROF ELEC ENG, UNIV CALIF, IRVINE, 80- *Concurrent Pos:* Indust consult, 67-; Nat Sci Fac fel, 64-65. *Honors & Awards:* Centennial Medal, Inst Elec & Electronics Engrs, 84. *Mem:* AAAS; Inst Elec & Electronics Engrs; Opers Res Soc Am; Sigma Xi; Am Soc Eng Educ. *Res:* Power systems; utility networks; operations research; failure analysis; contingency planning; engineering ethics. *Mailing Add:* Dept Elec & Comp Eng Univ Calif Irvine CA 92717

SCHIOLER, LISELOTTE JENSEN, b Copenhagen, Denmark, May 8, 50; US citizen. CRYSTAL CHEMISTRY. *Educ:* Ohio State Univ, BFA, 74, BS, 77; Mass Inst Technol ScD, 83. *Prof Exp:* Ceramic res engr, Army Mat Mech Res Ctr, Dept Defense, 79-85, prog mgr, Air Force Off Sci Res, 88-91; sr eng specialist, Aerojet TechSysts Co, 85-88; PRES, PROPOSAL RESOURCES, INC, 91- *Mem:* Am Ceramic Soc; Am Crystallog Asn. *Res:* Marketing of research ideas; preparing winning proposals for funding. *Mailing Add:* Proposal Resources Inc 2000 L St NW Suite 200 Washington DC 20036

SCHIPMA, PETER B, b Chicago, Ill, Oct 24, 41; m 62; c 3. INFORMATION SCIENCE. *Educ:* Ill Inst Technol, BS, 65, MS, 67. *Prof Exp:* Res asst physics, R R Donnelley & Sons Co, 62-66; from asst to assoc scientist info sci, ITT Res Inst, 67-70, res scientist, 70-72, mgr info sci, 72-84; PRES, IS GRUPE, INC, 84- *Concurrent Pos:* Adj assoc prof, Ill Inst Technol, 67-74; consult, WHO, 77-, Czechesloyaki, 78, Saudi Arabia, 81; instr, Trinity Christian Col, 77-86; chmn finance comt, Asn Info & Dissemination Ctrs, 78-79. *Mem:* Am Soc Info Sci; Asn Info & Dissemination Ctrs (secy-treas, 77). *Res:* Machine-readable data base design; application of video technology to information retrieval; artificial intelligence; CD-ROM. *Mailing Add:* ISG, Inc 948 Springer Dr Lombard IL 60148

SCHIPPER, ARTHUR LOUIS, JR, b Bryan, Tex, Apr 8, 40; m 64; c 2. PLANT PATHOLOGY, PLANT PHYSIOLOGY. *Educ:* Univ of the South, BS, 62; Univ Minn, St Paul, MS, 65, PhD(plant path), 68. *Prof Exp:* Plant physiologist, NCent Forest Exp Sta, USDA, 68-78, staff res plant pathologist, Wash Off, 79-84, asst sta dir, PNW Res Sta, Forest Serv, 84-90, ASSOC AREA DIR, NATLANTIC AREA, AGR RES SERV, USDA, 90- *Concurrent Pos:* Adj asst prof, Dept Plant Path, Univ Minn, 68-75, adj assoc prof, 75-78. *Mem:* Sigma Xi; Soc Am Foresters. *Res:* Biochemical and physiological changes in plants infected with plant parasitic fungi. *Mailing Add:* NAtlantic Area Agr Res Serv USDA 600 E Mermaid Lane Wyndmoor PA 19118

SCHIPPER, EDGAR, b Vienna, Austria, Sept 12, 20; nat US; m 51; c 2. ORGANIC CHEMISTRY. *Educ:* City Col New York, BS, 47; Univ Pa, MS, 48, PhD(chem), 51. *Prof Exp:* Pharmaceut chemist, Gold Leaf Pharmaceut Co, 41-42; assoc chemist, Ethicon, Inc, 51-61; mgr org & med chem, Shulton, Inc, 61-65; GROUP LEADER, ETHICON INC, 65- *Mem:* Am Chem Soc. *Res:* Medicinal organic chemistry; biomedical devices research. *Mailing Add:* 44 Nomahegan Ct Cranford NJ 07016

SCHIPPER, LEE (LEON JAY), b Santa Monica, Calif, Apr 7, 47; m 71; c 2. ENERGY ANALYSIS, ENERGY POLICY. *Educ:* Univ Calif, Berkeley, AB, 68, MA, 71, PhD(physics), 82. *Prof Exp:* LECTR ACOUST, SAN FRANCISCO CONSERV MUSIC, 71-; STAFF SCIENTIST ENERGY ANALYSIS, LAWRENCE BERKELEY LAB, UNIV CALIF, 77- *Concurrent Pos:* Specialist energy anal, Energy & Resources Group, Univ Calif, 74-77; mem Demand Panel, Study Nuclear & Alternative Energy Syst, Nat Acad Sci, 76-78; energy consult to various int orgn, 76-; guest researcher, Beijer Inst, Royal Swed Acad Sci, 77- *Mem:* Soc Heating & Air Conditioning Engrs, Sweden; AAAS; Int Asn Energy Economists. *Res:* Conservation; policy; economics of energy systems; energy use in developing countries; structure of clusters of galaxies; life and recordings of Wilhelm Furtwaengler, conductor; acoustics. *Mailing Add:* One Cyclotron Rd MS 90-4000 Univ Calif 2120 Oxford St Berkeley CA 94720

SCHIRBER, JAMES E, b Eureka, SDak, June 9, 31; m 55; c 7. SOLID STATE PHYSICS. *Educ:* Iowa State Univ, PhD(physics), 60. *Prof Exp:* Nat Acad Sci-Nat Res Coun fel, Bristol, 61-62; staff mem, 62-64, div supvr solid state physics, 64-68, MGR SOLID STATE RES DEPT, SANDIA LABS, 68- *Mem:* Fel Am Phys Soc; fel AAAS. *Res:* Low temperature high pressure metal physics; Fermi surface of metals under hydrostatic pressure; magnetism and superconductivity studies under pressure; high temperature superconductivity. *Mailing Add:* 7605 Spring NE Albuquerque NM 87110

SCHIRCH, LAVERNE GENE, b Chenoa, Ill, Aug 9, 36; m 58; c 3. BIOCHEMISTRY, ORGANIC CHEMISTRY. *Educ:* Bluffton Col, BS, 58; Univ Mich, PhD(biochem), 63. *Prof Exp:* From asst prof to prof chem, Bluffton Col, 63-78; ASSOC PROF BIOCHEM, MED COL VA, VA COMMONWEALTH UNIV, 78- *Concurrent Pos:* Res consult, Ente Nazionale Idrocarburi, Rome, 69-70. *Mem:* Am Chem Soc; Am Soc Biol Chemists. *Res:* Mechanism of action of serine and threonine aldolases, especially the role of pyridoxal phosphate in these enzymes. *Mailing Add:* Biochem Dept Va Commonwealth Univ PO Box 614 NCV Sta Richmond VA 23298

SCHIRMER, HELGA H, b Chemnitz, Ger, Oct 18, 27. MATHEMATICS. *Educ:* Univ Frankfurt, MSc, 53, Dr rer nat(math), 54. *Prof Exp:* Brit Coun res scholar math, Univ Oxford, 54-56; asst lectr, Univ Wales, 56-59; from asst prof to assoc prof, Univ NB, 59-66; assoc prof, 66-74, PROF MATH, CARLETON UNIV, 74- *Concurrent Pos:* Vis prof, Univ Calif, Los Angeles, 83, Delhi Univ, 86. *Mem:* Am Math Soc; Can Math Soc. *Res:* General and algebraic topology; theory of fixed points and coincidences; multifunctions. *Mailing Add:* Dept Math Carleton Univ Ottawa ON K1S 5B6 Can

SCHIROKY, GERHARD H, CHEMICAL VAPOR DEPOSITION, COMPOSITES. *Educ:* Friedrich-Alexander Univ, Ger, dipl, 78;Univ Mo, MS, 79; Univ Utah, PhD(mats sci & eng), 82. *Prof Exp:* Sr scientist, Gen Atomics, 82-86; scientist, 86, MGR MATS RES, LANXIDE CORP, 87- *Mem:* Am Ceramic Soc. *Res:* Ceramic matrix composites formed by the oxidation of molten metals; chemical vapor deposition; metal matrix composites; high temperature processing equipment. *Mailing Add:* Lanxide Corp 1300 Marrows Rd PO Box 6077 Newark DE 19714-6077

SCHISLA, ROBERT M, b Indianapolis, Ind, Mar 30, 30; m 53; c 5. ORGANIC CHEMISTRY. *Educ:* Purdue Univ, BS, 52, MS, 54, PhD(org chem), 57. *Prof Exp:* Asst, Purdue Univ, 52-54; res chemist, Monsanto Co, 57-61, sr res chemist, Monsanto Indust Chem Co, 61-65, res specialist, 65, group leader, 66-71, res specialist fine chem, 71-85; CONSULT, 85- *Res:* Synthetic organic chemistry related to functional fluids and fine chemicals; process development for fine chemicals. *Mailing Add:* 1333 Woodgate St Louis MO 63122

SCHISLER, LEE CHARLES, b Northampton, Pa, June 25, 28; m 51; c 2. MYCOLOGY. *Educ:* Pa State Univ, BS, 50, MS, 52, PhD(bot), 57. *Prof Exp:* Dir res, Butler Co Mushroom Farm, Inc, 57-64; prof, 64-88, EMER PROF PLANT PATH, PA STATE UNIV, 89- *Mem:* Am Soc Plant Physiol; Mycol Soc Am; Am Phytopath Soc; Sigma Xi; fel AAAS; Am Inst Biol Sci. *Res:* Physiology and pathology of cultivated mushrooms; fungus physiology and mycology. *Mailing Add:* 317 S Sparks St State College PA 16801

SCHISSLER, DONALD OWEN, physical chemistry, for more information see previous edition

SCHIVELL, JOHN FRANCIS, b Cleveland, Ohio, Sept 5, 42; m 66; c 2. PLASMA DIAGNOSTICS, IMAGE RECONSTRUCTION. *Educ:* Harvard Univ, AB, 63, AM, 64, PhD(physics), 68. *Prof Exp:* Physicist, Nat Accelerator Lab, 68-73; MEM TECH STAFF, PRINCETON PLASMA PHYSICS LAB, 73- *Mem:* Am Phys Soc. *Res:* Photoproduction of elementary particles; design and development of particle accelerators; design, construction, and basic measurements on tokamak plasma confinement devices; software systems for experiment control and analysis. *Mailing Add:* Princeton Plasma Physics Lab PO Box 451 Princeton NJ 08543

SCHJELDERUP, HASSEL CHARLES, b Vernon, BC, June 18, 26; US citizen; m 53; c 7. ENGINEERING MECHANICS. *Educ:* Univ BC, BASc, 49; Stanford Univ, MSc, 50, PhD(eng mech), 53. *Prof Exp:* Asst, Stanford Univ, 49-52; group leader struct eng, Northrop Aircraft Corp, 53-55; tech asst strength & dynamics, Douglas Aircraft Corp, 55-59; assoc dir eng, Nat Eng Sci Co, 59-64; dir eng, Dynamic Sci Corp, 64-65; mgr advan technol, Douglas Aircraft Corp, Long Island, 65-67, asst dir res, 67-68, dir res, 68-72, dep dir mat & process eng, 72-79, dir mat & process eng, 79-86; RETIRED. *Concurrent Pos:* Instr, Exten, Univ Calif, Los Angeles, 53-58; struct & dynamics consult, 53- *Mem:* Am Inst Aeronaut & Astronaut; Soc Advan Mat & Process Engrs. *Res:* Random vibrations; structural vibration and fatigue caused by jet engine noise; advanced composite materials. *Mailing Add:* 1630 West Dr San Marino CA 91108-2257

SCHLABACH, T(OM) D(ANIEL), b Cleveland, Ohio, July 4, 24; m 48; c 2. METALLURGY, NON-FUEL MINERAL RESOURCES. *Educ:* Baldwin-Wallace Col, BS, 48; Mich State Col, PhD(chem), 52. *Prof Exp:* Mem tech staff & res chemist, AT&T Bell Labs Inc, 52-59, supvr, 59-65, dept head metall, 65-89; RETIRED. *Mem:* Fel AAAS; fel Am Inst Chemists; fel Am Soc Metals Int; Minerals Metals & Mat Soc; Mat Res Soc. *Res:* Alloy development; materials conservation and substitution; high-temperature superconductors. *Mailing Add:* Four Adams Dr Whippany NJ 07981-2050

SCHLACHTER, ALFRED SIMON, b Cedar City, Utah, Feb 18, 42. ATOMIC PHYSICS. *Educ:* Univ Calif, Berkeley, AB, 63; Univ Wis, Madison, MA, 65, PhD(physics), 69. *Prof Exp:* Asst physics, Univ Wis, 63-68; prin res scientist, Honeywell Corp Res Ctr, 68-70; res assoc, Faculty Sci, Inst Fundamental Electronics, Univ Paris & Researcher, Saclay Nuclear Res Ctr, 71-75; PHYSICIST, LAWRENCE BERKELEY LAB, UNIV CALIF, 75-, SCI PROG COORDR, ADVAN LIGHT SOURCE, 89- *Concurrent Pos:* Nat Ctr Sci Res fel, Univ Paris, 71-72; Joliot-Curie Fel, Saclay Nuclear Res Ctr, 72-73; vis scientist, Justus-Liebig Univ, Giessen, WGer, 80-81; Alexander von Humboldt Found travelling fel, 80-81; adj assoc prof, NC State Univ, 86- *Mem:* Am Phys Soc; Optical Soc Am. *Res:* Atomic collisions; sources of negative ions; particle beams; multiply charged ion-atom collisions; fusion energy; synchrotron radiation. *Mailing Add:* Lawrence Berkeley Lab MS 46-161 Univ Calif Berkeley CA 94720

SCHLAEGER, RALPH, b Milwaukee, Wis, Nov 24, 21. RADIOLOGY. *Educ:* Univ Wis, BS, 42, MD, 45; Univ Pa, 48-49. *Prof Exp:* Asst instr radiol, Temple Univ, 50-52, instr, 52-54; from instr to assoc prof, 54-71, PROF CLIN RADIOL, COLUMBIA UNIV, 71- *Mem:* Radiol Soc NAm; fel Am Col Radiol; Soc Gastrointestinal Radiol; Am Asn Hist Med. *Res:* Radiology, especially the gastrointestinal tract. *Mailing Add:* Presby Hosp 622 W 168th St New York NY 10032

SCHLAFER, DONALD H, b Sidney, NJ, July 15, 48; m 80; c 2. PATHOPHYSIOLOGY. *Educ:* Cornell Univ, BS, 71, DVM, 74, MS, 75; Univ Ga, PhD(vet path), 82. *Prof Exp:* Assoc vet, Guilderland Animal Hosp, 75-77; relief vet, Ft Hill Animal Hosp, 79-80 & Aquebogue Vet Hosp, 80-82; dir, Bovine Health Res Ctr, 82-83 & 83-90, asst prof, 82-88, ASSOC PROF VET PATH, NY STATE COL VET MED, CORNELL UNIV, 88- *Concurrent Pos:* Lectr, Pedro Henriquez Urena Univ, Dominican Repub, 83; consult, Food & Drug Admin & indust, 83- *Mem:* Am Col Vet Pathologists; Am Col Vet Theriogenologists; Am Col Vet Microbiologists; Am Vet Med Asn; Soc Study Reprod; Soc Theriogenology. *Res:* Pathophysiology of diseases of pregnancy of domestic animals; transplacental infections by viruses, bacteria and protozoan; effects on the fetus and placenta by toxins; in vitro studies utilizing cultured trophoblast cells. *Mailing Add:* NY State Col Vet Med E205 Schurman Hall Cornell Univ Ithaca NY 14853

SCHLAFLY, ROGER, b Ill, Oct, 56. SOFTWARE SYSTEMS. *Educ:* Princeton Univ, BSE, 76; Univ Calif, PhD(math), 80. *Prof Exp:* Instr math, Univ Chicago, 80-83; programmer, Borland Inst, 85-90; PRES, REAL SOFTWARE, 90- *Mem:* Am Math Soc; Soc Indust Appl Math. *Res:* Optimization; differential geometry; computer software; encryption; finance. *Mailing Add:* PO Box 1680 Soquel CA 95073

SCHLAG, EDWARD WILLIAM, b Los Angeles, Calif, Jan 12, 32; m 55; c 3. PHYSICAL CHEMISTRY. *Educ:* Occidental Col, BS, 53; Univ Wash, PhD(phys chem), 58. *Prof Exp:* Res asst, Univ Wash, 54-58; Wissenschaftlicher asst, Inst Phys Chem, Univ Bonn, 58-59; res chemist, Yerkes Lab, E I du Pont de Nemours & Co, NY, 59-60; from asst prof to prof chem, Northwestern Univ, Evanston, 60-71; dean, fac chem, biol & geosci, 82-84, PROF PHYS CHEM, MUNICH TECH UNIV, 71- *Concurrent Pos:* Alfred P Sloan res fel, 65-67; selection comt, US Sr Award Prog, Alexander von Humboldt Found & Ger-Israeli Fel Prog, Minerva Found; mem, Ger-Israel Sci Prog Comt of Minerva, Nat Fulbright Comt Geo, Int Orgn Comt of the Int Congress Photochem, Gov Comt of Ger Phys Chem Soc; mem ed bd, Chem Physics, Chem Physics Letters, intern j of Mass Spectrometry and Ion Processes, j Physical Chem & Laser Chem. *Mem:* Fel Am Phys Soc; Bavarian Acad Sci. *Res:* Multiphoton ionization mass spectrometry; high resolution sub-Doppler molecular spectroscopy and dynamics; spectroscopy and kinetics molecular ions in a fast beam; dynamics of photoexcited states and van der Waals molecules; synchrotron radiation experiments on molecular ions, inner shell excitations. *Mailing Add:* Inst Phys Chem 8046 Garching Lichtenbergstr 4 Germany

SCHLAG, JOHN, b Liege, Belg, Feb 13, 27; US citizen; m 53; c 2. NEUROPHYSIOLOGY. *Educ:* Univ Liege, MD, 52. *Prof Exp:* Asst exp therapeut, Univ Liege, 53-61, lectr psychophysiol, 60-61; asst res anatomist, 61-63, from asst prof to assoc prof, 63-69, PROF ANAT, UNIV CALIF, LOS ANGELES, 69- *Concurrent Pos:* Fulbright grant, Univ Wash, 53-54; NIH res career develop award, 64-; secy, Belg Nat Ctr Anesthesiol, 59-61. *Honors & Awards:* Theophile Gluge Prize, Royal Acad Belg, 61. *Mem:* Am Physiol Soc; Fr Asn Physiol; Belg Asn Physiol; Int Brain Res Orgn; corresp mem Royal Acad Med Belg. *Res:* Cerebral control of motor functions; spontaneous activity of cerebral neurons; mechanisms of cortical evoked potentials. *Mailing Add:* Dept Anat 73-235 Bri Univ Calif 405 Hilgard Ave Los Angeles CA 90024

SCHLAGENHAUFF, REINHOLD EUGENE, b Amsterdam, Neth, Aug 14, 23; US citizen; m 55; c 2. NEUROLOGY. *Educ:* Univ Wurzburg, MD, 51. *Prof Exp:* Dir, EEG, Electromyography & Echoencephalography Dept, Meyer Mem Hosp, Buffalo, 63-81, assoc dir, Dept Neurol, 69-81; assoc prof neurol, State Univ NY Buffalo, 72-85, clin prof, 85-89; dir, Dept Neurol, Erie County Med Ctr, 81-86; RETIRED. *Concurrent Pos:* Consult neurol & EEG, Vet Admin Hosp, Buffalo, Buffalo Psychiat Ctr & Gowanda Psychiat Ctr, 68- & West Seneca Develop Ctr, 70-91. *Mem:* Acad Neurol; Am Med EEG Asn; Am Inst Ultrasonics in Med; AMA; Am Psychiat Asn. *Res:* Clinical neurology, electroencephalography, electromyography and ultrasound (Doppler-flow). *Mailing Add:* Dept Neurol 164 462 Grider St E State Univ NY 3435 Main St Buffalo NY 14214

SCHLAGER, GUNTHER, b New York, NY, Jan 14, 33; m 56; c 3. GENETICS. *Educ:* Univ Denver, BA, 56; Univ Kans, MA, 59, PhD(entom), 62. *Prof Exp:* USPHS fel, Hall Lab, Mammalian Genetics, Univ Kans, 61-62; assoc staff scientist, Jackson Lab, 62-65, staff scientist, 65-69; assoc prof, 69-72, chmn dept, 72-79, PROF SYSTS & ECOL, UNIV KANS, 72-, CHMN, DIV BIOL SCI, 79- *Concurrent Pos:* Lectr, Univ Maine, 64-69; vis prof genetics, Sch Med, Univ Hawaii, 75; mem genetic subgroup Hypertension Task Force, Nat Heart Lung & Blood Inst, NIH, 76-78; consult, Med Dept, Brookhaven Nat Lab, 77-80; fel Coun High Blood Pressure Res. *Mem:* Genetics Soc Am; Am Genetic Asn; Am Asn Univ Professors; Biometric Soc. *Res:* Quantitative genetics; genetics of blood pressure in rodents. *Mailing Add:* Dept Systs & Ecol Univ Kans Lawrence KS 66045

SCHLAGER, SEYMOUR I, tumor immunology, for more information see previous edition

SCHLAIKJER, CARL ROGER, b Boston, Mass, Mar 3, 40; m 70; c 2. ELECTROCHEMISTRY. *Educ:* Harvard Univ, AB, 61; Mass Inst Technol, PhD(inorg nuclear chem), 66. *Prof Exp:* Mem res staff, Arthur D Little, Inc, 65-67; mem staff, Lab Phys Sci, P R Mallory & Co, 68-75, staff scientist, 75-78; mgr res, Power Sources Ctr, GTE Labs Inc, 78-82; res fel, Duracell Res Ctr, 82-87; CHIEF SCIENTIST, BATTERY ENG, INC, 87- *Mem:* Am Chem Soc; Electrochem Soc; Sigma Xi. *Res:* Solid state electrolyte cells and systems; organic and inorganic electrolyte primary and secondary batteries; interaction of alkali and alkaline earth metals with organic and inorganic systems. *Mailing Add:* 105 Ridge Rd Concord MA 01742

SCHLAIN, DAVID, b Philadelphia, Pa, July 21, 10. CHEMICAL ENGINEERING. *Educ:* Univ Pa, BS, 32, MS, 37; Univ Md, PhD(chem eng), 51. *Prof Exp:* Metallurgist & chemist, US Bur Mines, 37-48, electrochemist, 48-52, chief, Galvanic Corrosion Sect, 52-55, supvry chem res engr, 55-58, proj coordr, 58-70, res supvr, 70-74, res chem engr, 74-81; RETIRED. *Mem:* Electrochem Soc; Am Inst Mining, Metall & Petrol Engrs; Nat Asn Corrosion Engrs; fel Am Inst Chemists; Am Chem Soc. *Res:* Electrowinning and electrorefining of metals; metallic corrosion; electrodeposition of coatings from molten salts and aqueous baths; effects of ultrasonics on metallurgical processes; electrometallurgy; crystallization. *Mailing Add:* PO Box 266 Greenbelt MD 20770-1717

SCHLAM, ELLIOTT, b New York, NY, Oct 7, 40; m 66; c 2. FLAT PANEL DISPLAYS, INTERACTIVE DISPLAY SYSTEMS. *Educ:* NY Univ, BEE, 61, MEE, 64, PhD(elec eng), 66; Fairleigh Dickinson Univ, MS, 74. *Prof Exp:* Design engr, RCA, 61-62, mem tech staff, 64; proj engr, Ecom, US Army, 68-73, team leader, 73-75, br chief, Eradcom, 75-85, div dir, Labcom, 85-87; PRES, ELLIOTT SCHLAM ASSOCS, 89- *Concurrent Pos:* Instr, George Wash Univ, 75-83, State Univ NY, Stony Brook, 79, Univ Calif, Los Angeles, 83-84 & Univ Wis, 83-85; assoc ed, Inst Elec & Electronics Engrs Trans Elec Devices, 78-80; lectr & course dir, Ctr for Prof Advan, 81-86; conf dir, Soc Photo-Optical Instrumentation Engrs, 83-86. *Mem:* Fel Soc Info Display; Soc Info Display; sr mem Inst Elec & Electronics Engrs. *Res:* Operation and fabrication techniques for thin film electroluminescent displays; high definition displays for commercial and military applications. *Mailing Add:* Elliott Schlam Assocs Four Mahoras Dr Wayside NJ 07712

SCHLAMEUS, HERMAN WADE, b Blanco, Tex, Nov 27, 37; m 84. ORGANIC CHEMISTRY. *Educ:* Southwest Tex State Univ, BS, 60. *Prof Exp:* Technician chem, Southwest Res Inst, 60-62, res chemist, 62-70, sr res chemist, 70-87, PRIN SCIENTIST, SOUTHWEST RES INST, 87- *Mem:* Sigma Xi; Controlled Release Soc. *Res:* Mainly concerned in research and development in the field of microencapsulation. *Mailing Add:* Southwest Res Inst 6220 Culebra Rd San Antonio TX 78228-0510

SCHLAMOWITZ, MAX, b New York, NY, Nov 13, 19; m 44; c 1. BIOCHEMISTRY, IMMUNOCHEMISTRY. *Educ:* City Col New York, BS, 40; Univ Mich, MS, 41, PhD(biochem), 46. *Prof Exp:* Res assoc, Manhattan Dist, Rochester Univ, 44-45; instr biochem, Univ Calif, 47-50; asst, Sloan-Kettering Inst Cancer Res, 51-54; assoc cancer res scientist, Roswell Park Mem Inst, 54-62; assoc prof microbiol, Baylor Col Med, 62-64; assoc biologist, Grad Sch Biomed Sci, Univ Tex, Houston, 64-67, assoc biochemist, 67-71, chief sect immunochem & immunol, 67-81, biochemist, Syst Cancer Ctr, Univ Tex M D Anderson Hosp & Tumor Inst, Houston, 71-85, assoc prof biochem, 67-78, prof biochem, Grad Sch Biomed Sci, Univ Tex, Houston, 79-85; RETIRED. *Concurrent Pos:* Markle Found fel, Univ Calif, 46-47; USPHS & Du Pont fels, Ohio State Univ, 50-51. *Mem:* AAAS; Am Soc Biol Chemists; Am Asn Immunol; NY Acad Sci; Reticuloendothelial Soc; Sigma Xi. *Res:* Carbohydrate chemistry; enzymes; ribonuclease; immunoglobulin chemistry and membrane transport; hormone receptors; phospho-glucomutase-phosphatase; pepsin; glycoproteins; Fc receptors. *Mailing Add:* 5503 Wigton Dr Houston TX 77096-4007

SCHLANGER, SEYMOUR OSCAR, b New York, NY, Sept 17, 27. GEOLOGY. *Educ:* Rutgers Univ, BSc, 50, MSc, 51; Johns Hopkins Univ, PhD(geol), 59. *Prof Exp:* Res asst, Rutgers Univ, 50-51; geologist, US Geol Surv, 51-59; prof, Petroleo Brasileiro, Brazil, 59-61; geologist, US Geol Surv, 61-62; from assoc prof to prof geol sci, Univ Calif, Riverside, 62-75; prof geol, State Univ Leiden & State Univ Utrecht, Neth, 75-77; prof marine geol, Univ Hawaii, 78-81; PROF GEOL, NORTHWESTERN UNIV, 81- *Concurrent Pos:* Vis prof, Tohuku Univ, Japan, 65-66; Guggenheim fel, Swiss Fed Inst Technol, 69-70. *Honors & Awards:* Pres Award, Am Asn Petrol Geologists, 81; F P Shepard Medal, Soc Econ Paleontologists & Mineralogists, 88. *Mem:* Fel Geol Soc Am; Int Asn Sedimentologists (treas, 75-77); Soc Econ Paleontologists & Mineralogists; Am Asn Petrol Geologists. *Res:* Geology of Pacific basin and coral reefs; geology of Europe; stratigraphy and sedimentary petrology; petroleum geology. *Mailing Add:* Dept Geol Northwestern Univ Evanston IL 60201

SCHLANT, ROBERT C, b El Paso, Tex, Apr 16, 29; m 80; c 3. INTERNAL MEDICINE, CARDIOLOGY. *Educ:* Vanderbilt Univ, BA, 48, MD, 51; Am Bd Internal Med, dipl, 58; Am Bd Cardiovasc Dis, dipl, 62. *Prof Exp:* House officer, Peter Bent Brigham Hosp, Mass, 51-52, from jr asst resident med to asst, 52-58; from asst prof to assoc prof, 58-67, PROF MED, SCH MED, EMORY UNIV, 67- *Concurrent Pos:* Res fel med, Harvard Med Sch, 56-58; mem, Subspecialty Bd Cardiovasc Dis, 71-; fel coun clin cardiol, Am Heart Asn. *Honors & Awards:* Distinguished Serv Award, Am Heart Asn. *Mem:* Fel Am Col Physicians; fel Am Col Cardiol; Am Fedn Clin Res; Asn Univ Cardiol; Am Physiol Soc. *Res:* Cardiovascular physiology. *Mailing Add:* Dept Med Emory Univ Sch Med 69 Butler St SE Atlanta GA 30303

SCHLAPFER, WERNER T, b Zurich, Switz, July 19, 35; US citizen; m 65; c 3. NEUROBIOLOGY, NEUROSCIENCES. *Educ:* Univ Calif, Berkeley, BA, 63, PhD(biophys), 69. *Prof Exp:* Instr med physics, Univ Calif, Berkeley, 69-70; asst res physiologist, Dept Psychiat, Univ Calif, San Diego, 70-72; res physiologist, psychiat serv, Vet Admin Hosp, San Diego, 72-80; CHIEF, WESTERN RES & DEVELOP OFF, VET ADMIN MED CTR, LIVERMORE, CALIF, 80-; DEPT BIOL/PHYSICS, OHLONE COL, FREMONT, CALIF, 82- *Concurrent Pos:* Lectr psychiat, Univ Calif, San Diego, 74-78, asst prof in residence, 78-80. *Mem:* AAAS; Soc Neurosci. *Res:* Mechanisms of the modulation of synaptic transmission, depression, facilitation, post-tetanic potentiation, heterosynaptic facilitation and heterosynaptic inhibition; presynaptic pharmacological regulation of neurotransmitter economics and release in molluscan central nervous systems. *Mailing Add:* Western Res & Develop Off Vet Admin Med Ctr Livermore CA 94550

SCHLARBAUM, SCOTT E, b Des Moines, Iowa, July 7, 51; m 80; c 1. FOREST TREE IMPROVEMENT. *Educ:* Colo State Univ, BS, 74, PhD(cytogenetics), 80; Univ Nebr, MS, 77. *Prof Exp:* Asst cytogenetics & tissue cult, Kans State Univ, 81-83; asst prof, 84-88, ASSOC PROF, UNIV TENN, 88- *Mem:* Am Soc Human Genetics. *Res:* Genetic improvement of softwood and hardwood trees; tree cytogenetics; somatic cell genetics in trees; chromosome transfer for gene mapping and tree improvement; cytotaxonomy and phylogeny of Coniferales. *Mailing Add:* Dept Forestry Wildlife & Fisheries Univ Tenn Knoxville TN 37996

SCHLATTER, JAMES CAMERON, b Madison, Wis, Feb 5, 45; m 67; c 2. CHEMICAL ENGINEERING, CHEMISTRY. *Educ:* Univ Wis, Madison, BS, 67; Stanford Univ, MS & PhD(chem eng), 71. *Prof Exp:* Sr res engr, Gen Motors Res Labs, 71-80; TECH DIR/ENG, CATALYTICA, 80- *Concurrent Pos:* Lectr, Stanford Univ, 83- *Mem:* Am Chem Soc; Am Inst Chem Engrs; Catalysis Soc. *Res:* Heterogeneous catalysis; automotive emission control. *Mailing Add:* 1718 Chitamook Ct Sunnyvale CA 94087

SCHLAUDECKER, GEORGE F(REDERICK), b Erie, Pa, Feb 10, 17; m 80; c 2. CHEMICAL ENGINEERING. *Educ:* Univ Notre Dame, BS, 38; Mass Inst Technol, MS, 39. *Prof Exp:* Develop engr, Am Locomotive Co, NY, 39; jr engr, E I du Pont de Nemours & Co, Del, 39-41, eng group leader, 41-45; secy-treas, Maumee Develop Co, 46-50, pres, 50-53, pres, Maumee Chem Co, 53-69; gen mgr, Sherwin-Williams Chem, 69-70, dir & group vpres, Sherwin-Williams Co, Cleveland, 70-78; pres, Erie Isles Assocs Inc, Port Clinton, Ohio, 79-83; RETIRED. *Concurrent Pos:* Asst prof, Univ Toledo, 47-49; dir, Indust Nucleonics Corp, 75-80, Energy Utilization Systs, 79-83 & Accuray Corp, 80-87. *Mem:* Am Inst Chem Engrs; Am Chem Soc. *Res:* Diffusional operations; organometallic reactions; organic chemical reaction rates; sublimation; precipitation and filtration; drying. *Mailing Add:* 23 Exmoor Toledo OH 43615-2175

SCHLAUG, ROBERT NOEL, b Jamaica, NY, Dec 21, 39; m 60; c 2. NUCLEAR ENGINEERING. *Educ:* Case Inst Technol, BS, 61; Univ Calif, Berkeley, PhD(nuclear eng), 65. *Prof Exp:* Staff mem high energy fluid dynamics, Gen Atomic Div, Gen Dynamics Corp, 65-68; staff mem radiation hydrodynamics, Systs, Sci & Software, Inc, 68-72; SR STAFF SCIENTIST RADIATION HYDRODYNAMICS SCI, APPLINS INT CORP, 72- *Mem:* Am Nuclear Soc; Sigma Xi. *Res:* Interactions of radiation and atomic particles with materials and development of computer methods for describing those interactions. *Mailing Add:* 3705 Sioux Ave San Diego CA 92117-5722

SCHLAX, T(IMOTHY) ROTH, b Kenosha, Wis, Mar 19, 38; m 65; c 4. ELECTRICAL ENGINEERING, COMPUTER TECHNOLOGY. *Educ:* Marquette Univ, BSEE, 60, MSEE, 63; Mass Inst Technol, PhD(solid state electronics), 68. *Prof Exp:* Develop engr, 72-75, PROJ MGR ENG STAFF, GEN MOTORS CORP, 75- *Honors & Awards:* 1975 SAE Vincent Bendix Automotive Electronics Award, Soc Automotive Engrs, 76. *Mem:* Inst Elec & Electronics Engrs; Soc Automotive Engrs. *Res:* Automotive control computer applications; custom computer approaches for fuel, ignition, and other engine control functions; advanced aids for refining control algorithms. *Mailing Add:* Gen Motor Corp GM Eng Staff GM Tech Ctr Warren MI 48090

SCHLECH, BARRY ARTHUR, b Bayonne, NJ, June 7, 44; m 63; c 5. PHARMACEUTICAL MICROBIOLOGY, ANTIMICROBIALS. *Educ:* Univ Tex, Austin, BA, 65, MA, 68, PhD(microbiol), 70. *Prof Exp:* Sr scientist, Alcon Labs Inc, 70, head microbiol, 70-76, corp microbiologist, 76-81, dir res & develop microbiol, 81-89, SR DIR RES & DEVELOP MICROBIOL, ALCON LABS, INC, 89- *Mem:* Am Soc Microbiol; Parenteral Drug Asn; Soc Indust Microbiol; Ocular Microbiol & Immunol Group; Pharmaceut Mfrs Asn Biol Sect. *Res:* Pharmaceutical microbiology; antibiotics; contact lens disinfection; opthalmic and cosmetic preservatives; sterilization; ocular infections; microbiological control; microbial limit testing; ultraviolet sterilization; biological indicators. *Mailing Add:* Alcon Lab Inc Mail Code R0-13 PO Box 6600 Ft Worth TX 76115

SCHLECHT, MATTHEW FRED, b Milwaukee, Wis, Nov 26, 53; m 86; c 1. MEDICINAL CHEMISTRY, AGROCHEMICALS. *Educ:* Univ Wis-Madison, BS, 75; Columbia Univ, PhD(chem), 80. *Prof Exp:* Res fel chem, Univ Calif, Berkeley, 80-82; asst prof chem, Polytech Univ, 82-88; SECT RES CHEMIST, DUPONT AGR PROD, 88- *Mem:* Am Chem Soc; AAAS; Am Soc Pharmacog. *Res:* Agricultural chemistry; organic synthetic methods; molecular modeling; oxidation reactions & mechanisms; medicinal chemistry. *Mailing Add:* DuPont Agr Prod Stine-Haskell Res Ctr PO Box 30 Newark DE 19714-0030

SCHLEE, FRANK HERMAN, b New York, NY, Apr 22, 35; m 58; c 3. NAVIGATION. *Educ:* Polytech Inst Brooklyn, BS, 56; Univ Mich, MS, 49, PhD(instrumentation eng), 63. *Prof Exp:* Engr, Sperry Gyroscope Co, 56-58; res assoc navig, Inst Sci & Technol, Univ Mich, 62-63; sr engr, 63-66, SR ENGR, INST NAVIG, FED SYSTS DIV, IBM CORP, 66- *Mem:* Am Inst Navig; Am Inst Aeronaut & Astronaut. *Res:* Space guidance; space and aircraft navigation, particularly using optimal filtering techniques. *Mailing Add:* 614 Ivory Foster Rd Owego NY 13827

SCHLEE, JOHN STEVENS, b Detroit, Mich, Sept 27, 28; div; c 1. GEOLOGY. *Educ:* Univ Mich, BS, 50; Univ Calif, Los Angeles, MA, 53; Johns Hopkins Univ, PhD(geol), 56. *Prof Exp:* Geologist, US Geol Surv, 56-58; asst prof geol, Univ Ga, 58-62; RES GEOLOGIST, US GEOL SURV, 62- *Concurrent Pos:* NSF grant, 60-62; distinguished lectr, Am Asn Petrol Geologists, 78-79. *Honors & Awards:* Meritorious Serv Award, Dept Interior, 83. *Mem:* AAAS; fel Geol Soc Am; Soc Econ Paleont & Mineral; Am Asn Petrol Geol. *Res:* Texture, composition and structures in sediments and sedimentary rocks; structure and stratigraphy of continental margins; bathymetry of Southern Lake Michigan; wide angle deep water side-scan sonar systems. *Mailing Add:* US Geol Surv Woods Hole MA 02543

SCHLEEF, DANIEL J, b Norfolk, Nebr, Sept 29, 27; m 56; c 2. MECHANICAL ENGINEERING. *Educ:* Univ Ark, BSME, 50; Kans State Univ, MS, 52; Purdue Univ, PhD(mech eng), 60. *Prof Exp:* Instr mech eng, Kans State Univ, 52-53; asst prof, Univ Cincinnati, 53-56; instr, Purdue Univ, 56-58; assoc prof, 59-61, head dept, 61-70, PROF MECH ENG, UNIV CINCINNATI, 61- *Concurrent Pos:* Mem, Year-in-Indust prog, E I du Pont de Nemours & Co, Del, 67-68. *Mem:* Am Soc Mech Engrs; Am Soc Eng Educ; Sigma Xi. *Res:* Thermodynamics; heat transfer; fluid mechanics; thermodynamic and transport properties of matter. *Mailing Add:* 471 Wood Ave Cincinnati OH 45220

SCHLEGEL, DAVID EDWARD, b Fresno, Calif, Sept 3, 27; m 48; c 4. PLANT PATHOLOGY. *Educ:* Ore State Univ, BS, 50; Univ Calif, PhD(plant path), 54. *Prof Exp:* Jr specialist, 53-54, instr plant path & jr plant pathologist, 54-56, asst prof & asst plant pathologist, 56-62, assoc prof & assoc plant pathologist, 62-69, Miller prof, 66-67, chmn, dept plant path, 70-76, actg dean, Col Natural Resources, 77-78, assoc dean res, 76-78, dean, Col Nat Resources, 78-85, PROF PLANT PATH, UNIV CALIF, BERKELEY, 69-, ASST DIR, AGR EXP STA 85- *Mem:* AAAS; Am Phytopath Soc. *Res:* Plant virology; plant pathology. *Mailing Add:* Div Agr & Natural Resources Univ Calif 300 Lakeside Dr 6th Floor Oakland CA 94612-3560

SCHLEGEL, DONALD LOUIS, b Dayton, Ohio, July 27, 34; m 52; c 5. PIPE JOINTING SYSTEMS, PIPE DESIGN. *Educ:* Univ Payton, BS, 52; Wright State Univ, MBA, 85. *Prof Exp:* Res engr, Univ Dayton Res Inst, 56-58; design engr, Alden Sticson & Assoc, 58-59, Bur Structr City Dayton, 59-65; struct engr, Felexible Co, 65-69; develop engr, 69-74, MGR RES & DEVELOP, PRICE BROS CO, 74- *Concurrent Pos:* Chmn, comt concrete consolidation, Am Concrete Inst, 80-85, subcomt use flyash in concrete, 79-86; mem comt admixtures, Am Concrete Inst, 75-, comt concrete coatings, 78- *Mem:* Am Concrete Inst. *Res:* Product and process improvements for pipe, including reinforced concrete, prestressed concrete and fiber reinforced plastic pipe. *Mailing Add:* 2035 Hamlet Dr Kettering OH 45440

SCHLEGEL, JAMES M, b Ogden, Utah, Aug 24, 37; m 62; c 1. PHYSICAL CHEMISTRY. *Educ:* Univ of the Pac, BS, 59; Iowa State Univ, PhD(phys chem), 62. *Prof Exp:* Res asst phys chem, Iowa State Univ, 59-62; asst prof, 62-69, assoc prof, 69-76, PROF PHYS CHEM, RUTGERS UNIV, NEWARK, 76- *Mem:* Am Chem Soc. *Res:* Stoichiometry and kinetics of reactions in fused salt media. *Mailing Add:* Dept Chem Newark Col Arts & Sci Rutgers Univ Newark NJ 07102

SCHLEGEL, JORGEN ULRIK, b Copenhagen, Denmark, July 18, 18; nat US; m 43; c 2. MEDICINE. *Educ:* Copenhagen Univ, MD, 45, PhD, 48; Tulane Univ, MD, 59. *Prof Exp:* Instr micros anat, Copenhagen Univ, 42-43, asst prof histol, 45-49; asst prof urol res & dir urol res lab, Univ Rochester, 49-57, asst prof urol surg, 57-59; assoc dean admin & clin affairs, 75-76, PROF UROL & HEAD DEPT, SCH MED, TULANE UNIV, 59-, CHIEF STAFF, UNIV HOSP, 77- *Concurrent Pos:* Rockefeller fel, Carnegie Inst, 48-49; asst resident, County Hosp, Copenhagen, Denmark, 44-45; asst resident, Strong Mem Hosp, Rochester, NY, 54-56, resident, 56-57; urologist in chief, Tulane Serv Charity Hosp, New Orleans, 59-; consult, Ochsner Found Hosp, 59-, Vet Admin Hosp, 60-, Touro Infirmary, 60-, Vet Admin Hosp, Alexandria, 61 & Lallie Kemp Charity Hosp, Independence, 63-; dir dept urol, Huey P Long Charity Hosp, Pineville, 64-; mem comt urol, Nat Acad Sci-Nat Res Coun, chmn ad hoc comt urol. *Mem:* Am Physiol Soc; Soc Exp Biol & Med; Am Asn Anat; Danish Med Asn. *Res:* Renal physiology. *Mailing Add:* Dept Urol Tulane Univ Sch Med 1430 Tulane Ave New Orleans LA 70112

SCHLEGEL, ROBERT ALLEN, b Chicago, Ill, Feb 17, 45; m 68. CELL BIOLOGY, BIOMEMBRANES. *Educ:* Univ Iowa, BS, 67; Harvard Univ, AM, 68, PhD(biochem & molecular biol), 71. *Prof Exp:* Fel, Walter & Eliza Hall Inst Med Res, Melbourne, 71-74; res asst prof immunol, Univ Utah, 74-76; from asst prof to assoc prof, 76-88, PROF MOLECULAR & CELL BIOL, PA STATE UNIV, 88-, DEPT HEAD, 91- *Concurrent Pos:* Prin investr res grants, NIH, Nat Cancer Inst, Am Cancer Soc & Am Heart Asn, 76-; estab investr, Am Heart Asn, 83-88. *Mem:* Am Soc Cell Biol; Int Cell Cycle Soc. *Res:* Introduction of macromolecules into eukaryotic cells; cell fusion; cell cycle regulation; membrane structure and function; hematopoietic cell surfaces. *Mailing Add:* Dept Molecular & Cell Biol Pa State Univ 101 S Frear University Park PA 16802

SCHLEGEL, ROBERT JOHN, b Ft Wayne, Ind, Dec 31, 27; m 49; c 3. PEDIATRICS, GENETICS. *Educ:* Univ Chicago, PhB, 49, MD, 55; Am Bd Pediat, dipl, 60. *Prof Exp:* From intern to chief resident, State Univ NY Upstate Med Ctr, 55-58; chief pediat serv, Walson Army Hosp, 59-60; asst chief pediat serv, Tripler Gen Hosp, 60-62, chief, 62-63; asst chief pediat serv, Walter Reed Gen Hosp & cytopathologist, Div Med, Walter Reed Army Inst Res, 66-69; assoc prof pediat, Sch Med, Stanford Univ, 69-71; PROF PEDIAT, SCH MED, UNIV CALIF, LOS ANGELES & PROF & CHAIRPERSON PEDIAT, CHARLES R DREW POSTGRAD MED SCH, 71- *Concurrent Pos:* Res fel human genetics & endocrinol, State Univ NY Upstate Med Ctr, 63-65 & Univ Georgetown, 66-67, clin prof, 67-; dir clin serv, Dept Pediat, Martin Luther King Gen Hosp, Los Angeles, 71-; mem, Nat Bd Med Examrs, Part III Clin Competence Comt. *Mem:* fel Am Acad Pediat; Endocrine Soc; Soc Pediat Res; Am Pediat Soc. *Res:* Peritoneal dialysis; human cytogenetics; cytogenetics of the embryo; embryogenesis of the gonads; sexual differentiation; biochemistry of phagocytosis, leucocyte chemotaxis and cell mediated immune responses; gene content and mode of action of genes of the x chromosome; gonadal proteins; dermatoglyphics; effects of infections on chromosomes. *Mailing Add:* 27838 Palos Verdes Dr E Rancho Palos Verdes CA 90274

SCHLEGELMILCH, REUBEN ORVILLE, b Greenbay, Wis, Mar 8, 16; m 43; c 4. MATHEMATICAL STATISTICS, PHYSICS. *Educ:* Univ Wis, BS, 38; Rutgers Univ, MS, 40; Mass Inst Technol, MS, 54. *Prof Exp:* Eng instr, Rugters Univ, 38-40, Cornell Univ, 40-41; res engr, Exp Sta, Univ Ill, 41-42;

proj engr & chief engr, Radar Lab, US Army/Air Materiel Command, Eatontown, NJ, 42-51; chief, Radar Lab/Electronic Warfare & Techniques Div, Rome Air Develop Ctr, USAF Air Systs Command, Rome, NY, 51-55, dir res & develop, 55-59; tech dir, Defense & Space Corp Hq, Westinghouse Elec Corp, Washington, DC, 59-63; mgr, Advan Technol & Missile Prog, Fed Systs Div, IBM, Owego, NY, 63-68; gen mgr & pres, Shilling Indust, 68-71; mgr, Preliminary Eng Design Directorate, US Army Advan Concepts Agency, Alexandria, Va, 71-74; mgr, Gun Fire Control Systs Develop, Naval Sea Systs Command, Washington, DC, 74-80; tech dir, Off Res & Develop, USCG Hq, Washington, DC, 80-86; CONSULT, 86- *Concurrent Pos:* Govt consult, Res & Develop Bd, Comt Electronics, Dept Defense, 49-54; Alfred P Sloan fel indust mgt, Sch Indust Mgt, Mass Inst Technol, 54-55; mem, Nat Prof Comt on Eng Mgt, Inst Elec & Electronic Engrs, 56-59, vchmn & chmn, Rome/Utica Sect, 56-59; indust consult, Guided Missile & Space Coun, Aerospace Indust Asn, 59-63; chmn, Southern Tier Empire Post, Am Defense Preparedness Asn, 67-68; mem, Univ Res Rev Bd, Dept Transp, 82-86, Small Bus Innovated Res Prog Rev Bd, 82-86; mem, Comt Visibility, Nat Transp Res Bd, 82-86; mem, Marine Facil Panel, US/Japan Coop Prog in Natural Resources, 82-86; mem, Tech Adv Comt, Great Lakes Comn, 82-86. *Mem:* Sr mem Inst Elec & Electronic Engrs; NY Acad Sci; Nat Soc Prof Engrs. *Res:* Electronics engineering; author of numerous technical articles and reports; recipient of one patent. *Mailing Add:* 8415 Frost Way Annandale VA 22003

SCHLEICH, THOMAS W, b Staten Island, NY, May 29, 38; m 62, 70. BIOCHEMISTRY, PHYSICAL CHEMISTRY. *Educ:* Cornell Univ, BS, 60; Rockefeller Univ, PhD(biochem), 66. *Prof Exp:* Res assoc biochem, Dartmouth Col, 66-67; res assoc chem, Univ Ore, 67-69; from asst prof to assoc prof, 69-79, PROF CHEM, UNIV CALIF, SANTA CRUZ, 79-; PROF PHARMACEUT CHEM, UNIV CALIF, SAN FRANCISCO, 86- *Concurrent Pos:* Helen Hay Whitney Found fel, 68-69; adj prof biol & chem, Univ Calif, Davis, 84- *Mem:* AAAS; Am Chem Soc; Biophys Soc; Am Soc Biol Chemists; Soc Magnetic Resonance Med; Asn Res Vision & Ophthal. *Res:* Biophysical chemistry; eye lens metabolism (cataractogenesis); in vivo nuclear magnetic resonance spectroscopy; lens protein structure and function. *Mailing Add:* Dept Chem & Biochem Univ Calif Santa Cruz CA 95064

SCHLEICHER, DAVID LAWRENCE, b Palmerton, Pa, July 22, 37; m 61; c 1. TECHNICAL WRITING. *Educ:* Pa State Univ, BS, 59, PhD(geol), 65; Calif Inst Technol, MS, 62. *Prof Exp:* GEOLOGIST, US GEOL SURV, 65- *Mem:* Fel Geol Soc Am. *Res:* Mechanics of tuffisite intrusion; seismicity induced by reservoirs in grabens; preparation of environmental impact statements. *Mailing Add:* US Geol Surv, Mail Stop 913 Box 25046 Fed Ctr Denver CO 80225

SCHLEICHER, JOSEPH BERNARD, b Nanticoke, Pa, Apr 4, 29; m 58; c 4. CELL BIOLOGY, VIROLOGY. *Educ:* Wilkes Col, BS, 51; Miami Univ, MS, 55; Kans State Univ, PhD(microbiol), 61. *Prof Exp:* Res asst immunol, Kans State Univ, 56-58; sr scientist, Develop Dept, Pitman-Moore Co, Ind, 58-62; head virus res sect, Alcon Labs, Inc, Tex, 62-63, head virus & microbiol res sect, 63-64; virologist, Virus Res Dept, Abbott Labs, Inc, 64-69, head antiviral tissue cult screening prog, 67-69, head res of biochem prod from tissue cult cells, Molecular Biol Dept, 69-71, res scientist, Biochem Develop Dept, 71-81 & Infectious Dis Immunol Diag Dept, 81-85, SR RES SCIENTIST, HEPATITIS/AIDS RES DEPT, ABBOTT LABS, INC, 85- *Concurrent Pos:* Consult tissue culture, Falcon Labs, Oxnard, Calif. *Mem:* AAAS; Am Soc Microbiol; Sigma Xi; NY Acad Sci; Tissue Cult Asn. *Res:* Research and development in virus vaccines for humans and animals, primarily in the area of respiratory diseases; antiviral chemotherapy and prophalaxis; interferon; tissue culture mass scale methodology and physiology; production of biochemicals by tissue culture cells; immunology; viral diagnostics. *Mailing Add:* 311 Green Bay Rd Lake Bluff IL 60044

SCHLEIDT, WOLFGANG MATTHIAS, ethology, bioacoustics, for more information see previous edition

SCHLEIF, FERBER ROBERT, b Oroville, Wash, Mar 6, 13; m 37; c 2. ELECTRICAL ENGINEERING. *Educ:* Wash State Univ, BS, 35. *Prof Exp:* Mem staff construct & oper eng, Coulee Dam, Wash, US Bur Reclamation, 36-48, syst planning work, Mo River Basin Proj, 48-50, oper & maintenance eng, 50-62, chief elec power br, 62-74; CONSULT ELEC POWER, 74- *Concurrent Pos:* Mem, NAm Power Systs Interconnection comt, Colo, 65, mem spec sessions, North-South Intertie Task Force, 65. *Mem:* Fel Inst Elec & Electronics Engrs. *Res:* Power system stabilization; hydraulic turbine governing; excitation control for stability; generator insulation test techniques; power system tests. *Mailing Add:* 800 S Fillmore St Denver CO 80209

SCHLEIF, ROBERT FERBER, b Wenatchee, Wash, Nov 22, 40; m 67; c 1. MOLECULAR BIOLOGY, BIOCHEMISTRY. *Educ:* Tufts Univ, BS, 63; Univ Calif, Berkeley, PhD(biophys), 67. *Prof Exp:* Helen Hay Whitney fel, Harvard Univ, 67-71; from asst prof to assoc prof, 71-81, PROF BIOCHEM, BRANDEIS UNIV, 81- *Concurrent Pos:* res grant, USPHS, 71- & NSF, 82-; ed, Proteins, assoc ed, J Molecular Biol. *Mem:* Am Soc Biol Chemists. *Res:* Regulatory mechanisms governing gene activity; genetic, physical and physiological studies; structure of proteins and nucleic acids. *Mailing Add:* Johns Hopkins Univ 34th & Charles Sts Waltham MA 02154

SCHLEIF, ROBERT H, b Watertown, Wis, Apr 20, 23; m 48; c 3. INDUSTRIAL PHARMACY. *Educ:* Univ Wis, BS, 47, PhD(pharm), 50. *Prof Exp:* Asst prof pharm, St Louis Col Pharm, 50-54, from assoc prof to prof, 54-62; res pharmacist, Nutrit Abbot Int, North Chicago, 62-65, mgr prof specifications & stability, 65-70, dir qual assurance, Consumer Div, Abbott Labs, 70-79, mgr qual assurance, 79-84; DIR MFG & QUAL ASSURANCE, GEN DERM CORP, 84- *Concurrent Pos:* Consult, Vet Admin Hosp. *Mem:* AAAS; Am Pharmaceut Asn; Am Soc Qual Control. *Res:* Arabic acid; ophthalmic solutions; antacids. *Mailing Add:* 2330 Grove Ave Waukegan IL 60085

SCHLEIFER, STEVEN JAY, b New York, NY, Mar 10, 50; m 71; c 4. PSYCHO-IMMUNOLOGY. *Educ:* Columbia Univ, BA, 71, Mt Sinai Sch Med, MD, 75. *Prof Exp:* Resident psychiat, Los Angeles County Hosp, Univ Southern Calif, 75-76; resident, 76-79, instr, 78-81, asst prof psychiat, Mt Sinai Sch Med, 82-87; ASSOC PROF PSYCHIAT, UNIV MED & DENT NJ, 87- *Concurrent Pos:* Prin investr, NIMH grants, 82-; ad hoc reviewer, Arch Gen Psychiat, Arch Internal Med, NIH & Alcoholism, Drug Abuse & Mental Health Asn study sects, & Brit Med Res Coun. *Mem:* Am Psychiat Asn; AAAS. *Res:* Effect of brain and behavior on the immune system; effects of bereavement and other life stresses on immunity; major depressive disorder and immunity; neuroendocrine mechanisms in stress effects on immunity; depression in patients with medical disorders; compliance; behavioral aspects of AIDS risk. *Mailing Add:* MSB-E501 Univ Med & Dent NJ 185 S Orange Ave Newark NJ 07103

SCHLEIGH, WILLIAM ROBERT, b Olean, NY, Feb 26, 41; m 62; c 2. ORGANIC CHEMISTRY. *Educ:* Clarkson Col Technol, BS, 62, PhD(org chem), 66. *Prof Exp:* Fel, Univ Wis, Madison, 65-67; RES CHEMIST, EASTMAN KODAK CO, 67- *Mem:* Am Chem Soc; Royal Soc Chem; Sigma Xi. *Res:* Heterocyclic chemistry; synthesis and reactions of heterocyclic compounds; organic reaction mechanisms; photochemistry of heterocyclic compounds. *Mailing Add:* PO Box 855 Clarkson NY 14430

SCHLEIMER, ROBERT P, b New York, NY, Apr 8, 52; c 2. IMMUNOLOGY, PHARMACOLOGY. *Educ:* Univ Calif, Davis, PhD(pharmacol), 80. *Prof Exp:* Asst prof, 82-87, ASSOC PROF MED, SCH MED, JOHNS HOPKINS UNIV, 87- *Res:* Immunopharmacology of inflammation. *Mailing Add:* Good Samaritan Hosp Unit Off 3A62 Sch Med Johns Hopkins Univ 301 Bayview Blvd Baltimore MD 21224

SCHLEIN, HERBERT, b New Haven, Conn, Nov 7, 27; m 52; c 2. ORGANIC CHEMISTRY. *Educ:* Harvard Univ, BS, 50; Boston Univ, PhD, 54. *Prof Exp:* Chemist, Children's Cancer Res Found, Boston, Mass, 53-58; chemist, Chem & Plastics div, Qm Res & Eng Command, US Dept Army, 58-60: head, Org Chem Sect, Explor Chem & Physics Div, Itek Corp, 60-63; vpres & dir res, Rahn Labs, 63-66; res group leader, Polaroid Corp, 66-85, sr scientist, 86-91; RETIRED. *Mem:* Am Chem Soc. *Res:* Diffusion transfer processes; product design; image-forming systems; electrophotography; photoconductors. *Mailing Add:* 106 Lothrop St Beverly MA 01915

SCHLEIN, PETER ELI, b New York, NY, Nov 18, 32; m 62; c 2. PHYSICS. *Educ:* Union Col, NY, BS, 54; Northwestern Univ, PhD, 59. *Prof Exp:* Res assoc physics, Johns Hopkins Univ, 59-61; res assoc, 61-62, from asst prof to assoc prof, 61-68, PROF PHYSICS, UNIV CALIF, LOS ANGELES, 68- *Concurrent Pos:* Vis physicist, Saclay, France, 63-64 & European Orgn Nuclear Res, 63-64 & 69-74; J S Guggenheim fel, 69-70; consult, Space Tech Labs, 66. *Mem:* Fel Am Phys Soc. *Res:* Experimental particle physics; properties of elementary particle interactions using electronic techniques. *Mailing Add:* Dept Physics 3-174 Knudsen Hall Univ Calif 405 Hilgard Ave Los Angeles CA 90024

SCHLEITER, THOMAS GERARD, b Evanston, Ill. SAFETY ENGINEERING, TECHNICAL APPRAISAL & INSPECTION. *Educ:* Univ Detroit, BME, 52; Univ Mich, MS, 54, MS, 55. *Prof Exp:* Nuclear engr reactor licensing, US Atomic Energy Comn, 56-61, reactor develop, 61-74; nuclear engr reactor develop, US Energy Res & Develop Admin, 74-77; nuclear engr facil develop, US Dept Energy, 77-85, team leader design & safety assessments, 85-88; NUCLEAR ENGR, ARGONNE NAT LAB, 88- *Concurrent Pos:* Instr Montgomery Col, 87- *Mem:* Am Nuclear Soc; Am Soc Mech Engrs. *Res:* Conceptual design of an advanced nuclear test reactor, as a thesis-type effort; managed design and safety reviews of major United States reactor facilities and other nuclear facilities. *Mailing Add:* 10011 Wedge Way Gaithersburg MD 20879

SCHLEMMER, FREDERICK CHARLES, II, b Watts Bar Dam, Tenn, Aug 10, 43; m 67; c 1. PHYSICAL OCEANOGRAPHY. *Educ:* US Naval Acad, BS, 65; Univ SFla, MA, 71; Tex A&M Univ, PhD(oceanog), 78. *Prof Exp:* Asst prof marine sci & asst to pres, Moody Col, 78-80; asst prof, 80-85, ASSOC PROF, MARINE SCI, TEX A&M UNIV, GALVESTON, 85- *Concurrent Pos:* Consult, Encyclopedia Britannica Film Rev Bd. *Mem:* Am Meteorol Soc; Am Geophys Union; Nat Geosci Hon Soc; Oceanog Soc. *Res:* Assessment of circulation patterns and hydrographic property distributions to evaluate pathways for water movement over large areas. *Mailing Add:* Tex A&M Univ PO Box 1675 Galveston TX 77553

SCHLEMPER, ELMER OTTO, b Pacific, Mo, Apr 13, 39; m 59; c 4. INORGANIC CHEMISTRY. *Educ:* Wash Univ, AB, 61, Univ Minn, PhD(inorg chem), 65. *Prof Exp:* Res assoc chem, Brookhaven Nat Lab, 64-66; from asst prof to assoc prof, 66-76, PROF INORG CHEM, UNIV MO, COLUMBIA, 76-, CHMN, CHEM DEPT, 90- *Honors & Awards:* Res Award, Sigma Xi. *Mem:* Am Chem Soc; Am Crystallog Asn; Chem Soc; Sigma Xi. *Res:* Structure of inorganic metal complexes and organometallics; hydrogen bonding; chemical bonding; develop of radiopharmaceuticals. *Mailing Add:* Dept Chem Univ Mo Columbia MO 65211

SCHLENDER, KEITH K, b Newton, Kans, Oct 3, 39; m 63; c 2. BIOCHEMISTRY, PHARMACOLOGY. *Educ:* Westmar Col, BA, 61; Mich State Univ, MS, 63, PhD(biochem), 66. *Prof Exp:* From asst prof to assoc prof, 69-81, PROF PHARMACOL & THERAPEUT, MED COL OHIO, 81-, DEAN GRAD SCH, 91- *Concurrent Pos:* NIH, res fel biochem, Univ Minn, Minneapolis, 66-69; NIH Res Career Develop Award, 78-83. *Mem:* Am Chem Soc; Am Heart Asn; Am Soc Pharmacol & Exp Therapeut; Am Soc Biochem & Molecular Biol; Sigma Xi. *Res:* Regulation of cellular processes by protein phosphorylation; protein kinases, protein phosphatases, glycogen metabolism. *Mailing Add:* Dept Pharmacol & Therapeut Med Col Ohio CS 10008 Toledo OH 43699-0008

SCHLENK, FRITZ, b Munich, Ger, 1909; nat US; m 40; c 2. BIOCHEMISTRY, MICROBIOLOGY. *Educ:* Univ Berlin, PhD(chem), 34. *Prof Exp:* Asst, Univ Stockholm, 34-37, res assoc, 37-40; asst prof biochem, Sch Med, Univ Tex, 40-43; assoc prof nutrit, Sch Med, prof biochem, Sch Dent & biochemist in charge, M D Anderson Hosp & Tumor Inst, 43-47; prof bact, Iowa State Univ, 47-54; sr biochemist, Argonne Nat Lab, 54-74; res prof biol, Univ Ill, Chicago, 75-90; RETIRED. *Concurrent Pos:* Res assoc prof, Univ Chicago, 54-74. *Mem:* Am Soc Biol Chem; Am Soc Microbiol; Am Acad Microbiol. *Res:* Enzymes, coenzymes intermediate metabolism; transmethylation; yeast cytology. *Mailing Add:* 3904 Forest Ave Downers Grove IL 60515

SCHLENK, HERMANN, b Jena, Ger, July 28, 14; nat US; m 46; c 2. ORGANIC CHEMISTRY, BIOCHEMISTRY. *Educ:* Univ Berlin, dipl, 36; Univ Munich, Dr rer nat(chem), 39. *Prof Exp:* Res chemist, Baden Anilin & Soda Works, Ger, 39-42; res assoc org chem, Univ Munich, 44-46; teaching asst & lectr, Univ Wuerzburg, 46-49; from asst prof to assoc prof biochem, Agr & Mech Col Tex, 49-52; from asst prof org chem to prof biochem, 53-85, asst dir, Inst, 67-75, EMER PROF, HORMEL INST, UNIV MINN, 85- *Mem:* Am Chem Soc; Am Soc Biol Chem; Am Oil Chem Soc. *Res:* Chemistry and biochemistry of lipids. *Mailing Add:* Hormel Inst 801 NE 16th Ave Austin MN 55912

SCHLENKER, EVELYN HEYMANN, b La Paz, Bolivia, May 30, 48; US citizen; m; c 1. RESPIRATORY PHYSIOLOGY. *Educ:* City Col New York, BS, 70; State Univ NY, Buffalo, MA, 73, PhD(biol), 76. *Prof Exp:* Vis asst physiol, Rochester Inst Technol, 76-77; fel respiration physiol, Univ Fla, 78-80; asst prof, 80-88, ASSOC PROF PHYSIOL, UNIV SDAK, 88- *Mem:* Am Soc Zoologist; Sigma Xi; Am Physiol Soc. *Res:* Control of respiration in animal models of respiratory-muscular diseases; factors affecting airway reactivity in human subjects, including air pollution, smoking and agricultural pollution. *Mailing Add:* Dept Physiol & Pharmacol Med Sch Univ SDak Vermillion SD 57069

SCHLENKER, ROBERT ALISON, b Rochester, NY, Oct 25, 40; m 68; c 2. RADIATION PHYSICS. *Educ:* Mass Inst Technol, SB, 62, PhD(nuclear physics), 68. *Prof Exp:* Appointee radium poisoning, Radioactivity Ctr, Mass Inst Technol, 68-69; asst physicist, 70-75, BIOPHYSICIST RADIATION DOSIMETRY, ARGONNE NAT LAB, 75-, ASSOC MGR, ENVIRON, SAFETY & HEALTH DEPT, 90- *Concurrent Pos:* Math instr, Col DuPage, 75-78; mem Sci Comt, Nat Coun Radiation Protection & Measurements, 57 & 78-, mem Task Group Prob Bone, 78-; assoc ed, Radiation Res, 83-; mem comt biol effects internally deposited radio nuclides, Nat Res Coun, 85-89. *Mem:* Am Asn Physicists Med; Health Physics Soc; Radiation Res Soc; AAAS; Soc Risk Anal. *Res:* Cellular radiation biology; radiation dosimetry; alpha, beta and gamma-ray spectrometry as applied to the study of the concentrations and effects of radioisotope in human and animal tissues, especially skeletal tissue; internal dosimetry of occupational radiation exposures. *Mailing Add:* Argonne Nat Lab 9700 S Cass Ave Argonne IL 60439

SCHLEPPNIK, ALFRED ADOLF, b Mar 2, 22; US citizen; m 49; c 1. SYNTHETIC ORGANIC CHEMISTRY, CHEMORECEPTION. *Educ:* Univ Vienna, DPhil(chem), 57. *Prof Exp:* Instr org chem, Univ Vienna, 56-57; res assoc, Wash Univ, 57-58; res chemist, Monsanto Co, 58-59, sr res chem, 59-63, from res specialist to sr res specialist, 63-76, sci fel, 76-89; RETIRED. *Mem:* AAAS; Sigma Xi. *Res:* Synthesis of flavor and fragrance aroma chemicals; molecular biochemistry of gustation and olfaction primarily at the level of peripheral stimulus-receptor interactions; induced selective reversible anosmia. *Mailing Add:* One Hanley Downs St Louis MO 63117

SCHLESINGER, ALLEN BRIAN, b New York, NY, Feb 18, 24; m 47; c 4. EMBRYOLOGY, ENVIRONMENTAL BIOLOGY. *Educ:* Univ Minn, BA, 49, MS, 51, PhD(zool), 57. *Prof Exp:* Instr, Creighton Univ, 52-54, from asst prof to assoc prof, 54-61, dir dept, 58-72, chmn dept, 77-85 & 88-90, PROF BIOL, CREIGHTON UNIV, 61- *Concurrent Pos:* Consult, Omaha Pub Power Dist. *Mem:* Soc Develop Biol; Am Soc Zool; Am Chem Soc; Am Asn Anat; Soc Nuclear Med; Am Fisheries Soc; Sigma Xi; fel AAAS. *Res:* Embryonic growth control; morphogenetic movement; environmental influences on development; effects of discharges of generating plants on river biota. *Mailing Add:* Dept Biol Creighton Univ 2500 California St Omaha NE 68178

SCHLESINGER, DAVID H, b New York, NY, Apr 28, 39; m 70; c 2. PEPTIDE SYNTHESIS, PROTEIN SEQUENCING. *Educ:* Columbia Col, BA, 62; Albany Med Col, MS, 65; Mt Sinai Med Sch, PhD(physiol & biophys), 72. *Prof Exp:* Res fel med, Mass Gen Hosp, Harvard Med Sch, 72-75, asst biochem, 75-77, instr med, Harvard Med Sch, 75-77; res assoc prof physiol & biophys, Univ Ill Med Ctr, 77-81; RES PROF MED BIOCHEM, NY UNIV MED CTR, 81- *Concurrent Pos:* Lectr, Fundecion Gen Mediterranea, 75; consult, Immunol Res Inst, 86-, Clin Technol Assoc, Inc, 87-; dir, Kaplan Cancer Ctr, NY Univ, co-dir, Neurosci Sect, Mental Health Clin Res Ctr. *Mem:* NY Acad Sci; Am Chem Soc; Protein Soc; Am Soc Biochem & Molecular Biol; Int Soc Immunopharmacol. *Res:* Structure determination and synthesis of calcium binding proteins; microsequencing and synthesis; neuroscience. *Mailing Add:* 150 Tennyson Dr Plainsboro NJ 08536

SCHLESINGER, EDWARD BRUCE, b Pittsburgh, Pa, Sept 6, 13; m 41; c 4. NEUROSURGERY. *Educ:* Univ Pa, BA, 34, MD, 38; Am Bd Neurol Surg, dipl, 49. *Prof Exp:* Res asst neurol, Col Physicians & Surgeons, Columbia Univ, 46-47, res assoc neurol surg, 47-49, assoc, 49-52, from asst prof clin neurol surg to prof clin neurol surg, 52-73, chmn dept, 73-80, Byron Stookey prof neurol surg, 73-80, BYRON STOOKEY EMER PROF, COL PHYSICIANS & SURGEONS, COLUMBIA UNIV, 80- *Concurrent Pos:* Teagle fel, Col Physicians & Surgeons, Columbia Univ, 46-47; jr asst neurologist, Presby Hosp, NY, 45-47, from asst attend neurol surgeon to attend neurol surgeon, 47-73; consult, Monmouth Mem Hosp, NJ, 47-49, Walter Reed Army Hosp, DC, 47-50 & Knickerbocker Hosp, NY, 47-49; attend neurol surgeon, Inst Crippled & Disabled, NY, 47-58 & White Plains Hosp, NY, 54-73; trustee, Wm J Matheson Found, 75, Int Ctr for the Disabled, 83 & Sharon Hosp, Conn; pres, med bd Presby Hosp, 76-79, consult neurosurg, 80-; chmn, Elsberg Fel Comt, NY Acad Med, 78- *Mem:* AAAS; Neurosurg Soc Am (vpres, 59-70, pres, 70-71); Am Asn Neurol Surg; Harvey Soc; fel NY Acad Sci; Soc Neurol Surgeons; NY Acad Med. *Res:* Use of radioisotopes in neurology; pharmacology and biogenetics of tumors of the central nervous system; genetic markers of neurological and orthopedic disorders. *Mailing Add:* 710 W 168th St New York NY 10032

SCHLESINGER, ERNEST CARL, b Hildesheim, Germany, Nov 25, 25; nat US; m 58; c 2. MATHEMATICAL ANALYSIS. *Educ:* Univ Wash, BS, 47, MA, 50; Harvard Univ, PhD, 55. *Prof Exp:* Instr philos, Univ Wash, 49-50; instr math, Yale Univ, 55-58; asst prof, Wesleyan Univ, 58-62; from asst prof to assoc prof, 62-73, PROF MATH, CONN COL, 73- *Concurrent Pos:* Fulbright lectr, Univ Col, Dublin, 68-69. *Mem:* Am Math Soc; Math Asn Am. *Res:* Functions of a complex variable. *Mailing Add:* Dept Math Box 5566 Conn Col 270 Mohegan Ave New London CT 06320-4196

SCHLESINGER, JAMES WILLIAM, b Salina, Kans, June 20, 31; m 52; c 4. MATHEMATICS. *Educ:* Mass Inst Technol, BS, 55, PhD(math), 64. *Prof Exp:* From instr to asst prof, 60-67, chmn dept, 69-73, ASSOC PROF MATH, TUFTS UNIV, 67- *Mem:* Am Math Soc. *Res:* Homotopy groups of spheres; semi-simplicial topology; homotopy theory. *Mailing Add:* Dept Math Tufts Univ Blomfield Pearson Hall Medford MA 02155

SCHLESINGER, JUDITH DIANE, b New York, NY. COMPILER DESIGN, ARTIFICIAL INTELLIGENCE. *Educ:* Brooklyn Col, BS, 67; Ohio State Univ, MS, 69; Johns Hopkins Univ, PhD(comput sci), 76. *Prof Exp:* Comput programmer, Honeywell Info Systs, 69-71; systs scientist, Mgt Adv Serv, 74-75; asst prof comput sci, Dept Math & Comput Sci, Univ Denver, 76-81; consult comput sci, JDS Consult Serv, 86-90; MEM RES STAFF, SUPERCOMPUT RES CTR, 90- *Concurrent Pos:* Adj assoc prof comput sci, Dept Math & Comput Sci, Univ Denver, 87- *Mem:* Asn Comput Mach; Inst Elec & Electronics Engrs Comput Soc; Am Asn Artificial Intel. *Res:* Computer design and development; artificial intelligence and expert systems; case and productivity tools; development of a metacompiler; natural language interface to an intelligent tutoring system. *Mailing Add:* PO Box 3926 Crofton MD 21114

SCHLESINGER, LEE, b Chicago, Ill, July 4, 26; m 54; c 3. ENGINEERING, SOILS. *Educ:* Ill Inst Technol, BS, 49, MS, 50. *Prof Exp:* Mat engr, Corps Eng, US Dept Army, 50; found & soils engr, Soil Testing Servs, Inc, 51-54; chief soils engr, Alfred Benesch & Assocs, 53-56; chief struct engr, Western-Knapp Eng Co, 56-60; proj mgr, Meissner Engrs, Inc, 60-63; PRES & DIR CONSTRUCT DIV, HOYER-SCHLESINGER-TURNER, INC, CHICAGO, 63- *Concurrent Pos:* Mem, Int Conf Soil Mech & Found Eng, 52; environ comnr, Village Morton Grove, Ill, 70- *Mem:* AAAS; Soc Am Mil Engrs; Am Soc Civil Engrs; Iron & Steel Asn; Am Concrete Inst; Sigma Xi. *Res:* Application of soil mechanics to civil engineering design; management concept of design-construction in heavy process industry. *Mailing Add:* 6417 Hoffman Terr Morton Grove IL 60053

SCHLESINGER, MARTIN D(AVID), b New York, NY, Aug 9, 14; m 45; c 3. CHEMICAL ENGINEERING. *Educ:* Univ Okla, BS, 41; NY Univ, MChE, 44. *Prof Exp:* Chem engr, M W Kellogg Co, 41-48; asst chief, Gas Synthesis Sect, US Bur Mines, 48-56, asst chief, Coal Hydrogenation Sect, 56-62, res coordr, 62-67, proj coordr process technol, 67-71, staff coordr, 71-74; dep res dir, Pittsburgh Energy Res Ctr, ERDA, 74-77; PRES, WALLINGFORD GROUP LTD, 77- *Concurrent Pos:* Consult coal conversion, UN & US Agency Int Develop. *Honors & Awards:* Distinguished Serv Award, Am Chem Soc, 89. *Mem:* Am Chem Soc; Am Inst Chem Engrs; Am Soc Pub Admin; AAAS. *Res:* Conversion of coal to synthetic fuels and chemicals; utilization of waste materials; research management. *Mailing Add:* Wallingford Group Ltd 4766 Wallingford St Pittsburgh PA 15213-1712

SCHLESINGER, MILTON J, b Wheeling, WVa, Nov 26, 27; m 55. BIOCHEMISTRY, MICROBIOLOGY. *Educ:* Yale Univ, BS, 51; Univ Rochester, MS, 53; Univ Mich, PhD(biochem), 59. *Prof Exp:* Res assoc, Univ Mich, 53-56 & 59-60; vis scientist, Int Ctr Chem Microbiol, Superior Inst Health, Italy, 60-61; res assoc biol, Mass Inst Technol, 61-64; from asst prof to assoc prof, 64-72, PROF MICROBIOL & IMMUNOL, SCH MED, WASH UNIV, 72- *Concurrent Pos:* Vis scientist, Imp Can Res Fund, London, 74-75; vis scholar, Harvard Univ, Cambridge, Mass, 89-90. *Mem:* Am Soc Biol Chemists; Am Chem Soc; Am Soc Microbiol; Am Soc Virol; AAAS. *Res:* Protein structure and function; molecular biology of animal viruses; protein-lipid interactions; heat-shock; ubiquitin; antivirals. *Mailing Add:* Dept Molecular Microbiol Sch Med Wash Univ St Louis MO 63110-1093

SCHLESINGER, MORDECHAY, b Budapest, Hungary, Sept 2, 31; m 57; c 2. CONDENSED MATTER, PHYSICS. *Educ:* Hebrew Univ, Jerusalem, Israel, MSc, 59, PhD(physics), 63. *Prof Exp:* NASA fel physics, Univ Pittsburgh, 63-65; from asst prof to assoc prof, Univ Western Ont, 65-68; PROF PHYSICS, UNIV WINDSOR, 68-, HEAD DEPT, 83- *Concurrent Pos:* Div ed, J Electrochem Soc, 79-90, assoc ed, 90- *Mem:* Am Phys Soc; Can Asn Physicists; Electrochem Soc; fel Inst Physics UK. *Res:* Crystal field studies; angular momentum algebra; thin films; magneto-optical and electrical properties of condensed media; electron microscopy; electrochemistry. *Mailing Add:* Dept Physics Univ Windsor Windsor ON N9B 3P4 Can

SCHLESINGER, R(OBERT) WALTER, b Hamburg, Ger, Mar 27, 13; US citizen; m 42; c 2. VIROLOGY. *Educ:* Univ Basel, MD, 37. *Prof Exp:* Fel bact & path, Rockefeller Inst Med Res, NY, 40-42, asst, 42-46; assoc res prof virol, Univ Pittsburgh, 46-47; assoc mem virol, Pub Health Res Inst, New York, NY, 47-55; prof & chmn microbiol, St Louis Univ Sch Med, 55-63; prof & chmn microbiol, Rutgers Med Sch, 63-83, EMER DISTINGUISHED PROF MOLECULAR GENETICS & MICROBIOL, ROBERT WOOD

JOHNSON MED SCH, UNIV MED & DENT NJ, 83- *Concurrent Pos:* Vis investr at numerous insts & orgn, 56-81; mem & chmn, numerous govt & insts rev & adv comt, 60-90; asst dean, Rutgers Med Sch, 63-67, acting dean, 70-71; ed, Virol, 63-83; prof, Rutgers Univ Grad Fac, 63-; Guggenheim fel, Guggenheim Found, 72-73. *Honors & Awards:* Selman Waksman Award, Am Soc Microbiol, 79; Humboldt Medal, Humboldt Found, Ger, 89; A von Graefe Medal, Berlin Med Soc, 81. *Mem:* Fel AAAS; Am Asn Immunologists; Am Soc Microbiol; Am Soc Virol; Am Asn Cancer Res; Soc Exp Biol & Med. *Res:* Various basic aspects of many different viruses from 1937 to the present; replication mechanisms, host adaptation, immunological and molecular aspects of arboviruses, picoruaviruses, herpesviruses, adenoviruses, influenza viruses, and factors determining pathogenesis. *Mailing Add:* Dept Molecular Genetics & Microbiol UMDNJ-Robert Wood Johnson Med Sch 675 Hoes Lane Piscataway NJ 08854-5635

SCHLESINGER, RICHARD B, b Mt Kisco, NY, Dec 19, 47; m; c 1. INHALATION TOXICOLOGY, PULMONARY PHYSIOLOGY. *Educ:* Queens Col, NY, BA, 68; NY Univ, MS, 71, PhD(biol), 75. *Prof Exp:* Assoc res scientist, 75-78, from asst prof to prof, 78-87, DIR, LAB PULMONARY BIOL & TOXICOL, DEPT ENVIRON MED, NY UNIV MED CTR, 86-, DIR, SYSTEMIC TOXICOL PROG, 88- *Concurrent Pos:* Consult, Sci Adv Bd, US Environ Protection Agency, 85-; mem, Panel Pulmonary Toxicol, Bd Environ Studies & Toxicol, Nat Res Coun, 86-89, Respiratory Modelling Group, Nat Coun Res & Planning, 84-; consult, US Environ Protection Agency, 87-; res career develop award, NIH, 83-88. *Honors & Awards:* Kenneth Morgareidge Award, 87. *Mem:* Am Thoracic Soc; Soc Toxicol; Am Indust Hyg Asn. *Res:* Analysis of the effects of ambient air pollutants upon the structure and physiology of the defense mechanisms of the lungs; development of lung disease due to defense mechanism dysfunction. *Mailing Add:* Dept Environ Med NY Univ Med Ctr 550 First Ave New York NY 10016

SCHLESINGER, RICHARD CARY, b Oberlin, Ohio, Apr 27, 40; m 61; c 2. SILVICULTURE. *Educ:* Middlebury Col, BA, 63; Yale Univ, MF, 65; State Univ NY Col Forestry, PhD(forest micrometeorol), 70. *Prof Exp:* Forester, Nat Forest Admin, 65-66, RES FORESTER, NCENT FOREST EXP STA, US FOREST SERV, 69- *Concurrent Pos:* Adj asst prof, Southern Ill Univ, 70-88. *Honors & Awards:* Walnut Res Award. *Mem:* Am Meteorol Soc; Soc Am Foresters; Sigma Xi. *Res:* Culture of black walnut with emphasis on the microenvironmental requirements; modeling of tree growth and of microenvironments. *Mailing Add:* 1-26 Agr Bldg Univ Mo Columbia MO 65202

SCHLESINGER, S PERRY, b New York, NY, Oct 9, 18; m 43, 79; c 2. MILLIMETER-SUBMILLIMETER WAVES, FREE ELECTRON LASERS. *Educ:* Mich State Univ, BA, 41; Union Col NY, MS, 50; Johns Hopkins Univ, DEng(elec eng), 57. *Prof Exp:* Prin investigator, NSF grants, Office Naval Reserve, Air Force Off Sci Res, 58-87, engr, Res Sect, Turban Generator Div, Gen Elec Co, 46-47; asst prof elec eng, Union Col NY, 47-50 & US Naval Acad, 50-53; res assoc microwaves, Radiation Lab, Johns Hopkins Univ, 53-56; from asst prof to prof, 56-87, EMER PROF ELEC ENG, COLUMBIA UNIV, 87- *Concurrent Pos:* Vis res assoc, Plasma Physics Lab, Princeton Univ, 62-63, consult, 63-64; pres, Faculties Assoc Consults Inc, 63-69; vis prof, Israel Inst Technol, 69-70, Tel Aviv Univ, 76-77, 83-84; consult, Naval Res Lab, 74-85; chmn, dept elec eng, Columbia Univ, 80-83. *Mem:* fel Inst Elec & Electronics Engrs; Am Phys Soc. *Res:* Plasma physics in general with emphasis on electromagnetic wave-plasma interaction; relativistic electron beam coherent sources of high power millimeter waves, in particular development of Raman free electron laser. *Mailing Add:* 400 W 119th St Apt 13-O New York NY 10027

SCHLESINGER, SONDRA, b Long Branch, NJ, July 10, 34; m 55. VIROLOGY, MICROBIOLOGY. *Educ:* Univ Mich, BS, 56, PhD(biochem), 60. *Prof Exp:* Res assoc microbiol, Mass Inst Technol, 61-64; from asst prof to assoc prof microbiol, 64- 76, PROF MOLECULAR MICROBIOL, SCH MED, WASH UNIV, 76- *Concurrent Pos:* Nat Found fel, Inst Superiore Sanita, Italy, 60-61; Am Cancer Soc grant, 64-65; USPHS grant, 65- *Mem:* Fedn Am Scientists Exp Biol; Am Soc Microbiol; Am Soc Virol. *Res:* Synthesis and structure of enveloped RNA viruses. *Mailing Add:* Dept Microbiol Box 8230 Wash Univ Sch Med St Louis MO 63110-1093

SCHLESINGER, STEWART IRWIN, b Chicago, Ill, Apr 22, 29; m 51; c 2. COMPUTER SCIENCE, MATHEMATICS. *Educ:* Ill Inst Technol, BS, 49, MS, 51, PhD(math), 55. *Prof Exp:* Staff mem, Los Alamos Sci Lab, 51-56; mgr math & comput, Aeronutronic Div, Ford Motor Co, 56-63; dir, Math & Comput Ctr, 63-69, gen mgr, Info Processing Div, 69-80, gen mgr, Mission Info Systs Div, 80-81, GEN MGR, SATELLITE CONTROL DIV, AEROSPACE CORP, 81- *Mem:* Soc Indust & Appl Math; Asn Comput Mach; Soc Comput Simulation (vpres, 78-79, pres, 79-82); Sigma Xi. *Res:* Real-time computing systems; applied mathematics, numerical analysis, data reduction, simulation, interactive computing, and management of software development, particularly involving large scale digital computers. *Mailing Add:* 12131 Skyway Dr Santa Ana CA 92705

SCHLESINGER, WILLIAM HARRISON, b Cleveland, Ohio, Apr 30, 50. PLANT ECOLOGY. *Educ:* Dartmouth Col, BA, 72; Cornell Univ, PhD(biol), 76. *Prof Exp:* Asst prof ecol, Univ Calif, Santa Barbara, 76-80; from asst prof to assoc prof, 80-88, PROF BOT, DUKE UNIV, 88- *Mem:* Ecol Soc Am; AAAS; Sigma Xi; Soil Sci Soc Am. *Res:* Ecosystem ecology including nutrient cycling in natural systems and global geochemical cycles; plant community structure. *Mailing Add:* Dept Bot Duke Univ Durham NC 27706

SCHLESSINGER, BERNARD S, b Mar 19, 30; m 52; c 3. INFORMATION SCIENCE. *Educ:* Roosevelt Univ, BS, 50; Miami Univ, Ohio, MS, 52; Univ Wis, PhD(phys chem), 55; Univ RI, MSLS, 75. *Prof Exp:* Res chemist, Am Can Co, 55-56; res supvr, USAF Sch Aviation Med, 56-58; indexer & dept head indexing, Chem Abstracts, 58-66; info scientist, Olin-Mathieson Chem Corp, 66-68; prof libr sci & asst dir div, Southern Conn State Col, 68-75; prof librarianship, Univ SC, 75-77; dean, Grad Libr Sch, Univ RI, 77-82; AT LIBR SCI, TEX WOMEN'S UNIV, DENTON. *Mem:* Am Chem Soc; Am Libr Asn; Spec Libr Asn. *Res:* Electrophoresis; cardiovascular disease diagnosis; indexing; abstracting; search strategy; library statistical data. *Mailing Add:* Libr Sci Sta Box 22905 Tex Women's Univ Denton TX 76204

SCHLESSINGER, DAVID, b Toronto, Ont, Sept 20, 36; US citizen; m 60; c 2. MOLECULAR BIOLOGY, MICROBIOLOGY. *Educ:* Univ Chicago, BA, 55, BS, 57; Harvard Univ, PhD(biochem), 60. *Prof Exp:* NSF fel, Pasteur Inst, Paris, France, 60-62; from instr to assoc prof, 62-72, PROF MICROBIOL, SCH MED, WASH UNIV, 72- *Concurrent Pos:* Macy fel, 81. *Mem:* Am Chem Soc; Am Soc Microbiol. *Res:* Cell physiology; biochemistry. *Mailing Add:* Dept Microbiol Wash Univ Sch Med Box 8230 St Louis MO 63110

SCHLESSINGER, GERT GUSTAV, b Karlsruhe, Ger, Mar 20, 33; nat US; m 59; c 2. WATER CHEMISTRY. *Educ:* City Col New York, BS, 53; Case Inst Technol, MS, 55; Univ Pa, PhD, 57. *Prof Exp:* Asst, Case Inst Technol, 53-55; asst instr chem, Univ Pa, 55-57; res assoc inorg chem, Univ Fla, 58-59; asst prof chem, Pace Col, 59-60; sr res chemist, Evans Res & Develop Corp, 60-61; assoc prof chem, Gannon Col, 61-66, Newark Col Eng, 66-68 & US Coast Guard Acad, 68-71; chief clin chemist, Fairfield Hills State Hosp, Conn, 71-73; SR ENVIRON CHEMIST, STATE OF CONN, HARTFORD, 73- *Mem:* Am Chem Soc; Am Inst Chemists; Royal Soc Chem; Water Pollution Control Fedn. *Res:* Coordination complexes; kinetics; application of inorganic reagents to organic syntheses; wastewater analysis. *Mailing Add:* Eight Norton Ct Norwich CT 06360

SCHLESSINGER, JOSEPH, b Mar 26, 45; m 70; c 2. RESEARCH ADMINISTRATION. *Educ:* Hebrew Univ, Jerusalem, BSc, 68, MSc, 69; Weizmann Inst Sci, Israel, PhD, 74. *Prof Exp:* Fel assoc, dept chem, Sch Appl & Eng Physics, Cornell Univ, 74-77; vis scientist, immunol br, Nat Cancer Inst, 77-78, sr scientist, dept chem immunol, Weizmann Inst Sci, Israel, 78-80, assoc prof, 80-83, prof, 83-84, dir, div molecular biol, Biotechnol Res Ctr, Meloy Lab Inc, Md, 85-86, dir Biotechnol Res Ctr, 86-87; res dir & adv, Rorer Biotechnol Inc, 87-90; CHMN & PROF DEPT PHARMACOL, NY UNIV MED CTR, 90- *Concurrent Pos:* Teaching asst, Hebrew Univ, Jerusalem, 68-69; Ruth & Leonard Simon Prof Cancer Res, dept chem immunol, Weizmann Inst Sci, 83- *Honors & Awards:* Hestrin Prize, Biochem Soc Israel, 83. *Mem:* European Molecular Biol Org. *Res:* Growth factor receptors. *Mailing Add:* Dept Pharmacol NY Univ Med Ctr 550 First Ave New York NY 10016

SCHLESSINGER, MICHAEL, b July 2, 37; m 58; c 3. ALGEBRA. *Educ:* Johns Hopkins Univ, BA, 59; Harvard Univ, PhD(math), 64. *Prof Exp:* Lectr math, Princeton Univ, 64-66; asst prof, Univ Calif, Berkeley, 66-73; assoc prof, 73-79, PROF MATH, UNIV NC, CHAPEL HILL, 79- *Concurrent Pos:* Res assoc, Inst Math, Pisa, 69 & Harvard Univ, 72; NSF res grants, 73-81. *Res:* Deformation theory in algebraic geometry, singularities. *Mailing Add:* Dept Math Univ NC Chapel Hill NC 27599

SCHLESSINGER, RICHARD H, b Greeley, Colo, Sept 20, 35. ORGANIC CHEMISTRY. *Educ:* Edinboro State Col, BSEd, 57; Ohio State Univ, PhD(org chem), 64. *Prof Exp:* Fel org chem, Harvard Univ, 64-65 & Columbia Univ, 65-66; from asst prof to assoc prof, 66-74, PROF ORG CHEM, UNIV ROCHESTER, 74- *Res:* Total synthesis of natural products; synthetic methods. *Mailing Add:* Dept Chem Univ Rochester Wilson Blvd Rochester NY 14627

SCHLEUSENER, RICHARD A, b Oxford, Nebr, May 6, 26; m 49; c 5. METEOROLOGY, ENGINEERING. *Educ:* Univ Nebr, BS, 49; Kans State Univ, MS, 56; Colo State Univ, PhD(irrig eng), 58. *Prof Exp:* Instr agr eng, Kans State Univ, 49-50; assoc prof civil eng, Colo State Univ, 58-64; dir inst atmospheric sci, 65-74, vpres & dean eng, 74-75, actg pres, 75-76, PRES, SDAK SCH MINES & TECHNOL, 76- *Mem:* Am Soc Civil Eng; Am Meteorol Soc; Am Soc Agr Eng; Am Geophys Union; Sigma Xi. *Res:* Development in weather modification. *Mailing Add:* 315 S Berry Pine Rd Rapid City SD 57702

SCHLEUSNER, JOHN WILLIAM, b Birmingham, Ala, Jan 16, 43; m 71. MATHEMATICS. *Educ:* Univ Ala, BS, 65, MA, 66, PhD(math), 69. *Prof Exp:* Asst prof, 69-74, ASSOC PROF MATH, WVA UNIV, 74- *Mem:* Am Math Soc. *Res:* Special functions; analysis. *Mailing Add:* Dept Math & Comput Sci Valdosta State Col Valdosta GA 31698

SCHLEYER, HEINZ, b Pforzheim, Ger, Oct 29, 27; m 63; c 1. BIOPHYSICS, ENZYMOLOGY. *Educ:* Karlsruhe Tech Univ, Diplom chem, 54, Dr rer nat(chem), 60. *Prof Exp:* Fel biophys, 61-64, res assoc, 64-69, ASST PROF BIOPHYS & BIOPHYS IN SURG, JOHNSON RES FOUND, SCH MED, UNIV PA, 69- *Mem:* Biophys Soc; Soc Ger Chem; Am Soc Biol Chem; AAAS. *Res:* Electron transfer systems of photosynthesis; photochemistry of bacteriochlorophyll; structure and function of the hemeprotein P-450 in steroid metabolism; chemical carcinogenesis; application of magnetic resonance and spectroscopic techniques to biologically important compounds. *Mailing Add:* Dept Biophysics 314 Meb Univ Pa Philadelphia PA 19104

SCHLEYER, PAUL VON RAGUÉ, b Cleveland, Ohio, Feb 27, 30; m 69; c 3. PHYSICAL ORGANIC CHEMISTRY. *Educ:* Princeton Univ, AB, 51; Harvard Univ, MA, 56, PhD(chem), 57. *Hon Degrees:* Dr, Univ Lyon, 71. *Prof Exp:* From instr to prof, Princeton Univ, 54-69, Eugene Higgins prof org chem, 69-76; INST CO-DIR & PROF, UNIV ERLANGEN-NURNBERG, WGER, 76- *Concurrent Pos:* A P Sloan res fel, 62-66; Guggenheim fel, 65-66; Fulbright res fel, Univ Munich, 65-66; vis & guest prof, Univ Colo, 63, Univ Wurzburg, 67, Univ Mich, 69, Univ Munich, 69 & 74-75, Carnegie-Mellon Univ, 69, Kyoto Univ, 70, Univ Munster, 71, Iowa State Univ, 72, Univ Geneva, 72, Univ Groningen, Neth, 72-73, Hebrew Univ Jerusalem, 73, Univ

Paris-Sud, 73, Univ Lausanne, Switz, 74, Univ Louvain, Belg, 74, Univ Liege, 74, Univ Western Ont, 78, Univ Copenhagen, 79-80 & Univ Utrecht, 82; consult, Hoffmann-La Roche, 71-72 & Hoechst AG, 80-; DuPont lectr, Clemson Univ, 71; Alexander von Humboldt Found sr US scientist award, 74-75; adj prof, Case Western Reserve Univ, 76-77 & Carnegie-Mellon Univ, 77-78; sr fel, Hydrocarbon Inst, Univ Southern Calif, 78-; co-ed, J Comput Chem, 80- *Honors & Awards:* Von Baeyer Medal, Ger Chem Soc, 86; Kahlbaum lectr, Univ Basel, 75; Ingersoll Mem lectr, Vanderbilt Univ, 83; J Musher Mem lectr, Israel, 85; J F Norris Award Phys Org Chem, Am Chem Soc, 87; Hersenberg Medal, World Asn Theoret Org Chemists, 87. *Mem:* Fel AAAS; Am Chem Soc; The Chem Soc; fel NY Acad Sci; fel Am Inst Chemists; fel Bavarian Acad Sci. *Res:* Bridged ring systems; adamantane and diamondoid molecules; structure, stability and rearrangements of carbonium ions; spectroscopy and hydrogen bonding; conformational analysis; theoretical calculations applied to organic intermediates and the exploration of new molecular structures; lithium and other electron deficient compounds. *Mailing Add:* Inst Org Chem Henkestrasse 42 Erlangen 8520 Germany

SCHLEYER, WALTER LEO, b Berlin, Ger, June 4, 19; nat US; m 51; c 2. BIOPHYSICAL CHEMISTRY. *Educ:* Rutgers Univ, AB, 48; Columbia Univ, AM, 50, PhD(phys chem), 52. *Prof Exp:* Res assoc, Col Physicians & Surgeons, Columbia Univ, 52-54; chemist, Res & Develop Dept, PQ Co, 54-63, tech field serv mgr, 63-66, mkt develop mgr, 66-67, commercial develop mgr, 67-69, tech serv mgr, 69-71, govt & indust rels mgr, 71-83; RETIRED. *Res:* Physical chemistry of biological processes; fundamental properties and industrial applications of alkali silicates; effects of silica and silicates on health and environment; predictive biomedical testing for regulatory purposes. *Mailing Add:* 8333 Seminole Blvd No 325 Seminole FL 34642-4358

SCHLEZINGER, NATHAN STANLEY, b Columbus, Ohio, June 2, 08; m 40; c 3. NEUROLOGY, PSYCHIATRY. *Educ:* Ohio State Univ, BA, 30; Jefferson Med Col, MD, 32; Columbia Univ, ScD(med), 38. *Prof Exp:* PROF CLIN NEUROL, JEFFERSON MED COL, 52- *Concurrent Pos:* Dir neuro-ophthal clin, Wills Eye Hosp, 39-; dir myasthenia gravis clin, Jefferson Hosp, 45-; consult, Vet Admin Hosp, Coatesville & Grandview Hosp, Sellersville. *Mem:* Am Neurol Asn; fel Am Acad Neurol; Asn Res Nerv & Ment Dis; Am Psychoanal Asn; fel Am Psychiat Asn. *Mailing Add:* Wyncote House Apt 813 25 Washington Ave Wyncote PA 19095-1414

SCHLICHT, RAYMOND CHARLES, b North Bergen, NJ, Oct 13, 27; m 48; c 5. SYNTHETIC ORGANIC CHEMISTRY. *Educ:* Cent Col, Iowa, BS, 48; Univ Maine, MS, 50; Ohio State Univ, PhD(org chem), 52. *Prof Exp:* Chemist, 52-62, res chemist, 62-70, sr res chemist, 70-84, res assoc, 84-87, SR RES ASSOC, TEXACO, INC, 87- *Honors & Awards:* Award for Innovation, Texaco Res Ctr, 84. *Mem:* Am Chem Soc. *Res:* Reaction chemistry of higher olefins; phosphorus chemistry; nitrogen chemistry; synthesis of lubricating oil additives and fluids; 46 United States patents. *Mailing Add:* Lyndon Rd RD 2 Fishkill NY 12524

SCHLICHTING, HAROLD EUGENE, JR, b Detroit, Mich, Mar 19, 26; m 49; c 6. PHYCOLOGY. *Educ:* Univ Mich, BS, 51; Mich State Univ, MS, 52, PhD(bot), 58. *Prof Exp:* Instr natural sci & bot, Mich State Univ, 52-54; fishery res biologist, US Fish & Wildlife Serv, 54-62; asst bot, Mich State Univ, 56-57; asst prof biol, Cent Mich Univ, 57-59; pvt res biologist, 59-60; assoc prof biol, NTex State Univ, 60-68; Fulbright-Hayes lectr, Univ Col Cork, 68-69; assoc prof bot, NC State Univ, 69-73; PRES, BIOCONTROL CO, INC, 73-; ADJ PROF BIOL, ST MARY'S COL, ORCHARD LAKE, 88- *Concurrent Pos:* Res grants, NIH, 59-68 & Sigma Xi, 63-64; vis prof, Univ Okla Biol Sta, 64, 66, 68, 70 & 72; NSF fel, Marine Lab, Duke Univ, 66; State of Tex res grant, 66-68; sea grant, 70-72; vis prof bot, Univ Minn Biol Sta, 73, 75, 76 & 77. *Mem:* Int Asn Aerobiol; Sigma Xi; Phycol Soc Am; Int Phycological Soc. *Res:* Dispersal of algae and protozoa; algal ecology; mass culturing of algae; biological monitoring; organic waste handling. *Mailing Add:* Box 43 Port Sanilac MI 48469

SCHLICK, SHULAMITH, b Yassy, Romania; US citizen; m; c 3. PHYSICAL CHEMISTRY, THERMODYNAMICS & MATERIAL PROPERTIES. *Educ:* Israel Inst Technol, BS, 55, Eng Dipl, 56, MS, 59, DS (phys chem), 63. *Prof Exp:* Guest scientist, Ford Res Labs, Dearborn, Mich, 65-67; sr lectr, Israel Inst Technol, Haifa, 70-73; sr res assoc magnetic resonance, Wayne State Univ, Detroit, 73-80; PROF POLYMER CHEM, UNIV DETROIT, 83- *Concurrent Pos:* Sabbatical, Ctr Study Nuclear Energy, Grenoble, France & Weitmann Inst Sci, Rehovot, Israel, 89-90. *Mem:* Am Chem Soc; Am Phys Soc; AAAS. *Res:* Multifrequency electron spin resonance (MESR) of polymers; morphology of ionomers, polymer blends and interpenetrating polymer networks; trapped electrons in organic crystals. *Mailing Add:* Dept Chem Univ Detroit Detroit MI 48221-9987

SCHLICKE, HEINZ M, b Dresden, Ger, Dec 13, 12; US citizen; m 39; c 2. ELECTRONICS ENGINEERING, APPLIED PHYSICS. *Educ:* Dresden Tech Univ, BS, 35, MS, 37, DSc(elec eng), 39. *Prof Exp:* Res engr, Telefunken, Ger, 38-40; dept head submarine commun, Naval Test Fields, 40-43, naval coun, High Command of Navy, 43-44, proj engr & spec consult, Paperclip scientist, Spec Devices Ctr, Off Naval Res, NY, 46-50; mgr, Electronics Labs, Allen-Bradley Co, 50-68, chief scientist, 68-75; CONSULT ENGR, 75- *Concurrent Pos:* Teaching electromagnetic compatibility courses, univs & indust; expert witness elec accident cases; US deleg, Sci Exchange US/USSR. *Honors & Awards:* Stoddart Award, Electromagnetic Compatibility Soc of Inst Elec & Electronics Engrs. *Mem:* Fel AAAS; fel Inst Elec & Electronics Engrs. *Res:* Electromagnetic interference and hazard control in civilian systems; author of 2 books and co-author of 4 books; over 60 scientific articles published; 20 patents. *Mailing Add:* 8220 N Poplar Dr Milwaukee WI 53217

SCHLIESSMANN, D(ONALD) J(OSEPH), b Colome, SDak, Aug 15, 17; m 44; c 4. ENVIRONMENTAL HEALTH. *Educ:* Univ Ill, BS, 41; Harvard Univ, MS, 49; Environ Eng Intersoc Bd, dipl, 70. *Prof Exp:* Mem staff malaria & typhus control, USPHS, 41-46, training off environ sanit & vector control, 47-48, engr & epidemiologist, 49-53, chief, Cumberland Field Sta, 54-57, dir & chief sanit eng, State Aids Sect, 57-61, dept chief, Tech Br, 61-63, chief aedes aegypti eradication br, 63-66, chief malaria eradication prog, Ga, 66-67; sanit engr, Pan Am Health Orgn, Washington, DC, 67-77; CONSULT ENGR, 77- *Concurrent Pos:* Consult comn enteric disease, US Armed Forces Epidemiol Bd, 51, mem comn environ sanit, 60-63, comn environ hyg, 63-72; consult, WHO, 58-63, 77, USAID, 78, APHA, 80, mem expert comt control enteric diseases, 63-68. *Mem:* Am Acad Environ Eng. *Res:* Public health engineering; epidemiology and control of diarrheal diseases; control of vectors and reservoirs of communicable diseases. *Mailing Add:* 3813 Savannah Sq E Atlanta GA 30340-4337

SCHLIMM, GERARD HENRY, b Baltimore, Md, May 26, 29; m 56; c 3. CIVIL ENGINEERING, STRUCTURES. *Educ:* Univ Md, BS, 57, PhD(structures), 70; NJ Inst Technol, MS, 60. *Prof Exp:* Mech engr compressors, Exxon Res & Eng Co, 57-60; instr civil eng, Univ Md, 60-62; asst prof mech eng, US Naval Acad, 62-66; DIR DIV ENG & PHYS SCI, EVE COL, JOHNS HOPKINS UNIV, 66- *Concurrent Pos:* Consult, Trident Eng Asn, 62-68 & Ellicott City Eng Co, 71-73; mem bd dirs continuing eng studies div, Am Soc Eng Educ, 72-75, mem eng manpower comt, 77-; mem, Gov's Sci Adv Coun, State Md, 75- *Mem:* Am Soc Civil Engrs; Am Soc Eng Educ (secy, 71-72); Nat Soc Prof Engrs; Sigma Xi. *Res:* Engineering manpower; continuing education for engineers. *Mailing Add:* 125 Croydon Rd Baltimore MD 21212

SCHLINGER, EVERT IRVING, b Los Angeles, Calif, Apr 17, 28; m 57; c 4. ENTOMOLOGY, SYSTEMATICS. *Educ:* Univ Calif, Berkeley, BS, 50; Univ Calif, Davis, PhD(insect taxon), 57. *Prof Exp:* Lab asst entom, Univ Calif, 46-50, asst, 50-54 & 55-56; prof collector, Calif Acad Sci, 54-55; jr entomologist, Univ Calif, Riverside, 56-57, asst entomologist, 57-61, assoc prof entom & assoc entomologist, 61-68, chmn dept, 68-69, chmn div entomol & parasitol, 75-76, chmn dept entom sci, 76-79, chmn dept, conserv & res studies, 82-83, prof, 69-86, EMER PROF ENTOM, UNIV CALIF, BERKELEY, 86- *Concurrent Pos:* Prof collector, Assocs Trop Biogeog, 53, 54; Guggenheim fel, 66-67; NSF award, 63-68. *Mem:* Fel AAAS; Entom Soc Am; Ecol Soc Am; Assoc Syst Collections; Sierra Club; Nature Conservancy; Wilderness Soc. *Res:* Insect (diptera) biosystematics; biogeography; insect ecosystems and land use management; spider ecology; arthropod non-target studies; biological control; biology of insect parasitoids. *Mailing Add:* Dept Entom Sci Univ Calif Berkeley CA 94720

SCHLINGER, W(ARREN) G(LEASON), b Los Angeles, Calif, May 29, 23; m 47; c 3. MECHANICAL ENGINEERING. *Educ:* Calif Inst Technol, BS, 44, MS, 46, PhD(chem & mech eng), 49. *Prof Exp:* Res fel, Calif Inst Technol, 49-53; chem engr, Texaco Inc, 53-57, sr chem engr, 57-60, supvr res, 60-68, dir, 68-70, mgr, Montebello Res Lab, 70-81, assoc dir gasification, Texaco Inc, 81-87; RETIRED. *Honors & Awards:* Tech Achievement Award, Am Inst Chem Engrs, 76; Chem Eng Pract Award, Am Inst Chem Engrs, 81; KFA Achievement Award, Elec Power Res Inst, 85. *Mem:* Nat Acad Eng; Am Chem Soc; fel Am Inst Chem Engrs; AAAS. *Res:* Fluid flow; heat transfer; coal gasification; hydrogen production and hydrogenation reactions; hydrocarbon gasification. *Mailing Add:* 3835 Shadow Grove Pasadena CA 91107

SCHLINK, F(REDERICK) J(OHN), b Peoria, Ill, Oct 16, 91; wid. PHYSICS, MECHANICAL ENGINEERING. *Educ:* Univ Ill, BS, 12, ME, 17. *Prof Exp:* Mech engr & physicist, Nat Bur Standards, 12-13, from lab asst to tech asst to the dir, 13-19; physicist in charge, Instruments-Control Dept, Firestone Tire & Rubber Co, Ohio, 19-20; mech engr & physicist, Western Elec Co, 20-22; asst secy, Am Standards Asn, 22-31; tech dir & ed in charge opers & publ, Consumers' Res, Inc, 29-82; RETIRED. *Concurrent Pos:* Lectr, Univ Tenn, 47-49, vis prof, 50; mem consumer adv coun, Underwriters' Labs. *Honors & Awards:* Longstreth Medal, Franklin Inst, 19. *Mem:* Fel Am Phys Soc; fel Am Soc Mech Engrs; Sigma Xi; Inst Elec & Electronics Engrs; Am Nat Standards Inst. *Res:* Measuring instruments and variant and hysteretic types of error; standards and specifications for products manufactured for ultimate consumer use; economics and technology of consumption goods; methods of test of appliances, materials and consumer-use products in general. *Mailing Add:* Rte 4 Box 209 Washington NJ 07882

SCHLIPF, JOHN STEWART, b Fargo, NDak, Oct 29, 48; m 70. MATHEMATICAL LOGIC. *Educ:* Carleton Col, BA, 70; Univ Wis-Madison, MA, 72, PhD(math), 75. *Prof Exp:* Instr math, Calif Inst Technol, 75-77; vis lectr math, Univ Ill, 77-79; asst prof math, St Mary's Col Md, 79-81; mem staff, Environ Control Inc, 81-83; asst prof math, 83-84, asst prof computer sci, 84-87, ASSOC PROF COMPUTER SCI, UNIV CINCINNATI, 87- *Mem:* Asn Symbolic Logic; Am Math Soc; Asn Comput Mach. *Res:* Logic programming and non-monotonic reasoning; computability and computational complexity; model theory. *Mailing Add:* Dept Computer Sci Univ Cincinnati Cincinnati OH 45221-0008

SCHLISELFELD, LOUIS HAROLD, b Chicago, Ill, Sept 15, 31. BIOCHEMISTRY. *Educ:* Univ Ill, Urbana, BS, 53, MS, 55; Vanderbilt Univ, PhD(biochem), 64. *Prof Exp:* Res assoc contractile proteins, Inst Muscle Dis, Inc, 66-74; asst res prof, dept biol chem, Univ Ill Med Ctr, Chicago, 74-85; RES AFFIL, DEPT NEUROSURG, ROSWELL PARK MEM INST, BUFFALO, 85- *Concurrent Pos:* USPHS fel biochem, Univ Wash, 64-66. *Mem:* Am Chem Soc; Am Soc Biol Chemists; Biophys Soc. *Res:* Enzymes and diseases involved in muscle glycogen breakdown; interaction of myosin and actin with adenosine triphosphate; mechanism of muscle contraction; metabolism of anti-cancer drugs in mouse leukemia cells. *Mailing Add:* Three Slate Creek Cr Apt Nine Cheektowaga NY 14227-2965

SCHLISSEL, ARTHUR, b Austria, July 7, 31; US citizen; div; c 2. MATHEMATICS. *Educ:* Brooklyn Col, BS, 54; NY Univ, MS, 58, PhD(math), 74. *Prof Exp:* Lectr math, Brooklyn Col, 57-60, Hunter Col, 60-63 & NY Univ, 63-64; asst prof, Fairleigh Dickinson Univ, 64-67 & Manhattan Col, 67-70; from instr to assoc prof, 70-80, chmn dept, 75-84,

PROF MATH, JOHN JAY COL CRIMINAL JUSTICE, CITY UNIV NEW YORK, 80- *Concurrent Pos:* Vis prof math, King-Kennedy Prog, Albert Einstein Med Sch, 68-69. *Mem:* Am Math Soc; Math Asn Am; Sigma Xi; Soc Indust Appl Math. *Res:* Asymptotic behavior of the solutions of ordinary and partial differential equations; development of analysis in the 19th and 20th century, with special reference to differential equations; computer science, with special reference to computer graphics and data bases. *Mailing Add:* 2679 21st St Brooklyn NY 11235

SCHLITT, DAN WEBB, b Lincoln, Nebr, Nov 2, 35; m 57; c 4. THEORETICAL HIGH ENERGY PHYSICS. *Educ:* Mass Inst Technol, BS, 57; Univ Wash, PhD(physics), 63. *Prof Exp:* Vis asst prof physics, Univ Md, 63-64; from asst prof to assoc prof, 64-77, PROF PHYSICS, UNIV NEBR, LINCOLN, 77- *Concurrent Pos:* NSF grant, 65-69; vis scientist, Inst Theoret Physics, State Univ Utrecht, 72-73. *Mem:* Am Phys Soc; Am Asn Physics Teachers; Soc Indust & Appl Math; Asn Comput Mach; Am Asn Univ Professors. *Res:* Elementary particle theory; mathematical physics; foundations of statistical mechanics; numerical analysis and computation. *Mailing Add:* Dept Physics & Astron City Col NY New York NY 10031

SCHLITT, WILLIAM JOSEPH, III, b Columbus, Ohio, June 12, 42. HYDROMETALLURGY, SOLUTION MINING. *Educ:* Carnegie Inst Technol, BS, 64; Pa State Univ, PhD(metall), 68. *Prof Exp:* Scientist, Kennecott Minerals Co, 68-75, sr scientist, 75-76, mgr, Hydrometall Dept, 77-81, prin prog mgr, Process Technol Group, 81-82, process staff mgr, 82-83; mgr technol, Mineral & Metal Indust Dept, Brown & Root, USA, 83-90; MGR TECHNOL, BROWN & ROOT BRAUN, 90- *Concurrent Pos:* mem, Oversight Comt Solution Mining Grant, NSF, 77-79, Oversight Comt Treatment Smelter Flue Dust Grant, Environ Protection Agency, 78-79; chmn, Mining & Explor Div, 87-88. *Mem:* Sigma Xi; Can Inst Mining & Metall; Soc Mining Engrs; Metall Soc. *Res:* Processes for extraction and refining of metal values contained in ores and other source materials including approaches involving hydrometallurgy and solution mining. *Mailing Add:* Brown & Root USA Inc PO Box 3 Houston TX 77001-0003

SCHLITTER, DUANE A, b Monona, Iowa, Apr 2, 42; m 63; c 2. MAMMALOGY. *Educ:* Wartburg Col, BA, 65; Univ Kans, MA, 69; Univ Md, PhD, 76. *Prof Exp:* Res & curatorial asst mammal, African Mammal Proj, Smithsonian Inst, 67-72; assoc cur mammal, 73-84, CUR MAMMALS, CARNEGIE MUS NATURAL HIST, 85- *Mem:* Am Soc Mammal (secy-treas, 77-80); Soc Syst Zool; Sigma Xi; AAAS; Zool Soc SAfrica. *Res:* Systematics, evolution, biogeography and ecology of mammals of Africa and Southwest Asia; relationship of diseases, ectoparasite vectors and mammal host in old world medical zoological problems; systematics and biogeography of mammals of eastern North America; conservation of endangered species of mammals in Africa and Southern Asia. *Mailing Add:* Carnegie Mus Natural Hist Annex 5800 Baum Blvd Pittsburgh PA 15206-3706

SCHLIWA, MANFRED, b Kulmbach, WGer, Dec 6, 45; m 78. CELL MOTILITY, ELECTRON MICROSCOPY. *Educ:* Univ Frankfurt, MS, 72, PhD(zool), 75. *Prof Exp:* Res assoc zool, Univ Frankfurt, WGer, 76-78; Heisenberg fel cell biol, Univ Colo, Boulder, 79-81; prof zool, Univ Calif, Berkeley, 81-90, dir, Electron Microscope Lab, 84-90. *Mem:* Am Soc Cell Biol; Ger Soc Cell Biol; Ger Zool Soc; Int Pigment Cell Soc. *Res:* Cell biology; subfield cell motility; structure and function of the cytoskeleton. *Mailing Add:* Inst Cell Biol Univ Munich Schillerstrasse D-8000 Munich 2 Germany

SCHLOEMANN, ERNST, b Borgholzhausen, Ger, Dec 13, 26; nat US; wid; c 3. MAGNETIC MATERIALS, MAGNETIC DEVICES. *Educ:* Univ Gottingen, BS, 51, MS, 53, PhD(theoret physics), 54. *Prof Exp:* Asst, Inst Theoret Physics, Univ Gottingen, 52-53; Fulbright fel solid state physics, Mass Inst Technol, 54-55; mem res staff, 55-60, proj dir, 60-64, SCI FEL, RAYTHEON CO, 64- *Concurrent Pos:* Vis assoc prof, Stanford Univ, 61-62 & Univ Hamburg, 66. *Mem:* Fel Am Phys Soc; fel Inst Elec & Electronics Engrs. *Res:* Solid state physics; magnetic phenomena; ferromagnetic resonance; lattice dynamics; thermal conductivity of solids; microwave physics and technology; statistical mechanics; environmental science; resource recovery from waste. *Mailing Add:* 38 Brook Rd Weston MA 02193

SCHLOEMER, ROBERT HENRY, b New York, NY, July 3, 46; m 69; c 2. VIROLOGY. *Educ:* Boston Col, BS, 68; Univ Va, MS, 72, PhD(biol), 73. *Prof Exp:* Fel, Sch Med, Univ Va, 73-75; ASST PROF MICROBIOL, SCH MED, IND UNIV, 75- *Mem:* Am Soc Microbiol; Sigma Xi. *Res:* Biochemical and biological analysis of viral membrane proteins; structure and assembly of viruses; mechanisms of virus persistence. *Mailing Add:* Dept Microbiol Ind Univ Sch Med 1120 South Dr Indianapolis IN 46223

SCHLOER, GERTRUDE M, b Milwaukee, Wis, May 16, 26. MICROBIOLOGY, MOLECULAR BIOLOGY. *Educ:* Marquette Univ, BS, 48, MS, 53; Univ Wis-Madison, PhD(virol), 65. *Prof Exp:* Res assoc virol, Univ Wis-Madison, 66-67; NIH fel, Inst Virol, Univ Giessen, 67-69; instr microbiol, Mt Sinai Sch Med, 69-70, assoc, 70-71, asst prof, 71-73; mem staff, Molecular Biol Lab & Plum Island Animal Dis Lab, USDA, 73-; RETIRED. *Mem:* AAAS; Am Soc Microbiol; Am Soc Virology. *Res:* Structure and genetics of influenza virus proteins; virulence and transmission of Newcastle Disease virus; epizootiology of adenovirus 127; structure and protein composition of malignant catarrhal fever virus and African swine fever virus. *Mailing Add:* 3060 Little Neck Rd Cutchogue NY 11935

SCHLOERB, PAUL RICHARD, b Buffalo, NY, Oct 22, 19; m 50; c 5. SURGERY, NUTRITION. *Educ:* Harvard Univ, AB, 41; Univ Rochester, MD, 44; Am Bd Surg, dipl, 52. *Prof Exp:* Asst surg, Peter Bent Brigham Hosp, Boston, 51; instr, Sch Med, Univ Rochester, 52; from asst prof to assoc prof, Med Ctr, Univ Kans, 52-64, res prof, 64-72, asst dean res, 70-72, dean res, 72-78, prof surg, 72-79; prof surg, Sch Med & Dent, Univ Rochester, 79-88; surgeon, Strong Mem Hosp, Rochester, 79-88; PROF SURG, UNIV KANS MED CTR, KANSAS CITY, KS, 88- *Concurrent Pos:* AEC fel med sci, Nat Res Coun, 48-49; USPHS career develop award, 62-67; consult, Vet Admin Hosps, Wichita, Kans, 55- & Kansas City, Mo, 59-79; adj prof surg, Sch Med & Dent, Univ Rochester, 88-90. *Mem:* AAAS; Am Physiol Soc; Soc Univ Surg; Am Surg Asn; fel Am Col Surgeons; Am Asn Cancer Res; Am Asn Surg Trauma; Am Soc Parenteral & Enteral Nutrit. *Res:* Postoperative care; surgical physiology; renal disorders; shock; water and electrolytes; transplantation; nutrition. *Mailing Add:* Univ Kans Dept Surg Rochester KS 66103

SCHLOM, JEFFREY, b Brooklyn, NY, June 22, 42; m 76; c 2. MOLECULAR BIOLOGY, BIOCHEMISTRY. *Educ:* Ohio State Univ, BS, 64; Adelphi Univ, MS, 66; Rutgers Univ, PhD(microbiol), 69. *Prof Exp:* Guest worker oncol, Nat Cancer Inst, 67-69; from instr to asst prof virol, Col Physicians & Surgeons, Columbia Univ, 69-73; chmn breast cancer virus segment, 73-77, head, Tumor Virus Detection Sect, 77-80,chief, Exp Oncol Sect , 80-82, CHIEF, LAB TUMOR IMMUNOL & B IOL, NAT CANCER INST, 82- *Concurrent Pos:* Adj prof, Grad Fac, George Washington Univ. *Honors & Awards:* Director's Award, NIH, 77 & 90; Leona Kopman Mem Award, 83; Rosenthal Found Award, Am Asn Cancer Res, 85. *Mem:* AAAS; Harvey Soc; NY Acad Sci; Am Asn Cancer Res; Tissue Cult Asn; Int Asn Breast Cancer Res; Int Asn Comp Res Leukemia Related Dis. *Res:* Tumor immunology; molecular biology; viral oncology. *Mailing Add:* Nat Cancer Inst Nat Inst Health Bldg 10 Rm 8B07 Bethesda MD 20892

SCHLOMIUK, DANA, b Bucharest, Romania, Jan 5, 37; Can citizen; m 58. MATHEMATICS. *Educ:* Univ Bucharest, dipl math, 58; McGill Univ, PhD, 67. *Prof Exp:* Fel, Univ Montreal, 67-68, res assoc, 69, from asst prof to assoc prof, 69-89, PROF MATH, UNIV MONTREAL, 89- *Concurrent Pos:* Invited prof, Univ Rome, 73 & 75; vis prof, Sci Univ Tokyo, Noda-Chiba, Japan, 86. *Mem:* Am Math Soc. *Res:* Dynamical systems; bifurcations of plane vectorfields with special emphasis on algebraic and global geometric aspects of the theory of polynomial vector fields. *Mailing Add:* Dept Math & Statist Univ Montreal PO Box 6128 Montreal PQ H3C 3J7 Can

SCHLOMIUK, NORBERT, b Cernauti, Rumania, Apr 23, 32; Can citizen; m 58. MATHEMATICS, HISTORY & PHILOSOPHY OF SCIENCE. *Educ:* Univ Bucharest, MA, 55; McGill Univ, PhD(math), 66. *Prof Exp:* Instr math, Univ Bucharest, 54-58; lectr, Dalhousie Univ, 61-62, asst prof, 62-63; res asst, Univ Calif, Berkeley, 63-64; lectr, McGill Univ, 64-66, asst prof, 66-68; asst prof, 68-72, ASSOC PROF MATH, UNIV MONTREAL, 72- *Concurrent Pos:* Vis prof, Math Inst, Univ Perugia, 71-72 & Math Res Inst, Swiss Fed Inst Technol, 71-72; vis res fel, Cornell Univ, 85, Kyoto Univ, 86. *Mem:* Am Math Soc; Math Soc France; Ital Math Union; Can Math Soc; NY Acad Sci. *Res:* Algebraic topology and homotopy theory; history of mathematics. *Mailing Add:* Dept Math Univ Montreal Montreal PQ H3C 3J7 Can

SCHLOSBERG, RICHARD HENRY, b New York, NY, May 23, 42; m 67; c 3. ORGANIC CHEMISTRY. *Educ:* City Univ New York, BS, 63; Mich State Univ, PhD(chem), 67. *Prof Exp:* Fel org chem, Case Western Reserve Univ, 67-69; asst prof chem, Univ Wis-Whitewater, 69-73; sr staff chemist coal sci, 73-80, res assoc fuels sci, Exxon Res & Eng Co, 80-84, RES ASSOC, EXXON CHEM CO, 84- *Concurrent Pos:* Ed, Chem of Coal Conversion & assoc ed, Liquid Fuels Technol, 82. *Mem:* Am Chem Soc; Am Inst Chem Engrs. *Res:* Chemistry of synthetic fuels; coal science; shale science; heavy oil science; thermal hydrocarbon chemistry; strong acid chemistry; friedel crafts chemistry; environmental chemistry. *Mailing Add:* Exxon Chem Co Rte 22 E Clinton Twp Annandale NJ 08801

SCHLOSS, JOHN VINTON, b St Louis, Mo, May 11, 51; m 72; c 2. ENZYMOLOGY. *Educ:* Univ Tulsa, BS, 73; Univ Tenn, PhD(biomed sci), 78. *Prof Exp:* Res fel, Univ Wis, 78-81; prin investr, 81-87, RES SUPVR, E I DU PONT DE NEMOURS & CO, 87- *Mem:* Am Chem Soc; Am Soc Biochem & Molecular Biol; Sigma Xi; AAAS; NY Acad Sci; Am Inst Chem. *Res:* Elucidation of enzymic reaction mechanisms; design and utilization of mechanism-based inhibitors of enzymes, primarily for enzymes of agronomic importance. *Mailing Add:* 2612 N Gate Rd Wilmington DE 19810

SCHLOSSER, HERBERT, b Brooklyn, NY, Nov 18, 29; m 60; c 2. CONDENSED MATTER THEORY, MOLECULAR BIOPHYSICS. *Educ:* Brooklyn Col, BS, 50; Polytech Univ New York, MS, 52; Carnegie Mellon Univ, PhD(physics), 60. *Prof Exp:* Proj supvr, Res Lab, Horizons Inc, 53-55; proj physicist, Carnegie Inst Technol, 59-60; sr physicist, Bayside Res Lab, Gen Tel & Electronics, 60-62; specialist physicist, Repub Aviation Corp, 62-63; asst prof, Polytech Inst Brooklyn, 63-68; assoc prof, 68-72, PROF PHYSICS, CLEVELAND STATE UNIV, 72- *Concurrent Pos:* Consult, Repub Aviation Co, 63; Fulbright-Hays Lectr, Univ Sao Paulo, 66-67; sr Weizmann res fel, Weizmann Inst Sci, Israel, 73-74; res assoc & vis prof, Dept Macromolecular Sci, Case Western Reserve Univ, 78-79. *Mem:* Am Phys Soc. *Res:* Universality theory, high pressure physics surface physics; condensed matter theory; electronic structure of macromolecules; molecular biophysics. *Mailing Add:* Dept Physics Cleveland State Univ Cleveland OH 44115

SCHLOSSER, JON A, b Houston, Tex, July 26, 37; m 64; c 2. MATHEMATICS, PHYSICS. *Educ:* Univ Tex, PhD(physics), 63. *Prof Exp:* Res assoc relativity theory, Univ Tex, 63-64; instr appl math, Univ Chicago, 64-66; asst prof math, La State Univ, Baton Rouge, 66-70; assoc prof, 72-78, PROF MATH, NMEX HIGHLANDS UNIV, 78- *Mem:* Opers Res Soc Am; Am Math Soc; Math Asn Am. *Res:* Functional analysis; relativity theory; operations research. *Mailing Add:* Dept Sci & Math NMex Highlands Univ Las Vegas NM 87701

SCHLOSSER, PHILIP A, nuclear engineering, radiation physics, for more information see previous edition

SCHLOSSMAN, IRWIN S, b New York, NY, July 2, 30; m 51; c 2. ORGANIC CHEMISTRY. *Educ:* City Col New York, BS, 51; Polytech Inst Brooklyn, MS, 59; Xavier Univ, Ohio, MBA, 71. *Prof Exp:* Chemist, Nopco Chem Co, NJ, 51-54; Gallowhur Chem Co, NY, 54-55; proj leader res &

develop, Halcon Int, Inc, NJ, 55-66; group leader res, Emery Industs, Inc, 66-85, chem systs coordr, 85-88, COORDR, SAFETY REGULATORY AFFAIRS, INFO SERV, QUANTUM CHEM CORP, 88- *Mem:* Am Chem Soc. *Res:* Liquid and vapor phase oxidation; catalysis; free radical chemistry; hydrogenation; polymer intermediates. *Mailing Add:* 8922 Cherry Blossom Lane Cincinnati OH 45231

SCHLOSSMAN, MITCHELL LLOYD, b Brooklyn, NY, Dec 30, 35; m 56; c 3. COSMETIC CHEMISTRY. *Educ:* NY Univ, BS, 56. *Prof Exp:* Group leader skin treat prod, Revlon Inc, 57-63; mgr res & develop, Leeming/Pacquin Div, Pfizer & Co, 64-69; dir tech opers, Paris Cosmetics Inc, 69-70; vpres res & develop, Prince Indust Ltd, 70-74; vpres mkt & res, Malmstrom Chem Div, Emery Indust Inc, 74-78; PRES, TEVCO, INC, 78-; PRES, PRESPERSE INC, 85- *Concurrent Pos:* Chmn, NY Chap, Soc Cosmetic Chemists, 66, nat dir, 67-68; lectr & instr, Continuing Educ Ctr, South Hackensack, NJ, 85- *Mem:* Fel Soc Cosmetic Chemists; Am Chem Soc; fel Am Inst Chemists. *Res:* Lanolin and cosmetic ester research; cosmetic product development; nail lacquers. *Mailing Add:* 454 Prospect Ave Unit 164 West Orange NJ 07052

SCHLOSSMAN, STUART FRANKLIN, b New York, NY, Apr 18, 35; m 58; c 2. IMMUNOLOGY, HEMATOLOGY. *Educ:* NY Univ, BA, 55, MD, 58. *Prof Exp:* Intern med, Ill Med Div, Bellevue Hosp, 58-59, asst resident, 59-60; res assoc, Lab Biochem, Nat Cancer Inst, 63-65; from instr to assoc prof, 65-77, PROF MED, HARVARD MED SCH, 77-, CHIEF DIV TUMOR IMMUNOL, DANA FARBER CANCER INST, 73- *Concurrent Pos:* Nat Found fel microbiol, Col Physicians & Surgeons, Columbia, 60-62; Ward hemat fel internal med, Sch Med, Wash Univ, 62-63; Guggenheim fel, 71; asst physician, Med Serv, Vanderbilt Clin, Presby Hosp, 60-62; clin instr med, Sch Med, George Washington Univ, 64-65; dir blood bank, Beth Israel Hosp, Mass, 65-66, assoc med, 65-67, from asst physician to assoc physician, 67-73; chief clin immunol, Beth Israel Hosp, Mass, 71-73; sr assoc med, Peter Bent Brigham Hosp, Mass, 76- *Mem:* Am Soc Hemat; Am Asn Immunol; Am Soc Clin Invest. *Res:* Internal medicine. *Mailing Add:* Dept Med Harvard Med Sch Dana Farber Cancer Inst 44 Binney St Boston MA 02115

SCHLOTFELDT, ROZELLA M, NURSING. *Prof Exp:* EMER DEAN, SCH NURSING, CASE WESTERN RESERVE UNIV. *Mem:* Inst Med-Nat Acad Sci. *Mailing Add:* Sch Nursing Case Western Reserve Univ University Circle Cleveland OH 44106

SCHLOTTHAUER, JOHN CARL, b Rochester, Minn, Oct 6, 30; m 64; c 2. PARASITOLOGY. *Educ:* Univ Minn, BS, 52, DVM, 54, PhD(vet parasitol), 65. *Prof Exp:* Gen practitioner, 54-55; scientist, Arctic Aeromed Lab, USAF, 55-57; instr vet diag, 57-58, instr vet parasitol, 59-65, from asst prof to assoc prof, 65-85, head microbiol-parasitol sect, 74-75, actg chmn dept vet biol, 75-76, PROF, COL VET MED, UNIV MINN, ST PAUL, 85- *Concurrent Pos:* Short term adv, Midwest Univs Consortium Int Activ/USAID Indonesian Higher Agr Educ Proj, 72. *Mem:* Am Vet Med Asn; Am Soc Parasitol; Am Asn Vet Parasitol; Wildlife Dis Asn. *Res:* Pathogenesis of canine dirofilariasis; ruminant helminthiasis; parasites of wildlife. *Mailing Add:* Col Vet Med Univ Minn, 1971 Commonwealth Ave St Paul MN 55108

SCHLOTTMANN, PEDRO U J, b Buenos Aires, Arg, Mar 28, 47; Ger citizen; m 77; c 1. HIGHLY CORRELATED ELECTRON SYSTEMS, LOW DIMENSIONAL MAGNETISM. *Educ:* Universidad de Cuyo, Bariloche, Licenciado, 70; Tech Univ Munich, Dr rer nat(physics), 73. *Hon Degrees:* Habilitation, Freie Univ Berlin, 78. *Prof Exp:* Postdoctoral researcher physics, Max-Planck Inst, Munich, 73-74; asst, Dept Physics, Freie Univ Berlin, 74-77, asst prof, 77-83; researcher, Inst Festkorperforschung KFA Julich, 82-86; prof, Temple Univ, Philadelphia, 85-90; PROF PHYSICS, DEPT PHYSICS, FLA STATE UNIV, 90- *Concurrent Pos:* Postdoctoral researcher, Univ Calif, Berkeley, 75-76; Heisenberg fel, DFG, Ger, 82-86; vis prof, Univ Gottingen, 85. *Mem:* Am Phys Soc. *Res:* Condensed matter theory and quantum statistical mechanics of systems with highly correlated states; heavy fermions, high temperature superconductors, narrow band phenomena and low dimensional magnetism and conductors. *Mailing Add:* Dept Physics Fla State Univ Tallahassee FL 32306

SCHLOUGH, JAMES SHERWYN, b Wheeler, Wis, Sept 14, 31; m 63; c 2. ANIMAL PHYSIOLOGY, ENDOCRINOLOGY. *Educ:* Wis State Univ-River Falls, BS, 60; Univ Wis, MS, 63, PhD(zool), 66. *Prof Exp:* From instr to assoc prof, 65-76, from asst dean to assoc dean, Col Letters & Sci, 69-74, PROF BIOL, UNIV WIS-WHITEWATER, 76-; CONSULT, NASCO INC, WIS, 67- *Mem:* AAAS; Soc Study Reproduction; Am Soc Mammal. *Res:* Early embryonic development in mammals; estrogen antagonism. *Mailing Add:* Dept Biol Univ Wis 800 W Main St Whitewater WI 53190

SCHLUB, ROBERT LOUIS, b Springfield, Ohio, Jan 22, 51; m 75; c 3. SUGARCANE PATHOLOGY. *Educ:* Ohio State Univ, BS, 73, MS, 75; Mich State Univ, PhD(plant path), 79. *Prof Exp:* Asst, Plant Dis Clinic, Ohio State Univ, 74; asst slide-tape teaching aids gen path, Mich State Univ, 76-78; fel, Soilborne Dis Lab, USDA, Beltsville, Md, 79-80; ASST PROF SUGARCANE DIS, DEPT PLANT PATH & CROP PHYSIOL, LA STATE UNIV, 80-; PRESCH EDUC ADMINR, OUR CHILDREN'S HOUSE, BATON ROUGE. *Mem:* Am Phytopath Soc. *Res:* Development of science and math educational materials for 4-9 year old children; etiology; epidemiology; environmental stress. *Mailing Add:* 9507 Highpoint Rd Baton Rouge LA 70810

SCHLUEDERBERG, ANN ELIZABETH SNIDER, b Detroit, Mich, May 31, 29; m 51; c 5. VIROLOGY, IMMUNOLOGY. *Educ:* Ohio State Univ, BS, 50; Johns Hopkins Univ, ScM, 54, ScD(microbiol), 59. *Prof Exp:* Instr med, Sch Med, Univ Md, 59-61; res assoc, Sch Med, Yale Univ, 62-69, sr res assoc epidemiol, 69-76; assoc prof epidemiol, Sch Hyg & Pub Health, Johns Hopkins Univ, 76-78; exec secy, Epidemiol Dis Control Study Sect, Div Res Grants, NIH, 78-87; virol prog officer, 87-90, CHIEF, VIROL BR, NAT INST ALLERGIES & INFECTIOUS DIS, NIH, 90- *Concurrent Pos:* USPHS fel, 61-62, USPHS grants, 61-78; Mem, Scholars Adv Panel, Fogarty Int Ctr, NIH, 80-84. *Mem:* Am Soc Virol; fel Infectious Dis Soc Am; Am Asn Immunol. *Res:* Measles and rubella virus characterization; evaluation of measles and rubella vaccines and vaccine regimens; viral immunity. *Mailing Add:* NIH Westwood Bldg Rm 736 Bethesda MD 20892

SCHLUETER, DONALD JEROME, b Oak Park, Ill, Nov 14, 31; c 1. NUCLEAR PHYSICS & STRUCTURE, SPECTROSCOPY & SPECTROMETRY. *Educ:* Northwestern Univ, BS, 53, MS, 57; Univ Kans, PhD(physics), 64. *Prof Exp:* From instr to asst prof, 62-68, ASSOC PROF PHYSICS, PURDUE UNIV, WEST, 68- *Concurrent Pos:* NSF consult, AID, India, 67; Midwest Univ Consortium Int Activities/ITM, Malaysia, 86-89. *Mem:* Am Asn Physics Teachers. *Res:* Optical interferometry; nuclear structure. *Mailing Add:* Dept Physics Purdue Univ West Lafayette IN 47907-1396

SCHLUETER, DONALD PAUL, b Milwaukee, Wis, July 24, 27; m 53; c 1. INTERNAL MEDICINE. *Educ:* Marquette Univ, BS, 51, MD, 59; Ga Inst Technol, MS, 56. *Prof Exp:* Chemist, E I du Pont de Nemours & Co, 51-52; instr chem, Ga Inst Technol, 53-55; intern med, Milwaukee County Hosp, Wis, 59-60, resident, 60-63; from instr to assoc prof, 64-75, PROF MED, MED COL WIS, 75-; CHIEF MED CHEST SERV, MILWAUKEE COUNTY HOSP, 68- *Concurrent Pos:* NIH res fel, 62-63; NIH res fel pulmonary physiol, Med Col Wis, 63-64; consult, Vet Admin Hosp, Wood, Wis, 63-; staff physician, Muirdale Sanatorium, 64-66, clin dir pulmonary dis, 66-68. *Mem:* Am Fedn Clin Res; Am Thoracic Soc; Am Col Chest Physicians; Am Med Asn; Am Occup Med Asn; Sigma Xi. *Res:* Inorganic paper chromatography; flame spectroscopy; pulmonary physiology-respiratory mechanics; hypersensitivity lung disease. *Mailing Add:* 6230 Fisher Lane Greendale WI 53129

SCHLUETER, EDGAR ALBERT, b Milwaukee, Wis, Sept 23, 18; m 57; c 3. ZOOLOGY, PARASITOLOGY. *Educ:* Univ NTex, BS, 42; Univ Wis, MS, 49, PhD, 62. *Prof Exp:* Instr biol & natural sci, Mich State Univ, 49-55 & 57-59; asst prof, Wis State Univ-Superior, 59-62; from asst prof to prof, Univ NTex, 62-83; RETIRED. *Mem:* Am Soc Parasitol; Am Micros Soc; Sigma Xi. *Res:* Host-parasite relationships; parasitology; biochemistry of diseases of parasitic origin. *Mailing Add:* 1105 Piping Rock Lane Denton TX 76205

SCHLUETER, MICHAEL ANDREAS, b Straubing, WGer, Feb 23, 45; m 72; c 1. SOLID STATE PHYSICS. *Educ:* Univ Karlsruhe, Ger, dipl, 69; Fed Polytech Inst, Lausanne, Switz, PhD(physics), 72. *Prof Exp:* Asst physics, Fed Polytech Inst, 72-73; fel physics, Univ Calif, Berkeley, 73-75; mem tech staff, 75-86, DEPT HEAD, BELL LABS, 86- *Honors & Awards:* Adler Award, Am Phys Soc, 90. *Mem:* Am Phys Soc. *Res:* Solid state theory; bandstructures; surfaces; defects; superconductors; semiconductors. *Mailing Add:* Bell Labs 1 D 446 Murray Hill NJ 07974

SCHLUETER, ROBERT J, b Chicago, Ill, Feb 28, 29; m 59; c 3. BIOCHEMISTRY, PHARMACEUTICAL. *Educ:* Valparaiso Univ, BA, 51; Northwestern Univ, PhD(biochem), 63. *Prof Exp:* Technician, Res & Develop biol Control, Armour & Co, 48-51, anal chemist, Res Control Lab, 51, biochemist, Pharmaceut Res Dept, 51-52, biochemist, Res Div, 54-60, sr res biochemist, 63-68, assoc res scientist, 68-76, prin scientist, 76-78, mgr biochem processes, 78-82, sr develop assoc, 82-84, mgr biochem develop, 84-88, SR RES FEL, ARMOUR PHARMACEUT CO, KANKAKEE, 88- *Concurrent Pos:* US Army Chem Corp, 52-54. *Mem:* Am Soc Pharmacol & Exp Therapeut; Am Soc Biol Chem & Molecular Biol; Soc Exp Biol & Med; Am Chem Soc; Endocrine Soc; Protein Soc. *Res:* Isolation, purification and characterization of biologically active natural products and synthetic polypeptides; analytical and physical biochemistry; enzymology; collagen chemistry; bioassay development; calcitonin; insulin; insulin-like growth factor; biologicals; ACTH; human plasma proteins. *Mailing Add:* Armour Pharmaceut Co Box 511 Kankakee IL 60901

SCHLUTER, MICHAEL, b Straubing, WGer, Feb 23, 45; m 72; c 1. SOLID STATE PHYSICS. *Educ:* Fed Inst Technol, PhD(physics), 73. *Prof Exp:* Tech staff mem, 75-86, DEPT HEAD RES, BELL LAB, 86- *Mem:* Am Phys. *Res:* Theory condensed matter in particular electronic structure metals and surfaces and interfaces of semiconductors. *Mailing Add:* AT&T Bell Lab 600 Mountain Ave Murray Hill NJ 07974

SCHLUTER, ROBERT ARVEL, b Salt Lake City, Utah, Aug 27, 24; div; c 1. ELEMENTARY PARTICLE PHYSICS. *Educ:* Univ Chicago, BS, 47, PhD(physics), 54. *Prof Exp:* Res assoc physics, Enrico Fermi Inst Nuclear Studies, Chicago, 54-55; from instr to asst prof, Mass Inst Technol, 55-60; PROF PHYSICS, NORTHWESTERN UNIV, 61- *Concurrent Pos:* Guest scientist, Brookhaven Nat Lab, 57-; vis physicist, Lawrence Radiation Lab, Univ Calif, 58; assoc scientist, Argonne Nat Lab, 60-72; vis scientist, Fermi Nat Accelerator Lab, 75- *Mem:* Sigmi Xi; Am Phys Soc; Nat Asn Scholars. *Res:* Interactions of fundamental particles; high energy and elementary particle physics; fluid film dynamics. *Mailing Add:* Dept Physics & Astron Northwestern Univ Evanston IL 60208

SCHMAIER, ALVIN HAROLD, b Neptune, NJ, Jan 6, 49; m 70; c 1. KININOGENS, CI INHIBITOR. *Educ:* Univ Va, BA, 70; Med Col Va, MD, 74; Am Bd Internal Med, dipl, 77, Hemat dipl, 80, Oncol dipl, 81. *Prof Exp:* Resident internal med, Temple Univ Sch Med, 74-77; fel hemat/oncol, Hosp Univ Pa, 77-79; fel thrombosis, Thrombosis Res Ctr, 79-80, from asst prof to assoc prof med, 80-91, PROF MED, TEMPLE UNIV SCH MED, 91- *Concurrent Pos:* Clin instr, Temple Univ Hosp, 79-80; dir, Temple Univ Hosp Coagulation Lab, 82-; clin investr award, NIH, 80, res career develop award, 87; mem Coun Thrombosis, Am Heart Asn. *Mem:* Am Fedn Clin Res; Am Soc Hemat; Int Soc Hemostasis & Thrombosis; Am Soc Clin Invest. *Res:* Investigating the mechanisms of regulation of cellular expression of the human kininogens, calpains and CI inhibitor in platelets, endothelial cells and cultured tumor cells by molecular, ligand binding and antigenic assay. *Mailing Add:* Hemat/Oncol Sect Thrombosis Res Ctr Temple Univ Sch Med 3400 N Broad St Philadelphia PA 19140

SCHMALBERGER, DONALD C, b Union City, NJ, Oct 24, 26; m 47. ASTROPHYSICS. *Educ:* Okla State Univ, BS, 58; Ind Univ, MA, 59, PhD(astrophys), 62. *Prof Exp:* From instr to asst prof astron, Univ Rochester, 62-67; asst prof astron, State Univ NY, Albany, 67-70, asst prof astron & space sci, 70-72, assoc prof astron & space sci, 72-77; RETIRED. *Mem:* AAAS; Am Astron Soc; fel Royal Astron Soc; NY Acad Sci; Int Astron Union. *Res:* Solar physics; theory of stellar atmospheres; physical structure of variable stars; turbulent energy transport in astrophysical media. *Mailing Add:* 75 Lenox Ave Albany NY 12203

SCHMALE, ARTHUR H, JR, b Lincoln, Nebr, Mar 14, 24; m 54; c 3. MEDICINE, PSYCHIATRY. *Educ:* Pa State Col, 45; Univ Md, MD, 51. *Prof Exp:* Med intern, Univ Hosp, Baltimore, Md, 51-52, asst resident med & psychiat, 52-53, asst resident psychiat, 53-54; asst resident med & psychiat, Strong Mem Hosp, 54-56; from instr to assoc prof med & psychiat, 56-73, PROF PSYCHIAT, SCH MED & DENT, UNIV ROCHESTER, 73- *Concurrent Pos:* NIMH Teaching fel psychiat, Sch Med, Univ Md, 53-54; Hochstetter fel med & psychiat, Sch Med & Dent, Univ Rochester, 54-55, USPHS res fel, 55-57, Markle scholar med sci, 57-62, Buswell fac fel, 60-; Found Fund Res Psychiat fel psychoanal training, 59-63. *Mem:* AAAS; Am Psychosom Soc; Am Asn Univ Prof. *Res:* Psychosomatic medicine; emotions; cancer. *Mailing Add:* Dept Psychiat Univ Rochester Sch Med & Dent Rochester NY 14642

SCHMALSTIEG, FRANK CRAWFORD, b Corpus Christi, Tex, Jan 30, 40. MOLECULAR MECHANICS OF HOST DEFENSE DEFECTS. *Educ:* Tex A&M Univ, PhD(phys-org chem); Univ Tex, Galveston, MD, 72. *Prof Exp:* From asst prof to assoc prof, 77-85, PROF ALLERGY-IMMUNOL, HUMAN BIOCHEM & GENETICS IMMUNOL, UNIV TEX MED BR, 86- *Mem:* Am Asn Immunologists; Reticuloendothelial Soc; Soc Pediat Res; Sigma Xi. *Mailing Add:* 1202 Harbor View Dr Galveston TX 77550-3114

SCHMALTZ, LLOYD JOHN, b Chicago, Ill, Apr 10, 29; m 52; c 4. GEOMORPHOLOGY, GLACIAL GEOLOGY. *Educ:* Augustana Col, AB, 53; Univ Mo, AM, 56, PhD(geol), 59. *Prof Exp:* Instr geol, Augustana Col, 54-55 & Univ Mo, 56-59; from asst prof to assoc prof, 59-66, head dept, 65-71, PROF GEOL, WESTERN MICH UNIV, 66-, CHMN, DEPT GEOL,74- *Mem:* Fel Geol Soc Am; Nat Asn Geol Teachers; Am Quaternary Asn. *Res:* Pediments in central Arizona; Pleistocene geology in southwestern Michigan. *Mailing Add:* Dept Geol Western Mich Univ Kalamazoo MI 49008

SCHMALZ, ALFRED CHANDLER, b Dedham, Mass, June 30, 24; m 47; c 2. ORGANIC CHEMISTRY. *Educ:* Bowdoin Col, AB, 47; Middlebury Col, MS, 51; Univ Va, PhD(org chem), 54. *Prof Exp:* Res chemist, Hercules Inc, Del, 54-61, Va, 61-62, group leader fiber develop, 62-63, res supvr fibers & film, 63-71, mgr, Prod Develop, 71-73, mgr, appl res, polymers-fibers, 73-78; RES SCIENTIST DEVELOP & FIBERS, HERCULES INC, 78- *Mem:* Am Chem Soc; Sigma Xi; Am Asn Textile Chem & Colorists. *Res:* Synthetic organic chemistry; paper chemicals; wet strength resins; water soluble polymers; antioxidants; light stabilizers; polyolefin fiber development; dyeing mechanisms. *Mailing Add:* 2594 Harvest Dr Conyers GA 30208-2406

SCHMALZ, PHILIP FREDERICK, b Buffalo, NY, Nov 15, 41; m 75; c 2. ELECTROPHYSIOLOGY. *Educ:* Northern Ill Univ, MS, 69. *Prof Exp:* Lead technician, 75-81, instr, 81-82, ASSOC PHYSIOL, MAYO FOUND, 82- *Mem:* Am Physiol Soc; Am Motility Soc. *Mailing Add:* Dept Physiol & Biophys Mayo Found Rochester MN 55905

SCHMALZ, ROBERT FOWLER, b Ann Arbor, Mich, May 29, 29; m 64; c 2. REEF ENVIRONMENT CHEMISTRY, EVAPORITE DEPOSITION. *Educ:* Harvard Col, AB, 51, AM, 54, PhD(geol), 59. *Prof Exp:* Asst marine sedimentation, Oceanog Inst, Woods Hole, 57-58; from asst prof to assoc prof, 58-68, head geol prog, 71-74, coordr, undergrad prog geol, 74-77, PROF GEOL, PA STATE UNIV, 69 - *Concurrent Pos:* Assoc ed, Sedimentol, 59-61. *Mem:* Am Asn Petrol Geologists; fel Geol Soc Am; Geochem Soc; Soc Econ Paleont & Mineral; fel AAAS. *Res:* Low temperature aqueous geochemistry, chemical oceanography and chemical sedimentation in reef and evaporite environments; chemistry of reduced marine basins and petroleum genesis; radioactive waste management. *Mailing Add:* 536 Deike Bldg University Park PA 16802

SCHMALZ, THOMAS G, b Springfield, Ill, July 12, 48. ELECTRONIC STRUCTURE OF ORGANIC POLYMERS. *Educ:* Mont State Univ, BS, 70; Univ Ill, Urbana, PhD(chem physics), 75. *Prof Exp:* Res assoc, James Franck Inst, Univ Chicago, 76-79; asst prof chem, Rice Univ, 79-81; asst prof, 81-85, ASSOC PROF MARINE SCI, TEX A&M, GALVESTON, 85- *Mem:* Am Phys Soc; Am Chem Soc. *Res:* Understanding large conjugated pi-electron systems such as recently discovered carbon cage molecules and electrically conductive carbon polymers; aromatic stabilization in these systems is examined with various AB initio and semiempirical tools; Theory of Aromaticity in conjugated systems. *Mailing Add:* Dept Marine Sci Tex A&M Univ PO Box 1675 Galveston TX 77553

SCHMALZER, DAVID KEITH, b Baltimore, Md, Aug 20, 42; m 65. CHEMICAL ENGINEERING. *Educ:* Johns Hopkins Univ, BES, 64, MS, 65; Univ Pittsburgh, PhD(chem eng), 69. *Prof Exp:* Assoc engr, Com Atomic Power Div, Westinghouse Elec Co, 67; proj engr, Gulf Res & Develop Co, Pa, 68-77; res mgr, Pittsburgh & Midway Coal Mining Corp, 77-80; mem staff, Gulf Mineral Resources Co, 80-; MEM STAFF, ARGONNE NAT LAB. *Mem:* Am Chem Soc; Am Inst Chem Engrs. *Res:* Hydrocarbon pyrolysis ranging from fundamental experimental work in small equipment at severe conditions to applied research and optimization of operating large scale industrial pyrolysis plants. *Mailing Add:* Argonne Nat Lab 9700 S Cass Ave Bldg 205 Argonne IL 60439

SCHMARS, WILLIAM THOMAS, b Lockport, Ill, Jan 10, 38. ELECTRON OPTICS, ELECTRICAL ENGINEERING. *Educ:* Univ Ill, BS, 61. *Prof Exp:* Res engr, 61-66, sr res engr, 66-68, SR ENGR, AUTONETICS DIV, ROCKWELL INT, 68- *Mem:* Nat Soc Prof Engrs. *Res:* Navigation instruments, especially photoelectric autocollimators, precision shaft angle encoders and vibrating string gyro and accelerometer; the electrochemical tiltmeter. *Mailing Add:* 1509 Beechwood Ave Fullerton CA 92635

SCHMECKENBECHER, ARNOLD F, b Allendorf, Ger, Feb 15, 20; US citizen; m 53; c 1. INORGANIC CHEMISTRY. *Educ:* Univ Heidelberg, dipl, 50; Univ Kiel, PhD(chem), 53. *Prof Exp:* Instr inorg chem, Univ Kiel, 53-54; chemist, Gen Aniline & Film Corp, NJ, 55-58; sr chemist, Remington Rand Univac, Pa, 58-60; assoc chemist, Components Div, IBM Corp, 60-61, staff chemist, 61-64, adv chemist, 64-69, sr chemist, 69-88; RETIRED. *Concurrent Pos:* Ger Res Asn scholar, 53-54. *Mem:* Am Chem Soc; Sigma Xi. *Res:* Magnetics materials, particularly materials for use in computer memories and phase locked oscillators; multilayered ceramic substrates for integrated circuit chips. *Mailing Add:* 228 Wilbur Blvd Poughkeepsie NY 12603

SCHMEDTJE, JOHN FREDERICK, b St Louis, Mo, July 9, 19; m 56; c 3. ANATOMY. *Educ:* Columbia Univ, AB, 41; Rutgers Univ, PhD(zool), 51. *Prof Exp:* Instr anat, Sch Med, St Louis Univ, 51-53, asst prof, 53-56; instr, Harvard Med Sch, 56-58; asst prof, Sch Med, Tufts Univ, 58-66; ASSOC PROF ANAT, SCH MED, IND UNIV, INDIANAPOLIS, 66- *Mem:* AAAS; Am Asn Anat; Electron Micros Soc Am; Histochem Soc; Am Soc Cell Biologists. *Res:* Immunocytochemistry; electron microscopy; cellular aspects of immune reactions in lymphatic tissue and epithelium. *Mailing Add:* Ind Univ Sch Med 635 Barnhill Dr Ind Univ Sch Med 635 Barnhill Dr Indianapolis IN 45223

SCHMEE, JOSEF, b Grieskirchen, Austria, Feb 13, 45; m 67. STATISTICS. *Educ:* Univ Com, Vienna, Magister, 68; Union Col, NY, MSc, 70, PhD(statist), 74. *Prof Exp:* Res asst sociol, Col Com, Vienna, 67-68; analyst finance, Gen Elec Co, 69; from asst prof to assoc prof, 72-80, dir, 80-86, PROF MGT, GRAD MGT INST, UNION COL, NY, 80- *Concurrent Pos:* Asst prof mgt, Univ Munich, 75; dir, Bur Health Mgt Standards, NY Dept Health, 78-79; Fulbright-Hays res scholar, Ger, 80-81; adj prof path, Albany Med Col, 81-; vis res fel, Gen Elec Res & Develop, 88-89. *Honors & Awards:* Wilcoxon Award, Am Soc Qual Control, 80, Brumbaugh Award, 81. *Mem:* Sigma Xi; Inst Math Statist; Am Statist Asn; Biomet Soc; fel Am Statist Asn. *Res:* Exact confidence intervals on mean, variance, percentiles and range of normal distribution with single censoring; sequential analysis t-test and estimation; semi-Markov models in health care systems; statistics in dentistry; censored data regression analysis; atherosclerosis in swine due to dietary effects. *Mailing Add:* Bailey Hall Grad Mgt Inst Union Col Schenectady NY 12308

SCHMEELK, JOHN FRANK, b Newark, NJ, July 19, 39; m 67. APPLIED MATHEMATICS. *Educ:* Seton Hall Univ, BS, 62; NY Univ, MS, 65; George Washington Univ, PhD(math), 76. *Prof Exp:* Instr math, Seton Hall Univ, 63-65 & NC State Univ, 65-66; teaching asst math, George Washington Univ, 71-73; instr math, Middlesex County Col, 74-75; ASST PROF MATH, VA COMMONWEALTH UNIV, 75- *Mem:* Am Math Soc; Math Asn Am; Soc Indust & Appl Math; Am Asn Univ Professors. *Res:* Development of an infinite-dimensional generalized function, for example, continuous linear functionals on test functions that are infinitely differentiable on an infinite dimensional vector space. *Mailing Add:* 1916 Sweetwater Lane Richmond VA 23229

SCHMEER, ARLINE CATHERINE, b Rochester, NY, Nov 14, 29. CELL BIOLOGY, SYNTHETIC ORGANIC & NATURAL PRODUCTS CHEMISTRY. *Educ:* Col St Mary, Ohio, BA, 51; Univ Notre Dame, MS, 61; Univ Colo Med & Grad Sch, Denver, PhD(cell biol), 69. *Hon Degrees:* DSc, Albertus Magnus Col, 74; DSc, State Univ NY, Potsdam, 90. *Prof Exp:* Chmn high sch sci dept, Ohio, 54-59 & NY, 59-63; asst prof biol & co-dir med res lab, Ohio Dominion Col, 63, chmn dept, 63-68, assoc prof biol & dir med res lab, 64-72, dir, St Thomas Inst Res Lab, 69-72; res scientist, Cancer Res Ctr & Hosp, Am Med Ctr, Denver, 72-73, dir med res fel prog & anticancer agents of marine origin, 72-82; DIR, MERCENENE CANCER RES INST, HOSP ST RAPHAEL, NEW HAVEN, CONN, 82- *Concurrent Pos:* NSF res fels, 59-64; USPHS Nat Cancer Inst & Am Cancer Soc grants; Nat Cancer Inst spec res fel; chmn biol educ sec schs, Archdiocese of New York; partic, NSF High Sch Biol Sci Curric Study & Comn Undergrad Educ Biol Sci Comt Col Biol Teacher Training, DC, 62-63; sr investr & mem cell biol specialty panel, Marine Biol Lab, Woods Hole, 62-, corp mem, 65-; partic, Int Cancer Cong, Tokyo, Japan, 66 & Houston, Tex, 70, Buenos Aires, Argentina, 78, Seattle, Wash, 82 & Budapest, Hungary, 86, Hamburg, Ger, 90; res scientist, Inst Med Biophys, Univ Würzburg & res prof, Univ Würzburg Med Sch, Ger, 69-70; vis scientist, Am Med Ctr Cancer Res; consult, Sch Trop Med, Univ Sydney; scholar & fel, Nat Cancer Inst, NIH, NSF, Med Sch, Univ Würzburg, WGer & Grad & Med Sch, Univ Colo; numerous grants, fel doctoral studies. *Honors & Awards:* St Joachim Award, Mercy Hosp, Watertown, NY. *Mem:* Am Soc Cell Biol; NY Acad Sci; Electron Micros Soc Am; Am Physiol Soc; fel Royal Micros Soc Eng. *Res:* Cellular biology; developmental drugs cancer; pharmacology toxicology and experimental therapeutics in use of growth inhibitors and biologically active moieties from naturally occurring products and effects of these products on abnormal growth such as cancer, AIDS, viral & bacterial activity; developmental therapeutics. *Mailing Add:* Hosp St Raphael Mercenene Cancer Res Inst Dir New Haven CT 06511

SCHMEHL, WILLARD REED, b Arlington, Nebr, Apr 16, 18; m 43; c 2. SOILS, INTERNATIONAL AGRICULTURE. *Educ:* Colo State Univ, BS, 40; Cornell Univ, PhD(soil sci), 48. *Prof Exp:* Asst, Cornell Univ, 40-42; supvr, Hercules Powder Co, 42-43; assoc prof & assoc agronomist, 48-56, prof agron & agronomist, 56-86, EMER PROF AGRON, COLO STATE UNIV, 86- *Concurrent Pos:* Proj assoc, Univ Wis, 54-55. *Mem:* Soil Sci Soc Am; Am Soc Agron; Am Soc Sugar Beet Technologists. *Res:* Soil acidity and fertility; availability of phosphates to plants; farming systems; sugar beet nutrition. *Mailing Add:* Dept Agron Colo State Univ Ft Collins CO 80523

SCHMEISSER, GERHARD, JR, b Baltimore, Md, Mar 27, 26; m 57. ORTHOPEDIC SURGERY. *Educ:* Princeton Univ, AB, 49; Johns Hopkins Univ, MD, 53. *Prof Exp:* From instr to assoc prof, 57-61, PROF ORTHOP SURG, SCH MED, JOHNS HOPKINS UNIV, 58- *Concurrent Pos:* Vis surgeon, Children's Hosp, 58-; orthop surgeon, Johns Hopkins Hosp, 58-; chief orthop surg, Baltimore City Hosps, 59-; consult, USPHS Hosp, 64-65. *Honors & Awards:* IR 100 Award, 71. *Mem:* Fel Am Acad Orthop Surg. *Res:* External power and control of limb prostheses and braces. *Mailing Add:* Dept Orthop Surg Johns Hopkins Univ 720 Rutland Ave Baltimore MD 21205

SCHMELING, SHEILA KAY, b Brookings, SDak, May 5, 49. WILDLIFE DISEASE. *Educ:* Univ Mass, Amherst, BS, 71; Colo State Univ, MS & DVM, 77. *Prof Exp:* WILDLIFE VET, NAT WILDLIFE HEALTH LAB, 79- *Mem:* Wildlife Disease Asn; Am Asn Wildlife Vets; Am Vet Med Asn; Sigma Xi. *Res:* Determination of the causes of mortality in free-flying raptors, especially the bald and golden eagle. *Mailing Add:* Box 196 Corozal Town Belize

SCHMELL, ELI DAVID, b Baltimore, Md, Jan 5, 50; m 71; c 2. IMMUNOCHEMISTRY, REPRODUCTIVE BIOLOGY. *Educ:* City Univ NY, BS, 71; Johns Hopkins Univ, PhD(cellular & molecular biol), 76. *Prof Exp:* Teaching fel biol, Johns Hopkins Univ, 76-79; staff fel reproductive biol, NIH, 79-81, sr staff fel, 81; sci officer biochem, 81-83, PROG MGR MOLECULAR BIOL, OFF NAVAL RES, 84- *Concurrent Pos:* Teacher asst, Johns Hopkins Med Sch, 74-75, res assoc, 80-81, adj asst prof, 82-83, adj assoc prof, 83-; vis scientist, Israel Inst Technol, 83 & Weizmann Inst Sci, 85; adminr, Mgt Res Progs in Biochem, Microbiol & Molecular Biol, US Navy. *Honors & Awards:* Forum Prize, Am Fertility Soc, 81. *Mem:* Am Soc Biol Chem; Am Soc Cell Biol; AAAS; Am Fertility Soc; Soc Study Reproduction. *Res:* Biochemical and immunochemical studies on the structure and function of human acetylcholinesterase. *Mailing Add:* Interpharm Labs Ltd Kiryat Weizmann Nesziona 76110 Israel

SCHMELLING, STEPHEN GORDON, b Kenosha, Wis, Mar 21, 40; m 66; c 2. HYDROLOGY & WATER RESOURCES. *Educ:* Mass Inst Technol, BS, 62; Univ Calif, Berkeley, PhD(physics), 67. *Prof Exp:* Res physicist, Lawrence Berkeley Lab, 67-70; asst prof physics, State Univ NY, Buffalo, 70-75; assoc prof physics, ECent Univ, 75-88; RES SCIENTIST, US ENVIRON PROTECTION AGENCY, RS KERR ENVIRON RES LAB, 80- *Concurrent Pos:* Sr lectr, Univ Ife, Ile-Ife, Nigeria, 74. *Mem:* Am Phys Soc; Am Geophys Union. *Res:* Subsurface contaminant transport particularly in fractured rock geological formations; remediation of ground-water contamination. *Mailing Add:* US Environ Protection Agency RS Kerr Environ Res Lab PO Box 1198 Ada OK 74820

SCHMELTZ, IRWIN, b New York, NY, Feb 26, 32; m 62; c 4. BIO-ORGANIC CHEMISTRY. *Educ:* City Col New York, BS, 53; Univ Utah, PhD(org biochem), 59. *Prof Exp:* Teaching fel chem, 55-59, fel biochem, Univ Utah, 59-60; Cigar Mfrs Asn sr res fel, Eastern Regional Lab, 60-62, res chemist tobacco invests, 62-65, head pyrolysis invests, tobacco lab, 65-70, head smoke invests, 70-71, lubricant invests, Animal Fat Prod Lab, USDA, 71-73; head bio-org chem, Div Environ Carcinogensis, Naylor-Dana Inst Dis Prev, Am Health Found, 73-79; tech fel, Hoffmann-LaRoche, Nutley, NJ, 79-87; DIR, US CUSTOMS LAB, NY, 87- *Concurrent Pos:* Consult, Nat Cancer Inst, 73- & Princeton Univ, 74- *Mem:* Am Oil Chemists Soc; Soc Environ Geochem & Health; AAAS; Am Chem Soc; Am Asn Cancer Res. *Res:* Chemical carcinogenesis, environmental chemistry; organic synthesis; biosynthesis of pteridines; chemical composition of tobacco and tobacco smoke; pyrolysis of organic compounds; products from animal fats; analytical organic chemistry. *Mailing Add:* Three Miriam Lane Monsey NY 10952

SCHMELZ, DAMIAN VINCENT, b Georgetown, Ind, May 7, 32. ECOLOGY. *Educ:* St Meinrad Col, BA, 58; Purdue Univ, Lafayette, MS, 64, PhD(ecol), 69. *Prof Exp:* Teacher high sch, Ind, 59-67; instr, 65-70, from asst prof to assoc prof, 70-75, PROF & ACAD DEAN, ST MEINRAD COL, 75- *Concurrent Pos:* Mem, Ind Natural Areas Surv, Purdue Univ, 67-69; mem, Ind Natural Resources Comn, 75- *Mem:* Ecol Soc Am; Sigma Xi. *Res:* Forest ecology. *Mailing Add:* Dept Biol St Meinrad Col St Meinrad IN 47577

SCHMERL, JAMES H, b Storrs, Conn, April 7, 40. LOGIC MODEL THEORY. *Educ:* Univ Calif, Berkeley, AB, 62, MA, 63, PhD(math), 70. *Prof Exp:* PROF MATH, UNIV CONN, STORRS, 72- *Mem:* Am Math Soc; Asn Symbol Logic. *Mailing Add:* Dept Math Univ Conn Storrs CT 06269

SCHMERLING, ERWIN ROBERT, b Vienna, Austria, July 28, 29; m 57; c 2. SPACE PHYSICS, DATA SYSTEMS. *Educ:* Cambridge Univ, BA, 50, MA, 54, PhD(radio physics), 58. *Prof Exp:* Vis asst prof elec eng, Pa State Univ. 55-57, from asst prof to assoc prof, 57-64; prog chief magnetospheric physics, 64-76, Off Space Sci, NASA Hq, 76-83; vis scholar, Stanford Univ, Calif, 83; asst dir space & earth sci, Goddard Space Flight Ctr, Greenbelt, Md, 84-86; CHIEF DATA SYST SCIENTIST, OFF SPACE SCI & APPLN NASA HQ, WASHINGTON, DC, 86- *Concurrent Pos:* Mem comns G & H, Int Sci Radio Union, 58, secy, US Comn III, 66-69, chmn, 69-72; mem, Wave Propagations Standards Comt, Inst Elec & Electronics Engrs, 70-; mem comt space res, 82, Adv Group Aerospace Res & Dev, 81-87. *Mem:* AAAS; fel Inst Elec & Electronics Engrs; Am Geophys Union. *Res:* Ionospheric and radio physics; physics of the ionosphere; radio wave propagation; electron densities and sounding of ionized regions from the ground and from space vehicles; atmospheric and space physics, environment of earth, planets and interplanetary space; scientific data systems; interactive computers for science education. *Mailing Add:* SC NASA Hq Washington DC 20546

SCHMERR, MARY JO F, b Dubuque, Iowa, Nov 4, 45; m 72; c 3. IMMUNOCHEMISTRY, MOLECULAR BIOLOGY. *Educ:* Clarke Col, Iowa, BA, 68; Iowa State Univ, PhD(biochem), 75. *Prof Exp:* RES CHEMIST BIOCHEM ANIMAL DIS, NAT ANIMAL DIS CTR, USDA, 75- *Mem:* Am Soc Microbiol; Am Chem Soc; Sigma Xi; NY Acad Sci. *Res:* Biochemistry and immunochemistry of animal diseases caused by viruses; study of the function of viral proteins. *Mailing Add:* Rte 4 Box 19 Woodward IA 50276

SCHMERTMANN, JOHN H(ENRY), b New York, NY, Dec 2, 28; m 56; c 4. GEOTECHNICAL ENGINEERING. *Educ:* Mass Inst Technol, BSCE, 50; Northwestern Univ, MS, 54, PhD(civil eng), 62. *Prof Exp:* Soils engr, Moran, Proctor, Mueser & Rutledge, NY, 51-54; geotech engr, US Army Corps Engrs, 54-56; from asst prof to prof civil eng, Univ Fla, 56-78; PRIN, SCHMERTMANN & CRAPPS CONSULT GEOTECH ENGRS, 78- *Concurrent Pos:* Prin investr, NSF res grants, 56-; NSF fel, Norweg Geotech Inst, Oslo, 62-63; Nat Res Coun Cross-Can lectr, 71; vis scientist, Nat Res Coun Can, 71-72. *Honors & Awards:* Collingwood Prize, Am Soc Civil Engrs, 56, Norman Medal, 71, State-of-the-Art Award, 77. *Mem:* Nat Acad Eng; Am Soc Civil Engrs; Am Soc Testing & Mat. *Res:* Soil mechanics and foundation engineering; consolidation and shear strength; methods for field exploration; special soil mechanics problems. *Mailing Add:* Schmertmann & Crapps Inc 4509 NW 23 Ave Suite 19 Gainesville FL 32606

SCHMICKEL, ROY DAVID, b Millville, NJ, Feb 9, 36; m 60; c 4. PEDIATRICS, GENETICS. *Educ:* Oberlin Col, BA, 57; Duke Univ, MD, 61; Am Bd Pediat, cert, 70. *Prof Exp:* Internship & jr residency, Duke Univ, 61-63; fel radiol health, USPHS, 63-65; sr residency, Johns Hopkins Univ, 65-66, fel, dept microbiol, 66-68, instr, 66-68; res assoc, human genetics, Univ Mich, 68-71, from asst prof to prof pediat, 68-81, from assoc prof to prof human genetics, 71-81; PROF HUMAN GENETICS & PEDIAT, SCH MED, UNIV PA, 81-, CHMN DEPT HUMAN GENETICS & DIR, HUMAN GENETICS CTR, 81-; PROF, WISTAR INST, PHILADELPHIA, 82- *Concurrent Pos:* Larry Silver res award, 57 & USPHS, 64; staff mem, Mott Children's Hosp, Ann Arbor, Mich, 68-81; dir, pediat genetics, Univ Hosp Mich, 72-81; vis prof, dept molecular biol, Univ Edinburgh, 75-76; mem, prenatal diag study sect, NIH, 74-76, genetics study sect, 78-82, chmn, 81-82, mem, dir adv coun, 82-; spec adv, President's Coun Ment Retardation, 76-77; mem bd trustees, Am Inst Ment Studies, NJ, 82-; sr physician, div genetics & teratology, dept med, Children's Hosp, Philadelphia, 82-; consult ed, Am J Ment Deficiency, 82-; mem, ment retardation res comt, Nat Inst Child Health & Human Develop, 83-, chmn, 84-; mem, permanent coun, Inst Basic Res in Develop Disabilities, Staten Island, NY, 84- *Mem:* Soc Pediat Res; Am Pediat Soc; Am Soc Human Genetics; Am Acad Pediat. *Res:* Chromosomal diseases and characterization of human DNA; human DNA and human chromosomes; isolation of human ribosomal genes. *Mailing Add:* Dept Human Genetics 195 Med Labs Bldg G3 Sch Med Univ Pa Philadelphia PA 19104

SCHMID, CARL WILLIAM, b Philadelphia, Pa. BIOPHYSICAL CHEMISTRY. *Educ:* Drexel Inst Technol, BS, 67; Univ Calif, Berkeley, PhD(chem), 71. *Prof Exp:* Chemist, Eastern Regional Res Labs, 63-67; res asst biophys chem, Univ Calif, Berkeley, 67-71; res fel, Calif Inst Technol, 71-73; ASSOC PROF CHEM, UNIV CALIF, DAVIS, 73- *Concurrent Pos:* Jane Coffin Childs Found fel, 71-73. *Mem:* AAAS; Am Chem Soc. *Res:* Determining the biological function of different DNA sequence classes which comprise the eukaryotic genome. *Mailing Add:* Dept Chem Univ Calif Davis CA 95616

SCHMID, FRANK RICHARD, b New York, NY, June 25, 24; m 54; c 9. MEDICINE. *Educ:* NY Univ, MD, 49. *Prof Exp:* From asst to resident to chief resident, Bellevue Hosp, NY, 50-51 & 52-54; asst med, Col Med, NY Univ, 54-57; from assoc to assoc prof, 57-69, PROF MED, MED SCH, NORTHWESTERN UNIV, CHICAGO, 69- *Concurrent Pos:* Fel, Arthritis & Rheumatism Found, 56-59; Markle scholar med sci, 60-65; trainee, Med Div, NIH, 54-56; asst vis physician, Bellevue Hosp, 54-57; attend physician, Vet Admin Res Hosp, 57-, Northwestern Mem Hosp, 61- & Cook County Hosp, 66-69; consult, Rehab Inst Chicago, 67- *Mem:* Am Rheumatism Asn; Am Fedn Clin Res; Cent Soc Clin Res. *Res:* Immunological considerations in rheumatic diseases. *Mailing Add:* 330 Greenbay Rd Glencoe IL 60022

SCHMID, FRANZ ANTON, b Hermersdorf, Czech, Sept 21, 22; US citizen; m 55; c 1. CANCER. *Educ:* Univ Munich, DVM, 51; Fordham Univ, MS, 54. *Prof Exp:* Res asst surg physiol, Sloan-Kettering Inst Cancer Res, 54-56; vet meat inspector, USDA, 56-58; res assoc, 58-84, assoc, 66-84, ASSOC MEM, EXP CANCER CHEMOTHER, WALKER LAB, SLOAN-KETTERING INST CANCER RES, 84- *Mem:* AAAS; Am Asn Cancer Res. *Res:* Experimental chemotherapy of cancer; chemical cancerigenesis; mouse genetics. *Mailing Add:* 122 Sterling Ave Harrison NY 10528

SCHMID, GEORGE HENRY, b Madison, Wis, Aug 6, 31; m 57, 87; c 2. ORGANIC CHEMISTRY. *Educ:* Univ Southern Calif, BS, 53, PhD(org chem), 61. *Prof Exp:* Fel, Harvard Univ, 61-63; from asst prof to assoc prof, 63-75, PROF CHEM, UNIV TORONTO, 75- *Mem:* Am Chem Soc. *Res:* The elucidation of the mechanism of electrophilic additions to unsaturated carbon-carbon bonds. *Mailing Add:* Dept Chem Univ Toronto Toronto ON M5S 1A1 Can

SCHMID, GERHARD MARTIN, b Ravensburg, Ger, Oct 26, 29; m 58; c 2. ELECTROCHEMISTRY. *Educ:* Innsbruck Univ, PhD(phys chem), 58. *Prof Exp:* Fel, Univ Tex, 58-62; temp asst prof, Univ Alta, 62-64; asst prof, 64-70, ASSOC PROF CHEM, UNIV FLA, 70- *Concurrent Pos:* Sr res chemist, Tracor, Inc, 60-62. *Mem:* Am Chem Soc; Electrochem Soc; Sigma Xi. *Res:* Structure of the electrical double layer; adsorption on solid electrodes; passivity of metals; corrosion and corrosion inhibition. *Mailing Add:* Chem Dept Univ Fla 348 Lei Bldg Gainesville FL 32611-2046

SCHMID, HARALD HEINRICH OTTO, b Graz, Austria, Dec 10, 35; m 77. BIOCHEMISTRY. *Educ:* Graz Univ, MS, 58, PhD, 64. *Prof Exp:* From res fel biochem to res assoc, 62-66, from asst prof to assoc prof, 66-74, actg dir, 85-86, PROF BIOCHEM, HORMEL INST, UNIV MINN, 74-, EXEC DIR, 87- *Concurrent Pos:* Ed, Chem Phys Lipids, 84- *Mem:* Am Chem Soc; Am Soc biochem & Molecular Biol. *Res:* Structure and biochemistry of natural products; metabolism of complex lipids in biomembranes; lipid peroxidation; lipid metabolism in ischemia and myocardial infarct. *Mailing Add:* Hormel Inst Univ Minn Austin MN 55912

SCHMID, JACK ROBERT, b Chicago, Ill, Oct 3, 24; m 48; c 4. PHARMACOLOGY, PHYSIOLOGY. *Educ:* Mich State Univ, BS, 52, MS, 54; Univ Ark, Little Rock, PhD(pharmacol), 67. *Prof Exp:* Assoc pharmacologist, Mead Johnson Res Ctr, Ind, 54-58; high sch teacher, Mich, 59-62; sr pharmacologist, Riker Res Labs, 3M Co, 66-73, pharmacologist specialist, 73-88; RETIRED. *Mem:* Int Soc Heart Res; NY Acad Sci; Am Chem Soc; Sigma Xi. *Res:* Cardiopulmonary effects of drugs. *Mailing Add:* 933 Copper Vista Dr Prescott AZ 86303-4801

SCHMID, JOHN CAROLUS, b Milwaukee, Wis, Apr 17, 20; m 48; c 3. APPLIED STATISTICS, EDUCATIONAL PSYCHOLOGY. *Educ:* Univ Wis, BS, 45, MS, 46, PhD(educ, statist), 49. *Prof Exp:* Instr math, Univ Wis Exten Div, 46-47; asst prof res, Mich State Univ, 49-52; res psychologist, Air Force Personnel & Training Res Ctr, 52-57; prof statist, Univ Ark, 57-66; prof, 66-84, emer prof res, Univ Northern Colo, 84-; RETIRED. *Concurrent Pos:* Ed, J Exp Educ, 63-84. *Mailing Add:* 1212 38th Ave Greeley CO 80634-2716

SCHMID, KARL, b Erlinsbach, Switz, July 23, 20; nat US; m 47; c 2. BIOCHEMISTRY. *Educ:* Swiss Fed Inst Technol, dipl rer nat, 43; Univ Basel, MA & PhD(biochem), 46. *Prof Exp:* Res assoc, Harvard Med Sch, 48-52; assoc biochemist, Lovett Mem Lab, Mass Gen Hosp, 52-63; assoc prof, 63-66, PROF BIOCHEM, SCH MED, BOSTON UNIV, 66- *Concurrent Pos:* Zurich Univ fel chem, Cambridge Univ, 47-48. *Mem:* Am Chem Soc; Am Soc Biol Chemists; Soc Complex Carbohydrates (pres, 80). *Res:* Isolation, purification, characterization, chemical structure and biological importance of human plasma proteins, especially glycoproteins; glycosaminoglycans of the tissues of various human organs. *Mailing Add:* Dept Biochem Sch Med Boston Univ Boston MA 02118

SCHMID, LAWRENCE ALFRED, b Philadelphia, Pa, Mar 11, 28; m 62. THEORETICAL PHYSICS. *Educ:* Univ Pa, BSEE, 49; Princeton Univ, MA, 51, PhD(physics), 53. *Prof Exp:* Asst prof physics, Mich State Univ, 53-59; physicist, Goddard Space Flight Ctr, NASA, 59-80; physicist, Nat Bur Standards, 80-85; GUEST SCIENTIST, NAT INST STANDARDS & TECHNOL, 85- *Mem:* AAAS; Am Phys Soc; Am Asn Physics Teachers; Soc Indust & Appl Math. *Res:* Fluid dynamics; thermodynamics; dynamic meteorology; relativity; variational formalism; group theory; surface tension phenomena. *Mailing Add:* 12 Maplewood Ct Greenbelt MD 20770

SCHMID, LOREN CLARK, b Ypsilanti, Mich, Feb 1, 31; m 54; c 3. ENERGY CONVERSION, TECHNOLOGY TRANSFER. *Educ:* Univ Mich, BS, 53, MS, 54, PhD(physics), 58. *Prof Exp:* Asst, Eng Res Inst, Univ Mich, 53-56; res assoc, Argonne Nat Lab, 56-58; physicist, Gen Elec Co, 58-65; res mgr, Battelle Mem Inst, 65-68, mgr, Reactor Physics Dept, 68-73, dir, Energy prog, 73-75, fusion technol prog mgr, 73-76, energy mission dir, 75-79, mgr planning, 79-83, res & technol appln, 83-85, PROG MGR, ADVAN ENERGY & TECHNOL TRANSFER, BATTELLE NORTHWEST, 85- *Concurrent Pos:* Actg assoc prof nuclear eng, Univ Wash, 68-74; coordr plutonium recycle short course, Joint Ctr Grad Study, Univ Wash, 73-77, affiliated prof, prog adv & coordr, Nuclear Eng, 74-; Far West Reg Coord, 83-88, nat chmn, Fed Lab Consortium Technol Transfer, 89- *Honors & Awards:* Harold Metcalf Award, Fed Lab Consortium Technol Transfer, 86. *Mem:* Sigma Xi; Fel Am Nuclear Soc; NY Acad Sci. *Res:* Fusion reactor technology; nuclear and reactor physics; analysis of radioactive-decay schemes; energy production and conservation; technology transfer. *Mailing Add:* Prog Mgr Pac Northwest Labs Battelle Northwest PO Box 999 Richland WA 99352

SCHMID, PETER, b Signau, Switz, Sept 5, 27; US citizen; m 54; c 2. MEDICAL SCIENCES, BIOCHEMISTRY. *Educ:* Winterthur Tech. Switz, BS, 52; Univ Calif, Berkeley, MS, 59; Univ Calif, San Francisco, PhD(biochem, pharmaceut chem), 64. *Prof Exp:* Investr org chem, Ciba Pharmaceut Co, Switz, 52-55; res assoc biochem, State Univ NY Upstate Med Ctr, 55-56; fel, Med Ctr, Univ Calif, San Francisco, 63; sr investr radiobiol, US Naval Radiol Defense Lab, 64-67; SR INVESTR & GROUP LEADER CUTANEOUS HAZARDS, LETTERMAN ARMY INST RES, 67- *Concurrent Pos:* Fac mentor, Columbia Pac Univ, San Rafael, Calif. *Mem:* AAAS; Am Chem Soc; NY Acad Sci; Dermal Clin Eval Soc; Sigma Xi; Am Mgt Asn. *Res:* Mechanism of cell replication and cell growth; biophysics of skin; medical research in dermatology to include biochemical, biophysical, morphometrics, biostatistics aspects, cosmethology, cell biology, non-invasive measurements, skin protection and dermal toxicology; contract review; budget and program development and management; toxicology, analytical and pharmaceutical chemistry; computer science; environmental science; research administration. *Mailing Add:* Letterman Army Inst Res San Francisco CA 94129-6800

SCHMID, RUDI, b Glarus, Switz, May 2, 22; nat US; m 49; c 2. INTERNAL MEDICINE, GASTROENTEROLOGY. *Educ:* Univ Zurich, MD, 47; Univ Minn, PhD, 54; Am Bd Internal Med, dipl, 57. *Prof Exp:* Intern, Univ Hosp, Univ Calif, San Francisco, 48-49; resident, Univ Hosp, Univ Minn, 49-51, instr med, Sch Med, 52-54; sr hematologist, Nat Inst Arthritis & Metab Dis, 55-57; assoc med, Harvard Med Sch, 57-59, asst prof, 59-62; prof, Univ Chicago, 62-66; PROF MED, UNIV CALIF, SAN FRANCISCO, 66- *Concurrent Pos:* USPHS spec res fel biochem, Columbia Univ, 54-55; asst physician, Thorndike Mem Lab, Boston City Hosp, 57-; consult, US Army Surgeon Gen & San Francisco Vet Admin Hosp. *Mem:* Nat Acad Sci; Am Acad Arts & Sci; Am Soc Exp Path; Am Soc Hemat; Am Soc Biol Chemists; AAAS; Am Soc Clin Invest. *Res:* Liver physiology and pathophysiology; hepatic enzymes and metabolism; porphyrins; bile pigments; jaundice; porphyria; liver diseases. *Mailing Add:* Dean's Off Sch Med Univ Calif 224-S San Francisco CA 94143-0410

SCHMID, WALTER EGID, b Philadelphia, Pa, Nov 24, 33; m 59; c 1. BOTANY. *Educ:* Univ Pa, AB, 55; Univ Wis, MS, 58, PhD(bot), 61. *Prof Exp:* Asst bot, Univ Wis, 55-61; jr res plant physiologist, Univ Calif, Davis, 61-62; from asst prof to assoc prof bot, 62-72, assoc dean grad sch, 70-72, PROF BOT, SOUTHERN ILL UNIV, CARBONDALE, 72- *Mem:* AAAS; Am Soc Plant Physiol; Bot Soc Am; Japanese Soc Plant Physiol; Sigma Xi. *Res:* Inorganic nutrition of plants. *Mailing Add:* Dept Plant Biol Southern Ill Univ Carbondale IL 62901

SCHMID, WERNER E(DUARD), b Waldershof, Ger, Feb 15, 27; nat US; m 63; c 2. HYDROLOGY & WATER RESOURCES. *Educ:* Munich Tech Univ, Dipl Ing, 53; Lehigh Univ, MSc, 55; Univ Vienna, Dr Techn Sc, 65. *Prof Exp:* Asst soil mech, Munich Tech Univ, 53; instr civil eng, Lafayette Col, 54-55, asst prof, 55-56; asst prof, 56-59, ASSOC PROF CIVIL ENG, PRINCETON UNIV, 59- *Concurrent Pos:* Pres, Technotron, Inc, 67-78; chmn & chief exec officer, Hibrospan, SAm, 85-, Geos, SAm, 88-; dir, Synergics, Inc, 87- *Honors & Awards:* Hogentogler Award, Am Soc Testing & Mat, 59. *Mem:* Am Soc Civil Engrs; Am Soc Testing & Mat; Am Soc Eng Educ; Ger Soc Soil Mech & Found Engrs. *Res:* Soil mechanics; alternate and integrated energy systems; construction technology and environmental impact. *Mailing Add:* Dept Civil Eng Princeton Univ Princeton NJ 08544

SCHMID, WILFRIED, b Hamburg, Ger, May 28, 43. MATHEMATICS. *Educ:* Princeton Univ, BA, 64; Univ Calif, Berkeley, MA, 66, PhD(math), 67; Marina D Bizzarri-Schmid, MD, 77. *Prof Exp:* Asst prof math, Univ Calif, Berkeley, 67-70; Sloan fel & vis assoc prof, Columbia Univ, 68-69; vis mem, Inst Advan Study, 69-70; prof math, Columbia Univ, 70-78; PROF MATH, HARVARD UNIV, 78- *Concurrent Pos:* Vis prof, Univ Bonn, 73-74; John Simon Guggenheim Mem fel, 75-76, 89-90; vis, Inst Advan Study, Princeton Univ, 75-76. *Honors & Awards:* Prix Scientifique de UAP, 86. *Res:* Representations of Lie groups; complex manifolds. *Mailing Add:* Dept Math Harvard Univ Cambridge MA 02138

SCHMID, WILLIAM DALE, b Santa Ana, Calif, Apr 21, 37; div; c 2. ANIMAL ECOLOGY, COMPARATIVE ANIMAL PHYSIOLOGY. *Educ:* Univ Minn, St Paul, BS, 59; Univ Minn, Minneapolis, PhD(zool), 62. *Prof Exp:* Asst prof biol, Univ NDak, 62-66; from asst prof to assoc prof, 66-74, PROF ZOOL, UNIV MINN, MINNEAPOLIS, 74-, DIR INST HERMONOGRAPHY, 70- *Mem:* AAAS; Ecol Soc Am; Am Soc Zoologists; Am Soc Mammalogists; Wildlife Dis Asn; Sigma Xi. *Res:* Amphibian water balance; vertebrate physiology; comparative animal physiology. *Mailing Add:* Dept Ecol & Behav Biol Univ Minn 108 Zool Bldg Minneapolis MN 55455

SCHMIDLE, CLAUDE JOSEPH, b Buffalo, NJ, June 14, 20; m 45; c 2. COATINGS TECHNOLOGY, RADIATION CURING. *Educ:* Univ Notre Dame, BS, 41, MS, 42, PhD(chem), 48. *Prof Exp:* Sr res scientist, Rohm and Haas Co, 48-62; sect head org & polymers, J T Baker Chem Co, 62-64; sect head coatings res, Gen Tire & Rubber Co, 64-73; supvr radiation curable coatings, Thiokol Corp, 74-78; mgr res, 79-81, PRIN SCIENTIST, CONGOLEUM CORP, 81- *Mem:* Am Chem Soc; Soc Plastics Engrs; Nat Asn Corrosion Engrs; Soc Mfg Engrs. *Res:* Aqueous and high solids coatings; radiation curable coatings and inks; acrylic monomers; acrylated urethanes; polyurethane and polyvinyl chloride coatings and foams. *Mailing Add:* 95 Jacobs Creek Rd West Trenton NJ 08628

SCHMIDLIN, ALBERTUS ERNEST, b Paterson, NJ, Apr 9, 17; m 43; c 3. TRANSPORTATION CONTAINER SAFETY, AEROSPACE HYDRAULICS & PNEUMATICS. *Educ:* Stevens Inst Technol, ME, 39, MS, 41, DSc(mech eng), 62. *Prof Exp:* Instr mech eng lab, Stevens Inst Technol, 39-41; develop engr, Walter Kidde & Co, Inc, 41-47, proj engr, 47-51, chief proj engr, 51-54, asst mgr develop dept, 54-56, mgr res dept, 56-62, assoc tech dir, 62-63; prin staff scientist & mgr fluidics dept, Singer-Gen Precision Inc, 63-69; indust consult, 69; sr res scientist, Fire Support Armament Ctr, Safety, Security & Survivability Br, US Army Armament Res Develop & Eng Ctr, 85-89; CONSULT ENGR, 90- *Concurrent Pos:* Fel, Am Soc Mech Engrs. *Honors & Awards:* Dedicated Serv Award, Am Soc Mech Engrs. *Mem:* Fel Am Soc Mech Engrs; Soc Automotive Engrs; Am Defense Preparedness Asn. *Res:* Fluid mechanics, heat transfer and mechanical design as related to safety, navigation, guidance and control of aerospace vehicles; internal flow in fluid components; gas dynamics; transient flow; fluid power transmission. *Mailing Add:* 28 Highview Rd Caldwell NJ 07006

SCHMIDLIN, FREDERICK W, b Maumee, Ohio, Aug 28, 25; m 59; c 3. SOLID STATE PHYSICS. *Educ:* Univ Toledo, BEEP, 50; Cornell Univ, PhD(physics), 56. *Prof Exp:* Asst, Cornell Univ, 50-54; mem tech staff, Ramo-Wooldridge Corp, 56-58; sr mem tech staff, Space Technol Labs, 58-60; sr scientist, Gen Technol Corp, 60-63; PRIN SCIENTIST, XEROX CORP, 63- *Mem:* AAAS; Soc Photog Scientists & Engrs; Am Phys Soc; Inst Elec & Electronics Engrs. *Res:* Theory of electrophotography; solid state theory; superconductivity; semiconductivity; photoconductivity; thin film devices; electrostatic printing. *Mailing Add:* Xerox Corp Webster Res Ctr Rochester NY 14644

SCHMIDLY, DAVID JAMES, b Lubbock, Tex, Dec 20, 43; m 66; c 2. MAMMALIAN SYSTEMATICS, NON-GAME WILDLIFE. *Educ:* Tex Tech Univ, BS, 66, MS, 68; Univ Ill, PhD(zool), 71. *Prof Exp:* From asst prof to assoc prof, 71-82, PROF WILDLIFE & FISHERIES SCI, TEX A&M UNIV, 82-, HEAD, DEPT WILDLIFE & FISHERIES, 86- *Concurrent Pos:* Consult, Wildlife Servs. *Mem:* Am Soc Mammalogists; Soc Conserv Biol; Sigma Xi; Southwestern Asn Naturalists. *Res:* Mammalian systematics, natural history and management with special emphasis on non-game mammals from the southwestern United States and northern Mexico; preservation management and utilization of biological collections. *Mailing Add:* Dept Wildlife & Fisheries Sci Tex A&M Univ College Station TX 77843

SCHMID-SCHOENBEIN, GEERT W, b Ebingen, WGer, Jan 1, 48; m 76; c 3. MICROCIRCULATION, BIOMECHANICS. *Educ:* Univ Calif, San Diego, MS, 73, PhD(bioeng), 76. *Prof Exp:* Sr staff assoc physiol, Columbia Univ, 76-79; from asst prof to assoc prof, 79-89, PROF BIOENG, UNIV CALIF, SAN DIEGO, 89- *Concurrent Pos:* Mem, Biomed Sci Grad Prog, Univ Calif, San Diego. *Honors & Awards:* Malphigi Award, Europ Soc Microcirculation, 80; Abott Award, 84; Melville Medal, Am Soc Mech Engrs,

90. *Mem:* Am Physiol Soc; Am Microcirculatory Soc; Biomed Eng Soc; Int Soc Biorheology; Europ Soc Mirocirculation; Am Heart Asn; Am Diabetes Asn; Am Acad Mech. *Res:* Biomechanics; microcirculation; bioengineering of cardiovascular diseases; math modeling. *Mailing Add:* Dept Appl Mech & Eng Sci-Bioeng Univ Calif San Diego B 109 La Jolla CA 92093-0412

SCHMIDT, ALAN FREDERICK, b Chicago, Ill, Mar 21, 25. LIQUIFIED NATURAL GAS TECHNOLOGY. *Educ:* Ill Inst Technol, BS, 51; Univ Colo, MS, 53. *Prof Exp:* Proj engr, Cryogenics Div, Nat Bur Standards, 52-63, consult staff, 63-76; CONSULT ENGR, ALAN F SCHMIDT, CONSULT-CRYOGENICS ENG, 76- *Mem:* Am Soc Mech Engrs. *Res:* Design and development of prototype land-based, aircraft-bourne and rocket-bourne cryogenic tankage and equipment; investigation of cryogenic fluid phenomena involving the pressurization, stratification, cooling, heating and flow of such fluids. *Mailing Add:* PO Box 3097 Pinewood Springs Lyons CO 80540

SCHMIDT, ALEXANDER MACKAY, internal medicine; deceased, see previous edition for last biography

SCHMIDT, ALFRED OTTO, b Mogilno, Ger, May 12, 06; nat US; m 41; c 1. MACHINE & TOOL DESIGN. *Educ:* Ilmenau Sch Eng, ME, 28; Univ Mich, MSE, 40, DSc(metal processing), 43. *Prof Exp:* Mech engr, Carl Zeiss Optical Works, 29-38; from instr to asst prof mech eng, Colo State Col, 40-42; assoc, Univ Ill, 42-43; res engr, Kearney & Trecker Corp, 43-61; res prof mech eng, Marquette Univ, 58-63; prof mach tool tech, Univ Roorkee, 64 & 81; prof, 64-71, EMER PROF INDUST ENG, PA STATE UNIV, 71- *Concurrent Pos:* Adv, UN Indust Develop Orgn, Israel, 67, Pakistan, 71, Kenya, 71, Brazil, 72, Srilanka, 77 & Argentina, 78 & 79; vis prof, Univ RI, 74-75 & Univ Wis-Milwaukee, 75-76, Korean Advan Inst Sci, 79, Univ Busan, Korea, 80; consult, Metal Indust Develop Centre, Taichung, Taiwan, Repub China, 89. *Honors & Awards:* Gold Medal, Soc Mfg Eng, 59. *Mem:* Am Soc Mech Engrs; Soc Mfg Engrs. *Res:* Metal processing; machine tool design and utilization; production; metal cutting calorimeter; carbide milling cutter; research surveys; laser applications. *Mailing Add:* 207 Hammond Bldg Pa State Univ University Park PA 16802

SCHMIDT, ANTHONY JOHN, b Winnipeg, Man, May 11, 27; m 55, 83; c 4. EMBRYOLOGY. *Educ:* Univ Wash, BA, 52, MSc, 54; Princeton Univ, PhD(biol), 57. *Prof Exp:* Instr biol, Princeton Univ, 57-58; from asst prof to prof anat, Univ Ill Med Ctr, 58-74; PROF ANAT & CHMN DEPT, PRESBY-ST LUKE'S MED CTR, RUSH MED COL, 75- *Mem:* AAAS; Am Asn Anat; Am Asn Clin Anat. *Res:* Chemistry of the cellular progression in organ development and regeneration; limb regeneration in amphibia; comparative anatomy; histology; neuroanatomy; experimental embryology; human gross anatomy. *Mailing Add:* Dept Anat Rush Med Col 600 S Paulina St Chicago IL 60612

SCHMIDT, ARTHUR GERARD, b Chicago, Ill, Jan 15, 44; m 72; c 5. COMPUTER MODELING OF PHYSICAL SYSTEMS. *Educ:* Depaul Univ, BS, 66; Univ Notre Dame, PhD, 74. *Prof Exp:* Instr physics, LaLumiere Sch, 72-73, physics, chem & biol, 73-74; fel, Oak Ridge Nat Lab, 74-76; instr physics, biol & chem, LaLumiere Sch, 76-78; asst prof physics, Lake Forest Col, 78-84; SR LECTR & DIR UNDERGRAD LABS, NORTHWESTERN UNIV, 84- *Concurrent Pos:* Guest fac fel, Notre Dame Univ, 76-78; instr physics, Ind Univ, South Bend, 77- *Mem:* Am Phys Soc; AAAS; Sigma Xi. *Res:* Measurement of triple admixtures in x-ray techniques in radioactive metals using xx angular correlation techniques; designing and fabrication of lecture demonstrations; writing progs for computers to simulate physical systems; nuclear physics. *Mailing Add:* CAS Dept Physics & Astron Northwestern Univ Evanston IL 60208

SCHMIDT, BARBARA A, ULTRASTRUCTURE, INSECT DEVELOPMENT. *Educ:* Northwestern Univ, PhD(biol), 76. *Prof Exp:* AT KANS STATE UNIV, 77- *Res:* Cell developmental biology. *Mailing Add:* Dept Teacher Prep Calif State Univ 6000 JST Sacramento CA 95819

SCHMIDT, BARNET MICHAEL, b New Milford, NJ, June 30, 58. DIGITAL SIGNAL PROCESSING, ADAPTIVE & STATISTICAL SIGNAL IDENTIFICATION. *Educ:* Stevens Inst Technol, BS, 80. *Prof Exp:* Res assoc chem, X-ray Diffraction Lab, Stevens Inst Technol, 79-80; sr engr, Timeplex Div, Open Transport Networks Eng Lab, Unisys Corp, 81-86; consult mem tech staff, Network Technol Lab, AT&T Bell Labs, 86-90, MEM TECH STAFF, NETWORK SYSTS ANALYSIS LAB, BELL COMMUN RES, INC, 90- *Concurrent Pos:* Sr mem tech staff, Div Commun Systs Develop, Computer Sci Corp Prof Consult Serv, 86-89. *Mem:* Inst Elec & Electronics Engrs; Asn Comput Mach. *Res:* Develop robust, fault-tolerant intelligent communications networks; apply the theory of digital signal processing and statistical signal identification to locating faults in transmission systems to insure the integrity of the international long distance network; one US patent. *Mailing Add:* 494 Reis Ave Oradell NJ 07649-2624

SCHMIDT, BERLIE LOUIS, b Treynor, Iowa, Oct 2, 32; m 54, 86; c 5. AGRONOMY, SOIL CONSERVATION. *Educ:* Iowa State Univ, BS, 54, MS. 59, PhD(agron), 62. *Prof Exp:* Soil scientist, Soil Conserv Serv, USDA, 54, 56-57; res assoc soil mgt, Iowa State Univ, 59-62; from asst prof to prof agron, Ohio State Univ, 62-87, assoc chmn, Ohio Agr Res & Develop Ctr, 69-75, chmn dept, 75-86; SOIL SCIENTIST, US DEPT AGR, COOP STATE RES SERV, 87- *Mem:* Am Soc Agron; Soil Sci Soc Am; Soil Conserv Soc Am; Int Soil Sci Soc; Sigma Xi. *Res:* Soil and water pollution from erosion and runoff; soil and water conservation; wind erosion control; water infiltration into soils; agronomic research administration. *Mailing Add:* US Dept Agr Coop State Res Serv/NRFSS Washington DC 20250-2200

SCHMIDT, BRUNO (FRANCIS), b Strawberry Point, Iowa, June 10, 42; m 64; c 2. PHYSICS. *Educ:* Cornell Col, BA, 64; Iowa State Univ, PhD(physics), 69. *Prof Exp:* Asst prof, 69-75, assoc prof, 75-81, prof physics, 81-84, PROF COMPUT SCI, SOUTHWEST MO STATE UNIV, 84- *Mem:* Asn Comput Mach; Sigma Xi. *Mailing Add:* Dept Physics Southwest Mo State Univ 901 S National Springfield MO 65804

SCHMIDT, CHARLES WILLIAM, b St Petersburg, Fla, Aug 11, 42; m 71; c 2. PHYSICS, ACCELERATOR TECHNOLOGY. *Educ:* Fla State Univ, BS, 64. *Prof Exp:* Sci asst physics, Argonne Nat Lab, 64-69; PHYSICIST ACCELERATOR PHYSICS, FERMI NAT ACCELERATOR LAB, 69-, HEAD, LINAC DEPT FERMILAB, 89- *Concurrent Pos:* Vis scientist, Ger Electron-Synchrotron, Hamburg, WGer, 84, Stanford, Calif, 86. *Mem:* Am Phys Soc; Sigma Xi. *Res:* Accelerator physics especially ion source development, magnetic measurements and operation. *Mailing Add:* 60 Oakwood Dr Naperville IL 60540

SCHMIDT, CLAUDE HENRI, b Geneva, Switz, May 6, 24; nat US; m 53; c 2. ENTOMOLOGY. *Educ:* Stanford Univ, BA, 48, MA, 50; Iowa State Univ, PhD(entom), 56. *Prof Exp:* Instr zool, Iowa State Univ, 55-56; med entomologist, USDA, 56-62; entomologist, Div Isotopes, Int Atomic Energy Agency, Vienna, Austria, 62-64; proj leader insect physiol & metab sect, Metab & Radiation Res Lab, 64-67, chief insects affecting man & animal res br, Entom Res Div, 67-72, dir, Dakotas-Alaska Area, 72-81, dir, Dakotas Area, N Cent Region, 81-82, dir metals & radiation res lab, 82-88, COLLABR, AGR RES SERV, USDA, 89- *Concurrent Pos:* Res consult, USAID, 68; consult, US Army Med Res & Develop Cmnd, 73-77, vol, 89- *Mem:* Fel AAAS; Am Mosquito Control Asn (pres, 81); Am Entom Soc; Am Inst Biol Sci; Am Chem Soc. *Res:* Application of radioisotopes and radiation in entomology; application of the sterile male technique. *Mailing Add:* 1827 Third St N Fargo ND 58102

SCHMIDT, CLIFFORD LEROY, b Los Angeles, Calif, Mar 27, 26; m 49; c 1. BOTANY, ECOLOGY. *Educ:* San Jose State Col, AB, 55, MA, 58; Stanford Univ, PhD(pop biol), 67. *Prof Exp:* Teacher, San Jose Unified Sch Dist, 56-60; asst prof natural sci, 60-65, from asst prof to prof biol, 65-90, EMER PROF BIOL, SAN JOSE STATE UNIV, 90- *Mem:* AAAS; Am Inst Biol Sci; Am Soc Plant Taxonomists. *Res:* Biosystematic studies involving chromosomal analysis of Ludwigia sect Dantia; human impact studies on alpine vegetation; taxonomic studies on Ceanothus; revegetation of disturbed sites. *Mailing Add:* 1745 Skyway St S Salem OR 97302

SCHMIDT, DAVID KELSO, b Lafayette, Ind, Mar 4, 43; c 3. AEROSPACE ENGINEERING, SYSTEMS ENGINEERING. *Educ:* Purdue Univ, BS, 65, PhD(eng), 72; Univ Southern Calif, MS, 68. *Prof Exp:* Engr & scientist missile design, McDonnell-Douglas Astronaut Corp, 65-69; vis asst prof aerospace eng, Purdue Univ, 72-73; res engr transp eng, Stanford Res Inst, 73-74; asst prof, 74-79, ASSOC PROF AEROSPACE ENG, PURDUE UNIV, 79- *Mem:* Am Inst Aeronaut & Astronaut; Opers Res Soc Am; Am Soc Eng Educ. *Res:* Systems analysis; optimization; control theory and applications; manual control; operations research; flight vehicle dynamics and control. *Mailing Add:* Sch Aeronaut & Astronaut Purdue Univ Grissom Hall West Lafayette IN 47907

SCHMIDT, DENNIS EARL, b Plymouth, Wis, Jan 23, 40; m 71; c 3. PSYCHOPHARMACOLOGY. *Educ:* Lakeland Col, BS, 62; Kans State Univ, PhD(biochem), 68. *Prof Exp:* Instr pharmacol, Mt Sinai Sch Med, 69-70; instr, 70-71, asst prof, 71-78, ASSOC PROF PHARMACOL, VANDERBILT UNIV, 79- *Concurrent Pos:* NIH fels, Cornell Med Col, 68 & Mt Sinai Sch Med, 68-69; Smith Kline & French fel, Vanderbilt Univ, 70-72. *Mem:* Am Soc Pharmacol & Exp Therapeut; Soc Neurosci; Neurochem Soc. *Res:* Investigation of central cholinergic mechanisms. *Mailing Add:* Dept Pharmacol Vanderbilt Univ 1601 23rd Ave S Nashville TN 37212

SCHMIDT, DONALD ARTHUR, b Wis, Jan 29, 22; m 51; c 3. VETERINARY PATHOLOGY, VETERINARY CLINICAL PATHOLOGY. *Educ:* Univ Wis, BS, 44; Mich State Univ, DVM, 47, PhD, 61; Univ Minn, MS, 50; Am Col Vet Pathologists, dipl. *Prof Exp:* Veterinarian, Chicago Zool Park, Ill, 50-53; instr vet path, Mich State Univ, 53-63, assoc prof path, 63-67, PROF VET PATH, UNIV MO, COLUMBIA, 67- *Mem:* Am Vet Med Asn; Am Col Vet Path. *Res:* Food producing animals. *Mailing Add:* Dept Vet Path Univ Mo W 213 Vet Med Bldg Columbia MO 65201

SCHMIDT, DONALD DEAN, b Highland. Ill, Dec 21, 42; m 64; c 2. PHYSICAL INORGANIC CHEMISTRY. *Educ:* Wabash Col, AB, 65; Ore State Univ, PhD(chem), 70. *Prof Exp:* Res chemist, Dow Chem Co, 70-84; RES SUPVR, AMOCO PROD CO, TULSA, OK, 84- *Concurrent Pos:* chmn, Tulsa sect, Am Chem Soc; chmn elect API Comt Drilling Fluids. *Mem:* Am Chem Soc; Soc Petrol Engrs. *Res:* Polymer chemistry; rheological behavior and colloidal properties of clays; coordination chemistry; surface chemistry; industrial chemistry; drilling fluids. *Mailing Add:* Amoco Prod Co PO Box 3385 Tulsa OK 74102

SCHMIDT, DONALD HENRY, b Rhinelander, Wis, July 20, 35; m 65; c 4. CARDIOVASCULAR DISEASE, INTERNAL MEDICINE. *Educ:* Univ Wis, BS, 57, MD, 60. *Prof Exp:* Asst med, Col Physicians & Surgeons, Columbia Univ, 66-67, instr, 67-68, asst prof clin med, 69-70, asst prof, 70-74; assoc prof, 74-77, PROF MED, UNIV WIS-MILWAUKEE, 77- *Concurrent Pos:* Asst vis prof, Harlem Hosp Ctr, 68, dir EKG, 70; asst dir, Cardiovasc Lab, Columbia Presby Hosp, New York, 70, asst attend physician, 71-, dir, 73-; head, cardiovasc sect & attend physician, Mt Sinai Med Ctr, 74- *Mem:* Nuclear Med Soc; Harvey Soc; Am Fedn Clin Res; Cent Soc Clin Res; Am Heart Asn. *Res:* Physiology of the coronary circulation; nuclear cardiology. *Mailing Add:* Mt Sinai Med Ctr Box 342 Milwaukee WI 53201

SCHMIDT, DONALD L, b Park Falls, Wis, Jan 29, 30; m 62; c 3. MATERIALS SCIENCE, CHEMISTRY. *Educ:* Wis State Univ, Superior, BSc, 52; Okla State Univ, MSc, 54. *Prof Exp:* Proj engr, Mat Lab, 54-57, mat engr, 57, aeronaut struct mat engr, 57-59, aeronaut struct mat res engr, 59-60, res mat engr mat cent, 60, supvry mat res engr directorate mat, 60-61, tech mgr, Thermal Proj Mat, Air Force Wright Aeronaut Labs, Wright-Patterson AFB, 61-87; mem, Advan Missile Mat Res Coun, US Dept Defense, 70-87; INT MAT CONSULT, 87- *Mem:* Sigma Xi; Soc Aerospace Mat & Process Engrs. *Res:* Thermal protection materials; high temperature materials sciences, including composites, plastics and glass. *Mailing Add:* 1092 Lipton Lane Dayton OH 45430

SCHMIDT, DONALD L, b Sept 10, 31; US citizen; m 63; c 3. ORGANIC POLYMER CHEMISTRY. *Educ:* Univ Utah, BS, 56, PhD(org chem), 62. *Prof Exp:* Res chemist, ARPA Lab, Dow Chem Co, 61-63, proj leader aluminum chem, 63-67, sr res chemist, SPL Lab, 67-68, sr res chemist, 68-78, res assoc, Phys Res Lab, 78-84, ASSOC SCIENTIST, DOW CHEM CO, 84- *Honors & Awards:* IR 100 Awards, Indust Res Mag, 69 & 72; A K Doolittle Award, Am Chem Soc, 75. *Mem:* AAAS; Am Chem Soc; Sigma Xi. *Res:* Metal hydride chemistry; inorganic polymer containing Al-O-P bonds; condensation polymerization involving cyclic sulfonium compounds; theory of surfactants and foams; membrane science. *Mailing Add:* Cent Res 1712 Dow Chem Co Midland MI 48640

SCHMIDT, DWIGHT LYMAN, b Fond du Lac, Wis, May 30, 26; m 58; c 2. GEOLOGY. *Educ:* Univ Wash, BS, 54, MS, 57, PhD(geol), 61. *Prof Exp:* GEOLOGIST, US GEOL SURV, 61- *Concurrent Pos:* Mem comt polar res, Nat Acad Sci. 68-70. *Mem:* AAAS; Geol Soc Am; Mineral Soc Am; Soc Econ Geol; Am Geophys Union. *Res:* Radioactive placer deposits and petrography of the Idaho batholith; geology of Pensacola and Lassiter Coast Mountains, Antarctica; Precambrian and Cenozoic geology of Saudi Arabia; geologic history of the Red Sea; geology of carbonate aquifers of southern Nevada. *Mailing Add:* US Geol Surv Bldg 25 Fed Ctr MS-913 Denver CO 80225

SCHMIDT, ECKART W, b Essen, Ger, Apr 16, 35; m 62; c 2. INDUSTRIAL CHEMISTRY, FUEL SCIENCE. *Educ:* Univ Marburg, BS, 58; Univ Tubingen, Dr rer nat(org chem, astron), 64. *Prof Exp:* Res chemist, Ger Res Inst Aero & Astronaut, 64-66; mgr chem res, 66-78, SR STAFF SCIENTIST, ROCKET RES CO, 66- *Concurrent Pos:* Consult, hazardous mat, 87- *Mem:* Am Inst Aeronaut & Astronaut; Planetary Soc. *Res:* High energy rocket propellants; explosives; fuel technology; astronautics; safety systems; extraterrestrial preparation of rocket propellants; planeto chemistry; energy storage and conversion. *Mailing Add:* Rocket Res Co PO Box 97009 Redmond WA 98073-9709

SCHMIDT, EDWARD GEORGE, b Cut Bank, Mont, Dec 13, 42; m 88; c 7. ASTRONOMICAL PHOTOMETRY VARIABLE STARS. *Educ:* Univ Chicago, BS, 65; Australian Nat Univ, PhD(astron), 70. *Prof Exp:* Res assoc astron, Univ Ariz, 70-72; sr res fel, Royal Greenwich Observ, 72-74; from asst prof to assoc prof, 74-80, PROF ASTRON, UNIV NEBR, LINCOLN, 80- *Mem:* Am Astron Soc; Int Astron Union; Royal Astron Soc. *Res:* Variable stars; application of CCD's to astronomical photometry; automation of instrumentation and data analysis; astronomical instrumentation. *Mailing Add:* Dept Physics & Astron Univ Nebr Lincoln NE 68588

SCHMIDT, EDWARD MATTHEWS, b Elgin, Ill, Mar 24, 33; m 59; c 2. BIOENGINEERING. *Educ:* Northwestern Univ, BSEE, 56; Purdue Univ, MSEE, 57, PhD, 65. *Prof Exp:* Asst, Argonne Nat Lab, 53-55; sr res engr, Borg-Warner Res Ctr, 57-61; from instr to assoc prof elec eng & vet anat, bioeng, Purdue Univ, West Lafayette, 61-72; SR STAFF SCIENTIST, LAB NEURAL CONTROL, NAT INST NEUROL, DIS & STROKE, 72- *Concurrent Pos:* Spec fel, Lab Neurol Control, NIH, Bethesda, 69-71. *Mem:* Inst Elec & Electronics Engrs; Soc Neurosci; Am Physiol Soc. *Res:* Biological control systems; neurophysiology; microelectrode study of cells in the motor cortex of the monkey and their relationship to limb movements; neuroprosthesis. *Mailing Add:* Lab Neural Control Nat Inst Neurol Dis Commun Dis & Stroke NIH Bldg 36 Rm 5A 31 Bethesda MD 20892

SCHMIDT, FRANCIS HENRY, b Cincinnati, Ohio, Aug 6, 41; m 68; c 1. PHYSICAL ORGANIC CHEMISTRY. *Educ:* Xavier Univ, BS, 63, MS, 65; Ind Univ, Bloomington, PhD(org chem), 74. *Prof Exp:* res chemist, 69-80, SR RES CHEMIST, E I DU PONT DE NEMOURS & CO, 80- *Mem:* Am Chem Soc. *Res:* The process development of finishes for natural and synthetic fibers and components for photo-sensitive systems. *Mailing Add:* 1632 Chippewa Ct Grove City OH 43123-2699

SCHMIDT, FRANK W(ILLIAM), b Madison, Wis, Mar 16, 29; m 54; c 4. MECHANICAL ENGINEERING. *Educ:* Univ Wis, BS, 50, PhD(mech eng), 59; Pa State Univ, MS, 52. *Prof Exp:* Engr, Kuchler-Huhn Corp, Philadelphia, 52-54; from instr to asst prof mech eng, Univ Wis, 56-62; from asst prof to assoc prof, 62-70, PROF MECH ENG, PA STATE UNIV, UNIVERSITY PARK, 70- *Concurrent Pos:* Nat Sci fel, Imp Col, Univ London, 65-66. *Mem:* Fel Am Soc Mech Engrs; Sigma Xi. *Res:* Fluid mechanics; heat and mass transfer. *Mailing Add:* Dept Mech Eng Pa State Univ University Park PA 16802

SCHMIDT, FRED HENRY, physics; deceased, see previous edition for last biography

SCHMIDT, FREDERICK ALLEN, b Cincinnati, Ohio, Dec 26, 30; m 51; c 4. MATERIAL SCIENCE, METALLURGY. *Educ:* Xavier Univ, BS, 51. *Prof Exp:* Jr chemist, 51-54, jr res assoc, 54-56, assoc, 56-59, assoc metallurgist, 59-71, metallurgist, 71-76, SR METALLURGIST, AMES LAB, IOWA STATE UNIV, 76-, DIR, MAT PREP CTR, 81- *Honors & Awards:* Serv Award, Am Inst Aeronaut & Astronaut; Significant Implication for Energy Related Technol in Metall & Ceramics, US Dept Energy, 87. *Mem:* Am Soc Metals; Am Inst Mining, Metall & Petrol Engrs; Am Inst Aeronaut & Astronaut. *Res:* Methods for preparing high purity refractory metals and thin film solar cells; electrotransport, thermotransport and diffusion of solutes in metals; characterization of refractory metals and alloys. *Mailing Add:* 121 Metals Develop Iowa State Univ Ames IA 50011-3020

SCHMIDT, GEORGE, b Budapest, Hungary, Aug 1, 26; US citizen; m 55; c 2. PHYSICS. *Educ:* Budapest Tech Univ, dipl eng, 50; Hungarian Acad Sci, PhD(physics), 56. *Hon Degrees:* MEng, Stevens Inst Technol, 66. *Prof Exp:* Res assoc physics, Cent Res Inst Physics, Hungary, 55-56; sr lectr, Israel Inst Technol, 57-58; res assoc, 58-59, from asst prof to assoc prof, 59-65, PROF PHYSICS, STEVENS INST TECHNOL, 65- *Concurrent Pos:* Consult, Grumman Aircraft Eng Co, 62, UK Atomic Energy Auth, 65-66 & French AEC, 66; vis prof, Univ Wis, 65, Univ Calif, Los Angeles, 72-73; consult, Exxon Corp, 71, Cornell Univ, 78-79, Appl Sci Inc, 81, Polytechnic Inst NY, 84 & Berkeley Assocs, 85; vis scientist, Polytech, France, 79-80; George Meade Bond Prof Physics & Eng Physics, 83. *Honors & Awards:* Ottens Res Award, 61. *Mem:* NY Acad Sci; fel Am Phys Soc. *Res:* Plasma physics; chaos theory. *Mailing Add:* Dept Physics Stevens Inst Technol Hoboken NJ 07030

SCHMIDT, GEORGE THOMAS, b Jersey City, NJ. SYSTEMS ENGINEERING. *Educ:* Mass Inst Technol, BS, 64, MS, 65, ScD(instrumentation), 70. *Prof Exp:* DIV LEADER, DECISION & CONTROL SYSTS DIRECTORATE, DRAPER LAB, 62- *Concurrent Pos:* Assoc prof systs elec & comput eng, Boston Univ, 71-74, adj prof, 65-70 & 75-82; mem adv group, Aerospace Res & Develop & Guidance & Control Panel, NATO, 87-90; lectr, Aeronaut & Astronaut, Mass Inst Technol, 90- *Mem:* Am Inst Aeronaut & Astronaut; Inst Elec & Electronics Engrs; Am Soc Eng Educ; Sigma Xi. *Res:* Control and estimation theory; guidance and navigation systems; guidance, navigation and control of systems engineering. *Mailing Add:* Div Leader MS No 2A Draper Lab Cambridge MA 02139

SCHMIDT, GERALD D, b Greeley, Colo, Mar 12, 34; m 58; c 2. ZOOLOGY, PARASITOLOGY. *Educ:* Colo State Col, AB, 60; Colo State Univ, MS, 62, PhD(zool), 64. *Prof Exp:* Instr zool, Univ Mont, 63-64; from asst prof to assoc prof, 64-72, PROF PARASITOL, UNIV NORTHERN COLO, 72-, PROF ZOOL, 74- *Concurrent Pos:* NATO sr fel sci, South Australian Mus, Adelaide, 70-71. *Mem:* Am Soc Parasitologists; Am Micros Soc; Wildlife Dis Asn; Am Soc Trop Med & Hyg; Int Filariasis Asn; Sigma Xi. *Res:* Morphology, biology and taxonomy of parasitic helminths. *Mailing Add:* Dept Biol Univ Northern Colo Greeley CO 80631

SCHMIDT, GILBERT CARL, b Cincinnati, Ohio, Apr 15, 21; m 66; c 3. PSYCHIATRY, SCIENCE ADMINISTRATION. *Educ:* Univ Cincinnati, BS, 46, MS, 49, PhD(biochem), 51. *Prof Exp:* Res assoc, Sperti, Inc, 46-48; asst prof chem, Col Pharm, Univ Cincinnati, 51-52, from assoc prof to prof pharmacog, 52-56, prof biol sci, 56-66, asst prof biochem, Col Med, 59-66; assoc prof biochem & asst prof pediat, Med Col, Univ Ala, 66-68; prof biochem, Northeast La State Univ, 68-70; prof pharmacog & biol, 70-81, PROF PHARMACUET SCI & ASST DEAN GRAD STUDIES & RES, COL PHARM, MED UNIV SC, 81- *Concurrent Pos:* USPHS fel chem, Col Pharm, Univ Cincinnati, 51-52, NSF fel, 52-53; dir res labs, Children's Hosp, Birmingham, Ala, 66-68. *Mem:* AAAS; Am Chem Soc; Am Soc Microbiol; Am Pharmaceut Asn; Am Inst Chemists; Soc Heterocyclic Chem. *Res:* Drug metabolism; biochemical pharmacology; azolesterases; synthesis isotopically labeled drugs; pineal gland metabolism; drug abuse; isolation and properties of natural products. *Mailing Add:* Col Pharm Med Univ SC Charleston SC 29425

SCHMIDT, GLEN HENRY, b Manning, Iowa, Jan 25, 31; m 52; c 5. DAIRY HUSBANDRY. *Educ:* Iowa State Univ, BS, 52; Cornell Univ, MS, 56, PhD(dairy sci), 58. *Prof Exp:* Asst animal sci, NY State Col Agr, Cornell Univ, 54-57, from instr to prof, 57-74, chmn dept, 74-84; prof dairy sci, 74-90, CHMN DEPT ANIMAL SCI, OHIO STATE UNIV, 91- *Honors & Awards:* MSD Ag Vet Award, Am Dairy Sci Asn. *Mem:* Am Dairy Sci Asn; fel AAAS; Coun Agr Sci & Technol. *Res:* Dairy management. *Mailing Add:* Ohio State Univ 2029 Fyffe Rd Columbus OH 43210

SCHMIDT, GLENN ROY, b Two Rivers, Wis, Feb 26, 43; m 66; c 1. MEAT SCIENCES. *Educ:* Univ Wis, BSc, 65, MSc, 68, PhD(animal sci), 69. *Prof Exp:* Vis scientist meat technol, Meat Indust Res Inst NZ, 69-70; vis scientist animal sci, Res Inst Animal Husb, Zeist, Neth, 70-71; from asst prof to assoc prof, Univ Ill, Urbana, 71-79; PROF ANIMAL SCI, COLO STATE UNIV, 79- *Mem:* Am Soc Animal Sci; Am Meat Sci Asn; Inst Food Technol; Sigma Xi. *Res:* Processed meat technology; swine physiology; muscle biochemistry. *Mailing Add:* Dept Animal Sci Colo State Univ Ft Collins CO 80523

SCHMIDT, GREGORY WAYNE, b Waterloo, Iowa, Mar 25, 47; m 70; c 1. CELL BIOLOGY, PHOTOSYNTHESIS. *Educ:* Grinnell Col, AB, 69; State Univ NY, Stony Brook, PhD(biol), 76. *Prof Exp:* Fel cell biol, Rockefeller Univ, 75-79; asst prof, 79-86, ASSOC PROF BOT, UNIV GA, 86- *Concurrent Pos:* Asst ed, Plant Physiol, 86- *Mem:* Am Soc Plant Physiologists; Am Soc Cell Biol; Int Soc Plant Molecular Biol. *Res:* Regulation of chloroplast biogenesis with emphasis on interactions between nuclear-cytoplasmic compartments and organelle in synthesis; maturation and assembly of subunits of photosynthetic complexes. *Mailing Add:* Dept Bot Univ Ga Athens GA 30602

SCHMIDT, HARTLAND H, b St Paul, Minn, Nov 22, 29; m 57; c 2. PHYSICAL CHEMISTRY. *Educ:* Univ Minn, BA, 51; Univ Calif, Berkeley, PhD(chem), 54. *Prof Exp:* From instr to assoc prof, Univ Calif, Davis & Univ Calif, Riverside, 54-68, PROF CHEM, UNIV CALIF, RIVERSIDE, 68- *Mem:* Am Chem Soc; Am Phys Soc; Sigma Xi. *Res:* Thermodynamics; statistical mechanics; critical phenomena. *Mailing Add:* Dept Chem Univ Calif Riverside CA 92521

SCHMIDT, HARVEY JOHN, JR, b Spokane, Wash, June 20, 41; m 69; c 1. MATHEMATICS. *Educ:* Lewis & Clark Col, BA, 63; Univ Ore, MA, 65, PhD(math), 69. *Prof Exp:* Asst prof, Ill State Univ, 69-74; asst prof, 74-76, ASSOC PROF MATH, LEWIS & CLARK COL, 76- *Concurrent Pos:* Vis fel math, Univ Warwick, 70-71. *Mem:* Am Math Soc; Math Asn Am. *Res:* Finite group theory; representation theory of finite groups. *Mailing Add:* Dept Math LC Box 111 Lewis & Clark Col 0615 SW Palatine Hill Rd Portland OR 97219

SCHMIDT, HELMUT, b Danzig, Ger, Feb 21, 28; m 55; c 3. PHYSICS, PARAPSYCHOLOGY. *Educ:* Univ Goettingen, MA, 53; Univ Cologne, PhD(physics), 54. *Prof Exp:* Asst prof physics, Univ Cologne, 54-55 & 58-59, docent, 60-63; vis lectr, Univ BC, 64-65; sr res physicist, Boeing Sci Res Lab, 66-69; res assoc, Inst Parapsychol, 69-70, dir, 70-73; RES ASSOC, MIND SCI FOUND, 74- *Concurrent Pos:* Nat Acad Sci fel, Univ Calif, Berkeley, 56-57; NATO exchange prof, Southern Methodist Univ, 62. *Mem:* Am Phys Soc; Parapsychol Asn. *Res:* Quantum physics; cosmology; solid state physics; study of parapsychological effects with modern electronic equipment. *Mailing Add:* 8301 Broadway No 100 San Antonio TX 78209-2006

SCHMIDT, JACK RUSSELL, b Milwaukee, Wis, July 23, 26; m 58. RESEARCH ADMINISTRATION. *Educ:* Univ Wis, BS, 48, MS, 50, PhD(med microbiol), 52. *Prof Exp:* Asst microbiol & immunol, Med Sch, Univ Wis, 49-52; virologist, Walter Reed Army Inst Res, 52-56; head, Viral Immunol Lab, Walter Reed Army Inst Res & Vet Admin Cent Lab Clin Path & Res, 56-57; head dept virol, US Naval Med Res Unit 3, 58-66, tech dir field facil, Ethiopia, 66-72, tech dir res & sci adv, Bur Med & Surg, Navy Dept, 72-74, dir progs & sci adv, Naval Med Res Develop Command, 74-83; chief, Int Coord Liaison Br, 84-88, actg dep dir, 88-89, DIR, SCHOLARS IN RESIDENCE PROG, FOGARTY INT CTR, NIH, 89- *Mem:* Am Soc Trop Med & Hyg; Royal Soc Trop Med & Hyg. *Res:* Ecology and epidemiology of arboviral infections; viral immunology; medical entomology and parasitology; tropical diseases. *Mailing Add:* Fogarty Int Ctr NIH Bldg 16 Rm 202 Bethesda MD 20892

SCHMIDT, JAMES L, pharmacology, industrial medicine, for more information see previous edition

SCHMIDT, JANE ANN, b Minneapolis, Minn, May 25, 51. ENZYMOLOGY, PROTEIN CHEMISTRY. *Educ:* Macalester Col, BA, 73; Iowa State Univ, PhD(biochem), 80. *Prof Exp:* Assoc, Dept Chem, Univ Iowa, 78-81; fel, Dept Chem, Univ Del, 81-86; assoc, Dept Biochem, Univ Minn, 86-88; SR RES & DEVELOP ASSOC, KALLESTAD DIAGNOSTICS, INC, 88- *Mem:* Am Chem Soc; AAAS; NY Acad Sci. *Res:* Modification of apo asparate aminotransferase with pyridoxal sulfate; modification of glutamate dehydrogenase with nucleotide affinity labels; kinetics and stereochemistry of decarboxylation reactions of serine hydroxymethylase; enzymology of signal transduction; immunodiagnostics research and development. *Mailing Add:* 430 Saratoga St S St Paul MN 55105-2545

SCHMIDT, JEAN M, b Waterloo, Iowa, June 5, 38. BACTERIOLOGY. *Educ:* Univ Iowa, BA, 59, MS, 61; Univ Calif, Berkeley, PhD(bact), 65. *Prof Exp:* NIH fel, Univ Edinburgh, 65-66; from asst prof to assoc prof, 66-79, PROF MICROBIOL, ARIZ STATE UNIV, 79- *Concurrent Pos:* NIH res grants, 67-70, 71-74; NSF res grant, 79-82. *Mem:* AAAS; Am Soc Microbiol; Brit Soc Gen Microbiol. *Res:* Cancer research: detection of new anticancer drugs using in vitro tumor cell assays; microbiol ultrastructure and differentiation. *Mailing Add:* Dept Microbiol Ariz State Univ Tempe AZ 85287

SCHMIDT, JEROME P, b Nortonville, Kans, Feb 1, 28; m 58. MEDICAL MICROBIOLOGY, BIOSAFETY. *Educ:* St Benedict's Col, Kans, BS, 49; Univ Kans, MA, 52; Univ NH, PhD(microbiol), 63. *Prof Exp:* Bacteriologist, Arctic Aeromed Lab, Fairbanks, Alaska, 56-61; chief infectious processes unit, 63-68, CHIEF MICROBIOL, USAF SCH AEROSPACE MED, 68- *Concurrent Pos:* Instr, Univ Alaska, 60-61; vis prof, Tex A&M Univ, 65- & NTex State Univ, 68-; mem Hibernation Info Exchange; chmn, Nat Acad Sci, subcomt Infectious Agts, 86-; mem, Am Bd Med Microbiol, 88- *Mem:* AAAS; Am Soc Microbiol; Am Biol Safety Asn (secy-treas, 85-); Soc Exp Biol & Med; fel Am Acad Microbiol; Sigma Xi. *Res:* Epidemiology and etiology of acute upper respiratory diseases; microbiological aspects of mammalian hibernation; effect of environmental factors on infectious processes. *Mailing Add:* USAF Sch Aerospace Med 6015 Woodwick San Antonio TX 78239

SCHMIDT, JOHN ALLEN, b Aberdeen, SDak, Dec 31, 40; m 68; c 1. PLASMA PHYSICS. *Educ:* SDak State Univ, BS, 62; Univ Wis, MS, 64, PhD(physics), 69. *Prof Exp:* Res assoc, Univ Wis, 69; res assoc physics, 69-78, co-head, Tokamac Fusion Test Reactor, 78-80, SR RES PHYSIST, PRINCETON UNIV, 78-, HEAD APPL PHYSICS DIV, PLASMA PHYSICS LAB, 80-, HEAD, BURNING PLASMA EXP PROJ. *Mem:* Fel Am Phys Soc. *Res:* Use of low plasma pressure toroidal magnetic field geometrics to confine plasmas for thermonuclear energy sources. *Mailing Add:* Plasma Physics Lab Princeton Univ Princeton NJ 08544

SCHMIDT, JOHN LANCASTER, b McPherson, Kans, Sept 19, 43; m 68; c 2. WILDLIFE ECOLOGY, WILDLIFE MANAGEMENT. *Educ:* Ottawa Univ, BA, 66; Colo State Univ, MS, 68, PhD(wildlife biol), 70. *Prof Exp:* Exten specialist wildlife, SDak State Univ, 70-72; exten prog leader, Colo State Univ, 72-75, asst dean natural resources, 75-78, assoc prof wildlife, 74-84; SR REGIONAL DIR, DUCKS UNLIMITED. *Concurrent Pos:* Consult, City Littleton, Colo, 73, Rogers Nagel Langhart Inc, 74-, Thorne Ecol Inst, 75-, Rocky Mountain Energy Co, Int Environ Consults & Dept Army. *Mem:* Wildlife Soc; Nat Wildlife Fedn; Nat Geog Soc. *Res:* Reintroduction of desert bighorn sheep in Colorado National Monument; feeding ecology at Cape Buffalo in Kruger National Park. *Mailing Add:* 1601 Cottonewood Pt Dr Ft Collins CO 80524

SCHMIDT, JOHN P, b Northhampton, Mass, Apr 17, 33; m 63; c 3. CHEMICAL ENGINEERING. *Educ:* Rensselaer Polytech Inst, BChE, 55; Mass Inst Technol, ScD(chem eng), 63. *Prof Exp:* Mem staff, Eng Dept, E I du Pont de Nemours & Co, Inc, 55-58; vpres res & develop, Halcon Int, Inc, 63-73; exec vpres, Oxirane Int, 73-80; sr vpres, Arco Chem, 80-85; CONSULT, 85- *Concurrent Pos:* Chmn bd, Acad Nat Sci Philadelphia. *Res:* Organic and petrochemical processing. *Mailing Add:* 11 Honey Lake Dr RD 2 Princeton NJ 08540

SCHMIDT, JOHN RICHARD, b Madison, Wis, July 3, 29; m 51; c 3. AGRICULTURAL ECONOMICS, OPERATIONS RESEARCH. *Educ:* Univ Wis, BS, 51, MS, 53; Univ Minn, PhD(agr econ), 60. *Prof Exp:* From asst prof to assoc prof, 56-65, chmn dept, 66-70, PROF AGR ECON, UNIV WIS-MADISON, 65-; DIR, NORTH CENT COMPUT INST, 81- *Concurrent Pos:* Consult, Am Farm Bur Fedn, 62, Bank Mex, 72-82, World Bank, 73-77 & Agr Develop Bank Iran, 74-75. *Mem:* Am Asn Agr Econ. *Res:* Economics of soil conservation; electronic farm records; computerized decision aids. *Mailing Add:* North Cent Comput Inst 666 WARF Off Bldg 610 Walnut St Madison WI 53706

SCHMIDT, JOHN THOMAS, b Louisville, Ky, Sept 25, 49; m 79; c 2. NEUROBIOLOGY, NEURAL CONNECTIONS. *Educ:* Univ Detroit, BS, 71; Univ Mich, PhD(biophysics & neurosci), 76. *Prof Exp:* Fel, Nat Inst Med Res, London, Eng, 76-77; fel, Anat Dept, Vanderbilt Univ, 77-80; asst prof, 80-85, ASSOC PROF BIOL SCI, SCH PUB HEALTH SCI, STATE UNIV NY, ALBANY, 85- *Concurrent Pos:* Prin investr grant, NIH, 81-; fel, Sloan Found, 81-85. *Mem:* Soc Neurosci; Asn Res Vision & Opthal. *Res:* Development and regeneration of retinotopic projections in the nervous system; role of activity in the stabilization of synaptic connections. *Mailing Add:* Dept Biol Sci State Univ NY 1400 Washington Ave Albany NY 12222

SCHMIDT, JOHN WESLEY, b Moundridge, Kans, Mar 13, 17; m 43; c 5. AGRONOMY. *Educ:* Tabor Col, BA, 47; Kans State Univ, MSc, 49; Univ Nebr, PhD(agron), 52. *Hon Degrees:* DSc, Kans State Univ, 84. *Prof Exp:* Assoc prof agron, Kans State Univ, 51-54; assoc agronomist, Univ Nebr, Lincoln, 54-62, prof agron, 62-80, Regents prof, 81-85; RETIRED. *Concurrent Pos:* NSF-US-Japan Coop Sci Prog vis scientist, Japan, 66. *Honors & Awards:* Crop Sci Award, 75. *Mem:* Fel Am Soc Agron; fel Crop Sci Soc Am; Am Genetic Asn. *Res:* Breeding, genetics and cytogenetics of wheat and related species and genera. *Mailing Add:* Dept Agron Univ Nebr Lincoln NE 68503-0915

SCHMIDT, JUSTIN ORVEL, b Rhinelander, Wis, Mar 23, 47; m 81; c 2. ENTOMOLOGY, BIOLOGY. *Educ:* Pa State Univ, BS, 69; Univ BC, MSc, 72; Univ Ga, PhD(entom), 77. *Prof Exp:* Fel biol, Univ NB, 77-78; res scientist, Univ Ga, 78-80; ENTOMOLOGIST & TOXINOLOGIST, BEE RES LAB, TUCSON, 80- *Concurrent Pos:* Adj prof, entom dept, Univ Ariz, Tucson, 82-84. *Mem:* Fel Royal Entom Soc; Entom Soc Am; Int Soc Toxinol; Int Soc Chem Ecol; AAAS; Animal Behav Soc; Sigma Xi. *Res:* Defensive behaviors of organisms; insect venoms; chemical ecology; insect physiology and biochemistry. *Mailing Add:* Carl Hayden Bee Lab 2000 E Allen Rd Tucson AZ 85719

SCHMIDT, KLAUS H, b Stuttgart, Ger, Oct 9, 28; m 55; c 1. RADIATION CHEMISTRY, PHYSICAL CHEMISTRY. *Educ:* Univ Tubingen, Dipl physics, 54; Univ Frankfurt, Dr phil nat(biophys), 60. *Prof Exp:* Res assoc biophys & radiation chem, Max Planck Inst Biophys, 60-63; resident res assoc radiation chem, 63-66, assoc physicist, 66-76, PHYSICIST, CHEM DIV, ARGONNE NAT LAB, 76- *Mem:* Radiation Res Soc. *Res:* Kinetics of radiation-induced chemical reactions with optical and electrical methods; developing techniques and equipment for radiation chemistry research; using radiation chemical techniques to study chemical mechanisms or structures. *Mailing Add:* 2815 SE 19th Pl Cape Coral FL 33904

SCHMIDT, KURT F, b New York, NY, Feb 25, 26; m 55; c 3. ANESTHESIOLOGY. *Educ:* Univ Munich, MD, 51. *Prof Exp:* Instr anesthesiol, Harvard Med Sch, 55; instr, Sch Med Yale Univ, 55-61, asst prof anesthesiol, 63-68; prof anesthesiol & chmn dept, Albany Med Col, 68-73; assoc prof anaesthesiol, Harvard Med Sch, 73-75; PROF ANESTHESIOL & CHMN DEPT, TUFTS-NEW ENG MED CTR, 75- *Concurrent Pos:* NIH spec fel neuropharmacol, 61-63. *Mem:* Am Soc Anesthesiol, AMA; NY Acad Sci; Int Anesthesia Res Soc; Asn Univ Anesthetists. *Res:* Neuropharmacology. *Mailing Add:* Dept Anesthesia Tufts-New Eng Med Ctr 171 Harrison Ave Boston MA 02111

SCHMIDT, LANNY D, b Waukegan, Ill, May 6, 38; m 62; c 2. CHEMICAL ENGINEERING, PHYSICAL CHEMISTRY. *Educ:* Wheaton Col, BS, 60; Univ Chicago, PhD(phys chem), 64. *Prof Exp:* Res assoc phys chem, Univ Chicago, 64-65; assoc prof, 65-69, PROF CHEM ENG, UNIV MINN, MINNEAPOLIS, 69- *Mem:* Am Phys Soc; Am Chem Soc; Am Vacuum Soc. *Res:* Surface chemistry and physics; adsorption; catalysis; electron microscopy; auger electron spectrometry; kinetics. *Mailing Add:* Dept Chem Eng & Mat Sci Univ Minn Minneapolis MN 55455

SCHMIDT, LEON HERBERT, experimental medicine; deceased, see previous edition for last biography

SCHMIDT, LOUIS VINCENT, aeronautical engineering, solid mechanics, for more information see previous edition

SCHMIDT, MAARTEN, b Groningen, Neth, Dec 28, 29; m 55; c 3. ASTRONOMY. *Educ:* Groningen, BSc, 49; Univ Leiden, PhD, 56. *Hon Degrees:* ScD, Yale, Univ, 66; Wesleyan Univ, 82. *Prof Exp:* Sci off, Leiden Observ, Neth, 53-59; assoc prof, Calif Inst Technol, 59-64, exec officer, 72-74, chmn, Div Physics, Math & Astron, 75-78, mem staff, Owens Valley Radio Observ, 70-78, prof astron, Calif Inst Technol, 64-81, mem staff, Hale Observ, 59-80, dir, 78-80, FRANCIS L MOSELEY PROF ASTRON ROBINSON LAB, CALIF INST TECHNOL, 81-; AT ROBINSON LAB, CALIF INST TECHNOL. *Concurrent Pos:* Carnegie fel, 56-58. *Honors & Awards:* Helen B Warner Prize, Am Astron Soc, 64; Watson Medal, Nat Acad Sci, 91. *Mem:* Foreign assoc Nat Acad Sci; Am Astron Soc; fel Am Acad Arts & Sci; Assoc Royal Astron Soc (London). *Res:* Structure, dynamics and evolution of the galaxy; radio astronomy; redshifts and cosmic distribution of quasars. *Mailing Add:* Dept Astron Calif Inst Technol 391 S Holliston Ave Pasadena CA 91125

SCHMIDT, MARK THOMAS, b New York, NY, Apr 17, 58. MUSEUM EDUCATION URBAN GEOLOGY. *Educ:* Ky State Univ, BS, 81 & MS, 86. *Prof Exp:* Instr earth sci, Cleveland Mus Nat Hist, 84-85, coordr, Sci Resource Ctr, 86-89; asst to dir educ, Am Geol Inst, 89-91; STAFF GEOLOGIST, WOODWARD-CLYDE CONSULT, 91- *Concurrent Pos:* Dir, Northeastern Ohio Sci & Eng Fair, 86-88; mem educ adv comt, Am Geol Inst, 87-88. *Mem:* Geol Soc Am; Nat Asn Geol Teachers; Nat Earth Sci Teachers Asn; Nat Sci Teachers Asn. *Res:* Rock types used for building and monuments; petrology of metasedimentary rocks from Taylor Valley, South Victoria Land, Antarctica. *Mailing Add:* 915 Aintree Park Dr Apt 204 Mayfield Village OH 44143

SCHMIDT, NATHALIE JOAN, virology, for more information see previous edition

SCHMIDT, NORBERT OTTO, b Highland Park, Mich, Nov 15, 25; m 53; c 5. CIVIL ENGINEERING, SOILS. *Educ:* US Mil Acad, BS, 49; Harvard Univ, MS, 55; Univ Ill, PhD(civil eng), 65. *Prof Exp:* From asst prof to assoc prof mil sci, Mo Sch Mines, 59-62; asst civil eng, Univ Ill, 63-65, from instr to asst prof, 65-66; from asst prof to assoc prof, 66-75, PROF CIVIL ENG, UNIV MO, ROLLA, 75- *Concurrent Pos:* Consult, tailings dams & recreational lake dams, 79- *Mem:* Am Soc Civil Engrs; Am Soc Testing & Mat; Nat Soc Prof Engrs; US Comn Large Dams. *Res:* Geotechnical engineering; dams; foundations; testing, dewatering and instrumentation. *Mailing Add:* HCR-32 Box 23 Rolla MO 65401

SCHMIDT, P(HILIP) S(TEPHEN), b Houston, Tex, Feb 26, 41; m 66; c 3. MECHANICAL ENGINEERING, HEAT TRANSFER. *Educ:* Mass Inst Technol, SB, 62; Stanford Univ, MS, 65, PhD(mech eng), 69. *Prof Exp:* Res engr, Bell Helicopter Co, 62-64; Woodrow Wilson teaching intern & assoc prof mech eng, Prairie View Agr & Mech Col, 68-70; from asst prof to prof mech eng, 70-90, DONALD J DOUGLAS PROF ENG, UNIV TEX, AUSTIN, 90- *Concurrent Pos:* Consult, Elec Power Res Inst & var corps. *Honors & Awards:* Ralph R Teetor Award, Soc Automotive Engrs, 72; Distinguished Serv Citation, Am Soc Mech Engrs, 86. *Mem:* Am Soc Eng Educ; Am Soc Mech Engrs. *Res:* Fluid mechanics; thermodynamics and industrial energy utilization. *Mailing Add:* Dept Mech Eng Univ Tex Austin TX 78712

SCHMIDT, PARBURY POLLEN, b Norwalk, Conn, Sept 4, 39; m 61; c 3. CHEMICAL PHYSICS. *Educ:* Kalamazoo Col, BA, 61; Wake Forest Col, MA, 64; Univ Mich, Ann Arbor, PhD(chem), 66. *Prof Exp:* NSF fel, Univ Col, Univ London, 66-67; NSF fel & vis res fel, Australian Nat Univ, 67-68; asst prof chem, Univ Ga, 68-70; from asst prof to assoc prof, 70-83, PROF CHEM, OAKLAND UNIV, 83- *Concurrent Pos:* Spec lectr gen studies sch, Australian Nat Univ, 68; Fulbright res scholar & vis prof, Southampton Univ, UK. *Mem:* Royal Soc Chem London; Am Chem Soc; Sigma Xi; Inst Navig; Electrochem Soc. *Res:* Theory of electron transfer reactions; theory of nerve impulse conduction; theory of nonradiative transitions in molecules. *Mailing Add:* Dept Chem Oakland Univ Rochester MI 48309

SCHMIDT, PAUL GARDNER, b Pasadena, Calif, June 9, 44; m 66; c 2. BIOPHYSICAL CHEMISTRY. *Educ:* Pomona Col, BA, 66; Stanford Univ, PhD(chem), 70. *Prof Exp:* Asst prof chem & biochem, Univ Ill, Urbana, 70-77; VPRES RES & DEVELOP, VESTAR RES INC, 87- *Concurrent Pos:* USPHS res grant, 70-; res career develop award, USPHS, 79-84. *Mem:* AAAS; Am Chem Soc; Biophys Soc; Am Soc Biol Chem. *Res:* Nuclear magnetic resonance; enzyme structure and function; transfer RNA structure. *Mailing Add:* 1730 Euclid San Marino CA 91108

SCHMIDT, PAUL J, b Cincinnati, Ohio, May 26, 43; m 69; c 2. ORGANIC CHEMISTRY. *Educ:* Xavier Univ, BS, 65; Univ Cincinnati, PhD(chem), 69. *Prof Exp:* Res chemist, 69-74, asst dir chem res, 74-75, dir chem res, 75-80, vpres res & develop, Hilton-Davis Div, Sterling Drug, Inc, 80-87; GEN MGR, ORG CHEM DIV, HILTON DAVIS CO, 88- *Mem:* Am Chem Soc. *Res:* Platinium complexes; organic synthetics. *Mailing Add:* 3590 Concerto Dr Sharonville Cincinnati OH 45241

SCHMIDT, PAUL JOSEPH, b New York, NY, Oct 22, 25; m 53; c 4. MEDICINE, CLINICAL PATHOLOGY. *Educ:* Fordham Univ, BS, 48; St Louis Univ, MS, 52; NY Univ, MD, 53; Am Bd Path, cert clin path, 64, cert blood banking, 73. *Prof Exp:* Intern, St Elizabeth's Hosp, Boston, 53-54; physician, Clin Ctr Blood Bank & chief blood bank sect, NIH, 55-60, resident, Clin Path Dept, 61-62, asst chief, 62-64, chief blood bank dept, 65-74; dir, 75-90, HEAD TRANSFUSION MED, SOUTHWEST FLA BLOOD BANK, INC, 91-; PROF PATH, COL MED, UNIV SFLA, 75- *Concurrent Pos:* From clin assoc prof to clin prof path, Sch Med, Georgetown Univ, 65-74. *Honors & Awards:* Silver Medal, Red Cross Spain; Emily Cooley Mem Award, Am Asn Blood Banks, 74. *Mem:* Col Am Path; Am Soc Clin Path; Int Soc Blood Transfusion; Am Soc Blood Banks (pres, 88). *Res:* Immunohematology; physiology of the formed elements of blood and the effects of storage on their viability; hepatitis; administration and education in blood banking and clinical pathology. *Mailing Add:* Southwest Fla Blood Bank PO Box 2125 Tampa FL 33601

SCHMIDT, PAUL WOODWARD, b Madison, Wis, May 8, 26; m 50; c 5. PHYSICS. *Educ:* Carleton Col, BA, 49; Univ Wis, MS, 50, PhD(physics), 53. *Prof Exp:* From asst prof to assoc prof, 53-66, PROF PHYSICS, UNIV MO, COLUMBIA, 66- *Mem:* Fel Am Phys Soc; Am Crystallog Asn; Sigma Xi; Mat Res Soc. *Res:* Small angle x-ray scattering, both theory and experiment; chemical physics; liquids; biophysics; colloids; fractals. *Mailing Add:* Dept Physics Univ Mo Columbia MO 65211

SCHMIDT, RAYMOND LEROY, b Tiffin, Ohio, July 7, 42; m 65; c 2. PETROLEUM ENGINEERING, ENHANCED OIL RECOVERY. *Educ:* Fla Presby Col, BS, 64; Emory Univ, PhD(phys chem), 67. *Prof Exp:* Instr chem, Emory Univ, 67-68; res fel chem eng, Calif Inst Technol, 68-70; from asst prof to assoc prof chem, Univ New Orleans, 70-78; sr res chemist, 78-84, SR RES ASSOC, CHEVRON OIL FIELD RES CO, 84-, SUP THERMAL RECOVERY, 86- *Concurrent Pos:* Fel Emory Univ, 67-68; vis assoc, Calif Inst Technol, 77-80. *Mem:* Am Chem Soc; Soc Petrol Engrs. *Res:* Laser scattering spectroscopy from fluid media; adsorption phenomena; surface chemistry of geologic materials; high pressure physical property measurements; thermal methods of enhanced oil recovery including tar sands; phase behavior of petroleum reservoir fluids. *Mailing Add:* 7751 Bowen Dr Whittier CA 90602

SCHMIDT, REESE BOISE, b Knoxville, Iowa, May 16, 13; m 49; c 5. ANALYTICAL CHEMISTRY. *Educ:* Cent Col, Iowa, BS, 34. *Prof Exp:* Technician exp canning, Calif Packing Corp, Ill, 39-40, foreman, 40-43; chemist, US Rubber Co, Iowa, 44-45; teacher pub sch, Iowa, 45-46; chemist, 46-62, mgr appl res, 62-69, mgr process & test, W A Sheaffer Pen Co, 70-78, CONSULT, SHEAFFER EATON, DIV TEXTRON INC, 78- *Mem:* Am Chem Soc. *Mailing Add:* 2109 Ave H Ft Madison IA 52627

SCHMIDT, RICHARD, b Pa, May 12, 25. CERAMICS ENGINEERING. *Educ:* Temple Univ, BS, 46; Am Univ, MA, 77. *Prof Exp:* Head, Metals Br, Bur Naval Weapons, 51-60; mgr mat res & develop, Naval Air Systs Command, Washington, DC, 80, dep dir, Aircraft Div, 80-83; MGR AEROSPACE & DEFENSE TECHNOL, ALCOA, 85- *Concurrent Pos:* Chmn, Aerospace Mat Div, Soc Automotive Engrs, 87. *Honors & Awards:* Von Karmen Mem Award, 79; Burgess Mem Award, Am Soc Metals. *Mem:* Fel Am Soc Metals; Soc Automotive Engrs. *Res:* Metal composites. *Mailing Add:* Alcoa Tech Ctr Alcoa Center PA 15069

SCHMIDT, RICHARD ARTHUR, b Elizabeth, NJ, Mar 18, 35; m 55; c 2. ECONOMIC GEOLOGY. *Educ:* Franklin & Marshall Col, BS, 57; Univ Wis-Madison, MS, 59, PhD(geol), 63. *Prof Exp:* Nat Acad Sci resident res assoc, NASA-Ames Res Ctr, 63-65; proj scientist, Aerospace Systs Div, Bendix Corp, 65-67; sr geologist, Stanford Res Inst, 67-74; TECH MGR FOSSIL FUEL RESOURCES, ELEC POWER RES INST, 74- *Mem:* Am Inst Mining, Metall & Petrol Engrs; fel Geol Soc Am. *Res:* Coal; surface mining; environmental assessment; resources management and planning. *Mailing Add:* 18342 E Hillcrest Ave Villa Park CA 92667

SCHMIDT, RICHARD EDWARD, b Detroit, Mich, Sept 3, 31; m 56; c 2. AGRONOMY. *Educ:* Pa State Univ, BS, 54, MS, 58; Va Polytech Inst, PhD(agron), 65. *Prof Exp:* Asst agron, Pa State Univ, 56-58; from instr to assoc prof, 58-86, PROF AGRON, VA POLYTECH INST & STATE UNIV, 86- *Concurrent Pos:* Consult, Weblite Corp, Va, 65- & US Mkt Group, 81- *Honors & Awards:* Fel, Crop Sci Soc Am. *Mem:* Am Soc Agron; fel Crop Sci Soc Am. *Res:* Turfgrass ecology, particularly environmental influences on the physiological affects of grasses. *Mailing Add:* 511 Cedar Orchard Dr Blacksburg VA 24060

SCHMIDT, RICHARD RALPH, b Milwaukee, Wis, Mar 28, 44; m 65; c 2. TERATOLOGY. *Educ:* Univ Wis-Madison, BA, 68; Med Col Wis, PhD(anat), 74. *Prof Exp:* INSTR GROSS ANAT, DANIEL BAUGH INST ANAT, JEFFERSON MED COL, 74- *Res:* Biochemical alterations in fetuses with multiple congenital skeletal malformations. *Mailing Add:* Dept Anat Jefferson Med Col 1020 Locust St Philadelphia PA 19107

SCHMIDT, ROBERT, b Ukraine, May 18, 27; US citizen; m 78; c 1. MECHANICS, ENGINEERING. *Educ:* Univ Colo, BS, 51, MS, 53; Univ Ill, PhD(civil eng), 56. *Prof Exp:* Asst prof mech, Univ Ill, 56-59; assoc prof, Univ Ariz, 59-63; chmn, Civil Eng Dept, 78-80, PROF ENG MECH, UNIV DETROIT, 63- *Concurrent Pos:* NSF res grants, 60-63, 64-67, 70-72 & 76-78; ed, Indust Math, Indust Math Soc, 68- *Honors & Awards:* First Gold Award, Indust Math Soc. *Mem:* Am Soc Civil Engrs; Am Soc Mech Engrs; Indust Math Soc (pres, 66-67 & 81-84); Am Acad Mech; Sigma Xi. *Res:* Theories of plates and shells, sandwich plates and shells and multilaminate plates and shells; nonlinear theories of arches and rods; elastic stability & postbuckling analysis; direct variational methods; biosophy. *Mailing Add:* Col Eng Univ Detroit 4001 W McNichols Rd Detroit MI 48221-9987

SCHMIDT, ROBERT GORDON, b Minneapolis, Minn, Nov 9, 24; m 52; c 2. GEOLOGY. *Educ:* Univ Wis, MS, 51. *Prof Exp:* Geologist, US Geol Surv, 51-90; RETIRED. *Mem:* Soc Econ Geologist; AAAS; Geol Soc Am; Asn Geosci Int Develop. *Res:* Satellite remote sensing for mineral exploration. *Mailing Add:* 3732 N Nelson St Arlington VA 22207-4836

SCHMIDT, ROBERT REINHART, b St Louis, Mo, Feb 18, 33; m 56; c 4. BIOCHEMISTRY. *Educ:* Va Polytech Inst, BS, 55, PhD(biochem), 61; Univ Md, MS, 57. *Prof Exp:* From asst prof to assoc prof, Va Polytech Inst & State Univ, 61-67, prof biochem, 67-80; MEM FAC, UNIV FLA, 80- *Concurrent Pos:* Res grants, NIH & NSF, 61- *Mem:* Am Soc Microbiol; Am Soc Biol Chemists; Am Soc Plant Physiol. *Res:* Use of synchronized cultures of microorganisms, plant and animal cells to study operation and control of metabolic pathways and enzymes located therein during cellular growth and division. *Mailing Add:* 6628 SW 100 Lane Gainsville FL 32608

SCHMIDT, ROBERT SHERWOOD, b Des Moines, Iowa, May 16, 28; m 51; c 3. ANURAN ACOUSTIC NEUROETHOLOGY. *Educ:* Ball State Teachers Col, BS, 50; Univ Mich, MS, 51; Univ Chicago, PhD(zool), 54. *Prof Exp:* From asst prof to assoc prof biol sci, Ill State Norm Univ, 54-61; asst prof otolaryngol, Univ Chicago, 62-65; PROF PHARMACOL, STRITCH SCH MED, LOYOLA UNIV, CHICAGO, 65- *Concurrent Pos:* Nat Inst Neurol Dis & Blindness career develop award, Univ Chicago, 63-65 & Stritch Sch Med, Univ Loyola Chicago, 66-70. *Mem:* AAAS; Animal Behav Soc; Am Soc Zool; Soc Neurosci. *Res:* Anuran acoustic neuroethology. *Mailing Add:* Pharmacol Bldg 135 Loyola Univ Stritch Sch Med Maywood IL 60153

SCHMIDT, ROBERT W, b Enid, Okla, Feb 16, 30; m 53; c 4. BIOCHEMISTRY. *Educ:* Bethel Col, AB, 52; Univ Okla, MSc, 55, PhD(chem), 60. *Prof Exp:* Instr math & sci high sch, Kans, 52-53; asst prof chem, Simpson Col, 58-61; assoc prof, 61-67, PROF CHEM, BETHEL COL, KANS, 67- *Concurrent Pos:* Assoc marine scientist, Va Inst Marine Sci, 69-70; vis prof, Biochem, State Univ Iowa Col Med, 76-77; vis prof, Chem Dept, Univ Okla, 83-84. *Mem:* Am Chem Soc; Am Sci Affil; Midwestern Asn Chem Teachers; Am Asn Univ Professors; Sigma Xi. *Res:* Natural products of plants. *Mailing Add:* Dept Chem Bethel Co North Newton KS 67117

SCHMIDT, ROBERT W, b Toledo, Ohio, July 22, 26; m 63; c 4. PATHOLOGY. *Educ:* Univ Toledo, BS, 50; Ohio State Univ, MD, 54. *Prof Exp:* Intern, 54-55, resident path, 55-59, from instr to assoc prof, 59-69, PROF PATH, UNIV HOSP, UNIV MICH, ANN ARBOR, 69-; SR ATTEND PATHOLOGIST, TOLEDO HOSP, OHIO, 90- *Concurrent Pos:* Dir path, Wayne County Gen Hosp, Eloise, Mich, 64-85. *Mem:* Am Soc Cytol. *Res:* Clinical and anatomical pathology; gynecologic pathology; exfoliative and fine needle aspirate cytology. *Mailing Add:* Dept Path Univ Mich Hosp Ann Arbor MI 48109

SCHMIDT, ROGER PAUL, b Abilene, Kans, Jan 16, 44; m 74. PARASITOLOGY, BIOLOGY. *Educ:* Univ Kans, BA, 66, MA, 72; Kans State Univ, PhD(parasitol), 78. *Prof Exp:* AT DEPT BIOL, COLUMBIA COL. *Mem:* Sigma Xi; AAAS; Am Inst Biol Sci. *Res:* Effects and interactions between pesticides and parasites in poultry. *Mailing Add:* Dept Biol Columbia Col 1301 Columbia Col Dr Columbia SC 29203

SCHMIDT, RONALD GROVER, b Bloomfield, NJ, Oct 13, 31; m 55; c 2. HYDROGEOLOGY. *Educ:* Columbia Univ, AB, 53, MA, 55; Univ Cincinnati, PhD(geol), 57; Am Inst Prof Geol & Calif Bd Regist, cert & regist geol. *Prof Exp:* Geologist, USAEC Contr NMex, 52; tech officer, Geol Surv Can, 53-54; asst geol, Columbia Univ, 54; instr Hunter Col, 55; asst, Univ Cincinnati, 55-57; geologist, Standard Oil Co, Tex, 56; asst prof, Univ Cincinnati, 57-63; pres & geol consult, Earth Sci Labs, Inc, 60-70; dir, Off Environ Studies, 70-74, Dir, Brehm Lab, 72-75, chmn, Dept Geol, 74-83, PROF GEOL & ENG, WRIGHT STATE UNIV, 70-, DIR, CTR GROUND WATER MGT, 86- *Mem:* AAAS; Am Geophys Union; Amer Water Res Asn; Geol Soc Am; Water Pollution Control Fed. *Mailing Add:* Dept Geol Wright State Univ Colonel Glenn Hwy Dayton OH 45435

SCHMIDT, RUTH A M, b Brooklyn, NY, Apr 22, 16. GEOLOGY, ENVIRONMENTAL SCIENCE. *Educ:* NY Univ, AB, 36; Columbia Univ, AM, 39, PhD(geol), 48. *Prof Exp:* Asst paleont, Columbia Univ, 39-42; geologist, US Geol Surv, 43-56, dist geologist, Alaska, 56-63; CONSULT GEOLOGIST & MICROPALEONTOLOGIST, 64- *Concurrent Pos:* Environ consult, Off Pipeline Coordr, Off of Gov, Alaska, 75-77; chmn, geol dept, Anchorage Community Col, Univ Alaska, 70-84. *Mem:* Fel AAAS; Fedn Am Scientists; fel Geol Soc Am; Am Inst Prof Geol; fel Arctic Inst NAm; Sigma Xi; Am Asn Petrol Geol. *Res:* Cretaceous and tertiary micropaleontology in Alaska; instructional television delivery of earth science to rural Alaska; general geology of Alaska. *Mailing Add:* 1402 W 11th Ave Anchorage AK 99501

SCHMIDT, STEPHEN PAUL, FESCUE TOXICITY, PROTEIN METABOLISM. *Educ:* Univ Wis, PhD(ruminant nutrit), 72. *Prof Exp:* ASSOC PROF ADVAN NUTRIT & INTRODUCTORY ANIMAL SCI, AUBURN UNIV, 76- *Mailing Add:* 409 Green St Auburn AL 36830

SCHMIDT, STEPHEN PAUL, b San Diego, Calif, Feb 8, 47; m 68; c 2. INSECT TOXICOLOGY. *Educ:* San Diego State Univ, BS, 73, MS, 75; Univ Calif, Riverside, PhD(entom), 79. *Prof Exp:* Res assoc entom, Okla State Univ, 79-82; RES BIOCHEMIST, RHONE POULENC AGR CO, 82- *Mem:* Entom Soc Am; AAAS. *Res:* Toxicology and biochemistry of novel pesticides in insects. *Mailing Add:* Rhone Poulenc Agr Co T W Alexander Dr PO Box 12014 Research Triangle Park NC 27709

SCHMIDT, STEVEN PAUL, IN VITRO FERTILIZATION, VASCULAR RESEARCH. *Prof Exp:* ASSOC DIR VASCULAR RES & DIR, IN VITRO FERTILIZATION LAB, VASCULAR RES LAB, AKRON CITY HOSP. *Mailing Add:* Dept 421 Abbott Park Bldg AP10 Abbott Labs Abbott Park IL 60064

SCHMIDT, THOMAS JOHN, b Mt Holly, NJ, Dec 13, 46. ENDOCRINOLOGY, CELLULAR PHYSIOLOGY. *Educ:* Univ Del, BA, 69; Cornell Univ, MS, 73, PhD(physiol), 76. *Prof Exp:* Biochem & endocrinol, Nat Cancer Inst, 76-79; sr fel, Fels Res Inst, 79-; ASST PROF, DEPT PHYSIOL & BIOPHYS, UNIV IOWA. *Honors & Awards:* Scholar, Leukemia Soc Am. *Mem:* Sigma Xi; Endocrine Soc; NY Acad Sci; Am Asn Cancer Res; Am Physiol Soc; Am Soc Biol Chemists. *Res:* Mode of action of steroid hormones; function of receptors, particularly for glucocorticoids, in normal and neoplastic cells. *Mailing Add:* Dept Physiol & Biophys Univ Iowa 5-432 Bowen Sci Bldg Iowa City IA 52242

SCHMIDT, THOMAS WILLIAM, b Evansville, Ind, Aug 9, 38; m 61; c 2. CHEMICAL PHYSICS. *Educ:* Univ Evansville, BA, 60; Univ Fla, MS, 63; Univ Tenn, PhD(chem), 67. *Prof Exp:* Sr res chem physicist, Phillips Petrol Co, 67-77, sect supvr, Eng Data, 77-82, br mgr, Alternate Energy, 82, br mgr, Planning & High Tech, 82-83, br mgr, Safety Div, 83-85, br mgr, fuels & lubes, crude oil, Environ Incineration Technol, 85-90, DIR ENVIRON PLANNING & TECHNOL, PHILLIPS PETROL CO, 90- *Honors & Awards:* IR-100 Award, Res & Develop Mag, 79. *Mem:* Am Phys Soc; Sigma Xi. *Res:* Molecular beams; mass spectroscopy; chemical kinetics; ultra high vacuum; vapor-liquid equilibrium. *Mailing Add:* Rte 1 Box 846 Ochelata OK 74051-9702

SCHMIDT, VICTOR A, b Brooklyn, NY, Nov 9, 36; m 68; c 2. GEOPHYSICS, PALEOMAGNETISM. *Educ:* Carnegie-Mellon Univ, BS, 58, MS, 60, PhD(physics), 66. *Prof Exp:* Instr physics, Carnegie-Mellon Univ, 66-67; NASA fel, 67-68, from asst prof to assoc prof, 68-86, PROF GEOPHYS, UNIV PITTSBURGH, 86- *Mem:* Am Geophys Union; fel Nat Speleol Soc. *Res:* Rock magnetism and paleomagnetism; geology and hydrology of caves in the Appalachians; mechanisms of thermoremanence. *Mailing Add:* Dept Geol & Planetary Sci Univ Pittsburgh EH 321 Pittsburgh PA 15260

SCHMIDT, VICTOR HUGO, b Portland, Ore, July 10, 30; m 58; c 4. SOLID STATE PHYSICS. *Educ:* Wash State Univ, BS, 51; Univ Wash, PhD(physics), 61. *Prof Exp:* Mech design engr, Gilfillan Bros, Inc, 53-54; assoc res engr, Boeing Airplane Co, 55-57; asst prof physics, Valparaiso Univ, 61-64; assoc prof, 64-73, PROF PHYSICS, MONT STATE UNIV, 73- *Mem:* Fel Am Phys Soc; Am Asn Physics Teachers; Sigma Xi; sr mem Inst Elec & Electronics Engrs. *Res:* Nuclear magnetic resonance, light scattering, dielectric and high pressure studies of ferroelectric phase transitions and proton glass; physical properties and applications of piezoelectric polymers; wind generation of electric power; liquid crystals. *Mailing Add:* Dept Physics Mont State Univ Bozeman MT 59717

SCHMIDT, VOLKMAR, b Heidelberg, Ger, Aug 27, 32; m 71; c 2. CARBONATE DIAGENESIS, SANDSTONE DIAGENESIS. *Educ:* Univ Heidelberg, BS, 56; Univ Kiel, PhD(geol), 61. *Prof Exp:* Sr res geologist, Mobil Oil Corp, Tex, 61-69, head, Geol Lab, Mobil Oil Can, Ltd, 68-76; mgr geol res, Calgary, 76-84, sr geol adv, 84-86, CONSULT, PETROL CAN, 86- *Mem:* Am Asn Petrol Geologists; Soc Econ Paleontologists & Mineralogists; Int Asn Sedimentologists; Can Soc Petrol Geologists. *Res:* Sandstone diagenesis; carbonate diagenesis; sedimentary processes; sediment geochemistry; petroleum reservoir petrography; evaporite facies and petrography; facies and paleoenvironmental studies; sedimentary petrography. *Mailing Add:* 4187 Varsity Rd NW Calgary AB T3B 2Y6 Can

SCHMIDT, WALDEMAR ADRIAN, SURGICAL PATHOLOGY, CYTOPATHOLOGY. *Educ:* Univ Ore, MD & PhD(human anat), 69. *Prof Exp:* ASSOC PROF PATH, MED SCH, UNIV TEX , HOUSTON, 77- *Res:* Carcinogenesis. *Mailing Add:* Dept Path & Lab Med Univ Tex Health Sci Ctr PO Box 20036 Houston TX 77225

SCHMIDT, WALTER HAROLD, b Gordon, Nebr, Sept 19, 35; m 59; c 4. AGRONOMY. *Educ:* Univ Nebr, Lincoln, BS, 57, MS, 60, PhD(crop prod), 65. *Prof Exp:* Asst county exten agent, Nebr Coop Exten Serv, fall 57; from asst prof to assoc prof, 65-76, PROF AGRON, OHIO COOP EXTEN SERV, OHIO STATE UNIV, 76- *Mem:* Am Soc Agron; Nat Asn Coop Agr Agents; Am Soc Sugar Beet Technol; Sigma Xi. *Res:* Crop production techniques for corn, forages, grain, soybeans, sugar beets, sunflowers and canola. *Mailing Add:* 1708 Oak Dr Fremont OH 43420

SCHMIDT, WERNER H(ANS), b Frankfurt, Ger, Sept 24, 14; nat US; m 41; c 2. CHEMICAL ENGINEERING. *Educ:* Tufts Univ, BS, 36. *Prof Exp:* Chemist foods, Johnson-Salisbury Co, 36; chemist edible oils, Lever Bros Co, 36-39, res chemist, 39-45, res supvr foods, 45-52, chief foods processing sect, 52-60, develop mgr, 60-64, develop mgr foods & toiletries, 64-73, dir develop foods, Foods Div, 73-78; RETIRED. *Concurrent Pos:* Consult, Lever Bros Foods Div, 78- *Mem:* Am Chem Soc; Am Oil Chem Soc; Inst Food Technologists. *Res:* Edible oil processing; shortening and margarine formulation and manufacture; catalytic hydrogenation; esterification. *Mailing Add:* 42 Liverpool Dr Yarmouth Port MA 02675

SCHMIDT, WILLIAM EDWARD, b Pittsburgh, Pa, Sept 7, 20; m 47; c 5. ANALYTICAL CHEMISTRY. *Educ:* George Washington Univ, BS, 43, MS, 50; Princeton Univ, MA & PhD(chem), 53. *Prof Exp:* Asst chem, George Washington Univ, 41-43, res assoc, Nat Defense Res Comt, 43-44, assoc, 46-50; asst anal chem, Princeton Univ, 50-53; from asst prof to assoc prof, 53-61, PROF CHEM, GEORGE WASHINGTON UNIV, 61- *Concurrent Pos:* Consult, US Vet Admin, 57-59; ed consult, Am Chem Soc, 65- *Mem:* AAAS; Am Chem Soc; Electrochem Soc; Am Inst Chemists. *Res:* Measurement of electrode potentials; mercury cathode electrolysis; electroanalysis; redox proteins; electroplating; radio tracers; reagent chemicals. *Mailing Add:* Dept Chem George Washington Univ Washington DC 20006

SCHMIDT, WOLFGANG M, b Vienna, Austria, Oct 3, 33; m 60; c 3. MATHEMATICS. *Educ:* Univ Vienna, PhD(math), 55. *Prof Exp:* Asst docent math, Univ Vienna, 55-56; instr, Univ Mont, 56-57; asst docent, Univ Vienna, 57-58; asst prof, Univ Mont, 58-59; asst docent, Univ Vienna, 59-60; asst prof, Univ Colo, 60-61; res assoc, Columbia Univ, 61-62; docent, Univ Vienna, 62-64; assoc prof, 64-65, PROF MATH, UNIV COLO, BOULDER, 65- *Concurrent Pos:* Univ Colo, Boulder fac fel, Univ Cambridge, 66-67; grant & invited address, Int Cong Mathematicians, Nice, 70; mem, Inst Advan Study, 70-71. *Honors & Awards:* Cole Prize Number Theory, Am Math Soc, 72. *Mem:* Am Math Soc; Austrian Math Soc. *Res:* Number theory, especially geometry of numbers and diophantine approximations. *Mailing Add:* Dept Math Ecot 4-41 Univ Colo Box 426 Boulder CO 80309

SCHMIDT, WYMAN CARL, b Ocheyedan, Iowa, Sept 9, 29; m 53; c 5. SILVICULTURE. *Educ:* Univ Mont, BS, 58, MS, 61, PhD, 80. *Prof Exp:* Forester, Black Hills Nat Forest, 59-60; from res forester to res silviculturist, 60-75, RES UNIT LEADER, INTERMOUNTAIN FOREST & RANGE EXP STA, FORESTRY SCI LAB, MONT STATE UNIV, 75- *Honors & Awards:* USDA Award; Sci Tech Award, Soc Am Foresters. *Mem:* Soc Am Foresters; Ecol Soc Am. *Res:* Autecological, synecological and silvicultural research in the coniferous forests of the northern Rocky Mountains, including forest regeneration, stand development, cone production, soil moisture, phenology and tree growth relationships; silviculture and forest ecology of subalpine forest ecosystems. *Mailing Add:* Forestry Sci Lab Mont State Univ Bozeman MT 59717

SCHMIDTKE, JON ROBERT, b Detroit, Mich, Mar 27, 43. IMMUNOLOGY. *Educ:* Mich State Univ, BS, 65; Univ Mich, MS, 67, PhD(immunol), 69. *Prof Exp:* Asst prof microbiol & surg, Univ Minn, Minneapolis, 72-76, assoc prof, 76-79; HEAD, DEPT IMMUNOL, ELI LILLY & CO, 79- *Concurrent Pos:* USPHS training grant, Scripps Clin & Res Found, La Jolla, Calif, 69-72. *Mem:* Am Asn Immunol; Am Soc Microbiol; Transplantation Soc; Am Soc Exp Path; Reticuloendothelial Soc. *Res:* Cellular immunology of human lymphocytes; macrophage function; modification of immunogenicity. *Mailing Add:* Dept Immunol Eli Lilly & Co Lilly Corp Ctr Indianapolis IN 46285

SCHMIDTKE, R(ICHARD) A(LLEN), b Benton Harbor, Mich, July 27, 25; m 48; c 2. MECHANICAL ENGINEERING. *Educ:* Univ Mich, BS, 48, MS, 49; Ill Inst Technol, PhD(mech eng), 53. *Prof Exp:* Instr mech eng, Ill Inst Technol, 49-53, asst prof, 53; mem tech staff, Melpar, Inc, Westinghouse Air Brake Co, 53, sr mem tech staff, 53-54, sr engr, 54-55, proj engr, 55-57, res br leader, 57-58, asst to vpres res & eng, 58-60, spec asst adv develop, 60; dir appl res, Gov Prod Div, Pratt & Whitney Aircraft Div, United Technol Corp, West Palm Beach, 60-70, sr prog mgr 70-76, vpres laser progs, 76-80, vpres engine progs, 80-82; RETIRED. *Mem:* Am Soc Mech Engrs; Am Astronaut Soc; Am Inst Aeronaut & Astronaut. *Res:* Free-piston, turbojet, ramjet and rocket engines; heat transfer; aerodynamics; thermodynamics, applied mathematics; marine propulsion; high energy lasers. *Mailing Add:* 372 Fairway N Tequesta FL 33469

SCHMIDT-KOENIG, KLAUS, b Heidelberg, Ger, Jan 21, 30; m 59; c 3. ZOOLOGY. *Educ:* Univ Freiburg, PhD(zool), 58. *Prof Exp:* Fel, Max Planck Inst Physiol of Behav, Ger, 55-57, mem staff, 58-63; zool, 59-71, adj assoc prof, 71-75, PROF ZOOL, DUKE UNIV, 75- *Concurrent Pos:* Pvt docent, Univ Gottingen, 63-71, appl prof, 71-75; prof zool, Univ Tubingen, 75- *Mem:* Sigma Xi; Am Orithol Union; Deutsche Ornithol-Gesellschaft (pres). *Res:* Animal orientation; biological rhythms; sensory physiology; biostatistics. *Mailing Add:* ABTF Verhaltensphys Beim Kupferhammer 8 Tubingen D-7400 Germany

SCHMIDT-NIELSEN, BODIL MIMI, b Copenhagen, Denmark, Nov 3, 18; nat US; m 39, 68; c 3. PHYSIOLOGY. *Educ:* Copenhagen Univ, DDS, 41, DOdont, 46, PhD, 55. *Hon Degrees:* DSc, Bates Col, 83. *Prof Exp:* Instr, Copenhagen Univ, 41-44, secy, res assoc & asst prof, 44-46; res assoc, Swarthmore Col, 46-48; res assoc, Stanford Univ, 48-49; res assoc, Col Med, Univ Cincinnati, 49-52, asst prof, 52; res assoc zool, Duke Univ, 52-54, sr res assoc, 54-57, assoc res prof, 57-61, assoc res prof zool & physiol, 61-64; prof biol, Case Western Reserve Univ, 64-71 chmn, dept biol, 70-71; res scientist, Mt Desert Island Biol Lab, 71-86; ADJ PROF, DEPT PHYSIOL, SCH MED, UNIV FLA, GAINSVILLE, 86- *Concurrent Pos:* Guggenheim fel, 52-53; established investr, Am Heart Asn, 54-62; trustee, Mt Desert Island Biol Lab, 55-69 & 76-, vpres, 79-81, dep dir, 79-, pres, 81-85; Bowditch lectr, 58; mem physiol training grant comt, Nat Inst Gen Med Sci, 67-71; adj prof, Case Western Univ, 71-75 & Brown Univ, 71-78; assoc ed, Am J Physiol, 76-81. *Honors & Awards:* NIH Career Award, 62-64. *Mem:* Fel AAAS; Am Physiol Soc (pres, 75-76); Soc Exp Biol & Med; Am Soc Nephrology; Am Soc Zoologists; Int Soc Nephrology; Int Soc Lymphology. *Res:* Biochemistry of saliva; water metabolism of desert animals; osmoregulation; comparative physiology of cellular volume and ion regulation; comparative renal physiology; physiology of the mammalian renal pelvis. *Mailing Add:* Mt Desert Island Biol Lab Salisbury Cove ME 04672

SCHMIDT-NIELSEN, KNUT, b Trondheim, Norway, Sept 24, 15; m 39; c 3. PHYSIOLOGY. *Educ:* Copenhagen Univ, Mag Sc, 41, PhD(zoophysiol), 46. *Hon Degrees:* DM, Lund UNiv, 85. *Prof Exp:* Res assoc, Swarthmore Col, 46-48; res assoc, Stanford Univ, 48-49; asst prof, Col Med, Univ Cincinnati, 49-52; prof, 52-63, JAMES B DUKE PROF PHYSIOL, DUKE UNIV, 63- *Concurrent Pos:* Docent, Univ Oslo, 47-49; Guggenheim fel, Univ Algeria, 53-54; consult, NSF, 57-61; trustee, Mt Desert Island Biol Lab, 58-61; sect ed, Am J Physiol & J Appl Physiol, 61-64; mem sci adv comt, New Eng Regional Primate Res Ctr, Harvard Med Sch, 62-66; regent's lectr, Univ Calif, 63; nat adv bd, Physiol Res Lab, Scripps Inst, Univ Calif, 63-69, chmn, 68-69; USPHS res career award, 64-85; mem subcomt environ physiol, US Nat Comt Int Biol Prog, 65-67; mem comt res utilization uncommon animals, Div Biol & Agr, Nat Acad Sci, 66-68; US Nat Comt Int Union Physiol Sci, 66-78, vchmn, 69-78; animal resources adv comt, NIH, 68; biomed eng adv comt, Duke Univ, 68-85; sect ed, Am J Physiol & J Appl Physiol, 61-64; ed, J Exp Biol, 75-79, 83-86; Wellcome vis prof, Univ SDak, 88- *Honors & Awards:* Brody Mem Lectr, Univ Mo, 62. *Mem:* Nat Acad Sci; fel AAAS; Am Acad Arts & Sci; Am Physiol Soc; fel NY Acad Sci; Royal Norweg Acad Arts & Sci; foreign mem Royal Soc London; French Acad Sci; Norweg Acad Sci; Royal Danish Acad; Int Union Physiol Soc (pres, 80-86). *Res:* Comparative physiology, respiration and oxygen supply; water metabolism and excretion; temperature regulation, physiology of desert animals. *Mailing Add:* Dept Zool Duke Univ Durham NC 27706

SCHMIEDER, ROBERT W, b Phoenix, Ariz, July 10, 41; m 63; c 3. ATOMIC PHYSICS, MARINE SCIENCES. *Educ:* Occidental Col, AB, 63; Calif Inst Technol, BS, 63; Columbia Univ, MA, 65, PhD(physics), 68. *Prof Exp:* Staff researcher, Lawrence Berkeley Lab, 69-73; MEM TECH STAFF, SANDIA NAT LABS, 72- *Concurrent Pos:* Exped leader, Cordell Expeds, 77-; instr, Univ Calif, Berkeley, 71-72; NATO Summer Inst, Cargese, France, 87; ed, Defense Res Rev, 86-; prog comm, Int Comb Symp, 82, 84 & 88. *Mem:* Am Phys Soc; Am Geophys Union; Am Inst Biol Sci. *Res:* Physics of highly ionized atoms; combustion physics and chemistry; marine biology and ecology. *Mailing Add:* Sandia Nat Labs Livermore CA 94551

SCHMIEDESHOFF, FREDERICK WILLIAM, b Brooklyn, NY, Mar 28, 25; wid; c 3. RESEARCH ADMINISTRATION, APPLIED MECHANICS. *Educ:* Rensselear Polytech Inst, BS, 48, MS, 53, PhD(mech), 66. *Prof Exp:* Engr chg physics, Beers & Heroy Co, NY, 48-52; physicist, W & L E Gurley Co, 52; res scientist, Rensselaer Polytech Inst, 52-56, asst prof mech, 56-59; chief theoret & exp labs, Watervliet Arsenal, 59-63, dir res, 63-77; CHIEF MECH, MAT BR, ARMY RES OFF, 77- *Concurrent Pos:* NATO consult, Adv Group Aeronaut, Res & Develop, London, 63; mem interface comt, Mat Adv Bd, Nat Acad Sci, 63-65, micromech comt, 64-65, designing with composites comt, 66-67; mem adv bd, Army Mat & Mech Res Ctr, Watertown, Mass, 70-75. *Mem:* AAAS; Sigma Xi. *Res:* Nonlinear heat conduction, thermal stresses, composite materials. *Mailing Add:* Rte 2 Box 204E Rockingham NC 28379

SCHMIEDESHOFF, GEORGE M, b Bridgeport, Conn, Nov 8, 55. SOLID STATE PHYSICS, THERMAL PHYSICS. *Educ:* Univ Bridgeport, BS, 79; Univ Mass, MS, 82, PhD(physics), 85. *Prof Exp:* Int Bus Mach postdoctoral fel physics, Mass Inst Tech, 85-87; lectr physics, Tufts Univ, 87; ASST PROF PHYSICS, BOWDOIN COL, 87- *Concurrent Pos:* Vis scientist, Francis Bitter Nat Magnet Lab, Mass Inst Technol, 87- *Mem:* Am Phys Soc; Am Asn Physics Teachers. *Res:* Experimental studies of magnetic properties of novel superconducting materials at low temperatures and in high magnetic fields. *Mailing Add:* Physics Dept Bowdoin Col Brunswick ME 04011

SCHMIEG, GLENN MELWOOD, b Detroit, Mich, Aug 25, 38; div; c 2. SCIENCE EDUCATION. *Educ:* Univ Mich, BSE, 60, MS, 62; Univ NC, PhD(physics), 67. *Prof Exp:* DISTINGUISHED LECTR SCI, OPPORTUNITIES IN SCI, INC, BEMIDJI, MINN, 91- *Concurrent Pos:* Expert witness. *Mem:* Am Asn Physics Teachers; Electrostatics Soc Am. *Res:* Classical field theory; electrostatics; mathematical physics. *Mailing Add:* 3224A N Oakland Milwaukee WI 53211

SCHMIEGEL, WALTER WERNER, b Chemnitz, Ger, Jan 13, 41; US citizen; m 71. RUBBER CHEMISTRY. *Educ:* Univ Mich, Ann Arbor, BS, 63; Dartmouth Col, AM, 65; Johns Hopkins Univ, PhD(chem), 70. *Prof Exp:* Res chemist, 69-80, res assoc, 80-88, SR RES ASSOC, E I DU PONT DE NEMOURS & CO, INC, 88- *Mem:* Am Chem Soc. *Res:* Synthetic elastomers; Ziegler catalysis; fluoroelastomer synthesis and reactivity; vulcanization chemistry; polymer nuclear magnetic resonance. *Mailing Add:* DuPont Co PPD ESL 353 Wilmington DE 19898

SCHMIR, GASTON L, b Metz, France, June 8, 33; US citizen; m 60; c 3. BIOCHEMISTRY, ORGANIC CHEMISTRY. *Educ:* Harvard Univ, AB, 54; Yale Univ, PhD(biochem), 58. *Prof Exp:* Asst scientist, USPHS, 58-60; from instr to assoc prof biochem, 60-69, from assoc prof to prof, 69-85, EMER PROF MOLECULAR BIOPHYS, YALE UNIV, 85- *Mem:* Am Chem Soc; Am Soc Biol Chemists. *Res:* Bio-organic reaction mechanisms; enzyme models. *Mailing Add:* Dept Molecular Biophys & Biochem Yale Univ New Haven CT 06510

SCHMISSEUR, WILSON EDWARD, b East St Louis, Ill, July 17, 42; m 69; c 2. FARM MANAGEMENT, PRODUCTION ECONOMICS. *Educ:* Univ Ill, BS, 64; Purdue Univ, MS, 66, PhD(agr econ), 73. *Prof Exp:* Res assoc, 71-79, asst prof, 79-81, ASSOC PROF AGR ECON, ORE STATE UNIV, 81- *Concurrent Pos:* Consult, Ethanol Int, Inc, 79. *Mem:* Am Asn Artificial Intel. *Res:* Economics of livestock production; expert systems for commercial livestock management. *Mailing Add:* Dept Agr & Resource Econ Ore State Univ Corvallis OR 97331-3601

SCHMIT, JOSEPH LAWRENCE, b Cold Springs, Minn, July 22, 33; m 56; c 6. PHYSICS, CRYSTAL GROWTH. *Educ:* St John's Univ, BA, 57. *Prof Exp:* RES SCIENTIST PHYSICS, SENSORS & SIGNAL PROCESSING LAB, 59- *Concurrent Pos:* Co-chair, MCT Workshop, 88. *Mem:* Am Phys Soc; AAAS; Fedn Am Scientists. *Res:* Growth and evaluation of HgCdTe suitable for infrared detectors; growth by Bridgman, by open tube slider LPE and metal-organic chemical vapor deposition; developed technique to measure composition, measurement of the band gap and calculation of the intrinsic carrier concentration of HgCdTe; currently growing HgCdTe by MOCVD on GaAs. *Mailing Add:* 3607 Farmington Rd Hopkins MN 55343

SCHMIT, LUCIEN A(NDRE), JR, b New York, NY, May 5, 28; m 51; c 1. STRUCTURAL SYNTHESIS, DESIGN OPTIMIZATION. *Educ:* Mass Inst Technol, SB, 49, SM, 50. *Prof Exp:* Struct engr, Grumman Aircraft Eng Corp, 51-53; res engr, Aeroelastic & Struct Res Lab, Mass Inst Technol, 54-58; from asst prof to prof struct, Case Western Reserve Univ, 58-69, Wilbert J Austin distinguished prof eng, 69-70, head div solid mech, struct & mech design, 66-70; chmn, Dept Mech & Struct, 76-79, PROF ENG & APPL SCI, UNIV CALIF, LOS ANGELES, 70- *Concurrent Pos:* Mem, Sci Adv Bd, USAF, 77-84. *Honors & Awards:* Walter L Huber Civil Eng Res Prize, 70; AIAA Struct Design Lect Award, 77; Struct Dynamics & Mat Award, Am Inst Aeronaut & Astronaut Struct, 79. *Mem:* Fel Am Soc Civil Engrs; fel Am Inst Aeronaut & Astronaut; Am Soc Mech Engrs; Nat Acad Eng 85-; fel Am Acad Mech. *Res:* Analysis and synthesis of structural systems; design optimization; finite element methods; nonlinear analysis; design methods for fiber composite structures. *Mailing Add:* 4531K Boelter Hall Univ Calif Los Angeles CA 90024

SCHMITENDORF, WILLIAM E, b Oak Park, Ill, Aug 6, 41; m 64; c 2. ENGINEERING. *Educ:* Purdue Univ, BS, 63, MS, 65, PhD(optimization tech), 68. *Prof Exp:* From asst prof to prof, Mech Eng, Northwestern Univ, 67-88; CHMN MECH ENG & AEROSPACE DEPT, UNIV CALIF, 88- *Concurrent Pos:* Assoc ed, Inst Elec & Electronics Engrs, Transactions Automatic Control, 80- & J Optimization Theory, 80- *Mem:* Inst Elec & Electronics Engrs. *Res:* Optimal control problems; zero-sum and nonzero-sum differential games; optimization problems with vector-valued criteria; controllability problems; minmox problems. *Mailing Add:* Dept Aerospace Univ Calif Irvine CA 92717

SCHMITT, CHARLES RUDOLPH, b New York, NY, Mar 31, 20; m 45; c 2. APPLIED CHEMISTRY. *Educ:* Queens Col, NY, BS, 42. *Prof Exp:* Supvr, Plum Brook Ord Works, Sandusky, Ohio, 42-43; pross mech, Spec Eng Detachment, US Army, Oak Ridge, Tenn, 44-45; tech engr, K-25 Plant, Union Carbide Corp, Oak Ridge, Tenn, 46-49, develop engr & specialist, 50-75, supvr develop, Y-12 Plant, 75-80; SR SCIENTIST, BECHTEL CORP, OAK RIDGE, TENN, 81- *Concurrent Pos:* Consult, Rust Eng Co, 70-75. *Mem:* Am Chem Soc; Nat Asn Corrosion Engrs; Am Nuclear Soc. *Res:* Polymerization of polyfurfuryl alcohol resins; dezincification of brass in sea water; metallurgy studies of high purity tungsten; uranium solubility and corrosion studies; water treatment for scale and corrosion control, treatment of chemical and radioactive wastes. *Mailing Add:* 110 Adelphi Rd Oak Ridge TN 37830

SCHMITT, DONALD PETER, b New Hampton, Iowa, Oct 29, 41; m 67; c 4. AGRICULTURE, BOTANY- PHYTOPATHOLOGY. *Educ:* Iowa State Univ, BS, 67, MS, 69, PhD(plant path), 71. *Prof Exp:* Plant pathologist & nematologist plant disease, Div Plant Industs, Tenn Dept Agr, 71-75; plant pathologist & nematologist soybeans, Dept Plant Path, NC State Univ, 75-90; PLANT PATHOLOGIST & NEMATOLOGIST, DEPT PLANT PATH, UNIV HAWAII, 90- *Mem:* Am Phytopath Soc; Soc Nematologists; Sigma Xi; Orgn Trop Am Nematologists. *Res:* Ecology of nematodes on tropical crops; epidemiology of diseases of tropical crops caused by nematodes. *Mailing Add:* Dept Plant Path Univ Hawaii Honolulu HI 96822

SCHMITT, ERICH, b Sandhausen, Ger, Jan 7, 28; m 56; c 1. COMPUTER SCIENCE. *Educ:* Univ Karlsruhe, Dipl Ing, 60, Dr Ing(elec eng), 64, Venia legendi, 67. *Prof Exp:* Dir res dept, Inst Info Processing, Univ Karlsruhe, 64-67; chief adv avionics, Bell Aerosysts Co, 67-68; ASSOC PROF COMPUT SCI, STATE UNIV NY, BUFFALO, 68- *Mem:* Asn Comput Mach; Inst Elec & Electronics Engrs. *Res:* Adaptive computing methods; pattern recognition; automatic design; theory of adaptive automata; information theory and coding. *Mailing Add:* Dept Elec/Comput Eng 248 Bell Hall State Univ NY N Campus Buffalo NY 14214

SCHMITT, FRANCIS OTTO, b St Louis, Mo, Nov 23, 03; m 27; c 3. MOLECULAR NEUROBIOLOGY. *Educ:* Wash Univ, AB, 24, PhD(physiol), 27. *Hon Degrees:* Numerous from US & foreign univs, 50-81. *Prof Exp:* Nat Res Coun fel chem, Univ Calif, Berkeley, 27-28, Univ London, 28-29 & Kaiser Wilhelm Inst, 29; from asst prof to prof zool, Wash Univ, 29-40, head dept, 40-41; prof biol, 41-55, head dept, 42-55, inst prof biol, 55-69, found scientist, neurosci res, 77-82, EMER INST PROF BIOL, MASS INST TECHNOL, 69- *Concurrent Pos:* Trustee, Mass Gen Hosp, 47-; mem, study sect morphol & genetics, NIH, 49-53, chmn, study sect biophys & biophys chem, 54-58; mem, Nat Adv Health Coun, 59-62; Gen Med Sci Coun, 69-71; bd sci consults, Sloan-Kettering Inst Cancer Res, 63-72; chmn neurosci res prog, 62-74, chmn res found, 62- *Honors & Awards:* Alsop Award, Am Leather Chem Asn, 47; Lasker Award, Am Pub Health Asn, 56; T Duckett Jones Award, Helen Hay Whitney Found, 63. *Mem:* Nat Acad Sci; Soc Develop Biol (treas, Soc Growth & Develop, 45-56, pres, 47); Electron Micros Soc Am (pres, 49); NY Acad Sci; Soc Neurosci; Sigma Xi. *Res:* Molecular biology; investigation of molecular organization of tissues, particularly of nerve, connective tissue and muscle, by biophysical and physical-chemical means; integration of all disciplinary levels in study of physical basis of brain function. *Mailing Add:* Dept Biol Rm 16-512 Mass Inst Technol Cambridge MA 62139

SCHMITT, GEORGE FREDERICK, JR, b Louisville, Ky, Nov 3, 39; m 65; c 2. EROSION OF MATERIALS DUE TO IMPINGEMENT, RESPONSE OF MATERIALS TO HIGH ENERGY LASERS. *Educ:* Univ Louisville, Ky, BChE, 62, MChE, 63; Ohio State Univ, MBA, 66; Air War Col, dipl strategy, 69. *Prof Exp:* Lt high temperature coatings, USAF, 63-66, proj eng erosion res mat, Mat Lab, 66-81, group leader coatings & protective mat, 81-83, tech area mgr laser hardened mat, 83-86, prog mgr space survivability, 86-90, actg chief plans & progs, 90-91, ASST DIR NONMETALLIC MAT, MAT LAB, USAF, 91- *Concurrent Pos:* Chmn, Comt G-2, Am Soc Testing & Mat, 77-80; mem, Comt Erosion in Energy Systs, Nat Mat Adv Bd, 79-80; rep, USAF & Govt Int Exchange Agreements, 83- *Honors & Awards:* Merit Award, Am Soc Testing Mat. *Mem:* Fel Soc Advan Mat & Process Engrs (pres, 81-82); Am Inst Aeronaut & Astronaut; fel Am Soc Testing & Mat (secy, 72-76); Am Chem Soc; Soc Photo-Optical Instrumentation Engrs. *Res:* Response of materials to impingement of rain, dust and ice at subsonic to hypersonic velocities, materials included plastics, composites, ceramics, elastomers and metals; published 30 technical reports and 60 articles. *Mailing Add:* 1500 Wardmier Dr Dayton OH 45459

SCHMITT, GEORGE JOSEPH, b Farmingdale, NY, June 21, 28; m 52; c 4. POLYMER CHEMISTRY. *Educ:* Polytech Inst Brooklyn, BS, 50; State Univ NY Col Forestry, Syracuse Univ, PhD(chem), 60. *Prof Exp:* Develop chemist, Am Cyanamid Co, 53-57; sr res chemist, cent res lab, Allied Chem Corp, 60-61, res supvr polymer chem, Allied Corp, 61-62, dir lab res, 62-64, asst dir cent res lab, 64-68, mgr polymer sci, Corp Res Lab, 68-80; DIR CORP STRUCT POLYMER LAB, ALLIED-SIGNAL CORP, 80- *Concurrent Pos:* mem bd dirs, Res & Develop Coun, NJ, 81-84. *Mem:* Am Chem Soc; Sigma Xi; Soc Advan Mat & Process Eng. *Res:* Free radical, ionic and condensation polymerization; fibers; polymer composites; membranes; electrically conducting polymers; biopolymers. *Mailing Add:* Chem Res Ctr Allied-Signal Corp Box 1021R Morristown NJ 07962

SCHMITT, HAROLD WILLIAM, b Sequin, Tex, Aug 11, 28; m 52; c 3. ATOMIC PHYSICS, ELEMENTAL ANALYSIS. *Educ:* Univ Tex, BA, 48, MA, 52, PhD(physics), 54. *Prof Exp:* Asst physics, Los Alamos Sci Lab, 52-54; physicist, Oak Ridge Nat Lab, 58-73, group leader, Physics of Fission Group, 60-73; pres, Environ Systs Corp, 73-81; FOUNDING PRES, ATOM SCI, INC, 81- *Concurrent Pos:* Founding pres & chmn bd dirs, Ortec, Inc, 60-64; guest scientist, Nuclear Res Ctr, Karlsruhe, Ger, 66-67; guest prof, Munich Tech Univ, 69 & Univ Frankfurt, 70. *Mem:* AAAS; fel Am Phys Soc; Sigma Xi; Am Chem Soc; Am Soc Testing & Mat. *Res:* Fission physics; neutron physics; nuclear reactions; accelerators; reactors; detectors; instrumentation; atomic physics; elemental analysis; energy and environmental sciences. *Mailing Add:* 121 Canterbury Rd Oak Ridge TN 37830

SCHMITT, JOHANNA, b Philadelphia, Pa, Mar 12, 53; m 83. PLANT POPULATION BIOLOGY, ECOLOGICAL GENETICS. *Educ:* Swarthmore Col, BA, 74; Stanford Univ, PhD(biol sci), 81. *Prof Exp:* Postdoctoral res assoc, Duke Univ, 81-82; asst prof, 82-87, ASSOC PROF BIOL, BROWN UNIV, 87- *Concurrent Pos:* Prin investr, NSF, 84-; assoc ed, Evolution, 90-92; coun mem, Soc Study Evolution, 90-92. *Mem:* Soc Study Evolution; Ecol Soc Am; Am Soc Naturalists; Bot Soc Am; Soc Conserv Biol; AAAS. *Res:* Plant population biology; ecological genetics; breeding system evolution; density-dependent phenomena; gene flow and population structure; evolutionary and ecological consequences of maternal effects. *Mailing Add:* Grad Prog Ecol & Evolutionary Biol Brown Univ Box G-W301 Providence RI 02912

SCHMITT, JOHN ARVID, JR, b Buffalo, NY, July 30, 25; m 47; c 2. MEDICAL MYCOLOGY. *Educ:* Univ Mich, BS, 49, MS, 50, PhD(mycol), 54. *Prof Exp:* Prof biol & head dept, Findlay Col, 53-54; asst prof, Univ Miss, 54-55; instr bot & plant path, 55-57, from asst prof to assoc prof, 57-69, assoc prof med microbiol, 66-71, chmn dept bot, 69-74, adj assoc prof med, Col Med, 67-74, prof bot, 69-86, head gen biol, 79-86, EMER PROF, OHIO STATE UNIV, 87- *Concurrent Pos:* Consult mycol, Merrell-Nat Labs & Allergy Labs Ohio, Inc, 65-80 & Philips Roxane Labs, 74-77. *Mem:* Mycol Soc Am; Am Soc Microbiol; Bot Soc Am; Med Mycol Soc Americas; Int Soc Human & Animal Mycol. *Res:* Medical mycology; zoopathogenic fungi, especially Candida albicans; paint mildew fungi; microbial succession in the establishment of paint mildew. *Mailing Add:* Bot Dept Ohio State Univ 1735 Neil Ave Columbus OH 43210

SCHMITT, JOHN LEIGH, b Newberry, Mich, July 30, 41; m 66; c 1. INSTRUMENTATION, ASTRONOMY. *Educ:* Mich Col Mining & Technol, BS, 63; Univ Mich, MS, 64, PhD(astron), 68. *Prof Exp:* Fel & part-time lectr astron, Univ Toronto, 68-69; asst prof physics, Southwestern at Memphis, 69-74; vis asst prof, Univ Mo, Rolla, 74-76, res asst prof physics, 76-85, res assoc prof physics & cloud physics, 85-90, ASSOC PROF PHYSICS, UNIV MO, ROLLA, 90- *Mem:* Am Astron Soc; Am Optic Soc; Am Chem Soc. *Res:* Vapor to liquid nucleation, cloud chambers and optical instrumentation. *Mailing Add:* Cloud & Aerosol Sci Lab Univ Mo Rolla MO 65401

SCHMITT, JOSEPH LAWRENCE, JR, b Cumberland, Md, Sept 22, 41; m 63; c 2. PHYSICAL CHEMISTRY. *Educ:* Shippensburg State Col, BS, 63; Bowling Green State Univ, MA, 67; Pa State Univ, PhD(fuel sci), 70. *Prof Exp:* Teacher pub sch, Pa, 63-67; res chemist, 70-74, sr res chemist, 74-77, proj leader, 77-85, Mgr, 85-86, DIR, AM CYANAMID CO, 86- *Mem:* Am Chem Soc; AAAS; Sigma Xi. *Res:* Heterogeneous catalysis; surface chemistry; carbon chemistry. *Mailing Add:* Four Settlers Rd Bethel CT 06801

SCHMITT, JOSEPH MICHAEL, b Louisville, Ky, Feb 9, 30; m 52; c 6. POLYMER CHEMISTRY. *Educ:* Univ Louisville, BS, 51, PhD(chem), 57. *Prof Exp:* Res chemist, 57-62, sr res chemist, 62-67, PRIN CHEMIST, AM CYNAMID CO, 67- *Mem:* Am Chem Soc. *Res:* Plastics; homopolymers, copolymers, multipolymer blends and properties; flocculants; polymers for water treatment. *Mailing Add:* PO Box 336 Ridgefield CT 06877

SCHMITT, KLAUS, b Rimbach, Ger, May 14, 40; US citizen; m 85; c 2. MATHEMATICS. *Educ:* St Olaf Col, BA, 62; Univ Nebr, MA, 64, PhD(math), 67. *Prof Exp:* Asst prof math, Nebr Wesleyan Univ, 66-67; from asst prof to assoc prof, 67-75, PROF MATH, UNIV UTAH, 75-, CHMN DEPT, 88- *Concurrent Pos:* Res grants, NASA, 67, NSF, 69-71 & US Army, 71-78, NSF, 78-; vis prof, Univ Wurzburg & Univ Karlsruhe, Ger, 73-74, Univ Bremen, T U Berlin, Univ Louvain, Univ Heidelberg; Alexander von Humboldt sr us scientist award, 78-79. *Mem:* AAAS; Math Asn Am; Am Math Soc. *Res:* Differential equations; nonlinear analysis; functional differential equations. *Mailing Add:* Dept Math Univ Utah Salt Lake City UT 84112

SCHMITT, NEIL MARTIN, b Pekin, Ill, Oct 25, 40; m 63; c 2. BIOMEDICAL ENGINEERING. *Educ:* Univ Ark, Fayetteville, BSEE, 63, MSEE, 64; Southern Methodist Univ, PhD(elec eng), 69. *Prof Exp:* Systs engr, IBM Corp, 66-67; engr, Tex Instruments, Inc, 67-70; assoc prof, 70-80, PROF ELEC ENG, UNIV ARK, FAYETTEVILLE, 80- *Concurrent Pos:* NSF res grant elec eng, Univ Ark, Fayetteville, 71-72. *Mem:* Inst Elec & Electronics Engrs; Biomed Eng Soc; Asn Advan Med Instrumentation; Am Soc Eng Educ. *Res:* Health care delivery systems; early detection of heart disease. *Mailing Add:* 1295 Joe Fred Stark Rd Fayetteville AR 72701

SCHMITT, OTTO HERBERT, b St Louis, Mo, Apr 6, 13; m 37. BIOPHYSICS, BIOMEDICAL ENGINEERING. *Educ:* Wash Univ, AB, 34, PhD(physics, zool), 37. *Prof Exp:* Nat Res Coun fel, Univ Col, London, 38, Sir Halley Stewart fel, 39; from instr to prof zool & physics, Univ Minn, Minneapolis, 39-80, prof biophys, 49-80, prof elec eng, 68-80, PROF BIOMED ENG, UNIV MINN, MINNEAPOLIS, 73- *Concurrent Pos:* Off investr, Nat Defense Res Comt Contract, 40-42; res engr, Columbia Univ, 42-43; supvr engr, Spec Devices Div, Airborne Instruments Lab, NY, 43-47; consult, USPHS & Inst Defense Anal; mem adv panel, Space Sci Bd Biol & Psychol, 58-61; chmn exec coun bioastronaut, Joint Armed Forces-Nat Acad Sci, 58-61. *Honors & Awards:* Lovelace Award, 60; Morlock Award, Inst Elec & Electronics Engrs, 63; Wetherill Medal, Franklin Inst, 72, Franklin Inst Medal, 84; Centennial Medal, Inst Elec & Electronics Engrs, 87. *Mem:* Nat Acad Eng; fel Am Phys Soc; Biophys Soc; Am Physiol Soc; Am Inst Aeronaut & Astronaut. *Res:* Nerve impulse mechanisms; tridimensional oscilloscopic displays; bivalent computers; biological tissue impedance analyses; direct current transformers; trigger circuits; electronic plethysmography; antenna radiation pattern measurements; stereovectorelectrocardiography; phase space displays; bioastronautics; biomimetics; electromagneto biology; technical optimization of biomedical communication and control systems; strand epidemiology; development of biometic sci and technol; personally portable whole life medical history; computerized electrosurgery. *Mailing Add:* 147 Pillsbury Dr Univ Minn Minneapolis MN 55455

SCHMITT, RAYMOND W, JR, b Pittsburgh, Pa, Mar 18, 50; m 81; c 3. PHYSICAL OCEANOGRAPHY. *Educ:* Carnegie-Mellon Univ, BS, 72; Univ RI, PhD(oceanog), 78. *Prof Exp:* Res assoc, Grad Sch Oceanog, Univ RI, 77-78; fel, 78-79, investr, 79-80, asst scientist, 80-84, ASSOC SCIENTIST, WOODS HOLE OCEANOG INST, 84- *Mem:* Am Geophys Union; AAAS. *Res:* Oceanic mixing and microstructure; double-diffusive convection (salt fingers); geophysical fluid dynamics. *Mailing Add:* Woods Hole Oceanog Inst Woods Hole MA 02543

SCHMITT, ROLAND WALTER, b Seguin, Tex, July 24, 23; m 51, 57; c 4. SOLID STATE PHYSICS. *Educ:* Univ Tex, BA & BS, 47, MA, 48; Rice Inst, PhD(physics), 51. *Hon Degrees:* Dr, Univ Pa, Worcester Polytech Inst & Union Col, 85 Lehigh Univ, 86 & Univ SC, 88. *Prof Exp:* Res assoc physics, Res Lab, Gen Elec Co, 51-57, mgr mat studies sect, 57-65; res assoc, Div Eng & Appl Physics, Grad Sch Pub Admin, Harvard Univ, 65; mgr, Metall & Ceramics Lab, Gen Elec Res & Develop Ctr, 66-68, res & develop mgr phys sci & eng, 68-74, energy sci & eng, from vpres to sr vpres corp res & develop, 78-86, sr vpres sci & technol, 86-88; PRES, RENSSELAER POLYTECH INST, 88- *Concurrent Pos:* Mem, Liaison Subcomt Mgt & Technol, Nat Acad Sci Adv Comt IIASA, Panel Condensed Matter, Physics Surv Comt & Comt on Surv Mat Sci & Eng; mem, Energy Adv Bd, Walt Disney Enterprises; past mem, adv bds, Univ Tex & Univ Va; chmn eval panel, Inst Basic Standards, Nat Bur Standards; mem, Nat Res Coun Solid State Sci Comt & Comt Nat Progs, Numerical Data Adv Bd; mem coun, Nat Acad Eng, 83-89; pres elect & mem bd dirs, Indust Res Inst; chmn, Nat Sci Bd, NSF, 84-88 & Coun Res & Technol; dir & mem bd dirs, Gen Signal Corp; dir, Coun Superconductivity

Am Competitiveness. *Honors & Awards:* Indust Res Inst Medalist Award, 89. *Mem:* Nat Acad Eng; fel Am Phys Soc; fel AAAS; fel Am Acad Arts & Sci; fel Inst Elec & Electronics Engrs; foreign mem Royal Swed Acad Eng Sci; foreign assoc Eng Acad Japan. *Mailing Add:* Rensselaer Polytech Inst 110 Eighth St Troy NY 12180-3590

SCHMITT, ROMAN A, b Johnsburg, Ill, Nov 13, 25; m 54; c 4. COSMOCHEMISTRY, GEOCHEMISTRY. *Educ:* Univ Chicago, MS, 50, PhD(nuclear chem), 53. *Prof Exp:* Instr nuclear chem, Univ Ill, 53-54, res assoc, 54-56; res scientist chem, Gen Atomic Div, Gen Dynamics Corp, 56-66; assoc prof, 66-69, PROF CHEM, ORE STATE UNIV, 69-, PROF GEOL & OCEANOG, 85- *Concurrent Pos:* Consult, NASA, 71-75. *Honors & Awards:* George P Merrill Award, Nat Acad Sci, 72. *Mem:* Fel AAAS; Geochem Soc; fel Meteoritical Soc. *Res:* Neutron activation analysis of rare earth elements and other elements in meteorites; terrestrial and lunar matter; cosmochemistry. *Mailing Add:* Dept Chem Ore State Univ Corvallis OR 97331

SCHMITTER, RUTH ELIZABETH, b Detroit, Mich. CELL BIOLOGY, PHYCOLOGY. *Educ:* Mich State Univ, BS, 64; Univ Edinburgh, MSc, 66; Harvard Univ, PhD(biol), 73. *Prof Exp:* Sr technician electron micros, AEC Plant Res Lab, Mich State Univ, 66-67; res fel biol, Harvard Univ, 73-74; Brown fel bot, Yale Univ, 74-75; asst prof biol, Univ Mass, Boston, 75-82; ASSOC PROF BIOL, ALBION COL, 82- *Concurrent Pos:* Fulbright scholar, Int Inst Educ, 64-66; vis assoc prof, Univ Okla, 88. *Mem:* Am Soc Cell Biol; Am Phycol Soc; Electron Micros Soc Am; Brit Phycol Soc; Int Soc Protozoologists; Hastings Ctr. *Res:* Cell ultrastructure, especially functional correlates of dinoflagellate fine structure, and organelle development; algal physiology, including algal nutrition; biochemistry of bioluminescence. *Mailing Add:* Dept Biol Albion Col Albion MI 49224-1899

SCHMITTHENNER, AUGUST FREDRICK, b Kotagiri, SIndia, Apr 16, 26; US citizen; m 54; c 2. PLANT PATHOLOGY. *Educ:* Gettysburg Col, BA, 49; Ohio State Univ, MSc, 51, PhD(bot), 53. *Prof Exp:* From instr to assoc prof, 52-66, PROF PLANT PATH, OHIO AGR RES & DEVELOP CTR & OHIO STATE UNIV, 66- *Mem:* Am Phytopath Soc; Sigma Xi. *Res:* Forage crop, soybean and root rot diseases; physiology of oomycetes and parasitism; photobiology; bean diseases. *Mailing Add:* 311 Elm Dr Wooster OH 44691

SCHMITZ, EUGENE H, b Wamego, Kans, Aug 13, 34; m 60; c 2. INVERTEBRATE ZOOLOGY. *Educ:* Univ Kans, AB, 56; Univ Colo, MA, 58, PhD(zool), 61. *Prof Exp:* Instr biol, Univ Colo, 59-60; from asst prof to assoc prof zool, La State Univ, 61-65; from asst prof to assoc prof, 65-74, PROF ZOOL, UNIV ARK, FAYETTEVILLE, 74- *Concurrent Pos:* Ed, Trans Am Micros Soc. *Mem:* Am Micros Soc; Sigma Xi; Crustacean Soc; Counc Biol Ed. *Res:* Invertebrate morphology. *Mailing Add:* Dept Biol Sci Univ Ark Fayetteville AR 72701

SCHMITZ, FRANCIS JOHN, b Raymond, Iowa, Jan 18, 32; m 61; c 3. NATURAL PRODUCTS CHEMISTRY, MARINE CHEMISTRY. *Educ:* Maryknoll Sem, BA, 54; Loras Col, BS, 58; Univ Calif, Berkeley, PhD, 61. *Prof Exp:* NIH fel, Stanford Univ, 61-62, NSF fel, 62-63; from asst prof to assoc prof, 63-71, PROF CHEM, UNIV OKLA, 71- *Mem:* Am Chem Soc; The Chem Soc. *Res:* Structure determination of natural products, emphasis on marine natural products; synthesis of natural products. *Mailing Add:* Dept Chem Univ Okla 620 Parrington Oval Norman OK 73019

SCHMITZ, GEORGE WILLIAM, b Minneapolis, Minn, Dec 15, 19. AGRONOMY. *Educ:* Univ Ariz, BS, 48; Ohio State Univ, MS, 50, PhD(soils), 52. *Prof Exp:* Sr agronomist, Zonolite Res Lab, 52-56; res agronomist, Calif Spray Chem Corp Div, Standard Oil Co, Calif, 56-60; asst prof agron, Fresno State Col, 60-66; assoc prof, 66-72, prof agron, 72-77, PROF PLANT & SOIL SCI, CALIF STATE POLYTECH UNIV, POMONA, 77- *Res:* Soil fertility; plant physiology. *Mailing Add:* Dept Plant & Soil Sci Calif State Polytech Univ 3801 W Temple Ave Pomona CA 91768

SCHMITZ, HAROLD GREGORY, b Helena, Mont, Aug 31, 43; m 66; c 4. ELECTRICAL ENGINEERING. *Educ:* Carroll Col, Mont, BA, 65; Mont State Univ, BS, 66, MS, 67, PhD(elec eng), 70. *Prof Exp:* Res engr, Mont State Univ, 69-70; prin investr comput technol, Honeywell Systs & Res Ctr, 70-76, sect chief, Comput Systs Technol, Honeywell, Inc, 76-89; vpres & gen mgr, Secure Comput Technol Corp, 89-91; CONSULT, 91- *Mem:* Inst Elec & Electronics Engrs. *Res:* Research and advanced development in the area of computer architecture and organization. *Mailing Add:* 1471 Bussard Ct Arden Hills MN 55112-3628

SCHMITZ, HENRY, b Vienna, Austria, Oct 2, 17; nat US; m 40; c 2. ORGANIC CHEMISTRY. *Educ:* NY Univ, BA, 47; Rutgers Univ, MS & PhD(org chem), 50. *Prof Exp:* Res chemist, J T Baker Chem Co, Vick Chem Co, 50-55; sr res scientist, Bristol Labs Div, Bristol-Myers Co, 55-81; RETIRED. *Concurrent Pos:* Adj prof, Onondaga Community Col, Syracuse, 80-88. *Mem:* Am Chem Soc; Sigma Xi. *Res:* Steroids; antibiotics; natural products. *Mailing Add:* 323 DeForest Rd Syracuse NY 13214

SCHMITZ, JOHN ALBERT, b Silverton, Ore, Oct 21, 40; m 71. VETERINARY PATHOLOGY & MICROBIOLOGY. *Educ:* Colo State Univ, DVM, 64; Univ Mo-Columbia, PhD(path), 71. *Prof Exp:* Res assoc path, Col Vet Med, Univ Mo, 68-71; asst prof, Univ Nebr, 71-72; assoc prof, Ore State Univ, 72-78, prof, 78-84, dir vet diag lab, 76-84, dir path serv, environ health sci ctr, 78-84; HEAD DEPT VET SCI, UNIV NEBR, LINCOLN, 84- *Mem:* Am Col Vet Pathologists; Int Acad Path; Am Vet Med Asn; AAAS; Am Asn Vet Lab Diagnosticians. *Res:* Infectious diseases of food-producing animals; congenital, infectious, nutritional, and toxicological conditions found in animals. *Mailing Add:* Dept Vet Sci Univ Nebr E Campus Lincoln NE 68583-0905

SCHMITZ, JOHN VINCENT, polymer chemistry, for more information see previous edition

SCHMITZ, KENNETH STANLEY, b St Louis, Mo, Sept 6, 43. BIOPHYSICAL CHEMISTRY. *Educ:* Greenville Col, BA, 66; Univ Wash, PhD(chem), 72. *Prof Exp:* NIH fel, Univ Wash, 72 & Stanford Univ, 72-73; asst prof chem, Fla Atlantic Univ, 73-75; from asst prof to assoc prof, 75-86, PROF CHEM, UNIV MO, KANSAS CITY, 86- *Concurrent Pos:* Univ Mo-Kansas City fac fel, 84; Am Soc Eng Educ fel, 85. *Honors & Awards:* Univ Kans City Fac Fel, 84. *Mem:* Int Asn Colloid & Interface Scientists; Am Chem Soc; Biophys Soc; Sigma Xi; Am Soc Biol Chem. *Res:* Conformational changes in biopolymers determined by quasielastic light scattering; theory of diffusion-controlled reactions; effect of cooperativity on binding isotherms; nonequilibrium thermodynamics. *Mailing Add:* Dept Chem Univ Mo Kansas City MO 64110

SCHMITZ, NORBERT LEWIS, b Green Bay, Wis, May 18, 21; m 50; c 4. ELECTRICAL ENGINEERING. *Educ:* Univ Wis, BS, 42, MS, 47, PhD, 51. *Prof Exp:* Instr, Eve Tech Div, Milwaukee Voc Sch, 44-45; instr elec eng, Marquette Univ, 45-46; from instr to prof, 47-83, EMER PROF ELEC ENG, UNIV WIS-MADISON, 83- *Concurrent Pos:* Elec engr, Cutler Hammer Inc, 42-46, consult, 46-47; consult, Gisholt Mach Co, 51, John Oster Mfg Co, 55-57, Sundstrand Aviation Co, 57-, Caterpillar Tractor Co, 61-70 & Marathon Elec Mfg Corp, 64-72. *Mem:* Inst Elec & Electronics Engrs; fel NY Acad Sci. *Res:* Electric machine theory and control; industrial control; power semiconductor applications. *Mailing Add:* 4717 Co Tr M Middleton WI 53562

SCHMITZ, ROBERT L, b Chicago, Ill, Mar 10, 14; c 5. CANCER. *Educ:* Univ Chicago, BS, 36, MD, 38; Am Bd Surg, dipl, 48. *Prof Exp:* From assoc clin prof to clin prof surg, Stritch Sch Med, Loyola Univ Chicago, 46-72; PROF SURG, UNIV ILL, 72- *Concurrent Pos:* Assoc attend surgeon, Mercy Hosp, Chicago, 46-58, sr attend surgeon, 58- *Mem:* Am Cancer Soc; AMA; fel Am Col Surg. *Res:* Surgical oncology. *Mailing Add:* Dept Surg M/C 958 Mercy Hosp & Med Ctr Chicago IL 60616

SCHMITZ, ROGER A(NTHONY), b Carlyle, Ill, Oct 22, 34; m 57; c 3. CHEMICAL ENGINEERING. *Educ:* Univ Ill, BS, 59; Univ Minn, PhD(chem eng), 62. *Prof Exp:* Instr chem eng, Univ Minn, 60-62; from asst prof to prof chem eng, Univ Ill, 62-79; chmn dept, 79-81, dean eng, 81-87, KEATING-CRAWFORD PROF CHEM ENG, UNIV NOTRE DAME, 79-, VPRES & ASSOC PROVOST, 87- *Concurrent Pos:* Guggenheim fel, 68-69. *Honors & Awards:* Colburn Award, Am Inst Chem Engrs, 70; Westinghouse Award, Am Soc Eng Educ, 77; Wilhelm Award, Am Inst Chem Engrs, 81. *Mem:* Nat Acad Eng; Am Inst Chem Engrs; Am Soc Eng Educ. *Res:* Dynamics of chemical reaction systems. *Mailing Add:* 202 Admin Bldg Univ Notre Dame Notre Dame IN 46556

SCHMITZ, WILLIAM JOSEPH, JR, b Houston, Tex, Dec 20, 37; m 59; c 4. PHYSICAL OCEANOGRAPHY. *Educ:* Univ Miami, ScB, 61, PhD(phys oceanog), 66. *Prof Exp:* Res aide, Univ Miami, 59-61; instr, Univ Miami, 64-66; fel, Nova Univ, 66-67; asst scientist, 67-71, assoc scientist, 71-79, SR SCIENTIST, WOODS HOLE OCEANOG INST, 79- *Mem:* Am Geophys Union. *Res:* Low-frequency ocean circulation. *Mailing Add:* Dept Oceanog Woods Hole Oceanog Inst Woods Hole MA 02543

SCHMITZ, WILLIAM ROBERT, b Wauwatosa, Wis, Jan 24, 24; m 51; c 2. LIMNOLOGY, FISH BIOLOGY. *Educ:* Univ Wis, BS, 51, MS, 53, PhD(zool), 58. *Prof Exp:* Asst zool, Univ Wis, 52-54, proj asst, 54-58, proj assoc, 58-59, instr biol, Ctr Syst, 59-60, asst prof, 60-66, chmn dept bot & zool, Ctr Syst, 67-70, assoc prof, 66-76, asst dir, Trout Lakes Res Sta, 67-77, prof zool, 76-89, EMER PROF CTR SYST, UNIV WIS, 89- *Mem:* Am Soc Limnol & Oceanog; Soc Int Limnol. *Res:* Hydrobiology; limnology, especially of ice-bound lakes; fisheries biology. *Mailing Add:* Univ Wis 518 S Seventh Ave Wausau WI 54401-5396

SCHMUCKER, DOUGLAS LEES, b McKeesport, Pa, Jan 22, 44. CELLULAR BIOLOGY, ANATOMY. *Educ:* Kenyon Col, AB, 65; Clark Univ, MA, 68, PhD(biol), 72. *Prof Exp:* Asst prof, 75-80, ASSOC PROF ANAT, UNIV CALIF, SAN FRANCISCO, 80-; RES BIOLOGIST, VET ADMIN MED CTR, SAN FRANCISCO, 73-, ASST CHIEF, CELL BIOL & AGING SECT, 80- *Concurrent Pos:* NIH grant, 72-73; fel anat, Nat Inst Aging grant, 77-81; Sigma Xi lectr, 81. *Mem:* Soc Exp Biol & Med; Am Soc Cell Biol; fel Geront Soc; Am Asn Anatomists; Am Aging Asn. *Res:* Age-related changes in cellular structure and function; mechanisms of hepatic bile secretion; mechanisms of drug actions; liver pathology; lipid metabolism; lipoprotein synthesis and secretion; drug metabolism. *Mailing Add:* Cell Biol Sect 151E Vet Admin Med Ctr 4150 Clement St San Francisco CA 94121

SCHMUCKLER, JOSEPH S, b Philadelphia, Pa, Feb 15, 27; m 50; c 4. SCIENCE EDUCATION. *Educ:* Univ Pa, BS, 52, MS, 54, EdD(chem educ), 68. *Prof Exp:* Instr sci educ, Univ Pa, 64-67; assoc prof sci, 68-73, PROF CHEM & SCI, TEMPLE UNIV, 73-, CHMN DEPT SCI EDUC, 69- *Concurrent Pos:* Chem Teacher, Haverford Twp High Sch, 53-68; partic, Chem Educ Mat Study Prog; consult, Sadtler Res Lab, 54-62; mem bd gov, Chem Educ Proj Corp, 64-67; pres, Chem Proj Corp, 80-81; prof chem educ, Tianjin Normal Univ, People's Rep China, 80- *Honors & Awards:* James Bryant Conant Award, 68; Benjamin Rush Medal, Chem Indust Coun, 68; Lindbach Award, 76. *Mem:* Am Chem Soc; Nat Sci Teachers Asn; Franklin Inst; fel Am Inst Chemists; Sigma Xi. *Res:* Chemistry; organic synthesis; science education at the secondary school level; chemistry education. *Mailing Add:* 864 Beechwood Rd Havertown PA 19083

SCHMUDE, KEITH E, b Rockford, Ill, Feb 10, 34; m 55; c 2. PHYSICAL CHEMISTRY. *Educ:* Carroll Col, BS, 55; Univ Rochester, PhD(chem), 59. *Prof Exp:* Assoc physicist, Armour Res Found, 59-61; from asst prof to assoc prof chem, Parsons Col, 61-64; res chemist, Dacron Res Lab, 64-68, sr res chemist, Textile Res Lab, 68-83, RES ASSOC, TEXTILE RES LAB, E I DU PONT DE NEMOURS & CO, INC, 83- *Mem:* AAAS. *Res:* Synthetic fibers; kinetics; radiation and nuclear chemistry. *Mailing Add:* Fibers Develop Ctr E I du Pont de Nemours & Co Inc Wilmington DE 19880-0702

SCHMUGGE, THOMAS JOSEPH, b Chicago, Ill, Oct 18, 37; m 61; c 4. REMOTE SENSING, EVAPOTRANSPIRATION. *Educ:* Ill Inst Technol, BS, 59; Univ Calif, Berkeley, PhD(physics), 65. *Prof Exp:* Asst prof physics, Trinity Col, Conn, 64-70; sr res assoc, Nat Acad Sci, 70-71; physicist, NASA-Goddard Space Flight Ctr, 71-86; PHYSICIST, HYDROL LAB, AGR RES SERV, USDA, BELTSVILLE, MD, 86- *Concurrent Pos:* Assoc ed, J Geophys Res, 79-83. *Mem:* AAAS; Am Geophys Union; Inst Elec & Electronics Engrs. *Res:* Magnetic resonance of rare earth ions; low temperature physics; remote sensing of the environment and interaction of electromagnetic waves with natural materials; microwave infrared emission from natural surfaces; soil moisture; soil physics; snow; hydrology; evapotranspiration and the atmospheric boundary layer. *Mailing Add:* Hydrol Lab-007 USDA-Agr Res Serv Beltsville MD 20705

SCHMUKLER, SEYMOUR, b Baltimore, Md, Oct 27, 25; m 57; c 2. ORGANIC CHEMISTRY, POLYMER CHEMISTRY. *Educ:* Johns Hopkins Univ, AB, 48; Columbia Univ, AM, 50, PhD(chem), 54. *Prof Exp:* Asst, Col Physicians & Surg, Columbia Univ, 53; res chemist, E I du Pont de Nemours & Co, 53-54; res chemist, Colgate-Palmolive Co, 54-57; develop chemist, Merck & Co, Inc, 57-59; res & develop chemist, Nopco Chem Co, 59-63; develop chemist, Gen Elec Co, Mass, 63-67; res assoc, Quantum-USI Chem Div, 67-88; CONSULT, 88- *Mem:* Am Chem Soc; Tech Asn Pulp & Paper Indust. *Res:* Molecular rearrangements; synthetic organic and polymer chemistry; polymer modification, evaluation and process development; polymer extrudable and coextrudable adhesives; flexible packaging, rigid packaging and peelable seals. *Mailing Add:* 445 S Elm St Palatine IL 60067

SCHMULBACH, CHARLES DAVID, b Belleville, Ill, Feb 2, 29; m 55; c 3. INORGANIC CHEMISTRY. *Educ:* Univ Ill, PhD, 58. *Prof Exp:* Asst prof chem, Pa State Univ, 58-65; assoc prof, 65-70, chmn dept, 75-82, PROF CHEM & BIOCHEM, SOUTHERN ILL UNIV, CARBONDALE, 70- *Mem:* Am Chem Soc. *Res:* Stabilization of uncommon oxidation states; electrochemical synthesis of inorganic compounds; homogeneous catalysis by transition metal complexes. *Mailing Add:* Dept Chem & Biochem Southern Ill Univ Carbondale IL 62901

SCHMULBACH, JAMES C, b New Athens, Ill, July 5, 31; c 2. FISH BIOLOGY. *Educ:* Southern Ill Univ, BA, 53, MA, 57; Iowa State Univ, PhD(fisheries biol), 59. *Prof Exp:* Asst, Southern Ill Univ, 55-57; asst, Iowa State Univ, 57-59; asst prof, 59-65, PROF BIOL, UNIV SDAK, 65- *Concurrent Pos:* Fel, Marine Lab, Miami, 63. *Mem:* AAAS; Am Fisheries Soc; Am Inst Biol Sci; Sigma Xi. *Res:* Limnology; bionomics of fishes; macrobenthos of lotic environments. *Mailing Add:* 922 Ridgecrest Dr Vermillion SD 57069

SCHMUTZ, ERVIN MARCELL, b St George, Utah, Oct 26, 15; m 36; c 1. RANGE MANAGEMENT, ECOLOGY. *Educ:* Utah State Univ, BS, 39, MS, 41; Univ Ariz, PhD(plant sci), 63. *Prof Exp:* Range exam, Agr Adjust Admin, 37, 39-40 & 41; sr fire guard, US Forest Serv, 38; range exam, Bur Animal Indust, 40; range conservationist, Soil Conserv Serv, USDA, 41-48 & 50-52, dist conservationist, 48-50, work unit conservationist, 52-55; res assoc weed control, 55-56, from instr to prof range mgt, 56-81, EMER PROF RANGE MGT, UNIV ARIZ, 82- *Concurrent Pos:* Res scientist, Ariz Agr Exp Sta; mem, Range Mgt Educ Coun; consult, range mgt. *Honors & Awards:* Commendation Award, Soil Conserv Soc Am, 74. *Mem:* Fel Soc Range Mgt; fel Soil Conserv Soc Am; Sigma Xi. *Res:* Range ecology, evaluation, conservation and management; reseeding, poisonous, allergenic and landscaping plants; book publisher. *Mailing Add:* Sch Renewable Natural Resources Univ Ariz Tucson AZ 85721-0001

SCHMUTZ, JOSEF KONRAD, b Ulm, WGermany, Oct 5, 50; Can citizen; m 72. RAPTOR BIOLOGY, WATERFOWL BIOLOGY. *Educ:* Univ Wis-Stevens Pt, BSc, 74; Univ Alta, MSc, 77; Queen's Univ, PhD(biol), 81. *Prof Exp:* Res assoc wildlife mgt, Arctic Inst NAm, 82-83; res assoc wildlife mgt, 83-84, asst prof biol, Dept Biol, 84-87, RES ASSOC, WILDLIFE BIOL, UNIV SASK, 87- *Concurrent Pos:* Scholar, Nat Res Coun, Govt Can, 77-79, Ont Grad Scholar, Govt Ont, 79-81; wildlife biologist, resource mgt, Webb Environ Serv Ltd, 82. *Honors & Awards:* J H Albertson Award, Univ Wis-Stevens Point, 73. *Mem:* AAAS; Am Ornith Union; Can Soc Zoologists; Cooper Ornith Soc; Wildlife Soc. *Res:* Evolutionary biology of birds; wildlife management. *Mailing Add:* Dept Biol Univ Sask Saskatoon SK S7N 0W0 Can

SCHNAAR, RONALD LEE, b Detroit, Mich, Nov 1, 50; m 72; c 3. MEMBRANE BIOCHEMISTRY, CELL-CELL INTERACTIONS. *Educ:* Univ Mich, BS, 72; Johns Hopkins Univ, PhD(biochem), 76. *Prof Exp:* Res fel biochem, Johns Hopkins Univ, 77; fel neurobiol, Nat Heart, Lung & Blood Inst, NIH, 77-79; asst prof pharmacol, 79-83, assoc prof pharmacol & neurosci, 84-90, PROF PHARMACOL & NEUROSCI, SCH MED, JOHNS HOPKINS UNIV, 90- *Concurrent Pos:* Fac Res Award, Am Cancer Soc, 84-89. *Mem:* Am Soc Cell Biol; Soc Neurosci. *Res:* The role of cell membrane carbohydrate (glycolipids and glycoproteins) in the control of cell-cell interactions; metabolism of cell surface carbohydrates during neuronal differentiation; neurotransmitter and neurotoxin mechanisms. *Mailing Add:* Dept Pharmacol Sch Med Johns Hopkins Univ 725 N Wolfe St Baltimore MD 21205

SCHNAARE, ROGER L, b Staunton, Ill, June 24, 38; m 60; c 3. PHARMACEUTICS, PARENTERALS. *Educ:* St Louis Col Pharm, BS, 60; Purdue Univ, MS, 63, PhD(pharm), 65. *Prof Exp:* Asst prof pharm, St Louis Col Pharm, 65-68; from asst prof to assoc prof, 68-77, PROF PHARM, PHILADELPHIA COL PHARM & SCI, 77- *Mem:* Am Pharmaceut Asn; Sigma Xi; Am Asn Pharmaceut Scientists. *Res:* Suspension, emulsion and parenteral dosage form design and development. *Mailing Add:* 230 Hutchinson Ave Haddonfield NJ 08033-3914

SCHNABEL, GEORGE JOSEPH, b Bornet, Tex, May 7, 16; m 40; c 2. NON-DESTRUCTIVE ANALYSIS, PIPING SYSTEMS STRESS ANALYSIS. *Educ:* Newark Col Eng, BSME, 61. *Prof Exp:* Tech, Pub Serv Elec & Gas Co, 39-47, draftsman, Eng Dept, 47-50, designer, 50-52, from asst engr to sr engr, 52-71, asst chief mech engr, 71-81, consult mech engr, 81-88; RETIRED. *Concurrent Pos:* Mem, Tech Adv Comt, Mat Properties Coun, 60-88 & A-1 Comt on Ferrous Mat, Am Soc Testing & Mat, 65-92; chmn, Steam Power Panel, Am Soc Mech Engrs-Am Soc Testing & Mat, 65-67, Edison Elec Inst Metall & Piping Task Force, 65-69, Metal Properties Coun, Nuclear Mat Comt, 68-74 & Nuclear Regulatory Comn Steam Generator Task Force, 72-78; consult, Elec Power Inst, Power Plant Mat, 70-82; elec utility adv, Elec Power Inst, 75-81; pres, NJ Soc Prof Engrs & Land Surveyors, 78-79. *Mem:* Fel Am Soc Mech Engrs; Mat Properties Coun; Am Soc Testing & Mat. *Res:* Materials in high temperature steam and nuclear power plants; evaluating extension of productive life of existing power plants; publications include economics, safety and reliability of power plants. *Mailing Add:* 70 Woodbridge Ave PO Box 568 Metuchen NJ 08840

SCHNABEL, ROBERT B, b New York, NY, Dec 18, 50; m 81; c 2. NUMERICAL COMPUTATION, PARALLEL LANGUAGES. *Educ:* Dartmouth Col, BA, 71; Cornell Univ, MS, 75, PhD(computer sci), 77. *Prof Exp:* From asst prof to assoc prof, 77-88, PROF COMPUTER SCI, UNIV COLO, BOULDER, 88-, CHAIR, 90- *Concurrent Pos:* Coun mem, Math Programming Soc, 85-88; vchair, Activ Group Optimization, Soc Indust & Appl Math, 86-88; assoc ed, Math Programming B, 88-, Soc Indust & Appl Math J Optimization, 90-, co-ed, Math Programming A, 89-; chair, Spec Interest Group Numerical Math, Asn Comput Mach, 89- *Mem:* Math Programming Soc. *Res:* Numerical computation; numerical solution of unconstrained and constrained optimization problems; systems of nonlinear equations and nonlinear least squares problems; parallel numerical languages and algorithms. *Mailing Add:* Dept Computer Sci Univ Colo Boulder CO 80309

SCHNABEL, TRUMAN GROSS, JR, b Philadelphia, Pa, Jan 5, 19; m 47; c 4. INTERNAL MEDICINE. *Educ:* Yale Univ, BS, 40; Univ Pa, MD, 43; Am Bd Internal Med, dipl, 52, recert, 74. *Prof Exp:* Intern med, Hosp Univ Pa, 44; asst resident, Mass Gen Hosp, 47-48; asst resident, Hosp Univ Pa, 48; instr physiol, 48-49, from asst instr to prof med, 49-73, vchmn dept med, 73-77, C MAHLON PROF MED, SCH MED, UNIV PA, 77-, STAFF PHYSICIAN CARDIOVASC SERV, 77- *Concurrent Pos:* Am Heart Asn fel, Sch Med, Univ Pa, 49-52, Markle scholar, 52-57; mem staff, Hosp Univ Pa, 52-; with Prof Lars Werko, St Erick's Hosp, Stockholm, Sweden, 55-56; asst ward chief, Philadelphia Gen Hosp, 56-59, ward chief, 59-73, coordr, Univ Pa Med Serv, 65-71, chief, 66-72; consult, Walston Gen Hosp, Ft Dix, NJ, 60-65; mem, Am Bd Internal Med, 63-72, secy-treas, 71-72; mem med educ adv comt, Rehab Serv Admin, HEW, Washington, DC, 68-71; mem clin res fel rev comt, Career Develop Rev Br, Div Res Grants, NIH, 68-71; D V Mattia lectr, Rutgers Med Sch, 75; Neuton Stern lectr & vis prof, Univ Tenn, 77. *Honors & Awards:* Alfred E Stengel Mem Award, Am Col Physicians, 78. *Mem:* Am Soc Clin Invest; Am Physiol Soc; AMA; Am Clin & Climat Asn (vpres, 68-69 & 76-77); Am Col Physicians (pres-elect, 73-74, pres, 74-75). *Res:* Cardiovascular physiology. *Mailing Add:* Dept Med-196 Gibson/G1 Univ Pa 3400 Spruce St Philadelphia PA 19104

SCHNABLE, GEORGE LUTHER, b Reading, Pa, Nov 26, 27; m 57; c 2. INORGANIC CHEMISTRY, MICROELECTRONICS. *Educ:* Albright Col, BS, 50; Univ Pa, MS, 51, PhD(chem), 53. *Prof Exp:* Asst chem, Univ Pa, 53; proj engr, Microelectronics Div, Philco Corp, 53-57, eng specialist, 57-59, eng group supvr, 59-62, head mat & processes develop group, 62-68, mgr advan mat & processes dept, 68-71; head process res group, RCA Labs, 71-79, head, Device Physics & Reliability Group, 79-87; HEAD DEVICE PHYSICS & RELIABILITY, DAVID SARNOFF RES CTR, 87- *Concurrent Pos:* Div ed, J Electrochem Soc, 79-91. *Mem:* Fel AAAS; Am Chem Soc; Electrochem Soc; fel Am Inst Chem; sr mem Inst Elec & Electronics Engrs. *Res:* Semiconductor devices; materials and processes for fabrication of transistors and integrated circuits; silicon chemistry and metallurgy. *Mailing Add:* David Sarnoff Res Ctr Princeton NJ 08543-5300

SCHNACK, LARRY G, b Harlan, Iowa, Mar 19, 37; m 55; c 4. ORGANIC CHEMISTRY. *Educ:* Iowa State Univ, BS, 58, PhD(org chem), 65. *Prof Exp:* Teacher, Minn High Sch, 58-61; asst prof, 65-69, assoc prof, 69-81, asst to vchancellor, 70-75, PROF ORG CHEM, UNIV WIS-EAU CLAIRE, 81-, ASST VCHANCELLOR ACAD AFFAIRS, 76- *Mem:* Am Chem Soc. *Res:* Stereochemistry and rearrangements. *Mailing Add:* Univ Wis Eau Claire WI 54701

SCHNAIBLE, H(AROLD) W(ILLIAM), b Lafayette, Ind, Apr 5, 25. CHEMICAL ENGINEERING. *Educ:* Purdue Univ, BS, 50, MS, 53, PhD(chem eng), 55. *Prof Exp:* Res engr, Gulf Res & Develop Co, 55-58; technologist, 59-63, SR RES ENGR, APPL RES LAB, US STEEL CORP, 63- *Mem:* Am Chem Soc; Am Inst Chem Engrs; Am Inst Mining, Metall & Petrol Engrs; Iron & Steel Soc; Sigma Xi. *Res:* Thermodynamics; kinetics, particularly hydrocarbon reactions; heat transfer; ingot solidification; computer simulation of processes. *Mailing Add:* PO Box 307 Darmont PA 15139

SCHNAPER, HAROLD WARREN, b Boston, Mass, Nov 11, 23; m 51; c 5. MEDICINE, CARDIOVASCULAR PHYSIOLOGY. *Educ:* Harvard Univ, AB, 45; La State Univ, cert, 44; Duke Univ, MD, 49. *Prof Exp:* Intern med, Boston City Hosps, 49-50; chief med, US Army 7th Evacuation Hosp, Ger, 51-53; resident, Mt Sinai Hosp, New York, 53-54; asst chief med serv, Vet Admin Hosp, DC, 54-60, chief internal med res, Vet Admin Cent Off, 60-64, assoc dir res serv, 64-66, actg dir, 66-67; exec vchmn dept med, 69-72, PROF MED & SR SCIENTIST, CARDIOVASC RES & TRAINING CTR, MED CTR, UNIV ALA, BIRMINGHAM, 69-, PROF PUB HEALTH & EPIDEMIOL & DIR DIV GERONT & GERIAT MED, 72-, DIR ALL UNIV CTR AGING, 76- *Concurrent Pos:* Fel neurol & dermat, Sch Med, Duke Univ, 49; fel cardiovasc dis, Sch Med, Georgetown Univ, 50-51; fel path, Mt Sinai Hosp, NY, 53; instr, Sch Med, Georgetown Univ, 54-58, asst prof, 58-66; attend physician, DC Gen Hosp, 54-66; vis prof, Mercy Hosp, Buffalo, NY, 59-66; co-dir cardiovasc res & training ctr, Med Ctr, Univ Ala, Birmingham, 66-70; assoc dir for heart & stroke, Ala Regional Med Prog, 69-76; chief, Vet Admin Cardiovasc Res Prog, 69-80. *Mem:* AAAS; AMA; fel Am Col Physicians; Am Fedn Clin Res; fel Geront Soc. *Res:* Clinical hypertension; aging mechanisms; multicenter clinical trials. *Mailing Add:* 3215 Sterling Rd Mountain Brook AL 35213

SCHNAPF, ABRAHAM, b New York, NY, Aug 1, 21; m 43; c 2. MECHANICAL ENGINEERING, SPACECRAFT SYSTEMS. *Educ:* City Col New York, BSME, 48; Drexel Univ, MSME, 53. *Prof Exp:* Develop engr aeronaut eng, Goodyear Aircraft Corp, Ohio, 48-50; leader develop eng, Airborne-Navig Systs, Defense Electronic Prod, Radio Corp Am, Camden, 50-55, mgr airborne weapon systs, 55-58, proj mgr Tiros, Astro-Electronics Div, 58-70, mgr prog mgt, 70-76, mgr satellite prog, 77-79, prin scientist, Astro-Electronics Div, RCA Corp, 79-82; PRES, AEROSPACE SYSTS ENG, 82- *Concurrent Pos:* Mem, Comn on Aerospace Applns, Nat Res Coun, 82- *Honors & Awards:* Ann Award, Am Soc Qual Control, 68; Dept Com Medal for Mgt, US Weather Satellite Prog, 85; Cert Appreciation Award, Dept Com. *Mem:* Nat Acad Sci; NY Acad Sci; fel Am Inst Aeronaut & Astronaut; Am Meteorol Soc; AAAS. *Res:* Development, design and testing of spacecraft, ground stations and field operations; management of satellite programs; remote sensing of earth from space. *Mailing Add:* Aerospace Systs Eng PO Box 160 Willingboro NJ 08046

SCHNAPPINGER, MELVIN GERHARDT, JR, b Baltimore, Md, Oct 29, 42; m 67; c 2. AGRONOMY. *Educ:* Univ Md, College Park, BS, 65, MS, 68; Va Polytech Inst & State Univ, PhD(agron), 70. *Prof Exp:* RES REP FIELD RES & DEVELOP, AGR DIV, CIBA-GEIGY CORP, 70- *Mem:* Am Soc Agron; Soil Sci Soc Am; Weed Sci Soc Am. *Res:* Field testing of herbicides, insecticides and micronutrient fertilizers. *Mailing Add:* Rte 3 Box 39 Centreville MD 21617

SCHNARE, PAUL STEWART, b Berlin, NH, Oct 16, 36; m 60; c 2. COMPUTER SCIENCE. *Educ:* Univ NH, BA, 60, MS, 61; Tulane Univ, La, PhD(math), 67. *Prof Exp:* Instr math, La State Univ, New Orleans, 61-66; asst prof, Univ Fla, 67-74; asst prof, Colby Col, 74-75 & Fordham Univ, 75-76; asst prof math, Univ Petrol & Minerals, Dhahran, Saudi Arabia, 76-80; ASSOC PROF MATH, EASTERN KY UNIV, 80- *Concurrent Pos:* NSF sci fac fel, Tulane Univ, 66-67. *Mem:* AAAS; London Math Soc; Math Asn Am; Am Math Soc; Nat Comput Graphics Asn; Asn Comput Mach. *Res:* Applied and computational mathematics; numerical methods; analytic inequalities; operations research. *Mailing Add:* Dept Math Sci Eastern Ky Univ Richmond KY 40475

SCHNATHORST, WILLIAM CHARLES, b Ft Dodge, Iowa, May 8, 29; m 51; c 3. PLANT PATHOLOGY. *Educ:* Univ Wyo, BS, 52, MS, 53; Univ Calif, PhD(plant path), 57. *Prof Exp:* Asst & lab instr plant physiol & bot, Univ Wyo, 52-53; asst plant path, exp sta, Univ Calif, Davis, 54-56, assoc, 56-85, lectr, 70; plant pathologist, Agr Res Serv, USDA, 56-85; PLANT PATHOLOGIST, UNIV CALIF, DAVIS, 85- *Mem:* Bot Soc Am; Mycol Soc Am; Am Phytopath Soc; Int Soc Plant Path; Sigma Xi. *Res:* Nature of disease resistance of plants; physiology of fungi; verticillium wilt; ecology of plant pathogens; diseases of field and tree crops. *Mailing Add:* 647 Cleveland Davis CA 95616

SCHNATTERLY, STEPHEN EUGENE, b Topeka, Kans, Oct 2, 38; m 63; c 2. SOLID STATE PHYSICS. *Educ:* Univ Wash, BS, 60, MS, 61; Univ Ill, PhD(physics), 65. *Prof Exp:* From instr to prof physics, Princeton Univ, 65-77; chmn, Physics Dept, 83-86, F H SMITH PROF PHYSICS, UNIV VA, 77- *Concurrent Pos:* Res Corp res grant, 66-67; prin investr, NSF grant, Univ Va, 77- *Mem:* Am Phys Soc. *Res:* Optical properties of solids; inelastic electron scattering spectroscopy; soft x-ray emission spectroscopy. *Mailing Add:* Physics Dept Univ Va Charlottesville VA 22901

SCHNECK, LARRY, b New York, NY, May 15, 26; m 59; c 3. PEDIATRIC NEUROLOGY, NEUROCHEMISTRY. *Educ:* NY Univ, BS, 49; Chicago Med Sch, MD, 53. *Prof Exp:* Resident pediat, Brooklyn Jewish Hosp, 54-56 & neurol, Bronx Munic Hosp, 57-60; asst prof neurol, 60-73, PROF NEUROL, STATE UNIV NY DOWNSTATE MED CTR, 74- *Concurrent Pos:* NIH fel, Albert Einstein Col Med, 67-70; dir neurol, Kingsbrook Jewish Med Ctr, 70-; dir, Albert Isaac Res Inst, 70-; attend physician, Vet Admin Hosp, Brooklyn, 71- *Mem:* Am Acad Pediat; Am Acad Neurol; Am Soc Neurochem; Int Soc Neurochem. *Res:* Neurochemistry and sphingolipidosis. *Mailing Add:* Kingsbrook Jewish Med Ctr 86 E 49th St-Neur Brooklyn NY 11203

SCHNECK, PAUL BENNETT, b New York, NY, Aug 15, 45; m 67; c 2. HIGH PERFORMANCE COMPUTER ARCHITECTURES, OPTIMIZING COMPILERS & LANGUAGES. *Educ:* Columbia Univ, BS, 65, MS, 66; NY Univ, PhD(computer sci), 79. *Prof Exp:* Mgr systs prog, Computer Appln, Inc, 67-69; sr computer scientist, Inst Space Studies, Goddard Space Flight Ctr, NASA, 69-76, asst dir for res, Mission & Data Opers, 76-79, asst to dir, Info Extraction Div, 80-81, asst dir computer & info sci, 81-83; head, Info Sci Div, Off Naval Res, 83-85; DIR, SUPERCOMPUT RES CTR, INST DEFENSE ANALYSIS, 85- *Concurrent Pos:* Adj prof, Computer Sci Dept, Univ Md, 81-82, adv bd mem, Computer Sci Bd, 89-; mem, Sci Supercomputer Subcomt, Inst Elec & Electronics Engrs/USAB, 82-90; panel mem, Off Technol Assessment, 83-86; distinguished visitor, Inst Elec & Electronics Engrs, 88-89. *Mem:* Sr mem Inst Elec & Electronics Engrs; Asn Comput Mach; Brit Computer Soc. *Res:* High performance computer architecture; optimizing compilers and languages for obtaining peak efficiency; algorithms directed towards parallel architectures. *Mailing Add:* Supercomput Res Ctr 17100 Science Dr Bowie MD 20715-4300

SCHNEEBERGER, EVELINE E, b The Hague, Holland, Oct 2, 34; US citizen. CELL BIOLOGY, ULTRASTRUCTURAL CYTOCHEMISTRY. *Educ:* Univ Colo, BA, 56, MD, 59. *Hon Degrees:* MA, Harvard Univ, 90. *Prof Exp:* Instr path, Harvard Med Sch, 67-68, assoc, 68-70, from asst prof to assoc prof, 70-88, PROF PATH, HARVARD MED SCH, 88- *Concurrent Pos:* Asst pathologist, Children's Hosp, Boston, Mass, 72-73, pathologist, 73-79; assoc pathologist, Mass Gen Hosp, Boston, 79- *Mem:* Fel AAAS; Sigma Xi; Am Soc Cell Biol; Am Asn Pathologists; Am Thoracic Soc; Microcirculatory Soc. *Res:* Cell biology of the lung; immunology of the lung. *Mailing Add:* Dept Path Mass Gen Hosp Cox Bldg 5 Boston MA 02114

SCHNEEMAN, BARBARA OLDS, b Seattle, Wash, Oct 3, 48; m 74; c 1. NUTRITION, FOOD SCIENCE. *Educ:* Univ Calif, Davis, BS, 70; Univ Calif, Berkeley, PhD(nutrit), 74. *Prof Exp:* Fel gastroenterol, Bruce Lyon Mem Res Lab, Children's Hosp, 74-76; from asst prof to assoc prof, 76-87, PROF, DEPT NUTRIT UNIV CALIF, DAVIS, 88, CHMN, 88- *Concurrent Pos:* NIH fel, 74-76, 77; mem, dietary guidelines adv comt, 90. *Honors & Awards:* Farma Int Fibre Prize, 89. *Mem:* Am Inst Nutrit; Am Physiol Soc; Soc Exp Biol & Med; AAAS; Inst Food Technologists. *Res:* Dietary regulation of digestion; impact of processed foods on nutrition and digestion. *Mailing Add:* Dept Nutrit Univ Calif Davis CA 95616

SCHNEEMEYER, LYNN F, b Baltimore, Md; m 76; c 2. SOLID STATE CHEMISTRY, HIGH TEMPERATURE SUPERCONDUCTORS. *Educ:* Col Notre Dame, Md, BA, 73; Cornell Univ, MS, 76, PhD(inorg chem), 78. *Prof Exp:* Postdoctoral res assoc chem, Mass Inst Technol, 78-80; mem tech staff, 80-87, DISTINGUISHED MEM TECH STAFF RES, AT&T BELL LABS, 87- *Concurrent Pos:* Chair inorg subdiv, North Jersey Am Chem Soc, 82-84; assoc ed, J Crystal Growth, 87- *Mem:* Am Chem Soc; Am Phys Soc; Electrochem Soc; Mat Res Soc; Am Asn Crystal Growth. *Res:* Preparation and characterization of interesting inorganic materials, typically in the form of single crystals; electrochemical, flux growth techniques and standard ceramic techniques have used to prepare new inorganic phases with interesting structures and physical properties. *Mailing Add:* AT&T Bell Labs Rm 1A-363 600 Mountain Ave Murray Hill NJ 07974-2070

SCHNEER, CECIL JACK, b Far Rockaway, NY, Jan 7, 23; m 43; c 2. GEOLOGY. *Educ:* Harvard Univ, AB, 43, AM, 50; Cornell Univ, PhD(geol), 54. *Prof Exp:* Mining geologist, Cerro de Pasco Co, SAm, 43-44; instr geol, Univ NH, 50; instr, Hamilton Col, 50-52; asst mineral, Cornell Univ, 52-54; from asst prof to prof, 54-87, EMER PROF GEOL & HIST SCI, UNIV NH, 87- *Concurrent Pos:* Pres, US Nat Comt on Hist Geol, 75-79; vpres, Int Comt Hist Geol, 76-84; assoc ed, Isis (Jour Hist Sci Soc), 77-80; nat lectr, Sigma Xi, 81-83. *Honors & Awards:* Hist Geol Award, Geol Soc Am, 85. *Mem:* Fel Geol Soc Am (chmn, hist geol div, 82); fel Mineral Soc Am; Hist Sci Soc; Nat Asn Geol Teachers; fel London Geol Soc; Hist Earth Sci Soc (pres, 86). *Res:* Dilational symmetry; snowflake morphology; history of science. *Mailing Add:* PO Box 181 Newfields NH 03856

SCHNEID, EDWARD JOSEPH, b Syracuse, NY, Apr 1, 40; m 67; c 1. NUCLEAR PHYSICS. *Educ:* LeMoyne Col, BS, 61; Univ Pittsburgh, PhD(physics), 66. *Prof Exp:* NSF res fel nuclear physics, Rutgers Univ, New Brunswick, 66-68; res scientist, 68-76, br head, 76-78, lab head, 78-84, sr lab head, 84-88, DIR, GRUMMAN AEROSPACE CORP, 88- *Mem:* Am Nuclear Soc; Inst Elec & Electronics Engrs; Am Phys Soc. *Res:* Ion beam analysis; advanced nuclear sensor development; gamma ray astronomy. *Mailing Add:* 28 Harvard Lane Commack NY 11735

SCHNEIDAU, JOHN DONALD, JR, b New Orleans, La, May 14, 13; m 43; c 3. MEDICAL MYCOLOGY. *Educ:* Loyola Univ, BS, 38; Tulane Univ, MS, 40, PhD(microbiol), 56. *Prof Exp:* From instr to assoc prof biol, Loyola Univ, 45-54; res assoc, Tulane Univ, 54-56, from asst prof to assoc prof, 56-73, prof microbiol, 73-78, prof immunol, 76-78, EMER PROF MICROBIOL & IMMUNOL, TULANE UNIV, 78- *Concurrent Pos:* Consult mycologist, Ochsner Found Hosp, New Orleans, La, 50-; lectr, ICA Prog Med Educ, Tulane-Colombia, 58-60. *Mem:* Am Soc Microbiol; Int Soc Human & Animal Mycol; Med Microbiol Soc Am. *Res:* Skin hypersensitivity to fungal antigens in systemic mycotic disease with particular emphasis on cross-reactivity among fungal skin-test antigens; taxonomy and biology of Nocardia and related Actinomycetales. *Mailing Add:* 1318 Jefferson Ave New Orleans LA 70115

SCHNEIDER, ALAN M(ICHAEL), b Milwaukee, Wis, Feb 28, 25; m 48; c 4. SYSTEMS & SIGNALS ENGINEERING, AUTOMATION. *Educ:* Villanova Univ, BEE, 45; Univ Wis, MS, 48; Mass Inst Technol, ScD(instrumentation), 57. *Prof Exp:* Res engr, Hughes Aircraft Co, 48-49; engr, AC Spark Plug Div, Gen Motors Corp, 50-53; res engr, Instrumentation Lab, Mass Inst Technol, 53-55; eng scientist, Airborne Systs Lab, Radio Corp Am, 57-59, sr eng scientist, Missile Electronics & Control Div, 59-61, mgr systs anal, Aerospace Systs Div, 61-65; assoc prof aerospace & mech eng sci, 65-68, PROF APPL MECH & ENG SCI, UNIV CALIF, SAN DIEGO, 68- *Concurrent Pos:* Consult, Gen Dynamics, Convair, 65, Aerospace Systs Div, Radio Corp Am, 65, Aerospace Corp, 65-68, Gen Micro-Electonics Div, Philco Corp, 66, TRW Systs, 67-70, Teledyne Ryan Aeronaut, 68-80, Naval Ocean Systs Ctr, 73-87, Air Pollution Technol Inc, 77, Linkabit Corp, 77-88, Pac Aerosyst, 83, Titan Systs, 83-86, Visutek, 83-84, Qualcomm, 85-86 & Teledyne Ryan Electronics, 85-89; vis assoc, Environ Qual Lab, Calif Inst Technol, 73; vis prof, Stanford Univ, 81; sr vis, Oxford Univ, 84. *Honors & Awards:* Samuel M Burka Award, 62. *Mem:* Assoc fel Am Inst Aeronaut & Astronaut; Inst Navig. *Res:* Vehicle navigation, guidance, and control; automatic and manual rendezvous guidance; inertial systems astrodynamics; systems theory and applications; modeling of physiological systems. *Mailing Add:* Dept Appl Mech & Eng Sci 0411 Univ Calif San Diego 9500 Gilman Dr La Jolla CA 92093-0411

SCHNEIDER, ALFRED, b Ger, Dec 17, 26 ; US citizen; m 50; c 3. NUCLEAR ENGINEERING, CHEMICAL ENGINEERING. *Educ:* Cooper Union, BChE, 51; Polytech Univ NY, PhD(chem eng), 58. *Prof Exp:* Chemist, US Testing Co, 51-52; develop proj mgr, Celanese Corp Am, 52-56; assoc chem engr, Argonne Nat Lab, 56-61; mgr nuclear res & develop, Martin Marietta Co, 61-64; mgr mat & processes, Nuclear Utility Serv, Washington, DC, 64-65; res assoc to dir nuclear technol, Allied Chem Corp, NJ, 65-71, dir nuclear technol, Allied-Gen Nuclear Serv, SC, 71-75; prof, 75-90, EMER PROF NUCLEAR ENG, GA INST TECHNOL, 90-; PRES, SCHNEIDER LABS, INC, GA, 90- *Concurrent Pos:* Consult, Allied-Gen Nuclear Serv, 75-83, NY State Energy Res & Develop Authority, 76- & Martin Marietta Energy Systs, Inc, 83-, Westinghouse Elec Co, 89- *Honors & Awards:* Antarctica Medal, USN, 63; Robert E Wilson Award, Am Inst Chem Engrs, 86. *Mem:* Am Chem Soc; Am Inst Chem Engrs; Am Nuclear Soc; AAAS. *Res:* Reprocessing of nuclear fuels; nuclear materials; radioactive waste management; nuclear power reactors; nuclear fuel cycles; isotope separation; energy systems. *Mailing Add:* Ga Inst Technol 5005 Hidden Branches Dr Atlanta GA 30338

SCHNEIDER, ALFRED MARCEL, b Vienna, Austria, Nov 7, 25; US citizen; m 53; c 2. MATHEMATICAL STATISTICS, OPERATIONS RESEARCH. *Educ:* Univ London, BSc, 48. *Prof Exp:* Res chemist, Vitamins, Ltd, Eng, 48-52; chem engr, Cyanamid Can, 53-55, exp statistician, Am Cyanamid Co, 55-57, leader math anal group, 58-61; mgr, Math Anal Dept, Dewey & Almy Chem Div, W R Grace & Co, 61-66, dir math sci, Tech Group, 66-71, dir opers res, 71-86; CONSULT, 86- *Mem:* Am Statist Asn; Opers Res Soc Am; Royal Soc Chem. *Res:* Experimental design; computer applications to chemistry and chemical engineering; simulation. *Mailing Add:* Seven Carleen Ct Summit NJ 07901

SCHNEIDER, ALLAN FRANK, b Chicago, Ill, Feb 7, 26; m 50; c 3. GLACIAL GEOLOGY. *Educ:* Beloit Col, BS, 48; Pa State Univ, MS, 51; Univ Minn, PhD(geol), 57. *Prof Exp:* Asst geol, Pa State Univ, 48-50, instr, 50-51; instr, Univ Minn, 51-54; from instr to asst prof, Wash State Univ, 54-59; geologist, Ind Geol Surv, 59-70, assoc map ed, 60-61, map & illus ed, 61-65; assoc prof, 70-80, PROF GEOL, UNIV WIS-PARKSIDE, 80- *Concurrent Pos:* Geologist, US Geol Surv, 49, Minn Geol Surv, 51-54 & Wis Geol Surv, 76 & 90; chair geol dept, Univ Wis-Parkside, 73-75, 80-83, 85-86; dist geologist, Lake Mich dist, Wis Dept Natural Resources, 86- *Mem:* Fel Geol Soc Am; Nat Asn Geol Teachers; Int Glaciol Soc; Am Quaternary Asn. *Res:* Geomorphology; glacial geology; sedimentary petrography; Pleistocene geology of Minnesota, Indiana and Wisconsin; late Quaternary history of Lake Michigan basin. *Mailing Add:* Dept Geol Univ Wis-Parkside Box 2000 Kenosha WI 53141-2000

SCHNEIDER, ALLAN STANFORD, b New York, NY, Sept 26, 40; m 68; c 2. NEUROPHARMACOLOGY, CELL REGULATORY BIOLOGY. *Educ:* Rensselaer Polytech Inst, BChemE, 61; Pa State Univ, MS, 63; Univ Calif, Berkeley, PhD(chem), 68. *Prof Exp:* Inst fel biomembranes, Weizman Inst Sci, Rehovot, Israel, 69-71; staff fel neurobiol, NIH, 71-73; from assoc to assoc mem cell regulation, Sloan Kettering Inst Cancer Res & from asst prof biochem to assoc prof biochem & cell biol, Cornell Univ Grad Sch Med Sci, 73-85, chmn biochem unit, 82-83; assoc prof, 85-86, PROF PHARMACOL & TOXICOL, ALBANY MED COL, NY, 87-; PROF BIOMED SCI, SCH PUB HEALTH, STATE UNIV NY, ALBANY, 88- *Concurrent Pos:* Vis res scholar, Norwegian Res Coun, Univ Bergen, Norway, 89; mem, Int Sci Adv Comt Chromaffic Cell Biol & Peripheral Catecholamines. *Mem:* Am Soc Biol Chemists; Biophys Soc; NY Acad Sci; AAAS; Soc Neurosci; Am Heart Asn. *Res:* Hormone and neurotransmitter secretion and action at cell surface receptors; nicotine addiction. *Mailing Add:* Dept Pharmacol & Toxicol Albany Med Col Albany NY 12208

SCHNEIDER, ARTHUR LEE, b St Louis, Mo, Feb 11, 39; m 64; c 2. PROTEIN CHEMISTRY. *Educ:* Univ Mo, Columbia, BS, 61, PhD(biochem), 66. *Prof Exp:* Res assoc biochem, Albert Einstein Col Med, Yeshiva Univ, 66-68; res assoc, Col Physicians & Surgeons, Columbia Univ, 69-70; assoc med, Albert Einstein Col Med, 71-73; GROUP LEADER CLIN CHEM, DADE, DIV AM HOSP SUPPLY CORP, 73- *Concurrent Pos:* USPHS fel, NIH, 67-69. *Mem:* Am Chem Soc; Am Asn Clin Chemists; AAAS. *Res:* Use of plasma proteins for clinical chemistry control materials; development of clinical chemistry control materials. *Mailing Add:* 526 Rte 303 Orangeburg NY 10962

SCHNEIDER, ARTHUR SANFORD, b Los Angeles, Calif, Mar 24, 29; m 50; c 3. PATHOLOGY, HEMATOLOGY. *Educ:* Univ Calif, Los Angeles, BS, 51; Chicago Med Sch, MD, 55. *Prof Exp:* Intern & resident, Vet Admin Hosp, Los Angeles, 55-59; chief med serv, USAF Hosp, Mather AFB, Calif, 59-61; instr med, Univ Calif, Los Angeles Med Sch, 61-64, asst prof med & path, 65-68; chief clin path, Vet Admin Hosp, Los Angeles, 62-68; dir clin path, City of Hope Med Ctr, 68-70, chmn clin path, 70-75; actg chief lab serv, 75-86, CHIEF, LAB HEMAT, VET ADMIN MED CTR, 86-; PROF & CHMN PATH, UNIV HEALTH SCI, CHICAGO MED SCH, 75- *Concurrent Pos:* Hemat trainee, Univ Calif, Los Angeles Sch Med, 61-62, res assoc, 62-66, asst clin prof med & path, 68-72, assoc clin prof, 72-75; attend specialist, Wadsworth Vet Admin Ctr, Los Angeles, 68-75. *Mem:* Asn Path Chmn; Am Soc Hemat; Am Soc Clin Pathologists; Acad Clin Lab Physicians & Scientists; Am Fedn Clin Res; Am Col Physicians; Col Am Pathologists; AMA; Asn Hematopath. *Res:* Inherited erythrocyte biochemical abnormalities; computer applications in laboratory medicine; immunofluorescent clinical chemistry analysis. *Mailing Add:* Dept Path Chicago Med Sch 3333 Green Bay Rd North Chicago IL 60064

SCHNEIDER, BARBARA G, IMMUNOCYTO CHEMISTRY, MEMBRANE PROTEIN BIOGENESIS. *Educ:* Univ Tex, MA, 75. *Prof Exp:* Res assoc, dept path, Yale Sch Med, 75-86, res coordr, 86-89; ASST PROF PATH, TEX HEALTH SCI CTR, 89. *Res:* Retinal cell biology. *Mailing Add:* Dept Path Tex Health Sci Str 7703 Floyd Curl Dr San Antonio TX 78284

SCHNEIDER, BARRY I, b Brooklyn, NY, Nov 16, 40; m 62; c 2. MOLECULAR PHYSICS, THEORETICAL CHEMISTRY. *Educ:* Brooklyn Col, BS, 62; Yale Univ, MS, 64; Univ Chicago, PhD(theoret chem), 68. *Prof Exp:* Fel chem, Univ Southern Calif, 68-69; mem tech staff, Gen Tel & Electronics Lab, 69-72; MEM TECH STAFF PHYSICS, LOS ALAMOS SCI LAB, 72- *Concurrent Pos:* Prog dir atomic, molecular & optical physics, NSF, 89-90. *Honors & Awards:* Fel, Am Phys Soc; Sr Scientist Humboldt Award, 86. *Mem:* Fel Am Phys Soc. *Res:* Scattering theory, photoionization; many-body theory; structure of molecules. *Mailing Add:* Theoret Div Los Alamos Sci Lab Los Alamos NM 87545

SCHNEIDER, BERNARD ARNOLD, b Washington, DC, June 8, 44; m 68; c 3. PLANT PHYSIOLOGY, AGRONOMY. *Educ:* Univ Md, College Park, BS, 66, MS, 68, PhD(forage physiol, biochem), 71. *Prof Exp:* Asst agron, Univ Md, 66-71; plant physiologist, Plant Biol Lab, Environ Protection Agency, 72-74, radioisotope safety officer, 74-77, supvry plant physiologist, 77-79, PLANT PHYSIOLOGIST & ENVIRON PROTECTION AGENCY, 79- *Concurrent Pos:* Rep, Plant Growth Regulator Soc Nomenclature Comt, Am Nat Standards Inst-K62, 74-; leader pesticide prod performance guidelines, Am Soc Testing & Mat, Terminol Subcomt, Chemigation Info Exchange Group, Environ Protection Agency. *Mem:* Am Soc Agron; Weed Sci Soc Am; Am Soc Hort Sci; Am Soc Testing & Mat; Am Chem Soc; Plant Growth Regulator Soc Am. *Res:* Develop methods for determining the biological effectiveness of algaecides, herbicides and plant regulators for public protection; author of handbook on toxicology and plant growth regulators; determine the fate and metabolism of pesticides in the environment using radioisotopes of pesticides; plant metabolism studies benefit risk assessments of pesticides; pesticide residue analyst; 30 publications. *Mailing Add:* Chem Br I HED (H7509C) 401 MST SW Washington DC 20460

SCHNEIDER, BRUCE ALTON, b Detroit, Mich, July 17, 41; m 67; c 2. AUDITORY PSYCHOPHYSICS, AUDITORY DEVELOPMENT. *Educ:* Univ Mich, Ann Arbor, BA, 63, Harvard Univ, PhD(psychol), 68. *Prof Exp:* Lectr psychol, Columbia Univ, 67-68, asst prof, 68-72, assoc prof, 72-74; assoc prof, 74-81, PROF PSYCHOL, UNIV TORONTO, 81- *Concurrent Pos:* Distinguished vis prof, Univ Alta, 81. *Mem:* Soc Math Psychologists. *Res:* Infant and adult auditory perception; how the ear processes sound and how the nature of this auditory processing system changes from infancy to adulthood. *Mailing Add:* Dept Psychol Erindale Col Univ Toronto 3359 Mississiauga Rd N Mississauga ON L5L 1C6 Can

SCHNEIDER, BRUCE E, b Sacramento, Calif, July 4, 50; m 75; c 3. BIOSTATISTICS. *Educ:* Brown Univ, BSc, 72; Villanova Univ, MS, 73; Temple Univ, PhD(appl statist), 77. *Prof Exp:* Group leader, Wyeth Labs, 72-75, supvr, 76-77, mgr biostatist, 77-81, assoc dir, biostatist & data systs, 81-85, dir clin info, 85-86, asst vpres clin opers, 87-90, VPRES CLIN OPERS, WYETH-AYERST RES, 90- *Concurrent Pos:* Steering comt, PMA Biostatist subsect, 84-86, chmn steering comt, 87-88, adv, 89- *Mem:* Am Statist Asn; Biometrics Soc. *Res:* Nonparametrics; statistical computing; pharmacokinetics. *Mailing Add:* Wyeth-Ayerst Res PO Box 8299 Philadelphia PA 19101

SCHNEIDER, BRUCE SOLOMON, b New York, NY, Feb 23, 42; m; c 3. HORMONES-DISEASES OF HYPOTHALAMIC-PITUITARY AXIS, HORMONE PRODUCTION BY TUMORS. *Educ:* Harvard Col, AB, 64; Harvard Med Sch, MD, 68. *Prof Exp:* Intern, MT Sinai Hosp, NY, 68-69, resident, 71-73; res assoc endocrinol, Solomon Benson Res Lab, Bronx Vet Admin Hosp, 73-75; asst prof & assoc physician, Rockefeller Univ, NY, 75-83; CHIEF ENDOCRINOL & METAB, LONG ISLAND JEWISH MED CTR, 83-; PROF MED, ALBERT EINSTEIN COL MED, 89- *Honors & Awards:* Career Scientist Award, Irma T Hirsch Trust, 81. *Mem:* Harvey Soc; Sigma Xi; Endocrine Soc; Fedn Am Socs Exp Biol; Am Diabetes Asn. *Res:* Physiology, biochemistry, molecular biology of neuropeptides; role of neuropeptides in nutritional homeostasis; expression of neuropeptides by brain and tumor cells. *Mailing Add:* Dept Endocrinol & Metab Long Island Jewish Med Ctr New Hyde Park NY 11042

SCHNEIDER, CARL STANLEY, b Baltimore, Md, Dec 20, 42; m 71; c 2. SOLID STATE PHYSICS. *Educ:* Johns Hopkins Univ, BA, 63; Mass Inst Technol, SM, 65, PhD(physics), 68. *Prof Exp:* From asst prof to assoc prof physics, 68-81, dir res, 86-89, PROF PHYSICS, US NAVAL ACAD, 81-, ASSOC DEAN, 89- *Concurrent Pos:* Naval Acad Res Coun grant, US Naval Acad-Nat Bur Standards, 69-75, NSF res grants neutron diffraction, 74-76; affil, David Taylor Res Ctr, 77-; pres, US Naval Acad, 80-81. *Mem:* Am Phys Soc; Sigma Xi (pres, 80-81); Am Asn Physics Teachers. *Res:* Neutron diffraction; prism refraction of thermal neutron for the determination of scattering amplitudes; nonlinear theory of magnetoelasticity; closed loop degaussing. *Mailing Add:* Res Off IIf US Naval Acad Annapolis MD 21402

SCHNEIDER, CHARLES L, obstetric physiology; deceased, see previous edition for last biography

SCHNEIDER, CRAIG WILLIAM, b Manchester, NH, Oct 23, 48; m 72; c 3. PHYCOLOGY. *Educ:* Gettysburg Col, BA, 70; Duke Univ, PhD(bot), 75. *Prof Exp:* From asst prof to assoc prof, 75-87, PROF BIOL, TRINITY COL, 87- *Mem:* Phycol Soc Am; Int Phycol Soc; Brit Phycol Soc; Sigma Xi. *Res:* Benthic algal studies on the Southeastern United States continental shelf and in Bermuda; benthic algal ecology in Connecticut; life-history cultural studies; red-algal morphological studies. *Mailing Add:* Dept Biol Trinity Col Hartford CT 06106

SCHNEIDER, DAVID EDWIN, b Philadelphia, Pa, Mar 16, 37; m 82. PHYSIOLOGICAL ECOLOGY, MARINE ECOLOGY. *Educ:* Bates Col, BS, 59; Duke Univ, PhD(zool), 67. *Prof Exp:* From instr to asst prof, 66-71, ASSOC PROF BIOL, WESTERN WASH UNIV, 71- *Mem:* AAAS; Am Soc Zoologists; Am Soc Limnol & Oceanog; Ecol Soc Am Sci. *Res:* Temperature and desiccation adaptations of marine intertidal animals; trophic relationships and physiological responses of Arctic marine species. *Mailing Add:* Dept Biol Western Wash Univ Bellingham WA 98225

SCHNEIDER, DENNIS RAY, b Sinton, Tex, June 10, 52; m 76; c 2. BIOREMEDIATION. *Educ:* Univ Tex, Austin, BA, 73, PhD(microbiol), 78. *Prof Exp:* Postdoctoral fel, Behringwerke AG Marburg/Lahn WGer, 78-79, Univ Mo Med Sch, 80-81; res microbiologist, New Eng Nuclear Dupont, 81-82; res & develop dir, Austin Biol Labs, 82-88; RES & DEVELOP DIR, MICRO-BAC INT, 88- *Concurrent Pos:* Adj assoc prof, Microbiol Dept, Univ Tex, Austin, 86-; res & develop dir, Green Sci Ltd, Taillaight, Rep Ireland, 90- *Mem:* Am Soc Microbiol; AAAS; Hazardous Mat Control Res Inst. *Res:* Development of microbial products for the treatment of environmentally important waste products; development of microbial products to improve oil and natural gas production. *Mailing Add:* Micro-Bac Int 9607 Gray Blvd Austin TX 78758

SCHNEIDER, DONALD LEONARD, b Muskegon, Mich, Jan 15, 41; m 79; c 1. BIOCHEMISTRY. *Educ:* Kalamazoo Col, BA, 63; Mich State Univ, PhD(biochem), 69. *Prof Exp:* Fel biochem, Cornell Univ, 69-71; res assoc biochem cytol, Rockefeller Univ, 71-72, asst prof, 72-73; asst prof biochem, Univ Mass, Amherst, 73-76; from asst prof to assoc prof, Dartmouth Med Sch, 77-90; HEALTH SCI ADMINR, NIH, 90- *Mem:* AAAS; Am Chem Soc; Am Soc Cell Biol; Am Soc Biochem & Molecular Biol; Am Inst Chemists; Leukocyte Biol Soc. *Res:* Lysosomes; proton pump ATPases; white blood cell defenses; sugar transport. *Mailing Add:* NIH/DRG/Spec Rev 5333 Westbard Ave Bethesda MD 20892

SCHNEIDER, DONALD LOUIS, b Ft Wayne, Ind, Apr 9, 19; m 41; c 2. BIOCHEMISTRY, NUTRITION. *Educ:* Evansville Col, BA, 52; Univ Ariz, MS, 60, PhD(biochem), 63. *Prof Exp:* Chemist, Mead Johnson Res Ctr, 52-58; res assoc biochem & nutrit, Ariz Agr Exp Sta, 58-62; sr scientist, Mead Johnson Res Ctr, 62-63, group leader, 63-68, prin investr, 68-73, sect leader nutrit, 73-76, prin res assoc, 76-81; RETIRED. *Mem:* AAAS; Am Inst Nutrit; NY Acad Sci; Am Inst Biol Sci; Am Inst Chemists. *Res:* Cyclopropenoid fatty acids, biochemical and physiological effects; lipid, cholesterol and bile salt metabolism; baby pig and infant nutrition. *Mailing Add:* Dept Nutrit Res Mead Johnson Res Ctr Evansville IN 47721

SCHNEIDER, E GAYLE, b St Louis, Mo, Aug 1, 46. STRUCTURE & FUNCTION OF MEMBRANE. *Educ:* Harvard Univ, PhD(biochem), 74. *Prof Exp:* ASST PROF BIOCHEM, MED CTR, UNIV NEBR, 79- *Mem:* Soc Develop Biol; Am Soc Cell Biol; Sigma Xi; AAAS. *Mailing Add:* Obstet-Gynec Dept Yale Univ Sch Med 333 Cedar St New Haven CT 06510

SCHNEIDER, EDWARD GREYER, b Indianapolis, Ind, Sept 2, 41; m 83; c 4. PHYSIOLOGY. *Educ:* DePauw Univ, BA, 63; Ind Univ, Indianapolis, PhD(physiol), 67. *Prof Exp:* Fel physiol, Univ Mo, 67-70, asst prof, 70-71; asst prof, Mayo Grad Sch Med, Univ Minn, 71-73; ASSOC PROF PHYSIOL, HEALTH SCI CTR, UNIV TENN, MEMPHIS, 73- *Concurrent Pos:* Estab investr, Am Heart Asn, 72-77; consult, Nova Pharm Corp, 86- *Mem:* Am Physiol Soc; Am Fedn Clin Res; Am Soc Nephrology; Sigma Xi; Int Soc Nephrology; Endocrin Soc. *Res:* Examination of the control of renal sodium excretion and the effects of alteration in fluid balance on the excretion of electrolytes by the kidney and the secretron of aldosterone by the adrenal. *Mailing Add:* Dept Physiol & Biophys Univ Tenn 894 Union Ave Memphis TN 38163

SCHNEIDER, EDWARD LEE, b Portland, Ore, Sept 14, 47. PLANT ANATOMY, POLLINATION BIOLOGY. *Educ:* Cent Wash Univ, BA, 69, MS, 71; Univ Calif, Santa Barbara, PhD(bot), 74. *Prof Exp:* From asst profto assoc prof, 74-84, PROF BOT, SOUTHWEST TEX STATE UNIV & CHMN DEPT BIOL, 84-, DEAN, SCH SCI, 89- *Concurrent Pos:* Mem, Int Comn Bee Bot, 77-; Prin investr, Nat Sci Foun, 81-84. *Mem:* Bot Soc Am; Sigma Xi; Int Asn Aquatic Vascular Plant Biologists. *Res:* Descriptive plant anatomy-morphology; evolution of flowering plants; reproductive biology of aquatic plants. *Mailing Add:* Sch Sci Southwest Tex State Univ San Marcos TX 78666

SCHNEIDER, EDWARD LEWIS, b New York, NY, June 22, 40. GERIATRICS, HUMAN GENETICS. *Educ:* Rensselaer Polytech Inst, BS, 61; Boston Univ, MD, 66. *Prof Exp:* Intern med, Cornell Univ-New York Hosp, 66-67, resident med, 67-68; res assoc, Lab Biol Viruses, Nat Inst Allergy & Infectious Dis, NIH, 68-70; res fel human genetics, Univ Calif Med Ctr, San Francisco, 70-73, prof med & biochem, 79-80; prog coordr, LCCP Res Ctr, 73-79, assoc dir to dep dir, Nat Inst Aging, NIH, 84-88; DEAN & EXEC DIR, ANDRUS GERONT CTR, 88- *Concurrent Pos:* Asst prof human genetics, Johns Hopkins Univ Sch Med, 73-76; adj prof biochem, George Washington Univ, 83-; clin prof med, Georgetown Univ, 85- *Honors & Awards:* Roche Award, Am Soc Clin Invest. *Mem:* Am Soc Cell Biol; Geront Soc; Am Soc Human Genetics; Tissue Cult Asn. *Res:* Studies on cellular aging utilizing human diploid cell cultures in vitro and animal systems to examine cell replication, nucleic acid metabolism and repair of DNA damage; molecular genetics. *Mailing Add:* Andrus Geront Ctr Univ Southern Calif University Park MC 0191 Los Angeles CA 90089-0191

SCHNEIDER, EDWIN KAHN, b Philadelphia, Pa, May 6, 48. METEOROLOGY. *Educ:* Harvard Col, AB, 70; Harvard Univ, MS, 73, PhD(appl physics), 76. *Prof Exp:* Res assoc, Mass Inst Technol, 74-77; NATO fel, Reading Univ, Eng, 77-78; RES FEL, HARVARD UNIV, 78- *Res:* Atmospheric general circulation and climate modelling; tropical meteorology. *Mailing Add:* 280 Harvard St No 4D Cambridge MA 02139

SCHNEIDER, ERIC DAVIS, b Wilmington, Del, Nov 21, 40; c 2. MARINE GEOLOGY, MARINE POLLUTION. *Educ:* Univ Del, BA, 62; Columbia Univ, MS, 65, PhD, 69. *Prof Exp:* Comdr, US Naval Oceanog Off, 67-69, dir, Global Ocean Floor & Anal Res Ctr, 68-71; dir, Off Spec Projs, Environ Protection Agency, 71-72, dir, Environ Res Lab, 72-79; spec asst to adminr, Nat Oceanic & Atmospheric Admin, 82-84; VIS PROF BIOL, CHES BIOL LAB, UNIV MD, 84- *Concurrent Pos:* Adj prof, Grad Sch Oceanog, Univ RI, 72-81; chmn, Working Group Effect Pollutants on Marine Organisms, US-USSR Joint Comt Coop Environ Protection, 72-76; consult, UN Econ Develop Admin, 84. *Res:* Evolutionary and ecological processes and their relationship to thermodynamic principles; global ocean pollution monitoring programs. *Mailing Add:* Cedal Hill Farm Box 135 RR-2 Prince Frederick MD 20678

SCHNEIDER, ERIC WEST, b Wilkes-Barre, Pa, Sept 1, 52; m 74; c 3. NUCLEAR CHEMISTRY. *Educ:* Rensselaer Polytech Inst, BS, 74; Univ Md, PhD(nuclear chem), 78. *Prof Exp:* Assoc res scientist, 78-81, sr res scientist, 81, staff res scientist, 81-87, SR STAFF RES SCIENTIST, RES LABS, GEN MOTORS CORP, 87- *Mem:* Am Chem Soc; Am Phys Soc; Sigma Xi. *Res:* Development of radioisotopic methods for industrial applications including, radiotracer methods for measuring wear, radioisotopic gauging and radiographic inspection of materials, and neutron activation and x-ray fluorescence elemental analysis. *Mailing Add:* Res Labs RCEL-215 Gen Motors Corp Warren MI 48090

SCHNEIDER, FRANK L, b New York, NY, May 26, 06; m 45; c 1. ANALYTICAL CHEMISTRY. *Educ:* Polytech Inst Brooklyn, BS, 28; NY Univ, MS, 30; Rutgers Univ, PhD(chem), 36. *Prof Exp:* Asst, Rutgers Univ, 33-34, instr, 34-37, exten div, 36-37; instr chem, Trinity Col, Conn, 37-39; from instr to prof, 39-72, EMER PROF CHEM, QUEENS COL, NY, 72- *Honors & Awards:* Benedetti-Pichler Award. *Mem:* Am Chem Soc; fel Am Inst Chemists; Sigma Xi; hon mem Austrian Microchem Soc. *Res:* Microchemistry; organic analysis; air and water pollution control. *Mailing Add:* Round Hill Lane Port Washington NY 11050

SCHNEIDER, FRED BARRY, b NY, Dec 7, 53. COMPUTER SCIENCE. *Educ:* Cornell Univ, BS, 75; State Univ NY, Stony Brook, MS, 77, PhD(comput sci), 78. *Prof Exp:* ASSOC PROF COMPUT SCI, CORNELL UNIV, 78- *Mem:* Am Asn Comput Mach; Inst Elec & Electronics Engrs. *Res:* Operating systems; programming languages; concurrency; software engineering. *Mailing Add:* Dept Comput Sci Cornell Univ Upson Hall Ithaca NY 14853

SCHNEIDER, FREDERICK HOWARD, b Detroit, Mich, Nov 19, 38; m 61; c 2. PHARMACOLOGY. *Educ:* Ariz State Univ, BS, 60, MS, 61; Yale Univ, PhD(autonomic pharmacol), 66. *Prof Exp:* Jr chemist, Merck Sharp & Dohme Res Labs, 61-62; asst prof pharmacol, Sch Med, Univ Colo, Denver, 67-73; assoc prof, Sch Med, Emory Univ, 73-75; pres, 75-86, CHIEF SCIENTIST OFFICER, BIOASSAY SYSTS CORP, CAMBRIDGE, MASS; DIR & SR VPRES, BOGART DELAFIELD FERRIER, 85- *Concurrent Pos:* NSF fel pharmacol, Oxford Univ, 66-67; estab investr, Am Heart Asn, 69-74. *Mem:* AAAS; Am Soc Pharmacol & Exp Therapeut; Tissue Cult Asn. *Res:* Effects of drugs on responses to sympathetic nerve stimulation; neurotransmitter secretion; lysosomal secretion mechanisms; nerve cells in tissue culture; toxicology; in vitro cytotoxicity. *Mailing Add:* Bogart Delafield Ferrier Yarmouthport MA 02675

SCHNEIDER, G MICHAEL, b Detroit, Mich, May 28, 45. SOFTWARE ENGINEERING. *Educ:* Univ Mich, BS, 66; Univ Wis, MSc, 68, PhD(comput sci), 74. *Prof Exp:* asst prof comput sci, Univ Minn, 74-; AT COMPUT-MATH DEPT, MACALESTER COL, ST PAUL, MINN. *Concurrent Pos:* Vis prof, Univ Calif, Berkeley, 79 & Imp Col, Univ London, 80. *Mem:* Asn Comput Mach; Inst Elec & Electronics Engrs. *Res:* Interface between the user and the computer system. *Mailing Add:* Comput Sci Dept Macalester Col 1600 Grand Ave St Paul MN 55105

SCHNEIDER, GARY, b Milwaukee, Wis, Feb 6, 34; m 56; c 2. FOREST ECOLOGY. *Educ:* Univ Mich, BS, 56, MF, 57; Mich State Univ, PhD(forest ecol), 63. *Prof Exp:* Asst dist ranger forest admin, US Forest Serv, 57; res asst forestry, Mich State Univ, 59-62; asst prof, Stephen F Austin State Univ, 62-65; from asst prof to prof forestry fisheries & wildlife, Mich State Univ, 65-77; PROF & HEAD DEPT FORESTRY, WILDLIFE & FISHERIES, UNIV TENN, 77-, ASST DEAN, COL AGR, 86- *Concurrent Pos:* Consult, AEC-Oak Ridge Assoc Univs, 66-72, US Forest Serv and various forest industries, 66- *Mem:* AAAS; fel Soc Am Foresters; Soil Sci Soc Am; Am Forestry Asn; Ecol Soc Am. *Res:* Production ecology; biomass and nutrient analysis of tree species; environmental factors influencing tree growth and development; nutrient cycle in the forest ecosystem; forest soil moisture relations. *Mailing Add:* Col Agr 125 Morgan Hall Univ Tenn Knoxville TN 37901

SCHNEIDER, GEORGE RONALD, b Chicago, Ill, Mar 6, 32. CHEMICAL ENGINEERING, CHEMISTRY. *Educ:* Ill Inst Technol, BS, 53; Mass Inst Technol, SM, 56, ScD(chem eng), 61. *Prof Exp:* Chem engr, Gen Elec Co, 53-54; MEM TECH STAFF, ADVAN PROGS DEPT, ROCKETDYNE DIV, ROCKWELL INT CORP, 61- *Mem:* Am Chem Soc; Am Inst Chem Engrs; Combustion Inst. *Res:* Chemical lasers and laser related diagnostics; chemical kinetics of combustion. *Mailing Add:* 3824 Calle Linda Vista Newbury Park CA 91320

SCHNEIDER, GERALD EDWARD, b Libertyville, Ill, Aug 20, 40; m 62, 77; c 4. NEUROSCIENCE. *Educ:* Wheaton Col, BSc, 63; Mass Inst Technol, PhD(psychol), 66. *Prof Exp:* Res assoc, 66-67, from asst prof to assoc prof, 67-78, PROF NEUROSCI, MASS INST TECHNOL, 78- *Concurrent Pos:* Prin investr, Nat Eye Inst, NIH, 70-; co-investr, Nat Inst Neurol Dis & Stroke, 75-78 & Nat Eye Inst, 78; mem, Biopsychol Study Sect, 78-82, Basic Sci Task Force, Long-term Res Strategies, Nat Inst Neurol & Commun Dis & Stroke, 78; Vision Res Rev Comt, Nat Eye Inst, 89- *Mem:* Soc Neurosci; Cajal Club. *Res:* Nervous system plasticity and regeneration in context of development; development, organization and function of the visual system. *Mailing Add:* Dept Brain & Cognitive Sci Mass Inst Technol 77 Massachusetts Ave Cambridge MA 02139

SCHNEIDER, HANS, b Vienna, Austria, Jan 24, 27; m; c 3. MATHEMATICS. *Educ:* Edinburgh Univ, MA, 48, PhD, 52. *Prof Exp:* Asst lectr, Queen's Univ, Belfast, 52-54, lectr, 54-59; from asst prof to assoc prof, 59-65, PROF MATH, UNIV WIS, 65-, JAMES JOSEPH SYLVESTER

PROF, 88- *Concurrent Pos:* Vis prof, Wash State Univ, Pullman, 56-57, Univ Calif, Santa Barbara, 64-65, Univ Tubingen, 70, Tech Univ Munich, 72 & 74, Univ Montreal, 77, Univ Wurzburg, 80-81, Technion, Lady Davis vis prof, Haifa, Israel, 85 & 86; ed, H Wielandt's collected works, Linear & Multilinear Algebra, 72-, J Algebraic & Discrete Methods, 79-87, ed-in-chief, Linear Algebra & Its Appl, 72-; assoc chmn math dept, Univ Wis, 65-66, chmn, 66-68, mem grad sch admin comt, 77-79; NSF res grants, 67- & 88- *Mem:* Am Math Soc; Math Asn Am; Soc Indust Appl Math. *Res:* Linear algebra. *Mailing Add:* Univ Wis 213 Van Vleck 480 Lincoln Dr Madison WI 53706

SCHNEIDER, HAROLD O, b Cincinnati, Ohio, Apr 8, 30; m 60; c 2. PHYSICS, MATHEMATICS. *Educ:* Univ Cincinnati, BS, 51, MS, 54, PhD(physics), 56. *Prof Exp:* Aeronaut res scientist, Lewis Res Ctr, NASA, 51-61; mathematician, Rand Develop Corp, Ohio, 61-62; staff mem, Lincoln Lab, Mass Inst Technol, 62-72; sr systs analyst, Dynamics Res Corp, Wilmington, Mass, 72-78; MEM STAFF, LOCKHEED CORP, SUNNYVALE, CALIF, 78- *Mem:* Am Inst Aeronaut & Astronaut; Sigma Xi; Soc Indust & Appl Math. *Res:* Spline method, general, analytical, optimal, nonlinear, parameter and state estimation/systems identification technique; 3-degrees-of-freedom (3DOF) application to aerodynamic modeling; multiple radar reentry estimation; systems identification with simulation, analytical covariance error analyses, extension to 6DOF and onboard instrumentation. *Mailing Add:* 855 Clara Palo Alto CA 94303

SCHNEIDER, HAROLD WILLIAM, b Rochester, NY, Oct 8, 43; m 69; c 3. TOPOLOGY, ACTUARIAL SCIENCE. *Educ:* Univ Rochester, AB, 65; Univ Chicago, MS, 66, PhD(math), 72. *Prof Exp:* Teaching asst math, Univ Chicago, 66-69; from instr to prof math, Roosevelt Univ, 69-86; assoc dir, Lincoln Nat Life Ins Co, Ft Wayne, 86-89; ASST ACTUARY, COLUMBUS LIFE INS CO, 90- *Mem:* Asn Comput Mach; Math Asn Am; Asn Soc Actuaries. *Res:* Differential topology; algebraic topology. *Mailing Add:* 256 N Cassington Rd Columbus OH 43209

SCHNEIDER, HENRY, b Los Angeles, Calif, July 25, 15; m 42; c 3. PLANT PATHOLOGY & ANATOMY. *Educ:* Univ Calif, Los Angeles, AB, 38; Univ Calif, Berkeley, MS, 39, PhD(plant path), 43. *Prof Exp:* Asst pathologist, Guayule Res Proj, USDA, 43-44, pathologist, Emergency Plant Dis Prev Proj, 44-45; assoc, Citrus Exp Sta, Univ Calif, 45-47, from asst plant pathologist to assoc plant pathologist, 47-60, plant pathologist, 60-86, lectr plant path, 59-86, EMER PLANT PATHOLOGIST & LECTR, UNIV CALIF, RIVERSIDE, 86- *Concurrent Pos:* Sabbatical leave, Univ Pretoria, 64. *Mem:* Am Phytopath Soc; Bot Soc Am; Biol Stain Comn (vpres, 81-85). *Res:* Pathological plant anatomy; graft transmissible diseases of trees. *Mailing Add:* 4391 Picacho Dr Riverside CA 92507

SCHNEIDER, HENRY, b Montreal, Que, Dec 5, 33; m 62; c 3. BIOCHEMISTRY, MICROBIOLOGY. *Educ:* Sir George Williams Univ, BSc, 55; Univ Western Ontario, MSc, 57; McGill Univ, PhD(phys chem), 63. *Prof Exp:* Res assoc, Cornell Univ, 62-64; res chemist, Miami Valley Labs, Procter & Gamble Co, 64-66; from asst res officer to assoc res officer, 66-78, SR RES OFFICER, NAT RES COUN CAN, 78- *Mem:* Chem Inst Can. *Res:* Pentose fermentation; microbiol process development; enzymology; polysaccharides. *Mailing Add:* Nat Res Coun Can Ottawa ON K1A 0R6 Can

SCHNEIDER, HENRY C, SR, medicine, for more information see previous edition

SCHNEIDER, HENRY JOSEPH, b York, Pa, Dec 21, 20; m 44; c 5. ORGANIC CHEMISTRY. *Educ:* La Salle Col, BA, 42; Temple Univ, MA, 48; Univ Wis, PhD(org chem), 51. *Prof Exp:* Chemist, Ugite Sales Corp, 42-43; res chemist, Rohm & Haas Co, 46-48; asst, Univ Wis, 48-51; res chemist, 51-56, mem staff sales develop, 56-60, mgr sales develop spec prod, 60-63, asst mgr, 63-68, asst mgr, indust chem, 68-69, mgr, 69-71, mkt develop mgr, Chem Div, 71-75, CORP DIR RES & DEVELOP, INDUST CHEM GROUP, ROHM & HAAS CO, PHILADELPHIA, 76- *Concurrent Pos:* Instr, La Salle Col, 52-56. *Mem:* Am Chem Soc; London Chem Soc; Com Develop Asn; Chem Mkt Res Asn. *Res:* Acetylene; hydrogen; high pressure reactions; ion exchange; acrylic monomers and polymers; petroleum additives; direction of broad scope. *Mailing Add:* 60 Meadow Brook Ave Hatboro PA 19040

SCHNEIDER, HENRY PETER, laboratory animal science, medical research; deceased, see previous edition for last biography

SCHNEIDER, HOWARD ALBERT, b Milwaukee, Wis, Dec 25, 12; m 37; c 2. NUTRITION, BIOCHEMISTRY. *Educ:* Univ Wis, BS, 34, MS, 36, PhD(biochem), 38. *Prof Exp:* Asst biochem, Univ Wis, 36-39; Rockefeller Found fel natural sci, Hosp, Rockefeller Inst, 39-40, from asst to assoc, 40-57, assoc prof nutrit & microbiol, 58-65; mem, Inst Biomed Res, Am Med Asn, 65-70, actg dir, 67-68, dep dir, 69-70; prof & dir, Inst Nutrit, 70-78, EMER PROF NUTRIT & BIOCHEM, SCH MED, UNIV NC, CHAPEL HILL, 78- *Concurrent Pos:* Chmn, Inst Lab Animal Resources, Nat Acad Sci-Nat Res Coun, 66-69; mem pub affairs comt, Fedn Am Soc Exp Biol, 68-74, chmn, 70-72; prof, NC State Univ, 70-78 & Univ NC, Greensboro, 70-78; chief ed, Nutrit Support Med Pract, 77-; vis distinguished scholar residence, Fredrik Wachmeister prof sci & eng, Va Mil Inst, 79. *Mem:* Fel AAAS; fel Am Inst Chemists; fel Am Inst Nutrit; Soc Nutrit Educ. *Res:* Nutrition; resistance to infection; discoverer of the pacifarins, a class of ecological ectocrines. *Mailing Add:* 228 Markham Dr Chapel Hill NC 27514

SCHNEIDER, HUBERT H, mathematical logic; deceased, see previous edition for last biography

SCHNEIDER, IMOGENE PAULINE, b Milwaukee, Wis, June 6, 34. PARASITOLOGY. *Educ:* Ohio State Univ, BS, 56, MS, 58; Univ Chicago, PhD(genetics), 61. *Prof Exp:* NSF fel, Univ Zurich, 61-62; res biologist, Yale Univ, 62-65; RES BIOLOGIST, WALTER REED ARMY INST RES, 65- *Mem:* AAAS; Am Soc Trop Med & Hyg; Am Soc Parasitol; Tissue Cult Asn. *Res:* Developmental genetics of Drosophila; insect tissue culture; nutritional requirements of malarial parasites in vitro. *Mailing Add:* Dept Entom Walter Reed Army Inst Res Washington DC 20307-5100

SCHNEIDER, IRWIN, b New York, NY, Aug 17, 32; m 59; c 2. SOLID STATE PHYSICS, OPTICS. *Educ:* Univ Ill, BS, 54, MS, 56; Univ Pa, PhD(physics), 63. *Prof Exp:* Fel physics, Lab Insulation Res, Mass Inst Technol, 63-64; RES PHYSICIST, US NAVAL RES LAB, 64- *Concurrent Pos:* Nat Acad Sci-Nat Res Coun fel, 64-65. *Mem:* Fel Am Phys Soc; fel Sigma Xi. *Res:* Color centers in alkali halide crystals; holography; color center lasers. *Mailing Add:* Code 6551 US Naval Res Lab Washington DC 20375

SCHNEIDER, JACOB DAVID, b St Louis, Mo, Apr 14, 46; m 72; c 2. EXPERIMENTAL ATOMIC PHYSICS, PROTOTYPE ENGINEERING. *Educ:* Univ Mo-Rolla, BS, 68; Kans State Univ, MS, 70. *Prof Exp:* Res asst, Los Alamos Sci Lab, 68-69; assoc physicist, Appl Physics Lab, Johns Hopkins Univ, 70-74; PHYSICIST, LOS ALAMOS SCI LAB, 74- *Mem:* Am Inst Physics. *Res:* Leading a team developing output beam diagnostics for a high-current high-quality ion accelerator; group leader of the accelerator experiments and injector development group (AT-10) in the accelerator technology division of the Los Alamos National Laboratory; specialty is the development and testing of high-current, high brightness H-ion injectors. *Mailing Add:* 675 Totavi Los Alamos NM 87544

SCHNEIDER, JAMES ROY, b Bellevue, Ky, Nov 28, 34; m 56; c 4. PHYSICS. *Educ:* Villa Madonna Col, AB, 56; Univ Cincinnati, MS, 59, PhD(physics), 65. *Prof Exp:* Instr physics, Villa Madonna Col, 59-63, asst prof, 63-64; from asst prof to assoc prof, 64-74, PROF PHYSICS, UNIV DAYTON, 74-, CHMN DEPT, 75- *Mem:* Am Phys Soc; Am Asn Physics Teachers; Sigma Xi. *Res:* Single crystals; x-ray techniques; optical properties of solids; laser interactions; visible and infrared radiation. *Mailing Add:* Dept Physics Univ Dayton 300 College Park Ave Dayton OH 45469

SCHNEIDER, JOHN ARTHUR, b Saginaw, Mich, Feb 27, 40; m 63; c 2. ORGANIC CHEMISTRY, POLYMER CHEMISTRY. *Educ:* Albion Col, AB, 62; Mass Inst Technol, PhD(org chem), 66. *Prof Exp:* Res chemist, Dow Chem Corp, 66-69, proj leader latex res, 69-72, develop specialist org chem, 72-73, group leader org chem develop, 73-77, NY dist sales mgr, 77-79, Chem & Metals Dept mgr for Brazil, 79-82, mgr new opportunity develop mkt res, 82-83, proj dir, comt opportunity develop, 83-85, TECH DIR ELECTRONIC PROD, DOW CHEM CORP, 85- *Concurrent Pos:* Instr, Delta Col, 67-69 & Cent Mich Univ, 74-77; vis indust scientist, Univ Wis-Superior, 67. *Mem:* Am Chem Soc (secy, 62-); Sigma Xi. *Res:* Cephalosporin C synthesis; bromine chloride; styrene-butadiene latexes; unsaturated polyesters; oxonium chemistry; polyamines; fire retardancy; peptide synthesis; electronic chemicals. *Mailing Add:* 2040 Dow Ctr Midland MI 48674

SCHNEIDER, JOHN H, b Eau Claire, Wis, Sept 29, 31; c 2. INFORMATION SCIENCE, BIOCHEMISTRY. *Educ:* Univ Wis, BS, 53, MS, 55, PhD(exp oncol), 58. *Prof Exp:* Asst prof biochem, Am Univ Beirut, 58-61 & Vanderbilt Univ, 61-62; ed-in-chief, Biol Abstr, 62-63; mem staff scientist-adminr training prog, NIH, 63-64, sci & prog info specialist, 64-67, sci & tech info officer, 67-73, dir, Int Cancer Res Data Bank Prog, 73-84, EXEC SECY, SPEC REV COMT, GRANTS REV BRANCH, NAT CANCER INST, 85- *Mem:* AAAS; Am Soc Info Sci. *Res:* Previously stressed development of large automated data bank for collection, processing and dissemination of all technical documents dealing with cancer and descriptions of all current cancer research projects; decimal classifications of biomedicine for use in program analysis; information retrieval and selective dissemination of information; design and development of computer systems for using hierarchical classifications in information systems; automating steps in processing research grants applications; biomedical research administration; cancer research. *Mailing Add:* Div Cancer Prev & Control Bldg EPN Rm 232 Nat Cancer Inst Bethesda MD 20892

SCHNEIDER, JOHN MATTHEW, b Coulterville, Ill, Apr 27, 35; m 60; c 3. ELECTROHYDRODYNAMICS, PHYSICAL ELECTRONICS. *Educ:* Univ Ill, BS, 59, MS, 60, PhD(elec eng), 64. *Prof Exp:* Instr elec eng, Univ Ill, 60-64, asst prof, 64-66; prof scientist res lab, Xerox Corp, 66-68, mgr res lab, 69-81; DIR, ADV TECHNOL LABS, MEAD DIGITAL SYSTS, DAYTON, OHIO, 81- *Res:* Electrostatics research dealing with the interaction of fields and liquids. *Mailing Add:* 3100 Research Blvd Dayton OH 45420

SCHNEIDER, JOSEPH, b Jersey City, NJ, June 25, 18. CHEMISTRY. *Educ:* Columbia Univ, BS, 39; NY Univ, MS, 47; Polytech Inst Brooklyn, PhD(org chem), 62. *Prof Exp:* From instr to asst prof chem, Long Island Univ, 46-55; lectr, Polytech Inst Brooklyn, 56-62; assoc prof, 63-70, PROF CHEM, ST FRANCIS COL, NY, 70- *Concurrent Pos:* Adj prof, Polytech Inst Brooklyn, 64-69; adj prof, Hunter Col, 70-71; US Off Educ fel, NY Univ, 71-72. *Mem:* AAAS; Am Chem Soc; Royal Soc Chem. *Res:* Enzyme model systems; decarboxylase models; reaction mechanisms; Diels-Alder reaction and its catalysis; formation of carbohydrates from formaldehyde; catalysis by metal ions. *Mailing Add:* Eight Cedarbrook Dr Somerset NJ 08873

SCHNEIDER, JURG ADOLF, b Basle, Switz, May 27, 20; nat US; m 46; c 4. PHARMACOLOGY. *Educ:* Univ Basle, MD, 45. *Prof Exp:* Resident surg, Hosp, Basle, Switz, 46-47; resident, Univ Hosp, Univ Basle, 48-51; res assoc cardiol, Sch Med, Univ Calif, 47-48; sr pharmacologist, Ciba Pharmaceut Prod, Inc, 52-54, dir physiol res, 54-57; dir macrobiol res dept, Chas Pfizer & Co, Inc, 57-62; dir pharmaceut res, E I du Pont de Nemours & Co, Inc, 62-72, dir prod lic, 72-85; EMERGENCY PHYSICIAN, DEPT EMERGENCY MED, MED CTR DEL, 71- *Concurrent Pos:* Res fel, Res Lab, Ciba, Ltd, 45-46; lectr, Col Physicians & Surgeons, Columbia Univ, 57-65, adj prof, 70-; consult, 85- *Mem:* Am Physiol Soc; Am Soc Pharmacol & Exp Therapeut; fel Am Col Clin Pharmacol & Chemother; fel Am Col Neuropsychopharmacol; NY Acad Sci; Am Col Emergency Physicians. *Res:* Central nervous system and cardiovascular pharmacology; research administration. *Mailing Add:* 520 Rothbury Rd Wilmington DE 19803

SCHNEIDER, KATHRYN CLAIRE (JOHNSON), b Wiltshire, Eng, March 5, 53; US citizen; m 75; c 1. ORNITHOLOGY, BEHAVIORAL ECOLOGY. *Educ:* Cornell Univ, BA, 75; Princeton Univ, MS, 77, PhD(biol), 79. *Prof Exp:* ASST PROF BIOL, UNIV RICHMOND, 79- *Concurrent Pos:* Marcy Brady Tucker Travel Award, Am Ornithologists Union, 79. *Mem:* AAAS; Am Ornithologists Union; Ecol Soc Am; Sigma Xi; Nat Audubon Soc. *Res:* Optimal foraging theory and its relationship to dominance hierarchies and predation in winter foraging flocks of birds. *Mailing Add:* Box 507 Claverack NY 12513

SCHNEIDER, KENNETH JOHN, b Denver, Colo, Nov 5, 26; m 55; c 3. CHEMICAL ENGINEERING. *Educ:* Colo Sch Mines, PRE, 50. *Prof Exp:* Process engr, Gen Elec Co, 50-58, develop engr, 58-61, sr engr, 61-65; Pac Northwest Lab, Battelle Mem Inst, 65-67, res assoc process develop & prog planning, 67-71; eng assoc, Process Develop & Prog Planning, Westinghouse-Hanford Co, 71-72; mgr process evalutions, 72-78, STAFF ENGR, PAC NORTHWEST LAB, BATTELLE MEM INST, 78- *Concurrent Pos:* Prin engr, Int Atomic Energy Agency, Vienna, Austria, 78-80. *Mem:* Am Inst Chem Engrs; Am Nuclear Soc. *Res:* Process and equipment development in reprocessing of spent nuclear fuels; development of methods for solidification of highly radioactive wastes for safe storage; project engineering; development program planning and management; evaluation of nuclear fuel cycles. *Mailing Add:* Dept Mech Eng Calif State Polytech Univ 3801 W Temple Ave Pomona CA 91768

SCHNEIDER, LAWRENCE KRUSE, b Portland, Ore, Dec 17, 36; m 61; c 3. ANATOMY. *Educ:* Univ Wash, BA, 60, PhD(cytochem), 66. *Prof Exp:* Instr biol struct, Univ Wash, 65-66; asst prof anat, Univ NDak, 66-68; asst prof anat & dir grad training, Dept Anat, Col Med, Univ Ariz, 68-73; assoc prof anat & head dept, 73-75, dir div biomed sci, 75-76, dir med admis & asst dean basic sci, 76-77, asst dean acad affairs, 77-78, PROF ANAT & HEAD DEPT, SCH MED SCI, UNIV NEV, RENO, 76- *Concurrent Pos:* Instnl res grant, Univ NDak, 66-68; Gen Res Support, NASA & Am Cancer Soc instnl res grants, Col Med, Univ Ariz, 68-72. *Mem:* Am Asn Anat; Am Soc Cell Biol; Sigma Xi. *Res:* Radioautography of DNA and RNA synthesis in chromosomes of mammalian lymphocytes in vitro; cell kinetics; effects of various agents on cell growth in tissue culture; cytogenetics. *Mailing Add:* Dept Anat Univ Nev Reno NV 89557

SCHNEIDER, LON S, GERIATRICS. *Educ:* Sarah Lawrence Col, AB, 74; Hahnemann Med Col, Md, 78. *Prof Exp:* ASSOC PROF PSYCHIAT, SCH MED, UNIV SOUTHERN CALIF, 83- *Mem:* Am Asn Geriat Psychiat (secy, 87-88). *Res:* Geriatric psychiatry; psychopharmacology, biological markers; Alzheimer's disease, depression. *Mailing Add:* Dept Psychiat Sch Med Univ Southern Calif 1934 Hospital Pl Los Angeles CA 90033

SCHNEIDER, MARC H, b Rochester, NY. WOOD SCIENCE & TECHNOLOGY. *Educ:* State Univ NY, BS, 65, MS, 67, PhD(wood & polymer sci), 78. *Prof Exp:* Fel wood coating, Paint Res Inst, 65-67; from asst prof to assoc prof, 67-80, PROF WOOD SCI & TECHNOL, UNIV NB, 80- *Mem:* Soc Wood Sci & Technol; Forest Prods Res Soc. *Res:* Wood-chemical interactions such as sorption, adhesion and interdiffusion; wood as fuel; wood-polymer composites. *Mailing Add:* Dept Forest Resources Univ NB Col Hill Box 4400 Fredericton NB E3B 5A3 Can

SCHNEIDER, MARTIN V, b Bern, Switz, Oct 20, 30; m 55. PHYSICS. *Educ:* Swiss Fed Inst Technol, MS, 55, PhD(physics), 59. *Prof Exp:* Res assoc microwave res lab, Swiss Fed Inst Technol, 60-62; mem tech staff radio res, 63-68, SUPVR RADIO RES DEPT, BELL LABS, 68- *Mem:* Am Phys Soc; Am Vacuum Soc; Inst Elec & Electronics Engrs; Sigma Xi. *Res:* Microwave active and passive devices and circuits; optical and thin film active devices; photodetectors. *Mailing Add:* 46 Line Rd Holmdel NJ 07733

SCHNEIDER, MAXYNE DOROTHY, b North Adams, Mass, Nov 4, 42. PHYSICAL CHEMISTRY. *Educ:* Col Our Lady of the Elms, AB, 65; Boston Col, PhD(chem), 75. *Prof Exp:* Teacher chem, Cathedral High Sch, Springfield, Mass, 65-69; instr, 73-81, asst prof chem, 81-83, ACAD DEAN, COL OUR LADY OF THE ELMS, 83- *Mem:* Am Chem Soc; Inst Theol Encounter with Sci & Technol; Am Asn Higher Educ. *Res:* Science education; science, technology, society, religion. *Mailing Add:* Col Our Lady of the Elms 291 Springfield St Chicopee MA 01013

SCHNEIDER, MEIER, b Worcester, Mass, Dec 8, 15; m 44; c 3. INDUSTRIAL HYGIENE, HAZARDOUS MATERIALS MANAGEMENT. *Educ:* Univ Rochester, NY, BA, 40; Calif State Univ, Northridge, MS, 73. *Prof Exp:* Sr chemist, Los Angeles County Air Pollution Control Dist, 48-52; res specialist, Los Angeles Div, Rockwell Int, 55-68; safety & indust hyg coordr, Lockheed-Calif Co, Burbank, 68-70; sr indust hyg engr, Environ Health Serv Prog, Occup Health Sect, Calif State Dept Health, 70-73; indust hyg engr, Dept Personnel, Med Serv Div, City of Los Angeles, 74-81; chief occup safety & health, Metropolitan Water Dist Southern Calif, 81-87; RETIRED. *Concurrent Pos:* Assoc & prof environ & occup health, Health Sci Dept, Calif State Univ, Northridge, 74-88; Calif indust hyg rep, Chem Agents Threshold Limit Value Comt, Am Conf Govt Indust Hygienists, Cincinnati, Ohio, 75-; field prof, Norton AFB Satellite Campus, Univ Southern Calif, 76-86; lectr, air pollution, occup & environ health & safety, indust hyg; expert witness, toxic tort & prod liability litigation; prof rep, Task Force Appl Competitive Technol-Hazardous Mat Technol, Calif Community Cols, 89-90. *Mem:* Am Indust Hyg Asn; Am Conf Govt Indust Hygienists; Sigma Xi. *Res:* Toxicology of industrial chemicals. *Mailing Add:* 1208 Point View St Los Angeles CA 90035-2621

SCHNEIDER, MICHAEL CHARLES, b Chicago, Ill, May 7, 29; m 54; c 2. GEOLOGY. *Educ:* Cornell Col, BA, 52; Miami Univ, MS, 56; Brigham Young Univ, PhD(geol), 67. *Prof Exp:* From instr to assoc prof geol, DePauw Univ, 59-68; prof geol, 68-76, PROF EARTH SCI, EDINBORO STATE COL, 76- *Mem:* AAAS; Geol Soc Am; Soc Econ Paleontologists & Mineralogists; Nat Asn Geol Teachers. *Res:* Water pollution; acid mine drainage; stratigraphic correlation and age dating; science education. *Mailing Add:* Dept Geol Sci Edinboro State Col Edinboro PA 16444

SCHNEIDER, MICHAEL J, b Saginaw, Mich, Apr 21, 38; m 67; c 1. PLANT PHYSIOLOGY, PHOTOBIOLOGY. *Educ:* Univ Mich, BS, 60; Univ Tenn, MS, 62; Univ Chicago, PhD(bot), 65. *Prof Exp:* Nat Acad Sci-Nat Res Coun fel, Plant Physiol Lab, USDA, Beltsville, Md, 65-67; USPHS fel bot, Univ Wis-Madison, 67-68; asst prof biol sci, Columbia Univ, 68-73; chmn dept natural sci, 75-80, 83-90, assoc provost, 90-91, PROF BIOL, UNIV MICH, DEARBORN, 77-, INTERIM PROVOST & VCHANCELLOR ACAD AFFAIRS, 91- *Concurrent Pos:* Dept Energy vis prof, Plant Res Lab, Mich State Univ, 80-81. *Honors & Awards:* Sigma Xi. *Mem:* AAAS; Am Soc Plant Physiol; Bot Soc Am; Am Soc Photobiol. *Res:* Physiology and biochemistry of plant growth and development. *Mailing Add:* Off Acad Affairs Univ Mich-Dearborn Dearborn MI 48128-1491

SCHNEIDER, MORRIS HENRY, b Sutton, Nebr, Nov 26, 23; m 52; c 1. INDUSTRIAL ENGINEERING. *Educ:* Univ Nebr, BS, 51 & 59; Kans State Univ, MS, 61; Okla State Univ, PhD(indust eng), 66. *Prof Exp:* Instr mech eng, Univ Nebr, 54-60; asst prof indust eng, Kans State Univ, 60-62 & Tex Tech Col, 62-64; assoc prof mech eng, 65-70, PROF INDUST ENG & CHMN DEPT, UNIV NEBR, LINCOLN, 70- *Concurrent Pos:* Am Soc Tool & Mfg Eng res grant, 66-67; NSF sci equip prog grant, 66-68. *Mem:* Am Soc Eng Educ; Am Inst Indust Engrs; Nat Soc Prof Engrs; Sigma Xi. *Res:* Production design and processes. *Mailing Add:* Dept Indust Eng 175 Nebraska Hall Univ Nebr Lincoln NE 68588

SCHNEIDER, NORMAN RICHARD, b Ellsworth, Kans, Mar 28, 43; m 68; c 1. TOXICOLOGY, DIAGNOSTIC MEDICINE. *Educ:* Kans State Univ, BS, 67, DVM, 68; Ohio State Univ, MSc, 72. *Prof Exp:* Chief vet serv, Goose AFB, Labrador, 68-70, Air Force Inst Technol fel, Ohio State Univ, 70-72; vet scientist, Armed Forces Radiobiol Res Inst, Md, 72-76; vet toxicologist, Aerospace Med Res Lab, Wright-Patterson AFB, Ohio, 76-79; ASSOC PROF & VET TOXICOLOGIST, DEPT VET SCI, UNIV NEBR, LINCOLN, 79- *Concurrent Pos:* Consult, Mid-Plains Poison Control Ctr, Nebr, 79-; assoc courtesy prof, Dept Pharmacodynamics & Toxicol, Col Pharm, Med Ctr, Univ Nebr, 82-85, Dept Pharmaceut Sci, 86-; vet prof rep, Nebr State Bd Health, 85-87. *Mem:* Am Bd Vet Toxicol (pres, 88-91); Am Vet Med Asn; Am Acad Vet & Comp Toxicol; Am Asn Vet Lab Diagnosticians; Asn Mil Surgeons US; Asn Off Anal Chemists. *Res:* Nitrite-nitrate pharmacokinetics and pathophysiology; maternal-fetal pharmacokinetics; mycotoxicoses related to fusariotoxins and ergot alkaloids, and chemical detoxification-decontamination; pesticide degradation in rendered animal byproducts; trace elements in nutritional-metabolic disorders. *Mailing Add:* Vet Diag CTR Dept Vet Sci Univ Nebr Lincoln NE 68583-0907

SCHNEIDER, PAUL, b New York, NY, May 3, 34; m 57; c 2. CLINICAL CHEMISTRY. *Educ:* City Col New York, BS, 56; Yale Univ, MS, 58; NY Univ, PhD(org chem), 61. *Prof Exp:* Res chemist, IIT Res Inst, 61-62; res scientist, Res Div, NY Univ, 62-64, asst prof chem, 64-65; res scientist, Diag Div, Chas Pfizer & Co, 65-68, mgr res & develop dept, 68-69; dir tech develop & eval, Technicon Instrument Corp, 69-75; dir sci rel, Bio-Dynamics/BMC, 75-81; DIR, REAGENT OPER METPATH, INC, 81- *Mem:* Fel Am Inst Chemists; NY Acad Sci; Am Chem Soc; Nat Registry Clin Chemists; Am Asn Clin Chemists. *Res:* Hemolytic disease of the newborn; bilirubin metabolism; causes of hypertension and their detection; chemical basis of mental disease; hyperlipidemea, automated analysis and calibration materials. *Mailing Add:* Metpath Inc One Malcolm Ave Teterboro NJ 07608

SCHNEIDER, PHILIP ALLEN DAVID, b St Louis, Mo, Oct 26, 38; m 63; c 6. INFORMATION SCIENCE, APPLIED STATISTICS. *Educ:* Cornell Univ, AB, 61; Duke Univ, PhD(methodology sci), 68. *Prof Exp:* Mathematician, Abbott Labs, 61-62; plans officer info sci, US Army Security Agency, 62-64; group leader, Defense Commun Agency, 64; consult, Dept Math, Duke Univ, 64-68; dir sci serv, US Army Syst Anal Group, 68-73; chief, Manpower Statist Div, 73-75, assoc dir workforce info, Bur Personnel Mgt Info Systs, US Civil Serv Comn, 75-80, ASST DIR WORKFORCE INFO, SR EXEC SERV, US OFF PERSONEL MGT, 80- *Concurrent Pos:* NDEA fel, 64-67, NSF fel, 67-68; assoc prof, Sch Gen Studies, Univ Va, 68-73 & Northern Va Community Col, 73-; prof logic, George Mason Univ, 86- *Mem:* Am Statist Asn; Am Philol Asn. *Mailing Add:* 8511 Browning Ct Annandale VA 22003

SCHNEIDER, PHILLIP WILLIAM, JR, b Corvallis, Ore, Sept 2, 44; m 69; c 2. FISHERIES, TOXICOLOGY. *Educ:* Ore State Univ, BS, 66, PhD(fisheries, pharmacol), 74; Univ Maine, MS, 71. *Prof Exp:* Fisheries biologist, US Environ Protection Agency, 68-71; fish & wildlife biologist, River Basins Studies, US Fish & Wildlife Serv, res toxicologist, 75-78, CHIEF, CHRONIC INVEST, TOXICOLOGIST, HASKELL LAB, E I DU PONT DE NEMOURS & CO, 78- *Mem:* Am Soc Testing & Mat; Am Fisheries Soc; AAAS. *Res:* Neuromuscular physiology, pharmacology and detoxification mechanisms in fish as compared with other vertebrates. *Mailing Add:* 8755 SW Woodside Portland OR 97225

SCHNEIDER, RALPH JACOB, b Oxford, Ohio, Sept 2, 22; m 45; c 3. ELECTRONICS, ELECTRICAL ENGINEERING. *Educ:* Clarkson Col Technol, BEE, 49. *Prof Exp:* Instrumentation engr, Bell Aircraft Corp, 49-54, preliminary design engr, 54-57; engr, Melpar Div, Westinghouse Air Brake Co, 57-58, proj engr, 58-60; electronics engr, Anal Serv, Inc, 60-85; CONSULT, 88- *Mem:* Sr mem Inst Elec & Electronics Engrs. *Res:* Evaluation of research aircraft and missile components such as storage batteries, critical mechanical components, and radar subsystems; analysis of airborne radar, communications and electronic penetration aids and fire control systems. *Mailing Add:* 718 N Overlook Dr Alexandria VA 22305-1222

SCHNEIDER, RICHARD THEODORE, b Munich, Ger, July 29, 27; m 50; c 2. PLASMA PHYSICS, SPECTROSCOPY. *Educ:* Univ Stuttgart, dipl, 58, PhD(physics), 61. *Prof Exp:* Res asst plasma physics, Inst High Temperature Res, Stuttgart, Ger, 59-61; sec chief, Allison Div, Gen Motors Corp, 61-65;

from assoc prof to prof nuclear eng, Univ Fla, 65-88; PRES, EYE RES LAB, 88- *Concurrent Pos:* Liason scientist br off, London, UK, Off US Naval Res, 74-75. *Honors & Awards:* Except Sci Achievement Medal, NASA, 75. *Mem:* Optical Soc Am; Soc Appl Spectros; Am Inst Aeronaut & Astronaut. *Res:* Plasma diagnostics using spectroscopic techniques; nuclear pumped lasers; infra red spectroscopy; uranium plasmas. *Mailing Add:* Eye Res Lab 1663 Technology Ave Alachua FL 32615

SCHNEIDER, ROBERT, b Brooklyn, NY, Apr 7, 21; m 46; c 3. HYDROGEOLOGY, HYDROGEOLOGY OF CONTAMINANTS IN GROUND WATER. *Educ:* Brooklyn Col, AB, 41. *Prof Exp:* Photogrammetrist, US Geol Surv, 41-43, geologist, 43-44, groundwater geologist, Tenn, 45-50, dist geologist, Minn, 50-60, asst to chief res sec, Groundwater Br, 60-62, actg chief res sect, 62-64, staff geologist, Gen Hydrol Br, 64-65, res geologist, 65-67, chief off radiohydrol, 67-71; water res scientist, Off Water Resources Res, US Dept Interior, 71-75; staff hydrologist, US Geol Surv, 75-84; CONSULT HYDROGEOLOGIST, 84- *Concurrent Pos:* Consult, Govt Brazil, 60 & Govt Israel, 62; ed bd, J Ground Water, 74-78. *Mem:* Geol Soc Am; Am Geophys Union; Nat Water Well Asn; Int Asn Hydrogeologists (secy-treas, 75-77); Am Inst Prof Geologists. *Res:* Thermal characteristics of aquifers; subsurface movement of hazardous wastes; geohydrology of glacial deposits. *Mailing Add:* 6212 N 31st St Arlington VA 22207

SCHNEIDER, ROBERT FOURNIER, b New York, NY, Feb 24, 33; div; c 3. CHEMICAL PHYSICS. *Educ:* Columbia Univ, AB, 54, MA, 56, PhD(chem), 59. *Prof Exp:* Res assoc chem, Brookhaven Nat Lab, 59-60; asst prof, 60-68, ASSOC PROF CHEM, STATE UNIV NY, STONY BROOK, 68-, ASSOC VPROVOST, 73- *Concurrent Pos:* Actg vprovost Comput & Commun, 86-88; vpres, NY State Educ Res Net. *Mem:* AAAS; Soc Res Admin; Nat Coun Univ Res Admin. *Res:* Research administration; computerization of administrative environments. *Mailing Add:* Off Res Admin State Univ NY Stony Brook NY 11794-4466

SCHNEIDER, ROBERT JULIUS, b Troy, NY, Mar 9, 39; m 61; c 2. ACCELERATOR PHYSICS, RADIATION PHYSICS. *Educ:* Oberlin Col, BA, 60; Wesleyan Univ, MA, 63; Tufts Univ, PhD(physics), 68. *Prof Exp:* Staff assoc physics, Columbia Univ, 68-73; res fel physics, Harvard Univ, 73-80; physicist, Gen Ionex Corp, 80-84; SR SCIENTIST, AM SCI & ENG, INC, 84- *Concurrent Pos:* Lectr physics, Harvard Univ, 79-80. *Mem:* Am Phys Soc; Am Asn Physicists Med; Sigma Xi. *Res:* Radiation applied to cancer therapy; accelerators; isotopes; nuclear physics. *Mailing Add:* Eight Braebrook Rd Acton MA 01720

SCHNEIDER, ROBERT W(ILLIAM), b Staten Island, NY, Dec 30, 25; m 46; c 3. MECHANICAL ENGINEERING. *Educ:* Lehigh Univ, BS, 48, MS, 49. *Prof Exp:* Design engr, Linde Co Div, Union Carbide Corp, 49-52, asst design & metall engr, 52-54; asst supv engr, Travelers Indemnity Co, 54-56, in chg boiler & pressure vessel design, 56-60; stress analyst, Oak Ridge Nat Lab, 60-62, asst supt, Inspection Eng Dept, 62-68; mgr eng, Energy Prod Group, Gulf & Western MFG Co, 68-82; CONSULT ENG, R W SCHNEIDER ASSOC, 82- *Mem:* Fel Am Soc Mech Engrs. *Res:* Pressure vessel design, especially evaluation of secondary or discontinuity stresses; plastic behavior of engineering materials; development of theory and design rules for analysis of reverse, raised face flanges and flat faced flanges in metal-to-metal contact; theoretical and experimental work on behavior of nozzles subjected to pressure and cyclic external loads. *Mailing Add:* 3918 Lincoln Pkwy W Allentown PA 18104

SCHNEIDER, RONALD E, b Akron, Ohio, Sept 14, 28; m 53; c 4. THEORETICAL PHYSICS, NUCLEAR PHYSICS. *Educ:* Univ Akron, BS, 51; Va Polytech Inst, MS, 53; John Carroll Univ, MS, 58; Case Inst Technol, PhD(physics), 64. *Prof Exp:* Engr, Goodrich Tire & Rubber Co, 53-54; sr engr, Goodyear Aerospace Corp, 56-58; instr math, John Carroll Univ, 58-60; asst prof, 62-64, ASSOC PROF PHYSICS, UNIV AKRON, 64- *Res:* Nuclear forces. *Mailing Add:* Dept Physics Univ Akron Akron OH 44325

SCHNEIDER, ROSE G, b Minsk, Russia, July 19, 08; US citizen; m 41; c 3. MEDICINE. *Educ:* Barnard Col, Columbia Univ, AB, 29; Harvard Med Sch, MA, 32; Cornell Med Col, PhD, 37. *Prof Exp:* Instr path, Univ Tex Med Br Galveston, 42-43; asst pathologist, Robert B Green Hosp, San Antonio, 44-45; res assoc, Tissue Cult Lab & Tissue Metab Res Lab, 48-62, asst res prof surg, 62-63, asst res prof pediat, 63-65, assoc res prof, 65-69, RES PROF PEDIAT & PROF HUMAN BIOL CHEM & GENETICS, UNIV TEX MED BR GALVESTON, 69- *Mem:* Int Soc Hemat; Soc Exp Biol & Med; Am Soc Human Genetics; Am Soc Hemat; Sigma Xi. *Res:* Abnormal hemaglobins. *Mailing Add:* Dept Pediat Univ Tex Med Br Galveston TX 77550

SCHNEIDER, SANDRA LEE, b Pueblo, Colo, July 10, 44; m 73. IMMUNOHEMATOLOGY, IMMUNOLOGY. *Educ:* Southern Colo State Col, BS & Univ Colo, MT, 66. *Prof Exp:* Med technologist immunohemat, Belle Bonfils Mem Blood Band, 66-68; supvr blood bank, East Tenn Baptist Hosp, 68-69; med technologist II hemat, Med Ctr, Univ Kans, 69-70; med technologist immunol, Knoxville Blood Bank & Reagents, 70-71; res scientist immunol & immunohemat, 73-90, RES ASSOC PROF, SOUTHWEST FOUND RES & EDUC, 90- *Concurrent Pos:* Tech transfusion consult, ETenn Baptist Hosp, Knoxville, Tenn, 68-69 & Knoxville Blood Ctr, 70-71. *Mem:* AAAS; Am Soc Primatologists; Am Soc Clin Pathologists; Am Asn Blood Banks; Am Soc Microbiol. *Res:* Design and development of techniques to determine alveolar macrophage and lymphocyte interactions, particularly in the baboon exposed to anti-cancer drugs, cigarette derived smoke and chemically defined environmental agents. *Mailing Add:* PO Box 3 Helotes TX 78023

SCHNEIDER, SOL, b New York, NY, Feb 24, 24; m 50; c 2. ELECTRONICS. *Educ:* Brooklyn Col, BA, 46; NY Univ, MS, 49. *Prof Exp:* Chief, Res Unit Power & Gas Tube Sect, Electronics Technol & Devices Lab, US Army, 48-55, asst sec chief, 55-56, chief, Plasma & Pulse Power Br, Electronics Technol & Devices Lab, 56-81; adj prof, Southeastern Ctr Elec Eng Educ, 81-87; CONSULT, 81- *Concurrent Pos:* Army mem, Power & Gas Tubes Panel, Adv Group Electrontubes, Dept Defense, 56-62, assoc army mem Adv Group Electron Devices, 74-81, Steering Comt, Pulsed Power Workshop, 76; chmn, Power Modulator Symp, Inst Elec & Electronics Engrs, 58-80, emer chmn, 81-, co-chmn, High Voltage Workshop, 88-89, adv exec comt, 90-, mem, exec comt Int Pulsed Power Conf, 76-80; mem, exec comt, Gaseous Electronics Comt, Am Phys Soc, 61- 66, Electron & Atomic Physics Div, 65-66, AF Panel Educ in Pulsed Power, 78-80, USN Pulsed Power Tech Adv Group, 78-80, Pulsed Power Tech Adv Panel, Strategic Defense Initiative, 84-; consult, Los Alamos Sci Lab, 80-81, Rockwell Int, 81, SRI Int, 83-91, Vitronics, 88-; consult, Pulse Power Ctr, US Army, 83-, tech studies, Res Off, Battelle, 81-88. *Honors & Awards:* Res & Develop Achievement Awards, 63 & 78, US Army & Secy Army Spec Act Award, 63; Bronze Medallion, Army Sci Conf, 78. *Mem:* Am Phys Soc (secy, 64); fel Inst Elec & Electronics Engrs; NY Acad Sci. *Res:* Modulators; pulse power; physical and gaseous electronics; plasma physics; electron tubes; lasers; modulators. *Mailing Add:* 100 Arrowwood Ct Red Bank NJ 07701

SCHNEIDER, STEPHEN HENRY, b New York, NY, Feb 11, 45; m 78; c 2. CLIMATOLOGY. *Educ:* Columbia Univ, BS, 66, MS, 67, PhD(mech eng), 71. *Prof Exp:* Nat Acad Sci-Nat Res Coun res assoc, Goddard Inst Space Studies, NASA, NY, 71-72; advan study prog fel, 72-73, sci & dep head, climate proj, 73-78, actg leader, Climate Sensitivity Group, 78-80, head, vis progs & dep dir, advanced study prog, 80-87, HEAD, INTERDISCIPLINARY CLIMATE SYSTS SEC, NAT CTR ATMOSPHERIC RES, 87- *Concurrent Pos:* mem, Comt Paleoclimatol & Climatic Change, Am Meteorol Soc, 73, mem global atmospheric res prog working group for numerical experimentation, 74-78; ed, Climatic Change, 75-; mem climate dynamics panel, Nat Acad Sci, 76-81; mem, Carter-Mondale Task Force on Sci Policy & Coun Sci & Technol for Develop, 76; Univ Corp Atmospheric Res Affil Prof, Lamont-Doherty Geol Observ, Columbia Univ, 76-; mem, Nat Acad Sci, Subcomt Resources & Environ, Int Inst Appl Systs Analysis, 78-80, Adv & Planning Comt, Social Indicators, Social Sci Res Coun, 79-, US Nat Climate Prog Adv Comt, 80-, Comt Pub Understanding Sci, NSF, 80-81, Sci Adv Comt, World Climate Studies Prog, UN Environ Prog, 80-; co-ed, Food-Climate Interactions, 81, Social Sci Res, An Interdisciplinary Appraisal, 82; ed, Climatic Change, 76- *Mem:* Fel AAAS; fel Scientists Inst Pub Info; US Asn Club Rome; Am Meteorol Soc; Am Geophys Union; Nat Coun Fedn Am Scientists. *Res:* Theoretical investigations of climatic changes arising from both natural and possible man-made causes; impact of human activities on climate; impact of climatic change on society; science policy; science popularization; effect of nuclear war on society and environment. *Mailing Add:* Nat Ctr Atmospheric Res PO Box 3000 Boulder CO 80307

SCHNEIDER, WALTER CARL, b Cedarburg, Wis, Sept 26, 19; m 42; c 3. NUCLEIC ACIDS, CYTOCHEMISTRY. *Educ:* Univ Wis, BS, 41, PhD(physiol chem), 45. *Prof Exp:* Res asst oncol, McArdle Mem Lab, Univ Wis, 41-45, instr, 47-48; res fel, Jane Coffin Childs Mem Fund, 45-47; chemist, Nat Cancer Inst, 48-62, head, Nucleic Acids Sect, 62-80; RETIRED. *Concurrent Pos:* Assoc ed, J Nat Cancer Inst, 52-54; res consult, George Washington Univ, 58-59. *Mem:* Sigma Xi; Am Soc Biol Chemists; Am Asn Cancer Res; Am Chem Soc. *Res:* Methods for nucleic acid analysis and for isolation of subcellular organelles such as nuclei and mitochondria from animal tissues; composition and function of subcellular organelles in normal and cancer tissues. *Mailing Add:* 15301 Barningham Ct Silver Spring MD 20906

SCHNEIDER, WILLIAM C, b New York, NY, Dec 24, 23; m 64; c 4. AEROSPACE ENGINEERING. *Educ:* Mass Inst Technol, SBAero, 49; Univ Va, MS, 52; Cath Univ, DEng(aero eng), 76. *Prof Exp:* Dep dir, Gemini, NASA, 63-65, mission dir, Gemini, 65-66 & Apollo, 66-68, prog dir Skylab, 68-74, dep assoc adminr, manned flight, 74-78, assoc adminr, tracking & data, 78-80; vpres, Computer Sci Corp, 80-90; CONSULT AEROSPACE, 90- *Concurrent Pos:* Mem, Aerospace Med Adv Comt, NASA, 83-, Life Sci Adv Comt, 84- & Space Sta Adv Comt, 86-, vpres finance, Alumni League, 86-; chmn, Honors & Awards Comt, Am Inst Aeronaut & Astronaut. *Honors & Awards:* Collier Trophy, 73. *Mem:* Int Astrol Asn; Am Astron Soc; Planetary Soc. *Mailing Add:* 11801 Clintwood Pl Silver Spring MD 20902

SCHNEIDER, WILLIAM CHARLES, b New Orleans, La, Jan 22, 40; m 68; c 3. SOLID MECHANICS, ENGINEERING MECHANICS. *Educ:* La State Univ, BS, 62; Univ Houston, MS, 68; Rice Univ, PhD(mech eng), 72. *Prof Exp:* Aerospace engr, 62-85, CHIEF, MECH DESIGN & ANALYSIS BR, NASA, 85- *Concurrent Pos:* Consult mech, 77-; eng consult, DiCaro & Assocs, 77-78; eng consult & vpres, Accident Anal Consult Engrs, 78- *Mem:* Sigma Xi. *Res:* Thermoelasticity; elasticity; fluid-filled porous elastic solids. *Mailing Add:* Johnson Space Ctr ES6 NASA Rd 1 Houston TX 77058

SCHNEIDER, WILLIAM GEORGE, b Wolseley, Sask, June 1, 15; m 40; c 2. PHYSICAL CHEMISTRY. *Educ:* Univ Sask, BSc, 37, MS, 39; McGill Univ, PhD(phys chem), 41. *Hon Degrees:* DSc, York Univ, 66, Mem Univ Newf, 68, Univ Moncton, 69, Univ Sask, 69, McMaster Univ, 69, Laval Univ, 69, Univ NB, 70, Univ Montreal, 70, McGill Univ, 70; LLD, Univ Alta, 68, Laurentian Univ, 68; DSc, Ottawa Univ, 78. *Prof Exp:* Royal Soc Can traveling fel, Harvard Univ, 41-43; res physicist, Woods Hole Oceanog Inst, 43-46; head gen phys chem sect, div pure chem, Nat Res Coun Can, 43-63, dir div pure chem, 63-65, vpres sci, 65-67, pres, 67-80; RES CONSULT, 80- *Concurrent Pos:* Pres, Int Union Pure & Appl Chem, 83-85. *Honors & Awards:* Medal, Chem Inst Can, 61; Henry Marshall Tory Medal, Royal Soc Can, 69; Montreal Medal, Chem Inst Can, 73; Order of Can, 77. *Mem:* Am Chem Soc; Am Phys Soc; fel Chem Inst Can; fel Royal Soc Can; fel Royal Soc London; Int Union Pure & Appl Chem. *Res:* Intermolecular forces; critical phenomena; ultrasonics; nuclear magnetic resonance; organic semiconductors. *Mailing Add:* 65 Whitemarl Dr Unit No 2 Ottawa ON K1L 8J9 Can

SCHNEIDER, WILLIAM PAUL, b Marietta, Ohio, Mar 25, 21; m 44; c 4. ORGANIC CHEMISTRY. *Educ:* Marietta Col, AB, 44; Univ Wis, MS, 46, PhD(chem), 50. *Prof Exp:* Instr chem, Marietta Col, 46-47, fel, Harvard Univ, 50-51; chemist, Upjohn Co, 51-83; RETIRED. *Mem:* Am Chem Soc. *Res:* Synthetic organic chemistry; steroids; prostaglandins; natural products. *Mailing Add:* 5539 Hwy 187 Anderson SC 29625-8961

SCHNEIDER, WOLFGANG JOHANN, b Vienna, Austria, Apr 5, 49; m 74; c 1. MOLECULAR BIOLOGY & BIOCHEMISTRY OF CELL SURFACE RECEPTORS. *Educ:* Tech Univ Vienna, Austria, BS, 70, MS, 73, PhD(biochem), 75; Univ Graz, Austria, Dozent(med biochem), 86. *Prof Exp:* Postdoctoral, Univ BC, 76-78; res assoc, Univ Tex Health Sci Ctr, Dallas, 78-80, asst prof molecular genetics, 81-85; assoc prof, 85-89, PROF BIOCHEM, UNIV ALTA, EDMONTON, 89- *Concurrent Pos:* Lectr, Univ Graz, Austria, 86-; mem, Lipid & Lipoprotein Res Group, 87-; ed, Biochim Biophys Acta, 89-; vis prof, Wihuri Res Inst, Helsinki, 90- *Honors & Awards:* Gabor Szasz Prize, Ger Soc Clin Invest, 87; Kurt Adam Prize, Ger Congress Soc, 87; Heinrich Weiland Prize, Heinrich Wieland Found, 91. *Mem:* Am Soc Cell Biol; Am Soc Biochem & Molecular Biol; Can Biochem Soc; Ger Soc Clin Invest; Austrian Atherosclerosis Soc. *Res:* Regulation of oocyte growth through receptor-mediated endocytosis; role of lipoprotein receptors and apolipoproteins in etiology of lipid disorders. *Mailing Add:* Lipid & Lipoprotein Res Group Univ Alta Edmonton AB T6G 2S2 Can

SCHNEIDER, WOLFGANG W, b Oberhausen, Ger, Mar 25, 35; m 63; c 1. HOMOGENEOUS CATALYSIS, HETEROGENEOUS CATALYSIS. *Educ:* Aachen Tech Univ, BS, 57, MS, 59; Max Planck Inst Coal Res, PhD(org chem), 62. *Prof Exp:* Asst metal-organics, Max Planck Inst Coals Res, 59-61, fel org chem, 61-62; res chemist, 63-65, sr res & develop chemist, 65-78, res & develop assoc, 78-80, SR RES & DEVELOP ASSOC, B F GOODRICH CO, 80- *Res:* Heterogenous and homogenous catalysis via transition metal pi complex; organic chemistry; polymerization; chemical engineering. *Mailing Add:* 3934 Harris Rd Cleveland OH 44147

SCHNEIDERMAN, HOWARD ALLEN, developmental biology, developmental genetics; deceased, see previous edition for last biography

SCHNEIDERMAN, JILL STEPHANIE, b New York, NY, May 13, 59. ENVIRONMENTAL SCIENCE. *Educ:* Yale Col, BS, 81; Harvard Univ, AM, 85; PhD(geol), 87. *Prof Exp:* ASST PROF GEOL, POMONA COL, 87- *Concurrent Pos:* Steele fel, Pomona Col, 91. *Mem:* Geol Soc Am; Nat Asn Geol Teachers; Coun Undergrad Res. *Mailing Add:* Dept Geol Pomona Col 609 N College Ave Claremont CA 91711

SCHNEIDERMAN, LAWRENCE J, b New York, NY, Mar 24, 32; m 56; c 4. INTERNAL MEDICINE. *Educ:* Yale Univ, BA, 53; Harvard Med Sch, MD, 57. *Prof Exp:* Intern path, Boston City Hosp, Mass, 57-58, intern med, Strong Mem Hosp, Rochester, NY, 58-59; clin assoc, Sect Human Genetics, Nat Inst Dent Res, 59-61; med resident, Sch Med, Stanford Univ, 62-64, from instr to asst prof med, 64-70; assoc prof community med, 70-80, PROF, SCH MED, UNIV CALIF, SAN DIEGO, 80- *Concurrent Pos:* Nat Heart Inst sr fel, Galton Lab, London, 61-62; vis scholar, Hastings Ctr; vis prof, Albert Einstein Sch Med & Montefiore Hosp, 80-81; vis prof, Dept Med Hist & Ethics, Univ Wash, 89. *Mem:* Am Col Physicians; Physicians for Social Responsibility; Am Soc Health & Human Values. *Res:* Medical ethics. *Mailing Add:* Dept Community/Family Med Sch Med Univ Calif San Diego La Jolla CA 92093

SCHNEIDERMAN, MARTIN HOWARD, b Brooklyn, NY, Dec 24, 41; m 72; c 1. RADIATION BIOLOGY, CELL BIOLOGY. *Educ:* Cornell Univ, BS, 63; Colo State Univ, MS, 67, PhD(physiol & biophys), 70. *Prof Exp:* Picker Found fel, Univ Fla, 70-72; res scientist, Battelle Mem Inst, 72-75; from asst porf to assoc prof radiation biol, Thomas Jefferson Univ, 75-84; courtesy assoc prof biol, Fla State Univ, 85-90; ASSOC PROF RADIOL, UNIV NEBR MED CTR, 90- *Concurrent Pos:* Mem exp combined modalities study group, Nat Cancer Inst, 76-78. *Mem:* Radiation Res Soc; Biophys Soc; Cell Kinetics Soc. *Res:* Effects of radiation, drugs and other challenges on the kinetics and survival of mammalian cells in culture; mechanisms related to observed changes. *Mailing Add:* Dept Radiol Univ Nebr Med Ctr Omaha NE 68198-1045

SCHNEIDERMAN, MARVIN ARTHUR, b New York, NY, Dec 25, 18; m 41; c 3. ENVIRONMENTAL HEALTH PUBLIC HEALTH & EPIDEMIOLOGY. *Educ:* City Col New York, BS, 39; Am Univ, MA, 53, PhD(math statist), 61. *Prof Exp:* Jr statistician, Bur Census, US Dept Commerce, DC, 40-41; statistician, US War Dept, 41-43, Wright-Patterson AFB, Ohio, 46-48; statistician, 48-60, assoc chief biomet br, 60-70, assoc dir field studies & statist, Nat cancer Inst, 70-80; prin scientist, Clement Assoc, 80-83; PRIN STAFF SCIENTIST, BD ENVIRON STUDIES & TOXICOL, NAT RES COUN-NAT ACAD SCI, 84- *Concurrent Pos:* Adj prof, Georgetown Univ; adj prof, Grad Sch Pub Health, Univ Pittsburgh, Uniformed Serv Univ of the health sci. *Honors & Awards:* Distinguished Serv Medal, US Dept Health & Human Serv. *Mem:* Fel AAAS; hon fel Am Statist Asn; Am Asn Cancer Res; fel Royal Statist Soc; Int Statist Inst; Soc Occup & Environ Health; Am Soc Prev Oncol. *Res:* Design of experiments; sequential analysis; risk analysis; environmental health. *Mailing Add:* 6503 E Halbert Rd Bethesda MD 20817

SCHNEIDERMAN, NEIL, b Brooklyn, NY, Feb 24, 37; m 60; c 3. PSYCHOPHYSIOLOGY. *Educ:* Brooklyn Col, BA, 60; Ind Univ, PhD(psychol), 64. *Prof Exp:* From asst prof to assoc prof, 65-74, PROF PSYCHOL, UNIV MIAMI, 74-, MEM STAFF, LAB QUANT BIOL, 65- *Concurrent Pos:* Asst, Physiol Inst, Univ Basel, 64-65; NSF res grant, 66-72; NIH training grant, 67- *Mem:* Psychonomic Soc; Am Psychol Asn; Sigma Xi. *Res:* Physiological psychology; psychopharmacology; conditioning; role of central nervous system in autonomic conditioning and cardiovascular regulation. *Mailing Add:* Psychol Dept PO Box 248185 Univ Miami Coral Gables FL 33124

SCHNEIDERWENT, MYRON OTTO, b Milwaukee, Wis, Jan 8, 35; m 54; c 2. PHYSICS, SCIENCE EDUCATION. *Educ:* Univ Wis-Stevens Point, 60; Western Mich Univ, MA, 63; Univ Miss, MSCS, 64; Univ Northern Colo, DEd(sci educ), 70. *Prof Exp:* Teacher, Muskegon Pub Schs, 60-63; instr physics, Interlochen Arts Acad, 64-67; PROF PHYSICS, UNIV WIS-SUPERIOR, 67- *Concurrent Pos:* Field consult, Harvard Proj Physics, 65-67, mgr, 67; dir proj AWARE, Univ Wis-Superior, 74-76; fac, Defense Equal Opportunity Mgt Inst. *Mem:* Am Asn Physics Teachers; Nat Sci Teachers Asn. *Res:* Classroom use of behavioral objectives; heavy metal uptake in aquatic flora due to coal leachate. *Mailing Add:* Dept Physics Univ Wis Superior WI 54880

SCHNEIDKRAUT, MARLOWE J, b New York, NY, Oct 27, 54. PROSTAGLANDINS. *Educ:* Albany Med Col, PhD(physiol), 81. *Prof Exp:* Fel, 81-84, res instr physiol, Sch Med, Georgetown Univ, 84-87; ASST PROF, SCH MED, WAYNE STATE UNIV, 87- *Mem:* Am Physiol Soc; NY Acad Sci; Am Fedn Clin Researchers; AAAS. *Mailing Add:* Internal Med Sch Med Wayne State Univ 421 E Canfield Elliman Bldg Rm 2222 Detroit MI 48201

SCHNEIR, MICHAEL LEWIS, b Chicago, Ill, Nov 17, 37; m 65; c 2. COLLAGEN METABOLISM, DIABETES COMPLICATIONS. *Educ:* Univ Ill, Urbana, BS, 59, Med Sch, Chicago, MS, 62, PhD(biochem), 66. *Prof Exp:* Fel, Dept Biochem, Tufts Sch Med, 65-66; fel, Dept Biochem, Univ Pittsburgh Sch Med, 66-67; asst prof, 67-70, ASSOC PROF BIOCHEM, UNIV SOUTHERN CALIF SCH DENT, 70- *Concurrent Pos:* Sabbatical leave, Univ Ala Dent Res Inst, 75. *Mem:* Int Asn Dent Res; Am Diabetes Asn; Am Chem Soc; Sigma Xi. *Res:* The effect of streptozotocin-induced diabetes on tissue collagen synthesis and degradaton with emphasis on procollagen catabolism in rat skin and oral tissues. *Mailing Add:* ACB 440 MC 1482 Univ Southern Calif Los Angeles CA 90089

SCHNEIWEISS, JEANNETTE W, b Corona, NY, Apr 14, 20; m 41; c 3. PHYSIOLOGY, BIOMETRICS. *Educ:* Brooklyn Col, BS, 58; NY Univ, MA, 61, PhD(biol), 63. *Prof Exp:* NY State Regents col teaching scholar, 59-61; teaching fel sci, NY Univ, 61-63; instr biol, Nassau Community Col, 63-64; asst prof, Adelphi Univ, 64-68; asst prof, 68-73, ASSOC PROF BIOL & PHYSIOL, HOFSTRA UNIV, 73- *Honors & Awards:* Founders Day Award, NY Univ, 62. *Mem:* Fel AAAS; Nat Asn Biol Teachers; Nat Sci Teachers Asn. *Res:* Determination of endocrinological effects of high fat diets in weaning, female and albino rats. *Mailing Add:* 217 Oakford St West Hempstead NY 11552

SCHNELL, GARY DEAN, b Lyons, Kans, July 30, 42; m 65; c 2. ZOOLOGY. *Educ:* Cent Mich Univ, BS, 64; Northern Ill Univ, MS, 66; Univ Kans, PhD(zool), 69. *Prof Exp:* Coordr biol, Origin & Struct of Ecosysts Integrated Res Prog, US Partic Int Biol Prog, 69-70; res assoc zool, Univ Tex, Austin, 69-70; from asst prof to assoc prof, 70-80. head cur, Stovall Mus Sci & Hist, 72-74, interim dir, 79-80. PROF ZOOL, UNIV OKLA, 83-, CUR BIRDS, OKLA MUS NAT HIST, 71-, HEAD CUR LIFE SCI, 74-, DIR OKLA BIOL SURV, 78- *Concurrent Pos:* Vis res assoc, Dept Biol & Ctr Evolution & Paleobiol, Univ Rochester, 77-78; ed, Syst Zool, Soc Syst Zool, 83-86; prog coordr, Am Ornith Union, 88- *Mem:* Fel AAAS; fel Am Ornith Union; Asn Syst Collections (secy, 83-86); Soc Study Evolution; Soc Syst Zool. *Res:* Systematic biology and ornithology; application of numerical techniques to the classification of organisms; evolutionary biology; environmental assessment. *Mailing Add:* Dept Zool Univ Okla Norman OK 73019

SCHNELL, GENE WHEELER, b Wapakoneta, Ohio, Jan 27, 24; m 46; c 3. MICROBIOLOGY, BIOCHEMISTRY. *Educ:* Ohio State Univ, BSc, 47, MSc, 48, PhD(microbiol), 57. *Prof Exp:* Microbiologist, Ft Detrick, Md, 49-57; sr scientist, Res Ctr, Mead Johnson & Co, 57-61, group leader, 61-63, sect leader, 63-69, prin investr, 69-73; mgr microbiol dept, Kraft Res & Develop, 73-79; prin consult, Bernard Wolnak & Assoc, 79-84, vpres, 84-86; RETIRED. *Concurrent Pos:* Consult biotechnol, 86-; tech adv sr environ employ prog, 86- *Mem:* AAAS; Am Soc Microbiol; Am Chem Soc. *Res:* Fermentation processes and products; enzymes; genetic engineering; molecular biology; plant genetics; tissue culture; monoclonal antibodies; diagnostics; pharmaceuticals; food ingredients; agricultural chemicals; market research; microbial physiology; biotechnology. *Mailing Add:* 1203 Shermer Rd Glenview IL 60025

SCHNELL, JAY HEIST, b Philadelphia, Pa, Nov 21, 32. RAPTORS, REMOTE SENSING INSTRUMENT DESIGN. *Educ:* Earlham Col, AB, 55; Univ Calif, Berkeley, MA, 57; Univ Ga, PhD(zool), 64. *Prof Exp:* Res asst ecol, Savannah River Plant, Univ Ga, 62-63, fel pop ecol, 64-65; fel radio tracking tech, Cedar Creek Radio-Tracking Sta, Univ Minn, Minneapolis, 65-69; res biologist, Tall Timbers Res Sta, Fla, 69-72; wildlife telemetry consult, Ill, 72-73; dir, Res Ranch, 73-74; dir, George Whittell Wildlife Preserve, Ariz, 74-76; CONSULT BIOL, 76- *Mem:* Raptor Res Found. *Res:* Population ecology of small mammals; behavior and ecology of birds of prey; radio-tracking techniques aiding studies in animal and bird behavior; black hawks nesting in Aravaipa Canyon, Arizona; designing, testing digital computer instruments used in eco-behavioral research. *Mailing Add:* Box 54 Klondyke Rural Sta Wilcox AZ 85643

SCHNELL, JEROME VINCENT, b Aitkin, Minn, July 19, 34. BIOCHEMISTRY. *Educ:* Col St Thomas, BS, 56; Univ Nebr, MS, 59, PhD(chem), 63. *Prof Exp:* Res assoc enzymol, Biol Div, Oak Ridge Nat Lab, 63-64; res assoc protozool, Med Ctr, Stanford Univ, 65-69; asst prof med microbiol & trop med, Univ Hawaii, Leahi, 70-75; res chemist, Nat Marine Fisheries Serv, Nat Oceanic & Atmospheric Admin, 75-81; assoc prof, Seattle Univ, 80-87, exec dir, Alcohol Studies Prog, 80-89; LECTR & CONSULT, PVT PRACT, 87- *Concurrent Pos:* USPHS fel, Oak Ridge Nat Lab, 64-65; instr, Alcohol Studies Prog, Seattle Univ, 78-87; expert witness, alcohol issues. *Res:* Biochemistry of malarial parasites; enzymology of amino acid and protein metabolism; xenobiotic metabolism in fish. *Mailing Add:* 3710 25th Pl W No 304 Seattle WA 98199

SCHNELL, LORNE ALBERT, pharmacy, for more information see previous edition

SCHNELL, ROBERT CRAIG, b Sturgis, SDak, Oct 14, 42; m 65; c 2. PHARMACOLOGY. *Educ:* SDak State Univ, BS, 65; Purdue Univ, MS, 67, PhD(pharmacol-toxicol), 69. *Prof Exp:* Asst prof pharmacol-toxicol, Wash State Univ, 71-72; assoc prof, Purdue Univ, 72-79; prof & chmn dept, Med Ctr, Univ Nebr, 79-85; DEAN GRAD STUDIES & RES, NDAK STATE UNIV, 85- *Concurrent Pos:* Mem bd trustees, Am Asn Accreditation Lab Animal Care, 76-85; Safe Drinking Water Comt, Nat Acad Sci, 79-82; assoc ed, Fundamental & Appl Toxicol, 86- *Honors & Awards:* Burroughs-Wellcome Toxicol Scholar Award, 83. *Mem:* Soc Toxicol; Am Soc Pharmacol & Exp Ther; Am Asn Pharmaceut Sci; Am Asn Col Pharm. *Res:* Drug metabolism; circadium rhythms; toxicology of heavy metals. *Mailing Add:* Dept Pharmaceut Sci NDak State Univ Old Main 202 Fargo ND 58105

SCHNELLE, K(ARL) B(ENJAMIN), JR, b Canton, Ohio, Dec 8, 30; m 54; c 2. AIR POLLUTION CONTROL, ATMOSPHERIC DIFFUSION MODELLING. *Educ:* Carnegie Inst Technol, BS, 52, MS, 57, PhD(chem eng), 59. *Prof Exp:* Chem engr, Columbia-Southern Chem Corp, 52-54; from asst prof to assoc prof chem eng, Vanderbilt Univ, 58-64; mgr ed & res, Instrument Soc Am, 64-66; assoc prof air resources eng, Vanderbilt Univ, 66-70, prof environ & air resources eng, 70-80, chmn, Dept Environ Eng & Policy Mgt & dir environ & water resources eng prog, 76-80, chmn dept, chem & environ eng, 80-88, PROF CHEM & ENVIRON ENG, VANDERBILT UNIV, 88- *Concurrent Pos:* Fulbright Chair, Univ Liege, Belg, Comn Int Exchange Scholars, 77. *Mem:* Am Soc Eng Educ; Instrument Soc Am; fel Am Inst Chem Engrs; Air Pollution Control Asn; Am Soc Environ Engrs. *Res:* Process dynamics and control; dynamic testing; instrumentation for pollution control; air pollution control; atmospheric diffusion modeling; control of sulfur oxides and nitrogen oxides in coal fired boilers. *Mailing Add:* Dept Chem Eng Vanderbilt Univ Nashville TN 37240

SCHNELLER, EUGENE S, b Cornwall, NY, Apr 9, 43; c 2. MEDICAL SOCIOLOGY, HEALTH POLICY. *Educ:* CW Post Col, BA, 68; New York Univ, PhD(sociol), 73. *Prof Exp:* Asst prof med sociol, Duke Univ Med Ctr, 72-75; assoc prof sociol, 75-78, assoc prof health admin, Union Col, 78-85; res scholar health policy, Columbia Univ, 83-84; PROF HEALTH ADMIN, ARIZ STATE UNIV, 85- *Concurrent Pos:* Accrediting comn on educ for Health Servs Admin; vis assoc prof, State Univ NY, Albany, 81-82 & Albany Med Col, 83-85. *Mem:* Am Sociol Asn; Am Pub Health Asn; Am Col Healthcare Execs. *Res:* Changes in health occupations and professions; the role of the physician in management; comparative health systems; AIDS policy; managed care systems. *Mailing Add:* Sch Health Admin & Policy Ariz State Univ Tempe AZ 85287

SCHNELLER, STEWART WRIGHT, b Louisville, Ky, Feb 27, 42; m 66; c 2. PHARMACEUTICAL CHEMISTRY. *Educ:* Univ Louisville, BS, 64, MS, 65; Ind Univ, Bloomington, PhD(org chem), 68. *Prof Exp:* NIH fel, Stanford Univ, 68-69, res assoc org chem, 69-70; res assoc, Univ Mass, 70-71; asst prof, Univ SFla, 71-75, asst chmn dept, 72-74, actg chmn dept 74-75, assoc prof, 75-78, PROF ORG CHEM, UNIV SFLA, 78-, CHMN, 86- *Concurrent Pos:* Petrol Res Fund-Am Chem Soc fel, Univ SFla, 71-74; NIH & Dept Army Support, 74- *Mem:* AAAS; Am Chem Soc; Royal Soc Chem; Int Soc Heterocyclic Chem (pres). *Res:* Heterocyclic synthetic methods; synthetic medicinal chemistry, nucleosides. *Mailing Add:* Dept Chem Univ SFla Tampa FL 33620-5250

SCHNEPFE, MARIAN MOELLER, b San Pedro de Macoris, Dominican Repub, Nov 15, 23; US citizen; m 54. INORGANIC CHEMISTRY, ANALYTICAL CHEMISTRY. *Educ:* George Washington Univ, BS, 53, MS, 60, PhD(chem), 66. *Prof Exp:* Analytical chemist, US Geol Surv, 54-80; RETIRED. *Concurrent Pos:* Consult, Indonesian Geol Surv Anal Labs, Bandung, Java, 81-82. *Mem:* Am Chem Soc; Sigma Xi. *Res:* Testing of natural materials with a view to their potential for fixation of some of the problem radio-nuclides; development of various analytical procedures; development of spectrophotometric and atomic absorption procedures for the determination of platinum, palladium, rhodium iridium and spectrophotometric procedures for the determination of antimony, arsenic, bromine, germanium, iodine, selenium and thallium in sub-microgram quantities. *Mailing Add:* Potomac Towers Apt 640 2001 Adams St Arlington VA 22201

SCHNEPP, OTTO, b Vienna, Austria, July 7, 25; m 50, 79; c 2. PHYSICAL CHEMISTRY, CHEMICAL PHYSICS. *Educ:* St John's Univ, China, BS, 47; Univ Calif, Berkeley, AB, 48, PhD(chem), 51. *Prof Exp:* Res asst chem, Univ Calif, Berkeley, 51-52; from lectr to sr lectr, Israel Inst Technol, 52-59, from assoc prof to prof, 59-65; PROF CHEM, UNIV SOUTHERN CALIF, 65-, DEPT CHMN, 89- *Concurrent Pos:* Res assoc, Duke Univ, 57-58; res physicist, Nat Bur Stand, DC, 58-59; sci counr, US Embassy, Beijing, 80-82. *Honors & Awards:* Assocs Award Creative Scholar & Res, Univ Southern Calif, 78; Super Honor Award, US Dept State, 82. *Mem:* Am Phys Soc; AAAS; Asn Asian Studies. *Res:* Molecular and solid state spectroscopy; coherent raman spectroscopy; lattice vibrations of molecular solids; circular dichroism and magnetic circular dichroism spectroscopy; vacuum ultraviolet spectroscopy; science and technology of China; science policy. *Mailing Add:* Dept Chem Univ Southern Calif Los Angeles CA 90089-1062

SCHNEPS, JACK, b New York, NY, Aug 18, 29; m 60; c 3. ELEMENTARY PARTICLE PHYSICS. *Educ:* NY Univ, BA, 51; Univ Wis, MS, 53, PhD(physics), 56. *Prof Exp:* From asst prof to assoc prof, 56-63, dept chmn, 80-89, PROF PHYSICS, TUFTS UNIV, 63- *Concurrent Pos:* NSF fel, 58-59; vis scientist, Europ Orgn Nuclear Res, 65-66; vis res fel, Univ Col, Univ London, 73-74; co-prin investr, 76-; vis res fel, Ecole Polytechnique, 82-83; vis prof, Technion, 89-90. *Mem:* Europ Phys Soc; fel Am Phys Soc; Sigma Xi. *Res:* Elementary particle research in neutrino and hadron interactions and proton decay. *Mailing Add:* Dept Physics Tufts Univ Medford MA 02155

SCHNETTLER, RICHARD ANSELM, b St Nazianz, Wis, May 3, 37; m 65; c 2. MEDICINAL CHEMISTRY, ORGANIC CHEMISTRY. *Educ:* Univ Wis, BS, 61; Univ Kans, PhD(med chem), 65. *Prof Exp:* Sr res chemist, Lakeside Labs Inc, 65-75; sr develop chemist, Merrell-Nat Labs, 75-81; SR RES CHEMIST, MERRELL DOW RES INST, 81- *Mem:* Am Chem Soc. *Res:* Reaction mechanisms; natural product synthesis; biogenesis of natural products; cardiovascular and psychopharmacologic agents. *Mailing Add:* 2110 Galbraith Rd Cincinnati OH 45215

SCHNEYER, CHARLOTTE A, b St Louis, Mo, Nov 21, 23; m 45. PHYSIOLOGY. *Educ:* Wash Univ, AB, 45; NY Univ, MS, 47, PhD(physiol), 52. *Prof Exp:* Asst zool, Wash Univ, 43-45; res assoc, 53-55, asst prof physiol, 62-65, PROF DENT, SCH DENT, UNIV ALA, BIRMINGHAM, 55-, PROF PHYSIOL & BIOPHYS, 65-, DIR, LAB EXOCRINE PHYSIOL, MED CTR, 76- *Concurrent Pos:* Undergrad teaching asst, Wash Univ, 42-45; teaching fel biol, NY Univ, 45-52 (fel res), & sch dent, Univ Ala, Birmingham, 42-43; mem adv coun gen med sci, NIH, 72-76; mem special grants review comt, Nat Inst Dent Res, 79-82. *Honors & Awards:* Hungarian Dent Soc Award. *Mem:* AAAS; Am Physiol Soc; Soc Exp Biol & Med. *Res:* Secretion of electrolytes and proteins by salivary glands; autonomic regulation of growth and development of salivary glands. *Mailing Add:* Lab Exocrine Physiol Dept Physiol Univ Ala Med Ctr Birmingham AL 35294

SCHNIEDERJANS, MARC JAMES, b Pocahontas, Ark, Oct 8, 50; m 71; c 3. GOAL PROGRAMMING. *Educ:* Univ Mo, St Louis, BSBA, 72; St Louis Univ, MBA, 74, PhD(mgt sci),78. *Prof Exp:* Dir bus admin, St Louis Univ, 74-78; asst prof decision sci, Univ Nebr, Omaha, 78-79, Univ Mo, St Louis, 79-80, Univ Hawaii, Hilo, 80-81; asst prof mgt sci, 81-85, ASSOC PROF MGT SCI, UNIV NEBR, LINCOLN, 85- *Concurrent Pos:* Lectr, St Louis Univ, 74-78; pres, Quickship Leasing Corp, 78-79; asst dir, Pac Bus Res Ctr, 80-81; chairperson dept bus & econ, Univ Hawaii, Hilo, 80-81; consult, Blue Hills Home Corp & Truck Transp Corp, 79, Ralston Purina Corp, 80, Union Diversified Enterprises, Inc, 83-84. *Mem:* Opers Res Soc Am; Inst Decision Sci; Inst Mgt Sci; Am Prod & Inventory Control Soc. *Res:* Applied mathematics and statistics in business, health care and education planning. *Mailing Add:* Dept Mgt Univ Neb Lincoln NE 68588

SCHNIEWIND, ARNO PETER, b Cologne, Ger, June 1, 29; US citizen; m; c 3. FOREST PRODUCTS. *Educ:* Univ Mich, BS, 53, MWT, 55, PhD(wood technol), 59. *Prof Exp:* Sr lab technician, 56, asst specialist wood sci & technol, 56-59, lectr forestry, 59-65, assoc prof, 66, PROF FORESTRY, UNIV CALIF, BERKELEY, 66- *Concurrent Pos:* NSF fel, 63-64. *Honors & Awards:* Wood Award, 59. *Mem:* Soc Wood Sci & Technol (pres, 73-74); Int Acad Wood Sci; Forest Prod Res Soc; Am Soc Testing & Mat; Am Inst Conserv Hist & Artistic Works. *Res:* Mechanical behavior and physical properties of wood. *Mailing Add:* Forest Prod Lab Univ Calif 1301 S 46th St Richmond CA 94804

SCHNITKER, DETMAR, b Wilhelmshaven, Ger, July 5, 37; m 64; c 2. MICROPALEONTOLOGY, PALEOCEANOGRAPHY. *Educ:* Univ NC, Chapel Hill, MS, 66; Univ Ill, Urbana, PhD(geol), 67. *Prof Exp:* Geologist, Soc Nat Petrol Aquitaine, 67-69; asst prof, 69-72, assoc prof, 72-79, PROF OCEANOG & GEOL SCI, UNIV MAINE, ORONO, 79- *Concurrent Pos:* Fulbright exchange scholar, 60-61; assoc dir, Submarine Geol & Geophys Prog, NSF, 80-81; co-chief scientist, Deep Sea Drilling Proj-Int Prog Ocean Drilling Leg, 81. *Mem:* Paleont Soc; Paleont Res Inst; Paleontologische Ges; Int Paleont Asn; Am Geophys Union. *Res:* Foraminiferal ecology; paleoecology; paleoceanography. *Mailing Add:* Dept Oceanog Univ Maine Orono ME 04469

SCHNITKER, JURGEN H, b Bremen, Ger, Mar 12, 58. SOLVATED ELECTRONS, LIQUID WATER. *Educ:* Univ Aachen, Ger, MS, 82, PhD(chem), 86. *Prof Exp:* Postdoctoral fel, Univ Tex, 86-88; tech staff, AT&T Bell Labs, 88-89; ASST PROF CHEM, UNIV MICH, 89- *Mem:* Am Chem Soc. *Res:* Computational chemistry: classical and quantum molecular dynamics simulations of liquids and chemical reactions in liquids. *Mailing Add:* Dept Chem Univ Mich Ann Arbor MI 48109-1055

SCHNITZER, BERTRAM, b Frankfurt, Ger, June 21, 29; m 59; c 3. HEMATOPATHOLOGY, IMMUNOPATHOLOGY. *Educ:* NY Univ, BA, 52; Univ Basle, Switzerland, MD, 58. *Prof Exp:* Resident physician path, Georgetown Univ Hosp, 59-63; pathologist, US Armed Forces Inst Pathol, 63-66; from instr path to assoc prof, 66-72, PROF PATH, UNIV MICH HOSP, 72-, DIR HEMAT, 76- *Concurrent Pos:* Consult, US Vet Admin Hosp; ed bd, Am J Clin Path, 83-, coun mem, Am Soc Clin Path, 88-, pres elect, Exec Comt, Soc Hematopath, 88; mem, path comt, Children Cancer Study Group, 74-; pres, Soc Hematopath, 90- *Mem:* Am Soc Clin Path; Am Soc Hemat; Int Acad Path; Int Soc Exp Hemat. *Res:* Immunophenotypic analysis in the diagnosis of lymphomas and leukemias by flow cytometry; DNA analysis. *Mailing Add:* Dept Path MSI M5242/0602 Univ Mich Med Ctr 1301 Catherine Ann Arbor MI 48109-0602

SCHNITZER, HOWARD J, b Newark, NJ, Nov 12, 34; m 66; c 2. THEORETICAL HIGH ENERGY PHYSICS, QUANTUM FIELD THEORY. *Educ:* Newark Col Eng, BS, 55; Univ Rochester, PhD(physics), 60. *Prof Exp:* Res assoc physics, Univ Rochester, 60-61; res assoc, 61-62, from asst prof to assoc prof, 62-68, PROF PHYSICS, BRANDEIS UNIV, 68- *Concurrent Pos:* Alfred P Sloan Found fel, 64-66; vis prof, Rockefeller Univ, 69-70; vis res assoc, Harvard Univ, 74-; assoc ed, Phys Review Lett, 78-80; chmn, dept physics, Brandeis Univ, 81-83; John S Guggenheim Found fel, 83-84. *Mem:* Fel Am Phys Soc; Sigma Xi. *Res:* Elementary particle theory; quantum field theory; string theory; conformal field theory. *Mailing Add:* Dept Physics Brandeis Univ Waltham MA 02254-9110

SCHNITZER, JAN EUGENEUSZ, b Pittsburgh, Pa, June 24, 57. BIOPHYSICS. *Educ:* Princeton Univ, BSE, 80; Univ Pittsburgh, MD, 85. *Prof Exp:* Assoc res scientist, Sch Med, Yale Univ, 85-90; ASST PROF PATH, PHYSIOL, CELL BIOL & MED, UNIV CALIF, SAN DIEGO, 90- *Concurrent Pos:* Prin investr, NIH grant, 89-94, Nat Am Heart Asn, 91-94. *Mem:* AAAS; Microcircuitry Soc; Am Soc Cell Biol. *Res:* Vascular endothial biology; capillary permcability; receptor-mediated trascytosis; biophysics of membrane transport; steric & electrostatic effects on incubrane transport; theoretical modeling. *Mailing Add:* Cellular & Molecular Med-0651 Univ Calif-San Diego 9500 Gilman Dr La Jolla CA 92093-0651

SCHNITZER, MORRIS, b Bochum, WGer, Feb 4, 22; Can citizen; m 48; c 1. SOIL CHEMISTRY, ORGANIC CHEMISTRY. *Educ:* McGill Univ, BSc, 51, MSc, 52, PhD(agr chem), 55. *Prof Exp:* Res scientist, Aluminum Labs Ltd, 55-56; PRIN RES SCIENTIST, LAND RESOURCE RES CTR, AGR CAN, 56- *Concurrent Pos:* Sabbatical fel, Imp Col, Univ London, 61-62. *Honors & Awards:* Soil Sci Award, Soil Sci Soc Am, 84. *Mem:* Fel Can Soc Soil Sci; Int Soc Soil Sci; fel Soil Sci Soc Am. *Res:* Chemical structure and reactions of humic substances in soils and waters; chemistry of soil nitrogen. *Mailing Add:* Land Resource Res Ctr Agr Can Ottawa ON K1A 0C6 Can

SCHNITZLEIN, HAROLD NORMAN, b Hannibal, Mo, Aug 29, 27; m 49; c 4. NEUROANATOMY. *Educ:* Westminster Col, Mo, AB, 50; St Louis Univ, MS, 52, PhD, 54. *Prof Exp:* From instr to prof anat, Sch Med, Univ Ala, Birmingham, 54-73; chmn dept, 73-78, PROF ANAT, COL MED, UNIV SFLA, 73-, PROF ANAT & RADIOL, 85-, PROF ANAT RADIOL & NEUROL, 88- *Concurrent Pos:* USPHS spec fel, 60-61. *Res:* Autonomic nervous system; comparative vertebrate neuroanatomy; imaging anatomy. *Mailing Add:* Dept Anat Univ SFla Box 6 Tampa FL 33612-4799

SCHNITZLER, RONALD MICHAEL, b Providence, RI, Jan 13, 39; m 68; c 2. PHYSIOLOGY, EDUCATIONAL ADMINISTRATION. *Educ:* Brown Univ, AB & ScB, 62; Univ Vt, MS, 64, PhD(physiol), 69. *Prof Exp:* Instr physiol, Med Col, Univ Vt, 69-70; Alexander von Humboldt scholar & res assoc, Microbiol Labs, Luisenhosp, Aachen, WGer, 70-71; res assoc physiol, Med Col, Univ Vt, 71-75; from asst prof to prof biol, Sci Dept, Bay Path Col, 75-85; DIR, MATH/SCI DIV, MATTATUCK COMMUNITY COL, 85- *Concurrent Pos:* NIH spec fel, 71-74. *Mem:* AAAS; Nat Sci Teachers Asn. *Res:* Physiology and pharmacology of neuromuscular transmission, especially mechanism of drug desensitization at the motor end plate; effects of ultrasound on biological tissue. *Mailing Add:* Div Math/Sci Mattatuck Community Col 750 Chase Pkwy Waterbury CT 06708

SCHNIZER, ARTHUR WALLACE, b Des Plaines, Ill, Jan 16, 23; m 47; c 3. ORGANIC CHEMISTRY. *Educ:* Baylor Univ, BS, 43; Northwestern Univ, PhD(chem), 51. *Prof Exp:* Develop chemist, Columbia Chem Div, Pittsburgh Plate Glass Co, 43-46; asst, Northwestern Univ, 47-50; res chemist, Celanese Corp, 50-52, res group leader, 52-55, res sect head, 55-61, dir chem res, 61-66, dir eng res, 66, mgr tech ctr, Celanese Chem Co, Tex, 66-69, dir chem & polymer res, Celanese Res Co, NJ, 69-71; tech dir, Day & Zimmermann, Inc, Philadelphia, 71-76; vpres res & develop, Bird & Son, Inc, 76-80; RETIRED. *Mem:* Am Chem Soc; AAAS; Am Inst Chem Engrs; Soc Chem Indust. *Res:* Reactions of alkyl sodiums, lithiums, magnesium bromides with methoxyl groups; reactions and synthesis of ketene; derivatives of oxygenated petrochemicals; asphalt roofing and plastics extrusion; organic synthesis and process development. *Mailing Add:* 333 University Dr Corpus Christi TX 78412-2741

SCHNOBRICH, WILLIAM COURTNEY, b St Paul, Minn, Nov 26, 30; m 56. ENGINEERING MECHANICS, STRUCTURAL ENGINEERING. *Educ:* Univ Ill, Urbana, BS, 53, MS, 55, PhD(thin shells), 62. *Prof Exp:* Res asst, 53-55, 58-62, from asst prof to assoc prof, 62-68, PROF CIVIL ENG, UNIV ILL, URBANA, 68- *Concurrent Pos:* Summer res engr, Space Technol Labs, Calif, 62 & 63; consult, John R Gullaksen, Struct Engr, Ill, 64-, Whitman Reguardt & Assocs, Greeley & Hansen Engrs, 76, Beaulieu Poulin, Robitaille & Assocs, 80, Klein & Hoffman, Argonne Nat Lab, 81, Kajima Corp, 84 & Lockwood Greene Engrs, 85; Humboldt sr US scientist award, Alexander von Humboldt-Found, 77 & 84. *Mem:* Am Soc Civil Engrs; Am Soc Mech Engrs; Int Asn Bridge & Struct Engrs; fel Am Concrete Inst; Int Asn Comput Mech; Int Asn Shell Struct. *Res:* Structural mechanics, particularly thin shell structures; nuclear reactor vessels; earthquake resistant design; cooling towers; finite element specialist. *Mailing Add:* 1419 Mayfair Rd Champaign IL 61821-5021

SCHNOES, HEINRICH KONSTANTIN, b Knetzgau, Ger, July 12, 39; m 69; c 1. NATURAL PRODUCTS, MASS SPECTROMETRY. *Educ:* Long Island Univ, BS, 61; Mass Inst Technol, PhD(org chem), 65. *Prof Exp:* Asst res chemist, Space Sci Lab, Univ Calif, Berkeley, 65-67; from asst prof to assoc prof, 67-74, PROF BIOCHEM, UNIV WIS-MADISON, 74- *Mem:* AAAS; Am Chem Soc; Am Soc Biol Chemists; Royal Soc Chem; Am Soc Mass Spectrometry. *Res:* Natural products chemistry and biochemistry; mass spectrometry and its application to structural and biochemical problems. *Mailing Add:* Dept Biochem Univ Wis Madison WI 53706

SCHNNBAUM, EDUARD, b Vienna, Austria, Sept 18, 23; Neth citizen; m 53. PHARMACOLOGY, PHYSIOLOGY. *Educ:* Univ Amsterdam, ChemCand, 50; McGill Univ, PhD(biochem), 55. *Prof Exp:* Res assoc physiol, 55-57, asst prof med res, 57-60, asst prof pharmacol, 60-63; head cent nerv syst pharmacol, 73-79, head gen pharmacol, 79-81, int coordr pharmacol res, Res & Develop Labs, Organon Int BV, OSS, Neth, 81-88; assoc prof pharmacol, 63-89, CONSULT PHARMACOLOGIST, UNIV TORONTO, 88- *Concurrent Pos:* Assoc med dir pharmacol, Ciba-Geigy Can Ltd, 68-73. *Mem:* AAAS; Can Physiol Soc; Can Pharmacol Soc; Am Physiol Soc; Dutch Soc Pharmacol; Brit Pharmacol Soc; Ger Pharmacol Soc. *Res:* Temperature regulation. *Mailing Add:* Peelkensweg 4 Venhorst 5428 NM Netherlands

SCHNOOR, JERALD L, b Davenport, Iowa, Aug 26, 50; m 72; c 2. WATER QUALITY MODELING, GLOBAL CLIMATE CHANGE. *Educ:* Iowa State Univ, BS, 72; Univ Tex, MS, 74, PhD(environ eng), 75. *Prof Exp:* NSF postdoctoral fel res, NSF-Manhattan Col, 76-77; from asst prof to assoc prof, 77-83, chmn, 85-90, PROF ENVIRON ENG TEACHING & RES, COL ENG, UNIV IOWA, 83-85, 90- *Concurrent Pos:* Vis prof, Swiss Fed Technol-Z06rich, 82 & 88; mem, Nat Res Coun Panel Lake Acidification, 83-84, Environ Protection Agency Global Climate Res Subcomt, 89-90; assoc ed, Water Resources Res, 85-87, Environ Sci & Technol, 91-; ed, Res J Water Pollution Control Fedn, 89-; prin investr, Hazardous Substances Res Ctr, 89-; co-dir, Ctr Global & Regional Environ Res, 90-; comt chair, Water Pollution Control Fedn, 89-90. *Honors & Awards:* Merit Award, Am Chem Soc, 80; Walter L Huber Res Prize, Am Soc Civil Engrs, 85. *Mem:* Am Soc Civil Engrs; Am Inst Chem Eng; Am Geophys Union; Am Chem Soc. *Res:* Water quality modeling; environmental engineering science; global climate change; lake eutrophication; pesticide fate and movement in the environment; author of two books and over 60 publications. *Mailing Add:* 1136 Eng Bldg Univ Iowa Iowa City IA 52242p

SCHNOPPER, HERBERT WILLIAM, b Brooklyn, NY, Mar 13, 33; m 69; c 2. ASTROPHYSICS, X-RAY ASTRONOMY. *Educ:* Rensselaer Polytech Inst, 54; Cornell Univ, MS, 58, PhD(physics), 62. *Prof Exp:* Sr scientist, Jet Propulsion Lab, 62-63; instr & res assoc physics, Cornell Univ, 63-66; from asst prof to assoc prof physics, Mass Inst Technol, 66-73, physicist, 73-74; physicist, Smithsonian Astrophys Observ & lectr, Dept Astron, Harvard Univ, 74-80; DIR, DANISH SPACE RES INST, 80- *Concurrent Pos:* Consult, Jet Propulsion Lab, 63-80 & Am Sci & Eng, 69-80; vis prof, Steward Observ, Univ Ariz, 70-73; guest physicist, Brookhaven Nat Lab, 71-80; consult, Quartz et Silice, Paris, 72-80; mem, Space Sci Comt, ESA, 80- & Europ Sci Found, 87- *Mem:* AAAS; Int Astron Union; Am Phys Soc; Am Astron Soc; NY Acad Sci; Sigma Xi; Europ Phys Soc. *Res:* Astrophysics; x-ray astronomy; diagnostics of high temperature plasmas; x-rays from heavy ion collisions; x-ray spectroscopy; x-ray optics. *Mailing Add:* Danish Space Res Inst Gl Lundtoftevej 7 DK 2800 Lyngby Denmark

SCHNUR, JOEL MARTIN, b Washington, DC, Feb 5, 45; m 71; c 1. CHEMICAL PHYSICS. *Educ:* Rutgers Univ, BS, 66; Georgetown Univ, PhD(phys chem), 71. *Prof Exp:* Head, molecular optics sect, 73-79, dept head, optic probes br, Naval Res Lab, 79-84, head, picosecond spectros sect, 79-84, dep coordr, Energetic Mat Progs, 80-85, head, Biol Molecular Eng Br, 84-89, DIR CTR BIOMOLECULAR SCI & ENG, 89- *Concurrent Pos:* Vpres, Concepts Unlimited, 71-; res assoc, Nat Res Coun, Naval Res Lab, 71-72; vis prof, Univ Paris, 83-; chmn, Gordon Conf on Thin Films, 90. *Mem:* Sigma Xi. *Res:* Picosecond spectroscopic techniques as applied to the elucidation of the kinetics of important ultra fast energetic materials reactions; time resolved techniques are also utilized to study electrohydrodynamics phemonmena at high pressures in lubricants as well as polymer morphology; development of receptor based biosensors, energy transtudion systems and hybrid devices; assess potential of self assembled and biologically derived microstructures for technological applications; study fundamentals of self organization in heterogeneous systems. *Mailing Add:* Code 6090 Naval Res Lab Washington DC 20375-5000

SCHNUR, RODNEY CAUGHREN, b Evanston, Ill, Dec 25, 45. MEDICINAL CHEMISTRY, SYNTHETIC ORGANIC CHEMISTRY. *Educ:* Williams Col, AB, 67; Pa State Univ, PhD(chem), 73. *Prof Exp:* NIH res fel chem, Stanford Univ, 73-74; RES SCIENTIST MED CHEM, PFIZER INC, 74- *Mem:* Am Chem Soc. *Res:* Design and synthesis of human pharmaceuticals. *Mailing Add:* Four Prospect St Mystic CT 06355-2312

SCHNUR, SIDNEY, b New York, NY, June 23, 10; m 44; c 1. MEDICINE. *Educ:* City Col New York, BS, 30, MS, 31; NY Univ, MD, 35; Am Bd Internal Med & Am Bd Cardiovasc Dis, dipl, 44. *Prof Exp:* Intern, Morrisania City Hosp, New York, 35-37; resident internal med, Kings County Hosp, 37-38; resident path, Jefferson Davis Hosp, Houston, 38-39; from clin asst prof to clin assoc prof med, Baylor Col Med, 46-62; coordr cardiol courses, Postgrad Sch Med, Univ Tex Grad Sch Biomed Sci, Houston, 53-75; clin prof, 62-76, EMER CLIN PROF MED, BAYLOR COL MED, 76-; EMER CLIN PROF MED, POSTGRAD SCH MED, UNIV TEX GRAD SCH BIOMED SCI, HOUSTON, 75-, MED SCH, 80- *Concurrent Pos:* Pvt pract internal med, Houston, 39-; assoc physician, Jefferson Davis Hosp, 40-, chief dept electrocardiol, 45-50, chief cardiac clin, 45-51, chief 4th div med, 58-60; consult, USPHS, 46-50, St Luke's Hosp, 54- & Polly Ryan Mem Hosp, Richmond, 57-59; assoc physician, Methodist Hosp, 46-53, attend physician, 54-66, consult cardiologist, 66-; attend specialist, Vet Admin Regional Off & Hosp, 46-61; clin asst prof med, Univ Tex Grad Sch Biomed Sci Houston, 50-52, clin assoc prof, 52-57; ed consult, Heart Bull, 51-70; contrib ed, Med Rec & Ann, 52-62; consult cardiologist, San Jacinto Mem Hosp, Baytown, 54-58; consult cardiologist & electrocardiologist, South Pac Hosp, 54-61; chief med & electrocardiol dept, St Joseph Hosp, Houston, 54-, pres med staff, 62-65; chief electrocardiol, Med Arts Hosp, 58-60, chief med, 63, pres staff, 62-63; mem coun clin cardiol, Am Heart Asn, 63; attend physician, Ben Taub Hosp, 64-; emer trustee, Houston Mus Natural Sci; clin prof med, Sch Med, Univ Tex Biomed Sci, Houston, 57-76. *Mem:* Am Heart Asn (vpres, 77); fel Am Col Physicians; emer fel Am Col Chest Physicians; emer fel Am Col Cardiol; sr mem Am Fedn Clin Res. *Res:* Clinical cardiology; internal medicine. *Mailing Add:* 2139 Sunset Blvd Houston TX 77005

SCHNURRENBERGER, PAUL ROBERT, veterinary public health; deceased, see previous edition for last biography

SCHOBER, CHARLES COLEMAN, b Shreveport, La, Nov 30, 24; m 47, 72; c 3. PSYCHIATRY, PSYCHOANALYSIS. *Educ:* La State Univ, Baton Rouge, BS, 46; La State Univ, New Orleans, MD, 49. *Prof Exp:* Intern, Philadelphia Gen Hosp, 49-51; resident psychiat & Nat Inst Ment Health residency training grant, Norristown State Hosp, Pa, 53-56, staff physician, 56-57; assoc clin dir, Pa Hosp Ment & Nerv Dis, 57-59; instr psychiat, Med Sch, Univ Pa, 58-62, assoc, 62-70, asst prof clin psychiat, 65-71; prof psychiat & head dept, Sch Med, La State Univ, Shreveport, 71-73; prof psychiat, Sch Med, St Louis Univ & mem fac, St Louis Psychoanal Inst, 73-78; PROF PSYCHIAT, MED SCH, LA STATE UNIV, 78-; PSYCHIAT SERV, CHARTER FOREST HOSP, SHREVEPORT, LA, 89- *Concurrent Pos:* Attend psychiatrist, Pa Hosp Inst, 63-68, sr attend psychiatrist, 68-71; chief psychiat serv, Vet Admin Hosp, Shreveport, La, Confederate Mem Hosp, 71-73 & Schumpert Med Ctr, 82-84; consult, Brentwood Neuropsychiat Hosp, 71-73; consult, Vet Admin Hosps, St Louis, Mo, 73-78; staff, Brentwood Psychiat Hosp, Schumpert Med Ctr & La State Univ Med Ctr Hosp, 78-; chief, Psychiat Serv, Schumpers Med Ctr, Shreveport, La, 82- 84; med & clin dir, Psychiat Serv, Willis Knighton Med Ctr, Shreveport, La, 86-89. *Mem:* fel Am Col Psychiat; fel Am Psychiat Asn; Am Psychoanal Asn; AMA. *Res:* Follow-up studies in psychotherapy of schizophrenia; evaluation of effectiveness of clinical teaching methods in psychiatric training of medical students and residents. *Mailing Add:* 1801 Fairfield Ave Suite 400 Shreveport LA 71101

SCHOBER, GLENN E, b Minneapolis, Minn, July 1, 38; m 66; c 2. MATHEMATICS. *Educ:* Univ Minn, BS, 60, PhD(math), 65. *Prof Exp:* Asst prof math, Univ Calif, San Diego, 65-66; from asst prof to assoc prof, 66-76, PROF MATH, IND UNIV, BLOOMINGTON, 76- *Concurrent Pos:* Vis assoc prof, Univ Md, 73-74; vis prof, Univ NC, 76; Distinguished vis prof, Tex Tech Univ, 83. *Mem:* Am Math Soc; Math Asn Am. *Res:* Complex function and potential theories. *Mailing Add:* Dept Math Ind Univ Bloomington IN 47405

SCHOBERT, HAROLD HARRIS, b Wilkes-Barre, Pa, Nov 13, 43; m 68; c 2. COAL CHEMISTRY. *Educ:* Bucknell Univ, BS, 65; Iowa State Univ, PhD(chem), 70. *Prof Exp:* Instr chem, Iowa State Univ, 70-72; res chemist, Deepsea Ventures Inc, 72-76; res chemist, Grand Forks Energy Tech Ctr, US Dept Energy, 76-78, supvr, anal chem, 78, mgr, anal res, 78-86; ASSOC PROF FUEL SCI, PA STATE UNIV, 86-, CHMN, 88- *Mem:* Am Chem Soc; Sigma Xi. *Res:* Studies of the chemistry of coal liquefaction and co-processing; studies of the structure of coal as it influences coal conversion processes; studies of the physical chemistry of coal ash slags; studies of thermal stability of jet fuels. *Mailing Add:* 209 Acad Projs Bldg Pa State Univ University Park PA 16802

SCHOCH, DANIEL ANTHONY, b Piqua, Ohio, July 25, 48; m 69; c 3. PRESS & DIE TESTING & ANALYSIS, PRESS VIBRATION SEVERITY & PRODUCTIVITY. *Educ:* Ohio State Univ, BS, 70. *Prof Exp:* Design engr, Minster Mach Co, 70-73, res proj engr, 73-77, advan eng supvr, 77-80, APPL RES MGR, MINSTER MACH CO, 80- *Concurrent Pos:* Secy & treas, Ohio Soc Prof Engrs, Midwest Chapter, 71-72; tech res comt, Precision Metal Forming Asn, 90-; adv bd, Ohio State Univ Eng Res Ctr, 90- *Mem:* Am Soc Mech Engrs; Nat Soc Prof Engrs. *Res:* Vibration severity; performance, reliability, part quality, productivity and accuracy of low and high speed mechanical metalforming presses and die tooling; granted six patents. *Mailing Add:* 47 Crestwood Dr Minster OH 45865

SCHOCHET, CLAUDE LEWIS, b Minneapolis, Minn, Aug 5, 44; m 67; c 3. MATHEMATICS. *Educ:* Univ Minn, BA, 65; Univ Chicago, MS, 67, PhD(math), 69. *Prof Exp:* Asst prof math, Roosevelt Univ, 69-70; amanuensis, Aarhus Univ, Denmark, 70-71; fel, Hebrew Univ, Jerusalem, 71-72; asst prof math, Ind Univ, Bloomington, 72-76; assoc prof, 76-80, PROF MATH, WAYNE STATE UNIV, 80-, ASSOC DEAN, COL LIB ARTS, 87- *Concurrent Pos:* Vis assoc prof, Univ Calif, Los Angeles, 79-80; vis prof, State Univ NY, Stoneybrook, 83-84; mem Math Sci Res Inst, Berkeley, 84-85; fac res award, Bd Gov, Wayne State Univ. *Mem:* Am Math Soc; Math Asn Am; Asn Women Math; London Math Soc. *Res:* Functional analysis; algebraic and differential topology. *Mailing Add:* Dept Math-FAB Wayne State Univ 5950 Cass Ave Detroit MI 48202

SCHOCHET, MELVIN LEO, b New York, NY, June 16, 24; m 51; c 3. PHYSICAL CHEMISTRY. *Educ:* City Col New York, BS, 44; NY Univ, PhD(chem), 54. *Prof Exp:* Instr chem, Brooklyn Col, 51-53; asst prof, Miss State Col, 54-57; from asst prof to assoc prof, 57-70, head dept, 74-80, PROF CHEM, BALDWIN-WALLACE COL, 70- *Mem:* AAAS; Am Chem Soc. *Res:* Kinetics of reactions in solutions; acid-base catalysis; corrosion. *Mailing Add:* 482 Eastland Rd Berea OH 44017

SCHOCHET, SYDNEY SIGFRIED, JR, b Chicago, Ill, Feb 7, 37; m 61; c 1. PATHOLOGY, NEUROPATHOLOGY. *Educ:* Tulane Univ, BS, 58, MD, 61, MS, 65. *Prof Exp:* Assoc prof pathol, Univ Tex Med Br, 73-79, prof, 79-81; PROF PATHOL, COL MED, WVA UNIV, 81- *Concurrent Pos:* Nat Inst Neurol Dis & Blindness spec fel, Armed Forces Inst Path, 66-67; prof path, Univ Okla, 79-81. *Mem:* Am Asn Neuropath; Am Asn Path & Bact; Am Asn Pathologists; Soc Exp Biol & Med; Int Acad Path; Sigma Xi. *Res:* Reactions of the neuron to injury; ultrastructural neuropathology; neuromuscular diseases. *Mailing Add:* Four Ridge Pl Morgantown WV 26505

SCHOCK, ROBERT NORMAN, b Monticello, NY, May 25, 39; m 59; c 3. GEOPHYSICS, HIGH PRESSURE PHYSICS. *Educ:* Colo Col, BSc, 61; Rensselaer Polytech Inst, MSc, 63, PhD(geophys), 66. *Prof Exp:* Res assoc, Univ Chicago, 66-68; sr res scientist high pressure physics, 68-72, group leader, 72-74, geosci sect leader, 74-76, earth sci div leader, 76-87, ENERGY PROG LEADER, LAWRENCE LIVERMORE NAT LAB, UNIV CALIF, 87- *Concurrent Pos:* Instr, Univ Chicago, 68 & Chabot Col, 69-71; sr Fulbright fel, Univ Bonn, 73; assoc ed, J Geophys Res, 78-80, J Physics & Chem Minerals, 84-87; vis res fel, Australian Nat Univ, 80-81; bd dir, Alameda County Flood Control & Water Conserv Dist, 84-86, chmn, 84-85; mem, Nat Res Coun, Continental Sci Drilling Comt, 84-86, Sci Adv Comt, Deep Observations & Sampling Earth's Continental Crust, Inc, 85-87, Solid Earth Sci panel, Energy Res Adv Bd, US Dept Energy, 85-87. *Mem:* Am Geophys Union; Sigma Xi. *Res:* High pressure physics; solid state processes; equation of state of solids; rock deformation. *Mailing Add:* Lawrence Livermore Nat Lab Univ Calif L-209 Livermore CA 94550

SCHODT, KATHLEEN PATRICIA, b Erie, Pa, Jan 27, 50. POLYMER SCIENCE. *Educ:* Case Western Reserve Univ, BS, 72, MS, 74, PhD(macromolecular sci), 77. *Prof Exp:* RES CHEMIST, E I DU PONT DE NEMOURS & CO, INC, 77- *Mem:* Am Chem Soc. *Res:* Physical and mechanical properties of polymers, especially elastomers; polymer blends. *Mailing Add:* Polymer Prod Chestnut Run Plaa PO Box 80713 Wilmington DE 19880-0713

SCHOEBERLE, DANIEL F, b Shipman, Ill, Aug 14, 31; m 65; c 2. ENGINEERING MECHANICS, MECHANICAL ENGINEERING. *Educ:* Univ Ill, Urbana, BS, 53, MS, 57, PhD(eng mech), 61. *Prof Exp:* Teaching fel mech eng, Univ Notre Dame, 56; asst mech engr, Argonne Nat Lab, 57-61; asst prof, 64-70, ASSOC PROF MAT ENG, UNIV ILL, CHICAGO CIRCLE, 70- *Res:* Nuclear reactor heat transfer; elastic wave propagation in solid bodies; thermoelasticity. *Mailing Add:* Dept Mech Eng M-C 251 Box 4348 Chicago IL 60680

SCHOEFER, ERNEST A(LEXANDER), b Brooklyn, NY, Sept 15, 08; m 37; c 3. METALLURGICAL ENGINEERING, MATHEMATICAL STATISTICS. *Educ:* Rensselaer Polytech Inst, CE, 32; Polytech Inst Brooklyn, MMetE, 57. *Prof Exp:* Statistician, Equity Corp, NY, 32-34; res engr, Repub Steel Corp, Ohio, 35-36; field engr, Distributors Group, Inc, NY, 36-37; secy, Alloy Casting Res Inst, 38-40, exec vpres, 40-70; METALL CONSULT, 70- *Concurrent Pos:* Mem, Tech Adv Comt High Alloy Castings, War Prod Bd, 42-45 & Nickel Conserv Panel, Metall Adv Bd, Nat Acad Sci, 52-53, High Alloys Comt, Welding Res Coun, 56- & Int Coun Alloy Phase Diagrams, 79-83. *Mem:* Am Soc Metals; Am Soc Testing & Mat; Am Inst Mining, Metall & Petrol Engrs. *Res:* Metallurgy of stainless steels and other heat and corrosion resistant casting alloys. *Mailing Add:* 43 Dinah Rock Rd PO Box 537 Shelter Island NY 11964

SCHOELLMANN, GUENTHER, b Stuttgart, Ger, Nov 17, 28; m 58; c 4. BIOCHEMISTRY, ORGANIC CHEMISTRY. *Educ:* Stuttgart Tech Univ, dipl, 55, PhD(org chem), 57. *Prof Exp:* Res assoc, 59-61, from instr to asst prof, 61-67, ASSOC PROF BIOCHEM, SCH MED, TULANE UNIV, 67-, ASSOC PROF OPHTHAL, 76- *Mem:* Am Chem Soc; NY Acad Sci; Ger Chem Soc; Ger Biol Soc. *Res:* Chemistry and function of proteins and amino acids; active site of enzymes; enzyme mechanism; peptide synthesis. *Mailing Add:* Tulane Med Ctr 1430 Tulane Ave New Orleans LA 70112

SCHOEMPERLEN, CLARENCE BENJAMIN, b Strathclair, Man, Mar 13, 13; m 39; c 3. MEDICINE. *Educ:* Univ Man, MD, 37, Royal Col Physicians & Surgeons Can, cert internal med, 46, FRCP(C), 72. *Prof Exp:* Demonstr, 46-49, lectr, 49-53, ASSOC PROF MED, UNIV MAN, 53-; ASST MED DIR, GREAT WEST LIFE ASSURANCE CO, 77- *Concurrent Pos:* Consult & endoscopist, Deer Lodge Vet Admin Hosp, 45-; assoc physician & bronchoesophagologist, Winnipeg Gen Hosp, 46-81, Children's Hosp, Winnipeg, 50-81 & Respiratory Dis Hosp; hon attend staff & consult, Health Sci Ctr, 81. *Mem:* Am Thoracic Soc; fel Am Col Physicians; sr mem & fel Am Col Chest Physicians; sr mem Am Broncho-Esophagol Asn; Can Med Asn. *Res:* Diseases of the chest; bronchoesophagology. *Mailing Add:* 351 Yale Ave Crescent & Wood Winnipeg MB R3M 0L5 Can

SCHOEN, FREDERICK J, b New York, NY, Feb 13, 46; m 75; c 1. CARDIOVASCULAR PATHOLOGY, BIOMATERIALS. *Educ:* Univ Mich, Ann Arbor, BSE, 66; Cornell Univ, PhD(mat sci), 70; Univ Miami, MD, 74. *Prof Exp:* Assoc scientist, Gulf Gen Atomic Co, San Diego, Calif, 70-72; asst prof, 80-85, ASSOC PROF PATH, HARVARD MED SCH, BOSTON, MASS, 85-; PATHOLOGIST/DIR, DEPT PATH, DIV CARDIAC PATH, BRIGHAM & WOMEN'S HOSP, 80-, DIR, AUTOPSY DIV, 88- *Concurrent Pos:* Mem, Spec Rev & Site Visit Comts, NIH, 83; lectr appl biol sci, Mass Inst Technol, 85; Lawrence J Henderson assoc prof health sci & technol, Div Health Sci & Technol, Harvard/Mass Inst Technol, 91- *Honors & Awards:* Ebert Prize, Am Pharmaceut Asn, 88. *Mem:* AAAS; Fedn Am Scientists; Am Soc Artificial Internal Organs; Int Acad Path; Soc Cardiovasc Path; Soc Biomat (pres, 89-90, treas, 86-90); Sigma Xi. *Res:* Clinicopathologic correlations in complications of cardiologic interventions; cardiovascular surgery and associated medical devices (including acute and chronic myocardial injury, angioplasty, endomyocardial biopsy, aortocoronary bypass, cardiac transplantation, heart valve prostheses, cardiac asst devices, and vascular grafts). *Mailing Add:* Dept Path Brigham & Women's Hosp 75 Francis St Boston MA 02115

SCHOEN, HERBERT M, b Long Island, NY, Oct 2, 28; m 51; c 4. CHEMICAL ENGINEERING. *Educ:* Syracuse Univ, BS, 52, MS, 53, PhD(chem eng), 57. *Prof Exp:* Res engr, Am Cyanamid Co, 57-60; dir res, Radiation Applns Inc, 60-62; dir contracts, Quantum Inc, 62-64; group leader, 64- 68, sr group leader, 68-69, res mgr, Birds Eye Div, 69-72, dir, Basic Sci, 72-78, dir fundamental res & phys sci, 78-79, corp scientist, 79-82, res fel, Gen Foods Corp, 82-86; PRIN, HMS ASSOC, 86- *Concurrent Pos:* Ed, Intersci Libr Chem Eng & Processing, 61-66; vis prof, Univ Tex, 67 & Cornell Univ, 68-69; Fulbright scholar, Tech Univ Braunschweig, Ger, 56-57; chmn, Food & Bioeng Div, Am Inst Chem Engr; consult, 85- *Mem:* Fel Am Inst Chem Engrs; Am Chem Soc; NY Acad Sci; Sigma Xi. *Res:* Crystallization, concentration and separation techniques; foam separation; food research; consumer psychology; frozen and refrigerated foods; research management; creative problem solving. *Mailing Add:* 73 Clay Hill Rd Stamford CT 06905

SCHOEN, JOHN WARREN, b Anacortes, Wash, Apr 17, 47; m 70; c 1. FOREST WILDLIFE ECOLOGY. *Educ:* Whitman Col, BA, 69; Univ Puget Sound, MS, 72; Univ Wash, PhD(wildlife ecol), 77. *Prof Exp:* Game biologist II, 76-77, GAME BIOLOGIST III, ALASKA DEPT FISH & GAME, 77- *Mem:* Wildlife Soc; Ecol Soc Am; Am Soc Mammalogists. *Res:* Wildlife habitat relationships principally black-tailed deer, mountain goats and brown bear; home range patterns and habitat selection to understand the value of old-growth forests as wildlife habitats. *Mailing Add:* Dept Fish & Game Wildlife Conserv 1300 College Rd Fairbanks AK 99701-1599

SCHOEN, KENNETH, b Bronxville, NY, Jan 18, 32; m 54; c 5. MATHEMATICS. *Educ:* Univ Conn, BA, 54; Yale Univ, AM, 55; Rensselaer Polytech Inst, 61; Univ Pittsburgh, PhD(math), 68. *Prof Exp:* Analytic engr, Hamilton Stand Div, United Technologies, 55-62; prof math, Wheeling Col, 62-66; instr, Univ Pittsburgh, 67-68; prof, Worcester Polytech Inst, 68-71; PROF MATH, WORCESTER STATE COL, 71- *Concurrent Pos:* Lectr, Univ Conn, 55-62; prof, WLiberty State Col, 63-66 & Clark Univ, 71-72. *Mem:* Soc Indust & Appl Math; Math Asn Am. *Res:* Numerical analysis. *Mailing Add:* 618 Salisbury St Worcester MA 01609

SCHOEN, KURT L, b Dec 14, 27; US citizen; m 56; c 3. CHEMISTRY. *Educ:* City Col New York, BS, 49; Polytech Inst Brooklyn, MS, 56. *Prof Exp:* Chemist, Felton Chem Co, 49-52; flavor chemist, H Kohnstamm & Co, Inc, 52-55; flavor chemist, 56-61, vpres & tech dir, 61-80, EXEC VPRES & TECH DIR, DAVID MICHAEL & CO, INC, 80- *Mem:* Am Chem Soc; Inst Food Technol; Soc Flavor Chemists. *Res:* Synthetic and natural flavorings; development and production of flavorings and synthetic food adjuncts to be used in comestibles. *Mailing Add:* 681 Meetinghouse Rd Elkins Park PA 19117

SCHOEN, MAX H, b New York, NY, Feb 4, 22; m 50; c 2. PUBLIC HEALTH. *Educ:* Univ Southern Calif, BS, 43, DDS, 43; Univ Calif, Los Angeles, MPH, 62, DrPH, 69. *Prof Exp:* Vis prof, Sch Dent Med, Univ Conn, 72; prof dent health serv, Sch Dent Med, Health Sci Ctr, State Univ NY, Stony Brook, 73-76, actg dean, 74-75, assoc dean clin affairs, 75-76; chmn sect, Sch Dent, Univ Calif, 76-82, asst dean acad affairs, 77-79, prof, 76-87, EMER PROF PREV DENT & PUB HEALTH, SCH DENT, UNIV CALIF, LOS ANGELES, 87- *Concurrent Pos:* Consult, Prov Man & BC, 74-75, Kaiser Found Health Plan, Southern Calif, 74-75, Blue Cross Southern Calif, 77-90 & Group Health Plan, Inc, St Paul, Minn, 78-80. *Honors & Awards:* Distinguished Serv Award, Am Asn Pub Health Dentists, 89; John W Knutson Distinguished Serv Award, Am Pub Health Asn, 90. *Mem:* Inst Med-Nat Acad Sci; fel Am Pub Health Asn; Am Dent Asn; Int Dent Fedn; Fedn Am Scientists. *Res:* Delivery of dental care, especially organization, economics and quality evaluation. *Mailing Add:* Sch Dent Univ Calif Ctr Health Sci Los Angeles CA 90024

SCHOEN, RICHARD ISAAC, b New Rochelle, NY, Aug 13, 27; m 51; c 2. SCIENCE ADMINISTRATION, MOLECULAR PHYSICS. *Educ:* Calif Inst Technol, BS, 49; Univ Southern Calif, MS, 54, PhD(physics) 60. *Prof Exp:* Asst prof physics, Mo Sch Mines, 55-59; asst, Univ Southern Calif, 59-60, res assoc, 60-61; staff mem, Boeing Sci Res Labs, 61-71; prog dir aeronomy, 71-73, prog mgr energy, 73-75, dep div dir energy & resources res, 75-78, SECT HEAD, APPL PHYS, MATH, BIOL SCI & ENG, DIV APPL RES, NAT SCI FOUND-RES APPLN NAT NEEDS, 78- *Mem:* AAAS; Am Phys Soc; Am Geophys Union; Sigma Xi. *Mailing Add:* Nat Sci Found 1800 G St Washington DC 20550

SCHOEN, RICHARD M(ELVIN), b Celina, Ohio, Oct 23, 50. MATHEMATTCAL DIFFERENTIAL GEOMETRY, PARTIAL DIFFERENTIAL EQUATIONS. *Educ:* Univ Dayton, BS, 72; Stanford Univ, PhD(math), 76. *Prof Exp:* Prof math, Univ Calif, Berkeley, 80-85 & Univ Calif, San Diego, 85-87; PROF MATH, STANFORD UNIV, 87- *Concurrent Pos:* Sloan Found fel, 79-81; MacArthur prize fel, MacArthur Found, 83-88. *Honors & Awards:* Böscher Prize, Am Math Soc, 89. *Mem:* Nat Acad Sci; Am Acad Arts & Sci; Am Math Soc; Math Asn Am. *Mailing Add:* Math Dept Stanford Univ Stanford CA 94305

SCHOEN, ROBERT, b New York, NY, Nov 2, 30; m 57; c 4. HYDROLOGY. *Educ:* Brooklyn Col, BS, 52; Univ Wyo, MA, 53, Harvard Univ, PhD(geol), 63. *Prof Exp:* HYDROLOGIST, US GEOL SURV, 62- *Concurrent Pos:* Asst chief off, Water Qual US Geol Surv, 75-90. *Mem:* Clay Minerals Soc; Geol Soc Am. *Res:* Application of solid-phase analysis to geochemical interpretation of water quality problems. *Mailing Add:* US Geol Surv 412 Nat Ctr Reston VA 22092

SCHOENBERG, BRUCE STUART, b New Brunswick, NJ, Nov 2, 42; m 73; c 2. NEUROLOGY, EPIDEMIOLOGY. *Educ:* Univ Pa, BA, 64; Yale Univ, MD, 68; Johns Hopkins Univ, MPH, 73; Univ Minn, MS, 76; Am Bd Psychiat & Neurol, cert neurol, 77. *Prof Exp:* Staff assoc cancer epidemiol, Nat Cancer Inst, 68-70; intern internal med, Mayo Grad Sch Med, 70-71; spec fel, Nat Inst Neurol Dis & Stroke at Mayo Clin & Johns Hopkins Univ, 71-74; HEAD SECT EPIDEMIOL, NAT INST NEUROL & COMMUN DIS & STROKE, 75- *Concurrent Pos:* Abstractor enzymol, Chem Abstr, 65-; lectr, Dept Epidemiol & Pub Health, Sch Med, Yale Univ, 68-; consult, Epidemiol Br, Nat Cancer Inst, 71-; clin instr neurol, Sch Med, Georgetown Univ, 72-75; instr neurol, med statist & epidemiol, Mayo Med Sch, Univ Minn, 74-; clin asst prof neurol, Sch Med, Georgetown Univ, 75-; vis scientist neurol, Mayo Clinic, 75-; mem biomet & epidemiol contract review comt, Nat Cancer Inst, 75-78; adv ed, J Neurol Sci, 77-; secy-gen, World Fedn Neurol Res Comt on Neuroepidemiol, 77- *Honors & Awards:* Roche Labs Award, 72; William C Menninger Award Res Neurol, 72; Award Res Hist Med, Mayo Found Hist Med Soc, 74; Mary Rooney Weigel Award, Int Acad Proctol Res Epidemiol, 74; Richmond Award, Am Acad Cerebral Palsy & Develop Med, 77. *Mem:* Fel AAAS; Am Acad Neurol; fel Royal Soc Med; Am Fedn Clin Res; Int Epidemiol Asn; Sigma Xi. *Res:* Chronic disease epidemiology, particularly the epidemiology of stroke and cancer; computer applications in medicine; medical history of the late nineteenth century. *Mailing Add:* 8520 Hazelwood Dr Bethesda MD 20814

SCHOENBERG, DANIEL ROBERT, b Chicago, Ill, Aug 14, 49; m 74. MOLECULAR ENDOCRINOLOGY. *Educ:* Univ Ill, Urbana, BS, 71; Univ Wis-Madison, PHD(oncol), 77. *Prof Exp:* Fel cell biol, Baylor Col Med, 78-80, instr, 80-81; ASSOC PROF PHARMACOL, UNIFORMED SERV UNIV HEALTH SCI, 81 - *Mem:* AAAS; Am Soc Cell Biol; Sigma Xi; Endocrine Soc. *Res:* Hormonal regulation of RNA stability; estrogen regulation of mammary tumor cell growth. *Mailing Add:* Dept Pharmacol Uniformed Serv Univ Health Sci 4301 Jones Bridge Rd Bethesda MD 20814

SCHOENBERG, LEONARD NORMAN, b Erie, Pa, Nov 29, 40; m 67; c 2. ELECTRONIC PACKAGING, PRINTED CIRCUIT BOARDS. *Educ:* Univ Rochester, BS, 62; Univ Mich, MS, 64, PhD(chem), 66. *Prof Exp:* MEM TECH STAFF, AT&T BELL LABS, 66- *Mem:* Int Electronics Packaging Soc; Am Chem Soc. *Res:* Coordination chemistry; printed circuit board fabrication; electrodeposition; electroless copper and nickel deposition; electroplating of nickel, copper, and gold; PWB design for manufacture; advanced electronic packaging technology. *Mailing Add:* Six Kathay Dr Livingston NJ 07039

SCHOENBERG, MARK, b New York, NY, Sept 3, 43; c 1. PHYSIOLOGY. *Educ:* Mass Inst Technol, SB, 64; NY Univ, MD, 68. *Prof Exp:* Intracurricular res fel, Mass Inst Technol-NY Univ, 66-67; intern pediat, Cleveland Metrop Gen Hosp, 68-69; resident internal med, Univ Chicago Hosps & Clins, 69-70; res assoc, 70-72, sr staff fel, 72-75, med officer res, Nat Inst Arthritis, Diabetes, Digestive & Kidney Dis, NIH, 75-86; MED OFFICER RES, NAT INST ARTHRITIS & MUSCULOSKELETAL & SKIN DIS, 86- *Concurrent Pos:* Res assoc, dept med, Univ Chicago, 69-70. *Honors & Awards:* Res Award in Med, Borden Inc, 68. *Mem:* Am Soc Clin Invest; Biophys Soc. *Res:* Muscle physiology. *Mailing Add:* Bldg 6 Rm 108 NIH Bethesda MD 20892

SCHOENBERG, THEODORE, b Brooklyn, NY, Aug 11, 39. CHEMICAL ENGINEERING. *Educ:* City Col New York, BChE, 60; Mass Inst Technol, SM, 62, ScD(chem eng), 65. *Prof Exp:* DIR OPERS, TEXTRON SPECIALTY MAT, 65- *Mem:* Am Inst Chem Engrs; Am Chem Soc; Nat Soc Prof Engrs. *Res:* Composite materials; high temperature chemical processes. *Mailing Add:* 120 Park Ave Medford MA 02155

SCHOENBERGER, JAMES A, b Cleveland, Ohio, July 16, 19; m 43; c 3. CLINICAL MEDICINE. *Educ:* Univ Chicago, BS, 42, MD, 43. *Prof Exp:* Asst med, Univ Chicago, 46; fel med, Univ Ill Med Ctr 49-50, from instr to asst prof, 50-60, from assoc clin prof to clin prof med, 60-71; PROF PREV MED & CHMN DEPT, RUSH-PRESBY-ST LUKES MED CTR, 75-, PRES, HOSP STAFF, 87- *Concurrent Pos:* Pres, Chicago Heart Asn, 74-76; chmn, dept prev med, Rush Med Col, 74- *Mem:* Fel Am Col Physicians; Am Fedn Clin Res; Sigma Xi; fel Am Col Cardiol; Cent Soc Clin Res; Am Heart Asm (pres, 80-81). *Res:* Renal function; capillary permeability; body water and electrolytes; therapy of hypertension; epidemiology and prevention of coronary heart disease; systems management of hypertension. *Mailing Add:* 104 Burr Ridge Club Dr Burr Ridge IL 60521

SCHOENBERGER, MICHAEL, b McKeesport, Pa, Jan 5, 40; m 64; c 3. SEISMIC SIGNAL ANALYSIS. *Educ:* Carnegie Inst Technol, BS, 61; Univ Ill, MS, 63, PhD(elec eng), 66. *Prof Exp:* From teaching asst to instr & res assoc elec eng & control theory, Univ Ill, 61-66; sr engr, Surface Div, Westinghouse Elec Corp, 66-67; sr res engr, Exxon Prod Res Co, 67-71, sr res specialist, 71-76, res assoc, 76-81, sr res assoc, 81-82, Explor Systs Div, geophys adv, 82-84, geophys scientist, Exxon USA, 84-85, sr res supvr, 85-87, res adv, 87-90, SR RES SUPVR, EXXON PROD RES CO, 90- *Concurrent Pos:* Spec ed proceedings, Inst Elec & Electronics Eng, 84; ed, Soc Explor Geophys, 87-89, Geophys, 88-89. *Mem:* Inst Elec & Electronics Eng; Europ Asn Explor Geophysicists; Soc Explor Geophys. *Res:* Geophysical data analysis; optimization and implementation of system parameters; seismic data enhancement. *Mailing Add:* Subsurface Imaging Div Exxon Prod Res Co Box 2189 Houston TX 77252-2189

SCHOENBERGER, ROBERT J(OSEPH), b Weissport, Pa, Mar 12, 38; m 61; c 2. ENVIRONMENTAL ENGINEERING. *Educ:* Drexel Univ, BSCE, 62, MS, 64, PhD(environ eng), 69. *Prof Exp:* Jr engr, Pa Power & Light Co, 57-60; sanit engr I, Pa Dept of Health, 62-63, engr-in-charge facil, 63-64; res assoc, Drexel Univ, 65-69, from asst prof to assoc prof environ eng, 69-78, assoc dir, Environ Studies Inst, 75-76; consult & vpres, Roy F Weston Co, 78-89; PRES, IMAGINEERING ASSOC, 89- *Concurrent Pos:* Pres, Imagineering Assocs, 75-78; dipl, Am Acad Environ Eng. *Mem:* Am Soc Civil Engrs; Am Soc Mech Engrs; Water Pollution Control Fedn; Air Pollution Control Asn. *Res:* Theoretical combustion of domestic solid wastes; biological and chemical treatment; thermal processing of wastes; resource recovery; hazardous waste management and control. *Mailing Add:* PO Box 51 Eagle PA 19480

SCHOENBORN, BENNO P, b Basel, Switz, May 2, 36; m 62. MOLECULAR BIOLOGY, BIOPHYSICS. *Educ:* Univ Calif, Los Angeles, BA, 58; Univ New South Wales, PhD(physics), 62. *Hon Degrees:* DSc, NJ Inst Technol, 82. *Prof Exp:* NIH fel molecular biol, Med Ctr, Univ Calif, San Francisco, 62-63, asst physicist, 63-64; res fel molecular biol, Cambridge Univ, 64-66; assoc prof pharmacol, Univ Calif, San Francisco, 66-67; from assoc biophysicist to biophysicist, 67-72, SR BIOPHYSICIST, BROOKHAVEN NAT LAB, 72-, ASSOC CHMN & HEAD CTR STRUCT BIOL, DEPT BIOL, 83-; ADJ PROF BIOCHEM, COLUMBIA UNIV MED SCH, 78- *Honors & Awards:* E O Lawrence Award, 80. *Mem:* Biophys Soc; assoc Australian Inst Physics; Am Crystallog Asn. *Res:* Molecular mechanism of anesthesia; x-ray and neutron scattering of biological structures; biophysics; hydrogen bonding & solvent structure. *Mailing Add:* Dept Biol Brookhaven Nat Lab Upton NY 11973

SCHOENBRUNN, ERWIN F(REDERICK), b Newark, NJ, July 15, 21; m 48; c 4. CHEMICAL ENGINEERING. *Educ:* Princeton Univ, BS, 47; Univ Pa, MS, 49. *Prof Exp:* Res engr, Sharples Corp, 47-51; proj mgr, Petrochem Dept, Nat Res Corp, 51-58; proj mgr, Res Labs, Escambia Chem Corp, Conn, 58-61; group leader, 61-66, head explor res & process develop, 66-68, SR CHEM ENGR, AM CYANAMID CO, STAMFORD, 68- *Mem:* Am Chem Soc; Am Inst Chem Engrs. *Res:* Ion exchange; hydrocarbon oxidation; process development; polymerization kinetics; catalysis. *Mailing Add:* 22 Christopher Rd Ridgefield CT 06877

SCHOENDORF, WILLIAM H(ARRIS), b New York, NY, Jan 21, 36; m 58; c 3. ELECTRICAL ENGINEERING. *Educ:* Mass Inst Technol, BSEE, 57; Univ Pa, MSEE, 58; Purdue Univ, PhD(elec eng), 63. *Prof Exp:* Instr elec eng, Purdue Univ, 58-62; assoc res engr, Conductron Corp, 63-65; radio astron observ, Univ Mich, Ann Arbor, 65-70; ASST GROUP LEADER, LINCOLN LAB, MASS INST TECHNOL, 70- *Mem:* Inst Elec & Electronics Engrs; Sigma Xi. *Res:* Scattering of electromagnetic waves by conductors, dielectrics, plasmas and random objects; wave propagation in plasmas and other random media; antenna theory; astrophysical processes; pattern recognition. *Mailing Add:* 16 Ledgewood Dr Bedford MA 01730

SCHOENE, NORBERTA WACHTER, b Pittsburg, Kans, July 9, 43. BIOCHEMISTRY, NUTRITION. *Educ:* Pittsburg State Univ, BA, 65; George Washington Univ, PhD(biochem), 71. *Prof Exp:* RES CHEMIST, LIPID NUTRIT LAB, NUTRIT INST, SCI & EDUC ADMIN-AGR RES, USDA, 71- *Mem:* Sigma Xi; AAAS. *Res:* Essential fatty acid and prostaglandin metabolism; phospholipases; effect of dietary essential fatty acids on platelet function and development of hypertension. *Mailing Add:* 6100 Westchester Park Dr Apt 1219 College Park MD 20740

SCHOENE, ROBERT B, b Columbus, Ohio, Dec 4, 46. RESPIRATORY PHYSIOLOGY, HIGH ALTITUDE PHYSIOLOGY. *Educ:* Columbia Univ, MD, 72. *Prof Exp:* Asst prof, 81-85, ASSOC PROF PHYSIOL, UNIV WASH, 85- *Mem:* Am Physiol Soc; Am Thoracic Soc; Am Col Chest Physicians. *Mailing Add:* Dept Med Univ Wash 325 9th Ave ZA 62 Seattle WA 98104

SCHOENER, AMY, biological oceanography, biogeography, for more information see previous edition

SCHOENER, EUGENE PAUL, b New York, NY, Oct 22, 43; m 65; c 2. NEUROPHARMACOLOGY, NEUROPHYSIOLOGY. *Educ:* City Col New York, BS, 64; Rutgers Univ, MS, 65, PhD(physiol), 70. *Prof Exp:* Res assoc neuropharmacol, Col Physicians & Surgeons, Columbia Univ, 73-74; asst prof, 74-78, ASSOC PROF PHARMACOL, SCH MED, WAYNE STATE UNIV, 78-, DIR, ADDICTION RES INST, 85- *Concurrent Pos:* NIH training grant, Col Physicians & Surgeons, Columbia Univ, 70-73; vis asst prof, State Univ NY Col Purchase, 72-73; career teacher substance abuse, Nat Inst Alcohol Abuse & Alcoholism, Wayne State Univ, 79-82. *Honors & Awards:* Award, Pharmaceut Mfg Asn Found, 75. *Mem:* Am Phys Soc; Soc Neurosci; Int Asn Study Pain; Res Soc Alcoholism; Asn Med Educ Res Substance Abuse. *Res:* Pharmacodynamics of alcohol and abused drugs; mechanisms of pain and analgesia; drug action on neural control systems. *Mailing Add:* Dept Pharmacol Sch Med Wayne State Univ Detroit MI 48201

SCHOENER, THOMAS WILLIAM, b Lancaster, Pa, Aug 9, 43; m 66, 86. ECOLOGY. *Educ:* Harvard Univ, BA, 65, PhD(biol), 69. *Prof Exp:* Jr fel, Harvard Univ, 69-71, from asst prof to assoc prof biol, 72-74; from assoc prof to prof zool, Univ Wash, 75-80; PROF ZOOL, UNIV CALIF, DAVIS, 80- *Concurrent Pos:* Ed, Theoret Pop Biol, 75-; ed bd, Oecologia, 83- *Honors & Awards:* MacArthur Prize, Ecol Soc Am. *Mem:* Nat Acad Sci; Am Soc Ichthyologists & Herpetologists; Am Soc Naturalists; Am Arachnol Soc; Ecol Soc Am; Soc Study Reptiles & Amphibians; Brit Ecol Soc; Herpetologists League; Am Soc Mammalogists; Am Ornith Union; Cooper Ornith Soc; Wilson Ornith Soc. *Res:* Feeding strategies; resource partitioning and the diversity of ecological communities; population dynamics; island ecology; biogeography; theoretical ecology; biology of lizards and spiders; food webs. *Mailing Add:* Dept Zool Univ Calif Davis CA 95616

SCHOENEWEISS, DONALD F, b Columbus, Ohio, July 27, 29; m 64; c 2. PLANT PATHOLOGY. *Educ:* Ohio State Univ, BA, 51, MSc, 53, PhD(bot), 58. *Prof Exp:* Asst plant path, Ohio Agr Exp Sta, 55-58; inspector plant parasitic nematodes, USDA, 58; from asst plant pathologist to assoc plant pathologist, State Natural Hist Surv, Ill, 58-72, plant pathologist, 72-90; RETIRED. *Mem:* Am Soc Hort Sci; Am Phytopath Soc; Sigma Xi. *Res:* Diseases of shade and ornamental trees and shrubs, especially nursery plants; influence of environmental stresses on disease susceptibility. *Mailing Add:* Bot Plant Path 172 Natural Resources Bldg 607 E Peabody Champaign IL 61820

SCHOENFELD, ALAN HENRY, b New York, NY, July 9, 47; m 70; c 1. PROBLEM SOLVING. *Educ:* Queens Col, NY, BA, 68; Stanford Univ, MS, 69, PhD(math), 73. *Prof Exp:* Lectr math, Univ Calif, Davis, 73-75, lectr sci educ, Berkeley, 75-78; asst & assoc prof, Hamilton Col, 78-81; assoc prof math & educ, Univ Rochester, 81-84; assoc prof, 84-87, PROF EDUC & MATH, UNIV CALIF, BERKELEY, 87- *Concurrent Pos:* Consult, Xerox Corp, 79; prin investr res grants, NSF, 79-, Spencer Found, 83-, Sloan Found, 84, 87-; vis lectr, Math Asn Am, 81-, chmn, Comt on Teaching Undergrad Math, 82-; mem math panel, Nat Bd Prof Teaching Standards, 90- *Honors & Awards:* Lester R Ford Award, Math Asn Am, 80. *Mem:* Math Asn Am; Am Educ Res Asn; Cognitive Sci Soc; AAAS; Nat Coun Teachers Math. *Res:* Psychology of mathematical problem solving. *Mailing Add:* Dept Educ & Math Univ Calif Berkeley CA 94720

SCHOENFELD, CY, b Brooklyn, NY, Nov 13, 39; m 76; c 2. MEDICAL & HEALTH SCIENCES, MOLECULAR BIOLOGY. *Educ:* Brooklyn Col, BS, 62; Long Island Univ, MSc, 65; NY Univ, PhD(biol), 72. *Prof Exp:* Tech lab, Margaret Sanger Res Bur, 62-63; lab supv, Infertil Clin, Belevue Hosp, 63-72; sr res tech lab path, Sch Med, NY Univ, 63-64, sr res tech lab prev med, 64-66, from asst res scientist lab to teaching asst prev med, 66-73, res asst prof urol, Urol Dept, 73-79, RES ASSOC PROF UROL, UROL DEPT, SCH MED, NY UNIV, 79- *Concurrent Pos:* Dir res, Fertil Lab Inc, New York, NY, 72-; consult, Universal Diag Lab, Brooklyn, NY, 74-; adj assoc prof, Obstet-Gynec Dept, Univ Med & Dent NJ, Newark, 86- *Mem:* Am Fertil Soc; Soc Study Reproduction; Am Soc Andrology; Am Soc Reproductive Immunol & Microbiol; Am Soc Microbiol; Am Soc Trop Med & Hyg; Sigma Xi. *Res:* Sperm physiology; hormone regulation in humans. *Mailing Add:* 137 E 36th St New York NY 10016

SCHOENFELD, DAVID ALAN, b Ft Monmouth, NJ, Apr 19, 45; m; c 3. BIOSTATISTICS. *Educ:* Reed Col, BA, 67; Univ Ore, MA, 68, PhD(math statist), 74. *Prof Exp:* Statistician, Sch Med, Univ Ore, 71-72; fel, dept statist, Stanford Univ, 74-75; res asst prof statist sci, State Univ NY Buffalo, 75-78; asst prof, 78-81, ASSOC PROF BIOSTATIST, HARVARD SCH PUB HEALTH, 81-, HARVARD MED SCH, 84- *Concurrent Pos:* Dir, Biostatist Ctr, Mass Gen Hosp. *Mem:* Biomet Soc; Am Statist Asn; Inst Math Statist. *Res:* Clinical trials of AIDS therapies with the AIDS Clinical Trials Group; development and application of statistical methodologies for clinical data; isotonic regression techniques for toxicology experiments. *Mailing Add:* Biostatist Ctr Mass Gen Hosp Fruit St Boston MA 02114

SCHOENFELD, LAWRENCE STEVEN, b New York, NY, Dec 22, 41; m 67; c 2. PSYCHIATRY. *Educ:* Ohio Wesleyan Univ, BA, 63; Univ Fla, MA, 65 & PhD(psychol), 67; Am Bd Prof Psychol, dipl, 76. *Prof Exp:* Sr asst scientist, USPHS, 67-69; PROF PSYCHIAT, UNIV TEX HEALTH SCI CTR, SAN ANTONIO, 69- *Concurrent Pos:* Exec dir & founder, Crisis Ctr San Antonio Area Inc, 72-79; consult, San Antonio Police Dept, 73-85, San Antonio Park Rangers, 75-, SW Res Inst, 78-90, San Antonio Fire Dept, 79-89, Univ Tex San Antonio Police, 79-, City New Baunfels Police Dept & City Seguin Police Dept, 82-; liaison officer, Tex Psychol Asn, 87-88. *Mem:* fel Am Psychol Asn; Int Asn Study Pain. *Res:* More than 75 publications involving training of psychologists, ethics, clinical psychology and chronic pain. *Mailing Add:* Univ Tex Health Sci Ctr 7703 Floyd Curl Dr San Antonio TX 78284-7792

SCHOENFELD, ROBERT GEORGE, b Topeka, Kans, Nov 29, 26; m 46; c 5. TOXICOLOGY. *Educ:* Univ Okla, BS, 49, MS, 56, PhD(biochem), 58; Am Bd Clin Chemists, dipl; Am Asn Clin Chemists, cert toxicol chem, 72. *Prof Exp:* Asst chemist, State Hwy Res Lab, Okla, 45-48; asst chief chemist, Wilson & Co, 49-51; clin biochemist, Vet Admin Hosp, 58-65; dir, Schoenfeld Clin Lab, Inc, 61-; AT LA MESA MED LAB, ALBUQUERQUE, NMEX. *Mem:* Am Chem Soc; fel Am Asn Clin Chemists; Am Acad Forensic Sci; fel Asn Off Racing Chemists (pres, 74-75 & 77-79); Sigma Xi. *Res:* Clinical chemistry, toxicology and methodology; drug and narcotic assays; GC-MS analyses. *Mailing Add:* 7204 Aztec Rd NE Albuquerque NM 87110-2252

SCHOENFELD, ROBERT LOUIS, b New York, NY, Apr 1, 20; m 90; c 4. ELECTRONICS, COMPUTER SCIENCE. *Educ:* NY Univ, BA, 42; Columbia Univ, BS, 44; Polytech Inst NY, MEE, 49, DEE, 56. *Prof Exp:* Electronic engr, Allied Lab Instrument Co, NY, 46-47; res assoc neurol, Col Physicians & Surg, Columbia Univ, 47-51; sr physicist, New York Dept Hosps, 51-52; res fel physics, Sloan-Kettering Inst, 52-57; asst prof electronics & comput sci, 57-63, assoc prof, 63-90, head Electronics & Comput Lab, 71-90, EMER PROF, ELECTRONICS & COMPUT SCI, ROCKEFELLER UNIV, 90- *Concurrent Pos:* From instr to assoc prof, Polytech Inst Brooklyn, 47-59, adj prof, 59-68, 77-83. *Honors & Awards:* Centennial Medal, Inst Elec & Electronic Engrs, 84. *Mem:* Fel Inst Elec & Electronics Engrs; Asn Comput Mach; Sigma Xi. *Res:* Electronic instrumentation for biophysics and neurophysiology; bioelectric signals; application of network and communication theory to physical systems; application of computer techniques in biology; microprocessor based instruments; laboratory applications of computer languages. *Mailing Add:* 500 E 63rd St New York NY 10021

SCHOENFELD, RONALD IRWIN, NEUROPHARMACOLOGY, BEHAVIORAL PHARMACOLOGY. *Educ:* Univ Chicago, PhD(pharmacol), 68. *Prof Exp:* Asst br chief, 77-91, DEPT DIR INTERMURAL RES, NIMH, 91- *Mailing Add:* Dept Dir Off Bldg 10 Rm 4N224 NIMH 5600 Fishers Lane Rockville MD 20857

SCHOENFELD, THEODORE MARK, b New York, NY, July 10, 07; m 46; c 1. SAFETY FROM HAZARDOUS CHEMICAL FLUID SPRAYOUTS, MATERIAL PERFORMANCE OF CHEMICALS. *Educ:* Col City New York, BS, 30. *Prof Exp:* Asst dir, Systs Dept, City New York, 33-42; admin officer, US Dept State, 44-46; chief indust engr, MGM Int Corp, 46-49; vpres & chief engr, Ramco Mfg Co, 74-89; pres, 50-74, PRES, T M SCHOENFELD SERV, 89- *Concurrent Pos:* Nat div dir, Inst Indust Engrs, 75; mem, Nat Ethics Comt, Am Inst Chemists, 86-, Ethics Comt & gov bd, NJ Inst Chemists, 88- *Mem:* Inst Indust Engrs; Am Soc Plastics Engrs; fel Am Inst Chemists. *Res:* Safety devices to prevent sprayouts of hazardous chemicals during production; inventor and developer of hazardous chemical safety shields for US military and major chemical processing plants. *Mailing Add:* 86C Empress Plaza Cranbury NJ 08512

SCHOENFIELD, LESLIE JACK, b Bronx, NY, Feb 20, 32; c 6. GASTROENTEROLOGY, INTERNAL MEDICINE. *Educ:* Temple Univ, BA, 52, MD, 56; Univ Minn, Minneapolis, PhD(med physiol), 64. *Prof Exp:* From instr to assoc prof internal med, Mayo Grad Sch Med, Univ Minn, 63-70; assoc prof, 71-72, PROF MED, UNIV CALIF, LOS ANGELES, 72-; DIR DEPT GASTROENTEROL, CEDARS-SINAI MED CTR, LOS ANGELES, 71- *Concurrent Pos:* NIH spec fel, Karolinska Inst, Stockholm, 65; res grant, Mayo Clin & Cedars-Sinai Med Ctr, 66-; res assoc, Fells Inst Gastrointestinal Res, Temple Univ, 63; consult internal med & gastroenterol, Mayo Clin, 63-70, assoc dir res unit & gastrointestinal training prog, 66-70; rev, Gen Med Study Sect, NIH, 70; mem res eval comt, Vet Admin, 70 & merit rev bd, 72; dir, Nat Coop Gallstone Study, 73-; vchmn, Nat Sci Adv Comt, Nat Found Ileitis & Colitis, Inc, 78-; chmn, AASLD Res Comt, 75- *Mem:* AAAS; AMA; Am Gastroenterol Asn; Am Asn Study Liver Dis; Int Asn Study Liver. *Res:* Bile flow and composition; hepatic conjugation and secretion; cholestasis biliary lipids and cholelithiasis; acute and chronic hepatitis. *Mailing Add:* Cedars-Sinai Med Ctr 8700 Beverly Blvd Los Angeles CA 90048-0750

SCHOENHALS, ROBERT JAMES, b Petoskey, Mich, Apr 29, 33; m 56; c 2. MECHANICAL ENGINEERING. *Educ:* Univ Mich, BSE, 56, MSE, 57, PhD(mech eng), 61. *Prof Exp:* Lab asst bearings, Res Labs Div, Gen Motors Co, 54, engr, AC Sparkplug Div, 55; teaching fel mech eng, Univ Mich, 56, res asst, 57-60; from asst prof to assoc prof, 60-68, PROF MECH ENG, PURDUE UNIV, WEST LAFAYETTE, 68- *Concurrent Pos:* Consult, Bendix Energy Controls Div, 63-69; vis prof, Ariz State Univ, 69-70; dir heat transfer prog, Eng Div, NSF, 73-74, energy res coordr, 74-75. *Mem:* AAAS; Am Soc Mech Engrs; Am Soc Eng Educ; Am Soc Heating, Refrig & Air-Conditioning Engrs; Sigma Xi. *Res:* Heat and mass transfer; fluid mechanics; dynamics and automatic control. *Mailing Add:* 2849 Barlow West Lafayette IN 47906

SCHOENHARD, DELBERT E, microbial genetics, for more information see previous edition

SCHOENHERR, ROMAN UHRICH, b St Henry, Ohio, Jan 2, 34; m 65; c 3. CHEMICAL ENGINEERING, PHYSICAL CHEMISTRY. *Educ:* Univ Dayton, BS, 56; Iowa State Univ, PhD(chem eng), 59. *Prof Exp:* Sr engr, 59-68, eng specialist, 68-72, sr eng specialist, 72-75, CORP ENG SCIENTIST, 3M CO, 75- *Mem:* Am Inst Chem Engrs. *Res:* Research and development in heat, mass and momentum transfer with emphasis on mathematical modeling. *Mailing Add:* 3296 Glen Oaks Ave White Bear Lake MN 55110

SCHOENHOLZ, WALTER KURT, b Recklinghausen, Germany, Jan 23, 23; US citizen; m 52; c 3. IMMUNOLOGY. *Educ:* Univ Calif, Berkeley, BSc, 56, MA, 59, PhD(med microbiol), 62. *Prof Exp:* Res assoc med microbiol, Med Ctr, Univ Calif, 60-62; asst prof microbiol, Univ NMex, 62-64; vis asst prof virol, Univ Ill, 64-65; assoc prof, 65-69, prof, 70-, EMER PROF MED MICROBIOL, CALIF STATE UNIV, HAYWARD. *Concurrent Pos:* Consult clin microbiol. *Res:* Immune mechanisms in infectious diseases; microbial latency; aids. *Mailing Add:* Dept Biol Sci Calif State Univ Hayward CA 94542

SCHOENIKE, ROLAND ERNEST, forest genetics, agriculture; deceased, see previous edition for last biography

SCHOENLY, KENNETH GEORGE, b Selma, Ala, Sept 20, 56. INSECT ECOLOGY, FORENSIC ENTOMOLOGY. *Educ:* Univ Tex, BS, 78, MS, 81; Univ NMex, PhD(biol), 89. *Prof Exp:* Res asst, Univ Tex, El Paso, 78, teaching asst biol, 79-81, vis lectr biol, 82; environ biologist, Univ Tex, El Paso & White Sands Nat Monument, 78-79; biol instr biol, Angelo State Univ, 81-84; grad teaching asst biol, Univ NMex, 85-89, asst cur entom, 89; POSTDOCTORAL ASSOC, ROCKEFELLER UNIV, 89- *Concurrent Pos:* Teaching assoc, Univ NMex, 85-89, adj asst prof, 89-; forensic entomologist, Ctr Medico Legal Res & Consult, NMex, 88- *Mem:* Am Soc Naturalists; Ecol Soc Am; Entom Soc Am; Am Acad Forensic Sci. *Res:* Community and ecosystem ecology; structure and dynamics of arthropod-dominated ecosystems including statistical analyses of successional communities and food webs; mathematical modelling of ecological systems. *Mailing Add:* Rockefeller Univ 1230 York Ave New York NY 10021-6399

SCHOENSTADT, ARTHUR LORING, b New York, NY, Nov 8, 42; m 64; c 2. APPLIED MATHEMATICS. *Educ:* Rensselaer Polytech Inst, BS, 64, MS, 65, PhD(math), 68. *Prof Exp:* PROF MATH, NAVAL POSTGRAD SCH, 70- *Honors & Awards:* Army Commendation Medal with Second Oak Leaf. *Mem:* Soc Indust & Appl Math; Sigma Xi. *Res:* Ordinary and partial differential equations; numerical methods. *Mailing Add:* Code MA/ZH Naval Postgrad Sch Monterey CA 93943

SCHOENWETTER, JAMES, b Chicago, Ill, Jan 2, 35; m 60; c 1. STRATIGRAPHIC POLYNOLOGY. *Educ:* Univ Chicago, BA, 55, BA, 56; Univ Ariz, MS, 60; Southern Ill Univ, PhD(anthropol), 67. *Prof Exp:* Res cur, Mus NMex, 63-67; from asst prof to assoc prof, 67-78, PROF ANTHROP, ARIZ STATE UNIV, 78- *Concurrent Pos:* Instr, NMex Highlands Univ, 65-66 & Eastern NMex Univ, 66-67; vis prof, Univ Liverpool, 82-83. *Mem:* Soc Am Archaeol; Am Anthropol Asn; Am Asn Stratig Polynologists. *Res:* Study of the pollen contained in sediment samples recovered from archaeological excavations to provide estimates of sample antiquity, paleoenvironmental reconstructions, and/or reconstructions of ancient cultural behaviors. *Mailing Add:* Dept Anthropol Ariz State Univ Tempe AZ 85287

SCHOENWOLF, GARY CHARLES, b Chicago, Ill, Nov 22, 49; m 71; c 2. MORPHOGENESIS, DEVELOPMENTAL BIOLOGY. *Educ:* Elmhurst Col, BA, 71; Univ Ill, Urbana Champaign, MS, 73, PhD(zool), 76. *Prof Exp:* Vis lectr embryol, Univ Ill, 76-77; fel anat, Sch Med, Univ NMex, 77-79; from asst prof to assoc prof, 79-89, PROF, ANAT & EMBRYOL, SCH MED, UNIV UTAH, 85- *Concurrent Pos:* Postdoctoral fel, Nat Res Serv Award - NIH, 78-79; assoc ed, The Anat Record, 81-; ed adv bd, Scanning Electron Micros, Inc, 81-86. *Honors & Awards:* Fogarty Award, NIH, 88. *Mem:* Am Soc Zoologists; Am Asn Anatomists; Soc Develop Biol; Am Soc Cell Biol; AAAS; Int Soc Develop Biol; Soc Neurosci; Teratology Soc. *Res:* Analysis of the mechanisms of morphogenesis, especially those mechanisms involved in the formation of the early rudiments of the nervous system (neurulation); gastrulation. *Mailing Add:* Dept Anat Sch Med Univ Utah Salt Lake City UT 84132

SCHOEPFER, ARTHUR E(RIC), b Chicago, Ill, Apr 28, 31; m 53; c 3. CHEMICAL ENGINEERING, CHEMISTRY. *Educ:* Univ Ill, BS, 52. *Prof Exp:* Process engr, Olin-Mathieson Chem Corp, 52-55, asst proj supvr, High Energy Fuel Div, 55-57, proj supvr, 57-59; develop engr, 59-61, sr develop engr, 61-65, group leader eng res, 65-77, prod mgr, 77-79, plant mgr, 79-85, DIR MFG, A E STALEY MFG CO, 85- *Mem:* Am Inst Chem Engrs. *Res:* High energy fuels; fused salt electrolysis; solvent extraction; economic evaluation; crystallization; fermentation; ion exchange; process development and design. *Mailing Add:* 4332 Leslie Lane Decatur IL 62526

SCHOEPFLE, GORDON MARCUS, b Louisville, Ky, June 11, 15; m 42; c 4. PHYSIOLOGY. *Educ:* DePauw Univ, AB, 37; Princeton Univ, AM, 39, PhD(biol), 40. *Prof Exp:* Asst physiol, Princeton Univ, 39-40; from instr to prof, Sch Med, Wash Univ, 41-69; prof, 69-85, EMER PROF NEUROBIOL IN PSYCHIAT, PHYSIOL & BIOPHYS, SCH MED, UNIV ALA, BIRMINGHAM, 85- *Mem:* AAAS; Soc Neurosci; Am Physiol Soc; Biophys Soc; Soc Exp Biol & Med; Sigma Xi. *Res:* Neurophysiology; muscle physiology; interfacial tensions; bioluminescence; mathematics of excitation. *Mailing Add:* 3704 Forest Run Rd Birmingham AL 35223

SCHOEPKE, HOLLIS GEORGE, b Kenosha, Wis, Feb 22, 29; m 54; c 2. PHARMACOLOGY. *Educ:* Univ Wis, BS, 51, PhD(pharmacol), 60. *Prof Exp:* Res asst, Univ Wis, 56-60; res pharmacologist, Abbott Labs, 60-65, head dept pharmacol, 65-69, dir div exp pharmacol, 69-71, dir prod planning & develop div, 71-73, vpres res & develop, Hosp Prod Div, 73-78; dir res & develop, Pharmaceut Div, E I du Pont de Nemours & Co, 78-80; sr vpres, Preclin Res & Develop, G D Scarle & Co, 80-87; VPRES RES & DEVELOP, ANAQUEST DIV, BOC GROUP, INC, 87- *Mem:* Am Soc Pharmacol & Exp Therapeut; Am Heart Asn; Sigma Xi. *Res:* Cardiovascular and autonomic pharmacology; cardiotonic drugs. *Mailing Add:* Two Galway Dr Mendham NJ 07945

SCHOESSLER, JOHN PAUL, b Denver, Colo, May 9, 42; m 64; c 5. OPTOMETRY, VISUAL PHYSIOLOGY. *Educ:* Ohio State Univ, BScOpt, 65, OD, 66, MSc, 68, PhD(physiol optics), 71. *Prof Exp:* From instr to prof optom & physiol optics, Ohio State Univ, 68-75, asst dean, Col Optom, 73-77. *Concurrent Pos:* Mem continuing educ comt, Am Optom Asn, 74- *Mem:* Am Acad Optom; Am Optom Asn; Asn Res Vision & Ophthal. *Res:* Corneal physiology; contact lenses; glaucoma detection; visual field defects. *Mailing Add:* Col Optom Ohio State Univ Col Med 320 W Tenth Ave Columbus OH 43210

SCHOETTGER, RICHARD A, b Arlington, Nebr, Oct 24, 32; m 54; c 2. FISH BIOLOGY, ENVIRONMENTAL BIOLOGY. *Educ:* Colo State Univ, BS, 54, MS, 59, PhD(zool), 66. *Prof Exp:* Physiologist, Fish Control Lab, 62-67, asst dir lab, 67-69, dir, Fish-Pesticide Res Lab, 69-80, DIR, NAT FISHERIES CONTAMINANT RES CTR, US FISH & WILDLIFE SERV, 80- *Concurrent Pos:* Res assoc, Dept Natural Resources, Univ Mo-Columbia, 71-; adv comt water qual criteria, Nat Acad Sci, 71-72; co-proj leaders, Effects of Pollutants on Aquatic Organisms & Ecosysts, Develop Water Qual Criteria, Assessment of Complex Anthropogenic Impacts on Ecosysts, Reservoirs & Rivers, US/USSR Environ Exchange Agreement, 79-,. *Honors & Awards:* Award of Excellence, Am Fisheries Soc, 86; Meritorious Serv Award, Dept Interior, 88. *Mem:* Am Fisheries Soc; Sigma Xi; Soc Environ Toxicol & Chem. *Res:* Toxicity, physiology and ecological effects of pesticides and other contaminants in aquatic organisms. *Mailing Add:* Nat Fisheries Contaminant Res Ctr US Fish & Wildlife Serv 4200 New Haven Rd Columbia MO 65201

SCHOFFSTALL, ALLEN M, b Harrisburg, Pa, Mar 20, 39; m 61; c 2. ORGANIC CHEMISTRY. *Educ:* Franklin & Marshall Col, BS, 60; State Univ NY Buffalo, PhD(org chem), 66. *Prof Exp:* NIH fel, Univ Ill, 66-67; from asst prof to prof chem, Univ Colo, Colo Springs, 67-87; AT DEPT CHEM, EMORY UNIV. *Concurrent Pos:* Consult, Kaman Sci Corp, Colo. *Mem:* AAAS; Am Chem Soc; Sigma Xi. *Res:* Nitrogen heterocyclic and organophosphorus chemistry; chemical evolution. *Mailing Add:* 1418 Mountview Lane Colorado Springs CO 80907

SCHOFIELD, DEREK, b Oldham, Eng, Feb 14, 28; m 55; c 3. PHYSICS. *Educ:* Univ Sheffield, BSc, 49, PhD(physics), 52. *Prof Exp:* Sci off, H M Underwater Detection Estab, Royal Navy, 52-55; sci off, Underwater Physics Sect, Naval Res Estab, 55-56, group leader transducer group, 56-58, head, Elec Sect, 59-61, head, Physics & Math Sect, 61-64, supt physics wing, 64-68, sci adv to vchief Defence Staff, Can Forces Hq, 68-72, chief, Defence Res Estab Atlantic, 72-77; dep chief res & develop labs, 77-83, CHIEF RES & DEVELOP, DEPT NAT DEFENCE, 83- *Res:* Underwater acoustics; electroacoustics. *Mailing Add:* Dept Nat Defence 305 Rideau St Ottawa ON K1A 0K6 Can

SCHOFIELD, EDMUND ACTON, JR, b Worcester, Mass, Nov 26, 38. ENVIRONMENTAL SCIENCES, CONSERVATION. *Educ:* Clark Univ, BA, 62, MA, 64; Ohio State Univ, PhD(bot), 72. *Prof Exp:* Tech ed, Battelle Mem Inst, 65-67; res asst, Dept Bot, Ohio State Univ, 67-71, res assoc plant ecol, 67-72, asst to dir, Inst Polar Studies, 72; NASA Nat Res Coun resident res assoc biol, Calif Inst Technol, 72-73; environ scientist, Ohio Dept Natural Resources, 73-76; dir res, Sierra Club, 76-77; staff ecologist, Inst Ecol, Butler Univ, 77-80; asst publ officer, Arnold Arboretum, Harvard Univ, 85-88. *Concurrent Pos:* Partic, US Antarctic Res Prog, Cape Hallett, Ross Island and southern Victoria Land, Antarctica, Clark Univ, 63-64 & Ohio State Univ, 67-68 & 68-69; res assoc, Inst Polar Studies, Ohio State Univ, 73-76; adj asst prof bot, Butler Univ, 78-80. *Res:* Ecology of Antarctic and Arctic lichens and blue-green algae; history of ecology; history of conservation movement. *Mailing Add:* 47 Elm St Scituate MA 02066

SCHOFIELD, JAMES ROY, b Spring, Tex, July 12, 23. ACADEMIC ADMINISTRATION. *Educ:* Baylor Univ, BS, 45, MD, 47. *Prof Exp:* From instr to asst prof anat, Baylor Col Med, 47-53, assoc prof admin med, 59-66, prof anat, 66-71, from asst dean to acad dean, 53-71; DIR DIV ACCREDITATION, ASN AM MED COLS, 71-, SECY, LIAISON COMT MED EDUC, 74- *Concurrent Pos:* Trustee, Baylor Med Found, 52-62, exec vpres, 52-56; Nat coordr, Med Educ Nat Defense, 55-58; consult, Surgeon Gen, US Army, 60 & Div Gen Med Sci, NIH, 61. *Mem:* Assoc Soc Exp Biol & Med; Asn Am Med Cols (asst secy, 59-66); AMA. *Res:* Medical education. *Mailing Add:* 700 Greenwood Ct Georgetown TX 78628

SCHOFIELD, RICHARD ALAN, b Royersford, Pa, June 30, 24; m 49; c 2. PATHOLOGY. *Educ:* Jefferson Med Col, MD, 48. *Prof Exp:* Instr path, Med Col & resident, Univ Hosp, Univ Ala, 50-51; instr & resident, Sch Med & Univ Hosp, Duke Univ, 53-55; pathologist & dir lab, Pottstown Mem Med Ctr, 57-89; RETIRED. *Mem:* Am Soc Clin Path; AMA; Col Am Path. *Res:* Mycology; hemoglobinopathies. *Mailing Add:* 1025 Briar Lane Pottstown PA 19464

SCHOFIELD, ROBERT EDWIN, b Milford, Nebr, June 1, 23; m 59; c 1. CULTURAL HISTORY, HISTORY OF CHEMISTRY. *Educ:* Princeton Univ, BA, 44; Univ Minn, MS, 48; Harvard Univ, PhD(hist sci), 55. *Prof Exp:* Res asst, Fercleve Corp, 44-45, Clinton Labs, 45-46; res assoc, Knolls Atomic Power Lab, Gen Elec, 48-51; from asst prof to assoc prof hist, Univ Kans, Lawrence, 55-60; assoc prof hist sci, Case Inst Technol, 60-72; Lynn Thorndike prof hist sci, Case Western Res Univ, 72-79; PROF HIST SCI & TECHNOL, IOWA STATE UNIV, 79- *Concurrent Pos:* Guggenheim fel, 59-60, 67-68; mem, Inst Advan Study, 67-68, 74-75; Sigma Xi nat lectr, 78-80; dir, Grad Prog Hist Technol & Sci, Iowa State Univ, Ames, 79; chair, Hist Physics Div, Am Phys Soc. *Honors & Awards:* Pfizer Prize, Hist Sci Soc, 64. *Mem:* Hist Sci Soc; fel Am Phys Soc; Soc Hist Technol; Royal Soc Arts; Int Acad Hist Sci; Am Soc Eighteenth Century Studies. *Res:* Eighteenth century natural philosophy; science, theology and society eighteenth century Britain life and work of Joseph Priestley; author of various publications. *Mailing Add:* Dept Hist Ross Hall Iowa State Univ Ames IA 50011

SCHOFIELD, WILFRED BORDEN, b NS, July 19, 27; m 56; c 3. BOTANY. *Educ:* Acadia Univ, BA, 50; Stanford Univ, MA, 56; Duke Univ, PhD, 60. *Hon Degrees:* DSc, Acadia Univ, 90. *Prof Exp:* Instr bot, Duke Univ, 58-60; from instr to assoc prof, 60-71, PROF BOT, UNIV BC, 71- *Honors & Awards:* G E Lawson Medal, Can Bot Asn, 86. *Mem:* Am Bryol & Lichenol Soc (vpres, 65-67, pres, 67-69); Int Asn Bryologists; Brit Bryol Soc; Nordic Bryol Soc. *Res:* Taxonomy, ecology, phytogeography and evolution of vascular plants and bryophytes. *Mailing Add:* Dept Bot Univ BC Vancouver BC V6T 2B1 Can

SCHOFIELD, WILLIAM, b Springfield, Mass, Apr 19, 21; m 46; c 2. PSYCHOTHERAPY, HEALTH PSYCHOLOGY. *Educ:* Springfield Col, BS, 42; Univ Minn, MA, 46 & PhD(psychol), 48. *Prof Exp:* From instr to assoc prof, 47-59, PROF PSYCHOL, UNIV MINN, 59- *Concurrent Pos:* Vis prof, psychol, Univ Wash, 60 & Univ Colo, 65; exec secy, Minn Psychol Asn, 54-59; mem, Med policy adv comt, Minn, 60-68, ment health serv rev comt, NIMH, 68-73, bd examrs psychologist, 83-86; bd dirs, Prof Exam Serv, NY, 76-81; instr & examr, USCG Aux, 67-; consult psychol, Vet Admin, Minneapolis, 53-, Episcopal Diocese of Minn, 69- *Mem:* Am Psychol Asn (secy-treas, 69-72); AAAS. *Res:* Psychodiagnostics and family history in mental illness; health psychology, professional issues in clinical psychology and social factors in psychotherapy. *Mailing Add:* Univ Minn Hosp Box 393 Mayo Minneapolis MN 55455

SCHOKNECHT, JEAN DONZE, b Urbana, Ill. MYCOLOGY, ULTRASTRUCTURE. *Educ:* Univ Ill, Urbana-Champaign, BS, 65, MS, 67, PhD(bot), 72. *Prof Exp:* Res assoc life sci, Univ Ill, 72-73; vis asst prof electron micros, 73-74, asst prof, 74-78, ASSOC PROF MYCOL & MICROBIOL, IND STATE UNIV, 78- *Concurrent Pos:* Adj mycologist, Ill Natural Hist Surv, 81, vis assoc prof vet microbiol, 84. *Mem:* Mycol Soc Am; Brit Mycol Soc; Brit Lichen Soc; Bot Soc Am; Sigma Xi; Electron Micros Soc Am; Med Mycol Soc Am; Am Micros Soc. *Res:* Cytology and development and systematics of the fungi; primarily the Ascomycetes and their imperfect stages and the Myxomycetes; Mycorrhizae; mineral translocation in cell develoment and between symbionts. *Mailing Add:* Sch Dent Ind State Univ 1121 W Michigan St B-1B Indianapolis IN 46202

SCHOLAR, ERIC M, b New York, NY, Aug 28, 39; m 65; c 3. METASTASIS, CHEMOTHERAPY. *Educ:* Rutgers Univ, BS, 61; Univ Ill, PhD(pharmacol), 68. *Prof Exp:* Res assoc biomed sci, 67-70, instr, 70-72, asst prof biochem pharmacol, Brown Univ, 72-75; asst prof, 75-82, ASSOC PROF PHARMACOL, MED SCH, UNIV NEBR, 82- *Mem:* Am Soc Exp Pharmacol & Therapeut; Am Asn Cancer Res; AAAS. *Res:* Mechanism of action of anti-neoplastic and immunosuppressive drugs; therapy of tumor metastasis; effect of nutrition on metastasis. *Mailing Add:* Dept Pharmacol Med Sch Univ Nebr Omaha NE 68198-6260

SCHOLBERG, HAROLD MILTON, chemistry; deceased, see previous edition for last biography

SCHOLER, CHARLES FREY, b Manhattan, Kans, May 31, 34; m 57; c 3. CONCRETE, HIGHWAYS & STREETS. *Educ:* Kans State Univ, BS, 56; Purdue Univ, MS, 57, PhD(civil eng mat), 65. *Prof Exp:* Engr, Burgwin & Martin Consult Engrs, 57-58; asst hwy engr, Riley County, Kans, 60; from instr to assoc prof, 60-89, PROF CIVIL ENG, PURDUE UNIV, WEST LAFAYETTE, 89- *Concurrent Pos:* Consult portland cement; dir, Highway Exten Res Project, Indiana Cities & Counties, Purdue Univ, West Lafayette; Concrete Materials Res Coun; mem bd dirs, Am Concrete Inst, 84-87. *Mem:* Am Soc Civil Engrs; Am Concrete Inst; Am Soc Testing & Mat; Am Pub Works Asn. *Res:* Construction materials, especially portland cement; concrete durability, physical and mechanical properties of concrete material and construction applications; low volume roads especially maintenance and construction; aggregates, soils and bituminous materials. *Mailing Add:* Sch Civil Eng Purdue Univ West Lafayette IN 47907

SCHOLES, CHARLES PATTERSON, b Auburn, NY, Oct 31, 42; m 66; c 2. BIOPHYSICS. *Educ:* Cornell Univ, AB, 64; Yale Univ, PhD(biophys), 69. *Prof Exp:* NSF fel, Oxford Univ, 69-70; NIH fel, Univ Calif, San Diego, 70-73; from asst prof to assoc prof, 73-84, PROF PHYSICS, STATE UNIV NY, ALBANY, 84- *Concurrent Pos:* NIH fel, 76-81; vis prof biochem & biophys, Chalmers Univ, Goteborg, Sweden, 85-86. *Mem:* Biophys Soc; Am Phys Soc; Am Chem Soc. *Res:* Study of biological molecules by techniques of electron paramagnetic resonance and electron nuclear double resonance. *Mailing Add:* Dept Physics State Univ NY 1400 Washington Ave Albany NY 12222

SCHOLES, NORMAN W, b Ogden, Utah, June 9, 30; m 50, 74; c 3. PHARMACOLOGY, PHYSIOLOGY. *Educ:* Univ Utah, BS, 53; Univ Calif, Los Angeles, MS, 56, PhD(pharmacol), 59; Univ Southern Calif, MS Ed, 74; Creighton Univ, BPh, & RPh, 84. *Prof Exp:* Res chemist, Wasatch Chem Corp, Utah, 53-54; Giannini-Bank Am fel, 59-61; res neuropharmacologist, City of Hope Med Ctr, Duarte, Calif, 60-64; asst prof pharmacol, Univ Calif, Davis, 64-68; prof pharmacol, Sch Med, Creighton Univ, 68-86; PHARM CONSULT SERV, INC, 86- *Concurrent Pos:* electronics for scientists, NSF, 65; educ health sci prof, United Soc Pharmacol Health Sci, 73; vis prof, Australian Nat Univ, Canberra, Australia, 76. *Mem:* AAAS; Soc Exp Biol & Med; Am Soc Pharmacol & Exp Therapeut. *Res:* Synaptic mechanisms in the central nervous system and their physiological significance; mode and site of action of drugs acting upon the central nervous system; neurochemical and electrophysiological correlates of learning; alcohol testing in humans, blood and breath ratio. *Mailing Add:* 11615 Calhoun Rd Omaha NE 68112

SCHOLES, SAMUEL RAY, JR, b Pittsburgh, Pa; m 44; c 2. CHEMISTRY. *Educ:* Alfred Univ, BS, 37; Yale Univ, PhD(phys chem), 40. *Prof Exp:* Asst quant anal, Yale Univ, 37-40; instr chem, Alfred Univ, 40-41 & phys chem, Tufts Col, 41-46; assoc prof, 46-55, chmn dept, 55-70, prof, 55-80, EMER PROF CHEM, ALFRED UNIV, 80- *Mem:* Am Chem Soc; Sigma Xi. *Res:* Properties of solutions of electrolytes; analysis of microgram quantities of fluorine; analysis of water in transformer oils. *Mailing Add:* 45 W University St Alfred NY 14802

SCHOLFIELD, CHARLES REXEL, b Morgan Co, Ill, Nov 9, 14. ANALYTICAL CHEMISTRY, ORGANIC CHEMISTRY. *Educ:* Ill Col, AB, 36; Univ Ill, AM, 41. *Prof Exp:* Sci aide chem, Regional Soybean Prod Lab, Northern Regional Res Ctr, USDA, 37-41, chemist, 41-42, res chemist, 42-81; RETIRED. *Mem:* Am Chem Soc; Am Oil Chemists Soc. *Res:* Chemistry and hydrogenation of fats and oils; liquid and gas chromatography. *Mailing Add:* 5921 N Cypress Dr Apt 1607 Peoria IL 61615-2628

SCHOLL, JAMES FRANCIS, b Albion, NY, Oct 4, 57. NON-LINEAR PHYSICS, SYMBOLIC COMPUTING. *Educ:* Univ Rochester, BS, 80, MS, 89; Univ Nev, Las Vegas, MS, 84. *Prof Exp:* Assoc scientist, Lockheed Eng & Mgt Serv Co, 83-85; SR RES ASSOC, ROCKWELL INT, 89- *Mem:* Optical Soc Am; Am Astron Soc; Am Math Soc; Soc Appl & Indust Math; Am Statist Asn. *Res:* Applied mathematical and computational methods problems in non-linear optics; symbolic computing; numerical mathematics. *Mailing Add:* Rockwell Int Sci Ctr 1049 Camino Dos Rios Thousand Oaks CA 91360

SCHOLL, MARIJA STROJNIK, b Ljubljana, Yugoslavia; US citizen; m 77; c 3. LASERS, INFRARED OPTICS. *Educ:* Univ Ariz, PhD(optical sci), 79. *Prof Exp:* res assoc optical sci, Univ Ariz, 75-78; sr staff mem optics, Rockwell Int, 78-79, mgr optics technol, 79-81; staff engr phosphor res, 81-87, SR SCIENTIST, JET PROPULSION LAB, 87- *Concurrent Pos:* Adj prof, Univ Southern Calif, 90. *Mem:* Sigma Xi; Am Phys Soc; Optical Soc Am; AAAS; Inst Elec & Electronic Engrs. *Res:* Optical engineering, optical image processing, laser diffraction, end to end simulation, high resolution displays and infrared target; development of coatings for high energy laser mirrors; fabrication of phosphors for color displays. *Mailing Add:* 721 E Chilton Dr Tempe AZ 85283

SCHOLL, PHILIP JON, b Madison, Wis, Jan 25, 45; m 75. LIVESTOCK ENTOMOLOGY, LIVESTOCK PARASITOLOGY. *Educ:* Univ Wis, Madison, BS, 70, MS, 73, PhD(entom), 78. *Prof Exp:* Res asst parasitol & entom, Univ Wis, 71-78, res assoc med entom, 78; RES ENTOMOLOGIST LIVESTOCK INSECTS, AGR RES SERV, USDA, 79- *Concurrent Pos:* Consult med entom, Univ Federal Rural do Rio de Janeiro, Brazil, 78; grad fac, Univ Nebr Syst, 79-82; asst prof, Univ Nebr, Lincoln, 79-82; US proj leader, US-Can Joint Cattle Grub Pilot Test Proj, 82-86; adj assoc prof, Montana State Univ, Bozeman, 82-86; prof consult, 5th Int Course Trop Myiasis, Rio de Janeiro, Brazil, 87; vis fac, NDak State Univ, Fargo, 90, Tex A&M Univ, College Station, 90; vis lectr, Univ Wis, Madison, 89-91. *Mem:* Sigma Xi; Entom Soc Am; Am Asn Vet Parasitologists. *Res:* Ecology and population parameters of blood-feeding and myiasis producing Diptera of veterinary and medical importance, especially livestock pests including gonotrophic age-grading, population sampling, seasonal distribution, and the effects of integrated control strategies on population characteristics. *Mailing Add:* Biting Fly-Cattle Grub Livestock Unit Livestock Insects Lab USDA PO Box 232 Kerrville TX 78029-0232

SCHOLLENBERGER, CHARLES SUNDY, b Wooster, Ohio, Aug 8, 22; m 49; c 2. POLYMER CHEMISTRY, ORGANIC CHEMISTRY. *Educ:* Col Wooster, AB, 43; Cornell Univ, PhD(org chem), 47. *Prof Exp:* Lab asst, Col Wooster, 42-43; lab asst, Cornell Univ, 44, asst org chem, 44-47; res chemist & sect leader, B F Goodrich Res & Develop Ctr, 47-75, res & develop fel, 75-84; RETIRED. *Concurrent Pos:* Chem analyst, Ohio Exp Sta, 42-44; polyurethane specialist & consult. *Honors & Awards:* Melvin K Mooney Distinguished Technol Award, Am Chem Soc, 90. *Mem:* Am Chem Soc; Polyurethane Mfg Asn. *Res:* Polyurethanes; stereo rubbers; polymers; environmental resistance. *Mailing Add:* 46 Hamden Dr Hudson OH 44236

SCHOLLER, JEAN, b Boston, Mass, Apr 14, 19. PHARMACOLOGY. *Educ:* Carnegie Inst Technol, BS, 41; Georgetown Univ, PhD, 52. *Prof Exp:* Res assoc path, Sch Med, Georgetown Univ, 48-52; assoc pharmacol, Sloan-Kettering Inst Cancer Res, 52-58, actg head dept, 55-56; asst prof, Sloan-Kettering Div, Med Col, Cornell Univ, 56-58; assoc cancer res, Christ Hosp Inst Med Res, Cincinnati, Ohio, 58-59; head dept pharmacol, John L Smith Mem Lab Cancer Res, Chas Pfizer & Co, Inc, 59-61; chmn dept exp therapeut, Stanford Res Inst, 61-68; prof pharmacol, Univ Tex, Austin, 68-77, dir lab comp pharmacol, 69-77; sr res toxicologist, Richmond Res Ctr, Stauffer Chem Co, 77-80; pres, Scholler Assocs, 80-90; RETIRED. *Mem:* AAAS; Am Soc Pharmacol & Exp Therapeut; Soc Exp Biol & Med; Am Asn Cancer Res; Soc Toxicol. *Res:* Pharmacology and toxicology; cancer chemotherapy and virology; drug metabolism. *Mailing Add:* 8913 Acorn Lane Santa Rosa CA 95409-6403

SCHOLNICK, FRANK, b Philadelphia, Pa, Apr 17, 25; m 57; c 2. ORGANIC CHEMISTRY. *Educ:* Temple Univ, BA, 46, MA, 48; Univ Pa, PhD(chem), 55. *Prof Exp:* Org chemist, E F Houghton & Co, 48-49 & Plastics & Coal Chems Div, Allied Chem Corp, 55-59; ORG RES CHEMIST, EASTERN REGIONAL RES CTR, USDA, 59- *Mem:* Am Chem Soc; Am Oil Chem Soc; Am Leather Chem Asn. *Res:* Polymers; coal tar chemistry; organic synthesis; detergents; coatings. *Mailing Add:* 2345 Pine Ridge Dr Lafayette Hill PA 19444

SCHOLNICK, STEVEN BRUCE, b New York, NY, June 4, 55. GENETICS, BIOLOGY. *Educ:* Columbia Univ, BA, 76; Cornell Univ, PhD(genetics), 82. *Prof Exp:* Res fel, Dept Biol Chem, Harvard Med Sch, 82-84; ASST PROF BIOL, CARNEGIE-MELLON UNIV, 85- *Res:* Analysis of developmental regulation of gene expression in Drosophila melanogaster by recombinant DNA techniques. *Mailing Add:* Dept Biol Sci Carnegie-Mellon Univ 5000 Forbes Ave Pittsburgh PA 15213

SCHOLTEN, PAUL DAVID, b Grand Haven, Mich, Apr 17, 49. SOLID STATE PHYSICS, THERMAL PHYSICS. *Educ:* Kalamazoo Col, BA, 71; Fla State Univ, PhD(physics), 76. *Prof Exp:* Res assoc, Tex A&M Univ, 76-77, vis asst prof, 77-78; asst prof, 78-82, ASSOC PROF PHYSICS, MIAMI UNIV, 82- *Mem:* Am Phys Soc; Sigma Xi. *Res:* Magnetism; Monte Carlo methods (computational physics). *Mailing Add:* Dept Physics Miami Univ Oxford OH 45056

SCHOLTENS, ROBERT GEORGE, b Grand Rapids, Mich, Feb 11, 29; m 52; c 3. EPIDEMIOLOGY, PARASITOLOGY. *Educ:* Mich State Univ, BS, 57, DVM, 59; Univ Ill, MS, 61; London Sch Hyg & Trop Med, dipl, 66. *Prof Exp:* Dir animal care, Biochem Res Found, 60-61; asst chief rabies, Nat Rabies Lab, Commun Dis Ctr, USPHS, 61-62, vet epidemiologist, 63-66, chief parasitic dis br, Ctr for Dis Control, 66-67, dep dir malaria prog, 67-73, dir vector biol

& control div, Ctr Dis Control, 72-76; assoc prof, Univ Tenn, 76-80, prof dept pathobiol, Col Vet Med, 80-85; dir, Training & Res, Int Livestock Ctr for Africa, 84-88; CONSULT, 88- *Concurrent Pos:* Mem subcomt animal dis surveillance, Animal Health Comt, Nat Acad Sci-Nat Res Coun, 64-; epidemiologist, London Sch Hyg & Trop Med, 65-66. *Mem:* Am Soc Trop Med & Hyg; Royal Soc Trop Med & Hyg; Am Soc Parasitol. *Res:* Epidemiology of parasitic diseases. *Mailing Add:* 5504 E Sunset Rd PO Box 1071 Knoxville TN 37914

SCHOLTES, WAYNE HENRY, b Clinton, Iowa, Dec 3, 17; m 41; c 3. SOIL SCIENCE. *Educ:* Iowa State Col, BS, 39, PhD(soil classification), 51; Duke Univ, MS, 40. *Prof Exp:* Jr soil scientist, Soil Conserv Serv, USDA, 41-45, assoc soil scientist, 45-46, soil scientist, Bur Plant Indust, 46-51; from asst prof to assoc prof, 51-55, PROF SOILS, IOWA STATE UNIV, 55-, PROF FORESTRY & DISTINGUISHED PROF AGR, 77- *Concurrent Pos:* Vis prof, Univ Ill, 58, Univ Ariz, 66 & 69 & San Carlos Univ Guatemala, 68 & 69; soils specialist, Fac Agron, Univ of the Repub, Uruguay, 63-65. *Mem:* Soil Sci Soc Am; Am Soc Agron; Soil Conserv Soc Am. *Res:* Soil classification and genesis. *Mailing Add:* 2430 Hamilton Dr Ames IA 50010

SCHOLTZ, ROBERT A, b Lebanon, Ohio, Jan 26, 36; m 62; c 2. ELECTRICAL ENGINEERING. *Educ:* Univ Cincinnati, EE, 58; Univ Southern Calif, MSEE, 60; Stanford Univ, PhD(elec eng), 64. *Prof Exp:* Mem tech staff, Hughes Aircraft Co, 58-63, staff engr, 63-68, sr staff engr, 68-78; res assoc, 63-65, from asst prof to assoc prof, 65-74, PROF ELEC ENG, UNIV SOUTHERN CALIF, 74- *Concurrent Pos:* Consult, Lincom, 76-80, Axiomatix, Inc, 79-87, JPL, 86, Technol Group, 87-89 & TRW, 89; vis colleague, Univ Hawaii, 69 & 78. *Honors & Awards:* Leonard G Abraham Award, Inst Elec & Electronics Engrs, 83 & Donald G Fink Award, 84. *Mem:* Fel Inst Elec & Electronics Engrs. *Res:* Communication and information theory; synchronization techniques; transmitter optimization and signal design; spread spectrum systems. *Mailing Add:* Dept Elec Eng Univ Southern Calif Los Angeles CA 90089-0272

SCHOLZ, CHRISTOPHER HENRY, b Pasadena, Calif, Feb 25, 43; div; c 2. GEOPHYSICS. *Educ:* Univ Nev, BS, 64; Mass Inst Technol, PhD(geol), 67. *Prof Exp:* Res fel, Seismol Lab, Calif Inst Technol, 67-68; res assoc, 68-71, assoc prof, Univ, 73-76, SR RES ASSOC GEOPHYS, LAMONT-DOHERTY GEOL OBSERV, COLUMBIA UNIV, 71-, PROF, 76- *Concurrent Pos:* Mem Nat Acad Sci Comn rock mech, 75-78 & Comn seismol; Sloan fel, 75-77 & Green fel, 80-81. *Mem:* Fel Am Geophys Union; Seismol Soc Am. *Res:* Mechanics of rock fracture and flow; earthquake mechanism and seismicity. *Mailing Add:* Dept Geol Sci Columbia Univ Broadway & W 116th New York NY 10027

SCHOLZ, DAN ROBERT, b Marysville, Kans, Sept 17, 20; m 34; c 1. MATHEMATICS. *Educ:* Southwest Tex State Teachers Col, BS, 41, MA, 42; Calif Inst Technol, MS, 43; Washington Univ, PhD(math), 51. *Prof Exp:* Asst math, Washington Univ, 48-50; asst prof, Southwestern La Inst, 51-52; from instr to assoc prof, 46-63, PROF MATH, LA STATE UNIV, BATON ROUGE, 63-, PROF MECH ENG, 77- *Mem:* Am Math Soc; Math Asn Am; Soc Indust & Appl Math. *Res:* Functions of a complex variable; numerical analysis. *Mailing Add:* 1245 Pickett Ave Baton Rouge LA 70808

SCHOLZ, EARL WALTER, b Marysville, Kans, Sept 24, 25; m 47; c 5. HORTICULTURE, PLANT PHYSIOLOGY. *Educ:* Kans State Col, BS, 50; Iowa State Univ, MS, 55, PhD(hort, plant physiol), 57. *Prof Exp:* Horticulturist, Agr Mkt Serv, USDA, 57-63; horticulturist, 63-80, ASSOC PROF HORT & FORESTRY, NDAK STATE UNIV, 80- *Mem:* Am Soc Hort Sci. *Res:* Vegetable and strawberry culture. *Mailing Add:* Dept Hort Forestry NDak State Univ Fargo ND 58102

SCHOLZ, JOHN JOSEPH, JR, b Parshall, NDak, June 11, 26; m 50; c 2. PHYSICAL CHEMISTRY. *Educ:* Univ NDak, BS, 48; Univ Ill, PhD, 55. *Prof Exp:* Res chemist, Minn Mining & Mfg Co, 53-55; res assoc, Univ Ill, 55-57; from asst prof to assoc prof, 57-68, vchmn dept, 74-81, PROF CHEM, UNIV NEBR, LINCOLN, 68- *Concurrent Pos:* Consult, Isco Inc, Lincoln, NE, 84- *Mem:* Am Chem Soc; Am Phys Soc. *Res:* Physical absorption; intermolecular forces. *Mailing Add:* Dept Chem Univ Nebr Lincoln NE 68508

SCHOLZ, LAWRENCE CHARLES, b New York, NY, Aug 8, 33; m 54; c 2. SOFTWARE ENGINEERING, SYSTEMS ENGINEERING. *Educ:* City Col New York, BEE, 54. *Prof Exp:* Engr electron tube design, Tube Div, RCA Corp, 54-60; res physicist plasma physics, IIT Res Inst, 60-65; group leader nuclear effects, Vitro Labs, 65-69; dir advan systs, Mantech, 69-70; mgr software eng, 70-80, mgr mission opers, satcom satellite proj, 80-84, div fel, 84-85, mgr systs eng, Space Sta Proj, Astro Electronics Div, RCA Corp, 85-87; div fel & mgr, SE&I, Space Sta Platform Proj, 87-89, DIV FEL, SCI & APPLICATIONS PROGS, GE ASTRO-SPACE DIV, 87- *Mem:* Sr mem Inst Elec & Electronics Engrs; Comput Soc; AAAS; Am Inst Aeronaut & Astronaut. *Res:* Systems engineering methodology and applications; software reliability; the relation between specification, implementation and testability; software organization for critical applications; spacecraft autonomy and fault tolerance. *Mailing Add:* 28 Old Salem Rd West Orange NJ 07052

SCHOLZ, PAUL DRUMMOND, b Sedro Woolley, Wash, Nov 11, 36; m 60; c 2. MECHANICAL ENGINEERING. *Educ:* Univ Wash, BS, 60; Northwestern Univ, MS, 65, PhD(mech eng, astronaut sci), 67. *Prof Exp:* Mech engr, Pasadena Annex, US Naval Ord Test Sta, 60-62; from asst prof to assoc prof, 67-78, PROF MECH ENG, UNIV IOWA, 78-, ASSOC DEAN, 79. *Mem:* AAAS; Am Soc Mech Eng; Am Soc Eng Educ; Sigma Xi. *Res:* Coagulation enhancement in aerosols; plasma non-equilibrium and spectroscopy; kinetic theory of aerosols; thermodynamics. *Mailing Add:* Eng Admin Univ Iowa Iowa City IA 52242

SCHOLZ, RICHARD W, b Ft Riley, Kans, Apr 1, 42. NUTRITIONAL BIOCHEMISTRY. *Educ:* Cornell Univ, BS; Purdue Univ, MS, 66, PhD(nutrit), 68. *Prof Exp:* Res asst nutrit, Purdue Univ, 64-65, NASA fel, 66-68; asst prof, 68-75, assoc prof, 75-81, PROF VET SCI, PA STATE UNIV, UNIVERSITY PARK, 81- *Mem:* Am Inst Nutrit. *Res:* Lung metabolism; metabolic adaptation to alterations in diet and other environmental factors. *Mailing Add:* Dept Vet Sci 103 Henning Bldg Pa State Univ University Park PA 16802

SCHOLZ, ROBERT GEORGE, b Chicago, Ill, July 3, 30; m 54; c 3. ANALYTICAL CHEMISTRY. *Educ:* Univ Ill, BS, 54; Purdue Univ, PhD(anal chem), 61. *Prof Exp:* Res chemist, Continental Can Co, 61-64 & IIT Res Inst, 64-71; mgr anal chem, Beatrice Foods, Chicago, 71-82; res chemist, Kendall Co, Barrinton, Ill, 82-83; tech support specialist, Varian Assocs, Park Ridge, Ill, 83-88; GROUP LEADER, WASTE MGT, INC, 88- *Concurrent Pos:* Lectr, Roosevelt Univ, 68-72. *Mem:* Am Chem Soc. *Res:* Technical support in atomic absorption and UV/VIS spectrophotometry; gas and liquid chromatography. *Mailing Add:* 376 Western Clarendon Hills IL 60514

SCHOLZ, WILFRIED, b Landau, Ger, Sept 14, 36; US citizen; m 66; c 2. PARTICLE SOLID INTERACTIONS, NUCLEAR & ATOMIC PHYSICS. *Educ:* Univ Freiburg, dipl physics, 60, PhD(physics), 62. *Prof Exp:* Asst nuclear physics, Univ Freiburg, 61-64; res assoc, Yale Univ, 64-67; asst prof physics, Univ Pa, 67-70; from asst prof to assoc prof, 70-82, PROF PHYSICS, STATE UNIV NY, ALBANY, 82- *Concurrent Pos:* Consult ed, Atomic Data & Nuclear Data Tables, 82-; consult, US Army Res, Develop & Eng Ctr, Watervliet, NY, 82. *Mem:* Am Phys Soc. *Res:* Nuclear spectroscopy and reactions; theory of nuclear structure; electron atom collisions; superconductivity; elastic properties of materials. *Mailing Add:* Dept Physics State Univ NY Albany NY 12222

SCHOMAKER, VERNER, b Nehawka, Nebr, June 22, 14; m 44; c 3. PHYSICAL CHEMISTRY, STRUCTURAL CHEMISTRY. *Educ:* Univ Nebr, BSc, 34, MSc, 35; Calif Inst Technol, PhD(chem), 38. *Prof Exp:* Hale fel chem, Calif Inst Technol, 38-40, sr fel, 40-45, from asst prof to prof, 45-58; chemist, Union Carbide Res Inst, 58-59, from asst dir to assoc dir, 59-65; prof, 65-84, chmn dept, 65-70, EMER PROF CHEM, UNIV WASH, 84- *Concurrent Pos:* Guggenheim Mem Found fel, 47-48; consult, Chem Div, Brookhaven Nat Lab, 48-49, Oak Ridge Nat Lab, 52-55, US Naval Ord Testing Sta, Calif, 55-58 & Union Carbide Corp, 57-58 & 65-80; mem ad hoc comt comput, Int Union Crystallog, 57, Nat Comt Crystallog, 61-64 & 71-74, subcomt molecular struct, Nat Acad Sci-Nat Res Coun, 61-63, adv comt math & phys sci, NSF, 67-69 & eval panel, Reactor Radiation Div, Nat Bur Stand, 74-77; vis assoc, chem, Calif Inst Technol, 84- *Honors & Awards:* Award, Am Chem Soc, 50. *Mem:* Fel AAAS; Am Crystallog Asn (pres, 62); fel NY Acad Sci. *Res:* Structural chemistry; determination of crystal and molecular structures by x-ray and electron diffraction; crystallographic computations; zeolite catalysis. *Mailing Add:* Dept Chem BG-10 Univ Wash Seattle WA 98195

SCHOMAN, CHARLES M, JR, b Rochester, NY, Dec 24, 24; m 46. RESEARCH & DEVELOPMENT. *Educ:* US Naval Acad, BS, 46; Rutgers Univ, MS, 57, PhD(sci), 60. *Prof Exp:* Equip engr, US Naval Supply Res & Develop Facil, 54, asst to tech dir, Res Div, 54-55, head Planning, Surv & Coordr Br, 55-57, asst tech dir, 57-58, asst tech dir & head eng planning & surv team, 58-59, sr staff tech consult & asst to officer in chg, 59-61, chief scientist, 61-64, tech dir, 64-66, head advan planning & systs anal, US Naval Ord Lab, 66-73; head advan planning, Naval Surface Weapons Ctr, 73-76; dir plans & progs, David W Taylor Naval Ships Res & Develop Ctr, 76-86, assoc tech dir, David Taylor Res Ctr, 86-89; SR ADV TO PRES, LSA, INC, 89- *Concurrent Pos:* sr exec assoc, Nat Conf Advan Res Conf Comt, 80-84, vpres, Nat War Col Alumni, 74-75, nat pres, Fed Prof Asn, 72-76. *Honors & Awards:* Jump Award, 58 & Isker Award, 62. *Mem:* Inst Food Technol; Fed Prof Exec Asn (nat pres, 72-76); Sr Execs Asn; Sigma Xi. *Res:* Scientific and engineering research management; technology transfer; food science. *Mailing Add:* 3600 Pimlico Pl Silver Spring MD 20906

SCHOMER, DONALD LEE, b Chicago, Ill, July 11, 46; m 70; c 4. NEUROPHYSIOLOGY. *Educ:* Mich State Univ, BS, 68; Univ Mich, MD, 72. *Prof Exp:* Instr neurol, Harvard Univ, 80-86; asst neurologist, 80-85, ASSOC NEUROLOGIST, BETH ISRAEL HOSP & CHILDRENS HOSP, BOSTON, 85-; ASST PROF NEUROL, HARVARD UNIV, 86- *Concurrent Pos:* Dir clin neurophysiol, Beth Israel Hosp, Boston, 80- & actg dir clin neurophysiol, Childrens Hosp, Boston, 86-89; vis physician, Clin Res Ctr, Mass Inst Technol, 85- & vis lectr neurol, 85-; prin investr, NIH grant, 86- *Mem:* Am Acad Neurol; Am Epilepsy Soc; fel Am EEG Soc; Am Acad Clin Neurophysiol. *Res:* The development of technology in neurophysiology; research approaches to epilepsy including experimental anticonvulsants and surgical techniques. *Mailing Add:* Dept Neurol Lab Clin Neurophys 330 Brookline Ave Boston MA 02215

SCHONBECK, NIELS DANIEL, b Baltimore, Md, Nov 1, 45; m 77; c 2. BIOENERGETICS, ENZYMOLOGY. *Educ:* Swarthmore Col, BA, 67; Univ Mich, Ann Arbor, PhD(biochem), 73. *Prof Exp:* Res technician biochem, Univ Mich, Ann Arbor, 69-71, fel, 73-74; Nat Cancer Inst fel, Univ Calif, Berkeley, 74-75, lectr biochem, 75-78; from asst prof to assoc prof, 78-85, PROF CHEM, METROP STATE COL, DENVER, 85- *Concurrent Pos:* Lectr health & med sci prog, Univ Calif, Berkeley, 75-; lectr, Div Natural Sci II, Univ Calif, Santa Cruz, 76; vis scientist, Nat Ctr Atmospheric Res, Boulder, Colo, 85-89; vchair, Rocky Flats Environ Monitoring Coun, 88-, mem, health adv panel, 90- *Mem:* AAAS; Sigma Xi; Union Concerned Scientists; Fedn Am Scientists. *Res:* Science and public policy issues concerning environmental and public health effects of nuclear weapons facilities; detection and amplification of atmospheric components (H2O2, SO2, NO2, NH3, HCHO) by enzyme coupling and cycling reactions; development of undergraduate programs in biochemistry. *Mailing Add:* Dept Chem Campus Box 52 PO Box 173362 Denver CO 80217-3362

SCHONBERG, RUSSELL GEORGE, b Minneapolis, Minn, Sept 15, 26; m 48; c 6. ENGINEERING, RESEARCH MANAGEMENT. *Educ:* Calif State Polytech Col, BS, 50. *Prof Exp:* Engr, US Air Force, McClellan Field, Calif, 50-51, Calif Res & Develop Co, 51-53 & Radiation Lab, Univ Calif, 53-55; engr, Varian Assocs, 55-58, proj engr, 58-60, mgr elec eng, 60-64, mgr lab opers, 64-68; vpres opers, SHM Nuclear Corp, 68-70; FOUNDER & PRES, SCHONBERG RADIATION CORP, 70- *Mem:* Sr mem Inst Elec & Electronics Engrs; Am Nuclear Soc; Am Soc Non-Destructive Testing. *Res:* Radiation research on polymerization; free radical chemistry and process techniques; development of new and improved radiation sources; design and development of miniature 3.5 million electron volt electron linear accelerator; development of real time x-ray imaging systems, automatic inspection devices and microprocessor controlled remote handling devices; development of 1.5 million electron volt and 6 million electron volt lightweight portable accelerators. *Mailing Add:* 3300 Keller St Santa Clara CA 95054

SCHONBRUNN, AGNES, b Budapest, Hungary, Oct 29, 48; Can citizen; m 74; c 2. RECEPTOR MECHANISMS, NEUROPEPTIDE ACTION. *Educ:* McGill Univ, BSc, 70; Brandeis Univ, PhD(biochem), 75. *Prof Exp:* Fel pharmacol, Harvard Sch Dent Med, 75-79; from asst prof to assoc prof physiol, Sch Pub Health, Harvard Univ, 79-87; assoc prof, 88-89, PROF PHARMACOL, MED SCH, UNIV TEX, 89- *Mem:* Am Soc Cell Biol; Endocrine Soc; Am Soc Biol Chemists; Soc Neurosci. *Res:* Structure, function and regulation of membrane receptors in eukaryotic cells; mechanisms of neuropeptide and neurotransmitter action. *Mailing Add:* Dept Pharmcol Univ Tex Health Sci Ctr PO Box 20708 Houston TX 77225

SCHONE, HARLAN EUGENE, b Bluffs, Ill, Feb 14, 32; m 56; c 3. PHYSICS. *Educ:* Univ Calif, PhD(physics), 61. *Prof Exp:* Physicist, Sci Res Lab, Boeing Airplane Co, 60-65; from asst prof to assoc prof, 65-74, actg grad dean, 70-71, PROF PHYSICS, COL WILLIAM & MARY, 74- *Mem:* Am Phys Soc. *Res:* Nuclear magnetic resonance; metal physics. *Mailing Add:* 209 Kingswood Dr Williamsburg VA 23185

SCHONEWALD-COX, CHRISTINE MICHELINE, b Paris, France; m 81; c 1. CONSERVATION OF PARKS, CONSERVATION BIOLOGY. *Educ:* Univ Calif, Davis, BA, 72; Univ MD, MS, 74, PhD(biol), 78. *Prof Exp:* Asst prof, biol, George Mason Univ, 78-79; biol, conserv, 78-82, coop unit coord, 82-87, RES SCIENTIST, CONSERV BIOL, US NAT PARK SERV, 82- *Concurrent Pos:* Adj prof, Div Environ Studies, Univ Calif, Davis, 82-, dept biol sci, Mont State Univ, 86-; res assoc, Ecol Inst, Univ Calif, Davis, 82-; mem, grad group ecol, fac, Univ Calif, Davis, 82- *Mem:* Evolution Soc; Sigma Xi; Am Inst Biol Sci. *Res:* Interdesciplinary synthesis to develop means of measuring effectiveness of boundaries in protecting parks and reserves; studies on mammalian carnivores and herbivores. *Mailing Add:* Inst Ecol Univ Calif Wickson Hall Davis CA 95616

SCHONFELD, EDWARD, b New York, NY, Oct 23, 30; m 76; c 2. MEDICAL ADHESIVES, CONTROLLED RELEASE GEL SYSTEMS. *Educ:* Brooklyn Col, BS, 52; Syracuse Univ, MS, 55, PhD(polymer & org chem), 57. *Prof Exp:* Chemist, Thiokol Chem Corp, 57-59; chemist & group leader, Nopco Chem Co, 59-64; chemist & lab mgr, Adhesives Prods Div, PPG Industs, 64-68; SR RES SCIENTIST, JOHNSON & JOHNSON, 68- *Concurrent Pos:* Instr, Brooklyn Col, 61-62, Bloomfield Col, 78-80; co-adj instr, New York City Tech Col, City Univ NY, 80-86 & Mercer County Community Col, 89- *Mem:* Am Chem Soc; Sigma xi. *Res:* Medical and surgical adhesives; pressure sensitive adhesives; sealants; urethane polymers; polymerization and characterization of polymers; bactericidal polymers; synthesis of quaternaries; UV and EB curing of polymers; gel dressings with controlled release properties for wound dressing applications; pharmaceutical development. *Mailing Add:* Ethicon Inc Div Johnson & Johnson PO Box 151 Somerville NJ 08876-0151

SCHONFELD, GUSTAV, b Mukacevo, Czech, May 8, 34; m 61; c 3. INTERNAL MEDICINE, METABOLISM. *Educ:* Wash Univ, BA, 56, MD, 60. *Prof Exp:* Asst prof internal med, Sch Med, Wash Univ, 68-70; assoc prof metab & human nutrit, Mass Inst Technol, 70-72; assoc prof, actg chmn dept, 83-87, PROF PREV MED & MED, SCH MED, WASH UNIV, 77-, KOUNTZ PROF MED, 87- *Concurrent Pos:* Asst dir, Mass Inst Technol Clin Res Ctr, 70-72; dir, Lipid Res Ctr, Sch Med, Wash Univ, 72-; mem, Adv Comt, Food & Drug Admin, 82-; NIH, Study Sect, 83- *Mem:* Endocrine Soc; Am Soc Clin Invest; Am Physiol Soc; Asn Am Physicians; Am Heart Asn; Am Soc Biochem Chemists. *Res:* Lipoproteins; hyperlipidemia; atherosclerosis; diabetes mellitus. *Mailing Add:* Lipid Res Ctr Box 8046 4566 Scott Ave St Louis MO 63110

SCHONFELD, HYMAN KOLMAN, b New York, NY, Aug 5, 19; m 44; c 2. PUBLIC HEALTH. *Educ:* Brooklyn Col, BA, 40; NY Univ, DDS, 43; Univ NC, MPH, 60, DrPH(epidemiol), 62. *Prof Exp:* Pvt dent pract, 47-56; dentist, Cent State Hosp, Petersburg, Va, 57-59; clin res assoc, Warner Lambert Res Inst, 62-64, biometrician, 64; sr res assoc med care, Dept Epidemiol & Pub Health, Sch Med, Yale Univ, 64-69, assoc prof pub health, Health Serv Admin, 69-74; sr staff officer, Nat Res Coun, Nat Acad Sci, 74-76; private consult, 76-81; staff mem, Dept Med & Surg, Vet Admin, 77-83; RETIRED. *Res:* Quality of medical and dental health care; evaluation of dental care systems. *Mailing Add:* 1116 Caddington Ave Silver Spring MD 20901

SCHONFELD, JONATHAN FURTH, quantum field theory, accelerator physics, for more information see previous edition

SCHONFELD, STEVEN EMANUEL, b New York, NY, Sept 3, 47; m 72; c 2. PERIODONTAL IMMUNOLOGY. *Educ:* State Univ NY Stony Brook, BS, 69; NY Univ, DDS, 73; Univ Southern Calif, PhD(cell & molecular biol), 76. *Prof Exp:* Asst prof, 76-82, ASSOC PROF PERIODONT, SCH DENT, UNIV SOUTHERN CALIF, 82-, CHMN DEPT, 86- *Concurrent Pos:* Consult dentist, Long Beach Vet Admin Hosp, 78- *Honors & Awards:* E H Hatton Award, Int Asn Dent Res, 75. *Mem:* AAAS; Int Asn Dent Res; Am Acad Periodont; Sigma Xi. *Res:* Host humoral immune responses in chronic oral disease; antigenic specificity of plasma cells in diseased human gingiva. *Mailing Add:* Dept Periodont Sch Dent Univ Southern California Los Angeles CA 90089

SCHONHOFF, THOMAS ARTHUR, b Quincy, Ill, July 11, 47; m 69; c 3. COMMUNICATIONS THEORY. *Educ:* Mass Inst Technol, BSEE, 69; Johns Hopkins Univ, MSEE, 72; Northeastern Univ, PhD(elec eng), 80. *Prof Exp:* Engr, Johns Hopkins Appl Physics Lab, 69-73; advan res engr, Gen Tel & Electron Sylvania, Inc, 73-78; mem tech staff, MITRE Corp, 78-80; MEM TECH STAFF, SPERRY RES CTR, 80- *Concurrent Pos:* Adj prof, Worcester Polytech Inst, 82- *Mem:* AAAS; Inst Elec & Electronics Engrs; Sigma Xi; Planetary Soc. *Res:* Using communication theory and principles to improve the performance of radio communications systems and digital magnetic recording systems. *Mailing Add:* 13 Heatherwood Dr Shrewsbury MA 01545

SCHONHORN, HAROLD, b New York, NY, Apr 2, 28; m 54; c 2. PHYSICAL CHEMISTRY, SURFACE CHEMISTRY. *Educ:* Brooklyn Col, BS, 50; Polytech Inst Brooklyn, PhD(phys chem), 59. *Prof Exp:* USAEC fel, Polytech Inst Brooklyn, 59-60; res scientist phys chem, Am Cyanamid Co, 60-61; mem tech staff surface chem, Bell Labs, Inc, 61-67, supvr surface chem, 67-84; VPRES, RES & DEVELOP LABS, KENDALL CO. *Honors & Awards:* Union Carbide Chem Prize, Am Chem Soc, 66. *Mem:* Am Chem Soc; Sigma Xi. *Res:* Surface chemistry and adhesion. *Mailing Add:* Res & Develop Labs Kendall Co 17 Hartwell Ave Lexington MA 02173

SCHONHORST, MELVIN HERMAN, b Slater, Iowa, Jan 21, 19; m 49; c 4. AGRONOMY, PLANT BREEDING. *Educ:* Iowa State Univ, BS, 51, MS, 53; Purdue Univ, PhD(agron, plant breeding), 58. *Prof Exp:* From asst agronomist to assoc agronomist, 56-64, agronomist plant breeder alfalfa improv, 56-83, prof, 64-83, EMER PROF PLANT SCI, UNIV ARIZ, 83- *Concurrent Pos:* Sabbatical leave, Mex Fed Alfalfa Breeding Prog, 76-77; consult int agr, 84-; mem, Nat Alfalfa Improv Conf. *Honors & Awards:* Pac Seedsmen Award, 72. *Mem:* Am Soc Agron; Crop Sci Soc Am. *Res:* Hybrid alfalfa; development of insect, nematode, and disease resistant varieties of alfalfa; pest resistance and tolerance to environmental stresses in alfalfa; also tolerance to high soil salinity, chloride concentration, high temperatures, increased nodulation and association with mycorrhizal fungi; use of organic mulches, mini catchments with limited rainfall for growing food crops in arid and semi-arid environments. *Mailing Add:* 6172 N Camino Almonte Tucson AZ 85718

SCHONING, ROBERT WHITNEY, b Seattle, Wash, Sept 29, 23; m 52, 82; c 4. FISHERIES. *Educ:* Univ Wash, BS, 44. *Prof Exp:* Res biologist, Ore Fish Comn, 47-52, in-chg Columbia River invests, 52-54, from asst dir to dir res, 54-58, from asst state fisheries dir to state fisheries dir, 58-71; dep dir, Nat Marine Fisheries Serv, US Dept Com, 71-73, dir, 73-78; vis prof, dept fisheries & wildlife, Ore State Univ, 78-82; sr policy adv, Nat Marine Fisheries Serv, US Dept Com, 82-86; RETIRED. *Concurrent Pos:* Mem fishing indust adv comt, US Dept State, 65-78; comnr, Int Pac Halibut Comn, 72-83 & Int NPac Fish Comn, 74-79; fish scientist, Am Fisheries Soc. *Honors & Awards:* Bronze Star Medal. *Mem:* Am Fisheries Soc; fel Am Inst Fishery Res Biol. *Res:* Research and management of fish and shellfish. *Mailing Add:* 622 NW Survista Corvallis OR 97330

SCHONSTEDT, ERICK O(SCAR), b Minneapolis, Minn, Sept 2, 17; m 57. MECHANICAL ENGINEERING. *Educ:* Univ Minn, BME & BBA, 41. *Hon Degrees:* LHD, Augustana Col, 87. *Prof Exp:* Mech engr, US Naval Ord Lab, 41-48, chief airborne magnetometer sect, 48-53; owner, Schonstedt Eng Co, 53-61; PRES, SCHONSTEDT INSTRUMENT CO, 61- *Concurrent Pos:* Dir, Augustana Col, Rock Island, Ill, 78-86. *Mem:* Am Geophys Union. *Res:* Invention of helical core magnetic field sensing element for use in rocket, satellite and laboratory magnetometers. *Mailing Add:* 1604 Greenbrier Ct Reston VA 22090

SCHONWALDER, CHRISTOPHER O, b Jan 31, 43; m; c 1. ENVIRONMENTAL HEALTH SCIENCES & TOXICOLOGY. *Educ:* Univ Vt, BS, 64; Pa State Univ, PhD(org chem), 68; Purdue Univ, MS, 73. *Prof Exp:* Res chemist, E I du Pont de Nemours & Co, Inc, 71-73; scientist admin, Off Assoc Comnr Sci, Food & Drug Admin, Rockville, 74-75; grants assoc, NIH, 75-76, prog adminr, Environ Toxicol Res Grants & Training Grants Prog, 76-84, prog adminr, Centers & Training Prog, 84-87, CHIEF, SCI PROG BR, DIV EXTRAMURAL RES & TRAINING, NAT INST ENVIRON HEALTH, NIH, 87- *Mem:* AAAS; Am Chem Soc; Soc Res Adminr. *Res:* Environmental health; toxicology; organic chemistry. *Mailing Add:* Nat Inst Environ Health Sci Sci Progs Br MD3103 104 T W Alexander Dr PO Box 12233 Research Triangle Park NC 27709

SCHOOLAR, JOSEPH CLAYTON, b Marks, Miss, Feb 28; m 60; c 5. PSYCHIATRY, PHARMACOLOGY. *Educ:* Univ Tenn, AB, 50, MS, 52; Univ Chicago, PhD, 57, MD, 60. *Prof Exp:* Asst zool, Univ Tenn, 50-52, instr biochem, Univ Tenn-AEC Lab, Oak Ridge, 53-54; res asst pharmacol, Univ Chicago, 54-57, res assoc, 57-60; resident & asst in psychiat, Baylor Col Med, 61-64, assoc prof psychiat & pharmacol, 63-74; asst dir, Tex Res Inst Ment Sci, 68-72, dir, 74-85; ASSOC PROF MENT SCI, UNIV TEX GRAD SCH BIOL SCI, HOUSTON, 68-; PROF PHARMACOL, BAYLOR COL MED, 74-, PROF PSYCHIAT, 75-, CHIEF DIV PSYCHOPHARMACOL, 85- *Mem:* AAAS; Am Psychiat Soc; AMA. *Res:* Psychopharmacology; drug abuse and addiction; effects of drugs on metabolic topography of the central nervous system; alcohol and substance abuse; adult and adolescent psychiatry; closed head injury. *Mailing Add:* Dept Pharmacol Rm 311D Baylor Col Med One Byalor Plaza Houston TX 77030

SCHOOLEY, ARTHUR THOMAS, b Plymouth, Pa, July 4, 32; m 55; c 3. CHEMICAL ENGINEERING, POLYMER CHEMISTRY. *Educ:* Carnegie Mellon Univ, BS, 54; Univ Akron, MS, 59. *Prof Exp:* Mat engr, Res Ctr, B F Goodrich Co, 54-56, res engr, 56-66, sr res engr, 66-68, res assoc, 68-79, sr res assoc, 79-89; RETIRED. *Mem:* Am Inst Chem Engrs. *Res:* Process modeling and simulation; microplants; process economics; process research. *Mailing Add:* 2015 Burlington Rd Akron OH 44313

SCHOOLEY, CAROLINE NAUS, b San Francisco, Calif, Feb 15, 32; m 53; c 3. ELECTRON MICROSCOPY. *Educ:* Univ Calif, Berkeley, BA, 53, MA, 58. *Prof Exp:* Res asst zool, 53-55, cancer res, 56-59, lab technician, 68-83, FACIL SUPVR, ELECTRON MICROSCOPE LAB, UNIV CALIF, BERKELEY, 83- *Concurrent Pos:* Dir Electron Micros Soc Am, 84-87. *Mem:* Electron Micros Soc Am; Royal Micros Soc; Am Soc Cell Biol; Am Women Sci; AAAS. *Res:* Biological ultrastructural research. *Mailing Add:* Electron Microscope Lab 26 Giannini Hall Univ Calif Berkeley CA 94720

SCHOOLEY, DAVID ALLAN, b Denver, Colo, Apr 17, 43; m 68; c 3. BIOLOGICAL CHEMISTRY, ANALYTICAL CHEMISTRY. *Educ:* NMex Highlands Univ, BSc, 63; Stanford Univ, PhD(org chem), 68. *Prof Exp:* Fel bio-inorg chem, Univ Fla, 68-69; fel bio-org chem, Columbia Univ, 69-71; sr biochemist, Zoecon Corp, 71-74, dir biochem res, 74-88; PROF BIOCHEM, UNIV NEV, RENO, 88- *Concurrent Pos:* NSF fel, Stanford Univ, 63-66; NIH fel, Columbia Univ, 69-70. *Honors & Awards:* Baxter, Burdick & Jackson Int Award, Agrochem Div, Am Chem Soc, 90. *Mem:* Am Chem Soc; Royal Soc Chem; Am Soc Biol Chem; AAAS; Protein Soc. *Res:* Biosynthesis and identification of insect juvenile hormones; insect hormone biochemistry; isolation and identification of neuropeptides in insects; analysis for hormones at physiological levels; intermediary metabolism and environmental chemistry; pesticide biochemistry. *Mailing Add:* Dept Biochem Univ Nev Reno NV 89557-0014

SCHOOLEY, JAMES FREDERICK, b Auburn, Ind, Aug 24, 31; m 53; c 7. THERMAL PHYSICS. *Educ:* Ind Univ, AB, 53, Univ Calif, Berkeley, MS, 55, PhD(nuclear chem), 61. *Prof Exp:* Physicist, Cryogenic Physics Sect, 60-74, chief, 74-82, PHYSICIST, TEMPERATURE DIV, NAT BUR STANDARDS, 82- *Concurrent Pos:* Nat Acad Sci-Nat Res Coun fel, 60-62. *Mem:* Am Phys Soc; Sigma Xi. *Res:* Study of temperature reference points based on superconductive transitions in pure metals. *Mailing Add:* 13700 Darnestown Rd Gaithersburg MD 20878

SCHOOLEY, JOHN C, b Chicago, Ill, Apr 24, 28; m 53; c 3. PHYSIOLOGY. *Educ:* Univ Calif, AB, 51, PhD, 57. *Prof Exp:* Physiologist, Donner Lab, PHYSIOLOGIST, LAWRENCE BERKELEY LAB, UNIV CALIF, BERKELEY, 71- *Mem:* AAAS; Am Physiol Soc; Soc Exp Biol & Med; Int Soc Hemat; Am Asn Immunologists; Am Soc Cell Biol. *Res:* Physiology of lymphoid tissue and bone; experimental hematology; production and differentiation of red blood cells; production of erythroprotein and action mechanism of the hormone in normal & pathphysiological conditions; role of stromal cells in modulating hematroporesis. *Mailing Add:* 3036 Hillegass Ave Berkeley CA 94705

SCHOOLEY, ROBERT T, Nov 10, 49; m 72; c 2. HERPES GROUP VIRUSES, IMMUNOLOGY. *Educ:* Johns Hopkins Univ, MD, 74. *Prof Exp:* ASSOC PROF MED, HARVARD MED SCH, 79- *Mem:* AAAS; Am Asn Immunologists; Infectious Dis Soc Am. *Res:* AIDS, immunology, antiviral chemotherapy. *Mailing Add:* Univ Colo Health Sci Ctr Box B178 4200 E Ninth Ave Denver CO 80262

SCHOOLMAN, HAROLD M, b Chicago, Ill, Jan 14, 24; m 59; c 2. MEDICINE. *Educ:* Univ Ill, MD, 50; Am Bd Internal Med, dipl, 57. *Prof Exp:* Resident med, Cook County Hosp, Ill, 51-53; assoc prof med, Cook County Grad Sch Med, 54-59; instr, Univ Ill, 57-59, clin asst prof, 59-65, assoc prof, 65-67; dir educ serv, Vet Admin Cent Off, Washington, DC, 67-70; spec asst to dir med prog develop & eval, 70-72, asst dep dir, Nat Libr Med, 72-77, DEP DIR RES & EDUC NAT LIBR MED, 77- *Concurrent Pos:* Fel hemat, Cook County Hosp, Ill, 53-54; NIH spec res fel, Div Med Sci, London Sch Trop Med, 59-60; assoc attend physician, Res & Educ Hosp & Cook County Hosp, 54-57; med educ, Cook County Hosp, 57-59, attend physician, 59, res assoc, Hektoen Inst Med Res, 54-59; chief, Hemat Res Labs, Vet Admin Hosp, Hines, Ill, 61-67 & biostatist res support ctr, 63-67; clin prof, Sch Med, Georgetown Univ, 70- *Mem:* AMA; fel Am Col Physicians; assoc Royal Soc Med. *Res:* Hematology; biostatistics. *Mailing Add:* Nat Lib Med 8600 Rockville Pike Bethesda MD 20894

SCHOON, DAVID JACOB, b Luverne, Minn, May 6, 43; m 69. CHEMICAL ENGINEERING, ELECTRONICS. *Educ:* Univ Minn, Minneapolis, BS, 65, PhD(chem eng), 69. *Prof Exp:* Sr chem engr, 3M CO, 69-86; RESEARCH SPECIALIST, PRINTWARE INC, 86- *Res:* Electrooptical object detection and counting systems; biomedical electronics. *Mailing Add:* Printware Inc 1385 Mendota Heights Rd Mendota Heights MN 55120

SCHOONMAKER, GEORGE RUSSELL, b Chicago, Ill, Dec 1, 16; m 44; c 2. GEOLOGY. *Educ:* Univ Chicago, BS, 38. *Prof Exp:* Asst inspector core drilling, Corps Engrs, US Dept Army, 39-40; geologist, Marathon Oil Co, 40-53, dist mgr, Can, 53-55, from asst mgr to explor mgr foreign dept, 55-60, explor mgr & vpres, Marathon Int Oil Co, 61-62, explor mgr, Marathon Oil Co, 62-80, vpres, 64-80; RETIRED. *Mem:* Am Asn Petrol Geol; Soc Explor Geophys. *Res:* General oil exploration. *Mailing Add:* 119 E Edgar Findlay OH 45840

SCHOONMAKER, RICHARD CLINTON, b Schenectady, NY, Dec 21, 30; m 56; c 4. CATALYSIS, MATERIAL PROPERTIES. *Educ:* Yale Univ, ChEng, 52; Cornell Univ, PhD(chem), 60. *Prof Exp:* Asst phys chem, Cornell Univ, 56-58; res physicist, Columbia Univ, 59-60; from asst prof to assoc prof, 60-69, chmn dept, 67-73 & 78-79, PROF CHEM, OBERLIN COL, 69- *Concurrent Pos:* NSF sci fac fel, Math Inst, Oxford Univ, 66-67; vis sr res fel, Dept Physics, Univ York, York Eng, 73-74; NSF prof develop fel; vis prof, Dept Chem, Univ Calif, Berkeley, 80-81; Max Planck fel, Fritz Haber Inst, Max Planck Soc, Berlin, 87-88. *Mem:* Am Phys Soc; Am Chem Soc; Sigma Xi. *Res:* Thermochemistry; thermochemical properties of gaseous molecules; phase equilibria; high temperature chemistry; mass spectrometry; thermodynamics and kinetics of vaporization and condensation processes; surface physics; molecular beams. *Mailing Add:* Dept Chem Kettering Lab Oberlin Col Oberlin OH 44074

SCHOOR, W PETER, b Frankfurt, WGer, June 4, 36; US citizen; m 70; c 1. BIOPHYSICAL CHEMISTRY, COMPARATIVE BIOCHEMISTRY. *Educ:* Auburn Univ, PhD(biochem), 66. *Prof Exp:* Fel, Inst Molecular Biophys, Fla State Univ, 66-68; asst prof pharmacol, Sch Med, La State Univ, 68-70; br chief, 70-80, MEM STAFF BIOCHEM, GULF BREEZE ENVIRON RES LAB, ENVIRON PROTECTION AGENCY, 80- *Concurrent Pos:* Adj prof, Univ WFla, 75- *Mem:* Am Chem Soc; Sigma Xi. *Res:* Biochemical molecular mechanisms; metabolism of chemical carcinogens by aquatic organisms; mechanism of osmoregulation and membrane transport. *Mailing Add:* Gulf Breeze Environ Res Lab Environ Protection Agency Gulf Breeze FL 32561

SCHOPF, JAMES WILLIAM, b Urbana, Ill, Sept 27, 41; m 65; c 1. PALEOBIOLOGY, ORGANIC GEOCHEMISTRY. *Educ:* Oberlin Col, AB, 63; Harvard Univ, AM, 65, PhD(biol), 68. *Prof Exp:* from asst prof to assoc prof, 68-73, vchair, dept earth & space sci, 82-83, dean, Div Honors, Col Lett & Sci, 83-85, PROF PALEOBIOL UNIV CALIF, LOS ANGELES, 73-, DIR, CTR STUDY EVOLUTION & ORIGIN OF LIFE, INST GEOPHYS & PLANETARY PHYSICS, 84- *Concurrent Pos:* Vis res chemist, Ames Res Ctr, NASA, 67; mem lunar sample preliminary exam team, NASA, 68-71, prin investr lunar samples, 69-74, mem, space sci adv comt, Space Prog Adv Coun, 69-82; mem, Inst Evolutionary & Environ Biol, 70-73 & Inst Geophys & Planetary Physics, 73-, Guggenheim fel, 73 & 88; assoc ed, Origins of Life, 73- & Paleobiol, 74-; Nat Acad Sci vis scientist, Soviet Union, 75; mem working groups Cambrian-Precambrian boundary, UNESCO Int Geol Correlation Prog, 75- & Precambrian biostratigraphy, 76-; mem terrestrial bodies sci working group, NASA, 75-76 & life sci comt, Space Prog Adv Coun, 76-; vis scientist, Bot Soc Am, China, 78-; vis prof, fac sci, Univ Nijmegen, Neth; Acad Chem vis res scientist, People's Repub China, 81 & 82. *Honors & Awards:* NY Bot Garden Award, Bot Soc Am, 66; Outstanding Paper Award, Soc Econ Paleontologists & Mineralogists, 71; Schuchert Award, Paleont Soc, 74; Alan T Waterman Award & Medal, NSF, 77; G Hawk Award, Univ Kans, 79; Golden Plate Award, Am Acad Achievement, 80; Mark Clark Thompson Medal, Nat Acad Sci, 86. *Mem:* Paleont Soc; Geol Soc Am; Bot Soc Am; Int Soc Study Origin Life (treas, 77-); Phycol Soc Am; Paleont Asn; Am Soc Microbiol; Am Philos Soc. *Res:* Precambrian paleobiology; optical and electron microscopy of fossil and modern microorganisms; origin of life; paleobotany; organic geochemistry of ancient sediments; interrelationships of atmospheric and biotic evolution. *Mailing Add:* Dept Earth & Space Sci Univ Calif Los Angeles CA 90024

SCHOPF, THOMAS JOSEPH MORTON, paleobiology; deceased, see previous edition for last biography

SCHOPLER, HARRY A, b Fuerth, Bavaria, Ger, Jan 5, 26; US citizen; m 49; c 2. BIOENGINEERING & BIOMEDICAL ENGINEERING. *Educ:* Univ Wis-Madison, BS, 50, Milwaukee, MS, 85. *Prof Exp:* Engr, Milwaukee Gas Co, Div Am Natural Resources, 50-56; regional mgr consult eng, Allstates Design & Develop Co, Inc, 56-82; PROF CHEM & PHYSICS, MILWAUKEE SCH ENG, 82- *Concurrent Pos:* Tech translator, Ger, 60-; lectr chem, Univ Wis-Milwaukee, 70-86. *Mem:* Fel Am Inst Chemists. *Res:* Applied research in the area of polymers and composites; instrumentation and equipment for biomedical applications such as percutaneous & transluminal angioplasty. *Mailing Add:* Milwaukee Sch Eng PO Box 644 Milwaukee WI 53201

SCHOPP, JOHN DAVID, b St Joseph, Mo, Oct 18, 27; m 57; c 2. ASTRONOMY. *Educ:* Northwestern Univ, BS, 49; Princeton Univ, PhD(astron), 54. *Prof Exp:* Asst, Princeton Univ, 49-53; res assoc, Northwestern Univ, 55; asst prof astron, Univ Mo, 55-62; from asst prof to assoc prof, 62-68, PROF ASTRON, SAN DIEGO STATE UNIV, 68- *Mem:* AAAS; Am Astron Soc. *Res:* Stellar spectroscopy and structure; photoelectric and photographic photometry; astronomy of binary stars. *Mailing Add:* Dept Astron San Diego State Univ 5300 Campanile Dr San Diego CA 92182

SCHOPP, ROBERT THOMAS, b Pontiac, Ill, Nov 5, 23; m 50; c 4. PHYSIOLOGY. *Educ:* Univ Ill, BS, 50, MS, 51; Univ Wis, PhD(physiol), 56. *Prof Exp:* Instr biol, St Norbert Col, 51-52; asst physiol, Univ Wis, 52-55, instr, 55-56; from instr to asst prof, Sch Med, Univ Colo, 56-67; assoc prof, Kirksville Col Osteop & Surg, 67-69; PROF PHYSIOL & CHMN DEPT, SCH DENT MED, SOUTHERN ILL UNIV, EDWARDSVILLE, 69- *Mem:* AAAS; Am Physiol Soc; Am Soc Pharmacol & Exp Therapeut; Am Inst Biol Sci; Am Soc Zoologists. *Res:* Neuromuscular research; autonomic nervous systems; reflex regulation of circulation and respiration. *Mailing Add:* 2934 E Exeter Tucson AZ 85716

SCHOPPER, HERWIG FRANZ, b Landskron, Czech, Feb 28, 24; Ger citizen; m 49; c 2. ELEMENTARY PARTICLE PHYSICS. *Educ:* Univ Hamburg, dipl, 49, PhD(physics & natural sci), 51. *Hon Degrees:* Dr, Univ Erlangen, 82, Univ Moscow, Univ Genève, Univ London, 89. *Prof Exp:* Assoc prof & lectr, Univ Erlangen, 54-57; prof & dir, Nuclear Phys Inst, Univ Karlsruhe, 57-60; res assoc, Europ Ctr Nuclear Res, Geneva, Switz, 66-67, head particle physics dept & mem directorate, 70-73, chmn intersecting storage ring comt, 73-76, mem sci policy comt, 79-80, dir gen, 81-88; RETIRED. *Concurrent Pos:* Res assoc, Tech Univ, Stockholm, 50-51, Cavendish Lab, Cambridge, Eng, 56-57 & Cornell Univ, 60-61; chmn, Sci Coun Nuclear Physics Ctr, Karlsruhe, 67-69; chmn directorate, German Electron-Synchrotron Particle Physics Lab, Hamburg, 73-80, & Asn Ger Nat Res Ctr, 78-80; prof, Univ Hamburg, 73-; mem expert comn, Ger Res Ministry, Max Planck Inst & var foreign insts; mem, Leopoldina Acad Sci, Halle & Joachim-Jungius Soc, Hamburg, Bayr Acad Sci, Munchen. *Honors & Awards:* Physics Award, Gottingen Acad Sci, 57; Carus Medal, Acad Leopoldina, Halle, 59; Ritter-von-Gerstner Medal, 78; Golden Plate Award, Am Acad Achievement, 84- *Mem:* Ger Phys Soc; Am Phys Soc; Europ Phys Soc. *Res:* Elementary particle physics; nuclear physics; optics; science and society; science and philosophie. *Mailing Add:* Europ Orgn Nuclear Res Geneva CH 1121 Switzerland

SCHOR, JOSEPH MARTIN, b New York, NY, Jan 10, 29; m 53; c 3. BIOCHEMISTRY. *Educ:* City Col New York, BS, 51; Fla State Univ, PhD(chem), 57. *Prof Exp:* Sr res chemist, Armour Pharmaceut Co, 57-59; res chemist, Lederle Labs, Am Cyanamid Co, 59-64; dir biochem, Endo Labs, E I du Pont de Nemours & Co, Inc, 64-77; VPRES SCI AFFAIRS, FOREST LABS, 77- *Mem:* Am Chem Soc; fel Am Inst Chemists; Int Soc Thrombosis & Hemostasis; Int Soc Hemat; AAAS; Sigma Xi. *Res:* Absorption, metabolism and disposition of drugs; blood clot lysis; blood coagulation; isolation and characterization of proteins; immunology; cardiovascular drugs; analgetic drugs; controlled release technology; central nervous system drugs. *Mailing Add:* 28 Meleny Rd Locust Valley NY 11560

SCHOR, NORBERTO AARON, b Cordoba, Arg, Dec 24, 29; m 58; c 2. PATHOLOGY. *Educ:* Nat Univ Litoral, Arg, MD, 55; Am Bd Path, dipl & cert anat path, 72. *Prof Exp:* Vis pathologist, Hosp Ramos Mejia, Buenos Aires, 55; asst prof histol, Nat Univ Litoral, 56-58; Brit Coun res fel, Postgrad Med Sch, Univ London, 58-59; Nat Res Coun Arg res fel, Dept Histol, Gothenburg Univ, Sweden, 59-60; from asst prof to assoc prof cell biol, Nat Univ Cordoba, 60-63; Guggenheim res assoc path, Stanford Univ, 64-67 & res fel, Univ Wis, 67-70; asst prof, 70-74, assoc prof, 74-79, PROF PATH, SCH MED, TULANE UNIV, 79- *Mem:* AAAS; Am Asn Pathologists & Bacteriologists; Histochem Soc; Int Acad Path. *Res:* Metabolic pathways of neoplastic cells; transfer of reducing equivalents in neoplastic cells; early activation and transformation of chemical carcinogens by lung and liver; response of lymphoid organs to the action of carcinogens. *Mailing Add:* Dept Path 6505 Hutch Tulane Univ Sch Med 1430 Tulane Ave New Orleans LA 70112

SCHOR, ROBERT, b New York, NY, Oct 25, 29; m 49; c 2. BIOPHYSICS, SOLID STATE PHYSICS. *Educ:* Mass Inst Technol, BS, 50; Univ Mich, MS, 52, PhD(physics), 58. *Prof Exp:* Asst, Univ Mich, 54-58; from instr to assoc prof, 58-80, PROF PHYSICS, UNIV CONN, 80- *Concurrent Pos:* Res fel, Inst Chemico-Phys Biol, Sorbonne, France, 66-67; vis scientist, Mass Inst Technol, 77-78. *Mem:* Am Phys Soc; Biophys Soc; Am Asn Physics Teachers; NY Acad Sci. *Res:* Structure and physical properties of fibrous proteins; chemical thermodynamics; crystal physics; phase transformations in magnetic systems; theory of diffusion coefficient of charged spherical macromolecules in solution. *Mailing Add:* Dept Physics Univ Conn Storrs CT 06268

SCHOR, STANLEY, b Philadelphia, Pa, Mar 3, 22; m 49; c 3. BIOSTATISTICS. *Educ:* Univ Pa, AB, 43, AM, 50, PhD(econ statist), 52. *Prof Exp:* Res assoc statist, Wharton Sch Com & Finance, Univ Pa, 50-55, from instr to assoc prof, 51-64, asst prof, Sch Med, 58-64, USPHS grant & dir nat periodic health exam res prog, 61-64; dir dept biostatist, AMA, 64-66; prof biomet & chmn dept, Med Sch, Temple Univ, 66-75; exec dir clin biostatist & res data syst, Merck & Co, 75-91; RETIRED. *Concurrent Pos:* Prof, Chicago Med Sch, 64-66; mem comt standard cert, USPHS, 64-66; mem task force on prescription drugs, US Dept HEW, 67-91; vis prof, Med Sch, Tel-Aviv Univ, 73-74; clin prof, Hahnemann Med Col, 74-85; adj prof, Med Sch, Temple Univ, 75-78. *Mem:* Fel Am Statist Asn; fel Am Pub Health Asn. *Res:* Application of biostatistics to solving of medical problems. *Mailing Add:* 536 Cedar Rd Lafayette Hill PA 19444

SCHORE, NEIL ERIC, b Newark, NJ, Mar 6, 48; m 78; c 2. ORGANIC CHEMISTRY, ORGANOMETALLIC CHEMISTRY. *Educ:* Univ Pa, BA, 69; Columbia Univ, PhD(org chem), 73. *Prof Exp:* Fel chem, Calif Inst Technol, 74-76; PROF CHEM, UNIV CALIF, DAVIS, 76- *Concurrent Pos:* Camille & Henry Dreyfus teacher scholar, 81-85. *Mem:* Am Chem Soc; NY Acad Sci; Sigma Xi. *Res:* Preparation and organic synthesis applications of new organometallic compounds; properties of compounds possessing intramolecular metal-metal interactions. *Mailing Add:* Dept Chem Univ Calif Davis CA 95616

SCHORER, CALVIN E, b Sauk City, Wis, June 29, 19; m 61; c 4. PSYCHIATRY. *Educ:* Univ Wis, BA, 39, MD, 55; Univ Chicago, MA, 42, PhD(eng), 48. *Prof Exp:* Intern, Detroit Receiving Hosp, 55-56; resident psychiat, 56-59, staff psychiatrist, 59-67, PROF, WAYNE STATE UNIV, 68-, DIR MED STAFF SERV PSYCHIAT, LAFAYETTE CLIN, 67-; PROF PSYCHIAT, SCH MED, WAYNE STATE UNIV, 68- *Mem:* Am Psychiat Asn; AMA. *Res:* Hypnosis; psychotherapy; teaching of psychiatry; psychopharmacology. *Mailing Add:* Psych/1434 Lafayette Clin Wayne State Univ 951 E Lafayette Detroit MI 48207

SCHORI, RICHARD M, b Tiskilwa, Ill, Oct 30, 38; m 79; c 1. NEURAL NETWORKS. *Educ:* Kenyon Col, BA, 60; Univ Iowa, MS, 62, PhD(math), 64. *Prof Exp:* From asst prof to prof math, La State Univ, Baton Rouge, 64-78; PROF MATH, ORE STATE UNIV, 78- *Concurrent Pos:* NSF res grant math, 68-78. *Mem:* Math Asn Am; Am Math Soc. *Res:* Inverse limits; hyperspaces and infinite dimensional topology; chaotic dynamical systems. *Mailing Add:* Math Dept Ore State Univ Corvallis OR 97331-4605

SCHORK, MICHAEL ANTHONY, b Elyria, Ohio, June 11, 36; m 85; c 7. BIOSTATISTICS. *Educ:* Univ Notre Dame, BA, 58, MS, 60; Univ Mich, MPH, 61, PhD(biostatist), 63. *Prof Exp:* From instr to assoc prof, 62-72, PROF BIOSTATIST, UNIV MICH, ANN ARBOR, 72- *Concurrent Pos:* Visitor, Inst Human Genetics, Univ Heidelberg, 69-70; consult, Col Am Pathologists, 74-84. *Mem:* Am Statist Asn; Biomet Soc. *Res:* Applications of biostatistical design and analysis techniques to biomedical problems. *Mailing Add:* Dept Biostatistics Univ Mich Sch Pub Health Ann Arbor MI 48109

SCHORNO, KARL STANLEY, b Berkeley, Calif, Nov 28, 39; m; c 2. ORGANIC GEOCHEMISTRY. *Educ:* Univ Calif, Berkeley, BA, 62, Okla State Univ, PhD(chem), 67. *Prof Exp:* Res chemist org chem, Univ Calif, 62; teaching asst, Okla State Univ, 62-67, fel, 67; fel med chem, Univ Kans, 67-68; res chemist, Phillips Petrol Co, 68-86; environ GC/MS sect leader, US Pollution Control, 86-87; res mgr geosci, US Dept Energy, 87-88; SR STAFF SCIENTIST, UNIV KANS, 88- *Concurrent Pos:* Consult, Univ Kans Ctr Bioanal Res, 86- *Mem:* Europ Geochem Soc; Geochem Soc Am; Am Chem Soc; Sigma Xi; NY Acad Sci; Am Soc Mass Spectrometry. *Res:* Medicinal chemistry in the study of drug design; physical organic chemistry, the study of mechanism of several reactions; organic geochemistry, the genesis of petroleum; environmental organic pollutants GC/MS, LC/MS in bioanalytical research. *Mailing Add:* 1036 College Blvd Lawrence KS 66049

SCHORR, HERBERT, b New York, NY, Jan 20, 36; m 62; c 3. ELECTRICAL ENGINEERING. *Educ:* City Col New York, BEE, 57; Princeton Univ, MA, 60, MS, 61, PhD(elec eng), 62. *Prof Exp:* Instr elec eng, Princeton Univ, 61-62; NSF fel math, Cambridge Univ, 62-63; asst prof elec eng, Columbia Univ, 63-64; mgr archit & prog, IBM Corp, Calif, 64-68, dir comput sci, NY, 68-72, vpres prod & serv planning, Advan Systs Develop Div, 73-75, mgr subsysts anal, Systs Prod Div, 75, mem corp tech comt, CHQ, 75-77, mgr systs technol, Res Div, 77-81, vpres systs, 80-84, group dir, Advan Systs, 84-88; EXEC DIR, INFO SCI INST, UNIV SOUTHERN CALIF, 88- *Concurrent Pos:* Adj asst prof, Columbia Univ, 64-65; lectr, Univ Calif, Berkeley, 65-70; res prof computer sci, Univ Southern Calif, 89- *Mem:* Asn Comput Mach; Inst Elec & Electronics Engrs. *Res:* Computer architecture and systems software. *Mailing Add:* USC-Information Sci Inst 4676 Admiralty Way 10th Floor Marina Del Rey CA 90292

SCHORR, LISBETH BAMBERGER, b Munich, Ger, Jan 20, 31; m 67; c 2. SOCIAL MEDICINE. *Educ:* Univ Calif, Berkeley, BA, 52. *Hon Degrees:* LHD, Wilkes Univ, 91. *Prof Exp:* Consult, med care, Int Union, United Automobile, Aerospace & Agr Implement Workers Am, 56-58; asst dir, Dept Social Sci, Am Fed Labor & Congress Indust Orgns, 58-65; consult, Dept Social Sci, Children's Defense Fund, 73-79; MEM, WORKING GROUP EARLY LIFE, DIV HEALTH POLICY RES & EDUC, HARVARD UNIV, 81-, LECTR SOCIAL MED, 84- *Concurrent Pos:* Mem, Comn Study Health Care Women, Am Col Obstetricians & Gynecologists, 70-73; mem adv bd, Nat Serv Study Proj, 83-84; mem bd dirs, City Lights, Washington, DC, 86-; mem nat adv comt, Healthy Tomorrows Partnership Children Prog, Bur Maternal & Child Health, Dept Health & Human Serv, 89-; mem, Task Force Children at Risk, United Way Am, 91- *Honors & Awards:* Dale Richmond Mem Award, Am Acad Pediat, 77. *Mem:* Inst Med-Nat Acad Sci. *Mailing Add:* Dept Social Med Harvard Univ Med Sch 25 Shattuck St Boston MA 02115

SCHORR, MARVIN GERALD, b New York, NY, Mar 10, 25; m 57; c 2. PHYSICS. *Educ:* Yale Univ, BS, 44, MS, 47, PhD(electromagnetics), 49. *Prof Exp:* Physicist, Tracerlab, Inc, 49-51; assoc tech dir, 51-57, exec vpres & treas, 57-62, pres, 62-88, CHMN, TECH OPERS, INC, 88- *Concurrent Pos:* Mem nuclear adv comt, Univ Lowell & AEC adv comt isotope & radiation develop, 64-66; chmn, Mass Technol Develop Corp, 74- *Mem:* Fel AAAS; Am Phys Soc; Opers Res Soc Am; Inst Elec & Electronics Engrs. *Res:* Wound ballistics; electromagnetic radiation; radioactivity and nuclear measurement; electronics; operations research. *Mailing Add:* 330 Beacon St Boston MA 02116

SCHOTLAND, RICHARD MORTON, b Irvington, NJ, Feb 11, 27; m 52; c 2. METEOROLOGY. *Educ:* Mass Inst Technol, SB, 48, SM, 50, ScD(meteorol), 52. *Prof Exp:* Asst, Mass Inst Technol, 50-52; res assoc, Oceanog Inst, Woods Hole, Mass, 52; from asst prof to prof meteorol, NY Univ, 52-73; PROF ATMOSPHERIC PHYSICS, UNIV ARIZ, 73- *Concurrent Pos:* Consult, Brookhaven Nat Lab, chmn, Group Laser Atmospheric Probing; mem, Army Basic Res Comt, Nat Res Coun, 74-, Adv Panel, Nat Ctr Atmospheric Res, 75- & adv panel, Wave Propagation Lab, Nat Oceanic & Atmospheric Admin, 78- *Mem:* Am Meteorol Soc; Am Geophys Union; Royal Meteorol Soc; Optical Soc Am; Sigma Xi. *Res:* Meteorological instrumentation; atmospheric physics; radiowave propagation; atmospheric radiation; remote sensing laser radar. *Mailing Add:* Dept Atmospheric Sci Univ Ariz Tucson AZ 85721

SCHOTT, FREDERICK W(ILLIAM), b Phoenix, Ariz, Oct 2, 19; m 46; c 1. ELECTRICAL ENGINEERING. *Educ:* San Diego State Col, AB, 40; Stanford Univ, EE, 43, PhD(elec eng), 48. *Prof Exp:* Jr engr, San Diego Gas & Elec Co, 43-44; instr physics, San Diego State Col, 46-47; from asst prof to assoc prof, 48-69, PROF ENG, UNIV CALIF, LOS ANGELES, 69- *Concurrent Pos:* Physicist, US Naval Electronics Lab, 49-50; res physicist, Hughes Aircraft Co, 56; consult, 78- *Mem:* Inst Elec & Electronics Engrs; Am Soc Eng Educ. *Res:* Rotating electric machines; applied electromagnetics. *Mailing Add:* 56-125B Engr IV Univ Calif Los Angeles CA 90024-1594

SCHOTT, GARRY LEE, b Detroit, Mich, Dec 20, 31; m 60; c 2. PHYSICAL CHEMISTRY. *Educ:* Univ Mich, BS, 52; Calif Inst Technol, PhD(chem), 56. *Prof Exp:* STAFF MEM, LOS ALAMOS NAT LAB, 56- *Mem:* Am Chem Soc; Am Phys Soc; Combustion Inst. *Res:* Shock wave processes in condensed matter and gases; chemical kinetics; gaseous combustion; detonation. *Mailing Add:* 120 Monte Vista Los Alamos NM 87544

SCHOTT, HANS, b Ger, Oct 25, 22; US citizen; m 58. PHYSICAL CHEMISTRY, PHARMACEUTICS. *Educ:* Univ Sao Paulo, 43; Univ Southern Calif, MS, 51; Univ Del, PhD(phys chem), 58. *Prof Exp:* Res chemist, Cent Lab, Matarazzo Industs, Brazil, 43-44, head chem lab, Viscose Rayon Plant, 44-47, plant adminstr, 49; res chemist, Thiokol Chem Corp, 47-48 & Textile Fibers Dept, Pioneering Res Div, E I du Pont de Nemours & Co, 51-56; res assoc, Film Div, Olin Mathieson Chem Corp, 58-60; sr res assoc phys chem sect, Res Ctr, Lever Bros Co, 61-67; res chemist, US Forest Prod Lab, 67-68; assoc prof, 69-72, PROF PHYS & COLLOID CHEM, SCH PHARM, TEMPLE UNIV, 72- *Concurrent Pos:* Geigy vis prof pharm, Univ Manchester, Eng, 85. *Mem:* Am Chem Soc; fel Acad Pharmaceut Sci. *Res:* Solubilization of cholesterol derivatives; effect of electrolytes on nonionic surfactants; colloidal properties and rheology of aqueous dispersions of drugs, bacteria and clays; physical chemistry of gelatin; surface and bulk properties of polymers; phase rule; colloid and polymer chemistry. *Mailing Add:* Temple Univ Sch Pharm 3307 N Broad St Philadelphia PA 19140

SCHOTT, JEFFREY HOWARD, b Angola, Ind, Feb 4, 47; m 66. CHEMICAL ENGINEERING, RESEARCH ADMINISTRATION. *Educ:* Univ Minn, BChemE, 70, MS, 74, PhD(chem eng), 78; Univ Chicago, MBA, 83. *Prof Exp:* Inst, unit opers, dept chem eng, Univ Minn, 72-75; res engr, Amoco Chem Corp, 75-78; dir res & develop, Eschem, Inc, subsid, Esmark, Inc, 78-84; exec dir, Tile Coun Am, Inc, 84-87; dir adhesive technol, Avery Dennison, 87-90; MGR, SIGMA CHEM CO, 90- *Mem:* Am Inst Chem Engrs; Am Chem Soc; Sigma Xi; Nat Soc Prof Engrs. *Res:* Application of chemical engineering principles to the development and production of adhesives, coatings, and related specialty chemicals. *Mailing Add:* PO Box 14508 St Louis MO 63178-9974

SCHOTTE, WILLIAM, b Burlington, Iowa, July 3, 27; m 50; c 3. CHEMICAL ENGINEERING. *Educ:* Columbia Univ, BS, 50, MS, 51, EngScD, 55. *Prof Exp:* Res assoc, Columbia Univ, 51-54; res engr, 54-58, res proj engr, 58-61, sr res engr, 61-69, sr res specialist, 69-78, SR RES ASSOC, ENG TECH LAB, EXP STA, E I DU PONT DE NEMOURS & CO, INC, 78- *Mem:* Am Chem Soc; Am Inst Chem Engrs. *Res:* Pollution abatement; chemical reactors; separation technology. *Mailing Add:* 2014 Wildwood Dr Woodland Park Wilmington DE 19805

SCHOTTEL, JANET L, MESSENGER RNA STABILITY, MICROBIAL PLANT INTERACTIONS. *Educ:* Wash Univ, St Louis, PhD(biol), 77. *Prof Exp:* Asst prof, 81-87, ASSOC PROF BIOCHEM, MOLECULAR BIOL & MICROBIOL, UNIV MINN, 87- *Mem:* Am Soc Microbiol; AAAS; Am Soc Biol Chemists. *Res:* Mechanism of MRNA turnover in Escherichia coli; mechanism of pathogenicity of stretomycin scabies in potato. *Mailing Add:* Dept Biochem Univ Minn 1479 Gortner Ave St Paul MN 55108

SCHOTTELIUS, BYRON ARTHUR, physiology; deceased, see previous edition for last biography

SCHOTTELIUS, DOROTHY DICKEY, b Lohrville, Iowa, Oct 9, 27; m 49. BIOCHEMISTRY. *Educ:* Univ Iowa, BA, 49; State Col Wash, MS, 51; Univ NC, PhD, 57. *Prof Exp:* Res asst biochem, Univ Iowa, 52-54; res asst, Univ NC, 54-57; res assoc radiation, 59-71, res assoc neurol & physiol, 71-75, ASST PROF NEUROL, UNIV IOWA, 75- *Mem:* Am Physiol Soc; Am Epilepsy Soc; Biophys Soc; Am Asn Clin Chem; Sigma Xi. *Res:* Neurochemistry; biochemistry of normal and diseased muscle; pharmacology of anticonvulsant drugs; protein and nucleic acid chemistry. *Mailing Add:* 1450 Grand Ave Iowa City IA 52246

SCHOTTMILLER, JOHN CHARLES, b Rochester, NY, Aug 6, 30; m 56; c 2. IMAGING MATERIALS, TECHNICAL MANAGEMENT. *Educ:* Univ Rochester, BA, 53; Syracuse Univ, PhD(chem), 58. *Prof Exp:* Asst, AEC, Syracuse Univ, 57-58; res chemist, Union Carbide Metals Co, 58-61; staff chemist, Components Div, Int Bus Mach Corp, 61-62; sr scientist, Spec Mat Mfg, Xerox Corp, 62-71, mgr photoreceptor develop & eng, 71-77, mgr process eng, Europ Opers, 77-79, mgr tech serv, 79-82, mgr phys & chem anal, 82-84, mgr qual & bus effectiveness, 84-89; PRES & TOTAL QUAL CONSULT, R M CONSULT, INC, 89- *Mem:* Sigma Xi; Am Soc Qual Control. *Res:* Photoconductivity; solid state and metallurgical chemistry; vacuum deposition; electrophotography; x-rays; microcircuitry; thermodynamics; phase diagrams; reactive metals; metal-metal oxide equilibria; analytical chemistry; total quality control. *Mailing Add:* 38 Alberta Dr Penfield NY 14526

SCHOTTSTAEDT, MARY GARDNER, psychiatry, for more information see previous edition

SCHOTTSTAEDT, WILLIAM WALTER, b Fresno, Calif, Mar 28, 17; m 47; c 4. MEDICINE, PUBLIC HEALTH. *Educ:* Univ Calif, BA, 47, MD, 48; Univ Mich, BMus, 40, MMus, 41. *Prof Exp:* From asst prof to assoc prof med, prev med & pub health, Sch Med, Univ Okla, 53-60, assoc prof psychiat, 56-60, prof prev med & pub health & chmn dept & consult prof psychiat, neurol & behav sci, 60-68, dean, Col Health, 68-73; dir Health Educ Ctr, 76-79, PROF PREV MED & COMMUNITY HEALTH, UNIV TEX MED BR, GALVESTON, 74-, ASSOC DEAN, CONTINUING EDUC, 79- *Concurrent Pos:* Commonwealth fel psychosom med, NY Hosp, 51-53. *Mem:* Am Psychosom Soc; AMA; Am Pub Health Asn. *Res:* Renal excretion. *Mailing Add:* 18 E Wildflower Dr Santa Fe NM 87501

SCHOTZ, LARRY, b Milwaukee, Wis, Dec 26, 49; m. ELECTRICAL ENGINEERING. *Educ:* Milwaukee Sch Eng, BS, 73. *Prof Exp:* Proj engr, Sherwood Electronics, 73-75; pres, Draco Labs, Inc, 75-80; PRES, L S RES, INC, 80- *Honors & Awards:* Eng Design Graphics Award, Am Soc Eng, 72. *Mem:* Inst Elec & Electronics Engrs; Audio Eng Soc. *Res:* Development of FM tuner controlled by a microprocessor; development of stereo TV decoder. *Mailing Add:* L S Res Inc N 52 W 6295 Mill St Cedarburg WI 53012

SCHOTZ, MICHAEL C, b New York, NY, June 30, 28; m 51; c 2. BIOCHEMISTRY. *Educ:* Marietta Col, BS, 47; Univ Southern Calif, MS, 50, PhD(biochem), 53. *Prof Exp:* Harvard Univ res fel cholesterol metab, Huntington Res Labs, Mass Gen Hosp, 53-54; USPHS officer, Res Div, Cleveland Clin Ohio, 55-57, asst staff mem, 57-60; BIOCHEMIST, LIPID RES LAB, VET ADMIN WADSWORTH MED CTR, 60-; PROF MED, UNIV CALIF, LOS ANGELES, 74- *Concurrent Pos:* Asst prof, Ctr Health Sci, Univ Calif, Los Angeles, 60-68, assoc prof, 68-72, adj assoc prof, 72-74; USPHS grant, 60. *Mem:* AAAS; Am Heart Soc; Am Chem Soc; Am Soc Biol Chem. *Res:* Role of lipoprotein lipase in the deposition of lipids in adipose tissue and heart tissue; determination of antibiotics using high-pressure liquid chromatography; the role of endotoxin in gram-negative septicemia. *Mailing Add:* Vet Admin Lipid Res Bldg 113 Rm 312 Wadsworth Med Ctr Wilshire & Sawtelle Blvds Los Angeles CA 90073

SCHOULTIES, CALVIN LEE, b Dayton, Ky, Nov 18, 43. PLANT PATHOLOGY. *Educ:* Univ Ky, BS, 65, PhD(plant path), 71. *Prof Exp:* Staff res assoc plant path, Univ Calif, Berkeley, 71-75; PLANT PATHOLOGIST, FLA DEPT AGR & CONSUMER SERV, 75- *Mem:* Sigma Xi; Am Phytopath Soc. *Res:* Soil microbiology and ecology; biological control of plant pathogens. *Mailing Add:* 212 Barre Hall Clemson Univ Clemson SC 29634-2775

SCHOULTZ, TURE WILLIAM, b Alhambra, Calif, June 6, 40; m 59; c 3. NEUROSCIENCES. *Educ:* Colo State Univ, BS, 65; Univ Colo, Boulder, MA, 67; Univ Colo Med Ctr, Denver, PhD(anat), 71. *Prof Exp:* Instr physiol, Med Sch, NY Univ, 71-72; asst mem neurobiol, Pub Health Res Inst, City of New York, 71-72; asst prof anat, Col Med, Univ Ark, Little Rock, 73-89, asst dean, 77-89; ASSOC DEAN, ACAD STUDENT AFFAIRS, ALA, 89- *Mem:* Soc Neurosci; Am Asn Anatomists. *Res:* The acute phase of mammalian spinal cord injury; determination of the role of catecholamines in trauma-induced progressive spinal cord destruction and determination of changes in dissolved oxygen concentration after injury. *Mailing Add:* 1617 Woods Pointe Circle Mobile AL 36609

SCHOWALTER, WILLIAM RAYMOND, b Milwaukee, Wis, Dec 15, 29; m 53; c 3. CHEMICAL ENGINEERING. *Educ:* Univ Wis, BS, 51; Univ Ill, MS, 53, PhD(chem eng), 57. *Prof Exp:* From asst prof to prof chem eng, princeton Univ, 57-89, actg chmn, Dept Chem Eng, 71, assoc dean, Sch Eng & Appl Sci, 72-77, chmn dept, 78-87; DEAN COL ENG, UNIV ILL URBANA-CHAMPAIGN, 90- *Concurrent Pos:* Mem adv bd for chem eng series, McGraw-Hill Book Co; vis sr fel, Brit Sci Res Coun, Cambridge, 70; mem, US Nat Comt Theoret & Appl Mech, 72-81, Nat Res Coun Comn Eng & Tech Systs, 83-; Sherman Fairchild Distinguished Scholar, Calif Inst Technol, 77-78. *Honors & Awards:* Lectureship Award, Am Soc Eng Educ, 71; William H Walker Award, Am Inst Chem Engrs, 82; Bingham Medal, Soc Rheology, 88. *Mem:* Nat Acad Eng; Am Chem Soc; Am Inst Chem Engrs; Soc Rheol (pres, 81-83). *Res:* Fluid mechanics; non-Newtonian flow; rheology. *Mailing Add:* 106 Eng Hall Univ Ill 1308 W Green St Urbana IL 61801

SCHOWEN, RICHARD LYLE, b Nitro, WVa, Aug 29, 34; m 63; c 2. MECHANISM CHEMISTRY. *Educ:* Univ Calif, Berkeley, BS, 58; Mass Inst Technol, PhD(org chem), 62. *Hon Degrees:* Dr rer nat, Martin Luther Univ, Halle-Wittenberg, Ger. *Prof Exp:* Res assoc, Mass Inst Technol, 62-63; from asst prof to prof, 63-77, SUMMERFIELD PROF CHEM, UNIV KANS, 77-, BIOCHEM, 86- *Concurrent Pos:* NIH res career develop award, 68-73; Syntex lectr, Can, 88. *Honors & Awards:* Dolph Simons Sr Award Biomed Res, 82. *Mem:* Fel AAAS; fel Am Inst Chemists; Am Chem Soc; Am Soc Biochem & Molecular Biol. *Res:* Reaction mechanisms; biodynamics; isotope effects; enzyme mechanisms. *Mailing Add:* RR 2 Box 222 Lawrence KS 66046

SCHOWENGERDT, FRANKLIN DEAN, b Bellflower, Mo, Mar 8, 36; m 62; c 2. ATOMIC PHYSICS. *Educ:* Univ Mo-Rolla, BS, 66, MS, 67, PhD(physics), 69. *Prof Exp:* Res assoc physics, Univ Nebr, 69-71, vis asst prof, 71-73; head physics dept, 76-89, assoc prof, 73-80, PROF PHYSICS, COLO SCH MINES, 80-, VPRES ACAD AFFAIRS, 90- *Mem:* Sigma Xi; Am Phys Soc. *Res:* Electron and ion collisions; ion energy loss spectroscopy; low energy electron spectroscopy; cloud physics applied to control of respirable coal dust. *Mailing Add:* Vpres Acad Affairs Colo Sch Mines Golden CO 80401

SCHOWENGERDT, ROBERT ALAN, b St Charles, Mo, Oct 10, 46; m 74; c 2. DIGITAL IMAGE PROCESSING, REMOTE SENSING. *Educ:* Univ Mo, BS, 68; Univ Ariz, PhD(optical sci), 75. *Prof Exp:* Res assoc, Optical Sci Ctr, 72-77, asst prof remote sensing, Off Arid Lands & Elec Eng, 77-84, ASSOC PROF ELEC & COMPUT ENG, ARID LANDS, OPTICAL SCI, UNIV ARIZ, 84- *Concurrent Pos:* Physical scientist, Earth Resources Observation Syst, US Geol Surv, Va, 75-80; Am Soc Eng Educ-NASA summer fac fel, Langley Res Ctr, NASA, 83; Fulbright sr scholar, Univ New South Wales, Canberra, Australia, 89. *Honors & Awards:* H J E Reid Award, NASA Langley Res Ctr, 83. *Mem:* Optical Soc Am; Am Soc Photogram; Inst Elec & Electronics Engrs. *Res:* Computer image enhancement; pattern recognition of satellite and aerial remote sensing images; sensor design and performance analysis; automated cartography; computer vision. *Mailing Add:* Dept Elec/Comput Eng Univ Ariz Tucson AZ 85721

SCHRACK, ROALD AMUNDSEN, b Ft Meade, Fla, Aug 26, 26; m 54; c 2. PHYSICS. *Educ:* Univ Calif, Los Angeles, BS, 49, MS, 50; Univ Md, PhD, 61. *Prof Exp:* Physicist, 49-56, NUCLEAR PHYSICIST, NAT BUR STANDARDS, 56- *Honors & Awards:* Silver Medal, Dept Com, 82; Indust Res 100 Award. *Mem:* Am Phys Soc; Inst Elec & Electronics Engrs. *Res:* Measurement of nuclear matter distribution by neutral meson photoproduction; nuclear structure physics; neutron cross sections; isotopic assay and distribution by resonance neutron radiography. *Mailing Add:* Div 846 Nat Inst Standards & Technol Gaithersburg DC 20899

SCHRADER, DAVID HAWLEY, b Syracuse, NY, Dec 9, 25; m 68; c 4. ELECTRICAL ENGINEERING. *Educ:* Univ Kans, BS, 51; Univ Wash, MS, 59, PhD(fading of radio waves), 63. *Prof Exp:* Assoc engr, Hazeltine Electronics Corp, 51-55; instr, Univ Wash, 55-63; PROF ELEC ENG, WASH STATE UNIV, 63- *Mem:* Inst Elec & Electronics Engrs; Sigma Xi. *Res:* Physics of the magnetosphere and the ionosphere-propagation of radio waves. *Mailing Add:* Rte 1 Box 232 Pullman WA 99163

SCHRADER, DAVID MARTIN, b Minneapolis, Minn, Sept 24, 32; m 55; c 3. PHYSICAL CHEMISTRY. *Educ:* Iowa State Univ, BS, 54; Univ Minn, Minneapolis, PhD(theoret chem), 62. *Prof Exp:* Res fel theoret chem, Columbia Univ, 61-62 & IBM Watson Lab, 62-63; asst prof phys chem, Univ Iowa, 63-67; asst prof chem, Univ Minn, Minneapolis, 67-68; PROF CHEM, MARQUETTE UNIV, 68- *Concurrent Pos:* Univ res fel & sr vis fel, Sci Res Coun, math dept, Univ Nottingham, UK. *Mem:* Am Phys Soc; Am Chem Soc. *Res:* Atomic and molecular quantum mechanics; positron annihilation. *Mailing Add:* 204 N 88th St Wauwatosa WI 53226

SCHRADER, DOROTHY VIRGINIA, b Boston, Mass, Jan 2, 21. MATHEMATICS. *Educ:* Mass State Col Bridgewater, BS, 42; Boston Col, MA, 46; Univ Wis, PhD(math, hist sci), 61. *Prof Exp:* Instr math, Col St Teresa, Minn, 46-52; asst prof, Dominican Col, Wis, 52-61; assoc prof, 61-64, chmn dept, 69-76, PROF MATH, SOUTHERN CONN STATE COL, 64- *Mem:* Math Asn Am; Hist Sci Soc; Sigma Xi. *Res:* History of medieval mathematics; mathematics education. *Mailing Add:* 7 Valleybrook Rd Branford CT 06405

SCHRADER, GEORGE FREDERICK, b Mattoon, Ill, July 21, 20; m 43; c 2. INDUSTRIAL ENGINEERING, RESEARCH ADMINISTRATION. *Educ:* Univ Ill, BS, 47, MS, 51, PhD(indust eng), 60. *Prof Exp:* Instr mech eng, Univ Ill, 47-51, asst prof indust eng, 53-61; prof, Okla State Univ, 61-62; prof & head dept, Kans State Univ, 62-66; dir tech serv, Univ Nebr, Lincoln, 66-67, dir indust res, 67-69; chmn dept indust eng & mgt systs, actg chmn dept eng math & comput systs, actg dir, Transp Systs Inst, 69-77, assoc dean, 77-86, assoc vpres res, 86-88, PROF INDUST ENG & MGT SYSTS, UNIV CENT FLA, 69-, EMER ASSOC DEAN, COL ENG, 88- *Concurrent Pos:* Consult, Joliet Ord & Ammunition Ctr, 53, Caterpillar Tractor Co, 54, Champion Paper & Fibre Co, 56-57, Ill Bell Tel Co, 59-60, Bayer & McElrath, Inc, 64-66 & Air Force Oper Anal Group, 63-71. *Mem:* Fel Inst Indust Engrs; Am Soc Eng Educ; Nat Soc Prof Engrs. *Res:* Manufacturing processes; management systems; research and development administration. *Mailing Add:* Univ Cent Fla Col Eng Box 25000 Orlando FL 32816

SCHRADER, KEITH WILLIAM, b Apr 22, 38; US citizen; div; c 2. MATHEMATICS. *Educ:* Univ Nebr, BS, 59, MS, 61, PhD(math), 66. *Prof Exp:* Engr electronic warfare, Sylvania Electronic Defense Lab, Gen Tel & Electronics Corp, 63-64; from asst prof to assoc prof, 66-78, chmn dept, 79-82 & 85-88, PROF MATH, UNIV MO, COLUMBIA, 78- *Concurrent Pos:* NASA study grant, 66-68; NSF study grant, 68-70. *Mem:* Am Math Soc; Math Asn Am; Soc Indust & Appl Math. *Res:* Differential equations; boundary value problems; oscillation; convergence theorems. *Mailing Add:* 202 Math Sci Univ Mo Columbia MO 65211

SCHRADER, LAWRENCE EDWIN, b Lancaster, Kans, Oct 22, 41; m 63, 81; c 3. PLANT BIOCHEMISTRY, PLANT PHYSIOLOGY. *Educ:* Kans State Univ, BS, 63; Univ Ill, Urbana, PhD(agron), 67. *Prof Exp:* Biochemist, US Army Med Res & Nutrit Lab, Fitzsimons Gen Hosp, 67-69; from asst prof to prof, agron, Univ Wis-Madison, 69-84; PROF & HEAD AGRON, UNIV ILL, 84- *Concurrent Pos:* Chief, Competitive Res Grants Off, USDA, Washington, DC, 80-81; vis prof, NC State Univ, 75-76. *Honors & Awards:* Soybean Res Award, Am Soybean Asn, 83. *Mem:* Am Chem Soc; Am Soc Agron; Crop Sci Soc Am; Am Soc Plant Physiol (pres, 87-88); fel AAAS; Sigma Xi. *Res:* Nitrate uptake and assimilation in higher plants; enzyme regulation; photosynthesis; translocation; carbon-nitrogen interactions. *Mailing Add:* Dept Agron Univ Ill 1102 S Goodwin Ave Urbana IL 61801

SCHRADER, LAWRENCE EDWIN, b Atchison, Kans, Oct 22, 41; m 63, 81; c 3. PHYSIOLOGICAL GENETICS. *Educ:* Kans State Univ, BS, 63; Univ Ill, PhD(agron), 67. *Prof Exp:* Res biochemist, US Army Med Res & Nutrit Lab, Denver, Colo, 67-69; from asst prof to prof, Univ Wis-Madison, 69-84; prof & head admin, Univ Ill, Urbana-Champaign, 84-89; DEAN, WASH STATE UNIV, 89- *Concurrent Pos:* Vis prof, NC State Univ, Raleigh, 75-76; mem, bd dirs, Crop Sci Soc Am, 77-79, AAAS, 91-94; chief, USDA Competitive Res Grants Off, Wash, 80-81; consult, Agracetus, Middleton, Wis, 81-89. *Mem:* Am Soc Plant Physiologists (secy, 83-85, pres, 87-88); fel AAAS; Am Chem Soc; fel Am Soc Agron; fel Crop Sci Soc Am. *Res:* Nitrogen and carbon metabolism of higher plants; physiological genetics; translocation and source-sink relations; effect of environmental stress on plant growth and metabolism. *Mailing Add:* Col Agr & Home Econ 421 Hulbert Hall Wash State Univ Pullman WA 99164-6242

SCHRADER, R(OBERT) J, b South Bend, Ind, Oct 27, 18; m 42; c 3. CHEMICAL ENGINEERING. *Educ:* Purdue Univ, BS & MS, 40; Mass Inst Technol, ScD(chem eng), 43. *Prof Exp:* Instr chem eng, Mass Inst Technol, 42-43; asst supt chem prod plant, Clinton Eng Works, 43-45, head dept chem eng, 45-46; sr engr, Tenn Eastman Corp, 46-52, in chg develop & process improv dept, 52-53, supt polyethylene dept, 53-60, from asst supt to gen supt, Plastics Div, 60-75, WORKS MGR, TEX EASTMAN CO, DIV EASTMAN KODAK CO, 75-, VPRES, 81- *Mem:* Am Chem Soc; Am Inst Chem Engrs; Soc Plastics Engrs. *Res:* Development and pilot plant work on new chemical processes; effect of pressure on enthalpy of hydrocarbons and their mixtures; synthetic resins and plastics; petroleum; natural gas and textile products. *Mailing Add:* 21 Palisades St Longview TX 75601

SCHRADER, WILLIAM THURBER, b Mineloa, NY, Oct 12, 43; m 67. ENDOCRINOLOGY, MOLECULAR BIOLOGY. *Educ:* Johns Hopkins Univ, BA, 64, PhD(biol), 69. *Prof Exp:* Res assoc obstet & gynec, Med Sch, Vanderbilt Univ, 71, asst prof, 71-72; from asst prof to assoc prof, 72-84, PROF CELL BIOL, BAYLOR COL MED, 84- *Mem:* Endocrine Soc. *Res:* Molecular mechanisms of hormone action; gene regulation in eukaryotic cells. *Mailing Add:* Dept Cell Biol Baylor Col Med One Baylor Plaza Houston TX 77030

SCHRADIE, JOSEPH, b Los Angeles, Calif, July 19, 33; m 60; c 3. PHARMACOGNOSY. *Educ:* Univ Southern Calif, PharmD, 57, MS, 61, PhD(pharmaceut chem, pharmacog), 66. *Prof Exp:* Asst prof, 66-70, ASSOC PROF PHARMACOG, COL PHARM, UNIV TOLEDO, 70-, CHMN, DEPT MED CHEM & PHARMACOG, 81- *Mem:* Am Pharmaceut Asn; Acad Pharmaceut Sci; NY Acad Sci; Am Soc Pharmacog. *Res:* Isolation of natural products from marine organisms, also their cultivation; biosynthesis and intermediary metabolism of carbohydrates in lower and higher plants; isolation of antibiotics; biological and phytochemical screens of higher plants. *Mailing Add:* Dept Pharmacog Univ Toledo Toledo OH 43606

SCHRADY, DAVID ALAN, b Akron, Ohio, Nov 11, 39; m 62; c 3. OPERATIONS RESEARCH. *Educ:* Case Inst Technol, BS, 61, MS, 63, PhD(opers res), 65. *Prof Exp:* Assoc dir, Off Naval Res, 70-71; asst prof opers res, 65-70, assoc prof, 71-74, chmn, Dept Opers Res & Admin Sci, 74-76, dean info & policy sci, 76-80, actg provost, 80-82, provost, 82-87, ACAD DEAN, NAVAL POSTGRAD SCH, 80-, PROF OPERS RES, 88- *Concurrent Pos:* Consult, Decision Studies Group, 67-69, Litton-Mellonics, 67-70 & Naval Supply Systs Command, 70-73, Singapore Ministry Defense, 83; hon treas, Int Fedn Opers Res Soc, 88-92. *Honors & Awards:* Wanner Award Mil Opers Res Soc, 84. *Mem:* Opers Res Soc Am (treas, 76-79, vpres, 82-83, pres, 83-84); Mil Opers Res Soc (pres, 78-79); Inst Mgt Sci. *Res:* Inventory control; command and control; logistics. *Mailing Add:* Naval Postgrad Sch Monterey CA 93943-5000

SCHRAER, HARALD, b Boston, Mass, June 10, 20; m 52; c 1. CELL BIOLOGY. *Educ:* Syracuse Univ, AB, 48, MA, 49; Cornell Univ, PhD(biol), 54. *Prof Exp:* Res assoc radiol, Albert Einstein Med Ctr, Pa, 52-56; res assoc physics, 56-58, sr res assoc biophys, 58-61, assoc prof, 61-67, PROF BIOPHYS, PA STATE UNIV, UNIVERSITY PARK, 67- *Concurrent Pos:* Vis scientist, NIH, 61; res assoc, Dept Anat, Med Sch, Harvard Univ, 67-68. *Mem:* AAAS; Am Physiol Soc; Soc Exp Biol & Med; Am Soc Cell Biol; Sigma Xi; Am Soc Bone & Mineral Res. *Res:* Structural-functional aspects of metal transport; electron microscopy; skeletal physiology; mineral physiology. *Mailing Add:* 9303 Box Springs Mtn Rd Moreno Valley CA 92387

SCHRAER, ROSEMARY, b Ilion, NY, Aug 1, 24; m 52; c 1. BIOCHEMISTRY, CELL BIOLOGY. *Educ:* Syracuse Univ, AB, 46, MS, 49, PhD(biochem), 53. *Prof Exp:* Res assoc biochem, Albert Einstein Med Ctr, 53-56; res asst biophys, Pa State Univ, 56-59, lectr, 59-60, from asst prof to assoc prof, 61-75, assoc dean res, Col Sci, 72-78, univ asst provost, 78-81, prof biochem, 75-85, univ assoc provost, 81-85; exec vchancellor, 86-87, CHANCELLOR, UNIV CALIF, RIVERSIDE, 87- *Concurrent Pos:* Vis fel, Lucy Cavendish Col, Cambridge Univ, Eng, 84-85, permanent fel, 87- *Mem:* Fel AAAS; Am Chem Soc; Am Soc Cell Biol; Am Physiol Soc; Am Soc Biol Chemists; Sigma Xi; Am Inst Chemists. *Res:* Cellular ultrastructure; calcium transport in cells and tissues; mechanisms of steroid hormone action. *Mailing Add:* Off Chancellor Univ Calif Riverside Riverside CA 92521

SCHRAG, JOHN L, b Siloam Springs, Ark, Apr 14, 37; m 60; c 1. POLYMER CHEMISTRY. *Educ:* Univ Omaha, BA, 59; Okla State Univ, MS, 61, PhD(physics), 67. *Prof Exp:* Proj assoc chem, 67-70, asst prof chem & eng, 70-71, asst prof chem, 71-75, ASSOC PROF CHEM, UNIV WIS-MADISON, 75- *Concurrent Pos:* Alfred P Sloan Found res fel, 73-74. *Mem:* Am Chem Soc; Am Phys Soc; Sigma Xi; AAAS; NY Acad Sci. *Res:* Optical and mechanical properties of macromolecules. *Mailing Add:* Dept Chem Univ Wis 1101 University Ave Madison WI 53706

SCHRAG, ROBERT L(EROY), b Moundridge, Kans, Nov 10, 24; m 66; c 2. ELECTROMAGNETIC FIELDS. *Educ:* Kans State Univ, BSEE, 45; Calif Inst Technol, MSEE, 46; Pa State Univ, PhD(ionosphere res), 54. *Prof Exp:* Res analyst, Douglas Aircraft Co, 46-48; instr elec eng, Pa State Univ, 48-53; mem tech staff, Bell Tel Labs, 54-57; PROF ELEC ENG, WICHITA STATE UNIV, 57- *Concurrent Pos:* NSF res grants, 62-64. *Mem:* Inst Elec & Electronics Engrs. *Res:* Electro-impulse deicing. *Mailing Add:* Dept Elec Eng Wallace Hall Box 44 Wichita State Univ 1845 Fairmount Ave Wichita KS 67208

SCHRAGE, F EUGENE, b Oak Park, Ill, July 13, 34; m 56; c 3. PLASTIC FILMS DEVELOPMENT, AUTOMOTIVE ANTIFREEZE DEVELOPMENT. *Educ:* Univ Ill, BA, 56; Northwestern Univ, MS, 57. *Prof Exp:* Group leader res & develop, Films Dept, Union Carbide, 63-69, tech mgr, Films & Packaging Div, 69-73, opers mgr, Home & Auto Div, 73-77, dir res & develop, 77-86; DIR, RES & DEVELOP, FIRST BRANDS CORP, 86- *Mem:* Soc Plastics Indust. *Res:* Development of plastic wrap and bags, antifreeze, automobile wax, and oil and gas additives. *Mailing Add:* 80 Scenic Hill Lane Monroe CT 06468

SCHRAGE, ROBERT W, b Brooklyn, NY, July 1, 25; m 65. PETROLEUM & CHEMICAL INDUSTRY ECONOMIC RESEARCH. *Educ:* Columbia Univ, BA, 46, BS & MS, 48, PhD(chem eng), 50. *Prof Exp:* Engr, East Coast Tech Serv Div, Esso Standard Oil Co, 50-58, sect head tech div, Mfg Dept, 58-61, sect head, Econ Coord Dept, Esso Standard Eastern Inc, 62-66, mgr econ & planning, Esso Eastern Chem Inc, 66-71, mgr corp affairs, Essochem Eastern Ltd, Hong Kong, 71-73, econ res adv, Exxon Chem Am, 73-86; RETIRED. *Mem:* Am Chem Soc. *Res:* Petroleum and chemical industry economics. *Mailing Add:* 1752 S Gessner Houston TX 77063

SCHRAGE, SAMUEL, b Vienna, Austria, Feb 1, 20; nat US; m 53; c 2. PHYSICAL CHEMISTRY, HISTORY OF SCIENCE. *Educ:* Dalhousie Univ, BSc, 44, MSc, 46; McGill Univ, PhD(phys chem), 51. *Prof Exp:* Instr phys chem, Dalhousie Univ, 45-46, lectr, 46-48; demonstr, McGill Univ, 48-51; res assoc, Univ Notre Dame, 51-52; from instr to asst prof, 52-59, dir, Univ Hons Progs, 70-82, ASSOC PROF PHYS CHEM, UNIV ILL, CHICAGO, 59- *Res:* Radiation chemistry; science education. *Mailing Add:* Dept Chem Box 4348 Univ Ill Chicago IL 60680

SCHRAM, ALFRED C, b Brussels, Belg, Sept 17, 30; US citizen; m 57; c 2. BIOCHEMISTRY. *Educ:* Polytech Inst Brooklyn, BS, 54; Univ Tex, Austin, MA, 56, PhD(chem), 58. *Prof Exp:* Res biochemist, St Barnabas Hosp, Minneapolis, Minn, 58-59; clin instr biochem, Southwestern Med Sch, Univ Tex, 59-62, clin asst prof, 62-65; assoc prof chem, 65-70, PROF CHEM, WTEX STATE UNIV, 70- *Concurrent Pos:* Res biochemist, Vet Admin Hosp, Dallas, 59-65; NIH grant cancer res, 63-65; abstractor, Chem Abstr, 63-; Welch Found res grant, 67-69. *Mem:* AAAS; Am Chem Soc; affil AMA. *Res:* Immunochemistry of synthetic antigens. *Mailing Add:* Dept Chem WTex State Univ Canyon TX 79016

SCHRAM, EUGENE P, b Milwaukee, Wis, Apr 19, 34; m 56; c 1. INORGANIC CHEMISTRY. *Educ:* Carroll Col, BS, 56; Purdue Univ, PhD(inorg chem), 62. *Prof Exp:* Res technician, Allis Chalmers Mfg Co, 54-55, res chemist, 56-58; res chemist, Carbon Prod Div, Union Carbide Corp, 62-64; asst prof, 64-69, ASSOC PROF CHEM, OHIO STATE UNIV, 69- *Mem:* AAAS; Am Chem Soc. *Res:* Chemistry of aluminum heterocycles and molecular compounds containing metal-metal bonds. *Mailing Add:* 4770 Teter Ct Columbus OH 43220

SCHRAM, FREDERICK R, b Chicago, Ill, Aug 11, 43; div; c 1. INVERTEBRATE ZOOLOGY, CARCINOLOGY. *Educ:* Loyola Univ, Ill, BS, 65; Univ Chicago, PhD(paleozool), 68. *Prof Exp:* Asst prof, 68-73, assoc prof zool, Eastern Ill Univ, 73-78; chief cur Earth & Marine Sci, Natural Hist Mus, San Diego, 78-87, cur paleont, 78-91. *Concurrent Pos:* Res assoc, Scripps Inst Oceanog; adj prof, San Diego State Univ; gen ed, Crustacean Issues, 83- *Mem:* Crustacean Soc; Paleont Soc; fel Linnean Soc London; Am Soc Zool; Crustacean Soc China. *Res:* Late Paleozoic history of the Malacostraca; comparative anatomy of crustaceans; arthropod relationships and evolution; morphology and systematics of remipede crustaceans; invertebrate morphology and evolution. *Mailing Add:* A-002 Marine Biol Scripps Inst Oceanog San Diego CA 92093

SCHRAMEL, ROBERT JOSEPH, b St Louis, Mo, Sept 6, 24; m 47; c 5. SURGERY. *Educ:* Tulane Univ, BS, 45, MD, 48. *Prof Exp:* From instr to assoc prof, 55-70, prof surg, 70-76, CLIN PROF SURG, SCH MED, TULANE UNIV, 76- *Concurrent Pos:* Prin investr, NIH, 57- *Mem:* AAAS; Am Asn Thoracic Surg; NY Acad Sci; Asn Hosp Med Educ; Asn Am Med Cols. *Res:* Pulmonary function in disease states; trauma; extracorporeal circulation; shock; hyperbaric oxygenation; surgical treatment of cardiovascular disease. *Mailing Add:* 2025 Gravier St No 613 New Orleans LA 70112

SCHRAMM, DAVID N, b St Louis, Mo, Oct 25, 45; m 86; c 2. THEORETICAL ASTROPHYSICS, NUCLEAR PHYSICS. *Educ:* Mass Inst Technol, BS, 67; Calif Inst Technol, PhD(physics), 71. *Prof Exp:* Res fel physics, Calif Inst Technol, 71-72; asst prof astron & physics, Univ Tex-Austin, 72-74; assoc prof astrophys, Univ Chicago, 74-76, actg chmn dept, 77, chmn Dept Astron & Astrophys, 78-84, PROF ASTRON & ASTROPHYS, ENRICO FERMI INST & COL, UNIV CHICAGO, 77-, PROF PHYS, UNIV CHICAGO, 77-, LOUIS BLOCK PROF PHYS SCI, 82-, PROF COMT CONCEPTUAL FOUND SCI, 84- *Concurrent Pos:* Consult, NSF Astrophys Curri, 74, Aerospace Corp, 74-, Lawrence Livermore Lab, 75-, Fermilab, 82-; lectr, Adler Planetarium, 76-; assoc ed, Am J Physics, 78-81; mem, Mayor Jane Byrne's Adv Comt High Technol, 81-83; mem sci adv comt, Demystifying Sci, Mus Sci & Indust, Chicago, Ill, 81-84; astrophys ed, Physics Reports, 81- & Univ Chicago Press, 81-; adj prof physics, Univ Utah, Salt Lake City, 81-; nat lectr, Sigma Xi, 83-85; distinguished vis scientist, Carnegie-Merlon Univ, 85; Alexander von Humboldt Award, 86-87; co-chair, Spec Comt Cosmology, NSF, 87-88; co-chmn, Theory & Lab Astrophys Panel, Astron & Astrophys Surv Comt, Nat Res Coun, 89-; sci assoc, Europ Orgn Nuclear Res, Geneva, Switz, 90; mem, Brit NAm comt, Am Phys Soc, 90- *Honors & Awards:* Robert J Trumpler Award, Astron Soc Pac, 74; Am Astron Soc Helen B Warner Prize, 78; Gunnar Kullen Mem Lectr, Lund, Sweden, 81; Richtmeyer Mem Award Lectr, Am Asn Physics Teachers, 84. *Mem:* Nat Acad Sci; Am Astron Soc; Int Astron Union; Sigma Xi; Meteoritical Soc; fel Am Phys Soc. *Res:* Theoretical studies of astrophysics with particular emphasis on: cosmology, the origin of the elements, cosmic rays, stellar evolution and supernova, neutrino astrophysics; nucleochronology, the early solar system and black holes; particle physics; author of over 300 published articles and numerous books. *Mailing Add:* Dept Astron Astrophys AAC-140 Univ Chicago 5640 S Ellis Chicago IL 60637

SCHRAMM, JOHN GILBERT, b Cincinnati, Ohio, Sept 15, 51; m 80; c 2. ELECTROCHEMICAL ENGINEERING, BIOCHEMICAL ENGINEERING. *Educ:* Univ Mich, Ann Arbor, BGS, 77. *Prof Exp:* Chemist, energy develop assoc, Gulf & Western, 78-82; sr instr assoc chem eng, dept chem eng, Col Eng, Univ Mich, 82-87; vpres oper, res & develop, Diamond Gen Develop Corp, 87-88; PRES BIO & ELECTRO CHEM ENG, NEXIAL INC, 88- *Mem:* AAAS; Am Inst Chem Engrs; Am Chem Soc. *Res:* Bio-sensors for use in biological or medical environments; electrochemical detectors for oxygen, hydrogen, pH, CO2 or bio-organic molecules of interest. *Mailing Add:* 1505 Pear St Ann Arbor MI 48105-1731

SCHRAMM, LEE CLYDE, b Portsmouth, Ohio, July 20, 34; m 64; c 3. PHARMACOGNOSY. *Educ:* Ohio State Univ, BSc, 57; Univ Conn, MS, 59, PhD(pharmacog), 62. *Prof Exp:* Asst prof, Col Pharm, Univ Minn, Minneapolis, 61-67; assoc prof pharmacog & head dept, Sch Pharm, Univ Ga, 67-81; SR REGIONAL MED ASSOC, SMITH KLINE & FRENCH LABS, 81- *Concurrent Pos:* Chmn pharmacog & natural prod sect, Acad Pharmaceut Sci, 79-80; chmn sect teachers biol sci, Am Asn Cols Pharm, 75-76. *Honors & Awards:* Kilmer Prize, Am Pharmaceut Asn, 57. *Mem:* Am Pharmaceut Asn; Soc Econ Bot; Ger Soc Drug Plant Res; Am Soc Pharmacog (treas, 75-81). *Res:* Phytochemistry, particularly plants and fungi with potential medicinal or toxic activity. *Mailing Add:* 266 Deerhill Dr Bogart GA 30622

SCHRAMM, MARTIN WILLIAM, JR, b Pittsburgh, Pa, Apr 21, 27; m 53; c 4. PETROLEUM GEOLOGY. *Educ:* Univ Pittsburgh, BS, 54, MS, 55; Univ Okla, PhD(geol), 63. *Prof Exp:* Explor geologist, Gulf Oil Corp, 55-57; consult petrol, A W McCoy Assocs, 57-59; proj supvr explor res, Cities Serv Oil Co, Okla, 59-69; mgr, Foreign Div, White Shield Explor Corp, 69-70 & explor & exploitation, White Shield Oil & Gas Corp, 70- 72, consult geologist, 72-74; exec vpres, Geoquest Int, Inc, 74-82; pres & chief exec officer, Seagull Int Explor, Inc, 82-85; PRES, SCHRAMM & ASSOC CONSULTS, 85- *Mem:* Am Asn Petrol Geologists; Am Inst Prof Geologists; fel Geol Soc Am. *Res:* Middle Ordovician stratigraphy and paleogeology; basin analysis; environments of deposition; seismic stratigraphy. *Mailing Add:* 1922 Anvil Dr Houston TX 77090

SCHRAMM, MARY ARTHUR, b Yankton Co, SDak, Mar 20, 32. NURSE ANESTHESIA. *Educ:* Mt Marty Col, BA, 65; Univ SDak, Vermillion, PhD(physiol), 77. *Prof Exp:* Instr anesthesia, Sch Nurse Anesthesia, Sacred Heart Hosp, 56-69, prog dir & dept head, 65-69; instr physiol, Univ SDak, Vermillion, 69-72; asst prof, 71-76, div head, health sci, 72-73 & 84-86, assoc prof & prog dir, 77-81, PROF ANESTHESIA, MT MARTY COL, YANKTON, SDAK, 81-, PROG DIR, 88- *Concurrent Pos:* Nursing serv dir & anesthetist, St Michael's Hosp, Tyndall, SDak, 56-; educ consult, St Scholastica Col, Duluth, 75, St Cloud Hosp, Minn, 76 & Jamaica Sch Nurse Anesthesia, Proj Hope, 81-84; biomed sci lectr, Am Asn Nurse Anesthetists, 78-81. *Honors & Awards:* James Award, 88. *Mem:* Am Asn Nurse Anesthetists; Am Asn Respiratory Care; AAAS; Am Nurses Asn. *Res:* Taurine in dog heart slices; respiratory and renal liver physiology; anesthesia research, practice and education. *Mailing Add:* Mt Marty Col Yankton SD 57078

SCHRAMM, RAYMOND EUGENE, b St Charles, Mo, Aug 11, 41. MATERIALS SCIENCE, PHYSICS. *Educ:* Regis Col, Colo, BS, 64; Mich Technol Univ, MS, 65. *Prof Exp:* Jr physicist nuclear physics, Ames Lab, Iowa State Univ, 65-66; PHYSICIST MAT, MAT RELIABILITY DIV, NAT INST STANDARDS & TECHNOL, 67- *Honors & Awards:* Bronze Medal, Dept Com, 90. *Mem:* Am Asn Physics Teachers; AAAS. *Res:* Microstructural properties of deformed metals; cryogenic mechanical properties of composites; ultrasonic non-destructive evaluation of metals using long-wavelength electromagnetic acoustic transducers. *Mailing Add:* Mat Reliability Div Nat Inst Standards & Technol Boulder CO 80303-3328

SCHRAMM, ROBERT FREDERICK, b Philadelphia, Pa, Feb 23, 42; m 68; c 2. CHEMISTRY. *Educ:* St Joseph's Col, Pa, BS, 64; Univ Pa, PhD(chem), 69. *Prof Exp:* Advan Res Projs Agency fel, Univ Pa, 69-70; asst prof to assoc prof, 70-75, PROF CHEM, EAST STROUDSBURG STATE COL, 75- *Concurrent Pos:* Lectr, St Joseph's Col, Pa, 70. *Mem:* AAAS; Am Chem Soc; Sigma Xi. *Res:* Synthesis and characterization of coordination complexes of palladium, platinum and gold; molecular orbital calculations; application of computers to scientific instruction. *Mailing Add:* RD 2 Box 570 East Stroudsburg PA 18301

SCHRAMM, ROBERT WILLIAM, b Wheeling, WVa, Nov 13, 34; m 67; c 2. PHYSICS. *Educ:* Liberty State Col, BS, 58; Univ WVa, MS, 59. *Prof Exp:* From instr to asst prof, 58-65, head dept, 66-71, ASSOC PROF PHYSICS, WEST LIBERTY STATE COL, 65- *Concurrent Pos:* Dir, Regional Sci Fair, 58-88; archivist, West Liberty State Col, 77- *Mem:* Am Asn Physics Teachers; Am Phys Soc. *Res:* Nuclear spectroscopy. *Mailing Add:* Dept Physics West Liberty State Col PO Box 296 West Liberty WV 26074

SCHRAMM, VERN LEE, b Howard, SDak, Nov 9, 41; m 64; c 2. BIOCHEMISTRY. *Educ:* SDak State Col, BS, 63; Harvard Univ, SM, 65; Australian Nat Univ, PhD(biochem), 69. *Prof Exp:* Nat Res Coun-NSF res assoc, NASA Ames Res Ctr, 69-71; from asst prof to assoc prof, 76-81, PROF BIOCHEM, SCH MED, TEMPLE UNIV, 81- *Concurrent Pos:* Mem biochem study sect, NIH, 81-85. *Mem:* AAAS; Am Chem Soc; Am Soc Biol Chem. *Res:* Regulation and mechanism of enzymes of nucleotide degradation and gluconeogenesis; nucleotide biosynthesis and degradation; metabolic controls. *Mailing Add:* Dept Biochem A Einsten Col Med 1300 Morris Park Ave Bronx NY 10461

SCHRANK, AULINE RAYMOND, b Hamilton, Tex, Aug 15, 15; m 42; c 2. PHYSIOLOGY, BIOPHYSICS. *Educ:* Tarleton Agr Col, AS, 37; Southwest Tex State Col, AB, 37; Univ Tex, PhD(physiol, biophys), 42. *Prof Exp:* From instr to assoc prof physiol, Univ Tex, Austin, 39-58, chmn, Dept Zool, 63-70, actg dean, 72-74, assoc dean, 74-78, dean, 78-80, prof, 58-, EMER PROF PHYSIOL, UNIV TEX, AUSTIN, 89- *Mem:* AAAS; Am Soc Plant Physiol; Soc Gen Physiol; Biophys Soc; Scand Soc Plant Physiol. *Res:* Bioelectrical fields in relation to growth phenomena and cell correlation; tropisms, regeneration and active transport. *Mailing Add:* 2502 Tower Dr Austin TX 78703

SCHRANK, GLEN EDWARD, b Omaha, Nebr, Aug 6, 26; m 47. PHYSICS. *Educ:* Univ Calif, Los Angeles, BA, 47, MA, 50, PhD, 53. *Prof Exp:* Res assoc physics, Univ Calif, Los Angeles, 53; from instr to asst prof, Princeton Univ, 53-61; res physicist, Lawrence Radiation Lab, Univ Calif, 61-63; ASSOC PROF PHYSICS, UNIV CALIF, SANTA BARBARA, 63- *Concurrent Pos:* Guest physicist, Brookhaven Nat Lab, 59-; consult, AEC; consult, Giannini Sci Corp, 59. *Mem:* Am Phys Soc; Ital Phys Soc. *Res:* Nuclear and elementary particle physics. *Mailing Add:* Dept Physics Univ Calif Santa Barbara CA 93106

SCHRANK, GORDON DABNEY, b San Angelo, Tex, Aug 11, 48; m 75; c 1. MICROBIOLOGY. *Educ:* Angelo State Univ, BS, 70; Univ Tex Med Br Galveston, PhD(microbiol), 74. *Prof Exp:* Med technologist bact & serol, Univ Tex Med Br Galveston, 74-75; instr microbiol, Ctr Health Sci, Univ Tenn, Memphis, 75-77, asst prof, 77-81; from asst prof to assoc prof, 81-87, PROF BIOL, ST CLOUD STATE UNIV, 87- *Concurrent Pos:* USDA Res Agreement Grant, Immunity to Brucellosis at Mucosal Surfaces, 78-81; USAF res fel, Isolation Tech for Legionellae, 84-85; Air Force Office Sponsored Res, Res Contract, 86-87; NSF Instrumentation Grants, 89-91, 90-92. *Mem:* Am Soc Microbiol; AAAS; Sigma Xi; Electron Micros Soc Am. *Res:* Area of host-parasite relationships in infectious diseases and how immunity may alter these relationships; plasmid fingerprinting in medically significant bacteria. *Mailing Add:* Dept Biol Sci MS 228 St Cloud State Univ 720 Fourth Ave S St Cloud MN 56301

SCHRANKEL, KENNETH REINHOLD, b Rice Lake, Wis, Mar 26, 45; m 70; c 2. TOXICOLOGY, ENVIRONMENTAL HEALTH. *Educ:* Wartburg Col, BS, 67; Ill State Univ, MS, 73, PhD(biol), 78. *Prof Exp:* Vis asst prof biol, Tex A&M Univ, 78-79; fel trainee, Univ Wis, 79-81; from res toxicologist to sr res toxicologist, 81-85, DIR, US FLAVOR & FRAGRANCE SAFETY ASSURANCE, INT FLAVORS & FRAGRANCES, 85- *Mem:* AAAS;

Sigma Xi; Soc Toxicol. *Res:* Mammalian toxicology; biochemical aspects of insect oogenesis; ultrastructural aspects of insect reproduction; dermatoxicology and phototoxicology of fragrance materials; safety and regulatory assessment of flavor and fragrance ingredients. *Mailing Add:* Corp Safety Assurance Int Flavors & Fragrances 1515 Highway 36 Union Branch NJ 07735

SCHRAUT, KENNETH CHARLES, b Hillsboro, Ill, May 19, 13; m 52; c 1. MATHEMATICS. *Educ:* Univ Ill, AB, 36; Univ Cincinnati, AM, 38, PhD(math), 40. *Prof Exp:* From instr to prof, 40-72, head dept math, 54-72, DISTINGUISHED SERV PROF MATH, UNIV DAYTON, 72- *Concurrent Pos:* Lectr, Wright Field Grad Ctr, Ohio State Univ, 48-52; dir, USAF Proj, Univ Dayton, 52-54; actg prof, Grad Sch, Univ Cincinnati, 58-60; pres, Honor Sem Metrop Dayton, Inc, 65-67, vchmn & bd dir, 85- *Mem:* Am Math Soc; Am Soc Eng Educ; Math Asn Am; Sigma Xi; Nat Asn Adv Health Prof. *Res:* Infinite series; mathematical analysis. *Mailing Add:* 412 Manchester Dr Dayton OH 45429

SCHRAUZER, GERHARD N, b Franzensbad, Czech, Mar 26, 32; m 57; c 4. INORGANIC CHEMISTRY. *Educ:* Univ Munich, BS, 53, MS, 54, PhD(chem), 56. *Prof Exp:* Res asst chem, Univ Munich, 55-57; res assoc, Monsanto Chem Co, 57-59; res asst chem, Univ Munich, 59-63, lectr inorg chem, 63-64; res supvr, Shell Develop Co, 64-66; PROF CHEM, UNIV CALIF, SAN DIEGO, 66- *Concurrent Pos:* Ed & Founder, Bioinorg Chem J, 70-; pres & founder, Int Asn Bioinorg Scientists, Inc. *Mem:* Am Chem Soc; NY Acad Sci; Soc Ger Chem. *Res:* Organometallic coordination and enzyme chemistry; homogeneous catalysis; trace element, vitamin and cancer research. *Mailing Add:* Dept Chem Univ Calif San Diego Box 109 La Jolla CA 92093

SCHRAY, KEITH JAMES, b Portland, Ore, Nov 25, 43; m 63; c 7. BIO-ORGANIC CHEMISTRY, IMMUNOASSAYS. *Educ:* Univ Portland, BS, 65; Pa State Univ, PhD(phys org chem), 70. *Prof Exp:* NIH fel, Inst Cancer Res, Philadelphia, 70-72; assoc prof, 72-80, PROF CHEM, LEHIGH UNIV, 80- *Mem:* Am Chem Soc. *Res:* Model and enzyme reaction mechanisms; anomerases; enzyme immunoassays; protein-surface binding. *Mailing Add:* Dept Chem Lehigh Univ Bethlehem PA 18015

SCHRAYER, GROVER J, JR, organic geochemistry, for more information see previous edition

SCHRECK, CARL BERNHARD, b San Francisco, Calif, May 18, 44; m 66; c 2. FISH BIOLOGY. *Educ:* Univ Calif, Berkeley, AB, 66; Colo State Univ, MS, 69, PhD(fish physiol), 72. *Prof Exp:* Asst prof, Va Polytech Inst & State Univ, 72-75; asst unit leader, 75-78, asst prof fisheries, 75-78, LEADER ORE COOP & ASSOC PROF FISHERIES, FISHERIES RES UNIT, 78- *Mem:* AAAS; Am Fisheries Soc; Am Inst Fishery Res Biologists; Am Soc Zoologists; Sigma Xi. *Res:* Biology of freshwater and anadromous fish with special emphasis on physiology, endocrinology, genetics and organism-environment interactions. *Mailing Add:* Ore Coop Fishery Res Unit Oregon State Univ Corvallis OR 97331

SCHRECK, JAMES O(TTO), b Houston, Tex, Nov 6, 37; m 66; c 2. ORGANIC CHEMISTRY. *Educ:* St Thomas Univ, Tex, BA, 59; Tex A&M Univ MS, 62, PhD(chem), 64. *Prof Exp:* Res assoc chem, Ga Inst Technol, 64-65; vis asst prof & fel, La State Univ, 65-66; from asst prof to assoc prof, 66-74, chmn dept, 75-81, PROF CHEM, UNIV NORTHERN COLO, 74- *Mem:* Am Chem Soc; Sigma Xi. *Res:* development of reduced scale procedures for the organic chemistry laboratory; chemical education. *Mailing Add:* Dept Chem Univ Northern Colo Greeley CO 80639

SCHRECKENBERG, MARY GERVASIA, b Paderborn, Germany, Jan 4, 16; US citizen. NEUROBIOLOGY, DEVELOPMENTAL BIOLOGY. *Educ:* Cath Univ Am, BS, 52, MS, 54, PhD(zool), 57. *Prof Exp:* Pres & dean, Tombrock Col, Paterson, NJ, 56-62; prof mod biol, Cheng Kung Nat Univ, Tainan, Taiwan, 63-70; prof neurobiol, Fairleigh Dickinson Univ, 74-86, DEPT BIOL, GEORGIAN COURT COL, NJ, 86- *Concurrent Pos:* Trustee, Tombrock Col, 72-74; fel, Columbia Univ, 70; NSF res grants, Univ Tex, Austin, 71; group leader, Eng Profs, Hua Zhang Univ Sci & Technol, Wuhan, China. *Mem:* Soc Neurosci; Soc Develop Biol; Soc Am Zoologists; Sigma Xi; AAAS. *Res:* Problems dealing with developmental neurobiology; brain opiates; neural plasticity. *Mailing Add:* Dept Biol Georgian Court Col Lakewood NJ 08701

SCHREIBEIS, WILLIAM J, b Pittsburgh, Pa, Oct 28, 29; m 80; c 4. INDUSTRIAL HYGIENE. *Educ:* Univ Pittsburgh, BS, 51, MPH, 56. *Prof Exp:* Sanit engr, USPHS, 51-54; pub health engr, City of Pittsburgh Dept Pub Health, 54-56; indust hyg engr, Indust Health Found, Carnegie-Mellon Univ, 56-65; indust hyg engr, Bell Labs, 65-84, mgr, indust hyg & safety, 85-88, prod safety, 88-89; CONSULT, 89- *Concurrent Pos:* Eng comt, Indust Health Found. *Mem:* Am Indust Hyg Asn; Air & Waste Mgt Asn; Brit Occup Hyg Soc; Semiconductor Safety Asn; Nat Fire Protection Asn; Air Pollution Control Asn. *Res:* Evaluation of environmental health and safety hazards. *Mailing Add:* 75 Dogwood Lane Berkeley Heights NJ 07922-2325

SCHREIBER, ALAN D, b Newark, NJ, Feb 26, 42. IMMUNOHEMATOLOGY, ONCOLOGY. *Educ:* Albert Einstein Col Med, MD, 67. *Prof Exp:* PROF MED, CANCER CTR, UNIV PA HOSP, 82- *Mem:* Am Soc Hemat; Am Asn Physicians; Am Asn Immunologists; Am Soc Clin Invest. *Mailing Add:* Med/7 Silver/G12 Univ Pa 3400 Spruce St Philadelphia PA 19104

SCHREIBER, B CHARLOTTE, b Brooklyn, NY, June 27, 31; m 50; c 2. SEDIMENTARY PETROLOGY, STRATIGRAPHY. *Educ:* Wash Univ, AB, 53; Rutgers Univ, New Brunswick, MS, 66; Rensselaer Polytech Inst, PhD(geol), 74. *Prof Exp:* Sedimentologist & oceanogr, Alpine Geophys Assoc, 66-68; instr geol, Lehman Col, City Univ New York, 68-69; instr, Barnard Col, Columbia Univ, 69-71; instr, Rensselaer Polytech Inst, 71-72, res asst, 72-74; from asst to assoc prof, 74-82, PROF GEOL, QUEENS COL NY, 82- *Concurrent Pos:* Teaching fel, Rutgers Univ, 63-66; consult sedimentologist, Johnson Soils Inc, 68-70; teaching fel, Rensselaer Polytech Inst, 71-72; NSF res fel, 72-74; NSF energy related res, Imperial Col, London, 76-77; assoc ed, Soc Econ & Petrol Geologists J, 77-; res assoc, Lamont-Doherty Geol Observ, Columbia Univ, 78-81, sr res assoc, 81- *Honors & Awards:* Levorsen Award, Am Asn Petrol Geologists, 75. *Mem:* Geol Soc Am; Soc Econ & Petrol Geologists; Paleont Res Found; fel Geol Soc London; Am Asn Petrol Geologists. *Res:* Origin and diagenesis of evaporites and associated carbonates; sedimentologic and stratigraphic sequences as developed on continental margins; deformation and tectonics of the Mediterranean; organic matter in evaporative environments. *Mailing Add:* Dept Geol Queens Col City Univ NY 65-30 Kissena Blvd Flushing NY 11367

SCHREIBER, DAVID LAURENCE, b Klamath Falls, Ore, Nov 15, 41; m 63; c 3. HYDRAULIC ENGINEERING, HYDROLOGIC ENGINEERING. *Educ:* Ore State Univ, BS, 63; Wash State Univ, MS, 65, PhD(eng sci), 70. *Prof Exp:* Res asst agr & civil eng, Wash State Univ, 63-67; asst prof civil eng, Univ Wyo, 67-68; instr, Wash State Univ, 68-69; res hydraul engr, Northwest Watershed Res Ctr, Agr Res Serv, USDA, 69-72; sr res engr, Water & Land Resources Dept, Pac Northwest Lab, Battelle Mem Inst, 72-74; hydraul engr, Hydrol-Meteorol Br, Off Nuclear Reactor Regulation, US Nuclear Regulatory Comn, 74-78; consult hydrol engr, 78-80; PRES, SCHREIBER CONSULT, INC, 80- *Concurrent Pos:* Groundwater dispersion honorarium, Argonne Nat Lab, 81. *Honors & Awards:* Robert E Horton Award, Am Geophys Union, 75. *Mem:* Am Geophys Union; Am Soc Agr Engrs; Am Soc Civil Engrs; Nat Soc Prof Engrs; Nat Water Well Asn; Am Water Resources Asn. *Res:* Hydraulics and hydrology of surface water and ground water; water resources planning and environmental impact analyses; radionuclide and pollutant dispersion in surface water and ground water; radioactive and hazardous waste management and disposal. *Mailing Add:* Grant Schreiber & Assoc 1000 W Hubbard Suite 220 Coeur d'Alene ID 83814

SCHREIBER, DAVID SEYFARTH, b Wilmington, Del, Apr 3, 36; m 65; c 2. SOLID STATE PHYSICS. *Educ:* Wabash Col, BA, 57; Cornell Univ, PhD(physics), 62. *Prof Exp:* NSF fel physics, Univ Calif, Berkeley, 62-63; asst prof, Northwestern Univ, 63-68; ASSOC PROF PHYSICS, UNIV ILL, CHICAGO, 68- *Concurrent Pos:* NSF res grant, 65-67. *Mem:* Am Phys Soc. *Res:* Study of magnetic properties, superconductivity and electronic band structure by means of nuclear magnetic resonance; alternative energy systems such as solar, wind and water. *Mailing Add:* Dept Physics MC 273 Univ Ill Box 4348 Chicago IL 60680

SCHREIBER, EDWARD, b Brooklyn, NY, Sept 11, 30; m 50; c 2. CERAMICS, GEOPHYSICS. *Educ:* State Univ NY BS, 56, PhD(ceramic sci), 63. *Prof Exp:* Res scientist, Am Standard Corp, 62-63; res scientist, Lamont Geol Observ, Columbia Univ, 63-65, res assoc, 65-68; assoc prof earth & environ sci, 68-71, PROF GEOL, QUEENS COL, NY, 72-, EXEC OFFICER UNDERGRAD SCHOLASTIC STANDARDS COMT, 87- *Concurrent Pos:* Lectr, Queens Col, NY, 65-66, adj asst prof, 66-67; vis res assoc, Lamont-Doherty Geol Observ, 68-71, vis sr res assoc, 72-; mem, Lunar Sample Rev Panel, 71-74 & lunar sample anal planning team, 74-77; assoc prog mgr geosci, Dept Energy, 81-83 & 85-86. *Mem:* AAAS; Am Ceramic Soc; Mineral Soc Am; Am Geophys Union; Geochem Soc; Sigma Xi; Soc Explor Geophysicists. *Res:* Study of physical properties of solids, equation of state, elastic properties and experimental mineralogy; inorganic materials at high pressures and temperatures; tectonophysics; submarine geology. *Mailing Add:* Dept Geol Queens Col Flushing NY 11367

SCHREIBER, ERIC CHRISTIAN, b Frankfurt, Germany, Aug 16, 21; US citizen; m 74; c 4. PHARMACOLOGY, ORGANIC CHEMISTRY. *Educ:* Polytech Inst Brooklyn, BS, 51, MS, 53; Univ Conn, PhD(pharmacol), 63. *Prof Exp:* Chemist, Nat Dairy Res Lab, 49-51; chemist, Chas Pfizer & Co, NY, 51-58, head radioisotope lab, Conn, 58-63; dept head drug metab, Wm S Merrell Co, 63-66; dir biopharmaceut res, E R Squibb & Sons, Inc, 66-67, dir drug metab, 67-75, assoc dir, Squibb Inst, 75-77, dir Squibb Int Res Ctr, Regensburg, WGer, 75-77; prof pharmacol & dir toxicol & drug metab, 77-83, adj prof med chem, Ctr Health Sci, Univ Tenn, 83-77; RETIRED. *Mem:* Emer fel Am Inst Chem; Am Pharmaceut Asn; Am Soc Pharmacol & Exp Therapeut; fel Acad Pharmaceut Sci. *Res:* Metabolism of drugs; pharmacokinetics; transfer of chemicals from mother to offspring, central nervous system deficits. *Mailing Add:* 830 Harrison Ferry Rd White Pine TN 37890-9550

SCHREIBER, HANS, b Quedlinburg, Ger, Feb 5, 44; m 69; c 3. IMMUNOLOGY OF CANCER, BIOLOGY OF CANCER. *Educ:* Univ Freiburg, MD, 69; Univ Chicago, PhD, 77. *Prof Exp:* Staff mem oncol, Oak Ridge Nat Lab, 70-74; res assoc oncol, 74-77, from asst prof to assoc prof, 77-85, PROF PATH, UNIV CHICAGO, 86- *Concurrent Pos:* Prin investr, 77-; prog dir, 85- *Mem:* Am Asn Cancer Res; Am Asn Path; Am Soc Cytol; Am Asn Immunol. *Res:* Regulation of immune responses to antigens, particularly to transplants and cancer cells; biology and genetics of tumor specific molecules; cancer immunology. *Mailing Add:* Univ Chicago Box 414 5841 S Maryland Ave Chicago IL 60637

SCHREIBER, HENRY DALE, b Lebanon, Pa, Nov 13, 48. PHYSICAL CHEMISTRY, MATERIALS SCIENCE. *Educ:* Lebanon Valley Col, BS, 70; Univ Wis-Madison, PhD(phys chem), 76. *Prof Exp:* Asst prof, 76-80, ASSOC PROF CHEM, VA MIL INST, 80- *Concurrent Pos:* Res asst, NASA Johnson Space Ctr, 73-76; dir, Ctr Glass Chem, 85- *Mem:* Am Chem Soc; Am Ceramic Soc; Am Geophys Union; Geochem Soc. *Res:* Oxidation-reduction processes of multivalent elements and gas solubility transport in silicate melts which simulate basaltic magmas as well as simple glass-forming systems; development of "high-tech" glass. *Mailing Add:* Dept Chem Va Mil Inst Lexington VA 24450

SCHREIBER, HENRY PETER, b Brunn, Czech, Nov 3, 26; nat Can; m 54; c 4. PHYSICAL CHEMISTRY. *Educ:* Univ Manitoba, BSc, 49, MSc, 50; Univ Toronto, PhD(phys chem), 53. *Prof Exp:* Fel phys chem, Nat Res Coun Can, 53-55; res chemist high polymer systs, Can Industs, Ltd, 55-74; PROF CHEM ENG, POLYTECH SCH MONTREAL, 74- *Honors & Awards:* Protective Coatings Award, Chem Inst Can, 77; Medaille Archambault & Alcan Prize, Asn Can-Francaise pour L'Advan Sci, 80. *Mem:* Soc Rheology; Chem Inst Can; Soc Plastics Eng. *Res:* Rheology of high polymer melts; properties of solutions of macromolecules; thermodynamics of liquids; structure of macromolecules. *Mailing Add:* Dept Chem Eng Ecole Polytech Cp 6079 Montreal PQ H3C 3A7 Can

SCHREIBER, JOSEPH FREDERICK, JR, b Baltimore, Md, June 2, 25; m 51; c 2. GEOLOGY. *Educ:* Johns Hopkins Univ, AB, 48, MA, 50; Univ Utah, PhD, 58. *Prof Exp:* Asst geol, Johns Hopkins Univ, 48-49, jr instr, 49-50; geologist, Chesapeake Bay Inst, Md, 51 & Calif Co, La, 54-55; asst prof geol, Okla State Univ, 55-59; assoc prof, 59-66, PROF GEOL, UNIV ARIZ, 66- *Concurrent Pos:* Jr geologist, Atlantic Refining Co, Tex, 49, 50 & Shell Oil Co, Wyo, 52. *Mem:* Soc Econ Paleontologists & Mineralogists; Am Asn Petrol Geol; Geol Soc Am; Int Asn Sedimentologists; Am Quaternary Asn. *Res:* Sedimentology; stratigraphy; marine geology. *Mailing Add:* Dept Geosci Univ Ariz Tucson AZ 85721

SCHREIBER, KURT CLARK, b Vienna, Austria, Feb 23, 22; nat US; m 51; c 3. PHYSICAL ORGANIC CHEMISTRY. *Educ:* City Col New York, BS, 44; Columbia Univ, AM, 47, PhD(chem), 49. *Prof Exp:* Asst, Columbia Univ, 46-49; res assoc chem, Univ Calif, Los Angeles, 49-51; from asst prof to assoc prof, 51-58, head dept, 58-72, assoc dean, Col Arts & Sci, 62-66, actg dean, grad sch, 81-84, PROF CHEM, DUQUESNE UNIV, 58- *Concurrent Pos:* Ed, Pa Acad Sci Newsletter, 79-86; dir, Pa Sci Talent Search, 78-88; pres, Pa Acad Sci, 88-90. *Honors & Awards:* Pittsburgh Award, Am Chem Soc, 85. *Mem:* Am Chem Soc; Soc Appl Spectros; AAAS. *Res:* Mechanism of organic reactions. *Mailing Add:* 1812 Wightman St Pittsburgh PA 15217

SCHREIBER, LAWRENCE, b Chicago, Ill, June 15, 31. PLANT PATHOLOGY. *Educ:* Northwestern Univ, BS, 53; Purdue Univ, MS, 59, PhD, 61. *Prof Exp:* PLANT PATHOLOGIST, SCI & EDUC ADMIN-AGR RES, USDA, 61- *Res:* Soil microbiology; vascular wilt diseases; general diseases of shade trees and ornamental plants. *Mailing Add:* Nursery Crops Res Lab USDA 359 Main Rd Delaware OH 43015

SCHREIBER, MARVIN MANDEL, b Springfield, Mass, Oct 17, 25; m 49; c 1. AGRONOMY, PLANT PHYSIOLOGY. *Educ:* Univ Mass, BS, 50; Univ Ariz, MS, 51; Cornell Univ, PhD(agron, plant physiol), 54. *Prof Exp:* Asst prof agron, Cornell Univ, 54-59; assoc prof, 59-73, PROF WEED SCI, PURDUE UNIV, 73- *Concurrent Pos:* Res agronomist, Agr Res Serv, USDA, 56- *Mem:* Fel AAAS; fel Am Soc Agron; fel Weed Sci Soc Am; Int Weed Sci Soc (pres, 79-81); Sigma Xi; Controlled Release Soc. *Res:* Phenology of weed species; integrated weed management; weed control in field crops; weed competition in field crops; microenvironment of weed competition; controlled release pesticides; weed science. *Mailing Add:* Dept Bot & Plant Path Purdue Univ West Lafayette IN 47907

SCHREIBER, MELVYN HIRSH, b Galveston, Tex, May 28, 31; m 77; c 4. RADIOLOGY. *Educ:* Univ Tex, BA, 53, MD, 55. *Prof Exp:* From instr to assoc prof, 59-67, PROF RADIOL, UNIV TEX MED BR GALVESTON, 67-, CHMN RADIOL, 76- *Concurrent Pos:* Markle Found scholar, 63-68; mem bd trustees, Am Bd Radiol. *Mem:* Asn Univ Radiol; fel Am Col Radiol; Am Roentgen Ray Soc; Radiol Soc NAm; Soc Chmn Acad Radiol Depts; Am Bd Radiol. *Res:* Cardiovascular and renal disease. *Mailing Add:* Dept Radiol Univ Tex Med Br Galveston TX 77550

SCHREIBER, PAUL J, b Buffalo, NY, June 23, 40; m 62; c 2. MOMENTUM SPACE WAVE FUNCTION. *Educ:* Univ Rochester, BS, 62; Univ Dayton, MS, 69; State Univ NY, Buffalo, PhD(physiol), 79. *Prof Exp:* PHYSICIST INFRARED DETECTORS & ELECTRO-OPTICS, AVIONICS LAB, ELECTRONICS TECHNOL DIV, AIR FORCE WRIGHT AERONAUT LAB, 62- *Mem:* Am Phys Soc. *Res:* Atomic structure of multi-electron atoms in momentum space; developing helium atom mometum space wave function; infrared heterodyne dectors and special properties of infrared detectors; helium compton profile; electro optics. *Mailing Add:* 2000 Staymar Dr Kettering OH 45440

SCHREIBER, RICHARD WILLIAM, b Lawrence, Mass, Apr 4, 17; m 47; c 1. BOTANY, CELL BIOLOGY. *Educ:* Univ NH, BS, 51, MS, 52; Univ Wis, PhD(bot), 56. *Prof Exp:* Instr biol, Exten Ctr, Univ Wis-Green Bay, 55-57; from asst prof to assoc prof bot, Univ NH, 57-67, prof, 67-87; RETIRED. *Mem:* AAAS; Sigma Xi; NY Acad Sci; Am Soc Cell Biol. *Res:* Metabolic autonomy and evolution of chloroplasts; structure and function of the kinetochore. *Mailing Add:* Cherry Lane Madbury NH 03824

SCHREIBER, ROBERT ALAN, b Brooklyn, NY, Feb 11, 40; m 67; c 2. BIOLOGICAL PSYCHOLOGY, BEHAVIORAL GENETICS. *Educ:* Univ NC, Chapel Hill, BA, 65; Univ Colo, Boulder, MA, 69, PhD(psychol), 70. *Prof Exp:* Fel dept biochem, Med Univ SC, 70-74; asst prof, 74-80, ASSOC PROF BIOCHEM, UNIV TENN CTR HEALTH SCI, 80- *Concurrent Pos:* NIMH fel, Med Univ SC, 72-73. *Mem:* Soc Neurosci; Behavior Genetics Asn; Int Soc Develop Psychobiol; Sigma Xi. *Res:* Central nervous system hyperreactivity, using mice susceptible to sound-induced seizures; brain energy reserves which are immediately available; physical dependence on ethanol. *Mailing Add:* 10910 Dunbrook Dr Houston TX 77070

SCHREIBER, SIDNEY S, b NY, May 1, 21; m 45; c 2. PHYSIOLOGY, NUCLEAR MEDICINE. *Educ:* City Col New York, BS, 41; NY Univ, MS, 45, MD, 49. *Prof Exp:* Instr physics, Townsend Harris, 41-42; instr biol, City Col New York, 42 & NY Univ, 45; instr histol, Hunter Col, 50; instr clin med, 59-64, from asst prof to assoc prof med, 64-74, PROF CLIN MED, SCH MED, NY UNIV, 74- *Concurrent Pos:* Clin asst, Mt Sinai Hosp, 52-58, sr clin asst, 58-; consult internist, Radioisotope Unit, Vet Admin Hosp, Bronx, sr consult, Dept Nuclear Med, 52- & New York, 56-; instr, Hunter Col, 60-61; assoc ed, Alcoholism J. *Honors & Awards:* Linder Surg Prize. *Mem:* AAAS; fel Am Col Physicians; NY Acad Sci; Int Study Group Res Cardiac Metab; Int Soc Heart Res; Sigma Xi; Am Col Nuclear Physicians. *Res:* Nerve regeneration; endocrinology; effect of hormones on tissue growth; effect of anti-folic acid substance on tissue growth; electrolyte metabolism of heart muscle; use of radioactive isotopes in physiology; cation exchange of heart muscle; protein metabolism in heart muscle in stress, hypertension, alcohol and ischemia. *Mailing Add:* Dept Nuclear Med NY Vet Admin Hosp New York NY 10016

SCHREIBER, THOMAS PAUL, b Detroit, Mich, Mar 5, 24; m 53; c 4. ELECTRON MICROSCOPY. *Educ:* Univ Notre Dame, BS, 46; Univ Mich, MS, 48. *Prof Exp:* Qual control engr, US Rubber Co, 46; jr physicist, Phys Instrumentation Dept, Gen Motors Corp, 47-49, from res physicist to sr res physicist, 49-59, staff res scientist, Anal Chem Dept, 59-87, sr staff res scientist & group leader microstruct characterization, 85-87, sr tech staff, 87-89; RETIRED. *Mem:* AAAS; Microbeam Anal Soc. *Res:* Electron probe microanalysis; scanning electron microscopy; surface analysis by electron spectroscopy; analytical electron microscopy. *Mailing Add:* 650 Country Club Dr St Clair Shores MI 48082-1063

SCHREIBER, WILLIAM F(RANCIS), b New York, NY, Sept 18, 25. ELECTRONICS. *Educ:* Columbia Univ, BS, 45, MS, 47; Harvard Univ, PhD(appl physics), 53. *Prof Exp:* Jr engr, Sylvania Elec Prod, Inc, 47-49; res assoc, Harvard Univ, 53; res physicist, Technicolor Corp, 53-59; from assoc prof, to profelec eng, Mass Inst Technol, 59-90, Bernard Gordon pfof, 79-83, dir, Advan TV Res Prog, 83-89; RETIRED. *Concurrent Pos:* Vis prof, Indian Inst Technol, Kanpur, 64-66; consult, Sylvania Elec Prod, Inc, 53, 60, Technicolor Corp, 59-62, Smithsonian Astrophys Observ, 61-64, Raytheon Corp, 62-64, ECRM, Inc, 68-78 & Addressograph-Multigraph, 78-; dir, Shintron Corp, 62- & ECRM, Inc, 68-78; vis prof, Univ Que, & Nat Inst Sci Res, 80-81; Gordon McKay fel, Harvard Univ, Charles Coffin fel; consult, Assoc Press, Electronic Image Systs, Inc, Shintron Co, Inc, Am Int, Inc, Scitex, Inc, Comugraphic, Inc, Harris, Inc & EFI, Inc. *Honors & Awards:* Honors Award, Tech Asn Graphic Arts, 83; David Sarnoff Gold Medal, Soc Motion Picture & TV Engrs, 90. *Mem:* Fel Inst Elec & Electronics Engrs; Sigma Xi. *Res:* Application of information theory to image transmission systems; image processing for the graphic arts; laser scanners; author of 12 publications. *Mailing Add:* Mass Inst Technol 36-677 Cambridge MA 02139

SCHREIBER, WILLIAM LEWIS, b New York, NY, Aug 10, 43; m 66; c 2. ORGANIC CHEMISTRY. *Educ:* Mass Inst Technol, BS, 65; Univ Rochester, PhD(chem), 70. *Prof Exp:* Res assoc org chem, Rockefeller Univ, 69-71; PROJ CHEMIST ORG SYNTHESIS, INT FLAVORS & FRAGRANCES INC, 71- *Mem:* Am Chem Soc; Sigma Xi. *Res:* Fragrance chemical synthesis; terpene chemistry. *Mailing Add:* Int Flavors & Fragrances Inc R&D 1515 Hwy 36 Union Beach NJ 07735

SCHREIBMAN, MARTIN PAUL, b New York, NY, Sept 18, 35; m 60; c 1. ZOOLOGY, COMPARATIVE ENDOCRINOLOGY. *Educ:* Brooklyn Col, BS, 56; NY Univ, MS, 59, PhD(biol), 62. *Prof Exp:* From instr to assoc prof, 62-72, PROF BIOL, BROOKLYN COL, 72-; RES ASSOC FISH ENDOCRINOL, NEW YORK AQUARIUM, NY ZOOL SOC, 66- *Concurrent Pos:* City Univ New York res grant, 77-81; NSF res grants, 65-70 & 77-80; res collabr, Brookhaven Nat Lab, 77-; NIH grants, 80- *Mem:* AAAS; Am Soc Zoologists; Endocrine Soc; Sigma Xi. *Res:* Comparative endocrinology of lower vertebrates, especially teleosts and relating to pituitary cytology and function; osmoregulation; melanogenesis; effects of hypophysectomy on endocrine functions; genetic control of sexual maturation and aging; hypothalamic-pituitary-gonadal axis. *Mailing Add:* Dept Biol Brooklyn Col Brooklyn NY 11210

SCHREIDER, BRUCE DAVID, b Denver, Colo, Mar 8, 46. CLINICAL PHARMACOLOGY, PHYSIOLOGY. *Educ:* Univ Chicago, BS, 68, MD & PhD(pharmacol & physiol), 75. *Prof Exp:* Res assoc toxicol, Dept Pharmacol & Physiol Sci, 75-81, asst prof, 81-90, ASSOC PROF, ANESTHESIA & CLIN PHARMACOL, UNIV CHICAGO, 90- *Mem:* AAAS; Sigma Xi. *Res:* Hepatic microsomal enzyme induction and inhibition by various therapeutic and environmental agents and the interactions between these agents. *Mailing Add:* 1417 Barry Chicago IL 60657

SCHREIER, ETHAN JOSHUA, b New York, NY, Sept 22, 43. ASTROPHYSICS, ASTRONOMY. *Educ:* City Univ New York, BS, 64; Mass Inst Technol, PhD(physics), 70. *Prof Exp:* Sr scientist, Am Sci & Eng, 70-73; physicist, Smithsonian Astrophys Observ, 73-81; chief data & oper scientist, 81-88, ASSOC DIR OPER, SPACE TELESCOPE SCI INST, 88- *Mem:* Am Astron Soc. *Res:* X-ray astronomy. *Mailing Add:* Space Telescope Sci Inst 3700 San Martin Dr Baltimore MD 21218

SCHREIER, HANSPETER, b Basel, Switzerland, Nov 3, 41; Can citizen; m 70; c 1. SOIL & WATER CHEMISTRY. *Educ:* Univ Colo, BA, 70; Univ Sheffield, MSc, 73; Univ BC, PhD(geomorphol), 76. *Prof Exp:* Res asst chem, Sandoz Pharmaceut, 61-64; fel terrain sci, 77-79, ASST PROF, DEPT SOIL SCI, WESTWATER RES CTR, UNIV BC, 79- *Mem:* Am Soc Photogram; Int Soc Soil Sci. *Res:* Water chemistry and sediments in river systems; terrain analysis, land evaluation, GIS geographic information systems; remote sensing of soils and terrain. *Mailing Add:* 2831 W 29th Vancouver BC V6L 1Y2 Can

SCHREIER, STEFAN, gas dynamics, mechanical engineering, for more information see previous edition

SCHREINER, ALBERT WILLIAM, b Cincinnati, Ohio, Feb 15, 26; m 53; c 1. INTERNAL MEDICINE, HEMATOLOGY. *Educ:* Univ Cincinnati, BS, 47, MD, 49. *Prof Exp:* Instr internal med, 55-59, asst clin prof, 59-62, assoc prof, 62-67, PROF MED, COL MED, UNIV CINCINNATI, 67-; DIR

DEPT INTERNAL MED, CHRIST HOSP, 68- *Concurrent Pos:* Fel hemat, Cincinnati Gen Hosp, Ohio, 51-52, attend physician, 57-, clinician, Outpatient Dept, 57-62, chief clinician, 62-65; clin investr leukemia res, Vet Admin Hosp, Cincinnati, 57-59, chief med serv, 59-68, consult, 68-; consult to med dir, Ohio Nat Life, 66-80; dir, dept internal med residency prog, Christ Hosp, 78-87. *Mem:* Am Col Physicians; Am Fedn Clin Res; NY Acad Sci; Am Soc Internal Med; Asn Prog Dirs Internal Med; Soc Res & Educ Primary Care Internal Med. *Res:* Experimental viral oncogenesis in rodents. *Mailing Add:* 8525 Given Rd Cincinnati OH 45243

SCHREINER, ANTON FRANZ, b Apr 29, 37; US citizen; m 66; c 1. INORGANIC CHEMISTRY. *Educ:* Univ Detroit, BS, 61, MS, 63; Univ Ill, PhD(inorg chem), 67. *Prof Exp:* Res assoc, Univ Ill, 67-68; from asst prof to assoc prof, 68-76, grad adminr, 74-77, PROF CHEM, NC STATE UNIV, 76-, CHMN ANALYTIC-INORG CHEM, 78- *Honors & Awards:* Outstanding Young Scientist Award, Sigma Xi, 73. *Mem:* Am Chem Soc; Sigma Xi; Royal Soc Chem. *Res:* Chemistry and magnetic circular dichroism of transition metals; magnetic circular luminescence; luminescence; crystal field, linear combination of atomic orbitals-molecular orbital, and normal coordinate theory applications; spectroscopy of inorganic materials; laser-optical semiconductor; 4d and 5d metal chemistry. *Mailing Add:* Dept Chem NC State Univ Box 8204 Raleigh NC 27695

SCHREINER, CEINWEN ANN, b Philadelphia, Pa, May 27, 43. IMMUNOLOGY, ANIMAL PATHOLOGY. *Educ:* Muhlenberg Col, BS, 65; Univ NH, MS, 67, PhD(genetics), 72. *Prof Exp:* Res assoc teratol, E R Squibb & Sons, Inc, 67-69; prin scientist mutagenicity, McNeil Labs, Inc, 72-79; supvr genetic toxicol, 79-87, MGR PATH-IMMUNOTOXICOL, MOBIL OIL CORP, 87-, MGR BIOCHEM TOXICOL, 88- *Concurrent Pos:* Vchmn, Gordon Conf Genetic Toxicol, 81, chmn, 83; sect ed, Cell Biol & Toxicol, 86-; mem coun, Environ Mutagen Soc, 86-89 & Am Col Toxicol, 86-89; fel, Acad Toxicol Sci, 87- *Mem:* Environ Mutagen Soc; Genetics Soc Am; Teratol Soc; Am Col Toxicol; Genetic Toxicol Asn (treas, 76-82). *Res:* Development of genetic toxicology immunotoxicology program for petroleum; management of pathological evaluation of dermal and inhalation toxicity studies, analytical chemistry, dermal sensitization, and research into mechanisms of toxicity. *Mailing Add:* Toxicol Div Mobil Oil Corp PO Box 1029 Princeton NJ 08940

SCHREINER, ERIK ANDREW, b Hammonton, NJ, Dec 7, 35; m 60; c 3. MATHEMATICS. *Educ:* Wayne State Univ, BA, 58, MA, 60, PhD(math), 64. *Prof Exp:* Asst prof, 63-68, assoc prof, 68-79, PROF MATH, WESTERN MICH UNIV, 79- *Concurrent Pos:* Vis asst prof, Univ Mass, 67-68. *Mem:* Am Math Soc; Math Asn Am. *Res:* Lattice theory. *Mailing Add:* Dept Math Western Mich Univ 6602 Evergreen Kalamazoo MI 49008

SCHREINER, FELIX, b Hamburg, Ger, Sept 28, 31; m 60, 69; c 2. PHYSICAL CHEMISTRY, INORGANIC CHEMISTRY. *Educ:* Univ Hamburg, Vordiplom, 54, dipl, 57, PhD(chem), 59. *Prof Exp:* Asst prof phys chem, Univ Kiel, 59-61; resident res assoc & fel, Argonne Nat Lab 61-63, asst chemist, 63-65, assoc chemist, 65-88; RETIRED. *Mem:* Am Chem Soc. *Res:* Thermodynamics; low-temperature calorimetry; noble gas chemistry; inorganic fluorine chemistry; energy research; nuclear waste management. *Mailing Add:* 19809 S 116th Ave Mokena IL 60448

SCHREINER, GEORGE E, b Buffalo, NY, Apr 26, 22; m 48; c 8. MEDICINE, PHYSIOLOGY. *Educ:* Canisius Col, AB, 43; Georgetown Univ, MD, 46; Am Bd Internal Med, dipl, 55. *Hon Degrees:* DHL, Canisius Col; DSc, Georgetown, 86; FRCPS(Glas). *Prof Exp:* Intern, Med Serv, Boston City Hosp, 46-47; asst physiol, Sch Med, NY Univ-Bellevue Med Ctr, 47-48, instr, 48-50; sr resident med, Mt Alto Hosp, 50-51; clin instr, 51-52, from instr to assoc prof, 52-61, prof med, 61-87, DISTINGUISHED PROF, GEORGETOWN UNIV, 87- *Concurrent Pos:* Fel physiol, NY Univ-Bellevue Med Ctr, 49-50; dir renal clin, Georgetown Univ Hosp, 52-59, nephrol div, 59-72 & clin study unit, 61-72; mem, Nat Res Coun, 60-63; chmn, Nat Drug Res Bd Comt for Armed Forces Inst Path Registry Adverse Reactions, 63, mem subcomt drug efficacy, 66; Secy-Gen, Int Cong Nephrology, 63-66; mem, Nat Kidney Found Sci Adv Bd & chmn, DC Chap, 63-68, pres, 68-70, chmn, Nat Adv Bd, 70-; mem, Nat Coun Regional Med Progs, Dept HEW; chmn comn res in nephrol, Vet Admin, 72-76, mem merit rev bd, nephrol; consult, NIH, Walter Reed Army Med Ctr, Vet Admin Hosp & Nat Naval Med Ctr; ed-in-chief, Trans, Soc Artificial Internal Organs, 55-85 & Nephron, Int Soc Nephrology, 63-72; nephrol registry, 78- *Honors & Awards:* Davidson Award, Nat Kidney Found, David Hume Award; President's Award; V Bologna Nettons Argento Award; Walton Lectr, Royal Col Physician & Surgeons (Glas). *Mem:* Am Fedn Clin Res (secy-treas, 58-61, pres elect, 61-62, pres, 62-63); Soc Artificial Internal Organs (secy-treas, 56-58, pres elect, 58-59, pres, 59-60); Am Soc Nephrology (secy, 66, pres, 71); AAAS; Int Soc Nephrology (pres-elect, 75, pres, 78-81); Am Soc Clin Invest; Assn Am Physicians; Clin & Climat Soc; Am Soc Hypertension; Am Soc Transplant Physicians; Renal Physicians Asn. *Res:* Clinical nephrology; internal medicine; renal physiology; hemodialysis; dialysis of poisons; renal homotransplantation, biopsy and pathology; pyelonephritis; nephrotic syndrome; ethics of nephrology; hemoperfusion. *Mailing Add:* Georgetown Univ Sch Med Washington DC 20007

SCHREINER, HEINZ RUPERT, b Apr 13, 30; US citizen; m; c 5. ENZYME & CLINICAL CHEMISTRY. *Educ:* Univ Nebr, BS, 51, MS, 54, PhD(phys & biol chem), 56. *Prof Exp:* Res chemist, Linde Div, Ocean Systs, Inc, 56-60, group leader, 60-63, res supvr, 63-68, dir res & develop, Ocean Systs, Inc, 68-71; prog mgr, Med Prod Div, Union Carbide Corp, 71-76, assoc dir corp res, 76-77, dir, 77-78, vpres technol, 78-80, dir, bus planning & prod develop, 80-81; vpres clin diagnostics, Technicon Instruments Corp, 81-89; VPRES REAGENT TECHNOL, MILES INC, 89- *Concurrent Pos:* Councilor, Am Chem Soc, 63-66. *Honors & Awards:* Albert R Behnke Jr Award, Undersea Med Soc, 89. *Mem:* Fel Aerospace Med Asn (vpres, 71); fel AAAS; Am Asn Clin Chem; Am Chem Soc; Am Inst Chemists; Am Soc Pharmacol & Exp Therapeut; Soc Nuclear Med; Undersea Med Soc (pres, 69); Sigma Xi. *Res:* Biotechnology; forty-five publications and patents in the Life Science Field. *Mailing Add:* RR3-Box 111 South Salem NY 10590

SCHREINER, PHILIP ALLEN, b Duluth, Minn, Sept 29, 43. EXPERIMENTAL HIGH ENERGY PHYSICS. *Educ:* Univ Calif, Los Angeles, BS, 65, MS, 66, PhD(physics), 70. *Prof Exp:* Fel, Argonne Nat Lab, 71-73, asst physicist high energy physics, 73-77, physicist, 77-81; MEM TECH STAFF, BELL LABS, 81- *Mem:* Am Phys Soc. *Res:* Neutrino interactions. *Mailing Add:* 1534 Chickasaw Dr Naperville IL 60563

SCHREINER, ROBERT NICOLAS, JR, b New York, NY, Jan 12, 35; m 56; c 6. TEST & EVALUATION, DESIGN & DEVELOPMENT. *Educ:* Capital Univ, BS, 56. *Prof Exp:* Engr, Fluid Dynamics Dept, Norair Div, Northrop Corp, Hawthorne, Calif, 56-59; mem tech staff, Dynamics Sect, TRW Inc, 59-64, staff engr, Dynamics Dept, 64-66, Eng Mech Lab, 66-69, Appl Mech Lab, 69-71, Res & Technol Opers, 71-72, head, Tracking Algorithm Design & Develop Sect, 72-78, subproj mgr, BETA Test Prog, 78-79, dept asst prog mgr, 79-81, staff engr, 81-82, design eng & develop, 82-83, field site test dir, Proj 9646, 83, dept asst prog mgr, 83-84, Proj 5810, 84-85, Proj 8524, 85-87, dept mgr test eng & integration, 87-90, MGR SYSTS DEVELOP STAFF & FIELD SERV, TRW, INC, 90- *Concurrent Pos:* Lectr mech eng & adv computer technol, var univs & indust firms, 65-72. *Res:* Author of eight technical articles published in professional journals on dynamics of solid propellants, mechanical shock and vibration, viscoelasticity, nuclear effects and advanced computer technology. *Mailing Add:* 30520 Via Rivera Rancho Palos Verdes CA 90274

SCHREIWEIS, DONALD OTTO, b Tacoma, Wash, July 27, 41. COMPARATIVE ANATOMY, VERTEBRATE EMBRYOLOGY. *Educ:* Univ Puget Sound, AB, 63, MS, 66; Wash State Univ, PhD(zool), 72. *Prof Exp:* Asst prof biol, Univ Puget Sound, 71-72; asst prof biol, Univ Nev, Las Vegas, 72-77; chairperson & assoc prof biol, Gonzaga Univ, 77-81; DIR PREPROFESSIONAL HEALTH STUDIES, ST LOUIS UNIV, 81- *Mem:* Am Soc Zoologists; Soc Syst Zool; Am Ornith Union; Am Soc Mammal; Am Soc Ichthyologists & Herpetologists; Sigma Xi. *Res:* Comparative myological studies of birds and bats using numerical methods; comparative vertebrate morphology; teratogenic effects of herbicides, pesticides and heavy metals on vertebrate development. *Mailing Add:* 548 Braebridge Rd Manchester MO 43011

SCHREMP, EDWARD JAY, b Newark, NJ, Aug 20, 12; m 47; c 1. SCIENCE & TECHNOLOGY POLICY, OCEANIC ENGINEERING. *Educ:* Mass Inst Technol, SB, 34, PhD(theoret physics), 37. *Prof Exp:* Teaching fel physics, Mass Inst Technol, 35-37, instr physics, 37; instr physics, Washington Univ, St Louis, 37-41; asst prof physics, Univ Cincinnati, 41-46; sci liasion officer, Off Asst Naval Attache Res, Am Embassy, London, 46-48; head, Theory Br, Nucleonics Div, US Naval Res Lab, Wash, 48-66, consult theoret physics, 66-70; CONSULT PHYSICIST, 70-; PRES, E J SCHREMP & CO, 75- *Concurrent Pos:* Mem staff, Radiation Lab, Mass Inst Technol, 41-46; off Sci Res & Develop, 41-46; vis scientist, theortet study div, Europ Orgn Nuclear Res, Geneva, Switz, 65-66, Int Ctr Theoret Physics, Trieste, Italy, 67, 68, 69, 70 & 71. *Mem:* Fel Am Phys Soc; NY Acad Sci; Inst Elec & Electronics Engrs; Inst Elec & Electronic Engrs Oceanic Eng Soc. *Res:* Oceanic engineering implications for national technology policy; ultimate quantum theoretical implications for fields and space-time geometry of simply transitive non-Abelian 4-parameter subgroups of the Poincare group, when examined in terms of the group-space of any such subgroup. *Mailing Add:* 226 S Fairfax St Alexandria VA 22314

SCHREMP, FREDERIC WILLIAM, b Utica, NY, Aug 14, 16; m 48, 58; c 4. PHYSICAL CHEMISTRY. *Educ:* Rensselaer Polytech Inst, BS, 42; Univ Wis, PhD(phys chem), 50. *Prof Exp:* Res elec engr, Am Steel & Wire Co, 42-44; res chemist, Oak Ridge Nat Lab, 44-45; sr res assoc, Chevron Oil Field Res Co, Standard Oil Co, Calif, 50-81; RETIRED. *Concurrent Pos:* Consult. *Mem:* Am Chem Soc; Nat Asn Corrosion Engrs. *Res:* Rheology of high polymers; oil field drilling fluid; corrosion. *Mailing Add:* 3225 Arbol Dr Fullerton CA 92635

SCHRENK, GEORGE L, b Seymour, Ind, Nov 28, 37; div. ENGINEERING. *Educ:* Ind Univ, BS, 55, MS, 56, PhD(theoret physics), 59. *Hon Degrees:* MA, Univ Pa, 71. *Prof Exp:* Sr res assoc, Inst Direct Energy Conversion, 63-65, chief plasma eng br, 64-66, asst prof eng, 65-68, assoc dir comput ctr, 66-69, assoc prof eng, Towne Sch Civil & Mech Eng, 68-80, assoc prof eng, Univ Pa, 70-80; ENG CONSULT, 59-; PRES COMP COMM, INC, 75- *Concurrent Pos:* Adj assoc prof eng, Univ Pa, 80-; fel, NSF, 58-60; expert witness telecommun indust. *Mem:* Am Phys Soc; Am Soc Mech Engrs; sr mem, Inst Elec & Electronics Engrs; Asn Comput Mach; Sigma Xi. *Res:* Electromagnetic propagation; telecommunications engineering; computational engineering physics; mathematical modeling and simulation. *Mailing Add:* Comp Comm Inc 900 Haddon Ave 4th Floor Collingswood NJ 08108

SCHRENK, WALTER JOHN, b Toledo, Ohio, Feb 12, 33. POLYMERIC OPTICAL THIN FILMS, MULTILAYER PACKAGING MATERIALS. *Educ:* Univ Mich, BS, 55. *Prof Exp:* SR RES SCIENTIST, DOW CHEM CO, 55- *Honors & Awards:* Fred O Conley Award for Plastics Eng, Soc Plastics Engrs, 87, Jack Barney Award for Contrib to Plastics, 87. *Mem:* Am Soc Mech Engrs; Soc Plastics Engrs; Soc Optical Eng. *Res:* Fabrication of polymers and polymer processing with emphasis on multilayer coextrusion of polymer films and sheets, orientation and enhanced properties by control of supramolecular structure thru processing. *Mailing Add:* Dow Chem Co 1702 Bldg Midland MI 48674

SCHRENK, WILLIAM GEORGE, b Hiawatha, Kans, July 13, 10; m 32; c 2. CHEMISTRY. *Educ:* Western Union Col, AB, 32; Kans State Col, MS, 36, PhD(chem), 45. *Prof Exp:* Teacher high schs, Iowa, 32-38; from instr to assoc prof, Kans State Univ, 38-51, asst chemist, Exp Sta, 43-51, prof chem & chemist, 51-76; RETIRED. *Mem:* AAAS; Am Chem Soc; Soc Appl Spectros. *Res:* Minor elements in plants; physical methods in chemical analysis; spectroscopy for analytical purposes; instrumental techniques in analysis; atomic absorption spectroscopy; flame emission spectroscopy. *Mailing Add:* Dept Chem Kans State Univ Manhattan KS 66506

SCHREUDER, GERARD FRITZ, b Medan, Indonesia, Apr 4, 37; Dutch citizen; m 61; c 2. FOREST ECONOMICS. *Educ:* State Agr Univ Wageningen, MS, 60; NC State Univ, MS, 67; Yale Univ, PhD(econ), 68. *Prof Exp:* Asst expert aerial photos, Orgn Am States, 61-64; asst prof opers res, Yale Univ, 67-70; assoc prof, 71-75, prof forestry, 75-77, PROF FORESTRY RESOURCES & DIR FOREST RESOURCES MGT STUDIES, UNIV WASH, 77- *Concurrent Pos:* Dir res, Univ Wash, 72-; mem comt renewable resources for indust mat, Nat Acad Sci, 75-76. *Mem:* AAAS; Neth Inst Agr Engrs; Sigma Xi. *Res:* Forest resource modelling; aerial photo interpretation as related to environmental impacts and animal damage; forest resource economics. *Mailing Add:* Univ Wash Bloedell Hall AR 10 Seattle WA 98195

SCHREURS, JAN W H, b Winterswijk, Neth, Feb 10, 32; m 56; c 4. PHYSICAL CHEMISTRY, MAGNETIC RESONANCE. *Educ:* Free Univ, Amsterdam, BSc, 53, MSc, 57, PhD(phys chem), 62. *Prof Exp:* Res assoc chem, Columbia Univ, 57-59; res fel, Nat Res Coun Can, 60-62; SR RES ASSOC PHYS PROPERTIES RES DEPT, CORNING INC, 62- *Mem:* Am Phys Soc; Sigma Xi. *Res:* Electron spin resonance and magnetic susceptibility of glasses and glass ceramics. *Mailing Add:* Phys Properties Res Dept Corning Inc Corning NY 14831

SCHREYER, JAMES MARLIN, b Asheville, NC, Dec 26, 15; m 51; c 3. CHEMISTRY. *Educ:* Univ NC, AB, 38; Ore State Col, PhD(inorg chem), 48. *Prof Exp:* Teacher, High Sch, NC, 38-41; inspector powder & explosives, Radford Ord Works, Va, 41-42; chem engr & area chief inspector, Badger Ord Works, Wis, 42-43; process engr, US Rubber Co, NC, 43-44; res engr, Sulphonics, Inc, Md, 44-45; asst, Ore State Col, 45-47; asst prof chem, Univ Ky, 48-50, prof, 50-51; sr chemist, Oak Ridge Nat Lab, 51-53; sr chemist, Union Carbide Nuclear Co, 53-78, head, Develop Dept, 78-80, sr tech adv, Union Carbide Corp, 80-82; sr mem staff, Oak Ridge Nat Lab, 82; RETIRED. *Mem:* Sigma Xi. *Res:* Properties of ferrates; preparation and analysis of potassium ferrate; ferrate oxidimetry; spectrophotometric studies cobalt-2 in alkaline solution; solubility of uranium-6 and -4 phosphates in phosphate solution; clean room technology; cleaned moon box and associated hardware for Apollo moon landings; solar energy technology; installation of solar water heaters. *Mailing Add:* 9100 Burchfield Dr Oak Ridge TN 37832

SCHREYER, RALPH COURTENAY, b Washington, DC, July 27, 19; m 44; c 2. ORGANIC POLYMER CHEMISTRY. *Educ:* Cath Univ Am, BA, 41; Purdue Univ, PhD(chem), 46. *Prof Exp:* Res chemist, Polychem Dept, Exp Sta, E I du Pont de Nemours & Co, Inc, Del, 46-60, tech assoc, Eastern Lab, Explosives Dept, NJ, 60-68, sr res chemist, Polymer Intermediates Dept, 68-73, staff chemist, 73-76, patent assoc, 76-78, patent assoc, Petrochem Dept, 78-84; RETIRED. *Concurrent Pos:* Patent agent, 76. *Mem:* Am Chem Soc. *Res:* Synthesis of flurocarbons; reactions of carbon monoxide and carbon monoxide and hydrogen with organic compounds; synthesis of nylon intermediates; polymerization of ethylene and fluoro-olefins. *Mailing Add:* 2522 Deep Wood Dr Wilmington DE 19810

SCHRIBER, STANLEY OWEN, b St Boniface, Man, July 20, 40; m 62; c 2. ACCELERATOR PHYSICS. *Educ:* Univ Man, BSc, 62, MSc, 63; McMaster Univ, PhD(physics), 67. *Prof Exp:* Asst res off, Chalk River Nuclear Labs, Atomic Energy Can Ltd, 66-70, assoc res off, 71-75, sr res off, 76-84; dep div leader, 84-87, DIV LEADER, LOS ALAMOS NAT LABS, 87- *Concurrent Pos:* Guest scientist, Los Alamos Nat Lab, 77 & 78-79 & Kek Lab High Energy Physics, Japan, 81; consult, Argonne Nat Lab, 80, Inst Fur Kernphysik, Karlspruhe, Ger, 80. *Mem:* Can Asn Physicists; Am Phys Soc; Inst Elec & Electronics Engrs. *Res:* Design, construction and testing of pulsed and continuous wave linear accelerations; charged particle beam dynamics; beam diagnostics; radio frequency systems and beam dynamics; practical uses of accelerator beams. *Mailing Add:* Accelerator Technol Div Los Alamos Nat Lab Los Alamos NM 87545

SCHRIBER, THOMAS J, b Flint, Mich, Oct 28, 35; m 67; c 3. SIMULATION MODELING, SIMULATION OUTPUT ANALYSIS. *Educ:* Univ Notre Dame, BS, 57; Univ Mich, MSE, 58, AM, 59, PhD(chem eng), 64. *Prof Exp:* Asst prof math, Eastern Mich Univ, 63-66; from asst prof to assoc prof, 66-72, PROF MGT SCI, UNIV MICH, ANN ARBOR, 72- *Concurrent Pos:* Consult, Ford Motor Co, 66, 71 & 74, Stanford Res Inst, 73, Int Tel & Tel, 74, CPC Int, 77 & 80, Occidental Petrol, 82, Gen Motors, 84-85 & Exxon, 84-85; vis scholar, Stanford Res Inst, 72-73; prin investr, Off Naval Res grant, 81-83. *Mem:* Inst Mgt Sci; fel Decisions Sci Inst; Soc Comput Simulation; Asn Comput Mach. *Res:* Computer applications in management science, especially discrete-event simulation, numerical methods and optimization. *Mailing Add:* Computer Info Systs Dept Univ Mich Ann Arbor MI 48109-1234

SCHRICKER, ROBERT LEE, b Davenport, Iowa, June 28, 28; m 56; c 3. VETERINARY MEDICINE, MICROBIOLOGY. *Educ:* Iowa State Univ, DVM, 52; Univ Ill, Urbana, MS, 58, PhD(vet sci), 61; Am Col Vet Microbiol, dipl. *Prof Exp:* Pvt pract, 52-54; res vet, US Army Biol Labs, 61-71; mem vet biol staff, USDA, 71-81; RETIRED. *Mem:* Am Vet Med Asn. *Res:* Precipitating antigens of leptospires; pathogenesis of zoonotic infectious diseases in primates; pathology and clinical biochemistry; veterinary biologics. *Mailing Add:* 1512 Rockcreek Dr Frederick MD 21702

SCHRIEFFER, JOHN ROBERT, b Oak Park, Ill, May 31, 31; m 60; c 3. THEORETICAL CONDENSED MATTER PHYSICS. *Educ:* Mass Inst Technol, BS, 53; Univ Ill, MS, 54, PhD(physics), 57. *Hon Degrees:* Dr rer nat, Munich Tech Univ, 68; Dr es Sci, Univ Geneva, 68; DSc, Univ Pa, 73, Univ Ill, 74, Univ Cincinnati, 77, Univ Tel Aviv, 86, Univ Ala, 90. *Prof Exp:* NSF fel, Univ Birmingham & Inst Theoret Physics, Univ Copenhagen, 57-58; asst prof physics, Univ Chicago, 57-60; from asst prof to assoc prof, Univ Ill, 59-62; prof physics, Univ Pa, 62-64, Mary Amanda Wood prof, 64-80; dir, Inst Theoret Physics, 84-89, PROF, UNIV CALIF, SANTA BARBARA, 80-, CHANCELLOR'S PROF, 84- *Concurrent Pos:* Guggenheim fel, 67-68; Andrew D White prof-at-lg, Cornell Univ, 67-73; Exxon fac fel, 79-; mem, Class I Mem Comt, Nat Acad Sci, 84-86, mem coun, 90-; mem, Space Sci & Applications Adv Comt, NASA, 88-89; fel, Los Alamos Nat Lab, 88-, dir, Advan Study Prog High Temperature Superconductivity Theory, 88-; consult, IBM, STI, Exxon. *Honors & Awards:* Nobel Prize in Physics, 72; Comstock Prize, Nat Acad Sci, 68; Oliver E Buckley Solid State Physics Prize, Am Phys Soc, 68; John Ericsson Medal, Am Soc Swed Engrs, 76; Nat Medal Sci, 84. *Mem:* Nat Acad Sci; Am Acad Arts & Sci; Am Philos Soc; fel Am Phys Soc; Danish Royal Acad Sci; Sigma Xi. *Res:* Theoretical solid state physics, especially superconductivity; surface physics; general theory of many body problem; magnetism; low dimensional conductors; nonlinear phenomena. *Mailing Add:* Inst Theoret Physics Univ Calif Santa Barbara CA 93106

SCHRIEMPF, JOHN THOMAS, b Sandusky, Ohio, July 6, 34; m 57; c 1. SOLID STATE PHYSICS. *Educ:* Carnegie Inst Technol, BS, 56, MS, 60, PhD(physics), 64. *Prof Exp:* Instr physics, Carnegie Inst Technol, 60-63; res physicist, US Naval Res Lab, 63-81, supvr, condensed matter & radiation sci div, 81-84; PHYS SCI INC, ALEXANDRIA, VA, 84- *Concurrent Pos:* Vis res scholar, Univ Calif, Irvine, 68-69. *Mem:* AAAS; Sigma Xi; fel Am Phys Soc; Am Asn Physics Teachers. *Res:* Transport properties, especially thermal conductivity of metals and alloys; transport properties of metals at high temperatures, especially in the liquid state; interaction between laser radiation and metallic systems. *Mailing Add:* Phys Sci Inc 635 Slaters Ln No G101 Alexandria VA 22314

SCHRIER, DENIS J, b Grand Rapids, Mich, Oct 29, 55; m 77; c 2. IMMUNOPHARMACOLOGY. *Educ:* Med Col Wis, PhD(path), 80. *Prof Exp:* Fel, Univ Mich, 80-82; ASSOC RES FEL, WARNER LAMBERT CO, 82- *Mem:* Am Thoracic Soc; Am Asn Path; Am Asn Immunol; fel Parker Francis Pulmonary; Soc Leukocyte Biol. *Res:* Direct multidisciplinary team which is involved in the identification of novel antiinflammatory and antiarthritic drugs; arachidonic acid metabolism; protein kinase c; cellular adhesion. *Mailing Add:* Warner Lambert Co 2800 Plymouth Rd Ann Arbor MI 48105

SCHRIER, EUGENE EDWIN, b New York, NY, May 24, 34; m 70. PHYSICAL BIOCHEMISTRY. *Educ:* Kenyon Col, AB, 55; Rensselaer Polytech Inst, PhD(phys chem), 61. *Prof Exp:* Vis instr chem, Kenyon Col, 60-61; res assoc, Cornell Univ, 61-63; from asst prof to assoc prof, State Univ NY, Binghamton, 63-74, prof chem, 74-80, chmn dept, 78-80, ADJ PROF CHEM, STATE UNIV NY, BINGHAMTON, 80- *Concurrent Pos:* NIH fel, 62-63; Mayo Found fel, 77-78. *Mem:* Am Soc Biol Chemists; Am Chem Soc. *Res:* Structure and thermodynamic properties of water and aqueous solutions; ion-nonelectrolyte interactions in solution; protein denaturation; biological calcification. *Mailing Add:* PO Box 3130 Guttenberg NJ 07093

SCHRIER, MELVIN HENRY, b Brooklyn, NY, Dec 13, 27; m 50; c 2. CHEMISTRY. *Educ:* Brooklyn Col, BS, 51. *Prof Exp:* Chemist, Pittsburgh Testing Lab, 51-53; chemist, Otto B May Inc, Newark, NJ, 53-59, head chemist, Anal Sect, 59-70, mgr, Anal Dept, 70-79; GROUP LEADER, US TESTING CO, HOBOKEN, NJ, 80- *Mem:* Am Chem Soc. *Res:* Analytical research in dyestuff and intermediates. *Mailing Add:* Dept Chem US Testing Co 1415 Park Ave Hoboken NJ 07030

SCHRIER, ROBERT WILLIAM, b Indianapolis, Ind, Feb 19, 36; m; c 5. NEPHROLOGY, RENAL DISORDERS. *Educ:* DePauw Univ, BA, 57; Ind Univ Sch Med, MD, 62; Am Bd Internal Med, dipl, 68. *Prof Exp:* Med intern, Marion County Gen Hosp, 62-63, med resident, Univ Wash Sch Med, 63-65, med asst, Harvard Univ Sch Med, 65-66, Endocrine-Metab res fel, Peter Brent Brigham Hosp, 65-66, res fel, Charing Cross Hosp Med Sch, London Univ, Eng, 67-68, consult, Walter Reed Gen Hosp & Army Inst Res, 66-67, 68-69; from asst to assoc prof med & Cardiovasc Res Inst, Univ Calif Med, Ctr, San Francisco, 69-72, assoc mem 70-72, assoc dir renal div, 71-72; PROF MED & HEAD DIV RENAL DIS, UNIV COLO SCH MED, 72-, CHMN DEPT MED, 76- *Concurrent Pos:* Fulbright scholar, Gutenberg Univ, Ger, 58, Guggenheim Fel, 86-87; NIH Hypertension Task Force, 75-76, Nephrol/Urol Survey Comt, 75-76, Consult, Heart, Blood & Lung Inst, NIH, 72-74; estab investr Am Heart Asn, 71-73; lectr, Int Cong Nephrol, Montreal, 78, Athens, 81, London, 87; Pfizer fel, clin Res Inst, Montreal, 86; Australian Nat Kidney Found Vis Prof, 83, Howard H Hiatt vis prof, Harvard Med Sch, 85, Mrs Ho Tam Kit Hing vis prof, Univ Hong Kong, 87; chmn, Health & Sci Affairs Comt, Nat Kidney Found, 84-86; hon prof, Univ Paris, 86-87 & Univ Beijing, 88; panel mem, Conf Geriat Assessment Methods Clin Decisionmaking, NIH, 87, chmn, End-Stage Renal Dis Comt, 90; ed, Advan Internal Med, 91. *Honors & Awards:* John Peters Distinguished Lectr, Yale Univ Sch Med, 79; Chandros lectr, Brit Renal Physicians Asn, 84; David M Hume Mem Award, Nat Kidney Found, 87; Sau Win-Lam Lectr, State Univ NY, 89; William Goldring Lectr, NY Univ, 89; John C Merrill Mem Lectr, Harvard Univ, 89; Eduardo Slatupolsky Lectr, Wash Univ, 89; Mayo Soley Award, Western Soc Clin Invest, 89. *Mem:* Inst Med-Nat Acad Sci; Int Soc Nephrology (treas, 81-90, vpres, 90-); Am Soc Nephrol (secy-treas, 79-81, pres, 83); Int Soc Physiol; fel Am Col Physicians; emer mem Am Soc Clin Invest (vpres, 80-81); Am Clin & Climatol Asn (vpres, 86); Asn Am Physicians; fel Am Col Clin Pharmacol; AAAS. *Res:* Kidney diseases; author of numerous technical publications. *Mailing Add:* Dept Med Div Renal Dis Univ Colo Health Sci Ctr 4200 E 9 Ave Box C280 Denver CO 80262

SCHRIER, STANLEY LEONARD, b New York, NY, Jan 2, 29; m 53; c 3. MEDICINE. *Educ:* Univ Colo, AB, 49; Johns Hopkins Univ, MD, 54; Am Bd Internal Med, dipl. *Prof Exp:* From instr to assoc prof, 59-72, PROF MED, MED CTR, STANFORD UNIV, 72-, HEAD DIV HEMAT, 68- *Concurrent Pos:* Markel scholar acad med, 61-66; consult, Palo Alto Vet Admin Hosp. *Mem:* Am Fedn Clin Res; Am Soc Hemat; Int Soc Hemat; Am Soc Clin Invest. *Res:* Metabolism and transport of red cell membranes; properties of erythrocyte membranes in health, disease and aging. *Mailing Add:* Div Hemat S-161 Stanford Univ Med Ctr 300 Pasteur Dr Stanford CA 94305

SCHRIESHEIM, ALAN, b NY, Mar 8, 30; m 53; c 2. RESEARCH ADMINISTRATION, ORGANIC CHEMISTRY. *Educ:* Polytech Inst Brooklyn, BS, 51; Pa State Univ, PhD(phy org chem), 54. *Prof Exp:* Res chemist, Nat Bur Standards, 54-56; res chemist, Exxon Res & Eng Co, 56-58, sr chemist, 58-59, res assoc, 59-63, sect head, 63-65, asst dir, 65-66, asst mgr, 66-69, dir chem sci lab, Corp Res Labs, 69-75, dir corp res labs, 75-78, gen mgr technol dept, 78-83; sr dep dir & chief oper officer, 83-84, DIR & CHIEF EXEC OFFICER, ARGONNE NAT LAB, 84- *Concurrent Pos:* Mem adv bd, Corp Vis Comt for Dept Chem, Mass Inst Technol, Chemtech & Stanford Energy Inst; co-chmn, Assembly Math & Phys Sci Comt on Chem Sci, Nat Res Coun, Indust Adv Comt-Rutgers Univ, Solid State Sci Adv Panel & Pure & Appl Chem Deleg to the People's Republic of China, Nat Acad Sci, 78. *Honors & Awards:* Petrol Chem Award, Am Chem Soc, 69; Karcher Silver Medalist Lectr. *Mem:* AAAS; Am Chem Soc; Sigma Xi; fel NY Acad Sci; fel Am Inst Chemists. *Res:* Kinetics and mechanism of acid and base catalyzed organic reactions including alkylation, isomerization, hydrogenation and polymerization. *Mailing Add:* Argonne Nat Lab 9700 S Cass Ave Argonne IL 60439-4832

SCHRIEVER, BERNARD ADOLF, b Ger, Sept 14, 10; nat US; m 38; c 3. AERONAUTICAL ENGINEERING. *Educ:* Tex Agr & Mech Col, BS, 31; Stanford Univ, MS, 42. *Hon Degrees:* DSc, Rider Col & Creighton Univ, 58, Adelphi Col, 59, Rollins Col, 61; LLD, Loyola Univ, Calif, 62, C W Post Col, Long Island, 65; DrAeroSci, Univ Mich, 62; DrEng, Polytech Inst Brooklyn, 62, PMC Cols, 66. *Prof Exp:* USAF, 32-66, chief staff, Far East Air Serv Command, 43-44, comdr adv echelon, 44-45, chief plans & policy div, Res & Develop, 45-49, asst opers, Develop Planning, 50-54, asst to comdr, Air Res & Develop Command, 54-59, comdr, Air Force Systs Command, 59-66; indust consult, 66-71; CONSULT, 71- *Concurrent Pos:* Dir, Am Med Int, Control Data Corp, Emerson Elec, Wackenhut Corp, Rockwell Int, Aerojet. *Honors & Awards:* Aviation Medal, Rome, 80; James Forrester Mem Award, 86. *Mem:* Nat Acad Eng; hon fel Inst Aeronaut & Astronaut. *Res:* Advanced technology in aerospace missions. *Mailing Add:* 2800 Shirlington Rd Suite 405A Arlington VA 22206

SCHRIEVER, RICHARD L, b Salt Lake City, Utah, Nov 22, 40; m 61; c 3. PHYSICS. *Educ:* Univ Utah, BS, 63, PhD(elec eng), 67. *Prof Exp:* Res engr, Lawrence Livermore Lab, 67-74; br chief, Laser Fusion Br, US Am Eng Comn, 74-76; dep dir, Off Inertial Fusion, Energy Res & Develop Admin, 76-77; dep dir, 77-81, DIR, OFF INERTIAL FUSION, US DEPT ENERGY, 81- *Mem:* Inst Elec & Electronics Engrs; Am Phys Soc. *Res:* Physics of laser and particle beam driven inertial fusion; application of inertial fusion to nuclear explosives and fusion power. *Mailing Add:* 4501 Dexter St Washington DC 20545

SCHRIRO, GEORGE R, b New York, NY, Sept 3, 21; m 48; c 4. APPLIED MATHEMATICS, AERODYNAMICS. *Educ:* NY Univ, BS, 47, MA, 50. *Prof Exp:* Night mgr, Western Union Tel Co, NY, 36-42; instr math, Wash Univ, St Louis, 48-49; teacher high sch, NJ, 49-51; chmn math dept, NY, 51-56; STRUCT ANALYST, DYNAMICIST & SR ENGR, GRUMMAN AEROSPACE CORP, 56- *Concurrent Pos:* Adj asst prof math, C W Post Col, Long Island Univ, 62-80, adj assoc prof, 80- *Mem:* Math Asn Am; Am Inst Aeronaut & Astronaut. *Res:* Stability and control; aeroelastic effects; dynamic analysis; stress and fatigue. *Mailing Add:* 121 Prospect St Farmingdale NY 11735

SCHROCK, GOULD FREDERICK, b Rockwood, Pa, Apr 23, 36; m 57; c 2. BOTANY, MYCOLOGY. *Educ:* Indiana Univ Pa, BS, 57, MEd, 61; Univ Chicago, PhD(bot), 64. *Prof Exp:* Joint high sch teacher, Ind, 58-61; assoc prof bot, Kutztown State Col, 64-68; PROF BOT, INDIANA UNIV PA, 68- *Mem:* Mycol Soc Am; Bot Soc Am; Int Soc Human & Animal Mycol; Am Inst Biol Sci; Am Hort Soc. *Res:* Factors influencing growth and development of human pathogenic fungi; succession of fungi in selected natural environments. *Mailing Add:* Weyandt 222 Indiana Univ Pa Indiana PA 15705-1090

SCHROCK, RICHARD ROYCE, b Berne, Ind, Jan 4, 45. ORGANOMETALLIC CHEMISTRY. *Educ:* Univ Calif, Riverside, AB, 67; Harvard Univ, PhD(chem), 71. *Prof Exp:* NSF fel, Cambridge Univ, 71-72; res chemist, Cent Res & Develop Dept, E I du Pont de Nemours & Co, Inc, 72-75; from asst prof to assoc prof chem, 75-80, PROF CHEM, MASS INST TECHNOL, 80- *Concurrent Pos:* A P Sloan fel, 76-80; Dreyfus teacher-scholar, 78-83. *Honors & Awards:* Organometallic Chem Award, Am Chem Soc, 85; Harrison Howe Award, Am Chem Soc, 91. *Mem:* Am Chem Soc; Am Acad Sci. *Res:* Synthetic and mechanistic organo-transition metal chemistry; homogeneous catalysis; early transition metal chemistry; metal-alkyl, metal-carbene and metal-carbyne complexes; reduction of carbon monoxide; olefin metathesis; olefin polymerization; acetylene metathesis. *Mailing Add:* Mass Inst Technol 6-331 77 Massachusetts Ave Cambridge MA 02139

SCHROCK, VIRGIL E(DWIN), b San Diego, Calif, Jan 22, 26; m 46; c 2. NUCLEAR POWER, NUCLEAR REACTOR SAFETY. *Educ:* Univ Wis, BS, 46, MS, 48; Univ Calif, ME, 52. *Prof Exp:* Instr mech eng, Univ Wis, 46-48; US Navy, Eng Off, USS O'Brien DD 725, 52-53; lectr mech eng, 48-51, asst prof, 54-60, assoc prof, 60-68, asst dean res, 68-74, PROF NUCLEAR ENG, UNIV CALIF, BERKELEY, 68- *Concurrent Pos:* Vis res fel, Ctr Info, Studies & Exp, Milan, Italy, 62-63 & 74-75; consult to indust & govt; tech ed, J Heat Transfer, Am Soc Mech Engrs, chmn, Heat Transfer Div, 78-79 & Nat Heat Transfer Conf Coord Comn, 80-81; chmn, Thermal Hydraulics Div, Am Nuclear Soc, 82-83; Japan Soc Prom Sci res fel, 84. *Honors & Awards:* Glenn Murphy Award, Am Soc Eng Educ, 83; Heat Trans Mem Award, Am Soc Mech Eng, 85, 50th Anniversary Award, 88. *Mem:* Fel Am Nuclear Soc; fel Am Soc Mech Engrs; Am Soc Eng Educ; Sigma Xi. *Res:* Thermodynamic and transport properties of fluids; heat transfer and fluid dynamics; boiling and two-phase flow; thermal design of nuclear power plants; environmental aspects of nuclear power; safety analysis of nuclear systems; resources conservation and planning. *Mailing Add:* Dept Nuclear Eng Univ Calif Berkeley CA 94720

SCHRODER, DAVID JOHN, b Edmonton, Alta, Oct 29, 41; m 65; c 2. FOOD SAFETY & QUALITY, ADMINISTRATION. *Educ:* Univ Alta, BSc, 64, MSc, 69; Univ Minn, PhD(food microbiol), 73. *Prof Exp:* Res scientist, Agr Can, 73-78; supvr, Alta Agr Food Res, 78-84; head, Alta Agr Food Processing Develop Ctr, 84-90; DIR, ALTA AGR FOOD LAB SERV, 90- *Concurrent Pos:* Adj prof, Univ Alta, 80-; exec, bd dirs, POS Pilot Plant Corp, 81-90; chmn, Nat Ann Conf, Can Inst Food Sci & Technol, 86; Prov rep, Can Comt Food, 88- *Mem:* Can Inst Food Sci & Technol, 89-90; Inst Food Technologists. *Res:* Food microbiology, safety, quality, product development and processing development. *Mailing Add:* Food Lab Serv 6909 116 St Edmonton AB T6H 4P2 Can

SCHRODER, DIETER K, b Lubeck, Ger, June 18, 35; US citizen; m 61; c 2. ELECTRICAL ENGINEERING, SOLID STATE PHYSICS. *Educ:* McGill Univ, BEng, 62 & MEng, 64; Univ Ill, PhD(elec eng), 68. *Prof Exp:* Sr engr, Westinghouse Res Labs, 68-72, fel eng, 72-76, adv engr, 76-79, mgr, 79-81; PROF ENG, ARIZ STATE UNIV, 81- *Concurrent Pos:* Vis engr, Inst Appl Solid State Physics, Freiburg, Ger, 78-79. *Mem:* Fel Inst Elec & Electronics Engrs; Electrochem Soc; Sigma Xi. *Res:* Solid state electronics, especially semi-conductor materials, defects, and material and device characterization. *Mailing Add:* Elec Eng Dept Ariz State Univ Tempe AZ 85287-5706

SCHRODER, GENE DAVID, b Atascadero, Calif, Oct 25, 44. ECOLOGY. *Educ:* Rice Univ, BA, 67, MA, 70; Univ NMex, PhD(ecol), 74. *Prof Exp:* Asst prof, 74-80, ASSOC PROF ECOL, SCH PUB HEALTH, UNIV TEX, HOUSTON, 80- *Concurrent Pos:* Adj assoc prof ecol, Rice Univ, Houston, 79- *Mem:* Ecol Soc Am; Am Soc Mammalogists. *Res:* Dynamics of species interactions with particular interests in competition among desert rodents and factors regulating urban rodent populations. *Mailing Add:* Dept Ecol Univ Tex Health Sci Ctr PO Box 20036 Houston TX 77225

SCHRODER, JACK SPALDING, b Atlanta, Ga, Jan 17, 17; m 38; c 5. INTERNAL MEDICINE. *Educ:* Georgetown Univ, AB, 37; Emory Univ, MD, 41; Am Bd Internal Med & Am Bd Gastroenterol, dipl. *Prof Exp:* From assoc prof to prof med, Sch Med, Emory Univ, 58-87; RETIRED. *Concurrent Pos:* Consult, Atlanta Vet Admin Hosp & Third Army Sr Surgeon; civilian consult. *Mem:* Am Gastroenterol Asn; fel Am Col Physicians. *Res:* Gastroenterology; clinical medicine. *Mailing Add:* 2600 Rivers Rd Atlanta GA 30305

SCHRODER, KLAUS, b Celle, Ger, Nov 1, 28; m 57; c 2. SOLID STATE PHYSICS, MATERIALS SCIENCE. *Educ:* Univ Marburg, Vordiplom, 51; Univ Gottingen, Dr rer nat, 54- *Prof Exp:* Res officer, Commonwealth Sci & Indust Res Orgn, Univ Melbourne, 55-58; res assoc mining & metall eng, Univ Ill, 58-60, res asst prof, 60-61; assoc prof metall eng, 61-68, PROF METALL, SYRACUSE UNIV, 68- *Mem:* Am Soc Metals; Am Phys Soc; Am Asn Univ Prof; Sigma Xi. *Res:* Magnetic memory; plastic properties of metals; specific heat; Hall and Seebeck effects of transition element alloys; optical properties of alloys; crack studies; magnetic thin films. *Mailing Add:* Dept Chem Eng & Mat Sci Syracuse Univ Syracuse NY 13210

SCHRODER, VINCENT NILS, b Chicago, Ill, Dec 8, 20; m 59; c 4. PLANT PHYSIOLOGY. *Educ:* Univ Ga, BSA, 48; Duke Univ, PhD(plant physiol), 56. *Prof Exp:* Asst prof, 55-69, ASSOC PROF AGRON, UNIV FLA, 69- *Mem:* Am Soc Agron; Am Soc Plant Physiologists. *Res:* Plant mineral nutrition; effects of environment on photosynthesis; soil temperature effects on plant growth. *Mailing Add:* 1630 NW 23rd St Gainesville FL 32605-3878

SCHRODER, WOLF-UDO, b Stralsund, Ger, May 25, 42; m 68. NUCLEAR CHEMISTRY. *Educ:* Univ Gottingen, Free Univ Berlin, dipl, 67; Tech Univ Darmstadt, PhD(nuclear physics), 71. *Prof Exp:* Res assoc nuclear physics, Univ Darmstadt, 68-75; res assoc, Univ Rochester, 75-79, sr res assoc, 79-80, prof nuclear sci, 81-83, assoc prof, 83-87, PROF CHEM, UNIV ROCHESTER, 87- *Concurrent Pos:* Fel, Ger Acad Exch Serv. *Mem:* Am Phys Soc; Am Chem Soc; Sigma Xi. *Res:* Heavy-ion and light-ion reactions; muon-induced reactions. *Mailing Add:* Dept Chem Univ Rochester Wilson Blvd Rochester NY 14627

SCHRODT, JAMES THOMAS, b Louisville, Ky, Oct 7, 37; m; c 2. CHEMICAL ENGINEERING, MATHEMATICS. *Educ:* Univ Louisville, BChE, 60, PhD(chem eng), 66; Villanova Univ, MChE, 62. *Prof Exp:* Jr engr, Tenn Eastman Co, 60; instr chem eng, Univ Louisville, 62-65; sr res engr, Tenn Eastman Co, 65-66; asst prof, 66-72, ASSOC PROF CHEM ENG, UNIV KY, 72- *Mem:* Am Chem Soc; Am Inst Chem Engrs. *Res:* Simultaneous heat and mass transfer; thermodynamics; electrodialysis. *Mailing Add:* Dept Chem Eng Univ of Ky Lexington KY 40506

SCHRODT, VERLE N(EWTON), b Muscatine, Iowa, Apr 26, 33; m 55; c 8. CHEMICAL ENGINEERING. *Educ:* Univ Ill, Urbana, BS, 55; Pa State Univ, MS, 58, PhD(chem eng), 61. *Prof Exp:* Supvr, Appl Sci Labs, Inc, 56-61; sr res engr, Monsanto Co, St Louis, 61-67, eng fel, 67-77, sr eng fel, 77-85; chief, Chem Eng Sci Div, Nat Bur Standards, Boulder, Co, 86-88; prof & dept head, Chem Eng, 88-89, PROF & ASST DEAN RES & GRAD STUDIES, UNIV ALA, TUSCALOOSA, ALA, 89- *Mem:* Am Inst Chem Engrs; Am Chem Soc. *Res:* Mathematical modeling of chemical and biological systems; design of agricultural growth facilities; image processing; x-ray analysis. *Mailing Add:* 1704 Hollow Lane Tuscaloosa AL 35406

SCHROEDER, ALFRED C(HRISTIAN), b West New Brighton, NY, Feb 28, 15; c 1. ELECTRICAL ENGINEERING, COLORIMETRY. *Educ:* Mass Inst Technol, BS & MS, 37. *Prof Exp:* Mem tech staff, David Sarnoff Res Ctr, RCA Corp, 37-80; RETIRED. *Honors & Awards:* David Sarnoff Gold Medal Award, Soc Motion Picture & TV Engrs, 65; Vladimir K Zworykin Award, Inst Elec & Electronics Engrs, 71-; Karl Ferdinand Braun Prize, Soc Info Display, 89. *Mem:* AAAS; fel Inst Elec & Electronics Engrs; Soc Motion Picture & TV Engrs; Optical Soc Am; Sigma Xi. *Res:* Sequential, simultaneous, and simultaneous subcarrier color television systems; tri-color tubes; mechanism of color vision. *Mailing Add:* 114 Pennswood Village Apt I Newtown PA 18940-0909

SCHROEDER, ALICE LOUISE, b Knoxville, Tenn, June 22, 41; m 66; c 2. GENETICS. *Educ:* Univ Colo, Boulder, BA, 63; Stanford Univ, PhD(biol, genetics), 70. *Prof Exp:* NIH fel, 69-70, lectr genetics, 71, asst prof, 71-78, ASSOC PROF GENETICS, WASH STATE UNIV, 78- *Concurrent Pos:* NIH res grants, 71-74, 78-80, 79-82, 85 & NSF, 85. *Mem:* AAAS; Genetics Soc Am; Am Women Sci. *Res:* DNA maintenance systems; recombination; genetics of radiation sensitivity; fungal, bacterial and viral genetics. *Mailing Add:* Prog Genetics & Cell Biol Wash State Univ Pullman WA 99164-4234

SCHROEDER, ALLEN C, animal physiology, comparative endocrinology; deceased, see previous edition for last biography

SCHROEDER, ANITA GAYLE, b Wichita, Kans. APPLIED STATISTICS. *Educ:* Baker Univ, BS, 66; Kans State Univ, MS, 68; Ore State Univ, PhD(statist), 72. *Prof Exp:* Proj dir, Ark Health Statist Ctr, 72-73; dir, Emergency Med Serv Data & Eval, Ark Health Systs Found, 73-74; pres, Schroeder & Assocs, 74-75; statistician, Westat, Inc, 75-84, dir, Social Sci Serv, 79-84; MEM STAFF, DEPT STATE, 84- *Concurrent Pos:* Asst prof biomet, Med Ctr, Univ Ark, 72-74. *Mem:* Sigma Xi; Am Statist Asn; Biomet Soc. *Res:* Surveys in social services and health; evaluation. *Mailing Add:* Am Embassy Islamabad PSC Box 29 APO New York NY 09614

SCHROEDER, DANIEL JOHN, b Manitowoc, Wis, Sept 23, 33; m 55; c 2. ASTRONOMY, OPTICS. *Educ:* Beloit Col, BS, 55; Univ Wis, MS, 57, PhD(physics), 60. *Prof Exp:* Res assoc astron, Univ Wis, 60-63; from asst prof physics to assoc prof physics & astron, 63-74, PROF PHYSICS & ASTRON, BELOIT COL, 74- *Concurrent Pos:* Telescope scientist, Space Telescope Proj, NASA. *Mem:* Am Astron Soc; Optical Soc Am; Int Astron Union. *Res:* Optical spectroscopy and instrumentation; astronomical optics and space astronomy. *Mailing Add:* Dept Physics Beloit Col Beloit WI 53511

SCHROEDER, DAVID HENRY, b Indianapolis, Ind, Feb 16, 40; m 62; c 2. BIOCHEMICAL PHARMACOLOGY. *Educ:* Purdue Univ, West Lafayette, BS, 62, MS, 66, PhD(biochem), 68. *Prof Exp:* Res assoc chem pharmacol, NIH, 68-70; RES BIOCHEMIST, WELLCOME RES LABS, 70- *Mem:* Am Chem Soc. *Res:* Development of analytical methods for detection and quantitation of drugs and metabolites; computerization of data; pharmacokinetics, bioavailability of drugs; use of computer spread sheet software for pharmacokinetics analyses. *Mailing Add:* Wellcome Res Lab-Pharmacokinetics & Drug Metab 3030 Cornwallis Rd Research Triangle Park NC 27709

SCHROEDER, DAVID J DEAN, b Hutchinson, Kans, Mar 21, 42; m 64; c 2. PERSONNEL RESEARCH, DRUG & ALCOHOL EFFECTS ON PERFORMANCE. *Educ:* Tabor Col, BA, 64; Kans State Teachers Col Emporia, MS, 67; Univ Okla, PhD(exp/soc psychol), 71. *Prof Exp:* Res psychologist, Civil Aeromed Inst, 70-72; clin psychol intern, Norfolk Regional Ctr & Northeast Mental Health Clin, Nebr, 72-73; clin psychologist, Vet Admin Hosp, Murfreesboro, Tenn, 73-75; adj asst prof psychol, Mid Tenn State Univ, 74-75; clin psychologist, Vet Admin Med Ctr, Topeka, Kans, 75-80; adj asst prof psychol, Washburn Univ, Topeka, Kans, 75-80; supvr, Clin Psychol Res Unit, Aviation Psychol Lab, 80-87, Clin Psychol Res Sect, Human Resources Br, 87-89, Field Performance Res Sect, Human Resources Res Div, 89-90, MGR, HUMAN FACTORS RES LAB, FED AVIATION ADMIN, 90- *Concurrent Pos:* Mem, comt, Aerospace Med Asn, 77-79, publicity comt, 80-83, arrangements comt, 77, regist comt, 77-, sci prog comt, 82-, aviation safety comt, 82-89, educ & training comt, 83-, sci & technol comt, 85, 86, 87, 88, long range planning comt, 86, 87, 88, human factors comt, 89-, chair-elect, Assoc Fel group, 81-82, chair, 82-83, chair, Poster Sessions subcomt, 85, 86, 87, deputy chair, sci prog comt, 89-90, chair, 90-91 & numerous other comts. *Mem:* Sigma Xi; fel Aerospace Med Asn; Am Psychol Asn; Psychonomic Soc. *Res:* The use of survey methodology to determine employee job satisfaction, well-being/stress, and reactions to job change/automation; assessment of factors associated with air traffic controller selection and performance; short and long term effects of alcohol and drugs on performance and the vestibular system; author & co-author of numerous publications. *Mailing Add:* Human Factors Res Lab Fed Aviation Admin Aeronaut Ctr Civil Aeromed Inst PO Box 25082 Oklahoma City OK 73125

SCHROEDER, DOLORES MARGARET, b New York, NY, July 30, 37; c 1. NEUROANATOMY. *Educ:* Notre Dame Col, BS, 58; John Carroll Univ, MS, 63; Case Western Reserve Univ, PhD(anat), 70. *Prof Exp:* Fel neurosurg, Med Sch, Univ Va, 70-72, from instr to asst prof, 72-75; assoc prof, 75-80, TENURED ASSOC PROF MED SCI, SCH MED, IND UNIV, BLOOMINGTON, 80- *Mem:* Soc Neurosci; Am Asn Anat; AAAS; Int Brain Res Orgn. *Res:* Comparative neuroanatomy; development and organization of brainstem and spinal cord. *Mailing Add:* Med Sci Prog Med Sch Univ Ind Bloomington IN 47401

SCHROEDER, DUANE DAVID, b Newton, Kans, Nov 4, 40; m 61; c 3. BIOCHEMISTRY. *Educ:* Bethel Col, AB, 62; Tulane Univ, PhD(biochem), 67. *Prof Exp:* Damon Runyon fel, Mass Inst Technol, 67-69; sr res biochemist, 69-71, biochem res supvr, 71-73, mgr biochem res, 73-80, assoc dir biochem res & develop, 80-83, assoc dir biochem res, 83-87, DIR, RES & DEVELOP PLANNING & ADMIN, CUTTER LABS INC, 87- *Concurrent Pos:* Res fel, Bayer AG, Ger, 75. *Mem:* AAAS. *Res:* Enzyme active sites and structure/function relationships; biologicals from plasma and recombinant DNA sources; hepatitis transmission; intravenous therapeutic immunoglobulins. *Mailing Add:* Cutter Labs Miles Inc PO Box 1986 Berkeley CA 94701

SCHROEDER, FRANK, JR, b Bartlesville, Okla, Sept 8, 27; m 49; c 2. NUCLEAR REACTOR SAFETY, REACTOR REGULATION. *Educ:* Univ Ill, BS, 49, MS, 51. *Prof Exp:* Asst physics, Univ Ill, 49-51; physicist, Atomic Energy Div, Phillips Petrol Co, 51-57, supvr Spert-3 reactor exps, 57-60, mgr, Spert Proj, 60-66, reactor safety prog officer, 66-68; dep dir div reactor licensing, US AEC, 68-72, asst dep tech rev directorate of licensing, 72-75; dep dir, Div Tech Rev, Off Nuclear Reactor Regulation, US Nuclear Regulatory Comn, 75-76, dep dir, Div Systs & Safety, 76-80, asst dir, Generic Projs, Div Safety Technol Off, 80-85, dep dir, Div Pressurized Water Reactor Licensing-B, 85-86, asst dir, Div Reactor Projs, 86-87; RETIRED. *Mem:* Fel Am Nuclear Soc. *Res:* Nuclear reactor safety and kinetics; reactor physics; nuclear engineering. *Mailing Add:* 802 S Belgrade Rd Silver Spring MD 20902

SCHROEDER, FRIEDHELM, b Kastorf, Ger, July 16, 47; US citizen; m 79. MEMBRANE LIPID ASYMMETRY, ATHEROSCLEROSIS. *Educ:* Univ Pittsburgh, BS, 70; Mich State Univ, PhD(biochem), 74. *Prof Exp:* NSF fel biochem, Mich State Univ, 70-74; Am Cancer Soc fel biol chem, Med Sch, Wash Univ, 74-76; ASST PROF PHARMACOL, SCH MED, UNIV MO, 76- *Concurrent Pos:* Prin investr grants, Am Heart Asn, 77-, Nat Cancer Inst, 78-, Pharmaceut Mfg Asn, 78-80, & Hereditary Dis Found, 80-; consult, Miles Res Labs, 76, Hemotropic Dis Group, 80-, & Hormel Inst, 81; mem, Am Heart Asn Arteriosclerosis Coun, 78- *Mem:* Am Soc Pharmacol & Exp Therapeut; Am Soc Biol Chemists; Soc Neurosci; Am Oil Chemists Soc. *Res:* Structure and function of lipids in membranous particles (plasma membranes and lipoproteins) from cancer cell, blood, brain, liver and skin fibroblasts; biochemical, biophysical (fluorescence and differential scanning calorimetry), and pharmacological methods. *Mailing Add:* Pharm Col Univ Cincinnati 3223 Eden Ave Cincinnati OH 45267

SCHROEDER, HANSJUERGEN ALFRED, b Lautawerk, Ger, Jan 21, 26; nat US; m 53. ORGANIC CHEMISTRY. *Educ:* Univ Berlin, BS, 49, MS, 50; Univ Freiburg, PhD(chem), 53. *Prof Exp:* Asst chem, Univ Berlin, 48-50, instr, 51; res assoc, 52-56, sr res chemist, 57-58, res specialist, 59-63, sect mgr, 64-69, venture mgr, 70-72, mgr res & develop, 73-74, DIR PROD RES, OLIN CORP, 75- *Mem:* Am Chem Soc; Ger Chem Soc. *Res:* Organic synthetic chemistry; nitrogen heterocycles; fluorine, boron and phosphorous compounds; pesticides; lubricants; high temperature polymers, biocides; pool chemicals; product development. *Mailing Add:* 609 Mix Ave Apt 3 Hamden CT 06514

SCHROEDER, HARTMUT RICHARD, b Hitzdorf, Ger, Aug 29, 42; US citizen; m 72; c 3. BIOCHEMISTRY. *Educ:* Youngstown Univ, BS, 66; Pa State Univ, MS, 70, PhD(biochem), 72. *Prof Exp:* Res assoc biochem, Mich State Univ, 72-74; from res scientist to sr res scientist biochem, Ames Div, Miles Lab, Inc, 74-85, sr staff scientist, 85-90; DIR IMMUNOL DEVELOP, EX OXEMIS INC, 90- *Concurrent Pos:* NAm ed, J Bioluminescence & Chemiluminescence, 86- *Mem:* Am Chem Soc; Am Soc Photobiol; Am Asn Clin Chem. *Res:* Bioluminescence, chemiluminescence and fluorescence, competitive protein binding reactions, enzyme assays, electrochemistry and immunochemistry; biosensors. *Mailing Add:* 12826 Castle Bend Dr San Antonio TX 78230

SCHROEDER, HERBERT AUGUST, b Cleveland, Ohio, Feb 26, 30; div; c 3. ORGANIC CHEMISTRY, WOOD & PULPING CHEMISTRY. *Educ:* Univ Idaho, BS, 52, MS, 54; Univ Hamburg, DSc(org chem), 60. *Prof Exp:* Res chemist, Forest Prod Lab, US Forest Serv, 61-63; asst prof forest prod chem, Forest Res Lab, Ore State Univ, 63-68; assoc prof, 68-79, PROF WOOD CHEM, FOREST & WOOD SCI, COLO STATE UNIV, 79- *Mem:* Am Chem Soc; Soc Wood Sci & Technol; Tech Asn Pulp & Paper Indust; Sigma Xi. *Res:* Chemistry of wood carbohydrates, wood polyphenolics and pulping processes; chemical treatment of wood; wood adhesives; biomass conversion to energy and chemicals. *Mailing Add:* Dept Forest & Wood Sci Colo State Univ Ft Collins CO 80523

SCHROEDER, HERMAN ELBERT, b Brooklyn, NY, July 6, 15; m 38; c 4. ORGANIC CHEMISTRY, POLYMER CHEMISTRY. *Educ:* Harvard Univ, AB, 36, AM, 37, PhD(chem), 39. *Prof Exp:* Res chemist, Exp Sta, E I du Pont de Nemours & Co, Inc, 38-45, res chemist, Jackson Lab, 45-46, group leader, 46-49, head miscellaneous dyes div, 49-51, asst dir lab, 51-57, asst dir res, Elastomer Chem Dept, 57-63, res dir, 63-65, dir res & develop, 65-80; PRES, SCHROEDER SCI SERV INC, 80- *Concurrent Pos:* Chmn res comt & trustee & vpres, Univ Del Res Found; sci consult, Metrop Mus Art, 80-84, Wintherthor Mus, 81-, Smithsonian, 84-88. *Honors & Awards:* Gen Award, Int Inst Rubber Producers; Goodyear Medal, Rubber Div, Am Chem Soc. *Mem:* Fel AAAS; Am Chem Soc; Soc Chem Indust. *Res:* Catalysis; resins; adhesives; polymers; rubber chemicals; color photography; vat dyes; pigments; application of dyes; fluorine chemicals; textile chemicals; elastomers; discovery and development of new elastomeric polymers and intermediates; fluoropolymers, art conservation. *Mailing Add:* No 74 Stonegates 4031 Kennett Pike Greenville DE 19807-2037

SCHROEDER, JOHN, b Pardan (Banat), Yugoslavia, Aug 31, 38; US citizen; m 64; c 2. HIGH PRESSURE PHYSICS. *Educ:* Univ Rochester, BS, 62, MS, 64; Cath Univ Am, PhD(physics), 74. *Prof Exp:* Res & develop officer, Atmospheric Effects Div, Defense Atomic Support Agency, USN, 67-70; res asst, Cath Univ Am, 70-74; physicist, Acoust Div, Naval Res Lab, 74-75; res assoc, dept chem, Univ Ill, 75-78; physicist, Corp Res & Develop, Gen Elec Co, 78-81; assoc prof, 82-90, PROF PHYSICS, RENSSELAER POLYTECH INST, 90- *Concurrent Pos:* Sr staff mem, Mats Res Lab, Univ Ill, 76-78; staff mem, Ctr Glass Sci & Technol, Rensselaer Polytech Inst, 85- *Mem:* Am Phys Soc; Optical Soc Am. *Res:* Brillouin, raman and rayleigh spectroscopy with emphasis on amorphous solids (glasses) and liquids; high pressure research; behavior of materials under extreme conditions of pressure and temperature; optical properties of glasses, fiber optics; magnetic properties of disordered solids; nonlinear optics; wave propagation in solids; photoluminescence; semiconductor microcrystallites in glass composites and colloids. *Mailing Add:* Dept Physics SC-1C18 Rensselaer Polytech Inst Troy NY 12180-3590

SCHROEDER, JOHN SPEER, b South Bend, Ind, May 6, 37. CARDIOLOGY. *Educ:* Univ Mich, MD, 62; Am Bd Internal Med, dipl, 69; Am Bd Cardiovasc Dis, dipl, 73. *Prof Exp:* Intern, 62-63, resident internal med, 65-67, asst prof med & cardiol, 70-77, ASSOC PROF MED & CARDIOL, MED CTR, STANFORD UNIV, 77-, DIR INTENSIVE CARDIAC CARE UNIT, 73- *Concurrent Pos:* Fel cardiol, Med Ctr, Stanford

Univ, 67-69. *Mem:* Am Fedn Clin Res; fel Am Col Cardiol; NY Acad Sci; fel Am Col Physicians. *Res:* Cardiac transplantation; coronary artery spasm; calcium antagonists; coronary artery disease. *Mailing Add:* Cardiol Div Cvre 261 Stanford Univ Hosp 300 Pasteur Dr Stanford CA 94305

SCHROEDER, JUEL PIERRE, b New England, NDak, Jan 23, 20; m 43, 77. ORGANIC CHEMISTRY. *Educ:* Univ NDak, BS, 41; Univ Wis, PhD(chem), 48. *Prof Exp:* Anal chemist, Org Chem Div, Monsanto Co, 42-43, res chemist, Plastics Div, 43-46; res chemist, Union Carbide, 48-58, asst dir res & develop, Plastics Div, 58-63; Robert A Welch fel, Univ Tex, 63-65; assoc prof chem, 65-68, prof, 68-80, EMER PROF CHEM, UNIV NC, GREENSBORO, 80- *Mem:* Am Chem Soc. *Res:* Liquid crystals; polymer chemistry; imidates and orthoesters. *Mailing Add:* 3016 Mayfield Way Michigan City IN 46360-1716

SCHROEDER, LAUREN ALFRED, b Long Prairie, Minn, Feb 24, 37; m 60; c 2. ECOLOGY. *Educ:* St Cloud State Col, BS, 60; Univ SDak, MA, 65, PhD(zool), 68. *Prof Exp:* From asst prof to assoc prof, 68-76, PROF BIOL, YOUNGSTOWN STATE UNIV, 76- *Mem:* AAAS; Am Inst Biol Sci; Ecol Soc Am; Entom Soc Am. *Res:* Ecological energetics of Lepidoptera especially as related to plant defense mechanisms and growth performance of larvae; population dynamics; ecosystem structure; PAH metabolism by aquatic vertebrates. *Mailing Add:* Dept Biol Youngstown OH 44555

SCHROEDER, LEE S, b Braddock, Pa, Apr 11, 38; m 57; c 4. ELEMENTARY PARTICLE PHYSICS, HIGH ENERGY NUCLEAR COLLISIONS. *Educ:* Drexel Inst, 61; Ind Univ, Bloomington, MS, 63, PhD(physics), 66. *Prof Exp:* Assoc physics, Iowa State Univ, 65-67, asst prof, 67-71; res physicist, 71-76, staff scientist, 76-83, SR STAFF SCIENTIST, LAWRENCE BERKELEY LAB, 83-, BEVALAC SCI DIR, 87- *Concurrent Pos:* Assoc, Ames Lab, AEC, 65-67, assoc physicist, 67-71; US Dept Energy, Nuclear Physics Div, 87-89. *Mem:* Am Phys Soc; Fel, Am Phys Soc. *Res:* Experimental elementary particle physics, particularly the use of bubble chamber, counters, and spark chambers to study the strong interactions of the elementary particles; high energy ions; streamer chambers; spectrometers to measure dilepton production and other kinematically forbidden processes. *Mailing Add:* 70-A-3307 Lawrence Berkeley Lab Berkeley CA 94720

SCHROEDER, LELAND ROY, organic chemistry, for more information see previous edition

SCHROEDER, LEON WILLIAM, b Guthrie, Okla, Jan 25, 21; m 42; c 2. ASTROPHYSICS, ASTRONOMY EDUCATION. *Educ:* Okla State Univ, BS, 47, MS, 48; Ind Univ, PhD(astrophys), 58. *Prof Exp:* From instr to prof, 47-84, EMER PROF PHYSICS, OKLA STATE UNIV, 84- *Concurrent Pos:* Guest investr, Dominion Astrophys Observ, 60; vis assoc prof, Northwestern Univ, 61-62; guest observer, Dyer Observ, 64; vis astronr, Kitt Peak Nat Observ, 71. *Mem:* Am Astron Soc. *Res:* Stellar spectrophotometry and photoelectric photometry; scanner energy distributions; education. *Mailing Add:* Rte 4 Box 640 Stillwater OK 74074-9679

SCHROEDER, LEROY WILLIAM, b Watertown, Wis, July 18, 43; m 67; c 1. PHYSICAL CHEMISTRY, BIOPHYSICS. *Educ:* Wartburg Col, BA, 64; Northwestern Univ, Evanston, PhD(phys chem), 69. *Prof Exp:* Nat Res Coun-Nat Bur Standards assoc, 69-71, res assoc, Am Dent Asn Res Div, 71-74, head dent crystallog, Am Dent Asn Health Found, Nat Bur Standards, 74-77; proj scientist, Div Chem & Physics, 77-80, RES CHEMIST, FOOD & DRUG ADMIN, 80- *Mem:* Am Chem Soc; Am Crystallog Asn. *Res:* Molecular structure and dynamics; thermodynamics; chemical processes; hydrogen bonding; diffusion in solids. *Mailing Add:* 23000 Timber Creek Lane Clarksburg MD 20871

SCHROEDER, MANFRED ROBERT, b Ahlen, Ger, July 12, 26; nat US; m 56; c 3. ACOUSTICS, NUMBER THEORY. *Educ:* Univ Goettingen, dipl, 51, Dr rer nat(physics), 54. *Prof Exp:* Sci asst microwaves & acoust, Univ Goettingen, 52-54; mem tech staff, Bell Tel Labs, Inc, 54-58, head acoust res, 58-63, dir, Acoust & Speech Res Lab, 63-64, dir, Acoust, Speech & Mech Res Lab, 64-69; PROF PHYSICS & DIR THIRD PHYSICS INST, UNIV GOETTINGEN, 69- *Concurrent Pos:* Ed, Speech & Speech Recognition, 85. *Honors & Awards:* Gold Medal, Audio Eng Soc, 72; W R G Baker Prize Award, Inst Elec & Electronics Engrs, 75; Sr Award Speech & Signal Processing, Inst Elec & Electronics Engrs, 79; Lord Rayleigh Gold Medal, Brit Inst Acoustics, 86; Gold Medal, Acoust Soc Am, 91. *Mem:* Nat Acad Eng; fel Acoust Soc Am; fel Inst Elec & Electronic Engrs; fel Audio Eng Soc; fel Am Acad Arts & Scis; Europ Phys Soc. *Res:* Speech synthesis and recognition; room acoustics; psycho-acoustics; electro-acoustics; coherent optics; spatial stochastic processes; digital signal processing; number theory; neural networks; chaos; author 2 books. *Mailing Add:* Univ Goettingen Third Physics Inst Buergerstrasse 42/44 Goettingen 34 Germany

SCHROEDER, MARK EDWIN, b Cincinnati, Ohio, July 31, 46; m 68; c 1. INSECT PHYSIOLOGY. *Educ:* Loyola Col, BS, 68; Purdue Univ, PhD(entom), 74. *Prof Exp:* Res fel insect physiol, Univ Calif, Riverside, 74-75; sr res entomologist, Shell Develop Co, 75-87; E I DU PONT DE NEMOURS CO, 88- *Mem:* AAAS; Entom Soc Am; Am Soc Zoologists; Am Chem Soc; Sigma Xi. *Res:* Design and execution of experiments to investigate, on a suborgan level, the mode of action of selected neurotoxins on the insect central nervous system. *Mailing Add:* 248 Mercer Mill Rd Landenberg PA 19350-9339

SCHROEDER, MELVIN CARROLL, b Saskatoon, Sask, July 19, 17; nat US; m 83; c 2. GEOLOGY. *Educ:* Wash State Univ, BS, 42, MS, 47, PhD(geol), 53. *Prof Exp:* Geologist, US Geol Surv, 49-54; from asst prof to assoc prof, 54-63, prof geol, 63-87, EMER PROF GEOL, TEX A&M UNIV, 87- *Concurrent Pos:* Counr, Geol Soc Am, 85-87; hon life mem, Sci Teachers Assoc Tex. *Mem:* Fel Geol Soc Am; Nat Asn Geol Teachers; Am Asn Petrol Geologists; hon mem Nat Sci Teachers Asn. *Res:* Ground water; radiohydrology; water contamination. *Mailing Add:* Dept Geol Tex A&M Univ College Station TX 77843-3115

SCHROEDER, MICHAEL ALLAN, b Little Falls, NY, Nov 13, 38. ORGANIC CHEMISTRY, PHYSICAL CHEMISTRY. *Educ:* Union Col, NY, BS, 61; Johns Hopkins Univ, PhD(chem), 68. *Prof Exp:* Res chemist, Naval Weapons Ctr, 67-68; RES CHEMIST, US ARMY BALLISTIC RES LABS, 68- *Concurrent Pos:* Resident res assoc, Nat Res Coun, 67-68. *Mem:* AAAS; Am Chem Soc; Am Defense Preparedness Asn; Sigma Xi. *Res:* Organic mechanisms; chemistry of organic nitro and nitroso compounds; heteroaromatic chemistry; explosive and propellant chemistry; laser spectroscopy; chemistry of high-nitrogen compounds; deamination chemistry. *Mailing Add:* Dir USA BRL Attn SLCBR-IB-I Aberdeen Proving Ground MD 21005

SCHROEDER, PAUL CLEMENS, b Brooklyn, NY, Aug 13, 38; m 66; c 2. EXPERIMENTAL ZOOLOGY. *Educ:* St Peter's Col, NJ, BS, 60; Stanford Univ, PhD(biol sci), 66. *Prof Exp:* NSF fel zool, Univ Calif, Berkeley, 66-67, USPHS fel, 67-68; from asst prof to assoc prof, 68-82, assoc chmn, 83-87, PROF ZOOL, 82-, CHMN, WASH STATE UNIV, 87- *Concurrent Pos:* Vis prof, Univ Southern Calif, 73; Alexander von Humboldt Found fel, Zool Inst, Univ Cologne, 74-75, Univ Mainz, 89, & Fogarty Int fel, NIH & Fulbright fel, Dept Anat, Univ Queensland, Brisbane, Australia, 82; prin investr, NIH, 74-77 & 79-82; ed, Marine Biol, 85-89, J Exp Zool, 86-89; vis scientist, Bioctr, Univ Basle, Switz, 89. *Mem:* Fel AAAS; Am Soc Cell Biol; Am Soc Zool; Am Inst Biol Sci; Coleopterists Soc; Int Soc Invert Reproduction (secy, 89-). *Res:* Hormonal control of developmental and reproductive processes primarily in polychaete worms, echinoderms and amphibians; ovulatory mechanisms in vertebrates to provide evolutionary perspective on mammalian ovary. *Mailing Add:* Dept Zool Wash State Univ Pullman WA 99164-4220

SCHROEDER, PETER A, b Dunedin, NZ, Dec 6, 28; m 53; c 2. LOW TEMPERATURE PHYSICS, ELECTRON TRANSPORT PROPERTIES. *Educ:* Univ Canterbury, MSc, 50; Bristol Univ, PhD(physics), 55. *Prof Exp:* Asst lectr physics, Univ Canterbury, 54-56, lectr, 56-59; fel, Nat Res Coun Can, 59-60, asst res officer, 60-61; from asst prof to assoc prof, 61-69, PROF PHYSICS, MICH STATE UNIV, 69- *Concurrent Pos:* Vis prof, Univ Sussex, Eng, 67, Univ Leeds, Eng, 74, Cath Univ Nijmegen, Neth, 82, Univ Paris-Sud, Orsay, France, 90, Max Planck Inst, High Magnetic Field Lab, Grenoble, France, 90. *Mem:* Fel Am Phys Soc. *Res:* Electron transport properties of metals, alloys and layered metallic systems; dielectric properties of clays. *Mailing Add:* Dept Physics & Astron Mich State Univ East Lansing MI 48824

SCHROEDER, ROBERT SAMUEL, b Chicago, Ill, July 9, 43; m 65; c 2. PESTICIDE CHEMISTRY, ORGANIC BIOCHEMISTRY. *Educ:* Iowa State Univ, BS, 64; Ind Univ, Bloomington, PhD(chem), 70. *Prof Exp:* Res chemist, 69-75, actg sect supvr, 74-75, sect supvr, Gulf Oil Chem Co, 76-80; mgr qual assurance, 82-87, admin asst, 87-89, TOXICOL SPECIALIST, MOBAY CORP, 90- *Mem:* Am Chem Soc; Soc Qual Assurance; Sigma Xi; Coun Agr Sci & Technol. *Res:* Pesticide metabolism and disposition in plants, animals, soil and water; development of residue methods for pesticides; application of instrumental analysis for structure determinations; direction of analytical biochemistry and environmental research; development of computer systems for data collection and reporting; dermal absorption and disposition of xenobiotics. *Mailing Add:* 17210 W 70th St Shawnee Mission KS 66217

SCHROEDER, RUDOLPH ALRUD, b Evansville, Minn, Oct 11, 23; m 66. PHYSICAL CHEMISTRY. *Educ:* NDak Agr Col, BS, 52, MS, 53; Univ Md, PhD(chem), 57. *Prof Exp:* Asst prof chem, Univ Ky, 57-58; from asst prof to assoc prof, Southwestern Univ, La, 58-74, prof chem, 74-; RETIRED. *Mem:* Am Chem Soc. *Res:* Infrared and Raman spectroscopy; hydrogen and interatomic bonding; bond energies; metal chelates; quantum mechanics. *Mailing Add:* 220 W Ardenwood Baton Rouge LA 70806

SCHROEDER, STEVEN A, b New York, NY, July 26, 39; m; c 2. INTERNAL MEDICINE. *Educ:* Stanford Univ, BA, 60; Harvard Univ, MD, 64; Am Bd Internal Med, dipl, 71 & 80. *Prof Exp:* Assoc prof health care sci & med, George Washington Med Ctr, 74-76; prof med & chief, Div Gen Internal Med, Univ Calif, San Francisco, 80-90; PRES, ROBERT WOOD JOHNSON FOUND & CLIN PROF MED, MED SCH, UNIV MED & DENT NJ, 91- *Concurrent Pos:* Vis prof, Dept Community Med, St Thomas's Hosp Med Sch, London, 82-83; mem, US Prospective Payment Comn, 83-88, chmn, 84-88; mem, Comt Implications For-Profit Enterprise Health Care, Inst Med, 83-85, Adv Panel Off Technol Assessment Study Physicians & Med Technol, 84-85, Prev Med & Pub Health Sect, Nat Bd Med Examrs, 86-90 & Health & Pub Policy Comt, Am Col Physicians, 87-89; ed, Western J Med, 86-90; chmn, Spec Study Qual Rev & Assurance Medicare, Inst Med, Nat Acad Sci, 87-90; Dozor vis prof, Ben Guiron Univ Negev, Israel, 87; Mack Lipkin vis prof, NY Hosp & Cornell Med Sch, 89; lectr, State Univ NY, Stony Brook, 91. *Honors & Awards:* Bartlett Mem Lectr, St Francis Mem Hosp, San Francisco, 90. *Mem:* Inst Med-Nat Acad Sci; Inst Health Policy Studies; fel Am Col Physicians; Am Fedn Clin Res; Am Pub Health Asn; Asn Am Physicians. *Res:* Auth over 100 publications. *Mailing Add:* Robert Wood Johnson Found PO Box 2316 Princeton NJ 08543-2316

SCHROEDER, THOMAS DEAN, b Reedsburg, Wis, May 2, 39; m 63; c 2. ANALYTICAL CHEMISTRY. *Educ:* Univ Wis-Platteville, BS, 65; Univ Iowa, MS, 68, PhD(anal chem), 69. *Prof Exp:* ASSOC PROF CHEM, SHIPPENSBURG STATE COL, 69- *Concurrent Pos:* Pa teaching fel, 81. *Mem:* Am Chem Soc; Sigma Xi. *Res:* Chemical instrumentation; gas chromatographic-mass spectrometry of biochemicals; construction of specific electrodes; flameless atomic absorption. *Mailing Add:* 9785 Forest Ridge Rd Shippensburg PA 17257

SCHROEDER, W(ILBURN) CARROLL, chemical engineering; deceased, see previous edition for last biography

SCHROEDER, WALTER ADOLPH, b Kansas City, Mo, Apr 30, 17; m 49; c 2. PROTEIN CHEMISTRY. *Educ:* Univ Nebr, BSc, 39, MA, 40; Calif Inst Technol, PhD(chem), 43. *Prof Exp:* Res fel, 42-46, sr res fel, 45-56, res assoc, 56-81, SR RES ASSOC CHEM, CALIF INST TECHNOL, 81- *Concurrent Pos:* Guggenheim fel, 59-60. *Res:* Carotenoids in plants; chromatographic separation of amino acids, peptides and proteins; propellants; organic chemistry; proteins; structure and function of hemoglobin and heme proteins. *Mailing Add:* Meadowbrook Apts 5325 Whetstone Rd Richmond VA 23234

SCHROEDER, WALTER ALBERT, b Jefferson City, Mo, Sept 9, 34; m 62; c 2. PRESETTLEMENT PRAIRIES. *Educ:* Univ Mo, Columbia, AB, 56; Univ Chicago, MA, 58. *Prof Exp:* Instr geog, Cent Mich Univ, 58-63; lectr, Univ Southern Ill, Carbondale, 63-64; INSTR GEOG, UNIV MO, COLUMBIA, 64-, CHMN DEPT & ASST PROF, 80- *Mem:* Asn Am Geogr; Am Name Soc. *Res:* Reconstruction of past environments in Missouri especially prairies and natural vegetation before European settlements; use by and impact of settlers from different cultures on the environment. *Mailing Add:* Dept Geog Univ Mo Columbia MO 65211

SCHROEDER, WARREN LEE, b Longview, Wash, Jan 3, 39; m 61; c 3. SOILS, CIVIL ENGINEERING. *Educ:* Wash State Univ, BS, 62, MS, 63; Univ Colo, Boulder, PhD(civil eng), 67. *Prof Exp:* From asst prof to assoc prof, 67-77, asst dean eng, 71-85, PROF CIVIL ENG, ORE STATE UNIV, 77-, ASSOC DEAN ENG, 85- *Concurrent Pos:* Staff engr, McDowell & Assocs, Consult Engrs, 66-67 &CH2M/Hill, Consult Engrs, 67-70; pres, Willamette Geotechnical, Inc, 78- *Honors & Awards:* Thomas Fitch Rowland Prize, Am Soc Civil Eng, 88. *Mem:* Am Soc Civil Engrs; Int Soc Soil Mech & Found Engr. *Res:* Deep foundations; retaining structures; behavior of submerged cohesionless soils; cofferdams and docks. *Mailing Add:* Col Eng Ore State Univ Corvallis OR 97331

SCHROEDER, WILLIAM, JR, b New York, NY, Apr 9, 27; m 58; c 3. ORGANIC CHEMISTRY. *Educ:* Purdue Univ, BS, 55, PhD(chem), 58. *Prof Exp:* Res assoc org chem, Upjohn Co, 58-65; dir res, Burdick & Jackson Labs, 65-66, vpres res, 66-75, secy, 68-75, vpres, 75-78, dir, Burdick & Jackson Labs, 75-86, pres, 78-86; vpres & gen mgr, Baxter Healthcare, 86-90; pres, Casadonte Res Labs, Muskegon, Mich, 88-90; RETIRED. *Mem:* Am Chem Soc. *Res:* Structures; natural products; carbohydrates; organic synthesis. *Mailing Add:* 3825 Harbor Point Rd Muskegon MI 49441

SCHROEDER, WILLIAM HENRY, b Breslau, Ger, Apr 27, 44; Can citizen; m 75; c 1. ATMOSPHERIC CHEMISTRY. *Educ:* Univ Alta, Calgary, BSc, 66; Univ Colo, PhD(chem), 71. *Prof Exp:* Res fel, Fresenius Inst, Wiesbaden, Fed Repub Ger, 71-72; head, Abstracting Sect, Environ Can, Air Pollution Control Directorate, Ottawa, 73-75, phys scientist, Technol Develop & Demonstr, Water Pollution Control Directorate, Burlington, 75-77, RES SCIENTIST ATMOSPHERIC CHEM, ATMOSPHERIC ENVIRON SERV, ENVIRON CAN, DOWNSVIEW, ONT, 77- *Concurrent Pos:* Sci liaison officer, Energy Recovery Demonstration Proj, Environ Can contract, St Lawrence Cement Co, Mississauga, Ont, 76-77; sci authority, Environ Can Contract, Barringer Res Inc, 80-82. *Mem:* Am Chem Soc; Chem Inst Can; Air & Waste Management Asn; Am Soc Testing & Mat; NY Acad Sci. *Res:* Atmospheric pathways (sources, transport, transformation and fate); characteristics (physico-chemical, toxicological) of toxic trace elements and organic substances; environmental analytical chemistry. *Mailing Add:* 90 Bedford Pk Ave Richmond Hill ON L4C 2N8 Can

SCHROEER, DIETRICH, b Berlin, Ger, Jan 24, 38; US citizen; m 64; c 2. ARMS-RACE ISSUES. *Educ:* Ohio State Univ, BSc, 60, PhD(physics), 65. *Prof Exp:* NATO fel, Munich Tech Univ, 65-66; asst prof, 66-73, assoc prof, 73-79, PROF PHYSICS, UNIV NC, CHAPEL HILL, 79- *Concurrent Pos:* Fulbright fel & Nat Endowment for Humanities fel, Munich, Ger, 72-73; res assoc, Int Inst Strategic Studies, London, 84-85. *Honors & Awards:* Am Inst Physics-US Steel Sci Writing Award, 72. *Mem:* AAAS; Am Asn Physics Teachers; Am Phys Soc; Soc Social Studies Sci; Arms Control Asn; Fedn Am Scientists. *Res:* Low energy nuclear spectroscopy; crystal-defect and radiation-damage studies by Mossbauer effect; science policy; arms-control issues. *Mailing Add:* Dept Physics & Astron Univ NC Chapel Hill NC 27599-3255

SCHROEER, JUERGEN MAX, b Berlin, Ger, Oct 2, 33; US citizen; m 64; c 4. MASS SPECTROMETRY, PHYSICS TEACHING. *Educ:* Ohio State Univ, BS & MS, 58; Cornell Univ, PhD(physics), 64. *Prof Exp:* Res assoc quantum electronics, Sch Elec Eng, Cornell Univ, 64-65; asst prof physics, Univ Wyo, 65-69; assoc prof, 69-74, PROF PHYSICS, ILL STATE UNIV, 74- *Concurrent Pos:* Vis prof & Fulbright travel grant, Univ Munster, 75; vis scientist, Mat Res Lab, Univ Ill, 83; vis prof, Ore Grad Ctr, 83-84 & Univ Ill, 85, 86, 90. *Mem:* Am Vacuum Soc; Am Soc Mass Spectrometry. *Res:* Surface physics; secondary ion mass spectrometry. *Mailing Add:* Dept Physics Ill State Univ Normal IL 61761-6901

SCHROEN, WALTER, b Munich, Ger, June 3, 30; m 67; c 4. INDUSTRIAL & MANUFACTURING ENGINEERING, TECHNICAL MANAGEMENT. *Educ:* Univ Munich, BS, 52, MS, 56; Clausthal Tech Univ, PhD(atomic physics), 62. *Prof Exp:* Physicist, Cent Res Labs, Siemens Corp, 53-55 & Semiconductor Div, 56-58; res asst physics, Clausthal Tech Univ, 58-62; sr scientist semiconductor res, Int Tel & Tel Semiconductors, Calif, 62-65; sect head, Semiconductor Res & Develop Labs, 65-75, semiconductor group process control, 75-79, MGR SEMICONDUCTOR ASSEMBLY & PACKAGING, TEX INSTRUMENTS INC, 80- *Concurrent Pos:* Ger co-rep, Int Seminar Nucelar Sci, Saclay, France, 59; fel, Tex Instruments Inc, 87; chmn, Electrochem Soc Symp, 75, Inst Elec & Electronics Engrs Reliab Physics Symp, 90. *Mem:* Am Phys Soc; Ger Phys Soc; Electrochem Soc; Inst Elec & Electronics Engrs. *Res:* Semiconductor and surface physics; physics of failure in electronics; analysis and modeling; thin films physics; superconductivity; physics of ionization; bipolar and metal-oxide semiconductor devices; semiconductor process control; semiconductor reliability; semiconductor packaging; multichip modules. *Mailing Add:* 6620 Churchill Way Dallas TX 75230

SCHROEPFER, GEORGE JOHN, JR, b St Paul, Minn, June 15, 32; c 5. BIOCHEMISTRY, CHEMISTRY. *Educ:* Univ Minn, BS, 55, MD, 57, PhD, 61. *Prof Exp:* Intern med, Univ Minn, 57-58, Nat Heart Inst res fel, 58-61, res assoc, 61-63, asst prof biochem, 63-64; asst prof, Univ Ill, Urbana, 64-67, from assoc prof to prof biochem & org chem, 67-72, dir sch basic med sci, 68-70; prof biochem & chem & chmn dept biochem, 72-83, RALPH & DOROTHY LOONEY PROF BIOCHEM, RICE UNIV, 83- *Concurrent Pos:* USPHS res career develop award, 62-64; fel, Harvard Univ, 62-63; fel, Coun Arteriosclerosis, Am Heart Asn, 64-; mem panel biochem nomenclature, Nat Acad Sci, 65-68; assoc ed, Lipids, 69-78; mem biochem training comt, NIH, 70-73. *Mem:* Fel AAAS; Am Chem Soc; Am Soc Biol Chem. *Res:* Sterol biosynthesis and metabolism; intermediary metabolism of lipids; stereochemistry and mechanism of enzymatic reactions. *Mailing Add:* Dept Biochem PO Box 1892 Rice Univ Houston TX 77251

SCHROER, BERNARD J, b Seymour, Ind, Oct 11, 41; m 63; c 2. SYSTEM SIMULATION. *Educ:* Western Mich Univ, BSE, 64; Univ Ala, MSE, 67; Okla State Univ, PhD(eng), 72. *Prof Exp:* Designer, Sandia Labs, 61-62; engr, Teledyne Brown Eng, 64-67; proj engr, Boeing, 67-70 & Computer Sci Corp, 70-71; DIR, JOHNSON RES CTR, UNIV ALA, HUNTSVILLE, 72-, PROF, COL ENGR, 88- *Concurrent Pos:* Mem, Gov Ala State Solar Comt, 76, Gov Cabinet, 82; mem bd dirs, Southern Solar Energy Ctr, 80-85; mem energy coun, Ala Dept Energy, 80- *Honors & Awards:* Energy Innovation Award, US Dept Energy, 84. *Mem:* Nat Soc Prof Engrs; Am Inst Indust Engrs; Sigma Xi; Robotics Int; Soc Comput Simulation. *Res:* System simulation of automated manufacturing-production systems. *Mailing Add:* 716 Owens Dr Huntsville AL 35801

SCHROER, RICHARD ALLEN, b Celina, Ohio, July 10, 44; div; c 2. CLINICAL PATHOLOGY. *Educ:* Kent State Univ, BS, 66, PhD(chem), 70. *Prof Exp:* Res assoc biochem, Univ Calif, Irvine-Calif Col Med, 71-73; dir biol, Nelson Res & Develop, 73-75; sr res biologist, Lederle Labs, 75-76, group leader clin chem-hemat, 76-88; MGR, CLIN PATH, MED RES DIV, AM CYANAMID CO, 89- *Concurrent Pos:* NIH fel, Univ Calif, Irvine-Calif Col Med, 71-73. *Mem:* AAAS; Am Asn Clin Chemists. *Res:* Toxicology; clinical pathology of laboratory animals. *Mailing Add:* 19 Strawberry Hill Lane West Nyack NY 10994

SCHROETER, GILBERT LOREN, b Reedley, Calif, May 24, 36; m 63; c 1. GENETICS, CYTOGENETICS. *Educ:* Fresno State Col, BA, 63; Univ Calif, Davis, PhD(genetics), 68. *Prof Exp:* Asst prof, 68-77, ASSOC PROF BIOL, TEX A&M UNIV, 77- *Mem:* AAAS; Genetics Soc Am. *Res:* Population cytology; chromosome evolution; cytotaxonomy of orthopteroid insects. *Mailing Add:* Dept Biol Tex A&M Univ College Station TX 77843

SCHROETER, SIEGFRIED HERMANN, chemistry; deceased, see previous edition for last biography

SCHROF, WILLIAM ERNST JOHN, b Cincinnati, Ohio, June 5, 31; m 58; c 5. ORGANIC POLYMER CHEMISTRY. *Educ:* Univ Cincinnati, AB, 58, PhD(org chem), 64. *Prof Exp:* Res chemist, 63-71, sr res chemist, 71-85, sr tech mkt rep, 85-87, SR INF SPEC, E I DU PONT DE NEMOURS & CO, INC, 87- *Concurrent Pos:* Patent searching, corp level. *Mem:* Am Chem Soc; Am Inst Chem; Sigma Xi. *Res:* Synthesis of anticancer compounds such as coumarins and furoquinolines; polymerization of polyamides for textile and industrial end uses; product development of textile and industrial yarns. *Mailing Add:* E I du Pont Barley Hill Plaza Bldg 14 Wilmington DE 19898

SCHROFF, PETER DAVID, b Munich, Ger, Apr 19, 26; US citizen; m 74; c 5. LIFE SCIENCES, CLINICAL PATHOLOGY. *Educ:* Royal Inst Sci, London, BSc, 44, MSc, 46; Mich State Univ, MS, 49, PhD(chem eng), 51; Ga State Univ, PhD(physiol), 70; Emory Univ, MD, 73. *Prof Exp:* Mgr pilot develop, Gen Elec, 51-53; chief develop engr, Fluor Corp, 53-60; dir qual, NAm Rockwell, 60-70; dir new prod, Coulter Diag, 75-80; dir prod, Ortho Diag, 80-82, DIR LIFE SCI, JANSSEN LIFE SCI, DIV JOHNSON & JOHNSON, 82- *Mem:* Am Soc Cell Biol; Am Soc Clin Path; Am Asn Clin Chem; Nat Soc Histotechnol; Am Asn Biochemists; Nat Soc Prof Engrs. *Res:* Immunology. *Mailing Add:* 956 Concord Way Neshanic Station NJ 08853-9684

SCHROGIE, JOHN JOSEPH, b Brooklyn, NY, Oct 24, 35; m 64; c 1. INTERNAL MEDICINE, CLINICAL PHARMACOLOGY. *Educ:* Boston Col, BS, 56; Yale Univ, MD, 60. *Prof Exp:* Intern, Univ Chicago Clins, 60-61; resident internal med, St Luke's Hosp, New York, NY, 61-63; asst chief, Coronary Heart Dis Sect, Heart Dis Control Br, USPHS, 63-65; dir div res & liaison, Bur Med, Food & Drug Admin, 67-70; chief fertil regulating methods eval br, Ctr Pop Res, Nat Inst Child Health & Human Develop, 70-73; dir clin pharmacol, Schering Corp, 73-77; asst clin prof med, Col Med & Dent, NJ, 73-77; sr dir clin pharmacol, Merck, Sharp & Dohme Res Labs, 77-80; PRES, PHILADELPHIA ASSOC CLIN TRIALS, 80- *Concurrent Pos:* Fel clin pharmacol, Johns Hopkins Univ, 65-67; chief, Div Clin Pharmacol & sr attend physician, St Michael's Med Ctr, Newark, NJ; adj assoc prof med & attend physician, Jefferson Med Col, Philadelphia, 77- *Mem:* Am Soc Pharmacol & Exp Therapeut; Am Soc Clin Pharmacol & Therapeut; Am Fedn Clin Res. *Res:* New drug development; drug metabolism. *Mailing Add:* Philadelphia Assoc Clin Trials 150 Radnor-Chester Rd Suite D-200 St Davids PA 19087

SCHROHENLOHER, RALPH EDWARD, b Cincinnati, Ohio, Aug 6, 33; m 60; c 2. PROTEIN CHEMISTRY, IMMUNOCHEMISTRY. *Educ:* Univ Cincinnati, BS, 55, PhD(biochem), 59. *Prof Exp:* Res chemist, Nat Cancer Inst, 58-60; asst prof arthritis res, Med Col Ala, 61-63; guest investr, Rockefeller Inst, 63-64; from asst prof to assoc prof, 64-87, PROF MED, SCH MED, UNIV ALA, BIRMINGHAM, 87- *Concurrent Pos:* Nat Inst Arthritis & Metab Dis grants, 65-78; Nat Cancer Inst res contract, 75-78. *Mem:* Am Col Rheumatology; AAAS; Am Chem Soc; Am Asn Immunol. *Res:* Immunoglobulin structure and function; autoantibodies and immune complexes in connective tissue diseases; structure of collagen. *Mailing Add:* Dept Med Univ Ala Sch Med Birmingham AL 35294

SCHROLL, GENE E, b North Manchester, Ind, June 29, 28; m 49; c 3. ORGANIC CHEMISTRY. *Educ:* Manchester Col, BA, 51; Purdue Univ, PhD(chem), 55. *Prof Exp:* Res chemist, Ethyl Corp, 55-61; res assoc, Celanese Coatings Co, 62-67; res assoc, 68-80, SUPVR, APPL SCI RES & DEVELOP DEPT, CINCINNATI MILACRON CO, 80- *Mem:* Am Chem Soc; Soc Plastic Engrs. *Res:* Protective coatings; plastics; organometallics; reaction mechanisms. *Mailing Add:* 10451 Londonderry Ct Cincinnati OH 45242

SCHROTER, STANISLAW GUSTAW, b Katowice, Poland, May 8, 17; US citizen; m 46; c 5. CHEMICAL ENGINEERING, INORGANIC CHEMISTRY. *Educ:* Polish Univ Col Eng, Dipl Eng, 49. *Prof Exp:* Tech officer, Steatite & Porcelain Prod Ltd, Eng, 49-52; engr, Can Radio Mfg Corp, Can, 52-56; sr engr, Raytheon Co, Quincy, 56-61, sr engr, 61-87; CONSULT, 87- *Concurrent Pos:* Consult, Georgetown Porcelain Ltd, 53-54. *Res:* Electron emissive materials; high temperature electrical insulation; ferromagnetic porcelains; potting and encapsulation; microelectronics. *Mailing Add:* 103 High New St Newton MA 02164

SCHROTH, MILTON NEIL, b Fullerton, Calif, June 25, 33; m 59; c 3. PLANT PATHOLOGY. *Educ:* Pomona Col, BA, 55; Univ Calif, Berkeley, PhD(plant path), 61. *Prof Exp:* From instr to assoc prof plant path, Univ Calif, 61-71, asst dean res, Col Agr Sci, 68-73, assoc dean res, Col Natural Resources & actg asst to vpres agr sci, 73-76, asst dir, Div Agri Sci, 76-79, PROF PLANT PATH, UNIV CALIF, BERKELEY, 71-, CHAIR, PLANT PATH DEPT, 89- *Honors & Awards:* Campbell Award, Am Inst Biol Sci, 64. *Mem:* Am Soc Microbioi; fel Am Phytopath Soc; Sigma Xi. *Res:* Root disease research, especially plant bacterial diseases; biological control. *Mailing Add:* Dept Plant Path Univ Calif Berkeley CA 94720

SCHROTT, HELMUT GUNTHER, b Wein, Austria, Jan 23, 37; m 61; c 2. ENDOCRINOLOGY, MEDICAL GENETICS. *Educ:* Western Reserve Univ, BA, 62; State Univ NY, Buffalo, MD, 66. *Prof Exp:* Intern & resident internal med, Buffalo Gen Hosp, 66-68; resident & fel internal med & endocrinol, Univ Utah Med Ctr, 68-70; fel med genetics, Univ Wash, 70-72; assoc consult med genetics, Mayo Clin, 72-73; ASSOC PROF INTERNAL & PREV MED, UNIV IOWA, 73- *Res:* Conduct clinical trials to determine the efficacy of intervening on disease states: hyperchol esterolemia, diabetes, menopause; conduct community health screening and intervention trials as related to hypercholesterolemia. *Mailing Add:* Lipid Res Clin Univ Iowa Westlawn 5-23 3117 Alpine Ct Iowa City IA 52242

SCHROY, JERRY M, b Dayton, Ohio, Nov 6, 39; m 69; c 3. CHEMICAL TRANSPORT & FATE, WORKPLACE EXPOSURE CONTROL. *Educ:* Univ Cincinnati, ChE, 63. *Prof Exp:* Engr, Detergents & Heavy Chem Sect, Inorg Chem Div, Monsanto, St Louis, Mo, 63-65, engr, Phosphorus Sect, Inorg Chem Process Eng, 65-67, supvr, Plant Tech Serv Group, Inorg Chem Phosphorus Technol Dept, Columbia, Tenn, 67-69, eng specialist, Monsanto Biodize Syst Inc, Long Island, NY, 69-71, eng specialist, Indust Water Pollution Control Dept, Monsanto Environ-Chem Systs, Inc, Chicago, Ill, 71-72, process eng mgr, 72-73, eng specialist, Water Pollution Control Dept, St Louis, Mo, 73-75, eng specialist, Environ Control Group CED, 75-76, prin eng specialist, 76-80, Monsanto fel, 81-85, Monsanto fel, Environ Technol Group MCC Opers, Monsanto Chem Co, 85-89, MONSANTO SR FEL, ENVIRON TECHNOL GROUP MCC OPERS, MONSANTO CHEM CO, ST LOUIS, MO, 89- *Concurrent Pos:* Mem, sci adv bd, Ctr Excellence Intermedia Transport, Univ Calif, Los Angeles, 84-87, tech adv bd, Cur Catastrophe Prev, NJ Dept Environ Protection, 86-87 & tech adv group, South Coast Air Qual Mgt Dist, Calif/Environ Protection Agency, 86-90; mem, task group, Am Inst Chem Engrs, 86- & tech rev panel, US Environ Protection Agency Coop Res Prog, 86-90; mem, Comt on Human Exposure Assessment, Nat Acad Sci/Nat Res Coun, 88-90; chmn, Design Inst for Phys Properties Res, Am Inst Chem Engrs, 90. *Mem:* Am Inst Chem Engrs; Water Pollution Control Fedn; Am Chem Soc; AAAS; Soc Risk Anal; Nat Asn Environ Professionals; Sigma Xi; Am Waste Mgt Asn; Am Acad Environ Engrs; Int Soc Exposure Anal. *Res:* Transport and fate of chemicals in the environment; physical chemical properties of 2, 3, 7, 8 Tetrachloropdioxin; mobility of low volatility chemicals in the soil; volatility/vaporization of chemicals from spills and wastewater systems. *Mailing Add:* Five Springlake Ct Ballwin MO 63011

SCHRUBEN, JOHANNA STENZEL, b Flushing, NY. MATHEMATICS, OPTICS. *Educ:* Queens Col, City Univ New York, BS, 64; Univ Mich, AM, 66, PhD(math), 68. *Prof Exp:* Instr math, Univ Minn, 68-69; sr scientist, Westinghouse Res & Develop Ctr, 69-82; assoc prof math sci, Univ Akron, 82-89; ASSOC PROF MATH SCI, UNIV HOUSTON, VICTORIA, 89- *Concurrent Pos:* NSF fel, Univ Mich, 65-67; mem nat comt, Int Comn Illum, 76-78. *Mem:* Am Math Soc; Soc Indust & Appl Math; Optical Soc Am; Illum Eng Soc NAm. *Res:* Partial differential equations; illumination design; thermal stresses in single-crystal silicon ribbon being grown for photovoltaic cells; attrition in fluidized beds. *Mailing Add:* Dept Math Sci Univ Houston 2302-C Red River Victoria TX 77901

SCHRUBEN, JOHN H, b Stockton, Kans, Jan 19, 26; m 48; c 8. BUILDING CONSTRUCTION, SYSTEMIZATION ARCHITECTURAL-ENGINEERING PRACTICE. *Educ:* Kans State Univ, BS, 48; Ill Inst Technol, MS, 53. *Prof Exp:* Staff engr, Standard Oil Co, Ind, 48-55; proj mgr, Skidmore, Owings & Merrill, 55-69; pres, Prod Systs Archit & Eng, 69-82; exec mgr, Am Inst Architects, 82-89; PERSONAL CONSULT, US DEPT STATE FBO/BDE, 89-91. *Concurrent Pos:* Master spec develop & consult, US Dept State, Foreign Bldg Oper, Bldg Design & Engr, Engr Support Br, Criteria & Specif, 87- *Mem:* Fel Am Inst Architects. *Res:* Principal developer, author, producer of master specification system. *Mailing Add:* 6200 Meadow Ct Rockville MD 20852

SCHRUM, MARY IRENE KNOLLER, b New York, NY, Apr 18, 26; m 69. BIOCHEMISTRY, INFORMATION SCIENCE. *Educ:* Col Notre Dame, Md, AB, 48; Georgetown Univ, MS, 56. *Prof Exp:* Sr org chemist, Crown Cork & Seal Co, 48-50; chemist, Dept Med, Johns Hopkins Univ, 50-52; biochemist cellular physiol & metab, Nat Heart Inst, NIH 52-55, supvr metab, 55-59, biochemist Nat Inst Arthritis & Metab Dis, 59-62, sci reference analyst, Div Res Grants, 62-63; chemist, Food & Drug Admin, Washington, DC, 63-66, head, Cent Retrieval Index Group, Sci Info Facility, 66-68, sr systs analyst, tech opers staff, 68-82; RETIRED. *Res:* Arteriosclerosis; physical-chemical studies of proteins; amino acid chemistry; information storage and retrieval of scientific information and data; research and development; chemical notations; automatic and electronic data processing. *Mailing Add:* 5528 Warwick Pl Chevy Chase MD 20815

SCHRUM, ROBERT WALLACE, b Hammond, Ind, Oct 19, 30; m 52; c 2. PETROLEUM CHEMISTRY. *Educ:* Univ Ill, BS, 52. *Prof Exp:* Chemist, Sinclair Res Labs, Inc, 52-55, asst sect leader engine oils, 55-59, asst sect leader fuels, 59-63, sect leader engine oils, 63-66; SECT MGR, RES DIV, ROHM & HAAS CO, 66- *Mem:* Soc Automotive Engr. *Res:* Fuel and lubricant additive research and development. *Mailing Add:* 1560 Temple Dr Maple Glen PA 19002-3317

SCHRUMPF, BARRY JAMES, b San Francisco, Calif, July 13, 43; m 72; c 1. RANGE ECOLOGY, REMOTE SENSING. *Educ:* Willamette Univ, BA, 66; Ore State Univ, MS, 68, PhD(rangeland resources), 75. *Prof Exp:* Actg dir, 73-75, dir, Environ Remote Appln Lab, 75-88, ASSOC PROF, ORE STATE UNIV, 80-, SEED CERT ASST, 88- *Res:* Multispectral, multiseasonal and multistage remote sensing for natural vegetation inventory and analysis. *Mailing Add:* Seed Cert Serv Ore State Univ 031 Crop Sci Bldg Corvallis OR 97331

SCHRYER, NORMAN LOREN, b Detroit, Mich, Jan 16, 43; m 65. APPLIED MATHEMATICS. *Educ:* Univ Mich, BS, 65, MS, 66, PhD(math), 69. *Prof Exp:* MEM TECH STAFF, BELL TEL LABS, 69- *Mem:* Asn Comput Mach; Soc Indust & Appl Math. *Res:* Numerical solution of elliptic and parabolic partial differential equations. *Mailing Add:* 122 Sulfrian New Providence NJ 07974

SCHRYVER, HERBERT FRANCIS, b New York, NY, Oct 15, 27; m 64. VETERINARY PHYSIOLOGY, VETERINARY PATHOLOGY. *Educ:* Cornell Univ, DVM, 54; Univ Pa, MS, 60, PhD(path), 64; Hofstra Col, BA, 61. *Prof Exp:* From instr to asst prof path, Univ Pa, 58-66; assoc prof path, NY, State Vet Col, Cornell Univ, 66-90; RETIRED. *Concurrent Pos:* Arthritis Found fel, 64-65. *Mem:* AAAS; Am Soc Cell Biol; Am Vet Med Asn; Int Acad Path; Am Inst Nutrit. *Res:* Mineral nutrition and metabolism; connective tissue physiology and pathology; bone diseases; diseases of domestic animals and wildlife. *Mailing Add:* Equine Res Prog NY State Vet Col Cornell Univ Ithaca NY 14850

SCHTEINGART, DAVID E, b Buenos Aires, Arg, Oct 17, 30; m 60; c 3. INTERNAL MEDICINE. *Educ:* Nat Col 6, Buenos Aires, BA, 47; Univ Buenos Aires, MD, 54. *Prof Exp:* Resident med, Hosp Nat Clin, Buenos Aires, 56-57; resident, 59-60, instr, 60-61 & 62-63, asst prof, 63-67, assoc prof, 67-73, PROF MED, UNIV HOSP, UNIV MICH, ANN ARBOR, 73- *Concurrent Pos:* Fel med, Mt Sinai Hosp, New York, 57-58; fel endocrinol, Maimonides Hosp, Brooklyn, 58-59 & Univ Hosp, Univ Mich, Ann Arbor, 61-62. *Mem:* Am Fedn Clin Res; NY Acad Sci; Endocrine Soc; Sigma Xi; Am Col Physicians. *Res:* Endocrinology; clinical abnormalities of adrenal cortical steroids; secretion and metabolism; obesity. *Mailing Add:* 5700 MSRB II Univ Mich Med Ctr Ann Arbor MI 48109

SCHUBACK, PHILIP, periodontics, oral pathology; deceased, see previous edition for last biography

SCHUBAUER, GALEN B, ENGINEERING. *Prof Exp:* RETIRED. *Mem:* Nat Acad Eng. *Mailing Add:* 10450 Lottsford Rd Unit 1211 Mitchellville MD 20721

SCHUBEL, JERRY ROBERT, b Bad Axe, Mich, Jan 26, 36; m 58; c 2. MARINE GEOLOGY. *Educ:* Alma Col, BS, 57; Harvard Univ, MAT, 59; Johns Hopkins Univ, PhD(oceanog), 68. *Prof Exp:* From asst res scientist to res scientist, Chesapeake Bay Inst, Johns Hopkins Univ, 67-74, adj res prof & assoc dir, 73-74; prof marine sci, State Univ NY, Stony Brook, 74-84, actg vprovost, Res & Grad Studies, 85-86, provost, 86-89, DIR, MARINE SCI RES CTR, STATE UNIV NY, STONY BROOK, 74-, DEAN & LEADING PROF MARINE SCI, 84- *Concurrent Pos:* Vis assoc prof, Univ Del, 69, lect prof, Univ Md, 69-71; vis prof, Franklin & Marshall Col, 70-71; sci dir & vpres, Hydrocon Inc, 71-74; mem, Univ-Nat Oceanog Lab Syst Adv Coun, 77-80, vchmn, 80; partic workshop, Sci Comt Ocean Res, Intergovt Oceanog Comn, 79, chmn, 80; panel chmn & partic workshop, Eng Found Conf Offshore Indust, Nat Oceanic & Atmospheric Admin, 79; mem oversight review team, Off Oceanog Facil, NSF, 80; mem bd trustees, Stony Brook Found, State Univ NY, 78-84; sr ed, Coastal Ocean Pollution Assessment, 80-84; mem sci work group, Nat Oceanic Satellite Syst, NASA, 80-86; chmn, Outer Continental Shelf Sci Comt, Minerals Mgt Serv, US Dept Interior, 85-86; hon prof, EChina Normal Univ, 85-; chmn bd dir, Marine Div Nat Asn, State Univ & Land Grant Col, 86-88, Nat Res Coun Marine Bd, 89- & NY Gov Task Force Coastal Resources, 90- *Mem:* Am Soc Limnol & Oceanog; AAAS; Nat Asn Geol Teachers; NY Acad Sci; Estuarine Res Fedn (vpres, 81-85, pres, 85-87). *Res:* Estuarine and shallow water sedimentation; suspended sediment transport; interactions of sediment and organisms; pollution; continental shelf sedimentation; marine geophysics; thermal ecology; coastal zone management; marine policy. *Mailing Add:* Marine Sci Res Ctr State Univ NY Stony Brook NY 11794-5000

SCHUBERT, BERNICE GIDUZ, b Boston, Mass, Oct 6, 13. PLANT TAXONOMY. *Educ:* Univ Mass, BS, 35; Radcliffe Col, MS, 37, PhD, 41. *Prof Exp:* Tech asst taxon bot, Gray Herbarium, Harvard Univ, 41-50; Guggenheim fel, 50-51; consult plant taxon, Pedobot Proj, Econ Coop Admin, Brussels, Belg, 51-52; plant taxonomist, New Crops Res Br, Agr Res Serv, USDA, 52-61; assoc cur, Harvard Univ, 62-69, lectr biol, 71, cur, Arnold Arboretum, 69-84, sr lectr biol, 75-84; RETIRED. *Concurrent Pos:* Ed, J Arnold Arboretum, 63-78. *Mem:* Am Soc Plant Taxon; Soc Econ Bot; Bot Soc Am; Am Inst Biol Sci; hon mem Soc Bot Mex. *Res:* Desmodium; American species of Dioscorea and Begonia. *Mailing Add:* Harvard Univ Herbaria 22 Divinity Ave Cambridge MA 02138

SCHUBERT, CEDRIC F, b Murray Bridge, Australia, Oct 20, 35; div; c 2. MATHEMATICS. *Educ:* Univ Adelaide, BSc, 57, Hons, 58, MSc, 60; Univ Toronto, PhD(math), 62. *Prof Exp:* Lectr math, Univ Toronto, 61-62; asst prof, Univ Calif, Los Angeles, 62-69; assoc prof, 69-77, PROF MATH, QUEEN'S UNIV, ONT, 77- *Mem:* Can Math Soc; AAAS; Am Math Soc. *Res:* Operator and function theoretic properties of linear elliptic partial differential equations. *Mailing Add:* Dept Math & Statist Queen's Univ Kingston ON K7L 3N6 Can

SCHUBERT, DANIEL SVEN PAUL, b Buffalo, NY, Sept 28, 35; m 69; c 1. PSYCHIATRY, PSYCHOLOGY. *Educ:* State Univ Buffalo, BA, 55, MD, 65; Univ Chicago, PhD(psychol), 69. *Prof Exp:* Resident psychiat, Yale Univ, 69-72; asst prof, 72-77, ASSOC PROF PSYCHIAT, CASE WESTERN RESERVE UNIV, 77- *Concurrent Pos:* Consult ed, J Creative Behav, 71-, J Psychiat Treat Eval, 79, ed, Int J Psychiat Med, 87- *Honors & Awards:* Fel Am Psychiat Asn. *Mem:* Am Psychol Asn; Am Psychiat Asn; Sigma Xi; Am Col Psychiat. *Res:* Psychobiology; psychosomatic medicine; variants of normal personality; measurement of personality. *Mailing Add:* Metrop Gen Hosp-Psychiat 3395 Scranton Rd Cleveland OH 44109

SCHUBERT, DAVID CRAWFORD, b Shillington, Pa, Jan 8, 25; m 49; c 3. LASERS. *Educ:* Lehigh Univ, BS, 49, MS, 50; Univ Md, PhD(physics), 61. *Prof Exp:* Asst physics, Lehigh Univ, 49-50; res physicist, Nat Bur Standards, 50-66; res physicist, 66-, SR ENGR, WESTINGHOUSE ELEC CORP. *Mem:* Am Phys Soc; Inst Elec & Electronics Engrs. *Res:* Physics of the free electron; electron optical study of gas flow at extremely low pressures; plasma physics; quantum optics. *Mailing Add:* 737 G St Pasadena MD 21122

SCHUBERT, EDWARD THOMAS, b Brooklyn, NY, Apr 24, 27; m 53; c 3. BIOCHEMISTRY. *Educ:* Fordham Univ, BS, 49, MS, 52, PhD(biochem), 59. *Prof Exp:* Lectr biochem, Fordham Univ, 56-59; res assoc, Med Col, Cornell Univ, 59-60, instr, 60-65, asst prof biochem, 65-88, ASSOC PROF, MED COL, CORNELL UNIV, 88- *Concurrent Pos:* Dir biochem, Pediat Ultramicro Chem Labs, New York Hosp, 66-76; asst dir, Clin Biochem, 76-88, dir clin biochem, 88- *Mem:* AAAS; Am Chem Soc; Am Asn Clin Chem; Sigma Xi. *Res:* Biochemistry of growth and development; clinical chemistry; enzymic changes in renal diseases. *Mailing Add:* Dept Biochem/Pediat/Path Cornell Univ Med Col 1300 York Ave New York NY 10021

SCHUBERT, GERALD, b New York, NY, Mar 2, 39; m 60; c 3. GEOPHYSICS, PLANETARY SCIENCES. *Educ:* Cornell Univ, BEngPhys & MAE, 61; Univ Calif, Berkeley, PhD(eng), 64. *Prof Exp:* Nat Acad Sci-Nat Res Coun res fel, Dept Appl Math & Theoret Physics, Univ Cambridge, 65-66; from asst prof planetary & space sci to assoc prof planetary physics, 66-74, PROF GEOPHYS & PLANETARY PHYSICS, UNIV CALIF, LOS ANGELES, 74- *Concurrent Pos:* Alexander von Humboldt fel & Fulbright grant, 69; John Simon Guggenheim Mem Found fel, 72. *Honors & Awards:* James B Macelwane Award, Am Geophys Union, 75. *Mem:* AAAS; Am Geophys Union. *Res:* Geophysical and astrophysical fluid dynamics; planetary physics. *Mailing Add:* Dept Earth & Space Sci Univ Calif Los Angeles CA 90024

SCHUBERT, JACK, TOXICOLOGY, CHELATION. *Educ:* Univ Chicago, PhD(phys chem), 44. *Prof Exp:* Prof chem, Univ Md, Baltimore County Campus, 80-86; prof, Mich State Univ, 86-, ADJ PROF BIOCHEM, 87- *Mailing Add:* 7012 Helena Ave West Olive MI 49460

SCHUBERT, JOHN ROCKWELL, b East Orange, NJ, Aug 1, 25; m 48; c 4. NUTRITIONAL BIOCHEMISTRY. *Educ:* Pa State Univ, BS, 48; Ore State Univ, MS, 51, PhD(nutrit), 56. *Prof Exp:* Res asst agr chem, Ore State Univ, 50; mem prod staff, Cutter Labs, Calif, 51; res asst agr chem, Ore State Univ, 51-57, asst prof, 57-63; exec secy nutrit study sect, Res Grants Rev Br, Div Res Grants, NIH, 63-85, referral off, 67-85; RETIRED. *Concurrent Pos:* Spec res fel, Exp Liver Dis Sect, Nat Inst Arthritis & Metab Dis, 60-61. *Mem:* Am Inst Nutrit; Am Soc Clin Nutrit. *Res:* Agricultural chemistry; animal nutrition; bacterial physiology; metabolic diseases of nutritional origin; science administration. *Mailing Add:* RR 1 Box 241 B Baker WV 26801

SCHUBERT, KAREL RALPH, b Urbana, Ill, Oct 12, 49; m 67; c 2. PLANT BIOCHEMISTRY, CELL BIOLOGY. *Educ:* WVa Univ, BS, 71; Univ Ill, MS, 73, PhD(biochem), 75. *Prof Exp:* Fel bot & plant path, Ore State Univ, 75-76; from asst prof to assoc prof biochem, Mich State Univ, 76-83; res mgr, Monsanto, 83-85; dir, Plant Genetic Resources Ctr, Mo Bot Garden, 88-89; asst dir, ctr plant sci & biotechnol, Wash Univ, 88-90; GEORGE LYNN CROSS PROF, UNIV OKLA, 90- *Concurrent Pos:* Adj assoc prof, Wash Univ, 83-85, vis assoc prof, 85-90, adj prof, 90-; consult, Mo Bot Garden, 87-89; pvt consult, 87- *Mem:* Am Soc Plant Physiologists; Am Chem Soc; Sigma Xi; Am Soc Biochem & Molecular Biol; AAAS; Int Soc Plant Molecular Biol. *Res:* Biochemical, physiological and molecular factors involved in the establishment of effective symbiotic association between leguminous and actinoriza plants and nitrogen-fixing bacteria including Rhizobium and Frankia; biochemical energetics; conservation of plant diversity; wheat tissue culture/transformation; mechanisms of plant defenses against insects; tropical genetic resources; research administration; resource management. *Mailing Add:* Bot/Microbiol Dept Univ Okla 770 Van Vleet Oval Norman OK 73019-0245

SCHUBERT, RUDOLF, b New York, NY, Sept 28, 40; m 63, 85; c 2. ELECTRICAL CONTACTS, GAS-METAL INTERACTIONS. *Educ:* Fairleigh Dickinson Univ, BS, 62, MS, 64; Univ Del, PhD(physics), 69. *Prof Exp:* Mem tech staff, Bell Tel Labs, 69-83; MEM TECH STAFF, BELL COMMUN RES, 84- *Concurrent Pos:* Prin investr, Bell Tel Labs, 69-83 & Bell Commun Res, 84-; instr, Am Vacuum Soc, 75-; comt chmn, Am Soc Testing & Mat & Inst Elec & Electronics Engrs. *Mem:* Am Vacuum Soc; Am Soc Testing & Mat; Inst Elec & Electronics Engrs Computer Soc. *Res:* Interaction of the environment with metal surfaces and specifically as to the corrosion of electrical contacts. *Mailing Add:* Bell Commun Res Inc PO Box 7040 Red Bank NJ 07701-7040

SCHUBERT, WALTER L, b New York, NY, Dec 17, 42; m 83. PHYSICS. *Educ:* Hofstra Univ, BA, 64. *Prof Exp:* CONSULT TECH WRITER, AVIONIC & COMMUN ELECTRONICS, DATA COMMUN TEACHING & DOC ASSOC, 87- *Mem:* Am Asn Physics Teachers; Soc Tech Commun. *Res:* Industrial commercial and defense electronics systems. *Mailing Add:* PO Box 20368 Cherokee Sta New York NY 10028-0053

SCHUBERT, WILLIAM K, b Cincinnati, Ohio, July 12, 26; m 48; c 4. PEDIATRICS. *Educ:* Univ Cincinnati, BS, 49, MD, 52; Am Bd Pediat, dipl, 57. *Prof Exp:* Intern, Med Ctr, Ind Univ, 52-53; resident pediat, Cincinnati Children's Hosp, 53-55; instr pediat, Med Ctr, Univ Cincinnati, 56-59, sr res assoc, 58-63, assoc prof, 63-69; dir clin res ctr, 63-76, assoc Chief Staff, 71-72, dir div gastroenterol & gastroenterol clin, 68-79, physician exec dir, 79-83, chief staff, 72-88, PRES & CHIEF EXEC OFFICE, MED CTR CHILDRENS HOSP, 83-; PROF PEDIAT, UNIV CINCINNATI, 69-, CHMN DEPT, 79- *Mem:* Soc Pediat Res; Am Fedn Clin Res; Am Asn Study Liver Dis; Am Gastroenterol Asn; Am Pediat Soc (Coun mem 86-); Assoc Med Sch Ped (Dept Chmn). *Res:* Gastroenterology; metabolism. *Mailing Add:* Children's Hosp Med Ctr Elland & Bethesda Aves Cincinnati OH 45229-2899

SCHUBERT, WOLFGANG MANFRED, b Hanover, Ger, Feb 16, 20; nat US; m 41, 64; c 2. ORGANIC CHEMISTRY. *Educ:* Univ Ill, BS, 41; Univ Minn, PhD(org chem), 47. *Prof Exp:* Res chemist, Am Cyanamid Co, Conn, 44-46; from instr to assoc prof, 47-58, PROF ORG CHEM, UNIV WASH, 58- *Concurrent Pos:* Guggenheim fel & Fulbright res scholar, 60-61. *Mem:* Am Chem Soc; Royal Soc Chem. *Res:* Mechanisms of organic chemical reactions; solvent effects; substituent effects; acid base catalysis. *Mailing Add:* Dept of Chem Univ of Wash Seattle WA 98195

SCHUBRING, NORMAN W(ILLIAM), b Port Hope, Mich, June 1, 24; m 48; c 1. MICROWAVES, ANTENNAS. *Educ:* Wayne State Univ, BSEE, 52, MSEE, 59. *Prof Exp:* Res asst, Willow Run Res Ctr, Univ Mich, 52; STAFF RES ENGR, GEN MOTORS RES LABS, 52- *Honors & Awards:* Sr Award, Inst Elec & Electronics Engrs, 58-59; Charles L McCuen Spec Achievement Award, Gen Motors Res Labs, 78. *Mem:* Inst Elec & Electronics Engrs; Soc Automotive Engrs. *Res:* Electronics; microwaves; automobile radar; antennas; guided missile countermeasures; instrumentation; ultrasonics; sonics; nondestructive testing; pyroelectrics, piezoelectrics and ferroelectrics; microwave sintering of cermaics. *Mailing Add:* Gen Motors Res Labs 30500 Mound Rd Warren MI 48090-9055

SCHUCANY, WILLIAM ROGER, b Dallas, Tex, Sept 7, 40; m 61; c 3. MATHEMATICAL STATISTICS, APPLIED STATISTICS. *Educ:* Univ Tex, BA, 63, MA, 65; Southern Methodist Univ, PhD(statist), 70. *Prof Exp:* Engr-scientist, Tracor, Inc, Tex, 63-66; sr engr, LTV Electrosysts, Inc, 66-68; lab mgr statist consult, 68-70, asst prof math statist, 70-74, assoc prof, 74-80, PROF STATIST, SOUTHERN METHODIST UNIV, 80- *Concurrent Pos:* Assoc ed, Commun Statist, 71-, J Educ Statist, 81-83, J Am Statist Asn, 83-86; consult, Medicus Corp, Louis, Bowles & Grace, Inc, Grove & Assocs & Mobil Oil. *Honors & Awards:* Res Award, Sigma Xi, 79. *Mem:* Fel Am Statist Asn; Inst Math Statist. *Res:* Extensions and applications of techniques for improvement of estimators and the construction of approximate statistical tests and interval estimation; nonparametric ranking statistics; computer simulation variance reductions; minimum distance, robustness and resampling plans. *Mailing Add:* Dept Statist Southern Methodist Univ Dallas TX 75275

SCHUCHER, REUBEN, b Zhitomir, Russia, June 11, 22; Can citizen; m 56; c 2. CLINICAL CHEMISTRY, BIOCHEMISTRY. *Educ:* McGill Univ, BSc, 49, PhD(biochem), 54; Univ Sask, MSc, 51. *Prof Exp:* Res asst biochem, McGill Univ, 51-55; biochemist & res assoc, Dept Med, 55-68, DIR DEPT CLIN CHEM, JEWISH GEN HOSP, 68- *Concurrent Pos:* Cancer Res Soc fel, Res Inst, Montreal Gen Hosp, 54-55; lectr, Dept Invest Med, McGill Univ, 59; treas & bd mem, Que Hosp Biochem Corp, 63-68; consult, Maimonides Hosp & Home for Aged, 66-, Jewish Convalescent Hosp, 67- & Med Data Sci Labs, 70-; proj dir, Lady Davis Inst Med Res, Jewish Gen Hosp, 68- *Mem:* NY Acad Sci; Chem Inst Can; Can Soc Clin Chemists (vpres, 67, pres, 68); Can Biochem Soc; Can Physiol Soc. *Res:* Synthesis of enzyme proteins by pancreas slices in vitro; metabolism of insulin by pancreas, especially synthesis and breakdown by pancreatic insulinase; insulin secretion in vivo; thyroid function tests in clinical laboratory. *Mailing Add:* 4928 Ponsard Ave Montreal PQ H3W 2A5 Can

SCHUCK, JAMES MICHAEL, b Chicago, Ill, Nov 5, 34; m 56; c 5. MAMMALIAN CELL CULTURE, PROTEIN CHEMISTRY. *Educ:* Univ Wis, BS, 56; Mass Inst Technol, PhD(org chem), 60. *Prof Exp:* Teaching asst, Mass Inst Technol, 57; res chemist, Res & Eng Div, Monsanto Chem Co, 60-61, sr res chemist, 61-64 & Cent Res Dept, 64-66, res group leader, Cent Res Dept, 66-68 & New Enterprise Div, 68-69, sr res group leader, New Enterprise Div, 69-77 & Corp Res & Develop, 77-78, res mgr biosynthesis, Corp Res Develop, 78-81, mgr explor res & develop, Environ Policy Staff, 81-83, mgr explor biomed res & develop & bio process & develop, 83-85, mgr res, biochem/cell cult, Cent Res Labs, 85-90, MGR RES, HEALTH SCI DEPT, MONSANTO CORP RES, MONSANTO CO, 90- *Honors & Awards:* Bausch & Lomb Sci Award, 52. *Mem:* AAAS; Sigma Xi. *Res:* Protein and synthetic organic chemistry; study of structure and function of mammalian growth hormones and analogs; blood protein fractionation; mammalian cell culture; structure and function of proteolytic enzymes; tumor angiogenesis and antiangiogenesis; development of large scale mammalian cell culture reactor systems. *Mailing Add:* Monsanto Corp Res Monsanto Co St Louis MO 63167

SCHUCKER, GERALD D, b McConnellstown, Pa, Oct 29, 36. ANALYTICAL CHEMISTRY. *Educ:* Juniata Col, BS, 58; Univ Mo, Rolla, MS, 67. *Prof Exp:* Chemist, Pa RR Test Dept, 58-62; res technician anal chem, Cornell Univ, 63-65; teaching asst, Univ Mo, Rolla, 66-67; SR RES CHEMIST, CORNING INC, 68- *Mem:* Am Chem Soc; Soc Appl Spectros. *Res:* Separation and preconcentration of trace amounts of elements and their determination by spectrophotometry; emission spectroscopy; atomic absorption; flame emission spectroscopy. *Mailing Add:* Corning Inc HP-ME-03-070 Corning NY 14831

SCHUCKER, ROBERT CHARLES, b Altoona, Pa, Jan 10, 45; m 68; c 2. PETROLEUM CHEMISTRY, CATALYSIS. *Educ:* Univ SC, BS, 68, MS, 70; Ga Inst Technol, PhD(chem eng), 74. *Prof Exp:* Res engr, Procter & Gamble Co, 74-77; mem, Exxon Corp Res Lab, 77-80; Esso Petrol Res Dept, Can, 85-86; staff engr, 80-84, RES ASSOC, EXXON RES & DEVELOP LABS, BATON ROUGE, LA, 87- *Mem:* Am Chem Soc; Am Inst Chem Engrs; NAm Membrane Soc; Soc Advan Mat & Process Eng. *Res:* Chemistry of heavy petroleum feedstocks with focus on macromolecular structure and reactivity; enhance yield of desirable product via catalytic reactions; separation of petroleum compounds for fuels refining. *Mailing Add:* Exxon Res & Develop Labs PO Box 2226 Baton Rouge LA 70821

SCHUDER, DONALD LLOYD, b Bartholomew Co, Ind, Mar 4, 22; m 45; c 2. ENTOMOLOGY. *Educ:* Purdue Univ, BSA, 48, MS, 49, PhD, 57. *Prof Exp:* From assoc prof to prof, 49-87, EMER PROF ENTOM, PURDUE UNIV, WEST LAFAYETTE, 88- *Concurrent Pos:* Ed, Ind Nursery News, 54-87. *Mem:* Entom Soc Am; Int Soc Arboriculture; Am Asn Nurserymen. *Res:* Coccidae; insect pests of ornamental trees, shrubs and Christmas trees. *Mailing Add:* Dept Entom Entom Hall Purdue Univ West Lafayette IN 47907

SCHUDER, JOHN CLAUDE, b Olney, Ill, Mar 2, 22; m 46; c 3. BIOPHYSICS. *Educ:* Univ Ill, BSEE, 43; Purdue Univ, MSEE, 51, PhD(elec eng), 54. *Prof Exp:* Instr elec eng, Purdue Univ, 49-54, asst prof, 54-56; assoc prof physics, Doane Col, 56-57; asst prof eng in surg res, Univ Pa, 59-60; from assoc prof to prof, 60-85, EMER PROF SURG, UNIV MO, COLUMBIA, 85- *Concurrent Pos:* Fel eng in surg res, Univ Pa, 57-59; res engr, Hosp Univ Pa, 57-60; estab investr, Am Heart Asn, 65-70. *Mem:* Inst Elec & Electronics Eng; Am Soc Artificial Internal Organs; Inst Elec & Electronics Engrs Eng in Med & Biol Soc. *Res:* Application of physics to medical problems, in particular, cardiac pacing, cardiac defibrillation, artificial hearts, electromagnetic energy transport through biological tissue. *Mailing Add:* Dept Surg Univ Mo Columbia MO 65212

SCHUE, JOHN R, b Gaylord, Minn, Feb 6, 32; m 57; c 4. ALGEBRA. *Educ:* Macalester Col, AB, 53; Mass Inst Technol, PhD(math), 59. *Prof Exp:* Instr math, Mass Inst Technol, 58-59; asst prof, Oberlin Col, 59-62; assoc prof, 62-69, PROF MATH, MACALESTER COL, 69- *Concurrent Pos:* NSF sci faculty fel, 68-69. *Honors & Awards:* Thomas Jefferson Award, 89. *Mem:* Am Math Soc; Math Asn Am. *Res:* Lie algebras and functional analysis. *Mailing Add:* Dept Math Macalester Col St Paul MN 55105

SCHUEL, HERBERT, b New York, NY, Apr 8, 35; m 62; c 2. CELL BIOLOGY, DEVELOPMENTAL BIOLOGY. *Educ:* NY Univ, BA, 56; Univ Pa, PhD(zool), 60. *Prof Exp:* Res assoc develop biol, Oceanog Inst, Fla State Univ, 60-61; res assoc chem, Northwestern Univ, 63-65; asst prof biol, Oakland Univ, 65-68; asst prof anat, Mt Sinai Sch Med, 68-72, assoc res prof, 72-73; assoc prof biochem, State Univ Downstate Med Ctr, 73-77; assoc prof, 77-89, PROF ANAT, STATE UNIV NY, BUFFALO, 89- *Concurrent Pos:* NIH fel, Oak Ridge Nat Lab, 61-63; NIH res career develop award, Oakland Univ & Mt Sinai Sch Med, 68-73; mem, Corp Marine Biol Labs; instr comt mem, Marine Biol Lab, 83-86; prin investr grants, Am Cancer Soc, 70-75, Pop Coun, 74-75, NIH, 69-72 & 83-86, NSF, 65-70 & 75-88 & Nat Inst Drug Abuse, 88-91; contribr to sci jour. *Mem:* AAAS; Am Physiol Soc; Am Asn Anat; Am Soc Cell Biol; Biophys Soc; Am Soc Zool; Sigma Xi; Soc Gen Physiologists; Soc Study Reprod; Am Inst Biol Sci. *Res:* Sub-cellular biochemistry; isolation and molecular characterization of sub-cellular organelles; elucidation of role in cellular functions; fertilization, prevention of polysperm secretion and cell division; acrosome reaction; cannabinoids. *Mailing Add:* Dept Anat Sci State Univ NY Buffalo NY 14214

SCHUELE, DONALD EDWARD, b Cleveland, Ohio, June 16, 34; m 56; c 6. SOLID STATE PHYSICS. *Educ:* John Carroll Univ, BS, 56, MS, 57; Case Inst Technol, PhD(physics), 63. *Prof Exp:* Instr math & physics, John Carroll Univ, 56-57; instr physics, Case Inst Technol, 57-59; from instr to assoc prof, Case Western Reserve Univ, 62-74, actg dean, 72-73, dean, 73-76, chmn dept, 76-78, vdean, 78-83, vpres, 83-84, dean, 84-86, dean sci, 88-89, PROF PHYSICS, CASE WESTERN RESERVE UNIV, 74-, MICHELSON PROF PHYSICS, 89- *Concurrent Pos:* Mem tech staff, Bell Tel Lab, 70-72; univ rep, Argonne Univ Assocs, 78-82; mem bd overseers, St Mary Sem, 74-81 & bd trustees, 82-; mem Alumni Coun, Case Alumni Asn, 86-; mem, bd trustees, Newman Found, 81- *Mem:* Am Inst Physics; Am Asn Physics Teachers; Sigma Xi. *Res:* Low frequency dielectric constant of ionic crystals, their pressure and temperature dependence; lattice dynamics; thermal expansion at low temperatures; elastic constants of single crystals; equation of state of solids; high pressure physics; electrical and mechanical properties of polymers and liquid crystals. *Mailing Add:* Dept Physics Case Western Reserve Univ 10900 Euclid Ave Cleveland OH 44106

SCHUELE, WILLIAM JOHN, b Philadelphia, Pa, Apr 27, 23; m 52; c 2. INORGANIC CHEMISTRY. *Educ:* Univ Pa, PhD(chem), 56. *Prof Exp:* Sr res chemist, Res Lab, Franklin Inst, 55-65; ADV CHEMIST, IBM CORP, 65- *Mem:* AAAS; Sigma Xi; Am Chem Soc; NY Acad Sci. *Res:* Fine particles; ferromagnetism; photolithography; semiconductor chemistry; chemical conservation. *Mailing Add:* 33 Clover St South Burlington VT 05403

SCHUELER, BRUNO OTTO GOTTFRIED, b Estcourt, SAfrica, Apr 21, 32; m 60; c 2. ORGANIC CHEMISTRY, CHEMICAL ENGINEERING. *Educ:* Univ Natal, BSc, 53, Hons, 54, MSc, 55, PhD(org chem), 57; Cambridge Univ, PhD(chem eng), 60. *Prof Exp:* Res engr process develop, Exp Sta, Del, 60-66, sr res engr, 66-69, asst div supt, Plastics Dept, Tex, 69-73, staff engr, 73-81, tech fel, plastics dept, 81-87, SR TECH FEL, DU PONT CHEMICALS, E I DU PONT DE NEMOURS & CO, INC, 87- *Res:* Indole alkaloids, especially voacangine and ibogaine; bubble formation at submerged orifices; effect of chemical structure on distribution coefficients; polymer manufacturing. *Mailing Add:* 1801 N 21 Orange TX 77630

SCHUELER, DONALD G(EORGE), b Harvard, Nebr, Oct 22, 40; c 2. ELECTRICAL ENGINEERING, ENERGY CONVERSION. *Educ:* Univ Nebr, Lincoln, BS, 62, MS, 63, PhD(elec eng), 69. *Prof Exp:* Staff mem electronics, Sandia Labs, 63-67; instr elec eng, Univ Nebr, 67-68; staff mem electronics, 69-70, supvr, Solid State Electronics Res, 70-75, supvr, Photovoltaic Projs, 75-81, MGR SOLAR ENERGY DEPT, SANDIA LABS, 81- *Mem:* Inst Elec & Electronics Engrs; Am Solar Energy Soc; Sigma Xi. *Res:* Solid state microwave devices; ferroelectric ceramic optical devices; energy conversion; photovoltaic devices; solar energy. *Mailing Add:* Dept 6220 Box 5800 Sandia Nat Lab Albuquerque NM 87185

SCHUELER, PAUL EDGAR, b New York, NY, Apr 18, 45; m 72; c 2. PHYSICAL ORGANIC CHEMISTRY. *Educ:* Univ Rochester, BS, 65; NY Univ, PhD(chem), 73. *Prof Exp:* Fel org chem, Rutgers Univ, 72-73; res assoc hot-atom chem, Brookhaven Nat Lab, 73-75; lectr chem, Rutgers Univ, New Brunswick, 75-77; from asst prof to assoc prof chem, Somerset County Col, 77-87; PROF CHEM, RARITAN VALLEY COMMUNITY COL, 87- *Concurrent Pos:* Coordr, NJ Master Fac Prog, 87-90. *Mem:* Am Chem Soc; AAAS; Sigma Xi; NY Acad Sci. *Res:* Mechanistic physical organic chemistry, with emphasis on reactive intermediates in polar aprotic solvents; chemistry education. *Mailing Add:* 426 Harvard Ave South Plainfield NJ 07080

SCHUELLEIN, ROBERT JOSEPH, b NY, Feb 22, 20. GENETICS. *Educ:* Univ Dayton, BS, 43; Univ Pittsburgh, MS, 48, PhD(genetics), 56. *Prof Exp:* Teacher parochial high sch, Pa, 43-49; teacher, Ohio Univ, 49-53; instr, Univ Dayton, 53, assoc prof biol, 59-64; mem staff, training grants & awards br, Nat Heart Inst, 64-65, exec secy grants assoc prog, Div Res Grants, NIH, 65-68, chief periodont dis & soft tissue prog, 68-70, chief soft tissue stomatol prog, 70-74, spec asst res manpower, Nat Inst Dent Res, 74-83; CONSULT SCI ADMIN, 84- *Mem:* Genetics Soc Am; Am Soc Human Genetics; Am Genetic Asn. *Res:* Genetics of Drosophila; population and human genetics; evolution and biometrical analysis. *Mailing Add:* 5626 Larmar Rd Bethesda MD 20816

SCHUERCH, CONRAD, b Boston, Mass, Aug 2, 18; m 48; c 4. SYNTHETIC ORGANIC & NATURAL PRODUCTS CHEMISTRY. *Educ:* Mass Inst Technol, BS, 40, PhD(org chem), 47. *Prof Exp:* Res assoc inorg war res, Mass Inst Technol, 42-43; chemist, Res Lab, Nat Adv Comt Aeronaut, Ohio, 45; sessional lectr, McGill Univ, 47-48, Hibbert fel, 48-49; from asst prof to prof, 49-78, chmn dept, 56-72, distinguished prof, 78-83, EMER DISTINGUISHED PROF CHEM, STATE UNIV NY COL ENVIRON SCI & FORESTRY, 83- *Concurrent Pos:* Guggenheim fel, 60-61; hon mem, Soc Fiber Sci & Technol, Japan, 90. *Honors & Awards:* Anselme Payen Award, Am Chem Soc, 72. *Mem:* Am Chem Soc; Tech Asn Pulp & Paper Indust. *Res:* Chemical synthesis of stereoregular polysaccharides; glycoside synthesis and synthetic carbohydrate antigens; solvent effects, accessibility phenomena and structural studies on lignin and wood; stereochemistry of vinyl polymerization; experimental pulping methods. *Mailing Add:* Dept Chem State Univ NY Col Environ Sci & Forestry Syracuse NY 13210

SCHUERMANN, ALLEN CLARK, JR, b Denver, Colo, Sept 28, 43; m 65; c 2. INDUSTRIAL ENGINEERING. *Educ:* Univ Kans, BA, 65; Wichita State Univ, MS, 68; Univ Ark, PhD(indust eng), 71. *Prof Exp:* Oper res analyst, Boeing Co, 65-69; sr oper res analyst, Boeing Computer Serv, 71; asst prof, 71-78, ASSOC PROF INDUST ENG, WICHITA STATE UNIV, 78-, CHMN DEPT, 79- *Concurrent Pos:* Prin investr, Rehab Serv Admin, 76-81; consult, Boeing Co, 72-73, Kans Gas & Elec Co, 78- & Kans Power & Light Co, 79- *Mem:* Am Inst Indust Engrs; Inst Mgt Sci; Oper Res Soc Am; Am Soc Eng Educ. *Res:* Modeling of the economic factors, incentives and disincentives which have an impact on the successful rehabilitation and employment of the severely physically disabled. *Mailing Add:* Dept Indust Eng Okla State Univ Stillwater OK 74078

SCHUESSLER, CARLOS FRANCIS, dentistry; deceased, see previous edition for last biography

SCHUESSLER, HANS A, b Mannheim, Ger, June 25, 33. ATOMIC PHYSICS. *Educ:* Univ Heidelberg, MS, 61, PhD(physics), 64. *Prof Exp:* Asst prof physics, Tech Univ, Berlin, 63-66; from res asst prof to res assoc prof, Univ Wash, 66-69; assoc prof, 69-81, PROF PHYSICS, TEX A&M UNIV, 81- *Concurrent Pos:* Mem, Nat Comt on Fundamental Constants, 82-88; founding mem, APS Topical Group on Precision Measurements & Fundamental Constants, 88- *Mem:* Am Phys Soc; Europ Phys Soc. *Res:* Radio frequency spectroscopy; optical pumping; ion storage; atomic clocks; lasers; photodissociation of ion molecules; polarized atomic beams; nuclear moments; level crossing; laser generated plasmas; on-line laser spectroscopy of short-lived isotopes; spectroscopy of highly-charged ions. *Mailing Add:* Dept Physics Tex A&M Univ College Station TX 77843-4242

SCHUETTE, EVAN H(ENRY), technical documentation, for more information see previous edition

SCHUETTE, OSWALD FRANCIS, JR, b Washington, DC, Aug 20, 21; m 47; c 3. PHYSICS. *Educ:* Georgetown Univ, BS, 43; Yale Univ, MS, 44, PhD(physics), 49. *Prof Exp:* Instr physics, Yale Univ, 43-44, asst instr, 46-48; assoc prof, Col of William & Mary, 48-53; Fulbright scholar, Ger, 53; sci liaison officer, Sci & Tech Unit, US Dept Navy, Ger, 54-58; mem staff, Nat Acad Sci, 59-60; dep spec asst space, Off Secy Defense, 60-63; head dept, 63-80, PROF PHYSICS, UNIV SC, 63- *Concurrent Pos:* Guest prof, Univ Vienna, Austria, 80. *Mem:* Am Phys Soc; Am Asn Physics Teachers. *Res:* Separation of isotopes; underwater sound; mass spectroscopy. *Mailing Add:* Dept Physics & Astron Univ SC Columbia SC 29208

SCHUETZ, ALLEN W, b Monroe, Wis, July 8, 36; m 65; c 2. EMBRYOLOGY, PHYSIOLOGY ANIMAL. *Educ:* Univ Wis, BS, 58, PhD(endocrinol), 65. *Prof Exp:* From instr to prof pop dynamics, obstet & gynec, 66-75, PROF POP DYNAMICS, SCH HYG & PUB HEALTH, JOHNS HOPKINS UNIV, 75- *Concurrent Pos:* Fels steroid biochem, Univ Minn, 64-65, 65-66; fel, Marine Biol Lab, Woods Hole, 65, corp mem, 80; Fogarty Found sr int fel, Cambridge Univ. *Mem:* Endocrine Soc; Cell Biol Soc; Soc Study Reprod; Am Soc Zoologists; Soc Develop Biol. *Res:* Cell-cell communication, cycle nucleotide-hormone interactions; specific role of gonadotrophic hormones in regulating gametogenesis and follicular maturation; control mechanisms in oocyte growth and meiotic maturation; nuclear-cytoplasmic interactions in fertilization and early development-cell cycle regulation. *Mailing Add:* Sch Hyg & Pub Health Johns Hopkins Univ Baltimore MD 21205

SCHUETZLE, DENNIS, b Sacramento, Calif, July 21, 42; m 68; c 2. ANALYTICAL CHEMISTRY. *Educ:* Calif State Univ, San Jose, BS, 65; Univ Wash, MS, 70, PhD(chem & eng), 72. *Prof Exp:* Technician, Stoner Labs, 64-65; res chemist, Stanford Res Inst, 65-68; res assoc, Univ Wash, 68-72, res assoc prof anal chem, 72-73; MGR ANALYTICAL SCI DEPT, FORD MOTOR CO, 73- *Concurrent Pos:* Consult, Calif Air Resources Bd, Environ Protection Agency, WHO, 70-75; R&D 100 award, 85 & 88. *Mem:* Am Chem Soc; Sigma Xi. *Res:* New analytical techniques for environmental monitoring, process monitoring, quality control and materials properties. *Mailing Add:* 5443 Crispin Way West Bloomfield MI 48323

SCHUFLE, JOSEPH ALBERT, b Akron, Ohio, Dec 21, 17; m 42; c 2. PHYSICAL CHEMISTRY, HISTORY OF SCIENCE. *Educ:* Univ Akron, BS, 38, MS, 42; Western Reserve Univ, PhD(chem), 48. *Prof Exp:* Chemist, Akron, Ohio, 39-42; instr math, Western Reserve Univ, 46-47; from asst prof to prof chem, NMex Inst Mining & Technol, 48-64; head dept chem & dir inst sci res, 64-70, prof, 64-83, EMER PROF CHEM, NMEX HIGHLANDS UNIV, 83- *Concurrent Pos:* Vis prof, Univ Col, Dublin, 61-62; vis scholar, Univ Uppsala, Sweden, 77. *Honors & Awards:* John Dustin Clark Medal, Am Chem Soc, 82. *Mem:* Fel AAAS; fel Am Inst Chemists; Am Chem Soc; Inst Chem Ireland; Hist Sci Soc; Am Soc Eighteenth Century Studies. *Res:* Biographies of Torbern Bergman, eighteenth century Swedish chemist and Juan José D'Elhuyar, eighteenth century Spanish and Colombian chemist. *Mailing Add:* 1301 Eighth St Las Vegas NM 87701

SCHUG, JOHN CHARLES, b New York, NY, Mar 31, 36; m 58; c 3. PHYSICAL CHEMISTRY. *Educ:* Cooper Union, BChE, 57; Univ Ill, MS, 58, PhD(chem), 60. *Prof Exp:* Res chemist, Gulf Res & Develop Co, 60-64; from asst prof to assoc prof, 64-73, PROF CHEM, VA POLYTECH INST & STATE UNIV, 73- *Concurrent Pos:* Consult, Philip Morris Res Ctr, 66- *Mem:* Am Chem Soc. *Res:* Quantum chemistry; high-resolution nuclear magnetic resonance; mass spectrometry. *Mailing Add:* Dept Chem Va Polytech Inst & State Univ Blacksburg VA 24061

SCHUG, KENNETH, b Easton, Pa, Aug 27, 24; m 48; c 3. INORGANIC CHEMISTRY. *Educ:* Stanford Univ, BA, 45; Univ Southern Calif, PhD, 55. *Prof Exp:* Res assoc chem, Univ Wis, 54-56; from instr to assoc prof, 56-75, chmn dept, 76-82 & 85-87 PROF CHEM, ILL INST TECHNOL, 75- *Concurrent Pos:* Consult, Argonne Nat Lab, 61-63; Fulbright res fel, Kyushu Univ, 64-65. *Mem:* AAAS; Am Chem Soc. *Res:* Inorganic, coordination and solution chemistry; inorganic mechanisms. *Mailing Add:* Dept Chem Ill Inst Technol Chicago IL 60616

SCHUGAR, HARVEY, b Pittsburgh, Pa, Dec 2, 36; m 63. INORGANIC CHEMISTRY, BIOINORGANIC CHEMISTRY. *Educ:* Carnegie Inst Technol, BS, 58; Columbia Univ, MA, 59, PhD(chem), 61. *Prof Exp:* Res chemist, Esso Res & Eng Co, 61-63; res chemist, Sci Design Co, 63-65; res assoc & lectr chem, Columbia Univ, 65-67; res assoc, Calif Inst Technol, 67-68; asst prof, 68-73, ASSOC PROF CHEM, RUTGERS UNIV, NEW BRUNSWICK, 73- *Mem:* Am Chem Soc. *Res:* Aquo chemistry of ferric complexes; metal ions in biological systems. *Mailing Add:* Dept Chem Box 939 Rutgers Univ Piscataway NJ 08854

SCHUH, FRANK J, RESEARCH ADMINISTRATION. *Educ:* Ohio State Univ, BS & MS, 56. *Prof Exp:* Field drilling engr, Atlantic Richfield Co, Oil & Gas Co, S La Dist, 57-58. res engr, Prod Res Dept, 58-62, staff drilling engr, Drilling Eng Group, 62-72, Drilling Technol Sect, 72-82, dir drilling & prod mech res, 82-85 & sr res adv drilling & opers res, Prod Res Ctr, 85-86; PRES, DRILLING TECHNOL, INC, 86-; COFOUNDER, SUPREME RESOURCES CORP, 88- *Concurrent Pos:* Chmn, Drilling Reprints Ser, Soc Petrol Engrs, 73, distinguished lectr, 81-82, nat dir region VI, 83-86, co-chmn, Drilling Reprint Ser Comt, 86-89; founder & first chmn, Drilling Eng Asn, 83-84, dir, 85-86; mem, Tech Eng & Develop Comt, Ocean Drilling Prog, NSF, 86-91. *Honors & Awards:* Petrol Eng Award, Nat Eng Asn, 80; Drilling Eng Award, Soc Petrol Engrs, 86. *Mem:* Nat Acad Eng; Soc Petrol Engrs; Am Petrol Inst; Drilling Eng Asn; Soc Independent Prof Earth Scientists. *Res:* Author of various publications; granted several patents. *Mailing Add:* Drilling Technol Inc 5808 Wavertree Suite 1000 Plano TX 75093

SCHUH, JAMES DONALD, b Chicago, Ill, Oct 9, 28; m 54; c 5. ANIMAL NUTRITION. *Educ:* Kans State Univ, BS, 53; Okla State Univ, MS, 57, PhD(animal nutrit), 60. *Prof Exp:* Exten dairy specialist, Univ Nev, 60-64; assoc prof, 64-72, PROF DAIRY SCI, UNIV ARIZ, 72- *Mem:* Am Dairy Sci Asn. *Res:* Dairy herd management; calf nutrition and immunity; water quality for dairy cattle; dairy heifer management. *Mailing Add:* Dept Animal Sci Univ Ariz Tucson AZ 85721

SCHUH, JOSEPH EDWARD, b Brooklyn, NY, Mar 26, 14. CYTOLOGY. *Educ:* Georgetown Univ, AB, 38; Woodstock Col, PhL, 39; Fordham Univ, MS, 41, PhD(biol), 51; Weston Col, STL, 46. *Prof Exp:* Instr biol, St Joseph's Col, Pa, 41-42; from asst prof to assoc prof, St Peters Col, NJ, 51-62, head dept, 51-62; sr lectr, Univ Lagos, 62-65; from assoc prof to prof, 65-84, head dept, 67-77, EMER PROF BIOL, ST PETERS COL, NJ, 84- *Concurrent Pos:* Vis res assoc, LeMoyne Col, 77-78. *Mem:* AAAS; Soc Protozool; Am Soc Zool; Am Asn Jesuit Sci; NY Acad Sci. *Res:* Development in Sciara, mosquitoes and Chaoborus; effects of colchicine on metamorphosis; use of tritiated thymidine in developmental and regeneration studies; limnoria, cytology and development. *Mailing Add:* Dept Biol St Peters Col Jersey City NJ 07306

SCHUH, MERLYN DUANE, b Avon, SDak, Feb 21, 45; m 69. PHYSICAL CHEMISTRY, BIOCHEMISTRY. *Educ:* Univ SDak, BA, 67; Ind Univ, Bloomington, PhD(phys chem), 71. *Prof Exp:* Asst prof phys chem & biochem, Middlebury Col, 71-75; from asst prof to assoc prof, 75-86, PROF PHYS CHEM & BIOCHEM, DAVIDSON COL, 86- *Honors & Awards:* Sci Fac Prof Develop Awardee, NSF, 81. *Mem:* Inter-Am Photochem Soc; Am Chem Soc; Sigma Xi; Am Soc Photobiol. *Res:* Lasers used to study the photophysics and spectroscopy of electronic excited state molecules; fluorescent and phosphorescent probes of proteins. *Mailing Add:* Dept Chem Davidson Col Davidson NC 28036-1749

SCHUH, RANDALL TOBIAS, b Corvallis, Ore, May 11, 43; m; c 1. SYSTEMATICS, BIOGEOGRAPHY. *Educ:* Ore State Univ, BS, 65; Mich State Univ, MS, 67; Univ Conn, PhD(entom), 71. *Prof Exp:* Chmn dept, 80-87, CUR ENTOM, AM MUS NATURAL HIST, 74- *Concurrent Pos:* Adj prof biol, City Univ New York, 78-; ed, J NY Entom Soc, 83-89, Cladistics, 91; adj prof entomol, Cornell Univ, 89- *Mem:* Entom Soc Am; Soc Syst Zool. *Res:* Taxonomy, phylogeny and biogeography of Hemiptera of the class Insecta, especially Miridae and semiaquatic families. *Mailing Add:* Dept Entom Am Mus Natural Hist Central Park W at 79th St New York NY 10024

SCHUHMANN, R(EINHARDT), JR, b Corpus Christi, Tex, Dec 16, 14; m 37; c 2. METALLURGICAL ENGINEERING. *Educ:* Univ Mo, Rolla, BS, 33; Mont Sch Mines, MS, 35; Mass Inst Technol, ScD, 38. *Prof Exp:* From instr to assoc prof metall, Mass Inst Technol, 37-54; prof & chmn div, 54-59, head sch metall eng, 59-64, Ross prof eng, 64-80, EMER PROF, PURDUE UNIV, WEST LAFAYETTE, 80- *Concurrent Pos:* Battelle vis prof, Ohio State Univ, 66-67; Kroll vis prof, Colo Sch Mines, Golden, 77. *Honors & Awards:* James Douglas Gold Medal, Am Inst Mining, Metall & Petrol Engrs, 70, Extractive Metall Sci Award, Metall Soc, 77; Extractive Metall Sci Award, Metall Soc, 59 & 77; Mineral Ind Educ Award, 75. *Mem:* Nat Acad Eng; fel Am Soc Metals; Am Chem Soc; Am Inst Mining, Metall & Petrol Engrs; fel Metall Soc; fel AAAS. *Res:* Applications of physical chemistry to metallurgical systems; thermodynamics of high temperature multicomponent systems; nonferrous extractive metallurgy. *Mailing Add:* Sch Mat Eng Purdue Univ West Lafayette IN 47907

SCHUHMANN, ROBERT EWALD, b El Paso, Tex, Sept 27, 24; m 53; c 1. PHYSIOLOGY, ENGINEERING. *Educ:* Univ Tex, Austin, BS, 49, MS, 52; Baylor Col Med, PhD(physiol), 69. *Prof Exp:* Instr eng, Univ Tex, Austin, 50-51; engr, Tenn Gas Transmission Co, 51-54; supvry engr, Trunkline Gas Co, 54-62; asst to chief engr, Bovay Engrs, Inc, 62-63, mgr dept mech eng, 63-64; sr res physiologist, Southwest Res Inst, 69-74; prof & dean, Univ Houston, Clear Lake, 74-78, prof physiol, Sch Natural & Appl Sci, 74-90; RETIRED. *Concurrent Pos:* Asst prof, Univ Tex Health Sci Ctr, San Antonio, 72-74; vis prof biomed eng, Baylor Col Med, 80-87. *Mem:* Am Physiol Soc; Sigma Xi; Am Heart Asn. *Res:* Cardiovascular and respiratory physiology and central nervous system control of respiration; transvalvular heart assist, left ventricular bypass without thoracotomy by ventricular catheterization; bioinstrumentation; biological control system theory; physiology of human stress; physiology of human aging. *Mailing Add:* 3810 Millbridge Houston TX 77059

SCHUIT, KENNETH EDWARD, b Ticonderoga, NY, Aug 28, 42; m 64; c 3. ANATOMY, CELL BIOLOGY. *Educ:* Wheaton Col, BS, 64; Univ Ill, MS, 66, PhD(cell biol), 69; Univ Va, MD, 74. *Prof Exp:* From instr to asst prof anat, Sch Med, Univ Va, 69-72, resident pediat, 74-76; fel pediat infectious dis, Univ NC, 76-78; ASST PROF PEDIAT, UNIV PITTSBURGH, 78- *Mem:* Am Soc Cell Biol; Am Soc Microbiol; Am Asn Anat. *Mailing Add:* Pediat/Children's Hosp 125 De Soto St Pittsburgh PA 15213

SCHUKNECHT, HAROLD FREDERICK, b Chancellor, SDak, Feb 10, 17; m 41; c 2. OTOLARYNGOLOGY. *Educ:* Univ SDak, SB, 38; Rush Med Col, MD, 40; Am Bd Otolaryngol, dipl, 49; FRCS(G); Harvard Univ, MA, 61. *Hon Degrees:* DSc, Univ SDak, 73. *Prof Exp:* Resident otolaryngol, Univ Chicago, 46-49, from instr to asst prof, Univ Chicago Sch Med, 49-53; assoc surgeon, Div Otolaryngol, Henry Ford Hosp, 53-61; chmn, dept otology & laryngol,61-84, Walter Augustus Le Compte prof otology & prof laryngol, 61-87, WALTER AUGUSTUS LE COMPTE EMER PROF, HARVARD MED SCH, 87-; EMER CHIEF, MASS EYE & EAR INFIRMARY, 84- *Concurrent Pos:* Chief otolaryngol, Mass Eye & Ear Infirmary, 61-84. *Honors & Awards:* Herbert Birkett Mem lectr; Hans Wilhelm Meyer lect, Copenhagen, 67; Joseph Toynbee Mem lect, Edinburgh, 68; Gavin Livingston lect, Edinburgh, 69; George Coats lect, Philadelphia,70; Sir William Wilde lect, Dublin, 71; Morris Bender lect, NY, 72; James Yearsley lect, London, 79; Louis H Clerf lect, Philadelphia, 81; Leroy Schall lect , Boston; George Kiperash lect, Los Altos, 83; Izabel Silverman lect, Toronto, 83; Frank

Lamberson lect, Ann Arbon, 84; Shambaugh Prize, 83. *Mem:* Am Otol Soc; Am Laryngol, Rhinol & Otol Soc; Am Acad Ophthal & Otolaryngol; fel Acoust Soc Am; AMA; Sigma Xi. *Res:* Physiology of hearing; otological surgery; ear pathology. *Mailing Add:* Mass Eye & Ear Infirmary 243 Charles St Boston MA 02114

SCHULDINER, SIGMUND, b Chicago, Ill, June 12, 13; m 46; c 3. PHYSICAL CHEMISTRY. *Educ:* NY Univ, BA, 38; Columbia Univ, MA, 39. *Prof Exp:* Phys sci aide, Glass Sect, Nat Bur Standards, 40-41; asst head metals sect, Norfolk Naval Shipyard, 41-45, head indust probs sect, 45-46; phys chemist, Corrosion Br, US Naval Res Lab, 46-48, head electrode mech sect, Electrochem Br, 48-71, head, Electrochem Br, 71-75; consult, Bur Mines, Avondale Res Ctr, 75-87; RETIRED. *Honors & Awards:* William Blum Award, Electrochem Soc, 60; Pure Sci Award, Sci Res Soc Am, 66. *Mem:* Am Chem Soc; Sigma Xi; Electrochem Soc. *Res:* Electrochemistry; kinetics of electrode processes; adsorption; gases in metals; catalysis; corrosion. *Mailing Add:* 12705 Prestwick Dr Ft Washington MD 20744

SCHULDT, MARCUS DALE, b Geneva, Ill, Aug 31, 30; m 50; c 2. COMPUTER SCIENCE. *Educ:* Aurora Col, BS, 60. *Prof Exp:* Draftsman, Geneva Kitchens, 46-51; engr, Burgess Norton Mfg Co, 55-60; oceanogr, US Coast & Geod Surv, 60-65 & Inst Oceanog, Environ Sci Serv Admin, 65-66; supvry res phys scientist, 66-76, environ scientist, 76-79, OCEANOGRAPHER, ENVIRON PROTECTION AGENCY, 79- *Res:* Application of automation; limnology; oceanography. *Mailing Add:* Box 598 Ely MN 55731

SCHULDT, SPENCER BURT, b St Paul, Minn, July 1, 30; m 57, 72; c 5. THEORETICAL PHYSICS, APPLIED MATHEMATICS. *Educ:* Univ Minn, BA, 51, MS, 57. *Prof Exp:* STAFF SCIENTIST, HONEYWELL CORP RES CTR, 57- *Mem:* Am Phys Soc. *Res:* Submicron physics; heat transfer; process control; mathematical programming methods; reliability physics. *Mailing Add:* 8830 Normandale Blvd Bloomington MN 55437

SCHULENBERG, JOHN WILLIAM, b Passaic, NJ, Sept 7, 30; m 52, 82; c 4. ORGANIC CHEMISTRY. *Educ:* Queens Col, NY, BS, 51; Columbia Univ, MA, 52, PhD(org chem), 56. *Prof Exp:* Res assoc, org chem, Sterling-Winthrop Res Inst, 56-60, sr res chemist, 60-89; RETIRED. *Mem:* Am Chem Soc. *Res:* Heterocyclic synthesis; medicinal chemistry. *Mailing Add:* 187 Adams St Delmar NY 12054

SCHULER, ALAN NORMAN, b Arlington, Mass, Dec 8, 49; m 71; c 2. POLYMER SCIENCE. *Educ:* Univ Mass, BS, 71, MS, 74, PhD(polymer sci & eng), 76. *Prof Exp:* Sr chemist emulsion polymer, Union Carbide Corp, 75-78; scientist polymerization, 78-79, sr scientist, 79-81, res group leader, Polaroid Corp, 81-84, SR RES GROUP LEADER, 84- *Mem:* Am Chem Soc. *Res:* Emulsion polymerization and polymer colloid characterization structure; properties correlations. *Mailing Add:* 15 Dee Rd Lexington MA 02173

SCHULER, GEORGE ALBERT, b Altoona, Pa, Sept 21, 33; m 61; c 3. FOOD SCIENCE, MICROBIOLOGY. *Educ:* Pa State Univ, BS, 59; Univ Tenn, MS, 66; Va Polytech Inst & State Univ, PhD(food sci), 70. *Prof Exp:* Mem staff sales & serv, Pa Farm Bur, 59-61; poultry expert, Windsor Poultry Serv Ltd, Australia, 61-63; mem staff sales & serv, Swift & Co, 63-64; teacher, Monroe County Sch Bd, 64-66; FOOD SCIENTIST, COOP EXTEN SERV, UNIV GA, 70- *Concurrent Pos:* Consult qual control & prod develop, Poultry Processors, Venezula, Columbia, Peru, Ecuador, Chile, EGer & Australia. *Mem:* Inst Food Technologists; World Poultry Sci Asn; Poultry Sci Asn; Sigma Xi. *Res:* Sanitation in and bacteriological surveys of food processing plants and food handling facilities; plans for food processing facilities; rabbit programs processing; quality audit/HACCP programs. *Mailing Add:* Dept Exten Food Sci Coop Exten Serv Univ Ga Athens GA 30602

SCHULER, MARTIN N, biochemistry; deceased, see previous edition for last biography

SCHULER, MATHIAS JOHN, b New York, NY, Apr 29, 18; m 42; c 4. DYEING TECHNOLOGY, COLOR SCIENCE. *Educ:* Brooklyn Col, BA, 38. *Prof Exp:* Asst, Mem Hosp, New York, 38-39; chemist, Continental Baking Co, 39-40; res chemist, Ansbacher Siegle Corp, 40-41; instr, Bd Educ, New York, 41-44; supvr, Kellex Corp, 44-45; res chemist, E I du Pont de Nemours & Co, Inc, 45-59, sr res chemist, 59-65, res assoc, 65-70, div head, 70-79, environ mgr, 79-82; RETIRED. *Honors & Awards:* Olney Medal, Am Asn Textile Chemists & Colorists. *Mem:* Am Chem Soc. *Res:* X-rays and chemical reactions; instrumentation; mass spectrometry; physics of interaction of light on dyes and pigments; color and color specification; mechanisms of dyeing hydrophobic fibers; organometallic compounds. *Mailing Add:* 102 Cyrus Ave Pitman NJ 08071

SCHULER, ROBERT FREDERICK, b New York, NY, Mar 17, 07; m 29; c 2. ORGANIC CHEMISTRY. *Educ:* Mass Inst Technol, BS, 28, MS, 29. *Hon Degrees:* ScD, St Andrews Univ, 58. *Prof Exp:* Jr res assoc, Am Petrol Inst, 28-29; res chemist, Standard Oil Co, NJ, 29-37; chief chemist & asst plant mgr, Jacqueline Cochran Cosmetics, Inc, 37-42, plant mgr, 42-45, dir res & mgr, J. C. Labs, NJ, 43-45; dir res & prod mgr, Cosmetic Div, Int Latex Corp, 45-50; tech dir, Prince Matchabelli, Inc, 50-58; sect head, Chesebrough-Pond's, Inc, 58-60; sr scientist, Lever Bros Co, 60-61; group mgr res & develop, Toiletries Div, Gillette Co, 61-72; CONSULT, 72- *Mem:* Am Chem Soc; Soc Cosmetic Chem. *Res:* Reactions of olefins to form alcohols; preparation of alkyl phenols from olefins for use as germicides; development and utilization of byproducts of petroleum industry; rubber and plastic products development; development and manufacture of antiseptic baby cosmetics; men's and women's cosmetics and toiletries. *Mailing Add:* 15 Trinity Terr Newton Center MA 02159

SCHULER, ROBERT HUGO, b Buffalo, NY, Jan 4, 26; m 52; c 5. PHOTOCHEMISTRY, RADIATION CHEMISTRY. *Educ:* Canisius Col, BS, 46; Univ Notre Dame, PhD(phys chem), 49. *Prof Exp:* Asst prof chem, Canisius Col, 49-53; from assoc chemist to chemist, Brookhaven Nat Lab, 53-56; staff fel & dir radiation res labs, Mellon Inst Sci, Carnegie-Mellon Univ, 56-76, prof chem, 67-76; PROF RADIATION CHEM & DIR RADIATION LAB, UNIV NOTRE DAME, 76-, ZAHM PROF, 85- *Concurrent Pos:* Mem adv comt, Mellon Inst Sci, 62-; mem adv comt, Radiation Chem Data Ctr, 65-76, chmn, 73-75; vis prof, Hebrew Univ, Israel, 80; Sir CV Raman prof, Univ Madras, India, 85-86; distinguished lectr, Univ Cordoba, Arg, 88. *Honors & Awards:* Notre Dame Centennial Award, 65. *Mem:* AAAS; Am Chem Soc; Am Phys Soc; Radiation Res Soc (pres, 75-76); Royal Soc Chem; Sigma Xi. *Res:* Radiation chemistry; kinetics of ionic reactions in the radiolysis of hydrocarbons; electron spin resonance and pulse radiolysis methods to study the nature and reaction kinetics of radiation produced radicals heavy particle radiation chemistry; Raman spectroscopy. *Mailing Add:* Radiation Lab Univ Notre Dame Notre Dame IN 46556-0768

SCHULER, RONALD THEODORE, b Manitowoc, Wis, Dec 26, 40; m 67; c 2. AGRICULTURAL ENGINEERING. *Educ:* Univ Wis-Madison, BS, 63, MS, 67, PhD(agr eng), 72. *Prof Exp:* Res asst agr eng, Univ Wis-Madison, 65-69, instr, 69-70; from asst prof to assoc prof, NDak State Univ, 70-76; assoc prof agr eng, Univ Minn, St Paul,76-81; prof agr eng & chmn dept, Univ Wis-Platteville, 81-84; PROF AGR ENGR, UNIV WIS-MADISON, 84- *Mem:* Am Soc Agr Engrs; Soil Conserv Soc Am; Sigma Xi; Am Soc Engr Educ; Am Asn Agr Sci. *Res:* Agricultural engineering instrumentation; reduced tillage for soil, water and energy conservation, sunflower seed drying; cattail harvesting; soil compaction. *Mailing Add:* 926 Arden Lane Madison WI 53711

SCHULER, RUDOLPH WILLIAM, b Stuttgart, Ger, Sept 2, 19; US citizen; m 44; c 2. CHEMICAL ENGINEERING. *Educ:* Purdue Univ, BS, 48, PhD, 51. *Prof Exp:* Res engr, Colgate-Palmolive-Peet Co, 50-51; res engr, Monsanto Co, 51-55, res group leader, 55-57, asst dir res, 57-64, dir eng & mat res, 64-70, dir technol, New Enterprise Div, 70-76, RES DIR, NEW ENTERPRISE DIV, MONSANTO RES CTR, MONSANTO CO, 76- *Mem:* Am Chem Soc; Am Inst Chem Engrs. *Res:* Reaction kinetics; reactor design; mass transfer. *Mailing Add:* 38 Shady Valley Dr River Bend Estates RR 2 Chesterfield MO 63017

SCHULER, VICTOR JOSEPH, b New York, NY, Mar 9, 33; m 56. ECOLOGICAL SCIENCES, ENVIRONMENTAL & MARINE CONSULTING. *Educ:* Cornell Univ, BS, 66, MS, 69. *Prof Exp:* Lab instr, Ithaca Col, 64-65; res asst, Cornell Univ, 66-68; sr res biologist, Ichthyological Assoc Inc, 67-68, vpres proj dir, 75-78, sr vpres, 78-83; pres, V J Schuler Assocs, Inc, 83-88; PRES, ENVIRON CONSULT SERV, INC, 88- *Mem:* Am Fisheries Soc; Am Inst Fishery Res Biologists. *Res:* Fisheries population studies; environmental impact studies; fish screening studies; ecological consultant, research program initiation and administration; engineering and design of water-screens for power plants; population-community studies of major East Coast estuaries and rivers; resign design; data analysis; expert testimony before regulating agencies; marine development wetlands and water quality responsibilities; marina setting, design, and permitting. *Mailing Add:* 100 S Cass St Middletown DE 19709

SCHULERT, ARTHUR ROBERT, b Gladwin, Mich, Feb 26, 22; m 49; c 7. CHEMISTRY. *Educ:* Wheaton Col, BS, 43; Princeton Univ, MA, 47; Univ Mich, PhD(biol chem), 51; Am Bd Health Physics, dipl, 63; Am Bd Clin Chem, dipl, 64. *Prof Exp:* Res asst, Manhattan Proj, Princeton Univ, 43-46, teaching asst chem, Univ, 46-47; instr chem & physics, Taylor Univ, 47-48; res asst, Dept Biol Chem, Univ Mich, 48-49, teaching asst, 49-51; New York City Pub Health Res Inst fel biol chem, Goldwater Mem Hosp, 51-53, Columbia Univ Res Serv res fel, 53-55, res assoc, Lamont Geol Observ, 55-61, actg dir geochem lab, 58-59; from asst prof to assoc prof biochem, Sch Med, Vanderbilt Univ, 61-70; pres, Environ Sci & Eng Corp, 70-88, CHIEF EXEC OFFICER, SCI CORP, 88- *Concurrent Pos:* Consult, Isotopes Inc, NJ, 57-61 & Interdept Comt for Nutrit for Nat Develop, 59-64; dir biochem div, US Naval Med Res Unit, Cairo, 61-64, head biochem dept, 64-66; mem sci adv comt, 2, 4, 5-T, Environ Protection Agency, 71. *Mem:* Fel Am Inst Chemists; Am Chem Soc; Am Inst Nutrit; Health Physics Soc; fel Am Asn Clin Chemists. *Res:* Role of trace elements in nutrition and disease; mineral metabolism; microanalytical techniques including low level radiochemistry; nuclear fallout, particularly the disposition of fission radioisotopes in the environment and man; drug metabolism and mechanism of action. *Mailing Add:* Environ Sci Corp Mt Juliet TN 37122-2602

SCHULKE, HERBERT ARDIS, JR, b New Ulm, Minn, Nov 12, 23; m 49; c 2. ELECTRONICS, COMMUNICATIONS. *Educ:* US Mil Acad, BS, 46; Univ Ill, Urbana, MS, 52, PhD(electronics), 54. *Prof Exp:* Assigned adv develop long range radio, Signal Res & Develop Labs, Signal Corps, US Army, Ft Monmouth, NJ, 54-56 & Signal Sch Regiment, 57-58, assoc prof elec eng, US Mil Acad, 58-61, chief of staff, Signal Res & Develop Labs, 61-63, commun-electronics proj officer, Adv Res Projs Agency, Vietnam, 64-65, mil asst tactical warfare, Off Dir Defense Res & Eng, Off Secy Defense, Washington, DC, 66-69, commanding officer, 29th signal group, US Army Strategic Commun Command, Thailand, 69-70, dep dir plans, Defense Commun Agency, Washington, DC, 70-76; gen mgr & exec dir, Inst Elec & Electronics Engrs, 75-77; dir telecommun, Chase Manhattan Bank, 77-84; RETIRED. *Concurrent Pos:* Pres, Aero-Tele-Com, Inc. *Mem:* Fel Inst Elec & Electronics Engrs; Sigma Xi. *Res:* Military electronics equipment research and development; management of all types of military research and development efforts. *Mailing Add:* 138 Borden Rd Middletown NJ 07748

SCHULKE, JAMES DARRELL, b Aurelia, Iowa, June 25, 32; m 56; c 3. PLANT GENETICS, AGRONOMY. *Educ:* Iowa State Univ, BS, 58; Univ Calif, MS, 60, PhD(genetics), 63. *Prof Exp:* PLANT BREEDER, SPRECKELS SUGAR CO, INC, 63- *Mem:* Am Soc Sugar Beet Technol; Crop Sci Soc Am; Am Inst Biol Sci; Sigma Xi. *Res:* Genetics and plant breeding of sugar beets. *Mailing Add:* Spreckels Sugar Co Inc PO Box 7428 Spreckels CA 93962

SCHULKIN, MORRIS, b Brooklyn, NY, Feb 6, 19; m 40, 64, 70; c 4. ACOUSTICS, REMOTE SENSING. *Educ:* Brooklyn Col, BA, 39; George Washington Univ, MS, 48; Cath Univ Am, PhD, 69. *Prof Exp:* Sci aide, US Weather Bur, 40-41; physicist & engr, Nat Bur Standards, 41-47; physicist, Naval Res Lab, US Dept Navy, 47-48; physicist & engr, Nat Bur Standards, 48-50; physicist, Underwater Sound Lab, 50-56; chief engr, Martin Co, 56-59; chief scientist, Marine Electronics Off, Avco Corp, 59-63; adv engr, Underseas Div, Westinghouse Elec Corp, 63-66; dir performance anal, Antisubmarine Warfare, Spec Proj Off, 66-67; res physicist, Naval Res Lab, 68; consult physicist, Naval Underwater Systs Ctr, 68-72; assoc sci & tech dir, Naval Oceanographic Off, 72-75; vpres, Mar Assoc, Inc, 75-76; assoc to dir & prin physicist, 76-85, SR FEL, APPL PHYSICS LAB, UNIV WASH, 86-; PRES & CONSULT, OCEAN ACOUST INC, 76- *Concurrent Pos:* Assoc, George Washington Univ & asst, Univ Md, 47-48; lectr, Drexel Inst, 59; instr, USDA Grad Sch, 61-65; consult, US Off Naval Res, 61 & 73-75; lectr ocean acoust & eng, Cath Univ Am, 73-77; adj prof ocean eng, Univ Miami, 86- *Mem:* Am Geophys Union; fel Acoust Soc Am; fel Inst Elec & Electronics Engrs. *Res:* Remote sensing; underwater sound; seismo-acoustics; wave propagation; oceanic variabilities. *Mailing Add:* 9325 Orchard Brook Dr Potomac MD 20854

SCHULKIND, MARTIN LEWIS, pediatrics, immunology, for more information see previous edition

SCHULL, WILLIAM JACKSON, b Louisiana, Mo, Mar 17, 22; m 46. HUMAN GENETICS. *Educ:* Marquette Univ, BS, 46, MS, 47; Ohio State Univ, PhD(genetics), 49. *Prof Exp:* Head dept genetics, Atomic Bomb Casualty Comn, Japan, 49-51; jr geneticist, Inst Human Biol, Univ Mich, 51-53, asst geneticist, 53-56, from asst prof to prof human genetics, Med Sch, 56-72, prof anthrop, 69-72; PROF HUMAN GENETICS, UNIV TEX GRAD SCH BIOMED SCI, HOUSTON, 72- *Concurrent Pos:* Vis fel, Australian Nat Univ, 69; consult, Atomic Bomb Casualty Comn, Japan, 54 & 56; consult, NIH, 56-, chmn genetics study sect, 69-72; dir, Child Health Surv, Japan, 59-60; vis prof, Univ Chicago, 63; Ger Res Asn guest prof, Univ Heidelberg, 70; vis prof, Univ Chile, 75; mem comt atomic casualties, Nat Res Coun, 51 & subcomt biol, Comt Dent, 51-55; mem comt on collab proj, Nat Inst Neurol Dis & Stroke, 57-; mem panel in genetic effects of radiation, WHO, 58- & panel of experts human heredity, 61-; mem nat adv comt radiation, USPHS, 60-64 & bd sci counsrs, Nat Inst Dent Res, 66-69; dir, Radiation Effects Res Found & head dept epidemiol & Japan, 78-80; adv, Nat Heart & Lung Inst; mem subcomt biol & med, AEC; mem human biol coun, Soc Study Human Biol. *Honors & Awards:* Centennial Award, Ohio State Univ, 70. *Mem:* AAAS; USMex Border Health Asn; hon mem Japanese Soc Human Genetics; hon mem, Peruvian Soc Human Genetics; hon mem Genetic Soc Chile; Sigma Xi. *Res:* Biometry. *Mailing Add:* Dept Med Genetics Univ Tex Health Sci Ctr PO Box 20334 Houston TX 77225

SCHULLER, IVAN KOHN, b Cluj, Rumania, June 8, 46; US citizen; m 74; c 2. PHYSICS. *Educ:* Univ Chile, Licenciado, 70; Northwestern Univ, MS, 72, PhD(physics), 76. *Prof Exp:* Res asst physics, Univ Chile, 66-70; res asst, Northwestern Univ, 70-74; sr res aide, Argonne Nat Lab, 74-76; adj asst prof, Univ Calif, Los Angeles, 76-78; sr physicist & group leader, Argonne Nat Lab, 78-87; PROF, UNIV CALIF, SAN DIEGO, 87- *Concurrent Pos:* Consult. *Honors & Awards:* Outstanding Sci Accomplishment in Solid State Physics, US Dept Energy, 87. *Mem:* Am Phys Soc; Sigma Xi; Inst Elec & Electronics Engrs; Soc Explor Geophysicists. *Res:* Low temperature physics; solid state physics; prospecting geophysics; superconducting electronics; microelectronics. *Mailing Add:* Physics Dept 0319 Univ Calif San Diego La Jolla CA 92093-0319

SCHULLERY, STEPHEN EDMUND, b Harrisonburg, Va, June 8, 43; m 69; c 1. LIPID MEMBRANES, VESICLE FUSION. *Educ:* Eastern Mich Univ, BA, 65; Cornell Univ, PhD(phys chem), 70. *Prof Exp:* From asst prof to assoc prof, 70-80, PROF CHEM, EASTERN MICH UNIV, 80- *Mem:* AAAS; Am Chem Soc; Biophys Soc. *Res:* Physical chemistry of biological macromolecules; structure and function of model biological cell membranes. *Mailing Add:* 2117 Collegewood Ypsilanti MI 48197

SCHULMAN, HAROLD, b Newark, NJ, Oct 26, 30; m 54; c 3. OBSTETRICS & GYNECOLOGY. *Educ:* Univ Fla, BS, 51; Emory Univ, MD, 55; Bd Maternal Fetal Med, cert, 75. *Prof Exp:* From instr to asst prof obstet & gynec, Temple Univ, 61-65; from asst prof to assoc prof, 65-71, prof obstet & gynec, Albert Einstein Col Med, 71-83; PROF OBSTET & GYNEC, STATE UNIV NY, STONY BROOK, 84- *Concurrent Pos:* Am Cancer Soc fel, 59-60; USPHS fel, 68. *Mem:* AAAS; Am Col Obstet & Gynec; Soc Gynec Invest; Am Gynec Obstet Soc. *Res:* Obstetric physiology. *Mailing Add:* 259 First St Mineola NY 11501

SCHULMAN, HERBERT MICHAEL, b New York, NY, Feb 7, 32. CELL BIOLOGY, BIOCHEMISTRY. *Educ:* Bard Col, BA, 55; Yale Univ, PhD(microbiol), 62. *Prof Exp:* Asst prof biol, Univ Calif, San Diego, 63-69; Actg dir, 88-90, STAFF INVESTR EXP HEMAT, LADY DAVIS INST, JEWISH GEN HOSP, 69- *Concurrent Pos:* NIH fel, Univ Calif, San Diego, 62-63 & res grant, 63-69; Med Res Coun & Nat Res Coun grants, Lady Davis Inst, Jewish Gen Hosp, 69-; vis prof, Univ Helsinki, 68-69 & Univ Wl, 85; Can Dept Agr grants; vis scientist, John Innes Inst, Norwich, Eng, 77-78; Nujfield Found Travel Grant, 77-78; consult nitrogen fixation res, Can Dept Agr, 76-; vis prof, Nat Univ Mex, 87; assoc prof med, McGill Univ. *Mem:* Can Soc Cell Biol; Int Soc Develop Biol. *Res:* Control of hemoglobin synthesis and development of erythrocytes; biochemistry of iron metabolism; synthesis of leg-hemoglobin; differentiation of root nodules; nitrogen fixation in the high arctic. *Mailing Add:* Lady Davis Inst Jewish Gen Hosp 3755 Cote St Catherine Rd Montreal PQ H3T 1E2 Can

SCHULMAN, HOWARD, b Holon, Israel, Feb 5, 49. NEUROSICENCES, BIOCHEMISTRY. *Educ:* Univ Calif, Los Angeles, BS, 71; Harvard Univ, PhD(biochem), 76. *Prof Exp:* Fel, Med Sch, Yale Univ, 76-78; ASSOC PROF PHARMACOL, SCH MED, STANFORD UNIV, 78- *Res:* Calcium and cyclic adenosine monophosphate-dependent protein phosphorylation in brain function; regulation of the cytoskeleton. *Mailing Add:* Dept Pharmacol R354 Stanford Univ Sch Med Stanford CA 94305-5332

SCHULMAN, IRVING, b New York, NY, Feb 17, 22; m 50; c 2. MEDICINE, PEDIATRICS. *Educ:* NY Univ, BA, 42, MD, 45. *Prof Exp:* Instr pediat, Sch Med, NY Univ, 49-50; from instr to assoc prof, Cornell Univ, 52-58; prof, Med Sch, Northwestern Univ, 58-61; prof & head dept, Univ Ill Col Med, 61-72; PROF PEDIAT & CHMN DEPT, SCH MED, STANFORD UNIV, 72- *Concurrent Pos:* USPHS res fel, Med Col, Cornell Univ, 50-52; consult, Nat Inst Child Health & Human Develop, 64-; ed-in-chief, Advances in Pediat. *Honors & Awards:* Mead-Johnson Award, 60. *Mem:* Soc Pediat Res (pres, 66); Am Pediat Soc; Am Soc Clin Invest; Am Acad Pediat; Am Soc Hemat. *Res:* Pediatric hematology; coagulation physiology. *Mailing Add:* Dept Pediat Stanford Univ Med Ctr S332 Stanford CA 94305

SCHULMAN, JAMES HERBERT, b Chicago, Ill, Nov 15, 15; m 40; c 3. SOLID STATE PHYSICS, OPTICAL PHYSICS. *Educ:* Mass Inst Technol, SB, 39, PhD(inorg chem), 42. *Prof Exp:* Instr, Suffolk Univ, 40-41; asst, Div Indust Coop, Mass Inst Technol, 41-44; sr engr, Sylvania Elec Prod, 46-53; head dielectrics br, US Naval Res Lab, 53-60, dep sci dir, Off Naval Res, London, 60-61, head dielectrics br, US Naval Res Lab, 62-64, chair mat sci, US Naval Res Lab, 64-65, supt optical physics div, 65-67, assoc dir res, US Naval Res Lab, 67-74, sci dir & chief scientist, London Br Off, 74-77, chair mat sci, US Naval Res Lab & actg tech dir, US Off Naval Res, 77-79; res prof, George Washington Univ, 79-81; CONSULT, 81- *Concurrent Pos:* Consult, Nat Mat Adv Bd, Nat Acad Sci-Nat Res Coun, 80-; US mem, Res Grants Prog Panel, NATO, 79-81; assoc ed, J Opt Soc Am, 71-80; consult, sci & tech mgt, 79-; mem, Panel on Reevaluation of Radiation Doses from Hiroshima & Nagasaki A-Bombs, Nat Acad Sci, 82-86; mem, Comt on US Army Thermoluminescent Dosimetry Syst, Nat Res Coun, 85-87, consult, Technol Issues Comt, Nat Acad Eng, 87-88. *Honors & Awards:* Sigma Xi Award, 57. *Mem:* Fel AAAS; fel Am Phys Soc; fel Optical Soc Am; Sigma Xi. *Res:* Luminescent materials; radiation sensitive solids; dosimetry; crystal chemistry and physics; glass; color centers; materials science. *Mailing Add:* 4615 N Park Ave Chevy Chase MD 20815-4509

SCHULMAN, JEROME LEWIS, b New York, NY, Nov 15, 25; m 49; c 3. PSYCHIATRY. *Educ:* Univ Rochester, BA, 46; Long Island Col Med, MD, 49. *Prof Exp:* Intern med, Jewish Hosp, Brooklyn, 49-50, resident pediat, 50-51 & 53-54; resident psychiat, Johns Hopkins Hosp, 54-57; asst prof pediat, psychiat & neurol, 57-65, PROF PEDIAT & PSYCHIAT & CHIEF PSYCHIAT, MED SCH, NORTHWESTERN UNIV, CHICAGO, 65- *Concurrent Pos:* Dir child guid & develop clins & attend pediatrician & psychiatrist, Children's Mem Hosp, 57-; consult, Joseph P Kennedy Jr Found, 58-; asst prof, Med Col, Cornell Univ, 68-69. *Mem:* Am Psychiat Asn; Am Asn Ment Deficiency; Am Pediat Soc; Soc Biol Psychiat; Am Acad Pediat. *Res:* Child psychiatry and development; mental retardation; child psychotherapy. *Mailing Add:* Dept Pediat Northwestern Univ 2300 Childrens Plaza Chicago IL 60614

SCHULMAN, JEROME M, b New York, NY, Oct 21, 38; m 65. CHEMICAL PHYSICS, MOLECULAR PHARMACOLOGY. *Educ:* Rensselaer Polytech Inst, BChE, 60; Columbia Univ, MA, 61, PhD(chem), 64. *Prof Exp:* Res assoc theoret chem, NY Univ, 64-66, asst prof chem, 66-67; sr res assoc, Yeshiva Univ, 67-68; from asst prof to assoc prof chem physics, Polytech Inst Brooklyn, 68-71; assoc prof chem, 71-74, PROF CHEM, QUEENS COL, NY, 74- *Concurrent Pos:* Alfred P Sloan fel, Polytech Inst Brooklyn & Queens Col, NY, 71-73. *Mem:* Am Chem Soc. *Res:* Quantum theory of atoms and molecules; perturbation theory; electromagnetic properties of atoms and molecules; theoretical organic chemistry; molecular pharmacology. *Mailing Add:* Dept Chem Queens Col Flushing NY 11367

SCHULMAN, JOSEPH DANIEL, b Brooklyn, NY, Dec 20, 41; m 64; c 2. HUMAN GENETICS, OBSTETRICS. *Educ:* Brooklyn Col, BA, 61; Harvard Univ, MD, 66. *Prof Exp:* Intern & resident pediat, Mass Gen Hosp, 66-68; clin assoc genetics, NIH, 68-70; resident obstet, Cornell Med Ctr, 70-73; fel develop biochem, Cambridge Univ, Eng, 73-74; head human genetics sect, 74-79, dir, IVF Inst, genetics training prog, 79-82, DIR, GENETICS & IVF INST, NIH, 84- *Concurrent Pos:* Prof obstet & gynec, George Washington Univ, Sch Med, 75-84, Med Col Va, 85- *Mem:* Soc Pediat Res; Soc Gynec Invest; Am Soc Human Genetics; Am Soc Clin Invest; Sigma Xi. *Res:* Human biochemistry, genetics and development; human genetic diseases; in vitro fertilization. *Mailing Add:* 9207 Aldershot Dr Bethesda MD 20034

SCHULMAN, LADONNE HEATON, b Jacksonville, Fla, May 13, 36; m 65. BIOCHEMISTRY, MOLECULAR BIOLOGY. *Educ:* Wheaton Col, Mass, BA, 57; Columbia Univ, MA, 60, PhD(chem), 64. *Prof Exp:* Res assoc biochem, Med Ctr, NY Univ, 66-68; from asst prof to assoc prof, 68-78, PROF DEVELOP BIOL & CANCER, ALBERT EINSTEIN COL MED, 78- *Concurrent Pos:* NIH fel biochem, Med Ctr, NY Univ, 64-66; NIH career develop award, 69-74; Am Cancer Soc fac res award, 74-77; mem, Study Sect Nucleic Acids & Protein Synthesis, Am Cancer Soc, 78-82; mem, Panel on Basic Biomed Sci Adv, Nat Res Coun, 79-80; mem, Study Sect Physiol Chem, NIH, 86-90. *Mem:* Am Chem Soc; Am Soc Biol Chem. *Res:* Protein-nucleic acid interactions; structure and function of transfer RNA; protein synthesis. *Mailing Add:* Dept Develop Biol & Cancer Albert Einstein Col Med 1300 Morris Park Ave Bronx NY 10461

SCHULMAN, LAWRENCE S, b Newark, NJ, Nov 21, 41; c 3. PHYSICS, STATISTICAL MECHANICS. *Educ:* Yeshiva Univ, BA, 63; Princeton Univ, PhD(physics), 67. *Prof Exp:* From asst prof to prof physics, Ind Univ, Bloomington, 67-78; assoc prof, 72-77, prof physics, israel inst technol, 77-; PROF PHYSICS & CHMN DEPT, CLARKSON UNIV, 85- *Concurrent Pos:* Consult, Los Alamos Sci Labs, 64, Lawrence Radiation Lab, Livermore, 65 & 68, IBM Corp, 75-; vis scientist, Israel Inst Technol, 70-71, Fr Atomic Energy Comn, Saclay; NATO fel, 70-71; vis prof, Univ Paris, Norweg Inst Technol, Trondheim; Donders prof, State Univ Utrecht. *Mem:* Am Phys Soc; Israel Phys Soc. *Res:* Mathematical physics; phase transitions; elementary particles; quantum mechanics; statistical physics. *Mailing Add:* Clarkson Univ Potsdam NY 13699-5820

SCHULMAN, MARVIN, b Brooklyn, NY, Nov 13, 27; m 54; c 4. ENGINEERING, LIFE SUPPORT EQUIPMENT. *Educ:* Brooklyn Col, BA, 54. *Prof Exp:* Electronic scientist, Mat Lab, Naval Air Develop Ctr, US Dept Navy, 50-54, supvry gen engr, Aircraft & Crew Systs Technol Directorate, 54-85; sr staff adv, Ketron Inc, 85-86; VPRES LME, INC, 87- *Concurrent Pos:* Mem occupant restraint systs comt, Soc Automotive Engrs; consult, human factors & systs eng, 85- *Mem:* Instrument Soc Am; SAFE. *Res:* Design and development of aircraft escape and fixed seating systems; restraint, occupant protective devices, energy management on impact and protection during high acceleration exposure. *Mailing Add:* LME Inc 444 Jacksonville Rd Warminster PA 18974

SCHULMAN, MARVIN DAVID, b New York, NY, Oct 6, 39; m 63; c 2. ENZYMOLOGY. *Educ:* Cornell Univ, BA, 61, PhD(biochem), 67. *Prof Exp:* Res & teaching fel, dept biochem, Sch Med, Case Western Reserve Univ, 67-70, instr, 70-73; res fel, 73-79, sr res fel, 79-86, SR RES INVESTR, MERCK SHARP & DOHME RES LABS, 86- *Mem:* Am Soc Biol Chemists; Am Chem Soc; Am Soc Microbiol. *Res:* Biochemistry and enzymology of secondary metabolites, particularly the avermectins; mechanism of action of anthelmintics. *Mailing Add:* Merck Sharp & Dohme Res Labs PO Box 2000 Rahway NJ 07065

SCHULMAN, SIDNEY, b Chicago, Ill, Mar 1, 23; m 45; c 3. NEUROLOGY. *Educ:* Univ Chicago, BS, 44, MD, 46. *Prof Exp:* From asst prof to assoc prof, 52-65, PROF NEUROL, UNIV CHICAGO, 65- *Mem:* Am Acad Neurol; Am Neurol Asn. *Res:* Neuropathology; clinical neurology; behavioral effects of experimental lesions in the thalamus. *Mailing Add:* Dept Biol Sci/Ch 405A Univ Chicago 1025 E 57th Chicago IL 60637

SCHULMAN, STEPHEN GREGORY, b Brooklyn, NY, June 11, 40; div; c 3. ANALYTICAL CHEMISTRY, PHOTOCHEMISTRY. *Educ:* City Col New York, BS, 61; Univ Ariz, PhD(chem), 67. *Prof Exp:* Asst chemist, Boyce Thompson Inst Plant Res, 62-64, phys chemist, 67-68; mem tech staff, Bellcomm Inc, 68-69; from asst prof to prof, 70-77, PROF PHARMACEUT, COL PHARM, UNIV FLA, 77- *Concurrent Pos:* Fel, dept chem, Univ Fla, 69-70; vis prof, Univ Ky, 76, State Univ Utrecht, 79, 81 & 83, Univ Stratheclyde, 85, Kumamoto Univ, 86-87, Acad Sinica, 87, Karl Franzens Univ, Graz, 89; comt revision, USP, 80-90; Food Chemicals Codex, Nat Acad Sci, 89-; consult, Futuretech, Inc. *Mem:* Am Chem Soc; Am Asn Pharmaceut Scientists; Sigma Xi. *Res:* Molecular electronic spectroscopy, mixed ligand chelates; analytical chemistry in biological fluids and living tissues; binding of drugs to proteins and nucleic acids; fast reaction kinetics; fluorescence optical sensors. *Mailing Add:* Col Pharm Univ Fla Gainesville FL 32610

SCHULSON, ERLAND MAXWELL, b Ladysmith, BC, May 28, 41; m 64; c 4. MATERIALS ENGINEERING, PHYSICAL METALLURGY. *Educ:* Univ BC, BASc, 64, PhD(metall eng), 68. *Hon Degrees:* MA, Dartmouth Col, 87. *Prof Exp:* Sr res fel, Univ Oxford, 68-69; res officer, Chalk River Nuclear Labs, Atomic Energy Can Ltd, 69-78; assoc prof, 78-84, PROF ENG, DARTMOUTH COL, 84-, DIR, ICE RES LAB, 83- *Concurrent Pos:* Prin investr, NSF, Dept Energy, NASA, ARO, ONR; consult, pvt indust & govt labs. *Mem:* Am Inst Mining, Metall & Petrol Engrs; Am Geophys Union; Mat Res Soc. *Res:* Materials science; relationship between the microstructure and the plastic flow and fracture of materials; scanning electron microscopy; mechanical properties of ice. *Mailing Add:* Thayer Sch Eng Dartmouth Col Hanover NH 03755

SCHULT, ROY LOUIS, b Brooklyn, NY, Aug 31, 34; m 58, 77; c 3. THEORETICAL PHYSICS. *Educ:* Univ Rochester, BS, 56; Cornell Univ, PhD(theoret physics), 61. *Prof Exp:* Res assoc, 61-63, res asst prof, 63-64, asst prof, 64-69, ASSOC PROF PHYSICS, UNIV ILL-URBANA, 69- *Concurrent Pos:* Vis assoc physicist, Brookhaven Nat Labs, 69-70; assoc mem, La Jolla Inst, Calif, 85- *Mem:* Am Phys Soc. *Res:* Weak interactions; theory of elementary particles; nonlinear systems. *Mailing Add:* Physics/237A Loomis Lab Univ Ill Urbana IL 61801

SCHULTE, ALFONS F, b Wulmeringhauser, W Ger. PROTEIN DYNAMICS, SPECTROSCOPY. *Educ:* Tech Univ Munich, dipl, 80 & PhD(physics), 85. *Prof Exp:* Res assoc, Tech Univ Munich, 85-86, Univ Ill, 86-87; VIS RES ASST PROF, UNIV ILL, URBANA. *Mem:* Am Phys Soc; Ger Phys Soc. *Res:* Dynamic structure and function of proteins; connections between proteins, glasses and spinglasses; ligand binding and relaxation in hemeproteins using time resolved spectroscopic techniques over a wide range of temperature and pressure; localization, superconductivity and correlation effects in amorphous metals. *Mailing Add:* 303 Tradewinds Dr No 4 San Jose CA 95123

SCHULTE, DANIEL HERMAN, b Osceola, Iowa, Aug 3, 29; m 55; c 3. ASTRONOMY, OPTICS. *Educ:* Phillips Univ, AB, 51; Univ Chicago, PhD(astron), 58. *Prof Exp:* Asst, Yerkes Observ, Univ Chicago, 51-56; optical engr, Perkin-Elmer Corp, 56-59; asst astronomer, Kitt Peak Observ, Ariz, Asn Univs Res in Astron, 59-65; staff physicist, Optical Design Dept, Itek Corp, 65-81; RES SCIENTIST, LOCKHEED PALO ALTO RES LABS, 81- *Concurrent Pos:* Consult, Haneman Assocs, Tex, 61 & Tropel, Inc, NY, 62-63. *Mem:* Am Astron Soc; Optical Soc Am; Int Astron Union. *Res:* Observational astronomy; astronomical spectroscopy and photometry; optical design; computer applications to astronomical problems. *Mailing Add:* 118 Mercy Mountain View CA 94041

SCHULTE, HARRY FRANK, b St Louis, Mo, Jan 1, 14; m 35; c 3. INDUSTRIAL HYGIENE. *Educ:* Wash Univ, BS, 34; Harvard Univ, MS, 46. *Prof Exp:* Res chemist, Shell Oil Co, 35-41; indust hyg engr, State Bd Health, Mo, 41-48; leader indust hyg group, Los Alamos Sci Lab, 48-74, prog coordr, Biomed & Environ Prog, Energy Off, 74-75, sci adv indust hyg group, 75-78, consult indust hyg, 78-85; RETIRED. *Concurrent Pos:* Lectr, Sch Med, Univ Kans, 42-48 & Harvard Sch Pub Health, 63-77; mem, Nat Coun Radiation Protection & measurements. *Honors & Awards:* Cummings Award in Indust Hyg, 72. *Mem:* Health Phys Soc; Am Indust Hyg Asn (pres, 63-64); Am Conf Govt Indust Hygienists; Am Acad Indust Hyg (pres, 72). *Res:* Evaluation and control of health hazards in industrial environment; dust; toxic chemicals; noise; air pollution. *Mailing Add:* 6515 Kathryn SE Albuquerque NM 87108

SCHULTE, HARRY JOHN, JR, b Newark, NJ, July 1, 25; m 49; c 3. EXPERIMENTAL PHYSICS. *Educ:* Rensselaer Polytech Inst, BS, 48; Univ Rochester, PhD(physics), 53. *Prof Exp:* Jr physicist, Univ Rochester, 48, asst, 48-52, instr, 52-53; res assoc physics, Univ Minn, 53-55; mem tech staff, Bell Tel Labs, Inc, 55-62 & Bellcomm, 62-64, mem tech staff, Bell Labs, Inc, 64-89; CONSULT, 89- *Res:* Instrumentation; digital, radio and optical communication techniques; undersea cable systems. *Mailing Add:* Ten Rustic Terr Fair Haven NJ 07704

SCHULTE, ROBERT LAWRENCE, b Covington, Ky, April 16, 45; m 70; c 1. NUCLEAR DETECTOR DEVELOPMENT, ION BEAM ANALYSIS OF MATERIALS. *Educ:* Thomas More Col, AB, 65; Univ Ky, MS, 67, PhD(physics), 71. *Prof Exp:* Res assoc, Univ Toronto, 71-74; LAB HEAD, GRUMMAN CORP, CORP RES CTR, 74- *Concurrent Pos:* Adj prof, Long Island Univ, C W Post Ctr, 82- *Mem:* Am Phys Soc; Am Nuclear Soc; Inst Elec & Electronics Engrs. *Res:* Effects of solute & trace element concentrations on the properties of materials using nuclear reaction analysis techniques; design & development of specialized nuclear detection systems for radioisotope measurements. *Mailing Add:* 19 Mitchell Rd Port Washington NY 11050

SCHULTE, WILLIAM JOHN, JR, b Stryker, Ohio, Nov 6, 28; m 61; c 7. SURGERY. *Educ:* Univ Toledo, BS, 52; Ohio State Univ, MD, 56; Am Bd Surg, dipl, 64. *Prof Exp:* Asst clin instr, 59-63, from instr to assoc prof, 63-77, PROF SURG, MED COL WIS, 77-; CHIEF SURG INTENSIVE CARE UNIT, VET ADMIN CTR, 70-, CHIEF SURG SERV, 80- *Concurrent Pos:* Gastrointestinal res fel, 64-65; attend staff, Milwaukee County Gen Hosp, 64- & Wood Vet Admin Hosp, 65-; Vet Admin clin investr, 65-68; staff attend, St Joseph's Hosp & Mt Sinai Hosp, Milwaukee, Wis; asst chief surg serv, Vet Admin Ctr, Milwaukee, 78-80. *Mem:* Am Fedn Clin Res; fel Am Col Surg; Am Gastroenterol Soc; Asn Acad Surg; Cent Surg Asn. *Res:* General surgery; parenteral nutrition; gastrointestinal motility; gastric and pancreatic physiology. *Mailing Add:* Dept Surg Vet Admin Ctr Milwaukee WI 53295

SCHULTER-ELLIS, FRANCES PIERCE, b Helena, Ala, Sept 22, 23; m 42, 77; c 2. ANATOMY, PHYSICAL ANTHROPOLOGY. *Educ:* Birmingham Southern Col, BS, 52; Emory Univ, MS, 54; George Washington Univ, PhD(anat), 72. *Prof Exp:* Instr biol, Chamblee High Sch, Ga, 60-61; instr zool, anat & physiol, Marjorie Webster Jr Col, 61-65; asst prof, 72-84, ASSOC PROF ANAT, SCH MED, UNIV MD, BALTIMORE, 84- *Concurrent Pos:* Res collabr, Div Phys Anthropol, Smithsonian Inst, 80- *Mem:* Human Biol Coun; Am Asn Phys Anthropologists; Am Acad Forensic Sci; NY Acad Sci; Am Asn Anatomists; Am Asn Univ Prof. *Res:* Craniometry, with special emphasis on temporal bone and cranial base variations, middle and inner ear disease, otitis media, human variation and asymmetry; bone aging, morphology of human hand skeleton; human skeleton race and sex identification. *Mailing Add:* 3465 Lockwood Dr Q76 Ft Collins CO 80525

SCHULTES, RICHARD EVANS, b Boston, Mass, Jan 12, 15; m 59; c 3. BOTANY. *Educ:* Harvard Univ, AB, 37, MA, 38, PhD(biol), 41; Nat Univ Colombia, 53. *Hon Degrees:* MH, Univ Nac Colombia, DSc, Mass Col Pharm. *Prof Exp:* Collabr, Inst Biol, Nat Univ Mex, 38, 39 & 41; hon res fel, Bot Mus, Harvard Univ, 41-54, cur, Ames Orchid Herbarium, 54-58, lectr econ bot, 58-70, dir Mus, 68-85; RETIRED, 85- *Concurrent Pos:* Nat Res Coun fel, Inst Natural Sci, Nat Univ Colombia, 41; cur econ bot, Bot Mus, 58-85, prof biol, 68-75, Mangelsdorf prof natural sci, 75-80, Jeffrey prof biol, 80-85; jungle explor & botanist, USDA, 43-54, collbr, Nat Sci Inst, Bogota, 41-; Guggenheim fel, 50; ed, Bot Mus Leaflets, Harvard Univ, 58-85 & Econ Bot, 63-80; adj prof pharmacog, Sch Pharm, Univ Ill, 75-; mem, Sci Adv Bd, Palm Oil Res Inst Malaysia, 79-83; collabr, Malaysian Rubber Res Inst, 90- *Honors & Awards:* Orden de la Victoria Regia, Colombian Govt; Gold Medal, World Wildlife Fund, 84; Gold Medal, Sigma Xi, 85; Lindbergh Award, 81; Tyler Prize Environ Achievements, 87. *Mem:* Nat Acad Sci; Col Acad Sci, Ecuador; Acad Sci, Argentina; fel Am Acad Arts & Sci; Linnean Soc; fel Am Col Neuropsychopharmacol. *Res:* Latin American ethnobotany, especially narcotics, medicines and poisons used by primitive peoples; orchid taxonomy; taxonomy of rubber plants. *Mailing Add:* Bot Mus Harvard Univ Oxford St Cambridge MA 02138

SCHULTHEIS, JAMES J, b Rochester, NY, Aug 22, 32; m 55; c 5. PHYSICS. *Educ:* John Carroll Univ, BS, 54; Univ Rochester, MS, 56. *Prof Exp:* Engr, Atomic Power Div, Westinghouse Elec Corp, 55-56; asst physicist, Argonne Nat Lab, 56-58; nuclear engr, 58-67, nuclear consult, 67-71, mgr mat systs reliability, Energy Res & Develop Admin, 71-77, mgr steam generator-coolant technol, 77-80, mgr, Ceramic Development Lab, 80-86, PROJ ENG, ADVAN STEAM GENERATOR, KNOLLS ATOMIC POWER LAB, DEPT ENERGY, 86- *Mem:* Am Nuclear Soc. *Res:* Radiological physics; reactor physics; advanced reactor engineering; heat transfer; reactor control; steam generators; materials engineering; ceramic engineering. *Mailing Add:* 1139 Fernwood Dr Schenectady NY 12309

SCHULTHEISS, PETER M(AX), b Munich, Ger, Oct 18, 24; nat US; m 59; c 3. ELECTRICAL ENGINEERING. *Educ:* Yale Univ, BE, 45, MEng, 48, PhD(elec eng), 52. *Prof Exp:* From instr to assoc prof, 48-64, PROF ENG & APPL SCI, YALE UNIV, 64- *Mem:* Inst Elec & Electronics Engrs; Acoust Soc Am. *Res:* Communication theory; automatic control; applied mathematics; underwater acoustics. *Mailing Add:* 229 Beeton Ctr Yale Univ New Haven CT 06520

SCHULTZ, ALBERT BARRY, b Philadelphia, Pa, Oct 10, 33; m 55; c 3. GERIATRIC BIOMECHANICS, MUSCULOSKELETAL BIOMECHANICS. *Educ:* Univ Rochester, BS, 55; Yale Univ, MEng, 59, PhD(mech eng), 62. *Prof Exp:* Lectr mech eng, Yale Univ, 61-62; asst prof,

Univ Del, 62-65; from asst prof to prof mat eng, Univ Ill, Chicago Circle, 65-83; VENNEMA PROF MECH ENG, UNIV MICH, ANN ARBOR, 83- *Concurrent Pos:* NIH spec res fel, Stockholm, 71-72; res career award, NIH, 75-80; assoc ed, J Biomech Eng, 76-82; vis prof, Sahlgren Hosp, Gothenburg, Sweden, 78-79; chmn, Bioeng Div, Am Soc Mech Engrs, 81-82; chmn, US Nat Comn Biomech, 82-85. *Honors & Awards:* Javits Neuroscientist Award, NIH, 85-92; Lissner Award, Am Soc Mech Engrs, 90. *Mem:* Am Soc Mech Engrs; Int Soc Study Lumbar Spine (pres, 81-82); Am Soc Biomech (pres, 82-83); Am Geriat Soc; Geront Soc Am. *Res:* Biomechanics; mobility impairment in the elderly; mechanical behavior of human spine. *Mailing Add:* MEAM Dept Univ Mich Ann Arbor MI 48109-2125

SCHULTZ, ALVIN LEROY, b Minneapolis, Minn, July 27, 21; c 4. INTERNAL MEDICINE, ENDOCRINOLOGY. *Educ:* Univ Minn, BA, 43, MD, 46, MS, 52. *Prof Exp:* Intern med, Ohio State Univ Hosp, 46-47; resident med, Univ Minn Hosps, 49-52; from instr to assoc prof, 52-65, chief med, Hennepin County Med Ctr, 65-88, prof med, 65-88, LECTR, MED SCH, UNIV MINN, MINNEAPOLIS, 52-; SR VPRES MED AFFAIRS, HEALTH ONE CORP, 88- *Concurrent Pos:* Assoc dir, Radioisotope Lab, Minneapolis Vet Admin Hosp, 52-53, asst chief med serv, 53-54, consult, Radioisotope Lab & Med Serv, 54-; attend physician, Endocrine Clin, Univ Minn Hosps, 52-60, chief clin, 60-65; dir radioisotope Lab, Methodist Hosp, 55-59; dir med educ & res, Mt Sinai Hosp, 59-65; gov for Minn, Am Col Physician, 84-87 & chmn elect, bd govenors, 86, chmn, 87, regent, 88-; chmn bd dir, Coun Med Specialty Soc, 88-90. *Honors & Awards:* Frances M Greenwalt Pharmacol Award, 75. *Mem:* Endocrine Soc; Cent Soc Clin Res; Am Fedn Clin Res; fel Am Col Physicians; Am Thyroid Asn. *Res:* Diseases of metabolism and endocrinology, especially of the thyroid gland; hormonal control of lipid metabolism; effects of thyroid function. *Mailing Add:* 5127 Irving Ave S Minneapolis MN 55419

SCHULTZ, ANDREW, JR, b Harrisburg, Pa, Aug 14, 13; m 39; c 2. ENGINEERING. *Educ:* Cornell Univ, BS, 36, PhD(admin eng), 41. *Prof Exp:* From instr to asst prof admin eng, Cornell Univ, 38-41, assoc prof, 46-50, prof & head dept, 50-61; vpres & dir res, Logistics Mgt Inst, Washington, DC, 61-63; from actg dean to dean, 63-72, actg dean, 78, Spencer T Olin Prof, 72-80, EMER PROF ENG, COL ENG, CORNELL UNIV, 80- *Concurrent Pos:* Dir, Chicago Pneumatic Tool Co, S I Handling Systs, Inc, Logistics Mgt Inst, Zurn Indust. *Mem:* Am Soc Eng Educ; Am Inst Indust Engrs. *Res:* Statistical applications in engineering; operations research; industrial engineering. *Mailing Add:* 22 Strawberry Hill Rd Ithaca NY 14850

SCHULTZ, ARNOLD MAX, b Altura, Minn, Sept 9, 20; m 49; c 5. PLANT ECOLOGY. *Educ:* Univ Minn, BSc, 41, MSc, 42; Univ Nebr, PhD(bot), 51. *Prof Exp:* Jr range conservationist, Soil Conserv Serv, USDA, 42; from asst specialist to specialist forestry, 49-66, ecologist, 66-77, PROF FORESTRY & RESOURCE MGT, AGR EXP STA, UNIV CALIF, BERKELEY, 77- *Mem:* AAAS; Soc Gen Systs Res; Ecol Soc Am. *Res:* Sampling and biometric techniques in range ecology; nutrient cycles and productivity in arctic tundra ecosystems; ecology of high mountain meadows; ecosystem management; interdisciplinary undergraduate programs. *Mailing Add:* Sch Forestry & Conserv Univ Calif 2120 Oxford St Berkeley CA 94720

SCHULTZ, ARTHUR GEORGE, b Chicago, Ill, Sept 14, 42; m 69. ORGANIC CHEMISTRY. *Educ:* Ill Inst Technol, BSc, 66; Univ Rochester, PhD(chem), 70. *Prof Exp:* Res fel chem, Columbia Univ, 70-72; asst prof chem, Cornell Univ, 72-78; assoc prof, 78-81, PROF CHEM, RENSSELAER POLYTECH INST, 81- *Mem:* Am Chem Soc. *Res:* Natural products synthesis; synthetic and mechanistic organic photochemistry; organo-sulfur chemistry; synthetic organic and natural products chemistry. *Mailing Add:* Dept Chem Rensselaer Polytech Inst Troy NY 12181

SCHULTZ, ARTHUR JAY, b Brooklyn, NY, July 15, 47; m 84. NEUTRON DIFFRACTION, INORGANIC CHEMISTRY. *Educ:* State Univ NY Stony Brook, BS, 69; Brown Univ, PhD(inorg chem), 73. *Prof Exp:* Res assoc chem, Univ Ill, Urbana, 73-76; res assoc, 76-78, asst chemist, 78-82, CHEMIST, ARGONNE NAT LAB, 82- *Mem:* Am Chem Soc; Am Crystallog Asn; AAAS. *Res:* Time-of-flight single-crystal neutron diffraction; transition metal coordination chemistry; structural studies by x-ray and neutron diffraction; one-dimensional conductors and organic superconductors; high-Tc superconductors. *Mailing Add:* Chem Div Argonne Nat Lab Argonne IL 60439

SCHULTZ, BLANCHE BEATRICE, b Palm, Pa, Aug 23, 20. MATHEMATICS. *Educ:* Ursinus Col, BS, 41; Univ Mich, MS, 49. *Prof Exp:* Teacher high sch, Pa, 41-42; from instr to prof math, Ursinus Col, 46-86, asst dean, 77-83; RETIRED. *Concurrent Pos:* Cryptographer & aerial navig instr, USN, 42-46. *Mem:* Am Math Soc; Math Asn Am. *Mailing Add:* 2354 Pleasant Ave Glenside PA 19038

SCHULTZ, CLIFFORD W, mineral processing, mineral economics, for more information see previous edition

SCHULTZ, CRAMER WILLIAM, b Laurel, Mont, May 2, 26; m 49; c 4. PHYSICS. *Educ:* Univ Calif, BA, 48; Univ Southern Calif, PhD(physics), 55. *Prof Exp:* Assoc prof, 53-64, PROF PHYSICS, CALIF STATE UNIV, LONG BEACH, 64- *Res:* Cryogenics; solid state physics. *Mailing Add:* Dept Physics Calif State Univ 1250 Bellflower Rd Long Beach CA 90840

SCHULTZ, DAVID HAROLD, b Park Falls, Wis, Sept 10, 42; m 65; c 2. COMPUTER SCIENCE, NUMERICAL ANALYSIS. *Educ:* Univ Wis-Madison, BS, 65, MS, 67, PhD(comput sci), 70. *Prof Exp:* Co-supvr data processing, Surv Res Lab, 65-70, res asst numerical anal, Comput Sci Dept, 68-76, asst prof math, 70-76, ASSOC PROF MATH, UNIV WIS-MILWAUKEE, 76- *Concurrent Pos:* Reviewer, Math Rev; consult, Argonne Nat Lab, 77-78. *Mem:* Soc Indust & Appl Math; Asn Comput Mach; Int Asn Math & Comput Simulation. *Res:* Numerical analysis; fluid flow problems; numerical solutions of differential equations. *Mailing Add:* Dept Math Univ Wis Milwaukee WI 53202

SCHULTZ, DONALD GENE, b Milwaukee, Wis, Aug 28, 28; m 53; c 3. ELECTRICAL ENGINEERING. *Educ:* Univ Santa Clara, BSEE, 52; Univ Calif, Los Angeles, MS, 55; Purdue Univ, PhD(automatic control), 62. *Prof Exp:* Assoc prof, 62-66, PROF ELEC ENG, UNIV ARIZ, 66-, HEAD SYSTS & INDUST ENGR DEPT, 74- *Concurrent Pos:* Consult, Los Alamos Sci Lab. *Mem:* Inst Elec & Electronics Engrs. *Res:* Automatic control; stability. *Mailing Add:* Dept Systems & Indust Eng Univ Ariz Tucson AZ 85721

SCHULTZ, DONALD PAUL, b Detroit, Mich, Feb 7, 30; m 51; c 4. ENVIRONMENTAL CONTAMINATION EVALUATION. *Educ:* Concordia Teachers Col, BS, 54; Auburn Univ, PhD(plant physiol), 67. *Prof Exp:* Prin & teacher, St Stephen Lutheran Sch, 54-62; fel, Pesticide Metab, Univ Mo, Columbia, 67-70; chemist herbicide metab, Fish Pesticide Res Lab, US Dept Interior, Mo, 70-71, chemist chief, Southeastern Fish Control Lab, Ga, 71-80; MEM STAFF, ENVIRON CONTAMINATION EVAL, FISH & WILDLIFE SERV, ATLANTA, 80- *Mem:* AAAS; Aquatic Plant Mat Soc; Sigma Xi. *Res:* Uptake and metabolism of pesticides; influence of pesticides on metabolic processes. *Mailing Add:* US Fish & Wildlife Serv R B Russell Bldg FWE/EC Atlanta GA 30303

SCHULTZ, DONALD RAYMOND, b North Tonawanda, NY, Nov 2, 18; m 42, 68; c 4. INORGANIC CHEMISTRY. *Educ:* Univ Mich, BS, 40, MS, 52, PhD(chem), 54. *Prof Exp:* Anal chemist, Pa Salt Mfg Co, 40; anal res chemist, McGean Chem Co, 40-42; res engr, Trojan Powder Co, 42-44; res engr, Inorg Res Brine Prods, Mich Chem Corp, 46-50; res engr, Boron Hydrides Eng Res Inst, Univ Mich, 51-54; sr patent liaison, Cent Res Dept, 3M Co, 54-82; RETIRED. *Concurrent Pos:* Lectr, Bethel Col, 65-66. *Mem:* AAAS; Am Chem Soc; Am Inst Chemists; Sigma Xi. *Res:* Preparation properties and uses of magnesia; inorganic bromides; thermography; coordination chemistry of copper, nickel, cobalt; patent literature; boron hydrides in liquid ammonia; vinyl polymerization with boron alkyls; photoconductivity. *Mailing Add:* 1002 Raven Rd Rogers AR 72756

SCHULTZ, DUANE ROBERT, b Bay City, Mich, June 24, 34; m 61; c 1. IMMUNOLOGY, PROTEIN CHEMISTRY. *Educ:* Univ Mich, BS, 57, MS, 60, PhD(microbiol), 64. *Prof Exp:* Staff immunologist, 66-71, ASSOC PROF MED, SCH MED, UNIV MIAMI, 71-; STAFF IMMUNOLOGIST, CORDIS LABS, 66- *Concurrent Pos:* NIH fel immunol, Walter Reed Army Inst Res, 64-66. *Res:* Isolation, purification and function of the nine components of complement. *Mailing Add:* Dept Med Sch Med Univ Miami PO Box 016960 R 120 Miami FL 33101

SCHULTZ, EDWARD, b Suffern, NY, Dec 4, 40; m 66; c 2. ANATOMY. *Educ:* Ithaca Col, BA, 62, BS, 65; Temple Univ, PhD(anat), 73. *Prof Exp:* Asst prof, 75-81, ASSOC PROF ANAT, UNIV WIS-MADISON, 81- *Concurrent Pos:* Muscular Dystrophy Asn Can fel, McGill Univ, 73-75. *Mem:* Am Asn Anatomists; AAAS; Am Soc Cell Biol. *Res:* Skeletal muscle regeneration; skeletal muscle growth; satellite cells; aging. *Mailing Add:* Dept Anat Univ Wis 353 Bardeen Lab Madison WI 53706

SCHULTZ, EDWIN ROBERT, b Detroit, Mich, July 24, 27; m 63; c 3. STABILITY & CONTROL, DESIGN & ANALYSIS VTOL AIRCRAFT. *Educ:* Wayne State Univ, BS, 51; State Univ NY, MS, 58. *Prof Exp:* Flight test engr, McDonnell Aircraft Corp, 51-52; res engr, Cornell Aero Lab, 52-59; mem tech staff, Space Technol Lab, 59-61; supvr, advan res, Kaman Aerospace Corp, 61-64; supvry opers analyst, USAF Hq, Europe, 64-76; TECH DIR, TECHNOL ASSESSMENT DIV, WL-TXA, WRIGHT PATTERSON AFB, 76- *Mem:* Am Inst Aeronaut & Astronaut. *Res:* Design and performance analyses of advanced conceptual aircraft for the United States Air Force; integrate emerging technologies into advanced aircraft configurations; development of life cycle cost modeling applied to future combat aircraft. *Mailing Add:* 5400 Lytle Rd Waynesville OH 45068

SCHULTZ, EVERETT HOYLE, JR, b Winston-Salem, NC, Sept 13, 27; m 55; c 4. MEDICINE, RADIOLOGY. *Educ:* Bowman Gray Sch Med, MD, 52. *Prof Exp:* Asst prof radiol, Univ Fla, 58-61; assoc prof, Univ NC, 61-67; radiologist, St Anthony's Hosp, 67-84; ASSOC PROF, MED COL, GA, 87- *Concurrent Pos:* Ed consult, Yearbk Cancer, 64-82. *Mem:* Am Roentgen Ray Soc; Radiol Soc NAm; fel Am Col Radiol. *Res:* Clinical research in human gastrointestinal diseases, particularly pancreatic diseases. *Mailing Add:* Radiol Med Col Ga Augusta GA 30912-2153

SCHULTZ, FRANKLIN ALFRED, b Whittier, Calif, Mar 27, 41; div; c 1. ANALYTICAL CHEMISTRY, ELECTROCHEMISTRY. *Educ:* Calif Inst Technol, BS, 63; Univ Calif, Riverside, PhD(chem), 67. *Prof Exp:* Res chemist, Beckman Instruments, Inc, Calif, 67-68; from asst prof to prof chem, Fla Atlantic Univ, 68-86; PROF CHEM, IND UNIV-PURDUE UNIV, INDIANAPOLIS, 86- *Mem:* AAAS; Am Chem Soc; Electrochem Soc. *Res:* Electrochemistry of transition metal complexes; redox chemistry of inorganic compounds as models for biological electron transfer; theory of electron transfer reaction. *Mailing Add:* Dept Chem Fla Atlantic Univ Boca Raton FL 33431

SCHULTZ, FRED HENRY, JR, b May 13, 09; m; c 3. GINSENG RESEARCH. *Educ:* Univ Colo, PhD(biochem), 38. *Prof Exp:* Dir Res Coord, Sandoz, Inc, 55-75; RETIRED. *Mem:* Am Soc Pharmacol & Exp Therapeut; NY Acad Sci; Sigma Xi; Am Asn Aging. *Mailing Add:* 3702 75th St Lubbock TX 79423-1204

SCHULTZ, FREDERICK HERMAN CARL, b Hanks, NDak, June 11, 21; m 49; c 3. PHYSICS. *Educ:* Univ NDak, PhB, 42; Univ Idaho, MS, 50; Wash State Univ, PhD(physics), 67. *Prof Exp:* Instr physics, Univ NDak, 42-44 & 46-48, NDak State Univ, 44 & Mont Sch Mines, 50-55; asst prof, Mont State Univ, 55-61; assoc prof, Minot State Col, 61-63; asst prof, Wash State Univ, 63-68; chmn dept, 68-77, PROF PHYSICS, UNIV WIS-EAU CLAIRE, 68- *Concurrent Pos:* Dir seismog sta, US Coast & Geod Surv, 55-61; energy consult, 77-; physicist, US Naval Ord Lab Corona, 57, 59, 61 & 63. *Mem:* Am

Asn Physics Teachers; Seismol Soc Am; Optical Soc Am; Sigma Xi; NY Acad Sci. *Res:* Seismology; small Montana earthquakes; interaction of polarized infrared radiation with materials and surfaces. *Mailing Add:* Dept Physics Univ Wis Eau Claire WI 54701

SCHULTZ, FREDERICK JOHN, b Davenport, Iowa, Oct 12, 29; m 55; c 4. ORGANIC CHEMISTRY, RESEARCH ADMINISTRATION. *Educ:* Augustana Col, Ill, BA, 52; DePauw Univ, MA, 56; Univ Iowa, PhD(chem), 60. *Prof Exp:* Res chemist, 59-62, sr res chemist, 62-65, prod develop mgr, 65-68, mgr res, Res Div, 68-75, dir, 75-78, VPRES RES & DEVELOP, LORILLARD INC, 78- *Mem:* AAAS; Am Chem Soc; NY Acad Sci; Am Inst Chemists. *Res:* Composition of tobacco and tobacco smoke; relation of composition to biological activity and organoleptic properties; selective filtration of tobacco smoke; analytical methods development; new products in areas of tobacco and food products. *Mailing Add:* Lorillard Res Ctr PO Box 21688 Lorillard Inc Greensboro NC 27420

SCHULTZ, GEORGE ADAM, b Phillipsburg, NJ, Apr 10, 32. ISOPOD CRUSTACEANS-ECOLOGY. *Educ:* Univ Chicago, BA, 53; Univ Mont, MA, 58; Duke Univ, PhD(zool), 64. *Prof Exp:* ASSOC PROF, JERSEY CITY STATE COL, 70- *Mem:* Crustacean Soc; Soc Syst Zool; Am Soc Zoologists; Am Asn Zool Nomenclature. *Res:* Isopod crustaceans, marine, freshwater and terrestrial. *Mailing Add:* 15 Smith St Hampton NJ 08827

SCHULTZ, GERALD EDWARD, b Red Wing, Minn, Sept 2, 36; div; c 1. VERTEBRATE PALEONTOLOGY & BIOSTRATIGRAPHY. *Educ:* Univ Minn, BS, 58, MS, 61; Univ Mich, PhD(geol), 66. *Prof Exp:* From asst prof to assoc prof geol, 64-74, PROF GEOL, WTEX STATE UNIV, 74- *Concurrent Pos:* NSF res grant Pleistocene vert, Tex Panhandle, 70-72. *Mem:* Soc Vert Paleont; Am Soc Mammal; Paleont Soc; Am Quaternary Asn; Sigma Xi. *Res:* Vertebrate paleontology, especially late Cenozoic vertebrates and stratigraphy of the High Plains; late Tertiary and Pleistocene microvertebrate faunas and paleoecology. *Mailing Add:* Dept Geosci WTex State Univ Canyon TX 79016

SCHULTZ, GILBERT ALLAN, b Camrose, Alta, Nov 25, 44; m 69; c 2. DEVELOPMENTAL BIOLOGY. *Educ:* Univ Alta, BSc, 65, MSc, 66; Univ Calgary, PhD(biol), 70. *Prof Exp:* Nat Res Coun Can fel, Weizmann Inst Sci, 70 & Med Ctr, Univ Colo, Denver, 71-72; from asst prof to assoc prof, 72-83, PROF MED BIOCHEM, FAC MED, UNIV CALGARY, 83- *Mem:* Can Soc Cell Biol; Am Soc Cell Biol; Soc Develop Biol; Soc Develop Biol. *Res:* Study of gene expression during early development of mammalian embryos. *Mailing Add:* Div Med Biochem Univ Calgary 3330 Hospital Dr NW Calgary AB T2N 4N1 Can

SCHULTZ, HARRY PERSHING, b Racine, Wis, Mar 9, 18; m 43; c 3. ORGANIC CHEMISTRY. *Educ:* Univ Wis, BS, 42, PhD(org chem), 46. *Prof Exp:* Res chemist, Nat Defense Res Comt, Univ Wis, 42-45 & Merck & Co, Inc, NJ, 46-47; from asst prof to prof chem, 47-84, chmn dept, 72-84, EMER PROF CHEM, UNIV MIAMI, 84- *Mem:* Am Chem Soc. *Res:* Synthesis organic nitrogen heterocycles; organic reduction; chemical topology; polypeptides. *Mailing Add:* 5835 SW 81st St South Miami FL 33143

SCHULTZ, HARRY WAYNE, b Burlington, Iowa, June 13, 30; m 55; c 3. PHARMACEUTICAL CHEMISTRY. *Educ:* Univ Iowa, BS, 52, MS, 57, PhD(pharmaceut chem), 59. *Prof Exp:* Asst prof, 59-66, ASSOC PROF PHARMACEUT CHEM, ORE STATE UNIV, 66- *Mem:* Am Pharmaceut Asn; Am Chem Soc. *Res:* Organic pharmaceutical chemistry; relationship of chemical structure to pharmacological activity. *Mailing Add:* 2137 NW Robin Hood Corvallis OR 97330

SCHULTZ, HILBERT KENNETH, b Butternut, Wis, Oct 27, 35; m 57; c 4. SYSTEMS ANALYSIS, COMPUTER SYSTEMS. *Educ:* Univ Wis, Oshkosh, BS, 59; Univ Wis-Madison, MS, 62, PhD(comput sci), 71. *Prof Exp:* Computer analyst supvr, AC Electronics, Inc, 59-61; consult, Info Syst & Modeling, 66-68; PROF MIS, UNIV WIS-OSHKOSH, 71- *Concurrent Pos:* Consult govt & indust, health care field, medicaid & retail, 75-85. *Mem:* Am Prod & Inventory Control Soc; Decisions Sci Inst. *Res:* Use of fourth generation languages; microcomputer applications; theoretical and practical analysis of transportation problems; experts systems development; systems analysis and design of computer systems for retail, construction, nursing homes and physicians. *Mailing Add:* Col Bus Admin Univ Wis Oshkosh WI 54901

SCHULTZ, HYMAN, b Brooklyn, NY, July 11, 31; m 57; c 2. COAL CHEMISTRY, TRACE ANALYSIS. *Educ:* Brooklyn Col, BS, 56; Pa State Univ, PhD(anal chem), 62. *Prof Exp:* Sr res engr, Rocketdyne Div, NAm Aviation, Inc, Calif, 62-67; scientist & head gas anal sect, Isotopes, Teledyne, Inc, NJ, 67-71; res supvr anal res & serv, Pittsburgh Energy Res Ctr, US Bur Mines, 71-75, res supvr anal res & serv, Pittsburgh Energy Res Ctr, 75-77, BR CHIEF ANALYSIS RES, PITTSBURGH ENERGY TECHNOL CTR, DEPT ENERGY, 77- *Concurrent Pos:* Organizing Comt, Pittsburgh Conf & Expo, 80- *Mem:* AAAS; Sigma Xi; Am Chem Soc; SACP. *Res:* Determination of trace materials in complex natural matrices; analysis of coal and the products of coal research; trace toxic materials in coal and their fate when coal is utilized; standardization of analytical methods for coal conversion materials. *Mailing Add:* Pittsburgh Energy Technol Ctr PO Box 10940 Pittsburgh PA 15236

SCHULTZ, IRWIN, b New York, NY, July 29, 29; m 55; c 5. MEDICINE. *Educ:* NY Univ, BA, 49, MD, 54; Harvard Univ, MSH, 60, ScD(trop med), 64. *Prof Exp:* Asst prof microbiol & med, Med Sch, Northwestern Univ, 61-64; asst prof med, Sch Med, Wash Univ, 65-69; ASSOC CLIN PROF MED, COL MED, ST LOUIS UNIV, 69- *Concurrent Pos:* Mem, Nat Inst Allergy & Infectious Dis, 59-61. *Mem:* AAAS; Am Fedn Clin Res; Am Col Physicians; Am Soc Microbiol. *Res:* Infectious diseases; epidemiology; pathogenesis of viral infections; host defense mechanisms in infectious disease; vaccine effectiveness. *Mailing Add:* 2865 Netherton Dr St Louis MO 63136

SCHULTZ, J(EROME) S(AMSON), b Brooklyn, NY, June 25, 33; m 55; c 3. CHEMICAL ENGINEERING, BIOENGINEERING. *Educ:* Columbia Univ, BS, 54, MS, 56; Univ Wis, PhD(biochem), 58. *Prof Exp:* Chem engr, Lederle Labs, Am Cyanamid Co, 58-59, in chg fermentation pilot plant, 59-61, group leader biochem res, 61-64; from asst prof to assoc prof, 64-70, chmn dept, 77-85, prof chem eng, Univ Mich, Ann Arbor, 70-87; DIR CTR BIOTECHNOL & BIOENG, UNIV PITTSBURGH, 87- *Concurrent Pos:* Res Career Develop Award, NIH, 70-75; sect head emerging technol, 85-86, dep dir; cross-disciplinary res, NSF, 86-87. *Mem:* Am Soc Artificial Internal Organs; AAAS; Am Chem Soc; Am Inst Chem Engrs. *Res:* Biochemical engineering; production of chemicals and pharmaceuticals by fermentation; kinetics; transport phenomena in membranes; compatibility of biomaterials; artificial organs; transport in blood and tissues; photochemical processes. *Mailing Add:* Ctr Biotechnol & Bioeng Univ Pittsburgh 911 Willaim Pitt Union Pittsburgh PA 15260

SCHULTZ, JACK C, b Chicago, Ill, Jan 4, 47. INSECT ECOLOGY, PLANT-INSECT ECOLOGY. *Educ:* Univ Chicago, AB, 69; Univ Wash, PhD(zool), 75. *Prof Exp:* Res instr, Dartmouth Col, 75-80, res asst prof dept biol sci, 81-83; asst prof, 83-89, ASSOC PROF, DEPT ENTOM, PESTICIDE RES LAB, PA STATE UNIV, 89- *Concurrent Pos:* Vis fel, Dept Entom, Cornell Univ, 78-79. *Mem:* Ecol Soc Am; Entom Soc Am; Soc Study Evolution. *Res:* Coevolutionary interactions among trees, insects and natural enemies; chemical and physiological responses of trees to insects; foraging and predator-avoidance behavior of insects; tropical ecology. *Mailing Add:* RD I Julian University Park PA 16802

SCHULTZ, JAMES EDWARD, b Sheboygan, Wis, Dec 25, 39; m 63; c 1. MATHEMATICS. *Educ:* Univ Wis-Madison, BS, 63; Ohio State Univ, MS, 67, PhD(math educ), 71. *Prof Exp:* Instr math, high sch, Wis, 63-68; admin assoc, 68-71, asst prof, 71-78, ASSOC PROF MATH, OHIO STATE UNIV, 78- *Concurrent Pos:* Vis asst prof math, Mich State Univ, 73-74; vis asst prof educ, Univ Chicago, 74-75. *Mem:* Math Asn Am. *Res:* Mathematics preparation of elementary teachers. *Mailing Add:* Dept Math 231 W 18th Ave No 118 Ohio State Univ Columbus OH 43210

SCHULTZ, JANE SCHWARTZ, b New York, NY, July 28, 32; m 55; c 3. IMMUNOGENETICS. *Educ:* Hunter Col, BA, 53; Columbia Univ, MS, 55; Univ Mich, MS, 67, PhD(human genetics), 70. *Prof Exp:* Res chemist, Gen Foods Corp, 54-55 & Forest Prod Lab, USDA, 55-58; sci teacher, Pearl River High Sch, NY, 58-59; res assoc, Univ Mich, Ann Arbor, 70-71 & 72-75, asst prof human genetics, 75-82, asst dean curriculum, 79-81, assoc prof human genetics, 82-83, asst dean student affairs, Univ Mich, Ann Arbor, 81-83; geneticist, Vet Admin Med Ctr, 72-83; chief genetics & transp biol br, Nat Inst Allergic & Infectious Dis, 83-88; DIR RES ADMIN HEALTH SCI, UNIV PITTSBURGH, 88- *Concurrent Pos:* Sr res investr, Dept Immunohaemetology, State Univ Leiden, 71-72; vet admin rep, Genetics Study Sect, NIH, 73-77, Nat Inst Gen Med Sci Coun, 77-81; chief, Div Prog Develop & Review, Vet Admin Res Serv, 76-79. *Honors & Awards:* Dirs Award, NIH; Spec Achievement Award, 88. *Mem:* Am Asn Immunologists; Genetics Soc Am; Am Soc Human Genetics; Am Asn Clin Histocompatibility Testing. *Res:* Elucidation of the immunological functions controlled by genetically determined transplantation antigens in mouse, rat and man. *Mailing Add:* Univ Pittsburgh-WPIC 3811 O'Hara St Pittsburg PA 15213

SCHULTZ, JAY WARD, b Detroit, Mich, Feb 9, 37; m 58; c 2. METALLURGY. *Educ:* Mich Technol Univ, BS, 58; Univ Mich, Ann Arbor, MSE, 60, PhD(metall eng), 65. *Prof Exp:* Asst res engr, Off Res Admin, Univ Mich, Ann Arbor, 60-64; res metallurgist alloy develop, Int Nickel Co, Inc, 64-65, res metallurgist corrosion, 66-70, supvr dry corrosion & nickel chem, 70-76, chem res mgr, 76-78; mgr res, 78-80, dir, 80-82, vpres, Inco Alloy Prod Co, 82-84. *Mem:* Nat Asn Corrosion Engrs; Am Soc Metals; Indust Res Inst. *Res:* Dry corrosion; alloy development; organometallic chemistry; electroplating, polymers; research management. *Mailing Add:* 11212 88th St Kenosha WI 53142

SCHULTZ, JEROLD M, b San Francisco, Calif, June 21, 35; m 60; c 4. MATERIALS SCIENCE. *Educ:* Univ Calif, Berkeley, BS, 58, MS, 59; Carnegie Inst Technol, PhD(metall), 65. *Prof Exp:* Intermediate engr mat res, Westinghouse Res Labs, 59-61; from asst prof to assoc prof metall, 65-73, PROF METALL, UNIV DEL, 73- *Concurrent Pos:* Vis asst prof, Stanford Univ, 68; vis prof, Univ Mainz, 74-75 & 82, Univ Sofia, 85; Humboldt sr scientist, Univ Saarbrucken, 77-78, 82-83 & Univ Bochum, 82-83; vis scientist, Du Pont Exp Sta, 85, Nat Chem Lab, India, 89-90. *Honors & Awards:* Humboldt Sr Scientist Award, 77; Kliment Ohridski Medal, Bulgaria, 86. *Mem:* Am Phys Soc; Polymer Processing Inst. *Res:* Polymeric materials; phase transformations; diffraction and scattering; composites; failure of materials. *Mailing Add:* Dept Chem Eng Univ Del Newark DE 19716

SCHULTZ, JOHN E, b Nowata, Okla, Mar 5, 36; m 55; c 3. ORGANIC CHEMISTRY. *Educ:* Westminster Col, Mo, BA, 58; Univ Ill, PhD, 63. *Prof Exp:* From asst prof to assoc prof, 64-70, chmn dept, 74-77, PROF CHEM, WESTMINSTER COL, MO, 70-, ASST ACAD DEAN, 77- *Mem:* Am Chem Soc; Sigma Xi. *Res:* Small ring carbocyclic compounds and free radical reactions; information retrieval; computers in education. *Mailing Add:* 2330 N Bluff St Fulton MO 65251

SCHULTZ, JOHN LAWRENCE, b Brooklyn, NY, June 22, 32. INFORMATION SCIENCE. *Educ:* St John's Univ, BS, 54; Univ Minn, PhD(chem), 59. *Prof Exp:* Asst inorg chem, Univ Minn, 54-56; res chemist, Pigments Dept, 57, info chemist, Patent Div, Textile Fibers Dept, 58-64, sr info specialist, Secy Dept, 64-73, sr info specialist, Info Systs Dept, 74-86 SR INFO CONSULT, HUMAN RESOURCES DEPT, E I DU PONT DE NEMOURS & CO INC, 86- *Res:* Solution calorimetry; heats of formation of metal ion complexes in aqueous solution; heats of ion exchange processes; storage and retrieval of chemical information. *Mailing Add:* Human Resources Dept Co Chem Inventory E I du Pont de Nemours & Co Wilmington DE 19898

SCHULTZ, JOHN RUSSELL, b Lanark, Ill, Apr 26, 08. GEOLOGY. *Educ:* Univ Ill, BA, 31; Northwestern Univ, MS, 33; Calif Inst Technol, PhD(geol), 37. *Prof Exp:* Asst geol, Northwestern Univ, 31-33; Nat Res Coun fel, Harvard Univ & Calif Inst Technol, 38; asst geologist, Standard Oil Co Calif, Saudi Arabia, 38-40; geologist, US Engrs Off, Miss, 41-43, geologist, Panama Canal, 46-47; assoc prof geol, Brown Univ, 47-49; chief geologist, US Waterways Exp Sta, 49-56; staff geologist, Harza Eng Co, 56-75, consult, 75-82; RETIRED. *Res:* Engineering geology; geology of dam sites; flood control projects; airfields. *Mailing Add:* 35 Mayflower Rd Winchester MA 01890

SCHULTZ, JOHN WILFRED, b Portland, Ore, Sept 15, 31. PHYSICAL CHEMISTRY. *Educ:* Ore State Col, BS, 53; Brown Univ, PhD(phys chem), 57. *Prof Exp:* Instr chem, Univ Wash, 56-58; from asst prof to assoc prof chem, Naval Postgrad Sch, 66-75; ASSOC PROF CHEM, NAVAL ACAD, 75- *Concurrent Pos:* Soc Appl Spectros; Coblentz Soc; Am Phys Soc; Am Chem Soc. *Res:* Molecular spectroscopy; infrared and Raman intensities; spectra of solids. *Mailing Add:* Dept Chem US Naval Acad Annapolis MD 21402

SCHULTZ, JONAS, b Brooklyn, NY, Mar 15, 35; m 58; c 3. PHYSICS. *Educ:* Columbia Univ, AB, 56, MA, 59, PhD(physics), 62. *Prof Exp:* Physicist, Nevis Cyclotron Labs, Columbia Univ, 61-63 & Lawrence Radiation Lab, Univ Calif, 63-66; assoc prof, 66-70, dean grad div, 73-76, PROF PHYSICS, UNIV CALIF, IRVINE, 70- *Concurrent Pos:* Assoc prog dir elem particle physics, NSF, 71-72. *Mem:* Am Phys Soc. *Res:* Elementary particle physics; studies of high energy phenomena. *Mailing Add:* Dept Physics Univ Calif Irvine CA 92717

SCHULTZ, LANE D, b Sellersville, Pa, Oct 20, 44. GEOPHYSICS, EDUCATION ADMINISTRATION. *Educ:* Franklin & Marshall Col, BA, 66; Lehigh Univ, MS, 72, PhD(geol), 74. *Prof Exp:* Instr, Dickinson Col, 70-71; teaching asst, Lehigh Univ, 71-74; geol sect supvr, Gilbert Assoc Inc, 74-81; mgr, geol, Western Geophys Corp, 81-84; mgr, geotech, ERT, 84-85; VPRES, OFF MGR, DUNN GEOSCI CORP, 85- *Mem:* Am Inst Prof Geol; Asn Eng Geol; Am Asn Petrol Geol; Am Soc Photogram. *Res:* Application of geology to engineered structure & characterization of subsurface geological & hydrogeological conditions. *Mailing Add:* Allan A Myers Inc PO Box 98 Worcester PA 19490

SCHULTZ, LINDA DALQUEST, b Yakima, Wash, Feb 23, 47; m 69; c 2. ANALYTICAL CHEMISTRY. *Educ:* Southern Methodist Univ, BA, 67, MS, 71; NTex State Univ, PhD(chem), 75; Registry Med Technologists, cert, 71. *Prof Exp:* Res technologist biochem, Univ Tex Southwestern Med Sch Dallas, 67-69; med technologist, Parkland Mem Hosp, 69-71; teaching asst chem, NTex State Univ, 71-74; res assoc sci, Howard Payne Univ, 75-76; fel, Tex Christian Univ, 76-78; ASSOC PROF, TARLETON STATE UNIV, 78- *Mem:* Am Chem Soc. *Res:* Liquid ammonia chemistry. *Mailing Add:* Rte 4 Box 187H Brownwood TX 76801

SCHULTZ, LORIS HENRY, b Mondovi, Wis, Feb 9, 19; m 49; c 3. DAIRY SCIENCE. *Educ:* Univ Wis, BS, 41, PhD, 49; Univ Minn, MS, 42. *Prof Exp:* From asst prof to prof animal husb, Cornell Univ, 49-57; PROF DAIRY SCI, UNIV WIS-MADISON, 57- *Mem:* Am Soc Animal Sci; Am Dairy Sci Asn (pres, 82); Nat Mastitis Coun (pres, 80). *Res:* Physiology of lactation; intermediary metabolism; metabolic disorders. *Mailing Add:* Dept Dairy Sci Univ Wis 266 Animal Sci Bldg Madison WI 53706

SCHULTZ, MARTIN C, b Philadelphia, Pa, Aug 29, 26; m 51; c 4. SPEECH & HEARING SCIENCES, AUDIOLOGY. *Educ:* Temple Univ, BA, 50; Univ Mich, MA, 52; Univ Iowa, PhD, 55. *Prof Exp:* Res assoc, Univ Iowa, 53-54; instr, Sch Speech, Northwestern Univ, 54-55; assoc otolaryngol, phys med & psychol, Univ Pa, 55-58, dir speech & hearing ctr, Univ Hosp, 55-58; supvr res lab, Cleveland Hearing & Speech Ctr, 58-60; asst prof, Univ Mich, 60-65; assoc prof speech, Ind Univ, Bloomington, 65-73; dir training in speech path & audiol, Develop Eval Clin, 72-77, dir, hearing & speech div, Children's Hosp Med Ctr, 73-85; prof commun dis, Emerson Col, 72-85; PROF, SOUTHERN ILL UNIV, 86- *Concurrent Pos:* Off Voc Rehab grant, Univ Pa, 55-58; NIH grant, Cleveland Hearing & Speech Ctr, 59-60; NIH grant, Univ Mich, 60-64; consult, State Dept Health, Pa, 57 & Woods Schs Except Children, 57-58; adj prof, Sch Educ, Boston Univ, 72-85; assoc otolaryngol, Harvard Med Sch, 74-86; res affil, Res Lab Electronics, Mass Inst Technol, 74-85; prin investr & res grant, US Dept Educ, 80-84. *Honors & Awards:* Editor's Award, J Speech & Hearing Dis, Am Speech & Hearing Asn, 74. *Mem:* AAAS; Acoust Soc Am; Am Speech & Hearing Asn; Int Soc Phonetic Sci; Sigma Xi. *Res:* Speech and hearing sciences; design methodology; clinical processes and models; hearing and language development in children. *Mailing Add:* Commun Dis & Sci Southern Ill Univ Carbondale IL 62901-6616

SCHULTZ, MARTIN H, b Boston, Mass, Dec 6, 40; m 65. COMPUTER SCIENCE. *Educ:* Calif Inst Technol, BS, 61; Harvard Univ, PhD(math), 65. *Prof Exp:* Asst prof math, Case Western Reserve Univ, 65-68; assoc prof, Calif Inst Technol, 68-70; PROF COMPUT SCI, YALE UNIV, 70-, CHMN DEPT, 74- *Mem:* Am Math Soc; Soc Indust & Appl Math; Asn Comput Mach. *Res:* Numerical analysis; computational complexity. *Mailing Add:* Dept Comput Sci 521 DI Yale Univ PO Box 2158 New Haven CT 06520

SCHULTZ, MYRON GILBERT, b New York, NY, Jan 6, 35; m 59; c 3. EPIDEMIOLOGY. *Educ:* NY State Vet Col, Cornell Univ, DVM, 58; Albany Med Col, MD, 62; London Sch Hyg & Trop Med, DCMT, 67. *Prof Exp:* Pvt pract vet med, 58-62; epidemic intel serv officer, 63-65, CHIEF PARASITIC DIS BR, NAT CTR DIS CONTROL, USPHS, 67- *Concurrent Pos:* Clin assoc prof prev med, Emory Univ, 67-, clin asst prof med, 71- *Mem:* Fel Am Col Physicians; Am Soc Trop Med & Hyg; Royal Soc Trop Med & Hyg. *Res:* Epidemiology, clinical tropical medicine; clinical drug evaluation. *Mailing Add:* Ctr Dis Control Atlanta GA 30306

SCHULTZ, PETER BERTHOLD, b Bucharest, Rumania, Oct 24, 46; m 88; c 2. ENTOMOLOGY. *Educ:* Univ Calif, Davis, BS, 68; Midwestern Univ, MS, 72; Va Polytech Inst & State Univ, PhD(entom), 78. *Prof Exp:* Instr entom, USAF, 69-73; regulatory inspector, Va Dept Agr, 73-78; entomologist, VA Truck & Ornamentals Res Sta, 78-85; ASSOC PROF ENTOM, VA POLYTECH INST & STATE UNIV, 85- *Mem:* Entom Soc Am. *Res:* Insect research on ornamental plants. *Mailing Add:* 1444 Diamond Springs Rd Virginia Beach VA 23455

SCHULTZ, PETER FRANK, b Oshkosh, Wis, Mar 23, 40; m 66; c 3. EXPERIMENTAL HIGH ENERGY PHYSICS. *Educ:* Univ Wis-Madison, BS, 62; Univ Ill, Urbana, MS, 64, PhD(physics), 69. *Prof Exp:* Res assoc physics, Univ Ill, Urbana, 69-72; res assoc, Argonne Nat Lab, 72-76, asst physicist, 76-81; MEM TECH STAFF, BELL TEL LAB, NAPERVILLE, ILL, 81- *Mem:* Am Phys Soc; Sigma Xi. *Res:* Elementary particle physics; acceleration technology; ion sources; telecommunications. *Mailing Add:* 501 Andrus Rd Downers Grove IL 60516

SCHULTZ, PETER HEWLETT, b New Haven, Conn, Jan 22, 44; m 67. PLANETARY GEOLOGY. *Educ:* Carleton Col, BA, 66; Univ Tex, PhD(astron), 72. *Prof Exp:* Resident res assoc, Nat Acad Sci-Nat Res Coun, Univ Santa Clara, 73-75, res assoc physics, NASA Ames Res Ctr, 75-76; sr staff scientist, Lunar & Planetary Inst, 76-84; ASSOC PROF & DIR NE PLANETARY DATA CTR, BROWN UNIV, 84- *Concurrent Pos:* Prin investr & assoc ed, Geophys Revs. *Mem:* Sigma Xi; Am Geophys Union; AAAS. *Res:* Morphology of impact craters on planets; impact cratering mechanics; degradational processes on planetary surfaces; volcanic modification of planetary surfaces; atmospheric effects on impact crater formation; lunar and Martian geologic history. *Mailing Add:* Dept Geol Sci Brown Univ Box 1846 Providence RI 02912

SCHULTZ, PHYLLIS W, b Connersville, Ind, Mar 9, 25; m 54. DEVELOPMENTAL BIOLOGY. *Educ:* Univ Cincinnati, BA, 47, MS, 50; Univ Wis, PhD(zool), 57. *Prof Exp:* Preparator zool, Univ Wis, 54-55; chemist med ctr, Univ Colo, 55-57; res assoc embryol, Univ Ore, 57-59; res assoc & vis instr, Med Ctr, 59-61, asst prof zool, 61-70, fac fel, 64-65, PROF BIOL & ASST DEAN NATURAL & PHYS SCI, MED CTR, UNIV COLO, DENVER, 71- *Concurrent Pos:* Res grants, USPHS, 58-64 & NSF, 64-66. *Mem:* Am Soc Cell Biol. *Res:* Effects of teratogenic agents or antimetabolites on protein formation, ultrastructure and cytochemistry of the chick and mammalian embryo and mammalian placenta. *Mailing Add:* 7243 Costilla St Littleton CO 80120

SCHULTZ, R JACK, b Caro, Mich, Aug 17, 29; m 57; c 2. POPULATION BIOLOGY, ICHTHYOLOGY. *Educ:* Mich State Univ, BS, 52, MS, 53; Univ Mich, PhD, 60. *Prof Exp:* Res assoc, Mus Zool, Univ Mich, 60-63; from asst prof to prof, 63-74, PROF BIOL, UNIV CONN, 75- *Concurrent Pos:* Ed, Copeia, Am Soc Ichthyologists & Herpetologists, 70-73; prog dir syst biol, NSF, 74-75; head ecol, Univ Conn, 79-85. *Mem:* Am Soc Ichthyologists & Herpetologists; Soc Study Evolution; Ecol Soc Am; Am Genetic Asn (pres, 88-89); Am Soc Naturalists. *Res:* Role of hybridization and polyploidy in the evolution and ecology of fishes; genetics of cancer in fishes. *Mailing Add:* Dept Ecol & Evolutionary Biol U-42 Univ Conn Storrs CT 06269-3042

SCHULTZ, RAY KARL, b Hereford, Pa, Aug 23, 37; m 64; c 4. CHEMISTRY. *Educ:* Muhlenberg Col, BS, 59; Lehigh Univ, MS, 61, PhD(rheology), 65. *Prof Exp:* Instr quant anal, Muhlenberg Col, 63; from asst prof to assoc prof, 65-82, PROF CHEM, URSINUS COL, 82- *Mem:* Am Chem Soc; Sigma Xi. *Res:* Equilibrium constants for formation of boratediol complexes; properties of polyelectrolytes; properties of poly(acrylic acid)-co-4-vinyl pyridine. *Mailing Add:* Dept Chem Ursinus Col Collegeville PA 19426

SCHULTZ, REINHARD EDWARD, b Chicago, Ill, Sept 13, 43; m 70; c 1. TOPOLOGY, MATHEMATICS. *Educ:* Univ Chicago, SB, 64, SM, 65, PhD(math), 68. *Prof Exp:* From instr to asst prof, 68-74, assoc prof, 74-80, PROF MATH, PURDUE UNIV, WEST LAFAYETTE, 80- *Concurrent Pos:* Purdue Res Found grant, 69, NSF res grant, 70-72. *Mem:* Am Math Soc. *Res:* Algebraic topology, differential topology; transformation groups. *Mailing Add:* Dept Math Purdue Univ West Lafayette IN 47907

SCHULTZ, RICHARD MICHAEL, b Philadelphia, Pa, Oct 28, 42; m 65; c 2. BIOCHEMISTRY, MOLECULAR BIOLOGY. *Educ:* State Univ NY Binghamton, BA, 64; Brandeis Univ, MA, 67, PhD(org chem), 69. *Prof Exp:* From asst prof to assoc prof, 71-84, PROF & CHMN BIOCHEM, STRITCH SCH MED, LOYOLA UNIV CHICAGO, 84- *Concurrent Pos:* NIH res fel biol chem, Harvard Med Sch, 69-71. *Mem:* Am Chem Soc; Am Soc Biol Chemists. *Res:* Mechanism of enzyme action, particularly the hydrolytic enzymes; cancer cell metastasis; peptide synthesis; coagulation enzyme purification and properties; protease gene regulation in cancer; association of aldehyde transition-state analogs to protease enzymes; molecular biology. *Mailing Add:* Dept Molecular & Cellular Biochem Loyola Univ Stritch Sch Med Maywood IL 60153

SCHULTZ, RICHARD MORRIS, b Malden, Mass, Mar 20, 49; m 79. PROTEIN PHOSPHORYLATION, MAMMALIAN DEVELOPMENT. *Educ:* Harvard Univ, PhD(biochem), 75. *Prof Exp:* PROF BIOL, UNIV PA, 78- *Mem:* AAAS; Am Soc Cell Biol; Soc Develop Biol; Soc Study Reproductive Biol. *Res:* Developmental biology. *Mailing Add:* Dept Biol Univ Pa Philadelphia PA 19104-6018

SCHULTZ, RICHARD OTTO, b Racine, Wis, Mar 19, 30; m 52, 90; c 3. MEDICINE, OPHTHALMOLOGY. *Educ:* Univ Wis, BA, 50, MSc, 54; MD, Albany Med Col, 56; Univ Iowa, MSc, 60. *Prof Exp:* Instr ophthal, Univ Iowa, 59-60; assoc, Proctor Found, Sch Med, Univ Calif, San Francisco, 63-64; assoc prof, 64-68, PROF OPHTHAL, MED COL WIS, 68-, CHMN DEPT, 64-; DIR OPHTHAL, MILWAUKEE REGIONAL MED CTR, 64- *Concurrent Pos:* NIH spec fel ophthal microbiol, Proctor Found Sch Med,

Univ Calif, San Francisco, 63-64; consult, US Vet Admin Hosp, Wood, Wis & Milwaukee Children's Hosp, 64-; Nat Adv Eye Coun, NIH. *Mem:* Fel Am Acad Ophthal; fel Am Col Surg; AMA; Asn Res Vision & Ophthal; Am Ophthal Soc; Assoc Univ Prof Ophthal. *Res:* Ocular microbiology; corneal disease; cataract surgery. *Mailing Add:* Dept Opthamol Med Col Wis Eye Inst 8700 W Wisconsin Ave Milwaukee WI 53226

SCHULTZ, ROBERT GEORGE, b Rahway, NJ, Jan 11, 33; m 58; c 3. PROCESS DEVELOPMENT. *Educ:* Mass Inst Technol, SB, 54; Univ Ill, PhD, 58; Northeast Mo State Univ, MA, 80. *Prof Exp:* Res specialist, Monsanto Co, 58-69, group leader, 69-85, mgr technol, 86-89, TECHNOLOGIST, MONSANTO CO, 89- *Mem:* Am Chem Soc. *Res:* Homogeneous catalysis; heterogeneous catalysis; process development. *Mailing Add:* Monsanto Co St Louis MO 63167

SCHULTZ, ROBERT JOHN, b Detroit, Mich, Apr 19, 44; m 67; c 2. MEDICINAL CHEMISTRY. *Educ:* Wayne State Univ, BSc, 66; Brown Univ, PhD(org chem), 71. *Prof Exp:* Res assoc, Univ Mich, 70-73; indust fel, Starks Assoc, Inc, 73, supvr, 73-75, prin investr, 75-79, co-prin investr, 79-80, PRIN INVESTR, STARKS C P, 80- *Mem:* Am Chem Soc; Sigma Xi. *Res:* Potential chemotherapeutic agents for anticancer screening programs. *Mailing Add:* 5649 Woodcrest Dr Edina MN 55424

SCHULTZ, ROBERT LOWELL, b Moscow, Idaho, Mar 18, 30; m 51; c 3. ANATOMY. *Educ:* Walla Walla Col, BA, 51, MA, 53; Univ Calif, Los Angeles, PhD(anat), 57. *Prof Exp:* Asst instr, 53-57, from instr to assoc prof, 57-74, PROF ANAT, SCH MED, LOMA LINDA UNIV, 74- *Concurrent Pos:* USPHS spec fel, Univ Calif, Los Angeles, 63-64. *Mem:* Electron Micros Soc Am; Am Asn Anat; Am Soc Cell Biol. *Res:* Electron microscopy and microanatomy of the nervous system. *Mailing Add:* Dept Anat Loma Linda Univ Sch Med Loma Linda CA 92350

SCHULTZ, RODNEY BRIAN, b Enid, Okla, Nov 2, 46; m 68; c 2. APPLIED PHYSICS, NUMERICAL SIMULATION. *Educ:* Okla State Univ, BS, 68; Univ Colo, MS, 71, PhD(astrophys), 74. *Prof Exp:* Asst solar physics, High Altitude Observ, 68-74; staff mem, Appl Theoret Physics Div, 74-88, GROUP LEADER, THERMONUCLEAR APPLNS GROUP, LOS ALAMOS NAT LAB, 88- *Concurrent Pos:* Los Alamos Liaison, Defense Intel Agency, Washington, DC, 86-87. *Honors & Awards:* Recognition Excellence Award, US Dept Energy, 90. *Mem:* Am Inst Aeronaut & Astronaut. *Res:* Thermonuclear weapons research. *Mailing Add:* 472 Grand Canyon Dr Los Alamos NM 87544

SCHULTZ, RONALD DAVID, b Freeland, Pa, Apr 21, 44; m 66; c 3. IMMUNOLOGY, VETERINARY VIROLOGY. *Educ:* Pa State Univ, BS, 66, MS, 67, PhD(microbiol), 70. *Prof Exp:* Res asst microbiol, Pa State Univ, 66-70; res assoc immunol, NY State Vet Col, 71-73, from asst prof to assoc prof immunol, Vet Virus Res Inst, Cornell Univ, 73-78, assoc dir, Dept Health Serv, Microbiol & Clin Lab, 73-78; prof, Dept Microbiol, Col Vet Med, Auburn Univ, 78-80; MEM FAC, NY STATE VET COL, CORNELL UNIV, 80- *Concurrent Pos:* Consult, Nat Cancer Inst, 72-78, Miles Lab, 75-, Corning Glass, 78- & Hybridoma Sci, 78-; res grants, NIH, USDA & Food & Drug Admin. *Mem:* Am Soc Microbiol; Conf Res Workers Animal Dis; Am Asn Vet Immunologists (pres, 75-80); US Animal Health Asn. *Res:* Developmental aspects of the immune response; cell-mediated immunity; immunoglobulins; clinical immunology; immunopathology; viral infections and the immune response; leukemia. *Mailing Add:* Sch Vet Med/Microbiol Univ Wisconsin Madison WI 53706

SCHULTZ, RONALD G(LEN), b Hammond, Ind, Nov 15, 31; m 56; c 3. ELECTRICAL ENGINEERING. *Educ:* Valparaiso Univ, BSEE, 53; Northwestern Univ, MS, 54; Univ Pittsburgh, PhD(elec eng), 59. *Prof Exp:* From instr to assoc prof elec eng, Univ Pittsburgh, 54-68; prof elec eng & chmn dept, 68-73, dean, Col Grad Studies, 73-81, assoc vpres acad affairs, 75-78, vprovost, 78-81, PROF ELEC ENG, CLEVELAND STATE UNIV, 81- *Concurrent Pos:* Consult, 59- *Mem:* Sigma Xi. *Res:* Computers and control systems; nonlinear control systems. *Mailing Add:* Elec Eng Dept Cleveland State Univ Cleveland OH 44115

SCHULTZ, RUSSELL THOMAS, AMYIOIDOSIS, EXPERIMENTAL ARTHRITIS. *Educ:* Univ Minn, MD, 53. *Prof Exp:* PROF MED, UNIV OKLA HEALTH SCI CTR, 73- *Res:* Experimental arthritis. *Mailing Add:* 921 NE 13th St Oklahoma City OK 73104

SCHULTZ, SHELDON, b New York, NY, Jan 21, 33; m 53; c 3. SOLID STATE PHYSICS. *Educ:* Stevens Inst Technol, ME, 54; Columbia Univ, PhD(physics), 59. *Prof Exp:* Res asst physics, Radiation Lab, Columbia Univ, 59-60; from asst prof to assoc prof, 60-71, PROF PHYSICS, UNIV CALIF, SAN DIEGO, 71-, DIR, CTR MAGNETIC RECORDING RES, 90- *Concurrent Pos:* Sloan Found fel, 62-64. *Mem:* Fel Am Phys Soc; Am Vacuum Soc; Inst Elec & Electronics Engrs; Mat Res Soc. *Res:* Solid state physics; magnetic resonance in metals; magnetic recording particles; surface magnetism; high temperature superconductors; electron paramagnetic resonance. *Mailing Add:* Dept Physics 0319 Univ Calif San Diego 9500 Gilman Dr La Jolla CA 92093-0319

SCHULTZ, STANLEY GEORGE, b Bayonne, NJ, Oct 26, 31; m 60; c 2. PHYSIOLOGY. *Educ:* Columbia Col, BA, 52; NY Univ, MD, 56. *Prof Exp:* Intern, Bellevue Hosp, NY, 56-57, resident internal med, 57-58; instr biophys, Harvard Med Sch, 64-65, assoc, 65-67; from assoc prof to prof physiol, Sch Med, Univ Pittsburgh, 70-79; PROF PHYSIOL & CHMN DEPT, MED SCH, UNIV TEX, HOUSTON, 79- *Concurrent Pos:* USPHS res fel cardiol, Lenox Hill Hosp, 58-59; Nat Acad Sci-Nat Res Coun res fel biophysics, Harvard Med Sch, 59-62; estab investr, Am Heart Asn, 64-69; UPSHS res career award, 69-72; res career develop award, NIH, 69-72; consult, NIH & Nat Bd Med Examrs; overseas fel, Churchill Col, Cambridge Univ, 76; ed, Am J Physiol, Physiol Rev & Ann Rev Physiol, Handbk Physiol, News in Physiol Sci. *Honors & Awards:* Hoffman-LaRoche Prize, Outstanding Contrib Gastroenterol. *Mem:* AAAS; Asn Am Physicians; Biophys Soc; Am Physiol Soc; Soc Gen Physiol; hon fel Am Soc Gynec-Obstet; Sigma Xi. *Res:* Membrane physiology; epithelial transport. *Mailing Add:* Dept Physiol & Cell Biol Univ Tex Med Sch Houston TX 77225

SCHULTZ, TERRY WAYNE, b Beloit, Wis, Feb 26, 46; m 68; c 1. TERATOGENESIS. *Educ:* Austin Peay State Univ, BS, 68; Univ Ark, MS, 72; Univ Tenn, PhD(zool), 75. *Prof Exp:* Fel, Biol Div, Oak Ridge Nat Lab, 75-77; asst prof histol & cell biol, Dept Biol, Pan Am Univ, 77-80; res assoc, Biol Div, Oak Ridge Nat Lab, 80-82; ASST PROF HISTOL, DEPT ANIMAL SCI, COL VET MED, UNIV TENN, 82-, ASSOC PROF. *Concurrent Pos:* Fac partic, Biomed & Environ Sci, Inst Lawrence Livermore Lab, 78; consult, Biol Div, Oak Ridge Nat Lab, 79. *Mem:* Am Micros Soc; Soc Environ Toxicol & Chem; Electron Micros Soc Am. *Res:* In vitro teratogenesis testing and screening using frog embryos; structure activity relationships of industrial chemicals and environmental toxicity; short-term cytotoxicity testing. *Mailing Add:* Dept Animal Sci Col Vet Med Univ Tenn PO Box 1071 Knoxville TN 37996

SCHULTZ, THEODORE DAVID, b Chicago, Ill, Jan 6, 29; m 57; c 2. SOLID STATE PHYSICS. *Educ:* Cornell Univ, BEngPhys, 51; Mass Inst Technol, PhD(physics), 56. *Prof Exp:* NSF fel math physics, Univ Birmingham, 56-58; res assoc physics, Univ Ill, 58-59, res asst prof, 59-60; PHYSICIST, WATSON RES CTR, IBM CORP, 60- *Concurrent Pos:* Adj prof, Syracuse Univ, 61-62; vis assoc prof, NY Univ, 64-65; vis prof, Univ Munich, 79-80. *Mem:* Fel Am Phys Soc. *Res:* Theory of solids. and quantum statistical mechanics; quantum field theory. *Mailing Add:* IBM Watson Res Ctr Rm 27-151 Yorktown Heights NY 10598

SCHULTZ, THEODORE JOHN, acoustics; deceased, see previous edition for last biography

SCHULTZ, THOMAS J, b Toledo, Ohio, July 22, 41; m 65; c 1. COMBUSTION, INCINERATION. *Educ:* Univ Toledo, BS, 64, MS, 65 & 68. *Prof Exp:* Proj engr, Midland Ross, 65-70, mgr chem eng, 70-80, asst dir develop, 80-88; DIR INDUST RES & DEVELOP, SURFACE COMBUSTION INC, 88- *Mem:* Am Inst Chem Engrs; Air & Waste Mgt Asn. *Res:* Reactions of natural gas; combustion catalytic oxidation; pyrolysis applied to waste disposal and recycling; heat transfer; several US patents. *Mailing Add:* 1268 Cass Maumee OH 43537

SCHULTZ, VINCENT, b Lakewood, Ohio, Mar 7, 22; m 48; c 3. ANIMAL ECOLOGY. *Educ:* Ohio State Univ, BSc, 46, MSc, 48, PhD(zool), 49; Va Polytech Inst, MSc, 54. *Prof Exp:* Wildlife biologist, US Fish & Wildlife Serv, 49-50; sr biologist, State Game & Fish Comn, Tenn, 50-52; asst prof wildlife mgt, Va Polytech Inst, 52-54; res fel biostatist, USPHS, Johns Hopkins Univ, 54-56; assoc prof biostatist & agr statistician, Univ Md, 56-59; ecologist environ sci br, Div Biol & Med, US AEC, 59-66; prof zool, Wash State Univ, 66-85; RETIRED. *Concurrent Pos:* Consult, Dept Energy, Nat Resource Coun, Sandia. *Res:* Application of statistical techniques to ecological research; radiation ecology. *Mailing Add:* NE 630 Oak Pullman WA 99163

SCHULTZ, WARREN WALTER, b Emporia, Kans, Sept 3, 41; m 79; c 3. TECHNICAL MANAGEMENT, SCIENCE POLICY. *Educ:* Kans State Univ, Emporia, BA, 64, MS, 66; Johns Hopkins Univ, ScD(virol), 72. *Prof Exp:* Teaching asst microbiol, Kans State Univ, Emporia, 64-66; res microbiologist, Naval Med Res Inst, 66-68, head, Div Pathobiol, 71-78; teaching asst pop biol, Johns Hopkins Univ, 69; prof chem & assoc chmn dept, US Naval Acad, 78-82; dep dir, biol sci div, Off Naval Res, 82-84, dep dir life sci, 84-86; BIOTECHNOL PROGS MGR, NAVAL RES LAB, 86- *Concurrent Pos:* Spec asst undersecy defense, Res & Advan Technol. *Mem:* Am Soc Microbiol; Asn Mil Surgeons; Soc Armed Forces Med Lab Scientists. *Mailing Add:* 4056 Cadle Creek Rd Edgewater MD 21037

SCHULTZ, WILLIAM C(ARL), b Sheboygan, Wis, July 30, 27; m 51; c 3. ELECTRICAL ENGINEERING. *Educ:* Univ Wis, BS, 52, MS, 53, PhD(elec eng), 58. *Prof Exp:* Asst, Univ Wis, 52-53, instr elec eng, 55-58; asst engr, Computer Lab, Allis-Chalmers Mfg Co, Wis, 53-55; asst engr, Cornell Aeronaut Lab, Inc, 58-70, head computer ctr, 70-75; ASSOC PROF TECHNOL, STATE UNIV COL NY, BUFFALO, 75- *Mem:* Inst Elec & Electronics Engrs; assoc fel Am Inst Aeronaut & Astronaut. *Res:* Computer sciences; computer facility management; administrative data processing; flight control systems; cockpit displays; human factors engineering. *Mailing Add:* State Univ NY 1300 Elmwood Ave Buffalo NY 14222

SCHULTZ, WILLIAM CLINTON, b Bainbridge, NY, Sept 19, 37; m 61; c 2. SYNTHETIC ORGANIC CHEMISTRY. *Educ:* Dartmouth Col, AB, 59; Rutgers Univ, PhD(org chem), 63. *Prof Exp:* SR DEVELOP CHEMIST, SYNTHETIC CHEM DIV, EASTMAN KODAK CO, 63- *Mem:* Am Chem Soc; Soc Photog Scientists & Engrs. *Res:* Development of economical manufacturing processes for specialty organic chemicals. *Mailing Add:* 67 Tulip Tree Lane Rochester NY 14617-2004

SCHULTZE, CHARLES L, b Alexandria, Va, Dec 12, 24. EDUCATION ADMINISTRATION. *Educ:* Georgetown Univ, BA, 48, MA, 50; Univ Md, PhD(econ), 60. *Prof Exp:* Lect & assoc prof econ, Ind Univ, 59-61; asst prof & prof econ, Univ Md, 61-88; RETIRED. *Concurrent Pos:* Asst dir, US Bur Budget, 62-64, sr fel, 68-77 & 81-87; sr fel, Brookings Inst, 68; chmn, Coun Econ Advsr, 77-80; distinguished vis prof res, Grad Sch Bus, Stanford Univ, 82-83; Lee Kuan Yew distinguished vis, Nat Univ Singapore, 85; dir econ studies prog, 87- *Mem:* Am Econ Asn; fel Nat Asn Bus Economists; Nat Acad Pub Admin. *Res:* Author of 11 books and 7 articles. *Mailing Add:* Brookings Inst 1775 Massachusetts Ave NW Washington DC 20036

SCHULTZE, HANS-PETER, b Swinemuende, Ger, Aug 13, 37; m 65; c 3. VERTEBRATE PALEONTOLOGY. *Educ:* Univ Freiburg, BSc, 58; Univ Tuebingen, MSc, 62, PhD(paleont), 65. *Prof Exp:* Fel, Ger Sci Found, Naturhistoriska Rikmuseet, Stockholm, 65-67; asst prof, Dept Paleont, Univ

Goettingen, 67-70; fel, Ger Acad Exchange, Am Mus Natural Hist, NY & Field Mus, Chicago, 70-71; from asst prof to assoc prof, Univ Goettingen, Ger, 71-78; from asst prof to assoc prof, 78-87, PROF DEPT SYST & ECOL, UNIV KANS, 87-, CUR, MUS NATURAL HIST, 78- *Concurrent Pos:* Chmn, dept syst & ecol, Univ Kans, 88-90. *Mem:* Paleont Soc, Ger; Soc Vertebrate Paleont; Sigma Xi; Paleont Asn Eng; Soc Syst Biol. *Res:* Morphology and evolution of fossil fishes and early tetrapods; histology of hard tissue; paleoenvironment of Paleozoic Lagerstätten. *Mailing Add:* Dept Syst & Ecol Univ Kans Lawrence KS 66045

SCHULTZE, LOTHAR WALTER, b Berlin, Ger, Dec 5, 20; US citizen; m 47; c 3. SCIENCE EDUCATION. *Educ:* State Univ NY Albany, BA, 42; Pa State Univ, MS, 52, DEd(higher educ), 55. *Prof Exp:* Proj engr, US Rubber Co, 42-45; assoc prof sci, State Univ NY Albany, 52-58, dir admis, 58-66; DIR INSTNL RES, STATE UNIV NY COL FREDONIA, 66-, EMER FAC. *Mem:* Am Chem Soc; Asn Inst Res. *Res:* Science education for non-science majors; non-academic predictors of college success; mobility of students in transfer. *Mailing Add:* State Univ NY Fredonia NY 14063

SCHULZ, ARTHUR R, b Brighton, Colo, Oct 9, 25; m 56. BIOCHEMISTRY, NUTRITION. *Educ:* Colo State Univ, BS, 50; Univ Calif, PhD, 56. *Prof Exp:* Res assoc biochem, Univ Minn, 57-58; asst prof, Okla State Univ, 58-61; asst prof chem, Colo State Univ, 61-62; biochemist, Vet Admin Hosp, 62-69; assoc prof, 69-80, PROF BIOCHEM, IND UNIV, 80- *Concurrent Pos:* NSF fel biochem, Swiss Fed Inst Technol, 56-57. *Res:* Steady-state enzyme kinetics, simulation of energy metabolism in the whole animal, biochemistry of exercise; metabolic control. *Mailing Add:* Dept Biochem Ind Univ Sch Med Indianapolis IN 46223

SCHULZ, CHARLES EMIL, b Blue Island, Ill. MOLECULAR BIOPHYSICS. *Educ:* Knox Col, BA, 72; Univ Ill, MS, 73, PhD(physics), 79. *Prof Exp:* Postdoctoral res assoc molecular biophys & biochem, Yale Univ, 79-81; ASSOC PROF PHYSICS, KNOX COL, 81- *Concurrent Pos:* Vis prof, Univ Iowa, 85 & Univ Ill, 90-91; prin investr, NIH grant, 87-90. *Mem:* Am Phys Soc; Am Asn Physics Teachers; Sigma Xi; Biophys Soc. *Res:* Spectroscopic studies of metalloproteins. *Mailing Add:* Dept Physics Knox Col Box K-74 Galesburg IL 61401

SCHULZ, DALE METHERD, b Fairfield, Ohio, Oct 20, 18; m 47; c 2. PATHOLOGY. *Educ:* Miami Univ, BA, 40; Wash Univ, MS, 42, MD, 49. *Prof Exp:* Res chemist, Tretolite Co, 42-45; from intern to resident path, Barnes Hosp, St Louis, Mo, 49-51; from instr to prof path, Sch Med, Ind Univ, Indianapolis, 52-85; RETIRED. *Concurrent Pos:* Fel, Med Ctr, Ind Univ, Indianapolis, 51-52; pathologist, Methodist Hosp, Indianapolis, 66-85. *Mem:* Am Asn Path & Bact; Int Acad Path. *Res:* Trace metals; fungus diseases; kidney diseases. *Mailing Add:* 9540 Copley Dr Indianapolis IN 46260

SCHULZ, DAVID ARTHUR, b Cleveland, Ohio, June 30, 34; m 57; c 5. MATERIALS SCIENCE, CERAMIC ENGINEERING. *Educ:* Ga Inst Technol, BCerE, 55; Univ Calif, Berkeley, MS, 57, PhD(eng sci), 61. *Prof Exp:* Develop engr, Niagara Develop Lab, Nat Carbon Co, Union Carbide Corp, 55-56; res engr, Inst Eng Res, Univ Calif, Berkeley, 59-60, engr, Inorg Mat Div, Lawrence Radiation Lab, 60-61; develop engr, Adv Mat Lab, Nat Carbon Co, 61-63, develop engr, Nuclear Prod Dept, Carbon Prod Div, 63-66, proj engr, Lawrenceburg Tech Opers, Tenn, 66-67, proj coordr, 67; proj engr, Union Carbide Corp, 68-73, staff engr, 73-76, sr engr, 76-79, sr res scientist, 79-81, sr group leader, Parma Tech Ctr, 81-86; sr res scientist, 86-90, SR RES ASSOC, AMOCO PERFORMANCE PROD, INC, ALPHARETTA, GA, 90- *Mem:* AAAS; Am Chem Soc; fel Am Ceramic Soc; Nat Inst Ceramic Engrs; Sigma Xi. *Res:* Process and product development relating to high-strength, high-modulus carbon fibers and carbon fiber reinforced composites. *Mailing Add:* 6429 Paradise Point Rd Flowery Branch GA 30542-3142

SCHULZ, DONALD NORMAN, b Buffalo, NY, May 24, 43; m 67; c 1. ORGANIC CHEMISTRY, POLYMER CHEMISTRY. *Educ:* State Univ NY Buffalo, BA, 65; Univ Mass, PhD(org chem), 71. *Prof Exp:* Res scientist, Cent Res Lab, Firestone Tire & Rubber Co, 71-75, group leader, org-polymer chem, 75-81; group head & res assoc water soluable polymers, 81-84, GROUP HEAD & SR RES ASSOC POLYMER SYNTHESIS, CORP RES, EXXON RES & ENG CO, 84- *Concurrent Pos:* Asst ed, Isotopics, 79, ed, 80-; assoc ed, Rubber Chem Technol, 84- *Mem:* Am Chem Soc; Am Inst Chemists; fel Am Inst Chem. *Res:* Synthetic and mechanistic organoantimony, organophosphorus and organometallic chemistry; anionic and cationic polymerizations; polymer synthesis and modification; polymer characterization. *Mailing Add:* 58 Balley Crest Rd Annandale NJ 08801

SCHULZ, HELMUT WILHELM, b Berlin, Ger, July 10, 12; US citizen; m 54; c 5. GENERATION OF ELECTRIC POWER FROM URBAN REFUSE & TOXIC WASTE, NOVEL PROCESS FOR ENHANCED OIL RECOVERY. *Educ:* Columbia Univ, BS, 33, ChE, 34, PhD(chem eng), 42. *Prof Exp:* Dir res & develop, Chem & Plastics, Union Carbide, 34-64, managing dir, Union Carbide Europ Res, 67-69; spec asst, Secy Defense, 64-67, US Comnr Educ, 70-71; sr res scientist, Columbia Univ, 72-83; PRES & CHIEF EXEC OFFICER, DYNECOLOGY INC, 74- *Concurrent Pos:* Chmn, Brandenburg Energy Corp, 79-; emer chmn, Global Energy Inc, 89- *Honors & Awards:* Atomic Energy Prize, US Dept Energy, 84. *Mem:* Fel Am Inst Chem Engrs; emer mem Am Chem Soc; NY Acad Sci. *Res:* Centrifugation process for enrichment of uranium isotopes; reinforced solid rocket motors for high accelleration missiles; laser catalysis; waste to energy conversion processes: simplex, bioplex and toxiplex. *Mailing Add:* 611 Harrison Ave Harrison NY 10528

SCHULZ, JAN IVAN, b Bratislava, Czech, Feb 3, 46; Can citizen; c 2. IMMUNOLOGY, INTERNAL MEDICINE. *Educ:* Univ Western Ont, MD, 70; FRCP Can, 74. *Prof Exp:* Res fel immunol, Montreal Gen Hosp Res Inst, 74-77 & Inst de Cancerologie et d'immunogenetique, France, 77-78; ASST PROF IMMUNOL & INTERNAL MED, MCGILL UNIV, 78- *Concurrent Pos:* Affil staff, Dept Immunol, Montreal Gen Hosp, 78-; mem staff, Dept Med, St Mary's Hosp, Montreal, 78-; assoc physician, Dept Med, Royal Victoria Hosp, Montreal, 79- *Mem:* Fel Am Col Physicians; Am Acad Allergy; Can Soc Allergy & Clin Immunol; Royal Soc Med. *Res:* Experimental and clinical immunotherapy of cancer; therapy of atopic diseases. *Mailing Add:* 687 Pine Ave Montreal PQ H3A 1A1 Can

SCHULZ, JEANETTE, b East Alton, Ill, Nov 30, 19. PEDIATRICS. *Educ:* Columbia Univ, AB, 49; Yale Univ, MD, 52; Am Bd Pediat, dipl. *Prof Exp:* From intern to resident pediat, Univ Minn Hosps, 52-55; res instr, Univ Calif, Los Angeles, 56-58, asst prof, 58-63; assoc prof pediat, Univ Ill Col Med, 63-76; Clin dir, Ill State Pediat Inst, 63-76; MED DIR, CHILD DEVELOP SERV, ALASKA, 76- *Concurrent Pos:* Fel pediat hemat, Sch Med, Univ Calif, Los Angeles, 55-56, Bank of Am-Giannini Found res fel, 56-58. *Mem:* Am Soc Hemat; Am Soc Human Genetics. *Res:* Hematology; congenital defects; cytogenetics. *Mailing Add:* 28731 Via del Sol Murrieta CA 92362

SCHULZ, JOHANN CHRISTOPH FRIEDRICH, b Göttingen, Ger, July 6, 20; nat US; m 58; c 2. ORGANIC CHEMISTRY. *Educ:* Mt Union Col, BS, 42; Syracuse Univ, PhD(chem), 59. *Prof Exp:* Asst, Syracuse Univ, 42-48; instr org & phys chem, Hobart Col, 48-50; asst prof, Wagner Col, 50-57 & Washington Col, 57-60; PROF ORG & PHYS CHEM, WAGNER COL, 60- *Mem:* AAAS; fel Am Inst Chemists; Am Chem Soc; NY Acad Sci; Sigma Xi. *Res:* Organic synthesis. *Mailing Add:* 103 W Rainbow Dr Bridgewater VA 22812

SCHULZ, JOHN C, b Peoria, Ill, Dec 4, 55. OPHTHAL. *Educ:* Univ WVa, PhD(pharmacol), 82, MD, 86. *Prof Exp:* Researcher, autonomic pharmacol, 79-82, RESIDENT OPHTHAL, UNIV LOUISVILLE, 87- *Mem:* Am Med Asn; Fedn Am Soc Exp Biol. *Res:* Currently engaged in residency training. *Mailing Add:* Dept Ophthal Univ Nebr Omaha NE 68198

SCHULZ, JOHN HAMPSHIRE, b New York, NY, Apr 10, 34; m 63; c 3. PAPER CHEMISTRY. *Educ:* Brooklyn Polytech Inst, BChemE, 55; Lawrence Univ, MS, 57, PhD(physics), 61. *Prof Exp:* Asst prof paper technol, Western Mich Univ, 61-63; mgr process develop, Paper & Bd Div, Continental Can Co, Ga, 63-69, tech dir, Hodge, 69-72, gen supt, 72-75; mgr res & develop, 75-80, TECH DIR, KRAFT DIV, ST REGIS PAPER CO, 80- *Mem:* Am Chem Soc; Tech Asn Pulp & Paper Indust. *Res:* Reaction of paper to stress; paper mill quality control; clay coating of paper; paper machine performance analysis. *Mailing Add:* 47 Demarest Mill Rd West Nyack NY 10994

SCHULZ, JOHN THEODORE, b Ames, Iowa, June 15, 29; m 53; c 5. ENTOMOLOGY. *Educ:* Iowa State Univ, PhD(entom), 57. *Prof Exp:* From asst prof to assoc prof entom, 57-67, actg chmn dept, 73-74, PROF ENTOM, NDAK STATE UNIV, 67-, CHMN DEPT, 74- *Mem:* AAAS; Entom Soc Am; Phytopath Soc; Sunflower Asn Am. *Res:* Insect transmission of plant diseases; economic entomology. *Mailing Add:* Dept Entom Box 5346 NDak State Univ Fargo ND 58105

SCHULZ, KARLO FRANCIS, physical chemistry; deceased, see previous edition for last biography

SCHULZ, LESLIE OLMSTEAD, b Milwaukee, Wis. NUTRITION, BIOCHEMISTRY. *Educ:* Univ NDak, BA, 74; NDak State Univ, MS, 77; Cornell Univ, PhD(nutrit biochem), 83. *Prof Exp:* Asst prof, 83-88, ASSOC PROF HEALTH SCI, UNIV WIS, MILWAUKEE, 88- *Concurrent Pos:* Mem, Nat Coun Against Health Fraud. *Mem:* Am Diabetes Asn; Am Dietetic Asn; Bangladesh Biochem Soc; Sigma Xi. *Res:* Metabolic regulation; hepatic glucose six phosphatase system and lipid metabolism in brown fat tissue; sports nutrition. *Mailing Add:* Dept Health Sci PO Box 413 Univ Wis Milwaukee WI 53201

SCHULZ, MICHAEL, b Petoskey, Mich, July 14, 43. SPACE PHYSICS, PLASMA PHYSICS. *Educ:* Mich State Univ, BS, 64; Mass Inst Technol, PhD(physics), 67. *Prof Exp:* Mem tech staff, Bell Tel Labs, 67-69; MEM TECH STAFF, AEROSPACE CORP, 69-, SR SCIENTIST, 82- *Concurrent Pos:* Secy, Magnetospheric Physics, Am Geophysics Union, 80-84; assoc ed, J Geophys Res, 76-78; ed, space res books, Am Geophys Union, 87- *Mem:* Am Phys Soc; Am Geophys Union. *Res:* Dynamics of partially ionized gases; theoretical plasma physics; adiabatic theory of charged particle motion; magnetospheric and radiation belt physics; solar wind; solar-terrestrial relationships. *Mailing Add:* Aerospace Corp Space Sci Lab PO Box 92957 Los Angeles CA 90009-2957

SCHULZ, NORMAN F, chemical & metallurgical engineering, for more information see previous edition

SCHULZ, RICHARD BURKART, b Philadelphia, Pa, May 21, 20; m 38; c 1. ELECTROMAGNETIC COMPATIBILITY. *Educ:* Univ Pa, BSEE, 42, MSEE, 51. *Prof Exp:* Res assoc, Univ Pa, 42-45; owner, Electro-Search, 47-55; prog develop coordr, Armour Res Found, 55-61; chief electro-interference, United Control Corp, 61-62; chief electro-compatibility, Boeing Corp, 62-70; staff eng, Southwest Res Inst, 70-74; scientific adv, ITT Res Inst, 74-8374; mgr electromagnetic compatibility, Xerox Corp, 83-87; ELECTROMAGNETIC COMPATIBILITY CONSULT, 87- *Concurrent Pos:* Comt electromagnetic compatibility, Soc Aeronaut Eng. *Honors & Awards:* L G Gumming Award, 80; Centennial Medal, Inst Elec & Electronic Engrs, R R S Stoddard Award, 88. *Mem:* Fel Inst Elec & Electronics Engrs Electromagnetic Soc (treas, 67, pres, 68); Soc Aeronaut Eng. *Res:* Electromagnetic shielding; shielding enclosures. *Mailing Add:* 2030 Cologne Dr Carrollton TX 75007

SCHULZ, ROBERT J, b Brooklyn, NY, Jan 12, 27; m 51; c 3. MEDICAL PHYSICS. *Educ:* Queens Col, NY, BS, 50; Cornell Univ, MS, 57; NY Univ, PhD, 67. *Prof Exp:* Asst physicist radiol physics, Mem Hosp, Sloan Kettering Inst, 52-56; asst prof, Albert Einstein Col Med, Yeshiva Univ, 56-70; PROF RADIOL PHYSICS, YALE UNIV, 70- *Concurrent Pos:* Attend physicist,

Montefiore Hosp, 57-68; chief physicist, Yale-New Haven Hosp. *Mem:* Asn Physicists Med; Am Col Radiol. *Res:* Applications of x-rays and radioactive materials to medical diagnostic and therapeutic problems. *Mailing Add:* Dept Therapeut Radiol 333 Cedar St Yale Univ New Haven CT 06510

SCHULZ, WALLACE WENDELL, b Basil Mills, Nebr, Feb 24, 26; m 47; c 2. INORGANIC CHEMISTRY. *Educ:* Univ Nev, BS, 49, MS, 50. *Prof Exp:* From res chemist to sr scientist, Hanford Atomic Prod Oper, Gen Elec Co, 50-65; sr res scientist, Battelle Northwest Labs, 65-69; staff chemist, Atlantic Richfield Hanford Co, 69-77, prin chemist, 77-80, CHIEF SCIENTIST, ROCKWELL HANFORD CO, 80- *Mem:* Am Chem Soc; Am Inst Mining, Metall & Petrol Engrs; Am Nuclear Soc; Sigma Xi. *Res:* Solvent extraction chemistry; uranium-plutonium separations processes; fission product separations processes; electrochemistry; nuclear waste management. *Mailing Add:* 727 Sweetleaf Dr Wilmington DE 19808

SCHULZ, WILLIAM, b Lakefield, Minn, Oct 14, 35; m 73; c 3. CHROMATOGRAPHY, MASS SPECTROMETRY. *Educ:* Mankato State Col, BS, 61, MS, 65; La State Univ, PhD(anal org chem), 75. *Prof Exp:* Instr chem, Wis State Univ-La Crosse, 65-68; asst prof, 68-76, assoc prof, 76-, PROF CHEM EASTERN KY UNIV. *Concurrent Pos:* Summer fac fel & consult, USAF, 86-90; vis prof chem, Univ Louisville, 84-85 & 89-90; vis assoc prof chem, Colo Sch Mines, 81-82. *Mem:* Am Chem Soc; Asn Off Anal Chemists. *Res:* Mechanism and products of jet fuel oxidation; products of flaming combustion and exhaust; trace environmental organic analysis. *Mailing Add:* Dept Chem Eastern Ky Univ Richmond KY 40475

SCHULZE, GENE EDWARD, b Louisville, Ky, May 8, 59. NEUROTOXICOLOGY, NEUROPHARMACOLOGY. *Educ:* Univ Ky, BS, 81, MS, 82, PhD(toxicol), 85; Am Bd Toxicol, cert & dipl, 89. *Prof Exp:* Res assoc toxicol, Tex A&M Univ, 85-86; vis scientist toxicol, Nat Ctr Toxicol Res, 86-88; staff scientist toxicol, Hazleton Labs Am Inc, 88-91, SR STAFF SCIENTIST, HAZLETON-WASH INC, 91- *Concurrent Pos:* Adj instr, dept pharmacol, Univ Ark Med Sci, 87-; adj asst prof, dept biol, Univ Ark, Pine Bluff, 88- *Honors & Awards:* First Ann Hazleton-Washington Award, 90. *Mem:* Soc Neurosci; Behav Toxicol Soc; Behav Pharmacol Soc; AAAS; NY Acad Sci. *Res:* Behavioral toxicity of drugs and pesticides; interaction between central nervous system and the immune system; the effect of drugs and chemicals on cognitive function using primate models of cognition. *Mailing Add:* 9200 Leesburg Turnpike Vienna VA 22182

SCHULZE, IRENE THERESA, b Washington, Mo, Feb 8, 29. VIROLOGY, BIOCHEMISTRY. *Educ:* St Louis Univ, BS, 52, PhD(microbiol), 62. *Prof Exp:* Res asst enzym & biochem, Pub Health Res Inst City New York, Inc, 65-68, assoc virol, 68-70; asst prof, 70-73, assoc prof, 73-80, PROF MICROBIOL, SCH MED, ST LOUIS UNIV, 80- *Concurrent Pos:* Fel microbiol, Vanderbilt Univ, 62-65; USPHS res grant, Sch Med, St Louis Univ, 71- *Mem:* AAAS; Am Soc Microbiol; Am Soc Biol Chemists. *Res:* Structure and chemical composition of large RNA-containing viruses, for example, influenza and oncogenic viruses; relationships between viral structure and biological activities; viral synthesis and virus-host relationships. *Mailing Add:* 7541 Westmoreland St Louis MO 63105

SCHULZE, KARL LUDWIG, b Aachen, Ger, Feb 28, 11; nat US; c 3. ENVIRONMENTAL ENGINEERING. *Educ:* Univ Berlin, Dr rer nat(bot), 39. *Prof Exp:* Fel, Kaiser-Wilhelm Inst Biol, Ger, 34-38; fel, Univ Wurzburg, 39, Univ Gottingen, 39-40 & Univ Berlin, 40-41; res dir yeast prod, paper & pulp indust, 41-45; asst prof, Munich Inst Technol, 46-49, pvt-dozent, 49-51; res dir waste treatment, Int Agr Prod, Inc, 51-55; res instr sanit eng, 55-57, from asst prof to assoc prof, 57-73, EMER ASSOC PROF SANIT ENG, MICH STATE UNIV, 73- *Concurrent Pos:* Consult, Fairfield Eng Co, Marion, Ohio. *Mem:* Am Inst Biol Sci; Water Pollution Control Fedn; Soc Indust Microbiol; Inst Advan Sanit Res Int; NY Acad Scis. *Res:* Microbiology of waste treatment; biological recovery of waste water; biotechnology; water pollution control. *Mailing Add:* 403 Kumquat Fairhope AL 36532

SCHULZE, WALTER ARTHUR, b Philadelphia, Pa, Dec 8, 43; m 71; c 1. MATERIALS SCIENCE, FERROELECTRICS. *Educ:* Pa State Univ, BS, 65, MS, 68, PhD(solid state sci), 73. *Prof Exp:* Res assoc, 74-80, SR RES ASSOC, MAT RES LAB, PA STATE UNIV, 80-; AT NY STATE COL CERAMICS, ALFRED UNIV. *Mem:* Am Ceramic Soc; Inst Elec & Electronics Engrs. *Res:* Preparation and characterization of ferroelectric materials and devices; electrical properties of piezoelectric and high dielectric constant ceramics. *Mailing Add:* NY State Col Ceramics Alfred Univ Main St Alfred NY 14802

SCHULZE, WILLIAM EUGENE, b Kiowa, Kans, Nov 27, 34. SYNTHETIC ORGANIC CHEMISTRY. *Educ:* Knox Col, Ill, AB, 56; Kans State Univ, MS, 62. *Prof Exp:* Chemist, Northern Lab, USDA, Ill, 56-57 & 59-60; Nat Defense Educ Act fel, Kans State Univ, 61-64; synthesis org chemist, Calif, 64-79, ORG SURFACTANT SYNTHESIS CHEMIST, EMERY INDUSTS, INC, SC, 79- *Mem:* AAAS; Am Chem Soc; Am Asn Cereal Chemists; Inst Food Technologists. *Res:* Cereal chemistry; products prepared from cereal grains and the modification of these products by processing methods, addition of surfactants and/or formula modification; organic surfactant synthesis; organic surface active materials. *Mailing Add:* PO Box 8263 West Chester OH 45069-8263

SCHUMACHER, BERTHOLD WALTER, b Karlsruhe, WGer, Apr 15, 21; US citizen; m 55; c 2. ENGINEERING-PHYSICS, HIGH POWER ELECTRON BEAM TECHNOLOGY. *Educ:* Univ Stuttgart, dipl, 50, Dr rer nat, 53. *Prof Exp:* Design engr, Electronics Indust, WGer, 53-54; res fel, Ont Res Found, Toronto, Can, 54-58, dir dept physics, 58-66; mgr electron beam technol, Westinghouse Res Labs, 66-77; prin engr res, Ford Motor Co, 77-87; CONSULT, 87- *Mem:* Am Phys Soc; Phys Soc Eng; Am Welding Soc; Ger Physics Soc. *Res:* Vacuum and electron beam physics and technology; electron beam attenuation, fluorescence and single-scatter probes for measuring gas parameters; atmospheric electron probe for x-ray analysis of matter; high power electron guns with beam transfer to the atmosphere for welding, cutting and upface hardening with electron beam and workpiece in air; theory of energy transport by electron, ion and laser beams and their interaction with matter; closed-loop control for welding processes with high energy beams, as well as for resistance welding. *Mailing Add:* 24635 Winona Dearborn MI 48124

SCHUMACHER, CLIFFORD RODNEY, b Waukegan, Ill, Dec 19, 29; m 50, 69; c 3. THEORETICAL PHYSICS, ELECTROMAGNETIC WAVE PHYSICS. *Educ:* Wayne State Univ, BS, 51; Cornell Univ, PhD(theoret physics), 62; Haskell Indian Jr Col, AA, 75. *Prof Exp:* Physicist, Air Develop Ctr, Wright-Patterson AFB, 51-54; mem tech staff, Bell Tel Labs, 54-57; mem, Inst Adv Study, 61-62; fels, NSF, Princeton Univ, 62-63 & Enrico Fermi, Chicago, 63-65; assoc prof physics, Pa State Univ, 65-70; vis assoc prof physics & vis fel, Lab Nuclear Studies, Cornell Univ, 70-72; adj prof physics & astron, Univ Kans, 73-81; PHYSICIST, DAVID TAYLOR NAVAL SHIP RES & DEVELOP CTR, 81- *Concurrent Pos:* Res assoc, Lab Nuclear Studies, Cornell Univ, 62 & 63; vis asst physicist, Brookhaven Nat Lab, 63; vis physicist, Argonne Nat Lab, 70; prof math, Sci Dept, Haskell Indian Jr Col, 72-81; vpres gov bd, Lawrence Indian Ctr, Inc, 73-75, pres 75-77 & treas, 77-79. *Mem:* Am Phys Soc; Inst Elec & Electronics Engrs; Optical Soc Am; Am Indian Sci & Eng Soc. *Res:* Theory of structure and interaction of elementary particles; analysis of high energy phenomena; microwave generators and electronics; electromagnetic interactions of the deuteron; meson resonances; American Indian traditions and science education; electromagnetic wave physics; radar research, electrodynamic similitude, and scale models. *Mailing Add:* Code 2840 Nonmetallics Div David Taylor Res Ctr Annapolis MD 21402-5067

SCHUMACHER, DIETMAR, b Yugoslavia, May 28, 42; US citizen; m 70. PETROLEUM GEOLOGY, PALEONTOLOGY. *Educ:* Univ Wis, Madison, BS, 64, MS, 67; Univ Mo, Columbia, PhD(geol), 72. *Prof Exp:* Asst prof geosci, Univ Ariz, 70-77; res geologist, 77-79, res supvr, 79-81, sr geol specialist, Phillips Petrol Co, 81-82; res assoc, 82-84, res mgr, 84-86, SR EXPLORATIONIST, PENNZOIL CO, 87- *Mem:* AAAS; Geol Soc Am; Paleont Soc; Am Asn Petrol Geologists; Sigma Xi. *Res:* Petroleum geology and geochemistry; micropaleontology and biostratigraphy; stratigraphy and sedimentation. *Mailing Add:* 622 Bendwood Dr Houston TX 77024

SCHUMACHER, GEBHARD FRIEDERICH B, b Osnabruck, WGer, June 13, 24; m 58; c 2. OBSTETRICS & GYNECOLOGY, IMMUNOLOGY. *Educ:* Univ Gottingen, MD, 51. *Prof Exp:* Intern gen med, Med Sch, Univ Tubingen, 51-52; resident in biochem, Max Planck Inst Biochem, 52-53 & Max Planck Inst Virus Res, 53-54; resident obstet & gynec, Med Sch, Univ Tubingen, 54-59, sci asst, 59-62, docent, 62 & 64-65; res assoc immunol, Inst Tuberc Res, Col Med, Univ Ill, 62-63; res assoc & asst prof obstet & gynec, Univ Chicago, 63-64; assoc prof obstet & gynec & asst prof biochem, Albany Med Col, 65-67; res scientist, Div Labs & Res, NY State Dept Health, 65-67; assoc prof, 67-73, PROF OBSTET & GYNEC, PRITZKER SCH MED, UNIV CHICAGO, 73-, CHIEF SECT REPRODUCTIVE BIOL, 71- *Concurrent Pos:* Ger Sci Found grant, 55-62; NIH grants; Ford Found funds; app fac mem, Div Comt on Immunol, 72-; consult & task force mem, World Health Orgn, Human Reprod Unit, 72-77; mem med adv bd, Int Fertil Res Prog, Triangle Park, NC, 77-; reviewer & ad hoc consult, Nat Inst Child Health & Human Develop, Bethesda, Md, 73-; prof, Div Comt Immunol, Univ Chicago Pritzker Sch Med, 74; prof, Biol Sci Col Div, Univ, Chicago, 82- *Mem:* Ger Soc Biol Chemists; Am Asn Pathologists; Soc Study Reprod; Am Soc Andrology; Europe Soc Immunol; fel Am Col Obstet & Gynec. *Res:* Biology of human reproduction; birth control; infertility; endocrinology-protein metabolism; serum proteins; immunology; inflammation and nonspecific resistance; trauma. *Mailing Add:* Dept Obstet & Gynec Sch Med Sect Reprod Biol Univ Chicago 5841 Maryland Ave Chicago IL 60637

SCHUMACHER, GEORGE ADAM, b Trenton, NJ, Sept 22, 12; m 41; c 4. MEDICINE NEUROLOGY. *Educ:* Pa State Univ, BS, 32; Cornell Univ, MD, 36. *Prof Exp:* From asst to assoc prof clin med neurol, Cornell Med Ctr, New York, 46-50; chmn div neurol, 50-68, prof neurol, 50-78, EMER PROF NEUROL, UNIV VT, 79- *Concurrent Pos:* From asst attend neurologist to dir neurol serv, Bellevue Hosp, New York, 46-50; mem med adv bd, Nat Multiple Sclerosis Soc, 49-, chmn, 64-66; attend neurologist, Med Ctr Hosp, VT, 50-78; consult, Hosps, NY & Vt, 50-78; mem prog proj comt, Nat Inst Neurol Dis & Blindness, 62-64; vis scientist, Arctic Health Res Lab, USPHS, 67-68; vis prof human ecol, Inst Arctic Biol, Univ Alaska, 67-68. *Mem:* Am Neurol Asn; Asn Res Nerv & Ment Dis; fel Am Acad Neurol; Am Med Asn. *Res:* Headache; pain; multiple sclerosis; spinal cord physiology. *Mailing Add:* 59 Bilodeau Ct Burlington VT 05401

SCHUMACHER, GEORGE JOHN, b Lindenwold, NJ, Dec 19, 24; m 49; c 3. PHYCOLOGY. *Educ:* Bucknell Univ, BS, 48; Cornell Univ, MS, 49, PhD(bot, vert zool), 53. *Prof Exp:* Instr, Cornell Univ, 52-53; from instr to assoc prof biol, 53-61, chmn dept, 59-66, PROF BIOL, STATE UNIV NY BINGHAMTON, 61- *Mem:* Am Phycol Soc; Int Phycol Soc. *Res:* Ecology and taxonomy of freshwater algae, populations and eutrophication. *Mailing Add:* Harpur Col State Univ NY Binghamton NY 13901

SCHUMACHER, H RALPH, JR, b Montreal, Que, Feb 14, 33; m 65; c 2. INTERNAL MEDICINE, RHEUMATOLOGY. *Educ:* Ursinus Col, Collegeville, Pa, BS, 55; Univ Pa, MD, 59. *Prof Exp:* Intern, Denver Gen Hosp, 59-60; resident, Wadsworth Vet Admin Hosp, Los Angeles, 60-62, fel rheumatol, 62-63; fel, Robert Brigham Hosp, Boston, 65-67; from asst prof to assoc prof, 67-79, PROF MED, SCH MED, UNIV PA, 79- *Concurrent Pos:* Staff physician & chief, Arthritis Sect, Vet Admin Hosp, 67-, dir, Rheumatol-Immunol Ctr, 77-; actg chief, Arthritis Sect, Sch Med, Univ Pa, 78-80 & 90- *Honors & Awards:* Hench Award, 85; Van Breeman Award, 89. *Mem:* Am Rheumatism Asn; Am Fedn Clin Res; Electron Micros Soc Am. *Res:* Pathogenic studies in the rheumatic diseases using light and electron microscopy, electron probe analysis, tissue culture and immunoelectron microscopy; crystal induced arthritis, rheumatoid arthritis; spondylarthropathies, rehabilitation and arthritis. *Mailing Add:* Arthritis Sect Univ Pa Hosp Philadelphia PA 19104

SCHUMACHER, IGNATIUS, b Munjor, Kans, Apr 1, 28; m 54; c 7. ORGANIC CHEMISTRY, MEDICINAL CHEMISTRY. *Educ:* Univ Kans, BS, 58; Univ Mich, PhD(med chem), 62. *Prof Exp:* Org researcher, 62-64, sr org chemist, 64-73, res specialist, 73-77, sr specialist, 77-82, fel, Monsanto Co, 82-86; PRES, IGNATUS SCHUMACHER CONSULT INC, 86- *Mem:* Am Chem Soc. *Mailing Add:* 15024 Clayron Rd St Louis MO 63011

SCHUMACHER, JOSEPH CHARLES, b Peru, Ill, Sept 15, 11; m 33, 85; c 4. INDUSTRIAL CHEMISTRY. *Educ:* Univ Southern Calif, AB, 46. *Prof Exp:* Res chemist & prod supvr, Carus Chem Co, Ill, 31-40; mem res develop & orgn staff, Fine Chem, Inc, Calif, 40-41, co-founder, vpres & dir res, Western Electrochem Co, 41-54; dir res, Am Potash & Chem Corp, 54-66, vpres rare earth div, 66-68; vpres electrochem, Vanadium & Rare Earth Div, Kerr-McGee Chem Corp, 68-72; pres, 72-77, CHMN OF THE BD, J C SCHUMACHER CO, 77- *Concurrent Pos:* Trustee, Whittier Col, 68-77. *Honors & Awards:* Gold Medal, Electrochem Technol, Electrochem Soc. *Mem:* Am Chem Soc; Electrochem Soc; Sigma Xi. *Res:* Industrial electrochemistry; chlorates, perchlorates and manganese compounds; organic chemistry; photochemicals; boron. *Mailing Add:* 2220 Ave of the Stars Apt 704 R Los Angeles CA 90067

SCHUMACHER, JOSEPH NICHOLAS, b Downers Grove, Ill, May 2, 28; m 50; c 9. ORGANIC CHEMISTRY. *Educ:* St Procopius Col, BS, 50; Ohio State Univ, MSc, 52, PhD(org chem), 54. *Prof Exp:* Asst res found, Ohio State Univ, 50-51, asst univ, 51-54; res chemist, R J Reynolds Tobacco Co, 54-68, group leader, 68-78, sect head, 78-82, master scientist, 82-87; RETIRED. *Mem:* Am Chem Soc. *Res:* Natural products; carbohydrates; chromatography. *Mailing Add:* 212 Cascade Ave Winston-Salem NC 27127

SCHUMACHER, RICHARD WILLIAM, b Chicago, Ill, 46; m 69; c 1. AGRICULTURAL & FOOD CHEMISTRY. *Educ:* Univ Ill, BS, 69; Va Polytech & State Univ, MS, 71; Univ Ky, PhD(plant physiol), 74. *Prof Exp:* Res leader, Agr Res Serv, USDA, 74-76; field res scientist, 76-79, MGR, AGR CHEM RES & DEVELOP, MONSANTO AGR PROD CO, 79- *Mem:* Sigma Xi. *Mailing Add:* 12 Heathercroft Ct Chesterfield MO 63166

SCHUMACHER, ROBERT E, b Heron Lake, Minn, Oct 14, 18; m 45; c 4. AQUATIC BIOLOGY, FISHERIES MANAGEMENT. *Educ:* Univ Minn, BS, 50. *Prof Exp:* Aquatic biologist, Fisheries Res Unit, State Dept Conserv, Minn, 50-57, res biologist, 57-65; dist fisheries mgr, Mont Dept Fish & Game, 65-71, regional fisheries mgr, 71-82; RETIRED. *Mem:* Am Fisheries Soc. *Res:* Fisheries management and research; aquatic ecology; trout population levels and dynamics. *Mailing Add:* 1227 Fifth St W Kalispell MT 59901

SCHUMACHER, ROBERT THORNTON, b Berkeley, Calif, Sept 29, 30; m 54; c 2. MUSICAL ACOUSTICS. *Educ:* Univ Nev, BS, 51; Univ Ill, MS, 53, PhD(physics), 55. *Prof Exp:* Instr physics, Univ Wash, 55-57; from asst prof to assoc prof, 57-66, PROF PHYSICS, CARNEGIE-MELLON UNIV, 66- *Concurrent Pos:* Sloan Found fel, 58-62, NSF sr fel, 65-66. *Mem:* Fel Am Phys Soc; AAAS. *Res:* Musical acoustics; measurements, theory, and computer simulations of oscillations of musical instruments, particularly bowed string instruments. *Mailing Add:* Dept Physics Carnegie-Mellon Univ 5000 Forbes Ave Pittsburgh PA 15213

SCHUMACHER, ROY JOSEPH, b Covington, Ky, Mar 15, 42; m 65; c 2. ORGANIC CHEMISTRY. *Educ:* Xavier Univ, Ohio, BS, 64, MS, 67; Univ Cincinnati, PhD(chem), 71. *Prof Exp:* Chemist, 71-75, res group leader, MC/B Mfg Chemists, 74-78, mkt mgr, 78-81, SR ANALYTICAL CHEMIST, MONSANTO RES CORP, 81- *Mem:* Am Chem Soc. *Res:* Ylid chemistry; organic synthesis and product development. *Mailing Add:* 3518 Grandview Ave Sharonville OH 45241

SCHUMACHER, WILLIAM JOHN, b Jersey City, NJ, Mar 11, 36; m 69; c 2. ENERGY ECONOMICS, ENERGY SUPPLY PLANNING. *Educ:* Cornell Univ, BChE, 58, PhD(chem eng), 64. *Prof Exp:* Prof chem eng, Nat Univ Trujillo, 65-70; DIR, ENERGY PRACT, SRI INT, 71- *Mem:* Am Inst Chem Engrs; Am Chem Soc. *Res:* Energy economics and planning. *Mailing Add:* SRI Int Menlo Park CA 94025

SCHUMAKER, JOHN ABRAHAM, b Marshall, Ill, July 24, 25. MATHEMATICS. *Educ:* Univ Ill, BS, 46, AM, 47; NY Univ, PhD(math educ), 59. *Prof Exp:* Instr math, Univ Ill, 46-47; from instr to asst prof, MacMurray Col, 47-53; instr, Grinnell Col, 53-55; from asst prof to assoc prof, Montclair State Col, 55-61; dir, NSF in-serv insts, 62-68, PROF MATH & CHMN DEPT, ROCKFORD COL, 61- *Concurrent Pos:* Vis prof, NSF Inst, Univ Vt, 62-73. *Honors & Awards:* Distinguished Serv Award, Math Asn Am. *Mem:* AAAS; Am Math Soc; Math Asn Am. *Res:* Number theory; history of mathematics; statistics. *Mailing Add:* 911 Woodridge Dr Rockford IL 61108-4012

SCHUMAKER, LARRY L, b Aberdeen, SDak, Nov 5, 39; m 63; c 1. MATHEMATICS. *Educ:* SDak Sch Mines & Technol, BS, 61; Stanford Univ, MS, 62, PhD(math), 66. *Prof Exp:* Lectr computer sci, Stanford Univ, 66; vis asst prof math, Math Res Ctr, Univ Wis-Madison, 66-68; from asst prof to prof math, Univ Tex, 68-79; prof math, Tex A&M Univ, College Station, 80-88; PROF MATH, VANDERBILT UNIV, 88- *Concurrent Pos:* Vis assoc prof, Math Res Ctr, Univ Wis, 73-74; vis prof math, Univ Munich, 74-75 & 89-90, Free Univ Berlin, Hahn Meitner Atomic Energy Inst, 78-79, Univ Sao Paulo, Brazil, 81 & Univ Würzburg, 87. *Honors & Awards:* Humbolt prize, 90. *Mem:* Am Math Soc; Soc Indust & Appl Math; Math Asn Am. *Res:* Classical approximation theory; total positivity; spline functions; investigating theoretical properties of spaces of multivariate splines and their applications to solving data fitting problems and in computer-aided design. *Mailing Add:* Dept Math Vanderbilt Univ Nashville TN 37235

SCHUMAKER, ROBERT LOUIS, b Sapulpa, Okla, July 28, 20; m 50; c 3. PHYSICS, MATHEMATICS. *Educ:* Tex Western Col, BS, 43; Univ Ariz, MS, 54. *Prof Exp:* From instr to asst prof physics & math, 46-61, asst dir, Schellenger Res Labs, 59-70, dir computer ctr, 67-70, dir admis, 70-77, physicist & head data anal sect, Schellenger Res Labs, 56-77, ASSOC PROF PHYSICS & MATH, UNIV TEX, EL PASO, 61- *Mem:* Am Meteorol Soc; Am Asn Physics Teachers; Sigma Xi. *Res:* Spectrography of high energy spark discharges; propagation and characteristics of atmospheric pressure oscillations, particularly ray-tracing techniques. *Mailing Add:* 3124 Aurora St El Paso TX 79930

SCHUMAKER, VERNE NORMAN, b McCloud, Calif, Sept 16, 29; m 51; c 2. PHYSICAL BIOCHEMISTRY. *Educ:* Univ Calif, AB, 52, PhD(biophys), 55. *Prof Exp:* Jr res biophysicist, Virus Lab, Univ Calif, 54-55, assoc res biophysicist, 55-56; Am Cancer Soc fel, Lab Animal Morphol, Brussels, Belg, 56-57; from asst prof to assoc prof biochem, Univ Pa, 57-65; assoc prof chem, 65-66, PROF CHEM, UNIV CALIF, LOS ANGELES, 66- *Concurrent Pos:* John Simon Guggenheim fel, 64-65. *Mem:* AAAS; Biophys Soc; Am Soc Biol Chemists. *Res:* Immunochemistry, structure and function of lipoproteins, chromatin structure; hydrodynamic theory and methodology. *Mailing Add:* Dept Chem Univ Calif Los Angeles CA 90024

SCHUMAN, GERALD E, b Sheridan, Wyo, July 5, 44; m 65; c 2. SOIL-PLANT RELATIONSHIPS, RECLAMATION OF DISTURBED LANDS. *Educ:* Univ Wyo, BS, 66; Univ Nev, Reno, MS, 69; Univ Nebr, PhD(agron), 74. *Prof Exp:* Soil scientist, USDA-Agr Res Serv, Reno, Nev, 66-69, Lincoln, Nebr, 69-75, Cheyenne, Wyo, 75-77, SOIL SCIENTIST & RES LEADER, USDA-AGR RES SERV, CHEYENNE, WYO, 77- *Concurrent Pos:* Adj prof plant soil & insect sci, Univ Wyo, Laramie, 75-; chmn energy resources div, Soil Conserv Soc Am, 79-81; fac affil dept agron, Colo State Univ, Ft Collins, 80-; chmn environ qual div, Am Soc Agron, 84-85. *Mem:* Fel Soil Conserv Soc Am; Soil & Water Conserv Soc; fel Am Soc Agron; Soc Range Mgt; Am Soc Surface Mining Reclamation; Int Soc Soil Sci. *Res:* Evaluation of soil-plant relationships on disturbed rangeland soils and highly erodible marginal soils; long term management of reclaimed lands. *Mailing Add:* USDA-Agr Res Serv High Plains Grasslands Res Sta 8408 Hildreth Rd Cheyenne WY 82009

SCHUMAN, LEONARD MICHAEL, b Cleveland, Ohio, Mar 4, 13; m 54; c 2. EPIDEMIOLOGY. *Educ:* Oberlin Col, AB, 34; Western Reserve Univ, MSc, 39, MD, 40; Am Bd Prev Med, dipl. *Prof Exp:* Chief div venereal dis control, State Dept Pub Health, Ill, 47-49, actg chief div commun dis, 49-50, dep dir in charge div prev med, 50-51, 53-54; epidemiologist, Cold Injury Res Team, US Dept Defense, Korea, 51-53; assoc prof pub health, 54-57, PROF EPIDEMIOL, SCH PUB HEALTH, UNIV MINN, MINNEAPOLIS, 57-, MAYO PROF PUB HEALTH, 83- *Concurrent Pos:* Rockefeller Found fel nutrit, Vanderbilt Univ, 46; vis lectr, Sch Med, Univ Ill, 47-54; lectr, Sch Nursing, Mem Hosp, Springfield, Ill, 47-54, guest lectr, 52-; comn officer, USPHS, 41-47; officer-in-chg, SE Nutrit Res Univ, USPHS, 45-47, consult, Commun Dis Ctr, 55-; chmn, Coun on Res, Am Col of Prev Med, 59-64, Epidemiol Sect, Am Pub Health Asn, 67; consult, Chronic Dis Div, 64-72, Div Radiol Health, 61-69, Nat Cancer Inst, 58-, Minn State Health Dept, 55-, Hennepin County Gen Hosp, 55-, USDA, 61-, Air Pollution Med Prog, 58 & Calif State Health Dept, 62; mem, Nat Adv Comt Gamma Globulin Eval, 52-53; adv comt polio vaccine field trials, Nat Found Infantile Paralysis, 52-55; adv polio vaccine eval ctr, Univ Mich, 53-55; mem, adv nat cancer control comt, NIH, 59-62, accident prev res study sect, 63-66; adv comt, Nat Coop Leukemia Study, 59-63; mem, adv field studies bd, Nat Adv Cancer Coun, 61-62; mem nat adv comt bio-effects radiation, USPHS, 66-69, task force smoking & health, 67-68; mem, Nat Adv Urban & Indust Health Coun, 68-69, Nat Adv Environ Control Coun, 69-71; mem comt health protection & dis prev adv, Secy, Health, Educ & Welfare, 69-71 & cancer res ctr rev comt, Nat Cancer Inst, 71-75; fel coun epidemiol, Am Heart Asn, chmn, Conf Chronic Dis Training Prog Dirs, 60-72, panel biomet & epidemiol, Nat Cancer Inst, 61-62, Nat Conf Res Methodology in Community Health & Prev Med, 62 & Surgeon Gen Adv Comt Smoking & Health, 62-64. task group on smoking, Nat Heart, Lung & Blood Inst, 78-79; task group on interstitial lung dis, Nat Heart, Lung & Blood Inst/NIH, 78-79; steering comt, Nutrit Res, Agency for Int Develop, US Dept of State, 78-82; policy bd, Mid-Career Med Fels, Bush Found, 78-84; mem, Prev Ctrs Grant Rev Comt, Ctr Dis Control, 86-, vchmn, Health & Environ Network Gov Bd, Freshwater Soc, 87-, bd dir, Smoke Free Generation, 86- *Mem:* Am Epidemiol Soc (vpres, 78); Am Thoracic Soc; Am Venereal Dis Asn; Asn Teachers Prev Med; NY Acad Sci; Am Asn Pub Health Physicians; Am Col Prev Med; Asn Military Surgeons US; Int Epidemiol Soc; Soc Epidemiol Res. *Res:* Epidemiology of communicable disease, non-communicable disease and cancer. *Mailing Add:* Epidemiol Div Sch Pub Health Univ Minn Box 1971 Mayo 420 Delaware St SE Minneapolis MN 55455

SCHUMAN, ROBERT PAUL, b Milwaukee, Wis, May 17, 19; m 49; c 3. NUCLEAR CHEMISTRY, PHYSICAL CHEMISTRY. *Educ:* Univ Denver, BS, 41; Ohio State Univ, MS, 44, PhD(chem), 46. *Prof Exp:* Chemist metall lab, Univ Chicago, 44-45, Knolls Atomic Power Lab, Gen Elec Co, 47-57, Atomic Energy Div, Phillips Petrol Co, 57-66 & Idaho Nuclear Corp, 66-69; assoc prof chem, Robert Col, Istanbul, 69-71 & Bogazici Univ, Turkey, 71-77; vis prof, Nuclear Eng, Iowa State Univ, 77-78; sr phys chemist, Idaho Nat Eng Lab, Allied Chem Corp, 78-79, Exxon Nuclear, 79-80 & EG&G Idaho, Inc, 80-84; RETIRED. *Concurrent Pos:* Fulbright grant, Sri Venkateswara Univ, India, 65-66. *Mem:* Am Nuclear Soc; Am Chem Soc; Am Phys Soc; Sigma Xi. *Mailing Add:* 751 Masters Dr Idaho Falls ID 83401

SCHUMAN, STANLEY HAROLD, b St Louis, Mo, Dec 29, 25; m 52; c 8. EPIDEMIOLOGY, PUBLIC HEALTH. *Educ:* Wash Univ, MD, 48; Univ Mich, MPH, 60, DrPH, 62; Am Bd Pediat, dipl, 60. *Prof Exp:* Clin instr pediat, Sch Med, Wash Univ, 54-59; from asst prof to prof epidemiol, Sch Pub Health, Univ Mich, Ann Arbor, 62-73; PROF EPIDEMIOL IN FAMILY PRACT COL MED, MED UNIV SC, 74-, PROF PEDIAT, 76-; MED DIR, AGROMED PROG, CLEMSON MED UNIV SC, 84- *Concurrent Pos:* Proj dir, SC Pesticide Study Ctr, Environ Protection Agency, 81-84; author, text Epidemiol, 86. *Mem:* Am Pub Health Asn; Am Acad Family Pract; Am Epidemiol Soc; Soc Epidemiol Res; Coun Agr Sci Technol; Sigma Xi. *Res:* Epidemiology of heat waves in United States' cities; human sweat studies in population survey; screening for cystic fibrosis; population surveys of injuries due to accidents; accident prevention; field trials with young drivers; epidemiology in family practice; computers in medicine; toxicology; cancer of the esophagus agricultural medicine. *Mailing Add:* Dept Family Pract Col Med Univ SC Charleston SC 29401

SCHUMAN, WILLIAM JOHN, JR, b Baltimore, Md, Jan 23, 30; m 52; c 3. MECHANICS. *Educ:* Univ Md, BS, 52; Pa State Univ, MS, 54, PhD(eng mech), 65. *Prof Exp:* Asst prof mech, US Air Force Inst Tech, 54-57; sr engr, Martin Co, Martin Marietta Corp, 57; aero res engr, Ballistic Res Lab, US Army, 57-62, res physicist, 62-83, physicist, Harry Diamond Labs, 83-85; sr res scientist, 85-89, VPRES, SI, DIV SPECTRUM 39, 89- *Mem:* Am Soc Mech Engrs; Am Acad Mech; Soc Advan Mat Processes. *Res:* Dynamic response of structures, including determination of high explosive blast parameters and details of loading; primary structures; cylindrical and conical shells; design of blast/thermal hardened electronic shelters. *Mailing Add:* SI Div Spectrum 39 PO Box 10970 Baltimore MD 21234-0970

SCHUMANN, EDWARD LEWIS, b Indianapolis, Ind, Aug 1, 23; m 47; c 3. ORGANIC CHEMISTRY, MEDICINAL CHEMISTRY. *Educ:* Ind Univ, BS, 43; Univ Mich, MS, 49, PhD(pharmaceut chem), 50. *Prof Exp:* Chemist, Linde Air Prods Co, 43-44; res chemist, Wm S Merrell Co, 49-57; res assoc, prod res & develop, Upjohn Co, 57-65, mgr clin drug regulatory affairs, 65-68, dir drug regulatory affairs, 68-85; RETIRED. *Concurrent Pos:* Consult, drug regulatory affairs, 85- *Mem:* AAAS; Am Fedn Clin Res; Am Col Clin Pharmacol & Chemother; Am Chem Soc; Sigma Xi. *Res:* Antispasmodics; histamine antagonists; central nervous system agents; cardiac-cardiovascular drugs; enzyme inhibitors; hormone antagonists; local anesthetics. *Mailing Add:* 809 Dukeshire Ave Kalamazoo MI 49008

SCHUMANN, THOMAS GERALD, b Los Angeles, Calif, Mar 15, 37. ELEMENTARY PARTICLE PHYSICS. *Educ:* Calif Inst Technol, BS, 58; Univ Calif, Berkeley, MA, 60, PhD(physics), 65. *Prof Exp:* Res assoc physics, Brookhaven Nat Lab, 65-67; asst prof, City Col New York, 67-71; lectr, 71-74, from asst prof to assoc prof, 74-85, PROF PHYSICS, CALIF POLYTECH STATE UNIV, SAN LUIS OBISPO, 85- *Mem:* Am Phys Soc. *Res:* Experimental high energy physics using bubble chambers. *Mailing Add:* Dept Physics Calif Polytech State Univ San Luis Obispo CA 93407

SCHUMER, DOUGLAS B, b Passaic, NJ, Mar 22, 51; m 76; c 2. ACOUSTO-OPTICS, COMMUNICATIONS. *Educ:* Carnegie-Mellon Univ, BS, 73, MS, 74; Rensselaer Polytech Inst, PhD(electroph & elec & syst eng), 77. *Prof Exp:* Mem tech staff, Bell Tel Labs, 77-80; mgr res, 80-83, dir res, 83-84, dir, 84-88, VPRES ENG & RES, OHAUS SCALE CORP, 88- *Mem:* Am Phys Soc; Optical Soc Am; Inst Physics Eng; Inst Elec & Electronics Engrs; NY Acad Sci; Soc Exp Mech. *Res:* Acousto-optics and surface acoustic waves for signal processing; data compression for graphics transmission; force-frequency quartz resonator sensors. *Mailing Add:* Ohaus Scale Corp 29 Hanover Rd Florham Park NJ 07932

SCHUMER, WILLIAM, b Chicago, Ill, June 29, 26; m 85; c 2. SURGERY, BIOCHEMISTRY. *Educ:* Ill Inst Technol, 44-45; Chicago Med Sch, MB, 49, MD, 50; Univ Ill, MS, 66; Am Bd Surg, dipl, 54. *Prof Exp:* Asst prof surg, Chicago Med Sch, 59-65, asst prof cardiovasc res, 64-65; dir, surg serv, Univ Calif, Davis, 65-67; from assoc prof to prof surg, Univ Ill, Vet Admin West Side Hosp, 67-75, chief surg, 67-75; PROF & CHMN, DEPT SURG, UNIV HEALTH SCI, CHICAGO MED SCH, 75-, DISTINGUISHED PROF BIOCHEM, 90- *Concurrent Pos:* Mem attend staff, Mt Sinai Hosp, Chicago, Ill, 62-65, asst chief surg, 63-64, chief dept surg, 90; mem attend staff, Vet Admin Hines Hosp, 60-63 & Vet Admin West Side Hosp, 63-64; dir surg serv, Sacramento County Med Ctr & Univ Calif, Davis, 65-67; mem, Am Bd Surg; mem med staff, Highland Park Hosp, Naval Reg Med Ctr, Great Lakes, 76- & St Mary's Nazareth Hosp Ctr, Chicago, 78; Morris L Parker Res Award, 76; chief surg serv, Va Med Ctr, NChicago, 75-79, chief gen surg sect, 79-82, attend, 82-88 & consult, 88- *Mem:* Am Col Surg; Am Physiol Soc; Shock Soc (pres, 78); Cent Surg Asn; Cent Soc Clin Res; Sigma Xi; Am Surg Asn; AAAS; Am Asn Univ Professors; AMA; Am Soc Contemporary Med & Surg; Am Soc Microbiol; Am Soc Biol Chemists; Am Trauma Soc; Asn Surg Educ; Asn Am Med Cols; Cent Soc Clin Res; Fedn Am Scientists; Fedn Am Socs Exp Biol; Intl Fedn Surg Cols. *Res:* Effect of trauma or shock on cell metabolism; correlation of cell biochemistry, physiology genetics and anatomy in humans and animals in shock. *Mailing Add:* Dept Surg Univ Health Sci Chicago Med Sch 3333 Green Bay Rd North Chicago IL 60064

SCHUMM, BROOKE, JR, b Shanghai, China, Oct 8, 31; US citizen; m 55; c 4. ELECTROCHEMICAL & CHEMICAL ENGINEERING. *Educ:* Rensselaer Polytech Inst, BS, 53; Univ Rochester, MS, 62, PhD(chem eng), 66. *Prof Exp:* Team leader chem warfare, US Army, 53-55; indust methods engr film processing, Eastman Kodak Co, 55-57, develop engr coatings, 57-58; fel chem engr, Univ Rochester, 58-62; sr technol assoc batteries, Union Carbide Corp, 62-89; PRES, EAGLE-CLIFFS, INC, 89- *Concurrent Pos:* Union Carbide fel, Univ Rochester, 60-62, sr tech assoc. *Mem:* Electrochem Soc; Am Inst Chemists. *Res:* Industrial coatings, fuel cell batteries, primary electrochemical cells and energy conversion. *Mailing Add:* Eagle Cliffs Inc 31220 Lake Rd Bay Village OH 44140

SCHUMM, DOROTHY ELAINE, b Dayton, Ohio, Aug 4, 43; m 87. BIOCHEMISTRY. *Educ:* Earlham Col, BA, 65; Univ Chicago, MS, 66, PhD(biochem), 69. *Prof Exp:* Res assoc biochem, Ben May Lab Cancer Res, Univ Chicago, 69-70; res assoc, 71-73, clin instr, 73-74, from instr to assoc prof biochem, 74-79, ASSOC PROF PHYSIOL CHEM, SCH MED, OHIO STATE UNIV, 79- *Mem:* AAAS; Am Asn Cancer Res; NY Acad Sci; Asn Women Sci; Am Soc Biochem & Molecular Biol. *Res:* Chemical carcinogenesis; oncogenesis; RNA synthesis and transport; aging. *Mailing Add:* Dept Physiol Chem Sch Med Ohio State Univ 333 W Tenth Ave Columbus OH 43210

SCHUMM, STANLEY ALFRED, b Kearny, NJ, Feb 22, 27; m 50; c 3. GEOMORPHOLOGY, WATERSHED MANAGEMENT. *Educ:* Upsala Col, AB, 50; Columbia Univ, PhD(geol), 55. *Prof Exp:* Geologist, US Geol Surv, 54-67; from assoc prof to prof geol, 67-86, assoc dean, 73-74, UNIV DISTINGUISHED PROF GEOL, COLO STATE UNIV, 86-; VPRES, WATER ENG & TECHNOL, 80- *Concurrent Pos:* Vis lectr, Univ Calif, 59-60; prof affil, Colo State Univ, 63; mem nat comt, Int Union Quaternary Res, 63-69; fel, Univ Sydney, 64-65; cor mem comn hillslope evolution & mem comn appl geomorphol, Int Geog Union, 65; vis geol scientist, Am Geol Inst, 66; vis scientist, Polish Acad Sci, 69; distinguished lectr, Univ Tex, 70; vis prof, Univ de los Andes, Venezuela, 72, Econ Geol Res Univ, Univ Witwatersrand, SAfrica, 75, Univ Canterbury, NZ, 84 & Univ Tsukuba, Japan, 85, Taiwan Nat Normal Univ, 88, Univ New South Wales, Australia, 88; prin investr res projs, Nat Sci Found, US Army Res Off, Nat Park Serv, Off Water Res & Technol, US Geol Surv & Colo Agr Exp Sta; co-investr, US Army CEngr, Bur Sports Fisheries & Wildlife & Fed Hwy Admin; consult, Reg Transp Dist, Denver, Cameron Eng Co, Atlantic Richfield Oil Co & DDI Explor, Paris, Integral Ltd, Medellin, Manawatu bd, NZ, corp engrs, Ohio River Dist, Vicksburg Dist & Sacramento Dist, US Dept Justice, Can Int Develop Agency, Atty Gen, State Colo. *Honors & Awards:* Horton Award, Am Geophys Union, 59; Kirk Bryan Award, Geol Soc Am, 79; David Linton Award, Brit Geomorphol Res Group, 82; G K Warren Prize, US Nat Acad Sci. *Mem:* AAAS; Geol Soc Am; Am Geophys Union; Am Soc Civil Engrs; Am Quaternary Asn. *Res:* Stream morphology; paleohydrology. *Mailing Add:* Dept Earth Resources Colo State Univ Ft Collins CO 80523

SCHUNDER, MARY COTHRAN, b Tulsa, Okla, Sept 28, 31; m 56; c 3. GROSS ANATOMY, CYTOHISTOCHEMISTRY. *Educ:* Tex Christian Univ, BA, 53, MA, 70; Baylor Univ, PhD(anat), 76. *Prof Exp:* Staff microbiol, Univ Dallas, 64-69; instr biol, Tex Christian Univ, 69-70; ASSOC PROF ANAT & CHMN DEPT, TEX COL OSTEOP MED, 70-; ASSOC GRAD FAC MEM BASIC HEALTH SCI, NTEX STATE UNIV, 78- *Concurrent Pos:* HEW training grant, Bur Health Prof, 72-75; consult anat, Nat Bd Examrs for Osteop Physicians & Surgeons, 75-; rep, Anat Bd State Tex, 70-, vpres, 78- *Mem:* Sigma Xi. *Res:* Calcium metabolism; histochemical localization of calcium and lipid in calcified tissue and gut. *Mailing Add:* 3212 Tanglewood Trail Ft Worth TX 76109

SCHUNK, ROBERT WALTER, b New York, NY. PLASMA TRANSPORT, NUMERICAL SIMULATIONS. *Educ:* NY Univ, BS, 65; Yale Univ, PhD(phys fluids), 70. *Prof Exp:* Inst Sci & Technol fel space physics, Univ Mich, 70-71; res assoc geophys, Yale Univ, 71-73; res assoc space physics, Univ Calif, San Diego, 73-76; assoc prof, 76-79, PROF PHYSICS, UTAH STATE UNIV, 79- *Concurrent Pos:* Assoc ed, J Geophys Res, 77-80; mem, Comt Solar Terrestrial Res, Geophys Res Bd, Nat Acad Sci, 79-; mem, Nat Ctr Atmospheric Res Comput Divisions Adv Panel, 80-; prin investr, Solar Terrestrial Theory Prog, 80- *Honors & Awards:* Governor's, Medal Sci & Technol, Utah, 88. *Mem:* Am Geophys Union; AAAS. *Res:* Planetary atmospheres, ionospheres and magnetospheres; solar wind. *Mailing Add:* Dept Physics Utah State Univ Logan UT 84322

SCHUNN, ROBERT ALLEN, b Martins Ferry, Ohio, July 15, 36; m 59; c 3. INORGANIC CHEMISTRY. *Educ:* Univ Ohio, BS, 58; Mass Inst Technol, PhD(inorg chem), 63. *Prof Exp:* Res chemist, 63-81, training supv, 81-83, PERSONNEL ADMINR, CENT RES DEPT, E I DU PONT DE NEMOURS & CO, INC, 83- *Mem:* Am Chem Soc. *Res:* Preparative inorganic chemistry, especially of transition metal compounds; heterogeneous catalysis; organometallic chemistry. *Mailing Add:* 19 Ski Hill Dr Bedminster NJ 07921

SCHUPF, NICOLE, b New York, NY, Jan 20, 43; m 68; c 1. PHYSIOLOGICAL PSYCHOLOGY. *Educ:* Bryn Mawr Col, BA, 64; NY Univ, PhD(psychol), 70. *Hon Degrees:* MPH, Univ Calif, Berkeley, 84. *Prof Exp:* Res assoc, Rockefeller Univ, 69-71; instr neurol, Med Sch, NY Univ, 74-77; from asst prof to assoc prof psychol, Manhattanville Col, 77-89; PROF, NY STATE INST BASIC RES, 88- *Concurrent Pos:* Adj asst prof neurol, Med Sch, NY Univ, 77-78; res assoc, Siergievsky Ctr, Columbia Univ, 84- *Mem:* AAAS; NY Acad Sci. *Res:* Neuroimmunology; neuroepidemiology. *Mailing Add:* 28 Schermerhorn St Brooklyn NY 10577

SCHUPP, GUY, b Blackwater, Mo, Oct 20, 33; m 62; c 2. MOSBAUER SCATTERING. *Educ:* Mo Valley Col, BS, 54; Iowa State Univ, PhD(physics), 62. *Prof Exp:* Res scientist, Indust Reactor Labs, US Rubber Co, NJ, 62-64 & Res Ctr, 64; from asst prof to assoc prof, 64-89, PROF PHYSICS, UNIV MO, COLUMBIA, 89- *Concurrent Pos:* Sabbatical, Argonne Nat Lab, 71-72; bd trustee, Missouri Valley Col, 88- *Mem:* AAAS; Am Phys Soc; Am Asn Physics Teachers. *Res:* Mössbauer scattering; nuclear spectroscopy; atomic excitation and ionization phenomena. *Mailing Add:* Dept Physics Univ Mo Columbia MO 65211

SCHUPP, ORION EDWIN, III, b Wilmington, Del, Jan 28, 32; c 3. ANALYTICAL CHEMISTRY. *Educ:* Univ Del, BS, 53; Ohio State Univ, PhD(chem), 58. *Prof Exp:* Res chemist, 58-63, sr res chemist, 63-69, res assoc anal chem, 69, LAB HEAD, EASTMAN KODAK CO, 69- *Mem:* Am Chem Soc. *Res:* Molecular weight and chemical composition of polymers, gel permeation chromatography; liquid chromatography; gas chromatography; polarography of complex ions for determination of stability constants. *Mailing Add:* 263 Lone Oak Ave Rochester NY 14616

SCHUPP, PAUL EUGENE, b Cleveland, Ohio, Mar 12, 37; m 66. MATHEMATICS. *Educ:* Case Western Reserve Univ, BA, 59; Univ Mich, Ann Arbor, MA, 61, PhD(math), 66. *Prof Exp:* Asst prof math, Univ Wis-Madison, 66-67; from asst prof to assoc prof, 67-75, assoc mem, Ctr Advan Study, 73-74, PROF MATH, UNIV ILL, URBANA, 75- *Concurrent Pos:* Vis mem, Courant Inst, 69-70; John Simon Guggenheim mem fel, 77-78. *Mem:* Am Math Soc; London Math Soc. *Res:* Theory of infinite groups and decision problems in algebraic systems; automate theory. *Mailing Add:* Dept Math 353 Atgeid Hall Univ Ill Urbana IL 61801

SCHUR, PETER HENRY, b Vienna, Austria, May 9, 33; US citizen; div; c 2. INTERNAL MEDICINE, IMMUNOLOGY. *Educ:* Yale Univ, BS, 55; Harvard Univ, MD, 58. *Prof Exp:* Assoc, 67-69, asst prof, 69-72, assoc prof, 72-77, PROF MED, HARVARD MED SCH, 78- *Concurrent Pos:* Helen Hay Whitney fel, Rockefeller Univ, 64-67. *Mem:* Am Soc Clin Invest; Am Asn Immunol; Am Col Physicians; Am Rheumatism Asn; Am Fedn Clin Res; Asn Am Phys. *Res:* Rheumatology; systemic lupus. *Mailing Add:* Brigham & Women's Hosp 75 Francis St Boston MA 02115

SCHURIG, JOHN EBERHARD, b Morristown, Tenn, Mar 16, 45. LICENSING. *Educ:* Univ Tenn, BS, 67, BS, 69, PhD(pharmacol), 73. *Prof Exp:* Sr res scientist pharmacol, Bristol Labs, 73-78 & antitumor biol, 78-90, asst dir, 80-82; assoc dir exp therapeut, 82-87 & anticancer res, 87-91, ASSOC DIR LICENSING, BRISTOL-MYERS SQUIBB PHARM GROUP, 91- *Mem:* Am Asn Cancer Res; Am Soc Pharmacol & Exp Therapeut; Licensing Exec Soc. *Res:* Anticancer therapies. *Mailing Add:* Dept 881 Bristol-Myers Squibb Co PO Box 5100 Wallingford CT 06492-7660

SCHURLE, ARLO WILLARD, b Clay Center, Kans, Oct 30, 43; m 63. MATHEMATICS. *Educ:* Univ Kans, BA, 64, MA, 66, PhD(math, topology), 67. *Prof Exp:* Asst prof math, Ind Univ, Bloomington, 67-71; assoc prof math, Univ NC, Charlotte, 71-78; Fulbright prof math, Univ Liberia, 78-80; ASSOC PROF MATH, UNIV NC, CHARLOTTE, 80- *Mem:* Math Asn Am; Am Math Soc. *Res:* Geometric topology, including decompositions of three-space; compactifications and dimension theory. *Mailing Add:* Rose Hulman Inst Technol Terre Haute IN 47803

SCHURMAN, GLENN AUGUST, b La Center, Wash, Sept 6, 22; m 44; c 3. AERODYNAMICS, GEOPHYSICS. *Educ:* State Col Wash, BS, 44; Calif Inst Technol, MS, 47, PhD(mech eng), 50. *Prof Exp:* Res engr, Nat Adv Comt Aeronaut, 44-47; from res engr to sr res engr, Calif Res Corp, 50-60, mgr producing res, 60-63, sr staff engr, 63-65, dist supt, Calif Co Div, Chevron Oil Co, 65-69; div prod supt, Standard Oil Co Tex, 69-71; asst gen mgr prod, Calco Div, Chevron Corp, 71-74, vpres & gen mgr prod, Denver, 74-75, managing dir, Chevron Petrol UK Ltd, 75-82, vpres prod, 82-87; RETIRED. *Concurrent Pos:* Distinguished lectr, Soc Petrol Engrs, 78. *Mem:* Nat Acad Eng; Soc Petrol Engrs; Am Soc Mech Engrs. *Res:* Auto-ignition of gases near heated surface; aerodynamics and fluid dynamics of gas turbines; oil production and geophysics. *Mailing Add:* 840 Powell St Apt 302 San Francisco CA 94108

SCHURMEIER, HARRIS MCINTOSH, b St Paul, Minn, July 4, 24; m 49; c 4. AERONAUTICAL ENGINEERING. *Educ:* Calif Inst Technol, BS, 45, MS, 48, AeroE, 49. *Prof Exp:* Sr res engr, Jet Propulsion Lab, Calif Inst Technol, 49-53, chief, Wind Tunnel Sect, 53-56, chief, Aerodyn Div, 56-58, dep prog mgr, Sergeant Missile, 58-59, mgr, Systs Div, 59-62, proj mgr, Ranger, 62-65, Mariner Mars 1969, 65-69, Voyager, 70-76, dep asst lab dir, Flight Projs, 69-76, asst lab dir, Energy & Technol Appln, 76-81, assoc dir, Defense & Civil Progs, 81-85; AEROSPACE CONSULT, 85- *Concurrent Pos:* Chmn, Supersonic Tunnel Asn, 54-56; mem, Res Steering Comt on Manned Space Flight, NASA, 59, Res Adv Comt Missile & Space Vehicle Aerodyn, 60-62, Comt Proj Mgt, 80-81; chmn, W M Keck Observ Proj Rev Bd, 86-, Galileo Proj Standing Rev Bd, 86-, Soaring Soc Am Tech Rev Bd, 89- *Honors & Awards:* Medal for Except Sci Achievement, NASA, 65, Exceptional Serv Medal, 69, Distinguished Serv Medal, 81; Astronaut Engr Award, Nat Space Club, 65 & 81; Von Karman Lectr, Am Inst Aeronaut & Astronaut, 75. *Mem:* Nat Acad Eng; Supersonic Tunnel Asn; fel Am Inst Aeronaut & Astronaut; AAAS; Sigma Xi. *Res:* Space exploration. *Mailing Add:* Jet Propulsion Lab Attn: Lynn Patterson Mail Stop 180-900 Pasadena CA 91109

SCHURR, AVITAL, b Jerusalem, Israel, Aug 23, 41; m 65; c 3. NEUROPHYSIOLOGY, NEUROCHEMISTRY. *Educ:* Hebrew Univ, Jerusalem, BSc, 67; Tel Aviv Univ, MSc, 70; Ben Gurion Univ, Beer Sheva, PhD(biochem pharmacol), 77. *Prof Exp:* Researcher biochem, Tel Aviv Univ, Israel, 67-70; researcher photosynthesis, Negev Inst Arid Zone Res Beer Sheva, Israel, 70-72; res assoc biomembranes, Ben Gurion Univ, Beer Sheva, Israel, 72-77; neurochemist neuropharmacol, Tex Res Inst Ment Sci, Houston, 77-79; res assoc neurobiol, Med Sch, Univ Tex, 79-81; asst prof anesthesiol, 81-86, ASSOC PROF ANETHESIOL, SCH MED, UNIV LOUISVILLE, KY, 86- *Concurrent Pos:* Lectr biochem, dept biol, Ben Gurion Univ, Beer Sheva, Israel, 73-77; lectr neuropharmacol, Baylor Col Med Grad Sch, Houston; actg dir res, dept anesthesiol, Sch Med, Univ Louisville, KY, 83-84, dir, anal lab, Anesthesia & Crit Care Res Unit, 83- *Mem:* Soc Neurosci; AAAS; NY Acad Sci; Am Soc Anesthesiologists; Int Anesthesial Res Soc; Int Soc Neurochem. *Res:* Studies of cerebral ischemia using an in vitro model for the better understanding of the consequences of stroke and brain injury; assessment of pharmacological agents for their potential as protectants against cerebral ischemia; studies on brain energy metabolism. *Mailing Add:* Dept Anesthesiol Univ Louisville Sch Med Louisville KY 40292

SCHURR, GARMOND GAYLORD, b Almont, NDak, Sept 7, 18; m 45; c 4. RESEARCH ADMINISTRATION. *Educ:* NDak State Univ, BS(chem), 40. *Prof Exp:* Group leader, Paint Res, Sherwin-Williams Co, 46-58, asst dir, 58-66, dir, 66-74, dir, Res Ctr, 74-79, tech advr, 79-82; CONSULT, 82- *Mem:* Am Chem Soc; Am Soc Testing & Mat; Nat Asn Corrosion Eng; Fedn Socs Coatings Technol. *Res:* Applications of polymeric resins in protective coatings; corrosion control by means of protective coatings. *Mailing Add:* 6220 W 127th Pl Palos Heights IL 60463-2314

SCHURR, JOHN MICHAEL, b Pittsfield, Mass, Nov 10, 37; m 58; c 2. PHYSICAL CHEMISTRY, MOLECULAR BIOPHYSICS. *Educ:* Yale Univ, BS, 59; Univ Calif, Berkeley, PhD(biophys), 64. *Prof Exp:* NIH fel chem, Univ Ore, 64-66; from asst prof to assoc prof, 66-78, PROF CHEM, UNIV WASH, 78- *Concurrent Pos:* Vis scientist, Swiss Fed Water Inst, 74; vis scholar, Kyoto Univ, Japan, 87. *Mem:* Biophys Soc; Am Phys Soc; Am Chem Soc. *Res:* Coherent dynamic light scattering, pulsed laser optical techniques and NMR relaxation; DNA dynamics, including deformational brownian motions; polyelectrolyte phenomena, diffusion, and electrophoresis; interaction of electromagnetic radiation with matter. *Mailing Add:* Dept Chem Univ Wash Seattle WA 98195

SCHURR, KARL M, b Logan Co, Ohio, Feb 28, 32; m 56; c 3. ENTOMOLOGY, POLLUTION BIOLOGY. *Educ:* Bowling Green State Univ, BA, 56, MA, 58; Univ Minn, PhD(entom, bot), 61. *Prof Exp:* Res asst, Univ Minn, 58-61; asst prof, 61-71, ASSOC PROF BIOL, BOWLING GREEN STATE UNIV, 71- *Concurrent Pos:* Grants, Fed Res, 58-61, NSF, 63-, US Dept HEW, 66-; consult Ohio State Univ, 63, Greeley & Hansen Co, Ill, 65, Columbus Labs, Battelle Mem Res Inst, 70- & Holmes County Pub Health Dept & Regional Planning Comn, 71-; res consult marine pollution, Col Law, Univ Toledo & J&S Steel Co; res consult maricult, toxicol & water resources, Auburn Univ; adj prof, Med Col Ohio. *Mem:* Int Asn Advan Earth & Environ Sci (vpres, 77-78); World Maricult Soc; AAAS; Entom Soc Am; Marine Technol Soc; Sigma Xi. *Res:* Invertebrate physiology; water resources; toxicology. *Mailing Add:* Dept Biol Bowling Green State Univ Bowling Green OH 43403

SCHURRER, AUGUSTA, b New York, NY, Oct 11, 25. MATHEMATICS. *Educ:* Hunter Col, AB, 45; Univ Wis, MA, 47, PhD(math), 52. *Prof Exp:* Computer, Off Sci Res & Develop, 45; PROF MATH, UNIV NORTHERN IOWA, 50- *Concurrent Pos:* NSF fac fel, Univ Mich, 57-58; mem panel suppl pub, Sch Math Study Group, 61-65. *Mem:* Am Math Soc; Math Asn Am. *Res:* Mathematical analysis; zeros of polynomials; mathematics education; teacher education. *Mailing Add:* 1224 20th St Cedar Falls IA 50613

SCHUSSLER, M(ORTIMER), metallurgical engineering, for more information see previous edition

SCHUSTEK, GEORGE W(ILLIAM), JR, b Oak Park, Ill, May 10, 15; m 44; c 3. CHEMICAL ENGINEERING. *Educ:* Univ Chicago, BS, 37, MBA, 51. *Prof Exp:* Res chemist, US Gypsum Co, 37-42; chem engr, Standard Oil Co (Ind), 46-48, group leader, 48-53, sect leader, Res Dept, 54-62, res supvr res & develop, Am Oil Co, 62-69, sr res scientist, 69-74; RETIRED. *Mem:* Am Chem Soc; Am Inst Chem Engrs. *Res:* Petroleum processes. *Mailing Add:* 1080 Prairieview Dr West Des Moines IA 50265-7207

SCHUSTER, CHARLES ROBERTS, JR, b Woodbury, NJ, Jan 24, 30; m 53; c 5. PSYCHOLOGY, PHARMACOLOGY. *Educ:* Gettysburg Col, AB, 51; Univ NMex, MS, 53; Univ Md, PhD(psychol), 62. *Prof Exp:* Asst instr endocrinol & res biologist, Temple Med Sch, 53-55; jr scientist, Smith, Kline & French Labs, 55-57; res assoc psychopharmacol, Univ Md, 61-63; asst prof pharmacol, Med Sch, Univ Mich, Ann Arbor, 63-69, lectr psychol, 66-69; assoc prof pharmacol & psychiat, Univ Chicago, 69-72, prof psychiat, pharmacol & physiol sci, 72-91; DIR, NAT INST DRUG ABUSE, 86- *Mem:* AAAS; Am Psychol Asn; Am Soc Pharmacol. *Res:* Psychological and physiological dependence on drugs; role of interoceptive processes in the control of behavior. *Mailing Add:* 5600 Fishers Lane Rockville MD 20857

SCHUSTER, DANIEL BRADLEY, psychiatry, for more information see previous edition

SCHUSTER, DAVID ISRAEL, b Brooklyn, NY, Aug 13, 35; m 62; c 1. ORGANIC CHEMISTRY, CHEMICAL DYNAMICS. *Educ:* Columbia Univ, BA, 56; Calif Inst Technol, PhD(chem, physics), 61. *Prof Exp:* Fel org photochem, Univ Wis, 60-61; from asst prof to assoc prof, 61-69, dir grad studies, Dept Chem, 74-78, PROF CHEM, NY UNIV, 69- *Concurrent Pos:* Alfred P Sloan fel, 67-69; vis fel, Royal Inst Gt Brit, 68-69; NSF sci fac fel, 75-76; vis prof, Yale Univ, 75-76. *Mem:* Am Soc Photobiol; Sigma Xi; Am Chem Soc; Inter-Am Photochem Soc; AAAS; Royal Inst Chem. *Res:* Mechanistic organic photochemistry; organic reaction mechanisms; spectroscopy and magnetic resonance; free radical chemistry; flash photolysis; biochemistry of Schizophrenia; mechanism of action of anti-psychotic drugs; characterization of dopamine receptors in the brain; photobiological processes; synthesis of prostaglandins. *Mailing Add:* Dept Chem NY Univ New York NY 10003

SCHUSTER, DAVID J, b Memphis, Tenn, Aug 29, 47; m 71; c 2. HOST PLANT RESISTANCE, PEST MANAGEMENT. *Educ:* Kans State Univ, BS, MS, 70; Okla State Univ, PhD(entom), 73. *Prof Exp:* Res assoc entom, Okla State Univ, 73-75; from asst prof to assoc prof, 81-85, PROF ENTOM, GULF COAST RES & EDUC CTR, UNIV FLA, BRADENTON, 85- *Honors & Awards:* Coun Mem Tomato Res Award, 81, 84, 90; Fla Fruit & Veg Assoc Ann Res Award, 85. *Mem:* Entom Soc Am; Sigma Xi. *Res:* Pest management of insect and mite pests of vegetables; interactions of insecticides, biological control and host plant resistance are emphasized. *Mailing Add:* Gulf Coast Res & Educ Ctr 5007 60th St E Bradenton FL 33508

SCHUSTER, DAVID MARTIN, materials science, metallurgical engineering, for more information see previous edition

SCHUSTER, EUGENE F, b Tintah, Minn, June 6, 41; m 63; c 6. NONPARAMETRIC STATISTICS, STATISTICAL COMPUTING. *Educ:* St John's Univ, BA, 63; Univ Ariz, MA, 65, PhD(math statist), 68. *Prof Exp:* Instr math, Univ Ariz, 68; sr analyst opers res, US Army Engr's Strategic Studies Group, 68-70; from asst prof to assoc prof math, 70-76, chmn dept, 79-81 & 83-86, PROF MATH, UNIV TEX, EL PASO, 76- *Concurrent Pos:* Vis assoc prof, Univ Ariz, 74-75, statist consult, hydrol dept, 75; statist consult, US Army Sci Serv, 81-83. *Mem:* Inst Math Statist; Math Asn Am; Am Statist Asn. *Res:* Large sample theory; nonparametric estimation of density, distribution and regression functions; statistical computing and stochastic simulation. *Mailing Add:* Dept Math Sci Univ Tex El Paso TX 79968-0514

SCHUSTER, FREDERICK LEE, b Brooklyn, NY, Jan 23, 34; m 60; c 2. PROTOZOOLOGY, ELECTRON MICROSCOPY. *Educ:* Brooklyn Col, BS, 56; Univ Calif, Berkeley, MA, 58, PhD(protozoan ultrastruct), 62. *Prof Exp:* Res specialist electron micros, Langley Porter Neuropsychiat Inst, 62-63; res assoc biol, Argonne Nat Lab, 63-66; from asst prof to assoc prof, 66-74, PROF BIOL, BROOKLYN COL, 74- *Mem:* Soc Protozool; Am Soc Cell Biol; Am Micros Soc; Am Soc Microbiol. *Res:* Pathogenic free-living Protozoa; Protozoan ultrastructure; morphogenesis of Protozoa. *Mailing Add:* Dept Biol Brooklyn Col Brooklyn NY 11210

SCHUSTER, GARY BENJAMIN, b New York, NY, Aug 6, 46; m 68; c 2. PHOTOCHEMISTRY. *Educ:* Clarkson Col Technol, BS, 68; Univ Rochester, PhD(chem), 71. *Prof Exp:* Res asst chem, Univ Rochester, 68-71; phys sci asst radiation chem, US Army, 71-73; res assoc chem, Columbia Univ, 73-75; from asst prof to assoc prof chem, 75-81, PROF CHEM, UNIV ILL, 81-. HEAD DEPT, 90- *Concurrent Pos:* NIH fel, 74-75; Alfred P Sloan fel, 77-79; John Simon Guggenheim fel, 85-86. *Mem:* Am Chem Soc; Am Soc Photobiol; Sigma Xi. *Res:* Energy partitioning in exothermic organic chemical reactions; chemical formation of electronically excited states. *Mailing Add:* Chem/261 Roger Adams Lab-Box 58 Univ Ill 1209 W Calif St Urbana IL 61801

SCHUSTER, GEORGE SHEAH, b Geneva, Ill, Sept 22, 40; m 63; c 1. MICROBIOLOGY, CELL BIOLOGY. *Educ:* Wash Univ, AB, 62; Northwestern Univ, DDS & MS, 66; Univ Rochester, PhD(microbiol), 70. *Prof Exp:* PROF MICROBIOL, CELL BIOL & MOLECULAR BIOL, SCH DENT & MED, MED COL GA, 70- *Mem:* AAAS; Tissue Cult Asn; Sigma Xi. *Res:* Viral carcinogenesis; metabolic functions of cells and the effects of virus infection on these; dental caries. *Mailing Add:* Dept Oral Biol Sch Dent Med Col Ga Augusta GA 30912

SCHUSTER, INGEBORG I M, b Frankfurt, Ger, Oct 30, 37: US citizen. PHYSICAL ORGANIC CHEMISTRY. *Educ:* Univ Pa, BA, 60; Carnegie Inst Technol, MS, 63, PhD(chem), 65. *Prof Exp:* Huff fel org chem, Bryn Mawr Col, 65-67; from asst prof to assoc prof chem, 73-83, PROF CHEM, PA STATE UNIV, OGONTZ CAMPUS, 83- *Mem:* Am Chem Soc. *Res:* Nuclear magnetic resonance spectroscopy elucidation of organic structures with emphasis on electronic effects. *Mailing Add:* Dept Chem Pa State Univ 1600 Woodland Rd Abington PA 19001

SCHUSTER, JAMES J(OHN), b Pottsville, Pa, Dec 13, 35; m 58; c 3. CIVIL ENGINEERING, TRANSPORTATION. *Educ:* Villanova Univ, BCE, 57, MCE, 61; Purdue Univ, PhD(civil eng), 64. *Prof Exp:* From instr to assoc prof, 58-70, PROF CIVIL ENG, VILLANOVA UNIV, 70-, DIR INST TRANSP STUDIES, 65- *Concurrent Pos:* Consult, Northern Calif Transit Demonstration Proj, 65-66; mem origin & destination comt, Hwy Res Bd, Nat Acad Sci-Nat Res Coun, 65- *Mem:* Inst Traffic Engrs; Am Soc Civil Engrs; Am Soc Eng Educ. *Res:* Origin-destination; vehicular trip prediction; modal split analysis; generation; distribution; assignment. *Mailing Add:* Dept Civil Eng Villanova Univ Villanova PA 19085

SCHUSTER, JOSEPH L, b Teague, Tex, May 21, 32; m 57; c 4. RANGE MANAGEMENT, ECOLOGY. *Educ:* Tex A&M Univ, BS, 54, PhD(range mgt), 62; Colo State Univ, MS, 59. *Prof Exp:* Range conservationist, US Soil Conserv Serv, 54-59 & US Forest Serv, 61-64; asst prof range mgt, Tex Tech Univ, 64-69, prof & chmn dept, 69-72; PROF RANGE SCI & HEAD DEPT, TEX A&M UNIV, 72- *Concurrent Pos:* Range mgt consult, Bur Land Mgt, 64. *Mem:* Soc Range Mgt (pres, 84); Soil Conserv Soc Am; Wildlife Soc. *Res:* Forest overstory-understory relations; research technique development; range improvement practices. *Mailing Add:* Dept Range Sci Tex A&M Univ College Station TX 77843

SCHUSTER, MICHAEL FRANK, b Mexia, Tex, May 29, 29; m 51; c 7. ECONOMIC ENTOMOLOGY. *Educ:* Tex A&M Col, BS, 55, MS, 61; Miss State Univ, PhD(entom), 71. *Prof Exp:* Asst entomologist, Tex Agr Exp Sta, Weslaco, Tex, 55-71; from asst prof to assoc prof entom, Miss Agr & Forest Exp Sta, 71-78; PROF ENTOM, AGR EXP STA, TEX A&M UNIV, 78-; ADJ PROF ENVIRON SCI, UNIV TEX, DALLAS, 82- *Concurrent Pos:* Consult entomologist, USAID, Brazil, 67; guest lectr, Biol Control, Trop Sch Agr, Cardinas Tabasco, Mex, 77; collabr, Centro de Investigaciones Agricoles del Gulfo Norte, INIA, Tampico, 78-; explor parasites of Lygusbugs, Egypt, Sudan, Kenya & Rep SAfrica, 81-85. *Mem:* Entom Soc Am; Orgn Biol Control; Sigma Xi. *Res:* Development of insect controls based on natural regulating factors, such as host plant resistance and biological control; resistance in plants is identified, evaluated and utilized; natural enemies are evaluated for effectiveness. *Mailing Add:* Agr Res & Exten Ctr 17560 Coit Rd Dallas TX 75252

SCHUSTER, ROBERT LEE, b Chehalis, Wash, Aug 29, 27; m 55; c 4. GEOLOGY & CIVIL ENGINEERING. *Educ:* State Col Wash, BS, 50; Ohio State Univ, MS, 52; Purdue Univ, MS, 58, PhD(civil eng), 60; Imp Col, London, dipl, 65. *Prof Exp:* Geologist, Snow, Ice & Permafrost Res Estab, Corps Engrs, US Army, 52-55; instr civil engr & eng geol, Purdue Univ, 56-60; from assoc prof to prof civil eng, Univ Colo, 60-67; prof civil eng & chmn dept, Univ Idaho, 67-74; chief, eng geol br, 74-79, GEOLOGIST, US GEOL SURV, 79- *Concurrent Pos:* NSF sci fac fel, Imp Col, London, 64-65; NATO sr fel sci, Imp Col, Univ London, 74; chmn, Joint Am Soc Civil Engrs-Geol Soc Am-Asn Eng Geol Comt on Eng Geol, 74-78; chmn eng geol comt, Transp Res Bd, Nat Res Coun-Nat Acad Sci, 77-81; chmn, 81 & mem exec comt, Eng Geol Div, Geol Soc Am, 82-86, chmn 84-85. *Honors & Awards:* Richard H Johns Distinguished Lectr, Geol Soc Am-Asn Eng Geol, 90; Distinguished Prac Award, Eng Geol Div, Geol Soc Am, 90. *Mem:* Asn Eng Geol; Geol Soc Am; Am Soc Civil Engrs; Geol Soc London; Int Asn Eng Geol. *Res:* Slope stability; soil and rock properties; engineering geology. *Mailing Add:* Geol Risk Assessment Br US Geol Surv MS 966 Box 25046 Denver CO 80225-0046

SCHUSTER, RUDOLF MATHIAS, b Altmuehldorf, Ger, Apr 8, 21; nat US; m 43; c 2. BOTANY. *Educ:* Cornell Univ, BSc, 45, MSc, 46; Univ Minn, PhD(entom, bot), 48. *Prof Exp:* Instr bot, Univ Minn, 48-50; asst prof, Univ Miss, 50-53 & Duke Univ, 53-56; asst prof bot & cur bryophyta, Univ Mich, 56-57; from assoc prof to prof bot, 57-90, dir herbarium, 64-70, EMER PROF BOT, UNIV MASS, AMHERST, 90- *Concurrent Pos:* NSF grants, 53-72 & 76-84; Guggenheim fel, 55-56 & 67; Fulbright prof, Univ Otago, NZ, 61-62. *Honors & Awards:* Award, Arctic Inst NAm, 60 & 66. *Mem:* Am Bryol & Lichenological Soc; Brit Bryol Soc. *Res:* Systematics, ecology and distribution patterns of North American, Arctic and Antipodal Hepaticae; phylogeny of the Archegoniates; plant geography; taxonomy of Mutillid wasps. *Mailing Add:* 821 W Calle Del Regald Green Valley AZ 85614-2805

SCHUSTER, SANFORD LEE, b Hastings, Nebr, Sept 28, 38; m 67; c 2. SOLID STATE PHYSICS. *Educ:* Univ Nebr, Lincoln, BS, 60, MS, 63, PhD(physics), 69. *Prof Exp:* From asst prof to assoc prof physics, 68-74, PROF PHYSICS, MANKATO STATE UNIV, 74- *Mem:* Am Phys Soc; Am Asn Physics Teachers; Am Crystallog Asn. *Res:* Thermal diffuse scattering of x-rays by metals. *Mailing Add:* Dept Physics Mankato State Univ Mankato MN 56001

SCHUSTER, SEYMOUR, b Bronx, NY, July 31, 26; m 54; c 2. MATHEMATICS. *Educ:* Pa State Univ, BA, 47, PhD(math), 53; Columbia Univ, AM, 48. *Prof Exp:* Asst math, Pa State Univ, 49-51, instr, 51-52; from instr to assoc prof, Polytech Inst Brooklyn, 53-58; assoc prof, Carleton Col, 58-63; res fel, Univ Minn, Minneapolis, 62-63, assoc prof math, 63-68; chmn dept, 73-76, PROF MATH, CARLETON COL, 68- *Concurrent Pos:* Fel, Univ Toronto, 52-53; vis assoc prof, Univ NC, Chapel Hill, 61; dir col geom proj, Minn Math Ctr, 64-74, Acad Year Inst for Col Teachers, 66-67; NSF sci fac fel, 70-71; vis scholar, Univ Calif, Santa Barbara, 70-71; guest scholar, Western Mich Univ, 76 & vis scholar, Western Mich Univ, 81; vis prof, Western Wash Univ, 83; guest scholar, Univ Ariz, 90; mem, Inst Combinatorics & Applications. *Mem:* Am Math Soc; Sigma Xi; Math Asn Am; Asn Women Math; Nat Asn Mathematcians. *Res:* Graph theory; projective and non-Euclidean geometry; mathematical film production. *Mailing Add:* Carleton Col Northfield MN 55057

SCHUSTER, TODD MERVYN, b June 27, 33; US citizen. BIOPHYSICAL CHEMISTRY. *Educ:* Wayne State Univ, AB, 58, MS, 60; Wash Univ, PhD(molecular biol), 63. *Prof Exp:* USPHS fel, Max Planck Inst, 63-66; asst prof biol, State Univ NY Buffalo, 66-70; assoc prof, 70-75, chmn dept, 77-81, PROF BIOL, UNIV CONN, 75-; CHIEF SCIENTIST, XENOGEN, INC, 81- *Concurrent Pos:* Consult, NIH, 71-75 & NSF, 85-88; McCollum-Pratt Prof, Johns Hopkins Univ, 79-80; NSF panel biol instrumentation, 85-, NSF panel biol facil & sci & technol sci, 87-; dir, Univ Conn Biotechnol Ctr. *Mem:* AAAS; Am Chem Soc; Am Soc Biol Chem; Biophys Soc; Am Soc Virol. *Res:* Physical biochemistry of proteins and protein-nucleic acid interactions; mechanisms of Ligand Binding to hemeproteins; self-assembly of viruses. *Mailing Add:* Dept Molecular & Cell Biol Box V-125 Univ Conn Storrs CT 06268

SCHUSTER, WILLIAM JOHN, b Elkhart, Ind, Mar 21, 48; c 1. ASTRONOMY. *Educ:* Case Western Reserv Univ, BS, 70; Univ Ariz, PhD(astron), 76. *Prof Exp:* INVESTR, INST ASTRON, MEX NAT UNIV, 73- *Mem:* Am Astron Soc; Astron Soc Pac. *Res:* Photometric astronomy; calibration of photometric indices; chemical compositions, evolutionary status and effective temperatures of stars; subdwarf stars, high velocity stars, Be stars, solar type stars. *Mailing Add:* PO Box 439027 San Diego CA 92143-9027

SCHUSTERMAN, RONALD JAY, b New York, NY, Sept 3, 32; m 57; c 3. MARINE BIOLOGY. *Educ:* Brooklyn Col, BA, 54; Fla State Univ, MA, 58, PhD(psychol), 61. *Prof Exp:* NSF res fel, Yerkes Labs Primate Biol, 61-62; asst prof psychol, San Fernando Valley State Col, 62-63; psychobiologist, Stanford Res Inst, 63-71, mgr animal behav, 69-71; PROF PSYCHOL & BIOL, CALIF STATE UNIV, HAYWARD, 72-; MARINE BIOLOGIST, INST MARINE SCI, UNIV CALIF, SANTA CRUZ, 85- *Concurrent Pos:* Prin investr, NSF grant, Stanford Res Inst, 63-70, Off Naval Res Contract, 67-71 & Calif State Univ, Hayward, 71-84, 85-; lectr psychol & biol, Calif State Univ, Hayward, 64-71. *Mem:* Fel AAAS; Animal Behav Soc; fel Am Psychol Asn. *Res:* Animal behavior and communication; animal cognition; biomarine mammals; animal psychophysics. *Mailing Add:* Dept of Psychol Calif State Univ Hayward CA 94542

SCHUT, HERMAN A, b Steenderen, Neth, Mar 23, 43; m 68; c 2. CARCINOGENESIS. *Educ:* McGill Univ, PhD(biochem), 74. *Prof Exp:* Asst prof, 80-84, ASSOC PROF PATH, MED COL OHIO, 84-; DIR, BILLSTEIN LIGAND LAB, 80- *Mem:* Am Asn Cancer Res; Soc Toxicol; Am Soc Pharmacol & Exp Therapeut. *Res:* Mechanisms of chemical carcinogenesis. *Mailing Add:* Dept Path Health Educ Bldg Rm 202 Med Col Ohio Toledo OH 43699

SCHUT, ROBERT N, b Hudsonville, Mich, Mar 6, 32; m 60; c 4. MEDICINAL CHEMISTRY. *Educ:* Hope Col, AB, 54; Mass Inst Technol, PhD(org chem), 58. *Prof Exp:* Res chemist, 59-62, group leader org chem, 62-65, sr res chemist, 65-66, sect head, 66-71, dir med chem dept, Miles Res Div, 71-75, DIR CHEM DEPT, CORP RES, MILES LABS, INC, 75- *Mem:* AAAS; Am Chem Soc; Royal Soc Chem; Sigma Xi; fel Am Inst Chemists. *Res:* Synthesis of compounds of potential therapeutic interest. *Mailing Add:* Corp Res Miles Labs Inc 400 Morgan Lane West Haven CT 06516

SCHUTT, DALE W, b Oak Park, Ill, Oct 1, 38; m; c 2. INSTRUMENTATION. *Educ:* Univ Ill, BS, 61; Univ Notre Dame, MS, 63. *Prof Exp:* Test equip design engr, Missile Div, Bendix Corp, 61-64; res asst, Univ Notre Dame, 64-66, from asst prof specialist to assoc prof specialist, 66-77; RES ASSOC & SUPVR PHYSICS DEPT, VA POLYTECH INST & STATE UNIV, 77- *Concurrent Pos:* Consult, Custom Electron Apparatus, Electro-optics. *Mem:* Inst Elec & Electronics Engrs. *Res:* Instrumentation used in basic research studies in the area of intermediate energy physics including detectors, wire chambers, data acquisition. *Mailing Add:* Dept Physics Va Polytech Inst & State Univ Blacksburg VA 24061

SCHUTT, PAUL FREDERICK, b Toledo, Ohio, Sept 1, 32; m 56; c 5. NUCLEAR REACTOR THEORY, RADIATION SAFETY. *Educ:* Ill Inst Tech, BS, 54; Univ Ariz, MS, 59. *Prof Exp:* Res scientist, Owens Corning Fiberglass, 54-55, US Army, 55-57; physicist, Babcock & Wilcox Co, 59-62, res & develop coordr, 62-66; prin concsult, Union Carbide Corp, 66-68; vpres, Nuclear Assurance Corp, 68-74, pres, 74-86, vchmn, 86-87; CHMN & CHIEF EXEC OFFICER, NUCLEAR FUEL SERV, 87- *Mem:* fel Am Nuclear Soc; Sigma Xi. *Res:* The development of remote, precise measurement systems to characterize spent nuclear fuel; the development of systems for the safe transport of spent fuel and highly radioactive materials; develop of advanced naval nuclear fuel. *Mailing Add:* 995 Windsor Trail Roswell GA 30075

SCHUTTA, JAMES THOMAS, b Milwaukee, Wis, Jan 11, 44; m 81; c 2. APPLYING STATISTICAL ANALYSIS OF DATA TO CONTROL PROCESSES, IDENTIFY NEEDS OF PRODUCT PERFORMANCE FROM CUSTOMER TO ENGINEERING. *Educ:* Milwaukee Sch Eng, BSEE, 78, MSEM, 88. *Prof Exp:* Technician, Western Elec, 58-62; engr & mgr, Johnson Controls Inc, 62-88; DIR, PMI FOOD EQUIP GROUP, 88- *Concurrent Pos:* Pres, E Troy Jaycees, 73-74; lectr, Milwaukee Sch Eng, 78-88, adv, Acad Adv Bd, 80-88; Milwaukee chamber, Chamber Com, 84-88; consult, Am Soc Qual Control, 86-87; mem, Comt Fire Testing, Nat Fire Prev Soc. *Res:* Communication techniques; ability to communicate technical information within a management setting. *Mailing Add:* 332 Chatham Dr Kettering OH 45429

SCHUTTE, A(UGUST) H(ENRY), b Boston, Mass, Dec 1, 07; m 35; c 2. CHEMICAL ENGINEERING. *Educ:* Dartmouth Col, AB, 29; Mass Inst Technol, MS(chem eng) & MS(chem eng pract), 30. *Prof Exp:* Asst dir & dir tech serv div, Standard Oil Co, Sumatra, 31-34; res & develop engr, W M Kellogg Co, 34-35; process design & Proj engr, Lummus Co, 35-42, proj engr, 42- 47, vpres 48-, process develop & eval, 48-57, mgr New York Div, 57-60; sr staff, Eng Div, Arthur D Little, Inc, 60-72, sr consult, Petro-Chem & Construct Proj, 72-; RETIRED. *Res:* Development of new devices; engineering and management consulting to developing countries. *Mailing Add:* Brookhaven at Lexington No 107A Lexington MA 02173

SCHUTTE, WILLIAM CALVIN, b Ponca, Nebr, May 14, 41; m 62; c 3. PHYSICAL CHEMISTRY. *Educ:* Wayne State Col, BAE, 62; Univ SDak, MNS, 67, PhD(phys chem), 72. *Prof Exp:* Teacher, South Sioux City Pub Sch, 62-67; ASST PROF CHEM, UNIV SDAK, 70- *Concurrent Pos:* Grant gen res fund, Univ SDak, 72-73 & 74-75, Exxon Educ Found, 76-77 & NSF, Acad Yr Proj, 77-78 & 78-79. *Mem:* Am Chem Soc; Sigma Xi. *Res:* Study of metal concentrations in walleye, paddlefish and morels. *Mailing Add:* 2912 Tipperary Idaho Falls ID 83401

SCHUTZ, BERNARD FREDERICK, b Paterson, NJ, Aug 11, 46; m 83; c 3. GENERAL RELATIVITY. *Educ:* Clarkson Col Technol, BSc, 67; Calif Inst Technol, PhD(physics), 72. *Prof Exp:* NSF fel physics, Univ Cambridge, 71-72; res fel physics, Yale Univ, 72-73, instr, 73-74; lectr astrophys, 74-76, reader in gen relativity, 76-84, PROF, UNIV WALES, COL CARDIFF, 84- *Concurrent Pos:* Vis prof, Wash Univ, St Louis, 83. *Mem:* Am Phys Soc; Royal Astron Soc; Sigma Xi; Int Astron Union; Soc Gen Relativity & Gravitation; Inst Physics. *Res:* General relativity and relativistic astrophysics; gravitational wave detection; numerical relativity. *Mailing Add:* Dept Physics & Astron Univ Wales Col Cardiff PO Box 913 Cardiff CF1 3TH Wales

SCHUTZ, BOB EWALD, b Brownfield, Tex, Sept 6, 40; m 68; c 2. ASTRODYNAMICS, REMOTE SENSING. *Educ:* Univ Tex, BS, 63, MS, 66, PhD(aerospace eng), 69. *Prof Exp:* Teaching asst, 65-69, asst prof, 69-77, assoc prof, 77-81, PROF AEROSPACE ENG, UNIV TEX, AUSTIN, 81- *Mem:* AAAS; Am Inst Aeronaut & Astronaut; Am Astron Soc; Am Geophys Union. *Res:* Rotation of the earth; estimation theory applied to orbit determination and geodynamics; applications of digital computers. *Mailing Add:* Dept Aerospace Eng Univ Tex Austin TX 78712

SCHUTZ, DONALD FRANK, b Orange, Tex, Sept 22, 34; m 58; c 2. GEOCHEMISTRY. *Educ:* Yale Univ, BS, 56, PhD(geol), 64; Rice Univ, MA, 58. *Prof Exp:* Instr geol, Kinkaid Sch, 58-60, res staff geologist, Yale Univ, 64; scientist, Isotopes, Inc, 64-65, dir, Proj Pinocchio, 65-66, mgr nuclear opers, 66-70, vpres & gen mgr, Westwood Labs, 70-75 PRES, TELEDYNE ISOTOPES, 75-, ENG GROUP EXEC, TELEDYNE, INC, 89- *Concurrent Pos:* Pres & Mem bd of dir, Am Asn Radon Scientists & Technologists, 86-89; mem radiation health sect, Am Pub Health Asn; mem environ radiation sect, Health Physics Soc; mem, environ sci div, Am Nuclear Soc. *Honors & Awards:* Cong Antarctic Serv Medal. *Mem:* Air & Waste Mgt Asn; Am Inst Mining, Metall & Petrol Engr, Soc Petrol Engrs; Geol Soc Am; Am Asn Petrol Geologists; Sigma Xi; Geochem Soc; Am Nuclear Soc; Scientists & Engrs Secure Energy. *Res:* Radiochemical tests for clandestine nuclear weapons tests; neutron activation analysis of trace elements in sea water; nuclear reactor environmental monitoring, isotope geochemistry, geochromometry, nuclear fuel analysis; radiotracer applications in enhanced oil recovery and refining; synfuel processes; radon surveys. *Mailing Add:* Teledyne Isotopes 50 Van Buren Ave Westwood NJ 07675

SCHUTZ, WILFRED M, b Eustis, Nebr, Jan 26, 30; m 57; c 3. COMPUTING, STATISTICS. *Educ:* Univ Nebr, BS, 57, MS, 59; NC State Univ, PhD(genetics, statist), 62. *Prof Exp:* Res geneticist, NC State Univ & Agr Res Serv, USDA, 62-68; dir comput, 85-87, PROF STATIST & HEAD, BIOMET & INFO SYSTS CTR, INST AGR & NATURAL RESOURCES, UNIV NEBR, LINCOLN, 68-, ASST VPRES & DIR, UNIV-WIDE COMPUT, 87- *Mem:* Am Soc Agron; Biomet Soc; Am Statist Asn. *Res:* Quantitative genetics, statistics and computing. *Mailing Add:* 231 Varner Hall Univ Nebr Lincoln NE 68583-0743

SCHUTZBACH, JOHN STEPHEN, b Pittsburgh, Pa, Mar 3, 41; m 62; c 4. BIOCHEMISTRY, MICROBIOLOGY. *Educ:* Edinboro State Col, BS, 63; Univ Pittsburgh, PhD(microbiol), 69. *Prof Exp:* Technician microbiol, Univ Pittsburgh, 65; res assoc, Med Col Wis, 71-72, asst prof biochem, 72-73; from asst prof to assoc prof, 73-88, PROF MICROBIOL, UNIV ALA, BIRMINGHAM, 88- *Concurrent Pos:* Fel, Med Col Wis, 69-71. *Mem:* Am Chem Soc; Am Soc Microbiol; Am Soc Biol Chemists. *Res:* Enzyme mechanisms; biosynthesis of macromolecules, particularly cell membrane constituents of mammalian cells; membrane structure and function. *Mailing Add:* Dept Microbiol Univ Ala Birmingham AL 35294

SCHUTZENHOFER, LUKE A, b East St Louis, Ill, Feb 14, 39; m 60; c 5. GAS DYNAMICS, ROTORDYNAMICS. *Educ:* St Louis Univ, BS, 60; Univ Ala, MSE, 70, PhD, 78. *Prof Exp:* Aerospace engr struct design, 60-62 aerospace engr struct vibrations, 62-64, aerospace engr unsteady gas dynamics, 64-81, BR CHIEF, SERVOMECH & SYSTS STABILITY BR, MARSHALL SPACE FLIGHT CTR, NASA, 81- *Res:* Structural vibrations; unsteady fluid and gas dynamics; applications of stochastic process theory; aero-structural interaction phenomena; statistical communication theory; stability theory; rotordynamics. *Mailing Add:* 1005 Toney Dr Huntsville AL 35802

SCHUTZMAN, ELIAS, b New York, NY, Jan 16, 25; m 60. ELECTRICAL ENGINEERING. *Educ:* NY Univ, BEE, 51, MEE, 53. *Prof Exp:* Res asst, NY Univ, 51-52, instr elec eng, 52-56, eng scientist, 56-59, asst dir, Grad Ctr, 59-66, adj assoc prof elec eng, 58-68; actg prog dir eng systs, 68, staff assoc, Elec Sci & Anal Prog, 68-72, prog dir, elec & commun prog, Eng Div, 72-85, PROG DIR, ENG RES CTR, NSF, 85- *Concurrent Pos:* Asst dir lab electrosci res & sr res scientist, NY Univ, 66-68. *Mem:* Am Soc Eng Educ; fel Inst Elec & Electronics Engrs; Sigma Xi. *Res:* Large scale computer communications networks; electronic circuits; digital systems; optical communication systems. *Mailing Add:* Div Cross Disciplinary Res NSF Washington DC 20550

SCHUUR, JERRY D, b Kalamazoo, Mich, Jan 14, 36; m 62; c 3. MATHEMATICS. *Educ:* Univ Mich, BS, 57, MA, 58, PhD(math), 63. *Prof Exp:* Engr, Boeing Airplane Co, 57; physicist, Cornell Aeronaut Lab, 59; teaching fel math, Univ Mich, 58-62, lectr, 62; from asst prof to assoc prof, 63-76, PROF MATH, MICH STATE UNIV, 76- *Concurrent Pos:* Off Naval Res fel & consult, 68-69; fel, Ital Nat Res Coun, 70; vis prof, Univ Florence, 77. *Mem:* Am Math Soc. *Res:* Ordinary differential equations. *Mailing Add:* Dept Math Mich State Univ East Lansing MI 48824

SCHUURMANN, FREDERICK JAMES, b East Grand Rapids, Mich, Jan 16, 40; m 64; c 2. MATHEMATICS. *Educ:* Calvin Col, BS, 62; Mich State Univ, MS, 63, PhD(math), 67. *Prof Exp:* Asst prof, 67-76, ASSOC PROF MATH, MIAMI UNIV, 76- *Concurrent Pos:* Vis res assoc, Aerospace Res Labs, Wright-Patterson AFB, Ohio, 73-74. *Mem:* Asn Comput Mach; Math Asn Am; Soc Indust & Appl Math. *Res:* Numerical analysis, especially computational problems of approximation theory; computational problems of multivariate statistical analysis. *Mailing Add:* Dept Math & Statist Miami Univ Oxford OH 45056

SCHUURMANS, DAVID MEINTE, b Ithaca, Mich, Apr 11, 28; m 51; c 2. INDUSTRIAL MICROBIOLOGY. *Educ:* Albion Col, BA, 49; Mich State Univ, MS, 51, PhD(bact), 54. *Prof Exp:* Jr bacteriologist, City Health Dept, Jackson, Mich, 49-50; bacteriologist, Div Antibiotics & Fermentation, Mich Dept Pub Health, 53-55, chief, Tissue Cult Unit, 57-61, chief, Antibiotic Screening Sect, 65-77 & Fermentation Sect, 68-77, chief, Div Antibiotics & Fermentation, 77-80, dep chief, biol prod prog, 80-84; RETIRED. *Mem:* Fedn Am Scientists. *Res:* Development of antitumor antibiotics; development and production of bacterial vaccines. *Mailing Add:* Pub Health Bur Dis Control Lab 3500 N Logan St Box 30035 Lansing MI 48909

SCHUURMANS, HENDRIK J L, b Malang, Indonesia, Dec 26, 28; US citizen; m 56, 74; c 5. POLYMER CHEMISTRY. *Educ:* State Univ Leiden, BS, 52, MS, 55, PhD(phys chem), 56. *Prof Exp:* Sr res chemist, Monsanto Co, Tex, 56-61; supvr, mgr & sr develop assoc, Mobil Chem Co, NJ & NY, 61-67; tech dir, Stein-Hall & Co, Inc, NY, 67-71; res dir, M&T Chem Inc, NJ, 71-74, dir new ventures, 74-75; dir res & develop, Plastics Div, ICI Am Inc, 75-82; PRES, SERVOCHEM INC, 82- *Concurrent Pos:* Consult, Med Br, Univ Tex, 58-59. *Mem:* Soc Plastics Engrs; Am Chem Soc; Photo Mkt Asn Int; Tech Asn Pulp & Paper Indust. *Res:* Natural and synthetic polymer evaluation; polymer synthesis, production and processing; thermodynamics and kinetics of rate processes; catalysis studies. *Mailing Add:* 615 Rossmore Rd Richmond VA 23225

SCHUYLER, ALFRED ERNEST, b Salamanca, NY, July 10, 35; m 68. BOTANY, TAXONOMY. *Educ:* Colgate Univ, AB, 57; Univ Mich, AM, 58, PhD(bot), 63. *Prof Exp:* Asst cur, 62-69, chmn bot dept, 62-75, ASSOC CUR, ACAD NATURAL SCI, PHILADELPHIA, 69- *Concurrent Pos:* Am Philos Soc traveling grants, 63-64, 66 & 72 & res grant, Franklin Inst, 70; vis prof, Mich State Univ, 66, Univ Mont Biol Sta, 78, 82, 85 & 87, Rutgers Univ, Camden, 81 & 82; vis lectr, Swarthmore Col, 67-70, 78 & 82; res assoc, Morris Arboretum, Univ Pa, 71-; ed, Bartonia, 71-; mem adv comt, Syst Resources Bot, NSF, 72-74; care & maintenance bot col, Acad Natural Sci, NSF, 72-77, assoc ed, Sci Publ, 78-82; chmn, Jessup-McHenry Comt, Acad Nat Sci, 88- *Mem:* Am Soc Plant Taxon; Bot Soc Am; Am Inst Biol Scientists. *Res:* Taxonomic research in the Cyperaceae; biology of aquatic vascular plants; environmental consulting; flora of the Delaware River system. *Mailing Add:* Dept Bot Acad Nat Sci 19th & Pkwy Philadelphia PA 19103

SCHUYTEMA, EUNICE CHAMBERS, b Rochester, NY, Feb 4, 29; m 54; c 1. MEDICAL MICROBIOLOGY. *Educ:* Cornell Univ, BS, 51, MS, 54; Univ Iowa, PhD, 56. *Prof Exp:* Res biochemist, Abbott Labs, 56-68; asst prof biol chem, Univ Ill Med Ctr, 68-71; Asst Dean Preclin Curric, Rush Med Col, 79-85; ASST PROF MICROBIOL, RUSH PRESBY ST LUKE'S MED CTR, 71- *Mem:* Am Soc Microbiol. *Res:* Nucleic acid chemistry; anaerobes; bacteriophage. *Mailing Add:* 4756 Crayton Ct Naples FL 33940

SCHUYTEN, JOHN, b Seattle, Wash, July 17, 14; m 39; c 3. METALLURGY. *Educ:* Univ Wash, BS, 36; Univ Calif, Berkeley, MS, 51. *Prof Exp:* Jr engr, Shell Chem Co, 36-39, metallurgist, 40-42, metallurgist, Shell Develop Co, 43-49; instr eng chem & metall, Contra Costa Col, 50-54; metallurgist, Volcan Foundry Co, 55-57, vpres, 57-69, pres, 70-86; METALL & FOUNDRY CONSULT, 86- *Concurrent Pos:* Dir, Iron Castings Soc, 74-78 & 80-82. *Mem:* Fel Am Soc Metals; Iron Castings Soc (secy, 79). *Res:* Hydrogen attack and stress corrosion of metals; development of ductile iron and alloys for heat, abrasion and corrosion resistance. *Mailing Add:* 260 Arlington Ave Berkeley CA 94707

SCHWAB, ARTHUR WILLIAM, b Minneapolis, Minn, July 17, 17; m 45; c 4. ORGANIC CHEMISTRY. *Educ:* Univ Minn, BChem, 41; Bradley Univ, PhD(chem), 52. *Prof Exp:* Chemist, WPoint Mfg Co, 41-42; RES CHEMIST, NORTHERN REGIONAL RES LAB, USDA, 42- *Mem:* Am Chem Soc; Am Oil Chem Soc. *Res:* Fundamental research on the chemistry of vegetable oils and modification of these oils for industrial utilization. *Mailing Add:* 2223 W Albany Ave Peoria IL 61604

SCHWAB, BERNARD, b Worcester, Mass, Dec 26, 26; m 57; c 3. CLINICAL MICROBIOLOGY. *Educ:* Clark Univ, AB, 49; Univ Mass, MS, 51; Nat Registry Microbiol, registered, 64, specialist med microbiol, 69, specialist food microbiol, 75. *Prof Exp:* Bacteriologist, Vt State Bur Labs, 52-62; bacteriologist in-chg lab, City of Kingston, NY, 62-64; chief bacteriologist, Erie Co Lab, Buffalo, 63-68; microbiologist in-chg diag & spec probs sect, 68-73, sr staff officer microbiol staff, Meat & Poultry Inspection Prog, Sci Serv Staff, Food & Safety Qual Serv, 73-77, sr staff officer microbiol div, 77-80, CHIEF, MED MICROBIOL BR, MICROBIOL DIV, SCI FOOD SAFETY & INSPECTION SERV, USDA, 81- *Mem:* Am Soc Microbiol; fel Am Asn Vet Lab Diagnosticians. *Res:* Public health microbiology; meat and poultry microbiology related to consumer protection programs; swine mycobacteriology, staphylococcus enterotin, antibiotics, species determination. *Mailing Add:* Bldg 322 Agr Res Ctr-E USDA Beltsville MD 20705

SCHWAB, ERNEST ROE, b Denver, Colo, July 19, 50; m 74. NEUROETHOLOGY, BIOACOUSTICS. *Educ:* Union Col, Lincoln, Nebr, BA, 76; Andrews Univ, Berrien Springs, Mich, MS, 82; Loma Linda Univ, Calif, PhD(biol), 89. *Prof Exp:* Grad asst physiol, Andrews Univ, Berrien Springs, Mich, 75-78 & Univ Notre Dame, Ind, 78-80; sci teacher health & biol, South Bend Jr Acad Ind, 80-81; grad asst res, Loma Linda Univ, Calif, 81-83, instr biol, Riverside, Calif, 83-90; ASST PROF BIOL, LA SIERRA UNIV, RIVERSIDE, CALIF, 90- *Concurrent Pos:* Vis res prof, Andrews Univ, Berrien Springs, Mich, 89. *Mem:* Sigma Xi; Entom Soc Am; NY Acad Sci; Am Soc Zoologists; Int Union Study Social Insects; Am Inst Biol Sci. *Res:* Role of individual nevrons in cricket courtship behavior as well as the modulation of that behavior by juvenile hormone. *Mailing Add:* 1461 Kevin Ave Redlands CA 92373

SCHWAB, FREDERIC LYON, b Brooklyn, NY, Jan 8, 40; m 65; c 4. SEDIMENTOLOGY. *Educ:* Dartmouth Col, AB, 61; Univ Wis, MS, 63; Harvard Univ, PhD(geol), 68. *Prof Exp:* From asst prof to assoc prof geol, 67-75, PROF GEOL, WASHINGTON & LEE UNIV, 75- *Concurrent Pos:* NSF sci fac fel, Univ Edinburgh, 71-72; consult ed, McGraw-Hill Dict Sci & Technol, Encycl Geol Sci & Encycl Energy; NATO sr scientist, Univ Grenoble, 77-78; mem adv bd, Petrol Res Fund, 85-; ed, Geosynclines. *Mem:* Int Asn Sedimentologists; Geol Soc London; Geol Soc Am; Soc Econ Mineralogists & Paleontologists. *Res:* Depositional environments, provenance and sedimentary tectonics of precambrian of the Blue Ridge; geochemistry of sedimentary rocks; sedimentation and tectonic history of the Cordilleran belt; sedimentation trends through time; Eocambrian Appalachian sediments; origin-early evolution Appalachian-Caledonide belt. *Mailing Add:* 916 Shenandoah Rd Lexington VA 24450

SCHWAB, FREDERICK CHARLES, b Meadville, Pa, Mar 1, 37; m 61; c 3. POLYMER CHEMISTRY. *Educ:* Cleveland State Univ, BChE, 61; Union Col, NY, MS, 66; Univ Akron, PhD(polymer sci), 69. *Prof Exp:* Chemist insulation develop, Gen Elec Co, 61-66; ASSOC CHEMIST POLYMERS, MOBIL CHEM CO, 69- *Mem:* Am Chem Soc. *Res:* Synthesis, characterization and mechanical properties of block polymers. *Mailing Add:* 34 Spear St Metuchen NJ 08840

SCHWAB, GLENN O(RVILLE), b Gridley, Kans, Dec 30, 19; m 51; c 3. AGRICULTURAL ENGINEERING. *Educ:* Kans State Univ, BS, 42; Iowa State Univ, MS, 47, PhD(agr eng, soils), 51. *Prof Exp:* From instr to prof agr eng, Iowa State Univ, 46-56; prof, 56-84, EMER PROF AGR ENG, OHIO STATE UNIV, 85- *Concurrent Pos:* Consult, Off State Exp Sta, USDA, 59-62, Sheladia Assoc, 86-88. *Honors & Awards:* Hancock Brick & Tile Drainage Eng Award, Am Soc Agr Engrs, 68, John Deere Gold Medal, 87. *Mem:* AAAS; life fel Am Soc Agr Engrs; Am Soc Eng Educ; AAAS; Soil Conserv Soc Am. *Res:* Agricultural drainage; irrigation; erosion and flood control; agricultural hydrology. *Mailing Add:* Dept Agr Eng Ohio State Univ 590 Woody Hayes Dr Columbus OH 43210

SCHWAB, HELMUT, b Nurnberg, Germany, Apr 3, 29; nat US; m 53; c 2. ANALYTICAL CHEMISTRY, PHYSICAL CHEMISTRY. *Educ:* Rutgers Univ, BS, 51, PhD(chem), 57. *Prof Exp:* Asst, Rutgers Univ, 53-54; res chemist, Nat Cash Register Co, 56-59, sect head, 59-72; mgr, mat res, Appleton Papers Inc, 73-90. *Mem:* Am Chem Soc; Soc Photograph Scientists & Engrs. *Res:* Heat and pressure sensitive recording media; color forming systems; photochromic systems; carbonless papers; color-blocked dyes; paper chemistry; organic chemistry; thermal papers. *Mailing Add:* 136 Crestview Dr Appleton WI 54915

SCHWAB, JOHN HARRIS, b Minn, Nov 20, 27; m 51; c 4. BACTERIOLOGY, IMMUNOLOGY. *Educ:* Univ Minn, BA, 49, MS, 50, PhD(bact), 53. *Prof Exp:* Asst bact, Univ Minn, 49-53; from instr to assoc prof, 53-67, PROF BACT & IMMUNOL, MED SCH, UNIV NC, CHAPEL HILL, 67- *Concurrent Pos:* NIH fel, Lister Inst, London, Eng, 60-61; mem, Med Res Coun Rheumatism Res Unit, Taplow, Eng, 68-69; Josiah Macy, Jr Found fac scholar, Radiobiol Inst, Rijswijk, Neth, 75-76; Fogarty Int fel exp immunother Inst Pasteur, Paris, 85-86. *Mem:* AAAS; Am Soc Microbiol; Am Asn Immunol. *Res:* Bacterial immunosuppressants; microbial factors in chronic inflammatory diseases; toxicity of bacterial cell walls; experimental models of rheumatic carditis and rheumatoid arthritis. *Mailing Add:* Dept Microbiol & Immunol Sch Med Univ NC CB 7290 Chapel Hill NC 27514

SCHWAB, JOHN J, b Cumberland, MD, Feb 10, 23; m 45; c 1. PSYCHIATRY, PSYCHOSOMATIC MEDICINE. *Educ:* Univ Ky, BS, 46; Univ Louisville, MD, 46; Univ Ill, MS, 49; Am Bd Internal Med, dipl, 55; Am Bd Psychiat & Neurol, dipl, 64. *Prof Exp:* From asst resident to resident med, Louis Gen Hosp, Ky, 49-50; internist & psychosomaticist, Holzer Clin, Gallipolis, Ohio, 54-59; resident psychiat, Col Med, Univ Fla, 59-61, from instr to asst prof, 62-65, assoc prof, 65-67, prof psychiat & med, 65-74, chief psychosom consult serv, 61-64, dir consult-liaison prog, 64-67, dir residency prog, 66-71; CHMN & PROF PSYCHIAT & BEHAV SCI, SCH MED, UNIV LOUISVILLE, 74- *Concurrent Pos:* Med fel, Col Med, Univ Ill, 48-49; fel psychosom med, Duke Univ, 51-52; NIMH career teacher award, 62-64; internist, Yokohama Army Hosp, Japan, 52-54; state trustee, Ment Health Fedn Ohio, 58-61; proj dir res grant, 65-68; prin investr, NIMH Res Grant, 69-73; chmn epidemiol studies rev comt, Ctr Epidemiol Studies, NIMH, 73-75; chmn coun res & develop, Am Psychiat Asn, 74-75; mem bd regents, Am Col Psychiat, 79-81 & bd dirs, Group Advan Psychiat, 85-87. *Honors & Awards:* Laughlin Award, Am Soc Physician Analysts. *Mem:* Fel Acad Psychosom Med (pres, 70-71); fel Am Asn Social Psychiat (pres, 71-73); AMA; fel Am Psychiat Asn; fel Am Col Psychiat; Group Advan Psychiat. *Res:* Applicability, both theoretical and practical, of psychiatric concepts to general medicine; establishing guidelines for the identification and management of medical patients whose illnesses are complicated by emotional distress; sociocultural aspects of mental illness; psychiatric epidemiology; risk for depression in the family. *Mailing Add:* Dept Psychiat & Behav Sci Sch Med Univ Louisville Louisville KY 40292-0001

SCHWAB, LINDA SUE, b St Louis, Mo, Oct 25, 51. MEDICINAL CHEMISTRY, NEUROCHEMISTRY. *Educ:* Wells Col, BA, 73; Univ Rochester, MS, 75, PhD(chem), 78. *Prof Exp:* Fel, Ctr Brain Res, Sch Med, Univ Rochester, 77-78, assoc neurochem, 78-82; lectr, 83-86, asst prof, 86-89, ASSOC PROF, DEPT CHEM, WELLS COL, AURORA, NY, 89- *Concurrent Pos:* Med chem & org chem div, Am Chem Soc. *Mem:* Am Chem Soc; Affil mem, Int Union Pure Appl Chem. *Res:* Chemistry of heterocycles; steroselective reactions. *Mailing Add:* Dept Chem Wells Col Aurora NY 13026

SCHWAB, MICHAEL, b New York, NY, Aug 9, 39; m 68; c 2. MATERIALS SCIENCE. *Educ:* Calif Inst Technol, BS, 61; Univ Calif, Berkeley, PhD(physics), 68. *Prof Exp:* Physicist, Lawrence Berkeley Lab, 68-69; PHYSICIST MAT SCI, LAWRENCE LIVERMORE LAB, 69- *Mem:* Am Phys Soc; AAAS. *Res:* Spectroscopy; nuclear and electron magnetic resonance; materials science; radiation damage; metallurgy. *Mailing Add:* 6215 Harwood Ave Oakland CA 94618

SCHWAB, PETER AUSTIN, polymer chemistry, for more information see previous edition

SCHWAB, ROBERT G, b Park Rapids, Minn, Mar 1, 32; m 51; c 3. PHYSIOLOGY. *Educ:* Univ Minn, BS, 58, MS, 65; Univ Ariz, PhD(zool), 66. *Prof Exp:* Asst prof, 64-74, ASSOC PROF WILDLIFE BIOL, COL AGR & ENVIRON SCI, UNIV CALIF, DAVIS, 74- *Mem:* Am Soc Mammalogists. *Res:* Environmental physiology; mammalogy. *Mailing Add:* Dept Wildlife & Fisheries Biol Univ Calif Davis CA 95616

SCHWAB, WALTER EDWIN, b Mexico City, Mex, Jan 4, 41; US citizen; m 64. NEUROBIOLOGY. *Educ:* Georgetown Univ, BS, 64; Va State Col, MS, 71; Univ Md, MS, 73, PhD(zool), 74. *Prof Exp:* Fel res assoc, Dept Develop & Cell Biol, Univ Calif, Irvine, 74-77; asst prof neurobiol, Dept Biol, 77-84, DIR COMP SERV, VA POLYTECH INST & STATE UNIV, 84- *Mem:* Am Soc Zoologists; AAAS; Am Inst Biol Sci. *Res:* Physiology and morphology of intracellular communication in primitive animals, primarily coelenterates. *Mailing Add:* Dept Comp Serv Col Vet Med Va Polytech Inst & State Univ Blacksburg VA 24061

SCHWABE, ARTHUR DAVID, b Varel, Ger, Feb 1, 24; US citizen; m 46. MEDICINE, GASTROENTEROLOGY. *Educ:* Univ Calif, Berkeley, AA, 51; Univ Chicago, MD, 56. *Prof Exp:* From intern to assoc resident, Univ Calif, Los Angeles, 56-59, chief resident, 60-61, from instr to assoc prof, 61-71, actg chmn dept, 72, vchmn dept, 72-74, prof med, Med Ctr, 71-89, chief gastroenterol, Dept Med, 67-88, EMER PROF MED, UNIV CALIF, LOS ANGELES, 89- *Concurrent Pos:* Ambrose & Gladys Bowyer fel med, Univ Calif, Los Angeles, 58 & 59, USPHS fel gastroenterol, 59-60; chief gastroenterol, Harbor Gen Hosp, Torrance, Calif, 62-67; consult, Vet Admin Ctr, Los Angeles, 64- *Mem:* Fel Am Col Physicians; NY Acad Sci; Western Asn Physicians; Am Gastroenterol Asn. *Res:* Intestinal fat absorption; familial Mediterranean Fever; intestinal neoplasia. *Mailing Add:* Dept Med Ctr Health Sci Univ Calif Los Angeles CA 90024-1684

SCHWABE, CALVIN WALTER, b Newark, NJ, Mar 15, 27; m 51; c 2. EPIDEMIOLOGY, PARASITOLOGY. *Educ:* Va Polytech Inst, BS, 48; Univ Hawaii, MS, 50; Auburn Univ, DVM, 54; Harvard Univ, MPH, 55, ScD(trop pub health), 56. *Prof Exp:* Assoc prof parasitol & chmn dept, Sch Med, Am Univ Beirut, 56-57, assoc prof parasitol & trop health & chmn dept trop health, Schs Pub Health & Med, 57-62, prof parasitol & epidemiol & asst dir, Sch Pub Health, 62-66; chmn dept epidemiol & prev med, Sch Vet Med, Univ Calif, Davis, 66-70, assoc dean, 70-71, prof epidemiol, Sch Med, 67-81, PROF EPIDEMIOL, SCH VET MED, UNIV CALIF, DAVIS, 66-; PROF EPIDEMIOL, SCH MED, UNIV CALIF, SAN FRANCISCO, 67- *Concurrent Pos:* USPHS res fel, Harvard Univ, 54-56 & Cambridge Univ,

72-73; Fulbright fel, Makerere Univ Col, Kenya, 61, Cambridge Univ, 72-73 & Univ Khartoum, 79-80; mem, WHO Secretariat, 64-66, consult, 60-; consult, Pan Am Health Orgn, 61-; adj prof, Agr History Ctr, Univ Calif Davis, 84- *Honors & Awards:* K F Meyer Goldheaded Cane Award. *Mem:* Am Soc Trop Med & Hyg; Am Soc Parasitol; Am Vet Med Asn. *Res:* Tropical public health and agriculture; veterinary medicine and human health; third world development; history of veterinary medicine. *Mailing Add:* Dept Epidemiol & Prev Med Univ Calif Davis CA 95616

SCHWABE, CHRISTIAN, b Flensburg, Ger, May 10, 30; US citizen. BIOLOGICAL CHEMISTRY. *Educ:* Univ Hamburg, DDS, 55; Univ Iowa, DDS, 60, MS, 63, PhD, 65. *Prof Exp:* Pvt pract, Ger, 55-56; res asst, Clinton Corn Processing Co, Iowa, 56-57; res asst stomatol, Col Dent, Univ Iowa, 57-60, instr biochem, Col Med, 63-65; instr biol chem, Sch Dent Med, Harvard Univ, 65, assoc, 66-67, asst prof, 67-71; assoc prof, 71-76, PROF BIOCHEM, MED UNIV SC, 76- *Concurrent Pos:* NIH career develop award, 66. *Mem:* Sigma Xi; Am Chem Soc; Am Soc Biol Chemists; Endocrine Soc; Int Asn Dent. Res. *Res:* Connective tissue; enzymology; protein chemistry; chemical endocrinology. *Mailing Add:* Biochem Dept Med Univ SC 171 Ashley Ave Charleston SC 29425

SCHWABER, JERROLD, b Evanston, Ill, May 24, 47. ANTIBODY DEFICIENCY DISEASES, MOLECULAR BASIS OF ANTIBODY DIVERSITY. *Educ:* Univ Chicago, BA, 69, PhD(biophys & theoret biol), 74. *Prof Exp:* Fel pediat, 74-78, instr pediat, 78-81, ASST PROF PATH, HARVARD MED SCH, 81- *Concurrent Pos:* Fel, Children's Hosp, Boston, 74-78, res assoc, 78- *Mem:* Am Asn Immunologists; Am Soc Cell Biol. *Res:* Antibody gene rearrangement and expression, especially the failure of gene rearrangement in antibody deficiency diseases; lymphocyte development and differentiation. *Mailing Add:* Harvard Med Sch Ctr Blood Res 800 Huntington Ave Boston MA 02115

SCHWAIGHOFER, JOSEPH, b Annaberg, Austria, Apr 16, 24. ENGINEERING MECHANICS. *Educ:* Graz Tech Inst, Dipl Ing, 51; Pa State Univ, MS, 57, PhD(eng mech), 58; Graz Univ, Dr Tech(civil eng), 65. *Prof Exp:* Asst prof eng mech, Pa State Univ, 58-59; from asst prof to assoc prof civil eng, 59-69, PROF CIVIL ENG, UNIV TORONTO, 69- *Concurrent Pos:* ASTEF French Govt res fel, Univ Nancy, 61. *Mem:* Soc Exp Stress Anal; Am Soc Testing & Mat. *Res:* Photoelasticity and experimental stress analysis; structural engineering. *Mailing Add:* Dept Civil Eng Univ Toronto Toronto ON M5S 1U1 Can

SCHWALB, MARVIN N, b New York, NY, May 23, 41; m 62; c 3. MYCOLOGY, MICROBIOLOGY. *Educ:* State Univ NY Buffalo, BA, 63, PhD(biol), 67. *Prof Exp:* Asst prof biol, Bridgewater State Col, 68-69; from instr to assoc prof, 69-78, PROF MICROBIOL, NJ MED SCH, UNIV MED & DENT, NJ, 78- *Concurrent Pos:* Fel, Brandeis Univ, 67-68. *Mem:* Am Soc Microbiol; Mycol Soc Am. *Res:* Biochemical and genetic regulation of development in higher fungi; recombinant DNA genetic systems in higher fungi. *Mailing Add:* Dept Microbiol NJ Med Sch Univ Med & Dent NJ 185 S Orange Ave Newark NJ 07103

SCHWALBE, LARRY ALLEN, b Austin, Tex, Feb 3, 45; m 68; c 2. SOLID STATE PHYSICS. *Educ:* Univ Ill, BS, 68, MS, 69, PhD(physics), 73. *Prof Exp:* Fel physics, Tech Univ Munich, 73-75; fel, 75-77, MEM STAFF PHYSICS, LOS ALAMOS NAT LAB, 77- *Res:* Theoretical and experimental studies of high explosives and explosively-driven metal systems, behavior of materials under conditions of high pressure and high strain rates. *Mailing Add:* Los Alamos Nat Lab PO Box 1663 Mail Stop F663 Los Alamos NM 87545

SCHWALL, DONALD V, b Nicolaus, Calif, July 2, 31; m 55; c 3. FOOD TECHNOLOGY, POULTRY SCIENCE. *Educ:* Calif State Polytech Col, BS, 53; Tex A&M Univ, MS, 60; Purdue Univ, PhD(food technol), 62. *Prof Exp:* Head poultry prod res & develop, Food Res Div, Armour & Co, 62-70; dir qual assurance & res & develop, 70-76, DIR RES & DEVELOP, VAN CAMP SEA FOOD DIV, RALSTON PURINA CO, 76- *Concurrent Pos:* Mem res coun, Poultry & Egg Inst Am, 65-, res award, 71. *Mem:* Inst Food Technologists; Poultry Sci Asn; World Poultry Sci Asn. *Res:* Direction and coordination of sea food research programs. *Mailing Add:* 5014 Walnut Apt B Kansas City MO 64112

SCHWALL, RICHARD JOSEPH, b Evanston, Ill, Oct 11, 49. DATABASE MANAGEMENT, APPLIED STATISTICS. *Educ:* Calif Inst Technol, BS, 71; Northwestern Univ, PhD(analytical chem), 77. *Prof Exp:* Prog mgr, Rockwell Int Environ Monitoring & Serv Ctr, 77-84; PRIN SCIENTIST, ABB ENVIRON SERV, 84- *Res:* Fast electrochemical measurements (kinetic and analytical) by FFT techniques; visibility degradation measurements and theory; statistical analysis of pollution data; data configuration management; fugitive emission modelling; speculative metaphysics. *Mailing Add:* ABB Environ Serv 4765 Calle Quetzal Camarillo CA 93012

SCHWALL, ROBERT E, b Detroit, Mich, May 31, 47. PHYSICS. *Educ:* St Mary's Univ Tex, BS, 68; Stanford Univ, PhD(appl physics), 73. *Prof Exp:* Mgr, Intermagnetics Gen Corp, 78-84; RES STAFF MEM, THOMAS J WATSON RES CTR, IBM CORP, 88-, MGR LITHOGRAPHIC OPTICS, 90- *Mem:* Am Phys Soc; Mat Res Soc. *Res:* Superconductivity; low temperature physics; semiconductor packaging; cryogenic electronics. *Mailing Add:* Thomas J Watson Res Ctr PO Box 218 IBM Corp Yorktown Heights NY 10598

SCHWALM, FRITZ EKKEHARDT, b Arolsen, Ger, Feb 17, 36; m 62; c 3. ZOOLOGY, DEVELOPMENTAL PHYSIOLOGY. *Educ:* Univ Marburg, PhD(zool), 64. *Prof Exp:* Lectr Ger, Folk Univ, Sweden, 59-60; res assoc exp embryol, Univ Marburg, 64-65; Anglo Am Corp SAfrica advan res fel electron micros, Univ Witwatersrand, 66-67; trainee, Univ Va, 68; res assoc embryol, Univ Notre Dame, 68-70; from asst prof to assoc prof biol, Ill State Univ, 70-82, actg chmn, 79-81; assoc prof, 82-87, PROF BIOL, TEX WOMENS UNIV, 87-, CHMN DEPT, 82- *Mem:* AAAS; Soc Develop Biol; Ger Zool Soc; Am Soc Zoologists. *Res:* Ultrastructure and biochemistry of oogenesis and embryogenesis in insects. *Mailing Add:* Dept Biol Tex Womans Univ Denton TX 76204

SCHWALM, MIZUHO K, b Tokyo, Japan, Sept 23, 40; m 77. SOLID STATE PHYSICS. *Educ:* Tokyo Gakugei Univ, BA, 64; Univ Wyo, MS, 70; Mont State Univ, PhD(physics), 78. *Prof Exp:* Asst instr, Univ Utah, 79-80; asst prof physics, Moorehead State Univ, 81-82; MEM STAFF, PHYSICS DEPT, UNIV NDAK, 82- *Mem:* Sigma Xi. *Res:* Methods for studying electronic structure and transport in disordered systems; computation of optical selection rules; modest size electron energy band calculations. *Mailing Add:* Dept Physics Univ NDak Box 8008 Grand Forks ND 58202

SCHWALM, WILLIAM A, b Portsmouth, NH, March 3, 47; m 77. SOLID STATE THEORY, GRAPH THEORY. *Educ:* Univ NH, BS, 69; Mont State Univ, PhD(physics), 78. *Prof Exp:* Fel physics, Univ Utah, 78-79, instr, 79-80; from asst prof to assoc prof, 80-90, PROF PHYSICS, UNIV NDAK, 90- *Concurrent Pos:* Vis asst prof, Univ Minn, Minneapolis, 83; vis assoc prof, Mont State Univ, 90-91. *Mem:* Am Asn Physics Teachers; Am Phys Soc; Math Asn Am; Sigma Xi. *Res:* Electronic structure of surfaces and thin films; transport properties in two dimensional systems; mathematical and computational methods; fractal lattices; transport in two dimensional systems; hierchical models applied to amorphous materials; spectral properties of families of infinite graphs. *Mailing Add:* Physics Dept Univ NDak Grand Forks ND 58202-8008

SCHWAN, HERMAN PAUL, b Aachen, Ger, Aug 7, 15; nat US; m 49; c 5. BIOPHYSICS, BIOMEDICAL ENGINEERING. *Educ:* Univ Frankfurt, PhD(biophys), 40, Dr habil, 46. *Hon Degrees:* DSc, NC, Univ Pa, 86. *Prof Exp:* Asst, Kaiser Wilhelm Biophys, 37-40, res assoc, 40-45, admin dir, 45-47; res specialist, US Naval Base, Pa, 47-50; asst prof physics in med, 50-60, asst prof phys med, 50-52, chmn biomed electronic eng, 61-73, prof bioeng, sch eng, 75-83, ASSOC PROF PHYS MED, SCH MED, UNIV PA, 52-, PROF ELEC ENG, MOORE SCH ELEC ENG & PROF ELEC ENG IN PHYS MED, SCH MED, 57-, EMER PROF BIOENG, SCH ENG, 83- *Concurrent Pos:* Asst prof, Univ Frankfurt, 46-55, vis prof, 62; vis MacKay prof, Univ Calif, Berkeley, 56; lectr, Johns Hopkins Univ, 62-66; sci mem, Max Planck Inst Biophys, 62-; consult, Gen Elec Co, 57-76, US Army & USN, 57-71 & NIH, 61-76; mem nat comt C95, Am Nat Stands Inst, 61- & Nat Acad Sci-Nat Res Coun, 67-73; mem nat adv coun environ health, Dept HEW, 69-71; Vis W W Clyde prof, Univ Utah, Salt Lake, 80; vis prof, Univ Wurzburg, Ger, 86-87. *Honors & Awards:* Cert of Commendation, Dept HEW, 66 & 72; Inst Elec & Electronics Engrs Awards, 53, 63 & 67; Rajewsky Prize for Biophys, 74; Edison Medal, Inst Elec & Electronics Engrs, 83 & Centennial Medal, 84; d'Arsonval Award, Bioelectromagnetics Soc, 85; Alexander von Humboldt Award, 80. *Mem:* Fel AAAS; Biophys Soc; Biomed Eng Soc; fel Inst Elec & Electronics Engrs; Nat Acad Eng; hon mem Ger Biophys Soc. *Res:* Impedance measurement techniques and electrodes for biological dielectric research; biophysics of nonionizing radiation; biomedical engineering; electrical and acoustical properties of tissues and cells; electrical properties of biological membranes and biopolymers; electrical engineering. *Mailing Add:* Dept Bioeng D3 Univ Pa Philadelphia PA 19104

SCHWAN, JUDITH A, b Apr 16, 25. CHEMICAL ENGINEERING. *Educ:* Univ Cincinnati, ChE, 48; Cornell Univ, MS, 50. *Prof Exp:* Res scientist, Eastman Kodak Co, 50-65, lab head, 65-68, asst div dir, 68-71, div dir, 71-75, asst dir res labs, 75-86, dir photog res labs, photog prod, 86-87; RETIRED. *Honors & Awards:* Herbert T Kalmus Mem Award, Soc Motion-Picture & TV Engrs, 79. *Mem:* Nat Acad Eng; fel Soc Motion-Picture & TV Engrs; Soc Photog Scientists & Engrs; Am Chem Soc. *Res:* Management, primarily in the area of photographic research and development. *Mailing Add:* 45 Park Ave Middleport NY 14105-1354

SCHWAN, THEODORE CARL, b Florida, Ohio, June 17, 18; m 44; c 4. ORGANIC CHEMISTRY, POLYMER CHEMISTRY. *Educ:* Valparaiso Univ, AB, 41; Univ Notre Dame, MS, 49, PhD(chem), 53. *Prof Exp:* Phys tester & chem analyst, US Rubber Co, Ind, 38-39, asst to res chemist, 41-42; chem analyst, Ind Steel Prods Co, 40-41; from instr to assoc prof chem, Valparaiso Univ, 48-62, chmn dept, 57-59, prof chem, 62-84; prof chem, Calif Lutheran Univ, Thousand Oaks, Calif, 84-90; EMER PROF CHEM, VALPARAISO UNIV, 90- *Concurrent Pos:* Res & develop chemist, Continental-Diamond Fibre Corp, Ind, 53-58; grant, Petrol Res Fund, Am Chem Soc, 59 & 63; Univ Ky prof, Univ Indonesia, Int Coop Admin, 59-61, chief of party, 61-62; UN Educ Sci & Cult Orgn vis prof, Haile Selassie Univ, 68-70; prof chem, Col Med & Med Sci, King Faisal Univ, Dammam, Saudi Arabia, 77-79, chmn dept, 78-79. *Mem:* emer mem Am Chem Soc. *Res:* Copolymerization of unsaturated organic compounds by free radical and heterogeneous catalyst systems; water as a fuel additive for internal combustion engines. *Mailing Add:* 154 McIntyre Ct Valparaiso IN 46383

SCHWAN, THOMAS JAMES, b Medina, NY, June 19, 34; m 61; c 3. MEDICINAL CHEMISTRY. *Educ:* St Bonaventure Univ, BS, 56, MS, 58; State Univ NY, Buffalo, PhD(org chem), 65. *Prof Exp:* Res fel med chem, State Univ NY, Buffalo, 64-65; sr res chemist, 65-68, unit leader org chem sect, 68-78, mgr, 78-80, dir, develop planning, 80-83, DIR, PROD DEVELOP, NORWICH-EATON PHARMACEUT, 83- *Mem:* Am Chem Soc. *Res:* Synthesis and transformations of nitrogen heterocycles; enzyme inhibitors; preparation of compounds of potential pharmacological activity; reaction mechanisms. *Mailing Add:* Eight Hillview Dr Norwich NY 13815

SCHWANDT, PETER, b Gottingen, Ger, Apr 7, 39; US citizen; m 63; c 1. NUCLEAR PHYSICS. *Educ:* Ind Univ, Bloomington, BS, 61; Univ Wis-Madison, MS, 63, PhD(physics), 67. *Prof Exp:* Res assoc nuclear physics, Univ Wis-Madison, 67-68 & Univ Colo, Boulder, 68-69; from asst prof to assoc prof, 69-80, PROF PHYSICS, IND UNIV, BLOOMINGTON, 80- *Mem:* Am Phys Soc; Sigma Xi. *Res:* Medium-energy physics; reaction mechanisms; spin dependent interactions; potential models for composite particle scattering; heavy-ion interactions. *Mailing Add:* Dept Physics Ind Univ Bloomington IN 47401

SCHWARCZ, ERVIN H, b Cleveland, Ohio, Aug 22, 24; m 48; c 4. PHYSICS. *Educ:* Ohio State Univ, BS, 45; Univ Mich, MS, 48, PhD(physics), 55. *Prof Exp:* Res physicist, Lawrence Radiation Lab, Univ Calif, 54-67; PROF PHYSICS, STANISLAUS STATE COL, 67- *Mem:* AAAS; Am Asn Physics Teachers; Am Phys Soc. *Res:* Nuclear structure and optical model analysis of elastic and quasi-elastic scattering. *Mailing Add:* Dept Physics Stanislaus State Col Turlock CA 95380

SCHWARCZ, HENRY PHILIP, b Chicago, Ill, July 22, 33; m 64; c 1. STABLE ISOTOPES, ARCHAEOMETRY. *Educ:* Univ Chicago, BA, 52; Calif Inst Technol, MSc, 55, PhD(geol), 60. *Prof Exp:* Res assoc isotopic geochem, Enrico Fermi Inst Nuclear Studies, Univ Chicago, 60-62; from asst prof to assoc prof, 62-72, chmn dept, 88-91, PROF GEOL, MCMASTER UNIV, 72- *Concurrent Pos:* Mem mineral, geochem & petrol subcomt, Nat Adv Coun Res-Geol Soc Can, 65-72; Fulbright fel, Nuclear Geol Lab, Univ Pisa, 68-69; mem subcomt isotopes & geochronology, Assoc Comt Geol & Geophys, Nat Res Coun Can, 72-74, mem subcomt meteorites, 77-; vis prof, Hebrew Univ Jerusalem, 75-76, 82-83; vis scientist, Res Lab Archeol, Oxford Univ, 78; assoc mem, dept anthrop, McMaster Univ, 88-; vis fel, Clare Hall, Univ Cambridge, 91-92; coun, Am Quaternary Asn, 90 & Acad Sci, Royal Soc Can. *Mem:* Fel Geol Soc Am; fel Geol Asn Can; Geochem Soc; fel Royal Soc Can; Am Quaternary Asn. *Res:* Stable isotope geochemistry; archeology; geochronology of cave deposits; disposal of radioactive waste. *Mailing Add:* Dept Geol McMaster Univ Hamilton ON L8S 4M1 Can

SCHWARK, WAYNE STANLEY, b Vita, Man, May 19, 42; m 63; c 2. PHARMACOLOGY, TOXICOLOGY. *Educ:* Univ Guelph, DVM, 65, MSc, 67; Univ Ottawa, PhD(pharmacol), 70. *Prof Exp:* Lectr physiol & pharmacol, Ont Vet Col, Univ Guelph, 65-67; biologist pharmacol div, Food & Drug Dir, Ont, 67-70, res scientist, 70-71; from asst prof to assoc prof, 72-87, PROF PHARMACOL, NY STATE COL VET MED, CORNELL UNIV, 87- *Concurrent Pos:* Vet consult med sch, Univ Ottawa, 69-70; vis lectr, NY State Vet Col, Cornell Univ, 71; consult, Food & Drug Admin, 77-; Fogarty Sr Int Fel, 84. *Mem:* Am Soc Vet Physiol & Pharmacol; Can Vet Med Asn; Soc Neurosci; Am Acad Vet Pharmacol & Therapeut; Am Epilepsy Soc. *Res:* Neurochemistry and neuropharmacology; neurochemical and neuropharmacological basis of epileptic disorders; clinical pharmacology in veternary medicine. *Mailing Add:* Dept Phamacol Cornell Univ Col Vet Med Ithaca NY 14853

SCHWARTING, ARTHUR ERNEST, b Waubay, SDak, June 8, 17; m 41; c 3. PHARMACOGNOSY, NATURAL PRODUCTS CHEMISTRY. *Educ:* SDak State Univ, BS, 40; Ohio State Univ, PhD(pharmacog), 43. *Prof Exp:* Assoc prof pharmacog, Univ Nebr, 43-49; dir, Am Found Pharmaceut Educ, 74-80; dean Sch Pharm, 70-80, prof pharmacog, 49-81, EMER PROF, UNIV CONN, 81- *Concurrent Pos:* Ed, Lloydia, 60-76; vis prof, Univ Munich, 68-69; consult drug efficacy, Food & Drug Admin, 75-83. *Honors & Awards:* Res Award, Am Pharmaceut Asn Fedn, 64. *Mem:* AAAS; Am Pharmaceut Asn; Am Soc Pharmacog; Acad Pharmaceut Sci. *Res:* Ergot culture; biosynthesis; alkaloid chemistry. *Mailing Add:* 330 Capstan Dr Cape Haze Placida FL 33946

SCHWARTING, GERALD ALLEN, b June 18, 46; m 70; c 2. GLYCOCONJUGATES, DEVELOPMENT. *Educ:* Univ Conn, BS, 69; Univ Munich, Ger, PhD(biochem), 74. *Prof Exp:* Postdoctoral fel, Albert Einstein Col Med, 74-77; asst biochemist, 78-82, assoc biochemist, 83-85, SR BIOCHEMIST, E K SHRIVER CTR, 86-, MEM, DEVELOP NEUROBIOL DIV, 90-; BIOCHEMIST NEUROL, MASS GEN HOSP, 89- *Concurrent Pos:* Asst biochem, Mass Gen Hosp, 79-84, assoc biochem, 85-88; res fel neurol, Harvard Univ, 83-88; asst prof neurosci, Neurol Dept, Harvard Med Sch, 89- *Mem:* Am Asn Immunologists; AAAS; Am Soc Biochem & Molecular Biol; Soc Neurosci; Soc Complex Carbohydrates; Asn Chemoreception Sci. *Res:* Expression of glycoconjugates during development of the mammalian nervous system. *Mailing Add:* Dept Biochem E K Shriver Ctr 200 Trapelo Rd Waltham MA 02254

SCHWARTZ, A(LBERT) TRUMAN, b Freeman, SDak, May 8, 34; m 58; c 2. CHEMICAL EDUCATION, HISTORY OF CHEMISTRY. *Educ:* Univ SDak, AB, 56; Oxford Univ, BA, 58, MA, 60; Mass Inst Technol, PhD(phys chem), 63. *Prof Exp:* Res chemist, Miami Valley Labs, Procter & Gamble Co, Ohio, 63-66; from asst prof to prof, 66-83, dean fac, 74-76, chmn dept, 80-86, DEWITT WALLACE PROF CHEM, MACALESTER COL, 83- *Concurrent Pos:* Asst, Mass Inst Technol, 58-63; NSF fels & grants, 59, 67, 72, 73, 79; mem, State & District Comt Selection Rhodes Scholar, 63-; Arthur Lee Haines lectr & vis scientist, Univ SDak, 65; Macalester fac for fel, Thermochem Lab, Univ Lund, 68; vis researcher, Univ Mass, 72-73; vis prof & NSF fel, Univ Wis-Madison, 79-80; lectr, Inst Chem Educ, Univ Wis-Madison, 84, 86 & Univ Calif-Berkeley, 85; dep dir, Div Teacher Prep & Enhancement, NSF, 86-87. *Honors & Awards:* Catalyst Award, Chem Mfrs Asn, 82; Thomas Jefferson Award, 84. *Mem:* Fel AAAS; Am Chem Soc; Nat Sci Teachers Asn. *Res:* Chemical education; history of chemistry; physicochemical properties of macromolecules, particularly conformation and aggregation of proteins; solution calorimetry; preparative electrophoresis; equilibrium properties of ion exchange resins. *Mailing Add:* Macalester Col St Paul MN 55105

SCHWARTZ, ABRAHAM, mathematics, for more information see previous edition

SCHWARTZ, ABRAHAM, b Rockville Centre, NY, Apr 4, 43. PHYSICAL CHEMISTRY, BIOMATERIALS. *Educ:* Bradley Univ, BA, 65; Case Inst Technol, MS, 67; Case Western Reserve Univ, PhD(phys chem), 69. *Prof Exp:* Biophysicist, Aerospace Med Res Lab, 69-73; instr mat sci, Cornell Univ, 73-77; sr chemist, Res Triangle Inst, 77-79, Becton Dickinson Res Ctr, 79-84; PRES, FLOW CYTOMETRY STANDARDS CORP,84- *Mem:* Am Phys Soc; Am Chem Soc. *Res:* Electron microscopy; structure and morphology of natural and synthetic polymers; thrombus formation and hard tissues, microbend synthesis, fluorescence standardization and flow cytometry. *Mailing Add:* PO Box 4344 Hato Rey PR 00919

SCHWARTZ, ALAN LEE, b Chicago, Ill, Dec 8, 41; m 65; c 2. MATHEMATICS. *Educ:* Mass Inst Technol, BS, 63; Univ Wis-Madison, MS, 64, PhD(math), 68. *Prof Exp:* Asst prof math, 68-74, ASSOC PROF MATH, UNIV MO, ST LOUIS, 74- *Concurrent Pos:* Indust consult. *Mem:* AAAS; Am Math Soc; Math Asn Am; Soc Indust Appl Math; Asn Comput Mach; Am Asn Univ Professors. *Res:* Harmonic analysis; integral transformations; orthogonal expansion; computational fluid dynamics. *Mailing Add:* Dept Math Univ Mo 8001 Natural Bridge Rd St Louis MO 63121

SCHWARTZ, ALAN WILLIAM, b New York, NY, Sept 9, 35; div; c 2. EXOBIOLOGY, CHEMICAL EVOLUTION. *Educ:* NY Univ, BA, 57; Fla State Univ, MS, 62, PhD(biochem), 65. *Prof Exp:* AEC fel, Biomed Res Group, Los Alamos Sci Lab, 65-67; NASA resident res assoc, Exobiol Div, Ames Res Ctr, NASA, 67-68; assoc prof exobiol, 68-70, PROF EXOBIOL, UNIV NIJMEGEN, 70- *Concurrent Pos:* Vis scientist, Salk Inst, Calif, 83-84. *Mem:* AAAS; Royal Neth Chem Soc; Geochem Soc; Am Soc Biol Chemists; Am Chem Soc. *Res:* Chemical evolution; polynucleotide chemistry. *Mailing Add:* Lab Exobiol Univ Nijmegen Toernooiveld Nijmegen 6525ED Netherlands

SCHWARTZ, ALBERT, b Cincinnati, Ohio, Sept 13, 23. SYSTEMATIC ZOOLOGY. *Educ:* Univ Cincinnati, BA, 44; Univ Miami, MS, 46; Univ Mich, PhD(mammal), 52. *Prof Exp:* Instr zool, Univ Miami, 46-48; cur vert zool, Charleston Mus, 52-56; from instr to asst prof biol, Albright Col, 56-60; independent res, 60-67; from assoc prof to prof biol, Miami-Dade Jr Col, 67-88; RETIRED. *Concurrent Pos:* NSF res grants, 57-61 & 69-72. *Mem:* Am Soc Ichthyologists & Herpetologists; Lepidopterists' Soc. *Res:* Mammalogy and herpetology; herpetofauna of West Indies; systematics and natural history of North American mammals, amphibians and reptiles; herpetogeography and systematics of West Indies. *Mailing Add:* 10000 SW 84th St Miami FL 33173

SCHWARTZ, ALBERT B, b Philadelphia, Pa, Dec 26, 22; m 63; c 1. TECHNICAL MANAGEMENT. *Educ:* Univ Pa, BS, 44. *Prof Exp:* Chem engr, Anthracite Industs Lab, 44-45; chem engr, Mobil Res & Develop Corp, 45-49, group leader, 49-56, supv technologist, 56-62, res assoc, 62-64, supvr, 64-70, group mgr, 70-72, sr scientist, 72-81, sect mgr, 81-84, mgr, Cent Res Div, 84-85, sci adv, 85-87; CONSULT, 87- *Mem:* Am Chem Soc; Catalysis Soc; Am Inst Chem Engrs. *Res:* Petroleum and petrochemical research and development, especially catalysis. *Mailing Add:* 1901 JFK Blvd 1204 Philadelphia PA 19103

SCHWARTZ, ALLAN JAMES, b New York, NY, Dec 8, 39; m 87. CLINICAL PSYCHOLOGY. *Educ:* Columbia Univ, BA, 61, MA, 66; Rensselaer Polytech Inst, MS, 67; Univ Rochester, PhD(psychol), 73. *Prof Exp:* Staff writer, Crowell-Collier Publ Co, 61-62; chmn sci dept, Riverdale Country Sch, 62-68; asst prof, 71-81, ASSOC PROF PSYCHIAT, SCH MED, UNIV ROCHESTER, 81-, ASSOC PROF PSYCHOL, COL ARTS & SCI, 81- *Concurrent Pos:* Psychologist, Strong Mem Hosp, 71-; mem & chmn, Ment Health Ann Prog Surv, Am Col Health Asn, 73-90; staff develop consult, Delphi House Drug Treatment Ctr, 73-75; group therapist & supvr, Rochester Ment Health Ctr, 73-76; chief, Ment Health Sect, Univ Health Serv, Sch Med, Univ Rochester, 81- *Mem:* Am Col Health Asn; Am Psychol Asn; Am Asn Sex Educr, Counr & Therapists; Int Transactional Anal Asn. *Res:* Epidemiology of mental disorders; non-verbal behavior, communication process and outcome in therapeutic, supervisory and consultative relationships; education in human sexuality and treatment of sexual dysfunction; college mental health and college student suicide. *Mailing Add:* Box 617 Strong Mem Hosp 260 Crittenden Blvd Rochester NY 14642

SCHWARTZ, ANTHONY, b New York, NY, July 30, 40; m 63; c 2. IMMUNOLOGY, VETERINARY MEDICINE. *Educ:* Cornell Univ, DVM, 63; Ohio State Univ, PhD(med microbiol), 72; Am Col Vet Surg, dipl, 71. *Prof Exp:* Small animal pract, Ft Hill Animal Hosp, NY, 63-66; res vet viral immunol, US Army, Ft Detrick, Md, 66-68; resident surg, New York Animal Med Ctr, 68-69; teaching assoc vet surg & res assoc med microbiol, Ohio State Univ, 69-72, asst prof & head small animal surg, 73; from asst prof to assoc prof comp med, Sch Med, Yale Univ, 73-79; assoc prof & head small animal surg, 79-83 , actg chmn surg, 81-82, PROF SURG, SCH VET MED, TUFTS UNIV, 83-, CHMN DEPT, 82-, ASSOC DEAN, 84- *Concurrent Pos:* Fel, Ohio State Univ, 71-72; consult, US Surg Corp, Norwalk, CT, 74-; Robert Wood Johnson Health Policy fel, Washington, DC, 88-89; mem, legis planning comt, Am Vet Med Asn, 89-; mem, bd regents, Am Col Vet Surg, 89-; chmn, animal welfare comt, Mass Vet Med Asn, 89- *Mem:* AAAS; Am Vet Med Asn; Am Asn Immunol; Am Col Vet Surg. *Res:* Cellular immunology; small animal general veterinary surgery; surgical device development. *Mailing Add:* Dept Surg Sch Vet Med Tufts Univ 200 Westboro Rd N Grafton MA 01536

SCHWARTZ, ANTHONY MAX, surface chemistry; deceased, see previous edition for last biography

SCHWARTZ, ARNOLD, b New York, NY, Mar 1, 29; m 56; c 2. PHARMACOLOGY, BIOCHEMISTRY. *Educ:* Brooklyn Col Pharm, BS, 51; Ohio State Univ, MS, 57; State Univ NY, PhD(pharmacol), 61. *Prof Exp:* From asst prof to assoc prof pharmacol, Baylor Col Med, 62-69, prof pharmacol & head div myocardial biol, 69-72, prof cell biophys & pharmacol & chmn dept cell biophys, 72-77; DIR DEPT PHARMACOL & CELL BIOPHYS, COL MED, UNIV CINCINNATI, 77- *Concurrent Pos:* Nat Heart Inst fel biochem, Inst Psychiat, Maudsley Hosp, Univ London, 60-61 & fel physiol, Univ Aarhus, 61-62; USPHS res career develop award, 64-74; NSF grant, 65-67; mem study sect CV-A, NIH, 72-76. *Mem:* Am Soc Cell Biol; Am Soc Pharmacol & Exp Therapeut; Brit Biochem Soc; Int Study Group Res in Cardiac Metab; Am Soc Biol Chem. *Res:* Mechanism of cardiac glycoside action on a biochemical level; etiology of congestive heart failure and ischemia. *Mailing Add:* Dept Cell Pharmacol & Cell Biophys/575 5210 Msb Univ Cincinnati Cincinnati OH 45221

SCHWARTZ, ARNOLD EDWARD, b Rochester, NY, Dec 15, 35; m 59; c 6. CIVIL ENGINEERING, SOIL MECHANICS. *Educ:* Univ Notre Dame, BSCE, 58, MSCE, 60; Ga Inst Technol, PhD(civil eng), 63. *Prof Exp:* From asst prof to assoc prof, 63-72, head dept civil eng, 67-69, dean grad sch & div univ res, 69-70, dean grad studies & univ res, 70-82, PROF CIVIL ENG, CLEMENSON UNIV, 72-, VPROVOST & DEAN GRAD SCH, 82- *Concurrent Pos:* Ford Found fel, 61-63; mem spec comt nuclear principles & appln, Hwy Res Bd, Nat Acad Sci-Nat Res Coun, 64-69; secy-treas, Conf Southern Grad Sch, 72-; mem bd dir, US Coun Grad Schs. *Mem:* Sigma Xi. *Res:* Soil mechanics and foundation engineering; highway materials and quality control testing; pavement design. *Mailing Add:* Grad Sch Clemson Univ 201 Sikes Hall Clemson SC 29634

SCHWARTZ, ARTHUR GERALD, b Baltimore, Md, Mar 13, 41; m 88; c 1. CELL BIOLOGY, CANCER. *Educ:* Johns Hopkins Univ, BA, 61; Harvard Univ, PhD(bact, immunol), 68. *Prof Exp:* from asst prof to assoc prof, 72-85, PROF MICROBIOL, FELS RES INST, MED SCH, TEMPLE UNIV, 85- *Concurrent Pos:* Jane Coffin Childs grant, Oxford Univ, 68-71 & Albert Einstein Col Med, 71-72. *Mem:* Am Asn Cancer Res; Geront Soc. *Res:* Chemical carcinogenesis in vitro; cancer chemoprevention. *Mailing Add:* Fels Res Inst Dept Microbiol Temple Univ Med Sch Philadelphia PA 19140

SCHWARTZ, ARTHUR HAROLD, b New York, NY, Apr 6, 36; div; c 1. PSYCHIATRY, ACADEMIC ADMINISTRATION. *Educ:* Columbia Univ, AB, 57; Harvard Med Sch, MD, 61. *Prof Exp:* Intern, Univ Ill Res & Educ Hosps, 61-62; Vet Admin fel, Med Ctr, Yale Univ, 62-64; USPHS fel, 64-65; dir psychiat, Dana Psychiat Clin, Yale-New Haven Hosp, 67-68; chief, In-Patient Serv, Conn Ment Health Ctr, 68-69, assoc dir, Gen Clin Div, 69-71, assoc psychiatrist-in-chief, 71-72, actg psychiatrist-in-chief, 72; assoc prof psychiat, Mt Sinai Sch Med, 72-81, dir, Ambulatory Serv, Dept Psychiat, 72-81; PROF PSYCHIAT, ROBERT WOOD JOHNSON MED SCH, UNIV MED & DENT NJ, 81- *Concurrent Pos:* Actg clin dir, Newport County Ment Health Clin, 66-67; asst prof psychiat, Sch Med, Yale Univ, 67-71, assoc prof clin psychiat, 71-72; consult, Hosp of St Raphael, New Haven, Conn, 70-71, Vet Admin Hosp, West Haven, Conn, 71-72 & Vet Admin Hosp, Bronx, NY, 72-81. *Mem:* AAAS; Am Psychoanal Asn; Am Psychopath Asn; fel NY Acad Med; Am Psychiat Asn. *Res:* Group process including group therapy and therapeutic communities; professional student values and attitudes; clinical psychiatry. *Mailing Add:* Dept Psychiat 675 Hoes Lane Piscataway NJ 08854

SCHWARTZ, BENJAMIN L, b Pittsburgh, Pa, Jan 11, 26; m 56; c 4. OPERATIONS RESEARCH. *Educ:* Carnegie Mellon Univ, BS, 46, MS, 47; Stanford Univ, PhD(oper res), 65; George Washington Univ, MBA, 81. *Prof Exp:* Asst prof math, Duquesne Univ, 50-53; asst div chief, Battelle Mem Inst, 53-58; suboff tech dir, Monterey Lab, 58-64; mem tech staff, Inst Defense Anal, 65-67; subdept head, Mitre Corp, 67-71; ADJ PROF, MARYMOUNT UNIV, VA, 85- & JOHNS HOPKINS UNIV, 88-; CONSULT, OPERS RES FED AGENCIES & PVT INDUST, 71- *Concurrent Pos:* Vis assoc prof, US Naval Postgrad Sch, 63-65; vis prof, George Washington Univ, 66-; prof, Am Univ, 69-, Georgetown Univ, 82-85; adj prof, Marymount Univ, Va, 85-, Johns Hopkins Univ, 88-90 & Averett Col, 91- *Mem:* Fel AAAS; Am Math Soc; Opers Res Soc Am; Asn Comput Mach; Math Asn Am. *Res:* Mathematics. *Mailing Add:* 216 Apple Blossom Ct Vienna VA 22181

SCHWARTZ, BERNARD, b Toronto, Ont, Nov 12, 27; nat US; m 54; c 4. OPHTHALMOLOGY, PHYSIOLOGY. *Educ:* Univ Toronto, MD, 51; Univ Iowa, MS, 53, PhD(physiol), 59; Am Bd Ophthal, dipl, 56. *Prof Exp:* Assoc prof ophthal, Col Med, State Univ NY Downstate Med Ctr, 58-68; prof ophthal, Sch Med, Tufts Univ & ophthalmologist-in-chief, Tufts-New Eng Med Ctr, 68-90; RETIRED. *Mem:* Asn Res Vision & Ophthal; fel Am Col Surg; fel Am Acad Ophthal & Otolaryngol; fel NY Acad Med; French Soc Ophthal. *Res:* Metabolism and permeability of the lens and cornea; physiology of intraocular fluid formation; patho-physiology of cataracts and glaucoma. *Mailing Add:* 180 Beacon St Boston MA 02111

SCHWARTZ, BERTRAM, b New York, NY, Nov 1, 24; m 48; c 2. SURFACE CHEMISTRY. *Educ:* NY Univ, BS, 49. *Prof Exp:* Mem tech staff, Interchem Corp, 51-52, Sylvania Elec Prod Co, 52-54 & Hughes Aircraft Co, 54-60; mem tech staff, 60-90, CONSULT AT&T BELL LABS, 90- *Honors & Awards:* Electronics Div Award, Electrochem Soc, 87. *Mem:* Am Phys Soc; Electrochem Soc. *Res:* Chemistry of solid surfaces; chemical etching of solids; semiconductor material preparation; semiconductor device fabrication techniques. *Mailing Add:* AT&T Bell Labs 600 Mountain Ave Rm 7-416 Murray Hill NJ 07974

SCHWARTZ, BRADFORD S, b Chicago, Ill, Mar 20, 52. HEMATOLOGY, BLOOD COAGULATION. *Educ:* Rush Univ, MD, 77. *Prof Exp:* ASST PROF MED, UNIV WIS, 83- *Mailing Add:* Dept Med-Hemat Univ Wis Sch Clin Ctr 1300 University Ave Madison WI 53706

SCHWARTZ, BRIAN B, b Brooklyn, NY, Apr 15, 38; m 61; c 2. THEORETICAL PHYSICS, SCIENCE EDUCATION. *Educ:* City Col NY, BS, 59; Brown Univ, PhD(physics), 63. *Prof Exp:* Teaching asst physics, Brown Univ, 59-61; res assoc, Rutgers Univ, 63-65; leader theoret physics group, Nat Magnet Lab, Mass Inst Technol, 65-77; dean, Sch Sci, Broolyn Col, 77-80, dean res, 78-83, vpres develop, 82-86, PROF PHYSICS, BROOKLYN COL, 87-; ASSOC EXEC SECY, AM PHYS SOC, 90- *Concurrent Pos:* Assoc prof physics, Mass Inst Technol, 69-74; co-dir, NATO Advan Study Inst Large Scale Appl Superconductivity, Entreve, Italy, 73 & Small Scale Devices, Gardone Riviera, Italy, 77; superconductor mat sci, Sintra, Portugal, 80; educ officer, Am Phys Soc, 87-90. *Mem:* Fel Am Phys Soc. *Res:* Low temperature physics; superconductivity; type II superconductors; Josephson junctions; response of ferromagnetic metals; many-body problem; scientific manpower projection and utilization; physics education for the science and non-science major; small business; high technology. *Mailing Add:* Am Phys Soc 335 E 45th St New York NY 10017

SCHWARTZ, CHARLES LEON, b New York, NY, June 9, 31; m 52; c 3. THEORETICAL PHYSICS. *Educ:* Mass Inst Technol, SB, 52, PhD(physics), 54. *Prof Exp:* Res assoc physics, Mass Inst Technol, 54-56; res assoc, Stanford Univ, 56-57, asst prof, 57-60; from asst prof to assoc prof, 60-67, PROF PHYSICS, UNIV CALIF, BERKELEY, 67- *Mem:* Am Phys Soc. *Res:* Theoretical studies of atoms, nuclei and elementary particles; interaction of science with human affairs. *Mailing Add:* Dept Physics Univ Calif Berkeley CA 94720

SCHWARTZ, COLIN JOHN, b Angaston, SAustralia, May 1, 31; div; c 4. CARDIO VASCULAR DISEASES, EXPERIMENTAL PATHOLOGY. *Educ:* Univ Adelaide, MB, BS, 54, MD, 59; FRACP; FRCP(C); FRCPath. *Prof Exp:* Resident med & surg, Royal Adelaide Hosp, Australia, 55; vice master, 57, actg master, Lincoln Univ Col, Univ Adelaide, 58; specialist pathologist, 62-67, head div med res, Inst Med Ved Sci, 67; prof path, Fac Med, McMaster Univ, 68-76; head dept cardiovasc dis res, Cleveland Clin Found, 76-78; PROF PATH, UNIV TEX HEALTH SCI CTR SAN ANTONIO, 78- *Concurrent Pos:* Mem Thrombosis Arteriosclerosis comt; C J Martin overseas res fel cardiovasc dis, Dept Med, Oxford Univ, 59-61; Med Res Coun Gt Brit fel, 61; Nat Heart Found Australia res grant, 62-66; NIH int fel cardiovasc path, C T Miller Hosp, Univ Minn, 67; Med Res Coun Can res grant, 68-; fel coun on arteriosclerosis, Am Heart Asn, 67, mem exec comt coun; dir, Southam Labs, Chedoke Hosps, 70-76; mem, Heart, Lung & Blood Res Rev Comt B; vpres, San Antonio Div, Am Heart Asn, 80-81; mem, vascular comt, Am Heart Asn; chmn, Gordon Res Conf. *Mem:* Am Asn Path; AMA; Path Soc Gt Brit & Ireland; Am Heart Asn; Int Acad Path. *Res:* Myocardial infarction and the etiology and pathogenesis of atherosclerosis and thrombosis; endothelial structure and function; lipid transport and metabolism; cellular biology; inflammation; macrophage biology, hemodynamics. *Mailing Add:* Dept Path 7703 Floyd Curl Dr San Antonio TX 78284

SCHWARTZ, DANIEL ALAN, b San Antonio, Tex, Oct 21, 42; m 69; c 2. X-RAY ASTRONOMY, COSMOLOGY. *Educ:* Washington Univ, BS, 63; Univ Calif, San Diego, MS, 66, PhD(physics), 69. *Prof Exp:* Lectr physics, Univ Calif, San Diego, 68-70, asst res physicist, 69-70; Nat Res Coun resident res assoc, Goddard Space Flight Ctr, NASA, 70-71; from staff scientist to sr scientist x-ray astron, Am Sci & Eng, Inc, 71-73; PHYSICIST, SMITHSONIAN ASTROPHYS OBSERV, 73- *Concurrent Pos:* Lectr astron dept, Harvard Univ, 81- *Honors & Awards:* Group Achievement Award, NASA, 78. *Mem:* Am Astron Soc; Am Phys Soc; Int Astron Union. *Res:* Experiment development and observation of extragalactic x-rays to study isotropy of the x-ray background and mechanism of source emission; development of x-ray imaging systems for spectral studies of cosmic x-rays. *Mailing Add:* Dept Astron Harvard Univ Cambridge MA 02138

SCHWARTZ, DANIEL EVAN, b Hollywood, Calif, Oct 6, 52; m 74; c 1. FLUVIAL SEDIMENTOLOGY, PETROLEUM ENGINEERING. *Educ:* Univ Calif, Berkeley, AB, 74; Univ Tex, Dallas, PhD(geol), 78. *Prof Exp:* Res engr prod, Shell Develop Co, 78-85; STAFF GEOL ENGR, DEVELOP GEOL, SHELL WESTERN E&P INC, 85- *Mem:* Geol Soc Am; Int Asn Sedimentologists; Soc Econ Paleontologists & Mineralogists. *Res:* Sedimentology of the braided-to-meandering transition zone of the Red River; applications of scanning electron microscopy to petroleum engineering; petrology and diagenesis of clastic and diatomite petroleum reservoirs; development of hydrocarbons from fractured reservoirs. *Mailing Add:* Shell Western E&P Inc PO Box 11164 Bakersfield CA 93389

SCHWARTZ, DANIEL M(AX), b San Francisco, Calif, Mar 22, 13; m 37, 85; c 2. MECHANICAL ENGINEERING, MANAGEMENT. *Educ:* Stanford Univ, AB, 33. *Prof Exp:* Mech engr, Pac Gear & Tool Works, Calif, 33-36 & Dept Water & Power, City of Los Angeles, 36-37; mech design engr, Falk Corp, Wis, 37-40; develop engr, Dravo Corp, Pa, 40-46; vpres in charge eng & dir, Eimco Corp, 46-63, sr vpres, 63-66, gen mgr, Tractor Div, 65-66; sr staff engr, Ground Vehicle Systs, Res & Develop Div, Lockheed Missiles & Space Co, 66-67, mgr construct equip syst, 67-72, prog mgr mil progs, 72-78; PRES, FOOTHILL ENG, INC, 78- *Concurrent Pos:* Expert witness, Patents, 86. *Mem:* Fel Am Soc Mech Engrs; Soc Automotive Engrs. *Res:* Government proposals, producibility engineering, engineering management; machine design and development of heavy duty equipment; heavy transmissions, clutches; shipyard cranes; mining and milling machinery; construction equipment; loaders and tractors; numerous patents in material handling equipment and power transmissions. *Mailing Add:* Foothill Eng 2190 Washington St Apt 1204 San Francisco CA 94109

SCHWARTZ, DONALD, b Scarsdale, NY, Dec 27, 27; m 50; c 4. CHEMISTRY. *Educ:* Univ Mo, BS, 49; Mont State Col, MS, 51; Pa State Univ, PhD(chem, fuel tech), 55. *Prof Exp:* Chemist, Gen Elect Co, 51-53; asst, Pa State Univ, 53-55; asst prof chem, Villanova Univ, 55-57; res chemist, Esso Res Lab, La, 57-58; asst prof chem, Moorhead State Col, 58-59; prof chem, NDak State Univ, 59-65; reg specialist, Cent Am, NSF-Am Chem Soc, 65-66; prog dir, NSF, 66-68; prof chem & assoc dean, Grad Sch, Memphis State Univ, 68-70; dean advan studies, Fla Atlantic Univ, 70-71; vpres acad affairs, 71-74, actg pres, State Univ NY Col Buffalo, 74; chancellor, Ind Univ-Purdue Univ, 74-78; chancellor, 78-83, PROF CHEM, UNIV COLO, 83- *Concurrent Pos:* Consult, Baroid Div, Nat Lead Co, 60-64; consult vpres, Mid-South Res Assocs, 68-70; bd dir, Ind Comn Humanities, 74-78, Penrose Cancer Hosp, 80-, Beth El Sch Nursing, 80- *Mem:* AAAS; Am Chem Soc. *Res:* Humic acids; desulfurization of coal; zirconium; science education; elucidation of structure of coal. *Mailing Add:* Dept Chem Univ Colo Colorado Springs CO 80933-7150

SCHWARTZ, DONALD ALAN, b Brooklyn, NY, Mar 5, 26; m 52; c 3. PSYCHIATRY. *Educ:* Case Western Reserve Univ, MD, 52. *Prof Exp:* From instr to assoc clin prof psychiat, Sch Med, Univ Calif, Los Angeles, 58-69; assoc prof psychiat, Sch Med, Univ Calif, Irvine, 69-71; from assoc prof to prof psychiat & chief adult psychiat, Neuropsychiat Inst, 71-74; MED COODR, BENJAMIN RUSH CTR, ST JOSEPH HOSP, ORANGE,

CALIF, 80- *Concurrent Pos:* Chief inpatient serv, Neuropsychiat Inst, Univ Calif, Los Angeles, 59-61; dep dir, Los Angeles County Dept Ment Health, 61-69; chief psychiat inpatient serv, Orange County Med Ctr, 69-71; adj prof psychiat, Sch Med, Univ Calif, Los Angeles, 74- *Mem:* AMA; fel Am Psychiat Asn. *Res:* Psychopathology; administrative medicine. *Mailing Add:* 17772 17th St Tustin CA 92680

SCHWARTZ, DORIS R, NURSING. *Prof Exp:* Assoc prof Sch Nursing, NY Hosp, Cornell Univ, 51-80; sr fel geriat grad studies, Sch Nursing, Univ Pa, 80-90; RETIRED. *Mem:* Inst Med-Nat Acad Sci. *Mailing Add:* Foulk Eways Apt 2110 Gwynedd PA 19436

SCHWARTZ, DREW, b Philadelphia, Pa, Nov 15, 19; wid; c 1. GENETICS. *Educ:* Pa State Col, 42; Columbia Univ, MA, 48, PhD(bot), 50. *Prof Exp:* Res assoc cytogenetics, Univ Ill, 50-51; sr biologist, Biol Div, Oak Ridge Nat Lab, 51-62; prof biol, Western Reserve Univ, 62-64; PROF GENETICS, IND UNIV, BLOOMINGTON, 64- *Mem:* AAAS; Genetics Soc Am. *Res:* Transposable elements in maize; methylation and gene regulation. *Mailing Add:* Dept Biol Ind Univ Bloomington IN 47401

SCHWARTZ, EDITH RICHMOND, b Karlsruhe, Ger; US citizen; c 3. BIOCHEMISTRY, ORTHOPEDICS. *Educ:* Columbia Univ, AB, 52, MA, 55; Cornell Univ, PhD(biochem), 64. *Prof Exp:* Res assoc biochem, Univ Tex, 66-69; vis asst prof, Univ Ill, 69-70; res assoc pediat, Sch Med, Univ Va, 70-71, res assoc orthop, 71-72, from asst prof to assoc prof orthop, 75-78; PROF ORTHOP SURG, SCH MED, TUFTS UNIV, 78- *Concurrent Pos:* Damon Runyon fel, Albert Einstein Col Med, 64-66. *Mem:* Orthop Res Soc; Fedn Am Soc Exp Biol; Rheumatism Asn. *Res:* Connective tissue research; osteoarthritis; sulfated proteoglycon metabolism in articular cartilage; human chondrocyte cultures. *Mailing Add:* 250 Beacon St Boston MA 02116-1203

SCHWARTZ, EDWARD, b Dec 25, 32; US citizen; m 58; c 3. PHARMACOLOGY. *Educ:* Philadelphia Col Pharm & Sci, BS, 55; Univ Pa, VMD, 59; Jefferson Med Col, PhD(pharmacol), 63; LaSalle Exten Univ, dipl comput programming, 69; Am Acad Toxicol Sci, dipl. *Prof Exp:* Sr toxicologist, Hoffman-LaRoche, Inc, 62-65; sr res assoc, Warner-Lambert Res Inst, 65-70, head dept toxicol, 70, dir dept toxicol, 70-77, assoc dir toxicol, 77; dir, 77-81, SR DIR PATH & TOXICOL, SCHERING-PLOUGH CORP, 81- *Mem:* Am Vet Med Asn; Soc Toxicol; Am Soc Pharmacol & Exp Therapeut; Europ Soc Study Drug Toxicity; Can Soc Toxicol; Am Col Toxicol. *Res:* Drug safety evaluation studies in animals. *Mailing Add:* Dept Toxicol 2003 Lower State Rd Doylestown PA 18902

SCHWARTZ, ELIAS, b New York, NY, Aug, 30, 35; m 60; c 2. PEDIATRICS, HEMATOLOGY. *Educ:* Columbia Col, AB, 56; Columbia Univ, MD, 60. *Hon Degrees:* MA, Univ Pa, 72. *Prof Exp:* Intern, Montefiore Hosp, NY, 60-61; residency, St Christopher's Hosp Children, Philadelphia, 61-63; chief, US Mil Serv, Offut AFB, Nebr, 63-65; fel hemat, Children's Hosp Med Ctr, Boston, 65-67; from asst prof pediat to assoc prof pediat, Jefferson Med Col, Philadelphia, 67-72, dir pediat hemat & oncol, 67-72; DIR DIV HEMAT, CHILDREN'S HOSP, PHILADELPHIA, 72-; PROF PEDIAT, UNIV PA, PHILADELPHIA, 72-, PROF HUMAN GENETICS, 79- *Concurrent Pos:* Instr, Univ Nebr, Omaha, 63-65; chmn, Gov's Comt Sickle Cell Dis, Pa, 74-77; vis sci, Inst Cancer Res, Philadelphia, 75-76, Weizmann Inst, Rehovot, Israel, 79. *Mem:* Am Soc Pediat Hemat & Oncol (vpres, 86-); Am Soc Clinic Invest; Soc Pediat Res; Am Pediat Res; Am Soc Hemat. *Res:* Molecular biology; hematologic, biochemical, genetic, and clinical studies of sickle cell disease, thalassemia and other hemoglobinopathies; molecular biology of magakaryocyte proteins. *Mailing Add:* 7703 West Ave Elkins Park PA 19117

SCHWARTZ, ELMER G(EORGE), b Pittsburgh, Pa, July 16, 27; m 47; c 5. NUCLEAR ENGINEERING, MATERIALS SCIENCE. *Educ:* US Merchant Marine Acad, BS, 50; Carnegie Inst Technol, MS, 60, PhD(nuclear eng), 64. *Prof Exp:* Develop engr, Bettis Atomic Power Lab, Westinghouse Elec Corp, 52-59, develop engr, Atomic Power Div, 60-61; assoc prof eng, 64-72, PROF ENG, UNIV SC, 72- *Concurrent Pos:* Consult, Carolinas-Va Nuclear Power Assocs, 64-68; res partic, Savannah River Lab, 66; vis assoc prof, Carnegie-Mellon Univ, 71-72; tech assoc, E R Johnson Assocs, Inc, 77-78. *Mem:* Am Soc Mech Engrs; Am Soc Metals. *Res:* Nuclear waste management; powder metallurgy compaction; mechanical properties of materials. *Mailing Add:* Dept Eng Univ SC Columbia SC 29208

SCHWARTZ, EMANUEL ELLIOT, b New York, NY, Oct 14, 23; m 46; c 2. MEDICINE, RADIOLOGY. *Educ:* NY Univ, BA, 46; State Univ NY, MD, 50. *Prof Exp:* Intern, Jewish Hosp Brooklyn, 50-51; resident radiol, Yale-New Haven Med Ctr, 51-54; instr, Sch Med, Univ Chicago, 54-55; asst radiother, Hosp Joint Dis, New York, 55-56; asst radiol, Albert Einstein Med Ctr, 57-61, assoc, 61-65; assoc prof & dir div radiation ther & nuclear med, Sch Med, Univ Va, 65-67; radiologist, Coatesville Hosp, 67-71; ASSOC PROF RADIOL, HAHNEMANN MED COL & HOSP, 71- *Concurrent Pos:* AEC fel, Argonne Cancer Res Hosp, Univ Chicago, 54-55; USPHS fel, Biol Div, Oak Ridge Nat Lab, 56-57; instr, Sch Med, Yale Univ, 53-54. *Mem:* Radiation Res Soc; Am Roentgen Ray Soc; Radiol Soc NAm; AMA; Am Asn Cancer Res. *Res:* Radiographic manifestations of chest disease, particularly in renal patients, and with complications of medical practice. *Mailing Add:* Dept Diag Radiol Hahnemann Med Col 230 N Broad St Philadelphia PA 19102

SCHWARTZ, ERNEST, b New York, NY, May 22, 24; m 51; c 4. ENDOCRINOLOGY, METABOLISM. *Educ:* Columbia Univ, BA, 45, MA, 50, MD, 51; Am Bd Internal Med, dipl, 57; Am Bd Endocrinol & Metab, dipl, 72. *Prof Exp:* Asst physics, Columbia Univ, 44-46, Jane Coffin Childs fel biochem, 48-49; asst biochem, Sloan-Kettering Inst, 45-46; resident med, Univ Calif Hosp, San Francisco, 51-53, Los Angeles Vet Admin Hosp, 53-54; capt, US Air Force, sect chief med, Wright-Patterson AFB Hosp, Ohio, 54-56; Jane Coffin Child fel endocrinol, Sloan-Kettering Inst, 56-57; from instr to asst prof, 57-68, ASSOC PROF MED, CORNELL UNIV, 68-; CHIEF METAB UNIT, VET ADMIN HOSP, BRONX, 57- *Concurrent Pos:* Asst, Sch Med, Univ Calif, Los Angeles & San Francisco, 52-54; co-dir, Metab Bone Unit, Hosp Spec Surg, NY, 76- *Mem:* Am Physiol Soc; Endocrine Soc; Soc Exp Biol & Med; Am Fedn Clin Res; fel Am Col Physicians; Am Soc Bone & Mineral Res. *Res:* Effects of thyroid hormone upon protein metabolism; interactions of growth hormone, estrogen and serum somatomedin in normals and in acromegaly; effects of high calcium intake upon radiocalcium kinetics; effects of sodium fluoride upon bone histomorphometry in osteoporosis. *Mailing Add:* Vet Admin Hosp 130 W Kingsbridge Rd Bronx NY 10468

SCHWARTZ, FRANK JOSEPH, b New Castle, Pa, Nov 20, 29. ICHTHYOLOGY, HERPETOLOGY. *Educ:* Univ Pittsburgh, BS, 50, MS, 52, PhD(ichthyol, ecol), 54. *Prof Exp:* Asst zool, Univ Pittsburgh, 50-55; asst prof, Univ WVa, 55-57; Md Dept Res & Educ biologist, Chesapeake Biol Lab, Univ Md, 57-61, res assoc prof, biol, 61-67, prof, 67; assoc prof, 68-71, PROF BIOL, INST MARINE SCI, UNIV NC, 71- *Concurrent Pos:* Assoc ed, Chesapeake Sci, 64-74, J Aquatic Organisms; ed, Trans Am Fisheries Soc, 66-68 & Copeia, Am Soc Ichthyol & Herpet, 68-72; ed, ASB Bull, 86- *Mem:* Int Acad Fishery Scientists; Int Oceanog Found; Am Fish Soc; Am Soc Ichthyologists & Herpetologists. *Res:* Taxonomy; distribution, ecology and life histories of marine and freshwater fishes; turtles; crayfishes. *Mailing Add:* Inst Marine Sci 3407 Arendell St Univ NC Morehead City NC 28557

SCHWARTZ, GERALD PETER, b Cleveland, Ohio, Mar 20, 38; m 63; c 3. BIOCHEMISTRY. *Educ:* John Carroll Univ, BS, 60; Univ Pittsburgh, PhD(biochem), 64. *Prof Exp:* Res assoc biochem, Med Dept, Brookhaven Nat Lab, 64-66, asst scientist, 66-68; ASST RES PROF BIOCHEM, MT SINAI SCH MED, 66- *Mem:* AAAS; Am Chem Soc. *Res:* Isolation of enzymes; study of enzyme action; synthesis of peptides of biological interest. *Mailing Add:* Dept Biochem Mt Sinai Sch Med Fifth Ave & 100th St New York NY 10029

SCHWARTZ, GERALDINE COGIN, b New York, NY, Apr 4, 23; m 43; c 2. PHYSICAL CHEMISTRY. *Educ:* Brooklyn Col, BA, 43; Columbia Univ, MA, 45, PhD(chem), 48. *Prof Exp:* Asst physics, SAM Labs, Manhattan Proj, Columbia Univ, 43; USPHS res fel, Tuberc Res Lab, 48 & Sloan Kettering Inst Cancer Res, 49, inst fel phys chem, 50-52; instr chem, Adelphi Col, 53 & Queens Col, NY, 53-54; asst prof, Bard Col, 58-62; chemist, 64-71, adv engr, Components Div, 71-78, SR ENGR, GEN TECHNOL DIV, IBM CORP, 78- *Mem:* Am Vacuum Soc; Sigma Xi; fel Electrochem Soc. *Res:* Physical chemistry of proteins; thin films; anodic oxidation reactive ion etching and dielectric films. *Mailing Add:* IBM Corp East Fishkill Facil Dept 206 Bldg 300-48A Hopewell Junction NY 12533

SCHWARTZ, HAROLD LEON, b Brooklyn, NY, Mar 14, 33; m 56; c 3. ENDOCRINOLOGY, BIOCHEMISTRY. *Educ:* Brooklyn Col, BS, 57; NY Univ, MS, 61, PhD(physiol), 64. *Prof Exp:* Res assoc med, State Univ NY Downstate Med Ctr, 57-64, instr, 64-66; biochemist, Endocrine Res Lab, Montefiore Hosp & Med Ctr & asst prof biochem, Albert Einstein Col Med, 67-76; ASSOC PROF MED, UNIV MINN COL MED, 76-, ASSOC DIR, DIV ENDOCRINOL & METAB, DEPT MED, 81- *Concurrent Pos:* USPHS fel, Nat Inst Med Res, London, Eng, 66-67. *Mem:* AAAS; NY Acad Sci; Am Thyroid Asn; Endocrine Soc. *Res:* Thyroid hormone biosynthesis and metabolism; mechanisms of hormone action; chemistry of thyroid proteins. *Mailing Add:* Univ Minn Col Med Mayo Bldg 420 Delaware St SE Minneapolis MN 55455

SCHWARTZ, HEINZ (GEORG), b Landsberg, Ger, Jan 15, 24; US citizen; m 50; c 1. PATHOLOGY. *Educ:* Rutgers Univ, New Brunswick, BS, 53, PhD(biochem), 57; Temple Univ, MD, 61. *Prof Exp:* Res asst biochem, Merck Inst Therapeut Res, Rahway, 50-57; intern, Abington Hosp, Pa, 61-62; resident path, Temple Univ Hosp, 62-65, instr, 65-66; asst prof & asst dir, 66-70, assoc prof path & assoc dir clin labs, 70-78, PROF PATH & DIR CLIN LABS, THOMAS JEFFERSON UNIV HOSP, 78- *Concurrent Pos:* Consult, Vet Admin Hosp, Coatesville, 67- & Children's Heart Hosp, 78- *Mem:* AMA; Am Asn Clin Path; Am Asn Clin Chem; Col Am Path; Sigma Xi. *Res:* Lipid analysis; immunoglobuline analysis. *Mailing Add:* Clin Lab Jefferson Med Col 11 Walnut Philadelphia PA 19107

SCHWARTZ, HENRY GERARD, JR, b St Louis, Mo, Aug 3, 38; m 60; c 2. ENVIRONMENTAL HEALTH ENGINEERING. *Educ:* Wash Univ, St Louis, BS, 61, MS, 62; Calif Inst Technol, PhD(environ health eng), 66. *Prof Exp:* Res fel environ health eng, Calif Inst Technol, 65-66; sr engr, 66-76, vpres & mgr, Environ Div, Sverdrup & Pracel & Assoc, Inc, 76-80, VPRES, CORP PRIN, SVERDRUP CORP, 78-, PRES, SVERDRUP ENVIRON, INC, 89- *Concurrent Pos:* Mgt adv group, US Environ Protection Agency, 81-83, 86-88; chmn, Water Pollution Control Fedn Res Found, 88- *Honors & Awards:* Edward Bartow Award, Am Chem Soc, 66; Arthur Sidney Bedell Award, Water Pollution Control Fedn, 76; Kappe Lectr, Am Acad Environ Engrs, 89. *Mem:* Nat Soc Prof Engrs; Am Soc Civil Engrs; Am Acad Environ Engrs; Water Pollution Control Fedn (pres, 85-86); Air Pollution Control Asn; Sigma Xi. *Res:* Adsorption and microbial degradation of pesticides in aqueous solutions; water recovery and reuse in space vehicles; water and air pollution control. *Mailing Add:* Sverdrup Corp 801 N 11th St Louis MO 63101

SCHWARTZ, HERBERT, b Limerick, Pa, Mar 8, 25; m 86; c 2. ORGANIC & BIOLOGICAL CHEMISTRY. *Educ:* Univ Freiburg, Ger, dipl, 55; Univ Utrecht, PhD(chem), 65. *Prof Exp:* Res chemist, Vineland Chem Co, 55-57; chemist anal chem, Food & Drug Admin, Washington, DC, 57-58; chemist biochem, Grad Hosp, Philadelphia, 58-59; res assoc, Inst Org Chem, Univ Utrecht, 59-65; res & develop consult chem, Biovivan Res Inst, 63-87, VPRES RES & DEVELOP, GRAN-TENESCO INC, PHILADELPHIA, PA, 88- *Concurrent Pos:* Adj prof chem, Cumberland County Col, 69-75 & Camden County Col, 77-78. *Mem:* AAAS. *Res:* Biologically active chemistry; investigation into physiologically induced human interactions based upon chronobiology and sociobiology; environmental chemistry; synthetic plant hormones. *Mailing Add:* 161 Rosenhayn Ave Bridgeton NJ 08302-1241

SCHWARTZ, HERBERT C, b New Haven, Conn, May 8, 26; m 58; c 3. MEDICINE, PEDIATRICS. *Educ:* Yale Univ, AB, 48; State Univ NY, MD, 52. *Prof Exp:* Intern med, Vet Admin Hosp, Newington, Conn, 52-53, intern pediat, Grace-New Haven Community Hosp, 53; med resident, Univ Serv, Kings County Hosp, 53-54 & Stanford Univ Hosp, 54-55; instr, Univ Utah, 58-60; from asst prof to assoc prof, 60-68, chmn dept, 69-71, PROF PEDIAT, SCH MED, STANFORD UNIV, 68- *Concurrent Pos:* Clin & res fel med, Univ Utah, 55-56, res fel, 56-57, res fel biochem, 57-58; Markle scholar acad med, 62. *Mem:* Am Fedn Clin Res; Soc Pediat Res; Am Soc Clin Invest; Am Pediat Soc; Am Soc Hemat. *Res:* Hemoglobin structure and synthesis in mammalian erythrocytes; effects of pregnancy on hemoglobin AIc in diabetic women; structure and function relationship of hemoglobins in deep diving mammals. *Mailing Add:* Pediat/Med Ctr S 348 Stanford Univ Stanford CA 94305

SCHWARTZ, HERBERT MARK, b Philadelphia, Pa, July 30, 48; m 74; c 2. NUCLEAR MAGNETIC RESONANCE SPECTROSCOPY, ANALYTICAL. *Educ:* Temple Univ, BA, 70; Univ Del, MA, 74 & PhD(physics), 78. *Prof Exp:* Postdoctoral assoc, nuclear magnetic resonance, Prof G Fasman Biochem Dept, Brandeis Univ, 77-78, Dr S Danyluk Div Biol & Med Res, Argonne Nat Lab, 78-81; DIR NUCLEAR MAGNETIC RESONANCE & SPECTROS, CHEM DEPT, RENSSELAER POLYTECH INST, 81- *Mem:* Am Phys Soc; Anal Lab Mgrs Asn; Sigma Xi. *Res:* Usage of nuclear magnetic resonance spectroscopy of various systems including nucleosides, drugs, polyamides and various types of biological tissues; development of methodologies for academic analytical laboratory management. *Mailing Add:* Chem Dept Rensselaer Polytech Inst Troy NY 12181

SCHWARTZ, HOWARD JULIUS, b New York, NY, Nov 24, 36; m 62; c 3. MEDICINE, ALLERGY. *Educ:* Brooklyn Col, BA, 56; Albert Einstein Col Med, MD, 60. *Prof Exp:* Res fel, Harvard Med Sch, 66-68; USPHS trainee, 68-71, asst prof, 71-74, asst clin prof, 74-77, assoc clin prof med, 77-86, CLIN PROF MED, SCH MED, CASE WESTERN RESERVE UNIV, 86- *Concurrent Pos:* Clin res fel allergy & immunol, Mass Gen Hosp, 66-68; assoc physician, Univ Hosps, Cleveland, 71-, chief allergy clin, 72-; consult, Hillcrest Hosp, Cleveland & Mt Sinai Hosp, Cleveland; chmn, Comt Insects, Am Acad Allergy. *Mem:* Am Asn Immunologists; fel Am Acad Allergy; fel Am Col Chest Physicians; Am Thoracic Soc; fel Am Col Allergy. *Res:* Allergic respiratory disease, including the interaction of rhinitis and asthma; insect allergy- bee, wasp, hornet and yellow jacket; immunology. *Mailing Add:* 1611 S Green Rd Cleveland OH 44121

SCHWARTZ, ILSA ROSLOW, b Brooklyn, NY, Aug 20, 41; m 64; c 2. NEUROANATOMY, AUDITORY SYSTEM. *Educ:* Vassar Col, AB, 62; Yale Univ, MS, 64, PhD(molecular biophys), 68. *Prof Exp:* Res assoc, Ctr Neural Sci, Ind Univ, Bloomington, 70-73, asst prof anat & physiol, 73-77; from asst prof to assoc prof surg, div Head & Neck Surg, Sch Med, Univ Calif, Los Angeles, 77-87; assoc prof surg, Sect Otolaryngol & Neuroanat, 87-89, PROF SURG, SECT OTOLARYNGOL, SCH MED, YALE UNIV, 89- *Concurrent Pos:* NIH fel & res fel neuroanat, Albert Einstein Col Med, 68-69; USPHS biomed sci res support grant, Ind Univ, Bloomington, 70-72, NIH res grants, 72; vis res anatomist, Sch Med, Univ Calif, Los Angeles, 76-77; mem, Commun Dis Rev Comt, Nat Inst Neurol & Commun Dis & Stroke, 81-83, chmn, 83-85; panel mem, NIH Consensus Devel Conf on Cochlear Implants, 88; Javits neuroscience investr award, Nat Inst Neurol Commun Dis & Strokes, 88; mem adv coun, Nat Inst Deafness & Other Commun Dis, 88-93; mem, Asn Res Otolaryngol Coun, 87-90; mem bd dir, Friends Nat Inst Deafness & Other Commun Dis, 90-91. *Mem:* Am Asn Anat; Asn Res Otolaryngol (pres, 90-91); Soc Neurosci; Women Neurosci; Asn Women Sci; Cajal Club. *Res:* Synaptic organization; synaptic development; structural and functional correlations of auditory and vestibular neural activity; autoradiographic and immunocytochemical studies of chemical properties of neurons in the auditory system. *Mailing Add:* Sect Otolaryngol Yale Univ Sch Med 333 Cedar St New Haven CT 06511

SCHWARTZ, IRA, b New York, NY, May 16, 47; m 68; c 3. BIOCHEMISTRY, MOLECULAR BIOLOGY. *Educ:* City Univ New York, BS, 68, PhD(biochem), 73. *Prof Exp:* Lectr, Dept Chem, City Col New York, 68-73; fel, Roche Inst Molecular Biol, 73-75; asst prof biochem, Univ Mass, 75-80; from asst prof to assoc prof biochem, 80-89, PROF BIOCHEM & MOLECULAR BIOL, NY MED COL, 89- *Concurrent Pos:* Sinsheimer fel, 81-84. *Mem:* AAAS; Am Chem Soc; Am Soc Microbiol; Am Soc Biochem & Molecular Biol. *Res:* Identification of the ribosomal components necessary for binding of nonribosomal protein factors; study of regulation of expression of genes for proteins involved in translation; molecular cloning. *Mailing Add:* Dept Biochem & Molecular Biol NY Med Col Valhalla NY 10595

SCHWARTZ, IRA A(RTHUR), b Brooklyn, NY, Mar 8, 15; m 56; c 2. METALLURGY. *Educ:* City Col New York, BS, BME, 42; Stevens Inst Technol, MS, 50; Assoc Appl Sci, electronics maj, 85. *Prof Exp:* Sr eng draftsman, Hull Design Div, NY Naval Shipyard, US Dept Army, 42-46, metallurgist, Mat Lab, 46-52, supvry metallurgist, 52-58, head, Wrought Metals & Radiographic Sect, 58-63, sr task leader, Naval Appl Sci Lab, 63-65, gen metallurgist, Off Chief Engrs, 65-80, res coordr metall & civil eng probs, 69-80; RETIRED. *Concurrent Pos:* Metall consult, Corp Coun, NY, 59-60 & Northern Va Community Col, 85-; consult welding metall & non-destructive testing, 80- *Mem:* Am Soc Metals; Am Foundrymen's Soc. *Res:* Foundry metallurgy involving casting of ferrous and nonferrous alloys by sand, shell and lost-wax methods; physical metallurgy of wrought metals, particularly heat treatment and notch-toughness; nondestructive testing radiography and ultrasonics; metallurgical failures; general metallurgical problems. *Mailing Add:* 8303 The Midway Annandale VA 22003

SCHWARTZ, IRVING LEON, b Cedarhurst, NY, Dec 25, 18; m 46; c 3. PHYSIOLOGY, MEDICINE. *Educ:* Columbia Col, AB, 39; NY Univ, MD, 43. *Prof Exp:* Intern, Third Med Div, Bellevue Hosp, New York, 43-44, asst resident, 46-47; from asst to assoc, Rockefeller Inst, 52-58; sr scientist & attend physician hosp, Med Res Ctr, Brookhaven Nat Lab, 58-61; Joseph Eichberg prof physiol & chmn dept, Col Med, Univ Cincinnati, 61-65; prof physiol & biophys, chmn dept physiol & dean grad sch biol sci, 65-80, Golden & Harold Lamport Distinguished Serv Prof-at-large & dir ctr Polypeptide & Membrane Res, 80-89, EMER DEAN, MT SINAI GRAD SCH BIOL SCI, CITY UNIV NEW YORK, 89- *Concurrent Pos:* NIH fel, Col Med, NY Univ, 47-50; Porter fel, Rockefeller Inst, 50-51, Am Heart Asn fel, 51-52; from asst physician to assoc physician, Rockefeller Hosp Inst, 52-58; res collabr, Med Res Ctr, Brookhaven Nat Lab, 61-; exec officer, Biomed Sci Doctoral Prog, City Univ New York, 68-70. *Honors & Awards:* Gibbs Mem Award, Rockefeller Inst, 50. *Mem:* Am Physiol Soc; Soc Exp Biol & Med; Am Soc Clin Invest; Biophys Soc; Am Fedn Clin Res; Endocrine Soc; Soc Neurosci. *Res:* Membrane and transport phenomena; mechanism of hormone action; conformation-structure-activity relationships of peptides and proteins. *Mailing Add:* Mt Sinai Med Ctr Box 932 New York NY 10029

SCHWARTZ, IRVING ROBERT, b New York, NY, May 7, 23; m 51; c 3. MEDICINE, HEMATOLOGY. *Educ:* NY Univ, AB, 47; State Univ NY, MD, 51. *Prof Exp:* Intern, Montefiore Hosp, Bronx, NY, 51-52; resident internal med, Bronx Vet Admin Hosp, 52-53 & Ohio State Univ Hosp, Columbus, 53-54; resident hemat, Cardeza Found, Jefferson Med Col, 54-55; from asst dir to dir Sacks Dept Hemat, Albert Einstein Med Ctr, 58-80; from asst prof to assoc prof, 65-86, PROF MED, TEMPLE UNIV, 86- *Concurrent Pos:* USPHS res fel, Jefferson Med Col, 55-57; Sacks res fel, Albert Einstein Med Ctr, 57-58; head, dir hemat, Albert Einstein Med Ctr, 80- *Mem:* Am Soc Hemat; AMA; Am Col Physicians; Transplantation Soc; NY Acad Sci. *Res:* Cancer chemotherapy; bone marrow transplantation; radiation effects; hemoglobinopathies; leukoagglutinins; bone marrow preservation. *Mailing Add:* Albert Einstein Med Ctr York & Tabor Rds Philadelphia PA 19141

SCHWARTZ, JACK, b New York, NY, May 4, 31; m 57; c 2. COMPUTER MODELING, IMAGE COLOR. *Educ:* City Col NY, BS, 53; Harvard Univ, AM, 54, PhD(physics), 58. *Prof Exp:* Res assoc, Brookhaven Nat Lab, 57-60; mem tech staff, RCA Labs, David Sarnoff Res Ctr, NJ, 60-64; mem tech staff, Sanders Assocs, Inc, 64-83; PRES, TOUCHSTONE TECHNOL, INC, 83- *Concurrent Pos:* Mem tech staff, Mitre Corp, 83-90. *Mem:* Am Phys Soc; NY Acad Sci; Sigma Xi; Soc Imaging Sci & Technol. *Res:* Seventeen US patents, one French patent; nonlinear optics; magnetic resonance phenomena; magnetism in thin ferromagnetic films; atomic beams; system analysis; operations research; solar energy; computer modeling of radar and communications systems; image color correction; seventeen US patents. *Mailing Add:* 147 Ridge St Arlington MA 02174-1733

SCHWARTZ, JACOB THEODORE, b New York, NY, Jan 9, 30; m 50, 89; c 2. FUNCTIONAL ANALYSIS. *Educ:* City Col New York, BS, 48; Yale Univ, MA, 49, PhD, 51. *Prof Exp:* From instr to asst prof math, Yale Univ, 52-57; assoc prof, 57-59, PROF MATH & COMPUT SCI, COURANT INST MATH SCI, NY UNIV, 59- *Concurrent Pos:* Dir, Info Sci & Tech Off, DARPA, 87-89. *Honors & Awards:* Wilbur Cross Medal, Yale. *Mem:* Nat Acad Sci. *Res:* Physics and functional analysis; physical mathematics; probability; computer science; robotics. *Mailing Add:* Dept Math NY Univ New York NY 10003

SCHWARTZ, JAMES F, b Milwaukee, Wis, Oct 28, 29; m 56; c 3. PEDIATRICS, NEUROLOGY. *Educ:* Swarthmore Col, BA, 51; Rochester Univ, MD, 55. *Prof Exp:* From asst prof to assoc prof, 63-70, PROF PEDIAT, SCH MED, EMORY UNIV, 70-, PROF NEUROL, 75- *Concurrent Pos:* Nat Inst Neurol Dis & Blindness spec fel pediat neurol, Columbia-Presby Med Ctr, 60-63. *Mem:* Fel Am Acad Pediat; fel Am Acad Neurol; Am Neurol Asn; Child Neurol Soc (pres, 75-76); Am Pediat Soc. *Res:* Neurological aspects of development; muscle and convulsive disorders in infancy and childhood. *Mailing Add:* Dept Pediat Emory Univ Sch Med Atlanta GA 30322

SCHWARTZ, JAMES WILLIAM, b Elmira, NY, Feb 11, 27; m 45; c 2. ENGINEERING. *Educ:* Cornell Univ, BE, 51, MS, 52. *Prof Exp:* Mem tech staff, RCA Labs, 52-58; vpres, Nat Video Corp, 61-64; vpres, Rauland Div, Zenith Rad, 69-79; pres, MD Systs, Inc, 80-86; CONSULT, 86- *Mem:* Sigma Xi; fel Inst Elec & Electronics Engrs. *Res:* Flat panel displays. *Mailing Add:* 2384 Stonebrook Dr Medford OR 97504

SCHWARTZ, JAY W(ILLIAM), b Scranton, Pa, Sept 28, 34; m 70; c 3. SPACE SYSTEMS ENGINEERING, TELECOMMUNICATIONS. *Educ:* Univ Pa, BS, 56; Yale Univ, MEng, 60, PhD(elec eng), 64. *Prof Exp:* Mem tech staff, Res & Eng Support Div, Inst Defense Anal, 63-67; vis assoc prof elec eng, Polytech Inst Brooklyn, 67-68; mech tech staff, Sci & Tech Div, Inst Defense Anal, Va, 68-72; consult to assoc, dir Res Space & Commun Sci & Technol, Naval Res Lab, 72-83; CHIEF SCIENTIST & VPRES, AVTEC SYSTS INC, ALEXANDRIA, VA, 83- *Concurrent Pos:* Mem ad hoc sci group, Tactical Satellite Commun, 67-68; chmn, Optical Commun Working group, Navy Laser Technol Prog Off, 72-76; adj assoc prof elec eng, George Washington Univ, 72-76; mem, Mil Man in Space Panel, 78-79. *Honors & Awards:* Res Publ Award, Naval Res Lab, 78, 80, & 84. *Mem:* Inst Elec & Electronics Engrs; Int Union Radio Sci. *Res:* Communication satellite systems; military communications; data processing in space vehicles; communication theory; military space systems. *Mailing Add:* Avtec Systs Inc 10530 Rosehaven St Fairfax VA 22030

SCHWARTZ, JEFFREY, b New York, NY, Jan 3, 45; m 70. ORGANOMETALLIC CHEMISTRY. *Educ:* Mass Inst Technol, SB, 66; Stanford Univ, PhD(chem), 70. *Prof Exp:* NIH fel, Columbia Univ, 70; ASST PROF CHEM, PRINCETON UNIV, 70-, PROF CHEM. *Mem:* Am Chem Soc; Royal Soc Chem. *Res:* Applications of organometallic chemistry to organic synthesis; novel organometallic complexes; intramolecular organometallic redox reactions. *Mailing Add:* Dept Chem Princeton Univ Princeton NJ 08540

SCHWARTZ, JEFFREY H, b Richmond, Va, Mar 6, 48. EVOLUTIONARY BIOLOGY, PHYSICAL ANTHROPOLOGY. *Educ:* Columbia Univ, BA, 69, MS, 73, PhD(phys anthrop), 74. *Prof Exp:* Adj lectr, Lehman Col, 73-74; from asst prof to assoc prof, 74-81, PROF PHYS ANTHROP, UNIV PITTSBURGH, 90- *Concurrent Pos:* Staff osteologist, Am Sch Oriental Res, 70-; res assoc, Carnegie Mus Natural Hist, 76- & Am Mus Nat Hist, 79- *Mem:* Am Asn Phys Anthrop; AAAS; Soc Syst Zool; Soc Vert Paleont; Soc Study Evolution; Sigma Xi. *Res:* Evolutionary theory and systematics; primate phylogeny and paleontology; human and faunal remains of the circum-Mediterranean and Near East; general physical anthropology and vertebrate paleontology. *Mailing Add:* Dept Anthrop Univ Pittsburgh Pittsburgh PA 15260

SCHWARTZ, JEFFREY LEE, b Far Rockaway, NY, Aug 19, 43. MICROBIAL BIOCHEMISTRY. *Educ:* Brooklyn Col Pharm, BS, 66; Univ Wis-Madison, MS, 68, PHD(pharm biochem), 71. *Prof Exp:* Res fel microbial biochem, Wesleyan Univ, 71-73; res fel, 73-74, RES INVESTR MICROBIAL BIOCHEM, SQUIBB INST MED RES, 74- *Mem:* Am Soc Microbiol; Am Chem Soc. *Res:* Investigating the production of novel antibiotics from various microorganisms and the development of novel and sensitive methods for antibiotic detection. *Mailing Add:* Dept Chem Princeton Univ Princeton NJ 08544

SCHWARTZ, JEROME LAWRENCE, b Birmingham, Ala, July 4, 38; m 64; c 2. CHEMISTRY, TECHNICAL MANAGEMENT. *Educ:* Univ SC, BS, 61, MS, 64; Univ Del, PhD(phys-org chem), 70. *Prof Exp:* Chemist polymer chem, E I du Pont de Nemours & Co Inc, 63-65, res chemist photog chem, 68-71, mkt rep clin instrumentation, 71-73, appl lab supvr clin chem, 73-75, mgr tech serv clin chem instrumentation, 76-78, sr methods res supvr, 78-79, res & develop mgr, 79-90; EXEC VPRES & CHIEF TECH OFFICER, OHMICRON CORP, NEWTON PA, 90- *Res:* Development of instrumentation and methodology for the automation of clinical chemistry tests; manufacturing of immunoassay test kits for environmental pesticides and residue testing. *Mailing Add:* 2817 Landon Dr Wilmington DE 19810-2212

SCHWARTZ, JESSICA, GROWTH HORMONE, ADIPOCYTES. *Educ:* Harvard Univ, PhD(physiol), 74. *Prof Exp:* asst prof, 69-87, ASSOC PROF PHYSIOL, MED SCH, UNIV MICH, 87- *Mem:* Am Diabetes Asn; Am Physiol Soc; Endocrine Soc. *Res:* Molecular and cellular mechanisms of growth factor action- regulation of gene expression. *Mailing Add:* Dept Physiol Med Sci Bldg Two Univ Mich Ann Arbor MI 48109

SCHWARTZ, JOAN POYNER, b Ont, Can, Aug 19, 43; m 67. NEUROBIOLOGY, NEUROPHARMACOLOGY. *Educ:* Cornell Univ, AB, 65; Harvard Univ, PhD(biol chem), 71. *Prof Exp:* Instr pharmacol, Dept Pharm, Rutgers Med Sch, 70-71; staff fel, Lab Neuroanat & Neurosci, Nat Inst Neurol & Commun Dis & Stroke, 72-76; GROUP HEAD MOLECULAR BIOL, LAB PRECLIN PHARM, NIMH, 76- *Concurrent Pos:* Mem, Neurol C Study Sect, 85- *Mem:* Am Soc Pharmaceut & Exp Therapeut; Soc Neurosci; Am Soc Neurochem; AAAS. *Res:* Nerve growth factor: regulation of synthesis and mechanism of actions; catecholamine-mediated regulation of gene expression via cyclic adenosine monophosphate levels. *Mailing Add:* LNRI NINCDS NIH Bldg 9 Rm 1W115 Bethesda MD 20892

SCHWARTZ, JOHN H, MEMBRANE TRANSPORT, RENAL METABOLISM. *Educ:* NY Univ, MD, 67. *Prof Exp:* ASSOC PROF MED, SCH MED, BOSTON UNIV, 83- *Res:* Renal physiology. *Mailing Add:* Thorndike Mem Lab Rm 406 Boston City Hosp 811 Harrison Ave Boston MA 02118

SCHWARTZ, JOHN T, b Hazelton, Pa, Aug 28, 26; m 56; c 6. OPHTHALMOLOGY, EPIDEMIOLOGY. *Educ:* Dartmouth Col, AB, 47; Univ Notre Dame, MS, 50; Jefferson Med Col, MD, 55; Harvard Univ, MPH, 63. *Prof Exp:* Intern med, Madison Gen Hosp, 55-56; head dept ophthal, Naval Base Dispensary, Norfolk, Va, 59-61; asst prof, Sch Med, Univ Mo, Columbia, 61-62; head sect ophthal field & develop res, Nat Inst Neurol Dis & Blindness, 63-68; dep chief dept ophthal, USPHS Hosp, Baltimore, 68-69; head sect ophthal field & develop res, Nat Eye Inst, 69-75; spec asst to dir, Div Hosp & Clin & chief, Dept Ophthal, 76-79, assoc dir med affairs, Bur Med Serv, Health Serv Admin, USPHS, 81-83; PRES, J T SCHWARTZ CONSULT INC, 83- *Concurrent Pos:* Clin asst prof ophthal, Sch Med, Univ Mo, 63-64; asst clin prof, George Washington Univ, 65-73; consult ophthal, USPHS Hosp, Baltimore, 69-; Nat Health Exam Surv, Nat Ctr Health Statist, 69-; guest worker, Geront Res Ctr, Nat Inst Aging, 79-80; consult ophthalmoepidemiol, Bur Med Devices, Food & Drug Admin, 80-81, med officer, 82-83. *Mem:* Am Acad Ophthal; Int Soc Twin Studies; Am Eye Study Club; Soc Epidemiol Res. *Res:* Epidemiologic and genetic investigations of etiology of ocular disorders and clinical practice ophthalmology. *Mailing Add:* 18000 Marden Lane Sandy Spring MD 20860

SCHWARTZ, JOSEPH BARRY, b Richmond, Va, June 11, 41; m 64; c 3. PHARMACEUTICAL CHEMISTRY. *Educ:* Med Col Va, BS, 63; Univ Mich, MS, 65, PhD(pharmaceut chem), 67. *Prof Exp:* Sr res pharmacist, Merck, Sharp & Dohme Res Labs, West Point, Pa, 67-72, res fel, 72-81; PROF PHARMACEUT & DIR INDUST PHARM RES, PHILADELPHIA COL PHARM & SCI, 81- *Concurrent Pos:* Consult, pharmaceut indust firms, 81-; chmn Pharmaceut Technol sect of Acad Pharmaceut Sci, 84-85; Linwood F Tice chair pharmaceut, Philadelphia Col Pharm & Sci, 87- *Mem:* AAAS; Am Pharmaceut Asn; fel Acad Pharmaceut Sci; fel Am Asn Pharmaceut Sci; Parentral Drug Asn; Controlled Release Soc; Am Asn Col Pharmacists; Sigma Xi. *Res:* Drug release form wax matrices; physical and chemical properties affecting drug availability; dosage form design and processing; controlled release. *Mailing Add:* Philadelphia Col Sci & Pharm 43rd St & Woodland Ave Philadelphia PA 19104

SCHWARTZ, JOSEPH ROBERT, b Chicago, Ill, Feb 17, 19; m 42; c 5. PHYSICAL CHEMISTRY, ORGANIC CHEMISTRY. *Educ:* Univ Chicago, BS, 41, MS & PhD(chem), 48. *Prof Exp:* Res assoc, Univ Calif, Los Angeles, 48-50 & Radioisotope Unit, Long Beach Vet Hosp, 50-53; from asst prof to assoc prof, 53-70, PROF CHEM, LOYOLA UNIV LOS ANGELES, 70-, CHMN DEPT, 74- *Mem:* AAAS; Am Chem Soc; Am Crystallog Asn; Sigma Xi. *Res:* Physical organic research; relation of mechanism to structure; color of C-nitroso compounds; determination of molecular interactions through x-ray crystallographic studies. *Mailing Add:* Dept Chem Loyola Marymount Univ Loyola Blvd W 80th St Los Angeles CA 90045

SCHWARTZ, JUDAH LEON, b Brooklyn, NY, July 13, 34; m 60; c 3. SCIENCE EDUCATION. *Educ:* Yeshiva Univ, BA, 54; Columbia Univ, AM, 57; NY Univ, PhD(physics), 63. *Prof Exp:* Asst physics, Columbia Univ, 54-57; reactor physicist, Am Mach & Foundry Co, 57; instr physics, Israel Inst Technol, 57-58; mem tech staff weapons physics, G C Dewey Corp, 58-61; sr physicist, Autometric Corp, 61-62 & G C Dewey Corp, 62-63; instr physics, NY Univ, 61-63; fel, Lawrence Radiation Lab, 63-64, physicist, 63-66; sr res scientist, Educ Res Ctr, 66-72, PROF ENGR SCI & EDUC, MASS INST TECHNOL, 73-; PROF EDUC, HARVARD UNIV, 85- *Concurrent Pos:* Consult Sci, Tech & World Affairs Prog, Carnegie Endowment for Int Peace, 63- & For Serv Inst, Dept State Educ Develop Ctr; assoc ed, Am J Physics, 72-; assoc ed, Int J Math Educ, 72-; hon res assoc, Dept Psychol, Harvard Univ, 74-75. *Mem:* AAAS; Am Phys Soc. *Res:* Science and mathematics education; computer graphics; computer generated films; cognitive psychology and the development of mathematical competence in children. *Mailing Add:* Sch Eng Mass Inst Technol 77 Massachusetts Ave Cambridge MA 02139

SCHWARTZ, LARRY L, b Fremont, Ohio, July 24, 35; m 69; c 2. PHYSICAL CHEMISTRY. *Educ:* Case Western Reserve Univ, AB, 57, MS, 59, PhD(phys chem), 63. *Prof Exp:* Res assoc nuclear chem, Argonne Nat Lab, 62-64; chemist, 64-76, dep dept head, Earth Sci Dept, 76-89, DEP ASSOC DIR ADMIN, CHEM & MAT SCI DEPT, LAWRENCE LIVERMORE LAB, 89- *Mem:* AAAS. *Res:* Chemical and radiochemical effects of nuclear explosions; nuclear test containment. *Mailing Add:* PO Box 808 Livermore CA 94550

SCHWARTZ, LAWRENCE B, b Oakridge, Tenn, July 1, 49. MASS CELLS. *Educ:* Wash Univ, MD & PhD(biochem), 76. *Prof Exp:* ASSOC PROF MED, MED COL VA, 83- *Mem:* Am Asn Immunol; Am Fedn Clin Res; Am Acad Allergy & Immunol. *Mailing Add:* Dept Internal Med Va Commonwealth Univ Box 263 Richmond VA 23298

SCHWARTZ, LEANDER JOSEPH, b Newton, Wis, Jan 10, 32; m 64; c 2. WASTE MANAGEMENT, RESOURCE RECOVERY. *Educ:* Wis State Univ, Platteville, BS, 57; Univ Wis, MS, 59, PhD(bot), 63. *Prof Exp:* Asst prof bot, Fox Valley Ctr, Univ Wis, 63-69, dean, 69-72; assoc prof environ sci, Univ Wis-Green Bay, 69-82, chmn dept biol, 72-75, dean natural sci, 85-88, PROF NATURAL & APPL SCI, UNIV WIS-GREEN BAY, 82- *Mem:* AAAS; Am Inst Biol Sci; Am Soc Microbiol. *Res:* Aquatic microbiology; resource recovery; anaerobic treatment of high strength liquid waste streams; application of sludges on agricultural lands. *Mailing Add:* Natural & Appl Sci Univ Wis Green Bay WI 54311-7001

SCHWARTZ, LELAND DWIGHT, b Enid, Okla, July 26, 25; m 54; c 2. POULTRY PATHOLOGY. *Educ:* Okla State Univ, DVM, 53; Univ Ga, MS, 63. *Prof Exp:* Vet, pvt pract, 53-55; vet, Hartsel Ranch, 55-56; inspection livestock, Animal & Plant Health Inspection Serv, USDA, 56-59 & poultry, 59-61; instr avian path, Univ Ga, 61-64; vet, exten, Pa State Univ, 64-84; SR AVIAN PATHOLOGIST, ANIMAL HEALTH DIAG LAB, COL VET MED, MICH STATE UNIV, 84- *Honors & Awards:* Game Bird Indust Award, 84; Wildlife Conserv Award, 84. *Mem:* Sigma Xi; Am Vet Med Asn; Am Asn Avian Pathologists; Am Asn Exped Vet. *Res:* Unidentified viral infections of commercial chickens. *Mailing Add:* Animal Health Diag Lab Col Vet Med PO Box 30076 Lansing MI 48909

SCHWARTZ, LEON JOSEPH, b New York, NY, Jan 28, 43; m 67; c 2. MATERIALS SCIENCE. *Educ:* City Col New York, BEng, 64; City Univ New York, MEng, 66, PhD(metall), 70. *Prof Exp:* Chemist, Photoconductors, 70-74, MAT SCIENTIST, PITNEY-BOWES, INC, STAMFORD, CONN, 74- *Mem:* Electrochem Soc; Soc Photog Scientists & Engrs. *Res:* Materials research on a phenomenalogical level; inorganics, especially metals, semiconductors, ceramics, intermetallic compounds; solid state diffusion and photoconductivity; materials, electronic, paper; graphic arts; design of experiments, human factors. *Mailing Add:* 30 Briarcliff Dr Monsey NY 10952

SCHWARTZ, LEONARD H, b New York, NY, Nov 25, 32; div; c 3. CHEMISTRY. *Educ:* City Col New York, BS, 54; NY Univ, PhD(chem), 61. *Prof Exp:* From asst prof to assoc prof, 63-71, PROF CHEM, CITY COL NEW YORK, 71- *Concurrent Pos:* Exec officer PhD prog chem, City Univ New York, 71-78. *Mem:* Am Chem Soc; Royal Soc Chem. *Res:* Organic chemistry. *Mailing Add:* Dept Chem City Col New York New York NY 10031

SCHWARTZ, LEONARD WILLIAM, b New York, NY, May 21, 43; m 75; c 2. FLOW BEHAVIOR OF COATINGS, POROUS MEDIA FLOWS. *Educ:* Cornell Univ, BEP, 65, MAE, 66; Stanford Univ, PhD(fluid mech), 72. *Prof Exp:* Aero engr, Lockheed Missiles & Space Co, 65-69; Nat Res Coun fel, Ames Res Ctr, NASA, 73-74; staff scientist, Flow Res Inc, 74-75; sr lectr appl math, Univ Adelaide, Australia, 75-80; sr staff mathematician, Res Lab, Exxon Corp, 80-86; PROF MECH ENG & MATH, UNIV DEL, 87- *Concurrent Pos:* Vis prof, Univ Calif, Berkeley, 77, Stanford Univ, 80, Rutgers Univ, 86-87; consult, Failure Anal Assocs, 80-81, Glidden Paints Div, ICI, 90-; ed, J Eng Math, 85-; selector, prin young investr panel, NSF, 85. *Mem:* Am Phys Soc; Soc Indust & Appl Math. *Res:* Fluid mechanics; flow in porous media; coating flows; naval hydrodynamics; aerodynamics; mathematical modelling of fluid flow and heat transfer; nonlinear problems. *Mailing Add:* Dept Mech Eng Univ Del Newark DE 19711

SCHWARTZ, LOWELL MELVIN, b Brooklyn, NY, Dec 1, 34. PHYSICAL CHEMISTRY. *Educ:* Mass Inst Technol, SB, 56, ScD(chem eng), 59; Calif Inst Technol, MS, 57. *Prof Exp:* Chem engr, Dept Appl Physics, Chr Michelsens Inst, Norway, 59-60 & Esso Res & Eng Co, Standard Oil Co NJ, 60-61; res assoc chem, Princeton Univ, 62-63; res instr, Dartmouth Col, 63-65; from asst prof to assoc prof chem, 65-76, PROF CHEM, UNIV MASS, BOSTON, 76- *Mem:* Royal Soc Chem. *Mailing Add:* Dept Chem Univ Mass-Harbor Campus Boston MA 02125

SCHWARTZ, LYLE H, b Chicago, Ill, Aug 2, 36; m 73; c 2. MATERIALS SCIENCE. *Educ:* Northwestern Univ, BSc, 59, PhD(mat sci), 64. *Prof Exp:* From asst prof to assoc prof, 64-72, asst chmn, dept mat sci & eng, 73-78, dir, mat res ctr, 79-84, prof mat sci, Northwestern Univ, Evanston, 72-84; DIR, MAT SCI & ENG LAB, NAT INST STANDARDS & TECHNOL, 84- *Concurrent Pos:* Consult, Solid State Sci Div, Argonne Nat Lab, 66-; visitor, Bell Tel Labs, 72-73; mem, Nat Res Coun panel to select NSF doctoral fels in eng, 74-77, chmn, 76 & 77. *Mem:* Am Phys Soc; Mat Res Soc; Am Crystallog Asn; Am Inst Mining, Metall & Petrol Engrs; fel Am Soc Metall. *Res:* X-ray and neutron diffraction; Mossbauer effect; spinodal alloys; alloy catalysts. *Mailing Add:* Mat Bldg Rm B309 Mat Sci & Eng Nat Inst Standards & Technol Gaithersburg MD 20899

SCHWARTZ, M(URRAY) A(RTHUR), b New York, NY, Nov 13, 20; m 44; c 2. MATERIALS SCIENCE ENGINEERING. *Educ:* Alfred Univ, BS, 43. *Prof Exp:* Ceramic engr, Bendix Aviation Corp, 43-44, Streator Drain Tile Co, 46-47, Fairchild Engine & Airplane Corp, 47-51 & Oak Ridge Nat Lab, 51; supvr ceramics, Aeronaut Res Lab, US Dept Air Force, 51-59; mgr, Aeronutronic Div, Ford Motor Co, 59-60; mat appln, United Tech Ctr, United Aircraft Corp, 60-65; asst dir ceramics div, Ill Inst Technol Res Inst, 65-71; coord engr, Dept Pub Works, Chicago, 71-72; supvr nonmetallic mat res, Tuscaloosa Res Lab, 72-74, staff ceramic eng, 74-90, MAT PROG MGR, US BUR MINES, 90-; DIR, MAT TECHNOL CONSULT INC. *Concurrent Pos:* Exec secy, Interagency Comt Mat, 75-79; ed, News From Washington, Bulletin Am Ceramic Soc, 76-86; mats policy analyst, White House Off Sci & Technol Policy, 82-84; deleg eng Affairs Coun, Am Asn Eng Soc, 85-; trustee, Fedn Mat Socs, 83- *Mem:* Fel Am Ceramic Soc (vpres, 81-82); fel Am Inst Chem; Am Soc Metals; Am Inst Ceramic Engrs; AAAS. *Res:* Nonmetallic materials; recycling of waste materials; utilization of mineral resources; industrial ceramics; new materials; basic behavior; refractories; whitewares; glass; aerospace, nuclear and electronic applications; environmental quality; energy conservation; construction materials. *Mailing Add:* Mat Technol Consult Inc 30 Orchard Way N Potomac MD 20854-6128

SCHWARTZ, MARSHALL ZANE, b Minneapolis, Minn, Sept 1, 45; m 71; c 2. GASTROINTESTINAL PHYSIOLOGY, ORGAN TRANSPLANTATION. *Educ:* Univ Minn, BS, 68, MD, 70. *Prof Exp:* Instr & asst surg, Children's Hosp Med Ctr, Med Sch, Harvard Univ, 78-79; from asst prof to assoc prof surg & pediat, Med Br, Univ Tex, 79-81, chief pediat & surg serv, Child Health Ctr, surgeon-in-chief, 80-83; assoc prof, 81-83, PROF SURG & PEDIAT, 86-, CHIEF PEDIAT SURG, UNIV CALIF, DAVIS, 83- *Concurrent Pos:* Prin investr, Basil O'Connor Starter Res Grant, March of Dimes, 81-86 & Young Investr Award, NIH, 82-85. *Honors & Awards:* Sam Segal Award, Univ Minn, 66, Boutell Award, 68; James W McLaughlin Award, Univ Tex, 83. *Mem:* Soc Univ Surgeons; Am Col Surgeons; Am Pediat Surg Asn; Am Acad Pediat; Soc Surg Alimentary Tract; Asn Acad Surg. *Res:* The relationship of gastrointestinal peptides to the function and adaptation of the small intestine; methods of identifying rejection following small intestine transplantation. *Mailing Add:* Dept Surg Univ Calif Med Ctr 4301 X St Rm 2310 Sacramento CA 95817

SCHWARTZ, MARTIN, plant physiology, biochemistry; deceased, see previous edition for last biography

SCHWARTZ, MARTIN ALAN, b New York, NY, July 5, 40; m 61; c 3. ORGANIC CHEMISTRY. *Educ:* Dartmouth Col, AB, 62; Stanford Univ, PhD(org chem), 66. *Prof Exp:* From asst prof to assoc prof, 66-76, chmn dept, 77-83, PROF CHEM, FLA STATE UNIV, 76- *Concurrent Pos:* Sr ed, J Org Chem, Am Chem Soc, 71- *Mem:* Am Chem Soc. *Res:* Synthesis of natural products. *Mailing Add:* Dept Chem Fla State Univ Tallahassee FL 32306

SCHWARTZ, MARTIN ALEXANDER, b Dec 31, 54; m 86; c 2. PHOTOCHEMICAL CROSSLINKING, CYTOPLASMIC PH. *Educ:* Stanford Univ, PhD(phys chem), 79. *Prof Exp:* Asst prof, 83-89, ASSOC PROF CELL BIOL & PHYSIOL, HARVARD UNIV MED SCH, 90- *Mem:* Am Soc Cell Biol. *Res:* Cytoskeleton-membrane interactions and signal transduction across the cell membrane; cell biology. *Mailing Add:* Dept Physiol Harvard Univ Med Sch 25 Shattuck St Boston MA 02115

SCHWARTZ, MAURICE EDWARD, b Laurinburg, NC, Sept 28, 39; m 65; c 2. THEORETICAL CHEMISTRY. *Educ:* Presby Col (SC), BS, 61; Vanderbilt Univ, PhD(chem), 66. *Prof Exp:* NSF & Ramsey fels, Oxford Univ, 66; res fel, Princeton Univ, 67-68; asst prof & sr scientist, Radiation Lab, 68-73, ASSOC PROF CHEM, UNIV NOTRE DAME, 73- *Concurrent Pos:* Vis prof, Univ Calif, Berkeley, 72; NATO sr fel sci, Univ Uppsala, Sweden & Oxford Univ, England, 73; assoc prog dir quantum chem, NSF, 76-77. *Mem:* Am Chem Soc; Am Phys Soc; Am Asn Univ Professors. *Res:* Quantum chemistry; photoelectron spectroscopy; radiation chemistry; surface chemistry and catalysis. *Mailing Add:* 725 Park Ave South Bend IN 46616

SCHWARTZ, MAURICE LEO, b Ft Worth, Tex, Sept 27, 25; m 50; c 5. GEOLOGICAL OCEANOGRAPHY, SCIENCE EDUCATION. *Educ:* Columbia Univ, BS, 63, MA, 64, PhD(geol), 66. *Prof Exp:* Lab instr geol, Columbia Univ, 63-66; lectr, Brooklyn Col, 64-68; from asst prof to assoc prof, 68-75, PROF GEOL & EDUC, WESTERN WASH UNIV, 75-, DEAN GRAD STUDIES & RES, 87-; PRES, COASTAL CONSULTS, INC, 81- *Concurrent Pos:* Fulbright-Hayes res scholar, Inst Oceanog & Fishing Res, Athens, Greece, 73-74; Nat Acad Sci specialist exchange prog visit to USSR, 78 & 86; actg grad dean, Western Wash Univ, 87- *Mem:* Geol Soc Am; Nat Asn Geol Teachers; Am Asn Univ Professors; Coastal Soc; Am Shore & Beach Preservation Asn. *Res:* Beach processes; sea level changes; earth science education; coastal archeology; barrier islands; artificial beach nourishment. *Mailing Add:* Dept Geol Western Wash Univ Bellingham WA 98225

SCHWARTZ, MELVIN, b New York, NY, Nov 2, 32; m 53; c 3. PHYSICS, DATA COMMUNICATION. *Educ:* Columbia Univ, AB, 53, PhD(physics), 58. *Prof Exp:* Res assoc, Brookhaven Nat Lab, 56-57, assoc physicist, 57-58; from asst prof to prof physics, Columbia Univ, 58-66; prof physics, Stanford Univ, 66-83; pres, Digital Pathways Inc, 70-91; ASSOC DIR, HIGH ENERGY & NUCLEAR PHYSICS, BROOKHAVEN NAT LAB, 91- *Concurrent Pos:* Sloan fel, 59-63; Guggenheim fel, Guggenheim Found, 68. *Honors & Awards:* Nobel Prize in Physics, 88; Prize, Am Phys Soc, 64. *Mem:* Nat Acad Sci; fel Am Phys Soc. *Res:* High energy experimental particle physics. *Mailing Add:* Brookhaven Nat Lab Bldg 510F Upton NY 11973

SCHWARTZ, MELVIN J, b Brooklyn, NY, Oct 8, 28; m 52; c 2. THEORETICAL PHYSICS. *Educ:* Brooklyn Col, BS, 51; State Univ Iowa, PhD(physics), 58. *Prof Exp:* Instr physics, Rutgers Univ, 56-57; lectr, Univ Minn, 57-59; res assoc, Syracuse Univ, 59-61; assoc prof, Adelphi Univ, 61-64; res scientist, NY Univ, 64-66; ASSOC PROF PHYSICS, ST JOHN'S UNIV, NY, 66- *Concurrent Pos:* Nat Sci Found res grant, 61-64; NASA res grant, 64-66. *Mem:* AAAS; Am Phys Soc; NY Acad Sci; Int Soc Gen Relativity & Gravitation; Sigma Xi. *Res:* Quantum field theory; correspondence principle quantization of electrodynamics; collisionless plasmas; general relativistic kinetic theory of plasmas; social effects of science. *Mailing Add:* Two Henry St Great Neck NY 11023

SCHWARTZ, MICHAEL AVERILL, b New York, NY, Aug 4, 30; div; c 2. PHARMACEUTICAL CHEMISTRY. *Educ:* Brooklyn Col Pharm, BS, 52; Columbia Univ, MS, 56; Univ Wis, PhD(pharm), 59. *Prof Exp:* Sr res scientist prod develop, Bristol Labs, Inc, 59-63; from asst prof to prof, State Univ NY, Buffalo, 63-78, 63-70, asst dean, 66-70, dean, Sch Pharm, 70-76; PROF PHARM & DEAN COL, UNIV FLA, 78- *Concurrent Pos:* USPHS grant, 64-69. *Mem:* Am Pharmaceut Asn; fel Am Asn Pharmaceut Sci; Am Chem Soc; Am Soc Hosp Pharmacists. *Res:* Pharmaceutics; chemistry of penicillins and drug allergy; models for enzymes; drug stability. *Mailing Add:* Dean Pharm Univ Fla Gainesville FL 52611

SCHWARTZ, MICHAEL H, b New Haven, Conn, June 16, 59. QUALITY ASSURANCE, AUDITING. *Educ:* George Washington Univ, BSEE, 81; Univ New Haven, MBA, 86. *Prof Exp:* Asst proj engr, Raytheon, 80-81; qual assurance engr, Pitney Bowes Inc, 81-84, staff qual engr, 84-85, proj mgr, 85-86, qual control mgr, 86-87, SR ENGR, PITNEY BOWES INC, 87- *Mem:* Inst Elec & Electronics Engrs; Am Soc Qual Control; Inst Interconnecting & Packaging Electronic Circuits. *Mailing Add:* Pitney Bowes Inc One Waterview Dr Shelton CT 06484

SCHWARTZ, MICHAEL MUNI, b Chicago, Ill, Aug 30, 43; m 70; c 1. INDUSTRIAL CHEMISTRY. *Educ:* Northwestern Univ, BA, 64; Fla State Univ, MS, 67, PhD(org chem), 70. *Prof Exp:* SR RES CHEMIST, AMOCO OIL CO, 70- *Mem:* Am Chem Soc. *Mailing Add:* 1138 Cheshire Naperville IL 60540-5704

SCHWARTZ, MISCHA, b New York, NY, Sept 21, 26; m 57, 70; c 1. COMPUTER & DIGITAL COMMUNICATIONS, TELECOMMUNICATIONS. *Educ:* Cooper Union, BEE, 47; Polytech Inst Brooklyn, MEE, 49; Harvard Univ, PhD(appl physics), 51. *Prof Exp:* Asst proj engr, Sperry Gyroscope Co, 47-49, proj engr, 49-52; from asst prof to prof elec eng, Polytech Inst Brooklyn, 52-74, head dept, 61-65; prof elec eng & comput sci, 74-88, dir, Ctr Telecommun Res, 85-88, CHARLES BATCHELOR PROF ENG & COMPUT SCI, COLUMBIA UNIV, 87- *Concurrent Pos:* NSF sci fac fel, Ecole Normale Superiore, Paris, 65-66; radiation physicist, Montefiore Hosp, 55-56; indust consult; chmn comn C, US Nat Comt, Int Union Radio Sci, 78-81; vis scientist, IBM Res, 80, Nippon Tel & Tel, 81 & NYNEX Corp, 86; dir, Inst Elec & Electronics Engrs, 78-79, pres Commun Soc, 84-85. *Honors & Awards:* Educ Medal, Inst Elec & Electronics Engrs, 83, Regional Award, 89; Gano Dunn Award, Cooper Union, 86. *Mem:* Fel AAAS; Asn Comput Mach; fel Inst Elec & Electronics Engrs; Sigma Xi; Am Asn Univ Professors. *Res:* Communication systems; computer communications. *Mailing Add:* Dept Elec Eng Columbia Univ 1206 SW Mudd New York NY 10027

SCHWARTZ, MORTON ALLEN, b Brooklyn, NY, Mar 12, 28; m 62. DRUG METABOLISM. *Educ:* City Col, NY, BS, 50; Univ Wis, MS, 52, PhD(biochem), 54. *Prof Exp:* Biochemist, Path Dept, Mercy Hosp, Pittsburgh, 54-56; res fel biophys, Sloan-Kettering Inst, 56-58; asst dir, Drug Metab Lab, 58-71, head sect drug disposition, 71-74, asst dir, 74-79, ASSOC DIR, DEPT BIOCHEM & DRUG METAB, HOFFMANN-LA ROCHE INC, 80. *Mem:* Am Chem Soc; Am Soc Pharmacol & Exp Therapeut; NY Acad Sci; Acad Am Pharmaceut Asn; AAAS. *Res:* Metabolism of drugs; relationship between drug metabolism and biological activity. *Mailing Add:* 35 Ingram Matawan NJ 07747

SCHWARTZ, MORTON DONALD, b Chicago, Ill, Oct 11, 36; m 58, 89; c 2. COMPUTERS, INSTRUMENTATION. *Educ:* Univ Calif, Los Angeles, BS, 58, MS, 60, PhD(eng), 64. *Prof Exp:* Sr engr, NAm Aviation, 64-66; sr scientist, TRW, 66-70; PROF ELEC ENG, CALIF STATE UNIV, LONG BEACH, 70- *Concurrent Pos:* Consult, St Mary Med Ctr, Long Beach, Calif, 70-, Hughes Aircraft Co, 78-; ed, J Clin Eng, 75-; mem, Bd Dir, Am Bd Clin Engrs, 75- *Mem:* Inst Elec & Electronic Engrs; Am Soc Eng Educ. *Res:* Computer applications in medicine including instrumentation for cardiac catheterization, pacemaker and firemen paramedic systems. *Mailing Add:* Computer Eng & Computer Sci Calif State Univ 1250 Bellflower Blvd Long Beach CA 90840

SCHWARTZ, MORTON K, b Wilkes-Barre, Pa, Oct 22, 25; m 66; c 2. BIOCHEMISTRY, CLINICAL CHEMISTRY. *Educ:* Lehigh Univ, BA, 48; Boston Univ, MA, 49, PhD(biochem), 52. *Prof Exp:* From instr to prof biochem, 54-84, prof, Develop Ther Clin Invest, 84-88, PROF, MOLECULAR PHARMACOL & THERAPEUT, SLOAN-KETTERING DIV, MED COL, CORNELL UNIV, 88- *Concurrent Pos:* Res fel, Sloan-Kettering Div, Med Col, Cornell Univ, 52-54; res fel, Sloan-Kettering Inst Cancer Res, 52-55; asst, Surg Metab Sect, Sloan-Kettering Inst Cancer Res, 55-56, assoc mem, Div Biochem, 57-, mem & lab head, Inst, 69-, assoc field coordr, Human Cancer, 75-81; dir clin res training, Mem Hosp Cancer & Allied Dis, 71-81, vpres lab affairs & dep gen dir, 77-81, attend clin chemist & chmn, Dept Clin Chem, 67- *Honors & Awards:* Van Slyke Award, Am Asn Clin Chem; Wiley Medal-Commissioners Citation, Food & Drug Admin. *Mem:* Am Soc Biochem & Molecular Biol; Am Chem Soc; Acad Clin Lab Phys & Scientists; Am Asn Clin Chem (pres, 75); Asn Clin Scientists; Am Asn Cancer Res; NY Acad Sci; Int Soc Clin Enzym (pres, 90). *Res:* Enzyme kinetics; serum enzymes; tumor markers; hormone receptor; automation. *Mailing Add:* Mem Sloan Kettering Cancer Ctr 1275 York Ave New York NY 10021

SCHWARTZ, NEENA BETTY, b Baltimore, Md, Dec 10, 26. PHYSIOLOGY, ENDOCRINOLOGY. *Educ:* Goucher Col, BA, 48; Northwestern Univ, MS, 50, PhD, 53. *Hon Degrees:* DSc, Goucher Col, 82. *Prof Exp:* From instr to assoc prof physiol, Col Med, Univ Ill, 53-57; dir biol lab, Inst Psychosom & Psychiat Res & Training, Michael Reese Hosp, 57-61; from assoc prof to prof physiol, Univ Ill Col Med, 61-70, prof neuroendocrinol, 70-73; asst dean, Col Med, Northwestern Univ, 68-70, prof physiol, Med Sch, 73-74, prof biol & chmn dept, 74-77, DEERING PROF NEUROBIOL & PHYSIOL, NORTHWESTERN UNIV, EVANSTON, 81- *Honors & Awards:* Williams Distinguished Serv Award, Endocrine Soc, 85. *Mem:* Endocrine Soc, (pres, 82-83); Am Physiol Soc; Soc Study Reproduction (pres, 77-78); fel AAAS. *Res:* Endocrinology; reproduction; environmental control of pituitary. *Mailing Add:* 1215 Chancellor St Evanston IL 60201

SCHWARTZ, NEWTON, b New York, NY, Aug 1, 25; m 49; c 2. SOLID STATE CHEMISTRY. *Educ:* City Col NY, BS, 47; Stevens Inst Technol, MS, 50; Univ Southern Calif, PhD(chem), 55. *Prof Exp:* Instr chem, Stevens Inst Technol, 47-49; res chemist, Maimonides Hosp, Brooklyn, 49-50; fel & lectr chem, Univ Southern Calif, 54-56; mem tech staff, Bell Tel Labs Inc, 56-66, dept head struct anal thin film mat, 66-70, supvr, Thin Film Mat Res Dept, 70-84; RETIRED. *Concurrent Pos:* Div ed, J Electrochem Soc. *Mem:* AAAS; Am Vacuum Soc; Am Chem Soc; Electrochem Soc. *Res:* Physical organic chemistry; free radical chemistry; formation and properties of thin metallic and inorganic films. *Mailing Add:* Village Green Bldg 8L Budd Lake NJ 07828

SCHWARTZ, NORMAN MARTIN, b New York, NY, Nov 9, 35; m 69; c 2. BIOLOGY. *Educ:* Hunter Col, BA, 56; Syracuse Univ, MS, 59; Univ Chicago, PhD(microbiol), 63. *Prof Exp:* NIH res fel microbiol, Yale Univ, 63-64; asst prof, 64-80, ASSOC PROF BIOL, CITY COL NEW YORK, 80- *Res:* Mutagenesis; microbial genetics. *Mailing Add:* Dept Biol City Col New York Convent Ave & 138th St New York NY 10031

SCHWARTZ, PAUL, b New York, NY, Sept 12, 48; m 71; c 2. DEVELOPMENT OF NEW ANTICANCER DRUGS. *Educ:* City Col NY, BS, 69, PhD(chem), 74. *Prof Exp:* Fel biophys, Mich State Univ, 74-75; res chemist, US Vet Admin Med Ctr, Charleston, SC, 75-80; SR RES SCIENTIST, ANDRULIS RES CORP, 80- *Mem:* Am Chem Soc. *Res:* Development of new anticancer drugs. *Mailing Add:* 14531 Woodcrest Dr Rockville MD 20853

SCHWARTZ, PAUL HENRY JR, b Baltimore, Md, Dec 19, 36; m 61; c 4. ENTOMOLOGY. *Educ:* Univ Md, BS, 59, MS, 61; Univ Fla, PhD(entom), 64. *Prof Exp:* Res entomologist, Agr Res Serv, USDA, 61-64; entomologist, USPHS, 64-65; res entomologist, 65-69, asst to chief, Fruit Insects Res Br, Entom Res Div, 69-72, staff specialist, 72-73, STAFF SCIENTIST, SCI & EDUC ADMIN-AGR RES, USDA, 73- *Mem:* Entom Soc Am. *Res:* Chemical and other methods for control of insect pests. *Mailing Add:* 8497 Heatherwold Dr Laurel MD 20723

SCHWARTZ, PAULINE MARY, b Philadelphia, Pa, Aug 8, 47; m 75. PHARMACOLOGY, BIOCHEMISTRY. *Educ:* Drexel Univ, BS, 70; Univ Mich, MS, 71, PhD(med chem), 75. *Prof Exp:* Teaching fel chem, Univ Mich, 70-71; res scholar cancer chemother, Los Angeles County-Univ Southern Calif Cancer Ctr, 75-76, cancer res training fel, 76; res assoc path, Sch Med, Univ Southern Calif, 76-77; fel, Dept Pharmacol, Yale Univ, 77-80, res assoc, 80-83; res assoc, E I Du Pont Pharmaceuticals, 83-84; RES ASSOC, DEPT DERMAT, YALE UNIV, 85- *Concurrent Pos:* Prin investr, Young Investrs Grant, Yale Univ; practr-in-residence, Dept Chem, Univ New Haven. *Honors & Awards:* Wilson R Earle Award, Nat Tissue Cult Asn, 74. *Mem:* Am Chem Soc; Am Soc Microbiol; Am Asn Cancer Res; Soc Invest Dermat. *Res:* Mechanism of action of anti hyperproliferative drugs; design of chemotherapy; nucleotide metabolism; epidermal differentiation; hyperproliferative skin diseases. *Mailing Add:* 101 Hammonasset Meadows Rd Madison CT 06443

SCHWARTZ, PETER LARRY, b Chicago, Ill, July 11, 40; m 64. CLINICAL BIOCHEMISTRY, MEDICAL EDUCATION. *Educ:* Univ Wis-Madison, BS, 62; Wash Univ, MD, 65. *Prof Exp:* lectr clin biochem, 71-77, SR LECTR CLIN BIOCHEM, MED SCH, UNIV OTAGO, NZ, 77- & SR LECTR, HIGHER EDUC DEVELOP CTR, 87- *Concurrent Pos:* Australian Res Grants Comt fel biochem, Monash Univ, Australia, 65-68; external exmnr, Sch Med Sci, Univ Sains, Malaysia, 84-86; vis prof, Med Sch, Univ Toronto, Can, 86. *Honors & Awards:* ANZAME-Smith, Kline & French Award, 89. *Mem:* Australasian & NZ Asn Med Educ. *Res:* Improved methods of medical and biochemical education; self-learning systems; computer-assisted instruction. *Mailing Add:* Dept Clin Biochem Univ Otago Med Sch Box 913 Dunedin New Zealand

SCHWARTZ, RALPH JEROME, b Chicago, Ill, Mar 14, 19; m 66. SPEECH PATHOLOGY, AUDIOLOGY. *Educ:* Northwestern Univ, BS, 41; Marquette Univ, MA, 48; Purdue Univ, PhD(speech path, audiol), 58. *Prof Exp:* Head speech clin, Am Red Cross, McPherson, Kans, 47-52; speech clinician, Inst Logopedics, 52-53; speech therapist & actg head speech & hearing dept, Children's Rehab Inst, Inc, Baltimore, Md, 53-55; asst prof logopedics, Univ Wichita & Inst Logopedics, 57-63; from asst prof to assoc prof speech, 63-71, assoc prof speech path & audiol, 71-89, EMER PROF, UNIV NORTHERN IOWA, 89- *Mem:* Am Speech & Hearing Asn; Acoust Soc Am; Int Asn Logopedics & Phoniatrics; Am Asn Phonetic Sci; Int Soc Phonetic Sci. *Res:* Voice; diagnosis and appraisal; phonology. *Mailing Add:* Dept Commun Dis Univ Northern Iowa Cedar Falls IA 50614

SCHWARTZ, RICHARD DEAN, b Hutchinson, Kans, Apr 17, 41; m 78. ASTRONOMY, ASTROPHYSICS. *Educ:* Kans State Univ, BS, 63; Union Theol Sem, MDiv, 66; Univ Wash, MS, 70, PhD(astron), 73. *Prof Exp:* Res asst atomic collision physics, Columbia Radiation Lab, Columbia Univ, 64-68; res teaching asst astron, Univ Wash, 68-73; astronomer pre-main sequence astron, Lick Observ, Univ Calif, Santa Cruz, 73-75; ASST PROF ASTRON, UNIV MO, 75- *Concurrent Pos:* Mem user's comt, Kitt Peak Nat Observ, 76-77; Copernicus prin investr, NASA grant, 76-78. *Mem:* Am Astron Soc; Int Astron Union. *Res:* Observational and theoretical pre-main sequence astronomy; T Tauri stars; Herbig-Haro objects; circumstellar dust shells; post-main sequence astronomy; white dwarfs; planetary nebulae. *Mailing Add:* Dept Physics Univ Mo 8001 Natural Bridge Rd St Louis MO 63121

SCHWARTZ, RICHARD F(REDERICK), b Albany, NY, May 31, 22; div; c 5. ELECTROMAGNETICS, ACOUSTICS. *Educ:* Rensselaer Polytech Inst, BEE, 43, MEE, 48; Univ Pa, PhD(elec eng), 59. *Prof Exp:* Asst physics, Rensselaer Polytech Inst, 46, asst elec eng, 46-48, instr, 48; develop engr, Adv Develop Sect, Radio Corp Am, NJ, 48-51; instr elec eng, Moore Sch Elec Eng, Univ Pa, 51-53, from assoc to res assoc, 53-59, from asst prof to assoc prof, 59-74, chmn grad group elec eng, 68-73; dept head, 73-79, PROF ELEC ENG, MICH TECHNOL UNIV, 73- *Concurrent Pos:* Consult, Marquette Univ, 78, Singer Corp, 80 & John Wiley & Sons, 81. *Mem:* AAAS; Am Soc Eng Educ; sr mem Inst Elec & Electronics Engrs; Nat Soc Prof Engrs; Acoust Soc Am. *Res:* Microwave theory and techniques; electromagnetic theory; antennas; communication engineering; electromagnetic compatability; musical acoustics; electroacoustics; electrical measurements. *Mailing Add:* Dept Elec Eng Watson Sch Eng-Appl Sci Tech State Univ NY Binghamton NY 13901

SCHWARTZ, RICHARD JOHN, b Waukesha, Wis, Aug 12, 35; m 57; c 8. SOLAR CELLS. *Educ:* Univ Wis, BSEE, 57; Mass Inst Technol, SMEE, 59, ScD(elec eng), 62. *Prof Exp:* Mem tech staff solid state develop, David Sarnoff Res Ctr, RCA Corp, 57-58; sr scientist, Energy Conversion, Inc, 61-64; assoc prof elec eng, 64-71, asst head sch, 72-83, PROF ELEC ENG, PURDUE UNIV, LAFAYETTE, 71-, HEAD SCH, 85- *Concurrent Pos:* Consult, RCA Corp, 65-77. *Mem:* Fel Inst Elec & Electronics Engrs. *Res:* Solid state devices; direct energy conversion; semiconducting materials. *Mailing Add:* Sch Elec Eng Purdue Univ West Lafayette IN 47907

SCHWARTZ, ROBERT, b New Haven, Conn, Dec 17, 22; m 47; c 3. PEDIATRICS. *Educ:* Yale Univ, BS, 43, MD, 47. *Prof Exp:* Intern, Children's Med Serv, Bellevue Hosp, 47-48; asst pediat, Col Med, NY Univ, 48-49, instr, 49; res fel, Harvard Med Sch, 49-51, res assoc med, 51-53, instr pediat, 53-54, assoc, 55-58; from assoc prof to prof pediat, Sch Med, Case Western Reserve Univ, 59-74; dir pediat metab & nutrit, RI Hosp, 74-87; PROF MED SCI, BROWN UNIV, 74- *Concurrent Pos:* Fel med, Children's Hosp, Boston, 49-51, NIH fel, 49-51; fel med, Thorndike Mem Lab, Boston City Hosp, 51-53; asst resident, Bellevue Hosp, 48-49, resident physician, 49; asst physician, Children's Hosp, 53-56, chief diabetic clin, 54-58, assoc physician & chief metab, 56-58; sr asst surgeon, Thorndike Mem Lab, Boston City Hosp, 51-53; res collabr, Brookhaven Nat Lab, 59-66; assoc pediatrician, Babies & Children's Hosp, Cleveland, 59-74; staff pediatrician, Metrop Gen Hosp, 59-74; dir dept pediat, Cleveland Metrop Gen Hosp, 67-74. *Mem:* AAAS; Am Soc Pediat Res; Am Fedn Clin Res; Am Acad Pediat; Pediat Soc. *Res:* Pediatrics metabolism; physiology. *Mailing Add:* Dept Pediat RI Hosp Providence RI 02902

SCHWARTZ, ROBERT BERNARD, b Brooklyn, NY, Sept 2, 29; m 56; c 2. NEUTRON DOSIMETRY, NUCLEAR PHYSICS. *Educ:* Union Univ, NY, BS, 51; Yale Univ, PhD(physics), 55. *Prof Exp:* Asst physicist, Brookhaven Nat Lab, 55-58; res assoc, Atomic Energy Res Estab, Harwell, Eng, 58-59; asst physicist, Brookhaven Nat Lab, 59-60; physicist, US Naval Res Lab, 60-62; PHYSICIST, US NAT BUR STANDARDS, 62- *Honors & Awards:* Silver Medal Award, Dept Com, 82. *Mem:* Am Phys Soc; Health Physics Soc. *Res:* Neutron spectroscopy by time-of-flight; neutron personnel dosimetry. *Mailing Add:* Nat Inst Standards & Technol Gaithersburg MD 20899

SCHWARTZ, ROBERT DAVID, b Brooklyn, NY, Apr 3, 41; div; c 2. INDUSTRIAL MICROBIOLOGY. *Educ:* Brooklyn Col, BS, 64; Long Island Univ, MS, 67; Rutgers Univ, PhD(microbial genetics), 69. *Prof Exp:* Technician metab res, Col Physicians & Surgeons, Columbia Univ, 64-65; med technologist, Middlesex Gen Hosp, New Brunswick, NJ, 68-69; res assoc oncogenic virol, Germfree Life Res Ctr, Life Sci Ctr, Nova Univ Advan Technol, 69-70; res biologist, Exxon Res & Eng Co, 70-76; proj scientist, Union Carbide Corp, 76-79; scientist, Food & Biotechnol Dept, Stauffer Chem Co, 79-87; SR DEVELOP SCIENTIST, FERMENTATION DEVELOP DEPT, ABBOTT LABS, 87- *Honors & Awards:* Porter Award, Soc Indust Microbiol, 88. *Mem:* AAAS; Am Soc Microbiol; Soc Indust Microbiol; Sigma Xi. *Res:* Fermentation process development; enzymatic hydroxylation; biotransformation; pollution control; single cell protein; microbial genetics; microbial energy production. *Mailing Add:* Fermentation Develop Dept 451 Abbott Labs North Chicago IL 60064-4000

SCHWARTZ, ROBERT JOHN, b Pittsburgh, Pa, Nov 14, 42; m 65, 76; c 1. NONWOVENS, MELT SPINNING OF FIBERS. *Educ:* Univ Md, BS, 64, PhD(chem eng), 68. *Prof Exp:* Res eng, Org Chem Div, FMC Corp, 67-69, sr res engr, 69-70; res engr, 70-72, sr res sci, 72-75, mgr prod develop, 75-78, dir process develop, 78-80 & technol develop, 80-82, vpres, non-wovens res & develop, 82-87, VPRES, LONG RANGE RES & DEVELOP, KIMBERLEY CLARK CORP, 87- *Mem:* Am Inst Chem Engrs. *Res:* Development of new filimant spinning and subsequent fabric/web formation technology with emphasis on increasing final product fabric attributes at the lowest cost; exploring novel approaches to fiber formation. *Mailing Add:* 2100 Habersham Marine Rd Cumming GA 30130

SCHWARTZ, ROBERT NELSON, b New Haven, Conn, Feb 4, 40; m 71; c 2. PHYSICAL CHEMISTRY. *Educ:* Univ Conn, BA, 62, MS, 65; Univ Colo, Boulder, PhD(chem), 69. *Prof Exp:* Fel chem, Univ Ill, Chicago, 69-70, from asst prof to assoc prof, 70-81; vis assoc prof chem, Univ Calif, Los Angeles, 79-81; mem tech staff, 81-87, SR STAFF PHYSICIST, HUGHES RES LAB, 87- *Concurrent Pos:* Vis scholar, Univ Calif, Los Angeles, 79-80; adj prof phys & astron, Calif State Univ, Northridge, 81- *Mem:* Sigma Xi; Am Phys Soc; NY Acad Sci. *Res:* Magnetic resonance; molecular structure and relaxation phenomena; magnetic and nonlinear optical properties of solids. *Mailing Add:* Hughes Res Labs 3011 Malibu Canyon Rd Malibu CA 90265

SCHWARTZ, ROBERT SAUL, b Brooklyn, NY, Apr 20, 42; m 65; c 2. ANALYTICAL CHEMISTRY. *Educ:* Brooklyn Col, BS, 63; City Univ New York, PhD(anal chem), 74. *Prof Exp:* Chemist pharmaceut res, Dept Health & Welfare, Food & Drug Admin, 74-75; RES CHEMIST ANALYTICAL RES, US CUSTOMS SERV, DEPT TREAS, 75- *Mem:* Am Chem Soc. *Res:* Analysis and detection of trace quantities of organic vapors; electroanalytical chemistry; instrumentation. *Mailing Add:* 19121 Brooke Grove Ct Gaithersburg MD 20879-2119

SCHWARTZ, ROBERT STEWART, b East Orange, NJ, Mar 14, 28; m 63; c 1. HEMATOLOGY. *Educ:* Seton Hall Col, BS, 50; NY Univ, MD, 54. *Prof Exp:* From intern to resident med, Montefiore Hosp, NY, 54-56; resident, New Haven Hosp, Conn, 56-57; clin fel hemat, 57-58, res fel, 58-60, from instr to assoc prof, 60-71, PROF MED, NEW ENG MED CTR, SCH MED, TUFTS UNIV, 71- *Concurrent Pos:* Dir blood bank, New Eng Med Ctr Hosp, 61-, chief clin immunol sect, 66-72; chief, Hemat-Oncol Div, 72-90. *Honors & Awards:* Stratton Prize, Am Soc Hemat; John Phillips Award, Am Col Physicians. *Mem:* Am Soc Clin Invest; Am Asn Immunol; Am Fedn Clin Res; Am Soc Hemat; Transplantation Soc. *Res:* Autoimmunity; experimental leukemia. *Mailing Add:* Dept Med New England Med Ctr 750 Washington St Boston MA 02111

SCHWARTZ, RUTH, b Berlin, Ger, Oct 9, 24. NUTRITION. *Educ:* Univ London, BSc, 47, PhD(nutrit biochem), 59. *Prof Exp:* Res asst biochem, Postgrad Med Sch, Univ London, 48-49; biochemist, Med Res Coun, Uganda, 50-56; res fel nutrit biochem, Med Sch, Wash Univ, 57-60; lectr nutrit, London Sch Hyg & Trop Med, 60-63; res assoc nutrit biochem, Mass Inst Technol, 63-65; asst nutrit, Univ Conn, 65-70; PROF NUTRIT, DIV NUTRIT SCI, CORNELL UNIV, 70- *Mem:* AAAS; Am Inst Nutrit; Brit Nutrit Soc; NY Acad Sci; Am Chem Soc. *Res:* Magnesium metabolism; stable isotopes as tracers; absorption and availability of minerals. *Mailing Add:* Cornell Univ N-205A Van Rensselaer Ithaca NY 14850

SCHWARTZ, SAMUEL, b Minneapolis, Minn, Apr 13, 16; m 37; c 9. MEDICAL RESEARCH. *Educ:* Univ Minn, BS, 38, MD, 43. *Prof Exp:* Intern, Univ Minn Hosps, 42-43; group leader res, Manhattan Proj, Univ Chicago, 43-46; from asst prof to assoc prof, 48-61, PROF EXP MED, UNIV MINN, MINNEAPOLIS, 61-; AT MINNEAPOLIS MED RES FOUND. *Concurrent Pos:* Commonwealth Fund fels, Univ Minn, Carlsberg Lab, Copenhagen & Karolinska Inst, Sweden, 46-48; USPHS res career award exp med, Nat Inst Gen Med Sci, 63-; vis prof, Hebrew Univ, Jerusalem, 61-62. *Mem:* Am Soc Biol Chem; Am Soc Clin Invest; Am Asn Cancer Res; Soc Exp Biol & Med. *Res:* Porphyrin and bile pigment metabolism; modification of radiosensitivity by metalloporphyrins. *Mailing Add:* Raptor Ctr Sch Vet Med 1920 Fitch Ave St Paul Campus St Paul MN 55108

SCHWARTZ, SAMUEL MEYER, b Winnipeg, Man, Feb 15, 29; m 54; c 3. MEDICINAL CHEMISTRY. *Educ:* Univ Man, BSc, 52; Univ Minn, PhD(med chem), 56. *Prof Exp:* Assoc prof pharmaceut chem, Sch Pharm, George Washington Univ, 56-64; scientist admin, NIH, 64-78, assoc dir sci rev, Div Res Grants, 78-83; SPEC ASST DIR, NAT INST ARTHRITIS, DIABETES & DEGENERATIVE KIDNEY DIS, 83- *Mem:* AAAS; Am Chem Soc. *Res:* Pharmacology; biochemistry. *Mailing Add:* 4620 N Park Ave No 708E Chevy Chase MD 20815

SCHWARTZ, SANFORD BERNARD, b Cheyenne, Wyo, Sept 9, 25; div; c 4. COMPUTER SCIENCES. *Educ:* Univ Colo, BS, 50, MA, 55, PhD, 59. *Prof Exp:* Physicist, Nat Bur Standards, 51-54; asst prof physics, Univ Nev, 58-59; assoc res scientist, Martin Co, 59-66; res scientist, Advan Res Labs, Douglas Aircraft Corp, 66-69, Douglas Advan Res Lab, 69-70, mgr systs programming, 70-80, SR SPECIALIST PLANNING, MCDONNELL DOUGLAS AUTOMATION CO, 80- *Mem:* Asn Comput Mach. *Res:* Physics of solar corona; astrophysical cross-sections; plasma physics; computer systems, especially measurement and evaluation of performance. *Mailing Add:* 302 Bluebird Canyon Laguna Beach CA 92651

SCHWARTZ, SEYMOUR I, b New York, NY, Jan 22, 28; m 49; c 3. THORACIC SURGERY. *Educ:* Univ Wis, BA, 47; NY Univ, MD, 50; Am Bd Surg & Am Bd Thoracic Surg, dipl. *Prof Exp:* Chief resident, Strong Mem Hosp, Rochester, NY, 56-57; from instr to assoc prof, 57-66, PROF SURG, UNIV ROCHESTER, 67-, DIR SURG RES, 62- *Concurrent Pos:* From asst surgeon to assoc surgeon, Strong Mem Hosp, Rochester, 57-63, sr assoc surgeon, 63-67, sr surgeon, 67-; Markle scholar acad med, 60-65; mem sci adv comt, Cent Clin Res Ctr, Roswell Park Mem Inst. *Mem:* Fel Am Col Surg; Soc Univ Surgeons; Am Surg Asn; AMA; Am Asn Thoracic Surg. *Res:* Portal hypertension; vascular surgery; platelets. *Mailing Add:* Dept Surg Univ Rochester Med Ctr 601 Elmwood Ave Rochester NY 14642

SCHWARTZ, SHELDON E, b New York, NY, July 21, 19; m 44; c 2. MEDICINE. *Educ:* Rensselaer Polytech Inst, BS, 40; NY Univ, MD, 43; Am Bd Internal Med, dipl, 52. *Prof Exp:* assoc prof clin med, Med Ctr, NY Univ, 47-; AT STATE UNIV NY UPSTATE MED CTR, SYRACUSE. *Concurrent Pos:* Chief arthritis clin, Bellevue Hosp; dir med, Hillcrest Hosp. *Mem:* Am Rheumatism Asn; fel Am Col Physicians. *Res:* Arthritis. *Mailing Add:* 218-65 99th Ave Queens Village NY 11429

SCHWARTZ, SORELL LEE, b Buffalo, NY, Sept 13, 37; m 63; c 2. PHARMACOLOGY. *Educ:* Univ Md, BS, 59; Med Col Va, PhD(pharmacol), 63. *Prof Exp:* Head pharmacol div, US Naval Med Res Inst, 66-68; PROF PHARMACOL, SCH MED, GEORGETOWN UNIV, 68-, SCI DIR, CTR ENVIRON HEALTH & HUMAN TOXICOL, 82- *Mem:* Soc Toxicol; Am Soc Pharmacol & Exp Therapeut; Soc Risk Anal; Am Col Toxicol; Am Acad Clin Toxicol. *Res:* Pharmacokinetics; exposure analysis; risk analysis; causal inference methods. *Mailing Add:* Dept Pharmacol Georgetown Univ Sch Med Washington DC 20007

SCHWARTZ, STANLEY ALLEN, b Newark, NJ, July 20, 41; m 65. IMMUNODEFICIENCY DISEASES, IMMUNOREGULATION. *Educ:* Rutgers Univ, AB, 63, MS, 65; Univ Calif, San Diego, PhD(cellular biol), 68; Albert Einstein Col Med, MD, 72. *Prof Exp:* Asst prof pediat & biol, Cornell Univ Med Col, 77-78; assoc prof, 78-83, PROF PEDIAT, MICROBIOL & IMMUNOL, & EPIDEMIOL, UNIV MICH, 83- *Concurrent Pos:* Asst attend physician pediat, Mem Hosp Cancer & Allied Dis & NY Hosp, 77-78; assoc, Sloan-Kettering Inst Cancer Res, 77-78. *Honors & Awards:* Meller Award, Mem Sloan-Kettering Cancer Ctr, 77; Res Career Develop Award, NIH, 83. *Mem:* Am Soc Clin Invest; fel Am Acad Allergy & Immunol; Am Asn Immunologists; Am Pediat Soc; Soc Pediat Res; Am Fedn Clin Res. *Res:* Analysis of the cellular and molecular immunopathogenic mechanisms underlying human immunodeficiency disorders including AIDS; immunoregulation of cellular cytotoxicity. *Mailing Add:* Dept Epidemiol Univ Mich Ann Arbor MI 48109-2029

SCHWARTZ, STEPHEN EUGENE, b St Louis, Mo, June 18, 41; m 80; c 2. ATMOSPHERIC CHEMISTRY & PHYSICS, PHYSICAL CHEMISTRY. *Educ:* Harvard Univ, AB, 63; Univ Calif, Berkeley, PhD(chem), 68. *Prof Exp:* Fulbright fel & Ramsay Mem fel. Cambridge Univ, 68-69; asst prof chem, State Univ NY, Stony Brook, 69-75; assoc chemist, 75-77 chemist, 77-90, SR CHEMIST, BROOKHAVEN NAT LAB, 90- *Concurrent Pos:* Ed, Advan in Environ Sci & Technol, 83; assoc ed, Atmospheric Environ, 84- & J Geophys Res, 86-89; mem, Comt on Atmospheric Chem, Am Meteorol Soc, 85-; mem comt atmospheric res, Nat Res Coun, 88- *Mem:* AAAS; Am Geophys Union; Am Chem Soc; Am Phys Soc; Am Meteorol Soc. *Res:* Physical chemistry; atmospheric chemistry; chemical kinetics; design, conduct and interpretation of measurements of trace atmospheric constituents; laboratory studies of gas and aqueous-phase kinetics; modeling gas-phase and heterogeneous atmospheric reactions; interpretation of residence times and scales of transport. *Mailing Add:* Dept Appl Sci Brookhaven Nat Lab Upton NY 11973

SCHWARTZ, STEPHEN MARK, b Boston, Mass, Jan 1, 42; m 64; c 2. PATHOLOGY, CARDIOVASCULAR DISEASES. *Educ:* Harvard Univ, AB, 63; Boston Univ, MD, 67; Univ Wash, PhD(path), 74. *Hon Degrees:* MD, Göteborg, Sweden, 89. *Prof Exp:* Asst dir labs, Long Beach Naval Regional Med Ctr, 73-74; from asst prof to assoc prof, 74-87, PROF PATH, UNIV WASH, 84- *Mem:* AAAS. *Res:* Structure, function and pathology of arterial endothelium; cell replication in endothelium and smooth muscle, hypertension. *Mailing Add:* Dept Path SJ60 Univ Wash I 420 Health Sci Bldg Seattle WA 98195

SCHWARTZ, STEVEN OTTO, b Hungary, July 6, 11; m 42; c 3. HEMATOLOGY. *Educ:* Northwestern Univ, BS, 32, MS, 35, MD, 36; Am Bd Internal Med, dipl, 42. *Prof Exp:* Prof internal med, Cook County Grad Sch Med, 39-68; asst prof med, Univ Ill, 42-47; prof hemat, Chicago Med Sch, 47-55; assoc prof, 55-59, prof, 59-79, EMER PROF MED, SCH MED, NORTHWESTERN UNIV, CHICAGO, 79- *Concurrent Pos:* Chief hemat clin, Mandel Clin & assoc hematologist, Michael Reese Hosp, 38-50; attend hematologist & dir hemat dept, Cook County Hosp, 39-68, hematologist, Hektoen Inst Med Res, 45-68; consult, Chicago State Hosp, 41-49, Highland Park Hosp, 50-84, Mother Cabrini Hosp, 51-84, Columbus Hosp, 55-84 & Hines Vet Admin Hosp, 56-84; assoc, Mt Sinai Hosp, 48-51; attend hematologist, West Side Vet Hosp, 55-56; sr attend physician, Northwestern Mem Hosp, 55-84. *Honors & Awards:* Solano Medal, Quincy Col, 66. *Mem:* Am Col Physicians; Int Soc Hemat; Am Soc Hemat; AMA; Am Soc Internal Med. *Mailing Add:* 610 Rice St Highland Park IL 60035

SCHWARTZ, STUART CARL, b New York, NY, July 12, 39; m 61; c 2. ELECTRICAL ENGINEERING. *Educ:* Mass Inst Technol, BS & MS, 61; Univ Mich, PhD(info & control), 66. *Prof Exp:* Res engr, Jet Propulsion Lab, Calif Inst Technol, 61-62; from asst prof to assoc prof elec eng, 66-74, assoc dean, Sch Eng & Appl Sci, 77-80, PROF ELEC ENG, PRINCETON UNIV, 74- *Concurrent Pos:* Ed, J Appl Math, 70; Guggenheim Mem Found fel, 72. *Mem:* Inst Math Statist; Inst Elec & Electronics Engrs; Soc Indust & Appl Math. *Res:* Application of probability and stochastic processes to problems in statistical communication and systems theory. *Mailing Add:* Dept Elec Eng/Eng Quadrangle Princeton Univ Princeton NJ 08540

SCHWARTZ, THEODORE BENONI, b Philadelphia, Pa, Feb 14, 18; m 48; c 6. MEDICINE. *Educ:* Franklin & Marshall Col, BS, 39; Johns Hopkins Univ, MD, 43; Am Bd Internal Med, dipl. *Prof Exp:* Intern med, Johns Hopkins Univ, 43-44; resident, Salt Lake Gen Hosp, 46-48; assoc, Duke Univ, 50-52, asst prof, 53-55; from assoc prof to prof, Col Med, Univ Ill, 55-70; CHMN DEPT INTERNAL MED, RUSH-PRESBY-ST LUKE'S MED CTR, 70-; PROF & CHMN DEPT, RUSH MED SCH, 71- *Concurrent Pos:* Damon Runyon sr clin fel, 40-52; fel, Duke Univ, 48-50; asst chief med serv, Vet Admin Hosp, Durham, NC, 53-55; dir endocrinol & metab, Presby-St Luke's Hosp, Chicago, Ill, 55-; ed, Yearbk Endocrinol, 64-; mem, Am Bd

Internal Med, 70-79 & Am Bd Med Specialties, 71-79. *Mem:* Endocrine Soc; Am Soc Clin Invest; Am Diabetes Asn; fel Am Col Physicians; Am Fedn Clin Res. *Res:* Endocrinology; protein metabolism. *Mailing Add:* 4820 Roberts Rd Boise ID 83705-2853

SCHWARTZ, THOMAS ALAN, b Plainfield, NJ, Jan 31, 51; m 79; c 3. GLASS, GLAZING & CURTAIN WALL CONSTRUCTION. *Educ:* Tufts Univ, BSCE, 73; Mass Inst Technol, MS, 77. *Prof Exp:* Field & lab engr, 73-75, sr engr, 77-79, sr staff engr, 80-82, assoc, 83-85, sr assoc, 86-89, prin, 89-90, PRIN & DEPT HEAD, SIMPSON GUMPERTZ & HEGER INC, 91- *Concurrent Pos:* Lectr, Nat Bur Standards, 77, 81, Boston Archit Ctr, 85 & Grad Sch Design, Harvard Univ, 85. *Mem:* Sigma Xi; Am Soc Civil Engrs; Am Soc Testing & Mat. *Res:* Building materials technology in connection with investigation of failures especially roofing, glass, windows, masonry and curtain walls. *Mailing Add:* Simpson Gumpertz & Heger Inc 297 Broadway Arlington MA 02174

SCHWARTZ, TOBIAS LOUIS, b Ft Wayne, Ind, Sept 8, 28; m 49; c 3. BIOPHYSICS, PHYSIOLOGY. *Educ:* City Col New York, BEE, 49; State Univ NY Buffalo, PhD(biophys), 66. *Prof Exp:* Design engr, Niagara Transformer Corp, NY, 54-58; USPHS fel lab neurophysiol, Col Physicians & Surgeons, Columbia Univ, 65-68; asst prof biol sci, 68-71, ASSOC PROF BIOL SCI, UNIV CONN, 71- *Concurrent Pos:* Nat Inst Neurol Dis & Stroke res grant, 69; master in res, Lab Cellular Neurobiol, Nat Ctr Sci Res, France, 74-75; mem & invited lectr, Marine Biol Lab Corp, Woods Hole, Mass. *Mem:* AAAS; Biophys Soc; Soc Gen Physiol; Soc Neurosci. *Res:* Diffusion phenomena in membranes; active transport; mechanisms of membrane excitability; membrane diffusion theory. *Mailing Add:* Dept Molecular/Cell Biol Univ Conn Storrs CT 06268

SCHWARTZ, WILLIAM BENJAMIN, b Montgomery, Ala, May 16, 22; c 3. INTERNAL MEDICINE. *Educ:* Duke Univ, MD, 45; Am Bd Internal Med, dipl, 56. *Prof Exp:* Intern & asst resident med, Univ Chicago Clins, 45-46; from instr to assoc prof, 50-58, chmn dept, 71-76, PROF MED, MED SCH, TUFTS UNIV, 58-, VANNEVAR BUSH PROF, 76- *Concurrent Pos:* Res fel, Harvard Med Sch, 48-50; fel, Boston Children's Hosp, 49-50; Markle scholar, 50-55; asst, Peter Bent Brigham Hosp, 48-50; from asst physician to sr physician, New Eng Med Ctr, 50-; estab investr, Am Heart Asn, 56-; chmn gen med study sect, NIH, 65-, chmn sci adv bd, Nat Kidney Found, 68-70; chmn health policy adv bd, Rand Corp, 70-, prin adv, Health Sci Prog, 77-; Endicott prof & physician in chief, New Eng Med Ctr Hosps, 71-76. *Mem:* Nat Inst Med; Am Soc Clin Invest; Am Acad Arts & Sci; Asn Am Physicians; Am Soc Nephrology (pres, 74-75). *Res:* Health care policy analysis. *Mailing Add:* Dept Med Tufts Univ Sch Med Boston MA 02111

SCHWARTZ, WILLIAM JOSEPH, b Philadelphia, Pa, Mar 28, 50; m 79; c 1. NEUROLOGY, CIRCADIAN RHYTHMS. *Educ:* Univ Calif, Irvine, BS, 71, San Francisco, MD, 74. *Prof Exp:* Med intern, Moffitt Hosps, Univ Calif, San Francisco, 74-75, neurol resident, 78-81; res assoc, NIMH, 75-78; instr neurol, Mass Gen Hosp, Harvard Med Sch, 81-82, asst prof, 82-86; assoc prof, 86-90, PROF NEUROL, MED SCH, UNIV MASS, 90- *Concurrent Pos:* Mem, NIH/ORG Neurol A Study Sect, 89-; vchmn, Gordon Conf Chronobiol, 91. *Mem:* Soc Neurosci; AAAS; Am Acad Neurol; Am Neurol Asn; Soc Res Biol Rhythms; Am Sleep Dis Asn. *Res:* Investigation of the neural regulation of circadian rhythms in mammals by the suprachiasmatic nuclei. *Mailing Add:* Dept Neurol Univ Mass Med Ctr 55 Lake Ave N Worcester MA 01655

SCHWARTZ, WILLIAM LEWIS, b Columbus, Ohio, Dec 11, 31; m 53; c 2. VETERINARY PATHOLOGY. *Educ:* Ohio State Univ, BSc, 53, DVM, 57; Tex A&M Univ, MS, 70. *Prof Exp:* Pvt pract, 57-60; dist vet, Ohio Dept Agr, 60-63, vet diagnostician, 63-64; asst prof diag vet med, Ga Coastal Plain Exp Sta, Univ Ga, 64-67; res assoc vet path & toxicol, 67, asst prof, 67-70, PATHOLOGIST, TEX VET MED DIAG LAB, TEX A&M UNIV, 70- *Concurrent Pos:* Consult mem health adv comt, Tex Specific Pathogen Free Swine Accrediting Agency, Inc, 71-75. *Mem:* Am Vet Med Asn; Am Asn Swine Practitioners; Am Asn Vet Lab Diagnosticians; Sigma Xi. *Res:* Diagnostic veterinary pathology and related fields; diseases of swine. *Mailing Add:* Tex Vet Med Diag Lab Drawer 3040 Col Stat TX 77841

SCHWARTZBACH, STEVEN DONALD, b Bronx, NY, May 24, 47; m 68; c 3. ALGAL PHYSIOLOGY, MOLECULAR BIOLOGY. *Educ:* State Univ NY, Buffalo, BA, 69; Brandeis Univ, PhD(biol), 75. *Prof Exp:* Fel molecular biol, Oak Ridge Nat Lab, 74-76; from asst prof to assoc prof, 76-87, PROF CELL BIOL, UNIV NEBR, LINCOLN, 88- *Mem:* Am Soc Plant Physiol; Japanese Soc Plant Physiol; AAAS. *Res:* Photoregulation of chloroplast development; organelle nucleic acids; regulation of protein synthesis. *Mailing Add:* 303 Lyman Hall Univ Nebr Lincoln NE 68588

SCHWARTZBART, HARRY, b Altoona, Pa, Jan 3, 23; m 53; c 3. WELDING, FAILURE ANALYSIS. *Educ:* Pa State Univ, BS, 43, MS, 48. *Prof Exp:* Metallurgist, Revere Copper & Brass, 43-44; aeronaut res scientist, Nat Adv Comt Aeronaut, 48-51; asst dir metals res, Res Inst, Ill Inst Technol, 51-68; dir mat eng, Rockwell Int, 68-83; mgr res, Aerojet Gen, 83; CONSULT, 83- *Mem:* Fel Am Soc Metals Int; Am Welding Soc; Int Soc Air Safety Investrs; Nat Forensic Ctr. *Res:* Mechanical metallurgy; welding; brazing; soldering; shape memory alloys; failure analysis; author of 75 publications. *Mailing Add:* 10951 Oklahoma Ave Chatsworth CA 91311

SCHWARTZBERG, HENRY G, b New York, NY, Oct 12, 25; m 55; c 3. CHEMICAL ENGINEERING, FOOD ENGINEERING. *Educ:* Cooper Union, BChE, 49; NY Univ, MChE, 59, PhD(chem eng), 66. *Prof Exp:* Chem engr, Chem & Radiol Labs, Army Chem Ctr, 50-53; res specialist process develop, Tech Ctr, Gen Foods Corp, 54-66; assoc prof chem eng, NY Univ, 66-73; PROF FOOD PROCESS ENG, UNIV MASS, 73- *Concurrent Pos:* Consult, Clairol Co, 66-67, Gen Foods Corp, 66-, Am Nat Res Cross, 67-74 & Devro, Inc, 68-77; vis lectr agr, Univ Netherland, 80; vis prof, ENSBANA Univ de Dijon, 84, Univ Nacional del Sur, Bahia Blanca, Argentina, 84; chmn, Food, Pharm & Biol Div, Am Inst Chem Engrs, 84; vis prof, Tech Res Ctr, Finland, 90. *Honors & Awards:* Food Engr Award, Dairy & Food Indust Supply Asn & Am Soc Agr Engrs, 85. *Mem:* Am Inst Chem Engrs; Inst Food Technol; Am Chem Soc; Int Inst Refrigeration. *Res:* Microwave heating; freeze concentration, food texturization by extrusion; membrane permeation; solid-liquid extraction; expression; freezing and thawing; evaporation; energy storage by brines; drying; preparative chromatography and puffing. *Mailing Add:* Dept Food-Sci Univ Mass Amherst MA 01003

SCHWARTZMAN, JOSEPH DAVID, b Washington, DC, Dec 9, 47; m 72; c 2. CLINICAL MICROBIOLOGY. *Educ:* Dartmouth Col, AB, 70; Dartmouth Med Sch, BMedSc, 72; Harvard Univ, MD, 74. *Prof Exp:* Resident physician path, Univ Colo Affil Hosps, 74-78; res fel path, Sch Med, Univ Colo, 76-78; teaching fel microbiol, Dartmouth Med Sch, 78-80; asst prof, 80-86, ASSOC PROF PATH, SCH MED, UNIV VA, 86-, ASSOC DIR, CLIN MICROBIOL LAB, UNIV VA HOSP, 80- *Concurrent Pos:* Prin investr, NIH grant. *Mem:* Am Soc Microbiol; Am Soc Trop Med & Hyg; Am Soc Parasitologists; Royal Soc Trop Med & Hyg; Soc Protozoologists; Infectious Dis Soc Am. *Res:* Cell biology of intracellular protozoan parasites; mechanism of parasite motility and host cell invasion of the coccidian Toxoplasma gondii. *Mailing Add:* Dept Path Box 168 Med Ctr Univ Va Charlottesville VA 22908

SCHWARTZMAN, LEON, b Brooklyn, NY, Feb 6, 31; m 54; c 2. MICROWAVE ENGINEERING, TECHNICAL MANAGEMENT. *Educ:* Polytech Inst Brooklyn, BEE, 58, MSEE, 63. *Prof Exp:* Engr, Sperry Gyroscope, 57-62, sr engr, 62-67, res sect head, 67-70, sr res sect head, 70-73, dept head, 73-82, mgr, ATR Prog, Sperry SFCS, 82-86, DIR PROG DEVELOP, UNISYS- SGSG, 86- *Concurrent Pos:* Radar panel, microwave expos, 67; chmn prog comt, Inst Elec & Electronics Engrs, 67-68, vchmn, 69-70, chmn, 70-71. *Mem:* Fel Inst Elec & Electrons Engrs (secy, 68-69); Am Defense Preparedness Asn. *Res:* Antennas and propogation; design and development of advanced microwave antennas; electronic scanning radars; solid state transmitters; gallium arsenide devices and digital signal processing. *Mailing Add:* 1475 Remsen Ave Brooklyn NY 11236

SCHWARTZMAN, ROBERT M, b New Haven, Conn, Nov 7, 26; m 60; c 3. DERMATOLOGY, IMMUNOLOGY. *Educ:* Univ Pa, VMD, 52; Univ Minn, MPH, 58, PhD(dermatopath), 59. *Prof Exp:* Instr vet med, Univ Minn, 53-59; from asst prof to assoc prof dermat, 59-67, PROF DERMAT, SCH VET MED, UNIV PA, 67-, ASST PROF COMP DERMAT, GRAD SCH MED, 62- *Honors & Awards:* Morris Animal Found award, 57-59; USPHS res career develop award, 63-72. *Mem:* Soc Invest Dermat; Am Soc Dermatopath; Am Soc Allergy; Am Vet Med Asn; Am Col Vet Dermat. *Res:* Veterinary and comparative dermatology. *Mailing Add:* 14 Wiltshire Rd Wynnewood PA 19096

SCHWARZ, ANTON, b Munich, Ger, May 26, 27; nat US; m 52; c 2. MEDICAL RESEARCH. *Educ:* Maximilian Univ, Ger, MD, 51. *Prof Exp:* Intern internal med, hosp, Munich, Ger, 51-52; intern, St John's Hosp, Long Island, 52-53; resident physician, Dobbs Ferry Hosp, 53-54; sr res assoc pediat, Children's Hosp Res Found, Col Med, Univ Cincinnati, 54-56; asst dir virus res, Res Div, Pitman-Moore Co, 56-63, dir virus res, Res Div, Pitman-Moore Div, 63-65, dir human health res & develop labs, 65-71, dir biol res & develop & biol labs, 71-75, med dir, Europ Area, Dow Chem Co, 75-77; dir corp med res, Int Region II, 77-80, DIR MED SCI, RES DIV, SCHERING-PLOUGH CORP, 81- *Honors & Awards:* Wolferine Frontiersman Award, 68. *Mem:* AAAS; Sigma Xi; AMA; NY Acad Sci; Soc Exp Biol & Med; Am Asn Immunol. *Res:* Medical sciences; research administration. *Mailing Add:* RR 1 Box 1069 Lakeview AR 72642-9638

SCHWARZ, CINDY BETH, b Bronx, NY, Sept 17, 58; m 87; c 1. ELEMENTARY PARTICLE PHYSICS. *Educ:* State Univ NY, BS, 80; Yale Univ, MPhil, 82, PhD(physics), 85. *Prof Exp:* ASST PROF PHYSICS, VASSAR COL, 85- *Concurrent Pos:* Consult & guest lectr, Int Bus Machines Corp, 89; guest researcher, Brookhaven Nat Lab, 89; prin investr ILI grant, NSF, 91- *Mem:* Am Phys Soc; Am Asn Physics Teachers; Sigma Xi; Am Asn Univ Professors. *Res:* Experimental particle physics. *Mailing Add:* Vassar Col Box 39 Poughkeepsie NY 12601

SCHWARZ, DIETRICH WALTER FRIEDRICH, b Stettin, Ger, Nov 22, 39. HEARING SCIENCE. *Educ:* Univ Freiburg, MD, 67, Dr, 69. *Prof Exp:* Res fel neurophysiol, Univ Toronto, 69-71, clin teadus otolaryngol, 71-75, from asst prof to assoc prof otolaryngol, 75- 83; assoc prof, 83-85, PROF OTOLARYNGOL, UNIV BC, 85- *Mem:* Soc Neurosci; Can Soc Physiol; Am Soc Physiol; Asn Res Otolaryngol. *Res:* Auditory and vestibular neurophysiology and neuroanatomy. *Mailing Add:* Rotary Hearing Ctr R153 Acute Care Unit 2211 Wesbrook Mall Vancouver BC V6T 2B5 Can

SCHWARZ, ECKHARD C A, b Luebeck, Ger, Nov 13, 30; US citizen; div; c 4. POLYMER SCIENCE. *Educ:* Univ Hamburg, Diplom, 56; McGill Univ, PhD(org chem), 62. *Prof Exp:* Chemist, E B Eddy Co, 57-59; sr res chemist, E I du Pont de Nemours & Co, 62-68, Kimberly-Clark Corp, 68-72 & E I du Pont de Nemours & Co, Ger, 72-73; dir res, Presto Prod, Inc, 73-75; PRES, BIAX-FIBERFILM CORP, 75- *Mem:* Am Chem Soc; Tech Asn Pulp & Paper Indust. *Res:* Research and development immodification of commodity polymers; design and development of fiber and film processes. *Mailing Add:* 115 N Park Ave Neenah WI 54956

SCHWARZ, FRANK, b Timisoara, Roumania, June 2, 24; US citizen; m 49; c 2. ELECTRICAL ENGINEERING, OPTICS. *Educ:* City Col New York, BEE, 50; Univ Conn, MEE, 61. *Prof Exp:* Develop engr, Spellman TV Co, 50 & Sigma Elec Co, 50-51; proj engr, Sorensen & Co, Inc, 51-53; proj & dept mgr & consult electrooptics, 53-69, MGR ADVAN DEVELOP DEPT, BARNES ENG CO, 69-; AT SOS INC, STANFORD, CT. *Mem:* Sr mem Inst Elec & Electronics Engrs; Optical Soc Am. *Res:* Infrared instruments and electrooptical systems, including infrared horizon sensors, radiometers, trackers, thermal imaging systems. *Mailing Add:* 156 Thunderhill Dr SOS Inc Stamford CT 06902

SCHWARZ, HANS JAKOB, b Leysin, Switz, Feb 3, 25; nat US; m 52; c 2. ORGANIC CHEMISTRY, BIOCHEMISTRY. *Educ:* Univ Basel, PhD(chem), 51. *Prof Exp:* Fel, Nat Res Coun Can, 51-53; res chemist, Res Div, Cleveland Clin, 53-54, res assoc, 54-55, asst mem staff, 55-58, mem staff, 58-59; biochem res, 59-63, head metab unit, 63-68, assoc head, Biochem Sect, 68-73, EXEC DIR, DRUG METAB DEPT SRI, SANDOZ PHARMACEUT DIV, SANDOZ, INC, 73- *Mem:* Am Chem Soc. *Res:* Isolation and structure of natural products; synthesis of polypeptides and labelled drugs; metabolism of drugs. *Mailing Add:* Drug Metab Sect Bldg 404 Sandoz Pharmaceut Rte 10 East Hanover NJ 07936

SCHWARZ, HAROLD A, b Nebr, Apr 1, 28; m 53; c 4. PHYSICAL CHEMISTRY. *Educ:* Univ Omaha, BA, 48; Notre Dame Univ, PhD(chem), 52. *Prof Exp:* Assoc chemist, 51-59, chemist, 59-66, SR CHEMIST, BROOKHAVEN NAT LAB, 66- *Mem:* Am Chem Soc; Radiation Res Soc. *Res:* Radiation and photochemistry. *Mailing Add:* Brookhaven Nat Lab Upton NY 11973

SCHWARZ, JOHN HENRY, b North Adams, Mass, Nov 22, 41; m 86. THEORETICAL PHYSICS, HIGH ENERGY PHYSICS. *Educ:* Harvard Univ, AB, 62; Univ Calif, Berkeley, PhD(physics), 66. *Prof Exp:* Instr physics, Princeton Univ, 66-68, lectr, 68-69, asst prof, 69-72; res assoc, 72-85, PROF THEORET PHYSICS, LAURITSEN LAB, CALIF INST TECHNOL, 85- *Concurrent Pos:* Guggenheim fel, 78-79; trustee, Azpen Ctr Physics. *Honors & Awards:* MacArthur Fel Award, 87- *Mem:* Fel Am Phys Soc. *Res:* Theoretical research in particle physics; supersymmetry; superstrings; conformal field theory. *Mailing Add:* Lauritsen Lab Calif Inst Technol Box 452-48 Pasadena CA 91125

SCHWARZ, JOHN ROBERT, b Passaic, NJ, Oct 8, 44. MARINE MICROBIOLOGY. *Educ:* Rensselaer Polytech Inst, BS, 67, PhD(biol), 72. *Prof Exp:* Res assoc, Univ Md, College Park, 72-75; from asst prof to assoc prof, 76-86, asst dean acad affairs, 78-79, vpres, 79-81, head, 83-87, PROF MICROBIOL, DEPT MARINE BIOL, MOODY COL, TEX A&M UNIV SYST, 86- *Mem:* Am Soc Microbiol; Soc Indust Microbiol; AAAS; Sigma Xi; Am Soc Limnol Oceanog. *Res:* Marine microbial ecology; biodegradation; microbial production on non-conservative gases in the marine environment. *Mailing Add:* Dept Marine Biol Tex A&M Univ PO Box 1675 Galveston TX 77553

SCHWARZ, JOHN SAMUEL PAUL, b Chicago, Ill, Mar 6, 32; m 56; c 1. INDUSTRIAL ORGANIC CHEMISTRY. *Educ:* Univ Ill, BS, 54; Univ Calif, PhD(org chem), 58. *Prof Exp:* Res chemist synthetic lubricants, Exxon Res & Eng Co, NJ, 57-58; sr res scientist pharmaceut res, Squibb Inst Med Res, 60-68; sr res chemist, Nease Chem Co, 68-72; PRES, PURE SYNTHETICS, INC, 73- *Mem:* Am Chem Soc; Inst Food Technologists. *Res:* Isolation, structure, stereochemistry of biologically-active natural products; synthesis and chemistry of tetracyclines and penicillins; acyclic isoimide-imide rearrangement; process research, development and production of flavor chemicals; heat capacity, atomic weight relationship. *Mailing Add:* 107 Highland Ave Ridgewood NJ 07450

SCHWARZ, KLAUS W, b Heidelberg, Ger, Mar 12, 38; US citizen; m 62; c 2. SUPERFLUIDS, TURBULENCE. *Educ:* Harvard Univ, BA, 60; Univ Chicago, MS, 62, PhD(physics), 67. *Prof Exp:* Asst prof physics, Univ Chicago, 69-76; res staff mem, 76-85, MGR, DYNAMICAL PHENOMENA GROUP, IBM WATSON RES CTR, 85- *Concurrent Pos:* Coun mem, Am Phys Soc. *Mem:* Fel Am Phys Soc. *Res:* Experimental and theoretical research in quantum liquids and fluid mechanics. *Mailing Add:* TJ Watson Res Ctr IBM PO Box 218 Yorktown Heights NY 10598

SCHWARZ, MARVIN, b Newark, NJ, Apr 7, 29; m 89; c 3. PHYSIOLOGICAL PSYCHOLOGY, EXPERIMENTAL PSYCHOLOGY. *Educ:* Lafayette Col, AB, 51, Yale Univ, MS, 52, PhD(psychol), 55. *Prof Exp:* Exp psychol, US Naval Sch Aviation Med, USNR, 55-58; Res asst prof, psychiat, Univ Iowa, 58-64; vprovost, 79-83, assoc prof, 64-66, PROF PSYCHOL, DEPT PSYCHOL, UNIV CINCINNATI, 66-, DIR GRAD STUDIES, 66-76 & 88- *Mem:* Am Psychol Asn; fel AAAS; NY Acad Sci; Sigma Xi. *Res:* Recording electrical activity of the brain and relating it to behavior; currently recording single cell activity in chronically implanted rabbits in relation to ingestion of different testing fluids. *Mailing Add:* Psychol Dept Univ Cincinnati Mail Location 376 Cincinnati OH 45221

SCHWARZ, MAURICE JACOB, b Northampton, Eng, Sept 13, 39; US citizen; m 65; c 2. ORGANIC CHEMISTRY. *Educ:* Univ Ore, BA, 62, PhD(chem), 65. *Prof Exp:* Develop chemist, Geigy Chem Corp, RI, 67-69, group leader develop, 69-71, develop mgr, Ciba-Geigy Facil, NJ, 71-75, dir chem develop, 75-78, dir prod, 78-83, VPRES, PHARMACEUT RES & DEVELOP, PHARMACEUT DIV, CIBA-GEIGY CORP, 83- *Concurrent Pos:* Mem, Pharmaceut Develop Subsect Steering Comt, 87-; mem bd dir, Res & Develop Coun; vchmn, Res & Develop Coun, NJ, 89- *Mem:* Am Chem Soc; Pharmaceut Mfrs Asn; Am Asn Pharmaceut Scientists; Am Pharmaceut Asn. *Res:* Process development and research; management. *Mailing Add:* 142 Van Houton Chatham NJ 07928

SCHWARZ, MEYER, b Amsterdam, Holland, Nov 6, 24; nat US; m 56; c 2. ORGANIC CHEMISTRY. *Educ:* Univ Geneva, BSc, 46, PhD(org chem), 50. *Prof Exp:* Res assoc chem, Fla State Univ, 50-52; res chemist, Sprague Elec Co, Mass, 52-56; chemist, Harry Diamond Labs, 56-64; CHEMIST, AGR ENVIRON QUAL INST, AGR RES SERV, USDA, 64- *Mem:* The Chem Soc; Am Chem Soc. *Res:* Organic synthesis and reaction mechanisms; dielectric materials; polymers; organic fluorine; phosphorus compounds; natural products as related to insect chemistry; insect hormones, pheromones, attractants and repellents. *Mailing Add:* 6612 Isle of Skye Dr Highland MD 20777

SCHWARZ, OTTO JOHN, b Chicago, Ill, Oct 19, 42; m 65; c 2. PLANT PHYSIOLOGY, BIOCHEMISTRY. *Educ:* Univ Fla, BSA, 64; NC State Univ, MS, 67, PhD(plant physiol), 70. *Prof Exp:* NIH fel, Biol Div, Oak Ridge Nat Lab, 69-71; asst prof, 71-77, ASSOC PROF PLANT PHYSIOL, UNIV TENN, KNOXVILLE, 77- *Concurrent Pos:* NIH biomed sci grant, Univ Tenn, Knoxville, 71-72; vis investr, Comp Animal Res Lab, Oak Ridge Assoc Univ, 79- *Mem:* Am Soc Plant Physiol. *Res:* Regulation of pyrimidine nucleoside phosphorylating enzymes; chemical regulation of secondary product formation in plants; paraquat induced oleoresin synthesis in Pinus; food chain transport of synfuels. *Mailing Add:* Dept Bot 611 Hesler Biol Bldg Univ Tenn Knoxville TN 37996

SCHWARZ, RALPH J, b Hamburg, Ger, June 13, 22; nat US; m 51; c 2. EDUCATIONAL ENGINEERING. *Educ:* Columbia Univ, BS, 43, MS, 44, PhD(elec eng), 49. *Prof Exp:* From asst to prof elec eng, 43-58, chmn dept elec eng, 58-65, 71-72, assoc dean acad affairs, Sch Eng & Appl Sci, 72-76, THAYER-LINDSLEY PROF ELEC ENG, COLUMBIA UNIV, 76-, VDEAN, 76- *Concurrent Pos:* Adv, Inst Int Educ, 51-70; vis assoc prof, Univ Calif, Los Angeles, 56; vis scientist, IBM Res Ctr, 69-70; trustee, Assoc Univs, Inc, 80-; dir, Armstrong Mem Res Found, 75- *Honors & Awards:* Centennial Medal, Inst Elec & Electronics Engrs, 84. *Mem:* Am Soc Eng Educ; fel Inst Elec & Electronics Engrs; AAAS. *Res:* Communication theory; system analysis; pattern recognition. *Mailing Add:* Sch Eng & Appl Sci Columbia Univ 500 W 120th St New York NY 10027

SCHWARZ, RICARDO, b Valdivia, Chile, July 5, 42; US citizen; m 66; c 2. SOLID STATE PHYSICS. *Educ:* Univ Chile, MS, 67; Univ Va, PhD(physics), 72. *Prof Exp:* Asst prof physics, Univ Chile, 66-68 & Univ Va, 72; vis asst prof, Univ Ill, Urbana, 73-75; physicist, Argonne Nat Labs, 75-85; STAFF MEM, CTR MAT SCI, LOS ALAMOS NAT LABS, 85- *Concurrent Pos:* Vis assoc, Keck Labs Mat Sci, Calif Inst Technol, 82-83. *Mem:* Am Phys Soc; Mat Res Soc; Am Inst Mining Metall & Petrol Engrs. *Res:* Mechanical properties of solids; ultrasonics; computer modelling of dislocation dynamics in alloys; solid state reactions in thin films; amorphous metallic alloys; dynamic compaction of powders. *Mailing Add:* Ctr Mat Sci Los Alamos Nat Labs Mail Stop K765 Los Alamos NM 87545

SCHWARZ, RICHARD HOWARD, b Easton, Pa, Jan 10, 31; m 55; c 4. OBSTETRICS & GYNECOLOGY. *Educ:* Jefferson Med Col, MD, 55; Am Bd Obstet & Gynec, dipl, 63, cert, 74. *Prof Exp:* Assoc obstet & gynec, Tulane Univ, 59-63; from instr to assoc prof, Sch Med, Univ Pa, 63-73, prof obstet & gynec, 73-78, dir, Jerrold R Giolding Div Fetal Med, 71-78; prof obstet & gynec & chmn dept, Downstate Med Ctr, 78-90, PROVOST & VPRES CLIN AFFAIRS, HSCB, STATE UNIV NY, 88- *Mem:* AAAS; fel Am Col Obstet & Gynec; Am Gynec & Obstet Soc; Infectious Dis Soc Obstet & Gynec (pres). *Res:* Perinatal and placental physiology; high risk obstetrics; diabetes; infectious disease. *Mailing Add:* Box 12 State Univ NY HSCB 450 Clarkson Ave Brooklyn NY 11203-2098

SCHWARZ, SIGMUND D, b Portland, Ore, Sept 27, 28; m; c 3. GEOPHYSICS, GEOLOGY. *Educ:* Ore State Univ, BSc, 52. *Prof Exp:* Photo-radar intelligence officer, USAF Strategic Air Command, Fairchild AFB, 52-54; geologist & geophysicist, Ore State Highway Dept, Salem, 54-58; pres, Geo Recon Inc, Seattle, 58-71; prin & sr geologist & geophysist, Shannon & Wilson, Inc, Seattle, 71-79; pres, Geo-Recon Int Ltd, 79-84; independent consult, 84-90; PRES, S D SCHWARZ & ASSOC, INC, 90- *Concurrent Pos:* Prin, Geo Recon Ore Ltd, Salem, 56-58; chmn, Wash State Sect, Asn Eng Geologists, 67. *Mem:* Am Inst Mining, Metall & Petrol Engrs; Asn Eng Geologists; Soc Explor Geophysicists; Europ Asn Explor Geophysicists; fel Geol Soc Am; Seismol Soc Am; Am Inst Prof Geologists. *Res:* Author of 20 publications in geophysics. *Mailing Add:* S D Schwarz & Assoc Inc PO Box 82-917 Kenmore WA 98028

SCHWARZ, STEVEN E, b Los Angeles, Calif, Jan 29, 39; m 63. ELECTRICAL ENGINEERING. *Educ:* Calif Inst Technol, BS, 59, MS, 61, PhD(elec eng), 64; Harvard Univ, AM, 62. *Prof Exp:* Mem tech staff, Hughes Res Labs, 62-64; from asst prof to assoc prof elec eng, 64-74, PROF ELEC ENG, UNIV CALIF, BERKELEY, 74- *Concurrent Pos:* Guggenheim fel, IBM Corp Res Lab, Zurich, 71-72; pres chair undergrad educ, Univ Calif, 90-93. *Mem:* Sr mem Inst Elec & Electronics Engrs. *Res:* Microwave circuits. *Mailing Add:* Dept Elec Eng Univ Calif 2120 Oxford St Berkeley CA 94720

SCHWARZ, THOMAS WERNER, pharmaceutical chemistry, pharmacy, for more information see previous edition

SCHWARZ, WILLIAM MERLIN, JR, b Hartford, Conn, Nov 13, 34; m 55; c 4. ELECTROCHEMISTRY, PHYSICAL CHEMISTRY. *Educ:* Pa State Univ, 56; Univ Wis, PhD(phys chem), 61. *Prof Exp:* Proj assoc, Univ Wis, 61-63; sr engr, Int Bus Mach Corp, 63-64; chemist, Nat Bur Stand, DC, 64-68; scientist, 68-70, SR SCIENTIST, XEROX CORP, 70- *Mem:* Am Chem Soc; Electrochem Soc; Am Inst Chemists. *Res:* Electrode kinetics; xerographic development; polarography; photoelectrophoresis; xerographic processes and materials. *Mailing Add:* 274 Southboro Dr Webster NY 14580

SCHWARZER, CARL G, b San Francisco, Calif, Apr 20, 17; m 37; c 1. ORGANIC CHEMISTRY. *Prof Exp:* Chemist, Shell Develop Co, 37-67; dir res & develop, Apogee Chem Co, 67-71; pres, Appl Resins & Technol, 71-73; mgr & chemist, Indust Tank, Inc, J & J Disposal, Inc, 74-76; waste mgt specialist, State Calif Health Serv, Hazardous Mat Mgt Sect, 76-80; TECH SPECIALIST AEROJET GEN ENVIRON STAFF & PROG MGR HAZARDOUS WASTE MAT, AEROJET ENERGY CONVERSION CO, 80- *Concurrent Pos:* Consult, US, Mex, Europe & China; mem bd dirs, World Asn Solid Waste Transfer & Exchange. *Mem:* Am Chem Soc; Am Civil Eng Asn. *Res:* Synthesis of organic and epoxy resins; hydrocarbon resin surface coatings; manufacture and applications of epoxy, peroxide, phenolic compounds and resins; industrial waste; disposal management in the environmental systems and technology; surveillance and management of hazardous waste materials; resource recovery and reuse of industrials; environmental chemistry disciplines. *Mailing Add:* 7760 Crystal Blvd Diamond Springs CA 95619-9625

SCHWARZER, THERESA FLYNN, b Troy, NY, Apr 14, 40; m 61; c 1. GEOLOGY, GEOCHEMISTRY. *Educ:* Rensselaer Polytech Inst, BS, 63, MS, 66, PhD(geol), 69. *Prof Exp:* Instr geol, State Univ NY Albany, 69; res fel remote sensing, Rice Univ, 69-72; sr res geologist, Exxon Prod Res Co, 72-74, res specialist, 74-76, sr res specialist, 76-78, sr explor geologist, Gulf Coast Div, Exxon, USA, 78-80, proj leader, Tex Offshore, 80-81, dist prod geologist, ETex Div, 81-83, sr supvr, Exxon Prod Res Co, 83-87, GEOL ADV, EXXON USA, 87- *Concurrent Pos:* Chairwoman women geoscientists comt, Am Geol Inst, 73-77. *Mem:* Geol Soc Am; Am Asn Petrol Geologists; Soc Explor Geophysicists; Geochem Soc. *Res:* Inorganic and organic geochemistry; remote sensing; multivariate statistical techniques; interpretation and integration of geophysical, geological and geochemical data for hydrocarbon exploration. *Mailing Add:* 17915 Echobend Spring TX 77379

SCHWARZSCHILD, ARTHUR ZEIGER, b New York, NY, Mar 24, 30; m 52; c 3. NUCLEAR PHYSICS. *Educ:* Columbia Univ, BA, 51, MA, 56, PhD(physics), 57. *Prof Exp:* Res assoc physics, Columbia Univ, 57-58; res assoc physics, 58-60, assoc physicist, 61-63, physicist, 63-70, dep chmn & head nuclear physics, 78-81, SR PHYSICIST, BROOKHAVEN NAT LAB, 70-, CHMN, PROG ADV COMN, TANDEM USERS GROUP, 75-, CHMN, PHYSICS DEPT, 81- *Concurrent Pos:* Consult, NY Univ, 64-80; NATO fel, 66-67; mem, Nuclear Sci Adv Comn, Dept Energy, NSF, 81-83; Argonne Univ Asn rev comt, Physics Div, Argonne Nat Lab, 81-83. *Mem:* Fel Am Phys Soc; AAAS; Sigma Xi; fel NY Acad Sci. *Res:* Heavy ion nuclear reactions; nuclear spectroscopy; measurements of electromagnetic transition probabilities for excited nuclear states; instrumentation for very short lifetime measurements. *Mailing Add:* 31 Howard St Patchogue Long Island NY 11772

SCHWARZSCHILD, MARTIN, b Potsdam, Ger, May 31, 12; nat US; m 45. STELLAR STRUCTURE & EVOLUTION, STELLAR DYNAMICS. *Educ:* Univ Gottingen, PhD(astron), 35. *Hon Degrees:* DSc, Swarthmore Col, 60, Columbia Univ, 73. *Prof Exp:* Nansen fel, Univ Oslo, 36-37; Littauer fel, Harvard Univ, 37-40; lectr astron, Columbia Univ, 40-44, asst prof, 44-47; prof astron, 47-51, Higgins prof, 51-79, EMER HIGGINS PROF ASTRON, PRINCETON UNIV, 79- *Concurrent Pos:* Vpres, Int Astron Union, 64-70. *Honors & Awards:* Newcomb Cleveland Prize, AAAS, 57; Draper Medal, Nat Acad Sci, 61; Eddington Medal, Royal Astron Soc, 63, Gold Medal, 69; Bruce Medal, Astron Soc Pac, 65; Rittenhouse Medal, Rittenhouse Astron Soc, 66; Albert A Michelson Award, Case Western Reserve Univ, 67; Dannie Heineman Prize, Acad Learning, Gottingen, 67; Prix Janssen Award, Soc Astron France, 70. *Mem:* Nat Acad Sci; Am Astron Soc (pres, 70-72); Int Astron Union; Am Philos Soc; AAAS. *Res:* Theory of stellar structure and evolution; stellar dynamics. *Mailing Add:* Princeton Univ Observ Peyton Hall Princeton NJ 08544

SCHWASSMAN, HORST OTTO, b Berlin, Ger, Aug 31, 22; m 60; c 1. BIOLOGY, PHYSIOLOGY. *Educ:* Univ Munich, Cand rer nat, 52; Univ Wis-Madison, PhD(zool), 62. *Prof Exp:* Lectr zool, Univ Wis-Madison, 62-63; USPHS fel, Univ Calif, Los Angeles, 63-65, asst res anatomist, 65-67; from asst res physiologist to assoc res physiologist, Scripps Inst Oceanog, 67-70; assoc prof psychol & biol, Dalhousie Univ, 70-72, prof, 72; assoc prof zool, 72-78, PROF ZOOL, UNIV FLA, 78- *Mem:* Asn Trop Biol; Am Soc Ichthyologists & Herpetologists. *Res:* Animal behavior; sensory physiology; visual system; circadian and biological rhythms; neurophysiology of vision in vertebrates. *Mailing Add:* Dept Zool Univ Fla Gainesville FL 32611

SCHWEBEL, SOLOMON LAWRENCE, b New York, NY, Oct 21, 16; m 49; c 2. THEORETICAL PHYSICS. *Educ:* City Col New York, BS, 37; NY Univ, MS, 47, PhD, 54. *Prof Exp:* Instr physics, City Col New York, 46-50; res scientist, Inst Math Sci, NY Univ, 52-54; instr physics, Brooklyn Col, 54-55; staff scientist, Missiles & Space Div, Lockheed Aircraft Corp, 55-61; assoc prof physics, Univ Cincinnati, 61-64; ASSOC PROF PHYSICS, BOSTON COL, 64- *Mem:* Am Phys Soc; Am Asn Physics Teachers. *Mailing Add:* Dept Physics Boston Col Chestnut Hill MA 02167

SCHWEBER, SILVAN SAMUEL, b Strasbourg, France, Apr 10, 28; nat US; m 65. THEORETICAL PHYSICS. *Educ:* City Col New York, BS, 47; Univ Pa, MS, 49; Princeton Univ, PhD(physics), 52. *Prof Exp:* Asst instr, Univ Pa, 47-49; instr, Princeton Univ, 51-52; NSF fel, Cornell Univ, 52-54; res physicist, Carnegie Inst Technol, 54-55; assoc prof, 55-61, chmn dept, 58-76, chmn sch sci, 62-68 & 73-74, PROF PHYSICS, BRANDEIS UNIV, 61-, PROF HIST SCI, 82- *Concurrent Pos:* Vis prof, Mass Inst Technol, 61-62 & 69-70 & Hebrew Univ, Jerusalem, 71-72; res assoc history sci, Harvard Univ, 77-86. *Mem:* Am Phys Soc; AAAS; Sigma Xi. *Res:* Field theory; statistical mechanics; history of science. *Mailing Add:* Dept Physics Brandeis Univ Col Arts & Sci Waltham MA 02154

SCHWEE, LEONARD JOSEPH, b Omaha, Nebr, Mar 19, 36; m 61; c 3. PHYSICS, ELECTRICAL ENGINEERING. *Educ:* Creighton Univ, BS, 60. *Prof Exp:* RES PHYSICIST MAGNETISM, NAVAL SURFACE WEAPONS CTR, 60- *Mem:* Inst Elec & Electronics Engrs. *Res:* Magnetic thin films (permalloy), and their application to devices such as magnetometers, recorders, and computer memories (the crosstie memory). *Mailing Add:* Naval Surface Weapons Ctr White Oak Lab 10901 New Hampshire Ave Silver Spring MD 20903

SCHWEGMANN, JACK CARL, b Denver, Colo, Nov 4, 25; m 47; c 2. ENTOMOLOGY. *Educ:* Tulane Univ, BS, 48; Univ Okla, MS, 50; La State Univ, PhD(plant path), 53. *Prof Exp:* Plant pathologist, Chalmette Works, Kaiser Aluminum & Chem Corp, 53-54, sr plant pathologist, 54-56, supvr air control, 56-58, supvr fume abatement, Metals Div, 58, coordr air control activ, 58-67, dir environ serv, 67-82, mgr air serv, corp environ affairs, 82-86; RETIRED. *Concurrent Pos:* Univ Calif Coop Exten Master Gardener Prog, Alameda County, 88- *Mem:* AAAS; Sigma Xi; Am Phytopath Soc; NY Acad Sci. *Res:* Air and stream pollution effects on plants and animals; diseases and insect pests of ornamental plants and vegetable crops. *Mailing Add:* 2001 Sandcreek Way Alameda CA 94501

SCHWEICKERT, RICHARD ALLAN, b Sonora, Calif, Feb 7, 46; m 67; c 2. GEOLOGY, TECTONICS. *Educ:* Stanford Univ, BS, 67, PhD(geol), 72. *Prof Exp:* Geologist, Texaco Inc, 71-72 & US Geol Surv, 73; asst prof geol, Calif State Col, Sonoma, 72, Calif State Univ, San Jose, 73 & Calif State Univ, San Francisco, 73; asst prof, 73-78, ASSOC PROF GEOL, COLUMBIA UNIV, 78- *Mem:* Geol Soc Am; Am Geophys Union; Soc Econ Paleontologists & Mineralogists. *Res:* Tectonics of orogenic belts and convergent plate boundaries; Paleozoic and Mesozoic tectonic evolution of the western cordillera of the United States; stratigraphy and structure of western Sierra Nevada; origin of melanges. *Mailing Add:* Mackay Sch Mines Univ Nev Reno NV 89557

SCHWEIGER, JAMES W, b Osage, Iowa, Oct 13, 29; c 2. DENTISTRY, PROSTHODONTICS. *Educ:* Univ Iowa, DDS, 54, MS, 57. *Prof Exp:* Instr prosthetic dent, Dent Sch, Univ Iowa, 57-58, asst prof dent technol, 58-59, asst prof otolaryngol & maxillofacial surg, Sch Med, 59-65, assoc prof, Sch Dent, 65-69; chief, Dent Serv & dir, Maxillofacial Prosthetic Ctr, Vet Admin Ctr, Wilmington, 70-83, coordr res, 81-83; chief, Dent Serv, Mem Sloan Kettering, 83-88; chmn, Div Prosthodontics, 87-91, CLIN PROF PROSTHODONTICS, COLUMBIA UNIV, 83- *Concurrent Pos:* Consult, Coun Dent Educ, Thomas Jefferson Univ Hosp, Wilmington Med Ctr Surg & Dent & Children's Hosp Philadelphia, Temple Univ; consult, Bronx Vet Hosp, NY, 88- *Mem:* Fel Am Col Prosthodont; assoc mem Am Acad Ophthal & Otolaryngol; fel Am Acad Maxillofacial Prosthodontics (pres, 81-82); Am Dent Asn; Am Bd Prosthodontics. *Res:* Development of facial plastics for maxillofacial prosthodontics. *Mailing Add:* PO Box 606 Lewes DE 19958

SCHWEIGER, MARVIN I, b Middletown, NY, Feb 10, 23; m 80; c 2. AERONAUTICAL ENGINEERING. *Educ:* Rensselaer Polytech Inst, BAeroE, 48. *Prof Exp:* Asst, Res Labs, Gen Elec Co, 47; res engr, Res Dept, United Aircraft Corp, 48-55, head propulsion sect, Res Labs, 55-61, chief propulsion, 61-62; vpres, Bowles Eng Corp, Md, 62-67; mgr, Aerothermo, 67-71, mgr preliminary design, 71-73, mgr, Internal Aero, 73-78, proj mgr contracted res, Columbus Aircraft Div, 78-80, tech dir res & technol, NAm Aircraft Opers, Rockwell Int Corp, 80-86; INDEPENDENT CONSULT, 86- *Mem:* Assoc fel Am Inst Aeronaut & Astronaut. *Res:* Aerodynamics; propulsion; control and aircraft. *Mailing Add:* 4765 Powderhorn Lane Westerville OH 43081

SCHWEIGERT, BERNARD SYLVESTER, food science; deceased, see previous edition for last biography

SCHWEIGHARDT, FRANK KENNETH, b Passaic, NJ, May 12, 44; m 68; c 2. MOLECULAR SPECTROSCOPY, FLUOROCARBON CHEMISTRY. *Educ:* Seton Hall Univ, BS, 66; Duquesne Univ, PhD(phys chem), 70. *Prof Exp:* Asst to dean pharmaceut chem, Sch Pharm, Duquesne Univ, 70- 71; fel Nat Res Coun, US Bur Mines, 71-72 res chemist, Dept Energy, 72-79; MGR & SR RES ASSOC, NEW PROD DEVELOP, AIR PROD & CHEM, INC, 79- *Concurrent Pos:* Chemist, Allegheny County Morgue, 69- 71; lectr, Chem Dept, Duquesne, 70; consult anal & forensic chem, 70-; postdoctoral fel, Nat Res Coun, 72; assoc ed, Pa Acad Sci J, 88- *Mem:* Am Chem Soc; Spectros Soc; Anal Soc; AAAS; Sigma Xi. *Res:* Development of fluon chemicals and their emulsions for bio-medical applications; development of high purity gases for trace analysis and the electronic's market. *Mailing Add:* IGD Specialty Gas Div Air Prod & Chem Inc 7201 Hamilton Blvd Allentown PA 18195-1501

SCHWEIKER, GEORGE CHRISTIAN, b Philadelphia, Pa, Feb 17, 24; m 50; c 4. CHEMISTRY, RESEARCH & DEVELOPMENT ADMINISTRATION. *Educ:* Temple Univ, AB, 49, AM, 52, PhD(chem), 53. *Prof Exp:* Adj prof chem, Drexel Univ, 50-53; res chemist, Hooker Chem Corp, 53-56, supvr polymer res, 56-57; mgr res, Velsicol Chem Corp, 57-60; mgr plastics res, Celanese Corp, 60-65; dir chem & polymers, Develop Div, Borg-Warner Corp, 65-71; vpres & dir res & develop, PQ Corp, 71-87; INDEPENDENT CONSULT, 87- *Concurrent Pos:* Mem, corp assoc comt, Am Chem Soc. *Mem:* Fel AAAS; fel Royal Soc Chem; Am Chem Soc; Indust Res Inst; Soc Chem Indust; NY Acad Sci. *Res:* Polymers; organic syntheses; rearrangements; industrial and agricultural chemicals; fire retardants; plastics and plastics additives; inorganic chemicals; research and development management. *Mailing Add:* 12518 Calle Tamega 127 San Diego CA 92128

SCHWEIKER, JERRY W, b Oshkosh, Wis, July 1, 31. ENGINEERING. *Educ:* Univ Ill, BS, 55, MS, 57, PhD(theoret & appl mech), 61. *Prof Exp:* Res asst theoret & appl mech, Univ Ill, 55-57, instr, 57-61; sr engr, McDonnell Aircraft Corp, 61-65; prof eng mech, St Louis Univ, 65-71; VPRES, ENG DYNAMICS INT, 71- *Mem:* Am Soc Eng Educ; Soc Exp Stress Anal. *Res:* Structural dynamics; noise control. *Mailing Add:* 627 Packford Dr Ballwin MO 63011

SCHWEIKERT, DANIEL GEORGE, b Bemidji, Minn, June 15, 37; m 61; c 3. COMPUTER AIDED DESIGN. *Educ:* Yale Univ, BE, 59; Brown Univ, ScM, 62, PhD(numerical anal), 66. *Prof Exp:* Res engr, Gen Dynamics/Elec Boat, 61-64; mem tech staff comput applns, Bell Tel Labs, 66-72, supvr, Bell Labs, 72-80; dir comput-aided design, United Technol Microelectron Ctr, 80-87, dir prog mgt & design methods, 87-88; DIR, CASE & TECH SERV, CADENCE DESIGN SYSTS, 88- *Concurrent Pos:* Mem, prog comt, 80-86, exec bd, Design Automation Conf, 86-, prog comt, Int Conf Comput Aided Design. *Mem:* Fel Inst Elec & Electronics Engrs; Asn Comput Mach. *Res:* Development of systems for the computer-aided design of integrated circuits. *Mailing Add:* Cadence Design Systs 555 River Oaks Pkwy Bldg 4 San Jose CA 95134

SCHWEIKERT, EMILE ALFRED, b Flawil, Switz, Sept 10, 39; m 65; c 3. ANALYTICAL CHEMISTRY. *Educ:* Univ Toulouse, Licensee in sci, 62; Univ Paris, Dr(anal chem), 64. *Prof Exp:* Res asst res ctr metall chem, Nat Ctr Sci Res, Vitry, France, 63-65; sci consult, Europ Nuclear Energy Agency, Orgn Econ Coop & Develop, Paris, 65; scientist, Swiss Govt Deleg, Atomic Energy Matters, Switz, 65-66; asst prof anal chem & asst res chemist, 66-70, assoc prof anal chem & assoc res chemist, 70-74, PROF CHEM, TEX A&M UNIV, 74-, DIR, CTR CHEM CHARACTERIZATION & ANALYSIS, 72- *Concurrent Pos:* Adj asst prof grad sch, Col Med, Baylor Univ, 68- *Honors & Awards:* George Hevesy Medal, 86. *Mem:* Am Chem Soc; NY Acad Sci; Swiss Asn Atomic Energy. *Res:* Analytical chemistry; administration of scientific affairs; secondary ion mass spectrometry; nuclear methods of analysis. *Mailing Add:* Ctr Chem Characterization & Analysis Tex A&M Univ College Station TX 77843-3144

SCHWEINLER, HAROLD CONSTANTINE, b Tacoma, Wash, June 1, 22; m 45. PHYSICS. *Educ:* Carnegie Inst Technol, BS, 43, MS, 44; Mass Inst Technol, PhD, 51. *Prof Exp:* Instr physics, Carnegie Inst Technol, 43-44; jr physicist, Metall Lab, Univ Chicago, 44-45; physicist, Oak Ridge Nat Lab, 45-48; asst prof physics, Mass Inst Technol, 51-52, mem solid state & molecular theory group, 52-54; physicist, Oak Ridge Nat Lab, 54-79; Ford Found Prof Physics, Univ Tenn, Knoxville, 63-84; RETIRED. *Concurrent Pos:* Lectr, Univ Tenn, Knoxville, 47-48 & 58-63. *Mem:* Am Phys Soc; Am Asn Physics Teachers. *Res:* Nuclear reactor theory; magnetic susceptibility; defects in solids; molecular potential energy curves; theory of solids; mathematical physics. *Mailing Add:* Cumberland Rd No R Rockwood TN 37854

SCHWEINSBERG, ALLEN ROSS, b Ellwood City, Pa, May 5, 42; m 69. MATHEMATICAL ANALYSIS. *Educ:* Univ Pittsburgh, BS, 62, MS, 65, PhD(math), 69. *Prof Exp:* Instr math, Univ Pittsburgh, 67-69; asst prof, 69-76, ASSOC PROF MATH, BUCKNELL UNIV, 76- *Mem:* Am Math Soc; Math Asn Am. *Res:* Operator theory. *Mailing Add:* Dept Math Bucknell Univ Lewisburg PA 17837

SCHWEISS, JOHN FRANCIS, b St Louis, Mo, June 25, 25; m 50; c 5. MEDICINE. *Educ:* St Louis Univ, MD, 48. *Prof Exp:* Asst surg, 52-57, from instr to assoc prof, 57-71, ASSOC PROF PEDIAT & PROF SURG, SCH MED, ST LOUIS UNIV, 71-, DIR SECT ANESTHESIOL, 62- *Concurrent Pos:* Resident, Presby Hosp, NY, 54-56; chief anesthesiol, Cardinal Glennon Mem Hosp, 56-; pvt pract. *Mem:* Am Soc Anesthesiol; AMA; Int Anesthesia Res Soc. *Res:* Anesthesiology. *Mailing Add:* 3635 Vista Ave St Louis MO 63110

SCHWEISTHAL, MICHAEL ROBERT, b Faribault, Minn, June 11, 36; m 65; c 5. FETAL ENDOCRINOLOGY, GROWTH & DEVELOPMENT. *Educ:* Luther Col, Iowa, BA, 58; Univ Minn, PhD(anat), 64. *Prof Exp:* Teaching asst anat, Univ Minn, 60-64; from asst prof to assoc prof, Med Col Va, 64-69; prof biol, ECarolina Univ, 70-71, prof anat & chmn dept, 71-77; assoc dean biomed sci, Oral Roberts Univ, 77, prof anat, 77- 88; CLIN PROF & SR LECTR, DEPT SURG, UNIV CALIF SCH MED, SAN DIEGO, 88- *Concurrent Pos:* NSF fel, 60; trainee anat, NIH, 60; lectr, Portsmouth Naval Hosp, 64-69; assoc prof anat & oral biol, Univ Ky, 69-70. *Mem:* AAAS; Am Asn Anatomists; Sigma Xi; Am Physiol Soc; Am Diabetes Asn; Am Asn Clin Anatomists. *Res:* Fetal alcohol syndrome; the effects of ethanol on the endocrine system of the maternal organism, the fetus and the neonate; the effects of ethanol on insulin and carbohydrate metabolism in a developing and growing system. *Mailing Add:* Dept Surg Univ Calif San Diego La Jolla CA 92093

SCHWEITZER, CARL EARLE, b Flint, Mich, Jan 19, 14:; m 41; c 2. ORGANIC CHEMISTRY, POLYMERS. *Educ:* Kalamazoo Col, AB, 36, MS, 37; Northwestern Univ, PhD(chem), 41. *Prof Exp:* Asst chem, Northwestern Univ, 37-40; res mgt, Polymer Prod Dept, E I du Pont de Nemours & Co Inc, 41-68, patents consult, 69-81; RETIRED. *Mem:* AAAS; Sigma Xi. *Res:* Polymers and plastics, particularly delrin acetal resin. *Mailing Add:* 205 Alapocas Dr Wilmington DE 19803

SCHWEITZER, DONALD GERALD, b New York, NY, Mar 25, 30; m 52; c 2. THERMODYNAMIC & MATERIALS PROPERTIES. *Educ:* City Col New York, BS, 51; Syracuse Univ, PhD(chem), 55. *Prof Exp:* Chemist, Dept Nuclear Energy, 55-85, head high temperature gas-cooled reactor safety div, 74-79, ASSOC CHMN & HEAD RADIOACTIVE WASTE MGT DIV, BROOKHAVEN NAT LAB, 79- *Mem:* Am Inst Mining, Metall & Petrol Eng. *Res:* Reactor safety; graphite research, chemistry of liquid metals; superconductivity; radiation damage; photochemistry; waste management. *Mailing Add:* Dept Nuclear Energy Bldg 701 Brookhaven Nat Lab Upton NY 11973

SCHWEITZER, EDMUND OSCAR, III, b Evanston, Ill, Oct 31, 47; m 77; c 3. ELECTRICAL ENGINEERING. *Educ:* Purdue Univ, BS, 68, MS, 71; Wash State Univ, PhD(elec eng), 77. *Prof Exp:* Elec engr radar syst, Nat Security Agency, 68-73; elec engr, Probe Syst Inc, 73-74; res asst elec power, Wash State Univ, 74-77; asst prof, Ohio Univ, 77-79; assoc prof elec power, Wash State Univ, 79-84; PRES, SCHWEITZER ENG LABS, 82- *Concurrent Pos:* Adj Prof, elect power, Wash State Univ, 84- *Mem:* Inst Elec & Electronics Engrs. *Res:* Application of microprocessors to electric power systems protection; revenue metering of electric energy; digital signal processing using microprocessors. *Mailing Add:* Schweitzer Eng Labs NE 2350 Hopkins Ct Pullman WA 99163

SCHWEITZER, GEORGE KEENE, b Poplar Bluff, Mo, Dec 5, 24; m 48; c 3. INORGANIC CHEMISTRY, PHILOSOPHY OF SCIENCE. *Educ:* Cent Col, Mo, BA, 45; Univ Ill, MS, 46, PhD(inorg chem), 48; Columbia Univ, MA, 59; NY Univ, PhD(philos of sci), 64. *Hon Degrees:* ScD, Cent Col, 64. *Prof Exp:* Asst chem, Cent Col, Mo, 43-45; from asst prof to assoc prof, 48-58, prof, 60-70, DISTINGUISHED PROF CHEM, UNIV TENN, KNOXVILLE, 70- *Concurrent Pos:* Fel, Columbia Univ, 59-61; vis prof, Meredith Col, 80 & Eastern Ill Univ, 82, 88; Am Chem Soc lectr. *Honors & Awards:* Staley lectr. *Mem:* Am Philos Asn; Am Chem Soc; Hist Sci Soc. *Res:* Templated coordination complex matrices; rare earth chelate fluorescence; ion exchanging polymers; history and philosophy of science; redox resins. *Mailing Add:* Dept Chem Univ Tenn Knoxville TN 37996-1600

SCHWEITZER, JEFFREY STEWART, b New York, NY, May 6, 46; m 70; c 1. NUCLEAR PHYSICS, NUCLEAR GEOPHYSICS. *Educ:* Carnegie Inst Technol, BS, 67; Purdue Univ, MS, 69, PhD(physics), 72. *Prof Exp:* Res fel nuclear physics, Calif Inst Technol, 72-74; sr res physicist, 74-87, SCI ADV, SCHLUMBERGER LTD, 88- *Concurrent Pos:* Reviewer, Sci Fac Prof Develop, NSF, 81; consult, Int Atomic Energy Agency, 85-; ed, Nuclear Geophys, 86- *Mem:* Am Phys Soc; Am Nuclear Soc; Inst Elec & Electronics Engrs; Soc Prop Well Log Analysts. *Res:* Experimental work in gamma-ray spectroscopy, neutron and gamma-ray interactions, low energy compound nuclear reactions and development of nuclear well logging tools for spectroscopic analysis. *Mailing Add:* Schlumberger-Doll Res Old Quarry Rd Ridgefield CT 06877-4108

SCHWEITZER, JOHN WILLIAM, b Covington, Ky, Apr 23, 41. MAGNETISM & SUPERCONDUCTIVITY IN STRONGLY CORRELATED SYSTEMS. *Educ:* Thomas More Col, AB, 60; Univ Cincinnati, MS, 62, PhD(physics), 66. *Prof Exp:* From asst prof to assoc prof, 66-78, PROF PHYSICS, UNIV IOWA, 78- *Mem:* Am Phys Soc. *Res:* Localized magnetic states; itinerant electron magnetism; valence fluctuations; statistical mechanics; solitons in molecular systems; superconductivity. *Mailing Add:* Dept Physics & Astron Univ Iowa Iowa City IA 52242

SCHWEITZER, LELAND RAY, b Merna, Nebr, July 23, 41; m 71; c 2. AGRONOMY, VEGETABLE CROPS. *Educ:* Ore State Univ, BS, 66, MS, 69; Mich State Univ, PhD(crop sci), 72. *Prof Exp:* Int Agr Ctr study fel, Off Seed Testing Sta, Wageningen, Neth, 72; SEED PHYSIOLOGIST, ASGROW SEED CO, 72- *Mem:* Am Soc Agron; Asn Off Seed Anal. *Res:* Seed physiology, technology and testing. *Mailing Add:* Asgro Seed Co PO Box 1235 Twin Falls ID 83303-1235

SCHWEITZER, MARK GLENN, b Mckeesport, Pa, July 12, 57; m 84, 91. CARBOHYDRATE CHEMISTRY. *Educ:* Thiel Col, BA, 79; Ohio State Univ, Columbus, PhD(chem), 84. *Prof Exp:* Prin res chemist, Lever Res, Inc, 84-85; scientist, Rohm & Haas Co, 85-89; DEPT MGR, BATTELLE, 89- *Mem:* Am Chem Soc. *Res:* Development of new analytical procedures for agricultural products of environmental concern; metabolism of agricultural products; residue analysis and method development. *Mailing Add:* 336 Mainsail Dr Westerville OH 43081-2743

SCHWEITZER, PAUL JEROME, b New York, NY, Jan 16, 41; div; c 3. OPERATIONS RESEARCH, APPLIED MATHEMATICS. *Educ:* Mass Inst Technol, BS(physics) & BS(math), 61, ScD(physics), 65. *Prof Exp:* Consult, Lincoln Lab, Mass Inst Technol, 64-65; staff mem, Weapons Systs Eval Div, Inst Defense Anal, 65-66, Jason Div, 66 & Weapons Systs Eval Div, 66-72; staff mem, IBM Thomas J Waston Res Ctr, 72-77; opers res area coordr, 85-90, PROF OPERS RES & COMPUT & INFO SYSTS, GRAD SCH MGT, UNIV ROCHESTER, 77- *Concurrent Pos:* Consult, Airborne Instruments Lab, NY, 62-63; mem staff, Am Univ, 69-70; vis mem fac, Israel Inst Technol, 70-72, 81-82; mem staff, Haifa Univ, 72; consult, Israel Aircraft Industs, 72, 83-84; consult, Exxon, 80-82; mem fac, Bengurion Univ, 83-84. *Mem:* Opers Res Soc Am; Inst Mgt Sci. *Res:* Military operations research; mathematical programming; stochastic processes; telecommunications and computer networks; aggregation methods; computer-integrated manufacturing. *Mailing Add:* Simon Sch Univ Rochester Rochester NY 14627

SCHWEIZER, ALBERT EDWARD, b Philadelphia, Pa, Mar 31, 43; m 61, 89; c 3. ZEOLITE CHEMISTRY. *Educ:* West Chester State Col, BS, 64; Rutgers Univ, MS, 68; Calif Inst Technol, PhD(chem), 74. *Prof Exp:* Sr res chemist, Cent Res Labs, Mobil Res & Develop Corp, 64-76; sr res chemist, indust chem res & develop, Air Prod & Chem, Inc, 76-80; sr res chemist, ICI Americas, 80-82; SR STAFF CHEMIST, EXXON RES & DEVELOP LABS, 82- *Mem:* Am Chem Soc; Am Ceramic Soc; Sigma Xi. *Res:* Synthesis, mechanistic studies and thermal properties; process development; catalysis; zeolites and transition metal chemistry. *Mailing Add:* 2069 W Magna Carta Baton Rouge LA 70815

SCHWEIZER, BERTHOLD, b Cologne, Ger, July 20, 29; nat US; m 61; c 1. MATHEMATICS. *Educ:* Mass Inst Technol, SB, 51; Ill Inst Technol, MS, 54, PhD(math), 56. *Prof Exp:* Instr math, Ill Inst Technol, 56-57; asst prof, San Diego State Col, 57-58; vis asst prof, Univ Calif, Los Angeles, 58-60; assoc prof, Univ Ariz, 60-65 & Univ Mass, Amherst, 65-68; prof, Univ Ariz, 68-70; PROF MATH, UNIV MASS, AMHERST, 70- *Mem:* Am Math Soc; Math Asn Am. *Res:* Probabilistic geometry; algebra of functions; functional equations. *Mailing Add:* Dept Math & Statist Univ Mass Amherst MA 01003

SCHWEIZER, EDWARD E, b Joliet, Ill, Apr 6, 33; m 53; c 2. AGRONOMY, WEED SCIENCE. *Educ:* Univ Ill, BS, 56, MS, 57; Purdue Univ, PhD(plant physiol), 62. *Prof Exp:* PLANT PHYSIOLOGIST, CROPS RES LAB, COLO STATE UNIV, USDA, 61- *Concurrent Pos:* mem, Coun Agr Sci & Technol. *Mem:* Fel Weed Sci Soc Am; Coun Agr Sci & Technol; Am Soc Sugar Beet Technol. *Res:* Weed research in sugar beets; integrated pest management; weed crop modeling. *Mailing Add:* Crops Res Lab 1701 Ctr Ave Ft Collins CO 80526

SCHWEIZER, EDWARD ERNEST, b Shanghai, China, Dec 7, 28; US citizen; m 68; c 7. ORGANIC CHEMISTRY. *Educ:* NDak State Univ, BS, 51, MS, 53; Mass Inst Technol, PhD(org chem), 56. *Prof Exp:* Sr chemist, Argos Establecimiento de Productos Colorantes, Arg, 51-52; sr res chemist, Minn Mining & Mfg Co, 56-59; res assoc org chem, Univ Minn, 59-60, instr, 60; asst prof chem, Hofstra Col, 60-61; from asst prof to assoc prof, 61-72, PROF CHEM, UNIV DEL, 72- *Honors & Awards:* Fulbright teaching award, Univ Madrid, 68-69. *Mem:* Am Chem Soc. *Res:* Reactions of phosphonium compounds; heterocyclics. *Mailing Add:* Dept Chem Univ Del Newark DE 19716

SCHWEIZER, FELIX, b Cologne, Ger, Sept 1, 27; US citizen; div; c 3. OPTICAL RADIATIONS, THEORETICAL PHYSICS. *Educ:* Rensselaer Polytech Inst, BS, 51, MS, 54; Univ Calif, Los Angeles, PhD(physics), 60. *Prof Exp:* Mem tech staff space physics, Space Technol Labs Inc, 60-62; scientist specialist, Jet Propulsion Lab, Calif Inst Technol, 62-64; asst prof physics, San Fernando Valley State Col, 64-67; physicist, Naval Weapons Sta, Seal Beach, Corona Site, 67-88; CONSULT, 88- *Concurrent Pos:* Mem, Working Group Infrared & Lasers, Joint Dept Defense-Metrol & Calibration Coord Group, 68-88. *Mem:* Am Phys Soc. *Res:* Nuclear physics; electromagnetic theory; radiometry; photometry; lasers; classical mechanics; anomalies in relativity and quantum mechanics. *Mailing Add:* 541 S Hepner Ave Covina CA 91723

SCHWEIZER, FRANCOIS, b Bern, Switz, Aug, 16, 42; m 75; c 4. OPTICAL ASTRONOMY, EXTRAGALACTIC RESEARCH. *Educ:* Univ Bern, Switz, licentiate, 68; Univ Calif, Berkeley, MA, 70, PhD(astron), 74. *Prof Exp:* Carnegie fel astron, Hale Observ, Pasadena, Calif, 74-75; staff astronr, Cerro Tololo InterAm Observ, La Serena, Chile, 76-81; STAFF ASTRONR, DEPT TERRESTRIAL MAGNETISM & ADJ STAFF MEM, OBSERVATORIES, CARNEGIE INST WASH, 81- *Mem:* Am Astron Soc; Int Astron Union; Swiss Astron Soc; Astron Soc Pac. *Res:* Optical studies of colliding and merging galaxies; structure and formation of elliptical galaxies; surface photometry and structure of spiral galaxies. *Mailing Add:* Dept Terrestrial Magnetism Carnegie Inst 5241 Broad Branch Rd NW Washington DC 20015

SCHWEIZER, MALVINA, b New Orleans, La, Jan 16, 06. ENDOCRINOLOGY. *Educ:* NY Univ, PhD(physiol), 33. *Prof Exp:* Spec asst dir, Nat Heart, Lung & Blood Inst, NIH, 76-89; RETIRED. *Mem:* AAAS; Am Physiol Soc. *Mailing Add:* Nat Heart Lung & Blood Inst NIH Bethesda MD 20205

SCHWEIZER, MARTIN PAUL, IN VIVO NUCLEAR MAGNETIC RESONANCE. *Educ:* Johns Hopkins Univ, PhD(phys biochem), 68. *Prof Exp:* PROF MED CHEM, UNIV UTAH, 78- *Res:* In vivo nuclear magnetic resonance studies of aging effects on metabolism and evaluation of pharmaceutical agents. *Mailing Add:* Dept Med Chem Univ Utah Salt Lake City UT 84112

SCHWELB, OTTO, b Budapest, Hungary, Mar 27, 31; Can citizen; m 66; c 3. INTEGRATED OPTICS, MICROWAVE DEVICES. *Educ:* Univ Tech Sci, Budapest, dipl, 54; McGill Univ, PhD(elec eng), 78. *Prof Exp:* Mem sci staff, Res & Develop Div, Northern Elec Co, 57-67; ASSOC PROF ELEC ENG, CONCORDIA UNIV, 67- *Concurrent Pos:* Lectr, Univ Ottawa, 62-66; consult, Ainslie Antenna Corp, 69-71, Mitec Electronics Ltd, 75-76, Can Elec Asn, 78-80, Com Dev Ltd, 85-87 & Can Marconi Co, 90. *Mem:* Inst Elec & Electronics Engrs; Optical Soc Am; Order Engrs Que; Int Soc Optical Eng. *Res:* Integrated optics; surface acoustic wave devices; microwave components; electromagnetic wave propagation in stratified and periodic media. *Mailing Add:* Dept Elec & Computer Eng Concordia Univ 1455 De Maisonneuve Blvd W Montreal PQ H3G 1M8 Can

SCHWELITZ, FAYE DOROTHY, b Milwaukee, Wis, June 17, 31. CELL BIOLOGY. *Educ:* Alverno Col, BA, 53; Purdue Univ, MS, 67, PhD(cell biol), 71. *Prof Exp:* NIH trainee, Purdue Univ, 69-71; asst prof biol, 71-78, ASSOC PROF BIOL, UNIV DAYTON, 78- *Mem:* Electron Micros Soc Am; Am Soc Plant Physiol; Biophys Soc; Am Soc Cell Biol; Sigma Xi. *Res:* Role of cyclic Adenine Monophosphatase in Euglena; photosynthetic mutants of Euglena; correlation of structural changes with functional and biochemical changes; chloroplast development. *Mailing Add:* 1049 Sherwood Dr Dayton OH 45406

SCHWENDEMAN, JOSEPH RAYMOND, JR, b Fargo, NDak, Dec 13, 30; m 52; c 3. PHYSICAL GEOGRAPHY, RESOURCE GEOGRAPHY. *Educ:* Univ Ky, BA, 56, MS, 57; Ind Univ, PhD(geog), 67. *Prof Exp:* Asst prof phys geog, Univ NDak, 62-63; staff instr, Ind Univ, 63-64; asst prof phys geog & chmn dept geog, Univ NDak, 64-66; chmn dept, 66-76, dean undergrad studies & prof geog, 76-83, assoc vpres acad planning & develop, 83-86, VPRES ADMIN AFFAIRS, EASTERN KY UNIV, 86- *Concurrent Pos:* Co-dir, Geog Studies & Res Ctr, 68- *Mem:* Asn Am Geog. *Res:* Climatology; regional potential studies; planners. *Mailing Add:* Vpres Admin Affairs Eastern Ky Univ Richmond KY 40475

SCHWENDEMAN, RICHARD HENRY, b Chicago, Ill, Aug 26, 29; m 53; c 3. PHYSICAL CHEMISTRY. *Educ:* Purdue Univ, BS, 51; Univ Mich, MS, 52, PhD, 56. *Prof Exp:* Res fel chem, Harvard Univ, 55-57; from asst prof to assoc prof, 57-69, PROF CHEM, MICH STATE UNIV, 69- *Mem:* Am Chem Soc; Am Phys Soc. *Res:* Determination of molecular structure and collisional relaxation rates by microwave spectroscopy and infrared laser spectroscopy; theoretical studies in molecular spectra and molecular dynamics. *Mailing Add:* Dept Chem/8 Chem Bldg Mich State Univ East Lansing MI 48824

SCHWENDIMAN, JOHN LEO, b Sugar City, Idaho, Sept 28, 09; m 34; c 6. SAND DUNE CONTROL, GRASS SEED PRODUCTION. *Educ:* Univ Idaho, BS, 35. *Prof Exp:* Jr agronomist, Soil Conserv Serv, USDA, 36-39, asst agronomist, 40-42, assoc agronomist, 42-45, agronomist, Plant Mat Ctr, 45-54, plant mat specialist, 54-76; PROF AGRON CONSULT, 76- *Concurrent Pos:* Chmn, Inland Empire Soil Conserv Soc, 65; mem bd, Parks & Recreation Comn, Pullman, Wash, 84-91. *Mem:* Fel Soc Range Mgt; Sigma Xi; Am Soc Agron; Soil Conserv Soc Am. *Res:* Improved grasses for use in soil conservation; improved plants and their conservation; author of numerous technical publications. *Mailing Add:* 1110 Juniper Way NE Pullman WA 99163

SCHWENK, FRED WALTER, b Dickinson, NDak, July 29, 38; m 63; c 2. PLANT PATHOLOGY, VIROLOGY. *Educ:* NDak State Univ, BS, 60, MS, 64; Univ Calif, Berkeley, PhD(plant path), 69. *Prof Exp:* Lab technician plant path, Univ Calif, Berkeley, 64-66; asst prof, 69-74, ASSOC PROF PLANT PATH, KANS STATE UNIV, 74- *Mem:* AAAS; Am Phytopath Soc. *Res:* Soybean pathology; teaching of undergraduate plant pathology. *Mailing Add:* Dept Plant Path Dickens Hall Kans State Univ Manhattan KS 66506

SCHWENKER, J(OHN) E(DWIN), b Bartlesville, Okla, June 27, 28; m 50; c 2. ELECTRICAL ENGINEERING. *Educ:* Univ Okla, BS, 52; Rutgers Univ, MS, 56. *Prof Exp:* Mem tech staff, 52-59, head logic technol res dept, 59-68, head data systs dept, 68-72, HEAD DATA APPLN ENG DEPT, BELL TEL LABS, INC, 72- *Mem:* AAAS; Inst Elec & Electronics Engrs. *Res:* Data communications; computers. *Mailing Add:* 61 Tulip Lane Colts Neck NJ 07722

SCHWENKER, ROBERT FREDERICK, JR, b Ann Arbor, Mich, July 3, 20; m 43; c 4. CELLULOSE CHEMISTRY, POLYMER CHEMISTRY. *Educ:* Univ Pa, AB, 48. *Prof Exp:* Asst biol, Wistar Inst, Univ Pa, 46-48; lab asst chem, Rutgers Univ, 48-49, asst instr, 49-51; chemist, Textile Res Inst, 51-55, group leader org chem, 55-60, assoc res dir, 60-66, dir chem & chem processing, 66; assoc dir res & develop, Personal Prod Co, Johnson & Johnson, 66-74, dir res & develop, 74-75, vpres res & develop, 75-82, mem bd dirs, 74-86, vpres sci technol, 82-86; RETIRED. *Mem:* Am Chem Soc; Fiber Soc; Tech Asn Pulp & Paper Indust; fel Am Inst Chemists; Sigma Xi. *Res:* Structure of cellulose derivatives; thermal degradation of high polymers; structure and properties of fibers; chemical specialties. *Mailing Add:* 92 Willow Run Lane Belle Mead NJ 08502

SCHWENSFEIR, ROBERT JAMES, JR, b Hartford, Conn, June 27, 34; m 67; c 3. NUCLEAR CRITICALITY SAFETY, CLAD METALS. *Educ:* Wesleyan Univ, BA, 56; Trinity Col, Conn, MS, 60; Brown Univ, PhD(physics), 66. *Prof Exp:* Exp physicist, Pratt & Whitney Aircraft Div, United Aircraft Corp, 56-60; res asst solid state physics, Brown Univ, 60-66; res assoc metal physics, Adv Mat Res & Develop Lab, Pratt & Whitney Aircraft Div, United Aircraft Corp, 66-68; asst prof physics, Bucknell Univ, 68-74; nuclear criticality safety engr, Naval Prod Div, United Nuclear Corp, 74-76, nuclear criticality safety specialist, 76-79; mgr Nuclear Safety & Mat, 79-82, MEM TECH STAFF, TEX INSTRUMENTS INC, 82- *Concurrent Pos:* Mem, Tech Adv Comt, Metals Properties Coun. *Mem:* AAAS; Am Asn Physics Teachers; Am Nuclear Soc; Am Phys Soc; Am Soc Metals; NY Acad Sci; Sigma Xi; Soc Advan Mat & Process Eng; Soc Automotive Engrs. *Res:* Nuclear criticality safety research. *Mailing Add:* 54 Marlise Dr Attleboro MA 02703

SCHWENTERLY, STANLEY WILLIAM, III, b Philadelphia, Pa, Aug 18, 45. CRYOGENICS, SUPERCONDUCTING MAGNETS. *Educ:* Yale Univ, BS, 67; Cornell Univ, PhD(physics), 73. *Prof Exp:* RES STAFF MEM PHYSICS, OAK RIDGE NAT LAB, 72- *Mem:* Am Phys Soc; Cryogenic Soc Am; Sigma Xi. *Res:* Research and development on cryogenic materials and equipment for application in nuclear fusion and other types of energy systems. *Mailing Add:* 1922 Plumb Creek Circle Knoxville TN 37932

SCHWENZ, RICHARD WILLIAM, b Portsmouth, Va, June 30, 55; m 81; c 2. PHYSICAL CHEMISTRY. *Educ:* Univ Colo, Boulder, BA, 77; Ohio State Univ, PhD(chem), 81. *Prof Exp:* Teaching fel, Univ Ill, Chicago, 81-82 & Northwestern Univ, 82-84; asst prof phys chem, Mundelein Col, 83-84; asst prof, 84-89, ASSOC PROF PHYS CHEM, UNIV NORTHERN COLO, 89- *Mem:* Am Chem Soc; Am Phys Soc; Sigma Xi. *Res:* Molecular spectroscopy of transition metal oxides and halides; infacing of laboratory instruments. *Mailing Add:* Dept Chem Univ Northern Colo Greeley CO 80639

SCHWENZER, KATHRYN SARAH, biochemistry, for more information see previous edition

SCHWEPPE, EARL JUSTIN, b Trenton, Mo, Sept 28, 27; m 48; c 3. COMPUTER SCIENCE, MATHEMATICS. *Educ:* Mo Valley Col, BS, 48; Univ Ill, MS, 51, PhD(math), 55. *Prof Exp:* Asst math, Univ Ill, 51-55; instr, Univ Nebr, 55-57; asst prof, Iowa State Univ, 57-61; mathematician, Dept Defense, 61-63; res asst prof computer sci & math, Univ Md, College Park, 63-65, assoc prof, 65-67; PROF COMPUTER SCI, SCH BUS, UNIV KANS, 67- *Concurrent Pos:* Consult, Fed Systs Div, IBM Corp, 65- *Mem:* Am Math Soc; Math Asn Am; Asn Comput Mach. *Res:* Abstract algebra; projective geometry; graph and network theory; automata theory; information structures; computer language design and translation; machine description and design; simulation; consequent prodecures; computer science curriculum development. *Mailing Add:* Dept Computer Sci Univ Kans 22nd St Lawrence KS 66045

SCHWEPPE, JOHN S, b Chicago, Ill, May 8, 17; m 43; c 3. MEDICINE. *Educ:* Harvard Univ, AB, 39; Northwestern Univ, MD, 43, MS, 47. *Prof Exp:* Fel internal med, Mayo Clin, 47-50; res assoc biochem, 55-60, assoc med, 60-62, from asst prof to assoc prof, 63-78, PROF MED & MOLECULAR BIOL, MED SCH, NORTHWESTERN UNIV, CHICAGO, 78- *Concurrent Pos:* Mem res dept, Northwestern Mem Hosp, 60-; fel coun arteriosclerosis, Am Heart Asn. *Mem:* Endocrine Soc; Am Fedn Clin Res; Am Asn Cancer Res; fel Am Col Physicians. *Res:* Cyclic adenosine monophosphate; protein kinases in endocrine and malignant tissues; control of gene expression; oncology; immunoconjugates in cancer therapy. *Mailing Add:* Rm 949W 845N Michigan Ave Chicago IL 60611

SCHWEPPE, JOSEPH L(OUIS), b Trenton, Mo, Jan 11, 21; m 42; c 3. ENGINEERING. *Educ:* Univ Mo, BS, 42, MS, 46; Univ Mich, PhD(chem eng), 50. *Prof Exp:* Jr chem engr, Tenn Valley Authority, Ala, 42-43; asst chem eng, Univ Mo, 46; chem engr, E I du Pont de Nemours & Co, Tex, 49-52; res engr, C F Braun & Co, Calif, 52-54, sr chem engr, 54-56, proj engr, 56-58; from assoc prof to prof mech eng, Univ Houston, 58-63, chmn dept, 59-63; pres, Houston Eng Res Corp, 60-90; DIR, J K CONTROL SYSTS, INC, 91- *Concurrent Pos:* Lectr, Univ Southern Calif, 54-56 & Univ Calif, Los Angeles, 55. *Mem:* Instrument Soc Am; Inst Elec & ElectronicS ENGRS; Am Soc Mech Engrs; Am Inst Chem Engrs. *Res:* Project engineering; process instrumentation; computer control. *Mailing Add:* 4987 Dumfries Dr Houston TX 77096

SCHWER, JOSEPH FRANCIS, b Georgetown, Ohio, July 25, 36; m 57; c 2. AGRONOMY, PLANT PHYSIOLOGY. *Educ:* Ohio State Univ, BSc, 57; Univ Ky, MSc, 59; Pa State Univ, PhD(genetics, breeding), 62. *Prof Exp:* Sr plant physiologist, Greenfield Labs, Eli Lilly & Co, Ind, 62-64, head Western Field Res, 65-66, head plant sci res, Lilly Res Labs Ltd, Eng, 66-69, plant sci field res-int, Eli Lilly & Co, 69-87; VPRES RES & DEVELOP, AGRIGENETICS CORP, 88- *Mem:* Am Soc Agron; Weed Sci Soc Am. *Res:* New pesticides and plant growth regulators. *Mailing Add:* 35575 Curtis Blvd Centre Plaza N Suite 300 Eastlake OH 44094

SCHWERDT, CARLTON EVERETT, b New York, NY, Jan 2, 17; m 45; c 3. VIROLOGY. *Educ:* Stanford Univ, AB, 39, MA, 40, PhD(biochem), 46. *Prof Exp:* Res assoc chem, Stanford Univ, 43-47; asst prof biochem, Sch Hyg & Pub Health, Johns Hopkins Univ, 47-50; from asst res biochemist to assoc res biochemist, Virus Lab, Univ Calif, 50-57; assoc prof med microbiol, 57-61, prof, 61-82, EMER PROF MED MICROBIOL, STANFORD UNIV, 82- *Mem:* Fel AAAS; Am Chem Soc; Soc Exp Biol & Med; Am Soc Biol Chemists; fel NY Acad Sci; Sigma Xi. *Res:* Physical and chemical characterization and biology of replication of animal viruses; crystallization of polio virus particles. *Mailing Add:* Dept Microbiol & Immunol Stanford Univ Stanford CA 94305-5402

SCHWERDTFEGER, CHARLES FREDERICK, b Philadelphia, Pa, July 20, 34; m 63; c 4. PHYSICS. *Educ:* Villanova Univ, BSc, 56; Univ Notre Dame, PhD(physics), 61. *Prof Exp:* Res assoc physics, Univ Basel, 61-62 & Ind Univ, 62-63; from asst prof to assoc prof, 63-73, PROF PHYSICS, UNIV BC, 73- *Mem:* Am Phys Soc; Can Asn Physicists. *Res:* Electronic properties of solids; electron spin resonance studies in semiconductors. *Mailing Add:* Dept Physics Univ BC 2075 Westbrook Mall Vancouver BC V6T 1W5 Can

SCHWERER, FREDERICK CARL, b Pittsburgh, Pa, Feb 1, 41; m 64; c 2. APPLIED PHYSICS, MATERIALS SCIENCE. *Educ:* Pa State Univ, BS, 62; Cornell Univ, PhD(appl physics), 67. *Prof Exp:* Scientist, Physics Div, US Steel Res Ctr, 67-76, assoc res consult, Basic Res, US Steel Res Lab, 76-85; SR TECH SPECIALIST, 85- *Concurrent Pos:* Prin investr, Apollo Lunar Sci Prog, 71-76. *Mem:* AAAS; Am Inst Mining, Metall & Petrol Engrs; Am Soc Nondestructive Testing; Am Soc Testing & Mat; Am Phys Soc; Inst Elec & Electronics Engrs. *Res:* Experimental and theoretical studies of transport and magnetic properties of metals and alloys; mixed metal oxides and industrial minerals; mathematical modeling of two-phase flow in porous media; theoretical and experimental studies of electromagnetically induced fluid flows in liquid metals. *Mailing Add:* 218 Adams St Export PA 15632

SCHWERI, MARGARET MARY, b Louisville, Ky, Aug 29, 46. RECEPTOR BINDING, BEHAVIORAL CHARACTERIZATION OF STIMULANT DRUGS IN ANIMALS. *Educ:* Marquette Univ, BS, 68; Univ Louisville, PhD(pharmacol), 80. *Prof Exp:* Chemist, Brown & Williamson Tobacco Corp, 68-72; res assoc, Biochem Dept, Univ Louisville, 73-76; staff fel, Lab Biorg Chem, Nat Inst Arthritis, Diabetes & Digestive & Kidney Dis, NIH, 80-83, sr staff fel, 83-84; asst prof, 84-90, ASSOC PROF PHARMACOL, SCH MED, MERCER UNIV, 90- *Concurrent Pos:* Prin investr, New Investr Res Award, Nat Inst Neurol & Commun Dis & Stroke, NIH, 85-88, biomed res support grant, NIH, 88-89, Small Instrument Prog, 88-89 & Acad Res Enhancement Award, 90-93; co-prin investr, Nat Inst Drug Abuse/RO1, 89-92. *Mem:* Soc Neurosci; Am Soc Pharmacol & Exp Therapeut; AAAS. *Res:* Characterization of the structure and function of the stimulant recognition site on the dopamine transport complex; development of affinity labels for this site to act as cocaine antagonists. *Mailing Add:* Mercer Univ Sch Med 1550 College St Macon GA 31207-0003

SCHWERING, FELIX, b Cologne, Ger, June 4, 30. ELECTRICAL ENGINEERING, THEORETICAL PHYSICS. *Educ:* Aachen Tech Univ, BS, 51, MS, 54, PhD(elec eng), 57. *Prof Exp:* Asst prof theoret physics, Aachen Tech Univ, 56-58; physicist, US Army Elec Lab, NJ, 58-61; proj leader radar res, Telefunken Co, Ulm, Ger, 61-64; RES PHYS SCIENTIST, US ARMY ELEC LAB, 64- *Concurrent Pos:* Vis prof, Rutgers Univ, 74, & NJ IT, 86-0010. *Honors & Awards:* IEEE Fel, 88. *Mem:* Inst Elec & Electronics Engrs; Am Geophys Union; Int Sci Radio Union. *Res:* Electromagnetic theory, particularly guided and free space propagation of electromagnetic waves; beam wave guides; antenna theory; periodic structures; diffraction and scatter theory; theoretical and electron optics; millimeter ware antennas and propagation. *Mailing Add:* US Army Commun Electronics Command Attn: AMSEL-RD-C3-TR-H Ft Monmouth NJ 07703-5202

SCHWERT, DONALD PETERS, b Wellsville, NY, Dec 12, 49; m; c 1. QUATERNARY GEOLOGY. *Educ:* Allegheny Col, BS, 72; State Univ NY, MS, 75; Univ Waterloo, PhD(earth sci), 78. *Prof Exp:* Asst prof, 78-86, ASSOC PROF GEOL, NDAK STATE UNIV, 86- *Concurrent Pos:* Fel, Univ Waterloo, 78- *Mem:* Geol Soc Am; Coleopterists Soc; Am Quaternary Asn. *Res:* Use of fossils to determine Quaternary distributions of insects. *Mailing Add:* Dept Geol NDak State Univ Fargo ND 58105

SCHWERT, GEORGE WILLIAM, b Denver, Colo, Jan 27, 19; m 43, 79; c 2. BIOCHEMISTRY. *Educ:* Carleton Col, BA, 40; Univ Minn, PhD(biochem), 43. *Prof Exp:* Asst agr biochem, Univ Minn, 41-42, instr, 42-43; biochemist, Sharp & Dohme, Inc, Pa, 43-44; instr & res assoc biochem, Duke Univ, 46-48, from asst prof to prof, 48-59; chmn dept, 59-74, prof, 59-85, EMER PROF BIOCHEM, COL MED, UNIV KY, 85- *Concurrent Pos:* Markle scholar, 49-54; consult, NIH, 59-64; assoc ed, J Molecular Cell Biol, 83-85. *Mem:* AAAS; Am Chem Soc; Am Soc Biol Chemists; Brit Biochem Soc; Sigma Xi. *Res:* Mechanisms of enzyme action; hydrolases; dehydrogenases; transaminases; relation of protein structure to biological activity. *Mailing Add:* 3316 Braemar Dr Lexington KY 40502

SCHWERZEL, ROBERT EDWARD, b Rockville Center, NY, Dec 14, 43; m 84; c 2. PHOTOCHEMISTRY, NONLINEAR OPTICAL MATERIALS. *Educ:* Va Polytech Inst & State Univ, BS, 65; Fla State Univ, PhD(phys org chem), 70. *Prof Exp:* Fel photochem, Cornell Univ, 70-71; fel magnetic resonance, Brown Univ, 71-72; res chemist org chem, Syva Res Inst, 72-73; res chemist photochem, 73-78, prin res scientist, 78-80, SR RES SCIENTIST PHOTOCHEM, BATTELLE MEM INST, 80- *Concurrent Pos:* Adj prof, Ohio State Univ, 82- & Bowling Green State Univ, 87- *Honors & Awards:* I R 100 Award, 80. *Mem:* Am Chem Soc; AAAS; Inter-Am Photochem Soc; Int Solar Energy Soc; Sigma Xi. *Res:* Novel applications of photochemistry and spectroscopy, including photochemical utilization of solar energy, optical data storage and processing, photoelectrochemical formation of fuels, development of improved laser dyes and exploratory studies on nonlinear optical materials; use of laser-produced x-ray for x-ray absorption fine structure spectroscopy. *Mailing Add:* Batelle Columbus Div 505 King Ave Columbus OH 43201-2693

SCHWERZLER, DENNIS DAVID, b St Louis, Mo, Dec 23, 44; m 67; c 2. MECHANICAL ENGINEERING. *Educ:* Univ Mo-Rolla, BS, 66; Purdue Univ, Lafayette, MS, 68, PhDmech eng), 71. *Prof Exp:* Assoc sr res engr, 71-74, DEVELOP ENGR, ENG STAFF, GEN MOTORS RES LABS, 74- *Mem:* Am Soc Mech Engrs. *Res:* Vehicle structural dynamics; application of finite element techniques to vehicle structures; experimental dynamic testing of vehicle structures. *Mailing Add:* Gen Motors Ctr CPC Eng Ctr L105 30003 Van Dyke Warren MI 48090-9060

SCHWETMAN, HERBERT DEWITT, b Waco, Tex, Aug 1, 11; m 39; c 3. ELECTRONICS, PHYSICS. *Educ:* Baylor Univ, BA, 32; Univ Tex, MA, 37, PhD(physics), 52; Harvard Univ, MS, 47. *Prof Exp:* Jr operator, Western Union Tel Co, 28-33; teacher pub schs, Tex, 34-41; instr electronics, Harvard Univ, 41-47; prof physics & chmn dept, Baylor Univ, 47-84; SR MEM TECHNOL STAFF, 84- *Mem:* Am Phys Soc; Am Asn Physics Teachers; sr mem Inst Elec & Electronics Engrs; Sigma Xi. *Res:* Mathematics; application of Laplace transforms to analysis of electric circuits and physical problems; electronic analog computers; mechanical harmonic analyzers and synthesizers; ultrasonic and microwave attenuation. *Mailing Add:* MCC 3500 W Balcones Ctr Dr Austin TX 78759

SCHWETTMAN, HARRY ALAN, b Cincinnati, Ohio, Aug 16, 36; m 58; c 3. PHYSICS. *Educ:* Yale Univ, BS, 58; Rice Univ, MA, 60, PhD(physics), 62. *Prof Exp:* Res assoc, 62-64, from asst prof to assoc prof, 64-77, PROF PHYSICS, STANFORD UNIV, 77- *Concurrent Pos:* Sloan res fel, 66-72. *Res:* Low temperature physics; development of superconducting accelerator; application of low temperature physics and nuclear physics to medicine. *Mailing Add:* Dept Physics/Hanson Labs Stanford Univ Stanford CA 94305

SCHWETZ, BERNARD ANTHONY, b Cadott, Wis, Nov 27, 40; m 62; c 2. TERATOLOGY, DEVELOPMENTAL TOXICOLOGY. *Educ:* Univ Wis-Stevens Point, BS, 62; Univ Minn, St Paul, DVM, 67; Univ Iowa, MS, 68, PhD(pharmacol), 70. *Prof Exp:* Toxicologist, Dow Chem Co, 70-82, dir toxicol res lab, 77-82; CHIEF, SYSTS TOXICOL BR, NAT INST ENVIRON HEALTH SCI, 82- *Honors & Awards:* Arnold J Lehman Award, Soc Toxicol, 91. *Mem:* Soc Toxicol; Behav Teratology Soc; Teratology Soc. *Res:* Reproduction; developmental toxicology; teratology. *Mailing Add:* Nat Inst Environ Health Sci PO Box 12233 Research Triangle Park NC 27709

SCHWIDERSKI, ERNST WALTER, b Satticken, Ger, Feb 24, 24; nat US; m 59; c 1. OCEAN TIDES, GEOPHYSICS. *Educ:* Karlsruhe Tech Univ, Dipl math, 52, Dr rer nat, 55. *Prof Exp:* Sci asst & instr math, Math Inst, Karlsruhe Tech Univ, 48-55; head traffic theory br, Stand Elektrik Co, Ger, 55-57; SR RES MATHEMATICIAN, DAHLGREN LAB, US NAVAL SURFACE WARFARE CTR, 58- *Concurrent Pos:* Prof lectr, Am Univ, 58-67; adj prof, Va Polytech Inst & State Univ, 68- *Honors & Awards:* John Adolphus Dahlgren Award. *Mem:* Am Phys Soc; Am Geophys Union. *Res:* Ordinary and partial differential equations; integral equations; real and complex analysis; numerical analysis; mathematical theory of inviscid and viscous fluid flow; ocean tides and currents; marine geodesy; geophysics. *Mailing Add:* Naval Surface Warfare Ctr Dahlgren VA 22448

SCHWIER, CHRIS EDWARD, b Neenah, Wis, Apr 11, 56. POLYMER STRUCTURE-PROPERTY RELATIONSHIPS, NEW PRODUCT DEVELOPMENT & COMMERCIALIZATION. *Educ:* Univ Wis-Madison, BS, 78; Mass Inst Technol, ScD(chem eng), 84. *Prof Exp:* RES SPECIALIST, MONSANTO CHEM CO, 84- *Mem:* Am Chem Soc; Soc Plastics Engrs. *Res:* Polymer structure-property relationships for ABS and nylon; rubber toughening of polymers; free radical and condensation polymerization kinetics; continuous polymerization reactor design; new product development and commercialization. *Mailing Add:* Monsanto Chem Co PO Box 12830 Pensacola FL 32575

SCHWIESOW, RONALD LEE, b Pittsburgh, Pa, May 22, 40; m 62; c 3. REMOTE SENSING, PHYSICAL OPTICS. *Educ:* Purdue Univ, BS, 62; Johns Hopkins Univ, PhD(physics), 68; Pa State Univ, MS, 74. *Prof Exp:* Jr instr physics, Johns Hopkins Univ, 62-66, res asst crystal spectros, 67-68; Nat Res Coun res fel atmospheric spectros, Res Labs, Environ Sci Serv Admin, 68-70; physicist, Environ Res Labs, Nat Oceanic & Atmospheric Admin, 70-84; RES ENGR, NAT CTR ATMOSPHERIC RES, 84- *Concurrent Pos:* Guest scientist, Rro14 Nat Lab, Denmark, 78-79; vis scientist, DLR-Oberpfaffenhofen, Ger, 82-83; comt laser atmospheric studies, 78-81, 87-90. *Mem:* Am Meteorol Soc; Sigma Xi; Optical Soc Am. *Res:* Remote measurement of meteorological parameters using lasers; micrometeorology of the boundary layer; inelastic scattering spectroscopy applied to environmental problems; Doppler laser wind instrumentation; cloud dynamics and microphysics; airborne instrumentation for meteorology. *Mailing Add:* Nat Ctr Atmospheric Res ATD/RAF PO Box 3000 Boulder CO 80307-3000

SCHWIMMER, SIGMUND, b Cleveland, Ohio, Sept 20, 17; wid; c 2. ENZYMOLOGY, FOOD BIOCHEMISTRY. *Educ:* George Washington Univ, BS, 41; Georgetown Univ, MS, 41, PhD(biochem), 43. *Prof Exp:* Jr chemist, USDA, Washington, DC, 41-43, from asst chemist to chemist, 43-58, prin chemist, Western Regional Res Lab, 58-65, chief chemist, 65-74, EMER CHIEF RES BIOCHEMIST, WESTERN REGIONAL RES LAB, USDA, 75- *Concurrent Pos:* NSF sr fel, Carlsberg Found Biol Inst & Royal Vet & Agr Col, Denmark, 58-59; res assoc, Calif Inst Technol, 63-65, head enzyme technol invests, 67-71; sr biochemist, UN Indust Develop Orgn Centre Indust Res, Haifa, Israel, 73-74; guest lectr, Dept Nutrit Sci, Univ Calif, Berkeley, 75-83; ed, J Food Biochem, 77-, Trends in Biotechnol, 83-89; consult, food enzym, 83-; guest ed, Trends in Biochem, 83-84; adj prof, Dept Nutrit Sci, Univ Calif, Berkeley, 84-88. *Mem:* Am Soc Biochem & Molecular Biol; Inst Food Technol. *Res:* Enzymology; plant and food biochemistry and biotechnology. *Mailing Add:* Western Regional Res Ctr USDA 800 Buchanan St Berkeley CA 94710

SCHWINCK, ILSE, b Hamburg, Ger, May 24, 23. DEVELOPMENTAL GENETICS. *Educ:* Univ Tuebingen, Dr rer nat(biol, biochem), 50. *Prof Exp:* Ger Res Asn fel insect physiol, Univ Munich, 51-54; Karl-Hescheler Found fel develop genetics, Univ Zurich, 54-56; res fel, Albert Einstein Col Med, 56; res fel, Col Med, NY Univ, 56-57; res fel, Columbia Univ, 57-58; guest investr develop genetics, Biol Lab, Cold Spring Harbor, 59-60; Ger Res Asn Grant & investr, Max Planck Inst, Mariense, Ger, 60-62; from asst prof to assoc prof biol, Univ Conn, 62-87; RETIRED. *Concurrent Pos:* Fulbright advan res fel, 56-58. *Mem:* Am Soc Cell Biol; Genetics Soc Am; Soc Develop Biol; NY Acad Sci; Swiss Genetics Soc. *Res:* Developmental and biochemical genetics; control of insect development. *Mailing Add:* Dept Molecular & Cell Biol Univ Conn Box U-125 Storrs CT 06268

SCHWIND, ROGER ALLEN, chemical engineering, physical chemistry, for more information see previous edition

SCHWINDEMAN, JAMES ANTHONY, b Cincinnati, Ohio, Oct 30, 55. HERBICIDE SYNTHESIS. *Educ:* Miami Univ, BS, 77; Ohio State Univ, PhD(org chem), 81. *Prof Exp:* Grad teaching asst org chem, Ohio State Univ, 77-78, assoc, 78-81; SR RES CHEMIST AGR CHEM, PPG INDUST, 81- *Mem:* Am Chem Soc; Royal Soc Chem. *Res:* Synthesis and evaluation of novel herbicidal compounds. *Mailing Add:* FMC Lithium Div PO Box 795 Bessemer City NC 28016

SCHWING, FRANKLIN BURTON, b Fairmont, WVa, Jan 6, 56. COASTAL PHYSICAL OCEANOGRAPHY, DYNAMICAL PHYSICAL OCEANOGRAPHY. *Educ:* Univ SC, BS, 78, MS, 81; Dalhousie Univ, PhD(phys oceanog), 89. *Prof Exp:* Res coordr, Skidaway Inst Oceanog, 81-85; PHYS OCEANOGR, PAC FISHERIES ENVIRON GROUP, NAT OCEANIC & ATMOSPHERIC ADMIN, NAT MARINE FISHERIES SERV, 88- *Mem:* Am Geophys Union; Am Meteorol Soc; Can Meteorol & Oceanog Soc; Oceanog Soc. *Res:* Dynamics of continental shelf, coastal and shallow water systems; role of circulation on transport of biological organisms; air-sea interactions; remote sensing of the marine environment. *Mailing Add:* Pac Fisheries Environ Group PO Box 831 Monterey CA 93942

SCHWING, GREGORY WAYNE, b Cincinnati, Ohio, Sept 30, 46; m 67; c 3. ORGANIC CHEMISTRY, AGRCHEMICALS. *Educ:* Purdue Univ, BS, 69; Univ Minn, PhD(organic chem), 72. *Prof Exp:* Res chemist agrchem, 72-77, sr res chemist, 77-78, RES SUPVR HERBICIDES, BIOCHEM DEPT, E I DU PONT DE NEMOURS & CO, INC, 78- *Mem:* Am Chem Soc. *Res:* Selective crop herbicides, industrial herbicides, plant growth modifiers, fungicides, insecticides and nematocides; synthesis and evaluation of new classes of compounds for agrichemical utility. *Mailing Add:* RD 1 Box 466 Lincoln Univ PA 19352

SCHWING, RICHARD C, b Buffalo, NY, Dec 8, 34; m 56; c 4. ENVIRONMENTAL POLICY, CHEMICAL ENGINEERING. *Educ:* Univ Mich, BS, 57, MS, 59, PhD(chem eng), 63. *Prof Exp:* From sr res engr to sr staff res engr, 63-87, PRIN RES ENGR, GEN MOTORS RES LABS, 87- *Honors & Awards:* John M Campbell Award. *Mem:* AAAS; Am Chem Soc; Int Asn Impact Assessment (pres, 84-85); Soc Risk Anal (pres, 88-). *Res:* Pollution control of internal combustion engines; measurement of corporate externalities; air pollution epidemiology regression analysis; risk-benefit, cost-benefit analyses; risk assessments; technology assessments; forecasting; human behavior and traffic safety. *Mailing Add:* Operating Sci Dept Gen Motors Res Labs Warren MI 48090-9055

SCHWINGER, JULIAN SEYMOUR, b New York, NY, Feb 12, 18; m 47. PHYSICS. *Educ:* Columbia Univ, AB, 36, PhD(physics), 39. *Hon Degrees:* DSc, Purdue Univ, 61, Harvard Univ, 62, Columbia Univ, 66. *Prof Exp:* Nat Res Coun fel, Columbia Univ, 39-40; res assoc, Univ Calif, 40-41; from instr to asst prof, Purdue Univ, 41-43; mem staff, Radiation Lab, Mass Inst Technol, 43-45; from assoc prof to prof physics, Harvard Univ, 45-66, Higgins prof, 66-75; prof, 75-, EMER PROF PHYSICS, UNIV CALIF, LOS ANGELES. *Concurrent Pos:* Mem staff, Metall Lab, Univ Chicago, 43; vis prof, Univ Calif, Los Angeles, 61. *Honors & Awards:* Nobel Prize in Physics, 65; C L Mayer Nature of Light Award, 49; Prize, Nat Acad Sci, 49; Einstein Award, 51; Nat Medal Sci in Physics, 64. *Mem:* Nat Acad Sci; AAAS; Am Phys Soc; Am Acad Arts & Sci; NY Acad Sci. *Res:* Nuclear and high energy physics; wave guide and electromagnetic theory; variational methods; quantum electrodynamics and field theory; foundations of quantum mechanics; many-particle systems; quantum gravitational theory. *Mailing Add:* Dept Physics Univ Calif Los Angeles CA 90024

SCHWINTZER, CHRISTA ROSE, US citizen; m 77; c 1. NITROGEN FIXATION, WETLAND ECOLOGY. *Educ:* Berea Col, BA, 62; Univ Mich, MA, 63, PhD(bot), 69. *Prof Exp:* Fel, Mo Bot Garden, 69-71; from asst prof to assoc prof ecol, Univ Wis-Green Bay, 71-78; Res Assoc, Harvard Univ Forest, 78-82; from asst prof to assoc prof, 82-89, PROF BOT, DEPT PLANT BIOL & PLANT, UNIV MAINE, 89- *Concurrent Pos:* Res assoc, Biol Sta, Univ Mich, 72-77. *Mem:* Ecol Soc Am; Brit Ecol Soc; Bot Soc Am; Am Soc Plant Physiologists. *Res:* Ecology of actinomycete-nodulated nitrogen fixing plants; ecology of northern bogs, swamps and fens emphasizing vegetation and nutrient status; physiological ecology of wetland plants. *Mailing Add:* Dept Plant Biol & Path Univ Maine Orono ME 04469

SCHWIRZKE, FRED, b Schwiebus, Ger, Aug 21, 27; m 58; c 3. PHYSICS. *Educ:* Karlsruhe Tech Univ, MS, 53, Dr rer nat(physics), 59. *Prof Exp:* Scientist, Max-Planck Inst Physics & Astrophys, 59-61; group supvr plasma physics res, Inst Plasma Physics, Munich, Ger, 61-62; staff mem plasma physics, Gen Atomic Div, Gen Dynamics Corp, Calif, 62-67; ASSOC PROF PHYSICS, NAVAL POSTGRAD SCH, 67- *Mem:* Am Phys Soc; Inst Elec & Electronics Engrs; Sigma Xi. *Res:* Plasma physics; controlled thermonuclear fusion; turbulence and anomalous diffusion of plasmas confined in magnetic fields; laser produced plasmas; self-generated magnetic fields; plasma diagnostics; impurities and plasma wall effects; plasma sheaths; ionization and charge exchange cross sections. *Mailing Add:* Dept Physics Naval Postgrad Sch Monterey CA 93940

SCHWITTERS, ROY FREDERICK, b Seattle, Wash, June 20, 44; m 65; c 3. PHYSICS. *Educ:* Mass Inst Technol, SB, 66, PhD(physics), 71. *Prof Exp:* Res assoc exp high energy physics, Stanford Univ, 71-74, from asst prof to assoc prof exp high energy physics, Stanford Linear Accelerator Ctr, 74-79; prof physics, Harvard Univ, 79-89; DIR, SUPERCONDUCTING SUPER COLLIDER LAB, 89- *Concurrent Pos:* Assoc ed, Annual Reviews of Nuclear & Particle Sci, 78-; div assoc ed, Phys Rev Letters, 79-84. *Honors & Awards:* Alan T Waterman Award, 80. *Mem:* Fel Am Phys Soc; fel AAAS. *Res:* Experimental high energy physics; development of large solid angle detection apparatus for use with high energy colliding beams; study of hadron production in electron-positron collisions. *Mailing Add:* Directorate Superconducting Super Collider Lab Dallas TX 75237

SCHWOEBEL, RICHARD LYNN, b New Rockford, NDak, Dec 26, 31; m 54; c 2. PHYSICS. *Educ:* Hamline Univ, BS, 53; Cornell Univ, PhD(eng physics), 62. *Prof Exp:* Sr engr, Gen Mills, Inc, 55-57; staff mem, Sandia Nat Labs, 62-65, supvr, 65-69, mgr Radiation & Surface Physics Res Dept, 78-82, DIR COMPONENTS, SANDIA NAT LABS, 88- *Concurrent Pos:* Vis prof, Cornell Univ, 71. *Mem:* Fel Am Phys Soc; sr mem Am Vacuum Soc; Sigma Xi; Mat Res Soc. *Res:* Oxidation of metals; defect nature and transport properties of oxides; microgravimetry; electron microscopy and diffraction; crystal growth processes; surface morphology; nuclear waste management; materials science engineering; research administration; technical management. *Mailing Add:* 8100 Osuna Rd NE Albuquerque NM 87109

SCHWOERER, F(RANK), b New York, NY, Sept 5, 22; m 49; c 1. MECHANICAL ENGINEERING. *Educ:* Webb Inst Naval Archit, BS, 44; Mass Inst Technol, MS, 47. *Prof Exp:* Develop engr, Aviation Gas Turbine Div, Westinghouse Elec Corp, 47-51, supvr compressor develop, 51-55, mgr adv develop, 55-57, adv engr, Bettis Atomic Power Lab, 57-59, mgr adv reactor develop, 59-64; mgr eng, NUS Corp, 64-65, tech dir, 65-66, vpres, 66-68; assoc, Pickard Lowe & Assocs, 68-75; tech dir, SNUPPS Proj, Nuclear Projs, Inc, 75-84; VPRES & TECH DIR, NEUTRON PROD INC, 84- *Mem:* Am Soc Mech Engrs; Am Nuclear Soc. *Res:* Nuclear reactor engineering, economics and project management. *Mailing Add:* Neutron Prods Inc P O Box 68 Dickerson MD 20842

SCHY, ALBERT ABE, b Przemysl, Poland, July 30, 20; m 57; c 4. AEROSPACE ENGINEERING, SYSTEMS CONTROL THEORY. *Educ:* Univ Chicago, BS, 42. *Prof Exp:* Aerospace engr control eng, 49-55, sect head, 55-60, asst br head, 60-66, br head, 66-78, asst br head control eng, 78-85, chief scientist, Guid & Control Div, 85-86, DISTINGUISHED RES ASSOC, NASA LANGLEY RES CTR, 86- *Mem:* Am Inst Aeronaut & Astronaut; Am Automatic Control Coun. *Res:* Dynamics and control of aerospace vehicles; computer aided control system design. *Mailing Add:* 722 Macon Rd Hampton VA 23665

SCHYVE, PAUL MILTON, b Rochester, NY, May 16, 44. PSYCHIATRY, MENTAL HEALTH ADMINISTRATION. *Educ:* Univ Rochester, BA, 66, MD, 70, dipl psychiat, 74. *Prof Exp:* Instr psychiat, Univ Rochester, 73-74; chief psychiat, USAF Regional Hosp, 74-75; staff psychiatrist, USAF Med Ctr, Wright-Patterson AFB, 75-76; unit chief, Ill State Phychiat Inst, 76-77, assoc dir, 77-79, clin dir, 79-82, dir, Lab Biol Psychiat, 82-83; assoc dir, Mental Illness, 83-85, DIR CLIN SERV, ILL DEPT MEN HEALTH & DEVELOP DISABILITIES 85- *Concurrent Pos:* Regional med consult, USAF Med Corps, 74-75; res assoc, Dept Psychiat, Univ Chicago, 76-80; asst prof, Dept Psychiat, Univ Chicago, 80-; attend mem, Dept Psychiat, Michael Reese Hosp, 80- *Mem:* Am Psychiat Asn. *Res:* Psychobiology of schizophrenia and affective illness; psychiatric diagnosis; mental health systems. *Mailing Add:* One Renaissance Blvd Oakbrook Terr IL 60181

SCIALDONE, JOHN JOSEPH, b Vitulazio, Italy, July 25, 26; US citizen; m 52; c 2. MECHANICAL ENGINEERING, AEROSPACE ENGINEERING. *Educ:* Univ Naples, dipl eng, 49, DE(mech & aerospace eng), 69; Carnegie Inst Technol, BS, 53; Univ Pittsburgh, MS, 60. *Prof Exp:* Pneumatic engr, Air Brake Div, Westinghouse Air Brake Co, 53-57, analyst, 57-62, sr analyst, Astronuclear Lab, Westinghouse Elec Corp, 62-64; asst mgr advan res, Test & Eval Div, 64-68, actg off head, 68-70, staff engr, 70-82, staff physicist, 82-87, HEAD POLYMERS SECT, GODDARD SPACE FLIGHT CTR, NASA, 87- *Concurrent Pos:* Adj prof physics & math, Capitol Col, Laurel, Md, 76; consult, NASA Spacecraft; prin investr, Int Sci Comts on Mat in Space Environ. *Mem:* Inst Environ Sci; Am Vacuum Soc; Soc Adv Mat & Process Eng; Am Inst Aeronaut & Astronaut. *Res:* Vacuum research and technology; space technologies; surface physics; kinetic theory; surface contamination; rarified gas dynamics; environmental testing; lubrication in space; instrumentation. *Mailing Add:* Head Polymers Sect Code 313 Goddard Space Flight Ctr NASA Greenbelt MD 20771

SCIAMANDA, ROBERT JOSEPH, b Erie Pa, Aug 11, 31; m 75; c 1. NIGHT VISION DEVICES, ELECTRO-OPTICS. *Educ:* St Bonaventure Univ, BA, 53; Cath Univ Am, MS, 60. *Prof Exp:* Chmn physics, Gannon Univ, 57-75; sr scientist, Idaho Nat Eng Lab, 75-80, Am Sterilizer Co, 80-88; ASSOC PROF, PHYSICS, EDINBORO UNIV PA, 88- *Mem:* Am Asn Physics Teachers; Am Asn Physicists Med. *Res:* Added color to night vision (patented); design of novel fiber optic image converters. *Mailing Add:* 3110 W 40th St Erie PA 16506

SCIAMMARELLA, CAESAR AUGUST, b Buenos Aires, Arg, Aug 22, 26; m 68; c 3. ENGINEERING MECHANICS. *Educ:* Univ Buenos Aires, Dipl Eng, 50; Ill Inst Technol, PhD(eng), 60. *Prof Exp:* Supvr design & stress anal, Hormigon Elastico Inversor, 51-53; spec assignment engr, Ducilo, Inc, 53-54; tech dir, Zofra, Inc, 55-56; sr researcher reactor eng, Arg Atomic Energy Comn, 56-57; assoc res engr, Ill Inst Technol, 58-59, instr, 59-60; prof eng sci & mech, Univ Fla, 61-67; prof appl mech & aerospace eng, Polytech Inst Brooklyn, 67-72; PROF APPL MECH, MECH ENG & AEROSPACE & DIR EXP STRESS LAB, ILL INST TECHNOL, 72- *Concurrent Pos:* Prof, Arg Army Eng Sch, 52-57 & Univ Buenos Aires, 56-57; lectr, Brit Sci Res Coun, 66; NSF vis lectr, Europe, 66; vis prof, Polytech Inst, Milan, Italy, 79, Univ Cagliari, Italy, 79, Polytech Inst Lausanne, Switz, 79, Univ Poitiers, France, 80. *Honors & Awards:* Award for Distinguished Res in Field of Eng, Sigma Xi, 66; Frocht Award, Soc Exp Mech, 80, Hetenyi Award, 82 & Lazan Award, 91. *Mem:* Fel Am Soc Mech Engrs; fel Soc Exp Mechanics; Am Astronaut Soc; Am Soc Testing & Mat; Optical Soc Am; Soc Photo-optical Instrumentation Engrs. *Res:* Experimental mechanics with particular emphasis in optical techniques; applications of experimental mechanics to the mechanics of materials; studies on the mechanism of failure of materials simple and composite. *Mailing Add:* Dept Mech & Aerospace Eng Ill Inst Technol Chicago IL 60616

SCIARINI, LOUIS J(OHN), b Branford, Conn, June 30, 15; m 53; c 2. ORGANIC CHEMISTRY. *Educ:* Pavia Univ, Italy, PhD(org chem), 39. *Prof Exp:* Asst chem microbiol, Fordham Univ, 40-45; res asst org chem, Yale Univ, 46-62, res assoc pharmacol, 63-82; RETIRED. *Mem:* AAAS; Am Chem Soc; Sigma Xi. *Res:* Mechanism of enzyme action; fermentation; wood-destroying fungi; chemical control of digitalis therapy; detoxication mechanisms; anti-viral and cancer chemotherapeutic agents. *Mailing Add:* 49 Spring Rock Rd Branford CT 06405

SCIARRA, DANIEL, b Sansevero, Italy, Aug 19, 18; nat US; m 46, 59, 72; c 6. MEDICINE. *Educ:* Harvard Univ, BA, 40, MD, 43; Am Bd Psychiat & Neurol, dipl, 49. *Prof Exp:* Instr neurol, Harvard Med Sch, 44-45; from instr to assoc prof, 47-67, PROF CLIN NEUROL, COL PHYSICIANS & SURGEONS, COLUMBIA UNIV, 67- *Concurrent Pos:* Attend neurologist, Neurol Inst. *Mem:* Asn Res Nerv & Ment Dis; Am Neurol Asn; Am Acad Neurol. *Res:* Epilepsy; multiple sclerosis; brain tumors. *Mailing Add:* Neurol Inst 710 W 168th St New York NY 10032

SCIARRA, JOHN J, b Brooklyn, Ny, Dec 28, 27; m 64; c 4. INDUSTRIAL PHARMACY. *Educ:* St John's Univ, NY, BS, 51; Duquesne Univ, MS, 53; Univ Md, PhD(pharm), 57. *Prof Exp:* Asst pharm, Duquesne Univ, 51-53; instr, Univ Md, 54-57; from asst prof to prof pharmaceut chem, St John's Univ, NY, 57-74, dir grad div, 66-73, asst dean col pharm & allied health professions, 72-73; prof indust pharm & dean, Brooklyn Col Pharm, Long Island Univ, 75-76, exec dean, Arnold & Marie Schwartz Col Pharm & Health Sci, 77-85, pres, Retail Drug Inst, 85-90, EMER PROF, LONG ISLAND UNIV, 90- *Concurrent Pos:* Consult, 60; mem, Nat Formulary Adv Comt & Adv Panel Pharmaceut, US Pharmacopeia Comt. *Honors & Awards:* Indust Pharm Award, 78. *Mem:* Am Pharmaceut Asn; fel Acad Pharmaceut Sci; fel Soc Cosmetic Chem (pres, 80). *Res:* Pharmacy; physical pharmacy; aerosol science and technology; particle size distribution. *Mailing Add:* Arnold & Marie Schwartz 75 DeKalb Ave University Plaza Brooklyn NY 11201

SCIARRA, JOHN J, b West Haven, Conn, Mar 4, 32; m 60; c 3. OBSTETRICS & GYNECOLOGY. *Educ:* Yale Univ, BS, 53; Columbia Univ, MD, 57, PhD(anat), 64. *Prof Exp:* Am Cancer Soc fel, 64-65; asst prof obstet & gynec, Col Physicians & Surgeons, Columbia Univ, 65-68; prof obstet & gynec & head dept, Med Sch, Univ Minn, Minneapolis, 68-73; PROF OBSTET & GYNEC & CHMN DEPT, SCH MED, NORTHWESTERN UNIV, 73- *Concurrent Pos:* NIH spec fel, 65-68; mem nat med comt, Planned Parenthood-World Pop, 71-74, 87-; mem exec bd, Int Fedn Gynec & Obstet, 86-; chmn, Am Col Obstet & Gynec, comt Int Affairs, 86- *Honors & Awards:* Carl G Hartman Award, Am Fertil Soc, 64. *Mem:* Am Col Obstet & Gynec; Am Fertil Soc; Am Col Surgeons; Am Asn Anatomists; Soc Gynec Invest. *Res:* Reproductive physiology and endocrinology. *Mailing Add:* Prentice Womens Hosp Med Ctr 333 E Superior St Chicago IL 60611

SCIARRONE, BARTLEY JOHN, b Jersey City, NJ, Nov 24, 26; m 65; c 1. PHARMACEUTICS. *Educ:* Rutgers Univ, BS, 52, MS, 55; Univ Wis, PhD(phys pharm), 60. *Prof Exp:* Instr pharmaceut chem, Col Pharm, Rutgers Univ, 54-55; asst, Univ Wis, 56-59; from asst prof to assoc prof, 58-69, asst dean, 81, assoc dean, 83, PROF PHARM, COL PHARM, RUTGERS UNIV, NEW BRUNSWICK, 69- *Mem:* Am Pharmaceut Asn. *Res:* Kinetics and mechanisms of interactions in pharmaceutical systems. *Mailing Add:* Sch Pharm Rutgers Univ New Brunswick NJ 08903

SCICLI, ALFONSO GUILLERMO, PHARMACOLOGY. *Educ:* Univ Buenos Aires, Arg, PhD(biochem), 63. *Prof Exp:* MEM SR STAFF, HYPERTENSION RES DIV, HENRY FORD HOSP, DETROIT, 73- *Mailing Add:* Hypertension Res Labs Henry Ford Hosp 2799 W Grand Blvd Detroit MI 48202

SCIDMORE, ALLAN K, b Grafton, NDak, Mar 11, 27; m 53; c 4. ELECTRICAL ENGINEERING. *Educ:* Univ NDak, BS, 51; Univ Wis, MS, 53, PhD(elec eng), 58. *Prof Exp:* Res asst chem eng, Univ Wis-Madison, 51-52, res fel elec eng, 53-54, res asst, 54-55, proj asst, 56-57, asst prof, 58-63, proj head, Digital Computer Lab, 57-59, assoc prof elec eng, 63-69, assoc dir, Univ Indust Res Prog, 76-80, assoc chmn dept, 82-86 & 89-90, PROF ELEC & COMPUTER ENG, UNIV WIS-MADISON, 69- *Concurrent Pos:* Dir, Nat Eng Consortium, 77-80. *Honors & Awards:* AT&T Found Award, ASEE; Polygon Award, outstanding instr, 77, 82, 83, 85, 88; Outstanding Counr Award, Inst Elec & Electronics Engrs; Benjamin Smith Reynolds Award. *Mem:* Inst Elec & Electronics Engrs; Am Soc Eng Educ. *Res:* Linear and digital circuit design and application. *Mailing Add:* Dept Elec & Computer Eng Univ Wis 1415 Johnson Dr Madison WI 53706

SCIDMORE, WRIGHT H(ARWOOD), b Saratoga Springs, NY, July 20, 25; m 51; c 3. OPTICS. *Educ:* Columbia Univ, BS, 50. *Prof Exp:* Staff mem, 50-58, chief, Optical Design Lab, Frankford Arsenal, US Army, 58-77; CONSULT OPTICAL DESIGN, SCIDMORE & SHEAN, 77- *Concurrent Pos:* Consult mil optical instruments, Int Sci Prog, 68-77. *Honors & Awards:* Karl Fairbanks Mem Award, Soc Photo-Optical Instrumentation Engrs, 73. *Mem:* Optical Soc Am; Sigma Xi. *Res:* Geometrical optics; lens design; military optical instruments. *Mailing Add:* Rte 8 Box 160A Brant Lake NY 12815

SCIFRES, CHARLES JOEL, b Foster, Okla, June 1, 41; m 61; c 2. RANGE SCIENCE, WEED SCIENCE. *Educ:* Okla State Univ, BS, 63, MS, 65; Univ Nebr, PhD(agron, bot), 68. *Prof Exp:* Asst res agronomist, Range Ecol, Agr Res Serv, USDA, 65-68; asst prof range mgt, 68-69, assoc prof range ecol & improvements, 69-76, PROF RANGE SCI, PROF RANGE ECOL & DEPT RANGE SCI, TEX A&M UNIV, 76- *Mem:* Weed Sci Soc Am; Soc Range Mgt; Sigma Xi. *Res:* Development of vegetation manipulation systems for rangeland resources management for maximum productivity of usable products from the resource while maintaining its ecological integrity; persistence and modes of dissipation of herbicides from the range ecosystem; life history of key range species and community dynamics following vegetation manipulation. *Mailing Add:* Dept Range Sci Tex A&M Univ College Station TX 77843

SCIFRES, DONALD RAY, b Lafayette, Ind, Sept 10, 46; m 69; c 5. PHYSICS, ELECTROOPTICS. *Educ:* Purdue Univ, Lafayette, BS, 68; Univ Ill, MS, 70, PhD(elec eng), 72. *Prof Exp:* Res scientist, res fel & mgr, Xerox Palo Alto Res Ctr, 72-83; PRES & CHIEF EXEC OFF, SPECTRA DIODE LABS, 83- *Honors & Awards:* Jack Morton Medal for Contrib to Electron Devices, Inst Elec Electronics Engrs 85. *Mem:* Am Phys Soc; fel Inst Elec & Electronics Engrs; fel Optical Soc Am. *Res:* Integrated optics and electro-optical devices; lasers. *Mailing Add:* Spectra Diode Labs 80 Rose Orchard Way San Jose CA 95134

SCIGLIANO, J MICHAEL, b Omaha, Nebr, Nov 22, 41; m 64; c 2. CHEMICAL & INDUSTRIAL ENGINEERING. *Educ:* Iowa State Univ, BS(chem eng) & BS(indust admin), 64, MS, 65; Wash Univ, DSc(chem eng), 71. *Prof Exp:* Process engr textile fibers, Monsanto Corp Eng, 65-68, sr engr polymers & petrochem, 68-72, process design supvr ABS resins, 72-75; process develop mgr nitration prod, 75-77, mgr process design chem group, Air Prod & Chem, 77-83; mgr, cent eng dept, 83-86, new ventures gen mgr, Johnson-Matthey, 87-90; CHIEF ENG, RODEL INC, 90- *Mem:* Am Inst Chem Engrs; Electrochem Soc. *Res:* Polymer solution thermodynamics; polymer characterization; polymer devolatilization; homogeneous catalysis; industrial equipment mortality; heterogeneous catalysis; fuel cells/electrochemistry. *Mailing Add:* 530 Tree Lane West Chester PA 19380

SCINTA, JAMES, b Buffalo, NY, Mar 15, 52. SPECIALTY CHEMICALS, SYNFUELS. *Educ:* Cornell Univ, BS, 73; State Univ NY at Buffalo, MS, 75, PhD(chem eng), 77. *Prof Exp:* Res engr, 77-79, sect supvr, Phillips Petrol Co, 79-86, plant mgr catalyst resources, 86-89, COMPLEX MGR, PHILLIPS 66 CO, 89- *Mem:* Am Inst Chem Eng; Am Chem Soc; Nat Soc Prof Eng. *Res:* Processes for recovering hydrocarbons from oil shale; catalyst manufacturing development; specialty chemical development. *Mailing Add:* Five Cobblestone Lane Borger TX 79007-8406

SCIORE, EDWARD, b July 13, 55. DATABASE SYSTEMS. *Educ:* Yale Univ, BA, 76; Princeton Univ, PhD(comput sci), 80. *Prof Exp:* ASST PROF COMPUTER SCI, STATE UNIV NY, STONY BROOK, 80- *Mem:* Asn Computer Mach. *Res:* Database systems, especially the semantics of data; database design methodologies; improved data description languages. *Mailing Add:* State Univ NY Stony Brook NY 11794-0001

SCIPIO, L(OUIS) ALBERT, II, b Juarez, Mex, Aug 22, 22; US citizen; m 42; c 3. SPACE SCIENCES. *Educ:* Tuskegee Inst, BS, 43; Univ Minn, BCE, 48, MS, 50, PhD, 54. *Prof Exp:* Instr archit & mech drawing, Tuskegee Inst, 46; struct engr, Long & Thorshov, 48-50; lectr aeronaut eng, Univ Minn, Minneapolis, 52-61; assoc prof mech, Howard Univ, 61-62; prof phys sci, Univ PR, 62-63; aerospace eng, Univ Pittsburgh, 63-67; prof aerospace eng & dir grad studies eng & archit, 67-70, prof space sci, 70-88, Emer Distinguished Univ Prof, Howard Univ, 88-; RETIRED. *Concurrent Pos:* Fulbright prof, Cairo Univ, 55-56; consult, NASA Knowledge Availability Systs Ctr, Univ Pittsburgh, John Wiley & Sons, Inc & Winzen Res Inc; mem, Army Sci Bd, 78-81; bd vis, Air Force Inst Technol, 79-82. *Honors & Awards:* Steinman Award, 58. *Mem:* AAAS; Soc Natural Philos; Am Phys Soc; Am Inst Aeronaut & Astronaut; Int Asn Bridge & Struct Engrs; fel Int Biog Asn. *Res:* Continuum mechanics; aerothermoelasticity; shell structures; viscoelasticity; US military history. *Mailing Add:* 12511 Montclair Dr Silver Spring MD 20904

SCISSON, SIDNEY EUGENE, engineering; deceased, see previous edition for last biography

SCITOVSKY, ANNE A, b Ludwigshafen, West Ger, Apr 17, 15; US citizen; m 42; c 1. HEALTH ECONOMICS. *Educ:* Barnard Col, BA, 37; Columbia Univ, MA, 41. *Prof Exp:* Economist, Bur Res & Statist, Social Security Bd, Washington, DC, 44-46; sr res assoc, 63-73, CHIEF, HEALTH ECON DEPT, RES INST, PALO ALTO MED FOUND, 73-; LECTR, INST HEALTH POLICY STUDIES, SCH MED, UNIV CALIF, SAN

FRANCISCO, 75- *Concurrent Pos:* Mem, Publ Adv Bd, Nat Ctr Health Serv Res & Develop, 69-71, consult 75-; mem, Comt Planning Study Costs Environ-related Health Effects, Inst Med, 79-80, President's Comn Study Ethical Probs Med & Biomed & Behav Res, 79-82, Adv Panel Life-Sustaining Technol & Elderly, Off Technol Assessment, 85-86, Coun Health Care Technol, Inst Med, 86-, Health Adv Comt, 87- & AIDS Adv Comt, 90- *Mem:* Inst Med-Nat Acad Sci. *Res:* Empirical studies of the medical care costs of specific illnesses, the effects of changing medical technology on medical care costs, the effect of coinsurance on the demand for physician services, the demand for health care services under prepaid and fee-for-service group practice, and medical care expenses in the last year of life; author of numerous technical publications. *Mailing Add:* 860 Bryant St Palo Alto CA 94301

SCIUCHETTI, LEO A, pharmacognosy; deceased, see previous edition for last biography

SCIULLI, FRANK J, b Philadelphia, Pa, Aug 22, 38; m 65; c 2. EXPERIMENTAL PHYSICS, ELEMENTARY PARTICLE PHYSICS. *Educ:* Univ Pa, AB, 60, MS, 61, PhD(K-meson decay), 65. *Prof Exp:* Res assoc particle physics, Univ Pa, 65-66; res fel particle physics, Calif Inst Technol, 66-68, from asst prof to prof, 69-81; PROF PHYSICS, COLUMBIA UNIV, 81- *Mem:* AAAS; fel, Am Phys Soc; Sigma Xi. *Res:* Weak interactions of elementary particles, particularly K-meson decays and neutrino interactions. *Mailing Add:* Four Deep Hollow Cose Box 38 Irvington NY 10533

SCIULLI, PAUL WILLIAM, b Pittsburgh, Pa, Aug 14, 47; m 73. BIOLOGICAL ANTHROPOLOGY, DENTAL ANTHROPOLOGY. *Educ:* Univ Pittsburgh, BA, 69, PhD(anthrop), 74. *Prof Exp:* Instr phys anthrop, Univ Pittsburgh, 73-74; vis asst prof, 74-76, ASST PROF PHYS ANTHROP, OHIO STATE UNIV, 76- *Concurrent Pos:* Consult breeding prog, SMI Chinchilla Farm, 75-78. *Honors & Awards:* res award, Col Soc & Behav Sci, Ohi State Univ, 75 & 79. *Mem:* Sigma Xi; Am Asn Phys Anthropologists; Am Asn Human Genetics. *Res:* Biocultural adaptations of prehistoric eastern woodland ameridians and genetic interactions in the production of coat color in the chinchilla. *Mailing Add:* Dept Anthrop Ohio State Univ Columbus OH 43210

SCIUMBATO, GABRIEL LON, b La Junta, Colo, Sept 12, 45; m 67; c 1. PLANT PATHOLOGY. *Educ:* Univ Eastern NMex, BA, 68; La State Univ, MS, 69, PhD(plant path), 73. *Prof Exp:* Technician plant path, La State Univ, 72-73; fel, Tex A&M Univ, 73-75; agronomist res & develop, US Borax Corp, 75-76; ASST PLANT PATHOLOGIST, MISS STATE UNIV, 76- *Mem:* Weed Sci Soc; Am Phytopath Soc. *Res:* Control of foliar cotton, soybean and rice diseases. *Mailing Add:* Delta Res & Exten Serv Old Leland Rd PO Box 097 Stoneville MS 38776

SCLAR, CHARLES BERTRAM, b Newark, NJ, Mar 16, 25; m 46; c 2. PETROLOGY, ORE DEPOSITS. *Educ:* City Col New York, BS, 46; Yale Univ, MS, 48, PhD(geol), 51. *Prof Exp:* Instr geol, Ohio State Univ, 49-51; res geologist, Battelle Mem Inst, 51-53, prin geologist, 53-57, asst consult, 57-62, res assoc, 62-65, assoc chief chem physics div & dir high-pressure res lab, 65-68; chmn dept, 76-85, prof geol, 68-90, EMER PROF GEOL, LEHIGH UNIV, 90- *Honors & Awards:* Ward Medalist in Geol, 46. *Mem:* Fel Geol Soc Am; Soc Econ Geol; Geochem Soc; fel Mineral Soc Am; Am Geophys Union. *Res:* Petrology, geochemistry, mineralogy, high-pressure synthesis, phase equilibria, and phase transformations; igneous and metamorphic petrology; shock metamorphism; shock effects in lunar minerals; mineral deposits and industrial applications of mineralogy. *Mailing Add:* Dept Geol Sci Williams Hall 31 Lehigh Univ Bethlehem PA 18015

SCLAR, NATHAN, b New York, NY, Mar 22, 20; m 51; c 4. IR DETECTOR & ARRAY, CRYOGENIC ELECTRONICS. *Educ:* NY Univ, BS, 48; Syracuse Univ, MS, 52 & PhD(physics), 67. *Prof Exp:* Res physicist, Naval Res Lab, 52-56; res engr, Dumont Labs, 56-58 & Avion Electronics, 58-59, staff engr, Dumont Labs, 62-64; mgr res, Nuclear Corp Am, 59-62; proj engr, sr scientist, mgr & prin scientist, Rockwell Int, 64-85; CONSULT, AEROJET ELECTRO SYSTS, 85- *Mem:* Fel Am Phys Soc; fel Inst Elec & Electronics Engrs; AAAS. *Res:* Solid state physics; design and development of electron devices and systems; detectors for nuclear, optical and IR radiation and the electronics to condition and to read out arrays of such detectors. *Mailing Add:* 23634 Decorah Rd Diamond Bar CA 91765

SCLATER, JOHN GEORGE, b Edinburgh, Scotland, June 17, 40; m 85. OCEANOGRAPHY, GEOPHYSICS. *Educ:* Univ Edinburgh, BSc, 62; Cambridge Univ, PhD(geophys), 66. *Prof Exp:* NSF res grant, Scripps Inst Oceanog, Univ Calif, San Diego, 65-67, asst geophys res, 67-72, lectr geol, Univ, 71-72; from assoc prof oceanog & marine geophys to prof marine geophys, Mass Inst Technol, 72-83; PROF GEOL, UNIV TEX, AUSTIN, 83- *Concurrent Pos:* Dir, Joint Prog Oceanog Woods Hole Oceanog Inst & Mass Inst Technol, 81-83; assoc dir, Inst Geophys, Univ Tex, Austin, 83-; Shell Co Distinguished Prof, 83-88. *Honors & Awards:* Rosenstiel Award in Oceanography, 79; Bucher Medal, Am Geophys Soc, 85. *Mem:* Nat Acad Sci; Fel Am Geophys Union; fel Geol Soc Am; Am Asn Prof Geologists; fel Royal Soc London. *Res:* Application of the theory of plate tectonics to the ocean environment. *Mailing Add:* Dept Geol Sci Univ Tex Austin TX 78712

SCLOVE, STANLEY LOUIS, b Charleston, WVa, Nov 25, 40; m 62, 90; c 2. STATISTICS. *Educ:* Dartmouth Col, AB, 62; Columbia Univ, PhD(math statist), 67. *Prof Exp:* Res assoc statist, Stanford Univ, 66-68; asst prof, Carnegie-Mellon Univ, 68-72; from assoc prof to prof math, 81-82, PROF INFO & DECISION SCI, UNIV ILL, CHICAGO, 82- *Concurrent Pos:* Consult, Alcoa Res Labs, Pa, 69; vis asst prof statist & educ, Stanford Univ, 71-72; vis assoc prof indust eng & mgt sci, Northwestern Univ, 80-81; expert witness, fed ct, Chicago, 85. *Mem:* Am Statist Asn; Inst Math Statist; Opers Res Soc Am; Classification Soc NAm. *Res:* Multivariate statistical analysis; regression analysis; time series analysis; fingerprint probabilities. *Mailing Add:* Dept Info & Decision Sci M/C 294 Col Bus Admin Univ Ill at Chicago Box 4348 Chicago IL 60680-4348

SCOBEY, ELLIS HURLBUT, b Kelso, Wash, Sept 15, 11; m 35; c 4. GEOLOGY. *Educ:* Cornell Col, AB, 33; Univ Iowa, MS, 35, PhD(geol), 38. *Prof Exp:* Asst geol, Univ Iowa, 35-38; asst geologist, Gulf Oil Corp, Ind, 38-44; geologist, Bay Petrol Corp, 44-47; dist geologist, Southern Minerals Corp, Tex, 47-51; chief geologist, Guy Mabee Drilling Co, 51-65 & Mabee Petrol Corp, 65-75; VPRES & TREAS, MCFARLAND & SCOBEY, INC, 75- *Mem:* Soc Econ Paleontologists & Mineralogists; Am Asn Petrol Geologists. *Res:* Sedimentation; stratigraphy and petroleum geology in Illinois Basin; petroleum geology in West Texas. *Mailing Add:* Two Chatham Ct Midland TX 79705

SCOBEY, ROBERT P, b Providence, RI, Sept 10, 38; m 60; c 3. PHYSIOLOGY, NEUROPHYSIOLOGY. *Educ:* Worcester Polytech Inst, BSEE, 60; Clark Univ, MA, 62; Johns Hopkins Univ, PhD(physiol), 68. *Prof Exp:* PROF NEUROL, SCH MED, UNIV CALIF, DAVIS, 77- *Res:* Neurophysiology of central nervous system; vision. *Mailing Add:* Dept Neurol/Physiol Univ Calif Sch Med Davis CA 95616

SCOBY, DONALD RAY, b Sabetha, Kans, Mar 18, 31; c 1. ENVIRONMENTAL BIOLOGY. *Educ:* Kans State Univ, BS, 57; Nebr State Teachers Col, MS, 60; NDak State Univ, PhD(bot, ecol), 68. *Prof Exp:* Instr & prin high sch, Kans, 57-60; instr high sch, Colo, 61-65, chmn sci dept, 65-66; instr gen biol & sci methods, NDak State Univ, 67-68, from asst prof to assoc prof biol, 68-78, prof, 78-89, EMER PROF BIOL & SCI EDUC, NDAK STATE UNIV, 89-; PROF BIOL, MOORHEAD STATE UNIV, 89- *Mem:* Asn Educ Sci Teachers; Ecol Soc Am; Sigma Xi; Nat Asn Biol Teachers. *Res:* Environmental education procedures; practices for environmental self sufficiency; application of ecological and biological principles to organic farming methods; science education. *Mailing Add:* Dept Bot Biol NDak State Univ Fargo ND 58105

SCOCCA, JOHN JOSEPH, b Philadelphia, Pa, Mar 23, 40; m 66; c 2. BIOCHEMISTRY, MOLECULAR BIOLOGY. *Educ:* Johns Hopkins Univ, BA, 61, PhD(biochem), 66. *Prof Exp:* NIH fel, McCollum-Pratt Inst, 66-68, asst prof, Univ, 68-72, assoc prof, 72-86, PROF BIOCHEM, SCH HYG & PUB HEALTH, JOHNS HOPKINS UNIV, 86- *Mem:* Am Soc Biol Chemists; Am Soc Microbiol. *Res:* Mechanism of genetic exchange in gram-negative bacteria; site specific recombination; DNA recognition mechanisms. *Mailing Add:* Sch Hyg & Pub Health Johns Hopkins Univ Baltimore MD 21205

SCOFIELD, DILLON FOSTER, b Norfolk, Va, Aug 10, 43; m 71. PHYSICS, MATERIALS SCIENCE. *Educ:* George Washington Univ, BS, 65, MS, 66, PhD(solid state physics), 69. *Prof Exp:* Independent systs analyst, 60-67; res analyst solid state & high energy physics, Foreign Technol Div, Air Force Systs Command, 67-70, res physicist solid state physics, Aerospace Res Labs, 70-71; Nat Res Coun assoc, Wright-Patterson AFB, 72-74; MEM STAFF PHOTO PROD DEPT, EXP STA, E I DU PONT DE NEMOURS & CO, INC, 74- *Concurrent Pos:* Pres, Appl Sci, Inc, 77- *Mem:* AAAS; Am Phys Soc; Inst Elec & Electronics Engrs. *Res:* Physical theory of photographic process, nonlinear mechanics, thin film fluid flow, mechanical and magnetic composites; high level languages for minicomputers; computer architecture; automated laboratory equipment; microprocessor system design. *Mailing Add:* 128 Country Flower Rd Newark DE 19711

SCOFIELD, GORDON L(LOYD), b Huron, SDak, Sept 29, 25; m 47; c 2. MECHANICAL ENGINEERING. *Educ:* Purdue Univ, BS, 46; Univ Mo, MS, 49; Univ Okla, PhD, 68. *Prof Exp:* Instr mech eng, SDak State Col, 46-47; asst, Univ Mo, Rolla, 47-48, from instr to prof, 48-69; PROF MECH ENG & ENG MECH & HEAD DEPT, MICH TECHNOL UNIV, 69- *Concurrent Pos:* Consult, Naval Ord Test Sta, Calif. *Mem:* Am Soc Mech Engrs; Soc Automotive Engrs (pres, 77); Am Soc Eng Educ; Am Inst Aeronaut & Astronaut. *Res:* Heat transfer and energy conversion. *Mailing Add:* SDSM&T Found 501 E St Joseph Rapid City SD 57701

SCOFIELD, HERBERT TEMPLE, b West Lafayette, Ind, May 16, 09; m 60. PLANT PHYSIOLOGY. *Educ:* Cornell Univ, AB, 30, PhD(plant physiol), 37. *Prof Exp:* Asst plant physiol, Cornell Univ, 31-36, instr, 36-37; instr bot, Ohio State Univ, 37-40, asst prof, 40-43; from asst prof to prof, 46-72, head dept, 50-63, EMER PROF BOT, NC STATE UNIV, 72- *Concurrent Pos:* Agent, Bur Entom & Plant Quarantine, USDA, 39; consult acad affairs to rector, Agrarian Univ, Peru, 59, 61-62, 63-65, 67-69, 70-72. *Res:* Tobacco physiology. *Mailing Add:* 501 E Whitaker Mill Rd No 203-C Raleigh NC 27608

SCOFIELD, JAMES HOWARD, b Gary, Ind, Oct 10, 33; m 59; c 3. THEORETICAL PHYSICS. *Educ:* Ind Univ, BS, 55, MS, 57, PhD(theoret physics), 60. *Prof Exp:* Res assoc theoret physics, Stanford Univ, 60-62; PHYSICIST, UNIV CALIF, LAWRENCE LIVERMORE LAB, 62- *Mem:* Fel Am Phys Soc. *Res:* Atomic structure calculations; interaction of electrons and x-rays with atoms. *Mailing Add:* Lawrence Livermore Lab Livermore CA 94551

SCOGGIN, JAMES F, JR, b Laurel, Miss, Aug 3, 21; m 48; c 3. PHYSICS, ELECTRICAL ENGINEERING. *Educ:* Miss State Univ, BS, 41; US Mil Acad, BS, 44; Johns Hopkins Univ, MA, 51; Univ Va, PhD(physics), 57; Indust Col Armed Forces, grad(nat security), 62. *Prof Exp:* Physicist, Signal Corps, US Army, 51-53 & 54-55 & Defense Atomic Support Agency, 58-61, engr, Adv Res Proj Agency, 62-65, proj mgr ground surveillance systs, Electronics Command, 66-68; from assoc prof to prof, 68-83, EMER PROF ELEC ENG, THE CITADEL, 83- *Honors & Awards:* Centennial Medal, Inst Elec & Electronic Engrs. *Mem:* Sr mem Inst Elec & Electronic Engrs; Am Nuclear Soc; Precision Measurements Asn; fel Radio Club Am. *Res:* Micrometeorology; cellular convection; tropical communications; surveillance systems; artillery sound ranging; nuclear energy; precision electrical measurements. *Mailing Add:* 1310 Pembrooke Dr Charleston SC 29407

SCOGGINS, JAMES R, b Aragon, Ga, Sept 22, 31; m 52; c 2. METEOROLOGY. *Educ:* Berry Col, AB, 52; Pa State Univ, BS, 54, MS, 60, PhD(meteorol), 66. *Prof Exp:* Mathematician, Lockheed Aircraft Corp, Ga, 57-58, meteorologist, Nuclear Lab, 59-60; meteorologist, Aerospace Environ Div, NASA Marshall Space Flight Ctr, 60-67; assoc prof, Col Geosci, Tex A&M Univ, 67-69, asst dean opers, 71-73, dir, Ctr Appl Geosci, 73-75, assoc dean res, 75-77, head dept meteorol, 80-90, PROF METEOROL, COL GEOSCI, TEX A&M UNIV, 69-, DIR, COOP INST APPL METEOROL STUDIES, 89- *Concurrent Pos:* Consult, Tex Dept Water Resources, 73- *Mem:* Fel Am Meteorol Soc. *Res:* Mesometeorology; applied meteorology. *Mailing Add:* Dept Meteorol Tex A&M Univ College Station TX 77843-3146

SCOGIN, RON LYNN, b Corpus Christi, Tex, Oct 6, 41; m 67; c 2. PLANT CHEMISTRY. *Educ:* Univ Tex, Austin, BA, 64, PhD(bot), 68. *Prof Exp:* Asst prof bot, Ohio Univ, 68-71; asst prof, 72-76, chmn dept, 79-85, ASSOC PROF BOT, CLAREMONT GRAD SCH, 76- *Concurrent Pos:* NSF fel, Univ Durham, Eng, 70-71. *Res:* Biochemical systematics and evolution; biochemical systematics of angiosperms. *Mailing Add:* Rancho Santa Ana Botanic Garden 1500 N College Claremont CA 91711

SCOLA, DANIEL ANTHONY, b Worcester, Mass, July 11, 29; m 53; c 4. CHEMISTRY, MATERIALS SCIENCE. *Educ:* Clark Univ, BA, 52; Williams Col, MA, 54; Univ Conn, PhD(org & phys chem), 59. *Prof Exp:* Lab asst chem, Williams Col, 52-54; res chemist, Durez Plastics Co, NY, 55; lab asst, Univ Conn, 55-57, asst instr, 57-58; sr res chemist, Monsanto Res Corp, Mass, 58-64, res group leader, 64-65; sr res scientist, United Aircraft Res Labs, Conn, 65-66; sr res engr, Res Labs, Norton Co, Mass, 66-67; sr res scientist & supvr org mat res, 66-74, SR MAT SCIENTIST, UNITED TECHNOL RES CTR, EAST HARTFORD, 74-; ADJ PROF POLYMER CHEM, UNIV CONN, 86- *Concurrent Pos:* Adj fac mem, Univ Hartford, 69-; dir & chmn relig educ comt, St Augustine Church, Glastonbury, 69-78; vis prof chem, Trinity Col, Hartford, Conn, 88-; founder, advan polymer compositures, Div Soc Plastics Engrs, 87, chmn, 90-91; adv bd mem, Polymer Sci Prog, Univ Conn, 88- *Honors & Awards:* Outstanding Achievement Award, Soc Plastics Engrs, 90. *Mem:* Am Chem Soc; Soc Plastics Engrs; Soc Advan Mat & Process Eng. *Res:* Materials research, especially fiber reinforced polymeric materials; advanced polymer composites, interface effects in composite materials; influence of environment on mechanical and thermal properties of composites; surface effects in adhesive bonding; synthesis of moisture resistant laminating resins, coatings and adhesives; synthesis and evaluation of high temperature polymeric materials for advanced composites and adhesives. *Mailing Add:* 83 Stone Post Rd Glastonbury CT 06033

SCOLARO, REGINALD JOSEPH, b Tampa, Fla, Oct 19, 39; m 66; c 2. INVERTEBRATE PALEONTOLOGY, PALEOECOLOGY. *Educ:* Univ Fla, BA, 60, BS, 62, MS, 64; Tulane Univ, PhD(paleont), 68. *Prof Exp:* Intern, Smithsonian Inst, 67-68; res fel, Univ Ga, 68-69; assoc prof, 69-79, chmn dept, 71-79, prof geol, Radford Col, 79; proj geologist, Gulf Oil Explor & Prod Co, 79-83; sr geologist I, 83-85, GEOLOGIST II, STANDARD OIL PETROL CO, 85- *Mem:* Geol Soc Am; Paleont Soc; Paleont Res Inst; Am Asn Petrol Geologists. *Res:* Taxonomy and paleoecology of Tertiary and recent Cheilostome Bryozoa. *Mailing Add:* Saudi Aramco-Dhahran Saudia Arabia Rte 1 Box 113 Wirtz VA 24184

SCOLES, GRAHAM JOHN, b Eng; Can citizen. PLANT BREEDING, CYTOGENETICS. *Educ:* Univ Reading, UK, BSc, 73; Univ Manitoba, MSc, 75, PhD(plant sci), 79. *Prof Exp:* ASSOC PROF CYTOGENETICS, UNIV SASK, CAN, 79- *Mem:* Can Genetics Soc; Am Soc Agron. *Res:* Interspecific hybridization and cross-compatibility in crop species. *Mailing Add:* Crop Sci Dept Univ Sask Saskatoon SK S7N 0W0 Can

SCOLLARD, DAVID MICHAEL, b NDak, July 26, 47; m 71; c 2. IMMUNOLOGY. *Educ:* St Olaf Col, BA, 69; Univ NDak, BS, 71; Univ Chicago, MD & PhD(path & immunol), 75. *Prof Exp:* Lectr path, Univ Hong Kong, 76-81; asst prof path, Univ Ill, 81-84; asst prof, 84-89, ASSOC PROF PATH, UNIV HAWAII, 89- *Concurrent Pos:* Asst prof path, Univ Chicago, 80; field dir, Chiang Mai-Ill Leprosy Res Proj, 81-84. *Mem:* Am Soc Leukocyte Biol; Am Soc Microbiol; Int Leprosy Asn; US & Can Acad Path. *Res:* Pathology and immunopathology of infection, with emphasis on mycobacterial diseases. *Mailing Add:* Dept Path Univ Hawaii Manoa HI 96822

SCOLMAN, THEODORE THOMAS, b Duluth, Minn, Oct 27, 26; m 55; c 2. NUCLEAR PHYSICS. *Educ:* Beloit Col, BS, 50; Univ Minn, PhD(physics), 56. *Prof Exp:* Staff mem physics, Weapon Div, 56-62, test div, 62-66, assoc group leader, 66-69, asst div leader physics, Test Div, 72-78, prog mgr, field testing, 78-86, TEST DIV LEADER & PROG DIR, LOS ALAMOS NAT LAB, UNIV CALIF, 86- *Mem:* Am Phys Soc. *Res:* Nuclear weapons test and development. *Mailing Add:* 198 El Viento Los Alamos NM 87545

SCOLNICK, EDWARD M, b Boston, Mass, Aug 8, 40; m 65; c 3. ONCOLOGY. *Educ:* Harvard Univ, AB, 61, MD, 65. *Prof Exp:* Intern internal med, Mass Gen Hosp, 65-66, asst residency, 66-67; res assoc USPHS Lab Biochem Genetics, Nat Heart Inst, NIH, 67-69, sr staff fel, 69-70 & viral lymphoma br, 70-71, spec adv, spec virus cancer prog, 73-78, mem, coord comt, 75-78, head molecular virol, virol sect & chief lab, tumor-virus genetics, 75-82; exec dir basic res, virus & cell biol res, Merck Sharp & Dohme Res Labs, 82-85, vpres, 83-84, sr vpres, 84-85, PRES, MERCK SHARP & DOHME RES LABS 85-, SR VPRES, MERCK & CO, 91- *Concurrent Pos:* Ed-in-chief, J Virol, 82-85; adj prof microbiol, Assoc Fac Sch Med, Univ Pa, 83-86; mem study sect, Prog Excellence Molecular Biol, Nat Heart, Lung & Blood Inst, 88. *Honors & Awards:* Arthur S Flemming Award, 76; Super Award, USPHS, 78; Eli Lilly Award, 80; Indust Res Inst Medal Award, 89. *Mem:* Nat Acad Sci; Am Soc Microbiologists; Am Soc Biol Chemists. *Res:* Author or co-author of over 170 publications. *Mailing Add:* Merck & Co PO Box 2000 Bldg 80K Rahway NJ 07065

SCOMMEGNA, ANTONIO, b Barletta, Italy, Aug 26, 31; US citizen; m 58; c 3. REPRODUCTIVE PHYSIOLOGY. *Educ:* Univ Bari, MD, 53; Am Bd Obstet & Gynec, dipl & cert reprod endocrinol, 74. *Prof Exp:* Intern, New Eng Hosp, Boston, Mass, 54-55; resident obstet & gynec, 56-59, fel, Dept Human Reproduction, 60-61, res assoc, 61, dir Sect Gynec & Endocrinol, 65-81, ATTEND PHYSICIAN, DEPT OBSTET & GYNEC, MICHAEL REESE HOSP & MED CTR, 61-, CHMN DEPT, 69-; PROF OBSTET & GYNEC, PRITZKER SCH MED, UNIV CHICAGO, 69- *Concurrent Pos:* Fullbright fel, 54-55; fel, Steroid Training Prog, Worcester Found, Exp Biol & Clark Univ, 64-65; assoc prof, Chicago Med Sch, 65-69; mem ad hoc review group contract proposals, Contraceptive Develop Br, Ctr Pop Res Nat Inst Child Health & Human Develop, 72-77; task force Ovum Transport & Inplantation, World Health Orgn, 72-73; examr, Am Bd Obstet & Gynec, 76-84, 86-; consult, IUD Core Adv Comt, Int Fertility Res Prog, 77-81; mem biomed adv comt, Pop Res Ctr, 78-85; mem scientific adv comt, Prog Appl Res Fertility Regulation, 83-87; mem, Obstet-Gynec Devices Panel, Ctr for Devices & Radiol Health, Food & Drug Admin, 85- *Honors & Awards:* Franklin Medal, Philadelphia Consortium for Pop Res, 74. *Mem:* Fel Am Col Obstet & Gynec; AMA; Am Fertil Soc; Soc Study Reproduction; hon mem Italian Soc Obstet & Gynec; Soc Gynec Invest; Am Gynec Obstet Soc. *Res:* Infertility and fertility; steroid chemistry; human reproduction; intrauterine contraceptive medication; hormones. *Mailing Add:* Dept Obstet & Gynec Michael Reese Hosp Lake Dr & 31st St Chicago IL 60616

SCOPP, IRWIN WALTER, b New York, NY, Dec 8, 09; m 42; c 1. ORAL MEDICINE, PERIODONTICS. *Educ:* City Col NY, BS, 30; Columbia Univ, DDS, 34; State Univ NY, cert pedag, 35. *Prof Exp:* Asst chief dent serv, Vet Admin Hosp, North Little Rock Ark, 44-49, chief dent serv, Vet Admin Regional Off, 49-54, CHIEF DENT SERV, VET ADMIN HOSP, NEW YORK, 54-; PROF PERIODONT, DIR DEPT MED & DIR CONTINUING EDUC, COL DENT, NY UNIV, 54- *Concurrent Pos:* Consult, Goldwater Hosp, New York, 66- *Honors & Awards:* Hershfeld Medal, Northeastern Soc Periodont; Samuel Charles Miller Mem Award, Am Acad Oral Med, 71. *Mem:* Am Acad Periodont; Am Acad Oral Med; fel Am Col Dent; fel Am Pub Health Asn; Sigma Xi (pres, 64-66). *Res:* Dentistry; dental medicine. *Mailing Add:* 110 Bleecker St New York NY 10012

SCORA, RAINER WALTER, b Mokre, Silesia, Poland, Dec 5, 28; US citizen; m 71; c 3. BOTANY. *Educ:* DePaul Univ, BS, 55; Univ Mich, MS, 58, PhD(bot), 64. *Prof Exp:* Master prep sch, Mich, 58-60; from asst botanist to assoc botanist, 64-75, PROF BOT, UNIV CALIF, RIVERSIDE, 75- *Concurrent Pos:* NSF res grant, 65-71. *Honors & Awards:* Cooley Award, Am Inst Biol Sci, 68. *Mem:* Phytochem Soc NAm; Bot Soc Am; Int Asn Bot Gardens; Int Orgn Biosyst; Int Soc Plant Taxon; Sigma Xi. *Res:* Biosystematics of the genus Monarda Persea of the subfamily Aurantiodeae; isozymes, phenolic, lipid and essential oil constituents of subfamily Aurantiodeae; history, origin and evolution of Citrus of Persea; phytochemistry of Parthemium, Asparagus, Chrysothamnus and Portulaca. *Mailing Add:* Dept Bot & Plant Sci Univ Calif Riverside CA 92521

SCORDELIS, ALEXANDER COSTICAS, b San Francisco, Calif, Sept 27, 23; m 48; c 2. STRUCTURAL ENGINEERING. *Educ:* Univ Calif, Berkeley, BS, 48; Mass Inst Technol, ScM, 49. *Prof Exp:* Struct designer, Pac Gas & Elec Co, 48; from instr to assoc prof, Univ Calif, Berkeley, 49-61, asst dean, Col Eng, 62-65, vchmn, Div Struct Eng & Struct Mech, 70-73, prof civil & struct eng, 62-90, EMER BYRON C & ELVIRA E NISHKIAN PROF STRUCT ENG, UNIV CALIF, BERKELEY, 90- *Concurrent Pos:* Engr, Bechtel Corp, San Francisco, 51-54; consult, eng firms & govt agencies. *Honors & Awards:* Moissieff Award, Am Soc Civil Engrs, 76 & 81, Howard Award, 89; Western Elec Award, Am Soc Eng Educ, 78; Axion Award, Hellenic Am Prof Soc, 79. *Mem:* Nat Acad Eng; hon fel Am Soc Civil Engrs; fel Am Concrete Inst. *Res:* Analysis and design of complex structural systems, especially reinforced and prestressed concrete shell and bridge structures. *Mailing Add:* Univ Calif 729 Davis Hall Berkeley CA 94720

SCORDILIS, STYLIANOS PANAGIOTIS, b Bridgeport, Conn, Nov 13, 48; m 83; c 2. PROTEIN BIOCHEMISTRY, CELL PHYSIOLOGY. *Educ:* Princeton Univ, AB, 69; State Univ NY, PhD(biophys), 75. *Prof Exp:* Fel, Muscular Dystrophy Asn Am, 75-78; asst prof, 78-84, ASSOC PROF BIOL, SMITH COL, 84- *Mem:* Am Soc Cell Biol; Biophys Soc; Am Chem Soc; NY Acad Sci. *Res:* Regulation of and energy transduction in muscle and non-muscle motility; contractile proteins and their interactions; role of fixed charge potentials in biological systems. *Mailing Add:* Dept Biol Sci Smith Col Northampton MA 01063

SCORNIK, JUAN CARLOS, IMMUNOLOGY. *Educ:* Univ Fla, MD. *Prof Exp:* ASSOC PROF IMMUNOL, COL MED, UNIV FLA. *Mailing Add:* Dept Path Col Med Univ Fla J Hillis Miller Ctr Gainesville FL 32610

SCORSONE, FRANCESCO G, b Palermo, Italy, June 12, 20; US citizen; m 45; c 2. MATHEMATICS. *Educ:* Univ Palermo, PhD(math), 45. *Prof Exp:* From asst prof to assoc prof math, Hartwick Col, 61-65; assoc prof, 65-66, PROF MATH, EASTERN KY UNIV, 66- *Mem:* Sigma Xi. *Res:* Differential equations; mathematical analysis. *Mailing Add:* 730 W Short St Lexington KY 40508

SCOTFORD, DAVID MATTESON, b Cleveland, Ohio, Jan 7, 21; m 47; c 4. PETROLOGY, STRUCTURAL GEOLOGY. *Educ:* Dartmouth Col, AB, 46; Johns Hopkins Univ, PhD(geol), 50. *Prof Exp:* Asst prof, 50-60, chmn dept, 60-79, PROF GEOL, MIAMI UNIV, 60- *Concurrent Pos:* Fulbright lectr grant, Turkey, 64-65; NSF travel grant, Brazil, 66; consult, Turkish Geol Surv, 65 & Shell Develop Co; fel, Univ Liverpool, 80-81. *Mem:* Fel Geol Soc Am; Mineral Soc Am. *Res:* Metamorphic petrology; structural geology; stratigraphy; shale mineralogy and petrography; feldspar structural and geochemistry studies related to geothermometry. *Mailing Add:* Dept Geol Miami Univ Oxford OH 45056

SCOTT, ALASTAIR IAN, b Glasgow, Scotland, Apr 10, 28; m 50; c 2. ORGANIC CHEMISTRY. *Educ:* Glasgow Univ, BSc, 49, PhD(chem), 52, DSc(chem), 63; DSc, Univ Coimbra, 90. *Hon Degrees:* MA, Yale Univ, 68. *Prof Exp:* Fel chem, Ohio State Univ, 52-53 & Birrbeck Col, Univ London, 54-56; fel, Glasgow Univ, 56-57, lectr, 57-62; prof, Univ BC, 62-65 & Univ Sussex, 65-68; prof chem, Yale Univ, 68-77; prof chem, Tex A&M Univ, 77-80; prof org chem, Univ Edinburgh, 80-82; DAVIDSON PROF SCI, TEX A&M UNIV, 82- *Concurrent Pos:* Roche Found fel, 63-65. *Honors & Awards:* Corday Morgan Medal, 64; E Guenther Award, Am Chem Soc, 75. *Mem:* Fel Royal Soc; Am Chem Soc; Chem Soc; Brit Biochem Soc; Swiss Chem Soc. *Res:* Chemistry and biochemistry of biologically significant molecules. *Mailing Add:* Dept Chem Tex A&M Univ College Station TX 77843

SCOTT, ALBERT DUNCAN, b Cupar, Sask, Nov 1, 21; m 47; c 3. SOIL CHEMISTRY. *Educ:* Univ Sask, BSA, 43; Cornell Univ, PhD(soils), 49. *Prof Exp:* From asst prof to assoc prof, 50-58, prof soils, 59-90, EMER PROF SOILS, IOWA STATE UNIV, 91- *Concurrent Pos:* Tech expert soil chem, Food & Agr Orgn, UN, Pakistan, 61-62; vis scientist, Commonwealth Sci & Indust Res Orgn, Adelaide, Australia, 68-69. *Mem:* Clay Minerals Soc; fel Am Soc Agron; fel Soil Sci Soc Am; Int Soc Soil Sci; Int Asn Study Clays. *Res:* Forms, reactions and plant availability of potassium in soils and minerals; clay mineralogy; mica weathering. *Mailing Add:* Dept Agron Iowa State Univ Ames IA 50011

SCOTT, ALLEN, b Louisa, Ky, Jan 22, 48; m 77. ORGANIC CHEMISTRY. *Educ:* Ohio Univ, BS, 69; Purdue Univ, PhD(org chem), 75. *Prof Exp:* Fel, Mass Inst Technol, 75-77; SCIENTIST ORG CHEM, UPJOHN CO, 77- *Concurrent Pos:* NIH Res Serv fel, 75-77. *Mem:* Am Chem Soc. *Res:* Synthesis of pharmaceutical compounds; development of synthetic methods and tools. *Mailing Add:* 5285 Buning Tree St Kalamazoo MI 49002

SCOTT, ALWYN C, b Worcester, Mass, Dec 25, 31; m 58, 81; c 3. SOLID STATE ELECTRONICS, APPLIED MATHEMATICS. *Educ:* Mass Inst Technol, BS, 52, MS, 58, ScD(elec eng), 61. *Prof Exp:* Engr, Sylvania Physics Lab, 52-54; from instr to assoc prof elec eng, Mass Inst Technol, 61-65; prof comput & elec eng, Univ Wis-Madison, 65-81; chmn, Ctr Nonlinear Studies, Los Alamos Nat Lab, 81-85; prof comput & elec eng, 84-87, PROF MATH, UNIV ARIZ, 87- *Concurrent Pos:* Guest prof, Univ Bern, 65-66; Belgian-Am Educ Found guest lectr, Univ Louvain, 66; researcher, Cybernet Lab, Univ Naples, Italy, 69-70; prof, Tech Univ Denmark, 86- *Res:* Experimental and theoretical aspects of solid state device theory and nonlinear wave propagation including semiconductors; superconductors, laser media, high density logic systems and neurophysics. *Mailing Add:* Dept Math Univ Ariz Tucson AZ 85721

SCOTT, ANDREW EDINGTON, b Newport, Scotland, Apr 27, 19; Can citizen; m 46; c 1. ORGANIC CHEMISTRY. *Prof Exp:* Res fel chem, Ont Res Found, 50-52; res chemist, Elec Reduction Co Can, Ltd, 53-55; res fel chem, Ont Res Found, 59-60; lectr, Bristol Univ, 60-62; from asst prof to prof, 63-79, head dept, 65-66, dean fac sci, 66-79, EMER PROF CHEM, UNIV WESTERN ONT, 79- *Mem:* Fel Can Inst Chem. *Res:* Lignin chemistry; chemistry of condensed phosphates. *Mailing Add:* 451 Westmount Dr London ON N6K 1X4 Can

SCOTT, ARTHUR FLOYD, b Dickson, Tenn, Jan 10, 44; m 66; c 2. VERTEBRATE ECOLOGY, HERPETOLOGY. *Educ:* Austin Peay State Col, BS, 65, MAEd, 67; Auburn Univ, PhD(zool), 76. *Prof Exp:* Instr biol, Univ SAla, 67-70; asst prof, Union Col, Ky, 74-77; field rep zool, Ky Nature Preserves Comn, 77-78; from asst prof to assoc prof, 78-90, PROF BIOL, AUSTIN PEAY STATE UNIV, 90- *Mem:* Soc Study Amphibians & Reptiles. *Res:* Ecology, natural history and distribution of amphibians and reptiles in Kentucky and Tennessee. *Mailing Add:* Dept Biol Austin Peay State Univ Clarksville TN 37044

SCOTT, BOBBY RANDOLPH, b US citizen. RADIATION BIOLOGY, RADIATION BIOPHYSICS. *Educ:* Southern Univ, BS, 66; Univ Ill, MS, 69, PhD(biophys), 74. *Prof Exp:* Postdoctoral partic, Univ Ill, 74-75; & Argonne Nat Lab, 75-77; BIOPHYSICIST, INHALATION TOXICOL RES INST, 77- *Concurrent Pos:* Mem, Early Effects Working Group, US Nuclear Regulatory Comn, 84- & Protracted Dose Working Group, US Defense Nuclear Agency, 86-; proj prin investr, Inhalation Toxicol Res Inst, 86-; mem sci comt 86, Nat Coun Radiation Protection & Measurements, 91. *Mem:* Radiation Res Soc; Health Physics Soc; Soc Indust & Appl Math; AAAS; Sigma Xi. *Res:* Developed predictive biomathematical models for stochastic and non-stochastic radiobiological effects including: models for effects of combined exposure to different radiations or exposure to radiation plus genotoxic chemical, dose-rate dependent models and microdosimetric models. *Mailing Add:* Inhalation Toxicol Res Inst PO Box 5890 Albuquerque NM 87185

SCOTT, BRUCE ALBERT, b Trenton, NJ, Feb 23, 40; m 62, 79; c 3. PHYSICAL CHEMISTRY, INORGANIC CHEMISTRY. *Educ:* Rutgers Univ, BS, 62; Pa State Univ, PhD(chem), 65. *Prof Exp:* Res chemist, Eastern Lab, E I du Pont de Nemours & Co, 66-67; mgr org solid state, 72-82, mgr, Chem Dynamics, 82-89, RES STAFF MEM, SOLID STATE CHEM, THOMAS J WATSON RES CTR, IBM CORP, 67-, SR MGR CHEM & MATS SCI, 89- *Mem:* Am Chem Soc; fel Am Phys Soc. *Res:* Electric and magnetic properties of solids; crystal chemistry; crystal growth; phase equilibria; kinetics and mechanisms of film growth; solid state transformations. *Mailing Add:* Thomas J Watson Res Ctr IBM Corp Box 218 Yorktown Heights NY 10598

SCOTT, BRUCE L, b Waco, Tex, Oct 8, 32; m 54; c 3. THEORETICAL NUCLEAR PHYSICS. *Educ:* Calif Inst Technol, BS, 53; Univ Ill, MS, 55; Univ Calif, Los Angeles, PhD(physics), 60. *Prof Exp:* Consult physics, Atomics Int Div, NAm Aviation Inc, 57-60; asst prof, Univ Southern Calif, 60-65; assoc prof, 65-68, PROF PHYSICS, CALIF STATE UNIV, LONG BEACH, 68- *Concurrent Pos:* Consult, TRW Systs, Inc, 61-72. *Mem:* Am Asn Physics Teachers; Am Phys Soc. *Res:* Nuclear many-body problem; nucleon-nucleon interaction; electron-hydrogen scattering; 3-body problem. *Mailing Add:* Dept Physics Calif State Univ 1250 Bellflower Blvd Long Beach CA 90840

SCOTT, CHARLES COVERT, b Sparta, Ill, Jan 18, 09; m 33; c 2. PHARMACOLOGY. *Educ:* Mo Valley Col, BS, 29; Univ Mo, BS, 33; Univ Chicago, PhD(physiol) & MD, 37. *Prof Exp:* Prof physiol, Chicago Col Osteop, 34-36; intern med, Billings Hosp, Chicago, 38; asst prof physiol, Sch Med, Univ Tex, 39-40; asst in med, Univ Chicago, 40-41; pharmacologist, Res Labs, Eli Lilly & Co, Ind, 41-47, head dept gen pharmacol, 47-48; internist, Inlow Clin, Ind, 48-50; dir clin pharmacol, Warner-Chilcott Labs, 50-54, pharmacol, 54-57, res, 57-58, vpres basic sci, 58-63, vpres sci affairs, 63-67, dir med regulatory document, Warner-Lambert Res Inst, 67-74; RETIRED. *Mem:* AAAS; Am Soc Pharmacol & Exp Therapeut; AMA; NY Acad Sci; Pharmacol Soc Can. *Res:* Gastrointestinal physiology; secondary shock; pharmacology of analgesic, cardiac, diuretic, antispasmodic and central nervous system drugs. *Mailing Add:* 19419 Spook Hill Rd Freeland MD 21053

SCOTT, CHARLES D(AVID), b Chaffee, Mo, Oct 24, 29; m 56; c 3. CHEMICAL ENGINEERING & BIOCHEMICAL ENGINEERING. *Educ:* Univ Mo, BS, 51; Univ Tenn, MS, 62, PhD(chem eng), 66. *Prof Exp:* Develop engr, Y-12 Plant, Nuclear Div, 53-57, develop engr, 57-67, group leader bio eng, Sect Chief Chem Technol Div, 70-76, assoc div dir, Oak Ridge Nat Lab, 76-83, res fel, 83-87, SR RES FEL, CHEM TECH DIV, OAK RIDGE NAT LAB, MARTIN-MARIETTA CORP, 87- *Concurrent Pos:* Vis lectr, Univ Tenn, 66-70, adj prof, 70- *Honors & Awards:* IR-100 Award, 71, 77, 78 & 79; Am Asn Clin Chem Award, 80; E O Lawrence Mem Award, Dept Energy, 80; Nathan W Dougherty Award, Univ Tenn, 87. *Mem:* AAAS; Am Inst Chem Engrs; Am Chem Soc; Nat Acad Engrs. *Res:* Separations technology, heterogeneous kinetics; biotechnology; energy production. *Mailing Add:* Bldg 4505 Oak Ridge Nat Lab PO Box 2008 Oak Ridge TN 37831

SCOTT, CHARLES EDWARD, b Philadelphia, Pa, Aug 26, 29; m 55; c 3. CHEMISTRY. *Educ:* St Joseph's Col, Pa, BS, 52; Univ Notre Dame, PhD(chem), 57. *Prof Exp:* Res chemist, US Rubber Co, NJ, 55-58; sect chief, Sun Oil Co, 58-63; asst dir carbon & elastomer res, Columbian Carbon Co, 63-72; asst dir petrochem res, 72-80, VPRES OXY RES & DEVELOP CO, 80- *Mem:* Am Chem Soc. *Res:* Petrochemical processing; carbon black development; new elastomers; petroleum refining; natural gas liquids. *Mailing Add:* Oxy Res & Develop Co PO Box 300 Tulsa OK 74102

SCOTT, CHARLES JAMES, b St Paul, Minn, Apr 16, 29; m 51; c 6. AERONAUTICAL ENGINEERING. *Educ:* Univ Minn, BS, 51, MS, 54, PhD, 64. *Prof Exp:* Scientist, Rosemount Aeronaut Labs, 52-65, ASSOC PROF MECH ENG, UNIV MINN, MINNEAPOLIS, 65- *Concurrent Pos:* NATO fel, Univ Naples, 65-66. *Mem:* Am Inst Aeronaut & Astronaut. *Res:* Aerothermodynamics. *Mailing Add:* Dept Mech Eng Univ Minn Minneapolis MN 55455

SCOTT, CHARLEY, b Meridian, Miss, June 10, 23; m 47; c 2. MECHANICAL ENGINEERING. *Educ:* Miss State Univ, BS, 44; Ga Inst Technol, MSME, 49; Purdue Univ, PhD(heat transfer, thermodyn), 53. *Prof Exp:* Instr drawing, Miss State Univ, 46-47; instr drawing & physics, Meridian Munic Jr Col, 47; instr mech eng, Univ WVa, 47-48; asst prof, Miss State Univ, 49-51, from assoc prof to prof, 53-63, res thermodynamicist, 61-63; from asst dean to dean grad sch, Univ Ala, Tuscaloosa, 63-76, prof mech eng, 63-66 & 69-84, actg dean, Grad Sch Libr Sci & planner, off instnl studies & serv, 70-71, dir, univ libr, 72, actg dean libr, 73, from asst acad vpres to assoc acad vpres, 73-84, actg co-vpres acad affairs & actg chief adminr, Univ Press, 77-78, actg dean, Capstone Col Nursing, 78-79, acad compliance officer, 80-84, actg dean, Sch Social Work, 81; dir athletics, Miss State Univ, 84-85, dir spec projs, 86-88, actg vpres acad affairs, 86, actg dean grad sch & actg assoc vpres acad affairs, 87-88; RETIRED. *Concurrent Pos:* Asst engr, Manhattan Proj, Oak Ridge Nat Lab, 44-46; fel univ admin, Ctr Study Higher Educ, Univ Mich, 62-63; dir instrn, Univ Ala, Huntsville, 63-66, prof, 66-69, dir acad affairs, 66-68, dir, Div Eng, 66-68, dir, Div Grad Studies, 68-69. *Mem:* Am Soc Eng Educ; fel Am Soc Mech Engrs. *Res:* Heat transfer; thermodynamics. *Mailing Add:* PO Box 3834 Mississippi State MS 39762-3834

SCOTT, DAN DRYDEN, b Petersburg, Tenn, Apr 1, 28; m 55; c 2. ANALYTICAL CHEMISTRY, SCIENCE EDUCATION. *Educ:* Middle Tenn State Univ, BS, 50; George Peabody Col, MA, 54, PhD(sci ed, chem), 63. *Prof Exp:* Teacher high sch, Tenn, 52-55; from instr to assoc prof, 55-65, PROF CHEM & PHYSICS, MID TENN STATE UNIV, 65-, CHMN, 81- *Concurrent Pos:* Consult chemist, Samsonite, Inc, 64- *Mem:* AAAS; Am Chem Soc; fel Am Inst Chem. *Res:* Curricula for beginning college chemistry; analysis of trace amounts of alkalai metals. *Mailing Add:* Dept Chem Box 68 Mid Tenn State Univ Murfreesboro TN 37131

SCOTT, DANA S, b Berkeley, Calif, Oct 11, 32; m 59; c 1. MATHEMATICS, COMPUTER SCIENCE. *Educ:* Univ Calif, Berkeley, BA, 54; Princeton Univ, PhD(math), 58. *Hon Degrees:* Dr, Rijksuniversiteit Utrecht, Neth, 86. *Prof Exp:* Bell Tel fel, Princeton Univ, 56-57, prof philos & math, 69-72; instr, Univ Chicago, 58-60; from asst prof to assoc prof math, Univ Calif, Berkeley, 60- 63; from assoc prof to prof logic & math, Stanford Univ, 63-69; prof math logic, Oxford Univ, 72-81; UNIV PROF COMPUTER SCI, MATH LOGIC & PHILOS, CARNEGIE-MELLON UNIV, 81-, HILLMAN PROF COMPUTER SCI, 89- *Concurrent Pos:* Miller Inst fel, Univ Calif, Berkeley, 60-61; Alfred P Sloan res fel, 63-65; vis prof math, Univ Amsterdam, 68-69; Guggenheim Found fel, 78-79; vis scientist, Xerox Palo Alto Res Ctr, 78-79. *Honors & Awards:* LeRoy P Steele Prize, Am Math Soc, 72; Turing Award, Asn Comput Mach, 76. *Mem:* Fel Nat Acad Sci; Asn Symbolic Logic; Am Philos Asn; Am Math Soc; Math Asn Am; Asn Comput Mach; fel Am Acad Arts & Sci; fel NY Acad Sci; fel AAAS; Finnish Acad Sci & Lett. *Res:* Author of various publications. *Mailing Add:* Sch Computer Sci Carnegie Mellon Univ 5000 Forbes Ave Pittsburgh PA 15213-3890

SCOTT, DAVID BYTOVETZSKI, b Providence, RI, May 8, 19; m 43, 65; c 4. MEDICAL & HEALTH SCIENCES. *Educ:* Brown Univ, AB, 39; Univ Md, DDS, 43; Univ Rochester, MS, 44. *Hon Degrees:* Dr, Col Med & Dent, NJ, Univ Louis Pasteur, France. *Prof Exp:* Carnegie fel, Univ Rochester, 43-44; mem staff, Nat Inst Dent Res, 44-56, chief lab histol & path, 56-65; Thomas J Hill distinguished prof phys biol, Sch Dent, Case Western Reserve Univ & prof anat, Sch Med, 65-75, dean, Sch Dent, 69-75; dir, Nat Inst Dent Res, NIH, 76-81; CONSULT, 82- *Concurrent Pos:* Cmndg Officer, USPHS, 44-65, asst surgeon gen, 76-82; mem gen res support prog adv comt, NIH, 72- *Honors & Awards:* Arthur S Flemming Award, 55; Award, Res in Mineralization, Int Asn Dent Res, 68; Birnberg Res Award, Columbia Univ, 83; Callahan Mem Award, 85. *Mem:* Am Dent Asn; Am Acad Forensic Sci; Electron Micros Soc Am; Am Col Dent; Int Asn Dent Res; Am Bd Forensic Odontol. *Res:* Physical biology; biological mineralization; histology and embryology of tooth structure by physical methods; dental caries; methods for age estimation in forensic odontology. *Mailing Add:* 10448 Wheatridge Dr Sun City AZ 85373

SCOTT, DAVID EVANS, b Los Angeles, Calif, June 27, 38; m 61; c 2. ANATOMY, NEUROENDOCRINOLOGY. *Educ:* Willamette Univ, BA, 60; Univ Southern Calif, MS, 65, PhD(anat), 67. *Prof Exp:* Instr neuroanat & histol, Med Sch, Univ Southern Calif, 65-67; from asst prof to assoc prof, 67-76, PROF ANAT, MED SCH, UNIV ROCHESTER, 76- *Concurrent Pos:* NIH grant, 68-; USPHS career develop award, 71-76; NSF grant, 78-80. *Mem:* Soc Neurosci; Am Asn Anat; Am Asn Neuropath; Electron Micros Soc Am. *Res:* Electron microscopy; brain-endocrine interaction. *Mailing Add:* Anat & Cell Biol EVa Med Sch PO Box 1980 Norfolk VA 23501

SCOTT, DAVID FREDERICK, b Watertown, Mass, Mar 18, 40; m 63; c 3. BIOCHEMISTRY. *Educ:* Northeastern Univ, BA, 63; Ind Univ, Indianapolis, PhD(biochem), 68. *Prof Exp:* USPHS fel oncol, Univ Wis, 68-71; asst prof, 71-80, ASSOC PROF CELL & MOLECULAR BIOL, MED COL GA, 80- *Mem:* Am Asn Dent Res; Am Asn Lab Animal Sci; Sigma Xi; Am Asn Cancer Res; Am Chem Soc; AAAS. *Res:* Intermediary metabolism; metabolic regulation; granulocytic function; biochemical oncology; enzymology; lipid metabolism and obesity; toxic shock syndrome. *Mailing Add:* Dept Cell & Molecular Biol Med Col Ga Augusta GA 30912

SCOTT, DAVID KNIGHT, b North Ronaldsay, Scotland, Mar 2, 40; m 66; c 3. NUCLEAR PHYSICS. *Educ:* Edinburgh Univ, BSc, 62; Oxford Univ, DPhil(nuclear physics), 67. *Prof Exp:* Res officer, Nuclear Physics Lab, Oxford Univ, 70-73, res fel nuclear physics, Balliol Col, 67-70, sr res fel, 70-73; physicist, Lawrence Berkeley Lab, Univ Calif, 73-75, sr scientist nuclear sci, 75-79; PROF PHYSICS & CHEM, NAT SUPERCONDUCTING CYCLOTRON LAB, MICH STATE UNIV, 80- *Concurrent Pos:* Assoc provost, 83-86, provost & vpres acad affairs, Mich State Univ, 86- *Mem:* Fel, Am Phys Soc; AAAS; Europ Phys Soc. *Res:* Study of nuclear structure and reaction mechanisms using heavy ion collisions; particularly interested in relation of high energy and low energy phenomena. *Mailing Add:* Off Provost Mich State Univ East Lansing MI 48824-1046

SCOTT, DAVID MAXWELL, b Glasgow, Scotland, Apr 30, 20; Can citizen; m 48; c 3. ZOOLOGY. *Educ:* McGill Univ, BSc, 42, MSc, 47, PhD, 49. *Prof Exp:* Lectr zool, McGill Univ, 48-51; from asst prof to prof, 51-85, EMER PROF ZOOL, UNIV WESTERN ONT, 85- *Mem:* Am Ornith Union. *Res:* Ornithology; reproductive biology of passerines, particularly of cowbirds. *Mailing Add:* Dept Zool Univ Western Ont London ON N6A 5B7 Can

SCOTT, DAVID PAUL, biophysics, biometrics, for more information see previous edition

SCOTT, DAVID ROBERT MAIN, b Toronto, Ont, Aug 30, 21; m 44; c 4. SILVICULTURE. *Educ:* Univ Va, BA, 42; Yale Univ, MF, 47, PhD(forestry), 50. *Prof Exp:* Res forester, Forestry Br, Dom Dept Resources & Develop, 50-51; forester, Div Res, Ont Dept Lands & Forests, 51-55; from asst prof to prof silvicult, Univ Wash, 55-88, assoc dean, Col Forestry, 64-69; RETIRED. *Concurrent Pos:* Lectr, Inst Forest Biol, NC State Col, 60. *Mem:* Soc Am Foresters; Ecol Soc Am; Can Inst Forestry. *Mailing Add:* 3915 48th Pl NE Seattle WA 98105

SCOTT, DAVID WILLIAM, b Trenton, NJ, Feb 4, 43; m 67; c 2. IMMUNOLOGY. *Educ:* Univ Chicago, MS, 64; Yale Univ, PhD(immunol), 69. *Prof Exp:* Jane Coffin Childs Mem Fund fel, Sch Path, Oxford Univ, 69-70; asst prof immunol, 71-74, assoc prof, 74-78, prof immunol, Duke Univ, 79-; DEANS PROF IMMUNOL CANCER CTR, UNIV ROCHESTER SCH MED & DENT. *Concurrent Pos:* Eleanor Roosevelt fel, 76-77 & 86. *Mem:* AAAS; Am Asn Immunol; Brit Soc Immunol. *Res:* Immunologic tolerance; differentiation of immunologic competence; cellular interactions among lymphocytes. *Mailing Add:* Cancer Ctr Dept Microbiol, Univ Rochester Sch Med & Dent Box 704 601 Elmwood Ave Rochester NY 14642

SCOTT, DON, b Brooklyn, NY, July 8, 25; m 46, 78; c 4. BIOTECHNOLOGY, FOOD SCIENCE. *Educ:* Cornell Univ, BS, 44, MS, 45; Univ Chicago, MBA, 70; Ill Inst Technol, PhD, 50. *Prof Exp:* Bacteriologist & biochemist, Vita-Zyme Labs, Ill, 45-46, tech dir, 51-54; instr bact, Ill Inst Technol, 46-50; res bacteriologist, Jackson Lab, E I du Pont de Nemours & Co, 50-51; vpres, Fermco Labs, Inc, 54-66, vpres & gen mgr, Fermco Div, G D Searle & Co, 66-72, pres, Searle Biochemics Div, 72-75, tech dir, New Ventures Div, 73-75; pres, Fermco Biochemics Inc, 75-85 & Fermco Develop Inc, 85-87; tech dir, Scott Biotechnology Inc, 87-90; RETIRED. *Mem:* Am Chem Soc; Am Asn Clin Chemists; Inst Food Technol; Am Inst Chem Engrs; Sigma Xi. *Res:* Microbial enzymes; food stabilization; clinical analytical procedures; research management; organizational structure. *Mailing Add:* Scott Biotechnol Inc PO Box 611 Elk Grove IL 60009-0611

SCOTT, DONALD, JR, b Oyster Bay, NY, Oct 16, 09; m 53; c 2. NEUROPHYSIOLOGY. *Educ:* Univ London, PhD(biophys), 39. *Prof Exp:* Assoc neurophysiol, Johnson Found, 39-48, assoc neurol, Sch Med, Univ Pa, 39-49, asst prof anat, 49-53, from asst prof to assoc prof, 53-77, prof, 77-81, EMER ASSOC PROF PHYSIOL, SCH MED, UNIV PA, 81- *Concurrent Pos:* Foreign investr award res, Turku Univ, 72. *Honors & Awards:* Claude Bernard Medal, Univ Paris. *Mem:* Am Physiol Soc; Brit Physiol Soc. *Res:* Influence on neural regeneration of proteosynthesis; identification and excitation of dentinal receptor in tooth; influence of heat, cold, pressure, anesthetics and microcirculation on receptor; role of metabolism and inorganic ions on receptor function; membrane physiology of receptor. *Mailing Add:* 214 Glenmore Rd Gladwin PA 19035

SCOTT, DONALD ALBERT, b Campville, NY. Nov 20, 17; m 46; c 2. ORGANIC CHEMISTRY. *Educ:* Cornell Col, AB, 39; Univ Ariz, MS, 41; Univ Iowa, PhD(org chem), 52. *Prof Exp:* Instr chem, Univ Ariz, 41-42 & Cornell Col, 42-43, 46-49; asst prof, Wash & Jefferson Col, 51-54; assoc prof, 54-57, chmn dept, 54-71, prof, 57-86, EMER PROF CHEM, DREW UNIV, 86- *Concurrent Pos:* NSF fac fel, Univ Calif, Los Angeles, 60-61, Univ Delft, 79. *Mem:* Am Chem Soc; Nat Sci Teachers Asn. *Res:* Structure of natural organic products; glycosides from bark; electrophilic substitution on carbon attached to sulfur. *Mailing Add:* Dept Chem Drew Univ Madison NJ 07940

SCOTT, DONALD CHARLES, b Washington, DC, Jan 6, 20; m 42; c 5. LIMNOLOGY, FISH BIOLOGY. *Educ:* Univ Mich, BS, 42; Ind Univ, PhD(zool), 47. *Prof Exp:* Asst, Ind Univ, 42-47; from instr to assoc prof zool, 47-65, chmn div biol sci, 67-72, prof, 65-80, EMER PROF ZOOL, UNIV GA, 80- *Concurrent Pos:* Biologist, USPHS, 51-52; staff assoc, Div Inst Prog, NSF, 64-65. *Mem:* Am Soc Ichthyologists & Herpetologists; Am Fisheries Soc; fel Am Inst Fishery Res Biol. *Res:* Ichthyology. *Mailing Add:* 225 Beech Creek Rd Athens GA 30606

SCOTT, DONALD HOWARD, b Indianapolis, Ind, July 11, 34; m 56; c 4. PHYTOPATHOLOGY. *Educ:* Purdue Univ, BS, 56; Univ Ill, MS, 64, PhD(plant path), 68. *Prof Exp:* Field rep grain dealers, 56-62; res asst plant path, Univ Ill, 62-64, asst exten plant pathologist, 64-68; assoc prof, 68-80, PROF & EXTEN PLANT PATHOLOGIST, PURDUE UNIV, LAFAYETTE, 80- *Concurrent Pos:* Consult, 78- *Mem:* Am Phytopath Soc; Sigma Xi. *Res:* Diseases and disease control of agronomic crops; effects of crop rotation and tillage practices on disease development; interaction of insects, nematodes, and disease on crop yield losses; chemical control of agronomic crop diseases. *Mailing Add:* 327 Lawn West Lafayette IN 47906

SCOTT, DONALD RAY, b Wichita Falls, Tex, Apr 27, 34; m 58; c 3. SPECTROSCOPY, CHEMOMETRICS. *Educ:* Univ Tex, Austin, BA, 56; Univ Houston, MS, 60, PhD(phys chem), 65. *Prof Exp:* Res chemist clay & anal chem, Texaco Res Lab, 56-61; instr & res fel spectros & quantum chem, dept chem, Univ Tex, Austin, 64-65; asst prof, Tex Tech Univ, Lubbock, 65-67; scientist polymer photo decomposition, Lockheed-Ga Mat Sci Lab, Atlanta, 67-71; sect chief environ anal chem, US Environ Protection Agency, Res Triangle Park, NC, 71-73; assoc prof spectros, dept chem, SDak Tech, Rapid City, 73-75; br chief environ anal chem, US Environ Protection Agency, Las Vegas, 75-78; dir & res scientist environ phys & anal chem, Inst Appl Sci, NTex State Univ, Denton, 78-80; res chemist & br chief, 80-81, SR SCI ADV, US ENVIRON PROTECTION AGENCY, RESEARCH TRIANGLE PARK, NC, 81- *Concurrent Pos:* Fel, Theoret Chem Group, Univ Tex, Austin, 64-65; consult, Intersoc Comt Heavy Metals, 74-75 & US Environ Protection Agency, 78-80; ed, US Environ Protection Agency ICP Newsletter, 76-78; adj prof, chem dept, NTex State Univ, 78-80; assoc ed, Chemometrics & Intel Lab Systs, 90- *Mem:* Am Chem Soc; Chemometrics Soc. *Res:* Electronic spectroscopy; applications of information theory and pattern recognition to spectral data; environmental analytical chemistry; expert systems for spectral analytical interpretations. *Mailing Add:* MD-78A US Environ Protection Agency Research Triangle Park NC 27711

SCOTT, DONALD S(TRONG), b Edmonton, Alta, Dec 17, 22; m 45; c 2. CHEMICAL ENGINEERING, PHYSICAL CHEMISTRY. *Educ:* Univ Alta, MSc, 46; Univ Ill, PhD(chem eng), 49. *Prof Exp:* Jr petrol engr, Imp Oil, Ltd, 44-45; chem engr, Nat Res Coun Can, 46-47; asst prof chem eng, Univ BC, 49-56, assoc prof, 56-64; chmn dept, 64-69, actg dean eng, 69-70, PROF CHEM ENG, UNIV WATERLOO, 64- *Concurrent Pos:* Vis prof, Univ Cambridge, 63-64, Imperial Col, London, 70, assoc dean eng, Univ Cambridge, 80-86. *Honors & Awards:* Plummer Medal, Eng Inst Can, 82; R S Jane award, Can Soc Chem Engrs. *Mem:* Fel Am Inst Chem Engrs; Am Chem Soc; fel Chem Inst Can; Can Soc Chem Engrs (vpres & pres, 70-72); Sigma Xi. *Res:* Three phase bubble column reaction with particular emphasis on hydrotreating of petroleum oils; reactor design; catalysis; extractive metallurgy; conversion of biomass to liquids by use of fluidized bed fast pyrolysis, thermal or catalytic; kinetics. *Mailing Add:* Dept Chem Eng Univ Waterloo Waterloo ON N2L 3G1 Can

SCOTT, DWIGHT BAKER MCNAIR, b Coldwater, Mich, May 5, 07; m 36; c 2. BIOCHEMISTRY. *Educ:* Vassar Col, AB, 29; Radcliffe Col, PhD(biochem), 36. *Prof Exp:* Biochemist, Thorndike Mem Lab, Boston City Hosp, 29-31; Huntington Mem Hosp, 33-34; asst biol chem, Harvard Med Sch, 34-36, Univ Col Hosp Med Sch, London, 36-38 & Pa Hosp Nervous & Ment Dis, 41-42; instr pediat, Johns Hopkins Univ, 42-43; instr chem, Wellesley Col, 43-45; assoc pediat & physiol chem, Children's Hosp, Philadelphia, 47-53; asst prof med & physiol chem, Sch Med, 54-57, physiol, 57-65, from asst prof to assoc prof biochem, 65-72, prof biochem, 72-76, PROF EMER, SCH VET MED, UNIV PA, 76- *Mem:* AAAS; Am Soc Microbiol; Am Soc Biol Chem; Am Chem Soc; Brit Biochem Soc; Am Asn Cancer Res; Sigma Xi. *Res:* Carbohydrate metabolism and cancer; the role of zinc in female reproductive system. *Mailing Add:* Logan Sq E No 1605 Two Franklin Town Blvd Philadelphia PA 19103

SCOTT, EARLE STANLEY, b Bellingham, Wash, Oct 16, 22; m 44; c 7. INORGANIC CHEMISTRY. *Educ:* Reed Col, BA, 49; Univ Ill, PhD(inorg chem), 52. *Prof Exp:* Instr chem, Univ Calif, 52-55; asst prof, Amherst Col, 55-60; vis prof, Earlham Col, 60-62; from assoc prof to prof chem, Ripon Col, 62-87, chmn dept, 78-80, May Bumby Severy distinguished serv prof, 82-87; RETIRED. *Mem:* Am Chem Soc. *Res:* Chemical education. *Mailing Add:* Dept Chem PO Box 248 Ripon Col Ripon WI 54971-1453

SCOTT, EDWARD JOSEPH, b Chicago, Ill, May 29, 13; m 40; c 1. MATHEMATICS. *Educ:* Maryville Col, Tenn, BA, 36; Vanderbilt Univ, MA, 37; Cornell Univ, PhD(math), 43. *Prof Exp:* Asst math, Univ Md, 37-39; instr, Lawrence Inst Technol, 39-43; instr, Cornell Univ, 43-46; from instr to assoc prof, 46-64, PROF MATH, UNIV ILL, URBANA, 64- *Res:* Partial differential equations; wave propagation. *Mailing Add:* 2105 Boudreau Dr Urbana IL 61801

SCOTT, EDWARD ROBERT DALTON, b Heswall, Eng, Mar 22, 47; m 80; c 2. COSMOCHEMISTRY, PLANETOLOGY. *Educ:* Univ Cambridge, BA, 68, PhD(mineral), 72. *Prof Exp:* Researcher geophys, Univ Calif, Los Angeles, 72-75; researcher mineral, Univ Cambridge, Eng, 75-78; res fel, Dept Terrestrial Magnetism, Carnegie Inst Wash, 78-80; sr res scientist, Inst Meteoritics, Univ NMex, 80-90; PROF, UNIV HAWAII, 90- *Concurrent Pos:* Counr, Meteorit Soc, 77-80; mem, NASA-NSF Antarctic Meteorite Working Group, 83-86; mem, NASA Lunar & Planetary Geosci Rev Panel, 85-87; assoc ed, J Geophys Res, 85-87; assoc ed, Proceedings Lunar Planetary Sci Conf, 87- *Mem:* Meteorit Soc; Am Geophys Union; Mineral Soc. *Res:* Origin and evolution of meteorites, asteroids, planets and solar nebula; composition, mineralogy and trace element distributions; analysis by electron probe and neutron activation analysis. *Mailing Add:* Planetary Geosci Div Dept Geol & Geophys Univ Hawaii Honolulu HI 96822

SCOTT, EDWARD W, geology; deceased, see previous edition for last biography

SCOTT, EION GEORGE, b Glasgow, Scotland, May 3, 31; US citizen; m 57; c 3. PLANT PHYSIOLOGY, PLANT BIOCHEMISTRY. *Educ:* Glasgow Univ, BSc, 54; Univ Calif, Davis, PhD(plant physiol), 58. *Prof Exp:* Nuffield fel plant physiol, Univ Col Swansea, Wales, 58-60; asst prof, Univ Southern Calif, 60-62; chmn dept hort, WVa Univ, 62-69, prof hort & horticulturist, 62-73, prof agr biochem, 69-73; consult scientist hort, Gen Elec Co, 73-75, tech dir, Controlled Environ Agr Oper, 75-79, mgr, 78-80; GEN MGR, CONTROLLED ENVIRON AGRON OPER, CONTROL DATA CORP, 80- *Mem:* AAAS; Am Soc Hort Sci. *Res:* Trace element metabolism; plant organ culture; plant environment interactions photobiology. *Mailing Add:* PO Box 183 Rte 2 Banks OR 97106

SCOTT, ELIZABETH LEONARD, mathematical statistics, biostatistics; deceased, see previous edition for last biography

SCOTT, ERIC JAMES YOUNG, b Gourock, Scotland, May 8, 24; nat US; m 59. PHYSICAL CHEMISTRY. *Educ:* Glasgow Univ, BSc, 45, PhD, 48. *Prof Exp:* Nat Res Coun Can fel photochem, 48-50; AEC fel & res assoc radiation chem, Univ Notre Dame, 50-52; sr res chemist, Mobil Res & Develop Corp, 52-75, assoc chemist, 75-84; RETIRED. *Mem:* Am Chem Soc; Sigma Xi. *Res:* Oxidation; kinetics of gas reactions; petroleum chemistry; chemical kinetics; homogeneous and heterogeneous catalysis; combustion. *Mailing Add:* 277 Nassau St Princeton NJ 08541

SCOTT, FRANCIS LESLIE, medicinal chemistry, organic chemistry, for more information see previous edition

SCOTT, FRANKLIN ROBERT, b Portland, Ore, Aug 23, 22; m 50; c 3. FUSION DEVELOPMENT, ENERGY RESEARCH MANAGEMENT. *Educ:* Reed Col, BA, 47; Ind Univ, MS, 49, PhD(physics), 52. *Prof Exp:* Res staff mem, Los Alamos Sci Lab, 51-57; asst dir fusion div, Gen Atomic, 57-67; prof physics, Univ Tenn, 67-73; chief, Open Systs Br, AEC-Energy Res & Develop Agency, 73-75; prog mgr, Elec Power Res Inst, 75-86; CONSULT, LOS LAMOS NAT LAB, 87- *Concurrent Pos:* Consult, Oak Ridge Nat Lab, 67-73, Dept Energy, 76- & Elec Power Res Inst, 87-; ed, Rev Sci Instruments, 70-71. *Mem:* Fel Am Phys Soc; Sigma Xi; AAAS; Inst Elec & Electronics Engrs. *Res:* Fusion; pulsed power; technical assessments; commercialization. *Mailing Add:* 4805 Terra Granada Dr Apt 2-B Walnut Creek CA 94595

SCOTT, FRASER WALLACE, b Montreal, Que, Nov 21, 46. METABOLISM, DIABETES. *Educ:* McGill Univ, BSc, 69, MSc, 72; Queen's Univ, PhD(biochem), 76. *Prof Exp:* Fel cancer res, Cancer Res Unit, Univ Alta, 76-77; RES SCIENTIST NUTRIT, DEPT NAT HEALTH & WELFARE, CAN, 77- *Mem:* Am Inst Nutrit; Can Soc Nutrit Sci. *Res:* Diet as a factor in the pathogenesis of insulin-dependent diabetes; nutrition and cancer; carbohydrate nutrition; diet and health. *Mailing Add:* Nutrit Res Div Health Protection Br Health & Welfare Tunney's Pasture Ottawa ON K1A 0L2 Can

SCOTT, FREDERICK ARTHUR, b Albany, NY, Mar 6, 25; m 51; c 2. ANALYTICAL CHEMISTRY. *Educ:* Rensselaer Polytech Inst, BS, 48, MS, 49, PhD, 52. *Prof Exp:* Sr scientist, Hanford Lab, Gen Elec Co, 52-65; chem div, Pac Northwest Labs, Battelle Mem Inst, 65-70; Wadco, 70-71; sr scientist, 71-80, MGR, WESTINGHOUSE HANFORD CO, 80- *Mem:* Am Chem Soc; Am Nuclear Soc; Sigma Xi. *Res:* Analytical instrument and methods development; sodium technology. *Mailing Add:* 1116 Wright Richland WA 99352

SCOTT, FREDRIC WINTHROP, b Greenfield, Mass, Nov 22, 35; m 57; c 3. VETERINARY VIROLOGY, FELINE MEDICINE. *Educ:* Univ Mass, BS, 58; Cornell Univ, DVM, 62, PhD(vet virol), 68; Am Col Vet Microbiol, dipl. *Prof Exp:* Vet, Rutland Vet Clinic, Vt, 62-64; res vet, Plum Island Animal Dis Lab, Agr Res Serv, 64-65; Nat Inst Allergy & Infectious Dis res fel virol Cornell Univ, 65-68, asst prof, 68-73, assoc prof, 73-78, PROF VET VIROL, COL VET MED, CORNELL UNIV, 78-, DIR, CORNELL FELINE HEALTH CTR, 74- *Concurrent Pos:* Prin investr res grant, State Agr Exp Sta, 69-79, USPHS Nat Inst Allergy & Infectious Dis, 70-73, Div Res Resources, USPHS, 74-77, Schering Corp, 84-86, 3M Corp, 87 & Morris Animal Found, 86-89; coinvestr res grant, USPHS Nat Inst Child Health & Human Develop, 71-74; prin investr res contract, Nat Inst Allergy & Infectious Dis, 75-81; mem, Int Working Teams on Small Enveloped RNA Viruses, Caliciviruses and Parvoviruses, WHO-Food & Agr Orgn Prog Comp Virol; fel, Mark L Morris Animal Found; mem adv comt, Ctr Vet Med, Food & Drug Admin, 84-86; mem, Am Vet Med Asn Coun Biol & Therapeut Agents, 86-88, chmn, 87-; Rhone Merieux res grant, 90-92. *Honors & Awards:* Res Award, Am Asn Feline Practitioners; Carnation Award for Excellence in Feline Med, Am Animal Hosp Asn, 90. *Mem:* Am Vet Med Asn; Conf Res Workers Animal Dis; Am Asn Feline Practitioners (pres, 76-78); Am Col Vet Microbiol; Am Animal Hosp Asn; Am Soc Virol; hon first fel Acad Feline Med. *Res:* Feline and bovine viral diseases; feline infectious pertionitis; feline immunodeficiency virus, feline panleukopenia; feline respiratory diseases; bovine winter dysentery; antiviral compounds; immunology. *Mailing Add:* Dept Microbiol Immunol Parasitol Col Vet Med Cornell Univ Ithaca NY 14853

SCOTT, GARLAND ELMO, JR, b Greensboro, NC, Nov 30, 38; m 61; c 2. CERAMICS. *Educ:* NC State Univ, BS, 61, MS, 64, PhD, 71. *Prof Exp:* Res asst solid state res, NC State Univ, 61-67; ceramist, Monsanto Co, 67-69; res ceramist, Gen Elec Co, 69-76, mgr, Lamp Div, 76-78; asst prof ceramics, 77-80, PROF MAT ENG, CALIF STATE POLYTECH INST, 80- *Concurrent Pos:* Adj prof, dept orthod, Loma Linda Univ. *Mem:* Am Ceramic Soc; Am Soc Metals; Nat Inst Ceramic Engrs. *Res:* Solid state sintering; behavior of glass with low silica content; fracture mechanics; failure analysis; injection molding. *Mailing Add:* 424 Adrian Ct Claremont CA 91711

SCOTT, GARY WALTER, b Topeka, Kans, Jan 19, 43; m 90; c 2. CHEMICAL PHYSICS. *Educ:* Calif Inst Technol, BS, 65; Univ Chicago, PhD(chem physics), 71. *Prof Exp:* NSF fel, Univ Pa, 71-72, NIH fel, 72-74; from asst prof to assoc prof, 74-85, PROF CHEM, UNIV CALIF, RIVERSIDE, 85- *Concurrent Pos:* Vis scholar, Wesleyan Univ, 80-81. *Mem:* Am Phys Soc. *Res:* Experimental chemical physics; spectroscopic studies of excited molecular states; development of short pulse lasers; applications of lasers to studies of the photophysics of n electron aromatics, fluorescent dyes, polymer photostabilizers, stabilized polymers, hydrogen-bonded molecules and photochemical hole burning. *Mailing Add:* Dept Chem Univ Calif Riverside CA 92521

SCOTT, GENE E, b Oberlin, Kans, June 11, 29; m 54; c 2. PLANT BREEDING. *Educ:* Kans State Univ, BS, 51, MS, 55, PhD(agron), 63. *Prof Exp:* Agent corn invests, USDA, 54-55, res agronomist, 55-72; RES LEADER, AGR RES SERV, USDA, MISS STATE UNIV, 72- *Mem:* Am Soc Agron; Crop Sci Soc Am; Am Phytopath Soc. *Res:* Insect and disease resistance in corn; aspects of corn improvement. *Mailing Add:* Dept Agron Box 5248 Miss State Univ Mississippi State MS 39762-5248

SCOTT, GEORGE CLIFFORD, b Shumway, Ill, Dec 6, 26; c 7. VETERINARY MEDICINE. *Educ:* Univ Ill, BS, 50, DVM, 52. *Prof Exp:* Vet Pract, 52-58; asst vet med dir, Smith Kline & French Labs, 58-64; dir res & develop, Vetco, Johnson & Johnson, 64-66 & Animal Div, Schering Corp, 66-67; dir, 67-74, vpres res & develop, 74-82, vpres sci & technol, 82-86, VPRES TECHNOL DEVELOP, ANIMAL HEALTH PROD DIV, SMITH-KLINE BECKMAN CORP, 86- *Mem:* Am Vet Med Asn; Soc Study Reprod; Am Dairy Sci Asn; Am Soc Animal Sci; Am Poultry Sci Asn. *Res:* Parasites of domestic animals, nutrition of ruminants and control of diseases in domestic animals. *Mailing Add:* 800 Hessian Circle West Chester PA 19382

SCOTT, GEORGE PRESCOTT, b Pittsfield, Mass, Sept 17, 21; m 47; c 5. PHILOSOPHY SCIENCE, ORGANIC CHEMISTRY. *Educ:* Worcester Polytech Inst, BS, 43; Univ Rochester, PhD(chem), 49. *Prof Exp:* From assoc prof to prof, 49-85, actg head dept, 59-60, EMER PROF CHEM, UNIV SDAK, VERMILLION, 85- *Concurrent Pos:* Res assoc, Univ Ill, 53-55; Fulbright lectr, UAR, 64-65; vis scholar, Univ Tex, Austin, 74, res fel, 85- *Mem:* Am Chem Soc. *Res:* Polymers; telomers; free radical kinetics; history and philosophy of science; chemical oscillations. *Mailing Add:* 31 Prentis Vermillion SD 57069

SCOTT, GEORGE TAYLOR, physiology; deceased, see previous edition for last biography

SCOTT, GEORGE WILLIAM, b Cape Charles, Va, Apr 2, 17; m 43; c 3. ORGANIC CHEMISTRY. *Educ:* Col of William & Mary, BS, 38; Univ Va, MS, 41, PhD(org chem), 42. *Prof Exp:* Res chemist, Jackson Lab, E I du Pont de Nemours & Co, Inc, 42-55, res chemist, Neoprene Works, 55-81; RETIRED. *Mem:* Am Chem Soc. *Res:* Polymerization of dienes; emulsion and colloidal chemistry. *Mailing Add:* 4706 Kittyhawk Way Louisville KY 40207

SCOTT, GERALD WILLIAM, b London, Eng, Jan 12, 31; Can citizen; m 55; c 5. GASTROINTESTINAL PHYSIOLOGY. *Educ:* Univ London, BS & MB, 55; Univ Minn, MS, 64; FRCS(C), 64; Am Bd Surg, dipl, 65. *Prof Exp:* Resident fel surg, Mayo Clinic, Minn, 60-64; teaching fel, Univ Alta, 64-65; surgeon, pvt pract, Calgary, Alta, 65-67; assoc prof, Univ Calgary, 68-73; chief surg, Charles Camsell Hosp, Edmonton, 73-78; PROF SURG, UNIV ALTA, 73-, DIR EXP SURG, 73-; DIR, SURG-MED RES INST, 73- *Concurrent Pos:* Fulbright scholar, Coun Inst Exchange Scholars, 60-64; examr, Royal Col Physicians & Surgeons Can, 72-80, exec secy, 79-; surgeon-consult, W W Cross Cancer Inst, 73; exec secy, Med Coun Can, 79-; vis prof, Univ Auckland, NZ, 79. *Honors & Awards:* Freyer Mem Lectr, Univ Col, Galway, Ireland, 83. *Mem:* Can Med Asn; Can Asn Clin Surgeons; Can Asn Gastroenterol; fel Am Col Surgeons. *Res:* Gallbladder motility and pathogenesis of gallstone disease. *Mailing Add:* Lady Minto Gulf Islands Hosp Univ Alta 1074B Dent-Pharm Bldg Edmonton AB T6G 2N8 Can

SCOTT, HAROLD GEORGE, b Williams, Ariz, Aug 20, 25; m 48; c 9. PUBLIC HEALTH & EPIDEMIOLOGY. *Educ:* Univ NMex, BS, 50, MS, 53, PhD, 57. *Prof Exp:* Entomologist, Med Field Serv Sch, Ft Sam Houston, US Army, Tex, 50-51, air res & develop Command, Kirtland AFB, USAF, NMex, 51-55, Commun Dis Ctr, USPHS, 55-67, Hosp, New Orleans, 67-69, Environ Health Serv, 69-71 & Environ Protection Agency, 71; prof trop med, 71-76, LECTR COMMUNITY MED, TULANE UNIV, 76- *Concurrent Pos:* Dep dir, Senegal River Valley Health Study, 77-80. *Mem:* Entom Soc Am; Asn Mil Surgeons US; Sigma Xi. *Res:* Tropical diseases and systematic entomology. *Mailing Add:* Four Pats Pl Metairie LA 70001

SCOTT, HENRY WILLIAM, JR, b Graham, NC, Aug 22, 16; m 42; c 4. SURGERY. *Educ:* Univ NC, AB, 37; Harvard Univ, MD, 41; Am Bd Surg, dipl, 48. *Hon Degrees:* DSc, Univ Aberdeen. *Prof Exp:* Asst surg, Harvard Med Sch, 43-44, Cushing fel neurosurg, 44-45; asst surg, Sch Med, Johns Hopkins Univ, 46, from instr to assoc prof, 47-51; prof surg & head dept, Sch Med & surgeon-in-chief, Univ Hosp, Vanderbilt Univ, 52-82, dir, Sect Surg Sci, 75-82, EMER PROF SURG & DIR SECT SURG SCI, VANDERBILT UNIV, 82- *Concurrent Pos:* Chief surg consult, Vet Admin Hosp, 52-; mem, Am Bd Surg, 56-62, vchmn, 61-62, rep to adv bd, Med Specialties; mem, Nat Bd Med Examr, 65-; mem study sect B, USPHS & chmn, 66-70; hon fel, Royal Australian Col Surg, 76, Swedish Surg Soc, 77; H William Scott Jr Chair Surg, Vanderbilt, 82. *Mem:* Soc Clin Surg (pres, 71); Soc Univ Surgeons (pres, 60); Am Surg Asn (treas, 58-65, pres, 73-74); Am Col Surgeons (treas, 67-, pres, 75-76); Soc Surg Alimentary Tract (pres, 70-71). *Res:* Physiology and physiopathology of cardiovascular diseases; gastrointestinal disorders; surgical aspects of cancer. *Mailing Add:* Dept Surg Vanderbilt Univ Med Ctr Nashville TN 37232

SCOTT, HERBERT ANDREW, b Marion, Va, Mar 29, 24; m 47; c 2. POLYMERIZATION, HEAT TRANSFER. *Educ:* Va Polytech Inst, BS, 44, MS, 47. *Prof Exp:* Chem engr, Tenn Eastman Co, 47-59, supt, Glycol Dept, 59-64 & Polymers Div, 65-68; plant mgr, Holston Defense Corp, Tenn, 68-71; dir systs develop, Tenn Eastman Co, 71-73, dir eng, 73-87; RETIRED. *Mem:* Am Inst Chem Engrs; Sigma Xi; Am Mgt Asn; Nat Soc Prof Engrs. *Res:* Hydrogenation; polyester polymers; polyurethane elastomers. *Mailing Add:* 4512 Chickasaw Rd Kingsport TN 37664

SCOTT, HOWARD ALLEN, b Ft Smith, Ark, Aug 12, 26; m 50; c 4. VIROLOGY, SEROLOGY. *Educ:* Memphis State Col, BS, 49; Univ Mont, MA, 54; Univ Calif, PhD(plant path), 59. *Prof Exp:* Asst specialist plant path, Univ Calif, 54-59; plant pathologist virol, Crops Res Div, USDA, Md, 59-67; assoc prof, 67-69, PROF PLANT PATH, UNIV ARK, FAYETTEVILLE, 69- *Mem:* Am Phytopath Soc; Soc Invert Path. *Res:* Purification; serological studies of plant and insect viruses; vectors of plant viruses. *Mailing Add:* Virol & Biocontrol Lab Rte 11 Univ Farms Fayetteville AR 72701

SCOTT, HUBERT DONOVAN, b Tarboro, NC, Apr 26, 44; m 68. SOIL SCIENCE. *Educ:* NC State Univ, BS, 66, MS, 68; Univ Ky, PhD(soil sci), 71. *Prof Exp:* Assoc prof, 71-80, PROF SOIL PHYSICS, UNIV ARK, FAYETTEVILLE, 80- *Mem:* Am Soc Agron; Soil Sci Soc Am. *Res:* Movement of water and water soluble substances in soil and their subsequent uptake by plants; geographic information systems. *Mailing Add:* Dept Agron Univ Ark Fayetteville AR 72701

SCOTT, HUGH LAWRENCE, JR, b Baltimore, Md, Jan 10, 44; m 66; c 2. BIOPHYSICS. *Educ:* Purdue Univ, BS, 65, PhD(physics), 70. *Prof Exp:* Res assoc physics, Univ Utah, 70-72; from asst prof to prof physics, 72-85, interim head dept, 85-86, PROF PHYSICS, OKLA STATE UNIV, 86-, HEAD DEPT, 90-; CONSULT, 86- *Concurrent Pos:* NSF grant, 74-83, 87-; Okla Water Resources Inst grant, 85. *Mem:* Am Phys Soc; Biophys Soc; AAAS. *Res:* Development and analysis of theoretical models, using statistical mechanics and computer simulation for lipid monolayer, lipid bilayer and biological membrane thermodynamic behavior; computer studies of CVD diamond film growth. *Mailing Add:* Dept Physics Okla State Univ Stillwater OK 74078

SCOTT, HUGH LOGAN, III, b Lexington, Ky, Oct 19, 40; m 63; c 2. EXPERIMENTAL NUCLEAR PHYSICS. *Educ:* Univ Ky, BS, 62, MS, 66, PhD(physics), 67. *Prof Exp:* Res fel physics, Bartol Res Found, 67-69; asst prof, Univ Ga, 69-76; MEM TECH STAFF, SANDIA LABS, 76- *Mem:* Am Phys Soc. *Res:* Nuclear spectroscopy; decay schemes; spin-parities of nuclear states; analog states and elemental analysis via proton-induced x-rays. *Mailing Add:* Sandia Nat Labs Div 9111 PO Box 5800 Albuquerque NM 87115

SCOTT, J(OHN) D(ONALD), Can citizen; m 56; c 2. GEOTECHNICAL ENGINEERING, OIL SANDS RESEARCH. *Educ:* Queen's Univ, Ont, BSc, 54; Univ Ill, Urbana, MSc, 58, PhD(soil mech), 64. *Prof Exp:* Engr, Hardy & Ripley, Consult, 54-57; res asst civil eng, Univ Ill, Urbana, 57-60; asst prof, Univ Waterloo, 60-64, assoc prof, 64-66; prof & chmn dept civil eng, Univ Ottawa, 66-78; sr geotech engr, R M Hardy & Assocs Ltd, 78-80; Aostra prof & chair, 80-85, PROF, UNIV ALTA 85. *Concurrent Pos:* Pvt consult, 57-78; Nat Res Coun res grant, Univ Waterloo, 61-65 & Univ Ottawa, 65-78; Ont Res Coun res grant, Univ Waterloo, 64-65; vis scientist, Nat Res Coun Can, 65-66; dir & consult engr, Fondex Ltd, 71-74; Ont Ministry Transp & Commun res grant, 75-78; Nat Res Coun Can res grant, 78-; adj res prof civil eng, Univ Alta, 78-; Aostra res grant, 80-; geotech engr consult, 80- *Mem:* Eng Inst Can; Can Geotech Soc (treas, 75-78); Can Inst Mining. *Res:* Slope stability; foundation performance; oil sand geotechnique; oil sand mining; tailings dams; oil sands in situ recovery. *Mailing Add:* Dept Civil Eng Univ Alta Edmonton AB T6G 2M7 Can

SCOTT, J(AMES) L(OUIS), b Memphis, Tenn, May 22, 29; m 53; c 2. MATERIALS SCIENCE, CERAMICS. *Educ:* Univ Tenn, BS, 52, MS, 54, PhD(metall), 57. *Prof Exp:* Instr chem eng, Univ Tenn, 53-56; metallurgist, 56-65, head ceramics lab, 65-74, mgr, fusion reactor mat prog, Metals & Ceramics Div, 74-84, RES STAFF MEM, OAK RIDGE NAT LAB, 84- *Concurrent Pos:* Mem, high temperature fuels comt, Atomic Energy Comn, 61-71 & int adv comt, First Int Conf Fusion Reactor Mat, Tokyo, Japan, 84; chmn, Special Purpose Mat Task Group, Dept Energy, 76-; gen chmn, Fifth Topical Meeting Technol Fusion Energy, Knoxville, Tenn, 83. *Mem:* Fel Am Nuclear Soc; fel Am Soc Metals; Am Ceramic Soc. *Res:* Fabrication and irradiation behavior of reactor materials. *Mailing Add:* X-10 Area Bldg 4508 Mail Stop 6098 Oak Ridge Nat Lab Oak Ridge TN 37831

SCOTT, JAMES ALAN, b Adrian, Mich, Aug 17, 43; m 65; c 1. ANALYTICAL CHEMISTRY, NYLON POLYMERS. *Educ:* Bowling Green State Univ, BS, 64, MS, 65. *Prof Exp:* From chemist to supvr, 65-73, asst chief chemist, Plastics Dept, 73-77, prod mgr, 78-85, SR MKT SPECIALIST, E I DU PONT DE NEMOURS & CO, INC, 86- *Mem:* Am Chem Soc; Soc Plastics Indust; Soc Electroplated Plastics; Am Mgt Asn. *Res:* Analytical chemistry of fluorocarbons and fluorocarbon polymers; atomic absorption spectrophotometry; flame emission spectrophotometry; thermal analysis; analysis automation; flame retardant thermoplastics; nylon polymers; polymer marketing and research. *Mailing Add:* Nine Jacqueline Dr Wellington Hills Hockessin DE 19707

SCOTT, JAMES FLOYD, b Beverly, NJ, May 4, 42; m 82; c 3. SOLID STATE PHYSICS. *Educ:* Harvard Univ, AB, 63; Ohio State Univ, PhD(physics), 66. *Prof Exp:* Res fel physics, Ohio State Univ, 66; mem tech staff, Bell Tel Labs, NJ, 66-72; PROF PHYSICS & ASTROPHYS, UNIV COLO, BOULDER, 72- *Concurrent Pos:* Sci Res Coun sr vis fel physics, Univ Edinburgh, 70-71 & Oxford Univ, 76-77; consult, Los Alamos Nat Labs, 74- & IBM, Zurich, 82; lectr, USSR Acad Sci, 77-81, Japan Soc Prom Sci, 80 & Chinese Acad Sci, 77-84. *Mem:* Fel Am Phys Soc; Sigma Xi. *Res:* Inelastic scattering of laser light from solid state excitations. *Mailing Add:* Dept Physics Campus Box 390 Univ Colo Boulder CO 80309

SCOTT, JAMES HENRY, b Marlboro, NY, Apr 19, 30; m 53; c 3. GEOPHYSICS, GEOLOGY. *Educ:* Union Col, NY, BS, 51. *Prof Exp:* Beers & Heroy, 51, Phillips Petrol Co, 51-54, USAEC, 54-62, US Geol Surv, Colo, 62-67 & US Bur Mines, 67-74; geophysicist, US Geol Surv, 74-86; CONSULT GEOPHYSICIST, 86- *Concurrent Pos:* Consult geophysicist, 86- *Mem:* Soc Explor Geophys; Soc Prof Well Log Analysts. *Res:* Well logging and surface geophysics for exploration and evaluation of mineral deposits and construction sites. *Mailing Add:* 12372 W Louisiana Ave Lakewood CO 80228

SCOTT, JAMES J, b Wiota, Wis, Apr 22, 28; m 47; c 5. MINING ENGINEERING. *Educ:* Mo Sch Mines, BS, 50; Univ Wis, MS, 59, PhD(mining eng), 62. *Prof Exp:* Mine engr, Bethlehem Steel Co, Pa, 50-53, mine foreman, 53-57; from instr to asst prof mining, Univ Wis, 57-63; assoc prof, Univ Mo, Rolla, 63-65, chmn depts mining & petrol, 70-76, prof mining, 65-80. *Concurrent Pos:* Gen mgr, Black River Mine, Marble Cliff Quarries Co, 67; asst dir mining res, US Bur Mines, 70-; adj prof mining eng, Univ Mo, Rolla, 80- *Mem:* Am Inst Mining, Metall & Petrol Engrs; Soc Exp Stress Anal; Can Inst Mining & Metall. *Res:* Field rock mechanics; mine operational problems; experimental stress analysis; photoelasticity; model studies; stress distribution problems; mine and research management. *Mailing Add:* HCR 33 Box 36 Rolla MO 65401

SCOTT, JAMES MICHAEL, b San Diego, Calif, Sept 20, 41; m 66; c 2. ENDANGERED SPECIES. *Educ:* San Diego State Univ, BS, 66, MA, 70; Ore State Univ, PhD(zool), 73. *Prof Exp:* Biol aide, US Bur Com Fisheries, 66-68; asst cur vertebrates, Nat Hist Mus, Ore State Univ, 69-73, researcher, Dept Fisheries & Wildlife, 73-74; biologist in charge, Mauna Loa Field Sta, US Fish & Wildlife Serv, 74-84, dir, Condor Field Sta, 84-86; LEADER, FISH & WILDLIFE RES UNIT, UNIV IDAHO, 86- *Concurrent Pos:* Instr ornithol, Malheur Environ Field Sta, Pac Univ, 72 & 73; leader, Maui Forest Bird Recovery Team, 75-79, Hawaii Forest Bird Recovery Team, 75-; mem, Am Ornithologists Union Conserv, 74-75 & 75-76, Sci Adv Bd, Nature Conserv Hawaii Forest Bird Proj, 81-; Richard M Nixon Scholar, Whittier Col; mem, Palila Recovery Team, 75-; mem, Nature Conservancy. *Mem:* Elective Am Ornithologists Union; Ecol Soc Am; The Wildlife Soc; Soc Conserv Biol; Inst Biol Sci. *Res:* Determining limiting factors for endangered species; devising methods for estimating bird numbers which are statistically sound and cost efficient; preserve design for native species and communities. *Mailing Add:* 1130 Kamiaken Moscow ID 83843

SCOTT, JAUNITA SIMONS, b Richland Co, SC, June 13, 36; m 59; c 3. BIOLOGY, ZOOLOGY. *Educ:* Livingstone Col, BS, 58; Atlanta Univ, MS, 62; Univ SC, EdD, 79. *Prof Exp:* Teacher biol & sci, Hopkins High Sch, SC, 58-60; instr biol, Morris Col, Sumter, SC, 64-65; from instr to assoc prof biol, Benedict Col, 63-80, dir, Biol Study Prog & Pre-Med Adv, 72-80, head, Dept Biol Sci & dir, Health Careers Proj, 81-87, PROF BIOL, BENEDICT COL, 81-, CHAIRPERSON, DIV MATH & NATURAL SCI, 87- *Concurrent Pos:* Prin investr, minority biomed support grant, NIH, 74-79 & 79-84; co-investr grant, NIH, 84-87; developer, Middle Sch Summer Lab Sci & Math Enrichment Prog, & dir, 84-; developer & dir, Middle Sch Sci Develop Prog for 5th & 6th grade teachers, 87-90; consult, SC State Dept Educ Progs, SC State Col, Orangeburg, 75. *Mem:* Am Inst Biol Sci. *Res:* Light and electron microscopic studies on development and regeneration in Rana pipiens as well as heavy metal pollutants on devlopment and ultrastructures in Rana pipiens (spring frog). *Mailing Add:* Benedict Col Harden & Blanding Sts Columbia SC 29204

SCOTT, JOHN CAMPBELL, b Edinburgh, Scotland, Oct 5, 49; m 75. PHYSICS, MATERIALS SCIENCE. *Educ:* Univ St Andrews, BSc, 71; Univ Pa, PhD(physics), 75. *Prof Exp:* Asst prof physics, Cornell Univ, 75-80; RES STAFF MEM, SAN JOSE RES LAB, IBM CORP, 80- *Mem:* AAAS; Am Phys Soc; Mat Res Soc. *Res:* Experimental solid state physics; magnetic properties of solids; low dimensional conductors; quasi-one-dimensional magnetic materials; electronic properties of polymers; photoconductivity. *Mailing Add:* IBM Almaden Res Ctr K34/803 650 Harry Rd San Jose CA 95120-6099

SCOTT, JOHN DELMOTH, b San Antonio, Tex, Aug 8, 44; m 67; c 3. OPTICS DESIGN FOR VUV THROUGH SOFT X-RAY, SYNCHROTRON RADIATION. *Educ:* Baylor Univ, BS, 67; NTex State Univ, PhD(chem), 74. *Prof Exp:* Teacher, John Marshall High Sch, 67-69; res fel, NTex State Univ, 75; from asst prof to assoc prof chem, Univ Mont, 78-88; res fel, 75-78, SR STAFF SCIENTIST, CTR ADVAN MICROSTRUCT & DEVICES, LA STATE UNIV, 88- *Concurrent Pos:* Res fels, Robert A Welch Found, 75 & Energy Res & Develop Admin, 75-78. *Mem:* Am Chem Soc; Optical Soc Am; Sigma Xi. *Res:* Experimental and theoretical investigation of excited electronic states of molecules, principally molecular Rydberg states; development of a synchrotron-radiation facility (storage ring). *Mailing Add:* Ctr Advan Microstruct & Devices La State Univ Baton Rouge LA 70803

SCOTT, JOHN E(DWARD), JR, b Portsmouth, Va, Nov 29, 27; m 52; c 4. AEROSPACE ENGINEERING. *Educ:* Va Polytech Inst, BS, 48; Purdue Univ, MS, 50; Princeton Univ, MA, 53, PhD(aeronaut eng), 59. *Prof Exp:* Instr mech eng, Va Polytech Inst, 48; asst, Purdue Univ, 48-50; sr scientist, Exp, Inc, Va, 51; asst, Princeton Univ, 52-54, actng tech dir, Proj Squid, 54-56; assoc prof aeronaut eng, 56-62, res dir, Astronaut Div, 56-60, head aerospace div, Res Labs Eng Sci, 60-67, chmn dept aerospace eng & eng physics, 72-77, chmn dept mech & aerospace eng, 80-82, PROF AEROSPACE ENG, UNIV VA, 62- *Concurrent Pos:* Dir Proj Squid, 62-67; liaison scientist, Br Off, Off Naval Res, London, 67-68; prog mgr res atomic interactions basic to macroscopic properties of cases, Univ Va, 68- *Mem:* AAAS; Am Phys Soc; Am Inst Aeronaut & Astronaut; Sigma Xi. *Res:* Gas dynamics; propulsion; astronautics; molecular physics. *Mailing Add:* Sch Eng & Appl Sci Univ Va PO Box 9025 Charlottesville VA 22906

SCOTT, JOHN FRANCIS, b New Orleans, La, July 29, 44; c 1. DNA REPLICATION, MOLECULAR GENETICS. *Educ:* Univ Calif, Berkeley, BS, 74; Stanford Univ, PhD(biochem), 79. *Prof Exp:* Asst molecular biologist & spec res fel, Molecular Biol Inst, Univ Calif, Los Angeles, 79-81; asst prof microbiol, Univ Ill, Urbana-Champaign, 81-87; ASSOC PROF BIOL, UNIV HAWAII, HILO, 86-, ASSOC PROF GENETICH, MANOA, HONOLULU, 89- *Concurrent Pos:* Prin investr, Inst Gen Med Sci res grants, NIH, 79-86, NSF res grant, 83-87; prog dir, NIH MBRS grant, 88- *Mem:* Am Soc Microbiol; Am Soc Cell Biol; AAAS; Sigma Xi; Am Chem Soc. *Res:* Mechanism of nuclear DNA replication in yeast; chromatin structure and function of yeast chromosomal replicators; further development of yeast as an organism useful for biotechnology and research. *Mailing Add:* Biol Dept Nat Sci Div Univ Hawaii 523 W Lanikaula St Hilo HI 96720-4091

SCOTT, JOHN MARSHALL WILLIAM, b Ipswich, Suffolk, UK, Nov 2, 30; div; c 3. CHEMISTRY. *Educ:* Univ London, BSc, 53, PhD(chem), 56. *Prof Exp:* Temp sci officer, Serv Electronics Res Lab, Harlow, UK, 55-57; Nat Res Coun Can fel chem, 57-59; temp lectr, Queen Mary Col, Univ London, 59-60; res chemist, Am Cyanamid Co, 60-62; assoc prof, 62-67, head dept, 68-69, 70-72, PROF CHEM, MEMORIAL UNIV, 67- *Concurrent Pos:* Hon vis prof, Univ Calgary, 69-70. *Res:* Organic mechanisms; physical and organic chemistry. *Mailing Add:* Dept Chem Mem Univ St John's NF A1B 3B7 Can

SCOTT, JOHN PAUL, b Kansas City, Mo, Dec 17, 09; m 33, 79; c 4. ZOOLOGY, PSYCHOLOGY. *Educ:* Univ Wyo, BA, 30; Oxford Univ, BA, 32; Univ Chicago, PhD(zool), 35. *Prof Exp:* Grad asst, Univ Chicago, 32-35; from assoc prof to prof zool, Wabash Col, 35-45, chmn dept, 35-45; res assoc & chmn div behav studies, Jackson Mem Lab, 45-57, sr staff scientist, 57-65, trustee, 46-49; res prof, 65-68, Ohio Regents prof psychol, 68-80, EMER REGENTS PROF PSYCHOL, BOWLING GREEN STATE UNIV, 80- *Concurrent Pos:* Vis prof, Univ Chicago, 58, Tufts Univ, 81-82; fel, Ctr Adv Study Behav Sci, 63-64. *Honors & Awards:* Jordan Prize, 47; Dobzhansky Award, Behav Genetics Asn, 87; Distinguished Animal Behaviorist, Animal Behav Soc, 90. *Mem:* AAAS; Am Soc Zool; Int Soc Develop Psychobiol (pres, 73-); Int Soc Res Aggression (pres, 73-74); Behav Genetics Asn (pres, 75-76); Animal Behav Soc. *Res:* Embryology and physiological genetics of the guinea pig; genetics and behavior of Drosophila; sociobiology; genetics and social behavior of dogs, mice and other mammals; development of behavior. *Mailing Add:* Dept Psychol Bowling Green State Univ Bowling Green OH 43403

SCOTT, JOHN STANLEY, b Hamilton, Ont, July 14, 29; m 56; c 2. GEOLOGY. *Educ:* McMaster Univ, BSc, 53; Univ Ill, Urbana, PhD(geol), 60. *Prof Exp:* Geologist, Photog Surv Corp Ltd, Can, 53-57; geologist, Geol Surv Can, 60-67; consult geologist, H G Acres & Co Ltd, Can, 67-69; res scientist, Geol surv Can, 69-74, dir, Terrain Sci Div, 74-87, dir gen, Geophys & Terrain Sci Br, 87-89, DIR GEN, SEDIMENTARY & CORDILLERAN GEOSCI BR, GEOL SURV CAN, 89- *Concurrent Pos:* Mem assoc comt geotech res, Nat Res Coun Can, 61-66 & 82-87; counr, Geol Soc Am, 87-89. *Mem:* Fel Geol Soc Am; fel Geol Asn Can; Can Geotech Soc; Sigma Xi. *Res:* Engineering geology; hydrogeology as related to construction; stability of slopes in overconsolidated shales; nuclear fuel waste management. *Mailing Add:* Geol Surv Can 601 Booth St Ottawa ON K1A 0E8 Can

SCOTT, JOHN W(ALTER), b Berkeley, Calif, May 27, 19; m 42; c 5. CHEMICAL ENGINEERING. *Educ:* Univ Calif, BS, 41, MS, 51. *Prof Exp:* Res chemist, Standard Oil Co, Calif, 46-56, sr res chemist, 56-57, supvr res engr, 57-59, supvr petrol process develop, 60-64, mgr petrol process res & develop div, 64-67, vpres process res, Chevron Res Co, 67-84. *Concurrent Pos:* Chmn & mem adv bd, Chem Eng Dept, Univ Calif, Berkeley, 74-81; consult, 84-; chmn res, Data Info Serv Comm, Am Petrol Inst, 71-73 & 77-80; awards comt, Am Inst of Chem Eng, 79-84; coun mem, Lawrence Hall Sci, 90- *Honors & Awards:* Award in Chem Eng Pract, Am Inst Chem Eng. *Mem:* Nat Acad Eng; fel Am Inst Chem Eng; fel AAAS; Am Chem Soc. *Res:* Physical chemistry; adsorption; synthetic fuels; processing; catalytic hydrogenation; hydrocracking; catalysis. *Mailing Add:* Box 668 Ross CA 94957

SCOTT, JOHN WARNER, b Rochester, NY, Sept 27, 48; m 75; c 2. HORTICULTURE. *Educ:* Mich State Univ, BS, 70, MS, 74; Ohio State Univ, PhD(hort), 78. *Prof Exp:* Res technician, Mich State Univ, 70-75; res assoc hort, Ohio State Univ, 75-78, asst prof, 78-81; ASST PROF VEG CROPS, UNIV FLA, 81- *Concurrent Pos:* Mem staff, Gulf Coast Res & Educ Ctr, 81-; consult, DNA Plant Technol, 85- *Mem:* Am Soc Hort Sci. *Res:* Breeding, genetics, and culture of vegetable crops, especially tomatoes. *Mailing Add:* GCREC 5007 60th St E Bradenton FL 34203

SCOTT, JOHN WATTS, JR, b Oct 5, 38; US citizen; m 66. NEUROANATOMY, NEUROPHYSIOLOGY. *Educ:* Ala Col, AB, 61; Univ Mich, Ann Arbor, PhD(psychol), 65. *Prof Exp:* NIMH fel, Rockefeller Univ, 65-67, asst prof physiol psychol, 67-69; asst prof, 69-76, PROF ANAT, SCH MED, EMORY UNIV, 76- *Concurrent Pos:* Nutrit Found grant, Emory Univ, 70-72, NSF grant, 71-73, 78-, NIMH res develop award, 71-76; Nat Inst Neurol & Commun Disorders & Stroke grant, 78-; mem, Behav & Neurosci Study Sect, NIH, 83-86 & Sensory Disorders & Lang Study Sect, 86-87. *Mem:* Am Asn Anatomists; Soc Neurosci. *Res:* Olfactory projections to the lateral hypothalamus, physiological properties of feedback circuits of the olfactory bulb, organization of the olfactory projection system. *Mailing Add:* Dept Anat Sch Med Emory Univ Atlanta GA 30322

SCOTT, JOHN WILSON, neurophysiology, for more information see previous edition

SCOTT, JOSEPH HURLONG, b Atlantic City, NJ, Dec 5, 34; m 76; c 5. PROCESS DEVELOPMENT, SOLID STATE DEVICE DESIGN. *Educ:* Lincoln Univ, Pa, AB, 57. *Prof Exp:* Eng, Solid State Div, RCA, 58, 67, mem tech staff David Sarnoff Res Ctr, 67-70, Head IC Tech, 70-74, dir, 74-79; dir, Res & Develop, Gen Instruments Corp, 79-82; dir, Monsanto Elec Materials, 82-84; pres & chief exec officer, Cadmemic Electronics, 84-85; ASSOC DIR, OLIN CHEM RES, 85- *Concurrent Pos:* Tech bd dir, Panel Vision Corp, 82-84; co-auth & consult, Nat Acad Sci, 83-84; consult, Rockwell Inst, 84-85. *Honors & Awards:* George C Marshall Space Flight Ctr, Nat Aeronaut & Space Admin, 71 & 72, Cert Recognition, 73. *Mem:* NY Acad Scis; Sigma Xi; Int Electronic Devices Soc; Am Chem Soc; Elec Chem Soc. *Res:* Solid State devices and materials while directing efforts in integrated circuit applications; design automation and materials. *Mailing Add:* 430 W Lake Ave 350 Knotter Dr PO Box 586 Guilford CT 06437

SCOTT, JOSEPH LEE, b Delano, Calif, Mar 18, 43; m 64; c 3. BOTANY, CYTOLOGY. *Educ:* Univ Calif, Santa Barbara, AB, 65, MA, 67; Univ Calif, Irvine, PhD(biol), 71. *Prof Exp:* From asst prof to assoc prof, 70-85, PROF BIOL, COL WILLIAM & MARY, 85- *Mem:* Phycol Soc Am; Int Phycol Soc; Am Soc Cell Biol. *Res:* Development and ultrastructure of algae, particularly cell division and reproductive differentiation in red algae. *Mailing Add:* Dept Biol Col William & Mary Williamsburg VA 23185

SCOTT, JUNE ROTHMAN, b New York, NY, Nov 28, 40; m 66. MICROBIAL GENETICS. *Educ:* Swarthmore Col, BA, 61; Mass Inst Technol, PhD(microbiol), 64. *Prof Exp:* Guest investr, Rockefeller Univ, 64-66, res assoc, 66-69; from asst prof to assoc prof, 69-81, PROF MICROBIOL, SCH MED, EMORY UNIV, 81- *Concurrent Pos:* Nat Cancer Inst fel, 65-66; div lectr, Am Soc Microbiol, 88, 90. *Mem:* Am Soc Microbiol; Genetics Soc Am. *Res:* Bacterial genetics; gene regulation; bacterial pathogensis. *Mailing Add:* Dept of Microbiol Sch of Med Emory Univ Atlanta GA 30322

SCOTT, KAREN CHRISTINE, b Newport Beach, Calif, Oct 6, 57. COMPARTMENTAL MODELING. *Educ:* Ohio State Univ, BS, 84, MS, 86, PhD(poultry sci), 89. *Prof Exp:* Res assoc, Ohio State Univ, 84-89; POSTDOCTORAL, AGR RES SERV WESTERN HUMAN NUTRIT RES CTR, USDA, 90- *Mem:* Assoc mem Am Inst Nutrit; Poultry Sci Asn; Am Soc Animal Sci. *Res:* Kinetic modeling of copper, zinc, and molybdenum metabolism in adult men; trace mineral analyses of menhaden fish tissues; fat-soluble vitamin analyses of fish meals and fish oils. *Mailing Add:* 219 Santa Cruz Dr Fairfield CA 94533

SCOTT, KENNETH ELSNER, b Webster, Mass, May 18, 26; m 52; c 4. MECHANICAL ENGINEERING. *Educ:* Worcester Polytech Inst, BS, 48, MS, 54. *Prof Exp:* From instr to assoc prof mech eng, Worcester Polytech Inst, 48-65, Alden prof eng & inst dir audiovisual develop, 72-75, dir instrnl TV, 72-90, PROF MECH ENG, WORCESTER POLYTECH INST, 65-, DIR COMPUTER AIDED DESIGN LAB, 81- *Honors & Awards:* Western Elec Fund Award, Am Soc Eng Educ, 73. *Mem:* Am Soc Mech Engrs; Am Soc Eng Educ; Sigma Xi. *Res:* Education; innovator of teaching methods; pioneer in use of individually prescribed instruction methods and use of audio-visuals in supporting these methods at Worcester Polytechnic Institute. *Mailing Add:* Dept Mech Eng Worcester Polytech Inst Worcester MA 01609

SCOTT, KENNETH RICHARD, b New York, NY, Apr 17, 34; m 52; c 2. MEDICINAL CHEMISTRY, ANALYTICAL CHEMISTRY. *Educ:* Howard Univ, BS, 56; Univ Buffalo, MSc, 59; Univ Md, PhD(pharm chem), 66. *Prof Exp:* Asst pharm, biochem & anal chem, Univ Buffalo, 56-59; from instr to assoc prof, 60-75, PROF ANALYTICAL CHEM & INORG PHARMACEUT CHEM, COL PHARM, HOWARD UNIV, 75-, ASST DEAN STUDENT AFFAIRS & RECRUITMENT, 71- *Concurrent Pos:* Consult, S F Durst Co, 66-67 & NIH, 67-68; proj dir sem recruitment & retention minority disadvantaged students health professions, Howard Univ, 71; consult, Student Nat Pharmaceut Asn, 72 & Student Health Manpower Conf, 72; actg chmn, Dept Biomed Chem, 76. *Mem:* Am Chem Soc; Am Asn Cols Pharm; Am Pharmaceut Asn; Nat Pharmaceut Asn; fel Am Inst Chemists; Sigma Xi. *Res:* Synthetic chemistry, spiranes, carbazoles, steroids and biological testing; analytical chemistry, newer techniques in the development of assay procedures of pharmaceutical preparations. *Mailing Add:* Dept Biomed Chem Howard Univ 2400 Sixth St NW NW Washington DC 20059

SCOTT, KENNETH WALTER, b Cleveland, Ohio, May 18, 25; m 67; c 4. POLYMER CHEMISTRY. *Educ:* Univ Mich, BS, 46; Princeton Univ. AM, 48, PhD(chem), 49. *Prof Exp:* Sr res chemist, Eastman Kodak Co, 49-55; res scientist, Goodyear Tire & Rubber Co, 55-57, sect head basic rubber res, 57-67, mgr, Basic Polymer Res, 66-73, New Prod Res, 74-77, Transp Prod Res, Res Div, 77-83, Tire & Process Sci, 83-87; VIS SCIENTIST, INST POLYMER SCI, UNIV AKRON, 87- *Mem:* Am Phys Soc; Am Chem Soc. *Res:* Viscoelastic behavior; chemistry and physics of high polymers. *Mailing Add:* 3030 Oakridge Dr Cuyahoga Falls OH 44224

SCOTT, KEVIN M, b Iowa City, Iowa, Aug 3, 35; div; c 2. GEOLOGY. *Educ:* Univ Calif, Los Angeles, BA, 57, MA, 60; Univ Wis, PhD(geol), 64. *Prof Exp:* Geologist, US Geol Surv, 59-60; proj assoc geol, Univ Wis, 61-64; NATO fel, Univ Edinburgh, 64-65; GEOLOGIST, WATER RESOURCES DIV, US GEOL SURV, 65- *Concurrent Pos:* Vis prof, Chinese Acad Sci, 90-93. *Honors & Awards:* Kirk Bryan Award, Geol Soc Am, 89. *Mem:* Am Asn Petrol Geol. *Res:* Sedimentology of marine and fluvial systems; sedimentary structures and their hydrodynamic interpretation; fluvial morphology; changes in sedimentologic parameters and mineralogy of sediments; environmental geomorphology; sedimentology and hazards assessment of lahars. *Mailing Add:* US Geol Surv 5400 MacArthur Blvd Vancouver WA 98661

SCOTT, L MAX, b Sweetwater, Tex, May 9, 34; m 56; c 3. HEALTH PHYSICS. *Educ:* Tex A&M, BS, 55; Purdue Univ, MS, 59, PhD(genetics), 61. *Prof Exp:* Internal dosimetry specialist, Union Carbide, 61-77; dir radiation health physics, Gulf Oil Corp, 77-85; ASST PROF NUCLEAR SCI & RADIATION SAFETY OFFICER, LA STATE UNIV, 85- *Concurrent Pos:* Prog chmn, Health Physics Soc, 87-89. *Mem:* Fel Health Physics Soc; Sigma Xi. *Res:* Radiation protection; exposure evaluation. *Mailing Add:* Ctr Energy Studies La State Univ Baton Rouge LA 70803

SCOTT, LAWRENCE TRESSLER, b Ann Arbor, Mich, June 11, 44; m 66; c 4. ORGANIC CHEMISTRY. *Educ:* Princeton Univ, AB, 66; Harvard Univ, PhD(org chem), 70. *Prof Exp:* Asst prof org chem, Univ Calif, Los Angeles, 70-75; from asst prof to prof, 75-85, FOUND PROF CHEM, UNIV NEV, RENO, 85- *Concurrent Pos:* Petrol Res Fund grant, 70-73, 75-77, 78-80 & 88-90, Res Corp grant, 74-75; NSF grant, 73-76, 79-85 & 85-88 & 88-91; NIH grant, 79-85 & 85-88; NATO sr scientist award, 81; sr scientist fel, Japan Soc Prom Sci; US-Israel Binat Found grant, 86-89; Dept Energy grant, 88-92; NATO grant, 88-90. *Mem:* Am Chem Soc. *Res:* Organic synthesis of chemically intriguing molecules; structural requirements for electron delocalization and the chemical consequences thereof; cyclic conjugation and homoconjugation; aromaticity and pericyclic reactions; cyclic polyacetylenes and nonplanar aromatic hydrocarbons; thermal rearrangements of aromatic compounds. *Mailing Add:* Dept Chem Univ Nev Reno NV 89557-0020

SCOTT, LAWRENCE VERNON, b Anthony, Kans, Jan 28, 17; m 45; c 3. VIROLOGY. *Educ:* Phillips Univ, BA, 40; Univ Okla, MS, 47; Johns Hopkins Univ, ScD, 50. *Prof Exp:* From asst prof to assoc prof bact, 50-58, PROF MICROBIOL, SCH MED, UNIV OKLA HEALTH SCI CTR, 58-, CHMN DEPT, 61- *Concurrent Pos:* Consult, St Anthony & Vet Admin Hosps. *Mem:* Sigma Xi; Am Soc Microbiol; fel Am Acad Microbiol; NY Acad Sci; Am Soc Trop Med & Hyg. *Res:* Viral diseases of man; influenza; herpes simplex; Rous sarcoma; arboviruses. *Mailing Add:* 4125 NW 61st Terr Oklahoma City OK 73112

SCOTT, LAWRENCE WILLIAM, b Manhattan, Kans, June 10, 24; m 49; c 3. FOOD SCIENCE, ANALYTICAL CHEMISTRY. *Educ:* Kans State Univ, BS, 51, MS, 54; Univ Mo, Columbia, PhD(food sci & nutrit), 70. *Prof Exp:* Res & qual control chemist, Gen Lab, Utah-Idaho Sugar Co, 54-61; ASSOC PROF CHEM, UNIV WIS, RIVER FALLS, 61- *Mem:* Inst Food Technologists; Am Chem Soc. *Res:* Design of new experiments and modification of old to present interesting science to non-science students; laboratory safety. *Mailing Add:* 521 E Maple Univ Wis River Falls WI 54022

SCOTT, LELAND LATHAM, b Elba, Ill, Mar 31, 19; m 46; c 3. MATHEMATICS. *Educ:* Southern Ill Univ, BS, 47; Univ Ill, MS, 48, PhD(math), 51. *Prof Exp:* Asst math, Univ Ill, 47-51; from asst prof to assoc prof, Univ Miss, 51-57; from assoc prof to prof, Southwestern at Memphis, 57-62; assoc prof, 62-64, PROF MATH, UNIV LOUISVILLE, 64-, PROF GRAD SCH, 69- *Concurrent Pos:* Ford Found fac fel, 55-56. *Mem:* Am Math Soc; Math Asn Am; Asn Symbolic Logic. *Mailing Add:* 4225 University Dr Fairfax VA 22030

SCOTT, LEONARD LEWY, JR, b Little Rock, Ark, Oct 17, 42; m 60; c 2. ALGEBRA. *Educ:* Vanderbilt Univ, BA, 64; Yale Univ, MA, 66, PhD(math), 68. *Prof Exp:* Instr math, Univ Chicago, 68-70; asst prof, Yale Univ, 70-71; from assoc prof to prof, 71-86, MCCDONNEL-BERNARD PROF MATH, UNIV VA, 87- *Concurrent Pos:* Mem ctr advan studies, Univ Va, 71-73; vis assoc prof, Univ Mich, 74-75; vis prof, Yale Univ, 78. *Mem:* Am Math Soc. *Res:* Finite permutation groups; representation theory; cohomology; algebraic groups. *Mailing Add:* Dept Math Univ Va Astro Bldg 207 Charlottesville VA 22903

SCOTT, LINUS ALBERT, b Jacksonville, Fla, June 8, 23; m 43; c 2. MECHANICAL ENGINEERING. *Educ:* Univ Fla, BME, 48, MSE, 51; Case Inst Technol, PhD(mech eng), 60. *Prof Exp:* From instr to asst prof mech eng, Univ Fla, 48-57; instr, Case Inst Technol, 57-58 & 59-60; assoc prof, Univ Fla, 60-62; prof, Univ Toledo, 63; PROF CHEM & MECH ENG & CHMN DEPT, UNIV SFLA, 64-, ASSOC DEAN ACAD AFFAIRS, 82- *Mem:* Am Soc Mech Engrs. *Res:* Instrumentation and automatic control of industrial operations. *Mailing Add:* Col Eng Univ SFla 4202 Fowler Ave Tampa FL 33620

SCOTT, MACK TOMMIE, b Grand Junction, Tenn, Dec 4, 31; m 58; c 2. ANIMAL SCIENCE, VETERINARY MEDICINE. *Educ:* Tenn State Univ, BS, 58; Tuskegee Inst, DVM, 62. *Prof Exp:* Instr clin vet med, Tuskegee Inst, 62-63; ASST PROF RES VET PHYSIOL & DIR EXP ANIMAL HOSP, MEHARRY MED COL, 63- *Mem:* Am Vet Med Asn; Am Asn Lab Animal Sci. *Res:* Laboratory animal nutrition and diseases; cellular physiology; experimental production of kernicterus and study of the pathogenesis of hemolytic anemia and jaundice in new born rabbits. *Mailing Add:* Animal Care Facil Meharry Med Col 1005 DB Todd Blvd Nashville TN 37208

SCOTT, MARION B(OARDMAN), b Ashland, Nebr, Dec 9, 12; m 42; c 2. CIVIL ENGINEERING. *Educ:* Univ Nebr, BSCE, 34; Purdue Univ, MSCE, 44. *Prof Exp:* Engr, State of Nebr Bur Rd & Irrig, 34-37; instr engr drawing, Purdue Univ, 37-41, instr civil eng, 42-44; engr, Third Locks Proj, CZ, 41-42; process engr & asst to head physics dept, Curtiss-Wright Res Lab, 44-46; from asst prof to assoc prof, 46-57, head struct eng, 62-64, asst dean, 64-68, PROF CIVIL ENG, PURDUE UNIV, 57-, ASSOC DEAN ENG, 68- *Mem:* AAAS; Am Soc Civil Engrs; Rwy Eng Asn; Soc Exp Stress Anal; Am Soc Eng Educ; Sigma Xi. *Res:* Analysis, stress measurement and performance of structural steel bridges, buildings and structural components. *Mailing Add:* 1500 N Grant St West Lafayette IN 47906

SCOTT, MARTHA RICHTER, b Dallas, Tex, July 8, 41. MARINE GEOCHEMISTRY. *Educ:* Rice Univ, BA, 63, PhD(geol), 66. *Prof Exp:* NSF fel geol, Yale Univ, 66-67; res assoc geol & oceanog. Fla State Univ, 67-69, 70-71; res assoc, 71-74, vis asst prof oceanog, 74-75, ASSOC PROF OCEANOG, TEX A&M UNIV, 80- *Mem:* Am Geophys Union; Am Geol Inst; Am Soc Oceanog; Geochem Soc; AAAS. *Res:* Interaction of land-derived materials with sea water; uranium series isotopes in sea water and sediments; adsorption chemistry in marine environment; incorporation of trace metals into ferromanganese deposits; chemistry of plutonium isotopes in the environment. *Mailing Add:* Dept Oceanog Tex A&M Univ Col Stat TX 77843

SCOTT, MARVIN WADE, b Clifton Forge, Sept 6, 36; m 62; c 2. BOTANY, MICROBIOLOGY. *Educ:* Hampden-Sydney Col, BS, 59; Va Polytech Inst & State Univ, PhD(bot), 68. *Prof Exp:* Chmn dept sci, Lynchburg Pub Sch Syst, Va, 59-60; res asst bot, Longwood Col, 60-62; instr biol, Hampden-Sydney Col, 62-63; assoc prof, 66-71, PROF BIOL, LONGWOOD COL, 71-, CHMN DEPT NATURAL SCI, 70- *Mem:* AAAS; Bot Soc Am. *Res:* Isolation and fermentation studies of Streptococcus lactis variety tardus; genetics and cytology studies of species of Illiamn. *Mailing Add:* Stevens Hall Longwood Col Farmville VA 23901

SCOTT, MARY JEAN, b Brooklyn, NY, Nov 8, 31; m 59; c 3. MEDICAL PHYSICS, NUCLEAR PHYSICS. *Educ:* St Lawrence Univ, BS, 52; Johns Hopkins Univ, PhD(physics), 58. *Prof Exp:* Vis asst physics, Brookhaven Nat Lab, 55, jr res assoc, 55-58; asst prof, Bryn Mawr Col, 58; vis res assoc nuclear physics, Atomic Energy Res Estab, Eng, 58-60; res assoc, Univ Witwatersrand, 60-61, lectr physics, 61-67; med physicist, Johannesburg Gen Hosp, 67-80; SR MED PHYSICIST, HILLBROW HOSP, JOHANNESBURG, 80- *Mem:* AAAS; Am Phys Soc; SAfrican Asn Physicists Med; SAfrican Inst Physics; SAfrican Soc Nuclear Med (secy-treas, 74-78); Am Asn Physics Teachers; Am Asn Physicists Med. *Res:* Proton polarization; radiobiology; effects of fractionated radiation therapy on cell populations and determination of cell parameters. *Mailing Add:* Dept Med Physics Hillbrow Hosp Joubert Park 2044 Johannesburg South Africa

SCOTT, MATTHEW P, b Boston, Mass, Jan 30, 53; m; c 1. MOLECULAR GENETICS, HOMEOTIC GENES. *Educ:* Mass Inst Technol, BS, 75, PhD(biol), 80. *Prof Exp:* Postdoctoral fel biol, Ind Univ, 80-83; from asst prof to assoc prof biol, Univ Colo, Boulder, 83-90; PROF DEVELOP BIOL, STANFORD UNIV SCH MED, 90- *Concurrent Pos:* Jr fac res award, Am Cancer Soc, 83-84; res career develop award, NIH, 84-89; Searle Scholar's Award, 85-89; Young Investr Award, Passano Found, 90; co-ed, Current Opinions in Genetics & Develop. *Mem:* Genetics Soc Am; Soc Develop Biol; Am Soc Cell Biol; Am Soc Biochem & Molecular Biol. *Res:* Research is focused on understanding how genes control animal development; molecular biology is combined with mutants that alter the structures of animals to determine how patterns are formed during embryogenesis. *Mailing Add:* Dept Develop Biol Stanford Univ Sch Med Beckman Ctr B 300 Stanford CA 94305-5427

SCOTT, MECKINLEY, b Shillong, India, Mar 1, 35; m 65; c 1. APPLIED MATHEMATICS, STATISTICS. *Educ:* Presidency Col Calcutta, India, BS, 55; Gauhati Univ, India, MS, 57; Univ NC, Chapel Hill, PhD(statist), 64. *Prof Exp:* Asst prof math, Colo Sch Mines, 64-65; from asst prof to prof math, Univ Ala, Tuscaloosa, 65-90; PROF MATH & CHAIRPERSON, WESTERN ILL UNIV, 91- *Mem:* Inst Math Statist; Oper Res Soc Am; Nat Coun Teachers Math; Math Asn Am. *Res:* Applied probability; theory of queues. *Mailing Add:* Math Dept Western Ill Univ Macomb IL 61455

SCOTT, MILTON LEONARD, b Tempe, Ariz, Feb 21, 15; m 38; c 2. NUTRITION. *Educ:* Univ Calif, AB, 37; Cornell Univ, PhD(nutrit), 45. *Prof Exp:* Vitamin chemist, Coop GLF Mills, Inc, Buffalo, NY, 37-42; fel, 42-44, res assoc, 44-45, from asst prof to prof nutrit, 45-76, Jacob Gould Schurman prof & chmn dept poultry sci, 76-79, JACOB GOULD SCHURMAN EMER PROF NUTRIT, CORNELL UNIV, 79- *Concurrent Pos:* Fel, Tech Univ Denmark, 61; consult, feed & pharmaceut industs. *Honors & Awards:* Borden Award, 65; NY Farmers Award, 71; Borden Award, Am Inst Nutrit, 77; Klaus Schwarz Award, 80; Earle W Crampton Award, 81. *Mem:* AAAS; Am Inst Nutrit; Am Poultry Sci Asn; Soc Exp Biol & Med. *Res:* Biochemistry and nutrition. *Mailing Add:* 16 Spruce St Ithaca NY 14850

SCOTT, NORMAN JACKSON, JR, b Santa Monica, Calif, Sept 30, 34; m 56; c 2. HERPETOLOGY, VERTEBRATE ECOLOGY. *Educ:* Humboldt State Col, BS, 56, MS, 62; Univ Southern Calif, PhD, 69. *Prof Exp:* Prof zool, Univ Costa Rica, 64-66; course coordr, Orgn Trop Studies, Miami, 66-70; asst prof biol sci, Univ Conn, 58-74; adj assoc prof biol, 75-90, BIOLOGIST, US FISH & WILDLIFE SERV, UNIV NMEX, 74-, ADJ ASSOC PROF BIOL, 90- *Mem:* Am Soc Ichthyologists & Herpetologists; Ecol Soc Am; Soc Syst Zool; Am Soc Zoologists; Asn Trop Biol; Soc Study Reptiles & Amphibians (pres, 87). *Res:* Herpetology; endangered species; tropical biology. *Mailing Add:* US Fish & Wildlife Serv Dept Biol Univ NMex Albuquerque NM 87131

SCOTT, NORMAN R(OSS), b Brooklyn, NY, May 15, 18; m 50; c 4. COMPUTER ENGINEERING. *Educ:* Mass Inst Technol, BS & MS, 41; Univ Ill, PhD(elec eng), 50. *Prof Exp:* Asst prof elec eng, Univ Ill, 46-50; from asst prof to prof & assoc dean, Col Eng, 51-68, dean, Dearborn Campus, 68-71, prof Elec Eng & Computer Sci, Univ Mich, Ann Arbor, 71-87; RETIRED. *Concurrent Pos:* Ed, Trans Electronic Computer, Inst Elec & Electronics Engrs, 61-65; mem, math & computer sci res adv comt, AEC, 62-65. *Mem:* Asn Computer Mach; fel Inst Elec & Electronics Engrs. *Res:* Engineering and logical design of electronic computers; arithmetic systems for computers. *Mailing Add:* Dept Elec Eng & Computer Sci Univ Mich Ann Arbor MI 48109

SCOTT, NORMAN ROY, b Spokane, Wash, Sept 6, 36; m 61; c 3. BIOENGINEERING & BIOMEDICAL ENGINEERING. *Educ:* Wash State Univ, BSAE, 58; Cornell Univ, PhD(agr eng), 62. *Prof Exp:* From asst prof to assoc prof agr eng, Cornell Univ, 62-76, chmn dept, 78-84, dir, Cornell Agr Exp Sta, 84-89, actg vpres comput info syst, 87-88, PROF AGR ENG, CORNELL UNIV, 76-, VPRES, RES & ADVAN STUDIES, 89- *Honors & Awards:* Henry Giese Award, Am Soc Agr Engrs. *Mem:* Nat Acad Engrs; fel Am Soc Agr Engrs; Am Soc Heat, Refrig & Air-Conditioning Engrs; Instrument Soc Am; NY Acad Sci; Sigma Xi; Am Soc Eng Educ. *Res:* Biological engineering study of poultry involving heat transfer and physiological responses; biomathematical modeling of animal systems; animal calorimetry; electronic instrumentation in biological measurements. *Mailing Add:* 314 Day Hall Cornell Univ Ithaca NY 14853-2801

SCOTT, PAUL BRUNSON, b Flint, Mich, Sept 8, 37; m 60; c 1. FLUID DYNAMICS, PHYSICS. *Educ:* Mass Inst Technol, SB & SM, 59, ScD(aeronaut eng), 65. *Prof Exp:* Mem staff vehicle anal, Space Tech Labs, 59-60; res staff molecular beam res, Mass Inst Technol, 65, asst prof aeronaut eng, 65-67; asst prof aerospace eng, Univ Southern Calif, 67-72; prin scientist, Xonics, Inc, 72-78; pres, Univ Consults, Inc, 78-80; vpres, Dinet Inc, 85-87; pres, Mammocare, Inc, 87-90; CONSULT, 90- *Concurrent Pos:* Ford Found fel. *Mem:* AAAS; Am Phys Soc. *Res:* Rarefied gas dynamics; molecular beams; intermolecular collisions; x-ray imaging; chemical kinetics; isotope separation, chemical and ultraviolet lasers; photovoltaics. *Mailing Add:* 17500 Lemarsh St Northridge CA 91325

SCOTT, PAUL G, b Newcastle-Upon-Lyme, Eng, June 11, 47; m 72; c 2. CONNECTIVE TISSUE BIOCHEMISTRY, PROTEIN STRUCTURE. *Educ:* Univ Liverpool, Eng, PhD(biochem), 73. *Prof Exp:* Assoc prof, 81-86, PROF ORAL BIOL, 86-, HON PROF EXP SURG, UNIV ALTA, 86- *Mem:* Am Chem Soc; Am Soc Biochem & Molecular Biol; Biochem Soc; Am Peptide Soc. *Res:* Collagen and proteoglycan structure and function. *Mailing Add:* Dept Oral Biol Univ Alta Edmonton AB T6G 2N8 Can

SCOTT, PETER CARLTON, b Seattle, Wash, Mar 20, 40. SCIENCE ADMINISTRATION, GRANT & CONTRACT ADMINISTRATION. *Educ:* Ore State Univ, BS, 63; Purdue Univ, Lafayette, PhD(plant physiol), 66. *Prof Exp:* Instr chem, Univ Calif, Santa Cruz, 66-67; instr chem, 69-72, chmn dept math & sci, 72-73, DIR DIV SCI & TECHNOL, LINN-BENTON COMMUNITY COL, 73- *Concurrent Pos:* Herman Frasch Found grant, Ore State Univ, 67-69; res assoc, Ore State Univ, 69- *Mem:* Nat Environ Training Asn; Am Chem Soc; Water Pollution Control Fedn; Am Pub Welfare Asn; Sigma Xi. *Res:* Physiology of plant growth regulators. *Mailing Add:* 3950 Me Hwy 20 Corvallis OR 97330

SCOTT, PETER DOUGLAS, b Kingston, Pa, Dec 13, 42; m 65; c 2. BIOMEDICAL ENGINEERING. *Educ:* Cornell Univ, BS, 65, MS, 68, PhD(elec eng), 71. *Prof Exp:* Asst prof, 70-77, ASSOC PROF ELEC ENG, STATE UNIV NY BUFFALO, 77- *Concurrent Pos:* Asst prof, Dept Biophys Sci, State Univ NY Buffalo, 75-; dir, Surg Res Computer Lab, Buffalo Gen Hosp, 75- *Mem:* Sigma Xi; Inst Elec & Electronics Engrs; Am Soc Eng Educ; Comt Social Responsibility in Eng. *Res:* Cardiac electrophysiology; automated intensive care unit monitoring; systems theory; cybernetics. *Mailing Add:* Elec/Computer Eng/136 Bell Hall State Univ NY at Buffalo-N Campus Buffalo NY 14260

SCOTT, PETER HAMILTON, b Providence, RI, Apr 6, 36; m 60; c 3. CHEMISTRY. *Educ:* Brown Univ, ScB, 60; Lehigh Univ, PhD(chem), 65. *Prof Exp:* Sr res chemist, Olin Res Ctr, Olin Mathieson Chem Corp, 65-67, sr res chemist II, 68-70; closure res mgr, Dewey & Almy Chem Div, W R Grace & Co, 70-75, asst dir res, 75-80, dir res, 80-83; dir technol, Georel Corp, 83-85; VPRES TECHNOL, COOLEY INC, 85- *Mem:* Am Chem Soc; Sci Res Soc Am; Soc Plastics Engrs; Am Asn Textile Chem & Colorists; Sigma Xi. *Res:* Reactions of nitrenes and nitrenelike intermediates; mechanisms of oxirane polymerization; organo-sulfur and heterocyclic chemistry; polyurethane chemistry, especially flame retardant polyurethanes; sealant compounds. *Mailing Add:* 33 Fales Ave Barrington RI 02806-1803

SCOTT, PETER JOHN, b Toronto, Can, July 2, 48. TAXONOMY. *Educ:* Univ Alta, BSc Hons, 70; Mem Univ Nfld, PhD(taxonomy), 73. *Prof Exp:* ASSOC PROF BOT & CUR HERBARIUM, MEM UNIV NFLD, 73- *Mem:* Sigma Xi; Am Soc Plant Taxonomists. *Res:* Flora of Newfoundland: origin, history and relationships with other floras. *Mailing Add:* Biol Dept Mem Univ Nfld St John's NF A1B 3X9 Can

SCOTT, PETER LESLIE, b San Francisco, Calif, May 14, 33; m 77; c 3. PHYSICS. *Educ:* Univ Calif, Berkeley, AB, 55, PhD(physics), 62; Univ Mich, MA, 57. *Prof Exp:* NSF fel, 62-63; asst prof physics, Stanford Univ, 63-66; ASSOC PROF PHYSICS, UNIV CALIF, SANTA CRUZ, 66- *Concurrent Pos:* Alfred P Sloan fel, 64-68; Fulbright-Hays fel, Galway, Ireland, 70-71, Danforth assoc, 79-85. *Mem:* Fel Am Phys Soc. *Res:* Magnetic resonance; solid state spectroscopy; nonlinear dynamics. *Mailing Add:* Nat Sci Div Univ Calif Santa Cruz CA 95064

SCOTT, PETER MICHAEL, b Blackpool, Eng, Aug 20, 38; UK & Can citizen; m 89; c 2. MYCOTOXINS. *Educ:* Cambridge Univ, BA, 59. PhD(org chem), 62, MA, 63. *Prof Exp:* NATO fel, Univ Calif, Berkeley, 62-64; fel, Univ BC, 64-65; RES SCIENTIST ANALYTICAL & ORG CHEM, CAN DEPT NAT HEALTH & WELFARE, 65- *Concurrent Pos:* Chmn, Joint Mycotoxins Comt, 76-, Asn Off Anal Chemists Gen Referee Mycotoxins, 82-; mem Mycotoxins Working Group, Int Union Pure & Appl Chem, 79-, adv bd, Mycotoxin Res, 84-, ed bd, J Food Protection, 84-92, Appl & Environ Microbiol, 86-93, Microbiol Aliments & Nutrit, 87- *Honors & Awards:* Harvey W Wiley Award, Asn Anal Chemist, 89. *Mem:* Chem Inst Can; Can Inst Food Sci & Technol; fel Asn Off Anal Chemists. *Res:* Mycotoxins and other fungal metabolites, isolation, identification and analysis in foodstuffs. *Mailing Add:* Health Protection Br Health & Welfare Can Ottawa ON K1A 0L2 Can

SCOTT, RALPH ASA, JR, b Sterling, Ill, July 23, 30; m 59; c 2. ANALYTICAL CHEMISTRY, RADIOCHEMISTRY. *Educ:* Univ Ill, BS, 52; Univ Okla, MS, 54; Tex A&M Univ, PhD(plant physiol, biochem, org chem), 57. *Prof Exp:* Radiochemist, Okla Res Inst, 52-53; res plant breeder, W Atlee Burpee Seed Co, Calif, 54-55; prin res chemist & dir waste eval proj, Int Minerals & Chem Corp, Fla, 57-58; res plant physiologist, Olin Mathieson Chem Corp, NY, 58; plant physiologist, Cotton Res Ctr, Crops Res Div, Agr Res Serv, USDA, Phoenix, Ariz, 58-61, sr res plant physiologist, Boll Weevil Res Lab, Starkville, Miss, 61-62; chief chemist, US Dept Defense, USAF, 6571st Aeromed Res Lab, Holloman AFB, NMex, 62-64, sr chemist, Adv Test Tech, Joint Chiefs Staff, Deseret Test Ctr, Ft Douglas, Utah, 65-66; chief div chem, Dept Pub Health, DC, 66-67; chief nationwide aquatic plant control prog, Off Chief Engrs, 67-69, chief phys scientist, Explosives Safety Bd, Dept Defense, 69-88; RETIRED. *Concurrent Pos:* Int sci adv, Secy State, 69-; personal rep Secy Defense, Chem Munition Safety, Okinawa; prog chmn, Chem Health & Safety Div, Am Chem Soc, 78-80 & 83-84, vchmn, 81-82 & chmn, 83-84. *Honors & Awards:* Patent Award & Eval Awards, US Govt, 65, 88. *Mem:* Am Chem Soc; Am Soc Plant Physiol; fel Am Inst Chemists. *Res:* Residue analysis, quality control and toxicology; air and water pollution analytical analysis control; allergies; carcinogenic chemicals mode of action; virological and microbiological chemistry as related to chemical ammunition. *Mailing Add:* PO Box 1104 Wickenburg AZ 85358

SCOTT, RALPH CARMEN, b Bethel, Ohio, June 7, 21; m 45; c 3. INTERNAL MEDICINE, CARDIOLOGY. *Educ:* Univ Cincinnati, BS, 42, MD, 45; Am Bd Internal Med, dipl; Am Bd Cardiovasc Dis, dipl. *Prof Exp:* Resident & asst path, Col Med, Univ Cincinnati, 48-49, from instr to assoc prof med, 50-68; dir cardiac clin, Cincinnati Gen Hosp, 65-75; prof, 68-90, EMER PROF MED, COL MED, UNIV CINCINNATI, 90- *Concurrent Pos:* Fel internal med & cardiol, Univ Cincinnati, 49-57; asst clinician internal med & cardiol, Cincinnati Gen Hosp, 50-51, clinician internal med, 52-, clinician cardiol, 52-55, asst chief clinician, 56-64, from asst attend physician to attend physician, 58-; attend physician, Vet Admin Hosp, 54-61, Holmes Hosp, 57- & Providence Hosp, Cincinnati, 71-; consult, USAF Hosp, Wright-Patterson AFB, 60-, Vet Admin Hosp, 62-85, Good Samaritan Hosp, 67-, Jewish Hosp, 68- & Children's Hosp, 68-85; fel coun clin cardiol, Am Heart Asn; staff, Med Ctr Univ Cincinnati. *Mem:* AMA; Int Cardiovasc Soc; fel Am Col Physicians; fel Am Col Chest Physicians; Am Heart Asn; Sigma Xi. *Res:* Electrocardiography. *Mailing Add:* 2955 Alpine Terr Cincinnati OH 45208

SCOTT, RALPH MASON, b Leemont, Va, Nov 23, 21; m 46; c 3. RADIOLOGY. *Educ:* Univ Va, BA, 47; Med Col Va, MD, 50. *Prof Exp:* Radiotherapist, Robert Packer Hosp, Sayre, Pa, 57-59; asst prof radiol, Sch Med, Univ Chicago, 59-60; assoc prof, Sch Med, Univ Louisville, 60-64, prof radiol, 64-77, dir radiation ther, 60-77; prof radiol & chmn, Dept Radiation Ther, Sch Med, Univ Md, 78-80; DIR, DEPT RADIATION MED, CHRIST HOSP, CINCINNATI, OHIO, 83-; CLIN PROF RADIOL, COL MED, UNIV CINCINNATI, 83- *Concurrent Pos:* pres, Am Bd Radiol, 72-74. *Mem:* Am Radium Soc; Am Col Radiol; Am Roentgen Ray Soc; Radiol Soc NAm; Asn Univ Radiol. *Res:* Clinical radiation therapy. *Mailing Add:* Christ Hosp Cincinnati OH 45219

SCOTT, RAYMOND PETER WILLIAM, b Erith, Eng, June 20, 24; m 46; c 2. PHYSICAL CHEMISTRY. *Educ:* Univ London, BSc, 46, DSc(chem), 58; FRIC, 58. *Prof Exp:* Lab mgr phys chem, Benzole Prod Res Labs, 50-60; div mgr phys chem, Unilever Res Labs, 60-69; DIR DEPT PHYS CHEM, HOFFMANN-LA ROCHE INC, 69- *Honors & Awards:* Chromatography Award, Am Chem Soc, 77; Tswett Medal Chromatography, Int Chromatography Symp, 78. *Mem:* Brit Chem Soc; Am Chem Soc. *Res:* Separations technology; gas chromatography; liquid chromatography; exclusion chromatography; gas chromatography/mass spectroscopy; liquid chromatography/mass spectroscopy; general physical chemical instrumentation; computer technology and data processing. *Mailing Add:* Two Sprucewood Lane Avon CT 06085

SCOTT, RICHARD ANTHONY, b Cork, Ireland, May 5, 36; m 67. APPLIED MATHEMATICS. *Educ:* Univ Col, Cork, BSc, 57, MSc, 59; Calif Inst Technol, PhD(eng sci), 64. *Prof Exp:* Lectr appl mech, Calif Inst Technol, 64-65, res fel, 65-67; from asst prof to assoc prof, 67-77, PROF ENG MECH, UNIV MICH, ANN ARBOR, 77-, ASSOC CHMN, DEPT MECH ENG & APPL MECH, 87- *Concurrent Pos:* Consult, Am Math Soc Math Rev, 68- *Mem:* Am Soc Mech Engrs; Am Acad Mech; Sigma Xi; Soc Indust & Appl Math. *Res:* Wave propagation in solids; elastic wave propagation; linear, nonlinear and random vibrations of solids; dynamics. *Mailing Add:* 104 Auto Lab Univ Mich Ann Arbor MI 48109

SCOTT, RICHARD ROYCE, b Fairfield, Ala, Mar 19, 33; m 69; c 1. FLUID MECHANICS. *Educ:* Univ Miss, BSME, 60; Univ Ala, PhD(eng mech), 68. *Prof Exp:* Proj engr, Procter & Gamble, Inc, 60-62; asst prof systs eng, Wright State Univ, 68-77; ASSOC PROF MECH ENG, UNIV SOUTHWESTERN LA, 75- *Res:* Supersonic fluidics; pollution control of automotive engines. *Mailing Add:* Dept Mech Eng Univ Southwestern La PO Box 44170 Lafayette LA 70504

SCOTT, RICHARD WALTER, b Modesto, Calif, June 30, 41; m 61; c 2. PLANT ECOLOGY, PLANT TAXONOMY. *Educ:* Univ Wyo, BS, 64, MS, 66; Univ Mich, MA, 69, PhD(plant ecol), 72. *Prof Exp:* NDEA fel plant ecol, Univ Mich, 66-69; asst prof biol, Albion Col, 69-75; instr biol, 75-86, PROF BIOL, CENT WYO COL, 87- *Mem:* Sigma Xi; Am Polar Soc. *Res:* Mountain ecosystems and plant community structure; Alpine plant communities of Western North America and human impact on wilderness environments; alpine flora of the Rocky Mountains. *Mailing Add:* Div Sci Health & Educ Cent Wyo Col Riverton WY 82501

SCOTT, ROBERT ALLEN, b Dixon, Ill, Apr 25, 53; m; c 1. BIOPHYSICAL CHEMISTRY, BIOINORGANIC CHEMISTRY. *Educ:* Univ Ill, Urbana, BS, 75; Calif Inst Technol, PhD(chem), 80. *Prof Exp:* NIH fel, Stanford Univ, 79-81; asst prof inorg chem, Univ Ill, Urbana, 81-87; ASSOC PROF CHEM & BIOCHEM, UNIV GA, ATHENS, 87- *Concurrent Pos:* Alfred P Sloan Res Fel, 86-88. *Honors & Awards:* Presidential young investr award, 85-90. *Mem:* Am Chem Soc; AAAS. *Res:* Inorganic and physical aspects of biologically important systems; kinetics of electron transfer in metalloenzymes and extended x-ray absorption fine structure studies of metalloproteins and models. *Mailing Add:* Dept Chem Sch Chem Sci Univ Ga Athens GA 30602

SCOTT, ROBERT BLACKBURN, b Wilmington, Del, July 22, 37; m 62. GEOLOGY. *Educ:* Univ Ala, BS, 60; Rice Univ, PhD(geol), 65. *Prof Exp:* Res geologist, Yale Univ, 65-67; asst prof geol, Fla State Univ, 67-71; assoc prof geol, Tex A&M Univ, 71-; AT US GEOL SURV. *Concurrent Pos:* Petrologist, Trans-Atlantic Geotraverse Proj, Nat Oceanic & Atmospheric Admin, 71- *Mem:* AAAS; Geochem Soc; Geol Soc Am; Mineral Soc Am; Am Geophys Union. *Res:* Marine volcanism; oceanic basalt geochemistry and petrology; seawater-basalt reactions and equilibria. *Mailing Add:* US Geol Surv MS 913 Denver Fed Ctr Denver CO 80225

SCOTT, ROBERT BLACKBURN, JR, b Greensboro, NC; m 36; c 4. ORGANIC CHEMISTRY. *Educ:* Pa State Univ, BS, 34, MS, 35; Univ Va, PhD(org chem), 49. *Prof Exp:* Chemist org chem, E I du Pont de Nemours & Co, 35-45; from assoc prof to prof, Univ Ala, 48-63; prof & chmn dept, 63-77, EMER PROF ORG CHEM, UNIV MISS, 77- *Concurrent Pos:* Consult, Nat Southern Prod, 50-63, gen, 77-; Ger translr, Buckeye Cellulose, 82-83; writing (ceramics & math), 83- *Mem:* Am Chem Soc. *Res:* Mechanisms; steric effects; reactions of aliphatic sulfonyl and sulfinyl compounds and 1, 3-glycols and diketones. *Mailing Add:* 406 Country Club Rd Oxford MS 38655

SCOTT, ROBERT BRADLEY, b Petersburg, Va, Nov 7, 33; m 58; c 3. GERIATRIC MEDICINE, HEMATOLOGY. *Educ:* Univ Richmond, BS, 54; Med Col Va, MD, 58. *Prof Exp:* From intern to resident internal med, Bellevue & Mem Hosps, New York, 58-61; fel biol, Mass Inst Technol, 63-64, res assoc, 64-65; from asst prof to assoc prof med, 65-74, assoc dean clin activities, 79-82, PROF MED, PATH & BIOCHEM, MED COL VA, VA COMMONWEALTH UNIV, 74- *Concurrent Pos:* Am Cancer Soc res scholar, 63-65; clin fel med, Mass Gen Hosp, 64-65; dir, Lab Hemat Res, 68-82; chief staff, Med Col Va Hosp, 79-82. *Mem:* Fel Am Col Physicians; Am Fedn Clin Res; Am Asn Cancer Res; Am Soc Hemat; Am Geriat Soc; Gerontological Soc Am. *Res:* Control of differentiation in normal and leukemic blood cell; cell biology; mechanism of cellular aging; geriatrics. *Mailing Add:* Dept Med Med Col Va Richmond VA 23298-0214

SCOTT, ROBERT EDWARD, b Crystal Springs, Miss, Oct 19, 22; m 45; c 3. ANALYTICAL CHEMISTRY. *Educ:* Univ Tex, BA, 43, MA, 44. *Prof Exp:* Anal chemist, New Prod Develop Lab, Gen Elec Co, Mass, 48-53; res anal chemist & group leader, Chem Div, Pittsburgh Plate Glass Co, 53-65, sr res supvr, Inorg Phys Res Dept, 65-67, SR RES SUPVR CHEM DIV, ANALYTICAL LABS, PPG INDUSTS, INC, 68- *Mem:* Am Chem Soc. *Res:* Wet chemical, ultraviolet-visible-infrared spectrophotometric, polarographic, x-ray, emission spectrographic and gas chromatographic analytical techniques applied to chlorosilanes, silicones, phenolics, autoxidation process hydrogen peroxide, chrome chemicals and chlorinated hydrocarbons. *Mailing Add:* 3214 Kensington Dr Corpus Christi TX 78414

SCOTT, ROBERT EUGENE, b Terre Haute, Ind, Oct 19, 41. EXPERIMENTAL PATHOLOGY, CANCER RESEARCH. *Educ:* Vanderbilt Univ, MD, 67. *Prof Exp:* PROF PATH & HEAD SECT EXP PATH, MAYO CLIN, 80- *Mem:* Am Asn Pathologist; Am Soc Cell Biol; Am Asn Cancer Res; Int Acad Path; Int Cell Cycle Soc; Int Soc Differentiation. *Mailing Add:* Dept Path Univ Tenn 800 Madison Ave Memphis TN 38163

SCOTT, ROBERT FOSTER, b Alberta, Can, June 23, 25; m 54; c 3. PATHOLOGY. *Educ:* Univ Alta, BSc, 49, MD, 51; FRCP, 58. *Prof Exp:* Asst prof path, Univ BC, 57-58, clin instr, 58-59; asst prof, 59-63, assoc prof, 63-66, PROF PATH, ALBANY MED COL, 66- *Concurrent Pos:* Fel, Nat Res Coun Can, 54-55, Life Insurance Med Res Found, 56-57; mem, Cardiovasc Study Sect, NIH, 64-68, Path Study Sect, NIH, 73-77; exec mem, Arterrosclerosis Coun, Am Heart Asn, 71-73; asst to dean acad affairs, Albany Med Col, 79-80, assoc to dean, 80- *Res:* Kinetics of arterial wall cells with regard to atherosclerosis utilizing experimental models. *Mailing Add:* Dept Path Albany Med Col 47 New Scotland Ave Albany NY 12208

SCOTT, ROBERT LANE, b Santa Rosa, Calif, Mar 20, 22; m 44; c 4. PHYSICAL CHEMISTRY. *Educ:* Harvard Univ, SB, 42; Princeton Univ, MA, 44, PhD(chem), 45. *Prof Exp:* Asst chem, Princeton Univ, 42-43, Manhattan Proj, 44-45; scientist, Manhattan Proj, Los Alamos Sci Lab, 45-46; Jewett fel, Univ Calif, 46-48; from asst prof to assoc prof, 48-60, chmn dept, 70-75, PROF PHYS CHEM, UNIV CALIF, LOS ANGELES, 60- *Concurrent Pos:* Guggenheim fel, 55; NSF sr fel, 61-62. *Honors & Awards:* Fulbright award, 68-69; Joel Henry Hildebrand Award, Am Chem Soc, 84. *Mem:* AAAS; Am Chem Soc; Am Phys Soc. *Res:* Statistical thermodynamics of liquids and solutions; high polymer solutions; fluorocarbon solutions; hydrocarbon solutions; critical phenomena; tricritical points; solubility and phase equilibria. *Mailing Add:* Dept Chem & Biochem Univ Calif Los Angeles CA 90024-1569

SCOTT, ROBERT NEAL, b Pawtucket, RI, Mar 8, 41; m 63; c 2. INDUSTRIAL CHEMISTRY. *Educ:* Brown Univ, ScB, 63; Northwestern Univ, PhD(chem), 68. *Prof Exp:* Sr res chemist polymer chem, 67-71, group leader org res & develop, 71-75, sect leader org & inorg res & develop, 75-78, sect mgr org & inorg res & develop, 78-82, mgr anal & res & develop admin, 82-84, ASSOC DIR RES, OLIN CORP, 84- *Mem:* Am Chem Soc; Am Soc Lubrication Engrs. *Res:* High temperature polymers; synthetic lubricants; surfactants textile chemicals; hydrazine applications and derivative chemistry; electronic chemicals. *Mailing Add:* Six Bruce Rd Wallingford CT 06492

SCOTT, ROBERT NELSON, b St John, NB, Apr 30, 33; m 52; c 5. REHABILITATION ENGINEERING, CLINICAL ENGINEERING. *Educ:* Univ NB, BSc, 55. *Hon Degrees:* DSc, Acadia Univ, 81. *Prof Exp:* From asst prof to assoc prof, 56-70, dir, Bioeng Inst, 65-90, PROF ELEC ENG, UNIV NB, 70- *Mem:* Sr mem Inst Elec & Electronics Engrs; fel Can Med & Biol Eng Soc; Asn Advan Med Instruments. *Res:* Biomedical engineering; myoelectric control of artifical limbs. *Mailing Add:* Dept Elec Eng Univ NB Col Hill PO Box 4400 Fredericton NB E3B 5A3 Can

SCOTT, ROBERT W, b Davenport, Iowa, June 7, 36; m 62; c 3. GEOLOGY, PALEONTOLOGY. *Educ:* Maryknoll Col, BA, 58; Univ Wyo, BA, 60, MA, 61; Univ Kans, PhD(geol), 67. *Prof Exp:* Asst geol, Univ Kans, 61-66; asst prof, Waynesburg Col, 66-70; asst prof, Univ Tex, Arlington, 70-74; sr res scientist, 74-77, res supvr, 77-83, SPEC RES ASSOC, AMOCO PROD CO, 83- *Mem:* Am Asn Petrol Geol; Paleont Soc; Soc Econ Paleont Mineral (secy/treas, 85-87); Sigma Xi. *Res:* Growth and evolution of Early Cretaceous bivalves and stratigraphy and sedimentary environments of Cretaceous rocks in the United States and Arabia. *Mailing Add:* Amoco Prod Co Res Ctr PO Box 3385 Tulsa OK 74102

SCOTT, ROLAND BOYD, b Houston, Tex, Apr 18, 09; m 35; c 3. ALLERGY, HEMATOLOGY. *Educ:* Howard Univ, Bs, 31, MD, 34. *Prof Exp:* From asst prof to prof, Col Med, 39-77, div chief, 45-49, head dept, 49-69, chmn dept, 69-73, distinguished prof pediat & child health, Howard Univ, 77-90, dir sickle cell dis, 71-90; RETIRED. *Concurrent Pos:* Mem bd dirs, Health Found DC, 65-72; consult, Head Start Prog, Dept Human Resources, Washington, DC, 67-70; vis prof, Children's Med Ctr & Dept Pediat, Harvard Med Sch, 70, Col Med, Univ Fla, 71 & Sch Med, Univ Nebr, 73; prof lectr child health & develop, Sch Med, George Washington Univ, 71-; mem adv comt, Nat Sickle Cell Dis Prog, NIH, 82-86. *Honors & Awards:* Jacobi Award, Am Acad Pediat, 85. *Mem:* AMA; AAAS; Am Soc Pediat Hemat & Oncol; Am Pediat Soc; Am Acad Allergy (vpres, 66-67); Am Acad Pediat. *Res:* Growth and development of children; allergic disorders in children; sickle cell disease. *Mailing Add:* 1723 Shepherd St NW Washington DC 20011

SCOTT, RONALD E, b Leslie, Sask, Mar 25, 21; US citizen; c 1. ELECTRICAL ENGINEERING. *Educ:* Univ Toronto, BASc, 43, MASc, 46; Mass Inst Technol, ScD(elec eng), 50. *Prof Exp:* Res assoc elec eng, Mass Inst Technol, 46-50, asst prof, 50-55; prof, Northeastern Univ, 55-60, dean eng, 60-67; dean eng, Col Petrol & Minerals, Dhahran, 67-78; proj head, Nat Inst Elec & Electronics, Algeria, 78-; PROF, NORTHEASTERN UNIV. *Mem:* Fel Inst Elec & Electronics Engrs; Am Soc Eng Educ; Soc Am Mil Engrs. *Res:* Circuit theory solid state microelectronics reliability and medical electronics. *Mailing Add:* Two Brook Rd No 5 Salem NH 03073-3612

SCOTT, RONALD F(RASER), b London, Eng, Apr 9, 29; m 59; c 3. SOIL MECHANICS. *Educ:* Glasgow Univ, BSc, 51; Mass Inst Technol, SM, 53, ScD(soil mech), 55. *Prof Exp:* Asst soil mech, Mass Inst Technol, 51-55; soil engr, Corps Engrs, US Dept Army, 55-57; div soil engr, Racey, MacCallum & Assocs, Ltd, Can, 57-58; from asst prof to prof civil eng, 58-87, DOTTIE & DICK HAYMAN PROF ENG, CALIF INST TECHNOL, 87- *Concurrent Pos:* Churchill fel, Churchill Col, Eng, 72; Guggenheim fel, 72; prin investr, Jet Propulsion Lab, NASA, mem, Soil Mech Team, Apollo Manned Lunar Missions & Phys Properties Team, NASA Viking; consult, pvt indust & govt agencies. *Honors & Awards:* Walter Huber Res Prize, Am Soc Civil Engrs, 69, Norman Medal, 72, Thomas A Middlebrooks Award, 82, Tersaghi Lectr, 83; Newcomb Cleveland Award, AAAS, 76; Rankine Lectr, Brit Geotech Soc, 87. *Mem:* Nat Acad Eng; Am Soc Civil Engrs; Am Geophys Union. *Res:* Soil engineering and mechanics; soil dynamics; earthquake engineering; author of 180 technical publications and four books. *Mailing Add:* 2752 N Santa Anita Ave Altadena CA 91001

SCOTT, RONALD MCLEAN, b Detroit, Mich, Apr 16, 33; m 57; c 3. BIOCHEMISTRY, TOXICOLOGY. *Educ:* Wayne State Univ, BS, 55; Univ Ill, PhD(biochem), 59. *Prof Exp:* From asst prof to assoc prof, 59-68, PROF CHEM, EASTERN MICH UNIV, 68- *Concurrent Pos:* Exchange prof, Coventry Col, Eng, 71-72; vis lectr, Warwick Univ, Eng, 72. *Mem:* Am Chem Soc; Sigma Xi; AAAS. *Res:* Hydrogen bonding in solution, solvation interactions; enzymology, immobilization and kinetics; chromatography and toxicology. *Mailing Add:* Dept Chem Eastern Mich Univ Ypsilanti MI 48197

SCOTT, ROY ALBERT, III, b Pottsville, Pa, Mar 22, 34; m 58; c 2. PHYSICAL CHEMISTRY, MOLECULAR BIOLOGY. *Educ:* Cornell Univ, AB, 58, PhD(chem), 64. *Prof Exp:* Asst prof chem, Cornell Univ, 63-66; asst prof biophys, Univ Hawaii, 66-68; ASSOC PROF BIOCHEM & MOLECULAR BIOL, OHIO STATE UNIV, 68- *Mem:* Am Chem Soc; Am Phys Soc. *Res:* Physical chemistry of biological macromolecules. *Mailing Add:* Dept Biochem Col Med Ohio State Univ 484 W 12th Ave Columbus OH 43210

SCOTT, SHERYL ANN, b Wilmington, Del, Feb 16, 49; m 84; c 2. DEVELOPMENTAL NEUROBIOLOGY. *Educ:* Duke Univ, BS, 71; Yale Univ, PhD(biol), 76. *Prof Exp:* Fel neurobiol, Med Ctr, McMaster Univ, 76-77 & Carnegie Inst Washington, 78-79; asst prof neurobiol, 79-84, ASSOC PROF NEUROBIOL & BEHAV, STATE UNIV NY STONYBROOK, 84- *Concurrent Pos:* Fels, Multiple Sclerosis Soc Can, 76-77, HEW Pub Health Serv, 77 & 78-, Grass Found, 78 & Alfred P Sloan Found, 82-83; Fogarty sr int fel, 88-89. *Mem:* Soc Neurosci; Int Soc Develop Neurosci. *Res:* Development of sensory inneravation patterns in the chick. *Mailing Add:* Dept Neurobiol & Behav State Univ NY Stony Brook NY 11794

SCOTT, STEVEN DONALD, b Ft Frances, Ont, June 4, 41; m 63; c 2. ECONOMIC GEOLOGY, MARINE GEOLOGY. *Educ:* Univ Western Ont, BSc, 63, MSc, 64; Pa State Univ, PhD(geochem), 68. *Prof Exp:* Res assoc geochem, Pa State Univ, 68-69; from asst prof to assoc prof, 69-79, assoc chmn, 80-84, PROF GEOCHEM, UNIV TORONTO, 79-, CHMN GEOL ENG, 88-, DIR, SESTIABANK MARINE GEOL RES LAB, 89- *Concurrent Pos:* Lectr, Massive sulfide short courses, Univ Toronto, Finland, SAfrica, Ger, Australia, China & Switz; consult, mining & marine indust, 70-; tech consult, television, radio, newspapers & magazines, 75-; vis prof, Australia, 76-77, 84, Japan, 77 & France, 85; Can leader, US-Japan-Can-Kuroko Res Group, 78-83; second chmn, Mineral Deposits Div, Geol Asn Can, 78-80 & 81-82; vpres, Ocean Explor & Mining Consult, 82-88; prin investr, Can Res Seafloor Hydrothermal Vents & Sulfide Deposits, 83-86; mem, Can Nat Comt Ocean Drilling Prog, 85-90, chmn, 88-90; subj-of-interest, PBS Prog Planet Earth; co-producer, videotape, Explorer Ridge, 85; gov, Earth Sci Sect, Royal Soc Can, 89-90; coun, Int Asn Genesis of Ore Deposits, 90-, Soc Geol Appl Gîtes Minéraux, 91- *Honors & Awards:* Waldemar Lindgren Citation, Soc Econ Geologists; Thayer-Lindsley; Bancroft Award, Royal Soc Can, 90. *Mem:* Mineral Soc Am; Mineral Asn Can; Geol Asn Can; Soc Econ Geol; Soc Mining Geologists Japan; Am Geophys Union; fel Royal Soc Can; Int Assoc Genesis Ore Deposits. *Res:* Modern seafloor hydrothermal systems and deposits; massive copper-zinc sulfide ores; marine geology and tectonics; physical geochemistry; oreforming processes by field and experimental methods; sulfide mineralogy; synthesis and crystal chemistry of metallic sulfides. *Mailing Add:* Dept Geol Earth Sci Ctr Univ Toronto Toronto ON M5S 3B1 Can

SCOTT, STEWART MELVIN, b Sherman, Tex, Oct 14, 26; m 60; c 1. CARDIOVASCULAR SURGERY, THORACIC SURGERY. *Educ:* Baylor Univ, MD, 51. *Prof Exp:* Asst chief cardiovasc surg, 59-62, ASSOC CHIEF STAFF FOR RES & CHIEF CARDIOVASC SURG SECT, VET ADMIN HOSP, 62-, CHIEF SURG, VET ADMIN HOSP, 82- *Concurrent Pos:* Clin instr, Woman's Med Col Pa, 52-56; assoc clin prof, Duke Univ Med Ctr, 68-; consult prof surg, 82- *Mem:* Fel Am Col Surgeons; Int Cardiovasc Soc; Am Asn Thoracic Surgeons; Soc Thoracic Surgeons; AMA. *Res:* Cardiac, pulmonary and vascular diseases; healing of vascular prostheses. *Mailing Add:* Vet Admin Hosp Asheville NC 28805

SCOTT, THOMAS A, physics; deceased, see previous edition for last biography

SCOTT, THOMAS WALTER, b Sewickley, Pa, Nov 10, 29; m 53; c 4. SOIL FERTILITY. *Educ:* Pa State Univ, BS, 52; Kans State Univ, MS, 56; Mich State Univ, PhD(soil sci), 59. *Prof Exp:* From asst prof to assoc prof, 59-74, PROF SOIL SCI, CORNELL UNIV, 74-, MEM DEPT AGRON, 77- *Concurrent Pos:* Travel grants from Cornell Univ & NY lime & fertilizer industs to Cambridge Univ, 66-67; USAID assignment to PR, 73-, temp assignment to Panama, 82-83. *Mem:* Am Soc Agron; Soil Sci Soc Am; Soil Conserv Soc Am. *Res:* Undergraduate teaching; animal waste management; soil fertility research and extension; soil fertility research on tropical soils; soil management. *Mailing Add:* Dept Soil Crop & Atmospheric Sci Bradfield Hall Cornell Univ Ithaca NY 14853

SCOTT, TOM E, b Cleveland, Ohio, Sept 10, 33; m 59; c 4. METALLURGY. *Educ:* Case Inst Technol, BS, 56, MS, 58, PhD(phys metall), 62. *Prof Exp:* Res metallurgist, Int Nickel Co, Ind, 61-63; asst prof mech metall, Iowa State Univ, 63-69, assoc prof metall, 69-72, prof metall, 72-81, sect chief, Ames Lab, 75-81; sr tech supvr, Aluminum Co Am, 81-82; from mgr, fabricating technol to mgr, Metal Matrix Composite Develop Dept, Agr Res Ctr Oper, 82-85; DIR, INST MATS PROCESSING, MICH TECH UNIV, 85- *Mem:* Am Soc Metals; Am Inst Mining, Metall & Petrol Engrs; Nat Asn Corrosion Engrs. *Res:* Hydrogen embrittlement; embrittlements; ferrous alloys; deformation; fracture mechanisms. *Mailing Add:* Appl Mat Ctr 535 N Exchange Ct Aurora IL 60504

SCOTT, TOM KECK, b St Louis, Mo, Aug 4, 31; div; c 4. PLANT PHYSIOLOGY. *Educ:* Pomona Col, BA, 54; Stanford Univ, MA, 59, PhD(biol), 61. *Prof Exp:* Teaching fel biol, 56-60, res asst, Stanford Univ, 57-61; res assoc, Princeton Univ, 61-63; from asst prof to assoc prof, Oberlin Col, 63-69; from asst prof to prof bot, Univ NC, 69-82, chmn biol curriculum, 70-75, chmn dept, 72-82, dir res serv, 85-90, PROF BIOL, UNIV NC, CHAPEL HILL, 82- *Concurrent Pos:* Vis res fel, Univ Nottingham, 67-68; Fulbright lectr, 72-73; assoc, Danforth Found, 73-; chmn space biol adv panel, NASA, 83-89, plant biol discipline working group, 88- *Mem:* Fel AAAS; Bot Soc Am; Am Soc Plant Physiol; Sigma Xi; Am Inst Biol Sci; Am Soc Gravitational & Space Biol; Nat Acad Sci. *Res:* Plant growth and development; auxin relationships and transport; growth regulator interactions; hormone physiology. *Mailing Add:* Dept Biol CB No 3280 Univ NC Chapel Hill NC 27514-3280

SCOTT, VERNE H(ARRY), b Salem, Ore, June 19, 24; m 48; c 3. CIVIL ENGINEERING, HYDROLOGY. *Educ:* Univ Mich, BS, 45, MS, 48; Colo State Univ, PhD(irrig eng), 59. *Prof Exp:* Res & irrig engr, hydrol, 46-47, assoc prof irrig, 48-63, chmn dept water sci & eng, 64-71, PROF WATER SCI & CIVIL ENG & HYDROLOGIST, UNIV CALIF, DAVIS, 63- *Concurrent Pos:* Dir, Campus Work-Learn Prog, Univ Calif, Davis, 63- *Mem:* Am Soc Eng Educ; Am Soc Civil Engrs; Am Soc Agr Engrs; Am Geophys Union. *Res:* Ground water and water resources. *Mailing Add:* Dept Land Air & Water Resources Univ Calif Davis CA 95616

SCOTT, W RICHARD, b Independence, Kans, 39. SOCIOLOGY OF MEDICINE. *Educ:* Duke Univ, MD, 64. *Prof Exp:* Intern, NY Hosp, 64-65, resident, 65-66; resident, Mass Gen Hosp, clin fel neurol, 71-74; STANFORD HOSP, 74-; PROF, DEPT SOCIOL, STANFORD UNIV. *Mailing Add:* 940 Lathrop Pl Stanford CA 94305

SCOTT, WALTER ALVIN, b Los Angeles, Calif, Feb 1, 43; m 70. BIOCHEMISTRY, CELL BIOLOGY. *Educ:* Calif Inst Technol, BS, 65; Univ Wis, PhD(physiol chem), 70. *Prof Exp:* Fel cell biol dept biophys & biochem, Univ Calif, San Francisco, 70-73; fel tumor virol, Dept Microbiol, Med Sch, Johns Hopkins Univ, 73-75; ASST PROF BIOCHEM, SCH MED, UNIV MIAMI, 75- *Concurrent Pos:* NSF fel, 70-71; Jane Coffin Childs Mem Fund fel, 71-72; NIH fel, 72-73 & 74-75; NIH grant, 75-78 & 78-81; NSF grant, 78-79. *Mem:* Am Soc Microbiol; AAAS; Sigma Xi. *Res:* Structure and function of Simian Virus 40 chromatin; recombination involved in integration and excision of SV40 from cell chromosomes; transformation of pancreatic islet cells by SV40 mutants. *Mailing Add:* 713 Navarre Ave Coral Gables FL 33134

SCOTT, WALTER NEIL, b Mar 2, 35; m 59; c 2. REGULATION OF HORMONAL TRANSPORT. *Educ:* Univ Louisville, MD, 60. *Prof Exp:* Assoc dean, Mt Sinai Sch Med, 76-82, prof physiol, 79-82; chmn dept, 82-87, PROF BIOL, NY UNIV, 82- *Mem:* Am Soc Biol Chem; Am Physiol Soc; fel NY Acad Sci (pres, 83); fel NY Acad Med. *Mailing Add:* Dept Biol NY Univ 1009 Main Bldg Washington Sq E New York NY 10003

SCOTT, WILLIAM ADDISON, III, b Indianapolis, Ind, Apr 27, 40; m 66; c 1. BIOCHEMISTRY, GENETICS. *Educ:* Univ Ill, Urbana, BS, 62; Calif Inst Technol, PhD(biochem), 67. *Prof Exp:* Guest investr biochem genetics & USPHS grant, 67-69, res assoc, 69-71, asst prof, 71-77, ASSOC PROF BIOCHEM GENETICS, DEPT CELL PHYSIOL & IMMUNOL, ROCKEFELLER UNIV, 77- *Mem:* AAAS; Sigma Xi; Harvey Soc. *Res:* Biochemical basis of morphology; membrane structure and function. *Mailing Add:* Dept Cell Physiol & Immunol Rockefeller Univ 1230 York Ave New York NY 10021

SCOTT, WILLIAM BEVERLEY, b Toronto, Ont, July 7, 17; m 42; c 2. ICHTHYOLOGY. *Educ:* Univ Toronto, BA, 42, PhD(zool), 50. *Hon Degrees:* DSc, Univ NB, 85. *Prof Exp:* Actg cur, Royal Ont Mus, 48-50, cur dept ichthyol & herpet, 50-76, assoc dir mus, 73-76; from assoc prof to prof zool, Univ Toronto, 63-83; exec dir, 76-82, SR SCIENTIST, HUNTSMAN MARINE LAB, ST ANDREWS, NB, 82-; EMER PROF ZOOL, UNIV TORONTO, 83- *Honors & Awards:* Centennial Medal, Govt Can, 67, Silver Jubilee Medal, 77. *Mem:* Am Soc Ichthyol & Herpet (pres, 73); Can Soc Zool; Am Fisheries Soc; Soc Systs Zool; Fel Am Soc Fish Res Biologists; fel Royal Soc Can; fel Royal Soc Arts. *Res:* Systematics and distribution of Canadian freshwater fishes and Northwest Atlantic fishes, particularly salmonids, myctophids and scombrids; food and feeding; commercial fisheries. *Mailing Add:* Huntsman Brandy Cove Marine Sci Ctr Brandy Cove Rd St Andrews NB E0G 2X0 Can

SCOTT, WILLIAM D(OANE), b Lakewood, Ohio, Feb 17, 31; m 59; c 2. CERAMICS ENGINEERING. *Educ:* Univ Ill, BS, 54; Univ Calif, Berkeley, MS, 59, PhD(eng sci), 61. *Prof Exp:* Res fel, Univ Leeds, 61-63; asst prof eng, Univ Calif, Berkeley & Lawrence Radiation Lab, 64-65; PROF ENG, UNIV WASH, 65- *Mem:* Fel Am Ceramic Soc. *Res:* Nucleation and growth of crystals in glass; mechanical properties of ionic solids and oxides; structure of grain boundaries. *Mailing Add:* Mat Sci & Eng Univ Wash Seattle WA 98195

SCOTT, WILLIAM EDWARD, b Middletown, Conn, Sept 6, 47; m 70; c 2. GEOLOGY. *Educ:* St Lawrence Univ, BS, 69; Univ Wash, MS, 71, PhD(geol), 74. *Prof Exp:* GEOLOGIST, US GEOL SURV, 74- *Mem:* AAAS; Geol Soc Am; Am Quaternary Asn; Am Geophys Union. *Res:* Assessment of volcanic hazards in the Pacific Northwest; late Quaternary eruptive histories of volcanoes in Oregon; Quaternary glaciation and climate change in the Pacific Northwest. *Mailing Add:* David A Johnston Cascades Volcano Observ 5400 MacArthur Blvd Vancouver WA 98661

SCOTT, WILLIAM EDWIN, b Rome, NY, Apr 8, 18; m 41; c 3. CHEMISTRY. *Educ:* Hamilton Col, NY, BS; Swiss Fed Inst Technol, DTechSc, 41. *Prof Exp:* Res chemist, 41-56, asst to dir res, 56-59, res coordr, 59-63, dir res admin, 63-69, DIR RES TECH SERVS, HOFFMAN-LA ROCHE, INC, 69-; PROF BUS ADMIN. *Mem:* Am Chem Soc; NY Acad Sci; The Chem Soc. *Res:* Structure and synthesis of organic medicinal compounds; purification and structure determination of natural products; synthesis of steroid analogs. *Mailing Add:* Three Allen North Caldwell NJ 07006

SCOTT, WILLIAM JAMES, JR, b New York, NY, Dec 8, 37; m 62; c 2. TERATOLOGY, DEVELOPMENTAL BIOLOGY. *Educ:* Univ Ga, DVM, 61; George Washington Univ, PhD(anat), 69. *Prof Exp:* Dir teratology, Woodard Res Corp, 64-68; Pharmaceut Mfrs Asn Found fel, 69-71, from asst prof res pediat to assoc prof res pediat, 71-78, PROF PEDIAT, CHILDREN'S HOSP RES FOUND, 78- *Concurrent Pos:* Vet consult, Sch Med, George Washington Univ, 64-69; mem, Human Embryol & Develop Study Sect, NIH, 82-86. *Honors & Awards:* Frank R Blood Award, Soc Toxicol, 77. *Mem:* Teratology Soc; Am Asn Lab Animal Sci. *Res:* Determination of the mechanisms by which environmental factors interfere with embryonic development to produce congenital malformations. *Mailing Add:* Div Basic Sci Res Children's Hosp Res Found Cincinnati OH 45229

SCOTT, WILLIAM RAYMOND, b Bloomingburg, Ohio, June 25, 19; m 45; c 3. ALGEBRA. *Educ:* Ohio State Univ, BA, 40, MA, 41, PhD, 47. *Prof Exp:* Res assoc math, Ohio State Univ, 47-48; instr, Univ Mich, 48-49; from asst prof to prof, Univ Kans, 49-65, actg chmn dept, 59-61; PROF MATH, UNIV UTAH, 65- *Concurrent Pos:* NSF fel, 55-56; managing ed, Rocky Mountain J of Math, 70- *Mem:* Am Math Soc; Math Asn Am; Sigma Xi. *Res:* Theory of groups. *Mailing Add:* 4774 Oak Terr Salt Lake City UT 84124

SCOTT, WILLIAM TAUSSIG, b Yonkers, NY, Mar 16, 16; m 42, 61; c 6. INTELLECTUAL BIOGRAPHY, ATMOSPHERIC PHYSICS. *Educ:* Swarthmore Col, BA, 37; Univ Mich, PhD(physics), 41. *Prof Exp:* Instr physics, Amherst Col, 41-44; asst prof math & physics, Deep Springs Jr Col, 44-45; from instr to prof physics, Smith Col, 45-61; prof, 61-81, EMER PROF PHYSICS, UNIV NEV, RENO, 81- *Concurrent Pos:* Res fel, Yale Univ, 59-60; vis prof, Univ Nev, 61-62, dir prog philos inquiry, 70-81; consult, Brookhaven Nat Lab, 47-53, Nat Bur Standards, 54-56; studio physicist, PSSC films, 58-59; res prof, Atmospheric Sci Ctr, Desert Res Inst, 64-; sr mem, Linacre Col, Oxford Univ, 69-70; mem anal comt, Higher Educ Progs on Sci, Technol & Human Values Res Proj, Univ Mich, 75. *Mem:* Fel Am Phys Soc; Am Asn Physics Teachers; Sigma Xi. *Res:* Biography of Michael Polanyi; philosophy of science; theory of multiple scattering of fast charged particles; interaction of science and religion; theory of cloud droplet growth by condensation and coalescence. *Mailing Add:* Dept Physics Univ Nev Reno NV 89557-0058

SCOTT, WILLIAM WALLACE, b Utica, NY, Oct 1, 20; m 42; c 4. MYCOLOGY. *Educ:* Univ Vt, AB, 48, MS, 50; Univ Mich, PhD(bot), 55. *Prof Exp:* Res asst plant path, Univ Vt, 48-50; from assoc prof to prof bot, Va Polytech Inst, 55-64; assoc prog dir, NSF, 64-66; dean, Madison Col, Va, 66-68, chmn dept bot, 68-75, assoc dean grad sch, 75-78, PROF BOT, EASTERN ILL UNIV, 68- *Concurrent Pos:* Vis prof, Univ Wis, 60. *Mem:* Fel AAAS; Bot Soc Am; Mycol Soc Am; Brit Mycol Soc. *Res:* Cryptogamic botany; aquatic fungi; fungus diseases of fish; marine microbiology. *Mailing Add:* Dept Bot Eastern Ill Univ Charleston IL 61920

SCOTT, WILLIAM WALLACE, b Kansas City, Kans, Jan 27, 13; m 36; c 1. PHYSIOLOGY. *Educ:* Univ Mo, AB, 34; Univ Chicago, PhD(physiol), 38, MD, 39. *Hon Degrees:* DSc, Univ Mo, 74. *Prof Exp:* Res assoc surg, Univ Chicago, 41-43, from instr to assoc prof urol, 43-46; prof, 46-74, David Hall McConnell Prof urol, 75-78, urologist-in-chg, Hopkins Hosp, 46-74, EMER PROF, SCH MED, JOHNS HOPKINS UNIV, 78- *Concurrent Pos:* Consult, US Naval Hosp, Bethesda, Clin Ctr, NIH & Walter Reed Gen Hosp, Washington, DC. *Honors & Awards:* Gold Medal, AMA, 40; Barringer Medal, Am Asn Genito Urinary Surgeons, 71, Keyes Medal, 79; Eugene Fuller Medal, Am Urol Asn, 80; Guiteral Award, Am Urol Asn, 83. *Mem:* Am Physiol Soc; Endocrine Soc; Am Asn Cancer Res; Am Asn Genito-Urinary Surg; Am Urol Asn. *Res:* Endocrine relations in prostatic disease; renal circulation. *Mailing Add:* Rm 500 Brady Urol Inst Johns Hopkins Hosp Baltimore MD 21205

SCOTTER, GEORGE WILBY, b Cardston, Alta, Jan 16, 33; m 59; c 2. ECOLOGY. *Educ:* Utah State Univ, BSc, 59, MSc, 62, PhD(range sci), 68. *Prof Exp:* Res scientist ecol, Can Wildlife Serv, 59-66; asst prof range ecol, Utah State Univ, 66-68; res scientist ecol, 68-77, prog leader parks res, 77-78, CHIEF CAN WILDLIFE SERV, 78- *Mem:* Ecol Soc Am; Soc Wildlife Mgt; Soc Range Mgt; Can Bot Asn. *Res:* Wildlife-range relationships and alpine research; wildland management and wilderness recreation research. *Mailing Add:* Rm 210 4999 98th Ave Edmonton AB T6B 2X3 Can

SCOUTEN, CHARLES GEORGE, b Atlanta, Ga, Nov 21, 40; c 64; c 2. ORGANIC CHEMISTRY. *Educ:* Univ Ga, Athens, BS, 68; Purdue Univ, Lafayette, PhD(org chem), 75. *Prof Exp:* Proj mgr, Xerox, 73-78; sr res chemist, Exxon Res & Eng, 78-87; SR SCIENTIST, AMOCO OIL RES & DEVELOP, 87- *Mem:* Am Chem Soc; AAAS. *Res:* Structure and reactivity of heavy fuel materials; new conversion processes for heavy fuels; colloid, interface and surface science. *Mailing Add:* Amoco Oil H-9 Box 3011 Naperville IL 60566-7011

SCOUTEN, WILLIAM HENRY, b Corning, NY, Feb 12, 42; m 65; c 6. ENZYMOLOGY, PROTEIN CHEMISTRY. *Educ:* Houghton Col, BA, 64; Univ Pittsburgh, PhD(biochem), 69. *Prof Exp:* NIH fel, State Univ NY, Stony Brook, 69-71; from asst prof to prof, dept chem, Bucknell Univ, 71-84; PROF & CHMN, DEPT CHEM, BAYLOR UNIV, 84- *Concurrent Pos:* Fulbright fel, 76; NSF fac develop fel, State Agr Univ, Wageningen & Cambridge Univ, UK, 78. *Mem:* Am Chem Soc; author of 47 publications and journal articles; Am Soc Biochem & Molecular Biol; Am Soc Biol Chemists. *Res:* DNA replication; multienzyme complexes; DNA-binding proteins; protein chemistry; solid state biochemistry; affinity chromatography; author of 20 publications and journal articles. *Mailing Add:* Dept Chem Baylor Univ Waco TX 76798

SCOVELL, WILLIAM MARTIN, b Wilkes-Barre, Pa, Jan 16, 44; m 65; c 4. BIOCHEMISTRY, BIOINORGANIC CHEMISTRY. *Educ:* Lebanon Valley Col, BS, 65; Univ Minn, Minneapolis, PhD(inorg chem), 69. *Prof Exp:* Teaching assoc gen chem, Univ Minn, 65-69, res assoc inorg chem, 69; researcher phys biochem, Princeton Univ, 69-70, instr phys chem, 70-71, gen chem, 71-72; vis asst prof chem, State Univ NY Buffalo, 72-74; assoc prof, 74-78, PROF CHEM, BOWLING GREEN STATE UNIV, 78- *Concurrent Pos:* Consult, NL Industs, Inc, 70- & Smith, Kline & French Labs, 75; NIH nat individual scholar, dept pathol, Fox Chase Cancer Ctr, 84-85; ed, J Chem Educ, feature entitled Concepts in Biochem. *Mem:* Am Chem Soc; AAAS; Sigma Xi (pres, 88-89). *Res:* Interaction of metal ions and complexes of therapeutic value nucleic acids and chromatin; DNA-protein interactions; use of vibrational spectroscopy, spectroscopy (Raman) in elucidating structure and conformation of biological macromolecules; studies of nucleosome structure; possible biological significance of the interaction of CIS-(NH3)2 P. *Mailing Add:* Dept Chem Hayes Hall Bowling Green State Univ Bowling Green OH 43403

SCOVILL, JOHN PAUL, b Fort Benning, Ga, Jan 16, 48; m 71; c 2. DRUG DESIGN. *Educ:* Cent Mich Univ, BSc, 70; Univ Mich, MSc, 73, PhD(med chem), 75. *Prof Exp:* Res chemist, Div Exp Therapeut, Walter Reed Army Inst Res, 75-84; staff officer, Mil Dis Hazards Res Progs, US Army Med Res & Develop Command, 85-86; chem instr, US Mil Acad, 86-87, from asst prof to assoc prof chem, 87-91; VIROL DIV, US ARMY MED RES INST INFECTIOUS DIS, FT DETRICK, MD, 91- *Concurrent Pos:* Vis Fulbright prof chem, Univ Zimbabne, 91- *Mem:* Am Chem Soc. *Res:* Design and preparation of potential chemotherapeutic agents; synthesis and mechanistic studies of organosulfur and organoselenium compounds, nitrogen heterocycles, and transition metal complexes. *Mailing Add:* Virol Div US Army Med Res Inst Infectious Dis Ft Detrick Frederick MD 21701-5011

SCOVILL, WILLIAM ALBERT, b Battle Creek, Mich, Nov 26, 40; wid; c 2. SURGERY. *Educ:* Univ Mich, BS, 63, MD, 66; Univ Ill, MS, 73. *Prof Exp:* Instr surg, Sch Med, Univ Ill, 72-74; from asst prof to assoc prof surg & physiol, Albany Med Col, 74-80; ASSOC PROF SURG, UNIV MD SCH MED, 80- *Concurrent Pos:* Dir, Trauma Serv, Albany Med Ctr, 74-80 & Burn Treat Ctr, 75-80; chief gastroentestinal surg, Univ Md Med Syst. *Mem:* Am Col Surgeons; Asn Acad Surg; Am Burn Asn; Reticuloendothelial Soc; Soc Surg Alimentary Tract; Soc Univ Surgeons. *Res:* Influence of trauma, burn injury or surgery on humoral and cellular aspects of reticuloendothelial host defense function. *Mailing Add:* Dept Gen Surg Univ Md Sch Med 22 Green St Baltimore MD 21201

SCOVILLE, RICHARD ARTHUR, b Torrington, Conn, Feb 14, 35. MATHEMATICS. *Educ:* Yale Univ, BA, 56, MA, 57, PhD(math), 62. *Prof Exp:* Asst prof, 61-74, ASSOC PROF MATH, DUKE UNIV, 74- *Concurrent Pos:* Lectr, Ehime Univ, Japan, 66-67. *Mem:* Am Math Soc. *Res:* Ergodic theory; measure-preserving transformations; sums of dependent random variables. *Mailing Add:* Dept Math Duke Univ 225 Physics Durham NC 27706

SCOW, KATE MARIE, b Washington, DC, Aug 22, 51. SOIL MICROBIAL ECOLOGY, ENVIRONMENTAL MICROBIOLOGY. *Educ:* Antioch Col, BS, 73; Cornell Univ, MS, 86, PhD(soil sci), 89. *Prof Exp:* Environ biologist, Arthur D Little, Inc, 77-82; ASST PROF SOIL MICROBIOL, UNIV CALIF, DAVIS, 89- *Mem:* Soil Sci Soc Am; Am Soc Microbiol; Soc Environ Toxicol & Chem; Sigma Xi. *Res:* Biodegradation of organic pollutants by microbial populations in soil and the subsurface; kinetics of microbial processes in soil; carbon cycling in agro-ecosystems. *Mailing Add:* Land Air & Water Resources Hoagland Hall Univ Calif Davis CA 95616

SCOW, ROBERT OLIVER, b Dos Cabezas, Ariz, Nov 17, 20; m 48, 77; c 4. PHYSIOLOGY, ENDOCRINOLOGY. *Educ:* Univ Calif, AB, 43, MA, 44, MD, 46. *Prof Exp:* Intern, San Francisco Hosp, 46 & Presby Hosp, NY, 47-48; from sr asst surgeon to surgeon, 48-59, MED DIR, NIH, 59-, CHIEF SECT ENDOCRINOL, LAB CELLULAR & DEVELOP BIOL, NAT INST DIABETES, DIGESTIVE & KIDNEY DIS, 61- *Concurrent Pos:* Guggenheim fel, 55; vis prof exp med & cancer res, Hebrew Univ-Hadassah Med Sch, Israel, 65-66; vis prof pediat, Univ Oulu, 75; vis scientist, Ctr Biochem & Molecular Biol, Nat Ctr Sci Res, France, 77. *Honors & Awards:* G Lyman Duff Lectr, Am Heart Asn, 74. *Mem:* Am Physiol Soc; Endocrine Soc; Am Asn Anatomists. *Res:* Hormonal influences on growth of striated musculature and bone; hormonal control of fat and carbohydrate metabolism; diabetes; perfusion of isolated organs; role of capillaries and lipoprotein lipase in metabolism of chylomicrons; hormonal regulation of lipoprotein lipase; fatty acid transport by lateral movement in cell membranes; discovery and purification of lingual lipase; genetic regulation of lipoprotein lipase and hepatic lipase. *Mailing Add:* Rm 137 Bldg 6 NIH Bethesda MD 20892

SCOZZIE, JAMES ANTHONY, b Erie, Pa, Nov 3, 43; m 70; c 3. ORGANIC CHEMISTRY. *Educ:* Gannon Col, AB, 65; Case Western Reserve Univ, MS, 68, PhD(chem), 70. *Prof Exp:* Jr res chemist, Cent Res Dept, Lord Corp, 65; res chemist, Diamond Shamrock Corp, 70-72, sr res chemist, 72-76, res supvr pharmaceut, 76-78, group leader agr chem, 78-81, assoc dir agr chem res, 81-83; dir agr chem res, 83-85, DIR CORP RES, SDS BIOTECH CORP, 85-; PRES, RICERCA, INC, OHIO 86- *Mem:* Am Chem Soc; Am Mgt Asn. *Res:* Structure and chemistry of peptide antibiotics; synthesis of biologically active compounds; pesticides; process studies of organic compounds; commercial evaluation; nutrition and animal health; herbicides; plant growth regulants; cardiovascular agents and antiinflammatory agents. *Mailing Add:* Ricerca Inc PO Box 1000 Painesville OH 44077

SCRABA, DOUGLAS G, b Blairmore, Alta, Apr 17, 40. MOLECULAR BIOLOGY, VIROLOGY. *Educ:* Univ Alta, BSc, 61, BEd, 63, PhD(biochem), 68. *Prof Exp:* Lectr chem, North Alta Inst Technol, 63-64; from asst prof to assoc prof, 70-80 PROF BIOCHEM, UNIV ALTA, 80- *Concurrent Pos:* Med Res Coun Can, centennial fel, Univ Geneva, 68-70. *Mem:* Can Biochem Soc; Am Soc Microbiol; Soc Gen Microbiol; Micros Soc Can; Am Soc Virol. *Res:* Structure and assembly of mammalian viruses. *Mailing Add:* Dept Biochem Univ Alta Edmonton AB T6G 2E1 Can

SCRANTON, BRUCE EDWARD, b Pittsfield, Ill, May 5, 46; m 68; c 5. MATHEMATICAL ANALYSIS, OPERATIONS RESEARCH. *Educ:* Northern Ill Univ, BS, 68; Purdue Univ, MS, 71, PhD(math), 74. *Prof Exp:* Sr assoc opers res, Daniel H Wagner, Assocs, 74-82; MGR, RESOURCE MGT TECHNOL, GEN ELEC, 82- *Mem:* Soc Indust & Appl Math; Asn Comput Mach; Inst Elec & Electronics Engrs. *Res:* Systems engineering and design of misson management systems for ground systems. *Mailing Add:* 32 Laurel Circle Malvern PA 19355

SCRANTON, MARY ISABELLE, b Atlanta, Ga, Feb 28, 50; m 81; c 1. MARINE BIOGEOCHEMISTRY. *Educ:* Mount Holyoke Col, BA, 72; Mass Inst Technol, PhD(oceanog), 77. *Prof Exp:* Nat Acad Sci-Nat Res Coun resident res assoc, US Naval Res Lab, 77-79; asst prof, 79-84, ASSOC PROF CHEM OCEANOG, MARINE SCI RES CTR, STATE UNIV NY, STONYBROOK, 84- *Mem:* Am Geophys Union; Am Soc Limnol & Oceanog; Sigma Xi; Am Soc Microbiol; AAAS. *Res:* Investigations of sources, sinks and distributions of reduced gases in the marine environment; interactions of biological processes and chemical cycles. *Mailing Add:* 101 Van Brunt Manor Rd East Satauket NY 11733

SCRIABINE, ALEXANDER, b Yelgava, Latvia, Oct 26, 26; nat US; m 64; c 2. PHARMACOLOGY. *Educ:* Cornell Univ, MS, 54; Univ Mainz, MD, 58. *Prof Exp:* Res asst pharmacol, Med Sch, Cornell Univ, 51-54; pharmacologist, Res Labs, Chas Pfizer & Co, Maywood, NJ, 54-56, sr pharmacologist, 59-61, supvr, 61-63, mgr, Chas Pfizer & Co, Groton, Conn, 63-66; chief pharmacologist, Div Cardiol, Philadelphia Gen Hosp, 66-67; sr res fel, Merck Inst Therapeut Res, 67-69, dir cardiovasc res, 69-72, sr dir pharmacol, 72-73, exec dir pharmacol, 73-78 assoc dir biol res, Wyeth Labs, Inc, 78-79; assoc prof pharmacol, Sch Med, Univ Pa, 75-89; DIR, MILES INST PRECLIN PHARMACOL, 79- *Concurrent Pos:* Mem coun thrombosis & coun hypertension, Am Heart Asn; ed, Pharmacol Antihypertensive Drugs, 80, New Cardiovasc Drugs, 85-87, Cardiovasc Drug Rev, 88- *Mem:* Am Soc Pharmacol & Exp Therapeut; Ger Pharmacol Soc; Am Soc Clin Pharmacol & Therapeut; AAAS; Int Soc Heart Res; NY Acad Sci; Am Hypertension Soc; Soc Neurosci. *Res:* Cardiovascular, central nervous system autonomic and renal pharmacology; antihypertensives, neuroprotectives, diuretics, antianginal agents; cardiotonics; pharmacology of tetrahydrozoline, benzthiazide, polythiazide, prazosin, clonidine, timolol, nimodipine, nisoldipine and nitrendipine; calcium channel antagonists. *Mailing Add:* Miles Inst Preclin Pharmacol 400 Morgan Lane West Haven CT 06516

SCRIBNER, BELDING HIBBARD, b Chicago, Ill, Jan 18, 21; m 42; c 4. NEPHROLOGY. *Educ:* Univ Calif, AB, 41; Stanford Univ, MD, 45; Univ Minn, MS, 51; Am Bd Internal Med, cert, 51; FRCP, 89. *Prof Exp:* From intern to resident med, San Francisco Hosp, 44-47; fel med, Mayo Found, Rochester, Minn, 47-50, asst staff mem, 50-51; from instr to assoc prof med, 51-62, head, Div Nephrology, 58-82, PROF MED, UNIV WASH, SEATTLE, 62- *Concurrent Pos:* Dir gen med res, Vet Admin Hosp, Seattle, 51-57; Markle scholar, Hammersmith Hosp, London, 57-58; mem, coun, Am Soc Nephrology, 74-79, Sci Adv Bd, Nat Kidney Found & Coun, Western Soc Clin Res. *Honors & Awards:* John Phillips Mem Award, Am Col Physicians, 73; David Hume Mem Award, 75; Mayo Solely Award, Western Soc Clin Res, 82; John Peters Award, Am Soc Nephrology, 86; Jean Hamburger Award, Int Soc Nephrology, 86. *Mem:* Inst Med-Nat Acad Sci; Am Soc Clin Invest; AMA; fel Am Col Physicians; Am Soc Nephrology (pres elect, 77, pres, 78-79); NY Acad Sci; Sigma Xi. *Res:* Fluid and electrolyte balance and kidney disease as pertaining to medicine; nephrology; dialysis; author of numerous technical publications. *Mailing Add:* Dept Med Div Nephrology Rm 11 Univ Wash Sch Med Seattle WA 98195

SCRIBNER, JOHN DAVID, organic chemistry, oncology; deceased, see previous edition for last biography

SCRIGNAR, CHESTER BRUNO, b Villa Park, Ill, Oct 15, 34; m 64. FORENSIC PSYCHIATRY, BEHAVIORAL MEDICINE. *Educ:* Ariz State Univ, Tempe, BA, 57; Tulane Univ Grad Sch, MS, 61; Tulane Univ Med Sch, New Orleans, 61. *Prof Exp:* ADJ PROF LAW & PSYCHIAT , TULANE UNIV SCH MED, DEPT PSYCHIAT & NEUROL, 72-, CLIN PROF , 79- *Concurrent Pos:* Adj prof anxiety treatment, Sch Social Work, 73-; adj prof, Xavier Univ, New Orleans, 84- *Honors & Awards:* John Herr Musser Mem Prize, 61; Milton G Erickson Award, 81. *Mem:* Am Psychiat Soc; Am Med Asn; Am Acad Psychiat & Law; Am Soc Clin Hypnosis; Am Soc Sex Educators; Asn Advan Behav Therapist. *Res:* Post-traumatic stress disorder and the psychological sequelae of trauma. *Mailing Add:* 2625 Gen Pershing New Orleans LA 70115

SCRIMGEOUR, KENNETH GRAY, b Vancouver, BC, Sept 10, 34; m 58; c 2. BIOCHEMISTRY. *Educ:* Univ BC, BA, 56, MSc, 57; Univ Wash, PhD(biochem), 61. *Prof Exp:* Res assoc biochem, Univ Wash, 61-62; from asst to assoc, Scripps Clin & Res Found, 62-67; assoc prof, 67-81, PROF BIOCHEM, UNIV TORONTO, 81- *Concurrent Pos:* Ed, Biochem & Cell Biol. *Mem:* Can Biochem Soc. *Res:* Enzymology; pteridine chemistry. *Mailing Add:* Dept Biochem Univ Toronto Toronto ON M5S 1A8 Can

SCRIMSHAW, NEVIN STEWART, b Milwaukee, Wis, Jan 20, 18; m 41; c 5. PUBLIC HEALTH. *Educ:* Ohio Wesleyan Univ, BA, 38; Harvard Univ, MA, 39, PhD(physiol), 41, MPH, 59; Univ Rochester, MD, 45; Am Bd Nutrit, cert, 64. *Hon Degrees:* DPS, Ohio Wesleyan Univ, 61; ScD, Univ Rochester, 74, Univ Tokushima, 79, Mahidol Univ, 82. *Prof Exp:* Instr embryol & comp anat, Ohio Wesleyan Univ, 41-42; intern, Gorgas Hosp, CZ, 43-46; asst resident physician obstet & gynec, Strong Mem Hosp & Genesee Hosp, Rochester, NY, 48-49; dir, Inst Nutrit Cent Am & Panama, Guatemala, 49-61; head, Dept Nutrit & Food Sci, Mass Inst Technol, 61-79, dir, Clin Res Ctr, 62-66 & 79-85, inst prof human nutrit, 76-88, EMER PROF HUMAN NUTRIT, MASS INST TECHNOL, 88- *Concurrent Pos:* Fel nutrit & endocrinol, Dept Vital Econ, Univ Rochester, 42-43, Rockefeller Found fel, Dept Obstet & Gynec, 46-47, Merck Nat Res Coun fel natural sci, 47-49; vis lectr trop pub health, Harvard Univ, 68-86; mem, Food & Nutrit Bd, Nat Acad Sci-Nat Res Coun, 67-77, mem exec comt, 68-75, mem, Comt Int Nutrit Progs, 64-72, chmn, 68-72, mem, Task Force Food-Health-Pop Prob, 73-76; mem adv comt med res, WHO, 71-78, chmn, 73-78; trustee, Rockefeller Found, 71-83; vpres, Int Union Nutrit Sci, 72-, pres, 78-81; mem, US Del Joint Comt, US-Japan Coop Med Sci Prog, 74-; chmn food & nutrit sect, Am Pub Health Asn, 63, mem res comt, 65, chmn ad hoc task force nutrit, 70; mem lectr prog, Inst Food Technol, 69-70, mem, Int Award Jury, 69-72, chmn, 74; mem expert work group on world hunger, UN Univ, 75; sr adv, World Hunger Prog, 75-; vis lectr, Harvard Sch Pub Health, 65-85; dir, Develop Studies Div, 85-86, Div Food & Nutrit Prog, 87-; res assoc, Nat Bur Econ Res, 86-; vis prof, Tufts Univ, 87-, Brown Univ, 88-; dir develop studies div, UN Univ. *Honors & Awards:* Osborne-Mendel Award, Am Inst Nutrit, 60; Int Award, Inst Food Technol, 69; Joseph Goldberger Award Clin Nutrit, AMA, 69; McCollum Award, Am Soc Clin Nutrit, 75; Bolton Carson Medal, Franklin Inst, 76; Bristol Meyers Award, 88; Feinstein Hunger Award, 91. *Mem:* Fel Am Col Nutrit; fel AAAS; fel Am Inst Nutrit; Am Col Prev Med; Am Physiol Soc. *Res:* Clinical and public nutrition; amino acid protein metabolism; nutrition and infection; diarrhea and malnutrition; single cell protein. *Mailing Add:* Harvard Ctr Pop Studies Nine Bow St Cambridge MA 02138

SCRIVEN, L E(DWARD), (II), b Battle Creek, Mich, Nov 4, 31; m 52; c 3. ENGINEERING SCIENCE. *Educ:* Univ Calif, Berkeley, BS, 52; Univ Del, MChE, 54, PhD(chem eng), 56. *Prof Exp:* Res engr, Shell Develop Co, 56-59; from asst prof to assoc prof, 59-89, assoc head dept chem eng, 74-77, FEL SUPERCOMPUT INST, UNIV, MINN, 89-, REGENTS PROF CHEM ENG & MAT SCI, 89- *Concurrent Pos:* Adv ed, Prentice-Hall, Inc; guest investr, Rockefeller Inst, 63; vis prof, Univ Pa, 67, Univ Fed Rio de Janeiro, 69 & Univ Witwatersand, 74 & Calif Inst Technol, 89; Guggenheim fel, 69-70; assoc ed, J Fluid Mech, 69-75; fac fel, Exxon Corp Res Labs, 84-89; mem bd, Chem Sci & Technol, 87-90, co-chair, 90-; Amundson Comt, Nat Res Coun, 84-88; sci adv, Packard Found, 88. *Honors & Awards:* Colburn Award, Am Inst Chem Engrs, 60, Walker Award, 77; Gibbs lectr, Am Math Soc, 86; Munphree Award, Am Chem Soc, 90. *Mem:* Nat Acad Eng; fel Am Inst Chem Engrs; Am Phys Soc; Soc Petrol Engrs; The Chem Soc; Soc Indust & Appl Math. *Res:* Interface, contact line and micellar physics; capillarity and small-scale free-surface flows; dynamic instability and pattern; finite element methods; coating process fundamentals. *Mailing Add:* Cardinal Point 2044 Cedar Lake Pkwy Minneapolis MN 55416

SCRIVER, CHARLES ROBERT, b Montreal, Que, Nov 7, 30; c 4. PEDIATRICS, GENETICS. *Educ:* McGill Univ, BA, 51, MD, CM, 55; FRCPS(C), 61. *Hon Degrees:* FRSC, 73. *Prof Exp:* From intern to jr asst resident med, Royal Victoria Hosp, Montreal, 55-56; jr asst resident, Montreal Children's Hosp, 57 & Med Ctr, Boston, Mass, 57-58; McLaughlin traveling fel, Univ Col, London Hosp, 58-60; chief resident pediat, Montreal Children's Hosp, 60-61; lectr pediat, 62-63, from asst prof to assoc prof, 63-69, PROF PEDIAT, MCGILL UNIV, 69-, PROF BIOL & HUMAN GENETICS, 77-; DIR, DEBELLE LAB BIOCHEM GENETICS, MONTREAL CHILDREN'S HOSP, 61- *Concurrent Pos:* Markle fel, 61-66; assoc physician, Dept Pediat Med, Montreal Children's Hosp, 63-65, physician, 65-; Med Res Coun assoc, 69- *Honors & Awards:* Wood Gold Medal, McGill Univ, 55; Royal Col Physicians & Surgeons, Can Medal, 61; F Mead Johnson Award, Am Acad Pediat, 68, Borden Award, 73; Borden Award, Nutrit Soc Can, 69; Allan Award, Am Soc Human Genetics, 78; Gardner Int Award, 79; Ross Award, Can Pediat Soc, 90. *Mem:* Am Acad Pediat; Am Soc Clin Invest; Am Soc Clin Nutrit; Am Soc Human Genetics; Am Pediat Soc; Sigma Xi; Asn Am Physicians. *Res:* Human genetics; membrane transport; amino acid metabolism; phosphate metabolism. *Mailing Add:* Debelle Lab Biochem/Genetics/Biol/Pediat Montreal Children's Hosp Res Inst 2300 Tupper St Montreal PQ H3H 1P3 Can

SCROGGIE, LUCY E, b Knoxville, Tenn, May 29, 35. ANALYTICAL CHEMISTRY. *Educ:* Univ Tenn, BS, 57, MS, 59; Univ Tex, PhD(anal chem), 61. *Prof Exp:* Anal chemist, Anal Chem Div, Oak Ridge Nat Lab, 61-66; anal chemist, Res & Develop Dept, Chem & Plastics Div, Union Carbide Corp, 66-69; anal chemist, Indust & Radiol Hyg Br, Div Environ Res & Develop, Tenn Valley Authority, 70-75, res chemist, Lab Br, Div Environ Planning, 75-80, mgt trainee, Off Natural Resources, 80-81, PROJ MGR, DIV SYST ENG, TENN VALLEY AUTHORITY, 81- *Mem:* Am Chem Soc; Am Conf Govt Indust Hygienists. *Res:* Analytical methods development; spectrophotometry; industrial hygiene chemistry; analytical chemistry applied to air and water quality and industrial hygiene. *Mailing Add:* Tenn Valley Authority 329 Evans Bldg Knoxville TN 37902-1499

SCROGGS, JAMES EDWARD, b Little Rock, Ark, Apr 20, 26; m 48; c 3. MATHEMATICS. *Educ:* Univ Ark, BA, 49; Univ Houston, MA, 54; Rice Inst, PhD(math), 57. *Prof Exp:* From asst prof to assoc prof math, Univ Ark, 57-62; asst prof, Univ Tex, 62-64; chmn dept, 66-79, PROF MATH, UNIV ARK, FAYETTEVILLE, 64- *Concurrent Pos:* Fulbright lectr, 80-81. *Mem:* Am Math Soc; Math Asn Am; Soc Indust & Appl Math. *Res:* Functional and complex analysis; Banach and Hilbert spaces. *Mailing Add:* Dept Math Univ Ark Fayetteville AR 72703

SCRUGGS, JACK G, b Cullman, Ala, Sept 9, 30; m 54; c 2. MAN MADE FIBERS, TEXTILES. *Educ:* Univ Mich, BS, 52; MS, 53; PhD(org chem), 56. *Prof Exp:* Group leader Monsanto Fibers Div, R&D, 56-66; VPRES TECH, PHILLIPS FIBERS CORP, 66- *Honors & Awards:* Borden Award, Univ Mich, Lenn & Fink Award. *Mem:* Am Chem Soc; AAAS; Sigma Xi. *Mailing Add:* Phillips Fibers Corp PO Box 66 Greenville SC 29611

SCUDAMORE, HAROLD HUNTER, b Wayne City, Ill, Dec 8, 15; m 42; c 4. MEDICINE. *Educ:* Mont State Col, BS, 37; Northwestern Univ, MS, 40, PhD(zool), 42, MD, 45; Am Bd Internal Med, dipl, 55; Am Bd Gastroenterol, dipl, 60. *Prof Exp:* Asst zool, Northwestern Univ, 37-39, asst sci, 39-42, instr, 42-43, asst pharmacol, 43; intern, Evanston Hosp, 45-46; fel med, Mayo Grad Sch Med, Univ Minn, 49-51, from instr to assoc prof, 52-71; CLIN ASSOC PROF MED, SCH MED, UNIV WIS, 71-; CHIEF GASTROENTEROL, MONROE CLIN, 71- *Concurrent Pos:* Mem staff, Mayo Clin, 51-71. *Mem:* AAAS; Am Gastroenterol Asn; fel Am Col Physicians; fel Am Col Gastroenterol; Am Soc Int Med; Am Med Asn. *Res:* Gastroenterology, especially malabsorption syndromes and diseases of the intestines; radioisotope studies; diseases of pancreas, stomach and postgastrectomy states. *Mailing Add:* 2612 Fourth St Monroe WI 53566

SCUDDAY, JAMES FRANKLIN, b Alpine, Tex, Sept 16, 29; m 50; c 3. VERTEBRATE BIOLOGY, WILDLIFE BIOLOGY. *Educ:* Sul Ross State Univ, BS, 52; Univ Idaho, MNS, 62; Tex A&M Univ, PhD(wildlife sci), 71. *Prof Exp:* Teacher independent sch dist, Tex, 52-54 & 56-61; instr biol, Sul Ross State Univ, 61-66; res asst wildlife sci, Tex A&M Univ, 67-69; from asst prof to assoc prof, 69-76, PROF BIOL, SUL ROSS STATE UNIV, 76- *Concurrent Pos:* NSF res partic syst bot, Okla State Univ, 64; pres, Chihuahuan Desert Res Inst, 84-; prin investr res proj. *Mem:* Herpet League; Am Soc Mammal; Soc Study Amphibians & Reptiles; Cooper Ornith Soc; Wildlife Soc. *Res:* Ecology and systematics of vertebrate animals; desert ecology. *Mailing Add:* Dept Biol Sul Ross State Univ Alpine TX 79832

SCUDDER, GEOFFREY GEORGE EDGAR, b Kent, Eng, Mar 18, 34; m 58. ENTOMOLOGY. *Educ:* Univ Wales, BSc, 55; Oxford Univ, DPhil(entom), 58. *Prof Exp:* Instr zool, 58-60, from asst prof to assoc prof, 60-68, PROF ZOOL, UNIV BC, 68-, CUR SPENCER ENTOM MUS, 58- *Concurrent Pos:* Royal Soc & Nuffield Found Commonwealth bursary, Imp Col, Univ London, 64-65. *Honors & Awards:* Gold Medal, Entom Soc Can, 75. *Mem:* Soc Study Evolution; Entom Soc Can; Can Soc Zool; fel Royal Entom Soc London; fel Royal Soc Can. *Res:* Systematics; evolution; entomology of Hemiptera; comparative morphology of insects; freshwater insect ecology and evolution. *Mailing Add:* Dept Zool Univ BC 2075 Westbrook Mall Vancouver BC V6T 1Z2 Can

SCUDDER, HARVEY ISRAEL, b Elmira, NY, Jan 2, 19; m 45; c 2. PUBLIC HEALTH, BIOLOGY. *Educ:* Cornell Univ, BS, 39, PhD(pub health), 53. *Prof Exp:* Asst entomologist, Boyce Thompson Inst Plant Res, 41; jr entomologist, USPHS, Fla, 42-43, state malaria control entomologist, Ala, 43-44, Carter Mem Lab, Ga, 44-46; chief div malaria control & sanit, Standard-Vacuum Oil Co, Sumatra, 47; tech coordr & officer in chg vector control field sta, State Dept Pub Health, Fresno, Calif, 51-54; res biologist, Tech Develop Labs, Commun Dis Ctr, USPHS, Savannah, Ga, 54-56, asst chief, Health Res Facil, Div Res Grants, NIH, 56-57, exec secy microbiol, virol & rickettsial study sect, 57-59; staff asst, Nat Cancer Inst, 59-61, chief virol res resources br, 61-62; asst to chief div res grants, NIH, 62; chief res training grants, Nat Inst Gen Med Sci, 62-65; manpower consult, Div Community Health Serv, USPHS, 65-66; head div biol & health sci, Calif State Univ, Hayward, 67-70, actg head div sci & math, 68-69, actg chmn dept biol sic, 70-71, coordr health sci, 71-72, prof microbiol, 67-80, EMER PROF MICROBIOL, CALIF STATE UNIV, HAYWARD, 80- *Concurrent Pos:* Mem microbiol fel rev panel, NIH, 58-60; mem, Moss Landing Marine Labs, 67-70, chmn policy bd, 69-70; mem, bd trustees, St Rose Hosp, Hayward, 69-83, chmn, 73-74, mem inst rev bd, 83-; trustee, Marine Ecol Inst, Redwood City, 71-, actg chmn, 74-78, chmn, 79-80 & 82-89; mem bd dir, E Bay Found Health Careers Educ, 72-85, chmn, 74-76 & 81-85; mem community adv comt, Fairmont Hosp, San Leandro, 74-83; chmn trustee corp bd, Calif Mosquito & Vector Control Asn, 90-91. *Mem:* AAAS; Am Mosquito Control Asn; Am Soc Trop Med & Hyg; Entom Soc Am; Am Soc Microbiol; Am Pub Health Asn; NY Acad Sci; Soc Vector Ecol; Sigma Xi. *Res:* Environmental sanitation and public health; health manpower; medical entomology; insect paleontology; decision processsing systems. *Mailing Add:* 7409 Hansen Dr Dublin CA 94568-2742

SCUDDER, HENRY J, III, b Brooklyn, NY, Sept 26, 35; c 4. INFORMATION SCIENCE. *Educ:* Cornell Univ, BEngPhys, 58, MEE, 60; Univ Calif, Berkeley, PhD(elec eng), 64. *Prof Exp:* Engr, Advan Electronics Ctr, Gen Elec Co, 58-60, info sci scientist, Res & Develop Ctr, 64-84; ASSOC PROF, DEPT MFG ENG, BOSTON UNIV, 85- *Concurrent Pos:* NSF fel, 62. *Mem:* Inst Elec & Electronics Engrs. *Res:* Information theory; pattern recognition; communication theory; data transmission and processing; learning machines; acoustic noise measurements; computer aided tomography; nondestructive evaluation. *Mailing Add:* Boston Univ Mfg Eng 64 Cummington St Boston MA 02215

SCUDDER, JACK DAVID, b Sao Paulo, Brazil, Sept 28, 47; US citizen; m 69; c 1. PLASMA PHYSICS. *Educ:* Williams Col, BA, 69; Univ Md, College Park, MS, 72, PhD(plasma physics), 75. *Prof Exp:* RES PHYSICIST SPACE PLASMA PHYSICS, NASA, GODDARD SPACE FLIGHT CTR, 69- *Concurrent Pos:* Max Planck fel, Max Planck Soc, 77. *Honors & Awards:* Mariner 10 Sci Award, NASA, 74; Int Sun Earth Explorer Team Award, NASA, 78. *Mem:* Am Geophys Union. *Res:* Kinetic physics of space magneto plasmas with emphasis on transport phenomena. *Mailing Add:* NASA Mail Code 692 Goddard Space Flight Ctr Greenbelt MD 20771

SCUDDER, JEFFREY ERIC, b Berkeley, Calif, Mar 20, 62; m 87; c 1. NUMERICAL ANALYSIS. *Educ:* Calif State Univ, BS(math) & BS(computer sci), 84, MS, 87. *Prof Exp:* Instr computer sci, Chapman Col, 86; instr math, Chabot Col, 87; lectr math & computer sci, Calif State Univ, Hayward, 85-88; instr computer sci, Univ Nebr, Omaha, 89-90; SOFTWARE ENGR, USAF STRATEGIC AIR COMMAND, 88- *Concurrent Pos:* Comt mem, Math Asn Am, 87- *Mem:* Math Asn Am; Armed Force Commun & Electronics Asn; Air Force Asn. *Res:* Mathematics applications in business; home computing applications. *Mailing Add:* 12439 Walker Dr Omaha NE 68123

SCUDDER, WALTER TREDWELL, b Elmira, NY, Aug 28, 20; div; c 3. WEED SCIENCE, VEGETABLE & FIELD CROPS. *Educ:* Cornell Univ, BS, 41, PhD(veg crops), 51; La State Univ, MS, 43. *Prof Exp:* Asst hort, La State Univ, 41-43; teacher, NY, 43-44; asst veg crops, Cornell Univ, 46-49; instr, 49-50; assoc horticulturist, SC Truck Exp Sta, 50-53; horticulturist, US Marine Corps, 53-55; assoc horticulturist, Univ Fla, 55-68, prof hort & horticulturist, Cent Fla Res & Educ Ctr, 68-86; RETIRED. *Concurrent Pos:* Coun, Agr Sci & Technol. *Mem:* Weed Sci Soc Am; Am Soc Hort Sci; Potato Asn Am; Am Soc Agron; Am Soybean Asn; Crop Sci Soc Am; Sigma Xi. *Res:* Chemical and biological weed control in vegetable and field crops; herbicide evaluation; persistence and degradation of herbicide residues in soil; weed species identification, terminology and distribution; biological nitrogen fixation in legumes. *Mailing Add:* 4001 S Sanford Ave Sanford FL 32773

SCULLY, ERIK PAUL, b Ossining, NY, Oct 22, 49; m; c 1. ZOOLOGY, POPULATION BIOLOGY. *Educ:* Fordham Univ, BS, 71; Univ RI, PhD(biol), 76. *Prof Exp:* Lectr ecol, Univ Calif, Irvine, 76-78; from instr to asst prof biol, 78-87, ASSOC PROF BIOL, TOWSON STATE UNIV, 87- *Concurrent Pos:* Vis scientist, Smithsonian Inst, 85-86. *Mem:* AAAS; Soc Study Evolution; Sigma Xi; Am Soc Naturalists. *Res:* Behavioral ecology and population biology of invertebrates, specifically marine invertebrates; mechanisms of resource utilization and intraspecific competition; use of the computer for instructional purposes. *Mailing Add:* Dept Biol Sci Towson State Univ Towson MD 21204

SCULLY, FRANK E, JR, b Brooklyn, NY, Mar 23, 47; m 71. CHEMISTRY. *Educ:* Spring Hill Col, BS, 68; Purdue Univ, PhD(chem), 73. *Prof Exp:* Instr org chem, Yale Univ, 73-75; asst prof chem sci, 75-80, ASSOC PROF CHEM, OLD DOMINION UNIV, 80- *Mem:* Sigma Xi. *Res:* Photosensitized oxygenations of cyclopropanes, vinylcyclopropanes, vinylcyclopropanols and norbornyl systems as a probe for an ionic mechanism of dioxetane formation; unsaturated alkoxide systems as an internal trap for intermediates in the reaction of singlet oxygen with olefins; use and effect of crown ethers in singlet oxygenations. *Mailing Add:* 1314 Brunswick Ave Norfolk VA 23508

SCULLY, MARLAN ORVIL, b Caspar, Wyo, Aug 3, 39; m 58; c 3. PHYSICS. *Educ:* Univ Wyo, BS, 61; Yale Univ, MS, 63, PhD(physics), 65. *Prof Exp:* Physicist, Gen Elec Co, 61-62; instr physics, Yale Univ, 65-67; from asst prof to assoc prof, Mass Inst Technol, 67-71; prof physics & optical sci, Univ Ariz, 70-80; DISTINGUISHED PROF, DEPT PHYSICS & ASTRON, UNIV NMEX, 80-; HEAD THEORY GROUP, MAX PLANCK INST QUANTUM OPTICS, 81- *Concurrent Pos:* Consult, United Aircraft Res Lab, 65-, Los Alamos Sci Lab, 70-, US Army, Redstone, 71-; Sci Appln Inc, 76- & Litton Indust, 76- mem, Joint Coun Quantum Electronics; John Simon Guggenheim & Alfred P Sloan fels; Humboldt fel. *Honors & Awards:* Adolph Lomb Medal, Optical Soc Am, 70. *Mem:* Fel AAAS; fel Am Phys Soc; fel Am Optical Soc. *Res:* Neutron and low temperature physics; laser physics; quantum statistical mechanics; solid state physics and quantum optics. *Mailing Add:* Rte 1 Box 8 Estancia NM 87016

SCULLY, ROBERT EDWARD, b Pittsfield, Mass, Aug 31, 21. PATHOLOGY. *Educ:* Col of the Holy Cross, AB, 41; Harvard Med Sch, MD, 44. *Prof Exp:* Asst clin prof, 59-63, assoc prof, 63-69, PROF PATH, HARVARD MED SCH, 70- *Concurrent Pos:* From assoc pathologist to pathologist, Mass Gen Hosp, 58- *Mem:* AMA; Int Soc Gynec Pathologists; Soc Gynec Oncol; Int Acad Path. *Res:* Gynecologic and testicular pathology and endocrinology. *Mailing Add:* Mass Gen Hosp Boston MA 02114

SCURRY, MURPHY TOWNSEND, b Houston, Tex, May 25, 33; m 55; c 2. MEDICINE, ENDOCRINOLOGY. *Educ:* Univ Tex, Austin, BA, 54; Univ Tex Med Br Galveston, MD, 58. *Prof Exp:* Rotating intern, Univ Pa, 59; resident med, Univ Mich, Ann Arbor, 61, NIH fel, 61-63; from asst prof to assoc prof, 66-85, CLIN PROF MED, UNIV TEX MED BR, GALVESTON, 85- *Concurrent Pos:* Consult, USPHS Hosp, Galveston, Tex, 70- *Mem:* Endocrine Soc; Am Diabetes Asn; Am Fedn Clin Res; Am Col Physicians. *Res:* Secretion of parathyroid hormone. *Mailing Add:* 1501 Broadway Galveston TX 77550

SCUSERIA, GUSTAVO ENRIQUE, b Buenos Aires, Arg, July 30, 56; m 82; c 1. COUPLED CLUSTER METHOD. *Educ:* Univ Buenos Aires, BSc, 79, PhD(physics), 83. *Prof Exp:* Grad asst, Dept Physics, FCEN, Univ Buenos Aires, 79-83, asst prof, 83-85; postdoctoral res assoc, Dept Chem, Univ Calif, Berkeley, 85-87; sr res assoc, Ctr Computational Quantum Chem, Univ Ga, 87-89; ASST PROF, DEPT CHEM, RICE UNIV, 89- *Mem:* Am Chem Soc; Am Phys Soc; AAAS. *Res:* Theoretical chemistry with an emphasis in computational quantum chemistry and development of new methods for molecular electronic structure. *Mailing Add:* Dept Chemistry Rice Univ, PO Box 1892 Houston TX 77251

SEABAUGH, PYRTLE W, b Millersville, Mo, Sept 14, 35; m 59; c 1. ANALYTICAL CHEMISTRY, APPLIED STATISTICS. *Educ:* Southeast Mo State Col, BS, 56; Iowa State Univ, PhD(inorg chem), 61; Univ Dayton, MBA, 71. *Prof Exp:* Fel struct chem, Univ Wis, 61-63; sr res chemist, Mound Lab, Monsanto Res Corp, 63-67, res specialist, 67-74, SR ANALYTICAL SPECIALIST, EG&G, 74- *Mem:* AAAS; Am Chem Soc; Am Crystallog Asn; Sigma Xi; Am Inst Physics. *Res:* Development of x-ray fluorescence and diffraction techniques; experimental design; interpretation of research and development data via applied statistics and numerical analysis; financial modeling; structural and pollution chemistry. *Mailing Add:* EG&G Mound Appl Tech PO Box 3000 Miamisburg OH 45343-3000

SEABLOOM, ROBERT W, b St Paul, Minn, Aug 14, 32; c 2. MAMMALOGY. *Educ:* Univ Minn, BA, 53, MS, 58, PhD(wildlife mgt), 63. *Prof Exp:* From asst prof to assoc prof biol, 61-75, PROF BIOL, UNIV NDAK, 75- *Concurrent Pos:* Vis scientist, Whiteshell Nuclear Res Estab, Atomic Energy Can Ltd, 71-72; fac, Itasca Biol Sta, 68, 70, 82 & 88; vis biologist, Univ Calif, Davis, 82-83. *Mem:* Am Soc Mammal; Wildlife Soc; Sigma Xi. *Res:* Vertebrate population ecology; mammalian ecology, life histories, and distributions; adrenal function in small mammal populations. *Mailing Add:* Dept Biol Univ NDak Grand Forks ND 58202

SEABORG, GLENN THEODORE, b Ishpeming, Mich, Apr 19, 12; m 42; c 6. NUCLEAR CHEMISTRY. *Educ:* Univ Calif, Los Angeles, AB, 34; Univ Calif, Berkeley, PhD(chem), 37. *Hon Degrees:* Forty-nine from US & foreign univs & cols, 51-78. *Prof Exp:* Res assoc chem, Univ Calif, Berkeley, 37-39, prof, 39-71, dir nuclear chem res, Lawrence Berkeley Lab, 46-58 & 72-75, assoc dir lab, 54-61, chancellor, 58-61, fac res lectr, 59, UNIV PROF CHEM, UNIV CALIF, BERKELEY, 71-, ASSOC DIR, LAWRENCE BERKELEY LAB, 72- *Concurrent Pos:* Sect chief metall lab, Univ Chicago, 42-46; mem gen adv comt, AEC, 46-50, mem hist adv comt, 58-61, chmn, AEC, 61-71; mem joint comn radioactivity, Int Coun Sci Unions, 46-56; mem, President's Sci Adv Comt, 59-61; mem, Pac Coast Comt, Am Coun Educ, 59-61; mem exec comt & chmn steering comt chem study, Chem Educ Mat Study, NSF, 59-74, mem bd, 60-61 & adv coun col chem, 62-67; mem, Fed Coun Sci & Technol, 61-71; mem, Fed Radiation Coun, 61-71; mem, Nat Aeronaut & Space Coun, 61-71; mem comn humanities, Am Coun Learned Socs & Coun Grad Schs, 62-65; mem sci adv comt, Pac Sci Ctr Found, 63-77; mem, Nat Coun Marine Res & Eng Develop, 66-71; mem sc adv bd, Robert Welch Found, 57-; trustee, Educ Serv, Inc, 61-67 & Pac Sci Ctr Found, 62-77; trustee, Sci Serv, 65-, pres, 66-; bd trustees, Swed Coun Am, 76- *Honors & Awards:* Nobel Prize in Chem, 51; Nat Medal of Sci, 91; Award Pure Chem, Am Chem Soc, 47, Nichols Medal, 48, Parsons Award, 64, Gibbs Medal, 66,

Marshall Madison Award, 72 & Priestly Medal, 79; Ericsson Gold Medal, Am Soc Swed Engrs, 48; Perkin Medal, Am Sect, Brit Soc Chem Indust, 57; Edison Found Award, 58; Enrico Fermi Award, AEC, 59; Priestley Mem Award, Dickinson Col, 60; Sci & Eng Award, Fedn Eng Socs, Drexel Inst, 62; Swed Am Yr, Vasa Order Am, 62; Franklin Medal, Franklin Inst, 63; Leif Erikson Award, Leif Erikson Found, 64; Arches of Sci Award, Pac Sci Ctr, 68; Chem Pioneer, Am Inst Chemists, 68, Gold Medal Award, 73; Prometheus Award, Nat Elec Mfrs Asn, 69; Oliver Townsend Award, Atomic Indust Forum, 71. *Mem:* Nat Acad Sci; fel AAAS (pres, 72); Am Chem Soc (pres, 76); fel Am Nuclear Soc; fel Am Phys Soc. *Res:* Heavy ion reactions; transuranium elements. *Mailing Add:* Lawrence Berkeley Lab Univ Calif Berkeley CA 94720

SEABORN, JAMES BYRD, b Panama City, Fla, Dec 15, 32; m 53; c 5. THEORETICAL NUCLEAR PHYSICS. *Educ:* Fla State Univ, BS, 60, MS, 62; Univ Va, PhD(nuclear theory), 65. *Prof Exp:* Asst prof physics, Univ Richmond, 65-66; res assoc, Univ Frankfurt, 66; asst prof, NTex State Univ, 67-69; vis lectr, Iowa State Univ, 69-70; assoc prof, 70-83, PROF PHYSICS, UNIV RICHMOND, 83-, CHMN DEPT, 82- *Concurrent Pos:* Consult, Lawrence Livermore Labs, 78-82; sabbatical leaves, Lawrence Livermore Labs, 78 & Univ Mainz, WGer, 85. *Res:* Nuclear structure studies; electromagnetic interactions in atomic nuclei. *Mailing Add:* Dept Physics Univ Richmond Richmond VA 23173

SEABROOK, WILLIAM DAVIDSON, b Ottawa, Ont, Apr 2, 35; m 60; c 2. INSECT NEUROPHYSIOLOGY. *Educ:* Carleton Univ, Can, BSc, 60, MSc, 64; Univ London, PhD(zool), 67. *Prof Exp:* Biologist, Govt Can, 60-62; PROF BIOL, UNIV NB, FREDERICTON, 67-, RES CONSULT, BIO-ENG INST, 67- *Concurrent Pos:* Dean grad studies, Univ NB, 84-89. *Mem:* Can Soc Zool; fel Can Entom Soc; Am Entom Soc. *Res:* Sensory physiology and behavior of insects, particularly chemical senses. *Mailing Add:* Dept Biol Univ NB Bag Serv No 45111 Fredericton NB E3B 5A3 Can

SEADER, J(UNIOR) D(EVERE), b San Francisco, Calif, Aug 16, 27; m 50, 61; c 8. APPLIED MATHEMATICS. *Educ:* Univ Calif, BS, 49, MS, 50; Univ Wis, PhD(chem eng), 52. *Prof Exp:* Instr chem eng, Univ Wis, 51-52; res engr, Calif Res Corp, Standard Oil Co, Calif, 52-57, group supvr chem process design, 57, supvr eng res, 58-59; sr res engr, Rocketdyne Div, NAm Aviation, Inc, 59-60, res specialist, 60-61, prin scientist, 60-65; prof chem eng, Univ Idaho, 65-66; chmn dept, 75-78, PROF CHEM ENG, UNIV UTAH, 66-, ADJ PROF FUELS ENG, 85- *Concurrent Pos:* Eve instr, eng exten, Univ Calif, 54-59; dir, Am Inst Chem Engrs, 83-85. *Honors & Awards:* 35th Ann Inst Lectr, Am Inst Chem Engrs, 83, Comput in Chem Eng Award, 88. *Mem:* Am Inst Aeronaut & Astronaut; fel Am Inst Chem Engrs, 83; Am Chem Soc. *Res:* Heat, mass and momentum transport; chemical kinetics; thermodynamics; physical properties; flammability; process synthesis and design; fuel processes; separation operations. *Mailing Add:* Dept Chem Eng Univ Utah Salt Lake City UT 84112

SEAGER, SPENCER LAWRENCE, b Ogden, Utah, Mar 10, 35; m 60; c 4. PHYSICAL CHEMISTRY. *Educ:* Univ Utah, BS, 57, PhD(phys chem), 62. *Prof Exp:* From asst prof to assoc prof chem, 61-69, PROF CHEM, WEBER STATE COL, 69-, CHMN DEPT, 68- *Mem:* Am Chem Soc; Sigma Xi; Nat Sci Teachers Asn. *Res:* Gas chromatography; gas phase diffusion. *Mailing Add:* Dept Chem 2503 Weber State Col Ogden UT 84408-2503

SEAGLE, EDGAR FRANKLIN, b Lincolnton, NC, June 27, 24; m 58; c 4. OCCUPATIONAL SAFETY & HEALTH. *Educ:* Univ NC, Chapel Hill, AB, 49; Univ Fla, BCE, 61; Univ NC, MSPH, 54; Univ Tex, DrPH(environ & occup health), 74; Am Acad Environ Engrs, dipl, 75. *Prof Exp:* Sanit consult, Div Epidemiol, NC State Bd Health, 54-56; chief, Indust Hyg Sect, Charlotte City Health Dept, NC, 56-59; engr, Div Radiol Health, USPHS, 61-66, chief, Prog Planning Off, 66-68, dir, Off Planning Strategy, 68-69, sr indust hyg engr, 69-75, dir occup safety, 75-78; asst dir, Fels Off Nat Acad Sci, 78-85; pub health engr, Md State Dept Health, 85-88; CONSULT ENGR, 78- *Honors & Awards:* Commendation Medal, USPHS. *Mem:* Am Soc Civil Engrs; Am Pub Health Asn. *Res:* Radiological health; industrial hygiene; occupational safety; sanitation. *Mailing Add:* 14108 Heathfield Ct Rockville MD 20853

SEAGLE, STAN R, ALLOYS. *Educ:* Purdue Univ, BS & MS. *Prof Exp:* VPRES RES & TECH DEVELOP, RMI TITANIUM CO, 79- *Mem:* Am Inst Mining Metall & Petrol Engrs; fel Am Soc Metals Int; Tech Asn Pulp & Paper Indust; Nat Asn Corrosion Engrs. *Res:* All aspects titanium research and development; numerous publications; seven US patents. *Mailing Add:* RMI Co 1000 Warren Ave Niles OH 44446

SEAGONDOLLAR, LEWIS WORTH, b Hoisington, Kans, Sept 30, 20; m 42; c 3. PHYSICS. *Educ:* Kans State Teachers Col, AB, 41; Univ Wis, PhM, 43, PhD(physics), 48. *Prof Exp:* Instr physics, Univ Kans, 47-48, from asst prof to prof, 48-65; head dept, 65-75, PROF PHYSICS, NC STATE UNIV, 65- *Concurrent Pos:* Civilian with Manhattan proj, Los Alamos Sci Lab, 59-60 & Hanford Lab, 62-64. *Mem:* Health Physics Soc; fel Am Phys Soc; Am Asn Physics Teachers. *Res:* Low energy nuclear physics; nuclear spectroscopy; Van de Graaff generators. *Mailing Add:* Dept Physics NC State Univ Raleigh NC 27695-8202

SEAGRAVE, JOHN DORRINGTON, b Bronxville, NY, Jan 23, 26; m 51; c 2. NUCLEAR PHYSICS, OPTICAL PHYSICS. *Educ:* Calif Inst Technol, BS, 46, MS, 48, PhD(physics), 51. *Prof Exp:* Mem staff physics, Los Alamos Nat Lab, 51-90; RETIRED. *Concurrent Pos:* Consult, Los Alamos Nat Lab, 90- *Mem:* Fel AAAS; Am Optical Soc; fel Am Phys Soc; Sigma Xi. *Res:* Optical physics and detectors; interactive image processing and pattern recognition; structure of very light nuclei and fast neutron scattering. *Mailing Add:* 3514 Arizona Los Alamos NM 87544

SEAGRAVE, RICHARD C(HARLES), b Westerly, RI, Dec 31, 35; m 59; c 1. CHEMICAL ENGINEERING. *Educ:* Univ RI, BS, 57; Iowa State Univ, MS, 59, PhD(chem eng), 61. *Prof Exp:* Asst prof chem eng, Univ Conn, 61-62; res fel, Calif Inst Technol, 62-63, asst prof, 63-66; assoc prof, 66-71, PROF CHEM ENG, IOWA STATE UNIV, 71- *Mem:* Am Inst Chem Engrs. *Res:* Transport phenomena; reactor dynamics; biomedical engineering. *Mailing Add:* Dept Chem Eng Iowa State Univ Ames IA 50011

SEALANDER, JOHN ARTHUR, JR, b Detroit Lakes, Minn, Dec 9, 17; m 47; c 3. ANIMAL PHYSIOLOGY. *Educ:* Luther Col, AB, 40; Mich State Univ, MS, 42; Univ Ill, PhD(zool), 49. *Prof Exp:* Asst zool, Mich State Univ, 40-42; asst zool & physiol, Univ Ill, 46-48; from asst prof to prof zool, Univ Ark, Fayetteville, 49-88; RETIRED. *Concurrent Pos:* USPHS spec fel, Inst Arctic Biol, Univ Alaska, 63-64; mem staff, Rocky Mt Biol Lab, 57; investr biol sta, Queen's Univ, Ont, 58; res assoc, Univ Ga, 68. *Mem:* AAAS; Am Soc Zool; Ecol Soc Am; Am Soc Mammal; Am Physiol Soc. *Res:* Comparative physiology; acclimatization of mammals to environmental temperature changes; natural history and ecology of mammals. *Mailing Add:* Dept Biol Sci Univ Ark Fayetteville AR 72703

SEALE, DIANNE B, b Birmingham, Ala, Apr 15, 45; m 73. ECOSYSTEMS, POPULATIONS. *Educ:* Wash Univ, St Louis, PhD, 73. *Prof Exp:* Fel, Nat Endowment Humanites, Pa State Univ, 73-74, instr environ sci, 73; res scientist, Ill Inst Technol, 74-75; res assoc, Dept Biol, Pa State Univ, 75-80; CTR SCIENTIST, CTR GREAT LAKES STUDIES, UNIV WIS, MILWAUKEE, 80-, FAC MEM BIOL SCI, 84- *Concurrent Pos:* Vis instr, Northwestern Univ, 74-75; prin investr, NSF grants, 75-78 & 78-80; Nat Oceanic & Atmospheric Admin Sea grants, 82-87; Environ Protection Agency grant, 84-86. *Mem:* Sigma Xi; Am Soc Limnol & Oceanog; Ecol Soc Am; Am Soc Ichthyologists & Herpetologists; Int Asn Great Lakes Res; Herpetologists' League. *Res:* Impact of suspension feeders on ecosystem processes; factors regulating suspension feeding dynamics; plant animal interactions; amphibian breeding behavior and larval community structure; nutrient release by amphibians and by Mysis relicta; phytoplankton ecology; animal energetics; microbiology. *Mailing Add:* Dept Biol Sci Univ Wis Box 413 Milwaukee WI 53201

SEALE, MARVIN ERNEST, b Edmonton, Alta, June 7, 22; m 53; c 4. ANIMAL BREEDING. *Educ:* Univ Alta, BSc, 48; Univ Minn, MS, 51, PhD(animal breeding), 65. *Prof Exp:* From asst prof to prof animal sci, Univ Man, 51-85, head dept, 73-80, assoc dean & dir, Glenlea Res Sta, 80-85; RETIRED. *Mem:* Am Soc Animal Sci; Can Soc Animal Prod; Genetics Soc Can. *Res:* Development of new breeds of livestock; inheritance of quantitative traits; evaluation of heterosis. *Mailing Add:* 195 Lyndel Dr Winnipeg MB R2H 1K5 Can

SEALE, RAYMOND ULRIC, b Snyder, Tex, Aug 19, 34; m 55; c 1. ANATOMY, EXPERIMENTAL EMBRYOLOGY. *Educ:* Eastern NMex Univ, BS, 56; Wash Univ, AM, 58; Univ Minn, PhD, 63. *Prof Exp:* Asst prof anat, Col Dent, Baylor Univ, 63-65; instr, Univ Tex Southwestern Med Sch, 65-66; asst prof, Sch Med, Univ Colo, Denver, 66-71; assoc prof, 71-80, PROF ANAT, RUSH MED COL, 80- *Concurrent Pos:* Vis instr, Southern Methodist Univ, 65-66; spec instr, Med Ctr, Baylor Univ, 63-66; AMA consult gross anat, Fac Med, Univ Saigon, 70. *Mem:* AAAS; NY Acad Sci; Am Asn Anat. *Res:* Developmental aspects of acquired immunological tolerance; morphogenetic movement in chick embryos in vitro; role of catecholamines in differentiation; mosaic and regulative capacity of organ primordia. *Mailing Add:* Dept Anat 1753 W Congress Rush-Presbyterian St Chicago IL 60612

SEALE, ROBERT L(EWIS), b Rosenberg, Tex, Mar 18, 28; m 47; c 4. NUCLEAR ENGINEERING. *Educ:* Univ Houston, BS, 47; Univ Tex, MS, 51, PhD(physics), 53. *Prof Exp:* Nuclear engr, Gen Dynamics Corp, Tex, 53-57, proj engr, 57-59, chief nuclear opers, 59-61; PROF NUCLEAR ENG, UNIV ARIZ, 61-, HEAD DEPT, 69- *Concurrent Pos:* Lectr, Southern Methodist Univ, 55-60; consult, Los Alamos Sci Lab, 61 & Sandia Corp, NMex, 66-; mem bd dir, Eng Coun Prof Develop, 72- *Mem:* Am Asn Physics Teachers; Am Phys Soc; Am Nuclear Soc; Nat Soc Prof Engrs; Sigma Xi. *Res:* Radiation shielding; nuclear reactor operations and safety; use of nuclear reactors. *Mailing Add:* 8815 Calle Bogota Tucson AZ 85715

SEALS, JONATHAN ROGER, INSULIN ACTION, GROWTH CONTROL. *Educ:* Wash Univ, St Louis, PhD(cell biol), 79. *Prof Exp:* ASST PROF BIOCHEM, SCH MED, UNIV MASS, 82- *Mailing Add:* Dept Protein Chem Cambridge Bio Sci Corp 365 Plantation St Worcester MA 01605

SEALS, RUPERT GRANT, b Shelbyville, Ky, Aug, 1932; m 54; c 4. DAIRY INDUSTRY. *Educ:* Fla Agr & Mech Univ, BS, 53; Univ Ky, MS, 56; Wash State Univ, PhD(dairy chem), 60. *Prof Exp:* Instr dairying, Fla Agr & Mech Univ, 54-55; res asst dairy sci, Wash State Univ, 55-59; assoc prof dairy mfg, Tenn Agr & Ind State, 59-64; res assoc, Iowa State Univ, 64-66, asst prof, 66-69; prof food chem & dean sch agr & home econ, Fla A&M Univ, 69-77; assoc dean & prof, 77-89, EMER PROF ANIMAL SCI, COL AGR, UNIV NEV, 89-; DIR, INT PROG, FLA A&M UNIV, 89- *Concurrent Pos:* Coordr spec prog, Coop State Res Serv, USBA, 74- *Mem:* Inst Food Technologists. *Res:* Amino acids in peanuts; milk proteins; lipid and flavor chemistry. *Mailing Add:* 1112 S Magnolia Apt J-101 Tallahassee FL 32301

SEALY, ROGER CLIVE, biophysics, for more information see previous edition

SEAMAN, EDNA, b Warsaw, Poland, July 2, 32; nat US; m 56; c 3. BIOLOGY. *Educ:* Brooklyn Col, BS, 56; Univ Ill, PhD(microbiol), 60. *Prof Exp:* Fel biochem, Brandeis, 60-68; asst prof biol, 68-73; assoc prof biol, 74-80, chmn dept, 78-80, ASSOC DEAN ARTS & SCI, UNIV MASS, HARBOR CAMPUS, BOSTON, 80- *Res:* Molecular biology; interrelations of DNA, RNA and proteins; bacterial transformations; synthesis of nucleic acids in subcellular systems. *Mailing Add:* Off Dean Col Arts & Sci Univ Mass Harbor Campus Boston MA 02125

SEAMAN, GEOFFREY VINCENT F, physical biochemistry, biomaterials, for more information see previous edition

SEAMAN, GREGORY G, b Alma, Mich, Jan 6, 38; m 80; c 3. ENVIRONMENTAL PHYSICS. *Educ:* Col Wooster, AB, 59; Yale Univ, MS, 60, PhD(physics), 65. *Prof Exp:* Appointee nuclear physics, Los Alamos Sci Lab, 64-66; fel, Rutgers Univ, 66-68; from asst prof to assoc prof nuclear physics, Kans State Univ, 68-76; prod specialist, 76-78, appln mgr, 79-80, proj mgr, 81-84, MGR SYSTS ENG, INSTRUMENTATION DIV, CANBERRA INDUSTS, 84- *Mem:* Am Phys Soc; Am Asn Physics Teachers. *Res:* Coulomb excitation; Ericson fluctuations; Doppler shift attenuation measurements of nuclear lifetimes; trace element analysis of foods by x-ray fluorescence. *Mailing Add:* One State St 45 Gracey Ave Meriden CT 06450

SEAMAN, LYNN, b De Queene, Ark, Aug 28, 33; m 89; c 4. CIVIL ENGINEERING, STRUCTURES. *Educ:* Univ Calif, Berkeley, BS, 59; Mass Inst Technol, PhD(civil eng), 61. *Prof Exp:* Civil engr, 61-67, PHYSICIST, SRI INT, 67- *Honors & Awards:* Alfred Sloan Scholarship; Gen Elec Fellowship. *Mem:* Am Soc Civil Engrs; Am Phys Soc. *Res:* Structural mechanics, shell buckling, structural dynamics; materials science, equations of state for soil and other porous media, crack growth and fracturing. *Mailing Add:* SRI Int 333 Ravenswood Ave Menlo Park CA 94025

SEAMAN, RONALD L, b Seaman, Ohio, Feb 10, 47; m 77; c 5. BIOENGINEERING & BIOMEDICAL ENGINEERING. *Educ:* Univ Cincinnati, BS, 70; Duke Univ, PhD(biomed eng), 75; Georgia Tech, MS(Mgt), 87. *Prof Exp:* Res assoc, Duke Univ, 74; res fel, Univ Tex Health Sci Ctr, Dallas, 75-76, instr physiol, 76-79; res engr, Ga Tech Res Inst, 79-82, sr reg engr, 82-86; ASSOC PROF, LA TECH UNIV, 86- *Concurrent Pos:* Res teaching asst, Duke Univ, 70-74; teaching asst, Univ Southern Calif, Santa Catalina Marine Biol Lab, 72; adj fac & adv bd, DeKalb Area Tech Voc Sch, 80-; adj fac, Oglethorpe Univ, 83-84; teacher, Col Mgt, Ga Inst Tech, 84. *Mem:* Inst Elec & Electronics Engrs; AAAS; Soc Neurosci; Bioelectromagnetics Soc; Am Soc Eng Educ; Soc Computer Simulation; NY Acad Scis. *Res:* Microwave biological effects, primarily neural systems; electromagnetic wave interactions with biological tissues; vestibular and auditory transduction; microwave exposure devices. *Mailing Add:* Dept Biomed Eng La Tech Univ Ruston LA 71272-0001

SEAMAN, WILLIAM B, b Chicago, Ill, Jan 5, 17; m 44; c 2. RADIOLOGY. *Educ:* Harvard Med Sch, MD, 41. *Prof Exp:* Instr radiol, Sch Med, Yale Univ, 48-49; from instr to prof, Sch Med, Wash Univ, 49-56; prof & chmn dept, 56-82, J PICKER EMER PROF RADIOL, COL PHYSICIANS & SURGEONS, COLUMBIA UNIV, 82- *Concurrent Pos:* Dir radiol serv, Presby Hosp, New York, 56-82; chmn comt radiol, Nat Acad Sci-Nat Res Coun, 69-73; mem bd chancellors, Am Col Radiol, 75-78. *Honors & Awards:* Gold Medal, Am Col Radiol, 83; W B Cannon Medal, Soc Gastrointestinal Radiol. *Mem:* Am Roentgen Ray Soc (pres, 73-74); Soc Gastrointestinal Radiol (pres, 73-74); Radiol Soc NAm; Asn Univ Radiol (pres, 56). *Res:* Diagnostic roentgenology. *Mailing Add:* Dept Radiol Columbia Univ Col Phys & Surg 622 W 168th St New York NY 10032

SEAMAN, WILLIAM E, b Washington, DC, May 22, 42. RHEUMATOLOGY. *Educ:* Harvard Univ, MD, 69. *Prof Exp:* ASSOC PROF MED, UNIV CALIF, SAN FRANCISCO, 76-; CHIEF ARTHRITIS & IMMUNOL, VET ADMIN MED CTR, SAN FRANCISCO, CALIF, 82- *Mem:* Am Asn Immunol; Am Fedn Clin Res; Am Soc Clin Invest. *Mailing Add:* Dept Med Univ Calif 4150 Clement St San Francisco CA 94121

SEAMAN, WILLIAM LLOYD, b Charlottetown, PEI, July 16, 34; m 59; c 3. PLANT PATHOLOGY. *Educ:* McGill Univ, BSc, 56; Univ Wis, PhD, 60. *Prof Exp:* Asst plant path, Univ Wis, 56-60; res officer, 60-73, RES SCIENTIST, CAN DEPT AGR, 73- *Mem:* Am Phytopath Soc; fel Can Phytopath Soc. *Res:* cereal grain diseases; disease survey. *Mailing Add:* Agr Can Plant Res Ctr K W Neatby Bldg CEF No 75 Ottawa ON K1A 0C6 Can

SEAMANS, DAVID A(LVIN), b Lawrence, Kans, June 13, 27; m 57; c 2. ELECTRICAL ENGINEERING, COMPUTER ENGINEERING. *Educ:* Univ Kans, BS, 50, MS, 56; Ore State Univ, PhD, 68. *Prof Exp:* Jr engr, Black & Veatch, Consult Engrs, 50-52; instr elec eng, Univ Kans, 53-54; from instr to asst prof, 54-63, ASSOC PROF ELEC ENG, WASH STATE UNIV, 63- *Mem:* Inst Elec & Electronics Engrs; Simulation Coun; Am Soc Eng Educ; Asn Comput Mach. *Res:* Analog and digital computer technology; artificial intelligence. *Mailing Add:* Dept Elec & Comput Eng Wash State Univ Pullman WA 99163

SEAMANS, R(OBERT) C(HANNING), JR, b Salem, Mass, Oct 30, 18; m 42; c 5. AERONAUTICAL ENGINEERING. *Educ:* Harvard Univ, BS, 39; Mass Inst Technol, MS, 42, ScD(instrumentation), 51. *Hon Degrees:* DSc, Rollins Col, 62, NY Univ, 67; DEng, Norwich Acad, 71, Notre Dame Univ, 74, Rensselaer Polytech Inst, 74, Univ Wyoming, 75, George Washington Univ, 75, Lehigh Univ, 76, Thomas Col, 80 & Curry Col, 82. *Prof Exp:* From Instr to assoc prof aeronaut eng, Mass Inst Technol, 41-55, staff engr instrumentation lab, 41-45, proj leader, 45-50, chief engr Proj Meteor, 50-55, dir flight control lab, 53-55; mgr airborne systs lab & chief systs engr, Airborne Systs Dept, Radio Corp Am, 55-58, chief engr, Missile Electronics & Controls Div, RCA Corp, 58-60; assoc adminr, NASA, 60-65, dep adminr, 65-68; Jerome Clarke Hunsaker prof aeronaut & astronaut, Mass Inst Technol, 68-69; Secy of the Air Force, 69-73; pres, Nat Acad Eng, 73-74; adminr, US ERDA, 74-77; Henry R Luce Prof Environ & Pub Policy, 77-84, dean eng, 78-81, SR LECTR AERONAUT, MASS INST TECHNOL, 84- *Concurrent Pos:* Mem subcomt automatic stabilization & control, Nat Adv Comt Aeronaut, 48-58 & group instrumentation & spec comt space technol, 58-59; consult, Sci Adv Bd, USAF, 57-59, mem, 59-62, assoc adv, 62-67; nat deleg, Adv Group Aerospace Res & Develop, NATO, 66-69; consult to adminr, NASA, 68-69; mem bd overseers, Harvard Univ, 68-74. *Honors & Awards:* Naval Ord Develop Award, 45; Lawrence Sperry Award, Am Inst Aeronaut & Astronaut, 51; NASA Distinguished Serv Medal, 65 & 69; Goddard Trophy, 68; Dept Air Force Exceptional Civilian Serv Award; Dept Defense Distinguished Pub Serv Medal; Gen Thomas D White USAF Space Trophy; Ralph Coats Roe Medal, Am Soc Mech Engrs. *Mem:* Nat Acad Eng (pres, 73-74); fel Am Astronaut Soc; fel Inst Elec & Electronics Engrs; hon fel Am Inst Aeronaut & Astronaut; AAAS. *Res:* Administration; instrumentation. *Mailing Add:* Mass Inst Technol Rm 33- 406 77 Massachusetts Ave Cambridge MA 02139

SEAMON, KENNETH BRUCE, HORMONE REGULATION, CALCIUM REGULATION. *Educ:* Carnegie-Mellon Univ, PhD(chem), 77. *Prof Exp:* Res chemist, Ctr Drugs & Biologics, 83-87, LAB CHIEF, LAB MOLECULAR PHARMACOL, DIV BIOCHEM & BIOPHYS, CTR BIOLOGICS EVAL & RES, FOOD & DRUG ADMIN, 87- *Mailing Add:* Food & Drug Admin 8800 Rockville Pike Bethesda MD 20892

SEAMON, ROBERT EDWARD, b Worcester, Mass, May 18, 39. NUCLEAR PHYSICS. *Educ:* Worcester Polytech Inst, BS, 61; Yale Univ, MS, 63, PhD(physics), 68. *Prof Exp:* Res staff mem nuclear reactor physics, 69-71, STAFF MEM NUCLEAR PHYSICS, LOS ALAMOS SCI LAB, 71- *Concurrent Pos:* Staff mem, Nuclear Data Sect, Int Atomic Energy Agency, Vienna, Austria, 77-78. *Mem:* Am Phys Soc; Am Nuclear Soc. *Res:* Evaluated nuclear data files (ENDF/B) and associated processing codes used in weapons calculations; phase-shift analysis of nucleon-nucleon scattering data; nucleon-nucleon potentials. *Mailing Add:* Box 421 Los Alamos NM 87544

SEANOR, DONALD A, b Gatley, Eng, June 10, 36; m 60; c 2. CHEMISTRY. *Educ:* Bristol Univ, BSc, 57, PhD(phys chem), 61. *Prof Exp:* Nat Res Coun Can fel catalysis, 61-63; from res chemist to sr res chemist, Chemstrand Res Ctr, Inc, Monsanto Co, NC, 63-67; scientist res labs div, Xerox Corp, 67-70, scientist info technol group, 70-71, sr scientist spec mat technol ctr, 71-81, prin scientist, 81-86, MGR, ELASTOMER TECHNOL, XEROX CORP, 86- *Concurrent Pos:* Guest lectr, NY State Col Environ Sci & Forestry, Syracuse, 75 & Mgt Res & Develop Course, Mitsloan Sch. *Mem:* Chem Soc; Am Chem Soc; Plastics & Rubber Asn. *Res:* Surface chemistry; solid state physics and chemistry; triboelectricity; polymers; photoconductivity and conduction in polymers; tribology; materials development; elastomer technology. *Mailing Add:* 264 Garnsey Rd Pittsford NY 14534

SEAPY, DAVE GLENN, b Santa Barbara, Calif, Aug 25, 56; m 84. PHYSICAL ORGANIC CHEMISTRY, PHOTOCHEMISTRY. *Educ:* Univ Calif Davis, BS, 78; Univ Colo Boulder, MS, 81, PhD(chem), 83. *Prof Exp:* Vis asst prof, 83-84, ASST PROF CHEM, FLA INST TECHNOL, 84- *Mem:* Am Chem Soc; Sigma Xi. *Res:* Organic photochemistry; the synthesis of rigid bridged ring compounds for the purpose of studying photochemical mechanisms; the process of intramolecular electron transfer. *Mailing Add:* 880 Roosevelt St Kingsburg CA 93631

SEAQUIST, ERNEST RAYMOND, b Vancouver, BC, Nov 19, 38; m 66; c 2. ASTRONOMY. *Educ:* Univ BC, BASc, 61; Univ Toronto, MA, 62, PhD(astron), 66. *Prof Exp:* Lectr, 65-66, from asst prof to assoc prof, 66-78, prof & assoc chmn, 78-88, CHMN, UNIV TORONTO, 88- *Mem:* Am Astron Soc; Can Astron Soc; Int Astron Union; Royal Astron Soc Can. *Res:* Galactic and extragalactic radio sources. *Mailing Add:* Dept Astron Univ Toronto Toronto ON M5S 1A7 Can

SEARCY, A(LAN) W(INN), b Covina, Calif, Oct 12, 25; m 45; c 3. MATERIALS SCIENCE, CHEMISTRY. *Educ:* Pomona Col, AB, 46; Univ Calif, PhD(chem), 50. *Prof Exp:* Asst chem, Univ Calif, 47-48, chemist, Radiation Lab, 48-49 & 50; from instr to asst prof, Purdue Univ, 49-54; from assoc prof to prof ceramic eng, 54-59, prof eng sci, 59-60, fac asst to chancellor, 63-64, vchancellor, 64-67, Miller res prof, 70-71, assoc dir, Lawrence Berkeley Lab & head Mat & Molecular Res Div, 80-84, PROF MAT SCI, UNIV CALIF, BERKELEY, 60- *Concurrent Pos:* Fulbright lectr, Balseiro Physics Inst, Arg, 60-61; consult, Los Alamos Sci Lab, Calif, 55-59 & Lawrence Radiation Lab, 56-61, assoc div head, Inorg Mat Div, 61-64; consult, adv res proj agency, US Dept Defense, 58-60; prin investr, Lawrence Berkeley Lab, 60-; mem, Nat Res Coun Comt on high temperature chem, 61-70; Guggenhiem fel, 67-68. *Mem:* Fel AAAS; fel Am Ceramic Soc; Am Chem Soc; Mat Res Soc. *Res:* Compositions of vapors at high temperatures and low pressures; thermodynamics of high temperature reactions; kinetics of gas-solid reactions and vaporization; thermodynamics of surfaces; thermodynamics and kinetics in temperature gradients. *Mailing Add:* Dept Mat Sci & Mineral Eng Univ Calif Berkeley CA 94720

SEARCY, CHARLES JACKSON, b Beaver, Okla, Feb 14, 35; m 61; c 2. MATHEMATICS. *Educ:* Panhandle A&M Col, BS, 57; Okla State Univ, MS, 63, EdD(math), 67. *Prof Exp:* Instr math, Cent State Col, Okla, 65-67; from asst prof to prof math, NMex Highlands Univ, 67-88; CONSULT, 88- *Mem:* Math Asn Am; Am Math Soc. *Res:* Topology; algebra. *Mailing Add:* Dept Math NMex Highlands Univ Las Vegas NM 87701

SEARCY, DENNIS GRANT, b Portland, Ore, Sept 25, 42; m 66; c 2. CELL PHYSIOLOGY, ARCHAE BACTERIA. *Educ:* Ore State Univ, BS, 64; Univ Calif, Los Angeles, PhD(zool), 68. *Prof Exp:* NIH trainee, Univ Calif, Los Angeles, 68-69; fel, Calif Inst Technol, 69-71; from asst prof to assoc prof, 71-88, PROF ZOOL, UNIV MASS, AMHERST, 88- *Concurrent Pos:* Vis scholar, Oxford, 78-79; Nat Ctr Sci Res fel, Nat Mus Natural History, Paris, 84-85. *Mem:* Am Soc Cell Biol; Int Soc Evolutionary Protistology. *Res:* Evolution; origin of eukaryotic cells; histones; chromatin structure; physiology of primitive organisms. *Mailing Add:* Dept Zool Univ Mass Amherst MA 01003

SEARIGHT, THOMAS KAY, b Vermillion, SDak, June 3, 29; m 54; c 2. GEOLOGY. *Educ:* Univ Mo, AB, 51, MA, 52; Univ Ill, PhD, 59. *Prof Exp:* Geologist, State Geol Surv, Mo, 52-54; from asst prof to assoc prof, 59-74, PROF GEOL, ILL STATE UNIV, 74- *Concurrent Pos:* Res affil, Ill State Geol Surv, 62- *Mem:* Geol Soc Am; Soc Econ Paleont & Mineral; Am Asn Petrol Geologists. *Res:* Pennsylvania stratigraphy and sedimentation of the mid-continent region. *Mailing Add:* Dept Geol Sch 246 Ill State Univ Normal IL 61761

SEARLE, CAMPBELL L(EACH), b Winnipeg, Man, July 24, 26; Can citizen; m 53; c 4. ELECTRONICS. *Educ:* Queen's Univ, Ont, BSc, 47; Mass Inst Technol, SM, 51. *Prof Exp:* Mem staff, Div Sponsored Res, Mass Inst Technol, 51-56, from instr to prof elec eng, 56-74; prof psychol & elec eng, Queen's Univ, Ont, 74-79; PROF ELEC ENG, MASS INST TECHNOL, 79- *Mem:* Inst Elec & Electronics Engrs; Acoust Soc Am. *Res:* Auditory and speech perception. *Mailing Add:* Dept Elec Eng & Comput Sci Mass Inst Technol 77 Massachusetts Ave 20 A 226 Cambridge MA 02139

SEARLE, GILBERT LESLIE, physiology, biochemistry, for more information see previous edition

SEARLE, GORDON WENTWORTH, b Providence, RI, Mar 9, 20; m 45; c 3. PHYSIOLOGY. *Educ:* Univ Ill, BS, 41, MD, 45; Northwestern Univ, MS, 49, PhD(physiol), 51. *Prof Exp:* Asst physiol, Northwestern Univ, 48-51; asst prof, Albany Med Col, Union, NY, 51-52; asst prof, 52-55, ASSOC PROF PHYSIOL, COL MED, UNIV IOWA, 55- *Concurrent Pos:* Vis res prof, Med Sch, Univ Newcastle, 64-65. *Mem:* AAAS; Am Physiol Soc; Soc Exp Biol & Med; Sigma Xi. *Res:* Intestinal absorption; bile secretion. *Mailing Add:* 345 Koser Ave Iowa City IA 52240

SEARLE, JOHN RANDOLPH, b Wilmington, Del, Jan 20, 47; m 73; c 2. BIOMEDICAL ENGINEERING. *Educ:* Wake Forest Univ, BS, 70; NC State Univ, BS, 70; Duke Univ, PhD(biomed eng), 75. *Prof Exp:* Biomed engr, Vet Admin, 75-77; ASST PROF BIOMED ENG, MED COL GA, 77- *Mem:* AAAS; Asn Advan Med Instrumentation; Inst Elec & Electronics Engrs; Eng Med & Biol. *Res:* Fetal heart rate analysis by micromputer; flow in collapsible vessels; high rate dilatometry of dental porcelains. *Mailing Add:* Dept Health Systs/Info Sci Med Col Ga CN-125 University Pl Augusta GA 30912

SEARLE, NORMA ZIZMER, b New York, NY, Jan 26, 25; m 49. WEATHERING & STABILIZATION OF POLYMERS. *Educ:* Hunter Col, BA, 46; NY Univ, PhD(phys chem), 59. *Prof Exp:* Control chemist, Purepac Pharmaceut Co, 46; res chemist, Montefiore Hosp, New York, 47-53; chemist, New York Dept Health, 53-54; chemist, Am Cyanamid Co, 57-69, sr res chemist, Am Cyanamid Co, 69-74, group head, Am Cyanamid Co, 74-82, CONSULT, PLASTICS & CHEM, 82- *Concurrent Pos:* Lectr, short courses on weathering of polymers & paints; consult, Atlas Elec Devices Co, Amoco Performance Prod, Borg Warner Chem, Shell Develop Co. *Mem:* Am Chem Soc; Am Soc Testing & Mat; Am Asn Textile Chemists & Colorists. *Res:* Application of photochemistry, spectroradiometry and spectroscopy (UV, VIS, NIR, IR, MS, Luminescence) techniques to development of light stable polymers and coatings including paint for space vehicles; material property evaluations by physical chemical and analytical techniques; developed techniques for determining wavelength sensitivity of polymers. *Mailing Add:* Grandview Gardens 106D Finderne Ave Bridgewater NJ 08807

SEARLE, ROGER, b Wilmington, Del, July 24, 36; m 61; c 2. ANALYTICAL CHEMISTRY. *Educ:* Oberlin Col, BA, 58; Univ Ill, PhD(org chem), 63. *Prof Exp:* Res chemist, Eastman Kodak Co, 63-88; RES CHEMIST, STERLING DRUG, 88- *Mem:* AAAS; Am Chem Soc; Sigma Xi. *Res:* Physical organic chemistry; organic photochemistry. *Mailing Add:* RD 2 Box 15 Averill Park NY 12018

SEARLE, SHAYLE ROBERT, b Wanganui NZ, Apr 26, 28; m 58; c 2. MATHEMATICAL STATISTICS. *Educ:* Victoria Univ Wellington, BA, 49, MA, 50; Cambridge Univ, dipl math stat, 53; Cornell Univ, PhD(animal breeding), 59. *Prof Exp:* Actuarial asst, Colonial Mutual Life Ins Co, NZ, 50-51; res statistician, NZ Dairy Bd, 53-62; res asst animal breeding, 56-58, res assoc, 58-59, from asst prof to assoc prof biol statist, 62-69, PROF BIOL STATIST, NY STATE COL AGR, CORNELL UNIV, 69- *Concurrent Pos:* Fulbright travel award, 56-59; short course teaching, George Washington Univ, 77- *Honors & Awards:* Alexander von Humboldt Sr US Scientist Award, 84. *Mem:* Fel Am Statist Asn; Biomet Soc; fel Royal Statist Soc; Int Statist Inst. *Res:* Matrix algebra; linear models; variance components. *Mailing Add:* Biomet Unit NY State Col Agr Cornell Univ 337 Warren Hall Ithaca NY 14850

SEARLE, WILLARD F, JR, b Columbus, Ohio, Jan 17, 24. OCEAN ENGINEERING, NAVAL & SALVAGE ENGINEERING. *Educ:* US Naval Acad, Annapolis, BS, 48; Mass Inst Technol, MS, 52. *Prof Exp:* Eng duty officer, USN, 47-56, head eng res, Exp Diving Unit, 56-64, supvr salvage, 64-69, proj mgr ship acquisition, 69-70; instr ocean eng, Marine Maritime Acad, 71-84; sr vis lectr, Mass Inst Technol, 71-90; RETIRED. *Concurrent Pos:* Mem, var studies, comts & panels, Marine Bd, Nat Res Coun, 70-84, var adv panels, Nat Oceanic & Atmospheric Admin, Ocean Thermal Energy Conversion, Dept Energy, UN Environ Prog & UN Relief Oper, Bangladesh; chmn, Standing Panel Salvage & Ocean Towing, Soc Naval Architects & Marine Engrs, Comt Diving & Salvage, Marine Technol Soc & MacKinnon Searle Consortium Ltd, 71- *Honors & Awards:* Harold E Saunders Award, Am Soc Naval Engrs, 85. *Mem:* Nat Acad Eng; Inst Nautical Architects; Soc Naval Architects & Marine Engrs; Am Soc Naval Engrs; Royal Inst Naval Engrs; Soc Am Mil Engrs. *Res:* Unique salvage and wreck clearance; deep search and recovery; explosive ordnance disposal; submarine seafloor cable laying; author of more than 100 technical publications. *Mailing Add:* 808 Timber Branch Pkwy Alexandria VA 22302

SEARLES, ARTHUR LANGLEY, b Nashua, NH, Aug 8, 20; m. ORGANIC CHEMISTRY. *Educ:* NY Univ, BA, 42, PhD(chem), 46. *Prof Exp:* Asst, Squibb Inst Med Res, 44; from instr to asst prof org chem, NY Univ, 46-56; assoc prof chem 56-73, chmn dept, 70-72, PROF CHEM, COL MT ST VINCENT, 74- *Concurrent Pos:* Consult, Roel-Cryston Corp, New York, 50-52, FMC Corp, Princeton, NJ, 55-57 & US Govt, 56; vis prof, Dept Chem, Hunter Col, 57 & NY Univ, 63, 64 & 66. *Honors & Awards:* Bene Merenti Medal, 81. *Mem:* Am Chem Soc; Royal Soc Chem; Sigma Xi. *Res:* Nitrogen heterocycles; organo-metallics; beta-ketoanilides. *Mailing Add:* Dept Chem Col Mt St Vincent Bronx NY 10471

SEARLES, RICHARD BROWNLEE, b Riverside, Calif, June 19, 36; m 57; c 3. PHYCOLOGY. *Educ:* Pomona Col, BA, 58; Univ Calif, Berkeley, PhD(bot), 65. *Prof Exp:* From asst prof to assoc prof, 65-83, PROF BOT, DUKE UNIV, 83- *Mem:* Phycol Soc Am; Int Phycol Soc (secy, 81-86). *Res:* Marine phycology; morphology; taxonomy and ecology of benthic marine algae. *Mailing Add:* Dept Bot Duke Univ Durham NC 27706

SEARLES, SCOTT, JR, b Minneapolis, Minn, Oct 15, 20; m 47, 59; c 5. ORGANIC CHEMISTRY. *Educ:* Univ Calif, Los Angeles, BA, 41, MA, 42; Univ Minn, PhD(org chem), 47. *Prof Exp:* Asst chem, Univ Minn, 42-43; res chemist, Am Cyanamid Co, Conn, 44-45; instr chem, Univ Minn, 46; instr, Univ Ill, 47-49; asst prof, Northwestern Univ, 49-52; assoc prof, Kans State Univ, 52-62, prof, 62-66; prof, 66-85, EMER PROF CHEM, UNIV MO, COLUMBIA, 85- *Concurrent Pos:* NSF sr fel, Calif Inst Technol & Cambridge Univ, 62-63. *Mem:* AAAS; Am Chem Soc. *Res:* Small ring heterocyclic compounds; rearrangements; reaction mechanisms; ionic catalysis. *Mailing Add:* 1007 Westwinds Ct Columbia MO 65203

SEARLS, CRAIG ALLEN, b Bremerton, Wash, Feb 27, 54. GENERAL ELECTROMAGNETIC & ACOUSTIC REMOTE SENSING APPLIED TO GEOPHYSICAL PROBLEMS. *Educ:* Univ Puget Sound, BS, 76; Univ Calif, Los Angeles, MS, 78, PhD(geophys & space physics), 81. *Prof Exp:* Sci intern, Northwest Asn Col & Univ Advan Sci, Pac Northwest Div, Batelle Mem Inst, 76; GEOPHYSICIST, WESTERN ELEC, SANDIA NAT LAB, 81- *Mem:* Am Geophys Union; AAAS; Soc Explor Geophysicists. *Res:* Design, monitoring, and analysis of geophysical experiments addressing a broad range of geophysical problems. *Mailing Add:* Sandia Nat Lab Org 9114 Albuquerque NM 87185-5800

SEARLS, JAMES COLLIER, b Mitchell, SDak, Aug 22, 26; m 47; c 4. ANATOMY. *Educ:* Cornell Col, BA, 50; Univ Iowa, DDS, 55, PhD(anat), 66. *Prof Exp:* Fel, Nat Inst Dent Res, 62-66; ASST PROF FAMILY DENT & ANAT, UNIV IOWA, 66- *Concurrent Pos:* Gen dent practr, 55-62. *Mem:* Am Asn Anatomists; Int Asn Dent Res. *Res:* Radioisotopic studies of the cartilagenous nasal septum and its role in maxillofacial growth. *Mailing Add:* Dept Oral Biol & Anat Univ Iowa Iowa City IA 52242

SEARLS, ROBERT L, b Madison, Wis, Oct 26, 31; m 61; c 4. BIOCHEMISTRY, EMBRYOLOGY. *Educ:* Univ Wis, BS, 53; Univ Calif, Berkeley, PhD(biochem), 60. *Prof Exp:* Fel embryol, Brandeis Univ, 60-63; asst prof biol, Univ Va, 63-68; assoc prof, 68-74, PROF BIOL, TEMPLE UNIV, 74- *Mem:* AAAS; Am Chem Soc; Soc Develop Biol; Int Soc Develop Biol. *Res:* Oxidative metabolism; chemical basis of morphogenesis and differentiation. *Mailing Add:* Dept Biol Temple Univ Philadelphia PA 19122

SEARS, CHARLES EDWARD, b Utica, Mich, Feb 3, 11; m 37; c 4. ECONOMIC GEOLOGY, ENGINEERING GEOLOGY. *Educ:* Va Polytech Inst, BS, 32, MS, 35; Colo Sch Mines, DSc, 53. *Prof Exp:* Instr mining eng & from asst prof to assoc prof geol, Va Polytech Inst & State Univ, 46-77, emer prof geol, 77, dir seismol observ, 61-77; RETIRED. *Concurrent Pos:* Consult geologist. *Mem:* Geol Soc Am; Am Geophys Union. *Res:* Hydrothermal alteration and mineralization at the Climax Molybdenum deposit, Climax, Colorado; geology and petrology; kimberlites; geothermal studies of Virginia areas. *Mailing Add:* 604 Airport Rd Blacksburg VA 24060

SEARS, CURTIS THORNTON, JR, b Wareham, Mass, Aug 3, 38; m 60; c 2. INORGANIC CHEMISTRY, ORGANOMETALLIC CHEMISTRY. *Educ:* WVa Wesleyan Col, AB, 61; Univ NC, PhD(chem), 66. *Prof Exp:* NATO fel, 66-67; asst prof chem, Univ SC, 67-71; asst prof, 71-77, ASSOC PROF CHEM, GA STATE UNIV, 77- *Honors & Awards:* O'Haus Award, Nat Sci Teachers Asn, 73. *Mem:* Am Chem Soc; Royal Soc Chem. *Res:* Chemistry of second and third row transition metals in low oxidation states. *Mailing Add:* Dept Chem Ga State Univ Univ Plaza Atlanta GA 30303

SEARS, DAVID ALAN, b Portland, Ore, Oct 20, 31; m 58; c 3. INTERNAL MEDICINE, HEMATOLOGY. *Educ:* Yale Univ, BS, 53; Univ Ore, MS, 58, MD, 59; Am Bd Internal Med, dipl, 66, cert hemat, 74. *Prof Exp:* Intern, Sch Med & Dent, Univ Rochester, 59-60, resident med, 60-62, trainee hemat, 62-63, asst prof med, 66-69; assoc prof, 69-77, prof med, 77-80, head div hemat, Univ Tex Med Sch San Antonio, 69-79; chief med serv, Harris County Hosp Dist, 80-90; Herman Brown Teaching prof & vchmn dept med, 80-90, PROF MED, BAYLOR COL MED, HOUSTON, 90- *Concurrent Pos:* Assoc physician, Strong Mem Hosp, Rochester, NY, 66-69; consult, Highland Hosp, Rochester, NY, 66-69 & Audie Murphy Vet Admin Hosp, San Antonio, 73-80. *Honors & Awards:* St George Medal, Am Cancer Soc, 88. *Mem:* Am Soc Hemat; Am Fedn Clin Res; Int Soc Hemat; fel Am Col Physicians. *Res:* Hemolytic disease; heme pigment metabolism; erythrocyte membrane; anemia. *Mailing Add:* Dept Med Baylor Col Med One Baylor Plaza Houston TX 77030

SEARS, DEREK WILLIAM GEORGE, b Maidstone, Eng, Dec 18, 48; m 71; c 3. METEORITES, THERMOLUMINESCENCE. *Educ:* Univ Kent, Eng, BSc, 70; Univ Col London, dipl, 71, Univ Leicester, PhD(geol astron), 74. *Prof Exp:* Res asst, Univ Manchester, 74-77; res fel, Univ Birmingham, 77-79; asst res chemist, Univ Calif, Los Angeles, 79-81; from asst prof to assoc prof chem, 81-88, PROF CHEM, UNIV ARK, 88- *Concurrent Pos:* Mem, NASA-NSF Meteorite Working Group, 85-89 & NASA Lunar & Planetary Geosci Rev Panel, 85-89; assoc ed, Geochemica Cosmochemica Acta, 89-, Meteoritics, 90- *Mem:* Fel Meteoritical Soc; Am Chem Soc. *Res:* Study of meteorites by thermoluminescence, caltrodoluminescence and related techniques; instrumental neutron activation; electron-microprobe analysis; thermodynamic modelling. *Mailing Add:* Dept Chem & Biochem Univ Ark Fayetteville AR 72701

SEARS, DONALD RICHARD, b Wilmington, Del, July 23, 28; c 1. ENVIRONMENTAL SCIENCE & ENGINEERING. *Educ:* Lawrence Col, BS, 50; Cornell Univ, PhD(phys chem), 58. *Prof Exp:* Asst chem, Cornell Univ, 50-52; chemist, Inst Paper Chem, Lawrence Col, 54; asst chem, Cornell

Univ, 54-57; res fel chem phys, Mellon Inst, 58-63; res staff mem, Oak Ridge Nat Lab, 63-68; develop specialist, Oak Ridge Y-12 Plant, 68-72; vis scientist, Nat Ctr Atmospheric Res, 72-73; dir, Air Pollution Eng Lab, Civil Eng Dept, WVa Univ, 73-75; res scientist, Lockheed Res & Eng Ctr, 75-79; PROJ MGR ENVIRON ENG, GRAND FORKS ENERGY TECHNOL CTR, 79- *Mem:* AAAS; Air Pollution Control Asn; Am Chem Soc; AAAS; Sigma Xi. *Res:* Characterization and control of particulate emissions; analysis of trace element and organic emissions from combustion of low rank western coals. *Mailing Add:* 4831 Vespucci Dr Sierra Vista AZ 85635-2334

SEARS, DUANE WILLIAM, b Denver, Colo, Mar 23, 46; m 69; c 2. IMMUNOLOGY, BIOCHEMISTRY. *Educ:* Colo Col, BS, 68; Columbia Univ, PhD(biophys chem), 74. *Prof Exp:* Fel, Albert Einstein Col Med, 74-77; ASST PROF IMMUNOL & BIOCHEM, UNIV CALIF, SANTA BARBARA, 77- *Mem:* Am Asn Immunologists; NY Acad Sci; AAAS; Am Chem Soc. *Res:* Structural analysis of major histocompatibility complex antigens; immunogenetic analysis of cytotoxic T lymphocyte reactivities; biochemical analysis of cytotoxic T lymphocyte target antigens; molecular analysis of K/NK cells. *Mailing Add:* Dept Biol Sci Univ Calif Santa Barbara CA 93106

SEARS, ERNEST ROBERT, b Rickreall, Ore, Oct 15, 10; m 36, 50; c 4. PLANT CYTOGENETICS. *Educ:* Ore State Col, BS, 32; Harvard Univ, AM, 34, PhD(biol), 36. *Hon Degrees:* DSc, Univ Gottingen, 70. *Prof Exp:* Geneticist, Agr Res Serv, USDA, 36-80, EMER PROF, UNIV MO, 81- *Concurrent Pos:* Fulbright res fel, Ger, 58; Hannaford res fel, Australia, 80; Einstein res fel, Israel, 80, Eng, 82; Michael res fel, Israel, 85. *Honors & Awards:* Stevenson Award, 51; Hoblitzelle Award, 58; Nat Agribus Sci Award, 81; Wolf Prize Agr, 86. *Mem:* Nat Acad Sci; Genetics Soc Am (pres, 78-79); Am Soc Agron; Bot Soc Am; Genetics Soc Can; Am Acad Arts & Sci. *Res:* Origin, evolution, cytogenetics of wheat. *Mailing Add:* 108 Curtis Hall Univ Mo Columbia MO 65211

SEARS, HAROLD FREDERICK, b Wilmington, Del, Feb 20, 47; m 68; c 2. BEHAVIORAL BIOLOGY. *Educ:* Northwestern Univ, BA, 69; Univ NC, PhD(zool), 76. *Prof Exp:* From instr to asst prof, 74-81, ASSOC PROF BIOL, UNIV SC, UNION, 81-, ASSOC DEAN, 84- *Mem:* AAAS; Sigma Xi; Animal Behav Soc; Am Ornithologists Union; Am Soc Zoologists; Wilson Ornithologists Soc; Cooper Ornithologists Soc. *Res:* Vertebrate communication and display behavior; evolution of display; invertebrate homing and orientation. *Mailing Add:* 3590 Glenn Springs Rd Spartanburg SC 29302

SEARS, J KERN, b Harper, Kans, May 13, 20; m 50; c 4. PLASTICS CHEMISTRY. *Educ:* Harding Col, BS, 42; Univ Mo, MA, 45, PhD(org chem), 47. *Prof Exp:* Asst instr, Univ Mo, 47; assoc prof, Harding Col, 47-51; res chemist, 51-71, res specialist, 71-84, consult technologist plasticizer applications, Monsanto Co, 84-85; PVT CHEM CONSULT, PLASTICIZERS. *Honors & Awards:* Contribution to Vinyl Plastics Award, Soc Plastics Engrs, 83. *Mem:* Soc Plastics Eng. *Res:* Organic chemistry; compatibility and solventability; resin modification by external additives; polymer testing and evaluation, stabilization of polyvinyl chloride, weathering of plasticized polyvinyl chloride; plasticizer chemistry. *Mailing Add:* 485 Hawthorne Ave St Louis MO 63119

SEARS, JACK WOOD, b Cordell, Okla, Aug 12, 18; m 43; c 3. GENETICS. *Educ:* Harding Col, BS, 40; Univ Tex, MA, 42, PhD(genetics), 44. *Prof Exp:* Instr zool, Univ Tex, 44-45; prof biol & head dept, Harding Univ, 45-86, EMER PROF & EMER HEAD, HARDING UNIV, 86- *Concurrent Pos:* Mem, Ark State Healing Arts Bd, 72- *Mem:* Fel AAAS; Genetics Soc Am; Am Inst Biol Sci; Am Fisheries Soc. *Res:* Cytogenetics and genetics of Drosophila; aquatic ecology; ecological relationships in Little Red River, Arkansas. *Mailing Add:* 920 E Market St Searcy AR 72143

SEARS, JOHN T, b LaCrosse, Wis, Nov 15, 38; m 71; c 4. CHEMICAL ENGINEERING. *Educ:* Univ Wis, BS, 60; Princeton Univ, PhD(chem eng), 65. *Prof Exp:* Asst chem engr, Nuclear Eng Dept, Brookhaven Nat Lab, 64-68; res engr, Esso Res & Eng Co, NJ, 68-69; from asst prof to assoc prof chem eng, WVa Univ, 69-81, prof, 81-; CHMN DEPT CHEM ENG, MONT STATE UNIV. *Mem:* Am Inst Chem Engrs; Am Nuclear Soc. *Res:* Radiation chemistry and processing; fluidized beds; air pollution control. *Mailing Add:* Dept Chem Eng Mont State Univ Bozeman MT 59717

SEARS, KARL DAVID, b Cedar Falls, Iowa, Mar 31, 41; m 65; c 2. CHEMISTRY. *Educ:* Iowa State Univ, BA, 63; Univ Wash, PhD(org chem), 68. *Prof Exp:* SR RES ASSOC, ITT RAYONIER INC, 68- *Mem:* Am Chem Soc. *Res:* Conversion of lignin by-product streams into new and improved chemical products. *Mailing Add:* ITT Rayonier Inc Shelton WA 98594

SEARS, LEO A, b Teaneck, NJ, Feb 10, 27; m 52; c 4. CHEMICAL ENGINEERING. *Educ:* Cornell Univ, BChE, 50. *Prof Exp:* Chem engr, 50-62, sr res engr, 62-66, supvr chem eng, 66-69, RES ASSOC, E I DU PONT DE NEMOURS & CO, INC, 69- *Mem:* Am Chem Soc. *Res:* Polymer synthesis and fabrication. *Mailing Add:* Five Stable Lane Wilmington DE 19803

SEARS, MARKHAM KARLI, b San Luis Obispo, Calif, May 3, 46; m 69; c 1. INSECT ECOLOGY. *Educ:* Univ Calif, Davis, BSc, 69, PhD(entom), 74. *Prof Exp:* From res assoc to teaching asst, 72-74, ASST PROF ENTOM, UNIV GUELPH, 75- *Concurrent Pos:* Mem subcomt woody ornamentals, flowers & turf, Ont Crop Protection Comt, 75- *Mem:* Sigma Xi; Entom Soc Am; Can Entom Soc. *Res:* Ecology of insects affecting turfgrasses and woody ornamentals; biology and systematics of immature Coleoptera. *Mailing Add:* Dept Environ Biol Univ Guelph Guelph ON N1G 2W1 Can

SEARS, MARVIN LLOYD, b New York, NY, Sept 16, 28; m 50; c 4. OPHTHALMOLOGY. *Educ:* Princeton Univ, AB, 49; Columbia Univ, MD, 53. *Prof Exp:* Intern, Columbia Bellevue Hosp, 53-54; from asst resident to chief resident, Wilmer Inst Ophthal, 54-61; PROF OPHTHAL & VISUAL SCI & CHMN DEPT, SCH MED, YALE UNIV, 61-, CHIEF, YALE NEW HAVEN MED CTR, 61- *Concurrent Pos:* Robert Weeks Kelly fel ophthal, 57-58; NIH trainee, 59-60; consult to Surgeon Gen, USPHS & mem visual sci study sect, Nat Inst Neurol Dis & Blindness, 62-66; mem bd sci coun, Nat Eye Inst, 70-; mem adv panel, US Pharmacopeia, 75-80. *Honors & Awards:* New Eng Ophthal Soc Award, 69; Friedenwald Award, Asn Res Vision & Ophthal, 77; Sanford R Gifford Mem Lectureship, 85; Alcon Res Inst Award, 85. *Mem:* AMA; Am Acad Ophthal & Otolaryngol; Am Ophthal Soc; Am Col Surgeons; Asn Res Vision & Ophthal. *Res:* Diseases of the eye. *Mailing Add:* 330 Cedar St New Haven CT 06510

SEARS, MILDRED BRADLEY, b New Castle, Pa, Feb 19, 33. INORGANIC CHEMISTRY, CHEMICAL ENGINEERING. *Educ:* Col Wooster, BA, 54; Univ Fla, PhD(inorg chem), 58. *Prof Exp:* RES STAFF MEM, OAK RIDGE NAT LAB, 58-70 & 72- *Mem:* Am Nuclear Soc; Am Chem Soc; Sigma Xi. *Res:* Engineering and environmental assessments of the nuclear fuel cycle, including waste treatment methods; chemistry of uranium and thorium carbides. *Mailing Add:* 130 Monticello Rd Oak Ridge TN 37830

SEARS, PAUL GREGORY, physical chemistry, for more information see previous edition

SEARS, RAYMOND ERIC JOHN, b Wellington, NZ, July 2, 34; m 62; c 4. NUCLEAR MAGNETIC RESONANCE. *Educ:* Univ Victoria, NZ, BSc, 57, MSc, 59; Univ Calif, Berkeley, PhD(physics), 66. *Prof Exp:* Res assoc chem, Mass Inst Technol, 66-67; asst prof, 67-80, ASSOC PROF PHYSICS, NTEX STATE UNIV, 80- *Mem:* AAAS; Am Phys Soc; Int Soc Magnetic Resonance. *Mailing Add:* Dept Physics Univ NTex Denton TX 76203

SEARS, RAYMOND WARRICK, JR, b Passaic, NJ, Mar 12, 35; m 58; c 3. RELIABILITY, MAINTAINABILITY. *Educ:* Cornell Univ, BEE, 58; Johns Hopkins Univ, MSE, 62, PhD(elec eng), 67. *Prof Exp:* Engr, Nat Security Agency, 61-67; mem tech staff, 67-71, SUPVR, AT&T BELL LABS, 71- *Concurrent Pos:* Asst prof, County Col Morris, 88- *Mem:* Inst Elec & Electronics Engrs. *Res:* Reliability and maintainability modeling and assessment of AT&T products produced for government and military use; high speed digital processing, signal processing, ocean acoustics, sonar systems engineering and business data systems. *Mailing Add:* 13 Garabrant St Bell Tele Lab One Whippany Rd Mendham NJ 07945

SEARS, RICHARD LANGLEY, b Boston, Mass, Mar 27, 31; m 73; c 1. ASTROPHYSICS. *Educ:* Harvard Univ, AB, 53; Ind Univ, MA, 55, PhD(astrophys), 58. *Prof Exp:* Vis fel astron, Princeton Univ, 58; instr, Ind Univ, 58-59; asst, Lick Observ, Univ Calif, 59-61; res fel physics, Calif Inst Technol, 60-61, sr res fel, 61-64; vis asst prof physics & astron, Vanderbilt Univ, 64-65; asst prof, 65-70, ASSOC PROF ASTRON, UNIV MICH, ANN ARBOR, 70- *Concurrent Pos:* Mem, Int Astron Union. *Mem:* AAAS; Am Astron Soc; Royal Astron Soc. *Res:* Stellar interiors and evolution; stellar photometry; theoretical astrophysics. *Mailing Add:* Dept Astron Univ Mich Ann Arbor MI 48109

SEARS, ROBERT F, JR, b Warren Co, Ky, June 13, 41; m 65; c 3. PHYSICS. *Educ:* Centre Col, BA, 63; Univ Colo, PhD(physics), 68. *Prof Exp:* From asst prof to assoc prof, 68-78, PROF PHYSICS, AUSTIN PEAY STATE UNIV, 78-, CHMN DEPT, 77- *Concurrent Pos:* Scientist 2, Oak Ridge Assoc Univ, 82. *Mem:* Am Asn Physics Teachers (treas, 90-); Astron Soc Pac; Nat Sci Teachers Asn. *Res:* Study of antiproton-proton interactions resulting in the production of a single pion; high energy physics; energy education. *Mailing Add:* Dept Physics Austin Peay State Univ Clarksville TN 37044-4608

SEARS, TIMOTHY STEPHEN, b Boston, Mass, Sept 29, 45; m 75. MARKET DEVELOPMENT, MARKET RESEARCH. *Educ:* Boston Univ, AB, 67; Univ Calif, Davis, CPhil, 70, PhD(chem), 73; Univ Calgary, MBA, 85. *Prof Exp:* Fel chem, Univ Calgary, 73-74, instr, 74-76, instr mgt, 86; process engr, Kaiser Aluminum & Chem Corp, 76-80; process res scientist, Petro Can, 80-83; mgr res & technol, Travis Chem, 84-85; MKT DEVELOP MGR, BAYMAG, 86- *Mem:* Am Chem Soc; Am Mgt Asn; Chem Mgt & Resource Asn. *Res:* Applications research for various industrial minerals in the pulp and paper, mining and chemicals industries. *Mailing Add:* Box 1323 Cochrane AB T0L 0W0 Can

SEARS, VARLEY FULLERTON, b Cadomin, Alta, Can, June 29, 37; m 63; c 2. SOLID STATE PHYSICS. *Educ:* Univ Toronto, BA, 59, MA, 60, PhD(physics), 63. *Prof Exp:* Asst, Oxford Univ, UK, 63-65; PHYSICIST, ATOMIC ENERGY CAN LTD, 65- *Mem:* Can Asn Physicists; fel Am Phys Soc. *Res:* Theory of thermal neutron scattering in condensed matter; neutron transport phenomena, diffraction, and optical effects. *Mailing Add:* Atomic Energy Can Ltd Chalk River ON K0J 1J0 Can

SEARS, WILLIAM R(EES), b Minneapolis, Minn, Mar 1, 13; m 36; c 2. AERODYNAMICS. *Educ:* Univ Minn, BAeroE, 34; Calif Inst Technol, PhD(aeronaut), 38. *Hon Degrees:* DSc, Univ Ariz, 87. *Prof Exp:* Asst aeronaut, Calif Inst Technol, 34-37, from instr to asst prof, 37-41; chief aerodynamicist, Northrop Aircraft Inc, Calif, 41-46; prof aeronaut eng & dir grad sch aeronaut eng, Cornell Univ, 46-64, dir ctr appl math, 63-67, J L Given prof eng, 63-74; prof, 74-85, EMER PROF AEROSPACE & MECH ENG, UNIV ARIZ, TUCSON, 85- *Concurrent Pos:* Consult, Calspan Corp; ed, J Aerospace Sci, Inst Aerospace Sci, Am Inst Aeronaut & Astronaut, 57-63. *Honors & Awards:* Vincent Bendix Award, 65; Ludwig Prandtl Ring, Deutsche Gesellschaft fur Luft- und Raumfahrt, 74; Reed Aeronaut Medal, Am Inst Aeronaut & Astronaut, 61, G Edward Pendray Award, 75; Am Soc Mech Engrs Medal, 89. *Mem:* Nat Acad Sci; Nat Acad Eng; fel Am Acad Arts & Sci; hon fel Am Inst Aeronaut & Astronaut; fel Int Acad Astronaut. *Res:* Fluid mechanics; wing and boundary layer theory; wind tunnels. *Mailing Add:* Dept Aerospace & Mech Eng Univ Ariz Tucson AZ 85721

SEASE, JOHN WILLIAM, b New Brunswick, NJ, Nov 10, 20; m 43; c 4. ELECTROCHEMISTRY. *Educ:* Princeton Univ, AB, 41; Calif Inst Technol, PhD(org chem), 46. *Prof Exp:* Asst inorg chem, Calif Inst Technol, 41-42, asst, Nat Defense Res Comt, 42-45; instr org chem, 46-48, from asst prof to prof chem, 48-88, E B Nye prof, 82-88, EMER PROF CHEM, WESLEYAN UNIV, 88- *Mem:* AAAS; Am Chem Soc. *Res:* Electrochemistry of organic compounds. *Mailing Add:* Dept Chem Wesleyan Univ Middletown CT 06457

SEATON, JACOB ALIF, b Wellington, Kans, Jan 2, 31; m 55; c 3. INORGANIC CHEMISTRY. *Educ:* Wichita State Univ, BS, 53, MS, 55; Univ Ill, PhD(inorg chem), 58. *Prof Exp:* Asst chemist chem eng div, Argonne Nat Lab, 57-58; proj engr polymer br mat lab, Wright Air Develop Ctr, 58-60; sr staff mem inorg res, Spencer Chem Co, 60-62; asst prof chem, Sam Houston State Col, 62-66; head dept chem, 66-78, chmn dept, 78-86, PROF CHEM, STEPHEN F AUSTIN STATE UNIV, 66- *Mem:* Fel Am Inst Chem; Am Chem Soc; Sigma Xi. *Res:* Non-aqueous solvents; inorganic polymers; transition and inner-transition metal compounds. *Mailing Add:* Dept Chem Stephen F Austin State Univ Box 13006 Nacogdoches TX 75962

SEATON, VAUGHN ALLEN, b Abilene, Kans, Oct 11, 28; m 54; c 2. VETERINARY PATHOLOGY. *Educ:* Kans State Univ, BS & DVM, 54; Iowa State Univ, MS, 57. *Prof Exp:* From instr to assoc prof, 55-64, PROF VET PATH & HEAD DEPT, COL VET MED, IOWA STATE UNIV, 64-, HEAD VET MED DIAG LAB, 74- *Concurrent Pos:* Mem, Conf Vet Lab Diagnosticians. *Mem:* Am Pub Health Asn; Am Vet Med Asn; Am Col Vet Toxicol. *Res:* Pulmonary adenomatosis in Iowa cattle; infectious diseases; veterinary toxicology. *Mailing Add:* Vet Diag Lab Col Vet Med Iowa State Univ Ames IA 50011

SEATON, WILLIAM HAFFORD, b Black Oak, Ark, Oct 22, 24; m 44; c 2. CHEMICAL ENGINEERING. *Educ:* Univ Ark, BSChE, 50; Ohio State Univ, MS, 55, PhD(chem eng), 58. *Prof Exp:* Chem engr, Monsanto Chem Co, 50-53; res assoc, Ohio State Res Found, 53-55; sr res engr, Tenn Eastman Co, 58-67, res assoc, 67-80; sr res assoc, Eastman Chem Co, 80-86; RETIRED. *Concurrent Pos:* Thermophys property data specialist. *Honors & Awards:* Dudley Medal, Am Soc Testing & Mat, 77. *Mem:* Fel Am Inst Chem Engrs; Nat Soc Prof Engrs; fel Am Soc Testing & Mat. *Res:* Chemical process data; unit operations research and development; energy hazard potential problems related to safety. *Mailing Add:* 1329 Belmeade Dr Kingsport TN 37664

SEATZ, LLOYD FRANK, b Winchester, Idaho, June 2, 19; m 49; c 1. SOIL FERTILITY. *Educ:* Univ Idaho, BS, 40; Univ Tenn, MS, 41; NC State Univ, PhD(agron), 49. *Prof Exp:* Asst agron, Univ Tenn, 40-41; asst, NC State Univ, 41-42 & 46-47; from asst prof to prof, 47-68, head dept, 61-84, Clyde B Austin distinguished prof, 68- 84, EMER PROF AGR, UNIV TENN, KNOXVILLE, 84- *Concurrent Pos:* Agronomist & asst chief soils & fertilizer res br, Tenn Valley Authority, 53-55. *Mem:* Am Soc Agron; Soil Sci Soc Am; Sigma Xi. *Res:* Phosphorus and trace element reactions and availability in soils; factors affecting crop response to fertilization. *Mailing Add:* 9729 Tunbridge Lane Knoxville TN 37922

SEAVER, SALLY S, b Marblehead, Mass, July 27, 46. CELL BIOLOGY OF SCALE-UP, ANTIBODIES FOR DIAGNOSTIC TESTS. *Educ:* Harvard Univ, AB, 67, AM, 68; Stanford Univ, PhD(phys chem), 73. *Prof Exp:* Jane Coffin Child fel biochem endocrinol, Inst Chemie Bulogique, Univ Louis Pasteur, Shasbourg, France, 73-75; asst prof molecular endocrinol, Dept Molecular Biol, Vanderbilt Univ, 75-81; sr scientist appln cell & molecular biol, Millipore Corp, 82-83, group leader, 83-85; ASSOC DIR RES & DEVELOP, IMMUNOL & CELL BIOL, HYGEIA SCI, 85- *Concurrent Pos:* Vis prof, Dept Biol, Univ Calif, San Diego, 81; consult ed, Immunol, 87-89, 91- *Mem:* Endocrine Soc; Am Soc Cell Biol; Am Chem Soc; Tissue Cult Asn; Sigma Xi. *Res:* Development of hybudenia lines and production (scale-up) of neono-clonal and polyclonal antibodies for diagnostic tests; development of equipment and reagents for biotechnology. *Mailing Add:* Hygeia Sci 330 Nevada St Newton MA 02160

SEAVEY, MARDEN HOMER, JR, b Preston, Cuba, Jan 12, 29; US citizen; m 55, 63; c 5. SOLID STATE PHYSICS. *Educ:* Harvard Univ, AB, 52; Northeastern Univ, MS, 56; Harvard Univ, PhD, 70. *Prof Exp:* Physicist, Air Force Cambridge Res Ctr, 52-55, Lincoln Lab, Mass Inst Technol, 55-62, Air Force Cambridge Res Labs, 62-70 & Philips Res Lab, Neth, 70-72; PRIN ENGR, EQUIP DIV, RAYTHEON CO, 72-; PRIN ENGR, DIGITAL EQUIP CO. *Mem:* Am Phys Soc; Inst Elec & Electronics Engrs. *Res:* Resonance phenomena and acoustic effects in ordered magnetic systems; laser gyroscopes; radiation effects in large scale integrated circuits. *Mailing Add:* 381 Cross St Carlisle MA 01741

SEAWRIGHT, JACK ARLYN, b Ware Shoals, SC, Sept 9, 41; m 62; c 2. ENTOMOLOGY, GENETICS. *Educ:* Clemson Univ, BS, 64, MS, 65; Univ Fla, PhD(entom), 69. *Prof Exp:* RES LEADER/ENTOMOLOGIST, MED & VET ENTOM LAB, USDA, 68- *Concurrent Pos:* Asst prof, 70-80, prof dept entom, Univ Fla, 80- *Mem:* Entom Soc Am; Am Mosquito Control Asn. *Res:* Genetics of insects with emphasis on genetic control mechanisms; toxicology of chemosterilants in insects. *Mailing Add:* USDA Agr Res Serv PO Box 14565 Gainesville FL 32604

SEAY, GLENN EMMETT, b Tahlequah, Okla, Mar 9, 26; m 59; c 3. PHYSICS. *Educ:* Univ Okla, BS, 50, MS, 53, PhD(physics), 57. *Prof Exp:* Mem staff, Los Alamos Sci Lab, Univ Calif, 47-62; div supvr, Sandia Corp, 62-64, dept mgr, 64-68; mgr, dept exp physics, Systs Sci & Software, La Jolla, Calif, 68-70, mgr, Systs & Software Div, 70-76; GROUP LEADER, LOS ALAMOS SCI LAB, UNIV CALIF, 76- *Mem:* Am Phys Soc. *Res:* Flash radiography; atomic spectroscopy; shock waves in gases and solids; detonation physics; initiation of detonation. *Mailing Add:* 101 San Ildefonso Rd Los Alamos NM 87544

SEAY, PATRICK H, b Lexington, SC, Jan 31, 20; m 48; c 3. PHARMACOLOGY. *Educ:* Princeton Univ, PhD(biol), 50. *Prof Exp:* EXEC DIR REGULATORY AFFAIRS, MCNEIL PHARMACEUT, 68-, SR REGULATORY AFFAIRS LAISON, 85- *Mem:* Am Soc Pharmacol & Exp Therapeut. *Mailing Add:* 325 Wenner Way Ft Washington PA 19034-2919

SEAY, THOMAS NASH, b Cincinnati, Ohio, Sept 29, 32; m 58; c 2. ENVIRONMENTAL SCIENCES, ENTOMOLOGY. *Educ:* Univ Fla, BSA, 55; Univ Ky, MSA, 63, PhD(biol sci), 67. *Prof Exp:* Asst prof, 66-70, ASSOC PROF BIOL, GEORGETOWN COL, 70-, DIR ENVIRON SCI, 74-, ACTG CHMN BIOL SCI, 84- *Mem:* Entom Soc Am; NAm Asn Environ Educ. *Res:* Environmental geography. *Mailing Add:* 1111 Chickasaw Terr Georgetown KY 40324

SEBALD, ANTHONY VINCENT, b US. ELECTRICAL ENGINEERING, SYSTEMS SCIENCE. *Educ:* Gannon Col, BEE, 63; Univ Ill, MSEE, 75, PhD(elec eng), 76. *Prof Exp:* Assoc engr, IBM Corp, 64-68; prof eng, Univ Catolica de Valparaiso, 69-72; ASST PROF SYST SCI, UNIV CALIF, SAN DIEGO, 76- *Mem:* Inst Elec & Electronics Engrs. *Res:* Energy and air pollution policy analysis; solar heating and cooling of buildings; estimation and control in systems which are incompletely specified. *Mailing Add:* Dept Appl Mech B-010 Univ Calif La Jolla CA 92037

SEBAN, RALPH A, b May 11, 17. HEAT TRANSFER THERMODYNAMICS. *Educ:* Univ Calif, Berkeley, PhD(mech eng), 48. *Prof Exp:* Prof mech eng, Univ Calif, Berkeley, 52-86; RETIRED. *Honors & Awards:* Max Jokob Award, Am Soc Mech Eng, 81. *Mem:* Nat Acad Eng; hon mem Am Soc Mech Engrs. *Mailing Add:* 5265 Harbord Dr Oakland CA 94618

SEBASTIAN, ANTHONY, b Youngstown, Ohio, July 11, 38; m 64. MEDICINE, NEPHROLOGY. *Educ:* Univ Calif, Los Angeles, BS, 60; Univ Calif, San Francisco, MD, 65. *Prof Exp:* From intern internal med to resident, Moffitt Hosp, 65-68, asst resident physician, 70-71, asst prof, 71-78, assoc prof, 78-84, PROF MED, UNIV CALIF, SAN FRANCISCO, 84- *Concurrent Pos:* Bank Am-Giannini Found fel renal dis, Univ Calif, San Francisco, 68-70. *Mem:* Am Soc Nephrol; Am Fedn Clin. *Res:* Am Soc Clin Invest; Int Soc Nephrol. Res: Renal and acid-base physiology and pathophysiology; renal acidosis; interrelationship of hydrogen ion and electrolyte transport in the kidney; renal tubular disorders. *Mailing Add:* Dept Med 1202 Moffitt Hosp Univ Calif Box 0126 San Francisco CA 94143

SEBASTIAN, FRANKLIN W, b Neenah, Wis, June 3, 20; m 48; c 3. ELECTRICAL ENGINEERING, CONSULTING. *Educ:* Ill Inst Technol, BSME, 45. *Prof Exp:* Staff engr mech, Kimberly Clark Paper Co, 47-49; engr mech, F H McGraw & Co, 49-52; design engr mech, Ingersoll Corp, 52-57; sales mgr sales, Cameron Mach Co, 63-67; chief engr mgt, Menasha Paper Co, 67-74; eng mgr admin, Phillip Morris Corp, 74-85; proj engr mech & elec, K V P Paper Co, 52-57, consult eng, K V P Paper Div James River, 85-88; RETIRED. *Concurrent Pos:* Consult engr & chmn bd, Artios Eng, 85- *Res:* New products and developed equipment and processes in the paper industry; developed new equipment to produce new products. *Mailing Add:* 2907 Sonora Kalamazoo MI 49004

SEBASTIAN, JOHN FRANCIS, b San Diego, Calif, Nov 20, 39; c 2. PHYSICAL ORGANIC CHEMISTRY, BIOCHEMISTRY. *Educ:* San Diego State Col, BS, 61; Univ Calif, Riverside, PhD(org chem), 65. *Prof Exp:* NIH fel enzyme catalysis, Northwestern Univ, 65-67; from asst prof chem to assoc prof, 67-81, PROF CHEM, MIAMI UNIV, 81- *Concurrent Pos:* Res Corp grant, 68- *Mem:* AAAS; Am Chem Soc; Royal Soc Chem; Sigma Xi. *Res:* Mechanisms of enzyme catalysis; enzyme model systems; nuclear magnetic resonance spectroscopy; organometallic and heterocyclic chemistry; applications of molecular orbital theory; DNA structure and function; protein-DNA interactions. *Mailing Add:* Dept Chem Miami Univ Oxford OH 45056

SEBASTIAN, LESLIE PAUL, b Tata, Hungary, July 11, 23; Can citizen; m 49; c 3. WOOD SCIENCE. *Educ:* Univ Forestry & Timber Indust, Hungary, ForEngr, 49; State Univ NY Col Forestry, Syracuse Univ, MSc, 65, PhD(wood sci), 71. *Prof Exp:* Demonstr forest utilization, Univ Forestry & Timber Indust, Hungary, 49-50, asst, 50-52, adj, 52-56; asst prof wood technol, Fac Forestry, Univ BC, 57-59; from asst prof to prof wood sci, Fac Forestry, Univ NB, 75-88; RETIRED. *Mem:* Int Asn Wood Anatomists; Forest Prod Res Soc; Soc Wood Sci & Technol; Tech Asn Pulp & Paper Indust; Can Inst Forestry; Sigma Xi. *Res:* Anatomical and physical properties of wood. *Mailing Add:* 116 Gorham Dr Fredericton NB E3A 2A7 Can

SEBASTIAN, RICHARD LEE, b Hutchinson, Kans, June 22, 42; m 64; c 2. SIGNAL PROCESSING, PATTERN RECOGNITION. *Educ:* Princeton Univ, AB, 64; Univ Md, Col Park, PhD(physics), 70. *Prof Exp:* Staff scientist, Ensco, Inc, Springfield, Va, 69-72, chief scientist, 72, div mgr, 72-74, vpres res, 74-83; chmn, Digital Optronics Corp, 84-87; PRES & CHMN, DIGITAL SIGNAL CORP, SPRINGFIELD, VA, 83- *Mem:* Am Phys Soc; Inst Elec & Electronics Engrs; Soc Explor Geophysics. *Res:* Signal processing and intelligent systems; geophysics including acoustic seismic and electromagnetic waves; source localization and classification and wave propagation. *Mailing Add:* Digital Signal Corp 8003 Forbes Pl No 100 Springfield VA 22151-2206

SEBBA, FELIX, surface chemistry, chemical engineering; deceased, see previous edition for last biography

SEBEK, OLDRICH KAREL, b Prague, Czech, July 3, 19; nat US; m 59; c 1. MICROBIAL PHYSIOLOGY, INDUSTRIAL MICROBIOLOGY. *Educ:* Charles Univ, Prague, DSc(natural sci) 46; Rutgers Univ, New Brunswick, ScD(natural sci), 49. *Prof Exp:* Asst microbiol, Charles Univ, 45-47; int fel fermentation, J E Seagram & Sons, Inc, 47-48; res assoc microbiol, Rutgers Univ, 48-49; fel chem & enzymol, Fordham Univ, 49-50; Muelhaupt scholar biol, Ohio State Univ, 50-52; sr scientist, Upjohn Co, 52-84; vis res scientist, 85-87, VIS RES PROF DEPT BIOL SCI, WESTERN MICH UNIV, KALAMAZOO, 88- *Concurrent Pos:* Abstractor, Chem Abstr, 50-70; vis scientist & res assoc Dept Biochem, Univ Calif, Berkeley, 66-67; ed, Appl Microbiol, 68-71; vis prof, Nat Polytech Inst, Mexico City, 73; US rep, Int Comn Appl Microbial Genetics, Int Asn Microbiol Socs, 74-82; partic foreign projs, Nat Res Coun, Nat Acad Sci, Czech, 80, Ger Dem Repub, 86, Indonesia, 86; ed, Appl Microbiol & Biotechol, 90- *Mem:* Am Soc Microbiol; Soc Indust Microbiol; fel Am Acad Microbiol. *Res:* Microbial metabolism and biochemistry; fermentations and biosynthesis; biotrans formation of microbial products, antibiotics, carotenoids, prostaglandins, pigments, amino acids, steroids and synthetic chemicals. *Mailing Add:* 1002 Short Rd Kalamazoo MI 49008-1138

SEBESTA, CHARLES FREDERICK, b North Braddock, Pa, Mar 6, 14; m 45; c 7. MATHEMATICS. *Educ:* Univ Pittsburgh, AB, 34, MA, 38, PhD, 56. *Prof Exp:* Teacher high sch, Pa, 34-43; instr math, Univ Pittsburgh, 46-54, asst prof, 54-56; assoc prof, 56-65, head dept, 56-74 & 78-79, PROF MATH, DUQUESNE UNIV, 65- *Mem:* Math Asn Am; Sigma Xi. *Res:* Abstract algebra; partial differential equations; analytic function theory. *Mailing Add:* 4344 E Barlind Dr Pittsburgh PA 15227

SEBETICH, MICHAEL J, b Nanty-Glo, Pa, Feb 25, 43. LIMNOLOGY, ECOLOGY. *Educ:* Duquesne Univ, BS, 65; Col William & Mary, MA, 69; Rutgers Univ, PhD(ecol, limnol), 72. *Prof Exp:* Asst prof, William Paterson Col, 72-73; aquatic biologist, US Geol Surv Water Resources Div, 73-77; ASSOC PROF BIOL, WILLIAM PATERSON COL, 77- *Concurrent Pos:* Ecol consult, 84- *Mem:* Ecol Soc Am; Am Soc Limnol & Oceanog; AAAS; Am Inst Biol Sci; Int Asn Theoret & Appl Limnol; Sigma Xi. *Res:* Nutrient cycling in freshwater ecosystems; lake management. *Mailing Add:* Dept Biol William Paterson Col Wayne NJ 07470

SEBO, STEPHEN ANDREW, b Budapest, Hungary, June 10, 34; m 68. ELECTRICAL ENGINEERING. *Educ:* Budapest Polytech Univ, MS, 57; Hungarian Acad Sci, PhD(elec eng), 66. *Prof Exp:* Elec engr, Budapest Elec Co, Hungary, 57-61; from asst prof to assoc prof elec power eng, Budapest Polytech Univ, 61-68; assoc prof elec eng, 68-74, PROF ELEC ENG, OHIO STATE UNIV, 74- *Concurrent Pos:* Consult engr, State Power Bd, Hungary, 61-64; Ford Found fel, 67-68; consult & res, electric utility co, 69-; res engr, Elec Power Res Inst, 75- *Mem:* Inst Elec & Electronics Engrs; Int Conf Large Elec Systs. *Res:* Electric power systems; high-voltage power transmission; power system analysis; electric power generation; high-voltage engineering; power system economics. *Mailing Add:* Dept Elec Eng Ohio State Univ 2015 Neil Ave Columbus OH 43210-1272

SEBORG, DALE EDWARD, b Madison, Wis, Mar 29, 41. CHEMICAL ENGINEERING. *Educ:* Univ Wis-Madison, BSc, 64; Princeton Univ, PhD(chem eng), 69. *Prof Exp:* Res asst chem eng, Princeton Univ, 64-68; from asst prof to prof chem & petrol eng, Univ Alta, 68-77; chmn dept, 78-81, PROF CHEM & NUCLEAR ENG, UNIV CALIF, SANTA BARBARA, 77- *Honors & Awards:* Meriam-Wiley Textbook Award, Am Soc Eng Educ, 90. *Mem:* Am Inst Chem Engrs; Inst Elec & Electronics Engrs. *Res:* Process control; computer control techniques; applied mathematics. *Mailing Add:* Dept Chem & Nuclear Eng Univ Calif Santa Barbara CA 93106

SEBRANEK, JOSEPH GEORGE, b Richland Ctr, Wis, Feb 22, 48; m 70; c 2. MEAT SCIENCES. *Educ:* Univ Wis, Platteville, BS, 70; Univ Wis, Madison, MS, 71, PhD(meat sci & food sci), 74. *Prof Exp:* Fel, Nat Cancer Inst, 74-75; asst prof meat sci, 75-79, assoc prof, 79-84, PROF ANIMAL SCI & FOOD TECH, IOWA STATE UNIV, 84- *Concurrent Pos:* Res chemist, USDA, 82-83. *Honors & Awards:* Meat Processing Award, Am Meat Asn. *Mem:* Inst Food Technologists; Am Soc Animal Sci; Am Meat Sci Asn; Am Chem Soc. *Res:* Meat processing, food additives, dehydration of processed meats, curing reactions, use of nitrite, color development and processed meat quality. *Mailing Add:* Animal Sci 101 Kildee Iowa State Univ Ames IA 50011

SEBREE, BRUCE RANDALL, b Marion, Kans, Feb 8, 56; m 73; c 4. FOOD SCIENCE, CHEMICAL ENGINEERING. *Educ:* Kans State Univ, BS, 78, MS, 81, PhD(cereal technol), 84. *Prof Exp:* Asst prod engr, Western Foods Corp, 78; res asst cereal sci, dept grain sci, Kans State Univ, 78-83; tech dir, 83-86, VPRES RES & QUAL CONTROL, FHISCHMANN-KURTH, 86- *Mem:* Am Asn Cereal Chemists; Am Soc Brewing Chemists; Master Brewers Asn Am; Inst Food Technologists; AAAS. *Res:* Malting of cereal grains including barley, wheat, oats and rye; analysis methodology associated with malting, cereal composition and malt process reactions. *Mailing Add:* Fleischmann-Kurth Malting Corp 2100 S 43rd St Milwaukee WI 53219

SECCO, ANTHONY SILVIO, b Antigonish, NS, Aug 10, 56; m 79. CRYSTALLOGRAPHY, DNA STRUCTURE. *Educ:* St Francis Xavier Univ, BSc, 78; Univ BC, PhD(chem), 82. *Prof Exp:* Fel, Univ Pa, 82-84; ASSOC PROF, UNIV MAN, 84- *Mem:* Am Crystallog Asn; Chem Inst Can. *Res:* Crystallographic investigations of novel inorganic compounds and biologically important molecules. *Mailing Add:* Dept Chem Univ Man Winnipeg MB R3T 2N2 Can

SECCO, ETALO ANTHONY, b Dominion, NS, Nov 8, 28; m 53; c 6. PHYSICAL CHEMISTRY. *Educ:* St Francis Xavier Univ, BSc, 49; Laval Univ, DSc, 53. *Prof Exp:* Instr gen chem, St Francis Xavier Univ, 49-50; instr phys chem, Laval Univ, 52-53; res assoc, Ind Univ, 53-55; from asst prof to assoc prof, 55-64, PROF CHEM, ST FRANCIS XAVIER UNIV, 64- *Concurrent Pos:* NATO overseas fel, Cavendish Lab, Cambridge Univ, 61-62. *Mem:* Am Chem Soc; NY Acad Sci; fel Chem Inst Can; Royal Soc Chem; fel Am Inst Chemists. *Res:* Kinetics of heterogeneous reactions temperatures; phase equilibria at high temperatures; radiotracers; structural problems; solid state decomposition kinetics; electrical conductivity studies; non-traditional glass preparation and studies; computer modeling of ameltropic solid solid solutions. *Mailing Add:* Dept Chem St Francis Xavier Univ Antigonish NS B2G 1C0 Can

SECHLER, DALE TRUMAN, b Pleasant Hope, Mo, Nov 30, 26; m 54; c 3. PLANT BREEDING. *Educ:* Univ Mo, BS, 50, MEd, 54, PhD(plant breeding), 60. *Prof Exp:* Teacher high schs, Mo, 50-55; instr field crops, Univ Mo, 55-60; asst prof agron, Univ Fla, 60-67; assoc prof, 67-75, PROF AGRON, UNIV MO, COLUMBIA, 75- *Mem:* Am Soc Agron. *Res:* Genetics and improvement of wheat and oats; grain crops production; plant breeding; international agronomy. *Mailing Add:* Dept Agron 106 Certis Univ Mo Columbia MO 65211

SECHRIST, CHALMERS FRANKLIN, JR, b Glen Rock, Pa, Aug 23, 30; m 57; c 2. AERONOMY, IONOSPHERIC PHYSICS. *Educ:* Johns Hopkins Univ, BE, 52; Pa State Univ, MS, 54, PhD(elec eng), 59. *Prof Exp:* From asst elec eng to instr, Pa State Univ, 52-55, asst, Ionosphere Res Lab, 55-59; sr engr, HRB-Singer, Inc, 59-63, staff engr, 63-65; from asst prof to assoc prof, 65-71, PROF ELEC ENG, UNIV ILL, URBANA-CHAMPAIGN, 71- *Concurrent Pos:* Mem, Educ Soc Admin Comt, Inst Elec & Electronics Engrs, 83-89, Educ Activ Bd, 90, vpres, Educ Soc, 90, pres, 91; prin investr on NSF proj to investigate upper atmosphere sodium layer, 77-84; assoc head, elec & comput eng dept, 84-86; asst dean of eng undergrad progs, 86- *Mem:* Inst Elec & Electronics Engrs; Am Geophys Union; Am Meteorol Soc; Am Soc Eng Educ. *Res:* Physics and chemistry of the lower ionosphere and the upper atmosphere sodium layer. *Mailing Add:* 207 Eng Hall Univ Ill 1308 W Green St Urbana IL 61801

SECHRIST, JOHN WILLIAM, b Manila, Philippines, Feb 22, 42; US citizen; m 64. NEUROANATOMY, DEVELOPMENTAL BIOLOGY. *Educ:* Wheaton Col, Ill, BS, 64; Univ Ill, PhD(anat), 67. *Prof Exp:* From instr to asst prof neuroanat, Sch Med, Univ Pittsburgh, 67-72; asst prof anat, Col Med, Univ Ariz, 72-76; lectr anat, Univ Ibadan, Nigeria, 76-79; assoc prof biol, Wheaton Col, 79-88; RES BIOLOGIST, UNIV CALIF IRVINE, 88- *Mem:* AAAS; Am Asn Anatomists; Am Soc Cell Biol. *Res:* Investigation of cytologic and metabolic changes during neuroblast differentiation and the retrograde reaction; comparative studies of earliest neurons in both vertebrates and invertebrates. *Mailing Add:* Develop Biol Ctr Univ Calif Irvine Irvine CA 92717

SECHRIST, LYNNE LUAN, b Milwaukee, Wis, Oct 26, 41; m 88; c 1. MICROBIOLOGY. *Educ:* Ohio Wesleyan Univ, BA, 63; Univ Wis-Madison, MS, 65; Ohio State Univ, PhD(bot), 69. *Prof Exp:* Asst prof biol, Ohio Dominican Col, 69-73; res fel, Univ Dayton, 74-76; from asst prof to assoc prof, State Univ NY Col Potsdam, 76-86, assoc dean Libr Studies, 86-90, PROF BIOL, STATE UNIV NY COL POTSDAM, 86-, SR ASST TO VPRES ACAD AFFAIRS, 90- *Mem:* Am Soc Microbiol; Soc Indust Microbiol; Sigma Xi. *Res:* Cellular slime molds; zygotic development, genetics and electron microscopy of Chlamydomonas; microbial metabolism and emulsification of hydrocarbons from freshwater ecosystems; antibiotic resistant bacteria. *Mailing Add:* Acad Serv Ctr State Univ NY Col Potsdam Potsdam NY 13676

SECHZER, JERI ALTNEU, b New York, NY, Nov 1, 30; m 48; c 3. PHYSIOLOGICAL PSYCHOLOGY, EARLY DEVELOPMENT. *Educ:* NY Univ, BS, 56; Univ Pa, MA, 61, PhD(psychol), 62. *Prof Exp:* Res fel physiol psychol, Sch Med, Univ Pa, 61-63, USPHS fel, 63-64; asst prof physiol psychol & anat, Col Med, Baylor Univ, 64-66; res scientist, Dept Rehab Med & Anat, NY Univ Med Ctr, 66-70; asst prof psychiat, 70-71, ASSOC PROF PSYCHIAT, MED COL, CORNELL UNIV, 71- *Concurrent Pos:* Mem, Univ Pa Medico Mission Algeria, 62; mem adv comt, NY Acad Sci, 73-, vchmn, 79-82, chmn, sect psychol, 82-84, ad hoc comt animal res, 77-87, chmn, 77-86, women in sci comt, 76-, co-prog chmn, 76-82; mem, Comt Animal Res & Experimentation, Am Psychol Asn, 80- 83, chmn, 82-83, bd sci affairs, Consultative Subcomt on Human Rights, ad hoc comt nonsexist res, 84-87; mem, Adv Panel Assessment of Alternatives to the Use of Animals in Testing, Res & Educ, Off Technol Assessments, US Cong, 84-85; mem, Hastings Ctr Task Force on Ethics of Animal Exp, 88-90; ad hoc consult, Asn Accreditation of Lab Animal Care, 88-91; mem, Comt Pain & Distress in Lab Animals, Nat Res Coun-Inst Lab Animal Resources, 89-91. *Honors & Awards:* Creative Talent Award Am, Inst Res, 63. *Mem:* Fel Am Psychol Asn; Am Asn Anatomists; Am Physiol Soc; Soc Neurosci; Asn Women Sci; AAAS; fel NY Acad Sci; Am Asn Univ Prof; Asn Univ Women; fel Am Psychol Soc; NY Acad Med. *Res:* Learning and memory; neurobiology and behavior; early development; behavioral toxicology; bioethics and animal research; bioethics: womens studies. *Mailing Add:* 180 East End Ave Apt 11D New York NY 10028

SECHZER, PHILIP HAIM, b New York, NY, Sept 13, 14; m 48; c 3. ANESTHESIOLOGY, MEDICO-LEGAL. *Educ:* City Col New York, BS, 34; NY Univ, MD, 38; Am Bd Anesthesiol, dipl, 47; FRCS(I). *Prof Exp:* Intern, Harlem Hosp, New York, 38-40; resident, Fordham Hosp, 41-42, dir anesthesiol & asst prof, Postgrad Med Sch, NY Univ, 55-56; asst prof anesthesiol, Sch Med & physician-anesthetist, Hosp Univ Pa, 56-64; from assoc prof to prof anesthesiol, Col Med, Baylor Univ, 64-66; PROF ANESTHESIOL, STATE UNIV NY DOWNSTATE MED CTR, 66-; EMER DIR, MAIMONIDES MED CTR, 87- *Concurrent Pos:* Chief anesthesiol, AAF Sch Aviation Med, Randolph Field, Tex, 46; dir anesthesiol, Fordham Hosp, NY, 47-55; Consult, USPHS Marine Hosp, Staten Island, NY, 48-82, Vet Admin Hosps, Philadelphia 60-63 & Houston, 64-66; dir anesthesiol, Seton Hosp, New York, 50-55; attend anesthesiologist, Vet Admin Hosps, New York, 55-56 & Philadelphia, 58-59; chief Baylor anesthesiol sect, Methodist Hosp, Houston; area consult, Vet Admin, DC, 64-; ed, Commun in Anesthesiol, 70-77; hon police surgeon, New York, 74-; mem, Nat Bd Acupuncture Med; ed, Bd Anesthesiol News, Med Malpractice Prev; dir anesthesiol, maimonides Med Ctr, 66-86, med dir 73-86, dir, Pain Ther Ctr, 72-86; Consult, EVP, 87-88, pres, 88-89. *Mem:* Fel Am Col Clin Pharmacol; fel Am Col Anesthesiol; Sigma Xi; fel Am Col Physicians; AMA; Am Soc Anesthesiologists; Soc Obstet Anesthesia & Perinatology; Soc Cardiovasc Anesthesiologists; Can Anaesthetists Soc; Int Asn Study Pain. *Res:* Circulatory and respiratory physiology; statistical methods and experimental design; evaluation of new drugs and anesthetic methods; medico-legal and ethical issues. *Mailing Add:* 180 East End Ave New York NY 10128

SECKEL, GUNTER RUDOLF, b Osnabruck, Ger, Nov 4, 23; nat US; m 65. OCEANOGRAPHY. *Educ:* Univ Wash, BS, 50, MS, 54. *Prof Exp:* Asst oceanog, Univ Wash, 50-53; oceanogr biol lab, Bur Com Fisheries, US Fish & Wildlife Serv, Hawaii, 53-63, supvry res oceanogr, 63, chief trade wind zone oceanog prog, 63-67, chief oceanog prog, 67-70; res oceanogr, PAC Environ group, Nat Marine Fisheries Serv, Nat Oceanic & Atmospheric Admin, 70-77, chief, 77-83; RETIRED. *Mem:* Am Geophys Union; Am Meteorol Soc; Sigma Xi. *Res:* Physical oceanography; climatic oceanography of Hawaiian waters; mechanisms producing seasonal and longer term changes in distribution of properties in north Pacific; structure of Pacific north equatorial current. *Mailing Add:* Nat Marine Fisheries Serv Box 831 Monterey CA 93942

SECKLER, BERNARD DAVID, b New York, NY, Feb 14, 25; m 53; c 2. MATHEMATICS. *Educ:* Brooklyn Col, BA, 45; Columbia Univ, MA, 48; NY Univ, PhD(appl math), 58. *Prof Exp:* Instr math, Long Island Univ, 48-53; lectr, Brooklyn Col, 47-54, instr, 57-58; asst appl math, NY Univ, 54-58; assoc prof math, Pratt Inst, 58-64; chmn dept, 68-72, PROF MATH, C W POST COL, LONG ISLAND UNIV, 64- *Mem:* Soc Indust & Appl Math. *Res:* Asymptotic expansions; geometrical and asymptotical solution of diffraction problems; Russian abstracting and translating of applied mathematics; Russian translating and translation editing of probability journal. *Mailing Add:* Dept Math C W Post Col Long Island Univ Greenvale NY 11548

SECOR, DONALD TERRY, JR, b Oil City, Pa, Nov 22, 34; m 59; c 3. APPALACHIAN TECTONICS, ROCK MECHANICS. *Educ:* Cornell Univ, BS, 57, MS, 59; Stanford Univ, PhD(geol), 62. *Prof Exp:* From asst prof to assoc prof, 62-81, chmn dept, 66-69, 77-81, PROF GEOL, UNIV SC, 81- *Mem:* Fel Geol Soc Am; Am Geophys Union; AAAS. *Res:* Mechanics of geological structures; tectonics of the Appalachian Mountains. *Mailing Add:* Dept Geol Sci Univ SC Columbia SC 29208

SECOR, JACK BEHRENT, b Indianapolis, Ind, Aug 18, 23. PHYSIOLOGICAL ECOLOGY. *Educ:* Butler Univ, BS, 48; Wash State Univ, PhD(bot), 57. *Prof Exp:* Botanist, US Geol Surv, 56-57; range conservationist, US Forest Serv, 57-58; Labatt fel bot, Univ Western Ont, 59-60; instr, Dept Natural Sci, Mich State Univ, 60-63; res asst biochem, Va Polytech Inst, 63-65; res scientist assoc, Univ Tex, Austin, 66-67; from asst prof to assoc prof biol, 67-77, PROF BIOL, EASTERN NMEX UNIV, 77- *Mem:* Ecol Soc Am. *Res:* Gypsumland ecosystems. *Mailing Add:* 1261 Willow Las Cruces NM 88001

SECOR, ROBERT M(ILLER), b New York, NY, Mar 21, 32; c 2. CHEMICAL ENGINEERING. *Educ:* NY Univ, BChE, 52; Yale Univ, DEng, 58. *Prof Exp:* Asst, Yale Univ, 52-55; res engr, Eastern Lab, 56-64, sr res engr, Eng Technol Lab, 64-74, RES ASSOC, E I DU PONT DE NEMOURS & CO, INC, 74- *Concurrent Pos:* Adj prof, Chem Eng Dept, Univ Del, 88- & Columbia Univ, 90-; Eastman Kodak fel, Yale Univ. *Mem:* Am Inst Chem Engrs. *Res:* Diffusion; mass transfer; chemical kinetics; applied mathematics; polymer processing. *Mailing Add:* Engineering Dept Exp Sta E I du Pont de Nemours & Co Inc Wilmington DE 19880-0304

SECORD, DAVID CARTWRIGHT, laboratory animal medicine, for more information see previous edition

SECORD, ROBERT N, b Newton, Mass, Dec 20, 20; m 44; c 4. CHEMICAL ENGINEERING. *Educ:* Mass Inst Technol, BS, 42, MS, 47. *Prof Exp:* Chem engr, 47-51, head appl res sect, New Prods Res Dept, 51-61, eng mgr, Oxides Div, 61-66, dir process develop, 66-70, res dir, 70-77, DIR, CAB-O-SIL RES & DEVELOP, CABOT CORP, 77- *Mem:* Am Chem Soc; Am Inst Chem Engrs. *Res:* Processes for high temperature chemical reactions. *Mailing Add:* Box 341 Sunapee NH 03782

SECOY, DIANE MARIE, b Kenton, Ohio, Oct 31, 38; m 70. VERTEBRATE BIOLOGY. *Educ:* Ohio State Univ, BS, 60, MS, 62; Univ Colo, PhD(herpet), 68. *Prof Exp:* Asst cur zool & paleont, Mus, Univ Colo, 67-68; from asst prof to assoc prof, 68-82, assoc dean, Grad Studies, 84-90, PROF BIOL, UNIV REGINA, 82- *Mem:* Sigma Xi; Am Soc Ichthyologists & Herpetologists; Herpetologists League; Soc Study Amphibians & Reptiles; Can Soc Zoologists; Am Soc Zoolgists; Can Soc Herpetologists. *Res:* Reptilian morphology; ecology of northern reptiles; history of pest control. *Mailing Add:* Dept Biol Univ Regina Regina SK S4S 0A2 Can

SECREST, BRUCE GILL, b Iowa City, Iowa, July 20, 36; m 64. MATHEMATICS, AERONAUTICAL ENGINEERING. *Educ:* Iowa State Univ, BS, 58, MS, 62, PhD(math), 64. *Prof Exp:* Assoc engr, Douglas Aircraft Co, Inc, 59; asst math, Iowa State Univ, 59-63, instr, 63-64; asst prof, Univ Nebr, 64-66; mem tech staff, Tex Instruments, Inc, 66-83; staff geophysicist, Standard Oil Prod Co, 83-88; STAFF GEOPHYSICIST, BP EXPLOR CO, 88- *Mem:* Inst Elec & Electronics Engrs. *Res:* Propagation of waves in fluid mechanics; methods of solution in applied mathematics. *Mailing Add:* BP Explor Co 5151 San Felipe PO Box 4587 Houston TX 77210

SECREST, DONALD H, b Akron, Ohio, Jan 3, 32; m 58; c 2. PHYSICAL CHEMISTRY. *Educ:* Univ Akron, BS, 55; Univ Wis, PhD(theoret chem), 61. *Prof Exp:* Instr phys chem, Univ Wis, 60-61; from asst prof to assoc prof, 61-82, PROF CHEM, UNIV ILL, 81- *Concurrent Pos:* Vis scientist, Max Planck Inst, Gottingen, Ger, 71-72; assoc ed, J Chem Physics, 77-79; Humboldt Award, Ger, 85-86. *Mem:* Fel Am Phys Soc; Soc Indust & Appl Math; Sigma Xi. *Res:* Atomic and molecular scattering problems; molecular structure of small systems, especially development of mathematical techniques for handling quantum mechanical problems; on-line application of computing machinery. *Mailing Add:* Dept Chem Univ Ill Urbana IL 61801

SECREST, EVERETT LEIGH, b Tioga, Tex, Jan 5, 28; m 48; c 2. RESEARCH ADMINISTRATION, SYSTEMS SCIENCE. *Educ:* NTex State Col, BS, 47, MS, 48; Mass Inst Technol, PhD(physics), 51. *Prof Exp:* From asst prof to assoc prof physics, NTex State Col, 51-54; chief nuclear physics, Convair Div, Gen Dynamics Corp, 54-57; sect chief proj physics atomic energy div, Babcock & Silcox Co, 57-58, asst mgr physics & math, 58-59; chief appl res, Gen Dynamics/Ft Worth, 59-63, chief scientist, 63-64; assoc dean eng grad studies & res, Univ Okla, 64-65; pres res found, 65-72, grad dean, 65-68, vchancellor advan studies & res, 68-72, CONTINENTAL NAT BANK PROF MGT SCI, TEX CHRISTIAN UNIV, 72-, VCHANCELLOR FINANCE & PLANNING, 81- *Concurrent Pos:* Consult, Gen Dynamics/Ft Worth, 52-54; guest Inst Syst Dynamics Group & Opers Resctr, Mass Inst Technol, 72. *Mem:* Am Phys Soc; Am Nuclear Soc; Am Soc Eng Educ; Soc Comput Simulation; Sigma Xi. *Res:* Management science with emphasis on applications of systems dynamics to complex social and economic structures; university administration. *Mailing Add:* 2415 Wabash Ft Worth TX 76109

SECRIST, JOHN ADAIR, III, b Vincennes, Ind, Sept, 26, 47; m 68; c 2. NUCLEOSIDE SYNTHESIS, DRUG SYNTHESIS. *Educ:* Univ Mich, BS, 68; Univ Ill, PhD(org chem), 72. *Prof Exp:* Fel, Harvard Univ, 72-73; asst prof chem, Ohio State Univ, 73-79; sr chemist, 79-80, head, Bioorg Sect, 80-84, from assoc dir to dir, org chem dept, 84-90, EXEC VPRES, SOUTHERN RES INST, 90- *Concurrent Pos:* Mem, Exam Comt, Org Chem Subcomt, Am Chem Soc, 75-90; adj scientist, Comprehensive Cancer Ctr, Univ Ala, Birmingham, 84-; mem, Am Chem Soc Nomenclature Comt, 87- *Mem:* Am Chem Soc; Chem Soc; AAAS; Int Soc Antiviral Res. *Res:* Synthetic organic chemistry; medicinal chemistry; chemotherapy and drug development. *Mailing Add:* Southern Res Inst PO Box 55305 Birmingham AL 35255-5305

SEDAR, ALBERT WILLIAM, b Cambridge, Mass, Dec 20, 22; m 53; c 4. MICROSCOPIC ANATOMY. *Educ:* Brown Univ, AB, 43, MS, 48; Univ Iowa, PhD(zool), 53. *Prof Exp:* Asst gen biol, Brown Univ, 46-48; asst histol, Univ Iowa, 50, asst cytol, 51-52; instr zool & histol, Syracuse Univ, 52-53; NIH res fel cytol, Rockefeller Inst, 53-55; from asst prof to assoc prof, 55-66, PROF ANAT, JEFFERSON MED COL, 67- *Concurrent Pos:* USPHS career develop award, 66-71. *Mem:* Am Soc Cell Biol; Electron Micros Soc Am; Am Asn Anatomists. *Res:* Histology; cytophysiology; electron microscopy; fine structure of cells; electron histochemistry of cells; ultrastructure of cells and tissues. *Mailing Add:* Dept Anat Jefferson Med Col Jefferson Hall Philadelphia PA 19107

SEDAT, JOHN WILLIAM, b Culver City, Calif, Aug 17, 42; m 75. MOLECULAR BIOLOGY, CHEMISTRY. *Educ:* Pasadena Col, BA, 63; Calif Inst Technol, PhD(biol), 70. *Prof Exp:* Fel, Helen Hay Whitney Found, 70-73; staff scientist, Lab Molecular Biol, Med Res Coun, 73-74; vis scientist, Dept Cell Biochem, Hadassah Med Sch, Jerusalem, Israel, 74-75; res assoc, Dept Radiobiol, Yale Univ Sch Med, 75-78; ASST PROF MOLECULAR BIOL, UNIV CALIF SCH MED, SAN FRANCISCO, 78- *Res:* Chromosome and interphase nuclear architecture. *Mailing Add:* Dept Biochem & Biophys Univ Calif 513 Parnassus Ave San Francisco CA 94143

SEDBERRY, JOSEPH E, JR, b Shreveport, La, Sept 18, 25; m 53; c 3. AGRONOMY, SOIL CHEMISTRY. *Educ:* Centenary Col, BS, 49; La State Univ, Baton Rouge, MS, 52, PhD(agron), 54. *Prof Exp:* Agronomist, NLa Exp Sta, La State Univ, Baton Rouge, 54-55; agronomist, Lion Oil Co, Monsanto Chem Co, 55-58; agronomist, Am Potash Inst, 58-63; prof agron, 63-88, EMER PROF, LA STATE UNIV, BATON ROUGE, 88-; PRES, JES INC, 80-; CHMN BD, STEEL FORGINGS, 80- *Concurrent Pos:* Ford Found consult, Latin Am, 65-66; res grants, Geigy Chem Co, Eagle Picher Co, Sherwin Williams Co & Am Cyanamid Chem Co, 67- *Mem:* Am Soc Agron; Soil Sci Soc Am. *Res:* Soil testing and fertility with major emphasis on effect of major, secondary and micronutrients on yield; chemical composition of food and fiber crops. *Mailing Add:* 172 Lee Dr Suite 3 Baton Rouge LA 70808

SEDELOW, SALLY YEATES, b Greenfield, Iowa, Aug 10, 31; m 58. COMPUTER SCIENCE, LINGUISTICS. *Educ:* Univ Iowa, BA, 53; Mt Holyoke Col, MA, 56; Bryn Mawr Col, PhD(Eng lit), 60. *Prof Exp:* Instr Eng, Smith Col, 59-60; asst prof, Parsons Col, 60-61 & Rockford Col, 61-62; human factors scientist, Systs Develop Corp, 62-64; asst prof eng, St Louis Univ, 64-66; assoc prof Eng & comput & info sci, Univ NC, Chapel Hill, 66-70; prof comput sci & ling, Univ Kans, 70-85, assoc dean col lib arts & sci, 79-85; PROF COMPUT SCI, UNIV ARK, LITTLE ROCK, 85- *Concurrent Pos:* Off Naval Res res grant automated lang, Univ NC, Chapel Hill & Univ Kans, 64-74; consult, Syst Develop Corp, 64-67; mem, Adv Panel, Instnl Comput Serv Sect, NSF, 68-70, Comput Applns Res Sect, 70-71, Adv Comt Comput Activities, 72-; vis scientist, NSF-Asn Comput Mach Vis Scientist Prog, 69-70; NSF grant, Univ Kans, 71-72; prog dir, Tech & Systs Prog, NSF, 74-76; chmn group comput in lang & lit, Modern Lang Asn Am, 77-78; mem US deleg comput-based natural-lang processing, USSR, 78; adj prof english, Univ Ark, Little Rock, 85-, adj prof electronics & instrumentation, Grad Inst Technol, 85-; adj prof, computer sci, Univ Ark, Fayetteville, 88- *Mem:* Asn Comput Mach; Asn Comput Ling; Am Soc Info Sci; Ling Soc Am; Modern Lang Asn Am; Western Social Sci Asn. *Res:* Computer-based language and literature analysis; stylistics; semantics; computing in the humanities. *Mailing Add:* PO Box 942 Heber Springs AR 72543-0942

SEDELOW, WALTER ALFRED, JR, b Ludlow, Mass, Apr 17, 28; m 58. COMPUTER SCIENCES, HISTORY & PHILOSOPHY OF SCIENCE. *Educ:* Amherst Col, BA, 47; Harvard Univ, MA, 51, PhD, 57. *Prof Exp:* Instr, Williams Col, 48-50; from instr to asst prof, Amherst Col, 54-60; assoc prof & chmn dept, Parsons Col, 60-61; Jane Addams assoc prof & chmn dept, Rockford Col, 61-62; human factors scientist, Syst Develop Corp, 62-64; assoc prof sociol, chmn dept sociol & anthrop & dir health orgn res prog, St Louis Univ, 64-66; from assoc prof to prof comput & info sci & sociol, Univ NC, Chapel Hill, 66-70, dean, Sch Libr Sci, 67-70; prof comput sci, sociol, & hist & philos of sci, Univ Kans, 70-85; PROF COMPUT SCI, UNIV ARK, LITTLE ROCK, 85- *Concurrent Pos:* Consult, Life Sci Div, McDonnell

Aircraft Corp, 64-66; res prof, Inst Res Soc Sci, Univ NC, Chapel Hill, prin investr, NASA res proj grant & Higher Educ Act fel prog, 68-70; mem comt info technol, Am Coun Learned Socs, 68-70; vis scientist, NSF-Asn Comput Mach Vis Scientist Prog, 69-70; prin investr, NSF study grant, Univ Kans & mem, adv comt proj alternative approaches mgt & financing univ comput ctrs, Univ Denver, 71-72; consult, Col Human Ecol, Mich State Univ, 73-74; dir, Networking Sci, Div Math & Comput Sci, NSF, 75 & 76; mem oversight comt, Info Networks Technol Assessment Proj, Columbus Labs, Battelle Mem Inst, 75-76; mem, Adv Panel Comput Applns, Nat Endowment Humanities, 75-76; Menninger Found fel interdisciplinary studies, 77-83; consult biotechnol resources, NIH, 78-; consult, Ketron Corp, 78-; adj prof, VA Med Sci, Col Med, 88-; prin investr artificial intel, Topog Modeling English Semantic Space, NSF, 85-; prin investr, Modelling Univ & Interdisciplinary Interdependence, Exxon Res Found grant, 85; adj prof electronics & instrumentation, Univ Ark, Grad Inst Technol, 85-, Col Med, Med Sci, 88-; vis fac, Topeka Inst Psychoanal, 81-82. *Mem:* AAAS; Asn Comput Mach; Am Soc Info Sci; fel Am Sociol Asn; Soc Gen Systs Res; Western Social Sci Asn. *Res:* Computer-assisted language analysis; information systems; public uses of computers, including education applications; human factors in computer based systems; computer to computer communication networks; sociology of science and technology; sociology of culture. *Mailing Add:* PO Box 942 Heber Springs AR 72543-0942

SEDENSKY, JAMES ANDREW, b Bridgeport, Conn, Aug 6, 36. CARDIOVASCULAR PHYSIOLOGY, PULMONARY PHYSIOLOGY. *Educ:* Fairfield Univ, BS, 58; Univ Tenn, Memphis, PhD(physiol, biophys), 66. *Prof Exp:* NSF trainee biomath, NC State Univ, 66-67, NIH trainee, 67-69; from instr to asst prof physiol, 69-76, ASSOC PROF PHYSIOL, SCH MED, WAYNE STATE UNIV, 76- *Concurrent Pos:* NIH fac educ develop award, 73. *Mem:* AAAS; Math Asn Am; Am Statist Asn; Am Physiol Soc; Am Thoracic Soc. *Res:* Computer simulation and statistical analysis of biomedical systems; electrical impedance plethysmography. *Mailing Add:* Dept Physiol 4126 Scott Wayne State Univ 540 E Canfield Detroit MI 48201

SEDGWICK, ROBERT T, b Rome City, Ind, Aug 2, 33; m 66; c 1. MECHANICS, MATERIALS SCIENCE. *Educ:* Tri-State Col, BSME, 59; Mich State Univ, MS, 60, PhD(mat sci), 65. *Prof Exp:* Asst instr mech, Mich State Univ, 60-65; scientist, Space Sci Lab, Gen Elec Co, Pa, 65-69; CONVENTIONAL MUNITIONS PROG MGR, SYSTS, SCI & SOFTWARE, 69- *Mem:* Soc Eng Sci. *Res:* Dislocation mechanics; continuum mechanics; elastic-plastic-hydrodynamic material flow; hypervelocity and ballistic impact studies; Eulerian and Lagrangian numerical code development; shaped charge and fragmentation munitions calculations; fuel-air explosives; conventional warhead design. *Mailing Add:* 1515 San Dieguito Dr Del Mar CA 92014

SEDLACEK, WILLIAM ADAM, b Glendive, Mont, Feb 22, 36; m 63; c 2. PLUTONIUM NDA & ACCOUNTABILITY. *Educ:* Univ Wyo, BS, 58, PhD(phys chem), 65. *Prof Exp:* Res chemist shale oil & petrol res sta, US Bur Mines, 59-60; NSF fel, Univ Fla, 65-66; STAFF MEM, LOS ALAMOS SCI LAB, UNIV CALIF, 66- *Mem:* Am Geophys Union; Sigma Xi. *Res:* Ternary fission, activation analysis; nuclear reactions, nuclear experimental techniques and environmental pollution; atmospheric dynamics; Aitken nuclei; trace elements, ozone, nondestructive assay of SNM and SNM accountability. *Mailing Add:* Los Alamos Sci Lab PO Box 1663 MS E513 Los Alamos NM 87545

SEDLAK, BONNIE JOY, cell biology, developmental biology, for more information see previous edition

SEDLAK, JOHN ANDREW, b Bridgeport, Conn, May 17, 34; m 87. ORGANIC CHEMISTRY. *Educ:* Wesleyan Univ, BA, 55; Tufts Univ, MS, 56; Ohio State Univ, PhD(org fluorine chem), 60. *Prof Exp:* From res chemist to sr res chemist, 60-84, PRIN RES CHEMIST, AM CYANAMID CO, 84- *Mem:* Am Chem Soc; Tech Asn Pulp & Paper Indust; Sigma Xi. *Res:* Fluorinated monomers; polynuclear aromatic hydrocarbons; polyelectrolytes; paper chemicals; isocyanates; urethanes; coatings crosslinkers; rubber chemicals; chemiluminescence. *Mailing Add:* Am Cyanamid Co 1937 W Main St PO Box 60 Stamford CT 06904-0060

SEDLET, JACOB, b Milwaukee, Wis, Apr 4, 22; m 44, 68; c 2. PHYSICAL CHEMISTRY. *Educ:* Univ Wis, BS, 45; Purdue Univ, PhD(chem), 51. *Prof Exp:* Asst chemist, Metall Lab, Univ Chicago, 44-46; asst, Purdue Univ, 46-48, asst instr, 48-50; CHEMIST, ARGONNE NAT LAB, 50- *Mem:* AAAS; Health Physics Soc; Am Chem Soc. *Res:* Analytical chemistry; radiochemistry; environmental chemistry. *Mailing Add:* Argonne Nat Lab 9700 S Cass Ave Argonne IL 60439

SEDMAN, YALE S, b Detroit, Mich, May 22, 29; m 55; c 3. ENTOMOLOGY. *Educ:* Ariz State Univ, BS, 50; Univ Utah, MS, 52; Univ Wis, PhD(entom), 55. *Prof Exp:* Asst entom, Univ Utah, 50-52; asst, Univ Wis, 52-54; from asst prof to assoc prof, 55-65, PROF BIOL SCI, WESTERN ILL UNIV, 65- *Mem:* Soc Syst Zool. *Res:* Systematics of Diptera, especially family Syrphidae. *Mailing Add:* Dept Biol Sci Western Ill Univ Adams St Macomb IL 61455

SEDNEY, R(AYMOND), fluid dynamics, applied mathematics; deceased, see previous edition for last biography

SEDOR, EDWARD ANDREW, b East Chicago, Ind, June 24, 39; m; c 3. ORGANIC CHEMISTRY. *Educ:* Lake Forest Col, Ill, BA, 61; Mich State Univ, PhD(org chem), 66. *Prof Exp:* Res chemist, Archer-Daniels-Midland Co, 65-67, sr res chemist, 67-73; mkt coordr, Ashland Chem Co, 73-74, group leader additive chem, 74-79; sect mgr, 79-87, assoc res dir, 87-90, DIR TOTAL QUAL, SHEREX CHEM CO, 90- *Concurrent Pos:* Assoc ed, J Am Oil Chemists Soc. *Mem:* Am Chem Soc; Am Oil Chemists's Soc. *Res:* Fundamental chemistry, processing technology and applications research of fatty derivatives including amines, alcohols, quaternaries andesters; applications areas-emulsifiers, surfactants; plasticizers, mining technology, fabric softeners. *Mailing Add:* 5710 Zetland Ct Dublin OH 43017-9488

SEDRA, ADEL S, b Egypt, Nov 2, 43; Can citizen; m 73; c 2. MICROELECTRONICS, CIRCUIT THEORY. *Educ:* Cairo Univ, BSc, 64; Univ Toronto, MASc, 68; Univ Toronto, PhD(elec eng), 69. *Prof Exp:* Instr & res engr, Cairo Univ, 64-66; from asst prof to assoc prof, 69-78, exec dir, Microelec Develop Ctr, 83-86, PROF ELEC ENG, UNIV TORONTO, 78-, CHMN DEPT, 86- *Concurrent Pos:* Consult, Elec Eng Consociates Ltd, 69-; Nat Sci & Eng Res Coun Can grant, Univ Toronto, 69-; assoc ed, Inst Elec & Electronics Engrs Trans Circuits & Systs, 81-83; ed, Circuits & Devices Mag, 85-86; mem bd dir, Dicon Systs Ltd; founding bd dirs, Info Technol Ctr. *Honors & Awards:* Darlington Award, Inst Elec & Electronics Engrs, 84, Cauer-Guillemin Award, 87; Frederick Emmons Terman Award, Am Soc Eng Educ, 88. *Mem:* Fel Inst Elec & Electronics Engrs. *Res:* Electronic circuit design; active network theory and design; active filters; analog and digital instrumentation; filter-theory & design; switched-capacitor networks; computer-aided design. *Mailing Add:* Dept Elec Eng Univ Toronto Toronto ON M5S 1A4 Can

SEDRANSK, JOSEPH HENRY, b New York, NY, Mar 3, 38; m 67; c 1. STATISTICS. *Educ:* Univ Pa, BS, 58; Harvard Univ, PhD(statist), 64. *Prof Exp:* From asst prof to assoc prof statist, Iowa State Univ, 63-69; from assoc prof to prof, Univ Wis-Madison, 69-74; prof statist sci, State Univ NY, Buffalo, 74-78; prof math & statist, State Univ NY, Albany, 78-85; PROF & CHMN STATIST, UNIV IOWA, 85- *Concurrent Pos:* Ed, J Am Statist Asn-Applns. *Mem:* Inst Math Statist; Fel Am Statist Asn; Sigma Xi. *Res:* Theory of sampling from finite populations; Bayesian methods; fisheries and forestry statistics. *Mailing Add:* Five Pine St Westmere Albany NY 12203

SEDRIKS, ARISTIDE JOHN, b Riga, Latvia, May 15, 38; US citizen; m 62; c 3. CORROSION SCIENCE, ELECTROCHEMISTRY. *Educ:* Univ Wales, BSc, 59, PhD(metall), 62. *Prof Exp:* Res scientist, Defense Standards Labs, Sydney, Australia, 62-65; head, metall dept, Res Inst Advan Studies, Martin-Marietta Labs, Baltimore, 65-71; head, corrosion sect, Int Nickel Co, Sterling Forest, NY, 71-79, res mgr, Wrightsville Beach, NC, 79-82; br head, Naval Res Lab, Washington, 82-85; PROG MGR, OFF NAVAL RES, ARLINGTON, VA, 85- *Concurrent Pos:* Mem, Publ Comt & Res Comt, Nat Asn Corrosion Engrs, 86- *Honors & Awards:* F N Speller Award, Nat Asn Corrosion Engrs, 89. *Mem:* Nat Asn Corrosion Engrs; Am Soc Metals; Metals Soc; Electrochem Soc. *Res:* Corrosion control by metallurgical modifications; stress corrosion cracking, hydrogen embrittlement, passivity and localized corrosion; coatings for corrosion control; author of two books. *Mailing Add:* Off Naval Res Code 1131 800 N Quincy St Arlington VA 22217-5000

SEE, JOHN BRUCE, pyrometallurgy, process metallurgy, for more information see previous edition

SEEBACH, J ARTHUR, JR, b Philadelphia, Pa, May 17, 38; m 59; c 1. MATHEMATICS. *Educ:* Gettysburg Col, BA, 59; Northwestern Univ, MA, 62, PhD(math), 68. *Prof Exp:* From asst prof to assoc prof, 65-87, PROF MATH, ST OLAF COL, 87- *Concurrent Pos:* Vis assoc prof, Swiss Fed Inst Technol, 71-72; assoc ed, Am Math Monthly, 70-85; ed, Math Mag, 76-80; ETS comt grad rec exam (math), 82-90, chair, 86-90. *Mem:* Am Math Soc; Math Asn Am; Asn Comput Mach. *Res:* Point set topology. *Mailing Add:* Dept Math St Olaf Col 1520 St Olaf Ave Northfield MN 55057-1098

SEEBASS, ALFRED RICHARD, III, b Denver, Colo, Mar 27, 36; m 58; c 2. AERODYNAMICS, APPLIED MATHEMATICS. *Educ:* Princeton Univ, BSE, 58, MSE, 61; Cornell Univ, PhD(aerospace eng), 62. *Prof Exp:* Fel, res & teaching asst, Cornell & Princeton Univs, 58-62; from asst prof to assoc prof, Grad Sch Aerospace Eng, Cornell Univ, 62-72, actg dir, Ctr Appl Math, 68-69, assoc dean, res & grad progs, Col Eng & prof mech & aerospace eng, 72-75; prof aerospace & mech eng & prof math, Univ Ariz, 75-81, actg chmn, prog appl math, 79-80; DEAN, COL ENG & APPL SCI, UNIV COLO, 81- *Concurrent Pos:* Consult, Inst Defense Anal, 64-65, Dept Transp, 67-71, Gen Appl Sci Labs, 68-72, Flow Res, Inc, 76, Boeing Co, 76 & 79, Lockheed Co, Ga, 79 & Calif, 79-80; prin investr, NASA grants, 66-82, Off Naval Res grant & Air Force Off Sci Res grant, 75-82 & US-Israel Bi-Nat Sci Fund grant, 79-81; staff mem, res div, Off Adv Res & Technol, NASA Hq, 66-67, mem, res & technol subcomt fluid mech, NASA, 70, res panel, 74-77 & adv coun, 81-83; consult, Comt SST-Sonic Boom, Nat Acad Sci, 67-71; adv, Interagency Aircraft Noise Abatement Prog, 68-70 & US deleg to Sonic Boom Panel, Int Civil Aviation Orgn, 69; fac assoc, Boeing Sci Res Labs, Seattle, 70; mem, Air Force Studies Bd, Nat Res Coun, 77-79 & Aeronaut & Space Eng Bd, 77-84, chmn & vchmn, 81-83, mem, Comn Eng & Tech Systs, 82-83 & subcomt fluids, Surv Comt Plasma Physics & Fluids, 83-84; assoc ed, Physics of Fluids, 78-80 & Am Inst Aeronaut & Astronaut J, 81-83, Progress in Aeronaut & Astronaut, 87-; vis prof aeronaut & astronaut & actg assoc dir, Aerospace & Energetics Res Prog, Univ Wash, 79; mem, Sci Adv Bd, USAF, 84-88. *Mem:* Nat Acad Eng; fel Am Inst Aeronaut & Astronaut; fel AAAS. *Res:* Aerodynamics and fluid mechanics; engineering education; geophysical fluid dynamics; computational fluid dynamics. *Mailing Add:* Col Eng & Appl Sci Univ Colo Boulder CO 80309

SEEBOHM, PAUL MINOR, b Cincinnati, Ohio, Jan 13, 16; m 42; c 1. MEDICINE. *Educ:* Univ Cincinnati, BA, 38, MD, 41; Am Bd Internal Med, cert allergy. *Prof Exp:* Asst resident physician, Cincinnati Gen Hosp, 46-48; resident allergy, Roosevelt Hosp, New York, 48-49; assoc internal med, 49-51, dir allergyclin, 49-70, from asst prof to prof internal med, 51-86, exec assoc dean col, 70-86, EMER PROF INTERNAL MED, COL MED, UNIV IOWA, 86-, CONSULT TO DEAN, COL MED, 86- *Concurrent Pos:* Mem spec med adv group, Vet Admin, 72-76; chmn allergenic extract rev panel, Food & Drug Admin, 74-84; mem, Iowa State Bd Health, 71-83, 76-83; mem bd dirs, Iowa Health Systs Agency, 76-80. *Mem:* Fel Am Col Physicians; Cent Soc Clin Res; fel Am Acad Allergy & Immunol (secy, pres, 66). *Res:* Pulmonary function in chronic respiratory disease. *Mailing Add:* Univ Iowa Col Med Iowa City IA 52242

SEEBURGER, GEORGE HAROLD, b Phillips, Wis, July 20, 35; m 59; c 2. SCIENCE EDUCATION, BIOLOGY. *Educ:* Wis State Univ, Stevens Point, BS, 57; Univ Ga, MEd, 62, EdD(sci educ), 64. *Prof Exp:* Teacher high sch, Wis, 57-60; from asst prof to assoc prof, 64-76, PROF BIOL, UNIV WIS-WHITEWATER, 76- *Mem:* Am Inst Biol Sci; Am Soc Mammal; Am Soc Ichthyologists & Herpetologists; Wis Soc Ornith; Wis Mycol Soc. *Res:* Preparation of high school biology teachers; ichthyology, fishes of Wisconsin; mammalogy. *Mailing Add:* Dept Biol 800 W Main St Univ Wis Whitewater WI 53190

SEED, HARRY BOLTON, geotechnical engineering, earthquake engineering; deceased, see previous edition for last biography

SEED, JOHN RICHARD, b Paterson, NJ, Apr 27, 37; m 59; c 2. MICROBIOLOGY, PARASITOLOGY. *Educ:* Lafayette Col, AB, 59; Yale Univ, PhD(microbiol), 63. *Prof Exp:* Res assoc biol, Haverford Col, 62-63; from asst prof to assoc prof, Tulane Univ, 65-73, prof, 73-74; prof biol & head dept biol, Tex A&M Univ, 74-80; prof & chair dept parasitol lab prac, 81-89, PROF DEPT EPIDEMIOL, SCH PUB HEALTH, UNIV NC, CHAPEL HILL, 90- *Concurrent Pos:* Am Cancer Soc fel, 62-63. *Honors & Awards:* Henry Baldwin Ward Medal, Am Soc Parasitol. *Mem:* Fel AAAS; Am Inst Biol Sci; Am Soc Microbiol; Soc Protozool; Am Soc Parasitol (pres-elect). *Res:* Immunological and physiological studies on parasitic protozoan infections, especially African trypanosomiasis. *Mailing Add:* Dept Epidemiol Univ NC Sch Pub Health Chapel Hill NC 27514

SEED, RANDOLPH WILLIAM, b Chicago, Ill, May 1, 33; m 68; c 5. SURGERY, BIOCHEMISTRY. *Educ:* Harvard Univ, BA, 54; Univ Chicago, MD, 60, PhD(biochem), 65. *Prof Exp:* Chmn dept surg, Grant Hosp, Chicago, 70-76; asst prof surg, Med Sch, Northwestern Univ, 71-87; ASST PROF SURG, RUSH UNIV MED SCH, 87- *Concurrent Pos:* Attend surg, Vet Admin Res Hosp, Chicago, 68- *Mem:* Am Thyroid Asn. *Res:* Human thyroid cancer; human thyroid tissue culture and biosynthesis of thyroglobulin. *Mailing Add:* 999 N Lakeshore Dr Chicago IL 60611

SEED, THOMAS MICHAEL, b Paterson, NJ, Dec 8, 45; m 68; c 2. HEMATOLOGY, RADIATION BIOLOGY. *Educ:* Univ Conn, BA, 68; Ohio State Univ, MS, 69, PhD(microbiol), 72. *Prof Exp:* Fel electron micros, Inst Path, Case Western Reserve Univ, 72-73; res assoc, Blood Res Lab, Am Nat Red Cross, 73-75; RES BIOLOGIST & GROUP LEADER, RADIATION HEMAT, DIV BIOL ENVIRON & MED RES, ARGONNE NAT LAB, 75- *Concurrent Pos:* Lectr, dept biol & physics, Cath Univ Am, 74-75; supvr, Electron Micros Ctr, Div Biol Environ & Med Res, Argonne Nat Lab, 75-; consult, Pharmaceut Res Div, Warner-Lambert/Parke-Davis, Detroit, 78-80; guest ed, Int J Scanning Electron Micros, 81-84. *Mem:* AAAS; Am Soc Microbiol; Int Soc Exp Hemat; Electron Micros Soc Am; Am Soc Radiation Res. *Res:* Structural and function studies of radiation-induced hematopathology; cellular mechanisms of preclinical phase leukemogenic processes; mechanistic studies on red cell destruction during infectious hemolytic anemias. *Mailing Add:* Div Biol Environ & Med Res Argonne Nat Lab 9700 S Cass Ave Argonne IL 60439

SEEDS, MICHAEL AUGUST, b Danville, Ill, Dec 14, 42; m; c 1. ASTRONOMY. *Educ:* Univ Ill, Urbana, BS, 65; Ind Univ, Bloomington, MS & PhD(astron), 70. *Prof Exp:* PROF ASTRON, FRANKLIN & MARSHALL COL, 70- *Mem:* AAAS; Am Astron Soc; Sigma Xi. *Res:* Photometry of short period variable stars and eclipsing binaries; narrow band photometry; undergraduate astronomy instruction, methods and materials; author of 2 books. *Mailing Add:* Dept Physics & Astron Franklin & Marshall Col PO Box 3003 Lancaster PA 17604-3003

SEEDS, NICHOLAS WARREN, b Circleville, Ohio, Dec 25, 42; m 64; c 2. NEUROSCIENCES. *Educ:* Univ NMex, BS, 64; Univ Iowa, PhD(biochem), 68. *Prof Exp:* NSF fel, NIH, 68-70; PROF BIOCHEM, BIOPHYS & GENETICS, MED CTR, UNIV COLO, DENVER, 70-, DIR, NEUROSCI CTR, 89- *Concurrent Pos:* Vis prof, Univ Heidelberg, Ger, 79-80. *Mem:* Am Soc Biol Chemists; AAAS; Am Soc Cell Biol; Soc Neurosci; Am Soc Neurochem. *Res:* Developmental neurobiology; microtubules. *Mailing Add:* Dept Biochem-Biophys & Genetics Univ Colo Med Sch Denver CO 80262

SEEFELDT, VERN DENNIS, b Lena, Wis. ANATOMY. *Educ:* Wis State Univ, La Crosse, BS, 55; Univ Wis, PhD(phys educ), 66. *Prof Exp:* Asst prof phys educ, Mich State Univ, 66-69; asst prof, Univ Wis-Madison, 69-71; assoc prof, 71-76, PROF PHYS EDUC, MICH STATE UNIV, 76- *Concurrent Pos:* Dir, Youth Sports Inst. *Mem:* Am Asn Health, Phys Educ & Recreation; Soc Study Human Biol. *Res:* Motor development; interrelationship of physical growth, motor development and academic achievement in pre-school and elementary aged children. *Mailing Add:* Phys Educ & Exercise Sci Mich State Univ 213 IM Sports Circle East Lansing MI 48824

SEEFELDT, WALDEMAR BERNHARD, b Milwaukee, Wis, Apr 4, 25; m 50; c 3. NUCLEAR ENGINEERING. *Educ:* Purdue Univ, BS, 47, MS, 48. *Prof Exp:* CHEM ENGR, ARGONNE NAT LAB, 48- *Mem:* Am Nuclear Soc; Am Inst Chem Engrs; Inst Nuclear Mat Mgt; AAAS; Res Soc Am. *Res:* Chemical processing of spent nuclear reactor fuels; methods of immobilizing highly radioactive nuclear waste; waste management systems. *Mailing Add:* 417 S Kensington Ave La Grange IL 60525

SEEFURTH, RANDALL N, b Milwaukee, Wis, Sept 19, 41; m 70; c 3. ELECTROCHEMISTRY, MOLTEN SALT CHEMISTRY. *Educ:* Univ Wis-Milwaukee, BS, 63. *Prof Exp:* Res chemist, Allison Div, 63-68, STAFF RES SCIENTIST, RES LABS, GEN MOTORS, 68- *Mem:* Electrochem Soc; Metall Soc. *Res:* High temperature, molten salt, chemical and electrochemical studies; lithium-alloy, sulfur and metal sulfide electrodes for application in molten salt or solid ionic conductor battery systems; chemical reduction of rare earth oxides and halides in molten salt media. *Mailing Add:* Phys Chem Dept 14 Tech Ctr Gen Motors Res Labs Warren MI 48090-9055

SEEGAL, RICHARD FIELD, b Newport, RI, Feb 13, 45; m 69; c 2. NEUROTOXICOLOGY. *Educ:* Brown Univ, AB, 66; Emory Univ, MA, 70; Univ Ga, PhD(physiol psychol), 72. *Prof Exp:* Fel endocrinol biobehav sci, Univ Conn, 72-74; res assoc neurochem, Dept Pharmacol, Mich State Univ, 77-78; res scientist virol, 74-80, RES SCIENTIST NEUROTOXIC, NY STATE DEPT HEALTH, 80- *Mem:* AAAS; Soc Neurosci. *Res:* Neurotoxicology of halogeneted hydrocarbons development of in vitro neurochemical test systems. *Mailing Add:* Wadsworth Ctr PO Box 509 Albany NY 12201-0509

SEEGER, CHARLES RONALD, b Columbus, Ohio, Jan 31, 31; m 61; c 2. EXPLORATION GEOPHYSICS, PLANETOLOGY. *Educ:* Ohio State Univ, BSc, 53; George Washington Univ, MS, 58; Univ Pittsburgh, PhD(earth & planetary sci), 66. *Prof Exp:* Analyst, Weapons Systs Eval Div, Inst Defense Anal, 58-60; earth sci analyst, Sci & Tech Intel Ctr, USN, 60-63; Nat Acad Sci-Nat Res Coun fel, NASA-Goddard Space Flight Ctr, 66-68; from asst prof to assoc prof, 68-77, PROF GEOL & GEOPHYS, WESTERN KY UNIV, 77- *Concurrent Pos:* Nat Acad Sci exchange scientist, EGermany, 86. *Mem:* Sigma Xi (pres, 83-84); Am Geophys Union; Geol Soc Am; Meteoritical Soc. *Res:* Geological and geophysical investigations of astroblemes, cryptoexplosion structures, volcanic and impact craters and other geological structures; tectonics of plate boundaries; lunar geology and geochemistry; planetology. *Mailing Add:* 630 Ironwood Dr Bowling Green KY 42103

SEEGER, PHILIP ANTHONY, b Evanston, Ill, Feb 19, 37; m 59; c 3. PHYSICS OF MATERIALS. *Educ:* Rice Inst, BA, 58; Calif Inst Technol, PhD(physics), 63. *Prof Exp:* Res fel low energy nuclear physics, Calif Inst Technol, 62-64; STAFF MEM MANUEL LUJAN NEUTRON SCATTERING CTR, LOS ALAMOS NAT LAB, 64- *Mem:* Am Phys Soc; Int Astron Union. *Res:* Calculation of stellar nucleosynthesis and the semiempirical atomic mass law; neutron time-of-flight experiments in nuclear physics, materials science and biophysics. *Mailing Add:* Los Alamos Natl Lab PO Box 1663 MS H805 Los Alamos NM 87545

SEEGER, RAYMOND JOHN, b Elizabeth, NJ, Sept 20, 06; m 29; c 2. QUANTUM THEORIES. *Educ:* Rutgers Univ, AB, 26; Yale, PhD(physics), 29. *Hon Degrees:* DSc, Kent State Univ, 58 & Univ Dubuque, 66. *Prof Exp:* Assoc prof physics, Presby Col, SC, 29-30; assoc prof physics, George Washington Univ & dir, Univ Chapel, 30-42; head, fundamental explosive res, USN, Bur Ord, 42-46, chief, Aeroballistic Res Dept, Naval Ord Lab, 46-52; spec asst to dir, math, physics & eng, NSF, 52-79; RETIRED. *Concurrent Pos:* Loomes fel, 43. *Honors & Awards:* Distinguished Serv Citation, Am Asn Physics Teachers, 56. *Mem:* Am Phys Soc; Am Physics Teachers; AAAS; Hist Sci Soc. *Res:* Quantum theories; co-editor and/or author of five books. *Mailing Add:* 4507 Wetherill Rd Bethesda MD 20816

SEEGER, ROBERT CHARLES, b Salem, Ore, May 9, 40; m 64; c 1. IMMUNOLOGY, PEDIATRICS. *Educ:* Willamette Univ, BA, 62; Univ Ore, MS & MD, 66. *Prof Exp:* Pediat intern, Med Sch, Univ Minn, 66-67, resident, 67-68; clin assoc immunol, NIH, 68-72; spec fel tumor immunol, Nat Cancer Inst, Univ Col, Univ London, 72-74; asst prof, 74-77, ASSOC PROF PEDIAT IMMUNOL, UNIV CALIF, LOS ANGELES, 77- *Concurrent Pos:* Res career develop award, Nat Cancer Inst, 75. *Mem:* Am Asn Cancer Res; Soc Pediat Res; Am Asn Immunologists. *Res:* Pediatric tumor immunology; childhood neuroblastoma with studies of tumor associated antigens; immune responses which kill tumor cells; chemo-immunotherapy in nude mice; human monocyte subsets. *Mailing Add:* Dept Pediat Med Sch Univ Calif Los Angeles CA 90024

SEEGERS, WALTER HENRY, b Fayette Co, Iowa, Jan 4, 10; m 35; c 1. PHYSIOLOGY, BIOCHEMISTRY. *Educ:* Univ Iowa, BA, 31, MS, 32, PhD(biochem), 34. *Hon Degrees:* ScD, Wartburg Col, 53 & Med Col Ohio, 78; MD, Justus Liebig Univ, Ger, 74. *Prof Exp:* Res assoc nutrit, Univ Iowa, 34-35, res assoc path, 37-42; res assoc nutrit & nitrogen metab, Antioch Col, 36-37; res, Parke, Davis & Co, Mich, 42-45; from assoc prof physiol to prof Sch Med, Wayne State Univ, 45-48, chmn dept physiol & pharmacol, 46-73, Traitel prof hemat, 64-80, chmn dept physiol, 73-80, SEEGERS PROF PHYSIOL & PHARMACOL, SCH MED, WAYNE STATE UNIV, 48- *Concurrent Pos:* Prof, Univ Detroit, 46-50; vis prof, Baylor Univ, 50; Nat Sci Coun vis prof, Nat Taiwan Univ, 75. Chmn panel blood coagulation, Nat Res Coun, 53-57; USPHS res grants, Univ Detroit; Sigma Xi res award, 57; Commonwealth Fund res award, 57-58; Henry Ford Hosp chmn, Nat Comt Platelet Conf, 60; mem, Ctr Health Educ, 61-80; mem, Mayor's Comt Rehab Narcotic Addicts, Detroit, 61-75; hon mem fac med, Univ Chile; mem bd regents, Wartburg Col, 66-80; Shirley Johnson Greenwalt mem lectr. *Honors & Awards:* Brown mem lectr, Am Heart Asn, 49; co-recipient Ward Burdick Award, Am Soc Clin Pathologists, 50, James F Mitchell Found Award, 69; Harvey lectr, 52; Reilly lectr, Notre Dame Univ, 59; John G Gibson II lectr, Columbia Univ, 60; Behringwerke lectr, Marburg/Lahn, Ger, 62; Beaumont lectr, Wayne County Med Soc, 66; H P Smith Award, Am Soc Clin Pathologists. *Mem:* Am Physiol Soc; Int Soc Hemat; fel Am Inst Chemists; fel Am Soc Clin Pharmacol & Therapeut; fel NY Acad Sci. *Res:* Nutrition; vitamins; protein metabolism; isolation and chemical characterization of plasma proteins; mechanisms of blood clotting; integrative physiology; hemostatic agents. *Mailing Add:* 2857 Ptarmigan No 5 Walnut Creek CA 94595

SEEGMILLER, DAVID W, b Nephi, Utah, Jan 6, 34; m 54; c 4. ELECTROCHEMISTRY. *Educ:* Brigham Young Univ, BS, 56, MS, 58; Univ Calif, Berkeley, PhD(nuclear chem), 63. *Prof Exp:* USAF, 58-, nuclear res officer, 58-60, from assoc prof to prof chem, USAF Acad, 62-76, dep dept head, 75-76, chief scientist, USAF Europ Off Aerospace Res & Develop, London, 76-78, dep dept head, 78-80, PROF & ACTG HEAD, DEPT CHEM & BIOL SCI, USAF ACAD, 80- *Concurrent Pos:* Res & Develop Award, USAF, 71. *Mem:* Am Chem Soc; Sigma Xi. *Res:* Electrochemistry and physical chemistry of fused salt systems; nuclear chemistry and reactions; thermochemical measurements; high energy-density batteries. *Mailing Add:* 5300 Van Christopher NE Albuquerque NM 87111

SEEGMILLER, JARVIS EDWIN, b St George, Utah, June 22, 20; m 50; c 4. BIOCHEMISTRY. *Educ:* Univ Utah, AB, 42; Univ Chicago, MD, 48. *Prof Exp:* Asst, US Bur Mines, Utah, 41; asst, Nat Defense Res Comt, Northwestern Tech Inst, 42-44, US Army, 44; asst med, Univ Chicago, 47-48; intern, Marburg Div, Johns Hopkins Hosp, 48-49; biochemist, Nat Inst Arthritis & Metab Dis, 49-51; res assoc, Thorndike Mem Lab, Harvard Med Sch, 52-53; vis investr, Pub Health Res Inst, NY, 53-54; chief sect human biochem genetics & asst sci dir, Nat Inst Arthritis & Metab Dis, 54-69; prof, Dept Med & dir, Div Rheumatology, 69-90, dir, Inst Res Aging, 83-90, EMER PROF, DEPT MED, & ASSOC DIR, SAM & ROSE STEIN INST RES AGING, UNIV CALIF, SAN DIEGO, 90- *Concurrent Pos:* Vis scientist, Univ Col Hosp Sch Med, London, 64-65; Harvey Soc lectr, 70; Macy scholar, Basel Inst Immunol, Switz & Sir William Dunn Sch Path, Oxford, Eng, 75-76; Guggenheim fel, Swiss Inst Exp Cancer Res, Lausanne, Switz, 82-83; John Simon Guggenheim Mem Found fel, 82; Fogarty Int fel, Dept Biochem, Oxford, Eng, 89. *Honors & Awards:* Gairdner Found Award, 68; Distinguished Serv Award, USPHS, 69; Geigy Award, 69; Philip Hench Award, 69; Balfourvis Lectr, Mayo Clin, 74; Malthe Lectr, Oslo, Norway, 75; Mayo Soley Award, Western Soc Clin Res, 79. *Mem:* Nat Acad Sci; hon mem Harvey Soc; Am Soc Biol Chemists; Am Rheumatism Asn; Am Fedn Clin Res; Am Soc Human Genetics; Am Soc Clin Invest; AAAS; Asn Am Physicians; Am Acad Arts & Sci. *Res:* Enzymology; intermediary carbohydrate and purine metabolism; biochemistry of liver disease; hereditary metabolic diseases; arthritis; human biochemical genetics; hereditary bases of arthritis; nature of the aging process; author of 336 publications. *Mailing Add:* Dept Med M-013 I Sch Med Univ Calif San Diego La Jolla CA 92093

SEEGMILLER, ROBERT EARL, b Salt Lake City, Utah, July 8, 43; m 63; c 8. DEVELOPMENTAL TOXICOLOGY. *Educ:* Univ Utah, BS, 65, MS, 67; McGill Univ, PhD(genetics), 70. *Prof Exp:* Res assoc develop biol, Univ Colo, 70-72; assoc prof, 72-80, PROF ZOOL, BRIGHAM YOUNG UNIV, 80- *Concurrent Pos:* Dipl, Pharmaceut Mfrs Asn Found, 70-72, res starter grant, 73; Basil O'Connor res starter grant, Nat Found March of Dimes, 74-76; vis res fel, Dept Pediat, Univ Chicago, 76-77 & Dept Pharmacol, Univ Wash, 87-88. *Mem:* Teratology Soc; AAAS; Int Soc Develop Biol. *Res:* Drug and gene induced defects of the endochondral skeleton in relation to limb, palate and lung development in laboratory animals; epidemiological assessment of birth defects in human populations. *Mailing Add:* Dept Zool Widb 593 Brigham Young Univ Provo UT 84602

SEEHRA, MOHINDAR SINGH, b W Pakistan, Feb 14, 40; m 63; c 2. SOLID STATE PHYSICS. *Educ:* Punjab Univ, India, BSc, 59; Aligarh Muslin Univ, MSc, 62; Univ Rochester, PhD(physics), 69. *Prof Exp:* Lab instr chem, Arya Col, India, 59-60; lectr physics, Jain Col, India, 62-63; from asst prof to assoc prof, 69-77, PROF PHYSICS, WVA UNIV, 77- *Concurrent Pos:* Alfred P Sloan Found res fel, 73-76; NSF & Dept of Energy res grants. *Mem:* Fel Am Phys Soc; Mat Res Soc. *Res:* Phase transitions and critical phenomena; magnetic resonance and spin-spin relaxation; magnetic optical and transport properties; properties of pyrite and other minerals in coal; catalysis and coal pyrolysis; magnetic properties of transition metal oxides, fluorides and sulfides using ESR, magnetic susceptibility, optical absorption and dielectric studies. *Mailing Add:* Dept Physics & Astron WVa Univ Morgantown WV 26506

SEELAND, DAVID ARTHUR, b St Paul, Minn, Nov 14, 36; m 61; c 2. GEOLOGY. *Educ:* Univ Minn, BA, 59, MS, 61; Univ Utah, PhD(geol), 68. *Prof Exp:* GEOLOGIST, US GEOL SURV, 67- *Concurrent Pos:* Geologist, US Geol Surv, 61-63. *Mem:* Geol Soc Am; Soc Econ Paleont & Mineral; Sigma Xi. *Res:* Sedimentology, particularly early Paleozoic marine shelf paleocurrents and depositional environments of Tertiary fluvial rocks; structure and stratigraphy of the northern Rocky Mountains. *Mailing Add:* US Geol Surv MS 919 Box 25046 Fed Ctr Denver CO 80225

SEELBACH, CHARLES WILLIAM, b Buffalo, NY, Dec 13, 23; m 46; c 3. POLYMER CHEMISTRY, INDUSTRIAL CHEMISTRY. *Educ:* Cornell Univ, AB, 48; Case Western Reserve Univ, MS, 51; Purdue Univ, PhD, 55. *Prof Exp:* Res chemist, Standard Oil Co, Ohio, 48-52; anal chemist, Purdue Univ, 52-55; res chemist, Esso Res & Eng Co, 55-57, asst sect head res lab, Esso Standard Oil Co, 57-58, sect head explor res, Esso Res & Eng Co, 58-61, chem coordr, Esso Int, Inc, 61-63, mgr polymers div, Esso Chem Co, Inc, NY, 63-66, mgr elastomers new investments & planning div, 66-67; dir indust chem develop, US Steel Corp, 67-69, dir hydrocarbon raw mat develop, 69-71, mgr com develop plastics dept, 71-77, mgr Com Develop Petrochem Develop Dept, USS Chem Div, 77-83; CONSULT, SUTRO MANAGEMENT ADVISORS, 83- *Mem:* AAAS; Am Oil Chem Soc; NY Acad Sci; Soc Plastics Indust; Com Develop Asn; Am Chem Soc. *Res:* Hydrocarbon stability; synthesis antioxidants; liquid thermal diffusion; separation of hydrocarbons and lipids; polymerization catalysis; plastic and elastomer synthesis; chemical intermediate syntheses; metal alkyl derivatives; olefin-diolefin derivatives; oxygenated derivatives. *Mailing Add:* PO Box 5590 Sun City Center FL 33571-5590

SEELEY, JOHN GEORGE, b North Bergen, NJ, Dec 21, 15; c 5. FLORICULTURE. *Educ:* Rutgers Univ, BSc, 37, MSc, 40; Cornell Univ, PhD(floricult), 48. *Prof Exp:* Asst, NJ Exp Sta, 37-40, garden supt, 40-41; res instr floricult, Cornell Univ, 41-43; asst agronomist bur plant indust, Soils & Agr Eng, USDA, Ga, 43-44; chemist in charge rubber mat lab, Wright Aeronaut Corp, NJ, 44-45; res instr floricult, Cornell Univ, 45-48, asst prof, 48-49; prof, Pa State Univ, 49-56; head dept, 56-70, PROF FLORICULT, CORNELL UNIV, 56- *Concurrent Pos:* Prof hort & D C Kiplinger chair floricult, Ohio State Univ, 84-85. *Honors & Awards:* Soc Am Florists Outstanding Res Award, 65; Leonard Vaughan Res Award, Am Soc Hort Sci, 50. *Mem:* Am Acad Florists; fel AAAS; fel Am Soc Hort Sci; Int Soc Hort Sci. *Res:* Nutrition and plant physiology; soils; soil aeration; light and temperature. *Mailing Add:* Dept Floricult Cornell Univ Ithaca NY 14853

SEELEY, ROBERT, b New York, NY, Dec 20, 36; m 67; c 4. RESEARCH ADMINISTRATION. *Educ:* Univ Ga, BSA, 60, MS, 63. *Prof Exp:* QUAL CONTROL MGR, MAPLE ISLAND, INC, 72- *Mem:* Inst Food Technologists. *Mailing Add:* Maple Island Inc Wanamingo MN 55983

SEELEY, ROBERT D, b Kansas City, Kans, May 15, 23; m 46; c 3. PHYSICAL CHEMISTRY. *Educ:* Tex A&I Univ, BS, 50, MS, 51; Wayne State Univ, PhD(phys chem), 59. *Prof Exp:* Asst instr chem, Tex A&I Univ, 50-51; instr gen sci & chem, McCook Jr Col, 51-52; asst instr chem, Wayne State Univ, 53-54; mat engr tire div, US Rubber Co, 54-59; sr chemist, Chemstrand Res Ctr, Inc, 59-62; staff mem high polymer res, Sandia Corp, 62-65; from assoc prof to prof, 65-83, EMER PROF CHEM, CENT MO STATE UNIV, 83- *Mem:* Sigma Xi. *Res:* Structural, surface and performance characteristics of high polymers; chemical sorption; reaction kinetics phenomena studies. *Mailing Add:* RR Three Box 176 Warrensburg MO 64093

SEELEY, ROBERT T, b Bryn Mawr, Pa, Feb 26, 32; m 58; c 4. MATHEMATICAL ANALYSIS. *Educ:* Haverford Col, BS, 53; Mass Inst Technol, PhD(math), 58. *Prof Exp:* Instr math, Harvey Mudd Col, 58-59, asst prof, 59-61; NATO fel, 61-62; from asst prof to assoc prof, Brandeis Univ, 63-67, prof, 67-72; PROF MATH, UNIV MASS, BOSTON, 72- *Concurrent Pos:* Sloan Found fel, 65-67. *Mem:* Am Math Soc; Math Asn Am. *Res:* Partial differential equations. *Mailing Add:* Dept Math Univ Mass Harbor Campus Boston MA 02125

SEELEY, ROD R, b Rupert, Idaho, Dec 29, 45; m 65; c 4. REPRODUCTIVE PHYSIOLOGY. *Educ:* Idaho State Univ, BS, 68; Utah State Univ, MS, 71, PhD(physiol), 73. *Prof Exp:* Asst prof, 73-77, ASSOC PROF PHYSIOL, IDAHO STATE UNIV, 77-, PROF BIOL, 80- *Concurrent Pos:* Chmn, Dept Biol Sci, Idaho State Univ. *Mem:* AAAS; Sigma Xi; Soc Study Reproduction. *Res:* The study of neural and endocrine mechanisms that regulate smooth-muscle motility in the testicular capsule and neural and endocrine mechanisms that control reproduction. *Mailing Add:* Dept Biol Idaho State Univ Pocatello ID 83209

SEELEY, SCHUYLER DRANNAN, b Huntington, Utah, Aug 5, 39; m 62; c 6. PLANT CHEMISTRY. *Educ:* Brigham Young Univ, BS, 64; Utah State Univ, MS, 67; Cornell Univ, PhD(pomol), 71. *Prof Exp:* Res assoc pomol, Cornell Univ, 71; from asst prof to assoc prof, 71-83, PROF PLANT SCI, UTAH STATE UNIV, 83- *Honors & Awards:* J H Gourley Medal, Am Soc Hort Sci, 75, 83. *Mem:* Am Soc Hort Sci; AAAS; Am Inst Chemists. *Res:* Hormonal physiology of fruit tree dormancy; mathematical modeling of chill units and growing degree hours for fruit tree physiodates; instrumental ultramicroanalysis of plant hormones. *Mailing Add:* 481 E 200 S Hyde Park UT 84318

SEELEY, THOMAS DYER, b Bellefonte, Pa, June 17, 52; m 79; c 2. BEHAVIORAL ECOLOGY, SOCIOBIOLOGY. *Educ:* Dartmouth Col, AB, 74; Harvard Univ, PhD (biol), 78. *Prof Exp:* Jr fel, Soc Fels, Harvard Univ, 78-80; from asst prof to assoc prof biol, Yale Univ, 80-86; asst prof, 86-88, ASSOC BIOL, CORNELL UNIV, 88- *Concurrent Pos:* Sci leader, Exped Thailand, Nat Geog Soc, 79-80; consult, Univ Oslo, Norway, 88. *Mem:* Animal Behav Soc; Int Union Study Social Insects; Int Bee Res Asn. *Res:* Physiological, behavioral, and ecological studies of the biology of social insects, especially the honeybee. *Mailing Add:* Sect Neurobiol & Behav Mudd Hall Cornell Univ Ithaca NY 14853

SEELIG, JAKOB WILLIAMS, b Brooklyn, NY, May 21, 37; m 65; c 3. OPERATIONS RESEARCH. *Educ:* Rensselaer Polytech Inst, BCE, 58, PhD(civil eng), 63; Columbia Univ, MS, 59; George Washington Univ, MEA, 72. *Prof Exp:* VPRES, ANALYTICAL SERV INC, 65- *Mem:* Opers Res Soc Am; Am Inst Aeronaut & Astronaut. *Res:* Acquisition of strategic, special operations and airlife systems; design of structures to withstand nuclear weapon effects. *Mailing Add:* 1215 Jefferson Davis Hwy Suite 800 Arlington VA 22202

SEELKE, RALPH WALTER, b Murfreesboro, Tenn, Nov 25, 51; m 77; c 3. MICROBIAL GENETICS. *Educ:* Clemson Univ, BS, 82; Univ Minn, PhD(microbiol), 82. *Prof Exp:* Fel cell biol, Mayo Clin, 81-83; asst prof microbiol, George Washington Univ, 83-85; ASST PROF MICROBIOL, UNIV WIS-MILWAUKEE, 85- *Mem:* Am Soc Microbiol. *Res:* Influence of chromosomal genes on plasmid DNA replication control in Escherichia coli; agrobacterium tumefaciens phage to assist its value in plant gene transfer. *Mailing Add:* Dept Biol Sci Univ Wis Milwaukee PO Box 413 Milwaukee WI 53201

SEELY, GILBERT RANDALL, b Bellingham, Wash, Jan 18, 29; m 56; c 4. PHYSICAL CHEMISTRY. *Educ:* Harvard Univ, AB, 50; Univ Calif, PhD(chem), 54. *Prof Exp:* Chemist, Gen Elec Co, NY, 53-54; fel boron chem, Univ Wash, Seattle, 56-57; chemist phys chem dept, Shell Develop Co, 57-62; investr, Battelle-C F Kettering Res Lab, 62, sr res scientist, 62-72, sr investr, 72-83, sr res scientist, 83-86; RES PROF, DEPT CHEM, ARIZ STATE UNIV, TEMPE, 86- *Mem:* Am Chem Soc; Inter-Am Photochem Soc; Am Soc Photobiol; Int Solar Energy Soc. *Res:* Photochemistry of chlorophyll and porphyrins; photochemistry and energy transfer in polymeric systems; environmental applications; model systems of photosynthesis. *Mailing Add:* 567 W Los Lagos Vista Mesa AZ 85210

SEELY, J RODMAN, b Willard, Utah, July 28, 27; m 55; c 3. PEDIATRICS. *Educ:* Univ Utah, BS, 50, MD, 52, PhD(biol chem), 64. *Prof Exp:* Intern pediat, Salt Lake Gen Hosp, 52-53, asst resident, 53-56; res instr, Col Med, Univ Utah, 57-58; asst prof, Sch Med, Univ Wash, 58-63; assoc prof, 64-76, prof, 76-79, CLIN PROF PEDIAT, SCH MED, UNIV OKLA, 79-; DIR, GENETICS DIAGNOSTIC CTR, PRESBY HOSP, OKLAHOMA CITY, 80- *Concurrent Pos:* Asst pediatrician, Salt Lake Gen Hosp, 55-58; dir premature ctr, Univ Hosp, Univ Wash, 60-63. *Mem:* AAAS; Am Acad Pediat; Endocrine Soc; NY Acad Sci; Am Soc Human Genetics. *Res:* Human biochemical genetics; cytogenetics; intrauterine diagnosis. *Mailing Add:* Genetics Diagnostics Ctr 711 Stanton Young Blvd Suite 405 Oklahoma City OK 73104

SEELY, JAMES ERVIN, b Marshalltown, Iowa, July 5, 54; m 78; c 1. POLYAMINE BIOSYNTHESIS. *Educ:* Univ SDak, Phd(biochem), 80. *Prof Exp:* Post-doctoral fel, Dept Physiol, Hershey Med Ctr, 81-84; res scientist, 84-89, SR RES SCIENTIST, DEPT BIOCHEM, PITMAN-MOORE, INC, 89- *Concurrent Pos:* Adj fac, Ind Univ Med Sch, 89- *Mem:* Am Soc Biochem & Molecular Biol; Biochem Soc; AAAS; Endocrine Soc. *Res:* Polypeptide growth factors; structure function studies on growth hormone and other polypeptide growth factors; endocrine growth regulation of livestock animals; folding recombinant proteins and polypeptides. *Mailing Add:* 7389 Troy Court Terre Haute IN 47802

SEELY, JUSTUS FRANDSEN, b Mt Pleasant, Utah, Feb 11, 41; m 65; c 3. STATISTICS, MATHEMATICS. *Educ:* Utah State Univ, BS, 63, MS, 65; Iowa State Univ, PhD(statist), 69. *Prof Exp:* Instr appl statist & comput sci, Utah State Univ, 64-65; PROF STATIST, ORE STATE UNIV, 69- *Mem:* Fel Inst Math Statist; fel Am Statist Asn; Int Statist Inst. *Res:* linear model theory; experimental design; mathematical statistcs. *Mailing Add:* Dept Statist Ore State Univ Kidder Hall 44 Corvallis OR 97331-4606

SEELY, SAMUEL, b New York, NY, May 7, 09; m 51; c 2. PHYSICS, ELECTRICAL ENGINEERING. *Educ:* Polytech Inst Brooklyn, EE, 31; Stevens Inst Technol, MS, 32; Columbia Univ, PhD(physics), 36. *Prof Exp:* Lectr physics, Polytech Inst Brooklyn, 35; from instr to asst prof elec eng, City Col New York, 36-46; assoc prof electronics, Grad Sch, US Naval Acad, 46-47; prof elec eng, Univ Syracuse, 47-56, chmn dept, 51-56; prof, Case Inst Technol, 56-64, head dept, 56-59; ed consult, 64-67; guest prof, Chalmers Univ Technol & Lund Inst Technol, Sweden, 67-68; vis prof elec eng, Univ Mass, Amherst, 68-69, coordr res & assoc grad dean, Grad Sch, 69-70; prof elec eng, Univ Conn, 71-72, Univ RI, 72-79; RETIRED. *Concurrent Pos:* Mem staff, Radiation Lab, Mass Inst Technol, 41-46; Fulbright lectr, Athens Polytech, 59-60; head eng sect, NSF, 61-63; vis prof, Johns Hopkins Univ, 65, City Col New York, 66 & Okla State Univ, 67. *Honors & Awards:* Silver Star, Order of Phoenix, 67. *Mem:* Fel Am Phys Soc; Am Soc Eng Educ; fel Inst Elec & Electronics Engrs. *Res:* Diamagnetism; nuclear physics; electronic circuits; microwave and longwave antennas; digital systems; author of 18 books on electrical engineering. *Mailing Add:* 37 Brainard Rd Westbrook CT 06498

SEEMAN, MARY VIOLETTE, b Lodz, Poland, Mar 24, 35; Can citizen; m 59; c 3. PSYCHIATRY. *Educ:* McGill Univ, BA, 55, MD, 60; FRCPCan, 68. *Prof Exp:* Res psychiatrist, Fullbourne Hosp, Cambridge, 65-67; psychiatrist, Toronto Western Hosp, 67-75; assoc prof, 75-80, PROF, UNIV TORONTO, 80-; CHIEF PSYCHIAT, MT SINAI HOSP, TORONTO, 85- *Concurrent Pos:* Chmn, Res Adv Bd, Can Friends of Schizophrenics; head, Active Treatment Clin, Clarke Inst Psychiat, 75- *Mem:* Am Col Psychiat; Can Med Asn. *Res:* Clinical research in schizophrenia. *Mailing Add:* Dept Psychiat Mt Sinai Hosp 600 University Ave Toronto ON M5G 1X5 Can

SEEMAN, NADRIAN CHARLES, b Chicago, Ill, Dec 16, 45. MOLECULAR BIOPHYSICS, X-RAY CRYSTALLOGRAPHY. *Educ:* Univ Chicago, BS, 66; Univ Pittsburgh, PhD(biochem, crystallog), 70. *Prof Exp:* Res assoc biol, Columbia Univ, 70-72; Damon Runyon Found fel, Mass Inst Technol, 72-73, NIH fel biophys, 73-76, res assoc biol, 76-77; from asst prof to assoc prof biol, State Univ NY, Albany, 77-88; PROF CHEM, NY UNIV, 88- *Concurrent Pos:* Consult, Lifecodes, Inc, 83-86 & Polyprobe; sr consult, Molecular Biophysics Technol, Inc, 83-87; res career develop award, 82-87. *Mem:* Am Crystallog Asn; Biophys Soc; Am Chem Soc; AAAS; NY Acad Sci; Am Soc Biochem & Molecular Biol. *Res:* Structure, dynamics, thermodynamics and applications of nucleic acid branched junctions; nucleic acid engineering. *Mailing Add:* Dept Chem NY Univ New York NY 10003

SEEMAN, PHILIP, b Winnipeg, Man, Feb 8, 34; m 59; c 3. NEUROPHARMACOLOGY, CELL BIOLOGY. *Educ:* McGill Univ, BSc, 55, MSc, 56, MD, 60; Rockefeller Univ, PhD(life sci), 66. *Prof Exp:* Intern med, Harper Hosp, Detroit, 60-61; from asst prof to assoc prof, 67-71, chmn dept, 77-87, PROF PHARMACOL, UNIV TORONTO, 71- *Concurrent Pos:* Med Res Coun Can fel, Univ Cambridge, 66-67; res awards, Clarke Inst Psychiat, Toronto, Can 75 & Clin Res Soc Toronto, 79; Upjohn award res, Can Pharmacol Soc, 80. *Honors & Awards:* Walter Murphey Lectr & Award, Can Soc Clin Pharmacol, 81; Lieber Prize, Nat Alliance Res Schizophrenia & Depression, 90; Stanley Dean Award, Am Col Psychiatrists, 91. *Mem:* AAAS; Am Soc Pharmacol & Exp Therapeut; Biophys Soc; Am Soc Cell Biol; Pharmacol Soc Can; Am Col Neuropsychopharmacol; fel Royal Soc Can. *Res:* Membrane biology; cell actions of anesthetics and tranquilizers; brain dopamine receptors; membrane ultrastructure; biology of schizophrenia. *Mailing Add:* Dept Pharmacol Univ Toronto Toronto ON M5S 2A8 Can

SEENEY, CHARLES EARL, b Jefferson City, Mo, Apr 2, 43; m 66; c 3. BUSINESS & PRODUCT DEVELOPMENT, MICROENCAPSULATION. *Educ:* Lincoln Univ, BS, 70; Univ Akron, MS, 74. *Prof Exp:* Polymer chemist polymer res, Calgon Corp, 74-76 & Ralston Purina Co, 76-78; supvr polymer res, IMC Corp, 78-80, mgr polymer res, 80-82, mgr macromolecular res, 82-84, dir chem res, 84-86, pres, Imcera Bioprod, 86-90; PROF MGR TECHNOL, KERR MCGEE CORP, 90- *Concurrent Pos:* Vis lectr, Urban League Beep Prog, 79-82; managing dir, ViCorp Exec, Inc, 89-; indust consult, Phenolic Thermoset Adhesives, 89-; gov appointee, Ind Health Policy Comt, 90. *Mem:* Am Chem Soc; Soc Plastics Engrs; Human Resource Asn. *Res:* Developing and linking technology to the marketplace; business enterprises from technological concepts in areas of biotechnology and polymer adhesives; awarded 16 patents. *Mailing Add:* 2504 Meadow View Edmond OK 73013

SEERLEY, ROBERT WAYNE, b Indianapolis, Ind, Oct 6, 30; m 51; c 2. ANIMAL SCIENCE, ANIMAL NUTRITION. *Educ:* Purdue Univ, BS, 52; Mich State Univ, MS, 57, PhD(animal nutrit), 60. *Prof Exp:* Exten swine specialist, Purdue Univ, 62-54; asst, Mich State Univ, 56-60; from asst prof to assoc prof animal sci, SDak State Univ, 60-67; assoc prof, 67-80, PROF ANIMAL SCI, UNIV GA, 80- *Concurrent Pos:* AID consult, Korea, 65. *Mem:* Am Soc Animal Sci. *Res:* Amino acids, minerals and vitamins in animal nutrition; building design and space requirements for animal environment. *Mailing Add:* Dept Animal Sci Livestock-Poultry Bldg 209 Univ Ga Athens GA 30601

SEERY, DANIEL J, b Philadelphia, Pa, Dec 17, 33; m 60; c 3. CHEMICAL KINETICS, FUEL SCIENCE. *Educ:* St John's Univ, BS, 55; Pa State Univ, MS, 58, PhD(fuel technol), 62. *Prof Exp:* Fel chem, Univ Minn, 62-64; res scientist, United Technol Res Ctr, 64-66, sr res scientist kinetics & heat transfer group, 66-77, prin scientist combustion sci sect, 78-81, mgr combustion sci, 81-90, SR PROG MGR, ENVIRON SCI, UNITED TECHNOL RES CTR, 90- *Mem:* Am Chem Soc; Am Phys Soc; Combustion Inst. *Res:* Chemical kinetics of combustion behind shock waves and in flames; dissociation of simple molecules; heterogeneous combustion in dust flames; catalytic combustion and combustion generated air pollution; coal devolatilization. *Mailing Add:* Silver Lane United Technol Res Ctr East Hartford CT 06108

SEERY, VIRGINIA LEE, b New York, NY, July 23, 34. SPECTROSCOPY, PHYSICAL METHODS. *Educ:* Seton Hill Col, AB, 56; Duke Univ, MA, 58; Univ Wash, PhD(biochem), 68. *Prof Exp:* Res assoc, Ore State Univ, 68-70; NIH trainee, Scripps Clin & Res Found, 70-72; asst prof biochem, Emory Univ, 72-79; assoc prof biochem, Western Ill Univ, 79-83; res chemist, USDA Eastern Reg Res Ctr, 83-87; ASSOC PROF BIOCHEM, PHILADELPHIA COL OSTEOP MED, 87- *Mem:* Am Chem Soc; AAAS. *Res:* Structure, function and interactions of proteins and enzymes; binding of small molecules to macromolecules; allosteric effectors; photoaffinity labeling; application of physical methods such as ultracentrifugation, fluorescence, circular dichroism and ultraviolet spectroscopy. *Mailing Add:* Philadelphia Col Osteop Med 4150 City Ave Philadelphia PA 11367-0904

SEESE, WILLIAM SHOBER, b Meyersdale, Pa, June 13, 32; m 58; c 2. BASIC CHEMISTRY. *Educ:* Univ NMex, BS, 54, MS, 59; Wash State Univ, PhD(chem), 65. *Prof Exp:* Instr chem, biol & math, Ft Lewis Col, Durango, Colo, 58-61; res biochemist, Int Minerals & Chem Corp, Wasco, Calif, 65-66; instr chem & pharmacol, Casper Col, Wyo, 66-87; PROF CHEM, ALICE LLOYD COL, PIPPA PASSES, KY, 89-91. *Concurrent Pos:* Assoc prof, King Fahd Univ Petrol & Minerals, Dhahran, Saudi Arabia, 73-76; asst prof, Univ NMex, Gallup, 76-77; Fulbright lectr, Omdurman Islamic Univ, Sudan, 87-88. *Mem:* Am Chem Soc. *Res:* Organic synthesis-heterocyclics, steriods, aromatics and carbohydrates. *Mailing Add:* 2915 Ridgecrest Dr Casper WY 82604

SEETHARAM, RAMNATH (RAM), b Kerala, India, Jan 9, 52; US citizen; m 81; c 2. PROTEIN CHEMISTRY, PEPTIDE CHEMISTRY. *Educ:* Univ Bombay, India, BSc, 72, MSc, 75; Indian Inst Sci, PhD(biochem), 81. *Prof Exp:* Res assoc protein chem, Rockefeller Univ, 82-84; sr res biochemist biochem, Monsanto Co, 84-88; sr res biochemist biochem, Du Pont Co, 88-91, SR RES BIOCHEMIST BIOCHEM, DU PONT MERCK PHARMACEUT CO, 91- *Mem:* Am Soc Biochem & Molecular Biol; Protein Soc. *Res:* Structure-function relationships in proteins; protein-protein interactions; protein folding; protein sequencing, peptide synthesis and chemical methods of analyzing proteins. *Mailing Add:* Du Pont Merck Pharmaceut Co Exp Sta E400/5249 Newark DE 19880-0400

SEEVERS, DELMAR OSWELL, b St John, Kans, June 26, 19; m 43; c 1. PHYSICS. *Educ:* Duke Univ, BS, 41, PhD, 51. *Prof Exp:* Physicist, Bur Ord, USN, 41-45; teaching assoc physics, Duke Univ, 46-51; res physicist, 51-55, sr res physicist, 55-56, res assoc, 56-60, SR RES ASSOC PHYSICS, CHEVRON RES CO, 60- *Mem:* Am Phys Soc. *Res:* Cosmic rays; nuclear and neutron physics; nuclear magnetic and electron paramagnetic resonance; physics of solid-liquid interfaces. *Mailing Add:* 832 Glenwood Circle Fullerton CA 92632

SEEVERS, ROBERT EDWARD, b Okanogan, Wash, Mar 18, 35; m 56; c 4. PHYSICAL CHEMISTRY. *Educ:* Portland State Col, BS, 63; Ore State Univ, PhD(chem), 68. *Prof Exp:* Sr analyst, Reynolds Metals Co, 60-61; chemist, Ore Steel Mills, 61-63; res asst chem, Ore State Univ, 66-67; asst prof, 67-74, assoc prof, 74-80, PROF CHEM, SOUTHERN ORE COL, 80- *Mem:* Am Chem Soc. *Res:* Solid state chemistry, especially electrical properties of alkali-halide crystals and semiconductors; quantum chemistry. *Mailing Add:* Dept Chem & Computer Sci Southern Ore Col 1250 Siskiyou Blvd Ashland OR 97520

SEFF, KARL, b Chicago, Ill, Jan 23, 38; m 83. ZEOLITE CRYSTALLOGRAPHY, INTRAZEOLITIC CHEMISTRY. *Educ:* Univ Calif, Berkeley, BS, 59; Mass Inst Technol, PhD(phys chem), 64. *Prof Exp:* Scholar chem, Mass Inst Technol, 64; scholar, Univ Calif, Los Angeles, 64-67, asst res chemist, 67-68; from asst prof to assoc prof, 68-75, PROF CHEM, UNIV HAWAII, 75- *Mem:* Am Crystallog Asn; Am Chem Soc. *Res:* Intrazeolitic chemistry and zeolite complex structure; structures of molecules of organic or biochemical interest; computer techniques. *Mailing Add:* Dept Chem Univ Hawaii Honolulu HI 96822

SEFF, PHILIP, b New York, NY, Oct 5, 23. FAULTING, EARLY MAN IN CALIFORNIA. *Educ:* Brooklyn Col, BS, 50; Univ Nebr, MS, 53; Univ Ariz, PhD(geol), 62. *Prof Exp:* Prof geol, Univ Redlands, 66-70; CONSULT, PHILIP SEFF & ASSOC, 70- *Concurrent Pos:* Res grant, NSF, 69; geol adv, Park Servs Petrified Forest Nat Park. *Mem:* Geol Soc Am; Sigma Xi. *Mailing Add:* 13020 South Lane Redlands CA 92373

SEFFL, RAYMOND JAMES, b Chicago, Ill, Sept 21, 27; m 51; c 2. ORGANIC CHEMISTRY. *Educ:* Univ Ill, BS, 50; Univ Colo, PhD(org chem), 54. *Prof Exp:* Jr res chemist, Velsicol Corp, 50-51; sr res chemist, M W Kellog Co div, Pullman, Inc, 54-57; chief chemist, Titan Chem Co, 57-58; LAB MGR, CHEM DIV, 3M CO, 58- *Mem:* Am Chem Soc. *Res:* Organic fluorine chemistry; synthesis and commerical product development. *Mailing Add:* 4941 Olson Terr N Lake Elmo MN 55042

SEGA, GARY ANDREW, b Cleveland, Ohio, Mar 23, 41; m 71; c 3. MOLECULAR GENETICS. *Educ:* Case Inst Technol, BS, 63; Univ Tex, Austin, MA, 66; La State Univ, Baton Rouge, PhD(genetics), 71. *Prof Exp:* Investr, 71-73, RES STAFF MEM MOLECULAR GENETICS, BIOL DIV, OAK RIDGE NAT LAB, 73- *Mem:* AAAS; Environ Mutagen Soc; Sigma Xi. *Res:* Molecular mechanisms of mutation induction in the mouse, including dosimetry of chemical mutagens in the germ cells and DNA repair. *Mailing Add:* 106 Clark Lane Oak Ridge TN 37830

SEGAL, ALAN H, b Pittsburgh, Pa, Dec 5, 21; m 48; c 2. DENTISTRY. *Educ:* Univ Pittsburgh, BS, 45, DDS, 46, MDS, 70. *Prof Exp:* Assoc prof dent anat, 59-70, prof grad periodont, 71-89, EMER PROF, SCH DENT, UNIV PITTSBURGH, 89- *Concurrent Pos:* Res grant, 63-66; mem, Coun Med TV. *Mem:* Am Dent Asn; Am Col Dentists; Am Acad Periodontists. *Res:* Dental anatomy; preclinical operative dentistry. *Mailing Add:* Passavant Prof Bldg 9102 Badcock Blvd SW 205 Pittsburgh PA 15237

SEGAL, ALEXANDER, b Novograd-Volynsky, USSR, Oct 10, 34; m 60; c 1. ACOUSTICS & ELECTROACOUSTICS. *Educ:* Polytech Inst Kiev, MS, 58; Inst Textile & Light Indust Leningrad, PhD(noise abatement), 73. *Prof Exp:* Sr elec engr, State Proj Inst Commun Kiev, 66-68; sr sci assoc, Sci Res Inst Labor Hyg, Kiev, 68-76; mem tech staff acoust, Wyle Lab, 78-79; environ mgt specialist, 79-86, ACOUT ENGR, COUNTY OF SAN DIEGO, 86- *Concurrent Pos:* Consult acoust eng, Country Kiev Noise Control Dept Kiev, 71-76 & San Diego, 80-; expert, Proj Inst Kievproject, 73-76; adj prof, San Diego State Univ, 82- *Mem:* Acoust Soc Am; Sigma Xi. *Res:* Community and industrial noise standards; effect of noise on worker productivity; aircraft acoustics; hearing protection; community and industrial noise abatement and control; environmental sciences. *Mailing Add:* 5222 Trojan Ave 316 San Diego CA 92115

SEGAL, ALVIN, b New York, NY, Mar 21, 29; m 58, 84, 89; c 2. BIOCHEMISTRY, CANCER. *Educ:* Long Island Univ, BS, 51 & 58; NY Univ, MS, 61, PhD(org chem), 65. *Prof Exp:* Pharmacist, Univ Hosp, New York, 58-59; org chemist, Ortho Pharmaceut Corp, 59-61; teaching fel chem, NY Univ, 61-63, univ fel, 63-64, USPHS fel environ med, Col Med, 64-66; asst prof pharmacog, Col Pharm, Univ Tenn, Memphis, 66-68; assoc res scientist environ med, 68-70, res scientist, 70-73, sr res scientist, 73-76, res assoc prof, 76-80, RES PROF ENVIRON MED, MED CTR, NY UNIV, 80- *Mem:* AAAS; Am Chem Soc; Am Asn Cancer Res. *Res:* Steroids; natural products chemistry; mechanisms of chemical carcinogenesis; major research has been in mechanisms of chemical carcinogenesis; minor research interests; steroids; natural products chemistry. *Mailing Add:* Dept Environ Med NY Univ Med Ctr 550 First Ave New York NY 10016

SEGAL, ARTHUR CHERNY, b Newark, NJ, July 22, 38; c 2. MATHEMATICS. *Educ:* Univ Fla, BS, 58; Univ Ariz, MS, 62; Tex Christian Univ, PhD(math), 66. *Prof Exp:* Physicist, ARO, Inc, Tenn, 58-59; asst prof math, Judson Col, 62-64, Univ Tex, Arlington, 65-66 & Tex A&M Univ, 66-67; chmn dept, 75-78, ASSOC PROF MATH, UNIV ALA, BIRMINGHAM, 67- *Mem:* Math Asn Am. *Res:* Biomathematics. *Mailing Add:* Dept Math-Biomath Univ Ala Univ Sta Birmingham AL 35294

SEGAL, BERNARD, b New York, NY, May 9, 36; m 62; c 2. ADDICTION STUDIES. *Educ:* City Col NY, BBA, 60, MSE, 63; Univ Okla, PhD(psychol), 67. *Prof Exp:* Asst prof psychol, Univ RI, 67-70; prof psychol, Murray State Univ, 70-77; PROF HEALTH SCI, UNIV ALASKA, ANCHORAGE, 77- *Concurrent Pos:* Police psychologist, Anchorage Police Dept, 90- *Mem:* Am Pub Health Asn; Am Soc Circumpolar Health. *Res:* Etiology, epidemiology, and psychsocial correlates of drug- taking behavior; biobehavioral factor in alcoholism. *Mailing Add:* Sch Health Sci 3211 Providence Dr Anchorage AK 99508

SEGAL, BERNARD L, b Montreal, Que, Feb 13, 29; m 63. INTERNAL MEDICINE, CARDIOLOGY. *Educ:* McGill Univ, BSc, 50, MD, CM, 55; Am Bd Internal Med, dipl, 63; Am Bd Cardiovasc Dis, dipl, 64. *Prof Exp:* Asst med, Sch Med, Johns Hopkins Univ, 56-57; clin asst, St George's Hosp, London, 59-60; assoc med, 61-62, assoc prof, 62-72, dir, Likoff Cardiovasc Inst, 80-86, PROF MED, HAHNEMANN MED COL & HOSP, 72-; DIR, PHILADELPHIA HEART INST, 86- *Concurrent Pos:* Teaching fel, Harvard Med Sch, 57-58 & Sch Med, Georgetown Univ, 58-59; USPHS fel, St George's Hosp, London, Eng, 59-60; Southeast Heart Asn Pa grants, 61-63 & 64-65; NIH grant, 62-64, res grant, 65-68; jr attend, Vet Admin Hosp & Hahnemann Med Col & Hosp, 61-; consult, 62-63; ed, Eng & Pract Med & Theory & Prac Auscultation; head, Auscultation Unit, Hahnemann Med Col & Hosp, 66-86; clin prof med Univ Pa. *Mem:* Fel Am Col Physicians; fel AMA; fel Am Col Cardiol; fel Am Col Chest Physicians; fel NY Acad Sci. *Res:* Atherosclerosis and coronary heart disease. *Mailing Add:* 1320 Race St Philadelphia PA 19102

SEGAL, BERNICE G, physical chemistry; deceased, see previous edition for last biography

SEGAL, DAVID MILLER, MOLECULAR & CELLULAR IMMUNOLOGY, CANCER RESEARCH. *Educ:* Johns Hopkins Univ, PhD(biochem), 66. *Prof Exp:* SR INVESTR, IMMUNOL BR, NAT CANCER INST, 75- *Mailing Add:* Immunol Br Nat Cancer Inst-NIH Bldg 10 Rm 4B17 Bethesda MD 20892

SEGAL, DAVID S, b Montreal, Que, Aug 7, 42; US citizen; m 63; c 3. NEUROPHARMACOLOGY, NEUROPSYCHOLOGY. *Educ:* Univ Calif, Santa Barbara, BA, 65; Univ Calif, Irvine, PhD(psychobiol), 70. *Prof Exp:* NIMH fel, 70-72, from asst prof to assoc prof, 72-78, PROF PSYCHIAT, UNIV CALIF, SAN DIEGO, 78- *Concurrent Pos:* Regional ed, Pharmacol, Biochem & Behav, 73-; NIMH res scientist develop award, 73-78. *Mem:* AAAS; Soc Neurosci. *Res:* Neurochemical substrates of arousal; drug-induced changes in brain biosynthetic enzymes in response to environmental changes; long-term effects of drugs on behavior and neurochemical mechanisms of adaptation. *Mailing Add:* Dept Psychiat M-003 Univ Calif San Diego Box 109 La Jolla CA 92093

SEGAL, EARL, b New York, NY, Dec 29, 23; m 54; c 2. ENVIRONMENTAL PHYSIOLOGY. *Educ:* Univ Southern Calif, BA, 49; Univ Calif, Los Angeles, MA, 53, PhD(zool), 55. *Prof Exp:* From asst prof to assoc prof biol, Kans State Teachers Col, 55-60; asst prof, Rice Univ, 60-63; assoc prof, 63-72, PROF BIOL, CALIF STATE UNIV, NORTHRIDGE, 72- *Concurrent Pos:* NSF lectr, NMex Highlands Univ, 58; lectr, Woods Hole Marine Biol Lab, 61; lectr, Marine Lab, Duke Univ, 62 & 63; coordr & lectr oceanog, US Naval Missile Range, Point Mugu, 65; Fulbright lectr, Penang, Malaysia, 71-72; US Info Serv lectr, SE Asia, 72; chmn bd govs, Southern Calif Ocean Studies Consortium, 76- *Mem:* Sigma Xi; Am Soc Zoologists; Marine Biol Asn UK. *Res:* Physiological ecology of poikilotherms; acclimation and physiological responses to stress. *Mailing Add:* PO Box 414 Holualoa HI 96725

SEGAL, GERALD A, b Pittsburgh, Pa, Dec 1, 34; m 79; c 6. THEORETICAL CHEMISTRY. *Educ:* Amherst Col, AB, 56; Carnegie Inst Technol, PhD(chem), 66. *Prof Exp:* NSF fel, Bristol Univ, 66-67; from asst prof to assoc prof, 67-75, PROF CHEM & CHMN DEPT, UNIV SOUTHERN CALIF, 75-, DEAN, COL LETT, ARTS & SCI, 89- *Concurrent Pos:* Sloan Found fel, 71-; sr Fulbright fel, France, 73-74. *Mem:* Am Phys Soc; Am Chem Soc. *Res:* Molecular orbital theory; theoretical spectroscopy. *Mailing Add:* Admin 200 Univ Southern Calif Los Angeles CA 90089-4012

SEGAL, HAROLD JACOB, b Winnipeg, Man, Mar 6, 41. PHARMACY ADMINISTRATION. *Educ:* Univ Man, BScPharm, 62; Purdue Univ, MS, 66, PhD(pharm admin), 68. *Prof Exp:* Pharmacist, Crescentwood Pharm Ltd, 62-65; res assoc & Can Dept Nat Health & Welfare fel, Comn Pharmaceut Serv, Toronto, 68-70; PROF PHARM, FAC PHARM, UNIV TORONTO, 70-, FAC MED, DIV COMMUNITY HEALTH, DEPT HEALTH ADMIN, 75- *Res:* Pharmaceutical and health product marketing; health care delivery systems planning; economics of health care. *Mailing Add:* Koffler Inst Pharm Mgt Univ Toronto Toronto ON M5S 1A1 Can

SEGAL, HAROLD LEWIS, b New York, NY, Nov 18, 24; m 45; c 2. BIOCHEMISTRY. *Educ:* Carnegie Inst Technol, BS, 47; Univ Minn, MS, 50, PhD(biochem), 52. *Prof Exp:* Instr, Univ Minn, 51-52; res assoc zool, Univ Calif, Los Angeles, 52-54; asst prof biochem, Univ Pittsburgh, 54-59; assoc prof pharmacol, St Louis Univ, 59-64; chmn dept, 64-67, grad dir, 80-84, PROF CELL & MOLECULAR BIOL, STATE UNIV NY, BUFFALO, 64- *Mem:* AAAS; Am Chem Soc; Am Soc Biol Chemists; NY Acad Sci. *Res:* Metabolic regulation. *Mailing Add:* Biol Dept State Univ New York Buffalo NY 14260

SEGAL, IRVING EZRA, b New York, NY, Sept 13, 18. MATHEMATICS, THEORETICAL PHYSICS. *Educ:* Princeton Univ, AB, 37; Yale Univ, PhD(math), 40. *Prof Exp:* Instr math, Harvard Univ, 41; res assoc, Princeton Univ, 42-43; asst, Inst Advan Study, 46, Guggenheim fel, 46-47; from asst prof to prof, Univ Chicago, 48-60; PROF MATH, MASS INST TECHNOL, 60- *Concurrent Pos:* Guggenheim fel, 51-52 & 67-68; vis assoc prof, Columbia Univ, 53-54; vis prof, Univ Paris, 65, State Univ Col Pisa, 72, Lund Univ & Univ Copenhagen, 71-72. *Honors & Awards:* Humboldt Award. *Mem:* Nat Acad Sci; Am Math Soc; Am Phys Soc; Am Astron Asn. *Res:* Harmonic analysis; operator rings in Hilbert space; analysis in infinite-dimensional spaces; quantum field and particle theory; astrophysics; nonlinear relativistic partial differential equations. *Mailing Add:* Rm 2-244 Mass Inst Technol Cambridge MA 02139

SEGAL, JACK, b Philadelphia, Pa, May 9, 34; m 55; c 2. TOPOLOGY. *Educ:* Univ Miami, BS, 55, MS, 57; Univ Ga, PhD(math), 60. *Prof Exp:* From instr to assoc prof, 60-70, chmn dept, 75-78, PROF MATH, UNIV WASH, 70- *Concurrent Pos:* NSF fel, 63-64; Fulbright fel, 69-70; exchange prof, NAS, 79-80. *Mem:* Am Math Soc; Math Asn Am. *Res:* Point-set topology; manifolds; dimension and fixed point theory; mapping; abstract spaces; inverse limit spaces; shape theory. *Mailing Add:* Dept Math Univ Wash Seattle WA 98195

SEGAL, MARK, b Montreal, Que, Sept 21, 35; m 59; c 3. PHARMACOLOGY, FOOD SCIENCE & TECHNOLOGY. *Educ:* McGill Univ, BSc, 56, MSc, 57, PhD(pharmacol), 60. *Prof Exp:* NIMH fel pharmacol, Univ Mich, 60-62; sr pharmacologist, Ayerst, McKenna & Harrison Ltd, 62-64; from asst prof to assoc prof pharmacol, Dalhousie Univ, 64-73; sr res pharmacologist, Psychiat Res Lab, Dept Psychiat, Hadassah Med Orgn, 74-83; dir, Psychopharmacokinetic Lab, Jerusalem Ment Health Ctr & dir res, Jerusalem Ctr for Drug Abuse Intervention, 83-84; PRES & DIR, SCI CONSULT AGENCY, JERUSALEM, 85- *Concurrent Pos:* vis prof pharmacol, Hadassah Med Sch, Hebrew Univ, Jerusalem, 71-72; vis prof pharmacol & exp therapeut, Sch Med, Boston Univ, 74; ed-in-chief, Revs Pure Appl Pharmaceut Sci; co-ed, Perspectives in Drug Abuse, 85- *Mem:* Am Soc Pharmacol & Exp Therapeut; Int Soc Biochem Pharmacol; Israel Soc Physiol Pharmacol. *Res:* Scientific writing and editing (medical topics, e.g., antioxidants and cancer, medicinal aspects of the Cannabinoids, and medicocosmetic science); mechanisms of tolerance and physical dependence; psychochemicals; interaction between tetrahydrocannabinols (THC) and other psychochemicals; pharmacology and biochemistry of drugs influencing sexual behavior; analgesic effectiveness of Cannabinoids; drug screening on a private basis; nutrition in relation to the provision of foods and diets for sufferers of cardiovascular diseases. *Mailing Add:* Sci Consult Agency 611/16 Adam St Maalot Moriah Jerusalem 93782 Israel

SEGAL, MOSHE, b Haifa, Israel, June 3, 34; m 61; c 3. OPERATIONS RESEARCH. *Educ:* Israel Inst Technol, BS, 56, Ing, 56; Johns Hopkins Univ, DEng, 61. *Prof Exp:* Mem tech staff, 61-64, supvr opers res methodology, 64-77, SUPVR OPERS RES TECH, BELL TEL LABS, INC, 77- *Concurrent Pos:* Vis scientist, Lady Davies fel, tech, Israel Inst Technol, 76-77. *Mem:* Opers Res Soc Am. *Res:* Queueing theory; communications networks; linear and non-linear mathematical programming; optimization techniques for the design of networks. *Mailing Add:* 47 Patridge Lane Tinton Falls NJ 07724

SEGAL, ROSALIND A, New York, NY, June 30, 58. NEUROLOGY. *Educ:* Rockefeller Univ, PhD(cell biol), 85; Cornell Univ, MD, 86. *Prof Exp:* Intern, 80-91, CLIN FEL NEUROL, BETH ISRAEL HOSP, BOSTON, 90- *Mem:* Am Soc Cell Biol. *Mailing Add:* Childrens Hosp Fagen 11 300 Longwood Ave Boston MA 02115

SEGAL, SANFORD LEONARD, b Troy, NY, Oct 11, 37; m 59; c 3. MATHEMATICS. *Educ:* Wesleyan Univ, BA, 58; Univ Colo, PhD(math), 63. *Prof Exp:* From instr to assoc prof math, 63-77, chmn dept, 79-87, PROF MATH, UNIV ROCHESTER, 77- *Concurrent Pos:* Fulbright res fel, Univ Vienna, 65-66; vis lectr, Univ Nottingham, Eng, 72-73; res grantee, Fed Univ, Rio de Janeiro, Brazil, 82; res grant, Alexander von Humboldt Found, WGer, 88. *Mem:* AAAS; Am Math Soc; Math Asn Am; Hist Sci Soc. *Res:* Elementary and analytic number theory; functional equations; complex functions of one variable; history of science, particularly 20th century german mathematics. *Mailing Add:* Dept Math Univ Rochester Rochester NY 14627

SEGAL, STANTON, b Camden, NJ, Sept 6, 27; m 56; c 2. MEDICINE, BIOCHEMISTRY. *Educ:* Princeton Univ, AB, 48; Harvard Med Sch, MD, 52. *Hon Degrees:* MA, Univ Pa, 71. *Prof Exp:* Res assoc, Sch Med, Univ Pa, 49-50; intern, Med Ctr, Cornell Univ, 52-53; resident med, Hosp Univ Pa, 53-54; clin assoc, NIH, 54-57; resident med, Hosp Univ Pa, 57-58; NIH sr investr, 58-65, chief sect diabetes & intermediary metab, Nat Inst Arthritis & Metab Dis, 65-66; PROF PEDIAT & CHIEF LAB MOLECULAR DIS & METAB, SCH MED, UNIV PA, 66-, PROF MED, 70-, ATTEND PHYSICIAN, HOSP, 70- *Concurrent Pos:* Vis scientist, Nat Inst Med Res, London, Eng, 63-64; mem metab study sect, NIH, 67-71, mem metab & diabetes training comt, Nat Inst Arthritis, Metab & Digestive Dis, 71-73; mem ment retardation comt, Nat Inst Child Health & Human Develop, 75-79; sr physician, Children's Hosp, Philadelphia. *Mem:* Asn Am Physicians; Am Soc Clin Invest; Endocrine Soc; Am Soc Biol Chemists; Brit Biochem Soc. *Res:* Intermediary metabolism; endocrinology; human genetics; inherited metabolic diseases. *Mailing Add:* Univ Pa 34th & Civic Ctr Blvd Children's Hosp Philadelphia PA 19104

SEGAL, WILLIAM, b Montreal, Que, Dec 22, 22; nat US; m 51; c 2. MICROBIOLOGY. *Educ:* McGill Univ, BS, 43; Univ Wis, MS, 48; Rutgers Univ, PhD(microbiol), 52. *Prof Exp:* Bacteriologist, Clin Bact Lab, Royal Victoria Hosp, 44-45; asst bacteriologist, Univ Wis, 47-48; vis investr, Pub Health Res Inst, NY, 52-57; asst prof microbiol, Sch Med, 58-66, assoc prof biol, 66-68, PROF BIOL, UNIV COLO, BOULDER, 68- *Concurrent Pos:* Vis scientist, Sch Med, Hebrew Univ, Israel, 68; NIH spec res fel, Nat Ctr Sci Res, France, 68-69. *Mem:* AAAS; Am Soc Microbiol; Brit Soc Gen Microbiol; Sigma Xi. *Res:* Intermediates in bacterial nitrogen-fixation; bacterial transformation of organic sulfur compounds; biochemistry and genetics of mycobacteria; biochemistry of tuberculous host-parasite interrelationship; pathogenicity and immunogenicity of tubercle bacillus; microbial ecology. *Mailing Add:* 540 N Lake Shore Dr Apt 506 Univ Colo CB 334 Chicago IL 60611

SEGALL, BENJAMIN, b New York, NY, July 23, 25; m 53; c 3. THEORETICAL PHYSICS. *Educ:* Brooklyn Col, BS, 48; Univ Ill, MS, 49, PhD(physics), 51. *Prof Exp:* Fel, Univ Ill, 51-52; fel, Copenhagen Inst Theoret Physics, Denmark, 52-53; res physicist, Radiation Lab, Univ Calif, 53-54; res physicist, Res & Develop Ctr, Gen Elec Co, NY, 55-68; PROF PHYSICS, CASE WESTERN RESERVE UNIV, 68- *Mem:* Fel Am Phys Soc. *Res:* Theoretical solid state and nuclear physics. *Mailing Add:* Dept Physics Case Western Reserve Univ 2040 Adelbert Rd Cleveland OH 44106

SEGALL, HAROLD NATHAN, clinical medicine; deceased, see previous edition for last biography

SEGALL, PAUL EDWARD, b New York, NY, Sept 28, 42. PROFOUND HYPOTHERMIA, AGE-DELAYING EFFECTS OF NUTRITIONAL RESTRICTION. *Educ:* State Univ NY, Stony Brook, BS, 63; Hofstra Univ, Hempstead, MA, 65; Univ Calif, Berkeley, PhD(physiol), 77. *Prof Exp:* Post-grad res physiologist, Dept Physiol Anat, Univ Calif, Berkeley, 78-80, asst res physiologist, 86; lectr human develop, Calif State Univ, Hayward, 80-81; res scientist, Biophys Res & Develop, Berkeley, Calif, 82-85; res dir, Cryomed Sci Inc, Bethesda, Md, 87-89, res scientist, 89-90; CHIEF EXEC OFFICER, BIO TIME INC, BERKELEY, CALIF, 90- *Concurrent Pos:* Lectr, Am Prog Bur, 87- *Mem:* Am Physiol Soc; Am Aging Asn. *Res:* Hypothermia; cryobiology; gerontology; author of scientific efforts to extend the human life span. *Mailing Add:* Bio Time Inc 1442 A Suite 474 Berkeley CA 94709

SEGALL, STANLEY, b Baltimore, Md, May 12, 30; m 54; c 2. ORGANIC CHEMISTRY, FOOD TECHNOLOGY. *Educ:* Northeastern Univ, SB, 53; Mass Inst Technol, 54-57, PhD(food tech), 57. *Prof Exp:* Chief chemist, Blue Seal Extract Co, Mass, 51-54; dir res, Rudd-Melikian, Inc, 57-68; ASSOC PROF BIOL & ENVIRON SCI, DREXEL UNIV, 68-, ASSOC PROF NUTRIT & FOOD & HEAD DEPT, 74- *Mem:* Am Chem Soc; Inst Food Technol; Sigma Xi. *Res:* Environmental and toxicological studies; taste and odor; flavor chemistry; food processing; public health aspects of automated industrial feeding. *Mailing Add:* 608 Rittenhouse Lane Wayne PA 19087

SEGALL, STEPHEN BARRETT, b Newark, NJ, Oct, 2, 42; m 65; c 4. PHYSICS. *Educ:* Columbia Col, BA, 64; Univ Md, MSc, 72, PhD(physics), 72. *Prof Exp:* PRIN SCIENTIST, FREE ELECTRON LASERS & DIR RES, KMS FUSION, INC, 72- *Res:* Laser-plasma interaction physics and free electron laser development. *Mailing Add:* 1349 King George Blvd Ann Arbor MI 48104

SEGAR, WILLIAM ELIAS, b Indianapolis, Ind, Dec 16, 23; m 54; c 2. PEDIATRICS. *Educ:* Ind Univ, BS, 44, MD, 47; Am Bd Pediat, dipl, 55. *Prof Exp:* Instr pediat, Yale Univ, 51-53; from asst prof to prof, Sch Med, Ind Univ, 55-67; prof, Mayo Grad Sch Med, 67-70; chmn dept, 74-85, PROF PEDIAT, UNIV WIS-MADISON, 70- *Mem:* AAAS; Soc Pediat Res; Am Pediat Soc; Am Fedn Clin Res; Asn Am Physicians. *Res:* Water and electrolyte metabolism; renal physiology. *Mailing Add:* Dept Pediat H4 458 Clin Sci Ctr Univ Wis Madison WI 53706

SEGARRA, JOSEPH M, neurology; deceased, see previous edition for last biography

SEGATTO, PETER RICHARD, b New York, NY, June 4, 28; m 54; c 3. ANALYTICAL CHEMISTRY, PHYSICAL CHEMISTRY. *Educ:* Univ Adelphi, AB, 53; Rutgers Univ, PhD(chem), 58. *Prof Exp:* Res chemist, 57-62, res chem assoc, 62-69, res assoc electronic mat & prod develop, 69-72, mgr tech prod dept, 72-77, portfolio mgr electronic/elec projs, 77-78, dir prod develop, 80-84, DIR TECH STAFF SERV, CORNING GLASS WORKS, 78-, DIR ELECTRONIC TECHNOL, 84- *Mem:* Am Chem Soc; Electrochem Soc. *Res:* Electrochemical analysis; thermodynamics; instrumental methods of analysis; instrumentation; electronic materials research; dielectrics; resister and magnetic materials; product development; vacuum technology; environmental chemistry; catalysis. *Mailing Add:* 1060 W Water St Elmira NY 14905

SEGEL, IRWIN HARVEY, b Staten Island, NY, Dec 29, 35; c 2. BIOCHEMISTRY. *Educ:* Rensselaer Polytech Inst, BS, 57; Univ Wis, MS, 60, PhD(biochem), 62. *Prof Exp:* NSF fel, 62-63; USPHS res fel, 63-64; from asst prof to assoc prof, 64-73, PROF BIOCHEM, UNIV CALIF, DAVIS, 73- *Mem:* AAAS; Am Soc Biol Chemists; Am Soc Plant Physiol; Am Chem Soc; Am Soc Microbiol. *Res:* Microbial biochemistry; enzymology and regulation of sulfur and nitrogen metabolism in microorganisms; membrane transport systems; enzyme kinetics; Author of a book,. *Mailing Add:* Dept Biochem & Biophys Univ Calif Davis CA 95616

SEGEL, L(EONARD), b Cincinnati, Ohio, Apr 16, 22; m 44; c 3. AUTOMOTIVE ENGINEERING. *Educ:* Univ Cincinnati, BSAE, 47; State Univ NY, Buffalo, MS, 53. *Prof Exp:* Res engr, Flight Res Dept, Cornell Aeronaut Lab, Inc, 47-56, prin engr, Vehicle Dynamics Dept, 56-57, head res sect, 57-60, asst dept head, 60-63, staff scientist, Appl Mech Dept, 63-66; from lectr to prof, 67-87, head, Eng Res Div, Transp Res Inst, EMER PROF MECH ENG, UNIV MICH, ANN ARBOR, 87- *Concurrent Pos:* Technion, Israel Inst Technol, 78, Tokyo Univ, 85; foriegn scholar grant, Int Asn Traffic & Safety Sci, 85; vis res fel, Japan Soc Prom Sci, 75. *Honors & Awards:* Crompton-Lanchester Medal, Brit Inst Mech Engrs, 58; Sci Contrib Prize, Soc Automotive Engrs Japan, 90. *Mem:* Fel Am Soc Mech Engrs; Am Inst Aeronaut & Astronaut; Human Factors Soc; Am Soc Testing & Mat; fel Inst Mech Engrs; Int Asn Vehicle Syst Dynamics, (pres, 87-). *Res:* Flight mechanics; vehicle stability and control; man-vehicle relationships; mobility of off-road vehicles; linear systems; tire mechanics; rotary-wing phenomena; tire-vehicle system dynamics. *Mailing Add:* Transp Res Inst Univ Mich Huron Pkwy & Baxter Rd Ann Arbor MI 48109-2150

SEGEL, LEE AARON, b Boston, Mass, Feb 5, 32; m 58; c 4. APPLIED MATHEMATICS, THEORETICAL BIOLOGY. *Educ:* Harvard Univ, AB, 53; Mass Inst Technol, PhD(math), 59. *Prof Exp:* Res fel, Aerodyn Div, Nat Phys Lab, Eng, 58-60; from asst prof to prof math, Rensselaer Polytech Inst, 60-73; res prof, Rensselaer Polytech Inst, 73-83; head dept, 73-78, dean, 78-89, PROF APPL MATH, WEIZMANN INST SCI, ISRAEL, 73- *Concurrent Pos:* Vis assoc prof, Mass Inst Technol, 63-64; vis assoc prof, Sch Med, Cornell Univ & vis scientist, Sloan-Kettering Inst, 68-69; Guggenheim fel & vis prof, Weizmann Inst Sci, 71-72; Vinton Hayes sr fel & vis prof appl math, Harvard Univ, 78-79; ed-in-chief, Bull Math Biol. *Mem:* Am Math Soc; Soc Indust & Appl Math; Israel Math Soc; fel AAAS; Soc Math Biol. *Res:* Theoretical biology, especially immunology neurobiology; general applied mathematics. *Mailing Add:* Dept Appl Math Weizmann Inst Sci Rehovot IL 76100 Israel

SEGEL, RALPH E, b New York, NY, Aug 29, 28; m 60; c 3. PHYSICS. *Educ:* Mass Inst Technol, SB, 48; Johns Hopkins Univ, PhD(physics). *Prof Exp:* Res assoc physics, Johns Hopkins Univ, 55; res assoc, Brookhaven Nat Lab, 55-56; physicist, USAF, 56-61; assoc physicist, Argonne Nat Lab, 61-70, sr physicist, 70-76; PROF PHYSICS, NORTHWESTERN UNIV, 66- *Concurrent Pos:* Sr res officer, Oxford Univ, 58-59; Humboldt Found award, Tech Univ Munich, 78. *Mem:* Fel Am Phys Soc. *Res:* Studies of the structure of the nucleus using nuclear reactions and other means. *Mailing Add:* Dept Physics Northwestern Univ 633 Clark St Evanston IL 60201

SEGEL, STANLEY LEWIS, b Philadelphia, Pa, Aug 23, 32; m 58; c 3. SOLID STATE PHYSICS. *Educ:* Allegheny Col, BS, 53; Univ Del, MS, 56; Iowa State Univ, PhD(physics, metall), 63. *Prof Exp:* Instr physics, Robert Col, 57-58; asst prof, Kalamazoo Col, 62-67; assoc prof, 67-76, PROF PHYSICS, QUEEN'S UNIV, ONT, 76-, ASSOC PROF ART, 76- *Mem:* Am Asn Physics Teachers; Phys Soc Japan. *Res:* Nuclear magnetic and quadrupole resonance. *Mailing Add:* Dept Physics Queen's Univ Kingston ON K7L 3N6 Can

SEGELKEN, JOHN MAURICE, b Baltimore, Md, Feb 17, 48; m 70. HARDWARE SYSTEMS, TECHNICAL MANAGEMENT. *Educ:* Univ Md, BS, 70; Purdue Univ, MS, 72. *Prof Exp:* Mem tech staff, 70-78, develop supvr, 79-89, RES SUPVR, AT&T BELL LABS, 89- *Concurrent Pos:* Session chmn & organizer, Electronic Components & Tech Conf, 88-, Nat Electronic Packaging Conf, 89-; chmn packaging prog comt, Electronics Components & Tech Conf, 89-; consult to chmn, Cong Report, Nat Res Coun, 89-90; elected bd gov, Inst Elec & Electronics Engrs Components, Hybrids & Mfg Technol. *Mem:* Sr mem Inst Elec & Electronics Engrs; Am Soc Mech Engrs; Int Soc Hybrid Microelectronics. *Res:* Electronic packaging: advanced packaging, multichip modules, system level packaging, 3 dimensional packaging, material science, physical design, thermal analysis, shock and vibration, reliability, assembly, connector technology, engineering economics, innovation, patents and engineering management. *Mailing Add:* Three Pepperidge Rd Morristown NJ 07960-2645

SEGELKEN, WARREN GEORGE, b Jamaica, NY, Mar 13, 26; m 57; c 3. MAGNETIC RESONANCE. *Educ:* Rutgers Univ, BS, 50, PhD(physics), 55. *Prof Exp:* Asst, Rutgers Univ, 50-52; res physicist, Gen Elec Co, 55-67; assoc prof, 67-72, PROF PHYSICS, ASHLAND COL, 72- *Mem:* Am Asn Physics Teachers; Sigma Xi; Am Phys Soc. *Res:* Solid state physics; nuclear and electronic paramagnetic resonance. *Mailing Add:* 711 Chestnut St Ashland OH 44805

SEGELMAN, ALVIN BURTON, b Boston, Mass, Sept 27, 31; m 72; c 2. PHARMACOGNOSY, PHYTOCHEMISTRY. *Educ:* Mass Col Pharm, BS, 54, MS, 67; Univ Pittsburgh, PhD(pharmacog), 71. *Prof Exp:* Chief pharmacist, Kenmore Prof Pharm, 54-61; dir pharmaceut serv, Bell Pharm Co, 61-65; fel, Mass Col Pharm, 65-67; instr pharmacog & microbiol, Univ Pittsburgh, 67-71; asst prof, 71-74, ASSOC PROF PHARMACOG, RUTGERS UNIV, 74- *Concurrent Pos:* Res dir, Cliniderm Labs, Boston, 57-65; consult pharmaceut, 72-; prin investr, Rutgers Photodynamic Ther Res Group, Fiber Optics Mat Res Prog, 88- *Mem:* AAAS; Am Soc Pharmacog; Acad Pharmaceut Sci; Am Asn Pharm Sci. *Res:* Isolation, purification and structure elucidation of biologically active natural products; design of phytochemical screening and isolation methods; alkaloids, antibiotics, anti-tumor and other biodynamic agents in terrestrial and marine plants and animals; microbiology; microbial transformations and fermentations; medicinal plant tissue culture and analytical toxicology. *Mailing Add:* Dept Pharmacog Rutgers Univ PO Box 789 New Brunswick NJ 08903

SEGELMAN, FLORENCE H, b Beaver, Pa, Apr 27, 41; m 72. ANALYTICAL BIOCHEMISTRY, CLINICAL CHEMISTRY. *Educ:* Allegheny Col, BS, 63; Univ Pittsburgh, PhD(pharmacog), 74. *Prof Exp:* Res scientist phytochem, Amazon Natural Drug Co, Peru, 66-67; teaching asst microbiol, Univ Pittsburgh, 70-71; res assoc pharmacog, Rutgers Univ, New Brunswick, 72-74; sect leader anal biochem, Carter-Wallace, Inc, Cranbury, NJ, 75-90. *Concurrent Pos:* Adj prof, Dept Pharmacog, Col Pharm, Rutgers Univ, 76-90. *Mem:* Am Soc Pharmacog; Am Chem Soc; Am Asn Clin Chem; Am Pharmaceut Asn. *Res:* Assay methodologies for determining biologically active compounds in biological fluids; natural products chemistry; isolation and determination of novel constituents of plants and animals. *Mailing Add:* 54 W 680 S Orem UT 84058-3123

SEGERS, RICHARD GEORGE, b Cincinnati, Ohio, July 4, 28; m 56; c 4. MATHEMATICS, OPERATIONS RESEARCH. *Educ:* Univ Dayton, BS, 50; Purdue Univ, MS, 52, PhD(math), 56. *Prof Exp:* Mem tech staff, Networks Dept, Bell Tel Labs, Inc, 55-59; sr mathematician, Vitro Labs, 59-63; res consult, Gen Precision Labs, 63-64; RES CONSULT, EXXON CORP, 64- *Concurrent Pos:* Vis assoc prof, Grad Sch Bus, Univ Chicago, 69-70; adj fac mem, Dept Math & Physics, Fairleigh Dickinson Univ, 71- *Mem:* Am Math Soc; Soc Indust & Appl Math; Sigma Xi. *Res:* Optimal control, statistics and game theory. *Mailing Add:* 18 Gunther St Mendham NJ 07945

SEGERSTROM, KENNETH, b Denver, Colo, Aug 01, 09. AERIAL MAPPING. *Educ:* Univ Denver, BA, 30; Harvard Univ, MS, 50. *Prof Exp:* Res geologist, US Geol Surv, 34-81; RETIRED. *Mem:* Geol Soc Am; Soc Econ Geologists. *Mailing Add:* 41 Morningside Dr Denver CO 80215

SEGHERS, BENONI HENDRIK, b Willebroek, Belgium, Dec 24, 44; Can citizen. ETHOLOGY, ICHTHYOLOGY. *Educ:* Univ BC, BSc, 67, PhD(zool), 73. *Prof Exp:* Lectr, Univ Man, 72-73; lectr, Univ Western Ont, 74-75, asst prof zool, 76-81; vis scientist, Univ Calgary, 81-82, res assoc, 83-89. *Concurrent Pos:* Fel, Univ Man, 72-74. *Mem:* Can Soc Zool; Animal Behav Soc; Int Soc Human Ethology; NAm Lake Mgt Soc; NAm Benthological Soc. *Res:* Behavior, ecology and evolution of fishes; biology of the fishes of Trinidad, West Indies; anti-predator adaptations in vertebrates; zoogeography; limnology; management of hydroelectric reservoirs. *Mailing Add:* Kananaskis Ctr Environ Res Univ Calgary Calgary AB T2N 1N4 Can

SEGLIE, ERNEST AUGUSTUS, b New York, NY, Aug 8, 45; m; c 2. NUCLEAR PHYSICS. *Educ:* Cooper Union, BS, 67; Univ Mass, PhD(nuclear physics), 73. *Prof Exp:* Assoc physics, Rensselaer Polytech Inst, 73-75; res assoc & lectr physics, Yale Univ, 75-79; res analyst & sci adv, Inst Defense Analyses, OSD/DOT, 79-88. *Honors & Awards:* Andrew Goodpastor Award Excellence Res, 87. *Mem:* AAAS; Int Test & Eval Asn. *Res:* Heavy ion physics, nuclear scattering and reaction theories and fusion reactions; defense analysis, operational testing and evaluation. *Mailing Add:* OSD/DOT & E Pentagon Rm 3E318 Washington DC 20301-1700

SEGLUND, JAMES ARNOLD, b Munising, Mich, Jan 31, 23; m 46; c 3. GEOLOGY. *Educ:* Univ Mich, BS, 48, MS, 49. *Prof Exp:* Dist geologist, Texaco Inc, 49-56; partner, Rodgers, Seglund & Shaw Assoc, 56-67; PRES, J A SEGLUND INC, 67- *Concurrent Pos:* Deleg, Am Asn Petrol Geologists; dir, Soc Independent Petrol Earth Scientist. *Mem:* Soc Geochemists. *Res:* Geological studies on the Michigan portion of the Mid-continent Rift System; geological studies on New Brunswick, Canada; gold deposits for Northern Michigan. *Mailing Add:* 4720 Southshore Dr Metairie LA 70002

SEGNER, EDMUND PETER, JR, b Austin, Tex, Mar 28, 28; m 52; c 5. STRUCTURAL ENGINEERING. *Educ:* Univ Tex, BS, 49, MS, 52; Tex A&M Univ, PhD(struct eng), 62. *Prof Exp:* Engr, United Gas Pipe Line Co, La, 49-50; sr struct engr, Gen Dynamics Corp, Tex, 51-52 & 53-54; engr, Forrest & Cotton, Inc, 53; from instr to assoc prof civil eng, Tex A&M Univ, 54-63; prof civil eng & struct group coordr, Univ Okla, 63-65; prof civil eng, Univ Ala, Tuscaloosa, 65-76, asst dean res & grad studies, 68-71, assoc dean eng, 71-76; assoc vpres Res & Grad Studies, Memphis State Univ, 76-80, prof civil eng & assoc vpres res, 76-90; CHMN & PROF, DEPT CIVIL ENG, UNIV ALA, BIRMINGHAM, 90- *Concurrent Pos:* Consult, 54- *Mem:* Am Soc Civil Engrs; Am Soc Eng Educ; Nat Soc Prof Engrs; Am Concrete Inst. *Res:* Design and analysis of structural steel and reinforced concrete structures; openings in flexural members; splices in tensile reinforcing bars; bond in reinforced concrete design and various flexural studies involving reinforced concrete and structural steel. *Mailing Add:* Dept Civil Eng Univ Ala Birmingham UAB Sta Birmingham AL 35294

SEGOVIA, ANTONIO, b Asuncion, Paraguay, Nov 3, 32; m 60, 81; c 3. GEOLOGY. *Educ:* Nat Univ Paraguay, BS, 54; Pa State Univ, PhD(geol), 63. *Prof Exp:* Geologist, Ministry Pub Works, Paraguay, 56; field geologist, Bolivian Gulf Oil Corp, 57; geologist, Photogeol Unit, Gulf Oil Corp, NJ, 57-58, consult, Western Explor Div, 58-60; adv photogeol, Nat Geol Surv, Colombia, 60-61; asst prof geol & head dept, Sch Petrol Eng, Eastern Univ Venezuela, 61-64 & Cent Univ Venezuela, 64-66; sr res scientist, Tulsa Res Ctr, Sinclair Oil Corp, 66-69; ASSOC PROF GEOL, UNIV MD, COLLEGE PARK, 69- *Concurrent Pos:* Consult, Skelly Oil Corp, 65-66, Berea Oil, 80-84 & World Bank, 83-84; pres, Geosysts Inc, 72-76; consult archaeol, Cath Univ Am, 73-75 & Am Univ, 74-75; vis res prof, Cent Univ Venezuela, 75 & Simon Bolivar Univ, Venezuela, 77; dir geol surv, PR, 76. *Mem:* AAAS; fel Geol Soc Am; Soc Petrol Engrs; Am Asn Petrol Geologists; Am Inst Mining, Metall & Petrol Engrs. *Res:* Photogeology; geomorphology; engineering geology; petroleum geology; structural geology; seismicity of regmites, especially deformation of Pleistocene surfaces; geological study of nuclear sites; South American geology. *Mailing Add:* Dept Geol Univ Md College Park MD 20742

SEGOVIA, JORGE, b Martinez, Arg, Mar 2, 34; m 63; c 2. COMMUNITY HEALTH, SOCIOMEDICAL SCIENCES. *Educ:* Univ Buenos Aires, MD, 59, MPH, 61. *Prof Exp:* Head health educ sect sociomed sci, Sch Pub Health, Univ Buenos Aires, 63-66; head med sociol sect, Ctr Educ Med Invest Clin Buenos Aires, 64-68; mem prof staff med sociol, Ctr Latino Am Admin Med Buenos Aires, 68-71; res assoc sociomed sci, Columbia Univ Sch Pub Health, 71-73; asst prof & consult med care prog, Health & Community, Univ Campinas, Brazil, 73-75; assoc prof, 76-83, PROF SOCIAL MED, MEM UNIV NFLD, 83- *Concurrent Pos:* Short term consult, Can Int Develop Agency, 84-85; mem adv comt, Can Pub Health Asn and occup health proj; counr, Int Health Section, Am Pub Health Asn, 84-87. *Mem:* Can Pub; Am Pub Health Asn; Am Sociol Asn; hon fel Royal Soc Health UK. *Res:* Health practices and medical care utilization; comparative health systems; health services research; health status measurement. *Mailing Add:* Div Community Med Mem Univ Nfld St John's NF A1B 3V6 Can

SEGRE, DIEGO, b Milano, Italy, Feb 3, 22; nat US; m 52; c 2. IMMUNOLOGY. *Educ:* Univ Milano, DVM, 47; Univ Nebr, MS, 54; Univ Wis, PhD(vet sci), 57. *Prof Exp:* Asst prof infectious dis, Univ Milano, 47-51; asst animal pathologist, Univ Nebr, 52-55; res asst vet sci, Univ Wis, 55-57, asst prof, 57-60; PROF VET PATHOBIOL, COL VET MED, UNIV ILL, URBANA, 60- *Concurrent Pos:* Nat Inst Aging, NIH, Bethesda, Md, 88-89. *Mem:* AAAS; Soc Exp Biol & Med; Am Asn Immunologists; Conf Res Workers Animal Dis; Am Col Vet Microbiol; Sigma Xi; Am Gerontol Soc. *Res:* Mechanisms of immunity and immunological tolerance; maturation and cytokinetics of the immune response; immunology of aging. *Mailing Add:* Col Vet Med Univ Ill 2001 S Lincoln Ave Urbana IL 61801

SEGRÉ, EMILIO GINO, nuclear physics, particle physics; deceased, see previous edition for last biography

SEGRE, GINO C, b Florence, Italy, Oct 4, 38; US citizen; m 62; c 2. THEORETICAL PHYSICS. *Educ:* Harvard Univ, AB, 59; Mass Inst Technol, PhD(physics), 63. *Prof Exp:* NSF fel physics, Europ Orgn Nuclear Res, Geneva, Switz, 63-65; res assoc, Lawrence Radiation Lab, Univ Calif, 65-69; assoc prof, 69-74, PROF, UNIV PA, 74- *Concurrent Pos:* A P Sloan Found fel, 63-71; Guggenheim fel, 75-76. *Mem:* Am Phys Soc. *Res:* Elementary particle physics. *Mailing Add:* Dept Physics 2n9a Drl E1 Univ Pa Philadelphia PA 19104

SEGRE, MARIANGELA BERTANI, b Milan, Italy, Oct 4, 27; US citizen; m 52; c 2. IMMUNOBIOLOGY. *Educ:* Univ Milan, Dr Sc(biol), 49. *Prof Exp:* Res assoc infectious dis, Col Vet Med, Univ Milan, 49-51; vis investr, Animal Dis Res Inst, Weybridge, Eng, 51; bacteriologist, Montecatini Corp, Milan, 51-52 & Nebr State Dept Health, 53-54; res assoc, 63-73, asst prof, 73-85, ASSOC PROF IMMUNOL, COL VET MED, UNIV ILL, URBANA, 85- *Concurrent Pos:* vis investr, Lab Immunol, NIH, 89. *Mem:* Am Asn Immunologists; Am Asn Microbiologists; Am Asn Vet Immunologists. *Res:* Mechanism of antibody formation; immunologic tolerance; immunologic aspects of aging; cytokinetics of the immune response; mechanisms of immunologic memory; immunological methods. *Mailing Add:* Vet Path 2001 S Lincoln Ave Univ Ill Urbana IL 61801

SEGREST, JERE PALMER, b Dothan, Ala, Aug 16, 40; m 66; c 3. PLASMA LIPOPROTEINS, PROTEIN DESIGN. *Educ:* Vanderbilt Univ, BA, 62, MD, 67, PhD(biochem), 69. *Prof Exp:* Resident path, Univ Hosp, Vanderbilt Univ, 68-70; assoc prof path, 74-80, from asst prof to assoc prof biochem, 74-82, asst prof microbiol, 75-79, PROF PATH, MED CTR, UNIV ALA, BIRMINGHAM, 80-, PROF BIOCHEM, 82- *Concurrent Pos:* Europ Molecular Biol Orgn fel, Nat Ctr Sci Res, Gif-sur-Yvette, France, 73; mem, NIH Study Sect Molecular Cytol, 78; mem, comt prof training, Am Chem Soc, 78-86; mem, Ed Affairs Comn, Am Soc Biochem & Molecular Biol, 81-87; vis scientist, Div Endocrinol Metab, Dept Med, Univ Wash, Seattle, 83; dir, Atherosclerosis Res Unit, Univ Ala, 83-, Atheroclerosis Detection & Prev Clin, 88-; prin investr, prog proj, Nat Heart Lung Blood Inst & mini prog proj, 85-87, prin investr on five reports of Invest Grant Rev Comn B, 88-; ed, Plasma Lipoproteins, 86; co organizer, Workshop Molecular Interactions Membranes, Ctr Physics, Aspen, Colo, 86; consult, Southern Biotechnol Assoc, Inc, 86; prof med, 84-; pres, Athero Tech, Inc, 87- *Mem:* Am Chem Soc; Am Chem Soc; Am Soc Cell Biol; Am Heart Asn; Am Soc Biochem & Molecular Biol. *Res:* Protein-lipid interactions in membrane and plasma lipoproteins; structure and function plasma high density lipoproteins; protein design; role of triplycinde-rich lipoprotein in atherogenesis; clinical detection, prevention and treatment of atherosclerosis; cell biology. *Mailing Add:* 3709 Forest Run Rd Birmingham AL 35223

SEGUIN, JEROME JOSEPH, b North Bay, Ont, Sept 27, 24; m 52; c 6. PHYSIOLOGY. *Educ:* Univ West Ont, BSc, 50, MSc, 52, PhD(physiol), 56. *Prof Exp:* Demonstr & asst, 50-57, lectr, 58-60, asst prof, 60-63, ASSOC PROF PHYSIOL, UNIV WESTERN ONT, 63- *Concurrent Pos:* McEachern sr med fel, Muscular Dystrophy Asn Can, 58-63; guest lectr, Inst Muscle Dis, Inc, NY, 62-63, Instituto Fisiologia Umana, Pisa, Italy, 71-72, Dept Physiol, Univ Adelaid, Australia, 86-87. *Mem:* Can Physiol Soc; Soc Neurosci; Int Asn Study Pain. *Res:* Muscle receptor physiology; neurophysiology of pain. *Mailing Add:* Dept Physiol Health Sci Ctr Univ Western Ont London ON N6A 5C1 Can

SEGUIN, LOUIS-ROCH, b Rigaud, Que, Apr 26, 20; m 50; c 4. FISHERIES. *Educ:* Col Bourget, BA, 43; Univ Laval, BSc, 47; Univ Montreal, MSc, 54. *Prof Exp:* Biologist, Laurentian Fish Hatchery, Que, 47-49; biologist & dir, Eastern Twp Fishery Sta, Que Dept Fish & Game, 50-61, chief biol, 62-63; prof fish culture, Univ Laval, 63; chief wildlife biologist, Can Int Paper Co, 63-71; exec dir, Que Fedn Camping & Caravaning, 71-72; head fishery sect, Appl Res Ctr Feeding Sci, Univ Que, Montreal, 72-76; prof aquacult, Col St Felicien, Quebec, 80-85; CONSULT FISHERY BIOLOGIST & MGR, JAMES BAY RESERVOIRS, QUE, 76- *Concurrent Pos:* Gen mgr, Que Outfitters Asn, 73-82; prof Aquacult, Col St Felicien, Quebec, 80-84. *Mem:* Aquacult Asn Can; Am Fishery Soc. *Res:* Trout culture; building and management of lakes; closed circuit for fish rearing. *Mailing Add:* 385 Sequin Rd Baldwin PQ J1A 2S4 Can

SEGUIN, MAURICE KRISHOLM, b Cedars, Que, May 30, 37. GEOPHYSICS, GEOLOGY. *Educ:* Univ Montreal, BA, 58, BSc, 62; McGill Univ, MSc, 63, PhD(geophys), 65. *Hon Degrees:* Degree, Royal Inst Technol, Stockholm, 68. *Prof Exp:* Geophysicist, Iron Ore Co, Can, 64-65; lectr, Royal Inst Technol, Stockholm, 65-68; assoc res, Soquem, Que, 68; from asst prof to assoc prof, 69-82, FULL PLEDGE PROF GEOPHYS, LAVAL UNIV, 82- *Concurrent Pos:* Mem, Comt Geol & Geophys, Nat Res Coun Can, 69. *Mem:* French-Can Asn Advan Sci; Europ Asn Explor Geophys; Europ Asn Geophys. *Res:* Applied geophysics; paleomagnetism; permafrost; geophysics. *Mailing Add:* Dept Geol Laval Univ Quebec PQ G1K 7P4 Can

SEGUIN, WARD RAYMOND, b Montpelier, Vt, Aug 28, 42; m 67; c 3. METEOROLOGY. *Educ:* Fla State Univ, BS, 65, MS, 67, PhD(meteorol), 72. *Prof Exp:* Fel meteorol, Univ Va, 72-73; RES SCIENTIST METEOROL, NAT OCEANIC & ATMOSPHERIC ADMIN, DEPT COM, 73-, DEP CHIEF, 88- *Honors & Awards:* Silver Medal, Dept Com. *Mem:* Am Meteorol Soc; Nat Weather Asn. *Res:* Study of energy and momentum transfers in the tropical marine atmospheric boundary layer. *Mailing Add:* SSMC No 2 Rm 10148 1325 E West Hwy Silver Spring MD 20902

SEGUNDO, JOSE PEDRO, b Montevideo, Uruguay, Oct 6, 22; c 5. NEUROPHYSIOLOGY. *Educ:* Univ Repub Uruguay, BS, 42, MD, 49. *Prof Exp:* Instr physiol, Univ Repub Uruguay, 50-57; head dept electrobiol, Inst Biol Sci, Montevideo, 57-60; instr anat, 53-55, PROF ANAT, HEALTH SCI CTR, UNIV CALIF, LOS ANGELES, 60- *Mem:* Am Physiol Soc; Biophys Soc; Soc Neurosci; LatinAm Ciencias Fisiologicas. *Res:* Functional organization of multisensory areas of the nervous system; interneuronal communication. *Mailing Add:* Dept Anat 73-235 Univ Calif 405 Hilgard Ave Los Angeles CA 90024

SEGURA, GONZALO, JR, b Cuba, Nov 25, 19; nat US. RADIOCHEMISTRY. *Educ:* Emory Univ, AB, 42, MS, 43. *Prof Exp:* Instr math, Emory Univ, 43-44; asst physics, Columbia Univ, 44; chemist, Tenn Eastman Corp, 45-46; instr chem, Finch Col, 46-51; radiochemist, Foster D Snell, Inc, 51-52, chief radiochemist, 52-59; res scientist, Philip Morris, Inc, 59-83; RETIRED. *Res:* Radiotracers; detergency; syntheses. *Mailing Add:* 3522 Grove Ave Richmond VA 23221

SEHE, CHARLES THEODORE, b Geneva, Ill, Feb 26, 23; m 53; c 5. DEVELOPMENTAL BIOLOGY, ENDOCRINOLOGY. *Educ:* N Cent Col, Ill, AB, 50; Univ Iowa, MS, 53, PhD(zool), 57. *Prof Exp:* Asst embryol, Univ Iowa, 52-57; asst prof endocrinol, Univ Ill, Urbana, 57-58; asst prof embryol & endocrinol, Univ Cincinnati, 58-61; assoc prof embryol & comp anat, N Cent Col, Ill, 61-64; res assoc prof develop biol & endocrinol, Stanford Med Ctr, 64-71; prof develop biol & endocrinol, 71-77, MEM BIOL FAC, MANKATO STATE COL, 77- *Concurrent Pos:* Resident res assoc, Argonne Nat Lab, 62-63; consult biol, Teacher Training Prog, Inst Nuclear Sci & Eng, Argonne Nat Lab, 64. *Mem:* AAAS; Am Soc Zool. *Res:* Developmental and secretory characteristics of the ultimobranchial body of vertebrates; hormonal factors in sexual behavioral development. *Mailing Add:* Dept Biol Mankato State Univ South Rd & Ellis Ave Mankato MN 56001

SEHGAL, LAKSHMAN R, b Hyderabad, AP, India, Feb 15, 42; US citizen; m 66; c 2. MICROBIOLOGY, IMMUNOLOGY. *Educ:* Sri Aurobindo Int Ctr Educ, BS, 62; Ill Inst Technol, Chicago, PhD(biol), 70. *Prof Exp:* Instr res, Med Sch, Univ Ill, 70-73, asst prof, 73-77; ASST RES PROF, PRITZKER SCH MED, UNIV CHICAGO, 80-; DIR SURG RES, MICHAEL REESE HOSP, 84- *Concurrent Pos:* Biochemist, dept surg, Cook County Hosp, 73-77; prin investr subcontract, Naval Res Labs, Washington, DC, 85- *Mem:* Asn Acad Surg; Am Asn Clin Chem; Asn Clin Scientists; Shock Soc. *Res:* Development of a hemoglobin based red cell substitute; manufacture of a polymerized pyridoxylated hemoglobin solution which has a normal oxygen carrying capacity. *Mailing Add:* Dept Surg Michael Reese Hosp Univ Chicago 5801 Ellis Ave Chicago IL 60637

SEHGAL, OM PARKASH, b Rawal Pindi, India, July 22, 32; m 62; c 2. PLANT PATHOLOGY, VIROLOGY. *Educ:* Univ Lucknow, MSc, 53; Univ Wis, PhD(plant path), 61. *Prof Exp:* Res asst virol, Indian Agr Res Inst, 55-57; res asst plant path, Univ Wis, 58-61; fel virol, Univ Ariz, 61-63; from asst prof to assoc prof, 63-78, PROF PLANT PATH & BIOL SCI, UNIV MO, COLUMBIA, 78- *Concurrent Pos:* Consult, UN/FAO, 90; pres, NCent Div, Am Phytopath Soc, 84-85. *Mem:* AAAS; Am Phytopath Soc; Soc Gen Microbiol; Am Soc Microbiol. *Res:* Viral structure and genetics. *Mailing Add:* Dept Plant Path Univ Mo Columbia MO 65211

SEHGAL, PRAVINKUMAR B, b Bombay, India, Sept 11, 49; m 72; c 2. VIROLOGY & IMMUNOLOGY, CYTOKINES. *Educ:* Seth G S Med Col, MB & BS, 73; Rockefeller Univ, PhD(virol & cell biol), 77. *Prof Exp:* Intern, King Edward Mem Hosp, 71-72; fel, 77-79, asst prof, 79-84, ASSOC PROF VIROL, ROCKEFELLER UNIV, 84- *Concurrent Pos:* Assoc ed, J Interferon Res, 79-, & Virol, 83-88; NIH res grant, 79-; estab investr, Am Heart Asn, 83; proj dir, Nat Found Cancer Res, 84- *Honors & Awards:* Irma T Hirschi Award, 81. *Mem:* AAAS; NY Acad Sci; Am Soc Microbiol; Am Soc Virol; Am Asn Immunol; Sigma Xi; Harvey Soc; Am Soc Molecular Biol & Biochem. *Res:* Interleukin-6 and other cytokines in health and disease; mechanisms of RNA transcription; plasma protein synthesis in acute infections; genetic sequence analysis. *Mailing Add:* Rockefeller Univ 1230 York Ave New York NY 10021

SEHGAL, PREM P, b Patiala, India, Nov 16, 34; m 61; c 1. PLANT PHYSIOLOGY. *Educ:* Univ Delhi, BSc, 54, MSc, 56; Harvard Univ, AM, 61; Duke Univ, PhD(bot), 64. *Prof Exp:* Asst prof bot, B R Col, Agra, 56-57; lectr, Ramjas Col, Delhi, 57-58; res asst, Duke Univ, 61-64, NSF res assoc, 64-65; NIH proj assoc biochem, Univ Wis, Madison, 65-66; asst prof, 66-69, assoc prof, 69-77, PROF BIOL, ECAROLINA UNIV, 77- *Concurrent Pos:* NC Bd Sci & Tech res grant, 67-70; NSF Cosip grant, 69-70. *Mem:* Am Soc Plant Physiol; Bot Soc Am; Brit Soc Exp Biol; Int Soc Plant Morphol; NY Acad Sci; Sigma Xi. *Res:* Interaction of hormones with nitrogen compounds, especially in chlorophyll production; biochemistry of plant tissue cultures with special reference to enzymatic changes; urease. *Mailing Add:* Dept Biol ECarolina Univ Greenville NC 27834

SEHGAL, S(ATYA) B(HUSHAN), soil mechanics, for more information see previous edition

SEHGAL, SURENDRA N, b Khushab, India, Feb 10, 32; Can citizen; m 61; c 2. MICROBIOLOGY, TUMOR BIOLOGY. *Educ:* Banaras Hindu Univ, BPharm, 52, MPharm, 53; Bristol Univ, PhD(microbiol), 57. *Prof Exp:* Asst prof pharm, Birla Col Pharm, 53-55; fel microbiol, Coun Sci & Indust Res, India, 55; Nat Res Coun Can fel, 58-60; sr scientist microbiol, Ayerst Labs Res Inc, 60-69, head microbial technol, 69-74, asst dir, dept microbiol, 74-83, asst dir, dept immunol, 83-88, asst dir, Dept Immunopharmacol, 88-89, SR RES FEL, WYETH-AYERST RES, 89- *Concurrent Pos:* Adj assoc prof microbiol, Concordia Univ, Montreal. *Mem:* Am Soc Microbiol; Chem Inst Can; Soc Indust Microbiol; AAAS. *Res:* Industrial microbiology; bioconversion of organic compounds; antibiotic fermentations; microbial chemistry; tumor biology; genetic toxicology; immunology; mechanisms involved in inumunosuppression and immunopotentiation; autoimmune disease. *Mailing Add:* Wyeth-Ayerst Res CN 8000 Princeton NJ 08543-8000

SEHGAL, SURINDER K, b Hoshiarpur, India, Apr 22, 38; m 66; c 2. ALGEBRA. *Educ:* Panjab Univ, India, BA, 57, MA, 59; Univ Notre Dame, PhD(math), 65. *Prof Exp:* Lectr math, DAV Col, Hoshiarpur, 59-61; teaching asst, Univ Notre Dame, 61-63; from teaching asst to asst prof, 63-71, assoc prof, 71-83, PROF MATH, OHIO STATE UNIV, 83- *Mem:* Am Math Soc; Math Asn Am. *Res:* Algebra; group theory. *Mailing Add:* Dept Math Ohio State Univ Columbus OH 43210

SEHMEL, GEORGE ALBERT, b Puyallup, Wash, Apr 8, 32; m 58; c 3. CHEMICAL ENGINEERING, AEROSOL PHYSICS. *Educ:* Univ Wash, Seattle, BS, 55, PhD(chem eng), 61; Univ Ill, MS, 56. *Prof Exp:* Engr, Gen Elec Co, NY, 57-58 & Wash, 61-63, sr engr, 63-83; staff engr, 65-83, SR RES ENGR, PAC NORTHWEST LABS, BATTELLE MEM INST, 83- *Concurrent Pos:* Lectr, Ctr Grad Study at Hanford, 64-69. *Mem:* Am Inst Chem Engrs. *Res:* Program management of aerosol physics research; aerosol particulate mass transfer behavior; meteorology; particulate deposition and resuspension; sampling of aerosols; evaluation of nuclear reactor fuel elements; heat transfer technology; smokes and obscurants; environmental assessments. *Mailing Add:* 2030 Howell Richland WA 99352

SEHON, ALEC, b Romania, Dec 18, 24; Can citizen; m 50; c 2. IMMUNOLOGY. *Educ:* Univ Manchester, BSc, 48, MSc, 50, PhD(phys chem), 51, DSc, 65. *Prof Exp:* Demonstr chem, Univ Manchester, 48-49 & Nat Res Labs, 51-52; res assoc chem, Calif Inst Technol & Inst Biochem, Univ Uppsala, 52-53; asst prof exp med, McGill Univ, 53-59, from asst prof to prof chem, 56-69, hon lectr biochem, 59-69; PROF IMMUNOL & HEAD DEPT, FAC MED, UNIV MAN, 69- *Concurrent Pos:* Co-dir, Div Immunochem & Allergy, Royal Victorian Hosp, Montreal, 53-59; biophys chemist, McGill Univ Clin, Montreal Gen Hosp, 60-69; Nat Res Coun Can sr res fel, John Simon Guggenheim Mem Found fel & res assoc, Harvard Univ, 63-64; mem, Res Grants Comt, Med Res Coun Can, 64-69; chmn, Gordon Res Conf Immunochem & Immunobiol, 66; dir, NATO Advan Studies Insts, Val Morin, Que, 68, Minaki, Ont, 70; RR Inst res award, Univ Man & Shering travel fel, Can Soc Clin Invest, 73; vis scientist, Walter & Eliza Hall Inst, Melbourne Univ Col, London & Med Res Coun Can, 73-74. *Mem:* Fel AAAS; Am Asn Immunologists; fel Am Acad Allergy; fel Am Col Allergists; Can Soc Immunologists (vpres, 67-69, pres, 69-71). *Res:* Antigen-antibody systems involved in common allergies; development of immunosuppressive therapeutic regimens; tumor immunology; immunodiagnostics. *Mailing Add:* Dept Immunol Fac Med Univ Man Winnipeg MB R3T 2N2 Can

SEIB, DAVID HENRY, b Exeter, Calif, Jan 23, 43; m 65; c 3. ELECTRICAL ENGINEERING. *Educ:* Calif Inst Technol, BS, 64; Stanford Univ, MS, 65, PhD(elec eng), 70. *Prof Exp:* Staff scientist solid state physics, tech staff, Aerospace Corp, 69-76; mem tech staff VII Elec Eng, Sci Ctr, 76-84, mgr, 84-88, MEM TECH STAFF, ROCKWELL INT, 88- *Mem:* Inst Elec & Electronics Engrs. *Res:* Semiconductor device research, specifically charge coupled devices and infrared detective arrays; integrated optics devices. *Mailing Add:* Rockwell Int 3370 Mira Loma Ave PO Box 3105 Mail Code 031-BC18 Anaheim CA 92803-3105

SEIB, PAUL A, b Poseyville, Ind, Jan 8, 36; m 58; c 2. ORGANIC CHEMISTRY, BIOCHEMISTRY. *Educ:* Purdue Univ, BS, 58, MS, 63, PhD(biochem), 65. *Prof Exp:* Asst prof org chem, Inst Paper Chem, 65-70; ASSOC PROF GRAIN SCI, KANS STATE UNIV, 70- *Mem:* Am Chem Soc; Am Asn Cereal Chemists; Inst Food Technologists. *Res:* Cereal chemistry; chemistry of vitamin C. *Mailing Add:* 836 Dondeed Manhattan KS 66502

SEIBEL, ERWIN, OCEANOGRAPHY, EDUCATIONAL ADMINISTRATION. *Educ:* City Univ New York, BS, 65; Univ Mich, MS, 66, PhD(oceanog), 72. *Prof Exp:* Logistics officer, US Army Corps Engrs, 67-69; master instr, US Army Engr Sch, 69-70, sect head, 70-71; assoc res oceanog & asst proj dir, Great Lakes Res Div, Univ Mich, Ann Arbor, 72-78, asst dir, Mich Sea Grant Prog, 75-78; prof oceanog & geol & dir, Tiburon Ctr Environ Studies, 78-81, prof & chmn dept geol sci, 81-88, DEAN UNDERGRAD STUDIES, SAN FRANCISCO STATE UNIV, 88- *Concurrent Pos:* Pres, San Franciso Bay chap, Marine Technol Soc, 82-83, exec secy oceans, 83; pres, San Francisco State Univ Chap, Sigma Xi, 82-84 & 90- *Mem:* Fel AAAS; Am Geophys Union; fel Geol Soc Am; Marine Technol Soc; Soc Econ Paleontologists & Mineralogists; Sigma Xi; fel Acad Sci. *Res:* Multidisciplinary approach to the solution of San Francisco Bay Area and adjacent Pacific Ocean environmental problems through monitoring and modeling the physical, economic social and cultural variables involved; investigation of the effect of nuclear power plant operation on the biological, chemical, geological and physical facets of the aquatic environment; study of the formation and breakup of ice using remote sensing techniques; dynamics of shoreline erosion and resultant sediment transport; enhancing retention rates of minority students. *Mailing Add:* Dean Undergrad Studies San Francisco State Univ 1600 Holloway Ave San Francisco CA 94132

SEIBEL, FREDERICK TRUMAN, b Corning, NY, May 30, 41; m 65; c 1. ARTIFICIAL INTELLIGENCE, COMPUTER SCIENCE. *Educ:* Yale Univ, BS, 63; Duke Univ, PhD(physics), 68. *Prof Exp:* Staff mem, Los Alamos Sci Lab, 68-77; sr prof tech staff, Princeton Plasma Physics Lab, Princeton Univ, 77-81; dir, Automation & Control, Syst & Advan Technol, Merck & Co, Inc, 81-85; DIR, COM ARTIFICIAL INTEL APPLNS, BOLT, BERANEL & NEWMAN LABS, 85- *Mem:* Sigma Xi. *Res:* Developing commercial applications of artificial intelligence technology. *Mailing Add:* 99 Larch Rd Cambridge MA 02138

SEIBEL, HUGO RUDOLF, b Radautz, Rumania, Nov 9, 37; m 64. ANATOMY, ELECTRON MICROSCOPY. *Educ:* Brooklyn Col, BS, 60; Univ Rochester, PhD(anat), 67. *Prof Exp:* Col sci asst & instr biol, Brooklyn Col, 60-62; instr anat, Univ Rochester, 66-67; from asst prof to assoc prof, 67-75, dir electron micros div, 67-86, PROF ANAT, MED COL VA, 75-, ASSOC DEAN MED, 84- *Concurrent Pos:* A D Williams grant, Med Col Va, 67-69, NIH grant, 68-71. *Mem:* AAAS; Soc Study Reproduction; Am Asn Anatomists; Pan-Am Asn Anatomists; Transplantation Soc; Sigma Xi. *Res:* Kidney and heart transplantation; electron microscopy and functional correlates of thyroid and pineal. *Mailing Add:* Dept Anat Med Col Va Box 709 Richmond VA 23298

SEIBEL, WERNER, b Krenau, WGer, Sept 27, 43; US citizen; m 67; c 2. GROSS ANATOMY, DENTAL RESEARCH. *Educ:* Brooklyn Col, BA, 65; Hofstra Univ, MA, 68; Med Col Va, Va Commonwealth Univ, PhD(anat), 73. *Prof Exp:* Asst anat, histol & neuroanat, Med Col Va, Va Commonwealth Univ, 68-72; instr, 72-73, asst prof, 73-77, ASSOC PROF DENT ANAT, SCH DENT, UNIV MD, 77- *Concurrent Pos:* NDEA fel, 68; vis prof, Bone Marrow Transplant Unit Dept Oncol, Johns Hopkins Sch Med. *Mem:* Am Asn Anatomists; Sigma Xi; Am Asn Dent Schs; Int Asn Dent Res; Am Asn Dent Res. *Res:* Mechanisms in the development of fibrotic tissue; functional activities of fibroblast cultures from dermis of rat model with chronic-graft-versus-host disease; pathogenesis of gingival overgrowth induced by phenytoin and cyclosporine-A; oral histology. *Mailing Add:* 114 Chestnut Hill Lane W Reisterstown MD 21136

SEIBER, JAMES N, b Hannibal, Mo, Sept 21, 40; m 67; c 2. ORGANIC CHEMISTRY. *Educ:* Bellarmine Col, Ky, AB, 61; Ariz State Univ, MS, 64; Utah State Univ, PhD(org chem), 66. *Prof Exp:* Res chemist, Dow Chem Co, 66-69; asst prof, 69-74, ASSOC PROF ENVIRON TOXICOL, UNIV CALIF, DAVIS, 74- *Concurrent Pos:* Vis scientist, Pesticides & Toxic Substances Effects Lab, US Environ Protection Agency, Fla, 73-74. *Mem:* AAAS; Am Chem Soc. *Res:* Isolation, structure determination, synthesis and reactions of biologically active chemicals, particularly pesticides, insect pheromones and plant-derived poisons; origin and fate of toxic chemicals in the environment; analytical chemistry of pesticides and pollutants. *Mailing Add:* Dept Environ Toxicol Univ Calif Davis CA 95616

SEIBERT, FLORENCE B, intravenous therapy, tuberculosis; deceased, see previous edition for last biography

SEIBERT, J A, b Dayton, Ohio, Oct 21, 53; m 87. RADIOLOGICAL QUALITY ASSURANCE TESTING. *Educ:* Univ Calif, Irvine, BS, 76, BA, 77, MS, 81 & PhD (radiol sci), 83. *Prof Exp:* ASST PROF RADIOL PHYSICS, UNIV CALIF, DAVIS, 83- *Concurrent Pos:* Lectr, Calif State Univ, Sacramento, 87-88. *Mem:* Am Asn Physicists Med; Radiol Soc NAm; Soc Photo-Optical & Instrumentation Engrs; Inst Elec & Electronics Engrs. *Res:* Medical image processing with goals of improving quantitative and qualitative features by removing degradations and artifacts occured during image acquisition process; radiation shielding specifications. *Mailing Add:* Diag Radiol Univ Calif Davis 2516 Stocton Blvd TICON II Sacramento CA 95817

SEIBERT, MICHAEL, b Lima, Peru, Nov 15, 44; US citizen; m 75; c 3. PHOTOSYNTHESIS, BIOTECHNOLOGY. *Educ:* Pa State Univ, University Park, BS, 66; Univ Pa, MS, 67, PhD(molecular biol & biophys), 71. *Prof Exp:* Scientist, Exp Sta, E I du Pont de Nemours & Co, 65-68; mem tech staff, GTE Labs, 71-77; sr scientist & task leader photobiol, 77-83, MGR, PHOTOCONVERSION RES BR, SOLAR ENERGY RES INST, 84-; RES PROF, DEPT BIOL, UNIV DENVER, 85- *Concurrent Pos:* mem, Proj 4, US/USSR Joint Working Group Microbiol, 79; chmn biotechnol & chem sci div, Am Solar Energy Soc, 80; res fel, Nat Ctr Sci Res, France, 83; vis scientist, Inst Phys & Chem Res, Japan, 87; vis prof, Moscow State Univ, USSR, 88; prog chmn, Int Energy Agency, 90. *Mem:* Biophys Soc; AAAS; Am Soc Photobiol; Am Soc Plant Physiologists; Am Solar Energy Soc; Int Asn Plant Tissue Cult. *Res:* Photosynthetic oxygen evolution; membrane surface chemistry; photobiological conversion of solar energy; primary photochemical process in photosynthesis; photomorphogenesis and cryopreservation of plant tis; horticultural applications of lighting. *Mailing Add:* Solar Energy Res Inst 1617 Cole Blvd Golden CO 80401

SEIBOLD, CAROL DUKE, b San Francisco, Calif, July 1, 43; m 64; c 1. PHOTOGRAPHIC CHEMISTRY. *Educ:* Creighton Univ, BS, 65; Univ Nebr, MS, 67, PhD(chem), 72. *Prof Exp:* Assoc prof chem, Bemidji State Col, 68-69; res chemist, Environ Res Corp, 72-73; MGR & SR CHEMIST, 3M CO, 73- *Mem:* Am Chem Soc; Soc Photog Scientists & Engrs. *Res:* Mechanisms and kinetics of photographic development. *Mailing Add:* Three Chicadee Lane St Paul MN 55127

SEIDAH, NABIL GEORGE, b Cairo, Egypt, Feb 1, 49; Can citizen; m 73; c 2. CLINICAL BIOCHEMISTRY, BIOPHYSICAL CHEMISTRY. *Educ:* Cairo Univ, BS, 69; Georgetown Univ, PhD(chem), 73. *Prof Exp:* Nat Res Coun Found fel, Dept Chem, Univ Montreal, 73-74, asst prof, 76-80, BIOCHEMIST, CLIN RES INST, SCH MED, UNIV MONTREAL, 74-, PROF, 80- *Honors & Awards:* Clarke Inst of Psychiat Award, Toronto, 77; Harold Piche Award, Montreal, 83. *Mem:* Med Res Coun Can. *Res:* Determination of amino acid sequence of polypeptide hormones both from pituitary gland and from tumor organs; maturation enzymes of pro-hormones and their genes. *Mailing Add:* Biochem Neuroendocrinology Lab Clin Res Inst 110 Pine Ave W Montreal PQ H2W 1R7 Can

SEIDE, PAUL, b Brooklyn, NY, July 22, 26; m 51; c 2. ENGINEERING MECHANICS. *Educ:* City Col New York, BCivEng, 46; Univ Va, MAeroE, 52; Stanford Univ, PhD(eng mech), 54. *Prof Exp:* Aeronaut res scientist, Nat Adv Comn Aeronaut, 46-52; res asst eng mech, Stanford Univ, 52-53; res engr, Northrup Aircraft, Inc, 53-55; head methods & theory sect, Space Technol Lab, 55-61; staff engr, Aerospace Corp, 61-65; PROF CIVIL ENG, UNIV SOUTHERN CALIF, 65- *Concurrent Pos:* NSF sr fel, 64-65; consult, Aerospace Corp, 65-68, Northrop Corp, 69, Norair Div, 72-77 & Rockwell Int, 82-85; Albert Alberman vis prof, Technion, Israel Inst Technol, 75; vis prof, Univ Sydney, Australia, 86, Univ Canterbury, NZ, 86. *Mem:* Am Soc Mech Engrs; Am Soc Civil Engrs; Am Acad Mech; Am Asn Univ Professors. *Res:* Stability of structures; nonlinear elasticity; shell analysis. *Mailing Add:* Dept Civil Eng Univ Southern Calif Los Angeles CA 90089-2531

SEIDEHAMEL, RICHARD JOSEPH, b Cleveland, Ohio, Dec 26, 40; m 63; c 3. PHARMACOLOGY, MEDICAL SCIENCE. *Educ:* Univ Toledo, BS, 63; Ohio State Univ, MSc, 65, PhD, 68. *Prof Exp:* Sr scientist, 69-71, sr investr, 71-74, sr res assoc, 74-80, prin res assoc, 80-83, ASSOC DIR, MEAD JOHNSON RES CTR, 83-; ASSOC DIR, CLIN STUDIES, BRISTOL-MYERS SQUIBB CO. *Mem:* Am Soc Pharmacol & Exp Therapeut; AAAS; NY Acad Sci; Am Soc Clin Pharmacol & Therapeut; Am Heart Asn. *Res:* Respiratory, cardiovascular, central nervous system, ocular and autonomic pharmacology. *Mailing Add:* Bristol Meyers Squibb Co 2400 W Lloyd Expressway Evansville IN 47721

SEIDEL, BARRY S(TANLEY), b Philadelphia, Pa, Aug 27, 32; m 53; c 2. FLUID MECHANICS. *Educ:* Univ Del, BSME, 53; Mass Inst Technol, SM, 56, ScD(mech eng), 59. *Prof Exp:* Design engr, Aviation Gas Turbine Div, Westinghouse Elec Co, 53-55; asst prof mech eng, Univ Del, 59-64; NSF res fel, Calif Inst Technol, 64-65; assoc prof, 65-69, PROF MECH & AEROSPACE ENG, UNIV DEL, 69- *Concurrent Pos:* Vis prof, Cambridge Univ, 76 & Mass Inst Technol, 84-85; consult, Northern Res & Eng Corp. *Mem:* Am Soc Mech Engrs; Am Inst Aeronaut & Astronaut. *Res:* Fluid mechanics of turbomachinery. *Mailing Add:* 244 Orchard Rd Newark DE 19711

SEIDEL, GEORGE ELIAS, JR, b Reading, Pa, July 13, 43; m 70; c 1. REPRODUCTIVE PHYSIOLOGY. *Educ:* Pa State Univ, University Park, BS, 65; Cornell Univ, MS, 68, PhD(physiol), 70. *Prof Exp:* NIH fel, Harvard Med Sch, 70-71; from asst prof to assoc prof, 71-83, PROF PHYSIOL, COLO STATE UNIV, 83- *Concurrent Pos:* Vis scientist, Yale Univ, 78-79, Whitehead Inst, 86-87; assoc ed, J Exp Zool, 79-82. *Honors & Awards:* Alexander von Humboldt Award & Nat Asn Animal Breeders Res Award, 83; Upjohn Physiol Award, Am Dairy Sci Asn, 86. *Mem:* Am Dairy Sci Asn; Am Soc Animal Sci; Sigma Xi; AAAS; Soc Study Reproduction; Soc Study Fertil; Int Embryo Transfer Soc (vpres, 78, pres, 79). *Res:* Superovulation and embryo transfer; in vitro fertilization; oogenesis; microsurgery to mammalian embryos; cryopreservation of embryos. *Mailing Add:* Animal Reproduction & Biotechnol Lab Colo State Univ Ft Collins CO 80523

SEIDEL, GEORGE MERLE, b Springfield, Mass, Aug 14, 30; m 54; c 3. SOLID STATE PHYSICS. *Educ:* Worcester Polytech Inst, BS, 52; Purdue Univ, MS, 55, PhD(physics), 58. *Prof Exp:* NSF fel, Univ Leiden, 58-59; res assoc & lectr physics, Harvard Univ, 59-62; from asst prof to assoc prof, 62-67, PROF PHYSICS, BROWN UNIV, 67- *Concurrent Pos:* Fulbright lectr, Atomic Ctr, Arg, 73-74. *Mem:* AAAS; Fedn Am Scientist. *Res:* Low temperature physics, electronic properties of metals; magnetism. *Mailing Add:* Dept Physics Brown Univ Brown Sta Providence RI 02912

SEIDEL, HENRY MURRAY, b Passaic, NJ, July 19, 22; m 45; c 3. PEDIATRICS. *Educ:* Johns Hopkins Univ, AB, 43, MD, 46. *Prof Exp:* From instr to asst prof pediat, 50-68, asst dean student affairs, 68-71, ASSOC PROF PEDIAT, SCH MED, JOHNS HOPKINS UNIV, 68-, ASSOC PROF MED CARE & HOSPS, 69-, ASSOC DEAN, SCH MED, 77- *Mem:* Fel Am Acad Pediat. *Res:* Malignancy; maternal attitudes; medical care staffing; delivery systems. *Mailing Add:* 6336 Sunny Spring Columbia MD 21044

SEIDEL, JOHN CHARLES, b Milwaukee, Wis, Sept 25, 33; m 63; c 1. BIOCHEMISTRY. *Educ:* Carroll Col, Wis, BS, 56; Univ Wis-Madison, MS, 59, PhD(biochem), 61. *Prof Exp:* From res asst to res assoc, Retina Found, 61-70; staff scientist, 70-73, SR STAFF SCIENTIST, BOSTON BIOMED RES INST, 73- *Concurrent Pos:* Res assoc, Harvard Univ, 68-69, assoc, 69- *Mem:* AAAS; Am Soc Biol Chemists; Am Chem Soc; Biophys Soc; NY Acad Sci. *Res:* Chemistry of muscle contraction; electron spin resonance. *Mailing Add:* Boston Biomed Res Inst 20 Staniford St Boston MA 02114

SEIDEL, MICHAEL EDWARD, b New York, NY, Jan 20, 45; m 70; c 2. VERTEBRATE ZOOLOGY, HERPETOLOGY. *Educ:* Univ Miami, BS, 67; NMex Highlands Univ, MS, 69; Univ NMex, PhD(biol), 73. *Prof Exp:* Instr, Univ NMex, 73-74; ASST PROF BIOL, MARSHALL UNIV, 74- *Mem:* Herpetologists League; Soc Study Amphibians & Reptiles; Am Soc Ichthyologists & Herpetologists. *Res:* Comparative physiology and ecology of reptiles; systematics of amphibians and reptiles. *Mailing Add:* Dept Biol Sci Marshall Univ Huntington WV 25701

SEIDEL, THOMAS EDWARD, b Altoona, Pa, Oct 8, 35; m 60; c 3. PHYSICS, SOLID STATE ELECTRONICS. *Educ:* St Joseph's Col, Pa, BS, 57; Univ Notre Dame, MS, 59; Stevens Inst Technol, PhD(physics), 65. *Prof Exp:* Engr, Semiconductor Div, RCA Corp, 59-60; mem staff, Sarnoff Labs, 61-62 & 65-66; mem staff, Bell Labs, 66-; SR VPRES, TECH BR, J C SHUMACHER CO. *Mem:* Electrochem Soc; Am Phys Soc; Inst Elec & Electronics Engrs. *Res:* Ion implantation phenomena and applications to semiconductor devices. *Mailing Add:* Seidel Consults 1165 Wales Pl Cardiff CA 92007

SEIDELL, BARBARA CASTENS, b Rockville Centre, NY, Jan 21, 57; m 80. EVAPORITE & CLASTIC SEDIMENTOLOGY. *Educ:* Buchnell Univ, BA, 79; Johns Hopkins Univ, PhD(sedimentology), 84. *Prof Exp:* ASST PROF GEOL, BRYN MAWR COL, 84- *Mem:* Geol Soc Am; Soc Econ Paleontologists & Mineralogists; Sigma Xi. *Res:* Compartive sedimentology; documentation of depositional processes (sedimentological, geochemical and hydrological) of modern arid evaporitic environments, in particular a non-marine closed basin; comparison and application of this information to the sylvan and Triassic regions of eastern North America. *Mailing Add:* Dept Geol Bryn Mawr Col Bryn Mawr PA 19010

SEIDELMANN, PAUL KENNETH, b Cincinnati, Ohio, June 15, 37; m 60; c 2. ASTRONOMY, CELESTIAL MECHANICS. *Educ:* Univ Cincinnati, EE, 60, MS, 62, PhD(dynamical astron), 68. *Prof Exp:* Astronr, US Naval Observ, 65-73, asst dir, 73-76, dir, Nautical Almanac Off, 76-90, DIR, ORBITAL MECH DEPT, US NAVAL OBSERV, 90- *Concurrent Pos:* Lectr, Cath Univ, 66; proj officer, Air Standards Coord Comt, 73-81; vis asst prof, Univ Md, 73 & 75, vis assoc prof, 77, 79, 81, 83, 85, 87 & 89; ed comn, Celest Mech, 76-80; secy, vchmn & chmn, Div Dynamical Astron, Am Astron Soc. *Mem:* AAAS; Am Astron Soc; Am Inst Navig (vpres, 78-79, pres, 79-80; Am Inst Aeronaut & Astronaut; Int Astron Union (vpres, 85-88, pres, 88-91); Royal Astron Soc. *Res:* Dynamical astronomy; planetary research; general planetary theories; celestial navigation; CCD astrometry. *Mailing Add:* US Naval Observ Washington DC 20392

SEIDEMAN, WALTER E, b Washington Co, Wis, Jan 21, 33; m 88; c 3. PROCESS & PRODUCT DEVELOPMENT, PLANT LAYOUT & DESIGN. *Educ:* Univ Wis, BS, 59; Univ Mo, MS, 62, PhD(food sci & nutrit), 66. *Prof Exp:* Inst food sci, Univ Mo, 63-65; food scientist res & develop, Wilson & Co, 65-70, asst mgr, Wilson-Sinclair, 70-71, mgr, 71-72 & Wilson & Co, Inc, 72-76; OWNER & DIR CONSULT FIRM, WALTER E SIEDEMAN, PHD & ASSOCS, 76. *Concurrent Pos:* Mem, Sci Adv Comt, Poultry & Egg Inst Am, 65-70, Nat Cheese Coun, 66-68, Am Meat Inst & Res & Develop Assocs Mil Food & Packaging Systs, Inc, 71-82; secy & treas, Okla Food Labs, Inc, 89- *Mem:* Inst Food Technologists; Am Meat Sci Asn. *Res:* Food product and process development specializing in futher processed meat, poultry and flour based products; technological control systems for processing; sanitation systems. *Mailing Add:* 921 NW 72nd Oklahoma City OK 73116

SEIDEN, DAVID, b New York, NY, Apr 14, 46; m 67; c 3. MORPHOLOGY, CELL BIOLOGY. *Educ:* City Univ New York, BS, 67; Temple Univ, PhD(anat), 71. *Prof Exp:* Teaching asst anat, Sch Med, Temple Univ, 68-70; instr, 71-73, asst prof anat, Rutgers Univ Med Sch, 73-79; ASSOC PROF ANAT, ROBERT WOOD JOHNSON MED SCH, 79-, ASSOC DEAN, 90- *Mem:* AAAS; Am Asn Anatomists; NY Acad Sci. *Res:* Electron microscopy and cytochemistry of skeletal muscle and cardiac muscle. *Mailing Add:* Dept Neurosci & Cell Biol Robert Wood Johnson Med Sch Piscataway NJ 08854

SEIDEN, HY, b New York, NY, Feb 12, 15. GEOLOGY. *Educ:* Univ Calif, Los Angeles, BA, 50, MA, 72. *Prof Exp:* Consult, geol sci, 75-80; RETIRED. *Mem:* Geol Soc Am; Am Asn Petrol Geol. *Mailing Add:* 1401 Ming Ave Bakersfield CA 93304

SEIDEN, LEWIS S, b Chicago, Ill, Aug 1, 34; m 62; c 2. PSYCHOPHARMACOLOGY. *Educ:* Univ Chicago, BA, 56, BS, 58, PhD(biopsychol), 62. *Prof Exp:* Res assoc pharmacol, 63-64, from instr to asst prof pharmacol & psychiat, 65-72, assoc prof, 72-77, PROF PHARMACOL & PSYCHIAT, UNIV CHICAGO, 77- *Concurrent Pos:* USPHS fels pharmacol, Gothenburg Univ, 62-63 & Stanford Univ, 64-65; USPHS res grant, 65-, res career develop award, 67-77; career res scientist, 77-82. *Mem:* Am Soc Pharmacol & Exp Therapeut; Am Psychol Asn; Soc Neurosci; Int Col Neuropsychopharmacol. *Res:* Relationships between behavior, drugs and biogenic amines in the brain. *Mailing Add:* Dept Pharmacol & Physiol Sci Ab 109 Univ Chicago 947 E 58th St Chicago IL 60637

SEIDEN, PHILIP EDWARD, b Troy, NY, Dec 25, 34; m 54; c 2. PHYSICS. *Educ:* Univ Chicago, AB, 54, BS, 55, MS, 56; Stanford Univ, PhD(physics), 60. *Prof Exp:* Asst betatron, Univ Chicago, 55-56; scientist solid state physics, Missiles & Space Div, Lockheed Aircraft Corp, 56-59; NSF fel magnetism, Univ Grenoble, 60; mem staff, 60-66, mgr coop phenomena group, 66-70, mgr physics group, Res Ctr, 70-72, dir phys dept, Res Ctr, 72-76, dir gen sci dept, 76-77, STAFF MEM, RES CTR, IBM CORP, NY, 78- *Concurrent Pos:* Vis prof, Ind Univ, Bloomington, 67-68; mem, Solid State Sci Panel, Nat Acad Sci, 70-; Lady Davis vis scientist, Technion Inst, Israel, 74-75. *Mem:* Am Astron Soc; AAAS; Sigma Xi; fel Am Phys Soc. *Res:* Superconductivity; metals; magnetism; ferromagnetism; electrical properties of organic materials; galactic structure; immunology; theoretical investigations in immunology by means of computer simulation of the immune system using a cellular automaton model. *Mailing Add:* Res Ctr IBM Corp PO Box 218 Yorktown Heights NY 10598

SEIDENBERG, ABRAHAM, mathematics; deceased, see previous edition for last biography

SEIDENFELD, JEROME, b Chomutov, Czech, Dec 31, 45; US citizen; m 71; c 1. CELL CYCLE REGULATION, POLYAMINE METABOLISM. *Educ:* Yeshiva Univ, BA, 67; Univ Chicago, MS, 69; Univ Calif, San Francisco, PhD(med chem), 79. *Prof Exp:* Res assoc, Ben May Lab Cancer Res, Univ Chicago, 79-81; asst prof pharmacol & cancer ctr, 81-87, ASSOC PROF, MED SCH, NORTHWESTERN UNIV, 87- *Concurrent Pos:* Mem res comt, Am Cancer Soc, Ill Div; adv bd, Leukemia Res Found. *Mem:* Am Asn Cancer Res; Am Soc Cell Biol; Am Soc Pharmacol & Exp Therapeut; Soc Anal Cytol; Sigma Xi; AAAS. *Res:* Growth control and cell cycle regulation; cancer chemotherapy; polyamine functions in cell proliferation. *Mailing Add:* Dept Pharmacol Cancer Ctr Northwestern Univ Med Sch 303 E Chicago Ave Chicago IL 60611

SEIDER, WARREN DAVID, b New York, NY, Oct 20, 41; m 65; c 2. CHEMICAL ENGINEERING, COMPUTER SCIENCE. *Educ:* Polytech Inst Brooklyn, BS, 62; Univ Mich, MS, 63, PhD(chem eng), 66. *Prof Exp:* Res assoc chem eng, Univ Mich, 66-67; from asst prof to assoc prof, 67-84, PROF CHEM ENG, UNIV PA, 84- *Concurrent Pos:* Mem, Comput Aids for Chem Engrs Educ Comt, 69-71, chmn, 71-; vis assoc prof, Mass Inst Technol, 74-75 & Denmark Tech Univ, 83; dir, Am Inst Chem Engrs, 84-86. *Mem:* Am Chem Soc; Am Inst Chem Engrs. *Res:* Process analysis, simulation, design and control; phase and chemical equilibria; chemical reaction systems; azeotropic distillation; heat and power integration; applied numerical methods. *Mailing Add:* Dept Chem Eng Univ Pa Philadelphia PA 19104-6393

SEIDERS, VICTOR MANN, b Chicago, Ill, Jan 7, 31; m 65; c 2. GEOLOGY. *Educ:* Franklin & Marshall Col, BS, 58; Princeton Univ, MA, 60, PhD(geol), 62. *Prof Exp:* GEOLOGIST, US GEOL SURV, 62- *Mem:* AAAS; Geol Soc Am. *Res:* Mapping of metamorphic and volcanic rocks in Venezuela, Puerto Rico, Virginia, New England and North Carolina; mapping of Franciscan rocks in Santa Lucia range, California. *Mailing Add:* US Geol Surv Mail Stop 75 345 Middlefield Rd Menlo Park CA 94025

SEIDL, FREDERICK GABRIEL PAUL, b New York, NY, June 12, 18; div; c 3. MATHEMATICAL PHYSICS, NUCLEAR PHYSICS. *Educ:* Univ Pa, PhD(physics), 43. *Prof Exp:* Assoc physicist, Metall Lab, Univ Chicago, 42-44, Los Alamos Sci Lab, 44-46, Argonne Nat Lab, 46-48 & Brookhaven Nat Lab, 48-55; physicist, Boeing Co, 55-59, Lawrence Radiation Lab, Univ Calif, 59-65 & Inst Space Studies, NASA, 65-71; physicist, Plasma Physics Lab, Princeton Univ, 72-86; RETIRED. *Mem:* Am Phys Soc. *Res:* Numerical calculations of tokamak plasmas, stellar collisions, and underground nuclear explosions; construction of neutron velocity selectors; neutron resonances; light-nuclear reactions; standard neutron source; cosmic-ray asymmetries. *Mailing Add:* 98 Sayer Dr Princeton NJ 08540

SEIDL, MILOS, b Budapest, Hungary, May 24, 23; m 61; c 1. PHYSICS. *Educ:* Prague Tech Univ, BSc, 47, PhD(phys electronics), 49, DSc(physics), 63. *Hon Degrees:* MEng, Stevens Inst Technol, 79. *Prof Exp:* Mem staff, Res Inst Vacuum Electronics, Prague, 49-53, group leader vacuum devices, 53-58; group leader plasma physics, Inst Plasma Physics, Prague, 59-68; vis scientist, Stanford Univ, 68-69; PROF PHYSICS, STEVENS INST TECHNOL, 69- *Concurrent Pos:* Lectr, Prague Tech Univ, 60-68; Int Atomic Energy Agency fel, Culham Lab, UK Atomic Energy Authority, 62-63; Jess Davis Mem Res Award, Stevens Inst Technol, 78. *Mem:* Am Vacuum Soc; AAAS; Sigma Xi; fel Am Phys Soc. *Res:* Surface physics; ion sources; charged particle optics; plasma production. *Mailing Add:* Dept Physics & Eng Physics Stevens Inst Technol Castle Point Hoboken NJ 07030

SEIDLER, RAMON JOHN, b Floral Park, NY, Aug 10, 41; m; c 3. MICROBIOLOGY. *Educ:* San Fernando Valley State Col, BA, 64; Univ Calif, Davis, PhD(microbiol), 68. *Prof Exp:* Asst prof biol, San Fernando Valley State Col, 67; fel, Univ Tex M D Anderson Hosp & Tumor Inst, 70; from asst prof to assoc prof microbiol, Ore State Univ, 70- 80; PROF & RES MICROBIOLOGIST, US ENVIRON PROTECTION AGENCY, 84- *Concurrent Pos:* USPHS fel, Univ Tex M D Anderson Hosp & Tumor Inst, 70-71. *Mem:* Am Soc Microbiol; Brit Soc Gen Microbiol; fel Am Acad Microbiology. *Res:* Molecular systematics; environmental biology; risk assessment methds for genetically engineered bacteria. *Mailing Add:* US Environ Protection Agency 200 SW 35th St Corvallis OR 97331

SEIDLER, ROSEMARY JOAN, b New Orleans, La, Oct 4, 39. ANALYTICAL CHEMISTRY, INORGANIC CHEMISTRY. *Educ:* Loyola Univ, BS, 61; Tulane Univ, PhD(anal chem), 66. *Prof Exp:* Asst prof, 66-73, ASSOC PROF CHEM, CENTENARY COL LA, 73- *Concurrent Pos:* Qual control chemist, O J Beauty Lotion, 78- *Mem:* Sigma Xi; Am Chem Soc. *Res:* Magnetic properties of alpha-amido acid metal complexes; metal complexes of azo dyes. *Mailing Add:* Dept Chem Centenary Col 2911 Centenary Blvd Shreveport LA 71104

SEIDMAN, DAVID N(ATHANIEL), b Brooklyn, NY, July 5, 38; m 73; c 3. MATERIALS SCIENCE. *Educ:* NY Univ, BS, 60, MS, 62; Univ Ill, PhD(phys metall), 65. *Prof Exp:* Res assoc mat sci, 64-66; from asst prof to prof mat sci & eng, Cornell Univ, 76-85; prof mat sci, Hebrew Univ, 83-85; PROF MAT SCI & ENG, NORTHWESTERN UNIV, 85- *Concurrent Pos:* Vis scientist, Israel Inst Technol, 69-70, CEN, Grenoble & CENET, Meylon, France, 81; Guggenheim fel, 72-73 & 80-81; vis prof, Tel-Aviv Univ, 76; Lady Davis vis prof, Hebrew Univ, 78 & 80-81; sci consult, Argonne Nat Lab, 85-90; Alexander von Humboldt sr fel, 88-89; vis sci, Univ Guettingen, 89; vis sci, Cen Saclay, 89. *Honors & Awards:* Robert Lansing Hardy Gold Medal, Am Inst Metall Engrs, 66. *Mem:* fel Am Phys Soc; Am Inst Mining, Metall & Petrol Engrs; Mats Res Soc. *Res:* Internal interfaces; point and line imperfections in metals, semiconductors and ceramics; radiation damage in metals; internal interfaces; analytical electron microscopy. *Mailing Add:* Dept Mat Sci & Eng Northwestern Univ 2145 Sheridan Rd Evanston IL 60208-3108

SEIDMAN, IRVING, b Brooklyn, NY, Oct 3, 30; m 56; c 3. PATHOLOGY. *Educ:* Univ Va, BA & MD, 51. *Prof Exp:* Asst prof, 61-66, ASSOC PROF PATH, MED CTR, NY UNIV, 66- *Concurrent Pos:* NIH fel path, Med Ctr, NY Univ, 58-60. *Mem:* Am Soc Exp Path; Am Asn Path & Bact. *Res:* Experimental oncology. *Mailing Add:* Dept Path NY Univ Med Ctr 550 First Ave New York NY 10016

SEIDMAN, JONATHAN G, b New York, NY, April 22, 50; m 73; c 2. IMMUNOGENETICS, HUMAN GENETICS. *Educ:* Harvard Univ, BA, 72; Univ Wis, PhD(molecular biol), 75. *Prof Exp:* Fel, Nat Inst Child Health & Human Develop, NIH, 75-79, staff fel, Lab Molecular Genetics, 79-81; from asst prof to assoc prof, 81-88, PROF, HARVARD MED SCH, 88- *Res:* Study of the immune system using molecular biologic and genetic techniques; study of the molecular basis of familial hypertrophic cardiomyopathy. *Mailing Add:* Dept Genetics Harvard Med Sch 20 Shattuck St Boston MA 02115

SEIDMAN, MARTIN, b Brooklyn, NY, June 20, 21; m 44; c 8. CARBOHYDRATE CHEMISTRY. *Educ:* Brooklyn Col, AB, 41; Okla State Univ, MS, 48; Univ Wis, PhD(biochem), 50. *Prof Exp:* Chemist, P J Schweitzer Co, 41-42; asst sci aide, US Dept Navy, 42-43; instr gen & org chem, Okla State Univ, 46; res asst biochem, Univ Wis, 48-50; res chemist, Visking Co Div, Union Carbide Corp, 50-55 & Salvo Chem Co, 55-57; group leader process res, A E Staley Mfg Co, 57-71, sr scientist, Fermentation Lab, 71-73, group mgr, Starch Syrups & Fermentation Lab, 73-85; GROUP MGR, M SEIDMAN ASSOCS, 85- *Mem:* AAAS; Am Chem Soc; Am Inst Chemists; Royal Soc Chem. *Res:* Syrup, starch and cellulose chemistry; fermentation; enzymes; process research. *Mailing Add:* 346 W Macon St Decatur IL 62522

SEIDMAN, STEPHEN BENJAMIN, b New York, NY, Apr 13, 44; m 69; c 2. COMBINATORICS & FINITE MATHEMATICS. *Educ:* City Col New York, BS, 64; Univ Mich, Ann Arbor, MA, 65, PhD(math), 69. *Prof Exp:* Asst prof math, NY Univ, 69-72; from asst prof to assoc prof math, George Mason Univ, 72-84, prof comput sci, 84-90; PROF COMPUTER SCI & ENG & DEPT HEAD, AUBURN UNIV, 90- *Concurrent Pos:* Vis scholar Anthrop, Ariz State Univ, 82-83. *Mem:* Asn Comput Mach; Inst Elec & Electronics Engrs; Comput Soc. *Res:* Parallel computation; software for parallel computation; design of programming environments. *Mailing Add:* Dept Computer Sci & Eng Auburn Univ Auburn AL 36849-5347

SEIDMAN, THOMAS ISRAEL, b New York, NY, Jan 7, 35; m 69; c 1. ANALYSIS & FUNCTIONAL ANALYSIS, SOLID STATE PHYSICS. *Educ:* Univ Chicago, AB, 52; Columbia Univ, MA, 53; NY Univ, MS, 54, PhD(math), 59. *Prof Exp:* Res asst, Courant Inst Math Sci, NY Univ, 55-58; mathematician, Lawrence Radiation Lab, Univ Calif, Livermore, 58-60, lectr math, Univ Calif, Los Angeles, 60-61; mem, Math Res Ctr, Univ Wis, 61-62; mathematician, Boeing Sci Res Lab, Wash, 62-64; assoc prof math, Wayne State Univ, 64-67 & Carnegie-Mellon Univ, 67-72; PROF MATH, UNIV MD, BALTIMORE COUNTY, 72- *Concurrent Pos:* Vis prof, Univ Nice, 80-81. *Mem:* Am Math Soc; Soc Indust & Appl Math. *Res:* Author of over 120 papers in control theory, especially boundary control for diffusion processes; computational methods for ill-posed problems; partial differential equations, especially semiconductor models; numerical analysis. *Mailing Add:* Dept Math & Statist Univ Md Baltimore County Catonsville MD 21228

SEIELSTAD, GEORGE A, b Detroit, Mich, Dec 8, 37; m 65; c 3. RADIO ASTRONOMY. *Educ:* Dartmouth Col, AB, 59; Calif Inst Technol, PhD(physics), 63. *Prof Exp:* Asst prof physics, Univ Alaska, 63-64; from res fel to sr res fel, Calif Inst Technol, 64-72, res assoc radio astron, 72-84, staff mem, Owens Valley Radio Observ, 66-84; ASST DIR & SCIENTIST, NAT RADIO ASTRON OBSERV, 84- *Concurrent Pos:* Docent, Chalmers Univ Technol, 69-70; fel, John Simon Guggenheim Mem Found, 69-70; vis assoc prof astron & elec eng, Univ Ill, Urbana-Champaign, 78. *Mem:* AAAS; Am Astron Soc; Astron Soc Pac; Int Astron Union; Int Sci Radio Union. *Res:* Interferometry; polarimetry; extragalactic radio sources; galactic magnetic field; supernovae; galactic nuclei. *Mailing Add:* Nat Radio Astron Observ Box Two Green Bank WV 24944-0002

SEIF, ROBERT DALE, b Cincinnati, Ohio, May 25, 27; m 50; c 4. BIOMETRY. *Educ:* Ohio State Univ, BS, 50, MS, 52; Cornell Univ, PhD, 57. *Prof Exp:* From asst prof biomet to assoc prof agron, 56-69, PROF AGRON, UNIV ILL, URBANA, 69- *Concurrent Pos:* HEW grant, Univ Minn, 65. *Mem:* Am Soc Agron; Crop Sci Soc Am; Biomet Soc; fel Nat Asn Col Teachers Agron. *Res:* Biological statistics; data processing applied to agriculture, especially agronomy and horticulture. *Mailing Add:* W-501 Turner Hall 1102 S Goodwin Univ Ill Urbana IL 61801

SEIFEN, ERNST, b Oct 28, 30; m; c 3. CARDIOTONIC AGENTS, CALCIUM AGENTS. *Educ:* Univ Saarbruecken, Germany, MD & PhD(shock res), 66. *Prof Exp:* PROF PHARMACOL, TOXICOL & ANESTHESIOL, UNIV ARK MED SCI, 70- *Mem:* NY Acad Sci; Am Soc Pharmacol & Exp Therapeut. *Mailing Add:* Dept Pharmacol & Toxicol Univ Ark Med Sci Slot 611 Little Rock AR 72205-5177

SEIFER, ARNOLD DAVID, b Newark, NJ, Apr 22, 40. APPLIED MATHEMATICS, SYSTEM ANALYSIS & DESIGN. *Educ:* Rensselaer Polytech Inst, BS, 62, MS, 64, PhD(math), 68. *Prof Exp:* Res specialist appl math, Elec Boat Div, Gen Dynamics Corp, 67-73; mathematician sr staff, Appl Phys Lab, Johns Hopkins Univ, 73-76; sr staff engr, Electronics & Space Div, Emerson Elec Co, 76-80; PRIN ENGR, EQUIP DEVELOP LABS, RAYTHEON CO, 80- *Mem:* Soc Indust & Appl Math; Inst Elec & Electronics Engrs; Sigma Xi. *Res:* Interdisciplinary analytical problems of radar system design; radar detection and tracking; analysis and systems design of fire control systems. *Mailing Add:* 66 Dinsmore Ave No 606 Framingham MA 01701

SEIFERT, GEORGE, b Jena, Ger, Mar 4, 21; nat US; m 48; c 2. APPLIED MATHEMATICS. *Educ:* State Univ NY, AB, 42; Cornell Univ, MA, 48, PhD(math), 50. *Prof Exp:* Asst prof math, Univ Nebr, 50-55; assoc prof, 55-62, PROF MATH, IOWA STATE UNIV, 62- *Concurrent Pos:* Mem staff, Res Inst Advan Study, 59-60. *Mem:* Am Math Soc; Math Asn Am; Soc Indust & Appl Math. *Res:* Nonlinear ordinary differential equations; volterra integral equations; functional differential equations. *Mailing Add:* 2526 Kellogg Ave Ames IA 50011

SEIFERT, JOSEF, b Prague, Czech, Sept 21, 42; US citizen; m 88; c 1. BIOCHEMICAL TOXICOLOGY, INSECTICIDE TOXICOLOGY. *Educ:* Prague Inst Chem Technol, MSc, 64, PhD(biochem toxicol), 73. *Prof Exp:* Res specialist biochem toxicol, Prague Inst Chem Technol, 72-77; biochemist, Pesticide Chem & Toxicol Lab, Univ Calif, Berkeley, 77-78, from asst to assoc res specialist, 78-85; ASSOC PROF PESTICIDE CHEM, DEPT ENVIRON BIOCHEM, UNIV HAWAII, 86- *Concurrent Pos:* Mem prog comt, Am Chem Soc. *Mem:* Am Chem Soc; Soc Toxicol. *Res:* Investigation of mechanisms of target and nontarget toxic effects of pesticides; insecticide neurotoxicity, teratogenicity, metabolism and analytical methods for residue analysis; author of 50 papers. *Mailing Add:* Dept Environ Biochem Univ Hawaii Honolulu HI 96822

SEIFERT, KARL E, b Orangeville, Ohio, Mar 16, 34; m 81; c 3. GEOLOGY. *Educ:* Bowling Green State Univ, BS, 56; Univ Wis, MS, 59, PhD(geol), 63. *Prof Exp:* Phys scientist, Geotech Br, Air Force Cambridge Res Labs, 61-65; from asst prof to assoc prof, 65-72, PROF GEOL, IOWA STATE UNIV, 72-, CHAIR, 88- *Mem:* Am Geophys Union; Geochem Soc; Geol Soc Am. *Res:* Trace element geochemistry of igneous and metamorphic rocks as a function of their origin, evolution and environment with an emphasis on layered stratiform complexes, anorthosite complexes and oceanic basalts. *Mailing Add:* Dept Geol Sci Iowa State Univ Ames IA 50011

SEIFERT, LAURENCE C, b Jersey City, NJ; m; c 4. COMMUNICATIONS. *Educ:* NJ Inst Technol, BS. *Prof Exp:* Var eng, mfg & prod planning positions, Western Elec Kearny Works, NJ, AT&T, dir eng, Oklahoma City Works & Merrimack Valley Works, North Andover, Mass, vpres mfg res & develop, Eng Res Ctr, Princeton, NJ, vpres eng, 57-89, VPRES, COMMUN PROD SOURCING & MFG, AT&T, 89- *Concurrent Pos:* Dir, Enhanced Training Opportunities Prog, AT&T & Int Brotherhood Elec Workers & Technol Int Purchasing Co; mem, Mfg Studies Bd, Nat Res Coun. *Mem:* Nat Acad Eng; Inst Elec & Electronics Engrs; Inst Indust Eng. *Mailing Add:* AT&T Co 55 Corporate Dr Bridgewater NJ 08807

SEIFERT, RALPH LOUIS, b Alma, Mich, Jan 4, 43; m 79. MATHEMATICS. *Educ:* Ind Univ, AB, 63; Univ Calif, Berkeley, MA, 66, PhD(math), 68. *Prof Exp:* Asst prof math, Ind Univ, Bloomington, 68-70; from asst prof to assoc prof, 70-90, PROF MATH, HANOVER COL, 90- *Concurrent Pos:* Writer, CEMREL-Comprehensive Sch Math Proj, 70-74. *Mem:* Am Math Soc; Math Asn Am; Sigma Xi. *Res:* Foundations of mathematics; cognition of quantitative concepts. *Mailing Add:* Dept Math Hanover Col Hanover IN 47243

SEIFERT, WILLIAM EDGAR, JR, b Bozeman, Mont, Dec 21, 48; m 69; c 4. ANALYTICAL BIOCHEMISTRY. *Educ:* Marietta Col, BS, 70; Purdue Univ, West Lafayette, MS, 73, PhD(biochem), 75. *Prof Exp:* Supvr, 75-76, sr res assoc, 76-77, res scientist, Anal Chem Ctr, 77-82, asst prof, 78-83, RES ASST PROF, DEPT BIOCHEM & MOLECULAR BIOL, MED SCH, UNIV TEX HEALTH SCI CTR, HOUSTON, 83-, ASST PROF, GRAD SCH BIOMED SCI, 79-, ASST DIR, ANALYTICAL CHEM CTR, 82- *Mem:* Am Soc Mass Spectrometry. *Res:* Biochemical applications of mass spectrometry; stable isotopes to study biochemical reactions; biochemical applications of x-ray energy spectrometry; investigation of neurological disfunction by mass spectrometry. *Mailing Add:* Univ Tex Health Sci Ctr Med Sch PO Box 20708 Houston TX 77225-0708

SEIFERT, WILLIAM W(ALTHER), b Troy, NY, Feb 22, 20; m 43; c 4. ELECTRICAL ENGINEERING, SYSTEMS ANALYSIS. *Educ:* Rensselaer Polytech Inst, BEE, 41; Mass Inst Technol, SM, 47, ScD(elec eng), 51. *Prof Exp:* Instr elec eng, Rensselaer Polytech Inst, 41-44; mem res staff, Field Sta, 44-45, from asst to res assoc, Dynamic Anal & Control Lab, 45-51, mem res staff, 51-57, from assoc prof to prof elec eng, 57-70, asst dir, Dynamic Anal Control Lab, 52-55, actg dir, 56, asst to dean eng, 59-62, asst dean, 62-67, dir proj transport, 64-70, PROF CIVIL ENG, MASS INST TECHNOL, 70- *Mem:* Sr mem Inst Elec & Electronics Eng. *Res:* Transportation, computation and control; engineering education; computer modelling of socioeconomic systems; transportation problems in developing countries; environmental problems. *Mailing Add:* 675 Massachusetts Ave Cambridge MA 02139

SEIFF, ALVIN, b Kansas City, Mo, Feb 26, 22; m 68; c 4. SPACE SCIENCE, PLANETARY ATMOSPHERES. *Educ:* Univ Mo, BS, 42. *Prof Exp:* Res engr, Tenn Valley Authority, 42-44; tech supvr mass spectros, Uranium Isotope Separation Plant, Oak Ridge, Tenn, 44-46; instr physics, Univ Tenn, 46-48; aeronaut res scientist & chief supersonic free-flight res br, Ames Res Ctr & Aeronaut Lab, 48-63, chief vehicle environ div, 63-72, staff scientist, Dir Off, 72-76, sr staff scientist, Space Sci Div, Ames Res Ctr, NASA, 76-86; SR RES ASSOC, DEPT METEOROL, SAN JOSE STATE UNIV, 87- *Concurrent Pos:* Mem adv comt basic res, NASA, 71-73, mem res coun, Off Advan Res & Technol, 71-74, mem sci steering group, Pioneer Venus Proj, 72-82; mem entry sci team, Viking Mission to Mars, 69-76; mem grad fac aerospace eng, Univ Kans, 79; mem proj sci group, Galileo Orbiter-Probe Mission to Jupiter, 79-; mem, comt planetary atmosphere, Comt Space Res, 82- & joint working group study team on outer planets, Nat Res Coun & Europ Sci Found, 83; mem, US team partic in Soviet-French Vega balloon mission on planet Venus, 83-86; co-investr, Titan Atmosphere Struct, Europ Space Agency Huygens Probe, 91-; mem, Sci Definition Team, US Mars Network Mission, 90- *Honors & Awards:* H Julian Allen Award, 82; Von Karman Medal & Lectr, Am Inst Aeronaut & Astronaut, 90- *Mem:* Assoc fel

Am Inst Aeronaut & Astronaut; Am Astron Soc; Am Geophys Soc; Planetary Soc. *Res:* Physical structure of atmospheres of Mars and Venus; in-situ measurements in the atmosphere of Jupiter; hypervelocity entry into atmospheres of earth and other planets; boundary layers and viscous flows; physical measurements from entry probes and balloons; dynamics of atmospheres of Mars and Venus. *Mailing Add:* Mail Stop 245-1 Ames Res Ctr Moffett Field CA 94035

SEIFFERT, STEPHEN LOCKHART, b Iowa, Nov 16, 42. NUCLEAR MATERIALS, METALLURGICAL PHYSICS. *Educ:* NMex State Univ, BS, 67; Univ Utah, MS, 72, PhD(mat sci), 74. *Prof Exp:* Physicist nuclear effects, White Sands Missile Range, 67; res asst, Univ Utah, 70-73; collaborator metal physics, Dept Res Fund, Centre d'Etude Nucleaires, Grenoble, France, 73-75; sr metallurgist nuclear mat, EG&G Idaho Inc, 76-79, scientist light water safety res, 79-81; prin staff mem, BDM Corp, 81-90; INDEPENDENT CONSULT, 90- *Concurrent Pos:* Affil prof metal, Idaho Nat Eng Lab Educ Prog, 77-81. *Mem:* Am Phys Soc; Am Inst Physics; Am Nuclear Soc; Am Soc Testing & Mat; Laser Inst Am; Am Inst Aeronaut & Astronaut. *Res:* Nuclear materials applications and reactor safety research; mechanical properties changes; radiation damage in metals; laser optical/thin films and metrology. *Mailing Add:* 9437 Thornton NE Albuquerque NM 87109

SEIFRIED, ADELE SUSAN CORBIN, b Chicago, Ill, Nov 16, 47; m 70; c 3. VIROLOGY, ELECTRON MICROSCOPY. *Educ:* Purdue Univ, BS, 68; Cornell Univ, MS, 70. *Prof Exp:* Teaching & res asst textile chem, Cornell Univ, 68-70, researcher food sci, 70-73; textile chemist, textile testing lab, Better Fabrics Testing Bur, 73-74; res chemist biochem & virol, 74-84, chemist over-the-counter drug eval, 84-86, CONSUMER SAFETY OFFICER, REGULATORY AFFAIRS, FOOD & DRUG ADMIN, 86- *Concurrent Pos:* Abstractor toxicol, Tracor Jitco, 82, Macro Systs, 83. *Mem:* AAAS; Am Soc Microbiol; Am Chem Soc; Regulatory Affairs Prof Soc. *Res:* Development of regulations for the marketing of prescription and over-the-counter drugs and biologics; biochemical basis of vaccine virus infections; phosphorylation, capping and polyadenylation of vaccine viruses and effects of interferon on their replication. *Mailing Add:* 9205 Quintana Dr Bethesda MD 20817

SEIFRIED, HAROLD EDWIN, b Suffern, NY, Apr 23, 46; m 70; c 3. BIOCHEMICAL PHARMACOLOGY, TOXICOLOGY. *Educ:* Univ Rochester, BS, 68; Cornell Univ, MS, 71, PhD(biochem), 73; Am Bd Toxicol, dipl, 80 & 84, 90; Am Bd Indust Hygiene, 86. *Prof Exp:* Sr chemist protein & lipid chem & cosmetic ingredient toxicity, Explor Prod Res, Avon Prod Inc, 73-74; Roche fel & guest worker carcinogen metab & enzymology activation deactivation, Nat Inst Arthritis, Diabetes, Digestive & Kidney Dis, NIH, 74-76, guest worker, transmethylation of fluorinated catecholamines & interaction with neurotransmitters, Nat Inst Arthritis, Metab & Digestive Dis, 77-88; biochem toxicologist indust hyg, occup & environ health, Stanford Res Inst, 76-77; prin toxicol, Indust Hygienist, Tech Resources Inc, 87-90; PROG DIR, NAT CANCER INST CHEM & PHYS CARCINOGENESIS BR, 90- *Concurrent Pos:* sr toxicologist & prin investr, Sci Policy in Toxicol, Tracor Jitco, Inc, 77-90; risk assessment, Clorox Co, Fuji, 88-90. *Mem:* Am Col Toxicol; Am Indust Hyg Asn; Am Chem Soc; Soc Toxicol; Int Soc Study Xenobiotics. *Res:* Chemical carcinogenesis; metabolism; toxicology; carcinogen and poly-cyclic aromatic hydrocarbon metabolism and binding to nucleic acid; enzymology of activation-deactivation; lipid biosynthesis; toxic substances relating to occupational exposure and carcinogenesis, industrial hygiene toxicology, asbestos analysis. *Mailing Add:* 9205 Quintana Dr Bethesda MD 20817

SEIFTER, ELI, b Cleveland, Ohio, Apr 17, 19; m 46; c 1. NUTRITION. *Educ:* Ohio State Univ, AB, 48; Univ Pa, PhD(bot), 53. *Prof Exp:* Asst instr bot, Univ Pa, 50-52; res biochemist, Monsanto Chem Co, 52-57; biochemist, Lond Island Jewish Hosp, NY, 57-62; asst prof, 62-72, ASSOC PROF BIOCHEM, ALBERT EINSTEIN COL MED, 72-, ASSOC PROF SURG, 76- *Mem:* AAAS; Am Chem Soc; Am Inst Nutrit; Sigma Xi. *Res:* Metabolic pathways; amino acids; metabolic effects. *Mailing Add:* 50 Morris Dr New Hyde Park NY 11044

SEIFTER, SAM, b Cleveland, Ohio, Dec 1, 16; m 43; c 2. BIOCHEMISTRY. *Educ:* Ohio State Univ, BA, 39; Western Reserve Univ, MS & PhD(biochem), 44. *Prof Exp:* Asst immunol, Western Reserve Univ, 40-44, instr, Med Sch, 44-45, sr instr immunochem, 45; asst prof biochem, Long Island Col Med, 45-49; assoc prof, Col Med, State Univ NY Downstate Med Ctr, 49-53; biochemist, Long Island Jewish Hosp, 54-56; from assoc prof to prof biochem, 56-90, from actg chmn to chmn, 75-76, EMER PROF BIOCHEM, ALBERT EINSTEIN COL MED, 90- *Mem:* Am Chem Soc; Am Soc Biol Chemists. *Res:* Protein chemistry; enzymology; immunochemistry; connective tissues. *Mailing Add:* Dept Biochem Albert Einstein Col Med New York NY 10461

SEIGEL, ARNOLD E(LLIOTT), b Washington, DC, July 16, 23; m 51; c 3. PHYSICS, MECHANICAL ENGINEERING. *Educ:* Univ Md, BS, 44; Mass Inst Technol, MS, 47; Univ Amsterdam, ScD, 52. *Prof Exp:* Mech engr, Signal Corps, US Dept Army, 45 & US Naval Res Lab, 45-46; air conditioning engr, William Brown Consult Engrs, 47; mech engr & physicist, US Naval Ord Lab, 48-61, chief Ballistics Dept, 61-74; DIR INSTRNL TV, UNIV MD, COLLEGE PARK, 75- *Concurrent Pos:* Lectr, Univ Md, 54-; consult, 58- *Mem:* Am Inst Physics; Am Inst Aeronaut & Astronaut; Am Phys Soc. *Res:* Gas dynamics; interior ballistics; thermodynamics; hydrodynamics; high speed flow; stress waves in solids; aeronautical engineering. *Mailing Add:* 3302 Pauline Dr Chevy Chase MD 20815

SEIGEL, RICHARD ALLYN, b Brooklyn, NY, Aug 2, 54; m 76. EVOLUTIONARY ECOLOGY, CONSERVATION BIOLOGY. *Educ:* Rutgers Univ, BA, 77; Univ Cent Fla, 79; Univ Kans, PhD(ecol & systematics), 84. *Prof Exp:* Postdoctoral assoc, Savannah River Ecol Lab, 84-87; asst prof, 87-91, ASSOC PROF ECOL, SOUTHEASTERN LA UNIV, 91- *Concurrent Pos:* Consult, Int Wildlife Coalition, 82-91, La Audubon Soc, 88-89; assoc ed, J Herpet, 89-91. *Mem:* Am Soc Naturalists; Soc Study Amphibians & Reptiles; Am Soc Ichthyologists & Herpetologists; Soc Conserv Biol. *Res:* Evolutionary ecology of amphibians and reptiles, with emphasis on evolution of life-history strategies; conservation biology, especially of non-game species. *Mailing Add:* Dept Biol Sci Southeast La Univ Box 814 Hammond LA 70402

SEIGER, HARVEY N, b New York, NY, June 20, 24; m 48; c 2. ELECTROCHEMICAL IMPREGNATION, SHUTTLE MECHANISMS. *Educ:* City Univ New York, BS, 49, MA, 52; Polytech Inst, PhD(chem), 62. *Prof Exp:* Dir res, Gulton Industs, 60-69; dir electrochem, Div Textron, Heliotek, 69-75; vpres res, Yardney Elec Corp, 75-79; consult, Harvey N Seiger Assocs, 79-83; PROG MGR, WESTINGHOUSE NAVAL SYSTS DIV, 83- *Concurrent Pos:* Mem, Nat Battery Adv Comt, 76-85. *Honors & Awards:* IR-100 Award, Indust Res, 67. *Mem:* Electrochem Soc. *Res:* Electrode phenomena with respect to energy storage; physical chemistry-kinetics; thermodynamics; irreversible processes-with batteries as the end product. *Mailing Add:* 6355 Woodhawk Mayfield Heights OH 44124

SEIGER, MARVIN BARR, b New York, NY, Nov 18, 26; m 60; c 3. GENETICS. *Educ:* Duquesne Univ, BS, 50; Univ Tex, MA, 53; Univ Calif, Los Angeles, MA, 59; Univ Toronto, PhD(genetics), 62. *Prof Exp:* Vis asst prof genetics, Purdue Univ, 62-63; res assoc, Univ Notre Dame, 63-64; NIH trainee biol, Univ Rochester, 64-65; from asst prof to assoc prof, 65-86, PROF GENETICS, WRIGHT STATE UNIV, 87- *Concurrent Pos:* Vis res prof, Univ Belgrade, Yugoslavia, 87; vis prof, Fed Univ, Brazil, 76-79. *Mem:* Soc Study Evolution; Genetics Soc Am; Am Soc Zool; Animal Behav Soc; Am Soc Naturalists; Sigma Xi. *Res:* Population dynamics; quantitative inheritance; behavior and ecological genetics. *Mailing Add:* Dept Biol Sci Wright State Univ Dayton OH 45435

SEIGLE, L(ESLIE) L(OUIS), b Dumbarton, Scotland, June 13, 17; m 50; c 3. PHYSICAL METALLURGY. *Educ:* Cooper Union, BChE, 41; Univ Pa, MS, 48; Mass Inst Technol, ScD, 52. *Prof Exp:* Res metallurgist, Int Nickel Co, 38-46; proj leader, Univ Pa, 46-48; instr metall, Drexel Inst Technol, 48; res asst labs, Gen Elec Corp, 48 & Mass Inst Technol, 48-51; head fundamental metall sect, Sylvania Elec Prod, Inc, Gen Tel & Electronics Corp, 51-56, mgr metall lab, Res Labs, 56-65; prof, 65-87, EMER PROF MAT SCI, STATE UNIV NY, STONY BROOK, 87- *Concurrent Pos:* Adj prof, NY Univ, 52-60; mem refractory metals panel, Mat Adv Bd, Nat Acad Sci, 55-65 & comt on coatings, 66-69; NY Univ exchange metall deleg, Moscow Steel Inst, USSR, 57; vis prof, Univ Pittsburgh, 64-65. *Mem:* Am Soc Metals; Am Inst Mining, Metall & Petrol Engrs. *Res:* Thermodynamics of alloys, diffusion and sintering of metals; protective coatings. *Mailing Add:* One Saywood Lane Stony Brook NY 11790

SEIGLER, DAVID STANLEY, b Wichita Falls, Tex, Sept 11, 40; m 61; c 2. BOTANY, ORGANIC CHEMISTRY. *Educ:* Southwestern State Col, Okla, BS, 61; Univ Okla, PhD(org chem), 67. *Prof Exp:* Assoc chem, Northern Regional Lab, USDA, 67-68; fel bot, Univ Tex, Austin, 68-70; asst prof, 70-76, assoc prof bot, 76-80, PROF PLANT BIOL, ENTOM, HORTICULT, UNIV ILL, URBANA, 85- *Concurrent Pos:* Fulbright Hays lectr, 76; Nat Acad Sci, exchange with Ger Democrat Rep, 80. *Mem:* Bot Soc Am; Am Chem Soc; Soc Econ Bot; Soc Chem Ecol. *Res:* Phytochemistry; study of secondary plant compounds; biochemical systematics; roles of plant secondary compounds in biological interactions; plant-fungus, plant-plant and plant-insert interactions. *Mailing Add:* Dept Plant Biol Univ Ill 505 S Goodwin Urbana IL 61801

SEIGLER, HILLIARD FOSTER, b Asheville, NC, Apr 19, 34; m 61; c 4. SURGERY, IMMUNOLOGY. *Educ:* Univ NC, BA, 56, MD, 60. *Prof Exp:* NIH res fel immunogenetics, 65-67, asst prof surg, 67-70, asst prof immunol, 69-70, assoc prof surg & immunol, 71-78, CO-PROG DIR, CLIN CANCER RES UNIT, MED CTR, DUKE UNIV, 70-, PROF SURG 78- *Concurrent Pos:* Clin investr, Vet Admin Hosp, Durham, NC, 67-70, chief surg, 71-72. *Honors & Awards:* Henry C Fordham Award, 62. *Mem:* Transplantation Soc; Am Col Surgeons; Int Primatol Soc; Asn Acad Surgeons; Soc Univ Surgeons. *Res:* Immunogenetics of transplantation; tumor immunology. *Mailing Add:* Dept Immunol & Surg 841 Med Ctr Duke Univ Box 3966 Durham NC 27706

SEIKEN, ARNOLD, b New York, NY, Feb 23, 28; m 55; c 3. MATHEMATICS. *Educ:* Syracuse Univ, BA, 51; Univ Mich, MA, 54, PhD(math), 63. *Prof Exp:* Instr math, Southern Ill Univ, 58-60; asst prof, Oakland Univ, 61-64; assoc prof, Univ RI, 64-67; ASSOC PROF MATH, UNION COL, 67-, CHMN DEPT, 68- *Mem:* Am Math Soc; Math Asn Am. *Res:* Differential geometry, particularly theory of connections. *Mailing Add:* Dept Math Bailey 2086 Union Col Schenectady NY 12308

SEIL, FREDRICK JOHN, b Nove Sove, Yugoslavia, Nov 9, 33; US citizen; m 55; c 2. NEUROLOGY. *Educ:* Oberlin Col, AB, 56; Stanford Univ, MD, 60. *Prof Exp:* Resident neurol, Stanford Univ, 61-64, fel, 64-65; fel neurol, Mt Sinai Hosp, NY & Albert Einstein Col Med, 65-66; asst prof neurol, Stanford Univ, 69-75; assoc prof, Health Sci Ctr, Univ Ore, 76-78; PROF NEUROL, ORE HEALTH SCI UNIV, 78- *Concurrent Pos:* Staff neurologist, Vet Admin Hosp, Palo Alto, 69-76; clin investr, Vet Admin Hosp, Portland, Ore, 76-79, staff neurologist, 79-81; dir, Vet Admin Off Regeneration Res Prog, 81- *Mem:* Int Brain Res Orgn; Soc Neurosci; Am Neurol Asn; Am Asn Neuropathologists; Int Soc Develop Neurosci. *Res:* Tissue culture studies of structure and function of the nervous system, of myelination and demyelination, of the pathophysiology of neurotoxic agents, and of neural development and plasticity. *Mailing Add:* Off Regeneration Res Prog Vet Admin Med Ctr Portland OR 97201

SEILER, DAVID GEORGE, b Green Bay, Wis, Dec 17, 40; m 63; c 2. SOLID STATE PHYSICS. *Educ:* Case Western Reserve Univ, BS, 63; Purdue Univ, Lafayette, MS, 65, PhD(physics), 69. *Prof Exp:* Physicist, Nat Bur Standards, 72-73; temp asst prof, 69-70, asst prof, 70-72, assoc prof physics, NTex State Univ, 73-; AT PROG MAT RES, NSF. *Mem:* Am Phys Soc; Optical Soc Am. *Res:* Semiconductors; energy band structures; scattering mechanisms. *Mailing Add:* Dept Physics NTex State Univ PO Box 5368 Dallas TX 76203

SEILER, FRITZ A, b Basel, Switz, Dec 20, 31; US citizen; m 64; c 3. HAZARDOUS MATERIAL MANAGEMENT. *Educ:* Kantonale Handelsschule, Basel, Switz, 51; Univ Basel, Switz, PhD (physics). *Prof Exp:* Res assoc, Univ Wis, Madison, 62-63; sci assoc physics, Univ Basel, Switz, 63-69, from lectr to asst prof, 75-80; SR SCIENTIST, LOVELACE INHALATION TOXICOL RES INST, 80- *Concurrent Pos:* Staff officer, Swiss Army Staff, 64-74; consult, Swiss Dept Defense, 68-74 & Lung Model Task Force, NCRP, 89; cmndg officer, Swiss Army Radiation Labs, 68-74; vis scientist, Lawrence Berkeley Labs, 74-75, consult, 76, guest scientist, 78; mem, Am Nat Standards Inst, 86-, mgt coun, 87- *Mem:* Fel Am Phys Soc; Health Physics Soc; Soc Risk Anal; Inst Nuclear Mats Mgt. *Res:* Risk assessment and management; dose effect relationships; risks of fusion and space nuclear reactors; uncertainty analysis; statistical sampling; health effects of one or more toxicants; route selection for hazardous material transports; methodologies risk analysis and management; risk comparison. *Mailing Add:* Int Technol Corp 5301 Central Ave NE Albuquerque NM 87108

SEILER, GERALD JOSEPH, b New Rockford, NDak, June 4, 49; m 74; c 2. PLANT BREEDING, GERM PLASM. *Educ:* NDak State Univ, BA, 71, MS, 73, PhD(bot), 80. *Prof Exp:* Fel bot & biol, NDak State Univ, 71-73; asst, State Biol Surv Kans, 73-74; res technician agron & crop physiol, 74-80, RES BOTANIST, AGR RES SERV, USDA, BUSHLAND, TEX, 80- *Concurrent Pos:* Cur wild sunflower germ plasm, Agr Res Serv, US Dept Agr, 80-; US coordr, Food & Agr Orgn, UN, Rome Italy, 81- *Mem:* Am Soc Plant Taxonomist; Agron Soc Am; Crop Sci Soc Am; Bot Soc Am; Sigma Xi. *Res:* Breeding of native species of sunflowers for specific characters of insect pest and disease resistance for commercial production; incompatibility of the species and development of techniques for interspecific crossing and physiology of stress and drought in wild species. *Mailing Add:* PO Box 5677 Fargo ND 58105

SEILER, STEVEN WING, b Glen Ridge, NJ, May 31, 50; m 72. PLASMA PHYSICS. *Educ:* Cornell Univ, BA, 72; Princeton Univ, MA, 74, PhD(physics), 77. *Prof Exp:* Res assoc neutron diagnostics, Plasma Physics Lab, Princeton Univ, 77-79; SR RES PHYSICIST, R&D ASSOCS, 79- *Mem:* Am Phys Soc; Sigma Xi; Inst Elec & Electronics Engrs. *Res:* Neutron and alpha particle diagnostics on fusion research tokamaks and micro instability research on Q-machines. *Mailing Add:* R&D Assocs 301A S West St Alexandria VA 22314

SEILHEIMER, JACK ARTHUR, b Kalamazoo, Mich, Nov 12, 35; m 54; c 4. ZOOLOGY, LIMNOLOGY. *Educ:* Western Mich Univ, BS, 60; Univ Louisville, PhD(zool), 63. *Prof Exp:* Instr limnol, Univ Louisville, 62; from instr to assoc prof, 63-73, PROF BIOL, SOUTHERN COLO STATE COL, 73- *Concurrent Pos:* Vis lectr, Colo-Wyo Acad Sci, 65 & 66; consult water pollution, Pueblo City County Health Dept, 66- *Mem:* AAAS; Am Soc Limnol & Oceanog; Ecol Soc Am; Sigma Xi. *Res:* Stream ecology; algology; plankton; ichthyology; pollution biology; radioecology. *Mailing Add:* One Remington Ct Pueblo CO 81008

SEILING, ALFRED WILLIAM, b Watseka, Ill, May 28, 36; m 57; c 4. SEMICONDUCTOR ASSEMBLY PROCESSES, FORMULATION OF HIGH PERFORMANCE EPOXY. *Educ:* Blackburn Col, BA, 57; Ind Univ, PhD(chem), 62. *Prof Exp:* Res chemist, Morton Chem, Div Morton Int, 61-64, tech serv supvr, 64-67, tech mgr electronics, 67-74, group mgr, 74-77, gen mkt electronic mat, 77-79, vpres gen mkt, 79-83, vpres mkt dynachem, 83-84; CONSULT, SEILING & ASSOCS, 84- *Res:* Development of materials and processes for assembly and packaging of integrated circuits and other electronic components; systems are fabricated from plastics, ceramics and metals by molding, photolethography and adhesive bonding processes; formulation of high performance epoxy, silicone and polyimide adhesives and molding compounds. *Mailing Add:* Two Windsor Rise Monterey CA 93940

SEIM, HENRY JEROME, b Granite Falls, Minn, Mar 20, 19; m 46; c 2. ANALYTICAL CHEMISTRY. *Educ:* St Olaf Col, BA, 41; Mont Sch Mines, MS, 43; Univ Wis, PhD(chem), 49. *Prof Exp:* Asst chem, Mont Sch Mines, 41-43, instr, 45-46; res & anal chemist, Boeing Airplane Co, 43-45; asst chem, Univ Wis, 46-49; from instr to assoc prof, Univ Nev, 49-62; mgr chem res, Res Div, Allis-Chambers, Wis, 62-67, dir, 67-69; mgr anal res & serv, Kaiser Aluminum & Chem Corp, 69-82, sr res assoc, Ctr Technol, 82-86; RETIRED. *Honors & Awards:* Award Merit, Am Soc Testing & Mat, 85. *Mem:* Am Chem Soc; Am Indust Hyg Asn; Am Soc Testing & Mat. *Res:* Instrumental analysis; ion exchange; electrochemistry. *Mailing Add:* 931 Val Aire Pl Walnut Creek CA 94596

SEINER, JEROME ALLAN, b Pittsburgh, Pa, Aug 21, 32; m 55; c 4. POLYMER COATINGS & PAINTS, METAL TREATMENTS & PRETREATMENTS. *Educ:* Carnegie Mellon Univ, BS, 54, BS, 60; Case Western Reserve Univ, MEND, 70. *Prof Exp:* Engr trainee to var positions, 54-84, DIR ADVAN RES, PPG INDUSTS, 84- *Concurrent Pos:* Lectr, Math & Econ, Carnegie Mellon Univ, 61-80; ed, I&EC Prod Res & Develop, Am Chem Soc J, 79-86, sr ed, I&EC Res, 86-; chmn, Chem & Physics Coatings & Films, Gordon Res Conf, 87 & Vis Comt, Mat Sci & Eng Dept, Univ Pittsburgh, 88-; elected pres, PPG Collegium, 90- *Mem:* Am Chem Soc; Federated Socs Coatings Technol. *Res:* Coatings and metal treatments; synthesis; formulation and testing of conventional and fire retardant materials. *Mailing Add:* PPG Industs Res & Develop Ctr PO Box 9 Pittsburgh PA 15101-0009

SEINER, JOHN MILTON, b Upper Darby, Pa, Feb 23, 44; m 68; c 4. AEROSPACE ENGINEERING, ACOUSTICS. *Educ:* Drexel Univ, BSME, 67; Pa State Univ, MSAE, 69, PhD(aeorspace eng), 74. *Prof Exp:* AEROSPACE ENGINEER JET NOISE, NASA LANGLEY RES CTR, 74- *Mem:* Sigma Xi. *Res:* Supersonic jet noise; nonlinear acoustics; physics of high speed turbulence; laser velocimetry and raman spectroscopy. *Mailing Add:* 209 Sheffield Rd Williamsburg VA 23185

SEINFELD, JOHN H, b Elmira, NY, Aug 3, 42. CHEMICAL ENGINEERING. *Educ:* Univ Rochester, BS, 64; Princeton Univ, PhD(chem eng), 67. *Prof Exp:* LOUIS E NOHL PROF CHEM ENG & CHMN, DIV ENG & APPL SCI, CALIF INST TECHNOL, 67- *Concurrent Pos:* Camille & Henry Dreyfus Found teacher-scholar grant, 72; Inst lectr, Am Inst Chem Engrs, 80. *Honors & Awards:* Donald P Eckman Award, Am Automatic Control Coun, 70; Curtis W McGraw Res Award, Am Soc Eng Educ, 76; Allan P Colburn Award, Am Inst Chem Engrs, 76; Pub Serv Award, NASA, 80; William H Walker Award, Am Inst Chem Engrs; George Westinghouse Award, Am Soc Eng Educ. *Mem:* Nat Acad Eng; Am Inst Chem Engrs; Am Chem Soc; Air Waste Mgmt Asn; Am Soc Eng Educ. *Res:* Air pollution. *Mailing Add:* Div Eng & Appl Sci Calif Inst Technol Pasadena CA 91125

SEIPEL, JOHN HOWARD, b Pittsburgh, Pa, Nov 9, 25; m 59; c 4. NEUROLOGY, LAW. *Educ:* Carnegie Inst Technol, BS, 46, MS, 47; Harvard Univ, MD, 54; Northwestern Univ, PhD(chem), 58; George Mason Univ, JD, 90. *Prof Exp:* Intern, Pa Hosp, Philadelphia, 54-55, resident surg, 55-56; res scientist, Nat Cancer Inst, 56-58; asst resident neurol, Mt Alto Vet Admin Hosp, Washington, DC, 58; chief resident, Georgetown Univ Hosp, 59 & DC Gen Hosp, 60; res fel, Georgetown Univ Hosp, 60-61, from clin instr to asst prof neurol, 61-79; dir neurol res, Friends Med Sci Res Ctr, Inc, 66-78; CHIEF MED OFFICER, DC DEPT CORRECTIONS, 79- *Concurrent Pos:* Neurol attending staff, Georgetown Univ Med Ctr, 61-79, Fairfax Hosp, Va, 61-79; Greenwalt fel neuroanat, NY Univ, 61; chief neurol lab, Georgetown Clin Res Inst, Fed Aviation Admin, 61-66, sr aviation med examr, Aviation Med Serv, 61-, consult neurol, 72-; consult neurol, Res Div, Md State Dept Ment Hyg & Spring Grove State Hosp, 66-78; staff neurologist, Neurol Serv, US Vet Admin Hosp, Washington, DC, 67-69, chief electrodiag sect, 68-69; dir neurol res, Md Psychiat Res Ctr, 69-78; attend neurologist, Nat Children's Rehab Ctr, Leesburg, Va, 72-73, Fauquier Hosp, Warrenton, 76-78; res assoc, Dept Psychiat, Md Psychiat Res Ctr, Baltimore, 77-78. *Honors & Awards:* S Weir Mitchell Award, Am Acad Neurol, 66. *Mem:* Fel Am Col Clin Pharmacol; Sigma Xi. *Res:* The application of physical science and technology to the solution of medical and legal problems, particularly in neurology; aerospace medicine; environmental sciences; pharmacology. *Mailing Add:* Maximum Security Facil PO Box 5200 Lorton VA 22199

SEIREG, ALI A, b Mahalla, Egypt, Oct 26, 27; US citizen; m 54; c 2. SYSTEMS DESIGN, ROBOTICS. *Educ:* Univ Cairo, BSc, 48; Univ Wis, PhD(mech eng), 54. *Prof Exp:* Lectr mech eng, Univ Cairo, 54-56; adv engr res & develop, Falk Corp, 56-59; prof theory appl mech, Marquette Univ, 59-65; PROF MECH ENG, UNIV WIS-MADISON, 65- *Concurrent Pos:* Consult, Falk Corp, 59-, Vet Admin Res, 64-, NSF, 80-82, & Jet Propulsion Lab, 81-82; ed, Comput in Mech Eng, 81- & SOMA, Eng Human Body, 86-; pres, Gear Res Inst, Am Soc Mech Engrs, 84- & chmn, Coun Eng, 85- *Honors & Awards:* George Washington Award, Am Soc Eng Educ, 70; Richard Mem Award, Am Soc Mech Engrs, 73; E P Connell Award, Am Gear Mfg Asn, 74. *Mem:* Am Soc Mech Engrs (sr vpres, 85-); Am Soc Eng Educ; Soc Exp Stress Anal; Am Gear Mfg Asn; hon mem Chinese Mech Eng Soc; int mem USSR Acad Sci. *Res:* Mechanical systems analysis; computer aided design; robotics; biomechanics; rehabilitation devices; underwater systems; gears and power transmission; friction lubrication and wear. *Mailing Add:* Dept Mech Eng Univ Fla Gainesville FL 32611

SEITCHIK, JEROLD ALAN, b Philadelphia, Pa, Jan 26, 35. SOLID STATE PHYSICS. *Educ:* Univ Del, BS, 56; Univ Pa, PhD(physics), 63. *Prof Exp:* Mem tech staff, Physics Res, Bell Tel Labs, 63-65; sr physicist, Univac Div, Sperry Rand Corp, 65-75; SR MEM TECH STAFF, TEX INSTRUMENTS, INC, 75- *Mem:* Am Phys Soc. *Res:* Nuclear magnetic resonance; transport properties of thin films; acoustic delay lines; magnetic bubble domains; bipolar device simulation. *Mailing Add:* 6927 Echo Bluff Dr Dallas TX 75248

SEITELMAN, LEON HAROLD, b New York, NY, May 27, 40; m 62; c 2. COMPUTER AIDED DESIGN, COMPUTER AIDED ENGINEERING. *Educ:* Cooper Union, BEE, 60; Univ Chicago, SM, 63; Brown Univ, PhD(appl math), 67. *Prof Exp:* Asst proj eng, 67-70, assoc res scientist, 70-73, SR APPL MATHEMATICIAN, PRATT & WHITNEY, 73- *Concurrent Pos:* Lectr, Trinity Col, Hartford, Conn, 68-70 & Vis Lectureship Prog, Soc Indust & Appl Math, 76-84; adj prof, Univ Conn, 75-76; contrib ed, Soc Indust & Appl Math, 78-, mem, Prog Comt, 82-84 & 88-90, Educ Comt, 86- & Joint Comt Employment Opportunities, 89-, chmn, Vis Lectureship Prog, 79-84, K-12 Panel, 87- & Joint Comt Employment Opportunities, 91; vis prof, Brown Univ, Providence, RI, 84-85; mem adv bd, Math Contest Modeling, 84- *Mem:* Soc Indust & Appl Math; Am Math Soc; Math Asn Am; AAAS; Sigma Xi; NY Acad Sci. *Res:* Development of curve and surface fitting procedures and associated programming systems for the efficient and accurate treatment of data throughout design, development and manufacturing. *Mailing Add:* 110 Cambridge Dr Glastonbury CT 06033

SEITZ, ANNA W, b Hong Kong, China, Dec 23, 41; m 69; c 3. DEVELOPMENTAL BIOLOGY. *Educ:* Wellesley Col, BA, 63; Univ Pa, PhD(anat), 80. *Prof Exp:* Fel biol, Sloan-Kettering Inst, 80-85; ASST PROF BIOL, UNIV MASS, 85- *Mem:* Soc Develop Biol; Am Soc Cell Biol; Europ Soc Develop Biol; Am Soc Microbiol. *Res:* Developmental genetics of mammalian embryos; genetic engineering of animals. *Mailing Add:* Dept Vet & Animal Sci Univ Mass Amherst MA 01003

SEITZ, EUGENE W, b Regina, Sask, Sept 27, 35; m 63, 85; c 3. MICROBIOL BIOCHEMISTRY, BACTERIOLOGY. *Educ:* Univ Sask, BSA, 57; Ore State Univ, MSc, 59, PhD(microbiol), 62. *Prof Exp:* Res asst bact, Ore State Univ, 57-59; res fel biochem, 59-62; Microl Res Ins, Res Br, Can Dept Agr, 62-63, res officer, 63-64, res scientist, Food Res Inst, 64-67, proj leader flavor res, 67-73, DIR, BIOL FLAVOR TECH, 80- *Concurrent Pos:* Field day dir, Univ Sask, 57; chmn, Ore SU chap, Phi Sigma Soc, 60-61; chmn, Am Chem Soc, Monmouth Co, 76-77. *Mem:* Am Chem Soc; Am Dairy Sci Asn; Can Soc Microbiol; Am Soc Microbiol; Sigma Xi; Soc Flavor Chemists. *Res:* Creation of and evaluation of natural flavors and fragrances; biogenesis of flavor compounds by micro-organisms, ie fermentation and microbiological enzyme process; design of microbiol and enzymatic processes. *Mailing Add:* 66 Pelican Rd Middletown NJ 07748

SEITZ, FREDERICK, b San Francisco, Calif, July 4, 11; m 35. PHYSICS. *Educ:* Stanford Univ, AB, 32; Princeton Univ, PhD(physics), 34. *Hon Degrees:* Twenty-nine from US & foreign univs & cols, 57-84. *Prof Exp:* Proctor fel, Princeton Univ, 34-35; from instr to asst prof physics, Univ Rochester, 35-37; res physicist, Gen Elec Co, 37-39; from asst prof to assoc prof physics, Randall Morgan Lab Physics, Univ Pa, 39-42; prof & head dept, Carnegie Inst Technol, 42-49; res prof, Univ Ill, 49-65, dir, Control Systs Lab, 51-52, tech dir, 52-57, head, Dept Physics, 57-64, dean, Grad Sch & vpres res, 64-65; pres, Nat Acad Sci, Washington, DC, 62-69; pres, Rockefeller Univ, 68-78; RETIRED. *Concurrent Pos:* Dir training prog atomic energy, Oak Ridge Nat Lab, 46-47; sci adv, NATO, 59-60; vpres, Int Union Pure & Appl Physics, 60-; consult educ comn inquiry, Ministry Educ, India, 64-; chmn, Sci Adv Coun, Ill, 64-; Midwest Sci Adv Comt, 65-; mem gov bd, Am Inst Physics, 54-, chmn gov bd, 54-59; mem, Naval Res Adv Comt, 55-, chmn, 60-62; mem, Defense Sci Bd, 58-72, vchmn, 61-62, chmn, 63-72; mem policy adv bd, Argonne Nat Lab, 58-; mem, President's Sci Adv Comt, 62-69; mem, President's Comt Nat Medal Sci, 62-, chmn, 62-63; mem statutory vis comt, Nat Bur Stand, 62-66; Grad Res Ctr Southwest, 63-66; liaison comt sci & technol, Libr of Cong, 63-; educ & adv bd, Guggenheim Mem Found, 65; bd dirs, Res Corp, Tex Instruments Inc, 71-, Akzona Inc, 73- & Ogden Corp, 77- bd trustees, Nutrit Found, Rockefeller Found, 64-77, Rockefeller Univ, 66-78, Res Corp, 66-, Princeton Univ, 68-72, Inst Int Educ, 71- & Woodrow Wilson Nat Fel Found, 72-, vchmn, Mem Sloan Kettering Cancer Ctr, 78-83; mem, Inst Defense Anal, 70-; mem, Nat Cancer Adv Bd, 72-74, 77-; mem, Belg Am Educ Found; mem bd overseers, mem Sloan Kettering Cancer Ctr, 83-; mem adv bd, Off Strategic Defense Initiative. *Honors & Awards:* Franklin Medal, Franklin Inst, 65; Hoover Medal, 68; Compton Award, Am Inst Physics, 70; Nat Medal Sci, 73; James Madison Award, Princeton Univ, 78; Vannevar Bush Award, NSF, 83; R Loveland Mem Award, Am Col Physicians, 83. *Mem:* Nat Acad Sci; Am Acad Arts & Sci; Am Crystallog Asn; Am Soc Metals; fel Am Phys Soc (pres, 61); Am Philos Soc; Optical Soc; Am Inst Mining; Metall & Petrol Engrs. *Res:* Theory of solids; nuclear physics. *Mailing Add:* Rockefeller Univ 1230 York Ave New York NY 10021

SEITZ, GARY M, b Santa Monica, Calif, May 10, 43. MATHEMATICS. *Prof Exp:* PROF MATH, UNIV ORE, 70- *Mailing Add:* Univ Ore Eugene OR 97403

SEITZ, LARRY MAX, b Hutchinson, Kans, June 30, 40; m 62; c 2. ANALYTICAL BIOCHEMISTRY, CEREAL CHEMISTRY. *Educ:* Kans State Univ, BS, 62; Univ Ill, PhD, 66. *Prof Exp:* Asst prof chem, Kans State Univ, 66-71; RES CHEMIST, US GRAIN MKT RES LAB, AGR RES SERV, USDA, 71- *Mem:* Am Asn Cereal Chemists; Am Chem Soc. *Res:* Mycotoxins; fungal metabolites; growth of fungi on cereal grains; composition of cereal grains. *Mailing Add:* 3008 Conrow Dr Manhattan KS 66502

SEITZ, S STANLEY, b Brooklyn, NY, June 13, 23; m 49; c 2. ELECTRONIC WARFARE, NAVIGATION. *Educ:* Pepperdine Univ, MBA, 74. *Prof Exp:* Asst gen mgr, CBS, 49-55; prog mgr, Arma, 55-58; corp dir indust eng, Am Electronics, 58-60; dir qual assurance, Sperry Gyroscope, 60-65; prog mgr, AIL, 65-69; dir mgt systs, Rockwell Int, 69-82; dir, Big Safari, Lockheed Aircraft, 82-88; SR PRIN SPECIALIST AVIONICS, MCDONNELL DOUGLAS, 88- *Concurrent Pos:* Sr lectr, Univ Calif Los Angeles, 82-83 & WCoast Univ, 75-82. *Mem:* Am Defense Preparedness Asn. *Res:* Effect of the cell organization structure on the design, development and production of electronic hardware. *Mailing Add:* Douglas Aircraft 3855 Lakewook Blvd Long Beach CA 90846

SEITZ, WENDELL L, b Mobeetie, Tex, Aug 3, 34; m 59; c 5. HIGH EXPLOSIVES. *Educ:* WTex State Univ, BS, 60; Univ Ariz, MS, 63. *Prof Exp:* Teaching asst physics, Univ Ariz, 60-62; staff mem, 62-85, sect leader, 85-90, ACTG DEP GROUP LEADER, LOS ALAMOS NAT LAB, 90- *Honors & Awards:* Award of Excellence, Dept Energy, 89. *Mem:* Am Phys Soc. *Res:* Performance and characterization of high explosives. *Mailing Add:* Los Alamos Nat Lab Group M9-MS P952 Los Alamos NM 87545

SEITZ, WESLEY DONALD, b Wapakoneta, Ohio, Sept 29, 40; m 64; c 2. AGRICULTURAL ECONOMICS. *Educ:* Ohio State Univ, BS, 62, MS, 64; Univ Calif, Berkeley, PhD(agr econ), 68. *Prof Exp:* Asst prof bus admin, 68-70, assoc prof agr econ, 68-74, assoc dir, Inst Environ Studies, 74-82 & head, Dept Agr Econ & Prof, 81-87, PROF AGR ECON, UNIV ILL, URBANA, 74- *Concurrent Pos:* Mem Ill Environ Agency Task Force on Agr Nonpoint Sources of Pollution, 77-78. Mem strip mined land reclamation task force, Argonne Nat Lab, 72-73; mem adv comt, Ill Environ Protection Agency on Appln of Sludge on Agr Land, 73-74; Off Water Res & Technol res grants, 73 & 74; chmn, Coun Agr Sci & Technol Task Force Rev Environ Protection Agency Proposed Guidelines for Registering Pesticides, 75; chmn, N Cent Regional Strategy Comt Natural Resources, 75-76; US Environ Protection Agency grants, 76 & 78; chmn, Comt Soil Resource Relation Surface Mining Coal, Nat Acad Sci, 78-79, mem, Comt Abandoned Mine Lands, 85-86. *Mem:* Am Agr Econ Asn; Am Econ Asn; Soil Conserv Soc Am; AAAS; Asn Resource & Environ Economists; Coun Agr Sci & Tech. *Res:* Alternative policies for the control of non-point sources of water pollution from agriculture; development of resource institutions. *Mailing Add:* 2203 Fletcher Urbana IL 61801

SEITZ, WILLIAM RUDOLF, b Orange, NJ, May 5, 43; m 69; c 2. ANALYTICAL CHEMISTRY. *Educ:* Princeton Univ, AB, 65; Mass Inst Technol, PhD(chem), 70. *Prof Exp:* Res chemist, Environ Protection Agency, 70-73; instr, Univ Ga, 73-75, asst prof chem, 75-76; from asst prof to assoc prof, 76-84, PROF CHEM, UNIV NH, 84- *Mem:* Am Chem Soc; Soc Appl Spectros. *Res:* Analytical applications of fluorescence; fiber optic sensors. *Mailing Add:* Dept Chem Univ NH Durham NH 03824

SEKA, WOLF, b Klagenfurt, Austria, May 15, 39; m 63; c 2. LASER PHYSICS, SPECTROSCOPY. *Educ:* Univ Tex, PhD(physics), 65. *Prof Exp:* Nat Res Coun fel, Univ BC, Can, 65-68; foreign scientist, French Atomic Energy Comn, France, 68-70; vis scientist, Europ Space Res Inst, 70-72; res scientist, Univ Bern, Switz, 72-76; SR SCIENTIST, LAB LASER ENERGETICS, UNIV ROCHESTER, NY, 76-, PROF, INST OPTICS, 86- *Mem:* Am Phys Soc. *Res:* High power solid state laser physics; yttrium-aluminum-garnet and glass oscillator development; highly efficient frequency conversion laser plasma interaction experiments; laser fusion. *Mailing Add:* Lab Laser Energetics 250 E River Rd Rochester NY 14627

SEKANINA, ZDENEK, b Mlada Boleslav, Czech, June 12, 36; m 66; c 1. PLANETARY SCIENCES, ASTROPHYSICS. *Educ:* Charles Univ, dipl physicist, 59, PhD(astron), 63. *Prof Exp:* Astronr, Stefanik Observ, Czech, 59-66 & Ctr Numerical Math, Charles Univ, 67-68; physicist, Smithsonian Astrophys Observ, 69-80; SR RES SCIENTIST, JET PROPULSION LAB, CALIF INST TECHNOL, 80- *Concurrent Pos:* Assoc, Harvard Col Observ, 69-80. *Honors & Awards:* Asteroid 1913 Named Sekanina, 76; Exceptional Sci Achievement Medal, NASA, 85. *Mem:* Int Astron Union; Am Astron Soc. *Res:* Physics and dynamics of comets; interplanetary particles; meteors. *Mailing Add:* Jet Propulsion Lab Calif Inst Technol 169-237 Earth & Space Sci Div 4800 Oak Grove Dr Pasadena CA 91109

SEKELLICK, MARGARET JEAN, b New Haven, Conn, Aug 15, 43. ANIMAL VIROLOGY. *Educ:* Univ Conn, BA, 65, MS, 67, PhD, 80. *Prof Exp:* Res asst develop genetics, 67-69, res asst virol, 69-81, res assoc biol, 81-84, ASST PROF IN RESIDENCE, UNIV CONN, 85- *Mem:* AAAS; Am Soc Microbiol; Soc Gen Microbiol; Int Soc Interferon Res. *Res:* Animal virus-host cell interactions; mechanisms of cell killing and persistent infection by viruses; mechanisms of induction and action of interferon; relationship of stress response to interferon induction and action. *Mailing Add:* Dept Molecular & Cell Biol U-44 Univ Conn Storrs CT 06268

SEKERKA, IVAN, b Prague, Czech, Dec 8, 27; Can citizen; m 58; c 1. ANALYTICAL CHEMISTRY, CORROSION. *Educ:* Charles Univ, Prague, BS, 50, MS, 52, PhD(phys chem), 55; Tech Univ, Ostrava, PhD(phys chem), 59. *Prof Exp:* Chemist anal chem, Nat Inst Mat Protection, 51-53, res scientist phys chem, 53-61; res mgr, Nat Inst Motor Vehicles, 62-68; res scientist phys chem, 69-82, HEAD ANALYTICAL CHEM RES SECT, NAT WATER RES INST, 83- *Res:* Electroanalytical chemistry; ion selective electrodes; water quality parameters; water chemistry; flow injection analysis. *Mailing Add:* 3537 Ravenswood Ct Burlington ON L7N 3L2 Can

SEKERKA, ROBERT FLOYD, b Wilkinsburg, Pa, Nov 27, 37; m 60; c 2. PHYSICS, METALLURGY. *Educ:* Univ Pittsburgh, BS, 60; Harvard Univ, AM, 61, PhD(physics), 66. *Prof Exp:* Tech metall, Westinghouse Res Labs, 55-58, sr scientist, 65-68, mgr theoret physics dept, 68, mgr mat growth & properties dept, 68-69; lectr, Carnegie-Mellon Univ, 65-66 & 67-69, assoc prof metall & mat sci, 69-72, prof, Dept Metall Eng & Mat Sci, 72-82, dept head, 76-82; PROF, PHYSICS & MATH, 82-, DEAN, MELLON COL SCI, CARNEGIE MELLON UNIV, 82- *Concurrent Pos:* Consult, Nat Inst Standards & Technol & Bell Tel Labs, 76-79, USRA/NASA, 75 & Europ Space Agency, 88; assoc ed, Metal Trans, 70-76, J Crystal Growth, 71- , ed board, Applied Microgravity Technol, 87-; mem space studies bd, Nat Res Coun, Nat Acad Sci, 88- *Honors & Awards:* A G Worthing Award, 60; Philip M McKenna Mem Award, 80. *Mem:* Am Phys Soc; Am Inst Mining, Metall & Petrol Engrs; Sigma Xi; AAAS; Am Asn Crystal Growth; fel Am Soc Metals; fel NSF. *Res:* Magnetism; solidification; crystals; applied mathematics; morphological stability; transport processes. *Mailing Add:* Off of the Dean Mellon Col Sci Carnegie Mellon Univ Pittsburgh PA 15213-3890

SEKHON, SANT SINGH, b Ludhiana, India, June 20, 31; m 62; c 2. ZOOLOGY, CYTOLOGY. *Educ:* Govt Col, Ludhiana, India, BSc, 52; Panjab Univ, India, BSc(hons), 54, MSc, 56; Univ Iowa, PhD(zool), 62. *Prof Exp:* Res assoc cytol, Univ Iowa, 62-67; res biologist, Vet Admin Hosp, Long Beach, Calif, 67-82; res anatomist, 67-80, ASSOC RES ANATOMIST, SCH MED, UNIV CALIF, LOS ANGELES, 80-; HEALTH SCI SPECIALIST, VET ADMIN HOSP, LONG BEACH, CALIF, 82- *Mem:* AAAS; Electron Micros Soc Am; Am Soc Cell Biol; Am Asn Anat; Sigma Xi. *Res:* Fine structure of germ cells; sense organs of insects; secretion; general cell structure; process of aging in the central nervous system. *Mailing Add:* Dept Path Vet Admin Hosp 5901 E Seventh St Long Beach CA 90822

SEKI, HAJIME, b Nishinomiya, Japan, Feb 11, 29; m 66; c 3. SURFACE VIBRATIONAL SPECTROSCOPY, TRIBOLOGY. *Educ:* Brown Univ, BS, 54; Univ Pa, PhD, 61. *Prof Exp:* Res asst physics, Metals Res Lab, Brown Univ, 53-54 & Univ Pa, 54-61; res staff mem, 61-67, proj mgr, 67-77, RES STAFF MEM, RES DIV, IBM CORP, 77- *Mem:* AAAS; Am Phys Soc; Phys Soc Japan; Am Chem Soc. *Res:* Solid state physics of semiconductors, junction, surface phenomenon, photoconductivity; cryogenics of liquid helium, superfluidity, superconductivity; ultrasonic attenuation in crystals; molecular solids; organic surface and interface physics; electrochemistry; surface & interfacial optical vibrational spectroscopy; tribology. *Mailing Add:* IBM Almaden Res Ctr K33-801 650 Harry Rd San Jose CA 95120-6099

SEKI, RYOICHI, b Toyama, Japan, Jan 13, 40; m 67. THEORETICAL NUCLEAR PHYSICS, ATOMIC PHYSICS. *Educ:* Waseda Univ, Japan, BS, 62; Northeastern Univ, MS, 64, PhD(physics), 68. *Prof Exp:* Res assoc physics & fel, Univ Denver, 67-68; fel, Univ Ga, 68-69; from asst prof to assoc prof, 69-76, PROF PHYSICS, CALIF STATE UNIV, NORTHRIDGE, 76- *Concurrent Pos:* Consult, Lawrence Radiation Lab, Univ Calif, Berkeley, 74-; vis assoc, Calif Inst Technol, 78- *Mem:* Am Phys Soc; Phys Soc Japan; Sigma Xi. *Res:* Theoretical intermediate energy physics, including interaction of mesons with nuclei and atoms. *Mailing Add:* Dept Physics & Astron Calif State Univ Northridge CA 91324

SEKIMOTO, TADAHIRO, b Nov 14, 26. ELECTRICAL ENGINEERING. *Educ:* Tokyo Univ, DEng, 62. *Prof Exp:* Staff, Nippon Elec Co, Ltd, 48-65, chief, Basic Res Dept, Commun Res Lab, 65; staff, Comsat, 65-67; mgr, Commun Res Lab, Cent Res Labs, Nippon Elec Co, Ltd, 67-72, gen mgr, Transmission Div, 72-74, mem bd dirs, 74-77, sr vpres & dir, 77-78, exec vpres & dir, 78-80, PRES NIPPON ELEC CO, LTD, 80- *Honors & Awards:*

Purple Ribbon Medal, His Majesty Emperor Japan, 82, Blue Ribbon Medal, 89; Edwin Haward Armstrong Achievement Award, Inst Elec & Electronics Engrs, 82. *Res:* Electrical engineering. *Mailing Add:* Nippon Elec Co Ltd 7-1 Shiba 5-chome Minato-Ku Tokyo 108-01 Japan

SEKINE, YASUJI, b Tokyo, Japan, Dec 7, 31; m 64; c 1. POWER SYSTEMS ENGINEERING, SYSTEMS THEORY. *Educ:* Univ Tokyo, BS, 54, MS, 56, Dr, 59. *Prof Exp:* Lectr, 59-60, assoc prof, 60-72, PROF ELEC ENG, UNIV TOKYO, 72- *Concurrent Pos:* Vis prof, Fed Inst Technol, Zurich, Switz, 87; pres, Power Systs Computation Conf, 90-92. *Honors & Awards:* Centennial Mem Award, Inst Elec Engrs, Japan, 88. *Mem:* Inst Elec Engrs Japan (pres, 89-90); Inst Elec & Electronics Engrs; Eng Acad Japan; Royal Swed Acad Eng Sci. *Res:* Electric power systems engineering; systems theory; mathematical programming; analysis, planning and operation of power systems. *Mailing Add:* Dept Elec Eng Univ Tokyo Hongo 7-3-1 Bunkyo-Ku Tokyo 113 Japan

SEKULA, BERNARD CHARLES, b Philadelphia, Pa, Dec 29, 51; m 77; c 3. LIPID BIOCHEMISTRY, YEAST FERMENTATION. *Educ:* Drexel Univ, BS, 74, MS, 76, PhD(biochem), 79. *Prof Exp:* Res asst, Drexel Univ, 74-76, teaching asst biochem, bot, life sci, 76-77, res asst, 77-79; trainee, Fels Res Inst, 79-81; sr microbiol chemist, 81-86, PRIN MICROBIOL CHEMIST, BEST FOODS RES & ENG CTR, 86- *Mem:* Sigma Xi; Am Oil Chemists' Soc. *Res:* Lipid biochemistry; yeast fermentations; control of microbial metabolic pathways; structure/function relationship of sterols and sterol metabolism in anaerobic yeast; fat replacers. *Mailing Add:* Best Foods Res & Eng Ctr 1120 Commerce Ave Union NJ 07083

SEKULA, STANLEY TED, b Niagara Falls, NY, Jan 30, 27; m 57. PHYSICS. *Educ:* Univ Buffalo, BA, 51; Cornell Univ, PhD(physics), 59. *Prof Exp:* PHYSICIST, OAK RIDGE NAT LAB, 58- *Concurrent Pos:* Vis prof, Mid East Tech Univ, Ankara, 63-64. *Mem:* Sigma Xi; Am Phys Soc. *Res:* Superconductivity; radiation damage in solids. *Mailing Add:* Solid State Div Bldg 3115 Oak Ridge Nat Lab PO Box 2008 Oak Ridge TN 37831

SEKULER, ROBERT W, b Elizabeth, NJ, May 5, 39; m 61; c 3. VISION. *Educ:* Brandeis Univ, BA, 60; Brown Univ, ScM, 63, PhD(psychol), 64. *Prof Exp:* Fel, Mass Inst Technol, 64-65; from asst prof to assoc prof psychol, Northwestern Univ, 65-73, prof neurobiol physiol, 81-84, assoc dean, Col Arts & Sci, 84-90, PROF PSYCHOL, NORTHWESTERN UNIV, 77-, PROF OPTHAL, MED SCH, 77-; LOUIS & FRANCES FALVAGE PROF PSYCHOL, BRANDEIS UNIV, 90- *Concurrent Pos:* Mem, Adv Panel Sensory Physiol, NSF, 74-77, Steering Comt Physiol Optics, Am Acad Optom, 81-; sr vpres, Optronix Corp, 79-82; consult, Nat Inst Aging, 81-90; chmn, Comt on Vision, Nat Acad Sci/Nat Res Coun, 83-85. *Mem:* Am Acad Optom; Soc Neurosci; Asn Res Vision & Ophthal; Optical Soc Am; Geront Soc. *Mailing Add:* 64 Strawberry Hill Rd Concord MA 01742-5502

SEKUTOWSKI, DENNIS G, b Hamtramck, Mich, Aug 14, 48; m 75; c 1. CATALYSIS, INORGANIC CHEMISTRY. *Educ:* Wayne State Univ, BS, 70; Univ Ill, PhD(inorg chem), 75. *Prof Exp:* Res chemist catalysis, Max Planck Inst Coal Res, Muelheim, WGer, 75-77; res assoc, Tex A&M Univ, 77-78; res chemist catalysis, Oxirane Int, 79-81; res chemist catalysis, Arco Chem Co, 81; sr res chemist, 81-85, RES ASSOC, PIGMENTS & ADDITIVES DIV, ENGELHARD CORP, 85- *Concurrent Pos:* Tech prog chmn, Gen Polymer Modifiers & Additives Div, Soc Plastic Engrs, 89-91. *Mem:* Am Chem Soc; Soc Plastics Engrs; Tech Asn Pulp & Paper Indust. *Res:* Homogeneous and heterogeneous catalysis, especially in relationship to the petrochemical industry; engineering plastic composites, especially relating to surface chemistry of reinforcing fillers; pigment technology related to coating applications. *Mailing Add:* Pigments & Additives Div Engelhard Corp Menlo Park CN 28 Edison NJ 08818

SELANDER, RICHARD BRENT, b Garfield, Utah, July 21, 27; m 50. ENTOMOLOGY. *Educ:* Univ Utah, BS, 50, MS, 51; Univ Ill, PhD, 54. *Prof Exp:* Instr biol, Univ Utah, 54; asst taxonomist, Ill Natural Hist Surv, 54-58; from asst prof to prof entom, 58-77, PROF GENETICS & DEVELOP, UNIV ILL, URBANA, 77- *Mem:* AAAS; Soc Syst Zool; Entom Soc Am. *Res:* Classification, phylogeny and biology of Meloidae; classification and phylogeny of Coleoptera. *Mailing Add:* Dept Entom 333 Morill Hall Univ Ill Urbana IL 61801

SELANDER, ROBERT KEITH, b Garfield, Utah, July 21, 27; m 51; c 2. BACTERIAL POPULATION GENETICS. *Educ:* Univ Utah, BS, 50, MS, 51; Univ Calif, Berkeley, PhD(zool), 56. *Prof Exp:* From instr to prof zool, Univ Tex, Austin, 56-74; prof biol, Univ Rochester, 74-87; EBERLY PROF BIOL, PENN STATE UNIV, 87- *Concurrent Pos:* Res fel, Am Mus Natural Hist, 60-61; Guggenheim fel, 65; Rand fel, 71. *Honors & Awards:* Walker Prize, 69; Painton Award, 70. *Mem:* Nat Acad Sci; Soc Study Evolution; fel Am Ornith Union; fel AAAS; foreign fel Linnaean Soc London; fel Am Acad Arts & Sci. *Res:* Population genetics; molecular evolution. *Mailing Add:* Dept Biol Pa State Univ University Park PA 16802

SELASSIE, CYNTHIA R, b Mombasa, Kenya, Aug 28, 51; US citizen; m 79; c 1. COMPUTER ASSISTED DRUG DESIGN, DRUG RESISTANCE. *Educ:* Mt St Mary's Col, BA, 74; Duke Univ, MA, 76; Univ Southern Calif, PhD(pharmaceut chem), 80. *Prof Exp:* Res assoc chem, Pomona Col, 80-86, adj assoc prof, 86-89, adj assoc prof, 89-90, ASSOC PROF CHEM TEACHING, POMONA COL, 90- *Mem:* Am Chem Soc; Am Pharmaceut Asn; Am Asn Cancer Res; Coun Undergrad Res; Asn Women Sci. *Res:* Computer assisted drug design particularly in the area of antifolates; applications of quantitative structure-activity relationship paradigm to resistance in bacteria and neoplastic cells; enzymatic reactions in nonaqueous solvents as models for drug-membrane associated enzymes. *Mailing Add:* Chem Dept Pomona Col 645 N College Ave Claremont CA 91711-6338

SELBERG, ATLE, b Langesund, Norway, June 14, 17; m 47; c 2. MATHEMATICS. *Educ:* Univ Oslo, PhD(math), 43. *Hon Degrees:* Dr, Univ Trondheim, Norway. *Prof Exp:* Res fel Math, Univ Oslo, 42-47; assoc prof, Syracuse Univ, 48-49; mem, 47-48, permanent mem, 49-51, PROF, INST ADVAN STUDY, 51- *Honors & Awards:* Fields Medal & Prize, Int Cong Mathematicians, 50; Wolf Prize, 86; Comdr W Star, Royal Norweg Order St Olav, 87. *Mem:* Royal Danish Soc Sci; Am Acad Arts & Sci; Norweg Acad Sci & Letters; Royal Norweg Soc Sci; Royal Swedish Acad Sci; Indian Nat Sci Acad; hon fel Tata Inst Fundamental Res. *Res:* Number theory; analysis. *Mailing Add:* Sch Math Inst Advan Studies One Olden Lane Princeton NJ 08540

SELBIN, JOEL, b Washington, DC, Aug 20, 31; m 55; c 4. INORGANIC CHEMISTRY. *Educ:* George Washington Univ, BS, 53; Univ Ill, PhD(chem), 57. *Prof Exp:* From asst prof to assoc prof, 57-67, dir grad studies, Chem Dept, 77-81, PROF CHEM, LA STATE UNIV, BATON ROUGE, 67- *Concurrent Pos:* Petrol Res Fund int fac award, Rome, Italy, 63-64; vis prof, Univ Calif, Berkeley, 72 & Harvard Univ, 82; Danforth Assoc. *Mem:* Am Chem Soc; fel AAAS; Sigma Xi. *Res:* Bioinorganic chemistry; physical chemical studies on complex inorganic compounds, mainly spectral properties of transition metal complexes. *Mailing Add:* Dept Chem La State Univ Baton Rouge LA 70803

SELBY, HENRY M, b US, Sept 20, 18; m 51; c 3. MEDICINE. *Educ:* La State Univ, MD, 43; Am Bd Radiol, dipl, 50. *Prof Exp:* ASST PROF CLIN RADIOL, MED COL, CORNELL UNIV, 51- *Concurrent Pos:* Asst attend radiologist, NY Hosp, 51 & James Ewing Hosp, 51-; assoc attend roentgenologist, Mem Hosp, 51-; dir radiol, Prev Med Inst, Strang Clin, 65- *Mem:* Am Radium Soc; Am Col Radiol. *Res:* Radiology. *Mailing Add:* 57 W 57th St New York NY 10019

SELBY, PAUL BRUCE, b Owatonna, Minn, Dec 5, 45; m 70; c 1. MAMMALIAN GENETICS. *Educ:* Westmar Col, BA, 67; Univ Tenn, PhD(biomed sci), 72. *Prof Exp:* Res assoc radiation genetics, Gesellschaft fur Strahlen-und Umweltforschung, Neuherberg, Ger, 72-75; RES ASSOC MUTAGENESIS, OAK RIDGE NAT LAB, 75- *Concurrent Pos:* Mem, Nat Res Coun Comt Biol Effects Ionizing Radiations, 77- & Genetic Effects Subcomt, 77- *Mem:* Genetic Soc Am; Environ Mutagen Soc. *Res:* Radiation and chemical mutagenesis in mice; study of induction and nature of dominant mutations that cause malformations of skeleton; induction of specific-locus mutations; cytogenetics; genetic risk estimation. *Mailing Add:* RR 3 No 352 Clinton TN 37716

SELDEN, DUDLEY BYRD, b Oakhill, WVa, Nov 24, 11; m 39. MATHEMATICS. *Educ:* Univ Richmond, BS, 32; Purdue Univ, MS, 61. *Prof Exp:* Asst dir, Ballistic Res Labs, Md, US Army, 46-47, chief proj officer, White Sands Proving Ground, NMex, 48-49, chief tech unit, Off Chief Ord, Washington, DC, 51-53, chief eng officer, Japan Ord Command, Oppama, Japan, 54-55, exec officer, Ballistic Missile Agency, Ala, 57-58, US Army Guided Missile Rep, Brit Ministry of Supply, Eng, 58-60; from instr to assoc prof, 61-74, EMER ASSOC PROF MATH, HAMPDEN-SYDNEY COL, 74- *Concurrent Pos:* NSF Col Mat Inst, Rutgers Univ, 66. *Res:* Teaching under graduate mathematics; development of military weapons, especially guided missiles. *Mailing Add:* 206 Agee St Farmville VA 23901

SELDEN, GEORGE, b Cleveland, Ohio, Oct 13, 15; c 2. ORGANIC CHEMISTRY, CHEMICAL ENGINEERING. *Educ:* Case Western Reserve Univ, BS, 36, MS, 39, PhD(org chem), 42. *Prof Exp:* Chemist, Upco Co, 36-39; resin chemist & head resin & varnish lab, Finishes Div, Interchem Corp, 41-47; factory mgr, Upco Co, USM Corp, 47-79, pres, 49-79, bus dir, Bostile Div, 79-81; pres, Selden Chem Consults Inc, 81-90; RETIRED. *Mem:* Fel AAAS; Am Chem Soc; Am Soc Testing & Mat; Am Concrete Inst; Fed Soc Coatings Technol. *Res:* Factors affecting the adhesion of paint films to metals. *Mailing Add:* Two Bratenahl Pl Apt 10D Cleveland OH 44108

SELDEN, ROBERT WENTWORTH, b Phoenix, Ariz, Aug 11, 36; m; c 1. DEFENSE APPLICATION OF SCIENCE. *Educ:* Pomona Col, BA, 58, Univ Wis, MS, 60, PhD(physics), 64. *Prof Exp:* Staff mem physics, Lawrence Livermore, Nat Lab, 67-73, group leader, Nuclear Exp Design Physics, 73-79, asst assoc dir nuclear exps, 78-79; div leader, Appl Theoret Physics Div, Los Alamos Nat Lab 79-83, deputy assoc dir strategic defense res, 83-84, assoc dir theoret & comput physics, 84-86, Dir, Ctr Nat Security Stud, 86-88; CHIEF SCIENTIST, USAF, 88- *Concurrent Pos:* Sci adv, US Dept Energy Team for Tech Discussion Nonproliferation, 76-77; President's Defensive Technols Adv Team, 83; Ballistic Missile Defense Technols Adv Panel, US Congress, 84-85; mem, USAF Sci Adv Bd, 84-88. *Honors & Awards:* Theodore von Karman Award, 89. *Mem:* Am Phys Soc; AAAS. *Res:* Technical management of basic and applied research in theoretical and computational physics; principal contributions in the areas of advanced computation, fusion energy and application of science to defense needs; development of advanced technologies in response to national security policies and implications of policy and strategy on technology development; mathematical and systems analysis of ballistic missile defense technologies. *Mailing Add:* HQ USAF/CCN The Pentagon Washington DC 20330-5040

SELDIN, DONALD WAYNE, b New York, NY, Oct 24, 20; m 42; c 3. INTERNAL MEDICINE. *Educ:* NY Univ, AB, 40; Yale Univ, MD, 43. *Prof Exp:* From instr to asst prof internal med, Yale Univ, 48-51; from asst prof to assoc prof, 51-52, chmn dept, 52-88, PROF INTERNAL MED, UNIV TEX HEALTH SCI CTR, DALLAS, 52- *Concurrent Pos:* Alexander Von Humboldt Sr US Scientist Award, 88-92. *Mem:* Inst Med-Nat Acad Sci; Am Soc Clin Invest; Asn Am Physicians; Am Col Physicians; Am Fedn Clin Res; Royal Soc Med. *Res:* Electrolyte and water metabolism; renal function; diabetes; adrenal gland. *Mailing Add:* Dept Internal Med Univ Tex Health Sci Ctr 5323 Harry Hines Blvd Dallas TX 75235

SELDIN, EMANUEL JUDAH, b Brooklyn, NY, Mar 20, 27; m 52; c 2. ENGINEERING PHYSICS. *Educ:* Brooklyn Col, BA, 49; Univ Wis, MS, 51; Univ Buffalo, PhD(physics), 58. *Prof Exp:* Asst physics, Univ Wis, 49-53 & Univ Buffalo, 53-57; physicist, Parma Tech Ctr, Carbon Prod Div, Union Carbide Corp, 57- 83; instr math, Lorain Co Community Col, 84-85; LECTR PHYSICS, BALDWIN-WALLACE COL, 85- *Concurrent Pos:* Lectr, Baldwin-Wallace Col, 64; instr physics, Cuyahoga Community Col, 85. *Mem:* Am Phys Soc. *Res:* Processing, mechanical and thermal properties of carbon and graphite. *Mailing Add:* 16384 Barriemore Ave Cleveland OH 44130

SELDIN, JONATHAN PAUL, b New York, NY, Jan 30, 42; m 79; c 1. COMBINATORY LOGIC & LAMBDA-CALCULUS, PROOF THEORY. *Educ:* Oberlin Col, BA, 64; Pa State Univ, MA, 66; Univ Amsterdam, Dr Math, 68. *Prof Exp:* Lectr pure math, Univ Col Swansea, Wales, 68-69; asst prof math, Southern Ill Univ, Carbondale, 69-81; ADJ ASSOC PROF MATH, CONCORDIA UNIV, MONTREAL, 81- *Concurrent Pos:* Researcher, Odyssey Res Assocs, 86-88. *Mem:* AAAS; Asn Symbolic Logic; Am Math Soc; Math Asn Am; Sigma Xi. *Res:* Combinatory logic. *Mailing Add:* Dept Math Concordia Univ 7141 Sherbrooke St W Montreal PQ H4B 1R6 Can

SELEGUE, JOHN PAUL, b Lorain, Ohio, Dec 31, 52; m 83; c 1. ORGANOMETALLIC CHEMISTRY, COORDINATION CHEMISTRY. *Educ:* Miami Univ, Oxford, Ohio, BS, 74; Mass Inst Technol, PhD(chem), 79. *Prof Exp:* Res assoc, Yale Univ, 78-80; ASST PROF CHEM, UNIV KY, LEXINGTON, 80-, ASSOC PROF, 86- *Concurrent Pos:* NSF fel, 74-77; Alexander von Humboldt fel, 87-88. *Mem:* Am Chem Soc; Sigma Xi. *Res:* Synthetic organotransition metal chemistry using spectroscopic and x-ray diffraction techniques; metallacumulene and carbide complexes; metal clusters; reactions of coordinated ligands. *Mailing Add:* Dept Chem Univ KY Lexington KY 40506-0055

SELF, GLENDON DANNA, b Waveland, Ark, Jan 1, 38; m 60. OPERATIONS RESEARCH, STATISTICS. *Educ:* Univ Ark, BS, 58, MS, 59; Okla State Univ, PhD(indust eng), 63; Univ Tex, JD, 79. *Prof Exp:* Qual control engr, Sandia Corp, 59-60, consult, 61-63; sr opers analyst, Gen Dynamics/Ft Worth, 63-64, proj opers analyst, 64-65; from asst prof to assoc prof opers res, Tex A&M Univ, 65-69; systs engr, 69-70, VPRES, ELECTRONIC DATA SYSTS CORP, 71- *Concurrent Pos:* Adj prof math, Tex Christian Col, 64-65; consult, Gulf Sch Suppl Educ Ctr, 67, Ctr Naval Anal, 68-69 & Nat Bur Standards, 68-; vis prof mech eng, Univ Tex, 78. *Mem:* Am Bar Asn; Opers Res Soc Am; Inst Mgt Sci. *Res:* Non-separable and multi-decision variable dynamic programming solutions; cost modeling of research and development; large-scale computer systems software development for the utilities; application of operations research contract law; artificial intelligence - parallel processing. *Mailing Add:* Electronic Data Systs Corp 7171 Forest Lane Dallas TX 75230

SELF, HAZZLE LAYFETTE, b Clairette, Tex, Aug 1, 20; m 43; c 4. ANIMAL SCIENCE. *Educ:* Agr & Mech Col, Tex, BS, 48; Tex Tech Col, MS, 50; Univ Wis, PhD(animal husb, genetics), 54. *Prof Exp:* Asst prof animal husb, Tarleton State Col, 48-52; assoc, Univ Wis, 53-54, asst prof, 54-59; assoc prof animal husb, Iowa State Univ, 59-61, prof, 61-, in chg exp farms, 60-; ALLEE RES CTR, IOWA STATE UNIV. *Honors & Awards:* Animal Mgt award, Am Soc Animal Soc, 78. *Mem:* AAAS; Am Soc Animal Sci; Am Forage & Grassland Coun. *Res:* Physiology of reproduction; breeding and artificial insemination; environmental effects on farm animals. *Mailing Add:* Rte 2 Box C-38 Hico TX 76457

SELF, STEPHEN, b London, Eng, Oct 26, 46. VOLCANOLOGY. *Educ:* Leeds Univ, BSc, 70; Imp Col Sci & Technol, PhD(geol), & DIC, 74. *Prof Exp:* Fel geol, Victoria Univ, NZ, 74-76; higher sci officer, Inst Geol Sci, UK, 76-77; res assoc, Goddard Inst Space Studies, NASA, & Dartmouth Col, 77-79; ASST PROF GEOL, ARIZ STATE UNIV, 80- *Concurrent Pos:* Vis scientist, Los Alamos Nat Lab, 78; vis prof, Mich Technol Univ, 79. *Mem:* Am Geophys Union; Geol Soc Am; Int Unquaternary Res; Int Asn Sedimentologists; NZ Geol Soc. *Res:* Quantitative volcanology; generation of volcanic rocks; quaternary geology; sedimentology. *Mailing Add:* Dept Geol Univ Tex Box 19069 Arlington TX 76019

SELFRIDGE, RALPH GORDON, b London, Eng, July 30, 27; m 82. MATHEMATICS, COMPUTER SCIENCES. *Educ:* Mass Inst Technol, BS, 47; Cornell Univ, MA, 53; Univ Ore, PhD(math), 53. *Prof Exp:* Mathematician, US Naval Ord Test Sta, Calif, 51-59; assoc prof math, Miami Univ, 59-61; assoc prof math, 61-72, dir comput ctr, 65-72, PROF COMPUT SCI, UNIV FLA, 72- *Mem:* Asn Comput Mach; Math Asn Am. *Res:* Numerical and harmonic analysis; computing technology; algorithmns for control of mechanisms, for graphics, and for simulation of automobile systems. *Mailing Add:* 300 CSE Univ Fla Gainesville FL 32611

SELGRADE, JAMES FRANCIS, b Washington, DC, Apr 25, 46; m 70; c 4. MATHEMATICS. *Educ:* Boston Col, BA, 68; Univ Wis-Madison, MS, 69, PhD(math), 73. *Prof Exp:* From asst prof to assoc prof, 73-87, PROF MATH, NC STATE UNIV, 87- *Mem:* Am Math Soc; Soc Indust & Appl Math; Soc Math Biol. *Res:* Qualitative theory of ordinary differential equations, global analysis and biomathematics. *Mailing Add:* 2905 Old Orchard Rd Raleigh NC 27607

SELIG, ERNEST THEODORE, b Harrisburg, Pa, Nov 25, 33; m 57; c 3. GEOTECHNICAL ENGINEERING, SOIL MECHANICS. *Educ:* Cornell Univ, BME, 57; Ill Inst Technol, MS, 60, PhD(civil eng), 64. *Prof Exp:* Res engr, Mech Div, IIT Res Inst, 57-66, mgr, Soil Mech Sect, 66-68; from assoc prof to prof civil eng, State Univ NY Buffalo, 68-78; PROF CIVIL ENG, UNIV MASS, 78- *Concurrent Pos:* Geotech eng consult, govt & private orgn, 68-; chmn, Soil Mech & Found Div Ill sect, Am Soc Civil Engrs, 64, pres, Buffalo sect, 73-74; chmn, Soil Dynamics Comt, Am Soc Testing & Mat, 66-74, Soil- Struct Interaction Comt, Transp Res Bd, 70-76, Soil & Rock Instrumentation Comt, 76-82, Soil Effects Comt, Soc Automotive Engrs, 66-69, Int Comt RR Geotechnol, 89-; ed, Geotech Eng J, 72-76, Geotech Testing J, 77-85; vis engr, Mass Inst Technol, 74-75; sr acad visitor, Oxford Univ, 86; vis prof, Nottingham, Eng, 86; mem, Comt Roadway and Ballast, Am Railway Eng Asn, 79- *Honors & Awards:* Gold Medal, Am Soc Mech Engrs; Charles E Dudley Award & Award of Merit, Am Soc Testing & Mat. *Mem:* Fel Am Soc Civil Engrs; fel Am Soc Testing & Mat; Transp Res Bd; Am Railway Eng Asn. *Res:* Dynamic behavior of soils; stress, strain and moisture instrumentation for soils; soil compaction and compaction equipment performance; soil-structure interaction; analysis of buried flexible and rigid culverts; behavior of railroad ballast and mechanics of track structure performance. *Mailing Add:* Marston Hall Dept Civil Eng Univ Mass Amherst MA 01003

SELIG, WALTER S, b Frankfurt am Main, Ger, Apr 13, 24; US citizen; div; c 3. ANALYTICAL CHEMISTRY. *Educ:* Roosevelt Univ, BS, 51; Miami Univ, MS, 52. *Prof Exp:* Chemist, R Lavin & Sons, Inc, Ill, 53-54, Simoniz Co, 54-59 & Sandia Corp, Calif, 59-60; chemist, 60-69, group leader org anal, 69-77, RESEARCHER, LAWRENCE LIVERMORE NAT LAB, UNIV CALIF, 77- *Concurrent Pos:* US AEC res & teaching fel & vis prof, Dept Org Chem, Hebrew Univ, Israel, 72-73. *Res:* Research and development of analytical methods for organic and inorganic materials; applications of ion-selective electrodes to organic and inorganic analysis; potentiometry. *Mailing Add:* Lawrence Livermore Lab Code L-310 PO Box 808 Livermore CA 94551

SELIGA, THOMAS A, b Hazleton, Pa, Dec 3, 37; m 63; c 2. ATMOSPHERIC SCIENCES, ELECTRICAL ENGINEERING. *Educ:* Case Inst Technol, BS, 59; Pa State Univ, MS, 61, PhD(elec eng), 65. *Prof Exp:* Instr elec eng, Pa State Univ, 61-65, asst prof, 65-69; prog dir aeronomy, NSF, 67-68; prof elec eng, Ohio State Univ, 69-85, dir atmospheric sci, 71-85; ASSOC DEAN GRAD STUDIES & RES & PROF ELEC ENG, COL ENG, PA STATE UNIV, 85- *Concurrent Pos:* Mem rep, Univ Corp Atmospheric Res, 73-85; consult, Environ Anal Asn Inc, 73- & 3M, 85- *Mem:* Inst Elec & Electronics Engrs; Am Geophys Union; AAAS. *Res:* Radar meteorology; radar polarimetry and ionospheric wave propagation; air pollution long range transport and effects; climatic variability; acid precipitation. *Mailing Add:* PO Box 85595 Seattle WA 98145

SELIGER, HOWARD HAROLD, b New York, NY, Dec 4, 24; m 44; c 2. PHYSICS, PHOTOBIOLOGY. *Educ:* City Col New York, BA, 43; Purdue Univ, MS, 48; Univ Md, PhD(physics), 54. *Prof Exp:* Asst instr physics, Purdue Univ, 48; prof leader radioactivity, Nat Bur Standards, 48-58; res assoc biophys, 58-63, assoc prof, 63-68, PROF BIOL, JOHNS HOPKINS UNIV, 68- *Concurrent Pos:* Guggenheim fel, 58-59; consult, Off Naval Res, 63-65; mem comt biol effects increased solar ultraviolet, Nat Acad Sci, 81. *Mem:* AAAS; fel Am Phys Soc; Am Soc Limnol & Oceanog; Am Soc Biol Chemists; Am Soc Photobiol (pres, 80-81). *Res:* Radioactivity standardization; bioluminescence; excited states of biological molecules; marine biology of bioluminescent dinoflagellates; photometry; estuarine ecology. *Mailing Add:* Dept Biol Johns Hopkins Univ 3400 N Charles St Baltimore MD 21218

SELIGER, WILLIAM GEORGE, b Chicago, Ill, Oct 8, 22; m 45; c 7. ANATOMY, DENTISTRY. *Educ:* Northwestern Univ, BS, 45, DDS, 46; Univ Wis, PhD(anat), 64. *Prof Exp:* Assoc prof anat, Colo State Univ, 64-68; prof anat & coordr, Sch Dent, Med Col Ga, 68-71; prof & chief oral med & chmn, Dept Anat, 71-81, PROF ANAT, SCH MED, TEX TECH UNIV, 71- *Concurrent Pos:* Nat Inst Dent Res fel, 62-64; NIH career res award, 66-68. *Mem:* Am Asn Anat; Int Asn Dent Res; World Asn Vet Anat; Am Asn Vet Anat. *Res:* Gross anatomy; histology; embryology; cytology; bone disease; tissue fluid movement through bone; periodontal disease; steroid secreting cells. *Mailing Add:* Dept Anat Tex Tech Univ Health Sci Ctr Lubbock TX 79409

SELIGMAN, GEORGE BENHAM, b Attica, NY, Apr 30, 27; m 59; c 2. MATHEMATICS. *Educ:* Univ Rochester, BA, 50; Yale Univ, MA, 51, PhD(math), 54. *Prof Exp:* Fine instr math, Princeton Univ, 54-56; from instr to assoc prof, 56-65, chmn dept, 74-77, PROF MATH, YALE UNIV, 65- *Concurrent Pos:* Fulbright lectr, Univ Munster, 58-59. *Mem:* Am Math Soc; Math Asn Am; Asn Women Math. *Res:* Lie algebras, especially semi-simple Lie algebras. *Mailing Add:* Dept Math Yale Univ New Haven CT 06520

SELIGMAN, ROBERT BERNARD, b Brooklyn, NY, Dec 30, 24; m 51; c 2. ORGANIC CHEMISTRY. *Educ:* Univ NC, BS, 48, PhD(org chem), 53. *Prof Exp:* Res chemist, 53-54, leader, Org Sect, 54-55, supvr, 55-57, asst mgr tobacco res, 57-58, develop, 58-59, mgr, 59-64, asst dir tobacco res & develop, 64-66, dir develop, 66-71, com develop, 71-76, vpres res & develop, 76-81, VPRES RES & DEVELOP, TOBACCO TAX GUIDE, PHILIP MORRIS USA, 81- *Mem:* Am Chem Soc; NY Acad Sci. *Res:* Synthetic tuberculostats; tobacco chemistry; consumer product development. *Mailing Add:* Philip Morris Res Ctr PO Box 26603 Richmond VA 23261

SELIGMAN, STEPHEN JACOB, b Brooklyn, NY, Feb 4, 31; m 85; c 2. INFECTIOUS DISEASES. *Educ:* Harvard Univ, AB, 52; NY Univ, MD, 56. *Prof Exp:* Fel infectious dis, Univ Calif, Los Angeles, 61-63, asst prof med, 63-68; assoc prof, 68-81, PROF MED, HEALTH SCI CTR, STATE UNIV NY, BROOKLYN, 81- *Concurrent Pos:* Infectious Dis Sect, Health Sci Ctr, State Univ NY, Brooklyn, 68-; consult, St Mary's Hosp Cath Med Ctr, Brooklyn, 71- *Mem:* AAAS; Infectious Dis Soc Am; Am Soc Microbiol; Soc Exp Biol & Med; Am Fedn Clin Res. *Res:* Three-dimensional structure of bacterial cell wall; methicillin resistant staphylococci; computer applications on bacteriologic data. *Mailing Add:* Infectious Dis Sect State Univ NY Health Sci Ctr Brooklyn NY 11203

SELIGMANN, BRUCE EDWARD, INFLAMMATION, HOST DEFENSE. *Educ:* Univ Md, PhD(biochem), 79. *Prof Exp:* SR STAFF SCIENTIST, CIBA GEIGY, 85- *Mailing Add:* IAR Lab Leukemia Res Ciba Geigy 556 Morris Ave Summit NJ 07901

SELIGSON, DAVID, b Philadelphia, Pa, Aug 12, 16; m 49; c 3. PATHOLOGY, BIOCHEMISTRY. *Educ:* Univ Md, BS, 40; Johns Hopkins Univ, ScD(biochem), 42; Univ Utah, MD, 46. *Hon Degrees:* MA, Yale Univ, 65. *Prof Exp:* Res biochemist, USDA, 42-43; chief hepatic & metab dis lab, Walter Reed Army Med Ctr, 43-45; USPHS fel, Univ Pa, 49-51, assoc prof clin chem in med, Grad Sch Med & dir div biochem, Grad Hosp, 53-58; assoc prof, 59-69, PROF MED & PATH, SCH MED, YALE UNIV, 69-, CHMN DEPT LAB MED, 71-, DIR CLIN LABS, YALE, NEW HAVEN MED CTR, 58- *Concurrent Pos:* Medici Publici & Med Alumni Asn fel, Col Med, Univ Utah, 66. *Honors & Awards:* Donald D VanSlyke Award, Am Asn Clin Chemists, 70, Ames Award, 71. *Mem:* Fel Am Soc Clin Pathologists; fel Col Am Pathologists; fel Am Col Physicians; Am Soc Clin Invest; Am Asn Clin Chemists (pres, 61-62); Sigma Xi. *Mailing Add:* Yale-New Haven Med Ctr 6022 CB 789 Howard Ave New Haven CT 06504

SELIGSON, FRANCES HESS, b Philadelphia, Pa, Sept 6, 49; m 75. REGULATORY AFFAIRS, FOOD SAFETY & TOXICOLOGY. *Educ:* Drexel Univ, BS, 71; Univ Calif, Berkeley, PhD(nutrit), 76. *Prof Exp:* Asst prof pub health nutrit, Univ NC, Chapel Hill, 76-77; scientist, Procter & Gamble Co, 77-87; MGR, HERSHEY FOODS CORP, 87- *Mem:* Am Dietetic Asn; Am Inst Nutrit; Inst Food Technologists. *Res:* Protein quality evaluation; calcium bioavailability; calcium-mineral interactions; blood lipid effects of sucrose polyesters and soy protein. *Mailing Add:* Hershey Foods Corp PO Box 805 Hershey PA 17033-0805

SELIGY, VERNER LESLIE, b Niagara-on-Lake, Ont, Sept 16, 40; m 66; c 3. MOLECULAR GENETICS, BIOTECHNOLOGY. *Educ:* Univ Toronto, BSc, 65, MSc, 66, PhD(molecular biol), 69. *Prof Exp:* Fel biochem, Nat Res Coun Can, 69-71, asst res officer chromatin struct, 71-75, assoc res officer molecular genetics, 76-82, head sect, 86-90, SR RES OFFICER MOLECULAR GENETICS, DIV BIOL SCI, NAT RES COUN CAN, 82-, MOLECULAR CELL BIOL, INST BIOL SCI, 90- *Concurrent Pos:* Group coordr molecular genetics, Div Biol Sci, Nat Res Coun Can, 82-85; adj prof biol, Carleton-Ottawa Univ Grad Ctr, 78-; assoc mem, Inst Biochem, Carleton Univ, 80-87 & Ottawa Univ, 87-89; bd dir, Plant Biotechnol, Univ Toronto, 87-90; sci adv, Indust Develop Off, Nat Res Coun, 80-, contracts res develop, 86- *Mem:* Genetics Soc Can; Can Biochem Soc; Genetics Soc Am. *Res:* Cloning, expression, regulation & engineering of genes coding for industrially important enzymes (glycosidases); regulatory proteins that control gene expression & cell development in yeast/fungi & insect/mammalian systems. *Mailing Add:* Inst Biol Sci Nat Res Coun Can 100 Sussex Dr Ottawa ON K1A 0R6 Can

SELIKOFF, IRVING JOHN, b New York, NY, Jan 15, 15; m 46. ENVIRONMENTAL MEDICINE, PUBLIC HEALTH. *Educ:* Columbia Univ, BS, 35; Royal Cols, Scotland, MD, 41; Am Bd Prev Med, dipl, 68. *Hon Degrees:* DSc, City Univ New York, 84. *Prof Exp:* Fel path, Mt Sinai Hosp, New York, 41; intern med, Newark Beth Israel Hosp, NJ, 43-44; resident, Sea View Hosp, New York, 44-47; physician, Paterson Clin, NJ, 47-68; dir, Environ Sci Lab, Mt Sinai Sch Med, 64-85, prof community med, 68-85, prof med, 70-85, EMER PROF MED & COMMUNITY MED, MT SINAI MED SCH, 85- *Concurrent Pos:* Consult, numerous govt agencies, 55-; ed-in-chief, Environ Res, 68- & Am J Indust Med, 80-; consult, Am Cancer Soc, 71- *Honors & Awards:* Lasker Award, Am Pub Health Asn, 55; Poiley Award & Medal, NY Acad Sci, 74; Haven Emerson Award, Pub Health Asn NY, 75; Nat Res Award, Am Cancer Soc, 77; Herman M Biggs Mem Award, 82; Buccieri La Ferla Int Award, 82. *Mem:* NY Acad Sci (pres, 69-70, gov, 70-); fel Am Pub Health Asn; fel Am Col Chest Physicians; Soc Occup & Environ Health (pres, 73-74). *Res:* Health effects of environmental factors, including environmental and occupational cancer. *Mailing Add:* 19 E 98th St Suite 10A New York NY 10029

SELIM, MOSTAFA AHMED, b Cairo, Egypt, June 11, 35; US citizen; m 64. OBSTETRICS & GYNECOLOGY, GYNECOLOGIC ONCOLOGY. *Educ:* Alexandria Univ, PNS, 54; Cairo Univ, MBBCH, 59. *Prof Exp:* Intern, Ahmed Maher Hosp, Egypt, 60; house officer, Royal Infirmary, UK, 61-62, Lister Hosp, 62-63 & Fairfield Gen Hosp, 63; residency, Womans Hosp, St Lukes Hosp Ctr, 64-66, chief resident, 66-67; fel pelvic cancer surg & res, Roswell Park Mem Inst, 67-68; pvt pract, Dar El Shiefa Hosp, Egypt, 69-70; inst, Case Western Reserve Univ, 70-71, sr instr, 71-72, from asst prof to assoc prof , 72-85, PROF, REPRODUCTIVE BIOL, CASE WESTERN RESERVE UNIV, 85- *Concurrent Pos:* Intern, St Vincents Hosp, NY, 63-64; dir, Div Gynec Oncol, Cleveland Metrop Gen Hosp, 70- & Div Gynec Serv, 72- *Mem:* Am Col Obstet & Gynec; Am Col Surgeons; Am Fertility Soc. *Res:* Improved methods of early diagnosis of gynecologic cancer and protocols for treatment with irradiation, chemotherapy and radical surgery; improve irradiation response by increasing the blood flow by chemical and physical factors. *Mailing Add:* Case Western Reserve Univ 3395 Scranton Rd Col St Thomas 2115 Summit Ave Cleveland OH 44109

SELIN, IVAN, b New York, NY, Mar 11, 37; m 57; c 2. RESOURCE MANAGEMENT. *Educ:* Yale Univ, BE, 57, MS, 58, PhD(elec eng), 60; Univ Paris, DrSci(math), 62. *Prof Exp:* Researcher, Rand Corp, 60-65; analyst & actg asst secy, Off Secy Defense, 65-70; chmn bd, Am Mgt Systs, Inc, 70-89; UNDER SECY MGT, DEPT STATE, 89-, CHMN-DESIGNATE, NUCLEAR REGULATORY COMN, 91- *Concurrent Pos:* Mem PSAC panel, Conventional Arms Transfers, 75; mem adv panel to Secy Defense B-1, 77; chmn mil econ adv panel to Dir CIA, 78-89. *Honors & Awards:* Distinguished Civilian Serv Medal, Secy Defense, 70. *Mem:* Sigma Xi. *Res:* Information theory, with applications to radar and to communications in a noisy environment. *Mailing Add:* 2905 32nd St NW Washington DC 20008

SELING, THEODORE VICTOR, b Lansing, Mich, Mar 27, 28; m 52; c 2. ELECTRICAL ENGINEERING, RADIO ASTRONOMY. *Educ:* Mich State Univ, BS, 49; Univ Mich, MSE, 60, PhD(elec eng), 69. *Prof Exp:* Engr, Pub Utilities Comn, State Mich, 49-50; ionosphere data anal elec engr, US Army Signal Corp, 50-52; proj engr, AC Spark Plug Div, Gen Motors Corp, 52-54, sr proj engr, 54-60 & Defense Systs Div, 60-62; assoc res engr, 62-69, res scientist, 69-82, CHIEF ENGR RADIO ASTRON, CALIF INST TECHNOL, 82- *Concurrent Pos:* Consult microwave receiving systs, Environ Res Inst-Mich, 73-74 & 81-82, Space Labs, 81-82. *Mem:* Inst Elec & Electronics Engrs; Am Astron Soc. *Res:* Radio astronomy instrumentation; centimeter & mm wave radiometers and associated electronic systems. *Mailing Add:* Dept Radio Astron 105-24 Calif Inst Technol Pasadena CA 91125

SELINGER, PATRICIA GRIFFITHS, b Cleveland, Ohio, Oct 15, 49. DATABASE MANAGEMENT SYSTEMS. *Educ:* Harvard Univ, AB, 71, MS, 72, PhD(appl math), 75. *Prof Exp:* Mem res staff, IBM Res, San Jose, 75-78, mgr, R Proj, 78-82, Off Systs Lab, 83 & Computer Sci Dept, 83-86, MGR, DATABASE TECHNOL INST, IBM RES, SAN JOSE, 86- *Concurrent Pos:* Vchair, Spec Interest Group Mgt Data, Asn Comput Mach, 83-85. *Honors & Awards:* Software Syst Award, Asn Comput Mach, 88. *Mem:* Asn Comput Mach. *Res:* Database management systems; relational language extensions; distributed data; query optimization; parallel query processing. *Mailing Add:* IBM Corp 650 Harry Rd San Jose CA 95120

SELINSKY, BARRY STEVEN, b Ashland, Pa, June 2, 58; m 81; c 2. BIOCHEMISTRY. *Educ:* Lebanon Valley Col, BS, 80; State Univ NY, Buffalo, PhD(biochem), 84. *Prof Exp:* Postdoctoral fel res, Dept Physiol, Duke Univ, 84-85; staff fel res, Nat Inst Environ Health Sci, 85-88; ASST PROF BIOCHEM, DEPT CHEM, VILLANOVA UNIV, 88- *Concurrent Pos:* Consult, Otsuka Electronics, 88-90. *Mem:* Biophys Soc; Am Chem Soc. *Res:* Protein-lipid interactions in biological membranes; biophysical measurements of the hepatic metabolism of fluorinated xenobiotics. *Mailing Add:* Dept Chem Villanova Univ Villanova PA 19085

SELKE, WILLIAM A, b Newburgh, NY, June 16, 22; m 52; c 3. CHEMICAL ENGINEERING. *Educ:* Mass Inst Technol, SB, 43, SM, 47; Yale Univ, DEng, 49. *Prof Exp:* Engr, State Water Comn Proj, Yale Univ, 47; assoc chem eng, Columbia Univ, 49-50, asst prof, 50-55, eng mgr, Atomic Energy Comn Proj, 54-55; dir fundamental res, Peter J Schweitzer, Inc, 55-57; dir res, Schweitzer Div, Kimberly-Clark Corp, 58-81, vpres res & develop, tech paper-spec products, 81-85, vpres technol assessment, 85-86; CONSULT, 86- *Concurrent Pos:* Engr, E I du Pont de Nemours & Co, 51; adj prof, Lenox Inst Water Technol, 87- *Mem:* AAAS; Am Chem Soc; Tech Asn Pulp & Paper Indust; Am Inst Chem Engrs; fel NY Acad Sci; Asn Environ Eng Prof. *Res:* Thermodynamics; ion exchange; dielectric materials; tobacco; paper making; sanitary and environmental engineering. *Mailing Add:* Meeting House Stockbridge MA 01262-0506

SELKER, MILTON LEONARD, b Detroit, Mich, Nov 2, 15; m 41; c 2. PHYSICAL ORGANIC CHEMISTRY. *Educ:* Western Reserve Univ, BS, 36, MA, 37, PhD(phys org chem), 40. *Prof Exp:* Lab asst qual anal, Western Reserve Univ, 35-36, org chem, 37; rubber res chemist, Bell Tel Labs, Inc, 40-46; engr, Kahn Co, 46-52 & Clevite Corp, 52-64, dir, Mech Res Div, 64-69; dir, Gould Mat Technol Lab, Gould Inc, 69-71, vpres, 71-75, vpres res & develop, 76-77, consult, 78-81; RETIRED. *Mem:* Am Chem Soc. *Res:* Physical organic chemistry; rubber-metal bearings; plating of metals; rubber recycling; metal-organic bonding; thin film technology. *Mailing Add:* 3175 Morley Rd Shaker Heights OH 44122

SELKIRK, JAMES KIRKWOOD, b New York, NY, Dec 3, 38; m 61; c 2. CHEMICAL CARCINOGENESIS, BIOCHEMISTRY. *Educ:* NY State Col Environ Sci Forestry, BS, 64; Syracuse Univ, BS, 64; Upstate Med Ctr Syracuse, PhD(biochem), 69. *Prof Exp:* Fel, McArdle Lab Cancer Res, Univ Wis, 69-72; staff fel, Chem Nat Cancer Inst, 72-75, sr staff fel, 74-75; group leader chem carcinogenesis, Oak Ridge Nat Lab, 75-78, sr scientist biol div, 78-85, chief Carcinogenesis & Toxicol Eval Br, 85-89, ASSOC DIR DIV TOXICOL RES & TESTING, OAK RIDGE NAT LAB, 89- *Concurrent Pos:* Lectr, Biomed Grad Sch, Univ Tenn, 76-80, sr lectr, 80-; assoc ed, Carcinogenesis, 79-85, Cancer Res, 82-85; mem, Breast Cancer Task Force, NIH, 80-81; mem, Comt Pyrene & Anogs, Nat Acad Sci, 81. *Mem:* Am Asn Cancer Res; Sigma Xi; NY Acad Sci; AAAS; Soc Toxicol. *Res:* Mechanism of action of chemical carcinogens in in vivo and in vitro systems; the enzymatic pathways involved and species variability. *Mailing Add:* Div Toxicol Res & Testing Natl Inst Environ Health Sci PO Box 12233 Toxicol Res & Testing Research Triangle Park NC 27709

SELKURT, EWALD ERDMAN, b Edmonton, Alta, Mar 13, 14; US citizen; m 41; c 2. PHYSIOLOGY. *Educ:* Univ Wis, BA, 37, MA, 39, PhD(zool), 41. *Prof Exp:* Asst zool & physiol, Univ Wis, 37-41; instr med physiol, Col Med, NY Univ, 41-44; from instr to assoc prof, Sch Med, Western Reserve Univ, 44-58; prof & chmn dept, 58-79, distinguished prof, 79-83, EMER PROF PHYSIOL, SCH MED, IND UNIV, INDIANAPOLIS, 83- *Concurrent Pos:* Mem subcomt shock, Comt Med & Surg, Nat Res Coun, 53-58; mem cardiovasc study sect, Nat Heart Inst, 63-67; Nat Sci fel, Univ Gottingen, 64-65. *Mem:* Fel Am Physiol Soc (pres-elect, 75); fel Soc Exp Biol & Med; fel Harvey Soc; fel Am Heart Asn; fel Am Soc Nephrology. *Res:* Physiology of circulation and kidney. *Mailing Add:* Summer Trace Retirement Community 12999 N Pennsylvania Apt E-249 Indianapolis IN 46032

SELL, GEORGE ROGER, b Milwaukee, Wis, Feb 7, 37; m 58; c 6. MATHEMATICS. *Educ:* Marquette Univ, BS, 57, MS, 58; Univ Mich, PhD(math), 62. *Prof Exp:* Benjamin Pierce instr math, Harvard Univ, 62-64; asst prof, Univ Minn, Minneapolis, 64-67; assoc prof, Univ Southern Calif, 67-68; assoc head, Sch Math, 70-71, assoc prof, 68-73, PROF MATH, SCH MATH, UNIV MINN, MINNEAPOLIS, 73- *Concurrent Pos:* Mathematician, Inst Defense Anal, NJ, 66; vis prof, Univ Florence, 71-72, Univ Palermo, 75, Tech Univ Warsaw, 75, Japan Soc Prom Sci, 77 & Australian Nat Univ, 79; prog dir, NSF, 77-78; co-founder & first assoc dir, Inst Math Appln, 81-87. *Mem:* Am Math Soc; Math Asn Am; Soc Indust & Appl Math. *Res:* Dynamical systems, ordinary and partial differential equations; applied mathematics. *Mailing Add:* Univ Math Univ Minn 206 Church St Minneapolis MN 55455

SELL, JEFFREY ALAN, b Anderson, Ind, Sept 18, 52; div; c 2. SPECTROSCOPY, CHEMICAL PHYSICS. *Educ:* Purdue Univ, BS, 74; Calif Inst Technol, PhD(chem), 79. *Prof Exp:* Assoc sr res scientist, 78-80, sr res scientist physics, 80-86, SR STAFF RES SCIENTIST & MGR, ADVAN MAT & PROCESSING, GEN MOTORS RES LABS, 86- *Concurrent Pos:* Lectr physics & math, Lawrence Inst Technol, 81- *Mem:* Optical Soc Am; Am Chem Soc; Am Phys Soc; Combustion Inst; Mat Res Soc. *Res:* Tunable diode laser spectroscopy; ultraviolet photoelectron spectroscopy; visible laser spectroscopy of rare earth crystals; thermodynamics; Raman spectroscopy; combustion diagnostics; laser deposition and ablation of materials; superconductivity; photothermal deflection spectroscopy; combustion ignition. *Mailing Add:* Dept Physics Gen Motors Res Lab Warren MI 48090-9055

SELL, JERRY LEE, b Adel, Iowa, Feb 6, 31; m 53; c 2. ANIMAL NUTRITION. *Educ:* Iowa State Univ, BS, 57, MS, 58, PhD(poultry nutrit), 60. *Prof Exp:* Assoc prof animal sci, Univ Man, 60-66; assoc prof animal sci, 66-68, prof animal nutrit, NDak State Univ, 68-76; prof animal nutrit, 76-86, C F CURTISS DISTINGUISHED PROF AGR, IOWA STATE UNIV, 86- *Concurrent Pos:* Mem sub-comt Poultry Nutrit, Nat Res Coun, US Nat Acad Sci, 81-84 & 89- *Honors & Awards:* Poultry Nutrit Res Award, Am Feed Mfrs, 78; Res Award, Nat Turkey Fed, 90. *Mem:* Sigma Xi; Am Poultry Sci Asn (pres, 86-87); World Poultry Sci Asn. *Res:* Energy efficiency of chickens and turkeys; functional development of gastro-intestinal tract of turkeys. *Mailing Add:* Dept Animal Sci Iowa State Univ Ames IA 50011

SELL, JOHN EDWARD, b Gainesville, Fla, June 14, 41. BIOCHEMISTRY, IMMUNOLOGY. *Educ:* Mich State Univ, BS, 63; Univ Wis-Madison, MS, 67; Univ Cincinnati, PhD(biochem), 71. *Prof Exp:* Res assoc microbiol, Univ Mich, Ann Arbor, 71-73, res assoc internal med, 73-76; RES ASSOC BIOCHEM, MICH STATE UNIV, 80- *Mem:* AAAS; Am Chem Soc. *Res:* Biochemistry of neonatal respiration; carcinofetal antigens; cellular fluorescence spectroscopy. *Mailing Add:* Dept Biochem Mich State Univ East Lansing MI 48824

SELL, KENNETH W, b Valley City, NDak, Apr 29, 31; m 50; c 4. BLOOD BANKING. *Educ:* Univ NDak, BA, 53; Harvard Med Sch, MD, 56; Cambridge Univ, PhD(immunopath), 68. *Prof Exp:* Intern & resident, Bethesda Naval Hosp, 56-59, mem pediat staff, 59-60; dir, Navy Tissue Bank, Md, 60-70; chmn, Dept Clin & Exp Immunol, Navy Med Res Inst, 70-77; sci dir, Nat Inst Allergy & Infectious Dis, NIH, 77-85; clin prof, Dept Pediat, Georgetown Sch Med, 77-85; PROF & CHMN, DEPT PATH & DIR, CANCER CTR, EMORY SCH MED, 85- *Concurrent Pos:* Command officer, Navy Med Res Inst, 74-77; lectr, Found Advan Educ Sci, NIH, 60-77; dir, Transplantation Serv, Nat Naval Med Ctr, 71-77; lectr, Uniformed Serv Univ Health Sci, 77-85. *Mem:* Am Asn Tissue Banks; Soc Cryobiol; Transplantation Soc; Am Acad Pediat; Am Col Path. *Res:* Clinical immunology and transplantation with contributions to immunoparasitology and immune regulation of responses to viral diseases; experimental and clinical study of immunosuppression for organ transplantation. *Mailing Add:* Dept Path Rm 703 WMB Emory Univ Atlanta GA 30322

SELL, NANCY JEAN, b Milwaukee, Wis, Jan 18, 45. CHEMICAL PHYSICS. *Educ:* Lawrence Univ, BA, 67; Northwestern Univ, MS, 68, PhD(chem physics), 72; Inst Paper Chem, MS, 86. *Prof Exp:* From asst prof to assoc prof, 71-82, chmn dept & coodr, grad prog environ sci, 81-83, PROF CHEM & PHYSICS, UNIV WIS-GREEN BAY, 77- *Concurrent Pos:* Consult, numerous industries & law firms; pres, N J Sell & Assocs SC, 84-; secy, Process Simulation Comt, Tech Asn Pulp & Paper Inst, 89- *Mem:* Am Chem Soc; Am Soc Testing & Mat; Sigma Xi. *Res:* Industrial resource recovery and subsequent industrial energy conservation; industrial pollution control; process simulation; pulp and paper engineering and technology. *Mailing Add:* 3244 Peterson Rd Green Bay WI 54311

SELL, SARAH H WOOD, b Birmingham, Ala, Mar 20, 13; m 52; c 2. PEDIATRICS, PUBLIC HEALTH. *Educ:* Berea Col, AB, 34; Vanderbilt Univ, MS, 38, MD, 48; Am Bd Pediat, dipl, 54. *Prof Exp:* Intern pediat, Vanderbilt Univ Hosp, 48-49; resident, Cincinnati Children's Hosp, 49-51; instr microbiol & pediat, Sch Med, La State Univ, 51-53; instr pediat, Sch Med, Tulane Univ, 53-54; from instr to prof, 54-78, EMER PROF PEDIAT, SCH MED, VANDERBILT UNIV, 78- *Concurrent Pos:* Res fel microbiol & pediat, Sch Med, La State Univ, 51-53; med consult, Tenn State Dept Pub Health; consult immunizations, Ctr Dis Control, 63-66; mem adv comt infectious dis, Nat Inst Allergy & Infectious Dis, 73-77, consult meningitis vaccine trial, Finland, 74. *Mem:* Am Pediat Soc; Am Acad Pediat; Am Col Chest Physicians; Infectious Dis Soc Am; Am Soc Microbiol; Am Asn Univ Prof. *Res:* Infectious diseases of infants and children; bacterial meningitis and sequelae, otitis media, respiratory infections and immunizations; life long emphasis on haemophilus, influenzae diseases and prevention; bacterial polysaccharide vaccines. *Mailing Add:* 3804 Woodlawn Dr Nashville TN 37215

SELL, STEWART, b Pittsburgh, Pa, Jan 20, 35; c 4. PATHOLOGY, IMMUNOLOGY. *Educ:* Col William & Mary, BS, 56; Univ Pittsburgh, MD, 60; Am Bd Path, dipl, 66, 83; Am Bd Lab Immunol, 81. *Prof Exp:* Intern & asst resident path, Mass Gen Hosp, 60-62; res assoc, NIH, 62-64; from instr to assoc prof, Sch Med, Univ Pittsburgh, 65-70; from assoc prof to prof path, Sch Med, Univ Calif, San Diego, 70-82; PROF PATH, PATH LAB MED, UNIV TEX HEALTH SCI CTR, HOUSTON, 82- *Concurrent Pos:* Nat Inst Allergy & Infectious Dis spec fel, Univ Birmingham, Eng, 64-65 & res career develop award, 65-70; mem adv path study sect B, NIH, 72-76. *Mem:* AAAS; Brit Soc Immunol; Am Asn Immunologists; NY Acad Sci; Am Soc Exp Path. *Res:* Immunology and pathology, lymphocyte receptors; alpha fetoprotein; chemical carcinogenesis; syphilis; viral pathogenesis; AIDS. *Mailing Add:* Dept Path & Lab Med Univ Tex Med Sch PO Box 20708 Houston TX 77025

SELLARS, JOHN R(ANDOLPH), b Ft Stanton, NMex, Mar 1, 25; m 50; c 3. AERONAUTICAL ENGINEERING. *Educ:* NMex State Univ, BS, 45; Univ Mich, MS, 50, PhD(aeronaut eng), 52. *Prof Exp:* Res assoc, Appl Physics Lab, Johns Hopkins Univ, 45-46; res assoc, Univ Mich, 46-52, asst prof, 52-55; mem tech staff, TRW Systs, 55-58, mgr, Aerodyn Dept, 58-61, dir Aerosci Lab, 61-66, mgr, Systs Labs Eng Oper, 66-69, mgr, res & technol opers, Appl Technol Div, TRW Systs, 69-81, vpres & gen mgr, Energy Technol Div, TRW Energy Develop Group, 81-86; RETIRED. *Mem:* Am Inst Aeronaut & Astronaut; Sigma Xi. *Res:* Reentry systems; stability of flow; heat transfer. *Mailing Add:* 128 Via Los Miradores Redondo Beach CA 90277

SELLAS, JAMES THOMAS, b Chicago, Ill, Dec 29, 24; m 51; c 3. ORGANIC CHEMISTRY. *Educ:* Univ Iowa, PhD(chem), 54. *Prof Exp:* Chemist, Stand Oil Co, 53-59, Aerojet-Gen Corp, Gen Tire & Rubber Co, 59-71 & Aerojet Solid Propulsion Co, 71-75, sr chem specialist, Aerojet Solid Propulsion Co, 75-78 & Aerojet Tech Syst Co, 78-85; RETIRED. *Concurrent Pos:* Consult, 85- *Mem:* Am Chem Soc; Sigma Xi. *Res:* Polymer chemistry; development of new and novel class of controllable high burning rate propellants for thrust vector control and controllable solid rocket application; expanding the technology of extinguishable solid propellants and the use of new oxidizers; expanding technology of ultra high burning rate propellants; development of high density-impulse propellants, and high combustion efficiency propellants; development of space storable propellants; additives developed and research carried out to determine methods to control both temperature and pressure sensitivity in the combustion of solid propellants. *Mailing Add:* 3708 Lynwood Way Sacramento CA 95864

SELLE, JAMES EDWARD, b Waukesha, Wis, Sept 1, 31; m 58; c 2. METALLURGICAL ENGINEERING, MATERIALS ENGINEERING. *Educ:* Univ Wis, BS, 55, MS, 56; Univ Cincinnati, PhD(metall eng), 67. *Prof Exp:* Res engr, Gen Motors Res Staff, Mich, 56-58 & Dayton Malleable Iron Co, Ohio, 58; sr res chemist, Mound Lab, Monsanto Res Corp, 58-67, group leader, 67-70, sr res specialist, 70-73; res staff mem, Oak Ridge Nat Lab, 74-80; ASSOC SCIENTIST, EG&G ROCKY FLATS, GOLDEN, COLO, 80- *Honors & Awards:* Wilson Award, Am Soc Metals, 72. *Mem:* Am Soc Metals. *Res:* Equilibrium diagrams; allotropic transformations; compatibility; nuclear reactor fuels studies; high temperature reactions; liquid metal corrosion; impurity effects on material properties. *Mailing Add:* 4755 W 101st Pl Westminster CO 80030

SELLE, WILBUR A, nuclear medicine, rehabilitative medicine; deceased, see previous edition for last biography

SELLECK, GEORGE WILBUR, b Sask, Can, Jan 28, 24; m 48; c 2. PLANT ECOLOGY, WEED SCIENCE. *Educ:* Univ Sask, BSA, 50, MSc, 53; Univ Wis, PhD(bot), 59. *Prof Exp:* Weed ecologist, Univ Sask, 53-55, instr ecol, 56, asst prof, 56-60; sr proj mgr, Monsanto Co, 60-63, mgr agr sales, Monsanto Europe, 63-66, mgr agr develop, 66-68, regional mgr mkt develop, 68-73, sr develop assoc, Monsanto Co, 73-75; PROF HORT & SUPT RES FARM, CORNELL UNIV, 75- *Mem:* AAAS; Ecol Soc Am; Weed Sci Soc Am; Can Soc Agron; Agr Inst Can. *Res:* Ecology of native vegetation; life history of perennial weeds; agricultural pesticides; weed control in horticultural crops; fertilization practice in potatoes and turf in relation to nitrate pollution in underground water. *Mailing Add:* Long Island Hort Res Lab 39 Sound Ave Riverhead NY 11901

SELLERS, ALFRED MAYER, b Philadelphia, Pa, Feb 23, 24; m 52; c 2. MEDICINE, CARDIOLOGY. *Educ:* Duke Univ, BS & MD, 51; Am Bd Internal Med, dipl, 58, recert, 75. *Hon Degrees:* MA, Univ Pa, 71. *Prof Exp:* Asst instr, 52-54, instr, 54-56, assoc, 56-59, asst prof, 59-66, chief hypertension clin, Univ Hosp, 61-71, ASSOC PROF MED, UNIV HOSP & MED SCH, UNIV PA, 66- *Concurrent Pos:* Attend physician, Vet Admin Hosp, Philadelphia, 67-; mem coun high blood pressure res & mem med adv bd & fel coun clin cardiol, Am Heart Asn. *Mem:* AAAS; fel Am Col Physicians; Am Fedn Clin Res; fel Am Col Cardiol; fel Am Col Chest Physicians; fel Am Col Clin Pharmacol. *Res:* Internal medicine; cardiology; hypertension. *Mailing Add:* Dept Med Univ Pa Hosp 3400 Spruce St Philadelphia PA 19104

SELLERS, ALVIN FERNER, b Somerset, Pa, Aug 9, 17; m 42; c 3. VETERINARY PHYSIOLOGY. *Educ:* Univ Pa, VMD, 39; Ohio State Univ, MS, 40; Univ Minn, PhD, 49. *Prof Exp:* Asst, Ohio State Univ, 39-40; instr animal physiol, Univ Minn, 40-42 & 46-49, assoc prof vet physiol, 49-54, prof vet physiol & pharmacol & head div, 54-60; prof physiol, 60-, EMER PROF, NY STATE VET COL, CORNELL UNIV. *Concurrent Pos:* Guggenheim fel, Physiol Lab, Cambridge & Rowett Res Inst, Scotland, 57-58. *Mem:* Am Physiol Soc; Soc Exp Biol & Med; Am Gastroenterol Asn. *Res:* Ruminant digestive tract; absorption; blood flow. *Mailing Add:* Dept Vet Med Cornell Univ Ithaca NY 14853

SELLERS, ALVIN LOUIS, b Philadelphia, Pa, Oct 16, 16; m 42; c 3. PHYSIOLOGY. *Educ:* Univ Calif, Los Angeles, BA, 40; Univ Calif, MD, 43. *Prof Exp:* Intern med, Univ Calif Hosp, 43-44; resident, Permanente Hosp, 44-46; res assoc, 50-70, SR RES ASSOC, CEDARS OF LEBANON HOSP, 70-; CLIN PROF MED, UNIV CALIF, LOS ANGELES, 79- *Concurrent Pos:* Nat Res Coun res fel, St Mary's Hosp Med Sch, Eng, 47-48; res fel, Cedars of Lebanon Hosp, Los Angeles, 48-49; assoc clin prof med, Univ Southern Calif, 56-73. *Mem:* Soc Exp Biol & Med; Am Physiol Soc; Am Heart Asn. *Res:* Physiology of the kidneys; hypertension. *Mailing Add:* 8635 W Third St Los Angeles CA 90048

SELLERS, CLETUS MILLER, JR, b Harrisonburg, Va, Sept 6, 44; m 70. ENVIRONMENTAL PHYSIOLOGY. *Educ:* Hampden-Sydney Col, BA, 66; James Madison Univ, MS, 70; Va Polytech Inst & State Univ, PhD(zool), 73. *Prof Exp:* Asst prof biol, 73-81, ASSOC PROF BIOL, JAMES MADISON UNIV, 81- *Mem:* Sigma Xi. *Res:* Development and implementation of biological monitoring for rapid detection and quantification of environmental toxicant effects. *Mailing Add:* Dept Biol James Madison Univ Harrisonburg VA 22807

SELLERS, DONALD ROSCOE, b Kansas City, Mo, June, 24, 46; m 68; c 2. TOXIC MATERIALS DETECTION, CLINICAL CHEMISTRY. *Educ:* Univ Mo, AB, 68, MS, 70, PhD(biochem), 72. *Prof Exp:* Head, Stress Biochem Lab, Wright Patterson AFB, Ohio, 72-77; sr biochemist, Midwest Res Inst, 77-83; BIOCHEMIST, ABBOTT LABS, 83- *Mem:* AAAS. *Res:* Development of methodologies and instrumentation that employ biological mechanisms to detect chemicals of interest in the environment and workplace. *Mailing Add:* Abbott Labs D9FA AP32 Abbott Park IL 60064

SELLERS, DOUGLAS EDWIN, b Santa Maria, Calif, Nov 25, 31; m 51; c 2. ANALYTICAL CHEMISTRY. *Educ:* Ft Hays Kans State Col, BS, 53; Kans State Univ, MS, 55, PhD(chem), 58. *Prof Exp:* Asst prof chem, Kans State Univ, 57-58 & Univ Southern Ill, 58-64; res specialist, Monsanto Res Corp, 64-66, group leader electrochem, explosives & gas anal, 66-69, anal mgr, 69-78; RETIRED. *Mem:* Am Chem Soc. *Res:* Instrumental methods of chemical analysis, particularly polarography and ultraviolet visible techniques. *Mailing Add:* RR 2 Box 35 Wakeeney KS 67672

SELLERS, EDWARD ALEXANDER, pharmacology; deceased, see previous edition for last biography

SELLERS, EDWARD MONCRIEFF, b Victoria, BC, June 12, 41. CLINICAL PHARMACOLOGY. *Educ:* Univ Toronto, MD, 65; Harvard Univ, PhD(pharmacol), 71; FRCP(C) & Am Bd Internal Med, dipl internal med, 72; FACP, 77. *Prof Exp:* Clin & res fel med, Mass Gen Hosp, 68-72; head, Clin Res Unit, Clin Inst Addiction Res Found, 73; dir, Div Clin Pharmacol, Clin Inst Addiction Res Found & Toronto Western Hosp, 75; assoc prof pharmacol & med, 76-80, PROF PHARMACOL & MED, UNIV TORONTO, 80- *Concurrent Pos:* Assoc ed, Drug Metab Rev, 74- *Honors & Awards:* Rawls Parlmer Award, Am Soc Clin Pharmacol & Exp Therapeut; Serv Award, Can Soc Clin Pharmacol. *Mem:* Am Soc Pharmacol & Exp Therapeut; Am Soc Clin Pharmacol & Therapeut; Can Soc Clin Pharmacol. *Res:* Alcohol and drug abuse; pharmacogenetics; drug metabolism and toxicology; drug treatment of alcohol and substance abuse. *Mailing Add:* Addiction Res Found 33 Russell St Toronto ON M5S 2S1 Can

SELLERS, ERNEST E(DWIN), b Manhattan, Kans, Aug 17, 25; m 49; c 4. ELECTRICAL ENGINEERING. *Educ:* Kans State Univ, BS, 48, MS, 49. *Prof Exp:* Res engr, Res Labs, Radio Corp Am, NJ, 49-51; instr elec eng, Kans State Univ, 51-52; res engr, 52-85, EMER ENGR, MANAGING DIR OFF, INST SCI & TECHNOL, UNIV MICH, ANN ARBOR, 85- *Concurrent Pos:* Sr engr, Tex Instruments Co, 59; head, Univ Mich Tech Inst Radar Lab, 63-66. *Mem:* AAAS; Inst Elec & Electronics Engrs; Instrument Soc Am. *Res:* Electronics systems analysis; air defense; combat surveillance; detection and control theory; radar; television; countermeasures; navigation; radio astronomy; microwave propagation and reflection; signal handling; data processing; data display; bio-medical engineering. *Mailing Add:* Univ Mich 503 Thompson Ave Ann Arbor MI 48109-0001

SELLERS, FRANCIS BACHMAN, b Washington, NC, Mar 22, 30; m 58; c 2. ATOMIC PHYSICS, NUCLEAR PHYSICS. *Educ:* Wake Forest Col, BS, 54; Univ Kans, PhD(physics), 60. *Prof Exp:* Sr physicist, Phys Res Dept, Allied Res Assocs, Inc, Mass, 60-62, head physics res dept, 62-64; head dept, 64-80, VPRES RADIATION PHYSICS, PANAMETRICS INC, 81- *Mem:* Am Phys Soc; Sigma Xi; Am Geophys Union; Am Meteorol Soc. *Res:* Interaction of nuclear particles with matter; production and measurement of x-radiation; measurement of atmospheric and extraterrestrial parameters; radiation detection techniques for rocket and satellite applications; atmospheric physics. *Mailing Add:* 541 Peakham Rd Sudbury MA 01776 Crescent St

SELLERS, FRANK JAMIESON, b Winnipeg, Man, Mar 10, 28; m 61; c 3. PEDIATRIC CARDIOLOGY. *Educ:* Univ Man, BSc, 54; Queen's Univ, Ont, MD & CM, 55. *Prof Exp:* From asst prof to assoc prof pediat, Univ Sask, 63-70; ASSOC PROF PEDIAT, UNIV OTTAWA, 70- *Mem:* Can Med Asn; Can Pediat Soc; Can Cardiovasc Soc; Can Soc Clin Invest. *Res:* Congenital heart disease; rheumatic fever. *Mailing Add:* Dept Pediat Univ Ottawa 452 Smyth Rd Ottawa ON K1N 8M5 Can

SELLERS, JOHN WILLIAM, b Wausau, Wis, Apr 13, 16; m 44; c 2. RUBBER, PLASTICS. *Educ:* Univ Ill, BS, 42; Ohio State Univ, PhD(chem), 49. *Prof Exp:* Tech dir foods, Food Mat Corp, 37-40; jr chem engr rubber, Firestone Tire & Rubber Co, 42-45, sr res chemist, 49-51; sr res chemist, Chem Div, Pittsburgh Plate Glass Co, 51-57, supvry org res, 57-58; head, Org Res Sect, Petrol Chems, Inc, 58-59; sr supvr org res, Chem Div, Pittsburgh Plate Glass Co, 59-63 & Rubber Chem Res, 63-65; tech adv to mgt & dir corp res, Tenneco Chem, Inc, 65-67; dir res & develop, Paterson Paper Co, 67-68, vpres tech, 68-70; mgr indust waste mgt, Procon, Inc, 70-71 & environ control, H J Heinz Co, 71-73; vpres opers, Interscience, Inc, 73-76, exec vpres, 77-82, pres & chief exec officer, 83-87; tech asst to sr vpres, Reeves/ Southeastern Corp, 87-89; RETIRED. *Concurrent Pos:* Consult, Interscience, Inc, 87- *Mem:* Sigma Xi; Am Chem Soc; Am Inst Chem Eng. *Res:* Organic synthesis; high polymers; plastics; vulcanization and reinforcement of elastomers; oxidation of hydrocarbons; kinetics and mechanisms; market development; management. *Mailing Add:* 8334 Fountain Ave Tampa FL 33615

SELLERS, PETER HOADLEY, b Philadelphia, Pa, Sept 12, 30; m 58; c 4. MATHEMATICS. *Educ:* Univ Pa, PhD(math), 65. *Prof Exp:* Programmer math, Johnson Found, Univ Pa, 58-61; master, Kangaru Sch, Embu, Kenya, 61-63; fel math, Johnson Found, Univ Pa, 65-66; res assoc, 66-72, ASSOC PROF MATH, ROCKEFELLER UNIV, 72-, SR RES ASSOC, 74- *Concurrent Pos:* Ed, Genomics; trustee, Col Atlantic, Bar Harbor, Maine. *Mem:* Am Math Soc; Math Asn Am; Soc Indust & Appl Math. *Res:* Combinatorial analysis; mathematics analysis of genetic sequences. *Mailing Add:* Lab Math Rockefeller Univ New York NY 10021-6399

SELLERS, THOMAS F, JR, b Atlanta, Ga, Apr 9, 27; m 49; c 3. PREVENTIVE MEDICINE, INFECTIOUS DISEASES. *Educ:* Emory Univ, BS, 47, MD, 50. *Prof Exp:* Res fel infectious dis, Sch Med, Emory Univ, 55-57 & Med Col, Cornell Univ, 57-58; asst prof med, Emory Univ, 58-60, prof prev med, Sch Med, 60-90, emer prof, 90-; RETIRED. *Concurrent Pos:* Mem adv comt health, Appalachian Regional Comn, 65-67; staff mem community med, Cent Middlesex Hosp, London, Eng, 74-75. *Mem:* Am Col Physicians; Am Pub Health Asn; Am Col Prev Med; Asn Teachers Prev Med. *Mailing Add:* 4875 Franklin Pond Rd Atlanta GA 30342

SELLGREN, KRISTEN, b San Diego, Calif, Nov 14, 55. INFRARED ASTRONOMY, INTERSTELLAR DUST. *Educ:* Univ Calif, San Diego, BA, 76; Calif Inst Technol, PhD(physics), 83. *Prof Exp:* Postdoctoral res assoc, Space Telescope Sci Inst, 83-84; asst astronr, Inst Astron, Univ Hawaii, 84-89, assoc astronr, 89-90; ASST PROF ASTRON, DEPT ASTRON, OHIO STATE UNIV, 90- *Honors & Awards:* Newton Lacy Pierce Prize, Am Astron Soc, 90. *Mem:* Am Astron Soc. *Res:* Infrared imaging and spectroscopy of the galactic center, interstellar dust and other astronomical sources. *Mailing Add:* Dept Astron Ohio State Univ 174 W 18th Ave Columbus OH 43210-1106

SELLIN, H A, b Creston, Iowa, Feb 21, 05; m; c 2. STRUCTUAL GEOLOGY. *Educ:* Augustana Col, BS, 27. *Prof Exp:* At Univ Colo, 28-29; mem, Ill State Geol Surv, 29-31; geologist & mgr, Mobil Oil, 36-70; RETIRED. *Mem:* AAAS; Am Asn Petrol Geologist; Geol Soc Am; Soc Econ Palent Mineralogists. *Mailing Add:* 14225 Palm Ridge Dr Sun City AZ 85351

SELLIN, IVAN ARMAND, b Everett, Wash, Aug 16, 39; m 62; c 2. ATOMIC PHYSICS. *Educ:* Harvard Univ, BS, 59; Univ Chicago, SM, 60, PhD(physics), 64. *Prof Exp:* Instr physics, Univ Chicago, 64-65; asst prof, NY Univ, 65-67; res physicist, Oak Ridge Nat Lab, 67-70; from assoc prof to prof, 70-83, DISTINGUISHED SERV PROF PHYSICS, UNIV TENN, KNOXVILLE, 83-, PROJ DIR, 70-; CONSULT, NSF, 80- *Concurrent Pos:* Lectr, Univ Chicago, 61-64; proj dir & consult, Oak Ridge Nat Lab, 70-; vis prof, Orgn Am States, 72; NSF grants; NASA grants, Off Naval Res Contracts, Dept Energy contracts, Univ Tenn, 72-82; mem adv comt atomic & molecular physics, Nat Acad Sci, 73-76, chmn, 80-83; chmn, Fourth Int Conf Beam-Foil Spectros, 75; guest prof, Swed Natural Sci Res Coun, 78; sr Fulbright Hays grant, Ger, 77; guest prof, Cent Atomico Bariloche, Arg, 81; counr coun PLA Am Phys Soc, 79-83; mem panel atomic, molecular & optical physics, NRC-NAS Natural Physics Soc, 83-84; Cecil & Ida Green hon chmn, Tex Christian Univ, 84; Allett fel, Univ Witwatersrand, 85; Humboldt Found award, 86-88. *Honors & Awards:* Sr US Scientist Award, Alexander von Humboldt Found, 77. *Mem:* Fel Am Phys Soc; Cosmos Club. *Res:* Physics of ion beams; structure and collisions of heavy ions; physics of highly ionized matter. *Mailing Add:* 1008 W Outer Dr Oak Ridge TN 37830

SELLINGER, OTTO ZIVKO, b Zagreb, Yugoslavia, Sept 14, 29; nat US; m 55; c 4. NEUROCHEMISTRY. *Educ:* Mass Inst Technol, SB, 54; Tulane Univ, PhD(biochem), 58. *Prof Exp:* NIH fel, Lab Physiol Chem, Univ Louvain, Belg, 58-59; NIH fel biochem, Ist Superiore Sanita, Rome, Italy, 59-60; asst prof biochem & med, Med Sch, Tulane Univ, 60-64; Fulbright vis prof biochem, Univ of the Repub, Montevideo, Uruguay, 64-65; assoc res pharmacologist, 65-68, RES SCIENTIST, MENT HEALTH RES INST, UNIV MICH, ANN ARBOR, 68- *Concurrent Pos:* Fogarty Int Scientist Award, Univ Claude Bernard, Villeurbanne, France, 81-82. *Mem:* Am Soc Neurochem; Int Soc Neurochem. *Res:* Neurochemistry; protein methylation in aging brain. *Mailing Add:* Lab Neurochem Ment Health Res Inst Univ Mich 205 Washtenaw Ann Arbor MI 48109-0720

SELLMER, GEORGE PARK, b Milwaukee, Wis, Mar 12, 18; m 43; c 2. ZOOLOGY. *Educ:* Upsala Col, AB, 48; Rutgers Univ, MS, 52, PhD(zool), 59. *Prof Exp:* From instr to assoc prof, 48-61, chmn dept, 58-79, PROF BIOL, UPSALA COL, 61- *Mem:* Soc Sci Study Sex; Am Asn Sex Educr, Councr & Therapists. *Res:* Anatomy and ecology of bivalve mollusks; human sexuality; biological control of insect pests. *Mailing Add:* Upsala Col Puder Hall Upsala Col Wirths Campus RD 3 Box 138A East Orange NJ 07019

SELLMYER, DAVID JULIAN, b Joliet, Ill, Sept 28, 38; m 62; c 3. SOLID STATE PHYSICS & MATERIALS SCIENCE ENGINEERING. *Educ:* Univ Ill, BS, 60; Mich State Univ, PhD(physics), 65. *Prof Exp:* From asst prof to assoc prof metall & mat sci, Ctr Mat Sci, Mass Inst Technol, 65-72; assoc prof, 72-75, chmn dept, 78-84, prof, 75-87, GEORGE HOLMES DISTINGUISHED PROF PHYSICS, UNIV NEBR, LINCOLN, 87-, DIR CTR MAT RES & ANALYSIS, 88- *Concurrent Pos:* Consult, USAF Cambridge Res Lab, Bedford, Mass, 71-72 & Dale Electron, Norfolk, Nebr, 79-; vis scientist, Nat Mag Lab, MIT-1105-80, Inst phys, Beijing, 85 & 86. *Mem:* AAAS; fel Am Phys Soc; Sigma Xi; Mat Resource Soc. *Res:* Electronic structure and magnetism in metallic compounds and alloys; physics of metallic glasses; amorphous magnetism; physics of thin films and multilayers; magnetic recording materials. *Mailing Add:* Behlen Lab Physics & Ctr Mat Res & Analysis Univ Nebr Lincoln NE 68588-0111

SELLNER, KEVIN GREGORY, b Albany, NY, Oct 11, 49. PLANKTON ECOLOGY. *Educ:* Clark Univ, BA, 71; Univ SC, MS, 73; Dalhousie Univ, PhD(oceanog), 78. *Prof Exp:* Res asst algal physiol, Univ SC, 71-73; asst cur dept limnol, 78-81, asst cur, 81-86, ASSOC CUR, BENEDICT ESTUARINE RES LAB, ACAD NATURAL SCI, 86- *Concurrent Pos:* Adj asst prof, Chesapeake Biol Lab, Univ Md, 84-87; adj res scientist, Chesapeake Bay Inst, Johns Hopkins Univ, 85-87. *Mem:* Am Soc Limnol & Oceanog; Phycol Soc Am; Estuarine Res Fedn. *Res:* Dynamics of carbon, nitrogen and oxygen in estuarine enviromnemts; the importance of plankton primary production in carbon, nitrogen and phosphorous flux to secondary producers, for example, bacteria and herbivorous zooplankton and fish; associated effects on carbon deposition and microbial oxidation in partially mixed estuaries. *Mailing Add:* Acad Natural Sci Benedict Estuarine Res Lab Benedict MD 20612

SELLO, STEPHEN, textile chemistry, for more information see previous edition

SELLS, BRUCE HOWARD, b Ottawa, Ont, Aug 15, 30; m 53; c 4. MOLECULAR BIOLOGY. *Educ:* Carleton Univ, BSc, 52; Queen's Univ, Ont, MA, 54; McGill Univ, PhD(biochem), 57. *Prof Exp:* Damon Runyon res fels, Lab Animal Morphol, Free Univ Brussels, 57-59 & State Serum Inst, Copenhagen, Denmark, 59-60; cancer res scientist, Roswell Park Mem Inst, 60-61; res assoc, Columbia Univ, 61-62; from asst prof to assoc prof, Lab Biochem, St Jude Children's Res Hosp & dept biochem, Univ Tenn, Memphis, 62-73, mem, Hosp, 68-72; prof & dir molecular biol, Med Sch, Mem Univ Nfld, 72-83, assoc dean basic med sci, 79-83; PROF & DEAN, COL BIOL SCI, UNIV GUELPH, CAN, 83- *Concurrent Pos:* Vis res scientist, Inst Animal Genetics, Univ Edinburgh, 69-70; mem, biochem grants comt, Med Res Coun Can, 73 , grants comt molecular biol & coun & centennial fel comt; assoc ed, Can J Biochem, 74; Killam sr res fel, Inst Molecular Biol, Univ Paris, 78-79; exchange scientist coop res prog, French Nat Ctr Sci Res, US Nat Res Coun; sci officer, Nat Cancer Inst, 79-81; chmn, comt on biotechnol develop grants, MRC, 83-85; sub-comt on biol phenomena, Nat Res Coun Can, 83-86; rapporteur, Microbiol & Biochem Div, Royal Soc Can, 85-87, convenor, 87-; chmn steering group, assoc comt on sci criteria environ qual, Nat Res Coun Can, 86; E W R Steacie Prize Awards comt, 86-88; Med Res Coun vis prof, Inst Pasteur, Paris, France, 89; Ayerst Award Selection comt, Can Biochem Soc, 90; mem, Life Sci Div Fel Rev comt, Acad Sci, Royal Soc Can, 90-; mem, standing comt, Genetic Basis Human Dis Network, Med Res Coun, 91. *Mem:* Can Biochem Soc (pres); Am Soc Biol Chemists; Am Soc Microbiol; Can Asn Univ Teachers; Am Soc Cell Biol; fel Royal Soc Can. *Res:* Nucleic acids; biosynthesis of ribosomes; studies on growth and differentiation; translational control. *Mailing Add:* Col Biol Sci Univ Guelph Guelph ON N1G 2W1 Can

SELLS, GARY DONNELL, b New Hartford, Iowa, Aug 5, 32; m 53; c 4. CELL BIOLOGY, GENERAL BIOLOGY. *Educ:* Univ Northern Iowa, BA, 54, MA, 59; Iowa State Univ, PhD(plant physiol), 65. *Prof Exp:* Teacher high schs, Iowa, 54-62; assoc prof, 65-71, PROF PHYSIOL, NORTHEAST MO STATE UNIV, 71- *Mem:* Nat Asn Biol; Am Inst Biol Sci; Sigma Xi. *Res:* Physiology; botany; developmental anatomy; mitochondrial research in plants & animals. *Mailing Add:* Six Grim Ct S Kirksville MO 63501

SELLS, JACKSON S(TUART), b Buffalo, NY, Dec 27, 20; m 41; c 2. ELECTRICAL ENGINEERING. *Educ:* Univ Miami, BS, 46, MS, 50; Purdue Univ, BSEE, 52. *Prof Exp:* Asst prof elec eng & physics, Univ Miami, 46-55, from assoc prof elec eng, 55-65, prof, 65-; RETIRED. *Concurrent Pos:* Consult, Mercy Hosp, Fla, 53-60 & City of Miami, 60-80. *Mem:* Am Soc Eng Educ; Illum Eng Soc; Inst Elec & Electronics Engrs; Am Sci Affil; Simulations Coun. *Res:* Servomechanisms; radiotelemetry; control and network theories. *Mailing Add:* 8215 Saragoza Ct Orlando FL 32819

SELLS, JEAN THURBER, b Butte, Nebr, May 24, 40; m 62; c 2. MATHEMATICS. *Educ:* Nebr Wesleyan Univ, AB, 61; Univ Minn, Minneapolis, MA, 63, PhD(math), 66. *Prof Exp:* Asst prof math, Tex A&I Univ, 66-67; asst prof, Univ Louisville, 67-70; from asst prof to assoc prof, Frostburg State Col, 70-72; assoc prof, Coker Col, 73-75; asst prof, Fordham Univ, 75-76; ASSOC PROF MATH, SACRED HEART UNIV, 76-, CHMN, DEPT MATH, 81- *Mem:* Am Math Soc; Math Asn Am; Nat Coun Teachers Math. *Mailing Add:* 36 September Lane Weston CT 06883

SELLS, ROBERT LEE, b Lancaster, Ohio, Oct 14, 25; m 47; c 2. PHYSICS. *Educ:* Univ Mich, BS, 48; Univ Notre Dame, PhD(physics), 53. *Prof Exp:* From asst prof to assoc prof physics, Rutgers Univ, 53-63; prof, 63-73, DISTINGUISHED TEACHING PROF PHYSICS & ASTRON & CHMN DEPT, STATE UNIV NY COL GENESEO, 73- *Mem:* Am Asn Physics Teachers; Am Phys Soc. *Res:* Theoretical physics; atomic and nuclear physics; solid state physics; teaching physics. *Mailing Add:* 207 Oak St Geneseo NY 14454

SELLSTEDT, JOHN H, b Minneapolis, Minn, June 11, 40; m 61; c 2. MEDICINAL CHEMISTRY. *Educ:* Univ Minn, BS, 62, PhD(org chem), 65. *Prof Exp:* Sr res chemist, Wyeth Labs, Inc, 65-69, group leader, 69-77, MGR RES ANALYTICAL CHEM, RES & DEVELOP LABS, WYETH LABS DIV, AM HOME PROD CORP, 77- *Res:* Antiallergy and cardiovascular research; mass spectroscopy; cephalosporins and penicillins. *Mailing Add:* Wyeth Lab Inc 64 Maple St Rouses Point NY 12979

SELMAN, ALAN L, b New York, NY, Apr 2, 41; m 63; c 2. THEORETICAL COMPUTER SCIENCE, MATHEMATICAL LOGIC. *Educ:* City Col Univ NY, BS, 62; Univ Calif, Berkeley, MA, 64; Pa State Univ, PhD(math), 70. *Prof Exp:* Instr math, Pa State Univ, 68-70; lectr, Carnegie-Mellon Univ, 70-72; asst prof, Fla State Univ, 72-77; assoc prof, 77-82, PROF COMPUTER SCI, IOWA STATE UNIV, 82- *Concurrent Pos:* Res Mathematician, Carnegie-Mellon Univ, 70-72; NSF grants, 75-87. *Mem:* Asn Symbolic Logic; Asn Comput Mach. *Res:* Low level automata based complexity; studies on whether certain computational problems can be feasibly computed versus whether they are intractable for practical computing. *Mailing Add:* Dept Computer Sci State Univ NY 226 Bell Hall Buffalo NY 14260

SELMAN, BRUCE R, ENERGY TRANSDUCTION, CHLOROPLAST ATPASE BIOGENESIS. *Educ:* Univ Rochester, PhD(biol), 73. *Prof Exp:* PROF BIOCHEM, UNIV WIS, 76- *Mailing Add:* Dept Biochem Univ Wis 420 Henry Mall Madison WI 53706

SELMAN, CHARLES MELVIN, b Brenham, Tex, Jan 18, 37; m 60; c 2. ORGANIC POLYMER CHEMISTRY, ORGANOMETALLIC CHEMISTRY. *Educ:* Southwestern Univ, BS, 59; N Tex State Univ, MS, 66, PhD(chem), 68. *Prof Exp:* Res chemist, Dow Chem Co, 60-63; sr polymerization chemist, 67-78, supvr polyolefins process, 79-82, MGR POLYOLEFINS, PHILLIPS PETROL CO, 83- *Mem:* Am Chem Soc; Soc Plastics Engrs. *Res:* Polymerization reactions of organic molecules catalyzed with organometallic compounds. *Mailing Add:* 5120 Parsons Dr Bartlesville OK 74006

SELMAN, KELLY, b Cleveland, Ohio, July 22, 42; m 84. CELL BIOLOGY, REPRODUCTIVE BIOLOGY. *Educ:* Univ Mich, BA, 64; Harvard Univ, MA, 65, PhD(biol), 72. *Prof Exp:* Instr biol, Simmons Col, 67-68 & Univ Va, 71-72; fel anat, Harvard Univ Med Sch, 72-74; asst prof, 74-79, ASSOC PROF ANAT, COL MED, UNIV FLA, 79- *Mem:* Am Soc Cell Biol; AAAS; Am Asn Anat; Am Soc Zoologists. *Res:* Oogenesis and fertilization. *Mailing Add:* Dept Anat & Cell Biol Univ Fla Col Med Gainesville FL 32610

SELMANOFF, MICHAEL KIDD, b Minneapolis, Minn, July 18, 49; m 75; c 2. REPRODUCTIVE NEUROENDOCRINOLOGY. *Educ:* Earlham Col, BA, 70; Univ Conn, PhD(neurobiol), 74. *Prof Exp:* Fel, Dept Obstet, Gynec & Reprod Sci, Reprod Endocrinol Ctr, Sch Med, Univ Calif, San Francisco, 74-77; asst prof, 77-82, ASSOC PROF, DEPT PHYSIOL, SCH MED, UNIV MD, 82- *Concurrent Pos:* Res career develop award, grants for 14 years, NIH, 81-86. *Mem:* Endocrine Soc; Soc Neurosci; Am Physiol Soc. *Res:* Role of tuberoinfundibular dopaminergic neurons in the release of prolactin and luteinizing hormone from the anterior pituitary gland. *Mailing Add:* Dept Physiol Sch Med Univ Md 655 W Baltimore St Baltimore MD 20201

SELOVE, WALTER, b Chicago, Ill, Sept 11, 21; m 55. PARTICLE PHYSICS. *Educ:* Univ Chicago, BS, 42, MS, 48, PhD(physics), 49. *Prof Exp:* Asst instr electronics, Univ Chicago, 42-43; mem staff, Radiation Lab, Mass Inst Technol, 43-45; from jr physicist to assoc physicist, Argonne Nat Lab, 47-50; from instr to asst prof physics, Harvard Univ, 50-56; assoc prof, 56-61, PROF PHYSICS, UNIV PA, 61- *Concurrent Pos:* Mem staff, Radiation Lab, Univ Calif, 53-54; NSF fel, 56; Guggenheim fel, 71-72. *Mem:* Fel Am Phys Soc; Sigma Xi. *Res:* Radar receivers; nuclear and particle physics. *Mailing Add:* Dept Physics Univ Pa Philadelphia PA 19104

SELOVER, JAMES CARROLL, b Los Angeles, Calif, Mar 25, 29; m 51; c 5. ORGANIC CHEMISTRY. *Educ:* Rutgers Univ, BS, 50; Stanford Univ, MS, 52, PhD(org chem), 53. *Prof Exp:* Res chemist, M W Kellogg Co, Pullman, Inc, 53-55; sr res chemist, Richfield Oil Corp, 55-59; tech dir new prods & quality control, Pilot Chem Co, 59; sr chem economist, Stanford Res Inst, 59-68, dir long range planning serv, 68-69; planning systs mgr, Bechtel Corp, 69-71, exec engr, 73-79; exec vpres, Lurgi Corp, 79-85; MGR CORP PLANNING, BECHTEL NAT INC, 71-, VPRES, 85- *Mem:* Am Chem Soc; Sigma Xi; Am Inst Chem Engrs. *Res:* Management consulting and corporate research planning. *Mailing Add:* Seven Maywood Lane Menlo Park CA 94025

SELOVER, THEODORE BRITTON, JR, b Cleveland, Ohio, Jan 13, 31; m 55; c 3. THERMODYNAMICS & MATERIAL PROPERTIES, CHEMICAL ENGINEERING. *Educ:* Brown Univ, ScB, 52; Western Reserve Univ, MS, 57. *Prof Exp:* Jr chemist, Standard Oil Co, Ohio, 52, chemist, 54-57, sr chemist & proj leader basic res, 57-60, tech specialist, 60-62, sr res chemist, 62-69, res assoc, 69-71, info specialist, 71-85; CONSULT CHEMIST, 85-; TECH DIR, AM INST CHEM ENGRS, DESIGN INST PHYS PROP DATA, 86- *Concurrent Pos:* Mem, Energy Soc Libr Bd, 79-, chmn, 85-87; mem tech comt, Design Inst Phys Prop Data, 80-, chmn, 86-; tech ed, Hemisphere Publ Co, 86- *Mem:* Am Chem Soc; Am Soc Testing & Mat; Am Phys Soc; Am Ceramic Soc; Sigma Xi; Am Inst Chem Eng. *Res:* Plasma chemistry; high temperature materials; fused salt batteries; technical editing of Russian translations of property data; aqueous capacitors; thermodynamic and transport property data; numerical database; information resources. *Mailing Add:* 3575 Traver Rd Shaker Heights OH 44122

SELSKY, MELVYN IRA, b Brooklyn, NY, June 25, 33; m 63; c 2. BOTANY. *Educ:* Brooklyn Col, BA, 54, MA, 56; Univ Ill, PhD(plant virol), 60. *Prof Exp:* Substitute teacher biol, Brooklyn Col, 54-56; asst bot, Univ Ill, 56-57 & plant virol, 57-59, res assoc, 60; from asst prof to assoc prof, 60-72, PROF BIOL, BROOKLYN COL, 72- *Concurrent Pos:* Nat Cancer Inst grant, 64-66; fel, Yale Univ, 69-70. *Mem:* AAAS; Am Soc Microbiol; NY Acad Sci. *Res:* Organelle transfer RNAs; chloroplast development in Euglena gracilis; blue-green algae. *Mailing Add:* Dept Biol Brooklyn Col Bedford Ave & Ave H Brooklyn NY 11210

SELTER, GERALD A, b Windsor, Ont, May 3, 40; m 63; c 3. ORGANIC CHEMISTRY. *Educ:* Wayne State Univ, BS, 62; Wash State Univ, PhD(chem), 66. *Prof Exp:* Fel, Univ Calif, Berkeley, 66-67 & 68; asst prof, 68-74, assoc prof, 74-80, PROF ORG CHEM, SAN JOSE STATE UNIV, 80- *Concurrent Pos:* Lectr, Univ Calif, Berkeley, 67-68. *Mem:* Am Chem Soc. *Res:* Reactivity of alpha-pentadienyl esters; solvolytic reactivity of allylic halides and esters, especially the mechanisms of the neighboring group participation in such reactivity. *Mailing Add:* Dept Chem San Jose State Univ Wash Square San Jose CA 95192

SELTIN, RICHARD JAMES, b Chicago, Ill, Nov 4, 27; m 53; c 3. VERTEBRATE PALEONTOLOGY. *Educ:* Univ Wyo, BS, 49; Univ Chicago, MS, 54, PhD(paleont), 56. *Prof Exp:* From instr to assoc prof, 56-69, chmn natural sci, 74-84, PROF NATURAL SCI, MICH STATE UNIV, 69- *Concurrent Pos:* Am Philos Soc grants, 58, 60, 62, 64, 66; consult, Mich State Univ, 57- *Mem:* Soc Vert Paleont; Soc Study Evolution; Sigma Xi; Soc Col Sci Teachers. *Res:* Primitive reptiles and amphibians. *Mailing Add:* 919 Collingwood East Lansing MI 48823

SELTMANN, HEINZ, b Frankfurt am Main, Ger, Sept 8, 24; nat US. PLANT PHYSIOLOGY. *Educ:* Drew Univ, BA, 49; Univ Chicago, MS, 50, PhD(bot), 53. *Prof Exp:* Asst, Univ Chicago, 50-53; asst prof bot, Barnard Col, Columbia Univ, 53-56; from asst prof to assoc prof bot, 56-74, PROF BOT & CROP SCI, NC STATE UNIV, 74-; PLANT PHYSIOLOGIST, USDA, 56- *Res:* Physiology of tobacco plant; plant growth regulators. *Mailing Add:* Box 7620 NC State Univ Raleigh NC 27695-7620

SELTSER, RAYMOND, b Boston, Mass, Dec 17, 23; m 46; c 2. PREVENTIVE MEDICINE, PUBLIC HEALTH. *Educ:* Boston Univ, MD, 47; Johns Hopkins Univ, MPH, 57; Am Bd Prev Med, dipl, 69. *Prof Exp:* Asst med, Sch Med, Boston Univ, 48-51; asst chief med info & intel br, Off Surgeon

Gen, US Dept Army, 53-56; epidemiologist, Div Int Health, USPHS, 56-57; from asst prof to assoc prof epidemiol, Johns Hopkins Univ, 57-66, from assoc prof to prof chronic dis, 63-69, prof epidemiol, 66-81, assoc dean, Sch Hyg & Pub Health, 67-81; dean, 81-87, emer prof & dean Pub Health, Univ Pittsburgh, 87-; assoc dir, Ctr Dis Control, 88-89, ASSOC DIR, SPEC POP RES, AGENCY HEALTH CARE POLICY & RES, PUB HEALTH SERV, 90- *Concurrent Pos:* Resident med & infectious dis, Mass Mem Hosp, 48-51; asst med, Sch Med, Harvard Univ, 50-51; consult, NIMH, 58-70, Nat Cancer Inst, 64-71, Fed Radiation Coun, 66-68, Nat Inst Environ Health Sci, 67-71, Bur Radiol Health, Dept HEW, 68-71 & Off Biomet, Nat Inst Neurol Dis & Stroke, 73-; fel coun epidemiol, Am Heart Asn, 65-, mem exec comt coun stroke, 69-72; vis prof, Med Col Pa, 66-69; mem coun pub health consult, Nat Sanitation Found, 67-69; secy, Asn Sch Pub Health, 69-71; mem bd overseers, Am J Epidemiol, 71-; chmn adv comt radiation registry physicians, Div Med Sci, Nat Acad Sci-Nat Res Coun, 72-; mem, Nat Coun Radiation Protection & Measurements, 73-; chmn biomet & epidemiol contract rev comt, Nat Cancer Inst, 73-; secy-treas, Am Bd Prev Med, 74-; chmn, Comt Human Volunteers, Johns Hopkins Univ, 77- *Mem:* Fel AAAS; fel Am Pub Health Asn; fel Am Col Prev Med; fel Am Heart Asn; Int Epidemiol Asn. *Res:* Streptococcal disease; poliomyelitis; hemorrhagic fever; influenza; ionizing radiation effects; cerebral vascular disease. *Mailing Add:* Agency Health Care Policy Res 2101 E Jefferson St Rockville MD 20852

SELTZER, BENJAMIN, b Philadelphia, Pa, Aug 5, 45; m 74; c 4. BEHAVIORAL NEUROLOGY, NEUROANATOMY. *Educ:* Univ Pa, AB, 65; Jefferson Med Col, MD, 69. *Prof Exp:* Intern, Boston City Hosp, 69-70, resident neurol, 70-73, asst neurologist, 73-75; from asst prof to assoc prof, neurol & psychait, Sch Med, Boston Univ, 78-88; PROF NEUROL & PSYCHIAT, SCH MED, TULANE, 88-, ADJ PROF ANAT, 88- *Concurrent Pos:* Clin fel neurol, Sch Med, Harvard Univ, 70-73, instr, 73-78, lectr, 78-88; assoc neurol, Beth Israel Hosp, Boston, 75-88; neurologist & clin investr, Geriat Res Ctr, Vet Admin Hosp, Bedford, Mass, 75-88, assoc dir, 84-88; dir, prog Behav Neurol & Clin Neurosci, Sch Med, Tulane Univ, 88- *Mem:* Fel Am Acad Neurol; fel Royal Soc Med; Soc Neurosci. *Res:* Anatomy of the cerebral cortex in the monkey: connections and architectonics of association areas; phenomenology and classification of organic mental disorders. *Mailing Add:* 1430 Tulane Ave New Orleans LA 70112

SELTZER, CARL COLEMAN, b Boston, Mass, June 1, 08; m 30; c 3. PHYSICAL ANTHROPOLOGY. *Educ:* Harvard Univ, AB, 29, PhD(phys anthrop), 33. *Prof Exp:* Anthropologist, Constitution Clinic, Presby Hosp, New York, 30-31; Nat Res Coun fel, Harvard Univ, 33-35, res asst, Fatigue Lab, 37-38, res assoc anthrop, 38-39, res assoc phys anthrop, 39-42, res fel, Peabody Mus, 42-63, res grant, Sch Pub Health, 42-47, anthropologist, Dept Hyg, 47-56, res assoc, 63-68, SR RES ASSOC BIOL ANTHROP, SCH PUB HEALTH, HARVARD UNIV, 68-, HON RES ASSOC PHYS ANTHROP, PEABODY MUS, 74- *Concurrent Pos:* Consult, Off Indian Affairs, US Dept Interior, 37-42; res assoc, Robert B Brigham Hosp, Boston, 40-; res assoc, Adolescent Unit, Children's Hosp, 57-; fel, Coun Epidemiol, Am Heart Asn, 64-; consult, Vet Admin Outpatient Clinic, Boston, 65- & Framingham Heart Study, Boston Univ, 71-73; vis prof nutrit, Tufts Univ, 79- *Mem:* AAAS; Am Asn Phys Anthrop; Am Anthrop Asn; NY Acad Sci; Soc Epidemiol Res. *Res:* Human constitution; constitutional medicine; growth and development; obesity. *Mailing Add:* Peabody Museum 11 Divinity Ave Cambridge MA 02138

SELTZER, EDWARD, b Chelsea, Mass, Oct 28, 11; m 41; c 3. CHEMICAL ENGINEERING, FOOD SCIENCE. *Educ:* Harvard Univ, BS, 33. *Prof Exp:* Res chemist & chem engr, Walter Baker Co Div, Gen Foods Corp, Mass, 34-39, res chemist & proj leader, Cent Res Lab, NJ, 39-42, from chem engr to head processing eng div, 41-46; chief res engr, Thomas J Lipton, Inc, NJ, 46-59, asst dir tech res, 59-69; prof, 69-84, EMER PROF FOOD PROCESS ENG, COOK COL, RUTGERS UNIV, NEW BRUNSWICK, 84- *Mem:* Am Chem Soc; Am Inst Chem Engr; fel, Inst Food Technologists; Res & Develop Assocs Mil Food & Packaging Systs. *Res:* Food research; process engineering; spray drying; dehydration; research administration; grant research. *Mailing Add:* 1175 Sussex Rd Teaneck NJ 07666

SELTZER, JAMES EDWARD, b Lebanon, Pa, Apr 15, 36. ELECTROMAGNETICS, MATHEMATICAL STATISTICS. *Educ:* US Mil Acad, BS, 58; Purdue Univ, MSE, 63, PhD(eng), 71. *Prof Exp:* res engr, Harry Diamond Labs, 67-; RETIRED. *Mem:* Inst Elec & Electronics Engrs; Union Radio Scientists. *Res:* Analysis and simulation of radar systems including backscatter from terrain and complex targets; evaluation of clutter effects. *Mailing Add:* 570 Mine Rd Lebanon PA 17042

SELTZER, JO LOUISE, b St Louis, Mo, July 10, 42; m 63; c 3. BIOCHEMISTRY. *Educ:* Wash Univ, AB, 63, PhD(pharmacol), 69. *Prof Exp:* Res assoc, Dept Pharmacol, 70-72, res instr biochem, 72-78, RES ASST PROF MED, DIV DERMAT, DEPT MED, SCH MED, WASH UNIV, 78- *Mem:* Sigma Xi; Am Soc Biol Chem; Tissue Cult Asn. *Res:* Neutral proteases of human skin including collagenase. *Mailing Add:* 8001 Cornell St Louis MO 63130

SELTZER, MARTIN S, b New York, NY, Apr 8, 37; m 60; c 3. SOLID STATE PHYSICS, PHYSICAL METALLURGY. *Educ:* NY Univ, BMetalEng, 58, Yale Univ, MEng, 60, DEng(metall), 62. *Prof Exp:* Res scientist, N V Philips Gloeilampenfabrieken, 62-63; fel, Dept Metall, Battelle Mem Inst, 63-77; MEM STAFF, PORTER, WRIGHT, MORRIS & ARTHUR, 77- *Mem:* AAAS; Am Ceramic Soc. *Res:* Self diffusion and mechanical properties of inorganic binary compounds. *Mailing Add:* 4860 Rustic Bridge Rd Columbus OH 43214

SELTZER, RAYMOND, b New York, NY, May 27, 35; m 66; c 2. POLYMER CHEMISTRY, ORGANIC CHEMISTRY. *Educ:* City Col New York, BS, 56; Purdue Univ, PhD(org chem), 61. *Prof Exp:* Res chemist, Eastman Kodak Co, 61-63 & M&T Chem, Inc, Am Can Co, NY, 63-68; assoc dir, 76-80, dir res & develop additives 80-88, RES ASSOC, PLASTICS & ADDITIVES DIV, RES DEPT, CIBA-GEIGY CHEM CO, 68-, VPRES RES, ADDITIVES DIV, 88- *Mem:* Am Chem Soc; Sigma Xi; NY Acad Sci; Soc Plastic Engrs. *Res:* Synthesis, characterization, evaluation and application of novel polymeric systems; developmental novel epoxy and other thermosetting resins; ultraviolet curable resins; high temperature stable polymers; film, coatings, adhesive, casting and composite applications; stabilizers for polymers; coatings; lubricants. *Mailing Add:* 11 Angus Lane New City NY 10956

SELTZER, SAMUEL, b Philadelphia, Pa, Feb 3, 14; m 46; c 1. DENTISTRY. *Educ:* Univ Pa, DDS, 37. *Prof Exp:* Assoc prof oral histol & path, Sch Dent, Univ Pa, 59-67; PROF ENDODONT, SCH DENT, TEMPLE UNIV, 67- *Mem:* AAAS; fel Am Col Dent; NY Acad Sci; Sigma Xi. *Res:* Biological aspects of dental pulp disease; injury of dental pulp and mechanisms of repair. *Mailing Add:* 1901 Kennedy Blvd Apt 1804 Philadelphia PA 19103

SELTZER, STANLEY, b New York, NY, Feb 25, 30; m 52; c 1. PHYSICAL CHEMISTRY, ORGANIC CHEMISTRY. *Educ:* City Col, BS, 50; Harvard Univ, AM, 56, PhD(chem), 58. *Prof Exp:* Phys chem analyst, M W Kellogg Co Div, Pullman, Inc, 50-52; from res assoc to chemist, 58-73, SR CHEMIST, BROOKHAVEN NAT LAB, 73- *Concurrent Pos:* Vis prof, Cornell Univ, 64-65; NIH spec fel, 69-70; vis prof, Brandeis Univ, 69-70; lectr, Columbia Univ, 69-73; vis prof, State Univ NY, Stony Brook, 74 & 77, adj prof, 78- *Mem:* Am Chem Soc; Biophys Soc; Am Soc Photobiol. *Res:* Mechanisms of reaction in organic and biochemical systems; kinetic isotope effects. *Mailing Add:* Dept Chem Brookhaven Nat Lab Upton NY 11973

SELTZER, STANLEY, b Brooklyn, NY, Aug 8, 25; m 51; c 4. CHEMICAL ENGINEERING. *Educ:* Cooper Union, BChE, 47; Univ Mich, MS, 48, PhD(chem eng), 51. *Prof Exp:* Chem engr, Chem Mat Dept, Mass, 51-60, process engr, Silicone Prods Dept, Waterford, 60-61, eng leader, 61-65, mgr process eng, 66-69, mgr intermediates mfg, 69-79, MGR ADVAN TECHNOL, GEN ELEC CO, 79- *Mem:* Am Chem Soc; Am Inst Chem Engrs. *Res:* Advanced incineration systems; fluidized bed reactions; process control computers. *Mailing Add:* 2481 McGovern Dr Schenectady NY 12309

SELUND, ROBERT B(ERNARD), b La Crosse, Wis, Jan 25, 07; m 34; c 1. CHEMICAL ENGINEERING. *Educ:* Univ Minn, BS, 30. *Prof Exp:* Chem engr, Standard Oil Co, 30-40, res group leader, 41-51, res sect leader, 52-61, tech serv sect leader, Am Oil Co, 61-77; RETIRED. *Mem:* Am Chem Soc; Am Inst Chem Engrs. *Res:* Asphalt; motor oils; wax; crude oil distillation and coking; petroleum refining. *Mailing Add:* 250 Seventh Ave S Naples FL 33940

SELVADURAI, A P S, b Matara, Sri Lanka, Sept 23, 42; Can citizen; m 72; c 4. MATERIALS SCIENCE ENGINEERING, ENVIRONMENTAL SCIENCES. *Educ:* Brighton Polytech, Sussex, Eng, dipl eng, 64; London Univ, Eng, DIC, 65; Stanford Univ, Calif, MS, 67; Univ Nottingham, Eng, PhD(theoret mech), 71, DSc, 86. *Prof Exp:* Lectr civil eng, Univ Aston, Birmingham, UK, 71-75; from asst prof to assoc prof, 75-81, PROF CIVIL ENG, CARLETON UNIV, OTTAWA, CAN, 81- *Concurrent Pos:* Staff res engr, Woodward Clyde Assocs, Oakland, Calif, 66-67; vis sr researcher, Bechtel Group, San Francisco, Calif, 81-82; consult, Atomic Energy Can Ltd, 83- & Ministry Transp Ont, 84-; dir, Am Acad Mech, 84-88 & 90-; vis prof, Dept Theoret Mech, Univ Nottingham, UK, 86; founding chmn, Eng Mech Div, Can Soc Civil Eng, 89; assoc ed, Can J Civil Eng, 90-; vis prof, Inst de Mecanique de Grenaoble, France, 91. *Honors & Awards:* King George VI Mem Fel, English Speaking Union of the Commonwealth, UK, 65; Davidson Dunton lectr, Carleton Univ, Ottawa, 87; Res Achievement Award, Carleton Univ, Ottawa, 90; Leipholz Medal, Can Soc Civil Eng, 91. *Mem:* Fel Eng Inst Can; fel Can Soc Civil Eng; fel Am Acad Mech; fel Inst Math & Its Appln. *Res:* Continuum mechanics and applied mathematics; finite elasticity; contact and interface problems; integral equations in mechanics; fracture and damage mechanics; composite materials mechanics; geomechanics; soil-structure interaction; nuclear waste management; hygrothermal processes in porous media; mechanics of layered systems; ice mechanics and ice structure interaction; environmental geomechanics. *Mailing Add:* Dept Civil Eng Carleton Univ Ottawa ON K1S 5B6 Can

SELVAKUMAR, CHETTYPALAYAM RAMANATHAN, b Karur, Tamil Nadu, India, Mar 17, 50; m 80; c 2. MICROELECTRONIC DEVICES, MICROELECTRONIC PROCESS TECHNOLOGY. *Educ:* Univ Madras, BE, 72; Indian Inst Technol Bombay, MTech, 74; Indian Inst Technol, Madras, PhD(elec eng), 85. *Prof Exp:* Proj assoc microelectronics, Indian Inst Technol, Madras, 78-84; postdoctoral fel, 85-87, ASST PROF MICROELECTRONICS, UNIV WATERLOO, 87- *Concurrent Pos:* Consult, Digital Equip Co, Ottawa & Gennum Corp, Ont, 90-; mem, Tech Prog Comt, Inst Elec & Electronics Engrs Bipolar Circuits & Technol Mgt, 90-; key assoc researcher, Info Technol Res Ctr, 91-; assoc researcher, Micronet, Fed Ctr Excellence, 91- *Mem:* Inst Elec & Electronics Engrs; Am Phys Soc. *Res:* Design, analytical/computer aided modeling and analysis of microelectronic devices and processes; bipolar transistors; polysilicon emitters; heterostructure devices; silicon-germanium alloys; photodetectors; microelectronic process technology; oxidation; ion-beam mixing; dc and ac electrical characterization of devices; microelectronic sensors. *Mailing Add:* Dept Elec & Computer Eng Univ Waterloo Waterloo ON N2L 3G1 Can

SELVERSTON, ALLEN ISRAEL, b Chicago, Ill, Jan 17, 36; m 63; c 4. NEUROPHYSIOLOGY, COMPARATIVE PHYSIOLOGY. *Educ:* Univ Ore, MA, 64, PhD(neurophysiol), 67. *Prof Exp:* Res assoc neurophysiol, Stanford Univ, 67-69; assoc prof neurophysiol, 74-76, assoc prof, 76-81, PROF BIOL, UNIV CALIF, SAN DIEGO, 81- *Concurrent Pos:* USPHS fel, 67-69; Alexander von Humboldt sr fel, 82-83. *Mem:* AAAS. *Res:* Animal behavior; neural mechanisms underlying behavior; integrative activity of invertebrate ganglia. *Mailing Add:* Dept Biol B 022 Univ Calif at San Diego Box 109 La Jolla CA 92093

SELVERSTONE, JANE ELIZABETH, b Cambridge, Mass, July 6, 56; m 84; c 2. METAMORPHIC PETROLOGY. *Educ:* Princeton Univ, AB, 78; Univ Colo, MS, 81; Mass Inst Technol, PhD (geol), 85. *Prof Exp:* Adj prof geol, Univ Colo, 85-86; asst prof, 86-89, ASSOC PROF GEOL, HARVARD UNIV, 90- *Concurrent Pos:* NSF presidential young investr award, 87; chair, Short Course Comt, Mineral Soc Am, 89-91. *Mem:* Geol Soc Am; Am Geophys Union; Mineral Soc Am; Asn Women Geoscientists. *Res:* Metamorphic petrology; application of petrologic techniques to interpretation of tectonic processes; determination of pressure-temperature-time-deformation paths of rocks; fluid-rock interactions in high-pressure rocks. *Mailing Add:* Dept Earth & Planetary Sci Harvard Univ Cambridge MA 02138

SELVESTER, RONALD H, cardiology, for more information see previous edition

SELVIDGE, HARNER, b Columbia, Mo, Oct 16, 10; m 33; c 4. ELECTRONICS. *Educ:* Mass Inst Technol, 32, MS, 33; Harvard Univ, SM, 34, SD(commun eng), 37. *Prof Exp:* Instr physics & commun eng, Harvard Univ, 35-38; assoc prof elec eng, Kans State Col, 38-41; sr engr, Carnegie Inst, 41-42; appl physics lab, Johns Hopkins Univ, 42-45; dir spec prods develop, Bendix Aviation Corp, 45-56, staff engr, 56-60; vpres & gen mgr, Meteorol Res, Inc, Calif, 60-69; CONSULT, ALTA ASSOCS, 69- *Concurrent Pos:* Mem, Harvard-Mass Inst Technol Eclipse Exped, Russia, 36; consult engr, Am Phenolic Corp, 38-42; assoc engr, Taylor Tube Co, Ill, 39-40; dir res, Fournier Inst, 42-45. *Mem:* Am Meteorol Soc; fel Inst Elec & Electronics Engrs; assoc fel Am Inst Aeronaut & Astronaut. *Res:* Antennas; propagation and transmission lines; vacuum tubes; proximity fuses; fire control radar; guided missile control systems; nucleonics; industrial instrumentation and controls; meteorological systems. *Mailing Add:* Alta Assocs Box 1128 Sedona AZ 86336

SELWITZ, CHARLES MYRON, b Springfield, Mass, July 20, 27; m 55; c 2. ORGANIC CHEMISTRY. *Educ:* Worcester Polytech Inst, BS, 49; Univ Cincinnati, PhD(chem), 53. *Prof Exp:* Asst inorg chem, Univ Cincinnati, 49-50; res chemist, Gulf Res & Develop Co, Pittsburgh, 53-55, group leader, 55-61, res chemist, 61-72, res assoc, 72-74, sect supvr, 74-76, dir synthetic chem, 76-82; SR CONSULT, GETTY CONSERV INST, MARINA DEL REY, CALIF, 83- *Concurrent Pos:* Mem, Int Comt Monuments & Sites. *Mem:* Am Chem Soc; Catalyst Soc; Mat Res Soc; Asn Preserv Technol. *Res:* Oxidation of organic compounds; chemical abstracts; free radical chemistry; organo-metallics; oxychlorination; phenols; coal chemistry; acetylene chemistry; high temperature polymers; conservation science; stone corrosion and consolidation; preservation of aged adobe; polymers for art conservation. *Mailing Add:* 3631 Surfwood Dr Malibu CA 90265

SELWYN, DONALD, audio & video engineering, for more information see previous edition

SELWYN, PHILIP ALAN, b New York, NY, Feb 12, 45; m 66; c 1. CHEMICAL PHYSICS, FLUID MECHANICS. *Educ:* Univ Rochester, BS, 65; Mass Inst Technol, PhD(chem physics), 70. *Prof Exp:* Res assoc, Mass Inst Technol, res staff mem geophys fluid mech, Inst Defense Anal, 70-76; prog mgr antisubmarine warfare & hydrodyn, Ocean Monitoring & Control Div, Tactical Technol Off, Defense Advan Res Projs Agency, 76-79; chief scientist, Fleet Ballistic Missile Submarine Security Technol Prog, Strategic Syst Prof Off, 79-82; spec asst vpres, Sci & Technol, Honeywell Inc, 82-83; DIR, OFF NAVAL TECHNOL, 83- *Mem:* Am Phys Soc; Am Geophys Union. *Res:* Nonequilibrium statistical mechanics; atmospheric and oceanic fluid mechanics; sensor technology; operations analysis; turbulence. *Mailing Add:* Off Naval Technol 800 N Quincy St Arlington VA 22217

SELZER, ARTHUR, b Lwow, Poland, July 3, 11; nat US; m 36; c 2. MEDICINE, CARDIOLOGY. *Educ:* Univ Lwow, MB, 35; Cracow Univ, MD, 36; Am Bd Internal Med, dipl, 43. *Prof Exp:* Asst, 41-42, clin instr, 42-47, from asst clin prof to clin prof, 47-76, EMER CLIN PROF MED, MED SCH, STANFORD UNIV, 76-; CLIN PROF MED, SCH MED, UNIV CALIF, SAN FRANCISCO, 60- *Concurrent Pos:* Chief cardiol & dir cardiopulmonary lab, Pac Med Ctr, 59-84; consult, Letterman Gen Hosp. *Mem:* Am Fedn Clin Res; distinguished fel Am Col Cardiol; master Am Col Physicians. *Res:* Clinical cardiovascular physiology; operable heart disease; coronary artery disease; heart failure; pharmacology of digitalis. *Mailing Add:* Pac Med Ctr Clay & Webster Sts San Francisco CA 94115

SELZER, MELVIN LAWRENCE, b New York, NY, Feb 3, 25; m 78; c 3. PSYCHIATRY. *Educ:* Tulane Univ, BS, 49, MD, 52; Am Bd Psychiat & Neurol, dipl, 59. *Prof Exp:* Intern, Univ Mich Hosp, 52-53; resident psychiat, Ypsilanti State Hosp, 54-57; assoc psychiatrist, Health Serv, Med Sch, Univ Mich, Ann Arbor, 57-59, from instr to prof psychiat, 59-78; CLIN PROF, DEPT PSYCHIAT, UNIV CALIF, SAN DIEGO, 81- *Concurrent Pos:* Examr, Am Bd Psychiat & Neurol, 62-68; mem comt alcohol & drugs, Nat Safety Coun, 63-72; resource consult, President's Comt Traffic Safety, 64-65; mem fac, Inst Continuing Legal Educ, Univ Mich-Wayne State Univ Law Schs, 65-70; mem criminal code revision comt, State Bar Mich, 66-69; ed referee, J Studies Alcohol, 67- & Am J Psychiat, 69-; mem fac, Pract Law Inst, NY, 68-71; consult, Fed Aviation Admin, 69, US Dept Transp, 69- & Archdiocese of Detroit, 72; consult ed, Life Threatening Behav, 70-78; res psychiatrist, Hwy Safety Res Inst, 70-75; mem res rev comt, Nat Inst Alcohol Abuse & Alcoholism, 76-78. *Mem:* Fel Am Psychiat Asn; Asn Am Med Cols; fel Am Col Psychiat. *Res:* Psychoanalytic therapy; student mental health; alcoholism and drug abuse; psychological aspects of traffic accidents. *Mailing Add:* 6967 Paseo Laredo La Jolla CA 92037

SEMAN, GABRIEL, b Budapest, Hungary, Sept 3, 25; m 52; c 5. VIROLOGY, CYTOLOGY. *Educ:* Univ Paris, BS, 46, MD, 54. *Prof Exp:* Gen practitioner, France, 54-61; res assoc, Res Ctr Cancerology & Radiobiol, Paris, 61-64; Eli Lilly res fel, Univ Tex M D Anderson Hosp & Tumor Inst Houston, 64-65; sr res assoc & head dept cytol & electron micros, Inst Cancerology & Immunogenetics, Villejuif, France, 65-68; ASSOC PROF VIROL & ASSOC VIROLOGIST, UNIV TEX M D ANDERSON HOSP & TUMOR INST HOUSTON, 68- *Mem:* AAAS; Am Asn Cancer Res; Tissue Cult Asn. *Res:* Ultrastructure cytology; normal and leukemic cytohematology; viral oncology; electron microscopy. *Mailing Add:* 7727 Sands Pt Houston TX 77036

SEMAN, GEORGE WILLIAM, b Pittsburgh, Pa, May 29, 40. ELECTRICAL ENGINEERING. *Educ:* Carnegie Inst Technol, BS, 62, MS, 64, PhD(elec eng), 66. *Prof Exp:* Res engr space syst, Avco Corp, 66-67, res engr, nuclear weapons effects, EG&G Inc, 67-68; sr res engr nuclear weapons effects, Ion Physics Corp, 68-69; RES MGR ELEC CABLES, GEN CABLE, PIRELLI CABLE CORP, 69-; VPRES RES CABLE TECHNOL LABS. *Mem:* Sr mem Inst Elec & Electronics Engrs. *Res:* New cable designs; evaluation of electrical and mechanical performance of extruded and laminar dielectric transmission and distribution cables. *Mailing Add:* Cable Technol Labs Triangle Rd Off Jersey Ave New Brunswick NJ 08903

SEMANCIK, JOSEPH STEPHEN, b Barton, Ohio, June 9, 38; m 63. PATHOLOGY, VIROLOGY. *Educ:* Western Reserve Univ, AB, 60; Purdue Univ, MS, 62, PhD(path), 64. *Prof Exp:* Asst plant pathologist & lectr plant path, Univ Calif, Riverside, 64-69; assoc prof plant path, Univ Nebr, Lincoln, 69-72; assoc prof, 72-74, chmn dept, 80-83, PROF PLANT PATH, UNIV CALIF, RIVERSIDE, 74- *Concurrent Pos:* USPHS grants, 65-68 & 69-73; assoc ed, Virol, 71-73 & Phytopathol, 74-76; NSF awards, 73-75, 75-78 & 78-81; Guggenheim fel, 78. *Honors & Awards:* Alexander von Humboldt Found Prize, 75. *Mem:* AAAS; fel Am Phytopath Soc; Sigma Xi. *Res:* Purification and characterization of pathogenic nucleic acids; cell biology; viroids; viroid-cell interactions. *Mailing Add:* Dept Plant Path Univ Calif Riverside CA 92521

SEMBA, KAZUE, b Tsuchiura, Japan, Jan 13, 49; m 72. PSYCHOBIOLOGY, NEUROSCIENCE. *Educ:* Tokyo Univ Educ, BEd, 71, MA, 73; Rutgers Univ, PhD(psychobiol), 79. *Prof Exp:* Res asst, Inst Aminal Behav, Rutgers Univ, 75-76, teaching asst physiol & exp psychol, Dept Psychol, 76-77; fel, Dept Vet Physiol & Pharmacol, Iowa State Univ, 79-80; res specialist, Col Med & Dent NJ, 80-81, instr, Rutgers Med Sch, Univ Med & Dent NJ, 81-84; RES ASSOC, UNIV BC, 84- *Concurrent Pos:* Grant-in-aid, Japan Soc NY, 78 & NJ State fel, 77-78. *Mem:* Soc Neurosci; Am Psychol Asn; Sigma Xi. *Res:* Central cholinergic systems. *Mailing Add:* 204-5925 Vine Vancouver BC V6M 4A3 Can

SEMEL, MAURIE, b NY, Jan 18, 23; m 50; c 3. ENTOMOLOGY. *Educ:* Cornell Univ, PhD(entom), 54. *Prof Exp:* Grad asst, 49-54, asst prof, 54-59, assoc prof, 59-88, EMER PROF ENTOM, LONG ISLAND HORT RES LAB, CORNELL UNIV, 88- *Concurrent Pos:* Vis sr scientist, Int Potato Ctr, Lima, Peru, 72-73; mem Adv Coun Agr, NY State Dept Agr & Mkt. *Mem:* Entom Soc Am; NY Acad Sci; Sigma Xi. *Res:* Control of insects affecting vegetable and ornamental crops. *Mailing Add:* 4356 Ranchwood Dr Bucyrus OH 44820

SEMELUK, GEORGE PETER, b Coleman, Alta, Apr 14, 24; m 49; c 3. PHYSICAL CHEMISTRY. *Educ:* Univ Alta, BSc, 47, MSc, 49; Ill Inst Technol, PhD(kinetics), 55; Cambridge Univ, PhD(kinetics), 60. *Prof Exp:* Res chemist, Lamp Div, Gen Elec Co, Ohio, 53-55; proj engr, Electrochem Labs, Inc, Okla, 55-58; assoc prof, 60-67, PROF CHEM, UNIV NB, 67- *Concurrent Pos:* Consult, Dow Chem Can, Ltd, 64- *Mem:* Fel AAAS; fel Chem Inst Can; Am Chem Soc; The Chem Soc. *Res:* Unimolecular decompositions; structure and chemistry of excited states; photosensitized decompositions. *Mailing Add:* 826 Windsor St Fredericton NB E3B 4G5 Can

SEMENIUK, FRED THEODOR, b Edmonton, Alta, Jan 3, 15; nat US; m 56; c 2. PHARMACEUTICAL CHEMISTRY. *Educ:* Univ Alta, BSc, 39; Purdue Univ, PhD(pharmaceut chem), 47. *Prof Exp:* Asst instr pharmaceut chem, Purdue Univ, 41-46; instr, Univ Wis, 46-47; from asst prof to prof pharmaceut chem, 47-80, EMER PROF, UNIV NC, CHAPEL HILL, 80- *Mem:* AAAS; Am Chem Soc; Am Pharmaceut Asn. *Res:* Organic medicinals; organic chemical nomenclature. *Mailing Add:* 1402 Mason Farm Rd Chapel Hill NC 27514

SEMENUK, NICK SARDEN, b Nestow, Alta, June 16, 37; US citizen; m 63; c 3. ORGANIC CHEMISTRY, INFORMATION SCIENCE. *Educ:* Univ Alta, BSc, 58; Purdue Univ, PhD(chem), 64. *Prof Exp:* Sr res chemist org div, Olin Mathieson Chem Co, 63-65, chem div, 65-69; info res scientist, Squibb Inst Med Res, New Brunswick, 69-71, dir sci info dept, 71-76, sect head, 76-87, dir biosci info, Princeton, 87-88; dir environ comp anal, 89, DIR MARKET INTEL, BRISTOL-MYERS SQUIBB, 90- *Mem:* AAAS; Am Chem Soc; Chem Inst Can; Drug Info Asn; NY Acad Sci. *Res:* Pharmaceutical chemistry; cyclic adenosine monophosphate. *Mailing Add:* 2871 Princeton Pike Lawrenceville NJ 08648

SEMERJIAN, HRATCH G, b Istanbul, Turkey, Oct 22, 43; US citizen; m 69, 86; c 2. COMBUSTION, LASER DIAGNOSTICS. *Educ:* Robert Col, Turkey, BS, 66; Brown Univ, MSc, 68, PhD(eng), 72. *Prof Exp:* Lectr chem, Univ Toronto, 72-73; res engr, Pratt & Whitney Aircraft, United Technol Corp, 73-77; group leader combustion, 77-87, CHIEF, PROCESS MEASUREMENTS DIV, CHEM SCI & TECHNOL LAB, NAT INST STANDARDS & TECHNOL, 87- *Concurrent Pos:* Res fel, Univ Toronto, 71-73. *Honors & Awards:* Silver Medal, Dept Com, 84. *Mem:* Am Inst Aeronaut & Astronaut; Am Soc Mech Engrs; Combustion Inst; Am Inst Chem Engrs; AAAS. *Res:* Laser diagnostics in reacting flows; combustion modelling and diagnostics; gas turbine combustion; particle sizing techniques; radiative heat transfer; spray combustion. *Mailing Add:* Nat Inst Standards & Technol Bldg 221 Rm B306 Gaithersburg MD 20899

SEMKEN, HOLMES ALFORD, JR, b Knoxville, Tenn, Jan 28, 35; m 57; c 2. VERTEBRATE PALEONTOLOGY. *Educ:* Univ Tex, BS, 58, MA, 60; Univ Mich, PhD(geol), 65. *Prof Exp:* Mus intern geol, Smithsonian Inst, 60-61; from asst prof to assoc prof, 65-73, PROF GEOL, UNIV IOWA, 73-, CHMN DEPT GEOL, 86- *Mem:* Geol Soc Am; Paleont Soc; Soc Vert Paleont; Am Soc Mammal; Am Quaternary Asn. *Res:* Paleoecology and biogeography of Pleistocene and Holocene mammals, especially rodents; archaeological geology, especially zooarchaeology; vertebrate paleontology. *Mailing Add:* Dept Geol Univ Iowa Iowa City IA 52242

SEMLER, CHARLES EDWARD, b Dayton, Ohio, Dec 27, 40; m 62; c 2. HIGH TEMPERATURE MINERALOGY, MATERIALS TESTING. *Educ:* Miami Univ, BA, 62, MS, 65; Ohio State Univ, PhD(mineral), 68. *Prof Exp:* Res engr ceramics, Ferro Corp, 64-65; sr res engr mat res, Monsanto Res Corp, 68-70; asst prof geol, Wash Univ, 70-71; sr res mineral refractories, Dresser Indust, 71-74; from asst prof to assoc prof, 74-86, dir refractories res ctr, 74-85, ADJ PROF & CONSULT, OHIO STATE UNIV, 86- *Concurrent Pos:* Abstractor, Am Chem Soc, 71-74; chmn Spalling Comt, Am Soc Testing & Mat, 75-; chmn refractories comt, Am Foundrymen's Soc, 76-85; ed comt, Interceram, WGer, 78-; NSF Exchange Scientist, India, 76; vis prof, Sydney, Australia, 81; fac consult, Sandia Nat Labs, Albuquerque, 82; guest worker, Nat Bur Standards, Gaithersburg, Md, 83 & 84; chmn, Refractories Div, Am Ceramic Soc, 84-85; fac fel, NASA-Lewis Res Ctr, Cleveland, 85. *Honors & Awards:* Cramer Award, 83. *Mem:* Fel Am Ceramic Soc; Am Soc Testing & Mat; Nat Inst Ceramic Eng; Brit Ceramic Soc; Can Ceramic Soc; Australian Ceramic Soc. *Res:* Refractories testing by destructive and non-destructive methods; high temperature phase equilibrium relations of oxide materials; microstructure of materials; thermal shock of materials; test development; in plant inspection and trouble shooting; refractories failure analysis; incinerator design and evaluation; technical writing, training seminars and video production. *Mailing Add:* Semler Mat Serv 4160 Mumford Ct Columbus OH 43220

SEMLITSCH, RAYMOND DONALD, evolutionary biology, behavioral ecology, for more information see previous edition

SEMLYEN, ADAM, b Gherla, Romania, Jan 10, 23; m; c 1. ELECTRICAL ENGINEERING. *Educ:* Timisoara Polytech Inst, Dipl Ing, 50; Iasi Polytech Inst, Dr Ing, 65. *Prof Exp:* Lectr power apparatus, Timisoara Polytech Inst, 50-58, from assoc prof to prof power systs, 58-69; vis assoc prof elec eng, 69-71, assoc prof, 71-74, PROF ELEC ENG, UNIV TORONTO, 74- *Concurrent Pos:* Engr, Power Sta Timisoara, Romania, 49-51; consult, Regional Power Authority, Timisoara, 60-69 & Elec Eng Consociates, 70-; Nat Res Coun Can grant, Univ Toronto, 71- *Mem:* Fel Inst Elec & Electronics Engrs; Int Conf Large Elec Systs. *Res:* Power system dynamics; power system optimization; switching transients in high voltage systems. *Mailing Add:* Dept Elec Eng Univ Toronto Toronto ON M5S 1A4 Can

SEMMELHACK, MARTIN F, b Appleton, Wis, Nov 19, 41; m 73; c 1. ORGANIC CHEMISTRY. *Educ:* Univ Wis, Madison, BS, 63; Harvard Univ, AM, 65, PhD(org chem), 67. *Prof Exp:* NIH fel, Stanford Univ, 67-68; from asst prof to prof org chem, Cornell Univ, 68-78; PROF ORG CHEM, PRINCETON UNIV, 78- *Concurrent Pos:* Fel, Alfred P Sloan Found, 72-74; teacher-scholar award, Camille & Henry Dreyfuss Found, 73-78; Guggenheim fel, 78-79. *Mem:* Am Chem Soc; The Chem Soc. *Res:* Synthesis of biologically active compounds; organometallic reagents and electrochemical techniques in organic synthesis. *Mailing Add:* Dept Chem Princeton Univ Princeton NY 08540

SEMMES, STEPHEN WILLIAM, b Savannah, Ga, May 26, 62. MATHEMATICS. *Educ:* Armstrong State Col, BS, 80; Wash Univ, PhD(math), 83. *Prof Exp:* PROF MATH, RICE UNIV, 87- *Mem:* Am Math Soc. *Mailing Add:* Dept Math Rice Univ PO Box 1892 TX 77251

SEMMLOW, JOHN LEONARD, b Chicago, Ill, Mar 12, 42. BIOENGINEERING, BIOMEDICAL ENGINEERING. *Educ:* Univ Ill, Champaign, BS, 64; Univ Ill Med Ctr, PhD(physiol), 70. *Prof Exp:* Sr engr, Motorola, Inc, 64-66; instr physiol optics, Univ Calif, Berkeley, 69-70; asst prof bioeng, Univ Ill, Chicago, 71-77; asst prof physiol, Med Sch, Rush Univ, Chicago, 71-77; asst prof elec eng, 77-80, ASSOC PROF SURG, MED SCH & ASSOC PROF BIO ENG, RUTGERS UNIV, 81- *Concurrent Pos:* NSF US/France Exchange fel, 85. *Mem:* Inst Elec & Electronics Engrs; Sigma Xi; NY Acad Sci; Biomed Eng Soc. *Res:* Eye movement and other physiological motor control systems; design and development of bioinstrumentation for noninvasive diagnosis, particularly in cardiology and neurology. *Mailing Add:* Dept Biomed Eng Rutgers Univ Piscataway NJ 08855

SEMON, MARK DAVID, b Milwaukee, Wis, Mar 27, 50. PHYSICS, MATHEMATICS. *Educ:* Colgate Univ, AB, 71; Univ Colo, MS, 73, PhD(physics), 76. *Prof Exp:* From asst prof to assoc prof, physics, Bates Col, 76-88; ASSOC PROF PHYSICS, AMHERST COL, 88- *Concurrent Pos:* Res asst, Los Alamos Sci Lab, 75 & 84-; asst ed, Am of Physics, 88- *Mem:* Am Phys Soc. *Res:* Quantum scattering theory; phase transitions and critical phenomena; quantum field theory; aharonov-bohm effect. *Mailing Add:* Dept Physics Amherst Col Amherst MA 01002

SEMON, WALDO LONSBURY, b Demopolis, Ala, Sept 10, 98; m 20; c 3. CHEMISTRY. *Educ:* Univ Wash, Seattle, BSChE, 20, PhD(chem), 23. *Hon Degrees:* ASLD, Univ Wash, Seattle, 46; DSc, Kent State Univ, 81. *Prof Exp:* Analytic chemist, Falkenburg Co, Wash, 18; instr chem, Univ Wash, Seattle, 20-26; res chemist & chem engr, B F Goodrich Co, 26-40, dir synthetic rubber res, 40-43, dir pioneering res, 43-54, dir polymer res, 54-58, dir explor res, 58-63; res prof chem, Kent State Univ, 63-72; RETIRED. *Concurrent Pos:* Chem engr, Everett Gas Works, Wash, 21; vpres & dir res, Hycar Chem Co, 40-42; vpres & dir, Farm Chemurgic Coun, 43-60; dir, Concepts Develop Inst, Ohio, 76-90. *Honors & Awards:* Modern Pioneer Award, Nat Asn Mfrs, 40; Goodyear Medal, Am Chem Soc, 45; 3rd Int Synthetic Rubber Gold Medal, London, 64; Elliot Cresson Gold Medal, Franklin Inst, 64; Am Inst Chem Eng Award, 64. *Mem:* Am Chem Soc; Am Inst Chem Eng. *Res:* Hydroxylamine and oximes; complexes of sulfur; rubber chemistry; antioxidants; synthetic copolymers, nitrile and deuterio rubber; rubber to metal adhesion; brass plating; polyvinyl chloride; Koroseal; butadiene; stereo-specific polymers; reactions of oxygen dissolved in water. *Mailing Add:* 21 Bard Dr No 206 Hudson OH 44236

SEMON, WARREN LLOYD, b Boise, Idaho, Jan 17, 21; m 45; c 4. MATHEMATICS. *Educ:* Univ Chicago, SB, 44; Harvard Univ, MS, 49, PhD(appl math), 54. *Prof Exp:* Instr math, Hobart Col, 46-47; res assoc comput lab, Harvard Univ, 49-55, lectr appl math & asst dir comput lab, 55-61; head appl math dept, Sperry Rand Res Ctr, Mass, 61-64; mgr comput & anal, Burroughs Res Ctr, Pa, 64-67; prof comput sci, Syracuse Univ, 67-85, dir systs & info sci, 68-76, dean, Sch Comput & Info Sci, 76-85, EMER PROF, SYRACUSE UNIV, 85- *Concurrent Pos:* Consult, USAF, 57-61 & Nat Security Agency, 58-59; ed-in-chief, Comput Soc of Inst of Elec & Electronics Engrs, 75- *Mem:* Asn Comput Mach; fel Inst Elec & Electronics Engrs; Math Asn Am; Sigma Xi. *Res:* Digital computer systems and applications; switching theory; automata theory. *Mailing Add:* c/o FMCA PO Box 44209 Cincinnati OH 54244

SEMONIN, RICHARD GERARD, b Akron, Ohio, June 25, 30; m 51; c 4. ATMOSPHERIC PHYSICS, PRECIPITATION CHEMISTRY. *Educ:* Univ Wash, Seattle, BSc, 55. *Prof Exp:* Res asst radar meteorol, 55-56, from res assoc to assoc prof sci, 56-65, prof, 65-71, asst head, 70-80, prin scientist, 71-86, CHIEF, ATMOSPHERIC SCI SECT 86- *Concurrent Pos:* NSF grants, 58-60 & 61-80; US Army Res & Develop Labs grant, 62-64; Dept Energy contract, 69-; Dept Interior contract, 71-; prof atmospheric sci, Univ Ill, 75- *Mem:* Fel AAAS; fel Am Meteorol Soc; Weather Modification Asn; Nat Weather Asn; Sigma Xi. *Res:* Microphysical processes necessary or attendant to the formation of clouds and precipitation; weather modification, controlled and inadvertent; causes and distribution of acid deposition. *Mailing Add:* Ill State Water Surv 2204 Griffith Dr Champaign IL 61820

SEMRAD, JOSEPH EDWARD, zoology, for more information see previous edition

SEMRAU, KONRAD (TROXEL), b Chico, Calif, June 5, 19. CHEMICAL ENGINEERING. *Educ:* Univ Calif, BS, 48, MS, 49. *Prof Exp:* Asst tech engr, Carbide & Carbon Chem Co Div, Union Carbide & Carbon Corp, 46; chem engr, 50-63, SR CHEM ENGR, SRI INT, 63- *Mem:* Am Inst Chem Engrs; Am Chem Soc; Sigma Xi; Air & Waste Management Asn. *Res:* Dust and mist collection; fine particle technology; air pollution control engineering; mass transfer. *Mailing Add:* SRI Int 333 Ravenswood Ave Menlo Park CA 94025

SEMTNER, ALBERT JULIUS, JR, b Oklahoma City, Okla, May 25, 41; m 69; c 2. OCEANOGRAPHY. *Educ:* Calif Inst Technol, BS, 63; Univ Calif, Los Angeles, MA, 65; Princeton Univ, PhD(geophys fluid dynamics), 73. *Prof Exp:* Lt comdr oceanog, Nat Oceanic & Atmospheric Admin, 68-73; adj asst prof meteorol, Univ Calif, Los Angeles, 73-76; scientist oceanog, Nat Ctr Atmospheric Res, 76-86; PROF OCEANOG, NAVAL POSTGRAD SCH, 86- *Concurrent Pos:* Consult, Rand Corp, 74-75. *Mem:* Am Geophys Union; Am Meteorol Soc; Oceanog Soc. *Res:* Numerical simulation of ocean circulation; prediction of climatic changes with coupled models of the atmosphere, the ocean and sea ice. *Mailing Add:* 3470 Edgefield Pl Carmel CA 93923

SEMTNER, PAUL JOSEPH, b Seminole, Okla, May 9, 45; m 70; c 3. ECONOMIC ENTOMOLOGY, INSECT ECOLOGY. *Educ:* Okla State Univ, BS, 67, MS, 70, PhD(entom), 72. *Prof Exp:* Res assoc entom, Okla State Univ, 72-73; instr, Connors State Col, 73-74; ASST PROF ENTOM, VA POLYTECH INST & STATE UNIV, 74- *Mem:* Entom Soc Am; Sigma Xi. *Res:* Pest management of tobacco insect pests and effects of the environment on their abundance. *Mailing Add:* Va Polytech Inst & State Univ Southern Piedmont Ctr PO Box 443 Blackstone VA 23824

SEMURA, JACK SADATOSHI, b Los Angeles, Calif, Sept 8, 41; m 67; c 1. STATISTICAL MECHANICS, THERMODYNAMICS. *Educ:* Univ Hawaii, BA, 63, MS, 65; Univ Wis-Madison, PhD(physics), 72. *Prof Exp:* Post doctoral res assoc, Univ Pittsburgh, 71-73; from asst prof to assoc prof, 73-82, PROF PHYSICS, PORTLAND STATE UNIV, 82- *Concurrent Pos:* Vis prof, Univ Wis-Madison, 80. *Honors & Awards:* Russell B Scott Award, Cryogenic Eng Soc, 85. *Mem:* Am Phys Soc; Sigma Xi; Am Asn Physics Teachers; AAAS. *Res:* Theoretical physics; statistical physics; maximum entropy analysis; thermodynamics; first order phase transitions; heat transfer in cryogenic fluids; physics of the environment. *Mailing Add:* Dept Physics Portland State Univ Portland OR 97207

SEN, AMAR KUMAR, b Calcutta, India, Mar 14, 27; m 56; c 2. PHYSIOLOGY. *Educ:* Univ Calcutta, MSc, 49, MB & BS, 55; Univ London, PhD(physiol), 60. *Prof Exp:* Instr physiol, Sch Med, Vanderbilt Univ, 60-61, asst prof phydiol, 63-66; sci pool officer, S K M Hosp, Calcutta, India, 62; dir clin res, Sandoz Ltd, Bombay, India, 62-63; assoc prof, 66-70, PROF PHARMACOL, FAC MED, UNIV TORONTO, 70-, COORD GRAD STUDIES, DEPT PHARMACOL, 84- *Concurrent Pos:* Consult pharmacologist, Addiction Res Found, Ont. *Mem:* NY Acad Sci; Can Biochem Soc; Can Pharmacol Soc; Am Physiol Soc; Soc Gen Physiol. *Res:* Transport of electrolytes and non-electrolytes across the cell membrane. *Mailing Add:* Dept Pharmacol Univ Toronto Fac Med Toronto ON M5S 1A8 Can

SEN, AMIYA K, b Calcutta, India, Dec 14, 30; m 63; c 1. ELECTRICAL ENGINEERING, PLASMA PHYSICS. *Educ:* Indian Inst Sci, Bangalore, dipl, 52; Mass Inst Technol, SM, 58, PhD(elec eng), 63. *Prof Exp:* Test engr elec eng, Gen Elec Co, 53-54, design anal engr, 54-55; teaching asst, Mass Inst Technol, 56-57, instr, 57-58; from instr to assoc prof, 58-74, PROF ELEC ENG, COLUMBIA UNIV, 74- *Concurrent Pos:* Various NSF, NASA and

Dept Energy grants, 65-; mem, Adv Subcomt, NSF, 80-82; vis prof, Technische Hogeschool, Eindhoven, Holland, 70, Nagoya Univ, Nagoya, Japan, 77, Univ Calif, Berkeley, 84. *Mem:* Fel Am Phys Soc; Am Geophys Union; Inst Elec & Electronics Engrs. *Res:* Energy conversion; plasma and space physics; magnetohydrodynamics; author or coauthor of over 55 scientific journal publications. *Mailing Add:* Dept Elec Eng Mudd Bldg Columbia Univ New York NY 10027

SEN, BUDDHADEV, b India, Aug 7, 23; nat US; m 56; c 2. PHYSICAL INORGANIC CHEMISTRY. *Educ:* Univ Calcutta, India, BSc, 45, MSc, 47, DPhil(sci), 54. *Prof Exp:* Lectr chem, City Col Calcutta, India, 48 & Jadavpur Univ, India, 48-56; fel, La State Univ, 54-56, from vis asst prof to asst prof, 56-60; vis asst prof, Univ Alta, 60-61; from asst prof to assoc prof, 61-72, PROF CHEM, LA STATE UNIV, BATON ROUGE, 72- *Concurrent Pos:* Petrol Res Fund int grant, Univ Col London, 66-67; chmn freshman chem prog, Dept Chem, La State Univ, 76. *Mem:* Am Chem Soc. *Res:* Coordination chemistry of transition and post-transition elements; thermodynamics of chelates; properties of mixed solvents. *Mailing Add:* 774 Baird Dr Baton Rouge LA 70803

SEN, DIPAK KUMAR, b Patna, India, Feb 28, 36. MATHEMATICAL PHYSICS. *Educ:* Patna Univ, India, MSc, 54; Univ Paris, DresSci(physics), 58. *Prof Exp:* From lectr to assoc prof, 58-77, PROF MATH, UNIV TORONTO, 77- *Mem:* Am Math Soc; Can Math Soc. *Res:* Relativity; cosmology. *Mailing Add:* Dept Math Univ Toronto 100 St George St St George Campus Toronto ON M5S 1A1 Can

SEN, GANES C, b Varanasi, India, Jan 17, 45; m 73; c 2. MOLECULAR BIOLOGY. *Educ:* Calcutta Univ, BSc, 65, MSc, 67; McMaster Univ, PhD(biochem), 74. *Prof Exp:* Fel molecular biol, Yale Univ, 74-76, res assoc, 76-78; ASST MEM MOLECULAR BIOL, SLOAN-KETTERING CANCER CTR, 78-, HEAD, LAB ONCOGENIC VIRUSES & INTERFERONS, 81-; STAFF MEM, MOLECULAR BIOL, CLEVELAND CLIN, 88- *Concurrent Pos:* Asst prof, Grad Sch Med Sci, Cornell Univ, 79-88. *Mem:* Am Soc Microbiol; Am Soc Virol; NY Acad Sci. *Res:* Molecular mechanisms of various actions of interferons including the actions on replication of and neoplastic transformation by RNA tumor viruses. *Mailing Add:* Dept Molecular Biol Cleveland Clin Found Res Inst Clin Ctr 9500 Euclid Ave NC2-103 Cleveland OH 44195-5285

SEN, MIHIR, b Calcutta, India, Jan 17, 47; m; c 2. HEAT TRANSFER, FLUID MECHANICS. *Educ:* Indian Inst Technol, Madras, India, BTech, 68; Mass Inst Technol, ScD, 75. *Prof Exp:* From asst prof to prof thermal & fluid sci, Nat Autonomous Univ Mex, Mexico City, 75-85; ASSOC PROF, AEROSPACE & MECH ENG, UNIV NOTRE DAME, 86- *Concurrent Pos:* Mech eng coordr, Nat Acad Eng-Mex, 80-83; consult, Dept Solar Energy, Nat Autonomous Univ Mex, 81-83, Inst Eng, 84-85, Sundstrand Heat Transfer, 89; vis prof, mech & aerospace eng, Cornell Univ, Ithaca, 85-86. *Honors & Awards:* Nat Researcher Award, Gov Mex, 84. *Mem:* Nat Acad Eng-Mex; Am Soc Mech Engrs; AAAS; Soc Indust & Appl Math; Am Phys Soc; Am Soc Eng Educ. *Res:* Heat transfer and fluid mechanics; mathematical modeling; numerical methods; heat exchangers; porous media flow; natural convection; thermosyphon loops; hydrodynamic stability; bifurcation theory and chaos; numerous technical and nontechnical publications. *Mailing Add:* 52833 Sporn Dr South Bend IN 46635

SEN, PABITRA NARAYAN, b Calcutta, India, Sept 5, 44; m 84; c 2. ELECTROMAGNETICS, ACOUSTICS. *Educ:* Calcutta Univ, BSc, 64, MSc, 67; Univ Chicago, PhD(physics), 72. *Prof Exp:* Res assoc physics, Mich State Univ, 72-73; mem res staff physics, Xerox Palo Alto Res Ctr, 73-76; sr scientist, Xonics Inc, 76-78; mem prof staff petrophys, 78-87, prog leader, 81-83, SCI ADV ROCK PHYSICS, SCHLUMBERGER-DOLL RES, 88- *Concurrent Pos:* Vis prof, Univ Provence, Marseille, 85; external examr, Brandeis, 85, Harvard Univ, 86 & Brooklyn Col, 88; adv comt, Int Sch Energy & Develop, Jamaica, 87, 2nd Int Conf Electronic Transport & Optical Properties of Inhomogeneous Media, Paris, 87-88; guest res fel, dept theoret physics, Oxford Univ, Eng, 88-89. *Mem:* Fel Am Phys Soc; Soc Petrol Engrs; Soc Explor Geophys; Soc Prof Well Log Analysts; Math Asn Am. *Res:* Electromagnetic, vibrational, structural and transport properties of inhomogeneous, amorphous, composite and porous media; statistical mechanics of weakly connected systems; rocks and geological materials; interfacial properties. *Mailing Add:* Schlumberger-Doll Res Old Quarry Rd Ridgefield CT 06877-4108

SEN, PRANAB KUMAR, b Calcutta, India, Nov 7, 37; m 63; c 2. BIOSTATISTICS, STATISTICS. *Educ:* Univ Calcutta, India, BSc, 55, MSc, 57, PhD(statist), 62. *Prof Exp:* Asst prof statist, Univ Calcutta, India, 61-64 & Univ Calif, Berkeley, 64-65; from asst prof to prof biostatist, 65-82, adj prof statist, 80-88, CARY C BOSHAME PROF BIOSTATIST, UNIV NC, CHAPEL HILL, 82-, PROF STATIST, 88- *Concurrent Pos:* USAF Systs Command contract statist anal, Univ NC, 71-78; Richard Merton guest prof, Univ Freiburg, Ger, 74-75; lect, math sth statist, Conf Bd Math Sci, NSF, 83- *Mem:* Fel Inst Math Statist; fel Am Statist Asn; Int Statist Inst. *Res:* Nonparametric methods; multivariate analysis; order statistics; weak convergence; nonparametric statistics and sequential procedures. *Mailing Add:* Dept Biostatist 400 Pub Health 20 In Univ NC Chapel Hill NC 27599-7400

SEN, TAPAS K, b Calcutta, India, Mar 1, 33; US citizen; m 66; c 2. PSYCHOLOGY, STATISTICS. *Educ:* Calcutta Univ, India, BS, 51, MSc 54; Johns Hopkins Univ, PhD(psychol), 63. *Prof Exp:* Res scholar statist, Indian Statist Inst, Calcutta, 55-59; assoc psychologist visual res, Appl Physics Lab, Johns Hopkins Univ, 60-63; mem tech staff speech res, Bell Labs, 63-72; dist mgr human resources planning/corp planning, 73-78, DIV MGR, HUMAN RESOURCES, AT&T, 79- *Concurrent Pos:* NSF fel, Johns Hopkins Univ, 59-63; chair, Coun Tech Group, Human Factors Soc, 72-77, Mayflower Group, 84-85; adv bd mem, Work Am Inst, 85- & Prog Technol, Pub Policy & Human Develop, Kennedy Sch, Harvard Univ, 89-91; mem, Tech Adv Comt Workforce Preparedness, Nat Planning Asn, 88-89. *Mem:* Am Psychol Asn; fel Human Factors Soc; Soc Indust & Orgn Psychol. *Res:* Organizational competency; understanding and developing models of effective organization and culture change using employee involvement and quality; macro-ergonomic and leadership dimensions are identified. *Mailing Add:* AT&T 100 Southgate Pkwy 3J-11 Morristown NJ 07046

SEN, UJJAL, b Calcutta, India, May 29, 57. SPECTROSCOPY & SPECTROMETRY, ELECTRONICS ENGINEERING. *Educ:* Indian Inst Technol, BSc, 77, Indian Inst Sci, BE, 80; State Univ NY, Buffalo, MS, 84. *Prof Exp:* Qual control engr, Siemens Ltd, 80-81; teaching asst elec eng, State Univ NY, Buffalo, 81-84, res asst, 81-87; res & develop engr, 87-89, chief res & develop engr, 89-90, RES & DEVELOP MGR, SPECTRONICS CORP, 90- *Concurrent Pos:* Scientist, Aerospace Mat Spec Nondestructive Testing, Soc Automotive Engrs, 88-; mem, Comt Nondestructive Testing, Am Soc Testing & Mat, 88- *Mem:* Inst Elec & Electronics Engrs; Soc Photo-Optical Instrumentation Engrs; Illum Eng Soc; Am Soc Testing & Mat; Am Soc Nondestructive Testing. *Res:* Optical diagnostics of electric discharges and gas plasmas; optical measurement and instrumentation; electro-optic devices and systems; pulsed power technology; electronic power control; nondestructive techniques and systems; author of numerous publications. *Mailing Add:* Spectronics Corp 956 Brush Hollow Rd Westbury NY 11590-0483

SENAGORE, ANTHONY J, b Nov 20, 58; US citizen; m 83; c 2. COLON & RECTAL SURGERY. *Educ:* Wayne State Univ, BSc, 80; Mich State Univ, MD, 81, MSc, 89. *Prof Exp:* ASST PROF, DEPT SURG, MICH STATE UNIV, 88-; SURGEON & DIR RES, FERGUSON HOSP, 89- *Mem:* AMA; assoc fel Am Col Surgeons; Asn Acad Surg; AAAS; Am Soc Colon & Rectal Surgeons; Am Physiol Soc. *Res:* Surgical techniques for the colon and rectum; recurrent colorectal carcinoma; colorectal anastomotic healing; author of 18 publications. *Mailing Add:* Ferguson Hosp 72 Sheldon Blvd SE Grand Rapids MI 49503

SENATORE, FORD FORTUNATO, b Brooklyn, NY, July, 14, 53; m 83. IMMOBILIZED ENZYME REACTOR DESIGN. *Educ:* Columbia Univ, BA, 75, MS, 77; Rutgers Univ, PhD(chem eng), 83. *Prof Exp:* ASST PROF CHEM ENG, TEX TECH UNIV, 83- *Concurrent Pos:* Prin investr, Am Heart Asn, 84-86; reviewer, Am Inst Chem Engrs J, 85- *Mem:* Sigma Xi; Am Inst Chem Engrs; Int Soc Artifical Organs. *Res:* Immobilized enzyme technology including multi-enzyme systems and multi-functional catalysts, is applied towards the design and modeling of bioreactors; fibrinolytic and thromboresistant small-caliber vascular prothesis for use in coronary bypass surgery and vascular reconstructions; mixed culture fermentation. *Mailing Add:* Dept Chem Eng Tex Tech Univ Lubbock TX 79409

SENAY, LEO CHARLES, JR, b Fall River, Mass, Jan 18, 27; m 51, 90; c 7. PHYSIOLOGY. *Educ:* Harvard Univ, AB, 49; State Univ Iowa, PhD(physiol), 57. *Prof Exp:* From instr to assoc prof, 57-68, PROF PHYSIOL, SCH MED, ST LOUIS UNIV, 68- *Concurrent Pos:* NIH career develop award, 65-70; Anglo-Am fel, SAfrica, 70-82; vis prof appl physiol, Univ Witwatersrand, 71; consult, NIH. *Mem:* AAAS; Am Physiol Soc; Am Col Sports Med; Sigma Xi. *Res:* Environmental physiology, particularly human responses to heat stress and exercise; body fluid dynamics; pharmacology of blood vessels. *Mailing Add:* Dept Physiol St Louis Univ Sch Med St Louis MO 63104

SENCER, DAVID JUDSON, b Grand Rapids, Mich, Nov 10, 24; m 51; c 3. MEDICINE. *Educ:* Univ Mich, MD, 51; Harvard Sch Pub Health, MPH, 58. *Prof Exp:* Med consult tuberculosis, USPHS, 55; med officer in charge, Muscogee County TB Field Res Facil, 55-59; prog officer, Bur State Serv, USPHS, Washington, DC, 59-60; asst chief, Ctr Dis Control, 60-64, dep chief, 64-66, dir, 66-77; sr vpres med & sci affairs, Becton Dickinson & Co, 77-81; COMNR, NEW YORK HEALTH DEPT, 82- *Concurrent Pos:* Mem, WHO Comt Int Surveillance Communicable Dis, 67-76; mem, US Deleg World Health Assembly, 68-76; mem, Int Comn Assessment Smallpox Eradication, 77; consult, Epidemiol Training Progs, WHO, India, 79. *Honors & Awards:* Rosenau Prize, Am Pub Health Asn, 76. *Mem:* Am Soc Trop Med & Hyg; Int Epidemiol Asn. *Mailing Add:* 1185 Coast View Dr Laguna Beach CA 92651

SENCIALL, IAN ROBERT, b Nottingham, Eng, June 2, 38; m 62; c 3. BIOCHEMISTRY. *Educ:* Univ London, BS, 61; Univ Birmingham, PhD(biochem), 67. *Prof Exp:* Org res chemist, Fisons Pest Control Ltd, Eng, 61-64; res biochemist, from lectr to assoc prof, 73-84, PROF BIOCHEM, SCH MED, MEM UNIV NFLD, 84- *Mem:* Can Biochem Soc; Endocrine Soc. *Res:* Studies on the biosynthesis and metabolism of steroid carboxylic acids by the human and by animals; steroid metabolism in essential hypertension; cytochrome P-450. *Mailing Add:* Sch Med Mem Univ Nfld St John's NF A1B 3V6 Can

SENCINDIVER, JOHN COE, b Martinsburg, WVa, Aug 21, 48; m 73; c 3. SOIL SCIENCE. *Educ:* WVa Univ, BS, 70, PhD(agron soil sci), 77. *Prof Exp:* NDEA fel, 71-74; soil scientist reclamation, Forest Serv, 75-78; asst prof, 78-82, ASSOC PROF SOIL SCI, WVA UNIV, 82- *Concurrent Pos:* Soil scientist soil surv, USDA Soil Conserv Serv, 71-75. *Mem:* Soil Sci Soc Am; Am Soc Agron; Soil Conserv Soc; Int Soc Soil Sci; Am Soc Surf Mining & Water Reclamation. *Res:* Soil genesis and classification; overburden and minesoil properties; surface mine reclamation. *Mailing Add:* Col Agr-Forestry WVa Univ Morgantown WV 26506

SENDA, MOTOTAKA, b Okayama, Japan, Feb 8, 54; m 87. MICROBIOLOGY. *Educ:* Okayama Univ, BS, 77, BS, 81, MS, 83; Osaka Univ, PhD, 88. *Prof Exp:* Jr scientist, Dept Microbiol & Chemother, Hoffman-La Roche Japan Res Ctr, 83-85, sr scientist, 86-89; POSTDOCTORAL FEL, DEPT MOLECULAR GENETICS & MICROBIOL, UNIV MED & DENT NJ, 89- *Concurrent Pos:* Vis fel, Dept Peptide Res, Hoffman-La Roche US Res Ctr, 85-86. *Mem:* Am Soc Biochem

& Molecular Biol. *Res:* Role of lipoprotein lipase in the metabolism of lipoprotein and apo-lipoprotein; interferon receptor and its role by the method of genomeii gene analysis and a signal transduction. *Mailing Add:* 32C Cedar Lane Highland Park NJ 08904

SENDELBACH, ANTON G, b Waumandee, Wis, Mar 23, 24; m 56; c 4. DAIRY SCIENCE, GENETICS. *Educ:* Wis State Univ, River Falls, BS, 54; Univ Wis, MS, 56, PhD(dairy sci), 60. *Prof Exp:* From instr to assoc prof, 56-75, PROF DAIRY SCI, UNIV WIS-MADISON, 75- *Mem:* Am Dairy Sci Asn; Sigma Xi. *Res:* Dairy sire and cow evaluation for dairy herd improvement. *Mailing Add:* Dept Dairy 278 Animal Sci Bldg Univ Wis Madison WI 53706

SENDERS, JOHN W, b Cambridge, Mass, Feb 26, 20; m; c 5. MAN-MACHINE SYSTEMS ANALYSIS. *Educ:* Harvard Univ, AB, 48, Univ Tilburg, Nether, PhD, 83. *Prof Exp:* Consult eng, 48-50; res psychologist, Aero Med Lab, Wright Patterson AFB, 50-56; head, Dept Psychol, Arctic Aero Med Lab, USAF, Fairbanks, 56-57; prin res scientist, Minneapolis-Honeywell Regulator Co, 57-62; prin scientist & consult, Bolt, Beranek & Newman, Inc, 62-70; vis prof, 73, prof, Univ Toronto, 74-85; RES PROF ENG & PYSCHOL, DEPT MECH ENG, UNIV MAINE, 81- *Concurrent Pos:* Mem, Comt Bio-Astronaut, Nat Acad Sci-Nat Res Coun, 59-60, Man-in-Space Comt, Space Sci Bd, 60-62, Comt Hearing, Bio-acoust & Bio-mech, 62-, Highway Safety Comt, Highway Res Bd, 65 & 72-, Comt Vision, 70-, Road Characteristics Comt, 72- & Indust Eng Grant Selection Comt, 74-77; lectr & sr res assoc psychol, Brandeis Univ, 65-72; sr lectr mech eng, Mass Inst Technol, 66-67; pres, Senders Assoc, Inc, 68-73; sr res assoc, Univ Calif, Santa Barbara, 79-80; adj prof, Univ Maine, Orono, 81-85, res prof, 85-; consult, various indust, govt agencies, res co & legal firms; pres, Human Factors North, Inc, 83-, chmn bd, 85-; pres, Emsco, Inc, 84-; mem, Sci Adv Bd, Div Life Sci, Air Force Off Sci Res, 85-; adj prof, dept indust eng, Univ Toronto, 85-; emer prof, Dept Indust Eng, Univ Toronto, 85- *Mem:* Sr mem Inst Elec & Electronics Engrs; fel Soc Eng Psychologists; fel Am Psychol Asn; fel AAAS; Psychonomic Soc; fel Human Factors Soc; Human Factors Asn Can. *Res:* Models of visual monitoring behavior; quantification of mental workload; nature and source of human error; design of electronic publications systems; human perceptual motor skill; human information processing. *Mailing Add:* Keneggy West Columbia Falls ME 04623

SENDLEIN, LYLE V A, b St Louis, Mo, May 11, 33; m 55; c 3. GEOLOGICAL ENGINEERING, GEOPHYSICS. *Educ:* Wash Univ, St Louis, BS, 58, AM, 60; Iowa State Univ, PhD(geol soil eng), 64. *Prof Exp:* From instr to prof geol & geophys, Iowa State Univ, 60-77; dir, Coal Extraction & Utilization Res Ctr, Southern Ill Univ, Carbondale, 77-; DIR, INST MINING & MINERALS RES, UNIV KY. *Concurrent Pos:* Consult, Atlantic Ref Co, Tex, 57 & Alpha Portland Cement Co, Mo, 58; consult engr, H M Reitz, 59-60 & US Gypsum Co, Ill, 60-77; asst div chief, Energy & Minerals Resources Res Inst, Iowa State Univ, 74-75; vis prof, Middle East Tech Univ, Ankara, Turkey. *Mem:* Geol Soc Am; Am Geophys Union; Soc Explor Geophys; Nat Water Well Asn; Sigma Xi. *Res:* Engineering geology related to urban and rural problems; ground water investigations of pollution from sanitary landfills, coal mines and other sources; mining and reclamation research in coal mines. *Mailing Add:* Ky Inst Mining Res Univ Ky 12 Porter Bldg Lexington KY 40506-0107

SENDROY, JULIUS, JR, physiological chemistry; deceased, see previous edition for last biography

SENEAR, ALLEN EUGENE, b Chicago, Ill, Nov 2, 19; m 48; c 4. ANALYTICAL CHEMISTRY. *Educ:* Williams Col, Mass, BA, 41; Calif Inst Technol, PhD(org chem), 46. *Prof Exp:* Asst radiochem, Williams Col, Mass, 40; asst chem, Calif Inst Technol, 41-43, Comt Med Res contract, 43-45; fel, Univ Ill, 46-47; from instr to asst prof chem, Univ Calif, 47-55; res engr, Boeing Co, 55-85; RETIRED. *Concurrent Pos:* Chair, Puget Sound Sect, Am Chem Soc, 86. *Mem:* Am Chem Soc; Soc Appl Spectros; AAAS. *Res:* Synthetic organic chemistry, especially monomer synthesis; polymer preparation. *Mailing Add:* 1446 92nd Ave NE Bellevue WA 98004

SENECA, HARRY, b Beirut, Lebanon, July 24, 09; m 48. INTERNAL MEDICINE, BACTERIOLOGY. *Educ:* Am Univ Beirut, MD, 33; Tulane Univ, MS, 43; Am Bd Internal Med, dipl. *Prof Exp:* Instr bact & parasitol, Med Sch, Am Univ Beirut, 33-37; from asst prof to assoc prof bact & parasitol & lectr trop dis, Royal Col Med, Iraq, 37-40; Rockefeller fel, Tulane Univ, 41; from instr to asst prof trop med, Tulane Univ, 42-43; ASSOC PROF UROL & COL PHYSICIANS & SURGEONS, COLUMBIA UNIV, 47-, RES ASSOC, 52- *Concurrent Pos:* Instr path, Med Sch, Am Univ Beirut, 35-36; dir, Govt Bact, Parasitol & Vaccine Insts, Baghdad, Iraq, 37-40; attend physician, Royal Hosp, Baghdad, 37-40, Charity Hosp, New Orleans, La, 41-44 & Columbia Presby Hosp, 47-; consult, Schering Corp, NJ, 46-51 & Chas Pfizer & Co, Inc, NY, 51-54; mem, Chagos Dis Found, Rio de Janeiro, Brazil, 73-; Gen Consult NJ Hospitals. *Honors & Awards:* Henderson Award, 64; Hugh Young Award, 76. *Mem:* Am Soc Trop Med & Hyg; Am Soc Parasitol; Am Geriat Soc; Am Soc Microbiol; fel Am Col Physicians. *Res:* Experimental medicine in relation to treatment and diagnosis; gastrointestinal diseases; bacterial and parasitic diseases; hemoflagellates; sepsis and septic shock; E histolytica; antibacterials and antibiotics; oxysteroids; antihistaminics; bacterial resistance; neoplastic diseases; pyelonephritis. *Mailing Add:* 401 Park Pl Ft Lee NJ 07024

SENECAL, GERARD, b Atwood, Kans, July 27, 29. PHYSICS. *Educ:* St Benedict's Col, Kans, BA, 51; Univ Mich, MA, 57; Kans State Univ, PhD(physics), 63. *Prof Exp:* Air Force Cambridge Res Labs res asst, Kans State Univ, 61-62; from instr to assoc prof physics, 62-72, actg chmn dept, 66-68 & chmn dept, 68-72, PRES, BENEDICTINE COL, 72- *Concurrent Pos:* Lectr-consult, Mobile Lab Prog, Oak Ridge Inst Nuclear Studies, 66-; proj dir, Undergrad Res Participation Grants St Benedict's Col, 66-67 & 68-69; NSF sci fac fel, Univ Calif, Berkeley, 71-72. *Mem:* Am Phys Soc; Am Asn Physics Teachers. *Res:* Electrical and magnetic properties of thin evaporated semiconducting films; electrical properties of polymer materials, particularly polyethylene crystals. *Mailing Add:* Off Pres Benedictine Col N Campus Second & Division Sts Atchison KS 66002

SENECAL, VANCE E(VAN), b Phillipston, Pa, Aug 16, 21; m 45; c 4. CHEMICAL ENGINEERING,. *Educ:* Slippery Rock State Teachers Col, BS, 43; Carnegie Inst Technol, BS, 47, MS, 48, DSc(chem eng), 51. *Prof Exp:* Develop engr, Pittsburgh Coke & Chem Co, 48; res engr, Exp Sta, 51-56, res proj engr, 56-59, res proj supvr, 59-60, res supvr, 60-62, res mgr, Polymer Tech, 62-66, develop mgr, 66-67, res mgr chem eng, 67-69, lab dir, Eng Technol Lab, 69-77, dir eng res, 77-83, VCHMN & EXEC DIR, COMT EDUC AID, E I DU PONT DE NEMOURS & CO, 83- *Mem:* Am Inst Chem Engrs; Am Chem Soc; Sigma Xi. *Res:* Solid deformation mechanics; liquid dynamics; heat transfer; fluid distribution; research and education administration. *Mailing Add:* 1309 Grayson Rd Welshire Wilmington DE 19803

SENECHAL, LESTER JOHN, b Chicago, Ill, June 10, 34; div; c 2. MATHEMATICS. *Educ:* Ill Inst Technol, BS, 56, MS, 58, PhD(math), 63. *Prof Exp:* Instr math, Ill Inst Technol, 59-60 & Univ Tenn, 60-62; asst prof, Univ Ariz, 63-65 & Univ Mass, Amherst, 66-68; assoc prof, 68-75, PROF MATH, MT HOLYOKE COL, 75- *Concurrent Pos:* Fulbright lectr, Brazil, 65; vis prof, Rijksuniversitijt te Groningen, Neth, 74-75 & Acad Sci USSR, 78-79; vis scholar, Univ Tex, Austin, 87-88; vis prof, Univ NC, Chapel Hill, 90-91. *Mem:* Am Math Soc; Neth-Am Acad Circle; Math Asn Am; Soc Indust & Appl Math. *Res:* Analysis and functional analysis; matrix theory; combinatorial algebra. *Mailing Add:* Dept Math Clapp Lab Mt Holyoke Col South Hadley MA 01075

SENECHAL, MARJORIE LEE, b St Louis, Mo, July 18, 39. MATHEMATICAL CRYSTALLOGRAPHY, DISCRETE GEOMETRY. *Educ:* Univ Chicago, BS, 60; Ill Inst Technol, MS, 62, PhD(math), 65. *Prof Exp:* Lectr, 66-67, asst prof, 67-74, assoc prof, 74-78, PROF MATH, SMITH COL, 78- *Concurrent Pos:* Sabbaticals, Univ Gronirgen, Netherlands, 74-75; exchange scientist, Inst Crystall Moscow, Nat Acad Sci, 78-79; mem, Adv Comt USSR Eastern Eruop, Comn Int Relations, Nat Acad Sci, 82-85. *Mem:* Am Math Soc; Sigma Xi. *Res:* Application of discrete geometry, including symmetry theory and the theory of tessellations to physical chemical and biological problems; history of the science of structure of matter. *Mailing Add:* Clark Sci Ctr Smith Col Northampton MA 01063

SENECHALLE, DAVID ALBERT, b Chicago, Ill, July 8, 40; m 63; c 1. MATHEMATICS. *Educ:* Univ Tex, BA, 65, PhD(functional anal), 67. *Prof Exp:* Res sci asst math, Univ Tex, Austin, 62-65; eng scientist, Tracor, Inc, 65-67; asst prof math, Univ Ga, 67-70; asst prof math, State Univ NY Col New Platz, 70-74, assoc prof, 74-78; SR SCIENTIST, TRACOR, INC, 78- *Mem:* Am Math Soc. *Res:* Functional analysis; electronic counter-measures. *Mailing Add:* 4803 Balcones Dr Austin TX 78731

SENFT, ALFRED WALTER, tropical medicine, parasitology; deceased, see previous edition for last biography

SENFT, JOHN FRANKLIN, b York, Pa, Apr 13, 33; m 58; c 2. FOREST PRODUCTS. *Educ:* Pa State Univ, BS, 55, MF, 59; Purdue Univ, PhD(wood tech), 67. *Prof Exp:* Instr forestry, Pa State Univ, 57-59; from instr forestry to asst prof wood sci, 59-71, ASSOC PROF WOOD SCI, PURDUE UNIV, WEST LAFAYETTE, 71- *Mem:* Forest Prod Res Soc; Soc Wood Sci & Technol; Am Soc Testing & Mat. *Res:* Mechanical properties of wood; stress rating of lumber, wood anatomy, and mechanical properties. *Mailing Add:* Dept Forestry & Nat Resources Purdue Univ West Lafayette IN 47907

SENFT, JOSEPH PHILIP, b York Co, Pa, Oct 2, 36; div; c 3. LAND RECLAMATION, SOLID WASTE MANAGEMENT. *Educ:* Juniata Col, BS, 59; State Univ NY Buffalo, MA, 61, PhD(biol), 65. *Prof Exp:* Fel physiol, Sch Med, Univ Md, 64-65, res assoc, 65-66, instr, 66-67; from asst prof to assoc prof, Rutgers Univ, New Brunswick, 67-72; vis prof, Juniata Col, 72-73, assoc prof, 73-77; res scientist, Rodale Res Ctr, 77-82; inst sci, Germantown Acad, 84-88; vis prof, Juniata Col, 88-89; RES ASSOC, PA STATE UNIV, 89-; CONSULT, LAND MGT DECISIONS, STATE COLLEGE, 89- *Concurrent Pos:* Instr sci, Morristown-Beard School, 83-84; mem corp, Marine Biol Lab, Woods Hole. *Mem:* AAAS; Am Asn Cereal Chemists; Am Physiol Soc; Biophys Soc; Soc Gen Physiol; Am Soc Surface Mining & Reclamation. *Res:* Nutritional quality of amaranth grain as possible food source; relationships between soil and plant nutrients for optimum nutritional quality of plant food products; conductance mechanisms in squid nerve axon membranes; active transport mechanisms in frog skin, lobster nerve, and rat gut; land reclamation; municipal, agricultural and industrial waste utilization. *Mailing Add:* 914-5 Southgate Dr State College PA 16801

SENFTLE, FRANK EDWARD, b Buffalo, NY, May 4, 21; m 49; c 6. BOREHOLE NUCLEAR GEOPHYSICS, MAGNETO CHEMISTRY. *Educ:* Univ Toronto, BS, 42, MA, 44, PhD(physics), 47. *Prof Exp:* Lectr chem, physics & math, St Michael's Col, Toronto, 40-46; physicist in charge radiation lab, Dept Mines & Resources, Ont, Can, 47-49; res assoc, Mass Inst Technol, 49-51; physicist in charge nucleonics group, US Geol Surv, 51-60, head solid state physics group, 60-65, head, Physics Lab, 65-89; ADJ PROF PHYSICS, HOWARD UNIV, 89- *Concurrent Pos:* Lectr, Ottawa Univ, 47-49; vis res prof, Howard Univ, 65-66 & 70-81; prin investr, Lunar Sci Prog, NASA, 70-73. *Honors & Awards:* NASA Award, 79. *Mem:* Am Phys Soc; Minerals & Geotech Logging Soc. *Res:* Development of instruments and methods in nuclear and solid state physics applied to geochemical processes; magnetic properties of crystalline materials. *Mailing Add:* Dept Physics Howard Univ Washington DC 20059

SENFTLEBER, FRED CARL, b Roslyn, NY, Nov 19, 48; m 71; c 2. ANALYTICAL CHEMISTRY. *Educ:* Univ Tampa, BS, 70; Southern Ill Univ, PhD(anal chem), 77. *Prof Exp:* Res assoc separation chem, Bucknell Univ, 75-76; instr anal & inorg chem, Providence Col, 76-77, asst prof, 77-78; asst prof anal chem, Murray State Univ, 78-; AT DEPT CHEM, JACKSONVILLE UNIV. *Mem:* Am Chem Soc. *Res:* Studies into changes in metabolic pathways brought about by various pathological conditions; studies involving the use of transition metal complexes in the catalysis of electrochemical reactions. *Mailing Add:* Dept Chem Jacksonville Univ 2800 University Blvd N Jacksonville FL 32211

SENGAR, DHARMENDRA PAL SINGH, b Aligarh, Uttar Pradesh, India, Jan 2, 41; Can citizen; m 71; c 3. HISTOCOMPATIBILITY ANTIGENS, TISSUE TYPING. *Educ:* Vikram Univ, Madhya Pradesh, India, BVSc & AH, 61; Agra Univ, Uttar Pradesh, India, MVSc & AH, 63; Univ Guelph, Ont, MSc, 67, PhD(immunogenetics), 69. *Prof Exp:* Asst prof reproductive physiol, Col Vet Med, Uttar Pradesh Agr Univ, India, 63-65; fel immunogenetics, dept surg, Univ Calif, Los Angeles, 69-72; fel oncol, dept med, Univ Ottawa, 72-73, asst prof med, 73- 79, from asst prof to assoc prof path & med, 79-90, PROF PATH & MED, UNIV OTTAWA, 90-; DIR RES & TECH, REGIONAL TISSUE TYPING LAB, OTTAWA GEN HOSP, ONT, 74- *Mem:* NY Acad Sci; Am Asn Immunologists; Transplantation Soc; Can Transplantation Soc. *Res:* Host immune responses in renal allograft recipients; association of histocompatibility antigens with disease states; mixed leukocyte response; lymphocyte subsets. *Mailing Add:* Dept Path Univ Ottawa Ottawa ON K1N 6N5 Can

SENGBUSCH, HOWARD GEORGE, b Buffalo, NY, Dec 14, 17; m 42; c 2. PARASITOLOGY. *Educ:* State Univ NY, BS, 39; Univ Buffalo, EdM, 47; NY Univ, MS & PhD(parasitol), 51. *Prof Exp:* Teacher pub sch, NY, 39-41; asst zool, Univ Buffalo, 46-47; instr gen biol, 51-52, from asst prof to assoc prof biol, 52-57, univ res found grant, 56 & 61, dir, Great Lakes Lab, 65-67, dean fac arts & sci, 65-70, res grant, 70, prof biol, 57-81, EMER PROF, STATE UNIV NY, BUFFALO, 81- *Concurrent Pos:* NSF grant, Mt Lake Biol Sta, Univ Va, 55-56; mem USAEC radiation safety prog, USPHS, Nev, 57, scientist dir, 80-; with Max Planck Inst, 57-58; NIH fel, La State Univ, Cent Am, 60; Fulbright fel, Cent Philippines Univ, 62-63; vis prof, Univ Mysore, 69-70; consult, Roswell Park Mem Inst, NY, 63-; fac exchange scholar, State Univ NY, 74-; res assoc entom, Bishop Mus, 77- & Buffalo Mus Sci, 81-85; adj prof biol, Inst Arthrodology & Parasitol, Ga Southern Univ, Statesboro, Ga, 83- *Mem:* Am Soc Parasitol; Entom Soc Am; fel Am Inst Biol Sci; Am Soc Acarology; Soc Syst Zool. *Res:* Physiology, ecology and taxonomy of soil mites; survey of dog heartworm, Dirofilaria immitis, in west New York; survey of toxoplasma antibodies in west New York. *Mailing Add:* 2112 Skyland Dr Tallahassee FL 32303

SENGE, GEORGE H, b Braunschweig, Ger, Oct 10, 37; US citizen. SEQUENTIAL MACHINES. *Educ:* Univ Calif, Berkeley, BA, 60; Univ Calif, Los Angeles, MA, 61, PhD(math), 65; Univ Wis-Madison, MS, 77. *Prof Exp:* Asst prof math, Univ Calif, San Diego, 65-68 & Univ Wis-Milwaukee, 68-74; asst prof math, Lawrence Univ, 75-76; SR SCIENTIST ENGR, HUGHES AIRCRAFT CO, 77- *Mem:* Am Math Soc; Math Asn Am. *Res:* Digital signal processing and synthetic aperture radar systems; information theory and digital communication systems; sequential machines. *Mailing Add:* 3437 Mountain View Ave Los Angeles CA 90066

SENGEL, RANDAL ALAN, b Liberal, Kans, Jan 8, 48; m 71; c 1. PSYCHOPATHOLOGY, SCHIZOPHRENIA. *Educ:* Univ Okla, BA, 70, MA, 73, PhD(human ecol), 76. *Prof Exp:* Res assoc behav sci, Vet Admin Hosp, 77-80, RES HEALTH SCIENTIST, VET ADMIN, 81- *Concurrent Pos:* Consult, Task Force Ecopsychiat Data Base, Am Psychiat Asn, 76-78; Nat Res Serv fel, 80. *Mem:* AAAS; Sigma Xi. *Res:* Cognitive and neuropsychological deficits in schizophrenia. *Mailing Add:* 737 S Lahoma Norman OK 73069

SENGER, CLYDE MERLE, b Portland, Ore, June 25, 29; m 51; c 6. MAMMALOGY, CAVE BIOLOGY. *Educ:* Reed Col, BA, 52; Purdue Univ, MS, 53; Utah State Univ, PhD, 58. *Prof Exp:* Co-op agent parasitol, Mont State Univ, 53-55 & Utah State Univ, 55-56; asst prof zool, Univ Mont, 57-63; assoc prof, 63-66, chairperson biol, 73-77, PROF BIOL, WESTERN WASH UNIV, 66- *Mem:* AAAS; Am Soc Mammal; Am Inst Biol Sci. *Res:* Cave biology-mycetophilid flies of lava tubes; taxonomy of fleas; ecology of bats. *Mailing Add:* Dept Biol Western Wash Univ Bellingham WA 98225-9060

SENGERS, JAN V, b Heiloo, Neth, May 27, 31; m 63; c 4. FLUID PHYSICS. *Educ:* Univ Amsterdam, Drs, 55, PhD(physics), 62. *Prof Exp:* Res asst physics, Van der Waals Lab, Univ Amsterdam, 53-55, res assoc, 55-63; physicist, Nat Bur Standards, 63-67; assoc prof, 68-74, PROF INST PHYS SCI & TECHNOL UNIV MD, COLLEGE PARK, 74- *Concurrent Pos:* Physicist, Nat Bur Standards, 68-; Jr Cornelis Gelderman Vis Prof, Delft Technol Univ, Neth, 74-75; corresp, Royal Dutch Acad Sci. *Honors & Awards:* Nat Bur Stand Awards, 66, 68, 69, 70 & 77. *Mem:* Fel AAAS; fel Am Phys Soc; Am Soc Mech Engrs; Am Inst Chem Engrs; Am Chem Soc. *Res:* Thermophysical properties of fluids; critical phenomena and fluctuation phenomena in gases and liquids. *Mailing Add:* Inst Phys Sci & Technol Univ Md College Park MD 20742

SENGERS, JOHANNA M H LEVELT, b Amsterdam, Neth, Mar 4, 29; m 63; c 4. CRITICAL PHENOMENA, FLUID MIXTURES. *Educ:* Univ Amsterdam, Drs, 54, PhD(physics), 58. *Prof Exp:* Wetenschappelijk ambtenaar, van der Waals Lab, Univ Amsterdam, 54-58; res assoc inst theoret chem, Univ Wis, 58-59; wetenschappelijk ambtenaar, Van Der Waals Lab, Univ Amsterdam, 59-63; physicist, Heat Div, Inst Basic Standards, Nat Bur Standards, 63-78, physicist & supvr, Thermophysics Div, Nat Eng Lab, 78-87; SR FEL, THERMOPHYSICS DIV, NAT INST STANDARDS & TECHNOL, 83- *Concurrent Pos:* Lectr, Cath Univ Louvain, 71; res assoc inst theor physics, Univ Amsterdam, 74-75; Regent's prof, Dept Chem, Univ Calif, Los Angeles, 82; chair, Working Group A, Int Asn Properties Steam, 86-90; chmn, Int Asn Properties Water & Steam, 91-92; corresp Neth Royal Acad Sci, 90. *Honors & Awards:* Silver Medal, Dept Com, 72, Gold Medal, 78; Wise Award, 85. *Mem:* Neth Phys Soc; Am Soc Mech Engrs; fel Am Phys Soc; Europ Phys Soc. *Res:* Thermodynamic properties of fluids and fluid mixtures; critical phenomena in fluids; equation of state, theoretical and experimental; supercritical aqueous systems. *Mailing Add:* Thermophysics Div Nat Inst Standards & Technol Gaithersburg MD 20899

SEN GUPTA, BARUN KUMAR, b Jamshedpur, India, July 31, 31; m 56; c 2. MICROPALEONTOLOGY, MARINE GEOLOGY. *Educ:* Calcutta Univ, BSc, 51, MSc, 54; Cornell Univ, MS, 61; Indian Inst Technol, Kharagpur, PhD(geol), 63. *Prof Exp:* From asst lectr to lectr geol, Indian Inst Technol, Kharagpur, 55-66; Nat Res Coun Can fel, Atlantic Oceanog Lab, Bedford Inst, 66-68, temporary res scientist, 68; asst prof geol, Univ Ga, 69-72, from assoc prof to prof, 72-79; PROF GEOL, LA STATE UNIV, 79- *Concurrent Pos:* NSF res grants, 69-72, 75-78, 82-86, 88-90, NATO res grant, 87-91; vis prof, Univ Fed Rio Grande Do Sul, Porto Alegre, Brazil, Univ Bordeaux I, France; pres, Cushman Found Foraminiferal Res, 87-88. *Mem:* AAAS; fel Geol Soc Am; fel Cushman Found Foraminiferal Res; Paleont Soc; Soc Econ Paleont Mineralogists. *Res:* Stratigraphy, ecology, and paleoecology of Cenozoic benthic foraminifera. *Mailing Add:* Dept Geol & Geophys La State Univ Baton Rouge LA 70803-4101

SENGUPTA, BHASKAR, b New Delhi, India, Feb 7, 44; m 70. OPERATIONS RESEARCH. *Educ:* Indian Inst Technol, BTech, 65; Columbia Univ, MS, 73, EngScD, 76. *Prof Exp:* Proj engr elec eng, Crompton Greaves Ltd, 65-69; syst engr comp sci, IBM World Trade Corp, 69-71, Serv Bur Co, 71-72; ASST PROF OPER RES, STATE UNIV NY, STONY BROOK, 76-; AT AT&T BELL LABS. *Concurrent Pos:* Consult, Turner Construct Co, 76- *Mem:* Inst Mgt Sci; Oper Res Soc Am. *Res:* Stochastic models in inventory, production and health care. *Mailing Add:* NEC Res Inst Four Independence Way Princeton NJ 08540

SENGUPTA, DIPAK L(AL), b Batisha, Bangladesh, Mar 1, 31; m 62; c 2. ELECTRONICS. *Educ:* Univ Calcutta, BSc, 50, MSc, 52; Univ Toronto, PhD(elec eng), 58. *Prof Exp:* Res fel electronics, Gordon McKay Lab Appl Physics, Harvard Univ, 59; res assoc, Radiation Lab, 59-61, assoc res physicist, 61-63; asst prof elec eng, Univ Toronto, 63-64; asst dir electronics, Cent Electronics Eng Res Inst Pilani, India, 64-65; lectr elec eng, Univ Mich, Ann Arbor, 65-66, assoc res eng, 65-68, res scientist & adj prof elec eng & comput sci, 68-86; PROF & CHMN, ELEC ENG & PHYSICS, UNIV DETROIT, 86- *Mem:* fel Inst Elec & Electronics Engrs; AAAS; Sigma Xi; Int Union Radio Sci. *Res:* Electromagnetic theory; antennas; plasma physics; interaction of electromagnetic waves and plasmas; acoustic and electromagnetic waves; electromagnetic interference. *Mailing Add:* Dept Elec Eng & Physics Univ Detroit 4001 W McNichols Rd Detroit MI 48221

SENGUPTA, GAUTAM, b Gaibandha, Bangladesh, Sept 1, 45; m 72; c 2. MECHANICAL & AEROSPACE ENGINEERING. *Educ:* Univ Calcutta, BSc Hons, 63; Indian Inst Technol, Kharagpur, BTech Hons, 66; Univ Southampton, PhD(struct acoust), 70. *Prof Exp:* Spec engr acoust fatigue, Eng Sci Data Univ, London, 70-71; resident res assoc struct dynamics, Langley Res Ctr, NASA, 72-73; sr specialist, 73-80, prin engr noise control, 73-85, LEAD ENGR, COMPUT ACOUST, BOEING CO, 86- *Concurrent Pos:* Reviewer, J Sound & Vibration, Shock & Vibration Digest, Am Inst Aeronaut & Astronaut J, 70-; resident res assoc, Langley Res Ctr, NASA-Nat Acad Sci, 72-73; dir educ, Pac Northwest Sect, Am Inst Aeronaut & Astronaut, 78-80. *Honors & Awards:* Outstanding Tech Contrib, Am Inst Aeronaut & Astronaut, 78. *Mem:* Fel Am Inst Aeronaut & Astronaut; Acoust Soc Am. *Res:* Vibration and noise control, especially cabin noise, airframe noise, aeroacoustics, sonic fatigue, wave propagation, periodic structures, matrix methods, correlation techniques and intrinsic structural tuning; computational fluid and structural dynamics; physics. *Mailing Add:* 13514 SE 181st Pl Boeing Co Renton WA 98058

SENGUPTA, SAILES KUMAR, b Bankura, India, Jan 1, 35; US citizen; m 69; c 2. OPERATIONS RESEARCH, SOFTWARE SYSTEMS. *Educ:* Univ Calcutta, BSc, 53, MSc, 56; Univ Calif, Berkeley, PhD(statist), 69. *Prof Exp:* Lectr math, WBengal Educ Serv, 57-62; teaching res asst statist, Univ Calif, Berkeley, 63-64 & 65-67; asst prof math, Univ Mo, Kansas City, 69-76; assoc prof, 76-81, PROF MATH, SDAK SCH MINES & TECHNOL, 81- *Concurrent Pos:* Fac res partic, Argonne Nat Lab; vis fel, Inst Math & Its Applns, Univ Minn, 87; fac res partic, NASA, Langley, ASD; sr Univ Corp Atmospheric Res fel, Naval Oceanog Atmospheric Res Lab. *Mem:* Sigma Xi; Am Statist Asn; Inst Elec & Electronics Engrs. *Res:* Statistical pattern recognition; satellite data analysis; computing, image understanding; multivariate statistical analysis of geochemical data. *Mailing Add:* Dept Math & Comput Sci SDak Sch Mines & Technol Rapid City SD 57701

SENGUPTA, SUMEDHA, b India, Feb 17, 43; US citizen; m 69; c 2. COMPUTER SCIENCE, STATISTICS. *Educ:* Patna Univ, BS, 62, MS, 64; Indian Inst Technol, PhD(statist), 68. *Prof Exp:* Lectr, Nat Col Bus, 76-78; res scientist statist res, Inst Atmospheric Sci, 78-80; sr res statistician, Res Inst Geochem, 80-81; statistician reliability & qual control, MPI-Control Data, 81-84; prin statistician, Defense Systs Div, Honeywell, 85-87; PRES STAT-TECH, CONSULT FIRM STATIST, MPS, 87- *Concurrent Pos:* Consult, Ace Elec Co, Kans, 71, Marion Lab, 76, Jackson County, Kans, 76, & to assoc adminr, Regional Hosp, SDak, 82-, MPI-Control Data, Honeywell, Nat Gould, Alexandria Exter. *Mem:* Sigma Xi; Am Statist Asn; Inst Math Statist; Grad Women Sci; Am Soc Qual Control. *Res:* Bayesian inference in life testing and reliability; statistical methodology in metereology and atmospheric science; quality and process control. *Mailing Add:* 7523 Fawn Ct Carmel CA 93923-9529

SENHAUSER, DONALD ALBERT, b Dover, Ohio, Jan 30, 27; m 61; c 2. PATHOLOGY, IMMUNOPATHOLOGY. *Educ:* Columbia Univ, AB, 48, MD, 51. *Prof Exp:* Intern, Roosevelt Hosp, New York, 51-52; asst resident & instr path, Columbia Presby Hosp, 55-56; asst resident, Cleveland Clin Found, Ohio, 56-58, sr resident, 58-59, clin assoc, 59-61, staff physician, 61-63; from assoc prof to prof path, Univ Mo, Columbia, 63-75, vchmn dept, 66-75, asst dean acad affairs, 69-71; PROF PATH & CHMN DEPT, COL MED, OHIO STATE UNIV, 75- *Concurrent Pos:* Fel path, Cleveland Clin Found, Ohio, 56-59; travelling fel immunol res, 60-61; consult, Study Group, WHO, 71- & long range planning comt, Nat Libr Med, 85-86; mem, bd gov, Col Am Path, 80-86; distinguished practitioner, Nat Acad Practise, 85- *Mem:* AAAS; Am Soc Clin Path; Col Am Path (pres-elect, 91); Int Acad Path; Asn

Am Pathologists. *Res:* Ultrastructural studies of the immunopathology of lymphocyte-target cell interaction; role of lipid molecules in cell membrane structural and antigenic integrity apheresis and immunosuppression; medical information science. *Mailing Add:* Dept Path Ohio State Univ Col Med Columbus OH 43210

SENICH, DONALD, b Cleveland, Ohio, Sept 5, 29; m 52; c 4. SOIL MECHANICS. *Educ:* Univ Notre Dame, BSME, 53; Iowa State Univ, MS, 61, PhD(soil mech), 66. *Prof Exp:* Resident engr, US Army, Frankfurt Am Main, WGer, 61-63, chief nuclear eng, Nuclear Power Field Off, Ft Belvoir, Va, 66-67, officer in chg, US Army Reactor, Ft Greely, Alaska, 67, chief construct engr, Mil Assistance Command, Vietnam, 68-69, staff scientist, Apollo Lunar Explor Off, NASA, 69-73; dir div, Energy Resources Res, 73-78, div dir, Integrated Basic Res, 78-79, dir div, Problem Focused Res, 79-81, DIV DIR, INDUST SCI & TECHNOL INNOVATION, NSF, 81- *Mem:* Sigma Xi. *Res:* Soil mechanics; physical and surface chemistry; development of fundamental knowledge to aid in the solution of major problems in the areas of discipline sciences. *Mailing Add:* NSF Washington DC 20550

SENIOR, BORIS, b Priluki, Russia, Apr 24, 23; US citizen; m 54; c 3. PEDIATRICS, ENDOCRINOLOGY. *Educ:* Univ Witwatersrand, Mb, BCh, 46; MRCP, 53, FRCP, 70. *Prof Exp:* Res asst, Univ Col Hosp, London, 51-54; Coun Sci & Indust Res sr bursar, Univ Cape Town, 54-55; asst prof pediat, Sch Med, Boston Univ, 62-63; from asst prof to assoc prof, 63-69, PROF PEDIAT, SCH MED, TUFTS UNIV, 70-, CHIEF PEDIAT ENDOCRINOL, NEW ENG MED CTR, 63- *Mem:* Endocrine Soc; Am Pediat Soc; Sigma Xi. *Res:* Disorders of carbohydrate metabolism; carbohydrate-lipid interrelationships. *Mailing Add:* 20 Ash St Boston MA 02111

SENIOR, JOHN BRIAN, b Cleveleys, Eng, Oct 11, 36. INORGANIC CHEMISTRY. *Educ:* Univ London, BSc, 58; McMaster Univ, PhD(chem), 62. *Prof Exp:* Dept Sci Indust Res-NATO fel chem, Birkbeck Col, London, 62-63; fel, Univ Western Ont, 63-64; asst prof chem, St Paul's Col, Man, 64-66; asst prof, 66-70, ASSOC PROF CHEM, UNIV SASK, 70- *Mem:* The Chem Soc; Chem Inst Can. *Res:* Chemistry of acidic nonaqueous solvent systems; halogen cations and oxycations. *Mailing Add:* Dept Chem & Chem Eng Univ Sask Saskatoon SK S7N 0W0 Can

SENIOR, JOHN ROBERT, b Philadelphia, Pa, July 17, 27; m 52; c 3. INTERNAL MEDICINE, GASTROENTEROLOGY. *Educ:* Pa State Univ, BS, 50; Univ Pa, MD, 54. *Prof Exp:* Instr med, Univ Pa, 57-59; res fel, Harvard Med Sch, 59-62; asst prof med & assoc biochem, Univ Pa, 62-68, assoc prof med, 68-78, assoc prof community med, 74-76, clin prof med, 78-80; vpres clin affairs, Sterling-Winthrop Res Inst, Rensselaer, NY, 81-; clin prof med, Albany Med Ctr, 81-85; at Dept Med, Albany Med Col, NY, 84. *Concurrent Pos:* Nat Inst Arthritis & Metab Dis res & training grants gastroenterol, Philadelphia Gen Hosp, 62-71; consult, US Naval Reg Med Ctr, Philadelphia, 66-, Nat Bd Med Examr & Am Bd Internal Med, 70-; dir clin invest, Presby-Univ Pa Med Ctr, 71-73; Carnegie Corp & Commonwealth Fund grant comput based exam, Philadelphia, 71-73; dir clin res Clin Res Ctr, Grad Hosp, Univ Pa, 73-74, dir, Emergency Serv & Spec Treatment Unit Alchol-Related Disorders & dir Off Eval, 74-79; dir regulatory proj, E R Squibb & Sons, NJ, 79-81; rear admiral, Med Corps, US Naval Reserve & dep fleet surgeon & comdr-in-chief, Pac Fleet, US Naval Health Sci Educ & Training Command, Bethesda, 80-; adj prof med, Univ Pa, 80-; consult, G H Besselaar Assoc, Princeton, NJ. *Mem:* Am Soc Clin Invest; Am Soc Clin Pharmacol & Therapeut; Am Asn Study Liver Dis (pres); Am Gastroenterol Asn; fel Am Col Physicians. *Res:* Lipid metabolism; biochemistry; cell physiology; computer science; medical information processing; hepatic and central nervous effects of ethanol; viral and drug-induced hepatitis; evaluation of clinical competence. *Mailing Add:* 54 Merbrook Lane Merion Station PA 19066

SENIOR, THOMAS BRYAN ALEXANDER, b Menston, Eng, June 26, 28; m 57; c 4. ELECTROMAGNETICS. *Educ:* Univ Manchester, BSc, 49, MSc, 50; Cambridge Univ, PhD(appl math), 54. *Prof Exp:* From sci officer to sr sci officer, Radar Res & Develop Estab, Ministry of Supply, Eng, 52-57; res assoc, Univ Ann Arbor, 57-58, from assoc res mathematician to res mathematician, 58-61, assoc dir, Radiation Lab, 61-75, dir, Radiation Lab, 75-87, PROF ELEC ENG, UNIV ANN ARBOR, 69-, ASSOC CHMN ELEC ENG & COMPUT SCI DEPT, 84-, ACTG CHMN, 87- *Concurrent Pos:* Ed, Radio Sci, 73-79; chmn, US Nat Comt, Int Union Radio Sci, 84, Comn B, 88-90. *Mem:* Fel Inst Elec & Electronics Engrs; Sigma Xi. *Res:* Theoretical problems in scattering and diffraction of electromagnetic and acoustical waves; radio wave propagation; optics. *Mailing Add:* Radiation Lab Elec Eng & Comput Sci Dept Univ Mich Ann Arbor MI 48109-2122

SENITZER, DAVID, b New York, NY, Oct 9, 44; m 66; c 3. IMMUNOLOGY, MICROBIOLOGY. *Educ:* City Col New York, BS, 66; La State Univ, PhD(microbiol), 69; Am Bd Med Lab Immunol, dipl. *Prof Exp:* Res assoc immunol, Col Physicians & Surgeons, Columbia Univ, 69-72; asst prof microbiol, Med Col Ohio, 72-77, assoc prof microbiol & path, 78-85; DIR TRANSPLANT IMMUNOL, MONTEFIORE MED CTR, 85- *Mem:* AAAS; Am Soc Microbiol; Am Asn Immunologists; Am Soc Clin Pathologists; Am Asn Clin Histocompatibility Testing. *Res:* Transplantation immunology; autoimmune disease mechanisms; regulation of the immune response. *Mailing Add:* Transplant Immunol Montefiore Med Ctr 111 E 210th St Bronx NY 10467

SENITZKY, BENJAMIN, b Vilno, Poland, Nov 15, 26; nat US; m 50; c 3. PHYSICS. *Educ:* Columbia Univ, BS, 48, PhD, 56. *Prof Exp:* Mem tech staff, Bell Tel Labs, Inc, 56-59; physicist, Tech Res Group, Inc, 59-66; assoc prof, 66-75, asst head dept, 74-78, PROF ELECTROPHYS, POLYTECH UNIV, 75- *Mem:* Am Phys Soc. *Res:* Atomic beam resonance techniques for study of hyperfine structure in atomic spectra; high field breakdown in semiconductors; millimeter wave amplification using resonance saturation effects; interaction of radiation and semiconductors. *Mailing Add:* 951 W Orange Grove Rd No 6-104 Tucson AZ 85704

SENITZKY, ISRAEL RALPH, b Vilna, Poland, Feb 28, 20; m 44; c 2. QUANTUM OPTICS. *Educ:* City Col, BS, 41; New York Univ, MS, 44; Columbia Univ, PhD(physics), 50. *Prof Exp:* Physicist, Inst Explor Res, US Army Electronics Command, 42-72; prof physics, Technion-Israel Inst Technol, 72-83; RES PROF PHYSICS, CTR LASER STUDIES, UNIV SOUTHERN CALIF, 85- *Concurrent Pos:* Vis prof physics, Univ Southern Calif, 83-84. *Mem:* Fel Am Phys Soc. *Res:* Quantum mechanics; quantum electronics; lasers; quantum optics; relationship between quantum theory and classical theory; resonance fluoresence; microwaves. *Mailing Add:* Dept Physics Univ Southern Calif Univ Park Los Angeles CA 90089-1112

SENKLER, GEORGE HENRY, JR, b Postville, Iowa, Oct 25, 45; m 72; c 2. PHYSICAL ORGANIC CHEMISTRY. *Educ:* Hamline Univ, BS, 67; Princeton Univ, MA, 72, PhD(chem), 75. *Prof Exp:* Staff res, chem, dye & pigments dept, 74-79, res supvr, 79-81, prod mgr, 81-85, SR RES CHEMIST, E I DUPONT DE NEMOURS, 74-, TECH MGR, 85- *Res:* Studies on the stereochemistry of organophosphorus, arsenic and sulfur compounds; development of industrially important processes; inorganic and organic pigment research. *Mailing Add:* Chem & Pigments Dept E I du Pont de Nemours & Co Wilmington DE 19898

SENKOWSKI, BERNARD ZIGMUND, b Dearborn, Mich, Feb 2, 27; m 52; c 2. PHARMACEUTICAL CHEMISTRY. *Educ:* Rutgers Univ, AB, 51, MS, 60, PhD(chem), 65. *Prof Exp:* Chemist, Hoffmann-La Roche Inc, 51-62, head anal res, 62-65, asst dir qual control, 65-70, dir qual control, 70, asst vpres, 71-77, dir pharmaceut opers, 72-77; vpres mgr & eng, Alcon Labs, 77-88; PRIN, BZS CONSULT SERV, 88- *Concurrent Pos:* Lectr, Rutgers Univ, 65-70; mem rev comt, US Pharmacopoeia, 70-75; mem comt vitamins, Nat Formulary. *Mem:* Fel AAAS; Sigma Xi; Am Chem Soc; fel Am Inst Chemists; fel Am Asn Pharmaceut Scientists. *Res:* Corporate manufacturing, domestic and international; facilities planning; GMP, OSHA regulatory compliance; analytical research; quality control; pharmaceutical operations. *Mailing Add:* 4366 Capra Way Ft Worth TX 76126-2237

SENKUS, MURRAY, b Redberry, Sask, Aug 31, 14; nat US; m 38; c 4. CHEMISTRY. *Educ:* Univ Sask, BSc, 34, MSc, 36; Univ Chicago, PhD(phys org chem), 38. *Prof Exp:* Instr chem, N Park Col, 37-38; chemist, Com Solvents Corp, 38-50; dir res & develop, Daubert Chem Corp, 50-51; dir chem res, 51-60, asst dir res, 60-64, dir res, 64-76, dir sci affairs, 76-79, CONSULT TOBACCO INST, R J REYNOLDS TOBACCO CO, 79- *Mem:* AAAS; NY Acad Sci; Am Chem Soc; Sigma Xi. *Res:* Tobacco and synthetic organic chemistry; insecticides; recovery of fermentation products; chemotherapeutic agents; chemistry of flavors. *Mailing Add:* 2516 Country Club Rd Winston-Salem NC 27104

SENN, MILTON J E, pediatrics, psychiatry; deceased, see previous edition for last biography

SENN, TAZE LEONARD, b Newberry, SC, Oct 16, 17; m 39; c 3. HORTICULTURE, SEAWEED. *Educ:* Clemson Univ, BS, 39; Univ Md, MS, 50, PhD(hort), 58. *Prof Exp:* Asst horticulturist, Clemson Col, 39-40, asst botanist, 40-41, assoc prof hort, 46-56; asst plant pathologist, Univ Tenn, 41-43 & 46; agriculturist, Eastern Lab, USDA, 56-58; prof, 58-81, EMER PROF HORT & HEAD DEPT, CLEMSON UNIV, 81- *Concurrent Pos:* Vpres, res & develop, Southeastern Resources Corp, 90. *Honors & Awards:* Gold Seal-Silver Seal, Nat Coun of State Garden Clubs. *Mem:* Fel AAAS; fel Am Soc Hort Sci; Sigma Xi; Am Soc Testing & Mat. *Res:* Active compounds of marine plants; physiological aspects of plant propagation; tobacco chemistry; horticultural physiology; post-harvest physiology of horticultural crops; horticultural therapy; humus chemistry. *Mailing Add:* 201 Strawberry Lane Clemson SC 29631

SENNE, JOSEPH HAROLD, JR, b St Louis, Mo, Nov 9, 19; m 46; c 1. CIVIL ENGINEERING. *Educ:* Wash Univ, BS, 48; Univ Mo, MS, 51; Iowa State Univ, PhD, 61. *Prof Exp:* Asst construct engr, Laclede Christy Clay Prod Co, Mo, 41-42; from instr to asst prof civil eng, Mo Sch Mines, 48-54; asst prof, Iowa State Univ, 54-63; PROF CIVIL ENG, UNIV MO, ROLLA, 63-, CHMN DEPT, 65- *Mem:* Am Soc Eng Educ; Am Soc Civil Engrs; Nat Soc Prof Engrs; Sigma Xi. *Res:* Welded wire fabric in reinforced concrete beams and pipe; structural design; orbital mechanics; satellite tracking. *Mailing Add:* Dept Civil Eng Univ Mo Rolla MO 65401

SENNELLO, LAWRENCE THOMAS, b New York City, NY, Feb 4, 37; m 62; c 4. PHARMACOKINETICS, DRUG ANALYSIS. *Educ:* Lake Forest Sch Mgt, Ill, MBA, 80; Univ Ill, Urbana, BS, 59, PhD(chem), 68. *Prof Exp:* Teaching asst chem, Univ Ill, Urbana, 66-67; group leader, Anal Chem Res, 69-70, SECT MGR, PHARMACOKINETICS & BIOPHARMACEUT, ABBOTT LABS, ABBOTT PARK, 71- *Mem:* Am Chem Soc; Sigma Xi; NY Acad Sci. *Res:* Analytical chemistry methodology. *Mailing Add:* Drug Metab Dept Abbott Labs Abbott Park IL 60064

SENNER, JOHN WILLIAM, evolutionary genetics, primatology, for more information see previous edition

SENOFF, CAESAR V, b Toronto, Ont, Apr 30, 39; m 68; c 1. INORGANIC CHEMISTRY. *Educ:* Univ Toronto, BSc, 61, MA, 63, PhD(chem), 65. *Prof Exp:* Fel chem, Clarkson Col Technol, 65-67 & Univ Tex, Austin, 67-68; asst prof, 68-73, ASSOC PROF CHEM, UNIV GUELPH, 73- *Mem:* Am Chem Soc. *Res:* Coordination chemistry of transition elements; organometallic chemistry. *Mailing Add:* Dept Chem Univ Guelph Guelph ON N1G 2W1 Can

SENOGLES, SUSAN ELIZABETH, b Latrobe, Pa. SIGNAL TRANSDUCTION. *Educ:* Univ Minn, BS, 77, PhD(biochem), 84. *Prof Exp:* Res assoc, Duke Univ, 84-89; ASST PROF, UNIV TENN, 89- *Mem:* Am Soc Biochem & Molecular Biol. *Res:* Transmembrane signal transduction; G protein specificity of effector coupling. *Mailing Add:* 858 Madison Ave G01 Memphis TN 38163

SENOZAN, NAIL MEHMET, b Istanbul, Turkey, Sept 13, 36; m 82. PHYSICAL CHEMISTRY. *Educ:* Brown Univ, ScB, 60; Univ Calif, Berkeley, PhD(chem), 65. *Prof Exp:* PROF CHEM, CALIF STATE UNIV, LONG BEACH, 64- *Concurrent Pos:* Fulbright prof, Izmir, Turkey. *Mem:* AAAS. *Res:* Biophysical chemistry; hemoglobin; hemocyanin. *Mailing Add:* Dept Chem Calif State Univ 1250 Bellflower Blvd Long Beach CA 90840

SENSEMAN, DAVID MICHAEL, b Dayton, Ohio, Dec 6, 48; m 81. NEUROPHYSIOLOGY, ANIMAL BEHAVIOR. *Educ:* Kent State Univ, BS, 71; Princeton Univ, PhD(biol), 77. *Prof Exp:* asst mem, Monell Chem Senses Ctr, 77-81, RES ASST PROF PHYSIOL, SCH DENT MED, UNIV PA, 78- *Concurrent Pos:* NIH fel, 76-78; res grant, Nat Inst Dent Res, 78-81. *Honors & Awards:* Merck Award, 71. *Mem:* AAAS; Biophys Soc. *Res:* Biophysics and neuropharmacology of exocrine and endocrine gland function. *Mailing Add:* Dept Life Sci Univ Tex 6700 N Fm 1604 W San Antonio TX 78285

SENSENY, PAUL EDWARD, b Chambersburg, Pa, June 14, 50; m 72; c 1. ENGINEERING MECHANICS, GEOPHYSICS. *Educ:* Univ Pa, BSME, 72; Brown Univ, ScM, 74, PhD(solid mech), 77. *Prof Exp:* Res asst, Brown Univ, 72-76; res engr, SRI Int, 76-79; mgr mat lab, Re/Spec Inc, 79-87, sr staff scientist, 87-89; CIVIL ENGR, DEFENSE NUCLEAR AGENCY, 89- *Mem:* Am Soc Mech Eng; Am Soc Civil Eng; Am Geophys Union; Soc Exp Stress Anal; Am Acad Mech; Soc Eng Sci; Am Soc Testing & Mat. *Res:* Development of constitutive equations for geologic materials that model the influence of pressure, temperature, loading rate, pore fluids and thermomechanical history on deformation and strength; deformation and damage of structures subjected to dynamic loads. *Mailing Add:* Defense Nuclear Agency Attn SPSD 6801 Telegraph Rd Alexandria VA 22310-3398

SENSIPER, S(AMUEL), b Elmira, NY, Apr 26, 19; m 50; c 3. MICROWAVE ENGINEERING, ANTENNA ENGINEERING. *Educ:* Mass Inst Technol, SB, 39, ScD, 51; Stanford Univ, EE, 41. *Prof Exp:* Asst, Stanford Univ, 39-41; from asst proj engr to sr proj engr & res sect head, Microwave Equip, Sperry Gyroscope Co, NY, 41-48, consult, 48-51; mem staff, Res Lab Electronics, Mass Inst Technol, 49-51; res engr & head, antenna sect, Microwave Lab, Hughes Aircraft Co, 51-53, head circuits & anal sect, Electron Tube Lab, 53-58, sr staff engr, Res Lab, 58-60; dir, Command & Control Labs, Space Electronics Corp, 60-63, assoc sr div mgr electronics opers, 63-64, dir eng, Electronic Systs, 64, mgr, Res & Ed Div, Space-Gen Corp, 64-67; mgr antenna syst lab, TRW Systs Group, 67-70; CONSULT ANTENNAS, ELECTROMAGNETICS & MICROWAVES, 70- *Concurrent Pos:* indust electronics fel, Mass Inst Technol, 47-49; Instr, Univ Southern Calif, 55-57 & 79-81. *Mem:* Fel AAAS; fel Inst Elec & Electronics Engrs; Sigma Xi; Sci Res Soc Am. *Res:* Systems theory and analysis, communications systems and components; electromagnetic theory; microwave electron tubes and devices; antennas and transmission components; microwave test equipment. *Mailing Add:* PO Box 3102 Culver City CA 90231-3102

SENTER, HARVEY, mathematics, for more information see previous edition

SENTERFIT, LAURENCE BENFRED, b Sarasota, Fla, July 30, 29; m 57; c 3. MICROBIOLOGY, IMMUNOLOGY. *Educ:* Univ Fla, BS, 49, MS, 50; Johns Hopkins Univ, ScD(pathobiol), 55; Am Bd Med Microbiol, dipl, 66. *Prof Exp:* Instr ophthal microbiol, Sch Med, Johns Hopkins Univ, 54-58; chief microbiologist & dir lab exp path, Charlotte Mem Hosp, NC, 58-64; res scientist, Chas Pfizer & Co, Inc, 64-66; assoc prof path & asst prof microbiol, Sch Med, St Louis Univ, 66-70; assoc prof microbiol, 70-84, PROF MICROBIOL & PATH, MED COl, CORNELL UNIV, 84-; DIR LAB MICROBIOL, NEW YORK HOSP, 70- *Concurrent Pos:* Attend microbiologist, Hosp for Spec Surg; attend microbiologist, Burke Rehab Ctr. *Mem:* Am Soc Microbiol; Am Asn Immunologists; Soc Exp Biol & Med; Am Venereal Dis Soc; Asn Clin Scientist; Int Org Mycoplasmology. *Res:* Immunology of parasitic diseases; vaccines for respiratory agents, particularly parainfluenza, respiratory syncytial virus and mycoplasma pneumonia; immunopathology and hypersensitivity; parasitology. *Mailing Add:* Dept Microbiol Cornell Univ Med Col 1300 York Ave New York NY 10021

SENTI, FREDERIC R(AYMOND), b Cawker City, Kans, Apr 29, 13; m 39; c 3. FOOD SAFETY ASSESSMENT. *Educ:* Kans State Univ, BS, 35, MS, 36; Johns Hopkins Univ, PhD(chem), 39. *Hon Degrees:* DSc, Kans State Univ, 72. *Prof Exp:* Asst chemist, Eastern Regional Res Lab, Bur Agr & Indust Chem, USDA, 41-42, assoc chemist, 42-45, chemist, 45-47, prin chemist, 47-48, chief, Anal Phys Chem & Physics Sect, Northern Utilization Res & Develop Div, Agr Res Serv, 49-54, Cereal Crops Sect, 54-59, dir, North Region Res Lab, 59-65, dept adminr Nutrit, Consumer & Indust Res, Agr Res Serv, USDA, 64-72, asst adminr, Nat Prog Staff, 72-74; res assoc, Feds Am Soc Exp Biol, 74-77, assoc dir, 77-84, sr sci consult, Life Sci Res Off, 84-86; CONSULT, 87- *Concurrent Pos:* Lectr physical chem, Grad Sch, Temple Univ, 46-48; consult, Agency Int Develop, Govt India, 66-68; mem, Panel World Food Supply, Pres Sci Adv Comt, 66-67; mem, Bd Trustees, Am Type Cult Collection, 62-67, chmn, 68-69; mem, Liaison Panel, Food Protection Comt, Nat Res Coun, 71-74; chief staff officer, Comt Processed Foods for Develop Countries & Domestic Food Distrib Prog, USDA, 64-74, vchmn, Task Group on Gen Recognized As Safe New Plant Varieties, Food & Drug Admin, 72-74. *Mem:* Am Chem Soc; Inst Food Technologists; fel AAAS; Am Crystallog Asn; Am Asn Cereal Chemists. *Res:* Structure and solution properties of starches and microbial polysaccharides; mycotoxins; cereal and oilseed chemistry and technology; development of processed foods for food assistance programs; safety evaluation of food additives. *Mailing Add:* 2601 N Pollard St Arlington VA 22207

SENTILLES, F DENNIS, JR, b Donaldsonville, La, Aug 7, 41; m 63; c 2. MATHEMATICS. *Educ:* Francis T Nicholls State Col, BS, 63; La State Univ, MS, 65, PhD(math), 67. *Prof Exp:* Instr math, La State Univ, 66-67; asst prof, 67-70, ASSOC PROF MATH, UNIV MO-COLUMBIA, 70- *Mem:* Am Math Soc. *Res:* Abstract functional analysis; application of functional analytic techniques to specific problems in function and measure spaces. *Mailing Add:* Dept Math Univ Mo 202 Math Sci Bldg Columbia MO 65211

SENTMAN, DAVIS DANIEL, b Iowa City, Iowa, Jan 19, 45. MAGNETOSPHERIC & SOLAR TERRESTRIAL PHYSICS. *Educ:* Univ Iowa, BA, 71, MS, 73, PhD(physics), 76. *Prof Exp:* Res assoc, Univ Iowa, 76-77; RES GEOPHYSICIST, INST GEOPHYS & PLANETARY PHYSICS, UNIV CALIF, LOS ANGELES, 78- *Mem:* Am Geophys Union; Am Phys Soc; Sigma Xi. *Res:* Physics of planetary magnetospheres; solar terrestrial interactions; terrestrial electromagnetic resonances. *Mailing Add:* Inst Geophys & Planetary Physics 405 Hilgard Ave Los Angeles CA 90024-1567

SENTMAN, LEE H(ANLEY), III, b Chicago, Ill, Jan 27, 37. AERONAUTICAL & ASTRONAUTICAL ENGINEERING. *Educ:* Univ Ill, BS, 58; Stanford Univ, PhD(aeronaut, astronaut), 65. *Prof Exp:* Sr dynamics engr, Lockheed Missiles & Space Co, Calif, 59-65; asst prof, 65-69, assoc prof, 69-79, PROF AERONAUT & ASTRONAUT ENG, UNIV ILL, URBANA, 79- *Concurrent Pos:* Vis prof aerospace eng, Univ Ariz, 71-72; consult, various aerospace companies. *Mem:* Am Inst Aeronaut & Astronaut; Am Phys Soc; Optical Soc Am. *Res:* Chemical lasers, unstable resonators; rotational nonequilibrium effects, vibrational relaxation of excited molecules, fluid dynamics; kinetic theory and statistical mechanics; supersonic combustion. *Mailing Add:* Dept Aeronaut & Astronaut Univ Ill 101 Trans Bldg 104 S Mathews Urbana IL 61801

SENTURIA, BEN HARLAN, otolaryngology; deceased, see previous edition for last biography

SENTURIA, JEROME B(ASIL), b San Antonio, Tex, Dec 2, 38; wid; c 2. COMPARATIVE PHYSIOLOGY, ENVIRONMENTAL PHYSIOLOGY. *Educ:* Univ Calif, Los Angeles, BA, 60; Rice Univ, MA, 63; Univ Tex, PhD(zool), 67. *Prof Exp:* Asst prof biol, 66-71, asst dean, Col Arts & Sci, 77-81 & 83-85, ASSOC PROF, BIOL, CLEVELAND STATE UNIV, 71-, ASSOC DEAN, COL ARTS & SCI, 89- *Concurrent Pos:* Nat Heart Inst fel, Heart Lab, Malmo, Sweden, 67-68. *Mem:* Soc Crybiol; Am Inst Biol Sci; Am Soc Zool; Am Physiol Soc; Am Heart Asn; Int Hibernation Soc. *Res:* Cardiovascular physiology in hibernating mammals; seasonal variation in the physiology of hibernating mammals. *Mailing Add:* Dept Biol Cleveland State Univ 1983 E 24th Cleveland OH 44115-2403

SENTURIA, STEPHEN DAVID, b Washington, DC, May 25, 40; m 61; c 2. SOLID STATE PHYSICS, ELECTRONICS. *Educ:* Harvard Univ, BA, 61; Mass Inst Technol, PhD(physics), 66. *Prof Exp:* Mem res staff, 66-67, asst prof, 67-70, ASSOC PROF ELEC ENG, CTR MAT SCI & ENG, MASS INST TECHNOL, 70- *Mem:* AAAS; Am Phys Soc; Inst Elec & Electronics Engrs; sr mem Instrument Soc Am. *Res:* Studies of crystalline, amorphous and polymeric semiconductors; the charge-flow transistor sensing device; electronic instrumentation. *Mailing Add:* 790 Boylston St Apt 19-F Boston MA 02199-7918

SENTZ, JAMES CURTIS, b Littlestown, Pa, Sept 24, 27; m 59; c 3. PLANT SCIENCE, EXPERIMENTAL STATISTICS. *Educ:* Pa State Univ, BS, 49; NC State Univ, MS, 51, PhD(agron), 53. *Prof Exp:* Asst statistician, NC State Univ, 52-54; res agronomist, Agr Res Serv, USDA, 54-57; from asst prof to assoc prof agron & plant genetics, Univ Minn, St Paul, 57-75, training officer, Int Agr Progs, 71-75; res adv, Tech Asst Bur, USAID, Washington, DC, 75-77; assoc prof & training officer, Col Agr, Univ Minn, St Paul, 77-80; RETIRED. *Concurrent Pos:* Cornell Univ Grad Educ Prog vis prof, Col Agr, Univ Philippines, 66-68. *Mem:* Am Soc Agron; Crop Sci Soc Am; Genetics Soc Am; Biomet Soc. *Res:* Population and quantitative genetics; Zea Mays and plant breeding; experimental statistics; grain legumes. *Mailing Add:* Int Agr Prog Univ Minn St Paul MN 55108

SENUM, GUNNAR IVAR, b Kristiansand, Norway, Nov 10, 48; US citizen. PHYSICAL CHEMISTRY. *Educ:* Brooklyn Col, BS, 70; State Univ NY, PhD(chem), 75. *Prof Exp:* Asst chemist, 76-78, assoc chemist, 78-81, CHEMIST, BROOKHAVEN NAT LAB, 81 - *Mem:* Am Chem Soc; Am Phys Soc; AAAS. *Res:* Atmospheric chemistry, analytical instrumentation for tracing and tagging using perfluorocarbons; trace analyses of ambient atmospheric constituents; development of new applications of perfluorocarbon tracers. *Mailing Add:* Environ Chem Div Bldg 426 Brookhaven Nat Lab Upton NY 11973

SENUS, WALTER JOSEPH, b Rome, NY, Feb 5, 46; m 68; c 4. SATELLITE GEODESY, ELECTRONIC SURVEYING. *Educ:* Syracuse Univ, BS, 67, MS, 71; Univ Hawaii, PhD(geodesy/geophys). *Prof Exp:* Staff physicist, Rome Air Develop Ctr, 68-72; electron engr, Res & Develop, USCG, 72-78; chief scientist, Defense Mapping Agency, 78-84; TECH DIR, ROME AIR DEVELOP CTR, 84- *Concurrent Pos:* Prof math, Northern Va Community Col, 74-; lectr math, George Mason Univ, 76-81 & State Univ NY, Utica-Rome; assoc prof, George Washington Univ, 76-83. *Mem:* Am Geophys Union. *Res:* The application of satellite technology to furthering the fields of geodesy and geophysics; defining the physical fields of the earth and other bodies in our solar system. *Mailing Add:* IR Div Rome Air Develop Ctr Griffiss AFB NY 13440

SENYK, GEORGE, b Kharkov, Ukraine, May 6, 26; US citizen; m 55; c 2. IMMUNOLOGY, MICROBIOLOGY. *Educ:* Univ Frankfurt, 46-49; Univ San Francisco, BS, 66; Univ Calif, San Francisco, PhD(microbiol), 71. *Prof Exp:* Lectr, San Francisco State Col, 70; fel biochem, Univ Calif, San Francisco, 71, trainee, Dept Clin Path & Lab Med & lectr med technol-clin microbiol, 72-73, asst prof clin path & lab med & training coordr curriculum med technol-clin microbiol, 73-77, asst dir, Grad Prog Clin Lab Sci-Clin Microbiol, 74-77, res microbiologist, Dept Microbiol, 77-82; ASSOC SCIENTIST, CETUS IMMUNE CORP, PALO ALTO, CA, 82- *Mem:* AAAS; Am Soc Microbiol; Am Asn Immunologists. *Res:* Role of cellular immunity in resistance to infections; in vitro correlates of cellular immunity; monoclonal antibodies. *Mailing Add:* Dept Immunol Cetus Corp 1400 53rd St Emeryville CA 94608

SENZEL, ALAN JOSEPH, b Los Angeles, Calif, May 26, 45; m 69; c 2. TOXICOLOGY, ANALYTICAL CHEMISTRY. *Educ:* Calif State Univ, Long Beach, BS, 67; Univ Calif, Los Angeles, MS, 69, PhD(anal chem), 70. *Prof Exp:* Assoc ed, Anal Chem & Mem Exec Comt, Div Anal Chem, Am Chem Soc, 70-74; anal methods ed, Asn Off Anal Chemists, 74-78; head info off, Chem Indust Inst Toxicol, 78-79; sr chemist, Del Green Assoc, Inc, Foster City, Calif, 81-84; dep mgr, Environ Systs, Environ Resources Mgt Inc, Exton, PA, 88-89; PROJ SCIENTIST, RESIDUE CHEM DEPT, AGR DIV, CIBA-GEIGY CORP, GREENSBORO, NC, 89- *Concurrent Pos:* Consult environ chem & eng, 78- *Honors & Awards:* Commendable Serv Award, US Food & Drug Admin, 78. *Mem:* AAAS; Am Chem Soc; Instrument Soc Am; Asn Off Anal Chemists; Soc Environ Chem & Toxicol; Soc Tech Commun. *Res:* Toxicology of bulk, commodity chemicals; analytical methodology for agricultural products, foods, feeds, beverages, drugs, pesticides, cosmetics, color additives, and other commodities important in public health; chemical literature and journal publication; air pollution measurement; hazardous waste analysis, treatment, storage and disposal; risk assessment of hazardous waste sites; pesticide registration. *Mailing Add:* 7704 Audubon Dr Raleigh NC 27615-3403

SEO, EDDIE TATSU, b Los Angeles, Calif, July 12, 35; m 64; c 2. BATTERY ENGINEERING. *Educ:* Univ Calif, Los Angeles, BS, 59; Univ Calif, Riverside, PhD(chem), 64. *Prof Exp:* Res assoc chem, Univ Kans, 64-65; mem tech staff, TRW Systs Group, 65-77; sr res assoc, Gates Corp, 77-90; SR SCIENTIST & ENGR, HUGHES AIRCRAFT CO, 90- *Mem:* Am Chem Soc; Electrochem Soc; Royal Soc Chem. *Res:* Electrochemical energy conversion and storage. *Mailing Add:* 7276 S Highland Dr Littleton CO 80120

SEO, STANLEY TOSHIO, b Honolulu, Hawaii, Mar 5, 28. CHEMISTRY. *Educ:* Univ Hawaii, BS, 50, MS, 52. *Prof Exp:* Chemist, 55-64, SUPVRY RES CHEMIST, HAWAII FRUIT & VEG RES LAB, AGR RES SERV, USDA, 64- *Mem:* AAAS; Am Chem Soc; Entom Soc Am; Am Inst Chemists. *Res:* Commodity treatment by fumigation, heat, low temperatures and use of gamma radiation; chemical factors affecting the infestation of fruit by fruitflies; radiation biology by use of gammaradiation; applications to sterile-insect technology for eradication and control of fruit flies; chemical factors pertaining to radiation biology. *Mailing Add:* 5404 Halapepe St Honolulu HI 98821

SEON, BEN KUK, b Fukuoka-Ken, Japan, May 5, 36; m 66; c 3. IMMUNOLOGY, CANCER. *Educ:* Osaka Univ, MS, 63, PhD(biochem), 66. *Prof Exp:* Fel biochem, Osaka Univ, 66-67; from cancer res scientist to sr cancer res scientist, Roswell Park Cancer Inst, 67-75, assoc cancer res scientist, 75-78, cancer res scientist V, 78-85, CANCER RES SCIENTIST VI, ROSWELL PARK CANCER INST, 85- *Mem:* AAAS; Sigma Xi; Am Asn Immunologists; Am Asn Cancer Res. *Res:* Utilization of hybridoma technology for cancer; application of immunoconjugates for cancer therapy. *Mailing Add:* Molecular Immunol Dept Roswell Park Cancer Inst Buffalo NY 14263

SEPERICH, GEORGE JOSEPH, b Chicago, Ill, Mar 18, 44; m 68; c 1. FOOD SCIENCE, MEAT CHEMISTRY. *Educ:* Loyola Univ, BS, 67; Mich State Univ, MS, 72, PhD(food sci), 76. *Prof Exp:* Dir educ prog, Mich Dept Labor, 75-76; dir, Sch Agr & Environ Resources, 82-88, ASSOC PROF AGR, DIV AGR, ARIZ STATE UNIV, 76- *Mem:* Inst Food Technologists; AAAS; Am Meat Sci Asn; Am Soc Animal Sci. *Res:* Muscle food constituents; muscle turnover and metabolism; isolation and characterization of food and muscle food proteins and muscle cellular physiology and biochemistry related to exercise and pathology. *Mailing Add:* Div Agr Col-Eng/Appl Sci Tempe AZ 85281

SEPINWALL, JERRY, b Montreal, Que, Dec 14, 40; m 63; c 3. PSYCHOPHARMACOLOGY, PHYSIOLOGICAL PSYCHOLOGY. *Educ:* McGill Univ, BA, 61; Cornell Univ, MA, 63; Univ Pa, PhD(psychol), 66. *Prof Exp:* Sr scientist, 66-71, res group chief, 71-75, RES SECT HEAD, HOFFMANN-LA ROCHE, INC, 75- *Concurrent Pos:* Instr, Brooklyn Col, 67 & Montclair State Col, 70. *Mem:* Soc Neurosci; Am Soc Pharmacol & Exp Therapeut; Am Psychol Asn; Behav Pharmacol Soc; NY Acad Sci. *Res:* Pharmacology of antianxiety agents; physiology and neurochemistry of motivation and learning. *Mailing Add:* Dept Pharmacol Hoffmann-La Roche Inc 340 Kingsland St Nutley NJ 07110

SEPKOSKI, J(OSEPH) J(OHN), JR, b Presque Isle, Maine, July 26, 48. PALEONTOLOGY, EVOLUTIONARY BIOLOGY. *Educ:* Univ Notre Dame, BS, 70; Harvard Univ, PhD(geol), 77. *Prof Exp:* From instr to asst prof geol, Univ Rochester, 74-78; from asst prof to assoc prof , 78-85, PROF GEOL, UNIV CHICAGO, 86-; RES ASSOC, FIELD MUS NATURAL HIST, 80- *Concurrent Pos:* Ed, Paleobiol, 83-86. *Honors & Awards:* Charles Schuchert Award, Paleont Soc, 83. *Mem:* AAAS; Paleont Soc; Sigma Xi; Soc Study Evolution; Soc Syst Zoologists. *Res:* Evolutionary paleobiology and paleoecology, with particular emphasis on diversification and distribution of marine invertebrates, especially in the early Paleozoic; Cambrian and Vendian paleontology and stratigraphy; geostatistics; mass extinction and periodicity of extinction. *Mailing Add:* Dept Geophys Sci Univ Chicago Chicago IL 60637

SEPKOSKI, JOSEPH JOHN, b Cleveland, Ohio, July 30, 21; m 47; c 5. ORGANIC CHEMISTRY. *Educ:* John Carroll Univ, BS, 43; Univ Notre Dame, MS, 50; Rutgers Univ, MBA, 56. *Prof Exp:* Analytic chemist, Cleveland Graphite Bronze Co, 42-43; prod chemist, Schering Corp, 50-53; develop chemist, Celanese Corp Am, 53-55; sr chemist, Chicopee Mfg Corp, 55-57, res supvr, 57-62; group leader, Thatcher Mfg Co, 62-63; sr res chemist, Indust Chem Div, Allied Chem Corp, 63-67 & Plastics Div, 67-68, group leader, 68-74, tech serv supvr spec chem, Plastics Div, 74-81, tech coordr, Allied Fibers & Plastics Co, 81-82, supvr tech support, Chem Sect, 82-86; RETIRED. *Mem:* Fel Am Inst Chem; Am Chem Soc. *Res:* Thermosetting resins, reinforced and non-reinforced; urethanes; low molecular weight polyethylene. *Mailing Add:* 323 West Shore Trail Sparta NJ 07871

SEPMEYER, L(UDWIG) W(ILLIAM), b East St Louis, Ill, Nov 6, 10; m 36; c 1. ELECTRICAL ENGINEERING. *Educ:* Univ Calif, BS, 33. *Prof Exp:* Consult engr, Calif, 34-41; asst, Univ Calif, Los Angeles, 34-37 & 40-41; engr, Lansing Mfg Co, Calif, 39-40; systs engr, Elec Res Prods Div, Western Elec Co, 41-42; engr, Div War Res, Univ Calif, 42-45 & Calif Inst Technol, 45; elec engr, US Naval Ord Test Sta, 45-51; engr, Rand Corp, 51-56 & Eng Systs Develop Corp, 56-63; CONSULT ENGR, 63- *Concurrent Pos:* Consult, 47- & US Naval Ord Test Sta, 51-57. *Mem:* Fel Acoust Soc Am; fel Audio Eng Soc; Am Soc Testing & Mat; Inst Noise Control Eng USA; Inst Elec & Electronics Engrs. *Res:* Architectural acoustics; noise control; acoustical instrumentation; electrical sound recording and reproduction. *Mailing Add:* 1862 Comstock Ave Los Angeles CA 90025

SEPPALA, LYNN G, b Watertown, SDak, May 21, 46; m 80. OPTICAL DESIGN. *Educ:* SDak State Univ, BS, 68; Univ Rochester, PhD(optics), 74. *Prof Exp:* Optical designer, ITEK Corp, 74-76; OPTICAL ENGR, LAWRENCE LIVERMORE LAB, 76- *Honors & Awards:* R&D 100 Award, 90. *Mem:* Optical Soc Am; Int Soc Optical Eng. *Mailing Add:* Lawrence Livermore Nat Lab L-491 Univ Calif PO Box 5508 Livermore CA 94550

SEPPALA-HOLTZMAN, DAVID N, b New York, NY, Aug 31, 50; m 83; c 1. HOMOTOPY THEORY, COHOMOLOGY OPERATIONS. *Educ:* State Univ NY, Stony Brook, BSc, 72; Oxford Univ, UK, MSc, 73, DPhil(math), 79. *Prof Exp:* Asst prof math, Cath Univ Nijmegen, 75-80; ASSOC PROF & CHMN DEPT MATH, ST JOSEPH'S COL, 81- *Mem:* NY Acad Sci; Am Math Soc; Math Asn Am; Nat Coun Teachers Math. *Res:* Algebraic topology; systems of higher order cohomology operations and their applications. *Mailing Add:* 25-27 120th St College Point NY 11354

SEPPI, EDWARD JOSEPH, b Price, Utah, Dec 16, 30; m 53; c 3. PHYSICS, MATHEMATICS. *Educ:* Brigham Young Univ, BS, 52; Univ Idaho, MS, 56; Calif Inst Technol, PhD(physics), 62. *Prof Exp:* Jr engr, Geneva Steel Co, 51; lab instr, Brigham Young Univ, 51-52; physicist, Hanford Labs, Gen Elec Co, 52-58; res asst physics, Calif Inst Technol, 58-60, Gen Elec fel, 60-62, univ fel, 62; staff physicist, Inst Defense Anal, 62-64; head res area physics group, Stanford Linear Accelerator Ctr, 64-66, res area dept head, 66-68, head exp facil dept, 68-74; mgr med diag inst, 74-76, eng mgr, 76-77, div mgr, 77-78, tech dir, 78-80, SR SCIENTIST, VARIAN ASSOCS, 80 - *Concurrent Pos:* Consult, Inst Defense Anal, 64-72. *Mem:* Am Phys Soc. *Res:* High energy, nuclear and solid state physics; lasers; medical diagnostic instrumentation; research administration. *Mailing Add:* 320 Dedalera Dr Portola Valley CA 94028

SEPSY, CHARLES FRANK, b Rochester, NY, May 19, 24; m 45; c 2. MECHANICAL ENGINEERING. *Educ:* Univ Tenn, Knoxville, BME, 49; Univ Rochester, MSc, 51. *Prof Exp:* Asst mech eng, Univ Rochester, 50-51; res assoc, Res Found, 51-56, from instr to assoc prof, 56-67, PROF MECH ENG, OHIO STATE UNIV, 67- *Concurrent Pos:* Consult, Owens-Corning Fiberglas, 58-, Bender & Assocs, Consult Engrs, 60-65 & Off Civil Defense, 63-; educ consult, NAm Heating, Air Conditioning Wholesalers Asn, 64-; consult, CVI Corp, 65- *Honors & Awards:* Carrier Award, Am Soc Heating, Refrig & Air-Conditioning Engrs, 67. *Mem:* Am Soc Heating, Refrig & Air-Conditioning Engrs; Am Soc Eng Educ; Inst Environ Sci. *Res:* Internal control of the environment for man and machines with respect to regulating temperature, humidity, contaminants, noise and distribution; system simulation and energy requirements of building environmental control systems. *Mailing Add:* 3675 Rushmore Dr Columbus OH 43210

SEPUCHA, ROBERT CHARLES, b Salem, Mass, June 12, 43; m 66; c 3. CHEMICAL PHYSICS, PHYSICAL OPTICS. *Educ:* Mass Inst Technol, SB, 65, SM, 67; Univ Calif, San Diego, PhD(eng physics), 71. *Prof Exp:* Sr res scientist, Aerodyne Res Inc, 71-76; physicist, High Energy Laser Syst Proj Off, US Army, 76-78; prof mgr, Space Defense Technol Div, Directed Energy Off, Defense Advan Res Proj Agency, 79-80, dep dir, 80-84; VPRES, SPACE TECHNOL, W J SCHAFER ASSOC INC, 84- *Mem:* Am Phys Soc; Optical Soc Am; Sigma Xi. *Res:* Advanced high energy laser technology programs which may head to eventual space-based weapon systems; high-power chemical lasers; acquisition tracking and precision pointing; large-apecture beam control. *Mailing Add:* 9790 Kedge Ct Vienna VA 22180

SEQUEIRA, JOEL AUGUST LOUIS, b Bombay, India, July 13, 47; US citizen; m 73; c 1. PHARMACOKINETICS, PHYSICAL PHARMACY. *Educ:* Univ Bombay, BPharm, 69; Columbia Univ, MS, 72; State Univ NY, Buffalo, PhD(pharmaceut), 76. *Prof Exp:* Sr res scientist, Pharmaceut Res & Develop, Johnson & Johnson Res, 76-78; sect leader, 78-87, ASSOC DIR, PHARMACEUT RES & DEVELOP, SCHERING CORP, 87- *Mem:* Acad Pharmaceut Sci; Am Pharmaceut Asn; Soc Invest Dermat; Am Asn Pharm Scientists. *Res:* Topical and transdermal dosage form design; pharmacokinetics; percutaneous absorption; aerosol dosage form design; oral control release dosage form design. *Mailing Add:* Schering Corp 2000 Galloping Hill Rd Kenilworth NJ 07033

SEQUEIRA, LUIS, b San Jose, Costa Rica, Sept 1, 27; m 54; c 4. PLANT PATHOLOGY. *Educ:* Harvard Univ, BA, 49, MA, 50, PhD(biol), 52. *Prof Exp:* Parker traveling fel from Harvard Univ, Biol Inst, Brazil, 52-53; from asst plant pathologist to dir, Coto Res Sta, United Fruit Co, Costa Rica, 53-60; res assoc plant path, NC State Univ, 60-61; from assoc prof to prof plant path, Univ Wis-Madison, 61-78, prof plant path & bact, 78-86, prof plant asn, 86-87, J C WALKER PROF PLANT PATH, UNIV WIS-MADISON, 88- *Concurrent Pos:* NSF sr fel, Univ Reading, 70-71; ed-in-chief, Phytopath, 79-81. *Mem:* Nat Acad Sci; fel Am Phytopath Soc (pres, 85-86); Bot Soc Am; Mycol Soc Am; Am Soc Plant Physiol. *Res:* Soil microbiology; root diseases; plant growth regulators; physiology of parasitism. *Mailing Add:* Dept Plant Path Univ Wis Madison WI 53706

SEQUIN, CARLO HEINRICH, b Winterthur, Switz, Oct 30, 41; m 68; c 2. COMPUTER AIDED DESIGN, COMPUTER GRAPHICS. *Educ:* Univ Basel, Switz, dipl, 65, PhD(physics), 69. *Prof Exp:* Mem tech staff MOS integrated circuits, Bell Tel Labs, 70-76; vis lectr logic design microprocessors, 76-77, PROF COMPUT SCI, UNIV CALIF, BERKELEY, 77- *Concurrent Pos:* Vchmn, CS Div, 80-83; consult, Xerox Computervision, Siemens. *Mem:* Fel Inst Elec & Electronics Engrs; Asn Comput Mach; fel Swiss Acad Eng Sci. *Res:* Computer-aided design of VLSI circuits; computer graphics and geometric modeling; computer architecture and multiprocessors. *Mailing Add:* Dept Elec Eng & Comput Sci Univ Calif Evans Hall Berkeley CA 94720

SERAD, GEORGE A, b Philadelphia, Pa, Feb 26, 39; m 67; c 2. CHEMICAL ENGINEERING. *Educ:* Drexel Inst Technol, BS, 61; Univ Pa, MS, 62, PhD(chem eng), 64. *Prof Exp:* Lab asst chem res, Scott Paper Co, Pa, 58 & 59; engr aide, Texaco Inc, NJ, 60 & 61; res engr, Celanese Res Co, 64-70, res eng, 70-73, tech group leader, 73-75, tech mgr, 75-83, TECH MGR SPECIALITIES OPERS, CELANESE FIBERS CO, 85- *Mem:* Am Inst Chem Engrs; Am Chem Soc. *Res:* Transport phenomena; electrochemistry; catalysis; fibers and polymers processing; product development. *Mailing Add:* 3008 Cutchin Dr Charlotte NC 28210

SERAFETINIDES, EUSTACE A, b Athens, Greece, June 4, 30; m 65. PSYCHIATRY. *Educ:* Nat Univ Athens, MD, 53; Royal Col Physicians & Surgeons, dipl psychol med, 59; Univ London, PhD(psychol med), 64. *Prof Exp:* Clin asst-res assoc, Maudsley Hosp, Inst Psychiat, Univ London, 57-65; assoc prof psychiat, neurol & behav sci, Sch Med, Univ Okla, 65-71; vis assoc prof psychiat, 71-72, PROF, DEPT PSYCHIAT & MEM, BRAIN RES INST, UNIV CALIF, LOS ANGELES, 72-; ASSOC CHIEF OF STAFF RES, VET ADMIN HOSP BRENTWOOD, LOS ANGELES, 72- *Concurrent Pos:* NIMH res scientist development award, 67-72. *Res:* Electro-clinical and psychopharmacological investigations of brain-behavior relationships; epilepsy; cerebral dominance consciousness and schizophrenia. *Mailing Add:* Dept Psychiat Univ Calif Los Angeles CA 90024

SERAFIN, FRANK G, b Passaic, NJ, Oct 8, 35. SURFACE CHEMISTRY, MATERIALS SCIENCE. *Educ:* Wesleyan Univ, Middletown, Conn, BA, 57; Northeastern Univ, Boston, MS, 68 & 76. *Prof Exp:* Chemist, 60-68, group leader, 68-72, res assoc, 72-82, SR RES ASSOC, W R GRACE & CO, CAMBRIDGE, 82- *Mem:* Am Chem Soc; Soc Appl Spectros; Royal Soc Chem; Sigma Xi. *Res:* Materials analysis and characterization including organic and inorganic compounds, mixtures and polymers. *Mailing Add:* 31 Livingston Dr Peabody MA 01960

SERAFIN, ROBERT JOSEPH, b Chicago, Ill, Apr 22, 36; m 61; c 4. RADAR ENGINEERING, RADAR METEOROLOGY. *Educ:* Univ Notre Dame, BS, 58; Northwestern Univ, MS, 61; Ill Inst Technol, PhD(elec eng), 72. *Prof Exp:* Engr elec eng radar, Hazeltine Res Corp, 60-62; res engr radar signal process, IIT Res Inst, 60-69, sr engr, 69-73, res assoc radar meteorol, 70-73; mgr field observ facil radar meteorol instrumentation, 73-81, dir, Atmospheric Tech Div, 81-89, NAT CTR ATMOSPHERIC RES, 89- *Mem:* Am Meteorol Soc; Inst Elec & Electronics Engrs; Sigma Xi. *Res:* Doppler radar signal processing theory and implementation; random signal theory; atmospheric turbulence, severe storms, tornadoes and damaging winds. *Mailing Add:* Nat Ctr Atmospheric Res PO Box 3000 Boulder CO 80303

SERAFINI, ANGELA, b Sassoferrato, Italy, July 27, 13; nat. BACTERIOLOGY, CYTOLOGY. *Educ:* Wayne State Univ, BA, 50; Univ Mich, MS, 56. *Prof Exp:* Med technologist, Wayne County Gen Hosp, Eloise, Mich, 33-35, sr bacteriologist, mycologist, parasitologist & instr, 44-55; sr bacteriologist, supvr med technologists & instr med technol, Grace Hosp, Detroit, 35-44; from asst res microbiologist to assoc res microbiologist, Parke, Davis & Co, 56-70; instr med technol, Sch Med, Wayne State Univ, 70-72, res asst, 72-78; RETIRED. *Mem:* Am Soc Microbiol; Am Soc Tissue Cult. *Res:* Respiratory human viruses; tumor viruses of animals and man; immunology; cytology. *Mailing Add:* 1255 Woodbridge Dr St Clair Shores MI 48080-3304

SERAFY, D KEITH, b St Petersburg, Fla, June 3, 47; m 86; c 1. ECHINODERM BIOLOGY, MARINE BENTHIC INVERTEBRATES. *Educ:* Univ SFla, BA, 69; Univ Maine, MS, 71, PhD(zool), 73. *Prof Exp:* Assoc res scientist marine sci, NY Ocean Sci Lab, 73-78 & Va Inst Marine Sci, 75-78; PROF BIOL & MARINE SCI, SOUTHAMPTON COL, LONG ISLAND UNIV, 78- *Concurrent Pos:* Asst prof marine sci, Col William & Mary, 75-78. *Mem:* Am Soc Zoologists; Sigma Xi; Union Concerned Scientists. *Res:* Systematics and ecology of the phylum Echinodermata (sea urchins, sand dollars, heart urchin, brittle stars and starfish); environmental surveys of benthic marine invertebrates in the north Atlantic between New York and Virginia. *Mailing Add:* Dept Biol Southampton Col Southampton NY 11968

SERAPHIN, BERNHARD OTTO, b Berlin, Ger, Nov 4, 23; nat US; m 58; c 2. SOLID STATE PHYSICS. *Educ:* Friedrich Schiller Univ, dipl, 50; Humboldt Univ, PhD(physics), 51. *Prof Exp:* Sci asst, Inst Solid State Res, Ger, 50-52; physicist, Res Lab, Siemens-Schuckertwerke Co, Co, 52-56; chief semiconductor team, Physics Lab, Brown Bovery Co, Switz, 56-59; head, Semiconductor Br, Michelson Lab, US Dept Navy, 59-70; PROF OPTICAL SCI, OPTICAL SCI CTR, UNIV ARIZ, 70- *Concurrent Pos:* Liaison scientist, Off Naval Res, London, 65-67; guest prof, Tech Univ Denmark, 69-70, Univ VI Paris, Univ Strasbourg, Univ Nice, 76-77, Int Ctr Theoret Phys, Trieste, Italy, 82-83; sr Fulbright scholar, Univ Split, Yugoslavia, 82. *Mem:* Fel Am Phys Soc. *Res:* Modulation spectroscopy; electroreflectance; energy band structure and optical properties of solids; material science aspects of solar energy conversion. *Mailing Add:* Optical Sci Ctr Univ Ariz Tucson AZ 85721

SERAT, WILLIAM FELKNER, b Chicago, Ill, Nov 14, 29; wid; c 3. AGRICULTURAL ENVIRONMENTAL CHEMISTRY. *Educ:* Univ Colo, AB, 51; Iowa State Univ, MS, 53; Univ Calif, Los Angeles, PhD(biochem), 58. *Prof Exp:* Res asst biochem, US Atomic Energy proj, 56-58; tech rep, Rohm and Haas Co, 58-62; res chemist, Calif State Dept Health, 62-69, coord & prin investr, Community Studies in Pesticides, 69-79; PRES, HARTSHORN CO, 79- *Concurrent Pos:* Lectr biol & biochem, St Mary's Col, Calif, 64-72. *Mem:* Am Chem Soc; Sigma Xi. *Res:* Environmental and health aspects of pesticides; physico-chemical aspects of environment. *Mailing Add:* 12 Camelford Ct Moraga CA 94556

SERBER, ROBERT, b Philadelphia, Pa, Mar 14, 09; m 33, 79; c 2. PHYSICS. *Educ:* Lehigh Univ, BS, 30; Univ Wis, PhD(physics), 34. *Hon Degrees:* DSc, Lehigh Univ, 72. *Prof Exp:* Nat Res Coun fel physics, Univ Calif, 34-36, res assoc, 36-38; from asst prof to assoc prof, Univ Ill, 38-45; prof, Univ Calif, 45-51; prof, 51-77, chmn dept, 75-77, EMER PROF PHYSICS, COLUMBIA UNIV, 77- *Concurrent Pos:* Res assoc, Metall Lab, Univ Chicago, 42-43; sr scientist, Los Alamos Sci Lab, Univ Calif, 43-45; mem, Atomic Bomb Group, Mariana, 45, dir physics measurements, Atomic Bomb Mission to Japan, 45; consult, Douglas Aircraft Co, 46-51 & Brookhaven Nat Lab, 51-; mem, Solvay Conf, Brussels, 48; sci policy comt, Stanford Linear Accelerator Ctr; high energy physics comt, Argonne Nat Lab; vis comt physics, Lehigh Univ; Guggenheim fel, 57; mem adv comts, Fermi Nat Lab, 67-72 & Los Alamos Meson Physics Facil, 78-82; trustee, Assoc Univs, 74-81. *Honors & Awards:* J Robert Oppenheimer Mem Prize, 72. *Mem:* Nat Acad Sci; fel Am Phys Soc (pres, 71); Am Acad Arts & Sci; Explorers Club. *Res:* Atomic and molecular structure; cosmic rays; quantum field theory; nuclear physics; particle accelerators; meson theory. *Mailing Add:* 450 Riverside Dr New York NY 10027

SERBIA, GEORGE WILLIAM, b New York, NY, May 28, 28; m 58; c 4. ASEPTIC PROCESSING & LOW ACID FOODS. *Educ:* Polytech Inst Brooklyn, BChemE, 54; Pepperdine Univ, MBA, 78. *Prof Exp:* Engr, Gen Foods Cent Labs, Hoboken, NJ, 54-60, admin asst, Jell-o Div, White Plains, NY, 60-61, group leader, Jell-o Div Labs, Tarrytown, NY, 62-66, tech res mgr, Gen Foods Int, Sao Paulo, Brazil, 66-68, sr group leader, Gen Foods Res, Tarrytown, NY, 68-70; assoc dir eng res & develop, Hunt-Wesson Foods, Fullerton, Calif, 70-85, dir prod & process develop, Hunt-Wesson/Beatrice Grocery Res & Develop Ctr Group, 85-89; RETIRED. *Concurrent Pos:* Adj prof, Chapman Col, 89- *Mem:* Inst Food Technol; Am Oil Chemists Soc; Nat Food Processors Asn. *Res:* Management of new product and improved product research and development; edible oils research and development; commercial development of aseptic pudding in plastic packaging and aseptic tomato products; commercial development of quick-cooking rice from benchtop to plant. *Mailing Add:* 248 S Malena Dr Orange CA 92669

SERBIN, LISA ALEXANDRA, b New York, NY, Sept 18, 46; US & Can citizen; m 71; c 2. PSYCHOLOGY, DEVELOPMENTAL PSYCHOLOGY. *Educ:* Reed Col, BA, 68; State Univ NY, Stony Brook, PhD(psychol), 72. *Prof Exp:* Asst prof psychol, State Univ NY, Binghamton, 73-78; assoc prof, 78-84, dir, Centre Res Human Develop, 81-91, PROF PSYCHOL, CONCORDIA UNIV, 84- *Mem:* Fel Am Psychol Asn; fel Can Psychol Asn; Soc Res Child Develop; Int Soc Study Behav Develop. *Res:* Early sex role development and the development of gender concepts in infants and preschoolers; inkergenetictional transfer of high risk status, a longitudinal study of outcomes for highly aggressive and/or withdrawn children. *Mailing Add:* Centre Res Human Develop Concordia Univ 1455 de Maisonneure Blvd W Montreal PQ H3G 1M8 Can

SERCARZ, ELI, b Bronx, NY, Feb 14, 34; m 55; c 4. IMMUNOLOGY, CELL BIOLOGY. *Educ:* San Diego State Col, BA, 55; Harvard Univ, MA, 56, PhD(bact, immunol), 60. *Prof Exp:* From asst prof to assoc prof microbiol, 63-70, PROF BACT, UNIV CALIF, LOS ANGELES, 70- *Concurrent Pos:* Fel bact, Harvard Med Sch, 60-62; fel virol, Mass Inst Technol, 62-63; grants, NSF, 63-69, NIH, 67-81 & Am Heart Asn, 69-72; consult, Eye Res Lab, Cedars-Sinai Med Ctr, 66-69; Guggenheim fel, Nat Inst Med Res, London, 70-71. *Mem:* Am Soc Microbiol; Am Asn Immunol; Nat Inst Med Res London. *Res:* Cell cooperation; iridium gene control; tritium cell specificity; immune unresponsiveness; idrotyke regulation. *Mailing Add:* Microbiol 5304 Life Sci Univ Calif 405 Hilgard Ave Los Angeles CA 90024

SERDAREVICH, BOGDAN, b Vel Bukovica, Yugoslavia, Dec 3, 21; m 64; c 2. BIOCHEMISTRY. *Educ:* Univ Zagreb, MSc, 55; Swiss Fed Inst Technol, PhD(org chem & biochem), 61. *Prof Exp:* Res assoc peptide chem, Univ Pittsburgh, 61-63; sr res assoc org chem, Swiss Fed Inst Technol, 63-64; asst prof lipid chem, Univ Western Ont, 64-75; PRES, SERDARY RES LABS, INC, 72- *Concurrent Pos:* USPHS fel, 61-63; res grants, Swiss Nat Res Coun, 63-64 & Med Res Coun Can, 64- *Mem:* Sr mem Chem Inst Can; NY Acad Sci; Am Oil Chemists Soc; Am Chem Soc. *Res:* Research and synthesis of lipids; lipid metabolism in relation to atherosclerosis and aging process; cell membranes structure and function. *Mailing Add:* Serdary Res Labs Inc 1643 Kathryn Dr London ON N6G 2R7 Can

SERDENGECTI, SEDAT, b Izmit, Turkey, July 28, 27; m 58; c 2. CONTROL & SYSTEMS ENGINEERING. *Educ:* Syracuse Univ, BS, 51; Calif Inst Technol, MS, 52, PhD(mech eng), 55. *Prof Exp:* Fel, Calif Inst Technol, 55-57; res engr, Turkish Gen Staff, 57-58 & Chevron Res Corp, Calif, 58-61; asst prof physics, 61-63, asst prof control systs eng, 63-65, assoc prof control systs, 65-71, PROF ENG, HARVEY MUDD COL, 71- *Concurrent Pos:* Consult, Metaflo Res Corp, Calif, 59-61, Chevron Res Corp, 61-64 & Gen Motors Corp, Mich, 65- *Res:* Applied mathematics and mechanics; rock physics; stress analysis; optimalizing control systems; computer science. *Mailing Add:* Dept Eng Harvey Mudd Col 12th & Columbia Claremont CA 91711

SERDUKE, FRANKLIN JAMES DAVID, b Berkeley, Calif, May 23, 42; m 63; c 2. NUCLEAR PHYSICS. *Educ:* San Francisco State Univ, AB, 64; Univ Calif, Davis, MA, 66, PhD(physics), 70. *Prof Exp:* Asst physicist, Physics Div, Argonne Nat Lab, 73-76; PHYSICIST, LAWRENCE LIVERMORE NAT LAB, 76- *Mem:* Am Phys Soc; Am Geophys Union. *Res:* Theoretical nuclear physics; nuclear shell theory; few body problems; nuclear matter; pion-nucleus interactions; numerical modeling; Speakeasy computer language. *Mailing Add:* Lawrence Livermore Nat Lab PO Box 808 L84 Livermore CA 94550

SERENE, JOSEPH WILLIAM, b Indiana, Pa, Apr 4, 47; m 69. LOW TEMPERATURE PHYSICS, SUPERFLUID FERMI LIQUIDS. *Educ:* Dartmouth Col, AB, 69; Cornell Univ, PhD(physics), 74. *Prof Exp:* Fel, Stanford Univ, 74-75; guest prof physics, Nordic Inst Theoret Atomic Physics, 75-76; asst prof, State Univ NY, Stony Brook, 76-79; asst prof appl physics, 79-84, NSF res physicist, 84-87, NSR RES PHYSICIST, YALE UNIV, 87- *Mem:* Am Phys Soc. *Res:* Theoretical condensed matter physics; superfluid helium 3, superconductivity and superfluids in neutron stars. *Mailing Add:* Naval Res Lab 4555 Overlook Ave SW Code 4690 Washington DC 20375-5000

SERFLING, ROBERT JOSEPH, b Kalamazoo, Mich, Jan 23, 39; m 61; c 2. MATHEMATICAL STATISTICS, PROBABILITY. *Educ:* Ga Inst Technol, BS, 63; Univ NC, Chapel Hill, PhD(math statist), 67. *Prof Exp:* Res statistician, Res Triangle Inst, 67; from asst prof to prof statist, Fla State Univ, 67-79; chmn, 82-85, 86-88, PROF MATH SCI, JOHNS HOPKINS UNIV, 79- *Concurrent Pos:* Statistician & chmn, Assessment Admin Rev Comn, State of Fla, 73-77; vis statist adv, NSF, 77-78; vis scholar, Limburgs Univ, Centrum, Belgium, 82; vis prof, Albert-Ludwigs Univ, Freiburg, WGer, 86. *Honors & Awards:* Humboldt Prize, Alexander von Humboldt Found, W Germany, 84. *Mem:* Fel Am Statist Asn; fel Inst Math Statist; Am Math Soc; Math Asn Am; Oper Res Soc Am; fel Int Statist Inst. *Res:* Probability theory; statistical inference; stochastic processes; operations research. *Mailing Add:* Dept Math Sci Johns Hopkins Univ 3400 N Charles St Baltimore MD 21218

SERFOZO, RICHARD FRANK, b Detroit, Mich, Mar 29, 39; m 62; c 1. OPERATIONS RESEARCH. *Educ:* Wayne State Univ, BS, 61; Univ Wash, MA, 65; Northwestern Univ, PhD(appl math), 69. *Prof Exp:* Prof engr, Ford Motor Co, 62; oper res analyst, Boeing Co, 62-68; vis asst prof oper res, Cornell Univ, 72; ASST PROF OPER RES, SYRACUSE UNIV, 69-72, 73- *Mem:* Am Math Soc; Oper Res Soc Am; Inst Mgt Sci. *Res:* Applied probability; semi-stationary processes; thinning of point processes, control of queues. *Mailing Add:* Dept Ind & Syst Engr Ga Inst Tech 225 North Ave NW Atlanta GA 30332

SERGEANT, DAVID ERNEST, b Hangchow, China, Jan 17, 27; Can citizen; m 54; c 3. BIOLOGY, ECOLOGY. *Educ:* Cambridge Univ, BSc, 48, MSc, 49, PhD(zool), 53. *Prof Exp:* Biologist fisheries biol, Fisheries Res Bd, Can Biol Sta, St John's, Nfld, 51-55 & Arctic Unit, Montreal, 55-65; biologist, Can Fisheries & Oceans Dept, Arctic Biol Sta, 65-87; RETIRED. *Concurrent Pos:* Adj prof, Renewable Resources Dept, Macdonald Col, McGill Univ. *Mem:* Soc Marine Mammal. *Res:* Study of life history, population dynamics of sea mammals and advice towards the management of endangered species. *Mailing Add:* 325 Main Rd Hudson PQ J0P 1H0 Can

SERGOVICH, FREDERICK RAYMOND, b Toronto, Ont, Nov 1, 33; m 60; c 2. CYTOLOGY, CYTOGENETICS. *Educ:* McMaster Univ, BA, 60; Univ Western Ont, PhD(anat), 64. *Prof Exp:* DIR CYTOGENETICS LAB, CHILDREN'S PSYCHIAT RES INST, 64- *Concurrent Pos:* Lectr, Med Sch, Univ Western Ont, 64-66, asst prof, 66-72; grants, Med Res Coun Can, 64-71, Ont Ment Health Fedn, 66-69 & 71-72 & Ont Cancer Treatment & Res Fedn, 69-72; consult, Victoria Hosp, London, Ont, 65- *Mem:* Am Soc Human Genetics; Can Asn Anat; Can Soc Cell Biol. *Res:* Human cytogenetics; correlation of clinical expression with chromosomes in lesion in mental retardation syndromes; fluorescence analysis of chromosomes in primary malignancies. *Mailing Add:* 103-280 Queens Ave London ON N6K 1X3 Can

SERIANNI, ANTHONY STEPHAN, b Chestnut Hill, Pa, Nov 18, 53. NMR SPECTROSCOPY, AB INITIO MOLECULAR ORBITAL CALCULATIONS. *Educ:* Albright Col, BS, 75; Mich State Univ, PhD(biochem), 80. *Prof Exp:* Res assoc biochem, Cornell Univ, 80-82; asst prof, 82-88, ASSOC PROF CHEM, UNIV NOTRE DAME, 88- *Concurrent Pos:* Pres & chief exec off, Omicron Biochems Inc, 82-; chmn-elect, Div Carbohydrate Chem, Am Chem Soc, 90- *Honors & Awards:* Horace S Isbell Award, Am Chem Soc, 88. *Mem:* Am Soc Biochem & Molecular Biol; AAAS; Int Soc Magnetic Resonance; NY Acad Sci. *Res:* Developing improved chemical/enzymic methods to introduce stable isotopes into carbohydrates and their derivatives; meclaristic studies of fundamental reactions of sugars in solution, carbohydrate conformation and in vivo biological metabolism. *Mailing Add:* Dept Chem Univ Notre Dame Notre Dame IN 46556

SERIF, GEORGE SAMUEL, b Saskatoon, Sask, Apr 5, 28; m 48; c 3. BIOCHEMISTRY. *Educ:* McMaster Univ, BSc, 51, MSc, 53, PhD(biochem), 56. *Prof Exp:* Mem tech staff, DuPont of Can, 52-54; Hoffmann-La Roche fel, agr biochem, Univ Minn, 56-57; mem staff, Scripps Clin & Res Found, Calif, 57-58; from asst prof to assoc prof biochem, Sch Med, Univ SDak, 58-62; from asst prof to assoc prof, 62-67, prof biochem & molecular biol & chmn fac, 67-71, PROF BIOCHEM & CHMN DEPT, OHIO STATE UNIV, 71- *Mem:* Am Soc Biol Chem; Am Chem Soc; Soc Exp Biol & Med; NY Acad Sci. *Res:* Mechanism of hormone action; carbohydrate metabolism; metabolism of fluoro derivatives; antimetabolites. *Mailing Add:* Dept Biochem Col Biol Sci Ohio State Univ 484 W 12th Ave Columbus OH 43210

SERIFF, AARON JAY, b Eagle Pass, Tex, Apr 26, 24; m 51; c 2. GEOPHYSICS. *Educ:* Univ Tex, BS, 44; Calif Inst Technol, PhD(physics), 51. *Prof Exp:* Instr math & physics, Univ Tex, 44-46; asst physics, Calif Inst Technol, 50; SR RES ASSOC, SHELL DEVELOP CO, 51- *Concurrent Pos:* Ed, Geophys, 73-75; adj prof, Rice Univ, 82-; fac assoc, Hanszen Col, 82- *Mem:* Am Geophys Union; Seismol Soc Am; Europ Asn Explor Geophys; hon mem Soc Explor Geophys. *Res:* Seismology; cosmic rays; exploration geophysics. *Mailing Add:* Shell Develop Co PO Box 481 Houston TX 77001

SERKES, KENNETH DEAN, b St Louis, Mo, Aug 18, 26; m 74; c 3. SURGERY. *Educ:* Yale Univ, BS, 50; Washington Univ, MD, 51; Am Bd Surg, dipl, 57. *Prof Exp:* Assoc surg, Jewish Hosp St Louis, 56-62, asst dir, 62-67; asst to dir, Albert Einstein Med Ctr, Pa, 67-68; assoc clin res, Baxter/Travenol Labs, Inc, 68-70, assoc dir clin res, 70-72, med dir, Artificial Organs Div, 72-83, vpres, Med Reg Affairs, Omnis Surg Div, 83-85, sr med adv, Baxter Labs, 85-87, MED CONSULT, 87- *Concurrent Pos:* From instr to asst prof, Sch Med, Wash Univ, 56-57; assoc prof, Sch Med, Temple Univ, 67-68; asst prof, Northwestern Univ, 68-87. *Mem:* Fel Am Col Surg; Am Soc Artificial Internal Organs; Int Soc Peritoneal Dialysis; Int Soc Nephrology; Am Col Surgeons; Cent Surg Asn. *Res:* Shock; blood volume; membrane and bubble oxygenation; artificial kidneys; organ preservation; general surgery; development of artificial kidneys; organ preservation; development of artificial organs; peritoneal dialysis research and development. *Mailing Add:* 860 Ladera Lane Montecito CA 93108

SERLIN, IRVING, b New York, NY, Apr 3, 23; m 50; c 2. POLYMER CHEMISTRY. *Educ:* City Col, BS, 43; Columbia Univ, MA, 47, PhD(org chem), 50. *Prof Exp:* Res chemist, Am Diet Aids, Inc, 43-44; lab instr org chem, Columbia Univ, 47-49, res assoc biochem, 51-53; res chemist, Cent Res Labs, Donut Corp Am, 50-51; assoc scientist, Brookhaven Nat Lab, 53-56; res chemist, Shawinigan Resins Corp, 56-65; res specialist, Monsanto Co, 65-70, sr res specialist, 70-85; CONSULT, 85- *Concurrent Pos:* Mem, Int Physiol Cong, Belg, 56. *Mem:* Am Chem Soc; Sigma Xi. *Res:* Organic chemistry; synthesis of new resins for reprographic systems; synthesis of high temperature resins; pressure sensitive adhesive basic research; computerized statistical data analysis of experimental designs. *Mailing Add:* 94 Hadley Springfield MA 01118

SERLIN, OSCAR, b New York, NY, Aug 10, 17; m 51; c 2. SURGERY. *Educ:* Dalhousie Univ, MD & CM, 41; Am Bd Surg, dipl, 52. *Prof Exp:* Chief surg serv, Vet Admin Hosp, Lebanon, 56-57 & Philadelphia, 57-67; chief surg serv, Vet Admin Hosp, East Orange, 67-75, chief staff, 75-88, CHIEF STAFF, VET ADMIN MED CTR, 88-; PROF SURG, COL MED NJ, COL MED & DENT, NJ, 71- *Concurrent Pos:* Clin prof, Sch Med, Temple Univ, 58-67. *Mem:* Fel Am Col Surg. *Res:* Chemotherapy in treatment of cancer. *Mailing Add:* Vet Admin Med Ctr Lyons NJ 07939

SERMOLINS, MARIS ANDRIS, b Liepaja, Latvia, Sept 4, 44; US citizen; m 76; c 1. RESEARCH ADMINISTRATION. *Educ:* City Col CUNY, BChE, 66; Rutgers Univ, MS, 71; Pace Univ, MBA(management), 74. *Prof Exp:* Mgr, pro develop Permacez Div, Johnson & Johnson, 66-77; vpres, res & develop, Arno Tapes Inc, Scholl Corp, 77-80; tech dir, Am Cyanamid - Eng Mat, 80-84, Essex Spec Pro-Essex Chem, 84-87; DIR PROD TECHNOL, SIKA CORP, 87- *Mem:* Soc Advan Mat & Process Eng; Am Concrete Inst; Am Soc Testing Mat; Trans Res Bd; Nat Asn Corrosion Engrs. *Res:* Polyurethane sealants; epoxy & acrylic resins and concrete; cement based products and admixtures used in construction and industrial markets. *Mailing Add:* 31 Willowbrook Rd Freehold NJ 07728

SERNAS, VALENTINAS A, b Klaipeda, Lithuania, Nov 3, 38; US citizen; m 62; c 2. MECHANICAL & AEROSPACE ENGINEERING. *Educ:* Univ Toronto, BASc, 61, MASc, 64, PhD(mech eng), 67. *Prof Exp:* From asst prof to assoc prof, 66-79, PROF MECH ENG, RUTGERS UNIV, NEW BRUNSWICK, 79- *Mem:* Am Soc Mech Engrs; Sigma Xi; Optical Soc Am. *Res:* Heat transfer; boiling and condensation; natural connection; optical methods applied to heat transfer. *Mailing Add:* Col Eng Rutgers Univ PO Box 909 New Brunswick NJ 08854

SERNE, ROGER JEFFREY, b Lakewood, Ohio, Mar 20, 46; m 68; c 2. GEOCHEMISTRY, ANALYTICAL CHEMISTRY. *Educ:* Univ Wash, BS, 69. *Prof Exp:* From res scientist, to sr res scientist, 69-78, sect mgr, 78, STAFF SCIENTIST EARTH SCI, BATTELLE NORTHWEST LABS, 78- *Mem:* Am Soc Agron; Soil Sci Soc Am; Am Geophys Union; Am Soc Testing & Mat. *Res:* Migration of hazardous wastes in geologic environment; adsorption mechanisms of trace constituents onto geomedia. *Mailing Add:* Battelle Northwest Labs PO Box 999 Sigma Five K6-81 Richland WA 99352

SERNKA, THOMAS JOHN, b Cleveland, Ohio, July 26, 41; div; c 1. MEDICAL PHYSIOLOGY, MEDICAL BIOPHYSICS. *Educ:* Oberlin Col, BA, 63; Harvard Univ, MA, 66; Univ Iowa, PhD(physiol, biophys), 69. *Prof Exp:* NIH fel, Univ Calif, San Francisco, 69-71; from instr to asst prof physiol, Univ Tex Med Sch Houston, 71-74; asst prof, 74-75, assoc prof physiol & biophys, med sch, La State Univ, Shreveport, 75-76; ASSOC PROF PHYSIOL & BIOPHYS, WRIGHT STATE UNIV, 76- *Mem:* Am Physiol Soc; Biophys Soc; Soc Exp Biol & Med. *Res:* Gastric mucosal transport and metabolism. *Mailing Add:* Dept Physiol & Biophys Wright State Univ Dayton OH 45401-0927

SEROVY, GEORGE K(ASPAR), b Cedar Rapids, Iowa, Aug 29, 26; m 71, 90; c 5. MECHANICAL ENGINEERING. *Educ:* Iowa State Univ, BS, 48, MS, 50, PhD(mech eng), 58. *Prof Exp:* Asst mech eng, Iowa State Univ, 48-49; aeronaut res scientist, Nat Adv Comt Aeronaut, 49-53; from asst prof to assoc prof mech eng, Iowa State Univ, 53-60, asst dir eng res inst, 68-72, asst to dean, Col Eng, 72-86, PROF MECH ENG, IOWA STATE UNIV, 60- *Concurrent Pos:* Mem, Nat Acad Sci propulsion adv panel, USAF Systs Command, 69-76, US mem NATO AGARD Propulsion & Energetics panel, 87-; Anson Marston distinguished prof, Iowa State Univ; tech ed, J turbomachinery & J Eng for Gas Turbines; Prof invite, EPFL, Fed Inst Technol, Lausanne, Switz, 87. *Mem:* Fel Am Soc Mech Engrs; Am Inst Aeronaut & Astronaut. *Res:* Fluid mechanics of turbomachinery; aircraft propulsion; internal flow; heat transfer. *Mailing Add:* Dept Mech Eng Iowa State Univ Ames IA 50011

SERR, FREDERICK E, b Minneapolis, Minn, July 7, 52. COMPUTER SCIENCES, DATA NETWORKS ANALYSIS. *Educ:* Mich State Univ, BS, 74; Stanford Univ, PhD(physics), 78. *Prof Exp:* Res assoc, Mich State Univ, 78-80; res assoc physics, Mass Inst Technol, 80-83; NETWORK ANALYST,

BBN COMMUNS, CAMBRIDGE, MASS, 88- *Concurrent Pos:* Vis scientist, Niels Bohr, Denmark, 80-81. *Mem:* Am Phys Soc. *Res:* Wide area data communications; Hatchetts Systems technology. *Mailing Add:* BBN Commun 150 Cambridge Park Dr Cambridge MA 02140

SERRIN, JAMES B, b Evanston, Ill, Nov 1, 26; m 52; c 3. PARTIAL DIFFERENTIAL EQUATIONS, CONTINUUM THERMOMECHANICS. *Educ:* Univ Ind, PhD(math), 51. *Hon Degrees:* DSc, Univ Sussex, 72. *Prof Exp:* Instr math, Mass Inst Technol, 52-54; from asst prof to prof math, 54-69, head, Sch Math, 64-65, PROF MATH & AERO ENG, UNIV MINN, MINNEAPOLIS, 60-, REGENT'S PROF MATH, 69- *Concurrent Pos:* Vis prof, Univ Chicago, 64 & 75 & Johns Hopkins Univ, 66; vis fel, Univ Sussex, 67, 72 & 76, Univ Modena, 88 & 90, Ga Inst Tech, 90. *Honors & Awards:* Birkhoff Prize, Am Math Soc, 73. *Mem:* Nat Acad Sci; Math Asn Am; Soc Natural Philos; fel AAAS; Am Math Soc; Am Acad Arts & Sci. *Res:* Partial differential equations; theoretical fluid mechanics; thermodynamics. *Mailing Add:* Dept Math Univ Minn Minneapolis MN 55455

SERTH, ROBERT WILLIAM, b Rochester, NY, Aug 30, 41; m 64; c 1. CHEMICAL ENGINEERING. *Educ:* Univ Rochester, BS, 63; State Univ NY, Buffalo, PhD(chem eng), 69; Univ Ariz, MA, 70. *Prof Exp:* Asst prof chem eng, Univ PR, 71-74; sr res engr energy & environ, Monsanto Res Corp, 74-77, res specialist, 77-78; ASSOC PROF CHEM ENG, TEX A&I UNIV, 78- *Mem:* Am Inst Chem Engrs; Air Pollution Control Asn; AAAS. *Res:* Fluid mechanics; rheology; applied mathematics; pollution control. *Mailing Add:* Dept Chem Eng Tex A&I Univ Kingsville TX 78363

SERVADIO, GILDO JOSEPH, b Ridgefield, Conn, Jan 27, 29; m 55; c 3. FOOD SCIENCE. *Educ:* Tufts Univ, BS, 52; Univ Mass, MS, 55, PhD(food technol), 61. *Prof Exp:* Proj leader, Pillsbury Co, 55-58; instr packaging, Univ Mass, 58-61; sr scientist, Mead Johnson & Co, 61-63; res mgr, Beech-Nut Life Savers, Inc, NY, 63-65, assoc dir res, 65-67; dir res & develop labs, Heublein Inc, Hartford, 67-75, vpres res & develop labs, 75-86; PRES, SERVALL INC, 86-; DIR, BARON RESOURCES INC, 86- *Concurrent Pos:* Mem, Wash Lab Comt, Nat Canners Asn, 65-86, Res & Develop Assocs, US Army Natick Labs, 65 & Distilled Spirits Inst, 69-86; chmn tech comt, Vinegar Inst, 70-86; chmn comt food protection, Grocery Mfrs Asn, 70-86. *Mem:* Inst Food Technol; Am Asn Cereal Chemists. *Res:* Food and beverage chemistry; food colorimetry; baking, infant nutrition and nutritional products technology; chemistry and processing of fruits, vegetables and formulated foods; technology of wines, beers and distilled spirits; research administration. *Mailing Add:* Ten Sagamore Dr Simsbury CT 06070

SERVAIS, RONALD ALBERT, b La Crosse, Wis, Apr 6, 42; m 64; c 2. AEROTHERMOCHEMISTRY, PROCESS MODELING. *Educ:* St Louis Univ, BS, 63, MS, 66; Wash Univ, DSc(chem eng), 69. *Prof Exp:* Assoc engr, McDonnell Aircraft Corp, 63-64, engr, 64-66; asst prof mech eng, Univ Mo-Rolla, 68-72; res scientist, USAF Mat Lab, 73-74; chmn, 74-81, prof chem eng, 81-86, CHMN, CHEM ENGR DEPT, UNIV DAYTON, 87- *Concurrent Pos:* Monsanto Enviro-Chem Systs, Inc, 72- & Gen Elec, 82-, Square D, 84-; partner, Creative Eng Consult & Assocs, 73- *Mem:* Am Inst Aeronaut & Astronaut; Am Inst Chem Engrs; Nat Soc Prof Engrs; Am Soc Eng Educ. *Res:* Numerical solution of equations describing reacting; viscous, conducting, and diffusing flow fields, including reentry physics and air pollution modeling; aerodynamics of automobiles and trucks; performance of materials exposed to intense radiant heating; computer modeling of manufacturing processing. *Mailing Add:* Advan Mfg Ctr Univ Dayton Dayton OH 45468-0190

SERVAITES, JEROME CASIMER, b Dayton, Ohio, Sept 15, 44. PLANT PHYSIOLOGY, BIOCHEMISTRY. *Educ:* Univ Dayton, BS, 68, MS, 72; Univ Ill, Urbana-Champaign, PhD(biol), 76. *Prof Exp:* Teaching asst, Dept Biol, Univ Dayton, 70-72; res asst, Dept Agron, Univ Ill, 72-75, res assoc, 76; res assoc, Dept Hort, Univ Wis, 76-77, fel agron, 77-78; plant physiologist res, Sci & Educ Admin-Agr Res, USDA, 78-80; MEM FAC, BIOL DEPT, VA POLYTECH INST & STATE UNIV, 80- *Mem:* AAAS; Am Soc Agron; Am Soc Plant Physiol. *Res:* Photosynthetic carbon metabolism and photorespiration in crop plants; biochemical mechanisms and processes limiting yield of crop plants. *Mailing Add:* Univ Dayton 300 College Park Dayton OH 45469

SERVE, MUNSON PAUL, b Medina, NY, Nov 26, 39; m 69. ORGANIC CHEMISTRY, BIOCHEMISTRY. *Educ:* Univ Notre Dame, BS, 61, PhD(org chem), 64. *Prof Exp:* Am Cancer Soc fel, Univ Chicago, 64-65; asst prof, 65-75, PROF CHEM, WRIGHT STATE UNIV, 75- *Concurrent Pos:* Vis scientist, NIH, 75-76 & Toxic Hazards Br, USAF, 80- *Mem:* Am Chem Soc; Sigma Xi. *Res:* Photochemistry; photolysis of benzotriazoles; shift reagents in organic structure determinations; chemical toxicology of enviromental pollutants. *Mailing Add:* Dept Org Biochem Wright State Univ Glenn Hwy Dayton OH 45431

SERVI, I(TALO) S(OLOMON), b Gallarate, Italy, Oct 3, 22; m 50; c 3. METALLURGY. *Educ:* Milan Univ, Dr, 46; Mass Inst Technol, MS, 49, ScD, 51. *Prof Exp:* Asst metall, Mass Inst Technol, 48-51; res metallurgist, Metals Res Labs, Union Carbide Metals Co, 51-56, staff metallurgist, 56-58, group mgr, 58-59; dir res, Metals Div, Kelsey-Hayes Co, 59-60, tech dir, 60-61; vpres & tech dir, Spec Metals, Inc, 61-62; staff scientist, Ledgemont Lab, Kennecott Copper Co, 62-65, asst dir res, 66-76, dir prod develop, 76-78, dir new venture technol, Lexington Develop Ctr, 78-81; CONSULT ENGR, 81- *Concurrent Pos:* Instr, Adult Educ Ctr, Niagara Falls, NY, 53-54, Northeastern Univ, Boston, Mass, 83-89. *Mem:* Fel Am Soc Metals Int; Am Inst Mining Metall & Petrol Engrs; fel Inst Metals; Soc Advan Mat Process Eng; Am Powder Metall Inst. *Res:* Physical and mechanical behavior of metals and alloys; marketing research. *Mailing Add:* Three Angier Rd Lexington MA 02173-1608

SERVI, LESLIE D, b Buffalo, NY, July 27, 55; m 82; c 2. APPLIED SCIENCES. *Educ:* Brown Univ, BS & MS, 77; Harvard Univ, MS, 78, PhD(appl sci), 81. *Prof Exp:* Mem tech staff, Dept Teletraffic Theory & Appln, Bell Labs, 81-83; sr mem, 83-87, PRIN MEM, TECH STAFF, GTE LABS, 87- *Concurrent Pos:* Chmn, invited sessions, Oper Res Soc Am & Inst Mgt Sci, 86-87; secy, Telecommun Col, Oper Res Soc Am, 87-89, coun mem, Appl Probability Col, 90; vis scientist, Mass Inst Technol & Harvard Univ, 89-90. *Mem:* Oper Res Soc Am; Inst Elec & Electronics Engrs. *Res:* Applied probability; communications; optimal control. *Mailing Add:* GTE Labs Inc 40 Sylvan Rd Waltham MA 02254

SERVIS, KENNETH L, b Indianapolis, Ind, July 27, 39; m 68; c 3. PHYSICAL CHEMISTRY, ORGANIC CHEMISTRY. *Educ:* Purdue Univ, BS, 61; Calif Inst Technol, PhD(chem), 65. *Prof Exp:* Am-Yugoslavia Cultural Exchange fel, Rudjer Boskovic Inst, Univ Zagreb, Yugoslavia, 64-65; from asst prof to assoc prof, 65-84, PROF CHEM, UNIV SOUTHERN CALIF, 84-, DEAN ACAD REC & REGISTR, 89- *Concurrent Pos:* Alfred P Sloan fel, 69-71; Guggenheim fel, 73-74; Fulbright Hays Award, Univ Zagreb, 74. *Mem:* Am Chem Soc; Int Soc Magnetic; Croatian Chem Soc; Am Asn Univ Prof. *Res:* Structure and reactivity of organic compounds; molecular rearrangements; nuclear magnetic resonance spectroscopy. *Mailing Add:* 14 Gaucho Dr Rolling Hills Estates CA 90274

SERVIS, ROBERT EUGENE, b Lansing, Mich, June 28, 41; m 69. BIO-ORGANIC CHEMISTRY. *Educ:* Univ Mich, BS, 63; NY Univ, MS, 66, PhD(chem), 69, MD, 74. *Prof Exp:* Res scientist, NY Univ, 69-74; RES SCIENTIST, DEPT MED RES, BLODGETT MED CTR, 74- *Mem:* AAAS; Am Chem Soc; NY Acad Sci; Am Inst Chemists. *Res:* Natural products; alkaloids; bioorganic chemistry; gas chromatography; mass spectrometry of biological materials. *Mailing Add:* 1564 E 40th St White Cloud MI 49349-9758

SERVOS, KURT, b Anrath, Ger, Dec 20, 28; nat US. MINERALOGY. *Educ:* Rutgers Univ, BS, 52; Yale Univ, MS, 54. *Prof Exp:* Sr cur geol, NY State Mus, 56-57; asst prof mineral, Stanford Univ, 57-60; independent geologist, 60-67; prof geol, Menlo Col, 67-90; RETIRED. *Concurrent Pos:* Geologist, Stanford Res Inst, 65-67. *Mem:* Mineral Soc Am; Am Crystallog Asn; Mineral Asn Can; Geol Soc Am; Am Geophys Union. *Res:* Crystallography. *Mailing Add:* 1281 Mills St No Nine Menlo Park CA 94025-3207

SERWAY, RAYMOND A, b Frankfort, NY, June 26, 36; m 59; c 4. SOLID STATE PHYSICS. *Educ:* Syracuse Univ, BA, 59; Univ Colo, MS, 61; Ill Inst Technol, PhD(physics), 67. *Prof Exp:* Teaching asst physics, Univ Colo, 59-61; res physicist, Rome Air Develop Ctr, 61-63; assoc physicist, IIT Res Inst, 63-67; from asst prof to prof physics, Clarkson Univ, 67-80; prof physics & head dept, 80-86, PROF PHYSICS, JAMES MADISON UNIV, 86- *Concurrent Pos:* Vis guest scientist, Argon Nat Lab, 73; vis guest prof, IBM Zurich Res Lab, Switz, 74. *Mem:* AAAS; Am Phys Soc; Am Asn Physics Teachers; Sigma Xi. *Res:* Thin-film solar cells and semiconducting solar cell materials exhibiting the photovoltaic effect; magnetic resonance spectroscopy; electron spin resonance absorption spectroscopy in inorganic systems; paramagnetic defect centers in irradiated solids; crystalline field splitting; structure of inorganic molecules. *Mailing Add:* Dept Physics James Madison Univ Harrisonburg VA 22807

SERWER, PHILIP, b Brooklyn, NY, Feb 5, 42; m 66; c 3. VIROLOGY, MACROMOLECULE FRACTIONATION. *Educ:* Univ Rochester, AB, 63; NY Med Col, MS, 68; Harvard Univ, PhD(biophysics), 73. *Prof Exp:* Fel res, Calif Inst Technol, 72-75, sr res fel, 75-76; asst prof, 76-81, assoc prof, 81-85, PROF BIOCHEM, UNIV TEX HEALTH SCI CTR, SAN ANTONIO, 85- *Concurrent Pos:* Sci Adv Bd, FMC Prod; Coun Electrophoresis Soc; Human Genome Study Sect, NIH. *Mem:* Am Soc Virologists; Electrophoresis Soc; Biophys Soc; Am Soc Biol Chemists; Electron Micros Soc Am. *Res:* Structure and assembly of viruses; structure and sieving of gels; fractionation of DNA; mutagenesis; detection and isolation of viruses and viral precursors. *Mailing Add:* Dept Biochem Univ Tex Health Sci Ctr 7703 Floyd Curl Dr San Antonio TX 78284-7760

SESHADRI, KALKUNTE S, b Jagalur, India, May 11, 24; m 51; c 2. PHYSICAL CHEMISTRY. *Educ:* Mysore Univ, BSc, 45, MSc, 47; Ore State Univ, PhD(phys chem), 60. *Prof Exp:* Lectr chem, Cent Col, Bangalore, 45-56; lectr phys chem, Maharani's Col, Mysore, 60; fel Nat Res Coun Can, 60-62; res assoc, Ohio State Univ, 62-66; res chemist, Gulf Res & Develop Co, 66-68, sr res chemist, 68-83; consult, Univ Pittsburgh, 83-85; SYST ANALYST, EG&G, WASH ANALYTICAL SERV CTR, MORGANTOWN OPER, 85- *Mem:* Am Chem Soc. *Res:* Infrared molecular structure determination, intensity measurement and band shape analysis; surface and catalytic studies by magnetic resonance and optical spectroscopic techniques; study of coal, coal liquids, shale oils, coaltar and polymer by proton, carbon-13, silicon-29, fluorine-19 and nitrogen-15 nuclear magnetic resonance spectroscopy and fourier transform infrared spectroscopy; high performance liquid chromatography, gas chromatography, supercritical fluid chromatography and mass spectroscopy of synthetic and natural fuels. *Mailing Add:* 1244 Parkview Dr Morgantown WV 26505-3245

SESHADRI, SENGADU RANGASWAMY, b Madras City, India, Oct 25, 28; m 59. APPLIED PHYSICS, ELECTRICAL ENGINEERING. *Educ:* Madras Univ, MA, 51; Indian Inst Sci, Bangalore, dipl, 53; Harvard Univ, PhD(appl physics), 59. *Prof Exp:* Lectr electronics, Madras Inst Tech, 53-55; res fel appl physics, Harvard Univ, 59-60 & 61-63; prin sci officer, Electronics Res & Develop Estab, Bangalore, 60-61; consult, Appl Res Lab, Sylvania Elec Prod Inc, Mass, 62-63, sr eng specialist, 63-67; PROF ELEC ENG, UNIV WIS-MADISON, 67- *Concurrent Pos:* Vis prof, Univ Toronto, 65; NSF sr fel, Calif Inst Technol, 70-71; vis scientist, G A Technologies Inc, 82-83. *Mem:* Sr mem Inst Elec & Electronics Engrs; Am Phys Soc; Am Opt Soc; Electromagnetics Soc. *Res:* Surface waves; antennas in anisotropic media; plasma instabilities; nonlinear waves; microsonics; magnetic waves devices; optical waveguide theory; metal surfaces. *Mailing Add:* Dept Elec & Comput Eng Univ Wis 1425 Johnson Dr Madison WI 53706

SESHADRI, VANAMAMALAI, b Madras, India, Apr 25, 28; m 49; c 2. MATHEMATICAL STATISTICS. *Educ:* Univ Madras, BA, 50, MA, 57; Okla State Univ, PhD, 61. *Prof Exp:* Teacher, Pub Schs, Ceylon, 50-54; asst lectr math, Mandalay Univ, 54-57; asst, Okla State Univ, 57-60; asst prof, Southern Methodist Univ, 60-62, asst prof math statist, 62-64; assoc prof math, 65-70, PROF MATH, McGILL UNIV, 70- *Mem:* Am Statist Asn; Inst Math Statist. *Res:* Statistical inference; distribution theory, characterization of distributions; applications of characterization to goodness of fit. *Mailing Add:* Dept Math McGill Univ 853 Sherbrooke St W Montreal PQ H3A 2M5 Can

SESHU, LILLY HANNAH, mathematics; deceased, see previous edition for last biography

SESONSKE, ALEXANDER, b Gloversville, NY, June 20, 21; m 52; c 2. NUCLEAR ENGINEERING, CHEMICAL ENGINEERING. *Educ:* Rensselaer Polytech Inst, BChE, 42; Univ Rochester, MS, 47; Univ Del, PhD(chem eng), 50. *Prof Exp:* Chem engr, Chem Construct Corp, 42-43; res assoc, S A M Labs, Columbia Univ, 43; chem engr, Houdaille-Hershey Corp, 43-45; res engr, Columbia Chem Div, Pittsburgh Plate Glass Co, 45-46; mem staff, Los Alamos Sci Lab, Univ Calif, 50-54; assoc prof, 54-58, prof, 58-86, EMER PROF, NUCLEAR & CHEM ENG, PURDUE UNIV, 86- *Concurrent Pos:* Mem staff, Los Alamos Sci Lab, 58-59, consult, 60-62; mem rev comt, Argonne Nat Lab, 65-68 & 75-81; consult, Oak Ridge Nuclear Lab, 62-65, United Nuclear Corp, 63 & Elec Power Res Inst, 73; independent consult, 86- *Honors & Awards:* Compton Award, Am Nuclear Soc, 87. *Mem:* Fel Am Nuclear Soc; Am Soc Eng Educ; Am Inst Chem Engrs. *Res:* Nuclear reactor engineering; heat transfer; nuclear fuel cycle analysis. *Mailing Add:* 16408 Felice Dr San Diego CA 92128-2804

SESSA, DAVID JOSEPH, b Hackensack, NJ, Mar 3, 38; m 86; c 6. BIOCHEMISTRY, ORGANIC CHEMISTRY. *Educ:* Tufts Univ, BS, 59. *Prof Exp:* Assoc chemist, Agr Res Serv, USDA, 63-68, RES CHEMIST, NAT CTR AGR UTILIZATION RES, 68- *Concurrent Pos:* Chmn, protein & co-prod sect, Am Oil Chemists Soc, 84-85, 85-86. *Mem:* Inst Food Technologists. *Res:* Thermal denaturation of soybean proteins; food and feed uses of soybean proteins; protein interactions. *Mailing Add:* Nat Ctr Agr Utilization Res 1815 N University St Peoria IL 61604

SESSA, GRAZIA L, b Italy; US citizen; m; c 2. BIOCHEMISTRY. *Educ:* Univ Rome, PhD(biol), 58. *Prof Exp:* Fel, Brit Cancer Campaign, Oxford, 60-63; Fulbright scholar, Res Inst Pub Health, New York, 63-64; res fel, Sch Med, NY Univ, 64-65, asst res scientist, 65-67, asst prof microbiol, 68-70; asst prof pharmacol, 70-75, SPEC TRAINING DIR LAB, MT SINAI HOSP, 75-; MEM STAFF, MERCK INT DIV, 81- *Concurrent Pos:* Dir, Clin Lab Training, Mt Sinai Hosp, 70-79; mem staff clin chem, Monmouth Med Ctr, Long Branch, NY, 79-81. *Res:* Interaction of drugs with brain membranes; functions of the blood-brain barrier. *Mailing Add:* 125 Alexander Ave Nutley NJ 07110

SESSION, JOHN JOE, immunogenetics, cytogenetics; deceased, see previous edition for last biography

SESSIONS, JOHN TURNER, JR, b Atlanta, Ga, July 8, 22; m 50. MEDICINE. *Educ:* Emory Univ, BS, 43, MD, 45; Am Bd Gastroenterol, dipl, 68. *Prof Exp:* Intern, Kings County Hosp, Brooklyn, 45-46; intern & resident, Grady Mem Hosp, Atlanta, 48-50; asst med, Sch Med, Boston Univ, 50-52; from asst prof to assoc prof, 52-64, PROF MED, SCH MED, UNIV NC, CHAPEL HILL, 64- *Concurrent Pos:* Res assoc, Evans Mem Hosp, Boston, 50-52; mem training comt gastroenterol & nutrit, Nat Inst Arthritis, Metab & Digestive Dis, 68-; ed, Viewpoints Digestive Dis, 68- *Mem:* AAAS; Am Col Physicians; Am Gastroenterol Asn; Am Fedn Clin Res; Am Asn Study Liver Dis; Am Clin & Climat Asn. *Res:* Biochemical and physiologic aspects of gastroenterology and hepatology. *Mailing Add:* Dept Med Clin Sci Bldg 229H Rm 324 Univ NC Sch Med Chapel Hill NC 27514

SESSLE, BARRY JOHN, b Sydney, Australia, May 28, 41; m 67; c 2. NEUROPHYSIOLOGY, ORAL BIOLOGY. *Educ:* Univ Sydney, BDS, 63, BSc & MDS, 65; Univ New South Wales, PhD(physiol), 69. *Prof Exp:* Fel physiol, Med Sch, Univ New South Wales, 65-68; vis assoc orofacial neurophysiol, Nat Inst Dent Res, 68-70; assoc prof dent, Fac Med, Univ Toronto, 71-76, chmn, Div Biol Sci, 77-85, assoc dean res, 85-90, PROF, FAC DENT & PROF PHYSIOL, FAC MED, UNIV TORONTO, 77-, DEAN RES, 90- *Concurrent Pos:* Mem, dent sci grants comt, Can Med Res Coun, 79-82; prin investr, Can Med Res Coun grants, 71-, NIH grants, 74-; vis prof, Can Med Res Coun, Univ BC, 80, Univ Alta, 85, Univ Montreal, 86, Laval Univ, 86, McGill Univ, 86; Japan Soc Prom Sci Fel, 80; pres, Can Asn Dent Res, 77-78; secy, Can Pain Soc, 82-87; pres, Neurosci Group, Int Asn Dent Res, 85-86. *Honors & Awards:* Oral Sci Award, Int Asn Dent Res, 76. *Mem:* Int Asn Study Pain; Soc Neurosci; Int Asn Dent Res; Can Physiol Soc. *Res:* Neural basis of facial and oropharyngeal function; general sensory and motor neurophysiology; perceptual and behavioral correlates; author of five books and 125 journal papers. *Mailing Add:* Fac Dent Univ Toronto Toronto ON M5G 1G6 Can

SESSLER, ANDREW M, b Brooklyn, NY, Dec 11, 28; m 51; c 3. THEORETICAL PHYSICS. *Educ:* Harvard Univ, BA, 49; Columbia Univ, MS, 51, PhD(physics), 53. *Prof Exp:* Asst physics, Columbia Univ, 49-52; NSF fel, Cornell Univ, 53-54; from asst prof to assoc prof physics, Ohio State Univ, 54-61; theoret physicist, 61-73, dir, 73-80, THEORET PHYSICIST, LAWRENCE BERKELEY LAB, UNIV CALIF, 80- *Concurrent Pos:* Mem high energy physics adv panel, US AEC, 69-72; mem comt on high energy physics, Argonne Univ, 71-73; chmn sci policy bd, Stanford Synchrotron Radiation Lab, 76-78; chmn advan fuels adv comt, Elec Power Res Inst, 78-81; chmn comt on Isabelle, Brookhaven Nat Lab, 80-82; mem review comt, Plasma Physics Lab, Princeton Univ, 81-85; mem comt concerned sci coun, 84-87; Japan Soc Prom Sci Fel, 85; mem Study Directed Energy Weapons, Am Phys Soc, 85-86; mem Fed Am Sci Coun, 79-82, 85-88, vchmn, 87-88, chmn, 88-; chmn panel pub affairs, Am Phys Soc, 88, vchmn, 87; mem comt appln physics, Am Phys Soc, 91-93; mem bd dir, AUI, 91-94; mem sci policy bd, SSRL, 91-96. *Honors & Awards:* E O Lawrence Award, 70; US Particle Accelerator Sch Prize, 88; Leland J Haworth Distinguished Scientist, Brookhaven Nat Lab, 91-92. *Mem:* Nat Acad Sci; fel Am Phys Soc; Sigma Xi; fel AAAS; sr mem Inst Elec & Electronics Engrs. *Res:* Theory of particle accelerators; plasma physics. *Mailing Add:* Lawrence Berkeley Lab Univ Calif Berkeley CA 94720

SESSLER, GERHARD MARTIN, b Rosenfeld, Ger, Feb 15, 31; m 61; c 3. ELECTROACOUSTIC TRANSDUCERS, ELECTRETS. *Educ:* Univ Munich, Ger, Vordiplom, 53; Univ Goettingen, Ger, Diplom, 57, Dr rer nat(physics), 59. *Prof Exp:* Mem tech staff, Bell Labs, Murray Hill, NJ, 59-65, supvr, 66-75; PROF ELEC ENG, TECH UNIV DARMSTADT, 75- *Concurrent Pos:* Consult, Bell Labs, Murray Hill, NJ, 75-88, 91-92, IBM Res Lab, San Jose, Calif, 84; vis prof, Tengji Univ, Shanghai, China, 81-87, consult prof, 87. *Honors & Awards:* Callinan Award, Electrochem Soc, 70; Dakin Award, Inst Elec & Electronics Engrs Dielectrics Soc, 86. *Mem:* Fel Inst Elec & Electronics Engrs; fel Acoust Soc Am; Am Phys Soc. *Res:* Electroacoustic sensors, particularly the capacitive and piezoelectric type; materials research, particularly in organic and inorganic electrets and in piezoelectric polymers. *Mailing Add:* Fichtestrasse 30 B Darmstadt D-6100 Germany

SESSLER, JOHN CHARLES, b Newark, NJ, Feb 14, 32; m 54; c 7. PHYSICS, OPERATIONS RESEARCH. *Educ:* Rutgers Univ, BA, 56; Univ Southern Calif, MA, 61; Georgetown Univ, PhD, 70. *Prof Exp:* Res engr, Rocketdyne Div, NAm Aviation, Inc, 56-60; assoc scientist, Opers Model Eval Group Off, Tech Opers, Inc, 60-62; opers analyst, Ctr Naval Anal, Univ Rochester, 62-72; sr prog analyst, Drug Abuse Coun, 72-78; EVAL COODR, USPHS, 78- *Concurrent Pos:* Marine Corps opers anal group rep, Staff, Commanding Gen, Fleet Marine Force, Atlantic, 64-65, Pac, 67-68; opers eval group rep, Comdr Attack Carrier Striking Force, Seventh Fleet, 70-71. *Mem:* Am Phys Soc; Opers Res Soc Am. *Res:* Analysis of control systems for large liquid propellant rocket engines; simulation of air battles; operations analysis; low energy nuclear physics; social program evaluation. *Mailing Add:* 110 E Walnut St Alexandria VA 22301

SESSLER, JOHN GEORGE, b Syracuse, NY, Apr 8, 20; m 53; c 5. MECHANICAL ENGINEERING, METALLURGICAL ENGINEERING. *Educ:* Syracuse Univ, BS, 50, MS, 62. *Prof Exp:* Res engr, Syracuse Univ, 50-60, sr res engr, 60-72; consult, 72-76, TECH CONSULT, ENGR PROD LIABILITY & ACCIDENT RECONSTRUCT, 76- *Honors & Awards:* NASA Minor Award, 68. *Mem:* Am Soc Testing & Mat; Am Inst Mining, Metall & Petrol Engrs; Am Soc Metals. *Res:* Mechanical behavior of metals and alloys, including creep, fatigue, notch behavior and fracture. *Mailing Add:* 121 Jean Ave Syracuse NY 13210

SESTANJ, KAZIMIR, b Zagreb, Yugoslavia, Nov 11, 27; Can citizen; wid. ORGANIC CHEMISTRY, MEDICINAL CHEMISTRY. *Educ:* Univ Zagreb, dipl, 55, PhD(org chem), 61. *Prof Exp:* Res chemist, Pliva Pharmaceut & Chem Works, Yugoslavia, 54-63; fel org photochem, Harvard Univ, 63-64; fel synthetic carcinostatics, Children's Cancer Res Found, Boston, Mass, 64-65; RES CHEMIST, AYERST LABS DIV, AYERST, MCKENNA & HARRISON LTD, 65-; SECT HEAD, CHEM DEPT, WYETH-AYERST RES, PRINCETON, NJ, 87- *Concurrent Pos:* Sect head chem dept, Ayerst, McKenna & Harrison Ltd, 80-87. *Mem:* Am Chem Soc; NY Acad Sci; Chem Inst Can; AAAS. *Res:* Synthetic organic chemistry; peptide synthesis; drug metabolism; chemistry of odors; synthetic pharmaceuticals; inhibitors aldose reductase. *Mailing Add:* Two Harper Rd Monmouth Junction NJ 08852

SETCHELL, JOHN STANFORD, JR, b Brooklyn, NY, Dec 4, 42; m 65; c 1. ELECTRICAL CONTACTS, COLOR SYSTEMS. *Educ:* Rensselaer Polytech Inst, BS, 63; Univ Ill, MS, 69. *Prof Exp:* Physicist mfg technol, Eastman Kodak Co, 69-73, sr physicist, 74-78, proj physicist mfg technol, 79-82, supvr engr copy prod, 83-84, proj engr electronic photog, 85-89, proj mgr printer prod, 90-91, COLOR SYSTS ENGR, EASTMAN KODAK CO, 91- *Mem:* Am Soc Testing & Mat; Am Sci Affil. *Res:* Physics of electrical contacts; electrophotographic process control; thermal diffusion dye transfer printing. *Mailing Add:* 376 English Rd Rochester NY 14616

SETH, BRIJ B, b Udaipur, India, Sept 4, 38; US citizen; m 65; c 2. STRUCTURE-PROPERTY CORRELATION, MATERIALS BEHAVIOR. *Educ:* Univ Rajasthan, India, BSc, 58; Indian Inst Sci, BE, 60; Univ Toronto, Can, MASc, 62 PhD(metall eng), 64. *Prof Exp:* Group leader res & develop, Atlas Steel Co, 64-69; adv engr, Westinghouse Elec Corp, 69-70, mgr mat develop, 70-74, mat egg, 74-79, L P Disc, 79-83 & diag, 83-86, consult engr, 84-91, MGR MAT & COMPUTER SYSTS, WESTINGHOUSE ELEC CORP, 91- *Concurrent Pos:* Lectr, Univ Toronto, 61-64; prin investr, numerous projs, 65-90; secy comt res, Can Inst Mining & Metall, 68-69; chmn, Power Activ Comt, Am Soc Metals, 79-81. *Mem:* Am Iron & Steel Inst; Inst Metals; fel Am Soc Metals; Can Inst Mining & Metall. *Res:* New materials and processes to enhance the reliability of steam turbines and application of quantitative techniques to optimize the use of materials; fracture mechanics; failure investigations. *Mailing Add:* 1641 Indian Dance Ct Maitland FL 32751

SETH, KAMAL KISHORE, b Lucknow, India, Mar 10, 33; m 62; c 3. NUCLEAR & PARTICLE PHYSICS. *Educ:* Univ Lucknow, BSc, 51 & 53, MSc, 54; Univ Pittsburgh, PhD(physics), 57. *Prof Exp:* Lectr, Univ Pittsburgh, 54-56; res assoc, Brookhaven Nat Lab, 56-57; res assoc, Duke Univ, 57-61; from asst prof to assoc prof, 61-74, PROF PHYSICS, NORTHWESTERN UNIV, EVANSTON, 74- *Concurrent Pos:* Vis scientist, Saclay Nuclear Res Ctr, France, 67; vis prof, Univ Tokyo, 71; consult, Oak Ridge Assocs Univs, Lewis Res Lab, NASA & Oak Ridge Nat Lab. *Mem:* Fel Am Phys Soc. *Res:* Nuclear structure, low and medium energy nuclear spectroscopy; neutron cross sections and charged particle induced reactions; pion spectroscopy; pion spectroscopy, antiproton induced reactions. *Mailing Add:* Dept Physics & Astron Northwestern Univ Evanston IL 60201

SETH, RAJINDER SINGH, b Lahore, India, June 11, 37; Can citizen; m 63; c 2. PULP & PAPER TECHNOLOGY. *Educ:* Panjab Univ, India, BSc, 57, MSc, 58; Univ Alberta, Can, PhD(physics), 69. *Prof Exp:* Lectr, DAV Col, India, 58-60, Govt Col, India, 60-64; sr res physicist, Consolidated Bathurst Inc, Can, 69-71; res scientist, 71-77, sr scientist & head, Fibre & Paper Physics Sect, Pointe Claire Lab, 77-86, HEAD, FIBRE & PROD QUAL SECT, PULP & PAPER RES INST CAN, VANCOUVER LAB, 87- *Honors & Awards:* Weldon Medal, Can Paper Asn, 84. *Mem:* Mat Res Soc; Tech Asn Pulp & Paper Indust; Can Pulp & Paper Asn; Am Phys Soc. *Res:* Structure and physical properties of wood pulp fibres and paper. *Mailing Add:* Pulp & Paper Res Inst Can 3800 Westbrook Mall Vancouver BC V6S 2L9 Can

SETH, SHARAD CHANDRA, b Madhya Pradesh, India, Nov 1, 42. COMPUTER SCIENCE, ELECTRICAL ENGINEERING. *Educ:* Univ Jabalpur, BE, 64; Indian Inst Technol, Kanpur, MTech, 66; Univ Ill, Urbana, PhD(elec eng), 70. *Prof Exp:* From asst prof to assoc prof, 72-83, PROF COMPUT SCI, UNIV NEBR, LINCOLN, 83- *Concurrent Pos:* Vis prof IIT Kanpur, India, 74-75, 82-83; consult, Bell Labs, GTE, RPI, NIH & US West. *Mem:* Asn Comput Mach; Inst Elec & Electronics Engrs. *Res:* Design and maintenance of reliable digital systems; document analysis of optically scanned images. *Mailing Add:* Dept Comput Sci Univ Nebr Lincoln NE 68588-0115

SETHARES, GEORGE C, b Hyannis, Mass, Oct 16, 30; m 52; c 4. MATHEMATICS. *Educ:* Boston Univ, BMus, 53; Univ Mass, MA, 59; Harvard Univ, PhD(math), 67. *Prof Exp:* Instr math, Boston Univ, 59-64; res mathematician, Air Force Cambridge Res Labs, 64-73; assoc prof, 73-80, PROF MATH, BRIDGEWATER STATE COL, 80- *Concurrent Pos:* Lectr, Northeastern Univ, 67-68. *Res:* Analytic function theory; Teichmüller mappings; network theory; automata theory; computer science. *Mailing Add:* Dept Math & Comput Sci Bridgewater State Col Junction Rt 18-28-104 Bridgewater MA 02324

SETHARES, JAMES C(OSTAS), b Hyannis, Mass, Dec 13, 28; m 73; c 3. SOLID STATE PHYSICS, ELECTRICAL ENGINEERING. *Educ:* Univ Mass, BSEE, 59; Mass Inst Technol, SMEE, 62. *Prof Exp:* Teaching asst elec eng, Mass Inst Technol, 59-62; RES PHYSICIST, MICROWAVE PHYSICS LAB, AIR FORCE CAMBRIDGE RES LABS, 62- *Concurrent Pos:* Lectr, Boston Univ, 60-63 & Lowell Technol Inst, 63-69; consult, NDE Asn Inc, 78- *Mem:* Inst Elec & Electronics Engrs. *Res:* Microwave magnetics research; liquid crystal displays for microwaves, millimeter waves and infrared. *Mailing Add:* NDE Asn Inc 131 Bedford St Burlington MA 01803

SETHER, LOWELL ALBERT, b Iola, Wis, Aug 5, 31; m 63; c 3. ANATOMY. *Educ:* Concordia Col, Minn, BA, 60; Univ NDak, MS, 62, PhD(anat), 64. *Prof Exp:* Asst prof, 64-74, ASSOC PROF ANAT, MED COL WIS, 74- *Concurrent Pos:* Consult, 25th ed, Dorlands Med Dictionary. *Res:* Cross sectional anatomy and imaging. *Mailing Add:* Dept Anat Med Col Wis 8701 Watertown Plank Rd Milwaukee WI 53226

SETHI, DHANWANT S, b Rawalpindi, WPakistan, Dec 13, 37; m 66; c 2. PHYSICAL CHEMISTRY. *Educ:* Delhi Col, BSc, 56; Hindu Col, MSc, 58; NY Univ, PhD(chem), 67. *Prof Exp:* Jr sci officer, AEC, India, 58-59; tech asst chem, Dir Gen Health Servs, India, 59-62; lectr, NY Univ, 66-67; NSF vis scientist, Nat Ctr Atmospheric Res, Colo, 68-69; asst res geophysicist, Inst Geophys, Univ Calif, Los Angeles, 69; PROF CHEM, UNIV BRIDGEPORT, 69-, CHMN DEPT, 80- *Concurrent Pos:* Res collab, Brookhaven Nat Lab, 72- *Mem:* AAAS; Am Chem Soc; Air Pollution Control Asn. *Res:* Gas phase kinetics, photochemistry, flash photolysis and kinetic absorption spectroscopy. *Mailing Add:* Dept Chem Univ Bridgeport 380 University Ave Bridgeport CT 06602

SETHI, ISHWAR KRISHAN, b India, Jan 30, 48; m 75; c 2. COMPUTER VISION, PATTERN RECOGNITION. *Educ:* Indian Inst Technol, Kharagpur, BTech, 69, MTech, 71, PhD(electronics), 78. *Prof Exp:* Lectr electronics, Indian Inst Technol, Kharagpur, 71-78, asst prof, 79-82; asst prof, 82-85, ASSOC PROF COMPUT SCI, WAYNE STATE UNIV, 85- *Concurrent Pos:* Consult, Comput Maintainance Corp India, Ltd, 82-, UN Develop Prog, 88 & Unisys, 90-; prin investr, NSF res grant, 85-86; vis prof, Indian Inst Technol, Delhi, India, 88; assoc ed, Pattern Recognition J, 90- *Mem:* Asn Comput Mach; Inst Elec & Electronics Engrs; Int Soc Neural Networks. *Res:* Dynamic scene analysis and object recognition; image processing; artificial neural networks; mobile robots. *Mailing Add:* Dept Comput Sci Wayne State Univ Detroit MI 48202

SETHI, JITENDER K, b Lahore, India, Oct 16, 39; US citizen; m 64; c 2. PHARMACY, IMMUNOLOGY. *Educ:* Punjab Univ, BS, 60, MS, 62; Univ Iowa, PhD(pharm), 72. *Prof Exp:* Prof serv rep pharmaceut, Pfizer India Ltd, 62-67; res & teaching asst pharm, Univ Iowa, 67-72; fel tumor immunol, Sloan Kettering Inst, NY, 72-74, res assoc immunodiagnosis, 74-81; sr scientist, 81-83, SR SCIENTIST, MED AFFAIRS DEPT, PARKE-DAVIS DIV, WARNER-LAMBERT CO, NJ, 83- *Concurrent Pos:* Instr, Cornell Univ Grad Sch Med Sci, 75-81. *Mem:* Am Asn Cancer Res; Tissue Cult Asn; Am Asn Immunolgists. *Res:* Research related to tumor antigens, particularly human sarcomas to determine and to clinically evaluate if these could be of immunodiagnostic value; role and specificity of complement in complement mediated antigen antibody reactions. *Mailing Add:* 76 Chilton St Bernardsville NJ 07924

SETHI, V SAGAR, anti-cancer drugs, clinical pharmacology, for more information see previous edition

SETHIAN, JOHN DASHO, b Washington, DC, Mar 20, 50; m 77. PLASMA PHYSICS. *Educ:* Princeton Univ, AB, 72; Cornell Univ, MS, 74, PhD(appl physics), 76. *Prof Exp:* Res assoc plasma physics, Univ Md, 76-77; RES PHYSICIST PLASMA PHYSICS, US NAVAL RES LAB, 77- *Mem:* Am Phys Soc. *Res:* Intense relativistic electron beams; electron beam induced CTR magnetic confinement systems, plasma heating and collective ion acceleration; generation and propagation of intense beams. *Mailing Add:* US Naval Res Lab Code 4762 Washington DC 20375

SETHNA, PATARASP R(USTOMJI), b Bombay, India, May 26, 23; nat US; m 54; c 4. MECHANICS, MATHEMATICS. *Educ:* Univ Bombay, BE, 45 & 46; Univ Mich, MSE, 48, PhD(eng mech), 53. *Prof Exp:* Instr, Univ Mich, 52-53; sr engr, Res Lab Div, Bendix Aviation Corp, 53-56; from asst prof to assoc prof, 56-63, PROF AERONAUT & ENG MECH, UNIV MINN, MINNEAPOLIS, 63-, HEAD DEPT, 66- *Concurrent Pos:* Vis prof, Brown Univ, 65-66; vis sr scientist, Univ Calif, Berkeley, 70; vis prof, Univ Warwick, 72; vis scholar, Stanford Univ, 77; comt mem, Nat Res Coun Comt Adv to Air Army Res Off, 78- *Mem:* Am Soc Mech Engrs; Am Inst Aeronaut & Astronaut. *Res:* Dynamical systems; nonlinear oscillation theory; applied mechanics. *Mailing Add:* Dept Aerospace Eng Univ Minn 107 Akerman Minneapolis MN 55455

SETHURAMAN, JAYARAM, b Hubli, India, Oct 3, 37; m 65; c 2. MATHEMATICAL STATISTICS, PROBABILITY. *Educ:* Madras Univ, MA, 58; Indian Statist Inst, Calcutta, PhD(statist), 62. *Prof Exp:* Lectr statist, Indian Statist Inst, 61-62; assoc prof, 65-68, PROF STATIST, FLA STATE UNIV, 68- *Concurrent Pos:* Fels, Univ NC, Chapel Hill, 62-63, Mich State Univ, 63-64 & Stanford Univ, 64-65; mem, Indian Statist Inst, 67-; US Army Res Off res grant, 72-82; vis prof statist, Univ Mich, 74-75; vis prof & actg head, Indian Statist Inst, Bangalore, 79-80; chmn, dept statist, Fla State Univ, 86-89. *Mem:* Fel Inst Math Statist; fel Am Statist Asn; Int Statist Inst; Sigma Xi. *Res:* Probability; stochastic processes. *Mailing Add:* Dept Statist Fla State Univ Tallahassee FL 32306-3303

SETHURAMAN, S, b Madras, India, Dec 16, 39; m 66; c 2. METEOROLOGY. *Educ:* Univ Roorkee, ME, 69; Colo State Univ, PhD(fluid mech), 72. *Prof Exp:* Lectr civil eng, Bakthavatsalam Polytech, Madras, 62-66; METEOROLOGIST, BROOKHAVEN NAT LAB, 72-, PROF METEROL, 82- *Mem:* Am Meteorol Soc; Am Geophys Union. *Res:* Geophysics, air-sea interaction, atmospheric turbulence, atmospheric diffusion, planetary boundary layer. *Mailing Add:* Dept Marine Earth & Atmospheric Sci NC State Univ Raleigh NC 27650

SETHY, VIMALA HIRALAL, NEUROPSYCHOPHARMACY. *Educ:* Univ Bombay, India, MD, 63, PhD(pharmacol), 67. *Prof Exp:* SR RES SCIENTIST, UPJOHN CO, 76- *Res:* Mechanism of action of antidepressants; physiology of the central cholinergic system. *Mailing Add:* Cent Nervous Syst Unit Upjohn Co Kalamazoo MI 49001

SETIAN, LEO, b Providence, RI, July 22, 30; m 57; c 5. ELECTRICAL ENGINEERING. *Educ:* Brown Univ, AB, 55; Univ RI, MS, 66; Mont State Univ, PhD(elec eng), 71. *Prof Exp:* Electronic engr, Underwater Sound Lab, 57-63 & Electronics Res Lab, 66-68; from asst prof to assoc prof, 70-79, PROF ELEC ENG, JOHN BROWN UNIV, 79- *Concurrent Pos:* Consult, Underwater Systems Ctr, New London, Conn. *Mem:* Am Sci Affiliation; Am Soc Eng Educ; Inst Elec & Electronics Engrs. *Res:* Moisture measurement in living foliage using an open-wire transmission line. *Mailing Add:* Dept Elec Eng John Brown Univ Siloam Springs AR 72761

SETLER, PAULETTE ELIZABETH, b Pittsburgh, Pa, Jan 1, 38. PHARMACOLOGY, PHYSIOLOGY. *Educ:* Seton Hill Col, BA, 59; Univ Pa, PhD(physiol), 70. *Prof Exp:* Res assoc pharmacol, McNeil Labs, 59-62; pharmacologist, Smith Kline & French Labs, 62-66; instr physiol, Sch Dent Med, Univ Pa, 68; supvr, Churchill Col, Cambridge Univ, 71-72; Smith Kline & French Labs, 72-81; McNeil Pharmaceut, 81-87; EXEC DIR, BIOL RES, JANSSEN RES FOUND, 87- *Concurrent Pos:* Res assoc, Physiol Lab, Cambridge Univ, 70-72. *Mem:* Assoc Am Physiol Soc; Soc Neurosci; AAAS. *Res:* Pharmacology of biogenic amines as studied behaviorally and biochemically and the interaction of amine neurotransmitters in regulation of behavior; the pharmacology of psychotropic drugs; control of ingestive behavior and body fluid balance. *Mailing Add:* Athena Neurosci 800F Gateway Blvd South San Francisco CA 94080

SETLIFF, EDSON CARMACK, b Indianola, Miss, Nov 3, 41; m 69; c 2. FOREST PATHOLOGY, FOREST MYCOLOGY. *Educ:* NC State Univ, BS, 63; Yale Univ, MF, 64; State Col Environ Sci & Forestry, Syracuse Univ, PhD(forest mycol), 70. *Prof Exp:* Res assoc, dept plant path, Univ Wis, 70-73; res scientist forest path, Cary Arboretum, NY Bot Garden, 73-77; res assoc, dept environ & forest biol, Col Environ Sci & Forestry, State Univ NY, Syracuse, 77-80; mem staff, Forintek Can Corp, Vancouver, BC, 80-85; SCH FORESTRY, LAKEHEAD UNIV, ONT, 85- *Concurrent Pos:* Adj assoc prof, Vassar Col, 75- & Univ BC. *Mem:* Am Phytopath Soc; Mycol Soc Am; Sigma Xi; Can Inst Forestry; Tech Asn Pulp & Paper Indust. *Res:* Taxonomy of tropical Polyporales; physiology and cultural morphology of wood decay fungi; mushroom culture; etiology of tree root diseases; tree disease diagnosis; cytology and fine structure of Polyporales; wood products pathology. *Mailing Add:* Sch Forestry Lakehead Univ Thunder Bay ON P7B 5E1 Can

SETLIFF, FRANK LAMAR, b Lake Charles, La, Sept 21, 38; m 62; c 2. ORGANIC CHEMISTRY. *Educ:* McNeese State Univ, BS, 60; Tulane Univ, MS, 62, PhD(org chem), 66. *Prof Exp:* Teaching asst org chem, Tulane Univ, 60-62, res asst, 64-66; asst prof, Little Rock Univ, 66-69; assoc prof, 69-74, chmn dept, 73-75, PROF CHEM, UNIV ARK, LITTLE ROCK, 74- *Mem:* Am Chem Soc; NY Acad Sci; Sigma Xi. *Res:* Small ring compounds; non-benzenoid aromatics; heterocyclic compounds; fluorination reactions. *Mailing Add:* Dept Chem Univ Ark 33rd University Ave Little Rock AR 72204

SETLOW, JANE KELLOCK, b New York, NY, Dec 17, 19; m 42; c 4. BIOPHYSICS. *Educ:* Swarthmore Col, BA, 40; Yale Univ, PhD, 60. *Prof Exp:* Asst biophys, Dept Radiol, Sch Med, Yale Univ, 59-60; biologist, Biol Div, Oak Ridge Nat Lab, 60-74; BIOLOGIST, BIOL DEPT, BROOKHAVEN NAT LAB, 74- *Concurrent Pos:* Mem, NIH Recombinant DNA Molecule Prog Adv Comt, 74-78, chmn, 78-80; ed J Genetic Eng, vols 1-13, 79-91. *Mem:* Biophys Soc (pres, 77). *Res:* Molecular biology; ultraviolet action spectra; photoreactivation; cellular repair mechanisms; bacterial recombination; mutagenesis. *Mailing Add:* Biol Dept Brookhaven Nat Lab Upton NY 11973

SETLOW, PETER, b New Haven, Conn, June 1, 44; m 65; c 2. BIOCHEMISTRY. *Educ:* Swarthmore Col, BA, 64; Brandeis Univ, PhD(biochem), 69. *Prof Exp:* NSF fel, Stanford Univ, 68-71; from asst prof to assoc prof 71-80, PROF BIOCHEM, UNIV CONN, 80- *Mem:* Am Soc Biol Chemists; Am Soc Microbiol. *Res:* Biochemical regulation of differentiation; bacterial sporulation and germination. *Mailing Add:* Dept Biochem Univ Conn Health Ctr Farmington CT 06030

SETLOW, RICHARD BURTON, b New York, NY, Jan 19, 21; m 42, 89; c 4. CANCER RESEARCH. *Educ:* Swarthmore Col, AB, 41; Yale Univ, PhD(physics), 47. *Hon Degrees:* DSc, York Univ, 85. *Prof Exp:* Asst physics, Med Sch, Yale Univ, 41-42, from asst instr to assoc prof, 42-61; biophysicist, Oak Ridge Nat Lab, 61-74, group leader biophys, 64-69; sr biophysicist, 74-86, actg assoc dir life sci, 84-86, chmn biol dept, 79-87, ASSOC DIR LIFE SCI, BROOKHAVEN NAT LABS, 86- *Concurrent Pos:* Prof biomed sci, Univ Tenn, Oak Ridge, 67-74, sci dir biophys & cell physiol, 69-84, dir, Grad Sch Biomed Sci, 72-74; adj prof biochem, State Univ NY, Stony Brook, 75- *Honors & Awards:* Finsen Medal, 80; Fermi Award, 89. *Mem:* Nat Acad Sci; Am Acad Arts & Sci; Biophys Soc; Am Soc Photobiol; Environ Mutagen Soc; Am Asn Cancer Res. *Res:* Far ultraviolet spectroscopy; ionizing and nonionizing radiation; molecular biophysics; action of light on proteins viruses and cells; nucleic acids; repair mechanisms; environmental carcinogenesis. *Mailing Add:* Biol Dept Brookhaven Nat Lab Upton NY 11973

SETLOW, VALERIE PETIT, b New Orleans, La, Jan 24, 50; m 76; c 2. AIDS POLICY DEVELOPMENT & ANALYSIS, PROGRAM MANAGEMENT & SUPERVISION. *Educ:* Xavier Univ, BS, 70; Johns Hopkins Univ, PhD(biol), 76. *Prof Exp:* Fel human genetics, Mt Sinai Hosp, NY, 76-77; fel, Nat Diabetes & Metab Inst, NIH, 77-79 & Nat Heart, Lung & Blood Inst, 79-81, asst prog dir, Diabetes Div, Nat Inst Diabetes & Digestive & Kidney Dis, 81-83, spec asst to dir, 83-86; sr analyst, Health Planning & Eval, 86-88; DIR POLICY, NAT AIDS PROG OFF, 88- *Concurrent Pos:* Prin investr, Sci Children Grant, Am Chem Soc, 87-90; consult, Elem Sch Sci, 89- *Honors & Awards:* Except Achievement Award, USPHS, 90. *Mem:* AAAS; Am Soc Biochem & Molecular Biol. *Res:* Gene regulation in viruses; public health policy analysis; development and management. *Mailing Add:* 200 Independence Ave SW Rm 738-G Washington DC 20201

SETO, BELINDA P L, b Canton, China, July 25, 48; US citizen; m 75; c 2. VIROLOGY. *Educ:* Univ Calif, Davis, BS, 70; Purdue Univ, PhD(microbiol), 74. *Prof Exp:* Staff fel, Nat Heart, Lung & Blood Inst, NIH, 74-80; RES CHEMIST, BUR BIOLOGICS, FOOD & DRUG ADMIN, 80- *Concurrent Pos:* Lectr, Howard Univ, 77. *Mem:* Am Soc Biol Chemists. *Res:* Human non-A and non-B hepatitis to isolate the soluble antigen associated with the hepatitis virus, to develop radioimmunoassay for testing, and structural studies of the viral DNA. *Mailing Add:* Westwood Bldg Rm 309 NIH Bethesda MD 20892

SETO, FRANK, b Los Angeles, Calif, Mar 12, 25; m 55; c 2. ZOOLOGY. *Educ:* Berea Col, BA, 49; Univ Wis, MS, 50, PhD(zool), 53. *Prof Exp:* Asst prof, Exten Div, Univ Wis, 53-56; asst prof zool, Berea Col, 56-62; asst prof, 64-69, assoc prof, 69-80, PROF ZOOL, UNIV OKLA, 81- *Concurrent Pos:* NSF res grant, Berea Col, 60-61; USPHS spec fel, Oak Ridge Nat Lab, 62-64, USPHS res grant, 75; Am Cancer Soc grant, 65-68. *Mem:* Nat Asn Biol Teachers; Int Soc Develop Comp Immunol; Am Soc Zoologists; Soc Develop Biologists; Soc Exp Hemat. *Res:* Developmental biology and immunology. *Mailing Add:* Dept Zool Univ Okla Norman OK 73019

SETO, JANE MEI-CHUN, b China, May 15, 27; US citizen; m 58; c 2. INTERNAL MEDICINE. *Educ:* Kwang-Ha Med Col, China, MD, 51; Queen's Univ Belfast, cert med, 57; Tulane Univ, cert trop med, 67. *Prof Exp:* Intern med, Regina Grey Nun's Hosp, Sask, 57-58; resident internal med, Providence Hosp, Seattle, 58-59; physician, Health Ctr, Univ Wash, 59-60; physician, Austin State Sch, Tex, 61-62; res asst cancer chemother, Univ Tex M D Anderson Hosp & Tumor Inst, 64-66; serologist, La State Bd Health, 67-69; instr internal med, Sch Med, Tulane Univ, 69-70, res assoc electrosci & biophys res group & NIH fel, Sch Eng, 69-82; RETIRED. *Mem:* Fel Royal Soc Health; AMA; Am Pub Health Asn. *Res:* Electromagnetic induced biological effects; dynamics of biological cells. *Mailing Add:* 4824 Purdue Dr Metairie LA 70003

SETO, JOSEPH TOBEY, b Tacoma, Wash, Aug 3, 24; m 59; c 2. ELECTRON MICROSCOPY, MOLECULAR BIOLOGY. *Educ:* Univ Minn, BS, 49; Univ Wis, MS, 55, PhD(bact), 57. *Prof Exp:* Asst bact, Sch Med, Univ Ill, 50-53; res assoc, Fermentation Div, Upjohn Co, Mich, 57; res virologist, Med Ctr, Univ Calif, Los Angeles, 58-59; asst prof biol, San Francisco State Univ, 59-60; from asst prof to assoc prof microbiol, 60-67, chmn dept, 64-75, prof, 67-88, EMER PROF MICROBIOL, CALIF STATE UNIV, LOS ANGELES, 88- *Concurrent Pos:* Consult & res virologist, US Naval Biol Lab, Univ Calif, 60 -63; guest prof, Inst Virol, Univ Giessen, 65-66, 72-73, 79-80 & 86-87; Humboldt Found award, 72; WHO res exchange worker, 72; Humboldt Found Reinvitation, 86. *Mem:* AAAS; Am Soc Microbiol; fel Am Acad Microbiol; Electron Micros Soc Am; Sigma Xi. *Res:* Characterization of orthomyxovirus and paramyxovirus glycoproteins; persistent infections of paramyxoviruses in tissue cultures; molecular basis of the pathogenesis of paramyxoviruses. *Mailing Add:* Dept Microbiol Calif State Univ Los Angeles CA 90032-6205

SETSER, CAROLE SUE, b Warrenton, Mo, Aug 26, 40; m 69; c 3. FOOD SCIENCE. *Educ:* Univ Mo, Columbia, BS, 62; Cornell Univ, MS, 64; Kans State Univ, PhD(foods, nutrit), 71. *Prof Exp:* Instr food sci, 64-66, asst to dean col home econ, 66-68, from asst prof to assoc prof, 74-86, PROF FOOD SCI, KANS STATE UNIV, 86- *Concurrent Pos:* Assoc ed, Cereal Chem, 88- *Mem:* Inst Food Technologists; Sigma Xi; Am Asn Cereal Chemists. *Res:* Sensory studies of foods; reduced calorie baked product development and sensory textural evaluation; NMR and DSC studies on sweetener functionality in foods. *Mailing Add:* Dept Foods & Nutrit Kans State Univ Manhattan KS 66506

SETSER, DONALD W, b Great Bend, Kans, Jan 2, 35; m 69; c 3. PHYSICAL CHEMISTRY. *Educ:* Kans State Univ, BS, 54, MS, 56; Univ Wash, PhD(phys chem), 61. *Prof Exp:* Fel phys chem, Univ Wash, 61-62; NSF fel, Cambridge Univ, 62-63; from asst prof to assoc prof, 63-69, PROF PHYS CHEM, KANS STATE UNIV, 70- *Concurrent Pos:* Alfred P Sloan fel, 68-70. *Mem:* Am Chem Soc; Am Phys Soc; Royal Soc Chem. *Res:* Chemical kinetics; spectroscopy of small molecules; energy transfer. *Mailing Add:* Dept Chem Kans State Univ Manhattan KS 66504

SETTERFIELD, GEORGE AMBROSE, cytology, for more information see previous edition

SETTERGREN, CARL DAVID, b Chicago, Ill, Dec 12, 35; m 65; c 1. FORESTRY, WATERSHED MANAGEMENT. *Educ:* Univ Mo, BSF, 58, MS, 60; Colo State Univ, PhD(watershed mgt), 67. *Prof Exp:* From instr ecol to assoc prof forest hydrol, 64-77, PROF FOREST HYDROL, UNIV MO, COLUMBIA, 77- *Mem:* Soc Am Foresters; Am Water Resources Asn. *Res:* Forest hydrology; forest influences. *Mailing Add:* Sch Forestry Univ Mo 1-30 Agr Bldg Columbia MO 65211

SETTERSTROM, CARL A(LBERT), b Brooklyn, NY, May 2, 15; m 40; c 2. CHEMICAL ENGINEERING. *Educ:* Polytech Inst, Brooklyn, BChE, 36. *Prof Exp:* Sci asst plant physiol, Boyce Thompson Inst Plant Res, 36-40; from tech rep to gen mgr, Textile Fibers Dept, Union Carbide Corp, 40-54; spec projs mgr, Chas Pfizer & Co, Inc, 54-58; asst dir, Res & Develop Div, Sun Oil Co, 58-59; asst to pres, Avisun Corp, 59, gen mgr mkt, 60; vpres, Rexall Chem Co, NJ, 60-68; MANAGING PARTNER, HARRINGTON RES CO, 68- *Concurrent Pos:* Pres, PNC Co, 70-79. *Mem:* Am Chem Soc; Commercial Develop Asn; AAAS; NY Acad Sci. *Res:* New product development; corporate acquisitions and evaluations; marketing of polymers; corporate profit strategies; plastics and fibers technology. *Mailing Add:* Harrington Res Co 323 Dogleg Dr Williamsburg VA 23188

SETTLE, FRANK ALEXANDER, JR, b Nashville, Tenn, Sept 19, 37; m; c 4. ANALYTICAL CHEMISTRY. *Educ:* Emory & Henry Col, BS, 60; Univ Tenn, PhD(chem), 64. *Prof Exp:* From asst prof to assoc prof, 64-74, PROF CHEM, VA MIL INST, 74- *Concurrent Pos:* NSF res fel, Va Polytech Inst & State Univ, 72-73; instr, Am Chem Soc Comput & Electronics Short Courses, 72-73 & 75-76; consult, Tenn Eastman Co, 77; instr, Microprocessor Short Courses, Va Mil Inst, 76 & 78; consult, Bendix Environ & Process Instruments Div, 79; proj dir, Sci Instrument Info & Curricula Proj, 80-; vis scientist, Ctr Anal Chem, Nat Bur Standards, Gaithersburg, Md, 87-88, 89 & 90; vis prof, USAF Acad, 91-92. *Honors & Awards:* Maury Res Award, 83. *Mem:* Am Chem Soc. *Res:* Computer interfacing of chemical instrumentation; computer-based information systems; expert systems for chemical analysis; laboratory automation; intelligent system. *Mailing Add:* Dept Chem Va Mil Inst Lexington VA 24450

SETTLE, RICHARD GREGG, b Ft Worth, Tex, Sept 4, 49; m 72. NEUROPSYCHOLOGY, PSYCHOPHYSICS. *Educ:* Colgate Univ, BA, 72; Univ Mo, Columbia, MA, 73, PhD(psychol), 75. *Prof Exp:* Fel psychol, Monell Chem Senses Ctr, 75-80, ASSOC RES DEPT SURG, UNIV PA, 80-; RES PHYSIOLOGIST, VET ADMIN MED CTR, 82- *Mem:* AAAS. *Res:* Structure and function of the hippocampus; neuroanatomical and physiological basis of learning and memory; genetic and environmental influences on alcoholism in humans and alcohol consumption in animals; taste sensation and perception. *Mailing Add:* Med Res Serv University & Woodland Aves Philadelphia PA 19109

SETTLE, WILBUR JEWELL, b Barren Co, Ky. BOTANY. *Educ:* Centre Col, Ky, AB, 62; Ohio State Univ, MSc, 65, PhD(bot), 69. *Prof Exp:* Asst prof biol, Bowling Green State Univ, 69-70; asst prof, 70-80, ASSOC PROF BIOL, STATE UNIV NY COL ONEONTA, 80- *Concurrent Pos:* Res Found State Univ NY fac res fel, 71. *Mem:* AAAS; Bot Soc Am; Am Soc Plant Taxonomists; Soc Econ Bot; Sigma Xi. *Res:* Plant biosystematics, especially the genus Blephilia. *Mailing Add:* Dept Biol State Univ NY Oneonta NY 13820-4015

SETTLEMIRE, CARL THOMAS, b Dayton, Ohio, July 14, 37; m 60; c 3. BIOCHEMISTRY. *Educ:* Ohio State Univ, BS, 59, MS, 61; NC State Univ, PhD(biochem), 67. *Prof Exp:* Instr nutrit, Ohio Agr Res & Develop Ctr, 61-62; NIH trainee biochem, NC State Univ, 62-66; NIH fel, Ohio State Univ, 66-69; asst prof biochem, 69-74, ASSOC PROF BIOL & CHEM, BOWDOIN COL, 74- *Concurrent Pos:* Mem, Nat Student Support Rev Comt, Dept HEW, 68-70. *Mem:* AAAS; Am Chem Soc. *Res:* Membrane biochemistry and ion transport in whole cells and in mitochondria; membrane changes in cystic fibrosis. *Mailing Add:* Dept Biol Bowdoin Col Brunswick ME 04011

SETTLEMYER, KENNETH THEODORE, b Arnold, Pa, Dec 19, 35; m 64; c 3. TAXONOMIC BOTANY, FLORISTICS. *Educ:* Pa State Univ, BS, 57, MEd, 61, DEd(biol sci), 71. *Prof Exp:* Instr biol, Freeport Area Joint Schs, Pa, 57-66; from asst prof to assoc prof, 66-72, chmn dept, 78-81 & 86-88, PROF BIOL SCI, LOCK HAVEN UNIV PA, 72- *Concurrent Pos:* Partic, NSF Inserv Inst, Univ Pittsburgh, 60-61; consult & reviewer, Choice; NSF grant, Pa State Univ, 63, 64 & 65. *Mem:* AAAS; Am Inst Biol Sci; Nat Asn Biol Teachers; Bot Soc Am; Am Soc Plant Taxonomists. *Res:* Taxonomy and floristics of woody plants. *Mailing Add:* Dept Biol Sci Lock Haven Univ Lock Haven PA 17745-2390

SETTLES, F STAN, b Oct, 1938; m; c 4. INDUSTRIAL ENGINEERING. *Educ:* LeTourneau Col, BS(indust eng) & BS(prod technol), 62; Ariz State Univ, MSE, 67, PhD(indust eng), 69. *Prof Exp:* Design engr, AiRes Mfg Co, 62-64, develop engr, 64-67, staff asst to mgr, 67-68, sr systs analyst, Mat Dept, 68-70, mat mgr aircraft propulsion engines, 70-74, mgr prod systs anal & design, Mat Dept, 74-78, mgr oper planning, Indust Eng Dept, 78-80, mgr indust eng, Garrett Pneumatic Systs Div, 80-83, mgr indust & mfg eng, Garrett Turbine Engine Co, 83-85, consult, Garrett/AiRes Mfg, Torrance,

Calif, 85, div vpres mfg opers, AiRes Mfg Co, 85- 87, corp dir indust & mfg eng, Garrett Corp, 87, DIR PLANNING, GARRETT ENGINE DIV, ALLIED-SIGNAL AEROSPACE, 88- *Concurrent Pos:* Inst dir, Opers Res Div, Inst Indust Engrs, 76-78, vpres, Systs Eng & Tech Div, 81-83. *Honors & Awards:* Opers Res Div Award, Am Inst Indust Engrs, 81. *Mem:* Nat Acad Eng; fel Am Inst Indust Engrs (pres, 87-88, treas, 90-); Opers Res Soc Am; Inst Mgt Sci; Soc Mfg Engrs; Sigma Xi; Am Mgt Asn. *Res:* Author of various publications. *Mailing Add:* 5508 S Marine Dr Tempe AZ 85283

SETTLES, GARY STUART, b Maryville, Tenn, Oct, 9, 49. TURBULENT & SEPARATED FLOWS. *Educ:* Univ Tenn, BS, 71; Princeton Univ, PhD(mech & aerospace eng), 76. *Prof Exp:* Res scientist, Princeton Combustion Labs, Div Flow Res Corp, 75-77; res scientist & lectr fluid mech, Mech & Aerospace Eng Dept, Princeton Univ, 77-83; assoc prof, 83-88, PROF & DIR, GAS DYNAMICS LAB, MECH ENG DEPT, PA STATE UNIV, 88- *Concurrent Pos:* Consult, Aerodyn Lab, Boeing Com Airplane Co, 80-81, IBM Corp, 82-83, United Technol Res Ctr, 82- *Mem:* Am Inst Aeronaut & Astronaut; Am Soc Mech Engrs; Optical Soc Am; Soc Photo-Optical Instrumentation Engrs; Am Phys Soc. *Res:* Fluid mechanics, specializing in experimental methods, high-speed flows, turbulent boundary layers, and shock waves; flow visualization techniques and optical flow diagnostics; scientific writing and photography. *Mailing Add:* RD 1 Box 313 Bellefonte PA 16823-9708

SETTLES, HARRY EMERSON, b Denver, Colo, Dec 19, 40; m 67; c 3. REGENERATION, ELECTROMYOGRAPHY. *Educ:* Wabash Col, AB, 63; Tulane Univ, PhD(anat), 67. *Prof Exp:* From instr to asst prof anat, NY Med Col, 69-78; asst prof, 78-79, ASSOC PROF ANAT, UNIV SDAK, 79- *Mem:* Sigma Xi; AAAS; Am Asn Anatomists; Am Soc Zoologists. *Res:* Experimental embryology; regeneration; electron microscopy. *Mailing Add:* Dept Anat Univ SDak Sch Med Vermillion SD 57069

SETTLES, RONALD DEAN, b Lawrence, Kans, Feb 12, 38; m 63; c 3. PHYSICS, MATHEMATICS. *Educ:* Va Polytech Inst, BS, 59, MS, 61; La State Univ, PhD(physics), 64. *Prof Exp:* SR STAFF PARTICLE PHYSICS, MAX PLANCK INST PHYSICS & ASTROPHYS, 64- *Mem:* Sigma Xi. *Res:* Elementary particle physics. *Mailing Add:* Max Planck Inst for Physics & Astrophys D-8000 Munich 40 Fohringer Ring 6 Germany

SETTOON, PATRICK DELANO, b Amite, La, Feb 15, 34; m 56; c 3. BIOCHEMISTRY, FOOD CHEMISTRY. *Educ:* Southeastern La Univ, BS, 57; La State Univ, Baton Rouge, MS, 61, PhD, 67. *Prof Exp:* Chemist, Union Carbide Corp, 57-58; from instr to assoc prof, Southeastern La Univ, 58-69, prof chem & chmn dept chem & physics, 69-80, dean, Col Sci & Technol, 80-86, provost & vpres, Acad & Student Affairs, 86-90; RETIRED. *Res:* Mechanisms involved in food decomposition; accessory growth factors for microorganisms. *Mailing Add:* Southeastern La Univ Sci & Technol 100 W Dakota Hammond LA 70401

SETZLER-HAMILTON, EILEEN MARIE, b Fremont, Ohio, Apr 28, 43; m 79. MARINE ECOLOGY, FISH BIOLOGY. *Educ:* Col of St Mary of Springs, BA, 65; Univ Del, MS, 69; Univ Ga, PhD(zool), 77. *Prof Exp:* Fisheries biologist, Southwest Fisheries Ctr, Nat Marine Fisheries Serv, Nat Oceanic & Atmospheric Admin, 74; res assoc, 75-87, sr res assoc, 87-89, ASSOC RES PROF FISH ECOL, CTR ENVIRON & ESTUARINE STUDIES, CHESAPEAKE BIOL LAB, UNIV MD, 89- *Mem:* Am Fisheries Soc; Sigma Xi; Estuarine Res Fedn; Atlantic Estuarine Res Soc (secy-treas, 78-79). *Res:* Fish ecology; population dynamics of estuarine species; estuarine food webs; larval and juvenile fish biology; roles of fish in energetics of estuarine nursery areas; striped bass biology. *Mailing Add:* Ctr Environ & Estuarine Studies Chesapeake Biol Lab Univ Md PO Box 38 Solomons MD 20688-0038

SEUBOLD, FRANK HENRY, JR, b Chicago, Ill, Nov 16, 22; m 42; c 2. ORGANIC CHEMISTRY, PUBLIC HEALTH ADMINISTRATION. *Educ:* Northwestern Univ, BS, 43; Univ Calif, Los Angeles, PhD(chem), 48. *Prof Exp:* Jr chemist, Shell Develop Co, 43-45, chemist, 47-52; asst prof chem, Northwestern Univ, 52-54; res chemist, Union Oil Co Calif, 54-59; sr chemist, Aerojet-Gen Corp Div, Gen Tire & Rubber Co, 59-63, tech dept mgr, Space-Gen Corp, Calif, 63-71; dir, Proj Mgt & Planning Div, Health Maintenance Orgn Serv, Health Serv & Ment Health Admin, Dept Health & Human Serv, 71-74, assoc bur dir, Health Serv Admin, 74-76, dir, Div Health Maintenance Orgns, Bur Med Serv, Health Serv Admin, 76-77, dir, Div Health Maintenance Develop, Off Asst Secy Health, 77-78, dep dir, Div Intramural Res, Nat Ctr Health Serv Res, 79-81, dir, Health Maintenance Orgns, Off Asst Secy Health, 80-87. *Mem:* Am Chem Soc. *Res:* Mechanisms of organic reactions free radical processes; oxidations of hydrocarbons; reaction of organic peroxides; polymerization; heterogeneous catalysis; solid rocket propellants; encapsulation; immunochemistry; biological and chemical detection systems and instrumentation; health care delivery systems; health maintenance organization; health services research. *Mailing Add:* 731 Gailen Ave Palo Alto CA 94301

SEUFERT, WOLF D, b Dusseldorf, Ger, Apr 17, 35; m 61; c 2. BIOPHYSICS, MOLECULAR BIOLOGY. *Educ:* Univ Dusseldorf, Dr Med, 60. *Hon Degrees:* Dsc, (physics), Univ Provence Marseille, 79. *Prof Exp:* Intern med, Hosp Nördlingen, Ger, 61-62; res assoc physiol, Univ Heidelberg, 63-64; med res scientist, Dept Basic Res, Eastern Pa Psychiat Inst, 64-65; guest investr biophys, Rockefeller Univ, 66-67; from asst prof to assoc prof, 67-75, PROF BIOPHYS, FAC MED, UNIV SHERBROOKE, 75- *Concurrent Pos:* Pres, Biophys Instrumentation Inc, Sherbrooke, Que, Can, 80- *Mem:* Biophys Soc. *Res:* Physico-chemical characteristics of biological membranes and membrane models; medical instrumentation; biomedical engineering; applied biophysics. *Mailing Add:* Dept Physiol & Biophys Univ Sherbrooke Fac Med Sherbrooke PQ J1H 5N4 Can

SEUFZER, PAUL RICHARD, b Cleveland, Ohio, Dec 23, 21; m 44. INORGANIC CHEMISTRY, RESEARCH ADMINISTRATION. *Educ:* Western Reserve Univ, BS, 43, MS, 49, PhD(inorg chem), 51. *Prof Exp:* Anal chemist, Harshaw Chem Co, 43-44; shift foreman, Tenn Eastman Corp & US Army, CEngrs, Tenn, 44-46; res chemist, Argonne Nat Lab, 51-53; supvr chem dept, 53-63, SUPT DEVELOP LAB, TECH DIV, GOODYEAR ATOMIC CORP, 63- *Mem:* AAAS; Am Chem Soc; Sigma Xi. *Res:* Physical inorganic chemistry of fluorine and uranium. *Mailing Add:* 15 Ridgeway Dr Chillicothe OH 45601

SEUGLING, EARL WILLIAM, JR, b Little Falls, NJ, Jan 6, 33; m 66; c 3. PHARMACY, CHEMICAL ENGINEERING. *Educ:* Rutgers Univ, BS, 55; Ohio State Univ, MSc, 57, PhD(pharm), 61. *Prof Exp:* Sr pharmaceut scientist, Vick Div Res, Richardson Merrell Inc, 60-63; res mgr, Res Dept, Strong Cobb Arner Inc, 63-65, dir res, 65-69; dir res, Block Drug Co, 69-78; CONSULT PHARMACEUT/DENT, COSMETIC, TOILETRIES DEVELOP, QUAL CONTROL/QUAL ASSURANCE PROD, 78-; SALES/TECH SERV, ALPINE AROMATICS INT, INC, 84- *Mem:* Am Pharmaceut Asn; Acad Pharmaceut Sci; Am Inst Chem Eng; Am Chem Soc; Soc Cosmetic Chem. *Res:* Ionic and adsorptive exchange reactions in pharmaceutical sciences; product and process development, specifically, improved granulation, compression and advanced tablet coating techniques; development of dental, pharmaceutical, cosmetic and toiletry products; research administration; technical management. *Mailing Add:* 44 Courtney Pl Palm Coast FL 32137-8126

SEUS, EDWARD J, b New York, NY, Jan 30, 35; m 57; c 4. ORGANIC CHEMISTRY, PHOTOGRAPHIC CHEMISTRY. *Educ:* Queens Col, NY, BS, 56; Univ Minn, PhD(org chem), 60. *Prof Exp:* Res chemist, Chem Div, 60-65, sr res chemist, Photo Mat Div, 65-72, tech assoc Film Emulsion Div, 72-75, SUPVR DEVELOP COLOR PRODS, EASTMAN KODAK CO, 75- *Honors & Awards:* R Max Goepp, Jr Mem Prize Chem, 56. *Mem:* Am Chem Soc; Soc Photograph Scientists & Engrs. *Res:* Synthesis of olefins via organophorous intermediates; chemistry of stilbenes; Vilsmeier formylation; preparation of photomaterials; development of color films. *Mailing Add:* Eastman Kodak Co Kodak Park Bldg Six Rochester NY 14650

SEVACHERIAN, VAHRAM, b Nov 25, 42; US citizen. ENTOMOLOGY, APPLIED STATISTICS. *Educ:* Univ Calif, Los Angeles, BA, 64; Calif State Univ, Los Angeles, MA, 66; Univ Calif, Riverside, PhD(entom), 70. *Prof Exp:* Asst zool, Calif State Univ, Los Angeles, 64-66; res asst entom, 66-70, lectr entom, asst res entomologist, lectr statist & asst res statistician, 71-72, asst prof, 72-81, ASSOC PROF STATIST & ENTOM, UNIV CALIF, RIVERSIDE, 81- *Mem:* Entom Soc Am; Ecol Soc Am; Entom Soc Can; Am Statist Soc; Sigma Xi. *Res:* Systems analysis; sampling techniques; statistical ecology; integrated pest management; economic thresholds in agriculture. *Mailing Add:* Dept Entom Univ Calif Riverside CA 92521

SEVALL, JACK SANDERS, b Jan 12, 46; US citizen. BIOCHEMISTRY. *Educ:* Willamette Univ, BA, 67; Purdue Univ, PhD(biochem), 71. *Prof Exp:* Lab instr biochem, Purdue Univ, 69-70; res assoc biol, Calif Inst Technol, 71-74; asst prof biochem, Tex Tech Univ, 74-80; sr scientist, Wadley Insts Molecular Med, 80-; ASSOC SCIENTIST, DEPT MOLECULAR & CELL BIOL, SOUTHWEST FOUND; RES SCIENTIST, HELICON FOUND. *Concurrent Pos:* Damon Runyon Mem Fund Cancer Res fel, Calif Inst Technol, 72-74; NSF res grant, 75, NIH res grants, 76-79 & 81-84, Roger A Welch res grant, 77-80. *Mem:* Soc Cell Biol; Biophys Soc; AAAS; Am Soc Biol Chemists. *Res:* Eukaryotic gene structure; nonhistone chromosomal protein role in gene structure and function. *Mailing Add:* Specialty Labs Inc 2211 Michigan Ave Santa Monica CA 90404

SEVENAIR, JOHN P, b Somerville, NJ, Oct 12, 43. SCIENCE EDUCATION. *Educ:* Mass Inst Technol, BS, 65; Univ Notre Dame, PhD(chem), 70. *Prof Exp:* Res assoc chem, Ga Inst Technol, 69-71 & Univ Ala, 71-72; teaching res fel, Tulane Univ, 72-74; asst prof, 74-80, assoc prof, 80-86, PROF CHEM, XAVIER UNIV, 86- *Mem:* Am Chem Soc; Nat Speleol Soc; Sigma Xi. *Res:* Chemistry of humic substances; science education of minorities. *Mailing Add:* Xavier Univ Box 100C New Orleans LA 70125

SEVENANTS, MICHAEL R, b Two Rivers, Wis, Mar 16, 38; m 62; c 3. FOOD CHEMISTRY. *Educ:* Univ Calif, Davis, BS, 61, PhD(agr chem), 65. *Prof Exp:* CHEMIST, PROCTER & GAMBLE CO, 65-, GROUP LEADER, 70- *Mem:* Am Chem Soc; Am Oil Chemists Soc; Sigma Xi. *Res:* Chemical identification and sensory correlation of natural flavorous components in foods. *Mailing Add:* 563 Beaufort Ct Cincinnati OH 45240

SEVER, DAVID MICHAEL, b Canton, Ohio, Feb 21, 48; m 69; c 2. HERPETOLOGY. *Educ:* Ohio Univ, BS, 70, MS, 71; Tulane Univ, PhD(biol), 74. *Prof Exp:* From asst prof to assoc prof, 74-87, PROF BIOL, ST MARY'S COL, IND, 87-, CHMN DEPT, 80- *Concurrent Pos:* Res grants, Highlands Biol Sta Ind, 73 & 75, Ind Acad Sci, 75, 82 & 86, Am Philos Soc, 77 & Ind Nat Res, 79, NSF, 86, 87 & 88. *Mem:* Am Inst Biol Sci; Am Soc Icthyologists & Herpetologists; Herpetologists League; Soc Study Amphibians & Reptiles; Sigma Xi. *Res:* Anatomy and evolution of the secondary sexual characters of salamanders; systematics, ecology, behavior and physiology of salamanders; author of 50 publications. *Mailing Add:* Dept Biol St Mary's Col Notre Dame IN 46556

SEVER, JOHN LOUIS, b Chicago, Ill, Apr 11, 32; m 56; c 3. PEDIATRICS, MICROBIOLOGY. *Educ:* Univ Chicago, BA, 51; Northwestern Univ, BS, 52, MS, 56, MD, PhD(microbiol), 57. *Prof Exp:* Instr microbiol, Med Sch, Northwestern Univ, 54-60; head sect infectious dis, Nat Inst Neurol Dis & Stroke, 60-71; assoc prof, 63-73, PROF PEDIAT, SCH MED, GEORGETOWN UNIV, 73-, PROF OBSTET & GYNEC, 80-; CHIEF INFECTIOUS DIS BR, NAT INST NEUROL & COMMUN DISORDERS & STROKE, 71- *Concurrent Pos:* Resident pediat, Chicago Children's Mem Hosp, Ill, 57-60; mem res & clin staff, Nat Children's Med Ctr, 65-; ed, J Teratology, 75-; rep, World Health Org Rotary Int, 84; Borden Award Med

Res. *Mem:* Am Soc Microbiol; AMA; Soc Pediat Res; Am Asn Immunologists; Am Epidemol Soc. *Res:* Infectious diseases; virology; perinatal infections; vaccines; chronic infections of central nervous system. *Mailing Add:* Dept Child Health Develop Obstet & Gynec George Washington Univ 111 Michigan Ave Washington DC 20010

SEVERIN, CHARLES HILARION, b Bendena, Kans, Sept 3, 96. BOTANY. *Educ:* Univ Chicago, BS, 25, MS, 27, PhD(bot), 30. *Hon Degrees:* LLD, St Mary's Col, Minn, 75. *Prof Exp:* High sch teacher, De La Salle Inst, 20-33; prof, 33-75, EMER PROF BIOL, ST MARY'S COL, MINN, 75- *Concurrent Pos:* Asst, Univ Chicago, 30; instr, De Paul Univ, 32-33. *Mem:* AAAS; Bot Soc Am; Ecol Soc Am; Am Genetic Asn; hon mem Nat Asn Biol Teachers (pres, 55). *Res:* Plant anatomy and ecology; morphology of botany; anatomy of roots of the lower monocotyls; anatomy of roots and seedlings of emergent spermatophytes. *Mailing Add:* Dept Biol St Mary's Col Winona MN 55987

SEVERIN, CHARLES MATTHEW, b Youngstown, Ohio, Dec 4, 48; m; c 2. NEUROANATOMY. *Educ:* St Louis Univ, BA, 70, MS, 72, PhD(anat), 75. *Prof Exp:* Asst anat, Sch Med, St Louis Univ, 73-74; ASSOC PROF ANAT, STATE UNIV NY, BUFFALO, 85- *Mem:* Sigma Xi; Soc Neurosci; Am Asn Clin Anat; Am Asn Anat. *Res:* Connections of the reticular formation; organization of the hippocampus; head injury. *Mailing Add:* Anat Sci 318 Farber Hall State Univ NY Health Sci Ctr 3435 Main St Buffalo NY 14214

SEVERIN, MATTHEW JOSEPH, b Omaha, Nebr, Aug 7, 33; m 58; c 4. MEDICAL MICROBIOLOGY. *Educ:* Creighton Univ, BS, 55, MS, 60; Univ Nebr, PhD(med microbiol), 68, JD, 86. *Prof Exp:* Bacteriologist, Clin Labs, Immanuel Hosp, Omaha, 58-60; asst prof biol, Univ Omaha, 61-63; dir labs, Omaha-Douglas County Health Dept, 63-73; dean students, Sch Med, 75-85, PROF MED MICROBIOL, CREIGHTON UNIV, 72-, PROF PREV MED & PUB HEALTH & EPIDEMIOL. *Concurrent Pos:* USPHS grants, 63-; lectr, Col St Mary, Nebr, 67-75; asst prof, Sch Med, Creighton Univ, 68- & Col Med, Univ Nebr, 72-77; asst prof, Col Med, Univ Nebr, 72-77; lectr, St Joseph's Sch Nursing, 72- *Mem:* Fel Am Soc Clin Scientists; Am Soc Microbiol; fel Am Pub Health Asn; Infectious Dis Soc Am; Am Soc Epidemiol. *Res:* Venereal disease agents and streptococcal disease agents and their chemotherapeutic sensitivities; AIDS and legal issues. *Mailing Add:* Dept Med Microbiol Creighton Univ Sch Med Omaha NE 68178

SEVERINGHAUS, CHARLES WILLIAM, b Ithaca, NY, Sept 3, 16; m 41, 71; c 3. WILDLIFE MANAGEMENT. *Educ:* Cornell Univ, BS, 39. *Prof Exp:* Conserv worker, State Conserv Dept, 39-41, asst game res investr, 41, game res investr, 41-59, dist game mgr, 41-44, leader deer mgt res, 44-56, leader big game mgt invests, 56-61, supv wildlife biologist, 61-74, prin wildlife biologist, 74-77; RETIRED. *Concurrent Pos:* Consult wildlife biologist, Cornell Univ, 77-86, pvt pract, 86- *Honors & Awards:* Achievement Award, NE Sect, Wildlife Soc, 51; Conserv Award, Am Motors Corp, Wildlife Soc, 62; Wildlife Conserv Award, Nat Wildlife Fed & Sears-Roebuck Found, 65. *Mem:* Wildlife Soc; Am Soc Mammalogists. *Res:* Game management; life history, population dynamics, ecology, reproduction and management of white-tailed deer and black bear. *Mailing Add:* 4665 Martin Rd Voorheesville NY 12186-2137

SEVERINGHAUS, JOHN WENDELL, b Madison, Wis, May 6, 22; m 48; c 4. MEDICINE. *Educ:* Haverford Col, BS, 43; Columbia Univ, MD, 49; Am Bd Anesthesiol, dipl, 58. *Hon Degrees:* Dr med, Univ Copenhagen, Denmark, 79. *Prof Exp:* Staff physicist, Radiation Lab, Mass Inst Technol, 43-45; staff physician, Sage Hosp, Ganado, Ariz, 51 & Embudo Hosp, NMex, 51; res assoc physiol, Univ Pa, 51-53; sr asst surgeon, Clin Ctr, USPHS, 53-56, surgeon, 56-58; from asst prof to assoc prof, 58-65, PROF ANESTHESIA, MED CTR, UNIV CALIF, SAN FRANCISCO, 65-, DIR ANESTHESIA RES, CARDIOVASC RES INST, 58- *Concurrent Pos:* Hon fel, Fac Anesthetists, Royal Col Surgeons, London, 75. *Mem:* Am Soc Anesthesiol; Am Physiol Soc; Am Soc Clin Invest; Asn Univ Anesthetists. *Res:* Pulmonary physiology; control of respiration; physiologic effects in anesthesia; electrodes for blood and tissue oxygen and carbon dioxide tension. *Mailing Add:* Dept Anesthesia HSE 1386 Univ Calif Med Ctr San Francisco CA 94143-0542

SEVERN, CHARLES B, b Breckenridge, Minn, Apr 11, 39; m 66; c 3. PEDIATRICS, NEONATOLOGY. *Educ:* Univ Minn, Minneapolis, BA, 62; Univ Mich, MS, 66, PhD(anat), 68; Univ Nebr, MD, 76. *Prof Exp:* Asst prof, 68-70, ASSOC PROF PATH & ANAT, COL MED, UNIV NEBR MED CTR, OMAHA, 70-, ASSOC PROF OBSTET & GYNEC & PEDIAT, 73- *Concurrent Pos:* Resident gen pediat, Univ Nebr, 76, chief resident, 78, fel neonatol, 78-79. *Mem:* Am Asn Anatomists; Teratology Soc. *Res:* Human embryology; perinatal pathology and development; fetal and newborn pathology. *Mailing Add:* Saint Alexis Med Ctr 900 E Broadway PO Box 5510 1658 Bismarck TX 58502

SEVERNS, MATTHEW LINCOLN, b Wilmington, Del, Nov 12, 52. OPHTHALMOLOGY, MEDICAL INSTRUMENTATION. *Educ:* Univ Del, BA, 76; Univ Va, ME, 78, PhD(biomed eng), 80. *Prof Exp:* Head field opers, Megonigal Electronics Inc, 75-76; staff engr, Rehab Eng Ctr, Univ Va, 77; asst prof computerized med, George Washington Univ Med Ctr, 80-82; head, Advan Automation Res Group, Am Red Cross Nat Labs, 82-86; dir res & develop, LKC Technologies, Inc, 86-91; RES ASSOC OPHTHAL, JOHNS HOPKINS UNIV MED CTR, 87-; VPRES RES & DEVELOP, ADVAN VISION SYSTS, INC, 91- *Concurrent Pos:* Consult biomed instrumentation, 76- *Mem:* Optical Soc Am; AAAS; Inst Elec & Electronics Engrs. *Res:* Techniques and instrumentation for diagnosis of eye diseases; signal processing and neural network analysis and classification of evoked potentials; rehabilitative technology for the visually impaired; design of medical electronic systems. *Mailing Add:* 3026 Mission Square Dr Fairfax VA 22031-1112

SEVERS, WALTER BRUCE, b Pittsburgh, Pa, June 10, 38; m 70; c 5. PHARMACOLOGY. *Educ:* Univ Pittsburgh, BS, 60, MS, 63, PhD(pharmacol), 65. *Prof Exp:* Instr pharmacol, Univ Pittsburgh, 64-65, USPHS fel, 65-66; USPHS fel, Lab Chem Pharmacol, Nat Heart Inst, 66-68; from asst prof to assoc prof, 68-77, PROF PHARMACOL, HERSHEY MED CTR, PA STATE UNIV, 77-, PROF NEUROSCI, 86- *Concurrent Pos:* Asst prof, Ohio Northern Univ, 65-66. *Honors & Awards:* I M Setchenov Medal, Acad Med Sci, USSR; Blue Medal for Sci, Bulgarian Union Sci Workers. *Mem:* Soc Neurosci; Am Soc Pharmacol & Exp Therapeut; Pavlovian Soc Am; Am Physiol Soc; Sigma Xi. *Res:* Hypertension; renin-angiotensin system; interaction between angiotensin and the central nervous system; salt/water balance; intracranial pressure control; neuropeptides. *Mailing Add:* Dept Pharmacol Pa State Univ Hershey Med Ctr Hershey PA 17033

SEVERSIKE, LEVERNE K, b Des Moines, Iowa, Nov 5, 36; m. AEROSPACE ENGINEERING. *Educ:* Iowa State Univ, BS, 58, MS, 61, PhD(aerospace eng), 64. *Prof Exp:* Res asst gas dynamics, Eng Exp Sta, 60-63, from instr to asst prof, 63-76, ASSOC PROF AEROSPACE ENG, IOWA STATE UNIV, 76- *Mem:* Am Inst Aeronaut & Astronaut; Am Astronaut Soc; Nat Space Soc; Planetary Soc; Sigma Xi. *Res:* Vehicle flight mechanics; flight and reentry trajectories; optimization techniques; gas dynamics; optimal controls. *Mailing Add:* Aerospace Engr Dept 328 Town Engr Iowa State Univ Ames IA 50011

SEVERSON, ARLEN RAYNOLD, b Clarkfield, Minn, Dec 5, 39; m 62; c 4. BIOLOGICAL STRUCTURE, CELL PHYSIOLOGY. *Educ:* Concordia Col, Moorhead, Minn, BA, 61; Univ NDak, MS, 63, PhD(anat), 65. *Prof Exp:* Res collabr histochem, Brookhaven Nat Lab, 65-67; asst prof anat, Sch Med, Ind Univ, Indianapolis, 67-72, res assoc orthop, 68-69; assoc prof, 72-79, PROF BIOMED ANAT, SCH MED, UNIV MINN, DULUTH, 79- *Concurrent Pos:* USPHS fel, 65-67, spec fel, 71; guest worker, Nat Inst Dent Res, Md, 71; res collabr hemat, Brookhaven Nat Lab, 76; assoc ed, Anat Rec, 78-; consult ed, Gerodont, 82- *Mem:* Am Asn Anatomists; Am Soc Bone & Mineral. *Res:* Histochemistry, biochemistry and cellular physiology of connective tissue, bone and teeth; proteoglycan and collagen metabolism; bone cell origin and metabolism; effects of hormones on connective tissue, bone and cartilage metabolism; osteoinduction. *Mailing Add:* Dept Anat & Cell Biol Sch Med Univ Minn Duluth MN 55812

SEVERSON, DAVID LESTER, BIOCHEMICAL PHARMACOLOGY. *Educ:* Univ BC, PhD(pharmacol), 72. *Prof Exp:* ASSOC PROF PHARMACOL & BIOCHEM, UNIV CALGARY, 76- *Res:* Metabolism of heart and blood vessels. *Mailing Add:* Dept Pharmacol Univ Calgary Fac Med 3330 Hosp Dr NW Calgary AB T2N 4N1 Can

SEVERSON, DONALD E(VERETT), b Minneapolis, Minn, Dec 16, 19; m 47; c 5. CHEMICAL ENGINEERING. *Educ:* Univ Minn, BChE, 41, PhD(chem eng), 58. *Prof Exp:* Res chem engr, Md Res Labs, Ford, Bacon, & Davis Inc, DC, 43-45; RETIRED. *Concurrent Pos:* Asst dir res contract, Univ NDak-US Army Qm Corps, 49-53; dir res contract, NSF grant, 57-59; dir res contracts, Univ NDak-Great Northern Rwy, 59-64 & 65-71; prin investr, Off Coal Res Proj Lignite, 72-; consult, Coal Res Lab, US Bur Mines, 64- *Mem:* Am Chem Soc; Am Inst Chem Engrs; Am Soc Eng Educ; Sigma Xi. *Res:* Food dehydration; mass transfer in evaporation; drying; coal research; gasification; high pressure technology. *Mailing Add:* 107 S Tanager Ct Louisville CO 80027

SEVERSON, HERBERT H, b Madison, Wis, May 27, 44; m 75; c 2. HEALTH PSYCHOLOGY, SCHOOL PSYCHOLOGY. *Educ:* Wis State Univ, Whitewater, BS, 66; Univ Wis-Madison, MS, 69, PhD(educ psychol), 73. *Prof Exp:* Sch psychologist, Madison Pub Sch-Wis, 69-72; asst prof psychol, Univ Northern Colo, 72-75; asst prof, 75-81, ASSOC PROF EDUC PSYCHOL, UNIV ORE, 81-; RES SCIENTIST & PRIN INVESTR, ORE RES INST, 79-; LIC CLIN PSYCHOLOGIST, ORE, 81- *Concurrent Pos:* Regional dir, Nat Asn Sch Psychologists, 79-81; adj-prof, Univ Ore, 83-; mem, Ore Sch Psychologist Asn; from dir to sci coordr, Ore Res Inst, 79-88. *Mem:* Fel Am Psychol Asn; Nat Asn Sch Psychologists; Asn Adv Behav Ther; Soc Behav Med; Soc Res Admin. *Res:* Smoking and smokeless tobacco prevention with adolescents; drug use prevention; chewing tobacco cessation; worksite health promotion; identification of at risk children; screening for behavior disorders; risk preception; health psychology; smoking cessation with pregnant women; passive smoking. *Mailing Add:* Ore Res Inst 1899 Williamette St, Ste Two Eugene OR 97401

SEVERSON, KEITH EDWARD, b Albert Lea, Minn, Dec 22, 36. RANGE SCIENCE, WILDLIFE ECOLOGY. *Educ:* Univ Minn, BA, 62; Univ Wyo, MS, 64, PhD(range mgt), 66. *Prof Exp:* Instr range ecol, SDak State Univ, 66-67, asst prof wildlife ecol, 67-70; range scientist, Rapid City, SDak, 70-77, wildlife biologist, 77-85, PROJ LEADER & RES WILDLIFE BIOLOGIST, ROCKY MOUNTAIN FOREST & RANGE EXP STA, US FOREST SERV, TEMPE, AZ, 85- *Mem:* Soc Range Mgt; Wildlife Soc. *Res:* Effects of grazing by livestock on wildlife habitat; nutrient cycling in pinyon-juniper woodlands; livestock distribution patterns and influences on riparian habitat; small mammals in pinyon-juniper habitats. *Mailing Add:* Rocky Mt Forest & Range Exp Sta Ariz State Univ Tempe AZ 85281

SEVERSON, ROLAND GEORGE, b Malta, Mont, Apr 1, 24; m 45; c 4. ORGANIC CHEMISTRY. *Educ:* Mont State Col, BS, 46; Purdue Univ, MS, 48, PhD(chem), 51. *Prof Exp:* Asst chem, Purdue Univ, 46-48, asst instr, 48-50; from instr to assoc prof, 50-58, prof, 58-88, chmn dept, 60-88, EMER PROF CHEM, UNIV NDAK, 88- *Mem:* Am Chem Soc. *Res:* Organometallics; synthesis of substituted organosilanes; organosilicon chemistry. *Mailing Add:* 705 Chestnut St Univ NDak Box 7185 Grand Forks ND 58201

SEVERSON, RONALD CHARLES, b Tracy, Minn, Nov 14, 45; m 69. GEOCHEMISTRY. *Educ:* Univ Minn, BS, 67, MS, 72, PhD(pedology), 74. *Prof Exp:* Soil scientist, USDA Soil Conserv Serv, Minn, 65-67; soil analyst, US Cold Regions Res & Eng Lab, NH, 68-69; res asst, Soil Sci Dept, Univ Minn, St Paul, 70-74; SOIL SCIENTIST GEOCHEM, BR GEOCHEM, ENVIRON GEOCHEM SECT, US GEOL SURV, 74- *Concurrent Pos:* Instr soil genesis & geog, Dept Plant & Earth Sci, Univ Wis-River Falls, 73. *Mem:* Am Soc Agron; Soil Sci Soc Am; Coun Agr Sci & Technol. *Res:* Spatial distribution of elements in soil materials and their changes with cultural activities. *Mailing Add:* 774 Independence St Denver CO 80215

SEVERUD, FRED N, b June 8, 1899; US citizen. STRUCTURAL ENGINEERING. *Educ:* Norwegian Inst Technol, dipl eng, 23. *Prof Exp:* CONSULT & FOUNDER, SEVERUD-SZEGEZDY, 27- *Honors & Awards:* Am Gold Medal, Am Inst Architects, 58; Franklin Award, Franklin Inst, 62; Ernest E Howard Award, Am Soc Civil Engrs, 64. *Mem:* Nat Acad Eng; Am Soc Civil Engrs. *Mailing Add:* 94 Maple Ave Rockaway NJ 07866

SEVIAN, WALTER ANDREW, b Copiague, NY, Sept 16, 40; m 64; c 3. APPLIED MATHEMATICS, DATA PROCESSING. *Educ:* State Univ NY, Stony Brook, BS, 62, MS, 64, PhD(appl math), 70. *Prof Exp:* Asst prof math, Southhampton Col, 70-71; systs engr energy systs anal, Brookhaven Nat Lab, 72-84. *Concurrent Pos:* Adj asst prof math, State Univ NY, Stony Brook, 74-75. *Mem:* Am Geophys Union; Asn Comput Users. *Res:* Computerized model/data couplings in environmental assessment of energy systems; free surface motion in groundwater hydrology; algorithm development for microcomputers. *Mailing Add:* 35 Central Ave Miller Place NY 11764

SEVIK, MAURICE, b Istanbul, Turkey, Mar 19, 23; US citizen; m 53; c 2. AEROSPACE ENGINEERING. *Educ:* Univ London, DIC, 46; Pa State Univ, PhD(eng mech), 63. *Prof Exp:* Mem staff, Bristol Aircraft Ltd & Hawker-Siddeley Ltd, 46-59; prof aerospace eng, Pa State Univ, 59-72 dir, Garfield Thomas Water Tunnel, 69-72; ASSOC TECH DIR ACOUST, DAVID TAYLOR RES CTR, 72- *Concurrent Pos:* Consult, IBM Corp, 64-65 & Off Res Anal, USAF, 65-66; vis prof, Cambridge Univ, 70, fel, Churchill Col, 70. *Mem:* Fel Am Soc Mech Engrs; Sigma Xi; fel Acoust Soc Am. *Res:* Acoustics. *Mailing Add:* 7817 Horsehoe Lane Potomac MD 20854

SEVILLA, MICHAEL DOUGLAS, b San Jose, Calif, Feb 16, 42; m 63; c 3. PHYSICAL CHEMISTRY, BIOPHYSICAL CHEMISTRY. *Educ:* San Jose State Col, BS, 63; Univ Wash, PhD(phys chem), 67. *Prof Exp:* Instr, Univ Wash, 67-68; res chemist, Atomics Int Div, NAm Rockwell Corp, 68-70; from asst prof to assoc prof, 70-83, PROF CHEM, OAKLAND UNIV, 83- *Concurrent Pos:* Assoc ed, Radiation Res J; res grants, Dept Energy, 73-91, USDOA, 82-84, NIH, 86-, Petrol Res Fund, 87-91. *Mem:* AAAS; Am Chem Soc; Sigma Xi; Radiation Res Soc; Am Soc Photobiol. *Res:* Electron spin resonance spectroscopy, particularly as an aid to understanding free radical mechanisms induced by radiation effects on biological molecules. *Mailing Add:* Dept Chem Oakland Univ Rochester MI 48309

SEVIN, E(UGENE), b Chicago, Ill, Jan 5, 28; m 51; c 3. ENGINEERING. *Educ:* Ill Inst Technol, BS, 49, PhD(eng mech), 58; Calif Inst Technol, MS, 51. *Prof Exp:* Mem tech staff, Ill Inst Technol, 51-65, dir, Eng Mech Div, 65-70; prof mech eng & head, Mech Eng Dept, Technion Israel Inst Technol, 70-74; chief, Strategic Struct Div, 74-80, asst dep dir, 80-86, DIR, OFFENSIVE & SPACE SYSTS, OFF DIR, DEFENSE RES & ENG, DEFENSE NUCLEAR AGENCY, 86- *Concurrent Pos:* Eve instr, Ill Inst Technol, 55-63, adj prof, 65-70; mem adv comts, Dept Defense Agencies, 55- *Mem:* Nat Acad Eng; Am Inst Aeronaut & Astronaut; Am Soc Mech Engrs. *Res:* Structural dynamics and shock isolation, with emphasis on numerical methods, and particular application to nuclear weapons effects and the design of hardened military construction; high-explosive testing techniques for simulating nuclear airblast; cratering and ground shock, and developed a number of new and improved hardened facilities concepts, including deep underground constructions and the first superhard silos design; numerical methods; computer applications of structural design; nuclear weapons effects; author of over 30 publications. *Mailing Add:* OUSDA/STNF/OSS Washington DC 20301

SEVOIAN, MARTIN, b Methuen, Mass, Mar 28, 19; m 54; c 2. VETERINARY MEDICINE. *Educ:* Univ Mass, BS, 49; Univ Pa, VMD, 53; Cornell Univ, MS, 54; Am Col Vet Microbiol, dipl. *Prof Exp:* Asst prof path, Cornell Univ, 54-55; PROF VET SCI, UNIV MASS, AMHERST, 55- *Mem:* Am Vet Med Asn. *Res:* Infectious diseases, especially neoplasms; biologic delivery systems using biodegradeable polymers; immunology. *Mailing Add:* Paige Labs Univ Mass Amherst MA 01002

SEVON, WILLIAM DAVID, III, b Andover, Ohio, July 22, 33; m 88; c 2. STRATIGRAPHY-SEDIMENTATION. *Educ:* Ohio Wesleyan Univ, BA, 55; Univ SDak, MA, 58; Univ Ill, PhD(geol), 61. *Prof Exp:* Lectr geol, Univ Canterbury, 61-65; GEOLOGIST, PA GEOL SURV, 65- *Mem:* AAAS; Geol Soc Am; Soc Econ Paleontologists & Mineralogists; Am Quaternary Asn; Sigma Xi. *Res:* Surficial geology of Pennsylvania Piedmont; Appalachian landscape development. *Mailing Add:* Pa Geol Surv PO Box 2357 Harrisburg PA 17105-2357

SEVY, ROGER WARREN, b Richfield, Utah, Nov 6, 23; m 48; c 2. PHARMACOLOGY. *Educ:* Univ Vt, MS, 48; Univ Ill, PhD(physiol), 51, MD, 54. *Prof Exp:* Asst physiol, Col Med, Univ Ill, 48-51, instr, 51-54; asst prof, 54-57, head dept, 57-73, dean, Sch Med, 73-78, PROF PHARMACOL, SCH MED, TEMPLE UNIV, 56- *Mem:* AAAS; Am Physiol Soc; Am Soc Pharmacol & Exp Therapeut; Endocrine Soc. *Res:* Cardiovascular pharmacology; vascular smooth muscle; hypertension; adrenal hormones and cardiovascular-renal function; pharmacology of platelets. *Mailing Add:* 242 Mather Rd Jenkintown PA 19046

SEWARD, FREDERICK DOWNING, b Goshen, NY, Dec 28, 31. SUPERNOVA REMNANTS, NEUTRON STARS. *Educ:* Princeton Univ, BA, 53; Univ Rochester, PhD(physics), 58. *Prof Exp:* Staff mem res, Lawrence Livermore Lab, 58-77; astrophysicist res, Smithsonian Inst, 77-86. *Concurrent Pos:* Vis fel, Univ Leicester, Eng, 76, Inst Astron, Cambridge, 82; consult, NASA, 75-86. *Mem:* Am Phys Soc; Am Astron Soc. *Res:* X-ray astronomy. *Mailing Add:* 158 Spencer Brook Rd Concord MA 01742

SEWARD, THOMAS PHILIP, III, b Brooklyn, NY, May 2, 39; m 65, 88; c 2. APPLIED PHYSICS, MATERIALS SCIENCE. *Educ:* Wesleyan Univ, BA, 61; Harvard Univ, MS, 63, PhD(appl physics), 68. *Prof Exp:* Res scientist, Corning Glass Works, 67-74, res supvr-mgr, 74-90, MGR, INT TECHNOL, CORNING, INC, 90- *Concurrent Pos:* Chmn glass div, Am Ceramic Soc, 88-89. *Mem:* Am Ceramic Soc; Optic Soc Am; Mat Res Soc. *Res:* Structure properties and composition of glass and glass ceramics; photochromic glass; interaction of light with glasses and ceramics. *Mailing Add:* Corning Inc Sullivan Park FR-2-11 Corning NY 14831

SEWARD, WILLIAM DAVIS, b Richmond, Va, Mar 14, 38; m 60; c 1. SOLID STATE PHYSICS. *Educ:* Univ Richmond, BS, 60; Cornell Univ, PhD(physics), 65. *Prof Exp:* Res assoc physics, Univ Ill, 65-67; asst prof, Univ Utah, 67-73; asst prof physics, Pomona Col, 73-76, assoc prof, 76-81; ENG STAFF, ENG RES ASSOCS, VIENNA, VA, 81- *Mem:* Am Phys Soc. *Res:* Solid helium; computer aided instruction, maintenance and testing. *Mailing Add:* Eng Res Assocs 1595 Springhill Rd Vienna VA 22182-2235

SEWELL, CURTIS, JR, b Iowa Park, Tex, Apr 14, 24; m 45; c 3. ELECTRONICS ENGINEERING, GEOLOGY COMPUTER HARDWARE. *Educ:* Hardin Col, 41-42; Tex Tech Col, 42-43; Univ NH, 43; Va Polytech Inst, 44. *Prof Exp:* Electronics engr, Los Alamos Sci Lab, 46-57; mgr eng div, Isotopes, Inc, NJ, 57-62; electronics engr, Lawrence Livermore Lab, 62-88; RETIRED. *Concurrent Pos:* Instr, Chabot Col, Livermore, Calif, 78- *Res:* Electronic circuit design; nuclear and laboratory instrumentation; digital computers; feedback amplifiers; microcomputers and systems for control and data processing. *Mailing Add:* 1625 Alviso Pl Livermore CA 94550

SEWELL, DUANE CAMPBELL, b Oakland, Calif, Aug 15, 18; m 43; c 1. PHYSICS. *Educ:* Col of the Pac, BA, 40. *Hon Degrees:* LLD, Univ of Pacific, 84. *Prof Exp:* Asst secy defense progs, US Dept Energy, 78-81, consult, 81-82; pvt consult, 82-89; Lab asst, Univ Calif, 40-41, physicist, Lawrence Berkeley Lab, 41-52, Lawrence Radiation Lab, 52-78, dir sci opns, 52-59, assoc dir support, 59-73, dep dir, 73-78, assoc dir at large, 89-90, DEP DIR, LAWRENCE LIVERMORE NAT LAB, UNIV CALIF, 90- *Concurrent Pos:* Mem, Gov Radio Defense Adv Comt, State of Calif, 60-; sci officer to gen adv comt, AEC, 63-68. *Honors & Awards:* US Atomic Energy Comn Citation, 71; Distinguished Assoc Award, US Energy Res & Develop Admin, 77; Distinguished Serv Medal, US Dept Energy, 81. *Mem:* Am Phys Soc; Am Nuclear Soc. *Res:* Uranium-235 mass spectrograph development; high energy neutron cross sections; magnetic field measurements. *Mailing Add:* 4265 Drake Ct Livermore CA 94550

SEWELL, FRANK ANDERSON, JR, b Atlanta, Ga, June 25, 34; m 58; c 1. ELECTRONIC PHYSICS. *Educ:* Vanderbilt Univ, BA, 56; Emory Univ, MA, 58; Brown Univ, PhD(physics), 66. *Prof Exp:* Instr physics, Emory Univ, 57-58; sr engr, Sperry Gyroscope Co, 58-60, res staff mem, Sperry Rand Corp, 65-73, mgr optoelectronics dept, 73-78, dir, semiconductor lab, Sperry Res Ctr, Sperry Rand Corp, 78-; SR SCIENTIST, T J WATSON RES CTR, IBM CORP. *Mem:* Sr mem Inst Elec & Electronics Engrs; Am Phys Soc; Soc Photog Scientists & Engrs. *Res:* Semiconductor device physics; electrophotography; high resolution x-ray and electron beam lithography. *Mailing Add:* T J Watson Res Ctr IBM Corp PO Box 218 Yorktown Heights NY 10598

SEWELL, HOMER B, b Red Bird, Mo, Aug 4, 20. ANIMAL NUTRITION. *Educ:* Univ Mo, BS, 53, MS, 63; Univ Ky, PhD(ruminate nutrit), 65. *Prof Exp:* County agt, 53-58, from asst prof to assoc prof, 58-72, PROF ANIMAL HUSB, EXTEN, UNIV MO, COLUMBIA, 72- *Mem:* Am Soc Animal Sci; Sigma Xi. *Res:* Beef cattle feeding; stability of vitamin A liver stores of ruminants. *Mailing Add:* S132 ASC Univ Mo Exten Columbia MO 65211

SEWELL, JOHN I, b Cedartown, Ga, Aug 28, 33; m 60; c 2. AGRICULTURAL ENGINEERING. *Educ:* Univ Ga, BSAE, 54; NC State Univ, MSAE, 58, PhD(agr eng), 62. *Prof Exp:* Instr agr eng, NC State Univ, 60-62; from asst prof to prof agr eng, Univ Tenn, 62-77, assoc head dept, 73-77, asst dean, 77-87, ASSOC DEAN, AGR EXP STA, UNIV TENN, KNOXVILLE, 87- *Concurrent Pos:* Mem, Tenn Air Pollution Control Bd, 90- *Mem:* Sr mem Am Soc Agr Engrs. *Res:* Land grading in alluvial lands for improved surface drainage; animal waste management; infrared aerial remote sensing; irrigation; pond sealing. *Mailing Add:* Assoc Dean Agr Exp Sta 103 Morgan Hall Univ Tenn Knoxville TN 37916

SEWELL, KENNETH GLENN, b Sherman, Tex, July 26, 33; m 54; c 2. PHYSICS. *Educ:* Okla State Univ, BS, 57; Southern Methodist Univ, MS, 60; Tex Christian Univ, PhD(physics), 64. *Prof Exp:* Aerodyn engr, Chance Vought Aircraft Corp, 57-60; res scientist, LTV Res Ctr, Ling-Tempco Vought Inc, 60-61, sr scientist, 64-67; assoc prof physics, Abilene Christian Col, 67-68; sr scientist, LTV Res Ctr Ling-Tempco Vought, Inc, Dallas, 68-69, tech dir Isoray, 69-70; dir res & develop, Varo Inc, 70-74, engr mgr, 74-78, gen mgr, Tex Div, 78-86; dir res & develop, Recon Optical, 86-89; ENG PROJECT MGR, GEN DYNAMICS, FT WORTH, 89- *Mem:* Am Phys Soc; Optical Soc Am; Inst Elec & Electronics Engrs. *Res:* Photoelectric devices; infrared, x-ray imaging, research and development management; nonlinear optics; atomic physics; electrooptics; infrared techniques. *Mailing Add:* 7661 La Bolsa Dallas TX 75248

SEWELL, RAYMOND F, b Seffner, Fla, Feb 20, 23; m 50; c 6. ANIMAL NUTRITION. *Educ:* Univ Fla, BSA, 49, MS, 50; Cornell Univ, PhD, 52. *Prof Exp:* Asst, Univ Fla, 50; from asst prof to assoc prof animal husb, Univ, Ga, 52-60, prof animal sci, 60-67; dir livestock res, Ralston Purina Co, 67-78; dir nutrit & tech serv, Cosby-Hodges Milling Co, 78-83; RETIRED. *Concurrent Pos:* Consult nutrit, 83- *Mem:* Am Soc Animal Sci; Am Dairy Sci Asn; Am Inst Nutrit. *Res:* Nutritional requirements of swine, dairy cattle, beef cattle, horses, laboratory species, catfish, rabbits, canine and feline. *Mailing Add:* 11 Oak Ridge Dr Pelham AL 35124

SEWELL, WINIFRED, b Newport, Wash, Aug 12, 17. INFORMATION SCIENCE, PHARMACY. *Educ:* Wash State Univ, AB, 38; Columbia Univ, BS, 40. *Hon Degrees:* DSc, Philadelphia Col Pharm & Sci, 79. *Prof Exp:* Jr asst librn, Columbia Univ, 40-42; librn, Wellcome Res Labs, NY, 42-46; sr librn, Squibb Inst Med Res, NJ, 46-61; med subj heading specialist, Nat Libr Med, 61-62, dep chief bibliog serv div, 62-64, head drug lit prog, 64-70; adj asst prof, Health Sci Ctr, Univ Md, Baltimore City, 70-85; CONSULT, 63- *Concurrent Pos:* Ed, Unlisted Drugs, 49-59 & 61-64; mem comn pharmaceut abstr, Int Fedn Pharm, 58-60; ad hoc comt on patent off steroid code, 59-60; comt current med terminology, AMA, 62-64; consult, Winthrop Labs, 63; comt mod methods handling chem info, Nat Acad Sci-Nat Res Coun, 64-67; adj lectr med lit, Univ Md, 69-; consult, Nat Asn Hosp Develop, Fed Libr Comt, Sheppard & Enoch Pratt Hosps, Biospherics, Nat Libr Med & Indust, acad & govt groups, 70-; ed, health affairs series of Gale Info Guides, 72-81; consult, Nat Health Planning Info Ctr, 75-81; Nat Libr Med res grant, 81-85. *Honors & Awards:* Eliot Prize, Med Libr Asn, 77. *Mem:* Drug Info Asn (vpres, 65-66, pres, 70-71); Spec Libr Asn (pres, 60-61); Am Libr Asn; Am Soc Info Sci; fel Med Libr Asn. *Res:* Coordination of chemical and biomedical terminology, especially in on-line information retrieval systems; library science; drug information centers; medical information transfer; online searching behavior of pharmacists and pathologists; bibliometrics of molecular biology. *Mailing Add:* 6513 76th Pl Cabin John MD 20818

SEXSMITH, DAVID RANDAL, b Niagara Falls, Ont, June 8, 33; nat US; m 82; c 3. ORGANIC CHEMISTRY. *Educ:* Kenyon Col, AB, 55; Univ Rochester, PhD(chem), 59. *Prof Exp:* Res chemist, Am Cyanamid Co, 58-65, group leader, 65-66; mgr res, Power Chem Div, 66-72, dir res, 72-74, vpres res, 74-80, VPRES TECHNOL, DREW CHEM CORP, DIV ASHLAND CHEM, 80- *Mem:* Am Chem Soc; Am Soc Testing & Mat; Tech Asn Pulp & Paper Indust; Nat Asn Corrosion Engr; Licensing Execs Soc. *Res:* Polymer application; water and waste treatment. *Mailing Add:* Drew Div Ashland Chem One Drew Chem Plaza Boonton NJ 07005

SEXSMITH, FREDERICK HAMILTON, b Ft Erie, Ont, Mar 30, 29; US citizen; m 64; c 2. CHEMISTRY. *Educ:* Queen's Univ, Ont, BA, 51, MA, 53; Princeton Univ, MA, 54, PhD(phys chem), 57. *Prof Exp:* Supvr, Chicopee Mfg Co, 56-62, dir res specialty chem, Refined Prod Co, 62-64 & Refined Prod Co Div, Millmaster-Onyx, 64-65; sect head, Ethicon Inc, 65-66; mgr Res & Develop Div, Hughson Chem Co Div, 66-80, dir res, Chem Prod Group, 80-84, div mgr, Elastomer Prod Div, 84-87, MGR, NEW BUS & TECHNOL, INT GROUP, LORD CORP, 87- *Concurrent Pos:* Textile Res Inst fel, Princeton Univ, 57. *Mem:* Am Chem Soc; Chem Inst Can; Fiber Soc; Am Asn Textile Chemists & Colorists; Am Inst Chemists. *Res:* Applied polymer chemistry; colloid science; surface chemistry; emulsion polymerization; textile finishes; adhesives for elastomers; polyurethane coatings; rubber chemicals. *Mailing Add:* Lord Chem Int Group 2000 W Grandview Blvd Erie PA 16514-0038

SEXSMITH, ROBERT G, b Regina, Can, Apr 13, 38; m 64; c 3. STRUCTURAL ENGINEERING. *Educ:* Univ BC, BASc, 61; Stanford Univ, MS, 63, Engr, 66, PhD(civil eng), 67. *Prof Exp:* Design engr, Phillips, Barratt & Partners, 63-64; from asst prof to assoc prof struct eng, Cornell Univ, 67-76; res scientist, Western Forest Prod Lab, 76-79; sr engr, 79-81, PRIN, BUCKLAND & TAYLOR LTD, 81- *Concurrent Pos:* Mem, Struct Div Exec Comt, Can Soc Civil Eng, 77- *Mem:* Am Concrete Inst; Am Soc Civil Engrs; Can Soc Civil Eng. *Res:* Application of probabilistic concepts to structural engineering; structural mechanics. *Mailing Add:* Buckland and Taylor Ltd 1591 Bowser North Vancouver BC V7P 2Y4 Can

SEXTON, ALAN WILLIAM, b Newark, NJ, Mar 25, 25; m 46; c 3. PHYSIOLOGY. *Educ:* Mont State Univ, BS, 50; Univ Mo, MA, 52, PhD(physiol), 55. *Prof Exp:* Instr physiol, Med Sch, Univ Mo, 54-55; instr human growth, Med Sch, Univ Colo, Denver Ctr, 55-59, physiologist, Child Res Coun, 55-62, asst prof physiol, Med Sch, 59-63, from asst prof to assoc prof phys med, 63-85; RETIRED. *Concurrent Pos:* Mgt consult, Nat Heart, Lung & Blood Inst. *Mem:* Am Physiol Soc; Soc Exp Biol & Med. *Res:* Physiology of muscle contraction; effects of hormones on muscle contractility. *Mailing Add:* 30163 Canterbury Circle Evergreen CO 80439

SEXTON, KEN, b Moscow, Idaho, Nov 6, 49. ENVIRONMENTAL RISK ASSESSMENT, TOXIC AIR POLLUTANTS. *Educ:* USAF Acad, BS, 72; Wash State Univ, MS, 77; Tex Tech Univ, MA, 79; Harvard Univ, PhD(environ health), 83. *Prof Exp:* Comput officer, USAF, 72-74; res asst, Wash State Univ, 75-77, environ engr chem eng, 77-79; environ engr, Acurex Corp, 79-80; prog dir, Calif Dept Health, 83-85; dir, Sci Rev, Health Effects Inst, 85-87, DIR, OFF HEALTH RES, US EPA, 87- *Mem:* Am Chem Soc; Am Pub Health Asn; Air Pollution Control Asn; NY Acad Sci; AAAS; Am Soc Heating, Refrig & Air-Conditioning Eng. *Res:* Human exposure to toxic air pollution, particularly health risk assessment; environmental policy analysis; indoor air quality in non-occupational environments. *Mailing Add:* Off Health Res US EPA 401 M St SW Wash DC 20460

SEXTON, MICHAEL RAY, energy systems, turbomachinery, for more information see previous edition

SEXTON, OWEN J, b Philadelphia, Pa, July 11, 26; m 52; c 4. ECOLOGY, VERTEBRATE ZOOLOGY. *Educ:* Oberlin Col, BA, 51; Univ Mich, MA, 53, PhD(vert natural hist), 57. *Prof Exp:* From instr to assoc prof, 55-68, exec officer dept, 64-66, PROF BIOL, WASH UNIV, 68- *Concurrent Pos:* NSF fel, Univ Chicago, 66-67; trop consult, UNESCO, 74-75; prof zool, Univ Mich Biol Sta, 75-83; vis res fel, Univ New Eng, NSW, Australia, 84. *Mem:* Ecol Soc Am; Am Soc Ichthyologists & Herpetologists. *Res:* Reproductive cycles in tropical vertebrates; predation in reference to color polymorphism and mimicry; habitat structure; tropical versus temperate biology; amphibian development; evolution of viviparity in reptiles; reptilian hibernation. *Mailing Add:* Dept Biol Wash Univ St Louis MO 63130

SEXTON, THOMAS JOHN, b Mt Holly, NJ, Aug 11, 42; m 64; c 3. CRYOBIOLOGY, ANIMAL PHYSIOLOGY. *Educ:* Del Valley Col, BS, 64; Univ NH, MS, 66; Pa State Univ, PhD(dairy sci), 72. *Prof Exp:* Res asst, Univ NH, 64-66 & Univ Conn, 66-67; res assoc, Pa State Univ, 67-71; res physiologist, 71-83, RES LEADER, AGR RES SERV, USDA, 83- *Concurrent Pos:* Adj prof, Auburn Univ, Ala, 84- *Honors & Awards:* Res Award, Poultry Sci Asn, 82. *Mem:* Poultry Sci Asn; World's Poultry Sci Asn; Sigma Xi; Soc Cryobiol. *Res:* Critical problems that limit the reproductive efficiency of the avian male; cryogenic preservation of semen, artificial insemination and isolation of female gamete. *Mailing Add:* Avian Physiol Lab B-262 Beltsville Agr Res Ctr USDA Beltsville MD 20705

SEXTRO, RICHARD GEORGE, b Odell, Nebr, Dec 31, 44; m 67; c 2. ENVIRONMENTAL SCIENCES. *Educ:* Carnegie Inst Technol, BS, 67; Univ Calif, Berkeley, MS, 69, PhD(nuclear chem), 73. *Prof Exp:* Fel nuclear chem, Lawrence Berkeley Lab, 74-75 & energy policy, NSF, 75-76; energy assessment, 76-82, STAFF SCIENTIST, INDOOR AIR QUAL, LAWRENCE BERKELEY LAB, 82- *Mem:* AAAS; Air Pollution Control Asn. *Res:* Measurement of indoor air pollutants, with emphasis on concentrations of radon and progeny in indoor air. *Mailing Add:* Lawrence Berkeley Lab Bldg 90-3058 Berkeley CA 94720

SEYA, TSUKASA, b Fukushima Prefecture, Oct 18, 50; m 90; c 2. COMPLEMENT RECEPTORS. *Educ:* Hokkaido Univ, Master D, 76, PhD, 84, MD, 87. *Prof Exp:* Resident clotting factors, Dept Internal Med, Sch Med, Hokkaido Univ, 76-79, res fel complement, Fac Pharmaceut Sci, 79-84; res assoc complement, Sch Med, Wash Univ, 84-87; ASSOC DIR COMPLEMENT, CTR ADULT DIS, OSAKA, 87- *Concurrent Pos:* Lectr, Osaka Univ. *Mem:* Am Asn Immunologists; NY Acad Sci. *Res:* Complement proteins, receptors, regulatory molecules and adhesion molecules participating in host-defense system which involves host lymphocytes phagocytes and cytokine networks. *Mailing Add:* Dept Immunol Ctr Adult Dis 1-3-3 Nakamichi Higashinari-ku Osaka 537 Japan

SEYB, LESLIE PHILIP, b Franklin, Iowa, May 11, 15; m 39; c 2. ENVIRONMENTAL CHEMISTRY. *Educ:* Coe Col, AB, 35; Univ Iowa, MS, 37, PhD(org chem), 39. *Prof Exp:* Chem asst, Patent Dept, Phillips Petrol Co, 39-42; res chemist & group leader, Diamond Alkali Co, 42-50, mgr res, 50-53, assoc dir res, 53-63; chief phys sci, Pac Northwest Water Lab, Environ Protection Agency, 64-67, asst prog dir eutrophication, 67-71, asst dir, 71-78; RETIRED. *Mem:* Am Chem Soc. *Res:* Fatty acids; wetting agents; textile bleaching; heterocyclic N-compounds; ethylene derivatives; N-chloro organic; chloro hydrocarbons; textile flameproofing; pesticides; xylene derivatives; metallo organics; acetylene derivatives; water pollution and industrial waste control; water and analytical chemistry. *Mailing Add:* 2960 NW Jackson St Corvallis OR 97330

SEYBERT, DAVID WAYNE, b Hazleton, Pa, May 2, 50; m; c 2. ENZYMOLOGY. *Educ:* Bloomsburg State Col, BA, 72; Cornell Univ, PhD(biochem), 76. *Prof Exp:* Res assoc, Med Ctr, Duke Univ, 76-79; asst prof, 79-82, ASSOC PROF CHEM, DUQUESNE UNIV, 82- *Concurrent Pos:* NIH prin investr, Duquesne Univ, 80- *Mem:* Am Chem Soc; AAAS; Sigma Xi. *Res:* Enzymology of biological oxidation, with particular emphasis on hemoproteins and cytochrome P-450 catalyzed steroid hydroxylations, lipid peroxidation. *Mailing Add:* Dept Chem Duquesne Univ Pittsburgh PA 15282

SEYBOLD, PAUL GRANT, b Camden, NJ, July 11, 37. STRUCTURE-PROPERTY RELATIONS, LUMINESCENCE. *Educ:* Cornell Univ, BA, 60; Harvard Univ, PhD(biophys), 68. *Prof Exp:* Postdoctoral fel theoret chem, Univ Uppsala, Sweden, 67-69; postdoctoral fel biochem, Univ Ill, 69-70; from asst prof to PROF CHEM, WRIGHT STATE UNIV, 70- *Concurrent Pos:* Vis scientist, Monell Chem Senses Ctr, 73, Univ Wash, 81; vis prof, Stockholm Univ, Sweden, 78-79, Univ Fla, 86; vis scholar, Univ Calif, San Diego, 86-87. *Mem:* Am Chem Soc; AAAS; Sigma Xi. *Res:* Development of molecular structure, property relationships for chemicals; molecular luminescence spectroscopy, especially room-temperature phosphorescence. *Mailing Add:* Dept Chem Wright State Univ Dayton OH 45435

SEYBOLD, VIRGINIA SUSAN (DICK), b Milwaukee, Wis, March 23, 51; m 72. NEUROSCIENCE, SPINAL CORD. *Educ:* Col William & Mary, BS, 72; Univ Minn, PhD(pharmacol), 77. *Prof Exp:* Instr, 78-80, ASSOC PROF NEUROANAT, DEPT ANAT, UNIV MINN, 80- *Mem:* Soc Neurosci. *Res:* Determination of transmitter-coded neuronal circuitry in the spinal cord using immunohistochemical, labeling via retrograde transport and autoradiographic methods with focus on pathways for pain, analgesia and autonomic function. *Mailing Add:* Anat 4-135 Univ Minn Jackson Hall 321 Church St SE Minneapolis MN 55455

SEYDEL, FRANK DAVID, b Davenport, Iowa, May 15, 44; m 70; c 2. BIOETHICS, BIOCHEMISTRY. *Educ:* Iowa Wesleyan Col, BS, 66; Iowa State Univ, PhD(biochem, cell biol), 73; Princeton Theol Sem, MDiv, 76; Am Bd Med Genetics, dipl, 90. *Prof Exp:* Asst prof chem, Univ Tenn, 76-80; assoc prof, Friends Univ, Wichita, 80-83; DIR, BIOCHEM GENETICS LAB, DEPT OBSTET/GYNEC, SCH MED, GEORGETOWN UNIV, 83- *Concurrent Pos:* Minister, Iowa Conf United Methodist Church, 74- Handicapped. *Mem:* Sigma Xi; Am Soc Human Genetics; Am Asn Ment Retardation. *Res:* Analysis of alfa feto-protein for prenatal diagnosis of neural tube defects. *Mailing Add:* 11134 Oak Leaf Dr Silver Spring MD 20901

SEYDEL, ROBERT E, b Davenport, Iowa, Aug 29, 42. MATHEMATICS. *Educ:* Iowa Wesleyan Col, BS, 65; Univ Iowa, PhD(math), 73. *Prof Exp:* Researcher, Lockheed Space Co, 79-85; sr engr, McDonald Douglas, 85-88; PROF MED, QUINCY COL, 88- *Res:* Finite projective planes. *Mailing Add:* 48 Spring Ct St Peters MO 63376

SEYER, JEROME MICHAEL, b Oran, Mo, Jan 2, 37; m 69; c 2. BIOCHEMISTRY. *Educ:* Univ Mo, Columbia, BS, 59, PhD(biochem), 66. *Prof Exp:* Orthop res fel, Mass Gen Hosp, Boston, 66-69; prof exp res asst biochem, Univ Mo, Columbia, 65-66; res assoc orthop res, Mass Gen Hosp, Boston, 66-71; res assoc orthop res, Childrens Hosp, Boston, 71-74; from asst prof to assoc prof, 74-84, PROF BIOCHEM, UNIV TENN, MEMPHIS, 84- *Concurrent Pos:* Res assoc, Harvard Med Sch, 66-; res chemist, Vet Admin Hosp, Memphis, 74-86, Carrer Scientist, 86- *Mem:* AAAS; Am Chem Soc; Orthop Res Soc. *Res:* Biological mechanism of calcification of bone and dental enamel and development of cartilage collagen; special emphasis of the two above areas along protein sequence analysis; diseases in connective tissues with special reference to fibrosis of lung, liver and scar tissue and amino acid sequence of collagen. *Mailing Add:* Vet Admin Hosp Res Serv 1030 Jefferson Ave Memphis TN 38104

SEYFERT, CARL K, JR, b Pecos, Tex, Feb 12, 38; m 60; c 2. STRUCTURAL GEOLOGY, PETROLOGY. *Educ:* Vanderbilt Univ, BA, 60; Stanford Univ, PhD(geol), 65. *Prof Exp:* From instr to asst prof geol, Queens Col, NY, 64-67; chmn dept, 67-72, assoc prof, 67-72, PROF GEOL, STATE UNIV NY COL BUFFALO, 72- *Concurrent Pos:* Geol Soc Am Penrose grant, Stanford Univ, 62-64, Shell grant fundamental res, 62-64; Sigma Xi grant, Queens Col, NY, 65-68; Res Found State Univ NY grant, State Univ NY Col Buffalo, 69-72. *Mem:* AAAS; Geol Soc Am; Am Geophys Union; Nat Asn Geol Teachers; Sigma Xi. *Res:* Igneous and metamorphic petrology and structural geology; compositional variation within granitic plutons; paleomagnetism and geotectonics; reconstruction of large scale movements of continents throughout geologic time. *Mailing Add:* Dept Geosci State Univ NY 1300 Elmwood Ave Buffalo NY 14222-1095

SEYFERTH, DIETMAR, b Chemnitz, Ger, Jan 11, 29; nat US; m 56; c 3. ORGANOMETALLIC CHEMISTRY. *Educ:* Univ Buffalo, BA, 51; Harvard Univ, MA, 53, PhD(inorg chem), 55. *Hon Degrees:* Dr, Univ Aix-Marseille III. *Prof Exp:* Res chemist, Dow Corning Corp, 55-56; res assoc, Harvard Univ, 56-57; from instr to assoc prof, 57-65, PROF CHEM, MASS INST TECHNOL, 65-, ROBERT T HASLAM & BRADLEY DEWEY PROF CHEM, 83- *Concurrent Pos:* Regional ed, J Organometal Chem, 63-81, coord ed, Organometal Chem Rev, 64-81; Guggenheim fel, 68; ed, Organometals, Am Chem Soc, 81- *Honors & Awards:* Frederic Stanley Kipping Award in Organosilicon Chem, 72; Distinguished Serv Award Advan Inorg Chem, Am Chem Soc; Sr Award, Alexander von Humboldt Found. *Mem:* Am Chem Soc; Royal Soc Chem; fel Ger Acad Scientists Leopoldina; fel AAAS; Ger Chem Soc. *Res:* Main group, especially silicon, lithium and mercury and transition metal organometallic chemistry; organophosphorus chemistry; organic synthesis; applications of organometallic chemistry to ceramics. *Mailing Add:* Dept Chem 4-382 77 Mass Ave Mass Inst Technol Cambridge MA 02139

SEYFRIED, THOMAS NEIL, b Flushing, NY, July 25, 46; m 73; c 3. NEUROBIOLOGY. *Educ:* St Francis Col, Maine, BA, 68; Ill State Univ, MS, 73; Univ Ill, PhD(genetics), 76. *Prof Exp:* Fel neurogenetics, Dept Neurol, Sch Med, Yale Univ, 76-79, asst prof, 79-84; assoc prof, 85-90, PROF BIOL, BOSTON COL, 90- *Mem:* Genetics Soc Am; AAAS; Am Soc Neurochem; Int Soc Neurochem; Soc Neurosci; Am Epilepsy Soc. *Res:* Cellular localization and function of brain gangliosides; developmental genetics of inherited epilepsy in mice; genetic control of brain myelinogenesis; cellular adhesion in brain tumors. *Mailing Add:* Dept Biol Boston Col Chestnut Hill MA 02167

SEYLER, CHARLES EUGENE, b Eustis Fla, June, 2, 48. PLASMA PHYSICS. *Educ:* Univ S Fla, BA, 70, MA, 72; Univ Iowa, PhD(physics), 75. *Prof Exp:* Res scientist plasma physics, Courant Inst Math Sci, NY Univ, 75-78; staff mem, Los Alamos Nat Lab, 78-81; ASST PROF PLASMA PHYSICS ELEC ENG, CORNELL UNIV, 81- *Concurrent Pos:* Vis staff mem, Los Alamos Nat Lab, 81- *Res:* Equilibrium, stability and transport of plasmas with applications to the development of controlled thermonuclear fusion reactors. *Mailing Add:* Phillips Hall Sch Elec Eng Cornell Univ Ithaca NY 14853

SEYLER, RICHARD G, b Du Bois, Pa, June 14, 33; m 64. NUCLEAR PHYSICS. *Educ:* Pa State Univ, BS, 55, MS, 59, PhD(physics), 61. *Prof Exp:* Vis asst prof, 61-63, from asst prof to assoc prof, 63-73, PROF PHYSICS, OHIO STATE UNIV, 73- *Mem:* Am Phys Soc. *Res:* Theoretical low-energy nuclear physics. *Mailing Add:* Dept Physics Ohio State Univ Columbus OH 43210

SEYMOUR, ALLYN H, b Seattle, Wash, Aug 1, 13; m 40; c 3. RADIATION ECOLOGY, FISH BIOLOGY. *Educ:* Univ Wash, BS, 37, PhD, 56. *Prof Exp:* Jr scientist, State Dept Fisheries, Wash, 40-41; asst scientist, Int Fisheries Comn, 42-47; res assoc & asst dir appl fisheries lab, Univ Wash, 48-56; marine biologist, Div Biol & Med, US AEC, DC, 56-58; asst dir lab radiation biol, 58-63, prof fisheries, Col Fisheries, 63-79 dir lab radiation ecol, 66-78, EMER PROF, COL FISHERIES, UNIV WASH, 79- *Concurrent Pos:* Chmn panel radioactivity in the marine environ, 68-71, & mem panel nuclear weapons effects, Nat Res Coun, 74-75; sci adv, Nat Coun Radiation Protection & Measurement, 87- *Mem:* Am Inst Fishery Res Biol; Health Physics Soc; Sigma Xi. *Res:* Biological distribution of radioisotopes; aquatic radioecology. *Mailing Add:* 5855 Oberlin NE Seattle WA 98105-2125

SEYMOUR, BRIAN RICHARD, b Chesterfield, UK, Sept 25, 44; US citizen; m 65; c 3. APPLIED MATHEMATICS, ACOUSTICS. *Educ:* Univ Manchester, UK, BSc, 65; Univ Nottingham, UK, PhD(theoret mech), 69. *Prof Exp:* Asst prof appl math, Lehigh Univ, Bethlehem, Pa, 69-70; asst prof math, NY Univ, 70-73; vis fel math, St Catherine's Col, Oxford, UK, 78-79; vis prof math, Monash & Melbourne Univ, Melbourne, Australia, 84-85; from asst prof to assoc prof, 73-81, actg dir appl math, Inst Appl Math, 86-88, PROF MATH, UNIV BC, VANCOUVER, 81-, DIR, INST APPL MATH, 88- *Concurrent Pos:* Sr fel, Sci Res Coun, UK, 78; sr Killam fel, Killam Found, Can, 84. *Mem:* Can Appl Math Soc; Can Meteorol & Oceanog Soc. *Res:* Analytical and numerical solutions of problems in fluid mechanics and elasticity; wave propagation in nonlinear and inhomogenesis materials. *Mailing Add:* Inst Appl Math Univ BC Vancouver BC V6T 1Z2

SEYMOUR, KEITH GOLDIN, b Fairfax, Mo, Jan 25, 22; m 43; c 2. PESTICIDE CHEMISTRY, FORMULATIONS. *Educ:* Iowa State Univ, BS, 43, MS, 50; Tex A&M Univ, PhD(soil chem), 54. *Prof Exp:* Chemist, Tex Div, Dow Chem Co, 54-59, res specialist, 59-65, group leader, Bioprod Dept, 65-71, res mgr, Agr Prod Dept, Dow Chem Co, 71-82; RETIRED. *Concurrent Pos:* Consult, UN Indust Develop Orgn, 83-84. *Mem:* Sigma Xi; Am Chem Soc. *Res:* Colloid, surface, agricultural and physical chemistry; pesticide formulations and application systems. *Mailing Add:* PO Box 115 Fairfax MO 64446

SEYMOUR, MICHAEL DENNIS, b St Cloud, Minn, May 17, 50; m 79; c 4. ANALYTICAL CHEMISTRY. *Educ:* St John's Univ, BA, 72; Univ Ariz, PhD(anal chem), 78. *Prof Exp:* Asst prof, 78-84, ASSOC PROF CHEM, HOPE COL, 85- *Concurrent Pos:* Vis scientist, Nat Ctr Atmospheric Res, 85-86. *Mem:* Am Chem Soc; Sigma Xi. *Res:* Chemical education; application of chemical amplifiers for environmental studies. *Mailing Add:* Dept Chem Hope Col Holland MI 49423

SEYMOUR, RAYMOND B(ENEDICT), b Brookline, Mass, July 26, 12; m 36; c 4. PLASTICS, ELASTOMERS. *Educ:* Univ NH, BS, 33, MS, 35; Univ Iowa, PhD(org chem), 37. *Prof Exp:* Instr chem, Univ NH, 33-35; instr org chem, Univ Iowa, 35-37; res chemist, Goodyear Tire & Rubber Co, 37-39; chief chemist, Atlas Minerals & Chem Co, 39-41; res group leader, Monsanto Co, 41-45; prof & dir res, Univ Chattanooga, 45-48; dir res, Johnson & Johnson, 48-49; pres & tech dir, Atlas Minerals & Chem Co, 49-55; pres & dir res, Loven Calif, 55-60; prof & chmn chem, Sul Ross State Univ, 60-64; prof & assoc chmn chem, Univ Houston, 64-76; DISTINGUISHED PROF POLYMER SCI, UNIV SOUTHERN MISS, 76- *Concurrent Pos:* Nat counr, Am Chem Soc, 46-48 & 74-76; nat dir, Am Asn Textile Chemists & Colorists, 47-49 & Soc Plastics Indust, 55-58; adj prof polymer sci, Los Angeles Trade Tech Col, 56-60; vis scientist, AAAS, 60-69, US Dept Com-AID, Bangladesh, 68, Australian Soc Coatings, Rubber & Plastics, 77 & Nat Acad Sci vis scientist, Univ Dacca, Bangladesh, 77; dir res, Inst NSF, 65; tech ed, Modern Plastics, 73-77; exchange prof, US Nat Acad Sci, Yugoslavia, 78; chair chem, TamKang Univ, Taipei, Taiwan, 78; chmn, Experimat Conf, Bordeux, France, 87. *Honors & Awards:* Charles H Herty Award, Am Chem Soc, 85; Chem Pioneer Award, Am Inst Chemists, 85; Int Plastics Award, Soc Plastics Engrs, 89; Carl Marvel Mem Lectr, Univ Rio de Janeiro, 89. *Mem:* Am Chem Soc; Soc Plastics Indust; Am Asn Textile Chemist & Colorists; Am Inst Chem Engrs; fel AAAS; fel Am Inst Chemists. *Res:* Invention of engineering plastics, high impact polystyrene, silica-filled thermoplastics, cultured marble, cellulose derivatives and ionomers. *Mailing Add:* Dept Polymer Sci Univ Southern Miss PO Box SS 10076 Hattiesburg MS 39406-0076

SEYMOUR, RICHARD JONES, b Harrisburg, Pa, Aug 28, 29; c 1. OCEANOGRAPHY. *Educ:* US Naval Acad, BS, 51; Univ Calif, San Diego, PhD(oceanog), 74. *Prof Exp:* Vpres, Wire Equip Mfg Co, Inc, 51-59; head rocket develop, Elkton Div, Thiokol Chem Corp, 59-62; chief engr, United Technol Div, United Aircraft Corp, 62-69; res asst oceanog, Scripps Inst Oceanog, Univ Calif, San Diego, 70-73; staff oceanogr, Calif Dept Boating & Waterways, 74-84; HEAD OCEAN ENG RES, SCRIPPS INST OCEANOG, UNIV CALIF, SAN DIEGO, 84- *Mem:* Am Soc Civil Engrs; Am Soc Mech Engrs. *Res:* Sediment transport; wave measurement and analysis; coastal processes. *Mailing Add:* Univ Calif 9500 Gilman Dr La Jolla CA 92093-0222

SEYMOUR, ROLAND LEE, b Palestine, Ill, Oct 12, 39. MYCOLOGY. *Educ:* Eastern Ill Univ, BS, 61; Va Polytech Inst & State Univ, PhD(bot), 65. *Prof Exp:* Asst bot, Eastern Ill Univ, 59-61; res asst, Va Polytech Inst & State Univ, 61-65, asst prof biol, 65-66; vis instr bot, Duke Univ, 66-67; asst prof biol, Univ Pittsburgh, 67-70; asst prof, 70-74, ASSOC PROF BOT, OHIO STATE UNIV, 74- *Concurrent Pos:* NSF grant, Univ Pittsburgh, 68-70; NSF grant, Ohio State Univ, 70-, Arctic Inst NAm grant, 72-74; consult mycologist, Bausch & Lomb, 72- *Mem:* Mycol Soc Am; Asn Trop Biol. *Res:* Aquatic mycology; role of aquatic fungi in tropical rainforest ecosystem; systematics and distribution of tropical aquatic phycomycetes; fungal parasites of mosquito and black fly of Mexico and West Africa. *Mailing Add:* Dept Bot 374 B12 Ohio State Univ 1735 Neil Ave Columbus OH 43210

SFAT, MICHAEL R(UDOLPH), b Timisoara, Rumania, Oct 28, 21; nat US; m 48; c 2. BIOTECHNOLOGY. *Educ:* Cornell Univ, BChE, 43, MChE, 47. *Prof Exp:* Res assoc chem eng, Cornell Univ, 43-44; from asst microbiologist to sr microbiologist, Merck & Co, Inc, 47-52; chem engr, Labs, Pabst Brewing Co, 52-54; res dir, Rahr Malting Co, 54-58, coordr res & develop, 58-60, vpres res & develop, 60-69; PRES, BIO-TECH RESOURCES INC, 69- *Mem:* Am Chem Soc; Am Soc Brewing Chem (pres, 74); Am Soc Microbiol; Am Inst Chem Engrs; Inst Food Technol; Am Asn Cereal Chem. *Res:* Industrial fermentations; malting; enzymes; bioengineering. *Mailing Add:* Bio-Tech Resources Inc 1035 S Seventh St Manitowoc WI 54220

SFERRA, PASQUALE RICHARD, b St Louis, Mo, Sept 2, 27; m 50; c 4. ENTOMOLOGY. *Educ:* Washington Univ, AB, 52; Rutgers Univ, MSc, 55, PhD(entom), 57. *Prof Exp:* Asst prof entom, Exp Sta, State Univ NY Col Agr, Cornell Univ, 56-62, asst prof biol, 62-63; assoc prof, 63-73, PROF BIOL, COL MT ST JOSEPH, 73- *Concurrent Pos:* Biol sci adv, US Environ Protection Agency. *Mem:* AAAS; Entom Soc Am; Am Inst Biol Sci; Sigma Xi. *Res:* Respiratory pacing in insects; insect toxicology; carbohydrate metabolism in insects; taxonomy of desert insects. *Mailing Add:* 5645 Candlelite Terr Cincinnati OH 45238

SFORZA, PASQUALE M, b New York, NY, Mar 5, 41; m 63; c 3. FLUID MECHANICS. *Educ:* Polytech Inst Brooklyn, BAeE, 61, MS, 62, PhD(astronaut), 65. *Prof Exp:* Res fel aerospace eng, Polytech Inst Brooklyn, 61-62, res asst, 62-63; res assoc aerospace eng, 63-65, from asst prof to assoc prof, 65-77, head dept mech & aerospace eng, 83-86, PROF AEROSPACE & MECH ENG, POLYTECH UNIV, 77-, HEAD DEPT AEROSPACE ENG, 88- *Concurrent Pos:* Pres, Flowpower, Inc, 78-; assoc ed, J Am Inst Aeronaut & Astronaut, 80-82, book review ed, 83- *Honors & Awards:* Technol Achievement Award, Am Inst Aeronaut & Astronaut, 77. *Mem:* Assoc fel Am Inst Aeronaut & Astronaut; Am Soc Mech Engrs; NY Acad Sci. *Res:* Theoretical and experimental fluid mechanics; high temperature energy transfer; hypersonic aerodynamics; wind engineering; energy conversion; airplane and engine design. *Mailing Add:* Polytech Univ Rte 110 Farmingdale NY 11735

SFORZINI, RICHARD HENRY, b Rochester, NY, July 25, 24; m 47; c 7. JET PROPULSION. *Educ:* US Mil Acad, BSc, 47; Mass Inst Technol, MechE, 54. *Prof Exp:* Instr ord, US Mil Acad, 54-56, asst prof, 56-57; proj dir missile systs, Res & Develop Div, Army Rocket & Guided Missile Agency, Redstone Arsenal, Ala, 58-59; engr, Huntsville Div, Thiokol Chem Corp, Ala, 59-62, mgr eng dept, 62-64, dir eng, Space Booster Div, Brunswick, Ga, 64-66; vis prof, 66-67, prof, 67-85, EMER PROF AEROSPACE ENG, AUBURN UNIV, 85- *Concurrent Pos:* Chmn solid rocket tech comt, Am Inst Aeronaut & Astronaut, 80-81. *Mem:* Assoc fel Am Inst Aeronaut & Astronaut. *Res:* Aircraft and missile propulsion systems, especially internal ballistics, combustion, ignition, swirling flow through nozzles and problems of very large solid-propellant rockets; aerodynamics. *Mailing Add:* 912 Cherokee Rd Auburn AL 36830-2723

SGOUTAS, DEMETRIOS SPIROS, b Thessaloniki, Greece, Sept 2, 29; US citizen; m 61; c 2. BIOCHEMISTRY, CLINICAL CHEMISTRY. *Educ:* Univ Thessaloniki, BS, 54; Univ Ill, Urbana, PhD, 63. *Prof Exp:* Res assoc food chem, Univ Ill, Urbana, 63-64, asst prof, 64-65; asst prof chem, Univ Thessaloniki, 65-66; asst prof food chem, Univ Ill, Urbana, 66-70; from asst prof to assoc prof path, 70-74, PROF PATH & LAB MED, MED SCH, EMORY UNIV, 75-, DIR, RADIOIMMUNOASSAY LAB, 73- *Concurrent Pos:* Prin investr, Chicago & Ill Heart Asn grants, 64-69; NIH grants, 67-; prof allied health professions, Emory Univ, 70- *Mem:* Am Chem Soc; Am Oil Chemists Soc; Am Soc Biol Chemists; Am Asn Clin Chemists. *Res:* Lipid metabolism as related to cardiovascular diseases. *Mailing Add:* Dept Path Emory Univ Med Sch 762 WMB Atlanta GA 30322

SGRO, J A, b New Haven, Conn, Nov 22, 37; m; c 2. HUMAN FACTORS. *Educ:* Lehigh Univ, MA, 61; Texas Christian Univ, PhD(psychol), 66. *Prof Exp:* From asst prof to prof psychol, Old Dominion Univ, 67-79; PROF & HEAD PSYCHOL DEPT, VA POLYTECH INST & STATE UNIV, 79- *Mem:* Am Psychol Asn; Psychonomic Soc. *Res:* Factors that contribute to leadership & following, dealing with sexual roles & perception that followers have of leaders & vice versa. *Mailing Add:* Dept Psychol Va Polytech Inst & State Univ 5088 Derring Hall Blacksburg VA 24061-0436

SHA, WILLIAM T, b Kiangsu, China, Sept 13, 28; US citizen; m 57; c 3. NUCLEAR ENGINEERING, NUCLEAR SCIENCE. *Educ:* Polytech Inst Brooklyn, BS, 58; Columbia Univ, DESc, 64. *Prof Exp:* Engr, Combustion Eng Inc, 57-60; fel scientist, Atomic Power Div, Westinghouse Elec Corp, 60-67; sr nuclear engr & dir anal thermal hydraul res prog, Mat & Components Technol Div, 67-80, SR CONSULT, THERMAL HYDRAUL, ARGONNE NAT LAB, 80- *Mem:* Fel Am Nuclear Soc. *Res:* Reactor dynamics; system stability; nuclear-thermal-hydraulic interaction calculation; multiphase fluid mechanics and heat transfer. *Mailing Add:* Mat & Components Technol Div Bldg 308 Argonne Nat Lab Argonne IL 60439

SHAAD, DOROTHY JEAN, b Newton, Mass, Aug 16, 09. OPHTHALMOLOGY, PSYCHOLOGY. *Educ:* Univ Kans, AB, 29; Bryn Mawr Col, MA, 30, PhD(exp psychol), 34; Univ Kans, MD, 44. *Prof Exp:* Asst, Howe Lab Ophthal, Harvard Med Sch, 31-32; instr, St Mary Col, Kans, 34; clin technician, Manhattan Eye, Ear & Throat Hosp, New York, 35-38; intern, Duke Univ Hosp, 44-45; assoc ophthal, Med Ctr, Univ Kans, 45-46, asst prof, 66-77, assoc res vision & ophthal, 71-77; RETIRED. *Concurrent Pos:* Practicing ophthalmologist, 45-69. *Mem:* AMA. *Res:* Light perception and dark adaptation; binocular vision. *Mailing Add:* 2322 W 51st St Shawnee Mission KS 66205-2010

SHA'AFI, RAMADAN ISSA, b Nabi-Rubien, Palestine, June 9, 38; Jordanian citizen; c 2. BIOPHYSICS, PHYSIOLOGY. *Educ:* Univ Ill, BS, 62, MS, 63, PhD(biophys), 65. *Prof Exp:* Fel, Harvard Med Sch, 65-67, instr biophys, 67-69; asst prof physiol, Am Univ Beirut, 69-72; from asst prof to assoc prof, 72-78, PROF PHYSIOL, UNIV CONN HEALTH CTR, 78- *Mem:* Biophys Soc; Am Physiol Soc; Soc Gen Physiol. *Res:* Mechanism of water and solute transport across mammalian red and white cells. *Mailing Add:* Dept Physiol Univ Conn Health Ctr Farmington CT 06032

SHAAK, GRAIG DENNIS, b Harrisburg, Pa, Oct 18, 42; m 75; c 1. SCIENCE ADMINISTRATION, PALEONTOLOGY. *Educ:* Shippensburg Univ, BS, 67; Ind Univ, MAT, 69; Univ Pittsburgh, PhD(geol), 72. *Prof Exp:* Asst cur paleontol, 72-78, chmn dept, 78-79, from asst dir to assoc dir, 79-86, actg dir, 86-87, ASSOC DIR, FLA MUS NATURAL HIST, 87- *Concurrent Pos:* Adj prof geol, dept nat sci, Fla State Mus, Univ Fla, 72-, diving officer, 74- *Mem:* Soc Econ Paleontologists & Mineralogists; Paleont Soc; Geol Soc Am; Paleont Res Inst; Int Paleont Asn; Am Asn Mus. *Res:* Diversity, structure, and evolution of shallow benthic marine communities; succession in late Pleistocene freshwater communities; echinoid evolution, biogeography and biometrics. *Mailing Add:* Fla Mus Natural Hist Univ Fla Gainesville FL 32611

SHAATH, NADIM ALI, b Jaffa, Palestine, Dec 10, 45; m 67; c 1. ORGANIC CHEMISTRY, MEDICINAL CHEMISTRY. *Educ:* Univ Alexandria, BSc, 67; Univ Minn, PhD(org chem), 73. *Prof Exp:* Teaching assoc chem, Univ Minn, 67-72, res specialist med chem, 72-75; from asst prof to assoc prof chem, State Univ NY, Purchase, 75-81; RES DIR, FELTON INT, 81- *Concurrent Pos:* Teaching assoc, Exten Div, Univ Minn, 70-73; prin investr multiple grants, 76- *Mem:* Am Chem Soc; AAAS. *Res:* Neuromuscular junction blocking or paralysing drugs; nuclear magnetic resonance shift reagents; mechanism of organic reactions; metabolism of narcotic stimulants; flavors and fragrances. *Mailing Add:* Felton Int 599 Johnson Ave Brooklyn NY 11237

SHABANA, AHMED ABDELRAOUF, b Domiat, Egypt. KINEMATICS DYNAMICS & CONTROL, COMPUTATIONAL MECHANICS. *Educ:* Cairo Univ, BSc, 74; Ain Shams Univ, MSc, 78; Univ Iowa, PhD (mech eng), 82. *Prof Exp:* Teaching asst mech eng, Ain Shams Univ, Egypt, 74-78; res asst mech eng, Univ Iowa, 79-82, fel, 82-83; asst prof, 83-88, ASSOC PROF MECH ENG, UNIV ILL, CHICAGO, 88- *Concurrent Pos:* Prin investr, Univ Ill Res Bd, 84-85. *Mem:* Am Soc Mech Engrs; Am Acad Mech. *Res:* Kinematics; dynamics and control of multibody systems; computational mechanics and computer aided design; scientific and technical papers on computational mechanics. *Mailing Add:* Dept Mech Eng Univ Ill Chicago PO Box 4348 Chicago IL 60680

SHABANOWITZ, HARRY, b Brooklyn, NY, Nov 11, 18; m 43; c 2. MATHEMATICS. *Educ:* City Col New York, BS, 49; Columbia Univ, MA, 50; Syracuse Univ, PhD(math educ), 74. *Prof Exp:* Sr engr, Westinghouse Elec Corp, 51-65 & Gen Elec Co, 65-66; from assoc prof to prof math, Elmira Col, 66-84; RETIRED. *Concurrent Pos:* Spec lectr, 57-65. *Mem:* Am Math Soc; NY Acad Sci. *Res:* Research and development of high sensitivity television camera tubes; theoretical and experimental investigation of the factors limiting television camera tube performance; study of high quantum efficiency photoemissive surfaces; secondary electron emission; electron optical geometry. *Mailing Add:* 205 Scenic Dr W Horseheads NY 14845

SHABAZZ, ABDULALIM, b Bessener, Ala, May 22, 27. INTEGRAL EQUATIONS, COMPLEX ANALYSIS. *Educ:* Lincoln Univ, AB, 49; Mass Inst Technol, MS, 51; Cornell Univ, PhD(math), 55. *Prof Exp:* Asst prof math, Tuskegee Inst, 56-57; assoc prof & chmn math dept, Atlanta Univ, 57-63; dir educ & minister, Masjid Muhammad, Washington, DC, 63-75, dir adult educ, Masjid Elijah Muhammad, Chicago, 75-79; dir adult educ & Imam, Masjid Wali Muhammad, Detroit, 79-82; from assoc prof to prof math, Umm Al Qura Univ, Makkah, 82-86; PROF MATH, CLARK ATLANTA UNIV, 86-, CHMN DEPT, 89- *Mem:* Sigma Xi; Am Math Soc; Am Soc Eng Educ; Math Asn Am; AAAS. *Res:* Eigen value problems for certain classes of hermitian forms and the analytic continuations of functions represented by certain power series, using padé approximants and the method of Borel, and the applications thereof to single-atom frequency-dependent polarizabilities. *Mailing Add:* Clark Atlanta Univ Box 196 223 James P Brawley Dr SW Atlanta GA 30314-4394

SHABEL, BARRIE STEVEN, b New York, NY, Aug 31, 38; m 62; c 1. MATERIALS CHARACTERIZATION, NONDESTRUCTIVE TESTING. *Educ:* Mass Inst Technol, SB, 59; Rensselaer Polytech Inst, MMetE, 61; Syracuse Univ, PhD(solid state sci), 67. *Prof Exp:* Mat engr, zirconium/columbium alloys, Knolls Atomic Power Lab, 59-60, mat engr radiation damage, 60-63; res engr, aluminum alloys, phys metall div, Alcoa Res Labs, 66-74, sr res engr, 74-75, sr res engr, mech & metalworking, eng properties & design div, 75-78, staff engr, mech & aluminum alloys, 80-84, SCI ASSOC, ALLOY TECHNOL DIV, ALCOA LABS, ALCOA TECH CTR, 84- *Honors & Awards:* IR-100 Award, 77. *Mem:* Am Soc Metals; The Metall Soc; Inst Mining, Metall & Petrol Engrs; Mat Res Soc; Sigma Xi. *Res:* Physical and mechanical metallurgy of aluminum alloys; mechanics of sheet metal forming, plasticity; application of statistics to industrial research; nondestructive testing. *Mailing Add:* 3464 Burnett Dr Murrysville Alcoa Center PA 15668

SHABICA, ANTHONY CHARLES, JR, b Meadville, Pa, Nov 20, 15; m 40; c 3. ORGANIC CHEMISTRY. *Educ:* Brown Univ, ScB, 38; Pa State Univ, MS, 39, PhD(org chem), 42. *Prof Exp:* Jr chemist, Calco Chem Co, 37 & 39; asst, Pa State Univ, 39-42; sr chemist, Merck & Co, Inc, 42-46; head develop dept, Ciba Pharmaceut Co, 46-48, dir develop res, 48-67, vpres develop & control, 67-81; ADJ PROF, COL VIRGIN ISLANDS, 81- *Concurrent Pos:* Mem bd, NJ Coun Res & Develop, 68-, chmn, 76-77; res assoc, Woods Hole Oceanog Inst. *Mem:* Am Pharmaceut Asn; Am Chem Soc; fel Am Inst Chemists; fel NY Acad Sci; Int Pharmaceut Fedn; Sigma Xi. *Res:* Heterocyclic chemistry; natural products; mechanism of the polymerization of olefins; steroidal sapogenins and related compounds alkaloids; process research, development and design. *Mailing Add:* Box 1631 Destin FL 32541

SHABICA, CHARLES WRIGHT, b Elizabeth, NJ, Jan 2, 43; m 67; c 3. COASTAL GEOLOGY. *Educ:* Brown Univ, AB, 65; Univ Chicago, PhD(geol), 71. *Prof Exp:* PROF EARTH SCI, NORTHEASTERN ILL UNIV, 80-, CHMN DEPT, 86-; PRES CHARLES SABICA & ASSOC, COASTAL CONSULTS, 85- *Concurrent Pos:* Coastal consult, City Highland Park, Loyola Univ & others, 84-; mem, Chicago Shoreline Protection Comn, 86-; designer shore protection structures, Great Lakes, 84-; vis prof, Col VI, 80-81; prin investr marine sci curric, NSF grant, 81-84, earth watch exped, 80- *Mem:* Am Asn Petrol Geologists; Sigma Xi; Am Meteorol Soc; AAAS; Am Shore & Beach Asn. *Res:* Coastal processes; coastal engineering. *Mailing Add:* 326 Ridge Winnetka IL 60093

SHACK, ROLAND VINCENT, b Chicago, Ill, Jan 15, 27; m 57; c 4. OPTICS. *Educ:* Univ Md, BS, 49; Am Univ, BA, 51; Univ London, PhD(physics), 65. *Prof Exp:* Physicist, Nat Bur Standards, 49-57 & Perkin-Elmer Corp, 57-64; res assoc, 64-65, assoc prof, 65-70, PROF OPTICS, UNIV ARIZ, 70- *Mem:* Optical Soc Am. *Res:* Optical image evaluation and testing; interferometry; systems analysis. *Mailing Add:* 6918 E Blue Lake Dr Tucson AZ 85715

SHACK, WILLIAM JOHN, b Pittsburgh, Pa, Jan 12, 43; m 75; c 1. APPLIED MECHANICS. *Educ:* Mass Inst Technol, BS, 64; Univ Calif, Berkeley, MS, 65, PhD(appl mech), 68. *Prof Exp:* From asst prof to assoc prof, Mass Inst Technol, 68-75; scientist, 76-85, SR SCIENTIST, ARGONNE NAT LAB, 85-, ASSOC DIR MAT & COMPONENTS TECH DIV, 89- *Mem:* Am Soc Mech Engrs. *Res:* Solid mechanics; fracture mechanics. *Mailing Add:* Argonne Nat Lab Argonne IL 60439

SHACKELFORD, CHARLES L(EWIS), b Wagoner, Okla, Oct 19, 18; m 45; c 1. ELECTRONICS. *Educ:* Okla State Univ, BS, 41; Univ Mo, MS, 42. *Prof Exp:* Repairman, Porum Tel Co, 35-40; asst, Univ Mo, 41-42; engr, Westinghouse Elec Corp, Pa, 42, NJ, 42-52; engr, Chatham Electronics Div, Tung-Sol Elec, Inc, 52-66, chief engr, Power Tube Div, NJ, 66-69; SR ENGR, ELECTRON TUBE DIV, INT TEL & TEL CORP, 69- *Mem:* Sr mem Inst Elec & Electronics Engrs. *Res:* Electron emission; conduction of electricity through gases and vapors. *Mailing Add:* 3916 Oakland Rd Bethlehem PA 18017

SHACKELFORD, ERNEST DABNEY, b Petersburg, Va, Aug 24, 26; m 51; c 5. BIOMEDICAL ENGINEERING, SUBSONIC AERODYNAMICS. *Educ:* Ga Inst Technol, BAE, 46; Med Col Va, MD, 52. *Prof Exp:* Intern, Univ Wis Hosp, Madison, 52-53; physician internal med, NC Mem Hosp, Chapel Hill, 53-55, res physician radiol, 63-67; pvt pract, Asheboro, NC, 55-63; radiologist, diag roentgenol, ultrasound & nuclear med, Randolph Mem Hosp, Inc, 67-90; ADJ FAC FLUID MECH, NC A&T STATE UNIV, 90- *Concurrent Pos:* Consult, Hospice of Randolph, NC, 84- *Mem:* Am Inst Aeronaut & Astronaut; Sigma Xi; Asn Cancer Res; Soc Indust & Appl Math. *Res:* Central nervous system causation factors in pulmonary edema; nuclear medicine; ultrasonology; radiology; fluid dynamics. *Mailing Add:* 203 Shannon Rd Asheboro NC 27203

SHACKELFORD, JAMES FLOYD, b Springfield, Miss, Sept 1, 44; m 71; c 1. MATERIALS SCIENCE ENGINEERING, CERAMICS ENGINEERING. *Educ:* Univ Wash, BS, 66, MS, 67; Univ Calif, PhD(mat sci & eng), 71. *Prof Exp:* Postdoctoral fel mat sci & eng, Univ Calif, Berkeley, 71 & McMaster Univ, Hamilton, Can, 72-73; from asst prof to assoc prof, 73-84, PROF MAT SCI & ENG, UNIV CALIF, DAVIS, 84-, ASSOC DEAN, COL ENG, 84- *Concurrent Pos:* Consult, var indust & legal clients, 72- & Lawrence Livermore Nat Lab, 77-; vis prof, Indian Inst Sci, Bangalore, 89 & Indian Inst Technol, Bombay, 89-90. *Mem:* Am Ceramic Soc; Am Soc Metals Int. *Res:* Author of over 50 publications including an introductory textbook; materials science and engineering; structure of noncrystalline solids; nondestructive testing; biomaterials. *Mailing Add:* Col Eng Univ Calif Davis CA 95616

SHACKELFORD, ROBERT G, b Atlanta, Ga, Oct 25, 36; m 62; c 2. ELECTROOPTICS. *Educ:* Ga Inst Technol, BEE, 59, MSEE, 62, MSPhys, 67. *Prof Exp:* Assoc aircraft engr, Lockheed-Ga Div, Lockheed Aircraft Corp, 59; res asst microwave eng, 59-62, from asst res engr to res engr, 62-68, sr res physicist, 68-69, assoc chief spec tech div, 69-75, chief, Electrooptics Div, 75-77, PRIN RES SCIENTIST & ASSOC DIR, ELECTROMAGNETICS LAB, ENG EXP STA, GA INST TECHNOL, 77- *Mem:* Inst Elec & Electronics Engrs; Sigma Xi. *Res:* Electromagnetic theory; dielectric properties of materials; laser design and development; atmospheric propagation in the infrared-visible spectrum. *Mailing Add:* 5833 W Fayetteville Rd College Park GA 30349

SHACKELFORD, SCOTT ADDISON, b Long Beach, Calif, Aug 11, 44; m 69; c 2. REACTION MECHANISMS, NOBLE GAS COMPOUND REACTIONS. *Educ:* Simpson Col, BA, 66; Northern Ariz Univ, MA, 68; Ariz State Univ, PhD(org chem), 73. *Prof Exp:* res chemist, Frank J Seiler Res Lab, 72-74, prin investr & div chief energetic chem, 74-77; Air Force exchange scientist, Inst Chem Antrieb & Verfahrenstech, WGer, 78-80; chief, Basic Chem Res Sect, Air Force Rocket Propulsion Lab, 80-84; chief, chem & energetics, Europ Off Aerospace Res & Develop, 84-87; SR SCIENTIST, FRANK J SEILER RES LAB, 87- *Concurrent Pos:* Lectr, dept chem, USAF Acad, 74-78, instr, 77, asst prof, 78-; secy, Joint Tech Coord Group, 75-77; prin investr org synthesis & thermochem mech, 81-84, propellant chem task mgr, 82-84; Nat Alliance Treaty Organ Advanced Study Inst Lectr, 89; mem Simpson Col Sci Adv Comt, 83-87, Jannaf Combustion Subcomt Panel, chmn on Propellant Combustion Chem, 90-91. *Mem:* Am Chem Soc. *Res:* Chemical mechanism elucidation with kinetic isotopic effects and deuterium labeling; selective organic compound xenon difluoride fluorination; deuterium isotope effects in condensed-phase decomposition and combustion processes; energetic aliphatic compound syntheses; several US patents. *Mailing Add:* 2134 Wildwood Dr Colorado Springs CO 80918

SHACKELFORD, WALTER MCDONALD, b Birmingham, Ala, Jan 8, 45; m 69; c 3. ANALYTICAL CHEMISTRY. *Educ:* Univ Miss, BS, 67; Ga Inst Technol, PhD(anal chem), 71. *Prof Exp:* Chemist, E I du Pont de Nemours & Co, Inc, Chattanooga Nylon Plant, 67; res assoc, Univ New Orleans, 73-74; res chemist, Athens Environ Res Lab, 74-85, SR SCIENTIST RES SYSTS, OFF INFO & RES MGT, US ENVIRON PROTECTION AGENCY, 85- *Concurrent Pos:* Chemist, Edgewood Arsenal, Md, 71-73. *Mem:* Sigma Xi; Am Chem Soc; Am Soc Mass Spectrometry. *Res:* Laboratory information management; computer systems; applications of mini-computers in analytical chemistry. *Mailing Add:* 103 Megan Ct Cary NC 27511-5877

SHACKLE, DALE RICHARD, b Caldwell, Ohio, Oct 4, 41; m 65; c 2. CHEMISTRY. *Educ:* Marietta Col, BS, 63; Ohio Univ, PhD(chem), 69. *Prof Exp:* Chemist process improv, Goodyear Atomic Corp, 63-65; teaching asst chem, Ohio Univ, 65-69; proj leader paper coatings, 69-74, sect head, 74-77, mgr process eng, 77-79, ASSOC DIR RES, MEAD CORP, 79-; VPRES, NEW SYSTS GROUP. *Res:* Specialty coatings for paper and the materials used in these coatings. *Mailing Add:* Meade Imaging 3020 Newmark Dr Miamisburg OH 45342

SHACKLEFORD, JOHN MURPHY, b Mobile, Ala, Dec 22, 29; m 58; c 2. ANATOMY. *Educ:* Spring Hill Col, BS, 57; Univ Ala, PhD(anat), 61. *Prof Exp:* From instr to assoc prof anat, Med Ctr, Univ Ala, 61-72, asst prof dent, 64-72; chmn dept, 72-80, PROF ANAT, UNIV S ALA, 72-, ASST DEAN ADMIS, COL MED, 87- *Concurrent Pos:* NIH res grants, 62-72. *Mem:* Am Asn Anatomists. *Res:* Cytochemistry and histophysiology of exocrine glands; electron microscopy of bones and teeth. *Mailing Add:* Off Admis Col Med Univ SAla 307 University Blvd Mobile AL 36688

SHACKLETT, ROBERT LEE, b Calif, Apr 5, 26; m 79; c 2. PHYSICS. *Educ:* Calif State Univ, Fresno, AB, 49; Calif Inst Technol, PhD(physics), 56. *Prof Exp:* From asst prof to assoc prof physics, 55-65, asst acad vpres, 67-68, asst dean sch grad studies, 68-75, actg dean sch grad studies, 75-76, prof, 65-79, EMER PROF PHYSICS, CALIF STATE UNIV, FRESNO, 79-; VPRES FOUND MIND-BEING RES, LOS ALTOS, CA, 87- *Concurrent Pos:* NSF fel, Univ Uppsala, 61-62; co-inventor, Digital Commun Syst, US Patent Off, 75- *Mem:* Am Phys Soc; Am Asn Physics Teachers; Int Soc Study Subtle Energies & Energy Med; Sigma Xi. *Res:* physics of consciousness. *Mailing Add:* PO Box 2128 Aptos CA 95001-2128

SHACKLETTE, LAWRENCE WAYNE, b New York, NY, Feb 26, 45; m 69; c 2. POLYMER PHYSICS, ELECTROCHEMISTRY. *Educ:* Brown Univ, BS, 67; Univ Ill, MS, 69, PhD(solid state physics), 72. *Prof Exp:* Assoc prof physics & electronics, Seton Hall Univ, 72-79; SR RES ASSOC POLYMER LAB, CORP TECHNOL, ALLIED-SIGNAL, INC, 79- *Mem:* Fel Am Phys Soc; Electrochem Soc. *Res:* Phase transitions and electronic transport in metals and ceramics; organic materials with special emphasis on electronic transport, spectroscopy and electrochemistry of conductive polymers; applications of conductive polymers as rechargeable batteries; electronic devices and electromagnetic shielding materials; polymer physics, blends, and composites. *Mailing Add:* Allied-Signal Inc PO Box 1021 Morristown NJ 07962

SHADDUCK, JOHN ALLEN, b Toledo, Ohio, Apr 22, 39; m 60; c 2. COMPARATIVE PATHOLOGY, VIROLOGY. *Educ:* Ohio State Univ, DVM, 63, MSc, 65, PhD(vet path), 67; Am Col Vet Pathologists, dipl. *Prof Exp:* Fel vet path, Ohio State Univ, 63-67, from asst prof to assoc prof vet path, 67-73; from assoc prof to prof comp path, Univ Tex Health Sci Ctr, Dallas, 73-80; PROF VET PATH & HEAD DEPT, COL VET MED, UNIV ILL, 80- *Concurrent Pos:* Fel comp virol & neuropath, Univ Munich, 67-68; consult indust, fed govt, Food & Drug Admin & WHO; prin investr grants & contracts; ed, Vet Path. *Mem:* AAAS; Am Vet Med Asn; Am Soc Exp Path; Am Col Vet Path; Sigma Xi. *Res:* Viral oncology; ophthalmic pathology; infectious diseases; host-parasite relationships; immunoregulatory events; comparative and functional aspects of inflammation. *Mailing Add:* Col Vet Med Tex A&M Univ College Station TX 77843

SHADDY, JAMES HENRY, b Everett, Wash, Aug 30, 38; m 66. ECOLOGY, ENTOMOLOGY. *Educ:* Okla State Univ, BS, 62, MS, 64; Mich State Univ, PhD(entom), 70. *Prof Exp:* Res technician, Mich State Univ, 67-68; PROF ECOL, NORTHEAST MO STATE UNIV, 69- *Mem:* Audubon Soc; AAAS; Entom Soc Am (secy/treas, 90-92); Nat Asn Col Teachers Agr; Sigma Xi. *Res:* Environmental assessment; computer assisted instruction; aquatic biology; old field ecology. *Mailing Add:* Sci Div Northeast Mo State Univ Kirksville MO 63501

SHADE, ELWOOD B, b Hollidaysburg, Pa, Aug 9, 13; div. FORESTRY, BIOLOGICAL SCIENCES. *Educ:* Juniata Col, BA, 35; Pa State Univ, BSF, 46, MF, 47. *Prof Exp:* Pub sch instr, Pa, 41-44 & Md, 44-45; forester, US Forest Serv, 47-55; park forester, City of Portland, Ore, 55-56; naturalist, Nat Park Serv, 56-57; from asst prof to assoc prof forestry, Univ Ark, Monticello, 60-81; RETIRED. *Mem:* Fel Soc Am Foresters. *Res:* Forest and outdoor recreation. *Mailing Add:* Dept Forestry Univ Ark Box 2528 Monticello AR 71655

SHADE, JOYCE ELIZABETH, b Louisville, Ky, Oct 30, 53. SYNTHETIC INORGANIC CHEMISTRY. *Educ:* Univ Louisville, BA, 75, PhD(chem), 80. *Prof Exp:* Asst chem, Univ Louisville, 75-80; res fel, Ohio State Univ, 80-82; ASST PROF CHEM, US NAVAL ACAD, 82- *Concurrent Pos:* Instr chem, Univ Louisville, 79; physics teacher, Presentation Acad, 77-79. *Mem:* Am Chem Soc; Sigma Xi. *Res:* Syntheses and characterization of cyclopentadienyl-type iron complexes; identification of diastereomeric isomers. *Mailing Add:* Dept Chem US Naval Acad Annapolis MD 21402

SHADE, RAY W(ALTON), b Souderton, Pa, Jan 11, 27; m 52; c 1. CHEMICAL ENGINEERING. *Educ:* Mass Inst Technol, SB, 49, SM, 51; Rensselaer Polytech Inst, PhD(chem eng), 64. *Prof Exp:* Chem engr, Knolls Atomic Power Lab, Gen Elec Co, 51-53 & Res & Develop Ctr, 53-66, mgr polymer processing, 66-68, mgr chem eng br, 68-70; assoc prof bio-environ eng, Rensselaer Polytech Inst, 70-77, chmn environ eng curriculum, 71-77; CHEM ENGR, GEN ELEC CO, 77- *Mem:* Am Chem Soc; Am Inst Chem Engrs. *Res:* Chemical and polymer processing; vacuum technology and thin vacuum deposited films; electroless plating. *Mailing Add:* 19 El Dorado Dr Clifton Park NY 12065

SHADE, ROBERT EUGENE, HYPERTENSION, RENAL PHYSIOLOGY. *Educ:* Ind Univ, PhD(physiol), 70. *Prof Exp:* ASSOC SCIENTIST, SOUTHWEST FOUND BIOMED RES, 83- *Mailing Add:* Dept Physiol & Biophys Southwest Found Biomed Res 7620 NW Loop 410 PO Box 28147 San Antonio TX 78228

SHADER, LESLIE ELWIN, b Ft Collins, Colo, July 18, 35; m 58; c 3. MATHEMATICS. *Educ:* Colo State Univ, BS, 57, MS, 61; Univ Colo, Boulder, PhD(math), 69. *Prof Exp:* Teacher math, High Sch, Colo, 57-58 & Wyo, 58-59; asst, Colo State Univ, 59-61; instr, Univ Wyo, 61-65; instr, Univ Colo, Boulder, 65-66 & 67-68; from instr to asst prof, 68-73, assoc prof, 73-81, PROF MATH, UNIV WYO, 81- *Mem:* Am Math Soc; Math Asn Am. *Res:* Polynomials over a finite field; matrix theory; combinatorics; number theory. *Mailing Add:* Dept Math Box 3036 Univ Wyo Laramie WY 82071

SHADER, RICHARD IRWIN, b Mt Vernon, NY, May 27, 35; m 58; c 3. PSYCHOPHARMACOLOGY, PSYCHOANALYSIS. *Educ:* Harvard Univ, Cambridge, Mass, BA, 56; Sch Med, New York Univ, MD, 60; Boston Psychoanal Soc Inst, MD, 70. *Prof Exp:* Intern, Greenwich Hosp, Conn, 60-61; resident psychiat, Mass Ment Health Ctr, 61-62 & 64-65, dir, psychopharmacol res lab, 68-79, dir training & educ, 75-77, dir continuing educ, 77-79; resident psychiat, NIMH, 62-64; asst prof psychiat, Med Sch, Harvard Univ, 68-70, assoc prof, 70-79; psychiatrist-in-chief, New England Med Ctr, 79-91; prof & chmn psychiat, 79-91, CLIN PROF PSYCHIAT, DENT MED, TUFTS UNIV, 85-, PROF PHARMACOL, SCH MED, 89-, CHMN PHARMACOL, 91- *Concurrent Pos:* J Michaels Merit Scholar, Boston Psychoanal Soc & Inst, 68-69; dir, Am Bd Psychiat & Neurol, 76-84, pres & dir, Am Bd Emergency Med, 80-90, exec comt, 85-90; bd dirs, Med Found, 80-87; ed-in-chief, J Clin Psychopharmacol, 80-; mem, Nat Adv Ment Health Coun, 84-87, Clin Affairs Coun, Vet Admin, Washington, DC, 85-; mem, Psychiatry Test Comn, Nat Bd Med Examrs, 87-, chair, 91-; mem, Med & Sci Adv Bd, Alzheimer's Dis & Related Disorders Asn, 88-; fel, Ctr Advan Study Behav Sci, Stanford, Calif, 90-91. *Honors & Awards:* Taylor Manor Hosp Psychiat Award, 80; Seymour Vestermark Award, Am Psychiat Asn, 88, 90. *Mem:* AMA; Am Psychiat Asn; Am Col Neuropsychopharmacol (pres, 90); Am Asn Chmn Departments Psychiat (pres, 85-86); Am Soc Clin Pharmacol & Therapeut; Asn Acad Psychiat; Am Bd Psychiat & Neurol (treas, 82-83, pres, 84). *Res:* Pharmacokinetic and pharmacodynamic factors which influence responses to psychoactive drugs, particular emphasis on aging, drug interactions and adverse drug reactions. *Mailing Add:* Dept Psychiat New Eng Med Ctr 750 Washington St Boston MA 02111

SHADOMY, SMITH, b Denver, Colo, Aug 29, 31; m 56; c 2. MEDICAL MICROBIOLOGY. *Educ:* Univ Calif, Los Angeles, BA, 56, PhD(microbiol), 63. *Prof Exp:* Instr biol, San Fernando Valley State Col, 61-62; asst prof infectious dis, 65-71, assoc prof, 71-77, PROF MED & MICROBIOL, DEPT MED, MED COL VA, VA COMMONWEALTH UNIV, 77- *Concurrent Pos:* Mem fac, Mil Nursing Pract & Res, Dept Nursing, Walter Reed Army Inst Res, 65-66 & Otolaryngol Basic Sci Course, Armed Forces Inst Path, 66. *Honors & Awards:* Meridian Award for Excellence Med Mycol, Med Mycol Soc of the Ams, 86; Selman A Wakesman hon lectr, Theobold Smith Soc, 88. *Mem:* Int Soc Human & Animal Mycol; Brit Mycol Soc; Am Soc Microbiol; Asn Mil Surgeons US; Infectious Dis Soc Am. *Res:* Chemotherapeutic evaluations; environmental microbiology; medical mycology; patient isolation; chemotherapeutic studies in bacterial and fungal diseases; epidemiology of mycotic infections. *Mailing Add:* 2320 Fillmore Circle Richmond VA 23235

SHADOWEN, HERBERT EDWIN, b Fredonia, Ky, Sept 11, 26; m 50; c 3. VERTEBRATE ZOOLOGY. *Educ:* Berea Col, BA, 50; Univ Ky, MS, 51; La State Univ, PhD(zool), 56. *Prof Exp:* Assoc prof zool, La Polytech Inst, 55-61; PROF BIOL, WESTERN KY UNIV, 61- *Mem:* Am Soc Mammal; Am Ornith Union. *Res:* Small-mammal population studies; taxonomy of amphibians and reptiles. *Mailing Add:* Dept Biol Western Ky Univ Bowling Green KY 42101

SHAEFFER, JOSEPH ROBERT, b New York, NY, June 3, 35; m 57; c 3. HEMATOLOGY. *Educ:* Mass Inst Technol, SB, 56; Univ Rochester, PhD(biophys), 62. *Prof Exp:* Res fel biochem, Univ Ky, 62-65; asst physicist, Univ Tex M D Anderson Hosp & Tumor Inst, 65-68, asst biologist, 68-71, assoc biologist, 71-76; vis assoc prof med, Peter Bent Brigham Hosp, Boston, 76-77; INVESTR, CTR BLOOD RES, 77-; ASSOC PROF MED BIOCHEM, HARVARD MED SCH, 78- *Concurrent Pos:* USPHS fel, 62-64. *Mem:* Biophys Soc; Am Soc Biochem & Molecular Biol; NY Acad Sci; Am Soc Hematol; Sigma Xi. *Res:* Proteolysis in human erythroid cells; human hemoglobin, structure, biosynthesis and degradation; thalassemia. *Mailing Add:* Ctr Blood Res Boston MA 02115

SHAEIWITZ, JOSEPH ALAN, b Brooklyn, NY, Oct 12, 52. DRUG DISSOLUTION, INTERFACIAL PHENOMENA. *Educ:* Univ Del, BS, 74; Carnegie-Mellon Univ, MS, 76, PhD(chem eng), 78. *Prof Exp:* Asst prof chem eng, Univ Ill, 78-84; ASSOC PROF CHEM ENG, W VA UNIV, 84- *Mem:* Am Inst Chem Engrs; Am Chem Soc; AAAS; Am Soc Eng Educ. *Res:* Diffusion and mass transfer; applications to drug dissolution and drug delivery; pharmaceutical applications of mass transfer; chemical engineering separation processes, especially biochemical separations. *Mailing Add:* Dept Chem Eng PO Box 6101 Morgantown WV 26506-6101

SHAER, ELIAS HANNA, b Beit-Jala, Israel, Aug 2, 41; US citizen; m 70; c 2. PHYSICAL & ANALYTICAL CHEMISTRY. *Educ:* Austin Peay State Univ, BS, 66; Univ Miss, MS, 70, PhD(phys chem), 77. *Prof Exp:* Lab instr chem, Univ Miss, 66-70; dir chem, Div Contractors, Vulcan Waterproofing Co, 70-73; lab instr chem, Univ Miss, 73-77; sr res chemist, 77-80, SCIENTIST RES & DEVELOP, DRACKETT CO, 80- *Mem:* Am Chem Soc. *Res:* Kinetics solvent effect on the rate of methyl radicals combination; thermodynamics of liquids and liquid mixtures; cleaning compositions and emulsions. *Mailing Add:* 5688 Bayberry Cincinnati OH 45342

SHAER, NORMAN ROBERT, b Boston, Mass, Mar 31, 37; m 58; c 2. COMPUTER SCIENCE, ELECTRICAL ENGINEERING. *Educ:* Tufts Univ, BSEE, 58; NY Univ, MEE, 60. *Prof Exp:* Mem tech staff, Bell Tel Labs, 58-63, supvr systs eng, 63-66, dept head electronic switching, 66-69; vpres bus info systs, Comput Systs Labs, 69-71; mgr comput sci, 72-74, dir develop, Mail Network Serv, 82-88, DIR, SYSTS APPL DEVELOP LAB, AT&T BELL TEL LABS, 88- *Mem:* AAAS; Inst Elec & Electronics Engrs. *Res:* Use of computers for large scale, real time systems; design and implementation. *Mailing Add:* AT&T Bell Labs Rm 1D401 Crawfords Corner Rd Holmdel NJ 07733

SHAEVEL, MORTON LEONARD, b Boston, Mass, June 7, 36; m 61; c 2. FOOD FORMULATION & PROCESSING, FROZEN FOOD STABILITY & QUALITY ASSURANCE SYSTEMS. *Educ:* Univ Mass, Amherst, BS. *Prof Exp:* Food technologist, B Manischewitz Co, 58-62; res chemist, Nabisco Brands, Inc, 62-64, res mgr, 64-68, dir new prod res, 68-73; vpres res & develop, Freezer Queen Foods, Inc, 73-82, vpres new bus develop, 82-84, gen mgr, 84-85, pres, 85-89; GEN MGR PROD DEVELOP & QUAL ASSURANCE SYSTS, SHAEVEL & ASSOCS, LTD, 89- *Concurrent Pos:* Chmn, Prepared Foods Comt, Am Frozen Food Inst, 75-77, res coun, 77-81, Qual Maintenance Comt, 81-85, mem bd dirs, 87-90; mem, Tech & Sci Issues Comt, Off Technol Assessment Food Stability TF, US Cong, 78-79 & Grocery Mfrs Am, 79-80. *Mem:* Inst Food Technologists; Am Soc Qual Control; Am Frozen Food Inst. *Res:* Conceived, developed and automated new product systems in commercial production facilities; state-of-the-art frozen food production, stability and durability; frozen food industry consultant. *Mailing Add:* PO Box 1565 Buffalo NY 14231-1565

SHAFAI, LOTFOLLAH, b Maraghen, Iran, Mar 17, 41; Can citizen; m 66; c 2. MICROWAVES, ELECTRONICS. *Educ:* Univ Tehran, BSc, 63; Univ Toronto, MSc, 66, PhD(elec eng), 69. *Prof Exp:* Lectr elec eng, Univ Man, 69-70, from asst prof to assoc prof, 70-76; vis scientist appl electromagnetic, Commun Res Ctr, Ottawa, 76-77; vis prof, Electromagnetic Inst, Tech Univ Denmark, 77-79; dir Inst Tech Develop, 85-88, head elec eng, 87-89, PROF ELEC ENG, UNIV MAN, 79-, APPL ELECTROMAGNETIC CHAIR, 89- *Honors & Awards:* Merit Award, Asn Prof Engrs, 83-84; Maxwell Premium Award, Inst Elec Engrs, 90. *Mem:* Fel Inst Elec & Electronics Engrs. *Res:* Antennas; electromagnetic scattering and diffraction; wave guides; computer solution of field problems; optics. *Mailing Add:* Dept Elec Eng Univ Man Winnipeg MB R3T 2N2 Can

SHAFER, A WILLIAM, b Great Bend, Kans, Nov 1, 27; m 50; c 2. MEDICINE. *Educ:* Univ Kans, BA, 50, MD, 54. *Prof Exp:* Assoc hemat, Scripps Clin & Res Found, 59-60, assoc mem, 60-66, head hemat, 64-66; from asst prof to prof med, Univ Okla, 66-73, from asst prof to prof lab med, 66-73; prof path & med, Col Med & dir clin labs, Med Ctr, Univ Ky, 73-75; DIR, SOUTHEASTERN MICH RED CROSS BLOOD CTR, 75- *Concurrent Pos:* Clin prof med & adj prof allied health, Wayne State Univ, 76- *Mem:* Fel Am Col Physicians; Am Bd Internal Med; Am Bd Path; Am Fedn Clin Res; fel Am Soc Clin Path. *Res:* Metabolism of normal, abnormal and stored erythrocytes. *Mailing Add:* Southeastern Mich Red Cross Blood Ctr Box 351 Detroit MI 48232

SHAFER, JULES ALAN, b New York, NY, Nov 21, 37; m 59; c 3. PROTEIN CHEMISTRY, ENZYMOLOGY. *Educ:* City Col New York, BChE, 59; Polytech Inst Brooklyn, PhD(chem), 63. *Prof Exp:* Fel chem, Polytech Inst Brooklyn, 62-63 & Harvard Univ, 63-64; from asst prof to assoc prof, 64-77, PROF BIOL CHEM, UNIV MICH, ANN ARBOR, 77- *Concurrent Pos:* NIH grants, 65-85. *Mem:* AAAS; Am Chem Soc; Am Soc Biol Chemists; NY Acad Sci; Int Soc Magnetic Resonance; Int Comt Thrombosis & Hemostasis; NY Acad Sci. *Res:* Mechanisms and models of enzyme action, especially thrombin insulin receptor kinase and enzymes requiring pyridoxal phosphate. *Mailing Add:* Merck Sharp & Dohme Res Labs Sunnytown Pike West Point PA 19486

SHAFER, M(ERRILL) W(ILBERT), b Grier City, Pa, July 25, 28; m 54; c 4. SOLID STATE CHEMISTRY, MATERIALS SCIENCE. *Educ:* Susquehanna Univ, BA, 52; Pa State Univ, MS, 54, PhD(ceramic tech), 56. *Prof Exp:* Asst, Pa State Univ, 54-56; res chemist, Res Lab, 56-72, MGR, SOLID STATE CHEM, IBM RES LAB, 72- *Mem:* Am Ceramic Soc; Am Phys Soc; Electrochem Soc; NY Acad Sci. *Res:* High temperature and solid state chemistry involving oxides, fluorides and sulfides; synthesis and evaluation of magnetic materials, particularly the rare earths; synthesis and evaluation of physical properties of electronic materials; especially magnetic, luminesent and semiconducting properties; flouride gasses and wide band gap insulators for electronic packaging. *Mailing Add:* 1312 N Park Ave Casa Grande AZ 85222

SHAFER, PAUL RICHARD, b Springfield, Ohio, June 17, 23; m 46; c 3. ORGANIC CHEMISTRY. *Educ:* Oberlin Col, BA, 47; Univ Wis, PhD(chem), 51. *Prof Exp:* Fel, Univ Ill, 51-52; from instr to prof chem, 52-88, EMER PROF CHEM, DARTMOUTH COL, 88- *Concurrent Pos:* NSF fel, Calif Inst Technol, 59-60. *Mem:* Am Chem Soc; Royal Soc Chem. *Res:* Natural products; reaction mechanisms; nuclear magnetic resonance of fast exchange reactions. *Mailing Add:* Rte 1 Box 277 Enfield NH 03748

SHAFER, RICHARD HOWARD, b New Rochelle, NY, July 20, 44; m 74; c 2. BIOPHYSICAL CHEMISTRY. *Educ:* Yale Univ, BA, 66; Harvard Univ, MA, 69, PhD(chem physics), 72. *Prof Exp:* Fel, Univ Calif, San Diego, 73-75; from asst prof to assoc prof, 75-87, PROF CHEM & PHARMACEUT CHEM, UNIV CALIF, SAN FRANCISCO, 87- *Concurrent Pos:* Fulbright scholar, Paris, France, 66-67. *Mem:* Am Chem Soc; Biophys Soc. *Res:* Physical chemistry of nucleic acids; drug-nucleic acid interactions and hydrodynamic properties of high polymers. *Mailing Add:* Dept Pharmaceut Chem Univ Calif San Francisco CA 94143-0446

SHAFER, ROBERT E, b San Francisco, Calif, June 2, 36; m 68. HIGH SPEED ELECTRONICS, ACCELERATOR DESIGN. *Educ:* Stanford Univ, BS, 58; Univ Calif, Berkeley, PhD(physics), 66. *Prof Exp:* Res assoc, Mass Inst Technol, 66-69; physicist res & develop, Fermi Nat Accelerator Lab, 69-86; PHYSICIST RES & DEVELOP, LOS ALAMOS NAT LAB, 86- *Mem:* Am Inst Physics; AAAS. *Res:* High energy physics experiments; designing and building particle accelerators. *Mailing Add:* MS H808 Los Alamos Nat Lab Los Alamos NM 87545

SHAFER, SHELDON JAY, b Passaic, NJ, June 17, 48. ORGANIC CHEMISTRY. *Educ:* Fairleigh Dickinson Univ, BS, 70; State Univ NY, Albany, PhD(org chem), 76. *Prof Exp:* Res fel org chem, Univ Calif, Santa Cruz, 76-77; develop chemist, Plastics Div, Gen Elec Co, 77-79, polymer chemist, 79-83, advan polymer chemist, 83-89, SR SCIENTIST, PLASTICS DIV, GEN ELEC CO, 89- *Mem:* Am Chem Soc. *Res:* Physical organic chemistry; reaction mechanisms; organic electron transfer reactions; organosulfur chemistry. *Mailing Add:* 62 Euclid Ave Pittsfield MA 01201-5924

SHAFER, STEPHEN JOEL, b Philadelphia, Pa, Dec 28, 39; m 70. BIOCHEMICAL & DEVELOPMENTAL GENETICS. *Educ:* Haverford Col, BA, 63; Univ Pa, MS, 65; Temple Univ, PhD(biochem genetics), 72. *Prof Exp:* Res asst genetics, Haverford Col, 65-66; res assoc biochem genetics, Temple Univ, 72-73; from asst prof to assoc prof, 73-85, PROF BIOL, DOWLING COL, 85- *Mem:* NY Acad Sci; AAAS; Am Soc Zool; Nat Asn Advan Health Prof. *Res:* Application of recombinant DNA techniques to mutants of drosophila melanogaster. *Mailing Add:* Dept Biol Dowling Col Oakdale NY 11769

SHAFER, STEPHEN QUENTIN, b Barrytown, NY, Dec 18, 44; m 66; c 3. EPIDEMIOLOGY, NEUROEPIDEMIOLOGY. *Educ:* Harvard Univ, BA, 66; Columbia Univ, MD, 70, MPH, 77, MA, 79; Am Bd Internal Med, dipl, 75; Am Bd Psychiat & Neurol, 85. *Prof Exp:* Intern & resident, Harlem Hosp Ctr, 70-72, clin fel neurol, 72-74, resident med, 74-75; clin scholar med, Johnson Clin Scholars Prog, 76-78, ASST PROF PUB HEALTH & NEUROL, SERGIEVSKY CTR, COLUMBIA UNIV, 78- *Concurrent Pos:* Asst attend physician, Harlem Hosp Ctr, 75-78; res assoc med, Columbia Col Physicians & Surgeons, 76-78; resident neurol, Columbia Univ, 81-84. *Mem:* Am Pub Health Asn; Int Soc Cardiol; Am Fedn Clin Res; Soc Epidemiol Res; Am Acad Neurol. *Res:* Cerebrovascular disease; chronic disease epidemiology; neuroepidemiology; epilepsy. *Mailing Add:* Harlem Hosp Ctr 506 Lenox Ave New York NY 10037

SHAFER, STEVEN RAY, b Troy, Ohio, Apr 5, 56; m 91. POLLUTANT-PARASITE INTERACTIONS, SOIL MICROBIOLOGY. *Educ:* Ohio State Univ, BS, 78, MS, 80; NC State Univ, PhD(plant path), 83. *Prof Exp:* RES PLANT PATHOLOGIST, AGR RES SERV, USDA, 83- *Concurrent Pos:* Assoc ed, Phytopath, Am Phytopath Soc, 88-90 & sr ed, 91-; mem, Panel on Monitoring & Managing Natural Resources, Nat Acad Sci, 90. *Mem:* Am Phytopath Soc; AAAS. *Res:* Effects of air pollutants on plants and associated soilborne microorganisms, including mycorrhizal fungi, rhizobia, nematodes, and rhizosphere-inhabiting bacteria. *Mailing Add:* USDA-NC State Univ Air Qual Res Prog 1509 Varsity Dr Raleigh NC 27606

SHAFER, THOMAS HOWARD, b Columbus, Ohio, Jan 23, 48; m 70; c 2. DEVELOPMENTAL BIOLOGY, SCIENCE EDUCATION. *Educ:* Duke Univ, BS, 70; Ohio State Univ, MS, 73, PhD(bot), 75. *Prof Exp:* Res assoc develop plant physiol, Ohio State Univ, 75-76; vis asst prof, Denison Univ, 76-77 & Univ Ill, 77-78; ASST PROF BIOL, UNIV NC, WILMINGTON, 78- *Mem:* AAAS. *Res:* Relation of plant hormones to the synthesis of gene products; gene activation in early embryogenesis. *Mailing Add:* Dept Biol Sci Univ NC 601 S College Rd Wilmington NC 28403

SHAFER, W SUE, b Alton, Ill, Oct 3, 41. ALCOHOL RESEARCH. *Educ:* Univ Wis, BS, 63; Univ Fla, PhD(develop biol), 68. *Prof Exp:* Fel, Univ Fla, 68-69; bus mgr, Soc Develop Biol, Inc, 70-74; prog adminr, Cellular & Molecular Basis Dis Prog, Nat Inst Gen Med Sci, 74-83, chief biomed eng & instrument develop, 78-83; chief, Off Prog Planning & Eval, Div Res Resources, NIH, 83-87; dep dir, Div Basic Res, Nat Inst Alcohol Abuse & Alcoholism, US Dept Health & Human Serv, 87-89; ASSOC DIR PROG ACTIV, NAT INST GEN MED SCI, NIH, 89- *Concurrent Pos:* Guest lectr, Kalamazoo Col, 69-72, vis lectr, 72-73. *Mem:* Soc Develop Biol; Am Soc Cell Biol. *Mailing Add:* Nat Inst Gen Med Sci NIH 938 Westwood Bldg Bethesda MD 20892

SHAFER, WILLIAM GENE, b Toledo, Ohio, Nov 15, 23; m 43. ORAL PATHOLOGY. *Educ:* Univ Toledo, BS, 47; Ohio State Univ, DDS, 47; Univ Rochester, MS, 50; Am Bd Oral Path, dipl, 52. *Prof Exp:* From instr to assoc prof, 50-59, chmn dept, 56-59, PROF ORAL PATH, SCH DENT, IND UNIV, INDIANAPOLIS, 59-, DENT CANCER COORDR, 52- *Concurrent Pos:* Consult, US Vet Admin, 53- & Surgeon Gen, USAF, 60-; secy-treas, Am Bd Oral Path, 61- *Mem:* AAAS; Soc Exp Biol & Med; Am Dent Asn; fel Am Acad Oral Path (vpres, 54, pres, 57); NY Acad Sci. *Res:* Experimental dental caries; salivary glands and endocrines; x-ray irradiation; experimental salivary gland tumors; tissue culture. *Mailing Add:* 3306 Kenilworth Dr Indianapolis IN 46208

SHAFFAR, SCOTT WILLIAM, b Waukeegan, Ill, Dec 1, 62. COMBUSTION RESEARCH, LASER DIAGNOSTICS. *Educ:* Calif Polytech Inst, BS, 84. *Prof Exp:* SR ENGR, NORTHROP CORP, 84- *Mem:* Am Inst Aeronaut & Astronaut; Am Asn Aerosol Res. *Res:* Combustion; laser based diagnostics; aerosol science; atomization; air pollution. *Mailing Add:* Northrop Corp M-S W12-GH 8900 E Washington Blvd Pico Rivera CA 90660

SHAFFER, BERNARD W(ILLIAM), b New York, NY, Aug 7, 24; m 47; c 2. ENGINEERING, RATIONAL ANALYSIS OF STRESS & MOTION. *Educ:* City Col New York, BME, 44; Case Inst Technol, MS, 47; Brown Univ, PhD(appl math), 51. *Prof Exp:* Aeronaut res scientist, Nat Adv Comt Aeronaut, Ohio, 44-47; res assoc, Grad Div, Appl Math, Brown Univ, 47-50; from asst prof to prof mech eng, NY Univ, 50-73, proj dir, Res Div, 51-73; PROF MECH & AEROSPACE ENG, POLYTECH UNIV, 73- *Concurrent Pos:* Spec lectr, Case Inst Technol, 46-47; res & consult for govt agencies & pvt enterprises, 50- *Honors & Awards:* Richards Mem Award Outstanding Achievement, Am Soc Mech Engrs, 68. *Mem:* Fel Am Soc Mech Engrs; assoc fel Am Inst Aeronaut & Astronaut. *Res:* Stress analysis; elasticity; plasticity; kinematics; mechanics of metal cutting; analysis of filament reinforced plastics. *Mailing Add:* 18 Bayside Dr Great Neck NY 11023

SHAFFER, CHARLES FRANKLIN, b Pittsburgh, Pa, Oct 9, 40; m 64; c 3. IMMUNOLOGY, TRANSPLANTATION BIOLOGY. *Educ:* Univ Pittsburgh, BS, 63, MS, 65; Univ Pa, PhD(biol), 68. *Prof Exp:* Pa scholar exp med, Dept Med Genetics, Sch Med, Univ Pa, 68-71; from asst prof to assoc prof biol, 71-85, PROF BIOL, WITTENBERG UNIV, 85- *Concurrent Pos:* Adj assoc prof microbiol & immunol, Med Sch, Wright State Univ, 77- *Mem:* Transplantation Soc; Am Asn Immunologists. *Res:* Immunologic and genetic aspects of tissue and organ transplantation. *Mailing Add:* Dept Biol Wittenberg Univ Springfield OH 45501

SHAFFER, CHARLES HENRY, JR, b Washington, DC, Nov 20, 13; m 40; c 2. MICROBIOLOGY. *Educ:* Univ Md, BS, 38. *Prof Exp:* Bacteriologist, HEW, 42-45 & Div Antibiotics, Food & Drug Admin, 45-63; asst supvr, USDA, 63-70; supvr microbiologist, US Environ Protection Agency, 70-78; RETIRED. *Concurrent Pos:* Consult, TRACOR-JITCO Sci Mgt Serv Inc, 78-79. *Mem:* Am Soc Microbiol; Am Soc Testing & Mat; Soc Indust Microbiol. *Res:* Development and testing of new disinfectants for tubercuocidal activity. *Mailing Add:* 622 Rollins Ave Rockville MD 20852

SHAFFER, CHARLES V(ERNON), b Melbourne, Fla, Feb 22, 22; m 44; c 3. ELECTRICAL ENGINEERING. *Educ:* Univ Fla, BEE, 44, MSE, 60; Stanford Univ, PhD(elec eng), 65. *Prof Exp:* Asst res prof electronic ord, 46-51, from assoc prof, to prof, 51-85, grad coord, 65-68, asst dean grad sch, 69-71, EMER PROF ELEC ENG, UNIV FLA, 85- *Concurrent Pos:* Consult, US Army Missile Command, 71-74; dir, Northeast Regional Data Ctr, State Univ Systs Fla, 72-74; mem tech staff, Bell Labs, Denver, Colo, 70-, Holmdel, NJ, 76 & IBM, Gen Systs Div, Boca Raton, Fla, 78. *Honors & Awards:* Award, NASA, 70; Centennial Medal, Inst Elec & Electronics Engrs, 84. *Mem:* Inst Elec & Electronics Engrs. *Res:* Network theory and design; computer communications; system science; microcomputers. *Mailing Add:* 3425 NW Eighth Ave Gainesville FL 32605

SHAFFER, DAVID, b Johannesburg, S Africa, Apr 20, 36; Brit citizen; c 2. CHILD PSYCHIATRY. *Educ:* Univ London, MB, BS, 61; MRCP, 64; FRCPsych, 81. *Prof Exp:* Lectr pediat, Univ Col Hosp, London, 64-65; registr psychiat, Maudsley Hosp, London, 65-69; sr registr & lectr child psychiat, Inst Psychiat, London, 69-74; sr lectr, NY State Psychiat Inst, 74-77, dir, Dept Child Psychiat, 77-; prof clin psychiat & pediat, Col Physicians & Surgeons, Columbia Univ, 77-; AT DEPT PSYCHOL, UNIV GA. *Concurrent Pos:* Von Ameringen fel, Found Fund Psychiat, Yale Univ, 67-68. *Mem:* Asn Child Psychiat & Psychol, London (treas, 75-77); Am Acad Child Psychiat; Am Psychiat Asn; Brit Pediat Asn; Royal Col Psychiat, London (secy, 75-77). *Res:* Relationships between brain damage and psychiatric disorder in children; suicide and depression in childhood; psychological aspects of enuresis; classification of child psychiatric disorders. *Mailing Add:* Dept Psychol Rm 401 Univ Ga Athens GA 30602

SHAFFER, DAVID BRUCE, b Berea, Ohio, Feb 17, 46; m 79. RADIO ASTRONOMY. *Educ:* Carnegie-Mellon Univ, BS, 68; Calif Inst Technol, PhD(astron), 74. *Prof Exp:* Teaching asst physics, Calif Inst Technol, 71-73; instr astron, Yale Univ, 73-75; asst scientist radio astron, Nat Radio Astron Observ, 75-77, assoc scientist, 77-79; staff scientist, Very Long Baseline Interferometer Group, Phoenix Corp, 79-82; CHIEF SCIENTIST, INTERFEROMETRICS, INC, 82- *Mem:* Am Astron Soc; Int Astron Union; Int Sci Radio Union; Inst Navig; Royal Astron Soc. *Res:* Long baseline radio interferometry of compact extragalactic radio sources, galaxies and quasars. *Mailing Add:* Interferometrics Inc 8150 Leesburg Pike Vienna VA 22182

SHAFFER, DOROTHY BROWNE, b Vienna, Austria, Feb 12, 23; nat US; m 43, 78; c 3. MATHEMATICS. *Educ:* Bryn Mawr Col, AB, 43; Radcliffe Col, MA, 45, PhD, 62. *Prof Exp:* Mem staff, Dynamic Anal & Control Lab, Mass Inst Technol, 45-47; asst, Harvard Univ, 48; consult, Corning Glass Works, 49-50; assoc mathematician, Cornell Aeronaut Lab, 52-56; staff mathematician, Dunlap & Assocs, Inc, 58-60; lectr, Univ Conn, Stamford Br, 63; from asst prof to assoc prof, 63-75, PROF MATH, FAIRFIELD UNIV, 75- *Concurrent Pos:* NSF fac fel, Courant Inst Math Sci, NY Univ, 69-70; Prof develop fel, Res Div, IBM Corp, 79; vis prof, Imp Col Sci & Technol, London 78, Univ Md, College Park & Univ Calif, La Jolla, 81 & 86. *Mem:* Am Math Soc; Math Asn Am; Sigma Xi; Asn Women Math; NY Acad Sci. *Res:* Conformal mapping and level curves of Green's Function; operations research; theoretical aerodynamics; polynomials; univalent functions; potential theory; special functions. *Mailing Add:* Dept Math Fairfield Univ Fairfield CT 06430

SHAFFER, DOUGLAS HOWERTH, b Danville, Pa, Oct 31, 28; m 52; c 2. MATHEMATICS, STATISTICS. *Educ:* Carnegie Inst Technol, BS, 50, MS, 51, PhD, 53. *Prof Exp:* Instr math, Carnegie Inst Technol, 53-54; res mathematician, Westinghouse Elec Corp, 54-63, fel mathematician, 63-65, adv mathematician, 65-75, consult mathematician, 75-79, mgr math, 79-90; CONSULT, 90- *Mem:* Am Statist Asn; Am Soc Qual Control. *Res:* Applied and statistical mathematics; design of experiments; mathematics education. *Mailing Add:* 150 Washington Pittsburgh PA 15218

SHAFFER, HARRY LEONARD, b Boston, Mass, Dec 15, 33; m 60; c 6. ELECTRICAL ENGINEERING. *Educ:* Northeastern Univ, BSEE, 62, MSEE, 64. *Prof Exp:* Sr technologist, GTE Sylvania Appl Res Lab, 64-70, eng specialist digital signal processing res, Eastern Div GTE Sylvania Electronics Systs Group, 70-87, GTE EDCD, 88- *Concurrent Pos:* NSF grant, 63-64. *Mem:* Inst Elec & Electronics Engrs. *Res:* Development of algorithms to digitally process speech and modem signals, in real-time on mini-computers and micro-processors; extensive study into algorithms for secure-speech transmission on communications channels; digital signal processing; currently involved with system configuration for electronic distribution of keys (aka key management). *Mailing Add:* 55 Howard Ave Lynnfield MA 01940

SHAFFER, JACQUELIN BRUNING, b Orange, Calif, Oct 12, 50; m 75. LUNG MOLECULAR BIOLOGY. *Educ:* San Diego State Univ, BS, 73; Univ SC, PhD(biol), 80. *Prof Exp:* Res assoc, Dept Genetics, NC State Univ, 80-85; RES SCIENTIST, LAB BIOCHEM & GENETIC TOXICOL, NY STATE DEPT HEALTH, 85-; ASST PROF, DEPT ENVIRON HEALTH & TOXICOL, GRAD SCH PUB HEALTH, STATE UNIV NY, ALBANY, 85- *Concurrent Pos:* Mem, Bd Dirs, Albany Res Inst, 89- & Gov Coun, NY State Lung Res Inst & Am Lung Asn NY State, 89- *Mem:* AAAS; Genetics Soc Am; Oxygen Soc. *Res:* Molecular mechanisms that control gene expression in higher eukaryotes; regulation of antioxidant enzyme expression in response to oxidative stress. *Mailing Add:* Wadsworth Ctr Labs & Res NY State Dept Health Albany NY 12201-0509

SHAFFER, JAY CHARLES, b Sunbury, Pa, July 21, 38; m 69; c 3. SYSTEMATIC ENTOMOLOGY. *Educ:* Bucknell Univ, BS, 61; Cornell Univ, PhD(syst entom), 67. *Prof Exp:* Vis res assoc entom, Smithsonian Inst, 66-67; from asst prof to assoc prof, 68-81, PROF BIOL, GEORGE MASON UNIV, 81- *Concurrent Pos:* Assoc ed, Biotropica, 72-; res assoc entom, Smithsonian Inst, 79- *Mem:* AAAS; Asn Trop Biol; Lepidop Soc. *Res:* Systematics and biology of the Pyralidae, Lepidoptera; origin, development and present status of the insect fauna of Aldabra atoll. *Mailing Add:* Dept Biol George Mason Univ Fairfax VA 22030-4444

SHAFFER, JOHN CLIFFORD, b Towanda, Pa, May 3, 38; m 66; c 2. PHYSICS. *Educ:* Franklin & Marshall Col, BS, 60; Univ Del, MS, 62, PhD(physics), 66. *Prof Exp:* Asst prof, 65-68, assoc prof, 68-75, PROF PHYSICS, NORTHERN ILL UNIV, 75-, CHMN DEPT, 73- *Mem:* AAAS; Am Phys Soc. *Res:* Optical properties of solids; luminescence; band structure of solids; disordered solids; surface physics. *Mailing Add:* Dept Physics Northern Ill Univ De Kalb IL 60115

SHAFFER, LAWRENCE BRUCE, b Delta, Ohio, Oct 14, 37; m 59; c 4. PHYSICS. *Educ:* Ohio State Univ, BSc, 59; Univ Wis, MSc, 60, PhD(physics), 64. *Prof Exp:* Asst prof physics, Hiram Col, 63-70, chmn dept, 64-70; PROF PHYSICS & CHMN DEPT, ANDERSON UNIV, IND, 70- *Concurrent Pos:* Res Corp grants, 64-; teaching equip grants, NSF, 65-67 & Kettering Found, 65-; NASA res grants, 66-69; guest scientist, Inst Solid State Res, Nuclear Res Ctr, Julich GmbH, WGer, 77-78, Oak Ridge Nat Lab, 70-75, 84-85 & 91-92, Brookhaven Nat Lab, 85, 86, 87, 88 & 89; Woodrow Wilson fel. *Mem:* AAAS; Am Phys Soc; Am Asn Physics Teachers; Am Crystallog Asn. *Res:* Thermodynamics of liquids and solutions using x-ray methods; measurement of dislocations and interstitial atom content in crystal surfaces by x-ray methods; diffuse scattering using synchrotron radiation. *Mailing Add:* Dept Physics Anderson Univ Anderson IN 46012-3462

SHAFFER, LOUIS RICHARD, b Sharon, Pa, Feb 7, 28; m 55; c 3. CONSTRUCTION ENGINEERING. *Educ:* Carnegie-Mellon Univ, BS, 50; Univ Ill, Urbana, MS, 57, PhD(systs civil eng), 61. *Prof Exp:* Asst to master mech, Nat Castings Co, Pa, 50-52, asst to dir eng, Sharon Steel Corp, 53-54; instr civil eng, 55-61, from asst prof to assoc prof, 61-65, PROF CIVIL ENG, UNIV ILL, URBANA, 65-; TECH DIR CONSTRUCT ENG RES LAB, US ARMY, 69- *Concurrent Pos:* Asst dir, Tech Dir Construct Eng Res Lab, US Army, 69-70 & dep dir, 70-76; coordr, Int Working Comn on Orgn & Mgt Construct, 74-; chmn, Tech Coun Res, Am Soc Civil Engrs, 76-77; chmn US Nat Comt, Int Coun Bldg Res, 77-83. *Honors & Awards:* Walter L Huber Res Prize, Am Soc Civil Engrs, 67, Construct Mgt Award, 78, Construct Res Award, 87. *Mem:* Am Soc Civil Engrs; Soc Am Mil Engrs. *Res:* Modern construction management with emphasis on technical innovation to increase productivity of management on all levels. *Mailing Add:* 2203 Pond St Urbana IL 61801

SHAFFER, MORRIS FRANK, b Revere, Mass, Feb 2, 10; m 43, 59; c 2. MEDICAL MICROBIOLOGY. *Educ:* Mass Inst Technol, SB, 30; Oxford Univ, PhD(bact), 34, DSc(microbiol), 73. *Prof Exp:* Asst biol & bact, Mass Inst Technol, 30-31; fel bact, Harvard Univ, 34-35, Nat Res Coun fel med, 35-36, asst bact, 36-38; res assoc div microbiol, Squibb Inst Med Res, NJ, 38-42; assoc bact & immunol, Harvard Med Sch, 42-43; from assoc prof to prof bact, Sch Med, Tulane Univ La, 43-73, chmn dept, 48-73; dean grad sch biomed sci, Col Med & Dent NJ, 73-77; coordr off res, Med Sch, La State Univ, 78-85; EMER PROF, TULANE UNIV, 78- *Concurrent Pos:* Sr bacteriologist, Mass Antitoxin & Vaccine Lab, 42-43; consult virus & rickettsia sect, Commun Dis Ctr, USPHS; consult, Tulane-Colombia Proj, Int Coop Admin; mem comn measles & mumps, Bd Prev Epidemic Dis in US Army; mem comn enteric infections, Armed Forces Epidemiol Bd; chmn microbiol & immunol study sect, NIH, chmn microbiol res training comt & mem comt int res fels; mem comt, Lederle Med Fac Awards; assoc ed, J Immunol, 55-59, J Bacteriol, 59-62 & Infection, 77-; chmn bd gov, Am Acad Microbiol, 66-67. *Mem:* Am Soc Exp Path; hon mem Am Soc Microbiol (vpres, 70-71, pres, 71-72); Soc Exp Biol & Med; Am Soc Trop Med & Hyg; Am Asn Immunol; Infectious Dis Soc Am. *Res:* Microbiology and immunology of human infections due to bacteria and viruses. *Mailing Add:* 5315 Camp St New Orleans LA 70115

SHAFFER, NELSON ROSS, b Galion, Ohio, Aug 18, 48; m 67, 80; c 6. OIL SHALES, CLAY MINERALOGY. *Educ:* Ohio State Univ, BSc, 72, MSc, 74. *Prof Exp:* Teaching asst geol, Ohio State Univ, 73-74, res asst geochem, 74; RES SCIENTIST GEOL, IND GEOL SERV, 74- *Concurrent Pos:* Res assoc, Glenn A Black Lab Archaeol, 87-; mem exec comt, IGCP, UN, 87-; prin invest, Basalt Waste Isolation proj, 85-87, US Dept Energy, 84-85; consult, 78- *Mem:* Fel Geol Soc Am; Soc Mining Engrs; Geochem Soc; Mineral Soc Am; Am Chem Soc; Am Asn Petrol Geologists. *Res:* Geochemistry and mineralogy of metalliferous black shales, limestone, phosphates, other sedimentary rocks, and ground water; stable isotope studies of sulfides, shales, archaelogical artifacts; mineralogy of meteorites, biominerals; industrial uses of minerals. *Mailing Add:* Indiana Geol Surv 611 N Walnut Grove Bloomington IN 47405

SHAFFER, PATRICIA MARIE, b Los Angeles, Calif, June 11, 28. BIOCHEMISTRY & MOLECULAR BIOLOGY. *Educ:* San Francisco Col Women, BA, 52; Stanford Univ, MS, 59; Univ Calif, San Diego, PhD(chem), 75. *Prof Exp:* Asst prof chem, San Diego Col Women, 59-68; student counr, 59-78, assoc prof, 71-81, PROF CHEM, UNIV SAN DIEGO, 81- *Concurrent Pos:* NATO fel, Univ Newcastle upon Tyne, Eng, 80-81; sabbatical leave, Dept Genetics, Univ Ga, 87-88. *Mem:* Am Soc Biochemists & Molecular Biologists; AAAS; Am Chem Soc. *Res:* Metabolism of primidine deoxyribonucleosides in Neurospora crassa and aspergillus nidulans; ammonium repressible enzymes; a-ketoglutarate dioxygenases. *Mailing Add:* Dept Chem Univ San Diego San Diego CA 92110

SHAFFER, ROBERT LYNN, b Long Beach, Calif, Dec 29, 29; m 58; c 1. MYCOLOGY, TAXONOMY. *Educ:* Kans State Col, BS, 51, MS, 52; Cornell Univ, PhD, 55. *Prof Exp:* From instr to asst prof bot, Univ Chicago, 55-60; from asst prof to prof bot, 60-81, dir, 75-86, CUR FUNGI, HERBARIUM, UNIV MICH, ANN ARBOR, 60-, WEHMEYER PROF FUNGAL TAXON, 81- *Mem:* Mycol Soc Am (secy-treas, 68-71, vpres, 71-72, pres, 73-74); NAm Mycol Asn; Int Asn Plant Taxon. *Res:* Taxonomy of fungi, especially Agaricales. *Mailing Add:* 4042a Nat Sci Biol Univ Mich Ann Arbor MI 48109-1048

SHAFFER, RUSSELL ALLEN, b Philadelphia, Pa, Nov 3, 33; m 58; c 4. THEORETICAL PHYSICS. *Educ:* Drexel Univ, BS, 56; Johns Hopkins Univ, PhD(physics), 62. *Prof Exp:* Res assoc physics, Vanderbilt Univ, 62-64; asst prof, 64-67, ASSOC PROF PHYSICS, LEHIGH UNIV, 67- *Mem:* Am Asn Physics Teachers. *Res:* Theory of elementary particle interactions; electroweak interactions; lepton physics. *Mailing Add:* Dept Physics Bldg 16 Lehigh Univ Bethlehem PA 18015

SHAFFER, THOMAS HILLARD, RESPIRATORY & DEVELOPMENTAL PHYSIOLOGY. *Educ:* Drexel Univ, PhD(appl mech), 72. *Prof Exp:* Asst prof physiol, Univ Pa Sch Med, 74-77; assoc prof, 77-87, PROF PHYSIOL & PEDIAT, TEMPLE UNIV, 87-, DIR RESPIRATORY, PHYSIOL SECT, 77- *Mem:* Am Thoracic Soc; Am Physiol Soc; AAAS; NY Acad Sci. *Res:* Biological development of the respiratory system. *Mailing Add:* Dept Physiol Sch Med Temple Univ 3400 N Broad St Philadelphia PA 19140

SHAFFER, WAVE H, b Fulton Co, Ohio, May 23, 09; m 57; c 3. PHYSICS. *Educ:* Hiram Col, BA, 33; Ohio State Univ, MA, 36, PhD(physics), 39. *Prof Exp:* Asst chem, Williams Col, 33-34; res fel physics, Univ Chicago, 39-40; asst, 36-38, from instr to prof, 40-76, EMER PROF PHYSICS, OHIO STATE UNIV, 76- *Concurrent Pos:* Res mem, Appl Physics Lab, Silver Spring, Md, 43-44 & 44-45; vis prof, Univ Colo, 64, Dartmouth Col, 65 & 66. *Mem:* Am Phys Soc; Am Asn Physics Teachers. *Res:* Theory of rotation-vibration, infrared, spectra of small polyatomic molecules; contact-transformation of quantum-mechanical hamiltonian of rotating-vibrating molecule. *Mailing Add:* 679 E Beaumont Rd Columbus OH 43214-2203

SHAFFNER, RICHARD OWEN, b Chicago, Ill, June 12, 38; div; c 2. GAS DISCHARGE PHYSICS. *Educ:* Ill Inst Technol, BSEE, 60, Mass Inst Technol, SMEE, 62; Case Western Reserve Univ, PhD(elec eng), 74. *Prof Exp:* Engr discharge lamps, Lighting Bus Group, Gen Elec Co, Ohio, 68-71, sr engr, 71-75, sr design engr, 75-78; sr staff scientist discharge plasmas, 78-79, mgr, Daymax Div, 80, SR PROJ ENGR, ILC TECHNOL INC, 81- *Mem:* Am Phys Soc; Inst Elec & Electronics Engrs; Sigma Xi. *Res:* Radiation from discharge plasmas and physical chemistry associated with discharge; metal halide arc lamp design. *Mailing Add:* 1711 Braddock Ct San Jose CA 95125

SHAFI, MOHAMMAD, b Peshawar, Pakistan, May 18, 37; m 63; c 3. SYSTEMS SERVICES, INFORMATION SYSTEMS. *Educ:* Islamia Col, Pakistan, BSc, 56; Peshawar Univ, MSc, 58; Georgetown Univ, MS, 61, PhD(physics), 63; Univ NMex, MA, 71. *Prof Exp:* Asst physics, Georgetown Univ, 58-63; instr, Swarthmore Col, 63-64; sr lectr, Univ Karachi, 64-65; lectr, Univ West Indies, Trinidad, 65-67; lectr & res assoc, Univ NMex, 67-68, asst prof, 68-69; consult, 69-71, sr res physicist, 71-72, sr systs analyst, 72-74, leading analyst, 74-76, mgr PSRO systs, 76-81, DIR SYSTS SERV, HANCOCK DIKEWOOD SERV, 81- *Concurrent Pos:* Res assoc, Univ NMex, 69-71, lectr, 71- *Mem:* AAAS; Am Phys Soc; Am Asn Physics Teachers; Am Soc Pub Admin; Am Mgt Asn. *Res:* Structure of diatomic molecules; theoretical and computer modeling of geomagnetic field; systems analysis; design, development and installation of management information systems for large programs; management of personnel and resources for operating large systems. *Mailing Add:* Eng Div Local Mkt Kuwait Nat Petrol Co PO Box 70 Safat 13001 Kuwait

SHAFIQ, SAIYID AHMAD, b Sitapur, India, Dec 29, 29; m 55; c 3. CELL BIOLOGY. *Educ:* Oxford Univ, DPhil(cytol), 54. *Prof Exp:* Lectr zool, Univ Dacca, 54-59; Fulbright fel anat, Univ Wash, 59-60; res assoc cell biol, Inst Muscle Dis, 60-63, asst mem, 63-69, assoc mem, 69-74; ASSOC PROF NEUROL, STATE UNIV NY DOWNSTATE MED CTR, 74- *Mem:* Am Asn Anatomists. *Res:* Cytology; muscle pathology; electron microscopy. *Mailing Add:* Dept Neurol Downstate Med Ctr State Univ NY 450 Clarkson Ave Brooklyn NY 11203

SHAFIT-ZAGARDO, BRIDGET, b Bronx, NY, May 6, 52; m 76; c 2. HUMAN GENETICS. *Educ:* C W Post Col, BS, 73; NY Univ, MS, 78; City Univ New York, MPhil, 80; Mt Sinai Sch Med, PhD(genetics), 81. *Prof Exp:* res assoc, dept cell biol, 81-84, ASST PROF, DEPT PATH, ALBERT EINSTEIN COL MED, 84- *Mem:* Am Soc Human Genetics; Am Soc Cell Biol. *Res:* Molecular biology of Alzheimer's disease; inherited disorders in man; brain-specific genes; human gene polymorphisms. *Mailing Add:* Dept Path Albert Einstein Med Col 1300 Morris Park Ave Bronx NY 10461

SHAFRITZ, DAVID ANDREW, b Philadelphia, Pa, Oct 5, 40; m 64; c 3. MOLECULAR BIOLOGY, MEDICINE. *Educ:* Univ Pa, AB, 62, MD, 66. *Prof Exp:* Res assoc, Molecular Dis Br, Sect Human Biochem, Nat Heart & Lung Inst, 68-71; from instr to asst prof med, Harvard Med Sch, 71-73; from asst prof to assoc prof, 73-81, PROF MED & CELL BIOL, ALBERT EINSTEIN COL MED, 81-, DIR, MARION BESSIN LIVER RES CTR, 85- *Concurrent Pos:* Nat Inst Arthritis, Metab & Digestive Dis spec res fel, Gastrointestinal Unit, Mass Gen Hosp, 71-73; Nat Inst Arthritis, Metab & Digestive Dis res career develop award, 75-80; assoc ed, Hepat, 81-86. *Honors & Awards:* Merrel lectr, Am Asn Study Liver Dis, 78; J Friedenvald Lectr, Univ Md, 83; Grace Kimball Lectr, Wilkes Col, 87; James Gibson Lectr, Univ Hong Kong, 87. *Mem:* AAAS; Am Asn Study Liver Dis; Int Asn Study Liver; Am Soc Biol Chemists; Am Soc Clin Invest; Harvey Soc; Asn Am Physicians; Am Gastroenterol Asn; Interurban Clin Club. *Res:* Mammalian protein synthesis and messenger RNA metabolism; eukaryotic gene regulation; liver regeneration and regulation of cellular differentiation, liver somatic gene therapy, hepatitis B virus infection, chronic liver disease and primary liver cancer; diseases of liver and intestine. *Mailing Add:* Albert Einstein Col Med 1300 Morris Park Ave Bronx NY 10461

SHAFROTH, STEPHEN MORRISON, b Denver, Colo, June 12, 26; m 54; c 3. ATOMIC PHYSICS. *Educ:* Harvard Univ, BA, 47; Johns Hopkins Univ, PhD(physics), 53. *Prof Exp:* From instr to asst prof physics, Northwestern Univ, 53-59; res physicist, France, 59 & Bartol Res Found, 60-67; from assoc prof to prof physics, Univ NC, Chapel Hill, 67-88; prof physics, Triangle Univs Nuclear Lab, Durham, 70-88; RESEARCHER, OAK RIDGE NAT LAB, 88- *Concurrent Pos:* Vis lectr, Temple Univ, 65; lectr & consult, US Naval Res Lab, 66-; adv bd, Nuclear Data Sheets, 66-78; assoc ed, Atomic Data & Nuclear Data Tables, 75-; ed, Scintillation Spectros of Gamma Radiation. *Mem:* Fel Am Phys Soc. *Res:* Experimental nuclear physics; high resolution x-ray spectroscopy following atomic excitation by heavy ions accelerated by a tandem Van de Graaff; high resolution projectile electron spectroscopy. *Mailing Add:* Dept Physics Univ NC Chapel Hill NC 25714

SHAFTAN, GERALD WITTES, b New York, NY, Apr 15, 26; m 49; c 2. SURGERY. *Educ:* Brown Univ, AB, 45; NY Univ, MD, 49. *Prof Exp:* Asst instr, State Univ NY Downstate Med Ctr, 56-57, from instr to assoc prof, 57-68, prof surg, 68-; STAFF MEM, BROOKDALE HOSP MED CTR, BROOKLYN. *Concurrent Pos:* Consult, Vet Admin Hosp, Brooklyn, 64-; chief surg serv, Kings County Hosp Ctr, Brooklyn, 72- *Mem:* Am Col Surgeons; Asn Acad Surg; Am Soc Surg Hand; Am Asn Surg Trauma; Sigma Xi. *Res:* Management of multiple trauma and shock; control of bleeding and healing of fractures. *Mailing Add:* Brookdale Hosp Med Ctr Linden Blvd at Brookdale Plaza Brooklyn NY 11212

SHAFTMAN, DAVID HARRY, b Philadelphia, Pa, Aug 27, 24; m 44, 75; c 4. MATHEMATICS, NUCLEAR ENGINEERING. *Educ:* Univ Chicago, BS, 48, MS, 49. *Prof Exp:* Asst mathematician, Naval Reactors Div & Reactor Eng Div, 50-55, assoc mathematician, Reactor Eng Div & Reactor Physics Div, 55-69, assoc mathematician, Appl Physics Div, 69-74, mathematician, Appl Physics Div, 74-77, MATHEMATICIAN, REACTOR ANALYSIS & SAFETY DIV, ARGONNE NAT LAB, 77- *Mem:* AAAS; Am Math Soc; Am Nuclear Soc. *Res:* Reactor design and development; theoretical reactor physics; theory and application of functional analysis. *Mailing Add:* 25 Oakwood Dr Naperville IL 60540

SHAGASS, CHARLES, b Montreal, Que, May 19, 20; nat US; m 42; c 3. PSYCHIATRY. *Educ:* McGill Univ, BA, 40, MD, CM, 49, dipl, 53; Univ Rochester, MS, 41. *Prof Exp:* Asst psychol, Allan Mem Inst Psychiat, McGill Univ, 45-48, asst & assoc dir res, Dept Psychol, 47-50, resident psychiat, 50-52, from lectr to asst prof psychiat, Univ & dir dept electrophysiol, Inst, 52-58; from assoc prof to prof psychiat, Univ Iowa, 58-66; dir, Temple Clin Serv, Eastern Pa Psychiat Inst, 66-82; dir, Temple Progs, Pa Psychiat Ctr, 82-90; prof psychiat, 66-, actg chair, Dept Psychiat, 86-90, EMER PROF PSYCHIAT, TEMPLE UNIV, 90- *Concurrent Pos:* Clin asst, Royal Victoria Hosp, 53-55, asst psychiatrist, 56-58; mem, B Study Sect, NIMH, 63-67, Extramural Res Adv Comt, 69-70 & Clin Prog Projs Study Sect, 70-74 & 79-82. *Honors & Awards:* Samuel Hamilton Award, Am Psychopath Asn, 75; Gold Medal Award, Soc Biol Psychiat, 77. *Mem:* Soc Biol Psychiat (pres, 74-75); Am EEG Soc; fel Am Psychiat Asn; Am Psychopath Asn (pres, 74-75); fel Am Col Neuropsychopharmacol; World Fedn Soc Biol Psychiat (pres 81-85). *Res:* Neurophysiological aspects of psychiatric illness, emotion and learning; psychosomatic relationships; objective tests for psychiatric diagnosis; experimental psychopathology; psychopharmacology. *Mailing Add:* Philadelphia Psychiat Ctr Ford Rd & Monument Ave Philadelphia PA 19131

SHAH, ASHOK CHANDULAL, b Palanpur, India, Apr 25, 39; m 66; c 1. PHARMACEUTICS. *Educ:* Univ Bombay, BSc (hons), 59, BSc, 61; Univ Wis, MS, 63, PhD(pharm), 65. *Prof Exp:* SR SCIENTIST, UPJOHN CO, 65- *Mem:* Am Pharmaceut Asn; fel Acad Pharmaceut Sci. *Res:* Chemical kinetics in solutions; physical properties of solids, including phase behavior, and surface, crystal, thermal, mechanical and dissolution properties; formulation of new solid drug dosage forms. *Mailing Add:* Dept Resource Servs Sir Sanford Fleming Col PO Box 8000 Lindsay ON K9V 4S6 Can

SHAH, ATUL A, b Ahmedabad, India, July 20, 40; m 64; c 2. REGULATORY AFFAIR, QUALITY CONTROL. *Educ:* Gujarat Univ, Ahmedabad, India, BS, 61; Univ Wis, Madison, MS, 63, PhD(phys pharm), 67. *Prof Exp:* Assoc res pharmacist, Parke, Davis & Co, 67-68; dir res & develop, Knoll Pharmaceut, 68-76; mgr anal res & develop, Div Am Home Prod, Whitehall Labs, 76-85; dir qual affairs, Lemmon Pharmaceut Co, 85-87; dir anal res & develop, Par Pharmaceut Co, 87-89; DIR SCI AFFAIRS, G & W LABS, INC, 89- *Mem:* Asn Am Pharmaceut Scientists. *Res:* Pharmaceutical and clinical development of new drug products. *Mailing Add:* Ten Washington Ct East Windsor NJ 08520

SHAH, BABUBHAI VADILAL, b Bombay, India, Feb 6, 35; US citizen; m 66; c 2. STATISTICS. *Educ:* Univ Bombay, BS, 55, MS, 57, PhD(statist), 60. *Prof Exp:* Res assoc statist, Iowa State Univ, 59-62; res analyst, Karamchand Premchand Ltd, India, 62-66; statistician, 66-68, mgr, 68-71, assoc dir, 72-75, CHIEF SCIENTIST, RES TRIANGLE INST, 76- *Concurrent Pos:* Adj prof biostatist, Univ NC, 79- *Mem:* Am Statist Asn; Soc Indust & Appl Math; Asn Comput Mach; Royal Statist Soc. *Res:* Data analysis in complex sample surveys, microsimulation models, optimization techniques. *Mailing Add:* Res Triangle Inst PO Box 12194 Research Triangle Park NC 27709

SHAH, BHAGWAN G, b Bombay, India, Mar 16, 24; Can citizen; m 55. ANIMAL SCIENCE & NUTRITION, AGRICULTURAL & FOOD CHEMISTRY. *Educ:* Bombay Univ, BSc, 45, MSc, 51; Univ Ill, Urbana, PhD(nutrit biochem), 64. *Prof Exp:* Fulbright res fel, US Educ Found, 60; RES SCIENTIST & SECT HEAD, MACRO NUTRIT & MINERALS, NUTRIT RES DIV, BUR NUTRIT SCI, HEALTH PROTECTION BR, HEALTH & WELFARE, CAN, 66- *Concurrent Pos:* Mem, Comt Rev Can Dietary Standard, 73-75; mem, Exec Bd, Can Soc Nutrit Sci, 79-82; reviewer, Nat Health Res & Develop Prog, grant appls, 80- & US Recommended Daily Allowance Comt, 85; assoc ed, Nutrit Reports Int. *Mem:* AAAS; Can Soc Nutrit Sci; Am Inst Nutrit; Int Union Nutrit Sci. *Res:* Role of calcium, phosphorus, fluorine in osteoporosis; role of minerals in cardiovascular disease; bioavailability from foods of mineral nutrients such as iron & zinc. *Mailing Add:* Three Brockington Cres Nepean ON K2G 4K6 Can

SHAH, BHUPENDRA K, b Visnagar, India, Dec 8, 35; US citizen; m 58; c 2. STATISTICS, BIOMETRICS. *Educ:* Univ Baroda, BS, 55, MS, 57; Yale Univ, PhD(statist), 68. *Prof Exp:* Lectr statist, Univ Baroda, 57-64; statistician, E I du Pont de Nemours & Co, 68-69; res math statistician, Info Sci Div, Rockland Res Inst, 69-70, assoc res scientist biomet, 70-83; adj prof, State Univ NY, Albany, 82-84; RES SCIENTIST, HELEN HAYES HOSP, 84- *Concurrent Pos:* Consult med statist, Nat Inst Neurol Dis & Strokes, Washington, DC, 69-71, Impact Nuclear Power Plants; NIMH res grant, 73; Fulbright scholar. *Mem:* Biomet Soc; Inst Math Statist; Sigma Xi; Am Statist Asn; Acad Pharm Sci; Soc Nuclear Med. *Res:* Application of statistics and computers to medical research; developing statistical theory to help establish bioequivalence of two pharmaceuticals. *Mailing Add:* 540 Highview Ave Pearl River NY 10965

SHAH, BHUPENDRA UMEDCHAND, b Bombay, India, June 15, 38; m 63; c 2. METALLURGY. *Educ:* Birla Eng Col, India, BEng, 59; Mich State Univ, MSc, 61, PhD(metall), 67; Univ Akron, MBA, 76. *Prof Exp:* Proj engr, Bausch & Lomb, NY, 62-63; RES SPECIALIST METALL, RES DIV, TIMKEN CO, 67- *Mem:* Am Inst Mining Metall & Petrol Engrs; Am Soc Metals; Am Foundrymen's Soc. *Res:* Development of new steelmaking processes and strand casting. *Mailing Add:* PO Box 35635 Canton OH 44735-5635

SHAH, DHARMISHTHA V, CLINICAL NUTRITION, NUTRITIONAL BIOCHEMISTRY. *Educ:* Univ Beroda, India, PhD(biochem), 62. *Prof Exp:* SR SCIENTIST, UNIV WIS-MADISON, 69- *Mailing Add:* c/o Dr V K Shah Univ Wis 107 Bacteriol Dept Madison WI 53706

SHAH, DINESH OCHHAVLAL, b Bombay, India, Mar 31, 38; m 68; c 2. BIOPHYSICS. *Educ:* Univ Bombay, BS, 59, MS, 61; Columbia Uni, PhD(biophys), 65. *Prof Exp:* Nat Res Coun-NASA resident res assoc, Ames Res Ctr, NASA, 67-68; res assoc surface chem, Lamont Geol Observ, Columbia Univ, 68-70; from asst prof to assoc prof, 70-75, PROF ANESTHESIOL, BIOPHYS & CHEM ENG, UNIV FLA, 75- *Concurrent Pos:* Consult, Barnes-Hind Pharmaceut, Inc, 72- & Sun Oil Co, 73- *Honors & Awards:* Excellence Teaching Award, Univ Fla, 72, President's Scholar Award & Outstanding Serv Award, 75; Best Paper Award, Int Cong Chem & Technol, India, 78. *Mem:* AAAS; Am Soc Eng Educ; Am Chem Soc; Am Inst Chem Engrs; Am Soc Anesthesiol. *Res:* Surface chemistry of biological systems and processes; lung surfactant; corneal surface; biomembranes; biomaterials; emulsification of fat; lipoproteins and mechanisms of anesthesia. *Mailing Add:* Dept Chem Eng Univ Fla Gainesville FL 32611

SHAH, GHULAM M, b Srinagar, India, May 22, 37; m 62; c 3. ANALYTICAL MATHEMATICS, APPLIED MATHEMATICS. *Educ:* Univ Jammu & Kashmir, India, BA, 56; Aligarh Muslim Univ, India, MA, LLB & dipl statist, 58; Univ Wis-Milwaukee, PhD(math), 66. *Prof Exp:* Lectr math, Amar Singh Col, India, 58-60; jr lectr, Regional Eng Col, Srinagar, 60-61, lectr, 61-63; asst, Univ Wis-Milwaukee, 63-66, asst prof, 66-70; assoc prof, 70-75, chmn dept, 73-80, PROF MATH, UNIV WIS-WAUKESHA, 75- *Res:* Zeros of polynomials in complex variables; differential equations and analytic function theory. *Mailing Add:* Dept Math Univ Wis Waukesha WI 53188

SHAH, HAMISH V, B Rajpipla, Gujarat, India, May 18, 53; US citizen; m; c 2. ANALYTICAL CHEMISTRY, INORGANIC CHEMISTRY. *Educ:* Saurashtra Univ, Rajkot, India, BS, 73, MS, 75. *Prof Exp:* Jr chemist org chem, IPCL, India, 75-77; chemist org chem, Kings Labs Inc, 78; chief chemist org chem, Columbia Organic Chemicals Co, 78-82; RES & DEVELOP, MGR ORG CHEM, CARBINAL CHEMCO, 82- *Mem:* Am Chem Soc. *Res:* Organotin compounds, their structure and reactivity; one patent. *Mailing Add:* 835 Knollwood Dr Columbia SC 29209

SHAH, HARESH C, b Godhra, India, Aug 7, 37; m 65. CIVIL ENGINEERING. *Educ:* Univ Poona, BS, 59; Stanford Univ, MS, 60, PhD(civil eng), 63. *Prof Exp:* Teaching asst civil eng, Stanford Univ, 61-62; actg asst prof, San Jose State Col, 62; from instr to assoc prof civil & mech eng, Univ Pa, 62-68; assoc prof, 68-73, dir, John A Blume Earthquake Eng Ctr, 76-85, PROF CIVIL ENG, STANFORD UNIV, 73-, CHMN DEPT, 85- *Concurrent Pos:* Consult, Local State & Fed Govt Agencies, 68-, UNESCO & var foreign govts, 68- *Mem:* Am Soc Civil Engrs; Earthquake Eng Res Inst; Am Concrete Inst; Seismol Soc Am. *Res:* Structural mechanics; application of theory of probability and statistics to civil engineering problems; earthquake engineering and risk analysis; innovations in engineering education. *Mailing Add:* Dept Technol North Harris County Col 2700 W Throne Dr Houston TX 77073

SHAH, HASMUKH N, b Radhanpur, India, Mar 25, 34; US citizen. POLYMER ENGINEERING. *Educ:* Univ Bombay, India, BS Hons, 55, BS, 57, MS, 60. *Hon Degrees:* Dipl plastics, SGer Plastic Inst. *Prof Exp:* Proj engr single use clin thermometers, Akzona Inc, Info-chem, 77-81 & insulin pump, Becton Dickinson Advan Bus Develop, 81-83; tech dir polymer powder coating, Plasti-coats & prints, 83-87; proj engr polymer catheter, Johnson & Johnson Interventional Syst, 89; RES POLYMERS, BRISTOL MYERS SQUIBB CONVATEC, 90- *Concurrent Pos:* Bd dirs, Med Div, Soc Plastics Engrs. *Mem:* Nat Acad Sci; Soc Plastics Engrs. *Res:* Develop new biodegradable polymer films; new adhesives; polymer powder coating; select appropriate materials for medical device. *Mailing Add:* Parklane Apt No 7G Ten Landing Lane New Brunswick NJ 08901

SHAH, ISHWARLAL D, b Rangoon, Burma, May 8, 35; m 63. METALLURGICAL ENGINEERING. *Educ:* Univ Bombay, BSc, 56; Univ Mo-Rolla, BS, 58; Purdue Univ, MS, 61; Stanford Univ, PhD(metall eng), 67. *Prof Exp:* Res asst metall eng, Purdue Univ, 58-61; res chemist, Delta Res, Ill, 61-63; res asst mineral eng, Stanford Univ, 63-67, res assoc, 67-68; res metallurgist, US Bur Mines, 68-84; RES ASST, DEAN WITTIER, 84- *Mem:* Am Inst Mining Metall & Petrol Engr; Am Soc Metals. *Res:* Fundamentals of roasting of copper sulphide ores; solubilities and diffusion of gases in metals at high temperatures; extractive metallurgy of copper-nickel sulfides, production of alumina, iron and steel making. *Mailing Add:* 3912 Grimes Lane Edina MN 55424

SHAH, JAGDEEP C, b Surat, India, Sept 3, 42; m 67; c 2. SOLID STATE PHYSICS. *Educ:* Univ Bombay, BSc, 62; Mass Inst Technol, PhD(solid state physics), 67. *Prof Exp:* Teaching asst physics, Rensselaer Polytech Inst, 62-63; res asst, Mass Inst Technol, 63-67; mem tech staff, 67-85, DISTINGUISHED MEM TECH STAFF, AT&T BELL LABS, 85- *Honors & Awards:* Humboldt Sr Scientist Award. *Mem:* Am Inst Physics; fel Am Phys Soc. *Res:* Optical, electrical, and transport properties of semiconductors; high intensity effects and non-equilibrium phenomena in semiconductors; electron-hole liquids in semiconductors, amorphous solids; ultrafast spectroscopy in semiconductors; semiconductor microstructures; physics of high speed electronic and optoelectronic devices. *Mailing Add:* Bell Labs Rm 4D-415 Holmdel NJ 07733

SHAH, JITENDRA J, organic chemistry , pharmaceutical chemistry, for more information see previous edition

SHAH, KANTI L, b Aligarh, India, Jan 6, 35; m 56; c 3. ENVIRONMENTAL SCIENCES, WATER RESOURCES. *Educ:* Aligarh Muslim Univ, India, BS, 55; Univ Kans, MS, 63; Univ Okla, PhD(environ sci), 69. *Prof Exp:* Asst engr, Irrig Dept, State of Uttar Pradesh, India, 55-59; city engr, Munic Corp, India, 59-60; lectr civil eng, M G Polytech, Hathras, India, 60-61; sanit engr, Kans State Dept Health, 63-67; res assoc, Univ Okla, 69; sanit engr, Pa Dept Health, 70; from asst prof to assoc prof, 70-76, PROF CIVIL ENG, OHIO NORTHERN UNIV, 76-, DISTINGUISHED CHAIR HOLDER, 76- *Concurrent Pos:* Consult water qual mgt, Pa Dept Health, 70. *Mem:* Sigma Xi; Am Soc Civil Engrs. *Res:* Interdisciplinary approach to environmental and water resources, especially wastewater and stream water quality. *Mailing Add:* Dept Civil Eng Ohio Northern Univ 500 S Main Ada OH 45810

SHAH, KEERTI V, b Ranpur, India, Nov 2, 28; m 67; c 2. VIROLOGY. *Educ:* B J Med Col, Poona, India, MBBS, 51; Johns Hopkins Univ, MPH, 57, DrPH, 63. *Prof Exp:* From intern to resident gen med, Sassoon Hosp, Poona, India, 51-53; asst res officer virol, Virus Res Ctr, Poona, 53-58, res officer, 58-61; res assoc pathobiol, 62-63, from asst prof to assoc prof, 63-74, PROF IMMUNOL & INFECTIOUS DIS, SCH HYG & PUB HEALTH, JOHNS HOPKINS UNIV, 74-, ASSOC CHMN, 85- *Mem:* AAAS; Am Soc Microbiol; Am Asn Immunologists; Am Soc Trop Med Hyg. *Res:* Biology of DNA tumor viruses. *Mailing Add:* Dept Immunol & Infectious Dis Johns Hopkins Univ Sch Hyg Baltimore MD 21205-2179

SHAH, MANESH J(AGMOHAN), b Bombay, India, July 9, 32; US citizen; m 60; c 2. CHEMICAL ENGINEERING, MATHEMATICS. *Educ:* Tech Inst Bombay, BS, 53, MS, 55; Univ Mich, MSChE, 57; Univ Calif, Berkeley, PhD(chem eng), 61. *Prof Exp:* Assoc engr phys chem, IBM Corp, 60-62, staff engr process control, 62-65, adv engr, 65-69, sr control systs engr, Data Processing Div, 69-72, sr engr, Gen Systs Div, 72-80, CONSULT ENGR, NAT ACCOUNTS DIV & STORAGE PROD DIV, IBM CORP, 81- *Concurrent Pos:* Adj prof, Univ PR, 78-79; sr lectr, Stanford Univ, 80-87. *Mem:* Am Inst Chem Engrs. *Res:* Process simulation and control; electric birefringence of colloids; heat transfer and fluid mechanics in non-Newtonian fluids; applied mathematics in chemical engineering; expert systems. *Mailing Add:* 1788 Frobisher Way San Jose CA 95124

SHAH, MIRZA MOHAMMED, b Delhi, India, Aug 11, 41; US citizen; m 77; c 2. MULTIPHASE HEAT TRANSFER & FLOW, HEATING VENTILATING & AIR CONDITIONING. *Educ:* Aligarh Muslim Univ, India, BSc 59 & 63; Calif State Univ, Los Angeles, MS, 73. *Prof Exp:* Scientist in chg airconditioning & refrig, Cent Mech Eng Res Inst, India, 65-68; res engr heat transfer, Tech Univ Norway, Trondheim, 68-69; mech engr, Bechtel Power Corp, 73 & United Engrs & Constructors, 74-76; sr engr mech eng, Gilbert-Commonwealth, 76-78; PRIN ENGR, EBASCO SERV INC, 78- *Concurrent Pos:* consult, Kirloskar Pneumatic Co, Poona, India, 68-69 & Gen Elec Co, 82-84. *Mem:* Am Soc Heating, Refrig & Air-Conditioning Engrs; Am Soc Mech Engrs. *Res:* Heat transfer in multiphase flow; general predictive techniques for heat transfer during boiling and condensation; critical heat flux; heat transfer in fluidized beds and other two component systems. *Mailing Add:* Ten Dahlia Lane Redding CT 06896-1421

SHAH, PRADEEP L, b Poona, India, Oct 17, 44; m 74; c 2. ELECTRICAL ENGINEERING, QUANTUM ELECTRONICS. *Educ:* Indian Inst Technol, BTech, 66; Rice Univ, PhD(elec eng), 70. *Prof Exp:* Supvr optics, Hycel Inc, 71-72; adj asst prof, Rice Univ, 72-73; mgr, metal oxide semiconductors/complementary metal oxide semiconductors technol, 73-80, complementary metal oxide semiconductors develop mgr, 80-83, EPROM DEVELOP MGR, TEX INSTRUMENTS, 83- *Concurrent Pos:* Res assoc, Rice Univ, 70-72. *Mem:* Inst Elec & Electronics Engrs. *Res:* Development of very large scale integrated circuits spanning areas of solid state physical electronics such as materials science, device physics and electronic circuit and system design and analysis. *Mailing Add:* Tex Instruments 1118 Horseshoe Dr Sugarland TX 77478

SHAH, RAMESH KESHAVLAL, b Bombay, India, Sept 23, 41; m 68; c 2. HEAT EXCHANGERS, HEAT TRANSFER. *Educ:* Gujarat Univ, India, BE, 63; Stanford Univ, MS, 64, Engr, 70, PhD(mech eng), 72. *Prof Exp:* Proj engr, Air Preheater, 64-66 & Avco Lycoming, 68-69; res engr, Gen Motors Corp, 71-76, tech dir res, 76-83, staff develop engr, 83-89, SR STAFF RES SCIENTIST, HARRISON DIV, GEN MOTORS CORP, 89- *Concurrent Pos:* Adj prof, State Univ NY, Buffalo, 78-; tech ed, J Heat Transfer, 81-86; chmn, Heat Transfer Div, Am Soc Mech Engrs, 85-86; chmn, First World Conf Exp Heat Transfer, Fluid Mech & Thermodyn, 87-88; tech prog chmn, Nat Heat Trans Conf, 88-89; ed-in-chief, Int J Exp Thermal & Fluid Sci, 88-; pres, Assembly World Confs Exp Heat Transfer, Fluid Mech & Thermodynamics, 88-; tech prog chmn, Nat Heat Transfer Conf, 88-89, co-chmn, 90- 91; invited to lectr & present short courses in 18 countries on five continents. *Honors & Awards:* Region III Tech Achievement Award, Am Soc Mech Engrs, 79, Valued Serv Award, 87; Outstanding Serv Award, Am Soc Mech Engrs J Heat Transfer, 86; Charles Russ Richards Mem Award, 89. *Mem:* Am Soc Mech Engrs; Soc Automotive Engrs; Asn Mech Sci Brazil. *Res:* Compact heat exchangers, process heat exchangers, and internal laminar and turbulent flow forced convection; heat transfer in manufacturing and materials processing. *Mailing Add:* Harrison Div Gen Motors Corp 200 Upper Mountain Rd Lockport NY 14094-1896

SHAH, RAMESH TRIKAMLAL, b Padra, India, Sept 13, 34; m 57; c 2. MECHANICAL ENGINEERING. *Educ:* Univ Baroda, BE, 56; Tech Univ Mech Eng, Magdeburg, Ger, DrIng(mech eng), 59. *Prof Exp:* Prof mech eng, Univ Baroda, 56-67 & Ill Inst Technol, 67-68; PROF AERONAUT & MECH ENG, CALIF STATE POLYTECH UNIV, SAN LUIS OBISPO, 68- *Concurrent Pos:* Consult, Jyoti Ltd, Sayaji Iron Works, Metrop Springs & Buganda Steel, Gaskets & Oil Seals, 60-67; consult, Maurey Mfg, Versionall Steel Press & Southern Calif Edison & O'Connar Eng, 67- *Mem:* Am Soc Mech Engrs; Am Soc Heating, Refrig & Air-Conditioning Engrs; Asn Ger Engrs; Indian Inst Engrs. *Res:* Mechanical design; stress analysis. *Mailing Add:* Dept Mech Eng Calif State Polytech Univ San Luis Obispo CA 93407

SHAH, SHANTILAL NATHUBHAI, b Dhulia, India, Aug 5, 30; m 56; c 3. BIOCHEMISTRY, NEUROCHEMISTRY. *Educ:* Univ Bombay, BS, 51; Univ Nagpur, BSc, 54, MSc, 56; Univ Ill, Urbana, PhD, 60. *Prof Exp:* Res asst, Univ Ill, Urbana, 56-60; Inst Metab Res res fel, Highland Hosp, Oakland, Calif, 60-62; asst res physiologist & Alameda County Heart Asn fel, Univ Calif, Berkeley, 62-65; lectr, Sardar Patel Univ, India, 65-67; res specialist, Sonoma State Hosp, 69-73; assoc res biochemist, 73-80, from res biochemist II to res biochemist III, 81-85, res biochemist IV, 85-88, RES BIOCHEMIST V, LANGLEY-PORTER PSYCHIAT INST, UNIV CALIF, SAN FRANCISCO, 88- *Concurrent Pos:* NIH grants, Sonoma State Hosp, Calif, 68- *Mem:* AAAS; Am Inst Nutrit; Biochem Soc; Int Soc Neurochem; Am Soc Neurochem; Am Soc Biol Chem; Soc Exp Biol. *Res:* Lipid metabolism of the central nervous system and other tissues. *Mailing Add:* Langley-Porter Neuropsychiat Inst Brain Behav Res Ctr Sonoma Develop Ctr Eldridge CA 95431-1493

SHAH, SHEILA, b Zanzibar, Tanzania, Mar 14, 45; US citizen; m 73; c 1. PATHOLOGY. *Educ:* Bombay Univ, MD, 68. *Prof Exp:* Intern med, KEM Hosp, Bombay, 68-69; resident, 70-74, asst prof & pathologist, 74-78, DIR HEMAT LAB PATH, WVA UNIV HOSP, 78- *Concurrent Pos:* Pathologist, Cancer & Leukemia Group B, 76- *Mem:* Int Acad Path. *Res:* Platelet function abnormalities in myeloproliferative disorders related to morphological abnormalities. *Mailing Add:* Heartland Hosp E 5325 Faraon St Joseph MO 64506

SHAH, SHIRISH, b Ahmedabad, India, May 24, 42; US citizen; m 73; c 1. ENVIRONMENTAL SCIENCES, EDUCATION ADMINISTRATION. *Educ:* Gujarat Univ, BSc, 62; Univ Del, Newark, PhD(phys chem), 68. *Prof Exp:* From asst prof to assoc prof sci, Chesapeake Co, 68-76; prog mgr, Md Dept Transp Res Prog, 81-82; coordr acad prog dev, 76-78, chairperson technol, 79-82, CHAIRPERSON COMPUT SYSTS & TECHNOL, COMMUNITY COL BALTIMORE, 82- *Concurrent Pos:* Adminr Marine & Food Sci Res Projs, Chesapeake Col, 72-76; educ mgr, Baltimore City's Manpower Proj, 80-81; educ consult, Baltimore City's Joint Apprenticeship Comt, 83-; mem, Baltimore City's Adult Educ Comt, 82- & Hazardous Chem Comn, 85- *Mem:* Nat Sci Teacher's Asn; Am Chem Soc; Nat Environ Training Asn; Inst Elec & Electronics Engrs; Nat Inst Chemists; Am Tech Educ Asn. *Res:* Radiation chemistry of aqueous solutions; food preservation techniques; curriculum development in marine science, food science, electronic communications, traffic engineering and waste water treatment. *Mailing Add:* 5605 Purlington Way Baltimore MD 21212

SHAH, SHIRISH A, b Bombay, India, Apr 26, 38; US citizen; m 66; c 3. QUALITY AFFAIRS, REGULATORY AFFAIRS. *Educ:* Univ Bombay, India, BPharm, 61; Univ Conn, MS, 64; Univ Iowa, PhD(pharmaceut), 75. *Prof Exp:* Sr scientist res & develop, Revlon Health Care, 63-67 & 75- 76; sect head formulations res, Pennwalt Pharmaceut Div, 67-72; asst mgr prod develop, Johnson & Johnson Baby Prod Co, 76-79; dir, Res & Tech Serv, Zenith Labs, 79-85; vpres develop & tech affairs, Lemmon Co, 85-87; dir prod develop, Ciba Consumer Pharmaceut, 88-89; MGR PROD DEVELOP, DUPONT-MERCK PHARMACEUT CO, 90- *Concurrent Pos:* Teaching asst, Col Pharm, Univ Iowa, 72-75. *Mem:* Am Asn Pharmaceut Scientists; Am Pharmaceut Asn; Am Chem Soc; Drug Info Asn; Controlled Release Soc. *Res:* In vitro dissolution kinetics of single and multi-drug systems and its correlation to in vivo bioavailability; preformulation and formulation development of conventional and controlled release pharmaceutical dosage forms. *Mailing Add:* 135 Jonathan Dr Stamford CT 06903

SHAH, SURENDRA P, b Bombay, India, Aug 30, 36; US citizen; m 62; c 2. CIVIL ENGINEERING, STRUCTURAL ENGINEERING. *Educ:* Col Eng, Bombay, India, BS, BVM, 59; Lehigh Univ, MS, 60; Cornell Univ, PhD(struct eng), 65. *Prof Exp:* Design engr, Modjeski & Masters, Pa, 60-62; res asst struct eng, Cornell Univ, 64-65; res assoc, 65-66; asst prof mat eng, Univ Ill, Chicago Circle, 66-73, prof civil eng & mat eng, 73-81; PROF CIVIL ENG, NORTHWESTERN UNIV, 81-, DIR, CTR CONCRETE & GEOMAT, 89-; DIR, SCI & TECHNOL CTR ADVAN CEMENT-BASED MAT, NSF, 89- *Concurrent Pos:* Develop engr, Res Labs, Portland Cement Asn, 66; consult, Corning Glass Works, 67, US Gypsum Co, 74-75 & Sci Mus Va, 76-77; vis assoc prof, Mass Inst Technol, 69-70; vis prof, Delft Univ

Technol, 76-77; NSF res grants, 78-; consult, Holderbark Mgt, Ltd, Switz; mem tech comt, Hwy Res Bd; mem ad-hoc comt, Nat Acad Sci; consult, Amoco Res; guest prof, Denmark Tech Univ, 84; vis prov, Univ Sidney, 87; vis sr NATO scientist, Paris, 86; Humboldt distinguished vis scientist, Ger, 90-95. *Honors & Awards:* Rilem Gold Medal, 80; Thompson Award, Am Soc Testing & Mat, 83; Vis Sr Scientist Award, NATO, 86. *Mem:* Fel Am Concrete Inst; Am Soc Civil Engrs; Prestress Concrete Inst; Am Soc Testing & Mat. *Res:* Relating macroscopic mechanical behavior of concrete to its microscopic properties; developing fiber-reinforced concrete; micromechanics of composite materials; fracture of brittle solids; application of ferrocement to low cost housing; properties of concrete; nondestructive testing; advanced cement-based materials; materials science and engineering. *Mailing Add:* Civil Eng Northwestern Univ Evanston IL 60208

SHAH, SWARUPCHAND MOHANLAL, b Deesa, India, Dec 30, 05; m 26; c 1. MATHEMATICAL ANALYSIS. *Educ:* Univ Bombay, BA, 27; Univ London, MA, 30, PhD(math), 42, DLitt(math), 51. *Hon Degrees:* DSc, Univ Kentucky, 79. *Prof Exp:* Sr lectr math, Muslim Univ, India, 30-47, reader, 47-53, prof & chmn dept, 53-58; vis prof, Univ Wis, 58-59, vis prof, Math Res Ctr, 59; vis prof, Northwestern Univ, 59-60 & Univ Kans, 60-66; prof, 66-76, EMER PROF MATH, UNIV KY, 76- *Concurrent Pos:* Ed, Math Student, 57-59; vis prof, Univ Tex, Arlington, 77; vis prof, Univ Brasilia, Brazil, 79. *Mem:* Am Math Soc; fel Nat Inst Sci India; fel Indian Acad Sci; fel Royal Soc Edinburgh; London Math Soc. *Res:* Theory of functions of a complex variable, particularly entire and meromorphic and univalent functions; Fourier analysis; difference equations; approximation theory; theory of numbers. *Mailing Add:* Dept Math Univ Ky Lexington KY 40506

SHAHABUDDIN, SYED, b Swat, Pakistan, Mar 2, 39; US citizen; m 67; c 2. MANAGEMENT INFORMATION SYSTEMS. *Educ:* Univ Peshawar, Pakistan, BA, 62; Univ Karachi, Pakistan, MBA, 65; Kent State Univ, MBA, 68; Univ Mo, Columbia, PhD(economet), 76. *Prof Exp:* Asst prof bus, Wis State Univ, 68-69, Talladega Col, 69-75 & Univ Notre Dame, 75-80; PROF BUS, CENT MICH UNIV, 80- *Concurrent Pos:* Consult, Off Prod Div, Int Bus Mach, 71-77; prof, Univ Wis-Madison, 76; Fulbright scholar, Univ Karachi, Pakistan, 89-90. *Mem:* Decision Sci Inst; Inst Mgt Sci; Am Statist Asn. *Res:* Forecasting; management information systems; manufacturing. *Mailing Add:* Dept Mgt Cent Mich Univ Mt Pleasant MI 48859

SHAHANI, KHEM MOTUMAL, b Hyderabad Sind, India, Sept 3, 23; nat US; m 54; c 4. FOOD TECHNOLOGY, BIOCHEMISTRY. *Educ:* Univ Bombay, BSc, 43; Univ Wis, PhD(dairy technol, biochem), 50. *Prof Exp:* Instr agr, King George V Agr Col, India, 43-45; fel dairy technol, Univ Ill, 50-51; bus consult chem, Int Bus Consults, India, 52-53; res assoc dairy technol, Ohio State Univ, 53-57; assoc prof, 57-61, PROF DAIRY & FOOD TECHNOL, UNIV NEBR, LINCOLN, 61- *Concurrent Pos:* Inst Food Technologists sci lectr; mem adv comt food hyg, WHO. *Honors & Awards:* Borden Award, 64; Gamma Sigma Delta Int Award Distinguished Serv to Agr, 66; Sigma Xi Outstanding Scientist Award, 77; Pfizer Award, 77; Nordica Int Award, 77; Dairy Res Found Award, 83. *Res:* Bioprocessing; food enzymes and their immobilzation and application; products; antibiotics in milk; mode of action of antibiotics and milk and microbial lipase; lysozymes; other enzymes; cultured dairy foods; lactase and proteases immobilization and their uses; infant foods, whey utilization and continuous alcohol fermentation by immobilized yeast. *Mailing Add:* Dept Food Sci Univ Nebr 115-E Campus Lincoln NE 68583-0919

SHAHBENDER, R(ABAH) A(BD-EL-RAHMAN), b Damascus, Syria, July 23, 24; m 54; c 3. ELECTRICAL ENGINEERING. *Educ:* Cairo Univ, BEE, 46; Wash Univ, St Louis, MSEE, 49; Univ Ill, PhD(elec eng), 51. *Prof Exp:* Engr, Anglo-Egyptian Oilfields, Ltd, Egypt, 46-48; sr res engr, Honeywell Controls Div, Pa, 51-55; develop engr, Radio Corp Am, 55-58, sr staff mem, Res Labs, 58-61, head digital device res, 61-72, head appl electronics res, 72-75, tech staff, RCA Labs, 75-87; CONSULT, 87- *Concurrent Pos:* Univ Ill fel, 50 & 51; chmn dept electronics physics, Evening Div, La Salle Col, 60-67. *Honors & Awards:* Indust Res-100 Award, 64 & 69; Mat Design Eng Award, 63; RCA Labs Outstanding Achievement Award, 60, 63, 73 & 82. *Mem:* AAAS; fel Inst Elec & Electronics Engrs; Sigma Xi. *Res:* Behavior of nonlinear automatic control systems and adaptive systems; nondestructive testing by means of ultrasonics; digital computer memory systems; digital video systems; kinescope displays; satellite communication systems. *Mailing Add:* 107 Autumn Hill RD Princeton NJ 08540-0432

SHAHEEN, DONALD G, b Trenton, NJ, Sept 5, 30; m 55; c 3. ANALYTICAL CHEMISTRY. *Educ:* Pa State Univ, BS, 53; NY Univ, MS, 58. *Prof Exp:* Chemist, Callery Chem Co, 55-59; sr chemist, Reaction Motors Div, Thiokol Chem Corp, 59-63; mgr, Chem Dept, Life Systs Div, Hazleton Labs, Inc, Va, 63-69; res mgr, Biospherics, Inc, 69-76; VPRES & TECH DIR, DEGESCH AM INC, 76- *Mem:* Am Chem Soc; Am Microchem Soc; Am Soc Testing & Mat. *Res:* Organic microanalyses; analytical method development; instrumental trace and functional group analyses; thermal stability; decomposition studies; phase diagram studies. *Mailing Add:* Degesch Am Inc 275 Triangle Dr Weyers Cave VA 24486

SHAHIDI, FEREIDOON, b Tehran, Iran, Apr 13, 51; Iranian & Can citizen; m 74; c 3. MEAT, OILSEEDS. *Educ:* Shiraz Univ, BSc, 73; McGill Univ, PhD(chem), 77. *Prof Exp:* Res fel org chem, McGill Univ, 77-78; res fel org & gen chem, 77-81, res assoc food eng, 81-84, adj prof food eng, Univ Toronto, 84-87; ASSOC PROF FOOD SCIE, MEM UNIV NFLD, 87- *Mem:* Am Chem Soc; Royal Chem Soc; Inst Food Technologists; Can Inst Food Sci & Technol; Can Meat Sci Asn. *Res:* New food processing systems and their chemical characteristics; solute-solvent interaction; flavor chemistry; food chemistry; meat and seafood research; seafoods aquaculture. *Mailing Add:* Dept Biochem Mem Univ Nfld St Johns NF A1B 3X9 Can

SHAHIDI, FREYDOON, b Tehran, Iran, June 19, 47; m 77; c 2. AUTOMORPHIC FORMS. *Educ:* Tehran Univ, BS, 69; Johns Hopkins Univ, PhD(math), 75. *Prof Exp:* Vis mem math, Inst Advan Study, 75-76; vis asst prof math, Ind Univ, Bloomington, 76-77; from asst prof to assoc prof, 77-86, PROF MATH, PURDUE UNIV, WEST LAFAYETTE, 86- *Concurrent Pos:* Prin investr, NSF, 79-; vis prof, Univ Toronto, Can, 81-82; vis mem math, Inst Advan Study, 83-84, 90-91. *Mem:* Am Math Soc. *Res:* Theory of automorphic forms, L-functions and group representations. *Mailing Add:* Dept Math Purdue Univ West Lafayette IN 47907

SHAHIDI, NASROLLAH THOMAS, b Meshed, Iran, Dec 11, 26; US citizen; c 3. HEMATOLOGY. *Educ:* Univ Montpellier, dipl, 47; Sorbonne, MD, 54. *Prof Exp:* Resident pediat, Hosp for Sick Children, Paris, France, 54-56; asst resident, Baltimore City Hosp, Md, 56-57; instr, Harvard Med Sch, 60-63; asst prof, Children's Hosp, Zurich, Switz, 64-66; dir pediat hemat, Children's Hosp, 70-77; assoc prof, 66-70, PROF PEDIAT, CTR HEALTH SCI, UNIV WIS-MADISON, 70- *Concurrent Pos:* Res fel, Harvard Med Sch, 57-60; asst physician, Children's Hosp Med Ctr, Boston, 60-63; vis investr, Swiss Nat Found, 64-66; hematologist, Children's Hosp, Wis, 66. *Mem:* Am Pediat Soc; Am Soc Hemat; Am Soc Pediat Res; affil Royal Soc Med. *Res:* Red cell 2, 3-diphosphoglycerate and oxygen transport; androgens and erythropoiesis; acquired and congenital thrombocytopenic purpura; red cell metabolism, glucose-6-phosphate dehydrogenase deficiency and drug-induced hemolytic anemias. *Mailing Add:* Dept Pediat K4-436 Csc Sch Clin Ctr Univ Wis Madison WI 53792

SHAHIN, JAMAL KHALIL, b Bethlehem, Jordan, Mar 5, 31; US citizen; m 58; c 3. MATHEMATICS. *Educ:* Univ Calif, Berkeley, BA, 60; Lehigh Univ, MS, 62, PhD(math), 65. *Prof Exp:* Asst math, Lehigh Univ, 60-62, from instr to asst prof, 62-66; assoc prof, 66-68, PROF MATH & CHMN DEPT, SALEM STATE COL, 68- *Mem:* Am Math Soc; Math Asn Am. *Res:* Differential geometry in affine space; Euclidean and Riemannian geometry. *Mailing Add:* Dept Math Salem State Col 352 Lafayette St Salem MA 01970

SHAHIN, MICHAEL M, b Isfahan, Iran, Sept 7, 32; m 58; c 2. PHYSICAL CHEMISTRY. *Educ:* Univ Birmingham, BSc, 55, PhD(phys chem), 58. *Prof Exp:* Nat Res Coun Can res fel phys chem, 58-60; res chemist, E I du Pont de Nemours & Co, 60-61; res assoc, Sch Med, Yale Univ, 61-63; sr scientist, Res Div, Xerox Corp, 63-68, lab mgr xerographic sci, 68-72 & imaging res, 72-73, ctr mgr, 73-78, mgr tech planning & corp res, 78-85, mgr matres, 85-89, MGR TECH PANELS, XEROX CORP, 89- *Mem:* Am Chem Soc; Am Phys Soc. *Res:* Photochemistry; gas-phase kinetics; chemical reactions in electrical discharges; radiation chemistry; gas chromatography; mass spectrometry; ion-molecule reactions; ionization phenomena; research management. *Mailing Add:* 12 Widewaters Lane Pittsford NY 14534

SHAHN, EZRA, b New York, NY, Nov 12, 33; m 83; c 1. MOLECULAR BIOLOGY, BIOPHYSICS. *Educ:* Bard Col, AB, 55; Univ Pa, PhD(molecular biol), 65. *Prof Exp:* Biophysicist, Off Math Res, Nat Inst Arthritis & Metab Dis, 58-60; fel bacteriophage, Wistar Inst, 65-66; asst prof biol, 66-70, ASSOC PROF BIOL, HUNTER COL, 71- *Concurrent Pos:* NSF res grant, 67-69, cause grant, 77-80. *Res:* Mathematical models of biological systems; mechanisms of genetic recombination of bacteriophage; effects of ultraviolet light on bacteriophage; thermodynamics of membrane function; science education. *Mailing Add:* Dept Biol Sci Hunter Col 695 Park Ave New York NY 10021

SHAHRIK, H ARTO, b Istanbul, Turkey, June 22, 23; US citizen; m 62; c 2. CYTOLOGY, ORAL BIOLOGY. *Educ:* Univ Istanbul, Lic es sc, 47, BDS, 50; Tufts Univ, DMD, 59. *Prof Exp:* Intern pediat dent & res, Forsyth Dent Ctr, Boston, Mass, 55-56, clin fel oral biol, 56-57; sr res fel histochem & cytochem & asst mem, Inst Stomatol Res, 59-63; assoc res specialist, Sci Resources Found, 63-64, assoc res clinician & assoc mem oral biol, 64-67, res specialist, head histochem & cytochem & mem oral biol, 67-76; ASST CLIN PROF, DEPT GEN DENT & ORAL DIAG, SCH DENT MED, TUFTS UNIV, 78- *Mem:* AAAS; Am Dent Asn. *Res:* Tissues of the oral cavity in health and disease, oxidative enzymes; keratinization; salivary fluids and cells; tobacco smoke toxicity on human oral and bronchial cells. *Mailing Add:* 193 Marrett Rd Lexington MA 02173

SHAHROKHI, FIROUZ, b Tehran, Iran, July 29, 38; US citizen; c 2. AEROSPACE ENGINEERING. *Educ:* Univ Okla, BSME, 61, PhD(mech eng), 66. *Prof Exp:* Design engr, Boeing Co, Wash, 61-62; res engr, Res Inst, Okla Univ, 62-65; asst prof, La State Univ, New Orleans, 65-66; from asst prof to assoc prof, 66-76, PROF AEROSPACE ENG & DIR SPACE INST, UNIV TENN, 76- *Mem:* Am Inst Aeronaut & Astronaut; Am Soc Mech Engrs. *Res:* Sensor technology radiating heat transfer; remote sensing for earth resources; radiative heat transfer in boundary layer flow. *Mailing Add:* PO Box 22027 Nashville TN 37202

SHAIKH, A(BDUL) FATTAH, b Sukkur, WPakistan, Aug 13, 37; m 67; c 3. STRUCTURAL ENGINEERING. *Educ:* Univ Karachi, BE, 60; Univ Hawaii, MS, 64; Univ Iowa, PhD(struct eng), 67. *Prof Exp:* Asst engr, Water & Power Develop Auth, WPakistan, 60-62; instr struct eng, Univ Iowa, 66-67; from asst prof to assoc prof, 67-81, PROF STRUCT ENG, UNIV WIS-MILWAUKEE, 81- *Honors & Awards:* Martin P Korn Award, Prestressed Concrete Inst, 71. *Mem:* Am Concrete Inst; Prestressed Concrete Inst; Am Soc Civil Engrs. *Res:* Deflections of concrete structures; connections in precast-prestressed concrete; behavior of concrete structures. *Mailing Add:* Dept Civil Eng Univ Wis Milwaukee WI 53201

SHAIKH, ZAHIR AHMAD, b Jullundur, Punjab, India, Mar 31, 45; m 75; c 3. TOXICOLOGY, METAL TOXICOLOGY. *Educ:* Univ Karachi, Pakistan, BSc, 65, MSc, 67; Dalhousie Univ, PhD(biochem), 72. *Prof Exp:* Res assoc environ health, Univ Okla, 72-73; sr fel pharmacol & toxicol, Univ Rochester, 73-75, asst prof, 75-81, assoc prof toxicol, 81-82; assoc prof, 82-86, PROF PHARMCOL & TOXICOL, UNIV RI, 86-, CHMN DEPT, 85- *Concurrent Pos:* Prin investr, NIH res grants, 75-, mem toxicol study sect,

85-89. *Mem:* Soc Toxicol; Am Soc Pharmacol & Exp Therapeut; Tissue Cult Asn; Soc Exp Biol & Med; NY Acad Sci; AAAS. *Res:* Heavy metal metabolism and toxicity; metallothionein and its role in modulating the effects of metal on liver, kidney, and testis; urinary metallothionein as a biological indicator of metal exposure. *Mailing Add:* Dept Pharmacol & Toxicol Univ RI Kingston RI 02881-0809

SHAIN, ALBERT LEOPOLD, b Brussels, Belg, Dec 7, 42; US citizen; m 66; c 2. PHYSICAL CHEMISTRY, POLYMER CHEMISTRY. *Educ:* Univ Calif, Los Angeles, BS, 65; Wash Univ, PhD(phys chem), 69. *Prof Exp:* Fel, Inst Phys Chem, Univ Amsterdan, 69-70 & Univ Calif, Los Angeles, 70-71; res assoc chem physics, Univ Del, 71-73; mem staff elaschem, Exp Sta, E I du Pont de Nemours & Co Inc, 73-80, mem staff, Polymer Prod Dept, 80-90; RETIRED. *Concurrent Pos:* Hon fel, Univ Siena, Italy. *Mem:* Am Chem Soc; Soc Plastics Engrs. *Res:* Magnetic resonance and optical spectroscopy of molecular excited states; polymer chemistry and physics, polymer flammability. *Mailing Add:* 1811 Bryce Dr Wilmington DE 19810-4508

SHAIN, IRVING, b Seattle, Wash, Jan 2, 26; m 47; c 4. ELECTROANALYTICAL CHEMISTRY. *Educ:* Univ Wash, BS, 49, PhD(chem), 52. *Prof Exp:* From instr to prof chem, Univ Wis-Madison, 52-75, chmn dept, 67-70, vchancellor, 70-75; prof chem, provost & vpres acad affairs, Univ Wash, 75-77; PROF CHEM & CHANCELLOR, UNIV WIS-MADISON, 77- *Concurrent Pos:* Dir, Olin Corp. *Mem:* Am Chem Soc; Int Soc Electrochem; Electrochem Soc; fel AAAS. *Res:* Instrumental analysis; polarography; kinetics of electrode reactions. *Mailing Add:* Olin Corp 120 Long Ridge Rd Stamford CT 06904

SHAIN, SYDNEY A, b Chicago, Ill, Aug 31, 40; m 67; c 2. BIOCHEMISTRY, ENDOCRINOLOGY. *Educ:* Univ Ill, BS, 62; Univ Calif, Berkeley, PhD(biochem), 68. *Prof Exp:* Res assoc, Sch Med, Univ Pa, 70-71; from asst to assoc found scientist, 71-76, FOUND SCIENTIST & CHMN DEPT CELLULAR & MOLECULAR BIOL, SOUTHWEST FOUND BIOMED RES, 77- *Concurrent Pos:* Pop Coun fel, Weizmann Inst Sci, Rehovoth, Israel, 68-70; res assoc, Audie L Murphy Mem Vet Hosp, 76-; adj assoc prof, Dept Physiol, Health Sci Ctr, Univ Tex, 78-81, adj prof, 81-; mem, Reproductive Endocrinol Study Sect, Asn Health Rec, 84. *Mem:* Endocrine Soc; Am Soc Andrology; Am Asn Cancer Res; Am Soc Biol Chemists; NY Acad Sci; Am Soc Cell Biol. *Res:* Pathophysiology of diseases of aging, normal and abnormal function in male accessories and genitalia; carcinogenesis in male accessories. *Mailing Add:* Tex Health Sci Ctr 7703 Floyd Curl Dr San Antonio TX 78284-7884

SHAIN, WILLIAM ARTHUR, b Louisville, Ky, Feb 16, 31; div; c 4. FORESTRY. *Educ:* Univ Ga, BSF, 53, MF, 56; Mich State Univ, PhD(forestry), 63. *Prof Exp:* Forester, Int Paper Col, 53; instr forestry, Miss State Univ, 56-59; asst prof forest mensuration, 61-70, assoc prof, 70-79, PROF DEPT FORESTRY, CLEMSON UNIV, 79- *Mem:* Soc Am Foresters; Am Soc Photogram. *Res:* Forest sampling techniques; aerial photogrammetry. *Mailing Add:* Dept Forestry Clemson Univ Clemson SC 29631

SHAININ, DORIAN, b San Francisco, Calif, Sept 26, 14; m 40; c 7. AERONAUTICAL ENGINEERING, QUALITY CONTROL. *Educ:* Mass Inst Technol, SB, 36. *Prof Exp:* Engr, Hamilton Standard Div, United Aircraft Corp, 36-43, chief inspector, 43-52; chief engr, Rath & Strong, Inc, 52-57, vpres & dir statist eng, 57-76; PRES, SHAININ CONSULTS, INC, 76- *Concurrent Pos:* Lectr, Univ Conn, 50-83, fac assoc, Sch Bus Admin, 64-; lectr, Am Mgt Asn, 51-78; consult med staff, Newington Children's Hosp, 57-; lectr, Assoc Bus Progs, London, Eng, 67-70 & Kwaliteitsdienst voor de Industrie, Rotterdam, Neth, 71-76. *Honors & Awards:* Brumbaugh Award, Am Soc Qual Control, 51, Edwards Medal, 69, E L Grant Award, 81, Shewhart Medal, 89. *Mem:* Fel AAAS; fel Am Soc Qual Control (exec secy, 53-54, vpres, 54-57); Am Statist Asn; Int Acad Qual. *Res:* Plant operating cost reduction; statistical quality control; reliability engineering; accelerated life testing; statistically designed experiments in research and development; creative and analytical methods of problem solving; product liability prevention; management and productivity improvement. *Mailing Add:* 35 Lakewood Circle S Manchester CT 06040-7018

SHAINOFF, JOHN RIEDEN, b Pittsburgh, Pa, Oct 9, 30; m 59; c 4. BIOCHEMISTRY. *Educ:* Univ Pittsburgh, BS, 51, MS, 54, PhD(biophys), 56. *Prof Exp:* Asst biophys, Univ Pittsburgh, 51-56; fel chem, Yale Univ, 56-57; res assoc, 57-59, mem asst staff, 60-63, mem staff, 63-75, actg dir atherosclerosis thrombosis res, 75-76, assoc dir, 76-78, DIR, THROMBOSIS RES SECT, RES DIV, CLEVELAND CLIN FOUND, 80- *Mem:* Am Soc Biol Chem; Biophys Soc; Am Heart Asn; Int Soc Thrombosis & Haemostasis; Am Asn Path. *Res:* Protein chemistry; blood coagulation; arteriosclerosis. *Mailing Add:* Res Inst Cleveland Clin Found Cleveland OH 44195

SHAIR, FREDRICK H, b Denver, Colo, May 25, 36; m 64; c 3. CHEMICAL ENGINEERING. *Educ:* Univ Ill, Urbana, BS, 57; Univ Calif, Berkeley, PhD(chem eng), 63. *Prof Exp:* Res engr, Space Sci Lab, Gen Elec Co, 61-65; from asst prof to assoc prof, 65-77, PROF CHEM ENG, CALIF INST TECHNOL, 77- *Mem:* Am Inst Chem Engrs; Am Phys Soc; Am Chem Soc; Sigma Xi. *Res:* Plasma chemistry; monequilibrium electrical discharges relating to chemical synthesis and separations; kinetics and transport associated with ecological systems; dispersion of pollutants; indoor air quality. *Mailing Add:* 2830 Glen Canyon Rd Altadena CA 91001

SHAIR, ROBERT C, b New York, NY, Aug 2, 25; m 49; c 2. CHEMICAL ENGINEERING, PHYSICAL CHEMISTRY. *Educ:* City Col New York, BChE, 47; Polytech Inst Brooklyn, MChE, 49, PhD(chem eng), 54. *Prof Exp:* Vpres res & develop, Gulton Industs, Inc, 59-71; mgr energy prod, Motorola, Inc, Ft Lauderdale, Fla, 71-79; mgr eng, Centec Corp, 79-83; MGR, PROG DEVELOP, WATER RESOURCES MGT DIV, BROWARD COUNTY, FLA, 83- *Mem:* Am Soc Civil Engrs. *Res:* Batteries; energy conversion; electrochemistry; power sources; energy engineering; environmental engineering; surface and groundwater management. *Mailing Add:* 4921 Sarazen Dr Hollywood FL 33021

SHAKA, ATHAN JAMES, b Newton, Mass, Apr 9, 58. DOUBLE RESONANCE, PULSE SEQUENCE DEVELOPMENT. *Educ:* Harvey Mudd Col, BS, 80; Oxford Univ, PhD(phys chem), 84. *Prof Exp:* Jr res fel chem, St Johns Col, Oxford, 84-85; Miller res fel phys chem, Univ Calif, Berkeley, 86-88; ASST PROF PHYS CHEM, UNIV CALIF, IRVINE, 88- *Concurrent Pos:* Dreyfus Found new fac award, 88; NSF presidential young investr, 89. *Mem:* Am Chem Soc; Am Phys Soc. *Res:* New techniques in nuclear magnetic resonance; multiple-pulse nuclear magnetic resonance; two dimensional and three dimensional nuclear magnetic resonance; spatial localization of nuclear magnetic resonance signals; nuclear magnetic resonance in solids; spectrometer hardware and software. *Mailing Add:* Chem Dept Univ Calif Irvine Irvine CA 92717

SHAKARJIAN, MICHAEL PETER, b Niagara Falls, NY, Nov 30, 55. IMMUNOPHARMACOLOGY, LEUKOCYTE BIOLOGY. *Educ:* State Univ NY, Buffalo, BS, 77, MA, 80; Va Commonwealth Univ, PhD(pharmacol), 89. *Prof Exp:* Nat Res Coun, Walter Reed Army Inst Res, 89-91; POSTDOCTORAL FEL, BRISTOL-MYERS SQUIBB PHARMACEUT RES INST, 91- *Concurrent Pos:* Vis scientist, Hosp Nuestra Senora del Mar, Inst Munic d'Investigatio Medica, 84. *Mem:* Fel Am Soc Pharmacol & Exp Therapeut; assoc mem Am Asn Immunologists; AAAS. *Res:* Investigation of the pathways involved in the activation of hematopoietic cells; importance of protein kinases and their substrates in the activation process. *Mailing Add:* Signal Transduction Lab Bristol-Myers Squibb Pharmaceut Res Inst Princeton NJ 08543-4000

SHAKE, ROY EUGENE, b Claremont, Ill, July 3, 32; m 57; c 6. PLANT ECOLOGY. *Educ:* Eastern Ill Univ, BS, 54; Univ Wis-Madison, MS, 56. *Prof Exp:* From instr to asst prof biol, Abilene Christian Col, 58-62; asst, Univ Fla, 62-64; prof, Polk Co Jr Col, 64-65; asst prof, 65-67, ASSOC PROF BIOL, ABILENE CHRISTIAN UNIV, 67- *Res:* Collecting and identifying reptiles and fish in West Central Texas; actinomycete activity in lake water; succession in lentic environments. *Mailing Add:* Dept Biol Abilene Christian Univ Box 8213 Abilene TX 79699

SHAKESHAFT, ROBIN, b Leamington Spa, Eng, June 3, 47. PHYSICS. *Educ:* Univ London, BSc, 68; Univ Nebr, PhD(physics), 72. *Prof Exp:* Res assoc, NY Univ, 72-75; from asst prof to assoc prof physics, Tex A&M Univ, 75-81; assoc prof, 81-84, PROF PHYSICS & CHAIR, UNIV SOUTHERN CALIF, 84- *Concurrent Pos:* Prin investr, NSF, 77-; vis asst prof physics, NY Univ, 78-79, vis assoc prof, 80-81. *Mem:* Am Phys Soc. *Res:* Theoretical atomic physics, in particular, scattering theory, interactions of atoms with intense fields. *Mailing Add:* Dept Physics Univ Southern Calif Los Angeles CA 90089-0484

SHAKHASHIRI, BASSAM ZEKIN, b Enfeh, Lebanon, Oct 9, 39, US citizen. CHEMISTRY. *Educ:* Boston Univ, AB, 60; Univ Md, College Park, MSc, 65, PhD(chem), 68. *Prof Exp:* Res assoc, Univ Ill, Urbana, 67-68, vis asst prof chem, 68-70; asst prof, 70-76, assoc prof, 76-80, PROF CHEM, UNIV WIS-MADISON, 80-; ASST DIR, NSF, 84- *Concurrent Pos:* Danforth Assoc, 80-; consult, Exxon Educ Found, Chicago Mus Sci & Indust, 81; chmn, Chem Educ Div, Am Chem Soc, 81. *Mem:* AAAS; Am Chem Soc; Royal Soc Chem; Nat Sci Teachers Asn. *Res:* Inorganic reaction mechanisms; innovations in undergraduate and graduate education in chemistry; lecture demonstrations. *Mailing Add:* Dept Chem Univ Wis Madison WI 53706

SHAKIN, CARL M, b New York, NY, Feb 17, 34; m 55; c 2. NUCLEAR PHYSICS. *Educ:* NY Univ, BS, 55; Harvard Univ, PhD(theoret physics), 61. *Prof Exp:* Instr physics, Mass Inst Technol, 60-63, NSF fel, 63-65, from asst prof to assoc prof, 65-70; assoc prof, Case Western Reserve Univ, 70-73; prof physics, 73-86, DISTINGUISHED PROF, BROOKLYN COL, 86- *Concurrent Pos:* Consult, Lawrence Radiation Lab & Brookhaven Nat Lab. *Mem:* Fel Am Phys Soc. *Res:* Nuclear theory; nuclear structure; nuclear reactions; intermediate energy physics. *Mailing Add:* Dept Physics Brooklyn Col Brooklyn NY 11210

SHAKLEE, JAMES BROOKER, b Salina, Kans, Mar 29, 45; m 65; c 2. FISH BIOLOGY, BIOCHEMICAL GENETICS. *Educ:* Colo State Univ, BS, 68, MS, 74; Yale Univ, MPhil, 70, PhD(biol), 72. *Prof Exp:* Res assoc develop genetics, Univ Ill, 72-73 & 74-75; asst prof zool, Univ Hawaii & asst marine biologist, Hawaii Inst Marine Biol, 75-81; sr res scientist, Div Fisheries Res, Commonwealth Sci & Indust Res Orgn, 81-85; BIOLOGIST, WASH DEPT FISHERIES, 85- *Concurrent Pos:* NIH fel, Univ Ill, 74-75. *Mem:* Soc Study Evolution; Am Soc Ichthyologists & Herpetologists; Am Soc Zoologists; Sigma Xi; Am Fisheries Soc. *Res:* Study of the population and evolutionary genetics of fishes primarily by analyzing their isozymes and other proteins. *Mailing Add:* Wash Dept Fisheries 115 Gen Admin Bldg Olympia WA 98504

SHAKUN, WALLACE, b New York, NY, July, 21, 34; m 58; c 3. FINANCIAL MANAGEMENT, THERMOECONOMICS. *Educ:* City Univ New York, BME, 58; Univ Vt, MS, 65; Univ Glasgow, PhD(math), 69; Univ Louisville, MBA, 76. *Prof Exp:* Advan design engr prod develop, Gen Elec Co, 60-66; sr res engr, Energy Mat Sci Lab, Eng Exp Sta, Ga Inst Technol, 80-87; vpres engr com prod, Modernfold, 78-80; SR RES ENGR, ENERGY MAT SCI LAB, ENG EXP STA, GA INST TECHNOL, 80-; DEAN TECHNOL ENG, CLAYTON STATE COL, 88- *Concurrent Pos:* Vis asst prof, Univ Vt, 60-65; vis asst prof, Univ Louisville, 72-78; pres, W W Shakun Consults, 80- *Mem:* Assoc fel Royal Aeronaut Soc. *Res:* Program development based on strategic planning requiring a multidisciplinary approach in order to ascertain the impact of new technologies on hardware development. *Mailing Add:* 420 Jefferson Circle NE Sandy Springs GA 30328

SHALABY, SHALABY W, b Dayrut, Egypt, Jan 3, 38; m 65; c 4. BIOMATERIALS, POLYMER SCIENCE. *Educ:* Ain Shams Univ, Egypt, BSc, 58; Univ Lowell, MS, 63, PhD(chem), 66, PhD(polymer sci), 67. *Prof Exp:* Asst forensic chem, Medico-Legal Dept, Egypt, 59-60; instr polymer sci, Univ Lowell, 65-67; instr math, Belvidere Sch, Mass, 66-67; lectr polymer sci,

Col Appl Sci, Egypt, 67-68; sr res chemist, Allied Chem Corp, NJ, 69-74; prin scientist, 74-78, mgr polymer res sect, 78-83, MGR, POLYMER TECHNOL DEPT, ETHICON, INC, 83-; MGR, JOHNSON & JOHNSON POLYMER CHEM CTR, 84- *Concurrent Pos:* Researcher, Nat Res Ctr, Cairo, 67-68; NASA res assoc, Old Dominion Univ, Va, 68-69. *Honors & Awards:* P B Hoffmann Award, 79. *Mem:* Sigma Xi; Am Chem Soc; NY Acad Sci; Am Phys Soc; Soc Plastics Engrs; Soc Biomat; Royal Soc Chem; Fiber Soc; Am Asn Pharm Scientists. *Res:* Synthesis and modification of macromolecules; study of structure-properties relationships and assessment of pertinent physical and structural parameters; structural design and development of new polymeric materials for biomedical applications; organometallic chemistry; thermal analysis; controlled drug delivery. *Mailing Add:* Dept Bioeng Clemson Univ 301 Rhodes Res Ctr Clemson SC 29634

SHALAWAY, SCOTT D, b Pottstown, Pa, Aug 13, 52; m 75. NONGAME WILDLIFE, HABITAT MANAGEMENT. *Educ:* Univ Del, BS, 74; Northern Ariz Univ, MS, 77; Mich State Univ, PhD(wildlife ecol), 79. *Prof Exp:* Resource specialist impact statements, Mich Dept Nat Resources, 80; vis asst prof biol, Dept Zool, 80-81, asst prof ornith, Univ Okla Biol Sta, 81, asst prof wildlife, Dept Zool, Okla State Univ, 81-84. *Concurrent Pos:* Mem, Animal Res Coun, Okla City Zoo, 80- *Mem:* Wildlife Soc; Am Ornithologists Union; Wilson Ornith Soc; Am Soc Mammalogists; Ecol Soc Am. *Res:* Breeding biology, behavior and habitat requirements of nongame and endangered species of wildlife; eastern bluebird biology; black-tailed prairie dog habitat impacts. *Mailing Add:* Rte 5 Box 76 Cameron WV 26033

SHALE, DAVID, b Christchurch, NZ, Mar 22, 32; US citizen; m 66; c 2. MATHEMATICS. *Educ:* Univ NZ, MSc, 53; Univ Chicago, PhD(math), 60. *Prof Exp:* Lectr math, Univ Toronto, 59-61; instr, Univ Calif, Berkeley, 61-62, asst prof, 61-64; from asst prof to assoc prof, 64-70, PROF MATH, UNIV PA, 70- *Concurrent Pos:* Temporary mem, Inst Advan Study, 63-64. *Mem:* Am Math Soc. *Res:* Abstract analysis and applications, especially to quantum theory. *Mailing Add:* Dept Math Univ Pa Philadelphia PA 19174

SHALEK, ROBERT JAMES, b Chicago, Ill, Apr 15, 22; m 51; c 6. BIOPHYSICS. *Educ:* Univ Ill, BA, 43; Southern Methodist Univ, MA, 48; Rice Inst, MA, 50, PhD(biophys), 53. *Prof Exp:* Instr radiol physics, Univ Tex M D Anderson Hosp & Tumor Inst, 50-53; USPHS fel, Univ London, 53-54; from asst physicist to assoc physicist, 54-60, PHYSICIST, UNIV TEX M D ANDERSON HOSP & TUMOR INST, 60-, PROF BIOPHYS, 65- *Concurrent Pos:* Consult, Oak Ridge Inst Nuclear Studies, 53-61. *Mem:* Am Phys Soc; Radiation Res Soc; Biophys Soc; Am Asn Physicists Med (pres, 66); Brit Inst Radiol; Sigma Xi. *Res:* Radiation chemistry; radiological physics. *Mailing Add:* Dept Physics M D Anderson Hosp Tex Med Ctr Houston TX 77030

SHALER, AMOS J(OHNSON), b Harrow, Eng, July 8, 17; US citizen; m 43; c 3. FAILURE METALLURGY, FORENSIC MATERIALS ENGINEERING. *Educ:* Mass Inst Technol, SB, 40, ScD(phys metall), 47. *Prof Exp:* Asst, New Consol Gold Fields, SAfrica, 40-42; heat treatment supt, Cent Ord Factory, 42; res & develop engr, C H Hirtzel & Co, Ltd, 42-43; tech dir, Indust Rys Equip Co, 43-45; mem staff, Div Indust Coop, Mass Inst Technol, 46-47, from asst prof to assoc prof phys metall, 47-53; prof metall & head dept, Pa State Univ, 53-60; PRES, AMOS J SHALER, INC, CONSULTS, 65-; VPRES, MGD, INC, 79- *Concurrent Pos:* Sci liaison officer, Off Naval Res, 50-51; dir, Belg-Am Educ Found, 58-; consult, 60-65; consult, President's Off Sci & Technol, 61-65; spec consult to asst secy gen, NATO, 69-70. *Honors & Awards:* Award, Am Inst Mining, Metall & Petrol Engrs, 51; Cert of Appreciation, US Off Naval Res, 51; Cert of Appreciation, US Mission to NATO, 70. *Mem:* Am Soc Metals; Int Oceanog Found; Nat Fire Protection Asn; Nat Acad Forensic Engrs. *Res:* Waste water recycling; chemical oceanography; water purification; failure prevention in buried pipes; powder metallurgy; refractory materials; new materials systems; new products and processes; economics of innovation; methods of forensic engineering. *Mailing Add:* 705 W Park Ave State College PA 16803

SHALIT, HAROLD, b Philadelphia, Pa, May 9, 19; m 42; c 2. ORGANIC CHEMISTRY. *Educ:* Univ Pa, AB, 41; Pa State Univ, MS, 43; Polytech Inst Brooklyn, PhD(org chem), 48. *Prof Exp:* Sr res assoc, Polytech Inst Brooklyn, 46-48; res chemist, Standard Oil Co Ind, 48-51; res chemist, Houdry Process Corp, 51-58, sect head explor & fuel cell res, 58-63; dir phys chem res, 63-72, mgr catalytic res, 72-76, SR RES ASSOC, ARCO CHEM DIV, ATLANTIC RICHFIELD CO, 76- *Concurrent Pos:* Adj prof chem, Drexel Univ, 74- *Mem:* Am Chem Soc; Catalysis Soc; Sigma Xi. *Res:* Physical, organic, petroleum and electro-organic chemistry; high pressure reactions; kinetics; catalysis. *Mailing Add:* 106 Meetinghouse Pond Wayne PA 19087-5513

SHALITA, ALAN REMI, b New York, NY, Mar 22, 36; m 60; c 2. DERMATOLOGY, BIOCHEMISTRY. *Educ:* Brown Univ, AB, 57; Free Univ Brussels, BS, 60; Bowman Gray Sch Med, MD, 64. *Hon Degrees:* DSc, Long Island Univ, 90. *Prof Exp:* Training grant fel dermat, Med Ctr, NY Univ, 68-70, from instr to asst prof dermat, 70-73; asst prof dermat, Col Physicians & Surgeons, Columbia Univ, 73-75; assoc prof, 75-79, head, Div Dermat, 75-80, prof med, Downstate Med Ctr, 79-80, PROF & CHMN, DEPT DERMAT, STATE UNIV NY HEALTH SCI CTR, BROOKLYN, 80- *Concurrent Pos:* USPHS spec fel biochem, Med Ctr, NY Univ, 70-72; dep dir div finance, Nat Prog Dermat, 70-; Dermat Found grant, 76. *Mem:* AAAS; NY Acad Sci; Soc Invest Dermat; Am Fedn Clin Res; fel Am Acad Dermat; Am Dermat Asn. *Res:* Factors involved in the pathogenesis of acne vulgaris, including cutaneous lipogenesis and microbial lipids and lipolytic enzymes. *Mailing Add:* Dept Dermat State Univ NY Health Sci Ctr Brooklyn NY 11203

SHALLCROSS, FRANK V(AN LOON), b Philadelphia, Pa, Nov 9, 32; m 56; c 2. PHYSICAL CHEMISTRY, SOLID STATE DEVICE TECHNOLOGY. *Educ:* Univ Pa, AB, 53; Brown Univ, PhD(chem), 58. *Prof Exp:* Chemist, M & C Nuclear, Inc, Mass, 57-58; mem tech staff, David Sarnoff Res Ctr, RCA Corp, 58-87, SRI INT, 87- *Mem:* AAAS; Am Chem Soc; Am Inst Chemists; Electrochem Soc. *Res:* Solid state physical chemistry; thin films; photoconductive materials; solid state image sensors; semiconductors; charge-coupled devices; integrated circuit technology; infrared imagers. *Mailing Add:* 12 Jeffrey Lane Princeton Junction NJ 08550

SHALLENBERGER, ROBERT SANDS, b Pittsburgh, Pa, Apr 11, 26; m 51; c 4. FOOD CHEMISTRY, CARBOHYDRATE CHEMISTRY. *Educ:* Univ Pittsburgh, BS, 51; Cornell Univ, MS, 53, PhD(biochem, hort, plant physiol), 55. *Prof Exp:* Assoc technologist chem, Gen Foods Corp, 55-56; asst prof biochem, 56-60, assoc prof food sci & technol, 60-66, PROF BIOCHEM, CORNELL UNIV, 66- *Mem:* Am Chem Soc. *Res:* Chemical reactions affecting color, flavor, texture and nutritive value in processed foods; carbohydrate structure and reactions. *Mailing Add:* Dept Food Sci & Technol NY State Agr Exp Sta Cornell Univ Geneva NY 14456

SHALLOWAY, DAVID IRWIN, b Miami, Fla, Apr 6, 48; m 85. ONCOPROTEINS, PROTEIN FOLDING. *Educ:* Mass Inst Technol, SB, 69, PhD(physics), 75; Stanford Univ, MS, 70. *Prof Exp:* Res fel molecular biol, Dana-Farber Cancer Inst, Harvard Med Sch, 77-81; from asst prof to assoc prof molecular biol, Pa State Univ, 82-90; res assoc physics, Lab Nuclear Studies, 75-77, GREATER PHILADELPHIA PROF BIOL SCI BIOCHEM, CORNELL UNIV, 90- *Concurrent Pos:* Mem, Pa State Biotechnol Inst, 87-90; vis assoc prof, Univ Calif, San Francisco, 88-89; prof, Dept Path, Vet Sch, Cornell Univ, 90- *Mem:* Am Soc Microbiol. *Res:* Molecular mechanisms of proto-oncoprotein signal transduction and their roles in carcinogenesis; theoretical prediction of protein structure; computer applications in molecular biology. *Mailing Add:* Biotechnol Bldg Rm 265 Cornell Univ Ithaca NY 14853

SHALOWITZ, ERWIN EMMANUEL, b Washington, DC, Feb 13, 24; m 52; c 3. CIVIL ENGINEERING. *Educ:* George Washington Univ, BCE, 47; Am Univ, MA, 54. *Prof Exp:* Consult waterfront struct, chief, Struct Res Eng & head, Defense Res, Navy Dept, Washington, DC, 48-59; CHIEF CONTRACT EVAL & ANALYSIS, GEN SERV ADMIN, WASHINGTON, DC, 59- *Concurrent Pos:* Adv atomic tests, Navy Dept, Wash, DC, 55-57, mem, Spec Weapons Effects Test Planning, 57-58; spec asst protective construct, proj mgr bldg systs, chief res, chief contract procedures & contract support, Gen Serv Admin, Wash, DC, 59-, chmn, Fire Safety Comt & Fallout Protection Comt, 59-61 & Bldg Eval Comt, 68-70; mem, Interagency Comt Housing Res & Bldg Technol, 68-70 & Nat Eval Bd Archit Eng Selection, 75-77; Gen Serv Admin rep, Procurement Policy Comt, Nat Acad Sci, 76-79. *Mem:* Soc Advan Mgt; fel Am Soc Civil Engrs; fel Am Biog Inst. *Res:* Atomic test reports; protective construction; civil engineering and water power; technical management; building research; contracting techniques and national contract organization staffing model. *Mailing Add:* 5603 Huntington Pkwy Bethesda MD 20814

SHALUCHA, BARBARA, b Springfield, Vt, Dec 9, 15. HORTICULTURE. *Educ:* Univ Vt, PhB, 37, MS, 38; Ohio State Univ, PhD(hort), 47. *Prof Exp:* Dow asst, Conn Col, 38-42, instr bot, 42-43; asst, Exp Sta, Ohio Univ, 43-45; asst cur pub educ, Brooklyn Bot Garden, 45-47; from instr to asst prof, 47-71, ASSOC PROF BOT, IND UNIV, BLOOMINGTON, 71- *Concurrent Pos:* Am Asn Univ Women & Nat Coun State Garden Clubs fel, Wye Col, Univ London, 54; vis prof, Cornell Univ, 70-71 & Royal Bot Garden, Kew, Eng; pres, Nat Civic Garden Ctrs, Inc, 75-77. *Mem:* Bot Soc Am; Am Soc Hort Sci; Am Hort Soc; Nat Sci Teachers Asn; Int Soc Hort Sci. *Res:* Extraction of auxins from plant tissues; environmental horticulture and horticulture education; civic garden centers. *Mailing Add:* Dept Bot Ind Univ Bloomington IN 47405

SHALVOY, RICHARD BARRY, b Norwalk, Conn, Apr 26, 49; m 72; c 2. SURFACE PHYSICS, ANALYTICAL CHEMISTRY. *Educ:* Rensselaer Polytech Inst, BS, 71; Brown Univ, ScM, 74, PhD(physics), 77. *Prof Exp:* Res fel, Univ Ky, 76-78, sr physicist electron spectros, 78-80; res chemist, Stauffer Chem Co, 80-85, sr res chemist anal res, 85-87; RES CHEMIST, OLIN CORP, 87- *Mem:* Am Vacuum Soc. *Res:* Characterization of catalysts and metals using electron spectroscopy; chemical bonding in semiconductors; analytical surface chemistry. *Mailing Add:* Olin Corp 350 Knotter Dr Cheshire CT 06410

SHAM, LU JEU, b Hong Kong, China, Apr 28, 38; m 65; c 2. THEORETICAL PHYSICS, SOLID STATE PHYSICS. *Educ:* Univ London, BSc, 60, Imp Col, ARCS, 60; Cambridge Univ, PhD(solid state physics), 63. *Prof Exp:* Physicist, Univ Calif, San Diego, 63-65, asst res physicist & lectr physics, 65-66; asst prof, Univ Calif, Irvine, 66-67; reader appl math, Queen Mary Col, Univ London, 67-68; assoc prof physics, 68-74, PROF PHYSICS, UNIV CALIF, SAN DIEGO, 74- *Concurrent Pos:* Vis prof, Max Planck Inst Solid State Res, Stuttgart, Ger, 78; Guggenheim fel, 83-84. *Honors & Awards:* Humboldt Found Award, 81. *Mem:* Fel Am Phys Soc. *Res:* Electronic properties in solids; theory of semiconductor heterostructures. *Mailing Add:* Dept Physics Univ Calif San Diego La Jolla CA 92093-0319

SHAMAN, PAUL, b Portland, Ore, Mar 30, 39; m 64; c 2. STATISTICS. *Educ:* Dartmouth Col, AB, 61; Columbia Univ, MA, 64, PhD(statist), 66. *Prof Exp:* From asst res scientist to assoc res scientist, NY Univ, 64-67; res assoc, Stanford Univ, 67-68; from asst prof to assoc prof statist, Carnegie-Mellon Univ, 68-77; assoc prof, 77-85, PROF STATIST, UNIV PA, 85- *Concurrent Pos:* Prog dir statist & probability, NSF, 84-85; managing ed, Inst Math Statist, 86- *Mem:* Am Statist Asn; Inst Math Statist. *Res:* Time series analysis. *Mailing Add:* Dept Statist Univ Pa Philadelphia PA 19104-6302

SHAMASH, YACOV A, b Jan 12, 50; m 76; c 2. CONTROL SYSTEMS, ROBOTICS. *Educ:* Imp Col, BSc, 70, PhD(control systs), 73. *Hon Degrees:* DIC, Imp Col, 73. *Prof Exp:* Lectr elec eng, Tel-Aviv Univ, 73-76; vis asst prof systs eng, Univ Pa, 76-77; prof & chair elec eng, Fla Atlantic Univ, 77-85; PROF & CHAIR ELEC ENG, WASH STATE UNIV, 85- *Concurrent Pos:* Dir, Ctr Analog/Digital Integrated Circuits, NSF, 89- & Keytronics, Inc, 90-;

bd gov, Inst Elec & Electronics Engrs Aerospace & Electronics Systs Soc, 91-93; mem, CAD/CAM Tech Comt, Am Inst Aeronaut & Astronaut, 85-89. *Mem:* Inst Elec & Electronics Engrs; Am Inst Aeronaut & Astronaut; Am Soc Eng Educ. *Res:* Control systems; robotics. *Mailing Add:* Elec Eng & Computer Sci Wash State Univ Pullman WA 99164

SHAMBAUGH, GEORGE E, III, b Boston, Mass, Dec 21, 31; m 56, 87; c 5. DEVELOPMENTAL BIOLOGY, NEUROBIOLOGY. *Educ:* Oberlin Col, BA, 54; Cornell Univ, MD, 58. *Prof Exp:* Gen med intern, Denver Gen Hosp Univ, 58-59; prev med adv, MAAG, Taiwan, 59-61; resident, Walter Reed Gen Hosp, 61-64; res internist, Walter Reed Army Med Ctr, 64-67; fel physiol chem, Univ Wis, 67-69; from asst prof to assoc prof, 69-81, PROF MED, MED SCH, NORTHWESTERN UNIV, 81- *Concurrent Pos:* Pres, Taipei Int Med Soc, 60-61; attend staff mem, Northwestern Mem Hosp, 69-; attend & chief endocrinol & metab, Vet Admin Lakeside Med Ctr, 74-; mem, Comt Nutrit Issues, Am Soc Clin Nutrit, 77-82; vis lectr, Sharam Zedek Med Ctr, Jerusalem, Israel, 90. *Mem:* Endocrine Soc; Cent Soc Clin Res; Am Thyroid Asn; Am Inst Nutrit; Am Soc Clin Nutrit; Am Physiol Soc. *Res:* Fetal fuels and growth factors in fetal and neonatal development in the rat with emphasis on metabolism and cellular mechanisms in fetal tissues; pyrimidine biosynthesis; neuroendocrine ontogeny; utilization of altered fuel mixtures by discrete tissues; modulation of cell replication by circulating factors in mother and fetus; neurochemistry. *Mailing Add:* Ctr Endocrinol Metab & Nutrit Northwestern Univ Med Sch 303 E Chicago Ave Chicago IL 60611

SHAMBAUGH, GEORGE FRANKLIN, b Columbus, Ohio, Nov 3, 28; m 53; c 3. ENTOMOLOGY. *Educ:* Wilmington Col, AB, 50; Ohio State Univ, MSc, 51, PhD(entom), 53. *Prof Exp:* Entomologist & asst chief pesticides br, Natick Qm Res & Eng Command, 55-62; assoc prof, 62-72, PROF ENTOM, OHIO AGR RES & DEVELOP CTR & OHIO STATE UNIV, 72- *Mem:* AAAS; Am Soc Zoologists; Am Entom Soc; Sigma Xi. *Res:* Electrophysiology of insect nerves and muscles; insect sense physiology; insect attractants and repellent; insect digestive enzymes. *Mailing Add:* 1574 Sunset Lane Wooster OH 44691

SHAMBELAN, CHARLES, b Philadelphia, Pa, Mar 16, 30; m 56; c 2. SYNTHETIC FIBERS. *Educ:* Temple Univ, BA, 51, MA, 55; Univ Pa, PhD(phys chem), 59. *Prof Exp:* Chemist, Frankford Arsenal, 51-55; res chemist, E I du Pont de Nemours & Co, Inc, 58-64, res supvr, 64, res assoc, 73, res fel, 77, sr res fel, 85-90; RETIRED. *Mem:* Am Chem Soc; Soc Advan Mat & Process Eng. *Res:* Polymers; nonwoven fabrics; synthetic fibers. *Mailing Add:* 3203 Summerset Rd Wilmington DE 19810

SHAMBERGER, RAYMOND J, b Munising, Mich, Aug 23, 34; m 59; c 6. BIOCHEMISTRY. *Educ:* Alma Col, BS, 56; Ore State Univ, MS, 60; Univ Miami, PhD(biochem), 63. *Prof Exp:* Asst, Ore State Univ, 57-59 & Univ Southern Calif, 63; dir res, Sutton Res Corp, Calif, 63-64; sr cancer res scientist, Roswell Park Mem Inst, 64-69; SECT HEAD ENZYM, CLEVELAND CLIN FOUND, 69- *Concurrent Pos:* Prof, Cleveland State Univ, 70-; mem comts nutrit & path, Fedn Am Socs Exp Biol & Med. *Mem:* Am Asn Cancer Res; Am Soc Clin Pathologists; Am Asn Clin Chemists. *Res:* Chemistry of trace metals; mechanisms of cancer formation; enzyme chemistry. *Mailing Add:* 9865 W Alpine Dr Kirkland OH 44094

SHAMBLIN, JAMES E, b Holdenville, Okla, Mar 24, 32; m 59; c 2. MECHANICAL ENGINEERING, INDUSTRIAL ENGINEERING. *Educ:* Univ Tex, BSME, 54, MSME, 62, PhD(mech eng), 64. *Prof Exp:* Test engr, Pratt & Whitney Aircraft Div, United Aircraft Corp, 54-55; res engr, Southwest Res Inst, 55-60; teaching asst mech eng, Univ Tex, 60-62, instr, 62-64; from asst prof to assoc prof, 64-69, PROF INDUST ENG, OKLA STATE UNIV, 69-, DIR, CTR LOCAL GOVT TECHNOL, 75- *Honors & Awards:* H B Maynard Innovative Achievement Award, Inst Indust Engrs; Chester F Carlson Award for Innovation in Eng Educ, Am Soc Eng Educ. *Mem:* Am Inst Indust Engrs; Am Soc Eng Educ; Am Pub Works Asn; Am Soc Civil Engrs. *Res:* Application of engineering and management technology to problems of local government. *Mailing Add:* 103 Crutchfield Hall Okla State Univ Stillwater OK 74078

SHAMBROOM, WILIAM DAVID, b Teaneck, NJ, Jan 18, 49. PARTICLE PHYSICS. *Educ:* Harvard Univ, AB & AM, 71, PhD(physics), 80. *Prof Exp:* Res assoc physics, 78-81, asst prof, 81-84, asst prof physics & comput sci, 84-85, ASST PROF COMPUT SCI, NORTHEASTERN UNIV, 85- *Mem:* Am Phys Soc; Am Asn Physics Teachers; Digital Equip Comput Users Soc; Asn Comput Mach; Inst Elec & Electronics Engrs. *Res:* Experimental elementary particle physics; methods of scientific computation; multiprocessor computing systems and architectures. *Mailing Add:* Prin Software Eng Wang Labs Inc One Industrial Ave M/S 014-590 Lowell MA 01851

SHAMBURGER, JOHN HERBERT, b Meridian, Miss, Nov 22, 25; m 48; c 4. ENGINEERING GEOLOGY. *Educ:* Univ Miss, BS, 49. *Prof Exp:* Civil engr, 49-51, geologist, 53-62, chief eng geol applications group, geotech lab, Waterways Exp Sta, US Army Corp Engrs, 62-86; CONSULT GEOLOGIST, 86- *Mem:* Asn Eng Geol; Am Soc Photogram; Soc Am Mil Engrs; Int Geog Union. *Res:* Engineering geologic site characterization; alluvial environment of deposition suitability for engineering requirements; groundwater containment at disposal sites; remote imagery interpretation methodology. *Mailing Add:* Eight Briarwood Pl Vicksburg MS 39180

SHAMES, DAVID MARSHALL, b Norfolk, Va, Dec 27, 39. NUCLEAR MEDICINE. *Educ:* Univ Va, BA, 61; Yale Univ, MD, 65. *Prof Exp:* Intern internal med, Yale-New Haven Hosp, 65-66; staff assoc kinetic anal metab systs, Math Res Br, NIH, 66-69; asst resident internal med, Johns Hopkins Hosp, 69-70; NIH fel, 70-71, asst prof, 71-75, assoc prof radiol, 75-80, ASSOC CLIN PROF, UNIV CALIF, SAN FRANCISCO, 80- *Concurrent Pos:* Nat Insts Gen Med Sci res career develop award, 72-77. *Mem:* Soc Nuclear Med; Am Fedn Clin Res. *Res:* Kinetic analysis of nuclear medicine tracer data using the computer, especially the cardiovascular, cerebrovascular and renal systems. *Mailing Add:* Providence Hosp PO Box 23020 Oakland CA 94809

SHAMES, IRVING H, b Boston, Mass, Oct 31, 23; m 54; c 2. ENGINEERING. *Educ:* Northeastern Univ, BS, 48; Harvard Univ, MS, 49; Univ Md, PhD, 53. *Prof Exp:* From instr to asst prof, Univ Md, 49-55; asst prof, Stevens Inst Technol, 55-57; prof eng sci & chmn dept, Pratt Inst, 57-62, actg chmn physics, 60-61; prof & head, State Univ NY, Buffalo, Div Interdisciplinary Studies & Res Eng, 62-70, fac prof eng & appl sci, 70-73, prof & chmn, Dept Eng Sci, Aerospace Eng & Nuclear Eng, 73, fac prof, Eng & Appl Sci, 79, DISTINGUISHED TEACHING PROF, STATE UNIV NY, BUFFALO, 80- *Concurrent Pos:* Vis lectr, Ord Lab, US Dept Navy, 52-55 & Res Lab, 53-55; vis prof, Mat Dept, Technion, Israel, 69, Mech Eng Dept, 75; prin investr, Esso grant, Sunyab grant, co-prin investr, NASA grant, NSF grant, 90. *Mem:* Am Soc Eng Educ; Sigma Xi. *Res:* Dynamics and mechanics; author of numerous publications and text books.. *Mailing Add:* Dept Civil Eng State Univ NY at Buffalo 222 Ketter Hall Amherst NY 14260

SHAMIR, ADI, b Tel Aviv, Israel, July 6, 52. COMPUTER SCIENCE. *Educ:* Tel Aviv Univ, BSc, 72; Weizmann Inst, MSc, 75, PhD(comput sci), 77. *Prof Exp:* Res asst, Univ Warwick, 76; instr, 77-78, ASST PROF COMPUT SCI, MASS INST TECHNOL, 78- *Res:* Combinatorics; algorithms; cryptography; semantics. *Mailing Add:* Appl Math The Weizmann Inst Rehovot 157 Israel

SHAMIS, SIDNEY S, b Norwalk, Conn, July 19, 20; m 42; c 2. ELECTRICAL ENGINEERING. *Educ:* Cooper Union, BSEE, 47; Stevens Inst Technol, MS, 50. *Prof Exp:* Instr advan commun eng, Army Air Force Off Electronics Sch, 42-44, course supvr, 44-46; electronics engr advan develop sect, Allen B Dumont Labs, Inc, 46-50, sect head, 50-52; from asst prof to prof elec eng, NY Univ, 52-73, assoc dean sch eng & sci, 72-73; prof elec eng & assoc dean eng, 73-81, dean eng, 81-83, ASSOC PROVOST, POLYTECH INST NY. *Concurrent Pos:* Arthur M Loew Found grant transistor techniques, 54-55; dir grad ctr at Bell Labs, 57-59, assoc dir lab electrosci res, 62-68; consult, Sprague Elec Co, 64-65, Eon Corp, 65-67 & Digital Device Corp, 65-67. *Mem:* Inst Elec & Electronics Engrs; Am Soc Eng Educ. *Res:* Electronic circuits; active networks; device technology. *Mailing Add:* Dept Eng Polytech Univ 333 Jay St Brooklyn NY 11201

SHAMMA, MAURICE, b Cairo, Egypt, Dec 14, 26; nat US; m 55; 55; c 1. NATURAL PRODUCTS CHEMISTRY. *Educ:* Berea Col, AB, 51; Univ Wis, PhD(chem), 55. *Prof Exp:* Fel, Wayne State Univ, 55-56; from asst prof to assoc prof, 56-66, PROF CHEM, PA STATE UNIV, UNIVERSITY PARK, 66- *Mem:* Am Chem Soc. *Res:* Isolation, characterization and synthesis of natural products, particularly alkaloids; synthesis of new nitrogen heterocycles. *Mailing Add:* Dept Chem Pa State Univ University Park PA 16802

SHAMOIAN, CHARLES ANTHONY, b Worcester, Mass, Oct 5, 31; m 61; c 2. PSYCHIATRY, BIOCHEMICAL PHARMACOLOGY. *Educ:* Clark Univ, AB, 54, MA, 56; Tufts Univ, PhD(physiol), 60, MD, 66. *Prof Exp:* Instr physiol, Med Sch, Tufts Univ, 61-62, res assoc pharmacol, 63-66; intern med, Bellevue Hosp, New York, 66-67; fel psychiat, Cornell Univ-NY Hosp, 67-70, from instr to assoc prof clin psychiat, 70-84, dir, Geriat Serv, Westchester Div, 79-89, PROF CLIN PSYCHIAT, CORNELL UNIV-NY HOSP, 84-, DIR, ACUTE TREAT SERV, WESTCHESTER DIV, 89- *Concurrent Pos:* Asst attend psychiatrist, Payne Whitney Psychiat Clin, 71-78, assoc attend psychiatrist, 78-; attend psychiatrist, NY Hosp, 84- *Mem:* AAAS; Am Physiol Soc; Am Psychopath Asn; Am Psychiat Asn; NY Acad Med; Am Asn Geriat Psychiat. *Res:* Catecholamine metabolism in affective illnesses; geriatric psychopharm. *Mailing Add:* New York Hosp Cornell 21 Bloomingdale Rd White Plains NY 10605

SHAMOO, ADIL E, b Baghdad, Iraq, Aug 1, 41; m 67; c 3. PHYSIOLOGY, BIOPHYSICS. *Educ:* Univ Baghdad, 62; Univ Louisville, MS, 66; City Univ New York, PhD(physiol, biophys), 70. *Prof Exp:* Instr physics, Univ Louisville, 65-68; from asst to assoc biophys, Mt Sinai Sch Med, 68-71, asst prof physiol & biophys, 71-73; from asst prof to assoc prof, Univ Rochester, 73-78; prof & chmn, 79-82, PROF, DEPT BIOL & CHEM, UNIV MD, BALTIMORE, 82- *Concurrent Pos:* Guest worker, Nat Inst Neurol Dis & Stroke, 72-73; guest prof, Max-Planck Inst Biophysics, Frankfurt, 77-78; ed-in-chief, Membrane Biochem, 77-, Accountability in Res Policies & Qual Assurance, 88- *Mem:* Am Asn Biol Chemists; NY Acad Sci; Biophys Soc; Am Physiol Soc; Soc Qual Assurance. *Res:* Physiology and biochemistry of membrane transport; membrane biochemistry-lipid membranes; quality assurance and ethics in research. *Mailing Add:* Biochem Dept Univ Md 660 W Redwood St Baltimore MD 21201

SHAMOS, MICHAEL IAN, b New York, NY, Apr 21, 47; m 73; c 2. COMPUTATIONAL GEOMETRY, COMPUTER LAW. *Educ:* Princeton Univ, AB, 68; Vassar Col, MA, 70; Am Univ, MS, 72; Yale Univ, MS, 73, MPhil, 74, PhD(comput sci), 78; Duquesne Univ, JD, 81. *Prof Exp:* Assoc engr comput sci, IBM Corp, 68-70; supvry programmer, Nat Cancer Inst, NIH, 70-72; teaching fel, Yale Univ, 72-75; asst prof math & comput sci, 75-90, ADJ SR RES COMPUT SCIENTIST, CARNEGIE-MELLON UNIV, 90- *Concurrent Pos:* Consult, various law firms; pres, Unus, Inc, 79-, Lunus Inc, 84-; assoc, Webb, Burden Ziesenheim & Webb, PC, 90- *Mem:* Asn Comput Mach; Math Asn Am; Nat Sci Teachers Asn; Sigma Xi; NY Acad Sci. *Res:* Theoretical computer science; graph theory; discrete mathematics; computational geometry; combinatorics; analysis of algorithms; computers and law. *Mailing Add:* 605 Devonshire St Pittsburgh PA 15213

SHAMOS, MORRIS HERBERT, b Cleveland, Ohio, Sept 1, 17; m 42; c 1. BIOPHYSICS, LABORATORY MEDICINE. *Educ:* NY Univ, AB, 41, MS, 43, PhD(physics), 48. *Prof Exp:* Sr vpres & chief sci officer, Technicon Corp, 75-83; PRES, M H SHAMOS & ASSOC, 83- *Concurrent Pos:* Consult, US AEC, 56-69, Nat Broadcasting Co, 57-65 & UN Info Serv, 58; chmn dept physics, Washington Sq Col, NY Univ, 57-70; sr vpres res & educ, Technicon Corp, 70-75; mem adv coun, NY Polytech Inst, 80-, fel, Polytechn Univ. *Mem:* Fel AAAS; Am Phys Soc; sr mem Inst Elec & Electronics Engrs; Nat Sci Teachers Asn (pres-elect, 66-67, pres, 67-68); NY Acad Sci (rec secy, 77-79, vpres, 80-81, pres, 82); Am Chem Soc; Sigma Xi. *Res:* Atomic and nuclear physics; cosmic rays; electron scattering; physical electronics; high energy physics; nuclear detectors and instrumentation; biophysics, electrical properties of hard tissues; biophysical theory of aging. *Mailing Add:* 3515 Henry Hudson Pkwy Bronx NY 10463

SHAMOUN, SIMON FRANCIS, b Habbaniya, Iraq, July 1, 46; Can citizen; m 79; c 3. PLANT PATHOLOGY, FOREST BIOTECHNOLOGY. *Educ:* Mosul Univ, Iraq, BSc, 72; NC State Univ, MSc, 79; Univ Ark, PhD(plant path), 88. *Prof Exp:* Agr engr forestry, Gen Directorate Forests, 72-74; lab instr forestry, Mosul Univ, Iraq, 74-76; res asst plant path, NC State Univ, 76-79 & Univ Ark, 79-88; RES SCIENTIST BIOCONTROL FOREST PESTS FOREST PATH, FORESTRY CAN, PAC FORESTRY CENTRE, CAN, 88- *Res:* Developing environmentally safe products for biological control of forest pests including weeds and diseases; forest tree diseases and biochemical analysis of fungi by means of biotechnology techniques. *Mailing Add:* Pac Forestry Centre Forestry Can 506 W Burnside Rd Victoria BC V8Z 1M5 Can

SHAMSIE, JALAL, b Delhi, India, Jan 29, 30; Can citizen; m 59; c 2. CHILD PSYCHIATRY. *Educ:* Punjab Univ, India, BSc, 47; Punjab Univ, Pakistan, MBBS, 53; FRCP(C), 62. *Prof Exp:* Dir child & child adolescent serv psychiat, Douglas Hosp, Montreal, Que, 61-71; asst prof, McGill Univ, 67-71; asst prof, 72-80, PROF PSYCHIAT, UNIV TORONT0, 80- *Concurrent Pos:* Dir Res & Educ, Thistletown Regional Ctr Children & Adolescents, Toronto, 72-80; consult child psychiat, Clarke Inst, 76- *Mem:* Royal Col Psychiatrists Gt Brit; Can Psychiat Asn. *Res:* Adolescent psychiatry; administrative psychiatry. *Mailing Add:* 32 Grovetree Rd Rexdale ON M9V 2Y2 Can

SHAMSUDDIN, ABULKALAM MOHAMMAD, b Comilla, Bangladesh, Mar 1, 48; m 71; c 1. CARCINOGENESIS, CHEMOPREVENTION. *Educ:* Dhaka Univ, Dhaka Med Col, MD, 72; Univ Md, Baltimore, Phd(carcinogenesis), 80. *Prof Exp:* Resident path, Baltimore City Hosp, 73-75; resident path, Univ Md Hosp, 75-77, instr path, Sch Med, Univ Md, 77-79, from asst prof to assoc prof, 80-88, PROF PATH, SCH MED, UNIV MD, 88- *Concurrent Pos:* Path fel, Johns Hopkins Hosp, 73-75; vis scientist, Cancer Inst Tokyo, 82 & Nat Cancer Ctr Res Inst, Tokyo, 85; reviewer & consult, Nat Cancer Inst, 83-; res fel, Univ Tokyo, Inst Med Sci, 89. *Mem:* AAAS; Am Asn Cancer Res; Am Asn Pathologists; Am Soc Cell Biol. *Res:* Carcinogenesis; tumor markers for early diagnosis of cancer; cancer prevention using dietary substances; new tests for cancer diagnosis. *Mailing Add:* Dept Path Univ Md Baltimore Ten S Pine St Baltimore MD 21201-1116

SHAN, ROBERT KUOCHENG, b Gaoan, China, Nov 9, 27; m 63; c 1. AQUATIC ECOLOGY. *Educ:* Taiwan Norm Univ, BS, 56; Univ BC, MS, 62; Ind Univ, Bloomington, PhD(zool), 67. *Prof Exp:* Asst fishery biol, Nat Taiwan Univ, 55-56, asst zool, 56-59; res assoc, Ind Univ, Bloomington, 67-69; from asst prof to assoc prof, 69-75, PROF BIOL, FAIRMONT STATE COL, 75- *Concurrent Pos:* Vis prof, Jinan Univ, China, 80 & 84. *Mem:* Am Inst Biol Sci; Am Soc Limnol & Oceanog; Ecol Soc Am. *Res:* Systematics and ecology of marine copepods; ecology and genetics of chydorid cladocerans. *Mailing Add:* Dept Biol Fairmont State Col Fairmont WV 26554

SHANAHAN, PATRICK, b Clyde, Ohio, Aug 4, 31; m 53; c 9. MATHEMATICS. *Educ:* Univ Notre Dame, BA, 53; Ind Univ, PhD(math), 57. *Prof Exp:* From instr to assoc prof, 57-67, PROF MATH, COL OF THE HOLY CROSS, 67- *Concurrent Pos:* NSF sci fac fel, Harvard Univ, 66-67. *Mem:* Am Math Soc; Math Asn Am; London Math Soc. *Res:* Differential topology; equivariant version of the Atiyah-singer index theorem, and its applications to gemetric and topological problems. *Mailing Add:* Dept Math Col Holy Cross College St Worcester MA 01610

SHANBERGE, JACOB N, b Milwaukee, Wis, Jan 14, 22; m 53; c 4. PATHOLOGY, HEMOSTASEOLOGY. *Educ:* Marquette Univ, BS, 42, MD, 44. *Prof Exp:* Assoc dir, Milwaukee Blood Ctr, 48; asst chief lab serv, Vet Admin Ctr Hosp, Wood, Wis, 52-55 & West Roxbury, Mass, 55-60; Nat Heart Inst spec res fel, Zurich, Switz, 60-61; assoc dir path, Michael Reese Hosp & Med Ctr, Chicago, 62-64; dir hemat labs & blood bank & assoc in path, Evanston Hosp, Ill, 64-69; dir dept path & lab med, Mt Sinai Med Ctr, 69-79; CHIEF COAGULATION & HEMOSTASIS, WILLIAM BEAUMONT HOSP, 79- *Concurrent Pos:* Res fel biochem, Sch Med, Marquette Univ, 48; assoc path, Peter Bent Brigham Hosp, Boston, 55-60. *Honors & Awards:* Murray Thelin Award, Nat Hemophilia Found, 77. *Mem:* Col Am Path; Am Asn Path; Int Soc Thrombosis & Haemostasis; Am Soc Hemat. *Res:* Blood coagulation. *Mailing Add:* 3601 W 13 Mile Rd Clin Labs William Beaumont Hosp Royal Oak MI 48072

SHAND, JULIAN BONHAM, JR, b Columbia, SC, Nov 6, 37; m 63; c 3. SOLID STATE PHYSICS. *Educ:* Univ SC, BS, 59; Univ NC, Chapel Hill, PhD(physics), 65. *Prof Exp:* Asst prof physics, Univ Ga, 64-67; DANA PROF PHYSICS & CHMN DEPT, BERRY COL, 67-, PROF COMPUT SCI, 73- *Mem:* Am Phys Soc; Am Asn Physics Teachers; Sigma Xi. *Res:* Electrons in metals; pseudopotentials; positron annihilation in solids. *Mailing Add:* 112 Parkway Dr Rome GA 30161

SHAND, MICHAEL LEE, b Stockton, Calif, July 2, 46; m 69; c 1. SOLID STATE PHYSICS, OPTICS. *Educ:* Princeton Univ, AB, 68; Univ Pa, MSc, 69, PhD(physics), 73. *Prof Exp:* Res assoc physics, Univ Paris, 73-74; vis asst prof, Ariz State Univ, 75-76; staff physicist, Allied Chem Corp, 76-80, sr res physicist, 80-85, MGR, LASER RES & DEVELOP, ALLIED SIGNAL INC, 85- *Concurrent Pos:* Res assoc fel, French Foreign Ministry, 73-74. *Mem:* Sigma Xi; Am Phys Soc; Inst Elec & Electronics Engrs; Optical Soc Am. *Res:* Raman scattering; quantum optics; nonlinear optics; laser physics. *Mailing Add:* Allied Signal Inc PO Box 1021R Morristown NJ 07960

SHANDS, HENRY LEE, b Madison, Wis, Aug 30, 35; m 62; c 3. PLANT GENETICS, PLANT BREEDING. *Educ:* Univ Wis, BS, 57; Purdue Univ, MS, 61, PhD(plant genetics), 63. *Prof Exp:* Asst prof agron, Purdue Univ-USAID Contract, Minas Gerais, Brazil, 63-65, asst prof plant genetics, Purdue Univ, Lafayette, 65-67; res agronomist, Dekalb Soft Wheat Res Ctr, Dekalb Agres, Inc, 67-86; NAT PROG LEADER, PLANT GERMPLASM, AGR RES SERV, USDA, 86- *Concurrent Pos:* Exec secy, Nat Plant Genetic Resources Bd. *Mem:* Fel AAAS; fel Am Soc Agron; fel Crop Sci Soc Am; Am Phytopath Soc; Genetics Soc Can. *Res:* Plant breeding through genetics and cytogenetics; disease resistance. *Mailing Add:* Bldg 005 Rm 140 USDA Agric Res Serv-Beltsville Agr Res Ctr-W Beltsville MD 20705-2350

SHANDS, JOSEPH WALTER, JR, b Jacksonville, Fla, Nov 1, 30; m 55; c 4. MEDICAL MICROBIOLOGY. *Educ:* Princeton Univ, AB, 52; Duke Univ, MD, 56. *Prof Exp:* Fel microbiol, Univ Fla, 61-64, from asst prof to prof immunol & med microbiol, Col Med, 67-76, prof med & chief, Div Infectious Dis, 76-91; CONSULT, 91- *Concurrent Pos:* Ed, J Infection & Immunity, 70-78, ed-in-chief, 79-; mem bacteriol & mycol study sect, NIH. 71-74. *Mem:* Am Soc Microbiologists; Reticuloendothelial Soc; Infectious Dis Soc Am; Am Asn Immunologists. *Res:* Endotoxin; coagulation; host-parasite relationships. *Mailing Add:* Dept Med Box J 277 M S B Univ Fla Gainesville FL 32610

SHANE, HAROLD D, b New York, NY, Jan 22, 36; m 62; c 2. MATHEMATICAL STATISTICS. *Educ:* Mass Inst Technol, SB, 57; NY Univ, MS, 62, PhD(math), 68. *Prof Exp:* Engr electronics, Elec Div, Daystrom Inc. 58-61; instr math, Sch Eng, Cooper Union, 62-68; from asst prof to assoc prof, 68-75, chmn dept, 71-85, PROF MATH, BARUCH COL, 76- *Mem:* Math Asn Am; Inst Math Statist. *Res:* Nonparametric statistical theory and methodology; inequalities for order statistics; mathematical applications to political science and management. *Mailing Add:* Dept Math Baruch Col New York NY 10010

SHANE, JOHN DENIS, b Gooding, Idaho, Aug 9, 52; m 76; c 4. PALEOPALYNOLOGY, ORGANIC MATTER ANALYSIS. *Educ:* Brigham Young Univ, BS, 77; Ariz State Univ, PhD(palynol), 82. *Prof Exp:* Geol field asst, US Geol Surv, 79, grad student appointee, 79-81; sr geologist, HQ Paleont, Explor Dept, Exxon Co, 81-83, RES GEOLOGIST, EXXON PROD RES CO, 83- *Mem:* Am Asn Stratig Palynologists; Int Orgn Paleobot; Int Asn Angiosperm Paleobot. *Res:* Organic matter analysis; Mesozoic palynology; fossil fungal spores. *Mailing Add:* Exxon Prod Res Co PO Box 2189 Houston TX 77001

SHANE, JOHN RICHARD, b San Diego, Calif, Sept 13, 36; m 59; c 2. MAGNETISM. *Educ:* Univ Maine. BS, 58; Mass Inst Technol, PhD(solid state physics). 63. *Prof Exp:* Mem res staff solid state physics, Sperry Rand Res Ctr, 63-68; asst prof, 68-70, ASSOC PROF PHYSICS, UNIV MASS, BOSTON, 70- *Mem:* Am Phys Soc. *Res:* Magnetic properties of matter; antiferromagnetism; antiferromagnetic and paramagnetic resonance; spin-lattice relaxation. *Mailing Add:* Dept Physics Univ Mass Harbor Campus Boston MA 02125

SHANE, PRESSON S, b Junction City, Kans, Feb 2, 20; m 50; c 4. CHEMICAL ENGINEERING. *Educ:* Univ Kans, BS, 41; Mass Inst Technol, MS, 46. *Prof Exp:* Tech supt, Heavy Water Dept, E I du Pont de Nemours & Co, Inc, 46-57; asst to gen mgr, McGean Chem Co, 57-58; dir, Solid Propellant Div, Atlantic Res Corp, 58-60, vpres, 60-63; partner, Columbia Assocs, 63-64; pres, Wash Technol Assocs, Inc, 64-67; PROF ENG ADMIN, GEORGE WASHINGTON UNIV, 68- *Mem:* Am Inst Aeronaut & Astronaut; Am Inst Chem Engrs. *Res:* Aerospace developments; management of advanced technology. *Mailing Add:* Double Mill R Easton MD 21601

SHANE, ROBERT S, b Chicago, Ill, Dec 8, 10; m 36; c 3. MATERIALS SCIENCE, TECHNOLOGY TRANSFER. *Educ:* Univ Chicago, BS, 30, PhD(chem), 33. *Prof Exp:* Res chemist, Nat Aniline Div, Allied Chem & Dye Corp, 34-35; chemist, Stein-Hall Mfg Co, 35-36; chemist, Fuel Antioxidants, Universal Oil Prod Co, 36; tech dir, Western Adhesives Co, 37-40; res chemist, Gelatin Prod Co, 41-42; plant supt, Amecco Chem, Inc, 42-43; group leader, Bausch & Lomb Optical Co, 43-45; owner, dry cleaning bus, 46-52; proj supvr govt contract res, Wyandotte Chem Corp, 52-54; asst dir new prod develop, Am Cyanamid Co, 54-55; mgr chem ceramics, Com Atomic Power Dept, Westinghouse Elec Corp, 55-57; nucleonics specialist, Bell Aircraft Corp, 57-58; consult engr, Light Mil Electronics Dept, Gen Elec Co, 58-64, res engr, Laminated Prod Dept, 64-66, systs specialist radiation effects, Spacecraft Dept, 66-67, mgr design rev, Reentry Systs, 67-69, mgr parts, Mat & Processes Eng, Space Systs Orgn, 69-70; staff scientist & consult, Nat Mat Adv Bd, Nat Acad Sci, 70-76; PRIN, SHANE ASSOCS, 76-; ED, MAT ENG, MARCEL DEKKER, INC, 78- *Concurrent Pos:* Guest instr, Pa State Univ, 56; consult, US Dept Defense, 76- *Honors & Awards:* Gold Key Award, Gen Elec Co, 63; Margaret Dana Award, Am Soc Testing & Mat, 83. *Mem:* Am Chem Soc; fel Am Soc Testing & Mat; fel Am Inst Chem Engrs; Am Soc Metals; AAAS. *Res:* Materials engineering; radiation effects; energy transmission; surface phenomena; plastics fabrication; physical chemistry; technology transfer and innovation. *Mailing Add:* 695 Venture Three 10701 S Ocean Dr Jensen Beach FL 34957

SHANE, SAMUEL JACOB, b Yarmouth, NS, May 17, 16; m 72; c 4. INTERNAL MEDICINE. *Educ:* Dalhousie Univ, BSc, 36, MD, CM, 40; FRCP(C). *Prof Exp:* From asst med dir to med dir, Point Edward Hosp, Sydney, NS, 49-57; from asst prof to assoc prof med, Dalhousie Univ, 57-68; assoc prof med, Fac Med, Univ Toronto, 68-81; dir cardiovasc unit, Sunnybrook Hosp, 68-81; consult cardiologist, Surrey Mem Hosp, Surrey, BC, 81-83; RETIRED. *Concurrent Pos:* Med dir tuberc div, Halifax Tuberc Hosp & Health Ctr, 57-64; cardiologist, Halifax Children's Hosp, 57-68; dir cardiac unit, Victoria Gen Hosp, 58-68; Can Tuberc Asn traveling fel, 60; consult, NS Rehab Ctr, 60-68 & Cardiol Halifax Infirmary, 64-68. *Mem:* Am Col Cardiol; Am Thoracic Soc; Am Col Physicians; Am Col Chest Physicians. *Res:* Diseases of chest and heart; cardiovascular hemodynamics and catheterization; clinical pharmacology. *Mailing Add:* 45 Huntingdale Blvd Penthouse 3 Scarborough ON M1W 2N8 Can

SHANEBROOK, J(OHN) RICHARD, b Syracuse, NY, July 10, 38; m 67; c 2. MECHANICAL ENGINEERING. *Educ:* Syracuse Univ, BME, 60, MME, 63, PhD(mech eng), 65. *Prof Exp:* From asst prof to assoc prof, 65-75, PROF MECH ENG, UNION COL, NY, 75-, dept chmn, 74-79. *Concurrent Pos:* NSF res grants, 67-72; Eng Found Res grant, 73-75, Cardiac Eng res grants, 75-81, Sloan Found, 83-88, R A Smith res grants, 89- *Mem:* Am Soc Mech Engrs. *Res:* Theoretical analysis of three-dimensional viscous flow fields, both laminar and turbulent; fluid dynamics of artificial heart valves, cardiac assist devices and biomedical catheters; energy conservation; societal issues related to nuclear technology; technological literacy for undergraduate liberal arts students. *Mailing Add:* Dept Mech Eng Steinmetz Hall Union Col Schenectady NY 12308

SHANEFIELD, DANIEL J, b Orange, NJ, Apr 29, 30; m 64; c 2. PHYSICAL CHEMISTRY. *Educ:* Rutgers Univ, BS, 56, PhD(phys chem). 62. *Prof Exp:* Sr tech specialist phys chem, ITT Fed Labs, 62-67; sr res chemist, AT&T Corp, 67-86; DISTINGUISHED PROF, CERAMICS ENG DEPT, RUTGERS UNIV, 86- *Concurrent Pos:* Assoc ed, J Am Ceramic Soc, 86- *Mem:* Am Chem Soc; Electrochem Soc; fel Am Inst Chemists; Am Ceramic Soc. *Res:* Integrated circuit packaging; ceramic tape casting; additives for ceramics. *Mailing Add:* Rutgers Univ Ceramics Engr Dept PO Box 909 Piscataway NJ 08855

SHANER, GREGORY ELLIS, b Portland, Ore, Dec 19, 42; m 64; c 2. PLANT PATHOLOGY. *Educ:* Ore State Univ, BS, 64, PhD(plant path), 68. *Prof Exp:* From asst prof to assoc prof, 68-81, head dir grad prog, bot & plant path, 82-87, PROF PLANT PATH, PURDUE UNIV, WEST LAFAYETTE, 81- *Concurrent Pos:* Sr ed, Phytopath, 82-84, editor-in-chief, 85-87. *Mem:* Fel Am Phytopath Soc; Crop Sci Soc Am; Am Soc Agron; Soc Econ Bot; Sigma Xi. *Res:* Plant disease epidemiology; development of improved varieties of wheat and oats; nature and genetics of disease resistance in small grains. *Mailing Add:* Dept Bot & Plant Path Purdue Univ West Lafayette IN 47907

SHANER, JOHN WESLEY, b Arlington, MA, June 3, 42; m 66; c 3. HIGH PRESSURE PHYSICS, VERY HIGH TEMPERATURE DENSE FLUIDS. *Educ:* Mass Inst Technol, BS, 64; Univ Calif Berkeley, PhD(physics), 69. *Prof Exp:* Assoc instr physics, Univ Utah Physics Dept, 70-72; group leader exp physics, Lawrence Livermore Lab, 72-78; GROUP LEADER, SHOCKWAVE PHYSICS, LOS ALAMOS NAT LAB, 78- *Mem:* Fel Am Phys Soc; Int Org High Pressure Sci & Technol (treas, 85-88). *Res:* Physics of materials at very high pressures and or very high temperatures; shock wave physics of explosives and inert materials; dense, non-ideal plasma physics. *Mailing Add:* 152 Piedra Loop Los Alamos NM 87544

SHANEYFELT, DUANE L, b Hastings, Nebr, Nov 16, 34; m 63; c 1. POLYMER CHEMISTRY, APPLIED CHEMISTRY. *Educ:* Hastings Col, BA, 56; Univ Nebr, MS. 59, PhD(org chem), 62. *Prof Exp:* Res chemist, Niagara Chem Div, FMC Corp, 61-64; fel, Coal Chem Res Proj, Mellon Inst Sci, 64-67; group leader, Res & Develop Div, Kraftco Corp, 67-72; sr chemist res & develop, 72-75, res supvr, 75-76, DIR RES & DEVELOP, MASURY-COLUMBIA CO, 77- *Mem:* Am Oil Chem Soc; fel Am Inst Chemists; Am Chem Soc. *Res:* Oxygen and nitrogen heterocyclic organic chemistry; organic sulfur compounds; agricultural chemistry; chemistry of coal tar derivatives; organic chemistry of ammonia; detergents, cleaners and degreasers, particularly maintenance specialties, emulsions, polymer coatings and systems; floor finishes, coatings and maintenance systems. *Mailing Add:* Misco Int 115 Messner Dr Wheeling IL 60090-7205

SHANFIELD, HENRY, b Toronto, Ont, May 17, 23; US citizen; m 50; c 2. PHYSICAL CHEMISTRY, CHEMICAL ENGINEERING. *Educ:* Univ Toronto, BASc, 46, MASc, 57, PhD(phys chem, chem eng), 51. *Prof Exp:* Sr res scientist, Nat Res Coun Can, 47-48; res engr, Esso Eng & Res Ctr, Standard Oil Co, 51-53; asst dir res, Paper-Mate Mfg Div, Gillette Co, 53-58; mgr chem lab, Aeronutronic Div, Philco-Ford Corp, Newport Beach, 58-68; dir res & eng, Polymetrics, Inc, 68-71; dir eng, Foremost Water Systs, 71; mgr water systs, KMS Technol Ctr, 71-74; ASSOC PROF CHEM, UNIV HOUSTON, 74- *Mem:* AAAS; Am Chem Soc; Am Inst Aeronaut & Astronaut; Combustion Inst; Sigma Xi. *Res:* Chemical kinetics; thermodynamics; electrochemistry; membrane transport phenomena; thin layer chromatography; semipermeable membranes. *Mailing Add:* Dept Chem Univ Houston 4800 Calhoun Rd Houston TX 77204-5641

SHANGRAW, RALPH F, b Rutland, Vt, June 11, 30; m 55; c 3. INDUSTRIAL PHARMACY. *Educ:* Mass Col Pharm, BS, 52, MS, 54; Univ Mich, PhD(pharmaceut chem), 58. *Prof Exp:* From asst prof to assoc prof, 58-70, PROF PHARMACEUT DEPT, SCH PHARM, UNIV MD, BALTIMORE, 70-, CHMN DEPT PHARMACEUT, 71- *Concurrent Pos:* Mem, XIX, XX, XI, XXII, XXIII, Comt, US Pharmacopeia, 70-95; distinguished scientist, Am Asn Pharmaceut Sci. *Mem:* Am Pharmaceut Asn; fel Acad Pharmaceut Sci; Soc Cosmetic Chemists; Am Asn Cols Pharm; fel Am Asn Pharmaceut Scientists. *Res:* Direct tablet compression; pharmaceutical excipients; nitroglycerin formulation; vitamins and nutritional supplements. *Mailing Add:* Univ Md Sch Pharm 20 N Pine St Baltimore MD 21201

SHANHOLTZ, VERNON ODELL, b Slanesville, WVa, Apr 22, 35; m 65; c 2. AGRICULTURAL ENGINEERING. *Educ:* WVa Univ, BS, 58, MS, 63; Va Polytech Inst & State Univ, PhD(civil eng), 70. *Prof Exp:* Hydraul engr, Agr Res Serv, 58-66; res instr, 66-70, asst prof, 70-78, ASSOC PROF SOIL & WATER CONSERV, VA POLYTECH INST & STATE UNIV, 78- *Mem:* Am Geophys Union; Am Soc Agr Engrs; Soil & Water Conserv Soc Am; Sigma Xi. *Res:* Modeling agricultural watershed systems. *Mailing Add:* 300 Dogwood Lane Christiansburg VA 24073

SHANHOLTZER, WESLEY LEE, b Cumberland, Md, Jan 25, 38; m 66; c 2. SOLID STATE PHYSICS. *Educ:* WVa Univ, BS, 62, MS, 64, PhD(physics), 68. *Prof Exp:* Asst prof, 66-73, assoc prof, 73-82, PROF PHYSICS, MARSHALL UNIV, 82- *Mem:* Sigma Xi; Am Asn Physics Teachers. *Res:* Electron spin resonance studies of conduction electrons paramagnetic susceptibility in lithium metal; electron spin resonance of doped semiconductor crystals. *Mailing Add:* Dept Physics Marshall Univ Huntington WV 25755

SHANK, BRENDA MAE BUCKHOLD, b Cleveland, Ohio, Sept 25, 39; m 69. CELL PHYSIOLOGY. *Educ:* Western Reserve Univ, BA, 61, PhD(biophys), 66; Rutgers Med Sch, MD, 76. *Prof Exp:* Res biophysicist, Lawrence Radiation Lab, Univ Calif, 66-68, NIH fel biophys, Donner Lab Med Physics, 68-69; asst prof radiol, Case Western Reserve Univ, 69; asst prof physiol, Rutgers Med Sch, 69-74; res & fel, Mem Sloan-Kettering Cancer Ctr, 76-80, asst attend, Mem Hosp, 80-85, assoc attend & assoc mem, 85-89; assoc prof Radiation Oncol Med, Cornell Univ Med Sch, 85-89; CHMN & PROF RADIATION ONCOL, MT SINAI SCH MED, 89-, DIR RADIATION ONCOL, MT SINAI HOSP, 89- *Mem:* AAAS; NY Acad Sci; Radiol Soc NAm; Radiation Res Soc; Am Soc Therapeut Radiol & Oncol; Am Soc Clin Oncol. *Res:* Osmotic adaptation in tissue-culture cells; flour beetle, tribolium confusum, regarding effects of radiation and weightlessness in biosatellite; electronic counting of erythrocytes; membrane properties and growth control of cultured cells; radiation kinetics; total body irradiation for marrow transplantation. *Mailing Add:* Dept Radiation Oncol Mt Sinai Med Ctr Box 1236 New York NY 10029-6574

SHANK, CHARLES PHILIP, b Pittsburgh, Pa, Feb 6, 41; m 63; c 4. POLYMER CHEMISTRY. *Educ:* Univ Dayton, BS, 63, MS, 65; Univ Akron, PhD(polymer sci), 68. *Prof Exp:* Res chemist, NCR Corp, 68-73; DEVELOP CHEMIST, PLASTICS DIV, GEN ELEC CO, 73-; AT DREXEL-HYSOL. *Mem:* Am Chem Soc. *Res:* Characterizations and use of polymers and copolymers, particularly impact modification of polymers and copolymer sequence distribution and its effect on properties. *Mailing Add:* Mallinckrodt Inc PO Box M Paris KY 40361

SHANK, CHARLES VERNON, b Mt Holly, NJ, July 12, 43. ELECTRICAL ENGINEERING. *Educ:* Univ Calif, Berkeley, BS, 65, MS, 66, PhD(elec eng), 69. *Prof Exp:* Mem tech staff, AT&T Bell Labs, 69-83, dir, Electronics Res Lab, 83-89; DIR, LAWRENCE BERKELEY LAB, 89- *Mem:* Nat Acad Sci; Nat Acad Eng. *Res:* Quantum electronics. *Mailing Add:* Bldg 50A Rm 4133 Cycletron Rd Berkeley CA 94720

SHANK, FRED R, b Harrisonburg, Va, Oct 11, 40; m 67; c 2. FOOD SCIENCE, NUTRITION. *Educ:* Univ Ky, BS, 62, MS, 64; Univ Md, PhD(nutrit), 69. *Prof Exp:* Res asst nutrit, Univ Ky, 62-64 & Univ Md, 64-68; biomed lab officer, USAF, Brooks AFB, San Antonio, Tex, 68-70, proj officer, Dietary Info Serv, Air Force Data Systs Design Ctr, Washington, DC, 70-71; nutritionist, Food & Nutrit Serv, US Dept Agr, Washington, DC, 71-74, chief, evaluation & tech serv br, Nutrit & Tech Serv Staff, 74-78; dep dir, div nutrit, 78-79, dep dir, off nutrit & food Sci, 79-86, DIR, OFF PHYS SCI, CTR FOOD SAFETY & APPL NUTRIT, FOOD & DRUG ADMIN, WASHINGTON, DC, 86- *Mem:* AAAS; Am Asn Cereal Chemists; Am Inst Nutrit; Am Soc Clin Nutrit; Inst Food Technologists; Sigma Xi. *Res:* Nutrient sufficiency, nutrient toxicity, assessment of nutritional status; effects of nutritional status, food consumption and food processing on human performance and disease. *Mailing Add:* HFF-1 Ctr Food Safety & Appl Nutrit 200 C St SW Washington DC 20204

SHANK, GEORGE DEANE, b Muncie, Ind, July 17, 40; m 62; c 2. NUMERICAL ANALYSIS. *Educ:* Purdue Univ, BSEE, 62; Univ Md, PhD(appl math), 68. *Prof Exp:* Math analyst, Comput Usage Co, 63-68; from res mathematician to advan res mathematician, 68-74, SR RES MATHEMATICIAN, DENVER RES CTR, MARATHON OIL CO, 74- *Mem:* Soc Indust & Appl Math; Am Math Soc; Asn Comput Mach; Sigma Xi. *Res:* Computational methods for solving problems of flow in porous media and other oil industry problems. *Mailing Add:* Marathon Oil Co PO Box 269 Littleton CO 80160

SHANK, HERBERT S, b Orange, NJ, Sept 25, 27. MATHEMATICS. *Educ:* Univ Chicago, BA, 49, MS, 52; Cornell Univ, PhD, 69. *Prof Exp:* Mathematician, Inst Syst Res, Univ Chicago, 54-59 & Labs Appl Scis, 59-65; mem prof staff, Ctr Naval Anal, 65-68; res assoc, Cornell Univ, 68-70; Nat Res Coun fel, 70-71, ASSOC PROF MATH, UNIV WATERLOO, 71- *Concurrent Pos:* Instr, Ill Inst Technol, 57-58; vis prof, Queen's Univ, 81; vis scientist, Cornell Univ, 81-82. *Mem:* Am Math Soc; Math Asn Am. *Res:* Graph theory; combinatorial mathematics; operations research; electrical network theory. *Mailing Add:* Dept Math & Statist Queens Univ Kingston ON K7L 3N6 Can

SHANK, KENNETH EUGENE, b Lancaster, Pa, Oct 26, 49; m 70; c 3. RADIOLOGICAL & ENVIRONMENTAL HEALTH. *Educ:* Elizabethtown Col, BS, 71; Purdue Univ, MS, 73, PhD(bionucleonics), 75. *Prof Exp:* Oper res anal, Defense Activ, 71-72; res assoc environ assessment, Oak Ridge Nat Lab, 75-80; mem staff, US Dept Energy, Oak Ridge, Tenn, 80-83; GROUP SUPVR, ENVIRON & CHEM, PA POWER & LIGHT, ALLENTOWN, 83- *Honors & Awards:* Glenn L Jenkins Award, Purdue Univ, 76. *Mem:* Health Physics Soc; Am Indust Hyg Asn. *Res:* Determining the environmental and health impacts of nuclear and nonnuclear energy systems. *Mailing Add:* 4115 Daisy Ct Orefield PA 18069

SHANK, LOWELL WILLIAM, b Hagerstown, Md, June 28, 39; m 63; c 2. FORENSIC CHEMISTRY. *Educ:* Goshen Col, BS, 61; Ohio State Univ, MSc, 64, PhD(anal chem), 66. *Prof Exp:* From asst prof to assoc prof, 66-82, dept head, 85-90, PROF ANALYTICAL CHEM, WESTERN KY UNIV, 82- *Mem:* Am Chem Soc. *Res:* Blood enzymes by electrophoresis; gunshot residue analysis. *Mailing Add:* Dept Chem Western Ky Univ Bowling Green KY 42101

SHANK, MAURICE E(DWIN), b New York, NY, Apr 22, 21; m 48; c 3. AERONAUTICAL PROPULSION, MECHANICAL ENGINEERING. *Educ:* Carnegie Inst Technol, BS, 42; Mass Inst Technol, ScD, 49. *Prof Exp:* Instr metall, Mass Inst Technol, 46-49, from asst prof to assoc prof mech eng, 49-60; dir advan mat, Pratt & Whitney, 60-70, mgr mat eng & res, 70-71, dir eng technol, Com Prods Div, Pratt & Whitney Aircraft Group, United Technologies Corp, 72-85, vpres, 85-87, AEROSPACE CONSULT, PRATT & WHITNEY OF CHINA, INC, 87- *Concurrent Pos:* Consult ed, McGraw-Hill Bk Co; mem res & technol adv coun, NASA, 73-77, mem adv comt aeronaut, 78-; mem comt res, NSF, 74-77; mem bd, Aeronaut & Space Eng, 89- *Mem:* Nat Acad Eng; fel Am Soc Mech Engrs; fel Am Inst Mining, Metall & Petrol Engrs; fel Am Soc Metals; fel Am Inst Aeronaut & Astronaut. *Res:* Advancing state-of-the-art technology in engine aerodynamic components; noise and emission reduction; fuel systems and controls; materials and structures to assure competitive engine performance weight and cost. *Mailing Add:* Two Enatai Dr Bellevue WA 98004

SHANK, PETER R, b Ithaca, NY, Feb 17, 46; m 70; c 2. VIROLOGY. *Educ:* Cornell Univ, BS, 68; Univ NC, Chapel Hill, PhD(virol), 73. *Prof Exp:* Fel, Univ Calif, San Francisco, 73-78; asst prof, 78-83, ASSOC PROF MED SCI, BROWN UNIV, 83- *Concurrent Pos:* Vis scientist, Lab molecular Virol, Nat Cancer Inst, NIH, 86-87. *Mem:* Am Soc Microbiol; AAAS; Soc Gen Microbiol; Am Soc Virol. *Res:* Molecular biology of retroviruses, respiratory syncytial virus and human immunodeficiency virus and role of oncogenes in hepatic cell carcinoma. *Mailing Add:* Div Biol & Med Brown Univ Providence RI 02912

SHANK, RICHARD PAUL, b Indianapolis, Ind, July 29, 41; m 67; c 1. NEUROCHEMISTRY, NEUROBIOLOGY. *Prof Exp:* Res physiologist, Vet Admin, 70-72; res trainee chem, Ind Univ, 72-74, res assoc, 74-75; asst prof, Sch Med, Temple Univ, 75-79; sr scientist, Franklin Inst, Philadelphia, Pa, 79-81 & Grad Hosp, 81-82; prin scientist, 82-86, res fel chem, McNeil Pharmaceut, 86-87; FEL BIOL RES, JANSSEN RES FOUND, 87- *Concurrent Pos:* Adj asst prof, Thomas Jefferson Med Col, 82-; adj assoc prof, Sch Med, Temple Univ, 82- *Mem:* Am Physiol Soc; Soc Neurosci; Am Soc Neurochem; Int Soc Neurochem. *Res:* Neurochemistry of amino acids; neurotransmitter receptors; antidepressants and their mechanisms of action; modulation of synaptic transmission. *Mailing Add:* Johnson Pharmaceut Res Inst Spring House PA 19477-0776

SHANK, ROBERT ELY, b Louisville, Ky, Sept 2, 14; m 42; c 3. NUTRITION. *Educ:* Westminster Col, Mo, AB, 35; Washington Univ, MD, 39. *Prof Exp:* From intern to house physician, Barnes Hosp, Mo, 39-41; resident physician, St Louis Isolation Hosp, 41; asst resident physician & asst, Rockefeller Inst Hosp, 41-46; assoc mem, Pub Health Res Inst, New York, 46-48; Danforth prof & head dept, 48-83, EMER PROF PREV MED, SCH MED, WASH UNIV, 83- *Concurrent Pos:* With nutrit surv, Nfld, 48-; spec consult, USPHS, 49-53; mem food & nutrit bd, Nat Res Coun, 50-71; mem adv comt metab, Surgeon Gen, US Dept Army, 56-60 & adv comt nutrit, 60-72; mem sci adv comt, Nat Vitamin Found, 58-61; mem human ecol study sect, NIH, 59-63 & nutrit study sect, 64-68; mem prof adv comt, Nat Found, 61-62 & Nat Adv Coun Child Health & Human Develop, 69-73; mem, Clin Appln & Prev Adv Comt, Nat Heart, Lung & Blood Inst, 76-80. *Honors & Awards:* Distinguished Serv Award, Am Heart Asn, 80. *Mem:* Am Soc Biol Chemists; Am Soc Clin Invest; Soc Exp Biol & Med; Asn Am Physicians; Fel Am Inst Nutrit. *Res:* Metabolism of progressive muscular dystrophy; cirrhosis of the liver; infectious hepatitis and homologous serum jaundice; appraisal of nutritional status; iron deficiency; relationship between nutrients and hormonal function. *Mailing Add:* 1325 Wilton Lane Kirkwood MO 63122

SHANKAR, HARI, b Aligarh, India, Nov 10, 30. MATHEMATICAL MODELING, HARMONIC FUNCTIONS. *Educ:* Aligarh Muslam Univ, MS, 56; Univ Cincinatti, MS, 63. *Prof Exp:* PROF MATH, OHIO UNIV, 63- *Mem:* Math Asn Am. *Res:* Growth properties, entire functions, harmonic functions & analytical functions. *Mailing Add:* Dept Math Ohio Univ Athens OH 45701

SHANKAR, RAMAMURTI, b New Delhi, India, Apr 28, 47; US citizen; m 76; c 4. PHYSICS. *Educ:* Indian Inst Sci, BTech, 69; Univ Calif, Berkeley, PhD(theoret phys), 74. *Prof Exp:* Jr fel, Harvard Soc fellows, 74-77; JW Gibbs instr, 77-79, from asst prof to assoc prof, 79-86, PROF, PHYSICS, YALE UNIV, 86- *Res:* Common problems in statistical mechanics and quantum field theory, statistical mechanics of homogeneous and random systems. *Mailing Add:* 55 Sloan Physics Labs Yale Univ New Haven CT 06511

SHANKEL, DELBERT MERRILL, b Plainview, Nebr, Aug 4, 27; m 58; c 3. MICROBIOLOGY. *Educ:* Walla Walla Col, BA, 50; Univ Tex, PhD(bact), 60. *Prof Exp:* Instr sci & math, Walla Walla Col, 50-51; instr chem, San Antonio Col, 54-55; res scientist bact, Univ Tex, 56-59; from asst prof to assoc prof, Univ Kans, 59-68, from assoc dean to actg dean arts & sci, 69-74, exec vchancellor, 74-80, actg chancellor, 80-81, PROF BACT, UNIV KANS, 68-, PROF MICROBIOL & SPEC COUNR TO CHANCELLOR, 81-, EXEC VCHANCELLOR, 90- *Concurrent Pos:* Consult, Cramer Chem Co, 59- & NCent Asn Cols & Sec Schs, 73-; NIH sr fel, Univ Edinburgh, 67-68; vis prof, Nat Inst Genetics, Japan, 88. *Mem:* AAAS; Am Soc Microbiologists; Genetics Soc Am; Radiation Res Soc; Environ Mutagen Soc; Sigma Xi. *Res:* Genetic effects of radiations and chemicals and interactions of repair processes; mutagenesis and antimutagenesis. *Mailing Add:* Dept Microbiol Univ Kans Lawrence KS 66045

SHANKLAND, DANIEL LESLIE, b San Diego, Calif, June 18, 24; m 55; c 3. NEUROPHYSIOLOGY, TOXICOLOGY. *Educ:* Colo State Univ, BS, 48; Univ Ill, MS, 52, PhD(entom), 56. *Prof Exp:* Salesman, Stauffer Chem Co, 52 & Farm Air Serv, 52-54; res rep, Stauffer Chem Co, 55-57; prof entom, Purdue Univ, West Lafayette, 57-76; prof entom & head dept, Miss State Univ, 76-80; prof entom & head dept, 80-86, DIR, CTR FOR ENVIRON TOXICOL, UNIV FLA, 86- *Mem:* AAAS; Entom Soc Am; Sigma Xi. *Res:* Physiology; neurophysiology, especially neurotoxic action of insecticides. *Mailing Add:* Dept Entom Univ Fla Gainsville FL 32611

SHANKLAND, RODNEY VEEDER, b Dixboro, Mich, Nov 12, 04; m 34; c 2. ORGANIC CHEMISTRY. *Educ:* Univ Mich, BS, 26, MS, 27, ScD(org chem), 30. *Prof Exp:* Res chemist, Standard Oil Ind, 30-37, group leader, 37-48, sect leader, 48-53, sr res assoc, 53-60, sr res assoc, Am Oil Co, 60-69; RETIRED. *Concurrent Pos:* Consult, 69- *Mem:* Am Chem Soc; AAAS; NY Acad Sci; Sigma Xi. *Res:* Synthetic lubricating oils; thermal and catalytic cracking and reforming; hydrocarbon conversion catalysts; information research. *Mailing Add:* 1307 Summit Pl Valparaiso IN 46383

SHANKLE, ROBERT JACK, b Ga, Sept 17, 23; m 52; c 2. DENTISTRY. *Educ:* Emory Univ, DDS, 48; Am Bd Endodontics, dipl. *Prof Exp:* Instr crown & bridge prothodont, Sch Dent, Emory Univ, 49-51; from assoc prof to prof oper dent, 51-66, dir admis, 64-75, prof endodont & chmn dept, 66-84, EMER PROF ENDODONT, SCH DENT, UNIV NC, CHAPEL HILL, 84- *Concurrent Pos:* Consult, Womack Army Hosp, Ft Knox, Ky, Ft Benning, Ga & Ft Dix, NJ; ed, NC Dent J, 72-77; dir pub rels & develop, 75- *Mem:* Am Dent Asn; fel Am Asn Endodont; fel Am Col Dent; fel Int Col Dent; fel Acad Gen Dent; fel Acad Dent Int. *Res:* Endodontics. *Mailing Add:* 104 Fox Ridge Rd Chapel Hill NC 27514

SHANKLIN, JAMES ROBERT, JR, b Bluefield, WVa, Dec 28, 41; m 66; c 2. SYNTHETIC ORGANIC CHEMISTRY, MEDICINAL CHEMISTRY. *Educ:* Yale Univ, BA, 64; Univ Va, PhD(chem), 72. *Prof Exp:* Teacher, Va Episcopal Sch, 64-66; fel, Wayne State Univ, 71-73; sr res chemist, 73-80, GROUP MGR CARDIOVASC SYNTHESIS, A H ROBINS CO, 80- *Concurrent Pos:* Adj fac mem, Va Commonwealth Univ, 73- *Mem:* Am Chem Soc; Sigma Xi. *Res:* Organosulfur chemistry; heterocyclic chemistry; medicinal chemistry specializing in the cardiovascular and central nervous system areas. *Mailing Add:* 8318 Brookfield Rd Richmond VA 23227

SHANKLIN, MILTON D(AVID), agricultural engineering; deceased, see previous edition for last biography

SHANKS, CARL HARMON, JR, b Martinsville, Ohio, May 20, 32; m 55; c 3. ENTOMOLOGY. *Educ:* Wilmington Col, BS, 54; Ohio State Univ, MSc, 55; Univ Wis, PhD(entom), 60. *Prof Exp:* ENTOMOLOGIST, WASH STATE UNIV, 59-, SUPT, SOUTHWESTERN WASH RES UNIT, 80- *Mem:* Entom Soc Am. *Res:* Biology and control of arthropod pests of small fruits and vegetables; plant resistance to insects and mites; biological control of weeds. *Mailing Add:* 106 Nashville Way Vancouver WA 98664

SHANKS, JAMES BATES, b Steubenville, Ohio, June 17, 17; m 43, 74; c 3. HORTICULTURE. *Educ:* Ohio State Univ, PhD, 49. *Prof Exp:* From assoc prof to emer prof hort, Univ Md, Col Park, 49-88; RETIRED. *Mem:* Fel Am Soc Hort Sci; Plant Growth Regulator Soc Am. *Res:* Greenhouse flowering crops. *Mailing Add:* 11340 Frances Dr Beltsville MD 20705

SHANKS, JAMES CLEMENTS, JR, b Detroit, Mich, Oct 15, 21; m 50; c 3. SPEECH PATHOLOGY. *Educ:* Mich State Univ, BA, 43; Univ Denver, MA, 49; Northwestern Univ, PhD(speech path), 57. *Prof Exp:* Instr speech, Iowa State Teachers Col, 49-50; asst prof speech path, Syracuse Univ, 52-55; from asst prof to assoc prof, 55-67, PROF SPEECH PATH & CLIN DIR SPEECH PATH SERV, SCH MED, IND UNIV, INDIANAPOLIS, 67- *Concurrent Pos:* Consult, New Castle State Hosp, 59. *Mem:* Am Speech & Hearing Asn; Am Cleft Palate Asn. *Res:* Speech disorders involving function of the larynx and valum manifested as deviations of voice quality and vocal resonance. *Mailing Add:* Dept Speech Path Ind Univ Med Ctr 702 Barnhill Dr Indianapolis IN 46202

SHANKS, ROGER D, b Libertyville, Ill, May 30, 51; m 71; c 2. DAIRY CATTLE BREEDING. *Educ:* Univ Ill, BS, 74; Iowa State Univ, MS, 77, PhD(animal sci), 79. *Prof Exp:* Asst prof, 79-84, ASSOC PROF GENETICS, UNIV ILL, 84- *Honors & Awards:* Agway Inc Young Scientist Award in Dairy Prod, Am Dairy Sci Asn, 84. *Mem:* Am Agr Econ Asn; Am Dairy Sci Asn; Am Genetic Asn; Biomet Soc; Genetics Soc Am; AAAS. *Res:* Genetic and economic aspects of dairy cattle improvement programs including evaluation, selection, and mating designs. *Mailing Add:* Dept Animal Sci Animal Sci Lab Univ Ill 1207 W Gregory Dr Urbana IL 61801

SHANKS, SUSAN JANE, b Toledo, Ohio. SPEECH LANGUAGE PATHOLOGY. *Educ:* Univ Toledo, BEd, 57; Bowling Green State Univ, MA, 60; La State Univ, PhD(speech path), 66. *Prof Exp:* Teacher, St Pius X Sch, 57-58; grad asst speech pathologist, Bowling State Univ, 58-59; speech pathologist, Samuel Gompers Rehab Ctr, 60-63; asst prof, Univ SFla, 67-68; asst prof, Stephen F Austin State Univ, 68-70; PROF SPEECH PATH, CALIF STATE UNIV, FRESNO, 70- *Mem:* Am Speech & Hearing Asn. *Res:* Voice disorders; language disorders in children and adults. *Mailing Add:* Dept Commun Disorders Calif State Univ Mail Stop No 80 Fresno CA 93740

SHANKS, WAYNE C, III, b Detroit, Mich, Aug 26, 47; m 70; c 2. GEOCHEMISTRY. *Educ:* Mich State Univ, BS, 69; La State Univ, MS, 71; Univ Southern Calif, PhD(geol), 76. *Prof Exp:* Teaching asst geol, La State Univ, 70-71; res asst, Univ Southern Calif, 71-75; asst prof geol, Univ Calif, Davis, 75-77; from asst prof Geol & Geophys, 77-80, to assoc prof Geol & Geophys, Univ Wis, Madison, 80-82; GEOLOGIST, US GEOLOGICAL SURVEY, RESTON VA, 82- *Concurrent Pos:* Instr geol, Pierce Col, Calif, 73-75; NSF traineeship, 74. *Mem:* Sigma Xi; Soc Econ Geol; Geochem Soc; Am, Geophys Union. *Res:* Geochemistry of submarine hydrothermal and geothermal ore fluids; origin of ore deposits and stable isotope geochemistry. *Mailing Add:* US Geol Surv 954 Nat Ctr Reston VA 22092

SHANMUGAM, KEELNATHAM THIRUNAVUKKARASU, b Keelnatham, India, Oct 15, 41; m 72; c 1. BACTERIAL PHYSIOLOGY, HYDROGEN METABOLISM. *Educ:* Annamalai Univ, India, BSc, 63; UP Agr Univ, India, MSc, 65; Univ Hawaii, PhD(microbiol), 69. *Prof Exp:* Assoc res microbiologist cell physiol, Univ Calif, Berkeley, 69-71; asst prof biol, Birla Inst Technol & Sci, India, 71-72; res chemist biochem, Univ Calif, San Diego, 73-75, asst res agronomist microbiol, Davis, 76-80; asst res scientist, 80-81, assoc prof, 81-89, PROF MICROBIOL, UNIV FLA, 89- *Mem:* Am Soc Microbiol. *Res:* Study of the mechanism of regulation of anaerobic processes using hydrogen metabolism in Escherichia coli as a model system; genetic alteration of nitrogen-fixing cyanobacteria for solar energy conversion to hydrogen and ammonia. *Mailing Add:* Dept Microbiol & Cell Sci 1059 McCarty Hall Univ Fla Gainesville FL 32611

SHANMUGAN, K SAM, b India, Jan 6, 43; US citizen; m 68; c 2. COMMUNICATION SYSTEMS, SIGNAL PROCESSING. *Educ:* Madras Univ, India, BE, 64; Indian Inst Sci, ME, 66; Okla State Univ, PhD(elec eng), 70. *Prof Exp:* Res assoc elec eng, Okla State Univ, 70-71 & Univ Kans, 71-73; assoc prof, Wichita State Univ, 73-78; mem tech staff, Bell Labs, NJ, 78-80; PROF ELEC ENG, UNIV KANS, 80- *Concurrent Pos:* Consult, Boeing Aircraft, Wichita, 74-75, United Telephone, 80, Tex Instruments, 83-84, Hughes Aircraft, 83-84 & TRW, 85-; assoc dir, Image Processing Lab, Univ Kans, 80-83, dir, Telecommunications Lab, 83- *Mem:* Inst Elec & Electronics Engrs. *Res:* Image processing; general systems theory; modeling and analysis of communication systems. *Mailing Add:* Elec Eng Dept Rm 224 Nichols Hall Univ Kans Lawrence KS 66045

SHANNON, BARRY THOMAS, b Philadelphia, Pa, Nov 2, 52; m 76; c 3. CLINICAL IMMUNOLOGY, FLOW CYTOMETRY. *Educ:* Ursinus Col, BS, 74; Pa State Univ, MS, 76; Wake Forest Univ, PhD(immunol), 80. *Prof Exp:* Fel immunol, Med Univ SC, 80-82; DIR, CLIN IMMUNOL LAB & FLOW CYTOMETRY RESOURCE CTR, COLUMBUS CHILDREN'S HOSP, 82-; CLIN ASSOC PROF IMMUNOL, DEPT PEDIAT & PATH, OHIO STATE UNIV, 83- *Concurrent Pos:* Prin investr, Children's Hosp Res Found & Ohio State Univ Comprehensive Cancer Ctr, 89-90,; co-investr, Muscular Dystrophy Asn, 88-90, NIH, 87-92; ed, Lab Immunol Newslett, 85- *Mem:* Am Asn Immunologists; Am Soc Microbiologists; Soc Anal Cytol. *Res:* Immunoregulatory aspects in Langerhans Cell Histiocytosis (LCH); DNA analysis and oncogene expression prognostic indicator of pediatric tumors; characterization of vasoactive intestinal polypeptide receptor on leukemia lympho-blasts; molecular biology. *Mailing Add:* Dept Lab Med Children's Hosp 700 Children's Dr Columbus OH 43205

SHANNON, CLAUDE ELWOOD, b Gaylord, Mich, Apr 30, 16; m 49; c 3. APPLIED MATHEMATICS. *Educ:* Univ Mich, BS, 36; Mass Inst Technol, MS & PhD(math), 40. *Hon Degrees:* MSc, Yale Univ, 54; DSc, Univ Mich, 61, Pittsburgh Univ, Princeton Univ, Northwestern Univ, Univ Edinburgh, Oxford Univ, Carnegie-Mellon Univ, Tufts Univ, Univ Pa. *Prof Exp:* Asst elec eng & math, Mass Inst Technol, 36-39; Nat Res Coun fel, Princeton Univ, 40, res mathematician, Nat Defense Res Comt, 40-41; res mathematician, Bell Tel Labs, Inc, 41-57; Donner prof sci, 58-80, PROF ELEC ENG, MASS INST TECHNOL, 57-, EMER DONNER PROF SCI, 80- *Concurrent Pos:* Fel, Ctr Advan Study in Behav Sci, 57-58; dir, Teledyne, Inc, 60-86; vis fel, All Souls Col, Oxford Univ, 78. *Honors & Awards:* Noble Award, 40; Morris Liebmann Mem Award, 49; Stuart Ballantine Medal, 55; Ballantine Medal, 55; Vanuxem Lectr, Princeton Univ, 58; Steinmetz Lectr, Univ Schenectady, 62; Medal of Honor, Inst Elec & Electronics Engrs, 66; Nat Medal Sci, 66; Shannon Lectr, Inst Elec & Electronics Engrs, 73; Chichele Lectr, Oxford Univ, 78; John Fritz Medal, 83. *Mem:* Nat Acad Sci; Nat Acad Eng; fel Inst Elec & Electronice Engrs; Am Acad Arts & Sci; Sigma Xi; Am Math Soc; Am Philos Soc; Royal Irish Acad. *Res:* Boolean algebra and switching circuits; communication theory; mathematical cryptography; computing machines. *Mailing Add:* Mass Inst Technol Five Cambridge St Winchester MA 01890

SHANNON, FREDERICK DALE, b Akron, Ohio, June 10, 31; m 57; c 1. ORGANIC CHEMISTRY, POLYMER CHEMISTRY. *Educ:* Univ Akron, BS, 53, MS, 59, PhD(chem), 64. *Prof Exp:* Res chemist, Inst Rubber Res, Univ Akron, 53-54 & 56-58; instr, 58-60, assoc prof, 60-62 & 64-65, dir summer sch, 71-73, acad dean, 73-85, PROF CHEM, HOUGHTON COL, 65- *Mem:* Am Chem Soc; Am Sci Affil; Am Asn Higher Educ; Am Conf Acad Deans. *Res:* Cyclic dienes; cationic polymerization; molecular structure determination. *Mailing Add:* Chmn Div Sci & Math Houghton Col Houghton NY 14744

SHANNON, IRIS R, NURSING. *Educ:* Fisk Univ, BSN, 48; Univ Chicago, MA, 54; Univ Ill, PhD(pub policy anal), 87. *Prof Exp:* Staff nurse, Chicago Bd Health, 48-50; instr, pub health nursing, Meharry Col, 51-56; teacher nursing, Head Start Chicago, 57-66; dir nursing, Mile Square Neighborhood Health Ctr, 66-69; co-dir, Nursing Asn, Rush-Presby-St Luke's Hosp, 71-76; chmn dept & assoc prof, Rush-Presby-St Luke's Med Ctr, 66-90, ASSOC PROF, DEPT COMMUNITY HEALTH NURSING, RUSH UNIV, 90- *Concurrent Pos:* Mem, Equal Health Opportunity Comt, Am Pub Health Asn, 69-70, Comt Nursing, Am Hosp Asn, 77-78, Nat Adv Coun Nurse Training, HEW, 78-81 & Task Force Credentialing Nursing, Am Nurse's Asn, 80-83; adj fac, Sch Pub Health, Univ NC, 77-; hon fel, Roy Soc Health, Eng, 89. *Honors & Awards:* Hildrus A Poindexter Distinguished Serv Award, Am Pub Health Asn, 81. *Mem:* Inst Med-Nat Acad Sci; Am Nurses Asn; fel Am Pub Health Asn (pres, 88-89). *Res:* Community health; author of numerous technical publications. *Mailing Add:* Dept Community Nursing Rush Col Nursing 600 S Paulina St Chicago IL 60617

SHANNON, JACK CORUM, b Halls, Tenn, Feb 27, 35; m 55; c 2. PLANT PHYSIOLOGY. *Educ:* Univ Tenn, BS, 58; Univ Ill, MS, 59, PhD(plant physiol), 62. *Prof Exp:* Res asst hort, Univ Ill, 58-60, res asst plant physiol, 60-62; asst prof, Purdue Univ, 62-63; plant physiologist, Crop Res Div, Agr Res Serv, USDA, 63-71; from assoc prof to prof hort, 71-87, PROF PLANT PHYSIOL, PA STATE UNIV, 87- *Concurrent Pos:* Chair, Intercol Grad Prog, Plant Physiol, Pa State Univ, 83- *Mem:* Am Soc Plant Physiologists; Am Soc Agron; Crop Sci Soc Am; Sigma Xi; Am Soc Hort Sci; Nat Sweet Corn Breeders Asn. *Res:* Study of sugar movement into Zea mays L kernels and the utilization of these sugars in starch biosynthesis. *Mailing Add:* 1455 Park Hills Ave State College PA 16803

SHANNON, JACK DEE, b McAlester, Okla, Oct 18, 43; m 79. METEOROLOGY. *Educ:* Univ Okla, BS, 65, MS, 72, PhD(meteorol), 75. *Prof Exp:* Asst meteorologist, 76-80, METEOROLOGIST, ARGONNE NAT LAB, 80-, DEP MGR ATMOSPHERIC PHYSICS PROG, 81- *Concurrent Pos:* Mem, Interagency Task Force on Acid Precipitation, 84-90, Am Meteorol Soc Comt on Atmospheric Chem, 85-88, Environ Protection Agency, Am Meteorol Soc Joint Steering Comt on Air Qual Modeling, 86-90. *Mem:* Am Meteorol Soc; Royal Meteorol Soc; AAAS; Am Geophys Union. *Res:* Numerical modeling of regional acid deposition; objective placement of sensors. *Mailing Add:* ERD Bldg 203 Argonne Nat Lab Argonne IL 60439

SHANNON, JAMES AUGUSTINE, b Hollis, NY, Aug 9, 04; m 33; c 2. PHYSIOLOGY. *Educ:* Col Holy Cross, AB, 25; NY Univ, MD, 29, PhD(physiol), 35. *Hon Degrees:* DSc, Col Holy Cross, 52, Duke Univ, 58, Providence Col, 58, Loyola Univ, Ill, 59, Cath Univ Am, 60, WVa Univ, 60, Univ Md, 65, NY Univ, 65 & Jefferson Med Col, 65; LLD, Univ Notre Dame, 57; LHD, Yeshiva Univ, 62; MD, Cath Univ Louvain, 64 & Karolinska Inst, Sweden, 64; DPhil, Rockefeller Univ, 75; plus many others. *Prof Exp:* Intern, Bellevue Hosp, New York, 29-31; asst, dept physiol, Col Med, NY Univ, 31-32, from instr to asst prof, 32-40, from asst prof to assoc prof, dept med, 40-46; dir, Squibb Inst Med Res, 46-49; assoc dir res, Nat Heart Inst, 49-52, assoc dir res, NIH, 52-55, dir, 55-68; dir grade & asst surgeon gen, Pub Health Serv, HEW, 52-68; scholar in residence, Nat Acad Sci, 68-70 & Nat Libr Med, NIH, 75-80; prof & spec asst to pres, 70-75, adj prof, 75-80, EMER PROF BIOMED SCI, ROCKEFELLER UNIV, 80- *Concurrent Pos:* Clin asst vis physician, NY Univ Med Div, Bellevue Hosp, New York, 31-32, asst vis physician, 38-42, assoc vis neuropsychiatrist, Third Psychiat Serv, 44; vis scientist, Physiol Lab, Cambridge Univ, 35, guest scientist & instr physiol, Marine Biol Lab, Woods Hole, 36-40; ed, J Soc Exp Biol & Med, 40; dir res serv, Goldwater Mem Hosp, 41-46, actg dir, NY Univ Med Div, 44-45; mem, Atebrine Conf, Malaria Conf & Subcomt Coord Malaria Studies, Nat Res Coun, 41-43; consult to Secy War, 43-46; mem, Bd Coord Malaria Studies & chmn, Panel Clin Testing, Comt Med Res, Off Sci Res & Develop, 43-46; chmn, Malaria Study Sect, NIH, 45-47, mem, 47-49; spec consult to surgeon gen, USPHS, 46-49; mem, Subcomt Malaria, Comt Med, Nat Res Coun, 51-56, Subcomt Shock, Comt Surg, 52-56, chmn, Malaria Panel, 53-56, mem, Panel Allocation Gamma Globulin, 53-54, exec comt, Div Med Sci, 53-60 & pub health serv rep, Div, 55-68; mem bd dirs, Gorgas Mem Inst Trop & Prev Med, Panama Canal Zone, 54-68; mem, Nat Comt, Int Union Physiol Sci, 55-62; mem, Expert Adv Panel Malaria, WHO, 56-66, Adv Comt Med Res, 59-63; mem, standing comt, Fed Coun Sci & Technol, 59-64, Dept HEW rep, 65-68; consult, President's Sci Adv Comt, 59-65 & Adv Comt Res, AID, 63-68; mem, Adv Comt Med Res, Pan-Am Health Orgn, 62-66; US deleg, US-Japan Coop Med Sci Comt, 65-68; mem, Bd Med, Nat Acad Sci, 67-68 & 70-71, Coun & Exec Comt, 70-74 & Exec Comt, Inst Med, 71-72; consult biomed sci, Merck, Sharpe & Dohme, 68-76 & adv, R J Reynolds Industs, Inc, 78-85; mem, Adv Comt Med Sci, AMA, 69-70. *Honors & Awards:* Sci Award, NY Univ, 58; Mendel Medal, Villanova Univ, 61; Pub Welfare Medal, Nat Acad Sci, 62; Distinguished Serv Medal, USPHS, 66; John M Russell Award, Markle Found, 66; Abraham Flexner Award, Asn Am Med Cols, 66; Alan Gregg Lect, 66; Nat Medal for Sci, 75; Int Award, Nat Soc Res Admin, 75; plus other awards from nat groups. *Mem:* Nat Acad Sci; Am Acad Arts & Sci; AAAS; Am Physiol Soc; Soc Exp Biol & Med. *Res:* Pharmacology; clinical investigation in chemotherapy; secretory mechanisms of the renal tubule; renal control of water and electrolyte balance; functional aspects of renal disease; measurement of glomerular filtration. *Mailing Add:* Carman Oaks B 125 3800 SW Carman Dr Lake Oswego OR 97035-2533

SHANNON, JERRY A, JR, b Meshoppen, Pa, Mar 31, 24; m 54; c 4. SCIENCE EDUCATION, BIOLOGY. *Educ:* Pa State Teachers Col, Mansfield, BS, 48; Peabody Col, MA, 49; Cornell Univ, EdD, 57. *Prof Exp:* Supvr biol, Demonstration Sch, Vanderbilt Univ, 49-50; high sch teacher, Ind, 50-51 & NY, 51-52; instr biol & conserv, Wis State Col, Oshkosh, 53; from instr to asst prof elem sch sci, Iowa State Teachers Col, 53-56; assoc prof biol & conserv, 56-64, PROF SCI EDUC & BIOL, STATE UNIV NY, COL ONEONTA 64- *Concurrent Pos:* Grants, State Univ NY, plant taxon, Univ Northern Iowa, 61, NSF, Univ NC, 65, biol educ, Univ Rochester, 68, sci educ, Univ Colo, 72. *Mem:* Nat Asn Biol Teachers; Conserv Educ Asn; Nat Sci Teachers Asn. *Res:* Science and environmental education; local flora. *Mailing Add:* 82 West St Oneonta NY 13820

SHANNON, LARRY J(OSEPH), chemical engineering; deceased, see previous edition for last biography

SHANNON, LELAND MARION, b Fullerton, Calif, Oct 24, 27; m 49; c 5. PLANT BIOCHEMISTRY. *Educ:* Univ Calif, Los Angeles, BS, 49, MS, 51; Rutgers Univ, PhD(plant physiol), 54. *Prof Exp:* Res assoc, Rutgers Univ, 51-54; from instr to assoc prof biochem, Univ Calif, Los Angeles, 54-68; PROF BIOCHEM, UNIV CALIF, RIVERSIDE, 68- , DEAN GRAD DIV. *Mem:* Am Soc Plant Physiologists; Am Soc Biol Chemists; Brit Biochem Soc. *Res:* Glycoproteins; protein chemistry; lectins; plant cell-cell interaction. *Mailing Add:* Dean Grad Div 900 University Ave Univ Calif Riverside CA 92521

SHANNON, ROBERT DAY, b Highland Park, Mich, Aug 28, 35; m 66; c 2. SOLID STATE CHEMISTRY, INORGANIC CHEMISTRY. *Educ:* Univ Ill, Champaign, BS, 57, MS, 59; Univ Calif, Berkeley, PhD(ceramic eng), 64. *Prof Exp:* Res assoc, Univ Ill, 59-60; RES CERAMIST, E I DU PONT DE NEMOURS & CO, INC, 64- *Concurrent Pos:* Res assoc, McMaster Univ, 71-72; assoc prof, Univ Grenoble, 72-73; vis scientist, Nat Ctr Sci Res, Inst Res, Catalyse, Lyon, France, 81-82. *Mem:* Fel Mineral Soc Am; Am Chem Soc. *Res:* Dielectric properties of oxides; synthesis of inorganic oxides; crystallography. *Mailing Add:* E I du Pont de Nemours & Co Inc Exp Sta 356 Wilmington DE 19880-0356

SHANNON, ROBERT RENNIE, b Mt Vernon, NY, Oct 3, 32; m 54; c 6. OPTICAL ENGINEERING. *Educ:* Univ Rochester, BS, 54, MA, 57. *Prof Exp:* Staff physicist, Itek Corp, 59-62, dept mgr optical design, 62-67, dir advan tech labs, 67-69; PROF OPTICAL SCI, UNIV ARIZ, 69-, DIR, OPTICAL SCI, 83- *Concurrent Pos:* Topical ed, Optical Soc Am, 75-78. *Honors & Awards:* Goddard Medal, Soc Photo-Optical Instrument Engrs, 81. *Mem:* Fel Optical Soc Am (pres, 85); fel Soc Photo-Optical Instrument Engrs (pres, 79-80). *Res:* Applied optics, especially lens design, image analysis, optical testing, laser application, and application of computers to engineering problems; atmospheric optics; massive optics design and fabrication. *Mailing Add:* Optical Sci Ctr Univ Ariz Tucson AZ 85721

SHANNON, SPENCER SWEET, JR, b Philadelphia, Pa, Apr 15, 27; div; c 4. EXPLORATION GEOCHEMISTRY, ORE GENESIS. *Educ:* Amherst Col, AB, 49; Yale Univ, MSc, 50; Univ Idaho, PhD(geol), 64. *Prof Exp:* Mineral economist, US Bur Mines, Washington, DC, 53-55; geologist, AEC, Rawlins, Wyo, 55-57; mining engr, US Bur Mines, Denver, Colo, 57-61; asst prof geol, Univ Idaho, 63-65; sr geologist, Minerals Div, Phillips Petrol Co, Utah, 65-66; res mgr, Navarro Explor Co Ltd, Windhoek, Namibia, 66-67; asst prof geol, Weber State Col, 67-68; assoc prof, Univ Tex, El Paso, 68-75; staff geologist, US Energy Res & Develop Admin, 75-78; staff mem, Los Alamos Nat Lab, N Mex, 78-85; prof geol, Adams State Col, 85-87. *Concurrent Pos:* Consult to var corp clients. *Mem:* Fel Geol Soc Am; Am Inst Prof Geologists; Soc Econ Geologists; fel Geol Soc SAfrica; Asn Explor Geochemists; Sigma Xi; AAAS. *Res:* Instruction, research, and participation in systematic exploration for mineral deposits using geology, geochemistry, and geophysics; examination and evaluation of mines and prospects; relation of volcanism to genesis of ore deposits. *Mailing Add:* PO Box 1251 Santa Fe NM 87504-1251

SHANNON, STANTON, b Phoenix, Ariz, Dec 28, 28; m 57; c 2. SOIL CHEMISTRY. *Educ:* Univ Ariz, BS, 52, MS, 53; Univ Calif, Davis, PhD(plant physiol), 61. *Prof Exp:* Sr lab technician, Citrus Exp Sta, Dept Soils & Plant Nutrit, Univ Calif, Riverside, 56-57; res asst plant anal, Univ Calif, Davis, 57-61; from asst prof to assoc prof veg crops, NY State Col Agr, Cornell Univ, 61-85; CONSULT, ROGERS BROS SEED CO, 85- *Mem:* Am Soc Plant Physiologists; Am Soc Hort Sci. *Res:* Post harvest physiology of vegetable crops; plant biochemistry; mineral element nutrition; cultural practices for mechanically harvested vegetables, water relations and photosynthetic efficiency. *Mailing Add:* 1413 White Rd Phelps NY 14532

SHANNON, WILBURN ALLEN, JR, b Springfield, Tenn, June 21, 41; m 64. MEDICAL RESEARCH, CYTOCHEMISTRY. *Educ:* Mid Tenn State Univ, BS, 64; Vanderbilt Univ, MA, 67, PhD(biol), 70. *Prof Exp:* Res assoc path, Harvard Med Sch & Mass Gen Hosp, 70-71; res assoc, Sch Med, Johns Hopkins Univ & Sinai Hosp Baltimore, 71-75; asst prof cell biol, Southwestern Med Sch, 74-83; dir, Res Morphology & Cytochem Unit, Gen Med Res Serv, Vet Admin Hosp, 75-83; RETIRED. *Concurrent Pos:* USPHS fel, Harvard Med Sch & Mass Gen Hosp, Boston, 70-71; Nat Cancer Inst fel, Sch Med, Johns Hopkins Univ & Sinai Hosp Baltimore, 71- *Mem:* Am Soc Cell Biol; Electron Micros Soc Am; Histochem Soc. *Res:* Ultracytochemistry, especially development of methods and application to normal and pathological tissue studies; endocrinology; mitochondriology. *Mailing Add:* 2001 Bryan Tower 30th Fl Dallas TX 75201

SHANNON, WILLIAM MICHAEL, b Mt Holly, NJ, Oct 11, 40; m 64; c 3. MICROBIOLOGY. *Educ:* Univ Ala, Tuscaloosa, BS, 62; Loyola Univ, La, MS, 65; Tulane Univ, PhD(microbiol), 69. *Prof Exp:* Radioisotope res technician, Vet Admin Hosp, Birmingham, Ala, 62-63; res asst endocrinol, Med Units, Univ Tenn, Memphis, 63-64; res asst virol, Loyola Univ, La, 64-65; res asst, Sch Med, Tulane Univ, 64-66, NASA fel, 66-69; res virologist, 69-71, sr virologist, 72-74, HEAD MICROBIOL, VIROL DIV, SOUTHERN RES INST, 75- *Mem:* AAAS; Am Soc Microbiol; NY Acad Sci; Am Asn Cancer Res; Brit Soc Gen Microbiol; Am Soc Virol. *Res:* Biochemistry of virus-infected cells; antiviral chemotherapy; retroviruses; herpesviruses; exotic RNA viruses; virus-host cell interactions; molecular virology. *Mailing Add:* 2212 Lime Rock Rd Birmingham AL 35216

SHANOR, LELAND, b Butler, Pa, July 21, 14; m 40; c 2. MYCOLOGY. *Educ:* Maryville Col, AB, 35; Univ NC, MA, 37, PhD(bot), 39. *Hon Degrees:* DSc, Ill Wesleyan Univ, 61. *Prof Exp:* Asst biol, Maryville Col, 33-35; asst bot, Univ NC, 35-39; instr, Clemson Col & Univ Ill, 39-43; pathologist, USDA, 43-44; res mycologist, George Washington Univ, 44-45; res assoc, Canal Zone Lab, Univ Pa, 46; from asst prof to prof bot, Univ Ill, 46-56, cur mycol collections, 48-56; prof bot & head dept biol sci, Fla State Univ, 56-62; dean div advan studies, Fla Inst Continuing Univ Studies, 62-64; sect head div sci personnel & educ, NSF, 64, div dir undergrad educ in sci, 65; chmn dept, 65-73, PROF BOT, UNIV FLA, 65- *Concurrent Pos:* Dir, Highlands Mus Natural Hist, Univ NC, 39; Guggenheim fel, 51-52; mem adv comt, Canal Zone Biol Area, 54-; trustee, Highlands Biol Sta, Univ NC, 57-, pres, 58-63; trustee, Fairchild Trop Garden, 66- *Mem:* AAAS; Bot Soc Am; Mycol Soc Am (secy-treas, 50-53, pres, 54); Sigma Xi. *Res:* Culture of entomogenous fungi; cytology of Saprolegniaceae; distribution, morphology and taxonomy of aquatic fungi. *Mailing Add:* Dept Bot Univ Fla 3171 MCC Bldg Gainesville FL 32601

SHANSKY, MICHAEL STEVEN, b Milwaukee, Wis, May 3, 43; m 68; c 1. VISUAL PHYSIOLOGY, PSYCHOLOGY. *Educ:* Marquette Univ, BA, 66; Syracuse Univ, PhD(psychol), 74. *Prof Exp:* Asst prof physiol optics, 71-77, ASSOC PROF VISUAL SCI, ILL COL OPTOM, 77-, CHMN DEPT VISUAL SCI & DIR RES, 72- *Concurrent Pos:* Guest lectr, DePaul Univ & Northeastern Ill Univ, 73- *Mem:* AAAS; Asn Res Vision & Ophthal. *Res:* Neurophysiology of vision; information processing in the vertebrate visual system; visual perception; polysensory integration. *Mailing Add:* Dept Optom Ferris State Col 901 S State St Big Rapids MI 49307

SHANTARAM, RAJAGOPAL, b Poona, India, Mar 29, 39; m 69. MATHEMATICAL STATISTICS. *Educ:* Ferguson Col, India, BS, 59; Univ Poona, MS, 61; Pa State Univ, PhD(math), 66. *Prof Exp:* Asst prof math & statist, State Univ NY, Stony Brook, 66-71; asst prof, 71-74, ASSOC PROF MATH & STATIST, UNIV MICH, FLINT, 74- *Mem:* Am Math Soc; Math Asn Am; Am Statist Asn. *Res:* Characteristic functions; limit distributions; mathematical statistics; mathematical modeling; operations research. *Mailing Add:* Dept Comput Sci Univ Mich Flint MI 48502

SHANTEAU, ROBERT MARSHALL, b Los Angeles, Calif, June 24, 47; m 80. HIGHWAY RESEARCH. *Educ:* San Jose State Univ, BS, 70; Univ Calif, Berkeley, MS, 76, PhD(civil eng), 80. *Prof Exp:* asst prof civil eng, Purdue Univ, 80-85; res engr, Ind Dept Hwys, 85-90; ENG DEPT TRAFFIC & ENG, MONTEREY, CALIF, 90- *Mem:* Opers Res Soc Am; Inst Transp Engrs; Am Soc Civil Engrs. *Res:* Application of mathematical and physical principles to problems in highway engineering and pavement management. *Mailing Add:* City Hall Monterey CA 93940

SHANTHIKUMAR, JEYAVEERASINGAM GEORGE, b Sri Lanka, July 1, 50; m 77; c 3. PRODUCTION SYSTEMS, STOCHASTIC MODELS. *Educ:* Univ Sri Lanka, BSc, 72; Univ Toronto, MASc, 77, PhD(indust eng), 79. *Prof Exp:* Asst lectr prod & mech eng, Univ Sri Lanka, 73-75; teaching asst indust eng, Univ Toronto, 75-79; asst prof, Syracuse Univ, 79-82; assoc prof, Univ Ariz, 82-84; asst prof mgt sci, 84-88, PROF, MGT SCI, UNIV CALIF, BERKELEY, 88- *Concurrent Pos:* Consult, Syracuse Res Inc, 82. *Mem:* Opers Res Soc Am; Inst Mgt Sci. *Res:* Develop models to understand the behavior to obtain performance measures and to design production systems; develop efficient techniques to derive solutions to stochastic and deterministic models of production systems. *Mailing Add:* Sch Bus Admin Univ Calif Berkeley CA 94720

SHANZER, STEFAN, b Krakow, Poland, June 3, 24; m 52; c 2. ELECTROENCEPHALOGRAPHY, EYE MOVEMENTS & VISION. *Educ:* Bologna Univ, Italy, MD, 51. *Prof Exp:* DIR NEUROL, BETH ISRAEL MED CTR, 77-; PROF CLIN NEUROL, MOUNT SINAI SCH MED, 77- *Mem:* Am Physiol Soc; Am Neurol Soc; Am Electroencephalographic Soc; Am Acad Neurol; Am Epilepsy Soc; Asn Res Nervous & Ment Dis. *Res:* Oculomotor system; electroencephalography. *Mailing Add:* Beth Israel Med Ctr First Ave & 16th St New York NY 10003

SHAPERE, DUDLEY, b Harlingen, Tex, May 27, 28; m 75; c 2. PHILOSOPHY OF SCIENCE, HISTORY OF SCIENCE. *Educ:* Harvard Univ, BA, 49, MA, 55, PhD(philos), 57. *Prof Exp:* Instr philos, Ohio State Univ, 57-60; from asst prof to prof, Univ Chicago, 60-72; prof, Univ Ill, Urbana-Champaign, 72-75; prof philos, Univ Md, College Park, 75-84; Z SMITH REYNOLDS PROF PHILOS & HIST SCI, WAKE FOREST UNIV, 84- *Concurrent Pos:* Vis assoc prof, Rockefeller Univ, 65-66; consult, Comn Undergrad Educ Biol Sci, 65-71; spec consult & prog dir, Hist & Philos Sci Prog, NSF, 66-75; vis prof, Harvard Univ, 68; NSF res grants, 70-72, 73-75, 76-78, 85-86; mem, Inst Advan Study, Princeton, NJ, 78-79, 81. *Mem:* Am Philos Asn; Philos Sci Asn; Hist Sci Soc; AAAS; Am Psychol Asn. *Res:* Characteristics of explanation; the relations of theory, observation, and experiment and the nature of change and innovation in the development of science. *Mailing Add:* Drawer 7229 Reynolds Sta Wake Forest Univ Winston-Salem NC 27109

SHAPERO, DONALD CAMPBELL, b Detroit, Mich, Apr 17, 42; m 69, 85; c 2. SCIENCE POLICY. *Educ:* Mass Inst Technol, BS, 64, PhD(physics), 70. *Prof Exp:* Fel theoret physics, Thomas J Watson Res Ctr, IBM Corp, 70-72; asst prof, Am Univ, 72-73 & Cath Univ, 73-75; exec dir energy res adv bd, US Dept Energy, 78-79; US-USSR exchange scientist, 73, sr staff officer, 75-78, spec asst prog coord, 79-82, EXEC DIR BD PHYSICS & ASTRON, NAT ACAD SCI, 82- *Concurrent Pos:* Mem, Nat Organizing Comt, 19th Gen Assembly, Int Union Pure & Appl Physics, 85-88, 20th Gen Assembly, 87-93 & Nat Organizing Comt, 20th Gen Assembly, Int Astron Union, 86-88. *Mem:* Nat Acad Sci; Am Phys Soc; Am Geophys Union; Int Astron Union; Am Astron Soc. *Res:* Many-body theory; quantum field theory; science policy on physical sciences. *Mailing Add:* Nat Acad Sci 2101 Constitution Ave Washington DC 20418

SHAPIRA, RAYMOND, b New Bedford, Mass, June 29, 28; m 56; c 3. BIOCHEMISTRY, NEUROCHEMISTRY. *Educ:* Univ NMex, BS, 50; Fla State Univ, PhD(chem), 54. *Prof Exp:* Res biochemist, Biol Div, Oak Ridge Nat Lab, 54-57; from asst prof to assoc prof, 58-71, PROF BIOCHEM, EMORY UNIV, 72- *Concurrent Pos:* Fulbright lectr, Weizmann Inst Sci, 67. *Mem:* AAAS; Am Chem Soc; Fedn Am Socs Exp Biol; Am Soc Biol Chemists; Am Soc Neurochem; Sigma Xi. *Res:* Studies on the structure, metabolism and function of myelin proteins; induction of tolerance in experimental allergic encephalomyelitis; studies on myelin from patients with multiple sclerosis. *Mailing Add:* Dept Biochem Emory Univ Woodruff Med Ctr Atlanta GA 30322

SHAPIRA, YAACOV, b Haifa, Israel, Jan 9, 38; m 73. PHYSICS. *Educ:* Brandeis Univ, BA, 60; Mass Inst Technol, PhD(physics), 64. *Prof Exp:* Mem res staff solid state physics, Nat Magnet Lab, 64-85, SR SCIENTIST, PHYSICS DEPT, MASS INST TECHNOL, 85- *Mem:* Am Phys Soc; AAAS. *Res:* Ultrasonic behavior of solids; Fermi surface; high field superconductors; magnetic phase transitions; magnetic semiconductors. *Mailing Add:* Dept Physics Univ Tufts Medford MA 02115

SHAPIRO, ALAN ELIHU, b New York, NY, Jan 6, 42; m 72. HISTORY OF SCIENCE. *Educ:* Polytech Inst Brooklyn, BS, 62, MS, 65; Yale Univ, MPhil, 69, PhD(hist sci), 70. *Prof Exp:* Instr physics, St John's Univ, NY, 65-66; NATO fel, Cambridge Univ, 70-71; asst prof hist sci, Oberlin Col, 71-72; from asst prof to assoc prof, 72-84, PROF HIST SCI, UNIV MINN, MINNEAPOLIS, 84-, DIR, PROG HIST SCI & TECHNOL, 89- *Concurrent Pos:* Assoc ed, Centaurus; mem coun, Hist Sci Soc, 77-80; mem-at-large, Sect L, AAAS, 84-87, nominating comt, 91-94; Guggenheim fel,

85-86; pres, Midwest Junto Hist Sci, 91-92. *Mem:* fel AAAS; corresp mem Int Acad Hist Sci; Hist Sci Soc; Sigma Xi. *Res:* History of physical sciences from the sixteenth to nineteenth centuries; Newton; optics; mechanics. *Mailing Add:* Univ Minn Sch Physics 116 Church St SE Minneapolis MN 55455

SHAPIRO, ALVIN, b Jersey City, NJ, Apr 20, 30; m 71; c 4. NUCLEAR SCIENCE & ENGINEERING. *Educ:* Polytech Inst Brooklyn, BME, 51; Univ Cincinnati, MS, 62, PhD(physics), 68. *Prof Exp:* Engr, Allis-Chalmers Mfg Co, 55-58 & Gen Elec Co, 58-62; from instr to assoc prof, 62-78, PROF NUCLEAR ENG, UNIV CINCINNATI, 78- *Concurrent Pos:* Consult, Mound Lab, Monsanto Res Corp, 70- *Mem:* Am Nuclear Soc; Am Phys Soc; Sigma Xi. *Res:* Methods of neutron and photon transport; nuclear reactor physics and radiation shielding; neutron dosimetry. *Mailing Add:* Dept Chem & Nuclear Eng Univ Cincinnati Cincinnati OH 45221

SHAPIRO, ALVIN PHILIP, b Nashville, Tenn, Dec 28, 20; m 51; c 2. INTERNAL MEDICINE, MEDICAL EDUCATION. *Educ:* Cornell Univ, AB, 41, Long Island Col Med, MD, 44. *Prof Exp:* Res fel, Cincinnati Gen Hosp, 48-49; instr internal med, Col Med, Univ Cincinnati, 49-51; asst prof, Univ Tex Southwestern Med Sch Dallas, 51-56; from asst prof to assoc prof, 56-60, assoc dean acad affairs, 71-75, vchmn med, 75-79, actg chmn, 77-79, PROF MED, SCH MED, UNIV PITTSBURGH, 67- *Concurrent Pos:* Asst clinician, Cincinnati Gen Hosp, 48-49, clinician & asst attend physician, 49-51; Fulbright vis prof, Univ Utrecht, 68; mem, Coun for High Blood Pressure, Am Heart Asn; attend physician, Parkland Mem & Vet Admin Hosps, Dallas, Tex, 51-56 & Presby Hosp, Vet Admin Hosp, Pittsburgh, 56-; attending physician, Shadyside Hosp, 86-, dir, Intern Med Training Prog & assoc chief med. *Honors & Awards:* Co-recipient, Albert Lasker Spec Pub Health Award, 80. *Mem:* Fel AAAS; Am Psychosom Soc (pres, 75); Acad Behav Med Res (pres, 88); Am Heart Asn; Am Soc Clin Invest. *Res:* Hypertension; experimental hypertension; psychosomatic disorders. *Mailing Add:* Dept Med Shadyside Hospital Pittsburgh PA 15232

SHAPIRO, ANATOLE MORRIS, b Syracuse, NY, June 28, 23; m 45; c 2. PHYSICS. *Educ:* Univ Buffalo, BA, 44; Cornell Univ, PhD(physics), 52. *Prof Exp:* Assoc physicist, Brookhaven Nat Lab, 52-54; res fel & lectr, Harvard Univ, 54-56, asst prof physics, 56-58; assoc prof, 58-65, PROF PHYSICS, BROWN UNIV, 65- *Mem:* Am Asn Physics Teachers; fel Am Phys Soc; Italian Phys Soc. *Res:* Nuclear physics; high energy particle interactions; proton-proton and pion-proton elastic and inelastic cross sections; production and decay of K-mesons and hyperons; hybrid systems for interactions above 100 GeV/c. *Mailing Add:* 10727 Gloxinia Dr Rockville MD 20852

SHAPIRO, ARTHUR MAURICE, b Baltimore, Md, Jan 6, 46; m 69; c 2. POPULATION BIOLOGY, BIOGEOGRAPHY. *Educ:* Univ Pa, BA, 66; Cornell Univ, PhD(entom), 70. *Prof Exp:* Asst prof biol, Richmond Col, NY, 70-72; from asst prof to assoc prof, 72-80, PROF & VCHMN ZOOL, UNIV CALIF, DAVIS, 81- *Mem:* Fel AAAS; Ecol Soc Am; Soc Study Evolution; fel Royal Entom Soc London; fel Explorers Club. *Res:* Genetics and ecology of colonizing species; adaptive strategies of weedy insects and plants; coevolution of insect-plant relationships; historical biogeography of the Andean region; phenology; biogeography and systematics of Pieridae and Hesperiidae. *Mailing Add:* Dept Zool Univ Calif Davis CA 95616

SHAPIRO, ASCHER H(ERMAN), b Brooklyn, NY, May 20, 16; m 85; c 3. FLUID DYNAMICS, BIOMEDICAL ENGINEERING. *Educ:* Mass Inst Technol, SB, 38, ScD(mech eng), 46. *Hon Degrees:* DSc, Univ Salford, Eng, 78, Israel, Inst Technol, 85. *Prof Exp:* Asst mech eng, Mass Inst Technol, 39-40, from instr to prof, 40-62, Ford prof eng, 62-75, chmn fac, 64-65, head dept mech eng, 65-74, inst prof, 75-86, EMER INST PROF, MASS INST TECHNOL, 86- *Concurrent Pos:* Consult mech engr, 38-; vis prof, Cambridge Univ, 55-56; Akroyd Stuart Mem lectr, Nottingham Univ, 56; mem subcomt, Nat Adv Comt Aeronaut, USAF; tech adv panel aeronaut, Off Secy Defense; chmn, Nat Comt Fluid Mech Films, 62-65 & 71-, mem, 65-; mem sci adv bd, USAF, 64-66; mem sci & pub policy comt, Nat Acad Sci, 73-77. *Honors & Awards:* Navy Ord Develop Award, 45; Richards Mem Award, 60 & Worcester Reed Warner Medal, 65, Am Soc Mech Engrs; Lamme Award, Am Soc Eng Educ, 77; Fluids Eng Award, Am Soc Mech Engrs, 81. *Mem:* Nat Acad Sci; Nat Acad Eng; fel Am Acad Arts & Sci; fel Am Soc Mech Engrs; fel Am Inst Aeronaut & Astronaut. *Res:* Fluid mechanics; supersonic flow of gases; gas turbine power plant; jet and rocket propulsion; dynamics and thermodynamics of compressible fluid flow; biomedical engineering; cardiovascular function; pulmonary function; centrifuges. *Mailing Add:* Dept Mech Eng Mass Inst Technol Cambridge MA 02139

SHAPIRO, BENNETT MICHAELS, b Philadelphia, Pa, July 14, 39; m 82; c 3. BIOCHEMISTRY. *Educ:* Dickinson Col, BS, 60; Jefferson Med Col, MD, 64. *Prof Exp:* Intern med, Hosp, Univ Pa, 64-65; res assoc, Lab Biochem, Nat Heart Inst, 65-68; vis scientist, Pasteur Inst, Paris, 68-69; chief sect cellular differentiation, Nat Heart & Lung Inst, 69-70; from assoc prof to prof biochem, 78-90, chmn dept, 85-90, EXEC VPRES, WORLDWIDE BASIC RES, MERCK SHARP & DOHME RES LABS, 90- *Concurrent Pos:* Guggenheim fel; Cell Biol Study Sect NIH, 80-83; bd of sci counr, NIH, 87- *Mem:* AAAS; Am Chem Soc; Soc Develop Biol; Am Soc Cell Biol; Am Soc Biol Chemists. *Res:* Regulation of cell behavior; ionic and enzymic regulation; role of the cell surface in fertilization and development. *Mailing Add:* Merck Sharp & Dohme Res Labs PO Box 2000 Rahway NJ 07065-0900

SHAPIRO, BERNARD, b Philadelphia, Pa, May 22, 25; m 49; c 3. NUCLEAR MEDICINE. *Educ:* Univ Pa, MD, 51; Am Bd Nuclear Med, dipl, 72. *Prof Exp:* From house officer to resident internal med, Hosp, Univ Pa, 51-53; Am Cancer Soc fel, Norweg Hydro Inst, Oslo, 53-54; Southeast Pa Heart Asn fel, Univ Pa, 54-55, Am Heart Asn fel, 55-56; head radioisotope lab, South Div, Albert Einstein Med Ctr, Philadelphia, 56-66, head radiation res lab, 56-90, head dept nuclear med, 66-90; CLIN PROF RADIOL, SCH MED, TEMPLE UNIV, 78- *Concurrent Pos:* Clin assoc prof radiol, Sch Med, Temple Univ, 67-78. *Mem:* AMA; Radiation Res Soc; Soc Nuclear Med; NY Acad Sci; Sigma Xi; fel Am Col Nuclear Physicians. *Res:* Clinical radioisotopes; radiation research. *Mailing Add:* A Einstein Med Ctr York & Tabor Rds Philadelphia PA 19141

SHAPIRO, BERNARD LYON, b Montreal, Que, June 16, 32; US citizen; m 60; c 2. ORGANIC CHEMISTRY, NUCLEAR MAGNETIC RESONANCE. *Educ:* McGill Univ, BSc, 52; Harvard Univ, AM, 54, PhD(org chem), 57. *Prof Exp:* Res fel chem, Harvard Univ, 56-58; res fel, Mellon Inst, 58-64; assoc prof, Ill Inst Technol, 64-68; admin officer, 68-71, PROF CHEM, TEX A&M UNIV, 68- *Concurrent Pos:* Ed, Ill Inst Technol Nuclear Magnetic Resonance Newslett, 58- & Nuclear Magnetic Resonance Abstracts Serv, Preston Tech Abstracts Co, 64-; consult, Res Div, W R Grace & Co, 64- & Shell Develop Co, 66. *Mem:* AAAS; Am Chem Soc; Am Phys Soc. *Res:* Organic chemistry; especially stereochemistry of haloketones; aliphatic fluorine compounds; organophosphorus compounds; nature of coupling constants, especially involving fluorine and phosphorous. *Mailing Add:* 966 Elsinore Ct Palo Alto CA 94303-3410

SHAPIRO, BERT IRWIN, b New York, NY, Jan 14, 41; m 62; c 2. NEUROPHYSIOLOGY, BIOPHYSICS. *Educ:* Swarthmore Col, BA, 62; Harvard Univ, MA, 65, PhD(biol), 67. *Prof Exp:* Instr biol & gen educ, 67-68, from asst prof to assoc prof biol, Harvard Univ, 68-76; MEM STAFF, NAT INST GEN MED SCI, NIH, 76- *Concurrent Pos:* Mem, Bermuda Biol Sta. *Mem:* Am Soc Cell Biol. *Res:* Neurophysiology, particularly of conduction mechanisms; action of pharmacological compounds, especially toxins; purification and structure of protein toxins, especially from invertebrates. *Mailing Add:* CMBD Prog Nat Inst Gen Med Sci Westwood Bldg Rm 903 Bethesda MD 20892

SHAPIRO, BURTON LEONARD, b New York, NY, Mar 29, 34; m 58; c 3. ORAL BIOLOGY, GENETICS. *Educ:* NY Univ, DDS, 58; Univ Minn, MS, 62, PhD(genetics), 66. *Prof Exp:* Teaching asst, Univ Minn, 61-62, instr, 62-66, assoc prof, 66-70, chmn, Dept Oral Biol, 68, dir grad studies, 71-75, prof, Ctr Humanistic Studies, Col Liberal Arts, 85-88, PROF ORAL PATH, SCH DENT & PROF, DEPT LAB MED & PATH, SCH MED, UNIV MINN, MINNEAPOLIS, 85- *Concurrent Pos:* Consult, lectr & cytologist, Minn Oral Cancer Detection, 63-66 & Wyo State Bd Health, 66-70; geneticist & oral pathologist, Cleft Palate-Maxillofacial Clin, Univ Minn, 65-69; mem, Grad Fac Dent & Genetics, oral biol Grad Sch, Univ Minn, Minneapolis, 66-, mem med staff, Univ Minn Hosp, 67-; chmn health sci policy and rev coun, 74-79; exec comt grad sch, 74-79; mem bd dirs, Minn Chapter Cystic Fibrosis Found, 75-83; spec vis prof, Japanese Min Educ, Sci & Cult, 83; vis scientist, Dept Biol, Univ Miami, 85, 86. *Mem:* Fel AAAS; Am Dent Asn; fel Am Acad Oral Path; Int Asn Dent Res; Am Soc Human Genetics; fel Am Assoc Oral Path; adv fel Am Cancer Soc. *Res:* Down syndrome; cystic fibrosis; genetics and oral disease; cleft palate microforms in American Indians; temporomandibular joint disorders; introduced concept of amplified developmental instability in down syndrome to explain its pathogenesis; ongoing tissue culture studies to determine basic defect in cystic fibrosis: discovered mitochondrial lesion and calcium abnormality. *Mailing Add:* Dept Oral Biol Univ Minn Minneapolis MN 55455

SHAPIRO, CAREN KNIGHT, b Berkeley, Calif, Apr 19, 45; m 72. MEDICAL MICROBIOLOGY, IMMUNOBIOLOGY. *Educ:* Univ Calif, Davis, AB, 67; Univ Wis, MS, 71, PhD(med microbiol), 72. *Prof Exp:* Res assoc, Ind Univ, 73-76; res affil, Roswell Park Mem Inst, 77; asst prof biol, 77-82, ASSOC PROF BIOL, D'YOUVILLE COL, NY, 82- *Mem:* Sigma Xi; NY Acad Sci; Am Soc Microbiol; AAAS; Asn Women Sci. *Mailing Add:* 142 Viscount Dr Williamsville NY 14221-1770

SHAPIRO, CHARLES SAUL, b Brooklyn, NY, June 16, 36; m 56; c 3. PHYSICS. *Educ:* Brooklyn Col, BS, 57; Syracuse Univ, MS, 60, PhD(physics), 65. *Prof Exp:* Assoc physicist, Int Bus Mach Corp, 57-59; teaching asst physics, Syracuse Univ, 59-60; res assoc reactor physics, Brookhaven Nat Lab, 61-64; staff mem physics, Int Bus Mach Corp, 64-67; from asst prof to assoc prof, 67-75, PROF PHYSICS, SAN FRANCISCO STATE UNIV, 75- *Concurrent Pos:* Consult, Lawrence Livermore Nat Lab, 84-; guest researcher, Stockholm Int Peace Res Inst, 73-74. *Mem:* Am Phys Soc; Am Nuclear Soc. *Res:* Radiation transport; shielding and dosimetry; reactor physics; environmental consequences of nuclear war; biogeochemical pathways of radionuclides; relations of science and humanities; science and society; science education; arms control and disarmament. *Mailing Add:* Dept Physics San Francisco State Univ 1600 Holloway Ave San Francisco CA 94132

SHAPIRO, DAVID, b New York, NY, July 20, 24; m 51. BEHAVIORAL MEDICINE, STRESS PSYCHOPHYSIOLOGY. *Educ:* Univ Ill, AB, 48; Univ Mich, AM, 50, PhD(psychol), 53. *Prof Exp:* Lectr, Dept Social Relations, 53-55, res assoc, Dept Psychiat, 56-63, assoc in psychol, 63-66, asst prof psychiat, 66-69, assoc prof psychol, 69-71, sr assoc psychiat, Harvard Med Sch, 71-74; PROF, DEPT PSYCHIAT, UNIV CALIF, LOS ANGELES, 74-, PROF, DEPT PSYCHOL, 75- *Concurrent Pos:* Consult, Nat Res Coun, 55, Dept Psychiat, Boston Univ, 56-60, Vet Admin, Dept Med Surg, 73-, John Wiley & Sons, 73, Ohio State Univ Pres, 73, Acad Press, 78-, New Eng Jour Med, 80-, NATO Sci Proj, 84-; prin investr, Off Naval Res Contract, 71-75; prin investr, Nat Heart, Lung & Blood Inst Res Grants, 74-; ed, Psychophysiol, 78-87. *Honors & Awards:* Res Recognition Award, Biofeedback Soc Am, 88; Distinguished Contrib Psychophysiol, Soc Psychophysiol Res, 88. *Mem:* Soc Psychophysiol Res (pres 75-76); fel Am Psychol Asn; fel AAAS; fel Soc Behav Med; Am Psychosom Soc; Acad Behav Med Res. *Res:* Human psychophysiological research on stress effects on autonomic response systems; role of behavior in hypertension and other cardiovascular disorders; published numerous articles in various publications. *Mailing Add:* Dept Psychiat Univ Calif 760 Westwood Plaza Los Angeles CA 90024

SHAPIRO, DAVID JORDON, b Brooklyn, NY, Apr 13, 46. BIOCHEMISTRY. *Educ:* Brooklyn Col, BS, 67; Purdue Univ, PhD(biochem), 72. *Prof Exp:* Helen Hay Whitney Found fel biochem, Med Sch, Stanford Univ, 72-73; biol, 73-74; asst prof, 74-78, ASSOC PROF BIOCHEM, UNIV ILL, URBANA, 78-, PROF. *Mem:* Am Chem Soc; Sigma Xi; Am Soc Biol Chemists; AAAS. *Res:* Control of gene expression and messenger RNA synthesis in animal cells; nucleic acid hybridization and use of immunologic techniques to isolate specific messenger RNAs in hormone dependent development. *Mailing Add:* Dept Biochem Univ Ill 1209 W California Urbana IL 61801

SHAPIRO, DAVID M, b New York, NY, June 9, 29; m 50; c 3. BIOCHEMISTRY. *Educ:* Queens Col, NY, BS, 51; Johns Hopkins Univ, PhD(biochem), 61. *Prof Exp:* Chemist, Sun Chem Co, 51-54; instr chem, Brooklyn Col, 54-55; chemist, Nopco Chem Co, 55-56; res assoc biol chem, Univ Mich, 61-64; asst prof biochem, Woman's Med Col Pa, 64-69; ASST PROF BIOCHEM, UNIV TEX HEALTH SCI CTR SAN ANTONIO, 69-, DIR STUDENT SERV, 73- *Concurrent Pos:* Fel, Inst Sci & Technol, Univ Mich, 61-62; NIH fel, 62-63. *Mem:* AAAS. *Res:* Photosynthesis and electron transport; protein synthesis in bacteriophage-infected systems. *Mailing Add:* Exec Dir Student Servs Univ Tex Health Sci Ctr 7703 Floyd Curl Dr San Antonio TX 78284-7701

SHAPIRO, DONALD M, b Pittsburgh, Pa, Nov 15, 35; m 56; c 3. APPLIED MATHEMATICS, COMPUTER SCIENCE. *Educ:* Univ Pittsburgh, BS, 56; Wash Univ, ScD(appl math), 66. *Prof Exp:* Reactor physicist, United Aircraft Corp, 56 & Internuclear Corp, 57; res assoc appl math, Washington Univ, 63-64; asst prof psychiat, Univ Mo, 64-66; asst prof appl math, Washington Univ, 66-67; PROF BIOSTATIST, NY MED COL, 67- *Mem:* AAAS; Asn Comput Mach; Soc Indust & Appl Math; Opers Res Soc Am; Data Processing Mgt Asn. *Res:* Use of computers and statistics in both research and hospital administration. *Mailing Add:* Dept Psychiat & Behavior Sci NY Med Col PO Box 31 Valhalla NY 10595

SHAPIRO, DOUGLAS YORK, b Houston, Tex, July 25, 41; m 75; c 1. BEHAVIORAL ECOLOGY. *Educ:* Harvard Univ, BA, 64; Case Western Reserve Univ, MD, 68; Cambridge Univ, PhD(animal behav), 77. *Prof Exp:* Med intern, New York Hosp & Cornell Med Ctr, 68-69; res assoc neurol, Nat Inst Neurol Dis & Stroke, NIH, 69-71; clin asst psychiat, W Suffolk Hosp, Eng, 73-75; from asst prof to assoc prof, 77-87, PROF ANIMAL BEHAV, DEPT MARINE SCI, UNIV PR, 87- *Concurrent Pos:* Vis prof pharmacol, Ponce Med Sch, PR, 78-80. *Mem:* Animal Behav Soc; Am Soc Ichthyologists & Herpetologists; Ecol Soc Am. *Res:* Behavioral aspects of socially controlled female-to-male sex reversal in coral reef fish; formation and development of fish social groups; ecological determinants of group size and spatial dispersion. *Mailing Add:* Dept Marine Sci Univ PR Mayaguez PR 00709

SHAPIRO, EDWARD K EDIK, b Kerch, Russia; US citizen. THERMAL PHYSICS, INORGANIC CHEMISTRY. *Educ:* Inst Mech Eng, Leningrad, MS, 66; Ioffe Inst Physics & Technol, Leningrad, PhD(solid state physics & mat sci), 73. *Prof Exp:* PRES, SEM TEK, 85-; SR ENGR, RAYTHEON CO, 89- *Concurrent Pos:* Lectr, Babson Col, Mass, 86-90 & Northeastern Univ, 87-88. *Mem:* Am Vacuum Soc; Mat Res Soc; Am Asn Crystal Growth; Am Inst Physics. *Res:* Technology development for microelectronics and optical application; direct energy conversion; display technology; reliability issues. *Mailing Add:* Sem Tek 11 Marshall Rd Lexington MA 02173

SHAPIRO, EDWIN SEYMOUR, b Los Angeles, Calif, June 20, 28; m 62; c 2. MATHEMATICS. *Educ:* Univ Calif, MA, 51; Univ Pittsburgh, PhD, 62. *Prof Exp:* Mathematician, US Naval Radiol Defense Lab, Calif, 51-69; prof, 69-78, PROF QUANT METHODS, UNIV SAN FRANCISCO, 78- *Mem:* Am Math Soc; Opers Res Soc Am; Math Asn Am. *Res:* Military operations; applied mathematics. *Mailing Add:* 3051 Atwater Dr Burlingame CA 94010

SHAPIRO, EUGENE, b New York, NY, Dec 26, 44; m 69. METALLURGICAL & MATERIALS ENGINEERING. *Educ:* Polytech Inst Brooklyn, BS, 65; Drexel Inst, MS, 67, PhD(mat eng), 69. *Prof Exp:* Res engr, 69-76, GROUP SUPVR, 76-, DIR, METAL RES LABS, OLIN CORP, 87- *Mem:* Am Soc Metals; Am Inst Mining Metall & Petrol; Am Soc Testing & Mat. *Res:* Fracture and ductility; hot working; formability; alloy development. *Mailing Add:* Metals Res Lab Olin Corp 91 Shelton Ave New Haven CT 06511

SHAPIRO, GILBERT, b Philadelphia, Pa, Mar 17, 34; m 58; c 3. PARTICLE PHYSICS. *Educ:* Univ Pa, BA, 55; Columbia Univ, MA, 57, PhD(physics), 59. *Prof Exp:* Res assoc physics, Nevis Cyclotron Lab, Columbia Univ, 59-61; physicist, Lawrence Radiation Lab, 61-63, from asst prof to assoc prof, 63-79, PROF PHYSICS, UNIV CALIF, BERKELEY, 79- *Concurrent Pos:* NSF sr fel, Saclay Nuclear Res Ctr, France, 65-66. *Mem:* Am Phys Soc. *Res:* Particle and muon physics; polarized proton targets. *Mailing Add:* Dept Physics Univ Calif Berkeley CA 94720

SHAPIRO, HARRY LIONEL, biological anthropology; deceased, see previous edition for last biography

SHAPIRO, HERBERT, b New York, NY, Nov 21, 07; m 46; c 2. BIOPHYSICS, NEUROPHYSIOLOGY. *Educ:* Columbia Univ, BS, 29, MA, 30; Princeton Univ, PhD(physiol), 34. *Prof Exp:* Med res scientist, Philadelphia State Hosp, 68-72; asst prof, Univ Pa; instr, Vassar Col & Hahnemann Med Col; CONSULT, 88- *Concurrent Pos:* Prin Investr, NIH grant, Univ Pa & Einstein Med Ctr; mem Radiation Lab, Mass Inst Technol; Guggenheim fel, Univ London; Nat Res Coun fel; mem, Marine Biol Lab, Woods Hole, Mass. *Mem:* Am Physiol Soc; fel AAAS. *Res:* Kinetics of cell division in marine eggs; bacterial luminescence; dynamic constant of molecular contraction; artificial parthenogenesis in rabbit egg; effect of temperature in man; nerve conduction; electrophysiology; tension at the surface of living cells; centrifuge microscope; delay lines in radar; contribution of papers to scientific journals. *Mailing Add:* 6025 N 13th St Philadelphia PA 19141

SHAPIRO, HERMAN SIMON, b New York, NY, Aug 29, 29. BIOCHEMISTRY, GENETICS. *Educ:* City Col New York, BS, 51; Columbia Univ, PhD(biochem), 57. *Prof Exp:* Asst biochem, Col Physicians & Surgeons, Columbia Univ, 51-52, res biochemist, 56-57, from res assoc to asst prof biochem, 58-69; ASSOC PROF BIOCHEM, COL MED NJ, 69- *Mem:* Brit Biochem Soc. *Res:* Elucidation of nucleotide sequences of the DNA of diverse cellular sources, especially the correlation of structure and function of these macromolecules. *Mailing Add:* Dept Biochem NJ Med Sch 100 Bergen St Newark NJ 07103

SHAPIRO, HOWARD MAURICE, b Brooklyn, NY, Nov 8, 41; m 64; c 2. MEDICINE, BIOENGINEERING. *Educ:* Harvard Univ, BA, 61; NY Univ, MD, 65. *Prof Exp:* Asst res scientist, NY Univ, 62-65; from intern to asst resident surg, Bellevue Hosp, 65-67; res assoc math statist & appl math, Biomet Br, Nat Cancer Inst, 67-70; sr staff fel, Baltimore Cancer Res Ctr, 70-71; resident surg, Col Med, Univ Ariz, 71-72; from asst dir med systs & instrumentation to dir clin res diag prod, G D Searle & Co, 72-76; asst to chmn, Cancer & Leukemia Group B, 76-77, chief cytokinetics, Sidney Farber Cancer Inst, 77-82, RES ASSOC, BETH ISRAEL HOSP, 82-, INVESTR, CTR BLOOD RES, 83-; PRES, HOWARD M SHAPIRO, MD, PC, 76- *Concurrent Pos:* Lectr path, Harvard Med Sch, 76-; vis prof path, Rush Med Col, 85-89. *Mem:* AAAS; Am Asn Cancer Res; Am Soc Clin Oncol; Am Soc Hemat; Inst Elec & Electronics Engrs. *Res:* Biomedical instrumentation and computing; oncology and hematology; analytical and theoretical biology; medical education and communication. *Mailing Add:* 283 Highland Ave West Newton MA 02165

SHAPIRO, IRVING, b Newark, NJ, Mar 3, 32; m 60; c 3. AUDIOLOGY. *Educ:* Western Mich Univ, BS, 58; Syracuse Univ, AM, 60; Western Reserve Univ, PhD(clin audiol), 64. *Prof Exp:* Instr otolaryngol, Med Sch, Univ Chicago, 64-66; audiologist, Vet Admin West Side Hosp, Chicago, 66-67; DIR COMMUN DISORDERS, HARBOR GEN HOSP, 67-; ASST PROF OTOLARYNGOL, MED SCH, UNIV CALIF, LOS ANGELES, 69-, LECTR HEAD & NECK SURG, 76- *Mem:* Am Speech & Hearing Asn; Am Audiol Soc; Int Soc Audiol; Sigma Xi. *Res:* Pediatric audiology; ototoxic drugs; systems for hearing aid evaluation. *Mailing Add:* 47 Juniper Lane W Hartford CT 06117

SHAPIRO, IRVING MEYER, b London, Eng, Oct 28, 37; m 65; c 1. BIOCHEMISTRY. *Educ:* London Hosp Med Col, LDSRCS BDS, 61; Univ Liverpool, MSc, 64; Univ London, PhD(biochem), 68. *Prof Exp:* Vis scientist biochem, Forsyth Dent Ctr, Boston, 64-66; from asst prof to assoc prof, 69-76, actg chmn dept, 75-80, PROF BIOCHEM, SCH DENT MED, UNIV PA, 76-, CHMN DEPT, 80- *Concurrent Pos:* Consult biochem, Koch Light Chem, Ltd, Eng, 67-70; res consult, Vet Admin Hosp, Philadelphia, 76- *Honors & Awards:* Basic Sci Award, Int Asn Dent Res, 74. *Mem:* AAAS; Biochem Soc; Royal Soc Med; Bone & Tooth Soc; Am Chem Soc. *Res:* Role of mitochondria in the initiation of the mineralization process; use of mineralized tissues in the diagnosis and treatment of lead poisoning. *Mailing Add:* Levy Oral Health 40-10 Locust St Philadelphia PA 19104-6002

SHAPIRO, IRWIN IRA, b New York, NY, Oct 10, 29; m 59; c 2. PHYSICS, ASTRONOMY. *Educ:* Cornell Univ, BA, 50; Harvard Univ, MS, 51, PhD(physics), 55. *Prof Exp:* Staff mem physics & astron, Lincoln Lab, Mass Inst Technol, 54-70, prof geophys & physics, 67-85; PROF ASTRON & PHYSICS, HARVARD UNIV, 82-, DIR, HARVARD-SMITHSONIAN CTR ASTROPHYS, 83-; SR SCIENTIST, SMITHSONIAN INST, 82- *Honors & Awards:* Michelson Medal, Franklin Inst, 75; B A Gould Prize, Nat Acad Sci, 79; Math & Phys Sci Award, NY Acad Sci, 82; Dannie Heineman Prize, Am Astron Soc, 83, Dirk Brouwer Award, 87. *Mem:* Nat Acad Sci; fel Am Phys Soc; fel Am Geophys Union; Am Astron Soc; Int Astron Union; Am Acad Arts & Sci. *Res:* Dynamical, radar and radio astronomy; experimental general relativity. *Mailing Add:* 17 Lantern Lane Lexington MA 02173

SHAPIRO, IRWIN LOUIS, b New York, NY, Jan 2, 32; m 63; c 2. BIOCHEMISTRY. *Educ:* Univ Iowa, BA, 53; Univ Del, MS, 59, PhD(chem), 62. *Prof Exp:* Fel biochem, Wistar Inst, Univ Pa, 61-63, assoc, 63-69; sr biochemist, J T Baker Chem Co, 69-73; ASST PROF CHEM, MONMOUTH COL, 74- *Concurrent Pos:* Instr, Sch Dent, Univ Pa, 63-65, asst prof, Sch Vet Med, 66-69; lectr, Philadelphia Col Pharm, 64-65. *Mem:* AAAS; Am Inst Nutrit; Am Asn Clin Chemists; Am Chem Soc; Am Oil Chemists Soc; Sigma Xi. *Res:* Triglyceride metabolism; cholesterol metabolism; nutritional biochemistry; inborn errors of metabolism; isotope applications in medicine; isotope instrumentation; clinical biochemistry. *Mailing Add:* Ten Dickson Rd Marlboro NJ 07746

SHAPIRO, JACK SOL, b Brooklyn, NY, Nov 3, 41; m 70; c 2. MATHEMATICS. *Educ:* Brooklyn Col, BS, 63; Yeshiva Univ, MA, 66, PhD(math), 70. *Prof Exp:* Asst prof, 70-78, ASSOC PROF MATH, BARUCH COL, CITY UNIV NEW YORK, 78- *Mem:* Am Math Soc; Math Asn Am. *Res:* Functional analysis and operator theory. *Mailing Add:* 1824 Ave S Brooklyn NY 11229

SHAPIRO, JACOB, b New York, NY, Sept 4, 25; m 48; c 2. BIOPHYSICS, RADIOLOGICAL HEALTH. *Educ:* City Col New York, BS, 44; Brown Univ, MS, 48; Univ Rochester, PhD(biophys), 54. *Prof Exp:* Instr physics, Univ RI, 46-47; tech adv nuclear energy, Opers Off, AEC, NY, 48-50; res assoc radiation biol, Univ Rochester, 53-55; supvr radiation anal, Elec Boat Div, Gen Dynamics Corp, 55-61; LECTR BIOPHYS IN ENVIRON HEALTH, SCH PUB HEALTH & UNIV HEALTH PHYSICIST, HARVARD UNIV, 61- *Mem:* Am Phys Soc; fel Health Physics Soc. *Res:* Radiation shielding, detection and dosimetry; health physics. *Mailing Add:* Harvard Univ Environ Heatlh 46 Oxford St Cambridge MA 02138

SHAPIRO, JAMES ALAN, b Chicago, Ill, May 18, 43; m 64; c 2. MICROBIOLOGY, MOLECULAR GENETICS. *Educ:* Harvard Col, BA, 64; Cambridge Univ, PhD(genetics), 68. *Prof Exp:* Fel, Pasteur Inst, Paris, 67-68 & Harvard Med Sch, 68-70; invited prof genetics, Univ Havana, 70-72; fel,

Brandeis Univ, 72-73; from asst prof to assoc prof, 73-82, PROF MICROBIOL, UNIV CHICAGO, 82- *Concurrent Pos:* Mem, Working Group Prod Useful Substances Microbiol Means, US/USSR Sci Exchange Prog, NSF, 75-78; NSF genetic biol panel, 81-84, Int Comn Indust Microorganisms, 82. *Mem:* Genetics Soc Am; Soc Gen Microbiol; Am Soc Microbiol; AAAS; Brit Genetical Soc; Sigma Xi. *Res:* Microbial hydrocarbon metabolism. *Mailing Add:* Dept Biochem & Molecular CLSC 861 Univ Chicago Chicago IL 60637

SHAPIRO, JEFFREY HOWARD, b New York, NY, Dec 27, 46; m 69; c 2. OPTICAL COMMUNICATIONS. *Educ:* Mass Inst Technol, SB, 67, SM, 68, EE, 69, PhD(elec eng), 70. *Prof Exp:* Asst prof elec eng, Case Western Reserve Univ, 70-73; assoc prof, 73-85, PROF ELEC ENG, MASS INST TECHNOL, 85-, ASSOC HEAD ELEC ENG, 89- *Concurrent Pos:* NSF res grants, 71-77, 81-91; consult, Lincoln Lab, Mass Inst Technol, 77- *Mem:* Inst Elec & Electronics Engrs; fel Optical Soc Am; Soc Photo-Optical Instrumentation Engrs. *Res:* Communication theory for optical systems; quantum noise reduction, coherent laser radars. *Mailing Add:* Dept Elec Eng & Comput Sci Mass Inst Technol Cambridge MA 02139

SHAPIRO, JEFFREY PAUL, b Monterey Park, Calif, Sept 1, 50; m 88; c 2. PLANT-INSECT INTERACTIONS, PROTEIN CHEMISTRY. *Educ:* San Diego State Univ, BS, 73, MS, 77; Cornell Univ, PhD(entom), 81. *Prof Exp:* Sr res biologist, Monsanto Co, 83-85; RES ENTOMOLOGIST, AGR RES SERV, USDA, 86- *Concurrent Pos:* Assoc ed, Arch Insect Biochem & Physiol, 86- *Mem:* Entom Soc Am; Am Soc Biochem & Molecular Biol; Phytochem Soc NAm. *Res:* Biochemistry; insect biochemistry; chemistry of insect proteins/lipoproteins interacting with plant compounds; phytochemicals affecting insects; plant enzymology and defenses against insects. *Mailing Add:* USDA Agr Res Serv 2120 Camden Rd Orlando FL 32803

SHAPIRO, JESSE MARSHALL, b Minneapolis, Minn, Nov 20, 29; m 51; c 3. MATHEMATICAL STATISTICS. *Educ:* Univ Minn, BA, 50, MA, 51, PhD(math), 54. *Prof Exp:* Asst math, Univ Minn, 51-53, instr, 53-54; from instr to prof, Ohio State Univ, 54-69; prof, Augsburg Col, 69-71; assoc prof math, 71-80, dean fac natural sci, 80-84, PROF MATH, BEN GURION UNIV OF THE NEGEV, ISRAEL, 80- *Concurrent Pos:* Vis prof, Univ Iowa, 76-77, Univ Southern Calif, 84-85. *Mem:* Am Math Soc. *Res:* Statistics; probability limit theorems. *Mailing Add:* Dept Math Ben Gurion Univ Negev Beersheba Israel

SHAPIRO, JOEL ALAN, b New York, NY, Feb 23, 42; m 65; c 2. THEORETICAL HIGH ENERGY PHYSICS. *Educ:* Brown Univ, ScB, 62; Cornell Univ, PhD(theoret physics), 67. *Prof Exp:* Vis res physicist, Univ Calif, Berkeley, 67-69; res assoc high energy physics, Univ Md, College Park, 69-71; from asst prof to assoc prof, 71-88, PROF, HIGH ENERGY PHYSICS, RUTGERS UNIV, 88- *Concurrent Pos:* Mem, Inst Advan Study, 86-87. *Mem:* Am Phys Soc. *Res:* Gauge field theories and strings. *Mailing Add:* Dept Physics Rutgers Univ New Brunswick NJ 08903

SHAPIRO, JOSEPH, b Montreal, Que, May 24, 29; m 52; c 2. LIMNOLOGY. *Educ:* McGill Univ, BSc, 50; Univ Sask, MSc, 52; Yale Univ, PhD(limnol), 57. *Prof Exp:* Res assoc limnol, Univ Wash, 56-58, res instr, 58-59; asst prof sanit eng, Johns Hopkins Univ, 59-64; assoc prof geol & geophys, 64-70, PROF GEOL & GEOPHYS, LIMNOL RES CTR, UNIV MINN, MINNEAPOLIS, 70-, ASSOC DIR, 64- *Mem:* Fel AAAS; Am Soc Limnol & Oceanog; Int Asn Theoret & Appl Limnol. *Res:* Chemical, physical and biological phenomena occurring in natural waters. *Mailing Add:* Ecol 109 Zool Bldg Univ Minn 318 Church St SE Minneapolis MN 55455-0302

SHAPIRO, LARRY JAY, b Chicago, Ill, July 6, 46; m 68; c 3. HUMAN GENETICS. *Educ:* Washington Univ, AB, 68, MD, 71. *Prof Exp:* Intern & resident pediat, St Louis Children's Hosp, 71-73; res assoc genetics, NIH, 73-75; from asst prof to assoc prof in residence pediat & genetics, 75-81, PROF PEDIAT & BIOL CHEM, SCH MED, UNIV CALIF, LOS ANGELES, 83- *Concurrent Pos:* Res career develop award, NIH, 80-85; pres, Soc Inherited Metabol Dis, 86-87; investr, Howard Hughes Med Inst, 87- *Honors & Awards:* E Mead Johnson Award, Am Acad Pediat, 82; Basil O'Connor Award, March of Dimes, 75. *Mem:* AAAS; Am Soc Human Genetics; Am Fedn Clin Res; fel Am Acad Pediat; Soc Inherited Metab Dis (pres, 86-87); Soc Pediat Res (pres elect, 90-91); Am Pediat Soc; Am Soc Clin Invest; Genetics Soc Am; Asn Am Physicians; Am Bd Med Genetics (vpres, 87-88, pres, 88-89). *Res:* Pediatrics, with emphasis on inborn errors of metabolism, disorders of sulfated steroid metabolism, molecular genetics, and biology of sex chromosomes. *Mailing Add:* Div Med Genetics Harbor UCLA Med Ctr 1000 W Carson St Sixth Fl Torrance CA 90509

SHAPIRO, LEE TOBEY, b Chicago, Ill, Dec 12, 43; m 70; c 2. ASTRONOMY. *Educ:* Carnegie Inst Technol, BS, 66; Northwestern Univ, MS, 68, PhD(astron), 74. *Prof Exp:* ASSOC PROF ASTRON & ASTROPHYS & DIR ABRAMS PLANETARIUM, MICH STATE UNIV, 74-; AT MOREHEAD PLANETARIUM. *Mem:* Am Astron Soc; Royal Astron Soc; Int Planetarium Soc; Am Asn Mus. *Res:* Late type giants and subgiants in close binaries; membership, size and configuration of the local group of galaxies; development and statistics of planetariums throughout the world. *Mailing Add:* Morehead Planetarium CB No 3480, Univ NC Chapel Hill NC 27599-3480

SHAPIRO, LEONARD DAVID, b San Francisco, Calif, Dec 4, 43; m 70; c 3. DATABASE MANAGEMENT SYSTEMS, EDUCATION. *Educ:* Reed Col, BA, 65; Yale Univ, PhD(math), 69. *Prof Exp:* From instr to asst prof math, Univ Minn, Minneapolis, 70-76, vis prof econ, 76-77; chmn, 77-85, prof, Dept Math Sci, NDak State Univ, Fargo, 79-87; PROF & HEAD COMPUT SCI, PORTLAND STATE UNIV, ORE, 87- *Concurrent Pos:* Vis scholar econ, 77, comput sci, Univ Calif, Berkeley, 83-84. *Mem:* Am Math Soc; Math Asn Am; Econometric Soc; Asn Comput Mach; Int Elec & Electronics Engrs Soc. *Res:* Topological dynamics; minimal sets; diophantine approximation; mathematical economics; database management systems; results on main memory and novel uses of database systems for artificial intelligence and operations. *Mailing Add:* Dept Comput Sci Portland State Univ PO Box 751 Portland OR 97207

SHAPIRO, LORNE, internal medicine, hematology; deceased, see previous edition for last biography

SHAPIRO, LUCILLE, b New York, NY, July 16, 40; m 60; c 1. MICROBIAL DEVELOPMENT. *Educ:* Brooklyn Col, BA, 62; Albert Einstein Col Med, PhD(molecular biol), 66. *Prof Exp:* Jane Coffin Childs fel biochem, Albert Einstein Col Med, 66-67, from asst prof to prof molecular biol, 67-77, chmn, Molecular Biol Dept, 78-85; prof & chmn, Molecular Biol Dept, Columbia Univ Col Physicians & Surgeons, 86-89; PROF & CHMN, DEVELOP BIOL DEPT, SCH MED, STANFORD UNIV, 89-, PROF GENETICS, 90- *Concurrent Pos:* Am Cancer Soc fac res assoc award, 68-71 & 71-76; Albert & Jane Nerken fel molecular biol, 70-76; mem, Study Sect Cell & Molecular Biol, NIH, 75, Study Sect Develop Biol, NSF, 76-77, Microbial Chem Study Sect, NIH, 78-80, Bd Sci Counr, NIAMKKD, 80-84, Gen Med Sci Coun, NIH, 81, adv bd, NSF Biol & Behav Sci Dir, 82 & 83-87, nat bd, Am Heart Asn, 84-87 & coun, Am Soc Biochem & Molecular Biol, 90-93; career scientist award, Hirschl Found, 76-81; co-chmn, adv bd, NSF Biol & Behav Sci Dir, 82 & 83-87; distinguished lectr, Carnegie Mellon Univ, 88, Woods Hole Arts & Sci, 89, Univ Colo, 89 & Southwestern Univ Tex, 90. *Honors & Awards:* De Witt Stetten Jr Lectr, NIH, 89. *Mem:* Inst Med-Nat Acad Sci; Sigma Xi; fel AAAS; Am Soc Biochem & Molcular Biol; Am Soc Microbiol; NY Acad Sci. *Res:* Unicellular differentiations; developmental biology. *Mailing Add:* Dept Develop Biol Beckman Ctr Stanford Univ Sch Med Stanford CA 94305-5427

SHAPIRO, MARK HOWARD, b Boston, Mass, Apr 18, 40; m 61; c 3. SURFACE PHYSICS, COMPUTER SIMULATION. *Educ:* Univ Calif, Berkeley, AB, 62; Univ Pa, MS, 63, PhD(physics), 66. *Prof Exp:* Res fel physics, Kellogg Radiation Lab, Calif Inst Technol, 66-68; res assoc, Nuclear Struct Res Lab, Univ Rochester, 68-70; from asst prof to assoc prof, 70-78, actg assoc dean, Sch Math, Sci & Eng, 85-86, PROF PHYSICS, CALIF STATE UNIV, FULLERTON, 78-, DEPT CHAIR, 89- *Concurrent Pos:* Res Corp grant & fac res grant, 71-72; vis assoc, Kellogg Radiation Lab, Calif Inst Technol, 76-; Calif Inst Technol President's venture fund grant, 77-78; US Geol Surv grant, 78-85; NSF grant, 84-87; actg dir, Off Fac Res & Develop, CSUF, 86-87; Rotator, Nat Sci Found, 87-88; vis scientist, Nat Inst Standards & Technol, 87- *Mem:* AAAS; Am Phys Soc; Am Asn Physics Teachers; Am Geophys Union; NY Acad Sci; Nat Coun Univ Res Admin. *Res:* Nuclear structure physics; nuclear reaction physics; laboratory nuclear astrophysics; applications of nuclear physics in geophysics; nuclear safety; ion-surface interactions. *Mailing Add:* Dept Physics Calif State Univ Fullerton CA 92634-9480

SHAPIRO, MARTIN, b New York, NY, Mar 18, 37; m 67; c 2. ENTOMOLOGY, MICROBIOLOGY. *Educ:* Brooklyn Col, AB, 58; Cornell Univ, MS, 61; Univ Calif, Berkeley, PhD(entom), 66. *Prof Exp:* Entomologist, USDA, Tex, 65-66; res entomologist, Int Minerals & Chem Corp, 66-70; USPHS fel insect path, Boyce Thompson Inst Plant Res, 70-72; entomologist, Maag & Easterbrooks, Inc, 72-73; mem staff, Res Unit Vector Path, Mem Univ Nfld, 73-75; mem staff, Gypsy Moth Methods Develop Lab, 75-85, AT INSECT PATH LAB, USDA, 85- *Concurrent Pos:* Consult, Int Minerals & Chem Corp, 70-73. *Mem:* Soc Invert Path; Entom Soc Am; Tissue Cult Asn; Int Orgn Biol Control. *Res:* Insect pathology and tissue culture; photobiology. *Mailing Add:* Bldg 011A Rm 230B Agr Res Ctr Beltsville MD 20705

SHAPIRO, MAURICE A, b Denver, Colo, June 4, 17; m 45; c 6. ENVIRONMENTAL HEALTH ENGINEERING, PUBLIC HEALTH. *Educ:* Johns Hopkins Univ, AB, 41; Univ Calif, Berkeley, MEng, 49; Am Acad Environ Engr, dipl. *Prof Exp:* Asst sanit engr, USPHS, 41-47, field officer, United Yugoslav Relief Fund Am, 47-48; eng res assoc, Am Pub Health Asn, 49-51; prof sanit eng, 51-69, prof 69-82, EMERITUS PROF ENVIRON HEALTH ENG, GRAD SCH PUB HEALTH, UNIV PITTSBURGH, 82- *Concurrent Pos:* Vis prof, Israel Inst Technol, 65-66; found dir, Westernport Bay Environ Study, Victoria, Australia, 73-74, vis mem sci fac, Univ Melbourne, 73-74; vis lectr, Monash Univ, 73-74, Wuhan Univ, 86. *Mem:* Am Soc Civil Engrs; Am Water Works Asn; Am Pub Health Asn; Am Pub Works Asn; Soc Environ & Occup Health. *Res:* Water and air pollution control; health aspects of energy conversion and water quality; radioactive wastes disposal; environmental health aspects of energy conversion, environmental health planning. *Mailing Add:* Grad Sch Pub Health Univ Pittsburgh Pittsburgh PA 15261

SHAPIRO, MAURICE MANDEL, b Jerusalem, Palestine, Nov 13, 15; US citizen; m 42; c 3. COSMIC RADIATION, NEUTRINO ASTROPHYSICS. *Educ:* Univ Chicago, SB, 36, SM, 40, PhD(physics), 42. *Prof Exp:* Assoc physicist, USN, 42-44; group leader, Los Alamos Lab, Univ Calif, 44-46; sr physicist, Oak Ridge Nat Lab & lectr, Nucleonics Sch, 46-49; chief scientist, Lab Cosmic Physics, Naval Res Lab, 49-82, dir, Nucleonics Div, 53-65; consult, NASA, 65-70; DIR, INT SCH COSMIC RAY ASTROPHYS, ERICE, ITALY, 77- *Concurrent Pos:* Chmn dept phys & biol sci, Austin Col, 38-41; lectr, George Washington Univ, 43-44, Reactor Dept, Erco Div, ACF Indust, Inc, 56-58; lectr, Grad Sch, Univ Md, 49-50 & 52-55, assoc prof, 50-51; Guggenheim fel & vis prof, Weizmann Inst, 62-63; mem comt emulsion exp, Space Sci Bd, Nat Acad Sci & mem panel x-ray & gamma ray astron, 65; mem vis comt, Bartol Res Found, Franklin Inst, 66-74; mem working group space biophys, Coun Europ, 70-; chmn div cosmic physics, Am Phys Soc, 71-72 & High Energy Astrophys Div, 82; chmn comt interdisciplinary res & consult panel cosmic radiation, US Nat Comt, Int Geophys Year; prin investr, Gemini, Skylab Cosmic-Ray Exp & Long Duration Exposure Facil, NASA, 63-82; deleg, int confs cosmic radiation, Int Union Pure & Appl Physics; deleg, int confs nuclear & high energy physics, nuclear photog & comt on space res; mem, Comn Honor for Celebration of Einstein Centenary, Acad Lincei, 77-79; mem steering comn, DUMAND Consortium; vis prof astron,

Northwestern Univ, 78, Univ Iowa & Univ Bonn, WGer, 81-84 & Univ Md, 85-; assoc ed, Phys Rev Lett, 78-84; Humboldt sr US scientist award, 81; vis scientist, Max Planck Inst Astrophys, Munich, 84-85; regents lectr, Univ Calif, Riverside, 85. *Honors & Awards:* Edison Lectr. *Mem:* Fel AAAS; fel Am Phys Soc; Sigma Xi; Am Technion Soc; Fedn Am Scientists; Am Astron Soc; Int Astron Union. *Res:* Cosmic rays, especially composition, origin, propagation and nuclear transformations; nuclear-emulsion techniques for high-energy physics and cosmic rays; charged sigma hyperons; physics of underwater explosions; neutron and fission physics; neutrino astrophysics; reactor design; piezoelectricity. *Mailing Add:* 205 Yoakum Parkway No 2-1720 Alexandria VA 22304

SHAPIRO, NATHAN, b Boston, Mass, July 16, 24; m 51; c 2. GENETICS. *Educ:* Univ Wis, BS, 49, MS, 50; Purdue Univ, PhD(genetics), 63. *Prof Exp:* Mycologist, Mat Testing Lab, NY Naval Shipyard, 51-53; res asst, Biol Dept, Brookhaven Nat Lab, 56-59; asst prof zool, Smith Col, 59-70; PROF BIOL, EASTERN CONN STATE COL, 70- *Res:* Radiation genetics; plant growth hormones. *Mailing Add:* Dept Biol Eastern Conn State Univ Willimantic CT 06226

SHAPIRO, PAUL JONATHON, b Baltimore, Md, Sept 10, 52; m 79; c 2. MECHANICAL ENGINEERING. *Educ:* Mass Inst Technol, BS, 73, MS, 75, PhD(fluid mech), 77. *Prof Exp:* Fel, Lady Davis Scholarship Found, 77; staff fel, NIH, 78-79; mem tech staff, Philips Lab, NAm Philips Corp, 79-86; CONSULT, 86- *Mem:* NY Acad Sci; AAAS; Sigma Xi. *Res:* Development of long life spaceborne mechanical refrigerator; x-ray diffraction studies of functioning muscle specimens; boundary-layer transition studies. *Mailing Add:* 123 Rockland Lane Spring Valley NY 10977-3125

SHAPIRO, PAUL ROBERT, b New Haven, Conn, Aug 2, 53. PHYSICS. *Educ:* Harvard Col, AB, 74; Harvard Univ, PhD(astron), 79. *Prof Exp:* Postdoctoral res fel astrophys, Inst Advan Study, 78-81; asst prof, 81-86, ASSOC PROF ASTRON, UNIV TEX, 86- *Concurrent Pos:* Alfred P Sloan Found res fel physics, 84-88. *Mem:* Am Astron Soc; Am Phys Soc; Sigma Xi. *Res:* Theoretical astrophysics research including the intergalactic medium, cosmology and the formation of galaxies and structure in the universe; interstellar medium; gas dynamics. *Mailing Add:* Dept Astron Univ Tex Austin TX 78712

SHAPIRO, PHILIP, b Brooklyn, NY, Sept 10, 23; wid; c 2. NUCLEAR PHYSICS, RADIATION EFFECTS. *Educ:* Brooklyn Col, BS, 48; Univ Iowa, PhD(physics), 53. *Prof Exp:* Asst physics, Univ Iowa, 48-53; physicist, Radiation Div, US Naval Res Lab, 53-66, Nuclear Physics Div, 66-71, from Cyclotron Br to Cyclotron Appln Br, 71-85; PHYSICIST, SFA INC, 85- *Mem:* AAAS; Am Phys Soc. *Res:* Neutron radiotherapy; dosimetry; nuclear reactions; radiation effects on microelectronics. *Mailing Add:* SFA Inc 1401 McCormick Dr Landover MD 20785

SHAPIRO, RALPH, b Malden, Mass, Nov 9, 22; m 45; c 3. DYNAMIC METEOROLOGY. *Educ:* Bridgewater Col, BS, 43; Mass Inst Technol, MS, 48, DSc(meteorol), 50. *Prof Exp:* Asst, Pressure Change Proj, Mass Inst Technol, 47-50; meteorologist, Planetary Atmospheres Proj, Lowell Observ, 50-51; proj scientist, Geophys Res Directorate, Hanscom AFB, 51-57, assoc chief, Meteorol Develop Lab & chief, Atmospheric Dynamics Br, Air Force Cambridge Res Ctr, 57-75, chief, Climatol & Dynamics Br, Air Force Geophys Lab, 75-80; SR SCIENTIST, S T SYSTS CORP, 80- *Concurrent Pos:* Consult ed, McGraw-Hill Encycl Sci & Technol, 73-80. *Mem:* AAAS; Am Geophys Union; fel Am Meteorol Soc; foreign mem Royal Meteorol Soc; Sigma Xi. *Res:* Mathematical modeling of large-scale atmospheric circulations; numerical weather prediction; numerical analysis of partial differential equations; digital filter design; statistical weather prediction; flux of solar radiation through the atmosphere. *Mailing Add:* S T Systs Corp 109 Massachusetts Ave Lexington MA 02173

SHAPIRO, RAYMOND E, b New York, NY, Oct 20, 27; m 64; c 2. CHEMISTRY. *Educ:* City Col New York, BS, 48; Ohio State Univ, PhD(chem), 52. *Prof Exp:* Chemist, Soil & Water Conserv Res Div, Agr Res Serv, USDA, 53-65; res chemist, Div Pharmacol, Intermediary Metab Br, US Food & Drug Admin, 65-70, sci coordr, Bur Foods, 70-73, sci coordr, Epidemiol Unit, 73-76; ASST DIR TOXICOL COORD, NAT INST ENVIRON HEALTH SCI, 76- *Concurrent Pos:* Guggenheim fel, Rothamsted Exp Sta, Herpenden, Eng, 62. *Mem:* Fel AAAS; Am Col Toxicol; Soc Environ Geochem & Health; Am Chem Soc; Soc Epidemiol Res. *Res:* Soil aeration; chemistry of submerged soils; kinetics of ion migration in soil systems; mycotoxins; metabolism; toxicology. *Mailing Add:* 126 Betty Rd New Hyde NY 11040

SHAPIRO, ROBERT, b New York, NY, Nov 28, 35; m 64; c 1. BIOCHEMISTRY, ORGANIC CHEMISTRY. *Educ:* City Col New York, BS, 56; Harvard Univ, AM, 57, PhD(chem), 59. *Prof Exp:* NATO fel chem, Cambridge Univ, 59-60; fel biochem, Sch Med, 60-61, from asst prof to assoc prof, 61-70, PROF CHEM, NY UNIV, 70- *Mem:* AAAS; Am Chem Soc; Am Soc Biochem & Molecular Biol. *Res:* Chemistry of nucleic acids; chemical mutagenesis and carcinogenesis; origin of life; human genome project. *Mailing Add:* Dept Chem NY Univ New York NY 10003

SHAPIRO, ROBERT ALLEN, b Long Branch, NJ, Aug 14, 30; m 53; c 3. INDUSTRIAL ENGINEERING. *Educ:* Okla State Univ, BS, 53, MS, 64, PhD(indust eng), 65. *Prof Exp:* Div prod engr, Shell Oil Co, 53-61; assoc prof, 64-72, asst to pres, 67-68, PROF INDUST ENG, & ASSOC VPRES ADMIN & FINANCE, UNIV OKLA, 72-, DIR SCH ENG, 65- *Concurrent Pos:* Consult, Okla Mgt Study Comt, 67, US Army & USPHS, 67- *Mem:* Am Inst Indust Engrs; Am Soc Petrol Engrs; Am Soc Eng Educ; Am Soc Qual Control. *Res:* Operations research, including recurrent processes, modern organization theory, econometric models of decisions under uncertainty and statistical quality control. *Mailing Add:* Rte 1 Box 114 B-1 Norman OK 73072

SHAPIRO, ROBERT HOWARD, b New Haven, Conn, July 18, 35; m 56; c 3. ORGANIC CHEMISTRY. *Educ:* Univ Conn, BS, 61; Stanford Univ, PhD(chem), 64. *Prof Exp:* NSF fel chem, Royal Vet & Agr Col, Denmark, 64-65; from asst prof to prof chem, Univ Colo, Boulder, 65-80; PROF & DEPT HEAD CHEM, JAMES MADISON UNIV, HARRISONBURG, VA, 80-, DEAN, COL LETT & SCI. *Concurrent Pos:* Consult, Criminal Lawyers, 66- *Mem:* Am Chem Soc; Royal Soc Chem. *Res:* Reaction mechanisms; mass spectrometry; environmental chemistry; natural products chemistry. *Mailing Add:* Col Lett & Sci James Madison Univ Harrisonburg VA 22801

SHAPIRO, RUBIN, b Chicago, Ill, Nov 7, 24; div; c 2. ANALYTICAL CHEMISTRY. *Educ:* Univ Ill, BS, 48; Univ Wis, PhD(anal chem), 53. *Prof Exp:* Res chemist, 53-57, supvr, 57-68, res assoc, 68-77, sr res assoc, 77-86, RES FEL, AM NAT CAN CO, 86. *Mem:* Am Chem Soc; Soc Appl Spectros. *Res:* Emission, mass and atomic absorption spectroscopy; gas chromatography; analytical chemistry. *Mailing Add:* 1215 N Waterman 1-J Arlington Heights IL 60004-5192

SHAPIRO, SAM, b New York, NY, Feb 12, 14; m 38; c 2. EPIDEMIOLOGY, BIOSTATISTICS. *Educ:* Brooklyn Col, BS, 33. *Prof Exp:* Chief, Natality Anal Br, Nat Off Vital Statist, USPHS, 47-54; sr study dir, Nat Opinion Res Ctr, 54-55; assoc dir, Div Res & Statist, Health Ins Plan of Greater New York, 55-59; vpres & dir, 59-73; dir, Health Serv Res & Develop Ctr, Johns Hopkins Med Insts, 73-83, prof, 73-85, EMER PROF HEALTH POLICY & MGT, SCH HYG & PUB HEALTH, JOHNS HOPKINS UNIV, 85- *Concurrent Pos:* Lectr pub health, Sch Pub Health & Admin Med, Columbia Univ, 61-80; adj prof community med, Mt Sinai Sch Med, 72-78; consult to var insts & orgns. *Mem:* Inst Med-Nat Acad Sci; fel Am Pub Health Asn; fel Am Statist Asn; fel Am Heart Asn; Am Epidemiol Soc; fel AAAS; Am Soc Prev Oncol; Asn Health Serv Res. *Res:* Evaluative research in organization; delivery; economics; quality of health care. *Mailing Add:* Sch Hyg & Pub Health Johns Hopkins Med Insts 624 N Broadway Baltimore MD 21205

SHAPIRO, SAMUEL S, b Brooklyn, NY, July 13, 30; m 56; c 2. APPLIED STATISTICS. *Educ:* City Col New York, BBA, 52; Columbia Univ, MS, 54; Rutgers Univ, MS, 61, PhD(statist), 64. *Prof Exp:* Statist qual control engr, Pittsburgh Plate Glass Co, 56-58; statistician, Res & Develop Ctr, Gen Elec Co, NY, 58-67; statistician, Prog Methodology, Inc, 67-72; PROF STATIST, FLA INT UNIV, 72- *Concurrent Pos:* Lectr, Union Col, 65-66; UN tech adv, Indian Statist Inst, 66. *Honors & Awards:* Jack Youdin Prize, 72. *Mem:* Inst Math Statist; Am Statist Asn; Sigma Xi; fel Am Statist Asn. *Res:* Testing for distributional assumptions. *Mailing Add:* Fla Int Univ Miami FL 33199

SHAPIRO, SANDOR SOLOMON, b Brooklyn, NY, July 26, 33; m 54; c 2. HEMATOLOGY. *Educ:* Harvard Univ, BA, 54, MD, 57. *Prof Exp:* Intern, Harvard Med Serv, Boston City Hosp, 57-58; asst surgeon, Div Biol Stand, NIH, USPHS, 58-60; asst resident, Boston City Hosp, 60-61; NIH spec fel, Mass Inst Technol, 61-64; from instr to assoc prof, 64-72, assoc dir, 78-85, PROF MED, CARDEZA FOUND, JEFFERSON MED COL, 72-, DIR, FOUND, 85- *Concurrent Pos:* Mem hemat study sect, NIH, 72-76 & 78-79; mem med adv coun, Nat Hemophilia Found, 73-75; chmn, Pa State Hemophilia Adv Comt, 74-76. *Mem:* Am Soc Clin Invest; Am Soc Hemat; Am Asn Immunologists; Int Soc Thrombosis & Hemostasis; Am Asn Physicians. *Res:* Hemostasis and thrombosis, prothrombin metabolism, hemophilia; lupus anticoagulants; endothelial cells. *Mailing Add:* Cardeza Found 1015 Walnut St Philadelphia PA 19107

SHAPIRO, SEYMOUR, b New York, NY, Feb 16, 24; m 47; c 1. BOTANY. *Educ:* Univ Mich, BS, 47, PhD(bot), 52. *Prof Exp:* Instr bot, Univ Mich, 50-51, res assoc, 51-52; res collabr biol, Brookhaven Nat Lab, 53, assoc botanist, NY, 53-61; assoc prof biol, Univ Ore, 61-64; head dept, 64-69, actg dean arts & sci, 69-70, actg head dept, 74-75, dean fac natural sci & math, 75-79, PROF BOT, UNIV MASS, AMHERST, 64- *Concurrent Pos:* Consult, NSF, 62-69, vis scientist, Inst Atomic Sci in Agr, Wageningen, Neth, 70-71; vis prof, Univ Col North Wales, UK, 74. *Mem:* AAAS; Am Soc Plant Physiologists; Am Soc Develop Biol; Bot Soc Am. *Res:* Regeneration; morphogenesis of higher plants. *Mailing Add:* Dept Bot Univ Mass Amherst MA 01003

SHAPIRO, SIDNEY, b Boston, Mass, Dec 4, 31; m 60; c 2. ELECTRONIC DEVICES, SUPERCONDUCTIVITY. *Educ:* Harvard Univ, AB, 53, AM, 55, PhD(appl physics), 59. *Prof Exp:* Res asst appl physics, Harvard Univ, 55-59; physicist, Arthur D Little, Inc, 59-64; mem tech staff, Bell Labs, Inc, NJ, 64-67; assoc prof, Univ Rochester, 67-73, assoc dean, Col Eng & Appl Sci, 74-79, chmn dept, 80-89, PROF ELEC ENG, UNIV ROCHESTER, 73- *Concurrent Pos:* Vis prof, Physics Lab I, Tech Univ Denmark, 72-73. *Mem:* Fel Inst Elec & Electronics Engrs. *Res:* Fast relaxation processes in devices; electron tunneling; Josephson effect; microwave phenomena and devices involving superconducting junctions and weak links. *Mailing Add:* Dept Elec Eng Univ Rochester Rochester NY 14627-0007

SHAPIRO, STANLEY, b Brooklyn, NY, Jan 3, 37; m 58; c 3. RESEARCH MANAGEMENT. *Educ:* City Col New York, BChe, 60; Rensselaer Polytech Inst, MS, 64; Lehigh Univ, PhD(metall), 66. *Prof Exp:* Res engr, Pratt & Whitney, 60-61; res engr, Res Lab, United Aircraft Corp, 61-64; instr & res asst, Lehigh Univ, 64-66; res sci supvr, Metals Res Labs, Olin Corp, 66-79; pres, Revere Res, Inc, 70-84; vpres res & develop, Nat Can Co, 84-87; vpres res & develop oper, Am Nat Can, 87-89; CONSULT, MERGERS & ACQUISITIONS, 89- *Mem:* Am Inst Metal Engrs; Am Soc Metals; AAAS; Am Soc Testing & Mats; Sigma Xi. *Res:* Non-ferrous metals research and development; alloy and process research. *Mailing Add:* 8818 N Kolmar Skokie IL 60076

SHAPIRO, STANLEY KALLICK, b Montreal, Que, Mar 20, 23; nat US; m 45. MICROBIOLOGY, BIOCHEMISTRY. *Educ:* McGill Univ, BSc, 44, MSc, 45; Univ Wis, PhD(microbiol), 49. *Prof Exp:* Asst prof bact, Iowa State Univ, 49-54; assoc biochemist, Argonne Nat Lab, 54-69; head dept, 74-85,

PROF BIOL SCI, UNIV ILL, CHICAGO, 69- *Mem:* AAAS; Am Soc Microbiol; Am Soc Biol Chemists; Sigma Xi. *Res:* Microbial physiology and biochemistry; sulfonium biochemistry, transmethylation, polyamine biosynthesis, sulfur amino acid metabolism. *Mailing Add:* Dept Biol Sci Univ Ill Box 4248 Chicago IL 60680

SHAPIRO, STANLEY SEYMOUR, b Brooklyn, NY, Sept 22, 40; m 65; c 3. DRUG DISCOVERY. *Educ:* Brooklyn Col, BS, 63; Univ Del, PhD(biochem), 66. *Prof Exp:* NIH fel molecular biol, Albert Einstein Col Med, 66-68; sr biochemist, Hoffmann-La Roche Inc, 68-76, res fel, 76-80, sr res fel, 80-85, DIR DERMAT RES, HOFFMANN-LA ROCHE INC, 85-, ASST VPRES, 90- *Concurrent Pos:* Mem coadj staff, Rutgers Univ, Newark, 69- *Mem:* AAAS; NY Acad Sci; Am Soc Biol Chemists; Am Chem Soc; Am Inst Nutrit; Soc Invest Dermat. *Res:* Biochemistry and cell biology of retinoids; identify and develop agents for the treatment of dermatological disorders. *Mailing Add:* Dermat Res Hoffmann La Roche Inc Nutley NJ 07110

SHAPIRO, STEPHEN D, b New York, NY, Feb 18, 41; m 69; c 1. COMPUTER & INFORMATION SCIENCE. *Educ:* Columbia Univ, BS, 63, MS, 64, PhD(digital syst), 67. *Prof Exp:* Mem tech staff comput software algorithms, Bell Tel Lab, 67-71; prof comput sci, Dept Elec Eng, Stevens Inst Technol, 74-80; PROF ELEC ENG, STATE UNIV NY, STONY BROOK, 80- *Concurrent Pos:* Consult info processing, var comn & indust orgn, 71-; consult dir comput sci educ prog, Bell Labs, 76-; NSF grant, 76-; chmn elec eng dept, Comput Sci Comn, 78- *Mem:* Sigma Xi; sr mem Inst Elec & Electronics Engrs; Asn Comput Mach. *Res:* Information processing; software engineering; picture processing; telecommunications. *Mailing Add:* Dept Elec Eng State Univ NY Stony Brook NY 11794

SHAPIRO, STEPHEN MICHAEL, b Pittsfield, Mass, June 21, 41; m 71; c 2. SOLID STATE PHYSICS. *Educ:* Union Col, BS, 63; Johns Hopkins Univ, PhD(physics), 69. *Prof Exp:* Res assoc, Physics Lab, Univ Paris, 69-70; res assoc, Brookhaven Nat Lab, 71-73; vis physicist, Riso Nat Lab, Denmark, 73-74; PHYSICIST, BROOKHAVEN NAT LAB, 74- *Mem:* Am Phys Soc. *Res:* Neutron scattering studies of phase transitions, magnetic phenomenon and solid electrolytes. *Mailing Add:* Dept Physics 510B Brookhaven Nat Lab Upton NY 11973

SHAPIRO, STEWART, b Springfield, Mass, Mar 21, 37; c 2. DENTISTRY, PUBLIC HEALTH. *Educ:* Boston Col, BA, 58; Tufts Univ, DMD, 62; Harvard Univ, MScH, 69; Century Univ, PhD, 81. *Prof Exp:* Res assoc oral physiol, Dent Sch, Tufts Univ, 62-63, instr oral physiol & oral diag, 64-65, staff mem, Forsyth Dent Ctr, 65-66; asst prof community dent, Sch Dent, Univ Md, 69-72; PROF FAMILY PRACT & COMMUNITY HEALTH, COLS MED & HEALTH, PROF COMMUNITY DENT, COL DENT & CHMN DIV COMMUNITY DENT, UNIV OKLA, 72- *Concurrent Pos:* Pvt pract dent, 62-68; lectr epidemiol, Sch Hyg & Pub Health, Johns Hopkins Univ, 69-70, assoc, 70-72; consult, Job Corps Prog, Dept Labor, 70-72, Vet Admin Hosp, Muskogee, Okla, 72- & Head Start, Div Dent Health, Dept HEW, 73- *Mem:* Am Dent Asn; Int Asn Dent Res; Am Pub Health Asn; Am Asn Dent Schs. *Res:* Prevention; gerontology; behavior. *Mailing Add:* Div Community Dent Univ Okla Sch Dent Oklahoma City OK 73190

SHAPIRO, STUART, b Brooklyn, NY, 1952. SECONDARY METABOLISM, BIOTECHNOLOGY. *Educ:* Washington Square Col, NY Univ, BA, 71; Univ Ill, Urbana-Champaign, MS, 76; Worcester Polytech Inst, PhD(biomed sci), 81. *Prof Exp:* Fel biol, Dalhousie Univ, 81-83; asst prof chem, Concordia Univ, 83-85; res, Bact Inst Armand-Frappier, 85-86; asst prof & res assoc biol, Dalhousie Univ, 86-87; mgr, Indust Microbiol Dept, Sigma Tau SpA, 88-91; MICROBIOL, ZAHNÄRTZ INST, UNIV ZURICH, 91- *Concurrent Pos:* Vis scientist, pharm, Univ Wis, Madison, 88. *Mem:* Soc Indust Microbiol; Swiss Soc Microbiol. *Res:* Microbial biochemistry; natural products chemistry; fermentation technology; stereochemistry of enzymic reactions. *Mailing Add:* Dept Oral Microbiol & Gen Immunol Zahnä Inst Univ Zurich Plattenstr 11 CH-8028 Zurich Switzerland

SHAPIRO, STUART CHARLES, b New York, NY, Dec 30, 44; m 72. COMPUTER SCIENCE, ARTIFICIAL INTELLIGENCE. *Educ:* Mass Inst Technol, SB, 66; Univ Wis, MS, 68, PhD(comput sci), 71. *Prof Exp:* From asst prof to assoc prof, Ind Univ, 71-78; from asst prof to assoc prof, 77-83, chmn dept, 84-90, PROF COMPUTER SCI, STATE UNIV, BUFFALO, 83-; RES SCIENTIST, NAT CTR GEOG INFO & ANAL, BUFFALO SITE, 89- *Concurrent Pos:* Teaching asst, Comput Sci Dept, Univ Wis, 66-67; consult, Ling Group, Rand Corp, 68-71, Anal & Simulation Inc, 85-87, Calspan-UB Res Ctr, Buffalo, 87- & Univ Southern Calif Info Sci Inst, 87-89; vis res asst prof, Comput Sci Dept, Univ Ill, Urbana, 74; actg chmn, Comput Sci Dept, State Univ NY, Buffalo, 78-79, NSF grant, 78-; external fel, Cognitive Sci Prog, Univ Rochester, 82-83; mem, Eval Panel NSF fels comput sci, Nat Res Coun, 83-85 & Rev Panel Res Prog Math & Info Sci Dir, Air Force Off Sci Res, 86-88; prin lectr & consult, Smart Systs Technol, McLean, Va, 83-85; prin investr, Rome Air Develop Ctr, Air Force Off Sci Res, 84-89 & Defense Adv Res Projs Agency, 87-89. *Mem:* Asn Comput Mach; Asn Comput Ling; Inst Elec & Electronic Engrs; Cognitive Sci Soc; Am Asn Artificial Intel; Sigma Xi. *Res:* Artificial intelligence; representation of knowledge, specifically, the semantics of semantic networks; reasoning; natural language processing; intelligent computer-human interfaces. *Mailing Add:* Dept Comput Sci State Univ NY 226 Bell Hall Buffalo NY 14260

SHAPIRO, STUART LOUIS, b New Haven, Conn, Dec 6, 47; m 71. ASTROPHYSICS. *Educ:* Harvard Univ, AB, 69; Princeton Univ, MA, 71, PhD(astrophys sci), 73. *Prof Exp:* Lab instr physics, Harvard Col, 67-69; teaching asst astron, Princeton Univ, 73; res assoc, 73-75, instr, 74-75, asst prof, 75-77, ASSOC PROF ASTRON, CTR RADIOPHYS & SPACE SCI, CORNELL UNIV, 77- *Mem:* Am Astron Soc; Int Astron Union; Sigma Xi. *Res:* Theoretical problems in relativistic astrophysics and high-energy astrophysics; black hole physics; the physics of compact objects, such as white dwarfs, neutron stars and black holes; x-ray astronomy; cosmology; dynamical astronomy. *Mailing Add:* Cornell Univ 426 Space Sci Bldg Ithaca NY 14850

SHAPIRO, VICTOR LENARD, b Chicago, Ill, Oct 16, 24; m 48; c 4. MATHEMATICAL ANALYSIS. *Educ:* Univ Chicago, BS, 47, MS, 49, PhD(math), 52. *Prof Exp:* Instr math, Ill Inst Technol, 48-52; from instr to prof, Rutgers Univ, 52-60; prof, Univ Ore, 60-64; fac res lectr, 78, PROF MATH, UNIV CALIF, RIVERSIDE, 64- *Concurrent Pos:* Asst, Univ Chicago, 51-52; mem, Inst Advan Study, 53-55 & 58-59, NSF fel, 54-55. *Mem:* Am Math Soc; Math Asn Am; Soc Indust Appl Math; AAAS. *Res:* Harmonic analysis; partial differential equations. *Mailing Add:* Dept Math Univ Calif Riverside CA 92521

SHAPIRO, WARREN BARRY, b Baltimore, Md, Nov 21, 41; m 66; c 2. HAIR COLOR, HAIR & SKIN CARE PRODUCTS DEVELOPMENT. *Educ:* Temple Univ, BS, 63, MS, 66, PhD(pharmacol chem), 70. *Prof Exp:* Res chemist herbicides, Amchem Prod Inc, 68-70; admin asst, Block Drug Co, Inc, 70-71, sr group leader, Block Drug Co, 71-73; mgr, Noxell Corp, 73-83; DIR HAIR & SKIN CARE, CLAIROL, INC, DIV BRISTOL MYERS, 83- *Mem:* Soc Cosmetic Chemists; Am Pharmaceut Asn; Acad Pharmaceut Sci. *Res:* Development of new hair and skin care products; development of over the counter drugs, oral care, analgesics, acne and antacids; development of household products. *Mailing Add:* 48 Grumman Ave Norwalk CT 06851

SHAPIRO, WILLIAM, b Newark, NJ, Dec 8, 27; m 51; c 3. INTERNAL MEDICINE, CARDIOLOGY. *Educ:* Duke Univ, AB, 47, MA, 48, MD, 54. *Prof Exp:* Intern, Mt Sinai Hosp, New York, 54-55; jr asst resident internal med, Duke Hosp, Durham, NC, 55-56, res fel cardiol, 56-57, sr asst resident, 57-58; from instr to asst prof, Med Col Va, 60-65; chief cardiovasc sect, Vet Admin Hosp, 68-81; from asst prof to assoc prof, 65-79, PROF INTERNAL MED, UNIV TEX HEALTH SCI CTR DALLAS, 79-; DIR, AUTOMATED ELECTROCARDIOGRAM SVCS, VETS ADMIN HOSP, DALLAS, 81- *Concurrent Pos:* Consult, Vet Admin Hosp, Dallas, 65-68 & Coronary Care Comt, Am Heart Asn, 68- *Mem:* Am Fedn Clin. *Res:* Am Physiol Soc; fel Am Col Physicians; Am Heart Asn; Am Col Cardiol. Res: Clinical cardiology; cardiovascular physiology. *Mailing Add:* Dallas Vet Admin Med Ctr 4500 S Lancaster Rd Dallas TX 75216

SHAPIRO, ZALMAN MORDECAI, b Canton, Ohio, May 12, 20; m 45; c 3. CHEMISTRY, PROCESS METALLURGY. *Educ:* Johns Hopkins Univ, AB, 42, MA, 45, PhD(phys chem), 48. *Prof Exp:* Res assoc, Johns Hopkins Univ, 42-46, jr instr chem, 46-48; sr scientist res labs, Westinghouse Elec Corp, 48-49, sr scientist, Bettis Atomic Power Div, 49-50, supv scientist, 50-53, mgr phys chem sect, 53-54, mgr chem metall sect, 54-55, asst mgr, Reactor Design Sub-Div, 55-57; pres, Nuclear Mat & Equip Corp, 57-70; exec asst to mgr, Breeder Reactor Div, Westinghouse Elec Corp, 71-73, dir, Fusion Power Systs Dept, 73-81 & Spec Proj, Nuclear Energy Syst, 81-83; CONSULT, ASSOC TECH CONSULTS, 83- *Concurrent Pos:* Chmn, Isotope & Radiation Enterprises, Ltd, 64-68; pres, Nuclear Decontamination Corp, Nuclear Mat Equip Corp & vpres, Arco Chem Co, 68-70; vpres, Kawecki Berylco Industs; dir, Diagnon Corp, Arco-Hanford Co, 68-70 & PSM Technologies, 88- *Mem:* AAAS; fel Am Nuclear Soc; Am Chem Soc; Am Soc Metals; Am Ceramic Soc; Sigma Xi. *Res:* Chemical erosion of steel; carbonyl chemistry; flame reactions of metallic halides; zirconium, hafnium, uranium and plutonium chemistry; metallurgy; fusion power systems technology. *Mailing Add:* 1045 Lyndhurst Dr Pittsburgh PA 15206

SHAPLEY, JAMES LOUIS, b Asotin, Wash, Mar 9, 20; m 47; c 3. AUDIOLOGY, SPEECH PATHOLOGY. *Educ:* Univ Wash, BA, 47, MA, 52; Univ Iowa, PhD(audiol), 54. *Prof Exp:* Assoc instr speech, Univ Wash, 47-51; chief audiologist, Houston Speech & Hearing Ctr, Tex, 54-56; from instr to asst prof audiol, Univ Iowa, 56-60; chief audiol & speech path, Vet Admin Hosp, Seattle, 60-86; clin assoc prof, 61-86, EMER CLIN ASSOC PROF, UNIV WASH, 86- *Mem:* Acoust Soc Am; Am Speech & Hearing Asn; Sigma Xi. *Res:* Psychoacoustics; clinical audiology. *Mailing Add:* 1253 23rd Ave E Seattle WA 98112

SHAPLEY, JOHN ROGER, b Manhattan, Kans, Apr 15, 46; m 70, 84; c 1. CATALYSIS. *Educ:* Univ Kans, BS, 67; Harvard Univ, PhD(chem), 72. *Prof Exp:* Assoc, Stanford Univ, 71-72; from asst prof to assoc prof, 72-79, PROF CHEM, UNIV ILL, URBANA, 79- *Concurrent Pos:* NSF fel, 71-72; A P Sloan Found fel, 78-80; teacher-scholar, Camille & Henry Dreyfus Found, 78-83. *Mem:* Am Chem Soc; Royal Soc Chem. *Res:* Organotransition metal chemistry; synthesis and characterization of novel compounds; metal clusters; catalysis; dynamic nuclear magnetic resonance. *Mailing Add:* 505 S Mathews Box 20 Univ Ill Urbana IL 61801

SHAPLEY, LLOYD STOWELL, b Cambridge, Mass, June 2, 23; m 55; c 2. GAME THEORY, MATHEMATICAL ECONOMICS. *Educ:* Harvard Univ, AB, 48; Princeton Univ, PhD(math), 53. *Hon Degrees:* PhD, Hebrew Univ Jerusalem, 86. *Prof Exp:* Mathematician, Rand Corp, 48-49; Henry B Fine instr math, Princeton Univ, 52-54; mathematician, Rand Corp, 54-55; sr res fel math, Calif Inst Technol, 55-56; mathematician, Rand Corp, 56-81; PROF MATH & ECON, UNIV CALIF, LOS ANGELES, 81- *Concurrent Pos:* Mem ed bd, Int J Game Theory, 70-, Math Prog, 71-, J Math Econ, 74- & Math Opers Res, 75- *Honors & Awards:* Von Neumann Theory Prize, Opers Res Soc Am, 81. *Mem:* Nat Acad Sci; Am Math Soc; Math Prog Soc; fel Economet Soc; fel Am Acad Arts & Sci. *Res:* Game theory and its application to economics and political science. *Mailing Add:* Dept Math & Econ Univ Calif Los Angeles Los Angeles CA 90024

SHAPLEY, ROBERT M, b New York, NY, Oct 7, 44; m 66; c 2. NEUROPHYSIOLOGY, BIOPHYSICS. *Educ:* Harvard Univ, AB, 65; Rockefeller Univ, PhD(biophys), 70. *Prof Exp:* Fel physiol, Northwestern Univ, 70-71 & Cambridge Univ, 71-72; from asst prof to assoc prof neurophysiol, Rockefeller Univ, 72-87; PROF PSYCHOL & BIOL, CTR NEURAL SCI, NY UNIV, 87- *Concurrent Pos:* assoc ed, J Gen Physiol, 83-, Visual Neurosci, 87-; MacArthur fel, 86; mem, Comt Vision, Nat Res Coun, 86-90; sensory physiol ed, Exp Brain Res, 90- *Mem:* Int Brain Res Orgn; Soc Neurosci; Asn Res Vision & Ophthal. *Res:* Visual neurophysiology; mathematical analysis of neural networks; visual perception. *Mailing Add:* Ctr Neural Sci New York Univ Six Washington Pl New York NY 10003

SHAPPIRIO, DAVID GORDON, b Washington, DC, June 18, 30; m 53; c 2. PHYSIOLOGY, CELL BIOLOGY. *Educ:* Univ Mich, BS, 51; Harvard Univ, AM, 53, PhD(biol), 55. *Prof Exp:* NSF fel, Molteno Inst, Cambridge Univ, 55-56; Nat Res Coun-Am Cancer Soc fel, Univ Louvain, 56-57; from instr to prof zool, Univ Mich, Ann Arbor, 57-75, assoc chmn, Div Biol Sci, 76-83, actg chmn, 76, 77, 78, & 80, actg chmn, Dept Cell & Molecular Biol, 75, PROF BIOL, UNIV MICH, ANN ARBOR, 75-, ARTHUR F THURNAU PROF, 89-, DIR, HONORS PROG, 83- *Concurrent Pos:* Lalor fel, Lalor Found, 53-55; Danforth assoc, Danforth Found, 68-; consult biol textbooks, res grant appl, NIH, NSF; grad training grants, NIH; vis lectr, Am Inst Biol Sci; consult, prof articles & books, various publishers. *Honors & Awards:* Bausch & Lomb Sci Award. *Mem:* Fel AAAS; Am Soc Cell Biol; Brit Soc Exp Biol; Am Soc Zoologists; Entom Soc Am; Am Inst Biol Sci; Xerces Soc. *Res:* Physiology and biochemistry of insect diapause; respiratory enzymology; cell biology; biology of wasps; undergraduate science education, especially honors/independent study experience. *Mailing Add:* Dept Biol Univ Mich 608 Soule Blvd Ann Arbor MI 48103-4625

SHAPTON, WILLIAM ROBERT, b Lansing, Mich, June 25, 41; m 63; c 2. VIBRATION, DESIGN. *Educ:* Mich State Univ, BS, 62, MS, 63; Univ Cincinnati, PhD(vibrations), 68. *Prof Exp:* Systs engr, Missile Div, Bendix Corp, 63-64, engr, Automation & Measurement Div, 64-65; assoc prof vibration, Univ Cincinnati, 65-79; PROF MECH ENG, MICH TECHNOL UNIV, 79- *Concurrent Pos:* Consult, Automation & Measurement Div, Bendix Corp, 65-68, Gen Elec Co, 69-70, Mound Lab, Monsanto Res Corp, 72-78, NASA, 76-82 & Cincinnati Milacron, 77-80; pres, Mich Technol Univ Senate, 86-89; mem bd dirs, Soc Automotive Engrs, 90-92. *Honors & Awards:* Medal of Honor, Soc Automotive Engrs, 89; Forest R McFarland Award, 86; Ralph Teetor Award, 79. *Mem:* Soc Automotive Engrs; Am Soc Mech Engrs; Am Soc Eng Educ. *Res:* Vibration, shock and measurement of mechanical systems; seismic response of structures; machine tool dynamics; experimental engineering; sound quality; noise, vibration and harshness. *Mailing Add:* Dept Mech Engr Mich Technol Univ Houghton MI 49931-1295

SHAR, ALBERT O, b Brooklyn, NY, Mar 11, 44. ALGEBRAIC TOPOLOGY, STATISTICS. *Educ:* Brandeis Univ, BA, 65; Fordham Univ, MS, 66; Univ Pa, PhD(math), 70. *Prof Exp:* Asst prof math, Univ Colo, 70-71; dir comput serv, Univ NH, 71-76 & 81-87, assoc prof, 76-81; EXEC DIR INFO TECHNOL, SCH MED, UNIV PA, 87- *Concurrent Pos:* Vis prof math, Res Inst Math, Zurich, Switz, 77-78. *Mem:* Am Math Soc. *Res:* Developing educational tools re: software & medical informational basis. *Mailing Add:* Dept Info Tech Univ Pa Sch Med 1R NEB Philadelphia PA 19104-6020

SHAR, LEONARD E, b SAfrica, 48; US citizen; m 86; c 1. HIGH SPEED COMPUTER ARCHITECUTURE. *Educ:* Univ Witwatersrand, BSc, 68; Stanford Univ, MS, 70, PhD(elec eng & comput sci), 72; Santa Clara Univ, MBA, 78. *Prof Exp:* Mgr, Hewlett Packard, 72-79; vpres, Elxsi Ltd, 79-88; PRIN, PANASYS INC, 89- *Concurrent Pos:* Vis lectr, Stanford Univ, 75. *Honors & Awards:* Archimedes Prize, Nat Sci Coun, 64. *Mem:* Inst Elec & Electronics Engrs; Asn Comput Mach. *Res:* All aspects of computer architecture & applications with emphasis on achieving highest performance. *Mailing Add:* 1101A Bay Laurel Dr Menlo Park CA 94025

SHARA, MICHAEL M, b Montreal, Que, Aug 12, 49; m 71; c 2. CATACLYSMIC BINARIES, WOLF-RAYET STARS. *Educ:* Univ Toronto, BSc, 71, MSc, 72; Tel-Aviv Univ, PhD(astrophys), 78. *Prof Exp:* Teaching fel, Univ Montreal, 78-80; vis asst prof, Ariz State Univ, 80-82; assoc astronr, 82-88, ASTRONR, SPACE TELESCOPE INST, 88- *Mem:* Am Astron Soc; Can Astron Soc; Int Astron Union. *Res:* Structure and evolution of cataclysmic binary stars; surveys of nearby galaxies for very luminous stars and hydrodynamical simulations of stellar collisions. *Mailing Add:* Space Telescope Sci Inst 3700 San Martin Dr Baltimore MD 21218

SHARAN, SHAILENDRA KISHORE, b Muzaffapur, Bihar, Nov 7, 47; Can citizen; m 72; c 2. STRUCTURE-FLUID INTERACTION, COMPUTATIONAL MECHANICS. *Educ:* Bilhar Univ, BSc, 69; Indian Inst Technol, Kanpur, MTech, 70; Queen's Univ, MSc, 75; Univ Waterloo, PhD(civil eng), 78. *Prof Exp:* Lectr civil eng, Indian Inst Technol, Delhi, 78; adj prof civil eng, Univ Estadual Maringa, Brazil, 79-81; from asst prof to assoc prof, 81-89, PROF CIVIL ENG, SCH ENG, LAURENTIAN UNIV, 89- *Concurrent Pos:* Vis fac, Indian Inst Technol, Delhi, 87-88. *Mem:* Can Soc Civil Eng; Am Soc Civil Engrs; Eng Inst Can; Can Inst Mining & Metall. *Res:* Seismic response analyses of structure-fluid systems such as dams and offshore structures; development of finite element techniques for infinite solid and fluid media; numerical modelling in geomechanics. *Mailing Add:* Sch Eng Laurentian Univ Sudbury ON P3E 2C6 Can

SHARAWY, MOHAMED, b Cairo, Egypt, Mar 13, 41; US citizen; m 65; c 3. DENTISTRY. *Educ:* Cairo Univ, PNS, 58, DDS, 62; Univ Rochester, PhD(anat), 70. *Prof Exp:* Intern oral surg, Sch Dent, Cairo Univ, 62, instr, 62-65; Fulbright fel, Inst Int Educ, 65-68; UAR scholar, Univ Rochester, 68-70; asst prof, Sch Dent, 70-73, assoc prof, 73-78, PROF ORAL BIOL & ANAT & COORD ANAT DENT, MED COL GA, 78- *Concurrent Pos:* Consult, electron micros, US Army. *Honors & Awards:* Outstanding Contrib to Dent Medal, Egyptian Dent Asn. *Mem:* Int Asn Dent Res; Am Asn Dent Res; An Asn Anatomists; Am Asn Dent Sch; Am Dent Asn; Am Acad Implant Prosthodontics. *Res:* Mechanism of bone induction, using demineralized bone powder; development of animal models for TMJ disc dislocation. *Mailing Add:* Dept Oral Biol Sch Dent Med Col Ga Augusta GA 30912

SHARBAUGH, AMANDUS HARRY, b Richmond, Va, Mar 28, 19; m 40; c 2. CHEMISTRY, PHYSICS. *Educ:* Western Reserve Univ, AB, 40; Brown Univ, PhD(chem), 43. *Prof Exp:* Res assoc, Res Lab, Gen Elec Co, 42-61, liaison scientist, 61-64, mgr dielec studies, 64-71, mgr, Plasma Physics Br, 71-80, sr consult, 80-83; PVT CONSULT, 83- *Concurrent Pos:* Fel physics, Union Univ, NY, 44-48; secy, Conf Elec Insulation, 54, vchmn, 55, chmn 56; mem adv comt, US Dept Defense, 56-57; mem conf elec insulation & dielec behav, Nat Acad Sci; US Adv to CIGRE; Holder world's record microwave communication. *Honors & Awards:* Potter Prize, Brown Univ; Dakin Award, Inst Elec & Electronics Engrs. *Mem:* Electrochem Soc; fel Inst Elec & Electronics Engrs. *Res:* Dielectric behavior; magnetron design; electronic breakdown and conduction; microwave spectroscopy. *Mailing Add:* 28 Hemlock Dr Clifton Park NY 12065

SHARBER, JAMES RANDALL, b Clarksville, Tenn, Aug 13, 41; m 74; c 1. SPACE PHYSICS, AURORAL MAGNETOSPHERIC PHYSICS. *Educ:* Murray State Col, BA, 63; Tex A&M Univ, PhD(physics), 72. *Prof Exp:* Res asst scientist, Div Atmospheric & Space Sci, Univ Tex, Dallas, 66-72; vis asst prof physics, US Naval Acad, 72-74; assoc prof physics & space sci, Fla Inst Technol, 74-84; STAFF SCIENTIST, DEPT SPACE SCI, SOUTHWEST RES INST, 84- *Concurrent Pos:* consult, Southwest Res Inst, 82-83. *Honors & Awards:* Consult, Boston Clin. *Mem:* Am Geophys Union; Am Asn Physics Teachers; Am Inst Physics. *Res:* Analysis of scientific data taken by earth satellites and ground observatories; descriptions of the electron and ion fluxes incident on the earth's auroral regions to determine the processes responsible for the entry of solar wind particles into the magnetosphere and their subsequent acceleration and precipitation to produce auroras and deposit energy into the atmosphere; instrumentation for particle measurements in the upper atmosphere and space. *Mailing Add:* 309 E Skyview San Antonio TX 78228

SHARE, GERALD HARVEY, b New York, NY, Oct 9, 40; m 80; c 2. GAMMA-RAY ASTRONOMY. *Educ:* Queens Col, NY, BS, 61; Univ Rochester, PhD(physics), 66. *Prof Exp:* Nat Acad Sci-Nat Res Coun resident res assoc cosmic radiation, 66-68, ASTROPHYSICIST, NAVAL RES LAB, 68- *Concurrent Pos:* Vis sr staff scientist, Astrophys Div, NASA Hq, 89-91. *Mem:* Am Phys Soc; Am Astron Soc; Sigma Xi. *Res:* Cosmic x-and-gamma radiation; investigation of solar and celestial gamma radiation; measurements of radioactive cobalt in supernova; diffuse Galactic continuum and lines from 26Al and positron annihilation; spectra of cosmic bursts; space background. *Mailing Add:* Naval Res Lab Code 4152 Washington DC 20375

SHARE, LEONARD, b Detroit, Mich, Oct 14, 27; m 49; c 3. PHYSIOLOGY. *Educ:* Brooklyn Col, AB, 47; Oberlin Col, AM, 48; Yale Univ, PhD(physiol), 51. *Prof Exp:* USPHS fel physiol, Sch Med, Western Reserve Univ, 51-52; from instr to prof physiol, 52-69; PROF PHYSIOL & BIOPHYS & CHMN DEPT, UNIV TENN CTR HEALTH SCI, MEMPHIS, 69- *Mem:* AAAS; Int Soc Neuroendocrinol; Endocrine Soc; Am Physiol Soc; Neurosci Soc. *Res:* Water and electrolyte metabolism; vasopressin secretion; metabolism. *Mailing Add:* Dept Physiol & Biophys Univ Tenn Ctr Health Sci Memphis TN 38163

SHARE, NORMAN N, b Montreal, Que, Dec 25, 30; m 56; c 1. NEUROPHARMACOLOGY. *Educ:* Univ Montreal, BPh, 58; McGill Univ, PhD(pharmacol), 62. *Prof Exp:* Fel, Columbia Univ, 62-64; chief pharmacologist, Charles E Frosst & Co, 64-65, mgr, Dept Pharmacol, 65-71, asst dir res biol, Dept Pharmacol, 71-79, dir pharmacol, 79-81, sr dir neuropsychopharmacol, Merck Inst Therapeut Res, 82-86; VPRES PHARMACOL, PAN LABS INT, INC, 86- *Mem:* Pharmacol Soc Can; Am Soc Pharmacol & Exp Therapeut; NY Acad Sci. *Res:* Chronic obstructive lung disease. *Mailing Add:* Pan Labs Int Inc 11804 North Creek Pkwy S Bothell WA 98011-8805

SHARER, ARCHIBALD WILSON, b Dayton, Ohio, Sept 19, 19; m 41; c 3. ZOOLOGY. *Educ:* Ohio State Univ, BS, 43; Univ Mich, MS, 48, PhD(zool), 59. *Prof Exp:* Asst zool, Univ Mich, 46-50; instr biol, Fla State Univ, 50-53; instr, Lake Forest Col, 53-56; instr zool, Duke Univ, 58-60; assoc prof biol, 60-63, chmn div sci, 63-77, PROF BIOL, NC WESLEYAN COL, 63- *Mem:* Am Arachnological Soc; Am Inst Biol Sci. *Res:* Behavior, ecology and natural history of vertebrates and invertebrates; herpetology; arachnology. *Mailing Add:* Dept Biol NC Wesleyan Col 3400 N Wesleyan Blvd Rocky Mt NC 27804

SHARER, CYRUS J, b Cleveland, Ohio, Mar 8, 22; m 55; c 2. ECONOMIC GEOGRAPHY, POPULATION. *Educ:* Univ Pa BS, 43, MA, 49; Univ Mich, PhD(geog), 55. *Prof Exp:* Instr ecol geog, Wharton Sch, Univ Pa, 48-50, lectr, 53-54; from instr to prof geog, Villanova Univ, 54-85, chmn dept, 62-81; RETIRED. *Mem:* Asn Am Geographers; Soc Hist Discoveries. *Res:* US and world patterns of iron and steel production; resource management and population change in the Bahama Islands and West Indies. *Mailing Add:* 505 E Lancaster Ave Apt 418 St Davids PA 19087

SHARF, DONALD JACK, b Detroit, Mich, Aug 4, 27; m 52; c 2. SPEECH & HEARING SCIENCES. *Educ:* Wayne State Univ, BA, 51, MA, 52; Univ Mich, PhD(speech sci), 58. *Prof Exp:* Asst ed, G & C Merriam Co, 57-61; asst prof speech, State Univ NY, Buffalo, 61-64; from asst prof to assoc prof, 64-73, PROF SPEECH SCI, UNIV MICH, ANN ARBOR, 73- *Mem:* Am Speech & Hearing Asn; Acoust Soc Am; Int Soc Phonetic Sci; Am Asn Phonetic Sci. *Res:* Acoustic and perceptual aspects of speech communication. *Mailing Add:* 3253 Lockridge Dr Ann Arbor MI 48104

SHARGEL, LEON DAVID, b Baltimore, Md, Nov 18, 41; m 67; c 2. PHARMACOLOGY, DRUG METABOLISM. *Educ:* Univ Md, BS, 63; George Washington Univ, PhD(pharmacol), 69. *Prof Exp:* Assoc res biologist, Sterling-Winthrop Res Inst, NY, 69-71, group leader, 72-75; asst prof, 75-77, ASSOC PROF, PHARM & PHARMACOL, COL PHARM & ALLIED HEALTH PROF, NORTHEASTERN UNIV, 77- *Concurrent Pos:* Mem, Drug Formulary Comn; consult, pharmaceut indust. *Mem:* AAAS; Am Soc Pharmacol & Exp Therapeut; Sigma Xi; Am Asn Pharmaceut Scientists; Am Pharmaceut Asn. *Res:* Physiological disposition of drugs, drug metabolism, biopharmaceutics; pharmacokinetics. *Mailing Add:* Dept Pharm Mass Col Pharm 179 Longwood Ave Boston MA 02115

SHARGOOL, PETER DOUGLAS, b Feb 25, 35; Can citizen; m 61; c 2. PLANT BIOCHEMISTRY. *Educ:* Univ London, BSc, 62; Univ Alta, MSc, 65, PhD(plant biochem), 68. *Prof Exp:* Res worker biochem, Wellcome Res Labs, Eng, 54-62; Nat Res Coun Can grant, 68-73, assoc prof, 73-79, PROF BIOCHEM, UNIV SASK, 79- *Concurrent Pos:* Med Res Coun grant, 75-77, Nat Sci & Engr Res Coun grant, 77-; Nat Res Coun res contract, 82-86. *Res:* Mechanisms regulating the biosynthesis of amino acids and other secondary products in plants. *Mailing Add:* Dept Biochem Health Sci Bldg Univ Sask Saskatoon SK S7N 0W0 Can

SHARITZ, REBECCA REYBURN, b Wytheville, Va, Aug 10, 44; m 76. ECOLOGY, BOTANY. *Educ:* Roanoke Col, BS, 66; Univ NC, Chapel Hill, PhD(bot), 70. *Prof Exp:* Asst prof biol, Saginaw Valley Col, 70-71; assoc res ecologist, 77-85, SR RES ECOLOGIST, SAVANNAH RIVER ECOL LAB, UNIV GA, 86-, HEAD, DIV WETLANDS ECOL, 88-, PROF BOT, 89- *Concurrent Pos:* Res assoc, Univ NC, 71; adj asst prof bot, 72-; plant ecologist, US Energy Res & Develop Admin, 75-76; acting div, Savannah River Ecol Lab, 87. *Mem:* Bot Soc Am; Am Inst Biol Sci; Ecol Soc Am (treas, 87-90, vpres 90-91); Int Asn Ecol (secy-gen, 90-). *Res:* Population dynamics of vascular plant species, structure and processes in swamp forest systems, response of wetland communities to environmental disturbance; structure and diversity of plant communities. *Mailing Add:* Savannah River Ecol Lab Drawer E Univ Ga Aiken SC 29802

SHARKAWI, MAHMOUD, b Cairo, Egypt, May 26, 35; m 70; c 4. PHARMACOLOGY. *Educ:* Cairo Univ, BPharm, 57; Univ Minn, Minneapolis, MSc, 61; Univ Calif, San Francisco, PhD(pharmacol), 64. *Prof Exp:* Lectr pharmacol, Cairo Univ, 64-65; res assoc, Univ Ill, 65-66; res assoc, Stanford Univ, 66-67; from instr to assoc prof, 68-80, PROF PHARMACOL, FAC MED, UNIV MONTREAL, 80- *Mem:* AAAS; Int Soc Biochem Pharmacol; Am Soc Pharmacol & Exp Therapeut; Pharmacol Soc Can. *Res:* Interactions between ethanol and centrally acting drugs; factors modifying ethanol elimination. *Mailing Add:* Dept Pharmacol Univ Montreal Fac Med Montreal PQ H3C 3J7 Can

SHARKEY, JOHN BERNARD, b Elizabeth, NJ, Sept 5, 40; m 63; c 3. INORGANIC CHEMISTRY, INSTRUMENTATION. *Educ:* NY Univ, BA, 64, MSc, 68, PhD(phys chem), 70. *Prof Exp:* Anal chemist, Engelhard Industs, NJ, 62-66; from asst prof to prof chem & phys sci, 70-90, chairperson, 77-90, ASSOC DEAN, PACE UNIV, 90- *Mem:* Am Chem Soc; AAAS; NY Acad Sci. *Res:* Polymorphism in inorganic compounds; HPLC analysis of drugs in serum; microscopy. *Mailing Add:* Dept Chem & Phys Sci Pace Univ New York NY 10038

SHARKEY, MARGARET MARY, CELL BIOLOGY. *Educ:* Fordham Univ, BS, 54, MS, 60; St John's Univ, NY, PhD(cell biol), 72. *Prof Exp:* Teaching asst cytol, histol & embryol, St John's Univ, NY, 69-70; asst prof cell biol, histol & embryol, St Thomas Aquinas Col, 70-72; SCI PROG DIR, TRAINING GRANTS, 88- *Mem:* Am Asn Cancer Res. *Res:* Administration of cell biology; genetics. *Mailing Add:* 1599 Clifton Rd Atlanta GA 30329

SHARKEY, MICHAEL JOSEPH, b Kitchener, Ont, Nov 2, 53; m 79; c 2. SYSTEMATICS, TAXONOMY. *Educ:* Univ Guelph, BSc, 77; McGill Univ, MSc, 81, PhD(entom), 84. *Prof Exp:* RES SCIENTIST, BIOSYST RES INST, AGR CAN, 82- *Concurrent Pos:* Ed, Ichnews, 84- *Mem:* Entom Soc Can; Int Hymenopterists Soc. *Res:* Taxonomy of Hymenoptera, especially Braconidae; biological control; systematic theory; quantitative aspects of phylogenetic reconstruction. *Mailing Add:* Biosyst Res Inst CEF Ottawa ON K1A 0C6 CAN

SHARKEY, THOMAS D, b Detroit, Mich, Jan 28, 53; m 74; c 1. ANALYTICAL GAS EXCHANGE, MODELING. *Educ:* Mich State Univ, BS, 74, PhD(bot & plant path), 80. *Prof Exp:* Fel, Res Sch Biol Sci, Australian Nat Univ, 80-82; asst res prof, Biol Sci Ctr, Desert Res Inst, 82-84, assoc res prof & assoc dir, 84-87; assoc prof, Biol Dept, Univ Nev, Reno, 86-87; from asst prof to assoc prof, Dept Bot, 87-91, PROF, DEPT BOT, UNIV WISMADISON, 91- *Concurrent Pos:* Vis, Dept Plant Biol, Carnegie Inst Wash, 80, Univ Göttingen, 85; mem, Prog Comt, Am Soc Plant Physiologists, 87-88; panel mem, Physiol Processes Prog, NSF, 88-91. *Mem:* AAAS; Am Soc Plant Physiologists; Australian Soc Plant Physiologists. *Res:* Biochemistry and biophysics of the exchange of gases between plant leaves and the atmosphere and environmental effects on plant-atmosphere interactions; photosynthesis and isoprene emission from plants. *Mailing Add:* Dept Bot Univ Wis Madison WI 53706

SHARKEY, WILLIAM HENRY, b Vinita, Okla, Oct 7, 16; m 42; c 2. PHYSICAL CHEMISTRY. *Educ:* Okla State Univ, BS, 37; Univ Ill, PhD(org chem), 41. *Prof Exp:* Res chemist, org res, Gen Elec Co, 39; res chemist, Du Pont Co, 40 & 41-53, res supvr chem, 53-80, admin assoc, 81; adj prof chem, Univ Miami, 84-87; RETIRED. *Mem:* Sigma Xi; Am Chem Soc; AAAS. *Res:* Organic synthesis; dye chemistry and dyeing materials; photooxidative processes; four-membered ring chemistry; fluorine chemistry including fluoropolymers; polymerization methods especially anionic and coordination polymerization. *Mailing Add:* 6001 Pelican Bay Blvd No 1502 Naples FL 33963-8168

SHARKOFF, EUGENE GIBB, b Washington, DC, Feb 1, 25; m 59; c 2. GENERAL PHYSICS, SCIENCE ADMINISTRATION. *Educ:* US Mil Acad, BMAS, 46; Mass Inst Technol, SM, 52, PhD(physics), 53. *Prof Exp:* Assoc prof physics, USAF Inst Technol, 63-66; dir, PPAAR Prog, Res Avionics, Princeton Univ, 66-69; var res & tech mgt, Picatinny Arsenal, USA, 69-80; instr physics, Isothermal Community Col, 84-88; pres, Rutherford County Habitat for Humanity, 89-90; RETIRED. *Honors & Awards:* Cert Adv Group Aerospace Res & Develop, NAtlantic Treaty Orgn, 59. *Mem:* Am Asn Physics Teachers; Sigma Xi. *Res:* Solid state thermal physics-measured thermal conductivity of magnesium at low temperature; structures with asymmetric loading; mechanics; one published manual. *Mailing Add:* Rte 1 Box 157 Union Mills NC 28167

SHARLOW, JOHN FRANCIS, b Potsdam, NY, Aug 28, 41; m 63; c 2. HARDWARE SYSTEMS. *Educ:* State Univ NY, Potsdam, BS, 63, MS, 65; Clarkson Col Technol, MS, 67; State Univ NY, Albany, EdD, 71. *Prof Exp:* Asst prof math, State Univ Agr & Technol Col, 65-69; PROF COMPUTER SCI & MATH, EASTERN CONN STATE UNIV, 69- *Mem:* Asn Comput Mach; Math Asn Am. *Res:* Mathematics education; computer science education. *Mailing Add:* 458 Beaumont Hwy Lebanon CT 06249

SHARMA, ARJUN D, b Bombay, India, June 2, 53; Can citizen; m 81; c 2. CARDIOLOGY. *Educ:* Univ Waterloo, BSc, 72; Univ Toronto, MD, 76. *Prof Exp:* Asst prof med, 83-88, assoc prof med & pharmacol, Univ Western Ont, 88-89; DIR, INTERVENTIONAL ELECTROPHYSIOL, SUTTER HOSP, 89- *Mem:* Am Col Cardiol; Am Col Physicians; Can Cardiovasc Soc. *Res:* Cardiac electrophysiology; andrenergic receptors in heart; adenosine effects on heart. *Mailing Add:* 3941 J St Suite 260N Sacramento CA 95819

SHARMA, BRAHMA DUTTA, b Khurja, India, Sept 20, 47: US citizen; c 3. DENTAL BIOMATERIALS RESEARCH. *Educ:* Agra Univ, India, BS, 66; Allahabad Univ, MS, 68; Queens Univ, Belfast, UK, PhD(chem), 71. *Prof Exp:* DIR RES, DENT MAT, COE LABS, INC, CHICAGO, 73- *Mem:* Am Chem Soc; AAAS; fel Acad Dent Mat; Int Asn Dent Res. *Res:* Dental biomaterial such as impression materials, resins & restoratives. *Mailing Add:* Coe Labs Inc 3737 W 127th St Chicago IL 60658

SHARMA, DINESH C, b Feb 1, 38. ENDOCRINOLOGY, SYSTEMS DESIGN. *Educ:* Kans State Univ, PhD, 61; Bombay Univ, India, DSc, 67; Univ Mich, MSE, 71. *Prof Exp:* HEALTH SCI ADMINR, NAT INST CHILD HEALTH & HUMAN DEVELOP, NIH, 72- *Concurrent Pos:* Regent's prof, Univ Calif. *Mem:* Inst Elec & Electronics Engrs; fel Royal Soc Chem. *Res:* Biochemistry of steroids; drug delivery systems. *Mailing Add:* 10111 Counselman Rd Potomac MD 20854

SHARMA, GHANSHYAM D, b Delhi, India, Feb 28, 31; m 61; c 1. MARINE GEOLOGY, GEOCHEMISTRY. *Educ:* Benares Hindu Univ, BSc, 52; Swiss Fed Inst Technol, Dipl, 58; Univ Mich, PhD(geol), 61. *Prof Exp:* Res engr, Sinclair Res, Inc, 61-63; from asst prof to prof, marine geol & geochem, Inst Marine Sci, 63-78, prof marine sci, Alaska Sea Grant, 78-84, DIR, PETROL DEVELOP LAB, UNIV ALASKA, FAIRBANKS, 84- *Honors & Awards:* Pres Award, Am Asn Petrol Geologists. *Mem:* Am Asn Petrol Geologists; Soc Econ Paleontologists & Mineralogists; Am Geophys Union; Soc Petrol Eng; Int Asn Sedimentologists. *Res:* Concentration and distribution of elements in marine water and sediments of arctic and subarctic in Alaska; study of diagenesis to reconstruct the paleoecology and environments of deposition; sediment transport and distribution on continental shelf. *Mailing Add:* Dept Petroleum & Mining Eng Univ Alaska Fairbanks AK 99701

SHARMA, GOPAL CHANDRA, b Churi-Ajitgarh, India, Apr 18, 32; US citizen; m 60; c 2. PHARMACOLOGY, CANCER. *Educ:* Agra Univ, BS, 53, DVM, 58; Univ Mo, Columbia, MS, 66, PhD(med pharmacol), 69. *Prof Exp:* Officer prev immunization, Dept Animal Husb, Govt India, 58-59, vet surgeon asst, 59-61; res asst genetics, Indian Coun Agr Res, 61-65; res asst med pharmacol, Univ Mo, Columbia, 66-67, Nat Cancer Inst spec proj fel, Med Sch, 67-69; asst prof pharmacol, Sch Med, Univ NDak, 69-70; sr res scientist cancer res, 70-73, dir plant res, Am Med Ctr, Denver, 73-77, PHARMACOLOGIST, PHARMACOL & TOXICOL BR, DIV VET MED RES, BUR VET MED, FOOD & DRUG ADMIN, HEW, 77- *Res:* Evaluation of anticancer drugs; effect of cancer on disposition of drugs in the body; drug toxicities and prevention; drug distribution and interaction; cyclic adenosine monophosphate; comparative pharmacology; drug interaction and analytical microdetection of drug and chemical residue in the tissue of food animals, effect of drugs on the metabolism of other drugs. *Mailing Add:* Two Holliben Ct Severna Park MD 21146

SHARMA, GOPAL KRISHAN, b Hoshiarpur, India, Mar 24, 37; m 66. ECOLOGY. *Educ:* Univ Mo, Columbia, BS, 61, MS, 62, PhD(bot), 67. *Prof Exp:* Teaching asst bot, Univ Mo, Columbia, 62-66; instr, Prestonsburg Community Col, Ky, 66-67, chmn div biol sci, 67-68; assoc prof, 68-76, PROF BIOL SCI, UNIV TENN, MARTIN, 76- *Concurrent Pos:* Instr, Cult Enrichment Prog, Prestonsburg, Ky, 66; NSF fel, Univ Mich, 68-69; res fel econ bot, Harvard Univ, 76-77. *Mem:* AAAS; Am Inst Biol Sci; Bot Soc Am; Ecol Soc Am; Sigma Xi. *Res:* Ecology and ethnobotany of Cannabis; cuticular features as indicators of environmental pollution. *Mailing Add:* Dept Biol Univ Tenn Martin TN 38238

SHARMA, GOVIND C, b Udaipur, India, Mar 3, 44; m 68; c 1. PLANT SCIENCE. *Educ:* Univ Udaipur, India, BS, 64; Univ Fla, MAgr, 65; Kans State Univ, PhD(veg crops), 70. *Prof Exp:* Res asst, Pesticide Lab, Univ Fla, 65-66; agr coordr, Peace Corps, 66; assoc prof, 70-73, PROF PLANT SCI & CHMN DEPT PLANT & SOIL SCI, ALA A&M UNIV, 73- *Concurrent Pos:* Consult int agr, USAID-sponsored projs; consult inst planning, area univs & cols. *Mem:* Am Soc Hort Sci; Am Soc Agron. *Res:* Plant breeding; tissue culture; host plant resistance and pesticide analysis. *Mailing Add:* PO Box 1208 Ala A&M Univ Normal AL 35762

SHARMA, HARI M, b Aligarh, India, Jan 16, 38; m; c 4. KIDNEY DISEASES. *Educ:* Lucknow Univ, India, MD, 61. *Prof Exp:* Dir, Anat Path Prog, 81-90, PROF PATH, OHIO STATE UNIV, 80-, DIR, CANCER PREV & NATURAL PROD RES, 90- *Concurrent Pos:* Ayurveda, Health Care Syst, India. *Mem:* Int Soc Nephrol; Int Acad Path; Am Col Pathologists. *Res:* Ayurveda; cancer; atherosclerosis; natural products. *Mailing Add:* Dept Path M 376 Starling-Loving Hall Ohio State Univ 320 W Tenth Ave Columbus OH 43210

SHARMA, JAGADISH, b Calcutta, India, Dec 15, 23; US citizen; m 51; c 2. SOLID STATE PHYSICS, ELECTRON SPECTROSCOPY. *Educ:* Calcutta Univ, BS, 44, MS, 47, PhD(physics), 53. *Prof Exp:* Nat Inst Res fel thermoluminescence, Khaira Physics Lab, Calcutta Univ, 53-54; res fel, Color

Ctr, Nat Res Coun, Div Elec Eng, Ottawa, Can, 55-57; lectr, dept physics, Indian Inst Technol, Kharagpur, 58-61, asst prof radiation damage, 65-67; res assoc, dept aerospace & mech sci, Princeton Univ, 61-64; res physicist, Div Explosives & Electron Spectros, Energetic Mat Res & Develop Comn, US Army, 67-80; PHYSICIST, NAVAL SURFACE WEAPONS CTR, 80- *Mem:* AAAS. *Res:* Fluorescence; thermoluminescence; color centers; radiation damage; solid state physics of explosives and propellants; Auger, uv, x-ray photoelectron spectroscopy of explosives and propellants. *Mailing Add:* Mat Eval Br Naval Surface Weapons Ctr White Oak Silver Springs MD 20910

SHARMA, JAGDEV MITTRA, b Punjab, India, June 28, 41; m 69. MICROBIOLOGY, VIROLOGY. *Educ:* Punjab Univ, India, BVSc, 61; Univ Calif, Davis, MS, 64, PhD(comp path), 67. *Prof Exp:* Jr specialist avian med, Univ Calif, Davis, 62-66; poultry pathologist, Wash State Univ, 67-71; vet med officer, Regional Poultry Res Lab, Sci & Educ Admin-Agr Res, USDA, 71-88; clin prof, dept path, Mich State Univ, 74-88; BENJAMIN POMEROY PROF, COL VET MED, UNIV MINN, 88- *Honors & Awards:* Upjohn Award, Am Asn Avian Pathologists, 82. *Mem:* Am Vet Med Asn; Am Asn Avian Pathologists; Int Asn Comp Res Leukemia & Related Dis; World Vet Poultry Asn. *Res:* Avian immunology; mechanisms of virus-induced neoplasms; infectious diseases. *Mailing Add:* Vet Pathobiol Col Vet Med Univ Minn St Paul MN 55108

SHARMA, MADAN LAL, b Faridkot, India, Aug 28, 34; Can citizen; m 68. ENTOMOLOGY, ECOLOGY. *Educ:* Univ Punjab, India, BSc, 54 & 56, MSc, 57; Univ Paris, DSc(entom), 65. *Prof Exp:* Scientist, Agr Dept, French Govt, 65-66; lectr, 66-68, from asst prof to assoc prof, 68-78, PROF ENTOM, UNIV SHERBROOKE, 78- *Concurrent Pos:* French Govt fel, Nat Inst Agron Res, 71; res fel, Quebec-France & France-Can Mission, Nat Inst Agron Res, France, 81. *Res:* Biology and ecology of Aphids and Coccids; bibliography of Aphidoidea. *Mailing Add:* 1510 Belvedere St Asct PQ J1K 1A1 Can

SHARMA, MANGALORE GOKULANAND, b Mulki, India, Nov 3, 27; US citizen; m 62; c 2. BIOMECHANICS, RHEOLOGY. *Educ:* Univ Mysore, India, BE, 52; Indian Inst Sci, DIISc, 54; Pa State Univ, PhD(eng mech), 60. *Prof Exp:* Design engr, Hindustan Aeronaut, 54-57; res asst eng mech, 57-60, asst prof, 60-64, assoc prof, 64-75, PROF ENG MECH, PA STATE UNIV, 75- *Concurrent Pos:* Vis prof, Indian Inst Sci, 67-68; consult, Gen Tire & Rubber Co, 64-68 & Fed Hgy Admin, 81- *Mem:* Am Soc Rheology; Am Soc Testing & Mat; Am Acad Mech; Am Soc Testing & Mat Comt Composite Mat; Sigma Xi. *Res:* Analysis of hemodynamic flow through blood vessels; fluid dynamical contribution to atherosclerosis; rheology of blood vessels; effect of hypertension on rheological properties; physical properties of biopolymers used in cardiac assist devices; mechanics of composite materials and their failure analysis; viscoplasticity of metals at elevated temperatures. *Mailing Add:* 948 S Sparks State Col PA 16801

SHARMA, MINOTI, b India, US citizen. SEPARATION TECHNIQUES NUCLEIC ACID SYNTHESIS, INSTRUMENTAL DEVELOPMENT. *Educ:* Tufts Univ, MS, 65; Southampton Univ, UK, PhD(org chem), 70. *Prof Exp:* Res asst prof bioenergetics, State Univ NY, Buffalo, 77-80; CANCER RES SCIENTIST BIOPHYS, ROSWELL PARK CANCER INST, 81-, ASST RES PROF, 84- *Mem:* Am Chem Soc; Soc Biol Chem; AAAS. *Res:* Developing new technology to assay DNA modification using laser-induced flourescence detection. *Mailing Add:* Dept Biophys Roswell Park Cancer Inst Buffalo NY 14263

SHARMA, MOHESWAR, b Jhanji, India; US citizen; m 62; c 2. ORGANIC CHEMISTRY. *Educ:* Calcutta Univ, BS, 53, MS, 56, PhD(chem), 62. *Prof Exp:* Fel chem, Tufts Univ, 63-65, Univ Alta, 65-67; vis lectr, Univ Southampton, 67-72; CANCER RES SCIENTIST CARBOHYDRATES, ROSWELL PARK MEM INST, 72- *Mem:* Am Chem Soc. *Res:* Carbohydrates; cell membranes. *Mailing Add:* 281 Coronation Dr Amherst NY 14226

SHARMA, OPENDRA K, b Dehradun, India, Sept 3, 41. MECHANISMS OF INTERFERON ACTION. *Educ:* Lucknow Univ, India, PhD(biochem), 66. *Prof Exp:* Sr scientist, 77-85, CHIEF, LAB MOLECULAR BIOL, CANCER RES CTR, AMA, 85- *Mem:* Am Soc Biol Chemists; Am Asn Cancer Res; Am Chem Soc. *Res:* Nucleotide analogs and anti-viral agents application of urinary toxicity of heavy metals and modified nucleosides in cancer diagnosis and cancer management. *Mailing Add:* NIH Bldg 31 5B-32 Bethesda MD 20892

SHARMA, RAGHUBIR PRASAD, b Bharatpur, India, Sept 9, 40; m 58; c 2. PHARMACOLOGY, TOXICOLOGY. *Educ:* Univ Rajasthan, BVSc, 59; Univ Minn, St Paul, PhD(pharmacol), 68; Am Bd Vet Toxicol, dipl, 74; Am Bd Toxicol, dipl, 81. *Prof Exp:* Clin vet, Govt Rajasthan, India, 59-60; instr pharmacol, Univ Rajasthan, 60-61; asst prof, Uttar Pradesh Agr Univ, India, 61-64; from asst prof to assoc prof, 69-79, PROF & TOXICOLOGIST ANIMAL, DAIRY & VET SCI, UTAH STATE UNIV, 79- *Concurrent Pos:* NIH grant, Utah State Univ, 69-72, NIMH res grant, 70-71; Food & Drug Admin res grant, 74-77, Nat Inst Environ Health Sci grant, 79-85, March of Dimes Found grant, 87-; mem, Am Conf Indust Hyg, 86- *Mem:* Am Soc Pharmacol & Exp Therapeut; Soc Toxicol. *Res:* Pharmacology and toxicology of selected chemicals; mechanisms, metabolism and biochemical alterations; neurochemical alterations and molecular interactions; immunotoxicology. *Mailing Add:* Toxicol Prog Dept Animal Dairy & Vet Sci Logan UT 84322-5600

SHARMA, RAM ASHREY, b Azamgarh, India, Sept 5, 43; m 61; c 3. MEDICINAL CHEMISTRY. *Educ:* Univ Gorakhpur, India, BS, 63, MS, 66; Univ Roorkee, PhD(chem), 70. *Prof Exp:* Asst prof chem, K N Govt Col Gyanpur, Varanasi, India, 66-68; res assoc, Coun Sci & Indust Res, New Delhi, 68-70; asst prof, Univ Roorkee, 70-71; CANCER RES SCIENTIST MED CHEM, ROSWELL PARK MEM INST, BUFFALO, 71- *Mem:* Am Chem Soc. *Res:* Synthesis and biological evaluation of metabolite analogs of purines and pyrimidines of potential medicinal interest. *Mailing Add:* 422 Carmen Rd Eggertsville NY 14226

SHARMA, RAM AUTAR, b Narnaul, Haryana, India, Aug 20, 27; US citizen. THERMODYNAMICS & MATERIAL PROPERTIES, METALLURGY ENGINEERING. *Educ:* Banaras Hindu Univ, India, MSc, 52; London Univ, UK, PhD(chem metall), 63. *Prof Exp:* Sr sci officer, Nat Metall Lab, Jamshedpur, India, 53-59; sr res fel, Univ Pa, Philadelphia, 63-66; res chemist, Argonne Nat Lab, Ill, 66-69; SR STAFF SCIENTIST, PHYS CHEM DEPT, GEN MOTORS RES LABS, 69- *Mem:* Electrochem Soc; Am Ceramic Soc. *Mailing Add:* Phys Chem Dept Gen Motors Res Labs Warren MI 48090-9055

SHARMA, RAM RATAN, b Jaipur, India, Oct 6, 36; m 67; c 1. PHYSICS, SOLID STATE PHYSICS. *Educ:* Maharaja's Col, Jaipur, BS, 58; Univ Bombay, MS, 62; Univ Calif, Riverside, MA, 64, PhD(physics), 65. *Prof Exp:* Lectr physics & chem, Indian Inst, Jaipur, 57-58; sci officer physics, Atomic Energy Estab, Bombay, 58-62; asst, Univ Calif, Riverside, 62-65; res assoc, Purdue Univ, West Lafayette, 65-68, Argonne Nat Lab, Argonne, Ill, 71-89; assoc prof, 68-72, PROF PHYSICS, UNIV ILL, CHICAGO, 72- *Concurrent Pos:* Advan Res Proj Agency fel, Purdue Univ, West Lafayette, 65-68; res assoc, Argonne Nat Lab, 71-; vis scientist, Atomic Energy Res Estab, Eng, 74; vis prof, Univ Liverpool, 75. *Mem:* Fel Am Phys Soc; Biophys Soc Am Phys Soc. *Res:* Solid state physics; properties of magnetic ions in solids and biological systems; bound-excitons and biexcitons in semiconductors, bound magnetic polarons in dilute magnetic semiconductors; electron-nuclear interactions in solids; shielding effects; polarizabilities of ions; surface color-centers; nuclear fusion; transport of drugs through blood-brain barrier; recrystallization of amorphous substances; epitaxial growth; high temperture superconductivity; muon spin resonance; author of one book. *Mailing Add:* Dept Physics Univ Ill Chicago Box 4348 Chicago IL 60680

SHARMA, RAMESH C, b Delhi, India, June 5, 31. NUCLEAR PHYSICS, SOLID STATE PHYSICS. *Educ:* Univ Delhi, BSc, 50, MSc, 52, MA, 53; Univ Toronto, PhD(nuclear physics), 59. *Prof Exp:* Lectr physics, Ramjas Col, Delhi, 55-56; res asst, Univ Toronto, 56-59; res fel, Ont Cancer Inst, Can, 59-60; lectr, Ramjas Col, Delhi, 60-61; lectr, Col Eng & Technol, Delhi, 61-62; asst prof, 62-69, ASSOC PROF PHYSICS, SIR GEORGE WILLIAMS CAMPUS, CONCORDIA UNIV, 69- *Concurrent Pos:* Nat Res Coun Can grant, 64-65. *Mem:* Am Soc Eng Educ; Can Asn Physicists; Am Phys Soc. *Res:* Nuclear structure, nuclear fission. *Mailing Add:* Dept Physics 1455 Boul De Maisonneuve W Montreal PQ H3G 1M8 Can

SHARMA, RAMESHWAR KUMAR, b Jammu, India, Dec 15, 35; m 65. MEDICINAL CHEMISTRY. *Educ:* Univ Jammu & Kashmir, BS, 53; Birla Inst Technol, India, BS, 58; Univ Conn, PhD(med chem), 63. *Prof Exp:* Assoc res chemist, Sterling Winthrop Res Inst, 63; lectr med chem, State Univ NY, Buffalo, 63-64; sr res scientist, Regional Res Labs, India, 64-65; res assoc steroids, Univ Miss, 65-67; fel biochem steroids, Worcester Found Exp Biol, 67-68; group leader chem & biochem, Thomas J Lipton, Inc, 68-69; asst prof biochem, med units, 69-78, PROF BIOCHEM CTR HEALTH SCI, UNIV TENN, MEMPHIS, 78- CHIEF STEROID BIOCHEMIST, VET ADMIN HOSP, MEMPHIS, 69- *Concurrent Pos:* Vis prof, McArdle Lab Cancer Res, Madison, Wi, 74-75. *Mem:* AAAS; Am Soc Biol Chemists; fel Acad Pharmaceut Sci; fel Am Inst Chemists; Am Chem Soc. *Res:* Regulatory mechanisms of hormones at cellular level in normal and tumor tissues; steroid biochemistry. *Mailing Add:* Dept Brain & Vasc Res Cleveland Clin Found 9500 Eculid Ave Cleveland OH 44106

SHARMA, RAN S, b Ruppura, Gujarat, India, June 9, 37; nat US; c 2. BIOSTATISTICS, MATHEMATICAL STATISTICS. *Educ:* Gujarat Univ, India, BA, 59, MA, 61; Univ Calif, Los Angeles, PhD(biostatist), 66. *Prof Exp:* Asst, Univ Calif, Los Angeles, 62-66; biostatistician, Riker Labs, 66; sr statistician, E R Squibb & Sons, Inc, 66-70; asst prof, Sch Med, Temple Univ, 68-74, assoc prof biomet & actg chmn dept, 74-80; DIR PRECLIN BIOSTATIST, ORTHO PHARMACEUT CORP, 80- *Concurrent Pos:* Res asst, Bur Econ & Statist, Gujarat State, 61-62; part-time biostatist consult, 63-65. *Mem:* Biomet Soc; Am Statist Asn. *Res:* Use of multivariate techniques in repeated measures and in related topics; non-parametrics; bioassay; repeated measures; pair comparisons. *Mailing Add:* Ortho Pharmaceutical Corp Rte 202--Res Bldg Raritan NJ 08869

SHARMA, SANSAR C, b Pirthipur, India, Mar 10, 38; m 70. NEUROBIOLOGY. *Educ:* Panjab Univ, India, BSc, 61, MSc, 62; Univ Edinburgh, PhD(physiol), 67. *Prof Exp:* Lectr, DAV Col, Ambala, India, 62-63; fel, Univ Edinburgh, 64-68; fel, Washington Univ, 68-72; assoc prof, 72-78, PROF OPHTHAL, NEW YORK MED COL, 78-, PROF ANAT, 81- *Concurrent Pos:* NSF grant, 74-75; Nat Eye Inst grant, 75- *Mem:* AAAS; Am Physiol Soc; Soc Neurosci; NY Acad Sci; Brit Soc Develop Biol. *Res:* Formation of specific nerve connections in the visual system; developmental neurophysiology of the spinal cord. *Mailing Add:* Dept Anat New York Med Col Valhalla NY 10595

SHARMA, SANTOSH DEVRAJ, b Kenya, Feb 24, 34. OBSTETRICS & GYNECOLOGY. *Educ:* B J Med Sch, Poona, India, MB & BS, 60; Am Bd Obstet & Gynec, dipl, 77. *Prof Exp:* From lectr to sr lectr obstet & gynec, Med Sch Makerere Univ, Kampala, Uganda, 67-72; asst prof obstet & gynec, Sch Med, Howard Univ, Washington, DC. 72-74; assoc prof, 74-78, PROF OBSTET & GYNEC, JOHN A BURN SCH MED, HONOLULU, 78- *Mem:* Fel Royal Col Obstetricians & Gynecologists; fel Am Col Obstetricians & Gynecologists; Am Soc Colposcopy & Cervical Path. *Res:* Clinical uses of prostagrandins in obstetrics and gynecology. *Mailing Add:* 1319 Punahou St No 801 Honolulu HI 96826

SHARMA, SHRI C, b Rewari, Haryana, June 18, 45; US citizen; m 75; c 3. MEDICAL FOODS, FOOD PROCESSING. *Educ:* Indian Inst Technol, Kharagpur, India, BTech Hons, 67; Univ Wis-Madison, MS, 71; Rutgers Univ, PhD, 71. *Prof Exp:* Postdoctoral fel food processing, Rutgers Univ, 73-75; sr food scientist, Best Foods CPC Int, 75-77, sect head, 77, prin scientist, 78; mgr mat res, PepsiCo Inc, Res & Technol Ctr, 79-80, assoc dir, 80-82; assoc dir chem res, Warner Lambert Co, 82-83, dir, 83-84, sr dir res

& develop, 84-85; vpres technol, Nutrit Technol Corp, 85-88; PRES, NUTRALAB INC, 88- *Concurrent Pos:* Consult, Sara Lee Corp, 85-87, Procter & Gamble Co, 88-89 & Frito-Lay, Inc, 91- *Mem:* Inst Food Technologists; Am Asn Cereal Chemists; Am Inst Nutrit; Am Soc Parental & Enteral Nutrit. *Res:* Development of novel food products; applied food technologies to pharmaceutical products; polymer technologies for development of food as well as pharmaceutical products; credited with commercialization of over 75 consumer products; 45 patents; author of 25 technical papers. *Mailing Add:* 5400 Indian Heights Dr Cincinnati OH 45243

SHARMA, SOMESH DATT, MOLECULAR PARASITOLOGY & IMMUNOLOGY. *Educ:* Univ Md, PhD(path), 76. *Prof Exp:* SR INVESTR, DEPT IMMUNOL & INFECTIOUS DIS RES INST, PALO ALTO MED FOUND, 79- *Mailing Add:* Dept Immunol Biospan Corp 440 Chesapeake Dr Redwood City CA 94063

SHARMA, SURESH C, b Bulandshahar, India, June 6, 45; US citizen; m 67; c 2. POSITRON ANNIHILATION, DIAMOND FILMS. *Educ:* Agra Univ, BSc, 65; Meerut Univ, MSc, 67; Brandeis Univ, PhD(physics), 76. *Prof Exp:* Lectr physics, M M H Col, Ghaziabad, 67; res asst nuclear physics, Univ Delhi, India, 68-69; from asst prof to assoc prof, 80-89, PROF PHYSICS, UNIV TEX, ARLINGTON, 90- *Concurrent Pos:* Assoc dir, Ctr Positron Studies, Univ Tex, Arlington, 80-82, dir, 83-; co-chmn, 6th Int Conf Positron Annihilation, 82; mem int adv comt, Int Workshops Positron & Positronium Chem, 85- *Honors & Awards:* Outstanding Res Contrib Award, Sigma Xi, 80. *Mem:* Am Phys Soc; Mat Res Soc. *Res:* Low energy positron annihilation to investigate positronium localization in density fluctuations, lattice defects, surfaces and fluids; deposition and electronic structure of diamond films. *Mailing Add:* Dept Physics Univ Tex Box 19059 Arlington TX 76019

SHARMA, UDHISHTRA DEVA, b Amritsar, India, Aug 16, 28; m 59; c 1. ANIMAL SCIENCE, BIOLOGY. *Educ:* Punjab Univ, BVSc, 48; Univ Ill, Urbana, MS, 54, PhD(reproductive physiol), 57. *Prof Exp:* Res asst animal genetics, Indian Vet Res Inst, 49-53; res assoc germ cell physiol, Am Found Biol Res, Madison, Wis, 57-58; prof animal genetics, Postgrad Col Animal Sci, Indian Vet Res Inst, 58-64; prof animal husb & head dept, Punjab Agr Univ, 64-65; vis prof anat, Med Ctr, Univ Ark, Little Rock, 65-66; PROF BIOL, ALA STATE UNIV, MONTGOMERY, 66- *Mem:* Am Vet Med Asn. *Res:* Physiology of reproduction, artificial insemination and preservation of bovine semen; animal sciences and veterinary medicine; biological sciences. *Mailing Add:* 4438 Eley Ct Montgomery AL 36106

SHARNOFF, MARK, b Cleveland, Ohio, July 26, 35; m 59; c 3. MOLECULAR PHYSICS, SOLID STATE PHYSICS. *Educ:* Univ Rochester, BS, 57; Harvard Univ, PhD(physics), 63. *Prof Exp:* Nat Acad Sci-Nat Res Coun resident res assoc, Nat Bur Stand, 63-65; from asst prof to assoc prof, 65-74, PROF PHYSICS, UNIV DEL, 74- *Concurrent Pos:* Regional ed, J Luminescence, 69-81; NIH spec fel & vis assoc prof biophys, Pa State Univ, 72-73. *Mem:* AAAS; Am Phys Soc; Biophys Soc; Am Chem Soc; Optical Soc Am. *Res:* Dynamic nuclear polarization; electron spin resonance of free radicals and transition metal ions; optical properties of coordination complexes; microwave-optical studies of isomerization and energy transport in organic systems; microdifferential holographic interferometry of activity patterns in biological tissues. *Mailing Add:* Dept Physics Univ Del Newark DE 19716

SHARON (SCHWADRON), YITZHAK YAAKOV, b Tel Aviv, Israel, Feb 29, 36; m 91. NUCLEAR STRUCTURE THEORY, NUCLEAR MODELS. *Educ:* Columbia Univ, AB, 58; Princeton Univ, MA, 60, PhD(physics), 66. *Prof Exp:* Teaching asst, dept physics, Columbia Univ, 57-58; consult, Phys Sci Study Comt, Educ Serv, Inc, 58-59; asst instruction, dept physics, Princeton Univ, 58-59, res asst, 59-60, asst instruction, 60-62, res asst, 63-65, asst, Inst Advan Study, 65-66; asst prof, Northeastern Univ, 66-72; assoc prof, 72-75, PROF PHYSICS, STOCKTON STATE COL, 75- *Concurrent Pos:* Physicist, Lawrence Radiation Lab, Berkeley, Calif, 68; res participant, Oak Ridge Nat Lab, Tenn, 69 & Nat Bur Standards, Washington, DC, 71; vis asst prof, Temple Univ, Philadelphia, 70-71; vis prof, lab nuclear physics, Univ Montreal, 70; vis fel, Princeton Univ, NJ, 80-82. *Mem:* Am Phys Soc; Am Asn Physics Teachers; Sigma Xi. *Res:* Nuclear theory, especially nuclear models and the connections between them; shell model, collective model, SU(3) models; projected wave functions; systematics of nuclear properties; physics education. *Mailing Add:* Dept Natural Sci Stockton State Col Pomona NJ 08240

SHARON, NEHAMA, b Tiberias, Israel, Mar 2, 29. VIROLOGY. *Educ:* Hebrew Univ, Jerusalem, PhD(immunol), 61. *Prof Exp:* ASSOC PROF PATH, SCH MED, NORTHWESTERN UNIV, 72- *Mem:* Am Soc Microbiol; Am Soc Clin Path; Sigma Xi. *Res:* Abnormal changes in serum protein. *Mailing Add:* Dept Path Evanston Hosp 2650 Ridge Ave Evanston IL 60201

SHARP, A C, JR, b Lorenzo, Tex, July 16, 32; m 55; c 3. MAGNETISM. *Educ:* Tex A&I Univ, BS, 57, MS, 58; Tex A&M Univ, PhD(physics), 65. *Prof Exp:* Instr physics, Univ Tex, Arlington, 58-61; asst prof, Tex Woman's Univ, 64-65; PROF PHYSICS, McMURRY COL, 65- *Concurrent Pos:* Fulbright exchange prof, Univ Liberia, 81-82. *Mem:* Am Phys Soc; Am Asn Physics Teachers; Optical Soc Am. *Res:* Polymer solution theory using polyisobutylene-n-alkane systems; magnetic properties of thin permalloy films. *Mailing Add:* Dept Physics McMurry Col Abilene TX 79697

SHARP, A(RNOLD) G(IDEON), b Worcester, Mass, May 16, 23. MECHANICAL ENGINEERING. *Educ:* Tufts Univ, BS, 45; Worcester Polytech Inst, MS, 53. *Prof Exp:* Instr mech eng, Worcester Polytech Inst, 46-53; asst prof civil eng, Univ Mass, 53-58; res engr, 58-63, res assoc, 63-78, RES SPECIALIST, WOODS HOLE OCEANOG INST, 78- *Concurrent Pos:* Lectr, Lincoln Col, Northeastern Univ, 60-61. *Mem:* Soc Exp Mech. *Res:* Applied mechanics; stress analysis; structural design; properties of materials; oceanographic instrumentation; deep-submergence vehicle design. *Mailing Add:* PO Box 434 Woods Hole MA 02543-0434

SHARP, AARON JOHN, b Plain City, Ohio, July 29, 04; m 29; c 5. PLANT GEOGRAPHY. *Educ:* Ohio Wesleyan Univ, AB, 27; Univ Okla, MS, 29; Ohio State Univ, PhD(bot), 38. *Hon Degrees:* DSc, Ohio Wesleyan Univ, 52. *Prof Exp:* From instr to prof bot, 29-65, distinguished serv prof, 65-74, EMER PROF BOT, UNIV TENN, 74- *Concurrent Pos:* Vis prof, WVa Univ, 39-41, Stanford Univ, 51, Biol Sta, Univ Mich, 54-57 & 59-64, Nat Univ Taiwan, 65, Biol Sta, Univ Minn, 71, Biol Sta, Univ Mont, 72, Biol Sta, Univ Va, 80 & Inst Exp Pedag, Univ Venezuela, 76; vis lectr, Am Inst Biol Sci, 67-70; consult, Time-Life Bks, 74, 75 & Nat Geog Bks, 85; fel, Guggenheim Found, 44, 46. *Mem:* Bot Soc Am (pres, 65); Am Soc Plant Taxonomists (pres, 61); Ecol Soc Am (vpres, 59); Am Bryol & Lichenol Soc (pres, 35); Int Soc Plant Taxonomists; Am Inst Biol Sci; AAAS (vpres, 63). *Res:* Relationships of modern flora of the world to their geological histories and modern climates; dependence of society on green plants. *Mailing Add:* Dept Bot Univ Tenn Knoxville TN 37996-1100

SHARP, ALLAN ROY, b Hamilton, Ont, Nov 3, 46; m 71; c 3. BIOPHYSICS, APPLIED PHYSICS. *Educ:* McMaster Univ, BSc, 67; Univ Waterloo, MSc, 69, PhD(physics), 73. *Prof Exp:* Fel, Univ Toronto, 72-74; lectr, Univ Natal, 74-75; from asst prof to assoc prof, 75-85, PROF PHYSICS, UNIV NB, 85- *Concurrent Pos:* Nat Sci Eng Res Coun Can res grant, 76- *Mem:* Can Asn Physicists. *Res:* Nuclear magnetic resonance studies of interactions of molecules with biopolymers, primarily water with cellulose; effects of these molecules on wood properties; materials science of wood. *Mailing Add:* Dept Physics Univ NB Coll Hill Box 4400 Fredericton NB E3B 5A3 Can

SHARP, DAVID HOWLAND, b Buffalo, NY, Oct 14, 38; m 65, 83; c 4. THEORETICAL PHYSICS. *Educ:* Princeton Univ, AB, 60; Calif Inst Technol, PhD(physics), 64. *Prof Exp:* NSF fel physics, Princeton Univ, 63-64, res assoc, 64-65; res fel, Munich Tech Univ, 65-66; res fel, Calif Inst Technol, 66; instr, Palmer Phys Lab, Princeton Univ, 66-67; asst prof physics, Univ Pa, 67-74; FEL, LOS ALAMOS SCI LAB, 74- *Concurrent Pos:* Consult, Jason Div, Inst Defense Anal, 60-66 & Lawrence Livermore Lab, Univ Calif, Livermore, 64-66; guest partic, Battelle Rencontres Math & Physics, 69; mem, Bd Hons Examrs, Swarthmore Col, 72; mem, J Robert Oppenheimer Mem Comt, 76-, chmn, 78. *Mem:* AAAS; Asn Math Physicists; Int Soc Gen Relativity & Gravitation; fel Am Phys Soc; Soc Indust & Appl Math. *Res:* Elementary particle physics; theory of gravitation; fluid dynamics, neural nets. *Mailing Add:* Univ Calif LASL-T8 Los Alamos NM 87545

SHARP, DEXTER BRIAN, b Chicago, Ill, June 14, 19; m 45; c 3. PESTICIDE CHEMISTRY, METABOLISM. *Educ:* Carleton Col, BA, 41; Univ Nebr, MA, 43, PhD(chem), 45. *Prof Exp:* Asst chem, Univ Nebr, 41-45; res chemist, Chem Dept, Exp Sta, E I du Pont de Nemours & Co, Inc, 45-46; Am Chem Soc fel & instr, Univ Minn, 46-47; from asst prof to assoc prof chem, Kans State Col, 47-51; res chemist, Monsanto Agr Co, 51-54, group leader, 54-68, mgr residue metab, 68-75, environ sci dir, 75-82, dir environ & formulations technol, 82-85; CONSULT PESTICIDES, DEXTER B SHARP, INC, 85- *Concurrent Pos:* Mem, comt registration pesticides, Nat Acad Sci, 68-, comt sci & regulatory issues underlying pesticide use patterns & agr innovation, 85-86; mem, Comn Pesticide Chem, Int Union Pure & Appl Chem, 84-88; consult, Stewart Pesticide Regist Assocs Inc, 88- *Mem:* Am Chem Soc; Weed Sci Soc Am; Int Union Pure & Appl Chem. *Res:* Mechanisms of organic reactions; oxidation; pesticides; gas-liquid chromatography; pesticide metabolism; plant and animals; residues; environmental fate; applicator and dietary exposure; pharmacokinetics; risk assessment; pesticide registration; residue in environmental sciences. *Mailing Add:* 13042 Weatherfield Dr St Louis MO 63146-3646

SHARP, EDWARD A, b Milwaukee, Wis, Oct 3, 20. MATHEMATICS, COMPUTER SCIENCE. *Educ:* St Louis Univ, AB, 43, AM, 56. *Prof Exp:* ASSOC PROF MATH, CREIGHTON UNIV, 57-, DIR COMPUT CTR, 66- *Mem:* Am Math Soc; London Math Soc; Soc Indust & Appl Math. *Mailing Add:* Dept Math Creighton Univ Omaha NE 68178

SHARP, EUGENE LESTER, b Spokane, Wash, Nov 19, 26; m 54; c 4. PLANT PATHOLOGY. *Educ:* Univ Idaho, BS, 49; Iowa State Col, MS, 51, PhD(plant path), 53. *Prof Exp:* Asst plant path, Iowa State Col, 49-53; asst plant path, biol lab, Chem Corps, US Dept Army, 53-55, res plant pathologist, 55-57; asst plant pathologist, 57-62, assoc plant pathologist, 62-67, PROF PLANT PATH, MONT STATE UNIV, 67-, HEAD DEPT, 73- *Concurrent Pos:* Bd dirs, Mont State Univ, 76-84. *Honors & Awards:* Sr Scientist Award, Alexander von Humboldt Found, 75; Fac Res Award, Sigma Xi, 70. *Mem:* Fel AAAS; Am Phytopath Soc. *Res:* Cereal pathology; genetics of pathogenicity and disease resistance with emphasis on cereal rusts; host-pathogen interactions from a genetic and environmental basis. *Mailing Add:* Dept Plant Path Mont State Univ Bozeman MT 59715

SHARP, GERALD DUANE, b Twin Falls, Idaho, July 28, 33; m 54; c 2. RUMINANT NUTRITION, LIVESTOCK EVALUATION. *Educ:* Univ Idaho, BS, 55, MS, 62; Wash State Univ, PhD(animal nutrit), 69. *Prof Exp:* Animal scientist, Caldwell Br Exp Sta, Univ Idaho, 62-64, Swine Herdsman, 64-66; asst dairy scientist, Wash State Univ, 69-70; asst prof, Fort Hays Kans State Col, 70-72, prof & farm supt, 72-76; assoc prof, 76-79, PROF & CHAIR, ANIMAL SCI DEPT, CALIF STATE POLYTECH UNIV, POMONA, 79- *Mem:* Sigma Xi; Am Soc Animal Sci; Coun Agr Sci & Technol. *Mailing Add:* 1010 Rosemary Lane La Verne CA 91750

SHARP, HENRY, JR, b Nashville, Tenn, Oct 14, 23; m 57; c 2. MATHEMATICS. *Educ:* Vanderbilt Univ, BE, 47; Duke Univ, AM, 50, PhD(math), 52. *Prof Exp:* Res assoc, Johns Hopkins Univ, 52-53; asst prof math, Ga Inst Technol, 53-56; from asst prof to prof math, Emory Univ, 58-83; PROF MATH, WASHINGTON & LEE UNIV, 83- *Concurrent Pos:* NSF fac fel, 64-65, res grant, 68-69. *Mem:* Am Math Soc; Math Asn Am. *Res:* Point-set topology, dimension theory and graph theory. *Mailing Add:* Dept Math Washington & Lee Univ Lexington VA 24450

SHARP, HOMER FRANKLIN, JR, b Lithonia, Ga, Sept 5, 36; m 61; c 2. ZOOLOGY, ECOLOGY. *Educ:* Emory Univ, BA, 59; Univ Ga, MS, 62, PhD(zool), 70. *Prof Exp:* Asst prof biol, LaGrange Col, 62-63; from instr to assoc prof, 63-79, PROF BIOL, OXFORD COL, EMORY UNIV, 79- *Mem:* Am Soc Mammalogists; Sigma Xi; AAAS. *Res:* Bioenergetics of populations; diversity of benthic communities; secondary succesion; wetland ecology. *Mailing Add:* Dept Biol Oxford Col Emory Univ Oxford GA 30267

SHARP, HUGH T, b New York, NY, Sept 6, 29; m 54; c 1. COST ESTIMATING, TECHNICAL PUBLISHING. *Educ:* Villanova Univ, BChE, 53. *Prof Exp:* Ed & mem sales & mkt staff, Chem Eng, McGraw-Hill Publ Co, 53-68, dir spec proj, Planning & Develop Dept, 68-70, advert sales mgr, 70-72; gen mgr, Sweet's Div, McGraw Hill Info Syst Co, 72-79 & Cost Info Syst Div, 79-80; vpres planning & develop, Sweet's Group, McGraw-Hill, Inc, 80-90; VPRES & GEN MGR, R S MEANS CO, INC, SOUTHERN CO, 90- *Concurrent Pos:* Vpres, Mfrs Agent Publ Co, Inc, 75-82. *Mem:* Am Inst Chem Engrs; Am Asn Cost Engrs; Info Indust Asn. *Mailing Add:* Four Van Wyck Montvale NJ 07645

SHARP, JAMES H, b Moosejaw, Sask, Apr 20, 34; c 4. PHYSICAL CHEMISTRY, CHEMICAL PHYSICS. *Educ:* Univ BC, BA, 57, MSc, 60; Univ Calif, Riverside, PhD(phys chem), 64. *Prof Exp:* Fel org solid state, Nat Res Coun Can, 63-64; scientist, Physics Res Lab, 64-68, mgr, Chem Physics Res Br, Res & Eng Sci Div, 68-71, mgr, Org Solid State Physics, Rochester Corp Res Ctr, 71-72, technol ctr mgr, Info Technol Group, 72-74, RES MGR, XEROX RES CENTRE CAN, XEROX CORP, 74- *Mem:* Am Phys Soc; Am Chem Soc; Chem Inst Can; Sigma Xi. *Res:* Photochemistry; photolyses and photophysics of organic materials; electron spin resonance; photoconductivity and transport phenomenon; optical sensitization and energy transfer; spectroscopy of the organic solid state. *Mailing Add:* 2660 Speakman Dr Mississauga ON L5K 2L1 Can

SHARP, JAMES JACK, b Glasgow, Scotland, July 22, 39; m 63; c 3. CIVIL ENGINEERING. *Educ:* Glasgow Univ, BSc, 61, MSc, 63; Univ Strathclyde, ARCST, 61, PhD(hydraul), 69. *Prof Exp:* Exec engr, Govt Malawi, 63-66; lectr civil eng, Univ Strathclyde, 66-70; PROF ENG, MEM UNIV NFLD, 70- *Mem:* Am Soc Civil Engrs; Brit Inst Civil Engrs. *Res:* Hydraulics; densimetric phenomena; energy dissipation. *Mailing Add:* Dept Eng Mem Univ Nfld Elizabeth Ave St Johns NF A1C 5S7 Can

SHARP, JOHN ARTHUR, physical chemistry, for more information see previous edition

SHARP, JOHN BUCKNER, b Maynardville, Tenn, Nov 5, 20; m 49; c 3. FORESTRY. *Educ:* Univ Tenn, BS, 43, MS, 45; Duke Univ, MF, 47; Harvard Univ, MPA, 50, DPA, 52. *Prof Exp:* Carnegie Corp fel, Harvard Univ, 50-52; dist forester, Agr Exten Serv, Univ Tenn, Knoxville, 47-49, assoc exten forester, 52-57, state exten forester, 57-74, PROF FORESTRY, UNIV TENN, KNOXVILLE, 74- *Mem:* Soc Am Foresters. *Res:* Farm forestry; forestry administration; holder of two US patents; author of one book. *Mailing Add:* 5052 Mountain Crest Dr Knoxville TN 37918

SHARP, JOHN GRAHAM, b Halifax, Eng, Feb 10, 46; m 73; c 2. EXPERIMENTAL HEMATOLOGY, CELL KINETICS. *Educ:* Univ Birmingham, Eng, BSc, 67, MSc, 68, PhD(exp hemat), 71. *Prof Exp:* Res fel anat, Univ Birmingham, Eng, 70-71; Damon Runyon res fel, Radiation Res Lab, Univ Iowa, 71-73, asst instr, dept anat, 72-73, instr radiol, 73; from asst prof to prof anat, 73-82, PROF ANAT & RADIOL, UNIV NEBR MED CTR, 82- *Concurrent Pos:* Yamagawa-Yoshida Mem Int Cancer fel, dept histopath, Univ London, 81-82; Paul Grange vis cancer res fel, dept anat, Monash Univ, Australia, 84- *Mem:* Int Soc Exp Hemat; Am Asn Immunologists; Am Asn Anatomists; Radiation Res Soc; Cell Kinetics Soc; AAAS. *Res:* Cell renewal systems; stromal cell regulation of stem cell proliferation and differentiation in bone marrow, thymus and intestine and its relevance to the development of cancers in these tissues; radioconjugate therapy of cancer. *Mailing Add:* Dept Anat Univ Nebr Med Ctr 600 S 42nd St Omaha NE 68198-6395

SHARP, JOHN MALCOLM, JR, b St Paul, Minn, Mar 11, 44; m 67; c 3. HYDROGEOLOGY. *Educ:* Univ Minn, BGeolE, 67; Univ Ill, MS & PhD(hydrogeol), 74. *Prof Exp:* Civil engr, USAF, 67-71; from asst to assoc prof geol, Univ Mo, Columbia, 74-82, chmn dept, 80-82; assoc prof, 82-84, C E Yager prof, 85-89, GULF FOUND CENTENNIAL PROF GEOL, UNIV TEX, 90- *Concurrent Pos:* Alexander von Humboldt fel, 81 & 83; chmn Hydrogeol Div, Geol Soc Am, 87-88, mem coun, 90-93. *Honors & Awards:* O E Meinzer Award, Geol Soc Am, 79. *Mem:* Am Inst Mining Engrs; Geol Soc Am; Am Geophys Union; Asn Groundwater Sci & Engrs; Int Water Resources Asn; Int Asn Hydrogeologists; Am Inst Hydrol (vpres, 88-91); Sigma Xi. *Res:* Energy transport in porous media; flood-plain hydrogeology; economics of water resource development; basinal hydrogeology; coastal subsidence; water resources of Trans-Pecos Texas and the Edwards aquifer. *Mailing Add:* Dept Geol Sci Univ Tex Austin TX 78713-7909

SHARP, JOHN ROLAND, b Joplin, Mo, Dec 5, 49; m 79; c 1. AQUATIC TOXICOLOGY, PHYSIOLOGICAL ECOLOGY. *Educ:* Southwest Mo State Col, BS, 71; Tex A&M Univ, PhD(biol), 79. *Prof Exp:* Res assoc, Tex A&M Univ, 78-79; ASST PROF BIOL, ZOOL & BIOL FISHES, INTRO AQUATIC TOXICOL, SOUTHEAST MO STATE UNIV, 80- *Concurrent Pos:* Instr marine ichthyol, Gulf Coast Res Lab, 76 & 81-82. *Mem:* Sigma Xi. *Res:* Individual and interactive effects of environmental stressors on the physiology and developmental biology of marine and freshwater fishes, particularly the embryo-larval life history stages; ecology of fishes. *Mailing Add:* Dept Biol Southeast Mo State Univ Cape Girardeau MO 63701

SHARP, JOHN T(HOMAS), b Dalhart, Tex, Nov 16, 24; m 49; c 3. RHEUMATOID ARTHRITIS. *Educ:* Columbia Univ, NY, MD, 47. *Prof Exp:* Div chief, rheumatology, Baylor Col Med, 62-76, prof med, 64-76; chief, Dept Med, Vet Admin Hosp, Danville, Ill, 76-80; prof med, Univ Ill, Urbana, 76-80 & Univ Colo, 80-86; dir rheumatology, Alpert Arthritis Ctr, Rose Hosp, 80-86; CLIN PROF MED, EMORY UNIV, 87- *Mem:* Am Col Rheumatology; Am Col Physicians; Am Asn Immunologists; emer mem Soc Am Microbiologists. *Res:* Radiologic assessment of disease progression in rheumatoid arthritis. *Mailing Add:* 712 E 18th St Tifton GA 31794

SHARP, JOHN TURNER, b Jamestown, NY, Jan 18, 27; m 49; c 5. INTERNAL MEDICINE. *Educ:* Univ Buffalo, MD, 49. *Prof Exp:* Intern med, Bellevue Hosp, New York, 49-50; resident internal med, Vet Admin Hosp, Buffalo, 52-54; Am Heart Asn res fel cardiopulmonary physiol & clin cardiol, Buffalo Gen Hosp & Sch Med, State Univ NY, Buffalo, 54-57; Am Heart Asn estab investr, 57-59; prof dir pulmonary dis serv, Vet Admin Hosp, Hines, 59-87; PROF MED, PHYSIOL & BIOPHYS UNIV S FLA, TAMPA, 87- *Concurrent Pos:* Nat Heart Inst res grants, 59-62 & 64-79; from assoc prof to prof med, Univ Ill Col Med 64-82. *Mem:* Am Soc Clin Invest; Am Col Physicians; fel Am Col Chest Physicians; fel Am Col Cardiol; Am Physiol Soc. *Res:* Cardiopulmonary physiology; respiratory physiology, particularly the mechanics of respiration and respiratory muscle function in normal and diseased man; human hemodynamics in health and disease; epidemiology of chronic respiratory diseases. *Mailing Add:* Pulmonary Sect Dept Med James A Haley Vet Admin Hosp 13000 Bruce D Downs Blvd Tampa FL 33612

SHARP, JONATHAN HAWLEY, b Bridgeton, NJ, Apr 29, 43; m 73; c 2. CHEMICAL OCEANOGRAPHY, BIOLOGICAL OCEANOGRAPHY. *Educ:* Lehigh Univ, BA, 65, MS, 67; Dalhousie Univ, PhD(oceanog), 72. *Prof Exp:* Fel oceanog, Scripps Inst Oceanog, Univ Calif, San Diego, 72-73; assoc prof, 73-87, PROF OCEANOG, COL MARINE STUDIES, UNIV DEL, 87- *Concurrent Pos:* Consult, Org Am States, 72-74; mem Md Environ Res Guidance Comt, 77-84; panel, Off Naval Res, 79-81; mgr, Del Estuary Res Proj, 81-86; chmn, Sci Tech Adv Comt, Del Estuary Prog, 89- *Mem:* AAAS; Am Soc Limnology & Oceanog; Am Chem Soc; Am Geophys Union; Oceanog Soc. *Res:* Biological chemistry of seawater, interaction between marine micro-organisms and chemistry of sea; dynamics of estuarine and coastal waters. *Mailing Add:* Col Marine Studies Univ Del Lewes DE 19958

SHARP, JOSEPH CECIL, b Salt Lake City, Utah, May 30, 34; m 56; c 2. NEUROBIOLOGY, EXPERIMENTAL PSYCHOLOGY. *Educ:* Univ Utah, BS, 57, MS, 58, PhD(psychol), 61. *Prof Exp:* Chief behav radiation, Walter Reed Army Inst Res, 60-67, chief exp psychol, 67-68; dep comn environ health, Dept Health, State NY, 69-70; dep dir neuropsychiat, Walter Reed Army Inst Res, 71-74; dep dir Life Sci, 74-85, DIR SPACE RES, NASA AMES RES CTR, 85- *Concurrent Pos:* Us space rep, UK, 87. *Mem:* Am Psychol Asn; Radiation Res Soc; Int Soc Chronobiol; Sigma Xi. *Res:* Central nervous system and radiation effects; space biology and medicine; experimental psychology, drug abuse and neuroendocrinology. *Mailing Add:* 615 Milverton Rd Los Altos CA 94022

SHARP, KENNETH GEORGE, b Dec 24, 43; US citizen; m 67; c 1. INORGANIC CHEMISTRY. *Educ:* Univ Calif, Riverside, BA, 65; Rice Univ, PhD(chem), 69. *Prof Exp:* Nat Res Coun-Nat Bur Stand fel, Nat Bur Stand, Washington, DC, 69-71; asst prof inorg chem, Univ Southern Calif, 71-80; assoc scientist, Dow Corning Corp, 80-90; SR RES ASSOC, POLYMERS DIV, E I DU PONT DE NEMOURS & CO, 90- *Mem:* Am Chem Soc. *Res:* Inorganic synthesis via energetic intermediates; high-temperature chemistry; silicon and fluorine chemistry. *Mailing Add:* E I du Pont de Nemours & Co PO Box 80323 Wilmington DE 19880-0323

SHARP, LEE AJAX, b Salt Lake City, Utah, Mar 27, 22; m 47; c 1. RANGE MANAGEMENT. *Educ:* Utah State Agr Col, BS, 48, MS, 49, PhD, 66. *Prof Exp:* From instr to assoc prof forestry & range mgt, 49-71, PROF RANGE MGT, UNIV IDAHO, 71-, CHAIRPERSON RES MGT, 77- *Mem:* Soc Range Mgt; Sigma Xi. *Res:* Salt desert shrub revegetation and evaluation; grazing management programs for seeded ranges. *Mailing Add:* Col Forestry Wildlife Univ Idaho Moscow ID 83844

SHARP, LOUIS JAMES, IV, b Washington, DC, Oct 13, 44; m 68; c 3. POLYMER CHEMISTRY. *Educ:* Univ Notre Dame, BS, 66; Calif Inst Technol, PhD(chem), 70. *Prof Exp:* Res fel, Radiation Lab, Univ Notre Dame, 69-72; asst prof chem, Marian Col, 72-74; dir resin res & develop, Lily Indust Coatings Inc, 74-80; group leader synthesis, Betz Labs, 80-81; mgr polymer chem, 81-87, DIR RES & ADMIN SERV, MIDLAND DEXTER CORP, 87- *Mem:* Am Chem Soc; Sigma Xi. *Res:* Synthesis and rheology of resins for coatings, including polyesters, acrylics, urea-formaldehyde resins, silicones, melamines, alkyds, water-based resins, high solids resins and ultraviolet cured resins. *Mailing Add:* Midland Dexter Corp E Water St Waukegan IL 60085

SHARP, PHILLIP ALLEN, b Falmouth, Ky, June 6, 44; m; c 3. MOLECULAR BIOLOGY, VIROLOGY. *Educ:* Union Col, BA, 66; Univ Ill, Urbana, PhD(chem), 69. *Prof Exp:* Fel biophys chem & molecular biol, Calif Inst Technol, 69-71; res staff virol & molecular biol, Cold Spring Harbor Lab, 71-74; assoc prof, 74-79, PROF DEPT BIOL, MASS INST TECHNOL, 79-, DIR, CTR CANCER RES, 85- *Concurrent Pos:* Consult, Biogen Ltd, 78-; lectr, Univ Ky, 79, Univ Chicago, 80, Loyola Univ Chicago, 81, Univ Ill, Urbana-Champaign, 84, Univ Calif, Los Angeles, 89, Purdue Univ, 89, Univ Calif, Berkeley, 90, NY Univ Med Ctr, 90, Southwestern Med Ctr, Univ Dallas, 90 & Univ Wis-Madison, 90. *Honors & Awards:* Fac Res Award, Am Cancer Soc, 74; Eli Lilly Award, 80; US Steel Award, Nat Acad Sci, 80; Alfred P Sloan Jr Prize, Gen Motors Res Found, 86; Albert Lasker Basic Med Res Award, 88. *Mem:* Nat Acad Sci; Inst Med-Nat Acad Sci; Am Soc Microbiol; fel AAAS; Am Chem Soc. *Res:* The molecular biology of gene expression in mammalian cells. *Mailing Add:* Ctr Cancer Res Rm E17-529B Mass Inst Technol Cambridge MA 02139

SHARP, RICHARD DANA, b Brooklyn, NY, June 17, 30; m 59; c 2. MAGNETOSPHERIC PHYSICS, IONOSPHERIC PHYSICS. *Educ:* Mass Inst Technol, BS, 52, PhD(physics), 56. *Prof Exp:* SR STAFF SCIENTIST, LOCKHEED PALO ALTO RES LAB, 56- *Mem:* Am Phys Soc; Am Geophys Union. *Res:* Hot plasma measurements in the magnetoshere; satellite experiments on auroral phenomena. *Mailing Add:* 261 Vista Verde Way Portola Valley CA 94025

SHARP, RICHARD LEE, b Kansas City, Mo, Sept 14, 35; m 64; c 1. CHEMICAL TOXICOLOGY. *Educ:* William Jewell Col, AB, 61; Purdue Univ, PhD(org chem), 66. *Prof Exp:* Prod chemist, 66-67, develop chemist, 67-69, supvr synthetic org chem, Synthetic Chem Div, 69-73, TECH ASSOC HEALTH & SAFETY LAB, EASTMAN KODAK CO, 73- *Mem:* Am Chem Soc; Am Indust Hyg Asn. *Res:* Directive effects in the hydroboration of functionally substituted olefins and related compounds; synthesis of fine organic chemicals; toxicity testing of organic chemicals. *Mailing Add:* Kodak Park Works Bldg 320 Eastman Kodak Co Health & Environ Lab Rochester NY 14652-3615

SHARP, ROBERT PHILLIP, b Oxnard, Calif, June 24, 11; m 38; c 2. GEOMORPHOLOGY. *Educ:* Calif Inst Technol, BS, 34, MS, 35; Harvard Univ, AM, 36, PhD(geol), 38. *Prof Exp:* Instr geol, Univ Ill, 38-43; prof, Univ Minn, 45-47; chmn div geol sci, 52-68, prof geomorphol, 47-79, EMER ROBERT P SHARP PROF GEOL, CALIF INST TECHNOL, 79- *Concurrent Pos:* Condon lectr, Univ Ore, 60. *Honors & Awards:* Kirk Bryan Award, 64 & Penrose Medal, Geol Soc Am, 78; Nat Medal Sci, 89. *Mem:* Nat Acad Sci; fel Geol Soc Am; Am Geophys Union; Glaciol Soc; fel Am Acad Arts & Sci. *Res:* Glaciology; glacial geology; arid region geomorphology; planetary surfaces. *Mailing Add:* Div Geol & Planetary Sci Calif Inst Technol Pasadena CA 91125

SHARP, ROBERT RICHARD, b Newport News, Va, May 15, 41; m 68; c 1. PHYSICAL CHEMISTRY. *Educ:* Case Western Reserve Univ, AB & MS, 65, PhD(chem), 68. *Prof Exp:* NSF fel, Oxford Univ, 67-69; from asst prof to assoc prof, 69-86, PROF CHEM, UNIV MICH, ANN ARBOR, 86- *Mem:* Am Chem Soc; AAAS; Sigma Xi. *Res:* Nuclear magnetic resonance of chemical and biological systems. *Mailing Add:* Dept Chem Univ Mich Ann Arbor MI 48109

SHARP, ROBERT THOMAS, theoretical physics, for more information see previous edition

SHARP, TERRY EARL, b Chicago, Ill, Nov 5, 35; m 58; c 3. CHEMICAL PHYSICS. *Educ:* Carnegie Inst Technol, BS, 57; Univ Calif, Berkeley, PhD(chem), 62. *Prof Exp:* Nat Res Coun res assoc mass spectrometry, Nat Bur Standards, 62-64; assoc res scientist, Lockheed Palo Alto Res Labs, 64-67, res scientist, 67-69; prin res scientist assoc, Ford Motor Co Sci Res Labs, 69-72; STAFF SCIENTIST, LOCKHEED PALO ALTO RES LABS, 72- *Mem:* Am Phys Soc; Am Chem Soc; Am Soc Mass Spectrometry; fel Am Inst Chem; Sigma Xi. *Res:* Mass spectrometry and ionization in gases; theory of energy transfer in gases; atomic cross sections; negative ion formation; laser-materials interactions; aging of polymers; decomposition of explosives. *Mailing Add:* Dept 93-50 Bldg 204 Lockheed Missile Space Co Palo Alto CA 94304

SHARP, THOMAS JOSEPH, b Hattiesburg, Miss, Oct 15, 44; m 70; c 2. ALGEBRA. *Educ:* Univ Southern Miss, BS, 65; Auburn Univ, MS, 66, PhD(math), 71. *Prof Exp:* Teaching asst math, Auburn Univ, 66-69; ASST PROF MATH, W GA COL, 69- *Mem:* Am Math Soc; Math Asn Am. *Res:* Characterization of projection-invariant subgroups of abelian groups and their relationship to fully-invariant subgroups; study of big subgroups and little homomorphisms. *Mailing Add:* Dept Math WGa Col Carrollton GA 30017

SHARP, WILLIAM BROOM ALEXANDER, b Glengarnock, Scotland, May 12, 42; m 67; c 3. CORROSION, MATERIALS SCIENCE. *Educ:* Univ Cambridge, BA, 64, MA, 68; Univ London, MSc, 68; Univ Ottawa, PhD(chem), 76. *Prof Exp:* Res officer high temperature oxidation res, Cent Elec Res Lab, Eng, 64-69; assoc scientist corrosion res, Pulp & Paper Res Inst, Can, 75-78; res engr, 78, sr res engr, 79-82, GROUP LEADER CORROSION & MAT, WESTVACO RES CTR, 83- *Concurrent Pos:* Chmn, Third Int Symp Corrosion in Pulp & Paper Indust, 80; ed, Corrosion Notebook column, Tech Asn Pulp & Paper Indust J, 80-; chmn, Tech Adv Coun, Mat Technol Inst, 87-88; chmn, Tech Adv Comt, NSF Eng Res Ctr, Mont State Univ, 90- *Honors & Awards:* Weldon Medal, Can Pulp & Paper Asn, 78; Eng Div Award & E H Neese Award, Tech Asn Pulp & Paper Indust, 85. *Mem:* Brit Inst Corrosion Sci & Technol; Brit Inst Metall; Electrochem Soc; Nat Asn Corrosion; fel Tech Asn Pulp & Paper Indust. *Res:* Mechanisms of corrosion in pulp and paper mill equipment; mechanisms and rates of propagation of localized corrosion of stainless steels. *Mailing Add:* 8524 Moon Glass Ct Columbia MD 20145

SHARP, WILLIAM EDWARD, III, b Del Norte, Colo, Feb 6, 40. EXPERIMENTAL SPACE PHYSICS, SPECTROSCOPY. *Educ:* William Jewell Col, BA, 62; Univ NH, MS, 65; Univ Colo, PhD(astro-geophys), 70. *Prof Exp:* Instr physics, Stetson Univ, 64-66; RES SCIENTIST, UNIV MICH, 70- *Concurrent Pos:* Lectr, Univ Mich, 72-; prin investr, NASA, 74-; assoc ed, J Geophys Res, 81-83; prog dir, NSF, 85-86, asst dir, Space Physics Res Lab, 86-90; vis prof physics, Univ Mich, Dearborn, 90- *Mem:* Am Geophys Union. *Res:* Chemical and physical processes occurring in the aurora and in atmospheric regions above 80 km. *Mailing Add:* Space Physics Res Lab 2455 Hayward Ann Arbor MI 48109

SHARP, WILLIAM R, b Akron, Ohio, Sept 13, 36. CELL BIOLOGY, MICROBIOLOGY. *Educ:* Univ Akron, BS, 63, MS, 64; Rutgers Univ, PhD(bot), 67. *Prof Exp:* NSF fel & NIH grant, Case Western Reserve Univ, 67-69; from asst to prof microbiol, Ohio State Univ, 69-78; dir, Pioneer Res, Campbell Soup Co, 78-81; exec vpres & res dir, DNA, Plant Technol Corp, 81-89; EXEC VPRES, DNA PHARMACEUT, INC, 89- *Concurrent Pos:* Fulbright Hays fel, Ctr Nuclear Energy in Agr, Univ Sao Paulo, 71, 73, & 74; vis prof, Orgn Am States, 72; foreign corresp, São Paulo Acad Sci, 86- *Mem:* AAAS. *Res:* Cellular aspects of plant genetics and breedings; propagation and genetic engineering; phytopharmaceuticals. *Mailing Add:* FDR Sta DNA Pharmaceut Inc PO Box 805 New York NY 10150-0805

SHARPE, CHARLES BRUCE, b Windsor, Ont, Apr 8, 26; nat US; m 54; c 5. ELECTRICAL ENGINEERING. *Educ:* Univ Mich, BS, 47, PhD(elec eng), 53; Mass Inst Technol, SM, 49. *Prof Exp:* Asst, High Voltage Lab, Mass Inst Technol, 47-49; res assoc, Willow Run Res Ctr, 49-50, res assoc electronics defense group, 51-53 & 55-60, from asst prof to assoc prof, 55-60, PROF ELEC ENG, UNIV MICH, ANN ARBOR, 61- *Mem:* Sr mem Inst Elec & Electronics Engrs. *Res:* Network synthesis; microwave circuit theory; theory and application of ferrites; microwave properties of ferroelectrics; synthesis of nonuniform transmission lines; electromagnetic sounding problems in geophysics. *Mailing Add:* Dept Elec Eng & Comput Sci Univ Mich Ann Arbor MI 48109

SHARPE, DAVID MCCURRY, b Orange, NJ, Jan 2, 38; m 61; c 2. PHYSICAL GEOGRAPHY. *Educ:* Syracuse Univ, BS, 60, MS, 63; Southern Ill Univ, Carbondale, PhD(geog), 68. *Prof Exp:* Jr forester, Forest Serv, USDA, Mt Hood Nat Forest, 60-61; FAC GEOG, SOUTHERN ILL UNIV, CARBONDALE, 66- *Concurrent Pos:* NSF grant, Div Environ Biol, Oak Ridge Nat Lab, 72-74. *Mem:* Asn Am Geog; AAAS; Ecol Soc Am. *Res:* Ecology and biogeography of regional forest ecosystems including impact of forest resource use and land use change; interaction between relict forest stands separated by nonforest land use; software development for environmental sciences. *Mailing Add:* Dept Geog Southern Ill Univ Carbondale IL 62901

SHARPE, GRANT WILLIAM, b Kentfield, Calif, May 15, 25; m 48; c 9. FORESTRY. *Educ:* Univ Wash, BS & MF, 51, PhD, 56. *Prof Exp:* From asst prof to assoc prof forestry, Univ Mich, Ann Arbor, 56-67; PROF WILDLAND RECREATION, COL FOREST RESOURCES, UNIV WASH, 67- *Concurrent Pos:* Soc Am Foresters vis scientist lectr, NSF; Pac Crest Trail Adv Coun, USDA, 78-81. *Mem:* Fel, Asn Interpretive Naturalists; Soc Am Foresters. *Res:* Recreational use of wild lands; park management. *Mailing Add:* Col Forest Resources Univ Wash Seattle WA 98195

SHARPE, LAWRENCE, b London, Eng, Dec 25, 30. PSYCHIATRY. *Educ:* Univ London, MB & BS, 54, dipl psychol med, 61. *Prof Exp:* House surgeon & physician, Highlands Hosp, London, Eng, 54-55; registr, Cell Barnes Hosp, St Alban's, 57; sr house officer & registr, Maudsley Hosp, London, 58-61; fel pediat & psychiat, Johns Hopkins Hosp, 62-63, instr pediat & psychiat, Johns Hopkins Univ, 63-65; asst prof psychiat, State Univ NY Downstate Med Ctr, 66-67; asst prof psychiat, Col Physicians & Surgeons, Columbia Univ, 67-76; MEM STAFF, DEPT NEUROPHARMACOL NIDA ADDICTION RES CTR, 77- *Concurrent Pos:* Consult, Md Children's Ctr, Boys Village of Md, Barrett Sch for Girls & McKim's Boys Haven, 63-65; psychiatrist, Res Found Ment Hyg, Inc, 67-; res psychiatrist biomet res, NY State Dept Ment Hyg, Inc, 67- *Mem:* Royal Col Psychiat; Royal Soc Med. *Mailing Add:* Dept Psych Columbia Univ Col Phys Surg 630 W 168th St New York NY 10027

SHARPE, LOUIS HAUGHTON, b Port Maria, Jamaica, WI, Jan 31, 27; m 68. ADHESION, ADHESIVES ENGINEERING. *Educ:* Va Polytech Inst, BS, 50; Mich State Univ, PhD(phys chem), 57. *Prof Exp:* Engr, State Hwy Dept, Mich, 54-55; mem tech staff, Bell Labs, 55-59, supvr, adhesives & surface chem, 59-65, surface chem appl res, 65-68, adhesives eng & develop, 68-85; CONSULT, 86- *Concurrent Pos:* Ed, J Adhesion, 69- *Honors & Awards:* Adhesives Award, Am Soc Testing & Mat, 68. *Mem:* Am Chem Soc; fel Am Inst Chem; fel Am Soc Testing & Mat; fel NY Acad Sci; Adhesion Soc (pres, 86-88). *Res:* Surface chemistry; adhesion; adhesives; mechanical properties of polymers. *Mailing Add:* 28 Red Maple Rd Sea Pines Plantation Hilton Head Island SC 29928

SHARPE, MICHAEL JOHN, b Sydney, Australia, Mar 15, 41; m 66; c 1. MATHEMATICS. *Educ:* Univ Tasmania, BSc, 63; Yale Univ, MA, 65, PhD(math), 67. *Prof Exp:* Asst prof, 67-73, assoc prof, 73-77, PROF MATH, UNIV CALIF, SAN DIEGO, 77- *Concurrent Pos:* Vis prof, Univ Paris, 73-74. *Mem:* Inst Math Statist; Am Math Soc. *Res:* Probability theory; Markov processes; continuous parameter martingales; potential theory. *Mailing Add:* Dept Math Univ Calif PO Box 109 La Jolla CA 92093

SHARPE, ROGER STANLEY, b Omaha, Nebr, Mar 31, 41; m 62; c 3. ECOLOGY. *Educ:* Munic Univ Omaha, BA, 63; Univ Nebr, MS, 65, PhD(zool), 68. *Prof Exp:* Asst prof, 68-72, ASSOC PROF BIOL, UNIV NEBR, OMAHA, 72- *Concurrent Pos:* Environ consult, numerous projs; vis prof, Charles Univ, Prague, Czech, 88, 90. *Mem:* Am Ornith Union; Cooper Ornith Soc; Wilson Ornith Soc. *Res:* Conservation biology; zoogeography of Great Plains avifauna; avian ecology and behavior. *Mailing Add:* Dept Biol Univ Nebr Omaha NE 68182

SHARPE, ROLAND LEONARD, b Shakopee, Minn, Dec 18, 23; m 46; c 3. DYNAMIC ANALYSIS OF STRUCTURES. *Educ:* Univ Mich, BSE, 47, MSE, 49. *Prof Exp:* Designer struct eng, Cummins & Barnard, Engrs, 47-48; asst prof struct eng, Univ Mich, 48-50; exec vpres & gen mgr, John A Blume & Assocs, Engrs, 50-73; chmn & chief exec officer, Eng Decision Anal Co, 74-87; PRES, ROLAND L SHARPE CONSULT STRUCT ENGR, 73-74 & 87- *Concurrent Pos:* Mem planning comn, City of Palo Alto, Calif, 55-60; mem bd dirs, Earthquake Eng Res Inst, 71-74; exec dir & managing dir, Appl Technol Coun, 73-82; prin investr, Guidelines for Seismic Design of Buildings, 73-78; chmn & chief exec officer, Calif Eng & Develop Co, 74-84; managing dir, Eng Decision Anal Co, GmbH, 74-82; chmn, Struct Div Exec Comt, Am Soc Civil Engrs, 83, Mgt Group B, 89-; mem, Nat Earthquake Hazard Reduction Prog Adv Comt, Fed Emergency Mgt Agency & US Geol Surv, 91. *Mem:* Am Concrete Inst; Am Soc Civil Engrs; Earthquake Eng Res Inst. *Res:* Development of building code requirements for seismic design of buildings; research on performance of structures subjected to earthquake ground motions. *Mailing Add:* 10320 Rolly Rd Los Altos CA 94024

SHARPE, THOMAS R, b Milwaukee, Wis, Nov 25, 44; m 68. HEALTH CARE ADMINISTRATION, STATISTICS. *Educ:* Univ Ill, BS, 70; Univ Miss, MS, 73, PhD(health care admin), 75. *Prof Exp:* Staff pharmacist, Northwestern Univ Hosp, 70-71; grad res asst health care, Admin Dept, 71-74, res asst prof, 74-80, RES INST PHARMACEUT SCI & ASST DIR ADMIN SCI RES, UNIV MISS, 80- *Concurrent Pos:* Charles R Walgreen Mem fel, Am Found Pharmaceut Educ, 72-74; consult, Nat Ctr Health Serv Res, 75-, Miss State Bd Health, 76- & Drug Info Designs Inc, 78- *Mem:* Am Pharmaceut Asn; Am Pub Health Asn; Am Soc Hosp Pharmacists; Am Sociol Asn; Am Heart Asn. *Res:* Drug post marketing surveillance; patient compliance; sociology of occupations and professions; epidemiology of hypertension; computer applications to pharmacy; social indicators; rural health, optimum delivery of comprehensive health care. *Mailing Add:* Res Inst Pharmaceut Sci Univ Miss University MS 38677

SHARPE, WILLIAM D, b Canton, Ohio, July 18, 27. PATHOLOGY, HISTORY OF MEDICINE. *Educ:* Univ Toronto, BA, 50; Univ Buffalo, MA, 53; Johns Hopkins Univ, MD, 58. *Prof Exp:* Intern med, Jersey City Med Ctr, 58-59; resident path, Pa Hosp, 59-63; asst prof, 63-68, CLIN ASSOC PROF PATH, COL MED & DENT NJ, 68- *Concurrent Pos:* Fel path, Pa Hosp, Philadelphia, 59-63; asst instr, Sch Med, Univ Pa, 60-63; dir labs, Cabrini Med Ctr, New York. *Mem:* Int Acad Path; Mediaeval Acad Am; Am Philol Asn; Am Asn Pathologists & Bacteriologists; Col Am Pathologists. *Res:* Chronic radiation intoxication in human beings; ancient, medieval and colonial American medical history. *Mailing Add:* Cabrini Med Ctr 227 E 19th St New York NY 10003

SHARPE, WILLIAM NORMAN, JR, b Pittsboro, NC, Apr 15, 38; m 59; c 2. MECHANICAL ENGINEERING. *Educ:* NC State Univ, BS, 60, MS, 61; Johns Hopkins Univ, PhD(mech), 66. *Prof Exp:* Instr mech, Johns Hopkins Univ, 65-66; asst prof mech, Mich State Univ, 66-70, assoc prof, 70-75, prof, 75-78; prof mech eng & chmn dept, La State Univ, 78-82; DECKER PROF, DEPT MECH ENG, JOHN S HOPKINS UNIV, 83- *Mem:* Am Soc Mech Engrs; Am Soc Eng Educ; Soc Exp Stress Anal; Am Soc Testing & Mat; Sigma Xi; fel Am Soc Mech Engr. *Res:* Experimental mechanics, especially strain measurement by laser interferometry; fatigue; fracture. *Mailing Add:* Dept Mech Eng Johns Hopkins Univ Baltimore MD 21218

SHARPLES, FRANCES ELLEN, b Brooklyn, NY, Feb 5, 50. ECOLOGY, MAMMALOGY. *Educ:* Barnard Col, AB, 72; Univ Calif, Davis, MA, 74, PhD(zool), 78. *Prof Exp:* Teaching asst zool & ecol, Univ Calif, Davis, 72-78; RES ASSOC ENVIRON IMPACT ASSESSMENT, OAK RIDGE NAT LAB, 78- *Mem:* AAAS; Ecol Soc Am; Am Soc Mammal; Sigma Xi. *Res:* Vertebrate terrestrial ecology; ecology and evolution of mammals; environmental impact assessment; endangered species status and preservation. *Mailing Add:* 2131 Everett Rd Knoxville TN 37932

SHARPLES, GEORGE CARROLL, b Toledo, Ohio, Dec 17, 18; m 44; c 2. HORTICULTURE. *Educ:* Univ Ariz, BS, 41, MS, 48. *Prof Exp:* Asst horticulturist, Agr Exp Sta, Univ Ariz, 48-58, res assoc hort, 58-62, from asst horticulturist to assoc horticulturist, 62-72, horticulturist, 72-83, emer prof, 83-; RETIRED. *Mem:* Am Soc Hort Sci; AAAS. *Res:* Grape nutrition and culture; citrus nutrition; lettuce seed physiology and precision planting. *Mailing Add:* 1353 Lemon Tempe AZ 85281

SHARPLESS, GEORGE ROBERT, BIOLOGY. *Educ:* Johns Hopkins Univ, ScD, 31. *Prof Exp:* Mgr qual control, Lederle Labs, 72; RETIRED. *Mailing Add:* 430 Orangeburg Rd Pearl River NY 10965

SHARPLESS, K BARRY, b Philadelphia, Pa, Apr 28, 41; m; c 3. HOMOGENEOUS CATALYSIS, ASYMMETRIC TRANSFORMATIONS. *Educ:* Dartmouth Col, BA, 63; Stanford Univ, PhD(chem), 68. *Prof Exp:* NIH fel chem, Stanford Univ, 68-69 & Harvard Univ, 69-70; from asst prof to assoc prof, Mass Inst Technol, 70-75, prof chem, 75-91; prof chem, Stanford Univ, 77-80; PROF CHEM, SCRIPPS RES INST, 90- *Honors & Awards:* Creative Work Org Synthesis Award, Am Chem Soc, 83, Arthur C Cope Scholar Award, 86; Paul Janssen Prize, Creative Org Synthesis, Belg, 86; Harrison Howe Award, Am Chem Soc, Remsen Award; Prelog Medal, Eidgenössiche Technische Hochschule, Z06rich; Rolf Sammet Prize, Johann Wolfgang Geothe Universität, Frankfurt-am-Main; Chem Pioneer Award, Am Inst Chemists. *Mem:* Nat Acad Sci; fel Am Acad Arts & Sci; fel AAAS. *Res:* Developing new homogeneous catalysts for the oxidation of organic compounds, using inorganic reagents to effect new transformations in organic chemistry, and asymmetric catalysis involving both early and late transition metal-mediated processes. *Mailing Add:* Dept Chem Scripps Res Inst LaJolla CA 92037

SHARPLESS, NANSIE SUE, biochemistry, neurochemistry; deceased, see previous edition for last biography

SHARPLESS, SETH KINMAN, b Anchorage, Alaska, May 21, 25; m 47; c 2. PHARMACOLOGY, PHYSIOLOGICAL PSYCHOLOGY. *Educ:* Univ Chicago, MA, 51; McGill Univ, PhD(psychol), 54. *Prof Exp:* Res assoc psychol, McGill Univ, 54-55; asst prof, Yale Univ, 55-57; asst prof pharmacol & sr res fel, Albert Einstein Col Med, 57-64, assoc prof pharmacol, 64-69; PROF PSYCHOL, UNIV COLO, BOULDER, 69- *Concurrent Pos:* NIH career develop fel, 61-69; mem rev comt psychol serv fel, NIMH, 71-, mem exp psychol study sect, NIH, 71- *Mem:* Am Psychol Asn; Philos Sci Asn; Asn Symbolic Logic; Am Physiol Soc; Am Soc Pharmacol & Exp Therapeut. *Res:* Neuropharmacology; physiological basis of behavior; drugs on central nervous system. *Mailing Add:* Dept Psychol Univ Colo Muen D052 Box 345 Boulder CO 80309

SHARPLESS, STEWART LANE, b Milwaukee, Wis, Mar 29, 26. ASTRONOMY. *Educ:* Univ Chicago, PhB, 48, PhD(astron), 52. *Prof Exp:* Carnegie fel, Mt Wilson & Palomar Observs, 52-53; astronomer, US Naval Observ, 53-64, dir, Astron & Astrophys Div, 63-64; PROF ASTRON & DIR C E KENNETH MEES OBSERV, UNIV ROCHESTER, 64- *Concurrent Pos:* Dir-at-large, AURA, Inc, 66-69. *Mem:* AAAS; Am Astron Soc; Int Astron Union. *Res:* Galactic structure; spectroscopy. *Mailing Add:* Dept Physics & Astron Univ Rochester Wilson Blvd Rochester NY 14627

SHARPLESS, THOMAS KITE, b Boston, Mass, May 23, 39; c 1. CYTOLOGY, BIOMEDICAL ENGINEERING. *Educ:* Haverford Col, AB, 62; Princeton Univ, PhD(biochem), 69. *Prof Exp:* Res assoc membrane biochem, Pub Health Res Inst of City of New York, Inc, 69-70; programmer, Hudson Comput Corp, 70-71; consult, 72-73; SYSTS ENGR, MEM HOSP CANCER & ALLIED DIS & ASSOC, SLOAN KETTERING INST, 73- *Concurrent Pos:* Res fel, USPHS, 69-70. *Mem:* Asn Comput Mach. *Res:* Automated, high-speed measurement and classification of single cells and cell populations by flow or pulse microphotometry; applications in cell immunology, cytogenetics and diagnostic cytopathology. *Mailing Add:* Mem Sloan-Kettering Cancer Ctr 1275 York Ave New York NY 10021

SHARPLEY, ROBERT CALDWELL, b Chicago, Ill, Jan 31, 46; m 66; c 2. APPLIED MATHEMATICS. *Educ:* Univ Tex, BA, 68, MA, 69, PhD(math), 72. *Prof Exp:* Vis asst prof math, La State Univ, 72; asst prof, Oakland Univ, 72-76; from asst prof to assoc prof, 76-82, PROF MATH, UNIV SC, 83- *Concurrent Pos:* Proj dir & fac fel, Oakland Univ, 73-74 & NSF res grant, 77-88; vis assoc prof math, McMaster Univ, 78-79; vis prof math, Univ Wyo, 86-87. *Mem:* Am Math Soc; Sigma Xi. *Res:* Interpolation of operators; Fourier analysis; functional analysis; partial differential equations; Sobolev and Besov spaces; singular integral operator; numerical analysis. *Mailing Add:* Dept Math Univ SC Columbia SC 29208

SHARRAH, PAUL CHESTER, b Jamesport, Mo, Oct 31, 14; m 36; c 2. PHYSICS. *Educ:* William Jewell Col, AB, 36; Univ Mo, PhD(physics), 42. *Prof Exp:* Instr physics & math, William Jewell Col, 38-40; from asst prof to prof, 42-82, chmn dept, 57-69, dir planetarium, 70-82, EMER PROF PHYSICS, UNIV ARK, FAYETTEVILLE, 82- *Concurrent Pos:* Physicist, US Naval Ord Lab, Washington, DC, 44-46, Neutron Res, Oak Ridge Nat Lab, Oak Ridge, TN, 54-55; instr & adv, 82-85. *Mem:* AAAS; Am Phys Soc; Am Asn Physics Teachers; Am Crystallog Asn; Int Soc Planetarium Educr; Sigma Xi. *Res:* X-ray and neutron diffraction by liquids. *Mailing Add:* Dept Physics Univ of Ark Fayetteville AR 72701

SHARRETT, A RICHEY, b Worcester, Mass, July 15, 37. PUBLIC HEALTH & EPIDEMIOLOGY. *Educ:* Oberlin Col, BA, 59; Univ Melbourne, Australia, MA, 61; Univ Pittsburgh, MD, 66; Johns Hopkins Sch Hyg & Pub Health, Dr PH(epidemiol), 79. *Prof Exp:* Med officer, 72-86, CHIEF, SOC & ENVIRON EPIDEMIOL BR, EPIDEMIOL & BIOMETRY PROG, NIH, 86- *Res:* Relationship of biochemical, physiological, life style and other environmental risk factors to atherosclerosis and the incidence of major cardiovascular disease. *Mailing Add:* Epidemiol & Biometry Prog Bldg Rm 2C08 Bethesda MD 20892

SHARROW, SUSAN O'HOTT, FLOW CYTOMETRY, T-CELL ONTOLOGY & DIFFERENTIATION. *Educ:* Mich State Univ, BS, 71. *Prof Exp:* CHEMIST, NAT CANCER INST, NIH, 71- *Res:* Expression of Class I antigens. *Mailing Add:* Immunol Br Rm 4B-17 Nat Cancer Inst-NIH Bldg Ten Bethesda MD 20892

SHARTS, CLAY MARCUS, b Long Beach, Calif, Feb 9, 31; m 52; c 4. FLUORINE CHEMISTRY, ORGANIC CHEMISTRY. *Educ:* Univ Calif, BS, 52; Calif Inst Technol, PhD(chem), 59. *Prof Exp:* Asst, Calif Inst Technol, 55-58; res chemist, Explosives Dept, E I du Pont de Nemours & Co, 58-62; assoc prof, 62-71, PROF CHEM, SAN DIEGO STATE UNIV, 71- *Concurrent Pos:* NSF sci fac fel, Univ Cologne, 71-72; consult, Technicon Instruments Corp, 73-83 & UNESCO, 80. *Mem:* AAAS; Am Chem Soc; NY Acad Sci. *Res:* Small-ring and organic fluorine compounds; spiropolymers; perfluorocarbons as oxygen carriers in artificial blood; perfluoroalhyl substituted steroids and sugars as surfactants for perfluorocarbons. *Mailing Add:* PO Box 15487 San Diego CA 92115-1487

SHASHA, DENNIS E, US citizen; c 1. KNOWLEDGE EXPLORATION, PARALLEL ALGORITHMS. *Educ:* Yale Univ, BS, 77; Syracuse Univ, MS, 80; Harvard Univ, PhD(appl math), 84. *Prof Exp:* Engr, Int Bus Mach Co, 77-80; ASSOC PROF COMPUT SCI, COURANT INST, NEW YORK UNIV, 84- *Concurrent Pos:* Expert witness, Cravath, Swaire & Moore; consult, AT&T; IBM fel. *Mem:* Asn Comput Mach. *Res:* Systems and theory for knowledge exploration; parallel algorithms for database systems; algorithms for tree-matching; puzzles; database management. *Mailing Add:* 251 Mercer St New York NY 10012

SHASHIDHARA, NAGALAPUR SASTRY, b Mysore, India, Aug 13, 40; m 71; c 1. APPLIED FLUID MECHANICS, HYDROTHERMAL ENGINEERING. *Educ:* Mysore Univ, BS, 61; Indian Inst Sci, MS, 64; Rutgers Univ, MS & PhD(fluid mech), 71. *Prof Exp:* Asst engr fluid mech, Simpson & Group Co, Bangalore, India, 61-62; sr sci officer hydrodynamics, Indian Inst Sci, 62-66; res asst fluid mech, Rutgers Univ & Princeton Univ, 66-71; SUPVR HYDROTHERMAL ENG, EBASCO SERV, INC, 71- *Concurrent Pos:* Res scientist rheology, E I du Pont de Nemours & Co, Inc, 67; consult fluid mech & other hydrothermal probs to many US & foreign elec utilities. *Mem:* Int Asn Hydraul Res; Am Geophys Union; Sigma Xi; Asn Sci Workers India (secy, 62-66). *Res:* Dispersion in air and water of thermal, radioactive and chemical effluents from industrial facilities and their effects on the environment. *Mailing Add:* 25 Suffolk Lane Princeton Junction NJ 08550

SHASHOUA, VICTOR E, b Kermanshah, Iran, Nov 15, 29; nat US; m 55; c 3. BIOCHEMISTRY. *Educ:* Univ London, BSc, 51; Loughborough Col Tech, Eng, dipl, 51; Univ Del, PhD(org chem), 56. *Prof Exp:* Res chemist, E I du Pont de Nemours & Co, Inc, 51-64, res assoc, 64-68; mem neurosci res prog, Mass Inst Technol, 64-68, res assoc biol, 68-70; asst prof, 70-75, ASSOC PROF BIOL CHEM, HARVARD MED SCH, 75-, ASSOC BIOCHEMIST, MCLEAN HOSP, 70- *Mem:* Am Chem Soc; Am Neurochem Soc; Int Soc Neurochem; Soc Neurosci; Am Soc Cell Biol. *Res:* Neurochemistry; behavior and biochemistry. *Mailing Add:* Biol Chem Harvard Med Sch McLean Hosp Belmont MA 02178

SHASTRY, B SRIRAM, b Akola, India, Nov 26, 50; m 75; c 2. HIGH TEMPERATURE SUPERCONDUCTIVITY. *Educ:* Tata Inst Fundamental Res, Bombay, PhD(physics), 76. *Prof Exp:* Lectr physics, Univ Hyderabad, India, 77-79; Royal Soc bursar, Imp Col, London, 79-80; postdoctoral instr physics, Univ Utah, Salt Lake City, 80-82; fel & reader, Tata Inst, Bombay, 82-89; MEM TECH STAFF, AT&T BELL LABS, 88- *Concurrent Pos:* Vis lectr physics, Princeton Univ, 87-88. *Mem:* Am Phys Soc; fel Indian Acad Sci. *Res:* Quantum spin systems; strongly correlated forms systems; high temperature superconductivity; magnetism. *Mailing Add:* AT&T Bell Labs 600 Mountain Ave Murray Hill NJ 07974-2070

SHATKIN, AARON JEFFREY, b Providence, RI, July 18, 34; m 57; c 1. BIOCHEMISTRY, VIROLOGY. *Educ:* Bowdoin Col, AB, 56; Rockefeller Univ, PhD(microbiol), 61. *Hon Degrees:* DSc, Bowdoin Col, 79. *Prof Exp:* Sr asst scientist virol, Cell Biol Sect, Nat Inst Health, 61-63, res biochemist, 63-68; assoc mem, Roche Inst Molecular Biol, 68-71, mem, 71-77, head, Lab Molecular Virol, 77-83, head, Dept Cell Biol, 77-86; PROF & DIR, CTR ADVAN BIOTECHNOL & MED, RUTGERS UNIV, 86-, PROF MOLECULAR BIOL & PROF MOLECULAR GENETICS, UNIV MED & DENT NJ, 86- *Concurrent Pos:* Vis prof, Georgetown Univ, 68; guest investr, Salk Inst, Calif, 68-69; mem molecular biol study sect, NSF, 71-74; instr, Cold Spring Harbor Lab, 72-74; ed, J Virology, 73-77, ed-in-chief, Molecular Cell Biol, 80-90; Inst Comt to Review Virus Cancer Prog, Nat Cancer Inst, 73, Comt Int Asilomar Conf Recombined DNA, 74, chair adv comt nuclear acid, Am Cancer Soc, 81-82, adv comt, NY Univ & Am Inst Biologists, 79, Einstein Med Col, 81, Jones Cell Sci Ctr, 82-, Brookhaven Nat Lab, 83-86, dept molecular biol, Princeton Univ, 86-, Wadsworth Lab, 86, Bowdoin Col Sci, 86-, Children Hosp Boston, 87, J Gamble Res Inst, 88-; vis prof, Univ PR, 78-80, Princeton Univ, 84-87; mem nuclear act comt, Am Chem Soc, 79-82; mem, Sci Adv Bd, Merck Sharp & Dohme Res Labs, 90-, Nat Res Adv Bd, Cleveland Clin Found, 90- *Honors & Awards:* US Steel Found Award Molecular Biol, 79. *Mem:* Nat Acad Sci; AAAS; Am Soc Biol Chem; Am Soc Microbiol; fel NY Acad Sci; Harvey Soc; Am Soc Cell Biol. *Res:* Structure and function of animal cells and viruses; biochemistry of virus replication. *Mailing Add:* Ctr Advan Biotechnol & Med 679 Hoes Lane Piscataway NJ 08854-5638

SHATTES, WALTER JOHN, b Oceanside, NY, May 27, 24; m 60; c 2. EXPERIMENTAL PHYSICS. *Educ:* Union Col, BS, 50; Rutgers Univ, PhD(physics), 56. *Prof Exp:* Res physicist, Tung-Sol Elec Co, 55-62; sr physicist, Airco, 62-89; RETIRED. *Mem:* Am Vacuum Soc; Am Phys Soc. *Res:* Vacuum technology; physical vapor deposition; superconductors; semiconductors; semiconductor devices and materials; thin films, cryogenics. *Mailing Add:* 70 Beverly Rd Bloomfield NJ 07003

SHATTUCK, THOMAS WAYNE, b Denver, Colo, Aug 10, 50; m 78. PHYSICAL CHEMISTRY. *Educ:* Lake Forest Col, BA, 72; Univ Calif, Berkeley, PhD(chem), 76. *Prof Exp:* ASST PROF CHEM, COLBY COL, 76- *Concurrent Pos:* Vis asst prof, Univ Maine, Orono, 78. *Mem:* Am Chem Soc; Am Phys Soc; Sigma Xi. *Res:* Solid state nuclear magnetic resonance; nuclear quadrupole resonance of transition metal complexes; liquid crystal phase transitions. *Mailing Add:* Dept Chem Colby Col Waterville ME 04901

SHATZ, STEPHEN S, b New York, NY, Apr 27, 37; c 2. MATHEMATICS. *Educ:* Harvard Univ, AB, 57, AM, 58, PhD(math), 62. *Prof Exp:* Instr math, Stanford Univ, 62-63, actg asst prof, 63-64; from asst prof to assoc prof, 64-69, PROF MATH, UNIV PA, 69- *Concurrent Pos:* Vis lectr, Haverford Col, 66; mem, Res Ctr, Physics & Math, Univ Pisa, 66-67; chmn, Dept Math, Univ Pa, 83-86; mem, MRSI, Berkeley, Calif, 86-87. *Mem:* Am Math Soc. *Res:* Algebraic geometry. *Mailing Add:* Dept Math Univ Pa Philadelphia PA 19104

SHAUB, WALTER M, b Mt Vernon, NY, Sept 21, 47; m 69; c 2. INFORMATION QUALITY ASSURANCE, CRITICAL ISSUES ANALYSIS. *Educ:* State Univ NY, BS, 69; Cornell Univ, PhD(phys chem), 75. *Prof Exp:* Post doctoral res assoc laser optics, US Naval Res Lab, 75-76, res chemist chem genetics, 76-81; asst prof chem, George Mason Univ, 76-78; res chemist chem kinetics, US Nat Bur Standards, 81-83, group leader res mgt, 83-86; TECH DIR INFO TRANSFER, US CONF MAYORS/CORRE, 86- *Concurrent Pos:* Tech dir, Coalition Resource Recovery & the Environ, 86-; consult, US Environ Protection Agency adv bd, 89, mem bd, 89-; spec ed, Elsevier, Sci of the Total Environ, 89-90. *Honors & Awards:* Medal, Am Inst Chemists, 69. *Mem:* AAAS; NY Acad Sci; Am Chem Soc; Air & Waste Mgt Asn; assoc mem Am Soc Mech Engrs. *Res:* Mathematical theories catalytic formation of dioxins and furans involving fly ash; physical and chemical properties of fly ash; writer/editor technical issues concerned with waste management. *Mailing Add:* US Conf Mayors 1620 Eye St NW Washington DC 20006

SHAUDYS, EDGAR T, b Washingtons Crossing, Pa, Nov 23, 28; m 52; c 4. AGRICULTURAL ECONOMICS. *Educ:* Wilmington Col, BS, 50; Ohio State Univ, MS, 52, PhD(agr econ), 54. *Prof Exp:* Teaching asst biol, Wilmington Col, 47-50; res asst, Ohio State Univ, 50-54, prof agr econ, 54-86; RETIRED. *Mem:* Am Soc Farm Mgrs & Rural Appraisers; Am Agr Econ Asn. *Res:* Farm management production and organization; land tenure and farm real estate taxation; cost of production studies for crops and livestock. *Mailing Add:* 1184 Fairview Ave Columbus OH 43212

SHAUGHNESSY, THOMAS PATRICK, b Dedham, Mass, July 12, 42; m 67; c 2. SEMICONDUCTOR PROCESSING, MICROLITHOGRAPHY. *Educ:* Boston Col, BS, 64, PhD(physics), 70. *Prof Exp:* Asst silicon photovoltaics, Boston Col, 72-75; group leader, Spire Corp, 75-79; mgr process eng, GCA Corp, Burlington, 79-85; PROCESS ENG DEPT, RATHION, 85- *Mem:* Am Vacuum Soc; Electrochem Soc; Inst Elec & Electronics Engrs. *Res:* Silicon integrated process development; photolithography; electron beam lithography; photoresist chemistry; reactive ion etch; ion implantation; anneal techniques; diffusion; depostions; device modeling and testing. *Mailing Add:* Five Foxrun Rd Bedford MA 01730

SHAULIS, NELSON JACOB, b Somerset, Pa, Sept 10, 13; m 41; c 2. VITICULTURE. *Educ:* Pa State Col, BS, 35, MS, 37; Cornell Univ, PhD(soils), 41. *Prof Exp:* Asst soil technol, Pa State Col, 35-37, asst county agent, 37-38, instr pomol, 38-44; asst soil conservationist, Soil Conserv Serv, US Dept Agr, 38-44; from asst prof to prof pomol, 44-67, prof viticult, 67-78, EMER PROF VITICULT, EXP STA, NY STATE COL AGR & LIFE SCI, CORNELL UNIV, 78- *Concurrent Pos:* Fulbright res scholar, Australia, 67-68; consult viticult, 79- *Honors & Awards:* Outstanding Achievement Award, E Sect, Am Soc Enol & Viticult, 84, Hon Res Lectr, 85; Award of Merit, Am Wine Soc, 90. *Mem:* Fel Am Soc Hort Sci; Am Soc Agron; Soil Sci Soc Am; hon mem Am Soc Enol & Viticult; Crop Sci Soc Am. *Res:* Vineyard sites; grapevine physiology; vineyard mechanization and management including mineral nutrition, rootstocks and canopy microclimate; define attributes of site, canopy and crop affecting attainment of viticultural goals. *Mailing Add:* Dept Hort Sci NY State Agr Exp Sta Geneva NY 14456

SHAVER, ALAN GARNET, b Brockville, Ont, Dec 17, 46; m 69; c 1. ORGANOMETALLIC CHEMISTRY, INORGANIC CHEMISTRY. *Educ:* Carleton Univ, BSc, 69; Mass Inst Technol, PhD(chem), 72. *Prof Exp:* Fel, Univ Western Ont, 72-75; from asst prof to assoc prof, 75-88, PROF CHEM, MCGILL UNIV, 89- *Mem:* Chem Inst Can; Am Chem Soc. *Res:* Synthesis and characterization of new types of complexes; special interest in optically active organometallic complexes and in complexes with catenated sulfur ligands. *Mailing Add:* Dept Chem McGill Univ Montreal PQ H3A 2K6 Can

SHAVER, EVELYN LOUISE, b London, Ont, Aug 21, 31. CYTOGENETICS, REPRODUCTIVE BIOLOGY. *Educ:* Univ Western Ont, BSc, 54, MSc, 58, PhD(anat), 68. *Prof Exp:* Res asst micro-anat, Univ Western Ont, 54-60; res technician biol, Atomic Energy of Can Ltd, 60-64; demonstr, 64-66, from lectr to assoc prof, 66-79, PROF ANAT, UNIV WESTERN ONT, 79- *Mem:* Am Asn Anat; Can Asn Anat; Can Soc Cell Biol; Soc Study Reproduction. *Res:* Factors influencing the chromosome complement of pre- and postimplantation embryos; invitro fertilization. *Mailing Add:* Dept Anat Univ Western Ont London ON N6A 5C1 Can

SHAVER, GAIUS ROBERT, b Pasadena, Calif, Aug 19, 49. PLANT ECOPHYSIOLOGY. *Educ:* Stanford Univ, BS, 72, AM, 72; Duke Univ, PhD(bot), 76. *Prof Exp:* Res assoc ecol, San Diego State Univ & Univ Alaska, 76-78; ASST SCIENTIST ECOL, ECOSYSTS CTR, MARINE BIOL LAB, 79- *Concurrent Pos:* Instr biol, San Diego State Univ, 77-78. *Mem:* Am Ecol Soc; Brit Ecol Soc; Soc Am Naturalists; AAAS. *Res:* Ecology of plants; ecology of arctic tundras; mineral nutrition, growth and ecophysiology of plants. *Mailing Add:* 50 Upland Ave Falmouth MA 02540

SHAVER, KENNETH JOHN, b Auburn, NY, Dec 18, 25; m 48; c 1. FOOD CHEMISTRY. *Educ:* Syracuse Univ, BS, 48, PhD(chem), 52. *Prof Exp:* RES CHEMIST, MONSANTO CO, 52-, GROUP LEADER, 64-, RES MGR, 78- *Mem:* Am Chem Soc; Int Asn Dent Res; Inst Food Technologists; Sigma Xi. *Res:* Food phosphates; food preservatives. *Mailing Add:* 32 Millbrook Lane St Louis MO 63122

SHAVER, ROBERT HAROLD, b North Henderson, Ill, Sept 8, 22; m 45; c 4. GEOLOGY & PALEONTOLOGY, STRATIGRAPHY & SEDIMENTATION. *Educ:* Univ Ill, BS, 47, MS, 49, PhD(geol), 51. *Prof Exp:* Asst geol, Univ Ill, 47-50; from asst prof to prof, Univ Miss, 51-56, chmn dept, 52-56; head geol sect, Ind Geol Surv, 56-86 & res & proj develop br, 86-87; assoc prof geol, Ind Univ, Bloomington, 56-64, asst chmn dept, 67-72, prof, 64-; RETIRED. *Concurrent Pos:* Co-ed J Paleont, 64-69; actg chmn, Ind Univ, Bloomington, 70. *Mem:* Fel Geol Soc Am; Paleont Soc; hon mem Soc Econ Paleontologists & Mineralogists (pres-elect, 75-76, pres, 76-77). *Res:* Silurian and Devonian stratigraphy, paleoecology and sedimentation; Silurian paleoecology (reef faunas); carboniferous paleontology (ostracoda). *Mailing Add:* 2012 Viva Dr Bloomington IN 47401

SHAVER, ROBERT JOHN, zoology, parasitology, for more information see previous edition

SHAVER, ROY ALLEN, b Rushville, Ill, Aug 4, 31; m 59; c 3. ORGANIC CHEMISTRY. *Educ:* Western Ill Univ, BS, 53, MS, 56. *Prof Exp:* From instr to assoc prof, 56-77, chmn dept, 77-79, PROF CHEM, UNIV WIS, PLATTEVILLE, 77- *Concurrent Pos:* Consult fuels, Shell Oil, 77-81. *Mem:* Am Chem Soc. *Res:* Nonbenzenoid aromaticity-electrophilic reactions of the tropylium ion; time predictions on cheese rancidification. *Mailing Add:* Dept Chem Univ Wis Platteville WI 53818

SHAVITT, ISAIAH, b Kutno, Poland, July 29, 25; US citizen; m 57; c 1. THEORETICAL CHEMISTRY, QUANTUM CHEMISTRY. *Educ:* Israel Inst Technol, BSc, 50, dipl eng, 51; Cambridge Univ, PhD(theoret chem), 57. *Prof Exp:* Instr chem, Israel Inst Technol, 53-54 & 56-57, lectr, 57-58; res assoc theoret chem, Naval Res Lab, Univ Wis, 58-59; asst prof chem, Brandeis Univ, 59-60; staff chemist, IBM Watson Sci Comput Lab, Columbia Univ, 60-62; sr lectr chem, Israel Inst Technol, 62-63, assoc prof, 63-67; staff mem theoret chem, Battelle Mem Inst, 67-81; PROF CHEM, OHIO STATE UNIV, 81- *Concurrent Pos:* Mem nat coun, Info Processing Asn Israel, 66-67; adj prof chem, Ohio State Univ, 68-81; vis scientist, Max Planck Inst Physics & Astrophys, Munich, Ger, 76; chmn, Sub-Div Theoret Chem, Am Chem Soc, 78-79; vis prof chem, Univ Wollongong, Australia, 79; fel, Japan Soc Promotion Sci, Inst Molecular Sci, Okazaki, Japan, 80; interim co-dir, Ohio Supercomput Ctr, 88. *Mem:* Am Chem Soc; Am Phys Soc; Int Acad Quantam Molec Sci. *Res:* Molecular quantum mechanics; computational methods. *Mailing Add:* Dept Chem Ohio State Univ Columbus OH 43210-1173

SHAW, A(LEXANDER) J(OHN), b Vancouver, BC, May 7, 20; m 47; c 2. CHEMICAL ENGINEERING, SONOCHEMISTRY. *Educ:* Univ BC, BASc, 44, dipl bus mgt, 62. *Prof Exp:* Shift supvr, BC Distillery, 44-46; from res engr to chief chemist & develop engr, Western Chem Industs, Ltd, 46-68;

chem engr, B H Levelton & Assocs Ltd, 68-90; SPEC PROJS, RES & DEVELOP LAB, CHATTERTON PETROCHEM CORP, 90- *Mem:* Sr mem Chem Inst Can. *Res:* Process and product development, environmental quality studies, waste management; vitamins; proteins, fats, oils; alpha-glyceryl ethers; nucleotides; production biochemicals and medicinal chemicals; air and water quality studies; hazardous chemical disposal; air and water pollution control; energy from biomass; sonochemistry in synthesis of aromatic chemicals. *Mailing Add:* 4427 W Fifth Ave Vancouver BC V6R 1S4 Can

SHAW, ALAN BOSWORTH, b Englewood, NJ, Mar 28, 22; m 82; c 2. GEOLOGY. *Educ:* Harvard Univ, AB, 46, AM, 49, PhD(geol), 49. *Prof Exp:* Asst prof geol, Univ Wyo, 49-55; area paleontologist, Shell Oil Co, Colo, 55-60; consult paleontologist, 60-61; res assoc, Amoco Prod Res, 61, res sect supvr paleont & palynol, 61-65, spec res assoc, Okla, 65-68, consult geologist, Denver Div Explor, 68-70, district geologist, 70-71, consult geologist, 71-76, chief paleontologist, 76-77, mgr geol, Chicago Br, 77-82, res consult, 82-85; RETIRED. *Concurrent Pos:* Consult, Amoco, Okla, 86-87. *Mem:* Paleont Soc (pres, 68); Soc Econ Paleont & Mineral; Am Asn Petrol Geol. *Res:* Invertebrate paleontology and carbonate stratigraphy. *Mailing Add:* 1315 Kamira Dr Kerrville TX 78028-8805

SHAW, BARBARA RAMSAY, b Newton, NJ. BIOPHYSICAL CHEMISTRY. *Educ:* Bryn Mawr Col, AB, 65; Univ Wash, MS, 67, PhD(phys chem), 73. *Prof Exp:* Res fel biophys chem, Ore State Univ, 73-75; asst prof 75-81, ASSOC PROF CHEM, DUKE UNIV, 82- *Honors & Awards:* Fac Career Develop Award, Am Cancer Soc, 86-90. *Mem:* AAAS; Fedn Am Soc Exp Biol. *Res:* Protein-nucleic acid interaction in chromatin and subunit structure of chromatin; association of collagen model polypeptides and thermodynamic analysis of their thermal transitions; DNA structure and mutagenesis. *Mailing Add:* Dept Chem Duke Univ Durham NC 27706

SHAW, BRENDA ROBERTS, b Dunkirk, NY, May 12, 56. ELECTROANALYTICAL CHEMISTRY, MODIFIED ELECTRODES. *Educ:* Earlham Col, AB, 77; Univ Ill, PhD(chem), 83. *Prof Exp:* Asst prof, 86-90, ASSOC PROF CHEM, UNIV CONN, 90- *Mem:* Am Chem Soc; Sigma Xi. *Res:* Nature and design of electrode surfaces; use of polymers and zeolites to support prospective electrocatalysts. *Mailing Add:* Dept Chem U-60 Univ Conn 215 Glenbrook Rd Storrs CT 06269-3060

SHAW, C FRANK, III, b Mt Vernon, NY, Mar 17, 44; div; c 2. INORGANIC BIOCHEMISTRY, PHARMACEUTICAL CHEMISTRY. *Educ:* Univ Del, BS, 66; Northwestern Univ, PhD(inorg chem), 70. *Prof Exp:* NSF grant, dept chem, Purdue Univ, Lafayette, 70-72; fel chem, McGill Univ, 72-74; from lectr to assoc prof, 74-86, chmn, 90-92, PROF CHEM, UNIV WIS-MILWAUKEE, 86- *Concurrent Pos:* Book rev ed, Can J Spectros, 73-77; indust consult. *Mem:* Am Chem Soc; Sigma Xi. *Res:* Inorganic biochemistry of essential metals; heavy metal drugs, environmental contaminants, gold, cadmium, zinc and metallo proteins. *Mailing Add:* Dept Chem Univ Wis Milwaukee WI 53201

SHAW, CHARLES ALDEN, b Detroit, Mich, June 8, 25; div; c 2. COMPUTER AIDED DESIGN, COMPUTER APPLICATIONS. *Educ:* Harvard Univ, BS, 45; Syracuse Univ, MSEE, 58. *Prof Exp:* Test & design engr, Gen Elec Co, Syracuse, NY, 47-51; chief engr, Onondaga Pottery Electronics, NY, 51-61; unit & subsect mgr, Gen Elec Semiconductor Prod Dept, Gen Elec Co, Paris, 61-66, consult to dir gen, Bull Gen Elec, 66-69, mgr comput aided design, Integrated Circuits Prod Dept, Syracuse, 69-71, proj mgr Solid State Appln Oper, 71-73, mgr, Comput Aided Design Ctr, 73- 77, mgr, Solid State Appln Oper, 78-81, dir, comput aided design, Intersil, Gen Elec, Cupertina, Calif, 81-88; TECH PROG MGR, CADENCE DESIGN SYSTS INC, SANTA CLARA, CALIF, 89- *Honors & Awards:* Recognition Award, Asn Comput Mach, 88. *Mem:* Asn Comput Mach; Inst Elec & Electronics Engrs. *Res:* Computer aided design for design of electronic circuits. *Mailing Add:* Cadence Design Systs Inc 2455 Augustine Dr Santa Clara CA 95054

SHAW, CHARLES BERGMAN, JR, b Dallas, Tex, June 7, 27; m 50, 66; c 2. PHYSICS, APPLIED MATHEMATICS. *Educ:* Calif Inst Technol, BS, 47; Univ Southern Calif, MS, 50, PhD(physics), 58. *Prof Exp:* Lab assoc physics, Univ Southern Calif, 47-49, lectr math, 52-54; res asst, Los Alamos Sci Lab, 49; physicist, Nat Bur Standards, 51; sr scientist, Missile Systs Div, Lockheed Aircraft Corp, 54-56; mem tech staff, Res Labs, Hughes Aircraft Co, 56-64; sr scientist, Electro-Optical Systs Inc, 64-66; sr tech specialist phys sci, Autonetics Div, NAm Aviation, Inc, 66-68, group scientist phys sci, 68-70, mem tech staff, 70-71, GROUP LEADER FLUID PHYSICS, SCI CTR, ROCKWELL INT, 71- *Concurrent Pos:* Vis prof, Loyola Univ, Calif, 60. *Mem:* Am Phys Soc; Mat Res Soc; Soc Indust & Appl Math; Inst Elec & Electronics Engrs; Am Welding Soc; Sigma Xi. *Res:* Plasma diagnostics; arc physics; physics of welding processes; computational physics; integral equations; improperly posed problems; reconstruction from projections; holographic interferometry; laser-generated plasmas. *Mailing Add:* IRC INGEL/EG&G Idaho Inc PO Box 1625 Idaho Falls ID 83415

SHAW, CHARLES GARDNER, b Springfield, Mass, Aug 12, 17; m 40; c 3. PHYTOPATHOLOGY, MYCOLOGY. *Educ:* Ohio Wesleyan Univ, BA, 38; Pa State Col, MS, 40; Univ Wis, PhD(bot, plant path), 47. *Prof Exp:* Instr, Bot Lab, Ohio Wesleyan Univ, 37-38; lab asst, Pa State Col, 38-40; lab asst, Univ Wis, 41-43, asst, 46-47; instr plant path & jr plant pathologist, Wash State Univ, 47-48, asst prof & asst plant pathologist, 48-51, assoc prof & assoc plant pathologist, 51-57, actg chmn dept, 60-61, chmn dept, 61-72, prof & plant pathologist, 57-84; RETIRED. *Concurrent Pos:* Consult, US Agency Int Develop Pakistan Agr Univ, 69-70 & 72; vis prof plant path, Dept Bot, Univ Auckland, NZ, 75; vis scientist, Govt New Zealand Div Indust & Sci Res, 81-82; chief of party, Wash State Univ-US AID, Univ Jordan, 77-79, Ministry Agr, Govt Jordan, 82-83. *Mem:* Mycol Soc Am; Am Phytopath Soc; Brit Mycol Soc; Int Soc Plant Path; Am Soc Plant Taxon. *Res:* Saprots of conifers; vectoring of wood decay fungi, delignification of wood; taxonomy of Peronosporaceae. *Mailing Add:* NW 325 Janet St Pullman WA 99163

SHAW, CHARLES GARDNER, III, b Colfax, Wash, June 30, 48; m 70; c 3. FOREST PATHOLOGY. *Educ:* Wash State Univ, BSc, 70; Ore State Univ, PhD(plant path), 75. *Prof Exp:* Scientist forest path, Forest Res Inst, NZ Forest Serv, 74-77; res plant pathologist, Forest Sci Lab, Juneau, Alaska, 77-86; PROJ LEADER, PEST IMPACT ASSESSMENT TECHNOL, FOREST SERV RES, FT COLLINS, CO, 86- *Concurrent Pos:* Path sect, Brit Forestry Comn, 82-83. *Mem:* Soc Am Foresters; Am Phytopath Soc; NZ Inst Foresters; Am Mycol Soc; Brit Mycol Soc. *Res:* Root diseases of forest trees, particularly Armillaria; tree growth responses to disease; dwarf mistletoes; tree declines of unknown cause; assessment of pest impacts on timbers and other resource values. *Mailing Add:* Rocky Mountain Res Sta 240 W Prospect St Ft Collins CO 80526

SHAW, CHENG-MEI, b Chang-Hua, Taiwan, Oct 24, 26; m 51; c 4. NEUROPATHOLOGY. *Educ:* Taihoku Gym, dipl, 46; Nat Taiwan Univ, MD, 50. *Prof Exp:* Resident surg, Nat Taiwan Univ Hosp, 50-53 & neuropsychiat, 53-54; fel neurosurg, Lahey Clin, 54-55; resident neurosurg, Col Med, Baylor Univ, 55-58; res fel path, Col Med, Baylor Univ, 58-60; from res instr to assoc prof, 60-74, PROF PATH, SCH MED, UNIV WASH, 74- *Concurrent Pos:* Nat Multiple Sclerosis Soc res fel, 60-63. *Mem:* Am Asn Neuropath; Int Acad Path; Am Asn Path; Sigma Xi. *Res:* Immunopathology; etiology and pathogenesis of experimental allergic encephalitis; mechanisms of immunological adjuvants; morphological studies of malformation of human central nervous system. *Mailing Add:* 6541 29th NE Seattle WA 98115

SHAW, DAVID ELLIOT, b Chicago, Ill, Mar 29, 51. COMPUTATIONAL FINANCE, PARALLEL COMPUTATION. *Educ:* Univ Calif, San Diego, BA, 72; Stanford Univ, MS, 74, PhD, 80. *Prof Exp:* Pres & chief exec officer, Stanford Systs Corp, 76-79; assoc prof computer sci & dir, Non-Von Supercomput Proj, Columbia Univ, 80-86; vpres, Morgan Stanley & Co, 86-88; managing gen partner, D E Shaw & Co, 88- *Concurrent Pos:* Chmn, New York City Mayor's Comt Technol & Pub Policy, 87. *Honors & Awards:* Fac Develop Award, IBM, 83. *Mem:* Asn Comput Mach; Inst Elec & Electronics Engrs. *Res:* Parallel computer architectures for artificial intelligence and other applications; computational finance. *Mailing Add:* D E Shaw & Co 251 Park Ave South 15th Fl New York NY 10010

SHAW, DAVID GEORGE, b Los Angeles, Calif, Apr 20, 45; m 69. CHEMICAL OCEANOGRAPHY, ORGANIC CHEMISTRY. *Educ:* Univ Calif, Los Angeles, BS, 67; Harvard Univ, AM, 69, PhD(chem), 71. *Prof Exp:* NSF fel, Harvard Univ, 71-72; res assoc chem oceanog, Marine Sci Inst, Univ Conn, 72-73; asst prof, 73-76, ASSOC PROF, INST MARINE SCI, UNIV ALASKA, 76- *Mem:* AAAS; Geochem Soc; Am Chem Soc; Am Soc Limnol & Oceanog. *Res:* Transport and reactivity of organic chemicals in marine systems; organic trace analysis; environmental quality. *Mailing Add:* 1601 Hans Way Fairbanks AK 99709

SHAW, DAVID HAROLD, b Grants Pass, Ore, Jan 18, 41; m 63; c 3. PHYSIOLOGY, PHARMACOLOGY. *Educ:* Univ Nebr, BS, 63, MS, 65; Univ of the Pac, PhD(physiol, pharmacol), 69. *Prof Exp:* Asst physiol, Univ Nebr, 63-64, asst & res assoc cell physiol, 64-65, instr physiol, 65-66; instr & res assoc physiol & pharmacol, Univ of the Pac, 67-69; from asst prof to assoc prof, 69-85, PROF & CHAIR ORAL BIOL, COL DENT, UNIV NEBR MED CTR, LINCOLN, 85- *Mem:* AAAS; Int Asn Dent Res; Am Asn Dent Res; Sigma Xi. *Res:* The effect of dental plaque extract on cultured cells; gingival retraction material toxicity; catecholamine response to dental treatment. *Mailing Add:* Dept of Oral Biol Univ Nebr Med Ctr Col Dent Lincoln NE 68583-0740

SHAW, DAVID T, b China, Mar 13, 38; m 61; c 2. ELECTROMECHANICAL ENGINEERING. *Educ:* Nat Taiwan Univ, BS, 59; Purdue Univ, MS, 61, PhD(nuclear eng), 64. *Prof Exp:* From asst prof to assoc prof, 64-74, co-dir, Ctr for Integrate Process Systs Technol, 86-88, PROF ELEC ENG, STATE UNIV NY BUFFALO, 74-, DIR LAB FOR POWER & ENVIRON STUDIES, 78-, EXEC DIR, NY STATE INST ON SUPERCONDUCTIVITY, 87- *Concurrent Pos:* NSF res initiation grant, 65-67; consult, Bell Aerosystem Co, NY, 66-69 & Jet Propulsion Lab, Calif, 68-; vis assoc, Dept Environ Eng Sci, Calif Inst Technol, 70-71; vis prof, Univ Paris, 76; consult, Atomic Energy Comn, Paris, 77- *Mem:* Am Nuclear Soc; Inst Elec & Electronics Engrs; Am Soc Mech Engrs; Am Soc Eng Educ; Am Inst Aeronaut & Astronaut. *Res:* Control of particulate emissions into the atmosphere; acoustic agglomeration; atmospheric nucleation processes; aerosol measurement techniques; aerosol physics; energy conversion; his research interests include high temperature synthesis of ultra fine particles; electromagnetic wave interactions with aerosols and superconductivity materials preparation and characterization. *Mailing Add:* Ten Ironwood Ct East Amherst NY 14051

SHAW, DENIS MARTIN, b St Annes, Eng, Aug 20, 23; m 46, 76; c 3. GEOCHEMISTRY. *Educ:* Cambridge Univ, BA, 43, MA, 48; Univ Chicago, PhD(geochem), 51. *Prof Exp:* From lectr to prof geol, McMaster Univ, 48-89, chmn dept, 53-59 & 62-66, dean grad studies, 78-84, EMER PROF GEOL, McMASTER UNIV, 89- *Concurrent Pos:* Vis prof, Ecole Nat Superieure de Geol, France, 59-60 & Inst Mineral, Univ Geneva, 66-67; exec ed, Geochimica et Cosmochimica Acta, 70-88. *Honors & Awards:* Miller Medal, Royal Soc Can, 81. *Mem:* Geochem Soc; fel Royal Soc Can; Geol Asn Can; Mineral Asn Can (pres, 64-65); Geol Soc Am; Meteoritical Soc. *Res:* Geochemistry; chemical mineralogy. *Mailing Add:* Dept Geol McMaster Univ Hamilton ON L8S 4M1 Can

SHAW, DEREK HUMPHREY, b Maidstone, Eng, Apr 27, 37; Can citizen; m 60; c 2. CARBOHYDRATE CHEMISTRY, MICROBIAL BIOCHEMISTRY. *Educ:* Univ Cape Town, BSc, 59, PhD(carbohydrate chem), 65. *Prof Exp:* Prod chemist, Seravac Labs Ltd, Cape Town, SAfrica, 60-63; Nat Res Coun Can fel biosci, Ottawa, Ont, 65-67; res scientist marine carbohydrates, Fisheries Res Bd Can, 67-72, div head fisheries technol, St John's, Nfld, 72-74; head, Marine Prod Div, Nfld Biol Sta, 75-76, head, Microbial Chem Sect, 76-86, HEAD FISH HEALTH & PARASITOL SECT,

NORTHWEST ATLANTIC FISHERIES CTR, 87- *Concurrent Pos:* Adj prof chem, Mem Univ Nfld. *Mem:* Fel Chem Inst Can. *Res:* Marine carbohydrates; gas chromatography of carbohydrate derivatives; microbial polysaccharides from bacteria pathogenic to fish; biochemistry of fish diseases. *Mailing Add:* Northwest Atlantic Fisheries Ctr PO Box 5667 St John's NF A1C 5X1 Can

SHAW, DON W, b Pecan Gap, Tex, Dec 26, 37; m 59; c 2. PHYSICAL CHEMISTRY, MATERIALS SCIENCE. *Educ:* East Tex State Univ, BS, 58; Baylor Univ, PhD(phys chem), 65. *Prof Exp:* Instr chem, Dallas Inst, 58-61; mem tech staff, Tex Instruments, Inc, 65-72, sr scientist, 72-77, head, Explor Mat Br, 78-79, DIR, MAT SCI LAB, TEX INSTRUMENTS, INC, 79- *Concurrent Pos:* Assoc ed, J Crystal Growth; div ed, J Electrochem Soc; vis comt chem, Univ Tex; adv bd matls, Processing Ctr, MIT; mem tech adv comt, Eng Res Ctr, Univ Ill. *Honors & Awards:* Electrochem Soc Electronics Div Award, 83. *Mem:* Nat Acad Eng; Electrochem Soc; Am Asn Crystal Growth; Am Chem Soc; Sigma Xi; Mats Res Soc. *Res:* Electronic materials, crystal growth mechanisms; kinetics of vapor phase epitaxial growth of semiconductors; materials for solid state microwave devices; preparation and properties of gallium arsenide. *Mailing Add:* 10009 Apple Creek Dr Dallas TX 75243

SHAW, EDGAR ALBERT GEORGE, b Middlesex, Eng, July 10, 21; Can citizen; m 45; c 2. PHYSICS. *Educ:* Univ London, BSc, 48, PhD(physics), 50. *Prof Exp:* Tech officer, UK Ministry Aircraft Prod & Brit Air Comn, US, 40-46; from asst res officer to prin res officer, 50-86, head acoust sect, 75-85, EMER RESEARCHER, INST MICROSTRUCT SCI, NAT RES COUN CAN, 86- *Concurrent Pos:* Lectr, Univ Ottawa, 58-73; mem, Comt Hearing, Bio-Acoust & Biomech, Nat Res Coun-Nat Acad Sci, 65-; chmn, Int Comn Acoust, 75-78; co-chmn, Int Comn Biol Effects of Noise, 88- *Honors & Awards:* Rayleigh Medal, Inst Acoust, Brit, 79. *Mem:* Fel Acoust Soc Am (pres, 73); fel Royal Soc Can; Brit Inst Physics; Can Asn Physicists; fel Inst Acoust Brit. *Res:* Electroacoustics; psychoacoustics; physiological acoustics; acoustic measurements; wave acoustics; urban and industrial noise; hearing and hearing protection; mechanical vibrations. *Mailing Add:* Inst Microstruct Sci Nat Res Coun Can Ottawa ON K1A 0R6 Can

SHAW, EDWARD IRWIN, b New York, NY, Jan 17, 27; m 56, 80; c 3. RADIATION BIOLOGY. *Educ:* Univ Mo, AB, 48, MA, 50; Univ Tenn, PhD(zool, radiation biol), 55. *Prof Exp:* Asst zool, Univ Mo, 48-51 & Univ Tenn, 52-55; res assoc biol div, Oak Ridge Nat Lab, 55-56; from asst prof to prof radiation biophys, 56-85, chmn dept, 68-78, PROF PHYSIOL & CELL BIOL, UNIV KANS, 80- *Concurrent Pos:* Instr, Univ Tenn, 55; consult, Menninger Found, Kans, 57-62 & AAAS, 71-75; dir, Radiation Protection, Energy Res Develop Admin Training Grant, 74-77. *Mem:* Fel AAAS; Radiation Res Soc; Health Physics Soc; Sigma Xi. *Res:* Radiation biophysics; radiation effects at the cell level; metabolism of radioactively labelled organic compounds as internal emitters; health physics; mutation-induction and recovery from pre-mutational lesions. *Mailing Add:* 1635 Mississippi St Lawrence KS 66044

SHAW, ELDEN K, b Brigham, Utah, Aug 20, 34; m 57; c 4. ELECTRICAL ENGINEERING. *Educ:* Utah State Univ, BS, 56; Stanford Univ, MS, 57, PhD(elec eng), 67. *Prof Exp:* Sr res engr, Litton Industs Tube Corp, 60-66; from asst prof to assoc prof, 66-77, PROF ELEC ENG, SAN JOSE STATE UNIV, 77- *Concurrent Pos:* Res asst, Stanford Univ, 64-66; consult, S F D Labs, Inc, 66-70. *Mem:* Inst Elec & Electronics Engrs. *Res:* Steady state plasma discharges; crossed field microwave tubes, electron optics. *Mailing Add:* Dean Eng Calif State Univ Fresno CA 93711

SHAW, ELLIOTT NATHAN, b Youngstown, Ohio, Apr 6, 20; m 55. BIOCHEMISTRY. *Educ:* Mass Inst Technol, SB, 41, PhD(chem), 43. *Prof Exp:* Res assoc, Squibb Inst Med Res, 43-48; vis investr biochem, Rockefeller Inst, 48-50, asst, 50-51, assoc, 51-57; from assoc prof to prof, Tulane Univ, 57-65; SR BIOCHEMIST, BROOKHAVEN NAT LAB, 65- *Concurrent Pos:* Fel, Nat Cancer Inst, 49-50. *Mem:* Am Chem Soc; Am Soc Biol Chem; Harvey Soc; NY Acad Sci. *Res:* Chemical structure and biological action; antimetabolites; enzyme active centers; proteolytic enzymes. *Mailing Add:* Friedrich Miescher Inst Postfach 2543 Basel FH-4002 Switzerland

SHAW, ELLSWORTH, b Chicago, Ill, Mar 7, 20; m 43; c 6. SOIL SCIENCE, AGRICULTURAL CHEMISTRY. *Educ:* Univ Chicago, BS, 41; Univ Ariz, PhD(soil chem), 49. *Prof Exp:* Res chemist, Testing Mat, Kellogg Switchboard & Supply Co, 41-42; plant indust sta, US Dept Agr, 49-52; consult agr, 53-70; PLANT & SOIL SCIENTIST & LAB & FIELD DIR, GAC PRODUCE, MEXICO, 70- *Concurrent Pos:* Res assoc agr chem & soils, Univ Ariz, 60; Independent agr consult, Ariz & Calif. *Mem:* Am Chem Soc; Am Soc Agron; fel Am Inst Chem; Sigma Xi. *Res:* Application of radioisotopes to soil and plant nutritional problems; effect of fumigants on soil; availability of soil zinc to plants; soil and fertilizer need to plants through measurement of nitrogen uptake. *Mailing Add:* 3602 E Flower Tucson AZ 85716

SHAW, ELWOOD R, b Clark Co, Ohio, Sept 29, 18; m 56; c 1. ANALYTICAL CHEMISTRY. *Educ:* Cedarville Col, AB, 40, BS, 41; Ohio State Univ, MSc, 56. *Prof Exp:* Instr math & chem, Cedarville Col, 40-41, prof chem, 46-54; teacher high schs, Ohio, 41-42, 45-46; asst prof chem & res chemist, Antioch Col, 54-59; from res assoc to sr res assoc chem, C F Kettering Res Lab, 59-84; RETIRED. *Mem:* Am Chem Soc. *Res:* Hydrothermal research; germanates and silicates; organic synthesis; substituted hydrazines; metalloporphins; photosynthesis. *Mailing Add:* 1102 Clifton Rd Xenia OH 45385

SHAW, EMIL GILBERT, b San Antonio, Tex, June 22, 22; m 50; c 3. BIOCHEMISTRY, NUTRITION. *Educ:* Southwest Tex State Col, BS, 47, MA, 48; Tex A&M Univ, PhD(bionutrit), 67. *Prof Exp:* Instr & prof biol, East Tex State Col, 48-50; chief lab serv, US Army Hosp, Ft Chaffee, 50-51; biochemist, Fourth Army Area Med Lab, Ft Sam Houston, 51-52, Surg Res Univ, Brooke Army Med Ctr, 52-53 & Tripler Gen Hosp, Honolulu, Hawaii, 53-55; chief of serv, Wilford Hall, US Air Force Hosp, Lackland Air Force Base, 55-56; liaison officer, Off Surgeon Gen, US Army Chem Ctr, Edgewood, Md, 56-58; dep chief, Environ Systs Div, US Air Force Sch Aerospace Med, 58-72, chief, 72-76; RETIRED. *Res:* Physiological effects of exposure to exotic atmospheres for prolonged periods upon the mineral metabolism of man and experimental animals. *Mailing Add:* 10606 West Ave San Antonio TX 78218

SHAW, EUGENE, b Burlington, Iowa, Aug 3, 25; m 56; c 4. MICROBIOLOGY, VIROLOGY. *Educ:* Northwestern Univ, BS, 49; Univ Iowa, MS, 55, PhD(microbiol), 63. *Prof Exp:* With Med Serv Corps, US Army, 51-70; SR SCIENTIST, VIROL DIV, ORTHO RES FOUND, 70- *Mem:* Am Soc Microbiol; Tissue Cult Asn; NY Acad Sci. *Res:* Enteroviruses; infectious hepatitis. *Mailing Add:* 35 Berkshire Dr Princeton Junction NJ 08550

SHAW, FREDERICK CARLETON, b Boston, Mass, July 8, 37; m 58; c 2. INVERTEBRATE PALEONTOLOGY. *Educ:* Harvard Univ, AB, 58, PhD(geol), 65; Univ Cincinnati, MS, 60. *Prof Exp:* From instr to asst prof geol, Mt Holyoke Col, 63-68; from asst prof to assoc prof, 68-78, chmn, Dept Geol & Geog, 72-81, PROF GEOL, LEHMAN COL, 78-, DEAN NATURAL & SOCIAL SCI, 81- *Concurrent Pos:* Geol Soc Am res grant, 65; instr & dir social sci training prog, Mt Hermon Sch, 65-66; NSF res grant, 68-71; prin investr, MBRS & MISIP training res grants , 85-; Smithsonian Foreign Currency Grant, 85- *Mem:* AAAS; Paleont Soc; Geol Soc Am; Sigma Xi. *Res:* Systematics and paleoecology of Paleozoic invertebrates; Ordovician biostratigraphy. *Mailing Add:* Dept Geol Herbert H Lehman Col Bronx NY 10468

SHAW, GAYLORD EDWARD, b London, Ont, Mar 10, 39; m 65; c 2. ANIMAL PHYSIOLOGY. *Educ:* Graceland Col, Iowa, BA, 62; Univ Ill, Urbana, MS, 66, PhD(vet med sci), 70. *Prof Exp:* Res asst pathol & hyg, Col Med, Univ Ill, 62-63; instr biol, Graceland Col, 63-64; res asst physiol & pharmacol, Col Vet Med, Univ Ill, 64-69, res asst vet biol structure, 69-70; ASSOC PROF ANAT & PHYSIOL MICROBIOL, KENT STATE UNIV, 70- *Concurrent Pos:* Lab dir, Caries Susceptibility Testing & Res Ctr, 78-82. *Mem:* AAAS; Sigma Xi. *Res:* Oral leukocyte infiltration and phagocytic activity; oral lactobacillus counts as a clinical parameter of caries activity. *Mailing Add:* 300 S Cherry St Lamoni IA 50140

SHAW, GEORGE, II, b Castro Valley, Calif, Oct 13, 59. ROBOTICS, EMBEDDED & REAL-TIME COMPUTER SYSTEMS. *Prof Exp:* Asst mgr & computer tech, Computer Systs Unlimited, San Lorenzo, Calif, 76; warehouse mgr, Tech Rep Assocs-TRA Sales, Hayward, Calif, 76-77; mgr & sr computer tech, Byte Shop Mountain View, Mountain View, Calif, 77; partner & mgr, Acropolis, Hayward, Calif, 81-82; programmer assoc, Am Inst Res, Palo Alto, Calif, 86-88; OWNER & CONSULT SOFTWARE ENGR, SHAW LABS LIMITED, HAYWARD, CALIF, 77- *Concurrent Pos:* Teaching, Basic Prog, Microcomputer Oper, 77-; consult, Microsysts Anal, Systs & Appln Prog, 77-; chairperson, Asilomar Forth Modification Lab, 82 & 85, publications comt, Forth Int Standards Team, 82-; dir, Forth Modification Lab, 87-88; founder & chairperson, NBay, Forth Interest Group, 88-90, ACM Special Interest Group, SIG Forth, 88-91. *Mem:* Asn Comput Mach; assoc mem Inst Elec & Electronics Engrs; assoc mem Inst Elec & Electronics Engrs Computer Soc. *Res:* Computer software; several inventions. *Mailing Add:* Shaw Labs PO Box 3471 Hayward CA 94540-3471

SHAW, GLENN EDMOND, b Butte, Mont, Dec 5, 38; m 57; c 5. ATMOSPHERIC PHYSICS. *Educ:* Mont State Univ, BS, 63; Univ Southern Calif, MS, 65; Univ Ariz, PhD, 71. *Prof Exp:* From asst prof to assoc prof, 71-79, PROF PHYSICS, GEOPHYS INST, UNIV ALASKA, 79- *Mem:* Am Meteorol Soc; Am Geophys Union; Royal Meteorol Soc; AAAS; Royal Inst. *Res:* Laser profiling of lower and upper atmosphere; atmospheric physics, especially cloud physics; physics of aerosols; nucleation; atmospheric chemistry. *Mailing Add:* Geophys Inst Univ of Alaska Fairbanks AK 99701

SHAW, GORDON LIONEL, b Atlantic City, NJ, Sept 20, 32; m 58; c 3. THEORETICAL PHYSICS. *Educ:* Case Inst Technol, BS, 54; Cornell Univ, PhD(theoret physics), 59. *Prof Exp:* Res assoc theoret physics, Ind Univ, 58-60 & Univ Calif, San Diego, 60-62; asst prof physics, Stanford Univ, 62-65; assoc prof, 65-68, PROF PHYSICS, UNIV CALIF, IRVINE, 68- *Mem:* Am Phys Soc. *Res:* Theory of strong interactions of elementary particles. *Mailing Add:* Dept Physics Univ Calif Irvine CA 92717

SHAW, HARRY, JR, b Miami, Fla, Feb 6, 27; m 49; c 2. APPLIED MATHEMATICS. *Educ:* Emory Univ, BA, 49; Univ Miami, MS, 51. *Prof Exp:* Instr math, Marion Mil Inst, 51-52; asst, Univ NC, 52-54 & Univ Md, 54-56; RETIRED. *Mem:* Asn Comput Mach; Sigma Xi. *Res:* Numerical analysis. *Mailing Add:* 6048 Stevens Forest Rd Columbia MD 21045

SHAW, HELEN LESTER ANDERSON, b Lexington, Ky, Oct 18, 36; m 88. AMINO ACIDS, TRACE MINERALS. *Educ:* Univ Ky, Lexington, BS, 58; Univ Wis-Madison, MS, 65, PhD(nutrit sci), 69. *Prof Exp:* Dietician, Roanoke Vet Admin Mem Hosp, 59-60 & Santa Barbara Cottage Hosp & Univ Calif, Santa Barbara, 60-63; from asst prof to prof, Dept Human Nutrit & Food, Univ Mo, Columbia, 69-88; PROF & CHAIR, DEPT FOOD & NUTRIT, UNIV NC, GREENSBORO, 89- *Concurrent Pos:* Chair, Grad Nutrit Area, Univ Mo, Columbia, 85-88. *Mem:* Am Inst Nutrit; Am Dietetic Asn; Am Soc Clin Nutrit; Soc Nutrit Educ; Am Home Econ Asn. *Res:* Amino acid requirements and metabolism; zinc bioavailability and requirement; zinc and copper interactions. *Mailing Add:* Dept Food & Nutrit Univ NC A4 Park Bldg Greensboro NC 27412-5001

SHAW, HENRY, b Paris, France, Oct 25, 34; US citizen. AIR POLLUTION CONTROL, REACTION KINETICS & CATALYSIS. *Educ:* City Col New York, BCHE, 57; Newark Col Eng, MSChE, 62; Rutgers Univ, PhD(phys chem), 67, MBA, 76. *Prof Exp:* Nuclear engr, Nuclear Design, Babcock &

Wilcox Co, 57-61, Nuclear Res, Mobil Oil Co, 61-65; instr radiation sci, Rutgers Univ, 65-67; mgr environ, Exxon Res & Eng Co, 67-86; PROF CHEM ENG, NJ INST TECHNOL, 86- *Concurrent Pos:* Mem, Workshop on Greenhouse Effect, AAAS/Dept Energy, 79; partic, Nat Comn Air Qual Workshop on Greenhouse Effect, 80; chmn res comt, Am Inst Chem Engrs, 83-86; chmn, Chem Technol Adv Comt, Oak Ridge Nat Lab, 84-86; mem, Study on Chem Eng Frontiers, Nat Res Coun, 85-86; chmn bd, Eng Found, 87-89. *Mem:* Fel Am Inst Chem Engrs; AAAS; Am Chem Soc. *Mailing Add:* NJ Inst Technol Newark NJ 07102

SHAW, HERBERT JOHN, b Seattle, Wash; c 3. PHYSICS. *Educ:* Univ Washington, BSEE, 41,MA, 43; Standford Univ, PhD (physics), 48. *Prof Exp:* Liaison scientist, US Off Naval Res, London, 68-69; PROF RES, STANFORD UNIV, 88- *Concurrent Pos:* Consult govt agencies & electronics firms. *Honors & Awards:* Morris N Liebmann Mem Award Inst Elec & Electronics Engrs, 78. *Mem:* Fel Inst Elec & Electronic Engrs; Nat Acad Eng. *Res:* Fiber optic device involving sensing and signal processing; microwave antennas, high power microwave tubes, solid state devices. *Mailing Add:* Edward L Ginzton Lab Stanford Univ Stanford CA 94305-4085

SHAW, HERBERT RICHARD, b San Mateo, Calif, Dec 7, 30; div; c 1. GEODYNAMICS. *Educ:* Univ Calif, Berkeley, PhD(geol), 59. *Prof Exp:* GEOLOGIST, IGNEOUS & GEOTHERMAL PROCESSES BR, US GEOL SURV, 59- *Concurrent Pos:* Vis prof, Univ Calif, Berkeley, 74-75, lectr, 76; Ernst Cloos Scholar, Johns Hopkins Univ, 78; freelance writer, 80-82. *Mem:* Fel Geol Soc Am; fel Am Geophys Union; AAAS; Sigma Xi; Mineral Soc Am. *Res:* Experimental, theoretical and field geologic invstigations relating to interpretation of terrestial evolution. *Mailing Add:* IGP Br 345 Middlefield Rd Mail Stop 910 Menlo Park CA 94025

SHAW, JAMES HARLAN, b Tyler, Tex, May 8, 46. WILDLIFE ECOLOGY, WILDLIFE RESEARCH. *Educ:* Stephen F Austin Univ, BS, 68; Yale Univ, MFS, 70, PhD(wildlife ecol), 75. *Prof Exp:* Teaching asst biol, Yale Univ, 73-74, teaching fel ecol, 74; asst prof wildlife ecol, 74-81, ASSOC PROF ZOOL, OKLA STATE UNIV, 81-, PROF WILDLIFE ECOL, 88- *Mem:* Wildlife Soc; Ecol Soc Am; Am Soc Mammalogists; Animal Behav Soc. *Res:* Ecology, behavior and management of large mammals; conservation education. *Mailing Add:* Rte 2 Box 1320 Perkins OK 74059

SHAW, JAMES HEADON, b Sharon, Ont, Jan 1, 18; nat US; m 43; c 2. DENTISTRY. *Educ:* McMaster Univ, BA, 39; Univ Wis, MS, 41, PhD(biochem), 43. *Hon Degrees:* MA, Harvard Univ, 55. *Prof Exp:* Res assoc nutrit, Univ Wis, 43-45; from instr to assoc, Harvard Univ, 45-48, asst prof dent med, 48-55, assoc prof biol chem, 55-65, dir training ctr, clin scholars Oral Biol, 72-84, prof nutrit, 65-84, EMER PROF NUTRIT, SCH DENT MED, HARVARD UNIV, 84. *Concurrent Pos:* Asst ed, Nutrit Rev, 46-89; consult, Forsyth Dent Infirmary Children, 47-63; mem comt dent, Med Sci Div, Nat Res Coun, 53-62, comt diet phosphate & dent caries, food & nutrit bd, Div Biol & Agr, 58-63; career investr, USPHS, 64-84. *Mem:* AAAS; Soc Exp Biol & Med; Am Inst Nutrit; Int Asn Dent Res. *Res:* Oral disease in rodents and subhuman primates; nutritional relationships to the development, calcification, metabolism and disease susceptibility of the oral structures. *Mailing Add:* 10 Stiles Terr Newton Center MA 02159-2317

SHAW, JAMES SCOTT, b Grand Junction, Colo, Oct 13, 42; c 1. ASTRONOMY. *Educ:* Yale Univ, AB, 64; Univ Pa, PhD(astron), 70. *Prof Exp:* Asst prof, 70-77, ASSOC PROF ASTRON, UNIV GA, 77- *Mem:* Sigma Xi; Int Astron Union; Am Astron Soc. *Res:* Eclipsing binary stars; intrinsic variable stars; photoelectric photometry. *Mailing Add:* Dept of Physics & Astron Univ of Ga Athens GA 30602

SHAW, JANE E, b Worcester, Eng, Feb 3, 39. PHYSIOLOGY, CLINICAL PHARMACOLOGY. *Educ:* Univ Birmingham, BS, 61, PhD(physiol), 64. *Prof Exp:* sr scientist, 70-72, PRIN SCIENTIST, ALZA RES, 72-, PRES, ALZA RES DIV, EXEC VPRES, ALZA CORP & CHMN BD, ALZA LTD, 85- *Mem:* AAAS; NY Acad Sci; Am Phys Soc. *Res:* Elucidation of the physiological role of the prostaglandins; mechanism of action of analeptics; mechanism of gastric secretion; physiology and pharmacology of skin. *Mailing Add:* Alza Corp 950 Page Mill Rd Palo Alto CA 94304

SHAW, JOHN ASKEW, b Raleigh, NC, Apr 20, 46; m 69; c 1. CHEMICAL ENGINEERING, ELECTRONICS ENGINEERING. *Educ:* NC State Univ, BS, 70. *Prof Exp:* Engr, Duke Power Co, 70-75; SR APPLN ENGR, ABB PROCESS AUTOMATION, 75- *Honors & Awards:* Kates Award, Instrument Soc Am, 84, Pond Award, 89. *Mem:* Instrument So Am; Am Inst Chem Engrs. *Res:* Process control; man-machine interface; process safety. *Mailing Add:* 374 Cromwell Dr Rochester NY 14610

SHAW, JOHN H, b Sheffield, Eng, Jan 25, 25; m 49; c 4. PHYSICS. *Educ:* Cambridge Univ, BA, 46, MA, 50, PhD(physics), 51. *Prof Exp:* From lectr to assoc prof, 53-64, PROF PHYSICS, OHIO STATE UNIV, 64- *Mem:* Am Meteorol Soc; fel Optical Soc Am; Royal Meteorol Soc. *Res:* Infrared spectroscopy; infrared studies of atmospheric gaseous constituents. *Mailing Add:* Dept Physics Ohio State Univ 174 W 18th Ave Columbus OH 43210

SHAW, JOHN THOMAS, b Philadelphia, Pa, Sept 12, 25; m 46; c 3. ORGANIC CHEMISTRY. *Educ:* Temple Univ, AB, 50, MA, 52, PhD, 54. *Prof Exp:* Res chemist, Org Chem Div, Am Cyanamid Co, 54-63; asst prof chem, Am Int Col, 63-65; assoc prof chem, 65-68, PROF CHEM, GROVE CITY COL, 68- *Concurrent Pos:* Res grant, Petrol Res Fund, 73-75, 76-77, 78-80, 81-86 & 89-91. *Mem:* Am Chem Soc; Sigma Xi; Int Soc Heterocyclic Chem. *Res:* Synthesis and reactions of nitrogen heterocycles. *Mailing Add:* 520 Woodland Ave Grove City PA 16127

SHAW, KENNETH C, b Cincinnati, Ohio, Sept 18, 32; m 55; c 4. ZOOLOGY. *Educ:* Univ Cincinnati, BS, 54; Univ Mich, MS, 58, PhD(zool), 66. *Prof Exp:* Asst zool, Univ Mich, 56-60, instr, 60-61; asst prof, 63-76, ASSOC PROF ZOOL, IOWA STATE UNIV, 76- *Mem:* Am Entomol Soc; Am Soc Zoologists; Orthopterist's Soc; Animal Behav Soc; Sigma Xi; Am Entom Soc. *Res:* Acoustical behavior of Homoptera and Orthoptera; nature and adaptive significance of the "chorusing" behavior of crickets and katydids. *Mailing Add:* Dept Zool Iowa State Univ Ames IA 50011

SHAW, KENNETH NOEL FRANCIS, b Vancouver, BC, Dec 4, 19; m 46; c 3. BIOCHEMISTRY. *Educ:* Univ BC, BA, 40, MA, 42; Iowa State Col, PhD(bio-org chem), 51. *Prof Exp:* Instr chem, Univ BC, 40-42; chemist, Inspection Bd UK & Can, 42-43; supvr pharmaceut prod, Ayerst, McKenna & Harrison, Ltd, 43-47; from instr to res asst, Iowa State Univ, 47-51; res assoc, Wash State Univ, 51-52; from res instr to asst res prof biochem, Col Med, Univ Utah, 52-57; sr res fel, Calif Inst Technol, 57-64; assoc prof, 64-77, PROF PEDIAT, SCH MED, UNIV SOUTHERN CALIF, 77-; DIR, METAB SECT, MED GENETICS DIV, CHILDRENS HOSP LOS ANGELES, 64- *Mem:* Soc Pediat Res; Am Soc Biol Chem; Am Chem Soc; Royal Soc Chem; Sigma Xi. *Res:* Chemistry and metabolism of amino acids; urinary phenol and indole acids; neural tumors; inherited metabolic disorders; mental diseases of biochemical origin. *Mailing Add:* 4900 Palm Dr La Canada Flintridge CA 91011

SHAW, LAWRANCE NEIL, b Hallock, Minn, Mar 15, 34; m 66. CROP TRANSPLANTER DEVELOPMENT, LAND LOCOMOTION. *Educ:* NDak State Univ, BS, 56; Purdue Univ, MS, 59; Ohio State Univ, PhD(agr eng), 69. *Prof Exp:* Grad assoc agr eng, Purdue Univ, 57-59; exten engr agr eng, Univ Maine, 59-67; grad assoc agr eng, Ohio State Univ, 67-69; from asst prof to assoc prof, 69-79, PROF AGR ENG, UNIV FLA, 79- *Concurrent Pos:* Vis prof, Nat Inst Agr Eng, Silsoe, Eng, 76 & Friedrich-Wilhelms Universitat, Bonn, Ger, 86. *Mem:* Am Soc Agr Engrs; Int Soc Horticult Sci. *Res:* Research and development of systems and equipment for the field production of vegetables, including machinery for soil tillage, crop planting, weed and pest control and harvesting; Four United States patents. *Mailing Add:* 8715 NW Fourth Pl Gainesville FL 32607

SHAW, LEONARD G, b Toledo, Ohio, Aug 15, 34; m 61; c 3. SYSTEMS ANALYSIS, ELECTRICAL ENGINEERING. *Educ:* Univ Pa, BS, 56; Stanford Univ, MS, 57, PhD(elec eng), 61. *Prof Exp:* Res asst, dept elec eng, Stanford Univ, 59-60; from asst prof to assoc prof, 60-75, PROF ELEC ENG, POLYTECH INST NY, 75-, DEPT HEAD, 82- *Concurrent Pos:* Vis prof, Tech Univ Eindhoven, Neth, 70; consult signal processing, Mgt Div, Sperry Syst, 73-; Nat Ctr Sci Res assoc, Automatic Control Lab, Univ Nantes, France, 76-77. *Mem:* Fel Inst Elec & Electronic Engrs; Am Soc Eng Educ; AAAS. *Res:* Image processing; stochastic control; spectral analysis; control of traffic and message queues; reliability. *Mailing Add:* Sch Elec Eng & Comput Sci Polytech Univ Brooklyn NY 11201

SHAW, LESLIE M J, b Newark, NJ, Feb 4, 41; m 67; c 4. PATHOLOGY, PHARMACOLOGY. *Educ:* LeMoyne Col, BS, 62; Upstate Med Ctr, State Univ NY, PhD(biochem), 68. *Prof Exp:* Res fel molecular biol, Johns Hopkins Univ, 68-70; res fel clin chem, 70-72, mem staff, 72-74, asst prof, 74-77, ASSOC PROF CLIN CHEM, DEPT PATH & LAB MED, & DIR, TOXICOL LAB, HOSP UNIV PA, 77-, PROF, 87- *Mem:* Am Chem Soc; Am Asn Clin Chem; AAAS; Nat Acad Clin Biochem; Am Asn Pathologists; Am Soc Clin Pharmacol & Therapeut. *Res:* Metabolism and mechanism of action of phosphorothioate radio and chemoprotectors in cancer patients and experimental animal models; clinical significance of unbound drug concentration in serum; pharmacokinetics of cyclosporine, pharmacologic activity of cylosporine metabolites. *Mailing Add:* Dept Path & Lab Med Univ Pa Hosp 3400 Spruce St Philadelphia PA 19104

SHAW, M(ELVIN) P, b Brooklyn, NY, Aug 16, 36; m 59, 87; c 2. SOLID STATE PHYSICS & ELECTRONICS, CLINICAL PSYCHOLOGY. *Educ:* Brooklyn Col, BS, 59; Case Inst Technol, MS, 63, PhD(physics), 65; Ctr Humanistic Studies, MA, 88. *Prof Exp:* Exp physicist & scientist-in-chg microwave physics, Res Labs, United Aircraft Corp, 64-70; PROF ELEC ENG & PHYSICS, WAYNE STATE UNIV, 70-; ADMIN DIR, ASN BIRMINGHAM, 88- *Concurrent Pos:* Lectr, Trinity Col, Conn, 65-66; adj asst prof, Rensselaer Polytech Inst, 68-70; vis lectr, Yale Univ, 69-70; consult, Energy Conversion Devices, 70-; adv, Nat Res Coun Comt, 75-78; dir, Res Inst Eng Sci, Wayne State Univ, 76-77, chmn, Dept Elec & Comput Eng, 82-85. *Mem:* Fel Am Phys Soc; sr mem Inst Elec & Electronic Engrs; Am Psychol Asn. *Res:* Amorphous solar cells; cyclotron resonance; electron spin resonance; superconductivity; solid state microwave sources; crystal growth and purification of metals and semiconductors; switching in crystalline and amorphous semiconductors; scientific creativity. *Mailing Add:* Dept Elec Eng Wayne State Univ Detroit MI 48202

SHAW, MARGARET ANN, b Louisville, Ky, Apr 29, 33. PHARMACY. *Educ:* Univ Ky, BS, 55; Univ Fla, PhD(pharm), 61. *Prof Exp:* Pharm practice, 55-58; asst prof pharm, Univ NC, 61-65; asst prof, 65-69, assoc prof, 69-80, PROF PHARM, COL PHARM, BUTLER UNIV, 80- *Mem:* Am Soc Hosp Pharmacists. *Res:* Isotonic solutions; aerosols. *Mailing Add:* Butler Univ Col Pharm 4600 Sunset Ave Indianapolis IN 46208

SHAW, MARGERY WAYNE, b Evansville, Ind, Feb 15, 23; div; c 1. HUMAN GENETICS, LEGAL GENETICS. *Educ:* Univ Ala, AB, 45; Columbia Univ, MA, 46; Univ Mich, MD, 57; Univ Houston, JD, 73. *Hon Degrees:* ScD, Univ Evansville, 77, Univ Southern Ind. *Prof Exp:* Instr zool, Univ Alaska, 51-53; from instr to assoc prof human genetics, Univ Mich, 58-67; assoc prof, 67-69, prof biol, Univ Tex M D Anderson Hosp & Tumor Inst, 69-75, prof genetics & Dir Med Genetics Ctr, Univ Tex Health Sci Ctr Houston, 71-88, EMER PROF, UNIV TEX HEALTH SCI CTR, 88- *Concurrent Pos:* Mem genetics study sect, NIH, 66-70, training comt, 70-74; mem med adv bd, Nat Genetics Found, 72- *Mem:* Am Soc Human Genetics (pres, 82); Am Soc Cell Biol; Tissue Cult Asn; Genetics Soc Am (pres, 77-78); Environ Mutagen Soc. *Res:* Inherited diseases; human chromosomes. *Mailing Add:* 2617 Pine Tree Dr Evansville IN 47711-2117

SHAW, MARY M, b Wash, DC, Sept 30, 43; m. SOFTWARE ENGINEERING, PROGRAMMING LANGUAGES. *Educ:* Rice Univ, BA, 65; Carnegie-Mellon Univ, PhD(computer sci), 72. *Prof Exp:* PROF COMPUTER SCI, CARNEGIE-MELLON UNIV, 72- *Concurrent Pos:* mem, Tech Comt Software Eng, Inst Elec & Electronic Engrs, Computer Soc, 81-92, Defense Sci Bd Task Force Software, 85-87, Working Group Syst Implementation Lang, Int Fedn Info Processing Soc, 85-92, Computer Sci & Telecommun Bd, Nat Res Coun, 86-92; assoc ed, Inst Elec & Electronic Engrs Software, 83-87; chief scientist, Software Eng Inst, Carnegie Mellon Univ, 84-88. *Mem:* Asn Comput Mach; fel Inst Elec & Electronic Engrs; Inst Elec & Electronic Engrs Computer Soc; NY Acad Sci; Sigma Xi. *Res:* Software architecture; programming language design; abstraction techniques for advanced programming; software engineering; computer science education. *Mailing Add:* Sch Computer Sci Carnegie-Mellon Univ 5000 Forbes Ave Pittsburgh PA 15213-3890

SHAW, MICHAEL, b Barbados, BWI, Feb 11, 24; m 48; c 4. PLANT PATHOLOGY, PLANT PHYSIOLOGY. *Educ:* McGill Univ, BSc, 46, MSc, 47, PhD(bot, plant path), 49. *Hon Degrees:* PhD, Univ Sask, 71; DSc, McGill Univ, 75. *Prof Exp:* Nat Res Coun Can fel, Bot Sch, Cambridge Univ, 49-50; from assoc prof to prof plant physiol, Univ Sask, 50-67, head dept biol, 61-67; dean agr sci, 67-75, vpres acad develop, 75-80, prof agr bot, 67-83, vpres & provost, 81-83, univ prof, 83-89, EMER UNIV PROF, UNIV BC, 89- *Concurrent Pos:* Vis researcher, Hort Res Labs, Univ Reading, 58-59; ed, Can J Bot, 64-79; vpres, Biol Coun Can, 71, pres, 72 & 87-89; mem adv comt biol, Nat Res Coun Can, 71-73; mem, Sci Coun Can, 76-82; mem, Natural Sci & Eng Res Coun Can, 78-80. *Honors & Awards:* Gold Medal, Can Soc Plant Physiol, 71; Flavelle Medal, Royal Soc Can, 76; Gold Medal, Biol Coun Can, 83. *Mem:* AAAS; Am Soc Plant Physiol; fel Royal Soc Can; Can Soc Plant Physiol; Can Bot Asn; fel NY Acad Sci; fel Am Phytopathlogy Soc. *Res:* Host-parasite relations of obligate plant parasites. *Mailing Add:* Dept Plant Sci Univ BC Vancouver BC V6T 1V3 Can

SHAW, MILTON C(LAYTON), b Philadelphia, Pa, May 27, 15; m 39; c 2. MECHANICAL ENGINEERING, TRIBOLOGY. *Educ:* Drexel Inst, BS, 38; Univ Cincinnati, MEngSc, 40, ScD(chem physics), 42. *Hon Degrees:* Dr, Cath Univ Louvain, 70. *Prof Exp:* Res engr, Cincinnati Milling Mach Co, 38-42; chief mat br, Nat Adv Comt Aeronaut, 42-46; from asst prof to prof mech eng, Mass Inst Technol, 46-61, head mach tool div, 46-61; head dept mech eng, Carnegie-Mellon Univ, 61-75, dir, Processing Res Inst, 70-71, univ prof, 74-77; PROF ENG, ARIZ STATE UNIV, 77- *Concurrent Pos:* Guggenheim fel, 56; Fulbright vis prof, Aachen Tech Univ, 57; vis prof, Univ Birmingham, 60, 61 & 64; Springer prof, Univ Calif, Berkeley, 72; distinguished guest prof, Ariz State Univ, 76; consult, various co, 46- *Honors & Awards:* Westinghouse Award, Am Soc Eng Educ, 56; Hersey Award, Am Soc Mech Engrs, Medalist, 85; Gold Medal, Am Soc Tool & Mfg Eng, 58; Wilson Award, Am Soc Metals, 71. *Mem:* Nat Acad Eng; hon mem Am Soc Mech Engrs; hon mem Am Soc Lubrication Engrs; hon mem Soc Mfg Eng; fel Am Acad Arts & Sci; hon mem Int Inst Prod Engrs, (pres, 61); fel Am Soc Metals. *Res:* Metal cutting; lubrication, friction and wear; behavior of materials; grinding research, brittle fracture. *Mailing Add:* Ariz State Univ ECG 247 Tempe AZ 85287-6106

SHAW, MONTGOMERY THROOP, b Ithaca, NY, Sept 11, 43; wid; c 1. POLYMER SCIENCE. *Educ:* Cornell Univ, BChE & MS, 66; Princeton Univ, MA, 68, PhD(chem), 70. *Prof Exp:* Chemist polymer thermodyn, Union Carbide Corp, 70-74, proj scientist polymer rheology, 74-76; assoc prof, 76-82, PROF, DEPT CHEM ENG, UNIV CONN, 82. *Mem:* Am Chem Soc; Soc Rheology (secy, 77-81); Sigma Xi; Inst Elec & Electronics Engrs; Soc Plastic Engrs; Am Phys Soc. *Res:* Research directed at relating the physical and chemical behavior high-polymers to the structure of the polymer, and developing the theory and experiments to substantiate these relationships. *Mailing Add:* Inst Mat Sci Univ Conn Storrs CT 06269-3136

SHAW, NOLAN GAIL, b Forsan, Tex, Oct 2, 29; m 67; c 5. GEOLOGY. *Educ:* Baylor Univ, AB, 51; Southern Methodist Univ, MS, 56; La State Univ, PhD(paleont), 66. *Prof Exp:* From asst prof to prof, 55-78, WILLIAM C WOOLF PROF CHAIR GEOL, CENTENARY COL LA, 78-, CHMN DEPT, 74- *Mem:* Am Asn Petrol Geologists; fel Geol Soc Am; Sigma Xi. *Res:* Paleontology. *Mailing Add:* 626 Lake Forbing Dr Shreveport LA 71006

SHAW, PAUL DALE, b Morton, Ill, Aug 12, 31; m 55; c 2. BIOCHEMISTRY. *Educ:* Bradley Univ, BS, 53; Univ Ill, PhD(biochem), 57. *Prof Exp:* Res fel chem, Harvard Univ, 58-60; from asst prof to assoc prof, 60-74, PROF BIOCHEM, UNIV ILL, URBANA, 74- *Mem:* Am Chem Soc; Am Soc Biochem & Molecular Biol; AAAS; Am Soc Microbiol; Am Phyopath Soc. *Res:* Chemistry and biochemistry of natural products; metabolism of microorganisms; bacterial molecular biology. *Mailing Add:* Dept of Plant Path Univ of Ill N 519 Turner Hall Urbana IL 61801

SHAW, PETER ROBERT, b Columbus, Ohio, June 8, 56; m 82; c 2. MARINE GEOPHYSICS, SEISMOLOGY. *Educ:* Mass Inst Technol, SB(physics) & SB(earth & planetary sci), 78; Scripps Inst Oceanog, PhD(geophys), 83. *Prof Exp:* Res asst, Scripps Inst Oceanog, 78-83; scholar, Dept Geol & Geophys, Woods Hole Oceanog Inst, 83-84, investr, 84-85, asst scientist, 85-89, ASSOC SCIENTIST, DEPT GEOL & GEOPHYS, WOODS HOLE OCEANOG INST, 89- *Mem:* Am Geophys Union. *Res:* Tectonics of mid-ocean ridges, especially faulting, flexure and magmatism; formation of abyssal hills; dynamical systems and faulting; seafloor topography; fracture zones; seismic structure of oceanic crust; inverse methods in seismology. *Mailing Add:* Dept Geol & Geophys Woods Hole Oceanog Inst Clark 243A Woods Hole MA 02543

SHAW, PHILIP EUGENE, b St Petersburg, Fla, July 21, 34; m 59; c 3. GENERAL CHEMISTRY. *Educ:* Duke Univ, BS, 56; Rice Univ, PhD(chem), 60. *Prof Exp:* Res assoc chem, Sterling-Winthrop Res Inst Div, Sterling Drug, Inc, 60-65; chemist, 65-72, RES LEADER, CITRUS AND SUBTROP PROD LAB, US DEPT AGR, 72- *Mem:* Am Chem Soc; Inst Food Technologists. *Res:* Isolation and identification of natural food flavors and insect attractants; chemistry of nutrients; development of analytical methodology. *Mailing Add:* Citrus & Subtrop Prods Lab US Dept of Agr Box 1909 Winter Haven FL 33883-1909

SHAW, RALPH ARTHUR, b Peoria, Ill, Dec 27, 30; m 55; c 2. METABOLISM, ENDOCRINOLOGY. *Educ:* Northwestern Univ, BS, 54, MD, 62; Purdue Univ, MS, 56, PhD(biochem), 58. *Prof Exp:* Consult, Children's Mem Hosp, 59-61, fel biochem, 61-62; intern, Evanston Hosp, Ill, 62-63; resident med, Hahnemann Hosp, 63-65, from sr instr med to assoc prof med & biochem, 66-72, dir clin chem lab, 70-71, PROF MED & RES PROF BIOCHEM, HAHNEMANN MED COL, 72-, ASSOC VPRES HEALTH AFFAIRS, 71- *Concurrent Pos:* Mem bd, Action for Brain Injured Children, Inc; fel biochem & med, Sch Med, Northwestern Univ, 58-61; USPHS fel, Hahnemann Hosp, 65-66. *Mem:* Am Fedn Clin Res; Am Diabetes Asn; Am Geriat Soc; NY Acad Sci; Am Schizophrenia Found. *Res:* Diabetes mellitus. *Mailing Add:* Dept Med 1840 Tall Oaks Rd Orwigsburg PA 17961-9540

SHAW, RICHARD FRANCIS, b Lawrence, Mass, Feb 22, 52; m; c 1. OCEANOGRAPHY. *Educ:* Univ Maine, PhD(oceanog), 81. *Prof Exp:* Res asst, Dept Oceanog, Univ Maine, 73-79; consult coordr, Human Sci Res Inc, 79-80; regional coordr & consult, Kathryn Chandler Assocs, Inc & Market Facts, Inc, 81; res assoc IV, coastal ecol lab, 81-85, interim dir, Coastal Fisheries Inst, 89-90, ASST PROF, COASTAL FISHERIES INST & DEPT MARINE SCI, CTR WETLAND RESOURCES, LA STATE UNIV, 85-, ASSOC PROF, DEPT OCEANOG & COASTAL SCI, 89-, DIR, 90- *Concurrent Pos:* Res assoc, Ira C Darling Ctr, Univ Maine, 79-80. *Mem:* Am Fisheries Soc; Am Soc Ichthyologists & Herpetologists; Am Soc Limnol & Oceanog; Estuarine Res Soc. *Res:* Ecology, population dynamics, growth rates, condition factors, age determination, recruitment, migration and transport of larval fishes. *Mailing Add:* Ctr Wetland Resources La State Univ Baton Rouge LA 70803

SHAW, RICHARD FRANKLIN, b Redlands, Calif, May 14, 24; m 44; c 4. THEORETICAL GERONTOLOGY, THEORETICAL EVOLUTIONARY GENETICS. *Educ:* Univ Calif, Berkeley, AB, 48, MA, 50, PhD(zool), 55. *Prof Exp:* Proj leader, Dept Animal Husb, Univ PR, 55-56; res fel, Dept Biostatist, Grad Sch Pub Health, Univ Pittsburgh, 56-57; asst prof genetics, Dept Prev Med, Univ Va, 57-63; res assoc, Dept Human Genetics, Univ Mich, 63-65; asst prof biol, Wayne State Univ, 65-68; asst prof pediat, Dalhousie Univ, Can, 68-72; assoc prof epidemiol, Fac Med, Univ Sherbrooke, Can, 72-74; res assoc, Univ Chicago, Sickle Cell Ctr, 74-77; proprietor, Shaw-Banfill Books, 78-85; adj assoc prof, Grad Sch, Howard Univ, 85-89; GUEST RES, DIV COMPUTER RES & TECHNOL, NIH, 88- *Concurrent Pos:* Assoc pediat, Wayne State Univ, 65-68, fac res fel, 67; researcher epidemiol, Can Govt, 72-74; consult, Howard Univ Cancer Ctr, 78-88; vis scientist, Nat Biomed Res Found, Georgetown, 85-86. *Mem:* AAAS; Am Soc Naturalists; Genetics Soc Am; Am Soc Human Genetics. *Res:* The causes of aging from the viewpoint of evolution and DNA structure of nuclear and mitochondrial genes; methods of computation in population genetics. *Mailing Add:* 10201 Grosvenor Pl No 522 Rockville MD 20852

SHAW, RICHARD GREGG, b Wilmington, Del, Nov 21, 29; m 57; c 3. POLYMER CHEMISTRY. *Educ:* Cornell Univ, AB, 51; Univ Pa, MS, 52; Ind Univ, PhD, 60. *Prof Exp:* Group leader opers div, Union Carbide Corp, 60-75, technol mgr wire & cable, Polyethylene Div, 75-77, mkt mgr wire & cable, Domestic Int Liaison, 77-78; CONSULT, 78- *Mem:* Am Chem Soc; Inst Elec & Electronics Engrs; Soc Photog Scientists & Engrs. *Res:* Polymer free radical chemistry; photosensitive systems; polymer property predictions; polymer synthesis and modification; engineered plastic materials design; physics of dielectrics. *Mailing Add:* 38 River St Somerville NJ 08876-5113

SHAW, RICHARD JOHN, b Fall River, Mass, Jan 15, 39; m 65; c 2. FLORICULTURE, HORTICULTURAL THERAPY. *Educ:* Univ RI, BS, 61; Univ Mo, MS, 63, PhD(hort), 66. *Prof Exp:* Asst prof hort, La State Univ, Baton Rouge, 66-70; asst prof, 70-76, ASSOC PROF PLANT & SOIL SCI, UNIV RI, 76- *Concurrent Pos:* Treas, New Eng Chap, Am Hort Ther Asn, 89- *Mem:* Am Soc Hort Sci; Am Hort Soc; Am Hort Ther Asn; Prof Plant Growers Asn. *Res:* Effects of environment, growth regulators, growing media and fertilizers on the production of greenhouse crops. *Mailing Add:* Dept Plant Sci Univ RI Kingston RI 02881-0804

SHAW, RICHARD JOSHUA, b Ogden, Utah, June 25, 23; m 47; c 3. BOTANY. *Educ:* Utah State Univ, BS, 48, MS, 50; Claremont Grad Sch, PhD(bot), 61. *Prof Exp:* From asst prof to assoc prof bot, 50-69, prof, 69-87, EMER PROF BIOL, UTAH STATE UNIV, 87- *Concurrent Pos:* Emer dir intermountain herbarium, Utah State Univ, 84. *Mem:* Am Soc Plant Taxon; Sigma Xi. *Res:* Taxonomic botany; biosystematics. *Mailing Add:* Dept Biol Utah State Univ Logan UT 84321

SHAW, RICHARD P(AUL), b Brooklyn, NY, June 23, 33; m 61; c 2. ENGINEERING, OCEANOGRAPHY. *Educ:* Polytech Inst Brooklyn, BS, 54; Columbia Univ, MS, 55, PhD(appl mech), 60. *Prof Exp:* Sr instr appl mech, Polytech Inst Brooklyn, 56-57; asst prof eng sci, Pratt Inst, 57-62; from assoc prof to prof eng sci, 62-80, PROF CIVIL ENG, STATE UNIV NY BUFFALO, 80-, PROF ENG, 80- *Concurrent Pos:* Nat Res Coun-Environ Sci Serv Admin fel, Joint Tsunami Res Effort, Univ Hawaii, 69-70, Nat Oceanog & Atmospheric Admin sr res assoc, 73-74; Am Coun Educ Govt exchange fel, Int Decade Ocean Explor, NSF, 77-78; vpres, Technol Systs Res, Inc; pres, Int Soc Innovative Numerical Analysis. *Honors & Awards:* Eminent Scientist Award, Comp Mech Inst, 88. *Mem:* Am Soc Civil Engrs; Am Geophys Union; Am Soc Mech Engrs; Acoust Soc Am; Marine Technol Soc; Int Soc Innovative Numerical Anal. *Res:* Numerical methods (boundary integral/element methods); wave motion (acoustic, elastic, water); ocean engineering and oceanography; structures and solid mechanics; oceanography; geophysical solid mechanics. *Mailing Add:* Dept Civil Eng State Univ NY Buffalo NY 14260

SHAW, ROBERT BLAINE, b Commerce, Tex, May 24, 49; m 76; c 1. AGROSTOLOGY, GRASS TAXONOMY. *Educ:* Southwest Tex State Univ, BS, 74; Tex A&M Univ, MS, 76, PhD(range sci), 79. *Prof Exp:* Res assoc, Range Sci Dept, Tex A&M Univ, 78-79; instr biol, Southwest Tex State Univ, 79-80; ASST PROF PLANT TAXON, SCH FORESTRY, UNIV FLA, 80- *Mem:* Soc Range Mgt; Am Soc Plant Taxonomists; Bot Soc Am. *Res:* Evolution, ecology, anatomy, morphology and systematics of the grasses; distribution, composition and evolution of the North American grasslands. *Mailing Add:* Colo State Univ Sch Forestry Ft Collins CO 80523

SHAW, ROBERT FLETCHER, b Montreal, Que, Feb 16, 10; m 35. ENGINEERING, EDUCATION ADMINISTRATION. *Educ:* BEngr, McGill Univ, 33; DSc, McMaster Univ, 67, McGill Univ, 85; DEng, Tech Univ, NS, 67. *Hon Degrees:* DEngr, Tech Univ, NS, 67, DSc, McMaster Univ, 67, Univ New Brunswick, 86, Mcgill Univ, 87. *Prof Exp:* From engr & estimator to asst to vpres, Found Co Can Ltd, 37-43, from shipyard mgr, Pictou, NS, to mgr eng & asst to pres, 43-58, exec vpres, Foundation Co Can Ltd, 58-62, pres & dir, 62-63; vprin admin, McGill Univ, 68-71; dep minister, Environ Can, 71-75; pres, Monenco Pipeline Consult Ltd, 75-78; CONSULT, MONENCO LTD, 78- *Concurrent Pos:* Vpres, dir & chief engr, Defense Construct Ltd, 51-52; with NATO, 52; past mem, Sci Coun Can & Nat Design Coun; mem Can Environ Adv Coun; dep comnr gen Expo '67, 63-68; past chmn Fed/Provincial Atlantic Fisheries Comt; past co-chmn BC Fed/ Provincial Fisheries Comt; spec adv, Minister Indust Develop, Nfld, 78-79 & Nfld & Labrador Hydro, 82-85. *Honors & Awards:* Companion of the Order of Can, 67; Govt Can Centennial Medal, 67; Julian C Smith Medal, Eng Inst Can, 67, Sir John Kennedy Medal, 79; Syst & Cybernet Award, Inst Elec & Electronics Engrs, 67; Gold Medal, Can Coun Prof Eng, 75. *Mem:* Eng Inst Can (pres, 75-76). *Mailing Add:* 3980 Cote des Neiges Rd Apt c29 Montreal PQ H3H 1W2 Can

SHAW, ROBERT HAROLD, b Madrid, Iowa, June 26, 19; m 45; c 3. AGRICULTURAL METEOROLOGY. *Educ:* Iowa State Univ, BS, 41, MS, 42, PhD(agr climat), 49. *Prof Exp:* From asst prof to prof, 48-80, DISTINGUISHED PROF, 80-, EMER PROF AGR CLIMAT, IOWA STATE UNIV, 86- *Concurrent Pos:* Mem comt meteorol and climat, Am Soc Agron, 60, 63-66, chmn, 65-66; mem working group biometeorol, Am Meteorol Soc, 63-66; assoc ed climat, Agron J, 64-66; mem comt agr meteorol, Am Meteorol Soc, 65-66; mem panel teaching climat, Am Asn Geographers, 66-67; mem panel natural resource sci, Comn Educ Agr and Nat Resources, Nat Res Coun, 66-68; mem working group, Instructions in Climat, World Meteorol Orgn, 67-68; mem comt meteorol and climat, Agr Bd, Nat Res Coun, 4 yrs. *Honors & Awards:* Outstanding Achievement in Biometeorol, Am Meteorol Soc. *Mem:* Fel AAAS; fel Am Soc Agron; Sigma Xi; fel Soil Sci Soc Am; Am Meteorol Soc. *Res:* Water use; weather statistics; evaluation of microclimate; crop-weather relationships; solar-wind energy climatology. *Mailing Add:* Climat-Meteorol Iowa State Univ Ames IA 50010

SHAW, ROBERT R, b Ft Fairfield, Maine, Mar 8, 39. CONSERVATION. *Educ:* Univ Maine, BS, 61; Xavier Univ, MBA, 83. *Prof Exp:* Asst state conservationist, Soil Conserv Serv, USDA, Vt, 74-76, state conservationist, 76-79, state conservationist, Ohio, 79-84, dir, Land Treatment Prog Div, 84-86, dir, Conserv Planning & Appl Div, 86, asst chief, 86, dep chief, Assess & Planning, 86-87, DEP CHIEF TECHNOL, SOIL CONSERV SERV, USDA, 87- *Mailing Add:* Soil Conserv Serv USDA PO Box 2890 Washington DC 20013-2890

SHAW, ROBERT REEVES, b Bokoshe, Okla, June 9, 36; m 62; c 4. MATERIALS SCIENCE, CERAMICS. *Educ:* Univ Wash, BS, 59, MS, 60; Mass Inst Technol, SM, 65, ScD(ceramics), 67. *Prof Exp:* Res engr, Boeing Corp, 59-60; res trainee, Gen Elec Res Lab, 60-62; res asst ceramics, Mass Inst Technol, 62-67; sr res physicist, Am Optical Corp, 67-77; prog supvr, Phys Sci Lab, P R Mallory & Co, Inc, 77-81; ADV ENGR, IBM CORP, 81- *Mem:* Am Ceramic Soc; Int Soc Optical Eng. *Res:* Glass structure and properties; composite materials; optical phenomena in crystal and glass systems; phase separation; electron microscopy; microstructure-controlled phenomena; battery reactions and systems; eletronic packaging technology. *Mailing Add:* Five Pine Ridge Rd Poughkeepsie NY 02603

SHAW, ROBERT WAYNE, b Davenport, Iowa, Sept 9, 47; m 69. SPECTROSCOPY. *Educ:* Iowa State Univ, BS, 69; Princeton Univ, PhD(chem), 74. *Prof Exp:* Res chemist, Indust Tape Lab, 3M Co, 68; fel Dept Chem, Univ Calif, Los Angeles, 73-75; RES CHEMIST, ANAL LYTICAL CHEM DIV, OAK RIDGE NAT LAB, 75- *Mem:* Am Chem Soc; AAAS; Sigma Xi; Optical Soc Am. *Res:* New spectroscopic methods and instrumentation for analytical and physical chemical research. *Mailing Add:* Analytical Chem Div Oak Ridge Nat Lab Box 2008 MS6142 Oak Ridge TN 37831

SHAW, ROBERT WAYNE, b Bartlesville, Okla, July 10, 49. BIOPHYSICAL CHEMISTRY. *Educ:* WVa Univ, BA, 71; Pa State Univ, PhD(biochem), 76. *Prof Exp:* Fel biochem, Inst Enzyme Res, Univ Wis, 76-81; ASST PROF CHEM, TEX TECH UNIV, 81- *Mem:* AAAS; Am Chem Soc; Am Soc Biol Chemists. *Res:* Mechanism of a variety of physiologically important metalloenzymes; electron paramagnetic resonance; optical spectroscopies coupled to rapid kinetic methods. *Mailing Add:* Dept Chem & Biochem PO Box 4260 Tex Tech Univ Lubbock TX 79409

SHAW, ROBERT WILLIAM, JR, b Ithaca, NY, Aug 10, 41; m 64; c 2. PHOTOVOLTAICS, SUPERCONDUCTIVITY. *Educ:* Cornell Univ, BEP, 64, MS, 64; Stanford Univ, PhD(appl physics), 68; Am Univ, MPA, 81. *Prof Exp:* Fel, Cavendish Lab, Cambridge Univ, 69; mem tech staff, Bell Labs, 69-72; assoc, Booz, Allen & Hamilton 72-74, res dir, 74-75, vpres, 75-79, sr vpres, 79-83; PRES, ARETE VENTURES, INC & UTECH VENTURE CAPITAL CORP, 83- *Mem:* AAAS; Am Phys Soc; Sigma Xi; Int Transactional Anal Asn; Asn Humanistic Psychol. *Res:* Consulting and financing of energy technology companies; author of many book chapters, articles and research papers. *Mailing Add:* 9405 Falls Bridge Ln Potomac MD 20854

SHAW, RODNEY, b Rotherham, UK, May 9, 37; m 63; c 2. OPTICS. *Educ:* Leeds Univ, BS, 58; Cambridge Univ, PhD(physics), 61. *Prof Exp:* Res scientist, Ilford Ltd, UK, 61-64; lectr appl physics, Hull Univ, UK, 64-67; prin lectr photog sci, Polytech Cent London, 67-70; res scientist, Ciba-Geigy Photochemie, Switz, 70-73; prin scientist physics, Xerox Corp, 73-81; res assoc, Res Labs, Eastman Kodak Co, 82-86; DIR, CTR FOR IMAGING SCI, ROCHESTER INST TECHNOL, 86- *Concurrent Pos:* Ed, J Imaging Sci, 77- *Mem:* Fel Optical Soc Am; fel Soc Photog Scientists & Engrs. *Res:* Application of image evaluation methodologies to unconventional imaging processes. *Mailing Add:* Ctr for Imaging Sci Rochester Inst Technol One Lomb Mem Dr PO Box 9887 Rochester NY 14623

SHAW, ROGER WALZ, b Chicago, Ill, May 29, 34; m 56; c 3. ELECTRONIC MATERIALS, CRYOGENICS. *Educ:* Univ Rochester, BS, 55; Univ Ill, MS, 56, PhD(physics), 59. *Prof Exp:* Asst physics, Univ Ill, 55-58, res assoc, 59-60; from asst prof to assoc prof, Rensselaer Polytech Inst, 60-67; SR RES SPECIALIST, MONSANTO ELECTRONIC MAT CO, 67-; sr res specialist, 67-89, FEL, MEMC ELECTRONIC MAT CO, 89- *Concurrent Pos:* Vis staff mem, Los Alamos Sci Lab, 66-67. *Mem:* Am Phys Soc; Electrochem Soc; Mat Res Soc. *Res:* Experimental physics of solids at low temperature; superconductivity; physics of magnetic thin films, particularly magnetic bubble materials; physics of semiconductors. *Mailing Add:* MEMC Electronic Mat Co PO Box eight St Peters MO 63376

SHAW, SPENCER, b New York, NY, Apr 28, 46. GASTROENTEROLOGY. *Educ:* Univ Rochester, MD, 70. *Prof Exp:* ASSOC PROF, MT SINAI SCH MED, 85-; ASST CHIEF SECT LIVER DIS & NUTRIT, VET ADMIN MED CTR, BRONX, 83- *Mem:* Am Fedn Clin Res; fel Am Col Physicians; Am Soc Clin Nutrit. *Mailing Add:* Bronx Vet Admin Hosp 130 W Kingsbridge Rd Bronx NY 10468

SHAW, STANLEY MINER, b Parkston, SDak, July 4, 35; m 62; c 3. NUCLEAR PHARMACY, MEDICAL RESEARCH. *Educ:* SDak State Univ, BS, 57, MS, 59; Purdue Univ, PhD(bionucleonics), 62. *Prof Exp:* Instr pharmaceut chem, SDak State Univ, 60-62; from asst prof to assoc prof, 62-71, PROF NUCLEAR PHARMACY, PURDUE UNIV, LAFAYETTE, 71-, DIV HEAD, 90-, ACTG HEAD, SCH HEALTH SCI, 90- *Concurrent Pos:* Lederle pharm fac awards, 62 & 65; res award, Parenteral Drug Asn, 70; mem, Bd Pharm Specialties, Specialty Coun Nuclear Pharm, 78-82; chmn, Sect Nuclear Pharm, Acad Pharm Pract, 79-80, historian, 81-85; mem, Founder's Award, Sect Nuclear Pharm, Acad Pharm Pract, 81-85. *Mem:* Am Pharmaceut Asn; fel Acad Pharm Pract & Mgt; Sigma Xi. *Res:* Use of radioactive isotopes for the diagnosis of disease states; therapeutic drugs and agents that interfere with diagnostic procedures. *Mailing Add:* Sch Pharm Purdue Univ W Lafayette IN 47907

SHAW, STEPHEN, b May 6, 48; c 2. IMMUNOGENETICS, CELL BIOLOGY. *Educ:* Harvard Univ, MD, 74. *Prof Exp:* SR INVESTR, NIH, 76- *Honors & Awards:* Commendation Medal, Pub Health Serv. *Mem:* Am Asn Immunologists; AAAS; Am Soc Histocompatibility & Immunogenetics. *Res:* Human T cell recognition. *Mailing Add:* NIH Bldg Ten Rm 4B17 Bethesda MD 20892

SHAW, VERNON REED, b Bellefontaine, Ohio, Apr 10, 37; m 66. ANALYTICAL CHEMISTRY. *Educ:* Wittenberg Univ, BS, 64; Univ Ill, MS, 66, PhD(chem), 68. *Prof Exp:* Asst, Univ Ill, 64-68; asst prof chem, Adrian Col, 68-73; asst prof, 73-77, ASSOC PROF CHEM, UNIV EVANSVILLE, 77- *Mem:* Am Chem Soc. *Res:* Gas chromatography of metal halide using fused salt liquid phases. *Mailing Add:* Dept Chem 1319 Greenfield Rd Evansville IN 47715-5140

SHAW, WALTER NORMAN, b Penns Grove, NJ, Dec 12, 23; m 52; c 3. BIOLOGICAL CHEMISTRY. *Educ:* Duke Univ, BA, 44; Univ Pa, PhD, 56. *Prof Exp:* Asst instr biochem, Univ Pa, 51-54, instr in res med, Univ Hosp, 56-58, assoc, 58-61; sr pharmacologist, 61-66, sr biochemist, 66-69, res scientist, 69-75, SR RES SCIENTIST, LILLY RES LABS, 75- *Mem:* AAAS; Am Diabetes Asn. *Res:* Hormonal control of lipid and carbohydrate metabolism; study of disease states in genetically determined disease states in laboratory animals; biological activity of various proteins prepared by recombinant methods. *Mailing Add:* Lilly Res Labs Lilly Corp Ctr Indianapolis IN 46285

SHAW, WARREN A(RTHUR), b Wichita, Kans, Mar 10, 25; m 50; c 2. STRUCTURAL & CIVIL ENGINEERING. *Educ:* Univ Kans, BS, 49, MS, 52; Univ Ill, Urbana, PhD(civil eng), 62. *Prof Exp:* Instr eng mech, Univ Kans, 49-52; res engr, 52-62, dir struct div, 62-72, HEAD CIVIL ENG DEPT, US NAVAL CIVIL ENG LAB, 72- *Mem:* Am Soc Civil Engrs; Am Concrete Inst; Sigma Xi. *Res:* Engineering mechanics; structural dynamics; blast resistance of structures and structural elements; nuclear weapons effects research. *Mailing Add:* 1835 Guava Ct Oxnard CA 93033

SHAW, WARREN CLEATON, agronomy, plant physiology; deceased, see previous edition for last biography

SHAW, WILFRID GARSIDE, b Cleveland, Ohio, May 30, 29; m 53; c 2. CATALYSIS, ENVIRONMENTAL SCIENCE. *Educ:* Oberlin Col, AB, 51; Univ Cincinnati, MS, 53, PhD(phys org chem), 57. *Prof Exp:* Sr chemist, Sohio Chem Co, 56-59, tech specialist, 59-62, sr res chemist, 62-71, res assoc, 71-74, sr res assoc, 75-82, dir res, 85-87, lab site mgr, 82-90; MGR RES & DEVELOP, BRIT PETROL AM, 88-; BR MGR RES & DEVELOP, BP AM, 88- *Mem:* Am Chem Soc; Catalysis Soc. *Res:* Heterogeneous and homogeneous catalysis; petrochemical processes and catalysts; environmental analysis and control; characterization of solids, liquid crystals; phototropy; molecular structure; petroleum processes. *Mailing Add:* B P Res 4440 Warrensville Ctr Cleveland OH 44128-2837

SHAW, WILLIAM S, b Glace Bay, NS, Oct 20, 24; m 50; c 6. GEOLOGY. *Educ:* St Francis Xavier Univ, BSc, 45; Mass Inst Technol, PhD(geol), 51. *Prof Exp:* Geologist, Geol Surv Can, 49-52; sr geologist, Dominion Oil Co Div, Standard Oil Co Calif, 52-56, div stratigrapher, Calif Co Div, 56-57; consult geologist, Rodgers, Seglund & Shaw, 57-68; prof geol & chmn dept, St Francis Xavier Univ, 68-90; PRES, CREIGNISH MINERALS LTD, 78- & SCOTIAROCK LTD, 89- *Concurrent Pos:* Mem bd dirs, Deuterium of Can, Ltd, 69-81; Dep, Minister Mines & Energy, NS, 79-80; mem, Vol Econ Planning Bd, NS, 69-80; mem NS Royal Comn Post-Sec Educ, 83-85; mem bd dirs, Eldorado Nuclear Ltd, 85-89. *Honors & Awards:* Distinguished Lectr, Can Inst Mining & Metall, 87; Cert Merit Can Eng Centennial Year, Asn Prof Engrs NS, 87. *Mem:* Am Asn Petrol Geol; Can Inst Mining & Metall; Geol Soc Am; fel Geol Asn Can; Can Soc Petrol Geologists; Mining Soc of NS. *Res:* Oil and gas exploration; geology of evaporites; mineral exploration including metallics, industrial minerals and groundwater. *Mailing Add:* Six Highland Dr Antigonish NS B2G 1N6 Can

SHAW, WILLIAM WESLEY, b Pittsburgh, Pa, May 12, 46; m 69; c 2. WILDLIFE CONSERVATION, RESOURCE MANAGEMENT. *Educ:* Univ Calif, Berkeley, BA, 68; Utah State Univ, MS, 71; Univ Mich, Ann Arbor, PhD(natural resources), 74. *Prof Exp:* PROF NATURAL RESOURCES, UNIV ARIZ, 74- *Concurrent Pos:* Int Wildlife & Nat Parks, Australia, Argentina, Egypt, Kuwait; Nat Res Coun Comt Fed Land Acquisition, 90-91. *Honors & Awards:* Leedy Nat Award for Urban Wildlife Conserv, 88. *Mem:* Wildlife Soc; Soc Conserv Biol. *Res:* Values of wildlife resources and management of wildlife for non-consumptive uses; urban wildlife conservation; wildlife and national parks in developing countries. *Mailing Add:* Sch Renewable Natural Resources Univ of Ariz Tucson AZ 85721

SHAWCROFT, ROY WAYNE, b LaJara, Colo, Aug 17, 38; m 67; c 2. SOIL PHYSICS, MICROMETEOROLOGY. *Educ:* Colo State Univ, BS, 61, MS, 65; Cornell Univ, PhD(soil sci), 70. *Prof Exp:* soil scientist, sci & educ admin-agr res, US Dept Agr, 61-65, 70-82; EXTENSION IRRIGATION AGRONOMIST, COL STATE UNIV, 82- *Mem:* Am Soc Agron; Soil Sci Soc Am. *Res:* Plant, soil, water and atmospheric relations in semi-arid region of central plains United States; dryland and irrigated agriculture; agronomic management of irrigation system, Northeastern Colo; general climatology and weather data analysis. *Mailing Add:* Cent Great Plains Res Sta Co-op Ext Serv Colo State Univ Akron CO 80720

SHAWCROSS, WILLIAM EDGERTON, b Norfolk, Va, Nov 29, 34. ECONOMIC BOTANY, ANTHROPOLOGY. *Educ:* Univ NC, Chapel Hill, AB, 59; Control Data Inst, cert comput prog, 71; Harvard Univ, cert nat sci, 81, MA, 82, 86. *Prof Exp:* Ed asst, Sky Publ Corp, Mass, 56-61, asst ed, 61-63, Mng ed, 64-87, PUBLISHER, SKY & TELESCOPE MAG, 87-, DIR & VPRES, 72-, PRES, 80- *Mem:* Am Astron Soc; fel AAAS. *Res:* Astronomy and archaeoastronomy; economic botany; anthropology. *Mailing Add:* 1105 Massachusetts Ave Apt 7A Cambridge MA 02138-5221

SHAWE, DANIEL REEVES, b Gardnerville, Nev, May 24, 25; m 51; c 3. GEOLOGY. *Educ:* Stanford Univ, BS, 49, MS, 50, PhD(geol), 53. *Prof Exp:* Chief br Rocky Mt mineral resources, 69-72, RES GEOLOGIST, US GEOL SURV, 51- *Mem:* Int Asn Genesis Ore Deposits; fel Geol Soc Am; Soc Econ Geol; Am Inst Mining, Metall & Petrol Eng; Soc Appln Geol Ore Deposits. *Res:* Ore deposits; geology uranium in sedimentary rocks; geology beryllium in volcanic rocks; resources fluorine in United States; structure in Great Basin; gold deposits in south-central Nevada. *Mailing Add:* 8920 W Second Ave Lakewood CO 80226

SHAWL, STEPHEN JACOBS, b San Francisco, Calif, June 18, 43; m 66; c 2. COOL VARIABLE STARS, GLOBULAR CLUSTERS. *Educ:* Univ Calif, Berkeley, AB, 65; Univ Tex, Austin, PhD(astron), 72. *Prof Exp:* From asst prof to assoc prof, 72-85, PROF PHYSICS & ASTRON, UNIV KANS, 85- *Mem:* Int Astron Union; Am Astron Soc. *Res:* cool variable stars; globular clusters. *Mailing Add:* Dept Physics & Astron Univ Kans 5073 MAL Lawrence KS 66045

SHAWYER, BRUCE L R, b Kirkcaldy, Scotland, May 12, 37; m 66; c 4. PURE MATHEMATICS. *Educ:* St Andrews Univ, BSc, 60, PhD(math), 63. *Prof Exp:* Asst lectr math, Univ Nottingham, 62-64, lectr, 64-66; from asst prof to prof prof math, Univ Western Ont, 66-85; PROF & HEAD MATH & STATIST, MEM UNIV NFLD, 85- *Mem:* Math Asn Am; Edinburgh Math Soc; London Math Soc; Can Math Soc; fel Inst Math & Appln. *Res:* Summability of series and integrals; approximation of series. *Mailing Add:* Dept Math & Statist Mem Univ Nfld St John's NF A1C 5S7 Can

SHAY, JERRY WILLIAM, b Dallas, Tex, Nov 7, 45; m 85; c 4. CELL BIOLOGY, SOMATIC CELL & MOLECULAR GENETICS. *Educ:* Univ Tex, Austin, BA, 66, MA, 69; Univ Kans, PhD(physiol & cell biol), 72. *Prof Exp:* Fel marine biol, Marine Biol Lab, 72; fel molecular cellular & develop biol, Univ Colo, 72-75; asst prof, 75-80, ASSOC PROF CELL BIOL, UNIV TEX SOUTHWESTERN MED CTR, 80- *Concurrent Pos:* NIH fel, 72-73; Muscular Dystrophy Asn fel, 73-75; res grants, NIH, Muscular Dystrophy Asn & Am Heart Asn, 75-80; course coordr med genetics, Southwestern Med Sch 76-87; mem exec comt, Cancer Ctr, 77-78; adj assoc prof, Univ Tex, 77-; adj staff mem, W Alton Jones Cell Sci Ctr, 77-78, course dir, 76-77; NIH res career develop award, 78-83; Nat Sci Found Res grant, 81-88; NIH res grant, 86-; Am Cancer Soc grant, 86-; mem, Human Genome Proj Study Sect, 89-90; mem, Mammalian Genetics Study Sect, NIH, 90- *Mem:* Am Soc Cell Biol; Tissue Culture Asn; Am Asn Cancer Res; Sigma Xi; Am Soc Microbiol; Int Soc Differentiation. *Res:* Somatic cell and molecular genetics of aging; molecular mechanisms of cell immortalization; mammalian mitochondrial genetics and molecular biology. *Mailing Add:* Dept Cell Biol 5323 Harry Hines Blvd Dallas TX 75235

SHAY, JOSEPH LEO, b Albany, NY, Mar 12, 42; m 63; c 2. SOLID STATE PHYSICS. *Educ:* Manhattan Col, BEE, 63; Stanford Univ, MS, 64, PhD(elec eng), 67. *Prof Exp:* Mem tech staff, 67-74, HEAD INFRARED PHYSICS & ELECTRONICS RES DEPT, BELL LABS, 74- *Mem:* Am Phys Soc. *Res:* Optical properties and electronic structure of semiconductors; properties of ordered ternary semiconductors; photovoltaic solar cells; electrochromic display devices. *Mailing Add:* 24 Longview Dr Holmdel NJ 07733

SHAYEGANI, MEHDI, b Rasht, Iran, Apr 2, 26; m 61; c 3. MEDICAL MICROBIOLOGY. *Educ:* Univ Tehran, PharmD, 53; Univ Pa, MS, 58, PhD(med microbiol), 61. *Prof Exp:* Dir, Mobile Unit Lab & Public Health Lab, Iran, 52-56; instr med microbiol, Med Sch, Univ Pa, 58-61; res assoc, Inst Microbiol, Rutgers Univ, 61-62; res assoc microbiol, Dept Commun Med, Sch Med, Univ Pa, 62-68, asst prof, Sch Med & Dept Path Biol, Vet Med, 68-72; DIR, BACTERIOL LABS, 73-, CHIEF LAB CLIN MICROBIOL, NY STATE DEPT HEALTH, ALBANY, 87-; ASSOC PROF, DEPT OBSTET & GYNECOL & DEPT MICROBIOL IMMUNOL, ALBANY MED SCH, 81-; ASSOC PROF, SCH PUB HEALTH, STATE UNIV NY, ALBANY, 85- *Concurrent Pos:* Prin investr grants, 61-; lectr, Sch Med Technol, Albany Med Ctr Hosp, 75-, Sch Pub Health, State Univ NY, Albany, 85- *Mem:* Fel Am Acad Microbiol; Am Soc Microbiol; Reticuloendothelial Soc; Sigma Xi. *Res:* Host-parasite relationship in selected bacteria; identification of virulent genes in Haemophilus ducreyi, the pathogens of sexually transmitted chancroid infections; Development of DNA probes and polymerase chain reaction for identification of H ducreyi in genital ulcers; molecular epidemiology for "fingerprinting" of bacteria involved in outbreaks. *Mailing Add:* 23 Fairway Ave Delmar NY 12054

SHAYKEWICH, CARL FRANCIS, b Winnipeg, Man, July 18, 41; m 67. SOIL PHYSICS. *Educ:* Univ Man, BSA, 63, MSc, 65; McGill Univ, PhD(soil physics), 68. *Prof Exp:* Asst prof, 67-74, ASSOC PROF SOIL PHYSICS, UNIV MAN, 74- *Mem:* Am Soc Agron; Can Soc Soil Sci. *Res:* Soil physics, especially water relations in soil-plant-atmosphere system. *Mailing Add:* Dept of Soil Sci Univ of Man Winnipeg MB R3T 2N2 Can

SHE, CHIAO-YAO, b Fukien, China, Aug 4, 36; m 64; c 2. QUANTUM ELECTRONICS, CONDENSED MATTER PHYSICS. *Educ:* Nat Taiwan Univ, BS, 57; NDak State Univ, MS, 61; Stanford Univ, PhD(elec eng), 64. *Prof Exp:* Asst prof elec eng, Univ Minn, 64-68; from asst prof to assoc prof, 68-74, PROF PHYSICS, COLO STATE UNIV, 74- *Concurrent Pos:* Vis prof, Naval Res Lab, Univ Md. *Mem:* Am Phys Soc; fel Optical Soc Am; Inst Elec & Electronics Engr. *Res:* Optical properties of solids; lidar application; laser spectroscopy; non-linear optical processes. *Mailing Add:* Dept of Physics Eng Bldg Colo State Univ Ft Collins CO 80523

SHEA, DANIEL FRANCIS, b Springfield, Mass, Aug 2, 37; m 66; c 2. MATHEMATICS. *Educ:* Am Int Col, BA, 59; Syracuse Univ, MS, 61, PhD(math), 66. *Prof Exp:* From asst prof to assoc prof, 65-72, PROF MATH, UNIV WIS-MADISON, 72- *Concurrent Pos:* Vis assoc prof, Purdue Univ, 70-71; vis prof, Calif Inst Technol, 74-75, Univ Hawaii, 78-79. *Mem:* Math Asn Am; Am Math Soc. *Res:* Functions of a complex variable; asymptotics; functional equations. *Mailing Add:* Dept Math Univ Wis Madison WI 53706

SHEA, FREDERICKA PALMER, b Pittsfield, Mass, June 13, 40. AIDS RELATE RESEARCH, TEEN RISK TAKING. *Educ:* Boston Univ, BS, 62, MS 63; Univ Mich, PhD(med care orgn), 86. *Prof Exp:* Staff nurse, Berkshire Med Ctr, 63-64; pub health nurse, Peace Corp, Togo, W Africa, 64-66; intensive care unit nurse, Berkshire Med Ctr, 66-77; asst prof, Wayne State Univ Col Nursing, 67-72, asst dir, 69-74, dir, Ctr Health Res, 86-90, ASSOC PROF NURSING, URBAN ENVIRON HEALTH, WAYNE STATE UNIV COL NURSING, 72- *Concurrent Pos:* Co-proj dir, AIDS Training for GM physicians & nurses, Gen Motors Corp, 88; prin investr, AIDS Training Prog for Health Care Providers, NIMH contract, 88-; co-prin investr, AIDS Educ for Children & Families, Nat Inst Child Health & Human Develop, 89- *Mem:* Am Nurses Asn. *Res:* Psychosocial aspects of AIDS, clients, families, health care providers, care delivery; adolescent risk-taking behavior; clinical decision making. *Mailing Add:* Wayne State Univ Col Nursing 5460 Cass Ave Detroit MI 48202

SHEA, JAMES H, b Eau Claire, Wis, Dec 20, 32; m 60; c 2. GEOLOGY. *Educ:* Univ Wis, BS, 58, MS, 60; Univ Ill, PhD(geol), 64. *Prof Exp:* Geologist, Texaco, Inc, 60-61; admin asst sec sch curric, Earth Sci Curric Proj, Am Geol Inst, 64-65, asst to dir, 65-66; asst prof geol, Univ Tenn, 66-67; assoc prof, 67-74, PROF GEOL, UNIV WIS-PARKSIDE, 74- *Mem:* AAAS; fel Geol Soc Am; Am Asn Petrol Geologists; Nat Asn Geol Teachers; Soc Econ Paleont & Mineral. *Res:* Philosophy of science; recent sediments; geological education; history of geology. *Mailing Add:* Dept Geol Univ Wis-Parkside Box 2000 Kenosha WI 53141

SHEA, JOHN RAYMOND MICHAEL, JR, b Burlington, Vt, Oct 9, 38; m 68. ANATOMY, HISTOLOGY. *Educ:* Rensselaer Polytech Inst, BS, 60; McGill Univ, MSc, 62, PhD(anat), 65; Univ Surrey, MSc, 75. *Prof Exp:* From sessional lectr to lectr anat, McGill Univ, 65-68; asst prof, 68-74, ASSOC PROF ANAT, JEFFERSON MED COL, 74- *Honors & Awards:* Christian R & Mary F Lindback Award, 70. *Mem:* Anat Soc Gr Brit & Ireland; Am Asn Anat. *Res:* Cytology; nuclear morphology; cytophotometric analysis. *Mailing Add:* Dept Anat Jefferson Med Col 1025 Walnut St Philadelphia PA 19107

SHEA, JOSEPH F(RANCIS), b New York, NY, Sept 5, 26; m 74; c 7. ENGINEERING MECHANICS. *Educ:* Univ Mich, BS, 49, MS, 50, PhD(eng mech), 55. *Prof Exp:* Instr eng mech, Univ Mich, 48-50 & 53-55; res mathematician, Bell Tel Labs, 50-53, engr, 55-59; dir advan systs, AC Spark Plug Div, Gen Motors Corp, 59-61; space prog dir, Space Tech Labs, Thompson Ramo Wooldridge, Inc, 61-62; dep dir systs eng, Off Manned Space Flight, NASA, 62-63, mgr Apollo Spacecraft Prog, Manned Spacecraft Ctr, 63-67, dep asst adminr manned space flight, Washington, DC, 67-68; vpres & gen mgr equip div, Raytheon Co, 68-69, sr vpres & gen mgr, 69-75, sr vpres & group exec, 75-81, sr vpres eng, 81-90; Hunsacker prof, 89, ADJ

PROF AERONAUT & ASTRONAUT, MASS INST TECHNOL, 90- *Honors & Awards:* Arthur S Flemming Award, 65. *Mem:* Nat Acad Eng; fel Am Astronaut Soc; fel Am Inst Aeronaut & Astronaut. *Res:* Guidance and navigation, both radio and inertial; systems engineering; space technology. *Mailing Add:* Mass Inst Technol Rm 33-213 Cambridge MA 02139

SHEA, MICHAEL FRANCIS, b Henderson, Ill, Sept 22, 33; m 61; c 7. ENGINEERING PHYSICS. *Educ:* Ill Benedictine Col, BS, 55; Univ Notre Dame, PhD(nuclear physics), 60. *Prof Exp:* Fel, Univ Notre Dame, 60-61; physicist, Midwestern Univs Res Asn, 61-64; res scientist space physics, Lockheed Palo Alto Res Labs, 64-67; assoc physicist, Argonne Nat Lab, 67-69; PHYSICIST, FERMI NAT ACCELERATOR LAB, 69- *Mem:* Am Phys Soc; Inst Elec & Electronics Eng; Comput Soc. *Res:* Bremstrahlung production; nuclear resonance fluorescence; space radiation measurements; satellite instrumentation; particle accelerator beam diagnostic instrumentation and measurements; computer control systems and microprocessor instrumentation. *Mailing Add:* Fermi Nat Accelerator Lab PO Box 500 Batavia IL 60510

SHEA, MICHAEL JOSEPH, b Eau Claire, Wis, Sept 4, 39; m 66; c 3. EXPERIMENTAL SOLID STATE PHYSICS. *Educ:* Marquette Univ, BS, 61; Univ Minn, Minneapolis, MS, 66; Bryn Mawr Col, PhD(physics), 69. *Prof Exp:* prof physics, 69-88, chmn dept, 74-79, CHMN DEPT, CALIF STATE UNIV, SACRAMENTO, 88- *Mem:* Am Asn Physics Teachers; Int Solar Energy Soc; Sigma Xi. *Res:* Photoproperties of lead monoxide; photoproperties of lead monoxide; holography; Building energy research. *Mailing Add:* Dept Physics Calif State Univ 6000 Jay St Sacramento CA 95819

SHEA, PHILIP JOSEPH, b Groton, NY, Dec 11, 21; m 49; c 3. PHARMACOLOGY, PHYSIOLOGY. *Educ:* Syracuse Univ, BS, 52, MS, 56. *Prof Exp:* Pharmacologist, Biochem Lab, Merrell Dow Res Inst, Indianapolis, Ind, 56-62, Human Health Labs, 62-72, Dow Pharmaceut, 72-81, res assoc, 81-88; RETIRED. *Mem:* NY Acad Sci; Am Soc Pharmacol & Exp Therapeut. *Res:* Cardiovascular-renal physiology and pharmacology; autonomic pharmacology; drug screening and evaluation. *Mailing Add:* Four Sycamore Rd Carmel IN 46032

SHEA, RICHARD FRANKLIN, b Boston, Mass, Sept 13, 03; m 30; c 3. ELECTRONICS, NUCLEAR INSTRUMENTATION. *Educ:* Mass Inst Technol, BS, 24. *Prof Exp:* Radio engr, Am Bosch Corp, 25-28, Amrad, Mass, 28-29, Kolster Radio Co, NJ, 29 & Atwater-Kent, Pa, 29-30; chief engr, Pilot Radio Corp, Mass, 30-31, Freed-Eisemann, NY, 32-34 & Fada Radio Co, 34-37; sect engr, Gen Elec Co, 37-50, mgr adv planning, Labs Dept, 50-54, res liaison, 54-55, consult engr, Knolls Atomic Power Lab, 55-63; consult, 63-80; RETIRED. *Concurrent Pos:* Consult, War Assets Bd; ed consult, John Wiley & Sons, Inc; ed-in-chief, Nuclear & Plasma Sci Soc, Inst Elec & Electronics Engrs. *Honors & Awards:* Nuclear & Plasma Sci Soc Spec Award, 73 & Richard F Shea Award, 86. *Mem:* Fel Inst Elec & Electronics Engrs. *Res:* Design of radio receivers; circuits and novel application of new materials; application of electronics in nuclear field; solid state circuits and applications. *Mailing Add:* 6501 17th Ave W I 105 Bradenton FL 34209

SHEA, STEPHEN MICHAEL, b Galway, Ireland, Apr 25, 26. PATHOLOGY. *Educ:* Nat Univ Ireland, BSc, 48, MB & BCh, 50, MSc, 51, MD, 59; Am Bd Path, dipl, 60. *Prof Exp:* Asst lectr pharmacol, Univ Col, Dublin, 53-56; resident path, Mallory Inst Path, Boston City Hosp, 56-59; asst prof, Univ Toronto, 59-61; from instr path to instr math biol, Harvard Med Sch, 61-64, from assoc to assoc prof path, 65-73; PROF PATH, ROBERT WARD JOHNSON MED SCH, UNIV MED & DENT NJ, 73- *Concurrent Pos:* NIH grants, 62, 73 & 86; assoc pathologist, Mass Gen Hosp, 72-73. *Mem:* Am Asn Path; Electron Micros Soc Am; Am Soc Cell Biol; Soc Math Biol; Biophys Soc; Microcirculatory Soc. *Res:* Morphometric and quantitative aspects of tissue and cellular structure; ultrastructure and microvascular permeability; glomerular blood flow and filtration. *Mailing Add:* Dept Path Robert Wood Johnson Med Sch Piscataway NJ 08854

SHEA, TIMOTHY EDWARD, b Newton, Mass, Aug 6, 1898; m 22; c 6. ELECTRICAL ENGINEERING. *Educ:* Mass Inst Technol, SB & SM, 19; Harvard Univ, SB, 19. *Hon Degrees:* ScD, Columbia Univ, 46; EngD, Case Inst Technol, 49. *Prof Exp:* Instr physics, Mass Inst Technol, 18-20; mem staff, Bell Tel Labs, New York, 21-39; dir war res, Columbia Univ, 41-45; pres, Teletype Corp, Chicago, 48-49; asst vpres, AT&T, New York, 50-52; vpres & gen mgr, Sandia Atomic Labs, Albuquerque, 52-54; vpres mfg, Western Elec Co, New York, 54-56, vpres personnel, 56-58, vpres eng & dir, 58-63; consult, Nat Acad Sci, 63-64; RETIRED. *Concurrent Pos:* Founder, Underwater Sound Lab, US Navy, 41; vpres, Bell Tel Labs, New York, 52-53, dir, 58-63; consult, US Navy, 55-; founder, Western Elec Eng Res Ctr, Princeton, NJ, 59; chmn undersea warfare comt, Nat Acad Sci, 64-72. *Mem:* Emer mem Nat Acad Eng; fel Inst Radio Engrs; fel Acoust Soc Am. *Res:* Submarine detection devices; transmission networks and wave filters. *Mailing Add:* W 90 Ridgewood Ave Paramus NJ 07652

SHEA, TIMOTHY GUY, b Elmhurst, Ill, Aug 22, 39; m 66; c 1. ENVIRONMENTAL ENGINEERING. *Educ:* Loyola Univ Los Angeles, BS, 62; Univ Calif, Berkeley, MS, 63, PhD(environ eng), 68. *Prof Exp:* VPRES ENG & SCI, TOUPS CORP, 87- *Honors & Awards:* Am Water Works Asn Qual Div Res Award, 71. *Mem:* Water Pollution Control Fedn; Int Asn Water Pollution Res (actg secy-treas, 67-69); Am Water Works Asn. *Res:* All aspects of environmental engineering related to water quality management, waste treatment, aquatic ecosystem modelling, resource allocation institutions, and storm water treatment. *Mailing Add:* 9922 Barnsbury Ct Fairfax Sta VA 22031

SHEAGREN, JOHN NEWCOMB, INTERNAL MEDICINE, INFECTIOUS DISEASE. *Educ:* Columbia Univ, MD, 62. *Prof Exp:* PROF & ASSOC CHMN DEPT INTERNAL MED & ASSOC DEAN, SCH MED, UNIV MICH, 83- *Mailing Add:* Vet Admin Med Ctr 2215 Fuller Rd Ann Arbor MI 48109-0600

SHEALY, CLYDE NORMAN, b Columbia, SC, Dec 4, 32; m 59; c 3. HOLISTIC MEDICINE, NEUROSURGERY. *Educ:* Duke Univ, BSc & MD, 56; Saybrook Inst, PhD(psychol), 77. *Hon Degrees:* DSc, Ryodoraku Inst. *Prof Exp:* Asst med, Sch Med, Duke Univ, 56-57; asst surg, Wash Univ, 57-58; teaching fel, Sch Med, Harvard Univ, 62-63; sr instr neurosurg, Sch Med, Western Reserve Univ, 63-66, asst prof, 66; chief, Dept Neurosurg, Gundersen Clin & Lutheran Hosp, LaCrosse, Wis, 66-71; clin assoc, Dept Psychol Univ Wis-Lacrosse, 71-82; CLIN & RES PROF PSYCHOL, FOREST INST PROF PSYCHOL, 87- *Concurrent Pos:* Asst clin prof neurosurg, Sch Med, Univ Wis, 67-74; assoc clin prof, Sch Med, Univ Minn, 70-75; sr dolorologist, Pain & Health Rehab Ctr; pres, Holos Inst Health. *Mem:* Am Holistic Med Asn (pres, 78-80); Am Asn Neurol Surgeons; AMA; Am Asn Study Pain; Am Asn Study Headache. *Res:* Holistic medicine, integration of body, mind, emotion, and spirit; psychophysiologic basis of stress; neurochemical aspects of pain and stress. *Mailing Add:* 1328 E Evergreen St Springfield MO 65803-4400

SHEALY, DAVID LEE, b Newberry, SC, Sept 16, 44; m 69; c 2. OPTICS. *Educ:* Univ Ga, BS, 66, PhD(physics): 73. *Prof Exp:* Syst analyst, Dept Physics, Univ Ga, 73; from asst prof to assoc prof, 73-84, PROF & CHMN PHYSICS, UNIV ALA, BIRMINGHAM, 84- *Concurrent Pos:* Consult, Dept Physics, Univ Ga, 75-, Motorola, 79-84. *Mem:* Optical Soc Am; Acoust Soc Am; Am Asn Physics Teachers; Am Phys Soc; Am Inst Aeronaut & Astronaut. *Res:* Formulation and implementation of new optical design techniques based on analytical expressions for the illuminance and the caustic surface of an optical system; x-ray/EUV optical instrumentation. *Mailing Add:* Dept Physics Univ Ala Birmingham Birmingham AL 35294

SHEALY, HARRY EVERETT, JR, b Columbia, SC, Oct 24, 42; m 65; c 2. BIOLOGY, BOTANY. *Educ:* Univ SC, BS, 65, MS, 71, PhD(biol), 72. *Prof Exp:* Fel plant sci, Univ Man, 72-73; ASSOC PROF BIOL, UNIV SC, 73- *Mem:* Bot Soc Am. *Res:* Biology of reproduction in seed plants; vascular plant systematics. *Mailing Add:* Dir Develop Univ SC 171 Univ Pkwy Aiken SC 29801

SHEALY, OTIS LESTER, b Little Mountain, SC, Oct 3, 23; m 50; c 4. FIBER MANUFACTURE, TEXTILES. *Educ:* Newberry Col, AB, 44; Univ NC, PhD(chem), 50. *Hon Degrees:* ScD, Newberry Col, 68. *Prof Exp:* Instr org chem, Univ NC, 48-49; res chemist, E I Du Pont de Nemours & Co, Inc, 50-52, res supvr, 52-55, res mgr, 55-59, prod develop mgr, 59-64, res dir, 64-66, tech dir, 66-83; CONSULT, 83- *Mem:* Fel Brit Textile Inst; Am Chem Soc; Fiber Soc. *Res:* Polyester and polyamide fibers engineering of sheet structures. *Mailing Add:* 109 Walnut Ridge Rd Wilmington DE 19807

SHEALY, Y(ODER) FULMER, b Chapin, SC, Feb 26, 23; m 50; c 3. SYNTHETIC ORGANIC CHEMISTRY, MEDICINAL CHEMISTRY. *Educ:* Univ SC, BS, 43; Univ Ill, PhD(chem), 49. *Prof Exp:* Chemist, Nat Defense Res Comt, 43-45; asst chem, Univ Ill, 45-47; Abbot Labs fel, Univ Minn, 49-50; res chemist, Upjohn Co, 50-56; asst prof chem, Univ SC, 56-57; sr chemist, 57-59, sect head, 59-66, div head, 66-90, DISTINGUISHED SCIENTIST, SOUTHERN RES INST, 90- *Mem:* AAAS; NY Acad Sci; Int Soc Heterocyclic Chem; Am Chem Soc; Am Pharmaceut Asn; Int Asn Vitamin & Nutrit Oncol; Am Asn Cancer Res. *Res:* Pyrimidines and purines; triazenes; steroids; anticancer agents; carbocyclic analogs of nucleosides; folic acid analogs; chloroethylating agents; antiviral agents; retinoids. *Mailing Add:* Southern Res Inst 2000 Ninth Ave S PO Box 55305 Birmingham AL 35255

SHEAR, CHARLES L, b Baltimore, Md, Apr 3, 53; m 77; c 3. PUBLIC HEALTH & EPIDEMIOLOGY. *Educ:* Univ Md, BS, 74; Tulane Univ, MPH, 76, PhD, 79. *Prof Exp:* Asst prof family med, Univ S Ala, 79-80; asst prof epidemiol, Sch Med, Univ Calif Irvine, 80-84 & La State Univ Sch Med, 84-86; RES SCI, MERCK, SHARP & DOHME RES LAB, 86- *Res:* Cardiovascular disease epidemiology. *Mailing Add:* Merck, Sharp & Dohme Res Lab Ten Sentry Parkway Blue Bell PA 19422

SHEAR, CHARLES ROBERT, b Chicago, Ill, Jan 20, 42; m 66. ANATOMY, NEUROBIOLOGY. *Educ:* Univ Ill, BS, 65; Columbia Univ, MA, 67, PhD(biol sci), 69. *Prof Exp:* Leverhulme Trust vis fel, Univ Hull, 69-70; from instr to assoc prof anat, Emory Univ, 70-76; mem fac dept anat, 76-85, ASSOC PROF ANAT, SCH MED, UNIV MD, BALTIMORE, 85- *Concurrent Pos:* Mem, Nat Adv Coun On Regeneration, Vet Admin. *Mem:* AAAS; Am Asn Anat; Am Soc Cell Biol. *Res:* Ultrastructural aspects of skeletal muscle growth, development and regeneration; cellular organization and function of the neural retina; electron microscopy. *Mailing Add:* Dept Anat Univ Md Sch Med Baltimore MD 21201

SHEAR, CORNELIUS BARRETT, b Vienna, Va, Sept 24, 12; m 35; c 3. PLANT PHYSIOLOGY. *Educ:* Univ Md, BS, 34, MS, 38. *Prof Exp:* Agent, Crops Res Div, Sci & Educ Admin, Agr Res, USDA, 33-35, asst sci aide, 35-39, from jr physiologist to sr plant physiologist, 39-63, prin res plant physiologist, 63-77; RETIRED. *Concurrent Pos:* Consult fruit nutrit, People's Repub China, 81, 83, 85. *Honors & Awards:* J H Gourley Award, 59. *Mem:* Int Soc Hort Sci; fel Am Soc Hort Sci. *Res:* Mineral nutrition of fruit trees and tissue analysis as means of determining their nutrient status; physiology of corking and relation of nutrition, especially calcium to quality in apples. *Mailing Add:* 4218 Kenny St Beltsville MD 20705

SHEAR, DAVID BEN, b Boston, Mass, Jan 26, 38; div; c 2. BIOPHYSICS. *Educ:* Swarthmore Col, BA, 59; Brandeis Univ, PhD(biophys), 66. *Prof Exp:* NIH fel, Univ Buffalo, 66-67; asst prof physics, Univ Ga, 67-69; asst prof, 67-73, ASSOC PROF BIOCHEM, UNIV MO-COLUMBIA, 73- *Mem:* Biophys Soc. *Res:* Applications of thermodynamics, kinetics and statistical mechanics to biology; bioenergetics; muscle contraction; photosynthesis; mathematical models in biology. *Mailing Add:* Biochem 322-A Chem Univ Mo Med Ctr M228 Med Sci Bldg Columbia MO 65212

SHEAR, LEROY, b Baltimore, Md, Feb 20, 33; m 55; c 2. INTERNAL MEDICINE, NEPHROLOGY. *Educ:* Johns Hopkins Univ, BA, 53; Univ Md, MD, 57; Am Bd Internal Med, dipl, 59. *Prof Exp:* Res fel nutrit & metab, Western Reserve Univ, 61-62; USPHS trainee, Cleveland Metrop Gen Hosp, 62-63, assoc physician, 66-69; sr instr med, Case Western Reserve Univ, 66-67, asst prof, 67-69; assoc prof, Sch Med & dir nephrol sect, Sch Med, Temple Univ, 69-73; dir renal sect, Med Ctr Western Mass, Springfield, 73-77; clin assoc prof, 73-76, CLIN PROF MED, SCH MED, TUFTS UNIV, 76-; MED DIR, WESTERN MASS KIDNEY CTR, SPRINGFIELD, 77- *Mem:* AAAS; Am Fedn Clin Res; Am Soc Artificial Internal Organs; Am Soc Nephrol; Soc Exp Biol & Med. *Res:* Metabolic aspects of renal and hepatic disease. *Mailing Add:* Baystate Med Ctr 66 Roe Ave Northampton MA 01060-1636

SHEAR, WILLIAM ALBERT, b Coudersport, Pa, July 5, 42; m 80. BIOLOGY. *Educ:* Col Wooster, BA, 63; Univ NMex, MS, 65; Harvard Univ, PhD(evolutionary biol), 71. *Prof Exp:* Asst prof biol, Concord Col, 70-74; assoc prof, 74-80, PROF BIOL, HAMPDEN-SYDNEY COL, 81- *Concurrent Pos:* Res assoc, Am Mus Natural Hist, 78- *Honors & Awards:* John Peter Mettauer Award, 80; Cabell Award, 85. *Mem:* Sigma Xi; Am Arachnol Soc. *Res:* Behavior, taxonomy and biogeography of arachnids and myriapods; early evolution of land animals; revisions of families and genera of Opiliones and Diplopoda, especially North American forms; web-building behavior of spiders; Devonian fossils of arachnids and myriapods. *Mailing Add:* Dept of Biol Hampden-Sydney Col Hampden-Sydney VA 23943

SHEARD, JOHN LEO, b Southbridge, Mass, Feb 24, 24; m 47; c 4. INORGANIC CHEMISTRY. *Educ:* Harvard Univ, AB, 45, AM, 47; Univ Minn, PhD(chem), 53. *Prof Exp:* Instr chem, Northeastern Univ, 46-47; res chemist, Electrochem Div, E I du Pont de Nemours, 52-70, staff scientist, 70-78, res assoc, Electronics Div, 78-85; RETIRED. *Mem:* Am Ceramic Soc; Sigma Xi. *Res:* Fabrication of multilayer capacitors; ferrocyanides and ferricyanides; oxygen fluorides; acrylonitrile and pulp bleaching; precious metal compositions for solid state circuitry. *Mailing Add:* 88 Sundow Trail Williamsville NY 14221

SHEARD, MICHAEL HENRY, b Manchester, Eng, Aug 23, 27; US citizen; m 52; c 3. PSYCHIATRY, NEUROPHARMACOLOGY. *Educ:* Univ Manchester, MB & ChB, 51, MD, 64; Am Bd Psychiat & Neurol, dipl, 61; Royal Col Physicians, DPM, 62. *Hon Degrees:* MA, Yale, 76. *Prof Exp:* From res assoc phychiat res, 64-78, PROF PSYCHAIT, YALE UNIV AT CONN MENTAL HEALTH CTR, 78- *Concurrent Pos:* NIMH fel psychait, Yale Univ, 62-64; Consult, Conn Dept Corrections, 58- *Mem:* Fel Am Psychiat Asn; fel Royal Col Physicians; Sigma Xi. *Res:* Psychopharmacology. *Mailing Add:* Conn Mental Health Ctr 34 Park St New Haven CT 06519

SHEARER, CHARLES M, b Ashland, Ohio, July 30, 31; m 58. ANALYTICAL ORGANIC CHEMISTRY, PHARMACEUTICAL ANALYSIS. *Educ:* ETenn State Col, BA, 53; Univ Detroit, MS, 64, PhD(chem), 68. *Prof Exp:* Chemist, Columbus Coated Fabrics, 54-55; res chemist, Parke, Davis & Co, 57-66; mgr, Wyeth Labs, 68-88, SR RES SCIENTIST, WYETH-AYERST RES, 88- *Mem:* Am Chem Soc. *Res:* Analysis and determination of rates of degradation of pharmaceuticals. *Mailing Add:* Six Bayview Dr St Albans VT 05478-5100

SHEARER, DUNCAN ALLAN, b Kamsack, Sask, Feb 15, 20; m 46; c 2. ANALYTICAL CHEMISTRY. *Educ:* Univ Sask, BA, 48, MA, 50; Univ Toronto, PhD(org chem), 54. *Prof Exp:* Res scientist, Can Dept Agr, 54-73, sr res scientist, 73-80; RETIRED. *Mem:* Fel Chem Inst Can; Spectros Soc Can. *Res:* Lignin; organo-mercury chemistry; analytical methods in agriculture; pheromones of the honeybee. *Mailing Add:* 1182 Gateway Rd Ottawa ON K2C 2W9 Can

SHEARER, EDMUND COOK, b Birmingham, Ala, May 20, 42; m 66; c 3. PHYSICAL CHEMISTRY. *Educ:* Ark Polytech Col, BS, 64; Univ Ark, PhD(chem), 69. *Prof Exp:* From asst prof to assoc prof, 68-79, PROF CHEM, FT HAYS STATE UNIV, 79- *Mem:* Am Chem Soc; Sigma Xi. *Res:* Reaction kinetics, atmospheric precipitation studies. *Mailing Add:* Dept of Chem Ft Hays State Univ Hays KS 67601

SHEARER, GREG OTIS, b Washington, DC, May 2, 47; m 68; c 5. PHYSICAL ORGANIC CHEMISTRY. *Educ:* Iowa State Univ, BS, 69; Creighton Univ, MS, 71; Univ Kans, PhD(med chem), 76. *Prof Exp:* ASST PROF CHEM, CREIGHTON UNIV, 75- *Mem:* Am Chem Soc. *Res:* Mechanistic studies of organic and biochemical reactions. *Mailing Add:* Dept of Chem Creighton Univ 2500 California St Omaha NE 68178-0002

SHEARER, J(ESSE) LOWEN, b Marengo, Ill, Apr 25, 21; m 44; c 3. MECHANICAL ENGINEERING. *Educ:* Ill Inst Technol, BS, 44; Mass Inst Technol, SM, 50, ME, 52, ScD, 54. *Prof Exp:* Eng trainee, Sundstrand Mach Tool Co, 39-44; asst mech eng, Mass Inst Technol, 49-51, from instr to assoc prof, 51-63, supvr automatic control systs div, 58-63; Rockwell prof eng, 63-74, dir systs & controls lab, 64-76, prof, 74-85, EMER PROF MECH ENG, PA STATE UNIV, 85- *Concurrent Pos:* Guest scientist, Swedish Inst Textile Res, 56-57; chmn component comt, Am Automatic Control Coun, 61-63, chmn tech comt components, Int Fedn Automatic Control, 66-69; vis prof, Tokyo Inst Technol, 69; guest scientist, Royal Inst Technol, Sweden, 70; tech ed, Am Soc Mech Engrs J Dynamic Systs, Measurement & Control, 76-84; vis scientist, Inst Hydraulic & Pneumatic Control Systs, R-WTH, Aachen, WGer, 77. *Honors & Awards:* Donald P Eckman Award, Instrument Soc Am, 65; Richards Mem Award, Am Soc Mech Engrs, 66; Rufus Oldenberger Award, 83. *Mem:* Fel Am Soc Mech Engrs. *Res:* Automatic control systems; mechanical, electrical and fluid control systems; design of engineering systems; materials recycling and resource recovery systems. *Mailing Add:* Dept Mech Eng Pa State Univ University Park PA 16802

SHEARER, JAMES WELLES, plasma physics, for more information see previous edition

SHEARER, MARCIA CATHRINE (EPPLE), b Akron, Ohio, Oct 27, 33; m 58. TAXONOMY OF ACTINOMYCETES, TISSUE CULTURE. *Educ:* Ohio State Univ, BS, 56; Wayne State Univ, Detroit, Mich, MS, 63. *Prof Exp:* Microbiologist, Parke, Davis & Co, 56-68; microbiologist, Smith Kline & French Labs, Div Smith Kline & French, 68-72, sr microbiologist, Div Smith Kline Corp, 72-80, assoc sr investr, Div Smith Kline Beckman, 80-85; sr scientist, 86-89, PRIN SCIENTIST, SCHERING CORP, 90- *Concurrent Pos:* Consult taxon, Var Corps, 85; mem exec bd, US Fedn Cult Collections, 90- *Mem:* Am Soc Microbiol; Soc Indust Microbiol; Am Inst Biol Sci; Sigma Xi; World Fedn Cult Collections; US Fedn Cult Collections. *Res:* Isolation of microorganisms from soils and other natural materials; knowledgeable culture collection curator with experience in preserving a wide variety of microorganisms; preparation of taxonomic descriptions of actinomycetes for patents and publications; tissue culture preparation for virology section. *Mailing Add:* 210 Green Hollow Dr Iselin NJ 08830

SHEARER, RAYMOND CHARLES, b Anaheim, Calif, June 3, 35; m 56; c 10. FORESTRY. *Educ:* Utah State Univ, BS, 57, MS, 59; Univ Mont, PhD, 85. *Prof Exp:* RES SILVICULTURIST, INTERMOUNTAIN RES STA, FOREST SERV, USDA, 57- *Mem:* Soc Am Foresters; Forest Hist Soc. *Res:* Regeneration of Larix occidentalis, Picea engelmannii and Pseudotsuga menziesii, including seed production, seed dissemination, germination, causes of seedling mortality and development; cutting methods and growth and mortality in western larch forests. *Mailing Add:* Forestry Sci Lab PO Box 8089 Missoula MT 59807-8089

SHEARER, THOMAS ROBERT, b South Bend, Ind, Aug 11, 42; m 64; c 2. BIOCHEMISTRY. *Educ:* Beloit Col, BA, 64; Univ Wis, MS, 67, PhD(biochem), 69. *Prof Exp:* Fel biochem, Univ Wis, 69; from asst prof to assoc prof, 69-80, PROF NUTRIT & DIR DIV, DENT & MED SCHS, ORE HEALTH SCI UNIV, 80- *Concurrent Pos:* Sabbatical biochem, Univ Wis, 76. *Mem:* Asn Res Vision & Opthal; Am Inst Nutrit; Int Asn Dent Res; Int Cong Eye Res. *Res:* Effects of toxic amounts of fluoride on intermediary metabolism in animals; selenium distribution in animals and biological materials; mechanism of selenium-overdose cataract; calpain enzymes. *Mailing Add:* Dept Biochem Ore Health Sci Univ 611 SW Campus Dr Portland OR 97201

SHEARER, WILLIAM MCCAGUE, b Zanesville, Ohio, June 24, 26; m 54; c 2. SPEECH & HEARING SCIENCE. *Educ:* Ind Univ, BA, 51; Western Mich Univ, MA, 54; Univ Denver, PhD(speech), 58. *Prof Exp:* Asst prof speech path, Minot State Col, 54-56; PROF COMMUN DISORDERS, NORTHERN ILL UNIV, 58- *Concurrent Pos:* Fel, Sch Med, Stanford Univ, 63-64; vis prof, Commun Sci Lab, Univ Fla, 68-69. *Mem:* Am Speech & Hearing Asn; Am Asn Phonetic Sci; Sigma Xi. *Res:* Anatomy of speech and hearing. *Mailing Add:* Dept Commun Disorders Northern Ill Univ De Kalb IL 60115

SHEARER, WILLIAM THOMAS, b Detroit, Mich, Aug 23, 37. ALLERGY, IMMUNOLOGY. *Educ:* Univ Detroit, BS, 60; Wayne State Univ, PhD(biochem), 66; Wash Univ, MD, 70; Am Bd Pediat, cert, 75; Am Bd Allergy & Immunol, 75. *Prof Exp:* Asst pediat, Wash Univ, 70-72, from instr to prof, 72-78; PROF PEDIAT, MICROBIOL & IMMUNOL, BAYLOR COL MED, 78-, HEAD SECT ALLERGY & IMMUNOL, DEPT PEDIAT, 78-; CHIEF, ALLERGY & IMMUNOL SERV, TEX CHILDREN'S HOSP, HOUSTON, 78- *Concurrent Pos:* Fel chem, Ind Univ, 66-67; res fel, USPHS, 72-74; res scholar award, Cystic Fibrosis Found, 74-77; dir div immunol, St Louis Children's Hosp, 74-76, dir div allergy & immunol, 76-78; from assoc pediatrician to pediatrician, Barnes Hosp, 76-78; fac res award, Am Cancer Soc, 77-79; dir, Allergy & Immunol Training Prog, Baylor Col Med, 79-; chmn, Allergy & Immunol Training Prog Dirs Comt, Am Acad Allergy & Immunol, 87-90; chmn, Sect Allergy & Immunol, Southern Med Asn, 90-91; exec comt mem, Sect Allergy & Immunol, Am Acad Pediat, 90-; dir, Am Bd Allergy & Immunol, 90-; numerous positions & comts, NIH, 88-91. *Mem:* Sigma Xi; Am Asn Immunologists; fel Am Acad Pediat; Soc Pediat Res; fel Am Acad Allergy; Am Soc Clin Invest. *Res:* Tumor immunology; interaction of antibody and complement with cell membrane antigens; immunodeficiency diseases of children; allergic disorders of children; cystic fibrosis; signal transduction across cell membranes HIV-1 infection and pediatric AIDS. *Mailing Add:* Dept Pediat Baylor Col Med One Baylor Plaza Houston TX 77030

SHEARIN, NANCY LOUISE, b Meridian, Miss, May 17, 38. SMOOTH MUSCLE, ION-SELECTIVE ELECTRODES. *Educ:* Millsaps Col, BS, 60; Tenn Tech Univ, MS, 71; Univ Wyo, PhD(physiol), 74. *Prof Exp:* Res fel, Mem Univ Nfld, 74-76 & Univ Alta, 76-79; res assoc, 79-82, RES ASST PROF, DEPT SURG, UNIV UTAH, 82- *Concurrent Pos:* Prin investr, NSF, 85- *Mem:* Biophys Soc; Am Physiol Soc; NY Acad Sci. *Res:* Intracellular ionic activity, motor function and functions of prostaglandins in smooth muscle. *Mailing Add:* Dept Surg Univ Utah 50 N Medical Dr Salt Lake City UT 84112

SHEARN, ALLEN DAVID, b Chicago, Ill, May 27, 42; m 64; c 3. DEVELOPMENTAL GENETICS. *Educ:* Univ Chicago, BA, 64; Calif Inst Technol, PhD(genetics), 69. *Prof Exp:* Helen Hay Whitney Found res fel molecular biophys & biochem, Yale Univ, 68-71; from asst prof to assoc prof, 71-81, PROF BIOL, JOHNS HOPKINS UNIV, 81- *Concurrent Pos:* Consult, Genetic Biol Adv Panel, NSF, 72-75. *Mem:* AAAS; Soc Develop Biol; Genetics Soc Am. *Res:* Applying the techniques of molecular biology to the study of the regulation of development in higher organisms. *Mailing Add:* Dept Biol Johns Hopkins 34th & Charles St Baltimore MD 21218

SHEARN, MARTIN ALVIN, b New York, NY, Dec 19, 23; m 51; c 3. INTERNAL MEDICINE, RHEUMATOLOGY. *Educ:* Ohio Univ, AB, 43; New York Med Col, MD, 49; Am Bd Internal Med, dipl, 56. *Prof Exp:* From intern to resident med, Bellevue Hosp, New York, 49-52, from clin asst vis physician to asst attend physician, 53-55; DIR MED EDUC, KAISER FOUND HOSP, OAKLAND, 56-; CLIN PROF MED, UNIV CALIF, SAN

FRANCISCO, 75- *Concurrent Pos:* Res fel cardiol, Sch Med, Stanford Univ, 52-53; instr, Med Sch, NY Univ, 53-55; from instr to asst clin prof med, Med Ctr, Univ Calif, San Francisco, 59-68, assoc prof med, Sch Med, 68-75, clin prof, 75-; consult, Vet Admin Hosp; attend physician, Highland Hosp, Oakland; chmn, Northern Calif Sect, Arthritis Found, 66; ed, Rheumatic Dis, Med Clin NAm; prof med, Univ Fed de Alagoas, Brazil. *Mem:* AMA; Am Heart Asn; fel Am Col Physicians; Am Fedn Clin Res. *Res:* Connective tissue disorders; rheumatic diseases; medical history; author of over 50 publications. *Mailing Add:* 1815 Arlington Ave El Cerrito CA 94530

SHEASLEY, WILLIAM DAVID, b Youngstown, Ohio, Oct 31, 46; m 69; c 2. PHYSICAL CHEMISTRY, POLYMER SCIENCE. *Educ:* Grove City Col, BS, 68; Ohio State Univ, MS, 70, PhD(phys chem), 72. *Prof Exp:* Fel chem lasers, Cornell Univ, 72-73; sr chemist, 73-79, RES SECT MGR, ROHM AND HAAS CO, 79- *Mem:* Am Chem Soc; Soc Rheology; Electron Micros Soc Am. *Res:* Analytical research and polymer characterization; particular emphasis on surface analysis electron spectroscopy for chemical analysis, morphological studies (electron microscopy and fluorescence techniques), and mechanical, rheological, and dielectric properties; adhesives and adhesion. *Mailing Add:* Rohm and Haas Co Spring House PA 19477

SHEATS, GEORGE FREDERIC, b Reno, Nev, Dec 19, 27; m 53; c 3. PHYSICAL CHEMISTRY. *Educ:* Univ Calif, BS, 51; Univ Rochester, PhD(phys chem), 55. *Prof Exp:* Res chemist, Am Cyanamid Co, 55-62; assoc prof chem, 62-69, chmn dept chem, 68-69, chmn dept comput sci, 70-75, PROF CHEM, STATE UNIV NY COL PLATTSBURGH, 69-, CHMN DEPT CHEM, 88- *Mem:* Am Chem Soc. *Res:* Absorption spectroscopy; microcalorimetry; ion chromatography; kinetics of aromatic nitration; luminescence of inorganic complexes; kinetics of biological interactions of calcium ion. *Mailing Add:* Dept Chem State Univ NY Col Plattsburg NY 12901

SHEATS, JOHN EUGENE, b Atlanta, Ga, Dec 20, 39; m 72; c 1. ORGANOMETALLIC CHEMISTRY, PHYSICAL ORGANIC CHEMISTRY. *Educ:* Duke Univ, BS, 61; Mass Inst Technol, PhD(chem), 66. *Prof Exp:* Asst prof chem, Bowdoin Col, 65-70; assoc prof, 70-78, PROF, RIDER COL, 78- *Honors & Awards:* Emmett Reid Award, Am Chem Soc, 84. *Mem:* Am Chem Soc; Sigma Xi. *Res:* Mechanism of decomposition of benzenediazonium ion; synthesis and properties of substituted Cobalticinium salts and other organo-transition metal compounds; preparation of organometallic polymers; biomedical applications of organometallic compounds; binuclear maganese complexes as models for photosynthetic oxygen evolution. *Mailing Add:* Dept Chem Rider Col Lawrenceville NJ 08648

SHECHMEISTER, ISAAC LEO, b Windaw, Latvia, June 11, 13; nat US; m 38; c 2. MEDICAL MICROBIOLOGY. *Educ:* Univ Calif, AB, 34, MA, 35, PhD(bact), 49; Am Bd Microbiol, dipl. *Prof Exp:* Asst bact, Univ Calif, 36-39, lectr epidemiol, Sch Pub Health, 46, prin bacteriologist, Infectious Dis Proj, 46-50; asst prof bact & immunol, Sch Med, Washington Univ, 50-52, asst prof microbiol, 52-53, assoc prof bact, Sch Dent, 53-57; assoc prof microbiol, 57-64, PROF MICROBIOL, SOUTHERN ILL UNIV, 64- *Concurrent Pos:* Consult, Radiol Defense Lab, US Dept Navy, Calif, 46-51; spec fel biophys, Statens Seruminstitut, Denmark, 66-67. *Mem:* Am Soc Microbiol; Am Asn Immunol. *Res:* Electron microscopy of antigen-antibody reactions; animal virology and immunology; dental caries. *Mailing Add:* Dept Microbiol Southern Ill Univ Carbondale IL 62901

SHECHTER, HAROLD, b New York, NY, July 12, 21. ORGANIC CHEMISTRY. *Educ:* Univ SC, BS, 41; Purdue Univ, PhD(chem), 46. *Prof Exp:* Asst chem, Purdue Univ, 41-42; from asst prof to assoc prof, 46-70, PROF CHEM, OHIO STATE UNIV, 70- *Mem:* Am Chem Soc. *Res:* Nitration of saturated hydrocarbons; mechanics of addition reactions of oxides of nitrogen; synthesis of polynitro compounds; kinetics of neutralization of pseudo acids; mechanisms of reactions of hydrazoic acid; homomorphic ring strain; chemistry of small ring compounds; alkylation of ambident ions; decomposition of carbenes. *Mailing Add:* Dept of Chem Ohio State Univ Columbus OH 43210

SHECHTER, LEON, b New York, NY, Dec 19, 12; m 37. ORGANIC CHEMISTRY, POLYMER CHEMISTRY. *Educ:* Univ SC, BS, 33, MS, 34; Univ Cincinnati, PhD(org chem), 37. *Prof Exp:* Asst, Univ Cincinnati, 34-37; res chemist, Union Carbide Plastics Co, 37-44, head coating resins res, 44-52, sect head plastics res, 52-56, asst dir res, 56-58, resident dir, 58-59, dir, 60-61, dir polymer res & develop, 61-63, dir appln res & develop, 63-64, vpres res & develop, 64-67, vpres res & develop chem & plastics, Union Carbide Corp, 67-74, vpres exploratory technol, Patents & Licensing Chem & Plastics, 74-77; CONSULT, 77- *Mem:* Am Chem Soc. *Res:* Alkyds; vinyl polymers; silicones; epoxy resins; phenolics; research and development management-long range research chemicals and plastics; patent management and licensing. *Mailing Add:* 22 Harvey Dr Summit NJ 07901

SHECHTER, YAAKOV, b Tel Aviv, Israel, Feb 11, 34; US citizen; m 59; c 2. HUMAN GENETICS, GENETIC COUNSELING. *Educ:* Univ Calif, Los Angeles, BSc, 59, PhD(plant sci), 65, cert med mycol, 67. *Prof Exp:* Res asst agr sci, Univ Calif, Los Angeles, 60-64, fel med mycol, Sch Med, 65-66; lectr bot, Univ Southern Calif, 66-67; res biochemist, Sch Med, Univ Calif, Los Angeles, 67-69; from asst prof to assoc prof, 69-76, PROF BIOL SCI, LEHMAN COL, 76- *Concurrent Pos:* Adj curator, NY Bot Garden, 69-; consult, Human Affairs Res Ctr, NY; vis assoc prof, Stein-Moore Lab, Rockefeller Univ, 75-76; lectr, Dept Pediat, Div Genetics, Col Physicians & Surgeons, Columbia Univ, 80- *Mem:* Sigma Xi; NY Acad Sci; Inst Soc, Ethics & Life Sci; Am Soc Human Genetics. *Res:* Human genetics; genetics of keratins; genetic counseling. *Mailing Add:* 165 E 72nd St Apt 149 New York NY 10021

SHEDD, DONALD POMROY, b New Haven, Conn, Aug 4, 22; m 46; c 4. SURGERY. *Educ:* Yale Univ, BS, 44, MD, 46. *Prof Exp:* Intern surg, Yale Univ, 46-47, from asst resident to resident, 50-53, from instr to assoc prof, 53-67; CHIEF DEPT HEAD & NECK SURG, ROSWELL PARK MEM INST, 67-; RES PROF, STATE UNIV NY BUFFALO, 70- *Concurrent Pos:* Harvey Cushing fel surg res, Yale Univ, 49-50; Markle scholar med sci, 53-58; mem head & neck cancer group, Nat Head & Neck Cancer Cadre, 73-76, head & neck cancer group, Organ Site Prog, Nat Cancer Inst, 86-88. *Mem:* Soc Univ Surg; Soc Head & Neck Surg (treas, 71-, pres, 76-77); Am Col Surg; Sigma Xi. *Res:* Oncology, particularly in head and neck cancer; physiology of deglutition; speech rehabilitation. *Mailing Add:* Dept Head & Neck Surg Roswell Park Mem Inst Buffalo NY 14263

SHEDLARSKI, JOSEPH GEORGE, JR, b Forty Fort, Pa, Mar 15, 39. MICROBIAL BIOCHEMISTRY. *Educ:* King's Col, Pa, BS, 61; St John's Univ, NY, MS, 63; Princeton Univ, MA, 66, PhD(biochem sci), 69; La State Univ, DDS, 81. *Prof Exp:* Instr biol, Col Misericordia, 63-64; asst prof biol, Univ New Orleans, 71-75, assoc prof, 76-77; PVT DENT PRACT, 81- *Concurrent Pos:* Can Nat Cancer Inst fel, McMaster Univ, 69-71; res award, Am Asn Dent, 81. *Mem:* Acad Gen Dent; Am Soc Geriat Dent; Am Dent Asn. *Res:* Microbial cell wall-sheath structure; biogenesis; regulation; sugar transport in bacteria. *Mailing Add:* 4409 Laudun St Metairie LA 70006

SHEDLER, GERALD STUART, b New York, NY, May 8, 39; m 68. OPERATIONS RESEARCH, COMPUTER SCIENCE. *Educ:* Amherst Col, BA, 61; Tufts Univ, MA, 64. *Prof Exp:* Asst scientist appl math, Res & Advan Develop Div, Avco Corp, 62-64; res staff mem comput sci, T J Watson Res Ctr, IBM Corp, 65-70; actg assoc prof oper res, Stanford Univ, 73-74; RES STAFF MEM COMPUT SCI, SAN JOSE RES LAB, IBM CORP, 70- *Concurrent Pos:* Actg assoc prof, Dept Oper Res, Stanford Univ, 75-76; vis lectr, Math Asn Am, 77-81; consult assoc propf dept oper res, Stanford Univ, 78-80; lectr dept info sci, Victoria Univ, NZ, 79. *Mem:* Asn Comput Mach. *Res:* Stochastic processes and their applications; discrete event simulation of stochastic systems; computer system performance analysis. *Mailing Add:* Dept Operations Res Stanford Univ Stanford CA 94305

SHEDLOCK, KAYE M, b Wash, DC, Mar 30, 51. SEISMOLOGY, TECTONICS. *Educ:* Univ Md, BS, 73; Johns Hopkins Univ, MS, 78; Mass Inst Technol, PhD(geophys), 86. *Prof Exp:* Mathematician, 78-87, geophysicist, 86-88, CHIEF, BR GEOL RISK ASSESSMENT, US GEOL SURV, 88- *Concurrent Pos:* Mem, Nat Earthquake Prediction Eval Coun, 90-, bd trustees, External Vis Comt Geophys Dept, Colo Sch Mines, 91-94 & bd dirs, Seismol Soc Am, 91-94; expert, NSF Presidential Young Investr Panel, 90-91. *Mem:* Seismol Soc Am; Am Geophys Union. *Res:* Seismotectonics of intraplate regions; subduction tectonics of the Pacific Northwest and the strike-slip San Andreas fault system. *Mailing Add:* US Geol Surv Denver Fed Ctr MS 966 Box 25046 Denver CO 80225

SHEDRICK, CARL F(RANKLIN), b South Bend, Ind, Aug 2, 20; m 46; c 3. CHEMICAL ENGINEERING. *Educ:* Purdue Univ, BS, 42; Columbia Univ, MSE, 48. *Prof Exp:* Res engr, E I du Pont de Nemours & Co, Inc, NJ, 42-50, tech economist, Polychem Dept, Del, 50-67, chem eng, Plastics Dept, E I du Pont de Nemours & Co, Inc, 67-85; RETIRED. *Mem:* Am Chem Soc; Am Inst Chem Engrs. *Res:* Process development of plastic materials; process and equipment design; applications research; economic evaluation. *Mailing Add:* 2535 Deepwood Dr Wilmington DE 19810

SHEEHAN, BERNARD STEPHEN, b Halifax, NS, July 25, 35; m 59; c 4. INFORMATION SYSTEMS, TELEMATICS. *Educ:* Tech Univ NS, BE, 57; Mass Inst Technol, SM, 61; Univ Conn, PhD(elec eng), 65. *Prof Exp:* Engr, Can Gen Elec Co, 57-58; lectr eng, St Mary's Univ, NS, 58-59; teaching asst, Mass Inst Technol, 59-61; instr, Univ Conn, 61-65; dean arts & sci, St Mary's Univ, NS, 65-67; asst to acad vpres, Univ Calgary, 67-69, dir, Off Instnl Res, 69-81, prof fac mgt, 81-90, assoc vpres, priorities & planning, 87-89, EMER PROF FAC MGT, UNIV CALGARY, 90-; ASSOC VPRES INFO & COMPUT SYSTS, UNIV BC, 90- *Concurrent Pos:* Res grants, Can Coun, 73-74 & Social Sci & Humanities Coun Can, 83-85; consult, Alta Advan Educ & Manpower, 75-77, Dept Commun, Can, 84-85; mem bd, Social Sci Fedn Can, 84-86; mem, Nat Adv Comt Educ Statist, 85-88; vchmn, Telematics, Can Higher Educ Res Network, 85-87. *Mem:* Distinguished mem Asn Instnl Res (vpres, 74-75, pres, 75-76); Can Soc Study Higher Educ (vpres, 82-83, pres, 83-84); distinguished mem Can Soc Study Higher Educ (vpres, 82-83, pres, 83-84). *Res:* Analysis in management, including institutional and system-wide problems in planning, resource allocation and decision processes; information technology. *Mailing Add:* 4186 Yuculta Crescent Vancouver BC V6N 3R5 Can

SHEEHAN, DANIEL MICHAEL, b Boston, Mass, Sept 5, 44; c 2. ENDOCRINE TOXICOLOGY, DEVELOPMENTAL TOXICOLOGY. *Educ:* Univ S Fla, Tampa, BA, 66 & MA, 68; Univ Tenn Oak Ridge Grad Sch Biomed Sci, PhD(biomed sci), 73. *Prof Exp:* Lab instr introd biol, Univ S Fla, Tampa, 66-67, lab coordr genetics, 67-68; Nat Defense Educ Act grad fel, Univ Tenn, Oak Ridge, 68-72; Oak Ridge Assoc Univ grad fel, 72-73; NIH trainee, Baylor Col Med, 73-74, fel, 74-75; res chemist, 75-78, RES BIOLOGIST, NAT CTR TOXICOL RES, 78-, CHIEF, DEVELOP MECHANISMS BR, 86- *Concurrent Pos:* Adj asst prof biochem, 76-84, adj assoc prof, 84-90, adj asst prof interdisciplinary toxicology & adj prof biochem, Univ Ark for Med Sci, 90-; Ed bd, proceedings of Soc Exp Biol & Med. *Mem:* Soc Toxicol; Teratol Soc; Endocrine Soc; Int Soc Study Xenobiotics; Soc Advan Contraception; Soc Exp Biol & Med. *Res:* Endocrine toxicology with emphasis on the developmental toxicity of estrogens including morphological and carcinogenic outcomes and tumor promotion by estrogens. *Mailing Add:* Div Reprod & Dev Toxicol Nat Ctr Toxicol Res Jefferson AR 72079

SHEEHAN, DESMOND, b Aldershot, Eng, Apr 8, 31; m 56; c 4. ORGANIC CHEMISTRY. *Educ:* Univ Reading, BSc, 55; Yale Univ, MS, 61, PhD(chem), 64. *Prof Exp:* Sci asst chem, Ministry Supply, Eng, 47-50, exp officer, 52 & 55-56; res chemist, Microcell Ltd, 56-58 & Am Cyanamid Co, Conn, 58-65; dir res, Techni-Chem Co, Conn, 65-70; sr res assoc, Allied Chem Corp, Morristown, 70-72, mgr org res, Corp Res Lab, 72-79, sr scientist, Corp Technol, Off Sci Technol, 80-86; RETIRED. *Mem:* Fel Am Chem Soc; The Royal Chem Soc. *Res:* Synthetic organic chemistry; reaction mechanism. *Mailing Add:* 850 Cleveland Rd PO Box 99 Bogart GA 30622

SHEEHAN, JAMES ELMER, materials science, for more information see previous edition

SHEEHAN, JOHN CLARK, b Battle Creek, Mich, Sept 23, 15; m 41; c 3. ORGANIC CHEMISTRY. *Educ:* Battle Creek Col, BS, 37; Univ Mich, MS, 38, PhD(org chem), 41. *Hon Degrees:* DSc, Univ Notre Dame, 63. *Prof Exp:* Res assoc, Nat Defense Res Comt Proj, Univ Mich, 41; res chemist, Merck & Co, Inc, NJ, 41-46; from asst prof to prof chem, 46-76, EMER PROF ORG CHEM, MASS INST TECHNOL, 76- *Concurrent Pos:* Sci liaison officer, Off Naval Res, 53-54; Swiss-Am Found lectr, 58; ed-in-chief, Org Syntheses. *Honors & Awards:* Am Chem Soc Awards, 51 & 59; Reilly lectr, Univ Notre Dame, 53; Swiss-Am Found lectr, 58; McGregory lectr, Colgate Univ, 58; Bachmann lectr, Univ Mich, 60; Dakin Mem lectr, Adelphi, 61; Medal, Synthetic Org Chem Mfrs Asn, 68. *Mem:* Nat Acad Sci; Am Acad Arts & Sci; Royal Soc Chem. *Res:* Synthetic penicillin; lactams; amino acids; peptides; alkaloids; steroids and the synthesis of high explosives. *Mailing Add:* Dept Chem Mass Inst Technol Cambridge MA 02139

SHEEHAN, JOHN FRANCIS, b Portsmouth, NH, July 28, 06; m 35. CLINICAL CYTOPATHOLOGY. *Educ:* Univ NH, BS, 28, MS, 30; Univ Iowa, PhD(biol), 45. *Prof Exp:* Asst zool, Univ NH, 28-30; from instr to assoc prof biol, 30-49, res assoc prof exfoliative cytol, 48-67, prof biol, 49-88, prof path, Sch Med, 67-88, prof gynec, 75-88, EMER PROF, PATH & GYNEC, CREIGHTON UNIV, 88- *Concurrent Pos:* Mem adj med staff, St Joseph Hosp, 72, chief, Cytol Lab, 78 -84. *Honors & Awards:* Sheehan Hall lectr, Creighton Univ, 84. *Mem:* Sigma Xi; Am Asn Anat; Am Soc Clin Pathologists; Am Soc Cytol; fel Am Soc Colposcopy & Cervical Pathol; Am Micros Soc; Am Asn Univ Profs; Am Inst Biol Sci. *Res:* Cytology cancer cell; exfoliative gynecologic cytology; colpomicroscopy; ultracentrifuge; cytoplasm. *Mailing Add:* 7300 Graceland Dr, No 307A Omaha NE 68134

SHEEHAN, THOMAS JOHN, b Brooklyn, NY, Apr 13, 24; m 50; c 3. ENVIRONMENTAL HORTICULTURE. *Educ:* Dartmouth Col, AB, 48; Cornell Univ, MS, 51, PhD(floricult, plant breeding & physiol), 52. *Prof Exp:* Asst floricult, Cornell Univ, 48-52; asst horticulturist, Exp Sta, Univ Ga, 52-54; asst ornamental horticulturist, Agr Exten Serv, 54-56 & Exp Sta, 56-63, from asst ornamental horticulturist to ornamental horticulturist, Exp Sta, 63-67, PROF ENVIRON HORT, UNIV FL, 67-, CHMN, 86- *Concurrent Pos:* Vis prof, Univ Hawaii, 62-63; consult floriculture, Food & Agr Orgn, UN, 71, 74-75 & 80, Jaflex, 71- & Cypress Gardens, 85- *Honors & Awards:* Silver Seal Award, Nat Fedn Garden Clubs, 80. *Mem:* Am Hort Soc; hon life mem Am Orchid Soc; Palm Soc; Fel Am Soc Hort Sci. *Res:* Nutrition and other cultural factors of orchids; photoperiod and photoperiod-temperature studies and growth tailoring compounds and their use with floricultural crops; Orchidaceae, Amryllidaceae and Zingiberaceae. *Mailing Add:* Dept Environ Hort Univ Fla Gainesville FL 32611

SHEEHAN, WILLIAM C, b Macon, Ga, Oct 31, 25; m 74; c 5. ORGANIC CHEMISTRY. *Educ:* Mercer Univ, AB, 49; Inst Textile Technol, MS, 51; Univ Tenn, PhD(org chem), 56. *Prof Exp:* Res chemist, Bibb Mfg Co, Ga, 49 & 51-53; instr chem, Univ Tenn, 55; res chemist, E I du Pont de Nemours & Co, Va, 56-59; head textile sect, Southern Res Inst, 59-62, asst head phys sci div, 62-64, head polymer div, 64-65; dir fiber res, Phillips Petrol Co, Okla, 65-70, dir res, 70-73, tech vpres, 73-79, vpres mkt, Phillips Fibers Corp, 79-86, mem bd dirs, 75-86; CONSULT, 87- *Concurrent Pos:* Mem comt textile functional finishing, Nat Res Coun, 65-74; fel, Inst Textile Technol & Univ Tenn. *Honors & Awards:* New Tech Prod Award, Indust Res, 63. *Mem:* Am Chem Soc; Am Mgt Asn; AAAS. *Res:* Fiber and polymer chemistry; textile finishing and auxiliaries. *Mailing Add:* Ten Skipper Keo-Wee Key Salem SC 29676

SHEEHAN, WILLIAM FRANCIS, b Chicago, Ill, Oct 19, 26; m 53; c 7. PHYSICAL CHEMISTRY, QUANTUM CHEMISTRY. *Educ:* Loyola Univ, Ill, BS, 48; Calif Inst Technol, PhD(chem), 52. *Prof Exp:* Chemist, Shell Develop Co, 52-55, chmn dept, 72-79; PROF CHEM, SANTA CLARA UNIV, 55- *Concurrent Pos:* Sabbatical leaves, Univ Sussex, Louis Pasteur Strasbourg Univ. *Mem:* Sigma Xi. *Res:* Quantum chemistry; structural chemistry; valence; thermodynamics; quantum chemistry. *Mailing Add:* Dept Chem Santa Clara Univ Santa Clara CA 95053

SHEEHE, PAUL ROBERT, b Buffalo, NY, Dec 8, 25; m 48; c 5. EPIDEMIOLOGY. *Educ:* Univ Buffalo, BSBA, 48, MBA, 54; Univ Pittsburgh, ScD(biostatist), 59. *Prof Exp:* Statistician, Erie County Health Dept, NY, 50-52; teaching fel statist, Univ Buffalo, 52-54; statistician, Pratt & Letchworth, 54-57; assoc biostatistician, Roswell Park Mem Inst, 59-65; assoc prof biostatist, 65-69, PROF PREV MED, STATE UNIV NY UPSTATE MED CTR, 69- *Mem:* Fel Am Col Epidemiol; Am Epidemiol Soc. *Res:* Statistics; epidemiol. *Mailing Add:* Dept Prev Med SUNY Health Sci Col Med 750 E Adams St Syracuse NY 13210

SHEEHY, THOMAS W, b Columbia, Pa, May 20, 21; m 44; c 4. MEDICINE, HEMATOLOGY. *Educ:* St Vincent Col, BS, 47; Syracuse Univ, MD, 51; Baylor Col Med, MS, 55; Am Bd Internal Med, dipl, 58. *Prof Exp:* From intern to resident med, Brooke Gen Hosp, US Army, 51-55, asst chief, Walter Reed Gen Hosp, 56-59, chief med div & hemat, Army Trop Res Lab, PR, 59-62, chief dept gastroenterol res, Walter Reed Army Inst Res, 62-65, med consult, Vietnam, 65-66, chief gen med & dir educ & med res, Walter Reed Gen Hosp, 66-67; prof med & assoc dir, div nutrit & clin res, Med Ctr, Univ Ala, Birmingham, 67-72, co-chmn dept med, 69-88; chief med serv, Vet Admin Med Ctr, Birmingham, 69-88; RETIRED. *Concurrent Pos:* Studentship, Walter Reed Army Inst Res, 55-56, fel hemat, 56-57; from asst prof to assoc prof, Sch Med, Univ PR, 59-62; assoc prof, George Washington Univ, 63-65; mem hemat study sect, NIH, 62-65 & 71-75. *Mem:* Am Fedn Clin Res; Am Soc Hemat; Asn Mil Surg US; fel Am Col Physicians; Am Gastroenterol Asn; Sigma Xi; Asn Am Physicians. *Res:* Study of folic acid metabolism, minimal daily requirements; gastroenterology, small bowel metabolism, absorption, function, enzymes; tropical disease, malaria, scrub typhus, tropical sprue. *Mailing Add:* Dept of Med Univ of Ala Med Ctr Birmingham AL 35294

SHEELER, JOHN B(RIGGS), b Anita, Iowa, Oct 25, 21; m 45; c 5. CHEMICAL ENGINEERING, CIVIL ENGINEERING. *Educ:* Iowa State Univ, BS, 50, PhD(chem & soil eng), 56. *Prof Exp:* Res assoc, Iowa State Univ, 50-56, asst prof civil eng, 56-59, assoc prof civil & chem eng, 59-83; RETIRED. *Concurrent Pos:* Mem, Hwy Res Bd, Nat Acad Sci-Nat Res Coun, 51. *Mem:* AAAS; Am Inst Chem Engrs; Am Chem Soc. *Res:* Soil mechanics, engineering and stabilization; physico-chemical phenomena in soils. *Mailing Add:* 505 Bel Aire Dr Marshalltown IA 50158

SHEELEY, EUGENE C, b Tiffin, Ohio, Jan 4, 33. AUDIOLOGY. *Educ:* Heidelberg Col, BA, 54; Western Reserve Univ, MA, 55; Univ Pittsburgh, PhD(audiol), 64. *Prof Exp:* Audiologist, Cincinnati Speech & Hearing Ctr, 55-60; asst res audiol, Univ Pittsburgh, 61 & 63-64; dir hearing test & child study ctr, NMex Sch Deaf, 64-67; PROF DEPT COMMUNICATIVE DISORDERS, UNIV ALA, TUSCALOOSA, 67- *Concurrent Pos:* Asst clin audiol, Univ Pittsburgh, 62; spec instr, Univ Eastern NMex, 65-67; consult, Partlow State Sch & Hosp, Tuscaloosa, 68-74, chmn commun skills comt, 74-82; consult hearing conserv progs, Ala Industs, 72-86. *Mem:* Acad Rehabilitative Audiol; Acoust Soc Am; Am Speech-Language-Hearing Asn; A G Bell Asn Deaf; Am Auditory Soc; Am Acad Audiol. *Res:* Central auditory masking; noise exposure. *Mailing Add:* Dept Commun Dis Univ Ala Box 870242 Tuscaloosa AL 35487-0242

SHEELEY, RICHARD MOATS, medicinal chemistry, food chemistry; deceased, see previous edition for last biography

SHEELY, W(ALLACE) F(RANKLYN), b Albany, NY, Nov 28, 31; m 55; c 2. METALLURGICAL ENGINEERING. *Educ:* Rensselaer Polytech Inst, BMetE, 53, MMetE, 56, PhD, 57. *Prof Exp:* Res assoc, Rensselaer Polytech Inst, 53-57; res metallurgist, Union Carbide Metals Co, 57-60; sr metallurgist, Div Res, USAEC, 60-68; sr staff engr, Chem & Metall Div, Pac Northwest Labs, Battelle Mem Inst, 68-69, assoc div mgr, 69-70; mgr appl res, Mat & Technol Dept, 70-76, mgr, Chem Eng Dept, 76-80, mgr Mat Technol, 80-82, ASST MGR TECHNOL, WESTINGHOUSE-HANFORD CO, 82- *Concurrent Pos:* Dept Com sci & technol fel, 65-66. *Mem:* Am Soc Metals; Am Nuclear Soc. *Res:* Plasticity and mechanical properties; nuclear metallurgy and chemistry. *Mailing Add:* 2900 S Garfield St Kennewick WA 99337

SHEEN, SHUH-JI, b Wookiang, China, Mar 21, 31; m 59; c 5. PLANT GENETICS. *Educ:* Chung Hsing Univ, Taiwan, BS, 53; NDak State Univ, MS, 58; Univ Minn, PhD(plant genetics), 62. *Prof Exp:* Asst agron, Chung Hsing Univ, Taiwan, 54-56; asst prof biol, Hanover Col, 62-66; from asst prof to assoc prof agron, 66-74, assoc prof, 74-79, PROF PLANT PATH, UNIV KY, 79- *Concurrent Pos:* NSF acad res exten grant, 65-66; USDA contract grants tobacco & health probs, 67-82. *Honors & Awards:* Hon Prof, Shandong Agr Univ, China, 87. *Mem:* AAAS; Am Chem Soc; Inst Food Technologists; Am Phytopath Soc; Soc Green Veg Res. *Res:* Plant biomass utilization with specific interest on the isolation and modification of leaf protein from tobacco and soybean; formulation of safer tobacco products. *Mailing Add:* Dept Plant Pathol Agr Sci Ctr N Univ Ky Agr Sci Ctr N Lexington KY 40506

SHEER, M(AXINE) LANA, b Brooklyn, NY, June 14, 45; m 78. PHYSICAL CHEMISTRY, POLYMER CHEMISTRY. *Educ:* Emory Univ, BS, 65, MS, 67, PhD(phys chem), 69. *Prof Exp:* Res chemist, Stauffer Chem Co, 69-73; res chemist, E I du Pont de Nemours & Co, Inc, 73-77, tech specialist, 77-79, prod specialist, 79-80, financial anal, 81-83, market res, 83-85, market develop, 85-90, ELEC PROGS MGR, E I DU PONT DE NEMOURS & CO, INC, 91- *Mem:* Soc Advan Mat Process Eng; Am Chem Soc; Sigma Xi. *Res:* Nuclear magnetic resonance; infrared and ultraviolet spectroscopy; computer applications to molecular spectral theory; Raman spectroscopy; applications of spectroscopy to polymeric materials; development of thermoplastic engineering materials; process development for injection molded engineering plastics; market and product management of engineering plastics; financial forecasting, advanced composites and engineering polymers market development. *Mailing Add:* E I du Pont de Nemours BMP 18 Wilmington DE 19880-0018

SHEERAN, PATRICK JEROME, b Meade Co, Ky, Aug 29, 42; m 63; c 4. PESTICIDE CHEMISTRY. *Educ:* Bellarmine Col, BS, 64; Univ Vt, PhD(chem), 68. *Prof Exp:* RES SUPERVR PROCESS DEVELOP, BIOCHEM DEPT, EXP STA, E I DU PONT DE NEMOURS & CO, INC, 68- *Res:* Organic synthesis involving 4-8 membered heterocycles; biologically active organic compounds. *Mailing Add:* Church Hill Rd Landenberg PA 19350

SHEERAN, STANLEY ROBERT, b Elizabeth, NJ, Nov 19, 16; m 43; c 3. CHEMICAL MANAGEMENT, MANUFACTURING MANAGEMENT. *Educ:* St Vincent Col, BS, 38; Univ Notre Dame, MS, 39, PhD(org chem), 41. *Prof Exp:* Chemist, E I du Pont de Nemours & Co, 41-42, prod supvr, 42-44, supvr prod develop, 44-45, supvr tech sales develop, 45-46, mgr tech sales & develop, Del, 46-51, mgr opers, NY, 51-52, mgr Midwest sales develop, 53-58, mgr chem develop, Del, 59-63; vpres mkt chem div, Archer Daniels Midland Co, Minn, 63-66; vpres, Tenneco Chem Inc, 66-68, sr vpres, 68-70, exec vpres, 70-71, vpres, 71-81; RETIRED. *Concurrent Pos:* Pres, Sheeran Assoc Inc, 81- *Mem:* AAAS; fel Am Inst Chem; Am Mgt Asn; Mfg Chem

Asn; Am Chem Soc. *Res:* Acetylene separation; polyvinyl alcohol and acetate resins; solubility and process of acetylene; cyanide chemistry; oxidation with peroxides; protective coating resins; plastics; industrial chemicals and specialities; paper and textile chemicals; management. *Mailing Add:* Box 1754 New York NY 10008

SHEERS, WILLIAM SADLER, b Pittsburgh, Pa, Oct 28, 48. MEDICAL PHYSICS. *Educ:* Washington & Jefferson Col, BA, 71; WVa Univ, MS, 75; Univ NMex, PhD(physics), 84. *Prof Exp:* Res fel, Baylor Col Med, 83-85; ASST PROF PHYSICS, WASHINGTON & JEFFERSON COL, 85- *Concurrent Pos:* Consult. *Mem:* Am Phys Soc. *Res:* Atomic/Molecular Physics; medical physics. *Mailing Add:* Dept Physics Washington & Jefferson Col Washington PA 15301

SHEETS, DONALD GUY, b Mt Vernon, NY, Nov 20, 22; m 50. CHEMISTRY. *Educ:* Cent Col, Mo, AB, 44; Univ Mich, MS, 47, PhD(pharmaceut chem), 50. *Prof Exp:* Lab asst chem, Cent Col, Mo, 43-44, instr physics, 44; res chemist, Monsanto Chem Co, 44-45; asst prof anal chem, Miss State Col, 49-50; prof gen chem, St Louis Col Pharm, 50-62; PROF GEN & PHYS CHEM, US NAVAL ACAD, 62- *Mem:* Am Chem Soc. *Res:* Reactions of phthalic anhydride; derivatives of thiophene; derivatives of thianaphthene. *Mailing Add:* Dept Indust Technol Lansing Community Col PO Box 40010 Lansing MI 48902

SHEETS, GEORGE HENKLE, b Washington Court House, Ohio, Apr 22, 15; m 41; c 2. PAPER CHEMISTRY, CHEMICAL ENGINEERING. *Educ:* Ohio State Univ, BChE, 37; Inst Paper Chem, MS, 39, PhD(chem), 41. *Prof Exp:* Develop engr res & develop, Mead Corp, 41-50, div mgr oper mgt, 50-60, managing dir, 60-61, exec vpres corp mgt, 62-80; RETIRED. *Mem:* Tech Asn Pulp & Paper Indust (pres, 69-70); Can Pulp & Paper Asn. *Res:* Pulping; papermaking. *Mailing Add:* 60 Harmon Terr Dayton OH 45419

SHEETS, HERMAN E(RNEST), b Dresden, Ger, Dec 24, 08; nat US; m 42; c 6. MECHANICAL & OCEAN ENGINEERING. *Educ:* Dresden Tech Univ, dipl, 34; Prague Tech Univ, DrTechSci(appl mech), 36. *Prof Exp:* Dir res, St Paul Eng & Mfg Co, Minn, 42-44; proj engr, Elliott Co, Pa, 44-46; eng mgr, Goodyear Aircraft Corp, Ohio, 46-53; dir res & develop, Elec Boat Div, Gen Dynamics Corp, 53-66, vpres eng & res, 66-69; prof ocean eng & chmn dept, 69-80, EMER PROF, UNIV RI, 80- *Concurrent Pos:* Mem marine bd, Nat Acad Eng; mem, Nat Acad Sci-Nat Acad Eng sci & eng adv comt to Nat Oceanic & Atmospheric Admin; consult engr, 80- *Mem:* Nat Acad Eng; fel Am Soc Mech Engrs; Soc Naval Archit & Marine Engrs; Am Soc Naval Engrs-Am Inst Aeronaut & Astronaut. *Res:* Ocean engineering systems; hydrodynamics. *Mailing Add:* 87 Neptune Dr Groton CT 06340

SHEETS, RALPH WALDO, b Point Cedar, Ark, Apr 18, 35; m; c 2. ENVIRONMENTAL CHEMISTRY. *Educ:* Henderson State Col, BS, 66; Univ Ark, MS, 69, PhD(phys chem), 71. *Prof Exp:* Fel surface chem, Ames Lab, AEC, 70-71; from asst prof to assoc prof, 71-82, PROF CHEM, SOUTHWEST MO STATE UNIV, 82- *Concurrent Pos:* Sci translr Russ, Consults Bur, Plenum Publ Corp, 71-, Faraday Press Inc, 71-72 & Allerton Press Inc, 72-76. *Mem:* Am Chem Soc; Am Nuclear Soc; Air & Waste Mgt Asn; Water Pollution Control Fedn; Sigma Xi. *Res:* Chemical characterization of natural water systems; indoor air pollution; infrared and ultraviolet spectroscopy of adsorbed species; atmospheric radioactivity. *Mailing Add:* Dept of Chem Southwest Mo State Univ Springfield MO 65804-0089

SHEETS, RAYMOND FRANKLIN, internal medicine; deceased, see previous edition for last biography

SHEETS, ROBERT CHESTER, b Marion, Ind, June 7, 37; m 59; c 3. METEOROLOGY. *Educ:* Ball State Teachers Col, BS, 61; Univ Okla, MS, 65, PhD(meteorol), 72. *Prof Exp:* Chief forecaster, US Air Force Air Weather Serv, Ft Knox, Ky, 61-64; asst meteorol, Univ Okla, 64-65; res meteorologist, Nat Oceanic & Atmospheric Admin, US Dept Com, 65-74, supvry meteorologist, Environ Res Lab, Nat Hurricane & Exp Meteorol Lab & actg chief Hurricane Group, 75-88; DIR, NAT HURRICANE CTR, 88- *Mem:* Am Meteorol Soc; Weather Modification Asn. *Res:* Experimental and theoretical research on the formation, motion, intensity, scale interactions and structure of hurricanes and other tropical storms; emphasis on hurricane modification schemes and evaluation techniques. *Mailing Add:* Nat Hurricane Ctr 1320 S Dixie Hwy Rm 631 Coral Gables FL 33146

SHEETS, THOMAS JACKSON, b Asheville, NC, Dec 11, 26; m 52; c 2. AGRICULTURAL CHEMISTRY, PESTICIDES. *Educ:* NC State Col, BS, 51, MS, 54; Univ Calif, PhD, 59. *Prof Exp:* Res instr, NC State Col, 51-54; res agronomist, Agr Res Serv, USDA, Calif, 54-59, plant physiologist, Delta Br Exp Sta, Miss, 59-60 & Md, 60-65; assoc prof entom & crop sci, 65-69, PROF ENTOM, CROP SCI & HORT SCI, NC STATE UNIV, 69-, TOXICOL, 89- *Concurrent Pos:* Ed, Weed Sci, 71-73; Ed-in-chief, Weed Sci Soc Am 74-78; dir, Pesticide Res Lab, 79. *Mem:* Coun Agr Sci & Technol; fel Weed Sci Am (vpres, 80, pres-elect, 81, pres, 82); Sigma Xi; Sod Weed Sci Soc. *Res:* Movement, persistence and modes of detoxification of pesticides; pesticide residues. *Mailing Add:* Pesticide Residue Res Lab NC State Univ PO Box 8604 Raleigh NC 27650

SHEETZ, DAVID P, b Colebrook, Pa, Dec 4, 26; m 46; c 3. PHYSICAL CHEMISTRY. *Educ:* Lebanon Valley Col, BS, 48; Univ Nebr, MS, 51, PhD(chem), 52. *Prof Exp:* Res chemist, 52-56, proj leader, 56-59, group leader, 59-65, asst lab dir, 65-66, lab dir, 66-67, asst dir res & develop, Midland Div, 67-71, dir res & develop, Mich Div, 71-78, tech dir, 78, vpres & dir, 78-80, corp dir res & develop, 80-85, SR VPRES & CHIEF SCIENTIST, DOW CHEM CO, 86- *Concurrent Pos:* Mem, Coun Chem Res. *Mem:* Am Chem Soc; Sigma Xi; Am Inst Chemists; Indust Res Inst; Soc Chem Indust. *Res:* Polymer, colloid and organic chemistry. *Mailing Add:* Dow Chem Co 1776 Bldg Midland MI 48674

SHEETZ, MICHAEL PATRICK, b Dec 11, 46; m; c 3. MUSCLE BIOCHEMISTRY, CELL MOTILITY. *Educ:* Calif Inst Technol, PhD(chem), 72. *Prof Exp:* PROF CELL BIOL, SCH MED, WASH UNIV, 85- *Concurrent Pos:* Univ Conn Health Ctr, 75-85. *Mem:* Am Soc Cell Biol; Biophys Soc. *Res:* Studying the molecular basis of intracellular organelle motility and the biophysical basis of motor enzyme function. *Mailing Add:* Dept Cell Biol Duke Univ Med Ctr 468 Sands Bldg Res Dr Box 3011 Durham NC 27710

SHEFER, JOSHUA, b Leipzig, Germany, Nov 1, 24; US citizen; m 50; c 3. ELECTRICAL ENGINEERING. *Educ:* Israel Inst Technol, BSc, 48; Univ London, PhD(elec eng), 55. *Prof Exp:* Res engr electronics res lab, Israeli Ministry Defence, 48-52 & 56-58; sci attache, Israeli Embassy, London, 58-60; res fel, Harvard Univ, 60-62; mem tech staff, Bell Tel Labs, 62-67; mem tech staff, RCA Labs, 67-86, STAFF ENGR, GE-ASTRO, 86- *Concurrent Pos:* Lectr microwave theory & technol, Israel Inst Technol, 57-58; consult, Foxbro Co, Mass, 61-62. *Mem:* Inst Elec & Electronics Engrs; Sigma Xi. *Res:* Electromagnetic theory; antenna and propagation studies; microwaves; mobile radio; display devices; television systems; satellite communications. *Mailing Add:* 223 Gallup Rd Princeton NJ 08540

SHEFER, SARAH, b Tel Aviv, Israel, Oct 6, 26; US citizen; m; c 3. LIPID METABOLISM, ARTERIOSCLEROSIS. *Educ:* London Univ, PhD(biochem), 56. *Prof Exp:* Assoc prof, 79-83, PROF DIGESTIVE DIS, MED SCH, UNIV MED & DENT NJ, 83- *Mem:* Am Heart Asn; Am Inst Nutrit; Am Soc Biol Chemists; AAAS. *Res:* Lipid metabolism. *Mailing Add:* Dept Med Div Digestive Dis & Nutrition MSB-H-534 Univ Med Dent NJ Med Sch 185 S Orange Ave Newark NJ 07103-2714

SHEFFER, ALBERT L, b Lewistown, Pa, Aug 7, 29; m 54; c 4. CLINICAL IMMUNOLOGY, ALLERGY. *Educ:* Franklin & Marshall Col, BS, 52; George Washington Univ, MD, 56. *Prof Exp:* Fel pulmonary dis, Grad Hosp & Henry Phipps Inst, Univ Pa, 57-58; res med, Grad Hosp, Univ Pa, Philadelphia, 56-60; fel appl immunol & allergy, Med Ctr, Temple Univ, 61; attend physician, Rockefeller Univ Hosp, 61-62; guest investr, Rockefeller Univ, 62; clin prof, Harvard Med Sch, 66; dir allergy clin, Beth Israel Hosp, Boston, 69-85; DIR ALLERGY CLIN, BRIGHAM & WOMEN'S HOSP, BOSTON, 71- *Concurrent Pos:* Allergy sect chief, New Eng Deaconess Hosp, Boston, 72-; asst allergy, Children's Hosp Med Ctr, Boston, 74-; consult, US Pharmacopeia, 75-; dir allergy training prog, Brigham & Women's Hosp, Boston, 76- *Mem:* Am Acad Allergy; Am Fed Clin Res. *Res:* Pathogenesis and treatment of asthma, anaphylaxis, urticaria, and angioedema. *Mailing Add:* 110 Francis St Boston MA 02215

SHEFFER, HOWARD EUGENE, b Schenectady, NY, Oct 3, 18; m 44; c 6. ORGANIC CHEMISTRY. *Educ:* Union Col, NY, BS, 39; Rensselaer Polytech Inst, MS, 40; Cornell Univ, PhD(org chem), 43. *Prof Exp:* Res chemist, Carbide & Carbon Chem Co, WVa, 43-45; from asst prof to prof chem, Union Col, NY, 45-83; RETIRED. *Concurrent Pos:* Consult, Schenectady Varnish Co, 47-; NSF fac fel, Univ Del, 61-62; Fulbright prof, 68-69, Neste Oil Found grant, Tech Univ Helsinki, 75-76. *Mem:* Am Chem Soc. *Res:* Organics; colloids; wire enamels; insulation varnishes; condensation polymerization; cationic polymerization; osmometry. *Mailing Add:* 6518 Draw Lane Sarasota FL 34238

SHEFFER, RICHARD DOUGLAS, b Portland, Ind, Apr 19, 42; m 69; c 1. CYTOGENETICS, BIOSYSTEMATICS. *Educ:* Purdue Univ, BS, 70; Univ Hawaii, PhD(hort), 74. *Prof Exp:* Res assoc, Univ Hawaii, 74-75; asst prof, Univ NB, 75-76; instr, Montclair State Col, 76-77; asst prof, 77-81, ASSOC PROF GENETICS, IND UNIV NORTHWEST, 81- *Concurrent Pos:* Res assoc, Mo Bot Garden, 79- *Mem:* AAAS; Am Genetics Asn; Am Bot Soc; Int Asn Plant Taxon; Am Genetics Soc. *Res:* Cytogenetics and cytotaxonomy of the genus anthurium. *Mailing Add:* Dept Biol Ind Univ Northwest Gary IN 46408

SHEFFI, YOSEF, b Jerusalem, Israel. TRANSPORTATION SYSTEMS, OPERATIONS RESEARCH. *Educ:* Israel Inst Technol, BSc, 75; Mass Inst Technol, SM, 77, PhD(transp), 78. *Prof Exp:* Asst prof, 78-80, ASSOC PROF TRANSP SYST, MASS INST TECHNOL, 80- *Concurrent Pos:* Sr oper res analyst, Transp Syst Ctr, US Dept Transp, 78-79; consult, 78-81. *Mem:* Transp Res Bd; Oper Res Soc Am; Inst Transp Engrs. *Res:* Transportation systems analysis; travel demand models; network analysis; performance of transportation facilities. *Mailing Add:* Dept Civil Eng Mass Inst Technol 77 Massachusetts Ave Cambridge MA 02139

SHEFFIELD, HARLEY GEORGE, b Detroit, Mich, Jan 10, 32; m 59; c 2. MEDICAL PARASITOLOGY, ELECTRON MICROSCOPY. *Educ:* Wayne State Univ, BS, 53, MS, 58; La State Univ, PhD(med parasitol), 62. *Prof Exp:* Res biologist, Parke, Davis & Co, 58-59; from scientist to sr scientist, 62-73, SCIENTIST DIR, MICROBIOL & INFECT DIS PROG, NAT INST ALLERGY & INFECTIOUS DIS, NIH, 73- *Mem:* Am Soc Parasitol; Am Soc Trop Med & Hyg; AAAS. *Res:* Electron microscopy of parasitic protozoa, especially toxoplasma and related organisms; electron microscopy of parasitic nematode intestine and other tissues. *Mailing Add:* MIDP Nat Inst Allergy & Infectious Dis NIH Rm 737 Westwood Bldg Bethesda MD 20892

SHEFFIELD, JOEL BENSON, b Brooklyn, NY, Dec 30, 42; m 65; c 1. CELL BIOLOGY, DEVELOPMENTAL BIOLOGY. *Educ:* Brandeis Univ, AB, 63; Univ Chicago, PhD(biol), 70. *Prof Exp:* Guest investr virol, Rockefeller Univ, 63-64; fel cell membranes, Dutch Cancer Inst, 70-71; asst mem virol, Inst Med Res, 71-77; from asst prof to assoc prof, 77-89, PROF BIOL, TEMPLE UNIV, 89-, DEPT CHAIR, 90- *Concurrent Pos:* Res fel, Int Agency Res Cancer, WHO, 70; Nat Cancer Inst fel, 71; fac assoc, Rutgers Univ, 74- *Mem:* AAAS; Am Soc Cell Biol; Asn Res Vision & Ophthal; Soc Neurosci. *Res:* Structure and biogenesis of cellular and viral membranes; retinal development. *Mailing Add:* Dept Biol Temple Univ Philadelphia PA 19122

SHEFFIELD, JOHN, b Purley, Eng, Dec 15, 36; m 64; c 2. PLASMA PHYSICS. *Educ:* London Univ, BSc, 58, MSc, 62, PhD(plasma physics), 66. *Prof Exp:* Exp officer plasma physics, Harwell Lab, UK Atomic Energy Authority, 58-61, Culham Lab, 61-66; asst prof plasma physics, Univ Tex, Austin, 66-71; prin sci officer fusion res, Culham Lab, UK Atomic Energy Authority, 71-77; assoc dir, 77-88, DIR, FUSION ENERGY DIV, OAK RIDGE NAT LAB, 88- *Honors & Awards:* Outstanding Achievement Award, Am Nuclear Soc, 90. *Mem:* Fel Am Phys Soc; Am Nuclear Soc. *Res:* Magnetic fusion; magnetic confinement schemes; diagnostics and technology for fusion. *Mailing Add:* Oak Ridge Nat Lab PO Box 2009 Oak Ridge TN 37831-8070

SHEFFIELD, L THOMAS, b Montgomery, Ala, Oct 25, 28; m 54; c 2. MEDICINE. *Educ:* Emory Univ, BA, 49; Univ Ala, MD, 54. *Prof Exp:* Intern, Michael Reese Hosp, Chicago, Ill, 54-55; asst med res, Univ Hosp & Vet Admin Hosp, Birmingham, Ala, 57-59; NIH cardiovasc res fel, Mass Mem Hosp & Sch Med, Boston Univ, 59-60 & Med Ctr, Univ Ala, Birmingham, 60-62; from instr to assoc prof, 62-73, PROF MED, UNIV ALA, BIRMINGHAM, 73-, DIR ECG LAB & EXERCISE LAB, 62- *Concurrent Pos:* Attend physician, Univ Hosp, Birmingham, 63-; mem, Am Heart Asn. *Mem:* AAAS; Am Fedn Clin Res; AMA; fel Am Col Cardiol. *Res:* Clinical cardiology; exercise electrocardiography; work physiology; computer-aided diagnosis. *Mailing Add:* Dept Med-Cardiol Sch Med Lyons Harrison Res Bldg No 318 Univ Ala Univ Sta Birmingham AL 35294

SHEFFIELD, RICHARD LEE, b Dayton, Ohio, Sept 22, 50; m 79; c 3. FREE-ELECTRON LASERS, ACCELERATOR PHYSICS. *Educ:* Wright State Univ, BS, 72; Mass Inst Technol, PhD(physics), 78. *Prof Exp:* Staff mem high energy, high-density physics, Los Alamos Nat Lab, 78-82, free electron laser technol, 82-85, dep group leader, 85-89, GROUP LEADER, ACCELERATOR THEORY & FREE ELECTRON LASER TECHNOL, LOS ALAMOS NAT LAB, 89- *Concurrent Pos:* Adj prof, Physics Dept, Wright State Univ, 86-; partner, Beam Energetics, 86-; consult, Felcorp, 87-89; lectr, US Accelerator, 89; prin investr, Advan FEL Initiative, Los Alamos Nat Lab, 90-; adv, UV/FEL Adv Panel, Brookhaven Nat Lab, 91- *Honors & Awards:* R&D 100 Award, R&D 100 Mag, 88; Strategic Defense Tech Achievements Award, Strategic Defense Preparedness Asn, 89. *Mem:* Am Phys Soc. *Res:* Advanced free-electron laser technology initiative; high-brightness electron linac for advanced free-electron laser. *Mailing Add:* MS H825 Los Alamos Nat Lab Los Alamos NM 87545

SHEFFIELD, ROY DEXTER, b Dorsey, Miss, Sept 5, 22; m 46. MATHEMATICS. *Educ:* Univ Miss, BA, 48, MA, 49; Univ Tenn, PhD(math), 56. *Prof Exp:* Asst prof math, Univ Miss, 51-56; sr nuclear engr, Gen Dynamics/Convair, 56-57; prof math, Univ Miss, 57-63; prof, Miss State Univ, 63-71; chmn dept math, Univ Miss, 71-77, prof, 71-80; RETIRED. *Concurrent Pos:* Consult, Gen Dynamics/Convair, 57-61. *Mem:* Am Math Soc; Math Asn Am. *Res:* Functional analysis; linear algebra; nonlinear programming. *Mailing Add:* 201 St Andrews Circle Oxford MS 38655

SHEFFIELD, WILLIAM JOHNSON, b Nashua, NH, May 9, 19; m 55; c 3. PHARMACY. *Educ:* Univ NC, BS, 42, MS, 49, PhD(pharm), 54. *Prof Exp:* From asst prof to assoc prof, 52-68, prof pharm, Col Pharm, 68-88, asst dean col, 56-58 & 68; RETIRED. *Mem:* AAAS; Acad Pharmaceut Sci; Am Pharmaceut Asn. *Mailing Add:* 1610 Blanchard Dr Round Rock TX 78681

SHEFFY, BEN EDWARD, b Luxemburg, Wis, Mar 12, 20; m 48; c 2. NUTRITION, MICROBIOLOGY. *Educ:* Univ Wis, BS, 48, MS, 50, PhD, 51. *Prof Exp:* CASPARY PROF NUTRIT & ASST DIR J A BAKER INST ANIMAL HEALTH, CORNELL UNIV, 51- *Concurrent Pos:* Guggenheim fel, Cambridge Univ, 59-60; NIH spec fel, Univ Munich, 66-67; nutrit consult, NY Zool Soc; consult, Joint Comn Rural Reconstruction, Republic China, 74-75. *Mem:* Brit Nutrit Soc; Am Asn Lab Animal Sci. *Res:* Nutrition and disease interrelationships. *Mailing Add:* Vet Baker Inst Animal Health Cornell Univ Ithaca NY 14853

SHEFT, IRVING, b Chicago, Ill, July 2, 19; m 44; c 2. ATMOSPHERIC CHEMISTRY. *Educ:* Univ Chicago, SB, 41, SM, 44. *Prof Exp:* Res asst, Univ Chicago, 41-42, Metall Lab, 42-46; SCIENTIST, ARGONNE NAT LAB, 46- *Mem:* Fel AAAS; Am Chem Soc; Sigma Xi. *Res:* Development of analytical procedure to separate atmospheric components to permit atmospheric modeling. *Mailing Add:* 411 N Lombard Ave Oak Park IL 60302

SHEFTER, ELI, b Philadelphia, Pa, Sept 10, 36; div; c 2. DRUG DELIVERY, PHARMACEUTICS. *Educ:* Temple Univ, BSc, 58; Univ Wis, PhD(phys pharm & chem), 63. *Prof Exp:* Nat Inst Gen Med Sci fel, 64-65; asst prof, Sch Pharm, State Univ NY Buffalo, 66-69, assoc prof, 69-81; mgr, Drug Delivery Systs, DuPont, 81-87; sr scientist, Pharmacol Res & Develop, Genentech Inc, 87-89; PROF PHARMACEUT, SCH PHARM, UNIV COLO, 90- *Concurrent Pos:* Pfeiffer fel, Am Found Pharmaceut Educ, 72; Norwegian Sci Found fel, 80-81. *Mem:* Fel Acad Pharmaceut Sci; AAAS; Am Crystallog Asn; Am Chem Soc; Am Asn Pharmaceut Scientists. *Res:* Correlations of structure and pharmacological activity; crystallographic studies on nucleic acid components and complexes of these fragments; phase transformations of solid pharmaceuticals; drug formulation; design of delivery systems for protein. *Mailing Add:* Sch Pharm 297 Univ Colo Boulder CO 80309-0297

SHEHADI, WILLIAM HENRY, b Providence, RI, June 30, 06; m 55; c 2. RADIOLOGY. *Educ:* Am Univ Beirut, DDS, 27, MD, 31; Am Bd Radiol, dipl, 41. *Prof Exp:* Instr anat, Schs Med & Dent, Am Univ Beirut, 31-33, asst radiologist, 32-39; asst prof radiol, Sch Med, Univ Vt, 41-42; dir, Mt Vernon Hosp, NY, 42-48; prof radiol & dir dept, New York Polyclin Med Sch, 48-60; dir dept, United Hosp, Port Chester, NY, 61-67; DIR, DEPT RADIOL, WESTCHESTER COUNTY MEDICAL CTR, VALHALLA, NY, 68-; PROF RADIOL, NEW YORK MED COL, 68- *Concurrent Pos:* Trainee radiol, Am Univ Beirut, 32-39; asst radiologist, Mary Fletcher Hosp, Burlington, Vt, 41-42; instr radiol, New York Med Col, 43-47; consult, St Joseph's Hosp, Yonkers, 43-71; trustee, Am Univ Beirut, 56-; consult, Food & Drug Admin, 75, Westchester County Med Ctr, Valhalla, NY, 76; chmn comt safety contrast media, Int Soc Radiol, 69. *Mem:* Am Roentgen Ray Soc; Radiol Soc NAm; AMA; fel Am Geog Soc; fel Am Col Radiol; Soc Nuclear Med. *Mailing Add:* 27 Byram Shore Rd Greenwich CT 06830

SHEIBLEY, FRED EASLY, b Cleveland, Ohio, Dec 18, 06. ORGANIC CHEMISTRY. *Educ:* Case Inst, BS, 31, PhD(org chem), 39; Univ Pa, MS, 37. *Prof Exp:* Asst org lab, Case Inst, 31-33; res chemist, Grasselli Chem Dept, E I DuPont de Nemours & Co, 33-35; asst instr, Univ Pa, 35-37; instr, Case Inst, 37-41; res chemist, Carborundum Co, NY, 41-43 & Battelle Mem Inst, 44-46; asst prof chem, Univ Ky, 46-47; res chemist, Gen Tire & Rubber Co, 47-49; chief chemist, Hefco Labs, Mich, 49-52; res chemist, B F Goodrich Chem Co, Ohio, 52-54 & Palmer Chem Co, 54-55; from asst prof to assoc prof chem, Cleveland State Univ, 55-72, emer assoc prof, 72-84; RETIRED. *Mem:* Am Chem Soc. *Res:* Synthesis of monomers; pure heterocyclic chemistry. *Mailing Add:* 1900 E 30 St Cleveland OH 44114

SHEID, BERTRUM, b Brooklyn, NY, Apr 19, 37. BIOCHEMISTRY. *Educ:* City Col New York, BS, 56; Brooklyn Col, MA, 60; Univ Conn, PhD(biochem), 65. *Prof Exp:* Fel biochem, Col Physicians & Surgeons, Columbia Univ, 65-67; asst prof path, Albert Einstein Col Med, 67-69; asst prof, 69-75, ASSOC PROF PHARMACOL, STATE UNIV NY, DOWNSTATE MED CTR, 75- *Mem:* AAAS; Am Chem Soc; Am Biol Scientists; NY Acad Sci; Sigma Xi. *Res:* Nucleic acid metabolism in normal and malignant tissues; experimental cancer chemotherapy. *Mailing Add:* Dept of Pharmacol State Univ NY Downstate Med Ctr Brooklyn NY 11203

SHEIKH, KAZIM, b Hyderabad, Pakistan, Sept 21, 36; Brit citizen; m 69; c 2. CANCER EPIDEMIOLOGY, ENVIRONMENTAL EPIDEMIOLOGY. *Educ:* Univ Karachi, Pakistan, MB & BS, 60; Royal Col Surgeons & Physicians, Eng, DIH, 73, FRCP. *Prof Exp:* Intern & resident, Nat Health Serv Gen Hosp, London & Southeast Eng, 60-70; med officer, Slough Indust Health Serv, Eng, 70-74; epidemiologist, Med Res Coun, UK, 74-78; med adv, BOC Ltd, London, 78-80; asst prof epidemiol, Sch Pub Health, Univ Mich, 81-87; RETIRED. *Concurrent Pos:* Consult prev med. *Mem:* Am Pub Health Asn; Am Heart Asn; Soc Epidemiologic Res; Am Occup Med Asn; Am Col Epidemiology; Am Col Prev Med; hon fel Royal Soc Physicians. *Res:* Medical care and community health research in stroke, and in chronic physical and mental disability; cancer epidemiology; occupational and general environmental epidemiology, and epidemiological methods. *Mailing Add:* 181 Smith Rd Charleston WV 25314

SHEINAUS, HAROLD, b New York, NY, Sept 5, 18; m 49; c 2. PHARMACY. *Educ:* City Col New York, BS, 39, Columbia Univ, BS, 49, MS, 51; Purdue Univ, PhD(pharm, pharmaceut chem), 55. *Prof Exp:* Instr pharm, Columbia Univ, 51-52; dir develop & process lab, Carroll Dunham Smith Pharmacol Co, 54-60; sr scientist, Warner-Lambert Res Inst, 61-63; dept head pharmaceut res & develop, 64-70, dir pharmaceut prod develop, 70-79, dir concept develop, Prod Div, Bristol-Myers Co, 79-81; RETIRED. *Concurrent Pos:* Vpres adv coun, County Off on Aging, 84-86; mem & chmn, Educ Comt, 83-91. *Mem:* Am Pharmaceut Asn; Acad Pharmaceut Sci; Sigma Xi. *Res:* Application of new developmental materials and techniques to pharmacy; aerosol pharmaceuticals; emulsions; sustained action formulations; preservatives; effervescent products; solubilization techniques. *Mailing Add:* 132 Wildwood Terr Watchung NJ 07060

SHEINESS, DIANA KAY, b Corpus Christi, Tex, Oct 1, 47. RETROVIRUSES, RETROVIRAL ONCOGENES. *Educ:* Univ Tex, Austin, BA, 67; Columbia Univ, MA, 73, PhD(cell biol), 74. *Prof Exp:* Fel genetics, Univ Edinburgh, Scotland, 74-75 & microbiol, Med Sch, Univ Calif, 76-81; ASST PROF BIOCHEM, LA STATE MED CTR, 81- *Mem:* Am Soc Microbiol; Sigma Xi. *Res:* Elucidating the function in normal cells of genes that have served as progenitors for retroviral oncogenes. *Mailing Add:* Dept Biochem La State Med Ctr 1901 Perdido St New Orleans LA 70112

SHEINGOLD, ABRAHAM, b New York, NY, Feb 17, 17; m 41; c 2. ELECTRONICS. *Educ:* City Col New York, BS, 36, MS, 37. *Prof Exp:* Instr high schs, NY, 36-43; instr elec commun, Mass Inst Technol, 43-46; from asst prof to assoc prof, 46-54, dean, Acad Admin, 77-82, PROF ELECTRONICS, US NAVAL POSTGRAD SCH, 54- *Mem:* Inst Elec & Electronics Engrs. *Mailing Add:* Dept Elec Eng Naval Postgrad Sch Code 615H Monterey CA 93940

SHEINGORN, MARK ELLIOT, b New York, NY, Dec 3, 44; m 72. MATHEMATICS. *Educ:* Dartmouth Col, AB, 65; Univ Wis-Madison, MS, 67, PhD(math), 70. *Prof Exp:* Nat Res Coun fel, Nat Bur Standards, 70-72; asst prof math, Hofstra Univ, 72-73; from asst prof to assoc prof, 73-82, PROF MATH, BARUCH COL, 82- *Concurrent Pos:* NASA fel, Univ Wis, 66-68; NSF grant, 73-84; vis mem, Inst Advan Study, Princeton, 74-75, 81 & 85-86; vis lectr, Univ Ill, Urbana, 78-79; vis prof, State Univ NY, Stony Brook, 82-83, Princeton Univ, 89-90; NAS exchange scientist, Polish Acad, Warsaw, 90. *Mem:* Am Math Soc. *Res:* Analytic number theory; Riemann surfaces. *Mailing Add:* 1200 Broadway New York NY 10001

SHEININ, ERIC BENJAMIN, b Chicago, Ill, Dec 23, 43; m 65, 76; c 5. PHARMACEUTICAL CHEMISTRY, ANALYTICAL CHEMISTRY. *Educ:* Univ Ill, Urbana, BS, 65; Univ Ill Med Ctr, PhD(pharmaceut chem), 71. *Prof Exp:* Res chemist anal & pharmaceut chem, 71-79, SUPVRY CHEMIST, FOOD & DRUG ADMIN, 79- *Concurrent Pos:* Adj prof chem, Montgomery Col, Rockville, Md. *Mem:* Am Chem Soc; Am Asn Pharm Sci; Asn Off Anal Chemists. *Res:* Use of nuclear magnetic resonance spectroscopy and mass spectrometry; development of methodology for the analysis of pharmaceutical preparations; laboratory evaluation of analytical methodology included in new drug applications. *Mailing Add:* Food & Drug Admin HFD-160 5600 Fishers Lane Rockville MD 20857

SHEININ, ROSE, b Toronto, Ont, May 18, 30; m 51; c 3. BIOCHEMISTRY, VIROLOGY. *Educ:* Univ Toronto, BSc, 51, MSc, 53, PhD(biochem), 56. *Hon Degrees:* DSc, Mt St Vincent Univ, 85; LHD & DSc, Acadia Univ, 87; DSc, Guelph Univ, 91. *Prof Exp:* Res assoc tumor virol, Ont Cancer Inst, 58-76; from asst prof to assoc prof med biophys, Univ Toronto, 67-75, chmn dept, 75-81, prof microbiol & parasitol, 75-81, prof microbiol, 81-89; PROF BIOL & VICE RECTOR ACAD, CONCORDIA UNIV, 89- *Concurrent Pos:* Fel, Brit Empire Cancer Campaign, 56-58; vis prof, Med Res Coun, 72, sci officer, 74-; Josiah Macy Jr fac scholar award, 81-82; vdean, Sch Grad Studies, Univ Toronto, 84-89. *Mem:* Can Biochem Soc (pres, 75); Can Soc Cell Biol (pres, 73); Am Soc Microbiol; fel Am Acad Microbiol; fel Royal Soc Can; Am Soc Virol. *Res:* Tumor virology; chromatin structure and replication; biochemical genetics; somatic cell genetics; women in medical sciences and medicine. *Mailing Add:* 7141 Sherbrooke St W Montreal PQ H4B 1R8 Can

SHEINSON, RONALD SWIREN, b Philadelphia, Pa, Dec 16, 42; m 68; c 2. PHYSICAL CHEMISTRY. *Educ:* Temple Univ, BA, 64; Mass Inst Technol, PhD(chem physics), 70. *Prof Exp:* RES CHEMIST, US NAVAL RES LAB, 70- *Mem:* Am Phys Soc; Sigma Xi; Am Chem Soc; Combustion Inst. *Res:* Combustion suppression; chemiluminescence; gas phase oxidation mechanisms; spectroscopy; electron paramagnetic resonance. *Mailing Add:* 809 N Belgrade Rd Silver Spring MD 20902-3245

SHEKELLE, RICHARD BARTEN, b Ventura, Calif, Mar 24, 33; m 52; c 5. EPIDEMIOLOGY. *Educ:* Univ Chicago, AB, 52, AM, 58, PhD(human develop), 62. *Prof Exp:* Res assoc psychiat, Univ Ill, Chicago, 59-61, res assoc prev med, 61-65, from instr to assoc prof, 65-74; from assoc prof to prof prev med, Rush-Presby-St Luke's Med Ctr, 74-83; PROF EPIDEMIOL, SCH PUB HEALTH, UNIV TEX, 83 - *Concurrent Pos:* Fel Am Heart Asn Coun Epidemiol. *Mem:* Soc Epidemiol Res; Am Epidemiol Soc. *Res:* Epidemiology of cardiovascular diseases and cancer. *Mailing Add:* Sch Pub Health Univ Tex Health Sci Ctr PO Box 20186 Houston TX 77225

SHELANSKI, MICHAEL L, b Philadelphia, Pa, Oct 5, 41; m 63; c 3. NEUROPATHOLOGY, NEUROSCIENCES. *Educ:* Univ Chicago, MD, 66, PhD(physiol), 67. *Prof Exp:* Intern path, Albert Einstein Col Med, 67-68, fel neuropath, 68-69, asst prof, 69-71; staff investr neurobiol, Lab Biochem Genetics, Nat Heart & Lung Inst, 71-73; Guggenheim fel, Inst Pasteur, Paris, 73-74; assoc prof neuropath, Harvard Med Sch, 74-78; prof pharmacol & chmn dept, Med Sch, NY Univ, 78-86; DELAFIELD PROF & CHMN, DEPT PATH, COL PHYSICIANS & SURGEONS, COLUMBIA UNIV, 87-, DIR, PATH SERV, PRESBYTERIAN HOSP, 87- *Concurrent Pos:* Nat Inst Neurol Dis & Stroke teacher-investr award, 70-71 & 73-74, asst prof in residence, Col, 71-73; mem, Neurol A Study Sect, NIH, 74-78 & Pharmacol Sci Study Sect, Nat Inst Gen Med Sci,86-; mem, Med & Sci Adv Bd, Alzheimers Dis & Related Dis Asn, 85-; chmn, Sci Adv Panel, NY Overhead Power Lines Proj, 81-86. *Mem:* Am Soc Cell Biol; Am Asn Neuropath; Am Soc Neurochem; Soc Neurosci. *Res:* Microtubule and neurofilaments; physical biochemistry of self-assembly; chemistry of senile and pre-senile dementias; neuronal differentiation. *Mailing Add:* Dept Path Columbia Univ 630 W 168th St New York NY 10032

SHELBURNE, JOHN DANIEL, b Washington, DC, Aug 27, 43; m 66; c 2. PATHOLOGY. *Educ:* Univ NC, Chapel Hill, AB, 66; Duke Univ, PhD(path), 71, MD, 72. *Prof Exp:* Intern, 72-73, asst prof, 73-78, ASSOC PROF PATH, MED CTR, DUKE UNIV, 78-; assoc dir, 73-77, DIR DIAGNOSTIC ELECTRON MICROS LAB, DUKE MED CTR & VET ADMIN HOSP, DURHAM, 77- *Concurrent Pos:* Assoc chief of staff res & develop & chief lab sci, VA Med Ctr, Durham, NC. *Mem:* Int Acad Path; Am Asn Pathologists; Electron Micros Soc Am; Microbeam Analysis Soc; Col Am Pathologists. *Res:* Lysosomes; autophagy; surgical pathology; x-ray microanalysis; ion microscopy. *Mailing Add:* Dept Path Box 3712 Duke Univ Med Ctr Durham NC 27710

SHELBY, JAMES ELBERT, b Memphis, Tenn, Mar 11, 43; div; c 1. GLASS & GLASS CERAMICS, GAS SOLIDS REACTIONS. *Educ:* Univ Mo, Rolla, BS, 65, MS, 67, PhD(ceramic eng), 68. *Prof Exp:* Staff mem, Sandia Nat Lab, Livermore, 68-82; PROF GLASS SCI, NY STATE COL CERAMICS, ALFRED UNIV, 82-, DIR GLASS SCI LAB, 88- *Concurrent Pos:* Chmn, Gordon Res Conf Glass, 85; consult glass, 82-; mem, Ceramic Educ Coun. *Honors & Awards:* Morey Award, Am Ceramic Soc, 75. *Mem:* Fel Am Ceramic Soc; Nat Inst Ceramic Engrs; Soc Glass Technol. *Res:* Glasses and glass-ceramics: diffusion, properties, phase separation, low melting glasses and solar energy; radiation effects, fluorides, heavy metal oxides, thermal analysis, hydroxyl, gas diffusion controlled reactions, optical properties and gases in glasses and melts; author of over 160 publications. *Mailing Add:* NY State Col Ceramics Alfred NY 14802

SHELBY, ROBERT MCKINNON, b Ogden, Utah, Aug 11, 50; m 90; c 4. OPTICAL PHYSICS, SPECTROSCOPY. *Educ:* Calif Inst Technol, BS, 72; Univ Calif, Berkeley, PhD(chem), 78. *Prof Exp:* Fel, 78-76, RES STAFF, ALMADEN RES CTR, IBM RES, 79- *Concurrent Pos:* Chmn, Western Spectros Asn, 87-88. *Mem:* Optical Soc Am; Am Phys Soc. *Res:* Laser physics; quantum optics; generation and application of non-classical light beams; dynamics of non-linear optical systems. *Mailing Add:* Dept K32802 IBM 650 Harry Rd San Jose CA 95120

SHELBY, T H, b Austin, Tex, Aug 12, 10. GEOLOGY. *Educ:* Univ Tex, BS, 33, MS, 34. *Prof Exp:* Staff geologist, Humble Oil Refining Co, Standard Oil Co, sr staff geologist, Exxon Corp, 34-74; RETIRED. *Concurrent Pos:* Geol consult, 74- *Mem:* Geol Soc Am; Am Asn Prof Geologists; AAAS. *Mailing Add:* 1402 W Second St Tyler TX 75701

SHELDEN, HAROLD RAYMOND, II, b Indianapolis, Ind, July 7, 42; m 65. ORGANIC CHEMISTRY. *Educ:* Loma Linda Univ, BA, 64; Univ Calif, Irvine, PhD(org chem), 69. *Prof Exp:* Asst prof, 69-74, assoc prof, 74-79, PROF CHEM, LOMA LINDA UNIV, LA SIERRA CAMPUS, 79-, CHMN DEPT, 80- *Mem:* Am Chem Soc. *Res:* Mechanisms of organic reactions. *Mailing Add:* Dept Chem Loma Linda Univ La Sierra Campus Riverside CA 92515

SHELDEN, ROBERT MERTEN, b Troy, Mont, Mar 23, 38; m 64; c 3. REPRODUCTIVE PHYSIOLOGY, EMBRYOLOGY. *Educ:* Univ Mont, BA, 64, PhD(zool), 68. *Prof Exp:* Asst prof biol, Moorhead State Col, 68-71; res assoc, Dept Obstet & Gynec, Ohio State Univ, 73-75; ASST PROF, DEPT OBSTET & GYNEC, ROBERT WOOD JOHNSON MED SCH, 76- *Concurrent Pos:* Minn State Col Bd res grant, 68-69; NIH fel reproductive endocrinol, Dept Obstet & Gynec, Ohio State Univ, 71-73. *Mem:* AAAS; Soc Study Reproduction. *Res:* Endocrine regulation of the uterine environment; immunogenic potentials of primate reproductive organs. *Mailing Add:* 12 Oxford New Brunswick NJ 08903

SHELDON, ANDREW LEE, b Greenfield, Mass, Apr 22, 38; m 90; c 2. ZOOLOGY, ECOLOGY. *Educ:* Colby Col, BA, 60; Cornell Univ, PhD(zool), 66. *Prof Exp:* Asst res zoologist, Sagehen Creek Field Sta, Univ Calif, 64-67; res assoc, Resources for the Future, Inc, 67-69; from asst prof to assoc prof, 69-77, PROF ZOOL, 78-, DIR WILDLIFE BIOL PROG, UNIV MONT, 90- *Concurrent Pos:* Vis scientist, Oak Ridge Nat Lab, 77-78, Savannah River Ecol Lab, 85-86. *Mem:* Am Soc Naturalists; Ecol Soc Am; NAm Benthological Soc; Am Fisheries Soc; Soc Conserv Biol. *Res:* Community structure and dynamics; comparative ecology; running water biology; biometrics; fishes, aquatic insects. *Mailing Add:* Div Biol Sci Univ Mont Missoula MT 59812

SHELDON, DONALD RUSSELL, science administration, for more information see previous edition

SHELDON, ELEANOR BERNERT, b Hartford, Conn, Mar 19, 20. MEDICAL ADMINISTRATION. *Educ:* Univ NC, AB, 42; Univ Chicago, PhD, 49. *Prof Exp:* Assoc dir, Chicago Community Inventory, Univ Chicago, 47-50; res assoc & lectr sociol, Univ Calif Los Angeles, 55-62; sociologist & exec assoc, Russell Sage Found, 61-72; mem bd dirs, UN Res Inst Social Develop, 73-79; trustee, Rockefeller Found, 78-85; trustee, Inst E&W Security Studies, Nat Opionion Res Ctr, 84-89; RETIRED. *Mem:* Inst Med-Nat Acad Sci. *Mailing Add:* 630 Park Ave New York NY 10021

SHELDON, ERIC, b Oct 24, 30; Brit citizen; m 59; c 1. THEORETICAL NUCLEAR PHYSICS, ASTROPHYSICS. *Educ:* Univ London, BSc, 51, Hons, 52, PhD(sci), 55, DSc(physics), 71. *Prof Exp:* Lectr & demonstr physics, Acton Tech Col, Eng, 52-55; assoc physicist, IBM Res Lab, Switz, 57-59; res assoc physics, Swiss Fed Inst Technol, 59-63, privat dozent, 63-64, prof, 64-69; vis prof, Univ Tex, Austin, 69-70; univ prof, 85-88, PROF PHYSICS, UNIV LOWELL, 70-85, 88- *Concurrent Pos:* NSF sr foreign scientist fel & vis prof, Univ Va, 68-69; chartered chemist, UK, 81-, chartered physicist, 85-; vis prof, Univ Oxford, UK, 89. *Mem:* Fel AAAS; fel Brit Inst Physics; fel Royal Soc Chem; Royal Inst Gt Brit; fel Am Phys Soc; fel Royal Astron Soc. *Res:* Theoretical nuclear physics involving nuclear reaction mechanism and nuclear structure studies in the low and intermediate energy range; astrophysics; cosmology; relativity. *Mailing Add:* Dept of Physics Univ of Lowell Lowell MA 01854-2881

SHELDON, HUNTINGTON, b New York, NY, Jan 14, 30; m 55; c 4. PATHOLOGY, MEDICINE. *Educ:* McGill Univ, BA, 51; Johns Hopkins Univ, MD, 56. *Prof Exp:* From asst to instr path, Johns Hopkins Univ, 56-59; from asst prof to assoc prof, 59-66, PROF PATH, MCGILL UNIV, 66-, STRATHCONA PROF, 80- *Concurrent Pos:* From intern to asst resident, Johns Hopkins Hosp, 56-59; vis prof, Harvard Univ, 72; co-ed, Intro to Study Dis, 77 & 80. *Mem:* Am Soc Cell Biol; Am Soc Exp Path; Biophys Soc; Int Acad Path. *Res:* Application of electron microscopy to problems in pathology. *Mailing Add:* PO Box 697 Shelburne VT 05482

SHELDON, JOHN WILLIAM, b Miami, Fla, Nov 21, 33; m 79; c 5. CHEMICAL PHYSICS, FLUID DYNAMICS. *Educ:* Purdue Univ, BS, 55, MS, 59; Tex A&M Univ, PhD(nuclear eng), 64. *Prof Exp:* Res engr, NASA, 55, head gaseous electronics sect, 64-66; from asst prof to assoc prof eng sci, Fla State Univ, 66-72; assoc prof phys sci, 72-76, chmn dept, 74-86, PROF PHYS, FLA INT UNIV, 76- *Mem:* Am Phys Soc; Am Vacuum Soc. *Res:* Atomic collision phenomena, measurement of collision cross sections by beam techniques; ionospheric flow and probe theory; electrical phenomena accompanying shock waves in two-phase flows; chemical kinetics of explosives. *Mailing Add:* Dept Phys Fla Int Univ Tamiami Trail Miami FL 33199

SHELDON, JOSEPH KENNETH, b Ogden, Utah, Nov 11, 43; m 65; c 2. INSECT ECOLOGY. *Educ:* Col Idaho, BS, 66; Univ Ill, PhD(entom), 72. *Prof Exp:* Assoc prof, 71-81, PROF BIOL, EASTERN COL, 81- *Concurrent Pos:* Fac mem, Au Sable Inst Environ Studies. *Mem:* AAAS; Ecol Soc Am; Am Entom Soc (vpres, 82, pres, 91); Am Sci Affil. *Res:* Biology and taxonomy of Chrysopidae; insect mimicry. *Mailing Add:* Dept Biol Eastern Col St Davids PA 19087

SHELDON, RICHARD P, b Tulsa, Okla, Oct 25, 23; m 66. GEOLOGY. *Educ:* Yale Univ, BS, 50; Stanford Univ, PhD, 56. *Prof Exp:* Geologist, US Geol Surv, 47-57; geologist, Lion Oil Co, 57-58; from geologist to asst chief geologist, US Geol Serv, 58-72, chief geologist, 72-77, res geologist, 77-82, GEOLOGIST, US MISSION, US GEOL SURV, SAUDI ARABIA, 82- *Concurrent Pos:* Consult, econ geologist, 82- *Mem:* AAAS; Geol Soc Am; Soc Econ Geol; Am Asn Petrol Geologists; Soc Econ Paleontologists & Mineralogists. *Res:* Sedimentary petrology; physical stratigraphy; sedimentary mineral deposits. *Mailing Add:* US Geol Surv APO New York NY 09697-7002

SHELDON, VICTOR LAWRENCE, b Maysville, Mo, Sept 24, 21; m 46; c 4. SOILS, FERTILIZERS. *Educ:* Univ Mo, BS, 43, MA, 48, PhD(soils, plant physiol), 50. *Prof Exp:* Asst prof soils, Univ Mo, 52-55; agronomist & consult, Olin Mathieson Chem Corp, 55-61; mgr agr serv, John Deere Chem Co, 61-66; mgr tech serv, Esso Chem Co, Standard Oil, NJ, 66-67; vpres mkt, Esso Pakistan Fertilizer Co, Karachi, 67-69; prof agr & chmn dept, Western Ill Univ, 69-81; PRES, VLS ASSOC, 76- *Concurrent Pos:* Leader deleg to

CENTO Agr Conf, US State Dept; consult, VLS Assoc, 72-76, Tenn Valley Authority, Int Inst Trop Agr & Int Fertilizer Develop Ctr. *Mem:* Soil Sci Soc Am; Am Soc Plant Physiol; Am Soc Agron; Sigma Xi. *Res:* Efficiency of phosphorus uptake from various compounds; influence of phosphorus on metabolism of plants; biodegradation of pesticides; fatty acids in soybean oil according to planting site and genotype; fertilizer marketing plan for Nigeria; management systems for fertilizer operaters. *Mailing Add:* 221 Woodchuck Lane Macomb IL 61455

SHELDON, WILLIAM GULLIVER, ecology, for more information see previous edition

SHELDON, WILLIAM ROBERT, b Ft Lauderdale, Fla, May 17, 27; m 79; c 3. PHYSICS. *Educ:* Univ Mo, BS, 50, MS, 56, PhD(physics), 60. *Prof Exp:* Sr physicist, Rocketdyne Div, NAm Aviation, Inc, 60; res specialist space physics, Aerospace Div, Boeing Co, 60-66; res scientist cosmic ray physics, Southwest Ctr Advan Studies, 66-68; assoc prof, 68-73, PROF PHYSICS, UNIV HOUSTON, 73-,. *Concurrent Pos:* Proj leader, Joint French-US Cosmic Ray Exped, Mont Blanc Tunnel; proj leader rocket and balloon measurements of x-rays and electric fields at Roberval and Ft Churchill, Can, Ft Yukon, Alaska, Kiruna, Sweden, Siple Sta, Antarctica and the Kerguelen Islands; leader US team, French-USSR ARAKS experiment, 74-75; dept chmn Dept Physics, Univ Houston, 73-; proj leader, Atmospheric Ozone Measurements, 87- *Mem:* Am Phys Soc; Am Geophys Union; Sigma Xi. *Res:* Cosmic rays; high energy muons; auroral particle precipitation; x-rays in the atmosphere; atmospheric ozone. *Mailing Add:* Dept Physics Univ Houston Houston TX 77004

SHELDRAKE, RAYMOND, JR, b Paterson, NJ, Sept 7, 23; m 54; c 4. HORTICULTURE. *Educ:* Rutgers Univ, BS, 49; Cornell Univ, MS, 50, PhD(veg crops, plant path, soils), 52. *Prof Exp:* Asst, NY Exp Sta, Geneva, 49-52; exten veg specialist, Univ Ga, 52-54; from asst prof to prof veg crops, Cornell Univ, 54-83. *Mem:* Am Soc Hort Sci; Potato Asn Am. *Res:* Critical temperatures for growth of higher plants; plastics and plant growing. *Mailing Add:* Sheldrake Res Ctr Pennsylvania Ave Trumansburg NY 14886

SHELDRICK, PETER, b Newark, NJ, Jan 22, 36; m 91; c 1. MOLECULAR BIOLOGY, VIROLOGY. *Educ:* Brown Univ, Providence, RI, BA, 58; Univ Calif, Berkeley, PhD(chem), 62. *Prof Exp:* Sr researcher, 68-77, DIR RES, NAT CTR SCI RES, FRANCE, 77- *Concurrent Pos:* Mem, Herpes Virus Study Group, Int Comt Taxon Viruses, 76-87. *Mem:* Europ Molecular Biol Orgn. *Res:* Structure and function of herpes virus genomes: DNA base sequence organization; transcription of RNA; in vivo polypeptide synthesis; enzymatic activities. *Mailing Add:* Inst Rech Sci Sur Le Cancer B P No 8 Villejuif Cedex 94802 France

SHELEF, LEORA AYA, b Haifa, Israel; US citizen; m 55; c 2. FOOD MICROBIOLOGY, FOOD SAFETY. *Educ:* Israel Inst Technol, BSc, 56, MSc, 59, DSc (food eng & biotechnol), 63. *Prof Exp:* Postdoctoral res assoc food rheology, Dept Agr Eng, Pa State Univ, 64-66; res assoc food microbiol, Dept Biol, Wayne State Univ, 67-71, from asst prof to assoc prof food sci, 71-79, actg chair, Dept Family & Consumers Res, 81-83, chair, 83-86, PROF FOOD SCI, WAYNE STATE UNIV, 79-, CHAIR, DEPT NUTRIT & FOOD SCI, 86- *Concurrent Pos:* Vis prof, Mercy Col, Detroit, 73 & 76 & Dept Environ Indust Health, Univ Mich, Ann Arbor, 79; NSF fac prof develop award, 78. *Honors & Awards:* Fulbright Sr Lectr, Tallinn Tech Univ, USSR, 90. *Mem:* Am Soc Microbiol; Int Asn Milk Food Environ Sanit; Inst Food Technologists. *Res:* Effect of indirect antimicrobials on food pathogens and spoilage microorganisms such as sodium lactate, lysozyme, and other naturally occurring substances; growth characteristics of Listeria monocytogenes in meat, fish, and similar foods; chemistry and nutrition of soy products. *Mailing Add:* Dept Nutrit & Food Sci Wayne State Univ Detroit MI 48202

SHELEF, MORDECAI, b Suvalki, Poland, June 28, 31; m 55; c 2. CATALYSIS, FUEL SCIENCE. *Educ:* Israel Inst Technol, BSc, 56, MSc, 59; Pa State Univ, PhD(fuel sci), 66. *Prof Exp:* Res scientist, Israel Mining Industs, Haifa, 56-63; res asst, Pa State Univ, 63-66; prin res scientist, 69-73, Ford Motor Co, staff scientist, 73-77, mgr fuels & lubricants dept, 77-81, mgr chem dept, 81-87, SR STAFF SCIENTIST, RES STAFF, FORD MOTOR CO, 87- *Mem:* Am Chem Soc; Catalysis Soc. *Res:* Development of new fuel sources; kinetics and mechanism of surface reactions; chemisorption; carbon gasification; mineral dressing; evaluation of energy systems. *Mailing Add:* Res Staff PO Box 2053 Dearborn MI 48121

SHELESNYAK, MOSES CHIAM, b Chicago, Ill, June 6, 09; m 42; c 2. BIODYNAMICS, PHYSIOLOGY OF REPRODUCTION. *Educ:* Univ Wis, BA, 30; Columbia Univ, PhD(anat), 33. *Prof Exp:* Asst endocrinol, Columbia Univ, 33-35; instr physiol & pharmacol, Chicago Med Sch, 35-36; Gen Educ Bd fel child develop, NY Asn Care Jewish Children, 36-38, dean boys, 38-40; Freidsam res fel endocrine pediat, Beth Israel Hosp, New York, 40-42; head ecol br, med sci div & actg head biophys br, US Off Naval Res, 46-49; dir, Arctic Inst NAm, DC, 49-50; sr scientist, Dept Exp Biol, Weizmann Inst Sci, 50-56, assoc, 56-57, prof endocrine & reproductive physiol, 58-61, prof biodynamics & head dept, 61-68; assoc dir, 67-68, dir interdisciplinary commun prog, Off Asst Secy Sci, 68-77, res assoc, Smithsonian Inst, 77-; EMER PROF. *Concurrent Pos:* Res assoc, Mt Sinai Hosp, 36-40; guest lectr, Columbia Univ, 37-38; vis prof geog, McGill Univ, 48; lectr, Johns Hopkins Univ, 49-50; hon consult, Panel Human Ecol Arid Zones, UNESCO, mem bd, 54-72; Marks fel & univ fel, Birmingham Univ, 57-58; consult & mem comn, Palais des Sci, World's Fair, Brussels, 58; mem Israel comt, Zool Sta, Naples; mem res comt, Int Planned Parenthood Fedn, 59-72; mem selection & adv comt, Int Training Prog Physiol Reproduction, Worcester, Mass, 59-72; vis prof, Col France, Paris, 60; mem neuroendocrinol panel, Int Brain Res Orgn, 61-77; mem expert adv comt human reproduction, WHO, 65-70; chmn bd, Interdisciplinary Commun Assocs, 69-77; dir, Int Prog Pop Analysis, 72-77; mem Israel comt, Int Biol Prog; consult, Am Physiol Soc, 77-84. *Honors & Awards:* Oliver Bird Prize, 58. *Mem:* Fel AAAS; fel Soc Res Child Develop; fel Arctic Inst NAm; Soc Exp Biol & Med; Am Physiol Soc; Soc Study Reproduction. *Res:* Reproduction and endocrine physiology; environmental physiology and human ecology; interdisciplinary communications; population dynamics; social biology. *Mailing Add:* 674 Chalk Hill Rd Solvang CA 93463

SHELINE, GLENN ELMER, b Flint, Mich, Mar 31, 18; m 48; c 1. MEDICINE. *Educ:* Univ Calif, BS, 39, PhD(physiol), 43, MD, 48. *Prof Exp:* Nat Res Coun fels, Univ Chicago, 49-50 & Univ Calif, San Francisco, 50-51; consult, US Naval Radiol Defense Lab, 53; from asst prof to prof radiol, Sch Med, Univ Calif, San Francisco, 55-88; RETIRED. *Concurrent Pos:* Commonwealth Fund fel, Royal Marsden Hosp, London, 57; consult, San Francisco Gen Hosp, 57-, San Francisco Vet Admin Hosp, 70 & AEC; dir radiation ther, Nat Cancer Inst. *Mem:* Soc Nuclear Med; Am Radium Soc; Radiation Oncol Roentgen Ray Soc; Am Thyroid Asn; Radiol Soc NAm. *Res:* Radiation therapy and effects; radiobiology. *Mailing Add:* Dept Radiation Oncol Univ Calif Sch Med 513 Parnassus Ave San Francisco CA 94143

SHELINE, RAYMOND KAY, b Port Clinton, Ohio, Mar 31, 22; m 51; c 7. NUCLEAR CHEMISTRY. *Educ:* Bethany Col, BS, 43; Univ Calif, PhD(chem), 49. *Prof Exp:* Asst, Bethany Col, 40-42; chemist, Manhattan Proj, Columbia Univ, 43-45; jr scientist, Los Alamos Sci Lab, Univ Calif, 45-46, asst, Univ, 46-49; instr, Inst Nuclear Studies, Univ Chicago, 49-51; res participant, Oak Ridge Inst Nuclear Studies, 51-52; assoc prof, 51-55, PROF CHEM, FLA STATE UNIV, 55-, PROF PHYSICS, 59-, ROBERT O LAWTON DISTINGUISHED PROF CHEM & PHYSICS, 67- *Concurrent Pos:* Res chemist, Merck Chem Co, 46; Fulbright res prof & Guggenheim fel, Niels Bohr Inst, Coepnhagen, Denmark, 55-56 & 57-58, Ford res prof, 57-58; Guggenheim fel, 64; Nordita prof, Univ Lund & Copenhagen Univ, 71-72; consult, NSF, Los Alamos Sci Lab, Univ Calif & Lawrence Livermore Nat Lab, 61-; Gillon lectureship, Nat Univ Zaire, Kinshasa, 76; res fel, Australian Nat Univ, 82-83; Fulbright prof, Univ Kinshasa, Zaire, 84- *Honors & Awards:* Niels Bohr Inst Silver Cup; Egyptian Nat Lectr, Eins Shams Univ, Cairo, 56; Am Inst Physics Citation, 63, Alexander von Humboldt sr scientist award, 76; Fla Award, Am Chem Soc, 80. *Mem:* Foreign mem Royal Danish Acad Sci & Lett; Am Chem Soc; fel Am Phys Soc. *Res:* Nuclear spectroscopy by decay scheme studies and Van de Graaff excitation; coulomb excitation; correlation of experimental data with nuclear models; muonic x-ray studies; octupole shapes in nuclei. *Mailing Add:* Dept Chem Fla State Univ Tallahassee FL 32306

SHELKIN, BARRY DAVID, b Brooklyn, NY, Oct 11, 28; m 52; c 3. GEOLOGY. *Educ:* Brooklyn Col, BS, 55. *Prof Exp:* Cartog aide, US Coast & Geod Surv, 55-56; geologist, Gulf Oil Corp, 56-60, Span Gulf Oil Co, 60-63 & Nigerian Gulf Oil Co, 63-64; sr photointerpreter, Data Anal Ctr, Itek Corp, 64-65; sr scientist, Autometric Oper, Raytheon Co, 65-69, head terrain sci sect, 69-70; chief, Terrain Sci Div, 70-73, chief, Support Div, 73-82, asst chief, Topog Reg Div, 82-89, ASST DIR, TECH INFO, DEFENSE NUCLEAR AGENCY, 90- *Mem:* Geol Soc Am; Am Asn Petrol Geologists; Int Soc Optical Eng. *Res:* Geological research, terrain, and environmental analysis through the medium of aerial photography; administration of cartographic contracts; technical information management. *Mailing Add:* 8206 Chancery Ct Alexandria VA 22308

SHELL, DONALD LEWIS, b Worth Twp, Sanilac Co, Mich, Mar 1, 24; m 46, 73; c 2. MATHEMATICS. *Educ:* Mich Technol Univ, BS, 44; Univ Cincinnati, MS, 51, PhD(math), 59. *Prof Exp:* Instr math, Mich Technol Univ, 46-49; mathematician, Gen Elec Co, 51-52, numerical analyst, 52-53, supvr systs anal & synthesis, 53-54, mgr comput tech develop, 54-56, mgr, Evendale Comput, 56-57, comput consult specialist, 57-59, mgr digital anal & comput, Knolls Atomic Power Lab, 60-61, eng math, adv tech lab, 61-63, mgr comput appln & processing telecommun & info processing opers, 63-66, mgr eng, Info Serv Dept, 66-68, mgr automation studies, Res & Develop Ctr, 68-69, mgr info servs tech planning, 69-71, mgr info servs qual assurance, 71-72; chmn bd & gen mgr, Robotics, Inc, 72-75; mgr file systs, Gen Elec Info Serv Co, 75-76, mgr technol systs, 76-78, mgr appln systs, 78-80, mgr Mark III Systs, 80- 84; RETIRED. *Mem:* Math Asn Am; Asn Comput Mach. *Res:* Numerical computation; applications of digital computers; sorting. *Mailing Add:* PO Box 1027 Lake Junaluska NC 28745

SHELL, EDDIE WAYNE, b Chapman, Ala, June 16, 30; m 53. FISH BIOLOGY. *Educ:* Auburn Univ, BS, 52, MS, 54; Cornell Univ, PhD(fishery biol), 59. *Prof Exp:* Asst fisheries, Auburn Univ, 52-54 & Cornell Univ, 56-58; asst fish culturist, 59-61, assoc prof fisheries, 61-70, PROF FISHERIES, AUBURN UNIV, 70-, HEAD DEPT, 73- *Concurrent Pos:* Dir Int Ctr Aquaculture, 73. *Mem:* AAAS; Am Fisheries Soc. *Mailing Add:* Dept Fisheries Auburn Univ Auburn AL 36849

SHELL, FRANCIS JOSEPH, b Medicine Lodge, Kans, Mar 27, 22; m 44; c 3. PHYSICAL CHEMISTRY. *Educ:* Ft Hays Kans State Col, AB & MS, 49; Univ Ky, PhD(phys chem), 53. *Prof Exp:* Res chemist, Phillips Petrol Co, 52-57, asst dir tech serv div, 57-66, mgr tech serv, 66-81, sr chem assoc, 81-85; RETIRED. *Concurrent Pos:* Chmn oil well cement comt, Am Petrol Inst, 76-78. *Mem:* Am Chem Soc; Am Petrol Inst; Soc Petrol Engrs. *Res:* Non-aqueous solutions; oil well cements; drilling fluids; fracturing; colloids. *Mailing Add:* 534 Crestland Dr Bartlesville OK 74006

SHELL, JOHN WELDON, b Waxahachie, Tex, Apr 20, 25; m 52; c 2. PHARMACEUTICAL CHEMISTRY. *Educ:* Univ Colo, BA, 49, BS, 53, PhD(pharmaceut chem), 54. *Prof Exp:* Asst chem, Univ Colo, 49-53; res assoc physics, Upjohn Co, 54-57, res scientist in prod res, 57-60, sr res scientist, 60-62; dir qual control, Allergan Pharmaceut, 62-64, dir res, 64-68; assoc dir, Inst Pharmaceut Chem Div, 68-71, dir ophthal res, 71-73, VPRES & DIR CLIN OPHTHAL, ALZA CORP, 73- *Concurrent Pos:* Vis grad lectr, Univ Southern Calif, 65-66; adj prof, 72-77, prof, Med Ctr, Univ Calif, San Francisco, 77- mem, Res Prevent Blindness, Inc. *Mem:* Am Pharmaceut Asn; fel Acad Pharmaceut Sci; Asn Res Vision & Ophthal. *Res:* Crystallography; x-ray analysis; biopharmaceutics; biochemistry; ocular pharmacology. *Mailing Add:* 952 Tournament Dr Burlingame CA 94010

SHELLABARGER, CLAIRE J, b College Corner, Ohio, Oct 23, 24; m 48; c 3. RADIOBIOLOGY, ENDOCRINOLOGY. *Educ:* Miami Univ, AB, 48; Ind Univ, MA, 49, PhD(zool), 52. *Prof Exp:* Asst, Ind Univ, 50-52; jr scientist, Brookhaven Nat Lab, 52-53, asst scientist, 53-54, assoc scientist, 54-57, scientist, 58-60, asst to chmn dept med, 59-60; scientist med dept, Brookhaven Nat Lab, 60-68; asst chmn med dept, 68-70, sr scientist & head radiobiol div, Brookhaven Nat Lab, Upton, 70-80; PROF PATH, STATE UNIV NY Stony Brook, 80- *Concurrent Pos:* Lectr, Adelphi Col, 56; USPHS fel, Nat Inst Med Res, Eng, 57-58 & Inst Cancer Res, London, 66-67. *Mem:* Am Soc Zool; Soc Exp Biol & Med; Am Soc Exp Path; Radiation Res Soc; Am Physiol Soc. *Res:* Radiation carcinogenesis. *Mailing Add:* Scientist Med Dept Brookhaven Nat Lab Upton NY 11973

SHELLENBARGER, ROBERT MARTIN, b Sacramento, Calif, June 25, 36; m 59; c 5. TECHNICAL MANAGEMENT. *Educ:* Col Pac, BS, 57; Univ NC, PhD(phys chem), 63. *Prof Exp:* Res chemist, 62-71, supvr, res & develop, 71-85, GROUP MGR, E I DU PONT DE NEMOURS & CO, INC, 85- *Res:* Structure and properties of synthetic fibers and nonwoven fabrics. *Mailing Add:* 3804 Valleybrook Dr Wilmington DE 19808

SHELLENBERGER, CARL H, b York, Pa, Sept 11, 35; m 84; c 3. PHYSIOLOGY, PHARMACOLOGY. *Educ:* Muhlenberg Col, BS, 58; State Univ NY, PhD(physiol), 68. *Prof Exp:* Res asst anesthesiol, Med Sch, Univ Pa, 59-61; mem staff pharmacol dept, Endo Labs, Inc, 67-70; sr clin scientist, Consumer Prod Div, Warner-Lambert Co, 70-73; asst dir, Sandoz, Inc, 73-75, assoc dir, 75-78, sr assoc dir clin res, 78-81; DIR CLIN AFFAIRS, WARNER-LAMBERT CO, 81- *Mem:* AAAS; NY Acad Sci; Am Acad Dermat. *Res:* Fibrinolysis; neuropharmacology; glucose and fat metabolism; adrenal gland metabolism; respiratory and analgesic pharmacology; clinical investigation of proprietary and ethical drugs; medical instrumentation. *Mailing Add:* Warner-Lambert Co Tabor Rd Morris Plains NJ 07950

SHELLENBERGER, DONALD J(AMES), b Altoona, Pa, Mar 30, 28; m 49; c 2. CHEMICAL ENGINEERING. *Educ:* Pa State Univ, BS, 50. *Prof Exp:* Asst indust hygienist, State Bur Indust Hyg, Pa, 50-51; process engr, Day & Zimmermann, Inc, 51-55; process engr, Jones & Laughlin Steel Corp, 55-56, sr process engr, 56-58, supvr chem eng servs, 58-64, supvr blast furnace res, 64-67, planning engr, 67-69, develop engr blast furnaces, 69-77; sr staff engr, Koppers Co, Inc, 77-81, asst dir tech opers, 81-83, dir, 83-84; prin engr, Kaiser Engrs, Inc, 83-88; MGR, PROCESS & DEVELOP, PAUL WORTH, INC, CARNEGIE, PA, 88- *Mem:* Am Inst Mining Metall & Petrol Engrs; Asn Iron & Steel Engrs. *Res:* Development of blast furnaces, ancillary equipment and iron ore agglomeration facilities; coordination of research and development in coal carbonization, byproducts, ore agglomerization and metallurgical processes. *Mailing Add:* 2730 Tischler Rd Bethel Park PA 15102

SHELLENBERGER, JOHN ALFRED, cereal chemistry; deceased, see previous edition for last biography

SHELLENBERGER, MELVIN KENT, b Pittsburg, Kans, Oct 29, 36; m 59; c 4. NEUROPHARMACOLOGY. *Educ:* Kans State Col Pittsburg, BS, 58; Univ Wash, MS, 62, PhD(pharmacol), 65. *Prof Exp:* Lab technician, Pharmacol Dept, Upjohn Co, 59-60; lab instr dent, med, pharm & pharmacol courses, Univ Wash, 60-65, lectr pharm & pharmacol courses, 62-65; from instr to assoc prof pharmacol, Univ Kans Med Ctr, 67-84; ADJ ASSOC PROF PHARMACOL & PHYS THER EDUC, 84-, VPRES, INT MED TECH CONSULTS, INC, 84- *Concurrent Pos:* Nat Inst Neurol Dis & Stroke fel pharmacol, Univ Mich, 65-66, NIMH trainee neuropsychopharmacol, 66-67; res assoc, Kans Ctr Ment Retardation & Human Develop, 71-; NIMH career develop res award, Univ Kans Med Ctr, 72-77. *Mem:* Int Soc Neurochem; Am Soc Pharmacol & Exp Therapeut; Soc Neurosci. *Res:* Correlating possible chemical mediators in the brain with electrical, physiological and behavioral events. *Mailing Add:* Parke-Davis Pharmaceut Res Div 2800 Plymouth Rd Ann Arbor MI 48105

SHELLENBERGER, PAUL ROBERT, b Dover, Pa, May 28, 35; m 57; c 2. DAIRY SCIENCE. *Educ:* Pa State Univ, BS, 57, MS, 59; Iowa State Univ, PhD(animal nutrit), 64. *Prof Exp:* Area dairy specialist, Agr Ext Serv, Tex A&M Univ, 64-66; actg assoc prof agr, Tarleton State Col, 66-67; from asst prof to assoc prof, 67-77, PROF DAIRY SCI, PA STATE UNIV, UNIVERSITY PARK, 77- *Concurrent Pos:* Mem, Coun Agr Sci & Technol; nat teacher fel award, Nat Asn Cols & Teachers Agr, 77. *Mem:* Nat Asn Col & Teachers Agr; Am Dairy Sci Asn; Coun for Agr Sci & Technol. *Mailing Add:* Dept Dairy-Animal 203 Borland Lab Penn State Univ 317 Henning Bldg University Park PA 16802

SHELLENBERGER, THOMAS E, b Havre, Mont, May 3, 32; m 53; c 3. BIOCHEMISTRY, TOXICOLOGY. *Educ:* Mont State Univ, BS, 54, MS, 55; Kans State Univ, PhD(biochem), 61. *Prof Exp:* Res asst chem, Mont State Univ, 54-55; asst instr, Kans State Univ, 55-60; biochemist, Stanford Res Inst, 60-66, mgr biochem toxicol labs, 66; chmn dept toxicol, Gulf South Res Inst, 66-72; actg dep dir, Nat Ctr Toxicol Res, Food & Drug Admin, 77-78, chief div comp pharmacol, 72-88; CONSULT. *Concurrent Pos:* Assoc prof biochem, Univ Ark, Little Rock, 74- *Mem:* AAAS; NY Acad Sci; Am Chem Soc; Soc Toxicol; Am Col Vet Toxicologists. *Res:* Metabolism of carcinogens; comparative endocrinology; reactions and mechanisms of organophosphates; hazards of pesticides to fish and wildlife; poultry nutrition, vitamins and protein. *Mailing Add:* PO Box 2537 Lorau MD 20708

SHELLEY, AUSTIN L(INN), b New Ross, Ind, Apr 9, 22; m 48; c 2. ELECTRICAL ENGINEERING. *Educ:* Univ Ky, BS, 47; Purdue Univ, MS, 52, PhD(elec eng), 58. *Prof Exp:* Instr elec eng, Miss State Col, 47-49; from instr to asst prof, 50-61, ASSOC PROF ELEC ENG, PURDUE UNIV, WEST LAFAYETTE, 61-, EXEC ASST TO HEAD SCH, 64- *Mem:* Illum Eng Soc; Inst Elec & Electronics Engrs; Am Soc Eng Educ; Sigma Xi. *Res:* Circuits; machinery; servomechanisms. *Mailing Add:* Sch of Elec Eng Purdue Univ Lafayette IN 47907

SHELLEY, EDWARD GEORGE, b Watford City, NDak, Jan 8, 33; m 52; c 2. SPACE PHYSICS. *Educ:* Ore State Univ, BS(physics) & BS(math), 59; Stanford Univ, MS, 61, PhD(nuclear physics), 67. *Prof Exp:* Res scientist, 59-73, staff scientist, 73-80, sr staff scientist, 80-84, PROJ LEADER, LOCKHEED RES LABS, 84- *Concurrent Pos:* Res assoc physics, Stanford Univ, 65-68. *Mem:* Am Geophys Union; Europ Geophys Soc. *Res:* Magnetospheric physics, primarily in area of satellite observations of space plasmas. *Mailing Add:* 2406 Villa Nuena Way Mountain View CA 94558

SHELLEY, EDWIN F(REEMAN), communications, systems engineering; deceased, see previous edition for last biography

SHELLEY, WALTER BROWN, b St Paul, Minn, Feb 6, 17; m 42, 80; c 5. DERMATOLOGY. *Educ:* Univ Minn, BS, 40, PhD(physiol), 41, MB & MD, 43. *Hon Degrees:* MD, Univ Uppsala, Sweden, 77. *Prof Exp:* Asst physiol, Univ Minn, 38-41; instr, Col St Thomas, 42-43; from instr to asst instr dermat, Univ Pa, 46-49; instr, Dartmouth Col, 49-50; from asst prof to prof dermat, Sch Med, Univ Pa, 50-80, chmn dept, 65-80; prof dermat, Peoria Sch Med, Univ Ill, 80-83; PROF DERMAT, DEPT MED, DIV DERMAT, MED COL OHIO, 83- *Concurrent Pos:* Pvt pract; chief clin, Univ Hosp, 51-56 & 65-66; regional consult, US Vet Admin, 55-59; mem comt cutaneous dis, Nat Res Coun, 55-59, mem coun, 61-64; Pollitzer lectr, NY Univ, 56; Rauschkolb Mem lectr, Univ Chicago, 57; Prosser White Oration, Univ London, 57; consult, Surgeon Gen, US Army, 58-61 & US Air Force, 58-61; mem comn cutaneous dis, Armed Forces Epidemiol Bd, 58-61, dep dir, 59-61; consult, Philadelphia Gen Hosp, 60-65, chief dermat serv, 65-67; mem & dir, Am Bd Dermat, 60-69, past pres; consult dermatologist, Children's Hosp Philadelphia, 65-80. *Honors & Awards:* Soc Cosmetic Chem Award, 55; Hellerstrom Medal, Karolinska Inst, Sweden, 71; Am Med Writers Asn Award, 73; Dohi Medalist, Nagoya, Japan, 81; Rose Hirschler Award, Women's Dermat Soc; Rothman Medal. *Mem:* Hon mem Soc Invest Dermat (pres, 76); Am Physiol Soc; hon mem Am Dermat Asn (pres, 76); hon mem Am Acad Dermat (pres, 72); Asn Prof Dermat. *Res:* Physiology of the skin, especially the eccrine and apocrine sweat gland, sebaceous gland and pruritus; allergic states. *Mailing Add:* Dept Dermat Div Dermat Med Col Ohio 3000 Arlington Ave Toledo OH 43699

SHELLEY, WILLIAM J, b Wichita, Kans, Mar 15, 22; m 54. CHEMICAL ENGINEERING. *Educ:* Univ Mich, BS, 48, MSE, 49. *Prof Exp:* Prod engr uranium div, Mallinckrodt Chem Works, 49-50, admin asst, 50-55, prod control mgr, 55-61, vpres & mgr, 61-67; asst to vpres, Kerr-McGee Corp, 67-71, dir & vpres nuclear licensing & regulations, Kerr-McGee Oil Industs, Kerr-McGee Corp, 71-84; RETIRED. *Mem:* Am Inst Chem Engrs; Am Chem Soc; Am Mgt Asn. *Mailing Add:* 42 Mayfair Dr Bella Vista AR 72714-5332

SHELLHAMER, DALE FRANCIS, b Tamaqua, Pa, Dec 4, 42; m 71; c 2. ORGANIC CHEMISTRY. *Educ:* Univ Calif, Irvine, BA, 69; Univ Calif, Santa Barbara, PhD(org chem), 74. *Prof Exp:* Assoc prof chem, 74-81, PROF CHEM, POINT LOMA COL, 81- *Concurrent Pos:* Am Heart Asn grant, 78. *Mem:* Am Chem Soc. *Res:* Electrophilic additions to alkenes, alkynes and dienes; physical organic properties of fluorinated hydrocarbons. *Mailing Add:* Dept Chem Point Loma Col 3900 Lomaland Dr San Diego CA 92106

SHELLHAMER, ROBERT HOWARD, anatomy, histochemistry; deceased, see previous edition for last biography

SHELLHAMMER, HOWARD STEPHEN, b Woodland, Calif, Aug 30, 35; m 56; c 1. MAMMALS, ECOLOGY. *Educ:* Univ Calif, Davis, BA, 57, PhD(zool), 61. *Prof Exp:* From asst prof to assoc prof, 61-70, PROF BIOL SCI, SAN JOSE STATE UNIV, 70- *Concurrent Pos:* Consult, H T Harvey & Assoc, Alviso, Calif. *Mem:* Am Soc Mammal; Animal Behav Soc; Wildlife Soc; Soc Conserv Biol. *Res:* Evolution and ecology of salt marsh harvest mice and other California mammals; ecology of large mammals; fire ecology; interactions with prescribed burning; behavior-ethology. *Mailing Add:* Dept Biol Sci San Jose State Univ One Washington Sq San Jose CA 95192-0100

SHELLOCK, FRANK G, b Glendale, Calif, Dec 16, 54. CARDIOVASCULAR PHYSIOLOGY. *Educ:* Columbia Pac Univ, PhD(physiol), 82. *Prof Exp:* RES SCIENTIST, CEDAR'S-SINAI MED CTR, LOS ANGELES, CALIF, 82- *Mailing Add:* Cedars-Sinai Med Ctr 8700 Beverly Blvd Los Angeles CA 90048

SHELLY, DENNIS C, b Chambersburg, Pa, Feb 16, 55; m 78; c 2. ANALYTICAL CHEMISTRY. *Educ:* Huntington Col, BS, 77; Tex A&M Univ, PhD(anal chem), 82. *Prof Exp:* Sr chemist, Lilly Res Labs, Eli Lilly & Co, 81-82; postdoctoral fel, Dept Chem, Ind Univ, 83-84; asst prof, Stevens Inst Technol, 84-90; ASST PROF, TEX TECH UNIV, 90- *Mem:* Am Chem Soc; Soc Appl Spectros; Sigma Xi. *Res:* Development of bioanalytical instrumentation with unique applicability to the rapid identification of bacteria and metabolic profiling of isolated tissue; behavior of polymers in microenvironments and on-line monitoring of polymer processing. *Mailing Add:* Dept Chem & Biochem Tex Tech Univ MS 1061 Lubbock TX 79409-1061

SHELLY, JAMES H, b Zanesville, Ohio, Nov 28, 32; m 56; c 3. MATHEMATICS, COMPUTER SCIENCE. *Educ:* Oberlin Col, BA, 54; Univ Ill, Urbana, AM, 56, PhD(math), 59. *Prof Exp:* Assoc engr, IBM Corp, 59-62, sr assoc engr, 62-64, staff engr, 64-66, proj engr, 66-67, adv engr, 67-73, sr engr, 73-85, mem sr tech staff, 85-87; vis lectr, 87-90, LECTR, NC STATE UNIV, 90- *Mem:* Asn Comput Mach; Soc Indust & Appl Math; Inst Elec & Electronic Engrs; Inst Elec & Electronic Engrs Comput Soc. *Res:* Processor and systems design and development; logical design; switching theory; combinatorial mathematics. *Mailing Add:* 1008 Bayfield Dr Raleigh NC 27606

SHELLY, JOHN RICHARD, b Sellersville, Pa, Jan 19, 49; m 74; c 1. FOREST PRODUCTS MANUFACTURING & WOOD BUILDING DESIGN. *Educ:* Pa State Univ, BS, 70; Univ Calif, Berkeley, MS, 77, PhD, 88. *Prof Exp:* Asst prof forest prod, Dept Forestry, Univ Ky, 81-86; res asst, Forest Prod Lab, 74-81, RES SCIENTIST, WOOD BLDG RES CTR, UNIV CALIF, BERKELEY, 88- *Concurrent Pos:* Extension specialist, Coop Extension, Univ Calif, 77-78. *Mem:* Forest Prod Res Soc; Soc Wood Sci & Technol; Sigma Xi. *Res:* Performance of wood products in structures; the theory of the flow of fluids through porous materials; principles and practice of wood drying methods, in particular energy efficiency in commercial lumber drying operations. *Mailing Add:* Forest Prod Lab 1301 S 46th St Richmond CA 94804

SHELOKOV, ALEXIS, b China, Oct 18, 19; nat US; m 47; c 1. VIROLOGY. *Educ:* Stanford Univ, AB, 43, MD, 48; Am Bd Microbiol, dipl; Am Bd Prev Med, dipl. *Prof Exp:* Physiologist, Climatic Res Lab, US War Dept, 43-44; res asst, Stanford Univ, 46-47; res asst med, Sch Med, Boston Univ, 48-49, instr, 48-50; med officer, Lab Infectious Dis, Nat Inst Allergy & Infectious Dis, NIH, 50-57, dir, Mid Am Res Unit, CZ, 57-61, chief lab trop virol, Nat Inst Allergy & Infectious Dis, 59-63, chief lab virol & rickettsiol, Div Biologics Standards, 63-68; prof microbiol & chmn dept, Univ Tex Health Sci Ctr San Antonio, 68-81; prof, 81-85, ADJ PROF, SCH HYG & PUB HEALTH, JOHNS HOPKINS UNIV, 85-; DIR VACCINE RES, SALK INST, 81- *Concurrent Pos:* House officer, Mass Mem Hosps, Boston, 47-50; asst pediat, Harvard Med Sch, 49-50; clin instr med, Georgetown Univ, 53-57; consult, DC Gen Hosp, 55-57, Gorgas, Coco Solo Hosps, CZ, 58-61, Pan Am Health Orgn, 58-63 & 71, Fogarty Int Ctr, NIH, 71 & 75 & Geog Med Br, Nat Inst Allergy & Infectious Dis, 72-76; mem sci adv bd, Gorgas Mem Inst, Panama, 59-72, 76-87 & mem bd dirs, 88-; exec coun, Am Comt Arthropod-Borne Viruses, 59-67; mem US Deleg Virus Dis, USSR, 61, WHO, 66; panel for arboviruses, Nat Inst Allergy & Infectious Dis, 62-66; ad hoc mem, Int Ctr Comt, NIH, 72-75; mem sci adv bd, WHO Serum Bank, Yale Univ, 64-68; chmn, US Deleg Hemorrhagic Fevers, USSR, 65 & 69; mem virol study sect, div res grants, NIH, 68-70; mem bd trustess, Am Type Cult Collection, 69-72; vis prof, Fac Med & Inst Hyg, Univ of the Republic, Uruguay, 71; bd sci counr, Nat Inst Dent Res, NIH, 71-75; mem viral dis panel, US-Japan Coop Med Sci Prog, 71-76; mem, Am Trop Med Deleg, China, 78; mem, Working Group Biol Weapon Control, Comn Int Security & Arms Control, Nat Acad Sci, 88-, core group, Expert Working Group Biol & Toxic Weapon Verification, FAS, 89 & Comn Microbiol Threats to Health, Inst Med, Nat Acad Sci, 90- *Honors & Awards:* Order of Rodolfo Robles, Guatemala, 59. *Mem:* Am Epidemiol Soc; Soc Exp Biol & Med; Am Soc Trop Med & Hyg; Am Asn Immunol; Infectious Dis Soc Am. *Res:* Epidemiology; preventive medicine; infectious diseases. *Mailing Add:* Salk Inst 7135 Minstrel Way Suite 203 Columbia MD 21045

SHELSON, W(ILLIAM), b Toronto, Ont, June 18, 22; m 58; c 2. ENERGY ANALYSIS, SYSTEM PLANNING. *Educ:* Univ Toronto, BASc, 44, PhD, 52; Pa State Univ, MS, 47. *Prof Exp:* Stress analyst, Curtiss-Wright Corp, 47-48; res assoc appl math, Brown Univ, 48-49; mech engr, Can Stand Asn, 50-51; res engr, Ont Hydro, 52-56, chmn, opers res group, 56-67, mgr opers res, 67-74, mgr fuel resources planning, 75-80, mgr energy resources planning, 80-83, CONSULT, ONT HYDRO, 84- *Concurrent Pos:* Special lectr, Univ Toronto, 56-57. *Mem:* Can Oper Res Soc. *Res:* Operations research; systems analysis; stress analysis; energy analysis and planning of electric power systems; evaluation and optimization of primary energy supply alternatives. *Mailing Add:* 53 Evanston Dr Downsview ON M3H 5P4 Can

SHELTON, DAMON CHARLES, b Richland, Ind, Apr 4, 22; m 43; c 2. BIOCHEMISTRY, NUTRITION. *Educ:* Purdue Univ, BSA, 47, MS, 49, PhD(agr biochem), 50. *Prof Exp:* Instr agr chem, Purdue Univ, 49-50, fel, 51-52; pvt bus, 52-53; from assoc prof to prof agr biochem, WVa Univ, 53-60, from assoc prof to prof med biochem, Sch Med, 60-67; res mgr, 67-75, res dir, Ralston Purina Co, 75-87; PRES, DAMON C SHELTON CONSULTS, INC, 87- *Concurrent Pos:* Assoc animal nutritonist, Ala Polytech Inst, 50-51. *Mem:* Am Chem Soc; Am Inst Nutrit; Am Soc Biochem & Molecular Biol; AAAS; Am Assoc Lab Analytical Sci; Can Assoc Lab Anal Sci. *Res:* Blood proteins; amino acids; antibiotics and vitamins in nutrition and pathology; microbiology; mineral and antibiotic interrelationships; lipid-protein interactions; biological transport-peptides and amino acids; nutrition management of research and special animals. *Mailing Add:* 9338 Lincoln Dr St Louis MO 63127

SHELTON, EMMA, b Urbana, Ill, June 10, 20. CELL BIOLOGY, ELECTRON MICROSCOPY. *Educ:* Brown Univ, PhD(biol), 49. *Prof Exp:* Jr biologist, Nat Cancer Inst, 44-46, res biologist, 49-78; secy, Am Soc Cell Biol, 78-81; RETIRED. *Concurrent Pos:* Vis biologist, Lab Electron Micros, Villejuif, France, 63-64; exec officer, Am Soc Cell Biol, 78-81. *Honors & Awards:* Superior Serv Award, USPHS, 78. *Mem:* Fel AAAS; Am Soc Cell Biol; Am Asn Path; Histochem Soc; Am Asn Cancer Res; Sigma Xi. *Res:* Fine structure of ribosomes, enzymes, immunoglobulins; electron microscopy of cell interactions in the immune response. *Mailing Add:* 8410 Westmont Terr Bethesda MD 20817-6813

SHELTON, FRANK HARVEY, b Flagstaff, Ariz, Oct 4, 24; m 48; c 3. NUCLEAR PHYSICS. *Educ:* Calif Inst Technol, BS, 49, MS, 50, PhD(physics), 52. *Prof Exp:* Res analyst, NAm Aviation, Inc, 50; mem staff, Sandia Corp, 51-55; tech dir, Armed Forces Spec Weapons Proj, US Dept Defense, 55-59; sr scientist, Nuclear Div, Kaman Aircraft Corp, 59-68, VPRES & CHIEF SCIENTIST, KAMAN SCI CORP, 68- *Concurrent Pos:* Mem subcomt civil defense, Nat Acad Sci, 57; mem sci adv group effects, Defense Nuclear Agency, Dept Defense, 74- *Mem:* Fel Am Phys Soc. *Res:* Military effects of nuclear weapons and missile applications; peaceful uses of nuclear detonations. *Mailing Add:* 1327 Culebra Ave Colorado Springs CO 80903

SHELTON, GEORGE CALVIN, b Tex, Apr 26, 23; m 49; c 2. VETERINARY PARASITOLOGY. *Educ:* Tex A&M Univ, DVM, 48; Auburn Univ, MS, 52; Univ Minn, PhD(vet microbiol), 65. *Prof Exp:* Asst prof vet bact & parasitol, Univ Mo, 49-51; res assoc, Auburn Univ, 52; from asst prof to prof vet bact & parasitol, Univ Mo-Columbia, 52-59, prof vet microbiol, Sch Vet Med, 59-73, assoc dean, Col Vet Med, 69-73, assoc dean acad affairs, 71-73; prof & emer dean, Col Vet Med, Tex A & M Univ, 73-88; RETIRED. *Concurrent Pos:* NSF fac fel, 61-62. *Mem:* Am Vet Med Asn; Am Soc Parasitol; Conf Res Workers Animal Dis. *Res:* Internal parasites of ruminants. *Mailing Add:* 7851 S Tomlin Hill Rd Columbia MO 65201

SHELTON, JAMES CHURCHILL, b Kansas City, Mo; m 63; c 4. MATERIALS SCIENCE, APPLIED PHYSICS. *Educ:* Cornell Univ, BEP, 63, PhD, 73. *Prof Exp:* Instr nuclear reactor eng, US Naval Nuclear Power Sch, 63-69; mem tech staff mat res & integrated optics, Bell Labs, 72-80; corp prod planning & mkt div, Western Elec Co, 80-85; MGR, ADVAN TECHNOL, AT&T NETWORK SYSTS, 85- *Concurrent Pos:* Grad fel, NSF. *Mem:* AAAS; Inst Elec & Electronic Engrs. *Res:* Communications and information network architectures, especially integrated and packet, photonic, broadband; technology assessment and planning; integrated optics, especially semiconductor laser sources, optical waveguides, switches, modulators, polarizers and detectors; surface physics, especially surface segregation, electron beam-solid interactions and diagnostics. *Mailing Add:* Two Winthrop Dr Holmdel NJ 07733

SHELTON, JAMES EDWARD, b Allais, Ky, Sept 26, 29; m 53; c 2. SOIL CHEMISTRY, SOIL FERTILITY. *Educ:* Univ Ky, BS, 53, MS, 57; NC State Univ, PhD(soils), 60. *Prof Exp:* From instr to asst prof, 59-78, ASSOC PROF SOILS, NC STATE UNIV, 78- *Mem:* Am Soc Agron; Soil Sci Soc Am; Int Soc Soil Sci; Am Soc Hort Sci. *Res:* Role of fertilizers in soil-plant relationships. *Mailing Add:* Mountain Hort Crops Res & Ext Ctr 2016 Fanning Br Rd Fletcher NC 28732

SHELTON, JAMES MAURICE, animal breeding, genetics, for more information see previous edition

SHELTON, JAMES REID, b Allerton, Iowa, Jan 16, 11; m 34; c 3. ORGANIC CHEMISTRY, POLYMER SCIENCE. *Educ:* Univ Iowa, BS, 33, MS, 34, PhD(org chem), 36. *Prof Exp:* Asst org chem, Univ Iowa, 35-36; instr chem, 36-41, from asst prof to assoc prof org chem, 41-48, dean grad studies, 66-67, prof org chem, 49-77, prof polymer sci, 68-77, EMER PROF CHEM & MACROMOL SCI, CASE WESTERN RESERVE UNIV, 77- *Honors & Awards:* Charles Goodyear Medal, Rubber Div, Am Chem Soc, 83. *Mem:* AAAS; Am Chem Soc. *Res:* Mechanism of oxidation and antioxidant action in rubber and related systems; mechanism of organic reactions; high polymers; organic sulfur compounds; reaction of free radicals with olefins; reactions of peroxides. *Mailing Add:* Dept Chem Case Western Reserve Univ Cleveland OH 44106

SHELTON, JOHN C, b Renovo, Pa, June 8, 37; m 62. ORGANIC CHEMISTRY. *Educ:* Lock Haven State Col, BS, 59; Cornell Univ, PhD(org chem), 64. *Prof Exp:* Instr chem, City Col San Francisco, 64-65; from asst prof to assoc prof org chem, 65-73, PROF ORG CHEM, CALIF STATE UNIV HAYWARD, 73- *Mem:* Am Chem Soc. *Res:* Highly strained bicyclic ring compounds. *Mailing Add:* Dept of Chem Calif State Univ Hayward CA 94542

SHELTON, JOHN WAYNE, b China Spring, Tex, Dec 28, 28; m 49; c 2. GEOLOGY. *Educ:* Baylor Univ, BA, 49; Univ Ill, MS, 51, PhD(geol), 53. *Prof Exp:* Asst, Univ Ill, 50-52; geologist, Shell Oil Co, 53-63; from asst prof to assoc prof, 63-70, prof geol, Okla State Univ, 70-88. *Concurrent Pos:* Consult, Continental Oil Co, 64- *Mem:* Geol Soc Am; Soc Econ Paleont & Mineral; Am Asn Petrol Geologists. *Res:* Sedimentation; structural geology. *Mailing Add:* Dept Geol Okla State Univ Stillwater OK 74078

SHELTON, KEITH RAY, b Chatham, Va, Jan 11, 41; m 65; c 3. BIOCHEMISTRY. *Educ:* Univ Va, BA, 63; Univ Ill, Urbana, PhD(biochem), 68. *Prof Exp:* Res assoc biochem, Rockefeller Univ, 67-69; Nat Res Coun Can fel, 69-70; from asst prof to assoc prof, 70-90, PROF BIOCHEM, MED COL VA, VA COMMON WEALTH UNIV, 90- *Mem:* Am Soc Cell Biol; Am Soc Biol Chemists; Am Chem Soc. *Res:* Molecular and cellular responses to toxic metals. *Mailing Add:* Dept Biochem & Molecular Biophys Va Commonwealth Univ Richmond VA 23298-0614

SHELTON, ROBERT NEAL, b Phoenix, Ariz, Oct 5, 48; m 69; c 3. SOLID STATE PHYSICS. *Educ:* Stanford Univ, BS, 70, MS, 73; Univ Calif, PhD(physics), 75. *Prof Exp:* From asst prof to prof physics, Iowa State Univ, 78-87; PROF PHYSICS, UNIV CALIF, DAVIS, 87-, VCHANCELLOR, RES ADMIN, 90- *Mem:* Am Phys Soc; Sigma Xi. *Res:* Experimental condensed matter physics- emphasis on novel materials; superconductivity; magnetism; correlated electron systems. *Mailing Add:* Univ Calif Davis CA 95616

SHELTON, ROBERT WAYNE, b Springfield, Ill, Dec 3, 23; m 46; c 3. ORGANIC CHEMISTRY. *Educ:* Ill Col, AB, 49; Univ Iowa, PhD(chem), 54. *Prof Exp:* Res chemist, E I du Pont de Nemours & Co, 53-56; from assoc prof to prof chem, 56-86, head dept, 58-66, EMER PROF, WESTERN ILL UNIV, 86- *Mem:* Am Chem Soc. *Res:* Synthesis; organophosphorus compounds. *Mailing Add:* 1103 Willow Ct Estes Park CO 80517-3441

SHELTON, RONALD M, b Shipman, Ill, July 11, 31; m 53; c 2. MATHEMATICS. *Educ:* Univ Ill, BS, 53, MS, 57, PhD(math educ), 65. *Prof Exp:* From asst prof to assoc prof, 60-61, PROF MATH, MILLIKIN UNIV, 71-, CHMN DEPT, 63- *Concurrent Pos:* Consult, Decatur Pub Schs, 67. *Mem:* Math Asn Am. *Res:* Teaching of mathematics at the college level. *Mailing Add:* Dept of Math Millikin Univ 1184 W Main St Decatur IL 62522

SHELTON, WILFORD NEIL, b Dalton, Ga, Dec 29, 35; m 65; c 3. ELECTRON PHYSICS. *Educ:* Univ Calif, Los Angeles, AB, 58; Fla State Univ, PhD(physics), 62. *Prof Exp:* From res asst to res assoc, 60-63, from actg asst prof to assoc prof, 63-76, PROF PHYSICS, FLA STATE UNIV, 76- *Mem:* Am Phys Soc. *Res:* Electron scattering on atoms and molecules; resonant charge exchange between atoms; electron-impact ionization of atoms; electron-photon angular correlations; states of nuclei; fourier transform infrared spectroscopy. *Mailing Add:* Dept Physics Fla State Univ Tallahassee FL 32306

SHELTON, WILLIAM LEE, b Tulsa, Okla, May 28, 39; m 65; c 1. FISHERIES. *Educ:* Okla State Univ, BS, 61, MS, 64; Univ Okla, PhD(zool), 72. *Prof Exp:* Fishery biologist, US Corps Engrs, Tulsa Dist, 67; ASST LEADER, ALA COOP FISHERY UNIT, US FISH & WILDLIFE SERV, AUBURN UNIV, 71- *Mem:* Am Soc Ichthyologists; Am Fisheries Soc; Sigma Xi. *Res:* Fishery biology in reservoirs and rivers; reproductive biology of fishes. *Mailing Add:* 1169 N Augusta Ave Camarillo CA 93010

SHELUPSKY, DAVID I, b New York, NY, Dec 9, 37; m 63. PHYSICS. *Educ:* City Col New York, BS, 59; Princeton Univ, MA, 61, PhD(physics), 65. *Prof Exp:* From instr to asst prof, 64-71, ASSOC PROF PHYSICS, CITY COL NEW YORK, 72- *Mem:* Am Math Soc; Am Phys Soc; Math Asn Am. *Res:* Axiomatic quantum field theory; algebraic methods in statistical mechanics. *Mailing Add:* Dept of Physics City Col of New York Convent at 138th St New York NY 10031

SHELVER, WILLIAM H, b Ortonville, Minn, May 31, 34; m 58. MEDICINAL CHEMISTRY, ORGANIC CHEMISTRY. *Educ:* NDak State Univ, BS, 56, MS, 57; Univ Va, PhD(org chem), 62. *Prof Exp:* From asst prof to assoc prof pharmaceut chem, 60-68, PROF PHARMACEUT SCI, NDAK STATE UNIV, 68-,. *Mem:* Am Chem Soc; Am Pharmaceut Asn; Sigma Xi. *Res:* Development of relationships between structure and biological activity by synthesis of new compounds; measurement of physical properties of new and existing compounds, especially in analgesic and hypotensive drugs. *Mailing Add:* Dept of Pharmaceut Chem/BTO NDak State Univ Fargo ND 58102

SHEMANCHUK, JOSEPH ALEXANDER, b Wostok, Alta, Apr 28, 27; m 51; c 2. VETERINARY ENTOMOLOGY. *Educ:* Univ Alta, BScAgr, 50, MSc, 58. *Prof Exp:* Entomologist med entom, Household Med Entom Unit, Can Dept Agr, Ottawa, 50-51 & Dominion Entom Lab, Saskatoon, 51-55; ENTOMOLOGIST VET-MED ENTOM, AGR CAN RES STA, LETHBRIDGE, 55- *Concurrent Pos:* Scientist exchange fel, Nat Res Coun Can & USSR Acad Sci, 71, Nat Res Coun Can & Czechoslovak Acad Sci, 80. *Mem:* Fel Entom Soc Can; Am Mosquito Control Asn; Can Soc Zool. *Res:* Behavior, culture and biological control of blood-sucking flies; development of repellents for protection of man and livestock; epidemiology of insect-borne diseases in livestock. *Mailing Add:* Agr Can Res Sta Lethbridge AB T1J 4B1 Can

SHEMANO, IRVING, b San Francisco, Calif, June 23, 28; m 56; c 2. PHARMACOLOGY. *Educ:* Univ Calif, AB, 50, MS, 51; Univ Man, PhD(pharmacol), 56. *Prof Exp:* Asst pharmacologist, Abbott Labs, 51-53; sr pharmacologist, Smith Kline & French Labs, 56-60; dir macrobiol res labs, Merrell-Nat Labs, 60-70, head dept immunol & endocrinol, 70-74, group dir, Clin Pharmacol Dept, 74-78; ASSOC DIR CLIN DEVELOP-ONCOL, ADRIA LABS, 78- *Mem:* Am Soc Pharmacol & Exp Therapeut; Am Soc Clin Pharmacol & Therapeut; Am Soc Clin Oncol. *Res:* Immunopharmacology; gastrointestinal pharmacology, cancer res. *Mailing Add:* Adria Lab Inc PO Box 16529 Columbus OH 43216

SHEMANSKY, DONALD EUGENE, b Moose Jaw, Sask, Apr 28, 36; m 67; c 4. ATMOSPHERIC PHYSICS. *Educ:* Univ Sask, BE, 58, MSc, 60, PhD(auroral physics), 66. *Prof Exp:* Physicist, Bristol Aero Industs, Ltd, 60-61; res asst, Univ Sask, 61-66; jr physicist, Kitt Peak Nat Observ, Ariz, 66-69; res asst prof physics, Univ Pittsburgh, 69-74; res assoc prof physics, Univ Mich, 74-78; res assoc, Univ Ariz, 78-79; sr res scientist, Tuscon Labs, Space Sci Inst, Univ Southern Calif, 79-84; SR RES SCIENTIST, LUNAR PLANETARY LAB, UNIV ARIZ, 84- *Concurrent Pos:* Assoc ed, J Geophys Res, 81-84; prin investr, NASA, NSF, DOE grants; co-investr, Voyager, Galileo, Cassini Spacecraft Experiments; comt mem, Nat Acad Sci Workshop, 85, NASA Planetary Atmospheres Rev Comts, NASA Saturn-Titan Voyager Workshop, 80, NASA Io Torus Plasma Workshop, 89, NSF CEDAR Workshops, NASA LEXSWG Atmospheric Sci Workshop, 90. *Honors & Awards:* Exceptional Sci Achievement Medal, NASA, 81, Group Achievement Award, Voyager Sci Instrument Develop, 81, Group Achievement Award, Voyager Sci Invest, 81, 86 & 90. *Mem:* Am Geophys Union; Am Astron Soc; Planetary Soc; Am Phys Soc. *Res:* Physics of Io plasma torus and Jupiter, Saturn, Uranus, Neptune and Titan atmosphere-magnetosphere interactions atmosphere-magnetosphere interactions; interstellar medium and cygnus loop study in extreme ultraviolet; theoretical physics of gas-surface interactions; earth atmosphere/magnetosphere; lunar, mercury atmosphere; laboratory astrophysics; atomic, molecular physics and chemistry; accretion disk theory. *Mailing Add:* 6655 E Calle De San Alberto Tucson AZ 85710

SHEMDIN, OMAR H, b Zakho, Iraq, Sept 12, 39; US citizen; m 65; c 3. OCEANOGRAPHIC ENGINEERING, OCEANOGRAPHY. *Educ:* Mass Inst Technol, BSc, 61, MSc, 62; Stanford Univ, PhD(eng), 66. *Prof Exp:* Asst prof civil eng, Stanford Univ, 66-67; from asst prof to assoc prof, 67-77, PROF COASTAL ENG, UNIV FLA, 77-; at Jet Propulsion Lab, Pasadena, Calif, 77-87, CHIEF SCI, OCEAN RES ENG, JET PROPULSION LAB, PASADENA, CALIF, 87- *Concurrent Pos:* Consult, Esso Res & Eng Co, 68-; consult res scientist, Naval Res Lab, 70-72, dir lab, 72-; consult res oceanographer, Nat Oceanic & Atmospheric Admin, 71-72. *Mem:* Am Geophys Union; Am Soc Civil Engrs; Am Meteorol Soc. *Res:* Air-sea interaction; coastal and near shore precesses. *Mailing Add:* 727 Georgian Rd La Canada CA 91011

SHEMENSKI, ROBERT MARTIN, b Martins Ferry, Ohio, June 17, 38; m 60; c 3. METALLURGICAL ENGINEERING. *Educ:* Univ Cincinnati, MetE, 61; Ohio State Univ, PhD(metall eng), 64. *Prof Exp:* Mem tech staff, Bell Tel Labs, 66-67; sr scientist, Battelle Mem Inst, 67-70; sr scientist, Fabric Develop Dept, 70-77, prin metall, 77-, MGR WIRE SCI & TECHNOL, GOODYEAR TIRE & RUBBER CO. *Mem:* Am Soc Metals; Am Inst Mining, Metall & Petrol Engrs; Nat Asn Corrosion Engrs. *Res:* Anger spectroscopy; secondary ion mass spectrometry; iron filamentary single crystals metallic dissolution kinetics; internal friction; properties of beryllium; scanning electron microscopy; failure analyses; fatigue; fiber reenforcement systems; x-ray photoelectron spectroscopy. *Mailing Add:* 204 Sutton Ave NE North Canton OH 44720

SHEMER, JACK EVVARD, b Phoenix, Ariz, Aug 22, 40; m 63; c 1. COMPUTER SCIENCE, ELECTRICAL ENGINEERING. *Educ:* Occidental Col, BA, 62; Ariz State Univ, MS, 65; Southern Methodist Univ, PhD(elec eng), 68. *Prof Exp:* Sr engr, Gen Elec Co, Phoenix, 62-67; sect mgr, Sci Data Systs, Santa Monica, Calif, 68-72; prin scientist & area mgr, Xerox Palo Alto Res Ctr, El Segundo, Calif, 72-76; vpres systs eng, Transaction Technol Inc, Citicorp, 76-79; chief exec officer, 79-87, chmn, 87-88, VCHMN, TERADATA CORP, 88- *Concurrent Pos:* Tech ed, Computer, 73-76. *Honors & Awards:* Spec Award, Comput Soc, 76. *Mem:* Inst Elec & Electronics Engrs; Asn Comput Mach. *Res:* Computer architecture; file and database management, performance analysis; queueing theory and applications; automatic control systems; digital image coding and display. *Mailing Add:* Teradata Corp 100 N Sepulveda Blvd 19th Floor El Segundo CA 90245

SHEMILT, L(ESLIE) W(EBSTER), b Souris, Man, Dec 25, 19; m 46; c 2. CHEMICAL ENGINEERING. *Educ:* Univ Toronto, BASc, 41, PhD(phys chem), 47; Univ Man, MSc, 46. *Prof Exp:* Lab supvr, Defence Industs, Ltd, 41-42, supvr tech dept & acid plant supvr, 42-44; lectr chem, Univ Man, 44-45; spec lectr, Univ Toronto, 45-47; from asst prof to prof, Univ BC, 47-60; prof chem eng & head dept, Univ NB, 61-69; dean fac eng, 69-79, prof, 69-87, EMER PROF CHEM ENG, MCMASTER UNIV, 87- *Concurrent Pos:* Vis prof, Univ Col, London, 59-60 & Ecole Polytech Lausanne, 75, Indian Inst Technol, Kanpur, 75, Madras, 75, Univ Sydney, 81 & spec vis prof, Yokohama Nat Univ, 87; chmn, NB Res & Productivity Coun, 62-69; sci adv, Prov NB, 64-69; chmn, Atlantic Prov Inter-Univ Comt on Sci, 66-69; mem, Nat Res Coun Can, 66-69; ed, Can J Chem Eng, 67-84, emer ed, 87-; chmn tech adv comt, Nuclear Fuel Waste Mgt, 79-; int ed, Chem Eng Res & Design, 84-89. *Honors & Awards:* T P Hoar Prize, Inst Corrosion Sci & Technol, 80; RS Jane Mem Lect Award, Can Soc Chem Eng, 85. *Mem:* Am Chem Soc; fel Am Inst Chem Engrs; hon fel Chem Inst Can (pres, 70-71); fel Eng Inst Can; Can Res Mgt Asn; fel Royal Soc Can; fel Can Acad Eng. *Res:* Fundamentals of corrosion; chemical engineering thermodynamics; industrial wastes; mass transfer; process dynamics. *Mailing Add:* Fac Eng McMaster Univ Hamilton ON L8S 4K1 Can

SHEMIN, DAVID, b New York, NY, Mar 18, 11; m 37, 63; c 2. BIOCHEMISTRY. *Educ:* City Col, BS, 32; Columbia Univ, AM, 33, PhD(biochem), 38. *Prof Exp:* Asst biochem, Col Physicians & Surg, Columbia Univ, 35-37, immunochem & virus res, 37-40, from instr to prof biochem, 40-68; PROF BIOCHEM, NORTHWESTERN UNIV, 68- *Concurrent Pos:* Lectr & scientist, Karolinska Inst, Stockholm & Swedish Med Res Coun, 47; Harvey lectr, 54; Guggenheim fel, 56 & 70; Commonwealth fel, 65; vis prof, Weizman Inst Sci; mem study sect, NIH, NSF; dep dir, Cancer Ctr, Northwestern Univ, 75-; Fogarth scholar, NIH, 80-83. *Honors & Awards:* Pasteur Medal. *Mem:* Nat Acad Sci; AAAS; Am Soc Biol Chem; Am Chem Soc; Am Acad Arts & Sci; hon mem Japanese Biochem Soc; hon mem Swiss Biochem Soc; Brit Biochem Soc. *Res:* Biosynthesis of porphyrins B12; porphyria enzymology; protein structure; control systems; microbiology; chlorophyll synthesis. *Mailing Add:* 33 Lawrence Farm Rd Woods Hole MA 02543

SHEN, BENJAMIN SHIH-PING, b Hangzhou, China, Sept 14, 31; nat US; m 71; c 2. ASTROPHYSICS, ENGINEERING SCIENCE. *Educ:* Assumption Col, AB, 54; Clark Univ, AM, 56; Univ Paris, DSc d'Etat (physics), 64. *Prof Exp:* Asst prof physics, State Univ NY Albany, 56-59; assoc prof space sci, aeronaut & astronaut, Sch Eng, NY Univ, 64-66; from assoc prof to prof astron & astrophys, Univ Pa, 66-72, chmn dept, 73-79, dir, Flower & Cook Observ, 73-79, assoc univ provost, 79-80, chmn, Coun Grad Deans, 79-81, Univ provost, 80-81, REESE W FLOWER PROF ASTRON & ASTROPHYS, UNIV PA, 72- *Concurrent Pos:* Consult, Space Sci Lab, Gen Elec Co, 61-68; guest staff mem, Brookhaven Nat Lab, 63-64, 65-70; gen chmn, Int Conf Spallation Nuclear Reactions & Their Appln , 75; assoc ed, Comments Astrophys, 79-85; mem, US Nat Sci Bd, 90- *Honors & Awards:* Vermeil Medal for Sci, Soc Encourgement of Progress, France, 78. *Mem:* Fel AAAS; fel Am Phys Soc; Int Astron Union; fel Royal Astron Soc. *Res:* Spallation nuclear reactions in astronomy and space- radiation shielding engineering; galaxies and quasars; science & technology policy. *Mailing Add:* Dept Astron & Astrophys Univ Pa Philadelphia PA 19104-6394

SHEN, CHE-KUN JAMES, b Taipei, Taiwan, 1949. BIOCHEMISTRY, MOLECULAR BIOLOGY. *Educ:* Nat Taiwan Univ, BS, 71; Univ Calif, Berkeley, PhD(biochem), 77. *Prof Exp:* Elec officer electronics & mech radar div, Navy Repub China, 71-73; NIH res fel biol, Calif Inst Technol, 78-80; asst prof, 81-83, ASSOC PROF GENETICS, UNIV CALIF, DAVIS, 83- *Mem:* Sigma Xi. *Res:* Molecular biology; human molecular genetics. *Mailing Add:* Dept Genetics Univ Calif Davis CA 95616

SHEN, CHIH-KANG, b Chekiang, China, Sept 21, 32; m 64; c 1. SOIL MECHANICS. *Educ:* Nat Taiwan Univ, BS, 56; Univ NH, MS, 60; Univ Calif, Berkeley, PhD(soil mech), 65. *Prof Exp:* Asst prof, Loyola Univ, Calif, 65-70; asst prof, 70-76, assoc prof, 76-80, PROF CIVIL ENG, UNIV CALIF, DAVIS, 80- *Concurrent Pos:* Res assoc, Univ Calif, Los Angeles, 66-; mem Hwy Res Bd, Nat Acad Sci-Nat Res Coun, 66- *Mem:* Am Soc Civil Engrs; Int Soc Soil Mech & Found Engrs. *Res:* Soil stabilization and compaction; flexible pavement design; shear strength of unsaturated soils. *Mailing Add:* Dept Genetics Univ Calif Davis CA 95616

SHEN, CHI-NENG, b Peiping, China, July 18, 17; m 47; c 2. CONTROL SYSTEMS, ENGINEERING SYSTEMS. *Educ:* Nat Tsing Hua Univ, China, BEng, 39; Univ Minn, MS, 50, PhD(eng), 54. *Prof Exp:* Instr mech eng, Univ Minn, 51-54; asst prof, Dartmouth Col, 54-58; from assoc prof to prof mech eng, 58-67, PROF ELEC & SYSTS ENG, RENSSELAER POLYTECH INST, 73- *Concurrent Pos:* Vis prof mech eng, Mass Inst Technol, 67-68; consult. *Mem:* Am Soc Mech Engrs; Am Nuclear Soc; Am Inst Aeronaut & Astronaut; Am Soc Eng Educ; Sigma Xi. *Res:* Martian vehicle navigation, including obstacle detection, terrain modeling and park selection; nuclear reactor stability for kinetics and two-phase flow phenomena; automatic controls, including nonlinear control systems and estimation theory; guidance and navigation. *Mailing Add:* Village Dr Troy NY 12180

SHEN, CHIN-WEN, b Shanghai, Oct 14, 24; m; c 3. OIL WELL DRILLING, ENHANCED OIL RECOVERY. *Educ:* Nat Chiao Tung Univ, BS, 46; Univ Tulsa, MS, 65; Tex A&M Univ, PhD(petrol eng), 69. *Prof Exp:* Mining engr coal mining, Fusin Colliery Co, China, 46-48; petrol engr drilling & production, Chinese Petrol Corp, 48-64; res scientist, Thermal Recovery, Getty Oil, Co, Houston, 69-84; RES ASSOC, THERMAL RECOVERY, HOUSTON EPTD, TEXACO, 84- *Mem:* Am Soc Petrol Engrs. *Res:* Steam displacement. *Mailing Add:* 11618 Cherry Knoll Houston TX 77077

SHEN, CHUNG YI, b Canton, China, May 23, 37; US citizen; m 82; c 3. APPLIED MATHEMATICS. *Educ:* Ore State Univ, BS, 60, MS, 63, PhD(math), 68. *Prof Exp:* PROF MATH, SIMON FRASER UNIV, 67- *Concurrent Pos:* Fel, Carnegie-Mellon Univ, 68-69; vis assoc prof, Nat Taiwan Univ, 74-75; consult, Northrop Corp, 85- *Mem:* Am Math Soc; Can Math Soc; Can Appl Math Soc. *Res:* Electromagnetic scattering; numerical methods. *Mailing Add:* R J Norton Co 21290 W Hillside Dr Topanga CA 90290

SHEN, CHUNG YU, b Dec 15, 21; m; c 3. FLUIDIZATION. *Educ:* Nat Southwestern Asn Univ, Kunming, China, BS, 42; Univ Louisville, MChE, 50; Univ Ill, PhD(chem eng), 54. *Prof Exp:* Sect chief, Cent Chem Works, Shanghai, China, 42-48; int fel, J E Seagram & Sons, Inc, 49-50; sr chem engr, Inst Indust Res, 50-52; sr fel, Monsanto Co, 54-86; CONSULT, SHEN & SHEN, INC, 86- *Concurrent Pos:* Mem, Chem Eng Prod Res Panel, 75-86; expert, People's Repub China. *Mem:* Fel Am Inst Chem Eng; Am Oil Chemist Soc; Am Res Soc; Sigma Xi; AAAS. *Res:* Process and product development; 65 US patents in phosphates, phosphonates, detergent processing, chelation and sequestion agents, various chemicals based on cyanides and fatty natural materials; 26 publications. *Mailing Add:* 12630 Conway Downs St Louis MO 63141

SHEN, COLIN YUNKANG, b Taichung, Taiwan. DOUBLE DIFFUSION CONNECTION, NUMERICAL MODELING. *Educ:* Univ Mass, BS, 71; Univ RI, PhD(phys oceanog), 77. *Prof Exp:* Postdoctoral, 78-81, sr oceanographer, Univ Wash, 82-83; OCEANOGRAPHER, NAVAL RES LAB, 84- *Concurrent Pos:* Vis scientist, Nat Ctr Atmospheric Res, 77-78. *Mem:* Am Geophys Union; Sigma Xi. *Res:* Double diffusive convection, fluid mixing, ocean internal waves, ocean eddies; specialize in numerical modeling of various fluid phenomena. *Mailing Add:* Atmospheric/Ocean Sensing Br Code 4220 Naval Res Lab Washington DC 20375

SHEN, D(AVID) W(EI) C(HI), b Shanghai, China, Jan 4, 20; nat US. ELECTRICAL ENGINEERING. *Educ:* Nat Tsing Hua Univ, China, BSc, 39; Univ London, PhD(elec eng), 48. *Prof Exp:* Lectr elec eng, Nat Univ Amoy, China, 40-44; with Marconi Wireless & Tel Co, Eng, 44-45; elec engr, Messrs Yangtse, Ltd, 46-48; sr lectr elec eng, Adelaide Univ, 50-53; vis asst prof, Univ Ill, 53-54, Mass Inst Technol, 54 & City Col New York, 55; from asst prof to assoc prof, 55-66, PROF ELEC ENG, MOORE SCH ELEC ENG, UNIV PA, 66- *Concurrent Pos:* Mem res staff, Stromberg-Carlson Div, Gen Dynamics Corp, NY; consult, CDC Control Serv, Inc, 55-; mem, Franklin Inst. *Mem:* Am Soc Eng Educ; fel Inst Elec & Electronics Engrs; fel Brit Inst Elec Eng; Tensor Soc; fel AAAS. *Res:* Electrical machinery; analogue computers; optimal and adaptive control theory; control in biological systems. *Mailing Add:* Moore Sch Elec Eng Univ Pa Philadelphia PA 19130

SHEN, HAO-MING, b Changzhou, China, 1933; m 59; c 3. ELECTROMAGNETIC PULSES, ELECTROMAGNETIC MEASUREMENT. *Educ:* Beijing Univ, China, Bachelor physics, 58; Chiaotung Univ, China, PhD(electronics eng), 66. *Prof Exp:* Teaching asst physics, Beijing Univ, 58-63; lectr, Harbin Civil Eng Col, 66-78, assoc prof elec eng, 78-79; res prof elec eng, Inst Electronics, Acad Sci, China, 83-86; RES ASSOC APPL PHYSICS, HARVARD UNIV, 86- *Concurrent Pos:* Mem, Comn E & var comts, Int Union Radio Sci, 85- *Mem:* Sr mem Inst Elec & Electronics Engrs Electromagnetic Compatibility Soc; sr mem Inst Elec & Electronics Engrs Microwave Theory & Technol Soc; Int Union Radio Sci. *Res:* Electromagnetic theory, antennas, microwaves, specifically in electromagnetic pulses, transient, electromagnetic compatibility. *Mailing Add:* Harvard Univ Nine Oxford St Cambridge MA 02138

SHEN, HSIEH WEN, b Peking, China, July 13, 31; m 56. HYDRAULICS. *Educ:* Univ Mich, BS, 53, MS, 54; Univ Calif, Berkeley, PhD(hydraul), 61. *Prof Exp:* Hydraul engr, US Army Corps Engrs, 55-56; struct engr, Giffels & Vallet, Inc, 56; res engr hydraul, Inst Eng Res, Univ Calif, 56-61; hydraul engr, Harza Eng Co, 61-63; assoc prof hydraul eng, 64-68, PROF HYDRAUL ENG, COLO STATE UNIV, FT COLLINS, 68- *Concurrent Pos:* Freeman fel, Am Soc Civil Engrs, 65-66; Guggenheim fel, 72; Humboldt Found sr res Award, Ger, 90-91. *Honors & Awards:* Horton Award, Am Geophys Union; Einstein Award, Am Soc Civil Engrs, 90-91. *Res:* Various aspects of fluvial hydraulics, including meandering, local scour, sediment transport under wind, stable channel shape and change of bed forms. *Mailing Add:* 412 O'Brien Hall Univ Calif Berkeley CA 94720

SHEN, KELVIN KEI-WEI, b Liao-pei, China, Sept 29, 41; m 68; c 2. POLYMER CHEMISTRY. *Educ:* Nat Taiwan Univ, BS, 64; Univ Mass, Amherst, MS, 66, PhD(chem), 68. *Prof Exp:* Guest scientist, Brookhaven Nat Lab, 66-68; res fel, Yale Univ, 68-69; instr chem, Drexel Univ, 69-70; asst prof, Calif State Univ, Los Angeles, 70-72; res chemist, US Borax Res Corp, 72-80, sr res chemist, 80-88; TECH SERV MGR, US BORAX, 88- *Concurrent Pos:* Petrol Res Fund grant, Calif State Univ, Los Angeles, 72. *Mem:* AAAS; Am Chem Soc; Soc Plastics Engrs. *Res:* Chemistry of small ring compounds; organometallic chemistry; fire retardants in plastics; x-ray crystallography; photochemistry; syntheses of herbicides; environmental studies of pesticides. *Mailing Add:* US Borax 3075 Wilshire Blvd Los Angeles CA 90010

SHEN, LESTER SHENG-WEI, b St Louis, Mo, Jul 8, 55. HEAT TRANSFER, THERMAL ENVIRONMENTAL ENGINEERING. *Educ:* Haverford Col, BS, 77; Ga Inst Technol, MSME, 79; Univ Minn, PhD(mech eng), 86. *Prof Exp:* RES ASSOC, UNDERGROUND SPACE CTR, UNIV MINN, 86- *Res:* Analysis of heat flow from residential building foundations; evaluation and implementation of weatherization procedures for low-income weatherization programs; numerical modelling of heat and moisture flow in soils. *Mailing Add:* Underground Space Ctr Univ Minn 790 CME Bdlg 500 Pillsbury Dr SE Minneapolis MN 55455

SHEN, LIANG CHI, b Chekiang, China, Mar 17, 39; US citizen; m 65; c 2. ELECTRICAL ENGINEERING. *Educ:* Nat Taiwan Univ, BS, 61; Harvard Univ, SM, 63, PhD(appl physics), 67. *Prof Exp:* Asst assoc prof, 67-77, chmn dept, 77-81, PROF ELEC ENG, UNIV HOUSTON, 77- *Concurrent Pos:* Consult, Gulf Oil, 81-82. *Mem:* Fel Inst Elec & Electronics Engrs; Am Asn Univ Profs; AAAS; Am Soc Eng Educ; Sigma Xi. *Res:* Antennas; microwaves; electromagnetic wave propagation in earth; well-logging. *Mailing Add:* Dept of Elec Eng Univ of Houston Houston TX 77004

SHEN, LINUS LIANG-NENE, b Chi-Kiang Prov, China, Aug 28, 41; US citizen; m; c 2. BIOCHEMISTRY, MICROBIOLOGY. *Educ:* Nat Taiwan Univ, Taipei, Taiwan, BS, 64; NC State Univ, MS, 69; Univ NC, Chapel Hill, PhD(biochem), 71. *Prof Exp:* From res asst to res assoc res Dept Biochem, Univ NC, Chapel Hill, 68-72, instr res & teaching, 72-75, asst prof res & teaching, 75; scientist II res, Corp Res, 75-81, sr scientist res, Pharmaceut Prod Div, 81-87, VOLWILER FEL RES, PHARMACEUT PROD DIV, ABBOTT LABS, 87- *Mem:* Am Soc Molecular Biol & Biochem. *Res:* Structure, function and inhibitors of microbial DNA topoisomerases; molecular mechanisms of the antibacterial activity of quinolones. *Mailing Add:* Antiinfectious Res Div Abbott Labs D47N AP9A Abbott Park IL 60064

SHEN, MEI-CHANG, b Shanghai, China, Oct 3, 31; m 64; c 2. APPLIED MATHEMATICS. *Educ:* Taiwan Univ, BSc, 54; Brown Univ, PhD(appl math), 63. *Prof Exp:* Res assoc appl math, Brown Univ, 62-63; vis mem, Courant Inst Math Sci, NY Univ, 63-65; from asst prof to assoc prof, 65-70, PROF APPL MATH, UNIV WIS-MADISON, 70- *Concurrent Pos:* Vis prof, Courant Inst Math Sci, NY Univ, 71-72; vis fel, Calif Inst Tech, 76. *Mem:* Am Math Soc; Soc Indust & Appl Math. *Res:* Asymptotic methods; biofluid-dynamics; nonlinear wave propagation; plasma dynamics; geophysical fluid dynamics. *Mailing Add:* Dept of Math Univ of Wis Madison WI 53706

SHEN, NIEN-TSU, b Taipei, Taiwan. STATISTICS, MATHEMATICS. *Educ:* Nat Taiwan Univ, BS, 75; Univ Calif, Santa Barbara, MA, 76; Purdue Univ, PhD(math), 82. *Prof Exp:* Vis lectr math, Univ Calif, Davis, 82-83; asst prof statist, Rider Col, NJ, 83-84; quant analytical eng mgr & statist process control, Nat Semiconductor Corp, 84-89; RELIABILITY & FAILURE ANALYSIS MGR, DIGITAL EQUIP CORP, 89- *Mem:* Am Soc Qual Control; Soc Indust & Appl Math; Chinese Qual Assurance Asn (pres); Am Statist Asn. *Res:* Applications of statistical methods in the semiconductor manufacturing environment; design of experiments, Taguchi methods. *Mailing Add:* Digital Equip Corp 10500 Ridgeview Ct MS UCF 1/200 Cupertino CA 95014-0715

SHEN, PETER KO-CHUN, b China, Oct 28, 38; US citizen; m 65; c 1. REACTOR PHYSICS. *Educ:* Nat Taiwan Univ, BS, 61; Univ Minn, MS, 65; Kans State Univ, PhD(nuclear eng), 70. *Prof Exp:* Res assoc nuclear shielding, Kans State Univ, 69-70; supv nuclear engr reactor physics & fuel mgt, Southern Calif Edison Co, 70-75; assoc prof reactor physics & fuel cycle, Joint Ctr Grad Study, Univ Wash, 75-79, dean, Joint Ctr Grad Study, 77-79, AFFIL PROF NUCLEAR ENG, UNIV WASH, 79-; TECHNICIAN & DIR, WASHINGTON PUB POWER SUPPLY SYST, 80- *Mem:* Am Nuclear Soc. *Res:* Nuclear fuel cycle core physics, thermal hydraulic and nuclear power plant design and operation. *Mailing Add:* Wash Pub Power Supply Syst PO Box 968 Richland WA 99352

SHEN, S(HAN) F(U), b Shanghai, China, Aug 31, 21; m 50; c 2. MECHANICAL ENGINEERING. *Educ:* Nat Cent Univ, China, BS, 41;. *Hon Degrees:* ScD, Mass Inst Technol, 49. *Prof Exp:* Res assoc math, Mass Inst Technol, 48-50; from asst prof to prof, Univ Md, 50-61; PROF AERONAUT ENG, GRAD SCH AERONAUT ENG, CORNELL UNIV, 61-, JOHN EDSON SWEET PROF ENG, 78- *Concurrent Pos:* Guggenheim fel, 57-58; vis prof, Univ Paris, 64-65 & 69-70, Tech Univ Vienna, 77 & Univ Tokyo, 84-85. *Honors & Awards:* US Sr Scientist Award, Humboldt Found, WGer, 84. *Mem:* Nat Acad Eng; Acad Sinica Repub China; Coresp mem Int Acad Astronaut. *Res:* Aerodynamics; rarefied gasdynamics; fluid mechanics. *Mailing Add:* Sibley Sch Mech & Aerospace Eng Cornell Univ Upson Hall Ithaca NY 14853

SHEN, SAMUEL YI-WEN, b Tientsin, China, Jan 4, 19; m 51; c 1. CHEMISTRY, PHYSICS. *Educ:* Yenching Univ, China, BSc, 41; Columbia Univ, MA, 51, EdD, 58. *Prof Exp:* Lectr chem, Rutgers Univ, 56-57; from asst prof to assoc prof, 57-64, PROF CHEM, LONG ISLAND UNIV, ZECKENDORF CAMPUS, 64- *Concurrent Pos:* Lectr, Polytech Inst New

York, 61- *Mem:* Am Chem Soc. *Res:* Chromatography; nonaqueous titrations; spectrophotometric analysis; chelate chemistry; electrochemistry. *Mailing Add:* Dept Chem Long Island Univ Brooklyn Ctr University Plaza Brooklyn NY 11201

SHEN, SHELDON SHIH-TA, b Shanghai, China, Nov 22, 47; US citizen; m 79; c 3. CELL PHYSIOLOGY, DEVELOPMENTAL BIOLOGY. *Educ:* Univ Mo, BS, 69; Univ Calif, Berkeley, PhD(physiol), 74. *Prof Exp:* Fel cancer biol, Univ Calif, Berkeley, 74-77, asst res zoologist cell biol, 77-79; asst prof, 79-83, assoc prof, 84-89, PROF ZOOL, IOWA STATE UNIV, 89- *Concurrent Pos:* NIH fel, 74-76; USPHS, NIH grant, 77-79, NSF grant, 80-92. *Mem:* Sigma Xi; Soc Develop Biol; Am Soc Cell Biol; AAAS; Am Soc Zoologists. *Res:* Ions co second messengers in signal transduction; cytoplasmic factors controlling chromosome diminution in Ascaris embryos; regulation of sea urchin egg activation. *Mailing Add:* Dept Zool Iowa State Univ Ames IA 50011

SHEN, SIN-YAN, b Singapore, Nov 12, 49; m 73; c 2. LOW TEMPERATURE PHYSICS. *Educ:* Univ Singapore, BS, 69; Ohio State Univ, MS, 70, PhD(low temp physics), 73. *Prof Exp:* Asst prof physics, Northwestern Univ, 74-77; scientist, 74-83, SR RES LEADER, ARGONNE NAT LAB, 83-; PROF, HARVARD UNIV, 88- *Concurrent Pos:* Adv, US Dept Energy, 77-, SUPCON Int, 86-, Nat Geog, 86, Int Energy Agency, 86-; ed-in-chief, World Resource Rev, 89-; chmn, Global Warming Sci & Policy Int Cong, 90 & 91. *Honors & Awards:* Nat Merit Scholar, 67; Fulbright Scholar, 69; Panel Mem, Nat Acad Sci, 86. *Mem:* Am Phys Soc; Chinese Music Soc NAm. *Res:* Surface second sound; superfluidity; surface phenomena in liquid helium; thermodynamics; hydrodynamics; acoustics; statistical mechanics; transport properties; Fermi liquids; quantum fluids; experimental ultra low temperature physics; cryogenics; fluid mechanics. *Mailing Add:* 2329 Charmingfare Woodridge IL 60517-2910

SHEN, TEK-MING, LASERS, MICROWAVES. *Educ:* Univ Hong Kong, BSc, 73; Univ Ca, Berkeley, PhD(physics),79. *Prof Exp:* Res asst, Univ Calif, 79-80; MEM TECHNICAL STAFF, AT&T, BELL LABS, MURRAY HILL, 80- *Mem:* Inst Elec & Electronics Engrs. *Res:* Studies of dynamic properties of multifrequency and single-frequency semiconductors lasers and their system performance. *Mailing Add:* Bell Lab 2C-307 600 Mountain Ave Murray Hill NJ 07974

SHEN, THOMAS T, b Chia-Xing, China, Aug 14, 26; US citizen; m 59; c 2. COMPUTER PROGRAMMING, MATHEMATICAL MODELING. *Educ:* St John's Univ, Shanghai, BSc, 48; Northwestern Univ, Evanston, MSc, 60; Rensselaer Polytech Inst, PhD(environ eng), 72. *Prof Exp:* Environ engr, Wash State Health Dept, 62-66; sr environ engr, NY State Health Dept, 66-70; SR RES SCIENTIST, NY STATE DEPT ENVIRON CONSERV, 70- *Concurrent Pos:* Chmn, NY State Coun, Am Soc Civil Engrs, 79-80, Air Pollution Control Comt, 87-89, Air & Radiation Comt, 89-90 & Air Pollution Control Comt, Am Acad Environ Engrs, 89-; adj fac, Div Environ Sci, Columbia Univ, 81-90; mem & consult, US Environ Protection Agency Sci Adv Bd, 87-; bd mem, NY State Asn Libr Bd & Chinese Am Acad & Prof Soc, 90- *Mem:* Am Soc Civil Engrs; Air & Waste Mgt Asn; Am Acad Environ Engrs. *Res:* Combustion emission control; measurement and monitoring techniques; volatile organic compounds emission assessment and control; multimedia pollution prevention methodologies. *Mailing Add:* 146 Fernbank Ave Delmar NY 12054

SHEN, TSUNG YING, b Peking, China, Sept 28, 24; nat US; m 53; c 6. ORGANIC CHEMISTRY. *Educ:* Nat Cent Univ, China, BSc, 46; Univ London, dipl, 48; Univ Manchester, PhD(org chem), 50, DSc, 78. *Prof Exp:* Fel, Ohio State Univ, 50-52; res assoc, Mass Inst Technol, 52-56; res fel synthetic org chem, 56-67, assoc dir, 67-69, dir synthetic chem res, 69-71, sr dir med chem, 71-74, exec dir med chem, 74-76, VPRES MEMBRANE & ARTHRITIS RES, MERCK SHARP & DOHME RES LABS, 76- *Concurrent Pos:* Vis chem prof, Univ Calif, Riverside, 73. *Honors & Awards:* Galileo Medal Sci Achievement, Univ Pisa, 76; Rene Descartes Silver Medal, Univ Paris, 77; Medal of Merit, Giornate Mediche Int del Collegium Biol Europa, 77; Burger Award, Am Chem Soc, 80. *Mem:* Am Chem Soc; NY Acad Sci; AAAS. *Res:* Medicinal chemistry; anti-inflammatory and immunopharmacological agents, viral and cancer chemotherapy; nucleosides, carbohydrate derivatives and membrane receptor regulators. *Mailing Add:* chem Dept Univ Va McCormick Rd Charlottesville VA 22901

SHEN, VINCENT Y, b US; m 69; c 2. COMPUTER SCIENCES, SOFTWARE ENGINEERING. *Educ:* Nat Taiwan Univ, BS, 64; Princeton Univ, PhD(elec eng), 69. *Prof Exp:* From asst prof to assoc prof comput sci, Purdue Univ, West Lafayette, 69-85; SR MEM TECH STAFF, MICROELECTRONIC & COMPUT TECHNOL RES, AUSTIN, TEX, 85- *Mem:* Asn Comput Mach; Inst Elec & Electronic Engrs. *Res:* Software engineering and software science; computer operating systems; programming languages; switching theory. *Mailing Add:* 9021 Lockleven Loop Austin TX 78750

SHEN, WEI-CHIANG, b Chekiang, China, May 3, 42; US citizen; m 68; c 2. CELL BIOLOGY, DRUG-PROTEIN CONJUGATES. *Educ:* Tunghai Univ, Taichung, Taiwan, BS, 65; Boston Univ, PhD(chem), 72. *Prof Exp:* Res fel biol chem, Harvard Med Sch, 72-73; res assoc biochem, Brandeis Univ, 73-76; from asst res prof to assoc res prof, Sch Med, Boston Univ, 76-83, assoc prof path & pharmacol, 83-87; ASSOC PROF PHARMACOL, UNIV SOUTHERN CALIF, SCH PHARM, 87- *Concurrent Pos:* Vis lectr biochem toxicol, Brandeis Univ, 80-87; Cancer Res Scholar Award, Mass Div, Am Cancer Soc, 82-85; vis prof, Coal Med Col, N China, 86. *Mem:* NY Acad Sci; Am Soc Cell Biol; Am Soc Pharmacol & Exp Therapeut; AAAS; Am Soc Biol Chemists. *Res:* Endocytosis and lysosomal degradation of macromolecules in mammalian cells; conjugates of drugs and monoclonal antibodies as potential tumor-targeting agents in cancer chemotherapy; development of immunoassay methods. *Mailing Add:* Sch Pharm Univ Southern Calif Los Angeles CA 90033

SHEN, WU-MIAN, b Shanghai, China, Aug 23, 42; m 74; c 1. PHOTOELECTROCHEMISTRY, CHARACTERIZATION OF SEMICONDUCTOR DEVICE. *Educ:* Shanghai Jiao-Tong Univ, BA, 64, ME, 81; City Univ NY, PhD(physics), 91. *Prof Exp:* Engr, Tianjin Electronic Wire & Cable Co, 64-78; lectr semiconductor physics, Dept Appl Chem, Shanghai Jiao-Tong Univ, 81-83; vis scholar, Dept Physics, 83-84, res asst, 84-91, RES ASSOC, APPL SCI INST, BROOKLYN COL, 91- *Mem:* Am Phys Soc; Electrochem Soc. *Res:* Photoelectrochemistry; solar energy conversion; optical and dielectric characterization of semiconductor with liquid junction or solid state junction; electrical insulation and dielectric phenomena; electrochemical deposition of high-temperature superconducting film. *Mailing Add:* Dept Physics Brooklyn Col City Univ New York Brooklyn NY 11210

SHEN, Y(UNG) C(HUNG), aeronautics, mathematics, for more information see previous edition

SHEN, YUAN-SHOU, b Peking, China, Dec 12, 21; m 48; c 2. PHYSICAL METALLURGY. *Educ:* Nat Southwest Assoc Univ, BS, 43; Ore State Univ, MS, 64, PhD(mat sci), 68. *Prof Exp:* Assoc prof metall, Cheng Kung Univ, Taiwan, 61-62; res metallurgist, Wah Chang Albany Corp, Ore, 65-67; sr res staff, P R Mallory & Co Inc, Mass, 67-78; res scientist, 68-80, SR RESEARCH SCIENTIST, ENGELHARD INDUSTS, PLAINVILLE, 80- *Mem:* Am Soc Metals; Metall Soc; Am Soc Testing & Mat. *Res:* Phase diagram; intermetallic compound; composite material; alloy development; electric contact material; silver alloy; powder metallurgy; brazing alloys. *Mailing Add:* Six Shackford Rd Reading MA 01867

SHEN, YUEN-RON, b Shanghai, China, Mar 25, 35; m 64. SOLID STATE PHYSICS. *Educ:* Nat Taiwan Univ, BS, 56; Stanford Univ, MS, 59; Harvard Univ, PhD(solid state physics), 63. *Prof Exp:* From asst prof to assoc prof, 64-70, PROF PHYSICS, UNIV CALIF, BERKELEY, 70- *Mem:* Fel Am Phys Soc. *Res:* Quantum electronics. *Mailing Add:* Dept of Physics Univ of Calif Berkeley CA 94720

SHEN, YVONNE FENG, b Hu-Nan, China, Sept 11, 42; US citizen; m 68; c 2. ORGANIC CHEMISTRY. *Educ:* Tunghai Univ, Taiwan, BS, 64; Univ Mass, PhD(org chem), 68. *Prof Exp:* Fel, Princeton Univ, 68-69, Drexel Univ, 69-70 & Univ Southern Calif, 70-71; jr res scientist, City Hope Med Ctr, 71-73; CHIEF CHEMIST, ORANGE COUNTY WATER DIST, 74- *Mem:* Am Chem Soc; Am Water Works Asn; Water Pollution Control Fedn. *Res:* Chemistry aspects of water and wastewater treatment. *Mailing Add:* 10500 Ellis Ave Fountain Valley CA 92647

SHENDRIKAR, ARUN D, b Gulberga, India, July 10, 38; m 66; c 2. ENVIRONMENTAL CHEMISTRY. *Educ:* Osmania Univ, India, BS, 57, MS, 61; Durham Univ, PhD(anal chem), 66. *Prof Exp:* Lectr chem, V V Sci Col, India, 61-62, lectr in-chg, 62-63; lectr, Osmania Univ, 62-63; asst prof, Environ Sci Inst, La State Univ, 66-73; sr fel aerosol, Nat Ctr Atmospheric Res, 73-74; sr res chemist, Oil Shale Corp, 74-76; tech dir, O A Labs, Indianapolis, 76-78; tech staff specialist, Meteorol Res Inc, Altadena, 78-80; sr chemist, Res Triangle Inst, Res Triangle Park, 80-82, inorg mgr, Compuchem Labs, 82-86. gen mgr, Beta Labs, 86-88; mgr anal serv, EIRA Inc, St Rose, La, 88-89; DIR ENVIRON AFFAIRS, LITHO INDUSTS, INC, 89- *Concurrent Pos:* Mem adv comt, Colo Air Pollution Control Comt, 74- *Mem:* Assoc mem Royal Inst Chem; Sigma Xi; Am Soc Testing & Mat; Am Chem Soc. *Res:* Interpretation of chemical data to investigate mechanisms of aerosol formation and modification; application of analytical chemistry for process performance evaluation; development of new source performance standards, artifacts and inhalable particulate matter sampling; design and management of environmental analytical laboratory; United States Environmental Protection Agency CLP project management; fixation of toxic wastes; plant and personnel safety; industrial hygiene and regulatory liaison. *Mailing Add:* 1011 St Helena Pl Apex NC 27502

SHENEFELT, RAY ELDON, b Spokane, Wash, Oct 2, 33; m 57; c 4. TERATOLOGY. *Educ:* Univ Wis-Madison, BS, 59, MD, 63. *Prof Exp:* Intern path, Univ Wis Hosps, 63-64; resident path, Univ Iowa Hosps, 64-66; fel path, Dartmouth Med Sch, 66-69; asst prof pediat & path, Cincinnati Childrens Hosp, 69-73; med officer teratol, Nat Ctr Toxicol Res, 74-79; INSTR PEDIAT & PATHOL, LE BONHEUR CHILDREN'S HOSP, 80- *Concurrent Pos:* Asst prof path, Med Sch, Univ Ark, 73-79. *Mem:* Teratol Soc; Pediat Path Club. *Res:* Pathologic sequence in development of malformations. *Mailing Add:* Le Bonheur Childrens Hosp 848 Adams Ave Memphis TN 38103

SHENEFELT, ROY DAVID, b Evanston, Ill, Jan 27, 09; m 32; c 2. ENTOMOLOGY. *Educ:* Spokane Col, AB, 32; State Col Wash, MS, 35, PhD(entom), 40. *Prof Exp:* From instr to asst prof zool, State Col Wash, 35-46; from asst prof to prof, 46-77, EMER PROF ENTOM FORESTRY, UNIV WIS-MADISON, 77- *Mem:* Entom Soc Am; Entom Soc Can. *Res:* Taxonomy of Braconidae; forest entomology; cacao insects. *Mailing Add:* 630 Oak St Oregon WI 53575

SHENEMAN, JACK MARSHALL, b Grand Rapids, Mich, Mar 26, 27; m 57; c 2. MICROBIOLOGY. *Educ:* Mich State Univ, BS, 52, MS, 54, PhD(microbiol), 57. *Prof Exp:* Asst microbiol, Mich State Univ, 54-57; res assoc, Wis Malting Co, 57-61, asst dir res, 61-63; sr food scientist, Eli Lilly & Co, 63-69; res microbiologist, Basic Veg Prod Inc, 69-75; FOOD & DRUG SCIENTIST, FOOD & DRUG BR, CALIF DEPT HEALTH SERV, 75- *Mem:* AAAS; Am Soc Microbiol; Inst Food Technologists; Asn Food & Drug Officials; Sigma Xi. *Res:* Industrial fermentations; food flavors and preservatives; microbiology of dehydrated vegetable products; new product and process development in vegetable dehydration; food hazard microorganisms; bacterial spores; microbiological quality control methods; food and drug regulation. *Mailing Add:* Food & Drug Br Calif Dept Health Serv 714 P St Rm 400 Sacramento CA 95814

SHENG, HWAI-PING, b Johore, Malaysia, July 18, 43; US citizen; m 77; c 1. MEDICAL PHYSIOLOGY, PEDIATRICS. *Educ:* Univ Singapore, BSc, 66, Hons, 67; Baylor Col Md, PhD(physiol), 71. *Prof Exp:* Lectr pharmacol, Univ Hong Kong, 71-74; asst prof physiol, 75-88, ASSOC PROF PEDIAT & MOLECULAR PHYSIOL & BIOPHYSICS, BAYLOR COL MED, 88- *Concurrent Pos:* Fulbright Fel, 67-70. *Mem:* Am Physiol Soc; Soc Exp Biol & Med; Am Inst Nutrit. *Res:* Growth, nutrition and body composition. *Mailing Add:* Dept Pediat Baylor Col Med Houston TX 77030

SHENG, PING, b Shanghai, China, Aug 5, 46; m 70; c 2. SOLID STATE PHYSICS. *Educ:* Calif Inst Technol, BS, 67; Princeton Univ, PhD(physics), 71. *Prof Exp:* Vis mem, Sch Natural Sci, Inst Advan Study, Princeton, NJ, 71-73; mem tech staff physics, RCA David Sarnoff Res Ctr, 73-80; group head, 80-86, SR RES ASSOC, CORP RES CTR, EXXON RES & ENG, 86- *Mem:* Fel Am Phys Soc; Optical Soc Am; Soc Explor Geophysicists. *Res:* Electrical transport in wave propagation ininhomogeneous systems; liquid crystals; structure and physical properties of random composites. *Mailing Add:* Corp Res Ctr Exxon Res & Eng Clinton Township Annadale NJ 08801

SHENG, YEA-YI PETER, b Shanghai, China, Aug 3, 46; m 70. FLUID MECHANICS, ENVIRONMENTAL SCIENCES. *Educ:* Nat Taiwan Univ, BS, 68; Case Western Reserve Univ, MS, 72, PhD(mech & aerospace eng), 75. *Prof Exp:* Res assoc environ, Case Western Reserve Univ, 75-77, sr res assoc, 77-78; assoc consult, Princeton, 78-79, consult meteorol & oceanog, 79-93, sr consult & mgr coastal oceanog, Aeronaut Res Assoc, Princeton, 84-86; assoc prof, 86-88, PROF, COASTAL & OCEANOG ENG, UNIV FLA, 88- *Mem:* Am Geophys Union; Am Soc Mech Engrs; Am Soc Civil Engrs. *Res:* Flow and dispersion of contaminants in coastal, estuarine, offshore and atmospheric environments; turbulent transport processes in stratified flows; turbulence modeling; computational fluid dynamics. *Mailing Add:* 9817 SW First Pl Gainesville FL 32607

SHENITZER, ABE, b Warsaw, Poland, Apr 2, 21; nat US; m 52; c 2. MATHEMATICS. *Educ:* Brooklyn Col, BA, 50; NY Univ, MSc, 51, PhD, 54. *Prof Exp:* Asst, Inst Math Sci, NY Univ, 52-55; mem staff, Bell Tel Labs, Inc, 55-56; from instr to asst prof math, Rutgers Univ, 56-58; assoc prof, Adelphi Univ, 58-63, prof, 63-69; PROF MATH, ARTS & EDUC, YORK UNIV, 69- *Mem:* Am Math Soc. *Res:* Group theory; differential equations; approximation theory. *Mailing Add:* Dept of Math York Univ 4700 Keele Downsview ON M3J 1P3 Can

SHENK, JOHN STONER, b Lancaster Co, Pa, July 25, 33; m 52; c 4. AGRONOMY, PLANT BREEDING. *Educ:* Pa State Univ, BS, 65; Mich State Univ, MS, 67, PhD, 69. *Prof Exp:* From asst prof to assoc prof, 70-81, PROF AGRON, PA STATE UNIV, 78- *Mem:* Am Soc Agron; Am Forage & Grassland Coun; Sigma Xi. *Res:* Plant breeding and genetics; development and utilization of chemical and bioassay technique for the production of new plant varieties with improved nutritional quality; chemical infrared and bioassay; computer analysis of infrared data. *Mailing Add:* RD 1 Box 109 Port Matilda PA 16870-9523

SHENK, WILLIAM E(DWIN), electrical engineering; deceased, see previous edition for last biography

SHENKEL, CLAUDE W, JR, b Lyons, Kans, Apr 29, 19; m 41; c 2. GEOLOGY. *Educ:* Kans State Col, BS, 41; Univ Colo, MS, 47, PhD(geol), 52. *Prof Exp:* From asst prof to prof geol, Kans State Univ, 49-87; RETIRED. *Concurrent Pos:* Consult, Spec Res Proj, East Venezuelan Basin, 56-57, res geol, Andes Mountains, SAm, 65, spec res geol, SAm, 65-66. *Mem:* Fel Geol Soc Am; Am Inst Prof Geol; Am Asn Petrol Geologists. *Res:* Regional stratigraphic analysis of San Juan and Paradox Basins; petroleum and subsurface geology. *Mailing Add:* 17807 Buntline Dr Sun City West AZ 85375

SHENKER, MARTIN, b New York, NY, Aug 12, 28; m 49; c 3. OPTICS. *Educ:* NY Univ, AB, 48, MS, 51. *Prof Exp:* Jr actuary, New York City Teachers' Retirement Bd, 48-49; inspector, 49-50, jr physicist, 50-51, optical designer, 51-57, asst chief optical designer, 57-60, chief optical designer, 60-73, VPRES OPTICAL DESIGN, FARRAND OPTICAL CO, 73- *Mem:* Fel Optical Soc Am; Soc Photo-Optical Instrument Engrs. *Res:* Optical design; geometrical optics. *Mailing Add:* Five Ormian Dr Pomona NY 10970

SHENKER, SCOTT JOSEPH, b Alexandria, Va, Jan 24, 56; m 80; c 1. GAME THEORY, COMPUTER METHODS. *Educ:* Brown Univ, ScB, 78; Univ Chicago, PhD(physics), 83. *Prof Exp:* Postdoctoral, Dept Physics, Cornell Univ, 83-84; MEM RES STAFF, PALO ALTO RES CTR, XEROX CTR, 84- *Mem:* Asn Comput Mach; Inst Elect & Electronic Engrs. *Res:* Analysis of computer network algorithms; analysis of resource allocation mechanisms from a game theoretic point-of-view. *Mailing Add:* Palo Alto Res Ctr Xerox Corp 3333 Coyote Hill Rd Palo Alto CA 94304

SHENKIN, HENRY A, b Philadelphia, Pa, June 25, 15; m 41; c 4. NEUROSURGERY. *Educ:* Univ Pa, AB, 35; Jefferson Med Col, MD, 39. *Prof Exp:* Charles Harrison Frazier traveling fel from Univ Pa to dept physiol & brain tumor registry, Sch Med, Yale Univ, 41-42; DIR NEUROSURG, EPISCOPAL HOSP, 58-; PROF NEUROSURG, MED COL PA, 74- *Concurrent Pos:* Assoc prof neurol surg, Div Grad Med, Univ Pa, 60-67; clin prof neurosurg, Sch Med, Temple Univ, 67- *Mem:* Am Asn Neurol Surg; Soc Neurol Surg; Am Col Surg. *Res:* Cerebral circulation and the metabolism of the neurosurgical patient. *Mailing Add:* Episcopal Hosp Front & Lehigh Ave Philadelphia PA 19125

SHENOI, BELLE ANANTHA, b Mysore, India, Dec 23, 29; m 61; c 2. ELECTRICAL ENGINEERING. *Educ:* Univ Madras, BS, 51; Indian Inst Sci, Bangalore, DIISc, 55; Univ Ill, MS, 58, PhD(elec eng), 62. *Prof Exp:* From teaching asst to instr elec eng, Univ Ill, 56-62; from asst prof to assoc prof, 62-67, fac res fel, 63, grad col grant, 63-67, PROF ELEC ENG, UNIV MINN, MINNEAPOLIS, 71- *Concurrent Pos:* NSF grant, 64-68. *Mem:* Fel Inst Elec & Electronics Engrs (pres, Circuits & Systs Soc, 75). *Res:* Digital filters and signal processing; theory and design of networks; analysis of systems. *Mailing Add:* Elec Eng Dept 134 Fawcett Hall Wright State Univ Dayton OH 45435

SHENOLIKAR, ASHOK KUMAR, b Hyderabad, India; US citizen; m 70; c 2. SOFTWARE SYSTEMS, COMPUTER SCIENCES. *Educ:* Osmania Univ, Hyderabad, India, BE, 60; Univ Okla, BS, 66, MS, 69. *Prof Exp:* Consult, 69-73; appln engr, Gibbs & Hills, Inc, 73-78; prin engr, Harris Corp, 78-86; SYST ARCHITECT, GRUMMAN DATA SYSTS, 86- *Mem:* Inst Elec & Electronic Engrs; Soc Mfg Engrs; Nat Computer Graphics Asn. *Res:* Technology evaluation and selection for advanced, integrated, large scale information systems; concept definition for strategic system architectures in multi vendor environments. *Mailing Add:* 22 Greene Dr Commack NY 11725

SHENOY, GOPAL K, India citizen; m 70; c 2. SYNCHROTRON RADIATION, MOSSBAUER SPECTROSCOPY. *Educ:* Univ Bombay, BS, 59, MS, 61, PhD(physics), 66. *Prof Exp:* sci assoc physics, Tech Univ Munich, 70-72; res assoc physics, Centre Recherche Nucleaire, Strausburg, 72-74; vis scientist, 74-76, physicist, 76-81, SR PHYSICIST, ARGONNE NAT LAB, 81-, ASSOC DIR ADV PHOTON SOURCE, 88- *Concurrent Pos:* Vis scientist, Tech Univ Helsinki, 71; group leader, Argonne Nat Lab, 84- *Mem:* Am Phys Soc; Mat Res Soc; AAAS; Sigma Xi. *Res:* Application of Mössbauer spectroscopy and synchrotron radiation to investigate properties of materials such as storage hydrides, rare earth, actinites and superconductors. *Mailing Add:* Bldg 360-APS Argonne Nat Lab Argonne IL 60439

SHEPANSKI, JOHN FRANCIS, b Rochester, NY, Mar 4, 54; m 76. ARTIFICIAL NEURAL SYSTEMS, PATTERN RECOGNITION. *Educ:* St Bonaventure Univ, BS, 76; Univ Rochester, MA, 79, PhD(physics), 82. *Prof Exp:* Res fel, Calif Inst Technol, 82-84; STAFF SCIENTIST, TRW INC, 84- *Mem:* Am Phys Soc; Int Neural Network Soc. *Res:* Artificial neural networks and their application to pattern recognition; biophysics (photosynthesis); chemical physics; laser spectroscopy; statistical mechanics. *Mailing Add:* 6627 Noble Ave Van Nuys CA 91405

SHEPARD, BUFORD MERLE, b Dexter, Ga, Apr 10, 42; m 60. ENTOMOLOGY, ECOLOGY. *Educ:* Mid Tenn State Univ, BS, 66; Univ Ga, MS, 68; Tex A&M Univ, PhD(entom), 71. *Prof Exp:* Entomologist, Hillsborough County Health Dept, 68-69; asst prof entom, Univ Fla, 71-72; from asst prof to prof entom, Clemson Univ, 72-83; entomologist, Int Rice Res Inst, 83-84, dept head entom, 85-89; AT COASTAL RES EDUC CTR. *Concurrent Pos:* Consult insect pest mgt, Brazil, Panama , Seychelles, Australia, Kenya & Costa Rica. *Mem:* Entom Soc Am; Philippine Asn Entomologists. *Res:* Biological control, insect ecology and integrated pest management; developing and evaluating integrated pest management programs in the United States, tropical Asia and South America. *Mailing Add:* Coastal Res Educ Ctr 1865 Savannah Hwy Charleston SC 29414

SHEPARD, DAVID C, b Montpelier, Vt, Mar 13, 29; m 52; c 2. BIOLOGY. *Educ:* Stanford Univ, AB, 51, PhD, 57. *Prof Exp:* From instr to assoc prof, 56-65, PROF BIOL, SAN DIEGO STATE UNIV, 65- *Mem:* Gerontol Soc; AAAS; Am Soc Cell Biol; Sigma Xi. *Res:* Cell growth and division; cell aging. *Mailing Add:* Dept of Biol San Diego State Univ San Diego CA 92182

SHEPARD, EDWIN REED, b Springfield, Ohio, June 24, 17; m 40; c 5. ORGANIC CHEMISTRY. *Educ:* Wittenberg Col, AB, 38; Ohio State Univ, PhD(org chem), 42. *Prof Exp:* Org chemist, Eli Lilly & Co, 42-51, head dept org chem res, 51-57, head dept appln res, 58-66, res assoc, Prod Develop Div, 66-72, res adv, 73-81; RETIRED. *Mem:* NY Acad Sci; AAAS; Am Chem Soc. *Res:* Chemotherapy, especially antibiotics; cardiovascular drugs; agricultural and food chemicals; product development through new drug application. *Mailing Add:* 2290 Demaret Dr Dunedin FL 34698

SHEPARD, HARVEY KENNETH, b Chicago, Ill, Sept 19, 38; c 2. NONLINEAR DYNAMICAL SYSTEMS, THEORETICAL HIGH ENERGY PHYSICS. *Educ:* Univ Ill, Urbana, BS, 60; Calif Inst Technol, MS, 62, PhD(physics), 66. *Prof Exp:* Consult, Rand Corp, Calif, 61-65; lectr & res fel high energy physics, Univ Calif, Santa Barbara, 65-67 & Univ Calif, Riverside, 67-69; from asst prof to assoc prof, 69-79, chmn, 85-88 PROF PHYSICS, UNIV NH, 79- *Concurrent Pos:* Mem, Inst Advan Study, 77-78; mem, Mass Inst Technol Ctr Theoret Physics, 84-85. *Mem:* Am Phys Soc; Sigma Xi. *Res:* Nonlinear dynamics; high energy theoretical physics; theoretical elementary particle physics; mathematical physics. *Mailing Add:* Dept of Physics Univ of NH Durham NH 03824

SHEPARD, JAMES F, b Rhinebeck, NY, Nov 16, 41; m 61; c 2. PLANT PATHOLOGY, PLANT VIROLOGY. *Educ:* Cornell Univ, BS, 59; Univ Calif, Davis, MS, 65, PhD(plant path), 67. *Prof Exp:* Res plant pathologist, Univ Calif, Davis, 67; from asst prof to prof plant path, Mont State Univ, 67-76; PROF PLANT PATH & HEAD DEPT, KANS STATE UNIV, 76- *Mem:* Am Phytopath Soc; Potato Asn Am. *Res:* Mesophyll protoplast culture and selection. *Mailing Add:* Hour Glass Travel 1701 First St E Bradenton FL 34208

SHEPARD, JOSEPH WILLIAM, b St Paul, Minn, Aug 16, 22; m 43; c 4. PHYSICAL CHEMISTRY. *Educ:* Wayne State Univ, BS, 43; Univ Mich, MS, 47, PhD(chem), 53. *Prof Exp:* Res chemist, Eng Div, Chrysler Corp, 47-49; dir, Imaging Res Lab, Minn Mining & Mfg Co, 52-68, tech dir Microfilm Prod Div, 68-80, tech dir Duplicating Prod Div, 75-78, Div Vpres, Eng Syst Div, 82-85; RETIRED. *Mem:* AAAS; Am Chem Soc; Soc Photog Scientists & Engrs; Sigma Xi. *Res:* Surface and crystal chemistry; colloids; solid state; photochemistry. *Mailing Add:* 9821 Admiral Dewey NE Albuquerque NM 87111

SHEPARD, KENNETH LEROY, b Hahira, Ga, Aug 5, 37; m 55; c 2. MEDICINAL CHEMISTRY. *Educ:* Stetson Univ, BS, 59; Univ NC, Chapel Hill, PhD(org chem), 63. *Prof Exp:* Mem staff, Chem Res & Develop Labs, Edgewood Arsenal, US Army, Md, 63-65; sr res chemist, Med Chem Dept, Merck, Sharp & Dohme Res Labs, West Point, 65-74, res fel, 74-75, sr res fel, 75-86, SR INVESTR, MED CHEM DEPT, MERCK, SHARP & DOHME RES LABS, WEST POINT, 86- *Mem:* AAAS; Am Chem Soc; Int Heterocyclic Chem Soc. *Res:* Synthesis of potential therapeutic agents. *Mailing Add:* PO Box 107 W Point PA 19486

SHEPARD, KENNETH WAYNE, b Columbus, Ohio, Jan 2, 41; m 62; c 2. ACCELERATOR PHYSICS. *Educ:* Univ Chicago, BS, 62; Dartmouth Col, MA, 64; Stanford Univ, PhD(physics), 70. *Prof Exp:* Res asst, Calif Inst Technol, 70-73, res assoc, 73-75; physicist, 75-86, SR PHYSICIST, ARGONNE NAT LAB, 86- *Mem:* Am Phys Soc. *Res:* Development of high-field, radio-frequency superconducting devices for use in particle accelerators; investigation of the RF properties of superconductors; development of superconducting devices. *Mailing Add:* 803 S Delphia Park Ridge IL 60068

SHEPARD, MARION L(AVERNE), b Owosso, Mich, Dec 20, 37; m 62; c 1. MATERIALS SCIENCE, METALLURGICAL ENGINEERING. *Educ:* Mich Tech, BS, 59; Iowa State Univ, MS, 60, PhD(metall), 65. *Prof Exp:* Anal engr, Pratt & Whitney Aircraft Div, United Aircraft Corp, 60-62, proj metallurgist, 65-67, sr proj metallurgist, 67; from asst prof to assoc prof, 67-78, PROF MECH ENG, DUKE UNIV, 78-, ASSOC DEAN, SCH ENG, 77- *Mem:* Am Soc Mech Engrs; Am Soc Eng Educ. *Res:* Phase equilibria in cryobiological systems; response of materials to chemical and thermal fields; precipitation hardening. *Mailing Add:* Dept Mech Eng & Mat Sci Duke Univ Durham NC 27706

SHEPARD, MAURICE CHARLES, b River Falls, Wis, Feb 29, 16; wid; c 3. MEDICAL BACTERIOLOGY. *Educ:* Univ Wis, BS, 39, MS, 40; Duke Univ, PhD(microbiol), 53. *Prof Exp:* Asst bact, Univ Wis, 39-40; instr, Univ Mass, 40-44; coordr field labs, Venereal Dis Div, USPHS, Washington, DC, 45-48, dir spec venereal dis res unit, NC, 49-52 & Ark, 52-53; chief div bact, US Naval Med Field Res Lab, 53-75, CONSULT PREV MED, NAVAL REGIONAL MED CTR, 75- *Mem:* Am Soc Microbiol; NY Acad Sci. *Res:* Biology of mycoplasma; ureaplasmas; etiology of nongonococcal urethritis. *Mailing Add:* 1008 River Jacksonville NC 28540

SHEPARD, PAUL FENTON, b Ann Arbor, Mich, Jan 27, 42; m 63; c 2. PARTICLE PHYSICS. *Educ:* Col William & Mary, BS, 63; Princeton Univ, MA, 66, PhD(physics), 69. *Prof Exp:* Adj asst prof physics, Univ Calif, Los Angeles, 69-73; from asst prof to assoc prof, 74-85, PROF PHYSICS & ASTRON, UNIV PITTSBURGH, 85- *Mem:* AAAS; Am Phys Soc. *Res:* High energy experimental particle physics; direct photon production and heavy quark spectroscopy; applications of solid state devices to high energy physics. *Mailing Add:* Dept Physics & Astron Univ Pittsburgh 100 Allen Hall Pittsburgh PA 15260

SHEPARD, RICHARD HANCE, b Tulsa, Okla, Jan 18, 22; m 45; c 3. MEDICINE. *Educ:* Washington & Lee Univ, AB, 43; Johns Hopkins Univ, MD, 46. *Prof Exp:* Intern, Johns Hopkins Hosp, 46-47; fel path, Univ Pa, 49-50; fel, Johns Hopkins Univ, 50-52; resident, Chest Serv, Columbia Div, Bellevue Hosp, 52-53; from instr to asst prof environ med, 53-57, from instr to asst prof med, 55-61, assoc prof physiol, 70-72, ASSOC PROF ENVIRON, SCH HYG & PUB HEALTH, JOHNS HOPKINS UNIV, 58-, ASSOC PROF MED, 61-, PROF PHYSIOL & BIOMED ENG, 72- *Concurrent Pos:* Res fel, Johns Hopkins Univ, 51-52, Nat Res Coun fel, 53-55. *Mem:* Am Thoracic Soc; Am Physiol Soc; Am Soc Clin Invest; fel NY Acad Sci. *Res:* Pulmonary diffusion and distribution in normal subjects and in patients with pulmonary diseases. *Mailing Add:* 1341 N Teal Ct Boulder CO 80303

SHEPARD, ROBERT ANDREWS, b Gaziantep, Turkey, Oct 22, 23; m 49; c 3. ORGANIC CHEMISTRY. *Educ:* Yale Univ, BS, 44, PhD(chem), 50. *Prof Exp:* Instr chem, Yale Univ, 46-50; from instr to assoc prof chem, Northeastern Univ, 50-58, chmn dept chem, 58-68, dean, Col Lib Arts, 68-76, prof chem, 58-86; RETIRED. *Concurrent Pos:* Actg dir, Marine Sci Inst, 79-82. *Mem:* Am Chem Soc. *Res:* Organic fluorine chemistry; marine chemistry; diazo compounds. *Mailing Add:* RFD 1 Box 125 Samoset Rd Boothbay ME 04537

SHEPARD, ROBERT STANLEY, b Washington, DC, June 19, 27; m 50; c 4. PHYSIOLOGY, PHARMACOLOGY. *Educ:* George Washington Univ, BS, 50, MS, 51; Univ Iowa, PhD(physiol), 55. *Prof Exp:* From spec instr to assoc prof, 55-70, PROF PHYSIOL, SCH MED, WAYNE STATE UNIV, 70- *Concurrent Pos:* Lectr, Mercy Col, Mich, 58-61 & Grace Hosp, 62-81. *Mem:* Am Physiol Soc. *Res:* Lathyrism; blood coagulation; muscle and cardiovascular physiology; teaching techniques. *Mailing Add:* Dept of Physiol Wayne State Univ Sch of Med Detroit MI 48202

SHEPARD, ROGER N, b Palo Alto, Calif, Jan 30, 29; m 52; c 3. COGNITIVE SCIENCE. *Educ:* Stanford Univ, BA, 51; Yale Univ, MS, 52, PhD(exp psychol), 55. *Prof Exp:* Res assoc, Naval Res Lab, 55-56; res fel, Harvard Univ, 56-58, prof psychol, 66-68, dir, Psychol Labs, 67-68; mem tech staff, Bell Tel Lab, 58-66, head dept, 63-66; prof psychol, Harvard Univ, 66-68; PROF PSYCHOL, STANFORD UNIV, 68- & RAY LYMAN WILBUR PROF SOCIAL SCI, 89- *Concurrent Pos:* John Guggenheim fel, 71-72; fel, Ctr Advan Study Behav Sci, 71-72; First Fowler Hamilton vis res fel, Christ Church, Oxford Univ, 87; William James fel, Am Psychol Soc, 89- *Honors & Awards:* Distinguished Sci Contrib Award, Am Psychol Assoc, 76; James McKeen Cattell Fund Award, 79, 80; Howard Crosby Warren Medal, Soc Exp Psychologists, 81; Behav Sci Award, NY Acad Sci, 87. *Mem:* Nat Acad Sci; fel AAAS; fel Am Acad Arts & Sci; fel Am Psychol Asn (pres, Div Exp Psychol, 80-81); Psychomet Soc (pres, 73-74); Soc Exp Psychologists. *Res:* Author of over 100 scientific papers and 3 books; originator of the experimental paradigm of "mental rotation"; originator of first method of nonmetric multidimensionol scaling; originator of the auditory illusion of endlessy ascending pitch; formulator of a proposed universal law of generalization. *Mailing Add:* Dept Psychol Bldg 420 Stanford Univ Stanford CA 94305-2130

SHEPARD, THOMAS H, b Milwaukee, Wis, May 22, 23; m 47; c 3. PEDIATRICS. *Educ:* Amherst Col, AB, 45; Univ Rochester, MD, 48. *Prof Exp:* Asst resident pediat, Albany Med Col, 49-50; chief resident, Univ Rochester, 50-52; fel endocrinol, Med Sch, Johns Hopkins Univ, 54-55; from instr to assoc prof, 55-68, PROF PEDIAT, SCH MED, UNIV WASH, 68-, HEAD CENT LAB HUMAN EMBRYOL, 62- *Concurrent Pos:* Res assoc embryol, Dept Anat & vis asst prof, Col Med, Univ Fla, 61-62; vis investr, Dept Embryol, Carnegie Inst Wash, 62 & fetal lab, Dept Pediat, Copenhagen Univ, 63. *Honors & Awards:* Joseph Warkany Lectr, 88. *Mem:* AAAS; Am Pediat Soc; Teratology Soc (pres, 68). *Res:* Human embryology and teratology; clinical pediatrics; development of the thyroid; embryo explantation and organ culture; metabolism of achondroplastic dwarfism; author of one book. *Mailing Add:* Dept Pediat Univ Wash Sch Med Seattle WA 98195

SHEPARDSON, JOHN U, b Winchendon, Mass, May 4, 20; m 42; c 3. ANALYTICAL CHEMISTRY. *Educ:* Univ Mass, BS, 42; Rensselaer Polytech Inst, MS, 48, PhD(analytical chem), 50. *Prof Exp:* Analyst, Lever Bros Co, 46-47; mgr control dept, Uranium Div, Mallinckrodt Chem Works, 50-59, mgr tech admin, 59-64, dir control, Winthrop Labs, Sterling Drug, 64-76; tech dir, Cis Radiopharmaceut Inc, 76-79; chief chemist, Astro Circuit Corp, 80-84; RETIRED. *Concurrent Pos:* Consult, 84-89. *Mem:* AAAS; Am Soc Qual Control; Am Chem Soc. *Res:* Organic analytical reagents; inorganic separations; organic analysis; pharmaceuticals. *Mailing Add:* 11 Hitchinpost Rd Chelmsford MA 01824-1919

SHEPHARD, RONALD W(ILLIAM), industrial engineering, for more information see previous edition

SHEPHARD, ROY JESSE, b London, Eng, May 8, 29; m 58; c 2. PHYSIOLOGY, MEDICINE. *Educ:* Univ London, BSc, 49, MB, BS, 52, PhD(sci), 54, MD, 59. *Hon Degrees:* DPE, Univ Gent. *Prof Exp:* Asst prof prev med, Univ Cincinnati, 56-58; sr sci officer, Chem Defence Res Estab, UK Ministry Defence, 58-59, prin sci officer, 59-64; PROF APPL PHYSIOL & PREV MED, UNIV TORONTO, 64-, PROF PHYSIOL, 66-, PROF MED, INST MED SCI, 68-, ASSOC PROF PHYS EDUC, SCH PHYS & HEALTH EDUC, 71-, DIR, SCH PHYS & HEALTH EDUC, 79- *Concurrent Pos:* Fulbright scholar, Univ Cincinnati, 56-58; consult, Toronto Rehab Ctr, 57-, Gage Inst Chest Dis, 71- & Univ Qué, Trois Rivières, 71-; vpres, Int Comt Phys Fitness Res; vis prof, Univ Paris, 85-86; ed in chief, Can J Sports Sci. *Honors & Awards:* Citation, Am Col Sports Med; Adolpho Abrahams Medal, Brit Asn Sport & Med. *Mem:* Brit Physiol Soc; Can Physiol Soc; Am Physiol Soc; Brit Med Res Soc; Am Asn Health, Phys Educ & Recreation; Am Col Sports Med (past pres); Can Asn Sports Sci (past pres); hon fel Belg Asn Sports Med. *Res:* Cardiorespiratory physiology with particular reference to endurance fitness, sport medicine and the environment. *Mailing Add:* Sch Phys & Health Educ Univ Toronto Toronto ON M5S 1A1 Can

SHEPHARD, WILLIAM DANKS, b Gary, Ind, July 8, 33; m 59. EXPERIMENTAL ELEMENTARY PARTICLE PHYSICS. *Educ:* Wesleyan Univ, BA, 54; Univ Wis, MS, 55, PhD, 62. *Prof Exp:* Asst physics, Univ Wis, 54-60; asst prof, Univ Ky, 60-63; from asst prof to assoc prof, 63-73, PROF PHYSICS, UNIV NOTRE DAME, 73- *Concurrent Pos:* Guest jr res asst, Brookhaven Nat Lab, 57-58, guest assoc physicist, 60-; Fulbright fel, Max Planck Inst Physics, 62-63; guest physicist & consult, Argonne Nat Lab, 63-; guest physicist, Nat Accelerator Lab, 71-; consult, Oak Ridge Nat Lab, 61-63; guest prof physics, Fac Math & Natural Sci, Univ Nijmegen, Netherlands, 75-76. *Mem:* Am Phys Soc; Sigma Xi. *Res:* Elementary particle physics; high energy interactions; experiments in hadron-hadron, photon-hadron and hadron-nucleus interactions involving heavy quark production and multiparticle final states. *Mailing Add:* Dept of Physics Univ of Notre Dame Notre Dame IN 46556

SHEPHERD, ALBERT PITT, JR, b Lexington, Miss, Dec 29, 43; m 65; c 1. PHYSIOLOGY. *Educ:* Millsaps Col, BS, 66; Univ Miss, PhD(physiol), 71. *Prof Exp:* Asst instr physiol, Univ Tex Med Sch Houston, 72-73; asst prof, Col Med, Univ Calif, Irvine, 73-74; asst prof, 74-77, PROF PHYSIOL, UNIV TEX HEALTH SCI CTR SAN ANTONIO, 77- *Concurrent Pos:* Mem Nat Bd Med Examiners; assoc ed, Am J Physiol. *Mem:* Am Physiol Soc; Microcirc Soc; Int Soc Oxygen Transport to Tissue; Am Heart Asn; Inst Elec & Electronics Engrs; Can Physiol Soc; Europ Microcirculatory Soc. *Res:* Intestinal circulation; control of microcirculation; biomedical instrumentation. *Mailing Add:* Dept of Physiol Univ of Tex Health Sci Ctr San Antonio TX 78284

SHEPHERD, BENJAMIN ARTHUR, b Woodville, Miss, Jan 28, 41; c 2. ZOOLOGY. *Educ:* Tougaloo Col, BA, 61; Atlanta Univ, MA, 63; Kans State Univ, PhD(zool), 70. *Prof Exp:* Instr biol, Tougaloo Col, 63-65; asst zool, Kans State Univ, 66-69; from instr to asst prof, 69-73, asst chmn dept zool, 76-78, actg chmn, 78, assoc prof, 73-79, PROF ZOOL, SOUTHERN ILL UNIV, CARBONDALE, 79-, ASSOC VPRES ACAD AFFAIRS, 79- *Mem:* Am Soc Zoologists; Soc Study Reproduction; AAAS; Am Asn Anatomists; Sigma Xi. *Res:* Epididymal histophysiology of the mammal including effects of androgens and age on the viability and fertilizing capacity of spermatazoa; influence of olfactory stimulation on estrous induction. *Mailing Add:* Dept of Zool Southern Ill Univ Carbondale IL 62901

SHEPHERD, D(ENNIS) G(RANVILLE), b Ilford, Eng, Oct 6, 12; nat US; m 39; c 3. MECHANICAL ENGINEERING, AERONAUTICAL & ASTRONAUTICAL ENGINEERING. *Educ:* Univ Mich, BSE, 34. *Prof Exp:* Engr, Power Jets, Ltd, Eng, 40-46; chief exp engr, Gas Turbine Div, A V Roe Can, 46-48; from asst prof to prof, 48-78, dir, 65-72, EMER PROF MECH ENG, CORNELL UNIV, 78- *Concurrent Pos:* Guggenheim fel, 54-55; Orgn Europ Econ Coop fel, 60-61; consult, Curtiss-Wright Corp, Bendix-Westinghouse Automotive Air Brake Co, Continental Engine Co & Carrier Corp. *Honors & Awards:* Worcester Reed Warner Medal, Am Soc Mech Engrs, 76. *Mem:* AAAS; fel Am Soc Mech Engrs; Am Wind Energy Asn. *Res:* Thermal engineering, especially turbomachinery, gas turbines, wind power and jet propulsion. *Mailing Add:* Sch Mech & Aerospace Eng Cornell Univ Ithaca NY 14853-7501

SHEPHERD, DAVID PRESTON, b Center, Tex, Aug 2, 40; m 88; c 1. RADIATION BIOLOGY, ZOOLOGY. *Educ:* Lamar Univ, BS, 63; Tex A&M Univ, MS, 67, PhD(biol), 70. *Prof Exp:* Instr zool, Tex A&M Univ, 69-70; from asst prof to assoc prof biol, 70-80, PROF BIOL, SOUTHEASTERN LA UNIV, 80- *Mem:* AAAS; Sigma Xi. *Res:* Radiation; physiology; ecology; herpetology; archeology; paleontology. *Mailing Add:* Southeastern La Univ State PO Box 829 Hammond LA 70402

SHEPHERD, FREEMAN DANIEL, JR, b Boston, Mass, June 7, 36; m 59; c 3. PHYSICS, ELECTRONICS. *Educ:* Mass Inst Technol, BS & MS, 59; Northeastern Univ, PhD(elec eng), 65. *Prof Exp:* Staff scientist, Air Force Cambridge Res Labs, 57-59, group leader, infrared devices, 66-76, br chief electronic devices, 66- 81, CHIEF ELECTRONIC DEVICE TECHNOL DIV, ROME LAB RL/ESE, HANSCOM AFB, 82- *Concurrent Pos:* Assoc mem adv group electron devices, Spec Devices Working Group, 70-76; mem, Working Group Basic Mech Radiation Effects, 71-76; Air Force rep, Comt Electromagnetic Detection Devices, Nat Mat Adv Bd-Nat Res Coun-Nat Acad Sci, 71-72; fel Rome Lab, 88; mem DOD Passive Sensing Steering Group, 88-91. *Honors & Awards:* Tech Achievement Award, US Air Force Systs Command, 70; Charles E Ryan Award, 78; Harry L Davis Award, 88; Harold E Brown Award, 89; Fed Lab Technol Transition Award, 90. *Mem:* Fel Inst Elec & Electronics Engrs; Sigma Xi; fel Soc Photo-Optical Instrumentation Engrs. *Res:* Infrared imaging and passive sensing; over 50 publications and seven patents. *Mailing Add:* 37 Berkeley Dr Chelmsford MA 01824

SHEPHERD, GORDON GREELEY, b Sask, Can, June 19, 31; m 53, 87; c 3. AERONOMY, SPACE PHYSICS. *Educ:* Univ Sask, BSc, 52, MSc, 53; Univ Toronto, PhD(molecular spectros), 56. *Prof Exp:* From asst prof to assoc prof physics, Univ Sask, 57-69; PROF PHYSICS, YORK UNIV, 69-; DIR, SOLAR TERRESTRIAL PHYSICS LAB, INST SPACE & TERRESTRIAL SCI, 88- *Concurrent Pos:* Mem ed adv bd, Planetary & Space Sci; Can Coun Killam Res fel, 91- *Mem:* Optical Soc Am; Can Asn Physicists; Am Geophys Union; fel Can Aeronaut & Space Inst; fel Royal Soc Can. *Res:* Interferometric spectroscopy of airglow and aurora from the ground; rockets and satellites; principal investigator for space shuttle and upper atmospheric research satellite investigations of upper atmospheric winds. *Mailing Add:* Centre for Res in Exp Space Sci York Univ Downsview ON M3J 1P3 Can

SHEPHERD, GORDON MURRAY, b Ames, Iowa, July 21, 33; m 59; c 3. NEUROPHYSIOLOGY. *Educ:* Iowa State Univ, BS, 55; Harvard Med Sch, MD, 59; Oxford Univ, PhD(neurophysiol), 62. *Prof Exp:* Res assoc biophys, NIH, 62-64; vis scientist neurophysiol, Karolinska Inst, Sweden, 64-66; asst prof physiol, 67-68, assoc prof, 69-79, PROF NEUROSCIENCE, SCH MED, YALE UNIV, 79- *Concurrent Pos:* USPHS fel, 59-62, spec fel, 64-66 & res grant, 66-; assoc fel, Retina Found, Boston, 66-67 & dept biol, Mass Inst Technol, 67; vis assoc prof, Neurol Inst, Univ Pa, 71-72; mem study sect, NIH, 75-79; chair, Nat Asn Chemorecep Sci, 81-82, comt in hist neuroscience, Soc Neuroscience, 85-; vis prof, Col France, 86-; physiol comt, Nat Bd Examrs, 84-88. *Honors & Awards:* R H Wright Award, 86; Freeman Award, 88. *Mem:* Soc Neurosci; Am Physiol Soc; Int Brain Res Orgn; Sigma Xi. *Res:* Synaptic organization of the olfactory system; transduction properties of sensors reactors; interactive properties of neuronal dendrites; mechanisms of cortical integration; computational neurology models. *Mailing Add:* Sect Neuroanat Med Sch Yale Univ 333 Cedar St New Haven CT 06510

SHEPHERD, HERNDON GUINN, clinical chemistry, toxicology, for more information see previous edition

SHEPHERD, HURLEY SIDNEY, b Oxford, NC, Dec 17, 50. PLANT MOLECULAR BIOLOGY, ORGANELLE GENETICS. *Educ:* Univ NC, BS, 73; Duke Univ, PhD(bot), 78. *Prof Exp:* Res assoc, Univ Calif, San Diego, 79-81; ASST PROF BOT & BIOCHEM, UNIV KANS, 81- *Mem:* Genetics Soc Am; Am Soc Plant Physiologists; Plant Molecular Biol Asn; AAAS; Sigma Xi. *Res:* Control of gene expression in plants due to environmental and developmental effects and interactions between the nuclear and organelle genomes. *Mailing Add:* 414 28th St New Orleans LA 70124

SHEPHERD, JAMES E, b Houston, Tex, May 29, 10. ENGINEERING. *Educ:* Univ Mo, BA, 32, MA, 33; Harvard Univ, MS, 35, DSc, 40. *Prof Exp:* Sci adv, Sperry Corp, 41-72; RETIRED. *Mem:* Fel Inst Elec & Electronic Engrs; fel AAAS; Am Phys Soc; Sigma Xi. *Mailing Add:* Box 27 210 Park Lane Concord MA 01742

SHEPHERD, JIMMIE GEORGE, b Lone Oak, Tex, Sept 2, 23; m 46; c 2. PHYSICS. *Educ:* North Tex State Univ, BS, 47, MA, 51. *Prof Exp:* Teacher high sch, Tex, 47-53; petrol engr, Sun Oil Co, 53-57; asst prof physics, Lamar State Col, 57-66, assoc prof physics, Lamar Univ, 66-87; RETIRED. *Concurrent Pos:* Sci teaching consult, Lamar Area Sch Study Coun, 66, 67 & 68. *Mem:* AAAS; Am Asn Physics Teachers. *Res:* Search for magnetic monopoles. *Mailing Add:* 55 County Rd 238 1/2 Cameron TX 76520

SHEPHERD, JOHN PATRICK GEORGE, b West Wickham, Eng, May 11, 37; m 63; c 2. SOLAR ENERGY. *Educ:* Univ London, BSc & ARCS, 58, PhD(electron physics) & dipl, Imp Col, 62. *Prof Exp:* Res asst electron physics, Imp Col, Univ London, 61-63; res assoc, Case Inst Technol, 63-65, asst prof, 65-66; sr sci officer, Royal Radar Estab, Eng, 66-69; assoc prof physics, 69-75, PROF PHYSICS, UNIV WIS-RIVER FALLS, 75- *Concurrent Pos:* Consult, 3M Co, 71- *Mem:* Am Phys Soc. *Res:* Band structure of solids; deHass van Alphen effect in metals and alloys, pseudo potential models of band structure of same; laser and nonlinear optics; band structure of semiconductors; solar energy; optical design; electro-chemical studies. *Mailing Add:* Dept Physics Univ Wis 372 W Charlotte St River Falls WI 54022

SHEPHERD, JOHN THOMPSON, b Northern Ireland, May 21, 19; m 45; c 2. PHYSIOLOGY. *Educ:* Queen's Univ Belfast, MB & BCh, 45, MCh, 48, MD, 51, DSc, 56. *Prof Exp:* House physician, Royal Victoria Hosp, Belfast, Northern Ireland, 45-46, extern surgeon, 46; lectr physiol, Queen's Univ Belfast, 47-53; Anglo-French Med Exchange bursary, 57; assoc prof, 57-62, dir res, Mayo Found, 69-77, prof physiol, Mayo Grad Sch Med, Univ Minn, 62-, dean Mayo Sch Med, 77-; Sigma Xi. *Concurrent Pos:* Leathem traveling fel & Fulbright scholar, 53-54; consult, Mayo Clin, 57- & Northern Ireland Hosps; chmn physiol sect, Mayo Found, 66-74. *Honors & Awards:* Carl J Wiggers Award, Am Physiol Soc, 78; Gold Heart Award, Asn Am Physicians, 78. *Mem:* Am Heart Asn (pres, 75-76). *Res:* Heart and peripheral circulation of man and animals in health and disease, especially hemodynamics. *Mailing Add:* Mayo Clin & Found Plummer 1043 Rochester MN 55905

SHEPHERD, JOSEPH EMMETT, b Joliet, Ill, Mar 7, 53; m 79; c 2. COMBUSTION, GAS DYNAMICS. *Educ:* Univ SFla, BS, 76; Calif Inst Technol, PhD(appl physics), 81. *Prof Exp:* MEM TECH STAFF, SANDIA NAT LABS, 80- *Mem:* Am Phys Soc; AAAS; Sigma Xi; Combustion Inst. *Res:* Modeling of rapid evaporation of superheated liquids; compressible flow in porous materials; gasdynamics of explosions; turbulent combustion of high-temperature gases. *Mailing Add:* Rensselaer Polytech AE & Mechanic Troy NY 12180-3590

SHEPHERD, JULIAN GRANVILLE, b Lutterworth, Eng, Dec 16, 42; US citizen; m 66, 84; c 1. INVERTEBRATE PHYSIOLOGY. *Educ:* Cornell Univ, BA, 64; Harvard Univ, PhD(biol), 72. *Prof Exp:* Res scientist insect reproduction, Int Ctr Insect Physiol & Ecol, 72-74; fel, Harvard Univ, 74-75; asst prof, 75-84, ASSOC PROF BIOL, STATE UNIV NY BINGHAMTON, 84- *Mem:* AAAS; Entom Soc Am; Soc for Invert Reproduction; Lepidopterist Soc; Sigma Xi. *Res:* Development and physiology of invertebrate, mainly insect, spermatozoa. *Mailing Add:* Dept Biol Sci SUNY-Binghamton Binghamton NY 13901

SHEPHERD, LINDA JEAN, b Philadelphia, Pa, May 25, 49. WOMEN IN SCIENCE, JUNGIAN PSYCHOLOGY. *Educ:* Millersville State Col, BA, 71; Pa State Univ, PhD(biochem), 76. *Prof Exp:* Mgr, Prod Dept, Worthington Diagnostics Corp, 76-81; mgr prod develop, Genetic Systs Corp, 81-87; SCI WRITER, 87- *Res:* Author in fields of science and spirituality, history and philosophy of science, and science policy and ethics. *Mailing Add:* 14985 256th Ave SE Issaquah WA 98027

SHEPHERD, MARK, JR, b Dallas, Tex, Jan 18, 23; m 45; c 3. ELECTRICAL ENGINEERING. *Educ:* Southern Methodist Univ, BS, 42; Univ Ill, Urbana, MS, 47. *Hon Degrees:* PhD, Southern Methodist Univ, 66, Rensselaer Polytech Inst, 79. *Prof Exp:* Test engr, Gen Elec Co, 42-43; engr, Farnsworth TV & Radio Corp, 47-48; proj engr, Geophys Serv Inc, 48-51, from asst chief engr to chief engr, Semiconductor Design, 52-54, asst vpres, Semiconductor-Components Div, 54-55, gen mgr, 54-61, vpres, 55- 61, exec vpres, 61-66, chief oper officer, 61-69, pres, 67-76, chief exec officer, 69-84, chmn, Tex Instruments, 76-88; RETIRED. *Mem:* Nat Acad Eng; Inst Elec & Electronics Engrs; Soc Explor Geophys. *Res:* Solid state control systems. *Mailing Add:* Tex Instruments Inc PO Box 655474 MS 407 Dallas TX 75265

SHEPHERD, RAYMOND EDWARD, b Joliet, Ill, Aug 21, 41; c 1. PHYSIOLOGICAL CHEMISTRY. *Educ:* Bethel Col, BS, 62; Univ Mont, MS, 68; Wash State Univ, PhD(exercise physiol), 74. *Prof Exp:* Instr biol, Graceville High Sch, 64-66; instr physiol, Dakota State Col, 68-70; NIH fel, Brown Univ, 74-76, from instr to asst prof physiol chem, 76-78; assoc prof exercise physiol, Univ Toledo, 78-80; asst prof, 80-84, ASSOC PROF PHYSIOL, LA STATE UNIV MED CTR, 84- *Concurrent Pos:* Fel, Nat Inst Arthritis Metab & Digestive Dis, 74. *Mem:* NY Acad Sci; AAAS; Am Col Sports Med; Am Physiol Soc; Sigma Xi. *Res:* Regulation of adenylate cyclase, protein kinase, and triglyceride lipase in fat cells by hormones and fatty acids. *Mailing Add:* Dept Physiol La State Univ Med Ctr New Orleans LA 70112

SHEPHERD, RAYMOND LEE, b Arkadelphia, Ark, Oct 13, 26; m 50; c 3. GENETICS, PLANT PATHOLOGY. *Educ:* Ouachita Baptist Col, 50; Univ Ark, MS, 60; Auburn Univ, PhD(plant breeding), 65. *Prof Exp:* Proprietor retail grocery bus, 54-57; res asst plant breeding, Univ Ark, 57-60; asst agron, 60-65, supvry res agronomist, Auburn Univ, 65-91; RETIRED. *Mem:* Am Soc Agron. *Res:* Genetics and breeding investigations on cotton, especially the development of basic breeding stocks which are resistant to nematodes and diseases. *Mailing Add:* 549 College Auburn AL 36830

SHEPHERD, REX E, b Greenville, Ohio, Nov 27, 45. REDOX REACTIONS, MODELING CATALYTIC REACTIVITIES. *Educ:* Purdue Univ, BS, 67; Stanford Univ, MS, 69, PhD(inorg chem), 71. *Prof Exp:* Fel, Dept Chem, Yale Univ, 71-72 & State Univ NY, Buffalo, 72-73; vis asst prof, Purdue Univ, 73-75; asst prof, 75-81, ASSOC PROF INORG ANAL, DEPT CHEM, UNIV PITTSBURGH, 81- *Concurrent Pos:* Consult, Maynard Metals, Inc, 74-75. *Mem:* Am Chem Soc. *Res:* Mechanistic inorganic chemistry as related to transition metal complexes, bioinorganic systems, catalytic activation of small molecules and photochemistry studied by physical chemical methods. *Mailing Add:* Dept Chem Univ Pittsburgh Pittsburgh PA 15260-0001

SHEPHERD, ROBERT JAMES, b Clinton, Okla, June 5, 30; m 78; c 3. PLANT PATHOLOGY, VIROLOGY. *Educ:* Okla State Univ, BS, 54, MS, 56; Univ Wis, PhD(plant path), 59. *Prof Exp:* Asst plant path, Okla State Univ, 54-55; asst, Univ Wis, 56-58, res assoc, 58-59, asst prof, 59-61; asst prof, Univ Calif, Davis, 61-65; assoc prof, Univ Ark, 65-66; from assoc prof to prof plant path, Univ Calif, Davis, 66-84, assoc plant pathologist, 66-72; PROF PLANT PATH, UNIV KY, 84- *Concurrent Pos:* Assoc ed, Phytopath, 65-66; ed, Virol, 71-74; chmn plant virus subcomt, Int Comt Taxonomy Viruses; assoc ed, J Gen Virol, 78-81; mem sci bd, Calgene, 80-85. *Honors & Awards:* Ruth Allen Award, Am Phytopath Soc, 81. *Mem:* Nat Acad Sci; Soc Gen Microbiol; fel Am Phytopath Soc. *Res:* Characterization and description of plant viruses; epidemiology and control of plant virus diseases; recombinant DNA vectors for plants. *Mailing Add:* Dept Plant Path Univ Ky S 305 Agr Sci Bldg Lexington KY 40546-0091

SHEPHERD, ROBIN, b York, Eng, June 18, 33; US citizen; m 86; c 2. FORENSIC ENGINEERING. *Educ:* Univ Leeds, UK, BS, 55, MS, 65; Univ Canterbury, NZ, PhD(civil eng), 71; Univ Leeds, UK, DSc, 73. *Prof Exp:* Asst eng, DeHavilland Aircraft Co, UK, 55-57, NZ Ministry Works, 58-59; fac mem struct, Univ Canterbury, NZ, 59-71; assoc prof struct, Univ Auckland, NZ, 72-79; dir, NZ Heavy Eng Res Asn, 79-80; PROF, STRUCT-EARTH CIVIL ENG, UNIV CALIF, IRVINE, 80- *Concurrent Pos:* Vis prof, Calif Inst Technol, 77; vis overseas scholar, St John's Col, Cambridge, UK, 84; Erskine fel, Univ Canterbury, NZ, 87; pres, Forensic Expert Adv Inc, Calif, 90- *Honors & Awards:* E R Cooper Medal, Royal Soc NZ, 72. *Mem:* Fel Am Soc Civil Engrs; fel Nat Acad Forensic Engrs; Earthquake Eng Res Inst; Seismogr Soc Am. *Res:* Application of structural dynamic analysis procedures to the prediction of earthquake-induced loads and movements in civil engineering structures including improved methods of seismic design. *Mailing Add:* Dept Civil Eng Univ Calif Irvine CA 92717

SHEPHERD, VIRGINIA L, CELL BIOLOGY, BIOCHEMISTRY. *Educ:* Univ Iowa, PhD(biochem), 75. *Prof Exp:* ASST PROF MED & BIOCHEM, UNIV TENN, MEMPHIS, 84- *Res:* Receptor modulation in macrophages. *Mailing Add:* Vet Admin Med Ctr Res 1301 24th Ave Nashville TN 37212

SHEPHERD, W(ILLIAM) G(ERALD), b Ft William, Ont, Aug 28, 11; nat US; m 36; c 3. ELECTRICAL ENGINEERING. *Educ:* Univ Minn, BS, 33, PhD(physics), 37. *Prof Exp:* Mem tech staff, Bell Tel Labs, NJ, 37-47; assoc dean inst technol, 54-56, head dept elec eng, 56-63, vpres acad admin, 63-73, prof elec eng, 47-79, dir, Space Sci Ctr, 74-79, EMER PROF ELEC ENG, UNIV MINN, MINNEAPOLIS, 79- *Concurrent Pos:* Consult, Bendix Aviation Corp, 49-63, Gen Elec Co, 55-63 & Control Data, 58-63; US Comn VII chmn, Int Sci Radio Union, 53-57 & Int Comn VII, 57-63; mem eng sci panel, NSF, 58-61, eng div adv comt, 64-69; mem adv group electron devices, Dept Defense, 58-68, chmn, 62-68; mem space technol adv comt, NASA, 64-69. *Honors & Awards:* Citation, Bur Ships, 47; Medal of Honor, Nat Electronics Conf, 65. *Mem:* Nat Acad Eng; fel Inst Elec & Electronics Engs (vpres, 65-66, pres, 66-67); Am Phys Soc. *Res:* Microwave electronics; physical electronics, especially electron emission. *Mailing Add:* 103 Shepherd Lab 100 Union St SE Minneapolis MN 55455

SHEPHERD, WILLIAM LLOYD, b Okla, Oct 2, 15; m 39; c 1. MATHEMATICS. *Educ:* Okla State Univ, BS, 38, MS, 41. *Prof Exp:* Instr math, Southwestern Col (Kans), 43-45; asst prof, Okla State Univ, 46-47; instr, Univ Tex, 47; instr, Univ Ore, 48-52; asst prof math & physics, Tex Western Col, 52-60; mathematician, White Sands Missile Range, US Dept Army, NMex, 60-73, res mathematician, Res Proj Off, Instrumentation Directorate, 73-80; RETIRED. *Concurrent Pos:* consult, 80-83. *Mem:* Am Math Soc. *Res:* Mathematics for missile range instrumentation problems; digital signal processing; number theoretic considerations in interferometric angle measurements. *Mailing Add:* 4829 Montreal Dr San Jose CA 95130

SHEPLEY, LAWRENCE CHARLES, b Washington, DC, Aug 11, 39. COSMOLOGY. *Educ:* Swarthmore Col, BA, 61; Princeton Univ, MA, 63, PhD(physics), 65. *Prof Exp:* Res physicist, Univ Calif, Berkeley, 65-67; asst prof, 67-70, ASSOC PROF PHYSICS, UNIV TEX, AUSTIN, 70-, ASSOC DIR, CTR RELATIVITY THEORY, 71- *Concurrent Pos:* NSF grant, 68-78; vchmn grad affairs, Physics Dept, Univ Tex, Austin, 71-73. *Mem:* Am Phys Soc; Am Asn Physics Teachers. *Res:* Cosmological models; equivalent lagrangians. *Mailing Add:* Dept Physics Univ Tex Austin TX 78712

SHEPP, ALLAN, b New York, NY, Apr 2, 28; m 48; c 2. LASER PHYSICS, IMAGING SCIENCE. *Educ:* Oberlin Col, BA, 48; Cornell Univ, PhD(phys chem), 53. *Prof Exp:* Fel div pure chem, Nat Res Coun Can, 53-55; chemist, Tech Opers, Inc, 55-69, dir chem, 65-69; sr scientist, Polaroid Corp, Cambridge, 69-85; DIR, BUS PLANNING/LASER PHYSICS, AVCO RES LAB, TEXTRON, 85- *Mem:* Am Phys Soc; sr mem Soc Photog Scientists & Engrs (exec vpres, 71-); Am Chem Soc. *Res:* Photographic theory; color imaging; lasers and optics. *Mailing Add:* Textron Defense Syst 2385 Revere Pkwy Bch Everett MA 02149

SHEPP, LAWRENCE ALAN, b Brooklyn, NY, Sept 9, 36; m 62; c 3. MATHEMATICS. *Educ:* Polytech Inst Brooklyn, BS, 58; Princeton Univ, MA, 60, PhD(math), 61. *Prof Exp:* Instr probability & statist, Univ Calif, Berkeley, 61-62; MEM TECH STAFF MATH, BELL LABS, 62- *Concurrent Pos:* Mem comt appl math, Nat Acad Sci, 74-77; vis scientist math dept, Mass Inst Technol, 75; adj prof, Columbia Univ & Neurol Inst, 74-; prof, Statist Dept, Stanford Univ, 84- *Honors & Awards:* Paul-Levy Prize, Inst Henri Poincare, Paris, 66 & 89; Distinguished Scientist Award, Inst Elec & Electronics Engrs, 82. *Mem:* Nat Acad Sci; Am Math Soc; Math Asn Am; AAAS; Inst Elec & Electronics Engrs; fel Inst Math Statist. *Res:* Probability and statistics; stochastic processes; Gaussian measure theory; analysis; asymptotics; random walk; limit theorems; computered tomography; medical imaging; reconstruction of pictures from projections. *Mailing Add:* Bell Labs 2C-374 Murray Hill NJ 07974

SHEPPARD, ALAN JONATHAN, b Parkersburg, WVa, Oct 11, 27; m 60. NUTRITIONAL BIOCHEMISTRY, ANIMAL NUTRITION. *Educ:* Ohio State Univ, BS, 51; Va Polytech, MS, 54; Univ Ill, MS, 56, PhD(animal nutrit), 59. *Prof Exp:* Agronomist, Northern Va Pasture Res Sta, 53; res asst animal sci, Agr Exp Sta, Univ Ill, Urbana, 59-60; chief lipid res sect, 60-70, RES CHEMIST, DIV NUTRIT, FOOD & DRUG ADMIN, 60-, CHIEF, FATS & ENERGY SECT, 70- *Concurrent Pos:* Instr grad sch, USDA, 60-63; actg chief macronutrient res br, Food & Drug Admin, 67; mem sub comn 9,12 di-cis-linoleic acid, Oils & Fats Sect, Int Union of Pure & Appl Chem, 74-78; adj prof chem, Am Univ, 77-; assoc prof, Human Nutrit Prog, Howard Univ, 77- *Mem:* Am Inst Nutrit; Am Oil Chem Soc; Am Soc Animal Sci; Am Dairy Sci Asn; Asn Off Analytical Chem. *Res:* Gas chromatography of vitamins; nutritional-metabolic studies in the lipid and fatty acid fields; interface between research and application to regulatory problems. *Mailing Add:* 10603 Vickers Dr Rte No 7 Vienna VA 22180

SHEPPARD, ALBERT PARKER, b Griffin, Ga, June 6, 36; m 78; c 3. ELECTRONICS, RESEARCH ADMINISTRATION. *Educ:* Oglethorpe Univ, BS, 58; Emory Univ, MS, 59; Duke Univ, PhD(elec eng), 65. *Prof Exp:* Instr physics, Univ Ala, 59-60; sr engr, Martin Co, Fla, 60-63; radio physicist, US Army Res Off, NC, 63-65; head, Spec Tech Br, Eng Exp Sta, 65-71, chief, Chem Sci & Mat Div, 71-72, assoc dean, Col Eng, 72-74, actg vpres res, 80, 88, PROF ELEC ENG, GA INST TECHNOL, 72-, ASSOC VPRES RES, 74- *Concurrent Pos:* Lectr, DeKalb Col, 67-71; pres, Microwave & Electronic Consults, Atlanta; Indust Res Inst Univ Comt; bd trustees, Univ Space Res Asn. *Mem:* Sr mem Inst Elec & Electronics Engrs; Am Soc Eng Educ; Nat Soc Prof Engrs. *Res:* Robotic applications; microwave engineering; microcomputer applications. *Mailing Add:* 1240 Jefferson Dr Norcross FL 33803

SHEPPARD, ASHER R, b Brooklyn, NY, Apr 4, 43; m 65; c 3. MEMBRANE BIOPHYSICS. *Educ:* Union Col, BS, 63; State Univ NY Buffalo, MS, 71, PhD(physics), 75. *Prof Exp:* Nat Inst Environ Health Sci fel environ biophys, Inst Environ Med, NY Univ Med Ctr, 74-76; Nat Inst Environ Health Sci fel environ biophys, Brain Res Inst, Univ Calif, Los Angeles, 76-78; RES PHYSICIST, PETTIS MEM VET HOSP, LOMA LINDA, CALIF, 78- *Concurrent Pos:* Asst res prof physiol, Loma Linda, Calif, 79-, Dept Neurosurg, 88-; sci adv to WHO, 80-88, Dept Energy, 80, Calif Dept Health Serv, 89-, Inst Elec & Electronic Engrs Comar, 88-, City Seattle, 84-86 & 88. *Mem:* AAAS; Am Phys Soc; Bioelectromagnetics Soc; Soc Neurosci; Biophys Soc; Bioelectrochem Soc. *Res:* Biological effects of electric and magnetic fields; physiological effects of fields at cell membrane surfaces and biophysical modeling of cell environment in fields; determination of health criteria for field exposures. *Mailing Add:* Res Serv-151 J L Pettis Mem Vet Hosp Loma Linda CA 92357

SHEPPARD, CHESTER STEPHEN, b Buffalo, NY, Sept 27, 27; m 52; c 2. ORGANIC CHEMISTRY. *Educ:* Canisius Col, BS, 52; Univ Pittsburgh, MLS, 55, PhD(org chem), 61. *Prof Exp:* Jr fel coal chem, Mellon Inst Indust Res, 52-61; res chemist, Lucidol Div, Wallace & Tiernan, Inc, 62-63; group leader res, 63-65, supvr res, 66-70, MGR RES, LUCIDOL DIV, PENNWALT CORP, 71- *Mem:* Am Chem Soc; Am Inst Chem; Royal Soc Chem; Sigma Xi. *Res:* Nitrogen chemistry; free radicals; peroxides; aliphatic azo chemistry; foamed polymers; polymerization; plasticizers; stabilizers; organic blowing agents; hydrazine chemistry. *Mailing Add:* 726 Parkhurst Kenmore NY 14223

SHEPPARD, DAVID E, b Chester, Pa, Jan 16, 38; m 60; c 3. BIOCHEMICAL GENETICS. *Educ:* Amherst Col, BA, 59; NIJohns Hopkins Univ, PhD(biochem genetics), 63. *Prof Exp:* Asst prof biol, Reed Col, 63-65; NIH fel, Univ Calif, Santa Barbara, 65-66; asst prof, 67-69, ASSOC PROF BIOL, UNIV DEL, 69- *Concurrent Pos:* NIH fel, Univ BC, 73. *Mem:* Genetics Soc Am; Am Soc Microbiol. *Res:* Transduction; feed-back inhibition; genetic control of protein synthesis; arabinose operon. *Mailing Add:* Dept Life Sci Univ Del Newark DE 19716

SHEPPARD, DAVID W, b Quincy, Mass, Dec 28, 27; m 52; c 2. PHYSICS. *Educ:* Gordon Col, BA, 52; Andover Newton Theol Sch, BD, 56; Brown Univ, MA, 61; Ohio State Univ, PhD(physics), 68; Univ Pittsburgh, MLS, 77. *Prof Exp:* Pastor, Congregational Church, Dunstable, Mass, 54-57 & Riverpoint Congregational Church, West Warwick, RI, 57-61; instr physics, Defiance Col, 61-64, asst prof, 64-65; chem dept physics, Thiel Col, 75-77, asst prof, 68-80; ASSOC PROF, WVA WESLEYAN COL, 80- *Mem:* Am Phys Soc; Am Asn Physics Teachers. *Res:* Nuclear magnetic resonance. *Mailing Add:* Dept Math & Comput Sci WVa Wesleyan Col PO Box 68 Buckhannon WV 26201

SHEPPARD, DEAN, b Bronx, NY, June 13, 49. PULMONARY MEDICINE. *Educ:* State Univ NY, Stony Brook, MD, 75. *Prof Exp:* Asst prof, 81-86, ASSOC PROF MED, UNIV CALIF, SAN FRANCISCO, 86- *Mem:* Am Physiol Soc; Am Thoracic Soc. *Mailing Add:* Dept Med Univ Calif Bldg One Rm 150 San Francisco Gen Hosp 1001 Potrero Ave San Francisco CA 94110

SHEPPARD, DONALD M(AX), b Port Arthur, Tex, Mar 21, 37; m 58; c 2. MECHANICAL ENGINEERING. *Educ:* Lamar State Univ, BS, 60; Tex A&M Univ, MS, 62; Ariz State Univ, PhD(mech eng), 69. *Prof Exp:* Design engr, Collins Radio Co, 60-61; asst mech eng, Tex A&M Univ, 61-62, Univ Ariz, 62-63 & Southwest Res Inst, 63-64; instr, Ariz State Univ, 64-68, NSF fel, 67-68; from asst prof to assoc prof eng sci, 69-77, asst chmn coastal & oceanog eng, 77-78, ASSOC PROF COASTAL & OCEANOG ENG, UNIV FLA, 77-, ACTG CHMN COASTAL & OCEANOG ENG, 78- *Concurrent Pos:* Instr, San Antonio Col, 63-64. *Res:* Stratified shear flows internal waves coastal hydraulics; nearshore sediment transport; geophysical fluid mechanics. *Mailing Add:* Dept Coastal & Oceanog Eng Univ Fla Col Eng Gainesville FL 32611

SHEPPARD, EMORY LAMAR, b Hendersonville, NC, Sept 18, 42; m 67; c 2. DIGITAL ELECTRONICS & MICROPROCESSORS, ELECTRONIC COMMUNICATIONS. *Educ:* Clemson Univ, BS, 67; NC State Univ, MS, 75. *Prof Exp:* Antenna engr, Radiation, Inc, 67-79; systs engr, Res Triangle Inst, 69-72; head dept EET, Spartanburg Tech Col, 73-75; assoc prof EET, Univ NC, Charlotte, 75-78; assoc prin engr, Harris Corp, 78-81; ASSOC PROF EET, CLEMSON UNIV, 81- *Concurrent Pos:* Prin investr, impedance matching lens radar proj, Appl Sci Assocs, Inc, 82, pulse detection presence impulse noise proj, Security Tag Systs, Inc, 83, passive subharmonic transponder proj, 84-85 & microstrip patch transponder proj, 85- *Mem:* Inst Elec & Electronics Engrs. *Res:* High gain low noise antennas; phased array radar; broadband autotracking feed systems; signal detection in the presence of noise; remote sensing; digital communications; microprocessor applications; physics. *Mailing Add:* Clemson Univ 224 Riggs Hall Clemson SC 29634-0915

SHEPPARD, ERWIN, b New York, NY, May 27, 21; m 43; c 2. PHYSICAL CHEMISTRY. *Educ:* City Col, BS, 42; Polytech Inst Brooklyn, PhD(chem), 51. *Prof Exp:* Res assoc phys biochem, Med Col, Cornell Univ, 49-57; staff chemist, 57-58, phys res mgr, S C Johnson & Son, Inc, 58-81; prog developer, Grad Sch, Univ Wis-Milwaukee, 82-85; RETIRED. *Mem:* Am Chem Soc; Electron Micros Soc Am; Soc Appl Spectros; Sigma Xi; AAAS. *Res:* Physical chemistry of polymeric and colloidal systems; organic coatings; surface chemistry; aerosol technology; electrokinetic and light scattering investigations; electron microscopy and corrosion studies. *Mailing Add:* 1108 N Milwaukee St Apt 145 Milwaukee WI 53202-3153

SHEPPARD, HERBERT, b New York, NY, June 14, 22; m 47; c 2. BIOCHEMISTRY. *Educ:* Cornell Univ, BS, 48; Univ Calif, PhD(biochem), 53. *Prof Exp:* Asst, Nutrit Lab, Cornell Univ, 48; assoc chemist, Biochem Res, Gen Foods Corp, Inc, 49; asst, Univ Calif, 49-50; assoc dir biochem pharmacol, Ciba Pharmaceut Co, NJ, 53-62, head, 62-67; sect head biochem pharmacol, Hoffmann-La Roche, Inc, 67-72, dir dept cell biol, 72-82, dir sci, Corp Lic Dept, 82-87; RETIRED. *Mem:* AAAS; Am Soc Pharmacol & Exp Therapeut; Am Chem Soc; Sigma Xi. *Res:* Cyclic adenosine monophosphate metabolism; metabolism of nervous tissue; steroid hormone action; lipid metabolism. *Mailing Add:* 33 Highwood Rd W Orange NJ 07052

SHEPPARD, JOHN CLARENCE, b San Pedro, Calif, June 4, 23; m 50; c 3. RADIOCHEMISTRY, INORGANIC CHEMISTRY. *Educ:* San Diego State Col, AB, 49; Washington Univ, MA, 54, PhD(chem), 55. *Prof Exp:* Chemist, US Naval Radiol Defense Lab, 49-51; asst, Washington Univ, 52-55; chemist, Hanford Labs, Gen Elec Co, 55-57; asst prof chem, San Diego State Col, 57-60; chemist, Hanford Labs, Gen Elec Co, 60-63; sr res scientist chem dept, Battelle Mem Inst, 65-71; assoc nuclear engr, Eng Res Div, 72-77, PROF CHEM ENG & ANTHROP, WASH STATE UNIV, 77- *Mem:* Am Chem Soc; AAAS; Sigma Xi. *Res:* Electron transfer reactions between ions in aqueous solution; solvent extraction of actinide elements; activation analysis; atmospheric chemistry; radiocarbon dating. *Mailing Add:* NE 1015 Alfred Lane Pullman WA 99163

SHEPPARD, JOHN RICHARD, b Minneapolis, Minn, Dec 29, 44; m 68; c 3. CELL BIOLOGY, BIOCHEMISTRY. *Educ:* Univ Minn, Minneapolis, BA, 65; Univ Colo, Boulder, PhD(chem), 69;. *Prof Exp:* NIH fel biochem, Princeton Univ, 69-70; asst prof neurol, Univ Colo, Denver, 70-72; from asst prof to assoc prof genetics, Univ Minn, Minneapolis, 72-78, asst dir, Dight Inst Human Genetics, 72-84, prof genetics, cell biol & med, 78-87, dir, Dight Inst Human Genetics, 84-87; vpres, Corp Finance, Kidder Peabody, NY, 87-88; VPRES, CORP FINANCE, VECTOR SECURITIES, DEERFIELD, IL, 88- *Concurrent Pos:* Sci adv, Nat Ataxia Found, 74-87 & Leukemia Task Force, 75-87; mem, adv comt biochem & carcinogenesis, Am Cancer Soc, 78-82 & cell physiol study sect, NIH, 79-83; Alexander von Humboldt fel, Heidelberg, Germany, 82. *Honors & Awards:* Jackson Mem Lectr, Med Sch, Northwestern Univ, 75. *Mem:* Am Asn Cancer Res; Am Soc Biol Chemists; Am Soc Cell Biol; Soc Neurosci; Tissue Cult Asn; Am Soc Pharmacol & Exp Therapeut. *Res:* The role of the plasma membrane in biological regulation, specifically cyclic nucleotide, plasma membrane enzymes and hormonal studies using cell cultures and nerve cells; Alzheimer's disease; Down's syndrome; cancer and tumor cell biology. *Mailing Add:* Vector Securities Int 1751 Lake Cook Rd Deerfield IL 60015

SHEPPARD, KEITH GEORGE, b London, Eng, June 28, 49; m 77; c 2. ELECTROCHEMICAL PROCESSES, CORROSION. *Educ:* Leeds Univ, Eng, BS, 71; Univ Birmingham, Eng, PhD(metall), 80. *Prof Exp:* Coordr, Stats (MR) Ltd, Birmingham, Eng, 72-73; exp officer, Metals & Alloys, Birmingham, Eng, 73-75; teaching fel, 79-80, res asst prof, 80-81, ASSOC PROF MAT & METALL ENG, STEVENS INST TECHNOL, NJ, 81- *Mem:* Electrochem Soc; Nat Asn Corrosion Engrs; Am Electroplaters & Surface Finishers Soc. *Res:* Structure and properties of electrodeposits, electroless deposits and deposits produced by laser-enhanced electrodeposition; corrosion and corrosion testing; electron microscopy; failure analysis; environmental degradation of engineering ceramics. *Mailing Add:* Dept Mat & Metall Eng Stevens Inst Technol Hoboken NJ 07030

SHEPPARD, LOUIS CLARKE, b Pine Bluff, Ark, May 28, 33; m 58; c 3. MEDICAL PRODUCTS LIABILITY, PATENT INFRINGEMENT. *Educ:* Univ Ark, BS, 57; Univ London, PhD(elec eng), 76. *Prof Exp:* Prof biomed eng & chmn dept, Univ Ala, Birmingham, 79-88, dir acad comput, Med Ctr, 81-83, prof elec eng & biostatist & biomath, 86-88; asst vpres res, 88-90, PROF PHYSIOL & BIOPHYS, UNIV TEX MED BR, 88-, ASSOC VPRES RES, 90- *Concurrent Pos:* Fac mem, Grad Sch, Univ Ala, Birmingham, 75-88, adj prof, Dept Biomed Eng, 89-; sr scientist, Cystic Fibrosis Res Ctr, 81-88; prof elec & computer eng, Univ Tex, Austin, 89-; adj prof, Dept Elec Eng, Cullen Col Eng, Univ Houston, 89-; mem sci staff, Shriners Burn Inst, Galveston, 89- *Honors & Awards:* Ayrton Premium, Inst Elec & Electronic Engrs. *Mem:* Am Inst Chem Engrs; sr mem Inst Elec & Electronic Engrs; sr mem Biomed Eng Soc; Brit Computer Soc; Sigma Xi. *Res:* Computer based systems for intensive care; automated blood and drug infusion for closed-loop, feedback control of cardiac and vascular pressures. *Mailing Add:* Univ Tex Med Branch 528 Admin Bldg Galveston TX 77550

SHEPPARD, MOSES MAURICE, b Hendersonville, NC, Sept 5, 28; m 51; c 2. SCIENCE EDUCATION. *Educ:* ECarolina Col, BS, 52, MA, 58; Ohio State Univ, PhD(sci ed, physics), 66. *Prof Exp:* Teacher pub schs, Norfolk County, 52-53; qual engr, Ford Motor Co, 53-54; teacher pub schs, Norfolk County, 54-58, sci coordr, 58-61; instr sci educ & math, Ohio State Univ, 62-63; from asst prof to, prof sci educ, 63-85, CHMN, DEPT SCI EDUC, E CAROLINA UNIV, 85- *Concurrent Pos:* Partic, NSF Acad Year Inst, Ohio State Univ, 61-62; fel earth sci, Southwest Ctr Advan Study, Tex, 68. *Mem:* Nat Sci Teachers Asn; Asn Educ Teachers Sci; Nat Asn Res Sci Teaching. *Res:* Science teaching on both the secondary school and college level. *Mailing Add:* Dept Sci Educ E Carolina Univ Greenville NC 27858

SHEPPARD, NORMAN F, JR, b Ipswich, Mass. BIOSENSORS, BIOMATERIALS. *Educ:* Mass Inst Technol, BS, 78, MS, 79, MS, 82, PhD(elec eng), 86. *Prof Exp:* Postdoctoral fel, chem eng, Mass Inst Technol, 86-89; ASST PROF BIOMED ENG, JOHNS HOPKINS UNIV SCH MED, 89- *Honors & Awards:* Presidential Young Investr, NSF, 90. *Mem:* Am Chem Soc; Inst Elec & Electronic Engrs; AAAS. *Res:* Development of biomedical sensors and devices using microfabrication technology; development of a miniature glucose sensor and devices for the controlled release of medications. *Mailing Add:* Dept Biomed Eng Johns Hopkins Univ Baltimore MD 21218

SHEPPARD, RICHARD A, b Lancaster, Pa, May 14, 30; m 57. GEOLOGY. *Educ:* Franklin & Marshall Col, BS, 56; Johns Hopkins Univ, PhD(geol), 60. *Prof Exp:* Geologist, 60-75, br chief, 76-79, RES GEOLOGIST, US GEOL SURV, 80- *Honors & Awards:* Spec Act Award, US Geol Surv, 66; Meritorious Serv Award, Dept Interior, 86. *Mem:* AAAS; Geol Soc Am; Mineral Soc Am; Clay Minerals Soc; Soc Econ Geol. *Res:* Geology of Cascade Mountains, especially petrology of Cenozoic volcanic rocks; distribution and genesis of zeolites in sedimentary rocks. *Mailing Add:* US Geol Surv Box 25046 Fed Ctr MS-917 Denver CO 80225

SHEPPARD, ROGER FLOYD, b Barberton, Ohio, July 25, 45; m 70. ENTOMOLOGY, INSECT BEHAVIOR. *Educ:* Ohio Univ, BS, 67; Ohio State Univ, MS, 73, PhD(entom), 76. *Prof Exp:* ASSOC PROF BIOL, CONCORD COL, 76 - *Mem:* Am Entom Soc. *Res:* Sexual behavior of insects. *Mailing Add:* Dept of Biol Concord Col Athens WV 24712

SHEPPARD, RONALD JOHN, b New Rochelle, NY, Apr 13, 39; m 63; c 2. MANAGEMENT DEVELOPMENT, BUSINESS & TECHNICAL ANALYSIS. *Educ:* Rensselaer Polytech Inst, BS, 61; Howard Univ, MS, 62, PhD(physics), 65; Rochester Inst Technol, MBA, 74. *Prof Exp:* Prin consult, Booz Allen Hamilton Inc, 66-71; prod mgr, Xerox Corp, 71-77; prod planning mgr, Ford Motor Co, 77-79; dir strategic anal, Gen Motors, 79-83; new bus develop mgr, Imperial Clevite, 83-84; dir, Ctr Bus & Indust, Univ Toledo, 84-90; DIR, OFF ECON DEVELOP & CONTINUING EDUC, CITY UNIV NY, 90- *Concurrent Pos:* NATO fel, Univ New Castle, 64; vis prof, Southern Univ, Bethune Cookman, 75 & 76; assoc prof, Empire State Col, 74-76; pres bd dir, Montessori Sch, Rochester, NY, 75-77; lectr, Rochester Inst Technol, 74-76, Univ Rochester, 75; NASA eval team, Denver, Colo, 81. *Mem:* Planetary Soc; Am Soc Training & Develop; Botany Club. *Res:* Technology assessment; research and development management; technology forecasting; creative process in science and art; issues in aerospace management and strategic defense. *Mailing Add:* City Univ NY 715 Ocean Terr Bldg A-110 Staten Island NY 10301

SHEPPARD, WALTER LEE, JR, b Philadelphia, Pa, June 23, 11; m 53; c 2. CORROSION ACID & ALKALI PROOF MASONRY MATERIALS. *Educ:* Cornell Univ, BChem, 32; Univ Pa, MS, 33. *Prof Exp:* From field sales to asst sales mgr & advert mgr, Atlas Mineral Prod, 38-48; chief engr & sales instr acid proof cements, Tanks & Linings, Ltd, Droitwich, Eng, 48-49; field sales, asst sales mgr & advert mgr, Electro-Chem Eng & Mfg Co, 49-66 & Interpace Inc, 66-68; nat acct mgr & field sales mgr, Pennwalt Corp, 68-76; PRES & CHIEF EXEC OFFICER, CCRM INC, 78- *Mem:* Nat Soc Prof Engrs; Am Acad Environ Engrs; Nat Asn Corrosion Engrs; Am Soc Testing & Mat. *Res:* Acid proof cements; all kinds of non-metallic structural and lining materials that are corrosion and chemical resistant; author of two textbooks and over 100 articles and technical papers on corrosion and chemical resistant materials. *Mailing Add:* 923 Old Manoa Rd Havertown PA 19083

SHEPPARD, WILLIAM JAMES, b Boston, Mass, Apr 10, 31; m 55; c 2. CHEMICAL MARKET STUDIES. *Educ:* Oberlin Col, AB, 52; Harvard Univ, MA, 54, PhD(chem), 59. *Prof Exp:* Instr chem, Swarthmore Col, 58-62, asst prof, 62-64; sr chem economist, 64-73, assoc sect mgr, 73-75, SR RES SCI, BATTELLE MEM INST, 75- *Mem:* Am Chem Soc; Sigma Xi. *Res:* Planning of research on chemicals and materials; economic and technical analysis of markets for chemical products and processes; fuel production and environmental control. *Mailing Add:* 505 King Ave Columbus OH 43201

SHEPPERD, WAYNE DELBERT, b Sterling, Colo, June 28, 47; m 70; c 2. FORESTRY, SILVICULTURE. *Educ:* Colo State Univ, BS, 70, MS, 74, PhD, 91. *Prof Exp:* Forestry res technician, Rocky Mountain Forest & Range Exp Sta, 70-76, forest silviculturist, Rio Grande Nat Forest, 76-78, RES FORESTER, ROCKY MOUNTAIN FOREST & RANGE EXP STA, US FOREST SERV, 78- *Concurrent Pos:* Fac affil, Colo State Univ. *Mem:* Sigma Xi; Nature Conservancy. *Res:* Silviculture of sub alpine forests; growth and yield prediction; silviculture of aspen. *Mailing Add:* 830 Wagonwheel Ft Collins CO 80521

SHEPPERSON, JACQUELINE RUTH, b Hopewell, Va, Feb 10, 35. PARASITOLOGY. *Educ:* Va State Univ, BS, 54; NC Cent Univ, Durham, MS, 56; Howard Univ, PhD(zool), 64. *Prof Exp:* Inst biol, Ft Valley State Col, 55-59; asst prof zool, Howard Univ, 64-65; assoc prof, 65-68, chmn Sci Dept, 67-79, PROF BIOL, WINSTON-SALEM UNIV, 68- *Res:* Survey studies of helminths in wild mammals; chemical relationships of host and parasite; physiological studies on plant parasitic nematodes. *Mailing Add:* Dept Natural Sci Box 13145 Winston-Salem State Univ Winston-Salem NC 27102

SHEPRO, DAVID, b Holyoke, Mass, Feb 14, 24; m 48; c 2. CARDIOVASCULAR PHYSIOLOGY. *Educ:* Clark Univ, BA, 48, MA, 50; Boston Univ, PhD, 58. *Prof Exp:* Instr biol, Simmons Col, 50-52, from asst prof to prof, 53-68; PROF BIOL, BOSTON UNIV, 68-, ASSOC PROF SURG, 71- *Concurrent Pos:* Mem corp, Marine Biol Lab, Woods Hole, 66-72, clerk of corp, 72-; ed-in-chief, J Microvascular Res. *Mem:* Microcirc Soc; Sigma Xi; Soc Gen Physiol; Am Physiol Soc; Am Soc Cell Biol. *Res:* Communication and interaction between endothelium and blood; calcium flux and release reaction in thrombocytes; homeostasis and the pulmonary microvasculature. *Mailing Add:* Dept Biol 2 Cummington St Boston Univ Boston MA 02215

SHEPS, CECIL GEORGE, b Winnipeg, Man, July 24, 13; nat US; m 37; c 1. PREVENTIVE MEDICINE. *Educ:* Univ Man, MD, 36; Yale Univ, MPH, 47; Am Bd Prev Med, dipl, 50. *Hon Degrees:* DSc, Chicago Med Sch, 70, Univ Man, 85; PhD Ben-Gurion Univ of Negev, 83. *Prof Exp:* Asst dep minister, Dept Pub Health, Sask, Can, 45-46; Rockefeller Found fel, Yale Univ, 46-47; assoc prof pub health admin, Univ NC, 47-51, res prof health planning, 52-53; lectr prev med, Harvard Med Sch, 54-58, clin prof, 58-60; prof, Grad Sch Pub Health, Univ Pittsburgh, 60-65; prof community med, Mt Sinai Sch Med, 66-68; dir Health Serv Res Ctr, 69-71, vchancellor health sci, 71-76, prof, 69-86, EMER PROF, SOCIAL MED, UNIV NC, CHAPEL HILL, 86- *Concurrent Pos:* Spec consult, Training Div Commun Dis Ctr, USPHS, 48-51; WHO traveling fel, 51; gen dir, Beth Israel Hosp, Boston, 53-60 & Beth Israel Med Ctr, New York, 65-68; mem & chmn Health Serv Study Sect, NIH, 55-62; mem, Nat Adv Comt Chronic Dis & Health of Aged, 57-61; nat planning comt, White House Conf Aging, 59-61; consult med affairs, Welfare Admin, Dept Health, Educ & Welfare, Washington, DC; chmn higher educ pub health, Millbank Mem Found Comn, 72-76. *Honors & Awards:* Miles Award Sci Achievement, Can Pub Health Asn, 70; Sedgewick Medal, Am Pub Health Asn, 90. *Mem:* Inst Med Nat Acad Sci; Am Pub Health Asn; Asn Teachers Prev Med. *Res:* Social medicine; medical and hospital administration. *Mailing Add:* Chase Hall CB7490 Univ NC Chapel Hill NC 27599-7490

SHER, ALVIN HARVEY, b St Louis, Mo, Aug 15, 41; m 63; c 3. NUCLEAR CHEMISTRY, SEMICONDUCTOR PHYSICS. *Educ:* Wash Univ, AB, 63, AM, 65; Simon Fraser Univ, PhD(nuclear chem), 67. *Prof Exp:* Res chemist, Nat Bur Standards, 68-72, sect chief semi- conductor processing, 72-75, asst chief, Electronic Technol Div, 75-77, dept dir, Ctr Electronics & Elec Eng, 78-83, chief, prog off, 84-87, asst dir, mgt info technol, 87-89; DIR PROG OFF, NAT INST STANDARDS & TECHNOL, 89- *Concurrent Pos:* Analyst, Nat Bur Standards Off Progs, 76. *Honors & Awards:* Silver Medal, Dept Com, 90. *Mem:* Inst Elec & Electronics Engrs. *Res:* Development of measurement methods relating to reliability of semiconductor microelectronic devices. *Mailing Add:* Rm A-1000 Admin Bldg Nat Inst Standards & Technol Gaithersburg MD 20899

SHER, ARDEN, b St Louis, Mo, July 5, 33; m 55; c 3. PHYSICS. *Educ:* Wash Univ, St Louis, BS, 55, PhD(physics), 59. *Prof Exp:* Res assoc physics, Wash Univ, St Louis, 59; NSF fel, Saclay Nuclear Res Ctr, France, 60; sr engr, Cent Res Labs, Varian Assocs, Calif, 61-67; assoc prof physics, Col William & Mary, 67-72, dir appl sci, 70-76, prof physics, 72-79; MEM STAFF, SRI INT, 79- *Concurrent Pos:* Consult, Langley Res Ctr, NASA, 68-69 & 72; consult prof mat sci, Stanford Univ, 82-, consult prof elec eng, 85- *Mem:* Fel Am Phys Soc. *Res:* Solid state physics. *Mailing Add:* 707 Crestview Dr San Carlos CA 94070

SHER, IRVING HAROLD, b Philadelphia, Pa, July 10, 24; m 49; c 2. INFORMATION SYSTEMS, INTELLIGENT SYSTEMS. *Educ:* Univ Pa, BA, 44; Johns Hopkins Univ, ScD(biochem), 53. *Prof Exp:* Res scientist, Mt Sinai Hosp, 53-54 & Nat Drug Co, Philadelphia, 54-57; sr info scientist, Smith, Kline, French Labs, 57-62; vpres res & develop, Inst Sci Info, 62-68 & Info Co Am, 68-70; vpres opers & prod, 3i Co, Philadelphia, 70-71; group mgr, Franklin Inst Res Labs, 72-74; mgr info systs, Univ City Sci Ctr, 74-78, DIR DEVELOP & QUAL CONTROL, INST SCI INFO, PHILADELPHIA, 78- *Honors & Awards:* McCollum Res Prize, Johns Hopkins Univ, 51. *Mem:* AAAS; Am Chem Soc; Am Soc Info Sci; Sigma Xi. *Res:* Information gathering, analysis, storage, indexing, retrieval and presentation; computer science. *Mailing Add:* 874 Brandon Lane Schwenksville PA 19473

SHER, JOANNA HOLLENBERG, b Winnipeg, Man, Can, May 23, 33; m 55; c 2. NEUROPATHOLOGY, MUSCLE PATHOLOGY. *Educ:* Univ Chicago, AB, 52, BS & MD, 56. *Prof Exp:* Intern med & surg, Kings County Hosp, 56-57; resident path, Kings County Hosp & State Univ NY, 57-58, fel, 60-62, NIH spec fel neuropath, 62-64, asst pathologist, 65-70, assoc prof path, 70-77, asst dean, 77-83, DIR NEUROPATH, HEALTH SCI CTR, KINGS COUNTY HOSP, 70-, PROF CLIN PATH, 77-, DISTINGUISHED SERV PROF, 87- *Concurrent Pos:* Consult, Long Island Col Hosp, 74-, Brookdale Hosp & Med Ctr, 74-, & Brooklyn Hosp, 75- *Mem:* Fel Am Soc Clin Pathologists; Int Acad Path; Am Asn Neuropathologists; Am Acad Neurol; AAAS; Sigma Xi. *Res:* Human skeletal muscle disease and development by the use of histochemical techniques and monoclonal antibodies to identify myosin isoforms. *Mailing Add:* Health Sci Ctr State Univ NY 450 Clarkson Ave Box 25 Brooklyn NY 11203

SHER, PAUL PHILLIP, b Oct 25, 39; m; c 3. CLINICAL CHEMISTRY, COMPUTER APPLICATION. *Educ:* Hobart Col, Geneva, NY, BS, 61; Washington Univ, St Louis, MD, 65. *Prof Exp:* DIR CLIN LABS, NY UNIV MED CTR, 81- *Concurrent Pos:* Ed, Lab Med. *Mem:* fel Col Am Pathologists; fel Asn Clin Scientists; fel Nat Acad Clin Biochem. *Mailing Add:* Clin Labs NY Univ Med Ctr 560 First Ave New York NY 10016-6402

SHER, RICHARD B, b Flint, Mich, Jan 21, 39; m 62; c 2. TOPOLOGY. *Educ:* Mich Technol Univ, BS, 60; Univ Utah, MS, 64, PhD(math), 66. *Prof Exp:* From asst prof to assoc prof math, Univ Ga, 66-74; head dept, 80-86, PROF MATH, UNIV NC, GREENSBORO, 74- *Mem:* Math Asn Am; Am Math Soc; Inst Advan Study. *Res:* Point set topology; piecewise linear topology; shape theory; infinite-dimensional manifolds; theory of retracts. *Mailing Add:* Dept Math Univ NC Greensboro NC 27412

SHER, RUDOLPH, b New York, NY, May 28, 23; m 52; c 2. NUCLEAR ENGINEERING. *Educ:* Cornell Univ, AB, 43; Univ Pa, PhD(physics), 51. *Prof Exp:* Staff mem radiation lab, Mass Inst Technol, 43-46; assoc physicist nuclear eng, Brookhaven Nat Lab, 51-61; assoc prof, 61-70, prof, 70-86, EMER PROF MECH ENG, STANFORD UNIV, 86- *Concurrent Pos:* Vis scientist, Comm a l'Energie Atomique, France, 58-59 & A B Atomenergi, Sweden, 68-69; consult, Brookhaven Nat Lab, 70-; ed, Progress in Nuclear Energy, 75-80; vis staff mem, Int Atomic Energy Agency, 78-79; consult, Elec Power Res Inst, 83- *Honors & Awards:* Fel, Am Nuclear Soc, 86. *Mem:* Am Nuclear Soc. *Res:* Nuclear reactor physics; nuclear data; nuclear safeguards; nuclear aerosols. *Mailing Add:* Dept Mech Eng Stanford Univ Stanford CA 94305

SHER, STEPHANIE ELLSWORTH, immunology, for more information see previous edition

SHERA, E BROOKS, b Oxford, Ohio, Aug 28, 35; m 56; c 2. NUCLEAR PHYSICS. *Educ:* Case Western Reserve Univ, BA, 56, PhD(physics), 62; Univ Chicago, MS, 58. *Prof Exp:* Instr physics, Case Western Reserve Univ, 61-62; resident res assoc, Argonne Nat Lab, 62-64; STAFF MEM, LOS ALAMOS SCI LAB, 64- *Mem:* Fel Am Phys Soc; AAAS. *Res:* Muonic x-ray studies; electron and proton scattering; neutron-capture gamma ray studies; low energy nuclear physics; nuclear structure; Biophysics; humangenome project (advanced techniques for DNA sequencing); muonic x-ray studies; electron scattering; neutron-capture gamma ray studies; low energy nuclear physics; nuclear structure. *Mailing Add:* Los Alamos Nat Lab D434 Los Alamos NM 87545

SHERALD, ALLEN FRANKLIN, b Frederick, Md, Nov 15, 42. DEVELOPMENTAL GENETICS. *Educ:* Frostburg State Col, BS, 64; Univ Va, PhD(biol), 73. *Prof Exp:* Teacher sci, Montgomery County Pub Schs, 64-66; res assoc genetics, Univ Va, 73-74; trainee genetics & biochem, Cornell Univ, 74-76; asst prof, 76-80, ASSOC PROF, GEORGE MASON UNIV, 80- *Mem:* AAAS; Genetics Soc Am. *Res:* Drosophila; biochemistry; sclerotization. *Mailing Add:* 9451 Lee Hwy 1209 Fairfax VA 22031

SHERBECK, L ADAIR, b New Norway, Alta, July 24, 22; m 43; c 3. POLYMER CHEMISTRY. *Educ:* Univ Alta, BSC, 47; McGill Univ, PhD(chem), 51. *Prof Exp:* From res chemist to sr res chemist, 51-59, supvr technol, 59-64, sr supvr nomex technol, 64-80, RES ASSOC, TEXTILE FIBERS DEPT, TECH DIV, E I DU PONT DE NEMOURS & CO, INC, 80- *Mem:* Am Chem Soc. *Res:* Wood and cellulose chemistry; addition and condensation polymerization; fiber production and properties. *Mailing Add:* 1826 Cherokee Rd Waynesboro VA 22980

SHERBERT, DONALD R, b Wausau, Wis, Feb 24, 35; m 81; c 2. MATHEMATICAL ANALYSIS. *Educ:* Univ Wis, BS, 57; Stanford Univ, PhD(math), 62. *Prof Exp:* From instr to asst prof, 62-68, dir undergrad prog, 80-87, ASSOC PROF MATH, UNIV ILL, URBANA, 68- *Concurrent Pos:* Mem math fac, Inst Teknologi Mara, Shah Alam, Malaysia, 87-88. *Mem:* Am Math Soc; Math Asn Am. *Res:* Functional analysis. *Mailing Add:* Dept Math Univ Ill 265 Altgeld Hall Urbana IL 61801

SHERBLOM, ANNE P, b New Haven, Conn, July 31, 49. GLYCOPROTEINS, IMMUNOSUPPRESSION. *Educ:* Bates Col, BS, 71; Dartmouth Col, PhD(chem), 75. *Prof Exp:* Vis asst prof, Dept Chem, Dartmouth Col, 76; res assoc, Dept Biol, Bowdoin Col, 76-77 & Okla State Univ, 77-80; asst prof, 80-86, ASSOC PROF, DEPT BIOCHEM, UNIV MAINE, ORONO, 86- *Concurrent Pos:* Vis scientist, NCI, NIH, 86-87. *Mem:* Am Chem Soc; Am Soc Biochem & Molecular Biol; Soc Complex Carbohydrates. *Res:* Isolation and characterization of tumor cell glycoproteins; role of oligosaccharides in immunoregulation. *Mailing Add:* Dept Biochem Hitchner Hall Univ Maine Orono ME 04469

SHERBON, JOHN WALTER, b Lewiston, Idaho, Oct 31, 33; m 57; c 2. DAIRY CHEMISTRY. *Educ:* Washington State Univ, BS, 55; Univ Minn, MS, 58, PhD(dairy sci), 63. *Prof Exp:* From asst prof to assoc prof, 63-77, PROF FOOD SCI, CORNELL UNIV, 77- *Mem:* Asn Off Anal Chem; Inst Food Technol; Am Dairy Sci Asn. *Res:* Physical state of fat in dairy and other food products; analysis for protein content of milk; protein gelation; chemical instrumentation in agricultural chemistry. *Mailing Add:* Dept of Food Sci Cornell Univ Ithaca NY 14853

SHERBOURNE, ARCHIBALD NORBERT, b Bombay, India, July 8, 29; Can citizen; m 59; c 6. STRUCTURAL ENGINEERING & MECHANICS. *Educ:* Univ London, BSc, 53; Lehigh Univ, BSCE, 55, MSCE, 57; Cambridge Univ, MA, 59, PhD(eng), 60; Univ London, DSc, 70. *Prof Exp:* Engr, Brit Rwys, 48-52; engr, Greater London Coun, 52-54; instr civil & mech eng, Lehigh Univ, 54-57; engr, US Steel Corp, Calif, 56; sr asst res eng, Cambridge Univ, 57-61; assoc prof, 61-63, chmn dept civil eng, 64-66, dean fac eng, 66-74, PROF CIVIL ENG, UNIV WATERLOO, 63- *Concurrent Pos:* Vis sr lectr, Univ Col, Univ London, 63-64; Orgn Econ Coop & Develop res fel, Swiss Fed Inst Technol, 64; vis prof, Univ West Indies, 69-70; Nat Res Coun sr res fel, Nat Lab Civil Eng, Portugal, 70; W Ger Acad Exchange fel, 75; NATO sr scientist fel & vis lectureship, Western Europe, 75-76; vis prof, Ecole Polytech Fed, Lausanne, Switz, 75-77, Mich Technol Univ, 80-81 & 81-83, Ocean Eng, Fla Atlantic Univ, 85, Archit & Civil Eng, NC AT&T State Univ, 87, Civil Eng, Kuwait Univ, 89 & Aerospace Eng, IISC, Bangalore, India, 90; consult/adv tech educ, Orgn Am States, Inter Am Develop Bank & Can Int Develop Agency; Gleddon sr vis fel, Univ Western Australia, 78. *Honors & Awards:* Eng Medal, Asn Prof Engrs, Ont, 75. *Mem:* Fel Brit Inst Struct Engrs; fel Royal Soc Arts; Int Asn Bridge & Struct Engrs; Am Acad Mech. *Res:* Design of steel structures; structural composites; plasticity. *Mailing Add:* Fac Eng, Dept Civil Eng Univ Waterloo Waterloo ON N2L 3G1 Can

SHERBURNE, JAMES AURIL, b Milo, Maine, Aug 8, 41; m 63; c 2. ECOLOGY. *Educ:* Univ Maine, BA, 67, MS, 69; Cornell Univ, PhD(ecol), 72. *Prof Exp:* Dir, Peace Corps Environ Prog, Smithsonian Inst, 75-78; unit leader, Coop Wildlife Res Unit, Univ Maine, 78-82; dir, African Oper Wildlife Found, Nairobi, 85-86; DIR, UNIV MAINE INT PROGS, 86- *Concurrent Pos:* Var int tech consultancies, Africa, Asia, Latin Am; sci adv, US Agency for Int Develop, US FIsh & Wildlife Serv Int Affairs, 82- *Mem:* AAAS; Sigma Xi; Ecol Soc Am; Int Union Conserv Nature & Natural Resources; Entom Soc Am. *Res:* Pesticide residues in vertebrates; chemical interactions between plants and vertebrates, particularly frugiverous birds and secondary chemicals, affecting behavior; developing wildlife and natural resource management projects internationally, particularly Africa, South America, and Asia; methods of plant dispersal by vertebrates; international policy studies. *Mailing Add:* Dept Wildlife Mgt Univ Maine Orono ME 04469

SHERBY, OLEG D(IMITRI), b Shanghai, China, Feb 9, 25; nat US; m 49; c 4. MATERIALS SCIENCE. *Educ:* Univ Calif, BS, 47, MS, 49, PhD(metall), 56; Univ Sheffield, DMet, 68. *Prof Exp:* Asst metall, Univ Calif, 47-49, res metallurgist, Inst Eng Res, 49-56; NSF fel, Univ Sheffield, 56-57; sci liaison officer metall, US Off Naval Res, Eng, 57-58; assoc prof, 58-62, PROF MAT SCI & ENGR, STANFORD UNIV, 62- *Concurrent Pos:* Consult, Los Alamos Nat Lab, 59-, Lockheed Palo Alto Res Lab, 61-64 & Lawrence Livermore Nat Lab, 65-; fel, Univ Paris, France, 67-68. *Honors & Awards:* Dudley Medal, Am Soc Testing & Mat, 58; Centennial Medal, Am Soc Mech Engrs, 80; Gold Medal, Am Soc Metals, 85; Yukawa Mem lect, Iron & Steel Inst, Japan, 88. *Mem:* Nat Acad Eng; Am Inst Mining, Metall & Petrol Engrs; fel Am Soc Metals; Sigma Xi. *Res:* Theoretical and experimental aspects of mechanical behavior of solids and diffusion in solids. *Mailing Add:* Dept Mat Sci & Eng Stanford Univ Stanford CA 94305

SHERCK, CHARLES KEITH, b Willard, Ohio, July 27, 22; m 46; c 2. FOOD SCIENCE. *Educ:* Miami Univ, BA, 47. *Prof Exp:* Chemist, Am Home Prod Corp, 47-50, mgr mfg, 50-54; mgr qual control, 54-59, mgr res opers, 59-64, mgr grocery prod res & develop, 64-69, dir corp res & develop, 70-77, dir, Res & Develop Facil & Tech Serv, 78-81, DIR RES & DEVELOP ADMIN, PILLSBURY CO, 81- *Concurrent Pos:* Bd dir, AACC, 73-74. *Mem:* Am Asn Cereal Chem; Inst Food Technol. *Res:* Management food research and development. *Mailing Add:* 4805 Markay Ridge Golden Valley MN 55427

SHERDEN, DAVID J, b Washington, DC, Oct 26, 40; m 63; c 2. PHYSICS. *Educ:* Univ San Francisco, BS, 62; Univ Chicago, MS, 64, PhD(physics), 70. *Prof Exp:* PHYSICIST, STANFORD LINEAR ACCELERATOR CTR, 69- *Res:* Experimental elementary particle physics; high energy electron and photon interactions. *Mailing Add:* Stanford Linear Accelerator Ctr PO Box 4349 Stanford CA 94305

SHEREBRIN, MARVIN HAROLD, b Winnipeg, Man, Mar 23, 37; m 66; c 3. BIOPHYSICS, BIOMEDICAL ENGINEERING. *Educ:* Univ Man, BSc, 60; Univ Western Ont, MSc, 63, PhD(biophys), 65. *Prof Exp:* Res asst biophys, 60-65, asst prof, 67-73, ASSOC PROF BIOPHYS, FAC MED, UNIV WESTERN ONT, 73- *Concurrent Pos:* Can Med Res Coun fel, Weizmann Inst Sci, 65-67, Can Med Res Coun res scholar, 67-72. *Mem:* Inst Elec & Electronics Eng; Can Med & Biol Eng Soc; Biophys Soc. *Res:* Mechanical properties of arteries; mechanochemistry of contractile systems; elastin and collagen: function and structure; analysis of peripheral pulse wave; electronics systems in hospitals. *Mailing Add:* Dept Med Biophys Univ Western Ont Fac Med London ON N6A 5C1 Can

SHERER, GLENN KEITH, b Allentown, Pa, May 11, 43; m 70; c 1. DEVELOPMENTAL BIOLOGY, CELL BIOLOGY. *Educ:* Muhlenberg Col, BS, 64; Temple Univ, PhD(biol), 72. *Prof Exp:* Res fel, NIH, 72-73, guest worker cell biol, 73; res assoc med, Col Physicians & Surgeons, Columbia Univ, 73-75; assoc med, Med Univ SC, 75-77, asst prof res med, 77-80; asst prof biol, Bowdoin Col, 80-84; ASSOC PROF BIOL & DEPT CHAIR, UNIV SAINT THOMAS, 84-, DIR, DIV SCI & MATH, 86- *Concurrent Pos:* NIH res fel, Col Physicians & Surgeons, Columbia Univ, 74-75 & Med Univ SC, 75-76. *Mem:* AAAS; Am Soc Cell Biol; Am Soc Zool; Soc Develop Biol; Tissue Cult Asn; Am Inst Biol Sci. *Res:* Epithelial-mesenchymal interactions in liver development; development of embryonic microvasculature; cell culture; tissue interaction in vertebrate organogenesis; embryonic development of the liver. *Mailing Add:* Dept Biol Univ St Thomas St Paul MN 55105

SHERER, JAMES PRESSLY, b Rock Hill, SC, Aug 25, 39; m 61; c 2. ORGANIC CHEMISTRY. *Educ:* Erskine Col, BA, 61; Duke Univ, MA, 63, PhD(org chem), 66. *Prof Exp:* SR RES ASSOC, E I DU PONT DE NEMOURS & CO, INC, 65- *Mem:* Am Chem Soc. *Res:* Synthesis and reactions of dicationoid aromatic systems containing one or more quaternary nitrogen atoms at bridgehead positions; research and development of synthetic fibers; synthesis of nitrogen heterocyclic compounds. *Mailing Add:* 2127 Gloucester Pl Wilmington NC 28403

SHERF, ARDEN FREDERICK, plant pathology; deceased, see previous edition for last biography

SHERGALIS, WILLIAM ANTHONY, b Hazleton, Pa, May 25, 41; m 64; c 2. ANALYTICAL & PHYSICAL CHEMISTRY. *Educ:* Univ Pa, BS, 62; Drexel Univ, MS, 64; Temple Univ, PhD(phys chem), 69. *Prof Exp:* Assoc prof chem, Widener Col, 68-78, chmn sci, 78-81; prof chem & acad dean, Cardinal Newman Col, 81-85; vpres acad affairs, Ohio Dominican Col, 85-90; DEAN, ARTS & SCI & PROF CHEM, KING'S COL, 90- *Mem:* Am Chem Soc; Sigma Xi; AAAS. *Res:* Electro-analytical techniques and their applications to chemical and biochemical systems; science education. *Mailing Add:* PO Box 5231 Wilkes-Barre PA 18710

SHERIDAN, DOUGLAS MAYNARD, b Faribault, Minn, Nov 9, 21; m 55; c 2. BASE & PRECIOUS METAL DEPOSITS, METALLIC MINERAL DEPOSITS. *Educ:* Carleton Col, BA, 43; Univ Minn, Minneapolis, MS(geol), 51. *Prof Exp:* GEOLOGIST, US GEOL SURVEY, 48- *Mem:* Mineral Soc Am; Soc Econ Geologists; Geol Soc Am. *Res:* Economic geology of precambrian base-metal sulfide deposits in Colorado; economic geology of metallic mineral deposits of all ages in Colorado. *Mailing Add:* 27334 Mildred Lane Evergreen CO 80439

SHERIDAN, JANE CONNOR, analytical chemistry, for more information see previous edition

SHERIDAN, JOHN FRANCIS, b Brooklyn, Ny, June 2, 49; m 72; c 4. IMMUNOLOGY, MICROBIOLOGY. *Educ:* Fordham Univ, BS, 72; Rutgers Univ, MS, 74, PhD(microbiol), 76. *Prof Exp:* Pub Health Serv fel div immunol, Med Ctr, Duke Univ, 76-78; res assoc immunol, Johns Hopkins Univ, 78-79, instr, 79-82, asst prof, div comp med, 82-84; asst prof med & dir, Bone Marrow Transplant Lab, 84-87, ASSOC PROF, SECT ORAL BIOL, DEPT MED MICROBIOL & IMMUNOL, OHIO STATE UNIV, 87- *Mem:* Am Soc Microbiol; AAAS; Am Asn Immunol; Am Soc Virol; Neurosci Soc. *Res:* Infection and immunity; cellular immunology; virology-herpesviruses, rotaviruses, influenza virus; neuroimmunology. *Mailing Add:* Box 192 Postle Hall Ohio State Univ 305 W 12th St Columbus OH 43210

SHERIDAN, JOHN JOSEPH, III, b Providence, RI, Feb 2, 45. CATALYSIS, MINERAL BENEFICIATION. *Educ:* Stevens Inst Technol, Hoboken, NJ, BE, 67; Univ Ill, Champaign, MS, 69, PhD(chem eng), 71. *Prof Exp:* Sr res engr, Process Res, Engelhard Minerals & Chem, Inc, 71-76; SR PRIN DEVELOP ENGR, PROCESS DESIGN, DEVELOP & JOINT VENTURE MGT, AIR PROD & CHEM, INC, 76- *Mem:* Am Inst Chem Engrs; Am Chem Soc. *Res:* Properties of reversible metal hydrides; hydrogenation chemistry, particularly unsaturated hydrocarbons over metallic surfaces; gas and metal interactions; gas separations via chemical complexation or absorption. *Mailing Add:* Air Prod Inc 7201 Hamilton Blvd Allentown PA 18195-1501

SHERIDAN, JOHN ROGER, b Helena, Mont, Sept 24, 33; m 59; c 2. ATOMIC PHYSICS, CHEMICAL PHYSICS. *Educ:* Reed Col, BA, 55; Univ Wash, PhD(physics), 64. *Prof Exp:* Res engr, Boeing Co, 55-60; res asst physics, Univ Wash, 60-63, assoc, 63-64; from asst prof to assoc prof, 64-71, head dept, 67-76 & 78-80, PROF PHYSICS, UNIV ALASKA, FAIRBANKS, 71- *Concurrent Pos:* Danforth assoc, 66-; vis scientist, Stanford Res Inst, 68-69; vis prof, Queen's Univ Belfast, 75-76; vis fac, Univ Calif, Irvine, 84-85. *Mem:* Sigma Xi; fel Am Phys Soc; Am Asn Physics Teachers; AAAS. *Res:* Use of photon coincidence techniques in laboratory studies of reactions of atoms, molecules; metastable and long-lived atomic, molecular reactions using molecular beams; optical calibration using photon coincidence; quenching and radiative lifetimes of excited atoms and molecules; development of software for computer-based instruction in physics. *Mailing Add:* Physics Dept Univ Alaska Fairbanks AK 99701

SHERIDAN, JUDSON DEAN, b Greeley, Colo, Nov 10, 40; m 63; c 2. CELL PHYSIOLOGY, CELL BIOLOGY. *Educ:* Hamline Univ, BS, 61; Oxford Univ, DPhil(neurophysiol), 65. *Prof Exp:* Res assoc neurophysiol & neuropharmacol, Neurophysiol Lab, Harvard Med Sch, 65-66, instr neurobiol, 66-68; from asst prof to assoc prof zool, Univ Minn, Minneapolis, 68-76, prof genetics & cell biol, 76-79, prof anat, Med Sch, 79-87, assoc dean, Grad Sch, 83-87; PROF PHYSIOL, BIOL SCI & PROVOST RES & GRAD DEAN, UNIV MO- COLUMBIA, 87- *Concurrent Pos:* USPHS fel, 65-68; assoc ed develop physiol, Develop Biol, 71-73; Nat Cancer Inst career develop award, 72; consult cell biol study sect, NSF, NIH, 72-; sr res fel, Dept Biochem, Univ Glasgow, Scotland, 74-75; Rhodes scholar, 62-65. *Mem:* AAAS; Am Soc Cell Biol; NY Acad Sci. *Res:* Cellular communication via specialized points of cell-to-cell contacts, especially during development and abnormal cell growth. *Mailing Add:* Grad Sch 202 Jesse Hall Univ Mo-Columbia Columbia MO 65211

SHERIDAN, MARK ALEXANDER, b Fullterton, Calif, Nov 8, 58; m 81; c 4. COMPARATIVE PHYSIOLOGY & BIOCHEMISTRY, COMPARATIVE ENDOCRINOLOGY. *Educ:* Humboldt State Univ, AB, 80, MA, 82; Univ Calif, Berkeley, PhD(zool), 85. *Prof Exp:* Post-grad res scientist, Univ Wash, 85; asst prof, 85-91, ASSOC PROF, NDAK STATE UNIV, 91- *Concurrent Pos:* Zool grad prog coordr, NDak State Univ, 86-90, dir, Regulatory Biosci Ctr, 91-; vis scientist, Nat Marine Fisheries Serv, Seattle, Wash, 86 & 88, Dept Biol, Humboldt State Univ, Arcata, Calif, 91 & Dept Biol, Hiroshima Univ, 91; consult, Biomed Instruments, Inc, Fullerton, Calif, 87; ad hoc ed, Am J Physiol, Aquacult, Aquatic Living Resources, Biol Bulletin, Can J Zool, Fish Physiol Biochem, Gen Comp Endocrinol, J Exp Zool & Lipids. *Mem:* AAAS; Am Fisheries Soc; Sigma Xi; Am Soc Zoologists. *Res:* Hormonal regulation of lipid metabolism and endocrine basis of metabolic alterations associated with salmon growth and development. *Mailing Add:* Dept Zool Stevens Hall NDak State Univ Fargo ND 58105

SHERIDAN, MICHAEL FRANCIS, b Springfield, Mass, Feb 20, 40; m 64; c 3. PETROLOGY, VOLCANOLOGY. *Educ:* Amherst Col, AB, 62; Stanford Univ, MS, 64, PhD(geol), 65. *Prof Exp:* Geologist, US Geol Surv, 64-74; instr geol, Amherst Col, 65-66; from asst prof to assoc prof, 66-76, PROF GEOL, ARIZ STATE UNIV, 76- *Concurrent Pos:* Consult geologist, 69-; NASA Surtsey Exped, 70; vis prof, Univ Tokyo, 72-73, Univ Pisa, 79-80 & Univ Calabria, 83; sr Fulbright-Hays scientist, Iceland, 78 & Antarctic Exped, 78. *Honors & Awards:* Castaing Award, 83. *Mem:* Geol Soc Am; Int Asn Volcanol & Chem Earth's Interior; Mineral Soc Am; Am Geophys Union. *Res:* Physical processes in volcanology; ash-flow tuff field relationships and mineralogy, volcanic risk, geothermal energy. *Mailing Add:* Dept Geol State Univ NY-Buffalo 4240 Ridge Lea Rd Buffalo NY 14260

SHERIDAN, MICHAEL N, b Augusta, Tex, Sept 5, 37; m 58; c 3. ANATOMY, ELECTRON MICROSCOPY. *Educ:* Stephen F Austin State Col, BS, 58; Med Col Va, PhD(anat), 63. *Prof Exp:* Instr anat, Med Col Va, 62-63, asst prof, 64-66; from asst prof to assoc prof anat, Sch Med & Dent, Univ Rochester, 66-77, prof, 77-80; MEM FAC, UNIFORMED SERV UNIV HEALTH SCI, 80- *Concurrent Pos:* NIH fel biochem, Inst Animal Physiol, Cambridge, Eng, 63-64. *Mem:* AAAS; Am Soc Cell Biol; Am Asn Anat. *Res:* Electron microscopy of steroid secretors; central nervous system. *Mailing Add:* Dept Anat Uniformed Serv Univ Health 4301 Jones Bridge Rd Bethesda MD 20814

SHERIDAN, PETER STERLING, b Portland, Maine, Mar 1, 44; m 67. INORGANIC CHEMISTRY. *Educ:* Kenyon Col, Ohio, BA, 66; Northwestern Univ, Evanston, PhD(chem), 71. *Prof Exp:* Lectr chem, Univ Kent, Canterbury, UK, 70-71, fel, 71-72; fel, Univ Southern Calif, 72-74; asst prof, State Univ NY Binghamton, 74-80; asst prof, 80-84, ASSOC PROF CHEM, COLGATE UNIV, 84- *Mem:* Am Chem Soc. *Res:* Photochemical reactions of transition metal complexes. *Mailing Add:* Dept Chem Colgate Univ Hamilton NY 13346-1399

SHERIDAN, PHILIP HENRY, b Washington, DC, June 29, 50; m 87; c 3. NEUROLOGY, DEVELOPMENTAL NEUROBIOLOGY. *Educ:* Yale Univ, BS, 72; Georgetown Univ, MD, 76. *Prof Exp:* Health scientist adminr, Epilepsy Br, 84-89, CHIEF, DEVELOP NEUROL BR, DIV CONVULSIVE DEVELOP & NEUROMUSCULAR DIS, NAT INST NEUROL DIS & STROKE, NIH, 89-; SR SURGEON, USPHS, 85- *Concurrent Pos:* Neurologist, Med Neurol Br, NIH Clin Ctr, 82-; attend physician, Va Dept Health Childrens Specialty Serv Child Neurol Clin, 82-; consult neurologist, Nat Naval Med Ctr, 84-; med monitor & proj officer, Antiepileptic Drug Develop Prog, NIH, 84-89; reviewer, J Am Med Asn, Annals Neurol, Pediat Neurol, Epilepsia, 89- *Mem:* Am Acad Neurol; Soc Neurosci; Child Neurol Soc; Am Acad Pediat; Am Epilepsy Soc; Am EEG Soc. *Res:* Complete involvement in the clinical research related to pediatric neurology, developmental neurobiology, epilepsy, and neuromuscular disorders. *Mailing Add:* NIH Fed Bldg Rm 8C-10 7550 Wisconsin Ave Bethesda MD 20892

SHERIDAN, RICHARD COLLINS, b Trotwood, Ohio, Aug 19, 29; m 57; c 4. CHEMISTRY. *Educ:* Murray State Col, BS, 56; Univ Ky, MS, 61. *Prof Exp:* Chemist, B F Goodrich Chem Co, 56-58; res chemist, Nat Fertilizer Develop Ctr, Tenn Valley Authority, 60-88; CHEM INSTR, UNIV NALA, 89- *Honors & Awards:* Wilson Dam Award, Am Chem Soc, 76. *Mem:* Am Chem Soc. *Res:* Synthesis of oxamide; pyrolysis of urea phosphate; preparation of ammonium polyphosphates; history of chemistry; preparation of new nitrogen-phosphorus fertilizer compounds; melamine phosphate; recovery of uranium from phosphate rock; ion chromatography; production of phosphoric acid. *Mailing Add:* 105 Terrace St Sheffield AL 35660

SHERIDAN, RICHARD P, b Detroit, Mich, Mar 10, 39; m 62; c 2. PLANT PHYSIOLOGY, BIOLOGY. *Educ:* Univ Ore, BA, 62, MA, 63, PhD(biol), 67. *Prof Exp:* NIH fel, Scripps Inst Oceanog, Univ Calif, San Diego, 67-68; asst prof, 68-74, ASSOC PROF BOT, UNIV MONT, 74- *Concurrent Pos:* NSF res grant, 69-71. *Mem:* Phycol Soc Am; Am Soc Plant Physiologists. *Res:* Algal physiology with emphasis on the photosynthetic mechanisms. *Mailing Add:* Dept Bot Univ Mont Missoula MT 59812

SHERIDAN, ROBERT E, b Hoboken, NJ, Oct 11, 40; m 66; c 3. GEOLOGY, OCEANOGRAPHY. *Educ:* Rutgers Univ, BA, 62; Columbia Univ, MA, 65, PhD(geol), 68. *Prof Exp:* Res asst, Lamont Geol Observ, NY, 62-68, res scientist, 68; from asst prof to prof geol, Univ Delaware, 68-86; PROF GEOL & GEOPHYSICS, RUTGERS UNIV, 86- *Concurrent Pos:* WAE appointment, US Geol Surv, 75-; coordr, Marine Geol & Geophys Prog, Col Marine Studies, Univ Del, 78-80; vis prof, Rutgers Univ, 79; chair, Joides Passive Margin Panel, 79-81. *Mem:* Geol Soc Am; Am Geophys Union; Am Asn Petrol Geol; Nat Maritime Hist Soc. *Res:* Geology and geophysics of the continental margin off eastern North America; stratigraphy of the Western North Atlantic; discovery of USS Monitor; recovery of oldest oceanic sediments; development of pulsation tectonic theory. *Mailing Add:* Dept Geol Sci Rutgers Univ New Brunswick NJ 08903

SHERIDAN, THOMAS BROWN, b Cincinnati, Ohio, Dec 23, 29; m 53; c 4. MAN-MACHINE SYSTEMS, HUMAN FACTORS. *Educ:* Purdue Univ, BS, 51; Univ Calif, Los Angeles, MS, 54; Mass Inst Technol, ScD, 59. *Hon Degrees:* Dr, Delft Univ Technol, Neth, 91. *Prof Exp:* Res asst eng, Univ Calif, Los Angeles, 53-54; res asst mech eng, 54-55, from instr to assoc prof mech eng, 55-70, PROF ENG & APPL PSYCHOL, MASS INST TECHNOL, 70- *Concurrent Pos:* Vis prof control eng, Delft Univ Technol, Neth, 72; mem NIH study sects accident prev & injury control, 73-77; mem US Cong Off Tech Assessment Comt Appropriate Technol, 81-82; mem NASA Tech Oversight Comn Flight Telerobot, 87-; chmn & mem Comt Human Factors, Nat Res Coun, 78-88; mem Defense Sci Bd Task Force Training, 87-88; mem Nuclear Regulatory Comn, Nuclear Safety Res Comn, 88-; mem, comn on com especially develop space facility, Nat Res Coun, 88- *Honors & Awards:* Paul Fitts Award, Human Factors Soc, 77; Centennial Medal, Inst Elec & Electronics Engrs, 82. *Mem:* Fel Human Factors Soc (pres, 90-91); fel Inst Elec & Electronic Engrs Systs Man & Cybernetics Soc (pres, 73-75). *Res:* Man-machine systems control and design for auto, aircraft and space vehicles, nuclear power, deep ocean science; robotics and human-operated remote manipulators; human-computer cooperation and decision-aiding. *Mailing Add:* 32 Sewall St Newton MA 02165

SHERIDAN, WILLIAM, b Cohoes, NY, Dec 1, 30; m 59; c 3. GENETICS. *Educ:* City Col New York, BA, 54; Univ Stockholm, Fil Lic, 62, Fil Dok(genetics), 68. *Prof Exp:* Instr genetics, Inst Genetics, Univ Stockholm, 60-61; res assoc, Lab Radiation Genetics, Sweden, 62-68; docent, Inst Genetics, Univ Stockholm, 68-73, actg dir inst, 70-71; head mammalian genetics sect, 73-88, STAFF MEM, OFF SR SCI ADV TO DIR, NAT INST ENVIRON HEALTH SCI, NIH, 88- *Concurrent Pos:* Asst ed, Mutation Res, 74- *Mem:* Genetics Soc Am; Environ Mutagen Soc. *Res:* Mammalian genetics; environmental mutagenesis; radiation genetics. *Mailing Add:* Nat Inst Environ Health Sci PO Box 12233 Research Triangle Park NC 27709

SHERIDAN, WILLIAM FRANCIS, b Lakeland, Fla, Dec 4, 36; m 64; c 2. GENETICS. *Educ:* Univ Fla, BSA, 58, MS, 60; Univ Ill, PhD(cell biol), 65. *Prof Exp:* Teaching asst bot, Univ Fla, 58-61; teaching asst, Univ Ill, 61-63, USPHS trainee biol, 63-65; instr, Yale Univ, 65-66, res fel sch med, 66-68; asst prof, Univ Mo-Columbia, 68-75; assoc prof, 75-80, PROF BIOL, UNIV NDAK, 80- *Concurrent Pos:* Fulbright res award, 77-78; vis prof, Dept Physiol, Carlsberg Res Lab, Copenhagen, 77-78, Purdue Univ, 88, Univ Oregon, 89. *Mem:* Genetics Soc Am; Soc Develop Biol. *Res:* Maize genetics and embryo development; maize morphogenesis; tissue culture of cereals and its genetic applications. *Mailing Add:* PO Box 8238 Grand Forks ND 58202

SHERIDON, NICHOLAS KEITH, b Detroit, Mich, Dec 8, 35; m 61; c 3. ELECTROOPTICS. *Educ:* Wayne State Univ, BS, 57, MS, 59. *Prof Exp:* Asst physicist systs div, Bendix Corp, 60-62, physicist res labs, 62, sr physicist, 64-67; sr physicist, Xerox Webster Res Ctr, Xerox Corp, 67-71, scientist, 71-72, sr scientist, 72-75, prin scientist, Palo Alto Res Ctr, 75-80, res fel, 80-88, SR RES FEL, XEROX CORP, 88- *Mem:* Optical Soc Am; Inst Elec & Electronic Eng; Soc Photog Instrumentation Eng; Soc Photog Scientists & Engrs. *Res:* Electrography, non-impact printing; display devices, electronic imaging; electrophotography; optical and acoustical holography; optical data processing; electron physics; surface chemistry; gaseous electronics. *Mailing Add:* Xerox Corp Palo Alto Res Ctr 3333 Coyote Hill Rd Palo Alto CA 94304

SHERIFF, ROBERT EDWARD, b Mansfield, Ohio, Apr 19, 22; m 45; c 6. EXPLORATION GEOPHYSICS. *Educ:* Wittenberg Univ, AB, 43; Ohio State Univ, MS, 47, PhD(physics), 50. *Prof Exp:* Physicist, Manhattan Proj, 44-46; from geophysicist to chief geophysicist, Standard Oil Co, Calif, 50-75; sr vpres, Seiscom-Delta Inc, 75-80; PROF, UNIV HOUSTON, 80- *Concurrent Pos:* Adj prof geophysics, Univ Houston, 73- *Honors & Awards:* Kauffman Gold Medal, Soc Explor Geophysicists, 69. *Mem:* Hon mem Soc Explor Geophysicists (1st vpres, 72-73); Europ Asn Explor Geophysicists; Sigma Xi; Am Asn Petrol Geologists; AAAS. *Res:* Techniques of geophysical interpretation, including data acquisition and data processing techniques. *Mailing Add:* Dept Geol Univ Houston 4800 Calhoun Rd Houston TX 77024-5503

SHERINS, RICHARD J, b Brooklyn, NY, July 6, 37; m 60; c 3. ENDOCRINOLOGY, ANDROLOGY. *Educ:* Univ Calif, Los Angeles, BA, 59, San Francisco, MD, 63. *Prof Exp:* Fel endocrinol, Univ Wash, 67-69; fel, Nat Cancer Inst, 69-70, sr investr endocrinol, Endocrinol & reproduction res br, 70-77, DEVELOP ENDOCRINOL BR, NAT INST CHILD HEALTH & HUMAN DEVELOP, NIH, 77- *Concurrent Pos:* Assoc clin prof obstet & gynec, Sch Med, George Washington Univ, 75- *Mem:* Am Fedn Clin Res; Endocrine Soc; Am Col Physicians; Am Fertility Soc; Am Soc Andrology (treas, 79-81, pres, 82-83). *Res:* Disorders of male reproduction; control of pituitary gonadotropin secretion. *Mailing Add:* NIH Nat Inst Child Health Bethesda MD 20892

SHERK, FRANK ARTHUR, b Stayner, Ont, May 20, 32; m 54; c 4. MATHEMATICS. *Educ:* McMaster Univ, BA, 54, MSc, 55; Univ Toronto, PhD(math), 57. *Prof Exp:* Lectr math, 57-61, asst prof, 61-65, assoc prof, 65-81, PROF MATH, UNIV TORONTO, 81- *Mem:* Am Math Soc; Can Math Cong. *Res:* Projective geometry; regular maps; discrete groups. *Mailing Add:* Dept Math Univ Toronto Toronto ON M5S 1A1 Can

SHERMA, JOSEPH A, b Newark, NJ, Mar 2, 34; m 61; c 2. ANALYTICAL CHEMISTRY. *Educ:* Upsala Col, BS, 55; Rutgers Univ, PhD(anal chem), 58. *Prof Exp:* From assoc prof to prof, 58-82, CHARLES A DANA PROF ANAL CHEM, LAFAYETTE COL, 82-, HEAD CHEM DEPT, 85- *Concurrent Pos:* researcher, Perrine Primate Lab, Environ Protection Agency, 72; vis researcher, Argonne Nat Lab, Iowa State Univ, Syracuse Univ Res Corp, Hosp Univ Pa, Waters Assocs & Whatman Inc; ed, J Asn Official Anal Chemists. *Mem:* Am Inst Chem; Am Chem Soc; Soc Appl Spectros; Sigma Xi. *Res:* Ion exchange and solution chromatography; analytical separations; pesticide analysis; quantitative thin-layer chromatography; food additive determination; clinical analysis; author of over 350 publications. *Mailing Add:* Dept Chem Lafayette Col Easton PA 18042

SHERMAN, ADRIA ROTHMAN, b 1950; m; c 2. TRACE ELEMENT NUTRITION, IRON IMMUNITY. *Educ:* Pa State Univ, PhD(nutrit), 77. *Prof Exp:* Assoc prof nutrit, Univ Ill, Urbana, 84-87; PROF NUTRIT, RUTGERS UNIV, 87- *Mem:* Am Inst Nutrit; Am Soc Clin Nutrit. *Res:* Nutritional immunology; iron; developmental nutrition. *Mailing Add:* Dept Nutrit Sci Rutgers Univ Thompson Hall Cook Col New Brunswick NJ 08903

SHERMAN, ALBERT HERMAN, b Philadelphia, Pa, Mar 5, 21; m 44; c 3. ORGANIC CHEMISTRY. *Educ:* Rutgers Univ, BS, 42, MS, 50, PhD(chem), 53. *Prof Exp:* Chemist, Rare Chems Div, Nopco Chem Co, 47-49; asst anal chem, Rutgers Univ, 49-52; sr chemist, Nitrogen Div, Allied Chem Corp, 52-57; res chemist, Atlas Powder Co, 57-71 & ICI Am, Inc, 71-80. *Mem:* Am Chem Soc. *Res:* Surfactants; detergents; textile additives; organic synthesis; cosmetic chemistry. *Mailing Add:* 20100 W Country Club Dr No 504 Miami FL 33180-1633

SHERMAN, ALFRED ISAAC, b Toronto, Ont, Sept 4, 20; US citizen; m 44; c 3. OBSTETRICS & GYNECOLOGY. *Educ:* Univ Toronto, MD, 43; Am Bd Obstet & Gynec, dipl; Bd Gynec Oncol, dipl, 74. *Prof Exp:* From sr intern to asst resident obstet & gynec, Hamilton Gen Hosp, Ont, 44-45; fel path, Univ Rochester, 46-47; intern, St Louis Maternity & Barnes Hosps, 47-48, asst resident, 49-50; fel obstet & gynec, Sch Med, Washington Univ, 50-51, from instr to prof obstet & gynec, 51-65; PROF OBSTET & GYNEC, SCH MED, WAYNE STATE UNIV, 67-; DIR, DEPT OBSTET & GYNEC RES & EDUC, SINAI HOSP, DETROIT, 75- *Concurrent Pos:* DIR, DEPT OBSTET & GYNEC ONCOL, BEAUMONT HOSP, ROYAL OAK, 75-; resident, Barnard Hosp, 50-51; consult & vis physician, St Louis City Hosp; consult, Mallinckrodt Inst Radiol, Homer G Phillips, St Luke's & Jewish Hosps; civilian consult, US Air Force; co-dir cytogenetics lab, John Hartford Found, Mo, 62-67. *Mem:* Radium Soc; Am Col Obstet & Gynec; Endocrine Soc; Soc Gynec Invest; Soc Gynec Oncol. *Mailing Add:* 6767 W Outer Dr Detroit MI 48235

SHERMAN, ANTHONY MICHAEL, b Barberton, Ohio, Mar 3, 40; m 63; c 4. POLYMER SYNTHESIS & CHARACTERIZATION, POLYMER PROCESSING. *Educ:* Univ Akron, BS, 65, MS, 68, PhD(polymer sci), 75. *Prof Exp:* Chemist, Goodyear Res, 63-66; sr chemist, B F Goodrich Chem Co, 66-77; sr res chemist, Mobil Chem Co, 77-85; RES MGR, ALCOLAC, 85- *Concurrent Pos:* Lectr, John Carroll Univ, Cleveland, 75-76. *Mem:* Am Chem Soc. *Res:* Specialty monomers and derived specialty polymers. *Mailing Add:* Alcolac Inc 3440 Fairfield Rd Baltimore MD 21226-1593

SHERMAN, ARTHUR, b Brooklyn, NY, Oct 15, 31; m 56; c 3. PLASMA PHYSICS. *Educ:* Polytech Inst Brooklyn, BME, 53; Princeton Univ, MSE, 58; Univ Pa, PhD(eng), 65. *Prof Exp:* Res engr, Gen Elec Corp, 53-68; pres, Med Diag Ctrs, 68-73; staff engr, RCA Corp, 73-79; vpres eng, Combustion Power Corp, 79-81; sr engr, Appl Mat, Inc, 81-83; SR SCIENTIST, VARIAN CORP, 83- *Concurrent Pos:* Lectr, Univ Pa, 65-68 & Univ Calif, Berkeley, 82- *Mem:* Am Phys Soc; Electrochem Soc; Am Vacuum Soc. *Res:* Chemical vapor deposition; plasma physics; magnetohydrodynamics; fluid mechanics. *Mailing Add:* 438 Chaucer St Palo Alto CA 94301

SHERMAN, BURTON STUART, b Brooklyn, NY, Nov 12, 30; m 52; c 4. ANATOMY. *Educ:* NY Univ, BA, 51, MS, 56; State Univ NY Downstate Med Ctr, PhD(anat), 60. *Prof Exp:* Res asst leukemia, Sloan-Kettering Inst Cancer Res, 51-52; res chemist, Jewish Hosp Brooklyn, NY, 52-56, res assoc biochem, 59-60; asst, State Univ NY, Downstate Med Ctr, 56-59, from instr to asst prof, 60-71, assoc prof, 71-85, EMER ASSOC PROF ANAT, STATE UNIV NY DOWNSTATE MED CTR, 85-; PROF ANAT & EMBRYOL, SCH HEALTH SCI, TOURO COL, 85-, EMER DEAN, 89- *Concurrent Pos:* Vis scientist, Strangeways Res Lab, Eng, 64-65; dean, Sch Health Sci, Touro Col, 85-89. *Mem:* Am Asn Anat; AAAS. *Res:* Leukemia research; osteogenesis; calcification of collagen and macromolecules; tissue and organ culture; vitamin A metabolism, biochemistry and effects on growth and proliferation of tissues; developmental anatomy. *Mailing Add:* 184 Elmwood St Valley Stream NY 11581

SHERMAN, BYRON WESLEY, b St Louis, Mo, Sept 20, 35; m 55; c 4. ELECTRICAL ENGINEERING. *Educ:* Univ Mo, BS, 57, MS, 59, PhD(elec eng), 66. *Prof Exp:* Asst instr elec eng, Univ Mo, 57-59; eng consult, Columbia Pictures Corp, Calif, 60-61 & Wells Eng, 61-62; electronics engr, McDonnell Aircraft Corp, 62-63; instr, 63-66, asst prof elec eng, 66-81, PROF, UNIV MO-COLUMBIA, 81- *Concurrent Pos:* Eng consult, Audience Studies, Inc, Calif, 66-67; summer fac fel, NASA Manned Spacecraft Ctr, 67-68, consult, 68-; consult, Lockheed Electronics Corp, Tex, 67-68 & Inst Bioeng Res, Univ Mo, 68-75. *Mem:* Inst Elec & Electronics Engrs. *Res:* Electric field instrumentation. *Mailing Add:* Dept Elec Eng Univ Mo Columbia MO 65211

SHERMAN, CHARLES HENRY, b Fall River, Mass, Dec 16, 28; m 51; c 2. PHYSICS. *Educ:* Mass Inst Technol, BS, 50; Univ Conn, MS, 57, PhD(physics), 62. *Prof Exp:* Physicist, Tracerlab, Inc, 50-53; physicist, US Navy Underwater Sound Lab, 53-63; assoc dir res, Parke Math Labs, Inc, Mass, 63-71; head appl res br, Transducer Div, New London Lab, Naval Underwater Systs Ctr, 71-74, head Transducer & Arrays Div, 74-81, prog mgr, 81-88; RETIRED. *Concurrent Pos:* Adj prof dept ocean eng, Univ RI, 75- *Mem:* Fel Acoust Soc Am; Am Phys Soc; Sigma Xi. *Res:* Acoustics; theoretical physics; solid state physics. *Mailing Add:* 22 Champlin Dr Westerly RI 02891

SHERMAN, D(ONALD) R, b Cleveland, Ohio, Aug 2, 35; m 60; c 3. STRUCTURAL ENGINEERING. *Educ:* Case Inst Technol, BS, 57, MS, 60; Univ Ill, PhD(struct eng), 64. *Prof Exp:* Engr, Pittsburgh Testing Lab, 57-58; design engr, Aerojet-Gen Corp, 63; engr, Esso Res & Eng Co, 64-66; assoc prof, 66-72, PROF STRUCT, UNIV WIS-MILWAUKEE, 72-, DIR, STRUCT LAB, 87- *Mem:* Am Soc Civil Engrs; Soc Exp Stress Anal; Struct Stability Res Coun. *Res:* Stability of structural steel columns; tubular structures; equipment design for seismic loads. *Mailing Add:* Dept Civil Eng Box 784 Univ Wis Milwaukee WI 53201

SHERMAN, DAVID MICHAEL, b Redwood City, Calif, Aug 2, 56; m 81; c 1. MINERAL PHYSICS, MINERAL SPECTROSCOPY. *Educ:* Univ Calif, Santa Cruz, BA(chem) & BS(earth sci), 80; Mass Inst Technol, PhD(geochem), 84. *Prof Exp:* GEOLOGIST, MINERAL & GEOCHEM, US GEOL SURV, 84- *Mem:* Mineral Soc Am; Am Geophys Union. *Res:* Applications of quantum chemistry and spectroscopy, optical, infrared and Mossbauer, to problems in mineralogy and geochemistry; electronic structures of transition metal oxides and silicates. *Mailing Add:* US Geol Surv Mail Stop 964 Box 25046 DFC Denver W Bldg 2 Rm 127 Denver CO 80225

SHERMAN, DOROTHY HELEN, speech pathology, statistics; deceased, see previous edition for last biography

SHERMAN, EDWARD, b New York, NY, Feb 8, 19; m 45; c 2. TECHNICAL MANAGEMENT, ORGANIC CHEMISTRY. *Educ:* Univ Ill, BS, 40; Lehigh Univ, MS, 47, PhD(chem), 49; Northwestern Univ, MBA, 70. *Prof Exp:* Asst org chem lab, Univ Ala, 40-41; jr inspector, US Food & Drug Admin, 41-42; asst oil chem, Lehigh Univ, 46-47; res assoc high nitrogen compounds, Ill Inst Technol, 49-50; group leader chem res, Quaker Oats Co, 50-64, sect leader, 64-71, coordr tech & admin serv, 71-72, asst dir chem res & develop, 72-78, mgr spec projs, 78-80, chem prod mgr, 80-83; RETIRED. *Concurrent Pos:* Consult, 83- *Mem:* AAAS; Am Chem Soc; Polyurethane Mfrs Asn. *Res:* Chemical product development and marketing; environmental sciences; industrial hygiene; furans; tetrahydrofurans; lactones; lactams; aromatics; ring cleavage; polymers; photochemistry; levulinic acid; cycloaliphatics; nitrogen heterocycles; nitroaminoguanidine; nitroguanyl azide; nitroaminotetrazole; pharmaceuticals; oil oxidation; hemin and hemochromogens; resins. *Mailing Add:* 17723 Tiffany Trace Dr Boca Raton FL 33487-1225

SHERMAN, FRED, b Minneapolis, Minn, May 21, 32; m 58; c 3. GENETICS, BIOPHYSICS. *Educ:* Univ Minn, BA, 53; Univ Calif, Berkeley, PhD(biophys), 58. *Prof Exp:* Fel genetics, Univ Wash, 59-60 & Lab Physiol Genetics, France, 60-61; from sr instr to prof, 61-81, prof, 71-81, WILSON PROF RADIATION BIOL & BIOPHYS, SCH MED & DENT, UNIV ROCHESTER, 81- *Concurrent Pos:* Instr, Cold Spring Harbor Lab, 70-87. *Mem:* Nat Acad Sci; AAAS; Genetics Soc Am; Am Soc Microbiol; Biophys Soc. *Res:* Mutational alteration of yeast cytochrome C; yeast genetics; cytoplasmic inheritance; cytochrome deficient mutants of yeast; amino acid changes in cytochrome C; DNA changes in yeast genes. *Mailing Add:* Sch Med & Dent Univ Rochester Rochester NY 14642

SHERMAN, FREDERICK GEORGE, b McGregor, Mich, Apr 16, 15; m 42; c 3. BIOLOGY. *Educ:* Univ Tulsa, BS, 38; Northwestern Univ, PhD(physiol), 42. *Prof Exp:* Fel, Sch Med, Wash Univ, St Louis, 46; from instr to prof biol, Brown Univ, 46-60; prof zool & chmn dept, 60-68, prof biol, 68-82, EMER PROF, SYRACUSE UNIV, 82- *Concurrent Pos:* Fel, Picker Found, 51-52; res collabr, Brookhaven Nat Lab, 53-56 & 58-68; vis scientist, NIH, Bethesda, 57, spec fel, 67-68. *Mem:* Am Physiol Soc; Soc Gen Physiol(secy, 57-59); Sigma Xi. *Res:* Biochemical changes associated with aging; regulation of protein synthesis. *Mailing Add:* 106 Dewitt Rd Syracuse NY 13214

SHERMAN, FREDERICK S, b San Diego, Calif, Apr 14, 28; m 53; c 2. MECHANICS. *Educ:* Harvard Univ, BS, 49; Univ Calif, MS, 50, PhD(mech eng), 54. *Prof Exp:* Instr mech eng, Univ Calif, 54-56; aeronaut res engr mech br, Off Naval Res, US Dept Navy, Washington, DC, 56-58; asst prof mech eng, Univ Calif, 58-59, assoc prof, 59-65, prof aeronaut sci, 65-70, asst dean, Col Eng, 73-80, prof mech eng, 70-91, EMER PROF, UNIV CALIF, BERKELEY, 91- *Concurrent Pos:* Mem fluid mech subcomt, Nat Adv Comt Aeronaut, 57-58. *Mem:* Am Inst Phys. *Res:* Fluid mechanics; mixing in stratified fluid flows; non-linear instability; free convection; unsteady flow separation. *Mailing Add:* Dept of Mech Eng Univ of Calif Berkeley CA 94720

SHERMAN, GARY JOSEPH, b Bellaire, Ohio, Dec 18, 41; m 64; c 3. MATHEMATICS. *Educ:* Bowling Green State Univ, BS, 63, MA, 68; Ind Univ, Bloomington, PhD(math), 71. *Prof Exp:* Asst prof, 71-78, PROF MATH, ROSE-HULMAN INST TECHNOL, 78- *Mem:* Am Math Soc; Math Asn Am. *Res:* Group theory; partially ordered groups. *Mailing Add:* Dept Math Rose-Hulman Inst Technol Terre Haute IN 47803

SHERMAN, GEORGE CHARLES, b Pasadena, Calif, Mar 4, 38; m 59; c 5. PHYSICAL OPTICS. *Educ:* Stanford Univ, BS, 60; Univ Calif, Los Angeles, MS, 65, PhD(meteorol), 69. *Prof Exp:* Mem tech staff, Aerospace Corp, 65-68; res assoc optics, Dept Physics & Astron & Inst Optics, Univ Rochester, 69-70, asst prof optics, Inst Optics, 70-76; staff electro-optic scientist, Itek Corp, 76-79; mem prof staff, Schlumberger-Doll Res Ctr, 79-; AT MISSION RES CORP. *Mem:* Optical Soc Am. *Res:* Radiation, propagation, diffraction and scattering of waves; atmospheric optics; oil exploration. *Mailing Add:* Mission Res Corp 735 State St PO Drawer 719 Santa Barbara CA 93102

SHERMAN, GERALD PHILIP, b Philadelphia, Pa, Mar 20, 40; m 66; c 2. PHARMACOLOGY, PHARMACY. *Educ:* Philadelphia Col Pharm & Sci, BSc, 63, MSc, 65, PhD(pharmacol), 67. *Prof Exp:* NIH fel, Univ Pittsburgh, 67-68; asst prof physiol & pharmacol, Univ Pittsburgh, 68-69; asst prof mat med, 69-70, Univ Ky, from asst prof to assoc prof clin pharm, 70-78; dir undergrad studies, 78-80, PROF PHARMACOL, UNIV TOLEDO, 80-, CHMN, DEPT PHARMACOL, 81- *Mem:* Am Asn Col Pharm; Am Soc Clin Pharmacol & Therapeut. *Res:* Clinical drug efficacy studies; autonomic and cardiovascular pharmacology; sports medicine. *Mailing Add:* Col of Pharm Univ of Toledo Toledo OH 43606

SHERMAN, GORDON R, b Menomonee, Mich, Feb 24, 28; m 51; c 2. COMPUTER SCIENCES, OPERATIONS RESEARCH. *Educ:* Iowa State Univ, BS, 53; Stanford Univ, MS, 54; Purdue Univ, PhD(math), 60. *Prof Exp:* Assoc prof math, 60-69, prof, 69-73, head dept comput sci, 70-73, PROF MATH & COMPUT SCI, UNIV TENN, KNOXVILLE, 73-, DIR, COMPUT CTR, 60- *Concurrent Pos:* NASA res grant, 62-70, grant data entry systs, 74-75; NSF grant, 63-64 & 69-71; prog dir, NSF, 71-72; chmn, Knoxville/Knox County Comput Adv Comt, 75. *Mem:* Asn Comput Mach; Soc Indust & Appl Math; Opers Res Soc Am; Am Statist Asn; Sigma Xi; fel Brit Comput Soc; Data Processing Mgt Asn. *Res:* Optimization of discrete functions; application of digital computers; computer installation management. *Mailing Add:* 200 SMC Univ Tenn Knoxville TN 37996-0520

SHERMAN, HAROLD, b Newark, NJ, Oct 19, 21; m 43; c 2. PHYSICS, NUCLEAR OIL WELL LOGGING. *Educ:* Brooklyn Col, AB, 42; NY Univ, PhD, 56; Pace Univ, JD, 86. *Prof Exp:* Physicist, Signal Corps, US Dept Army, Ohio, 42-44; electronics engr, Fada Radio, NY, 44; physicist, Premier Crystal Labs, 44-47; asst physics, NY Univ, 48-49, res assoc, 52-56; instr, St Peter's Col, 49-51; sr engr, A B Dumont Labs, NJ, 51-52; sr scientist, Avco Corp, Conn, 56; sr res proj physicist, Schlumberger-Doll Res Ctr, 56-83; RETIRED. *Mem:* Am Phys Soc; Sigma Xi. *Res:* Gas discharges; nuclear instrumentation; nuclear well logging. *Mailing Add:* 24 Webster Rd Ridgefield CT 06877

SHERMAN, HARRY LOGAN, b Anniston, Ala, Dec 5, 27; m 55; c 4. BOTANY, ECOLOGY. *Educ:* Jacksonville State Univ, BS, 55; Univ Tenn, Knoxville, MS, 58; Vanderbilt Univ, PhD(biol), 69. *Prof Exp:* Instr & res asst bot, Univ Tenn, Knoxville, 57-59; from asst prof to assoc prof biol, 61-69, PROF BIOL SCI & HEAD DEPT, MISS UNIV WOMEN, 69- *Mem:* Am Inst Biol Sci; Am Soc Plant Taxon; Int Soc Plant Taxon. *Res:* Biosystematics, including cyto- and chemo-taxonomy and reproductive biology; taxonomy. *Mailing Add:* 421 S Tenth St Columbus MS 39701

SHERMAN, HERBERT, electrical engineering; deceased, see previous edition for last biography

SHERMAN, IRWIN WILLIAM, b New York, NY, Feb 12, 33; m 66; c 2. ZOOLOGY, PARASITOLOGY. *Educ:* City Col New York, BS, 54; Northwestern Univ, MS, 59, PhD(biol), 60. *Prof Exp:* Asst protozool, Univ Fla, 54; lab technician, US Army, 54-56; teacher, Yonkers Bd Educ, 56-57; asst parasitol, Northwestern Univ, 57-60; NIH fel, Rockefeller Inst, 60-62; from asst prof to assoc prof, 62-73, chmn, Dept Biol, 74-79, dean, Col Natural & Agr Sci, 81-88, PROF ZOOL, UNIV CALIF, RIVERSIDE, 73-,. *Concurrent Pos:* Guggenheim Mem Found fel, Carlsberg Found, Copenhagen, 67; eve lectr biol, City Col New York, 57, 60-62; spec NIH fel, Nat Inst Med Res, Mill Hill, Eng, 73-74; mem trop med & parasitol study sect, NIH, 70-74 & 79-80; chmn ad hoc, Study Group Parasitic Dis, Dept Army, 77-78; mem steering comt Malaria Chemother, World Health Orgn, 78-86. *Honors & Awards:* Ward Medal Biol, 54; Wellcome Trust Lectr, 88. *Mem:* AAAS; Soc Protozool; Am Soc Parasitol; Am Soc Tropical Med & Hygiene. *Res:* Biochemisry and cell biology of malaria; malaria immunity. *Mailing Add:* Dept of Biol Univ of Calif Riverside CA 92521

SHERMAN, JAMES H, b Detroit, Mich, Mar 14, 36; m 65; c 2. PHYSIOLOGY. *Educ:* Univ Mich, BS, 57; Cornell Univ, PhD(cell physiol), 63. *Prof Exp:* From res assoc to asst prof, 63-71, asst dir, Off Allied Health Educ, Dept Postgrad Med, 72-77, ASSOC PROF PHYSIOL, MED SCH, UNIV MICH, ANN ARBOR, 71- *Mem:* Biophys Soc; NY Acad Sci; Am Physiol Soc; Sigma Xi. *Res:* Cell membrane permeability; intracellular pH; active transport of amino acids in tumor cells; surface properties of cell membranes; co-auth Human Physiology the Mechanisms of Body Function. *Mailing Add:* Dept Physiol Med Sch Univ Mich Ann Arbor MI 48104

SHERMAN, JEROME KALMAN, b Brooklyn, NY, Aug 14, 25; m 52; c 3. ANATOMY, CRYOBIOLOGY. *Educ:* Brown Univ, AB, 47; Western Reserve Univ, MS, 49; Univ Iowa, PhD(zool), 54. *Prof Exp:* Asst biol, Western Reserve Univ, 47-49; from asst to res assoc urol, Univ Iowa, 49-54; res assoc, Am Found Biol Res, 54-58; from asst prof to assoc prof, 59-66, PROF ANAT, MED COL, UNIV ARK MED SCI, LITTLE ROCK, 67- *Concurrent Pos:* Consult, Am Breeders Serv, 55-56, Winrock Farm, 59-60, Idant Corp, 72-73 & Dow Chem Co, 77-; mem adv bd, Am Type Cult Collections, 73-77; Lederle med fac award, 61-63; Fulbright sr res award, Univ Munich, Ger, 65-66; spec chair prof, Nat Chung-Hsin Univ, Taiwan, 73-74; Nat Sci Award, Taiwan, 73-74. *Mem:* Am Asn Anat; Soc Cryobiol; Sigma Xi; Am Asn Tissue Banks; Soc Exp Biol & Med. *Res:* Effects of cooling, freezing and rewarming on protoplasm of various cells and tissues; low temperature preservation of living cells, especially mammalian gametes; fertility and sterility; cytology; ultrastructural and biochemical cryoinjury of cellular organelles; cryobanking of human semen and cryosurvival of organisms which cause sexually transmitted diseases during crybanking of semen; male infertility. *Mailing Add:* Dept Anat Med Col Univ Ark for Med Sci Little Rock AR 72205

SHERMAN, JOHN EDWIN, b Brooklyn, NY, Jan 18, 22; m 45; c 3. SUPERCOMPUTERS TECHNICAL COMPUTING. *Educ:* Hofstra Col, BA, 50. *Prof Exp:* Physicst, US Naval Air Missile Test Ctr, 51-53; design engr, McDonnell Aircraft Corp, 53-54; math analyst analog comput, Lockheed Missile & Space Co, 54-56, group engr, 56-57, sect leader, 57-59, dept mgr hybrid comput, 59-66, div mgr, 66-69, div mgr data processing, 69-75, div mgr sci comput, 75-80, dir, Tech Comput Serv, 80-88; CONSULT, LARGE COMPUTER SYSTEMS, 88- *Concurrent Pos:* Pres, Simulation Coun, Inc, 59-60, dir, 60-66, dir pub, 64-74; dir, Am Fedn Info Processing Socs, 62-72. *Mem:* Inst Elec & Electronics Engrs; Soc Comput Simulation. *Res:* Analog and hybrid computing; digital computing; data processing. *Mailing Add:* 1647 Grant Rd Mountain View CA 94040

SHERMAN, JOHN FOORD, b Oneonta, NY, Sept 4, 19; m 44; c 2. PHARMACOLOGY. *Educ:* Union Univ, NY, BS, 49; Yale Univ, PhD(pharmacol), 53. *Hon Degrees:* ScD, Albany Col Pharm, 70. *Prof Exp:* Pharmacologist, Lab Trop Dis, Nat Microbiol Inst, 53-56, dep chief extramural progs, Nat Inst Arthritis & Metab Dis, 56-61, assoc dir extramural progs, Nat Inst Neurol Dis & Blindness, 61-62, Nat Inst Arthritis & Metab Dis, 62-63 & NIH, 64-68, dep dir, NIH, 68-74; vpres, Asn Am Med Col, 74-87, exec vpres, 87-90; RETIRED. *Mem:* Inst Med-Nat Acad Sci; Sigma Xi; AAAS; Nat Multiple Sclerosis Soc; Nat Asn Biomed Res. *Res:* Pharmacology of the central nervous system; chemotherapy; medical research and education administration. *Mailing Add:* Asn Am Med Cols One Dupont Circle NW Washington DC 20036

SHERMAN, JOHN WALTER, b Auburn, NY, Aug 19, 45; m 70, 80. PALEOLIMNOLOGY. *Educ:* Hamilton Col, BA, 67; Univ Vt, MS, 72; Univ Del, PhD(geol), 76. *Prof Exp:* Sr scientist, Acad Natural Sci Philadelphia, 75- *Mem:* Am Soc Limnol & Oceanog; Sigma Xi; NAm Lakes Mgt Soc. *Res:* Paleoecological studies involving the use of diatoms as indicators of past water quality; ecological requirements of diatom communities. *Mailing Add:* Div Environ Res 19th & The Parkway Philadelphia PA 19103

SHERMAN, JOSEPH E, b Chicago, Ill, July 27, 19; m 45; c 3. MICROBIOLOGY, FOOD TECHNOLOGY. *Educ:* Univ Ill, BS, 39. *Prof Exp:* Bacteriologist, Vet Admin Hosp, Wis, 39-41; res labs, Swift & Co, Ill, 46-60; group leader microbiol res ctr, Kraftco Corp, 60-69; tech dir, Runyon Lab, 69-73; qual assurance mgr, Schulze & Burch Biscuit Co, 73-81; CONSULT, SHERMAN FOOD LAB, 82- *Mem:* Am Soc Microbiol; Inst Food Technologists; Am Soc Qual Control; Soc Indust Microbiol. *Res:* Medical and forensic microbiology. *Mailing Add:* 3925 N Triumvera Dr Apt 12B Glenview IL 60025

SHERMAN, KENNETH, b Boston, Mass, Oct 6, 32; m 58; c 3. BIOLOGICAL OCEANOGRAPHY. *Educ:* Suffolk Univ, BS, 54; Univ RI, MS, 60; Morski Inst Ryback, DSc, 78. *Hon Degrees:* DSc, Suffolk Univ, 79. *Prof Exp:* Instr conserv educ, Mass Audubon Soc, 54-55; fishery aide, US Bur Com Fisheries, Mass, 55-56; teacher high sch, Mass, 59-60; fishery res biologist zooplankton ecol, US Bur Com Fisheries, Hawaii, 60-63, Maine, 63-71; coord, Marine Resources Monitoring Assessing & Prediction Prog, 71-73, chief resource assessment div, US Dept Com, Nat Oceanic & Atmospheric Admin, Washington, DC, 73-75, LAB DIR & CHIEF, MARINE ECOSYST BR, US DEPT COM, NAT OCEANIC & ATMOSPHERIC ADMIN, NAT MARINE FISHERIES SERV, NARRAGANSETT, RI, 75- *Concurrent Pos:* Mem biol oceanog comt, Int Coun Explor Sea, 72-87; US proj officer, Plankton Sorting Ctr, Szczecin, Poland, 73-; adj prof, Grad Sch Oceanog, Univ RI, 80- *Mem:* AAAS; Am Soc Limnol & Oceanog; Ecol Soc Am; Am Soc Zool; Am Soc Ichthyol & Herpet. *Res:* Tuna oceanography; zooplankton ecology; Atlantic herring biology; distribution and abundance of epipelagic marine copepods; estuarine ecology; taxonomy of marine copepods; predator prey relationships of pelagic fishes; plankton in ecosystems; productivity of living marine resources. *Mailing Add:* Am Red Cross Blood Res Old Georgetown Rd Bethesda MD 20814

SHERMAN, KENNETH ELIOT, virology, for more information see previous edition

SHERMAN, LARRY RAY, b Easton, Pa, June 26, 34; m 66. PESTICIDE CHEMISTRY. *Educ:* Lafayette Col, BSc, 56; Utah State Univ, MSc, 61; Univ Wyo, PhD(anal chem), 69. *Prof Exp:* res assoc, NC A&T State Univ, Greensboro, 69-71, assoc prof chem, 70-74; dir res, Hillyard Chem Co, St Joseph, Mo, 74-76; asst prof chem, Univ Miss, 76-78; asst prof, Univ Akron, Ohio, 78-81; ASST PROF CHEM, UNIV SCRANTON, PA, 81- *Concurrent Pos:* Engr, Gen Elec Co, Cleveland, Ohio, 56-58; chem anal, Weyerhauser Co, Seattle, Wash, 62-64; instr chem, Northern Ill Univ, DeKalb, 64-66; NASA fel, Univ Wyo, Laramie, 67-69; res assoc, NC A&T State Univ, 69-71; NSF fel, Pa State Univ, 72; res assoc, Univ Miss, 76-78; fel, Univ Akron, Ohio, 80, vis prof, 87-88; NASA fel, Lewis Res Ctr, Cleveland, Ohio, 84; guest prof, Univ Dortmund, WGer, 86; fac fel, Brooks Air Force Base, San Antonio, Tex, 88; NATO fel, Vrije Univ, Brussels, Belg, 89. *Mem:* Sr mem Sigma Xi; fel Royal Soc Chem; Am Chem Soc; Controlled Release Soc. *Res:* Organotin chemistry: analyses, toxicology and in vitro and in vivo biological and antitumor activity of di and tri-alkyl/aryl tin compounds; author of approximately 50 publications. *Mailing Add:* Dept Chem Univ Scranton Scranton PA 18510-4626

SHERMAN, LAURENCE A, b Cambridge, Mass, Jan 1, 35; m 65; c 2. PATHOLOGY, HEALTH SCIENCES. *Educ:* Univ Chicago, BA & BS, 56; Albany Med Col, MD, 64. *Prof Exp:* Intern med, Boston City Hosp, Mass, 64-65; resident, Univ Calif, Los Angeles, 65-66; fel enzym, Sch Med, Wash Univ, 66-68, from instr to prof med path, 69-83; assoc dir, 73-82, CHIEF MED SERV, MO-ILL REGIONAL RED CROSS BLOOD PROG, 82-; CLIN PROF PATH & MED, MED SCH, WASH UNIV, ST LOUIS, 83- *Concurrent Pos:* Dir coagulation lab & chief vascular div, Jewish Hosp, St Louis, 69-73, assoc dir blood bank, 70-73; dir blood bank, Barnes Hosp, St Louis, 73-82, co-dir hemostasis lab, 78-80; dir blood banking, training prog, Nat Heart, Lung & Blood Inst, Wash Univ, 77-85; chmn sci prog, pres elect, 87-88, Am Asn Blood Banks, 79-84, bd dirs, 80-; mem, comt apheresis, Am Red Cross, 85-, dir coun, 86-87. *Mem:* Am Fedn Clin Res; Am Soc Hemat; Am Asn Pathologists; Am Soc Clin Invest; Am Heart Asn; fel Am Soc Clin Path; fel Am Col Physicians; Am Asn Blood Banks (treas, 84). *Res:* Thrombosis and metabolism of coagulation moieties, particularly fibrinogen and its derivatives; transfusions medicine. *Mailing Add:* Northwestern Univ Med Sch 303 W Chicago Ave Chicago IL 60611

SHERMAN, LINDA ARLENE, b Brooklyn, NY, Feb 27, 50; m 78; c 2. IMMUNOLOGY, BIOCHEMISTRY. *Educ:* Barnard Col, AB, 71; Mass Inst Technol, PhD(biol), 76. *Prof Exp:* Postdoctoral immunol, Albert Einstein Sch Med, 76-77; postdoctoral immunol, Harvard Med Sch, 77-78; asst mem, 78-85, ASSOC MEM IMMUNOL, RES INST SCRIPPS CLIN, 85- *Concurrent Pos:* Consult & panel reviewer, NSF, 85-89; consult synbiotics, 88-90; sect ed, J Immunol, 90- *Mem:* Am Asn Immunologists; AAAS. *Res:* Molecules involved in cytolytic T lymphocyte recognition of alloantigens and foreign antigens. *Mailing Add:* Dept Immunol IMM16 Scripps Res Inst 10666 N Torrey Pines Rd La Jolla CA 92037

SHERMAN, LOUIS ALLEN, b Chicago, Ill, Dec 16, 43; m 69; c 2. PHOTOSYNTHESIS, MEMBRANE STRUCTURE. *Educ:* Univ Chicago, BS, 65, PhD(biophysics), 70. *Prof Exp:* Fel photosynthesis, Cornell Univ, 70-72; from asst prof to assoc prof, 72-82, dir biol sci, 85-89, PROF BIOL SCI, UNIV MO, 82-; PROF & HEAD BIOL SCI, PURDUE UNIV, 89- *Concurrent Pos:* Fulbright res fel, Univ Leiden, Netherlands, 79-80. *Mem:* Biophys Soc; Am Soc Microbiol; Am Soc Photobiol; AAAS; Plant Physiol. *Res:* Photosynthesis and the structure of photosynthetic membranes; isolation of photosynthetic mutants in cyanobacteria; isolation of photosynthetic membrane components; cloning of photosynthesis genes on specially designed hybrid plasmids. *Mailing Add:* Dept Biol Sci Purdue Univ Lilly Hall West Lafayette IN 47907

SHERMAN, MALCOLM J, b Chicago, Ill, July 28, 39; m 63; c 2. MATHEMATICS. *Educ:* Univ Chicago, SB & SM, 60; Univ Calif, Berkeley, PhD(math), 64. *Prof Exp:* Asst prof math, Univ Calif, Los Angeles, 64-68; asst prof, 68-70, ASSOC PROF MATH, STATE UNIV NY, ALBANY, 70- *Concurrent Pos:* Mem, US Comn Civil Rights, 84-85. *Mem:* Am Math Soc; Math Asn Am; Am Statist Assoc. *Res:* Applied statistics and functional analysis (operator theory). *Mailing Add:* Dept Math & Statist State Univ NY Albany Albany NY 12222

SHERMAN, MARTIN, b Newark, NJ, Nov 21, 20; m 43, 75; c 2. ENTOMOLOGY, INSECTICIDE TOXICOLOGY. *Educ:* Rutgers Univ, BSc, 41, MSc, 43; Cornell Univ, PhD(insect toxicol), 48. *Prof Exp:* Res asst entom, Cornell Univ, 45-48; entomologist, Beech-Nut Packing Co, 48-49; asst prof entom, Univ & asst entomologist, Agr Exp Sta, 49-52, assoc prof & assoc entomologist, 52-58, prof entom & entomologist, 58-85, EMER ENTOMOLOGIST & PROF, UNIV HAWAII, 85- *Concurrent Pos:* Fulbright scholar, Univ Tokyo, 56-57 & State Entom Lab & Royal Vet & Agr Col, Denmark, 66; vis prof, Rutgers Univ, 73; mem gov bd, Entom Soc Am, 74-77. *Mem:* Am Chem Soc; Soc Toxicol; Soc Environ Toxicol & Chem; Int Soc Study Xenobiotics; fel Am Inst Chemists. *Res:* Comparative vertebrate and insect toxicology; insecticide residue analysis; insecticide formulation; metabolism of insecticides. *Mailing Add:* 1121 Koloa St Honolulu HI 96816

SHERMAN, MERRY RUBIN, b New York, NY, May 14, 40. ENDOCRINE BIOCHEMISTRY. *Educ:* Wellesley Col, BA, 61; Univ Calif, Berkeley, MA, 63, PhD(biophysics), 66. *Prof Exp:* NIH fel, Weizmann Inst, 66-67 & Nat Inst Dent Res, 68-69; res assoc, Sloan-Kettering Inst, 71-76, assoc mem, 76-86; PROF BIOCHEM, RUTGERS UNIV, 86- *Concurrent Pos:* Vis investr, Cardiovascular Res Inst, Univ Calif, San Francisco, 75-76; assoc prof biochem, Grad Sch Med Sci, Cornell Univ, 77-86. *Mem:* Am Soc Biol Chemists; Endocrine Soc; Am Asn Cancer Res. *Res:* Steroid hormone receptors; proteolytic enzymes. *Mailing Add:* Dept Biol Sci Rutgers Univ 101 Warren St Newark NJ 07102

SHERMAN, MICHAEL IAN, b Montreal, Que, Sept 27, 44; m 68. DEVELOPMENTAL BIOLOGY & ONCOLOGY, VIROLOGY. *Educ:* McGill Univ, BS, 65; State Univ NY Stony Brook, PhD(molecular biol), 69. *Prof Exp:* Fel, dept path, Univ Oxford, 69-70 & dept zool, 70-71; from asst mem to mem, Roche Inst Molecular Biol, Roche Res Ctr, 71-86, dir cell biol, 86-90; PRES & CHIEF EXEC OFFICER, PHARMAGENICS, INC, 90- *Concurrent Pos:* Mem rev panel, NASA, 78; mem, Human Embryol Develop Study Sect, NIH, 79-81; adj assoc prof human genetics, Columbia Col Physicians & Surgeons, 80-86. *Mem:* Am Asn Cancer Res. *Res:* Cancer; virology. *Mailing Add:* 314 Forest Ave Glen Ridge NJ 07028

SHERMAN, NORMAN K, b Kingston, Ont, Jan 28, 35; m 60; c 5. LASER-DRIVEN ACCELERATORS, NUCLEAR & RADIATION PHYSICS. *Educ:* Queen's Univ, Ont, BSc, 57, MSc, 59, PhD(physics), 62. *Prof Exp:* Lectr physics, Royal Mil Col, Ont, 59-60; lectr, Queen's Univ, Ont, 61-62; foreign fel, CEN Saclay, France, 62-64; NATO fel nuclear physics, Univ Paris, 62-64 & Yale Univ, 64-65; asst prof physics, McGill Univ, 65-68; assoc res officer, Nat Res Coun Can, 68-75, x-rays & nuclear radiations, 68-84, physicist laser & plasma physics, Div Physics, 84-88, consortium adv, 88-90, SR RES OFFICER, NAT RES COUN CAN, 75-, PLANNING & PROG DEVELOP, SCI AFFAIRS OFF, 90- *Concurrent Pos:* Res assoc, Accelerator Lab, Univ Sask 65; user zero gradient synchrotron, Argonne Nat Lab, 66-68; ed, Youth Sci News, 73-76; mem ad hoc comt synchrotron radiation, Nat Res Coun, 76-78; actg head, Electron Linac Lab, Nat Res Coun, 77-80; secy, comt intermediate physics, NSERC, 78-81; proj mgr, Can Synchrotron Radiation Fac, 78-; secy, Adv Bd TRIUMF, Nat Res Coun Can, 78-81, comt high energy physics, 78-84, comt subatomic physics, 84-85; foreign collabr, Saclay linear accelerator, 82; external user, superconducting Linear Accelerator, MUSL, Univ Ill, Champaign, 83; agreement adminr, Can Audio Res Consortium, 88-; facilitator, Solid State Optoelectronics Consortium Can, 88-90; adv, Simulated Mfg Res Consortium, 88-90; secy, proj Athena Steering Comt, 89-; working party, sci merit, KAON, 90, cost benefit, 90; adminr, Nat Res Coun, Queen's Univ contrib agreement Sudbury Neutrino Observ, Carleton Univ agreement high energy physics, 91- *Honors & Awards:* Farrington Daniels Award Achievement Radiation Dosimetry, Am Asn Physicists Med, 75. *Mem:* AAAS; Am Phys Soc; NY Acad Sci; Can Asn Physicists; Youth Sci Found (pres, 76-78); Inst Elec & Electronic Engrs. *Res:* Superconducting nuclear particle detector; magnesium photofission; lithium phototritons; proton scattering; shielding; radiation-damage reversal; de-excitation neutrons; bremsstrahlung spectra, radiators, filters, depth dose; photoneutron fine structure by time-of-flight; photoneutron angular distributions; photon total absorption; electron pair cross section of uranium; liquid-deuterium gamma-ray spectrometer; linearly polarized gamma rays; laser photocathodes; laser driven accelerators; laser produced gamma rays; picosecond optical damage to metals; two-temperature heat transport in metals on picosecond time scale; electron-to-lattice coupling constant in metals, synchrotron radiation, photonuclear reactions. *Mailing Add:* Physics Div Nat Res Coun Ottawa ON K1A 0R6 Can

SHERMAN, PATSY O'CONNELL, b Minneapolis, Minn, Sept 15, 30; m 53; c 2. CORPORATE TECHNICAL EDUCATION & DEVELOPMENT, APPLIED CHEMISTRY. *Educ:* Gustavus Adolphus Col, BA, 52. *Prof Exp:* Chemist cent res dept, 52-57, chemist chem div, 57-67, res specialist, 67-70, sr res specialist, 70-73, res mgr chem resources div, 73-81, MGR TECH DEVELOP, MINN MINING & MFG CO, 82- *Mem:* Am Chem Soc; Am Soc for Eng Educ; Am Soc Training & Develop. *Res:* Fluorine-containing polymers; oil and water repellent textile treatments. *Mailing Add:* 9300 11th Ave S Minneapolis MN 55420

SHERMAN, PAUL DWIGHT, JR, b San Diego, Calif, May 18, 42; m 63; c 2. INDUSTRIAL ORGANIC CHEMISTRY. *Educ:* Univ NH, BS, 64; Brown Univ, PhD(org chem), 70. *Prof Exp:* Group leader & technol mgr, Union Carbide Corp, 76-82, info systs mgr, 82-84, bus mgr, 84-86, group leader, 86-89, ASSOC DIR, UNION CARBIDE CORP, 89- *Mem:* Am Chem Soc. *Res:* Application of analytical techniques for the solution of industrial process problems; organic-analytical chemistry; hydroformylation process development; technical management. *Mailing Add:* 1014 Knob Way South Charleston WV 25309

SHERMAN, PAUL WILLARD, b July 6, 49; US citizen; m 81; c 2. EVOLUTIONARY BIOLOGY, ANIMAL BEHAVIOR. *Educ:* Stanford Univ, BA, 71; Univ Mich, MS, 74, PhD(biol), 76. *Prof Exp:* Miller fel zool, Univ Calif, Berkeley, 76-78, asst prof psychol, 78-80; from asst prof to assoc prof, 80-91, PROF NEUROBIOL & BEHAV, CORNELL UNIV, 91- *Concurrent Pos:* Guggenheim fel, 85. *Honors & Awards:* A B Howell Award, Cooper Ornith Soc, 74; A M Jackson Award, Am Soc Mammalogists, 77; Clark Award, Cornell Univ, 84. *Mem:* Animal Behav Soc; Am Soc Naturalists. *Res:* Evolution of social behavior. *Mailing Add:* Neurobiol & Behav Dept Cornell Univ Ithaca NY 14850

SHERMAN, PHILIP MARTIN, b Norwalk, Conn, July 10, 30; m 55; c 3. STRATEGIC PLANNING, COMPUTER SCIENCE. *Educ:* Cornell Univ, BEPhys, 52; Yale Univ, MEE, 56, PhD(elec eng), 59. *Prof Exp:* Engr, Sperry Gyroscope Co, NY, 52-55; instr elec eng, Yale Univ, 57-59; engr, Bell Tel Labs, 59-63, supvr, 63-67, dept head, 67-69; mgr info systs, Webster Res Ctr, 69-78, mgr technol planning, 78-84; COMPUTER CONSULT, SHERMAN DATA SYSTS, 84- *Mem:* Inst Elec & Electronic Engrs; Asn Comput Mach. *Res:* Computer programming and analysis; systems analysis; computer language studies; computer data management. *Mailing Add:* Sherman Data Systems 471 Claybourne Rd Rochester NY 14618

SHERMAN, ROBERT GEORGE, b Charlevoix, Mich, Mar 22, 42; m 65; c 2. NEUROPHYSIOLOGY, ZOOLOGY. *Educ:* Alma Col, BS, 64; Mich State Univ, MS, 67, PhD(zool), 69. *Prof Exp:* USPHS fel, Univ Toronto, 69-70, spec fel, 70-71; from asst prof to assoc prof physiol, Clark Univ, 71-78; PROF & CHMN ZOOL, MIAMI UNIV, 78- *Concurrent Pos:* Consult, NY Bd Reagents, 84-85, Rev Doctoral Progs Biol; mem, Nat Res Coun, Biomed Panel Rev Nat Sci Found, fel applications, 85-87. *Mem:* Am Soc Zool; Soc Gen Physiologists; Soc Neurosci; Electron Microscopy Soc Am; Bermuda Biol Sta. *Res:* Structure and function of synapses in arthropods; ultrastructure of arthropod muscle; cardiac physiology of arthropods; development of nerve and muscle. *Mailing Add:* Dept Zool Miami Univ Oxford OH 45056

SHERMAN, ROBERT HOWARD, b Chicago, Ill, Nov 18, 29; m; c 3. PHYSICAL CHEMISTRY, CRYOGENICS. *Educ:* Ill Inst Technol, BS, 51; Univ Calif, PhD(chem), 55. *Prof Exp:* STAFF MEM, LOS ALAMOS NAT LAB, 55- *Concurrent Pos:* Consult, Argonne Nat Lab, 59-60. *Mem:* Am Chem Soc; Am Phys Soc. *Res:* Thermodynamics, especially at low temperatures; liquid helium; critical point phenomena; hydrogen isotope technology (tritium); isotope separation; Raman spectroscopy. *Mailing Add:* Group MST-3 MS C348 Los Alamos Nat Lab PO Box 1663 Los Alamos NM 87545

SHERMAN, ROBERT JAMES, b Bristow, Iowa, July 28, 40; m 62; c 1. BOTANY, ECOLOGY. *Educ:* Coe Col, BA, 62; Ore State Univ, MS, 66, PhD(bot), 68. *Prof Exp:* Asst prof biol, Univ Colo, Colorado Springs, 68-70; chmn dept, 75-78, asst prof, 70-80, actg dean, Sch Nat Sci, 90-91, PROF BIOL, SONOMA STATE UNIV, 80- *Concurrent Pos:* Consult grasslands biome, Int Biol Prog, 69-70; NSF grants; Dept Health, Educ & Welfare grant. *Mem:* Ecol Soc Am. *Res:* Structure and pattern of Pinus ponderosa forests; fire ecology; oak woodlands ecology. *Mailing Add:* Dept Biol Sonoma State Univ Rohnert Park CA 94928

SHERMAN, ROGER TALBOT, b Chicago, Ill, Sept 30, 23; m 52; c 5. SURGERY. *Educ:* Kenyon Col, AB, 46; Univ Cincinnati, MD, 48; Am Bd Surg, dipl, 57. *Prof Exp:* Fel path, St Luke's Hosp, Chicago, 49-50; asst resident surgeon, Cincinnati Gen Hosp, 50-55, instr surg, Col Med, Univ Cincinnati, 55-56; from asst prof to prof, Col Med, Univ Tenn, Memphis, 59-72; prof surg & chmn dept, Col Med, Univ S Fla, 72-82; PROF SURG, SCH MED, EMORY UNIV, ATLANTA, GA, 83-; CHIEF SURG, GRADY MEM HOSP, ATLANTA. *Honors & Awards:* Curtis P Artz Trauma Soc Award. *Mem:* Am Surg Asn; Am Asn Surg of Trauma. *Res:* Surgical infections, shock and trauma. *Mailing Add:* Dept Surg Emory Univ Sch Med 69 Butler St SE 312 Glenn Bldg Atlanta GA 30303

SHERMAN, RONALD, b Philadelphia, Pa, Sept 5, 41; wid; c 2. ELECTRICAL ENGINEERING, APPLIED MATHEMATICS. *Educ:* Univ Pa, BSEE, 62, PhD(elec eng), 65; Columbia, MBA, 82. *Prof Exp:* Mem tech staff radar systs, 65-67, mem tech staff ocean systs, 67-70, supvr electromagnetic pulse eng & design principles group, 70-76, supvr systs anal & planning group, 76-80, SUPVR FINANCIAL MODELING STUDIES GROUP, BELL LABS, 80- *Concurrent Pos:* Vis sr lectr, Stevens Inst Technol, 69-71, assoc prof elec eng, 71-77, prof, 77-80. *Mem:* Inst Elec & Electronics Engrs. *Res:* Electromagnetics; effect of pulsed fields on communications systems; estimation and control theory; adaptive tracking; statistics; clustering algorithm development; financial model development. *Mailing Add:* Bell Tel Labs Rm Mh Sc-110 600 Mountain Ave Murray Hill NJ 07974

SHERMAN, SAMUEL MURRAY, b Pittsburgh, Pa, Jan 4, 44; m 69. NEUROSCIENCES, OPHTHALMOLOGY. *Educ:* Calif Inst Technol, BS, 65; Univ Pa, PhD(neuroanat), 69. *Hon Degrees:* MA, Univ Oxford, 85. *Prof Exp:* USPHS fel, Australian Nat Univ, 70-72; from asst prof to prof physiol, Sch Med, Univ Va, 72-78; prof anat, Dept Anat Sci, 79-80, PROF NEUROBIOL & BEHAV, STATE UNIV NY, STONY BROOK, 80- *Concurrent Pos:* NSF res grant, 73-; USPHS res grant, 75-; USPHS res career develop award, 75-80; mem, Visual Sci B Study Sect, NIH, 75-79; AB Sloan fel, 77-80; Newton-Abraham vis prof, Univ Oxford, 85-86; mem & chair Behav & Neurosci Study Sect 1, NIH, 89- *Mem:* Soc Neurosci; Am Physiol Soc; Am Asn Anat; Asn Res Vision & Ophthal; AAAS. *Res:* Functional organization of the mammalian visual system. *Mailing Add:* Dept Neurobiol & Behav State Univ NY Stony Brook NY 11794

SHERMAN, SEYMOUR, systems design, systems science, for more information see previous edition

SHERMAN, THOMAS FAIRCHILD, b Ithaca, NY, May 25, 34; m 70; c 5. BIOLOGY. *Educ:* Oberlin Col, AB, 56; Oxford Univ, DPhil(biochem), 60. *Prof Exp:* Vis asst prof zool, Oberlin Col, 60-61; res fel hist sci, Yale Univ, 61-62; vis asst prof zool, Pomona Col, 62-65; res fel math biol, Harvard Univ, 65-66; from asst prof to assoc prof, 66-79, PROF BIOL, OBERLIN COL, 79- *Res:* Vascular branching; connective tissue transport; history of science. *Mailing Add:* Dept Biol Oberlin Col Oberlin OH 44074

SHERMAN, THOMAS LAWRENCE, b Los Angeles, Calif, Nov 17, 37. MATHEMATICS. *Educ:* Univ Calif, Los Angeles, AB, 59; Univ Utah, MS, 61, PhD(math), 63. *Prof Exp:* Mem, US Army Math Res Ctr, Univ Wis, 63-64; from asst prof to assoc prof, 64-74, PROF MATH, ARIZ STATE UNIV, 74- *Mem:* Am Math Soc; Math Asn Am; Soc Indust & Appl Math. *Res:* Ordinary differential equations. *Mailing Add:* Dept Math Ariz State Univ Tempe AZ 85287

SHERMAN, THOMAS OAKLEY, b Brooklyn, NY, May 6, 39; m 61; c 3. MATHEMATICS. *Educ:* Mass Inst Technol, BS, 60, PhD(math), 64. *Prof Exp:* Mem staff math, Inst Advan Study, 64-65; asst prof, Brandeis Univ, 65-69; asst prof, 69-70, ASSOC PROF MATH, NORTHEASTERN UNIV, 70- *Res:* Lie groups; harmonic analysis; numerical analysis. *Mailing Add:* Dept Math Northeastern Univ 360 Huntington Ave Boston MA 02115

SHERMAN, WARREN V, b London, Eng, Jan 7, 37; m 63; c 2. RADIATION CHEMISTRY, PHOTOBIOLOGY. *Educ:* Univ London, BSc, 58, PhD(org chem), 61. *Prof Exp:* Fulbright res fel chem, Brandeis Univ, 61-63; res fel, US Army Natick Lab, 63-64 & Israel Atomic Energy Comn, 64-66; res fel radiation lab, Univ Notre Dame, 66-68; assoc prof, 68-74, PROF CHEM, CHICAGO STATE UNIV, 74- *Concurrent Pos:* NSF fac develop grant, 79-80; vis prof physics, Ill Inst Technol, 79-80. *Mem:* Am Chem Soc; Royal

Soc Chem; fel Europ Molecular Biol Orgn; Royal Inst Chem; Am Soc Photobiol. *Res:* Photochemistry and high-energy radiation chemistry of organic and biological compounds; visual pigments. *Mailing Add:* Dept Chemistry Chicago State Univ Chicago IL 60628

SHERMAN, WAYNE BUSH, b Lena, Miss, Feb 25, 40; m 63; c 1. PLANT BREEDING, HORTICULTURE. *Educ:* Miss State Univ, BS, 61, MS, 63; Purdue Univ, PhD(hort), 66. *Prof Exp:* Asst prof hort, 66-72, assoc prof, 72-78, PROF HORT, UNIV FLA, 78- *Mem:* Am Soc Hort Sci; Am Pomol Soc. *Res:* Fruit crops. *Mailing Add:* Dept Hort Univ Fla Gainesville FL 32611

SHERMAN, WILLIAM REESE, b Seattle, Wash, Jan 18, 28; m 51; c 2. BIOCHEMISTRY, MASS SPECTROMETRY. *Educ:* Columbia Univ, AB, 51; Univ Ill, PhD(org chem), 55. *Prof Exp:* Sr res chemist, Abbott Labs, 55-59, group leader, 59-61; res asst, 61-63, res asst prof, 63-69, assoc prof, 69-75, PROF BIOCHEM, DEPT PSYCHIAT, SCH MED, WASHINGTON UNIV, 75-, PROF BIOL CHEM, DEPT BIOL CHEM & MOLECULAR BIOPHYS, 78- *Mem:* Am Chem Soc; Int Soc Neurochem; Am Soc Neurochem; Am Soc Mass Spectrometry; Fedn Am Socs Exp Biol. *Res:* Biochemistry of the inositols and the phosphoinositides with emphasis on the effects of lithium on the metabolism of these substances; biochemical applications of mass spectrometry. *Mailing Add:* Dept of Psychiat Washington Univ Sch of Med St Louis MO 63110

SHERMAN, ZACHARY, b New York, NY, Oct 26, 22; m 47; c 2. STRUCTURAL & MECHANICAL ENGINEERING, AEROSPACE ENGINEERING. *Educ:* City Col New York, BCE, 43; Polytech Inst Brooklyn, MCE, 53, PhD(mech, struct), 69; Stevens Inst Technol, MME, 68. *Prof Exp:* Stress analyst, Gen Dynamics, Calif & Tex, 43-45; sr stress analyst, Repub Aviation Corp, 45-47; struct designer, Cent RR of NJ, 48-49; struct engr, Parsons, Brinckerhoff, Hall & MacDonald, New York, 49-51; designer-in-chg, F L Ehasz, 51-52; engr-in-chg, Loewy-Hydropress Co, 52-54; from assoc prof to prof civil eng, Univ Miss, 54-59; prin engr, Repub Aviation Corp, 59-62; in-chg stress anal lab, Stevens Inst Technol, 62-67; lectr civil eng, City Col New York, 67-69; assoc prof aerospace eng, Pa State Univ, 69-73; CONSULT ENGR, 73-; DESIGNATED ENG REP, FED AVIATION ADMIN, 86- *Concurrent Pos:* Consult engr, S S Kenworthy Eng & Concrete Eng Co, Tenn, 54-58; independent consult engr, indust & ins, 58-; NSF int travel grant to Int Aeronaut Fedn Cong, 71; adj prof civil eng, Sch Eng, Cooper Union, 78-; adj prof math, Pace Univ, 78- *Mem:* Fel Am Soc Civil Engrs; Am Inst Aeronaut & Astronaut; Sigma Xi; NY Acad Sci. *Res:* Dynamics and vibrations; optimization; tornado resistant construction; design; elasticity. *Mailing Add:* 25 Neptune Blvd Apt 7H Long Beach NY 11561

SHEROCKMAN, ANDREW ANTOLCIK, inorganic chemistry; deceased, see previous edition for last biography

SHERR, ALLAN ELLIS, b Detroit, Mich, July 26, 26; m 55; c 3. ORGANIC CHEMISTRY, TOXICOLOGY. *Educ:* Wayne State Univ, BS, 48, MS, 51, PhD(chem), 56. *Prof Exp:* Chief chemist, Radioactive Prods, Inc, Mich, 51; res chemist, Conn, 55-57, tech rep new prod develop dept, NY, 57-58, res chemist, 58-60, group leader res div, Conn, 60-66, group leader, 67-77, Am Cyanamid Co; tech dir, Glendale Optical Co, 71-77; coordr toxic chem regist, Cent Res Div, 77-79, APPROVALS MGR, TOXICOL DEPT, CHEMICALS GROUP, AM CYANAMID CO, 79- *Concurrent Pos:* Mem comt safe use of lasers & masers, Am Nat Standards Inst; chief US del, Inst Standards Orgn Comt Eye Safety, 74-80; sci adv, Glendale Optical Co, 77- *Mem:* AAAS; Soc Plastics Eng; Sigma Xi; Indust Safety Equipment Asn; Am Chem Soc. *Res:* Synthetic organic chemistry; preparation and properties of polymers; additives to modify properties of polymers; personal safety equipment as respirators, safety spectacles, laser goggles, hearing protectors and head protection; optical plastics; flame retardance; toxicology; national and international health and safety regulations. *Mailing Add:* 9 Wychwood Way Warren NJ 07060

SHERR, BARRY FREDERICK, b New York, NY, Mar 9, 44; m 79; c 2. AQUATIC MICROBIAL ECOLOGY, PROTOZOOLOGY. *Educ:* Kans Wesleyan Univ, BA, 65; Univ Kans, MA, 68; Univ Ga, PhD(zool), 77. *Prof Exp:* Res assoc microbiol, Univ Ga, 77-79; res assoc aquatic ecol, Israel Oceanog & Limnol Res, Ltd, 79-81; res assoc, Univ Ga Marine Inst, 82-84, asst marine scientist, 84-87, assoc marine scientist microbiol ecol, 87-90; ASSOC PROF OCEANOG, ORE STATE UNIV, 90- *Concurrent Pos:* prin investr, NSF, 83-; vis scientist overseas labs. *Mem:* Am Soc Microbiol; Am Soc Limnol & Oceanog; Soc Protozoologists; Estuarine Res Fedn. *Res:* Flow of carbon-energy and nutrient cycling in marine pelagic food webs; roles of planktonic protozoa in facilitating these processes via their trophic interactions with components of the microbial community and with metazooplankton. *Mailing Add:* Col Oceanog Ore State Univ Ocean Admin Bldg 104 Corvallis OR 97331

SHERR, EVELYN BROWN, b Dublin, Ga, Dec 19, 46; m 79; c 2. MICROBIAL ECOLOGY, AQUATIC FOOD WEBS. *Educ:* Emory Univ, BS, 69; Duke Univ, PhD(zool), 74. *Prof Exp:* Res assoc microbiol, Univ Ga, 74-75, res assoc marine ecol, 75-77, asst marine scientist, 77-82, assoc marine scientist, Marine Inst, 82-90; ASSOC PROF, COL OCEANOG, ORE STATE UNIV, 90- *Concurrent Pos:* Prin investr NSF, 76-91; vis investr, Kinneret Limnological Lab, Israel, 79-81; ed adv, Marine Ecol Prog Ser, 86. *Mem:* Am Soc Limnol & Oceanog; Am Soc Microbiol; Soc Protozoologists; Estuarine Res Fedn. *Res:* Pathways of carbon flow in aquatic food webs; feeding and growth rates of pelagic ciliates and flagellates; ecological roles of heterotrophic protozoa. *Mailing Add:* Col Oceanog Ore State Univ Oceanog Admin Bldg 104 Corvallis OR 97331

SHERR, RUBBY, b Long Branch, NJ, Sept 14, 13; m 36; c 2. NUCLEAR PHYSICS. *Educ:* NY Univ, BA, 34; Princeton Univ, PhD(physics), 38. *Prof Exp:* Asst radioactivity, Harvard Univ, 38-39, instr nuclear physics, 39-42; staff mem radiation lab, Mass Inst Technol, 42-44; staff mem, Manhattan Proj, Los Alamos Sci Lab, Univ Calif, 44-46; from asst prof to prof, 46-82, EMER PROF PHYSICS, PRINCETON UNIV, 82- *Mem:* Fel Am Phys Soc; Sigma Xi. *Res:* Radar systems; radioactivity; nuclear structure. *Mailing Add:* Dept Physics Princeton Univ-Jadwin Hall Princeton NJ 08540

SHERR, STANLEY I, b Washington, DC, Oct 17, 34; m 63; c 3. BIOCHEMISTRY. *Educ:* George Washington Univ, BS, 57, MS, 58, PhD(biochem), 64. *Prof Exp:* Res assoc biochem, Harvard Univ, 63-65; asst prof, 65-71, ASSOC PROF BIOCHEM, UNIV MED & DENT NJ, 72- *Mem:* AAAS; Am Soc Microbiol. *Res:* Pulmonary fat embolism and its relation to trauma; lipid metabolism in bacteria and fungi; intestinal absorption and metabolism of lipids. *Mailing Add:* Dept Biochem Col Med & Dent NJ Med Sch 100 Bergen Newark NJ 07103

SHERRARD, JOSEPH HOLMES, b Waynesboro, Va, June 12, 42; m 64; c 2. ENVIRONMENTAL ENGINEERING. *Educ:* Va Mil Inst, BS, 64; Sacramento State Col, MS, 69; Univ Calif, Davis, PhD(civil eng), 71. *Prof Exp:* Jr civil engr, Calif Div Hwy, 64-65; Zurn Industs fel, Cornell Univ, 71-72; asst prof bioeng, Okla State Univ, 72-74; from assoc prof to prof civil eng, Va Polytech Inst & State Univ, 82-89; PROF & HEAD CIVIL ENG, MISS STATE UNIV, 89- *Honors & Awards:* Fulbright Lectr, Ecuador, 80, 88-89; Walter L Huber Res Prize, 87; Wesley W Horner Award, 90. *Mem:* Am Soc Civil Engrs; Am Water Works Asn; Water Pollution Control Fedn. *Res:* Biological and chemical wastewater treatment. *Mailing Add:* Dept Civil Eng Miss State Univ Starkville MS 39762

SHERREN, ANNE TERRY, b Atlanta, Ga, July 1, 36; m 66. ANALYTICAL CHEMISTRY. *Educ:* Agnes Scott Col, BA, 57; Univ Fla, PhD(chem), 61. *Prof Exp:* Instr chem, Tex Woman's Univ, 61-63, asst prof, 63-66; assoc prof, 66-76, chmn dept, 75-78, 81-84 & 88-90, chmn sci div, 83-87, PROF CHEM, NORTH CENT COL, ILL, 76- *Mem:* AAAS; Am Chem Soc; Am Inst Chem; Coblentz Soc. *Res:* Turbidity measurements; technetium chemistry; neutron activation analysis; electrochemistry; absorption spectrophotometry. *Mailing Add:* Dept Chem NCent Col Box 3063 Naperville IL 60566

SHERRER, ROBERT E(UGENE), b Abilene, Kans, Aug 20, 23; m 56; c 2. ENGINEERING MECHANICS. *Educ:* Univ Kans, BS, 48; Univ Wis, MS, 53, PhD(eng mech), 58. *Prof Exp:* Engr, Allis-Chalmers Mfg Co, 48-52; instr eng mech, Univ Wis, 52-57, asst prof, 57-60; assoc prof, 60-, EMER PROF MECH ENG, UNIV WASH. *Concurrent Pos:* Consult, Forest Prod Lab, Madison, Wis, 56-58 & Boeing Co, 60- *Mem:* Am Soc Eng Educ. *Res:* Fluid and solid mechanics; mechanical vibration; structural analysis; materials; dynamics. *Mailing Add:* Dept Mech Eng Univ Wash Seattle WA 98195

SHERRICK, CARL EDWIN, b Carnegie, Pa, Oct 28, 24; m 54; c 3. PSYCHOLOGY. *Educ:* Carnegie Inst Technol, BS, 48; Univ Va, MA, 50 & PhD(psychol), 52. *Prof Exp:* Asst prof psychol, Wash Univ, St Louis, 53-59; res assoc, Cent Inst for the Deaf, 59-61, Univ Va, 61-62; res psychologist, Princeton Univ, 62-70, SR RES PSYCHOLOGIST, 70- *Concurrent Pos:* Prin investr psychol, Princeton Univ, 72-; lectr psychol, Princeton Univ, 74-; mem, NANCDS Coun, NIH, 86-89, NADCD Coun, 89-91. *Mem:* Am Psychol Asn; Acoustical Soc Am; Am Pyschol Soc. *Res:* The capacity of the skin for processing information to identify the receptive systems; adaptation of devices that transform visual and auditory signals for the blind and or deaf. *Mailing Add:* Psychol Green Hall Princeton Univ Princeton NJ 08544-1010

SHERRICK, JOSEPH C, b Monmouth, Ill, June 4, 17; m 44, 84; c 3. PATHOLOGY. *Educ:* Monmouth Col, Ill, BA, 37; Harvard Med Sch, MD, 41. *Prof Exp:* Demonstr path, Sch Med, Western Reserve Univ, 43-44; fel, 51-52, instr, 53-54, assoc, 54-55, from asst prof to assoc prof, 56-67, PROF PATH, MED SCH, NORTHWESTERN UNIV, CHICAGO, 67-; asst prof, Univ Ill Col Med, 56-59. *Concurrent Pos:* Ill Div Am Cancer Soc grant, 63-67. *Mem:* AAAS; Asn Clin Sci (pres, 64); Col Am Path; Am Soc Clin Path(pres 86); Am Asn Path & Bact. *Res:* Cancer research; clinical pathology. *Mailing Add:* Dept Path Rm P-232 Northwestern Univ Med Sch 303 E Superior St Chicago IL 60611

SHERRILL, BETTE CECILE BENHAM, b Vernon, Tex, Aug 7, 44; m 63; c 2. BIOCHEMISTRY, SCIENCE ADMINISTRATION. *Educ:* NMex State Univ, BS, 66; Tex Christian Univ, PhD(phys chem), 73. *Prof Exp:* Asst scientist rocket telemetry, Phys Sci Lab, White Sands Missile Range, 62-66; res fel membrane transp, Dept Internal Med, Univ Tex Health Sci Ctr, 73-75, instr, 75-77, asst prof cholesterol metab & membrane transp, 77-81; ASST PROF LIPOPROTEIN METAB & MEMBRANE TRANSP, DEPTS MED & BIOCHEM, BAYLOR COL MED, 81-; EXEC DIR, INST BIOSCI & BIOENGR, RICE UNIV, 87- *Concurrent Pos:* Res asst chem, NMex State Univ, 66-67; res fel, Robert A Welch Res Found, 73 & NIH grant, 73-75; consult, Sherrill Eng Consults, Inc, Irving, Tex, 78-; consult & pres, Sherrill Environ Consult Inc, Houston, 81-; estab investr, Am Heart Asn, 81-84. *Mem:* Fel Am Heart Asn; Am Chem Soc; Am Asn Applied Sci; Sigma Xi; fel Am Inst Chemists. *Res:* Membrane transport; hepatic transport kinetics of intestinal and serum lipoproteins; thermodynamic properties of lipids; liquid diffusion studies. *Mailing Add:* PO Box 153568 Irving TX 75015-3568

SHERRILL, J(OSEPH) C(YRIL), b Philadelphia, Pa, Nov 18, 17; m 42; c 3. CHEMICAL ENGINEERING. *Educ:* Pa State Univ, BS, 41, MS, 47, PhD(chem), 52. *Prof Exp:* Asst petrol res, Pa State Univ, 41-47, instr textile chem, 47-52; prof detergency res & asst dean, Tex Woman's Univ, 52-57; sect head detergency eval, Armour & Co, 57-58, mgr soap res, 58-59, sales mgr pvt brand detergents, Armour Grocery Prod Co, 59-69; mkt mgr, Darrill Industs, Inc, 69-71; PRES, SHERRILL ASSOCS, INC, 71- *Mem:* Am Inst Chemists; Am Inst Chem Engrs. *Res:* Detergency evaluation; forces bonding soils to surfaces, especially fabric surfaces; ultrasonic methods for evaluation of detergency. *Mailing Add:* 2360 Maple Rd Homewood IL 60430

SHERRILL, MAX DOUGLAS, b Hickory, NC, Jan 2, 30; m 55; c 5. SOLID STATE PHYSICS. *Educ:* Univ NC, BS, 52, PhD(physics), 61. *Prof Exp:* Physicist, Gen Elec Res & Develop Ctr, 60-67; assoc prof physics, 67-75, PROF PHYSICS, CLEMSON UNIV, 75- *Mem:* Am Phys Soc. *Res:* Electrical and magnetic properties of metals, particularly superconductivity. *Mailing Add:* Dept Physics & Astron 201 Sikes Hall Clemson Univ Clemson SC 29631

SHERRILL, WILLIAM MANNING, b San Antonio, Tex, Feb 23, 36; m 58; c 3. RADIOPHYSICS. *Educ:* Univ Tex, BA & BS, 57; Rice Univ, MS, 59. *Prof Exp:* Anal engr, Pratt & Whitney Aircraft Div, United Aircraft Corp, Conn, 58-59; res engr, 59-63, sr res engr, 63-66, mgr intercept & direction finding res, 66-71, asst dir dept appl electromagnetics, 71-74, DIR DEPT RADIO LOCATION SCI, SOUTHWEST RES INST, 74- *Mem:* Inst Elec & Electronics Eng; Am Astron Soc; Sigma Xi; AAAS. *Res:* Ionospheric propagation and mode angular spectra; radio location research; high frequency radio direction finding; solar system radio astronomy. *Mailing Add:* Southwest Res Inst 6220 Culebra Rd San Antonio TX 78228

SHERRIS, JOHN C, b Colchester, Eng, Mar 8, 21; m 44; c 2. MICROBIOLOGY. *Educ:* Univ London, MRCS & LRCP, 44, MB & BS, 48, MD, 50; Am Bd Med Microbiol, cert, 61; Am Bd Path, cert, med microbiol, 66; FRCPath, 68. *Hon Degrees:* Dr med, Karolinska Inst, Sweden, 75. *Prof Exp:* House surgeon & physician med, King Edward VII Hosp, Windsor, Eng, 44-45; trainee path & microbiol, Stoke Mandeville Hosp, 45-48, sr registr, 48-50; sr registr, Radcliffe Infirmary, Oxford, 50-52; lectr bact, Univ Manchester, 53-56, sr lectr, 56-59; from assoc prof to prof, 59-86, chmn dept, 70-80, EMER PROF, SCH MED, UNIV WASH, 86- *Concurrent Pos:* Chmn, Am Bd Med Microbiol, 71-73; vchmn Am Acad Microbiol, 76-78. *Honors & Awards:* Becton- Dickinson Award, Am Soc Microbiol, 78. *Mem:* Fel Am Acad Microbiol; Am Soc Microbiol (pres 82-83). *Res:* Clinical microbiology; chemotherapy; pathogenesis of infection. *Mailing Add:* Dept Microbiol & Immunol Univ Wash SC-42 Seattle WA 98195

SHERRITT, GRANT WILSON, b Hunter, NDak, Mar 27, 23; m 52; c 3. ANIMAL SCIENCE. *Educ:* Iowa State Univ, BS, 48; Univ Ill, MS, 49; Pa State Univ, PhD(animal husb), 61. *Prof Exp:* Asst animal sci, Univ Ill, 48-49; instr animal husb, 49-61, from asst prof to assoc prof animal sci, Pa State Univ, 81-84; RETIRED. *Concurrent Pos:* Mem comt, Nat Swine Indust, 63. *Mem:* AAAS; Am Soc Animal Sci. *Res:* Crossbreeding and selection experiments in swine breeding; swine management studies, especially as related to the sow. *Mailing Add:* 131 E Lytle Ave State College PA 16801

SHERROD, LLOYD B, b Goodland, Kans, Mar 5, 31; m 63; c 2. ANIMAL NUTRITION. *Educ:* SDak State Univ, BS, 58; Univ Ark, MS, 60; Okla State Univ, PhD(animal nutrit), 64. *Prof Exp:* Asst animal scientist, Univ Hawaii, 64-67; assoc prof animal sci, Res Ctr, Tex Tech Univ, 67-73, prof, 73-79; prof chem, Frank Phillips Col, Borger, Tex, 79-88; RETIRED. *Mem:* AAAS; Am Inst Biol Sci; Am Soc Agron; Am Dairy Sci Asn; Am Soc Animal Sci. *Res:* Ruminant animal nutrition research with emphasis on factors influencing ration component digestibility, nutrient utilization and retention and nutrient requirements. *Mailing Add:* PO Box 1017 Panhandle TX 79068

SHERROD, THEODORE ROOSEVELT, b Ala, July 29, 15; m 41; c 1. PHARMACOLOGY. *Educ:* Talladega Col, AB, 38; Univ Chicago, MS, 41; Univ Ill, PhD(pharmacol), 45, MD, 49. *Prof Exp:* From instr to assoc prof, 45-58, PROF PHARMACOL, UNIV ILL COL MED, 58- *Mem:* Am Soc Pharmacol & Exp Therapeut; Soc Exp Biol & Med. *Res:* Cardiovascular and renal pharmacology. *Mailing Add:* 901 S Wolcott Ave Chicago IL 60612

SHERRY, ALLAN DEAN, b Viroqua, Wis, Oct 13, 45; m 82; c 2. BIOINORGANIC CHEMISTRY. *Educ:* Wis State Univ, LaCrosse, BS, 67; Kans State Univ, PhD(inorg chem), 71. *Prof Exp:* Fel bioinorg chem, NIH, 71-72; from asst prof to assoc prof, 72-82, PROF CHEM, UNIV TEX, DALLAS, 82-, HEAD, PROG CHEM, 79- *Concurrent Pos:* Sr fel, NIH, 83-84. *Mem:* Am Chem Soc. *Res:* Aqueous lanthanide chemistry; lanthanides as probes of calcium sites in proteins; fluorescence and nuclear magnetic resonance spectroscopy of metalloproteins; biological nuclear magnetic resonance spectroscopy. *Mailing Add:* Dept Chem Univ Tex PO Box 830688 Richardson TX 75083-0688

SHERRY, CLIFFORD JOSEPH, b Chicago, Ill, Jan 16, 43; m 69; c 3. NEUROPHYSIOLOGY, PSYCHOPHARMACOLOGY. *Educ:* Roosevelt Univ, BS, 68; Ill Inst Technol, MS, 74, PhD, 76. *Prof Exp:* Res assoc pharmacol, Med Sch, Univ Ill, 74-75; from instr to asst prof biol, Tex A&M Univ, 75-82; biofeedback therapist, Biofeedback & Stress Mgt Consult, 83-87; SR SCIENTIST, SYSTS RES LABS, 89- *Concurrent Pos:* Writer, Word Plus, 83- *Mem:* Neurosci Soc. *Res:* Neurophysiology, especially the relationship between single neuron activity and behavior; behavioral teratology; psychopharmacology, especially hallucinogens and drugs of abuse; psychobiology of reproductive and sexual behavior; biobehavioral effects of electromagnetic fields and potentials. *Mailing Add:* 1406 Shrivanek Dr Bryan TX 77802

SHERRY, EDWIN J, b Jamaica, NY, Nov 10, 33; m 65; c 2. MATHEMATICS. *Educ:* Fordham Univ, AB, 57, MA, 59 & 61; Yeshiva Univ, PhD(math), 64. *Prof Exp:* Instr math & physics, Brooklyn Prep Sch, 58-59; res asst, Univ NMex, 63-64; mem tech staff, Sandia Labs, 65-67; MEM TECH STAFF, JET PROPULSION LAB, CALIF INST TECHNOL, 67- *Concurrent Pos:* Vis lectr, Univ NMex, 65-67. *Mem:* Am Math Soc; NY Acad Sci. *Res:* Partial differential equations; operations research and analysis of the planetary quarantine problem; development of a computer model to simulate the microbial accumulation on a spacecraft during assembly; the problem of assaying low levels of microbial contamination. *Mailing Add:* Jet Propulsion Lab 4800 Oak Grove Pasadena CA 91103

SHERRY, HOWARD S, b New York, NY, Nov 18, 30; m 55; c 4. PHYSICAL INORGANIC CHEMISTRY, CHEMICAL ENGINEERING. *Educ:* NY Univ, BSChE, 57; State Univ NY Buffalo, MA, 62, PhD(phys chem), 63. *Prof Exp:* Res chem engr, Union Carbide Metals Co, 57-59; Union Carbide fel, 62-63; sr res chem appl res & develop div, Mobil Res & Develop Corp, 63-65; vis lectr chem, Univ Colo, 65-66; sr res chem cent res div, Mobil Res & Develop Corp, 66-69, res assoc, 69-71, res assoc appl res div, 71-77; tech mgr, Res & Develop Div, 77-80, gen mgr zeolites, 80-90, VPRES ZEOLITES & CATALYSTS, PQ CORP, 90- *Concurrent Pos:* Mem panel rare earths, Nat Acad Sci, 69. *Mem:* Am Chem Soc; Am Inst Chem Engrs; Mineral Soc Am; Catalysis Soc; Int Zeolite Asn; British Zeolite Asn. *Res:* Catalysis; ion exchange in zeolites; zeolite science and technology and catalysis; rare earth chemistry. *Mailing Add:* 416 S Cranford Rd Cherry Hill NJ 08003

SHERRY, JOHN M, b Munice, Ind, Oct 11, 13. METALLURGICAL ENGINEERING. *Educ:* Ball State Univ, BS, 35. *Prof Exp:* Asst chief spectrogr, Aluminum Res Lab, 41-44; founder & pres, John M Sherry Labs, 47-69; metallurgist, Ont Corp, 69-80; RETIRED. *Mem:* Fel Am Soc Metals. *Mailing Add:* 101 N Riley Rd Muncie IN 47304

SHERRY, PETER BURUM, physical chemistry; deceased, see previous edition for last biography

SHERRY, SOL, b New York, NY, Dec 8, 16; m 46; c 2. MEDICINE. *Educ:* NY Univ, BA, 35, MD, 39. *Hon Degrees:* DSc, Temple Univ, 80. *Prof Exp:* Fel med, Col Med, NY Univ, 39-41; intern & resident, Bellevue Hosp, 41-42 & 46; asst, NY Univ, 46-47, from instr to asst prof, 47-51; asst prof, Univ Cincinnati, 51-54; from assoc prof to prof med, Sch Med, Wash Univ, 54-68; prof internal med & chmn div med, 68-84, dir, Thrombosis Res Ctr, 71-80, dean & distinguished prof, 84-86, DISTINGUISHED EMER PROF, SCH MED, TEMPLE UNIV, 87- *Concurrent Pos:* Dir, May Inst Med Res, Cincinnati, Ohio, 51-54 & med serv, Jewish Hosp, St Louis, 54-58; chmn coun thrombosis, Am Heart Asn & Int Soc Cardiol; chmn subcomt clin invest, Int Comt Haemostasis & Thrombosis; former chmn thrombosis, thrombolytic agents & gen clin res, NIH & task force on thrombosis, Nat Acad Sci-Nat Res Coun; selections comt for career res, Vet Admin; emer consult, US Army & Develop Command; NIH career res award. *Honors & Awards:* Robert Grant Medal, Inst Soc Thrombosis & Haemostasis, Gold Medal Award; John Phillips Medal, Am Col Physicians; Founders Med, Am Col Cardiologist; Sci Achievement Award, Am Heart Asn. *Mem:* Am Soc Clin Invest; Asn Am Physicians; Int Soc Cardiol; Int Soc Thrombosis & Haemostasis (pres); Asn Prof Med (pres); Am Physiol Soc. *Res:* Thrombosis research; fibrinolysis, coagulation and hemostasis. *Mailing Add:* Fourth Floor FSB Temple Univ Sch Med Philadelphia PA 19140

SHERSHIN, ANTHONY CONNORS, b Clifton, NJ, Oct 16, 39; m 68; c 3. OPERATIONS RESEARCH, MATHEMATICS. *Educ:* Georgetown Univ, AB, 61; Univ Fla, MS, 63, PhD(math), 67; Fla Int Univ, MSM, 84. *Prof Exp:* Opers res analytical autonetics div, NAm Rockwell Corp, 63-64; asst prof math, Univ SFla, 67-72; asst prof, 72-74, ASSOC PROF MATH, FLA INT UNIV, 74- *Mem:* Am Math Asn. *Res:* Graph theory; networks; finance & economics. *Mailing Add:* Dept Math Fla Int Univ Miami FL 33199

SHERTZER, HOWARD GRANT, b New York, NY, Oct 9, 45; m 68; c 2. TOXICOLOGY, CANCER RESEARCH. *Educ:* Univ Mich, Ann Arbor, BS, 67; Univ Calif, Los Angeles, PhD(cell biol), 73. *Prof Exp:* Fel biochem, Cornell Univ, 73-75; asst prof cell biol, Tex A&M Univ, 75-79; ASSOC PROF TOXICOL, MED CTR, UNIV CINCINNATI, 79- *Concurrent Pos:* Vis prof, Cornell Univ, 76-; prin investr, Nat Inst Environ Health Sci, 79- & Nat Cancer Inst, 84-; vis prof, Karolinska Inst, Stockholm, Sweden, 89. *Mem:* Am Soc Pharmacol & Exp Therapeut; Int Soc Study Xenobiotics; Soc Toxicol. *Res:* Phenomena and mechanisms associated with chemical protection from environmental toxins, mutagens and carcinogens; chemoprotective dietary constituents, in particular, indole compounds and derivatives. *Mailing Add:* Kettering Lab Univ Cincinnati Med Ctr Cincinnati OH 45267-0056

SHERVAIS, JOHN WALTER, b Philadelphia, Pa, Mar 9, 48; m; c 2. GEOCHEMISTRY, VOLCANOLOGY. *Educ:* San Jose State Univ, Calif, BSc, 71; Univ Calif, Santa Barbara, PhD(petrol), 79. *Prof Exp:* Fac assoc petrol & mineral, Univ Calif, Santa Barbara, 75-79, lectr petrol, 81-82; NATO postdoctoral res, Confederate Tech Acad, Zurich, 79-80; res assoc lunar petrol, Univ Tenn, Knoxville, 82-84; asst prof, 84-88, ASSOC PROF IGNEOUS PETROL, UNIV SC, 88- *Concurrent Pos:* Regent's fel, Univ Calif, 76-77; prin investr, NASA, 85- & NSF, 85-; assoc ed, Geol Soc Am Bull, 89- *Mem:* Geol Soc Am; Am Geophys Union. *Res:* Petrology and geochemistry of Earth's mantle; evolution of basic and intermediate magma systems; origin and significance of ophiolites, ocean island basalts, arc volcanism, flood basalts, and lunar mare volcanism. *Mailing Add:* Dept Geol Univ SC Columbia SC 29208

SHERWIN, ALLAN LEONARD, b Montreal, Jan 29, 32; m 64; c 1. NEUROLOGY, IMMUNOCHEMISTRY. *Educ:* McGill Univ, BSc, 53, MD & CM, 57, PhD(immunochem), 65; FRCPS(C), 64. *Prof Exp:* Life Ins Med Res Fund fel, 60-61, lectr neurol, 62-65, asst prof neurol & asst neurologist, 65-71, ASSOC PROF NEUROL & NEUROLOGIST, McGILL UNIV, 72- *Concurrent Pos:* Markle scholar acad med, 62-67. *Mem:* Am Acad Neurol; Can Neurol Soc; Can Med Asn. *Res:* Clinical neuropharmacology; antiepileptic drugs. *Mailing Add:* Dept Neurol Montreal Neurol Hosp 3801 University St Montreal PQ H3A 2B4 Can

SHERWIN, MARTIN BARRY, b New York, NY, July 27, 38; m 64; c 3. GENERAL CHEMISTRY, RESEARCH ADMINISTRATION. *Educ:* City Col New York, BChE, 60; Polytech Inst Brooklyn, MS, 63; City Univ New York, PhD(chem eng), 67. *Prof Exp:* Jr process engr, Sci Design Co, 60-62; process develop engr, Halcon Int, Inc, 62-64; lectr chem eng, City Col New York, 64-66; staff engr, Chem Systs, Inc, 66-68, mgr process develop, 68-69, dir res & develop, 69-71, vpres res & develop, 71-78, managing dir, Chem Systs Int, Ltd, 79-80; vpres eng res, 80-87, EXEC VPRES, W R GRACE &

CO, 87- , GEN MGR, GRACE MEMBRANE SYSTS, 87- *Mem:* Fel Am Inst Chem Engrs; Am Chem Soc. *Res:* Development of new and improved processes and products; chemical engineering. *Mailing Add:* W R Grace & Co 7379 Rt 32 Columbia MD 21044

SHERWIN, RUSSELL P, b New London, Conn, Mar 11, 24; m 48; c 3. PATHOLOGY. *Educ:* Boston Univ, MD, 48. *Prof Exp:* Instr path, Georgetown Univ, 49-50; from instr to asst prof, Boston Univ, 53-62; asst prof, 62-63, Hastings assoc prof, 63-69, HASTINGS PROF PATH, SCH MED, UNIV SOUTHERN CALIF, 69- *Concurrent Pos:* Asst, Harvard Univ, 56-62; mem path ref panel asbestosis & neoplasia, Int Union Against Cancer, 65-; sci consult oncol, Vet Admin, Washington, DC, 71-74; mem subcomt exp biol, Nat Cancer Inst, 72-75; sci consult, Am Cancer Soc; mem res grants, Am Lung Asn, Air Qual Health Adv Comt State Calif. *Mem:* Am Soc Exp Path; Am Asn Cancer Res; Int Acad Path; AAAS; Sigma Xi. *Res:* Air quality standards and pollutant effects; cancer, specifically breast, lung and adrenal gland cancer; diseases of the lung, especially emphysema and fibrosis; histoculture; pathobiology; tissue culture; electromicroscopy; cinemicrography; histochemistry. *Mailing Add:* 2025 Zonal Ave Los Angeles CA 90033

SHERWOOD, A GILBERT, b Lloydminster, Sask, June 17, 30; m 53; c 2. PHYSICAL CHEMISTRY. *Educ:* Univ Man, BSc, 53, MSc, 58; Univ Alta, PhD(photochem), 64. *Prof Exp:* Fel, Univ Alta, 64-65; Nat Res Coun fel, 65-66; asst prof, 66-77, ASSOC PROF CHEM, SIMON FRASER UNIV, 77- *Mem:* Chem Inst Can. *Res:* Gas phase photochemistry; free radical kinetics; photoelectrochemistry; electrochemical storage of solar energy. *Mailing Add:* Dept of Chem Simon Fraser Univ Burnaby BC V5A 1S6 Can

SHERWOOD, ALBERT E(DWARD), b New Haven, Conn, Sept 19, 30; m 53; c 2. MICROCOMPUTER APPLICATIONS. *Educ:* Mass Inst Technol, SB & SM, 57; Univ Calif, Berkeley, PhD(chem eng), 64. *Prof Exp:* Chem engr, Esso Res & Eng Co, 57-58; res assoc molecular physics, Univ Md, 64-65; chem engr, Lawrence Livermore Nat Lab, Univ Calif, 65-85; CONSULT, 85- *Mem:* Am Chem Soc; Am Nuclear Soc; Am Inst Chem Engrs; Sigma Xi. *Res:* Energy and mass transport; industrial applications of nuclear explosions; equations of state; mathematical modeling of in-situ coal gasification; tritium technology; radioactive gas containment and capture systems. *Mailing Add:* 3060 Miranda Ave Alamo CA 94507

SHERWOOD, ARTHUR ROBERT, b Berkeley, Calif, Sept 6, 36; m 84; c 2. PLASMA PHYSICS. *Educ:* Pomona Col, BA, 58; Univ Calif, Berkeley, PhD(physics), 67. *Prof Exp:* Physicist, Lawrence Radiation Lab, 67, STAFF MEM PHYSICS, LOS ALAMOS SCI LAB, UNIV CALIF, 67- *Honors & Awards:* Fel, Am Phys Soc. *Mem:* Am Phys Soc. *Res:* Experimental plasma physics, especially as related to the controlled thermonuclear reactor program. *Mailing Add:* 4332 Sycamore St Los Alamos NM 87544

SHERWOOD, BRUCE ARNE, b Laporte, Ind, Dec 12, 38; m 59; c 3. PHYSICS, SYSTEMS DESIGN & SYSTEMS SCIENCE. *Educ:* Purdue Univ, BS, 60; Univ Chicago, MS, 63, PhD(physics), 67. *Prof Exp:* Asst prof physics, Calif Inst Technol, 66-69; from asst prof to assoc prof, Comput Based Educ Res Lab & Dept Physics, Univ Ill, Urbana-Champaign, 69-83, assoc prof, Dept Ling, 82-83; ASSOC DIR, CTR DESIGN EDUC COMPUT & INFO TECH CTR, CARNEGIE MELLON UNIV, 85-, PROF, DEPT PHYSICS, 85- *Concurrent Pos:* Secy, Comn Int Confer Univ, Universal Esperanto Asn, Rotterdam, Neth, 80- *Mem:* Am Phys Soc; Am Asn Physics Teachers; AAAS; Ling Soc Am; Esperanto Studies Asn Am; Int Acad Esperanto; Fel Asn Develop Comput-Based Instruct, 87. *Res:* Design and development of computer-based education systems; computer-based physics teaching. *Mailing Add:* Dept Physics Carnegie Mellon Univ Pittsburgh PA 15213

SHERWOOD, JESSE EUGENE, b Pittsburgh, Pa, Dec 27, 22; div; c 1. ATOMIC PHYSICS. *Educ:* Univ Pittsburgh, BS, 43; Univ Md, MS, 52; Univ Ky, PhD(physics), 73. *Prof Exp:* Physicist, Nat Bur Standards, 49-53, Oak Ridge Nat Lab, 53-61, Lawrence Radiation LAB, 61-64; asst prof physics, Univ Miss, 64-66; staff physicist, Oak Ridge Tech Enterprises, 66-69; lectr physic, Univ Ky, 69-73; sr res officer, Nuclear Physics Res Unit Res Unit, Univ of Witwatersrand, Johannesburg, SAfrica, 73; vis asst prof, NC State Univ, Raleigh, 75-76; from asst prof to assoc prof physics, Valdosta State Col, 76-80; ASSOC PROF PHYSICS, MIDWESTERN STATE UNIV, 80- *Concurrent Pos:* Assoc prof, Univ Tenn, Martin, 82-88. *Mem:* Am Phys Soc. *Res:* Measurement of nuclear moments; magnetic behavior of the neutron; polarized ion sources; polarization of photoneutrons; measurement of fast neutron polarization from reactions; production and properties of negative ions. *Mailing Add:* 4126 NW 43rd Terr Gainesville FL 32606

SHERWOOD, JOHN L, b Shreveport, La, Feb 24, 52; m 74. PLANT VIROLOGY. *Educ:* Col William & Mary, BS, 74; Univ Md, MS, 77; Univ Wis-Madison, PhD(plant path), 81. *Prof Exp:* Res assoc, 81-82, asst prof, 82-87, ASSOC PROF PLANT PATH, OKLA STATE UNIV, 87- *Mem:* Am Phytopath Soc; Sigma Xi. *Res:* Virus diseases of cereals, peanut and vegetables; use of monoclonal antibodies to address agricultural problems; mechanisms of viral cross-protection. *Mailing Add:* 715 Dryden Circle Stillwater OK 74074

SHERWOOD, LOUIS MAIER, b New York, NY, Mar 1, 37; m 66; c 2. ENDOCRINOLOGY, METABOLISM. *Educ:* Johns Hopkins Univ, AB, 57; Columbia Univ, MD, 61. *Prof Exp:* NIH trainee endocrinol & metab, Col Physicians & Surgeons, Columbia Univ, 66-68; assoc med, Harvard Med Sch, 68-69, from asst prof to assoc prof, 69-72; prof med, Div Biol Sci Pritzker Sch Med, Univ Chicago, 72-; physician-in-chief & chmn dept med, Michael Reese Hosp, 72-; AT DEPT MED, YESHIVA UNIV, NEW YORK. *Concurrent Pos:* Chief endocrine unit, Beth Israel Hosp, Boston, Mass, 68-72, assoc physician, 71-72; attend physician, West Roxbury Vet Admin Hosp, 71-72; trustee, Michael Reese Hosp & Med Ctr, 74-77; mem gen med B study sect, NIH, 75-79; Macy Found fel, vis scientist, Weizmann Inst, Israel, 78-79. *Mem:* Endocrine Soc; Am Soc Biol Chem; Am Fedn Clin Res; Cent Soc Clin Res; Am Soc Clin Invest. *Res:* Protein chemistry, correlations between structure and function; polypeptide hormones-parathyroid hormones; human placental lactogen; correlations of structure-function and factors regulating synthesis and secretion. *Mailing Add:* Dept Med Albert Einstein Col Med 1300 Morris Park New York NY 10461

SHERWOOD, ROBERT TINSLEY, b West Orange, NJ, Feb 21, 29; m 55; c 3. PLANT PHYSIOLOGY. *Educ:* Cornell Univ, BS, 52, MS, 54; Univ Wis, PhD, 58. *Prof Exp:* Res asst prof plant path, NC State Univ, 58-69, prof, 69-71; PLANT PATHOLOGIST, USDA, 58- *Concurrent Pos:* Adj prof, Pa State Univ, 71- *Mem:* Am Phytopath Soc; Am Soc Agron. *Res:* Physiology of parasitism; forage crop disease; fungal diseases of plants; regulation of apomictic seed formation. *Mailing Add:* US Pasture Res Lab USDA Pa State Univ University Park PA 16802

SHERWOOD, WILLIAM CULLEN, b Washington, DC, Feb 8, 32; m 58; c 3. GEOCHEMISTRY. *Educ:* Univ Va, BA, 54, MA, 58; Lehigh Univ, PhD(geol), 61. *Prof Exp:* Mat res analyst, Va Hwy Res Coun, 61-67; asst prof geol, Univ Va, 67-70, asst prof environ sci, 70-72; assoc prof geol, 72-73, PROF GEOL, JAMES MADISON UNIV, VA, 73- *Concurrent Pos:* Mem subcomt on fundamentals of binder-aggregate adhesion, Hwy Res Bd, Nat Acad Sci-Nat Res Coun, 65-67; lectr, Univ Va, 66-67; fac consult, Va Hwy Res Coun. *Mem:* Geochem Soc; Geol Soc Am; Nat Asn Geol Teachers; Am Inst Prof Geologists. *Res:* Erosion and sedimentation control; land use planning; geochemistry of weathering and soils; chemistry of natural waters; environmental geology. *Mailing Add:* 120 Ott St Harrisonburg VA 22801

SHERWOOD-PIKE, MARTHA ALLEN, b Eugene, Ore, Nov 8, 48; m 83; c 1. PALEOMYCOLOGY, MYCOLOGY. *Educ:* Univ Ore, BA, 70; Cornell Univ, PhD(mycol), 77. *Prof Exp:* Cryptogamic botanist, Farlow Herbarium, Harvard Univ, 77-79; res asst mycol, Commonwealth Mycol Inst, Kew, 79-80; RES ASSOC PALEONT, GEOL DEPT, UNIV ORE, 81- *Mem:* Mycol Soc Am. *Res:* Taxonomy and paleoecology of fossil fungi; morphology and taxonomy of living Ascomycetes including lichens; monographic studies in the Phacidiales and Ostropales. *Mailing Add:* 38563 Wendling Rd Marcola OR 97454

SHERYLL, RICHARD PERRY, b Pa, June 18, 56. OCEANOGRAPHIC INSTRUMENTATION, BIOLOGICAL OCEANOGRAPHIC. *Educ:* State Univ NY, BS, 82. *Prof Exp:* Res asst mat analysis, Metrop Mus Art, 81-84; RES & DEVELOP ENGR ELECTRONICS, CYCLOPS RES & DEVELOP, 84- *Concurrent Pos:* Res scientist, City Col, 82-84; res & develop engr, World Wide Wind, 84-85, Seaspec Assocs, 85-86 & VIAC Inc, 88-90. *Mem:* AAAS; NY Acad Sci. *Res:* Pure and applied research as it pertains to oceanographic and marine studies; design and development of instrumentation for such studies; author of several publications; granted various patents. *Mailing Add:* Cyclops Res & Develop 340 W 87th St 3A New York NY 10024

SHESHTAWY, ADEL A, b Zefta, Egypt, Feb 1, 40; US citizen; m 69; c 2. PETROLEUM ENGINEERING, EARTH SCIENCE. *Educ:* Cairo Univ, BS, 63; NMex Inst Mining & Technol, MSc, 74. *Prof Exp:* Resident drilling engr, Co Oriental Des Petrol Egypt, 63-66; petrol eng supvr contractor, Phillips Petrol Co, Egypt, 66-68; dist mgr, Sontrach, Algeria, 68-72; drilling supvr petrol eng, Mobil Explor & Producing Serv, 74; sr engr, Amoco Int Oil Co, 74-75; asst prof, NMex Inst Mining & Technol, 75-76; ASSOC PROF PETROL ENG, OKLA UNIV, 76-; PRES, INT PETROL ENG CORP, 78- *Mem:* Soc Petrol Eng; Am Petrol Inst; Int Asn Oil Well Drilling Contractors. *Res:* Deep oil and gas well drilling optimization; offshore operations; computer application in oil well drilling. *Mailing Add:* 2814 Walnut Rd Norman OK 73072

SHESKIN, THEODORE JEROME, b New York, NY, June 11, 40. NUMERICAL SOLUTION MARKOV CHAINS, MATRIX INVERSION. *Educ:* Mass Inst Technol, BS, 62; Syracuse Univ, MS, 65; Pa State Univ, PhD(indust eng), 74. *Prof Exp:* Test equip engr, Int Bus Mach Corp, 62-64; logic design engr, Burroughs Corp, 63-68; digital systs engr, Digital Info Devices, 68-71; from asst prof to assoc prof, 74-88, PROF INDUST ENG, CLEVELAND STATE UNIV, 88- *Concurrent Pos:* Summer fac fel, Nat Aeronaut & Space Admin, 79-91. *Mem:* Sr mem Inst Indust Engrs; Am Soc Eng Educ; Am Asn Univ Professors. *Res:* Systems analysis of space stations and satellites; development of new methodologies for the numerical solution of markov chains and the inversion of matrices. *Mailing Add:* Indust Eng Dept Cleveland State Univ Cleveland OH 44115

SHESTAKOV, ALEKSEI ILYICH, b Yugoslavia, Feb 8, 49; US citizen. NUMERICAL ANALYSIS, COMPUTATIONAL PHYSICS. *Educ:* Univ Calif, Berkeley, BS, 70, MA, 73, PhD(appl math), 75. *Prof Exp:* PHYSICIST MAGNETIC FUSION ENERGY, LAWRENCE LIVERMORE NAT LAB, UNIV CALIF, 76- *Mem:* Am Math Soc; Am Phys Soc. *Res:* Developing numerical models for use in problems arising in magnetic fusion energy; calculations; solution of linear systems. *Mailing Add:* L-561 Lawrence Livermore Nat Lab PO Box 5509 L-471 Livermore CA 94550

SHETH, ATUL C, b Bombay, India, Dec 2, 41; US citizen; m 65; c 2. ENERGY CONVERSION, POLLUTION CONTROL. *Educ:* Univ Bombay, India, BChemEng, 64; Northwestern Univ, MS, 69, PhD(chem eng), 73. *Prof Exp:* Shift supvr, Esso Standard Eastern Inc, 64-67; chem operator, Riker Labs, 67; process engr, Armour Indust Chem Co, 69; clerk, Charlotte Charles Inc, 71-72; fel chem eng, Argonne Nat Lab, 72-74, chem engr, 74-80; chem engr, Exxon Res & Eng Co, 80-84; assoc prof & mgr anal & eng lab serv group, 84-90, ASSOC PROF & MGR POLLUTION CONTROL & ENVIRON SECT, DEPT CHEM ENG, SPACE INST, UNIV TENN, TULLAHOMA, 90- *Concurrent Pos:* Tech leader seed reprocessing team, Dept Energy & NASA, 78-80; Murphy fel, 67-71. *Honors & Awards:* Anil P Desai Prize, 60. *Mem:* Am Inst Chem Engrs; Sigma Xi; Air & Waste Mgt Asn. *Res:* Alternate energy fields such as nuclear reactors, high-temp electric batteries, magnetohydrodynamics and coal liquefaction;

catalytic coal gasification process with Exxon; seed and sorbent regeneration; pollution control & trace element study; hazardous waste management. *Mailing Add:* Dept Chem Eng Space Inst Univ Tenn Tullahoma TN 37388-8897

SHETH, BHOGILAL, b Bombay, India, Sept 18, 31; US citizen. PHARMACEUTICAL TECHNOLOGY. *Educ:* Gujerat Univ, India, BPharm, 52; Univ Mich, MS, 55, PhD(pharmaceut chem), 61. *Prof Exp:* Sect head prod develop, Alcon Labs, 60-64; group leader pharm res, Warner-Lambert Co, 64-68; assoc prof, 68-74, prof pharmaceut, Col Pharm, Univ Tenn, 74-78; area dir, Vicks Health Care Div Res & Develop, 78-87; PROF PHARMACEUT, COL PHARM, UNIV TENN, 87- *Concurrent Pos:* Consult, var pharmaceut cos, 68-78, 87- *Mem:* Am Pharmaceut Asn; Acad Pharmaceut Sci; Am Soc Hosp Pharmacists; Am Asn Pharm Sci. *Res:* Pharmaceutical research and development; surface chemistry applications to dosage forms; rheology; drug dissolutions; product formulations. *Mailing Add:* Dept Pharmaceut Col Pharm Univ Tenn Memphis TN 38163

SHETH, KETANKUMAR K, b Gondal, Gujarat, India, Jan 21, 59; m. FLUID DYNAMICS, TURBO MACHINERY. *Educ:* Sardar Patel Univ, India, BE, 80; Tex A&M Univ, MS, 85. *Prof Exp:* Maintenance engr, Tata Chemicals Ltd, India, 80-82; res assoc, Tex A&M Univ, 83-85; sr assoc engr, 85-87, staff engr, 87-89, ENG SUPVR, CENTRILIFT, 89- *Mem:* Am Soc Mech Engrs; Soc Petrol Engrs. *Res:* Design, prototype testing, development of centrifugal pumps; plant maintenance. *Mailing Add:* 8810 S 73rd E Ave Tulsa OK 74133

SHETLAR, DAVID JOHN, b Columbus, Ohio, June 30, 46; m 68; c 1. ENTOMOLOGY, LANDSCAPE PEST MANAGEMENT. *Educ:* Univ Okla, BS, 69, MS, 75; Pa State Univ, PhD(entomol), 77. *Prof Exp:* Instr & cur, 75-76, asst prof teaching & res, Dept Entomol, Pa State Univ, 77-83; res scientist, Chemlawn Servs Res & Develop, Delaware, Ohio, 84-90; ASST PROF EXTEN & RES, DEPT ENTOMOL, OHIO STATE UNIV, 90- *Mem:* Sigma Xi; Entomol Soc Am. *Res:* Pest management in ornamental plants, turf and Christmas trees. *Mailing Add:* Dept Entomol Ohio State Univ 1991 Kenny Rd Columbus OH 43210-1090

SHETLAR, MARTIN DAVID, b Wichita, Kans, Aug 28, 38; m 66; c 1. PHOTOCHEMISTRY, PHOTOBIOLOGY. *Educ:* Kans State Univ, BS, 60; Univ Calif, Berkeley, PhD(chem), 65. *Prof Exp:* Biophysicist, Donner Lab, Univ Calif, Berkeley, 65-68; from asst prof to assoc prof, 68-84, PROF CHEM & PHARMACEUT CHEM, SCH PHARM, UNIV CALIF, SAN FRANCISCO, 84- *Concurrent Pos:* AEC fel, 66-68. *Mem:* AAAS; Am Chem Soc; Am Soc Photobiol; Int-Am Photochem Soc; Sigma Xi; European Soc Photobiol. *Res:* Photochemistry and free radical chemistry in nucleic acid-protein systems; organic photochemistry; photochemical kinetics. *Mailing Add:* Sch Pharm Univ Calif San Francisco CA 94143

SHETLAR, MARVIN ROY, b Bayard, Kans, Apr 3, 18; m 40; c 3. NUTRITION. *Educ:* Kans State Univ, BS, 40; Ohio State Univ, MS, 43, PhD(biol chem), 46. *Prof Exp:* Asst milling indust, Kans State Univ, 38-40; asst, Ohio State Univ, 41-42, instr agr chem, 44-46; Am Cancer Soc fel, Sch Med, Univ Okla, 46-47, res assoc, 47-50, from asst prof to prof biochem, 53-64, chmn dept, 64, res prof, 64-66; prof, Univ Tex Med Br, Galveston, 66-73; assoc chmn dept biochem, 72-75, PROF BIOCHEM & DERMAT, HEALTH SCI CTR, TEX TECH UNIV, 77-,. *Concurrent Pos:* NIH sr res fel, 57-62; vis prof, Univ Calif-Univ Airlangga Proj Med Educ Indonesia, 64-66 & Col Med, Univ King Faisal, Dammam, Saudi Arabia, 79-81; mem, Am Bd Clin Chem; adj prof nutrit, Tex Tech Univ, 85- *Mem:* Am Chem Soc; Am Inst Chem; Soc Exp Biol & Med; Am Asn Biol Chem; Am Asn Clin Chem; Sigma Xi. *Res:* Animal glycoproteins and mucopolysaccharides; wound healing; fetal alcohol syndrome. *Mailing Add:* Dept Nutrit Tex Tech Univ Lubbock TX 79409-1162

SHETLER, ANTOINETTE (TONI), b Ont; US citizen; c 3. COMPUTER PROGRAMMING TOOLS, COMPUTER SYSTEM PERFORMANCE. *Educ:* Univ Wash, BA, 62. *Prof Exp:* Sci programmer, Boeing Co, 65-66; systs programmer, Pac Northwest Bell, 66-68; systs programmer, Rand Corp, 68-71, assoc dept head, 73-76; sr prog staff mem, Xerox Data Systs, 71-73, computer scientist, Star Proj, 76-82; staff engr, 82-85, ENG MGR, SYSTS DIV, TRW, 86- *Concurrent Pos:* Dep chair conf bd, Asn Comput Mach, 80-86, mem, Facil Planning Comt, 88-90. *Mem:* Asn Comput Mach; Inst Elec & Electronics Engrs. *Res:* Operating systems and computer programming tools; analysis and improvement of the software engineering process; computer system performance; development of technical conferences. *Mailing Add:* Systs Div TRW FVA6-3444 PO Box 10400 Fairfax VA 22031

SHETLER, STANWYN GERALD, b Johnstown, Pa, Oct 11, 33; m 63; c 2. PLANT TAXONOMY. *Educ:* Cornell Univ, BS, 55, MS, 58; Univ Mich, PhD, 79. *Prof Exp:* Asst cur, 62-63, assoc cur, 63-81, dir, Flora NAm Prog, 71-80, asst dir progs, 84-86, CUR, DEPT BOT, SMITHSONIAN INST, NAT MUS NAT HIST, 81-, ACTG DEP DIR, 86- *Concurrent Pos:* Pres, Audubon Naturalist Soc Cent Atlantic States, Inc, 74-77; mem bd dirs, Piedmont Environ Coun, Va, 84-86; mem of nat comt, Man & Biosphere Prog, 88- *Mem:* AAAS; Am Soc Plant Taxon; Bot Soc Am; Am Inst Biol Sci; Arctic Inst NAm. *Res:* Taxonomy and ecology of Campanula; flora and vegetation of the Arctic, especially Alaska; history of Russian botany; biological conservation. *Mailing Add:* Dept Bot Smithsonian Inst Nat Mus Nat Hist 166 Washington DC 20560

SHETTERLY, DONIVAN MAX, b Des Moines, Iowa, Oct 29, 46; m 66; c 2. HEAT TRANSFER, GLASS SCIENCE. *Educ:* Cent Col, BA, 68; Purdue Univ, MS, 70. *Prof Exp:* Assoc physicist, Owens-Illinois Inc, Toledo, 70-74, physicist, 74-80, sr physicist, glass technol sect, 80-85; eng mgr, Warren Tech Assocs, 85- 86; PHYSICIST, GLASS TECH INC, 86- *Mem:* Am Ceramic Soc; Sigma Xi; Soc Mfg Engrs. *Res:* Heat transfer; glass forming; phase separation in glass; glass homogeneity and process control. *Mailing Add:* 3033 Gallatin Rd Toledo OH 43606

SHETTLE, ERIC PAYSON, b New York, NY, Nov 23, 43; m 65; c 2. ATMOSPHERIC PHYSICS, ATMOSPHERIC AEROSOLS. *Educ:* Johns Hopkins Univ, BA, 65; Univ Wis-Madison, MA, 67. *Prof Exp:* Teaching & res asst, dept physics, Univ Wis, 66-68, res asst, Dept Meteorol, 68-72; instr & res assoc, Dept Physics, Univ Fla, 72-73; PHYSICIST, AIR FORCE GEOPHYS LAB, 73- *Mem:* Optical Soc Am; Am Meteorol Soc; Am Geophys Union; AAAS. *Res:* The optical and infrared properties of the atmosphere, especially the atmospheric aerosols and their effects on radiative transfer of light in scattering atmospheres; atmospheric remote sensing, lidor applications. *Mailing Add:* 5504 Uppingham St Chevy Chase MD 20815

SHEVACH, ETHAN MENAHEM, b Brookline, Mass, Oct 16, 43; m 67; c 2. CELLULAR IMMUNOLOGY, IMMUNOGENETICS. *Educ:* Boston Univ, AB & MD, 67. *Prof Exp:* Intern med, Bronx Munic Hosp Ctr, NY, 67-68, asst resident, 68-69; clin assoc immunol, lab clin invest, 69-71, sr staff fel, 71-72, lab immunol, 72-73, sr investr immunol, Lab Immunol, Nat Inst Allergy & Infectious Dis, NIH, 73-87, HEAD, CELLULAR IMMUNOL SECT, LAB IMMUNOL, BETHESDA, MD, 87- *Concurrent Pos:* Ed-in-chief, J Immunol, 87- *Mem:* Am Asn Immunologists; Am Soc Clin Invest; Am Fedn Clin Res; Asn Am Physicians; Coun Biol Ed. *Res:* Immunogenetics, basic mechanisms that control immunocompetent cell interactions; mechanisms of lymphocyte activation, cell surface antigens involved in T cell triggering and growth factor receptors. *Mailing Add:* Lab Immunol Nat Inst Allergy & Infect Dis NIH Bldg 10 Rm 11N315 Bethesda MD 20892

SHEVACK, HILDA N, b Brooklyn, NY, Apr 12, 34; div. MECHANICAL ENGINEERING. *Educ:* City Col NY, BSME, 59. *Prof Exp:* Asst to chief engr, Starret TV Corp, NY, 50-52; INTEGRATED LOGISTICS SUPPORT MGR, GOVT SYSTS DIV, GEN INSTRUMENT CORP, HICKSVILLE, NY, 52- *Concurrent Pos:* Dir & chmn, Publ Comts, Soc Logistics Engrs, 68- *Mem:* Fel Soc Logistics Engrs; Inst Elec & Electronics Engrs; Am soc Mech Engrs; Am Inst Indust Engrs; Am Defense Preparedness Asn; Soc Women Engrs. *Mailing Add:* 8672 18th Ave Brooklyn NY 11214-3702

SHEVEL, WILBERT LEE, b Monessen, Pa, Oct 26, 32; m 54; c 4. MAGNETICS, SYSTEMS DEVELOPMENT. *Educ:* Carnegie-Mellon Univ, BS, 54, MS, 55, PhD(elec eng), 60. *Prof Exp:* Vpres & asst gen mgr, Motorola, 73-74; vpres & gen mgr, Rockwell Int, 74-76; pres, Omex, 76-80 & Barrington, 80-82; vpres, Burroughs Corp, 82-84 & Unisys Defense Systs, 84-88; PRES, PARAMAX, UNISYS CO, 88- *Concurrent Pos:* Mem, Long Range Planning Comt, Inst Elec & Electronics Engrs, 83-85 & bd dirs, Aerospace Industs Asn Can, 90- *Mem:* Fel Inst Elec & Electronics Engrs; Inst Elec & Electronics Engrs Magnetics Soc (pres, 67-68); Sigma Xi. *Res:* Impact on information systems of technological advances; integration levels, speed, automated assembly, software, security, storage; cooperative computing and processing of information. *Mailing Add:* 12279 Fairway Pointe Row San Diego CA 92128

SHEVIAK, CHARLES JOHN, b Chicago, Ill, May 31, 47; m 68. PLANT SYSTEMATICS. *Educ:* Univ Ill, Urbana, BS, 70, MS, 72; Harvard Univ, PhD(plant syst), 76. *Prof Exp:* Dir, Ill Endangered Plants Proj, Natural Land Inst, 77-78; CUR BOT, NY STATE MUS, 78- *Mem:* AAAS; Am Soc Plant Taxonomists; Sigma Xi. *Res:* Systematics, ecology and biogeography of North American orchids; biogeography, especially forest-grassland relationships; evolution of colonizing species; northeastern floristics. *Mailing Add:* Cur of Bot NY State Mus Albany NY 12234

SHEVLIN, PHILIP BERNARD, b Mineola, NY, June 28, 39. ORGANIC CHEMISTRY. *Educ:* Lafayette Col, BS, 61; Yale Univ, MS, 63, PhD(chem), 66. *Prof Exp:* Res assoc chem, Brookhaven Nat Lab, 65-66 & 68-70; from asst prof to assoc prof, 70-79, PROF CHEM, AUBURN UNIV, 80- *Mem:* Am Chem Soc. *Res:* Nuclear magnetic resonance studies of chemical kinetics; liquid and gas photochemistry; chemistry of atomic carbon. *Mailing Add:* Dept of Chem Auburn Univ Auburn AL 36830

SHEW, DELBERT CRAIG, b Canton, Ohio, Dec 29, 40; m 67. MASS SPECTROMETRY. *Educ:* Hanover Col, AB, 62; Ind Univ, MS, 66; Univ Ark, PhD(chem), 69. *Prof Exp:* Res chemist, E I du Pont de Nemours & Co, 68-71; assoc prof math, Washington Tech Inst, 71-72; RES CHEMIST, R S KERR ENVIRON RES LAB, ENVIRON PROTECTION AGENCY, 72- *Concurrent Pos:* Consult legal aspects environ pollution, various law firms, 74- *Mem:* Am Soc Mass Spectrometry. *Res:* Isolation of organic pollutants from ground water and identification using electron impact and chemical ionization mass spectrometry. *Mailing Add:* PO Box 1373 Ada OK 74820

SHEWAN, WILLIAM, b Chicago, Ill, May 24, 14; m 49; c 6. ELECTRICAL ENGINEERING, MATHEMATICS. *Educ:* Valparaiso Univ, BS, 50; Univ Notre Dame, MS, 52; Purdue Univ, PhD(elec eng), 66. *Prof Exp:* Chief electronic technician, US Navy, 42-45; jr eng, Northern Ind Pub Serv Co, 45-46; instr electronics, Valparaiso Tech Univ, 46-50, from instr to assoc prof, 52-57, chmn dept, 57-76, actg dean, Col Eng, 78-80, prof, 80-84, EMER PROF ELEC ENG, VALPARAISO UNIV, 84- *Concurrent Pos:* Consult, res & develop, US Navy Crane, 73-76; vis prof, Northern Western Univ Tech Ctr, Evanston, 67. *Mem:* Sr mem Instrument Soc Am; sr mem Inst Elec & Electronics Engrs; Am Soc Eng Educ; Sigma Xi. *Res:* Nonlinear circuit analysis; variable speed drives for rotating machines-solid state devices; numerical methods in systems engineering. *Mailing Add:* 2154 Ransom Rd Valparaiso IN 46383

SHEWCHUN, JOHN, b Toronto, Ont, Aug 14, 38; m 63; c 1. SOLID STATE PHYSICS & ELECTRONICS. *Educ:* Univ Toronto, BASc, 60; Univ Waterloo, MASc, 61, PhD(elec eng), 63. *Prof Exp:* Fel mat sci, Brown Univ, 64-65; mem tech staff solid state physics, RCA Labs, 65-66; asst prof mat sci, McMaster Univ, 66-68, actg chmn dept, 67-69, assoc prof eng physics, 68-87, chmn dept, 69-87; PROF &N CHIEF SCIENTIST, SOLAR POWER DEVELOP INT LTD, 88- *Concurrent Pos:* Spec lectr, City Col New York, 66-67; E W R Steacie Mem fel, Nat Res Coun Can, 72-74; exchange scientist,

Soviet Acad Sci, 72. *Mem:* AAAS; Am Phys Soc; Inst Elec & Electronics Engrs; Am Vacuum Soc; Can Asn Physicists. *Res:* Metal-oxide-semiconductor and semiconductor-oxide semiconductor structures; tunneling devices and spectroscopy; ion implantation; epitaxial growth of semiconductor films; ellipsometry; semiconductor lasers; device physics and transport in thin films; microelectronics; large scale integrated circuits; solar cells. *Mailing Add:* 3219 Healthfield Dr Burlington ON L7M 1E2 Can

SHEWEN, PATRICIA ELLEN, b Stratford, Ont, Mar 21, 49; m 72; c 2. IMMUNOLOGY, MICROBIOLOGY. *Educ:* Univ Guelph, BSc, 71, DVM, 75, MSc, 79, PhD(immunol), 82. *Prof Exp:* Vet, Magilvary Vet Hosp, Toronto, 75-77; asst prof immunol, 82-86, ASSOC PROF IMMUNOL, UNIV GUELPH, 86- *Mem:* Am Asn Immunologists; Can Soc Immunol; Am Asn Vet Immunologists; Am Soc Microbiol; Can Vet Med Asn. *Res:* Respiratory disease focused on the immunologic aspects of pneumonic pasteurellosis and the virulence mechanisms of Pasteurella haemolytica. *Mailing Add:* Dept Vet Microbiol & Immunol Univ Guelph Guelph ON N1G 2X5 Can

SHEWMAKER, JAMES EDWARD, b Paragould, Ark, Jan 22, 22; m 46; c 5. ENVIRONMENTAL EARTH & MARINE SCIENCES. *Educ:* Harding Univ, BS, 44; Univ Nebr, MS, 49, PhD(phys chem), 51. *Prof Exp:* Res chemist, Exxon Res & Eng Co, 51-58, sr chemist, 58-61, res assoc, 62-83; RETIRED. *Mem:* Am Chem Soc; NY Acad Sci. *Res:* Surfactants and detergents; biodegradability; demulsification; oil-field chemicals; radiation chemistry; trace metal contaminants in petroleum; pollution control; cleaning cargo spaces of oil tankers. *Mailing Add:* 1370 S Martine Ave Scotch Plains NJ 07076

SHEWMON, PAUL G(RIFFITH), b Rochelle, Ill, Apr 18, 30; m 52; c 3. MATERIALS SCIENCE ENGINEERING. *Educ:* Univ Ill, BSc, 52; Carnegie Inst Technol, MS & PhD(metall eng), 55. *Prof Exp:* Res engr, Res Lab, Westinghouse Elec Corp, 55-58; from asst prof to prof metall eng, Carnegie Inst Technol, 58-67; assoc dir, Metall Div, Argonne Nat Lab, 67-68, dir, Mat Sci Div, 69-73; dir, Div Mat Res, NSF, 73-75; chmn dept, 75-83, PROF METALL ENG, OHIO STATE UNIV, 75- *Concurrent Pos:* NSF fel, 64-65; mem, adv com reactor safeguards, US Nuclear Regulatory Comn, 77-, chmn, 82; Alexander von Humboldt Sr Scientist Award, 84. *Honors & Awards:* Alfred Noble Prize, 60; Howe Medal, Am Soc Metals, 77; Mathewson Gold Medal, 82. *Mem:* Nat Acad Eng; Am Inst Mining, Metall & Petrol Engrs; Am Nuclear Soc; fel Am Soc Metals; fel AAAS; fel Metall Soc. *Res:* Physical metallurgy; nuclear materials; kinetics of reactions in solids. *Mailing Add:* Dept Mat Sci & Eng Ohio State Univ Columbus OH 43210

SHI, YUN YUAN, b Nanking, China, Sept 26, 32; m 60; c 3. AERONAUTICS, APPLIED MATHEMATICS. *Educ:* Nat Taiwan Univ, BSc, 55; Brown Univ, MSc, 58; Calif Inst Technol, PhD(aeronaut, math), 63. *Prof Exp:* Res asst eng physics, Brown Univ, 57-58; scholar & teaching asst, Calif Inst Technol, 58-62; res specialist, Missile & Space Syst Div, Douglas Aircraft Co, 62-63; asst prof eng & sci, Carnegie Inst Technol, 63-65; sr scientist, McDonnell Douglas Astronaut Co, 65-78, prin scientist, 78-80, staff scientist res & develop, 80-87, SR STAFF MGR, ADVAN TECHNOL CTR, MCDONNELL DOUGLAS SPACE SYST CO, 87- *Mem:* AAAS; fel Am Inst Aeronaut & Astronaut; sr mem Inst Elec & Electronics Engrs. *Res:* Wave propagation in anelastic solids; fluid mechanics; magneto-hydrodynamics; nonlinear oscillations; singular perturbation methods; astronomical science; electromagnetic wave scattering from turbulent plasma; optimal control and estimation; digital signal processing; guidance, control and navigation of launch and space vehicles; Kalman filtering and real time tracking; sensor and space experiments; optimizations and nonlinear programming. *Mailing Add:* 6792 Sunview Dr Huntington Beach CA 92647

SHIAO, DANIEL DA-FONG, b Kiangsi, China, Apr 6, 37; m 74; c 3. PHYSICAL CHEMISTRY, PHYSICAL BIOCHEMISTRY. *Educ:* Nat Taiwan Univ, BS, 58; NMex Highlands Univ, MS, 63; Univ Minn, PhD(phys chem), 68. *Prof Exp:* Fel, Yale Univ, 68-70; sr res chemist, 70-79, RES ASSOC, EASTMAN KODAK, CO, 79- *Concurrent Pos:* Consult, UNESCO, 87. *Mem:* Am Chem Soc; Soc Photog Sci & Eng. *Res:* Physical chemistry at interfaces of silver halide grains; mathematical modeling of diffusion processes and heterogeneous catalysis; physical-chemical measurements pertaining to product design; product performance optimization via statistical analysis and experimental design. *Mailing Add:* 1009 Whalen Rd Penfield NY 14526

SHIAU, YIH-FU, b Chang-Hwa, Taiwan, Jan 12, 42; m; c 2. LIPID METABOLISM, G-I PHYSIOLOGY. *Educ:* Taipei Med Col, MD, 66; George Washington Univ, PhD(physiol), 71. *Hon Degrees:* MA, Univ Pa, 83. *Prof Exp:* Chief, G-I Sect, Vet Admin Med Ctr, 82-89; ASSOC PROF MED, UNIV PA, 83- *Concurrent Pos:* Assoc ed, Digestive Dis & Sci, 82-87. *Mem:* Am Physiol Soc; Am Soc Gastrointestinal Endoscopy; Am Gastroenterol Asn; Am Fed Clin Res; AAAS. *Res:* Lipid absorption and lipid metabolism; bile salt absorption and bile salt secretion. *Mailing Add:* Dept Med G-I Sect 111-GI Vet Admin Med Ctr Woodland Ave Philadelphia PA 19104

SHIBATA, EDWARD ISAMU, b Gallup, NMex, Mar 1, 42; m 73. EXPERIMENTAL HIGH ENERGY PHYSICS. *Educ:* Mass Inst Technol, SB, 64, PhD(physics), 70. *Prof Exp:* Res assoc physics, Northeastern Univ, 70-72; from asst prof to assoc prof, 72-84, PROF PHYSICS, PURDUE UNIV, WEST LAFAYETTE, 84- *Mem:* Am Phys Soc; Am Asn Phys Teachers. *Res:* Meson and baryon spectroscopy; electron-positron annihilations at high energy. *Mailing Add:* Dept Physics Purdue Univ West Lafayette IN 47907

SHIBATA, SHOJI, b Kyoto, Japan, Nov 12, 27; m 58; c 2. PHARMACOLOGY. *Educ:* Nara Med Col, Japan, MD, 52; Kyoto Univ, PhD(pharmacol), 57. *Prof Exp:* Japanese Govt fel, 54-57; instr pharmacol, Sch Med, Kyoto Univ, 57-59; asst prof, Sch Med, Univ Miss, 63-66; assoc prof, 67-69, PROF PHARMACOL, SCH MED, UNIV HAWAII, MANOA, 70- *Concurrent Pos:* Instr & lectr, Sch Med, Univ Southern Calif, 60-61; assoc prof pharmacol & chmn dept, Col Pharm, Kyoto Univ, 62-63. *Mem:* Am Soc Pharmacol & Exp Therapeut; Am Soc Physiol; Sigma Xi; Int Soc Heart Res, Am Sect. *Res:* Cardiovascular pharmacology and natural products. *Mailing Add:* Dept of Pharmacol Univ of Hawaii Sch of Med Honolulu HI 96822

SHIBIB, M AYMAN, b Damascus, Syria, Feb 14, 53; US citizen; m 82; c 2. HIGH VOLTAGE INTEGRATED CIRCUITS. *Educ:* Am Univ Beirut, BS, 75; Univ Fla, Gainesville, MS, 76, PhD(elec eng), 79. *Prof Exp:* Grad res asst microelectronics, elec eng dept, Univ Fla, 76-78, grad res assoc, 78-79, vis asst prof, 79-80; mem, 80-87, DISTINGUISHED MEM TECH STAFF, HIGH VOLTAGE INTEGRATED CIRCUITS DEPT, AT&T BELL LABS, 87- *Concurrent Pos:* Vchmn & chmn electron device group, Inst Elec & Electronics Eng, Lehigh Valley Sect, 84-88, vchmn exec comt, 87-88 & chmn, 88-89; sect chmn, Inst Elec & Electronic Engrs, ISPSD, 91; guest ed, Trans Electron Devices, 91. *Mem:* Sr mem Inst Elec & Electronics Engrs; Am Phys Soc; Electrochem Soc; Sigma Xi; NY Acad Sci; Electrochem Soc. *Res:* Development of silicon device physics and technology development; high voltage integrated circuits used in telecommunication systems; fundamental limitations of performance of silicon solar cells and bipolar transistors. *Mailing Add:* Seven Tewkesbury Dr Wyomissing Hills PA 19610

SHIBKO, SAMUEL ISSAC, b Bargoed, Wales, Oct 2, 27. BIOCHEMISTRY. *Educ:* Univ Birmingham, BSc, 54; Imp Col, dipl, 58, Univ London, PhD(biochem), 58. *Prof Exp:* Res assoc, McCollum Pratt Inst, Johns Hopkins Univ, 58-60; asst biochemist dept food sci & technol, Univ Calif, Davis, 60-64; instr nutrit & food sci, Mass Inst Technol, 64-65, asst prof, 65-67; rev scientist, 67-72, spec asst to dir toxicol div, 72-76, actg chief SPAL, 76-77, CHIEF CONTAMINANTS & NATURAL TOXICANTS EVAL BR, FOOD & DRUG ADMIN, 77- *Concurrent Pos:* Gen referee toxicol tests, Asn Off Anal Chemists, 74- *Mem:* AAAS; Soc Exp Biol & Med; NY Acad Sci; Soc Toxicol; Soc Environ Geochem & Health. *Res:* Food toxicology; evaluation of safety of food additives, and hazards associated with contamination of the food supply with heavy metals, industrial chemicals and natural toxicants; techniques for establishing safety of food chemicals. *Mailing Add:* 6634 31 St Pl NW Washington DC 20015

SHIBLES, RICHARD MARWOOD, b Brooks, Maine, Feb 12, 33; m 64; c 1. CROP PHYSIOLOGY, PHOTOSYNTHESIS. *Educ:* Univ Maine, BS, 56; Cornell Univ, MS, 58, PhD(agronomy), 61. *Prof Exp:* From asst prof agron to assoc prof agron, 60-69, PROF AGRON, IOWA STATE UNIV, 69- *Concurrent Pos:* Lectr fel, Japanese Soc Promo Sci, 88. *Mem:* Fel AAAS; fel Am Soc Agron; fel Crop Sci Soc Am; Am Soc Plant Physiol; Am Asn Univ Profs. *Res:* Genotypic and environmental aspects of plant physiology as they relate to crop productivity; photosynthetic and respiratory metabolism. *Mailing Add:* 2117 North Dakota Ave Ames IA 50011

SHIBLEY, JOHN LUKE, b Gentry, Ark, Apr 12, 19; m 48; c 4. ZOOLOGY. *Educ:* Univ Okla, BS, 41; Univ Ga, MS, 49, PhD(zool), 56. *Prof Exp:* Instr zool, Univ Ga, 49-50; from assoc prof to prof biol, La Grange Col, 50-86; RETIRED. *Res:* Population survival of Euglena with ultraviolet irradiation treatment. *Mailing Add:* 1313 Vernon Rd La Grange GA 30240

SHICHI, HITOSHI, b Nagoya, Japan, Dec 20, 32; m 62; c 2. BIOCHEMISTRY, BIOPHYSICS. *Educ:* Nagoya Univ, BS, 55, MS, 57; Univ Calif, Berkeley, PhD, 62. *Prof Exp:* Res biochemist, Univ Calif, Berkeley, 62; asst prof biochem, Nagoya Univ, 62-63; asst prof, Univ Tokyo, 63-67; res chemist, Nat Inst Neurol Dis & Stroke, 67-69, res chemist, Lab Vision Res, Nat Eye Inst, 69-81; prof & asst dir, Eye Res Inst, Oakland Univ, Rochester, 81-88; PROF OPHTHAL & DIR RES, WAYNE STATE UNIV, KRESGE INST, 88- *Concurrent Pos:* Fulbright scholarship, 57-62. *Honors & Awards:* US-Japan Eye Res Exchange Prog Award, 76. *Mem:* Biophys Soc; Japanese Biochem Soc; Am Soc Photobiol; Am Soc Biol Chem; Asn Res Vision & Ophthal; Am Chem Soc; AAAS. *Res:* Drug metabolism; biochemistry of the visual process. *Mailing Add:* Dept Ophthal Wayne State Univ 47117 St Ottawan Detroit MI 48201

SHICHMAN, D(ANIEL), b Brooklyn, NY, Aug 20, 28; m 49; c 3. MECHANICAL ENGINEERING. *Educ:* Univ Mich, BS, 50; Stevens Inst Technol, MS, 61. *Prof Exp:* Develop engr, Nylon Div, E I du Pont de Nemours & Co, Del, 50-51, develop engr, Atomic Energy Div, Ind, 51-53; res engr, Mech Eng Dept, Res Ctr, Uniroyal, Inc, 53-60, mgr fiber eng res, 60-66, mgr eng res, 66-78; dir res & develop, NRM Corp Div, Condec Corp, 78-80. *Mem:* Sigma Xi; Am Chem Soc; Nat Soc Prof Engrs (vpres, 74-76). *Res:* Process equipment development for chemical, rubber and plastic industry; sprayed metals; synthetic fiber and automation; process equipment for rubber and plastics industry. *Mailing Add:* 20 Copper Kettle Rd Trumbull CT 06611

SHICK, PHILIP E(DWIN), b Kendallville, Ind, Jan 7, 18; m 63; c 2. PULP CHEMICAL RECOVERY FURNACE DESIGN & OPERATION. *Educ:* Harvard Univ, SB, 39; Lawrence Col, MS, 41, PhD(phys chem), 43. *Prof Exp:* Res chemist, Masonite Corp, Miss, 43-45; res proj leader, WVa Pulp & Paper Co, 45-48; res supvr, Mead Corp, 48-56, tech asst to vpres res, 56; tech dir mill div, Owens-Ill Glass Co, 56-60, tech dir forest prod div, 60-63, actg dir res & eng, 63-65, dir res, 65-66, sr res scientist, 66-82, CONSULT, FOREST PROD DIV, OWENS-ILL, INC, 82- *Mem:* Can Pulp & Paper Asn; Am Chem Soc; fel Tech Asn Pulp & Paper Indust. *Res:* High temperature thermodynamics; chemistry of lignin; pulping and papermaking; neutral sulfite semichemical recovery. *Mailing Add:* 3628 Cavalear Dr Toledo OH 43606-1145

SHICKLUNA, JOHN C, soil fertility; deceased, see previous edition for last biography

SHIDA, MITSUZO, b Hamamatsu, Japan, Oct 2, 35; m 59; c 2. CHEMISTRY. *Educ:* Kyoto Univ, BS, 58; Polytech Inst Brooklyn, PhD(chem), 64. *Prof Exp:* Sr res chemist, W R Grace & Co, 63-66; sr res scientist, Plastic Div, Allied Chem Corp, 66; mgr polymer physics, Chemplex Co, Rolling Meadows, 66-70, polymer res dept, 70-86; PRES, MS INTERTECH, 86- *Mem:* Am Chem Soc; Soc Rheol; Chem Soc Japan. *Res:* Polymer physical chemistry, especially solution properties; morphological and rheological aspects of polymer science. *Mailing Add:* 68 Timberlake Dr Barrington IL 60010

SHIDELER, GERALD LEE, b Detroit, Mich, May 19, 38; m 63; c 2. SEDIMENTOLOGY, MARINE GEOLOGY. *Educ:* Mich State Univ, BS, 63; Univ Ill, MS, 65; Univ Wis, PhD(geol), 68. *Prof Exp:* Explor geologist, Mobil Oil Co, 63; teaching asst geol, Univ Ill, Urbana, 63-65; teaching asst, Univ Wis, Madison, 65-66, res asst, 66-68; asst prof, Old Dom Univ, 68-72, assoc prof geol & oceanog, 72-74; marine geologist, Off Marine Geol, US Geol Surv, Corpus Christi, Tex, 74-86; ASSOC PROG COORDR, OFF REGIONAL GEOL, US GEOL SURV, DENVER, 86- *Concurrent Pos:* Nat Acad Sci exchange scientist, Poland. *Mem:* AAAS; fel Geol Soc Am; Soc Econ Paleont & Mineral; Sigma Xi. *Res:* Modern sedimentation of coastal regions and continental shelves; Paleozoic and Cenozoic sedimentation and stratigraphy. *Mailing Add:* US Geol Surv Fed Ctr PO Box 25046 Denver CO 80225-0046

SHIDELER, ROBERT WEAVER, b Joliet, Ill, Apr 25, 13; m 48; c 2. BIOCHEMISTRY. *Educ:* Goshen Col, AB, 34; Univ Chicago, MS, 41; Univ Tex, PhD, 56. *Prof Exp:* Teacher high sch, NDak, 35-36, Ind, 36-40; instr sci, Independence Jr Col, Iowa, 41-42; instr, US Army Air Force Training Sch, Wis, 42-44; assoc prof chem, 46-50, prof chem & chmn dept chem, 54-77, chmn natural sci area, 66-77, EMER HAROLD & LUCY CABE DISTINGUISHED PROF CHEM, HENDRIX COL, 78- *Mem:* AAAS; Am Chem Soc; NY Acad Sci. *Res:* Methods in investigating intermediary metabolism; individual differences in mineral metabolism. *Mailing Add:* 1601 Quail Creek Dr Conway AR 72032

SHIEH, CHING-CHYUAN, b Pingtung, Taiwan, Repub of China, Oct 20, 50; US citizen; m 77; c 3. TELECOMMUNICATION FACILITY NETWORK OPERATIONS, NETWORK OPERATIONS SYSTEMS DESIGN. *Educ:* Nat Taiwan Univ, BS, 72; Univ Maine, MS, 76; Univ Pa, PhD(elec eng), 84. *Prof Exp:* Digital prod mgr data commun telemetry, Sonex Philmont Electronics Inc, 80-83; mem tech staff micro-syst, AT&T Bell Labs, 84-85, mem tech staff computer syst, AT&T Info Systs, 85-86, MEM TECH STAFF, TELECOMMUN, AT&T BELL LABS, 87- *Concurrent Pos:* Lectr, Nat Taiwan Univ, 74-75; syst consult syst design, DazoTron Electronics Assoc, 82-83. *Mem:* Inst Elec & Electronics Engrs. *Res:* Mechanization of telecommunication transport network provisioning and maintenance; integrated operation systems, expert systems and user interface systems architecture design for automated transport network operations; network core data integrity, central and local database synchronization; real-time automatic control system and data communication system design; high-performance microprocessor based single board computer development. *Mailing Add:* Seven Apple Manor Lane East Brunswick NJ 08816

SHIEH, HANG SHAN, microbiology, for more information see previous edition

SHIEH, JOHN SHUNEN, b Shanghai, China, Jan 10, 46; Can citizen; m 76; c 2. COMPUTER SCIENCE. *Educ:* Univ Sci & Technol China, BSc, 67; Simon Fraser Univ, PhD(computer sci), 86. *Prof Exp:* Technician, Wangching Standard Space Parts Factory, 67-76; lectr computer sci, Cent China Univ Sci & Technol, 76-81; res scientist, Simon Fraser Univ, 87-88; ASST PROF COMPUTER SCI, DEPT COMPUTER SCI, MEM UNIV NFLD, 88- *Concurrent Pos:* Sr res scientist, Nordco Ltd, 88-91. *Res:* Mobile robot vision, scene analysis and object recognition; knowledge based system design. *Mailing Add:* Dept Computer Sci Mem Univ Nfld St Johns NF A1C 5S7

SHIEH, KENNETH KUANG-ZEN, b Keelung, Taiwan, Feb 15, 36; US citizen; m 66; c 2. INDUSTRIAL MICROBIOLOGY. *Educ:* Taiwan Prov Col Agr, BS, 61; Ill Inst Technol, MS, 65, PhD(microbiol), 69. *Prof Exp:* Sr res microbiologist, 68-75, RES ASSOC, CENT RES DEPT, CORN PRODS SECT, ANHEUSER-BUSCH, INC, 75- *Mem:* Am Soc Microbiol. *Res:* Production of enzymes from a microbial origin for transformation of one kind of carbohydrate to another. *Mailing Add:* 501 Prince Way Ct Manchester MO 63011

SHIEH, LEANG-SAN, b Tainan, China, Jan 10, 34; US citizen; m 62; c 2. DIGITAL CONTROL SYSTEMS, NETWORK THEORY. *Educ:* Nat Taiwan Univ, BS, 58; Univ Houston, MS, 68, PhD(elec eng), 70. *Prof Exp:* Design engr, Taiwan Florescent Lamp Co, 60-61; power systs engr, Taiwan Elec Power Co, 61-65; from asst prof to assoc prof, 71-78, PROF CONTROL SYSTS, UNIV HOUSTON, 78- *Concurrent Pos:* Vis prof, Inst Univ Polit, Venezuela, 76-77; vis scientist, Battelle Columbus Lab, 77 & Math Res Ctr, Univ Wis-Madison, 79. *Mem:* Inst Elec & Electronics Engrs; Am Soc Elec Engrs. *Res:* Model reduction, identification, realization and adaptive control of multivariable control systems. *Mailing Add:* Dept Elec Eng Univ Houston-Univ Park Houston TX 77004

SHIEH, PAULINUS SHEE-SHAN, b Nanchang, China, Apr 5, 31; m 60; c 3. NUCLEAR ENGINEERING, PHYSICS. *Educ:* Nat Taiwan Univ, BS, 55; Univ SC, MS, 58; NC State Univ, PhD(nuclear eng), 69. *Prof Exp:* Asst prof physics & chmn dept, King's Col, Pa, 60-63; instr, NC State Univ, 63-69; from asst prof to assoc prof nuclear eng, Miss State Univ, 69-79; sr nuclear engr, C P & L, 79-81; SUPVR NUCLEAR ANAL, H L & P, TEX, 79- *Concurrent Pos:* Vis prof, Pinstech, Pakistan, 76-77. *Mem:* AAAS; Am Nuclear Soc; Am Soc Eng Educ. *Res:* Neutron transport theory; radiation measurements; nuclear power. *Mailing Add:* 771 Seacliff Dr Houston TX 77062

SHIEH, YUCH-NING, b Chan-hua, Taiwan, Feb 15, 40; m 66; c 2. GEOCHEMISTRY, PETROLOGY. *Educ:* Nat Taiwan Univ, BS, 62; Calif Inst Technol, PhD(geochem), 69. *Prof Exp:* Fel geochem, McMaster Univ, Can, 68-72; asst prof, 72-79, assoc prof, 79-87, PROF GEOCHEM, PURDUE UNIV, 87- *Concurrent Pos:* Vis scientist, Inst Earth Sci, Acad Sinica, Taipei, China, 80-81, vis prof, 89-90; assoc ed, Isotope Geosci, 82- *Mem:* Geochem Soc; Am Geophys Union; Geol Soc China. *Res:* Oxygen, carbon and hydrogen isotopes in igneous and metamorphic rocks; sulfur isotopes in ore deposits and coals; stable isotope studies of active geothermal systems. *Mailing Add:* Dept Earth & Atmospheric Sci Purdue Univ West Lafayette IN 47907

SHIELD, RICHARD THORPE, b Swalwell, Eng, July 9, 29; m 58; c 2. MECHANICS. *Educ:* Univ Durham, BSc, 49, PhD(appl math), 52. *Prof Exp:* Res assoc appl math, Brown Univ, 51-53; sr res fel, A R E Ft Halstead, Eng, 53-55; from asst prof to prof, Brown Univ, 55-65; prof appl mech, Calif Inst Technol, 65-70; head dept, 70-84, PROF THEORET & APPL MECH, UNIV ILL, URBANA, 70- *Concurrent Pos:* Guggenheim mem fel, Univ Durham, 61-62; co-ed, J Appl Math & Physics, 65-, assoc ed, J Appl Mech, 79-; Alcoa vis prof, Univ Pittsburgh, 70-71; mem comt, US Nat Theoret & Appl Mech, 81-86; distinguished scholar, Calif Inst Technol, 82; Russell Severance Springer vis prof, Univ Calif, Berkeley, 84. *Mem:* Fel AAAS; fel Am Acad Mech (pres, 77-78); fel Am Soc Mech Engrs; fel Soc Eng Sci; Sigma Xi. *Res:* Elasticity, plasticity; stability theory. *Mailing Add:* 40 Lake Park Rd Champaign IL 61820

SHIELDS, ALLEN LOWELL, mathematics; deceased, see previous edition for last biography

SHIELDS, BRUCE MACLEAN, b Wilkinsburg, Pa, Sept 27, 22; m 51; c 2. METALLURGICAL ENGINEERING, METALLURGY. *Educ:* Carnegie Inst Technol, BS, 44; Mass Inst Technol, MS, 52. *Prof Exp:* Metall observer, Homestead Works, Carnegie Ill Steel, 42-43 & 46-47; res technologist metall, South Works, US Steel Corp, 47-50; res asst, Mass Inst Technol, 50-51; chief develop metallurgist, Duquesne Works, 51-54, chief process metallurgist, 54-56, asst chief metallurgist, 56-57, chief metallurgist, 57-60, chief metallurgist, South Works, 60-65, mgr process metall, 65-68, mgr tubular prod metall, 68-72, gen mgr process metall, 72-75, gen mgr customer tech serv, 75-78, dir metall eng, 78-84, MGT CONSULT, US STEEL CORP, 84- *Concurrent Pos:* Consult, fields of qual assurance, iron & steel making, ferrous metall. *Mem:* Fel Am Soc Metals; Am Inst Mining, Metall & Petrol Engrs; Am Iron & Steel Inst; Am Petrol Inst. *Res:* Physical chemical reactions of iron and steelmaking, including both the equilibrium relationships and the reaction kinetics; transformation characteristics of alloy steel systems. *Mailing Add:* 104 Altadena Dr Pittsburgh PA 15228

SHIELDS, DENNIS, b London, Eng, Sept 9, 48; m 75. CELL BIOLOGY, BIOCHEMISTRY. *Educ:* Univ York, Eng, BA, 71; Nat Inst Med Res, London, PhD(biochem), 74. *Prof Exp:* Fel cell biol, Rockefeller Univ, 74-77; ASST PROF ANAT, ALBERT EINSTEIN COL MED, 78-, ASSOC PROF. *Honors & Awards:* Solomon A Berson Award, Am Diabetes Asn, 78. *Mem:* Am Soc Cell Biol; Biochem Soc; Endocrine Soc. *Res:* Biosynthesis and subcellular compartmentation of secretory and membrane proteins; pancreatic islet hormone biosynthesis; post-translational modifications of proteins. *Mailing Add:* Dept Anat Albert Einstein Col Med 1300 Morris Park Ave Bronx NY 10461

SHIELDS, FLETCHER DOUGLAS, b Nashville, Tenn, Oct 27, 26; m 48; c 4. PHYSICS. *Educ:* Tenn Polytech Inst, BS, 47; Vanderbilt Univ, MS, 48, PhD(physics), 56. *Prof Exp:* Res physicst, Carbide & Carbon Chem Corp, 48-49; from asst prof to assoc prof physics, Middle Tenn State Col, 49-59; assoc dean lib arts, 76-79, PROF PHYSICS, UNIV MISS, 59- *Mem:* Fel Acoust Soc Am; Sigma Xi. *Res:* Thermal relaxation processes in gas by means of sound absorption measurements; energy and mometum accommodation of gas molecules on solid surfaces. *Mailing Add:* Dept Physics Univ Miss University MS 38677

SHIELDS, GEORGE SEAMON, b Bombay, NY, Oct 18, 25; m 48; c 5. INTERNAL MEDICINE, COMPUTER MEDICINE. *Educ:* Mass Inst Technol, SB, 48; Cornell Univ, MD, 52. *Prof Exp:* Intern, Bellevue Hosp, New York, 52-53, resident, 53-54; USPHS res fel hemat, State Univ NY Downstate Med Ctr, 54-56; resident, Salt Lake County Gen Hosp, 56-57; Am Cancer Soc fel biochem & hemat, Col Med, Univ Utah, 57-60, instr internal med, 59-61; from asst prof to assoc prof, Col Med, Univ Cincinnati, 61-70; dir med systs, Good Samaritan Hosp, 69-75; VPRES, MICRO-MED, INC, 80- *Concurrent Pos:* Med dir, Winton Hills Med Ctr, 73-75, Wesley Hall Home, 75-, Riverview Home, 78- & Three Rivers Home, 78-; pres, Sycamore Prof Asn Inc, 71- *Mem:* Biomed Eng Soc; Soc Comput Med; Soc Advan Med Systs; Am Fedn Clin Res; Am Chem Soc. *Res:* Biochemistry, protein, enzymology, nucleic acids, trace metals, copper; hematology, anemias, leukemias, pathogenesis; computer applications. *Mailing Add:* 4903 Vine St Cincinnati OH 45217

SHIELDS, GERALD FRANCIS, b Anaconda, Mont, Nov 9, 43; m 67; c 2. CYTOGENETICS, MOLECULAR EVOLUTION. *Educ:* Carroll Col, BA, 66; Cent Wash State, MS, 70; Univ Toronto, PhD(zool), 74. *Prof Exp:* Teacher biol, Billings Cent High Sch, 66-68; NSF res asst, Cent Wash State Col, 68-70; teaching asst, Dept Zool, Univ Toronto, 70-74; ASST PROF ZOOL, INST ARCTIC BIOL, UNIV ALASKA, 75- *Concurrent Pos:* Nat Res Coun fel, 72-74. *Mem:* Am Ornith Union; Soc Study Syst Zool; Soc Study Evolution; Can Soc Genetics & Cytol; Sigma Xi. *Res:* Cytogenetics and chromosomal evolution in insects and vertebrates; DNA evolution in vertebrates; rates of chromosome and DNA change. *Mailing Add:* Zool Dept Univ Alaska Fairbanks AK 99701

SHIELDS, HOWARD WILLIAM, b Tomotla, NC, May 19, 31; m 60; c 3. SOLID STATE PHYSICS. *Educ:* Univ NC, BS, 52; Pa State Univ, MS, 53; Duke Univ, PhD(physics), 56. *Prof Exp:* Res assoc physics, Duke Univ, 56-57; from asst prof to assoc prof, 58-66, PROF PHYSICS, WAKE FOREST UNIV, 66- *Concurrent Pos:* Sloan Found res fel, 63-65; vis prof, Yale Univ, 68-69. *Mem:* Fel Am Phys Soc; Radiation Res Soc; Sigma Xi. *Res:* Electron spin resonance; studies made on irradiation damage in organic solids; phase transitions, and high T superconductors. *Mailing Add:* 3380 Sledd Ct Winston-Salem NC 27106

SHIELDS, JAMES EDWIN, b Marion, Ind, July 29, 34; m 62; c 1. BIOCHEMISTRY, ENTOMOLOGY. *Educ:* DePauw Univ, AB, 56; Univ Calif, Berkeley, PhD(biochem), 61. *Prof Exp:* USPHS fel, Univ Zurich, 60-62; asst prof chem, Case Western Reserve Univ, 62-68; sr biochemist, 68-76, RES SCIENTIST, LILLY RES LABS, ELI LILLY & CO, 77-; FOUNDER & PRES, AMARYLLIS RES INST, 78- *Concurrent Pos:* Adj prof chem, Indiana Univ & Purdue Univ, 79- *Mem:* Am Chem Soc; Lepidopterists Soc; AAAS; Sigma Xi. *Res:* Peptide and protein chemistry; horticulture. *Mailing Add:* Dept MC 797 Bldg 88-4 Eli Lilly & Co Indianapolis IN 46285-0002

SHIELDS, JIMMIE LEE, b St Louis, Mo, Aug 9, 34; m 65; c 2. ENVIRONMENTAL PHYSIOLOGY. *Educ:* Cent Methodist Col, AB, 56; Univ Mo, AM, 58, PhD(physiol), 62. *Prof Exp:* Instr physiol, Univ Mo, 62-64, asst prof, 64-65; res physiologist, Fitzsimons Gen Hosp, Denver, 65-68; mem staff, 68-74, asst dir health info progs, 74-77, ASST DIR PREV EDUC & CONTROL & ACTG DEP DIR, DIV HEART & VASCULAR DIS, NAT HEART, LUNG & BLOOD INST, 78- *Concurrent Pos:* Res assoc, Space Sci Res Ctr, Univ Mo, 64-65, res consult, 65-68. *Mem:* Am Physiol Soc; Sigma Xi. *Res:* Altitude and arctic, metabolic and performance aspects of environmental physiology. *Mailing Add:* 9125 Aldershot Dr Bethesda MD 20817

SHIELDS, JOAN ESTHER, b Cambridge, Mass, Oct 11, 34. ORGANIC CHEMISTRY. *Educ:* Regis Col, Mass, AB, 56; Tufts Univ, MS, 58; Boston Col, PhD(org chem), 66. *Prof Exp:* Instr chem, Regis Col, Mass, 58-63; fel, Max Planck Inst Coal Res, 66-68; asst prof, 68-71, ASSOC PROF CHEM, C W POST COL, LONG ISLAND UNIV, 71- *Mem:* Am Chem Soc; Sigma Xi. *Res:* Organometallic chemistry; organosulfur and nitrogen heterocyclics and photochemical cycloadditions. *Mailing Add:* Dept of Chem C W Post Col Greenvale NY 11548

SHIELDS, LORA MANEUM, b Choctaw, Okla, Mar 13, 12; m 31; c 1. BIOLOGY. *Educ:* Univ NMex, BS, 40, MS, 42; Unv Iowa, PhD(bot), 47. *Prof Exp:* Assoc prof biol, NMex Highlands Univ, 47-54, prof biol & head dept, 51-78, dir, Environ Health Div, 72-78; RESEARCHER & VIS PROF, NAVAJO COMMUNITY COL, SHIPROCK, NMEX, 78- *Concurrent Pos:* Res grants, AEC, NIH, NSF, Sigma Xi & Squibb & Searle, March Dimes Birth Defects Found, Minority Biomed Res Support. *Mem:* AAAS; Ecol Soc Am (vpres, 62-63). *Res:* Plant nitrogen sources and leaf nitrogen content; soil algae; nuclear effects on vegetation; serum lipids; hypoglycemic principles in plants; serum lipids in Spanish and Anglo-Americans; birth anomalies in the Navajo uranium district; streptococcal diseases among the Navajo; lead-210 in Navajo teeth in the Shiprock uranium area. *Mailing Add:* 4825 W Ninth St Greeley CO 80634

SHIELDS, LORAN DONALD, b San Diego, Calif, Sept 18, 36; m 57; c 4. ANALYTICAL CHEMISTRY, ACADEMIC ADMINISTRATION. *Educ:* Univ Calif, Riverside, BA, 59; Univ Calif, Los Angeles, PhD(chem), 64. *Prof Exp:* From asst prof to assoc prof chem, Calif State Univ, Fullerton, 63-67, chmn fac coun, 67, vpres admin, 67-70, actg pres, 70-71, prof chem, 67-, pres, 71-; PRES, SOUTHERN METHODIST UNIV. *Concurrent Pos:* Consult, Calif State Senate for Legis in Support of Res in Calif State Cols, 68-69 & NSF, 70; mem exec comt, Coun of Pres, Calif State Univ & Cols Syst, 73-, chmn, 75-76; resource leader, Res Corp Conf for Pub Univ Sci Dept Chairs, Ala, 74; mem, Nat Sci Bd, 74-80, mem budget comt, Progs Comt, Sci Educ Adv Comt, 74- & Nat Comt on Coop Educ; exec dir, Calif Coun Sci & Technol. *Mem:* Am Inst Chemists; Sigma Xi; AAAS; Am Chem Soc; Am Col Pub Rels Asn. *Res:* Transition metal coordination chemistry; instrumental methods of analytical chemistry. *Mailing Add:* 1746 Tattenham Rd Leucadia CA 92024

SHIELDS, PAUL CALVIN, b South Haven, Mich, Nov 10, 33; div; c 6. PURE MATHEMATICS. *Educ:* Colo Col, AB, 56; Yale Univ, MA, 58, PhD(math), 59. *Prof Exp:* CLE Moore instr math, Mass Inst Technol, 59-61; asst prof, Boston Univ, 61-63; asst prof, Wayne State Univ, 63-69; vis scholar & res assoc, Stanford Univ, 70-73; vis lectr, Univ Warwick, Coventry Eng, 73-74; assoc prof, 74-76, PROF MATH, UNIV TOLEDO, 76- *Concurrent Pos:* Res assoc, Willow Run Lab, Univ Mich, 60; vis prof elec eng, Cornell Univ, 78, Stanford Univ, 80-82; Fulbright Scholar, Math Inst, Budapest, Hungary, 85; prof stat, Univ Toronto, Can, 85-86. *Mem:* Am Math Soc; Sigma Xi. *Res:* Ergodic theory; information theory; statistical mechanics; operator theory; linear algebra. *Mailing Add:* Dept Math Univ Toledo Toledo OH 43606

SHIELDS, ROBERT JAMES, b Philadelphia, Pa, Apr 23, 34; m 58. PARASITOLOGY. *Educ:* East Stroudsburg State Col, BS, 55; Ohio State Univ, MS, 59, PhD(zool), 62. *Prof Exp:* Instr zool, Ohio State Univ, 62-63; instr biol, 63-65, asst prof, 65-70. assoc prof, 70-76, chmn dept, 78-81, PROF BIOL, CITY COL NEW YORK, 77- *Mem:* Am Soc Parasitol; Am Soc Zoologists; Am Micros Soc. *Res:* Life-histories, ecology and physiology of marine and fresh-water copepods parasitic on fishes. *Mailing Add:* Zool Dept City Col NY New York NY 10031

SHIELDS, ROBERT PIERCE, b Nashville, Tenn, June 11, 32; m 60; c 2. VETERINARY PATHOLOGY, BIOCHEMISTRY. *Educ:* Auburn Univ, DVM, 56, MVS, 59; Univ Ark, Little Rock, MS, 66; Mich State Univ, PhD(path), 72. *Prof Exp:* Instr path, Auburn Univ, 56-59; pathologist, Ralston Purina Co, 59-62; resident pathologist, Med Ctr, Univ Ark, Little Rock, 65-66; assoc prof path, Auburn Univ, 66-72; assoc prof path, Div Comp Med, Col Vet Med & Dir, Animal Resource Diag Lab, Univ Fla, 72-90; RETIRED. *Concurrent Pos:* NIH fel chem path, Univ Ark, 62-65, spec fel path, Mich State Univ, 70-72. *Mem:* Am Col Vet Pathologists; Am Vet Med Asn. *Res:* Biochemistry and pathology associated with neuromuscular diseases; nutritional pathology; creatine metabolism. *Mailing Add:* 3959 NW 29th Lane Univ of Fla Col Vet Med Gainesville FL 32606

SHIELDS, THOMAS WILLIAM, b Ambridge, Pa, Aug 17, 22; m 48; c 3. THORACIC SURGERY. *Educ:* Kenyon Col, BA, 43; Temple Univ, MD, 47; Am Bd Surg, dipl, 55; Bd Thoracic Surg, dipl, 56. *Hon Degrees:* DSc, Kenyon, 78. *Prof Exp:* Intern, Allegheny Gen Hosp, Pittsburgh, 47-48; resident, New Eng Deaconess Hosp, Boston, 48-49, Passavant Mem Hosp, Chicago, 49-50 & Vet Admin Res Hosp, Chicago, 54-55; sr resident, Chicago Munic Tuberc Sanitarium, 55-56; instr, Northwestern Univ, Chicago, 56-57, assoc, 57-62, assoc prof, 64-68, chief surg serv, Vet Admin Lakeside Hosp, 68-86, PROF SURG, NORTHWESTERN UNIV, CHICAGO, 68- *Concurrent Pos:* Allan B Kanavel fel surg, Passavant Mem Hosp, Chicago, 49-50; thoracic surgeon, Vet Admin Res Hosp, 56-57, mem staff thoracic surgeons, 57-; assoc surg staff, Passavant Mem Hosp, 56-57, attend surgeon, 58-72; attend thoracic surgeon, Chicago Munic Tuberc Sanitarium, 56-68 & attend surgeon, Northwestern Mem Hosp, 72-80. *Mem:* Western Surg Asn; Am Col Surg; Am Asn Thoracic Surg; Soc Thoracic Surg; Cent Surg Asn. *Res:* Clinical cancer research; editor and contributor to surgical, thoracic surgical, and cancer literature. *Mailing Add:* Northwestern Mem Hosp Suite 201 250 E Superior St Chicago IL 60611

SHIER, DOUGLAS ROBERT, b Cleveland, Ohio, Oct 16, 46. OPERATIONS RESEARCH, APPLIED MATHEMATICAL MODELING. *Educ:* Harvard Univ, AB, 68; London Sch Econ, PhD(oper res), 73. *Prof Exp:* Res statistician, Ctr Dis Control, USPHS, 68-70; res assoc, Nat Bur Standards, 73-74; asst prof quant methods, Univ Ill, 74-75; mathematician, Nat Bur Standards, 75-80; prof math sci, Clemson Univ, 80-87; PROF OPER RES, COL WILLIAM & MARY, 88- *Mem:* Am Statist Asn; Math Asn Am; Opers Res Soc Am. *Res:* Mathematical and statistical modeling; network and location problems; environmental sciences. *Mailing Add:* Dept Math Col William & Mary Williamsburg VA 23187-8795

SHIER, WAYNE THOMAS, b Harriston, Ont, Dec 1, 43; m 69; c 3. TOXINS, SINGLE CELL PROTEIN. *Educ:* Univ Waterloo, BSc, 66; Univ Ill, Urbana, MS, 68, PhD(chem), 70. *Prof Exp:* Res assoc biochem, Salk Inst, 70-72, asst res prof, 72-80; assoc prof, dept pharmaceut cell biol, 80-85, PROF, DEPT MED CHEM, UNIV MINN, 85- *Concurrent Pos:* Am Cancer Soc Dernham jr fel, 70-71; consult, Tokyo Tanabe Co, Ltd, 71 & Marcel Dekker, Inc, 79-; USPHS res grant, 72-80, 85-88; Cystic Fibrosis Found res grant, 78-80; NSF res grant, 80-85, 85-87 & 87-90; Am Cancer Soc res grant, 83-86; ed, J Toxicol-Toxins Rev, 82- *Mem:* AAAS; Am Soc Biol Chemists; Am Soc Biochem & Molecular Biol; Am Chem Soc; Soc Toxicol; Am Soc Cell Biol; Int Soc Toxinol. *Res:* Structure and synthesis of glycoproteins; biochemistry of cell membranes; phospholipases; cystic fibrosis; cyclic nucleotides; lipids; cytotoxic mechanisms; regulation of prostaglandin synthesis; calcium homeostasis; antiviral agents; single cell protein. *Mailing Add:* Dept Med Chem Col Pharm Health Sci Unit F Univ Minn Minneapolis MN 55455

SHIFFMAN, BERNARD, b New York, NY, June 23, 42; m 65; c 2. COMPLEX VARIABLES. *Educ:* Mass Inst Technol, BS, 64; Univ Calif, Berkeley, PhD(math), 68. *Prof Exp:* Moore instr math, Mass Inst Technol, 68-70; asst prof, Yale Univ, 70-73; assoc prof, 73-77, PROF MATH, JOHNS HOPKINS UNIV, 77-, CHAIR, DEPT MATH, 90- *Concurrent Pos:* Mem, Inst Advan Study, 75; res fel, Alfred P Sloan Found, 73-75; vis prof, Univ Kaiserslautern, 77; lectr Inst Math, Acadeimia Sinica, Beijing, 78, Nordic Summer Sch, Joensuu, Finland, 81; mem, Inst des Hautes Etudes Scientifiques, 79; vis prof, Univ Paris, 81 & 85; ed, Forum Math, 88-; assoc ed, Am J Math, 90- *Mem:* Am Math Soc. *Res:* Meromorphic mappings; value distribution theory; complex manifolds. *Mailing Add:* Dept of Math Johns Hopkins Univ Baltimore MD 21218

SHIFFMAN, CARL ABRAHAM, b Boston, Mass, Nov 14, 30; m 56; c 2. PHYSICS. *Educ:* Mass Inst Technol, BS, 52; Oxford Univ, DrPhil, 56. *Prof Exp:* Physicist, Nat Bur Standards, 56-60 & Mass Inst Technol, 60-67; PROF PHYSICS, NORTHEASTERN UNIV, 67- *Res:* Low temperature physics; superconductivity. *Mailing Add:* Dept Physics Dana Bldg Rm 103 Northeastern Univ Boston MA 02115

SHIFFMAN, MAX, b New York, NY, Oct 30, 14; div; c 2. MINIMAX THEORY AERODYNAMICS. *Educ:* City Col New York, BS, 35; NY Univ, MS, 36, PhD(math), 38. *Prof Exp:* Instr math, St John's Univ, NY, 38-39 & City Col New York 38-42; res mathematician, US Army, US Off Naval Res & Appl Math Panel, US Off Sci Res & Develop, NY Univ, 41-48, assoc prof math, 46-49; prof math, Stanford Univ, 49-66; prof math, Calif State Univ, Hayward, 67-81; MATHEMATICIAN & OWNER, MATHEMATICO, 70- *Concurrent Pos:* Res mathematician, Rand Co, Santa Monica, Calif, 51; consult mathematician, George Washington Univ, 58-61, US Gov't, 61-62. *Mem:* Math Asn Am; Am Math Soc; Soc Indust & Appl Math. *Res:* Minimal surfaces; stationary extremals; groups; conformal mapping; potential theory; complex variable; calculus of variations; hydrodynamics; aerodynamics; partial differential equations; games; probability; topological and linear space methods in analysis; non-measurable and partially measurable sets. *Mailing Add:* 16913 Meekland Ave No 7 Hayward CA 94541

SHIFFMAN, MORRIS A, b New York, NY, Oct 12, 22; m 50; c 2. ENVIRONMENTAL HEALTH. *Educ:* Middlesex Univ, DVM, 44; Univ Mich, MPH, 45; Nat Vet Sch, Alfort, France, DVet, 49; Univ Pa, MGA, 57, PhD, 67. *Prof Exp:* Qual control supvr, Gen Ice Cream Corp, 45; sr veterinarian, UN Relief & Rehab Admin, 46-47; food & drug sanit supvr, Milwaukee Health Dept, 49-53; chief milk & food sanit, Phila Dept Pub Health, 53-64; assoc prof environ sanit, 64-69, PROF ENVIRON HEALTH,

UNIV NC, CHAPEL HILL, 69- *Concurrent Pos:* Mem subcomt food sanit, Nat Res Coun, 59-62, chmn, 62-65; mem food estab sanit adv comt, USPHS, 60-63; adv panel zoonoses, WHO, 61-72; mem comt sanit eng & environ, Nat Acad Sci-Nat Res Coun, 62-65; mem adv panel food hyg, WHO, 72- & Sci Adv Comt, Pan-Am Zoonoses Ctr, 72- *Mem:* AAAS: Am Soc Pub Admin; Sigma Xi. *Res:* Administration of environmental health programs; environmental health policy; project-impact studies in developing countries. *Mailing Add:* 201 Ridgecrest Dr Chapel Hill NC 27514

SHIFLET, THOMAS NEAL, b Marysville, Tex, July 25, 30; m 53; c 2. RANGE MANAGEMENT, ECOLOGY. *Educ:* Tex A&M Univ, BS, 51; Univ Calif, Berkeley, MS, 67; Univ Nebr, PhD(range mgt), 72. *Prof Exp:* Area range mgt conservationist, Soil Conserv Serv, USDA, 57- 60, state range conservationist, 60-67, staff range conservationist, 67-70, regional range conservationist, 70-74, chief range conservationist, 74-75, dir, Div Ecol Sci, 75-87; RETIRED. *Concurrent Pos:* Range conservationist, Soil Conserv Serv, USDA, 53-54, work unit conservationist, 54-57. *Mem:* Soc Range Mgt; Soil Conserv Soc Am; Coun Agr Sci & Technol; Nat Asn Conserv Districts. *Mailing Add:* 4859 S Crescent Ave Springfield MO 65804

SHIFLETT, LILBURN THOMAS, b Adamsville, Ala, Feb 4, 21; m 42; c 2. MATHEMATICS. *Educ:* Univ N Ala, BS, 46; George Peabody Col, MA, 48, PhD(math), 63. *Prof Exp:* Teacher high sch, Ark, 46-48; from instr to assoc prof, Southwest Mo State Univ, 48-58, dir div honors, 65-67, prof math, 58-85, head Dept Math, 68-85; RETIRED. *Res:* Geometry; statistics. *Mailing Add:* 829 S Weller Springfield MO 65804-0089

SHIFLETT, RAY CALVIN, b Levensworth, Wash, Dec 3, 39; m 59; c 2. MATHEMATICS. *Educ:* Eastern Wash State Col, BA, 63; Ore State Univ, MS, 65, PhD(math), 67. *Prof Exp:* NSF res asst, Ore State Univ, 66-67; asst prof math, Wells Col, 67-75, chmn dept, 69-75; assoc prof, 75-80, PROF MATH, CALIF STATE UNIV, FULLERTON, 80- *Mem:* Math Asn Am; Am Math Soc. *Res:* Analysis, especially measure theory, doubly stochastic measures and Markov operators. *Mailing Add:* 2013 E Union Ave Fullerton CA 92631

SHIFRINE, MOSHE, microbiology, for more information see previous edition

SHIGEISHI, RONALD A, b Vancouver, BC, Apr 5, 39; m 68. PHYSICAL CHEMISTRY. *Educ:* Univ Toronto, BSc, 61; Queen's Univ, PhD(chem), 65. *Prof Exp:* Nat Res Coun fel, 65-67; ASST PROF CHEM, CARLETON UNIV, 67- *Concurrent Pos:* Nat Res Coun res grants, 67-69. *Res:* Surface chemistry of gas-metal systems. *Mailing Add:* Dept of Chem Carleton Univ Ottawa ON K1S 5B6 Can

SHIGLEY, JOSEPH E(DWARD), b Delphi, Ind, Apr 10, 09; m 34; c 2. MECHANICAL ENGINEERING. *Educ:* Purdue Univ, BS, 31 & 32; Univ Mich, MS, 46. *Prof Exp:* From assoc prof to prof, 57-78, EMER PROF MECH ENG, UNIV MICH, ANN ARBOR, 78- *Concurrent Pos:* Res writer & consult. *Honors & Awards:* Mechanisms Award, Am Soc Mech Engrs, 74, Worcester Reed Warner Medal, 77, Mach Design Award, 85. *Mem:* Fel Am Soc Mech Engrs. *Res:* Kinematic and dynamic analysis and computer simulation of mechanical systems; mechanical design. *Mailing Add:* 125 Timber Trail Pinecone Beach Roscommon MI 48653

SHIGO, ALEX LLOYD, b Duquesne, Pa, May 8, 30; m 54; c 2. PLANT PATHOLOGY. *Educ:* Waynesburg Col, BS, 56; Univ WVa, MS, 58, PhD(plant path), 59. *Prof Exp:* Plant pathologist, 59-74, CHIEF PLANT PATHOLOGIST & LEADER, PIONEER PROJ, US FOREST SERV, 74- *Concurrent Pos:* Lectr, Univ Maine; adj prof, Univ NH. *Honors & Awards:* New Eng Logger Award Outstanding Res, 73; Ciba-Giegy Award, Am Phytopath Soc, 75. *Mem:* Mycol Soc Am; Am Phytopath Soc. *Res:* Decay and discoloration in living trees; physiology of wood-inhabiting fungi; diseases of northern hardwoods; fungi parasitic on other fungi; myco-parasites. *Mailing Add:* Four Denbow Rd Durham NH 03824

SHIH, ARNOLD SHANG-TEH, b Shanghai, China, May 17, 43; US citizen; m 69; c 2. PHYSICS. *Educ:* Univ Calif, Berkeley, AB, 65; Columbia Univ, PhD(physics), 72. *Prof Exp:* Physicist surface chem, Nat Bur Standards, 72-75; RES PHYSICIST SURFACE PHYSICS, NAVAL RES LAB, 75- *Mem:* Am Phys Soc; Am Vacuum Soc. *Res:* Electronic and structural properties at the surface of solids; Van der Waals forces between atoms and solid surfaces. *Mailing Add:* Naval Res Lab Code 6840 4555 Overlook Ave Washington DC 20375-5000

SHIH, CHANG-TAI, b Xiamen, China, Aug 15, 34; Can citizen; m 63; c 2. SYSTEMATICS, BIOLOGICAL OCEANOGRAPHY. *Educ:* Nat Taiwan Univ, BSc, 58; McGill Univ, PhD(marine sci), 66. *Prof Exp:* Asst prof biol, Lakehead Univ, 66-67; scientist res group & cur zool, 67-79, CUR CRUSTACEANS, CAN MUS NATURE, 79- *Concurrent Pos:* Chmn, Second Int Conf on Copepoda. *Mem:* Am Soc Limnol & Oceanog; Ecol Soc Am; Soc Study Evolution; Sigma Xi; Plankton Soc Japan; Can Soc Zool; World Asn Copepodologists; Chinese Soc Limnol & Oceanog; Chinese Soc Zool. *Res:* Systematics and zoogeography of Hyperiidea and Calanoida; taxonomy and ecology of marine zooplankton of northern waters of northern hemisphere. *Mailing Add:* Nat Mus Natural Sci Ottawa ON K1P 6P4 Can

SHIH, CHING-YUAN G, b Taiwan, China, May 25, 34; m 67; c 2. PLANT PHYSIOLOGY, CROP BREEDING. *Educ:* Chung Hsing Univ, Taiwan, BS, 58; Nat Taiwan Univ, MS, 62; Univ Calif, Davis, PhD(plant physiol), 68. *Prof Exp:* Fel veg crops, Univ Calif, Davis, 68-69; res assoc, 69-71, adj asst prof, 78-81, ADJ ASSOC PROF BOT, UNIV IOWA, 81-, DIR, UNIV SCANNING ELECTRON MICROS LAB, 71- *Mem:* AAAS; Electron Micros Soc Am; Am Soc Cell Biol. *Res:* Structure of phloem cells in relation to translocation; plant hormones in relation to nucleic acid synthesis; virus-host-relationship in cultured cells; scanning electron microscopy in biology; biological microsculptures. *Mailing Add:* Windmill 25278 McIntyre Laguna Hills CA 92653

SHIH, CORNELIUS CHUNG-SHENG, b Rukuan, Formosa, Nov 15, 31; m 60; c 4. FLUID MECHANICS. *Educ:* Nat Taiwan Univ, BS, 54; Mich State Univ, MS, 57, PhD(fluid mech), 59. *Prof Exp:* Res asst fluid mech, Mich State Univ, 56-59; assoc prof civil eng, Auburn Univ, 59-65; PROF ENG MECH, UNIV ALA, HUNTSVILLE, 65-, HEAD FLUID DYNAMICS LAB, RES INST, 69- *Mem:* Am Soc Civil Engrs; Eng Soc Japan. *Res:* Hydraulic open-channel flows, secondary flows; fluid amplifiers, vortex motions. *Mailing Add:* Dept Mech Eng Univ Ala 4701 University Dr Huntsville AL 35899

SHIH, FREDERICK F, b China, Dec 11, 36; US citizen; m 77; c 1. FOOD SCIENCE & TECHNOLOGY. *Educ:* Fort Hays State Univ, MS, 66; La State Univ, PhD(chem), 76. *Prof Exp:* Assoc prof chem, Maryville Col, 65-76; RES CHEMIST, USDA, 76- *Mem:* Am Chem Soc; Sigma Xi. *Res:* Textile research on the improvement of cotton fabrics with chemical treatments; chemical and enzymatide modification of oil seed proteins for the improvement of functional properties. *Mailing Add:* SRRC-USDA 4624 David Dr Kenner LA 70065

SHIH, HANSEN S T, b Shanghai, China, April 15, 42. LASER PHYSICS, NONLINEAR OPTICS. *Educ:* Mass Inst Technol, BS, 65, MS, 68; Harvard Univ, PhD(appl physics), 70. *Prof Exp:* Res assoc, Dept Physics & Optical Sci Ctr, Univ Ariz, Tucson, 71-75; res physicist, Naval Res Lab, 75-76; sr res physicist, Tech Ctr, Libby-Omens-Ford, 77-79; TECH SPECIALIST & PROJ MGR, XEROX CORP, 79- *Mem:* Am Phys Soc. *Res:* Laser physics; nonlinear optics; acoustics; process modelling of physical processes of industrial interests, including vacuum film deposition, color, fluid dynamics and acoustics. *Mailing Add:* 828 Bradford Circle Lynn Haven FL 32444

SHIH, HSIANG, b Chungking, China, Nov 11, 43; m 70. POLYMER CHEMISTRY. *Educ:* Nat Taiwan Univ, BS, 65; Yale Univ, MPh, 68, PhD(chem), 69. *Prof Exp:* Fel chem, Stanford Univ, 69-70; res chemist, 70-75, SR RES CHEMIST, PIONEERING RES LAB, TEXTILE FIBERS DEPT, E I DU PONT DE NEMOURS & CO, 75- *Mem:* Am Chem Soc. *Res:* Polymer chemistry and its industrial applications; thermodynamics and other physical chemistry fields. *Mailing Add:* Exp Sta PO Box 80-302 Wilmington DE 19880-0302

SHIH, HSIO CHANG, b Wuchang, Hupei, China, Apr 1, 37; m 71. PHYSICS. *Educ:* Ohio Univ, BSME, 58; Ill Inst Technol, MS, 62, PhD(physics), 68. *Prof Exp:* Mech designer, Precision Transformer Corp, 58-60; asst prof physics, Ill Inst Technol, 68; lectr, 67-68, actg chmn dept, 71, ASSOC PROF PHYSICS, ROOSEVELT UNIV, 68- *Concurrent Pos:* Chmn dept physics & eng sci, 73-77. *Mem:* Am Phys Soc; Am Asn Physics Teachers. *Res:* Electrodynamics. *Mailing Add:* 240 N Linden Westmont IL 60559

SHIH, JAMES WAIKUO, b China, July 24, 41; US citizen; m 68; c 2. VIROLOGY, IMMUNOCHEMISTRY. *Educ:* Chung Hsing Univ, Taiwan, BS, 64; Vanderbuilt Univ, PhD, 70. *Prof Exp:* Mem staff, Molecular Anat Prog, Oak Ridge Nat Lab, 71-78; assoc prof microbiol, Div Molecular Virol & Immunol, Georgetown Univ, 78-80; res microbiologist, Hepatitis Br, Div Blood & Prod, Bur Biologics, Food & Drug Admin, 80-83; SR INVESTR, DEPT TRANFUSION MED, NIH, 83- *Concurrent Pos:* Res assoc pharm, Sch Pharm, Univ Wis, 70; vis fel biochem, Nat Heart & Lung Inst, NIH, 70-71. *Mem:* Am Soc Microbiol; Am Asn Immunologist. *Res:* Identification of antigenic systems of viruses and characterization of structure-function relationship by molecular and immunochemical procedures; specific application in clinical and diagnostic research of viral hepatitis and infectious disease pathogens. *Mailing Add:* Transfusion Transmitted Virus Lab Clin Ctr Nat Inst Health Bethesda MD 20892

SHIH, JASON CHIA-HSING, b Hunan, China, Oct 8, 39; US citizen; m 67; c 2. BIOTECHNOLOGY. *Educ:* Nat Taiwan Univ, BS, 63, MS, 66; Cornell Univ, PhD(biochem), 73. *Prof Exp:* Lectr chem, Tunghai Univ, Taiwan, 66-69; res asst nutrit, Cornell Univ, 69-73; res assoc biochem, Univ Ill, 73-75; sr res assoc poultry sci, Cornell Univ, 75-76; asst prof, 76-80, ASSOC PROF POULTRY SCI, NC STATE UNIV, 80- *Concurrent Pos:* Adv, NC-China Coun Adv Bd, 79-; vis lectr, Ministry Agr, China, 82; vis fel, Univ Col, Cardiff, Wales, 83; vis prof, Shenyang Agr Col, China, 85 & Nat Taiwan Univ, 86; UNDP specialist to China, 87. *Mem:* Poultry Sci Asn; Am Inst Nutrit; Am Soc Microbiol; Soc Chinese Bioscientists Am (secy-treas, 85-86). *Res:* Biotechnology of anaerobic digestion; study of basic microbiology, product utilization and environmental benefits of the bioprocess of anaerobic fermentation of poultry waste; experimental atherosclerosis; study of quail atherosclerosis as a model for pathogenesis; prevention and therapy of atherosclerosis in humans. *Mailing Add:* Dept Poultry Sci NC State Univ PO Box 7608 Raleigh NC 27695-7608

SHIH, JEAN CHEN, b Yunan, China, Jan 29, 42; US citizen; m 69; c 2. BIOCHEMISTRY, PHARMACOLOGY. *Educ:* Nat Taiwan Univ, Taipei, BS, 64; Univ Calif, Riverside, PhD(biochem), 68. *Prof Exp:* Scholar biochem, Dept Biol Chem & Psychiat, Sch Med, Univ Calif, Los Angeles, 68-70, asst res biochemist, 70-74; from asst prof to assoc prof, 74-85, prof biochem, 85-88, BOYD & ELSIE WELIN PROF BIOCHEM, SCH PHARM, UNIV SOUTHERN CALIF, 88- *Concurrent Pos:* Vis assoc prof pharmacol, Dept Pharmacol, Sch Med, Univ Calif, Los Angeles, 81; mem, Psychopath & Clin Biol Res Rev Comt, NIMH, Dept Health & Human Serv, 84-88; mem, Cellular Neurobiol & Psychopharmacol Subcomt, Neurosci Res Rev Comt Dept Health & Human Serv, Nat Inst Mental Health, 90-94; res scientist Award, Nat Inst Mental Health, 90-94; Wellcome vis prof, Burroughs Wellcome Fund & Fed Am Soc Exp Biol. *Mem:* Am Soc Biol Chemists; Am Soc Neurochem; Soc Neurosci; Am Asn Col Pharm; AAAS. *Res:* Molecular mechanism of neurotransmission with emphasis on the structure and function relationship of membrane proteins in central nervous systems; regulation of enzymes and receptors, and their implicaiton in aging and in disease states. *Mailing Add:* Sch Pharm Univ Southern Calif 1985 Zonal Ave Los Angeles CA 90033

SHIH, KWANG KUO, b China, Oct 13, 32; m 64; c 2. SOLID STATE DEVICES & PHYSICS. *Educ:* Nat Taiwan Univ, BS, 53; Purdue Univ, MS, 56; Va Polytech Inst, MS, 59; Stanford Univ, MS, 61, PhD(elec eng), 66. *Prof Exp:* Engr, Sunbeam Corp, 56-57; engr, Hewlett Packard Co, 61-62; res asst, Stanford Univ, 64-66; RES STAFF MEM RES CTR, IBM CORP, 66- *Mem:* Fel Am Inst Chemists; Am Vacuum Soc; Am Phys Soc; Inst Elec & Electronic Engrs; Sigma Xi. *Res:* Electroluminescence; growth of III-V ternary compounds, defects in silicon and the physics of semiconductors; thin film devices; reactive evaporation; sputtering, material and devices including metal, resistine materials, insulating materials and magnetic materials. *Mailing Add:* IBM TJ Watson Res Ctr PO Box 218 Yorktown Heights NY 10598

SHIH, THOMAS Y, b Taipei, Taiwan, July 10, 39; m 68; c 2. CANCER RESEARCH, CELL BIOLOGY. *Educ:* Nat Taiwan Univ, MD, 65; Calif Inst Technol, PhD(biochem, chem), 69. *Prof Exp:* Asst biol, Calif Inst Technol, 66-69; res assoc biochem, Brandeis Univ, 69-71; vis fel molecular biol, 71-72, vis assoc, Lab Biol Viruses, 73-82, MEM STAFF, LAB MOLECULAR ONCOL, NAT CANCER INST, NIH, 82-, HEAD ONCOGENE BIOCHEM GROUP, 85- *Mem:* AAAS; Am Chem Soc; Am Soc Microbiol. *Res:* Oncogenes and cancer; molecular biology of tumor viruses; molecular mechanisms of malignant transformation; genetic engineering; protein biochemistry. *Mailing Add:* 6820 Marbury Rd Bethesda MD 20817

SHIH, TSUNG-MING ANTHONY, b Taipei, Taiwan, Oct 8, 44; US citizen; m 70; c 2. NEUROPHARMACOLOGY, TOXICOLOGY. *Educ:* Kaohsiung Med Col, Taiwan, BS, 67; Univ Pittsburgh, PhD(pharmacol), 74. *Prof Exp:* Teaching asst, Columbia Univ, 68-69; teaching asst, Dept Pharmacol, Univ Pittsburgh, 69-71, NIH trainee, 72-74, res asst III, Dept Psychiat, 79-76, res assoc, 76-78; pharmacologist, Biomed Lab, Chem Systs Lab, 78-80, PHARMACOLOGIST, US ARMY MED RES INST CHEM DEF, ABERDEEN PROVING GROUND, MD, 80- *Concurrent Pos:* Fel, Western Psychiat Inst & Clin, Univ Pittsburgh, 74-76, dean Chinese Lang Sch Baltimore, 82-84, prin, 84-87. *Mem:* Sigma Xi; Am Chem Soc; Am Soc Neurochem; Soc Toxicol; Soc Neurosci. *Res:* Central neuropharmacological mechanisms of action of anticholinesterases and their treatment compounds; transport of choline from plasma, via choline in the brain, to acetylcholine in the brain; central neurotransmitter system dynamics and interactions; mechanism of action of anticonvulsants in organophosphorus compound poisoning. *Mailing Add:* US Army Med Res Inst Chem Defense Bldg E-3100 Aberdeen Proving Ground MD 21010-5425

SHIH, VIVIAN EAN, b China, Dec 27, 34; US citizen; m 65; c 2. BIOCHEMICAL GENETICS. *Educ:* Col Med, Nat Taiwan Univ, MD, 58. *Prof Exp:* Dir amino acid lab, Joseph P Kennedy Jr Mem Labs, 67-, asst neurol, 68-75, ASSOC PROF NEUROL, MASS GEN HOSP & HARVARD UNIV MED SCH, 75-, PEDIATRICIAN, MASS GEN HOSP, 88- *Concurrent Pos:* Consult & co-prin investr, Mass Metab Disorders Prog, Mass Dept Pub Health, 67-80; consult amino acid metab dis, Walter E Fernald State Sch, Waltham, Mass, 68- & Wrentham State Sch, Wrentham, Mass, 68-70; consult pediat, Cambridge Hosp, Cambridge, Mass, 69-; asst prof neurol, Harvard Med Sch, 70-75; assoc, Ctr Human Genetics, 71-80. *Honors & Awards:* Javits Neurosci Investr Award. *Mem:* Soc Pediat Res; Am Acad Pediat; Am Asn Human Genetics; Am Pediat Soc; Soc Inherited Metabolic Disorders. *Res:* Biochemical genetics and hereditary metabolic disorders. *Mailing Add:* Mass Gen Hosp Dept Neurol, 32 Fruit St Boston MA 02114

SHIH, WEI JEN, b Taiwan, Oct 22, 36; US citizen; m 68; c 2. NUCLEAR MEDICINE. *Educ:* Nat Defense Med Ctr, MD, 63; Am Nuclear Med Bd, 82- *Prof Exp:* Fel nuclear med, Am Nuclear Bd Examiners, 76-79, fel pathol,79-81; asst prof nuclear med, 81-86, ASSOC PROF NUCLEAR MED, UNIV KY MED CTR, 86-; CHIEF NUCLEAR MED, VET ADMIN MED CTR, LEXINGTON, 90- *Concurrent Pos:* Attending physician nuclear med, Univ Ky Med Ctr, 81-, Vet Admin Med Ctr, Lexington, 81-, asst chief, 81-90; Fulbright scholar, WGer. *Mem:* Soc Nuclear Med; Radiol Soc N Am; Am Roentgenology Soc. *Res:* Imaging study using radionuclear tracers in humans and animals; recently interested in brain and lung imaging using new radiopharmaceutical 1-123 IMP or HIPDM. *Mailing Add:* Dept Diag Radiol Univ Ky Med Ctr Lexington KY 40536

SHILEPSKY, ARNOLD CHARLES, b Norwalk, Conn, Dec 10, 44; m 68; c 2. MATHEMATICS, COMPUTER SCIENCES. *Educ:* Wesleyan Univ, AB, 66; Univ Wis-Madison, PhD(math), 71. *Prof Exp:* Asst prof, Ark State Univ, 71-74; from asst prof to assoc prof, 74-85, chmn, div phys & math sci, 78-81 & 83-85, PROF MATH & HERBERT E IVES PROF SCI, WELLS COL, 85- *Concurrent Pos:* Proj dir, Exxon Comput Literacy grant, 81-82; Pew vis prof, Cornell Univ, 88. *Mem:* Math Asn Am; Am Math Soc; Asn Women Math; Asn Computer Mach. *Res:* Geometric topology; properties of embeddings in Euclidean spaces; mathematics and computer science education. *Mailing Add:* Dept of Math Wells Col Aurora NY 13026

SHILLADY, DONALD DOUGLAS, b Norristown, Pa, Aug 27, 37; m 68; c 3. THEORETICAL CHEMISTRY. *Educ:* Drexel Univ, BS, 62; Princeton Univ, MA, 65; Univ Va, PhD(chem), 70. *Prof Exp:* Fel chem, Univ Va, 69-70; PROF CHEM, ACAD DIV, VA COMMONWEALTH UNIV, 70- *Mem:* Am Chem Soc; Sigma Xi. *Res:* Quantum chemistry, ab initio and semiempirical computational methods applied to chemical bonding and interpretation of circular dichroism and magnetic circular dichroism of lanthanides in polymers. *Mailing Add:* Dept Chem Va Commonwealth Univ 1003 W Main St Richmond VA 23284

SHILLING, WILBUR LEO, b St Joseph, Mo, Aug 30, 21; wid; c 4. ORGANIC CHEMISTRY, POLYMER CHEMISTRY. *Educ:* Univ Mo, AB, 42; Univ Notre Dame, MS, 47; Ohio State Univ, PhD(chem), 49. *Prof Exp:* Res chemist, Mallinckrodt Chem Works, 42-46; sr res chemist, Crown Zellerbach Corp, 49-69, supvr polymer-fiber res, 69-72, res assoc, 72-77, proj mgr, Pioneering Res, 78-81; RETIRED. *Mem:* Am Chem Soc. *Res:* Cellulose and carbohydrates; lignin and lignans; consumer paper products; applied polymer research; advanced technology assessment. *Mailing Add:* 10615 SW Highland Dr Tigard OR 97224

SHILLINGTON, JAMES KEITH, b Clarion, Iowa, Nov 4, 21. ORGANIC CHEMISTRY. *Educ:* Iowa State Col, BS, 43; Cornell Univ, PhD(chem), 52. *Prof Exp:* Instr chem, Evansville Col, 46-48 & Amherst Col, 52-53; from asst prof to assoc prof, 53-66, PROF CHEM, WASHINGTON & LEE UNIV, 66- *Concurrent Pos:* NSF res grant, 56-57. *Mem:* Am Chem Soc. *Res:* Macro carbon rings; optical resolution; resolving agents. *Mailing Add:* Dept of Chem Washington & Lee Univ Lexington VA 24450

SHILLITOE, EDWARD JOHN, b Hull, Eng, Sept 13, 47; m 77; c 3. ORAL CANCER. *Educ:* Univ London, BDS, 71, PhD(immunol), 76. *Prof Exp:* Fel virol, Dept Microbiol, Hershey Med Ctr, Pa State Univ, 76-78; asst prof oral path, Dent Sch, Univ Calif, San Francisco, 78-83; PROF & CHAIR, DEPT MICROBIOL, DENT BR, UNIV TEX, HOUSTON, 83- *Res:* Etiology of oral cancer, utilizing virological methods. *Mailing Add:* Microbiol Univ Tex Health Sci PO Box 20068 Houston TX 77225

SHILMAN, AVNER, b Tel Aviv, Israel, Aug 28, 23; US citizen; m 72. ANALYTICAL CHEMISTRY, PHYSICAL CHEMISTRY. *Educ:* Columbia Univ, MS, 53, MA, 57; Polytech Inst Brooklyn, PhD(chem), 61. *Prof Exp:* Mgr pharm, Shilman Pharmacy, 45-48 & 50-51; fel, Polytech Inst Brooklyn, 61-63; from asst prof to assoc prof chem, 63-68, PROF CHEM, NEWARK COL ENG, 68- *Mem:* Am Chem Soc. *Res:* Analytical methods and physical chemical principles of chromatographic and electrophoresis processes. *Mailing Add:* 150 East 18th St Apt 11D New York NY 10003-2444

SHILS, MAURICE EDWARD, b Atlantic City, NJ, Dec 31, 14; m 39; c 2. CLINICAL NUTRITION. *Educ:* Johns Hopkins Univ, BA, 37, ScD(nutrit, biochem), 40; NY Univ MD, 58. *Prof Exp:* Asst biochem, Sch Hyg & Pub Health, Johns Hopkins Univ, 40-41, instr, 41-42; asst biochemist, Edgewood Arsenal, US Dept Army, 42-43, food technologist, Off Qm Gen, 43-45; exec secy subcomt nutrit & indust fatigue, Nat Res Coun, 45-46; instr nutrit & indust hyg, Sch Pub Health, Columbia Univ, 46-49, asst prof nutrit, 49-54; res assoc, Sloan-Kettering Inst, 57-59, head, Surg Metab Lab, 59 & metab lab, 61, assoc mem, 60, asst prof biochem, Sloan-Kettering Div, 59-62,; from asst prof to prof, 62-85, EMER PROF MED, MED COL, CORNELL UNIV, 85- *Concurrent Pos:* Assoc attend physician, Mem Hosp, New York, 67-72, attend physician, 72-85, dir clin nutrit, 77-85,; exec secy, 76-85, Comm Pub Health, NY Acad Med, consult clin nutrit, 85-, res scholar, 85- *Honors & Awards:* Goldberger Award Clinical Nutrition, AMA, 83; Rhoads Lectr, Aspen, 87. *Mem:* AAAS; fel Am Inst Nutrit; Am Soc Clin Nutrit (pres, 85-86); Soc Exp Biol & Med; Harvey Soc; Am Col Physicians; Sigma Xi. *Res:* Clinical nutrition research; nutrition and metabolism; trace elements in man; intravenous nutrition. *Mailing Add:* Bowman Gray Sch Med 300 S Hawthorne Rd Winston-Salem NC 27103

SHILSTONE, JAMES MAXWELL, JR, b New Orleans, La, Feb 19, 55; m 82; c 1. QUALITY CONTROL, SOFTWARE DEVELOPMENT. *Educ:* Rice Univ, BA, 77. *Prof Exp:* VPRES, SHILSTONE & ASSOCS, 77- & SHILSTONE SOFTWARE, 85- *Mem:* Am Concrete Inst; Am Soc Concrete Construct. *Res:* Development of computer programs for concrete technical management relating to quality control and production. *Mailing Add:* 8577 Manderville Dallas TX 75231-1001

SHIM, BENJAMIN KIN CHONG, b Honolulu, Hawaii, May 21, 29; m 53; c 5. NICKEL CATALYSTS, RESEARCH & DEVELOPMENT. *Educ:* Univ Rochester, BS, 52; Northwestern Univ, PhD(phys chem), 56. *Prof Exp:* Chemist, Esso Res & Eng Co, 56-62; sr chemist, Lord Corp, 62-68; sr res chemist, Calsicat Div, Mallinckrodt, Inc, 68-86; tech dir, Chem Surveys, Inc, 86-87; CONSULT, 87- *Mem:* Am Chem Soc. *Res:* Heterogeneous catalysts-preparation; characterization and process research and development. *Mailing Add:* 901 Hartt Rd Erie PA 16505-3209

SHIM, JUNG P, b Korea, Sept 3, 47; m 80; c 2. MANAGEMENT SCIENCE, INFORMATION TECHNOLOGY. *Educ:* Yeungnam Univ, BBA, 71; Seoul Nat Univ, MBA, 73; Univ Nebr, Lincoln, PhD(mgt sci), 83. *Prof Exp:* Grad asst mgt sci, Mgt & Info Systs, Univ Nebr, 78-82; asst prof mgt sci, Mgt & Info Systs, Univ Wis-LaCrosse, 82-84; vis prof, Decision Support Systs, Ga State Univ, 90-91; assoc prof, 84-89, PROF MGT SCI, MGT & INFO SYSTS, MISS STATE UNIV, 89- *Concurrent Pos:* Invited lectr, Numerous univs & Res Insts, Korea, 85-; vis prof, Ga State Univ, 90-91; lectr, US Army Masters Bus Admin Prog, 91. *Mem:* Inst Mgt Sci; Acad Mgt. *Res:* Management Information Systems/DDS; expert systems; hypertext/hypermedia; multiple criteria decision making; coauthored several books and software packages. *Mailing Add:* Dept Mgt & Info Systs Miss State Univ Mississippi State MS 39762

SHIMABUKURO, FRED ICHIRO, b Honolulu, Hawaii, Sept 3, 32; m 67; c 1. RADIO ASTRONOMY. *Educ:* Mass Inst Technol, BS, 55, MS, 56; Calif Inst Technol, PhD(elec eng), 62. *Prof Exp:* Mem tech staff, Hughes Aircraft Co, 56-58; mem tech staff, Electronics Res Lab, 62-74, STAFF SCIENTIST, AEROSPACE CORP, 74- *Mem:* Am Astron Soc; Inst Elec & Electronics Engrs. *Res:* Solar radio astronomy; millimeter-wave propagation. *Mailing Add:* Aerospace Corp PO Box 92957 Los Angeles CA 90009

SHIMABUKURO, RICHARD HIDEO, b Hakalau, Hawaii, Sept 20, 33; m 65; c 3. PLANT PHYSIOLOGY. *Educ:* Univ Hawaii, BS, 56; Univ Minn, MS, 62, PhD(plant physiol), 64. *Prof Exp:* RES PLANT PHYSIOLOGIST, BIOSCI RES LAB, USDA, 64-; EXP CONSULT, INT ATOMIC ENERGY AGENCY, SUDAN, 84- *Concurrent Pos:* Expert consult, Int Atomic Energy Agency to the Sudan, 84; distinguished vis scholar, Adelaide Univ, Australia, 90. *Honors & Awards:* Foreign Res Scientist Award, Japanese Govt, 73; Outstanding Res Award, Weed Sci Soc Am, 88. *Mem:* Fel AAAS; Am Soc Plant Physiol; Weed Sci Soc Am; Scand Soc Plant Physiologists. *Res:* Metabolism of chemical pesticides in plants; mechanism of action of herbicidal chemicals in plants and their metabolism in different plant organs. *Mailing Add:* Biosci Res Lab State Univ Sta Fargo ND 58105

SHIMADA, KATSUNORI, b Tokyo, Japan, Mar 12, 22; nat US; m 54; c 2. ELECTRONICS ENGINEERING. *Educ:* Univ Tokyo, BS, 45; Univ Minn, MS, 54, PhD(elec eng), 58. *Prof Exp:* Engr, Tokyo Shibaura Elec Co, Japan, 45-49; instr elec eng, Univ Minn, 54-58; assoc prof, Univ Wash, 58-64; resident res appointee, Jet Propulsion Lab, Calif Inst Technol, 64-65, res group supvr, 65-88; CONSULT, NASDA LA, 88- *Concurrent Pos:* Consult, Boeing Co, Wash, 61-63; vis prof, Inst Space & Aeronaut Sci, Tokyo, Japan, 73. *Honors & Awards:* NASA Cert of Recognition, 73, 77, 78, 81, 83, 84, 88, 89. *Mem:* Inst Elec & Electronics Engrs. *Res:* Celential Sensors. *Mailing Add:* 3840 Edgeview Dr Pasadena CA 91107

SHIMAMOTO, YOSHIO, b Honolulu, Hawaii, Oct 4, 24; m 55; c 2. THEORETICAL PHYSICS, MATHEMATICS. *Educ:* Univ Hawaii, AB, 48; Harvard Univ, AM, 51; Univ Rochester, PhD(physics), 54. *Prof Exp:* Assoc physicist, Brookhaven Nat Lab, 54-58, physicist, 58-64, chmn dept appl math, 64-75, sr scientist, 64-; RETIRED. *Concurrent Pos:* Vis res prof digital comput lab, Univ Ill, Urbana, 64, consult dept comput sci, 67-; mem math & comput sci res adv comt, US AEC, 65-72, chmn, 69-71; adj prof dept math statist, Columbia Univ, 71-72; vis prof math inst, Hanover Tech Univ, 72-73; assoc ed, J Comput Physics, 75-77; pvt consult. *Honors & Awards:* Sr Scientist Award, Alexander von Humboldt-Stiftung, 72. *Res:* Reactor and particle physics; computer design; graph theory. *Mailing Add:* 158 Donegan Ave East Patchogue NY 11772

SHIMAMURA, TETSUO, b Yokohama, Japan, Feb 18, 34; m 60; c 2. PATHOLOGY. *Educ:* Yokohama Munic Univ, MD, 59. *Prof Exp:* Intern med, US Army Med Command, Japan, 59-60; intern, Bexar County Hosp, 60-61; residency in path, Sch Med, Washington Univ, 61-64; resident, Methodist Hosp, Houston, Tex, 64-65; res asst, Baylor Sch Med, & resident res assoc, Vet Admin Hosp, Houston, 65-66; asst prof, Baylor Sch Med, 66; asst prof, Univ SDak, 67-68; from asst prof to assoc prof path, Rutgers Med Sch, 68-75, prof, 75-; AT DEPT CHEM, UNIV CALIF. *Mem:* Int Acad Path; Am Soc Nephrol; Int Soc Nephrol. *Res:* Renal pathology; experimental amyloidosis and hydronephrosis; experimental chronic renal disease; nutrition and glomerular diseases. *Mailing Add:* Robert Wood Johnson Med Sch 675 Hoes Lane Piscataway NJ 08854

SHIMAN, ROSS, b Washington, DC, May 22, 38; m 64; c 2. BIOCHEMISTRY. *Educ:* Columbia Col, BA, 60; Univ Calif, Berkeley, PhD(biochem), 65. *Prof Exp:* Res chemist, NIH, 65-68; PROF BIOL CHEM, M S HERSHEY MED CTR , PA STATE UNIV, 69- *Concurrent Pos:* NIH fel, 66-67. *Mem:* Am Chem Soc; Am Soc Cell Biol; Am Soc Biol Chemists. *Res:* Mechanism of phenylalanine hydroxylase action; regulation of enzyme expression and activity in mammalian cells. *Mailing Add:* Dept Biol Chem Hershey Med Ctr Pa State Univ Hershey PA 17033

SHIMANUKI, HACHIRO, b Kahului, Hawaii, July 25, 34; m 58, 83; c 3. INSECT PATHOLOGY, APICULTURE. *Educ:* Univ Hawaii, BA, 56; Iowa State Univ, PhD(bact), 63. *Prof Exp:* Res microbiologist, 63-66 invest leader, Bioenviron Bee Lab, 66-72, microbiologist, 72-75, lab chief, Bioenviron Bee Lab,; res leader, 85-87, RES MICROBIOL, BENEFICIAL INSECTS LAB, AGR RES SERV, USDA 87- *Concurrent Pos:* Ed, Apidologie, 82- *Honors & Awards:* J I Hambleton Mem Award, 78; First Apicult Res Award, Apiary Insp Am, 78. *Mem:* Am Soc Microbiol; Entom Soc Am; Soc Invert Path; Am Beekeeping Fedn; Sigma Xi; Am Honey Prod; Am Asn Prof Apicult. *Res:* Diseases of honey bees; computer simulation of honey bee populations; honey bee nutrition; parasitic bee mutes. *Mailing Add:* Bee Res Lab Bldg 476 Arc-East Beltsville MD 20705

SHIMAOKA, KATSUTARO, b Nara, Japan, Sept 4, 31; m 56; c 2. ONCOLOGY, THYROIDOLOGY. *Educ:* Keio Univ, Japan, MD, 55. *Prof Exp:* Intern, St Luke's Hosp, Denver, Colo, 56-57; resident med, Louisville Gen Hosp, Ky, 57-58; resident med, Roswell Park Mem Inst, Buffalo, NY, 58-59, fel 59-61; res asst, Univ Col Hosp Med Sch, London, 61-63; sr res assoc, Roswell Park Mem Inst, 63-65, sr cancer res scientist, 65-67, cancer res internist I, 67-69, cancer res internist II, 69-79, assoc chief cancer clinician, 79-86; chief Nagasaki Lab, 87-89, ASSOC CHIEF RES, RADIATION EFFECTS RES FOUND, 89- *Concurrent Pos:* From res asst prof to res assoc prof, State Univ NY, Buffalo, 71-81, res prof med, 81-, res assoc prof physiol, 72-; chief Endocrinol Clin, E J Meyer Mem Hosp, Buffalo, 74; mem consult staff med, Erie Co Med Ctr, Buffalo, 75-86; attending & consult physician, Vet Admin Med Ctr, Buffalo, 79-; lectr, Nagasaki Univ Sch Med, 87-; lectr, Res Inst Environ Med, Nagoya Univ, 88-89. *Mem:* Endocrine Soc; Am Asn Cancer Res; Soc Nuclear Med; Am Soc Clin Oncol; Am Thyroid Asn; Am Soc Bone Mineral Res. *Res:* Thyroid and iodine metabolism; radiation induced cancer; cancer chemotherapy; lymphoma and leukemia; parathyroid and calcium metabolism. *Mailing Add:* Radiation Effects Res Found 1-8-6 Nakagawa Nagasaki 850 Japan

SHIMIZU, C SUSAN, biochemistry, physiology; deceased, see previous edition for last biography

SHIMIZU, HIROSHI, b Kyoto, Japan, Aug 21, 24; m 51; c 2. OTOLARYNGOLOGY, AUDIOLOGY. *Educ:* Kyoto Prefectural Univ Med, MD, 49, MScD, 55. *Prof Exp:* Asst otolaryngol, Kyoto Prefectural Univ Med, 50-56, instr, 56-57; clin fel, White Mem Hosp, Los Angeles, 57-58; resident otolaryngol, Hosp, 58; res fel audiol, Johns Hopkins Univ, 58-60; instr, Kyoto Prefectural Univ Med, 60-63; asst prof, 63-67, ASSOC PROF OTOLARYNGOL, SCH MED, JOHNS HOPKINS UNIV, 67-, DIR HEARING & SPEECH CTR, 76- *Concurrent Pos:* NIH res grants, 65-68 & 67-70. *Mem:* Am Auditory Soc; Am Speech & Hearing Asn; Japan Soc Otolaryngol; Japan Soc Audiol; Acoust Soc Am. *Res:* Auditory evoked potentials in both normal listeners and patients. *Mailing Add:* 120 Othoridge Rd Lutherville MD 21093

SHIMIZU, NOBUMICHI, b Tokyo, Japan, Feb 4, 40; m 65; c 2. TRACE ELEMENT GEOCHEMISTRY, GEOCHEMICAL KINETICS. *Educ:* Univ Tokyo, BSc, 63, MSc, 65, DSc, 68. *Prof Exp:* Instr geol, Univ Tokyo, 68-75; res assoc geochem, Carnegie Inst Wash, 71-74; assoc prof geochem, Universite de Paris 6, 74-78; sr res sci geochem, 78-88, VIS PROF GEOCHEM, MASS INST TECHNOL, 88-; SR SCI GEOCHEM, WOODS HOLE OCEANOG INST, 88- *Mem:* Geochem Soc; Am Geophys Union. *Res:* Distribution of trace elements among minerals; kinetics of geochemical processes; geochemical evolution of the mantle. *Mailing Add:* Dept Geol & Geophys Clark 102B Woods Hole Oceanog Inst Woods Hole MA 02543

SHIMIZU, NOBUYOSHI, b Osaka-city, Japan, Aug 10, 41; m 67; c 1. CELL BIOLOGY, HUMAN GENETICS. *Educ:* Nagoya Univ, BA, 65; Inst Molecular Biol, MSc, 67, PhD(molecular biol), 70. *Prof Exp:* Res assoc, Inst Molecular Biol, 70-71; res biologist, Univ Calif, 71-74 & Yale Univ, 74-76; from asst prof to assoc prof, 77-82, PROF RES & TEACHING, UNIV ARIZ, 82- *Concurrent Pos:* Res biologist, Yale Univ, 74-75, vis asst prof, 77; jr fac res award, Am Cancer Soc, 78-80. *Mem:* NY Acad Sci; AAAS; Tissue Cult Asn; Am Soc Cell Biol; Am Soc Biol Chemists. *Res:* Genetic control of mammalian cell surface functions with emphasis on receptor-mediated hormonal signal transfer mechanisms and malignant transformation; parasexual approaches to human genetics and chromosome mapping. *Mailing Add:* Dept Molecular & Cellular Biol BSW 308 Univ Ariz Tucson AZ 85721

SHIMIZU, YUZURU, b Gifu, Japan, Jan 17, 35; m 63; c 2. NATURAL PRODUCTS CHEMISTRY, PHARMACOGNOSY. *Educ:* Hokkaido Univ, BS, 58, MS, 60, PhD(pharm sci), 63. *Prof Exp:* Scientist, Worcester Found Exp Biol, 63-64; res assoc dept chem, Univ Ga, 64-65; instr pharm sci, Hokkaido Univ, 65-69; asst prof pharmacog, 69-73, assoc prof, 73-77, PROF PHARMACOG, UNIV RI, 77- *Concurrent Pos:* Water resources grant, 71-75; Dept Health, Educ & Welfare grant, 74-; chmn marine natural prod sect, Gordon Res Conf, 75- *Honors & Awards:* Award, Matsunaga Sci Found, Japan, 69. *Mem:* Am Chem Soc; Am Soc Pharmacog; Pharmaceut Soc Japan; Sigma Xi. *Res:* Isolation, structural elucidation and synthesis of natural products, especially of marine origins; marine pharmacognosy. *Mailing Add:* Col Pharm Univ RI Kingston RI 02881

SHIMKIN, MICHAEL BORIS, cancer; deceased, see previous edition for last biography

SHIMM, ROBERT A, b New York, NY, Jan 30, 26; m 58; c 3. INTERNAL MEDICINE. *Educ:* Columbia Univ, AB, 45, MD, 48; Am Bd Internal Med, dipl, 56, recert, 77. *Prof Exp:* Intern med, Presby Hosp, New York, 49-50, jr asst resident, 50-51; sr asst resident, Duke Hosp, Durham, NC, 51-52; NIH trainee cardiol, Mt Sinai Hosp, New York, 52-53; resident med, Bellevue Hosp, 55-56; instr, 56-59, assoc, 59-62, from asst prof to assoc prof med, 62-76, CLIN PROF MED, ALBERT EINSTEIN COL MED, 76- *Mem:* Am Soc Internal Med; Am Col Physicians. *Res:* Medical administration; comprehensive ambulatory private health care. *Mailing Add:* 1180 Morris Park Ave Bronx NY 10461

SHIMONY, ABNER, b Columbus, Ohio, Mar 10, 28; m 51; c 2. THEORETICAL PHYSICS, PHILOSOPHY OF SCIENCE. *Educ:* Yale Univ, BA, 48, PhD(philos), 53; Univ Chicago, MA, 50; Princeton Univ, PhD(physics), 62. *Prof Exp:* Instr philos, Yale Univ, 52-53; from asst prof to assoc prof, Mass Inst Technol, 59-68; assoc prof, 68-73, PROF PHYSICS & PHILOS, BOSTON UNIV, 73- *Concurrent Pos:* Sr NSF fel, 66-67; fel, Am Coun Learned Soc, 67 & Guggenheim Found fel, 72-73; Luce prof cosmol, Mt Holyoke Col, 81; Nat Endowment Humanities, 91. *Mem:* Am Philos Soc; Am Phys Soc; fel Am Acad Arts & Sci. *Res:* Foundations of quantum mechanics; foundations of statistical mechanics; philosophy of physics; naturalistic epistemology; inductive logic. *Mailing Add:* Dept Physics Boston Univ 111 Cummington St Boston MA 02215

SHIMOTAKE, HIROSHI, b Dec 6, 28; US citizen; m 60; c 3. CHEMICAL ENGINEERING. *Educ:* Nihon Univ, Tokyo, BS, 51; Northwestern Univ, MS, 57, PhD(chem eng), 60. *Prof Exp:* Res engr, Whirlpool Res Labs, 60-63; chem engr & group leader, Argonne Nat Lab, 63-85; sr res engr, 85-87, MGR, AMOCO LASER CO, 88- *Concurrent Pos:* Adj prof chem eng, Case Western Reserve Univ, 85- *Honors & Awards:* IR-100 Award, 68. *Mem:* Am Chem Soc; Electrochem Soc Japan; Japanese Soc Chem Engrs; Electrochem Soc. *Res:* Energy conversion and storage; high energy electrochemical cells and batteries. *Mailing Add:* 726 Franklin St Hinsdale IL 60521

SHIMP, NEIL FREDERICK, b Akron, Ohio, Aug 19, 27; m 49; c 2. ANALYTICAL CHEMISTRY. *Educ:* Mich State Univ, BS, 50, MS, 51; Rutgers Univ, PhD, 56. *Prof Exp:* Res chemist citrus exp sta, Univ Fla, 51-52; asst prof, Rutgers Univ, 56-57; assoc chemist, Ill State Geol Surv, 57-63, chemist head analytical chem sect, 63-73, prin chemist, 73-88; RETIRED. *Honors & Awards:* R A Glenn Award, Am Soc Testing & Mat. *Mem:* Am Chem Soc; Soc Appl Spectros; Am Soc Test & Mat. *Res:* Geochemistry; trace elements; instrumental analysis; coal chemistry; environmental geology; spectrochemistry. *Mailing Add:* 2900 Oak Ct Spring Lake MI 49456

SHIMURA, GORO, b Hamamatsu, Japan, Feb 23, 30; m 59; c 2. MATHEMATICS. *Educ:* Univ Tokyo, BS, 52, DSc(math), 58. *Prof Exp:* Asst prof math, Univ Tokyo, 57-61; prof, Univ Osaka, 61-64; vis prof, 62-64, PROF MATH, PRINCETON UNIV, 64 - *Concurrent Pos:* Res mem, Nat Ctr Sci Res, Paris, France, 57-58; mem, Inst Advan Study, 58-59, 67, 70-71 & 74-75; John Simon Guggenheim fel, 70-71. *Honors & Awards:* Cole Prize in Number Theory, Am Math Soc, 77. *Mem:* Am Math Soc; Math Soc Japan. *Res:* Number theory; automorphic functions; algebraic geometry. *Mailing Add:* Fine Hall Princeton Univ Princeton NJ 08544

SHIN, ERNEST EUN-HO, b Chindo, Korea, Dec 31, 35; US citizen; m 63; c 4. DENSITY MATRIX THEORY, RECIPROCITY PRINCIPLE. *Educ:* Carneigie Inst Technol, BS, 57; Harvard Univ, Am, 60, PhD(physics), 61. *Prof Exp:* Teaching fel physics, Harvard Univ, 58-60; res assoc physics,

Advan Res Div, Arthur D Little, Inc, 60-62; res assoc, Nat Magnet Lab, Mass Inst Technol, 62-66; prof physics, Univ Miami, 66-72; dir, Nieman Inst, 72-74; chmn & chief exec officer, Yulsan Am, Inc, 74-89; DIR, CHESTNUT HILL INST, 84-, CHMN & CHIEF EXEC OFFICER, CHESTNUT HILL GROUP, 89- *Concurrent Pos:* Vis scientist, Korea Atomic Energy Res Inst, 73-74. *Mem:* Am Physical Soc; NY Acad Sci; AAAS; Int Platform Asn. *Res:* Quantum theory of optical and electrical properties of solids; nonlocal field theory and symmetry properties of elementary particles; nonlinear optical properties of electrons. *Mailing Add:* Chestnut Hill Inst PO Box 3510 Napa CA 94558

SHIN, HYUNG KYU, b Kochang, Korea, Sept 3, 33; m 63; c 3. PHYSICAL CHEMISTRY, THEORETICAL CHEMISTRY. *Educ:* Univ Utah, BS, 59, PhD(phys chem), 61. *Prof Exp:* Res chemist, Nat Bur Standards, Washington, DC, 61-63; fel theoret chem, Cornell Univ. 63-65; from asst prof to assoc prof 65-67, assoc prof 67-70, chmn dept, 76-80, PROF PHYS CHEM, UNIV NEV, RENO, 70- *Concurrent Pos:* Petrol Res Fund grant, 65-67; Air Force Off Sci Res grant, 67-78; vis res prof, Univ Calif, Berkeley, 80; Petrol Res Fund grant, 81-84; univ found prof, Univ Nev, Reno, 84-87, fel, Ctr Advan Study, 84- *Honors & Awards:* Award, Sigma Xi Soc, 61. *Mem:* Am Phys Soc; Sigma Xi; Am Chem Soc. *Res:* Theory of inelastic collisions; theory of non-equilibrium rate processes; dynamics of weakly bound complexes. *Mailing Add:* Dept Chem Univ Nev Reno NV 89557-0020

SHIN, KJU HI, b Pusan, Korea, Nov 11, 29; US citizen; c 3. CHEMISTRY. *Educ:* Seoul Nat Univ, BS, 52; Univ Frankfurt, MS, 58, PhD(chem), 60. *Prof Exp:* Fel & res assoc, Cornell Univ, 60-62; Nat Res Coun Can fel, 62-63; staff chemist res labs, UniRoyal Co, Can, 63-66; res chemist, 66-75, RES ADV, ETHYL CORP, 75- *Mem:* Am Chem Soc; Korean Chem Soc; Korean Scientists & Engrs Am. *Res:* Organic synthesis and catalysis; antioxidants; oxidation; synthesis of fine chemicals. *Mailing Add:* Tech Ctr 8000 GSRI Ave Baton Rouge LA 70808

SHIN, MOON L, b Seoul, Korea, Feb 1, 38. NEUROIMMUNOLOGY. *Educ:* Korean Univ, MD, 62. *Prof Exp:* Assoc prof, 78-85, PROF, DEPT PATH, UNIV MD, BALTIMORE, 85- *Mem:* Am Asn Immunologists; Am Asn Pathologists. *Mailing Add:* Sch Med Univ Ten S Pine St Baltimore MD 21201

SHIN, MYUNG SOO, b Seoul, Korea, May 5, 30; m 60; c 3. RADIOLOGY. *Educ:* Seoul Nat Univ, MD, 56, PhD(radiol), 67. *Prof Exp:* Instr radiol, Sch Med, Seoul Nat Univ, 63-67, asst prof, 67-68; from asst prof to assoc prof, 69-75, PROF RADIOL, SCH MED, UNIV ALA, BIRMINGHAM, 75- *Concurrent Pos:* Consult, Radiol Serv, Vet Admin Hosp, Birmingham, 68- *Mem:* AMA; fel Am Col Radiol; Radiol Soc NAm; Am Roentgen Ray Soc; Asn Univ Radiol; Soc Thoracic Radiol; fel Am Col Chest Physicians. *Res:* Chest imaging: computed tomography and MRI. *Mailing Add:* Dept Diag Radiol Univ Ala Hosp Birmingham AL 35233

SHIN, SEUNG-IL, b Wonju City, Korea, Nov 10, 38. GENETICS, CELL BIOLOGY. *Educ:* Brandeis Univ, BA, 64, PhD(biochem), 69. *Prof Exp:* Dutch Orgn Sci Res fel, State Univ Leiden, 69-70; mem, Basel Int Immunol, 70-72; asst prof, 72-77, assoc prof, 77-81, prof genetics, Albert Einstein Col Med, 82-85; PRES, EUGENE TECH INT, 84- *Concurrent Pos:* Fac res award, Am Cancer Soc, 76-81; mem, Scientific Adv Comt, Damon Runyon-Walter Winchell Cancer Fund, 77-82; vis prof microbiol, Seoul Nat Univ, Korea, 79. *Mem:* AAAS; Genetics Soc Am; Am Soc Cell Biol; Am Asn Cancer Res; Am Diabetes Asn. *Res:* Molecular genetics of somatic cells; tumor biology; etiologic mechanisms of insulin-dependent diabetes mellitus; research administration. *Mailing Add:* Eugene Tech Int Inc Four Pearl Ct Allendale NJ 07401

SHIN, SOO H, b Chonnam, Korea, Feb 5, 40; US citizen; m 73; c 2. II-VI SEMICONDUCTORS, INFRARED DETECTORS. *Educ:* Chosun Univ, Korea, BS, 63; Yonsei Univ, Korea, MS, 65; Purdue Univ, PhD(phys chem), 72. *Prof Exp:* Lectr chem, Yonsei Univ, Seoul, Korea, 65-66; asst, Sogang Univ, Seoul, Korea, 66-68; res assoc physics, Yeshiva Univ, NY, 73-78; MEM TECH STAFF, ROCKWELL INT SCI CTR, 78- *Concurrent Pos:* Prin investr long wavelength avalanche photodiode, Rockwell Int Sci Ctr, 88- *Mem:* Am Chem Soc; Am Phys Soc. *Res:* Epitaxial thin film growth and characterization of semiconductors for infrared detector and opto-electronic device applications; author of over 80 scientific and technical publications; awarded two patents. *Mailing Add:* 2997 Canna St Thousand Oaks CA 91360

SHIN, SUK-HAN, b Seoul, Korea, Aug 28, 30; US citizen. OPERATIONS RESEARCH. *Educ:* Seoul Nat Univ, Korea, BA, 54; Clark Univ, Mass, MA, 67; Univ Pittsburgh, PhD(geog), 75. *Prof Exp:* Teacher geog, Jim Mynny Girl's High Sch, 54-62; PROF, GEOG, EASTERN WASH UNIV, 69- *Concurrent Pos:* Asst planner, Western Pa Regional Planning Comt, 70; res fel, Korean Res Inst Human Settlement, 80; environ consult, Eng-Sci, Inc, 81; vis prof, Seoul Nat Univ, Korea, 81 & 87; invited lectr, Environ Sci Tycining, 83; exchange prof, Dungguk Univ, Seoul. *Mem:* Asn Am Geogr; Northwest Sci Asn. *Res:* Changing environmental perception in relation to level of economic development and social value system; impacts of industrialization on socio-economic environment and amenity resources. *Mailing Add:* 522 Short St Cheney WA 99004

SHIN, YONG AE IM, b Seoul, Korea, Aug 2, 32; m 61; c 3. INORGANIC CHEMISTRY, MOLECULAR BIOLOGY. *Educ:* Tift Col, BA, 56; Ohio State Univ, MSc, 58, PhD(chem), 60. *Prof Exp:* Res fel chem, Univ Ill, 61-62; res assoc chem, Ohio State Univ, 62-64; fel, 65-67, RES CHEMIST, GERONT RES CTR, NIH, 67- *Concurrent Pos:* mem, Health Sci Admin, Nat Inst Aging, Nat Inst Diabetes & Digestive Dis & Kidney, Nat Inst Gen Med Sci. *Mem:* Biophys Soc; Am Soc Biol Chemists; Korean Scientists & Engrs Am; fel Am Inst Chemists. *Res:* Stereospecificity in coordination compounds; the role of metals in nucleic acids and proteins; structure and function of nucleic acids and nucleoproteins; aging; polymorphism of nucleic acids. *Mailing Add:* Geront Res Ctr Nat Inst Aging/NIH Baltimore MD 21224

SHIN, YONG-MOO, b Seoul, Korea, June 14, 31; m 56; c 2. NUCLEAR PHYSICS. *Educ:* Yonsei Univ, Korea, BS, 57; Univ Pa, MS, 60, PhD(physics), 63. *Prof Exp:* From res assoc to asst prof, Univ Tex, 63-65; from asst prof to assoc prof, 65-75, PROF PHYSICS & DIR ACCELERATOR LAB, UNIV SASK, 75- *Mem:* Can Asn Physics; Can Nuclear Soc; Am Phys Soc. *Res:* Nuclear structure and reaction mechanism. *Mailing Add:* Dept Physics Univ Sask Saskatoon SK S7N 0W0 Can

SHINBROT, MARVIN, mathematics; deceased, see previous edition for last biography

SHINDALA, ADNAN, b Mosul, Iraq, July 1, 37; m 64; c 3. SANITARY ENGINEERING. *Educ:* Univ Baghdad, BSc, 58; Va Polytech Inst, MSc, 61, PhD(civil eng), 64. *Prof Exp:* Asst resident engr, Govt Iraq, 59-60; asst prof civil eng, Lehigh Univ, 64-65; lectr, Univ Baghdad, 65-67; from asst prof to assoc prof sanit eng, 67-77, PROF ENVIRON ENG, MISS STATE UNIV, 77-; PRIN, COOK COGGIN ENGRS, INC, 76- *Concurrent Pos:* vpres, Cook Coggin Engrs, Inc, Tupelo, MS; Herrin-Hess outstanding prof civil eng, 89. *Mem:* Am Soc Civil Engrs; Water Pollution Control Fedn; Am Water Works Asn; Am Water Resources Asn; Asn Environ Eng Professors. *Res:* Water resources engineering; water quality modeling; water and waste treatment. *Mailing Add:* PO Drawer CE Mississippi State Univ Mississippi State MS 39762

SHINE, ANDREW J(OSEPH), b Cass, WVa, Jan 3, 22; m 46; c 5. MECHANICAL ENGINEERING. *Educ:* Rensselaer Polytech Inst, BME, 46, MME, 47; Ohio State Univ, PhD(mech eng), 57. *Prof Exp:* Asst, Univ Minn, 47-48; instr & asst, Rensselaer Polytech Inst, 48-49; from asst prof to assoc prof, 49-58, PROF MECH ENG & HEAD DEPT, US AIR FORCE INST TECHNOL, 58- *Mem:* Am Soc Mech Engrs; Am Soc Eng Educ. *Res:* Heat transfer; fluid flow. *Mailing Add:* 2065 SR 235 Xenia OH 45385

SHINE, ANNETTE DUDEK, b Toledo, Ohio, Jan 20, 54; m 81; c 2. CHEMICAL ENGINEERING. *Educ:* Wash Univ, BS & AB, 76; Case Western Res Univ, MSE, 79; Mass Inst Technol, PhD(chem eng), 83. *Prof Exp:* Res scientist, Eastman Kodak Co, 82-85; asst prof chem eng, Colo Sch Mines, 86-88; ASST PROF CHEM ENG, DEPT CHEM ENG, UNIV DEL, 89- *Mem:* Am Inst Chem Engrs; Am Chem Soc; Soc Women Engrs; Soc Rheology. *Res:* Relationship between structure and processing of polymeric materials, especially liquid crystalline polymers and polymer blends and composites. *Mailing Add:* Dept Chem Eng Univ Del Newark DE 19716

SHINE, DANIEL PHILLIP, b Chicago, Ill, Aug 10, 34. COSMETIC CHEMISTRY. *Educ:* Xavier Univ, BS, 55, MS, 57; Univ Akron, PhD(chem), 61. *Prof Exp:* Instr chem, Villa Madonna Col, 61-64; RES CHEMIST, ANDREW JERGENS CO, 64- *Res:* Analysis of cosmetics, fats, oils and waxes. *Mailing Add:* Andrew Jergens Co 2535 Spring Grove Ave Cincinnati OH 45214

SHINE, HENRY JOSEPH, b London, Eng, Jan 4, 23; m 53; c 2. ORGANIC CHEMISTRY. *Educ:* London Univ, BSc, 44, PhD(chem), 47. *Prof Exp:* Chemist, Shell Develop Co, Eng, 44-45; res fel org chem, Iowa State Univ, 48-49; res fel, Calif Inst Technol, 49-51; res chemist, US Rubber Co, 51-54; from asst prof to prof chem, 54-68, chmn dept, 69-75, PAUL WHITFIELD HORN PROF CHEM, TEX TECH UNIV, 68- *Concurrent Pos:* Distinguished Sr US Scientist Award, Alexander von Humboldt Found, 86-87. *Mem:* AAAS; Am Chem Soc; Royal Soc Chem; Sigma Xi. *Res:* Reaction mechanisms; aromatic rearrangements; organosulfur chemistry; ion radical reactions; electron spin spectroscopy; heavy-atom kinetic isotope effects. *Mailing Add:* Dept Chem & Biochem Tex Tech Univ Lubbock TX 79409-4260

SHINE, KENNETH I, b Worchester, Mass. MEDICAL ADMINISTRATION. *Educ:* Harvard Univ, MD, 61. *Prof Exp:* Assoc med, Beth Israel Hosp, Boston, 61; intern, Mass Gen Hosp, 61-62, res, 62-63 & 65-66, fel cardiologist, 66-68; assoc med, Beth Israel Hosp, Boston, 61-71; from asst prof to prof, 71-79, chief cardiol, 79-86, DEAN, SCH MED, UNIV CALIF, LOS ANGELES, 86- *Concurrent Pos:* Surgeon, USPHS, 63-65; instr med, Harvard, 68. *Mem:* Nat Acad Sci; Inst Med-Nat Acad Sci. *Mailing Add:* Sch Med Univ Calif Los Angeles CA 90024

SHINE, M CARL, JR, b Newton Center, Mass, Feb 27, 37. CORROSION, MOS-SEMICONDUCTOR PHYSICS. *Educ:* Univ Pa, BS, 59, PhD(metal), 68; Mass Inst Technol, MS, 61, Engrs, 62. *Prof Exp:* Staff engr, Thomas J Watson Labs, IBM, 68-70, IBM, E Fishkill, NY, 70-73; sr scientist, TRW Philadelphia Labs, 73-75; prin metallurgist, Fischer & Porter, Warminster, Pa, 76-79; sr scientist, P R Mallory, Burlington, Mass, 79-80; sr metallurgist, Raytheon Corp, Waltham, Mass, 80-81; prin engr, Andover, Mass, 81-88, CONSULT ENGR, DIGITAL EQUIP, CUPERTINO, CALIF, 88- *Concurrent Pos:* Consult, Fischer & Porter Corp, Warminster, Pa, 75-76. *Mem:* Mat Res Soc. *Res:* Metallurgical kinetics; thermal fatigue, creep, superplasticity in solder joints; MOS semiconductor physics; fast states, concentration profiling analysis, dielectric trapping analysis; corrosion, electromigration, dielectric relaxation; metal/plastic adhesion. *Mailing Add:* 11592 Bridge Park Ct Cupertino CA 95014

SHINE, ROBERT JOHN, b Orange, NJ, May 21, 41; m 67; c 3. ORGANIC CHEMISTRY, COMPUTER SCIENCE. *Educ:* Seton Hall Univ, BS, 62; Pa State Univ, PhD(org chem), 66; Stevens Inst Techonol, MSc, 85. *Prof Exp:* Teaching asst, Pa State Univ, 62; res chemist, Walter Reed Army Inst Res, 67-69; asst prof chem, Trinity Col, DC, 69-71; from asst prof to assoc prof, 71-74, PROF CHEM, RAMAPO COL, NJ, 74-, PROF COMPUT SCI, 85- *Mem:* Am Chem Soc; Sigma Xi. *Res:* Structural determination of alkaloids; chemistry of organic sulfur and organic selenium compounds. *Mailing Add:* Sch of Theoret & Appl Sci Ramapo Col Mahwah NJ 07430-1680

SHINE, TIMOTHY D, b New York, NY, June 6, 39. ORGANIC CHEMISTRY. *Educ:* Merrimack Col, BS, 60; Univ Conn, PhD(org chem), 67. *Prof Exp:* Asst inst chem, Univ Conn, 62-66; res assoc, Univ Mich, 66-67; asst prof, 67-71, assoc prof, 71-77, PROF CHEM, 77-, CHAIR CHEM DEPT, CENT CONN STATE UNIV, 84- *Mem:* Sigma Xi; Am Chem Soc. *Res:* Acyl and alkoxy group migrations between oxygen and nitrogen in oo-aminophenols; sulfur and nitrogen in o-aminothiophenols. *Mailing Add:* Dept Chem Cent Conn State Univ New Britain CT 06050

SHINE, WILLIAM MORTON, b St Louis, Mo, Nov 24, 12; m 46; c 4. ORGANIC CHEMISTRY. *Educ:* Wash Univ, St Louis, BS, 34; Univ Ill, MS, 37. *Prof Exp:* Res chemist med sch, Wash Univ, St Louis, 34-36; res chemist, Univ Ill, 36-37; chemist, Pfanstiehl Chem Co, 37-41; sect leader res & chem develop plastics dept, Gen Elec Co, 41-45; mkt develop chemist, Gen Aniline & Film Corp, 45-50; dir mkt develop, Arnold Hoffman & Co, Inc, 50-53; dir develop dept, Celanese Corp Am, 53-55, techno-com dir cent tech dept, 55-59, vpres, Celanese Develop Co, 59-63; PRES, WILLIAM M SHINE CONSULT SERV, 63- *Mem:* Am Chem Soc; Soc Plastics Eng; Chem Develop Asn; Chem Mkt Res Asn; fel Am Inst Chem; Soc Chem Indust; NY Acad Sci. *Res:* Organic synthesis; polymer chemistry; international technology; marketing research; corporate planning; technical-economics evaluations; acquisition studies. *Mailing Add:* William M Shine Consult Serv PO Box 2069 Heritage Village Southbury CT 06488

SHINEFIELD, HENRY R, b Paterson, NJ, Oct 11, 25; m; c 4. MEDICINE, PEDIATRICS. *Educ:* Columbia Univ, AB, 44, MD, 48; Am Bd Pediat, dipl. *Prof Exp:* From asst prof to assoc prof pediat, Med Col, Cornell Univ, 59-65; chief, 65-68, EMER CHIEF PEDIAT, KAISER FOUND HOSP, 68-; CLIN PROF, SCH MED, UNIV CALIF, SAN FRANCISCO, 68- *Concurrent Pos:* Nat Found res fel, 59-61; Lederle med fac award, 61-63; assoc clin prof, Univ Calif, San Francisco, 66-68; chief pediat, Permanente Med Group, San Francisco; mem bact & mycol study sect, Res Rev Br, NIH, 70-74; co-dir, Kaiser-Permanente Vaccine Res Ctr, 65- *Mem:* Inst Med-Nat Acad Sci; fel Am Acad Pediat; Soc Pediat Res; Infectious Dis Soc Am; Am Pediat Soc; Am Bd Pediat. *Res:* Infectious diseases; epidemiology; medical care. *Mailing Add:* 2200 O'Farrell St San Francisco CA 94115

SHINEMAN, RICHARD SHUBERT, b Albany, NY, May 21, 24. INORGANIC CHEMISTRY. *Educ:* Cornell Univ, AB, 45; Syracuse Univ, MS, 50; Ohio State Univ, PhD(inorg chem), 57. *Prof Exp:* From inst to asst prof chem, Purdue Univ, 59-62; PROF CHEM, STATE UNIV NY COL, OSWEGO, 62- *Concurrent Pos:* Chmn dept chem, State Univ NY Col, Oswego, 62-67. *Mem:* AAAS; Am Chem Soc. *Res:* Inorganic nitrogen chemistry; x-ray crystallography. *Mailing Add:* 308 Washington Blvd Oswego NY 13126-1727

SHINER, EDWARD ARNOLD, b Chicago, Ill, Feb 18, 24; m 51; c 3. ORGANIC CHEMISTRY. *Educ:* Northwestern Univ, BS, 47; Univ Wis, PhD(chem), 51. *Prof Exp:* Instr chem, Univ Wis-Milwaukee, 47-48; res chemist, Food Prod Div, 51-60, mgr food casing develop, 60-65, tech mgr, 65-69, asst dir res & develop, 69-73, DIR RES & DEVELOP, UNION CARBIDE CORP, 73- *Mem:* Am Chem Soc. *Res:* Cellulose; lignin; polymer science. *Mailing Add:* 400 N Linden Ave Oak Park IL 60302

SHINER, VERNON JACK, JR, b Laredo, Tex, Aug 11, 25; m 46; c 3. PHYSICAL ORGANIC CHEMISTRY. *Educ:* Tex Western Col, BS, 47; Cornell Univ, PhD(chem), 50. *Prof Exp:* Res assoc org chem, State Agr Exp Sta, NY, 47; Fulbright scholar, London Univ, 50-51; Du Pont fel, Harvard Univ, 51-52; from instr to assoc prof chem, 52-60, chmn dept, 62-67, dean, Col Arts & Sci, 73-78, PROF CHEM, IND UNIV, BLOOMINGTON, 60-, CHMN DEPT, 82- *Mem:* Am Chem Soc; Royal Soc Chem; Sigma Xi. *Res:* Kinetics and mechanisms of organic reactions; deuterium isotope rate effects. *Mailing Add:* Dept Chem Ind Univ Bloomington IN 47401

SHING, YUEN WAN, b Kunming, China, Apr 29, 45; m 73; c 2. GROWTH FACTORS, HORMONE RECEPTORS. *Educ:* Taiwan Univ, BSc, 66; Univ Kans, PhD(biochem), 74. *Prof Exp:* Res assoc growth factors, State Univ NY, Albany, 74-76; res assoc receptors, Univ Wis-Madison, 76-80; asst prof biochem, Southeastern Mass Univ, North Dartmouth, 80-81; teaching fel biochem, 82-84, ASST PROF SURG & BIOCHEM, MED SCH, HARVARD UNIV, 84-; RES ASSOC, CHILDREN'S HOSP, BOSTON, 80- *Mem:* Am Soc Biol Chemists; Am Soc Cell Biol; NY Acad Sci; AAAS. *Res:* Characterization and purification of tumor-derived angiogenic factors; biochemical and physiological properties of polypeptide growth factors, especially those derived from milk; molecular biology of hormone receptors. *Mailing Add:* Enders Bldg 1086 Children's Hosp 300 Longwood Ave Boston MA 02115

SHING, YUH-HAN, b Anhwei, China, June 18, 41; Can citizen; m 67; c 3. THIN FILM MATERIALS, DEVICE PHYSICS. *Educ:* Taiwan Normal Univ, BSc, 63; Univ Calgary, MSc, 69, PhD(physics), 72. *Prof Exp:* Nat Res Coun fel physics, McGill Univ, 72-74, reader & res assoc, 74-76, reader, 76-79; mem res staff solar energy, Xerox Res Ctr, Can, 79-80; SR SCIENTIST MAT, ARCO SOLAR INDUST, 80- *Mem:* Am Phys Soc; Inst Elec & Electronics Engrs; Electrochem Soc. *Res:* Synthesis and characterization of energy conversion materials; electronic and optical properties of thin film; material synthesis techniques for amorphous and crystalline semiconductors using sputtering, glow-discharge and melt-spinning. *Mailing Add:* Jet Propulsion Lab MS302-306 Calif Inst Technol 4800 Oak Grove Dr Pasadena CA 91109

SHINGLETON, HUGH MAURICE, b Stantonsburg, NC, Oct 11, 31; c 3. OBSTETRICS & GYNECOLOGY, ONCOLOGY. *Educ:* Duke Univ, AB, 54, MD, 57; Am Bd Obstet & Gynec, dipl. *Prof Exp:* Intern, Jefferson Med Col Hosp, 57-58; asst resident obstet & gynec, NC Mem Hosp, Chapel Hill, 60-61 & Margaret Hague Maternity Hosp, Jersey City, NJ, 62; resident, NC Mem Hosp, Chapel Hill, 62-63, chief resident, 63-64; from instr to asst prof obstet & gynec, Sch Med, Univ NC, Chapel Hill, 64-69, asst prof path, 68-69; assoc prof, 69-74, asst prof, 69-81, PROF OBSTET & GYNEC, SCH MED, UNIV ALA, BIRMINGHAM, 74-, ASSOC PROF PATH, 81-, CHMN DEPT, MED CTR, 78- *Concurrent Pos:* Am Cancer Soc fel, Mem Hosp, Chapel Hill, NC, 62-63; Nat Cancer Inst spec fel, Col Physicians & Surgeons, Columbia Univ, 66-67. *Mem:* Am Col Obstet & Gynec; Am Col Surgeons; AMA; Soc Gynec Oncol (secy-treas, 81); Am Gynec Soc. *Res:* Gynecologic oncology; use of electron microscope and clinical research. *Mailing Add:* Dept of Obstet & Gynec Univ of Ala Sch of Med Birmingham AL 35294

SHININGER, TERRY LYNN, developmental biology, plant physiology, for more information see previous edition

SHINKAI, ICHIRO, b Japan, Dec 4, 41; m 66; c 2. ORGANIC & ANALYTICAL CHEMISTRY. *Educ:* Doshisha Univ, BSc, 64, MSc, 66; Kyushu Univ, PhD(org chem), 71. *Prof Exp:* Res asst & lectr org chem, Kyushu Univ, 66-72; res assoc dept chem, Univ Ala, 72-76; DIR PROCESS CHEM, MERCK, SHARP & DOHME RES LABS, MERCK & CO, INC, 76- *Mem:* Am Chem Soc; Japan Chem Soc; Int Union Pure & Appl Chem. *Res:* Synthetic organic chemistry; reaction mechanisms of heterocycles; reactive intermediate. *Mailing Add:* Merck Sharp & Dohme Res Labs PO Box 2000 R801-200 Rahway NJ 07065

SHINKMAN, PAUL G, b New York, NY, June 18, 36; m 69. PHYSIOLOGICAL PSYCHOLOGY. *Educ:* Harvard Univ, AB, 58; Univ Mich, AM, 62, PhD(psychol), 62. *Prof Exp:* Instr psychol, Univ Mich, 61-62, NIMH fel, Brain Res Lab, 64-66; vis asst prof psychobiol, Univ Calif, Irvine, 66-67; from asst prof to assoc prof psychol & neurobiol, 67-77, PROF PSYCHOL & NEUROBIOL, UNIV NC, CHAPEL HILL, 77-, DIR EXP PSYCHOL PROG, 75- *Mem:* Am Psychol Asn; Psychonomic Soc; Soc Neurosci. *Res:* Central nervous system and behavior. *Mailing Add:* Dept Psychol 201 Davie Hall 013a Univ NC Chapel Hill NC 27514

SHINN, DENNIS BURTON, b Keene, NH, Sept 2, 39; m 60; c 3. INORGANIC CHEMISTRY. *Educ:* Univ NH, BS, 61, MS, 64; Mich State Univ, PhD(chem), 68. *Prof Exp:* Adv develop engr, 68-70, engr in charge chem appln, 70-72, prog mgr high intensity discharge mat, 72-78, PROG MGR MAT ENG LAB, SYLVANIA LIGHTING CTR, 78- *Mem:* Am Chem Soc; Am Crystallog Asn; Am Ceramic Soc; Sigma Xi. *Res:* Synthesis and properties of solid state inorganic materials. *Mailing Add:* 28 Colrain Rd Topsfield MA 01983

SHINN, JOSEPH HANCOCK, b Atlantic City, NJ, Jan 4, 38; m 75; c 2. METEOROLOGY, POLLUTION ECOLOGY. *Educ:* Del Valley Col, BS, 59; Cornell Univ, MS, 62; Univ Wis-Madison, PhD(meteorol), 71. *Prof Exp:* Phys sci aide microclimate, Agr Res Serv, USDA, 59-62; proj asst meteorol, Univ Wis-Madison, 62-67; res meteorologist, US Army Electronics Command, Ft Huachuca, 67-70 & White Sands Missile Range, 70-73; METEOROLOGIST POLLUTANT EFFECTS, ENVIRON SCI DIV, LAWRENCE LIVERMORE LAB, 73- *Concurrent Pos:* Chmn, US Army Electronics Command Res Bd, 71-72; adv tactical environ support study, US Army Intel Sch, 72-73; br chief automatic meteorol systs, US Army Atmospheric Sci Lab, 72-73; dep sect chief environ sci div, Lawrence Livermore Lab, 77-84, sect leader, 85- *Mem:* Sigma Xi; Am Meteorol Soc; Air & Waste Mgt Asn. *Res:* Dynamics of the atmospheric boundary layer; inhalation exposure and suspension of toxic particles; processes of deposition of gases and particles on vegetation; forest meteorology; air pollution meteorology. *Mailing Add:* Lawrence Livermore Lab L-524 PO Box 5507 Livermore CA 94550-9516

SHINNAR, REUEL, b Vienna, Austria, Sept 15, 23; US citizen; m 48; c 2. CHEMICAL ENGINEERING. *Educ:* Israel Inst Technol, BSc, 45; Columbia Univ, ScD(chem eng), 57. *Prof Exp:* Chem engr indust, 45-54; assoc prof chem eng, Israel Inst Technol, 58-62; res assoc aeronaut eng, Princeton Univ, 62-64; prof, 64-79, DISTINGUISHED PROF CHEM ENG, CITY COL NEW YORK, 79- *Concurrent Pos:* Consult chem & petrol indust, 64- *Honors & Awards:* Wilhelm Mem lectr, Princeton Univ, 85; Kelly lectr, Purdue Univ, 91. *Mem:* Nat Acad Eng; AAAS; NY Acad Sci; Am Inst Chem Engrs; Am Chem Soc; Am Inst Aeronaut & Astronaut. *Res:* Process dynamics and control; process design and economics; industrial economics; chemical reactor design. *Mailing Add:* Dept Chem Eng City Col New York New York NY 10031

SHINNERS, CARL W, b Milwaukee, Wis, Aug 13, 28; m 54; c 3. PHYSICS. *Educ:* Marquette Univ, PhB, 52; La State Univ, MS, 60, PhD(nuclear spectros), 65. *Prof Exp:* Instr physics, La State Univ, 59-63; assoc prof, 65-67, chmn dept, 67-71, PROF PHYSICS, UNIV WIS-WHITEWATER, 67- *Concurrent Pos:* Wis State res grants, 68 & 70; NSF grants, 70 & 71. *Mem:* AAAS; Am Phys Soc; Am Asn Physics Teachers. *Res:* Beta and gamma ray spectroscopy; nuclear structures; solar energy and energy education. *Mailing Add:* Dept Physics Univ Wis 800 Main St W Whitewater WI 53190

SHINNERS, STANLEY MARVIN, b New York, NY, May 9, 33; m 56; c 3. ELECTRICAL ENGINEERING, EDUCATION. *Educ:* City Col NY, BEE, 54; Columbia Univ, MS, 59. *Prof Exp:* Engr, Western Elec Co, 53-55; staff engr, Electronics Div, Otis Elevator Co, 55-56; proj engr, Polarad Electronics Corp, 56-57 & Consol Avionic Corp, 57-58; SR RES SECT HEAD, UNISYS CORP, 58- *Concurrent Pos:* Adj prof, Polytech Inst Brooklyn, 59-71, Cooper Union, 66-73 & 79- & NY Inst Technol, 74- *Mem:* Fel Inst Elec & Electronics Engrs; Am Soc Eng Educ. *Res:* Control systems and systems engineering. *Mailing Add:* 28 Sagamore Way North Jericho NY 11753

SHINNICK-GALLAGHER, PATRICIA L, b Chicago, Ill, July 28, 47; m 74; c 2. NEUROPHARMACOLOGY, NEUROPHYSIOLOGY. *Educ:* Univ Ill, BS, 70; Loyola Univ Chicago, PhD(pharmacol), 74. *Prof Exp:* Pharmacist, Nosek Apothecary, 71-73; res assoc neurophysiol, Sch Med, Loyola Univ Chicago, 74-75; instr, 75-76, asst prof, 76-81, ASSOC PROF PHARMACOL, UNIV TEX MED BR, GALVESTON, 81- *Mem:* AAAS; Am Pharmaceut Asn; Am Soc Pharmacol Exp Therapy; Sigma Xi. *Res:*

Pharmacological and physiological dissection of reflex pathways in the isolated spinal cord; analysis of drug action on ganglionic and neuromuscular transmission. *Mailing Add:* Dept of Pharmacol & Toxicol Univ of Tex Med Br Galveston TX 77550

SHINOHARA, MAKOTO, b Naha, Japan, Jan 30, 37; m 64; c 2. POLYMER CHEMISTRY, POLYMER SCIENCE. *Educ:* Tokyo Inst Technol, BSc, 60, MSc, 62; State Univ NY Col Forestry, PhD(phys chem), 69; Syracuse Univ, PhD(phys chem), 69. *Prof Exp:* Res Found fel, State Univ NY Col Forestry, Syracuse Univ, 64-69; proj chemist, Dow Corning Corp, 69-74, sr proj chemist, 74-75; res assoc, Int Playtex, Inc, 75-78; mgr molding compound res & develop, Morton Chem Co, 78-84; tech mgr, polyset, Dynachem Corp, Morton-Thiokol Inc, 84-87; TECH DIR, FURANE PROD, CIBA-GEIGY CORP, 87- *Mem:* Am Chem Soc; Am Geog Soc; Sigma Xi; AAAS; NY Acad Sci. *Res:* Structure-mechanical, electrical and physico-chemical property relation in thermo-plastic and thermosetting polymers; polymer composites and toughening; polymer characterization; mechanisms and kinetics of polymerization, polyaddition, polycondensation and ring-opening polymerization; water soluble polymers and gels; organo-silicone polymers. *Mailing Add:* 21202 Georgetown Dr Saugus CA 91350

SHINOZUKA, HISASHI, ONCOLOGY. *Educ:* McGill Univ, Can, PhD(exp path), 63. *Prof Exp:* PROF PATH, UNIV PITTSBURGH, 79- *Mailing Add:* Dept Path Scaife Hall Rm 750 Univ Pittsburgh Sch Med Pittsburgh PA 15261

SHINOZUKA, MASANOBU, b Tokyo, Japan, Dec 23, 30; m 54; c 3. CIVIL ENGINEERING, ENGINEERING MECHANICS. *Educ:* Kyoto Univ, BS, 53, MS, 55; Columbia Univ, PhD(civil eng), 60. *Prof Exp:* Res asst, Columbia Univ, 59-61, from asst prof to prof, 61-77, Renwick prof civil eng, 77-88; SOLLENBERGER PROF CIVIL ENG, PRINCETON UNIV, 88- *Concurrent Pos:* US coordr, US-Japan Joint Seminars; res analyst, US Air Force, 67-68; consult, Jet Propulsion Lab, 68- & Kawasaki Heavy Indust, Kobe, Japan, 71; res struct engr, Naval Civil Eng Lab, 70; consult, US Air Force, Flight Dynamics Lab, Wright-Patterson AFB, Ohio, US Army Mat & Mech Res Ctr, Watertown, Mass, US Navy Strategic Eng Surv Off, Bethesda, Md, US Atomic Energy Comn, US Naval Civil Eng Lab, Port Hueneme, Calif, US Nuclear Res Coun, Adv Comt Reactor Safeguards, Washington, DC; Gen Dynamics, Ft Worth Div, Tex, Northrop Corp, Aircraft Div, Hawthorne, Calif, Rockwell Int, Los Angles, Eng Decision Anal Co, Inc, Palo Alto, US Nuclear Regulatory Comn, Wash, Lawrence Livermore Lab, Livermore, Calif, Kawasaki Heavy Indust, Ltd, Kobe, Japan, Shimizu Construct Co, Tokyo, Japan; dir, Nat Ctr Earthquake Eng Res, State Univ NY Buffalo, 90- *Honors & Awards:* Walter L Huber Civil Eng Res Prize, Am Soc Civil Engrs, 72, Am Freudenthal Medal, Nathan M Newmark Medal, 85; Moisseiff Award, 88. *Mem:* Nat Acad Eng; Am Soc Civil Engrs; Am Soc Mech Engrs; Am Inst Aeronaut & Astronaut. *Res:* Structural reliability analysis; random vibration; inelasticity; structural analysis. *Mailing Add:* E232 Eng Quad Princeton NJ 08540

SHINSKEY, FRANCIS GREGWAY, b North Tonawanda, NY, Oct 29, 31; m 58; c 8. PH CONTROL, DISTILLATION CONTROL. *Educ:* Univ Notre Dame, BSc, 52. *Prof Exp:* Process engr, E I du Pont Co, 54-55; process engr, Olin-Mathieson Chem Co, 55-57, instrument engr, 57-60; syst engr, Foxboro Co, 60-68, control systs consult, 68-72, sr systs consult, 72-83, chief consult, 83-90, RES FEL, FOXBORO CO, 90- *Concurrent Pos:* Bristol fel, Foxboro Co, 82. *Honors & Awards:* Sprange Appln Award, Instrument Soc Am, 77, Eckman Educ Award, 83, Sperry Founders Award, 88. *Mem:* Fel Instrument Soc Am. *Res:* Evaluation of feedback controller performance; design of a self-turning high-performance model-based controller. *Mailing Add:* 251 Main St Foxboro MA 02035

SHIONO, RYONOSUKE, b Kobe, Japan, Nov 12, 23; m 58; c 3. CRYSTALLOGRAPHY. *Educ:* Osaka Univ, MSc, 45, DSc(physics), 60. *Prof Exp:* Instr physics, Osaka Univ, 49-56; res assoc lectr, 56-61, asst res prof, 61-65, assoc res prof crystallog, 66-69, ASSOC PROF CRYSTALLOG, UNIV PITTSBURGH, 69- *Concurrent Pos:* Vis prof, Univ Sao Paulo, 69. *Mem:* Am Crystallog Asn; Royal Soc Chem; Chem Soc Japan; Phys Soc Japan. *Res:* Crystal structure analysis and application of computer in crystallography. *Mailing Add:* Dept Crystallog Univ Pittsburgh Pittsburgh PA 15260

SHIOTA, TETSUO, b Los Angeles, Calif, Jan 1, 23; m 48; c 3. BIOCHEMISTRY. *Educ:* Roosevelt Col, BS, 48; Univ Ill, MS, 50, PhD(bact), 53. *Prof Exp:* Res assoc, NIH, Bethesda, Md, 53-60, sr scientist microbiol, 60-67; assoc prof, 67-71, PROF MICROBIOL, UNIV ALA, BIRMINGHAM, 71-, ASSOC PROF BIOCHEM, 69- *Concurrent Pos:* Sr scientist, Cancer Res & Training Prog, Univ Ala, Birmingham, 72- *Mem:* Am Soc Microbiol; Am Soc Biol Chemists. *Res:* Biochemistry of pteridines and folic acid compounds in bacteria and mammalian cells. *Mailing Add:* 2040 13th Ave W No 24 Seattle WA 98119-2757

SHIOYAMA, TOD KAY, b Seattle, Wash, Aug 16, 51. ANALYTICAL CHEMISTRY, INORGANIC CHEMISTRY. *Educ:* Western Wash State Col, BS, 73; Wash State Univ, PhD(anal), 78. *Prof Exp:* CHEMIST RES, PHILLIPS PETROL CO, 78- *Mem:* Am Chem Soc. *Res:* Catalysis; kinetics. *Mailing Add:* 2824 Montecello Bartlesville OK 74006

SHIP, IRWIN I, oral medicine; deceased, see previous edition for last biography

SHIPCHANDLER, MOHAMMED TYEBJI, b Surat, India, May 19, 41; m 71; c 2. MEDICINAL CHEMISTRY, ORGANIC CHEMISTRY. *Educ:* Univ Bombay, BSc, 62, BSc, 64; Univ Minn, Minneapolis, PhD(med chem), 69. *Prof Exp:* NIH fel & res assoc med chem, Univ Kans, 68-70; NIH fel & res assoc natural prod chem, Col Pharm, Ohio State Univ, 70-72; instr, Columbus Tech Inst, Ohio, 72-73; res chemist, Com Solvents Corp, Terre Haute, Ind, 73-80; SR SCIENTIST, ABBOTT LABS, 80- *Mem:* Am Chem Soc; Sigma Xi; Am Asn Clin Chem. *Res:* Synthesis of medicinal agents and natural products; immuno assay development. *Mailing Add:* Dept D-93C AP-20 Dia Div Abbott Labs Abbott Parks IL 60064

SHIPE, EMERSON RUSSELL, b Knoxville, Tenn, July 28, 47; m 76; c 3. AGRONOMY, PLANT BREEDING. *Educ:* Univ Tenn, BS, 69; Western Ky Univ, MS, 70; Va Polytech Inst & State Univ, PhD(agron), 78. *Prof Exp:* Teaching asst, Western Ky Univ, 69-70; agriculturalist, US Peace Corps, Cent Am, 73-75; teaching asst, Va Polytech Inst & State Univ, 75-78; asst prof soil & crop sci, Tex Agr Exp Sta, 78-80; from asst prof to assoc prof, 80-89, PROF AGRON & SOILS, CLEMSON UNIV, 89- *Mem:* Am Soc Agron; Crop Sci Soc Am; Sigma Xi. *Res:* Development of soybean germplasm, cultivars with improved nematode and insect resistance, higher seed yields, and adaptation to southeast soil and climatic conditions. *Mailing Add:* Dept Agron & Soils Clemson Univ Clemson SC 29634-0359

SHIPE, WILLIAM FRANKLIN, b Middletown, Va, Mar 8, 20; m 48; c 2. FOOD SCIENCE. *Educ:* Va Polytech Inst, BS, 41; Cornell Univ, PhD(dairy chem), 49. *Prof Exp:* Instr dairy mfg, Va Polytech Inst, 45-46; from asst to assoc prof dairy indust, 46-60, prof Food Sci, 61-88, EMER PROF, CORNELL UNIV, 88- *Concurrent Pos:* Res assoc, NC State Col, 56; travel fel, Cornell Univ, 62; res consult, Dept Agr & Mkt, 63; Nat Inst Res Dairying fel, Reading, Eng, 70 & 79. *Mem:* Am Chem Soc; Am Dairy Sci Asn; Int Food Technologists. *Res:* Enzymatic changes in food products; flavor and texture of foods; nutritional quality of foods. *Mailing Add:* Dept Food Sci Cornell Univ Ithaca NY 14850

SHIPINSKI, JOHN, b Wisconsin Rapids, Wis, Aug 25, 32; m 66. MECHANICAL ENGINEERING. *Educ:* Univ Wis, BSME, 60, MSME, 63, PhD(mech eng), 67. *Prof Exp:* Engr, Deere & Co, 67-75 & Chicago Pneumatic, 75-76; ENGR, WARNER ELEC CO, 76- *Honors & Awards:* Dugald Clerk Prize, Inst Mech Engrs, London, 72. *Mem:* Soc Automotive Engrs; Am Soc Mech Engrs; Am Soc Agr Engrs. *Res:* Electromagnetic clutches and brakes; clutches and brakes; diesel engines; energy technology; supercharging of engines. *Mailing Add:* 4265 Crestline Dr Ann Arbor MI 48103

SHIPKOWITZ, NATHAN L, b Chicago, Ill, Mar 29, 25; m 56; c 4. MICROBIOLOGY. *Educ:* Univ Ill, BS, 49, MS, 50; Mich State Univ, PhD(bact, pub health), 52. *Prof Exp:* Asst prof vet sci, Univ Mass, 52-54; asst res bacteriologist, Hooper Found, Med Ctr, Univ Calif, San Francisco, 54-58; bacteriologist & virologist, Path Dept, Good Samaritan Hosp, Portland, Ore, 59-62; sr res microbiologist, 63-75, ASSOC RES FEL, ABBOTT LABS, 76- *Mem:* AAAS; Am Soc Microbiol; Sigma Xi. *Res:* microbiology. *Mailing Add:* Abbott Labs Dept 47T Abbot Park IL 60064-3500

SHIPLEY, EDWARD NICHOLAS, b Baltimore, Md, Jan 26, 34; m 63; c 3. DATABASE TECHNOLOGY, COMMUNICATION NETWORKS. *Educ:* Johns Hopkins Univ, AB, 54, PhD(physics), 58. *Prof Exp:* From instr to asst prof physics, Northwestern Univ, 58-63; mem tech staff, Bellcomm, Inc, Washington, DC, 63-72; MEM TECH STAFF, BELL LABS, 72- *Concurrent Pos:* Consult, Argonne Nat Lab, 59-63; distinguished mem tech staff, Bell Labs, 81. *Mem:* Am Phys Soc. *Res:* Low energy nuclear reactions; lifetimes of excited nuclear states; hyperfragment decay modes; K meson reactions; lunar surface mechanical properties; Martian atmospheric phenomena; maintenance and operations of telephone switching systems; database administration; numbering for telecommunication services; systems planning and development. *Mailing Add:* AT&T Bell Labs Rm 2J-603 Crawford Corners Rd Holmdel NJ 07733

SHIPLEY, GEORGE GRAHAM, b London, Eng, Nov 18, 37. BIOPHYSICS, BIOCHEMISTRY. *Educ:* Univ Nottingham, Eng, BSc, 59, PhD(phys chem), 63 & DSc, 84. *Prof Exp:* Scientist biophys, Unilever Res Lab, Eng, 63-69, sect leader, 69-71; from asst res prof to assoc res prof med, 71-82, from asst prof to assoc prof biochem, 73-82, PROF BIOCHEM & RES PROF MED, SCH MED, BOSTON UNIV, 82- *Mem:* AAAS; Am Chem Soc; Am Crystallog Asn; Am Heart Asn; Am Soc Biol Chemists; Biophys Soc. *Res:* Structure and function of biological lipids, cell membranes and serum lipoproteins and their relationship to pathological processes, notably atherosclerosis and hyperlipidemia. *Mailing Add:* Biophys Inst Boston Univ 80 E Concord St Boston MA 02118-2307

SHIPLEY, JAMES PARISH, JR, b Clovis, NMex, Jan 3, 45; m 62; c 3. SCIENCE POLICY, TECHNICAL MANAGEMENT. *Educ:* NMex State Univ, BS, 66; Univ NMex, MS, 69, PhD(elec eng), 73. *Prof Exp:* Staff mem electronics, Los Alamos Nat Lab, 66-73, nuclear sci & solar energy, 73-76, systs sci, 76-78, group leader safeguards systs, 78-82, prog mgr Safeguards & Security, Low-Intensity Conflict, 82-87; sr adv to ambassador, Nonproliferation Policy, Dept State, 87-89; prog mgr Arms Reduction Treaty Verification, 89-91, PROG MGR ENVIRON MGT TECHNOL DEVELOP, LOS ALAMOS NAT LAB, 91-; OWNER & PRES, JP SYSTS, INC. *Mem:* Inst Elec & Electronic Engrs; AAAS; Inst Nuclear Mat Mgt; Am Soc Indust Security; Am Nuclear Soc. *Res:* Nuclear safeguards systems; systems science; statistical decision theory; science and technology policy; environmental management; arms control. *Mailing Add:* 1646 Camino Uva Los Alamos NM 87544

SHIPLEY, MICHAEL THOMAS, b Kansas City, Mo, Apr 22, 41; m 76; c 2. NEUROANATOMY, NEUROPHYSIOLOGY. *Educ:* Univ Mo, Kansas City, BA, 67; Mass Inst Technol, PhD(neurosci), 72. *Prof Exp:* Fel anat, Univ Aarhus, 72-74; asst prof, Univ Lausanne, 74-78; asst prof cell biol & anat, Med Sch, Northwestern Univ, 78-; PROF ANAT DEPT, UNIV CINCINNATI. *Concurrent Pos:* Prin investr, NIH grants, 82-, mem study sect, 83-89; Woodrow Wilson fel. *Mem:* Soc Neurosci; Am Asn Anatomists; AAAS; Am Chem Soc. *Res:* Neuroanatomy and physiology of sensory-lumbic interactions in cerebral cortex and neuroanatomy; development of olfactory nervous system; image analysis. *Mailing Add:* Dept Anat Univ Cincinnati Cincinnati OH 45221

SHIPLEY, REGINALD A, b Dayton, Ohio, Oct 15, 05; m 35; c 3. TRACERS FOR IN VIVO KINETICS. *Educ:* Otterbein Col, BS, 27; Western Reserve Univ, MD, 31. *Prof Exp:* Staff physician, Univ Hosps Cleveland, 38-74; from instr to prof, 38-74, EMER PROF MED, SCH MED, CASE WESTERN RESERVE UNIV, 74- *Concurrent Pos:* Chief nuclear med, Vet Admin Hosp, Cleveland, 50-74, chief staff res admin, 65-75; mem study sect endocrinol, NIH, 57-63. *Mem:* AAAS; Am Physiol Soc; Am Soc Clin Invest; Endocrine Soc; Soc Exp Biol & Med. *Res:* Physiology and pathophysiology of the heart and the endocrine glands; glucose metabolism; radioactive tracer applications in physiology and pathophysiology. *Mailing Add:* 35 Lyman St Easthampton MA 01027

SHIPLEY, THORNE, b New York, NY, Apr 11, 27; m 71; c 2. PSYCHOLOGY, OPHTHALMOLOGY. *Educ:* Johns Hopkins Univ, BA, 49; New Sch Social Res, MA, 53; NY Univ, PhD(psychol), 55. *Prof Exp:* Instr psychol, Long Island Univ, 53-55; res psychol, Am Optical Co, 55-58; NIH spec fel, Imp Col, Univ London, 58-59 & Fac Med, Univ Paris, 59-60; assoc prof visual sci, Med Sch & assoc prof neuropsychol, Sch Arts & Sci, 60-77, PROF VISUAL SCI, MED SCH & PROF NEUROPSYCHOL, SCH ARTS & SCI, UNIV MIAMI, 77- *Concurrent Pos:* Founding ed, Vision Res, 60-78; dir, Inst Advan Study Sci & Humanities, 80- *Mem:* AAAS; Am Psychol Asn; Soc Neurosci; Soc Social Responsibility in Sci; Optical Soc Am; fel World Acad Arts & Sci. *Res:* Theoretical psychology; sensory communication; sense function in children; cognition; communication and learning disabilities; history and philosophy of science. *Mailing Add:* Dept of Ophthal Univ of Miami Med Sch Miami FL 33101

SHIPMAN, C(HARLES) WILLIAM, b Phillipsburg, NJ, Aug 29, 24; m 46; c 3. COMBUSTION, THERMODYNAMICS. *Educ:* Mass Inst Technol, SB, 48, SM, 49, ScD(chem eng), 52. *Prof Exp:* Instr chem eng, Mass Inst Technol, 49-50, asst combustion res, 50-52, res assoc, 55-58; asst prof chem eng, Univ Del, 52-55; from asst prof to prof, Worcester Polytech Inst, 58-74, dean grad studies, 71-74; asst dir, Corp Res Dept, Cabot Corp, 78-80, engr, 74-86, mgr, Carbon Black Res & Develop, 80-86; CONSULT, 86- *Concurrent Pos:* Consult, Avco Corp, 58-65, United Aircraft Corp, 65-74 & Kennecott Corp, 68-74; dir, Combustion Inst, 78-90. *Honors & Awards:* Silver Combustion Medal, 64. *Mem:* Am Chem Soc; Am Inst Chem Engrs; Combustion Inst. *Res:* Thermodynamics; combustion; mass transfer; fine particles technology. *Mailing Add:* PO Box 32 Prospect Harbor ME 04669-0032

SHIPMAN, CHARLES, JR, b Ventura, Calif, Nov 1, 34; m 73; c 2. VIROLOGY. *Educ:* Univ Calif, Los Angeles, AB, 56; Calif State Univ, Fresno, MA, 63; Ind Univ, PhD(microbiol), 66. *Prof Exp:* Assoc res microbiologist, Parke-Davis & Co, 66-68; asst prof microbiol, Sch Med, 68-75, from asst prof to assoc prof oral biol, Sch Dent, 68-84, ASSOC PROF MICROBIOL, MED SCH, UNIV MICH, ANN ARBOR, 75-, PROF BIOL & MATS SCI, SCH DENT, 84- *Concurrent Pos:* Consult, Coun Dent Educ, 72-73; vis assoc prof mol biol biochem, Univ Calif, Irvine, 78; coun dent therapeut, Am Dent Asn, 83-; co-chmn, virol div, Inter-Am Soc Chemother, 84-87. *Mem:* Am Soc Microbiol; Tissue Cult Asn; Am Soc Virol; Soc Gen Microbiol; Am Asn Dent Schs. *Res:* Mechanism of action of nucleoside antibiotics and 2-acetylpyridine thiosemicarbazones on human herpes viruses and human immunodeficiency virus. *Mailing Add:* Dept Biol & Mats Sci Univ Mich Dent Sch Ann Arbor MI 48109-1078

SHIPMAN, HAROLD R, b Rock Rapids, Iowa, Feb 20, 11; m 38; c 2. SANITARY ENGINEERING. *Educ:* Univ Minn, BS, 37, MS, 48; Am Acad Environ Engrs, dipl. *Prof Exp:* Dir div rural sanit, dir div hotels, resorts & restaurants, regional engr & dist engr coord, Minn State Dept Health, 37-50; sanit engr, Int Red Cross, Korea, 51; sanit engr, WHO, Turkey, 52-54 & Egypt, 54-58; sanit engr & chief br environ sanit, Pan-Am Health Orgn, 58-62; water & wastes adv, 62-76, chief, Water Supply Sect, 63-76, CONSULT, ASIAN DEVELOP BANK, WORLD BANK, 77- *Concurrent Pos:* Adv, Govt Turkey, 52-54, Govt Egypt, 54-58 & numerous eng firms, 76-90; chmn subcomt water resources, UN, 71-72; mem, Columbia sem water resources. *Honors & Awards:* Centennial Fel Award, Johns Hopkins Univ, 76. *Mem:* Am Soc Civil Engrs; Nat Soc Prof Engrs; Am Water Works Asn; Am Pub Health Asn; Inter-Am Asn Sanit Engrs; Int Water Supply Asn; Water Pollution Control Asn. *Res:* Management, financing, appraisal and design of water supply and sewerage systems internationally. *Mailing Add:* 7108 Edgevale St Chevy Chase MD 20815

SHIPMAN, HARRY LONGFELLOW, b Hartford, Conn, Feb 20, 48; m 70; c 2. ASTROPHYSICS, PHYSICS. *Educ:* Harvard Univ, BA, 69; Calif Inst Technol, MS, 70, PhD(astron), 71. *Prof Exp:* J W Gibbs instr astron, Yale Univ, 71-73; asst prof physics, Univ Mo, St Louis, 73-74; from asst prof to assoc prof, 74-81, PROF PHYSICS, UNIV DEL, 81- *Concurrent Pos:* Guest investr, Kitt Peak Nat Observ, 72-74 & var satellite progs, NASA, 74-; astronomer, McDonnell Planetarium, 73-74; prin investr grants, NSF, 74-, Res Corp, 74-76, Univ Del Res Found, 75-76 & NASA, 76-79, 81, & 83-; John Simon Guggenheim Mem fel, 80-81. *Mem:* Sigma Xi; Am Astron Soc; AAAS; Am Asn Physics Teachers; Astron Soc Pac; Int Astron Union. *Res:* Analysis of stellar spectra via stellar-atmosphere calculations; white-dwarf stars and other final stages of stellar evolution; theoretical astrophysics. *Mailing Add:* Dept Physics & Astron Univ Del Newark DE 19711

SHIPMAN, JERRY, b Elamville, Ala, Feb 20, 43. DISCRIMINANT ANALYSIS. *Educ:* Ala A&M Univ, BS, 63; Western Wash Univ, MS, 66; Pa State Univ, PhD(math), 73. *Prof Exp:* Assoc prof & chmn, dept physics & math, 73-78, PROF & CHMN, DEPT MATH, ALA A&M UNIV, 78- *Concurrent Pos:* Assoc mathematician, Northrop Space Labs, 63-64. *Mem:* Math Asn Am; Soc Indust & Appl Math. *Mailing Add:* Dept Math Ala A&M Univ Box 326 Normal AL 35762

SHIPMAN, LESTER LYNN, b Topeka, Kans, Mar 28, 47; m 69. INTELLIGENT SYSTEMS. *Educ:* Washburn Univ, BA & BS, 69; Univ Kans, PhD(chem), 72. *Prof Exp:* Fel chem, Cornell Univ, 72-74; appointee, Argonne Nat Lab, 74-75, res assoc chem, 75-76, asst chemist, 76-80, chemist, 80-81; res scientist, 81-85, sr specialist, 85-87, CONSULT, E I DUPONT DE NEMOURS & CO, 87- *Concurrent Pos:* Consult, Norwich Pharmacal Co, 73-75. *Mem:* Am Asn Artifical Intelligence. *Res:* Ab initio molecular quantum mechanics; conformational and intermolecular potential energy functions; primary events of photosynthesis; structure-activity relationships; theoretical biophysical chemistry; molecular aspects of chemical carcinogenesis; mechanisms of energy transfer; theory of excitons in molecular aggregates; expert systems; intelligent scheduling systems. *Mailing Add:* Info Syst Dept E I du Pont de Nemours & Co Inc Wilmington DE 19898

SHIPMAN, ROBERT DEAN, b Moundsville, WVa, May 12, 21; m 46; c 2. FOREST ECOLOGY, SILVICULTURE. *Educ:* Univ Mich, BSF & MF, 42; Mich State Univ, PhD(forestry), 52. *Prof Exp:* Asst forest soils, Childs-Walcott Forest, Conn, 42; munic park forester, Oglebay Park, Wheeling, WVa, 47; agr aide, US Forest Serv, 49, mem staff, 52-58; asst forest res, Mich State Univ, 50-51; assoc prof forestry, Clemson Univ, 58-63; assoc prof forest ecol, 63-75, PROF FOREST ECOL, SCH FOREST RESOURCES, PA STATE UNIV, 75- *Concurrent Pos:* Moderator Northeast Weed Control Conf, NY, 64; mem, Soc Am Foresters Nat Task Force on Herbicides, 75-76. *Mem:* Soil Sci Soc Am; Soc Am Foresters; Ecol Soc Am; Weed Sci Soc Am; Sigma Xi. *Res:* Tree physiology and soils; silvics and herbicides. *Mailing Add:* Sch Forest Resources 207 Ferguson Bldg Pa State Univ University Park PA 16802

SHIPMAN, ROSS LOVELACE, b Jackson, Miss, Nov 20, 26; m 48; c 1. GEOLOGY. *Educ:* Univ Miss, BA, 50. *Prof Exp:* Geologist, Miss State Geol Surv, 49-50; jr geologist, Humble Oil & Refining Co, 50-51, dist geologist, 51-55; petrol consult, 55-67; asst exec dir, Am Geol Inst, 67-71; res prog mgr, Bur Econ Geol & Div Natural Resources & Environ, 71-75, assoc dir admin, Marine Sci Inst, 75-79, assoc vpres, Res Admin, Univ Tex, Austin, 79-85; pres & CEO Live Oak Energy, 85-87; PETROL INVEST CONS, 87- *Concurrent Pos:* Tex Coastal & Marine Coun; Cons Intl Bound & Wtr Comn; AIPG; Dir U Tx Indus Assoc Prog. *Honors & Awards:* Hon Mem Am Inst Prof Geol; Distinguished Serv Award, Am Asn Petrol Geol. *Mem:* Fel Geol Soc Am; Am Asn Petrol Geol; Am Inst Prof Geol; Fel Geol Soc (London); Soc Indep Prof Earth Sci. *Res:* Petroleum exploration and production; mining exploration; environmental geology; marine geophysics. *Mailing Add:* 1807 Polo Rd Austin TX 78703

SHIPP, JOSEPH CALVIN, b Northport, Ala, Feb 10, 27; m 62; c 1. MEDICINE. *Educ:* Univ Ala, BS, 48; Columbia Univ, MD, 52. *Prof Exp:* Intern & asst resident med, Presby Hosp, New York, 52-54; Nat Res Coun res fel, Harvard Med Sch, 54-56; res fel biochem, Oxford Univ, 58-59; instr med, Harvard Med Sch, 59-60; from asst prof to assoc prof med, Med Col, Univ Fla, 60-68, prof & dir clin res ctr, 68-70, dir diabetes res & training prog, 64-70; PROF MED & CHMN DEPT, COL MED, UNIV NEBR MED CTR, 70- *Concurrent Pos:* Sr asst resident, Peter Bent Brigham Hosp, 56-57, chief resident physician, 57-58, jr assoc, 59-60; sr investr, Boston Med Found, 59-60; Markle scholar, 61; res fel biochem, Univ Munich, 65-66. *Mem:* Am Diabetes Asn; Am Fedn Clin Res. *Res:* Endocrinology; diabetes; acting of hormones at the cellular level. *Mailing Add:* Valley Med Ctr 445 S Cedar Ave Fresno CA 93702

SHIPP, OLIVER ELMO, b Big Creek, Miss, June 13, 28; m 52; c 3. ENTOMOLOGY, PLANT PHYSIOLOGY. *Educ:* Miss State Univ, BS, 52, MS, 58; Tex A&M Univ, PhD(entom), 63. *Prof Exp:* Asst county agent, Miss Exten Serv, 54-56; asst prof cotton insect res, Tex A&M Univ, 60-63; res biologist, 63-68, MEM STAFF FIELD RES, CHEMAGRO CORP, 68- *Mem:* Entom Soc Am; Southern Weed Sci Soc; Sigma Xi. *Res:* Tenacity of insecticide residues on field crops; penetration characteristics of plant tissue by insecticides; insect control; rearing techniques for laboratory insect cultures; herbicide evaluation and development. *Mailing Add:* 218 Valleywood Dr East Collierville TN 38017

SHIPP, RAYMOND FRANCIS, b Hay Springs, Nebr, June 1, 31; m 66; c 3. AGRONOMY. *Educ:* Univ Nebr, BS, 53, MS, 58; Pa State Univ, PhD(agron), 62. *Prof Exp:* Soil scientist, Pa State Dept Health, 62-66 & US Dept Interior Bur Reclamation, 66-72; exten agronomist, 72-91, ASSOC PROF AGRON EXTEN, PA STATE UNIV, 78- *Mem:* Fel AAAS; Soil Sci Soc Am; Sigma Xi. *Res:* Use of sewage sludge by-products from municipal waste water treatment plants for the enhancement of crop production. *Mailing Add:* Dept Agron Pa State Univ 116 ASI Bldg University Park PA 16802

SHIPP, ROBERT LEWIS, b Tallahassee, Fla, Aug 22, 42; m 64; c 3. ICHTHYOLOGY. *Educ:* Spring Hill Col, BS, 64; Fla State Univ, MS, 66, PhD(biol), 70. *Prof Exp:* Instr biol, Fla A&M Univ, 68-70; from asst prof to assoc prof, 71-78, PROF BIOL, UNIV S ALA, 79- *Concurrent Pos:* Ed, Northeast Gulf Sci, 77-, Systematic Zoology, 87-90. *Mem:* Am Soc Ichthyologists & Herpetologists; Am Fisheries Soc; Soc Syst Zool. *Res:* Marine zoogeography; fish systematics and phylogeny; artificial reef development; development of Guatemalan fisheries; river ecology. *Mailing Add:* Dept of Biol Univ Sala 307 Univ Blvd Mobile AL 36688

SHIPP, WILLIAM STANLEY, b Little Rock, Ark, July 11, 39; m 62; c 1. BIOPHYSICS. *Educ:* Univ Chicago, PhD(biophys), 65. *Prof Exp:* Nat Acad Sci-Nat Res Coun res fel bact genetics, Genetics Inst, Univ Cologne, 65-66; Am Cancer Soc res fel, Max Planck Inst Biol, Ger, 66-67; asst prof, Brown Univ, 67-72, from asst provost to assoc provost, 80-83, assoc prof med sci, 72, res inst dir, Inst Res Info & Scholar, 83-90; RETIRED. *Mem:* Inst Elec & Electronic Engrs Comput Soc. *Res:* Digital signal processing; local area networks; operating systems; operating systems. *Mailing Add:* 155 Georgia Providence RI 02912

SHIPPY, DAVID JAMES, b Oelwein, Iowa, July 26, 31; m 54; c 2. ENGINEERING MECHANICS. *Educ:* Iowa State Univ, BS, 53, MS, 54, PhD(theoret & appl mech), 63. *Prof Exp:* Aerophysics engr, Gen Dynamics Corp, 54-56; instr physics & eng, Graceland Col, 56-64; from asst prof to assoc prof eng mech, 64-78, NSF grant, 66-67, PROF ENG MECH, UNIV KY, 78- *Concurrent Pos:* Instr, Iowa State Univ, 60-61; Air Force Off Sci Res grant, 75-78; NSF grant, 81-84. *Mem:* Am Soc Eng Educ. *Res:* Numerical methods in solid mechanics, especially the Boundary Integral Equation Method. *Mailing Add:* Dept Eng Mech Anderson Hall Univ Ky Lexington KY 40506

SHIPSEY, EDWARD JOSEPH, b New York, NY, Aug 22, 38; c 2. PHYSICAL CHEMISTRY. *Educ:* Stanford Univ, BS, 60; Ohio State Univ, PhD(phys chem), 67. *Prof Exp:* res assoc physics, Univ Tex, Austin, 72-90; TEX RR COMN, 90- *Res:* Theory and computation of atomic and molecular collisions. *Mailing Add:* Tex RR Comn 2801 Wheless Dr Austin TX 78712

SHIRAKI, KEIZO, HUMAN PHYSIOLOGY, ENVIRONMENTAL PHYSIOLOGY. *Educ:* Kyoto Pref Sch Med, MD, 61, PhD(physiol), 67. *Prof Exp:* Sr instr res, Dept Physiol, Kyoto Prefectural Univ Med, Japan, 67-68; asst prof res, Inst Arctic Biol, Univ Alaska, 68-69 & Dept Nutrit, Tokushima Univ Sch Med, Japan, 70-78; PROF RES & CHMN DEPT PHYSIOL, SCH MED, UNIV OCCUP & ENVIRON HEALTH, KITAKYUSHA, JAPAN, 78- *Mem:* Am Physiol Soc; Am Col Sports Med; Undersea Med Soc; NY Acad Sci; Physiol Soc Japan; Japanese Soc Nutrit & Food Sci. *Res:* Environmental physiology of humans by applying humoral and neural analysis technique; adaption mechanisms of humans to extreme environments, high altitude, high pressure, hot, and cold. *Mailing Add:* Dept Physiol Sch Med Univ Occup & Environ Health 1-1 Iseigaoka Yahata-nishiku Kitakyushu 807 Japan

SHIRANE, GEN, b Nishinomiya, Japan, May 15, 24; m 50; c 2. SOLID STATE PHYSICS. *Educ:* Univ Tokyo, BE, 47, DSc(physics), 54. *Prof Exp:* Res assoc physics, Tokyo Inst Technol, Japan, 48-52; res assoc, Pa State Univ, 52-55, asst prof, 55-56; assoc physicist, Brookhaven Nat Lab, 56-57; res physicist, Res Labs, Westinghouse Elec Corp, 57-58, adv physicist, 59-63; physicist, 63-68, SR PHYSICIST, BROOKHAVEN NAT LAB, 68- *Honors & Awards:* Buckley Prize, Am Physics Soc, 73; Warren Award, Am Crystallog Asn, 73; Humboldt Award, 85. *Mem:* Nat Acad Sci; Phys Soc Japan; Am Phys Soc. *Res:* Neutron scattering; magnetism; lattice dynamics. *Mailing Add:* Dept Physics Brookhaven Nat Lab Upton NY 11973

SHIRAZI, MOSTAFA AYAT, b Najaf, Iraq, Sept 27, 32; US citizen; m 61; c 3. FLUID DYNAMICS. *Educ:* Calif State Polytech Col, BS, 59; Univ Wash, MS, 61; Univ Ill, Urbana, PhD(mech eng), 67. *Prof Exp:* Assoc res engr, Boeing Co, Wash, 61; sr res engr, Hercules Inc, Md, 67-69; res mech engr, 69-80, SR RES SCIENTIST, CORVALLIS ENVIRON RES LAB, ENVIRON PROTECTION AGENCY, 80- *Mem:* Am Soc Mech Engrs. *Res:* Analysis of jet diffusion for the prediction of heated plume behavior in large bodies of water; jet diffusion; ecosystems modeling and analysis. *Mailing Add:* 2025 NW Jackson Creek Dr Corvallis OR 97330

SHIREMAN, RACHEL BAKER, b Springhill, La, Aug 3, 40; m 61; c 4. NUTRITION, BIOCHEMISTRY. *Educ:* La State Univ, BS, 62; Iowa State Univ, MS, 66; Univ Fla, PhD(biochem), 78. *Prof Exp:* Instr biol, Univ Southwest La, 67-72; res asst med, 78-79, asst prof, 79-84, ASSOC PROF NUTRIT, UNIV FLA, 84- *Mem:* Soc Exp Biol & Med; Am Oil Chemists Soc; Am Soc Biol Chem; AAAS. *Res:* Relationships between the structure, composition and function of the plasma lipoproteins; low density lipoprotein. *Mailing Add:* Dept Food Sci & Human Nutrit Univ Florida Gainesville FL 32611

SHIREN, NORMAN S, b New York, NY, Feb 7, 25; m 47; c 2. QUANTUM ACOUSTICS, QUANTUM OPTICS. *Educ:* Tufts Univ, BS, 45; Stanford Univ, PhD(physics), 56. *Prof Exp:* Staff scientist, Hudson Lab, Columbia Univ, 51-55; staff mem, Gen Elec Res Ctr, 55-61; staff mem/mgr, T J Watson Res Ctr, IBM, 61-87, consult, 87-90; RETIRED. *Concurrent Pos:* Guggenheim fel, Clarendon Lab, Oxford Univ, UK, 69-70; Sci Res Coun vis fel, Lancaster Univ, UK, 82-83. *Mem:* Fel AAAS (pres, 83); fel Am Phys Soc. *Res:* Quantum acoustics, quantum optics, microwave measurements and nonlinear acoustics at microwave frequencies-primarily wave interactions; discovered backward wave phonon echoes, acoustic phase conjugation, holographic echo storage; theory of kapitza conductance; measured microwave properties of superconductors and organic conductors. *Mailing Add:* 20 Twin Ridges Rd Ossining NY 10562

SHIRER, DONALD LEROY, b Cleveland, Ohio, May 10, 31. COMPUTER SCIENCES, ENGINEERING PHYSICS. *Educ:* Case Western Reserve Univ, BS, 52; Ohio State Univ, MSc, 53, PhD(physics), 57. *Prof Exp:* Res assoc, Res Found, Ohio State Univ, 57; from asst prof to prof physics, Valparaiso Univ, 57-78; adj prof elec eng, Univ Ariz, Tucson, 80-81, dir, Computer-Based Instr Lab, 81-88; LAB DIR, YALE UNIV, 88- *Concurrent Pos:* Consult, Argonne Nat Lab, 58-65; NSF fel, Univ Ill, 71-72; assoc ed, Am J Physics, 71-76; vis prof elec eng, Univ Ariz, 78-79. *Mem:* Am Phys Soc; Am Asn Physics Teachers; Asn Develop Comput-Based Instruction Systs; Inst Elec & Electronics Engrs; Am Soc Eng Educ. *Res:* Educational uses of computers; speech and music synthesis; acoustics. *Mailing Add:* 40 Wall St Wallingford CT 06492

SHIRER, HAMPTON WHITING, b Newton, Mass, Aug 8, 24; m 47; c 5. PHYSIOLOGY. *Educ:* Washburn Univ, BS, 45; Univ Kans, MD, 48. *Prof Exp:* Intern, Med Ctr, Univ Kans, 48-49, USPHS res fel, 49-51, instr surg, 53-54, from instr to asst prof physiol, 54-61; head biophys group biol sci & systs dept, Gen Motors Defense Res Labs, 61-63; asst prof physiol, Univ Mich, 63-64; assoc prof physiol, Dept Comp Biochem & Physiol & Dept Elec Eng, 64-67, prof elec eng, physiol & cell biol, 67-73, PROF PHYSIOL & CELL BIOL, UNIV KANS, 73- *Concurrent Pos:* Lederle Med Fac Award, 56-59. *Mem:* AAAS; Biophys Soc; Am Meteorol Soc; Inst Elec & Electronics Engrs; Sigma Xi. *Res:* Bioelectricity; cardiovascular regulation; physiological instrumentation; biotelemetry. *Mailing Add:* Dept of Physiol & Cell Biol Univ of Kans Lawrence KS 66044

SHIRES, GEORGE THOMAS, b Waco, Tex, Nov 22, 25; m 48; c 3. SURGERY. *Educ:* Univ Tex, BS, 44, MD, 48; Am Bd Surg, cert, 56. *Prof Exp:* From asst prof to prof surg & chmn dept, Southwest Med Sch, Univ Tex, Dallas, 57-74; chmn & prof surg, Sch Med, Univ Wash, 74-75; Lewis Atterbury Stimson prof & chmn, Dept Surg, Med Col, Cornell Univ, 75-91, Stephen & Suzanne Weiss dean & provost med affairs, 87-91; PROF SURG & CHMN DEPT, HEALTH SCI CTR, TEX TECH UNIV, 91- *Concurrent Pos:* Surgeon-in-chief, Parkland Mem Hosp, Dallas, 60-74, Harborview Med Ctr & Univ Hosp, Seattle, 74-75 & NY Hosp, 75-91; consult, Surgeon Gen, Nat Inst Gen Med Sci, 65- & var hosps, 74; mem coun, Am Surg Asn, 69-74 & 80-, Ad Hoc External Exam Rev Comt, Asn Am Med Cols, 80-, Adv Comt & Work Groups Health Policy Agenda, AMA, 83-; assoc ed-in-chief, Infections Surg, 81-89, sr contrib ed, 82-89; ed, Surg, Gynec & Obstet, 82-; chief surg, Univ Med Ctr, Lubbock, Tex, 91- *Honors & Awards:* Numerous Named Lectr, 71-90; Distinguished Achievement Award, Am Trauma Soc, 81, Curtis P Artz Mem Award, 84; Sheen Award, 85; Harvey Stuart Allen Distinguished Serv Award, Am Burn Asn, 88. *Mem:* Inst Med-Nat Acad Sci; Am Surg Asn (secy, 69-74, pres, 79-80); Am Col Surg (pres, 82); Soc Univ Surg; Soc Int Surg; Am Asn Surg Trauma; AMA; Am Burn Asn. *Res:* Surgical trauma; author of numerous technical publications. *Mailing Add:* Health Sci Ctr Tex Tech Univ 3601 Fourth St Lubbock TX 79430

SHIRES, THOMAS KAY, b Buffalo, NY, July 12, 35; m 61; c 3. CELL BIOLOGY. *Educ:* Colgate Univ, BA, 57; Univ Okla, MS, 61, PhD(cell biol), 65. *Prof Exp:* Res lectr depts anat & urol, Sch Med, Univ Okla, 65-68; res fel, McArdle Cancer Res Inst, Univ Wis-Madison, 68-72; from asst prof to assoc prof, 72-81, PROF PHARMACOL, COL MED, UNIV IOWA, 81- *Concurrent Pos:* Vis prof, Univ Wis-Madison, 81-82. *Mem:* Am Asn Cancer Res; NY Acad Sci; Am Soc Pharmacol & Exp Therapeut; Am Soc Cell Biol; Am Asn Pathologists; Am Ornithologists Union. *Res:* Glycation of proteins; toxicology of glucose; transcytosis of proteins. *Mailing Add:* Dept Pharmacol Col Med Univ Iowa Iowa City IA 52240

SHIRK, AMY EMIKO, b Honolulu, Hawaii, July 14, 46. INORGANIC CHEMISTRY. *Educ:* Univ Hawaii, BS, 68; Northwestern Univ, PhD(chem), 73. *Prof Exp:* Res assoc mat sci, Northwestern Univ, 73-74; res assoc chem, Ill Inst Technol, 74-88. *Concurrent Pos:* Mem tech staff chem, Bell Labs, 77-78. *Mem:* Am Chem Soc; Sigma Xi. *Res:* Infrared and Raman spectroscopy; molecular structure; inorganic syntheses. *Mailing Add:* Naval Res Lab 4555 Overlook Dr Washington DC 20375-5000

SHIRK, B(RIAN) THOMAS, b Schoeneck, Pa, Sept 20, 41; m 66; c 4. SOLID STATE SCIENCE, ELECTRICAL ENGINEERING. *Educ:* Pa State Univ, BS, 63, PhD(solid state sci), 68. *Prof Exp:* Res scientist, Stackpole Carbon Co, 68-72; tech dir, Lydall Magnetics Co, 72-75; PRES, HOOSIER MAGNETICS, INC, 75- *Mem:* Inst Elec & Electronics Engrs; Am Ceramic Soc. *Res:* Magnetic oxides and glasses; single crystal growth and characterization; research and development of ferrites. *Mailing Add:* 5217 River Ridge Circle Sylvania OH 43560

SHIRK, JAMES SILER, b Chambersburg, Pa, Mar 7, 40; m 72. SPECTROSCOPY, NON-LINEAR OPTICS. *Educ:* Col Wooster, BA, 62; Univ Calif, Berkeley, PhD(chem), 66. *Prof Exp:* Res assoc chem, Imp Col, Univ London, 66-67; Nat Acad Sci-Nat Res Coun res assoc fel spectros, Nat Bur Standards, 67-69; from asst prof to prof chem, Ill Inst Technol, 69-87; RES CHEMIST, NAVAL RES LAB, 87- *Concurrent Pos:* Mem tech staff, Bell Labs, Murray Hill, NJ, 77-78. *Mem:* Am Chem Soc; Am Phys Soc; Sigma Xi. *Res:* Infrared and ultraviolet spectroscopy; matrix isolation studies; lasers; instrumentation; non linear optical materials and processes. *Mailing Add:* Code 6551 Naval Res Lab Washington DC 20375

SHIRK, PAUL DAVID, b Waterloo, Iowa, June 7, 48. MOLECULAR GENETICS, MOLECULAR ENDOCRINOLOGY. *Educ:* Univ Northern Iowa, BA, 70; Tex A&M Univ, MS, 75, PhD(zool), 78. *Prof Exp:* Res assoc, Univ Ore, 78-79, NIH fel, 79-81; asst prof, Ore State Univ, 81-84; RES PHYSIOLOGIST, AGR RES SERV, USDA, 85- *Mem:* Am Soc Zoologists; Soc Develop Biol; Sigma Xi. *Res:* Hormonal control of yolk polypeptide expression and control of reproduction in stored product insect pests. *Mailing Add:* Agr Res Serv USDA PO Box 14565 Gainesville FL 32604-2516

SHIRK, RICHARD JAY, b Tyrone, Pa, Feb 10, 30; m 51; c 4. BACTERIOLOGY, DATA PROCESSING. *Educ:* Pa State Univ, BS, 51. *Prof Exp:* Bacteriologist fermentation process develop, Heyden Chem Corp, 51-53; bacteriologist, 53-55, res bacteriologist food res group, 55-64 & bact chemother group, 64-74, sr tech systs analyst, 74-83, SUPERVISING ANALYST, AM CYANAMID CO, 83- *Mem:* Data Processing Mgt Asn; Asn Comput Machinery. *Res:* Antibiotic fermentations; food technology; antioxidants; food preservatives; food coatings; bioassays; experimental infections; chemotherapy of animal diseases; statistical analysis; computer programming; computerized biological screening systems; research and development computer applications; research administration systems. *Mailing Add:* Agr Res Ctr Am Cyanamid Co PO Box 400 Princeton NJ 08540

SHIRKEY, HARRY CAMERON, b Cincinnati, Ohio, July 2, 16; m 58; c 3. PEDIATRICS. *Educ:* Univ Cincinnati, BS, 39, MD, 45, DSc(pharm), 76; Am Bd Pediat, dipl, 52; Am Bd Clin Toxicol, dipl, 76. *Prof Exp:* Assoc prof pharm, Col Pharm, Univ Cincinnati, 40-41; pharmacist, Children's Hosp, 41-42; asst pharmacol, Col Med, Univ Cincinnati, 43-46 & 48-53, instr pediat, 53-57, asst clin prof, 57-60; prof pediat, Med Col Ala, 60-68; prof pediat & chmn dept & prof pharmacol, Sch Med, Univ Hawaii, 68-71; prof pediat & chmn dept, Med Ctr, Tulane Univ, 71-77; RETIRED. *Concurrent Pos:* Resident, Children's Hosp, Cincinnati, 48-51; assoc prof pharmacol, Col Pharm, Univ Cincinnati, 48-60; mem rev comt & chmn panel pediat, US Pharmacopeia, 52-; dir pediat, Cincinnati Gen Hosp, 53-60; chmn admis comt, Nat Formulary, 60-; med & admin dir, Children's Hosp, 60-68; dir, Jefferson County Poison Control Ctr, 60-68; mem staff, Univ Hosp, 60-68; consult, Baptist Hosp, Crippled Children's Hosp & Clin & St Vincent Hosp, 60-68; prof pharmacol, Samford Univ, 60-68; med & exec dir, Kauikeolani Children's

Hosp, 68-71; dir pediat serv, Charity Hosp New Orleans, 71-77; mem adv comt drug efficacy study, Nat Res Coun-Nat Acad Sci, mem drug res bd, 71-74. *Mem:* Fel Am Acad Pediat; Am Pediat Soc; Am Soc Pharmacol & Exp Therapeut. *Mailing Add:* 1216 Paxton Ave Cincinnati OH 45208

SHIRLEY, AARON, b Gluckstadt, Miss, Jan 3, 33; m; c 4. PEDIATRICS. *Educ:* Tougaloo Col, BS, 55; McHarry Med Col, MD, 59; Univ Miss, MD, 68. *Prof Exp:* Intern, Herbert Hosp, Tenn, 59-60; gen practice, Vicksburg, 60-65; PROJ DIR, JACKSON-HINDS COMPREHENSIVE HEALTH CTR, 70- *Concurrent Pos:* Fac med, Tufts Univ Med, Mass, 68-73 & Univ Miss Med Sch, 70-; head start consult, Am Acad Pediat, 69-74; mem bd dirs, Field Found, 73-89; mem adv bd, Rural Pract Proj, Robert Wood Johnson Found, 74-78; mem, Select Panel Prom Child Health, Washington, DC, 79-81; mem coun, Inst Med-Nat Acad Sci, 88- *Mem:* Inst Med-Nat Acad Sci. *Mailing Add:* Jackson-Hinds Comprehensive Health Ctr 4433 Medgar Evers Blvd Jackson MS 39213

SHIRLEY, BARBARA ANNE, b Muskogee, Okla, Oct 15, 36. REPRODUCTIVE PHYSIOLOGY, ENDOCRINOLOGY. *Educ:* Okla Baptist Univ, BA, 56; Univ Okla, MS, 61, PhD(zool), 64. *Prof Exp:* Asst, Univ Okla, 58-62; from asst prof to assoc prof zool, 64-79, PROF ZOOL, UNIV TULSA, 79- *Mem:* AAAS; Soc Study Reproduction; Am Physiol Soc. *Res:* Reproduction; embryo culture. *Mailing Add:* Dept Biol Sci Univ Tulsa Tulsa OK 74104

SHIRLEY, DAVID ALLEN, organic chemistry; deceased, see previous edition for last biography

SHIRLEY, DAVID ARTHUR, b North Conway, NH, Mar 30, 34; m 56; c 5. CHEMICAL PHYSICS. *Educ:* Univ Maine, BS, 55; Univ Calif, Berkeley, PhD(chem), 59. *Hon Degrees:* ScD, Univ Maine, 78; Dr Rer Nat, FU Berlin, 87. *Prof Exp:* Fel chem, Lawrence Radiation Lab, Univ Calif, 58-59; lectr chem, Univ Calif, Berkeley, 59-60, from asst prof to assoc prof, 60-67, from vchmn to chmn dept, 68-75, dir, Lawrence Berkeley Lab, 80-89, PROF CHEM, UNIV CALIF, BERKELEY, 67- *Concurrent Pos:* NSF fels, Oxford Univ, 66-67 & Free Univ Berlin, 70; assoc ed, J Chem Physics, 74-76; assoc dir, Lab & head, Mat & Molecular Res Div, Lawrence Berkeley Lab, 75-80. *Honors & Awards:* Ernest O Lawrence Award, USAEC, 72. *Mem:* Nat Acad Sci; AAAS; Am Chem Soc; Am Phys Soc; Fedn Am Scientists; Am Acad Arts & Sci. *Res:* Electron spectroscopy of atoms, molecules and solids with emphasis on surfaces and many-electron effects. *Mailing Add:* Dept Chem Univ Calif Berkeley CA 94720

SHIRLEY, FRANK CONNARD, b Minneapolis, Minn, Dec 18, 33; wid; c 4. FOREST MANAGEMENT, FOREST ECONOMICS. *Educ:* Cornell Univ, AB, 55; State Univ NY Col Forestry, Syracuse, MF, 60; Univ Mich, Ann Arbor, AM, PhD(forestry), 69. *Prof Exp:* Forester, US Forest Serv, 60-64; asst prof forest econ, Colo State Univ, 69-72; consult forest mgt policy & econ, 72-74; forest economist, 74-85, OPERS RES ANALYST, ST REGIS PAPER CO, 85-; PRES, ASPEN FOREST CONSERV SYSTS, 85- *Mem:* Soc Am Foresters; Am Forestry Asn. *Res:* Forest land management in an affluent society. *Mailing Add:* 18219 S Vaughn Rd KPN Vaughn WA 98394

SHIRLEY, HERSCHEL VINCENT, JR, b Alpharetta, Ga, Aug 29, 23; m 51; c 3. ANIMAL GENETICS, ANIMAL PHYSIOLOGY. *Educ:* Univ Ga, BSA, 49, MSA, 51; Univ Ill, PhD(animal genetics & physiol), 55. *Prof Exp:* Assoc prof & assoc poultry geneticist, Agr Exp Sta, Univ Tenn, 55-75, prof animal physiol & poultry geneticist, 75-90; RETIRED. *Mem:* Poultry Sci Asn; Am Genetic Asn. *Res:* Endocrinology of reproduction and stress physiology. *Mailing Add:* 232 Essex Dr Knoxville TN 37922

SHIRLEY, RAY LOUIS, b Berkeley Co, WVa, Dec 11, 12; m 43, 78; c 5. ANIMAL NUTRITION. *Educ:* WVa Univ, BS, 37, MS, 39; Mich State Univ, PhD(biochem), 49. *Prof Exp:* Asst agr chem, WVa Univ, 37-39; asst biochem, Mich State Univ, 39-41, asst prof, 41-42; res chemist, Hercules Powder Co, 42-47; asst prof biochem, Mich State Univ, 47-49; prof animal nutrit & biochemist, Univ Fla, 49-51; prof chem, Shepherd Col, 51-53; prof, 53-82, EMER PROF ANIMAL SCI & ANIMAL NUTRITIONIST, UNIV FLA, 82- *Honors & Awards:* Gustav Bohstedt Award, Am Soc Animal Sci, 75. *Mem:* AAAS; Am Chem Soc; Soc Exp Biol & Med; fel Am Soc Animal Sci; Am Inst Nutrit. *Res:* Nitrogen compounds and minerals, isotopes, enzymes, energy in diets. *Mailing Add:* 1523 NW 11th Rd Gainesville FL 32605

SHIRLEY, ROBERT LOUIS, b Fairview Village, Ohio, Jan 11, 33; m 55; c 3. ORGANIC CHEMISTRY. *Educ:* Col Wooster, BA, 55; Ohio State Univ, PhD(org chem), 60. *Prof Exp:* Res chemist, Jefferson Chem Co, Inc, 60-63, mem staff mkt develop, 63-70; mgr mkt develop, Ott Chem Co, 70-73; sales mgr, Story Chem Corp, 73-77; tech sales rep, Polymer Chem Div Upjohn, 77-85; sr sales specialist, Dow Chem Co, 85-86; OWNER/PRES, TECK-TOOL SUPPLY, 86- *Mem:* Soc Plastics Engrs; Soc Automotive Engrs. *Res:* Phosgene chemistry; isocyanates, specialty organic chemicals. *Mailing Add:* 725 Lyncott St North Muskegon MI 49445-2836

SHIRLEY, THOMAS CLIFTON, b Falfurrias, Tex, Sept 17, 47; m 79. CRUSTACEAN BIOLOGY, BENTHIC ECOLOGY. *Educ:* Tex A&I Univ, BS, 69, MS, 74; La State Univ, PhD(zool), 82. *Prof Exp:* Instr biol & invertebrate zool, La State Univ, 78-82; res scientist, Auke Bay Labs, Nat Marine Fisheries Serv, 82; asst prof biol & fisheries, Univ Alaska, Juneau, 82-87; assoc prof biol & fisheries, Univ Alaska Southeast, 87-88; ASSOC PROF MARINE ECOL & FISHERIES, JUNEAU CTR FISHERIES & OCEAN SCI, UNIV ALASKA, FAIRBANKS, 88- *Concurrent Pos:* Mem plan team, NPac Fisheries Mgt Coun, 87-; vis assoc prof, Marine Lab, Duke Univ, 88-89. *Mem:* AAAS; Am Soc Limnol & Oceanog; Am Soc Zoologists; Ecol Soc Am; Crustacean Soc. *Res:* Early life history of decapod crustaceans, benthic ecology, benthic-pelagic coupling, meiofauna ecology, bioenergetics and physiology of marine invertebrates. *Mailing Add:* Juneau Ctr Fisheries & Ocean Sci Univ Alaska 11120 Glacier Hwy Juneau AK 99801

SHIRN, GEORGE AARON, b Williamsport, Pa, June 30, 21; m 46; c 2. PHYSICS. *Educ:* Columbia Univ, BS, 46; Rensselaer Polytech Inst, MS, 50, PhD(physics), 54. *Prof Exp:* Sr scientist, Res & Develop Ctr, Sprague Elec Co, 54-87; RETIRED. *Concurrent Pos:* Adj prof physics, N Adams State Col, Mass, 68- *Mem:* Am Phys Soc; Electrochem Soc; Am Vacuum Soc. *Res:* Solid state physics; semiconductors; oxides; metals; thin films. *Mailing Add:* 37 Jamieson Heights Williamstown MA 01267-2001

SHIRTS, RANDALL BRENT, b Mt Pleasant, Utah, Apr 28, 50; m 74; c 6. INTRAMOLECULAR VIBRATIONAL & ROTATIONAL DYNAMICS, SEMICLASSICAL METHODS. *Educ:* Brigham Young Univ, BS, 73; Harvard Univ, AM, 78, PhD(chem physics), 79. *Prof Exp:* Res assoc, Joint Inst, Lab Astrophys & Dept Chem, Univ Colo, 79-81; asst prof chem, Georgetown Univ, 81-82; asst prof chem, Univ Utah, 82-87; SCI SPECIALIST, IDAHO NAT ENG LAB, 87- *Concurrent Pos:* Vis scientist, Los Alamos Nat Lab, 86. *Mem:* Am Chem Soc; Am Phys Soc; Sigma Xi. *Res:* Spectroscopy and dynamics of intramolecular vibrational-rotational motion; laser-molecule interactions. *Mailing Add:* EG&G Idaho Inc PO Box 1625 Idaho Falls ID 83415-2208

SHIU, ROBERT P C, MOLECULAR BIOLOGY, BREAST CANCER. *Educ:* McGill Univ, PhD(endocrinol), 74. *Prof Exp:* PROF ENDOCRINOL & CELL BIOL, UNIV MAN, 77- *Mailing Add:* Dept Physiol Fac Med Univ Man 770 Bannatyne Ave Winnipeg MB R3T 2N2 Can

SHIUE, CHYNG-YANN, b Tainan, Taiwan, Dec 15, 41; m 67; c 2. ORGANIC CHEMISTRY, MEDICINAL CHEMISTRY. *Educ:* Taiwan Normal Univ, BSc, 65; Brown Univ, PhD(chem), 70. *Prof Exp:* Asst org chem, Inst Chem, Acad Sinica, 64-65; res asst, Brown Univ, 66-70; res assoc, Univ Ky, 70-72; res assoc biochem pharmacol, 72-74; instr biochem, Brown Univ, 74-76; chemist, Brookhaven Nat Lab, 76-89; PROF RADIOL & PHARMACOL, DIR CTR METAB IMAGING, CREIGHTON UNIV, 89- *Concurrent Pos:* Mem, Ad Hoc Tech Rev Group, Nat Cancer Inst, 87, 88 & 91, US Dept Energy, 87, 89 & 90. *Mem:* Am Chem Soc; Soc Nuclear Med; NY Acad Sci. *Res:* Synthesis of radiopharmaceuticals and other biologically active compounds; nuclear medicine. *Mailing Add:* Ctr Metab Imaging Creighton Univ Omaha NE 68108

SHIVANANDAN, KANDIAH, b Parit Buntar, Malaya, Aug 22, 29; US citizen; m 62; c 2. ASTROPHYSICS, COSMOLOGY. *Educ:* Univ Melbourne, BSc, 58; Univ Toronto, MA, 59; Cath Univ Am, PhD(physics), 69. *Prof Exp:* Tech asst, Weapons Res Estab, Melbourne, Australia, 50-57; radiation physicist, Australian Atomic Energy Comn, 57-58; res physicist, Mass Inst Technol, 59-64; PHYSICIST, NAVAL RES LAB, 65- *Concurrent Pos:* Consult mem, Fed Radiation Coun, 60-63 & Int Atomic Energy Asn, Vienna, 60-65; mem, US Nuclear Weapons Study Comt, 63-; adv mem, US Arms Control & Disarmament Agency, 66, Europ Space Res Orgn, Paris, 68, Defense Sci Bd, 88 & Sci adv bd, NAtlantic Treaty Orgn, 85. *Mem:* Am Astron Soc; fel Royal Astron Soc; Europ Phys Soc; NY Acad Sci; Indian Inst Physics. *Res:* Biological effects of radiation from nuclear weapon tests; relativistic astrophysics and experimental infrared astronomy in relation to cosmology; infrared and submillimeter astronomy. *Mailing Add:* Naval Res Lab Code 4200 Washington DC 20375

SHIVE, DONALD WAYNE, b Hanover, Pa, June 24, 42; m 71; c 2. ANALYTICAL CHEMISTRY. *Educ:* Pa State Univ, BS, 64; Mass Inst Technol, PhD(chem), 69. *Prof Exp:* From asst prof, to assoc prof, 69-78, PROF CHEM, MUHLENBERG COL, 78- *Concurrent Pos:* Consult, J T Baker Chem Co, 79-, Microchagnostics Inc, 85- *Honors & Awards:* Lindback Award, 78. *Mem:* Am Chem Soc. *Res:* Electrochemistry; chromatography. *Mailing Add:* 2204 Huckleberry Rd Allentown PA 18104-1344

SHIVE, JOHN BENJAMINE, JR, plant physiology, botany; deceased, see previous edition for last biography

SHIVE, PETER NORTHROP, b Plainfield, NJ, July 2, 41; m 64. GEOPHYSICS. *Educ:* Wesleyan Univ, BA, 64; Stanford Univ, PhD(geophys), 68. *Prof Exp:* Res assoc geophys, Stanford Univ, 68-69; asst prof geol, 69-72, assoc prof geol & physics, 72-76, PROF GEOL & PHYSICS, UNIV WYO, 76- *Mem:* AAAS; Am Geophys Union; Soc Terrestrial Magnetism & Elec; Soc Explor Geophys. *Res:* Rock magnetism; paleomagnetism; time series analysis; exploration seismology. *Mailing Add:* Dept Geol Univ Wyo Box 3006 Laramie WY 82071

SHIVE, ROBERT ALLEN, JR, b Dallas, Tex, Oct 27, 42; m 64; c 3. MATHEMATICS. *Educ:* Southern Methodist Univ, BA, 64, MS, 66; Iowa State Univ, PhD(math), 69. *Prof Exp:* Instr math, Iowa State Univ, 67-69; asst prof, 69-74, assoc prof, 74-79, PROF MATH & ASSOC DEAN, MILLSAPS COL, 79-, DIR INFO SYSTS, 81- *Concurrent Pos:* Consult comput; adm intern, Am Coun Educ, 78-79. *Mem:* Am Math Soc; Math Asn Am. *Res:* Integration theory; computers in education. *Mailing Add:* Dept Comput Millsaps Col Jackson MS 39210

SHIVE, WILLIAM, AMINO ACIDS, CONTROL MECHANISMS. *Educ:* Univ Tex, PhD(chem), 41. *Prof Exp:* RJ WILLIAMS PROF CHEM, UNIV TEX, 45- *Res:* Control mechanisms. *Mailing Add:* Dept Chem Univ Tex Austin TX 78712

SHIVELY, CARL E, b Laurelton, Pa, June 8, 36; m 58; c 3. BACTERIOLOGY. *Educ:* Bloomsburg State Col, BS, 58; Bucknell Univ, MS, 61; St Bonaventure Univ, PhD(biol), 68. *Prof Exp:* Instr bact, Bucknell Univ, 61; asst prof bact & genetics, State Univ NY Col Cortland, 63-65; asst prof bact & genetics, 68-71, assoc prof, 71-77, head, div biol sci, 77-83, PROF BACT & BIOCHEM, ALFRED UNIV, 78- *Honors & Awards:* Outstanding Scientist Award, Eastman Kodak. *Mem:* Am Soc Microbiol; Sigma Xi; Am Soc Enol & Viticulture. *Res:* Biochemistry and fermentation chemistry; production of ethanol on a continuous basis by immobilized Zymomonas mobilis s c in an immobilized cell reactor; immobilizing material, foam glass; protein chemistry of wines. *Mailing Add:* Dept Biol Alfred Univ Alfred NY 14802

SHIVELY, CHARLES DEAN, b Dyersburg, Tenn, Feb 4, 44; m 66; c 2. PHARMACEUTICAL CHEMISTRY. *Educ:* Purdue Univ, BS, 67, PhD(indust & phys pharm), 72. *Prof Exp:* Sr scientist pharm develop, Alcon Labs, Inc, 71-75, prod mkt mgr, 75-76, res sect head, 76-77; MGR PHARM DEVELOP, COOPER LABS, INC, 77- *Concurrent Pos:* Lectr, Calif Bd Pharm Continuing Educ, 77-; adv bd mem, Nat Eye Res Found, 77- *Mem:* Am Pharmaceut Asn; Nat Eye Res Found; Sigma Xi. *Res:* Pharmaceutical dosage form development; pharmaceutical formulation; physical pharmacy; surface chemistry. *Mailing Add:* 18816 N 94th Ave Peoria AZ 85345

SHIVELY, FRANK THOMAS, b Cuyahoga Falls, Ohio, Oct 31, 34; m 69. PHYSICAL & ATMOSPHERIC OPTICS. *Educ:* Oberlin Col, BA, 54; Yale Univ, MS, 57, PhD(physics), 61. *Prof Exp:* Asst physicist, Physics Dept, Yale Univ, 60-61; res physicist, Ctr Nuclear Studies, France, 61-63; res assoc, Lawrence Radiation Lab, Univ Calif, 63-66; sr researcher, Inst Nuclear Physics, Univ Paris, 66-68; mem staff, Los Alamos Meson Physics Facil, Univ Calif, 69-74; staff scientist & consult, Lawrence Berkeley Lab, Univ Calif, & Univ Calif, Los Angeles, 74-77; consult & sr scientist, Sci Simulation, Inc, 78-81; PRIN SCIENTIST, ORION INT, 82- *Concurrent Pos:* Adj instr, Physics Dept, Univ Calif, Berkeley, 64; vis lectr, Colo Col, 75; dir consult, C D Sci Consults, 75-78; lectr, Honors Prog, Univ NMex, 76; vis prof physics, Univ Calif, Los Angeles, 77. *Mem:* Am Phys Soc; Sigma Xi; Optical Soc Am. *Res:* Strong, electromagnetic and weak interactions of particles and nuclei; particle beam propagation, physical optics; atmospheric physics & remote sensing theory; simulation methodology; author or coauthor of over 50 publications. *Mailing Add:* 1094 Governor Dempsey Dr Santa Fe NM 87501

SHIVELY, JAMES NELSON, b Moran, Kans, Feb 9, 25; m 53; c 3. VETERINARY PATHOLOGY. *Educ:* Kans State Univ, DVM, 46; Johns Hopkins Univ, MPH, 53; Univ Rochester, MS, 56; Colo State Univ, PhD, 71. *Prof Exp:* Officer in charge vet med, Army Med Lab, 46-52; res vet, Agr Res Prog, Univ Tenn, AEC, 54-55; chief vet virol, Div Vet Med, Walter Reed Army Inst Res, 56-60; vet, Res Br, Div Radiol Health, USPHS, 60-62, Radiol Health Lab, Colo, 62-68 & Path Sect, Div Biol Effects, Bur Radiol Health, Md, 68-70; assoc prof ultrastructural path, NY State Vet Col, Cornell Univ, 71-75; PROF VET SCI, UNIV ARIZ, 75- *Mem:* Am Vet Med Asn; Electron Micros Soc Am; Int Acad Path; Sigma Xi. *Res:* Pathology; electron microscopy. *Mailing Add:* Dept Vet Sci Univ Ariz Tucson AZ 85721

SHIVELY, JESSUP MACLEAN, b Monrovia, Ind, Nov 9, 35; m 59; c 2. BIOCHEMISTRY, MICROBIOLOGY. *Educ:* Purdue Univ, BS, 57, MS, 59, PhD(microbiol), 62. *Prof Exp:* Instr microbiol, Purdue Univ, 61-62; from asst prof to assoc prof, Univ Nebr, Lincoln, 62-70; chmn biochem sect, Biol Div, 70-71, assoc prof, 70-73, head dept, 71-82, PROF BIOCHEM, CLEMSON UNIV, 73- *Concurrent Pos:* NSF fel, Scripps Inst Oceanog, 65; fel, Inst Enzyme Res, Univ Wis-Madison, 68-69, Dept Physiol Chem, Med Sch, Johns Hopkins Univ, 69-70, Med Div, Oak Ridge Assoc Univs, 71, Wash State Univ, 76; guest prof, Univ Hamburg, Fed Repub Ger, 81-82, St Louis Univ, 82. *Mem:* AAAS; Am Soc Microbiol; Brit Soc Gen Microbiol; Am Soc Biol Chemists. *Res:* Lipids and membranes of bacteria; ribulose bisphoephate carboxylase/oxygenase, carboxysomes, sulfur oxidation, autotrophic microbes. *Mailing Add:* Dept Biol Sci Clemson Univ Clemson SC 29631-1903

SHIVELY, JOHN ADRIAN, b Rossville, Ind, Oct 29, 22; m 45; c 4. PATHOLOGY, HEMATOLOGY. *Educ:* Ind Univ, BA, 44, MD, 46. *Prof Exp:* Lab asst physiol, Ind Univ, 43-44; clin pathologist, Clin Hosp, Bluffton, Ind, 52-54; asst prof path, Sch Med, Ind Univ, 54-57; pathologist, Manatee Mem Hosp, Bradenton, Fla, 57-62; assoc prof path, Col Med, Univ Ky, 62-63; pathologist, Univ Tex M D Anderson Hosp & Tumor Inst, 63-68; prof path, Sch Med, Univ Mo-Columbia, 68-71; chmn dept, Univ Tenn, Memphis, 71-76, prof path, 71-83, vchancellor acad affairs, Ctr Health Sci, 76-83; med dir, Smith Kline Bio-Sci Labs, Tampa, 83-88; clin prof, 83-88, PROF PATH, COL MED, UNIV SFLA, 88- *Mem:* AAAS; Col Am Pathologists; Am Soc Clin Path; Am Col Physicians; Am Asn Blood Banks (pres, 67-68); Am Soc Hematol. *Res:* Bone marrow failure; platelet physiology; oncology; medical education. *Mailing Add:* 12901 N 30th St Box 11 Tampa FL 33612

SHIVELY, JOHN ERNEST, b Chicago, Ill, Aug 25, 46; m 67; c 2. BIOCHEMISTRY, IMMUNOLOGY. *Educ:* Univ Ill, Urbana, BS, 68, MS, 69, PhD(biochem), 75. *Prof Exp:* SCIENTIST IMMUNOCHEM, BECKMAN RES INST CITY OF HOPE, 75- *Mem:* Am Soc Biochem & Molecular Biol; Protein Soc. *Res:* Microsequence studies on carcinoembryonic antigen and human plasma fibronectin; protein structural analysis. *Mailing Add:* 1657 Wilson Ave Arcadia CA 91006

SHIVELY, RALPH LELAND, b Mt Morris, Ill, Nov 22, 21; m 50; c 2. MATHEMATICS. *Educ:* Univ Mich, BSE, 47, MA, 48, PhD(math), 54. *Prof Exp:* From instr to asst prof math, Western Reserve Univ, 51-55, from asst prof to assoc prof, 56-61; assoc prof, Manchester Col, 55-56 & Col, 61-64; sr mathematician, Oak Ridge Nat Lab, 64-65; prof math & chmn dept, 65-87, EMER PROF MATH, LAKE FOREST COL, 87- *Concurrent Pos:* NSF fac fel, Univ Calif, Berkeley, 60-61; vis res fel, Comput Lab, Oxford Univ, 71-72 & Univ Chicago, 83. *Mem:* Am Math Soc; Math Asn Am; Sigma Xi. *Res:* Special functions of classical analysis. *Mailing Add:* Dept Math Lake Forest Col Lake Forest IL 60045

SHIVER, JOHN W, b Jacksonville, Fla, Aug 31, 57. EXPERIMENTAL IMMUNOLOGY. *Educ:* Wofford Col, BS, 78; Univ Fla, PhD(chem), 85. *Prof Exp:* Postdoctoral res assoc, Purdue Univ, 85-88; guest res fel, 88-90, SR STAFF RES FEL, NIH, 90- *Mem:* Am Soc Biochem & Molecular Biol; Am Asn Immunologists. *Mailing Add:* Exp Immunol Br Nat Cancer Inst NIH Bldg 10 Rm 4B17 Bethesda MD 20892

SHIVERICK, KATHLEEN THOMAS, b Burlington, Vt, Dec 1, 43; c 1. BIOCHEMICAL PHARMACOLOGY, PHYSIOLOGY. *Educ:* Univ Vt, BS, 65, PhD(physiol), 74. *Prof Exp:* Res asst, Univ Vt, 65-68; fel endocrinol, McGill Univ, 74-76, res assoc pharmacol, 76-78; ASST PROF PHARMACOL, UNIV FLA, 78- *Concurrent Pos:* Fel, Am Lung Asn, 74-76; res assoc, Roche Develop Pharm Unit, McGill Univ, 76-78. *Res:* Endocrine pharmacology. *Mailing Add:* Dept Pharmacol & Therapeut Univ Fla JHMHC Box J-267 Gainesville FL 32610

SHIVERS, CHARLES ALEX, b Goodlettsville, Tenn, Sept 16, 32; m 55; c 3. REPRODUCTIVE BIOLOGY, DEVELOPMENTAL BIOLOGY. *Educ:* George Peabody Col, BS, 55, MA, 56; Mich State Univ, PhD(zool), 61. *Prof Exp:* Instr chem, Cumberland Univ, 56-58, res asst develop, Mich State Univ, 59-60; USPHS fel, Fla State Univ, 61-63; from asst prof to assoc prof zool, 63-71, PROF ZOOL, UNIV TENN, KNOXVILLE, 71- *Concurrent Pos:* Dir, IVF Lab, E Tenn Baptist Hosp, 84-87 & Ft Sauders Hosp, 88-89; vis prof, Cambridge Univ, Catholic Univ & Monash Univ. *Mem:* AAAS; Am Soc Zoologists. *Res:* Immunochemical studies on fertilization mechanisms. *Mailing Add:* Dept Zool F207 Walter Sci Bldg Univ of Tenn Knoxville TN 37996

SHIVERS, RICHARD RAY, b Salina, Kans, Dec 18, 40; c 2. ZOOLOGY, CELL BIOLOGY. *Educ:* Univ Kans, AB, 63, MA, 65, PhD(zool), 67. *Prof Exp:* from asst prof to assoc prof, 67-86, PROF ZOOL, UNIV WESTERN ONT, 86- *Mem:* Am Soc Cell Biol; Am Asn Anat; Can Asn Anat; Soc Neurosci; Pan-Am Asn Anat. *Res:* Electron microscopic anatomy of invertebrate nervous systems; freeze-fracture of nerve regeneration; neurosecretory portions of the crayfish optic ganglia; reptile blood-brain interface, ultrastructure and freeze-fracture of vertebrate blood-brain barrier; freeze-fracture and ultrastructure of human brain tumor vascular systems; culture of brain microvessel endothelial cells. *Mailing Add:* Dept Zool Univ Western Ont London ON N6A 5B8 Can

SHIZGAL, HARRY M, b Montreal, Que, June 23, 38; m 60; c 3. CRITICAL CARE MEDICINE. *Educ:* McGill Univ, BSc, 59, MD, 63, FRCPS(C). *Prof Exp:* Intern, Royal Victoria Hosp, 63-64, surg resident, 64-65 & 67-69, res fel, 65-67, asst surgeon, 71-77, assoc surgeon, 77-84; fel, Univ Calif, San Francisco, 70-71; from asst prof to assoc prof, 71-80, PROF SURG, MCGILL UNIV, 80-; SR SURGEON & DIR RES INST, ROYAL VICTORIA HOSP, 84- *Mem:* Fel Am Col Surg; Soc Univ Surgeons; Am Surg Asn; Am Soc Clin Nutrit; Am Soc Parenteral & Enteral Nutrit. *Res:* Nutritional support of the hospitalized critially ill patient. *Mailing Add:* 687 Pine Ave W Montreal PQ H3A 1A1 Can

SHKAROFSKY, ISSIE PETER, b Montreal, Que, July 4, 31; m 57; c 4. PLASMA & FUSION PHYSICS, MICROWAVE ELECTRONICS. *Educ:* McGill Univ, BSc, 52, MSc, 53, PhD(physics), 57. *Prof Exp:* Res & develop fel, Res & Develop Labs, RCA Ltd, Ste Anne de Bellevue, 57-76; res & develop fel, 77-88, DIR FUSION TECHNOL DIV, MPB TECHNOL INC, DORVAL, QUE, 88- *Mem:* Can Asn Physicists; fel Am Phys Soc; Am Geophys Union. *Res:* Plasma kinetics and waves; fusion; lasers and microwaves; propagation; turbulence; space and ionosphere; radar. *Mailing Add:* 1959 Clinton Ave Montreal PQ H3S 1L2 Can

SHKLAR, GERALD, b Montreal, Que, Dec 2, 24; nat US; m 48; c 3. ORAL PATHOLOGY. *Educ:* McGill Univ, BSc, 47, DDS, 49; Tufts Univ, MS, 52; Am Bd Oral Path, dipl; Am Bd Periodont, dipl. *Hon Degrees:* MA, Harvard Univ, 71. *Prof Exp:* Res assoc oral path & periodont, Sch Dent Med, Tufts Univ, 51-52, from instr to assoc prof, 52-60, assoc prof oral path & assoc res prof periodont, 60-61, prof oral path & chmn dept, 61-71; CHARLES A BRACKETT PROF ORAL PATH & CHMN DEPT ORAL MED & ORAL PATH, SCH DENT MED, HARVARD UNIV, 71- *Concurrent Pos:* Nat Res Coun Can dent res fel, 51-52; dir cancer training prog, Sch Dent Med, Tufts Univ, 52-71, prof dent hist, 60-63, lectr social dent, 63-71, lectr oral path, Forsyth Sch Dent Hygienists, 53-71, res prof periodont, 60-71; oral pathologist, Boston City Hosp, 61-; consult, Mass Gen Hosp, Peter Bent Brigham & Women's Hosp & Children's Hosp Med Ctr. *Mem:* Am Dent Asn; Am Acad Oral Path; Am Acad Periodont; fel Int Col Dentists; fel Am Col Dentists. *Res:* Tumors of mouth and jaws; experimental pathology of salivary glands and periodontal tissues; pathology and physiology of bone; experimental carcinogensis; histochemistry of oral diseases; ultrastructure oral tissues; radiation biology. *Mailing Add:* 7 Chauncy Lane Cambridge MA 02138

SHKLOV, NATHAN, b Winnipeg, Man, Aug 3, 18; m 51; c 1. COMBINATORICS & FINITE MATHEMATICS. *Educ:* Univ Man, BA, 40; Univ Toronto, MA, 49. *Prof Exp:* Spec lectr math, Univ Sask, 51-52, from instr to assoc prof, 52-67, chmn comput ctr, 62-67; PROF MATH, UNIV WINDSOR, 67- *Concurrent Pos:* Consult, Defence Res Bd Can, 56-60; ed-in-chief, Can Comput Sci Asn, 66-71. *Mem:* Math Asn Am; Inst Math Statist; Statist Soc Can (pres, 72-74). *Res:* Non-associative algebras; statistical design of experiments; quadrature formulas. *Mailing Add:* Dept of Math Univ of Windsor Windsor ON N9B 3P4 Can

SHKOLNIKOV, MOISEY B, b USSR; US citizen; c 1. FINITE ELEMENT METHOD TECHNOLOGY, STRUCTURAL MECHANICS. *Educ:* Maritime Univ, USSR, MS, 50; Automotive Univ, USSR, PhD(mech eng), 61; Supreme Certifying Comn, USSR, DSc, 74. *Prof Exp:* Sr scientist, Combat Eng Res Inst, Moscow, 61-66; head, Res Lab, NAMI, Major Res Inst USSR Automotive Indust, 66-81; sr engr, Gramman Flexible Corp, Ohio, 81-85; STAFF RES ENGR, GEN MOTORS CORP, MICH, 85- *Res:* Nonlinear finite element technology; simulation of complex automotive structures and occupant behavior under impact loads in collision using super computers. *Mailing Add:* 1408 S Bates Birmingham MI 48009

SHLAFER, MARSHAL, CARDIAC ISCHEMIA, OXYGEN RADICALS. *Educ:* Med Col Ga, PhD(pharmacol), 74. *Prof Exp:* ASSOC PROF PHARMACOL & SURG, UNIV MICH MED CTR, 77- *Mailing Add:* Dept Pharmacol Med Sci Bldg 1 Box 035 Univ Mich Med Sch Ann Arbor MI 48109-0010

SHLANTA, ALEXIS, b Scranton, Pa, June 17, 37; m 65, 83; c 3. ATMOSPHERIC PHYSICS. *Educ:* Univ Tex, El Paso, BS, 62, MS, 65; NMex Inst Mining & Technol, PhD(atmospheric physics), 72. *Prof Exp:* Res engr space physics & radar, NAm Aviation, Inc, 62-63; instr physics, Buena Vista Col, 66-67; physicist atmospheric physics, 73-84, PROG MGR, NAVAL WEAPONS CTR, CHINA LAKE, 84- *Concurrent Pos:* Fel, Nat

Oceanic & Atmospheric Admin, 72-73. *Mem:* Am Meteorol Soc; Fed Mgrs Asn; fel Naval Weapons Ctr. *Res:* The determination and quantification of the effects of atmospheric conditions on the performance of tactical weapon systems; currently Phoenix Missile Program Technical Manager. *Mailing Add:* PO Box 815 Ridgecrest CA 93556

SHLEIEN, BERNARD, b New York, NY, Feb 5, 34; m 60; c 2. HEALTH PHYSICS. *Educ:* Univ Southern Calif, PharmD, 57; Harvard Univ, MS, 63; Am Bd Health Physics, dipl, 66; Am Bd Health Physics, cert health physicist. *Prof Exp:* Staff pharmacist, USPHS Hosp, Seattle, 57-59, anal chemist, Med Supply Depot, 59-60; asst to chief radiol intel fallout studies, Robert A Taft Sanit Eng Ctr, Bur Radiol Health, 60-62, chief dosimetry, Northeastern Radiol Health Lab, 63-70, chief environ radiation, 68-70; co-proj officer, Fed Radiation Coun-Nat Acad Sci Risk Eval, Dept Health, Educ & Welfare, 70-72, tech adv, Off Bur Dir, 72-78, asst dir sci affairs, Bur Radiol Health, Food & Drug Admin, 78-84; head, Isotopes & Radiol Health Inst Environ Health, Tel Aviv Univ, 84-86; ISRAEL MINISTRY HEALTH, 86-; PRES, SCINTA, INC, 90- *Concurrent Pos:* Mem Intersoc Comn Ambient Air Sampling, chmn radioactive substances, 66-72; mem temporary staff, Fed Radiation Coun-Environ Protection Agency spec studies group, 70-; pres, Nucleon Lectr Assocs. *Honors & Awards:* Commendation Medal, US Pub Health Sect. *Mem:* AAAS; Health Physics Soc; Am Indust Hyg Asn; fel Am Pub Health Asn. *Res:* Radiation hazards; radioactive aerosols; population doses; radiation carcinogenesis, and standards setting; energy planning for radiation accidents. *Mailing Add:* 2421 Homestead Dr Silver Spring MD 20902

SHLEVIN, HAROLD H, b Providence, RI, Sept 15, 49; m; c 1. ELECTROPHYSIOLOGY, OPHTHALMOLOGY. *Educ:* Univ Rochester, PhD(physiol), 77. *Prof Exp:* Mgr res, 83-84, dir, Sci Info Syst, 84-86, exec dir, Strategy & Bus Develop, 88-89, DIR, STRATEGY, SCI & TECHNOL, CIBA-GEIGY, SUMMIT, NJ, 87-; VPRES, RES, PRODUCT & BUS DEVELOP, CIBA-GEIGY/CIBA VISION OPHTHALMICS, ATLANTA, GA, 90- *Concurrent Pos:* Postdoctoral, Mayo Clinic. *Mem:* Am Soc Pharmacol & Exp Therapeut; Asn Res Vision & Ophthal; Am Physiol Soc; Biophys Soc. *Res:* Ophthalmology. *Mailing Add:* CIBA Vision Ophthalmics 2910 Amwiler Ct Atlanta GA 30360

SHMAEFSKY, BRIAN ROBERT, b Brooklyn, NY, July 13, 56; m 86; c 2. SCIENCE & TECHNOLOGY PUBLIC EDUCATION, GENETICS ETHICS EDUCATION. *Educ:* City Univ New York, BS, 79; Southern Ill Univ, Edwardsville, MS, 81, EdD(sci educ), 87. *Prof Exp:* Chemist, Sigma Chem Co, 82-85; instr biol educ, Southern Ill Univ, Edwardsville, 85-88; ASST PROF & CHMN BIOL, NORTHWESTERN OKLA STATE UNIV, 88- *Concurrent Pos:* Vis instr, McKendree Col, Ill, 82-88, St Louis Community Col, 88; pres, Sci Ed Sect, Okla Acad Sci, 88-89 & 90-91, vpres, 89-90; contrib ed, Nat Sci Teachers Asn, 88-; consult, sci educ, sponsored by Soc Col Sci Teaching, 89-; book reviewer, J Biopharm, 90- *Mem:* Nat Sci Teachers Asn; Nat Biol Teachers Asn; Sigma Xi; Nat Asn Sci Technol & Soc. *Res:* Genetic ethics-the social attitudes and origins; conducting a study introducing simple scientific research into elementary classes in US and Canada. *Mailing Add:* Biol Dept Northwestern Okla State Univ Alva OK 73717

SHMOYS, JERRY, b Warsaw, Poland, Oct 6, 23; US citizen; m 53; c 2. ELECTROPHYSICS. *Educ:* Cooper Union, BEE, 45; NY Univ, PhD(physics), 52. *Prof Exp:* Jr electronic engr, Marine Div, Bendix Aviation Corp, 45; asst proj engr, Res Labs, Sperry Gyroscope Co, 45-46; res asst, Physics Dept, NY Univ, 46-49, math res group, 49-52, res assoc, Inst Math Sci, 52-55; from instr to asst prof elec eng, 55-60, assoc prof electrophys, 60-68, prof electrophysics, 68-80, PROF ELEC ENG, POLYTECH UNIV, 80- *Concurrent Pos:* Mem US Comn B, Int Sci Radio Union, 66-; Nat Acad Sci-Nat Res Coun sr res fel, Ames Res Ctr, NASA, 67-68. *Mem:* Sr mem Inst Elec & Electronics Engrs; Am Phys Soc. *Res:* Diffraction and propagation of electromagnetic waves; antenna theory; plasma research. *Mailing Add:* Polytech Inst NY 18 Crandon St Melville NY 11747

SHNEOUR, ELIE ALEXIS, b Paris, France, Dec 11, 25; nat US; m 90; c 2. BIOCHEMISTRY, NEUROSCIENCES. *Educ:* Columbia Univ, BA, 47; Univ Calif, Berkeley, MA, 55, Univ Calif, Los Angeles, PhD(biol chem), 58. *Hon Degrees:* DSc, Bard Col, 68. *Prof Exp:* Asst dir res & develop, J A E Color Works, NY, 48-50; sr res technician, Univ Calif, Berkeley, 50-53, asst biochem, 53-55, asst, Univ Calif, Los Angeles, 55-58; Am Heart Asn res fel, Univ Calif, Berkeley, 58-62; res assoc genetics, Stanford Univ, 62-65; assoc prof molecular & genetic biol, Univ Utah, 65-69; vis res neurochemist, Div Neurosci, City of Hope Nat Med Ctr, Duarte, Calif, 69-71; dir res, Calbiochem, 71-74; PRES, BIOSYST INSTS, INC & DIR, BIOSYSTS RES INST, 74- *Concurrent Pos:* Consult, Melpar Inc, 64-65, Gen Elec Co, 65- & NAm Aviation, Inc, 66-70 & other maj US & Foreign Corps, 74-; exec secy biol & explor Mars study group, Nat Acad Sci, 64-65; chmn regional biosci res coun manned earth orbiting missions, Am Inst Biol Sci, 66-69; chmn sci adv prog, Am Soc Biol Chemists, 73-78, mem sci & pub policy comt, 74-76; mem sci adv coun, Cousteau Soc, Inc, 77-; mem bd dir, San Diego Biomed Inst, 81-86; chmn, Sci Adv Bd, Quadroma, Inc, 81-84; mem, Sci Adv Bd, Advan Polymer Systs Inc. *Mem:* Am Chem Soc; Int Soc Neurochem; NY Acad Sci; Am Soc Biochem & Molecular Biol; Am Soc Neurosci; Int Soc Neurochem. *Res:* Pharmacology and biochemistry of natural substances; developmental neurochemistry; information processing by biological systems; research administration; science communications; science policy. *Mailing Add:* Biosysts Res Inst 700 Front St CDM-608 San Diego CA 92101-6009

SHNIDER, BRUCE I, b Ludzk, Poland, Jan 20, 20; nat US; m 42; c 3. INTERNAL MEDICINE, ONCOLOGY. *Educ:* Wilson Teachers Col, BS, 41; Georgetown Univ, MD, 48; Am Bd Internal Med, dipl, 55. *Prof Exp:* Intern, DC Gen Hosp, 48-49, from jr asst resident to chief resident, Georgetown Univ Hosp, 49-52, from instr to assoc prof med, 52-66, from asst prof to assoc prof med & pharmacol, 61-66, cancer coordr, 60-66, from asst dean to assoc dean, 61-79, prog dir clin cancer training, 70-75, dir, Breast Cancer Detection Demonstration Prog, 76-79, prof med & pharmacol, sch med, 66-85, EMER PROF, SCH MED, GEORGETOWN UNIV, 86- *Concurrent Pos:* Dir Tumor serv & cancer chemother res prog, Georgetown Univ Med Div, DC Gen Hosp, 57-, chief vis physician, 57-59, exec officer, 59-60, coord teaching activ, 61-; prin investr, East Coop Group Solid Tumor Chemother, 57-, chmn, 61-71; dir div oncol, Dept Med, Georgetown Univ, 61-80; consult, oncol clin & vis physician, Georgetown Univ Med Ctr, Holy Cross Hosp, Wash Hosp Ctr & Mt Alto Vet Hosp; consult, Clin Ctr, NIH; vis prof oncol, Sch Med, Tel Aviv Univ, 72-73 & 89-90, Sch Med, Hebrew Univ, 79-80, Ben Gurion Univ, Fac Health Sci, 85. *Mem:* Am Soc Clin Oncol; fel Col Clin Pharmacol & Chemother; Am Asn Cancer Res; fel Am Col Physicians; Am Soc Clin Pharmacol & Therapeut. *Res:* Medical oncology; clinical pharmacology; chemotherapy of cancer. *Mailing Add:* 610 Sisson St Silver Spring MD 20902

SHNIDER, RUTH WOLKOW, b Louisville, Ky, June 15, 15; m 47. APPLIED MATHEMATICS, PHYSICS. *Educ:* Wellesley Col, AB, 34; Univ Chicago, SM, 37. *Prof Exp:* Instr electronics, Univ Chicago, 42-43; instr physics, Ind Univ, 43-44; electronics engr to nuclear physicist, US Naval Res Lab, 44-52, from nuclear physicist to res physicist, US Naval Radiol Defense Lab, 52-69; prin res physicist & sr systs analyst, URS Res Co, San Mateo, Calif, 69-76; sr analyst, Ctr Planning & Res Inc, Palo Alto, Calif, 76-80; consult, 80-85; RETIRED. *Concurrent Pos:* Dirs adv coun, San Francisco Exploratorium. *Mem:* AAAS; Oper Res Soc Am. *Res:* Analysis of test results and correlation of experimental and theoretical analysis of nuclear weapon design to provide predictive estimates of effects of nuclear bursts and development of safety criteria for military and civilian personnels; development of procedures and criteria for evaluation of the efficiency of environmental monitoring networks. *Mailing Add:* 2745 Summit Dr Burlingame CA 94010-6039

SHNIDER, SOL M, b Yorkton, Sask, June 13, 29; US citizen. MEDICINE. *Educ:* Univ Man, BSc & MD, 53; Am Bd Anesthesiol, dipl. *Prof Exp:* Intern, Winnipeg Gen Hosp, Man, 52-53; gen pract, Sask, 53-57; resident, Presby Hosp, New York, 57-59; instr anesthesiol, Col Physicians & Surgeons, Columbia Univ, 59, assoc, 60-62; from asst prof to assoc prof, 62-72, PROF ANESTHESIA, OBSTET & GYNEC, SCH MED, UNIV CALIF, SAN FRANCISCO, 72-, VCHMN DEPT ANESTHESIA, 73- *Concurrent Pos:* Mem, World Fedn Socs Anesthesiol; asst anesthesiologist, Presby Hosp, New York, 59-62; lectr, numerous insts, 71-90; vis prof, Dept Anaesthesia & Obstet & Gynecol, Univ Hosp Wales, Welsh Nat Sch Med, Heath Park, Cardiff, 84, vis prof, Dept Anaesthesia, Univ Bristol, Charing Cross Hosp & Queen Charlotte's Maternity Home, London, Nuffield Dept Anaesthesia, Univ Oxford; distinguished vis prof, Osaka City Univ, Osaka, Japan, 86, John J Bonica distinguished vis prof, Seattle, Wash, 86. *Mem:* AMA; Am Soc Anesthesiol; fel Am Col Anesthesiol; hon fel, Col Anaesthetists, Royal Col Surgeons of Eng. *Res:* Obstetrical anesthesia; impact of maternal anesthesia on fetus and newborn. *Mailing Add:* Dept of Anesthesia Univ of Calif Med Ctr San Francisco CA 94143

SHNITKA, THEODOR KHYAM, b Calgary, Alta, Nov 21, 27; m 87. PATHOLOGY, CELL BIOLOGY. *Educ:* Univ Alta, BSc, 48, MSc, 52, MD, 53; Royal Col Physicians & Surgeons Can, cert path, 58; FRCP(C), 72. *Prof Exp:* From instr to assoc prof, 54-67, prof path, Univ Alta, 67-87, dir electron micros unit, dept path, 68-87, chmn fac med, 80-87, EMER PROF, 87- *Concurrent Pos:* Jr intern, Univ Alta Hosp, 53-54, from asst resident to resident path, 54-59; fel, Sch Med, Johns Hopkins Univ, 59-60; Can Cancer Soc McEachern Mem fel, 59-60. *Mem:* Electron Micros Soc Am; Histochem Soc; Am Soc Cell Biol; Am Soc Exp Path; Int Acad Path. *Res:* Experimental and molecular pathology, utilizing techniques of enzyme cytochemistry and electron microscopy; cell structure and function; diagnostic electron microscopy. *Mailing Add:* 12010 87th Ave Edmonton AB T6G 0Y7 Can

SHOAF, CHARLES JEFFERSON, b Roanoke, Va, July 22, 30; m 58; c 2. TEXTILE CHEMISTRY. *Educ:* Va Mil Inst, BS, 52; Purdue Univ, MS, 54, PhD(chem), 57; Del Law Sch, JD, 80. *Prof Exp:* Res chemist, E I du Pont de Nemours & Co, 57; instr chem, US Air Force Inst Technol, 57-59, asst prof, 59; res chemist, 59-63, from patent chemist to sr patent chemist, 63-74, patent agent, 74-80, patent atty, 80-87, SR COUN, LEGAL DEPT, E I DU PONT DE NEMOURS & CO, INC, 87- *Mem:* Am Chem Soc; Sigma Xi; Am Intellectual Law Asn. *Res:* Synthetic textile fibers. *Mailing Add:* E I du Pont de Nemours & Co Inc Legal Dept Wilmington DE 19898

SHOAF, MARY LA SALLE, b Milwaukee, Wis, Feb 29, 32; m 58; c 2. PHYSICAL CHEMISTRY. *Educ:* Cardinal Stritch Col, BA, 53; Univ Calif, MS, 55; Purdue Univ, PhD(phys chem), 60. *Prof Exp:* Phys chemist, US Dept Air Force, 58-59; assoc prof, ECarolina Univ, 61-63; mem sci fac, Tatnall Sch, Del, 64-66; from assoc prof to prof physics, West Chester State Col, 66-75, dean grad studies, 75-77; asst dir, Princeton Plasma Physics Lab, 79-89; CONSULT, TECH MGT SERV, 89- *Concurrent Pos:* Mem, Nuclear Safety Res Rev Comt, Nat Res Coun. *Mem:* Fel Am Inst Chemists; Am Chem Soc; fel Am Phys Soc (dep exec secy, 73-79); fel AAAS. *Res:* Ion-induced fission reactions; determination of thermodynamic data for transition ions; technical management. *Mailing Add:* 1113 Independence Dr West Chester PA 19382

SHOBE, L(OUIS) RAYMON, b Waverly, Kans, Apr 22, 13; m 36; c 3. MATHEMATICS, ENGINEERING MECHANICS. *Educ:* Emporia State Univ, BS, 36; Kans State Univ, MS, 40. *Prof Exp:* Lab asst physics, Emporia State Univ, 35-36; teacher high schs, Kans, 36-38; instr math, Kans State Univ, 38-40 & Univ Kans, 40-41; instr math & mech, Gen Motors Inst, 41-46; assoc prof math, Bemidji State Teachers Col, 46-47; assoc prof civil eng & eng mech, 47-56, prof, 57-78, EMER PROF ENG SCI & MECH, UNIV TENN, KNOXVILLE, 78- *Concurrent Pos:* Consult, Oak Ridge Nat Lab, 59- *Mem:* Fel Am Soc Civil Engrs; Nat Soc Prof Engrs; Am Soc Eng Educ. *Res:* Solid mechanics; stress analysis. *Mailing Add:* 1432 Tugaloo Dr Knoxville TN 37919

SHOBER, ROBERT ANTHONY, b St Louis, Mo, Oct 18, 48; m 76; c 2. SUPERCOMPUTERS, NUMERICAL ANALYSIS. *Educ:* Univ Fla, BSNES, 70, MS, 72; Mass Inst Technol, PhD(nuclear eng), 76. *Prof Exp:* Engr safety physics, Combustion Eng, Inc, 72-74; numerical methods develop, Agronne Nat Lab, 76-79; mem tech staff, 79-82, SUPVR, BELL LABS, 82- *Concurrent Pos:* Consult, Nat Eval Systs, Inc, 76; mem comput benchmarks probs comt, Am Nuclear Soc, 77- *Mem:* Am Nuclear Soc. *Res:* Supercomputers; parallel/vector processing; efficient numerical calculation. *Mailing Add:* AT&T Bell Labs Crawfords Corner Rd Holmdel NJ 07733

SHOBERT, ERLE IRWIN, II, b DuBois, Pa, Nov 19, 13; m 39; c 2. SOLID STATE, GRAPHITE MATERIALS TRIBOLOGY. *Educ:* Susquehanna Univ, AB, 35; Princeton Univ, MA, 39. *Hon Degrees:* DSc, Susquehanna Univ, 57. *Prof Exp:* Res engr, 39-46, eng dir, 46-54, res dir 54-72, vpres res, 72-78, CONSULT, STACKPOLE CORP, 78- *Concurrent Pos:* Bd dirs, Chem Corp, 62-82, chmn, 80-82; bd dirs, Am Soc Testing Mats, 64-72; mem bd dir, Susque hanna Univ, 65-, chmn, 78-86. *Honors & Awards:* Holm Award, 73 & Armington Award, Inst Elec & Electronics Engrs, 86. *Mem:* Fel Inst Elec & Electronics Engrs; fel Am Soc Testing Mats (pres, 71-72). *Res:* Solid state, friction, commutation, contacts, electrical machines, carbon and graphite materials tribology. *Mailing Add:* PO Box 343 St Marys PA 15857

SHOCH, DAVID EUGENE, b Warsaw, Poland, June 10, 18; nat US; m 45; c 2. OPHTHALMOLOGY. *Educ:* City Col New York, BS, 38; Northwestern Univ, MS, 39, PhD(biochem), 43, BM, 45, MD, 46. *Prof Exp:* Intern, Cook County Hosp, 45-46; res fel, Med Sch, Northwestern Univ, 48-50; resident ophthal, Cook County Hosp, 50-52; clin asst, 52-56, assoc, 56-60, asst prof, 60-66, prof ophthal & chmn dept, 66-83, MAGERSTADT PROF, MED SCH, NORTHWESTERN UNIV, CHICAGO, 83- *Concurrent Pos:* Mem Am Bd Ophthal, 69-80, & chmn, 79; chmn coun, Am Opthal Soc, 85, pres-elect, 88. *Mem:* AMA; Asn Res Vision & Ophthal; Am Col Surgeons; Am Acad Ophthal (pres, 81); Am Ophthal Soc. *Res:* Biochemistry and diseases of the crystalline lens. *Mailing Add:* 303 E Chicago Ave Chicago IL 60611

SHOCH, JOHN F, b Evanston, Ill. COMPUTER COMMUNICATIONS, DISTRIBUTED SYSTEMS. *Educ:* Stanford Univ, BA, 71, MS, 77, PhD(comput sci), 79. *Prof Exp:* Mem res staff, Xerox Palo Res Ctr, Xerox Corp, 71-80, exec asst to pres & dir, Corp Policy Comt, 80-81, vpres, Xerox Off Systs Div, 82-83, pres, 83-85; PARTNER, ASSET MGT CO, 85- *Concurrent Pos:* Vis fac mem, Comput Sci Dept, Stanford Univ, 78; mem, Nat Conf Lawyers & Scientists, AAAS, 91- *Mem:* Asn Comput Mach; Inst Elec & Electronic Engrs. *Res:* Computer communication and distributed systems; local computer networks; internetwork communication; packet radio; protocol design. *Mailing Add:* Asset Mgt Co 2275 E Bayshore Rd Palo Alto CA 94303

SHOCHET, MELVYN JAY, b Philadelphia, Pa, Oct 31, 44; m 67; c 2. ELEMENTARY PARTICLE PHYSICS. *Educ:* Univ Pa, BA, 66; Princeton Univ, MA & PhD(physics), 72. *Prof Exp:* Res assoc, Enrico Fermi Inst, 72-73, from instr to assoc prof, 73-85, PROF PHYSICS, UNIV CHICAGO, 85- *Concurrent Pos:* Co-spokeswoman, CDF Collab, 88- *Mem:* Am Phys Soc; AAAS. *Res:* Experimental elementary particle physics. *Mailing Add:* Univ of Chicago 5630 S Ellis Ave Chicago IL 60637

SHOCK, CLINTON C, b Los Angeles, Calif, June 11, 44; m 66; c 3. AGRICULTURAL RESEARCH. *Educ:* Univ Calif-Berkeley, BA, 66; Univ Calif-Davis, MS, 73, PhD(plant physiol), 82. *Prof Exp:* Asst prof, La State Univ, 82-84; ASSOC PROF & SUPT, MALHEUR EXP STA, ORE STATE UNIV, 84- *Concurrent Pos:* Proj leader revegation res, IRI Res Inst, Brazil, 73-75, actg gen mgr, 78, mgr & supt, IRI Exp Sta, 75-78; res asst & grad student, Univ Calif, Davis, 78-82; mem, Am Forage & Grassland Coun. *Mem:* Am Soc Agron; AAAS; Am Soc Plant Physiol; Crop Sci Soc Am; Soil Sci Soc Am; Soc Range Mgt; Nitrogen Fixing Tree Asn; Am Potato Asn; Nat Onion Asn; Sigma Xi. *Res:* Soil and irrigation management effects on potato production and quality; variety evaluation and management of cereals and forages for eastern Oregon; performance of soybean genotypes grown at four locations in Oregon. *Mailing Add:* Malheur Exp Sta Ore State Univ 595 Onion Ave Ontario OR 97914

SHOCK, D'ARCY ADRIANCE, b Fowler, Colo, June 13, 11; m 55; c 4. PHYSICAL CHEMISTRY. *Educ:* Colo Col, BS, 33; Univ Tex, MA, 46. *Prof Exp:* Chemist, Dow Chem Co, Mich, 33-36 & McGean Chem Co, 36-42; chief chemist, Int Minerals & Chem Corp, 42-44; instr chem, Univ Tex, 44-45; field correlator, Nat Gas Asn, 46-47, res scientist, 47-49; sr res chemist, Prod Res Lab, Continental Oil Co, 49-51, res group leader chem & metal sect, Prod Res Div, 51-56, mgr, Cent Res Div, 56-75, mgr, Mining Res Div, 75-78; CONSULT SOLUTION MINING & SLURRY TRANSPORT, 78- *Concurrent Pos:* Comt Waste Isolation Pilot Plant, Nat Acad Sci, 76-90. *Mem:* Am Chem Soc; Nat Asn Corrosion Eng; Am Inst Mining, Metall & Petrol Eng; distinguished mem Soc Mining Engrs. *Res:* Corrosion of iron and steel in oil production and processing; anodic passivation; manufacture, transportation and storage of cryogenic gases; hydraulic fracturing technology; solution mining; waste disposal; solids pipelining; mining and milling of uranium and copper ores; coal mining research. *Mailing Add:* 233 Virginia Ponca City OK 74601

SHOCK, NATHAN WETHERILL, physiology, psychology; deceased, see previous edition for last biography

SHOCK, ROBERT CHARLES, b Dayton, Ohio, July 18, 40; m 64; c 2. ALGEBRA, OPERATIONS RESEARCH. *Educ:* Bowling Green State Univ, BS, 62; Univ Ariz, MA, 64; Univ NC, Chapel Hill, PhD(math), 68. *Prof Exp:* From asst prof to assoc prof math, Southern Ill Univ, 68-78; PROF MATH & CHAIRPERSON DEPT, EAST CAROLINA UNIV, 78- *Mem:* Am Math Soc. *Res:* Non-commutative ring theory; linear programming. *Mailing Add:* Dept Comput Sci Wright State Univ Dayton OH 45435

SHOCKEY, DONALD ALBERT, b New Kensington, Pa, July 26, 41; m 68; c 1. MATERIALS SCIENCE. *Educ:* Grove City Col, BS, 63; Carnegie Inst Technol, MS, 65; Carnegie-Mellon Univ, PhD(metall, mat sci), 68. *Prof Exp:* Scientist, Ernst-Mach Inst Freiburg, Ger, 68-71; PHYSICIST, STANFORD RES INST INT, 71- *Res:* Fracture behavior of materials; fracture mechanics; response of materials to high rate loading; effects of environment, radiation, and microstructure on plasticity and fracture. *Mailing Add:* 467 Claremont Way Menlo Park CA 94025

SHOCKEY, WILLIAM LEE, b Frostburg, Md, Apr 7, 53; m 79; c 3. DAIRY CATTLE, RUMINANT NUTRITION. *Educ:* WVa Univ, AB, 75, PhD(agr biochem), 79. *Prof Exp:* Postdoctoral res assoc, Ohio Agr Res & Develop Ctr, 79-81, res scientist, 90-91; res dairy scientist, USDA Agr Res Serv, 81-90; DAIRY CONSULT, PURINA MILLS, INC, 91- *Concurrent Pos:* Adj asst prof, Ohio State Univ, 82-; rep Am Forage & Grassland Coun, Am Dairy Sci Asn, 88- *Mem:* Am Dairy Sci Asn; Am Soc Animal Sci; Am Inst Nutrit; Am Registry Prof Animal Scientists; Am Soc Agron; Am Forage & Grassland Coun. *Res:* Chemistry and microbiology of hay crop silage fermentations and related changes in fermentation; utilization of silage by lactating dairy cows; contribution of inorganic forage components to the silage fermentation. *Mailing Add:* Purina Mills Inc PO Box 606 Milford IN 46542

SHOCKLEY, DOLORES COOPER, b Clarksdale, Miss, Apr 21, 30; m 57; c 4. PHARMACOLOGY. *Educ:* La State Univ, BS, 51; Purdue Univ, MS, 53, PhD(pharmacol), 55. *Prof Exp:* Asst pharmacol, Purdue Univ, 51-53; asst prof, 55-67, ASSOC PROF PHARMACOL, MEHARRY MED COL, 67- *Concurrent Pos:* Fulbright fel, Copenhagen Univ, 55-56; vis asst prof, Albert Einstein Col Med, 59-62; Lederle fac award, 63-66. *Mem:* AAAS; Am Pharmaceut Asn. *Res:* Measurement of non-narcotic analgesics; effect of drugs on stress conditions; effect of hormones on connective tissues; nutrition effects and drug action. *Mailing Add:* 4141 W Hamilton Rd Nashville TN 37218

SHOCKLEY, GILBERT R, b Mo, Sept 20, 19; m 43; c 2. CHEMICAL ENGINEERING. *Educ:* Univ Mo, BS, 42 & CEng, 60. *Hon Degrees:* DEng, Univ Mo, 70DE, 70. *Prof Exp:* Design chem engr, Monsanto Chem Co, 42-46; design & supvry engr, Wood Res Inst, 46-47; design chem engr, Goslin Birmingham Mfg Co, 47-49; asst mgr, Filter Div, Eimco Corp, 49-53; vpres, Metals Div, Olin Mathieson Chem Corp, 53-61; gen dir, Prod Develop Div, Reynolds Metals Co, 61-74, gen mgr opers serv, Mill Prod Div, 74-85, exec vpres, Reynolds Res Corp, 66-74; RETIRED. *Mem:* NY Acad Sci; Am Chem Soc; Am Inst Chem Engrs; Am Soc Metals; Soc Automotive Engrs. *Mailing Add:* 207 Nottingham Rd Richmond VA 23221

SHOCKLEY, JAMES EDGAR, b Richmond, Va, Dec 26, 31; div; c 3. MATHEMATICS. *Educ:* Univ NC, AB, 57, AM, 59, PhD(math), 62. *Prof Exp:* Asst prof math, Col William & Mary, 61-64; assoc prof, Univ Wyo, 64-66; ASSOC PROF MATH, VA POLYTECH INST & STATE UNIV, 66- *Mem:* Am Math Soc; Math Asn Am. *Res:* Elementary number theory. *Mailing Add:* Dept Math Va Polytech Inst & State Univ Blacksburg VA 24061

SHOCKLEY, THOMAS D(EWEY), JR, b Haynesville, La, Nov 2, 23; m 47; c 2. ELECTRICAL ENGINEERING. *Educ:* La State Univ, BS, 50, MS, 52; Ga Inst Technol, PhD(elec eng), 63. *Prof Exp:* Instr eng, La State Univ, 50-53; aerophys eng, Gen Dynamics/Ft Worth, 53-56; res engr, Ga Inst Technol, 56-58, asst prof elec eng, 58-63; assoc prof, Univ Ala, 63-64; prof, Univ Okla, 64-67; chmn dept, Memphis State Univ, 67-78; pres, SSC, Inc, 78-; AT DEPT ELEC ENG, MEMPHIS STATE UNIV. *Mem:* Inst Elec & Electronics Engrs; Am Soc Eng Educ; Nat Fire Protection Asn; Nat Soc Prof Engrs. *Res:* Microwave and antenna systems; computer systems. *Mailing Add:* 1526 Poplar Estates Pkwy Germantown TN 38138

SHOCKLEY, THOMAS E, b Rock Island, Tenn, Mar 15, 29; m 57; c 4. BACTERIOLOGY. *Educ:* Fisk Univ, BA, 49; Ohio State Univ, MSc, 52, PhD, 54. *Prof Exp:* Asst, Ohio State Univ, 50-52; asst prof, Meharry Med Col, 54-59; Rockefeller Found fel, 59-60; Am Cancer scholar, 60-61; assoc prof, 62-67, vchmn dept, 67-71, PROF MICROBIOL, MEHARRY MED COL, 67-, CHMN DEPT, 71- *Concurrent Pos:* Vis investr, Rockefeller Univ, 59-61. *Mem:* Am Soc Microbiol. *Res:* Microbial physiology and genetics; molecular biology; medical microbiology. *Mailing Add:* Dept Microbiol Meharry Med Col 1005 18th Ave N Nashville TN 37208

SHOCKLEY, W(OODLAND) G(RAY), b Crisfield, Md, June 3, 14; m 39; c 2. CIVIL ENGINEERING, SOIL MECHANICS. *Educ:* Antioch Col, BS, 36. *Prof Exp:* Engr, Soils & Pavements Sect, Little Rock Dist, US Corps Engrs, 38-46, chief, Bituminous Sect, US Army Engrs, Waterways Exp Sta, 46-47, asst chief, Embankment & Found Br, 47-52, asst chief, Flexible Pavement Br, 52-53, chief, Embankment & Found Br, 53-58, asst chief, Soils Div, 58-63, chief, Mobility & Environ Systs Lab, 63-78, prog mgr mil eng, 78-80; CONSULT ENGR, 80- *Concurrent Pos:* Mem task group tech panel adv comt, Dept Housing & Urban Develop. *Honors & Awards:* Award of Merit, Am Soc Testing & Mat. *Mem:* Hon mem Am Soc Civil Engrs; Nat Soc Prof Engrs; Am Soc Testing & Mat. *Res:* Mobility research. *Mailing Add:* 326 Lake Hill Dr Vicksburg MS 39180

SHOCKLEY, WILLIAM, physics; deceased, see previous edition for last biography

SHOCKMAN, GERALD DAVID, b Mt Clemens, Mich, Dec 22, 25; m 49; c 2. MICROBIOLOGY, BIOCHEMISTRY. *Educ:* Cornell Univ, BS, 46; Rutgers Univ, PhD(microbiol), 50. *Hon Degrees:* Docteur honoris causa, Univ de l'Etat a Lièg. *Prof Exp:* Fel, Rutgers Univ, New Brunswick, 47-50; res assoc, Univ Pa, 50-51; res fel, Inst Cancer Res, 51-60; assoc prof, 60-66, chmn, Dept Microbiol & Immunol, 74-90, PROF MICROBIOL, SCH MED, TEMPLE UNIV, 66- *Concurrent Pos:* Am Cancer Soc-Brit Empire Cancer Campaign exchange fel, 54-55; NIH res career develop award, 65-70; vis scientist, Lab Enzymol, Nat Ctr Sci Res, Gif-sur-Yvette, France, 68-69; prof,

Univ Liege, 71-72. *Mem:* Am Soc Microbiol; Am Acad Microbiol; Am Soc Biol Chemists; AAAS; Sigma Xi. *Res:* Physiology and biochemistry of the surface structures (walls, membranes); secreted polymers of bacteria; synthesis, nature, action and significance of autolysins and peptidoglycan hydrolases. *Mailing Add:* Dept Microbiol & Immunol Temple Univ Sch Med 3400 N Broad St Philadelphia PA 19140

SHODELL, MICHAEL J, b New York, NY, Mar 9, 41; m 68; c 2. CELL BIOLOGY. *Educ:* NY State Univ Stony Brook, BS, 62; Univ Calif, Berkeley, PhD(molecular biol), 68. *Prof Exp:* Vis prof tissue cult, Ctr Invest Polytechnic, Mex, 68; Damon Runyon fel cancer res, Imp Cancer Res Fund, London, 69, Am Cancer Soc fel, 69-71, head, Dept Cell Proliferation Studies, 71-75; asst prof, 75-80, prof biol, C W Post Col, 80-83; PROF BIOL, LONG ISLAND UNIV, 83- *Concurrent Pos:* Contrib ed, Science, 84-86; dir, Banbury Ctr Cold Spring Harbor Lab, 82-86. *Res:* Regulation of growth in normal and neoplastic cells. *Mailing Add:* One Briarcliff Dr Port Washington NY 11050

SHOEMAKER, CARLYLE EDWARD, b Columbus, Ohio, Feb 19, 23; m 48; c 3. INORGAINC CHEMISTRY. *Educ:* Ohio State Univ BChE, 43; Univ Ill, MS, 46, PhD(inorg chem), 49. *Prof Exp:* Asst smoke munitions res, Univ Ill, 44-45, asst anal rubber res, 45-46 & asst inorg chem, 46-49; res chemist atomic energy, Monsanto Chem Co, 49-54; from sr res & develop chemist to group leader inorg & chem eng develop, J T Baker Chem Co, 54-60; mem tech staff, Bell Tel Labs, Inc, 60-64; res engr, Bethlehem Steel Corp, 64-72, sr res engr, 72-77; proj mgr, Elec Power Res Inst, 78-89; RETIRED. *Mem:* Nat Asn Corrosion Engrs; Am Chem Soc. *Res:* Analytical and preparative inorganic chemistry; surface chemistry of metals; corrosion; metallurgy; high temperature aqueous chemistry. *Mailing Add:* 3561 Amber Dr San Jose CA 95117

SHOEMAKER, CHRISTINE ANNETTE, b Berkeley, Calif, July 2, 44. PEST MANAGEMENT & WATER QUALITY MODELING. *Educ:* Univ Calif, BS, 66; Univ Southern Calif, MS, 69, PhD(math), 71. *Prof Exp:* Vis asst syst ecologist entom, Univ Calif, Berkeley, 76-77; from asst prof to assoc prof, 72-85, prof environ eng & chmn dept, Cornell Univ, 85-88; STUDY GROUNDWATER CONTAMINATION, NAT ACAD SCI, 84- *Concurrent Pos:* Panel mem, Study Pest Control, Nat Acad Sci, 72-73; co-organizer, Conf Resource Mgt, NATO, Parma, Italy, 78, Conf Energy & Environ, Oper Res Soc Am, 86; mem, expert panel pest mgt, Food & Agr Orgn, UN, 84-; mem, US Nat Comt Sci Coun Problems Environ. *Mem:* Opers Res Soc Am; Entom Soc Am; Am Geophys Union; Biometric Soc; Entom Soc Can. *Res:* Application of operations research techniques to environmental problems, especially pesticide use and water quality management; mathematical modeling of pest management, ecosystems, pollution of groundwater and sewage networks; optimization techniques and statistical analysis of environmental data; acid rain. *Mailing Add:* Dept Environ Eng 218 Hollister Hall Cornell Univ Ithaca NY 14853

SHOEMAKER, CLARA BRINK, b Rolde, Netherlands, June 20, 21; US citizen; m 55; c 1. CRYSTALLOGRAPHY. *Educ:* State Univ Leiden, Doctoraal, 46, PhD(chem), 50. *Prof Exp:* Instr inorg chem, State Univ Leiden, 46-53; res assoc, Mass Inst Technol, 53-55; res assoc biochem lab, Harvard Med Sch, 55-56 & Mass Inst Technol, 58-70; res assoc, 70-75, res assoc prof, 75-82, sr res prof, 82-84, EMER PROF CHEM, ORE STATE UNIV, 84- *Concurrent Pos:* Int Fedn Univ Women fel, Oxford Univ, 50-51; proj supvr chem, Boston Univ, 63-64; mem comt alloy phases, Metall Soc, 69-79, struct reports, Int Union Crystallog, 70-90; mem crystallographic data comt, Am Crystallog Asn, 75-78, Fankuchen Award Comt, 76. *Mem:* Am Crystallog Asn. *Res:* X-ray crystallography; crystal structures of metals, alloys (including alloys related to quasicrystals) and organometallic compounds. *Mailing Add:* Dept of Chem Ore State Univ Corvallis OR 97331-4003

SHOEMAKER, DALE, CANCER RESEARCH. *Prof Exp:* BR CHIEF, REGULATORY AFFAIRS BR, NAT CANCER INST, NIH, 85- *Mailing Add:* NIH Nat Cancer Inst Regulatory Affairs Br 6130 Executive Blvd Bethesda MD 20892

SHOEMAKER, DAVID POWELL, b Kooskia, Idaho, May 12, 20; m 55; c 1. SOLID STATE CHEMISTRY, STRUCTURAL CHEMISTRY. *Educ:* Reed Col, BA, 42; Calif Inst Technol, PhD(chem), 47. *Prof Exp:* Nat Defense Res Comt projs, Calif Inst Technol, 43-46, sr res fel, 48-51; Guggenheim fel, Inst Theoret Physics, Copenhagen, Denmark, 47-48; from asst prof to prof chem, Mass Inst Technol, 51-70; chmn dept chem, 70-80, prof, 70-84, EMER PROF, DEPT CHEM, ORE STATE UNIV, 84- *Concurrent Pos:* Consult res labs, Exxon Co USA, 57-86; co-ed, Acta Crystallographica, Int Union Crystallog, 64-69; chmn, US Nat Comt Crystallog, 67-69; vis scientist, Nat Ctr Sci Res, Grenoble, France, 67 & 78-79; mem vis comt, Chem Dept, Brookhaven Nat Lab, 74-79, chmn, 79; mem eval panel, Div Mat Sci, Nat Bur Standards, 78-80. *Honors & Awards:* Vollum Award, Reed Col, 86. *Mem:* AAAS; Am Acad Arts & Sci; Am Chem Soc; Am Phys Soc; Am Crystallog Asn (pres, 70). *Res:* Chemical crystallography; x-ray diffraction; metals and alloys; zeolites and catalytic materials. *Mailing Add:* Dept Chem Ore State Univ Corvallis OR 97331

SHOEMAKER, EDWARD MILTON, b Wilkinsburg, Pa, Jan 5, 29; m 70. APPLIED MATHEMATICS. *Educ:* Carnegie Inst Technol, MS, 51, PhD(math), 55. *Prof Exp:* Res engr, Chance Vought Aircraft, Inc, 55-56; res mathematician, Boeing Airplane Co, 56-60; assoc prof theoret & appl mech, Univ Ill, Urbana, 60-65; PROF MATH, SIMON FRASER UNIV, 65- *Concurrent Pos:* Consult, Sandia Labs, 62-70. *Mem:* Am Geophys Union; Int Soc Glaciol. *Res:* Glaciology; tectonophysics; solid mechanics. *Mailing Add:* Dept Math Simon Fraser Univ Vancouver BC V5A 1S6 Can

SHOEMAKER, EUGENE MERLE, b Los Angeles, Calif, Apr 28, 28; m 51; c 3. ASTRONOMY. *Educ:* Calif Inst Technol, BS, 47, MS, 48; Princeton Univ, MS, 54, PhD, 60. *Hon Degrees:* DSc, Ariz State Col, 65, Temple Univ, 67 & Univ Ariz, 84. *Prof Exp:* Chief br astrogeol, US Geol Surv, 61-65, chief scientist, Ctr Astrogeol, 66-68; chmn div geol & planetary sci, Calif Inst Tehcnol, 69-72, prof geol, 69-85; GEOLOGIST, US GEOL SURV, 48- *Concurrent Pos:* Co-investr, TV Exp, Proj Ranger, 61-65, prin investr, Proj Surveyor, 63-68; vis prof, Calif Inst Technol, 62, res assoc, 64-68; actg dir manned space sci div, NASA, 63; prin investr field geol exp, Proj Apollo, NASA, 66-70 & co-investr, Proj Voyager, 78-90. *Honors & Awards:* Wetherill Medal, Franklin Inst, 65; NASA Medal, 67; A S Fleming Award, Geol Soc Am, 66, Day Medal, 82, Gilbert Award, 83; Barringer Award, Meteoritical Soc, 84, Leonard Medal, 85; Kuiper Prize, Am Astron Soc, 84. *Mem:* Nat Acad Sci; Mineral Soc Am; Soc Econ Geol; Geochem Soc; Am Asn Petrol Geol; Geol Soc Am; Am Geophys Union; Meteoritical Soc; Am Astron Soc. *Res:* Geology of the Colorado Plateau; meteorite impact and nuclear explosion craters; geology of the moon and of satellites of the outer planets; paleomagnetism; comets and planet-crossing asteroids; origin and impact history of the earth. *Mailing Add:* US Geol Survey 2255 N Gemini Dr Flagstaff AZ 86001

SHOEMAKER, FRANK CRAWFORD, b Ogden, Utah, Mar 26, 22; m 44; c 2. EXPERIMENTAL HIGH ENERGY PHYSICS. *Educ:* Whitman Col, AB, 43; Univ Wis, PhD(physics), 49. *Hon Degrees:* DSc, Whitman Col, 78. *Prof Exp:* Mem staff radiation lab, Mass Inst Technol, 43-45; asst, Univ Wis, 46-49, instr physics, 49-50; prof physics, 62-89, MEM FAC, PRINCETON UNIV, 50-, EMER PROF PHYSICS, 89- *Concurrent Pos:* Vis scientist, Rutherford High Energy Lab, Eng, 65-66; head main ring sect, Nat Accelerator Lab, 68-69. *Mem:* Fel Am Phys Soc. *Res:* High energy accelerator design and construction; elementary particle physics. *Mailing Add:* Dept Physics Princeton Univ Princeton NJ 08544

SHOEMAKER, GRADUS LAWRENCE, b Zeeland, Mich, Jan 18, 21; m 52; c 2. ORGANIC CHEMISTRY. *Educ:* Hope Col, AB, 44; Univ Ill, MS, 47, PhD(org chem), 49. *Prof Exp:* Asst org chem, Univ Ill, 46-47; res fel, Rutgers Univ, 49; from asst prof to assoc prof chem, 49-65, actg chmn dept, 63-64, chmn, 65-67, chmn div natural sci, 67-80, prof, 65-87, EMER PROF CHEM, UNIV LOUISVILLE, 88- *Mem:* Am Chem Soc. *Res:* Reactions of nitroparaffins; adaptation of the chemistry laboratory for handicapped students; synthesis of alkylboranes and alkylboronic acids. *Mailing Add:* Dept Chem Univ of Louisville Louisville KY 40292

SHOEMAKER, JOHN DANIEL, JR, b Lawton, Okla, May 9, 39; m 72; c 3. CHEMISTRY. *Educ:* Univ Okla, BS, 60, MS, 62; Univ Kans, PhD(org chem), 67; La State Univ, Baton Rouge, MBA, 72. *Prof Exp:* Chemist, Esso Res Labs, Humble Oil & Ref Co, La, 66-70; chemist, Union Camp Corp, 72-78; from res assoc to sr res assoc, 78-90, PRIN SCIENTIST, ERLING RIIS LAB, INT PAPER CO, 90- *Mem:* Tech Asn Pulp & Paper Indust; Am Chem Soc; Sigma Xi. *Res:* Chemistry of pulp bleaching; preparation of petroleum refining catalysts; Kraft pulping process; wood chemistry. *Mailing Add:* Int Paper Co Box 2787 Mobile AL 36652

SHOEMAKER, PAUL BECK, b Bridgeton, NJ, Feb 26, 41; m 64; c 2. EXTENSION, EPIDEMIOLOGY. *Educ:* Rutgers Univ, BS, 63, MS, 65; Cornell Univ, PhD(plant path), 71. *Prof Exp:* From asst prof to assoc prof, 70-80, PROF PLANT PATH, NC STATE UNIV, 80- *Concurrent Pos:* Mem, Coun Agr Sci & Technol. *Mem:* Am Phytopath Soc; Sigma Xi. *Res:* Epidemiology and control strategies for diseases of vegetables and burley tobacco; practical control of tomato Verticillium wilt, early blight and bacterial canker and tobacco blue mold and Thielaviopsis root rot. *Mailing Add:* Mountain Hort Crops Res Sta 2016 Fanning Bridge Rd Fletcher NC 28732-9628

SHOEMAKER, RICHARD LEE, b Grand Rapids, Mich, Dec 31, 44; m 68; c 3. CHEMICAL PHYSICS, QUANTUM OPTICS. *Educ:* Calvin Col, BS, 66; Univ Ill, PhD(phys chem), 71. *Prof Exp:* Vis scientist molecular physics, IBM Res Div, San Jose Calif, 71-72; from asst prof to assoc prof, 72-81, PROF OPTICAL SCI, OPTICAL SCI CTR, UNIV ARIZ, 81- *Concurrent Pos:* Alfred P Sloan fel, 76-80. *Mem:* Optical Soc Am; Soc Photo-Optical Inst Engrs; Inst Elec & Electronics Engrs. *Res:* Coherent optical transient effects; nonlinear spectroscopic techniques; laboratory microcomputers; multiprocessor computers. *Mailing Add:* Optical Sci Ctr Univ of Ariz Tucson AZ 85721

SHOEMAKER, RICHARD LEONARD, b Cullman, Ala, Sept 28, 31; m; c 4. PHYSIOLOGY. *Educ:* Auburn Univ, BS, 54, MS, 59; Univ Ala, Birmingham, PhD, 67. *Prof Exp:* Asst animal nutrit, Auburn Univ, 51-54; supvr animal care, 60-66, dir animal care dept, 64-67, assoc prof, 67-73, PROF PHYSIOL & BIOPHYS, MED CTR, UNIV ALA, BIRMINGHAM, 73- *Concurrent Pos:* Consult meteorologist, Eastern Airlines, 59-60. *Mem:* AAAS; Am Physiol Soc; Biophys Soc. *Res:* Electrophysiology and membrane transport in regards to the mechanisms of ion transport across cell membranes; alternation of membrane potential in vascular smooth muscle cells in hypertensive animals; alterations in the mechanisms of ion transport in cystic fibrosis. *Mailing Add:* Dept Physiol & Biophysics Univ Ala Birmingham Birmingham AL 35294

SHOEMAKER, RICHARD NELSON, b Allentown, Pa, Nov 21, 21; m 47; c 2. MEDICAL EDUCATION. *Educ:* Lehigh Univ, PhD(microbiol), 50. *Prof Exp:* Instr biol sci, Lehigh Univ, 47-49; dir explor res, Pfizer, Inc, 52-55, tech mgr, 55-70; coordr med educ, Greater Del Valley Regional Med Prog, 70-73; DIR EDUC, MERCY HOSP, SCRANTON, PA, 73- *Concurrent Pos:* Med educ consult, Lackawanna County Med Soc, Pa, 71-73. *Res:* Microbiological conversions; antibiotics; vitamins; steroidal chemicals. *Mailing Add:* Mercy Hosp Scranton PA 18501

SHOEMAKER, RICHARD W, b Toledo, Ohio, Nov 8, 18; m 45; c 4. MATHEMATICS. *Educ:* Univ Toledo, BS, 40, MS, 42; Univ Chicago, cert, 43; Univ Mich, MA, 49, PhD(educ math), 54. *Prof Exp:* From asst prof to assoc prof, Univ Toledo, 46-58, chmn dept, 58-63, prof math, 58-; RETIRED. *Mem:* Math Asn Am. *Mailing Add:* 2426 Meadowwood Dr Toledo OH 43606

SHOEMAKER, ROBERT ALAN, b Toronto, Ont, July 9, 28; m 50; c 4. MYCOLOGY. *Educ:* Univ Guelph, BSA, 50, MSA, 52; Cornell Univ, PhD(mycol), 55. *Prof Exp:* Asst bot, Univ Guelph, 50-52; asst plant path, Cornell Univ, 52-55; asst mycologist, Plant Res Inst, Can Dept Agr, 55-56, from assoc mycologist to sr mycologist, 56-67, head mycol sect, Biosystematics Res Inst, 67-; RETIRED. *Concurrent Pos:* With inst specialized bot, Swiss Fed Inst Technol, 61-62. *Mem:* Mycol Soc Am; Can Bot Asn; Can Phytopath Soc Res. *Res:* Taxonomy of pyrenomycetes. *Mailing Add:* 1414-1375 Prince Wales Dr Ottawa ON K2C 3L5 Can

SHOEMAKER, ROBERT HAROLD, b Buffalo, NY, Sept 15, 21; c 2. METALLURGY. *Educ:* Ohio Univ, BS, 43. *Prof Exp:* CHMN & CHIEF EXEC OFFICER, KOLENE CORP. *Concurrent Pos:* Mem bd dirs, Am Soc Metals, 66 & Indust Heating Equip Asn, 79; chmn, Legis Action Comt, Indust Heating Equip Asn, 78-81. *Mem:* Hon mem Am Soc Metals Int; Indust Heating Equip Asn; Am Iron & Steel Inst; Soc Automotive Engrs. *Res:* Metal conditioning salt bath technology; author of numerous technical publications. *Mailing Add:* Kolene Corp 12890 Westwood Ave Detroit MI 48223

SHOEMAKER, ROY H(OPKINS), civil & hydraulic engineering; deceased, see previous edition for last biography

SHOEMAKER, VAUGHAN HURST, b Chicago, Ill, Apr 4, 38; m 82; c 2. COMPARATIVE PHYSIOLOGY, PHYSIOLOGICAL ECOLOGY. *Educ:* Earlham Col, AB, 59; Univ Mich, MA, 61, PhD(zool), 64. *Prof Exp:* Instr zool, Univ Mich, 64-65; from asst prof to assoc prof zool, 65-75, chmn dept biol, 82-88, PROF ZOOL, UNIV CALIF, RIVERSIDE, 75- *Concurrent Pos:* NIH fel, 65; NSF res grants, 66-92; mem adv panel, Pop Biol & Physiol Ecol, 79-82. *Mem:* Fel AAAS; Am Soc Zool; Am Inst Biol Sci; Am Soc Ichthyologists & Herpetologists. *Res:* Water and electrolyte metabolism, energetics, and thermal relations in terrestrial vertebrates. *Mailing Add:* Dept Biol Univ Calif Riverside CA 92521

SHOEMAKER, WILLIAM C, b Chicago, Ill, Feb 27, 23; m 53; c 4. SURGERY. *Educ:* Univ Calif, AB, 44, MD, 46. *Prof Exp:* Res fel surg, Harvard Med Sch, 56-59; from asst prof to prof, Chicago Med Sch, 59-69; prof, Mt Sinai Sch Med, 69-74, chief, Div Surg Metab, 71-74; PROF SURG, SCH MED, UNIV CALIF, LOS ANGELES, 74-; CHMN, DEPT EMERGENCY MED, KING/DREW MED CTR, 91- *Concurrent Pos:* Dir dept surg res, Hektoen Inst Med Res, Cook County, Ill, 59-68; mem prog-proj rev comt B, Nat Heart Inst, 68-70; mem comt shock, Nat Res Coun-Nat Acad Sci, 69-71; chief third surg serv, Cook County Hosp, Ill; Nat Heart Inst res career award; chief, Acute Care Ctr, Harbor Gen Hosp, 74-85. *Mem:* AAAS; Am Physiol Soc; Soc Exp Biol & Med. *Res:* Regional hemodynamics and metabolism; hemorrhagic shock; hepatic physiology; electrolyte metabolism. *Mailing Add:* 12021 S Wilmington Ave Los Angeles CA 90059

SHOEMAN, DON WALTER, b Tracy, Minn, Feb 24, 41; m 67; c 1. PHARMACOLOGY. *Educ:* Macalester Col, BA, 65; Univ Minn, Minneapolis, PhD(pharmacol), 71. *Prof Exp:* Instr pharmacol, Univ Kans Med Ctr, Kansas City, 70-74, asst prof, 74-80; asst prof & chmn dept physiol & pharmacol, New Eng Col Osteopathic Med, 80-85. *Mem:* Am Soc Pharmacol & Exp Therapeut. *Res:* Drug metabolism and clinical pharmacology. *Mailing Add:* 177 Demar Ave Shoreview MN 55112

SHOENFIELD, JOSEPH ROBERT, b Detroit, Mich, May 1, 27. MATHEMATICS. *Educ:* Univ Mich, BS, 49, MS, 51, PhD(math), 52. *Prof Exp:* From instr to assoc prof, 52-65, PROF MATH, DUKE UNIV, 66- *Concurrent Pos:* NSF fel, 56-57. *Mem:* Am Math Soc; Asn Symbolic Logic. *Res:* Mathematical logic. *Mailing Add:* Dept Math Duke Univ Durham NC 27706

SHOFFNER, JAMES PRIEST, b New Madrid, Mo, Jan 14, 28; m 56; c 3. ORGANIC CHEMISTRY. *Educ:* Lincoln Univ, Mo, BS, 51; DePaul Univ, MS, 56; Univ Ill, Chicago, PhD(org Chem), 65. *Prof Exp:* Res chemist, Corn Prod Co, 55-61; RES SPECIALIST, UNIVERSAL OIL PRODS INC, 63- *Mem:* Am Chem Soc; AAAS. *Res:* Synthesis of aromatic amines; rubber vulcanization; nuclear magnetic resonance spectroscopy; imine exchange reactions; corrosion inhibitors. *Mailing Add:* 296 Parkchester Rd Elk Grove Village IL 60007

SHOFFNER, ROBERT NURMAN, b Junction City, Kans, Mar 2, 16; m 38; c 3. POULTRY GENETICS. *Educ:* Kans State Col, BS, 40, Univ Minn, MS, 42, PhD(animal genetics), 46. *Prof Exp:* From asst to prof, 40-86, actg head dept, 65-66, EMER PROF POULTRY HUSB, UNIV MINN, ST PAUL, 86- *Concurrent Pos:* Vis prof, Iowa State Univ, 57 & Univ Tex, Houston, 69; Fulbright scholar, Univ Queensland, 62. *Honors & Awards:* Merck Res Award, 82. *Mem:* Fel AAAS; fel Poultry Sci Asn (vpres, 64 & 65, pres, 66); Am Genetic Asn; Genetics Soc Am; World Poultry Sci Asn. *Res:* Avian genetics and cytogenetics; chromosome methodology, cytotaxonomy, quantitative genetics and molecular genetics. *Mailing Add:* Dept Animal Sci Univ Minn 1404 Gortner St Paul MN 55101

SHOGER, ROSS L, b Aurora, Ill, Jan 14, 30; m 57; c 3. ZOOLOGY, DEVELOPMENTAL BIOLOGY. *Educ:* NCent Col, BA, 51; Purdue Univ, MS, 53; Univ Minn, PhD(zool), 59. *Prof Exp:* From instr to assoc prof, 59-69, chmn dept, 69-73, PROF BIOL, CARLETON COL, 69- *Concurrent Pos:* NSF sci fac fel, Waseda Univ, Japan, 68-69. *Mem:* AAAS; Soc Develop Biol; Am Soc Zool. *Res:* Experimental embryology development; node regression studies. *Mailing Add:* Dept Biol Carleton Col Northfield MN 55057

SHOGREN, MERLE DENNIS, b Lindsborg, Kans, Nov 20, 26; m 74; c 4. CEREAL CHEMISTRY, BIOCHEMISTRY. *Educ:* Bethany Col, BS, 51; Kans State Univ, MS, 54. *Prof Exp:* RES FOOD TECHNOLOGIST CEREAL CHEM, USDA, 54- *Mem:* Sigma Xi; Am Asn Cereal Chemists; Am Chem Soc. *Res:* Improving breadmaking qualities of United States wheat (new varieties); improving nutritional qualities of bread-wheat flour. *Mailing Add:* US Grain Mkt Res 1515 College Ave Manhattan KS 66502

SHOHET, JUDA LEON, b Chicago, Ill, June 26, 37; m 69; c 3. PLASMA PHYSICS, ELECTRICAL ENGINEERING. *Educ:* Purdue Univ, BS, 58; Carnegie Inst Technol, MS, 60, PhD(elec eng), 61. *Prof Exp:* Asst prof elec eng, Johns Hopkins Univ, 61-66; assoc prof, 66-71, PROF ELEC ENG, UNIV WIS-MADISON, 71- *Concurrent Pos:* Consult, US Army Res & Develop Labs, Va, 61-63, Westinghouse Elec Corp, 63-66, 81, Trane Co, 68 & Argonne Nat Lab, 68-78; mem staff, Ctr Nuclear Study, Saclay, France, 69-70, Los Alamos Sci Lab, 77-, McGraw Edison Co, 79-80 & Nicolet Instrument Corp, 82- *Honors & Awards:* Frederick Emmons Terman Award, Am Soc Eng Educ, 77; Centennial Medal, Inst Elect & Electronics Engrs, 84 & Merit Award, 78. *Mem:* AAAS; fel Inst Elec & Electronics Engrs; Sigma Xi; fel Am Phys Soc. *Res:* Fusion; quantum electronics; waves and instabilities in plasmas; mathematical models of biological systems; electromagnetic theory and microwaves; plasma processing and technology. *Mailing Add:* Dept of Elec & Comput Eng Univ of Wis Madison WI 53706

SHOKEIR, MOHAMED HASSAN KAMEL, b Mansoura, Egypt, July 2, 38: Can citizen; m 68; c 2. IMMUNOLOGY. *Educ:* Cairo Univ, MB & BCh, 60, DCh, 63 & 64; Univ Mich, Ann Arbor, MS, 65, PhD(human genetics), 69. *Prof Exp:* Intern med, Cairo Univ Hosps, 60-62, resident orthop surg, 62-64; Fulbright scholar & trainee human & med genetics, Univ Mich, 64-66, Fulbright res scholar, 66-69; from asst prof to assoc prof pediat & Queen Elizabeth II scientist, Univ Sask, 69-72; assoc prof & Queen Elizabeth II scientist, Univ Man, 72-75; DIR, DIV MED GENETICS, DEPT PEDIAT, UNIV SASK, 75-, PROF PEDIAT, 77-, HEAD DEPT, 79- *Concurrent Pos:* Vis prof, Univ Alta, 76, Univ Fla, 80, Wayne State Univ, 82, Univ Calgary, 84, Garyounis Univ, Banghazi, Libya, 85, Arab Med Univ, Banghazi, 85 & Univ Mich, 86. *Mem:* Am Soc Human Genetics; Am Pediat Soc; Can Soc Clin Invest; Soc Pediat Res; fel Can Col Med Geneticists. *Res:* Genetics, biochemistry and immunology of copper containing enzymes and other metallo-proteins; neurobiology, especially of Huntington's chorea, Wilson's disease and parkinsonism; congenital malformations in man; genetic polymorphisms of proteins in man. *Mailing Add:* Dept Pediat Univ Sask Saskatoon SK S7N 0X0 Can

SHOLANDER, MARLOW, b Topeka, Kans, Mar 13, 15; m 40; c 3. MATHEMATICS. *Educ:* Univ Kans, AB & AM, 40; Brown Univ, PhD(math), 49. *Prof Exp:* Instr math, Univ Kans, 38-40 & Brown Univ, 40-45; from instr to assoc prof, Wash Univ, 46-54; from assoc prof to prof, Carnegie Inst Technol, 54-60; prof, 60-80, EMER PROF MATH, CASE WESTERN RESERVE UNIV, 80- *Mem:* Am Math Soc; Math Asn Am. *Res:* Mean values and medians; foundations; convex plates and lattices. *Mailing Add:* 2921 Warrington Rd Shaker Heights OH 44120-2422

SHOLDT, LESTER LANCE, b Greeley, Colo, Aug 18, 38; m 62; c 2. MILITARY ENTOMOLOGY, MEDICAL ENTOMOLOGY. *Educ:* Colo State Univ, BS, 62, PhD(entom), 78; Univ Hawaii, MS, 64. *Prof Exp:* Res asst, Cotton Insect Lab, USDA, Ariz, 65-66; asst opers officer, Dis Vector Control Ctr, Calif, 66-67; div entomologist, 1st Marine Div, Vietnam, 68; head, Dept Entom, Prev Med Unit #2, Va, 69-72, Naval Med Res Unit #5, Ethiopia, 73-76; officer in charge, Dis Vector Ecol & Control Ctr, Fla, 78-81; head, Dis Vector Control Sect, Pre Med Div, Bur Med & Surg, 81-84; Indust Col Armed Forces, 84-85; ASSOC PROF, UNIFORMED SERV UNIV. *Concurrent Pos:* Consult, Trop Dis Res Orgn, Africa, Asia, Central Am & S Am. *Honors & Awards:* Outstanding Med & Vet Entomologist Award, Am Registry Prof Entomologists, 90. *Mem:* Entom Soc Am; Am Mosquito Control Asn; Asn Military Surgeons of US; Sigma Xi. *Res:* Epidemiology and control of human lice and louse-borne diseases; insect repellents for the protection of human subjects; malaria prevention and control; vector control in post-disaster situations. *Mailing Add:* Uniformed Serv Univ Health Sci 4301 Jones Bridge Rd Bethesda MD 20814

SHOLL, HOWARD ALFRED, b Northampton, Mass, Oct 14, 38; m 60; c 2. COMPUTER SCIENCES. *Educ:* Worcester Polytech Inst, BS, 60, MS, 63; Univ Conn, PhD(comput sci), 70. *Prof Exp:* Engr, US Army Signal Res & Develop Lab, 60-61; asst elec eng, Worcester Polytech Inst, 61-63; sr engr, Sylvania Elec Co, 63-66; from instr to assoc prof, 66-85, PROF ELEC ENG & COMPUT SCI, UNIV CONN, 85-, DIR, BOOTH RES CTR, 86- *Concurrent Pos:* Leverhulme vis fel, Univ Edinburgh, 73-74; mem, Task Force Software Eng, Digital Syst Eval Comt, 75; Fulbright sr res fel, Tech Univ Munich, 82-83. *Mem:* Inst Elec & Electronics Engrs; Asn Comput Mach. *Res:* Digital systems design; engineering and analysis of real time, distributed, computer systems. *Mailing Add:* Booth Res Ctr Univ Conn Box U-31 Storrs CT 06268

SHOMAKER, JOHN WAYNE, b Pueblo, Colo, Apr 18, 42; m 78; c 6. GROUNDWATER FLOW MODELLING. *Educ:* Univ NMex, BS, 63 & MS, 65; St John's Col, MA, 84; Univ Birmingham, MSc, 85. *Prof Exp:* Hydrologist, Water Res Div, US Geol Surv, 65-69; geologist, NMex Bur Mines & Mineral Resource, 69-73; consult geologist, 73-86; PRES, JOHN W SHOMAKER INC, 86- *Concurrent Pos:* Adj assoc prof geol, Univ NMex, 76- *Mem:* Geol Soc Am; Am Asn Petrol Geologists; Am Inst Hydrol; Am Inst Prof Geologists (secy-treas, 76); Int Asn Hydrogeologists; Asn Ground-Water Scientists & Engrs. *Res:* Ground-water flow modelling; water supply and water rights studies; ground water contamination studies. *Mailing Add:* 2703 Bradford Pkwy NE Suite D Albuquerque NM 87107

SHOMAY, DAVID, b Brooklyn, NY, Aug 31, 24. ZOOLOGY. *Educ:* Long Island Univ, BS, 48; Univ Ill, MS, 49, PhD(zool), 55. *Prof Exp:* From asst to instr, 49-56, from instr to asst prof, 56-63, ASSOC PROF ZOOL, UNIV ILL, CHICAGO CIRCLE, 63- *Mem:* AAAS; Soc Vert Paleont; Am Soc Zoologists. *Res:* Comparative anatomy; invertebrate zoology; structure and evolution of nervous system. *Mailing Add:* Dept Biol Sci Univ Ill Box 4348 MC-066 Chicago IL 60680

SHOMBERT, DONALD JAMES, b Pittsburgh, Pa, Oct 31, 28; m 55; c 4. PHYSICAL CHEMISTRY. *Educ:* Univ Pittsburgh, BS, 53, PhD(chem), 58. *Prof Exp:* Res assoc phys chem, Res Labs, Merck & Co, Inc, NJ, 58-61; mgr surface physics & chem, CBS Labs, Conn, 61-62; asst prof to assoc prof chem, 62-87, EMER PROF, DOUGLASS COL, RUTGERS UNIV, NEW BURNSWICK, 87- *Mem:* Am Chem Soc; Inst Elec & Electronic Eng. *Res:* Semiconductor materials; thin-film depostion and device technology; physical measurements and electronic instrumentation. *Mailing Add:* 199 Chaucer Dr Berkeley Heights NJ 07922

SHOMURA, RICHARD SUNAO, fish biology, for more information see previous edition

SHON, FREDERICK JOHN, b Pleasantville, NY, July 24, 26; m 46; c 1. NUCLEAR PHYSICS. *Educ:* Columbia Univ, BS, 46. *Prof Exp:* Jr engr, Publicker Alcohol Co, 46-47; proj engr, Thermoid Co, 47-48; opers physicist, Mound Lab, 48-52; reactor opers supvr, Lawrence Radiation Lab, Univ Calif, 52-61, lectr nuclear eng, 56-61; chief reactor oper & supvr licensing br, Div Licensing & Regulations, US AEC, 61-62, chief reactor & criticality safety br, Div Oper Safety, 63-67, asst dir nuclear facilities, 67-72; DEP CHIEF JUDGE, TECH, ATOMIC SAFETY & LICENSING BD PANEL, US NUCLEAR REGULATORY COMN, 72- *Concurrent Pos:* Radiation chemist, Atomics Int Div, NAm Aviation, Inc, 51-52; consult, Aerojet-Gen Nucleonics Div, Gen Tire & Rubber Co, 58- & US AEC, 59-; physicist, Lawrence Radiation Lab, Univ Calif, 62-63; mem, Int Atomic Energy Agency Safety Adv Mission to the Spanish Junta De Energia Nuclear, 71-72. *Mem:* Am Nuclear Soc. *Res:* Nuclear reactor design and operation; neutron physics; radiation detection, measurement and safety. *Mailing Add:* US Nuclear Regulatory Comn Washington DC 20555

SHONE, ROBERT L, b Gary, Ind, July 28, 37; m 61; c 2. ORGANIC CHEMISTRY, MEDICINAL CHEMISTRY. *Educ:* Ind Univ, BS, 59; Mich State Univ, MS, 61, PhD(org chem), 65. *Prof Exp:* Res chemist, Swift & Co, 65-66; res fel, Ill Inst Technol, 66-67; sr investr, 67-71, RES SCIENTIST, G D SEARLE & CO, 71- *Mem:* Am Chem Soc; The Chem Soc. *Res:* Anti-viral and antihypertensive drugs; nucleic acids; enzyme inhibitors of nucleic acid metabolites; synthesis of nucleosides, amino acids and carbohydrates; antiallergy drugs; synthesis of pyrones; synthesis of gastric and secretory prostaglendins. *Mailing Add:* 1441 Joan Dr Palatine IL 60067

SHONE, ROBERT TILDEN, b Rochester, NY, July 22, 28; m 54; c 2. PHOTOGRAMMETRY, FORESTRY. *Educ:* Syracuse Univ, BS, 49, MS, 54. *Prof Exp:* Photogram aide mapping, Army Map Serv, 52-53; sect mgr photogram prod sales, Bausch & Lomb Optical Co, 54-59; mem tech staff, Ramo-Wooldridge Div, Thompson Ramo Wooldridge, Inc, 59-61; sr staff engr, Librascope Div, Gen Precision Inc, 61-63; dept head photogram instrument res & develop, Bausch & Lomb, Inc, 63-67, tech dir, Spec Prod Div, 67-73, vpres spec prod develop, 73-81, dir res & develop, Sci Optical Prod Div, 81-85; DIR ENG, IMAGE INTERPRETATION SYSTS, INC, 87- *Mem:* Can Inst Survrs; Am Soc Photogram. *Res:* Instrument accuracy studies; mapping instrument automation; system analysis; analytical photogrammetry; photographic interpretation; instrument development management. *Mailing Add:* 56 Round Trail Pittsford NY 14534

SHONEBARGER, F(RANCIS) J(OSEPH), b Sugar Grove, Ohio, Nov 2, 24; m 51; c 4. CERAMIC ENGINEERING. *Educ:* Ohio State Univ, BCerE & MS, 51, PhD, 61. *Prof Exp:* Glass technologist, Gen Elec Co, 54-61; glass technologist, Anchor Hocking Corp, 61-69, mgr explor res, 69-76; INSTR CERAMIC ENG, HOCKING TECH COL, 76- *Mem:* Am Ceramic Soc; Brit Soc Glass Technol. *Res:* Glasses to imbed electroluminescent phosphors; infrared glasses; glass surfaces, composition, physical properties and refractory reactions. *Mailing Add:* Dept Engr Hocking Tech Col Rm 331 Nelsonville OH 45764

SHONICK, WILLIAM, b Poland, Oct 3, 19; US citizen; m 41; c 1. BIOSTATISTICS, PUBLIC HEALTH. *Educ:* City Col New York, BS, 42; George Wash Univ, MA, 48; Univ Calif, Los Angelos, PhD(biostatist), 67. *Prof Exp:* Jr acct, Wm Janis CPA, New York, 42; statistician-economist, Off Price Admin, 43; bus agent, Local 203, United Fed Workers of Am-CIO, 44-45; teacher social studies, Montgomery County Sch Syst, Md, 45-51; pvt bus, 52-55; budget & statist anal, Fedn Jewish Philanthropies, New York, 55-61; coordr biostatist, Rehab Res & Training Ctr, Sch Med, Univ Southern Calif, 68-69; from asst prof to assoc prof, 69-80, PROF PUB HEALTH, SCH PUB HEALTH, UNIV CALIF, LOS ANGELES, 80- *Concurrent Pos:* NIH fel biostatist, Univ Calif, Los Angeles, 68; asst prof community med, Med Sch, Univ Southern Calif, 68-69. *Mem:* Am Statist Asn; Am Pub Health Asn; Opers Res Soc Am. *Res:* Governmental policies and health services; health policy formulation and planning methods, delivery. *Mailing Add:* 1244 Beverly Green Dr Los Angeles CA 90035

SHONK, CARL ELLSWORTH, b Plymouth, Pa, Nov 11, 22; m 51; c 2. BIOCHEMISTRY. *Educ:* Bucknell Univ, BS, 48; MS, 49; Rutgers Univ, PhD(biochem, physiol), 62. *Prof Exp:* Chemist, E I du Pont de Nemours & Co, Pa, 48; biochemist, Merck & Co, Inc, 49-66; ASSOC PROF CHEM, CENT MICH UNIV, 66- *Mem:* AAAS; Am Chem Soc; NY Acad Sci; Biochem Soc; Am Inst Chemists. *Res:* Enzyme chemistry; cancer biochemistry; analytical biochemical methods. *Mailing Add:* 8680 S Vandecar Rd Shepherd MI 48883-9549

SHONKWILER, RONALD WESLEY, b Chicago, Ill, Feb 20, 42. MATHEMATICS. *Educ:* Calif State Polytech Col, Kellogg-Voorhis, BS, 64; Univ Colo, Boulder, MS, 67, PhD(math), 70. *Prof Exp:* Aerospace engr, US Naval Ord Lab, 64-65; asst prof, 70-76, ASSOC PROF MATH, GA INST TECHNOL, 76- *Mem:* Soc Math Biol; Am Math Soc; Soc Indust & Appl Math. *Res:* Operator theory; mathematic biology. *Mailing Add:* 329 Robin Hood Rd NE Atlanta GA 30309

SHONLE, JOHN IRWIN, b Indianapolis, Ind, Oct 1, 33; m 71; c 6. MUSICAL ACOUSTICS. *Educ:* Wesleyan Univ, BA, 55; Univ Calif, Berkeley, MA, 57, PhD(physics), 61. *Prof Exp:* From asst prof to assoc prof physics, Reed Col, 60-67; assoc prof physics & astrophys, Univ Colo, Denver, 67-73, prof physics, 73-79; CONSULT, 79- *Concurrent Pos:* NSF sci fac fel, 66-67. *Mem:* AAAS; Am Asn Physics Teachers; Acoust Soc Am. *Res:* Physics teaching methods; psychoacoustics of music; environmental physics. *Mailing Add:* Two Village Brook Rd Yarmouth ME 04096

SHONS, ALAN R, SURGERY. *Prof Exp:* DIR PLASTIC SURG, CASE WESTERN RESERVE UNIV, UNIV HOSPS, 85- *Mailing Add:* Dept Surg Case Western Reserve Univ Univ Hosps 2074 Abington Rd Cleveland OH 44106

SHONTZ, CHARLES JACK, b Sewickley, Pa, Jan 2, 26; m 52. ANIMAL ECOLOGY, HUMAN ECOLOGY. *Educ:* Ind State Col, Pa, BS, 49; Univ Pittsburgh, MS, 53, PhD(zool), 62. *Prof Exp:* Teacher high sch, Pa, 49-55; Fulbright lectr sci, Kambawza Col, Burma, 55-56; teacher high sch, Pa, 56-57; asst prof biol, Clarion State Col, 57-59, from assoc prof to prof biol & physiol, 59-84, head dept, 59-62, dean acad serv, 64-84, assoc vpres acad affairs, 78-84; RETIRED. *Concurrent Pos:* NSF fac fel, 61-62; assoc vpres acad affairs & dean summer sessions, Univ Pa. *Mem:* Am Soc Ichthyol & Herpet; Wilderness Soc; Am Nature Study Soc. *Res:* Effects of environment on the evolution of populations of fishes, especially the family Cyprinidae. *Mailing Add:* Box 11 Clarion PA 16214

SHONTZ, JOHN PAUL, b Meadville, Pa, Oct 11, 40; m 67; c 2. PLANT ECOLOGY. *Educ:* Edinboro State Col, BS, 62; Miami Univ, MA, 64; Duke Univ, PhD(bot), 67. *Prof Exp:* Teacher pub sch, Pa, 62; instr biol sci, Mt Holyoke Col, 67-68, asst prof, 68-74; from asst prof to assoc prof, 74-82, chmn, 79-89, PROF BIOL, GRAND VALLEY STATE UNIV, 82- *Mem:* AAAS; Bot Soc Am; Ecol Soc Am; Sigma Xi. *Res:* Ecology of desert annuals; plant species interaction; seed germination ecology. *Mailing Add:* Dept Biol Grand Valley State Univ Allendale MI 49401

SHONTZ, NANCY NICKERSON, b Pittsburgh, Pa, July 9, 42; m 67; c 2. HUMAN GENETICS. *Educ:* Smith Col, AB, 64, PhD(biol), 69; Duke Univ, MA, 66. *Prof Exp:* Instr biol, ECarolina Univ, 66-67; lectr zool, Univ Mass, 67-68; asst prof bot, Holyoke Community Col, 69-74; adj fac, 74-85, asst prof, 85-89, ASSOC PROF BIOL, GRAND VALLEY STATE UNIV, 89- *Mem:* AAAS; Sigma Xi. *Res:* Electrophoresis of salamander proteins; seed germination studies; ecotypic variation; bacterial strain identification using plasmids. *Mailing Add:* Biol Dept Grand Valley State Univ Allendale MI 49401

SHOOK, BRENDA LEE, b Newport Beach, Calif, Nov 30, 52. DEVELOPMENTAL NEUROANATOMY, PRENATAL DEVELOPMENT PHYSIOLOGY. *Educ:* Calif State Univ, Stanislaus, BA, 75, MA, 76; Brandeis Univ, PhD(physiol psychol), 82. *Prof Exp:* Asst res anatomist, Brandeis Univ, 79-82; NIH res fel & vis scholar, Univ Calif, Davis, 82-85 & vis lectr, 83-84, ASSOC RES ANATOMIST, UNIV CALIF, LOS ANGELES, 85- *Mem:* Soc Neurosci; Int Brain Res Orgn; NY Acad Sci; Am Asn Adv Sci. *Res:* Phenomena of recovery of behavioral and physiological function after early brain damage; human mental retardation which results from prenatal and early postnatal damage, by disease or trauma to the developing central nervous system; anatomy and physiology of oculomotor system in primates. *Mailing Add:* 4810 Hollow Corner Rd Culver City CA 90230

SHOOK, CLIFTON ARNOLD, b Lamont, Alta, Oct 10, 34; m 59; c 4. FLUID & FLUID PARTICLE MECHANICS. *Educ:* Univ Alta, BSc, 56; Univ London, PhD(chem eng), 60. *Prof Exp:* From asst prof to assoc prof, 60-71, PROF CHEM ENG, UNIV SASK, 71- *Mem:* Chem Inst Can. *Res:* Fluid mechanics; heat transfer; mass transfer; fluid-particle systems; rheology; pipeline flow of suspensions. *Mailing Add:* Dept Chem Eng Univ Sask Saskatoon SK S7N 0W0 Can

SHOOK, THOMAS EUGENE, b Pasadena, Calif, Mar 10, 28; m 58; c 3. CHEMISTRY. *Educ:* Tex Tech Univ, BS, 51. *Prof Exp:* Res asst biochem, Ft Detrick, Md, 52-53, chief biochem br, 53-71, environ coordr, Pine Bluff Arsenal, 72-82, CHIEF DEVELOP & TECHNOL DIV, PINE BLUFF ARSENAL, US ARMY, 71- *Mem:* Am Chem Soc; Am Inst Chemists; Sigma Xi; Am Statist Asn. *Res:* Physical-engineering sciences; statistics; production development; process evaluation; research management. *Mailing Add:* 1716 Alberta Dr Little Rock AR 72207

SHOOK, WILLIAM BEATTIE, b Columbus, Ohio, Oct 3, 28; m 50; c 4. CERAMICS ENGINEERING. *Educ:* Ohio State Univ, BCerE, 53, PhD(ceramics eng), 61. *Prof Exp:* Res asst ceramics res, Eng Exp Sta, Ohio State Univ, 50-53, res assoc, 53-55, supvr building res, 55-57, dir ceramics res, 57-63; vis prof, Indian Inst Technol, Kanpur, 63-65; from asst prof to prof ceramics eng & chmn dept, Ohio State Univ, 65-82; RETIRED. *Honors & Awards:* Cramer Award, Am Ceramic Soc, 81. *Mem:* Fel Am Ceramic Soc; Am Soc Eng Educ; Am Ord Asn; Nat Inst Ceramic Engrs; Am Soc Nondestruct Testing. *Res:* Brittle failure mechanisms in impact testing, especially influence of elastic properties, density and geometry of test specimens on the response system; viscosity; measurement and interpretation in melting and crystallizing at non-equilibrium; nondestructive testing of ceramics. *Mailing Add:* 111 Glencoe Rd Columbus OH 43214

SHOOLERY, JAMES NELSON, b Worland, Wyo, June 25, 25; m 51, 71; c 3. PHYSICAL CHEMISTRY. *Educ:* Univ Calif, BS, 48; Calif Inst Technol, PhD(chem), 52. *Prof Exp:* Dir appln lab, Varian Assocs, 52-62, mkt mgr anal inst div, Calif, 62-69; independent consult, 69-72; sr appln chemist, Varian Assoc, 72-90; RETIRED. *Honors & Awards:* Sargent Award, 64; Anachem Award, 82. *Mem:* Am Chem Soc. *Res:* Microwave spectroscopy; chemical effects in nuclear magnetic resonance. *Mailing Add:* 2301 Bowdoin St Palo Alto CA 94306

SHOOMAN, MARTIN L, b Trenton, NJ, Feb 24, 34; m 62; c 2. COMPUTER & ELECTRICAL ENGINEERING. *Educ:* Mass Inst Technol, SB & SM, 56; Polytech Inst Brooklyn, DEE, 61. *Prof Exp:* Teaching asst elec eng, Mass Inst Technol, 55-56; mem staff, Res & Develop Group, Sperry Gyroscope Co, NY, 56-58; from instr to assoc prof elec eng, 58-74, dir, Div Comput Sic, 81-84, PROF, ELEC ENG & COMPUT SCI, POLYTECH INST NY, FARMINGDALE, 74- *Concurrent Pos:* Consult govt & indust, 59-; vis assoc prof, Mass Inst Technol, 71. *Mem:* Fel Inst Elec & Electronics Engrs; Asn Comput Mach. *Res:* Reliability theory and application to computer systems; reliability of electronic and mechanical systems; software engineering. *Mailing Add:* Dept Elec Eng Polytech Univ Long Island Ctr Rte 110 Farmingdale NY 11735

SHOOP, C ROBERT, b Chicago, Ill, Aug 12, 35; div; c 2. ECOLOGY. *Educ:* Southern Ill Univ, BA, 57; Tulane Univ, MS, 59, PhD(zool, bot), 63. *Prof Exp:* Instr zool & physiol, Wellesley Col, 62-64, asst prof biol sci, 64-69,; assoc prof zool, 69-74, PROF ZOOL, UNIV RI, 74- *Concurrent Pos:* US AEC res contract, 65-75; dir, Inst Environ Biol, 70-72; dir NIH training grant environ physiol, 70-72; collabr, Nat Park Serv, 78-; contracts, Dept Interior, 78-82 & Nat Marine Fisheries Serv, 80-81, 82-83 & 87-; re awards, Wellesley Col, 62-66. *Honors & Awards:* Stoye Prize, Am Soc Ichthyol & Herpet, 60. *Mem:* AAAS; Am Inst Biol Sci; Am Soc Ichthyol & Herpet; Am Soc Mammal; Animal Behavior Soc. *Res:* Behavior and ecology of vertebrates; radiobiology. *Mailing Add:* Dept Zool Univ RI Kingston RI 02881

SHOOSMITH, JOHN NORMAN, b London, Eng, Oct 9, 34; US citizen; m 66; c 3. APPLIED MATHEMATICS, COMPUTER SCIENCE. *Educ:* Queen's Univ, Ont, BSc, 56; Col William & Mary, MS, 67; Univ Va, PhD(appl math), 73. *Prof Exp:* Comput specialist, Avro Aircraft Corp, Can, 56-59; aerospace technologist comput, Manned Spacecraft Ctr, 59-64, aerospace technologist, Gemini Prog, 64-65, HEAD COMPUT APPL, LANGLEY RES CTR, NASA, 65- *Concurrent Pos:* Asst prof lectr, George Washington Univ, 74- *Mem:* Asn Comput Mach; Am Inst Aeronaut & Astronaut; Soc Indust & Appl Math. *Res:* Numerical analysis, specifically high-order accurate numerical solutions to bondary-value problems of ordinary and partial differential equations. *Mailing Add:* 105 Cambridge Lane Williamsburg VA 23185

SHOOTER, ERIC MANVERS, b Mansfield, Eng, Apr 18, 24; m 49; c 1. BIOCHEMISTRY. *Educ:* Cambridge Univ, BA, 45, MA, 49, PhD(chem), 50; Univ London, DSc(biochem), 64; Univ Cambridge, ScD(biochem), 86. *Prof Exp:* Fel chem, Univ Wis, 49-50; sr scientist biochem, Brewing Indust Res Found, Eng, 50-53; lectr, Univ Col, Univ London, 53-64; assoc prof genetics, 64-68, prof genetics & biochem, 68-75, prof neurobiol & chmn dept, 75-87, PROF NEUROBIOL, SCH MED, STANFORD UNIV, 87- *Concurrent Pos:* USPHS int fel, Stanford Univ, 61-62. *Honors & Awards:* Fel, Royal Soc, 88; Wakeman Award, 88. *Mem:* Foreign assoc Inst Med Nat Acad Sci; Brit Biophys Soc; Am Soc Biol Chemists; Am Soc Neurochem; Int Soc Neurochem; Soc Neurosci; Brit Biochem Soc. *Res:* Physical chemistry of proteins; structure of normal and abnormal hemoglobins; genetic control of protein synthesis, replication of DNA; molecular neurobiology; nerve growth factor. *Mailing Add:* Dept Neurobiol-Fairchild Sch Med Stanford Univ Stanford CA 94305-5401

SHOOTER, JACK ALLEN, b Austin, Tex, June 16, 40; m 77; c 4. ACOUSTICS, COMPUTER SCIENCE. *Educ:* Univ Tex, Austin, BS, 63. *Prof Exp:* res scientist assoc V acoust, 63-88, SPECIAL RES ASSOC, APPL RES LABS, UNIV TEX, AUSTIN, 80- *Mem:* Assoc mem Acoust Soc Am; Sigma Xi. *Res:* Underwater acoustics; physical acoustics; signal processing. *Mailing Add:* 11305 January Austin TX 78713-8029

SHOPE, RICHARD EDWIN, JR, b Philadelphia, Pa, Sept 4, 26; m 61; c 6. VIROLOGY, IMMUNOLOGY. *Educ:* Williams Col, BA, 47; Univ Wis, BS, 49; Cornell Univ, DVM, 59; Univ Minn, PhD(microbiol), 64. *Prof Exp:* Asst prof, 59-70, ASSOC PROF VET MED & MICROBIOL, UNIV MINN, ST PAUL, 70- *Concurrent Pos:* NIH career develop award, 64-68. *Mem:* Am Vet Med Asn; US Animal Health Asn; Am Asn Vet Clinicians. *Res:* Enteric and respiratory viral disease of domestic animals; mammalian leukemias and other tumors; immune tolerance and autoimmune diseases of domestic animals. *Mailing Add:* Dept Vet Biol Col Vet Med Univ Minn St Paul MN 55108

SHOPE, ROBERT ELLIS, b Princeton, NJ, Feb 21, 29; m 58; c 4. VIROLOGY. *Educ:* Cornell Univ, BA, 51, MD, 54. *Prof Exp:* Intern, Grace-New Haven Community Hosp, Conn, 54-55, asst resident internal med, 57-58; mem staff virus labs, Rockefeller Found, NY, 58-59 & Belem Virus Lab, 59-65; from asst prof to assoc prof, 65-75, PROF EPIDEMIOL, SCH MED, YALE UNIV, 75- *Res:* Arboviruses. *Mailing Add:* Yale Arbovirus Res Unit 60 Col St Box 3333 New Haven CT 06510

SHOPP, GEORGE MILTON, JR, b Harrisburg, Pa, May 21, 55. IMMUNOTOXICOLOGY, MARINE MAMMAL TOXICOLOGY. *Educ:* Bucknell Univ, BS, 77; Med Col Va, PhD(toxicol), 84. *Prof Exp:* Postdoctoral fel, Inhalation Toxicol Res Inst, 84-86; ASSOC SCIENTIST, LOVELACE MED FOUND, 86- *Concurrent Pos:* Clin asst prof, Col Pharm, Univ NM, 86-, adj asst prof, Dept Biol, 88-; adj res scientist, Marine Environ Res Inst, 90-; ad hoc proposal reviewer, Nat Oceanog Atmospheric Admin, 91-; ad hoc proposal reviewer, Health Effects Inst, 90; prin investr, Nat Inst Environ Health Sci, 89- *Mem:* Am Col Toxicol; Soc Toxicol; Am Asn Immunologists. *Res:* Toxic effects of environmental chemicals and drugs on immune function in laboratory animals, marine mammals and humans. *Mailing Add:* Lovelace Med Found 2425 Ridgecrest Dr SE Albuquerque NM 87108

SHOPSIS, CHARLES S, IN VITRO TOXICOLOGY. *Educ:* City Univ NY, PhD(biochem), 74. *Prof Exp:* ASST PROF BIOCHEM, ROCKEFELLER UNIV, 82- *Res:* Cell membrane function. *Mailing Add:* Dept Chem Adelphi Univ Garden City NY 11530

SHOR, AARON LOUIS, b New York, NY, Jan 13, 24; m 60. VETERINARY MEDICINE, ANIMAL NUTRITION. *Educ:* Cornell Univ, BS, 47, DVM, 53; Univ Del, MS, 49. *Prof Exp:* Field investr animal dis, Farm & Home Div, Am Cyanamid Co, 55-57, ruminant specialist, 57, field investr, Agr Div, 57-60, mgr, Clin Develop Lab & Poultry Prog, 63-77, regist coordr, Agr Div, 77-80; mgr clin develop, Smith Kline Animal Health Prod, 80-85; mgr, Reg Affairs & Mfg Qual Assurance, 85-90; CONSULT, 90- *Concurrent Pos:* Adj prof, Trenton State Col, 73-75. *Mem:* Am Vet Med Asn; Am Soc Animal Sci; Am Dairy Sci Asn; Poultry Sci Asn; Indust Vet Asn (secy, 65-70, pres, 71-72); Am Asn Avian Pathologists. *Res:* Development of drugs to prevent or treat disease or improve production efficiency of animals. *Mailing Add:* 32 Hopemont Dr Mt Laurel NJ 08054-4513

SHOR, ARTHUR JOSEPH, b New York, NY, June 10, 23; m 52; c 2. INORGANIC CHEMISTRY, CHEMICAL ENGINEERING. *Educ:* City Col New York, BChE, 43; Univ Tenn, MS, 64, PhD(chem), 67. *Prof Exp:* Assoc chem engr, Argonne Nat Lab, 46-56; res chem engr, IIT Res Inst, 56-57; mem res staff chem, Oak Ridge Nat Lab, 58-89; RES ENGR, WEITZMANN INST, 90- *Mem:* Am Chem Soc; Sigma Xi. *Res:* Fused salt phase studies; effects of reactor irradiation on nuclear fuels and fertile materials; development of reverse osmosis membranes and apparatus for cleanup of waste and brackish waters. *Mailing Add:* PO Box 10191 Tel-Aviv 61101 Israel

SHOR, GEORGE G, JR, b New York, NY, June 8, 23; m 50; c 3. MARINE GEOPHYSICS. *Educ:* Calif Inst Technol, BS, 44, MS, 48, PhD(seismol), 54. *Prof Exp:* Seismol party chief, Seismic Explor, Inc, 48-51; res asst, Calif Inst Technol, 51-53; from asst res geophysicist to res geophysicist, 53-69, prof marine geophys & sea grant prog mgr, 69-73, ASSOC DIR, SCRIPPS INST OCEANOG, UNIV CALIF, SAN DIEGO, 68-, EMER PROF GEOPHYS, 90- *Mem:* AAAS; Soc Explor Geophys; fel Geol Soc Am; fel Am Geophys Union. *Res:* Marine geophysics; structure, origin and properties of ocean floor; marine technology. *Mailing Add:* Scripps Inst Oceanog Univ Calif La Jolla CA 92093-0210

SHOR, STEVEN MICHAEL, b New York, NY, Apr 5, 44; m 70. CHEMICAL ENGINEERING. *Educ:* Univ Mass, Amherst, BS, 65; Northwestern Univ, MS, 67; Iowa State Univ, PhD(chem eng), 70. *Prof Exp:* Sr chem engr, 70-88, DIV SCIENTIST, 3M CO, 88- *Mem:* Am Inst Chem Engrs. *Res:* Small particle technology, especially as applied to grain size distribution involved in crystallization and precipitation processes; effects of various parameters on nucleation and growth kinetics. *Mailing Add:* 9177 Edinburgh Lane St Paul MN 55125

SHORB, ALAN MCKEAN, b Baltimore, Md, July 18, 38; m 60; c 3. MATHEMATICAL PROGRAMMING. *Educ:* Swarthmore Col, BA, 60; Cornell Univ, MA, 65; Univ Minn, Minneapolis, PhD(math), 69. *Prof Exp:* Instr, State Univ NY Binghamton, 64-66; from asst prof to assoc prof math, Naval Postgrad Sch, 68-75; SYSTS ANALYST, DEVELOP ANAL ASSOCS, 76- *Concurrent Pos:* Res mathematician, Nat Bur Standards, 59-66; mathematician, Naval Electronics Lab Ctr, San Diego, 72-73. *Mem:* Soc Indust & Appl Math; Sigma Xi. *Res:* Computer simulation of social systems. *Mailing Add:* 58 Northgate Rd Wellesley MA 02181

SHORE, BRUCE WALTER, b Visalia, Calif, Feb 27, 35; c 3. ATOMIC PHYSICS. *Educ:* Col of Pacific, BS, 56; Mass Inst Technol, PhD(nuclear chem), 60. *Prof Exp:* Res chemist, Shell Oil Co, Calif, 56; res scientist, US Naval Radiol Defense Lab, 57; instr physics, Suffolk Univ, 57-60; analyst develop planning, Anal Serv, Inc, Va, 60-62; lectr astron & res fel astrophys, Harvard Col Observ, 62-68; assoc prof physics, Kans State Univ, 68-72; PHYSICIST, LAWRENCE LIVERMORE LAB, 72- *Concurrent Pos:* Sci Res Coun fel, Imp Col, Univ London, 70-71; vis scientist, Imp Col London, 83-84 & Max Planck Inst for Quantum Optics, 84. *Mem:* Am Phys Soc; Int Astron Union. *Res:* Atomic structure and theoretical spectroscopy; photon physics; quantum optics. *Mailing Add:* Lawrence Livermore Nat Lab Livermore CA 94550

SHORE, DAVID, physics, aeronautical engineering, for more information see previous edition

SHORE, FERDINAND JOHN, b Brooklyn, NY, Sept 23, 19; m 46; c 5. NUCLEAR PHYSICS. *Educ:* Queens Col, NY, BS, 41; Wesleyan Univ, MS, 43; Univ Ill, PhD(physics), 52. *Prof Exp:* Res physicist photom of pyrotech, Wesleyan Univ, 41-45, res physicist piezoelec, 45-46; asst physics, Univ Ill, 46-49, asst nuclear physics, 49-52; assoc physicist, Brookhaven Nat Lab, 52-60; assoc prof, 60-65, PROF PHYSICS, QUEENS COL, NY, 65- *Concurrent Pos:* Consult, Brookhaven Nat Lab, 61-; consult, Nat Coun Radiation Protection & Measurements, 62-70, consociate mem, 76- *Mem:* Sigma Xi; AAAS; Am Asn Physics Teachers; Am Phys Soc. *Res:* Pyrotechnics; piezoelectricity; decay schemes; reactor shielding; neutron cross sections; cryogenics; energy system analysis. *Mailing Add:* 77 Southern Blvd East Patchogue NY 11772

SHORE, FRED L, b Bakersfield, Calif, Sept 3, 42; m 64; c 2. AGRICULTURAL & FOREST SCIENCES. *Educ:* Fresno State Col, BS, 64; Ariz State Univ, Phd(chem), 71. *Prof Exp:* From asst prof to prof chem, Jackson State Univ, 70-82; supvr, Lockheed Eng & Man Serv Co, 82-85; sect head, 85-87, TECH DIR & DEPT HEAD, RADIAN CORP, 87- *Mem:* Am Chem Soc. *Res:* Environmental analytical chemistry method development with GC/MS and GC specialties. *Mailing Add:* c/o Radian Corp PO Box 201088 Austin TX 78720-1088

SHORE, GORDON CHARLES, ORGANELLE BIOGENESIS, PROTEIN TARGETING. *Educ:* McGill Univ, PhD(biochem), 74. *Prof Exp:* PROF BIOCHEM, MCGILL UNIV, 86- *Mailing Add:* Dept Biochem McGill Univ 3655 Drummond St Montreal PQ H3G 1Y6 Can

SHORE, HERBERT BARRY, b Brooklyn, NY, Nov 18, 39. THEORETICAL SOLID STATE PHYSICS. *Educ:* Mass Inst Technol, BS, 61; Univ Calif, Berkeley, PhD(physics), 66. *Prof Exp:* Asst res physicist, Univ Calif, San Diego, 66-67, asst prof physics, 67-75; assoc prof, 75-79, PROF PHYSICS, SAN DIEGO STATE UNIV, 79- *Mem:* Am Phys Soc; AAAS. *Res:* Theory of paraelectric resonance; resonance in biological systems; electron-hole liquid; theory of electron gas; impurities in semiconductors; metal-insulator transition. *Mailing Add:* Dept of Physics San Diego State Univ San Diego CA 92182

SHORE, JAMES H, b Winston-Salem, NC, Apr 6, 40; m 63; c 2. PSYCHIATRY. *Educ:* Duke Univ, MD, 65. *Prof Exp:* Chief, Portland Area Indian Health Serv, 69-73; from assoc prof to prof psychiat, Med Sch, Univ Ore, 73-85, dir, community psychiat training prog, Health Sci Ctr, 73-75, chmn dept psychiat, 75-85; PROF & CHMN, DEPT PSYCHIAT, UNIV COLO HEALTH SCI CTR, 85-; SUPT, CO PSYCHIAT HOSP, 85- *Concurrent Pos:* Chmn ment health res comt, Health Prog Systs Ctr, Indian Health Serv, Tucson, Ariz, 70-73; consult, Psychiat Educ Br, Div Manpower & Training Progs, NIMH, 74-80, Am Indian Ment Health Res & Develop Ctr, Denver, Colo, 85-; mem dirs adv bd, Ore Ment Health Div, Salem, 75 & Ore Bd Med Examiners, 78-85; Found Fund researcher psychiat, 79; dir, Am bd psychiat, Neurol, 87-, chair psychiat residency rev comt, 89- *Honors & Awards:* Commendation Medal, Dept Health, Educ & Welfare, 72. *Mem:* Fel Am Psychiat Asn; Am Asn Chmn Dept Psychiat; fel Am Col Psychiatrists. *Res:* Psychiatric epidemiology; psychiatric education; suicidology; civil commitment; transcultural psychiatry; impaired physicians; stress disorders. *Mailing Add:* Health Sci Ctr Univ Colo 4200 E 9th Denver CO 80262

SHORE, JOHN EDWARD, b Slough, Gt Brit, Sept 2, 46; US citizen; m 69; c 1. INFORMATION THEORY, SPEECH PROCESSING. *Educ:* Yale Univ, BS, 68; Univ Md, PhD(theoret physics), 74. *Prof Exp:* Res Scientist Physics & Comput Sci, Naval Res Lab, 68-; AT DEPT ENG, GEORGE WASHINGTON UNIV. *Concurrent Pos:* Res publ award, Naval Res Lab, 71, 76 & 78; prof lectr comput sci, George Washington Univ, 78-; adj lectr elec engr, Univ Md, 80- *Mem:* Asn Comput Mach; Inst Elec & Electronics Engrs; Am Phys Soc. *Res:* Information theory, especially the foundations and applications of maximum entropy and related techniques; software engineering; programming language design; speech processing. *Mailing Add:* 906E Capital St NE Washington DC 20002

SHORE, JOSEPH D, b New York, NY, Apr 2, 34; m 68; c 2. ENZYMOLOGY, PHYSICAL BIOCHEMISTRY. *Educ:* Cornell Univ, BS, 55; Univ Mass, MS, 57; Rutgers Univ, PhD(biochem), 63. *Prof Exp:* Muscular Dystrophy Asn fel, Nobel Med Inst, Stockholm, Sweden, 64-66; sr staff investr dept biochem, 66-79, HEAD DIV BIOCHEM RES, HENRY FORD HOSP, 79-, DIR RES, 82- *Concurrent Pos:* Adj prof, Med Sch, Wayne State Univ; Am Heart Asn Coun Thrombosis. *Mem:* Am Chem Soc; Am Soc Biol Chem & Molecular Biol; Biophys Soc; Nat Coun Univ Res Admin. *Res:* Blood coagulation; transient kinetics and fluorescence techniques. *Mailing Add:* Dept Biochem Henry Ford Hosp 2799 W Grand Blvd Detroit MI 48202

SHORE, LAURENCE STUART, b Philadelphia, Pa, Mar 16, 44; m 67; c 4. ENDOCRINOLOGY. *Educ:* Yeshiva Univ, BA, 65; Hahnemann Med Sch, MA, 69, PhD(physiol), 72. *Prof Exp:* Postdoctoral, Med Sch, Wash Univ, 72-73; instr physiol, Dept Obstet-Gynec & Physiol, Temple Univ, Philadelphia, 73-76; INVESTR, DEPT HORMONE RES, KIMRON VET INST, BET DAGAN, 78- *Concurrent Pos:* Vis scientist, Dept Obstet-Gynec, Hosp Univ Pa, 84-85. *Mem:* Am Physiol Soc; Israel Endocrine Soc. *Res:* Hormones present in food and water and their effects on animals; hyperestrogenism and premature puberty in cattle; prolaped oviduct and salpingitis in poultry. *Mailing Add:* Dept Hormone Res Kimron Vet Inst PO Box 12 Bet Dagon 50200 Israel

SHORE, MILES FREDERICK, b Chicago, Ill, May 26, 29; m 53; c 3. PSYCHIATRY. *Educ:* Univ Chicago, AB, 48; Harvard Univ, BA, 50, MD, 54; Am Bd Psychiat & Neurol, dipl, 60. *Prof Exp:* Intern, Univ Ill Res & Educ Hosp, 55; resident psychiat, Mass Ment Health Ctr, 55-56 & Beth Israel Hosp, Boston, 59-61; instr, Harvard Med Sch, 64-65; from asst prof to prof psychiat, Sch Med, Tufts Univ, 71-75, assoc dean community affairs, 72-75, dir ment health ctr, 68-75; BULLARD PROF PSYCHIAT, HARVARD MED SCH, 75-; SR PROG CONSULT & DIR, PROG FOR CHRONICALLY MENT ILL, ROBERT WOOD JOHNSON FOUND, 85- *Concurrent Pos:* Chmn bd trustees, Boston Psychoanal Soc & Inst, 70-73; dir community & ambulatory med, New Eng Med Ctr Hosp, 72-75; supt & area dir, Mass Ment Health Ctr, 75- *Honors & Awards:* Admn Psychiat Award, Am Psychiat Asn. *Mem:* Fel Am Psychiat Asn; Group Advan Psychiat; Am Col Psychiatrists. *Res:* Community psychiatry; psychoanalysis; psychohistory. *Mailing Add:* 62 Meadowbrook Rd Needham MA 02192

SHORE, MORIS LAWRENCE, b Russia, Dec 7, 27; US citizen; m 58; c 3. METABOLIC KINETICS, RADIOBIOLOGY. *Educ:* Southwestern Univ, Memphis, BA, 50; Univ Tenn, PhD(physiol), 54. *Prof Exp:* Staff mem Biophysics Br, US Navy Radiol Defense Lab, US Navy Med Serv Corps, 54-61; asst prof physiol, Marquette Univ, 61-62; chief Biophysics & asst officer in command, Res Br Lab, 62-67, chief, Physiol & Biophysics Lab, 67-69, chief, Exp Studies Br, 69-70, DIR, DIV BIOL EFFECTS, BUR RADIOL HEALTH, DEPT HEALTH & HUMAN SERV, FOOD & DRUG ADMIN, 70- *Concurrent Pos:* Chief, Physiol Sect, Res Serv, Wood Vet Admin Ctr, 61-62. *Mem:* Radiation Res Soc; Health Physics Soc; Res Soc; NY Acad Sci. *Res:* Tracer kinetics; phospholipid metabolism; experimental atherosclerosis; reticulo-endothelial system function; effects of ionizing and nonionizing radiation; development of regulatory and voluntary health protection standards and guidelines. *Mailing Add:* 12411 Kemp Mill Rd Silver Spring MD 20902

SHORE, NOMIE ABRAHAM, b Chicago, Ill, Oct 2, 23; div; c 2. PEDIATRICS, HEMATOLOGY. *Educ:* Univ Calif, Los Angeles, BA, 47; Univ Southern Calif, MD, 53. *Prof Exp:* Intern med, Los Angeles County Gen Hosp, 52-53; resident pediat, Children's Hosp Los Angeles, 53-55, fel hemat, 55-56; asst clin prof pediat, Univ Calif, Los Angeles, 57-60; asst prof, 61-68, ASSOC PROF PEDIAT, SCH MED, UNIV SOUTHERN CALIF, 68-; ASSOC HEMATOLOGIST, CHILDREN'S HOSP, LOS ANGELES, 61- *Concurrent Pos:* Consult pediat hemat, St John's Hosp, Santa Monica, 61-; mem hon staff, Santa Monica Hosp, 61- *Mem:* AMA; Am Soc Hemat; Am Acad Pediat; Am Asn Cancer Res. *Res:* Evaluating the effects of chemotherapeutic agents in treatment of leukemia and other neoplastic diseases; erythropoietin physiology; bone marrow stem cell kinetics. *Mailing Add:* Children's Hosp of Los Angeles PO Box 54700 Los Angeles CA 90054

SHORE, RICHARD A, b Boston, Mass, Aug 18, 46; m 69; c 2. RECURSION THEORY. *Educ:* Hebrew Col, BJEd, 66; Harvard Univ, AB, 68; Mass Inst Technol, PhD(math), 72. *Prof Exp:* Instr, Univ Chicago, 72-74; from asst prof to assoc prof, 74-83, PROF MATH, CORNELL UNIV, 83- *Concurrent Pos:* Asst prof, Univ Ill, 77; vis assoc prof, Univ Conn, Storrs, 79 & Mass Inst Technol, 80; vis prof, Hebrew Univ, Jerusalem, 82-83; ed, J Symbolic Logic, 84- *Mem:* Am Math Soc; Asn Symbolic Logic. *Res:* Computability theory: degrees of difficulty of computability, recursively enumerable sets and degrees; generalizations of recursion theory and applications and effective mathematics. *Mailing Add:* Dept Math White Hall Cornell Univ Ithaca NY 14853

SHORE, ROY E, b Mina, NY, Oct 30, 40; m 62; c 6. ENVIRONMENTAL EPIDEMIOLOGY, RADIATION EPIDEMIOLOGY. *Educ:* Houghton Col, BA, 62; Syracuse Univ, MA, 66, PhD(psychol), 67. *Hon Degrees:* Dr, Columbia Univ, 82. *Prof Exp:* fel psychol, Educ Testing Serv, Princeton, 67-69; res scientist environ med, 69-76, from asst prof to assoc prof, 76-87, PROF ENVIRON MED, NY UNIV MED CTR, 87- *Concurrent Pos:* Comt on fed res on health effects of ionizing radiation, Nat Acad Sci, 80-81; adv comt health & environ res, US Dept Energy, 83-86; bd sci counr, Div Cancer Etiol, Nat Cancer Inst, 84- *Mem:* Fel Am Col Epidemiol; Nat Coun Radiation & Measurements; Soc Epidemiol Res; Biometric Soc. *Res:* Cancer in relation to ionizing radiation exposures, diet, occupational exposures and biomarkers. *Mailing Add:* Dept Environ Med NY Univ Med Ctr 341 E 25th St Rm 208 New York NY 10010-2598

SHORE, SAMUEL DAVID, b Lewistown, Pa, Nov 9, 37; m 64; c 2. MATHEMATICS. *Educ:* Juniata Col, BS, 59; Pa State Univ, MA, 61, PhD(gen topology), 64. *Prof Exp:* Instr math, Pa State Univ, 64-65; asst prof, 65-70, ASSOC PROF MATH, UNIV NH, 70- *Mem:* Am Math Soc; Math Asn Am. *Res:* Spaces of continuous functions; compactifications and extensions; ordered spaces. *Mailing Add:* Dept of Math Univ of NH Durham NH 03824

SHORE, SHELDON GERALD, b Chicago, Ill, May 8, 30. INORGANIC CHEMISTRY. *Educ:* Univ Ill, BS, 51; Univ Mich, MS, 54, PhD(chem), 57. *Prof Exp:* Instr chem, Univ Mich, 56-57; from asst prof to assoc prof, 57-62, PROF CHEM, OHIO STATE UNIV, 65- *Honors & Awards:* Reilley Lectr, Inorg Chem, Univ Notre Dame, 82; Morley Award & Medal, Cleveland Sect, Am Chem Soc, 89, Columbus Sect Award, 90. *Mem:* Am Chem Soc; Royal Soc Chem. *Res:* Synthesis and study of transition metal and non-metal cluster systems: polynuclear metal carbonyl hydrides, metallaboranes, and boron hydrides. *Mailing Add:* Dept Chem Ohio State Univ Columbus OH 43210

SHORE, STEVEN NEIL, b New York, NY, July, 16, 53; m 74. THEORETICAL ASTROPHYSICS. *Educ:* State Univ NY, Stony Brook, MSc, 74; Univ Toronto, PhD(astron), 78. *Prof Exp:* Res assoc astron, Columbia Univ, 78-79, lectr, 79; assoc prof & dir, Astrophysics Res Ctr, 85-89; ASST PROF ASTRON, CASE WESTERN RESERVE UNIV, 79-; ASTROPHYSICIST, COMPUTER SCI CORP, GODDARD SPACE FLIGHT CTR, 89- *Concurrent Pos:* Shapley lectr astron, Am Astron Soc, 80-; vis asst prof astron, Ohio State Univ, 81; vis prof, Ecloe Normale Superiure. *Mem:* Am Astron Soc; Sigma Xi; Hist Sci Soc; Int Astron Union. *Res:* Magnetic fields in stellar structure and atmospheres; chemical evolution of the galaxy; ultraviolet spectroscopy; history of science. *Mailing Add:* Code 681 Goddard Space Flight Ctr Greenbelt MD 20771

SHORE, VIRGIE GUINN, b Lavaca, Ark, Oct 20, 28; m 52. BIOCHEMISTRY, PHYSIOLOGY. *Educ:* Univ Calif, AB, 50, PhD(biochem), 55. *Prof Exp:* Res asst physiol, Sch Med, Wash Univ, 57-58, from res instr to res asst prof, 58-61, asst prof, 61-63; BIOCHEMIST, BIO-MED DIV, LAWRENCE LIVERMORE LAB, UNIV CALIF, 63- *Concurrent Pos:* Mem, Metab Study Sect, NIH & exec comt coun on arteriosclerosis, Am Heart Asn. *Mem:* Am Physiol Soc. *Res:* Resonance energy transfer; structure of lipoproteins and membranes. *Mailing Add:* Bio-Med Div Univ Calif Lawrence Livermore Lab Livermore CA 94550

SHORER, PHILIP, theoretical atomic physics; deceased, see previous edition for last biography

SHORES, DAVID ARTHUR, b Towanda, Pa, Jan 10, 41; m 63; c 3. METALLURGICAL CHEMISTRY. *Educ:* Pa State Univ, BS, 62, MS, 64, PhD(mat sci), 67. *Prof Exp:* Fel metall, Ohio State Univ, 68-70; metallurgist, Large Steam Turbine Generator Div, 70-74, Metallurgist Hot Corroston, Corp Res & Develop Ctr, Gen Elec Co, 74-82; ASSOC PROF, DEPT CHEM ENG & MAT SCI, UNIV MINN, 82- *Mem:* Electrochem Soc; Am Ceramics Soc. *Res:* High temperature oxidation; thermochemistry and electrochemistry of molten salts; high temperature fuel cells. *Mailing Add:* Univ Minn 221 Church St SE Minneapolis MN 55455

SHORES, THOMAS STEPHEN, b Kansas City, Kans, May 28, 42; m 68. MATHEMATICS. *Educ:* Univ Kans, BA, 64, MA, 65, PhD(math), 68. *Prof Exp:* Assoc prof, 68-77, dept vchmn, 76-79, PROF MATH, UNIV NEBR, LINCOLN, 77-, ACTG CHMN, 81- *Mem:* Am Math Soc; Math Asn Am. *Res:* Generalized solvable and nilpotent groups; structure theory for modules and commutative ring theory. *Mailing Add:* Dept Math Univ Nebr Lincoln NE 68588

SHOREY-KUTSCHKE, ROSE-ANN, NUTRITION. *Prof Exp:* DIV HEAD NUTRIT & PROF NUTRIT, UNIV TEX, 71- *Mailing Add:* Dept Grad Nutrit Univ Tex GEA 115 Austin TX 78712

SHORR, BERNARD, b New York, NY, July 5, 28; m 58. OPERATIONS RESEARCH. *Educ:* City Col NY, BA, 50; NY Univ, MS, 51, PhD, 70. *Prof Exp:* Meteorologist, Gen Elec Co, Wash, 51-55; staff mem div sponsored res, Mass Inst Technol, 56-58; res assoc opers res, 58-63, asst dir res, 63-66, assoc dir, 66-70, SECOND VPRES, CORP RES DIV, TRAVELERS INS CO, 70- *Concurrent Pos:* Lectr, Univ Conn, 70-77. *Mem:* Fel AAAS; Opers Res Soc Am; Inst Mgt Sci. *Res:* Management; financial and economic analysis. *Mailing Add:* PO Box 1024 Avon CT 06001

SHORT, BYRON ELLIOTT, b Putnam, Tex, Dec 29, 01; m 37; c 2. MECHANICAL ENGINEERING. *Educ:* Univ Tex, BS, 26, MS, 30; Cornell Univ, MME, 36, PhD(mech eng), 39. *Prof Exp:* From instr to prof mech eng, 26-73, chmn dept, 45-47, 51-53, actg dean, 48-49, in charge heat eng lab, 30-64, EMER PROF MECH ENG, UNIV TEX, AUSTIN, 73- *Concurrent Pos:* Consult, Defense Res Lab, Tex, 45-63, Tex Gulf Sulphur Co, 49-54, Oak Ridge Nat Lab, 56-59, Atomics Int, 61-63 & US Army Eng Corp, 64-68. *Mem:* Nat Soc Prof Engrs; Am Soc Mech Engrs; Am Soc Eng Educ; Am Soc Heating, Refrig & Air-Conditioning Engrs. *Res:* Fluid flow; refrigeration; heat transfer. *Mailing Add:* 502 E 32nd St Austin TX 78705-3105

SHORT, CHARLES ROBERT, b Rochester, NY, Nov 7, 38; m 64; c 2. PHARMACOLOGY. *Educ:* Ohio State Univ, DVM, 63, MS, 65; Univ Mo-Columbia, PhD, 69. *Prof Exp:* From instr to assoc prof pharmacol, Sch Med, Univ Mo-Columbia, 65-75; PROF VET PHARMACOL & TOXICOL, SCH VET MED, LA STATE UNIV, BATON ROUGE, 75-, VET DIAG TOXICOLOGIST, 76- *Concurrent Pos:* Spec fel med educ, Univ Southern Calif, 74. *Mem:* AAAS; NY Acad Sci; Am Soc Pharmacol & Exp Therapeutics; Am Col Vet Toxicologists. *Res:* Drug disposition pharmacology in the fetus and neonate; chemical carcinogenesis. *Mailing Add:* Dept Vet Physiol 2436 Vet Med La State Univ Baton Rouge LA 70803

SHORT, DONALD RAY, JR, b Camp McCoy, Wis, Sept 13, 44; m 78; c 2. MATHEMATICS. *Educ:* Univ Calif, Los Angeles, BA, 65; Ore State Univ, PhD(math), 69. *Prof Exp:* From asst prof to assoc prof, 69-75, PROF MATH & DEAN, COL SCI, SAN DIEGO STATE UNIV, 75- *Mem:* Sigma Xi; Am Math Soc; Soc Indust & Appl Math. *Res:* Algebraic topology; cohomology theory; sheaf theory; spectral sequences; branched immersions. *Mailing Add:* Col Sci San Diego State Univ San Diego CA 92182

SHORT, EVERETT C, JR, b Monett, Mo, Dec 27, 31; c 3. BIOCHEMISTRY. *Educ:* Kent State Univ, BS, 58; Colo State Univ, DVM, 62; Univ Minn, PhD(biochem), 68. *Prof Exp:* From instr to assoc prof, Col Vet Med, Univ Minn, St Paul, 64-73, prof biochem & assoc dean, 73-78; PROF & HEAD PHYSIOL SCI, COL VET MED, OKLA STATE UNIV, STILLWATER, 80- *Mem:* Am Soc Microbiol; Am Vet Med Asn; Soc Exp Biol Med; Soc Environ Toxicol & Chem. *Res:* Effects of toxic substances on aquatic organisms; enteric diseases of baby pigs. *Mailing Add:* Dept Physiol Scis Okla State Univ Stillwater OK 74078

SHORT, FRANKLIN WILLARD, b Charleston, WVa, Feb 24, 28; m 57; c 2. MEDICINAL CHEMISTRY, CLINICAL DRUG DEVELOPMENT. *Educ:* Univ Buffalo, BA, 48; Columbia Univ, PhD(org chem), 52. *Prof Exp:* Jr chemist, Nat Aniline Div, Allied Chem Corp, 47-48; from assoc res chemist to sr res chemist, Parke, Davis & Co, 52-67, assoc lab dir orgn chem, 67-70, sect dir chem dept, 70-78, clin scientist, clin res dept, Warner-Lambert-Parke-Davis Pharmaceut Res Div, 78-86; SR SCI ADV, DAVCO MFG CORP, 86- *Mem:* Am Chem Soc. *Res:* Synthesis of antiparasitic, antiinflammatory, analgetic, antifungal, antibacterial pulmonary/allergy and gastrointestinal agents; clinical development of antiinflammatory and antibacterial agents; development of water purification and automotive products. *Mailing Add:* 3431 Clover Dr Saline MI 48176-9539

SHORT, HENRY LAUGHTON, b Penn Yan, NY, Apr 6, 34; m 62; c 2. ECOLOGY, WILDLIFE BIOLOGY. *Educ:* Swarthmore Col, BA, 56; Johns Hopkins Univ, MS, 59; Mich State Univ, PhD(fisheries, wildlife), 62. *Prof Exp:* Res asst vert ecol, Johns Hopkins Univ, 57-58; res asst fisheries & wildlife, Mich State Univ, 59-61; asst prof & mem grad fac forest recreation & wildlife, Colo State Univ, 61-63, wildlife nutritionist, Colo Coop Wildlife Res Unit, 61-63; wildlife biologist, Southern Forest Exp Sta, Forest Serv, USDA, 64-73, wildlife biologist, Rocky Mountain Forest & Range Exp Sta, 73-77; TERRESTRIAL ECOLOGIST, WESTERN ENERGY & LAND USE TEAM, FISH & WILDLIFE SERV, US DEPT INTERIOR, 77- *Concurrent Pos:* Cooperator, Int Biol Prog, 65; mem, Grad Fac Forestry, Stephen F Austin State Univ, 64-73 & Wildlife Sci, Tex A&M Univ, 69-73. *Mem:* Ecol Soc Am; Wildlife Soc; Am Soc Mammal; Am Soc Animal Sci. *Res:* Ecology and life history of migratory bats; anatomy, digestive physiology and nutrition of deer; determination of physiology and nutrition of deer; determination of physiological requirements of wild animals; forage quality for wild animals; predicted and determined quality of wildlife habitat. *Mailing Add:* 7909 Yancey Dr Falls Church VA 22042

SHORT, JAMES HAROLD, b Leavenworth, Kans, July 9, 28; m 73; c 1. PHARMACOLOGY. *Educ:* Stanford Univ, BS, 50; Univ Kans, PhD(pharmaceut chem), 54. *Prof Exp:* Res chemist med chem, Abbott Labs, 56-71; res assoc, 72, asst prof pharmacol, Univ Louisville, 73-76; supvr chem, Adria Labs, 76-84; REV CHEMIST, FOOD & DRUG ADMIN, 84- *Mem:* Am Chem Soc. *Mailing Add:* 19117 Rhodes Way Gaithersburg MD 20879-2153

SHORT, JAMES N, b Dayton, Ohio, Nov 14, 22; m 45; c 4. POLYMER CHEMISTRY, RESEARCH ADMINISTRATION. *Educ:* Univ Cincinnati, BChE, 45, MS, 47, ScD, 49. *Prof Exp:* Res chemist, Warren-Teed Labs, Ohio, 49-51; res chemist, Phillips Petrol Co, 51-55, mgr solution polymerization sect, 55-59, rubber synthesis br, 59-66, rubber & carbon black processes br, 66-69 & chem processes br, 69-72, mgr plastics develop br, Phillips Petrol Co, 72-83; RETIRED. *Mem:* Am Chem Soc; Soc Plastics Eng. *Res:* Stereospecific polymerization; polyolefins; synthetic rubber engineering plastics; fibers; technical direction; polymers. *Mailing Add:* 2360 SE Windsor Way Bartlesville OK 74006

SHORT, JOHN ALBERT, b Pittsburgh, Pa, Feb 2, 36; m 62; c 2. CELL BIOLOGY, HISTOLOGY. *Educ:* Univ Pittsburgh, BS, 57, PhD(anat & cell biol), 72. *Prof Exp:* Res asst, Univ Pittsburgh, 63-67, res assoc, 71-75, res asst prof, 75-76, asst prof anat & histol, 76-82, res assoc prof, 82-86, from assoc prof to prof, 86-91; assoc prof, Slippery Rock Univ, Pa, 87-89; ASSOC PROF ANAT, DUQUESNE UNIV, 91- *Concurrent Pos:* Res biochemist, Vet Admin Med Ctr, 71-80, co-investr, 75-80. *Mem:* Fedn Am Scientists; Am Asn Anatomists; Am Soc Cell Biol; NY Acad Sci. *Res:* Control of mammalian cell proliferation in vivo, including the effects of the thyroid hormones and glucocorticoids on this process; science education. *Mailing Add:* 5753 Phillips Rd Gibsonia PA 15044

SHORT, JOHN LAWSON, b Dorking, UK, May 22, 46; m 75; c 2. MEMBRANE TECHNOLOGY, DAIRY PROCESSING. *Educ:* Univ Manchester Inst Sci & Technol, BSc, 67; Univ London, MSc, 71; Univ Col, London, dipl (biochem eng), 71. *Prof Exp:* Biotechnologist, Chem Indust Basel Geigy AG, Switz, 71-75; sect head, NZ Dairy Res Inst, 75-79; sr engr, Corning Glass Works, NY, 79-80; int mkt mgr, Romicon-Rohm & Haas, Mass, 80-85; mgr process technol, 85-87, DIR COM DEVELOP, KOCH MEMBRANE SYSTS, MASS, 87- *Concurrent Pos:* Mobil Environ grant, Mobil Oil NZ Ltd, 78. *Mem:* Inst Food Technol; Am Inst Chem Engrs; Inst Chem Engrs UK. *Res:* Application of membrane separation technology (ultrafiltration, reverse osmosis, electrodialysis) to food, pharmaceutical, industrial and other process streams. *Mailing Add:* 850 Main St Wilmington MA 01887-3388

SHORT, LESTER LE ROY, b Port Chester, NY, May 29, 33; m 55, 78; c 2. ORNITHOLOGY. *Educ:* Cornell Univ, BS, 55, PhD(vert zool), 59. *Prof Exp:* Asst vert zool, Cornell Univ, 54-59; instr biol, Adelphi Univ, 60-62, asst prof, 62; Chapman fel, Am Mus Natural Hist, 62-63; chief bird sect, Bird & Mammal Labs, US Fish & Wildlife Serv, 63-66; assoc cur, 66-68, chmn, 80-87, LAMONT CUR BIRDS, AM MUS NATURAL HIST, 68-, CHMN, ORNITH DEPT, 80- *Concurrent Pos:* Hon cur NAm birds, Smithsonian Inst, 63-66; adj prof, City Univ NY, 70- *Mem:* Soc Study Evolution; Am Ornith Union; Cooper Ornith Soc; Soc Syst Zool; Royal Australasian Ornith Union. *Res:* Systematic and evolutionary zoology; speciation; hybridization; taxonomy and classification of birds; avian ethology and ecology; zoogeography; ornithology. *Mailing Add:* Dept Ornith Am Mus Natural Hist New York NY 10024

SHORT, MICHAEL ARTHUR, b London, Eng, Aug 15, 30; US citizen; m 57; c 4. X-RAY ANALYSIS, X-RAY PHYSICS. *Educ:* Univ Bristol, Eng, BSc, 52, MSc, 57; Pa State Univ, PhD(x-ray diffraction), 61. *Prof Exp:* Scientist, Gen Elec Co, Eng, 54-57; mem tech staff, Bell Telephone Lab, NJ, 61-64; engr, Assoc Elec Indust, 64-67; staff scientist, Ford Motor Co, Dearborn, Mich, 67-80; res assoc, Occidental Res Corp, 80-85; SR STAFF SCIENTIST, SRS TECHNOL, CALIF, 85- *Mem:* Am Crystallog Asn; Am Phys Soc; Am Chem Soc; Am Soc Metals; Microbeam Analytical Soc. *Res:* Development of improved instrumentation and analytical techniques for X-ray diffraction, X-ray fluorescence, X-ray scattering, and electron microprobe analysis. *Mailing Add:* 24832 Weyburn Dr Laguna Hills CA 92653-4310

SHORT, NICHOLAS MARTIN, b St Louis, Mo, July 18, 27; m 61; c 1. GEOLOGY. *Educ:* St Louis Univ, BS, 51; Wash Univ, MA, 54; Mass Inst Technol, PhD(geol), 58. *Prof Exp:* Instr geol, Univ Mo, 54-55; geologist, Gulf Res & Develop Co, Pa, 57-59; geologist-physicist, Lawrence Radiation Lab, Univ Calif, 59-64; from asst prof to assoc prof geol, Univ Houston, 64-67; Nat Acad Sci res assoc, Planetology Br, NASA, 67-69, res geologist, Earth Resources Prog, 69-77, dir training, Regional Appln Prog, 77-81, res scientist, Geophys Br, Goddard Space Flight Ctr, 81-88; PROF GEOL & GEOG, BLOOMSBURG STATE UNIV, 88- *Mem:* Fel Geol Soc Am. *Res:* Geochemistry; astrogeology; shock effects in meteorite craters and underground nuclear explosion sites; remote sensing. *Mailing Add:* 1105 Cherry Hill Rd Bloomsburg PA 17815

SHORT, ROBERT ALLEN, b Dayton, Wash, Nov 7, 27; m 49; c 7. ELECTRICAL ENGINEERING, COMPUTER SCIENCE. *Educ:* Ore State Univ, BS, 49, BA, 52; Stevens Inst Technol, MS, 56; Stanford Univ, PhD(elec eng), 61. *Prof Exp:* Mem tech staff, Bell Tel Labs, 52-56; sr res engr, Stanford Res Inst, 56-66; chmn dept comput sci, 72-78, PROF ELEC ENG & COMPUT SCI, ORE STATE UNIV, 66- *Concurrent Pos:* Lectr, Santa Clara Univ, 65-66; mem gov bd, Inst Elec & Electronics Comput Soc, 70-72, 74-76, ed-in-chief, 71-75, spec tech ed, 77-78; ed, Trans on Comput, Inst Elec & Electronic Engrs, 71-75. *Honors & Awards:* Computer Soc Spec Awards, Inst Elec & Electronic Engrs, 76 & 79. *Mem:* AAAS; Inst Elec & Electronic Engrs; Asn Comput Mach; Am Soc Cybernetics. *Res:* Information systems, computer science, logic design, switching theory, automata, fault-tolerant computing, coding theory, teaching effectiveness. *Mailing Add:* Dept Elec Eng Ore State Univ Corvallis OR 97331

SHORT, ROBERT BROWN, b Changsha, China, Feb 28, 20; US citizen; m 47; c 3. PARASITOLOGY. *Educ:* Maryville Col, BA, 41; Univ Va, MS, 45; Univ Mich, PhD(zool), 50. *Prof Exp:* Instr math, Sewanee Mil Acad, 41-43; instr math & biol, Va Episcopal Sch, 43-44; from asst prof to assoc prof biol sci, 50-57, PROF BIOL SCI, FLA STATE UNIV, 57- *Concurrent Pos:* Mem study sect trop med & parasitol, NIH, 65-69; NIH spec res fel, Tulane Univ, La, 70-71. *Mem:* Am Soc Parasitol (vpres, 77, pres, 82); Am Soc Trop Med & Hyg; AAAS; Am Micros Soc. *Res:* Biology and cytogenetics of schistosomes and other trematodes. *Mailing Add:* Dept of Biol Sci Fla State Univ Tallahassee FL 32306

SHORT, ROLLAND WILLIAM PHILLIP, carbohydrate chemistry; deceased, see previous edition for last biography

SHORT, SARAH HARVEY, b Little Falls, NY, Sept 22, 24; m 46; c 3. SPORTS NUTRITION, NUTRITION & BIOCHEMISTRY. *Educ:* Syracuse Univ, BS, 46, PhD(nutrit), 70, EdD(instrnl technol), 75; State Univ NY, Upstate Med Ctr, MS, 66. *Prof Exp:* Researcher chem, Bristol Labs, 46-50; asst prof chem, State Univ NY, Upstate Med Ctr, 63-80; PROF NUTRIT, SYRACUSE UNIV, 66- *Concurrent Pos:* Bd adv, Am Coun Sci & Health, 78-; Int Cong Individualized Instr, 75-76; mem bd sci counr, USDA. *Mem:* Am Dietetic Asn; Soc Nutrit Educ; Asn Educ Commun & Technol; Inst Food Technologists; Am Col Sports Med. *Res:* Nutrition education at university and medical school including developing and evaluating self instruction units using computer assisted instruction, audiovisual media and rate controlled speech; evaluation of trained athletes' diet using computer analysis; TV/radio nutrition communications. *Mailing Add:* Dept Nutrit & Food Mgt Syracuse Univ Syracuse NY 13244-1250

SHORT, TED H, b Wauseon, Ohio, Mar 13, 42; m 68; c 2. AGRICULTURAL ENGINEERING, HORTICULTURAL ENGINEERING. *Educ:* Ohio State Univ, BS, 65, MS, PhD(agr eng), 69. *Prof Exp:* Res assoc, Ohio State Univ, 65-69; from asst prof to assoc prof, 69-82, PROF AGR ENGR, OHIO AGR RES & DEVELOP CTR, 82- *Concurrent Pos:* Consult, TVA Waste Heat Prog, 75-76 & Bechtel Corp Waste Heat Study, 78; Greenhouse res exchange, Holland, 81; host, Int Greenhouse Energy Symp, 83; solar greenhouse proj, Food & Agr Org UN Develop Prog, Antalya, Turkey, 85. *Honors & Awards:* Concept of the Year, Am Soc Agr Engrs, 77. *Mem:* Sigma Xi; Solar Energy Soc; Int Soc Hort Sci; Am Soc Agr Engrs. *Res:* Energy conservation for greenhouses; mechanization of greenhouse growing systems; solar ponds for heating greenhouses and rural residences; hydroponic greenhouse production systems; computer climate control for greenhouses; passive solar greenhouse heating systems. *Mailing Add:* Dept Agr Eng Ohio Agr Res & Develop Ctr Wooster OH 44691

SHORT, W(ILLIAM) LEIGH, b Calgary, Alta, Jan 30, 35; c 5. CHEMICAL & ENVIRONMENTAL ENGINEERING. *Educ:* Univ Alta, BSc, 56, MSc, 57; Univ Mich, PhD(chem eng), 62. *Prof Exp:* Proj engr, Edmonton Works, Can Industs Ltd, 57-59; res engr, Chevron Res Co, Calif, 62-67; from asst prof to prof chem eng, Univ Mass, Amherst, 67-79, assoc head dept, 69-76, head dept, 76-79; mgr, Houston Eng Div, 79-80, vpres & dir, Environ Eng Div, Environ Res & Technol Inc, 80-85; sr prog mgr, Radian Corp, Herdon, VA, 85-87; VPRES, WOODWARD CLYDE CONSULTS, WAYNE, NJ, 87- *Concurrent Pos:* Pub Health Serv res grant & Co Dir Air Pollution training grant, 69-; consult, Kenics Corp, M W Kellogg & Arthur D Little, 69-, Environ Protection Agency, 70- & Gen Acct Off, 78; mem sci adv bd, Environ Protection Agency, 76-80; mem, SBIR, RFA, Res Grants Rev Panel, 80- *Mem:* Am Inst Chem Engrs; Am Chem Soc; Air Pollution Control Asn; AAAS. *Res:* Chemical engineering applications in air and water pollution control; thermodynamics; hazardous waste treatment technologies. *Mailing Add:* Two Johnson Dr Convent Station NJ 07961

SHORT, WALLACE W(ALTER), b Ogdensburg, NY, May 21, 30; m 58; c 4. FLUID DYNAMICS, HIGH ENERGY LASERS. *Educ:* Mo Sch Mines, BS, 51; Calif Inst Technol, MS, 53, PhD(chem & elec eng), 58. *Prof Exp:* Prod supvr, Merck & Co, Inc, 51-52; res engr, Rocketdyne Div, N Am Aviation, Inc, 53; from res scientist to staff scientist, Convair Div, Gen Dynamics Corp, 58-62; mem tech staff, Gen Res Corp, 62-74; pres, Appl Technol Assocs, 75-90. *Concurrent Pos:* Consult, Inst Defense Analysis, Washington, DC, 64-70; bd dirs, NMex Mus Nat Hist, 88- *Res:* Theoretical and experimental heat and mass transfer, especially evaporation, ablation and boundary layer flow; rocket propulsion, especially chemical and electrical devices; reentry phenomena; hypersonic wakes; radar scattering by plasmas; ballistic missile defense; high energy laser controls. *Mailing Add:* 8716 Rio Grande Blvd NW Albuquerque NM 87114

SHORT, WILLIAM ARTHUR, b West Chester, Pa, Feb 18, 25; m 50; c 2. ORGANIC CHEMISTRY, BIOCHEMISTRY. *Educ:* Furman Univ, BS, 50; Univ SC, MS, 52; Univ Ala, MS, 57, PhD(biochem), 61. *Prof Exp:* Res asst org chem & biochem, Southern Res Inst, 52-61; prof chem & chmn div natural sci & math, 61-77, PROF CHEM & CHMN DEPT, ATHENS STATE COL, ALA, 77- *Mem:* Am Chem Soc; fel Am Inst Chemists. *Res:* Carbohydrate chemistry. *Mailing Add:* Dept Chem Athens State Col Athens AL 35611-3589

SHORTELL, STEPHEN M, b New London, Wis, Nov 9, 44. HOSPITAL PHYSICIAN RELATIONSHIPS. *Educ:* Univ Notre Dame, BBA, 66; Univ Calif, Los Angeles, MPH, 68; Univ Chicago, MBA, 70, PhD(behav sci), 72. *Prof Exp:* Res asst, Nat Opinion Res Ctr, 69, instr/res assoc, Ctr Health Admin Studies, 70-72, actg dir, grad prog in Hosp Admin, Univ Chicago, 73-74; from asst prf to assoc prof 74-79, prof, Sch Pub Health & Comm Med, Dept Health Serv, Univ Wash, 79-82; A C BUEHLER DISTINGUISHED PROF, HOSP & HEALTH SERV MGT, NORTHWESTERN UNIV, 82- *Concurrent Pos:* Consult, VA, Robert Wood Found, Henry Kaiser Found; asst prof Health Servs Orgn, Univ Chicago, 72-74; adj asst prof, Dept Sociol, Univ Washington, 75-76, sch Pub Health & Community Med, Dept Health Serv, doctoral prog dir, 76- 78; prof sociol, Dept Sociol, Northwestern Univ, 82; prof community med, Dept Community Health & Preventive Med, Sch Med, Nrthwestern Univ. *Mem:* Inst Med, Nat Acad Sci; Am Pub Health Asn. *Res:* A national study of the quality of care in Intensive Care Units between Jan 1988-Dec 1990; numerous publications in various journals. *Mailing Add:* J L Kellogg Grad Sch Mgt Leverone Hall Northwestern Univ Evanston IL 60208-2007

SHORTER, DANIEL ALBERT, b Goltry, Okla, May 20, 27; m 46; c 4. ENTOMOLOGY, ZOOLOGY. *Educ:* Northwestern Okla State Univ, BS, 49; Okla State Univ, MS, 60, PhD(entom), 66. *Prof Exp:* Instr pub schs, Kans, 49-58; dir admin, 72-75, PROF BIOL, NORTHWESTERN OKLA ST UNIV, 60-72, 75- *Mem:* Am Soc Mammal; Entom Soc Am. *Res:* Syrphidae of Oklahoma; ecology of the beaver. *Mailing Add:* Dept of Biol Northwestern Okla State Univ Alva OK 73717

SHORTER, ROY GERRARD, b London, Eng, Jan 11, 25; US citizen; m 48; c 2. EXPERIMENTAL MEDICINE. *Educ:* Univ London, MB, BS, 48, MD, 52. *Prof Exp:* Fulbright travel award, 58-59; consult physician, Sect Path, 61-66, Sect Surg Res, 66-68 & Sect Tissue & Org Transplantation, 68-71, consult physician exp med, Sect Anat Path, 71-75, PROF PATH, MAYO MED SCH 74-, PROF MED, 75- *Concurrent Pos:* Consult physician, Sect Med Path & Dept Med, Mayo Clin & Found, 75- *Mem:* Am Asn Path & Bact; Am Gastroenterol Asn; Soc Exp Biol & Med; Brit Asn Clin Path; Brit Soc Gastroenterol; Sigma Xi; fel Royal Col Path; fel Am Col Physicians; fel Royal Col Physicians, London. *Res:* Immunology of idiopathic inflammatory bowel disease. *Mailing Add:* Sect Med Path Mayo Clin 200 First St SW Rochester MN 55905

SHORTESS, DAVID KEEN, b Baltimore, Md, July 29, 30; m 49; c 4. GENETICS, PLANT PHYSIOLOGY. *Educ:* Lycoming Col, BA, 52; Pa State Univ, MEd, 59, PhD(genetics), 66. *Prof Exp:* Teacher high sch, Pa, 55-60; asst prof biol, Bloomsburg State Col, 61-63; actg head dept, 67-68, head dept, 68-72, PROF BIOL, N MEX INST MINING & TECHNOL, 66- *Concurrent Pos:* Fulbright-Hays lectr, Univ Jordan, 77-78. *Mem:* AAAS; Am Genetic Asn; Sigma Xi. *Res:* Genetics and physiology of pollination and seed germination. *Mailing Add:* Dept Biol NMex Inst Mining & Technol Socorro NM 87801

SHORTLE, WALTER CHARLES, b Laconia, NH, Apr 26, 45; m 66; c 4. PLANT PATHOLOGY. *Educ:* Univ NH, BS, 68, MS, 70; NC State Univ, PhD(plant path), 74. *Prof Exp:* Res asst plant path, Univ NH, 68-70 & NC State Univ, 70-74; res plant pathologist, 74-90, SUPVRY RES PLANT PATHOLOGIST, NORTHEASTERN FOREST EXP STA, US FOREST SERV, 90- *Concurrent Pos:* Adj prof, Univ NH, 76-88 & Univ Maine, 78-83. *Mem:* Am Phytopath Soc; Am Chem Soc; Int Asn Wood Anatomists. *Res:* Basic research in biochemistry and physiology of diseases which result in decay of wood in living trees and disease defense mechanisms. *Mailing Add:* Northeastern Forest Exp Sta Box 640 Durham NH 03824

SHORTLIFFE, EDWARD HANCE, b Edmonton, Alta, Aug 28, 47; m 70; c 2. MEDICAL COMPUTER SCIENCE. *Educ:* Stanford Univ, PhD(med info sci), 75, MD, 76. *Prof Exp:* Intern, Mass Gen Hosp, 76-77; resident, 77-79, PROF MED COMPUT SCI, STANFORD UNIV MED SCH, 90- *Mem:* Inst Med-Nat Acad Sci; Am Col Physicians; Amer Soc Clin Invest; Amer Col Med Informatics. *Res:* Medical artificial intelligence. *Mailing Add:* Dept Med MSOB X-215 Stanford Med Ctr Stanford CA 94305-5479

SHORTRIDGE, ROBERT GLENN, JR, b Los Angeles, Calif, Aug 19, 45. CHEMICAL KINETICS. *Educ:* Loyola Univ, Los Angeles, BS, 67; Univ Calif, Irvine, PhD(chem), 71. *Prof Exp:* Res chem, Pa State Univ, 71-73; res assoc, Naval Res Lab, 73-75; asst res chemist, Dept Chem & Statewide Air Pollution Res Ctr, Univ Calif, Riverside, 75-77; MEM STAFF, BELL AEROS TEXTRON, BUFFALO, 77- *Concurrent Pos:* Res chemist, Naval Weapons Support Ctr, Crane, IN, 82- *Mem:* Am Chem Soc; Sigma Xi. *Res:* Research and development of military pyrotechnics. *Mailing Add:* RR5 Box 218C Bloomfield IN 47424

SHORTRIDGE, ROBERT WILLIAM, b Newport News, Va, Sept 1, 18; m 47; c 4. SCIENCE COMMUNICATIONS, TECHNOLOGY COMMUNICATIONS. *Educ:* Wabash Col, AB, 38; Ohio State Univ, PhD(org chem), 43. *Prof Exp:* Res chemist, Monsanto Chem Co, Ohio, 43-45; assoc chemist, Midwest Res Inst, 45-48; res chemist, Commercial Solvents Corp, 48-51; sr chemist, Midwest Res Inst, 53-57, head phys chem sect, 57-63, head org chem sect & asst dir chem div, 63-68; dir, Tech Info Ctr, Univ Mo, 68-83; RETIRED. *Mem:* Sigma Xi. *Res:* Scientific and technological information transfer; non-metallic materials technology. *Mailing Add:* 4409 W 78th St Prairie Village KS 66208

SHOSTAK, STANLEY, b Brooklyn, NY, Nov 3, 38; div; c 2. DEVELOPMENTAL BIOLOGY, PRIMATOLOGY. *Educ:* Cornell Univ, BA, 59; Brown Univ, ScM, 61, PhD(biol), 64. *Prof Exp:* NIH fel, Western Reserve Univ, 64-65; asst prof, 65-70, ASSOC PROF BIOL, UNIV PITTSBURGH, 70- *Mem:* Soc Develop Biol; Int Soc Develop Biol. *Res:* Author of one book on developmental biology. *Mailing Add:* Dept Biol Sci Univ Pittsburgh Pittsburgh PA 15260

SHOTLAND, EDWIN, b Rulzheim, Ger, Dec 18, 08; nat US; m 46; c 1. PHYSICS. *Educ:* Univ Munich, BS, 31, MS, 32; Univ Heidelberg, Dr phil nat, 34. *Prof Exp:* Phys engr, Kurman Electronic Co, NY, 41-42, Kompolite Co, 42 & Kurman Electronic Co, NY, 45-46; sr proj anal engr, Chance Vought Aircraft, Inc, Conn, 46-48, Tex, 48-50; sr staff mem & physicist, Johns Hopkins Univ, 50-55, res proj supvr, 55-57, prin staff mem, Appl Physics Lab, 58-86; CONSULT APPL, 86- *Mem:* Am Phys Soc. *Res:* Dynamics of aircraft, missiles and artificial satellites; aeroelasticity of airframes; information theory and communication engineering; missile guidance and radar intelligence; investigation of communication by radio millimeter waves. *Mailing Add:* 418 E Indian Spring Dr Silver Spring MD 20901

SHOTT, LEONARD D, b Twin Falls, Idaho, June 10, 34; m 58; c 3. VETERINARY PATHOLOGY. *Educ:* Colo A&M Col, BS, 57; Colo State Univ, DVM, 59, PhD(vet path), 67; Am Col Vet Pathologists, dipl. *Prof Exp:* Vet, Hawthorne Vet Clin, 59-63; NIH trainee path, Colo State Univ, 63-67; vet pathologist, Hazleton Labs, Inc, 67-71; head dept, 71-81, sr head, 81-84, DIR PATH, SYNTEX RES, 85- *Mem:* Am Col Vet Pathologists; Am Vet Med Asn; Int Acad Path. *Res:* Drug safety evaluation for human and veterinary pharmaceuticals; studies utilizing common laboratory animals and sophisticated laboratory techniques. *Mailing Add:* Syntex Res 3401 Hillview Ave Palo Alto CA 94304

SHOTTAFER, JAMES EDWARD, b Utica, NY, Dec 13, 30; m 53; c 2. MATERIALS SCIENCE, WOOD TECHNOLOGY. *Educ:* State Univ NY Col Forestry, Syracuse Univ, BS, 54, MS, 56; Mich State Univ, PhD(wood sci), 64. *Prof Exp:* Design group leader, Mat Design, United Aircraft Corp,

59-61; res group leader wood, Brunswick Corp, 61-62, res group leader physics, 62-64; assoc prof, 64-70, PROF WOOD TECHNOL, SCH FOREST RESOURCES, UNIV MAINE, ORONO, 70- *Concurrent Pos:* Consult, 64- *Mem:* Soc Wood Sci & Technol; Forest Prod Res Soc; Am Soc Testing & Mat. *Res:* Materials science and technology, especially on wood, adhesives, nonmetallic materials; study of adhesion timber physics, surface phenomena; materials processing technology; operations analysis; research management. *Mailing Add:* Forest Prod Lab Col Forest Resources Univ Maine Orono ME 04473

SHOTTS, ADOLPH CALVERAN, b Rush Springs, Okla, Dec 28, 25; m 54; c 2. ORGANIC CHEMISTRY. *Educ:* Cent State Col, Okla, BS, 50. *Prof Exp:* Teacher & head sci & math, High Sch, NMex, 50-52; from asst res chemist to assoc res chemist, Continental Oil Co, 52-58; res chemist, Petrol Chem, Inc, S8-60; sect supvr anal chem, Cities Serv Res & Develop Co, 60-62; tech staff asst, Columbian Carbon Co, 62-64, asst to dir, Lake Charles Chem Res Ctr, 64-67, admin mgr, Technol & Planning Div, NJ, 67-68, mgr bus serv, Cities Serv Res & Develop Co, 68-74, prod mgr, Petrochem Sales Dept, Cities Serv Oil Co, 74-76 & proj coord, Admin Div, Chem Group, Cities Serv Co, 76-86; RETIRED. *Concurrent Pos:* Mgr environ & safety affairs, Columbian Chem Co, 80-86; environ & safety consult, 86- *Mem:* Am Chem Soc. *Res:* Organic chemical research as applied to petrochemicals; hydrocarbon oxidations. *Mailing Add:* 801 W Knollwood Broken Arrow OK 74011

SHOTTS, EMMETT BOOKER, JR, b Jasper, Ala, Sept 23, 31; m 56; c 2. MEDICAL MICROBIOLOGY, VETERINARY MICROBIOLOGY. *Educ:* Univ Ala, BS, 52; Med Col Ala, cert, 53; Univ Ga, MS, 58, PhD, 66. *Prof Exp:* Asst microbiol & prev med, Sch Vet Med, Univ Ga, 56-57, res assoc path & parasitol, Southeastern Coop Deer Dis Study, 57-58; med bacteriologist & epidemic intel serv officer, Vet Pub Health Lab, Commun Dis Ctr, USPHS, 59-62, res microbiologist, Rabies Invest Lab, 62-64; asst prof path & parasitol, 66-68, chief, Clin Microbiol Lab, 66-78, from asst prof to assoc prof, 69-76, PROF MED MICROBIOL, COL VET MED, UNIV GA, 76-, MEM GRAD FAC, 67-, CONSULT, MICROBIOL LAB, 78- *Concurrent Pos:* Consult microbiol, Southeastern Comp Wildlife Dis Study, 66-; specialist, Pub Health & Med Lab Microbiol, Am Acad Microbiol. *Mem:* Am Soc Microbiol; Wildlife Dis Asn; Am Fisheries Soc; Int Asn Aquatic Animal Med; Conf Res Workers Animal Dis. *Res:* Leptospira, serology, culture, isolation and identification; food borne diseases; diagnostic bacteriology; fluorescent antibody applications; virus-helminth interrelationships; zoonoses; diseases of fresh and salt water fish; Aeromonas, Edwardsiella and Flexibacter. *Mailing Add:* Dept of Med Microbiol Col Vet Med Univ of Ga Athens GA 30602

SHOTWELL, ODETTE LOUISE, b Denver, Colo, May 4, 22. ORGANIC CHEMISTRY. *Educ:* Mont State Col, BS, 44; Univ Ill, MS, 46, PhD(org chem), 48. *Prof Exp:* Asst inorg chem, Univ Ill, 44-48; chemist, Northern Regional Res Lab, Bur Agr & Indust Chem, USDA, 48-52, chemist, Agr Res Serv, 53-77, res leader mycotoxin anal & chem res, 75-84, res leader mycotoxin res, Northern Region Res Ctr, 85-89; RETIRED. *Concurrent Pos:* Consult, Bur Vet Med, FDA, 81-86 & Can Health & Welfare Dept, 83-89; mem, Comt Protection Against Trichothecene Mycotoxins, Nat Res Coun-Nat Acad Sci, 82-83; adv, Int Found Sci, Stockholm, Sweden, 83-89; collabr, Northern Ctr Agr Utilization Res, Agr Res Serv, USDA, 90- *Honors & Awards:* Harvey W Wiley Award, Am Oil Chem Soc, 82. *Mem:* AAAS; Am Chem Soc; Am Oil Chem Soc; Am Asn Cereal Chemists; fel Asn Off Anal Chemists (pres-elect, 87-88). *Res:* Synthetic organic chemistry; chemistry of natural products including isolation purification and characterization; antibiotics; microbial insecticides; mycotoxins. *Mailing Add:* Northern Regional Res Serv Lab Agr Res Serv USDA Peoria IL 61604

SHOTWELL, THOMAS KNIGHT, b Hillsboro, Tex, May 31, 34; m 55; c 1. PROTOCOLS FOR STUDY OF HEALTH RELATED PRODUCTS. *Educ:* Tex A&M Univ, BS, 55, MEd, 58; La State Univ, PhD(agr educ), 65. *Prof Exp:* Teacher biol & gen sci, Allen Acad, 58-60; dept head biol, Allen Col, 60-65; regulatory mgr pharmaceut, Salsbury Labs, Inc, 66-71 & Zoecon Corp, 71-74; PRES, SHOTWELL & CARR, INC, 74- *Concurrent Pos:* Assoc ed, J Clin Res & Drug Regulatory Affairs, 87- *Honors & Awards:* Distinguished Serv Award, Am Soc Agr Consults, 78 & 80. *Mem:* AAAS; NY Acad Sci; Am Asn Indust Vet; Inst Religion Age Sci; US Animal Health Asn. *Res:* Strategies to assure new drug and new animal drug studies meet requirements for approval to market products; guiding research and development programs. *Mailing Add:* 3003 LBJ Freeway No 100 Dallas TX 75234

SHOTZBERGER, GREGORY STEVEN, b Lewistown, Pa, Jan 17, 48. CARDIOVASCULAR PHARMACOLOGY, PROJECT MANAGEMENT. *Educ:* Pa State Univ, BS, 69; State Univ NY, Buffalo, PhD(pharmacol), 74. *Prof Exp:* Res pharmacologist, E I DU Pont De Nemours & Co, Inc, 73-77, sr res pharmacologist, 77-81, proj mgr, 81-83, mgr info & technol anal, 83-88, mgr licensing technol, 88-90; MGR TECHNOL ASSESSMENT, DU PONT MERCK PHARMACEUT CO, 91- *Mem:* Am Col Clin Pharmacol; NY Acad Sci; Inflammation Res Asn; Drug Info Asn; Am Soc Clin Pharmacol & Ther. *Res:* Cardiovascular and gastrointestinal pharmacology; local anesthetics; antiarrhythmic drugs; new drug development; strategic planning. *Mailing Add:* Du Pont Merck Pharmaceut Co PO Box 80025 Wilmington DE 19880-0025

SHOUB, EARLE PHELPS, b Washington, DC, July 19, 15; m 62; c 2. OCCUPATIONAL HEALTH, ACCIDENT PREVENTION. *Educ:* Polytech Univ, BS, 37. *Prof Exp:* Chemist, Hygrade Food Prod Corp, New York City, 40-41 & Nat Bur Standards, 41-43; at US Bur Mines, 43-70, chief, Div Accident Prev & Health, Washington, DC, 63-70; dep dir, Appalachian Lab Occup Respiratory Dis, Nat Inst Occup Safety & Health, Morgantown, WVa, 70-79, Div Safety Res, 77-79; mgr occup safety & consult indust environ , Safety Prod Div, 79, CONSULT, AM OPTICAL CORP, SOUTHBRIDGE, MASS, 79-; CONSULT, 80- *Concurrent Pos:* Prof, Col Mineral & Energy Resources, Univ WVa, 70-79, assoc prof, dept anesthesiol, Med Ctr, 77-82; consult occup health & safety, 80-; dir, Int Soc Respiratory Protection. *Honors & Awards:* Gold Medal, US Dept Interior, 59. *Mem:* Fel Am Inst Chemists; Am Indust Hyg Asn; Am Inst Mining, Metall & Petrol Engrs; Int Soc Respiratory Protection (pres); Am Nat Standards Inst; Am Soc Testing & Mat; Am Soc Safety Engrs; Nat Fire Protection Asn; Nat Soc Prof Engrs; Sigma Xi. *Res:* Occupational health and safety research and development in mineral industries; improved respiratory protective devices. *Mailing Add:* 5850 Meridian Rd Apt 202C Gibsonia PA 15044-9605

SHOUBRIDGE, ERIC ALAN, b Toronto, Ont, Apr 2, 51; m 91. GENE EXPRESSION REGULATION, NUCLEAR MAGNETIC RESONANCE. *Educ:* McGill Univ, BSc, 74, MSc 77; Univ BC, PhD(zool), 81. *Prof Exp:* Asst prof, 85-91, ASSOC PROF NEUROL, MONTREAL NEUROL INST, MCGILL UNIV, 91- *Mem:* AAAS; Soc Magnetic Resonance Med. *Res:* Molecular genetic analysis of the mitochondrial genome and its role in human neurological disease. *Mailing Add:* Montreal Neurol Inst 3801 University St Montreal PQ H3A 2B4 Can

SHOUGH, HERBERT RICHARD, b Springfield, Ohio, Jan 7, 42; m 63; c 3. PHARMACOGNOSY, MEDICINAL CHEMISTRY. *Educ:* Univ Tenn, Memphis, BS, 64, PhD(pharm sci), 68. *Prof Exp:* From asst prof to assoc prof pharmacog, Col Pharm, Univ Utah, 68-78; assoc prof, 78-80, asst dean, 78-83, actg dean, 82, interim dean, 83-84, PROF HEALTH SCI CTR, COL PHARM, UNIV OKLA, 80-, ASSOC DEAN, 83- *Concurrent Pos:* Consult, Palmer Chem & Equip Co, Ga, 65-68; res comt grant, Univ Utah, 68-72; Am Cancer Soc inst grant, 73-74; Univ Utah res comt grant, 75-; Smith-Kline Corp grant, 81-83. *Mem:* Am Asn Pharmaceut Scientists; Am Asn Cols Pharm; Acad Pharmaceut Sci; Am Pharmaceut Asn; Am Soc Pharmacog. *Res:* Ergot alkaloid chemistry and biochemistry; pharmacy education. *Mailing Add:* Health Sci Ctr Univ of Okla Col of Pharm Oklahoma City OK 73190

SHOUKAS, ARTIN ANDREW, CARDIOVASCULAR SURGERY, CONTROL SERIES. *Educ:* Case Western Reserve Univ, PhD(biomed eng), 72. *Prof Exp:* ASSOC PROF CARDIOVASC CONTROL, SCH MED, JOHNS HOPKINS UNIV, 72- *Res:* Cardiovascular system; systems analysis. *Mailing Add:* Dept Biomed Eng Sch Med Johns Hopkins Univ 720 Rutland Ave Baltimore MD 21205

SHOUMAN, A(HMAD) R(AAFAT), b Egypt, Aug 8, 29; m 60; c 5. MECHANICAL ENGINEERING. *Educ:* Cairo Univ, BS, 50; Univ Iowa, MS, 54, PhD(mech eng), 56. *Prof Exp:* Instr, Cairo Univ, 50-53; asst prof mech eng, Univ Wash, 56-60; assoc prof, 60-65, PROF MECH ENG, N MEX STATE UNIV, 65- *Concurrent Pos:* Consult, Boeing Co, 59-63, ARO Inc, 64, AiResearch Mfg Co, 66, NASA, 68, Serv Technol Corp, & E I du Pont de Nemours & Co, Inc, 69-70; vis prof, Laval Univ, 66; consult Nat Acad Sci-Nat Res Coun sr fel, Marshall Space Flight Ctr, NASA, 67. *Mem:* Fel AAAS; Am Soc Mech Engrs; Am Soc Eng Educ. *Res:* Thermodynamics; compressible fluids; gas turbines and heat transfer. *Mailing Add:* 1006 Bloomdale St Las Cruces NM 88005

SHOUP, CHARLES SAMUEL, JR, b Nashville, Tenn, Dec 10, 35; m 58; c 3. PHYSICAL CHEMISTRY. *Educ:* Princeton Univ, AB, 57; Univ Tenn, MS, 61, PhD(phys chem), 62. *Prof Exp:* Chemist, Oak Ridge Nat Lab, 57; prod specialist, Indust Prod Div, Goodyear Tire & Rubber Co, 57-58; chemist, Oak Ridge Nat Lab, 62-67; mgr spec projs, Electronics Div, Union Carbide Corp, NY, 67-68; vpres, Bell & Howell Schs, Inc, 68-69; mgr technol planning, Cabot Corp, 69-70, dir corp res, 70-73, vpres & gen mgr, E-A-R Corp, 73-81, gen mgr, E-A-R Div, Cabot Corp, 81-87, vpres, 84-87; pres, Alphaflex Industs, Inc, 87-88; PRES & MEM BD DIRS, CEMKOTE CORP, 88- *Concurrent Pos:* Mem steering comt tech physics, Am Inst Physics, 70-73; vpres & gen mgr, Nat Res Corp, 70-73; bd trustees, Indust Safety Equip Asn, 79-82; pres, Noise Control Prod & Mat Asn, 82-84. *Mem:* AAAS; Am Inst Physics; fel Am Inst Chemists; Sigma Xi. *Res:* Infrared spectroscopy; surface chemistry; molecular force fields and structure; infrared spectra of adsorbed species; irreversible thermodynamics; technology transfer; technological innovation; new venture management; hearing protection. *Mailing Add:* 13019 Andover Dr Carmel IN 46032

SHOUP, JANE REARICK, b Kansas City, Mo, June 19, 41; m 62; c 2. ZOOLOGY. *Educ:* Univ Rochester, AB, 62; Univ Chicago, PhD(zool), 65. *Prof Exp:* Res assoc zool, Univ Chicago, 65-66; from asst prof to assoc prof biol, 66-82, head dept, 74-82, PROF BIOL, PURDUE UNIV, CALUMET CAMPUS, 82- *Concurrent Pos:* Bd mem nat abortion rights action league, 75-78. *Mem:* Bot Soc Am; Am Soc Cell Biol; AAAS; Inst Soc Ethics & Life Sci; Midwest Co Biol Teachers. *Res:* Fine structural aspects of the genetic control of development. *Mailing Add:* Dept Biol Sci Purdue Univ Calumet 2233 171 St Hammond IN 46323

SHOUP, ROBERT D, b Sinking Spring, Pa, Mar 14, 33; m 87; c 1. COLLOIDAL OR FINE PARTICLE PROCESSING. *Educ:* Albright Col, BS, 60; Univ Pittsburgh, PhD(inorg chem), 64. *Prof Exp:* Res chemist, W R Grace & Co, 64-68; sr res chemist, 68-75, RES SUPVR, CORNING GLASS WORKS, 75- *Mem:* Am Chem Soc; Am Ceramic Soc. *Res:* Glass, ceramics, optical fibers by sol-gel techniques; colloidal chemistry of nuclear fuels, uranium dioxide, uranium carbide and uranium nitride; silicate materials research; catalysts and support systems for environmental pollution control; boron hydride chemistry; synthetic flourohectorites by hydrothermal reactions. *Mailing Add:* Corning Glass Works Sullivan Park-FR-05-1 Corning NY 14831

SHOUP, TERRY EMERSON, b Troy, Ohio, July 20, 44; m 66. MECHANICAL ENGINEERING. *Educ:* Ohio State Univ, BS, 66, MS, 67, PhD(mech eng), 69. *Prof Exp:* Res asst mech eng, Ohio State Univ, 65-66, teaching asst, 67; teaching assoc, Ohio State Univ, 69; from asst prof to assoc prof, Rutgers Univ, 69-75; assoc prof mech eng, Univ Houston, 75-; AT DEAN COL ENG, FLA ATLANTIC UNIV. *Concurrent Pos:* Ed-in-chief, Mechanism & Mach Theory, 77- *Mem:* Am Soc Mech Engrs; Am Soc Eng Educ. *Res:* Mechanisms; kinematic synthesis and analysis of linkages, machine design; dynamic analysis of machines and machine control systems; computer-aided design techniques. *Mailing Add:* Dean Eng Santa Clara Univ Santa Clara CA 95053

SHOVE, GENE C(LERE), b Havensville, Kans, Feb 18, 27; m 49; c 3. AGRICULTURAL ENGINEERING. *Educ:* Kans State Univ, BS, 52, MS, 53; Iowa State Univ, PhD(agr eng, theoret & appl mech), 59. *Prof Exp:* Asst agr eng, Kans State Univ, 52-53; asst, Iowa State Univ, 53-55, 56-58, ext agr engr, 55-56; assoc prof, 58-72, PROF AGR ENG, UNIV ILL, URBANA, 72- *Concurrent Pos:* Eng aid, USDA, 52-53. *Honors & Awards:* Paul A Funk Recognition Award, Col Agr, Univ Ill, 80. *Mem:* Am Soc Agr Engrs; Am soc Eng Educ. *Res:* Crop drying and storage; feed and materials handling; farm building design and use; application of solar energy to grain drying. *Mailing Add:* Dept Agr Eng Univ Ill 1304 W Pennsylvania Ave Urbana IL 61801

SHOVLIN, FRANCIS EDWARD, b Jamaica, NY, Oct 13, 29; m 57; c 5. ENDODONTICS, MICROBIOLOGY. *Educ:* City Col New York, BS, 57; Seton Hall Col, DDS, 61, MS, 65. *Prof Exp:* Nat Inst Dent Res fel, Seton Hall Col, 62-65; pvt pract endodont, 65-72; chmn dept, 72-84, assoc dean res, 84-89, PROF ENDODONT, COL MED & DENT NJ, 72- *Concurrent Pos:* Nat Inst Dent Res fels, 66-72 & 74-77; Omicron Kappa Upsilon, NJ Col Med & Dent, 71. *Mem:* Int Asn Dent Res; Am Soc Microbiol. *Res:* Microbiology of dental caries; cell wall components of cariogenic bacteria; intracellular polyphosphate storage in bacteria; phosphoprotein used by lactobacilli; species identification of lactobacillus. *Mailing Add:* Dept Endodont Col Med & Dent NJ Newark NJ 07103-2423

SHOW, IVAN TRISTAN, b Belleville, Ill, May 12, 43. SYSTEMS ECOLOGY, MATHEMATICAL & NUMERICAL MODELING. *Educ:* Univ Southern Miss, BMEd, 66, MS, 73; Tex A&M Univ, PhD(oceanog & statist), 77. *Prof Exp:* Sr oceanographer, Sci Appl Inc, 77-80, div mgr, 79-80; independent consult, 80-90. *Concurrent Pos:* Res assoc, Gulf Coast Res Lab, 72-, Hubbs-Sea World Res Inst, 79-90, US Environ Protection Agency, 83-90; consult, US Navy, 79-90, Nat Oceanic & Atmospheric Admin, 80-90, USEPA, 83-90. *Mem:* AAAS; Am Soc Limnol & Oceanog; Am Soc Naturalists; Biometrics Soc; N Am Soc Phlebology. *Res:* Theoretical and applied systems ecology and biostatistics; effects of natural and man-made perturbations on ecosystem structure and function; statistical epidemiology. *Mailing Add:* 1604 Calle Plumerias Encinitas CA 92024

SHOWALTER, DONALD LEE, b Louisville, Ky, Jan 22, 43; m 64; c 3. RADIOCHEMISTRY. *Educ:* Eastern Ky Univ, BS, 64; Univ Ky, PhD(chem), 70. *Prof Exp:* NASA grant radiochem anal extraterrestrial samples, under Dr Roman A Schmitt, Ore State Univ, 70-71; asst prof chem, Univ Wis-Stevens Point, 71-73; asst prof sci, Iowa Western Community Col, 73-76; assoc prof, 76-83, PROF CHEM, UNIV WIS-STEVENS PT, 83- *Concurrent Pos:* Consult radiation protection, Wis State Health Lab, 80-; lab demonstr, World of Chem; vis scientist award, East Conn Sect, Am Chem Soc, 90. *Mem:* Am Chem Soc; Sigma Xi. *Res:* Neutron activation applied to geochemical analysis; radiochemical solutions to problems of chemical analysis; environ monitoring of radioactivity. *Mailing Add:* Dept Chem Univ Wis Stevens Pt WI 54881

SHOWALTER, HOWARD DANIEL HOLLIS, b Broadway, Va, Feb 22, 48; m 73; c 2. MEDICINAL CHEMISTRY. *Educ:* Univ Va, BA, 70; Ohio State Univ, PhD(nat prod chem), 74. *Prof Exp:* Fel org chem, Rice Univ, 74-76; res scientist, 76-80, sr scientist, 80-83, res assoc, 83-86, SR RES ASSOC MED CHEM, WARNER LAMBERT/PARKE DAVIS PHARMACEUT RES, 86- *Honors & Awards:* Excellence in Indust Chem Res Award, Am Chem Soc, 83. *Mem:* Am Chem Soc; Int Soc Heterocyclic Chemistry; Am Asn Cancer Res. *Res:* Total synthesis of organic compounds of therapeutic significance, especially anticancer agents; new synthetic methodology; heterocyclic synthesis. *Mailing Add:* 3578 Lamplighter Dr Ann Arbor MI 48103-1702

SHOWALTER, KENNETH, b Boulder, Colo, Apr 9, 49; m 70; c 2. CHEMICAL REACTION KINETICS. *Educ:* Ft Lewis Col, BS, 71; Univ Colo, PhD(chem), 75. *Prof Exp:* Res assoc chem, Univ Ore, 75-77, vis asst prof, 77-78; asst prof, 78-82, assoc prof, 82-86, PROF CHEM, WVA UNIV, 86- *Mem:* Am Chem Soc. *Res:* Chemical waves; multiple stationary states in pumped chemical systems; oscillatory chemical reactions. *Mailing Add:* Dept Chem WVa Univ Morgantown WV 26506-6045

SHOWALTER, ROBERT KENNETH, b Middlebury, Ind, Feb 17, 16; m 43; c 1. HORTICULTURE, FOOD SCIENCE. *Educ:* DePauw Univ, AB, 38; Purdue Univ, MS, 40. *Prof Exp:* Asst hort, Purdue Univ, 38-43; asst chem, Allison Div, Gen Motors Corp, 43; chemist, US Rubber Co, 43-45; from assoc prof to prof hort, Exp Sta, 45-81, EMER PROF, HORT & FOOD SCI, UNIV FLA, 81- *Mem:* Fel AAAS; fel Am Soc Hort Sci. *Res:* Quality maintenance and evaluation of vegetables during handling, transportation and marketing; effects of mechanization of harvesting and handling on market quality of vegetables. *Mailing Add:* Dept Veg Crops Univ Fla Gainesville FL 32611

SHOWELL, JOHN SHELDON, b Camden, NJ, Oct 29, 25; m 51, 71; c 2. ORGANIC CHEMISTRY. *Educ:* Calif Inst Technol, BS, 46, MS, 47; Univ Minn, PhD(org chem), 51. *Prof Exp:* Fel org chem, Univ Ill, Urbana, 51-53; asst prof, Rutgers Univ, 53-55; res fel, Columbia Univ, 55-57; from res chemist to sr res chemist, Agr Res Serv, USDA, 57-66; assoc prog dir org chem, 66-68, assoc prog dir synthetic chem, 68-72, prog dir synthetic org & natural prod chem, 72-87, PROG DIR ORG SYNTHESIS, NAT SCI FOUND, 87- *Mem:* Am Chem Soc; Royal Soc Chem. *Res:* Cyclopropane chemistry; monomer and polymer synthesis; lipid chemistry; organic synthesis; chemical applications of computers; x-ray crystallography. *Mailing Add:* Chem Div NSF 1800 G St NW Washington DC 20550

SHOWERS, MARY JANE C, b Iowa City, Iowa, June 30, 20. ANATOMY. *Educ:* Univ Chicago, BSc, 43; Univ Mich, MSc, 49, PhD(neuroanat), 57. *Prof Exp:* USPHS fel, Kresge Found, Mich, 47-58; staff nurse, Geneva Community Hosp, Ill, 41-42 & Chicago Mem Hosp, 42-43; instr sci, Sch Nursing, Christ Hosp, Ohio, 44-55, dir educ prog, 48-55; assoc prof biol, Our Lady Cincinnati Col, 58-62; asst prof anat, Col Med, Univ Ky, 62-64; from assoc prof to prof anat, Hahnemann Med Col, 64-73, head sect neuroanat, 68-73; prof anat, Philadelphia Col Osteop Med, 73-78; prof anat, 78-84, FAC, UNIV CINCINNATI MED SCH, 84- *Concurrent Pos:* Gelston fel med res, Anat Inst, Norway, 58; spec lectr, Rutgers Univ, 64-69; consult, Sch Nursing, Christ Hosp, 55-61 & Sch Nursing, Deaconess Hosp, 61-65; mem comp vert neuroanat comt, Study Sect, NIH, 63-67; adj prof med educ & anat, Med Col, Univ Cincinnati, 84- *Honors & Awards:* Lindback Award, 70. *Mem:* AAAS; NY Acad Sci; Animal Behav Soc; Am Asn Neuropath; Am Asn Anat. *Res:* Additional motor areas of brain; comparative anatomy of the vertebrate nervous system. *Mailing Add:* Dept Anat & Biol Med Col Univ Cincinnati Cincinnati OH 45267

SHOWERS, RALPH M(ORRIS), b Plainfield, NJ, Aug 7, 18; m 44; c 3. ELECTRONICS, COMMUNICATIONS. *Educ:* Univ Pa, BS, 39, MS, 41, PhD(eng), 50. *Prof Exp:* Lab asst, Farnsworth Radio & Tel Co, 39; testing engr, Gen Elec Co, Pa & NY, 40-41; lab asst, 41-43, instr elec eng, 42-43, res engr & lab supvr, 43-45, asst prof & proj supvr, 45-53, assoc prof, 53-58, PROF ELEC ENG, UNIV PA, 59- *Concurrent Pos:* Chmn, SC A Measurements, Int Spec Comt Radio Interference (CISPR), 62-69, vpres, 73-79, chmn, 80-85; chmn, Am Nat Standards Comt, 63, Electromagnetic Compatibility Soc, 68-; vpres, US Nat Comt Int Electro-tech Comn, 75- *Honors & Awards:* Richard R Stoddart Award, Electromagnetic Compatibility Soc, Inst Elec & Electronics Engrs, 79, Charles Proteus Steinmetz Award, 82, Centennial Medal, 85. *Mem:* Fel Inst Elec & Electronics Engrs; Opers Res Soc Am; Am Soc Eng Educ; Am Asn Univ Profs. *Res:* Electrical engineering; radio interference; solid-state electronics; electromagnetic compatibility between equipments and systems. *Mailing Add:* Moore Sch Elec Eng Univ Pa Philadelphia PA 19104-6390

SHOWERS, WILLIAM BROZE, JR, b St Joseph, Mo, Nov 9, 31; m 63; c 4. INSECT ECOLOGY, BIOCLIMATOLOGY. *Educ:* Univ Ariz, BS, 58; La State Univ, MS, 66; Iowa State Univ, PhD(entomol), 70. *Prof Exp:* Entomologist, rice insects, Baton Rouge, 63-66, res, corn insects, 66-70, PROJ LEADER, ECOL CORN INSECTS, AGR RES SERV, USDA, ANKENY, IOWA, 70-; PROF DEPT ENTOMOL, IOWA STATE UNIV, AMES, 82- *Concurrent Pos:* From asst prof to assoc prof, 74-82, Iowa State Univ, Ames, 74-82. *Honors & Awards:* Cert Merit, USDA, 83 & 90. *Mem:* Am Registry Prof Entomol; Entomol Soc Am; Sigma Xi. *Res:* Evaluating non-chemical insect pest suppression strategies and investigating extrinsic factors that progam noctuid moths to mate and/or migrate. *Mailing Add:* Corn Insects Res Lab Iowa State Univ Res Farm Box 45B USDA-ARS Ankeny IA 50021

SHOWS, THOMAS BYRON, b Brookhaven, Miss, May 4, 38; m 59; c 2. HUMAN GENETICS, CELL GENETICS. *Educ:* San Diego State Univ, BA, 61; Univ Mich, MS, 63, PhD(biochem genetics) hon, 67. *Prof Exp:* Head, Biochem Genetics Sect, 75-79, assoc chief, Dept Exp Biol, 79-80, RES PROF BIOL, ROSWELL PARK CANCER INST, STATE UNIV NY, BUFFALO, 78-, DIR DEPT HUMAN GENETICS, 80- *Concurrent Pos:* USPHS fel, Yale Univ, 67-69. *Honors & Awards:* Newcomb Cleveland Prize Sci, AAAS, 86 & 87. *Mem:* Sigma Xi; Genetics Soc Am; Am Soc Human Genetics; Am Soc Cell Biol; fel AAAS; Am Asn Cancer Res. *Res:* Human genetics: gene mapping and control of gene expression; biochemical and somatic cell genetics. *Mailing Add:* Dir Dept Human Genetics Roswell Park Cancer Inst Elm & Carlton Buffalo NY 14263

SHOZDA, RAYMOND JOHN, b Pittsburgh, Pa, Sept 5, 31; m 59; c 4. HETEROGENOUS CATALYSIS, PROCESS DEVELOPMENT. *Educ:* Carnegie-Mellon Univ, BS, 53, MS, 56, PhD(chem), 57. *Prof Exp:* CHEMIST, CENT RES & DEVELOP DEPT, E I DU PONT DE NEMOURS & CO, INC, 57- *Mem:* Am Carbon Soc; Sigma Xi; NAm Thermal Anal Soc; Catalysis Soc; NY Acad Sci. *Res:* Industrial chemistry; process chemistry at high pressures and temperatures; intermediates synthesis; materials of construction. *Mailing Add:* Cent Res & Develop Dept E I du Pont de Nemours & Co PO Box 80251 Wilmington DE 19898

SHPIZ, JOSEPH M, elementary particle physics, for more information see previous edition

SHRADER, JOHN STANLEY, b Yakima, Wash, Apr 17, 22; m 56; c 2. SCIENCE EDUCATION. *Educ:* Univ Wash, BS, 47, MA, 51, EdD, 57. *Prof Exp:* Asst zool, Univ Wash, 47-48; teacher pub schs, Wash, 48-55; asst zool, Univ Wash, 55-56; teacher pub schs, Wash, 56-57; from assoc prof to prof sci educ, Cent Wash State Col, 57-63; interim prof, Univ Fla, 63-64; prof sci educ, Cent Wash Univ, 64-; RETIRED. *Concurrent Pos:* NSF grant dir, Earth Sci Inst, 70-71. *Mem:* Nat Asn Res Sci Teaching; Asn Educ Teachers Sci; Nat Sci Teachers Asn; Northwest Sci Asn. *Res:* Instructional problems of beginning secondary science teachers in the Pacific Northwest; understanding of college chemistry by intermediate grade pupils; analysis of middle and junior high school teaching; false explorations in science. *Mailing Add:* 14812 NE 12th St Bellevue WA 98007

SHRADER, KENNETH RAY, pharmacy administration; deceased, see previous edition for last biography

SHRADER, WILLIAM D, b Bellflower, Mo, Oct 26, 12; m 35; c 3. SOILS. *Educ:* Univ Mo, BS, 35, MA, 41; Iowa State Univ, PhD(soils), 53. *Prof Exp:* Soil scientist soil conserv serv, USDA, 35-37; instr soils, Univ Mo, 37-42; soil scientist, US Forest Serv, 42-45; soil correlator, US Bur Plant Indust, Soils & Agr Eng, 45-52; from asst prof to prof, 52-81, EMER PROF SOILS, IOWA STATE UNIV, 81- *Concurrent Pos:* Consult, Govt Iran, 58-59 & Thailand, 62-; prof soils, fac agron & party chief, Iowa State Mission to Uruguay, 66. *Mem:* Am Soc Agron; Soil Sci Soc Am; Soil Conserv Soc Am. *Res:* Interpretation of soil properties in terms of plant growth. *Mailing Add:* Rt 2 Box 191 Hermann MO 65041

SHRADER, WILLIAM WHITNEY, b Foochow, China, Oct 17, 30; US citizen. RADAR SYSTEMS RADAR. *Educ:* Univ Mass, BS, 53; Northeastern Univ, MS, 61. *Prof Exp:* Res engr, Boeing Airplane Co, 53-56; systs engr, 56-70, CONSULT SCIENTIST, RAYTHEON CO, 70- *Mem:* Fel Inst Elec & Electronics Engrs. *Res:* Expert in surface based radar systems particularly large phase array radars and Moving Target Indication (MTI) radars; author of numerous papers. *Mailing Add:* Raytheon Co Boston Post Rd Wayland MA 01778

SHRAGER, PETER GEORGE, b Brooklyn, NY, Apr 18, 41; m 66. PHYSIOLOGY, BIOPHYSICS. *Educ:* Columbia Col, AB, 62; Columbia Univ, BS, 63; Univ Calif, Berkeley, PhD(biophys), 69. *Prof Exp:* NIH fel, Med Ctr, Duke Univ, 69-71; asst prof, 71-76, ASSOC PROF PHYSIOL, SCH MED & DENT, UNIV ROCHESTER, 76- *Mem:* Am Physiol Soc; Soc Neurosci; Soc Gen Physiol; Biophys Soc. *Res:* Biophysics and biochemistry of cell membranes; molecular basis of excitation; conduction in demyelinated nerve; neuron-neuroglia interactions. *Mailing Add:* Dept Physiol Box 642 Univ Rochester Med Ctr 601 Elwood Ave Rochester NY 14642

SHRAGO, EARL, b Omaha, Nebr, Apr 9, 28; m 55; c 4. BIOCHEMISTRY, MEDICINE. *Educ:* Univ Omaha, BA, 49; Univ Nebr, MD, 52. *Prof Exp:* From instr to asst prof med, 59-67, assoc prof, 67-71, PROF NUTRIT SCI, SCH MED, UNIV WIS-MADISON, 71-, PROF HEALTH SCI MED, 77-, PROF ASSOC, ENZYME INST, 61- *Concurrent Pos:* USPHS fel, 59-61. *Res:* Mechanisms of hormonal and metabolic control. *Mailing Add:* Dept Med-430 B Univ Wis Med Sch-Nutrit Sci 1415 Linden Dr Madison WI 53706

SHRAKE, ANDREW, b Cleveland, Ohio, Nov 9, 41. CHEMISTRY. *Educ:* Princeton Univ, BA, 64; Yale Univ, PhD(phys chem), 69. *Prof Exp:* NATO postdoctoral fel, Dept Chem, Univ Cambridge, Eng, 69-70; NIH postdoctoral fel biochem, Univ Ariz, 71-74; sr staff fel, Nat Heart, Lung & Blood Inst, NIH, 74-80; RES CHEMIST, CTR BIOLOGICS, FOOD & DRUG ADMIN, 80- *Mem:* Am Chem Soc; Biophys Soc; Am Soc Biochem & Molecular Biol. *Mailing Add:* Ctr Biologics Food & Drug Admin 8800 Rockville Pike Bldg 21 Rm 222 Bethesda MD 20892

SHRAUNER, BARBARA ABRAHAM, b Morristown, NJ, June 21, 34; m 65; c 2. PLASMA PHYSICS, III-V SEMICONDUCTOR TRANSPORT. *Educ:* Univ Colo, BA, 56; Harvard Univ, AM, 57, PhD(physics), 62. *Prof Exp:* Researcher statist mech & plasma physics, Free Univ Brussels, 62-64; resident res assoc plasma & space physics, Ames Res Ctr, NASA, 64-65; from asst prof to assoc prof, 66-77, PROF ELEC ENG, WASH UNIV, 77- *Concurrent Pos:* Am Asn Univ Women fel, 62-63; Air Force grant, 63-64; vis scientist, Los Alamos Sci Lab, 75-76, consult, 79 & collabr, 84; vis scientist, Lawrence Berkeley Lab, 85. *Mem:* Am Phys Soc; Am Geophys Union; Am Asn Univ Profs; Sigma Xi. *Res:* Theoretical problems in plasma physics, including kinetic theory of plasmas and space plasmas; lie group point transformations of nonlinear equations; electron transport in III-V semiconductors for sub millimeter device; discharge plasmas in plasma etching. *Mailing Add:* Dept Elec Eng Wash Univ St Louis MO 63130

SHRAUNER, JAMES ELY, b Dodge City, Kans, Mar 10, 33; m 65; c 2. PHYSICS. *Educ:* Univ Kans, BS, 56; Columbia Univ, MA, 60; Univ Chicago, PhD(physics), 63. *Prof Exp:* Res asst biophys, Radio Res Lab, Columbia Univ, 56-58 & theoret physics, Enrico Fermi Inst Nuclear Res, Univ Chicago, 60-63; res assoc physics, Inst Theoret Physics, Stanford Univ, 63-65; from asst prof to assoc prof, 65-77, PROF PHYSICS, WASH UNIV, 77- *Concurrent Pos:* Vis scientist, Los Alamos Sci Lab, 75-76; assoc scientist, Ames Lab, Dept of Energy, 77-85; trustee, Univs Res Asn, 79-85; vis scientist, superconducting super collider cent design group, Lawrence Berkeley Lab, Univ Calif, Berkeley, 85-86, consult, 86- *Mem:* Fel Am Phys Soc; Fedn Am Sci. *Res:* Theoretical physics; quantum field and elementary particle theories. *Mailing Add:* Dept of Physics Wash Univ St Louis MO 63130

SHRAWDER, ELSIE JUNE, b Norristown, Pa, Nov 15, 38. CLINICAL BIOCHEMISTRY, DNA PROBES. *Educ:* Thiel Col, BA, 60; Univ Notre Dame, MS, 66, PhD(chem), 70. *Prof Exp:* Asst control chemist, Miles Labs, Inc, 60-63, assoc res biochemist, Ames Co, 63-68, res biochemist basic & appl res, 70-72; mem staff, Abbott Labs, Chicago, 77-84; VPRES OPERS, MOLECULAR BIO-SYSTEMS, 84- *Mem:* Am Chem Soc; Am Asn Clin Chemists. *Res:* Enzymes, especially their structure and function; enzymes and their isoenzymes and their relation to diagnostic enzymology; DNA probes and their applications to clinical diagnostics; diagnostic test product development. *Mailing Add:* Molecular Bio-Systems 11180 Roselle Suite A San Diego CA 92121

SHREEVE, JEAN'NE MARIE, b Deer Lodge, Mont, July 2, 33. INORGANIC CHEMISTRY. *Educ:* Univ Mont, BA, 53; Univ Minn, MS, 56; Univ Wash, PhD(inorg chem), 61. *Hon Degrees:* DSc, Univ Mont, 82. *Prof Exp:* Teaching asst chem, Univ Minn, 53-55; asst, Univ Wash, 57-61; from asst prof to assoc prof, 61-67, head dept, 73-87, PROF CHEM, 67-, ASSOC VPRES RES & DEAN COL GRAD STUDIES, UNIV IDAHO, 87- *Concurrent Pos:* Fel, Cambridge Univ, 67-68; NSF fel, 67-68; hon US Ramsey fel, 67-68; Alfred P Sloan Found fel, 70-72; mem chem res eval panel, Air Force Off Sci Res, 72-76 & petrol res fund adv bd, Am Chem Soc, 75-78, bd dir, 85-; vis prof, Univ Bristol, 77; Alexander von Humboldt sr scientist award, 78; guest prof, Univ Göttingen, 78; mem adv comt chem, NSF, 78-82; mem chem div, Argonne Univ Asn Rev Comt, Argonne Nat Lab, 80-86; secy, chair-elect & chair, AAAS (sec C), 84-89, bd dir, 91- *Honors & Awards:* Garvan Medal, Am Chem Soc, 72; Fluorine Award, Am Chem Soc, 78. *Mem:* Fel AAAS; Am Chem Soc; Royal Soc Chem; Am Inst Chemists. *Res:* Synthesis of inorganic and organic fluorine-containing compounds. *Mailing Add:* Univ Res Off Univ Idaho Moscow ID 83843

SHREEVE, WALTON WALLACE, b 1921; m 45; c 4. MEDICINE. *Educ:* DePauw Univ, BA, 43; Ind Univ, MD, 44; Western Reserve Univ, PhD(biochem), 51. *Prof Exp:* Lab instr biochem, Sch Med, Western Reserve Univ, 46-47, sr instr, 51-52; head, Radioisotope Lab, US Naval Hosp, Oakland, Calif, 52-54; scientist & assoc physician, Biochem Div, Med Res Ctr, Brookhaven Nat Lab, 54-64, sr scientist & attend physician, 64-73; PROF MED, STATE UNIV NY, STONY BROOK, 73-, PROF RADIOL, 81-; CHIEF NUCLEAR MED, VET ADMIN MED CTR, NORTHPORT, 73-, DIR, SCH NUCLEAR MED TECHNOL, VET ADMIN MED CTR, 75-, DIR, NUCLEAR MED RESIDENCY PROG, 76-, DEP DIR, NUCLEAR MED, VET ADMIN REGION 1, 86- *Concurrent Pos:* Res physician, Radiosotope Unit, Vet Admin Hosp, Cleveland, Ohio, 50-52; guest prof & NIH spec fel, Karolinska Inst, Sweden, 67; consult, Nassau County Med Ctr, 69- & WHO, India, 71; vis staff mem, Los Alamos Sci Lab, NMex, 72; guest scientist, Inst Med, Nuclear Res Ctr, Juelich, Fed Repub Ger, 81-82. *Mem:* Am Soc Biol Chemists; Endocrine Soc; Am Diabetes Asn; Soc Nuclear Med; NY Acad Sci. *Res:* Intermediary metabolism of carbohydrates, fats and amino acids; clinical applications of radioactive and stable nuclides in diabetes, liver disease and other endocrine, metabolic or nutritional disorders; isotope tracer methodology and uses in nuclear medicine. *Mailing Add:* Nuclear Med Serv Vet Admin Med Ctr Northport NY 11768

SHREFFLER, DONALD CECIL, b Kankakee, Ill, Apr 29, 33; m 57; c 2. IMMUNOGENETICS. *Educ:* Univ Ill, Urbana, BS, 54, MS, 58; Calif Inst Technol, PhD(genetics), 62. *Prof Exp:* Res assoc, 61-64, from asst prof to prof genetics, Med Sch, Univ Mich, Ann Arbor, 64-75; chmn dept, 77-84, PROF GENETICS, SCH MED, WASH UNIV, 75- *Concurrent Pos:* USPHS res career develop award, 66-75; mem, Immunobiol Study Sect, NIH, 70-74, Comt Maintenance Genetic Stocks, Genetics Soc Am, 78-82, Bd Sci Counselors, Nat Inst Allergy & Infectious Dis, 81-85 & Inst Lab Animal Resources Coun, Nat Res Coun, 83-86. *Mem:* Nat Acad Sci; Inst Med-Nat Acad Sci; Am Soc Human Genetics; Transplantation Soc; Am Asn Immunol (pres, 87-88); Genetics Soc Am. *Res:* Mammalian biochemical genetics and immunogenetics; genetic control of variants in serum proteins and cellular antigens. *Mailing Add:* Dept Genetics Sch Med Wash Univ St Louis MO 63110

SHREFFLER, JACK HENRY, b Melrose Park, Ill, May 26, 44; m 70; c 2. METEOROLOGY, AIR POLLUTION. *Educ:* Univ Wis, BS, 65, MS, 67; Ore State Univ, PhD(oceanog), 75. *Prof Exp:* Aerospace engr, Manned Spacecraft Ctr, NASA, 67-70; phys scientist, Nat Oceanic & Atmospheric Admin, 75-80, supvry phys scientist, Air Resources Lab, 80-83; dep dir, Atmospheric Sci Res Lab, 83-88, DIR, CHEM PROCESSES DIV, ATMOSPHERIC RES & EXPOSURE ASSESSMENT LAB, US ENVIRON PROTECTION AGENCY, 88- *Mem:* Am Meteorol Soc; Sigma Xi. *Res:* Numerical modeling related to air pollution meteorology; statistical analysis of air pollution data. *Mailing Add:* 117 Lynwood Pl Chapel Hill NC 27514

SHRENSEL, J(ULIUS), b Newark, NJ, Feb 6, 22; m 47; c 2. ECONOMIC & DECISION ANALYSIS. *Educ:* Newark Col Eng, BS, 44; Stevens Inst Technol, MS, 49. *Prof Exp:* Engr, Baker & Co, Inc, 44-46; instr physics, Newark Col Eng, 46-51; res engr, Allied Chem & Dye Corp, 51-60, res engr, Nat Aniline Div, Allied Chem Dye Corp, 60-63, supvr fiber process eng, Fibers Div, 63-65, proj mgr Mid E, Int Div, 65-69, mgr tech econ sect, eng dept, Specialty Chem Div, 69-71, mgr Bus Analysis Sect Planning Dept Allied Chem Co, Allied Chem Corp, 71-86; CONSULT, 86- *Mem:* Am Chem Soc. *Res:* Plastics extrusion; equipment design; production of synthetic fibers; decision analysis; strategic planning. *Mailing Add:* 97 Laurel Dr Springfield NJ 07081

SHREVE, DAVID CARR, b Lafayette, Ind, May 25, 42. MATHEMATICS. *Educ:* NC State Univ, BS, 64; Rice Univ, PhD(math), 69. *Prof Exp:* Instr, Univ Minn, Minneapolis, 68-69, asst prof, 69-72; asst prof, Univ Wis-Milwaukee, 72-76, assoc prof, 76-81; RES MATHEMATICIAN, CITIES SERV CO, 81- *Mem:* Am Math Soc; Math Asn Am; Soc Indust & Appl Math. *Res:* Numerical analysis and Fourier analysis. *Mailing Add:* 319 Concord No 7 El Segundo CA 90245

SHREVE, GEORGE WILCOX, b Cincinnati, Ohio, Jan 11, 13; m 33; c 2. PHYSICAL CHEMISTRY. *Educ:* Stanford Univ, AB, 35, PhD(chem), 46. *Prof Exp:* Chemist, B Cribari & Sons, Calif, 35-36, Paraffine Co, 37 & Standard Oil Co, Calif, 37-38; res dir, Pac Can Co, Calif, 39-40; res group leader, Permanente Corp, 45; res chemist, Gen Elec Co, NY, 45-46; asst prof chem, Kenyon Col, 46-49; sr phys chemist, Stanford Res Inst, 49-53; res chemist, Monsanto Chem Co, 53-54; chief chemist, Hewlett-Packard Co, 54-57; RES & DEVELOP CONTRACTOR & CONSULT, 57- *Concurrent Pos:* Res Corp grant. *Mem:* AAAS; Am Chem Soc; NY Acad Sci. *Res:* Colloidal properties of non-alkali soaps; chromatography; phase systems of sodium soaps; surface potentials; adsorption; electrochemical cells; alcoholic fermentation, preservation and spoilage of food; protective linings for food containers; catalysis; lignin derivatives; air pollution; solid state chemistry; materials of electronics. *Mailing Add:* 20 Berenda Way Menlo Park CA 94028

SHREVE, LOY WILLIAM, b Smoke Hole, WVa, Oct 8, 26; m 51; c 3. FORESTRY, HORTICULTURE. *Educ:* WVa Univ, BSF, 51; Kans State Univ, MS, 67,PhD(hort), 72. *Prof Exp:* Serv forester, Ky Div Forestry, 54-57, asst dist forester, 57-59, dist forester, 59-63; exten forester fire control, Dept Hort & Forestry, Kans State Univ, 64-68, exten forester tree improv, Dept State & Exten Forestry, 68-76; area exten horticulturist, 76-81, EXTEN HORTICULTURIST, DEPT HORT, TEX A&M UNIV, 81- *Concurrent Pos:* Mem, tree improv comt, Walnut Coun, 72-79, crop adv comt, Juglans, 85-; exchange scientist, Romania & Hungary, 80. *Res:* Development of practical and economical methods for production of genetic duplicates of forest tree species and horticultural varieties of trees and shrubs; breeding of walnuts, pecans and poplars; plant adaptation to site; plant exploration and collection; high density orchards; dwarfing fruit and nut trees. *Mailing Add:* PO Drawer 1849 Uvalde TX 78802-1849

SHREVE, RONALD LEE, b Los Angeles, Calif, Oct 18, 30; m 62; c 1. GEOMORPHOLOGY, GLACIOLOGY. *Educ:* Calif Inst Technol, BS, 52, PhD(geol), 59. *Prof Exp:* Instr geol, Calif Inst Technol, 57-58; from instr to assoc prof geol & geophysics, 58-69, PROF GEOL & GEOPHYS, UNIV CALIF, LOS ANGELES, 69- *Concurrent Pos:* NSF fel, Swiss Fed Inst Technol, 58-59; hon res fel geol, Harvard Univ, 65-66, hon res assoc, 71-72; vis assoc prof, Univ Minn, 68; vis prof & Crosby lectr, Mass Inst Technol, 71-72; Sherman Fairchild Distinguished Scholar, Calif Inst Technol, 78-79; distinguished vis prof, Univ Wash, 85; distinguished vis scientist, Cascades Volcano Observ, 87. *Honors & Awards:* Kirk Bryan Award, Geol Soc Am, 69; Oualline lectr, Univ Texas, 88. *Mem:* AAAS; Geol Soc Am; Am Geophys Union; Int Glaciol Soc. *Res:* Geomorphology; glaciology; physical geology; geophysics. *Mailing Add:* Dept Earth & Space Sci Univ Calif Los Angeles CA 90024-1567

SHRIER, ADAM LOUIS, b Warsaw, Poland, Mar 26, 38; US citizen; m 61; c 4. COMMERCIALIZATION OF TECHNOLOGY, TECHNOLOGY POLICY. *Educ:* Columbia Univ, BS, 59; Mass Inst Technol, SM, 60; Yale Univ, DEng, 65; Fordham Univ, JD, 76. *Prof Exp:* Sr tech staff, Exxon Res & Eng Co, 63-72; environ coordr, Exxon Int Co, 72-73, div mgr, 83-86; venture mgr, Exxon Enterprises Inc, 74-81; hq consult, Exxon Corp, 81-82; mgr policy & planning, Exxon Co, Int, 86-88; MANAGING PARTNER, SPECIALTY TECHNOL ASSOCS, 88- *Concurrent Pos:* Vis scholar, Dept Chem Eng, Cambridge Univ, 65-66; lectr, Dept Chem Eng, Columbia Univ, 67-69; mem, Indust Adv Bd, Int Energy Agency OECD, 83-88; adv, Int Res Ctr Energy & Econ Develop, Univ Colo, 84-88, Energy & Environ Policy Ctr, Harvard Univ, 86-88, Int Energy Prog, Johns Hopkins Univ, 86-88; sr adv, Global Energy Forum, 88-; sr assoc, Cambridge Energy Res Assocs, 88- *Mem:* Am Inst Chem Engrs. *Res:* Commercialization of technology in the fields of energy supply and use; environmental conservation, and the manufacture of specialty chemicals and materials; formulation and implementation of energy, environmental, and technology policy; start up of technology-based business. *Mailing Add:* 543 Park St Upper Montclair NJ 07043

SHRIER, STEFAN, b Mexico City, Mex, Nov 7, 42; US citizen. MACHINE INTELLIGENCE, NUMERICAL ANALYSIS. *Educ:* Columbia Univ, BS, 64, MS, 66; Brown Univ, PhD(appl math), 77. *Prof Exp:* Teaching fel appl math, Brown Univ, 71-72; chmn comput sci, Wellesley Col, 72-75; res asst appl math, Brown Univ, 75-76, specialist, Comput Lab, 76-77; sr engr, Booz, Allen & Hamilton, Inc, 77-79; dir, Softech, Inc, 79-80; mem res staff, Syst Planning Corp, 80-; TECH DIR, GRUMMAN-CTEC, INC. *Concurrent Pos:* Dir, Acad Comput Serv, Wellesley Col, 73-75; secy, New Eng Regional Comput Prog, 73-75; assoc prof & lectr statist, George Washington Univ, 81- *Mem:* Asn Comput Mach; Inst Elec & Electronics Engrs; Pattern Recognition Soc; Sigma Xi; Soc Indust & Appl Math. *Res:* Numerical analysis; operations research; pattern theory; machine intelligence; software engineering. *Mailing Add:* PO Box 19139 Alexandria VA 22320

SHRIFT, ALEX, b New York, NY, Apr 19, 23; m 55; c 3. PLANT PHYSIOLOGY. *Educ:* Brooklyn Col, AB, 44; Columbia Univ, AM, 48, PhD(plant physiol), 53. *Prof Exp:* Asst plant physiol, Columbia Univ, 46-51 & 52-53; instr pharmacog & plant physiol, Sch Pharm, Univ Calif, San Francisco, 53-55; asst prof bot, Univ Pa, 55-60; assoc res scientist, Lab Comp Biol, Kaiser Found Res Inst, 60-66; assoc prof, 66-69, PROF BIOL, STATE UNIV NY BINGHAMTON, 69- *Concurrent Pos:* NIH spec res fel bot & microbiol, Univ Col, Univ London, 72-73. *Mem:* Am Inst Biol Scientists; Am Soc Plant Physiol; Sigma Xi; Am Asn Univ Professors. *Res:* Sulfur and selenium metabolism in plants, animals and microorganisms. *Mailing Add:* Dept Biol State Univ NY PO Box 6000 Binghamton NY 13902-6000

SHRIGLEY, ROBERT LEROY, b Zanesville, Ohio, Apr 10, 29; m 52; c 3. SCIENCE EDUCATION. *Educ:* Ohio Univ, BS, 53, ME, 54; Pa State Univ, DEd, 68. *Prof Exp:* Asst prof educ, Ohio Univ, 55-63; sci adv, Kano Teachers Col, Nigeria, 63-65; assoc prof, 66-80, PROF SCI EDUC, PA STATE UNIV, 80- *Mem:* Nat Asn Res Sci Teaching; Sch Sci & Math Asn; Nat Sci Teachers Asn. *Res:* Attitude modification theory, preservice and in-service elementary teachers toward science. *Mailing Add:* Dept Curric & Instr Pa State Univ Main Campus University Park PA 16802

SHRIME, GEORGE P, b Fakeha, Lebanon, Nov 27, 40. ELECTRICAL ENGINEERING. *Educ:* Am Univ Beirut, BS, 62; Northwestern Univ, MS, 63, PhD(elec eng), 65. *Prof Exp:* Asst prof elec eng, Univ Hawaii, 65-66; sr engr, Tex Instruments, Inc, 65-69, br mgr, Semi-Conductor Circuits Div, 69-87; PRES, EMI INC, 88- *Res:* Computer control of industrial processes; semi-conductor manufacturing; automatic control; deltamodulation. *Mailing Add:* 9611 Mill Trail Dr Dallas TX 75238

SHRIMPTON, DOUGLAS MALCOLM, b Nuneaton Warks, Eng, Mar 29, 35; Can citizen; m 56; c 4. PLANT BIOCHEMISTRY. *Educ:* Univ BC, BA, 57, MA, 58; Univ Chicago, PhD(plant biochem), 61. *Prof Exp:* Nat Res Coun Can fel, 61-63; enzyme chemist, Can Packers Ltd, Ont, 63-65; res officer biochem, 65-74, RES SCIENTIST, CAN FORESTRY SERV, 74- *Mem:* Can Soc Plant Physiol. *Res:* Formation of heartwood and wound response tissues in conifers; enzyme preparations and methods of preservation. *Mailing Add:* Forestry Can Pac & Yukon Region Pac Forestry Ctr 506 W Burnside Victoria BC V8Z 1M5 Can

SHRINER, DAVID SYLVA, b Spokane, Wash, July 20, 45; m 68; c 1. PHYTOPATHOLOGY, FOREST ECOLOGY. *Educ:* Univ Idaho, BS, 67; Pa State Univ, MS, 69; NC State Univ, PhD(plant path), 74. *Prof Exp:* Res ecol, Environ Sci Div, 74-82, RES GROUP LEADER, ENVIRON SCI DIV, OAK RIDGE NAT LAB, 82- *Mem:* Soc Am Foresters; Air Pollution Control Asn; AAAS; Sigma Xi. *Res:* Effects of air pollutants on terrestrial ecosystems; biogeochemical cycling of pollutants; stress physiology of plants; plant host-parasite interactions; assessment of regional-scale environmental problems. *Mailing Add:* Environ Sci Div Oak Ridge Nat Lab PO Box 2008 Oak Ridge TN 37831-6038

SHRINER, JOHN FRANKLIN, JR, b Montgomery, Ala, Feb 6, 57. PROTON RESONANCES, STATISTICAL PROPERTIES OF NUCLEI. *Educ:* Univ South, BS, 78; Duke Univ, MA, 80, PhD(physics), 83. *Prof Exp:* Res assoc, Duke Univ, 83; assoc res physicist, Yale Univ, 83-85; ASST PROF PHYSICS, TENN TECHNOL UNIV, 85- *Mem:* Am Phys Soc; Sigma Xi; AAAS. *Res:* Nuclear spectroscopy of unbound levels studied with proton resonances; applications to statistical theories of nuclei. *Mailing Add:* Tenn Technol Univ Box 5051 Tenn Technol Univ Cookeville TN 38505

SHRINER, RALPH LLOYD, b St Louis, Mo, Oct 9, 99; c 1. ORGANIC CHEMISTRY. *Educ:* Wash Univ, BS, 21; Univ Ill, MS, 23, PhD(org chem), 25. *Prof Exp:* Instr chem, Wash Univ, 21-22; res assoc & asst prof, NY Agr Exp Sta, Geneva, 25-27; from asst prof to prof chem, Univ Ill, 27-41; prof & chmn dept, Ind Univ, 41-46; prof org chem, Univ Iowa, 47-63, head dept chem, 52-62; VIS PROF CHEM, SOUTHERN METHODIST UNIV, 63- *Concurrent Pos:* Mem chem panel cancer chemother, Nat Serv Ctr, Nat Cancer Inst, 59-62, chmn, 61-62; ed-in-chief, Chem Rev, Am Chem Soc, 50-66 & Org Syntheses Cumulative Indices, 71-76. *Mem:* AAAS; Am Chem Soc; Sigma Xi. *Res:* Syntheses and structure of organic compounds; anthocyanins and flavylium salts; lignin model compounds; synthetic drugs and stereoisomerism; organic chemical identification methods. *Mailing Add:* 2709 Hanover St University Park Dallas TX 75225

SHRIVASTAVA, PRAKASH NARAYAN, b Narsingpur, India, Sept 5, 40; m 68; c 3. RADIOLOGICAL PHYSICS. *Educ:* Univ Nagpur, BSc, 58, MSc, 61; Univ Tex, Austin, PhD(nuclear physics), 66; Am Bd Health Physics, cert, 75; Am Bd Radiol, cert, therapeut radiol physics, 75. *Prof Exp:* Res assoc nuclear physics, Ctr Nuclear Studies, Univ Tex, Austin, 66-68; Ont Cancer Inst fel, Princess Margaret Hosp, Toronto, Ont, 68-69; attending radiol physicist, Allegheny Gen Hosp, Pittsburgh, PA, 69-73, dir Div Physics, 74-89; PROF, DEPT RADIATION ONCOL, UNIV SOUTHERN CALIF LOS ANGELES, 89-; CHIEF PHYSICIST RADIATION ONCOL, LOS ANGELES COUNTY, UNIV SOUTHERN CALIF MED CTR, 89- *Concurrent Pos:* Dir, Mideast Ctr Radiol Physics, 74-86, Med Physics & Eng Res Proj, Allegheny-Singer Res Inst, 76-89, Computational Resource Group, 83-89, Nat Hyperthermia Physics Ctr, 83-89 & Allegheny-Singer Dosimetry Calibration Lab, 83-89; sr lectr eng, Carnegie-Mellon Univ Med, 82-89; clin prof, Dept Human Oncol, Univ Wis-Madison, 84-89; third Dr Padam Singh mem lectr, Indian Cong Radiation & Oncol, 84. *Honors & Awards:* Gupta Gold Medal, 58; Thamma Silver Medal, 61; Paranjpe Gold Medal, 61. *Mem:* Am Asn Physicists in Med; Am Col Radiol; Am Pub Health Asn; NY Acad Sci; Soc Nuclear Med; Am Soc Therapeut Radiologists; Health Physics Soc; Radiation Res Soc. *Res:* Therapeutic and diagnostic radiologic physics; hyperthermia for cancer therapy and radiation effects; radiation measurement and 3-D dosimetry; biological effects of heat and radiation. *Mailing Add:* Dept Radiation Oncol LAC-USC Med Ctr Rm OPD 1P17 1200 N State St Los Angeles CA 90033

SHRIVER, BRUCE DOUGLAS, b Buffalo, NY, Oct 18, 40; m 63; c 4. COMPUTER SCIENCE. *Educ:* Calif State Polytech Univ, BS, 63; W Coast Univ, MS, 68; State Univ NY Buffalo, PhD(comput sci), 71. *Prof Exp:* Res engr, Millard D Shriver Co Inc, 63-68; res asst comput sci, State Univ NY Buffalo, 68; NSF fel, 69-71; vis lectr, Aarhus Univ, 71-73; from assoc prof to prof, Univ Southwestern La, 73-84, Alfred Lamson prof comput sci, 81-84, vpres res, 89-90; dept group mgr software technol, IBM T J Watson, 84-88; dir, Pac Res Inst Info Systs, Univ Hawaii, 88-89; DIR RES, D N BROWN ASSOCS, TARRYTOWN, NY, 90- *Concurrent Pos:* Consult educ, IBM Corp, 74-; proj investr, NATO grant, 74-75; ed, Inst Elec & Electronic Engrs Software, 83-87, computer, 87-91. *Mem:* Asn Comput Mach; fel Inst Elec & Electronic Engrs; Am Math Asn; Soc Indust & Appl Math; Inst Elec & Electronic Engrs Computer Soc (pres-elect, 91). *Res:* Parallel architectures, programming languages, algorithms and implementation technology; multi-paradigm design and programming suites; object-oriented systems technology; artificial neural networks and architectures; dataflow languages and architectures; computer supported cooperative work. *Mailing Add:* 17 Bethea Dr Ossining NY 10562-1620

SHRIVER, DAVID A, b Syracuse, NY, May 29, 42; m 64; c 2. GASTROINTESTINAL PHARMACOLOGY. *Educ:* Purdue Univ, BS, 66; Univ Iowa, MS, 68, PhD(pharmacol), 70. *Prof Exp:* Res scientist, Wyeth Labs, Div Am Home Prods, 70-77; RES MGR GASTROINTESTINAL/AUTONOMICS/CNS PHARMACOL, ORTHO PHARMACEUT CORP, DIV JOHNSON & JOHNSON, 77- *Mem:* NY Acad Sci; Am Soc Pharmacol & Exp Theory; Gastrointestinal Res Group; Sigma Xi. *Res:* Pharmacology of the gastrointestinal tract, central nervous system and cardiovascular system. *Mailing Add:* R W Johnson Pharmaceut Res Inst Welsh & McKean Rds Spring House PA 19477-0776

SHRIVER, DUWARD F, b Glendale, Calif, Nov 20, 34; m 57; c 2. INORGANIC CHEMISTRY. *Educ:* Univ Calif, Berkeley, BS, 58; Univ Mich, PhD(chem), 61. *Prof Exp:* Chemist, Univ Calif Radiation Lab, Livermore, 58; from instr to prof, 61-87, 71-87, MORRISON PROF CHEM, NORTHWESTERN UNIV, 87- *Concurrent Pos:* Mem, Mat Res Ctr & Ipatieff Catalysis Ctr, Northwestern Univ; Alfred P Sloan Found res fel, 67-69; vis prof, Univ Tokyo, 77 & Univ Western Ontario, 79; pres, Inorg Syntheses, Inc, 81-83; Guggenheim fel, 83-84; consult, Los Alamos Nat Lab, 84-, Gen Motors, 88-, Medtronic, 90- *Honors & Awards:* Distinguished Serv Award, Am Chem Soc, 87; Medal, Mat Res Soc, 90. *Mem:* AAAS; Am Chem Soc; Royal Soc Chem; Electrochem Soc; Mat Res Soc. *Res:* Synthesis and physical investigation of organometallics, metal cluster compounds, solid-state superionic conductors, complexes and hydrides; infrared and Raman spectroscopy of inorganic and bioinorganic systems, homogenous and heterogenous catalysis. *Mailing Add:* Dept of Chem Northwestern Univ Evanston IL 60208

SHRIVER, JOHN WILLIAM, b Fairmont, WVa, Aug 9, 49; m 80. MUSCLE CONTRACTION, NUCLEAR MAGNETIC RESONANCE. *Educ:* WVa Univ, BA, 71; Case Western Reserve Univ, PhD(chem), 77. *Prof Exp:* Fel biochem, Univ Alta, 77-81; ASST PROF MED BIOCHEM, SCH MED & ASST PROF CHEM, SOUTHERN ILL UNIV, 81- *Mem:* Biophys Soc; Sigma Xi. *Res:* Energetics of conformational state changes in myosin associated with energy transduction in muscle contraction; use of nuclear magnetic resonance in biophysical problems. *Mailing Add:* Dept Chem & Biochem Southern Ill Univ Carbondale IL 62901

SHRIVER, JOYCE ELIZABETH, b Quincy, Ill, Sept 14, 37. ANATOMY. *Educ:* William Jewell Col, AB, 59; Univ Kans, PhD(anat), 65. *Prof Exp:* Nat Inst Neurol Dis & Stroke fel, Col Physicians & Surgeons, Columbia Univ, 64-68; asst prof, 68-71, asst dean student affairs, 76-81, ASSOC PROF ANAT, MT SINAI SCH MED, 71-, MEM FAC, MT SINAI GRAD SCH BIOL SCI, 71-, ASSOC DEAN STUDENT AFFAIRS, 81- *Mem:* AAAS; Am Asn Anat; Am Soc Zoologists; Int Primatol Soc; Soc Neurosci; Sigma Xi. *Res:* Comparative and experimental neurology; study of integration of sensory and motor pathways. *Mailing Add:* Dept Anat Mt Sinai Sch Med New York NY 10029

SHRIVER, M KATHLEEN, VIROLOGY, MONOCLONAL ANTIBODIES. *Educ:* Univ Wash, PhD(biochem), 78. *Prof Exp:* PROG MGR VIROL, GENETICS SYST CORP, 82- *Mailing Add:* Genetics Syst Corp PO Box 97016 Redmond WA 98073-9716

SHROCK, ROBERT RAKES, b Wawpecong, Ind, Aug 27, 04; m 33; c 2. PALEONTOLOGY, SEDIMENTOLOGY. *Educ:* Ind Univ, AB, 25, AM, 26, PhD(geol), 28. *Hon Degrees:* ScD, Ind Univ, 71. *Prof Exp:* Asst geol, Ind Univ, 23-26; asst geol, Univ Wis, 28-29, from instr to asst prof, 29-37; from asst prof to prof, 37-70, chmn dept geol, 49-65, EMER PROF GEOL, MASS INST TECHNOL, 70- *Concurrent Pos:* Sr lectr, Mass Inst Technol, 70-75. *Honors & Awards:* W H Twenhofel Medal, Soc Econ Paleontologists & Mineralogists, 76. *Mem:* Geol Soc Am; Paleont Soc (treas, 38-41); Soc Econ Paleontologists & Mineralogists (pres, 57); Am Asn Petrol Geologists; Nat Asn Geol Teachers (pres, 59). *Res:* Sedimentology; history of geology; index fossils; fossil invertebrates. *Mailing Add:* 18 Loring Rd Lexington MA 02173

SHRODE, ROBERT RAY, b Louisville, Colo, Oct 23, 19. ANIMAL BREEDING. *Educ:* Colo State Univ, BS, 43; Iowa State Univ, MS, 45, PhD(animal breeding), 49. *Prof Exp:* From assoc prof to prof genetics, Agr & Mech Col, Tex, 48-58; geneticist, Wm H Miner Agr Res Inst, NY, 58-60; pop geneticist, De Kalb Agr Asn, Inc, 60-66; PROF ANIMAL SCI, UNIV TENN, KNOXVILLE, 66- *Mem:* AAAS; Am Soc Animal Sci; Biometric Soc; Am Genetic Asn; Nat Asn Cols & Teachers Agr. *Res:* Quantitative genetics of beef cattle, swine, sheep and flour beetles; general biometrical genetics; computer applications; beef cattle breeding. *Mailing Add:* Dept Animal Sci Univ Tenn Knoxville TN 37901-1071

SHRODER, JOHN FORD, JR, b Troy, NY, July 5, 39; m 83. GEOLOGY, GEOMORPHOLOGY. *Educ:* Union Col, BS, 61; Univ Mass, Amherst, MS, 63; Univ Utah, PhD(geol), 67. *Prof Exp:* Instr geol, Westminster Col, 66; lectr geol & geog, Univ Malawi, 67-69; asst prof geol, 69-74, assoc prof geog & geol, 74-78, PROF GEOG & GEOL, UNIV NEBR, OMAHA, 78- *Concurrent Pos:* Grants, Univ Malawi, 68-69, NSF, 70, 73-74, 79, Univ Nebr, Omaha, 70-71, 80, 85, Fulbright, 78, 84 & Smithsonian, 84 & 86, US AID, 88, Nat Park Serv, 89-90, Kiewit Found, 90-91. *Mem:* AAAS; Geol Soc Am; Int Asn Quaternary Res; Sigma Xi; Am Asn Geogrs. *Res:* Mass wasting; periglacial geomorphology; tropical geomorphology. *Mailing Add:* Dept of Geog & Geol Univ of Nebr Omaha Omaha NE 68182

SHROFF, ARVIN PRANLAL, b Surat, India, July 2, 33; US citizen; m 62; c 2. PHARMACEUTICAL CHEMISTRY. *Educ:* Univ Baroda, BS, 54; Duquesne Univ, MS, 58; Univ Md, PhD(pharmaceut chem), 62. *Prof Exp:* Lectr, Univ Col, Md, 61-63, fel, Univ, 62-63; sr scientist & group leader analysis res, Ortho Res Found, 63-74; chemist, 74-75, chief Prod Surveillance Br, 75-81, dir, Div Field Sci, 81-86, actg dep dir, 86-88, DEP DIR, OFF REGIONAL OPERS, FOOD & DRUG ADMIN, 88- *Honors & Awards:* Philip B Hoffman Award, Johnson & Johnson, 72. *Mem:* Am Chem Soc; Am Pharmaceut Asn; Am Microchem Soc; Soc Appl Spectros. *Res:* Steroids, alkaloids and heterocyclics; biotransformation, bioavailability and bioequivalence; pharmaceutical analyses and stability; chromatography; spectroscopy. *Mailing Add:* 3902 Bel Pre Rd Silver Spring MD 20906

SHROFF, RAMESH N, b Jambusar, India, Apr 27, 37; m 65; c 2. POLYMER PHYSICS, POLYMER RHEOLOGY. *Educ:* St Xavier's Col, India, BS, 59; Lehigh Univ, MS, 61, PhD(chem), 66. *Prof Exp:* Sr res physicist, Goodyear Tire & Rubber Co, Ohio, 65-68; sr res scientist, 68-77, asst mgr polymer res, 77-81, res assoc, 81-86, assoc scientist, 86-89, RES SCIENTIST, NORCHEM, 90- *Mem:* Am Chem Soc; Soc Rheology; Soc Plastics Engrs. *Res:* Rheology; extrusion; screw and die design; molecular weight and distribution; long-chain branching; morphology; short-chain branching and distribution; new product development; coextrusion. *Mailing Add:* Quantum Chem-USI Div 3100 Golf Rd Rolling Meadows IL 60008-4070

SHRONTZ, JOHN WILLIAM, b Newark, Ohio, Mar 6, 16; m 42, 70; c 4. RUBBER COMPOUNDING, RUBBER & PETROLEUM NEW PRODUCT DEVELOPMENT. *Educ:* Denison Univ, Granville, Ohio, BSc, 38. *Hon Degrees:* PhD, Am Inst Chemists, 77. *Prof Exp:* Chemist qual control, Pharis Tire & Rubber Co, Newark, Ohio, 40-43, chemist res & develop, 44-48; tech rep rubber & plastics chem sales, Harwick Chem Co, Akron, Ohio, 48-60, br mgr, 60-66, asst to vpres, 66-69; asst to pres & dir pub rels, Edgington Oil Co, Long Beach, Calif, 69-77. *Mem:* Emer mem Am Chem Soc; fel Am Inst Chemists; sr mem Soc Plastics Engrs; AAAS. *Res:* Synthetic rubber and guayule; rubber chemicals for specialized compounding; low temperature plasticizer esters of ethylene glycol. *Mailing Add:* 3261 Orangewood Ave Los Alamitos CA 90720

SHROPSHIRE, WALTER, JR, b Washington, DC, Sept 4, 32; m 58; c 3. BIOPHYSICS, PLANT PHYSIOLOGY. *Educ:* George Washington Univ, BS, 54, MS, 56, PhD(plant physiol, photobiol), 58; Wesley Theol Sem, MDiv, 90. *Prof Exp:* Plant physiologist, Astrophys Observ, Smithsonian Inst, 54-57; res fel biophys, Calif Inst Technol, 57-59; physicist, Div Radiation & Organisms, Smithsonian Inst, 59-64, asst dir, Radiation Biol Lab, 64-83, asst dir, Environ Res Ctr, 83-86, acting dir, 86; DIR, OMEGA LAB, 86- *Concurrent Pos:* Prof lectr, George Washington Univ, 63-86; Smithsonian res award, 65-67; consult, Nat Acad Sci pre & postdoctoral fel award panels, 65-71; guest prof, Univ Freiburg, 68-69; mem coun, Am Asn Advan Sci, 67-77, mem comt coun affairs, 75-78; mem US Nat comt photobiol, Nat Res Coun, 73-75; chmn Am sect, Int Solar Energy Soc, 75-76; guest prof, Univ Zurich, Switzerland, 85-86; adj prof, Wesley Theol Sem, 90- *Honors & Awards:* Special Act Award, Smithsonian Inst, 68. *Mem:* AAAS; Am Soc Plant Physiol; Biophys Soc; Int Solar Energy Soc; Am Soc Photobiol (pres, 84-85); fel Explorers' Club. *Res:* Photobiology; action and transmission spectra; photomorphogenesis; seed germination; spectral distribution of solar radiation; cell physiology; phototropism and light growth responses of fungi. *Mailing Add:* Omega Lab PO Box 5126 Timonium MD 21093

SHRUM, JOHN W, b Jeannette, Pa, Apr 30, 25; m 47; c 3. GEOLOGY, SCIENCE EDUCATION. *Educ:* Pa State Univ, BS, 48; Bowling Green State Univ, MEd, 59; Ohio State Univ, PhD(earth sci, sci educ), 63. *Prof Exp:* Teacher pub schs, Ohio, 56-59; from instr to assoc prof geol & sci educ, Ohio State Univ, 60-68; prof sci educ, 67-87, assoc dir, Biosci Teaching Ctr, 77-82, assoc dean educ, 84-87, EMER PROF SCI EDUC, UNIV GA, 87- *Concurrent Pos:* Dir teacher prep, Earth Sci Curric Proj, Colo, 64-66, Ga Sci Teacher Proj, 68-75; mem panel teacher prep, Coun Educ Geol Sci, 64-67; mem inst eval panel, NSF, 64-75; field reader, US Off Educ Res Proposals, 65-72; chmn dept sci educ, Univ Ga, 67-74. *Mem:* Nat Sci Teachers Asn; Nat Asn Res Sci Teaching; Nat Asn Geol Teachers. *Res:* Evaluation of science instruction; photomacrography; biology science education; course development in geology. *Mailing Add:* 195 Dogwood Dr Athens GA 30606-4603

SHRYOCK, A JERRY, b Canton, Ill, Apr 16, 30; m 50; c 2. MATHEMATICS. *Educ:* Bradley Univ, BS, 50; Ill State Univ, MS, 55; Univ Iowa, PhD(math educ), 62. *Prof Exp:* Mem fac math, 55-72, chmn dept, 70-76, PROF MATH, WESTERN ILL UNIV, 72- *Mem:* Math Asn Am. *Res:* Logic and its applications. *Mailing Add:* 70 Lake Michael Dr Macomb IL 61455

SHRYOCK, GERALD DUANE, b Sharon, Okla, Jan 24, 33; m 59; c 4. ORGANIC CHEMISTRY. *Educ:* Northwestern State Col, BS, 56; Okla State Univ, MS, 59; Univ of the Pac, PhD(chem), 66. *Prof Exp:* Instr chem, Murray State Col, 58-59 & Imp Ethiopian Col, 59-61; instr chem, Murray State Col, 61-62; from asst prof to assoc prof, 63-67, PROF CHEM, BLACK HILLS STATE COL, 67-, CHMN DIV SCI & MATH, 68- *Mem:* Am Chem Soc. *Res:* Carbohydrate chemistry. *Mailing Add:* 2 Lourie Lane Spearfish SD 57783-1144

SHTRIKMAN, SHMUEL, b Brisk, Poland, Oct 21, 30; Israeli citizen; m 55; c 3. PURE & APPLIED PHYSICS. *Educ:* Technion, Israel Inst Technol, BSc, 53, Diplomaed Engr, 54, DSc, 58. *Prof Exp:* Researcher, Electronics Dept, Weizmann Inst Sci, 54-64, from assoc prof to prof, 64-71, dept head, 81-82 & 86-87, SAMUEL SEBBA PROF PURE & APPL PHYSICS, ELECTRONICS DEPT, WEIZMANN INST SCI, 72- *Concurrent Pos:* Vis prof, Physics Dept, Univ Pa, 64-65; vis res prof, Imp Col Sci & Technol, London, 71-72; adj prof, Physics Dept, Univ Calif, San Diego, 85- *Honors & Awards:* Michael Landau Prize, Mifal Hapayis Found, Israel, 75; Rothschild Prize, Yad Hanadiv Found, Israel, 84. *Mem:* Fel Inst Elec & Electronics Engrs; Am Phys Soc. *Res:* Basic research and mission oriented research and developments in pure and applied physics. *Mailing Add:* Dept Electronics Weizmann Inst Sci Rehovot 76100 Israel

SHU, FRANK H, b Kunming, China, June 2, 43; m. ASTROPHYSICS. *Educ:* Mass Inst Technol, BS, 63; Harvard Univ, PhD(astron), 68. *Prof Exp:* From asst prof to assoc prof earth & space sci, State Univ NY, Stony Brook, 68-73; assoc prof, 73-76, chmn, Astron Dept, 84-88, PROF ASTRON, UNIV CALIF, BERKELEY, 76- *Concurrent Pos:* Alfred P Sloan Found fel, 72-74; vis scientist, Kapteyn Astron Inst, Groningen, Neth, 73; lectr, Harlow Shapley Vis Prof Prog, Am Astron Soc, 75-79, 81-84, counr, 82-85, chmn, Mem Questionnair Subcomt, 89-; mem, Brouwer Award Comt, Div Dynamical Astron, Am Astron Soc, 84-87, chmn, 86-87; mem, Class Mem Comt, Nat Acad Sci, 89, US Nat Comt, Int Astron Union, 90- *Honors & Awards:* Bok Prize, Harvard Univ, 72; Warner Prize, Am Astron Soc, 77. *Mem:* Nat Acad Sci; Int Astron Union; Am Astron Soc (vpres, 88-92); Sigma Xi. *Res:* Author of various publications. *Mailing Add:* Astron Dept Univ Calif Berkeley CA 94720

SHU, LARRY STEVEN, b Kuala Lumpur, Malaysia, Mar 8, 36; US citizen; m 65; c 2. BUILDING SCIENCE & TECHNOLOGY. *Educ:* Taiwan Cheng Kung Univ, BS, 58; Brown Univ, MS, 61, PhD(eng), 66. *Prof Exp:* Res scientist fiber-reinforced composites, Space Sci Lab, Gen Elec Co, 65-69; sr res chemist, Celanese Res Co, Celanese Corp, 69-70; mat scientist, W R Grace & Co, 71-72, sr group leader, 72-82, res sect mgr, Bldg Sci & Technol, 83-89, DIR RES, W R GRACE & CO, 89- *Mem:* Am Soc Mech Engrs; Am Soc Testing & Mat; NY Acad Sci; AAAS. *Res:* Fiber-reinforced composites; materials science and engineering; building science and technology. *Mailing Add:* Res Lab Construct Prod Div W R Grace & Co 62 Whittemore Ave Cambridge MA 02140

SHUB, MICHAEL I, b Brooklyn, NY, Aug 17, 43; m 88; c 1. MATHEMATICS, COMPUTER SCIENCE THEORY. *Educ:* Columbia Col, AB, 64; Univ Calif, Berkeley, MA, 66, PhD(math), 67. *Prof Exp:* Lectr & asst prof math, Brandeis Univ, 67-71, from asst prof to assoc prof math, Univ Calif, Santa Cruz, 71-73; assoc prof, 73-75, prof math, Queens Col, NY, 75-86; RES STAFF MEM IBM, T J WATSON RES CTR, 86- *Concurrent*

Pos: NATO fel, 69; Sloan res fel, 72. *Mem:* Am Math Soc; Soc Indust & Appl Math; fel NY Acad Sci. *Res:* Orbit structure of discrete and continuous differentiable dynamical systems; geometric theory of computational complexity. *Mailing Add:* IBM T J Watson Res Ctr Yorktown Heights NY 10598-0218

SHUBE, EUGENE E, b New York, NY, Jan 26, 27; m 48; c 3. ELECTROMECHANICAL ACTUATION, ELECTRO MECHANICAL SYSTEMS. *Educ:* City Col New York, BME, 46; Stevens Inst Technol, MS, 51. *Prof Exp:* Self-employed, Metro Cooling Systs Inc, 46-51; asst proj engr, Wright Aeronaut Div, Curtiss Wright Corp, 51-54; chief preliminary design, Stratos Div, Fairchild Stratos Corp, 54-61, chief engr, 66-71; asst to pres, Dorne & Margolin Inc, 61-76; VPRES ENG, G E C AEROSPACE INC, UK, 71- *Concurrent Pos:* Teacher related tech subjects, NY Bd Educ, 48-51. *Mem:* Soc Automotive Engrs. *Mailing Add:* 564 Ridge Rd Elmont NY 11003

SHUBECK, PAUL PETER, b Elizabeth, NJ, Oct 21, 26; m 53; c 2. ECOLOGY, ENTOMOLOGY. *Educ:* Seton Hall Univ, BS, 50; Montclair State Col, AM, 55; Rutgers Univ, PhD(zool), 67. *Prof Exp:* Instr biol, Thomas Jefferson High Sch, 51-60; guid counsr, Battin High Sch, 61-67; from asst prof to assoc prof, 67-77, chmn dept, 76-79, PROF BIOL, MONTCLAIR STATE COL, 77- *Concurrent Pos:* Vis prof entom, Rutgers Univ, 80. *Mem:* Sigma Xi; Entom Soc Am; N Am BenthＯlog Soc; Coleopterists Soc. *Res:* Insect ecology and behavior; orientation of carrion beetles to carrion; phenology and flight activity of carrion beetles. *Mailing Add:* Dept of Biol Montclair State Col Upper Montclair NJ 07043

SHUBERT, BRUNO OTTO, b Ostrava, Czech, Apr 15, 34; m 60; c 1. MATHEMATICS, OPERATIONS RESEARCH. *Educ:* Czech Tech Univ, MS, 60; Charles Univ, Prague, PhD(probability, statist), 65; Stanford Univ, PhD(elec eng), 68. *Prof Exp:* Res assoc appl probability, Inst Info Theory & Automation, Czech Acad Sci, 64-68; vis asst prof math, Morehouse Col, 68; vis asst prof elec eng, Univ Colo, Boulder, 68-69; asst prof, 69-73, ASSOC PROF OPERS RES, NAVAL POSTGRAD SCH, 73- *Mem:* Inst Math Statist; Am Math Soc; Math Asn Am. *Res:* Stochastic models, theory of games, statistical decisions and learning. *Mailing Add:* Dept Opers Res-Systs Anal Naval Postgrad Sch Code 0223 Monterey CA 93943

SHUBERT, L ELLIOT, b St Louis, Mo, May 16, 43; c 2. PHYCOLOGY. *Educ:* Univ Mo-Kansas City, BS, 66; Univ Conn, PhD(phycol), 73. *Prof Exp:* asst prof, 73-78, ASSOC PROF BIOL, UNIV NDAK, 78- *Concurrent Pos:* Prin-investr, Dept of Interior Bur Reclamation grant, Univ NDak, 74-78; co-investr, Proj Reclamation, 75-80; co-dir, Inst for Energy & Coal Develop for Educators, 77-81; hon vis prof, Univ Durham, Eng, 80-81; sr assoc, Off Instrnl Develop, Univ NDak, 82-83; res assoc, USDA/Agr Res Serv, Human Nutrit Res Ctr, 83-85; fac lect ser, Univ NDak, 86; vis scientist, NASA, Kennedy Space Ctr, 86- *Mem:* Am Soc Limnol & Oceanog; Phycol Soc Am; Sigma Xi; Brit Phycol Soc; Int Phycol Soc. *Res:* Freshwater algae-aquatic and soil; ecology, physiology and nutrition of algae; algal bioassays; algal succession on disturbed and natural soils; soil microcosms; periphyton ecology of prairie lakes. *Mailing Add:* Dept Biol Univ ND PO Box 8238 Grand Forks ND 58202-8238

SHUBIK, PHILIPPE, b London, Eng, Apr 28, 21; US citizen; m 64; c 3. PATHOLOGY, ONCOLOGY. *Educ:* Oxford Univ, BMBCh, 43, DPhil, 49, DM, 71. *Prof Exp:* Demonstr path, Sir William Dunn Sch Path, Oxford Univ, 47-49; instr & biologist, Med Sch, Northwestern Univ, 49-50; cancer coordr, Chicago Med Sch, 50-53, prof oncol & dir dept, 53-68; Eppley Prof Oncol & Path & dir, Eppley Inst, Col Med, Univ Nebr Med Ctr, 68-80; SR RES FEL, GREEN COL, OXFORD, ENG, 80- *Concurrent Pos:* Mem morphol study sect, NIH, 58-59, cell biol study sect, 58-60 & path study sect, 60-62; expert adv panel, WHO, 59-; Nat Adv Cancer Coun, 62-66 & sr mem, Nat Cancer Adv Bd, 70-; co-managing ed, Cancer Letters, 75-; pres, Toxicol Forum, Inc, Washington, DC, 74-; toxicologist, Eppley Inst, Univ Nebr, dir, Eppley Inst for Res in Cancer. *Honors & Awards:* Co-recipient of Ernest W Bertner Mem Award, 78. *Mem:* Am Soc Path & Bact; Am Soc Exp Path; Am Asn Cancer Res; Am Soc Prev Oncol; Soc Toxicol. *Res:* Experimental pathology; chemical carcinogenesis; environmental and industrial cancer; toxicology; tumor biology. *Mailing Add:* Green Col Univ Oxford Oxford 0X2 6HG England

SHUBKIN, RONALD LEE, b New York, NY, Aug 27, 40; m 67; c 2. SYNTHETIC LUBRICANTS. *Educ:* Univ NC, BS, 62; Univ Wis, PhD(inorg chem), 67; Mich State Univ, MBA, 81. *Prof Exp:* Res fel, Queen Mary Col, London, 66-67; from res chemist to sr res chemist, 67-79, res assoc, 79-81, econ eval, 82-83, res supvr, 81-88, RES MGR, ETHYL CORP, 89- *Concurrent Pos:* Instr, Wayne County Community Col; assoc prof, Oakland Community Col. *Mem:* Am Chem Soc; Soc Tribologists & Lubrication Engrs. *Res:* Synthetic lubricants; industrial chemistry; organic chemistry. *Mailing Add:* Ethyl Corp 8000 GSRI Ave Baton Rouge LA 70820-7497

SHUCHAT, ALAN HOWARD, b Brooklyn, NY, Oct 6, 42; m 85; c 1. MATHEMATICAL MODELING. *Educ:* Mass Inst Technol, SB, 63; Univ Mich, Ann Arbor, MS, 65, PhD(math), 69. *Prof Exp:* Asst prof math, Univ Toledo, 69-71 & Mt Holyoke Col, 71-74; from asst prof to assoc prof, 74-83, PROF MATH, WELLESLEY COL, 83-, ASSOC DEAN, 89- *Concurrent Pos:* Fac fel, Transp Systs Ctr, 81; reviewer, Math Revs, 72-; sci fac prof develop grant, NSF, 79-80, improv lab instrumentation grant, 89-91; vis scientist, Mass Inst Technol, 84; translr, Russ math; vis lectr, Math Asn Am. *Mem:* Am Math Soc; Math Asn Am; Opers Res Soc Am. *Res:* Mathematical models in operations research; curriculum development in mathematics and technology. *Mailing Add:* Dept Math Wellesley Col Wellesley MA 02181

SHUCK, FRANK O, b Glasgow, Mont, Feb 19, 36; m 56; c 2. CHEMICAL ENGINEERING. *Educ:* Carnegie Inst Technol, BS, 58, MS, 60, PhD(chem eng), 62. *Prof Exp:* Asst prof, 62-77, ASSOC PROF CHEM ENG, IOWA STATE UNIV, 77- *Mem:* Am Inst Chem Engrs. *Res:* Diffusion in binary and multicomponent liquids and liquid metals; mass transfer in liquid systems. *Mailing Add:* Dept Eng McNeese State Univ Lake Charles LA 70609

SHUCK, JOHN WINFIELD, b Cumberland, Md, Apr 9, 40; m 64; c 1. MATHEMATICS, EDUCATION. *Educ:* Mass Inst Technol, BS, 63; Tufts Univ, MS, 68; Northeastern Univ, PhD(math), 69. *Prof Exp:* Asst prof math, Univ Mich, 69-70 & Univ Rochester, 70-77; from asst prof to assoc prof math, 77-90, dept head, 83-88, PROF URSINUS COL, 90- *Concurrent Pos:* NSF teaching fel, 68-69. *Mem:* Am Math Soc; Math Asn Am; Sigma Xi; Coun Undergrad Res. *Res:* Algebraic number theory; non-Archimedean analysis and its applications to number theory; Diophantine equations. *Mailing Add:* Dept Math Ursinus Col Collegeville PA 19426

SHUCK, LOWELL ZANE, b Bluefield, WVa, Oct 23, 36. MECHANICAL ENGINEERING, PETROLEUM ENGINEERING. *Educ:* WVa Inst Technol, BSME, 58; WVa Univ, MSME, 65, PhD (theoret & appl mech-biomech), 70. *Prof Exp:* Sales engr, WVa Armature Co, 58-59; from instr to assoc prof mech eng, WVa Inst Technol, 59-69, chmn dept, 65-69; res mech engr & proj leader, Morgantown Energy Res Ctr, 70-76; prof mech eng & mech & assoc dir Eng Exp Sta, WVa Univ, 76-80; PRES, TECHNOL DEVELOP INC, 80- *Concurrent Pos:* Consult, Railcar Div, Food Mach Corp, WVa, 65-66; NSF faculty fel & res eng, 68-70; res mech engr, Morgantown Energy Technol Ctr, 76-; Governor's appointee to WVa Coal & Energy Res Adv Comt, 77-81; sci adv to WVa Gov Jay Rockefeller IV, 78-81; adj prof, Col Eng, WVa Univ, 80- *Honors & Awards:* Mat Testing Award, Am Soc Testing & Mat, 70; Ralph James Award, Am Soc Mech Engrs, 80. *Mem:* Am Soc Eng Educ; Am Soc Mech Engrs; Soc Petrol Engrs; Instrument Soc Am; Nat Soc Prof Engrs; Sigma Xi. *Res:* Vibrations; acoustics; data acquisition; metrology; biomechanics; rheology; theoretical and experimental stress analysis; design and development of transducers; instrumentation; biomechanics; petroleum, natural gas and coal in situ recovery technology research and development including theoretical, laboratory and field projects. *Mailing Add:* 401 Highview Place Morgantown WV 26505

SHUDDE, REX HAWKINS, b Santa Monica, Calif, Dec 25, 29; m 58; c 2. NUCLEAR CHEMISTRY. *Educ:* Univ Calif, Los Angeles, BS & AB, 52, Berkeley, PhD(chem), 56. *Prof Exp:* Sr res chemist, Atomics Int Div, NAm Aviation, Inc, 56-61, supvr reactor code develop, 61, supvr numerical applns, 61-62; assoc prof, opers anal, Naval Postgrad Sch, 62-76, assoc prof opers res, 76-90; RETIRED. *Mem:* Opers Res Soc Am; Asn Comput Mach; Am Phys Soc. *Res:* Mathematics; electronics; operations research; mathematical programming; computer systems; numerical analysis. *Mailing Add:* 27105 Arriba Way Carmel CA 93923

SHUE, ROBERT SIDNEY, b Burlington, NC, May 24, 43; m 67; c 2. POLYMER CHEMISTRY, ORGANOMETALLIC CHEMISTRY. *Educ:* Univ NC, Chapel Hill, AB, 64, PhD(org chem), 68. *Prof Exp:* Teaching asst org chem, Univ NC, Chapel Hill, 67-68; res chemist, Res Ctr, Phillips Petrol Co, 68-72, sr res chemist, 72-77, group leader, 77-78; mgr mkt develop, Phillips Eng Plastics, Phillips Chem Co, 78-80, group leader, Res Ctr, 80-82, planning & budgeting specialist, 82-84, mkt develop specialist, 84-87, SR PATENT DEVELOP CHEMIST, PHILLIPS PETROL CO, 87- *Mem:* Soc Advan Mat & Process Eng; Soc Plastic Engrs. *Res:* Homogeneous transition metal catalyzed organic reactions; reaction mechanisms; synthesis, characterization and structure-property relationships of macromolecules. *Mailing Add:* 3508 SE Oakdale Dr Bartlesville OK 74006

SHUEY, MERLIN ARTHUR, b Pottsville, Pa, Sept 6, 36; m 60, 80; c 3. MANNED VEHICLE ENVIRONMENTAL CONTROL SYSTEMS, THERMAL CONTROL SYSTEMS. *Educ:* Drexel Univ, BSME, 60; Rensselaer Polytech Univ, MSME, 65; Univ Conn, MBA, 71. *Prof Exp:* Exp engr, Hamilton Standard, Div United Technol Corp, 60-61, engr preliminary design tech mkt, 61-64, sr engr, 64-65, asst preliminary design engr tech mkt & group leader, 65-73 & 74-76, asst prog mgr, 73-74, sr mkt engr, Prod Line Tech Mkt, 76-78, prod mkt mgr, 78-88, prog mgr & dep prog mgr, 88-89, PROF MGR BUS DEVELOP, HAMILTON STANDARD, DIV OF UNITED TECHNOL CORP, 89- *Concurrent Pos:* Mem bd, Inst Advan Studies in Life Support, 90-, chmn annual mkt, 91. *Mem:* Am Inst Aeronaut & Astronaut. *Res:* Support of life in enclosed environment, primarily manned spacecraft; author of numerous papers and publications; support of plants in closed environment. *Mailing Add:* 11 Woodduck Lane Tariffville CT 06081

SHUEY, R(ICHARD) L(YMAN), b Chicago, Ill, May 7, 20; m 44; c 2. ELECTRONICS, INFORMATION SCIENCE. *Educ:* Univ Mich, BS(eng physics) & BS(eng math), 42; Univ Calif, MS, 47, PhD(elec eng), 50. *Prof Exp:* Engr, Radiation Lab, Univ Calif, 46-50; res assoc res lab, 50-55, mgr info studies sect, 55-65, mgr info studies br, Res & Develop Ctr, 65-75, staff consult, Res & Develop Ctr, Gen Elec Co, 75-84; ADJ PROF, RENSSELAER POLYTECH INST, 87- *Concurrent Pos:* Adj prof, Rensselaer Polytech Inst, 53-59. *Mem:* Inst Elec & Electronics Engrs; Asn Comput Mach; Soc Mfg Engrs; AAAS. *Res:* Information theory; communications systems; computers; distributed computer and information systems. *Mailing Add:* 2338 Rosendale Rd Schenectady NY 12309

SHUEY, WILLIAM CARPENTER, b Emporia, Kans, July 1, 24; m 43; c 4. CEREAL CHEMISTRY. *Educ:* Univ Wichita, BS, 48; NDak State Univ, MS, 67, PhD(cereal technol), 70. *Prof Exp:* Exp miller & baker, Gen Mills, Inc, 48-51, in chg exp milling & phys dough test sect, 51-62; res cereal food technologist in chg hard red spring & durum wheat qual lab, Agr Res Serv, USDA, 62-77, ADJ PROF CEREAL CHEM & TECHNOL, AGR EXP STA, NDAK STATE UNIV, 77- *Honors & Awards:* Carl Wilhelm Brabender Award, Am Asn Cereal Chem, 70. *Mem:* AAAS; Am Asn Cereal Chem. *Res:* Chemical composition of wheat and flour, their physical properties, finished products, influence of nutrition, temperature, disease, and other environmental factors on quality of wheat. *Mailing Add:* 5142 26th Ave S Gulfport FL 33707

SHUFORD, RICHARD JOSEPH, b Hobart, Okla, Dec 20, 44. ORGANIC POLYMER CHEMISTRY. *Educ:* Stetson Univ, BS, 66; Southern Ill Univ, Carbondale, PhD(org chem), 71. *Prof Exp:* RES CHEMIST POLYMERS, ORG MAT LAB, ARMY MAT & MECH RES CTR, 71- *Mem:* Am Chem

Soc; Sigma Xi; Catalysis Soc; Am Soc Nondestructive Testing; Soc Adv Mat & Process Eng. *Res:* Piezoelectric and pyroelectric polymers; polymer morphology; nondestructive evaluation of composites; fabrication and characterization of fiber reinforced composites; develop and evaluate quality control and cure monitoring techniques for composites. *Mailing Add:* Army Mat & Mech Res Ctr Watertown MA 02172

SHUGARMAN, PETER MELVIN, b Duluth, Minn, July 28, 27. PLANT PHYSIOLOGY, BIOCHEMISTRY. *Educ:* Univ Calif, Los Angeles, PhD(chlorophyll biosynthesis), 66. *Prof Exp:* Lab technician leukocyte metab, Dept of Med, 51-58 & chlorella physiol, Dept Bot & Plant Biochem, 58-66; asst prof biol sci, 66-70, asst dean student affairs, 73-78, from asst dean to assoc dean, 78-86, assoc dean student affairs to assoc dean nat sci, 86-89, DIR, INTERDISCIPLINARY MAJ, UNIV CALIF, LOS ANGELES, 90-, ASSOC PROF CELL PHYSIOL, UNIV SOUTHERN CALIF, 70- *Mem:* AAAS; Am Soc Plant Physiol; Am Inst Biol Sci. *Res:* Control of chlorophyll biosynthesis in Chlorella. *Mailing Add:* Dept Biol Sci Univ Southern Calif Los Angeles CA 90089-0371

SHUGARS, JONAS P, b Liberty, Ky, Feb 8, 34; m 60; c 2. PLANT SCIENCE, SOIL SCIENCE. *Educ:* Univ Ky, BSA, 55, MS, 57; Univ Tenn, Knoxville, PhD(agr plant & soil sci), 70. *Prof Exp:* Teacher high sch, Ky, 59-60; ASSOC PROF HORT, BEREA COL, 60- *Mem:* Am Soc Hort Sci. *Res:* Plant nutrition; flower physiology of plants; pollen study related to allergic reactions. *Mailing Add:* Dept Agr Berea Col Berea KY 40404

SHUGART, CECIL G, b Ennis, Tex, Oct 13, 30; m 55; c 2. NUCLEAR PHYSICS, POLYMER PHYSICS. *Educ:* North Tex State Univ, BA, 57; Univ Tex, Austin, MA, 61, PhD(nuclear physics), 68. *Prof Exp:* Staff asst physics, Southwestern Bell Tel Co, 57-58; res scientist, Defense Res Lab, Univ Tex, Austin, 58-61; assoc engr, Develop Lab, Int Bus Mach Corp, 61-62; asst prof physics & chmn dept, Hardin-Simmons Univ, 62-65; asst prof physics, Southwestern Univ, 65-66; res assoc nuclear physics, Univ Tex, Austin, 67-68; dir soc physics studies, Am Inst Physics, 68-70; assoc prof, 70-73, prof physics & head dept, Northeast La Univ, 73-77; PROF PHYSICS & CHMN DEPT, MEMPHIS STATE UNIV, 77- *Concurrent Pos:* Sci fac fel, NSF, 66-67; vis scientist, Am Asn Physics Teachers, 68-70. *Mem:* Fel AAAS; Am Phys Soc; Am Asn Physics Teachers. *Res:* Nuclear and polymer physics; atmospheric electricity; magnetics; physics education. *Mailing Add:* Dept of Physics Memphis State Univ Memphis TN 38152

SHUGART, HERMAN HENRY, JR, b El Dorado, Ark, Jan 19, 44; m 66; c 2. ECOLOGY, ZOOLOGY. *Educ:* Univ Ark, BS, 66, MS, 68, PhD(zool), 71. *Prof Exp:* ECOLOGIST, OAK RIDGE NAT LAB, 71- *Concurrent Pos:* Lectureship, asst to assoc prof, Dept Bot, Univ Tenn, 71- *Mem:* Am Ornith Union; Ecol Soc Am; AAAS. *Res:* Systems analysis in ecology; theoretical ecology; synecology; niche theory, mathematical ecology, use of multivariate statistics in ecology. *Mailing Add:* Dept Envir Sci Clark Hall Univ Va Charlottesville VA 22903

SHUGART, HOWARD ALAN, b Orange, Calif, Sept 21, 31; m 71. PHYSICS. *Educ:* Calif Inst Technol, BS, 53; Univ Calif, MA, 55, PhD, 57. *Prof Exp:* Teaching asst, Univ Calif, Berkeley, 53-56, assoc, 57, lectr, 57-58, from actg asst prof to assoc prof, 58-67, group leader, Lawrence Berkeley Lab, 64-79, vchmn dept physics, 68-70, 79-87, PROF PHYSICS, UNIV CALIF, BERKELEY, 67-, VCHMN DEPT, 88- *Concurrent Pos:* Consult, Gen Dynamics/Convair, 60-61; mem comt nuclear constants, Nat Res Coun, 60-63. *Mem:* Fel Am Phys Soc; fel Nat Speleol Soc; Sigma Xi. *Res:* Atomic and molecular beams; low energy nuclear physics; atomic and nuclear properties, including lifetimes, hyperfine structure, spins and static multipole moments. *Mailing Add:* Dept Physics Univ Calif Berkeley CA 94720

SHUGART, JACK ISAAC, veterinary & medical entomology, pesticide development, for more information see previous edition

SHUGART, LEE RALEIGH, b Corbin, Ky, Dec 23, 31; m 52, 84; c 3. BIOCHEMISTRY, ENVIRONMENTAL GENOTOXICITY. *Educ:* East Tenn State Univ, BS, 51; Univ Tenn, MS, 62, PhD(microbiol), 65. *Prof Exp:* NIH fel biol, 65-67, BIOCHEMIST, ENVIRON SCI, OAK RIDGE NAT LAB, 67- *Mem:* Soc Environ Toxicol & Chem; Am Chem Soc; Am Soc Biochem & Molecular Biol; Sigma Xi. *Res:* Biochemical measurement of damage to DNA; molecular mechanisms of environmental genotoxicity; interaction of proteins with nucleic acids; isolation and characterization of nucleic acids. *Mailing Add:* Environ Sci Div Oak Ridge Nat Lab Box 2008 Oak Ridge TN 37831-6036

SHUKLA, ATUL J, b Dar-Es-Salaam, Tanzania, May 31, 57; m 82; c 2. PHARMACEUTICS, CONTROLLED RELEASE TECHNOLOGY. *Educ:* Univ Bombay, India, BS, 79; Univ Ga, MS, 82, PhD(pharmaceut), 85. *Prof Exp:* Asst prof pharaceut, Duguesne Univ, 85-89; ASST PROF PHARMACEUT, DEPT PHARMACEUT, COL PHARM, UNIV TENN, MEMPHIS, 89- *Concurrent Pos:* Invited speaker, Tanzania, Pharmaceut Coun, 88, Journees Galenique, Gatlefosse Corp, France, 88; prin investr grants, Gatlefosse Corp, 88-90, Edward Mendell Corp, 91- *Mem:* Am Asn Pharmaceut Scientists; Controlled Release Soc; NAm Thermal Analysis Soc. *Res:* Design and formulation of controlled release drug delivery systems, such as microcapsules; biodegradable injectable drug delivery systems; oral controlled release drug delivery systems; transdermal drug delivery systems and evaluation of excipients for their use in the manufacturing of tablets. *Mailing Add:* Dept Pharmaceut Col Pharmacy Univ Tenn 26 S Dunlap Rm 214 Memphis TN 38163

SHUKLA, KAMAL KANT, b India, Jan 1, 42. MUSCLE BIOCHEMISTRY. *Educ:* Agra Univ, India, BS, 61; Banaras Hindu Univ, MS, 63; State Univ NY Stony Brook, PhD(physiol & biophysics), 77. *Prof Exp:* Asst prof physics, K N Govt Col, India, 63; jr sci officer radiation physics, Inst Nuclear Med, India, 64-70; med assoc, Brookhaven Nat Lab, 71-73; lectr, State Univ NY, Stony Brook, 78-79, asst prof muscle biochem, 79-; ASSOC PROG DIR, BIOPHYS PROG, NSF. *Concurrent Pos:* Sr sci officer, Inst Nuclear Med, India, 70-81. *Res:* Muscle biochemistry, in particular the mechanism of adenosine triphosphate hydrolysis by acto-myosin and its relation to contractions. *Mailing Add:* NSF 1800 G St NW Washington DC 20550

SHUKLA, SHIVENDRA DUTT, b Mirzapur, India, Oct 11, 51; m 70; c 3. EMBRANES & RECEPTOR FUNCTIONS. *Educ:* Banaras Hindu Univ, India, BSc, 68, MSc, 70; Univ Liverpool, Eng, PhD(biochem), 77. *Prof Exp:* Res fel, Univ Birmingham, Eng, 76-80; res asst prof biochem, Univ Tex Health Ctr, 80-84; ASSOC PROF PHARMACOL, SCH MED, UNIV MO, 84-, DIR GRAD STUDIES, 89- *Concurrent Pos:* Lectr, Banaras Hindu Univ, 71-73; prin investr, Am Heart Asn, 81-84 & NIH, 84-; res career develop award, NIH, 89-94. *Mem:* AAAS; Am Heart Asn; Am Soc Biochem & Molecular Biol; Am Soc Pharmacol & Exp Therapeut. *Res:* Biochemical and pharmacological investigations on the inositol lipid turnover and transmembrane signalling in cells. *Mailing Add:* Dept Pharmacol Univ Mo Sch Med Columbia MO 65212

SHULDINER, PAUL W(ILLIAM), b New York, NY, June 19, 30; m 51; c 7. CIVIL ENGINEERING. *Educ:* Univ Ill, Urbana, BSCE, 51, MSCE, 53; Univ Calif, Berkeley, DrEng(transp), 61. *Prof Exp:* Instr civil eng, Ohio Northern Univ, 53-54, asst prof, 54-55; asst prof, Northwestern Univ, 60-63, assoc prof, 63-65; consult transp planning, Off Under Secy Transp, 65-66, sr transp engr, Off High Speed Ground Transp, US Dept Transp, 66-67, chief transp systs planning div, 68-69; fed exec fel, Brookings Inst, 70; dep dir nat transp planning study, Nat Acad Sci, 70-71; PROF CIVIL ENG & REGIONAL PLANNING, UNIV MASS, AMHERST, 71- *Concurrent Pos:* Adv, Northeastern Ill Planning Comn, 64-65; mem Hwy Res Bd, Nat Acad Sci-Nat Res Coun. *Honors & Awards:* Walter L Huber Res Prize, Am Soc Civil Engrs, 66. *Mem:* Am Soc Civil Engrs. *Res:* Transportation systems engineering; urban and regional planning. *Mailing Add:* Dept of Civil Eng Univ of Mass Amherst MA 01003

SHULER, CHARLES F, b Jamesville, Wis, Feb 17, 53. BIOLOGY. *Educ:* Univ Wis-Madison, BS, 75; Harvard Univ, DMD, 79; Univ Chicago, PhD(path), 84. *Prof Exp:* Asst prof dent, Ohio State Univ, 84-89; ASST PROF DENT, UNIV SOUTHERN CALIF, 89- *Mem:* Am Soc Cell Biol; AAAS; Am Soc Dent Res. *Mailing Add:* Ctr Craniofacial Molecular Biol Univ Southern Calif 2250 Alcazar St CSA First Floor Los Angeles CA 90033

SHULER, CRAIG EDWARD, b Wichita, Kans, Aug 27, 38; m 60; c 4. FOREST PRODUCTS, WOOD SCIENCE & TECHNOLOGY. *Educ:* Colo State Univ, BS, 60, MS, 66, PhD(wood sci), 69. *Prof Exp:* Instr wood technol, Colo State Univ, 67-68; asst prof wood technol, Sch Forest Resources, Univ Maine, 69-75, assoc prof, 75-79; ASSOC PROF WOOD SCI & TECHNOL, COLO STATE UNIV, 79- *Mem:* Soc Wood Sci & Technol; Forest Prod Res Soc; Am Soc Testing & Mat; Sigma Xi. *Res:* Timber mechanics; particle board production and use; timber physics; residue utilization. *Mailing Add:* 2413 Constitution Ft Collins CO 80526

SHULER, KURT EGON, b Nuremberg, Ger, July 10, 22; nat US; m 44. THEORETICAL CHEMISTRY, CHEMICAL PHYSICS. *Educ:* Ga Inst Technol, BS, 42; Cath Univ, PhD(theoret chem), 49. *Prof Exp:* AEC fel, Appl Physics Lab, Johns Hopkins Univ, 49-51, sr staff mem, 51-55; mem sr staff, Nat Bur Stand, 55-58, consult to chief, Heat Div, 58-60, consult to dir, 60-61, sr res fel, 63-68; chmn dept, 68-70, 84-87, PROF CHEM, UNIV CALIF, SAN DIEGO, 68- *Concurrent Pos:* Consult, Advan Res Proj Agency, Dept Defense, 61-74, spec asst to vpres res, Inst Defense Analysis, 61-63; vis prof chem, Univ Calif, San Diego, 66-67; mem adv panel, Chem Sect, NSF, 73-75; Solvay Found fel, Solvay Inst, Univ Brussels, Belg, 75. *Honors & Awards:* Gold Medal, US Dept Com, 68. *Mem:* Fel AAAS; fel Am Inst Chemists; Am Chem Soc; fel Am Phys Soc. *Res:* Statistical mechanics; nonlinear phenomena and processes; stochastic processes. *Mailing Add:* Dept Chem 3-040 Univ Calif San Diego La Jolla CA 92093

SHULER, MICHAEL LOUIS, b Joliet, Ill, Jan 2, 47; m 72; c 4. BIOTECHNOLOGY. *Educ:* Univ Notre Dame, BS, 69; Univ Minn, PhD(chem eng), 73. *Prof Exp:* From asst prof to assoc prof, 74-83, PROF CHEM ENG, CORNELL UNIV, 84- *Concurrent Pos:* Vis scholar, Univ Wash, 80-81; ed-in chief, Biotechnology Progress, 85-88; vis prof, Univ Wis, 88-89. *Honors & Awards:* Colburn Lectr, Univ Del, 82; Marvin Johnson Award, Am Chem Soc, 86; Food, Pharm & Bioeng Award, Am Inst Chem Eng, 89. *Mem:* Nat Acad Eng; Am Chem Soc; Am Soc Microbiol; Am Soc Pharmacog; Am Inst Chem Engrs. *Res:* Biochemical engineering; mathematical models of individual cells; plant cell cultures; immobilized cell reactors; bioreactors for genetically-modified cells; interaction of heavy metals and biofilms; extractive fermentations; insect cell tissue culture; pharmadynamics. *Mailing Add:* Sch of Chem Eng Cornell Univ Ithaca NY 14853

SHULER, PATRICK JAMES, b Joliet, Ill, Dec 11, 48; m 80. PETROLEUM & CHEM ENGINEERING. *Educ:* Univ Notre Dame, BS, 71; Univ Colo, MS, 74, PhD(chem eng), 78. *Prof Exp:* RES ENGR PETROL ENG, CHEVRON OIL FIELD RES CO, DIV STANDARD OIL CALIF, 78- *Mem:* Sigma Xi; Am Inst Chem Engrs; Soc Petrol Engrs. *Res:* Enhanced oil recovery by chemical flooding. *Mailing Add:* Chevron Oil Field Res Co PO Box 446 La Habra CA 90633-0446

SHULER, ROBERT LEE, b West Columbia, SC, Mar 18, 26. SURFACE CHEMISTRY. *Educ:* Guilford Col, BS, 50; Georgetown Univ, MS, 60, PhD(chem), 69. *Prof Exp:* Chemist aeronaut fuels res, 54-61, res chemist biochem, 61-69, RES CHEMIST SURFACE CHEM, LAB CHEM PHYSICS, NAVAL RES LAB, 69- *Mem:* Am Chem Soc; Sigma Xi. *Res:* Study of the behavior of various types of polymers spread as monomolecular films on aqueous and nonaqueous liquids. *Mailing Add:* Apt 1512 5840 Cameron Run Terr Alexandria VA 22303

SHULL, CHARLES MORELL, JR, b Connellsville, Pa, Apr 8, 22; m 52; c 5. PHYSICAL CHEMISTRY, ANALYTICAL CHEMISTRY. *Educ:* Univ Tulsa, BS, 48; Univ Utah, MA, 50, PhD(chem), 53. *Prof Exp:* Instr chem, Univ Tulsa, 48-49; chemist, Newmont Explor, Ltd, 53-56; res metallurgist eng labs, Litton Eng Labs, Inc, 56-58; asst prof chem, Colo Sch Mines, 59-64, assoc prof, 64-66; tech writer curric develop, Educ Develop Ctr, 66-69; assoc prof, San Diego State Univ, 69-72, chmn dept, 74-80, prof natural sci, 72-83, ACTG ASSOC DEAN, IMPERIAL VALLEY CAMPUS, SAN DIEGO STATE UNIV, 80- *Concurrent Pos:* Consult curric res & writing, Ed Servs Inc, 66-67. *Honors & Awards:* Fulbright, Uruguay SA, 87. *Mem:* Am Soc Eng Educ; Am Chem Soc; Royal Soc Chem; Am Inst Mining, Metall & Petrol Eng. *Res:* Reaction rates; formation constants and molecular structure of complex ions; curriculum development in science education. *Mailing Add:* 415 N State St Apt 305 Bellingham WA 98225

SHULL, CLIFFORD GLENWOOD, b Pittsburgh, Pa, Sept 23, 15; m 41; c 3. SOLID STATE PHYSICS. *Educ:* Carnegie Inst Technol, BS, 37; NY UNiv, PhD(nuclear physics), 41. *Prof Exp:* Asst, NY Univ, 37-41; res physicist, Tex Co, 41-46; from prin physicist to chief physicist, Oak Ridge Nat Lab, 46-55; prof, 55-86, EMER PROF PHYSICS, MASS INST TECHNOL, 86- *Concurrent Pos:* Humboldt Sr Scientist Award, 80. *Honors & Awards:* Buckley Prize, Am Phys Soc, 56. *Mem:* Nat Acad Sci; AAAS; fel Am Acad Arts & Sci; Am Phys Soc; Sigma Xi. *Res:* Solid state and neutron physics. *Mailing Add:* Four Wingate Rd Levington MA 02173

SHULL, DON LOUIS, b Bridgewater, Va, Aug 2, 35; m 59; c 2. PHYSICAL CHEMISTRY. *Educ:* Bridgewater Col, BA, 56; Univ Va, PhD(phys chem), 61. *Prof Exp:* Res chemist, Hercules Inc, 61-64; sr chemist, Texaco Exp Inc, 64-69, group leader, Texaco, Inc, 69-75; dir labs, Commonwealth Labs, Inc, 76-77; dir res & sci advisor, Va Gen Assembly, 77-80; EXEC DIR, VA FUEL CONVERSION AUTHORITY, 81- *Mem:* Am Chem Soc; Sigma Xi. *Res:* Molecular spectroscopy; surface and colloid chemistry; chemical kinetics; analytical and testing; political science; information science. *Mailing Add:* 100 Hillview St Bridgewater VA 22812

SHULL, FRANKLIN BUCKLEY, b Ann Arbor, Mich, Apr 23, 18; m 47; c 4. PHYSICS. *Educ:* Univ Mich, AB, 39, MA, 40, PhD(physics), 48. *Prof Exp:* Assoc physicist, US Naval Ord Lab, 41-46; from asst prof to assoc prof, 48-63, prof, 63-88, EMER PROF PHYSICS, WASH UNIV, 88- *Mem:* Am Phys Soc. *Res:* Nuclear physics, especially beta-decay, nuclear reactions and polarization of nucleons. *Mailing Add:* Dept of Physics Wash Univ St Louis MO 63130

SHULL, HARRISON, b Princeton, NJ, Aug 17, 23; m 48, 62; c 4. QUANTUM CHEMISTRY, SCIENCE POLICY. *Educ:* Princeton Univ, AB, 43; Univ Calif, Berkeley, PhD(phys chem), 48. *Prof Exp:* Assoc chemist, US Naval Res Lab, Washington, DC, 43-45; Nat Res Coun fel, Univ Chicago, 48-49; assoc scientist, Ames Lab, AEC, 49-54; asst prof phys chem, Iowa State Univ, 49-55; PROVOST & ACAD DEAN, NAVAL POSTGRAD SCH, MONTEREY, CA, 88-; vpres acad affairs, provost & prof chem, Rensselaer Polytech Inst, 79-82; chancellor, 82-85, PROF CHEM, UNIV COLO, BOULDER, 82- *Concurrent Pos:* Guggenheim Found fel, 54-55 & Sloan res fel, 56-58; asst dir res, Quantum Chem Group, Sweden, 58-59; mem, comt awards, div chem & chem technol, 59-67, chmn, 63-67, mem, comt phys chem, 63-66, mem, panel surv chem, Westheimer Comt, 64-65, chmn, Comn Human Resources, 77-81, mem, comt adv Off Naval Res chem, Nat Res Coun, 85-; mem, adv panel chem, NSF, 64-67, chmn, 66-67, sr fel, 68-69, adv comn res, 74-76; consult, Off Sci Info Serv, 65-70; mem, chem vis comt, Brookhaven Nat Lab, 67-69, chmn, 69-70; mem, comt sci & pub policy, Nat Acad Sci, 69-72, mem coun & exec comt, 71-74, Naval Studies Bd, 73-79; mem, adv comt, Chem Abstr Serv, 72-75; trustee, Assoc Univs, Inc, 73-74; dir, Storage Technol Corp, 83-; mem, bd trustees, Argonne Univs Asn, 70-75 & Inst Defense Analysis, 84-; mem, exec panel, Chief Naval Opers, 84-89. *Mem:* Nat Acad Sci; AAAS; fel Am Phys Soc; fel Am Acad Arts & Sci (vpres, 75-79, 80-83); Am Chem Soc; foreign mem Royal Swedish Acad Sci; foreign mem Royal Uppsala Acad Arts & Sci. *Res:* Quantum chemistry; theoretical and experimental molecular spectroscopy and structure. *Mailing Add:* Code 01 Naval Postgraduate School Monterey CA 93943-5000

SHULL, JAMES JAY, b Chester, Pa, Sept 1, 29; m 52; c 2. MICROBIOLOGY, BIOCHEMISTRY. *Educ:* Pa State Col, BS, 51, MS, 55; Pa State Univ, PhD(microbiol), 63. *Prof Exp:* Res supvr microbiol, Wilmot Castle Co, 59-61; instr, Syracuse Univ, 61-63; sr res assoc, Am Sterilizer Co, 63-65; mgr biol & med sci lab, Gen Elec Co, 65-69, exp consult aerospace biol, 69-71, tech dir microbial protein prod, 72-74; vpres res & strategic planning, Ariz Feeds, 74-80; oper mgr, Vega Biotechnol, Inc, 80-82; VPRES RES, SHULCON INDUSTS, INC, 82-; VPRES MFG, VALPAR INT CORP, 86- *Mem:* Am Soc Microbiol; AAAS; NY Acad Sci. *Res:* Gastrointestinal disease and antidiarrheal agents; animal nutrition and health; production of microbial protein; energy production through biological systems. *Mailing Add:* 4351 E Saranac Dr Tucson AZ 85718

SHULL, PETER OTTO, JR, b Summit, NJ, July 24, 54. SUPERNOVA REMNANTS, ASTROPHYSICAL JETS. *Educ:* Princeton Univ, AB, 76; Rice Univ, MS, 79, PhD(astron), 81. *Prof Exp:* Res assoc, Max Planck Inst Astron, 82-83 & physics dept, Ariz State Univ, 84; asst prof, 84-89, ASSOC PROF ASTRON, PHYSICS DEPT, OKLA STATE UNIV, 89- *Concurrent Pos:* Vis scientist, Max Planck Inst Astron, 85, 88; Sci & Eng Res Coun vis fel, Univ Manchester, 83, 85, 88. *Mem:* Am Astron Soc; Sigma Xi. *Res:* Observational and theoretical astrophysics; structure and evolution of supernova remnants; Crab Nebula's jet; expansion of the Cygnus Loop; interactions of supernova progenitor stars with the interstellar medium. *Mailing Add:* Dept Physics Okla State Univ Stillwater OK 74078-0444

SHULLS, WELLS ALEXANDER, b Hudson, Wis, Oct 2, 16; m 42; c 1. MICROBIOLOGY. *Educ:* Mich State Univ, BS & MS, 39; Wayne State Univ, PhD, 57. *Prof Exp:* Assoc microbiologist, Parke, Davis & Co, 44-49; asst prof bact, Col Pharm, Detroit Inst Technol, 49-57; prof microbiol, Univ Colo, 57-82; RETIRED. *Concurrent Pos:* Res assoc, Col Med, Wayne State Univ, 53-57. *Mem:* AAAS; Am Soc Microbiol. *Res:* Methane bacteria in soils and succession of bacteria in alpine soils in plant decay. *Mailing Add:* 2315 Forest Ave Boulder CO 80304

SHULMAN, CARL, b Chelsea, Mass, Jan 12, 17. ELECTRICAL ENGINEERING, OPTICS. *Educ:* Mass Inst Technol, BS, 38, MS, 39; Princeton Univ, MA, 48, PhD(physics), 57. *Prof Exp:* Res engr, Submarine Signal Co, Mass, 39-40 & Res Labs, Radio Corp Am, 40-55; assoc prof, 55-70, PROF ELEC ENG, CITY COL NEW YORK, 70- *Mem:* AAAS; Am Phys Soc; Inst Elec & Electronics Engrs; NY Acad Sci. *Res:* High frequency electronics; solid state physics; electromagnetic theory. *Mailing Add:* Sch Technol City Col NY New York NY 10031

SHULMAN, DAVID-DIMA, b Kremenchug, USSR, July 9, 58; Israeli citizen. SEMICONDUCTOR DEVICES, DESIGN OF BIPOLAR & CMOS CIRCUITS. *Educ:* Israel Inst Technol, Haifa, BS, 83, MS, 86. *Prof Exp:* Res asst elec eng, Israel Inst Technol, Haifa, 83-85; develop engr circuit design, Tadiran, Commun Div, Israel, 86-88; res asst elec eng, Univ Waterloo, Ont, Can, 88-89; RES ASST ELEC ENG, UNIV BC, CAN, 89- *Mem:* Electrochem Soc. *Res:* Modeling and processing of gallium arsenide devices; investigation of parasitic effects in gallium arsenide integrated circuits; design and analysis of monolithic silicon circuits. *Mailing Add:* Dept Elec Eng Univ BC 2356 Main Mall Vancouver BC V6T 1W5 Can

SHULMAN, GEORGE, b West New York, NJ, Sept 3, 14; m 43; c 4. INDUSTRIAL ORGANIC CHEMISTRY. *Educ:* City Col New York, BS, 36. *Prof Exp:* Res chemist, Insl-X co, 37-39; res chemist, 39-40, plant chemist, 40-51, tech dir, 51-64, VPRES, PFISTER CHEM INC, 64- *Mem:* Am Chem Soc; Math Asn Am; NY Acad Sci. *Res:* Organic synthesis; dye intermediates; thickeners for liquid hydrocarbons; plastics and plasticizers; textile chemicals. *Mailing Add:* 715 Winthrop Rd Teaneck NJ 07666-2266

SHULMAN, HAROLD, b Newark, NJ, Feb 12, 25; m 58; c 3. MATHEMATICS. *Educ:* George Washington Univ, BS, 48; Johns Hopkins Univ, MA, 51; NY Univ, PhD(math), 58. *Prof Exp:* Tutor, Queens Col, NY, 52-54; assoc res scientist, Inst Math Sci, NY Univ, 54-60; sr analyst, Comput Usage Co, 60-62; prin comput engr, Repub Aviation Corp, 62-64; asst prof math, Hunter Col, 64-68; ASST PROF MATH, LEHMAN COL, 68- *Concurrent Pos:* Consult, Corps Engrs, Dept Defense, 57-59. *Mem:* Am Math Soc; Math Asn Am. *Res:* Numerical analysis; applied mathematics; statistics. *Mailing Add:* 144-43 Jewel Ave Flushing NY 11367

SHULMAN, HERBERT BYRON, b Weehawken, NJ, June 29, 47; m 73; c 2. COMPUTER PERFORMANCE ANALYSIS, SYSTEMS ENGINEERING. *Educ:* Cornell Univ, AB, 68; Univ Calif, Berkeley, PhD(math), 72. *Prof Exp:* Instr math, Yale Univ, 72-74 & Univ Pa, 74-75; asst prof math, Belfer Grad Sch Sci, Yeshiva Univ, 75-78; mem tech staff, 78-82, SUPVR, BELL TELEPHONE LABS, 82- *Concurrent Pos:* NSF grant, 72-78. *Res:* Queuing theory; traffic engineering. *Mailing Add:* 521 Marl Rd Colts Neck NJ 07722

SHULMAN, HERMAN L, b New York, NY, Feb 24, 22; m 42; c 2. CHEMICAL ENGINEERING, EDUCATIONAL ADMINISTRATION. *Educ:* City Univ New York, BChE, 42; Univ Pa, MS, 48, PhD(chem eng), 50. *Hon Degrees:* DSc, Clarkson Univ, 87. *Prof Exp:* Chem engr, Gen Motors Corp, 42-43, Barrett Div, Allied Chem & Dye Corp, 43-46; res chem engr, Publicker Industs, Inc, 46-47; from asst prof to prof, Clarkson Univ, 48-77, assoc dir div res, 54-59, dir, 59-77, chmn dept, 59-64, dean grad sch, 64-77, vpres, 68-77, dean sch eng, 68-85, provost, 77-85, exec vpres, 85-87, EMER PROF & PROVOST, CLARKSON UNIV, 88- *Mem:* Am Soc Eng Educ; Am Chem Soc; fel Am Inst Chem Engrs; Nat Soc Prof Engrs; Sigma Xi. *Res:* Mass transfer absorption; packed columns; ion exchange; gas-bubble columns; organic processes; flowmeters. *Mailing Add:* Clarkson Univ Potsdam NY 13699-5500

SHULMAN, IRA ANDREW, b Los Angeles, Calif, Feb 26, 49; m 72; c 2. TRANSFUSION MEDICINE. *Educ:* Univ Calif, Los Angeles, BA, 71; Univ Southern Calif, MD, 75. *Prof Exp:* Blood Bank Dir Med, St Mary Med Ctr, 80-81 TRANS FUSION MED, 81-; BLOOD BANK DIR & ASSOC PROF TRANS FUSION MED, UNIV SOUTHERN CALIFMED CTR, 81- *Concurrent Pos:* Co-chmn, teleconf planning group, Am Soc Clin Pathol, 87-, coun transfusion med, 86-; pres, Coun Blood Bank dir, 85-; bd dir, Calif Blood Bank Soc, 88-, chmn sci prog comt, 85-88. *Mem:* Am Assoc Blook Banks; Am Soc Clin Pathol; Col Am Pathol. *Res:* Investigating the clinical significance of red blood cell alloantibodies and auto antibodies. *Mailing Add:* 3488 Mc Laughlin Ave Los Angeles CA 90066

SHULMAN, JONES A(LVIN), b Baltimore, Md, Sept 5, 36; m 58; c 2. INTERNAL MEDICINE. *Educ:* Univ Md, MD, 60; Am Bd Internal Med, dipl, 68. *Prof Exp:* Intern internal med, Univ Md Hosp, Baltimore, 60-61; asst resident med, Univ Wash, 61-62, R G Petersdorf fel infectious dis & med, 62-64, instr & chief resident med, Univ Wash Hosp, 66-67; from asst prof to assoc prof infectious dis, 70-74, coordr clin curric, 70-76, dir Div Infectious Dis, 72-84, assoc dean clin educ,76-86, PROF MED & PROF PREV MED & COMMUNITY HEALTH, EMORY UNIV, 74- *Concurrent Pos:* CHIEF MED, CRAWFORD LONG HOSP, EMORY UNIV. *Mem:* Infectious Dis Soc Am. *Res:* Epidemiology of hospital infections; antibiotics, clinical pharmacology and efficacy studies. *Mailing Add:* Dept of Med Div Infectious Dis Sch Med Emory Univ Atlanta GA 30303

SHULMAN, LAWRENCE EDWARD, b Boston, Mass, July 25, 19; m 59; c 2. INTERNAL MEDICINE. *Educ:* Harvard Univ, AB, 41; Yale Univ, PhD(pub health), 45, MD, 49. *Prof Exp:* Res assoc, John B Pierce Found, Conn, 42-45; intern med, Johns Hopkins Hosp, 49-50, asst resident, 52-53, physician, 53; dir, Connective Tissue Div, Dept Med, 54-75, from instr to asst prof, 53-63, ASSOC PROF MED, SCH MED, JOHNS HOPKINS UNIV, 63-; dir, Div Arthritis, Musculoskeletal & Skin Dis, Nat Inst Arthritis,

Diabetes & Digestive & Kidney Dis, 76-86, DIR, NAT INST ARTHRITIS & MUSCULOSKELETAL & SKIN DIS, 86- *Concurrent Pos:* Fel, Sch Med, Johns Hopkins Univ, 50-52; sr investr, Arthritis & Rheumatism Found, 57. *Honors & Awards:* Heberden Medal Res, 76; Superior Serv Award, Pub Health Serv, 85. *Mem:* AAAS; Am Rheumatism Asn (pres, 74-75); Asn Am Med Cols; Am Fedn Clin Res; NY Acad Sci; Pan Am League Against Rheumatism (pres, 82-86). *Res:* Connective tissue disorders, including rheumatic diseases, especially collagen disorders; immunologic and epidemiologic studies. *Mailing Add:* NIH Bldg 31 Rm 4C-32 Bethesda MD 20892

SHULMAN, MORTON, b Chicago, Ill, July 7, 33; m 55; c 3. ANESTHESIOLOGY. *Educ:* Univ Ill, BS, 55, MD, 58. *Prof Exp:* Asst, 59-61, res fel, 61-62, from instr to asst prof, 63-70, clin prof anesthesiol, 81, ASSOC PROF ANESTHESIOL, UNIV ILL COL MED, 70- *Concurrent Pos:* Attend anesthesiologist, Ill Masonic Med Ctr, 66-89, Shriners Hosp Crippled Children, Chicago, 78; consult, Col Vet Med, Univ Ill, 68-70; vis prof, Univ Louisville, 69, Med Col Wis, 70, Mayo Clin, 78 & Brigham Women's Hosp, Harvard Univ, 81; sr attend anesthesiologist, Rush Presby St Lukes Med Ctr, Chicago, Ill, 89; prof anesthesiol, Rush Med Col, 89. *Mem:* Am Soc Anesthesiol; Int Anesthesia Res Soc. *Res:* Local anesthetics and analgesics; pain control. *Mailing Add:* 1115 Thorn Tree Lane Highland Park IL 60035

SHULMAN, N RAPHAEL, b Hartford, Conn, June 23, 25; m; c 3. CLINICAL HEMATOLOGY. *Educ:* Johns Hopkins & Harvard Univs, AB, 43; Johns Hopkins Univ, MD, 47; Am Bd Internal Med, dipl, 55. *Prof Exp:* Intern, Boston City Hosp, Harvard Med Serv, 47-48; asst resident med, Mem Hosp, Cornell Med Serv, NY, 48-49, res med, 49-50, chief res med, 50-51; med corps, USN, Naval Med Res Inst, Hemat Div & Nat Naval Med Ctr, Dept Med, Bethesda, Md, 51-55, chief, Hemat Div, Naval Med Res Inst & hematologist, Nat Naval Med Ctr, 55-57; chief hemat res, Metab Dis Br, Nat Inst Arthritis & Metab Dis, 57-62, actg dep sci dir, Intramural Res, 89-90, CHIEF, CLIN HEMATOL BR, NAT INST DIABETES & DIGESTIVE & KIDNEY DIS, NIH, BETHESDA, 62- *Concurrent Pos:* USN rep, Nat Res Coun Comt, Blood & Blood Prod, 51-57; instr clin med, George Washington Univ, 51-57, assoc clin prof med, 57-61, prof, 61-; mem, Hemat Study Sect, NIH, 55-57 & 61-67, adv comt Nat Inst Arthritis & Metab Dis Extramural Hemat Prog, 64-65; postdoctoral clin res fel rev panel, NIH, 57-63; chmn, Clin Res Comt Med Bd, Clin Ctr, NIH, 62-64, mem, 59-65, Exp Adv Panel Immunol, WHO, 69-; actg dir, Nat Inst Arthritis & Metab Dis, 64-65; reviewer, J Clin Invest, Immunol, J Exp Med, NEng J Med, Am J Med, Annals Internal Med, Blood, Metab, Endocrinol, Transfusion; actg dep sci dir, Intramural Res, Nat Inst Dibetes & Digestive & Kidney Dis, NIH, 89- *Mem:* Am Soc Clin Invest; Asn Am Physicians; Am Soc Hemat. *Res:* Hematology; metabolic diseases; patents. *Mailing Add:* Nat Inst Diabetes Digestive & Kidney Dis NIH Clin Hemat Br Bldg 10 Rm 8C101 Bethesda MD 20892

SHULMAN, ROBERT GERSON, b New York, NY, Mar 3, 24; m 85; c 3. CHEMICAL PHYSICS. *Educ:* Columbia Univ, AB, 43, AM, 47, PhD(chem), 49. *Prof Exp:* AEC fel, Calif Inst Technol, 49-50; head semiconductor res, Hughes Aircraft Co, 50-53; mem tech staff, Bell Labs, Inc, 53-79; PROF, DEPT MOLEC BIOPHYS & BIOCHEM, YALE UNIV, 79- *Concurrent Pos:* Guggenheim fel, Lab Molecular Biol, Med Res Coun Eng, 61; vis prof, Ecole Normale Superieur Univ, Paris, 62; vis lectr, Princeton Univ, 71-72. *Honors & Awards:* Rask Oersted lectr, Copenhagen Univ, 59; Appleton lectr, Brown Univ, 65; Reilly lectr, Univ Notre Dame, 69. *Mem:* Nat Acad Sci; Biophys Soc; Inst Med. *Res:* Microwave spectroscopy; semiconductors; nuclear magnetic resonance; molecular orbital theory of transition metal complexes; radiation damage to DNA; metalloenzymes; phage genetics; paramagnetic metal ion complexes of nucleic acids; high resolution nuclear magnetic resonance of hemoglobin and tRNA NMR of metabolism in vivo. *Mailing Add:* Dept Molec Biophys & Biochem Yale Univ PO Box 3333 New Haven CT 06510

SHULMAN, ROBERT JAY, b Newark, NJ, Jan 6, 50. GASTROENTEROLOGY, PEDIATRICS. *Educ:* Emory Univ, BA, 72; Chicago Med Sch, MD, 76. *Prof Exp:* Pediat residency, Univ Mich, 76-79; pediat fel, Baylor Col Med, 79-81, instr gastroenterol & nutrit, 81-82, asst prof, 82-90, ASSOC PROF GASTROENTEROL & NUTRIT, BAYLOR COL MED, 90- *Concurrent Pos:* Dir nutrit support team, Tex Childrens Hosp, 82- *Mem:* Soc Pediat Res; Am Gastroent; Am Acad Pediats; Am Soc Parenteral & Enteral Nutrit; NAm Soc Pediat. *Res:* Growth and repair of the small intestine; factors involved in normal growth of the intestine; repair of the damaged or surgically shortened intestine and the effects of these on carbohydrate digestion and resorption. *Mailing Add:* Dept Pediat Baylor Col Med One Baylor Plaza Houston TX 77030

SHULMAN, SETH DAVID, b Lynn, Mass, Mar 11, 43; m 69; c 3. X-RAY ASTRONOMY. *Educ:* Harvard Univ, BA, 63; Columbia Univ, PhD(physics), 70. *Prof Exp:* Res physicist, E O Hulburt Ctr Space Res, 70-81, CONSULT X-RAY ASTRON, NAVAL RES LAB, 81-; PRES, BIOSCAN, INC, 80- *Mem:* Am Astron Soc; Am Phys Soc; Am Chem Soc. *Res:* X-ray astronomy and studies of the interstellar medium; nuclear radiation detectors. *Mailing Add:* Bioscan Inc 4402 Que St NW Washington DC 20007

SHULMAN, SIDNEY, b Baltimore, Md, Aug 22, 23; m 45, 68; c 6. IMMUNOLOGY. *Educ:* George Washington Univ, BS, 44; Univ Wis, PhD(chem), 49. *Prof Exp:* Assoc, Allegany Ballistics Lab, US Army Ballistics Missile Agency, Md, 44-46; asst chem, Univ Wis, 46-48, proj assoc, 49-52, assoc immunochem, Sch Med, State Univ NY Buffalo, 52-54, asst prof, 54-58, assoc prof immunochem & biophys, 58-65, prof immunochem, 65-68, chmn dept microbiol, 68-69, PROF MICROBIOL, NEW YORK MED COL, 68-, DIR SPERM ANTIBODY LAB, 70-, RES PROF UROL, 73-,. *Concurrent Pos:* Lederle med fac fel, 54-57; USPHS sr res fel, 58-62; NIH res career award, 63-68; consult immunol, Buffalo Vet Admin Hosp, 64-; Fulbright travel award, 65; Commonwealth Fund travel award, 65; vpres, Int Coord Comt Immunol Reproduction, 67-; assoc ed, Cryobiol, 69-; chmn workshop on immunoreproduction, 1st Int Cong Immunol, 71; assoc ed, Contraception, 73-; assoc ed, Int J Fertil; mem WHO task force immunol methods fertil regulation, 75-; res prof obstet & gynec, New York Col Med, 74- *Mem:* Int Soc Immunol Reproduction (vpres, 75); Am Chem Soc; assoc fel Am Col Obstet & Gynec; Am Soc Microbiol; Am Acad Allergy. *Res:* Tissue proteins; autoantibodies; cryobiology; urogenital tract antigens and enzymes; immunology of reproduction and infertility. *Mailing Add:* Sperm Antibody Lab Fertility Antibody Diag 945 West End Ave Suite 1D New York NY 10025

SHULMAN, SOL, b Smorgon, White Russia, Nov 6, 29; US citizen; m 53; c 4. ORGANIC CHEMISTRY, POLYMER CHEMISTRY. *Educ:* Univ Wash, BS, 52; Univ Wis, MS, 54; NDak State Univ, PhD(chem), 63. *Prof Exp:* Teaching asst chem, Univ Wis, 53-54; res chemist, Archer-Daniels-Midland Co, Minn, 54-59; from instr to asst prof, NDak State Univ, 59-65; from assoc prof to prof chem, Moorhead State Col, 65-69, chmn dept, 66-69; head dept, 69-75, PROF CHEM, ILL STATE UNIV, 69- *Concurrent Pos:* Sr res assoc, Rice Univ, 75. *Mem:* Fel AAAS; Am Chem Soc; fel Am Inst Chemists. *Res:* Chemistry of lipids; derivatives of fats and oils, polyurethanes and synthesis. *Mailing Add:* Dept of Chem Ill State Univ Normal IL 61761

SHULMAN, STANFORD TAYLOR, b Kalamazoo, Mich, May 13, 42; m 64; c 3. PEDIATRICS, INFECTIOUS DISEASES. *Educ:* Univ Cincinnati, BS, 63; Univ Chicago, MD, 67. *Prof Exp:* Intern-resident pediat, Univ Chicago, 67-69, chief resident, 69-70; fel, Univ Fla, 70-73, asst prof, 73-75, assoc prof pediat inf dis & immunol, 75-79; actg chmn, Dept Pediat, 81-83, PROF PEDIAT, MED SCH, NORTHWESTERN UNIV, 79- *Concurrent Pos:* Fel immunol, Inst Child Health, London, Eng, 70; chief infectious dis, Children's Mem Hosp, Chicago, 79-; chmn, comt rheumatic fever & infective endocariditis, Am Heart Asn, 81-; vis prof, Univ Minn, Yale Univ. *Mem:* Soc Pediat Res; Am Acad Pediat; Infectious Dis Soc Am; Am Asn Immunol; AAAS. *Res:* Pathogenesis of rheumatic fever; kawasaki disease: etiology, pathogenesis and therapy; clinical immunology. *Mailing Add:* Dept Pediat Northwestern Univ Children's Mem Hosp 2300 Children's Plaza IL 60614

SHULMAN, YECHIEL, b Tel Aviv, Israel, Jan 28, 30; nat US; m 50; c 3. ENGINEERING, MANAGEMENT. *Educ:* Mass Inst Technol, SB(aeronaut eng), SB(bus & eng admin) & SM, 54, ScD(aeronaut & astronaut eng), 59; Univ Chicago, MBA, 73. *Prof Exp:* Res engr, Aeroelastic & Struct Res Lab, Mass Inst Technol, 54-56, asst, 57-59; asst prof mech eng, Tech Inst, Northwestern Univ, 59-62, assoc prof mech eng & astronaut, 62-67; res consult, Anocut Eng Co, 67-68, vpres advan eng, 69-72, dir electronic systs group, 70-72; vpres corp planning, Alden Press, Inc, John Blair & Co, 72-84; pres, MMT Environ, Inc, 84-86; adj prof mech eng, 86-89, H W SWEATT CHAIR & DIR, CTR DEVELOP TECHNOL LEADERSHIP, UNIV MINN, 90-, PROF MECH ENG, 90- *Concurrent Pos:* Consult, Am Mach & Foundry Co, 60-62 & Res Div, Gen Am Transp Corp, 62-67; mem res & develop comt, Nat Mach Tool Builders Asn, 68-72. *Mem:* Am Inst Aeronaut & Astronaut; Soc Mfg Engrs; Am Soc Eng Educ; Am Soc Mech Engrs; Graphic Commun Asn. *Res:* Electrochemical machining; digital control systems for process machinery; laser metrology, aerothermoelasticity; shell structures; astrodynamics and optimization; structural dynamics; flight mechanics; management of technology. *Mailing Add:* 1201 Yale Pl No 1504 Minneapolis MN 55403-1959

SHULT, ERNEST E, b Tonica, Ill, Sept 29, 33; m 57; c 3. ALGEBRA, COMBINATORICS. *Educ:* Southern Ill Univ, BA, 58, MA, 60; Univ Ill, PhD(math), 64. *Prof Exp:* Res assoc genetics yeast, Biol Res Lab, Southern Ill Univ, 54-57, chief theoretician, 58-61, instr math, 63-64, asst prof, 64-65; NSF fel, Univ Chicago, 65-66; from assoc prof to prof math, Southern Ill Univ, 66-70; prof, Univ Fla, 70-74; DISTINGUISHED REGENTS PROF MATH, KANS STATE UNIV, 74- *Concurrent Pos:* Mem, Inst Advan Study, 68-69; Alexander von Humboldt Sr Am Scientist award. *Honors & Awards:* Leo Kaplan Res Prize, Sigma Xi, 70. *Mem:* Sigma Xi; Am Math Soc. *Res:* Abstract algebra; finite geometry; microbial genetics, especially yeast genetics; theory of finite groups combinatorics. *Mailing Add:* Dept of Math Kans State Univ Manhattan KS 66506

SHULTIS, J KENNETH, b Toronto, Ont, Aug 22, 41; m 67; c 2. RADIATION SHIELDING & PROTECTION, RADIATION TRANSPORT. *Educ:* Univ Toronto, BASc, 64; Univ Mich, MSc, 65, PhD(nuclear sci & eng), 68. *Prof Exp:* PROF NUCLEAR ENG, KANS STATE UNIV, 69- *Concurrent Pos:* Sci officer, Univ Groningen, Neth, 68-69; Black & Veatch distinguished prof, Kans State Univ, 78; von Humboldt res fel, Univ Karlsruhe, 80; guest prof, Univ Karlsruhe, Ger, 80-81; consult, many elec utilities & eng firms; prin investr, over 30 fed & state grants; mem, Exec Comt, RP&S Div, Am Nuclear Soc, 86-90. *Honors & Awards:* Western Elec Award, Am Soc Eng Educ, 79 & Glenn Murphy Award, 81. *Mem:* Am Soc Eng Educ. *Res:* Numerical and computer analyses; neutron transport theory; radiative transfer; combustion modeling; remote sensing; radiation shielding; reactor physics; author of over 100 publications. *Mailing Add:* Dept Nuclear Eng Ward Hall Kans State Univ Manhattan KS 66506

SHULTS, WILBUR DOTRY, II, b Atlanta, Ga, Nov 24, 29; m 50; c 3. ANALYTICAL CHEMISTRY. *Educ:* Emory Univ, AB, 50, MS, 51; Ind Univ, PhD(chem), 66. *Prof Exp:* From jr chemist to assoc chemist, 51-55, from chemist to group leader, 57-67, asst div dir, 67-71, assoc div dir, 71-76, DIV DIR ANAL CHEM, OAK RIDGE NAT LAB, 76- *Mem:* Am Chem Soc; AAAS; Sigma Xi. *Res:* Instrumental analysis. *Mailing Add:* 1011 W Outer Dr Oak Ridge TN 37830

SHULTZ, ALLAN R, b Huntington, Ind, Jan 8, 26; m 49; c 4. PHYSICAL CHEMISTRY, POLYMER PHYSICS. *Educ:* Manchester Col, AB, 48; Cornell Univ, PhD(chem), 53. *Prof Exp:* Res assoc chem, Mass Inst Technol, 52-54; res chemist, Minn Mining & Mfg Co, 54-63; RES CHEMIST, GEN ELEC RES & DEVELOP CTR, SCHENECTADY, 63- *Mem:* AAAS; Am Chem Soc; NY Acad Sci. *Res:* Polymer chemistry and physics; radiation chemistry of polymers; thermodynamics; polymer structure/property relationships. *Mailing Add:* 111 Acorn Dr Scotia NY 12302

SHULTZ, CHARLES H, b Lancaster, Pa, May 29, 36. VOLCANIC GEOLOGY, SCIENCE EDITING. *Educ:* Franklin & Marshall Col, BS, 58; Ohio State Univ, PhD(petrol), 62. *Prof Exp:* Geologist, Humble Oil & Refinery Co, 62-64; asst prof geol, Ohio State Univ, 64-70; assoc prof, 70-72, PROF GEOL, SLIPPERY ROCK UNIV, 72- *Concurrent Pos:* Ed, Geol Pa, 83- *Mem:* Fel Geol Soc Am; Am Geophys Union; AAAS; Int Asn Volcanic & Chem Earth Interiors; Sigma Xi. *Res:* Tertiary volcanic petrology; tectonics and metamorphism of the Appalachian Mountains; volcanology of Antarctica. *Mailing Add:* Dept Geol Slippery Rock Univ Slippery Rock PA 16057

SHULTZ, CLIFFORD GLEN, b Wichita, Kans, Sept 9, 24; m 50; c 3. SCIENCE ADMINISTRATION, ANALYTICAL CHEMISTRY. *Educ:* McPherson Col, AB, 49; Univ NMex, PhD(chem), 57. *Prof Exp:* Group leader, anal res, Mich Chem Corp, 57-60; assoc prof chem, Evansville Col, Ind, 60-66; PRES, NAT LABS, INC, 69- *Concurrent Pos:* Mem coun, Am Chem Soc, 63-64. *Mem:* Am Chem Soc; Am Inst Chemists; Nat Soc Prof Engrs; Water Pollution Control Fedn; Air & Waste Mgt Asn. *Res:* Destruction of hazardous chemical and biological wastes by thermal reduction with molten aluminum; recovery of valuable materials from waste products; extraction of hazardous materials from contaminated soils and sludge; five US patents. *Mailing Add:* 1701 Glendale Evansville IN 47712

SHULTZ, FRED TOWNSEND, b Grinnell, Iowa, Mar 5, 23; m 61; c 4. GENETICS. *Educ:* Stanford Univ, AB, 47; Univ Calif, Berkeley, PhD(genetics), 52. *Prof Exp:* Asst poultry husb, 49-52, jr res geneticist, 52-53, RES ASSOC POULTRY HUSB, UNIV CALIF, 53-; DIR, BIOL FRONTIERS INST, 60- *Concurrent Pos:* Pres, Animal Breeding Consult, 52- *Honors & Awards:* Res Prize, Am Poultry Sci Asn, 53. *Mem:* Am Poultry Sci Asn; Genetics Soc Am; Soc Study Evolution; World Poultry Sci Asn; Nat Shellfisheries Asn. *Res:* Aquaculture; animal breeding; marine biology. *Mailing Add:* PO Box 313 Sonoma CA 95476

SHULTZ, LEILA MCREYNOLDS, b Bartlesville, Okla, April 20, 46; c 1. PLANT SYSTEMATICS. *Educ:* Univ Tulsa, BS, 69; Univ Colo, MA, 75; Claremont Grad Sch, PhD(bot), 83. *Prof Exp:* CURATOR, INTERMOUNTAIN HERBARIUM, UTAH STATE UNIV, 78-, ADJ ASST PROF BIOL DEPT, 86- *Concurrent Pos:* Teacher biol & french, Elwood High Sch, Kans, 69-70; res asst Herbarium, Univ Colo, 71-73; asst curator, Intermountain Herbarium, Utah State Univ, 73-78; consult, Bur Land Mgt, 76-78, US Forest Serv, 78-81, US Fish & Wildlife Serv, 78-; instr, Nat Wildlife Fedn, 73-88; ed, Flora NAm Proj, 86-; vis scholar, Harvard Univ, 88-89; vis researcher, Univ Calif, Los Angeles, 88-89; coun mem, Am Soc Plant Taxonomist; syst rep, Bot Soc. *Mem:* Am Soc Plant Taxonomist; Bot Soc Am; Int Asn Plant Taxonomists. *Res:* Systematics of Artemisia; ecology and evolutionary specializations in leaf anatomy. *Mailing Add:* Dept Biol Utah State Univ Logan UT 84322-5500

SHULTZ, LEONARD DONALD, b Boston, Mass, Apr 16, 45; m 69; c 2. CANCER. *Educ:* Northeastern Univ, BA, 67; Univ Mass, PhD(microbiol), 72. *Prof Exp:* Res asst cancer, Sch Med, Tufts Univ, 67-68; teaching asst microbiol, Univ Mass, 68-70, lectr immunol, 70-71; trainee, 72-74, res assoc cancer, 74-76, assoc staff scientist, 76-79, STAFF SCIENTIST, JACKSON LAB, 80- *Concurrent Pos:* Prin investr grants, Nat Cancer Inst & Nat Inst Allergy & Infectious Dis. *Mem:* Am Soc Microbiol; Am Asn Immunologists. *Res:* The study of lymphoid cell differentiation and immunoregulatory mechanisms in tumorigenesis; immuno-deficiency diseases; autoimmunity; AIDS. *Mailing Add:* Jackson Lab Bar Harbor ME 04609

SHULTZ, TERRY D, b Caldwell, Idaho, July 26, 47. NUTRITION ASSESSMENT, VITAMIN B-SIX METABOLISM. *Educ:* Ore State Univ, PhD(human nutrit biochem), 80. *Prof Exp:* ASST PROF BIOCHEM & NUTRIT, SCH MED, LOMA LINDA UNIV, 82- *Mem:* Am Inst Nutrit; Endocrine Soc; Sigma Xi; Am Soc Bone & Mineral Res. *Mailing Add:* Dept Food Sci & Human Nutrit Wash State Univ Pullman WA 99164-6376

SHULTZ, WALTER, b Philadelphia, Pa, Nov 11, 31; m 58; c 3. BIOPHARMACEUTICS. *Educ:* Temple Univ, BS, 53; Philadelphia Col Pharm, MS, 54, PhD(pharmaceut chem), 61. *Prof Exp:* Develop chemist, 60-68, group leader pharmaceut prod develop, 68-78, SR RES SCIENTIST, LEDERLE LABS DIV, AM CYANAMID CO, 78- *Mem:* AAAS; Am Chem Soc; Am Pharmaceut Asn; Acad Pharmaceut Sci. *Res:* Physical pharmacy; synthesis of steroid derivatives. *Mailing Add:* Pharmaceut Prod Develop Sect Lederle Labs Pearl River NY 10965

SHUM, ANNIE WAICHING, b US. COMPUTER SCIENCE, MATHEMATICAL MODELING. *Educ:* Univ Calif, Berkeley, BA, 72; Harvard Univ, MSc, 73, PhD(comput sci), 76. *Prof Exp:* Res staff comput sci, IBM Corp, 77-78; ASST PROF COMPUT SCI, DIV APPL SCI, HARVARD UNIV, 78- *Concurrent Pos:* Consult, BGS Systs Inc, 78-; res grant div appl sci, Harvard Univ, 78-79. *Mem:* Asn Comput Mach; Sigma Xi; Inst Elec & Electronics Engrs; Math Soc. *Res:* Queueing theory; mathematical models; operating system design and performance evaluation of computer systems; efficient computational algorithms of combinatoric problems. *Mailing Add:* BGS Systs Inc, 1 Univ Office Pk 29 Sawyer Rd Waltham MA 02254

SHUM, ARCHIE CHUE, b Hong Kong, Aug 13, 42; US citizen; m 73. CLINICAL MICROBIOLOGY, BIOCHEMISTRY. *Educ:* Idaho State Univ, BS, 68, MS, 70; Univ Iowa, PhD(microbiol), 73. *Prof Exp:* Asst prof microbiol, Calif State Univ, Los Angeles, 73-76; clin microbiologist, St John's Hosp, 77-89; intern, clin labs, 76-77, MICROBIOLOGIST, UNIV CALIF, LOS ANGELES, 88- *Concurrent Pos:* Consult, Santa Paula Hosp, Santa Paula, Calif & Pleasant Valley Hosp, Camarillo, Calif, 77. *Mem:* Am Soc Microbiol; AAAS; Sigma Xi. *Res:* Antimicrobial susceptibility testing on anaerobic bacteria of clinical significance; bacteriological studies on bowhead whales; cellular immunology. *Mailing Add:* 1420 N M St Oxnard CA 93030

SHUM, WAN-KYNG LIU, biochemistry, endocrinology, for more information see previous edition

SHUMACKER, HARRIS B, JR, b Laurel, Miss, May 20, 08; m 33; c 2. SURGERY. *Educ:* Univ Tenn, Chattanooga, BS, 27; Vanderbilt Univ, MA, 28; Johns Hopkins Univ, MD, 32; Am Bd Surg, dipl, 46; Am Bd Thoracic Surg, dipl, 67. *Hon Degrees:* DSc, Ind Univ, 85. *Prof Exp:* Instr surg, Sch Med, Yale Univ, 37-38; instr, Johns Hopkins Univ, 38-41, asst prof, 41-46; assoc prof, Sch Med, Yale Univ, 46-48; prof & chmn dept, 48-70, EMER DISTINGUISHED PROF SURG, IND UNIV, 78-; prof & sr adv, Dept Surg, 81-88, DISTINGUISHED PROF SURG, UNIFORMED SERVS UNIV HEALTH SCI, 88- *Concurrent Pos:* Consult, Surgeon Gen, US Army, 50-55, mem adv comt, Environ Med, 57-61; surg study sect, US Pub Health Dept, 50-56 & Therapeut Eval Comt, 69-73; vpres, Int Soc Surg, 57-59, pres, NAm Chap, 56-58; vchmn, Am Bd Surg, 59-60; chmn, surg sect, AMA, 60-61; vpres, Pan-Pac Surg Asn, 69-72; vpres, Int Surg Group, 74-75, pres, 75-76; mem adv comt, Off Naval Res, 48-53; hon fel, Societa Italiana di Chirwegia, Polish Surg Soc & Royal Col Surgeons, England. *Honors & Awards:* Roswell Park Medal, 68; Curtis Medal, 70; Distinguished Serv Medal, Am Col Surgeons, 68 & Uniformed Serv Univ, Dept of Defense. *Mem:* Fel Am Col Surgeons; Am Surg Asn (1st vpres, 60-61, secy, 64-68); Soc Clin Surg (pres, 60-62); Soc Univ Surg (pres, 50-51); Soc Vascular Surg (treas, 47-53, pres, 58-59); Am Heart Asn; Coun Cardiovasc Asn (vpres). *Mailing Add:* Suite 400 St Vincent Prof Bldg 8402 Harcourt Rd Indianapolis IN 46260

SHUMAKER, JOHN BENJAMIN, JR, b Ames, Iowa, Jan 17, 26; m 54; c 1. OPTICAL PHYSICS. *Educ:* Iowa State Col, BS, 49; Yale Univ, PhD(chem), 52. *Prof Exp:* Analyst opers res, Opers Eval Group, Mass Inst Technol, 52-55; res physicist, Nat Bur Standards, 55-86; RETIRED. *Mem:* Optical Soc Am. *Res:* Plasma physics; plasma spectroscopy; spectral radiometry; physical optics. *Mailing Add:* 905 Montrose Rd Rockville MD 20852

SHUMAKER, ROBERT C, b Fort Wayne, In, June 06, 31; m 53; c 3. BASIN ANALYSIS, REMOTE SENSING. *Educ:* Brown Univ, AB, 53; Cornell Univ, MS, 57 & PhD(geol), 60. *Prof Exp:* Staff mem geol, Exxon, 60-72; PROF GEOL, IV VA UNIV, 72- *Concurrent Pos:* Dir, Appalachian Petrol Geol Symp, 72- & Appalachian Basin Indust Assoc, 81-; consult major grants, DOE & Gas Res Inst. *Mem:* Fel Geol Soc Am; Sigma Xi; Am Asn Petrol Geologist. *Res:* Study of structural styles in geology; their expression and distribution in paleozoic sedimentary basins; fracture permeability in tight reservoirs such as the Devonian shale Appalachian basin. *Mailing Add:* Dept Geol & Geog WVa Univ 425 Whitehall Morgantown WV 26506

SHUMAN, BERTRAM MARVIN, b Boston, Mass, May 2, 31; m 59; c 2. GEOPHYSICS, SPACE PHYSICS. *Educ:* Harvard Col, AB, 52; Boston Univ, AM, 68. *Prof Exp:* Res physicist geophys, Air Force Geophys Lab, 53-87; mem tech staff, W J Schafer Assoc, 87-88. *Mem:* Am Geophys Union. *Res:* Magnetic field measurements in space using rocket and satellite-borne sensors; active control of spacecraft charging. *Mailing Add:* 78 Hill St Lexington MA 02173

SHUMAN, CHARLES ROSS, b Harrisburg, Pa, Sept 18, 18; m 44; c 2. INTERNAL MEDICINE. *Educ:* Gettysburg Col, AB, 40; Temple Univ, MD, 43, MS, 49. *Hon Degrees:* DSc, Gettysburg Col, 73. *Prof Exp:* From instr to assoc prof, 49-63, clin prof, 63-66, PROF MED, SCH MED, TEMPLE UNIV, 66- *Concurrent Pos:* Consult, Vet Admin Hosp & Philadelphia Gen Hosp, 57-68; mem bd trustees, Am Diabetes Asn, 76-; pres, Philadelphia County Med Soc, 77- *Mem:* Am Col Physicians; Am Diabetes Asn; Am Fedn Clin Res; AMA; Sigma Xi. *Res:* Metabolism of carbohydrate, fat and protein; oral antiabetic agents; nutrition; disease. *Mailing Add:* 1111 Delene Rd Jenkintown PA 19046

SHUMAN, FRED LEON, JR, agricultural & biological engineering, for more information see previous edition

SHUMAN, LARRY MYERS, b Harrisburg, Pa, Apr 3, 44; m 70; c 2. SOIL CHEMISTRY. *Educ:* Pa State Univ, BS, 66, MS, 68, PhD(agron), 70. *Prof Exp:* Asst prof, 72-79, ASSOC PROF AGRON, UNIV GA, 79- *Concurrent Pos:* Assoc ed, Soil Sci Soc Am J, 86-91; mem, Coun Agr Sci & Technol. *Mem:* Am Soc Agron; Soil Sci Soc Am; Soc Environ Geochem & Health. *Res:* Influence of soil properties on the retention and release of microelements to plants; chemical forms of microelements in soil; chemistry of soil aluminum and manganese. *Mailing Add:* Ga Exp Sta Griffin GA 30223-1797

SHUMAN, MARK S, b Yakima, Wash, July 29, 36; m 63; c 3. ENVIRONMENTAL CHEMISTRY, ELECTROANALYTICAL CHEMISTRY. *Educ:* Wash State Univ, BS, 59; Univ Wis, PhD(chem), 66. *Prof Exp:* Asst prof chem, Tex Christian Univ, 66-69 & Whitman Col, 69-70; asst prof, 70-75, assoc prof, 75-80, PROF ENVIRON SCI & ENG, UNIV NC, CHAPEL HILL, 80- *Mem:* Am Chem Soc. *Res:* Transport of trace inorganics in natural water systems; trace metal-organic associations in natural water; electroanalytical chemistry. *Mailing Add:* Dept of Environ Sci Univ of NC Chapel Hill NC 27599-7400

SHUMATE, KENNETH MCCLELLAN, b Houston, Tex, Nov 9, 36; m 66; c 2. ORGANIC CHEMISTRY, SCIENCE EDUCATION. *Educ:* Baylor Univ, BS, 58, MS, 63; Univ Tex, PhD(org chem), 66. *Prof Exp:* Res chemist, Petro-Tex Chem Corp, 66-67; from asst prof to assoc prof, 67-75, chmn dept, 73-86, PROF CHEM, SAN ANTONIO COL, 75-, VPRES ACAD AFFAIRS, 86- *Mem:* Am Chem Soc. *Res:* Thermal and photochemical reactions of conjugated medium ring dienes; low molecular weight polymers of butadiene. *Mailing Add:* Dept Chem San Antonio Col 1300 San Pedro Ave San Antonio TX 78284

SHUMATE, PAUL WILLIAM, JR, b Philadelphia, Pa, July 15, 41; m 64; c 1. FIBER OPTICS, LIGHTWAVE DEVICES. *Educ:* Col William & Mary, BS, 63; Univ Va, PhD(physics), 68. *Prof Exp:* Asst prof physics, Univ Va, 68-69; mem tech staff, Bell Labs, 69-75, supvr, 75-83, dist mgr, 83-86, DIV MGR, BELL COMMUN RES, 86- *Concurrent Pos:* Ed-in-chief, Transactions on Magnetics, Inst Elec & Electronics Engrs, 74-79 & Photonics

Technol Lett, 88- *Mem:* Am Phys Soc; fel Inst Elec & Electronic Engrs; Sigma Xi. *Res:* Magnetics; magnetic bubble domain materials and device applications; solid-state electronics and integrated circuits; semiconductor lasers, optical fibers and device reliability; optical networks. *Mailing Add:* Bell Commun Res 445 South St Morristown NJ 07960-1910

SHUMATE, STARLING EVERETT, II, b Martinsville, Va, Aug 20, 47. BIOCHEMICAL ENGINEERING. *Educ:* Va Polytech Inst & State Univ, BS, 70; Univ Tenn, MS, 74, PhD(chem eng), 75. *Prof Exp:* Res engr bioeng, 74-76, leader bioeng res group, 76-81, mgr biotechnol & environ prof, Oak Ridge Nat Lab, 81-82; VPRES, RES & DEVELOP, ENGENICS, INC, 82- *Concurrent Pos:* Lectr, Dept Chem, Metall & Polymer Engr, Univ Tenn, 80- *Mem:* Am Inst Chem Engrs; Am Chem Soc; Sigma Xi. *Res:* Bioengineering; chemical engineering science, especially reaction kinetics and mass transfer; separation processes; chemical and biochemical reactor design. *Mailing Add:* 3760 Haven Dr Menlo Park CA 94025

SHUMRICK, DONALD A, b Newark, NJ, Mar 8, 25; c 9. OTORHINOLARYNGOLOGY. *Educ:* Seton Hall Univ, BS, 49; Univ Minn, Minneapolis, MS, 52, MD, 57; Am Bd Otolaryngol, dipl, 64. *Prof Exp:* Teaching asst physiol, Univ Minn, Minneapolis, 52-57; intern surg, San Francisco City & County Hosp, 57-58, resident gen surg, 58-59; resident, instr otolaryngol & NIH fel, Washington Univ, 59-63; asst prof, Univ Iowa, 63-66; PROF OTOLARYNGOL & MAXILLOFACIAL SURG & PROF & CHMN, MED CTR, UNIV CINCINNATI, 66- *Concurrent Pos:* Consult, Study Sect, NIH, 70- *Mem:* Am Acad Otolaryngal-Head & Neck Surg; Am Acad Facial Plastic & Reconstructive Surg; Am Soc Head & Neck Surgeons; Am Acad Ophthal & Otolaryngol; Pan Am Asn Oto-Rhino-Laryngol & Broncho-Esophagol; Soc Univ Otolaryngol (secy, 70-); Am Col Surg. *Res:* Head and neck cancer; maxillofacial surgery. *Mailing Add:* Dept Otolaryngol & Maxillofacial Surg Univ Cincinnati Med Ctr 231 Bethesda Ave ML No 528 Cincinnati OH 45267

SHUMWAY, CLARE NELSON, (JR), b Painted Post, NY, Oct 28, 25; m 55; c 2. PEDIATRICS, HEMATOLOGY. *Educ:* Univ Buffalo, MD, 48; Am Bd Pediat, dipl, 53, cert pediat hemat-oncol, 74. *Prof Exp:* Intern med, Buffalo Gen Hosp, 48-49; resident pediat, Buffalo Children's Hosp, 49-52; instr, Sch Med & Dent, Univ Rochester, 52-57; assoc prof, Sch Med, Univ Buffalo, 57-64; prof pediat, Med Col Va, 64-72; dir pediat, Harrisburg Polyclin Hosp, 72-77; DIR HEALTH SERV, GETTYSBURG COL, 77- *Concurrent Pos:* Am Cancer Soc fel, Univ Rochester, 52-53, USPHS res fel, 55-57; dir hemat, Buffalo Children's Hosp, 57-64; spec res fel, Univ Wash, 69-70. *Mem:* AAAS; Am Acad Pediat; Am Pediat Soc; NY Acad Sci; Am Soc Hemat. *Res:* The role of bacterial hemolysin in the pathogenesis of pneumococcal infections. *Mailing Add:* 20 Byers Rd RD 2 Dillsburg PA 17019

SHUMWAY, LEWIS KAY, b Salt Lake City, Utah, Dec 3, 34; m 58; c 7. PLANT GENETICS. *Educ:* Brigham Young Univ, BS, 60, MS, 62; Purdue Univ, PhD(plant genetics), 65. *Prof Exp:* Res botanist, Univ Calif, Davis, 65-66; asst res botanist, Univ Calif, Berkeley, 66-67; from asst prof to assoc prof genetics & bot, Wash State Univ, 67-77; ASSOC DEAN, SAN JUAN CAMPUS, COL EASTERN UTAH, 77- *Mem:* AAAS. *Res:* Plant cell ultrastructure; chloroplast inheritance, development and ultrastructure; protoplasts. *Mailing Add:* San Juan Campus Col Eastern Utah 639 W 100 S Blanding UT 84511

SHUMWAY, RICHARD PHIL, b Taylor, Ariz, Aug 21, 21; m 43; c 6. ANIMAL PHYSIOLOGY, ANIMAL HUSBANDRY. *Educ:* Utah State Univ, BS, 47, PhD, 59; Univ Minn, MS, 49. *Prof Exp:* Asst prof agr, Utah State Univ, 47-48; from asst prof to assoc prof, 49-63, prof animal sci, Brigham Young Univ, 63-87; RETIRED. *Concurrent Pos:* Chmn dept animal sci, Brigham Young Univ, 63-74. *Mem:* Am Soc Animal Sci. *Res:* Animal science and nutrition. *Mailing Add:* Dept of Animal Sci Brigham Young Univ Provo UT 84602

SHUMWAY, SANDRA ELISABETH, b Taunton, Mass, Mar 29, 52. SHELLFISH BIOLOGY. *Educ:* Long Island Univ, BS, 74; Univ Col N Wales, PhD(marine biol), 76. *Prof Exp:* Postdoctoral fel, Portobello Marine Lab, NZ, 78-79; res asst, State Univ NY, Stony Brook, 80-82; SCIENTIST, STATE MAINE, 83- *Concurrent Pos:* Adj grad fac mem, Univ Maine, Orono, 83-; adj prin investr, Bigelow Lab Ocean Sci, 84-; ed, J Shellfish Res, 87- *Mem:* Nat Shellfish Asn (pres, 91-92); Asn Women Sci; Am Malacological Union; Am Soc Zoologists; Marine Biol Asn UK; Coun Biol Ed. *Res:* Physiological ecology of marine invertebrates, primarily respiratory and osmoregulatory physiology; toxic algal blooms and their effects on shellfish. *Mailing Add:* Dept Marine Resources Univ Maine West Boothbay Harbor ME 04575

SHUNG, K KIRK, b China, June 2, 45; US citizen; m 71; c 3. BIOMEDICAL ULTRASOUND, BIOINSTRUMENTATION. *Educ:* Nat Cheng-Kung Univ, Taiwan, BSEE, 68; Univ Mo, MSEE, 70; Univ Wash, PhD(elec eng), 75. *Prof Exp:* Res assoc, Ctr Bioeng, Univ Wash, 70-75; res fel, Providence Med Ctr, Seattle, Wash, 75-76, res engr, 76-79; from asst prof to assoc prof bioeng, 79-89, acting head dept, 87-88, PROF BIOENG, PA STATE UNIV, 89- *Concurrent Pos:* Prin investr NSF, 77-81 NIH, 79-; NIH diagnostic radiology. *Honors & Awards:* Early Career Achievement Award, Eng Med & Biol Soc, Inst Elec & Electronic Engrs, 85. *Mem:* Inst Elec & Electronic Engrs; Am Inst Ultrasound Med; assoc mem Acoust Soc Am. *Res:* Ultrasonic imaging and tissue characterization; diagnostic imaging. *Mailing Add:* 233 Halowell Bldg Pa State Univ University Park PA 16802

SHUPE, DEAN STANLEY, b Clarion, Iowa, July 7, 37; m 62; c 2. TECHNICAL WRITING. *Educ:* Iowa State Univ, BS, 60; Stanford Univ, MS, 61; Mass Inst Technol, ScD(mech eng), 69. *Prof Exp:* Process engr, Procter & Gamble Co, 61-62; from instr to prof, 63-89, EMER PROF MECH ENG, UNIV CINCINNATI, 89-; PRIN, SHUPE & ASSOCS, 89- *Concurrent Pos:* Prin, Eng & Mgt Assocs, 77-89; exec vpres, Cincinnati Industs, Inc. *Honors & Awards:* Ralph R Teetor Award, Soc Automotive Engrs, 80. *Mem:* Am Soc Mech Engrs; Am Soc Heating, Refrig & Air Conditioning Engrs. *Res:* Energy conservation and energy sciences; applied heat transfer and thermodynamics; technical writer, including textbooks and software manuals. *Mailing Add:* Shupe & Assocs 10304 Gunpowder Rd Florence KY 41042

SHUPE, JAMES LEGRANDE, b Ogden, Utah, Nov 5, 18; m 57; c 2. VETERINARY MEDICINE. *Educ:* Utah State Univ, BS, 48; Cornell Univ, DVM, 52. *Prof Exp:* From asst prof to prof vet med, Utah State Univ, 52-61; res vet animal dis & parasite res div, Agr Res Serv, USDA, 61-66; head Dept Vet Sci, 73-76, prof vet med, 66-80, PROF ANIMAL DAIRY & VET SCI, UTAH STATE UNIV, 80- *Concurrent Pos:* Resident path, Armed Forces Inst Path, Walter Reed Med Ctr, 57-58; chmn subcomt fluorosis, Nat Res Coun. *Mem:* Am Vet Med Asn; Am Col Vet Toxicol (pres, 72); Am Acad Clin Toxicol; Pan-Am Med Asn; Int Acad Path; Sigma Xi. *Res:* Toxicology; pathology. *Mailing Add:* Vet Dept Utah State Univ Logan UT 84322-4815

SHUPE, JOHN W(ALLACE), b Liberal, Kans, Mar 30, 24; m 44; c 4. CIVIL ENGINEERING, RENEWABLE ENERGY. *Educ:* Kans State Univ, BS, 48; Univ Calif, MS, 51; Purdue Univ, PhD(civil eng), 58. *Prof Exp:* Instr appl mech, Kans State Univ, 48-49, asst prof, 51-53; lectr civil eng, Univ Calif, 49-51; struct engr, Convair Div, Gen Dynamics Corp, 53-54; assoc prof appl mech, Kans State Univ, 54-65, assoc dean eng, 60-65; dean eng, Univ Hawaii, 65-80, dir, Hawaii Natural Energy Inst, 80-83; DIR, PAC SITE OFF, US DEPT ENERGY, 83- *Concurrent Pos:* Mem environ coun, State of Hawaii; dir, Hawaii Geothermal Proj; chmn, Governor's Comt Alt Energy; sci adv to asst secy for energy technol, US Dept Energy, 77-78. *Honors & Awards:* Templin Award, Am Soc Testing & Mat, 60. *Mem:* Am Soc Civil Engrs; Am Soc Eng Educ; Nat Soc Prof Engrs; Inst Solar Energy Soc; Geothermal Resources Coun. *Res:* Geothermal and solar energy. *Mailing Add:* 1629 Wilder Ave Honolulu HI 96822

SHUPE, ROBERT EUGENE, b Sparks, Kans, Sept 6, 34; m 57; c 3. RADIATION PHYSICS, RADIOBIOLOGY. *Educ:* Wayne State Col, BS, 62; Purdue Univ, West Lafayette, MS, 67, PhD(radiol physics), 70. *Prof Exp:* Instr radiol physics, Southwestern Radiol Health Lab, USPHS, Las Vegas, 69-70; ASST PROF RADIATION BIOL, MED CTR, IND UNIV, INDIANAPOLIS, 70- *Concurrent Pos:* Mem, Mid-West Radiation Protection, Inc, 71-; consult radiation physics. *Mem:* AAAS; Radiation Res Soc; Sigma Xi; Bioelectromagnetic Soc. *Res:* Hyperthermia research combined with ionizing radiation and chemotherapy in the treatment of malignant diseases. *Mailing Add:* Radiation-Oncol Cl Ind Univ Sch Med 535 Barnhill Dr Indianapolis IN 46223

SHUR, BARRY DAVID, b Elizabeth, NJ, Jan 3, 50; m 71; c 1. DEVELOPMENTAL BIOLOGY. *Educ:* Marietta Col, BS, 71; Johns Hopkins Univ, PhD(biol), 76. *Prof Exp:* Fel develop genetics, Mem Sloan-Kettering Cancer Ctr, 76-78; ASST PROF ANAT, HEALTH CTR, UNIV CONN, 78- *Concurrent Pos:* Helen Hay Whitney Found fel, 76-78. *Mem:* Am Soc Zoologists; Am Soc Anatomists; AAAS. *Res:* Cell surface biochemistry of normal and mutant morphogenesis. *Mailing Add:* Anderson Hosp & Tumor Inst 1515 Holcombe Blvd Box 117 Houston TX 77030

SHUR, MICHAEL, b Kamensk-Uralski, USSR, Nov 13, 42; US citizen; c 2. ELECTRICAL ENGINEERING, SOLID STATE PHYSICS. *Educ:* Leningrad Electrotech Inst, MSEE, 65; A F Ioffe Inst Physics & Technol, PhD(physics), 67. *Prof Exp:* Researcher, A F Ioffe Inst Physics & Technol, Leningrad, USSR, 65-76; res assoc, Wayne State Univ, 76-77, asst prof elec eng, 77-78; asst prof, Oakland Univ, Rochester, Mich, 78-79; from assoc prof to prof elec eng, Univ Minn, 79-89; JOHN MARSHALL MONEY PROF, UNIV VA, 89- *Concurrent Pos:* Vis res assoc, Cornell Univ, 76-80; consult, Honeywell, Xerox & Gen Elec; mem, Ctr Advan Studies, Univ Va, 89- *Mem:* Fel Inst Elec & Electronic Engrs; Am Phys Soc. *Res:* Semiconductor devices and integrated circuits; compound semiconductor devices; ballistic devices and amorphous silicon devices. *Mailing Add:* Dept Elec Eng Univ Va Charlottesville VA 22903-2442

SHURBET, DESKIN HUNT, JR, b Lockney, Tex, Aug 27, 25; m 58; c 3. SEISMOLOGY. *Educ:* Univ Tex, BS, 50, MA, 51. *Prof Exp:* Dir Bermuda-Columbia Seismograph Sta, Lamont Geol Observ, 51-56; PROF GEOL & DIR SEISMOL OBSERV, TEX TECH UNIV, 56- *Mem:* Fel AAAS; Seismol Soc Am; Am Geophys Union; Soc Explor Geophys. *Res:* Earthquake seismology. *Mailing Add:* Dept Geosci Tex Tech Univ Lubbock TX 79406

SHURE, DONALD JOSEPH, b Washington, DC, July 21, 39; m 65; c 2. ECOSYSTEMS PROCESSES. *Educ:* Western Md Col, BA, 61; Rutgers Univ, MS, 66, PhD(zool), 69. *Prof Exp:* Teaching asst zool, Rutgers Univ, 64-67, NSF fel ecol, 68-69; asst prof, 69-74, ASSOC PROF BIOL, EMORY UNIV, 74- *Concurrent Pos:* Consult, Allied Gen Nuclear Serv, 70-80; dir grad studies, Dept Biol, Emory Univ, 78-81. *Mem:* Ecol Soc Am; Am Soc Mammalogists; Am Inst Biol Sci; AAAS. *Res:* Perturbation effects on ecosystem structure and function; nutrient cycling, decomposition, consumer dynamics and plant-animal interactions in systems undergoing natural succession. *Mailing Add:* Dept Biol Emory Univ 1364 Clifton Rd Atlanta GA 30322

SHURE, FRED C(HARLES), b New York, NY, Feb 26, 34; m 63; c 3. PHYSICS, NUCLEAR ENGINEERING. *Educ:* Harvard Col, AB, 55; Univ Mich, MS, 57, PhD(physics), 61. *Prof Exp:* Instr physics, Univ Mich, 59-61; assoc res physicist, Conductron Corp, 61-62; from lectr to asst prof, 62-65, ASSOC PROF NUCLEAR ENG, UNIV MICH, ANN ARBOR, 65- *Concurrent Pos:* Physicist plasma physics lab, Princeton Univ, 63-64; pres ESZ Assocs, Inc. *Mem:* Am Phys Soc; Am Nuclear Soc. *Res:* Transport theory; reactor theory; plasma physics; applied mathematics. *Mailing Add:* 1127 Brooks Ann Arbor MI 48103

SHURE, KALMAN, b Brooklyn, NY, Mar 14, 25; m 51; c 2. NUCLEAR SCIENCE. *Educ:* Brooklyn Col, AB, 45; Mass Inst Technol, PhD(physics), 51. *Prof Exp:* Gen phys scientist, US Air Force Cambridge Res Ctr, 49; sr scientist, Bettis Atomic Power Div, 51-54, supvry scientist, 54-65, adv scientist, 65-73, CONSULT, BETTIS ATOMIC POWER LAB, WESTINGHOUSE ELEC CORP, 73- *Mem:* Am Phys Soc; fel Am Nuclear Soc. *Res:* Shielding, penetration of gamma rays and neutrons in materials; decay energies of radioactive isotopes. *Mailing Add:* 5612 Woodmont St Pittsburgh PA 15217

SHURMAN, MICHAEL MENDELSOHN, b St Louis, Mo, Aug 4, 21. ASTRONOMY. *Educ:* Univ Wis, BA, 43, MA, 46, PhD(physics), 51. *Prof Exp:* Instr physics, Exten, Univ Wis, 46-48; physicist, Los Alamos Sci Lab, 52-55; from asst prof to prof physics, 55-83, assoc dean sci, 62-65, EMER PROF PHYSICS, UNIV WIS-MILWAUKEE, 83- *Mem:* Am Asn Physics Teachers. *Mailing Add:* PO Box 32188 Jerusalem Israel

SHURTLEFF, DAVID B, b Fall River, Mass, July 1, 30; m 52; c 3. PEDIATRICS. *Educ:* Tufts Univ, MD, 55. *Prof Exp:* Intern pediat, Mass Gen Hosp, Boston, 55-56, asst resident, 56-57; chief resident, Children's Orthop Hosp, Seattle, Wash, 57-58; from instr to assoc prof, 60-71, PROF PEDIAT, SCH MED, UNIV WASH, 71- *Concurrent Pos:* Teaching fel, Harvard Med Sch, 55-57; assoc, Sch Med, Univ Wash, 57-58; sr consult, US Army, Madigan Hosp, 61-85; Nat Found March of Dimes fel, Welsh Nat Sch Med, Univ Wales, 69-70, vis prof & consult, 69-71; Ross award res, Western Soc Pediat Res. *Honors & Awards:* Casey Holter Lectr, Hydrocepholus & Spina Befida, Int Soc Res; Gallagher Lectr, Soc Adolescent Med. *Mem:* Am Acad Pediat; Soc Pediat Res; Am Pediat Soc. *Res:* Congenital defects; clinical study of epidemiology and cerebrospinal fluid dynamics and ecology of children with hydrocephalus and meningomyelocele. *Mailing Add:* Dept Pediat RD 20 Univ Wash Sch Med Seattle WA 98195

SHURTLEFF, MALCOLM C, JR, b Fall River, Mass, June 24, 22; m 50; c 3. PHYTOPATHOLOGY. *Educ:* Univ RI, BS, 43; Univ Minn, MS, 50, PhD(plant path), 53. *Prof Exp:* Asst, Univ Minn, 47-50; instr bot, Univ RI, 50-51, asst res prof plant path & asst exten prof plant path & entom, 51-54; asst prof plant path & exten plant pathologist, Iowa State Univ, 54-58, assoc prof bot & plant path, 58-61; assoc prof plant path, 61-65, PROF PLANT PATH, UNIV ILL, URBANA, 65-, EXTEN PLANT PATHOLOGIST, 61- *Concurrent Pos:* Chmn exten comt, Int Soc Plant Path, 78-80; chief ed, Plant Dis, 79-82. *Honors & Awards:* Adventurers in Agr Sci Award Distinction, IX Int Cong Plant Protection, 79; Distinguished Serv Award, USDA, 86. *Mem:* AAAS; fel Am Phytopath Soc; Bot Soc Am. *Res:* Fungicides; turf, field crop and ornamental diseases. *Mailing Add:* 2707 Holcomb Dr Urbana IL 61801

SHURVELL, HERBERT FRANCIS, b London, Eng, Sept 3, 34; m 60; c 3. SPECTROCHEMISTRY. *Educ:* Univ Exeter, BSc, 59; Univ BC, MSc, 62, PhD(chem), 64. *Hon Degrees:* DSc, Univ Exeter, 81. *Prof Exp:* Res attache, Nat Ctr Sci Res, Fac Sci, Marseille, France, 64-65; from asst prof to assoc prof chem, 65-77, PROF CHEM, QUEEN'S UNIV, ONT, 77- *Concurrent Pos:* Res fel, Univ Queensland, 72-73; vis prof, Sao Paulo, Brazil, 79 &Univ Queensland, 78, 81 & 87; res ed, Thornton Res Ctr, Shell Res Ltd, Chester, UK, 87-88. *Honors & Awards:* Bicentennial Medal, Ont, 84. *Mem:* Fel Chem Inst Can; Spectros Soc Can (pres, 78-79); Royal Soc Chem. *Res:* Infrared and Raman spectroscopy. *Mailing Add:* Dept Chem Queen's Univ Kingston ON K7L 3N6 Can

SHUSHAN, MORRIS, b New Orleans, La, June 13, 07; m 35; c 1. ALLERGY FOOD, WELLNESS MEDECINE. *Educ:* Tulane Univ, BS, 27, MD, 29. *Prof Exp:* Researcher, med, Touro Infirmary, 29-31; instr, dept med, La State Univ Sch Med, 31-38, clin assoc prof med psychol & psychiat, 38-74; PVT PRACT, 35- *Concurrent Pos:* Bd mem, Jefferson Coun Aging; chief med, 78th Sta , 42-46; chief, Sr Citizen health Maintenance Study, Jefferson Parish, 75-, consult, 76-; chmn, E Jefferson, 78-79. *Mem:* Nat Coun Aging; Am Soc Holistic; Inst Noetic Sci; Am Asn Retired Persons. *Mailing Add:* River Ridge Holistic Ctr 271 Sauve Rd River Ridge LA 70123

SHUSHAN, SAM, b Bronx, NY, July 6, 22; m 46; c 3. LICHENOLOGY, MYCOLOGY. *Educ:* City Col New York, BS, 43; Rutgers Univ, MS, 47, PhD(bot), 49. *Prof Exp:* From instr to assoc prof, 49-72, PROF BOT, UNIV COLO, BOULDER, 72- *Mem:* AAAS; Bot Soc Am; Am Bryol & Lichenological Soc; Mycol Soc Am; Phycol Soc Am; Sigma Xi. *Res:* Developmental plant anatomy; lichen taxonomy. *Mailing Add:* Biol Dept Campus Box 334 EPO Univ of Colo Boulder CO 80309

SHUSKUS, ALEXANDER J, b Hartford, Conn, June 15, 29; m 55; c 3. SOLID STATE PHYSICS. *Educ:* Univ Conn, BA, 50; Univ Ala, MS, 57; Univ Conn, PhD(physics), 61. *Prof Exp:* Engr, Hart Mfg Co, 50-51; engr, Pratt & Whitney Aircraft Div, United Aircraft Corp, 53-55, physicist, Res Labs, 61-65, group leader microwave physics, 65-68, mgr microelectronics lab, 68-71, SR RES CONSULT, RES CTR, UNITED TECHNOLOGIES CORP, 71- *Mem:* Inst Elec & Electronics Engrs; Am Phys Soc. *Res:* Electron spin resonance studies; radiation effects in solids; microwave properties of solids; semiconductor physics; thin films; photovoltaics; ion implantation. *Mailing Add:* United Technologies Corp Res Ctr Silver Lane West Hartford CT 06107

SHUSMAN, T(EVIS), chemical engineering,polymer chemistry; deceased, see previous edition for last biography

SHUSTER, CARL NATHANIEL, JR, b Randolph, Vt, Nov 16, 19; m 44; c 5. AQUATIC ECOLOGY, INVERTEBRATE ZOOLOGY. *Educ:* Rutgers Univ, BSc, 42, MSc, 48; NY Univ, PhD(biol), 55. *Prof Exp:* Instr zool, Rutgers Univ, 49-54, lectr & demonstr sci, Univ Col, 53-55, lectr zool, 54-55; asst prof biol sci & dir marine labs, Univ Del, 55-63; dir Northeast Marine Health Serv Lab, USPHS, 63-69, ecologist, Bur Water Hyg, 69-71; br chief water progs, Environ Protection Agency, 71-72; asst adv environ qual, Ecol Systs Anal, 72-74, actg adv environ qual, 74-75, ecol systs analyst, Fed Energy Regulatory Comn, 75-84; RETIRED. *Concurrent Pos:* Mem, US Nat Mus Smithsonian-Bredin Caribbean exped, 58; adj prof zool & oceanogr, Univ RI, 63-70; adj prof biol oceanogr, Va Inst Sch Marine Sci, Williams & Mary, 79- *Mem:* Fel AAAS; Am Soc Limnol & Oceanog; Ecol Soc Am; fel NY Acad Sci; Am Soc Zoologists. *Res:* Authority on ecology of limulidae; estuarine ecology, especially of arthropods and mollusks; evaluation of environmental impacts on aquatic ecosystems from federal actions, particularly interrelations within river basins, electrical power systems and water use management. *Mailing Add:* 3733 N 25th St Arlington VA 22207-5011

SHUSTER, CHARLES W, GRAM NEGATIVE TOXINS, BACTERIAL INVASION MECHANISMS. *Educ:* Univ Ill, PhD(biochem), 58. *Prof Exp:* ASSOC PROF, MOLECULAR BIOL & MICROBIOL, SCH MED, CASE WESTERN RESERVE UNIV, 64- *Mailing Add:* Dept Molecular Biol & Microbiol Sch Med Case Western Reserve Univ Cleveland OH 44106

SHUSTER, JOSEPH, b Montreal, Que, Jan 29, 37; m 64; c 2. CANCER, IMMUNOLOGY. *Educ:* McGill Univ, BS, 58; Univ Alta, MD, 62; Univ Calif, PhD(immunol), 68. *Prof Exp:* From asst prof to assoc prof med, 68-78, assoc dir, McGill Cancer Ctr, 78-81, PROF MED, MCGILL UNIV, 78-; SCI DIR, MONTREAL GEN HOSP RES INST, 81-, DIR CLIN CHEM, 86- *Concurrent Pos:* Med Res Coun Can scholar, 68-73; from asst phys to assoc phys, Montreal Gen Hosp, 68-77, sr physician, 77-, dir, Div Clin Immunol & Allergy, 80-; clin res assoc, Nat Cancer Inst Can, 74-80; fac med, Univ Uruguay, 83; dir, Div Clin Immunol & Allergy, Montreal Childrens Hosp, 90. *Mem:* NY Acad Sci; Can Soc Immunol; Can Soc Oncol; Can Soc Clin Invest; Asn Am Immunologists; Am Soc Cancer Res; Clin Immunol Soc; Am Acad Allergy. *Res:* Identification and characterization of human tumor antigens, particularly alpha fetoprotein, carcinoembryonic antigen and tumor modified histocompatibility antigens; AIDS and hemophilia. *Mailing Add:* Div Clin Immunol & Allergy Mont Gen Hosp 1650 Cedar Ave Rm 7135 Montreal PQ H3G 1A4 Can

SHUSTER, KENNETH ASHTON, b Trenton, NJ, Apr 3, 46; m 69; c 1. ENVIRONMENTAL SYSTEMS & TECHNOLOGY. *Educ:* Rutgers Univ, BS & BA, 69; Xavier Univ, MBA, 72. *Prof Exp:* Staff engr solid waste collection, Dept HEW, USPHS, 69-70; staff engr, 70-71, sect chief solid waste systs, 71-74, PROG MGR LAND DISPOSAL, US ENVIRON PROTECTION AGENCY, 74- *Concurrent Pos:* Staff comt safety standards solid waste equip, Am Nat Standards Inst, 74-77; adv staff task force solid waste, Nat Comn Prod, 72-73; chmn, Environ Protection Agency Disposal Regs Work Group, 76-79. *Honors & Awards:* Bronze Medal, US Environ Protection Agency, 77, Silver Medal, 79. *Mem:* Am Pub Works Asn. *Res:* Environmental and economic analyses technologies and regulations of solid waste management systems, particularly land disposal of hazardous and non-hazardous wastes, resource conservation and recovery and waste storage and collection systems. *Mailing Add:* Off Solid Waste WH-564 US Environ Protection Agency Washington DC 20460

SHUSTER, LOUIS, b Wysock, Poland, Apr 17, 29; nat US; m 59; c 2. PHARMACOLOGY, NEUROCHEMISTRY. *Educ:* Univ BC, BA, 50; Johns Hopkins Univ, PhD(biochem), 54. *Prof Exp:* Nat Res Coun Can overseas fel, Nat Inst Med Res, Eng, 54-55; vis scientist, Nat Cancer Inst, 55-58; from asst prof to assoc prof pharmacol, Sch Med, Tufts Univ, 58-70, assoc prof biochem, 67-70, actg chmn pharmacol, 87-91, PROF BIOCHEM & PHARMACOL, SCH MED, TUFTS UNIV, 70- *Mem:* Am Chem Soc; Am Soc Biol Chemists; Brit Biochem Soc; Am Soc Pharmacol & Exp Therapeut; Soc Neurosci. *Res:* Mechanisms of drug action; liver damage from drugs; addiction to narcotics and stimulants; pharmacogenetics. *Mailing Add:* Dept Pharmacol Tufts Univ Sch Med Boston MA 02111

SHUSTER, ROBERT C, b Brooklyn, NY, Dec 15, 32; m 58; c 2. BIOCHEMISTRY, MOLECULAR BIOLOGY. *Educ:* Brooklyn Col, BA, 53; Purdue Univ, MS, 59; Albany Med Col, PhD(biochem), 63. *Prof Exp:* Res assoc biochem, Sch Med, Yale Univ, 63-66; staff fel, NIH, 66-68; asst prof, 68-74, ASSOC PROF BIOCHEM, EMORY UNIV, 74- *Mem:* Am Soc Biol Chemists. *Res:* Studies of the human insulin receptor gene in disorders affecting glucose homeostasis. *Mailing Add:* Dept of Biochem Emory Univ Atlanta GA 30322

SHUTER, ELI RONALD, b New York, NY, June 16, 35; m 58; c 4. NEUROLOGY, NEUROCHEMISTRY. *Educ:* Cornell Univ, AB, 56; Wash Univ, MD, 60; Am Bd Psychiat & Neurol, dipl, 71. *Prof Exp:* Intern med, NY Hosp, 60-61; asst resident neurol, Mass Gen Hosp, 61-62; resident neurol, Cleveland Metrop Gen Hosp, 64-65 & neuropath, 65-66; Nat Inst Neurol Dis & Stroke spec res fel, Wash Univ, 66-69; asst prof neurol, Sch Med, St Louis Univ, 69-75, asst clin prof, 75-79; ASST PROF CLIN NEUROL, SCH MED, WASHINGTON UNIV, 79- *Concurrent Pos:* Teaching fel, Harvard Univ, 61-62; consult, Vet Admin Hosps, St Louis, 70-77; USPHS res grant, St Louis Univ, 70-75; mem active staff, Christian Hosp Northeast-Northwest, 75-; asst neurologist, Barnes Hosp, 79-; consult staff, St Anthony's Hosp, 80-, Alton Mem Hosp, 81-, Jewish Hosp, 82- & DePaul Community Health Ctr, 87- *Mem:* Asn Res Nerv & Ment Dis; Am Acad Neurol; Soc Neurosci; Am Asn Study Headache; NY Acad Sci. *Res:* Biochemical changes in neuropathologic conditions; biochemical changes during development of the central nervous system; metabolism of gangliosides; hexosaminidases in the nervous system; pseudobulbar palsy; headache. *Mailing Add:* 11155 Dunn Rd St Louis MO 63136

SHUTER, WILLIAM LESLIE HAZLEWOOD, b Rangoon, Burma, Jan 17, 36; m 63; c 2. RADIO ASTRONOMY. *Educ:* Rhodes Univ, SAfrica, BSc, 57, MSc, 59; Univ Manchester, PhD(physics), 63. *Prof Exp:* Lectr physics, Rhodes Univ, SAfrica, 63-65; from asst prof to assoc prof, 65-79, PROF PHYSICS, UNIV BC, 79- *Mem:* Can Astron Soc; Am Astron Soc; Royal Astron Soc. *Res:* Millimeter wave astronomy; interstellar gas; radio astronomical spectroscopy; galactic structure; dark matter. *Mailing Add:* Dept Physics Univ BC Vancouver BC V6T 2A6 Can

SHUTSKE, GREGORY MICHAEL, US citizen. MEDICINAL CHEMISTRY, DRUG DESIGN. *Educ:* Rose-Hulman Inst Technol, BS, 71; Ind Univ, PhD(org chem), 75. *Prof Exp:* Sr res chemist, Hoechst-Roussel Pharmaceut, 75-78, res assoc, 78-81, sr res assoc, 81-89, PRIN RES SCIENTIST, HOECHST-RUSSELL PHARMACEUT, 89- *Concurrent Pos:* Exchange scientist, Hoechst AG, Frankfurt, WGer, 80-81. *Mem:* Am Chem Soc; Int Soc Heterocyclic Chem. *Res:* Synthesis of heterocyclic compounds of medicinal interest; drugs affecting the central nervous system; determination of organic structures by nuclear magnetic resonance methods. *Mailing Add:* Hoechst-Roussel Pharmaceut Inc Rte 202-206N PO Box 2500 Somerville NJ 08876-1258

SHUTZE, JOHN V, b Hale, Colo, Apr 21, 24; m 47; c 2. POULTRY NUTRITION. *Educ:* Colo State Univ, BS, 55, MS, 57; Wash State Univ, PhD(poultry nutrit), 64. *Prof Exp:* Rancher, Imperial, Nebr, 49-52; instr poultry sci, Wash State Univ, 57-63; exten poultryman, Pa State Univ, 63-65; exten poultryman & assoc poultry scientist, Colo State Univ, 65-70; head exten poultry sci dept & exten poultry scientist, 70-85; EMER PROF, UNIV GA, 85- *Concurrent Pos:* Consult poultry scientist & nutritionist, 85- *Honors & Awards:* Pfizer Exten Poultry Sci Award, 77. *Mem:* AAAS; Poultry Sci Asn; World Poultry Sci Asn. *Res:* Effect of pesticides on growth and reproduction; effect of polychlorinated biphenyls on hatchability and tissue residue; effect of calcium sources and additives on eggshell quality; effect of calorie protein ratio and energy sources on the thiamine requirement in chicks; fatty acid metabolism in laying hens. *Mailing Add:* 115 Witherspoon Rd Athens GA 30606

SHUVAL, HILLEL ISAIAH, b Washington, DC, July 16, 26; m 52; c 3. ENVIRONMENTAL ENGINEERING, ENVIRONMENTAL HEALTH. *Educ:* Univ Mo, BSCE, 48; Univ Mich, MPH, 52. *Prof Exp:* Dep chief sanit engr, Ministry Health, Israel, 49-57, chief sanit engr, 58-65; assoc prof environ health, 65-75, dir, Div Environ Sci, 72-88, LUNENFELD-KUNIN PROF ENVIRON HEALTH & DIR, ENVIRON HEALTH LAB, HEBREW UNIV JERUSALEM, 76- *Concurrent Pos:* Design engr, Metcalf & Eddy Consult Eng, 55-56; Nat Water Coun, Ministry Agr, Israel, 57-68; Expert Adv Panel, Environ Health Div, WHO, 60-; vis prof, Sch Pub Health, Univ Mich, 61-62 & 78-79, Harvard Univ, 83-84 & Mass Inst Technol, 88-89; Sci Adv Bd, WAPORA-Environ Consult Engrs, 74-82; Reactor Safety Comn, Israel AEC, Environ Eng Consult, World Bank & Consult WHO, United Nations Environ Prog, 76-; chmn, Israel Soc Ecol & Environ Qual, 81-84; dipl, Am Acad Environ Engrs. *Honors & Awards:* Distinguished Foreign Lectr, Asn Environ Eng Profs, 74; Presidential Citation, US Nat Asn Environ Health, 62; Israel Environ Qual Prize, 83. *Mem:* Int Asn Water Pollution Res & Control (vpres, 68-70); fel Am Pub Health Asn; Am Soc Civil Engrs; Int Acad Environ Safety; Water Pollution Control Fedn; Sigma Xi; Am Acad Environ Engrs. *Res:* Environmental engineering and health; water quality management; wastewater renovation and reuse; viruses in water; marine pollution; water microbiology; low cost waste treatment; environmental health in developing countries; international water resources conflict resolution. *Mailing Add:* Div Environ Sci Hebrew Univ Jerusalem Jerusalem 91904 Israel

SHVARTZ, ESAR, b Tel-Aviv, Israel, Aug 1, 35; US citizen; m 71; c 2. SPORTS MEDICINE, HUMAN FACTORS IN EQUIPMENT DESIGN. *Educ:* Univ Calif, Los Angeles, BA, 60 MS, 62; Univ SC, PhD (phys educ), 65. *Prof Exp:* Asst prof phys educ, Washington State Univ, 65-66; asst prof phys educ, Ind State Univ, 66-67; dir, Environ Physiol Lab, Negen Inst Arid Zone Res, Beer Sheva, Israel, 67-70; sr scientist environ physiol, Heller Inst Med Res, Sheba Med Ctr, Israel, 70-76; sr res assoc environ physiol, NASA-Ames Res Ctr, Moffett Field, Calif, 76-77; sr engr & scientist human factors, Douglas Aircraft Co, 78-84; SYST SAFETY ENGR, SYST SAFETY, ROCKWELL INT, 84- *Concurrent Pos:* Res adv, Ben-Gurion Univ, Israel, 67-73; vis prof, Tel-Aviv Univ, 70-73; dir, Res in Fitness, Med Corps, Israel Defence Forces, 70-76; vis scientist, Human Sci, Capital Chamber of Mines of SAfrica, Johannesburg, 73-74. *Honors & Awards:* Environ Sci Award, Aerospace Med Asn, 82. *Mem:* Am Physiol Soc; fel Am Col Sports Med; Human Factors Soc; assoc fel Aerospace Med Asn. *Res:* Crew safety in space flights; environmental physiology, mainly effects of heat, cold, altitude on man; industrial safety; physical fitness and training. *Mailing Add:* Rockwell Int MC AD-60 12214 Lakewood Blvd Downey CA 90241

SHWE, HLA, b Rangoon, Burma, May 2, 34; US citizen; m 62; c 3. HIGH ENERGY PHYSICS, NUCLEAR PHYSICS. *Educ:* Univ Calif, Berkeley, AB, 58, MA, 59, PhD(physics), 62. *Prof Exp:* US AEC fel, Lawrence Radiation Lab, 62; from asst prof to assoc prof physics, Ripon Col, 63-69; chmn dept, 69-74, dean fac sci, 74-79, dean, Sch Arts & Sci, 79-83, PROF PHYSICS, EAST STROUDSBURG UNIV. 69- *Concurrent Pos:* US AEC & Assoc Cols Midwest fel, Argonne Nat Lab, 66-67; consult, Argonne Nat Lab, 68-72, Oak Ridge Nat Lab, 70-72 & Lawrence Berkeley Lab, 72-76. *Mem:* Am Asn Univ Prof; Sigma Xi; NY Acad Sci; Am Phys Soc; Am Asn Physics Teachers. *Res:* Particle physics; neutron physics, especially in the area of cross section work; heavy-ion nuclear physics; high-energy heavy ions. *Mailing Add:* Dept Physics East Stroudsburg Univ East Stroudsburg PA 18301

SHYAMSUNDER, ERRAMILLI, b Pune, India, Feb 18, 57. PHYSICS, BIOPHYSICS. *Educ:* Fergusson Col, BSc, 77; Indian Inst Technol, MSc, 79; Univ Ill, PhD(physics), 86. *Prof Exp:* Res asst physics, Univ Ill, 79-86; res assoc, 86-87, instr, 87-89, ASST PROF PHYSICS, PRINCETON UNIV, 89- *Mem:* Am Phys Soc; Biophys Soc. *Res:* Effects of high pressure on biological systems; understanding of the physics underlying complex biological molecules. *Mailing Add:* Dept Physics Joseph Henry Lab PO Box 708 Princeton NJ 08540

SHYKIND, DAVID, US citizen. SOLID STATE NMR, OPTICALLY DETECTED NMR. *Educ:* Yale Univ, BS, 83; Univ Calif Berkeley, PhD(physics), 89. *Prof Exp:* Postdoctoral res assoc, Calif Inst Technol, 89-90; RES PHYSICIST, QUANTUM MAGNETICS, INC, 90- *Concurrent Pos:* Prin investr, Defense Advanced Res Proj Agency, 91; vis res assoc chem, Calif Inst Technol, 91-92. *Mem:* Am Phys Soc. *Res:* Applying nuclear magnetic resonance spectroscopy and magneto-optics to problems in materials science and non- destructive testing and evaluation; electromagnetism. *Mailing Add:* Quantum Magnetics Inc 11578 Sorrento Valley Rd No 30 San Diego CA 92121

SHYKIND, EDWIN B, b Los Angeles, Calif, Oct 10, 31; m 57; c 3. MARINE GEOLOGY, SCIENCE POLICY. *Educ:* Northwestern Univ, BS, 53; Univ Chicago, SM, 55, PhD(geol), 56. *Prof Exp:* Instr geol, Wright Jr Col, 54-55; res engr, Montaine Corp, Ill, 56-57; asst prof earth sci, Northern Ill Univ, 57-62; chief earth sci br & spec asst to dir sci inform exchange, Smithsonian Inst, 62-64; assoc staff dir interagency comt oceanog, Fed Coun Sci & Technol, Exec Off of Pres, 64-67, actg exec secy, 67, exec secy interagency comt marine res educ & fac, Nat Coun Marine Resources & Eng Develop, 67-69; staff dir marine sci affairs staff, Off Oceanogr, Dept Navy, 69; sr staff mem, Nat Coun Marine Resources & Eng Develop, Exec Off of Pres, 69-71; dir, Environ Affairs Div, 71-77, dir, Off Bux & Policy Anal, Bus Domestic Com, 71-79, sr tech adv, Off Regulatory Policy, 79-82, SCI ADV TRADE ADMIN, US DEPT COM, 82- *Mem:* AAAS; Am Asn Petrol Geol. *Res:* Sedimentation; hydrodynamics; scientific information; ocean engineering; environmental affairs. *Mailing Add:* Trade Develop 14 & Constitution Ave NW Rm 4043 Trade Develop Washington DC 20230

SHYNE, J(OHN) C(ORNELIUS), b Detroit, Mich, Nov 26, 25; m 47; c 5. METALLURGY. *Educ:* Univ Mich, BS(math) & BS(metall), 51, MS, 52, PhD(metall), 58. *Prof Exp:* Res engr, Ford Motor Co, 52-59; proj dir mat res, Mueller Brass Co, 59-60; from asst prof to assoc prof mat sci, 60-66, chmn dept mat sci & eng, 71-75, PROF MAT SCI, STANFORD UNIV, 66- *Concurrent Pos:* Head, Metall & Mat Sect, NSF, 75-77. *Mem:* Am Soc Metals; Am Inst Mining, Metall & Petrol Engrs. *Res:* Internal friction, crystalline defects and phase transformations in metals; relation of structure to properties in metals; metallurgical failure analysis. *Mailing Add:* 4195 Dake Ave Palo Alto CA 94306

SHYSH, ALEC, b Vilna, Alta, Apr 2, 36; m 61; c 1. BIONUCLEONICS, RADIOPHARMACY. *Educ:* Univ Alta, BSc, 58, MSc, 68, PhD(bionucleonics), 70. *Prof Exp:* Teaching asst pharm, 65-70, from asst prof to assoc prof bionucleonics, 70-84, PROF BIONUCLEONICS & RADIOPHARM, UNIV ALTA, 84- *Mem:* Can Pharmaceut Asn; Soc Nuclear Med. *Res:* Development, quality control and application of radiopharmaceuticals as diagnostic scanning agents. *Mailing Add:* Fac Pharm & Pharmaceut Sci Univ Alta Edmonton AB T6G 2N8 Can

SIAKOTOS, ARISTOTLE N, b Dedham, Mass, July 19, 28; m 72; c 4. BIOCHEMISTRY. *Educ:* Univ Mass, BS, 52, MS, 54; Cornell Univ, PhD(entom), 58. *Prof Exp:* Res asst entom, Cornell Univ, 54-56; entomologist, Med Res Labs, US Army Chem Ctr, 58-62, biochemist, 62-68; from asst prof to assoc prof, 68-78, PROF PATH, MED CTR, IND UNIV, INDIANAPOLIS, 78- *Concurrent Pos:* Nat Retinitis Pigmentosa sr res fel, 74-76. *Mem:* Am Chem Soc; Am Oil Chem Soc; Am Soc Neurochem; Am Soc Neurosci; Am Soc Exp Path. *Res:* Subcellular particulates of the central nervous system; drug induced changes in brain; lipid composition and metabolism in subcellular particles; retinal degeneration; biochemistry of the eye; vision; ophthalmology; biochemical pathology; atypical slow virus diseases; lipopigments; aldehyde metabolism. *Mailing Add:* Dept Path Ind Univ Med Ctr 1100 W Michigan St Indianapolis IN 46223

SIANO, DONALD BRUCE, b Sewickley, Pa, June 30, 42; m 72. PHYSICAL CHEMISTRY, BIOPHYSICS. *Educ:* Kent State Univ, BS, 60; Iowa State Univ, MS, 68, PhD(biophys), 78. *Prof Exp:* Res assoc chem, Columbia Univ, 76-78; res physicist emulsion sci, Exxon Res & Eng Co, 78-86, RES PHYSICIST, POLYMER SCI, EXXON CHEM CO, 86- *Mem:* AAAS; Am Chem Soc. *Res:* Rayleigh and dynamic light scattering; biopolymers; microemulsions; enhanced oil recovery; polymer blends. *Mailing Add:* Exxon Chem Co 1900 E Linden Ave Linden NJ 07036

SIAPNO, WILLIAM DAVID, b Norfolk, Va, Aug 29, 26; div; c 3. MARINE GEOLOGY, EXPLORATION. *Educ:* Va Polytech Inst, BS, 51; Univ Colo, MS, 53. *Prof Exp:* Geol eng, US AEC, 53-55, asst chief airborne explor, 56-59; sr res engr, Space Div, NAm Aviation, 62-64, chief geologist, Ocean Systs, 65-68; chief geologist, Deepsea Ventures, 68-71, chief scientist, 72-76, dir marine sci, 77-86; CONSULT, MARINE SCI, 87- *Concurrent Pos:* Mem, Sea Sat Users Comt, NASA, 77-79 & Sci Appln Adv Comt, 83-86; lectr, NATO Advan Res Inst, 85, UN, Law of Sea, 87; vis scientist, Vinogradov Exped, USSR, 86; vis lectr, USSR Acad Sci, 87, Am corresp, 90-; team leader, UN Southeast Asia Eval, 89. *Honors & Awards:* Deap Ocean Gastropod named in honor, Amaea Siapnoi, 77. *Mem:* Sigma Xi; Geol Soc Am; Int Marine Minerals Soc; fel Explorers Club. *Res:* Global geoscientific programs, emphasis on remote sensing and evaluation of mineral resources. *Mailing Add:* Box B96 Ordinary VA 23131

SIAS, CHARLES B, chemical engineering, for more information see previous edition

SIAS, FRED R, JR, b Jacksonville, Fla, Aug 17, 31; m 61; c 2. PHYSIOLOGICAL CONTROL SYSTEMS, ROBOTICS. *Educ:* Univ Fla, BS, 54, MS, 59; Univ Miss, PhD(physiol & biophys), 70. *Prof Exp:* Asst prof, Sch Med, Univ Miss, 70-72; asst prof, Sch Info & Comput Sci, Ga Inst Technol, 74-76; ASSOC PROF ELEC & COMPUT ENG, CLEMSON UNIV, 76- *Concurrent Pos:* Prin investr, several res grants & contracts. *Mem:* Inst Elec & Electronic Engrs; Am Physiol Soc; Biomed Eng Soc; Soc Comput Simulation; Am Soc Eng Educ. *Res:* Microcomputer applications in medicine, mobile robotics; mobile robotic. *Mailing Add:* Dept Elec & Comput Eng Riggs Hall Clemson Univ Clemson SC 29634-0915

SIAS, FREDERICK RALPH, b Grand Rapids, Mich, June 1, 05; m 30; c 2. ELECTRICAL & MECHANICAL ENGINEERING. *Educ:* Univ Fla, BS, 28. *Hon Degrees:* EE, Univ Fla, 59. *Prof Exp:* Sales engr, Weston Elec Instrument, 28-29; teacher manual arts, Andrew Jackson High Sch, Jacksonville, Fla, 29-31; owner, Sias Labs, Jacksonville & Orlando, 31-32; instrument maintenance, Pan Am Airways, 32-40; aircraft inst design & electrical inst design, Gen Electric Co, Lynn, Mass, 40-57, consult engr, Rectifier Dept, Va, 57-58, Philadelphia, Pa, 58-69, Space Div, Valley Forge, Pa, 69-70; RETIRED. *Mem:* Inst Elec & Electronics Engrs; Inst Elec & Electronics Engrs Instrumentation & Measurement Soc. *Res:* Electromechanical instrumentation and measurements pertaining to aircraft indicating instruments, nuclear propulsion systems and semiconductor manufacturing; 19 US patents. *Mailing Add:* 8232 Lake Lucy Dr Orlando FL 32818

SIATKOWSKI, RONALD E, b Newark, NJ, Oct 30, 50. CIRCULAR DICHROISM SPECTROSCOPY, MOLECULAR ELECTRONIC DEVICES. *Educ:* Pa State Univ, BS, 72; Temple Univ, MA, 78, PhD(biophys chem), 85. *Prof Exp:* Teaching asst, Biol Dept, Temple Univ, 76-78, teaching assoc, Chem Dept, 78-85; Naval res assoc, Nuclear Magnetic Resonance Spectroscopy Lab, 85-87, consult/res biol & Naval res lab scientist, Chem Warfare Defense, 87-88; ASST PROF, CHEM DEPT, US NAVAL ACAD, 88- *Concurrent Pos:* Res scientist, ImClone Systs Inc, New York, NY, 89, McNeil Consumer Prod Co, Ft Washington, Pa, 90, Purdue-Frederick Pharmaceut, Yonkers, NY, 91. *Mem:* Sigma Xi; fel Am Inst Chemists; NY Acad Sci; Am Chem Soc; Am Phys Soc; Int Union Pure & Appl Chem. *Res:* Molecular devices; theoretical and experimental circular dichroism spectroscopy; solid-state nuclear magnetic resonance spectroscopy; dioxirane chemistry; molecular modeling and computer assisted analysis. *Mailing Add:* Chem Dept US Naval Acad Annapolis MD 21402

SIAU, JOHN FINN, b Detroit, Mich, Mar 30, 21; div; c 2. WOOD SCIENCE, CHEMICAL ENGINEERING. *Educ:* Mich State Univ, BS, 43; State Univ NY Col Environ Sci & Forestry, MS, 66, PhD(wood sci), 68. *Prof Exp:* Engr loudspeaker mfr, Utah Radio Prod Div, Newport Steel Corp, 46-48; mfg & designing engr vacuum tube, Gen Elec Co, 48-58; physics teacher, Paul Smiths Col, 58-68; from assoc prof to prof, 68-85, EMER PROF WOOD SCI, COL ENVIRON SCI & FORESTRY, STATE UNIV NY, 85- *Concurrent Pos:* Proj dir, NSF, 69-77 & 83-86. *Honors & Awards:* Wood Award, Forest Prod Res Soc, 68. *Mem:* Forest Prod Res Soc; Soc Wood Sci & Technol; Int Asn Wood Anatomists. *Res:* Flow of liquids and gases through wood; non-isothermal moisture diffusion in wood; wood drying; wood preservation. *Mailing Add:* PO Box 41 Keene NY 12942

SIBAL, LOUIS RICHARD, b Chicago, Ill, Aug 6, 27; m 64; c 2. MICROBIOLOGY. *Educ:* Univ Ill, BS, 49; Univ Colo, MS, 54, PhD(microbiol), 57. *Prof Exp:* Assoc prof microbiol, Col Med, Univ Ill, 57-65; spec fel, Nat Cancer Inst, 65-66, res microbiologist, 66-71, dep assoc dir, 71-76, actg assoc dir viral oncol, 76-80, assoc dir, Div Cancer Cause & Prev, 80-82, EXTRAMURAL PROG PROCEDURES OFF, NAT CANCER INST, 82- *Mem:* AAAS; Am Asn Cancer Res; Am Soc Microbiol; Am Asn Immunol. *Res:* Viral oncology; immunologic aspects of virus-induced cancer of animals and man. *Mailing Add:* NIH Bldg 1 Rm 314 Bethesda MD 20203

SIBBACH, WILLIAM ROBERT, b Chicago, Ill, Mar 9, 27; m 50; c 2. EXTRUSION COATING, ADHESIVE LAMINATING. *Educ:* Northwestern Univ, Evanston, PhD(chem & math), 55. *Prof Exp:* Chemist, Kraft Foods Co, 49-55 & Tee-Pak, Inc, 55-58; int adminr, Pillsbury Co, 58-59; tech dir & plant mgr, Champion Paper Co, 59-65; sr scientist, Chicopee Div, Johnson & Johnson, 65-68; tech dir, Ludlow Corp, 68-78; TECH DIR, LAMINATING & COATING CO, JEFFERSON SMURFIT CORP, 78- *Concurrent Pos:* Chmn, Polymers, Laminations & Coatings Div, Tech Asn Pulp & Paper Indust, 83-85. *Mem:* Tech Asn Pulp & Paper Indust. *Res:* New packaging structure for food and pharmaceuticals; awarded six patents. *Mailing Add:* 4400 N Wildwood Ct Hoffman Estates IL 60195

SIBBALD, IAN RAMSAY, b Eng, Sept 20, 31; nat Can; m 55; c 4. NUTRITION. *Educ:* Univ Leeds, BSc, 53; Univ Alta, MSc, 55, PhD(animal nutrit), 57. *Hon Degrees:* DSc, Leeds Univ, 82. *Prof Exp:* Asst nutrit, Macdonald Col, McGill Univ, 57; asst prof, Ont Agr Col, 57-63; group leader, Animal Prod Res, John Labatt Ltd, 63-70, mgr food res sect, 70-72; prin res scientist, Animal Res Ctr, Agr Can, Ottawa, 72-89; RETIRED. *Honors & Awards:* Borden Award, Nutrit Soc Can, 60; Tom Newman Mem Int Award, 77; Am Feed Mfgs Award, Poultry Sci Asn, 79. *Mem:* Poultry Sci Asn; World's Poultry Sci Asn; Nutrit Soc Can. *Res:* Utilization of energy and nitrogen by monogastrics; various aspects of poultry nutrition; feeding stuff evaluation. *Mailing Add:* Box 291 Small Mountain ON K0E 1W0 Can

SIBBALD, WILLIAM JOHN, b London, Ont, June 28, 46; m; c 5. CRITICAL CARE. *Educ:* Univ Western Ont, MD, 70. *Prof Exp:* AT PROG CTR, CRITICAL CARE TRAUMA UNIT, VICTORIA HOSP CORP, 79-; PROF MED & SURG, UNIV WESTERN ONT, 85- *Res:* Technology evaluation; clinical-basis research; sepsis-multiple organ failure. *Mailing Add:* Dept Med Crit Care Victoria Hosp London ON N6A 4G5 Can

SIBENER, STEVEN JAY, b Brooklyn, NY, Apr, 3, 54; m. SURFACE CHEMISTRY, MOLECULAR BEAM SCATTERING. *Educ:* Univ Rochester, BA & ScB, 75; Univ Calif, Berkeley, MS, 77, PhD(chem), 79. *Prof Exp:* Res fel, Bell Labs, 79-80; from asst prof to assoc prof, 85-89, PROF, DEPT CHEM & JAMES FRANCK INST, UNIV CHICAGO, 89- *Concurrent Pos:* Alfred P Sloan res fel, 83-87. *Honors & Awards:* Marlow Medal, Faraday Div, Royal Soc Chem, 88. *Mem:* Am Phys Soc; Royal Soc Chem; Sigma Xi; AAAS; Am Chem Soc. *Res:* Molecular beam, laser spectroscopic, and ultra-high vacuum surface characterization techniques; gas-surface interaction potentials; chemisorption; physisorption; heterogeneous catalysis; two-dimensional phase transitions; surface structure; surface phonons; semiconductor reconstruction; epitaxial film growth. *Mailing Add:* James Franck Inst Univ Chicago 5640 S Ellis Ave Chicago IL 60637

SIBERT, ELBERT ERNEST, b DeQueen, Ark, Oct 7, 41; m 63; c 2. COMPUTER SCIENCE, MATHEMATICS. *Educ:* Rice Univ, BA, 63, PhD(math), 67. *Prof Exp:* Asst prof comput sci, Rice Univ, 67-70; assoc prof, 70-77, PROF COMPUT & INFO SCI, SYRACUSE UNIV, 77- *Concurrent Pos:* Develop engr, Schlumberger Well Serv, 69-70, consult, 68-81. *Mem:* Asn Comput Mach; Comput Soc; AAAS. *Res:* Computational logic; logic programming; parallel computing. *Mailing Add:* Box 31 Oran NY 13125

SIBERT, JOHN RICKARD, b Glendale, Calif, Dec 3, 40; Can citizen; m; c 1. MARINE ECOLOGY, STATISTICS. *Educ:* Univ Pac, BA, 62; Columbia Univ, PhD(zool), 68. *Prof Exp:* Fel oceanog, Univ BC, 68-69, teaching fel bot, 69-70; res scientist marine ecol, Pac Biol Sta, 70-82; sr fisheries scientist, 82-84, coordr, Tuna & Billfish Prog, S Pacific Comn, 84-87. *Mem:* AAAS; Am Soc Limnol & Oceanog. *Res:* Sources and fates of detritus in aquatic ecosystems; productivity of estuarine meiofauna; mathematical modelling of estuarine ecosystems; statistical descriptions of community structure; population dynamics; fisheries management. *Mailing Add:* Box 279 Errington BC V0R 1V0 Can

SIBILA, KENNETH FRANCIS, b Canton, Ohio, Dec 2, 11; m 40; c 3. ELECTRICAL ENGINEERING. *Educ:* Case Inst Technol, BSEE, 35, MSEE, 37. *Prof Exp:* Engr, Ohio Crankshaft Co, 37-40; from instr to assoc prof elec eng, 40-47, prof & head dept, 47-68, dir electronic systs eng, 68-77, EMER PROF ELEC ENG, UNIV AKRON, 77- *Mem:* Am Soc Eng Educ; sr mem Inst Elec & Electronics Engrs; Nat Asn Educ Broadcasters; Sigma Xi. *Res:* Electronic systems used in educational TV instruction. *Mailing Add:* 1929 Silver Lake Ave Cuyahoga Falls OH 44223

SIBILIA, JOHN PHILIP, b Newark, NJ, Mar 12, 33; m 58; c 3. CHEMICAL PHYSICS. *Educ:* Rutgers Univ, BA, 53; Univ Md, PhD(chem), 58. *Prof Exp:* Res chemist, US Rubber Co, 59-61; res chemist, Allied Corp, 61-62, group leader, 62-67, res supvr, 67-77, mgr chem physics dept, 77-83, DIR ANAL SCI LAB, ALLIED-SIGNAL INC, 83- *Concurrent Pos:* Lectr spectros & molecular workshop, Fairleigh Dickinson Univ, 65-66, instr polymer sci, 72-78; vchmn, Res & Develop Coun NJ, 85, chmn, 87-89; lectr, Polytechnic Inst, 91. *Mem:* Am Chem Soc; Am Soc Testing & Mat; Soc Appl Spectros. *Res:* Molecular structure of organic compounds, morphology of polymers; analysis of materials through spectroscopic; x-ray diffration; microscopy, nuclear magnetic resonance and thermal analytical techniques; material structure-property relationships and pollution analysis. *Mailing Add:* 12 Balmoral Dr Livingston NJ 07039

SIBLEY, CAROL HOPKINS, b Freeport, NY, Oct 9, 43; m 66; c 2. DEVELOPMENTAL GENETICS, IMMUNOGENETICS. *Educ:* Univ Rochester, BA, 65, MS, 69; Univ Calif, San Francisco, PhD(biochem), 74. *Prof Exp:* Fel, Calif Inst Technol, 74-76; asst prof, 76-83, ASSOC PROF GENETICS, UNIV WASH, 83- *Concurrent Pos:* Vis prof, Calif Inst Technol, 81; Fogarty Int fel, Free Univ Brussels, 84-85; assoc ed, J Immunol. *Mem:* Am Asn Immunologists; AAAS; Asn Women Sci; Am Asn Microbiologists. *Res:* Control of B lymphocyte development using transcription, processing and synthesis of the antibody genes as a model for control of developmental processes; chloroquine resistance in plasmodium. *Mailing Add:* SK-50 Dept Genetics Univ Wash Seattle WA 98195

SIBLEY, CHARLES GALD, b Fresno, Calif, Aug 7, 17; m 42; c 3. SYSTEMATIC ZOOLOGY, ORNITHOLOGY. *Educ:* Univ Calif, Berkeley, AB, 40 & PhD(zool), 48. *Prof Exp:* Instr zool, Univ Kans, 48-49; asst prof, San Jose State Col, Calif, 49-53; from assoc prof to prof zool, Cornell Univ, 53-65; from prof to emer prof biol & William R Coe prof ornith, Yale Univ, 65-86, cur birds, Peabody Mus Natural Hist, 65-86; DEAN'S PROF SCI & PROF BIOL, SAN FRANCISCO STATE UNIV, 86- *Concurrent Pos:* Guggenheim fel, 59-60; dir, Peabody Mus Natural Hist, Yale Univ, 70-76. *Honors & Awards:* Brewster Mem Award, Am Ornith Union, 71; Daniel Giraud Elliot Medal, Nat Acad Sci, 88. *Mem:* Nat Acad Sci; Soc Syst Zool; Am Soc Naturalists; Am Ornith Union (treas, 53-62, pres, 86-88); Inst Ornith Cong (pres, 86-90); AAAS. *Res:* Fossil birds; geographic variation, speciation and interspecific hybridization in wild populations of birds; biochemical and molecular techniques applied to proteins and DNA to reconstruct evolutionary history of birds and mammals; rates of genomic evolution; dating of divergence events; published 130 titles mostly pertaining to molecular evolution. *Mailing Add:* 95 Seafirth Rd Tiburon CA 94920

SIBLEY, DUNCAN FAWCETT, b Newton, Mass, Mar 9, 46; m 68. SEDIMENTARY PETROLOGY. *Educ:* Lafayette Col, BA, 68; Rutgers Univ, MS, 71; Univ Okla, PhD(geol), 75. *Prof Exp:* Asst prof, 74-80, ASSOC PROF GEOL, MICH STATE UNIV, 80- *Mem:* Soc Econ Paleontologists & Mineralogists. *Res:* Understanding of diagenesis in sandstone and carbonate rocks with specific interest in the origin and evolution of porosity. *Mailing Add:* Dept of Geol Mich State Univ East Lansing MI 48824

SIBLEY, LUCY ROY, b Fayetteville, Ark, June 10, 34; m 56; c 3. TEXTILE FABRIC PSEUDOMORPHS, RADIOCARBON DATING OF TEXTILES. *Educ:* Auburn Univ, BS, 56, MS, 58; Univ Mo, PhD(hist textiles), 81. *Prof Exp:* Instr textiles, La State Univ, 58-61, Univ Puget Sound, 69; lectr home economics, St Mary Col, 74-80, dir, Life Planning Ctr, 78-80; asst prof textile hist, Univ Ga, 80-84; assoc prof, 84-85, CHMN DEPT TEXTILES & CLOTHING, OHIO STATE UNIV, 85- *Concurrent Pos:* Prin investr, Am Philos Soc grant, 81-83, Ohio State Univ seed grant, 85-86; co-prin investr, Werner-Gren Found Anthrop Res, 84-86. *Mem:* Archaeol Inst Am. *Res:* Direct forms of archaeological fabric evidence, including textile fabric pseudomorphs, the application of small sample radiocarbon dating to Coptic textiles; developing of a sampling methodology for ancient textiles; ancient textile technology. *Mailing Add:* 7831 Maplecreek Ct Powell OH 43065

SIBLEY, WILLIAM ARTHUR, b Ft Worth, Tex, Nov 22, 32; m 57; c 3. SOLID STATE PHYSICS. *Educ:* Univ Okla, BS, 56, MS, 58, PhD(physics), 60. *Prof Exp:* Res physicist, Nuclear Res Estab, Julich & Inst Metal Physics, Aachen Tech Univ, 60-61; res physicist, Oak Ridge Nat Lab, 61-70; chmn dept physics, 70-78, PROF PHYSICS, OKLA STATE UNIV, 70-, ASST VPRES RES, 78- *Mem:* Fel Am Phys Soc; Sigma Xi. *Res:* Optical and mechanical properties of both irradiated and unirradiated materials; light scattering; optical absorption and luminescence of laser materials. *Mailing Add:* NSF Rm 1225 Washington DC 20550

SIBLEY, WILLIAM AUSTIN, b Miami, Okla, Jan 25, 25; m 54; c 4. NEUROLOGY. *Educ:* Yale Univ, BS, 45, MD, 48. *Prof Exp:* Asst resident neurologist, Presby Hosp, New York, 51 & 53-55, resident neurologist, 56; asst neurol, Col Physicians & Surgeons, Columbia Univ, 55-56; from asst prof to assoc prof, Sch Med, Western Reserve Univ, 56-67, actg dir div, 59-61; dir dept neurol, 67-82, PROF, COL MED, UNIV ARIZ, 67- *Concurrent Pos:* Consult neurologist, Vet Admin & Benjamin Rose Hosps, Cleveland, 58-; physician-in-chg neurol, Univ Hosps, 59-61. *Mem:* Am Neurol Asn (vpres, 79-80); fel Am Acad Neurol (vpres, 85-87); Asn Res Nerv & Ment Dis; Int Fed Multiple Sclerosis Soc (exec comt). *Res:* Multiple sclerosis: treatment, cause and triggering factors. *Mailing Add:* Dept Neurol Univ Ariz Health Sci Ctr Tucson AZ 85724

SIBOO, RUSSELL, b Trinidad, Wis, Mar 21, 30; Can citizen; m 66; c 2. IMMUNOLOGY, MICROBIOLOGY. *Educ:* McMaster Univ, BA, 58; Univ Toronto, MSc, 62; McGill Univ, PhD(immunol), 64. *Prof Exp:* Med Res Coun Can fel immunol, Univ Lund, 64-65; asst prof, Univ Sask, 65-67; asst prof, Univ Ottawa, 67-69; asst prof, 69-73, ASSOC PROF IMMUNOL, MCGILL UNIV, 73- *Concurrent Pos:* Head, Diagnostic Immunol Lab, Kuwait Univ, 83-85. *Mem:* Can Soc Immunol; Can Soc Microbiol. *Res:* Complement synthesis; complement synthesis; membrane receptors; CRP; fluorescence immunoassay; oral spirochete infection; DR-typing in multiple sclerosis and diabetes; &-chain disease. *Mailing Add:* Dept of Microbiol & Immunol McGill Univ Sherbrooke St W Montreal PQ H3A 2M5 Can

SIBUL, LEON HENRY, b Voru, Estonia, Aug 30, 32; US citizen; m 61; c 2. APPLIED MATHEMATICS, UNDERWATER ACOUSTICS. *Educ:* George Washington Univ, BEE, 60; NY Univ, MEE, 63; Pa State Univ, PhD(appl math & elec eng), 68. *Prof Exp:* Mem tech staff syst design, Bell Tel Labs, 60-64; SR SCIENTIST & PROF ACOUST, STOCHASTIC PROCESSES, ADAPTIVE SYSTS, APPL RES LAB, PA STATE UNIV, UNIVERSITY PARK, 64- *Concurrent Pos:* Consult, Nat Acad Sci, Nat Res Coun, 72-73; assoc ed, Trans Electronic & Aerospace Systs, Inst Elec & Electronic Engrs, 80- *Mem:* Soc Indust & Appl Math; Inst Elec & Electronic Engrs. *Res:* Adaptive systems and application of adaptive algorithms to array processing with applications to sonar, radar and seismic signal processing; broadband ambiguity function and signal design; system optimization theory; stochastic system theory. *Mailing Add:* Appl Res Lab Pa State Univ PO Box 30 University Park PA 16802

SIBULKIN, MERWIN, b New York, NY, Aug 20, 26; m 49; c 2. FLUID DYNAMICS. *Educ:* NY Univ, BS, 48; Univ Md, MS, 53; Calif Inst Technol, AeroE, 56. *Prof Exp:* Aeronaut res scientist, Nat Adv Comt Aeronaut, 48-51; res engr, US Naval Ord Lab, 51-53; res engr, Jet Propulsion Lab, Calif Inst Technol, 53-56; staff scientist, Sci Res Lab, Convair Div, Gen Dynamics Corp, 56-63; assoc prof eng, 63-66, PROF ENG, BROWN UNIV, 66- *Mem:* Am Soc Mech Engrs; Combustion Inst. *Res:* Combustion and fire research; heat transfer; viscous flow. *Mailing Add:* Div of Eng Brown Univ Providence RI 02912

SIBUYA, YASUTAKA, b Maizuru, Japan, Oct 16, 30; m 62; c 4. MATHEMATICS. *Educ:* Univ Tokyo, BS, 53, MS, 55; Univ Calif, Los Angeles, PhD(math), 59; Univ Tokyo, DS(math), 61. *Prof Exp:* Res assoc, Mass Inst Technol, 59-60; temp mem, Courant Inst, NY Univ, 60-61; asst prof math, Ochanomizu Univ, Tokyo, 62-63; assoc prof, 63-65, PROF MATH, UNIV MINN, 65- *Concurrent Pos:* Vis prof, Math Res Ctr, Univ Wis-Madison, 72-73. *Mem:* Math Soc Japan; Am Math Soc; Soc Indust & Appl Math. *Res:* Field of analytic theory of ordinary differential equations. *Mailing Add:* Sch of Math Univ of Minn Minneapolis MN 55455

SICA, LOUIS, b Miami, Fla, June 14, 35; m 78. UNCONVENTIONAL IMAGING, OPTICAL PHYSICS. *Educ:* Fla State Univ, BA, 58; Johns Hopkins Univ, PhD(physics), 66. *Prof Exp:* Res asst infrared spectros, Lab Astrophys & Phys Meteorol, Johns Hopkins Univ, 63-67; RES PHYSICIST OPTICAL PHYSICS, NAVAL RES LAB, 67- *Concurrent Pos:* Surveyor res topics mod optics, Off Naval Res, 75-76. *Mem:* Optical Soc Am. *Res:* Unconventional imaging optical coherence effects; nonlinear propagation; interferometry; space variant imaging, image- sharpness criteria. *Mailing Add:* Code 6530 Naval Res Lab Washington DC 20375

SICARD, RAYMOND EDWARD, b Lawrence, Mass, Apr 18, 48; div; c 1. DEVELOPMENTAL PHYSIOLOGY. *Educ:* Merrimack Col, AB, 69; Univ RI, MS, 72, PhD(biol sci), 75. *Prof Exp:* Jr bacteriologist anal, Mass Dept Pub Health, 69; hemat tech anal, New Eng Deaconess Hosp, 70-73; res assoc, Dept Biol, Amherst Col, 74; res fel, Shriners Burns Inst, Mass Gen Hosp & Harvard Med Sch, 75-76; asst prof, Dept Biol, Boston Col, 76-83; res assoc, Dept Biol, Amherst Col, 83-84; res proj develop asst, Dept Pediat, RI Hosp, 84-91; ASST PROF, DEPT SURG, UNIV MINN, 91- *Concurrent Pos:* Grad teaching asst, Dept Zool, Univ RI, 69-71 & 73-74; lectr, Div Pharm & Allied Health, Northeastern Univ, 75-76; lectr, Dept Biol & Continuing Educ, Regis Col, 83-84; grant, RI Found, 86; tech consult, Dept Surg, Univ Minn, 91- *Mem:* Am Inst Biol Sci; NY Acad Sci; Int Soc Develop & Comp Immunol; Int Soc Develop Biologists; Soc Develop Biol. *Res:* Nature of physiological (neural, endocrine, and immunological) regulation of the developmental events occurring during the process of forelimb regeneration in amphibians. *Mailing Add:* Dept Surg Box 120 UMHC Univ Minn Sch Med Minneapolis MN 55455

SICCAMA, THOMAS G, b Philadelphia, Pa, July 6, 36; m 62; c 1. ECOLOGY. *Educ:* Univ Vt, BS, 62, MS, 63, PhD(bot), 68. *Prof Exp:* Assoc in res ecol, Yale Univ, 67-69, asst prof, 69-80, lectr forest ecol, 77-80. *Mem:* AAAS; Ecol Soc Am; Torrey Bot Club. *Res:* Ecosystem analysis; computer applications in ecology; heavy metals cycling in natural ecosystems. *Mailing Add:* Yale Forestry Sch 107 Gml PO Box 2021 New Haven CT 06520

SICH, JEFFREY JOHN, b Youngstown, Ohio, Oct 8, 54. IMMUNOMODULATION, NEUTROPHIL FUNCTION. *Educ:* Davidson Col, NC, BS, 77; Univ Cincinnati, MS, 81, PhD(microbiol), 83. *Prof Exp:* Asst prof biol, Denison Univ, 83-84; asst prof biol, Univ Tampa, 84-86; ASST PROF BIOL, YOUNGSTOWN STATE UNIV, 86- *Concurrent Pos:* Res biologist, Nat Inst Arthritis, Diabetes, Digestive & Kidney Dis, NIH, 85; mem bd educ & training, Am Soc Microbiol, 90- *Mem:* Am Soc Microbiol; Sigma Xi. *Res:* Immunomodulation of macrophage and neutrophil function; effects of immunopotentiators; immunosuppressive effects associated with tumor growth. *Mailing Add:* Youngstown State Univ 410 Wick Ave Youngstown OH 44555

SICHEL, ENID KEIL, b Burlington, Vt, May 14, 46. PHYSICS, MATERIALS SCIENCE ENGINEERING. *Educ:* Smith Col, Mass, AB, 67; Rutgers Univ, New Brunswick, PhD(physics), 71. *Prof Exp:* Vis scientist, Nat Magnet Lab, Mass Inst Technol, 70-71; res assoc, Rutgers Univ, New Brunswick, 71-73; mem tech staff, David Sarnoff Res Ctr, RCA Labs, 73-79; mem staff, GTE Advan Technol Lab, 79-85; CONSULT, 88- *Concurrent Pos:* Consult electro-optics, Tufts Univ, 86; vis prof elec eng, Mass Inst Technol, 86-88; prog dir, NSF, 88-89. *Mem:* AAAS; Am Phys Soc; Mat Res Soc; Inst Elec & Electronic Engrs. *Res:* Thermophysical properties; transport and optical properties of thin solid films; conducting polymers; amorphous semi-conductors; carbon black-polymer composites. *Mailing Add:* Box 236 Lincoln MA 01773

SICHEL, JOHN MARTIN, b Montreal, Que, Dec 2, 43; m 67; c 3. MOLECULAR ORBITAL CALCULATIONS. *Educ:* McGill Univ, BS, 64, PhD(quantum chem), 68. *Prof Exp:* Fel theoret chem, Univ Bristol, 67-69; res assoc chem & physics, Ctr Res Atoms Molecules, Laval Univ, 69-70, asst prof physics, 70-72; from asst prof to assoc prof, 72-83, PROF CHEM, UNIV MONCTON, 83- *Concurrent Pos:* Vis prof, Univ Montreal, 81-82. *Mem:* Chem Inst Can; Can Asn Physicists; Fr Can Asn Advan Sci. *Res:* Molecular orbital calculations on molecules and surfaces. *Mailing Add:* Dept Chem & Biochem Univ Moncton Moncton NB E1A 3E9 Can

SICHEL, MARTIN, b Stuttgart, Ger, Sept 1, 28; m 52; c 3. AEROSPACE ENGINEERING, FLUID DYNAMICS. *Educ:* Rensselaer Polytech Inst, BME, 50, MME, 51; Princeton Univ, PhD(aerospace eng), 61. *Prof Exp:* Develop engr thermal power, Gen Elec Co, 51-54; res aide aerospace, Princeton Univ, 58-61; from asst prof to assoc prof, 61-68, PROF AEROSPACE ENG, UNIV MICH, ANN ARBOR, 68- *Mem:* Am Phys Soc; fel Am Inst Aeronaut & Astronaut; Am Soc Eng Educ; Combustion Inst. *Res:* Fluid dynamics; shock wave structure; boundary layer theory; flow with chemical reactions; hypersonic flow; detonations. *Mailing Add:* Dept Aerospace Eng Univ Mich Ann Arbor MI 48109-2140

SICHLER, JIRI JAN, b Prague, Czech, Dec 30, 41; m 68. MATHEMATICS. *Educ:* Charles Univ Prague, MSc, 66, PhD(math), 68. *Prof Exp:* Nat Res Coun Can fel, 69-70, asst prof math, 70-75, PROF MATH, UNIV MAN, 75- *Concurrent Pos:* Nat Res Coun Can grants, 70-92. *Mem:* Am Math Soc; Can Math Soc. *Res:* Algebra; general mathematical systems; category theory. *Mailing Add:* Dept of Math Univ of Man Winnipeg MB R3T 2N2 Can

SICILIAN, JAMES MICHAEL, b Bronx, NY, May 25, 47. NUMERICAL ANALYSIS. *Educ:* Mass Inst Technol, BS, 69; Stanford Univ, MS, 70, PhD(nuclear eng), 73. *Prof Exp:* Res analyst, Savannah River Lab, 73-76; asst group leader, Los Alamos Sci Lab, 76-80; SR SCIENTIST, FLOW SCI INC, 80- *Concurrent Pos:* Special fel, Nuclear Sci & Eng, AEC, 69-72. *Mem:* Sigma Xi; AAAS; Am Soc Mech Engrs; Am Inst Aeronaut & Astronaut. *Res:* Application of numerical simulation methods to the analysis and solution of problems in hydrodynamics with emphasis on the development of convenient, efficient programs. *Mailing Add:* Flow Sci Inc PO Box 933 Los Alamos NM 87544

SICILIANO, EDWARD RONALD, b Brooklyn, NY, Aug, 13, 48; m 75. NUCLEAR SCATTERING THEORY, REACTION THEORY. *Educ:* Univ Conn, BA, 70; Ind Univ, MS, 73, PhD(physics), 76. *Prof Exp:* Fel, Dept Physics, Case Western Reserve Univ, 76-78 & Meson Physics Div, Los Alamos Nat Lab, 78-80; asst prof physics, Univ Colo, 80-83; asst prof physics, Univ Ga, 83-86; STAFF MEM, LOS ALAMOS NAT LAB, 86- *Concurrent Pos:* Vis staff mem, Meson Physics Div, Los Alamos Nat Lab, 80-86. *Mem:* Am Phys Soc; Am Asn Physics Teachers; Sigma Xi. *Res:* Nuclear structure, scattering and reaction theories; electron, pion, proton, and kaon induced reactions. *Mailing Add:* Group T-2 MS-B243 Theory Div Los Alamos Nat Lab Los Alamos NM 87545

SICILIANO, MICHAEL J, b Brooklyn, NY, May 12, 37; m 61; c 3. BIOCHEMICAL GENETICS. *Educ:* St Peter's Col, BS, 59; Long Island Univ, MS, 62; NY Univ, PhD(biol), 70. *Prof Exp:* From instr to assoc prof biol, Long Island Univ, 61-72; asst prof, 72-76, ASSOC PROF BIOL & HEAD, DEPT GENETICS, UNIV TEX SYST CANCER CTR, M D ANDERSON HOSP & TUMOR INST, 76- *Concurrent Pos:* Fel, Dept Biol, M D Anderson Hosp & Tumor Inst, 70-72; consult, Tex Epidemiol Studies Prog, 77- *Mem:* Fel Genetics Soc Am; Am Soc Cell Biol; Am Soc Ichthyologists & Herpetologists; Environ Mutagen Soc. *Res:* Control of gene expression in normal and neoplastic cells; somatic cell, animal model and human tissue materials used to study the genetics of the control of enzyme phenotypes. *Mailing Add:* 12402 Barry Knoll Houston TX 77024

SICILIO, FRED, b Italy, Aug 10, 20; nat US; m 44; c 4. INORGANIC CHEMISTRY, RADIOCHEMISTRY. *Educ:* Centenary Col, BS, 51; Vanderbilt Univ, MA, 53, PhD(chem), 56. *Prof Exp:* Asst chemist, Springhill Paper Co, 49-51; asst, Vanderbilt Univ, 53-56; sr nuclear engr, Convair Div, Gen Dynamics Corp, Tex, 56-58; res assoc prof chem, assoc prof chem eng & head, Radioisotopes Lab, Ga Inst Technol, 58-61; from assoc prof to prof chem, 61-85, EMER PROF, TEX A&M UNIV, 85- *Mem:* Am Chem Soc. *Res:* Radioisotope separations and purifications; radiation chemistry of organic substances; studies on free radicals in solution. *Mailing Add:* Dept Chem Tex A&M Univ College Station TX 77843

SICK, LOWELL VICTOR, b Elmira, NY, Apr 28, 44; m 72. ORGANOMETALLIC CHEMISTRY. *Educ:* Univ Conn, BS, 65; NC State Univ, PhD(zool biochem), 70. *Prof Exp:* Res asst zool, Nat Marine Fisheries Serv, 66-67; res assoc zool, Duke Univ, 69-70; asst prof zool nutrit, Univ Ga, Skidaway Inst, 70-75; asst prof, 75-80, ADJ ASST PROF ZOOL BIOCHEM, UNIV DEL, 80- *Concurrent Pos:* Res fel, Univ Del, 76. *Mem:* Am Soc Limnol & Oceanog; Am Chem Soc; Atlantic Estuarine Res Soc; World Mariculture Soc. *Res:* Nutritional and biochemical studies of crustacea and mollusks; specific interest in trace metal physiology. *Mailing Add:* NMFS-Charleston Lab Box 12607 Charleston SC 29412

SICK, THOMAS J, BRAIN METABOLISM, NEUROPHYSIOLOGY. *Educ:* Tulane Univ, PhD(physiol), 79. *Prof Exp:* ASSOC PROF NEUROL, SCH MED, UNIV MIAMI, 81- *Mailing Add:* Dept Neurol Sch Med Univ Miami Miami FL 33101

SICKA, RICHARD WALTER, b Cleveland, Ohio, May 28, 38. PHYSICAL CHEMISTRY, POLYMER SCIENCE. *Educ:* Case Inst Technol, BS, 59, MS, 61. *Prof Exp:* Sr res assoc, Horizons Inc, 64-66, proj supvr, 66-69, group leader, 60-74; sr res scientist, 76-81, ASSOC SCIENTIST, CENT RES, FIRESTONE TIRE & RUBBER CO, 81-; CONSULT, R SICKA & ASSOCS, 74- *Mem:* Am Chem Soc; Soc Rheology; Polymer Processing Soc. *Res:* Applied physical chemistry; polymer structure property relationships; biomedical applications of polymers; inorganic polymers; polyphosphazene polymers; composite materials; polymer processing; rheology; mixing and extrusion; coextrusion. *Mailing Add:* 3207 Boston Rd Brecksville OH 44141-3315

SICKAFUS, EDWARD N, b St Louis, Mo, Mar 7, 31; m 53; c 2. SOLID STATE PHYSICS. *Educ:* Mo Sch Mines, BS, 55, MS, 56; Univ Va, PhD(physics), 60. *Prof Exp:* Vis lectr, Sweet Briar Col, 59-60; res physicist, Denver Res Inst, 60-67, actg head physics div, 61-62, from asst prof to assoc prof physics, Univ Denver, 62-67; prin res scientist surface sci, 67-80, MGR, ELECTRONIC MAT & DEVICES DEPT, FORD SCI LAB, 80- *Mem:* Am Phys Soc; Am Vacuum Soc; Sigma Xi. *Res:* Electron spectroscopy and crystal physics related to surface physics; secondary electron cascade theory; auger electron spectroscopy; low energy electron diffraction; crystal defect state and growth; dynamic transmission analysis of x-rays and electrons; theoretical calculations of interstitial geometry; smart sensors. *Mailing Add:* 27981 Elba Dr Grosse Ile MI 48138

SICKO-GOAD, LINDA MAY, b Highland Park, Mich, Sept 21, 48; m 74; c 1. PHYCOLOGY. *Educ:* Wayne State Univ, BS, 70; City Univ New York, PhD(biol), 74. *Prof Exp:* Technician elec micros, Mich Cancer Found, 69-70; adj lectr biol, Herbert H Lehman Col, City Univ New York, 70-72 & Bronx Community Col, 72, lab technician elec micros, Herbert H Lehman Col, 72-74; res assoc, 74-76, asst res scientist, 76-78, assoc res scientist, 78-84, RES SCIENTIST, GREAT LAKES RES DIV, UNIV MICH, 84-, MARINE SUPT, 86- *Mem:* Phycol Soc Am; Electron Micros Soc Am; Am Soc Limnol & Oceanog; Int Asn Stereology. *Res:* Algal ultrastructure, lipid metabolism in phytoplankton and quantitative electron microscopy and ecological applications. *Mailing Add:* Ctr Great Lakes & Aquatic Sci Inst Sci & Technol Bldg Univ Mich Ann Arbor MI 48109-2099

SICOTTE, RAYMOND L, b Waltham, Mass, Feb 21, 39; m 64; c 3. PRODUCT DESIGNER MICROWAVE. *Educ:* Northeastern Univ, BSEE, 62; Univ Conn, MS; 64. *Prof Exp:* Staff mem, Mass Inst Technol, Lincoln Lab, 64-69; mem technol staff, Comsat Lab, 69-75; lab mgr, Fairchild Indust, 75-79; pres, 79-90, CHMN, AM MICROWAVE CORP, 90- *Mem:* Sr mem Inst Elec & Electronic Engrs. *Res:* Development of microwave solid state circuits for communications and radar systems; published on wide band varactor upconverters and C-M wave impatt amplifiers; solid state microwave switch and attenuator and detector log video amplifier products. *Mailing Add:* 9512 Faith Lane Damascus MD 20872

SICOTTE, YVON, b Montreal, Que, Oct 12, 30; m 59; c 2. PHYSICAL CHEMISTRY. *Educ:* Univ Montreal, BSc, 54, MS, 56, PhD(chem), 59. *Prof Exp:* From asst prof to assoc prof, 58-74, PROF CHEM, UNIV MONTREAL, 74- *Concurrent Pos:* NATO fel, Res Ctr Macromolecules, Strasbourg, France, 59-60; guest prof phys chem, Univ Bordeaux, 66-67. *Mem:* Chem Inst Can; Fr-Can Asn Advan Sci. *Res:* Light scattering by pure liquids and solutions; molecular anisotropy and dielectric polarization at optical frequencies, correlations of molecular orientations in liquids; polymer chemistry and polymer characterization in solution. *Mailing Add:* Dept Chem Univ Montreal Montreal PQ H3C 3J7 Can

SICULAR, GEORGE M, b New York, NY, Aug 15, 21; m 48; c 2. HYDRAULICS, HYDROLOGY. *Educ:* Cooper Union, BS, 49; Columbia Univ, MS, 53; Stanford Univ, Engr, 71. *Prof Exp:* Lectr civil eng, City Col, 49-54; PROF CIVIL ENG, SAN JOSE STATE UNIV, 54- *Concurrent Pos:* Consult, 55-; vis prof, Univ Roorkee, 63-64; Ford Found-Univ Wis adv, Univ Singapore, 67-69. *Mem:* Fel Am Soc Civil Engrs; Am Geophys Union. *Res:* Hydraulics; planning in water resources. *Mailing Add:* 387 Canon Del Sol Dr La Selva Beach CA 95076

SIDA, DEREK WILLIAM, b Barking, Eng, Nov 4, 26; m 54; c 3. APPLIED MATHEMATICS, ASTRONOMY. *Educ:* Univ London, BSc, 51, MSc, 52, PhD(math), 55. *Prof Exp:* Asst lectr math, Univ Leeds, 55-57, lectr, 57-59; sr lectr, Univ Otago, NZ, 59-62; assoc prof, St Patricks Col, 62-71, chmn dept, 68-71, dean fac arts, 71-73, prof math, Carleton Univ, 68; RETIRED. *Mem:* Royal Astron Soc London; Royal Astron Soc Can; Int Astron Union. *Res:* Stellar dynamics; cosmology; history of astronomy; population dynamics in biology. *Mailing Add:* 51 Grosvenor Dr Ottawa ON K1S 4S1 Can

SIDBURY, JAMES BUREN, JR, b Wilmington, NC, Jan 13, 22; m 53; c 5. PEDIATRICS. *Educ:* Yale Univ, BS, 44; Columbia Univ, MD, 47. *Prof Exp:* Intern med, Roosevelt Hosp, NY, 47-48, asst resident, 48-49; intern pediat, Johns Hopkins Hosp, 49-50; asst resident, Univ Hosps Cleveland, Ohio, 50-51; asst & instr, Sch Med, Emory Univ, 51-53; from instr to asst prof, Johns Hopkins Univ, 54-61; assoc prof, 61-68, prof & dir clin res unit, Sch Med, Duke Univ, 68-75; sci dir, 75-82, SR SCIENTIST, NAT INST CHILD HEALTH & HUMAN DEVELOP, NIH, 82- *Concurrent Pos:* Fel, Sch Med, Johns Hopkins Univ, 54-57. *Mem:* AAAS; Soc Pediat Res; Am Pediat Soc; Am Soc Human Genetics; Am Acad Pediat. *Res:* Biochemical genetics. *Mailing Add:* Bldg 10 Rm 8C429 NIH Bethesda MD 20892

SIDDALL, ERNEST, b Halifax, Eng, Dec 10, 19; Can citizen; m 45; c 3. SAFETY RESEARCH, SOCIO-ECONOMIC IMPACT STUDIES. *Educ:* Univ London, Eng, BSc, 39; Banff Sch Advan Mgt, Alta, dipl, 67. *Hon Degrees:* DEng, Univ Waterloo, Ont, 91. *Prof Exp:* Engr tel technol, Post Off Tel Syst, Brit, 39-49; sr scientist blast instrumentation, Ministry of Supply, Brit, 49-51; engr flight simulators, Can Aviation Electronics Ltd, 52-54; div head, nuclear plant design, Atomic Energy Can Ltd, 54-76 & 79-84; MEM, INST RISK RES, UNIV WATERLOO, ONT, 86-, ADJ PROF, DEPT SYSTS DESIGN ENG, 89- *Concurrent Pos:* Major radio technol, Royal Corps Signals, Brit Army, 40-46; assoc consult, Canatom Inc, Toronto, 77-79. *Honors & Awards:* W B Lewis Medal, Can Nuclear Asn, 82. *Mem:* Fel Can Acad Eng. *Res:* Fundamentals of safety in modern society; the total socio-economic impact of technology. *Mailing Add:* 946 Porcupine Ave Mississauga ON L5H 3K5 Can

SIDDALL, THOMAS HENRY, III, b Sumter, SC, Oct 4, 22. CHEMISTRY. *Educ:* Univ NC, AB, 42; Univ Chicago, MS, 48; Duke Univ, PhD(chem), 51. *Prof Exp:* Chemist, E I Du Pont De Nemours & Co, 50-54, res supvr chem, 54-63, res assoc, 63-69; PROF CHEM, UNIV NEW ORLEANS, 69- *Mem:* Am Chem Soc. *Res:* Chemistry of actinide elements and fission products; role of structure in determining behavior of organic extractants toward actinides and fission products; molecular dynamics; isomerism; nuclear magnetic resonance. *Mailing Add:* Dept Chem La State Univ New Orleans LA 70122

SIDDELL, DERRECK, b Consett, Co Durham, Eng, Aug 11, 42; Can & Brit citizen; m 69; c 2. STEELMAKING, NUCLEAR ENGINEERING. *Educ:* Lanchester Col Technol, Coventry Eng, Assoc Inst Metallurgists, 65; Univ Surrey, Eng, PhD(metall), 68. *Prof Exp:* Apprentice metallurgist, Consett Iron Co, 58-65; researcher, Univ Surrey, 65-68; res engr, Atlas Steels, Welland, Ont, 69-73; struct mat engr, Nuclear Fuel Handling, 73-78, supvr qual assurance dept, 78-80, MGR, METALS BR, GE CAN, 80- *Concurrent Pos:* Mem comt, Am Soc Mat, 80-84, chmn, 90-91; mem res bd, Can Welding Inst, 82-; vis prof, Univ Montreal, 86- *Mem:* Fel Inst Metals UK; Am Soc Mat. *Res:* Development of improved processes and designs to improve efficiency of electric motors from appliance size to major motors and large generators. *Mailing Add:* Eng Lab GE Can 107 Park St N Peterborough ON K9J 7B5 Can

SIDDER, GARY BRIAN, b Detroit, Mich, Nov 27, 54; m 81; c 2. METALLIC MINERAL DEPOSITS, IGNEOUS ROCKS. *Educ:* Colo State Univ, BS, 76; Univ Ore, MS, 81; Ore State Univ, PhD(geol), 85. *Prof Exp:* Geologist, Amarillo Oil Co, Pioneer Nuclear, Inc, 76-78, Houston Oil & Minerals Corp, 79 & Western Mining Corp Ltd, 80; teaching asst econ geol, Univ Ore, 80; teaching asst petrography, Ore State Univ, 80-84; field asst, 75, GEOLOGIST, US GEOL SURV, 85- *Mem:* Soc Econ Geologists; Geol Soc Am; Am Geophys Union. *Res:* Geology, geochemistry, petrology and origin of mineral deposits in the Guayana shield of Venezuela and in the southeast Missouri iron province of the midcontinent of the United States. *Mailing Add:* US Geol Surv DFC MS-905 Box 25046 Denver CO 80225

SIDDIQEE, MUHAMMAD WAHEEDUDDIN, b Pakistan, Aug 23, 31; US citizen; m 61; c 2. TRANSPORTATION SYSTEMS. *Educ:* Panjab Univ, BA, 51, BSc, 55; Univ Tenn, MS, 60; Univ Minn, PhD(control sci), 67. *Prof Exp:* Elec engr, Siemens Pakistan, 55-56 & Siemens Schuckertwerke, WGer, 56-58; sr elec engr, Siemens Pakistan, 58-59, exec engr, 60-62; res engr, 67-69, sr res engr, 69-74, staff scientist & mgr transp, Sri Int, 74-82; INFO SYSTS SPECIALIST, LOCKHEED, 82- *Mem:* Inst Elec Engrs, Pakistan. *Res:* Advanced transportation systems; air traffic analysis; railroad analysis; urban transportation systems and electric power systems and information systems. *Mailing Add:* 1733 Banff Dr Sunnyvale CA 94087

SIDDIQUE, IRTAZA H, b Budaun, India, July 4, 29; m 54; c 2. VETERINARY MEDICINE, MICROBIOLOGY & PUBLIC HEALTH. *Educ:* Bihar Vet Col, Patna, GBVC, 50; Univ Minn, St Paul, MS, 61, PhD(vet med), 63; Univ Ala, Birmingham, MPH, 78. *Prof Exp:* Vet, Govt Uttar Pradesh, India, 50-59; from asst prof to assoc prof, 64-71, PROF MICROBIOL, SCH VET MED, TUSKEGEE INST, 71-, PROF PUB HEALTH, 78- *Concurrent Pos:* Proj dir, USPHS grant, 64-69, proj dir, 69-; dir, NSF grant, 65-75; dir, USDA-Coop State Res Serv grant, 74-; chmn S-92 tech comt, Southern Regional Proj, 77-78; dir, USDA-CSRS grant, 83-88, MBRS grant, 84-87. *Mem:* Conf Res Workers Animal Dis; Am Vet Med Asn; Am Soc Microbiol; NY Acad Sci; Sigma Xi. *Res:* Listeriosis; immunofluorescent techniques and electron microscopy; cattle abortion and bacterial infections and pathogenesis. *Mailing Add:* Sch Vet Med Tuskegee Inst Tuskegee AL 36088

SIDDIQUI, ASLAM RASHEED, b British India, Dec 12, 46; US citizen; m 73; c 3. NUCLEAR MEDICINE. *Educ:* Chittagong Med Col, EPakistan, MD, 69. *Prof Exp:* From asst prof to assoc prof, 76-84, PROF RADIOL, SCH MED, IND UNIV, 84- *Concurrent Pos:* Consult radiol, Vet Admin Hosp, Indianapolis & staff physician radiol, Wishard Mem Hosp, Indianapolis, 76- *Mem:* Soc Nuclear Med; AMA; Radiol Soc NAm; Am Col Physicians; AAAS. *Res:* Nuclear medicine studies in pediatric oncology; nuclear thyroidology. *Mailing Add:* Riley Hosp 702 Barnhill Dr Dept Radiol Indianapolis IN 46202-5200

SIDDIQUI, HABIB, b Khulna, Bangladesh, July 14, 53; m. MATHEMATICAL MODELING, COMPUTATIONAL FLUID DYNAMICS. *Educ:* Bangladesh Univ Eng & Technol, BScEngg, 77; Univ Saskatchewan, Can, MSc, 80; Univ Calif, Santa Barbara, MS, 83; Univ Southern Calif, Los Angeles, PhD(chem eng), 89. *Prof Exp:* Consult engr, Key Connections Int, Long Beach, 86; RES ENGR, ROHM & HAAS CO, BRISTOL, 89- *Concurrent Pos:* Teaching asst, Dept Chem Eng, Univ Saskatchewan, 78-80; teaching asst, Dept Chem & Nuclear Eng, Univ Calif, 80-81, res asst, 81-82; computer programmer, Pioneer Sci & Technologies, Inc, Los Angeles, 83; grad asst, Dept Chem Eng, Univ Southern Calif, 83-89. *Res:* Transport phenomena; mixing and dispersion of phases; flow through porous media; thermal hydraulics; corrosion science; computational methods. *Mailing Add:* Rohm & Haas Co PO Box 219 Bristol PA 19007

SIDDIQUI, IQBAL RAFAT, b India, Jan 28, 31; m 56. ORGANIC CHEMISTRY. *Educ:* Univ Sind, Pakistan, BSc, 48, MSc, 50; Univ Birmingham, PhD(chem), 57. *Hon Degrees:* DSc, Univ Birmingham, 69. *Prof Exp:* Lectr chem, Govt Col, Lahore, Pakistan, 51-52; Sugar Res Found fel, Univ London, 57-58; Nat Res Coun Can fel, 58-60; Harold Hibbert Mem fels, McGill Univ, 61-62; GROUP LEADER STRUCT & SENSORY TEAM, FOOD RES INST, CAN DEPT AGR, 62- *Mem:* Fel Royal Soc Chem; fel Chem Inst Can; Am Chem Soc. *Res:* Structural, synthetic and analytical studies of carbohydrates; carbohydrates of honey; polysaccharide of bacteria, fungi, rapeseed, alginates and tobacco; dietary fibers from Canadian vegetables. *Mailing Add:* Cent Exp Farm 1153 Trent St Ottawa ON K1Z 8J5 Can

SIDDIQUI, M A Q, b Hyderabad, India, Feb 10, 37. GENE EXPRESSION OF CARDIAC MUSCLE. *Educ:* Univ Houston, PhD(biol sci), 67. *Prof Exp:* Assoc mem, 79-86, MEM BIOCHEM, ROCHE INST MOLECULAR BIOL, 86- *Mailing Add:* SUNY Health Sci Ctr 450 Clarkson Ave Box 5 Brooklyn NY 11203

SIDDIQUI, MOHAMMED MOINUDDIN, b Hyderabad, India, Apr 19, 28; m 57; c 5. MATHEMATICAL STATISTICS. *Educ:* Univ Osmania, India, MA, 48; Am Univ, MA, 54; Univ NC, PhD(math, statist), 57. *Prof Exp:* Lectr statist, Univ Panjab, Pakistan, 49-53, sr lectr, 58-59; res asst, Univ NC, 54-57; math statistician, Nat Bur Standards, 57-58 & 59-64; PROF MATH STATIST, COLO STATE UNIV, 64- *Concurrent Pos:* Consult, US Forest Serv, 84-85. *Mem:* Fel AAAS; Am Statist Asn; Inst Math Statist; Am Geog Soc; Int Statist Inst. *Res:* Distribution theory; stationary time series; order statistics and stochastic processes. *Mailing Add:* Henningson Durham & Richardson Gulf Hosp Proj PO Box 1480 Dammam 31431 Saudi Arabia

SIDDIQUI, WAHEED HASAN, b Bijnor, India, Aug 2, 39; US citizen; m 69; c 2. TERATOLOGY, PHARMACOKENETICS. *Educ:* Agra Univ, India, BSc, 60; Sind Univ, Pakistan, MSc, 64; Carleton Univ, Ont, PhD(biol), 71; Univ Western Ont, MEng, 77. *Prof Exp:* Sr biologist, Environ Control Consult, Ltd & Natural Mus Nat Sci, Can, 72-76; res scientist toxicol, Health Protection Br, Dept Health & Welfare, Govt Can, 77-79; ASSOC SCIENTIST, DOW CORNING CORP, MICH, 79- *Mem:* Soc Toxicol; Soc Toxicol Can; Sigma Xi; Teratology Soc. *Res:* Safety evaluation of new chemicals; long and short-term effects of chemicals in animal models; risk assessment for human exposure; metabolism and pharmacokinetics; developmental toxicology. *Mailing Add:* Dow Corning Corp 2200 W Saltzburg Rd Midland MI 48686-0994

SIDDIQUI, WASIM A, b India, May 10, 34; m 65; c 2. PARASITOLOGY, TROPICAL MEDICINE. *Educ:* Aligarh Muslim Univ, India, BSc, 52, MSc, 54; Univ Calif, Berkeley, PhD(zool), 61. *Prof Exp:* Lectr zool, Aligarh Muslim Univ, 54-57; res asst, Univ Calif, Berkeley, 57-61; lectr, Aligarh Muslim Univ, 61-62; NIH res fel parasitol, Rockefeller Univ, 62-65; instr trop med, Sch Med, Stanford Univ, 65-69; assoc prof, 69-73, PROF TROP MED & MICROBIOL, SCH MED, UNIV HAWAII, MANOA, 73- *Mem:* Soc Protozool; Am Soc Parasitol; Am Soc Trop Med & Hyg; Int Soc Immunopharmacol. *Res:* Protozoa, especially parasitic protozoans; amoebiasis and malaria. *Mailing Add:* Dept of Med Microbiol & Trop Med Univ of Hawaii Sch of Med Honolulu HI 96822

SIDEBOTTOM, OMAR M(ARION), b Forest City, Ill, Mar 13, 19; m 55; c 2. MECHANICS. *Educ:* Univ Ill, BS, 42, MS, 43. *Prof Exp:* Asst, Univ Ill, Urbana, 41-44, res assoc, 44-46, from instr to assoc prof, 46-57, prof theoret & appl mech, 57-82; RETIRED. *Mem:* Am Soc Mech Engrs; Am Soc Eng Educ; Am Soc Testing & Mat; Soc Exp Stress Analysis. *Res:* Creep analysis of load carrying members; creep buckling of columns; multiaxial creep; inelasticity; load carrying capacity of members which have been inelastically deformed; experimental evaluation of assumptions made in theory. *Mailing Add:* 7415 Highland Groove Dr Lakeland FL 33809

SIDEL, VICTOR WILLIAM, b Trenton, NJ, July 7, 31; m 56; c 2. COMMUNITY HEALTH, PREVENTIVE MEDICINE. *Educ:* Princeton Univ, AB, 53; Harvard Med Sch, MD, 57. *Prof Exp:* From intern to jr asst resident med, Peter Bent Brigham Hosp, 57-59; clin assoc, Nat Heart Inst, 59-61; sr asst res, Peter Bent Brigham Hosp, 61-62; instr biophys, Harvard Med Sch, 62-64, assoc prev med, 64-68, asst prof med, 68-69; prof community health, 69-74, chmn dept social med, 69-74, DISTINGUISHED UNIV PROF SOCIAL MED, MONTEFIORE MED CTR, ALBERT EINSTEIN COL MED, 74- *Concurrent Pos:* Am Heart Asn advan res fel, 62-64; consult physician, Child Health Div, Children's Hosp Med Ctr, Boston, 63-67; Med Found, Inc res fel, 64-68; asst med & chief prev & community med units, Mass Gen Hosp, 64-69; Milbank Mem Fund fac fel, 64-71; consult, USPHS, 64-70, WHO, 69, 74 & 77 & Int J Health Serv, 71-79; vis prof community health & social med, City Col New York, 73-; vis prof, Scand Sch Pub Health, 75-88; attend physician, NCent Bronx Hosp, 76-; adj prof pub health, Cornell Univ Med Col, 87- *Mem:* Am Pub Health Asn (pres, 84-85); NY Acad Med; Asn Teachers Prev Med; Physicians Soc Responsibility (pres, 87-88); Royal Soc Health Gt Brit. *Res:* International health care comparisons; health care delivery; medical ethics; health policy. *Mailing Add:* Dept Epidemiol Soc Med Montefiore Med Ctr Albert Einstein Col Med 111 E 210th St Bronx NY 10467

SIDELL, BRUCE DAVID, b Manchester, NH, Mar 20, 48; m 70; c 3. COMPARATIVE PHYSIOLOGY, FISH BIOLOGY. *Educ:* Boston Univ, AB, 70; Univ Ill, MS, 72, PhD(physiol), 75. *Prof Exp:* Asst res scientist aquatic biol, Chesapeake Bay Inst, Johns Hopkins Univ, 75-77; from asst prof to assoc prof zool, 82-88, assoc dean res, Col Sci, 89-90, PROF ZOOL, UNIV MAINE, 88- *Concurrent Pos:* Mem res comt Maine affil, Am Heart Asn, 82-86; ed adv, Marine Ecol Prog Ser, 83-90; assoc ed, J Exp Zool, 85-88. *Mem:* Am Soc Zoologists; Sigma Xi; Am Physiol Soc. *Res:* Physiological and biochemical adaptations of aquatic ectotherms; physiology of fishes. *Mailing Add:* Dept Zool Univ Maine Orono ME 04469

SIDELL, FREDERICK R, b Marietta, Ohio, July 27, 34. CHEMISTRY. *Educ:* Marietta Col, BS, 56; NY Univ, MD, 60. *Prof Exp:* DIR, MED MGT CHEM CASUALTIES COURSE & SR MED OFFICER, US ARMY MED RES INST CHEM DEFENSE, 64- *Mem:* Am Soc Pharmacol & Exp Therapeut; Am Soc Clin Pharmacol. *Mailing Add:* US Army Med Res Inst Chem Defense Aberdeen Proving Ground MD 21010

SIDEN, EDWARD JOEL, b Miami Beach, Fla, Aug 5, 47; m 84. MOLECULAR IMMUNOLOGY, B CELL DEVELOPMENT. *Educ:* Brandeis Univ, BA, 69; Univ Calif, San Diego, PhD(biol), 75. *Prof Exp:* Teaching fel bacteriophage, Univ Calif, San Diego, 75-76 & molecular immunol, Mass Inst Technol, 76-80; asst prof immunol & med microbiol, Col Med, Univ Fla, 80-87; ASST PROF MED, MT SINAI MED SCH, NEW YORK, 87- *Mem:* Am Soc Microbiol; Am Asn Immunologists. *Res:* Immortalizing immature hemopoietic cells using Abelson murine leukemia virus to study early stages of B cell differentiation. *Mailing Add:* Col Med Jhahc Univ Fla Box J 266 Gainsville FL 32610

SIDEROPOULOS, ARIS S, b Thessaloniki, Greece, Jan 30, 37; US citizen. PHOTOBIOLOGY, ENVIRONMENTAL POLLUTANTS. *Educ:* Concordia Col, Minn, BS, 60; NDak State Univ, MS, 62; Univ Kans, PhD(molecular genetics), 67. *Prof Exp:* Fel microbiol genetics, Palo Alto Res Found, 67-70; res assoc, Univ Tex, Austin, 70-71; asst prof, Med Col Pa, 71-75; PROF MICROBIOL, DUQUESNE UNIV, 75- *Concurrent Pos:* Chief, Microbiol Lab, Fairmont Dairy Co, Minn & instr bacteriol, NDak State Univ, 62-63. *Mem:* Am Soc Microbiol; Sigma Xi; AAAS. *Res:* Cellular repair of potentially mutagenic damage to establish and utilize an experimental method in evalution of mutagenic effects of chemicals on ultraviolet irradiated living cells. *Mailing Add:* Dept Biol Sci Duquesne Univ Pittsburgh PA 15282

SIDES, GARY DONALD, b Tuscaloosa, Ala, Oct 5, 47; c 3. INSTRUMENT DEVELOPMENT, ENVIRONMENTAL MONITORING. *Educ:* Univ Ala, BS, 69; Univ Fla, MS, 71, PhD(phys chem), 75. *Prof Exp:* Res scientist, Aerospace Res Labs, 71-75; res assoc prof, Wright State Univ, 75-77; sr scientist, Chem Defense Div, Southern Res Inst, head, 78-86; PRES, CMS RES CORP, 86- *Mem:* Am Soc Mass Spectrometry; Am Chem Soc. *Res:* Automated analytical instruments for the detection of nanogram levels of toxic organophosphorus and organosulfur compounds. *Mailing Add:* 1732 Tahiti Lane Alabaster AL 35007

SIDES, PAUL JOSEPH, b Birmingham, Ala, Jan 23, 51; m 87. ELECTROCHEMICAL ENGINEERING, SEMICONDUCTOR PROCESSING. *Educ:* Univ Utah, BS, 73; Univ Calif, Berkeley, PhD(chem eng), 81. *Prof Exp:* Asst prof, 81-86, ASSOC PROF CHEM ENG, CARNEGIE MELLON UNIV, 86- *Mem:* Am Inst Chem Engrs; Am Inst Mining, Metall & Petrol Engrs; Electrochem Soc. *Res:* Investigation of electrolytic gas evolution; electrochemical engineering of primary aluminum production; organometallic vapor phase epitaxy of mercury cadmium tellwride. *Mailing Add:* Dept Chem Eng Carnegie Mellon Univ Pittsburgh PA 15213

SIDHU, BHAG SINGH, b Ludhiana, India, Apr 30, 29; m 54; c 2. PLANT GENETICS, PLANT SCIENCE. *Educ:* Punjab Univ, India, BSc, 51, MSc, 53; Cornell Univ, PhD(genetics, plant sci), 60. *Prof Exp:* Chmn dept bot, Govt Col, Rupar, India, 54-57; chmn dept agr & biol, Govt Col, Faridkot, India, 61-62, prin-pres, 63; Nat Res Coun Can fel, McGill Univ, 64-65; assoc prof biol, Winston-Salem State Col, 65-66; UN adv agron & biol, UNESCO, UN Develop Prog, Manila, P I, 66-68; PROF BIOL, WINSTON-SALEM STATE UNIV, 69- *Concurrent Pos:* Res award, Asn Cols Agr, Philippines, 67; NSF fel, NC State Univ, 70. *Mem:* AAAS; Am Inst Biol Sci; Am Soc Agron; Crop Sci Soc Am. *Res:* Biosystematics and germ plasm screening of international field food crops materials; radiation genetics and international seed production and distribution. *Mailing Add:* Dept Life Sci Winston-Salem State Univ Winston-Salem NC 27110

SIDHU, DEEPINDER PAL, b Chachrari, India, May 13, 44; m 69; c 4. THEORETICAL PHYSICS, COMPUTER SCIENCE. *Educ:* Univ Kans, BE, 66; State Univ NY Stony Brook, PhD(theoret physics), 73, MS, 79. *Prof Exp:* Res assoc physics, Rutgers Univ, New Brunswick, 73-75; asst physicist, Brookhaven Nat Lab, 75-77, assoc physicist, 77-80; mem tech staff, Mitre Corp, 80-82; mgr, Secure Distrib Systs Dept, Dept Computer Sci, Iowa State Univ, 84-88; PROF COMPUT SCI, UNIV MD, 88- *Res:* Unified gauge theories of strong, electromagnetic and weak interactions; computer communication networks. *Mailing Add:* Dept Computer Sci Univ Md Catonsville MD 21228

SIDHU, GURMEL SINGH, b Pasla, India, May 23, 40; Can citizen; m 79; c 2. GENETICS, PLANT PATHOLOGY. *Educ:* Punjab Univ, BSc, 58, MSc, 60; Univ BC, PhD(genetics), 72. *Prof Exp:* Res assoc plant genetics, Punjab Agr Univ, 62-64, lectr genetics cytogenetics, 64-66; res assoc plant sci, Univ BC, 66-67; fel, Simon Fraser Univ, 72-76, res scientist genetics & path, 76-80; asst prof plant pathol, Univ Nebr, 80-86; RES SCIENTIST & PROF, CALIF STATE UNIV, 86- *Concurrent Pos:* Vis prof, Punjab Agr Univ, 76-77; Genetics Soc Am Travel grant, 78 & 83; assoc ed, Phytopath, 82-86. *Mem:* Genetics Soc Can; Genetics Soc Am; Am Soc Phytopath; Can Phytopath Soc; Indian Soc Crop Improvement. *Res:* Genetics of host-parasite interactions; genetics of plant disease complexes; breeding for disease resistance; fungal genetics; molecular biology of host-parasite interaction. *Mailing Add:* Dept Biol California State-Fresno Fresno CA 93740

SIDI, HENRI, b Provadia, Bulgaria, Mar 13, 19; US citizen; m 56; c 2. ORGANIC CHEMISTRY, POLYMER CHEMISTRY. *Educ:* Univ Toulouse, ChE, 41, PhD(org chem), 48. *Prof Exp:* Asst anal chem, Univ Toulouse, 46-47; sr chemist, Poudrerie Nationale, Toulouse, France, 47-48; asst dir dyestuffs, Francolor, Oissel, 48-50; fel, Rutgers Univ, 50-51; sr chemist, Heyden Chem Corp, 51-61; group leader, Heyden Div, Garfield, 61-71, SR SCIENTIST, INTERMEDIATES DIV, TENNECO CHEM, INC, PISCATAWAY, 71- *Mem:* AAAS; Am Chem Soc; Chem Soc France. *Res:* Polymerization of formaldehyde; fire retardant chemicals; amino-alcohols; pesticides. *Mailing Add:* 156 Victoria Ave Paramus NJ 07652-1923

SIDIE, JAMES MICHAEL, b Elizabeth, NJ, May 22, 41; m 66; c 1. NEUROETHOLOGY, COMPARATIVE PHYSIOLOGY. *Educ:* Univ Notre Dame, BS, 64, MS, 67, PhD(biol), 70. *Prof Exp:* Pub Health Serv fel, Ind Univ, 69-71; vis lectr, Princeton Univ, 71-72; res assoc, Univ Ore, 72-74, vis asst prof, 74-76; ASST PROF PHYSIOL/NEUROBIOL, STATE UNIV NY, BUFFALO, 76- *Concurrent Pos:* NSF fel, Woods Hole, 71. *Mem:* AAAS; Am Soc Zoologists; Animal Behav Soc; Sigma Xi. *Res:* Neuroethological studies of communication in honeybees, Apis mellifera; computer assisted analyses of neural spike trains in nerve/muscle systems. *Mailing Add:* Dept Biol Urisinus Col Collegeville PA 19426

SIDKY, YOUNAN ABDEL MALIK, b Khartoum, Sudan, Feb 9, 28; div; c 2. IMMUNOLOGY, TUMOR BIOLOGY. *Educ:* Cairo Univ, BSc, 50, MSc, 55; Univ Marburg, PhD(zool), 56. *Prof Exp:* Demonstr zool, Cairo Univ, 50-59, lectr, 59-65; fel, Univ Alta, 69-72; res assoc, Univ Wis-Madison, 65-69, proj res assoc, 72-76, assoc scientist, 76-87, sr scientist, 87-90, RES ASSOC PROF, UNIV WIS-MADISON, 90- *Mem:* Am Asn Immunologists; Am Asn Cancer Res; Int Soc Interferon Res. *Res:* Parathyroid glands in reptiles; effect of steroids on mouse thymus development; development of immunity in turtles; hibernation and immunity in hamsters; inhibitory effect of brown fat on the immune response; tumor and lymphocyte-induced angiogenesis; effects of interferons and cytotoxic drugs on tumor growth. *Mailing Add:* Cancer Ctr Med Col Wis 8701 Watertown Plank Rd Milwaukee WI 53226

SIDLE, ROY CARL, b Quakertown, Pa, Oct 31, 48; m 84; c 2. LANDSLIDES, CUMULATIVE EFFECTS. *Educ:* Univ Ariz, BS, 70, MS, 72; Pa State Univ, PhD(soil sci), 76. *Prof Exp:* Hydrologist, Wright Water Engrs, Inc, 72; res asst soil sci, Pa State Univ, 72-76; res soil scientist, Agr Res Serv, USDA, 76-78; asst prof watershed sci, dept forest eng, Ore State Univ, 78-80; RES SOIL SCIENTIST & HYDROLOGIST, FORESTRY SCI LAB, FOREST SERV, USDA, 80- *Concurrent Pos:* Res fel, Japan, 91. *Mem:* Soil Sci Soc Am; Am Soc Agron; Sigma Xi; Int Soc Soil Sci; Am Geophys Union; Am Water Resources Asn. *Res:* Cumulative effects of land management practices; modeling slope stability; processes of erosion and sediment transport; role of subsurface flow on landslide initiation; natural hazards. *Mailing Add:* 860 N 1200 Logan E Logan UT 84321

SIDLER, JACK D, b Rochester, Pa, Sept 19, 39; m 58; c 3. ORGANIC CHEMISTRY. *Educ:* Geneva Col, BS, 61; State Univ NY Buffalo, PhD(chem), 66. *Prof Exp:* Teaching asst, State Univ NY Buffalo, 61-62; asst prof chem, Geneva Col, 65-67; assoc prof, 67-72, PROF CHEM, MANSFIELD UNIV, 72- *Mem:* Am Chem Soc. *Res:* Organic synthesis; organolithium chemistry; organic reaction mechanisms. *Mailing Add:* Dept Chem Mansfield Univ Mansfield PA 16933

SIDMAN, RICHARD LEON, b Boston, Mass, Sept 19, 28; m 50, 74; c 2. NEUROPATHOLOGY, DEVELOPMENTAL GENETICS. *Educ:* Harvard Univ, AB, 49, MD, 53. *Prof Exp:* Intern med, Boston City Hosp, Mass, 53-54; Moseley traveling fel, Strangeways Res Lab, Cambridge Univ & Dept Human Anat, Oxford Univ, 54-55; asst resident neurol, Mass Gen Hosp, Boston, 55-56; assoc to prof neuropath, Harvard Med Sch, 63-68; chief, Div Neurosci, Children's Hosp, Boston, 72-88; BULLARD PROF NEUROPATH HARVARD NED SCH, 69- *Concurrent Pos:* Asst, Mass Gen Hosp, 59-69. *Honors & Awards:* Stearns mem lectr, Albert Einstein Col Med, 58. *Mem:* Nat Acad Sci; Am Asn Anat; Tissue Cult Asn (secy, 64-70); Am Asn Neuropath; Soc Neurosci; Int Soc Develop Neurosci; Histochem Soc; Am Acad Arts & Sci. *Res:* Developmental genetics of the normal and diseased mammalian brain. *Mailing Add:* Div Neurogenetics New Eng Regional Primate Res Ctr One Pine Hill Dr Southboro MA 01772

SIDNEY, STUART JAY, b New Haven, Conn, June 8, 41; m 65; c 4. MATHEMATICAL ANAYLSIS. *Educ:* Yale Univ, BA, 62; Harvard Univ, MA, 63, PhD(math), 66. *Prof Exp:* Instr math, Yale Univ, 66-68, asst prof, 68-71; assoc prof, 71-80, PROF, UNIV CONN, 80- *Concurrent Pos:* Vis prof, Univ Grenoble, 71-72 & 78-79. *Mem:* Am Math Soc; Math Asn Am. *Res:* Analysis, especially uniform algebras and Banach spaces; geometry. *Mailing Add:* Dept Math Univ Conn Storrs CT 06269-3009

SIDOTI, DANIEL ROBERT, b North Bergen, NJ, Jan 17, 21; m 51; c 1. NUTRITION. *Educ:* Union Col, Schenectady, BA, 47; Stevens Inst Technol, MS, 59. *Prof Exp:* Qual control chemist, Robert A Johnston Co Inc, 40-42; group leader, Gen Foods Corp, 46-66; sect head, Monsanto Co, 66-69; res mgr, Anheuser-Busch Co Inc, 69-91; RETIRED. *Concurrent Pos:* Prod mgr, Carter-Wallace, 69; counr, Inst Food Technologists, 78-91. *Res:* New product and process development; functional oil applications in food products; fat and oil substitutes; high intensity sweeteners; new functional ingredients in food products; awarded six patents. *Mailing Add:* 500 Wellshire Ct Ballwin MO 63011

SIDRAN, MIRIAM, b Washington, DC, May 25, 20. SOLID STATE PHYSICS. *Educ:* Brooklyn Col, BA, 42; Columbia Univ, MA, 49; NY Univ, PhD(physics), 56. *Prof Exp:* Instr physics & chem, Brothers Col, Drew Univ, 46-47; instr physics, Adelphi Col, 47-49; asst solid state physics, NY Univ, 50-55, Nat Carbon Co fel, 55-58; sr physicist, Balco Res Labs, NJ, 55; asst prof chem & physics, Staten Island Community Col, State Univ NY, 58-59; res scientist, Grumman Aircraft Eng Corp, 59-67; prof physics & dep chmn dept, NY Inst Technol, New York Campus, 67-72; prof, 72-90, chmn, 83-89, EMER PROF PHYSICS, BARUCH COL, 90- *Concurrent Pos:* NSF sci fac fel, Nat Marine Fisheries Serv, Nat Oceanic & Atmospheric Admin, 71-72. *Honors & Awards:* NY Univ Founder's Day Award, 56. *Mem:* Assoc fel Am Inst Aeronaut & Astronaut; NY Acad Sci; Soc Women Engrs; Sigma Xi; Am Asn Physics Teachers. *Res:* Microwave spectrometry; optical spectrometry; infrared astronomy; lunar luminescence; lunar surface studies; radiation dosimetry; rotational energy levels of asymmetric molecules; remote sensing of sea surface temperature. *Mailing Add:* Box 502 Baruch College 17 Lexington Ave New York NY 10010

SIDRANSKY, HERSCHEL, b Pensacola, Fla, Oct 17, 25; m 52; c 2. PATHOLOGY. *Educ:* Tulane Univ, BS, 48, MD, 53, MS, 58; Am Bd Path, dipl, 58. *Prof Exp:* Intern, Charity Hosp La, New Orleans, 53-54, vis asst pathologist, 54-58; instr path, Tulane Univ, 54-58; med officer, Nat Cancer Inst, 58-61; prof path, Sch Med, Univ Pittsburgh, 61-72; prof path & chmn dept, Col Med, Univ SFla, 72-77; PROF PATH & CHMN DEPT, GEORGE WASHINGTON UNIV MED CTR, 77- *Concurrent Pos:* Consult, Div Biologics, Stand Contract Comt, NIH, 66-67, mem path study sect res grants rev br, Div Res Grants, 68-72, mem nutrit study sect, 73-77; vis scientist, Weizmann Inst Sci, 67-68, Eleanor Roosevelt Int Cancer fel travel award & USPHS spec res fel, 67-68; mem bd, Am Registry Path, 79-87; adv comt, Life Sci Res Off, FASEB, 82-88; mem, Intersoc Comt Path Info, 82-83, chmn, 83- *Mem:* Soc Exp Biol & Med; Am Inst Nutrit; Sigma Xi; AAAS; Int Acad Path; Am Soc Clin Nutrit; Am Asn Cancer Res; Am Asn Path; NY Acad Sci. *Res:* Chemical pathology of nutritional deficiencies; experimental liver tumor igenesis; tryptophar metabolism. *Mailing Add:* Dept Path George Washington Univ Med Ctr 2300 Eye St NW Washington DC 20037

SIDWELL, ROBERT WILLIAM, b Huntington Park, Calif, Mar 17, 37; m 57; c 6. VIROLOGY, CHEMOTHERAPY. *Educ:* Brigham Young Univ, BS, 58; Univ Utah, MS, 61, PhD(microbiol), 63. *Prof Exp:* Head serol & rickettsial res, Univ Utah, 58-63; head virus research, Univ Utah & Dugway Proving Grounds, 63; sr virologist, Chemother Dept, Southern Res Inst, 63-66, head virus sect, 66-69; head dept virol, 69-72, chemother div, 72-75, dir, ICN Pharmaceut Nucleic Acid Res Inst, 75-77; res prof, depts biol & animal, dairy & vet sci, 77-87, PROF VIROL, UTAH STATE UNIV, 87- *Concurrent Pos:* Prin investr, NIH, 63-69, 87-, Thrasher Res Fund, 79-84 & US Army Med Res & Develop Command, 85-; asst prof, Dept Microbiol, Univ Ala, Birmingham, 64-69; rev ed, J Chemother, J Antiviral Res, Antimicrobial Agts & Chemother, 69-; mem, tech adv comt, Thrasher Res Fund Found, 84-89; chmn, Basic Res Subcomt, Div AIDS, NIH & mem, AIDS Liaison Subcomt; D W Thorne res award, Utah State Univ, 87; bd trustees, Inter-Am Soc Chemother, 90- *Mem:* Am Asn Immunologists; Soc Exp Biol & Med; Am Soc Microbiol; Am Soc Virol; Inter-Am Soc Chemother (secy, 86-89); Int Soc Antiviral Res; Int Soc Chemother; Nat Asn Col Teachers Agr; Sigma Xi. *Res:* Basic and applied research on viral disease chemotherapy and immunotherapy. *Mailing Add:* Dept Animal Dairy & Vet Sci Utah State Univ Logan UT 84322-5600

SIE, CHARLES H, b Shanghai, China, Sept 12, 34; US citizen; m 58; c 3. ELECTRICAL ENGINEERING, MATERIAL SCIENCE. *Educ:* Manhattan Col, BS, 57; Drexel Univ, MS, 60; Iowa State Univ, PhD(elec eng), 69. *Prof Exp:* Elec engr, Radio Corp Am, NJ, 57-63; mem res staff, Watson Res Ctr, IBM Corp, NY, 66-69; mgr, Memory Device Develop, Energy Conversion Devices, Inc, 69-74; mgr, Component Lab, Burroughs Corp, 74-77; mgr component & packaging technol, 77-84, MGR, ELECTRONIC DESIGN & TECHNOL, XEROX CORP, 84- *Mem:* Inst Elec & Electronics Engrs; Am Phys Soc. *Res:* Semiconductor and ferromagnetic memory devices; amorphous chalcogenide semiconductor; electrical circuit design; component and reliability engineering; software management. *Mailing Add:* Mgr Elec Design & Technol Xerox Corp 801-046 1350 Jefferson Rd Rochester NY 14623

SIE, EDWARD HSIEN CHOH, b Shanghai, China, Jan 17, 25; US citizen; wid; c 1. BIOCHEMISTRY. *Educ:* Univ Nanking, BS, 46; Ill Inst Technol, MS, 51; Princeton Univ, PhD(biochem), 57. *Prof Exp:* Int trainee, Joseph E Seagram & Sons, Inc, Ky, 47-48; biochemist trainee, E R Squibb & Sons, Inc, NY, 51-52; jr biochemist, Ethicon Inc, 52-54; res assoc biochem, Princeton Univ, 57-59; res assoc microbial chem, Mt Sinai Hosp, New York, 59-62; sr chemist, Space Div, NAm Aviation, Inc, 62-63; res specialist, 63-66, sr tech specialist, Autonetics Div, NAm Rockwell Corp, 66-68; res supvr, Biomed Div, Gillette Res Inst, 68-70; sr scientist, 70-73, sales mgr, Eastern region US, 73-78, sr develop scientist, Carlsbad opers, Beckman Instruments Inc, 78-83; mgr, tech mkt, Carlsbad Opers, Smith, Kline, Beckman, 84-85; SR SCIENTIST, MICROBICS CORP, 85- *Mem:* AAAS; Am Chem Soc; NY Acad Sci. *Res:* Industrial enzymology and fermentation; chemistry of bioluminescence and its application; growth and nutrition of thermophilic bacteria; keratin of hair and skin; action of keratinase; clinical enzymology and diagnostic reagents. *Mailing Add:* Microbics Corp 2232 Rutherford Rd Carlsbad CA 92008-8883

SIEBEIN, GARY WALTER, b New York, NY, Jan 3, 51; m 84; c 6. ARCHITECTURAL ACOUSTICS, AUDITORIUM DESIGN & ANALYSIS. *Educ:* Rensselaer Polytech Inst, BS, 72, BArch, 78; Univ Fla, MArch, 80. *Prof Exp:* Health facil planner, Philips Med Systs, Inc, 73-75; proj coordr, Elec Eng Servs, Norwalk, Conn, 75-77; asst prof, 80-85, ASSOC PROF ARCHIT, ARCHIT TECHNOL RES CTR, UNIV FLA 85-, DIR,. *Concurrent Pos:* Consult, archit acoust, 82-; prin investr, NSF, 83-86; vis lectr, various univs, Registration Inst, 84- *Honors & Awards:* Progressive Archit Citation for Appl Res Acoustics; Educ Hon Citation, Am Inst Archit. *Mem:* Acoust Soc Am; Soc Bldg Sci Educrs; Am Soc Heating, Refrig & Air Conditioning Engrs. *Res:* Architectural acoustical modeling; development of acoustical design criteria; analysis of auditoria and concert halls; electric lighting quality and day lighting; thermal performance evaluation of buildings. *Mailing Add:* Dept Archit Univ Fla 231 Arch Gainesville FL 32611

SIEBEL, M(ATHIAS) P(AUL) L, b Witten, Ger, Mar 6, 24; nat US; m 60. MECHANICAL ENGINEERING. *Educ:* Bristol Univ, BS, 49, PhD, 52. *Prof Exp:* In charge res & develop & asst works mgr, Tube Investments, Ltd, Eng, 53-57; res assoc, Columbia Univ, 58-59; gen mgr, Pressure Equip Div, Pall Corp, 59-64; vpres & mgr opers, Radiation Dynamics, Inc, NY, 64-65; dep dir mfg eng lab, Michoud Assembly Facil, NASA, 65-68, dir process eng lab, 68-74, staff scientist, Marshall Space Flight Ctr, 74-79, mgr, 79-87; CONSULT, 87-; ASSOC DEAN, COL ENG, UNIV NEW ORLEANS, 89- *Mem:* Sigma Xi. *Res:* Industrial and research management; strength of materials; space sciences. *Mailing Add:* 5204 Janice Ave Kenner LA 70065

SIEBELING, RONALD JON, b Oostburg, Wis, Nov 10, 37; m 59; c 4. MICROBIOLOGY. *Educ:* Hope Col, AB, 60; Univ Ariz, MS, 62, PhD(microbiol), 66. *Prof Exp:* Asst prof, 66-74, ASSOC PROF MICROBIOL, LA STATE UNIV, BATON ROUGE, 74- *Mem:* Am Soc Microbiol. *Res:* Immunobiology; immune suppression and antigenic competition. *Mailing Add:* Dept Microbiol 676 Life Sci La State Univ Baton Rouge LA 70803-1715

SIEBENS, ARTHUR ALEXANDRE, b Altanta, Ga, July 13, 21; m 48; c 6. PHYSIOLOGY, REHABILITATION MEDICINE. *Educ:* Oberlin Col AB, 43; Johns Hopkins Univ, MD, 47. *Prof Exp:* Asst physiol, Sch Med, Johns Hopkins Univ, 45-48; asst prof physiol & pharmacol, Long Island Col Med, State Univ NY Downstate Med Ctr, 48-52, from assoc prof physiol to prof, 54-58; prof physiol, Sch Med & dir respiratory & rehab ctr, Hosps, Univ Wis-Madison, 58-71; prof rehab med & surg, 71-76, RICHARD BENNETT DARNALL PROF REHAB MED & PROF SURG, JOHNS HOPKINS UNIV, 77-, CHIEF DIV REHAB MED, UNIV & DIR REHAB MED, HOSP, 71- *Concurrent Pos:* Mem physiol study sect, USPHS, 49-64 & prog-proj comt, Nat Heart Inst, 65-69; dir dept rehab med, Good Samaritan Hosp, Baltimore, Md, 71-; consult, Vet Admin Hosp, Madison, Wis. *Mem:* AAAS; Am Physiol Soc; fel Am Acad Phys Med & Rehab; Am Cong Rehab Med. *Res:* Respiratory, cardiovascular and nervous system physiology; swallowing impairment; orthotics. *Mailing Add:* 617 W 40th St Baltimore MD 21211

SIEBENTRITT, CARL R, JR, b Jersey City, NJ, Aug 14, 22; m 57; c 6. HEALTH PHYSICS. *Educ:* Univ Cincinnati, BSME, 47, MS, 49. *Prof Exp:* Physicist, Instrument Div, Keleket X-ray Corp, 49-51; supvr nuclear instrument develop, Cincinnati Div, Bendix Corp, 51-62; dir, Nucleonics Div, Defense Civil Preparedness Agency, Dept Defense, 62-73, staff dir, Detection & Countermeasures Div, 73-79; Chief, Radiol Defense Br & Chmn Interagency Subcomt Offsite Emergency Instrumentation, Fed Emergency Mgt Agency, 79-88, RES PHYS SCIENTIST, FED EMERGENCY MGT AGENCY, 88- *Concurrent Pos:* Chmn, Interagency Task Force for Radiol Emergencies Involving Fixed Nuclear Facil, 73-81. *Mem:* Health Physics Soc. *Res:* Instrumentation and systems for detection and measurement of radioactivity; nuclear instrumentation for radiological emergency response; radiation damage to insulators. *Mailing Add:* Fed Emergency Mgt Agency Rte 601 Bldg 217 PO Box 129 Berryville VA 22611-0129

SIEBER, FRITZ, b Erstfeld, Switz, Jan 28, 46; c 2. CANCER RESEARCH, BLOOD PRODUCTS. *Educ:* Swiss Fed Inst Technol, PhD(biochem & cell biol), 76. *Prof Exp:* Asst prof med & oncol, Sch Med, Johns Hopkins Univ, 75-85; assoc prof, 85-90, PROF PEDIAT, MED COL WIS, 90- *Honors & Awards:* Frederick Stohlman Mem Award, 87. *Res:* Autologous bone marrow transplantation, photodynamic therapy; antiviral agents. *Mailing Add:* Dept Pediat Med Col Wis 8701 Watertown Plank Rd Milwaukee WI 53226

SIEBER, JAMES LEO, b New Glasgow, NS, Nov 3, 36; US citizen. MATHEMATICS, DISCRETE STRUCTURES. *Educ:* Shippensburg State Col, BS, 58; Pa State Univ, MA, 61, PhD(math), 63. *Prof Exp:* From asst prof to assoc prof, 63-67, chmn dept, 64-85, PROF MATH & COMPUT SCI, SHIPPENSBURG STATE COL, 67- *Concurrent Pos:* Actg dir comput ctr, Shippensburg State Col, 67-68; dir, NSF Inst, 69-71. *Mem:* Math Asn Am; Asn Comput Mach. *Res:* Abstract spaces in general topology, syntopogenous spaces, quasi-uniform spaces and quasi-proximity spaces; computer science and mathematics administration. *Mailing Add:* Dept of Math & Comput Sci Shippensburg Univ Shippensburg PA 17257

SIEBER-FABRO, SUSAN M, b Hattiesburg, Miss, May 18, 42; m 71. PHARMACOLOGY. *Educ:* Univ Va, BS, 64; George Washington Univ, MS, 69, PhD(pharmacol), 70. *Prof Exp:* Staff fel pharmacol, 71-76, pharmacologist, Lab Chem Pharmacol, 76- 84, DEP DIR, DIV CANCER ETIOL, NAT CANCER INST, 84- *Mem:* Am Asn Cancer Res; Teratology Soc; AAAS; Am Soc Pharmacol & Exp Therapeut. *Res:* Developmental pharmacology; drug disposition; carcinogenesis; toxicology. *Mailing Add:* Div Cancer Etiol Nat Cancer Inst Bldg 31 Rm 11A03 Bethesda MD 20892

SIEBERT, ALAN ROGER, b Cleveland, Ohio, July 16, 30; m 54; c 3. PHYSICAL CHEMISTRY, POLYMER CHEMISTRY. *Educ:* Fenn Col, BChemEng, 53; Western Reserve Univ, MS, 54, PhD(chem), 57. *Prof Exp:* Sr res chemist, Res Ctr, 56-68, res assoc, 68-69, sect leader specialty elastomer, 69-73, proj tech mgr, Res & Develop Ctr, 73-77, prod mgr reactive liquid polymers, 77-78, sr prod mgr new prod, Chem Div, 78-80, MKT MGR & RES & DEVELOP MGR, REACTIVE LIQUID POLYMERS, CHEM GROUP, B F GOODRICH CO, 80- *Mem:* Am Chem Soc; Soc Aerospace Mat & Process Engrs. *Res:* Relation between structure and properties of polymers; emulsion and solution polymerization; polymerization and characterization of reactive liquid polymers; polymerization and characterization of rubber and plastic latexes; impact resistant; thermo setting and thermoplastic resins. *Mailing Add:* 3999 E Meadow Lane Cleveland OH 44122

SIEBERT, DONALD ROBERT, b Oak Ridge, Tenn, July 6, 46; m 68; c 2. SOLID STATE LASERS, OPTICS. *Educ:* Union Col, NY, BS, 68; Columbia Univ, New York, PhD(phys chem), 73. *Prof Exp:* Asst prof chem, Drew Univ, Madison, NJ, 74-80; res chemist, Photon Chem Dept, Allied Corp, 78-79, res physicist electrooptical prod, 80-81, proj engr electrooptical prod, 81-82, proj leader sci lasers, 82-83, sr res physicist, 83-84, res assoc & prog mgr, Laser Res & Develop Dept, 84-90; PROGRAM MGR, LASER SYSTS, ALLIED-SIGNAL AEROSPACE CO, 90- *Mem:* Soc Photo-Optical Instrumentation Engrs; Am Phys Soc; Optical Soc Am. *Res:* Design and development of scientific solid state lasers and laser systems; molecular spectroscopy; thermal lensing; applications of lasers to chemical, physical and environmental problems. *Mailing Add:* 178 Hillcrest Ave Morris Township NJ 07960

SIEBERT, ELEANOR DANTZLER, b Birmingham, Ala, July 18, 41; m 67; c 2. PHYSICAL CHEMISTRY, SCIENCE EDUCATION. *Educ:* Duke Univ, BA, 63; Univ Calif, Los Angeles, PhD(chem), 69. *Prof Exp:* Res chemist, Allied Chem Corp, 63-65; mem fac, Dept Chem, Univ Calif, Los Angeles, 70-71; mem fac, Dept Phys Sci, Westlake Sch, 71-73; assoc prof, 74-83, PROF PHYS SCI, MT ST MARY'S COL, 83- *Concurrent Pos:* Proj dir, NSF CAUSE Prog, 76-79; staff res, Univ Calif, Los Angeles, 83-; proj dir, NSF/CSIP Prog, 87-89; prin invest, NIH-MBRS Prog, 86- *Mem:* AAAS; Am Chem Soc; Sigma Xi; Soc Col Sci Teachers (sec-treas, 86-89). *Res:* Nucleation and growth in phase separation; equation of state data; corresponding states; thermodynamic measurements; molecular potentials; individualized learning. *Mailing Add:* 3863 Marcasel Ave Los Angeles CA 90066

SIEBERT, JEROME BERNARD, b Fresno, Calif, Dec 12, 38; m 60; c 2. AGRICULTURAL ECONOMICS. *Educ:* Univ Calif, Davis, BS, 60; Univ Calif, Berkeley, PhD(agr econ), 64. *Prof Exp:* Asst undersecy admin, USDA, 69-70, exec asst admin, Consumer & Mkt Serv, 70-71, asst secy, 71-72; economist agr econ, 66-69, assoc dir admin, Univ Calif Coop Exten, 72-80, dir, 80-88, ECONOMIST DEPT AGR RES ECON, UNIV CALIF BERKELEY, 88- *Concurrent Pos:* Consult to dir, Calif Dept Food & Agr, 67-69; alt, Walnut Control Bd, USDA, 67-69, mem, Joint USDA & Nat Asn State Univ & Land Grant Cols Comt Educ & chmn, Independent Study & Exten Educ Comt, USDA Grad Sch, 71-72; dir, Tri Valley Growers Inc, 72-75; mem, Blue Ribbon Comt Agr, State Calif, Lt Gov Off, 73-74; dir, Calif Farm Bur, 80-88; chmn, Walnut Mkt Bd, 82- *Mem:* Am Agr Econ Asn; Am Econ Asn. *Res:* Agricultural marketing and policy; environmental policy. *Mailing Add:* 11 Wandel Dr Moraga CA 94556

SIEBERT, JOHN, b Aurora, Ill, Mar 12, 40; m 61; c 2. ORGANIC CHEMISTRY, BUSINESS ADMINISTRATION. *Educ:* Procopius Col, BS, 62; Wichita State Univ, MS, 63; Univ Mo, PhD(org chem), 67. *Prof Exp:* Res chemist, Procter & Gamble Co, 67-70; dir household prod res & develop, Merck-Calgon Consumer Prod Div, 70-74; dir new prod develop, Gillette Toiletries Div, 74-77; DIR RES & DEVELOP, AMWAY CORP, 77- *Mem:* Am Chem Soc; Indust Res Inst; Sigma Xi; Soc Cosmetic Chemists. *Res:* Novel steroid chemistry; synthesis and nuclear magnetic resonance analysis of sterically hindered alkyl and aromatic quaternary carbon olefins. *Mailing Add:* Miles Inc Consumer Health Care PO Box 340 Elkhart IN 46515-9914

SIEBERT, KARL JOSEPH, b Harrisburg, Pa, Oct 29, 45; m 70; c 2. MULTIVARIATE ANALYSIS, CHEMOMETRICS. *Educ:* Pa State Univ, BS, 67, MS, 68, PhD(biochem), 70. *Prof Exp:* Synthetic chemist, Appl Sci Labs, Inc, 70; res assoc, Stroh Brewery Co, 71, head res & develop sect, 71-76, mgr res & develop lab, 76-82, dir res, 82-90; PROF & CHMN, FOOD SCI & TECHNOL DEPT, CORNELL UNIV, 90- *Concurrent Pos:* Mem, tech comt, Am Soc Brewing Chemists, 83-89, chmn, 86-88; bd visitors, Oakland Univ Biol Dept, Rochester, Mich, 85-89; vpres, Strohtech, 86-89; assoc dir, Cornell Inst Food Sci, 90-, mem bd dirs, Cornell Res Found, 90-; int consult brewing & food sci, 90- *Honors & Awards:* Nat Sci Found Fel, 69; Master Brew Pres Award, 86, 90. *Mem:* Am Chem Soc; Am Soc Brewing Chemists; Master Brewers Asn Am; Inst Food Technologists; Int Chemometrics Soc. *Res:* Beverage properties and processing; fermentation; process monitoring and controlling instrumentation; computer programming; laboratory automation; laboratory management; experiment design and statistical analysis; development and optimization of analytical methods, processes, and fermentation operations, using statistical experiment designs, mathematical modelling and multivariate analysis techniques; use of pattern recognition for cultivar identification, adulteration detection; extensive beverage experience. *Mailing Add:* Dept Food Sci & Technol Cornell Univ NYSAES Geneva NY 14456

SIEBERT, W(ILLIAM) M(CCONWAY), b Pittsburgh, Pa, Nov 19, 25; m 49; c 4. COMMUNICATIONS. *Educ:* Mass Inst Technol, SB, 46, ScD, 52. *Prof Exp:* Jr res engr, Res Labs, Westinghouse Elec Corp, 46-47; asst & instr elec eng, 47, from asst prof to prof elec eng, 52-84, mem staff & group leader, Lincoln Lab, 53-55, FORD PROF ENG, MASS INST TECHNOL, 84- *Concurrent Pos:* Mem, security resources panel, Gaither Comt, 56. *Honors & Awards:* Pioneer Award, Inst Elect & Electronics Engrs Aerospace & Electronic Syst Soc, 88. *Mem:* Fel Inst Elec & Electronics Engrs; Acoust Soc Am. *Res:* Statistical communication theory; radar theory; electrical communications; communications biophysics. *Mailing Add:* Dept of Elec Eng & Comput Sci Mass Inst Technol Cambridge MA 02139

SIEBES, MARIA, b Dinslaken, Ger, Feb 14, 58. BLOOD FLOW IN DISEASED ARTERIES, MEDICAL IMAGE PROCESSING. *Educ:* Fachhochschule Giessen, Ger, Dipl-Ing, 81; Univ Southern Calif, MS, 84, PhD(biomed eng), 89. *Prof Exp:* Res engr, Kerckhoff Klinik, Max Planck Inst Physiol & Clin Res, Bad Nauheim, Ger, 81-83; mem tech staff, Caltech, Biomed Image Processing Group, Jet Propulsion Lab, 83-89; ASST PROF, DEPT BIOMED ENG, UNIV IOWA, IOWA CITY, 89- *Honors & Awards:* Honor for Meritorious Res, AMA, 88. *Mem:* Assoc mem Sigma Xi; Am Soc Mech Engrs; Inst Elec & Electronics Engrs. *Res:* Hemodynamics of coronary artery stenoses, including arterial wall mechanics; quantitative image processing of coronary angiograms to obtain functional information regarding coronary physiology; mathematical modeling. *Mailing Add:* Dept Biomed Eng Univ Iowa Iowa City IA 52242

SIEBRAND, WILLEM, b Ijsselmuiden, Netherlands, Aug 12, 32; m 61; c 2. CHEMICAL PHYSICS, THEORETICAL CHEMISTRY. *Educ:* Univ Amsterdam, Drs, 60, Dr(phys chem), 63. *Prof Exp:* From asst res officer to assoc res officer, 63-72, sr res officer, 72-83, PRIN RES OFFICER THEORET CHEM, NAT RES COUN CAN, 83- *Concurrent Pos:* Ed, Can J Chem, 85-; corresp, Royal Dutch Soc Arts & Sci. *Mem:* Fel Royal Soc Can. *Res:* Electronic and spectroscopic properties of molecular crystals; vibrational-electronic coupling in molecules; radiationless transitions; hydrogen tunneling. *Mailing Add:* 752 Lonsdale Rd Ottawa ON K1K 0K2 Can

SIEBRING, BARTELD RICHARD, b George, Iowa, Sept 24, 24. INORGANIC CHEMISTRY. *Educ:* Macalester Col, BA, 47; Univ Minn, BS, 48, MS, 49; Syracuse Univ, PhD, 53. *Prof Exp:* Instr chem, Worthington Jr Col, 48-50; prof, Jamestown Col, 53; from asst prof to assoc prof, 53-62, PROF CHEM, UNIV WIS-MILWAUKEE, 62- *Concurrent Pos:* Vis prof, US Military Acad, West Point, 81-82 & 88-90. *Mem:* Am Chem Soc; Hist Sci Soc; Nat Asn Res Sci Teaching. *Res:* History of chemistry; coordination compounds; identification and description of areas of excellence in the training of scientists, especially the study of institutions and professors; modern trends and issues of chemical education; identification of personal attributes of leading personalities in the history of chemistry, with emphasis on possibly common characteristics. *Mailing Add:* Dept Chem Univ Wis-Milwaukee Milwaukee WI 53201

SIEBURTH, JOHN MCNEILL, b Calgary, Alta, Sept 1, 27; nat US; m 50; c 5. MARINE MICOROBIOLOGY, ATMOSPHERIC CHEMISTRY & PHYSICS. *Educ:* Univ BC, BSA, 49; Wash State Univ, MS, 51; Univ Minn, PhD(bact), 54. *Prof Exp:* Assoc prof vet sci, Va Polytech Inst & State Univ, 55-60; researcher biol oceanog, 60-61, PROF OCEANOG, UNIV RI, 66-, PROF MICROBIOL, 68- *Concurrent Pos:* Vis prof, Marine Biol Lab, Woods Hole, Mass, 73-74 & Norweg Inst Seaweed Res, Trondheim, 66-67; antarctic microbiologist, Int Geophys Yr, Arg Navy, 57-59. *Mem:* Am Soc Microbiol; Am Soc Limnol & Oceanog; Phycol Soc Am; Sigma Xi; AAAS; Soc Protozoologists; Brit Phycol Soc. *Res:* Physics, chemistry and microbiology of the oxic-anoxic transition zone of anoxic marine basins; microbial production and consumption of greenhouse gases in the upper ocean; role of bacteria, microalgae and protozoa in marine microbial food webs. *Mailing Add:* Grad Sch Oceanog Univ Rhode Island Bay Campus Narragansett RI 02882-1197

SIECKHAUS, JOHN FRANCIS, b St Louis, Mo, Sept 23, 39; m 63; c 3. INDUSTRIAL CHEMISTRY. *Educ:* Rockhurst Col, AB, 61; St Louis Univ, MS, 65, PhD(chem), 67. *Prof Exp:* Sr chemist & group leader, 66-80, MGR, LIFE SCI RES & DEVELOP DEPT, OLIN CORP, 80- *Honors & Awards:* IR-100 Award, Indust Res Mag, 71. *Mem:* Am Chem Soc; Sigma Xi. *Res:* Biotechnology development related to agriculture, energy and chemical production. *Mailing Add:* 89 Brooklawn Dr Milford CT 06460-2806

SIECKMANN, DONNA G, b Austin, Tex, Mar 26, 46; m 84. CELLULAR IMMUNOLOGY, IDIOTYPIC REGULATION. *Educ:* Univ Nebr, BS, 69; Mont State Univ, MS, 70; Univ Calif, San Diego, PhD(biol), 75. *Prof Exp:* Fel, Lab Immunol, Nat Inst Allergy & Infectious Dis, NIH, 75-78, staff fel, 78-79; MICROBIOLOGIST, INFECTIOUS DIS DEPT, NAVAL MED RES INST, 79- *Mem:* Am Soc Microbiol; Am Asn Immunologists; NY Acad Sci; Sigma Xi; AAAS; Am Soc Trop Med & Hyg. *Res:* B lymphocyte activation by anti-immunoglobulin antibody; anti-idiotypic vaccines; regulation of immune responsiveness by anti-idiotypic antibody. *Mailing Add:* Naval Med Res Inst Mail Stop 32 Bethesda MD 20889-5055

SIECKMANN, EVERETT FREDERICK, b Tobias, Nebr, Dec 31, 28; m 51; c 2. THEORETICAL PHYSICS, EXPERIMENTAL SOLID STATE PHYSICS. *Educ:* Doane Col, BA, 50; Fla State Univ, MS, 52; Cornell Univ, PhD(theoret solid state physics), 60. *Prof Exp:* Asst chem, Fla State Univ, 50-52; asst physics, Cornell Univ, 52-57; asst prof, Univ Ky, 57-62; assoc prof, 62-67, PROF PHYSICS, UNIV IDAHO, 67- *Concurrent Pos:* Consult, Librascope Div, Gen Precision, Inc, Calif, 61 & US Air Force, Holloman AFB, NMex, 62-63; vis prof physics, Univ Calif, Berkeley, 76, Jet Propulsion Lab, 80. *Mem:* Am Asn Physics Teachers; Am Phys Soc. *Res:* Impurity states in ionic crystals; single crystals of alkaline earth oxides; electrochemistry; applied mathematics; theoretical calculations on electron hole drops in Ge; optical dispersion of crystals containing color centers; heat storage by alkaliflourides. *Mailing Add:* 509 S Polk Moscow ID 83843

SIEDLE, ALLEN R, b Pittsburgh, Pa. INORGANIC CHEMISTRY. *Educ:* Ind Univ, PhD(chem), 73. *Prof Exp:* Fel chem, Nat Bur Standards, 73-75, res chemist, 75-77; res chemist, 77-85, SR RES SPECIALIST, SCI RES LAB, 3M CENT RES LAB, 85- *Res:* Synthetic inorganic chemistry; catalytic and surface chemistry; materials science-ceramics. *Mailing Add:* Sci Res Lab 3M Cent Res Lab 201-2E 3M Ct St Paul MN 55101-1428

SIEDLER, ARTHUR JAMES, b Milwaukee, Wis, Mar 17, 27; m 76; c 6. FOOD SCIENCE, NUTRITION. *Educ:* Univ Wis, BS, 51; Univ Chicago, MS, 56, PhD(biochem), 59. *Prof Exp:* Asst biochem, Am Meat Inst Found, 51-53, from asst biochemist to biochemist, 53-64, chief, Div Biochem & Nutrit, 59-64; group leader spec biol, Norwich Pharmacal Co, 64-65, chief physiol, 65-69, biochem, 69-72; PROF FOOD SCI & HEAD DEPT, UNIV ILL, URBANA, 72- *Concurrent Pos:* Instr & asst prof, Univ Chicago, 59-64. *Mem:* AAAS; Inst Food Technologists; Am Chem Soc; Am Inst Nutrit; Am Heart Asn; Sigma Xi. *Res:* Antimicrobial agent; biological activity napthoquinones; chemotherapeutic agents; food additives. *Mailing Add:* Dept Food Sci Univ Ill 1304 W Pennsylvania Urbana IL 61801

SIEDOW, JAMES N, b Chicago, Ill, Sept 21, 47. PLANT PHYSIOLOGY. *Educ:* Ind Univ, PhD(plant biochem), 72. *Prof Exp:* ASSOC PROF BOT, DUKE UNIV, 81- *Mailing Add:* Dept Bot Duke Univ Durham NC 27706

SIEDSCHLAG, KARL GLENN, JR, b Akron, Ohio, Oct 28, 20; m 41; c 1. ORGANIC CHEMISTRY. *Educ:* Western Reserve Univ, BS, 46, MS, 48, PhD(chem), 49. *Prof Exp:* Asst prof chem, Univ WVa, 49-51; res chemist, Benger Lab, E I du Pont de Nemours & Co, 51-54, res chemist, Patent Div, Textile Fibers Dept, 54-74, patent assoc, 74-79; RETIRED. *Mem:* Am Chem Soc; Sigma Xi. *Res:* Condensation polymer fibers; inorganic fibers. *Mailing Add:* Box 78 Winesburg OH 44690

SIEFKEN, HUGH EDWARD, b Warsaw, Ind, Apr 13, 40; m 62; c 2. EXPERIMENTAL NUCLEAR PHYSICS. *Educ:* Greenville Col, BA, 62; Univ Kans, MS, 65, PhD(physics), 68. *Prof Exp:* Res fel nuclear physics, Univ BC, 68-69; from asst prof to assoc prof, 69-77, PROF PHYSICS, GREENVILLE COL, 77- *Concurrent Pos:* Vis scientist, Nuclear Res Ctr, Univ Alta, 76-77, 86-87, McDonnell Douglas Astronaut Co, Neutral Particle Beam Integrated Exp. *Mem:* Am Phys Soc; Sigma Xi; Am Asn Physics Teachers. *Res:* low energy nuclear physics; gamma ray spectroscopy and ion bombardment of solids & gases; ion optics & in source design. *Mailing Add:* Physics Dept Greenville Col Greenville IL 62246

SIEFKEN, MARK WILLIAM, b Hankinson, NDak, Oct 3, 39; m 61; c 3. ORGANIC POLYMER CHEMISTRY. *Educ:* NDak State Univ, BS, 61; Univ Wis-Madison, PhD(org chem), 67. *Prof Exp:* Capt chem, Med Res Lab, Edgewood Arsenal, Md, 67-69; sr chemist, Cent Res Lab, 3M Co, 69-73, res scientist recreation & athletic prod, 73-74, supvr res & develop indust mineral prod div, Minn, 74-77; res mgr metals, detergents, surfactants, cleaning & sanitizing compounds, Diversey Chem, 77-78; tech vpres, Diversey Corp, 78-81; div mgr, 81-90, NAT SALES MGR, TEXO PULP & PAPER CORP, 90- *Concurrent Pos:* Fulbright grant, Tech Univ, Stuttgart, Ger. *Mem:* Am Chem Soc. *Res:* Tricyclic hydrocarbon rearrangements; polymer synthesis and development; new product research and development. *Mailing Add:* 11282 Terwilliger's Valley Lane Cincinnati OH 45249

SIEFKER, JOSEPH ROY, b Brownstown, Ind, Mar 14, 33; m 56; c 2. ANALYTICAL CHEMISTRY, INORGANIC CHEMISTRY. *Educ:* Wabash Col, AB, 55; Univ NDak, MS, 57; Ind Univ, PhD(chem), 60. *Prof Exp:* Res chemist, E I du Pont de Nemours & Co, 57; instr chem, St Louis Univ, 60-62; from asst prof to assoc prof, 62-71, PROF CHEM, IND STATE UNIV, TERRE HAUTE, 71- *Concurrent Pos:* Consult, Sporlan Valve Co, St Louis, 61 & OA Labs, Indianapolis, 70-; vis asst prof, Ind Univ, 62 & 64. *Mem:* Am Chem Soc; Sigma Xi. *Res:* Water pollution; coordination compounds and complex ions; spectrophotometry; polarography; non-aqueous solvents; redox titrants; electroanalytical chemistry. *Mailing Add:* Dept Chem Ind State Univ Terre Haute IN 47809-0001

SIEG, ALBERT LOUIS, b Chicago, Ill, Mar 25, 30; m 55; c 3. ORGANIC CHEMISTRY. *Educ:* Univ Ill, BS, 51; Univ Rochester, PhD, 55; Harvard Grad Sch Bus, PMD, 72. *Prof Exp:* Anal chemist, Eastman Kodak Co, 54-55, develop engr, 55-59, sr develop engr, 59-66, supvr chem testing, Paper Div, 67-69, supvr process develop, 69-71, supvr emulsion control, 71-72, corp coordr instant photog, 72-76, asst dir, 74-76, dir, Paper Serv Div, 76-78, asst mgr paper mfg, 78-80, mgr, 80-81, dir photog strategic planning, 81-84, vpres, 81, GEN MGR JAPANESE REGION, PRES & CHIEF EXEC OFFICER, EASTMAN KODAK CO, JAPAN, 84- *Concurrent Pos:* Lectr, Univ Rochester, 59-65, sr lectr, 65-69. *Mem:* AAAS; fel Am Inst Chemists; Am Chem Soc; Soc Photog Sci & Eng. *Res:* Synthesis of natural products; photographic chemistry; chemistry of gelatins; lithographic chemistry. *Mailing Add:* Eastman Kodak Co 343 State St Rochester NY 14650

SIEGAL, BERNARD, b Brooklyn, NY, Nov 11, 24; m 48; c 3. PHARMACEUTICAL SCIENCES. *Educ:* Yeshiva Univ, BA, 45; Rutgers Univ, PhD(pharmaceut), 68. *Prof Exp:* Chemist, Estro Chem Co, 49-53; from chemist to lab dir, Prod Div, Bristol Meyers, 53-69; dir res & develop, Toiletries Div, Gillette Co, 69-77; DIR RES & DEVELOP, H V SHUSTER, INC, 77- *Mem:* Am Chem Soc; Am Pharmaceut Asn; Acad Pharmaceut Sci; Soc Cosmetic Chem; Am Asn Textile Chemists & Colorists. *Res:* Over-the-counter pharmacy, health aids and allied products; product research and development; safety and claim substantiation; regulatory compliance. *Mailing Add:* H V Shuster Inc 5 Hayward St Quincy MA 02171

SIEGAL, BURTON L, b Chicago, Ill, Sept 27, 31. MECHANISMS, PATENT AVOIDANCE. *Educ:* Univ Ill, BS, 53. *Prof Exp:* Torpedo designer, US Naval Ord, 53-54; chief engr, Gen Aluminum Corp, 54-55; prod designer, Chicago Aerial Indust, 55-58; chief designer, Emil J Paider Co, 58-59; PRES, CONSULT, BUDD ENG CORP, 59- *Concurrent Pos:* Bd mem, Math, Eng & Sci Adv Bd, Niles TWP, 75-79; lectr, Am Soc Mech Engrs, 91. *Mem:* Soc Mech Engrs. *Res:* Expert conception, design, engineering and placing into production of products for annual quantities from five to a million; approximately 90 patents in over 30 industries. *Mailing Add:* 8707 Skokie Blvd Skokie IL 60077

SIEGAL, FREDERICK PAUL, b New York, NY, Sept 24, 39; m 66; c 2. IMMUNOLOGY, INTERNAL MEDICINE. *Educ:* Cornell Univ, AB, 61; Columbia Univ, MD, 65. *Prof Exp:* From intern to resident internal med, Mt Sinai Hosp, New York, 65-67; asst prev med officer, Walter Reed Army Med Ctr, 67-69; resident internal med, Mt Sinai Hosp, New York, 69-70, clin asst physician, 70-73; assoc immunol, Mem Sloan-Kettering Cancer Ctr, 73-79, assoc mem, 77-78, head, Lab Human Lymphocyte Differentiation, Sloan-Kettering Inst, 75-78; asst prof med, Cornell Univ Col Med, 75-78; ASSOC PROF MED & DIR, DIV CLIN IMMUNOL, MT SINAI SCH MED, NEW YORK, 78- *Concurrent Pos:* Helen Hay Whitney Found fel, Rockefeller Univ, 70-73, asst res physician, Rockefeller Univ Hosp, 70-73, vis assoc physician, 73-74; adj asst prof, Rockefeller Univ, 73-74; asst attend physician, Mem Hosp, New York, 73-78; asst prof biol, Sloan-Kettering Div, Cornell Univ, 74-78; assoc sci, Mem Sloan-Kettering Cancer Ctr, 78- *Mem:* AAAS; Am Asn Immunologists; Am Fedn Clin Res; NY Acad Sci. *Res:* Clinical immunology; immunodeficiency diseases; development of lymphoid cells; cell surfaces. *Mailing Add:* Long Island Jewish Med Ctr 270-05 76th Ave New Hyde Park NY 11042

SIEGAL, GENE PHILIP, b Bronx, NY, Nov 16, 48; m 72; c 2. TUMOR INVASION & METASTASIS, NEOPLASMS OF BONE. *Educ:* Adelphi Univ, BA, 70; Univ Louisville, MD, 74; Univ Minn, PhD(exp path), 79; Am Bd Path, dipl, 78. *Prof Exp:* Intern path, Mayo Grad Sch Med, Mayo Clinic, 74-75, resident, 75-76, res fel, 76-77, sr resident, 77-78, chief resident, dept anat path, 78-79; res assoc exp path & biochem, Lab Pathophysiol, Nat Cancer Inst, NIH, 79-81; med specialist & fel surg path, dept lab med & path, Div Surg Path, Univ Minn, Minneapolis, 81-82; from asst prof to assoc prof path, Univ NC, Chapel Hill, 82-90; assoc dir, surg path, NC Mem Hosp, Chapel Hill, 88-90; PROF PATH, UNIV ALA BIRMINGHAM, 90- *Concurrent Pos:* Instr path, Mayo Med Sch, Rochester, Minn, 76-79; clin fel, Am Cancer Soc, 81-82, jr fac fel, 83-86; attend pathologist, Univ NC Hosp, Chapel Hill, 82-90, dir histopath labs, 84-90 & dir spec procedures lab, 88-90; mem, Lineberger Cancer Res Ctr, Univ NC, Chapel Hill, 83-90; sr scientist, Comp Cancer Ctr Ala, 90-; dir anat path, Univ Ala Birmingham Hosp, 90- *Mem:* Am Asn Pathologists; US-Can Acad Path; Am Asn Cancer Res; AAAS; NY Acad Sci; AMA; Sigma Xi. *Res:* Experimental tumor invasion and metastasis; immunohistochemistry of solid tumors; human uterine carcinogenesis; neoplasms of bone and related conditions. *Mailing Add:* Dept Path UAB Sta Univ Ala Birmingham Birmingham AL 35294

SIEGART, WILLIAM RAYMOND, b Paterson, NJ, July 28, 31; m 55; c 3. ORGANIC CHEMISTRY. *Educ:* Gettysburg Col, BA, 53; Univ Pa, MS, 55, PhD(org chem), 57. *Prof Exp:* Asst instr chem, Univ Pa, 53-54, res assoc org chem, 54-56; chemist, 58-59, sr chemist, 59-62, res chemist, 62-67, group leader, 67-69, asst supvr petrochem res, 69-71, sr technologist, Mfg Div, Res & Technol Dept, 71-72, sr technologist prog planning & coord, Res & Technol Dept, 72-76, staff coordr, 76-78, sr staff coordr, Strategic Planning Dept, 78-80, sr coordr, 80-82, assoc dir altenate energy & resources dept, 82-88, SR STAFF CONSULT, TEXACO INC, 88-; VPRES, TEXACO SYNGAS, INC, 84- *Mem:* Am Chem Soc; Sigma Xi; NY Acad Sci; AAAS. *Res:* Petrochemicals; products and processes; lubricant additives and products; long range research planning; coal gasification; corporate planning; alternate energy. *Mailing Add:* 91 Round Hill Rd Poughkeepsie NY 12603

SIEGBAHN, KAI MANNE BORJE, b Lund, Sweden, Apr 20, 18. PHYSICS. *Prof Exp:* Prof physics, Royal Inst Technol, Stockholm, 51-54; prof, 54-83, EMER PROF PHYSICS, UNIV UPPSALA, 83- *Honors & Awards:* Nobel Prize in Physics, 81; Lindblom Prize, 45; Bjorken Prize, 55 & 77; Harrison Howse Award, 73; Maurice F Hasler Award, 75. *Mem:* Royal Swed Acad Sci; Royal Soc Arts & Sci; Papal Acad Sci. *Mailing Add:* Inst Physics Univ Uppsala Box 530s Upsala 751 21 Sweden

SIEGEL, ALBERT, b New York, NY, Aug 20, 24; m 47; c 4. MOLECULAR GENETICS, PLANT VIROLOGY. *Educ:* Cornell Univ, BA, 47; Calif Inst Technol, PhD(genetics), 51. *Prof Exp:* USPHS fel, Univ Calif, Los Angeles, 51-53, res assoc bot, 53-59; prof agr biochem, Univ Ariz, 59-72; chmn dept, 72-74, PROF BIOL, WAYNE STATE UNIV, 72- *Concurrent Pos:* NSF fel, 65-66, prog dir, 67-68; Fel award, Am Phytopath Soc. *Mem:* AAAS; Am Soc Cell Biol; Am Soc Microbiol; Genetics Soc Am; Am Phytopath Soc. *Res:* Nucleic acids; mechanism of plant virus replication; organization of plant nuclear and organelle genomes. *Mailing Add:* Dept Biol Wayne State Univ Detroit MI 48202

SIEGEL, ALLAN, b New York, NY, June 18, 39. NEUROBIOLOGY, NEUROANATOMY. *Educ:* City Col New York, BS, 61; State Univ NY Buffalo, PhD(psychol), 66. *Prof Exp:* USPHS fel, Sch Med, Yale Univ, 65-67; instr anat, 67-69, asst prof, 69-73, assoc prof anat & neurosci, 73-77, PROF NEUROSCI, NJ MED SCH, 77- *Concurrent Pos:* Consult neurol & res, East Orange Vet Admin Hosp, 72-; mem Neurol Sci II Study Sect, Nat Inst Neurol & Commun Dis & Stroke, 84-; vis prof, Cornell Univ Med Sch, 85. *Mem:* AAAS; Am Asn Anat; Soc Neurosci; NY Acad Sci; Int Soc Res Aggression; Int Soc Res Emotions. *Res:* Experimental psychology; anatomy and neurophysiology of limbic system; biology of aggressive behavior. *Mailing Add:* 100 Bergen St NJ Col Med Newark NJ 07103

SIEGEL, ALVIN, b New York, NY, Aug 29, 31; m 59; c 4. CHEMICAL OCEANOGRAPHY. *Educ:* City Col New York, BS, 53; Rutgers Univ, PhD(phys chem), 62. *Prof Exp:* Asst scientist, Woods Hole Oceanog Inst, 61-67; assoc prof, 67-71, dir, Natural Sci Div, 80-86, PROF CHEM, SOUTHAMPTON COL, LONG ISLAND UNIV, 71-, DIR, MARINE SCI PROG, 73-, PROF MARINE SCI, 75-, ACAD DEAN, 89- *Mem:* AAAS; Am Soc Limnol & Oceanog; Am Chem Soc. *Res:* Polyelectrolytes; speciation of metal ions in natural waters; organic bonding to metal ions; extraction of polar organics from sea water. *Mailing Add:* Southampton Col Long Island Univ Southampton NY 11968

SIEGEL, ANDREW FRANCIS, b Cambridge, Mass, Jan 6, 50. GEOMETRICAL PROBABILITY, BIOSTATISTICS. *Educ:* Boston Univ, AB, 73; Stanford Univ, MS, 75, PhD(statist), 77. *Prof Exp:* Asst prof, Univ Wis-Madison, 77-79; ASST PROF STATIST, PRINCETON UNIV, 79- *Concurrent Pos:* Vis res assoc, Dept Paleobiol, Smithsonian Inst, 78; res fel, Dept Biostatist, Harvard Univ, 78-79; vis staff mem, Statist Group, Los Alamos Nat Lab, 79; consult, Statist Group, Bell Telephone Labs, 79-80; co-prin investr, US Army Res Off, 79-; vis scholar, Dept Statist, Stanford Univ, 80, Univ Wash, Seattle, 81. *Mem:* Am Statist Asn; Inst Math Statist; Biomet Soc; Royal Statist Soc; AAAS. *Res:* Geometric probability; statistical robustness; pattern matching; morphology; probability distributions; biological and medical applications of statistics. *Mailing Add:* Dept Statist Fine Hall Princeton Univ Princeton NJ 08540

SIEGEL, ARMAND, b New York, NY, Oct 10, 14; m 43; c 3. STATISTICAL MECHANICS, BIOMATHEMATICS. *Educ:* NY Univ, AB, 36; Univ Pa, AM, 44; Mass Inst Technol, PhD(physics), 49. *Prof Exp:* Asst physics, Univ Pa, 42-43, instr, 43-44; instr elec commun, Radar Sch, Mass Inst Technol, 44-45, res assoc physics, 52-53; instr, Worcester Polytech Inst, 49-50; from instr to prof, 50-80, EMER PROF PHYSICS, BOSTON UNIV, 80-, PROF PSYCHIAT PHYSICS, SCH MED, 75- *Concurrent Pos:* Guggenheim fel, Univ Mich, 57-58. *Mem:* Am Phys Soc. *Res:* Relativistic nucleon-nucleon interactions; differential space formulation of quantum mechanics; kinetic theory; stochastic processes; mathematical theory of turbulence; electroencephalography of petit-mal epilepsy; stochastic aspects of the origin of the electroencephalogram. *Mailing Add:* 56 Marshall St Brookline MA 02146

SIEGEL, BARBARA ZENZ, b Detroit, Mich, July 22, 31; m 50; c 4. BIOLOGY. *Educ:* Univ Chicago, AB, 60; Columbia Univ, MA, 63; Yale Univ, PhD(biol), 66. *Prof Exp:* Purchasing liaison, Army Chem Corp, Ft Detrick, Md, 50-52; res assoc human genetics, Sch Med, NY Univ, 62-63; res staff mem biol, Yale Univ, 66-67; from asst prof to assoc prof microbiol, Univ Hawaii, 67-75, dir biol prog, 71-75, researcher Biomed Res Ctr & assoc prof Current Res & Develop, 75-76, interim dir Res Admin & interim dean Grad Sch, 79-82, res prof Pac Biomed Res Ctr, 76-, dir, Pesticide Hazard Assessment Proj, 83-87; PROF & SR RESEARCHER ENVIRON HEALTH SCI, SCH PUB HEALTH & ASSOC DEAN ACAD AFFAIRS, UNIV HAWAII, 88. *Concurrent Pos:* Mem, Boston Univ-NASA exped, Iceland & Surtsey, 70 & Nat Geog Soc-Hawaii Found & Cottrell Found expeds, Iceland & Surtsey, 72; Fulbright-Hays sr res scholar, Univ Belgrade, Marine Sta, Montenegro, Yugoslavia & Univ Heidelberg, 73-74; sr NATO fel, Nordic Volcanol Inst, Iceland, 75; res assoc, Volcani Agr Ctr, Israel; proj leader, US Antarctic Res Prog, NSF, McMurdo, Antarctic, 78-79; vis prof bot & geol, Univ BC, 82; vis prof, Weizmann Inst, Israel, 86-87. *Honors & Awards:* vis researcher Inst of Biophysics, PISA award, Nat Res Coun, Italy, 87-88; vis researcher Inst of Biophysics, PISA award, Nat Res Coun, Italy, 87-88; fulbright Hay Sr Res scholar, Oalu Univ, Finland, 88-89. *Mem:* Am Chem Soc; Am Genetic Asn; Genetics Soc Am; Am Soc Plant Physiologists; Am Inst Biol Sci; Int Soc Chem Ecol. *Res:* Biochemical mechanisms in growth and development; biological oxidations, biochemical interactions in eco-systems; volcanic gas emissions on biological systems; ecology of heavy metals; impact of alternate energy sources (geothermal, wind, ocean thermal energy conversion, etc) on the environment. *Mailing Add:* Univ Hawaii Biomed D104J Sch Pub Health Honolulu HI 96822

SIEGEL, BARRY ALAN, b Nashville, Tenn, Dec 30, 44; div; c 2. NUCLEAR MEDICINE, RADIOLOGY. *Educ:* Washington Univ, AB, 66, MD, 69. *Prof Exp:* Intern med, Barnes Hosp, St Louis, 69-70; resident nuclear med & radiol, Edward Mallinckrodt Inst Radiol, 70-73; assoc prof med, 80-83, from asst prof to assoc prof, 73-79, PROF RADIOL, SCH MED, WASH UNIV, 79-, PROF MED, 83-; DIR NUCLEAR MED, EDWARD MALLINCKRODT INST RADIOL, 73- *Concurrent Pos:* asst prof radiol, Sch Med, Johns Hopkins Univ, 74-76; chief, Radiol Sci Div, Armed Forces Radiobiol Res Inst, Defense Nuclear Agency, 74-76; mem radioactive pharmaceut adv comt, Food & Drug Admin, 74-77, 81-85, chmn, 82-85 & task force short-lived radionuclides appln nuclear med, 75-76; mem task force nuclear med, Energy Res & Develop Admin, 75-76; mem adv panel radiopharmaceut, US Pharmacopeia, 76-; consult ed bd, J Nuclear Med, 76-81; mem adv comt on med radioisotopes, Los Alamos Sci Lab, 76-79; assoc ed, Radiol, 80-; mem, Residency Rev Comt Nuclear Med, 84-85, Mo Low Level Radioactive Waste Adv Comt, 84-86; trustee, Am Bd Nuclear Med, 85-90; asst ed, Am J Roentgenol, 87-; chmn, US Nuclear Regulatory Comn Adv Comt Med Uses Isotopes, 90- *Honors & Awards:* Comnr Spec Citation, Food & Drug Admin, 88. *Mem:* Fel Am Col Nuclear Physicians; fel Am Col Radiol; Radiol Soc NAm; fel Am Col Physicians; Soc Nuclear Med; AMA. *Res:* Clinical research with positron-emitting radionuclides and coincidence axial tomographic detection systems; radioisotopic detection of vascular thrombosis; radioisotopic evaluation of cancer. *Mailing Add:* Edward Mallinckrodt Inst Radiol 510 Kingshighway Blvd St Louis MO 63110-1076

SIEGEL, BENJAMIN MORTON, b Superior, Wis, Mar 26, 16; m 44; c 3. ION BEAMS, IMAGE PROCESSING. *Educ:* Mass Inst Technol, BS, 38, PhD(phys chem), 40. *Prof Exp:* Vis fel, Calif Inst Technol, 40-41; res assoc, Heat Res Lab, Nat Defense Res Comt Proj, Mass Inst Technol, 41-42 & 44-46 & US Bur Ships Proj, 48-49; Nat Defense Res Comt Proj, Harvard Univ, 42-44; Weizmann Inst Sci assoc & chg electron micros lab, Polytech Inst Brooklyn, 46-48; assoc prof, 49-59, PROF APPL & ENG PHYSICS, CORNELL UNIV, 59- *Concurrent Pos:* Vis prof, Hebrew Univ, Israel, 62-63; vis fel, Salk Inst Biol Sci, 71. *Honors & Awards:* Distinguished Award for contrib electron micros, Electron Micros Soc Am, 82. *Mem:* AAAS; Sigma Xi; Am Vacuum Soc; Am Phys Soc; Electron Micros Soc Am (pres, 73). *Res:* High brightness field ionization sources; electron and ion optics; field ion probe systems for nanometer ion beam lithography; computer image processing of high resolution electron microscope images. *Mailing Add:* Clark Hall Cornell Univ Ithaca NY 14853

SIEGEL, BENJAMIN VINCENT, b New York, NY, Dec 14, 13; m 43; c 3. VIROLOGY. *Educ:* Univ Ga, BS, 34; Columbia Univ, MA, 37; Stanford Univ, PhD(bact, exp path), 50; Am Bd Microbiol, dipl. *Prof Exp:* Teacher pub schs, Calif, 39-42; chmn biol & phys sci high sch, 46-48; asst bact physiol, Stanford Univ, 49-50, instr bact & exp path, 50-52, Nat Found Infantile Paralysis fel, 52-53; fel med phys & virol, Univ Calif, Berkeley, 53-54, asst res virologist, 54-56; assoc res virologist, Sch Med, San Francisco & res microbiologist, Donner Radiation Lab, Berkeley, 56-61, lectr microbiol, 59-60; PROF PATH, MED SCH, UNIV ORE, PORTLAND, 61- *Mem:* Fel Am Acad Microbiol; fel NY Acad Sci; Am Soc Exp Path; Int Acad Path; fel AAAS. *Res:* Tumor virology; experimental pathology; immunology. *Mailing Add:* 3900 SW Pendleton St Portland OR 97221

SIEGEL, BERNARD, b New York, NY, Oct 14, 28; m 54; c 2. PHYSICAL CHEMISTRY, INORGANIC CHEMISTRY. *Educ:* Brooklyn Col, BS, 48; Polytech Inst Brooklyn, PhD(phys chem), 53. *Prof Exp:* Res chemist, Shell Oil Co, 52-55; head phys chem br, Naval Propellant Plant, US Off Naval Res, 55-58; res chemist, Gen Corp, 58-61; sect head propellant chem, Aerospace Corp, Calif, 61-70; CONSULT, 70- *Mem:* Am Chem Soc; NY Acad Sci; The Chem Soc. *Res:* High temperature chemistry, chemistry; inorganic syntheses, especially hydrides and fluorides; energetics of chemical propellants. *Mailing Add:* Anthrop Bldg 110 Rm 111M Stanford Univ Stanford CA 94305

SIEGEL, BROCK MARTIN, b Binghamton, NY, Aug 25, 47; m 78; c 3. ORGANIC CHEMISTRY, BIOORGANIC CHEMISTRY. *Educ:* Syracuse Univ, BS, 69; Univ Ill, PhD(chem), 74. *Prof Exp:* NIH fel, Columbia Univ, 74-76; asst prof chem, Univ Minn, Minneapolis, 76-80; res mgr, Henkel Corp, 80-89; RES & DEVELOP MGR, MILLIPORE CORP, 89- *Concurrent Pos:* DuPont fac fel, Univ Minn, 76-77, NIH grant, 78-79. *Mem:* AAAS; Am Chem Soc; NY Acad Sci; Sigma Xi. *Res:* Kinetics mechanisms; synthetic and physical-organic chemistry; biomimetic enzyme catalysts; emerging technologies; oleochemical lipid specialties; aromatic substitution chemistry; nucleic acid and peptide synthesis; carbohydrate chemistry. *Mailing Add:* 2936 Santa Marta Ct Santa Rosa CA 95405

SIEGEL, CAROLE ETHEL, b US, Sept 29, 36; m 57; c 2. BIOSTATISTICS. *Educ:* NY Univ, BA, 57, MS, 59, PhD(math), 63. *Prof Exp:* Prin res scientist math, 65-73, HEAD, EPIDEMIOL & HEALTH SERV RES LAB, NATHAN KLINE INST, 73- *Concurrent Pos:* Adj asst prof math, NY Univ, 65-68 & 69, res prof, Dept Psychiat, 79-; adj prof, Fairleigh Dickinson Univ, 68; consult, World Health Orgn, 84-; prin investr, Nat Inst Alchol Abuse & Alcoholism, 78-81 & Nat Ctr Health Serv Res, 79-81, Nat Inst Mental Health, 88. *Mem:* Inst Math Statist; Asn Women Math. *Res:* The development and application of statistical and quantitative models for examining mental health issues covering: epidemiology, health services and economics; mathematical and statistical approaches to mental health data; epidemiology; health services research. *Mailing Add:* Statist Sci & Epidemiol Div Nathan Kline Inst Orangeburg NY 10962

SIEGEL, CHARLES DAVID, b New York, NY, Sept 28, 30. BIOLOGY. *Educ:* NY Univ, BA, 53, MS, 55, PhD(biol), 60. *Prof Exp:* Instr biol sci, Sch Com, NY Univ, 56-59; from instr to assoc prof, 59-77, PROF BIOL, WASH SQUARE COL, 77-, ASST DEAN, 70- *Mem:* AAAS; Am Asn Anat; Am Soc Zoologists; Am Soc Hemat; fel NY Acad Sci. *Res:* Physiologic hematology; mechanism involved in the formation and destruction of blood and bone marrow cellular elements. *Mailing Add:* Dept Biol Washington Sq Campus NY Univ New York NY 10003

SIEGEL, CLIFFORD M(YRON), b Apple River, Ill, Apr 15, 21; m 46; c 4. ELECTRICAL NETWORK THEORY, COMPUTER AIDED INSTRUCTION. *Educ:* Marquette Univ, BEE, 47; Univ NH, MSEE, 49; Univ Wis, PhD(elec eng), 51. *Prof Exp:* Engr, Wis Elec Power Co, 41-43; instr elec eng, Marquette Univ, 47; asst, Univ NH, 47-48, instr, 48-49; asst, Univ Wis, 49-51; from asst prof to assoc prof, 51-62, acting chmn dept, 67-69, PROF ELEC ENG, UNIV VA, 62- *Mem:* Am Soc Eng Educ; Inst Elec & Electronics Engrs. *Res:* Methods for electrical engineering education; computer-assisted instruction. *Mailing Add:* Dept of Elec Eng Univ of Va Charlottesville VA 22901

SIEGEL, EDWARD, b New York, NY, Aug 1, 19; m 44. BIOPHYSICS. *Educ:* City Col New York, BS, 41; Univ Calif, Berkeley, PhD(biophys), 66. *Prof Exp:* Physicist-in-chg optical res, Universal Camera Corp, NY, 41-44; physicist, Frankford Arsenal, US Dept Army, 45, biophysicist, Aero-Med Lab, Wright Field, 45-46; res assoc, Col Eng, Rutgers Univ, 47-48; physicist-in-chg, Med Physics Lab, Montefiore Hosp, 48-62; assoc prof radiol physics, Sch Med, Stanford Univ, 66-70; actg dir nuclear med sect, Sch Med, Univ Mo-Columbia, 71-73, prof radiol sci & med, 70-76, prof biol sci, 75-76, dir nuclear med sect, 73-76; prof radiol & radiation sci & dir, Dept Radiol & Radiation Sci, Sch Med, Vanderbilt Univ, 76-83, prof physics, Col Arts & Sci, 79-88; ADJ PROF RADIOL, DEPT RADIOL, UNIV CALIF, SAN FRANCISCO, 88- *Concurrent Pos:* Consult physicist, Radioisotope Dept, Newark Beth Israel Hosp, NJ, 50-55 & Radioisotope Dept, Lebanon Hosp, NY, 53-62; guest scientist, Donner Lab, Univ Calif, 66-70; consult physicist, Palo Alto Vet Admin Hosp, 67-70; consult adv comt human uses radioactive mat, Bur Radiol Health, State of Calif, 67-70; consult physicist, USPHS Hosp, San Francisco, 68-70 & Vet Admin Hosp, Columbia, Mo, 72-76; ed, Med Physics, 79-81; radiation safety officer, Vet Admin Hosp, San Francisco, 88- *Mem:* Endocrine Soc; Am Asn Physicists Med; Radiation Res Soc; Radiol Soc NAm; Brit Hosp Physicists Asn; Soc Nuclear Med. *Res:* Applications of radioisotopes to biology and medicine; radiation physics; thyroid physiology and thyroid cancer; cell culture; radiation dosimetry; nuclear medicine; high resolution radioautography; radiation biology; biological effects of radiation and ultrasound. *Mailing Add:* Two Milland Ct Mill Valley CA 94941

SIEGEL, EDWARD T, b Lawrence, Mass, Aug 3, 34; m 59; c 3. VETERINARY ENDOCRINOLOGY. *Educ:* Univ Pa, VMD, 58; Jefferson Med Col, PhD(physiol), 64. *Prof Exp:* Scientist, Worcester Found Exp Biol, 61-63; asst prof biochem in med, 63-68, assoc prof med, 68-73, PROF MED, UNIV PA, 73- *Concurrent Pos:* NIH res fel, 60-63, res grant, 64-66; Inst Coop Res grants, Univ Pa & Am Cancer Soc, 63-64; Seeing Eye, Inc grant, 68-71. *Mem:* Am Vet Med Asn; Am Col Vet Internal Med. *Res:* Hormonal synthesis in the canine Sertoli cell tumor; steroid biosynthesis and catabolism in domestic animals; clinical endocrinology in domestic animals. *Mailing Add:* Two Milland Court Mill Valley CA 94941

SIEGEL, ELI CHARLES, b Newark, NJ, July 25, 38; m 86; c 1. MICROBIAL GENETICS. *Educ:* Rutgers Univ, BA, 60, PhD(microbiol), 66. *Prof Exp:* Fel biochem, Albert Einstein Col Med, 66-68; from asst prof to assoc prof, 68-82, PROF BIOL, TUFTS UNIV, 82- *Concurrent Pos:* USPHS fel, 67-68. *Mem:* AAAS; Am Soc Microbiol; Genetics Soc Am; Am Asn Univ Prof. *Res:* Mutator genes; DNA repair; bacterial genetics. *Mailing Add:* Dept Biol Tufts Univ Medford MA 02155

SIEGEL, ELLIOT ROBERT, b New York, NY, May 31, 42; m 67; c 2. COMMUNICATIONS SCIENCE, INFORMATION SCIENCE. *Educ:* Brooklyn Col, BA, 64; Mich State Univ, MA, 66, PhD(commun), 69. *Prof Exp:* Res scientist commun, Human Sci Res, Inc, 69-70; res assoc commun & info sci, Off Commun, Am Psychol Asn, 70-72, mgr & exec ed, 72-74 & sci affairs officer, 75-76; info scientist commun, Lister Hill Nat Ctr Biomed Commun, 76-82, spec asst opers res, off dir, 82-87, ASST DIR PLAN & EVAL, NAT LIBR MED, 87- *Concurrent Pos:* Assoc ed, Am Psychologist J, 75-76; mem dissemination comt, Fed Coun Educ Res & Develop, 78-83; mem adv comt med appl res, NIH, 83-, eval res rev comt, 84-90 & technol transfer comt, 89-; mem, Fed Libr & Info Ctr Comn, 85-88; secy info, comput & commun, AAAS, 85- *Honors & Awards:* Award Merit, NIH, 88. *Mem:* AAAS; Am Soc Info Sci; Am Psychol Asn; Int Commun Asn; Am Col Med Informatics. *Res:* Information transfer and knowledge utilization in the health sciences; medical informatics; information systems research, development and evaluation; scientific and technical communication processes; science policy research. *Mailing Add:* Nat Libr Med Bethesda MD 20894

SIEGEL, FRANK LEONARD, b Brooklyn, NY, Apr 15, 31; m 62; c 2. BIOCHEMISTRY. *Educ:* Reed Col, BA, 53; Univ Tex, PhD(chem), 60. *Prof Exp:* Asst chem, Univ Tex, 54, Clayton Found fel biochem, 60-64; from instr to asst prof, 64-71, assoc prof pediat & physiol chem, Sch Med, 71-75, PROF PEDIAT & PHYSIOL CHEM, NEUROCHEM SECT, WAISMAN CTR, UNIV WIS-MADISON, 75- *Concurrent Pos:* Res scientist, NIMH Develop Comt. *Honors & Awards:* Javits Neurosci Investr Award, 87. *Mem:* AAAS; Am Chem Soc; Am Soc Biol Chemists; Am Soc Neurochem; Soc Neurosci. *Res:* Protein methylation; calmodulin, glutathione S-transferases. *Mailing Add:* Waisman Ctr Neurochem Lab Univ Wis Madison WI 53706

SIEGEL, FREDERIC RICHARD, b Chelsea, Mass, Feb 8, 32; m 62; c 2. GEOCHEMISTRY. *Educ:* Harvard Univ, BA, 54; Univ Kans, MS, 58, PhD(geol), 61. *Prof Exp:* Prof & researcher geochem & sedimentology, Inst Miguel Lillo, Nat Univ Tucuman, 61-63; div head geochem, Kans Geol Surv, 63-65, res assoc, 64-65; assoc prof, 65-69, chmn geol dept, 76-86, PROF GEOCHEM, GEORGE WASHINGTON UNIV, 69- *Concurrent Pos:* Vis prof, Univ Buenos Aires, 63; lectr, Univ Kans, 65; res assoc, Smithsonian Inst, 67-; Fulbright scholar, Facultad de Minas, Medellin, Colombia, 70; consult, World Bank, 79, UN Develop Prog, 80, US Dept State, 88-90, minerals indust, 82 & 84; ed, United Nations Educ Sci Cultural Orgn, volume geochemistry, 79; assoc ed, exploration geochemistry, 72-76. *Mem:* AAAS; Asn Explor Geochem; Geochem Soc; Soc Econ Paleont & Mineral; Soc Environ Geochem & Health; Int Assoc Geochemists & Cosmochemists. *Res:* Geochemical prospecting; suspended sediment mineralogy and geochemistry; environmental geology and geochemistry; geologic hazards in land-use planning; trace element distribution delta systems and in modern marine sediments; clay mineralogy; thermoluminescence in oil exploration; carbonate geochemistry. *Mailing Add:* Dept Geol George Washington Univ Washington DC 20052

SIEGEL, GEORGE JACOB, b Bronx, NY, Aug 6, 36; m 57; c 3. NEUROLOGY, NEUROBIOLOGY. *Educ:* Yeshiva Col, BA, 57; Univ Miami, MD, 61. *Prof Exp:* Resident neurol, Mt Sinai Hosp, New York, 62-65; res assoc neurochem, Nat Inst Neurol Dis & Stroke, 65-68; from asst prof to assoc prof physiol & neurol, Mt Sinai Sch Med, 68-73; assoc prof, 73-75, CHIEF NEUROL CHEM LAB MED SCH, UNIV MICH, ANN ARBOR, 73-, PROF NEUROL, 75- *Concurrent Pos:* Assoc attend neurologist, Mt Sinai Hosp & chmn neurosci integrated curric comt. *Mem:* Am Neurol Asn; Int Soc Neurochem; Am Asn Neuropath; Am Acad Neurol; Harvey Soc. *Res:* Biology of active transport and neural membrane development. *Mailing Add:* Dept Neurol Univ Mich Med Sch 1103 E Huron St Ann Arbor MI 48109

SIEGEL, GEORGES G, b Port-au-Prince, Haiti, Oct 11, 30; US citizen; m 64; c 3. THERMODYNAMICS & MATERIAL PROPERTIES. *Educ:* Rensselaer Polytech Inst, BChE, 53; Yale Univ, MS, 60, PhD(chem), 65. *Prof Exp:* Asst chem, Rensselaer Polytech Inst, 53-55; asst prof, Univ PR, 55-59; res asst, Yale Univ, 59-65; sr res engr, US Steel Corp, 65-71; from asst to assoc prof, 71-76, PROF CHEM, UNIV PR, MAYAGUEZ, 76- *Concurrent Pos:* Vis prof, Univ Leuven, Belgium, 88-89. *Mem:* Sigma Xi; Am Chem Soc. *Res:* Change of volume for ionic reactions in solution; H-bond interactions in non-electrolyte solutions by dipolar, calorimetric and spectrophotometric measurements. *Mailing Add:* Box 5391 Mayaguez PR 00709-5391

SIEGEL, HENRY, b New York, NY, Oct 16, 10; m 34; c 2. PATHOLOGY. *Educ:* City Col New York, BS, 33; NY Univ, MD, 37; Am Bd Path, dipl. *Hon Degrees:* ScD, Mercy Col, 78. *Prof Exp:* Asst to toxicologist, Off Chief Med Examr, New York, 28-33; asst med examr, 47-50; assoc prof path, Col Med, State Univ NY Downstate Med Ctr, 50-55; from asst med exam to exec dep chief, Off Chief Med Examr, NY, 55-70; PROF PATH, NY MED COL, 70- *Concurrent Pos:* Res pathologist, Merck Inst Therapeut Res, NJ, 41-43; assoc med, Long Island Col Med, 43-44; asst pathologist, Lenox Hill Hosp, 48-50; chief pathologist, Kings County Hosp, 50-55; assoc prof, Albert Einstein Col Med, 55-58; pathologist, Bronx Munic Hosp Ctr, 55-58; pathologist & dir labs, Grand Cent Hosp, 58-63; res specialist, Rutgers Univ, 58; pathologist-med examr, Westchester County, NY, 70-78, consult, Labs & Res Dept, 78- *Mem:* Fel Am Soc Clin Path; Am Asn Path & Bact; fel Col Am Path; NY Acad Sci. *Res:* Toxicology; morphologic effects of chemical compounds; forensic and experimental pathology. *Mailing Add:* Scarborough Manor PO Box 307 Scarborough NY 10510

SIEGEL, HERBERT, b New York, NY, Apr 23, 25; m 56; c 2. INFORMATION SCIENCE, ORGANIC CHEMISTRY. *Educ:* Ind Univ, BS, 47; WVa Univ, PhD(org chem), 56. *Prof Exp:* Chemist, Ohio-Apex, Inc, WVa, 48-50; assoc prof chem, Waynesburg Col, 55-58; asst ed, Chem Abstr Serv, 59-61, asst dept head appl org ed, 62-63, dept head, 64-66, spec projs mgr, 66-69, mgr org ed anal dept, 69-72, asst mgr chem technol dept, 72-73, asst mgr org chem dept, 73-75; info chemist, Int Occup Safety & Health Info Ctr, 75-79, head, 80-85; RETIRED. *Concurrent Pos:* Res fel & lectr, Bedford Col, Univ London, 58-59. *Mem:* Am Chem Soc; Am Soc Info Sci. *Res:* Biphenyl stereochemistry; chemical information storage and retrieval; abstracting and indexing. *Mailing Add:* BP 52 01212 Ferney-Voltaire Cedex France

SIEGEL, HERBERT S, b Mt Vernon, NY, Aug 29, 26; m 48; c 3. PHYSIOLOGY. *Educ:* Pa State Univ, BS, 50, MS, 57, PhD, 59. *Prof Exp:* Asst poultry husb, Pa State Univ, 55-57; from asst prof to assoc prof, Va Polytech Inst & State Univ, 58-64; res physiologist, Agr Res Serv, USDA, 64-84; HEAD DEPT POULTRY SCI, PA STATE UNIV, 84-, MEM GRAD FAC PHYSIOL, 84- *Concurrent Pos:* Adj mem grad fac vet physiol & pharmacol, Univ Ga, 66-84; Fulbright-Hays res award, Neth, 80-81; ed-in-chief, Poultry Sci, 80-86. *Honors & Awards:* Res Award, Poultry Sci Asn, 61. *Mem:* AAAS; Am Soc Zoologists; Poultry Sci Asn; fel NY Acad Sci; Soc Exp Biol & Med. *Res:* Avian physiology; environmental physiology, stress, adrenals and the immune systems; pesticide residues. *Mailing Add:* Dept Poultry Sci Pa State Univ University Park PA 16802

SIEGEL, HOWARD JAY, b Newark, NJ, Jan 16, 50; m 72. COMPUTER ARCHITECTURE. *Educ:* Mass Inst Technol, BS(mgt) & BS(elec eng), 72; Princeton Univ, MA, 74, MSE, 74, PhD(elec eng), 77. *Prof Exp:* Researcher compiler design, Mass Inst Technol, 70, info systs, 71; researcher & teaching asst elec eng & comput sci, Princeton Univ, 72-76; asst prof elec eng, 76-81, res staff, LARS, 79-81, assoc prof, 81-85, PROF ELEC ENG, PURDUE UNIV, 85-, COORDR, PARALLEL PROCESSING LAB, 89- *Concurrent Pos:* Consult, TRW, Huntsville, 79, Xerox Corp, Rochester, 79, Gen Motors Res Lab, Dearborn, 80, Arvin/Calspan, Advan Technol Ctr, Buffalo, 81-82, Dynamic Computer Architect Inc, Lincoln, 82, IBM Fed Systs Div, Manassas, Va, 83-87, Hewlett-Packard, Ft Collins, Colo, 84, Westinghouse Elec Corp, Baltimore, 84, KLA Instruments Corp, Santa Clara, 85, Ball Aerospace, Boulder, 85, MCC, Austin, Tex, 86, Citicorp/TTI, Santa Monica, 86, Gen Dynamics, Fort Worth, 86, NCR Corp, Minneapolis, 88, Sandia Nat Labs, Livermore, Ca, 88-89, NCR, San Diego, 90 & Cray Res Inc, Mendota Heights, Minn, 90; numerous res grants, 77-89; chmn, Computer Soc Tech Comt Computer Architect, Inst Elec & Electronic Engrs, 82; chmn, Assoc Comput Mach Spec Interest Group Computer Architect, 83-85, co-chmn Int Conf Parallel Processing, 83; res proj leader, Supercomput Res Ctr, Lanham, Md, 87-88. *Honors & Awards:* Cert Appreciation, Inst Elec & Electronic Engrs, 83, 84; Serv Award, Am Comput Mach, 86. *Mem:* Asn Comput Mach; fel Inst Elec & Electronic Engrs; Sigma Xi. *Res:* Development of large-scale parallel and distributed multimicrocomputer systems for image and speech processing, including system hardware and software and study of parallel algorithms; design and analysis of interconnection networks for parallel machines; modeling of parallel processing systems; co-authored over 130 technical papers. *Mailing Add:* Sch Elec Eng Purdue Univ West Lafayette IN 47907

SIEGEL, IRVING, b Brooklyn, NY, June 15, 24; m 50; c 2. SOLID STATE PHYSICS, CHEMICAL PHYSICS. *Educ:* Univ Iowa, BA, 50; Ill Inst Technol, MS, 61; Univ Toledo, PhD, 70. *Prof Exp:* Physicist, Semiconductor Div, Battelle Mem Inst, 51-55, Physics Res Div, IIT Res Inst, 55-59 & Semiconductor & Mat Div, RCA, 59-60; res physicist fundamental res sect, Owens-Ill Tech Ctr, Ohio, 60-70; asst prof physics, Calif Polytech State Univ, 71-76; RES PHYSICIST, NORTHROP CORP, 79- *Concurrent Pos:* Adj lectr, Univ Toledo, 61-70. *Mem:* Am Phys Soc; Inst Elec & Electronics Engrs; Magnetics Soc. *Res:* Electron paramagnetic resonance; ferromagnetic resonance; magnetic materials and magnetism; teaching. *Mailing Add:* 2071 Hope St San Luis Obispo CA 93401

SIEGEL, IRWIN MICHAEL, b New York, NY, Apr 18, 30; m 56; c 1. PHYSIOLOGY, GENETICS. *Educ:* City Col New York, BS, 51; Columbia Univ, MS, 54, MA, 58, PhD(vision), 60. *Prof Exp:* From asst prof to assoc prof, 60-73, PROF EXP OPHTHAL, MED CTR, NY UNIV, 73- *Mem:* AAAS; Asn Res Vision & Ophthal; Sigma Xi. *Res:* Vision physiology; ophthalmic genetics. *Mailing Add:* Dept Ophthal Med Ctr NY Univ 560 First Ave New York NY 10016

SIEGEL, IVENS AARON, b Bay Shore, NY, Jan 28, 32; m 59; c 3. PHARMACOLOGY, ORAL BIOLOGY. *Educ:* Columbus Univ, BS, 53; Univ Kans, MS, 58; Univ Cincinnati, PhD(pharmacol), 62. *Prof Exp:* From instr to asst prof pharmacol, State Univ NY Buffalo, 62-68; from assoc prof to prof pharmacol & oral biol, Univ Wash, 68-79, chmn, Dept Oral Biol, 76-79; PROF & CHMN PHARMACOL, UNIV ILL-URBANA, 79- *Mem:* Int Asn Dent Res; Am Soc Pharmacol & Exp Therapeut. *Res:* Ion transport, transport in salivary glands; physiology and pharmacology of salivary glands; drug transport across the oral mucosa. *Mailing Add:* 190 Med Sci Bldg 506 S Matthews Ave Urbana IL 61801

SIEGEL, JACK MORTON, b Sioux City, Iowa, June 11, 22; m 46; c 3. BIOCHEMISTRY. *Educ:* Univ Calif, Los Angeles, AB, 44; Wash Univ, PhD(chem), 50. *Prof Exp:* Assoc chemist radiochem, Oak Ridge Nat Lab, 44-46; asst chem, Wash Univ, 46-48, instr, Univ Col, 49-50; asst prof biochem, Sch Med, Univ Ark, 50-55; tech dir, Pharmacia P-L Biochem, Inc, 55-70, vpres, 70-80, sr vpres, 80-87, pres, 87-89; RETIRED. *Mem:* Am Chem Soc; AAAS. *Res:* Uranium fission products; photosynthetic bacteria, nucleotides and coenzymes. *Mailing Add:* 8815 N Rexleigh Dr Milwaukee WI 53217

SIEGEL, JAY PHILIP, b New York, NY, May 18, 52; m 82; c 2. CELLULAR IMMUNOLOGY, CYTOKINES. *Educ:* Calif Inst Technol, BS, 73; Stanford Univ, MD, 77. *Prof Exp:* Resident internal med, Univ Calif, San Francisco, 77-80; fel infectious dis, Sch Med, Stanford Univ, 80-82; sr staff fel immunol, Div Virol, 82-86, sr investr, Ctr Biol Evol & Res, 86-88, LAB CHIEF, LAB IMMUNOL, DIV CYTOKINE BIOL, CTR BIOL EVAL & RES, FOOD & DRUG ADMIN, 88- *Concurrent Pos:* Fel immunol & infectious dis, Palo Alto Med Found, 80-82; surgeon, USPHS, 86-87, sr surgeon, 87-; attend physician, Div Infectious Dis, Nat Naval Med Ctr, Bethesda, 89- *Honors & Awards:* Physician Recognition Award, AMA, 89. *Mem:* Am Asn Immunologists; fel Am Col Physicians; Am Fedn Clin Res; AAAS. *Res:* Regulation of cellular immune responses; cytotoxic responses; roles of cytokines in the activation and differentiation of cytotoxic lymphocytes. *Mailing Add:* NIH Bldg 29A 2B24 HFB 820 Bethesda MD 20892

SIEGEL, JEFFRY A, b New York, NY. NUCLEAR MEDICINE. *Educ:* Univ Cincinnati, BS, 73, MS(chem), 76, MS(radiol physics), 77; Univ Calif, Los Angeles, PhD(med physics), 81. *Prof Exp:* ASST PROF NUCLEAR MED PHYSICS, TEMPLE UNIV HOSP, 81- *Mem:* Soc Nuclear Med; Asn Physicists Med; Inst Elec & Electronics Engrs. *Res:* Quantitative nuclear medicine; nuclear cardiology; image processing of digital data. *Mailing Add:* Dept Radiation Oncol Cooper Hosp Univ Med Ctr One Cooper Plaza Camden NJ 08103

SIEGEL, JOHN H, b Baltimore, Md, Dec 12, 32; m 56; c 3. SURGERY, PHYSIOLOGY. *Educ:* Cornell Univ, BA, 53; Johns Hopkins Univ, MD, 57; Am Bd Surg, dipl, 66. *Prof Exp:* Intern surg, Grace-New Haven Community Hosp, 57-58; Cardiovasc fel, Dept Surg, Yale Univ, 58-59 & Lab Cardiovasc Physiol, Nat Heart Inst, 59-61; dir cardiovasc physiol lab, Dept Surg, Sch Med, Univ Mich, 62-65; instr surg, Albert Einstein Col Med, 65-66, assoc, 66-67, from asst prof to assoc prof, 67-72; prof surg & biophys, State Univ NY, Buffalo, 72-; PROF SURG & DEP DIR, MD INST EMERGENCY MED SERV SYSTS, UNIV MD, 82- *Concurrent Pos:* USPHS trainee acad surg, Dept Surg, Sch Med, Univ Mich, 62-65; Health Res Coun City of New York career scientist award, 66-71; prin investr, Nat Heart Inst grants, 62-65, 66-72 & Nat Inst Gen Med Sci grant, 69-71; dir renal transplantation serv & assoc dir clin res ctr-acute, Albert Einstein Col Med, 67-72; asst vis surgeon, Bronx Munic Hosp, 65-67, from assoc attend surgeon to attend surgeon, 67-72; attend, Hosp, Albert Einstein Col Med, 66-72; chief dept surg, Buffalo Gen Hosp, 72-82. *Mem:* Am Physiol Soc; fel Am Col Surgeons; Am Asn Surg of Trauma; Am Surg Asn; Int Cardiovasc Soc. *Res:* General and vascular surgery; computer science; physiologic evaluation of the critically ill. *Mailing Add:* Md Inst Emergency Med Serv Systs Univ Md 22 S Green St Baltimore MD 21201

SIEGEL, JONATHAN HOWARD, b Chicago, Ill, Mar 2, 51; m 78; c 3. SENSORY NEUROBIOLOGY. *Educ:* Univ Ark, Fayetteville, BS, 73; Wash Univ, St Louis, PhD(physiol & biophys), 78. *Prof Exp:* Res asst, Wash Univ, St Louis, 74-78, res fel, 78-81; ASSOC PROF AUDIOL, NORTHWESTERN UNIV, 81- *Mem:* Acoust Soc Am; Soc Neurosci; Asn Res Otolaryngol; Sigma Xi. *Res:* Synaptic transmission between the sensory receptor cells of the mammalian cochlea and the afferent neurons; active cochlear mechanical processes; receptor and nerve cell physiology; cell biology. *Mailing Add:* Northwestern Univ 2299 Sheridan Rd Evanston IL 60208

SIEGEL, LAWRENCE SHELDON, b Fargo, NDak, Oct 13, 10; m 42; c 1. MEDICINE. *Educ:* Univ NDak, BS, 33; Rush Med Col, MD, 36. *Prof Exp:* From asst prof to prof, 54-70, EMER PROF PEDIAT, SCH MED, LOMA LINDA UNIV, 70-; ASSOC PROF, SCH MED, UNIV SOUTHERN CALIF, 64- *Concurrent Pos:* Sr attend pediatrician, Los Angeles City Hosp, 50-; from assoc attend pediatrician to attend pediatrician, Cedars of Lebanon Hosp, 50-67, sr attend pediatrician, 69- *Mem:* AMA; Am Acad Pediatricians. *Res:* Pediatrics. *Mailing Add:* 6010 Wilshire Blvd Los Angeles CA 90036

SIEGEL, LESTER AARON, b New York, NY, Sept 25, 25. PHYSICS. *Educ:* Mass Inst Technol, SB, 45, PhD(physics), 48. *Prof Exp:* Instr physics, Mass Inst Technol, 48-50; from res physicist to sr res physicist, Am Cyanamid Co, 50- 61, group leader, 61-69, sr res physicist, 69-86; RETIRED. *Mem:* Am Phys Soc; Am Chem Soc; Am Crystallog Asn. *Res:* X-ray diffraction. *Mailing Add:* 44 Strawberry Hill Ave Apt 10E Stamford CT 06902

SIEGEL, LEWIS MELVIN, b Baltimore, Md, Aug 7, 41; m 60; c 3. BIOCHEMISTRY. *Educ:* Johns Hopkins Univ, BA, 61, PhD(biol), 65. *Prof Exp:* Res assoc biol, Brookhaven Nat Lab, 65; res assoc, 66-68, PROF BIOCHEM, SCH MED, DUKE UNIV, 68- *Concurrent Pos:* Res chemist, Vet Admin Hosp, Durham, 68- *Mem:* Am Soc Biol Chemists; Am Chem Soc. *Res:* Mechanisms of electron transport in metalloflavoproteins; multi-electron reductions; sulfur and nitrogen metabolism. *Mailing Add:* Dept of Biochem Duke Univ Sch of Med Durham NC 27706

SIEGEL, MALCOLM RICHARD, b New Haven, Conn, Nov 5, 32; m 62; c 2. PLANT PATHOLOGY, TOXICOLOGY. *Educ:* Univ Conn, BS, 55; Univ Del, MS, 59; Univ Md, PhD(bot), 63. *Prof Exp:* From asst prof to assoc prof, 66-73, PROF PLANT PATH, UNIV KY, 73-, PROF, CTR TOXICOL, 75- *Concurrent Pos:* Assoc ed, Phytopathology, 73-76; assoc ed, Pesticide Biochem & Physiol, 78-87; sabbatical, Inst Org Chem, 75; mem staff, USDA, 79. *Mem:* Am Phytopath Soc; Sigma Xi. *Res:* Action and metabolic fate of fungicides; epidemiology, chemical and biological control of plant pathogens. *Mailing Add:* Dept Plant Path Univ Ky Lexington KY 40506

SIEGEL, MARTHA J, b New York, NY, Nov 5, 39; m 62; c 2. MATHEMATICS. *Educ:* Russell Sage Col, BA, 60; Univ Rochester, MA, 63, PhD(math), 69. *Prof Exp:* Asst prof, Goucher Col, 67-71; assoc prof, 71-77, PROF MATH, TOWSON STATE UNIV, 77- *Concurrent Pos:* Fel, Sch Hyg & Pub Health, Johns Hopkins Univ, 77-78. *Mem:* Math Asn Am; Am Math Soc; Soc Indust & Appl Math. *Res:* Birth and death processes; collegiate mathematics curriculum reform. *Mailing Add:* Dept of Math Towson State Univ Towson MD 21204

SIEGEL, MARVIN I, b Brooklyn, NY, July 11, 46; m 87; c 1. BIOCHEMISTRY. *Educ:* Johns Hopkins Univ, PhD(biochem), 73. *Prof Exp:* Res scientist, Burroughs Corp, 75-82; assoc dir, biol res, Schering Plough Co, 82-85, dir biol res, 85-88, SR DIR BIOL RES, SCHERING PLOUGH CO, 88- *Mem:* Am Soc Pharmacol & Exp Therapeut. *Res:* Allergy; inflamation; immunology. *Mailing Add:* 165 Rummel Rd Milford NJ 08848-9504

SIEGEL, MAURICE L, b New York, NY, Aug 7, 27; m 59. ORGANIC CHEMISTRY. *Educ:* City Col New York, BS, 49; NY Univ, MS, 51, PhD, 58. *Prof Exp:* Dir res, Caryl Richards Co, 57-67; vpres, Faberge, Inc, 67-78, dir res, 67-87, exec vpres, 78-87; CONSULT, 87- *Concurrent Pos:* Adj asst prof, NY Univ, 59-67. *Mem:* AAAS; Soc Cosmetic Chem; Am Chem Soc. *Res:* Hair technology. *Mailing Add:* 15 Sierra Ct Hillsdale NJ 07642

SIEGEL, MELVIN WALTER, b New York, NY, May 26, 41; m 68; c 1. SENSORS, INTELLIGENT SYSTEMS. *Educ:* Cornell Univ, BA, 62; Univ Colo, MS, 67, PhD(physics), 70. *Prof Exp:* Instr physics & math, Achimota Col, Ghana, 62-64; res assoc physics, Joint Inst Lab Astrophysics, Univ Colo, 70; sr scientist, Univ Va, 70-72, lectr, 71-72; asst prof physics, State Univ NY Buffalo, 72-74; physicist, Extranuclear Labs, Inc, 74-82, dir, res & develop, 78-82; SR RES SCIENTIST, ROBOTICS INST, CARNEGIE MELLON UNIV, 82- *Honors & Awards:* IR-100 Awards, 78, 79 & 86. *Mem:* AAAS; Am Phys Soc; Am Soc Mass Spectros; Am Asn Artificial Intelligence; Inst Elec & Electronics Engrs; Soc Photo Optical Instrumentation Engrs. *Res:* Ionization phenomena; mass spectrometry; intelligent sensors and their applications in artificial intelligence based systems control; ion optics; instrumentation, robotics, sensors and artificial intelligence. *Mailing Add:* Robotics Inst Carnegie Mellon Univ DH 3304 Pittsburgh PA 15213

SIEGEL, MICHAEL ELLIOT, b New York, NY, May 13, 42; m 66; c 2. NUCLEAR MEDICINE, RADIOLOGY. *Educ:* Cornell Univ, BA, 64; Chicago Med Sch, MD, 68. *Prof Exp:* NIH fel diag radiol, Temple Univ Med Ctr, 70-71; NIH fel nuclear med, Johns Hopkins Univ, 71-73, asst prof radiol & environ health, 73-76; asst clin prof radiol, George Washington Univ, 75-77; ASSOC PROF RADIOPHARM, SCH PHARM & ASSOC PROF RADIOL & MED, SCH MED, UNIV SOUTHERN CALIF, 76- *Concurrent Pos:* Radiologist, Johns Hopkins Univ, 71-76; consult nuclear med, Ann Arundel Hosp, Annapolis, Md, 74-76; dir, Dept Nuclear Med, Orthop Hosp, Los Angeles, 76-; dir, Dept Nuclear Med & sr attend physician, Los Angeles County-Univ Southern Calif Med Ctr, 76-; dir nuclear med, Kenneth Norris Cancer Hosp & res Ctr, Los Angeles. *Honors & Awards:* Silver Medal, Soc Nuclear Med, 74 & 75. *Mem:* Soc Nuclear Med; Am Col Nuclear Physicians; Radiol Soc NAm; Asn Univ Radiologists; Am Col Nuclear Med. *Res:* Development of new applications of radioisotopes for prognostic and diagnostic evaluation of vascular disease, both cardiac and peripheral; diagnosis and therapy of malignancies. *Mailing Add:* Los Angeles County-Univ Southern Calif Med Ctr 1200 N State St Los Angeles CA 90033

SIEGEL, MICHAEL IAN, b Brooklyn, NY, Nov 24, 42; m 84. PHYSICAL ANTHROPOLOGY, PRIMATOLOGY. *Educ:* Queens Col, NY, BA, 67; City Univ, New York, PhD(phys anthrop), 71. *Prof Exp:* Lectr phys anthrop, Hunter Col, 67-69; instr, Adelphi Univ, 69-71; from asst prof to assoc prof phys anthrop, 71-82, orthod, Col Dent, 75-76, asst prof, 76-78, assoc prof anat & cell biol, 78-80, PROF PHYSICS ANTHROP, UNIV PITTSBURGH, 82- *Concurrent Pos:* Vis scientist, Lab Exp Med & Surg Primates, 70-79; adj lectr prev dent, NY Univ, 70-79; sr res, Cleff Palate Ctr, 75-, NIH grants, 77, 80 & 85. *Mem:* Fel AAAS; Am Asn Anat; Am Asn Phys Anthropologists; Am Soc Mammalogists. *Res:* Experimental morphology and functional anatomy; growth and development; cleft palate models; middle ear disease; 3D computers. *Mailing Add:* Dept of Anthrop Univ of Pittsburgh Pittsburgh PA 15260

SIEGEL, MORRIS, b US, Mar 2, 04; m 33; c 2. PREVENTIVE MEDICINE, EPIDEMIOLOGY. *Educ:* City Col New York, BA, 24; NY Univ, MD, 28; Johns Hopkins Univ, MPH, 39. *Prof Exp:* Intern, Bellevue Hosp, New York, 29 & 31; resident physician, Sea View Hosp, 32-34; res assoc, New York Health Dept, 35-41; assoc, Pub Health Res Inst New York, 41-47; health officer, New York Health Dept, 47-52, chief poliomeylitis div, 49-50; from assoc prof to prof, 52-74, EMER PROF PREV MED, STATE UNIV NY DOWNSTATE MED CTR, 74- *Concurrent Pos:* Vis lectr, Sch Pub Health, Harvard Univ, 59-65. *Mem:* Harvey Soc; Epidemiol Soc; NY Acad Med; Sigma Xi. *Res:* Public health, preventive medicine and epidemiology of acute and chronic diseases. *Mailing Add:* 345 E 69th St New York NY 10021

SIEGEL, NORMAN JOSEPH, b Houston, Tex, Mar 8, 43; m 67; c 2. PEDIATRIC NEPHROLOGY, PATHOPHYSIOLOGY. *Educ:* Tulane Univ, BA, 64; Univ Tex, Galveston, MA & MD, 68. *Prof Exp:* Fel nephrology, Sch Med, Yale Univ, 70-72, asst prof pediat, 72-75, from asst prof to assoc prof pediat & med, 75-82, PROF PEDIAT & MED, SCH MED, YALE UNIV, 82-, VCHMN PEDIAT, 76- *Concurrent Pos:* Chmn, Sub-Bd Pediat Nephrology, Am Bd Pediat, 85-89 & coun Pediat Nephrology & Urol, Nat Kidney Found, 87-; mem, Sci Adv Comt, Nat Kidney Found, 88-; assoc ed, Am J Kidney Dis, 90-; distinguished vis prof, Nat Kidney Found Western Tenn, 90. *Honors & Awards:* Spec Recognition Award, Soc Pediat Res, 85. *Mem:* Am Soc Pediat Nephrol (pres, 88-89); Soc Pediat Res; Am Soc Pediat. *Res:* Recovery of the kidney from injury; pathophysiology, cellular and molecular mechanisms responsible for the restoration and regeneration of renal tubular epithelium following ischemia, toxins and ureteral obstructions. *Mailing Add:* Dept Pediat Yale Univ Sch Med 333 Cedar St New Haven CT 06510

SIEGEL, PAUL BENJAMIN, b Hartford, Conn, Nov 19, 32; m 57; c 3. GENETICS. *Educ:* Univ Conn, BS, 53; Kans State Univ, MS, 54, PhD(genetics), 57. *Prof Exp:* From asst prof to prof, 57-75, UNIV DISTINGUISHED PROF POULTRY SCI, VA POLYTECH INST & STATE UNIV, 75- *Mem:* AAAS; Animal Behav Soc; Am Genetic Asn; Poultry Sci Asn. *Res:* Genetic aspects of behavior; population genetics. *Mailing Add:* Dept Poultry Sci Va Polytech Inst & State Univ Blacksburg VA 24061

SIEGEL, RICHARD C, b New York, NY, Jan 25, 52; m 75; c 2. IMMUNOCHEMISTRY, ANALYTICAL BIOCHEMISTRY. *Educ:* Boston Univ, AB, 74, Tufts Univ, PhD(biochem), 80. *Prof Exp:* Postdoctoral chem & immunol, Univ Calif Los Angeles, 79-81; sr scientist res & develop, Technicon Instruments Corp, 81-83; group leader res & develop, Cytogen Corp, 83-88; ASSOC DIR BIOPHARMACEUT RES & DEVELOP, CENTOCOR CORP, 88- *Mem:* Am Chem Soc; Fedn Am Socs Exp Biol; AAAS. *Res:* Therapeutic monoclonal antibodies; development of manufacturing methods, protein characterization and assay development. *Mailing Add:* Centocor Inc 244 Great Valley Pkwy Malvern PA 19355

SIEGEL, RICHARD W(HITE), b Cambridge, Mass, May 21, 37; m 62; c 2. MATERIALS SCIENCE, METAL PHYSICS. *Educ:* Williams Col, AB, 58; Univ Ill, Urbana, MS, 60, PhD(metall), 65. *Prof Exp:* Res assoc , Dept Mats Sci & Eng, Cornell Univ, 64-66; from asst prof to assoc prof, Dept Mat Sci, State Univ NY, Stony Brook, 66-75; group leader metal physics, Mat Sci Div, Argonne Nat Lab, 74-82, group leader defects metals, Mat Sci & Tech Div, 82-86, res prog mgr, 80-88, RES SCIENTIST, MAT SCI DIV, ARGONNE NAT LAB, 74- *Concurrent Pos:* Guest prof, Max Planck Inst for Metal Res, Stuttgart, Ger, 72-73; adj prof, Dept Mat Sci, State Univ NY Stony Brook, 75-76; vis prof, Mat Eng Dept, Ben-Gurion Univ Negev, Beer-Sheva, Israel, 80-81, Dept Nuclear Physics, Univ Madras, India, 82 & 85; assoc ed, Mat Letters, 81-; Nat Mat Adv Bd Comt on Mat with Submicron-Sized Microstruct, 86-89; Dept Energy Panel on Clusters & Cluster-Assembled Mat, 88; founder, dir & consult, Nanophase Technologies Corp, 89- *Mem:* Am Phys Soc; Metall Soc; Inst Physics; AAAS; Mat Res Soc. *Res:* Properties and interactions of atomic defects in metals and alloys-diffusion; electron microscopy; positron annihilation spectroscopy; synthesis, processing, characterization, and properties of nanophase materials including metals and ceramics. *Mailing Add:* Mat Sci Div Argonne Nat Lab Argonne IL 60439

SIEGEL, RICHARD WEIL, genetics, for more information see previous edition

SIEGEL, ROBERT, b Cleveland, Ohio, July 10, 27; m 51; c 2. MECHANICAL ENGINEERING, HEAT TRANSFER. *Educ:* Case Inst Technol, BS, 50, MS, 51; Mass Inst Technol, ScD(mech eng), 53. *Prof Exp:* Asst mech eng, Case Inst Technol, 50-51; fluid dynamics, Mass Inst Technol, 52-53; res engr, Gen Eng Lab, Gen Elec Co, 53-54, res engr, Knolls Atomic Power Lab, 54-55; RES ENGR, NASA, 55- *Concurrent Pos:* Assoc tech ed, J Heat Transfer, 73-83; assoc tech ed, J Thermophysics & Heat Transfer, 86- *Honors & Awards:* Heat Transfer Mem Award, Am Soc Mech Engrs, 70. *Mem:* Fel Am Soc Mech Engrs; fel Am Inst Aeronaut & Astronaut. *Res:* Heat transfer and fluid dynamics theory; forced and free convection; transient fluid flow and heat convection; thermal radiation exchange, boiling and solidification. *Mailing Add:* Off Chief Scientist NASA Lewis Res Ctr 21000 Brookpark Rd Cleveland OH 44135

SIEGEL, ROBERT TED, b Springfield, Mass, June 10, 28; m 51; c 5. PARTICLE PHYSICS. *Educ:* Carnegie Inst Technol, BS, 48, MS, 50, DSc, 52. *Prof Exp:* Resident physicist, Carnegie Inst Technol, 52-54, from asst prof to assoc prof physics, 54-63; dean grad studies, Col William & Mary, 64-67, dir, Space Radiation Effects Lab, Newport News, Va, 67-78, prof physics, 63-69, W F C FERGISON PROF PHYSICS, COL WILLIAM & MARY, 79- *Mem:* Fel Am Phys Soc; AAAS. *Res:* Elementary particle physics; weak interactions; muon physics. *Mailing Add:* Dept Physics Col William & Mary Williamsburg PA 23185

SIEGEL, SAMUEL, b Lake Mills, Wis, Feb 15, 17; m 48; c 2. ORGANIC CHEMISTRY. *Educ:* Univ Calif, BS, 38; Univ Calif, Los Angeles, MA, 40, PhD(chem), 42. *Prof Exp:* Res assoc, Northwestern Univ, 42-43; res assoc chem warfare agents & insect repellants, Nat Defense Res Comt Proj, Harvard Univ, 43-45; asst prof chem, Ill Inst Technol, 46-51; assoc prof, 51-57, chmn dept, 57-63, prof chem, 57-86, univ prof, 86-87, EMER PROF, UNIV ARK, FAYETTEVILLE, 87- *Concurrent Pos:* Consult, Universal Oil Prod Co, 56-62; Am Chem Soc-Petrol Res Fund Int fac award, 63-64; vis prof, Queen's Univ, Belfast, 63-64, Res Inst Catalysis, Hokkaido, Japan, 78, Northwestern Univ, 84; sr fel, Japan Soc Promotion of Sci, 77; exchange fel, Inst Isotopes, Hungarian Acad Sci, 83. *Mem:* AAAS; Am Chem Soc; Royal Soc Chem; Catalysis Soc; NY Acad Sci. *Res:* Stereochemistry; quantitative structure-reactivity relationships; mechanism of heterogeneous and homogeneous catalytic hydrogenation. *Mailing Add:* Dept Chem & Biochem Univ Ark Fayetteville AR 72701

SIEGEL, SANFORD MARVIN, b Kansas City, Mo, Sept 3, 28; m 50; c 4. ENVIRONMENTAL CHEMISTRY, BIOGEOCHEMISTRY. *Educ:* Univ Chicago, SM, 50, PhD(biol), 53. *Prof Exp:* Asst bot, Univ Chicago, 49-50; biochemist, Chem Corps Biol Lab, US Dept Army, Md, 50-52; Am Cancer Soc res fel, Calif Inst Technol, 53-54; asst prof biol & instr phys sci, Univ Tampa, 54-55; asst prof biol, Univ Rochester, 55-58; group leader phys biochem, Union Carbide Res Inst, 58-67; PROF BOT, UNIV HAWAII, 67-, CHMN DEPT, 80- *Concurrent Pos:* Guggenheim fel, 57-58; NASA grants, 63-67, 67-76, 67-74; mem, Environ Biol Adv Panel, NASA, 68-71; Res Corp Cottrell res grant toxic volcanic emissions, Iceland & Hawaii, 72-73, Nat Geog Soc grant, 72; sr fel, Weizmann Inst Sci, 74; pub responsibility officer, Am Inst Biol Sci, State Hawaii, 72-76; mem staff, Hawaii Natural Energy Inst, 78-87; environ coordr, Hawaii Natural Energy Inst, 85-87; guest scientist, Academia Sinica, 84-85; vis prof, fac Agr, Hebrew Univ,Rehout,Israel, 86-87; vis scientist, Biophys Inst, Res Coun Italy. *Mem:* AAAS; Am Chem Soc; Scand Soc Plant Physiologists; Phytochem Soc; NY Acad Sci. *Res:* Planetary biology and biogeochemistry of mercury and other heavy elements; physiology of stress; biology of extreme environments; exobiology; salinity and salt tolerance; membrane physiology; planetary biology; environmental impacts of energy development; author of over 260 publications. *Mailing Add:* Dept Bot Univ Hawaii Honolulu HI 96822

SIEGEL, SEYMOUR, b New York, NY, Oct 19, 32; m 55; c 2. CHEMICAL PHYSICS, ACADEMIC ADMINISTRATION. *Educ:* Brooklyn Col, BS, 54; Harvard Univ, MA, 56, PhD(phys chem), 59. *Prof Exp:* Mem sr staff, Appl Physics Lab, Johns Hopkins Univ, 58-59; chem physicist, Aerojet-Gen Corp, 59-61; head chem physics dept, Mat Sci Lab, Aerospace Corp, 61-73, dir, Chem & Phys Lab, 73-80; dir explor res, Occidental Res Corp, 81-83; assoc vchancellor res progs, Univ Calif, Los Angeles, 84-89; PROF TECHNOL MGT, SCH BUS & MGT, PEPPERDINE UNIV, 89- *Concurrent Pos:* Adj prof chem, Univ Calif, Los Angeles. *Mem:* Am Phys Soc; Am Chem Soc; AAAS. *Res:* Free radical and excited molecule reactions; space environmental effects on materials; forensic material analysis; thin film chemistry and physics; photochemistry; surface kinetics and catalysis; technology transfer; organizational management; technology management. *Mailing Add:* 11432 Bolas St Los Angeles CA 90049

SIEGEL, SHELDON, b New York, NY, May 10, 32; m 54; c 3. PHARMACEUTICAL CHEMISTRY. *Educ:* Long Island Univ, BS, 53; Columbia Univ, MS, 57; Fairleigh Dickinson Univ, MBA, 69. *Prof Exp:* Teaching asst, Columbia Univ, 55-57; sr chemist, Merck & Co, 57-63; mgr res planning, 63-69, prod mgr mkt, 69-70, dir, Am Chicle Develop, 70-72, vpres res & develop, Am Chicle, 72-79, vpres, Res & Develop Consumer Prod, 79-81, VPRES WORLDWIDE TECHNOL, CONSUMER PROD, WARNER-LAMBERT CO, 81- *Concurrent Pos:* Mem bd dir, Nat Confectioners Asn US; mem, Exec Comt, Nat Asn Chewing Gum Mfg. *Mem:* Nat Asn Chewing Gum Mfrs; Am Pharmaceut Asn; Am Chem Soc; Int Food Technologists. *Res:* Areas of confectionery products and proprietary drugs. *Mailing Add:* Leaf Inc Bannockburn 2345 Waukegan Rd Deerfield IL 60015-1592

SIEGEL, SIDNEY, b New York, NY, Jan 10, 12; m 37; c 4. PHYSICS OF SOLIDS, NUCLEAR TECHNOLOGY. *Educ:* Columbia Univ, AB, 32, PhD(physics), 36. *Prof Exp:* Asst physics, Columbia Univ, 33-38; res engr, Res Labs, Westinghouse Elec Corp, 38-44, sect mgr, 44-46, mgr, Physics Dept, Atomic Power Div, 49-50; chief physicist, Oak Ridge Nat Lab, 46-49; assoc dir, Atomic Res Dept, NAm Aviation, Inc, 50-55, tech dir, Atomic Int Div, 55-60, vpres, Atomic Int Div, NAm Rockwell Corp, 60-72; dep assoc dir, Oak Ridge Nat Lab, 72-74; CONSULT, ADVAN ENERGY SYSTS, 74- *Concurrent Pos:* Lectr, Univ Pittsburgh, 38-40; res assoc, Calif Inst Technol, 51. *Mem:* Fel Am Nuclear Soc (vpres, 65, pres, 66); fel Am Phys Soc; Sigma Xi. *Res:* Solid state physics; ferromagnetism; radiation effects; nuclear reactor development; energy economics. *Mailing Add:* 722 Jacon Way Pacific Palisades CA 90272

SIEGEL, STANLEY, b Waterloo, Iowa, Feb 5, 15; m 40. CRYSTALLOGRAPHY. *Educ:* Univ Chicago, BS, 36, MS, 38, PhD(physics), 41. *Prof Exp:* Asst, Univ Chicago, 39-40 & Duke Univ, 40-41; assoc physicist, Nat Bur Standards, 41-44; physicist, Armour Res Found, 44-50; PHYSICIST, ARGONNE NAT LAB, 50- *Mem:* Am Phys Soc; Am Crystallog Asn. *Res:* X-ray diffraction; crystallography; crystal physics. *Mailing Add:* 442 Bunning Dr Downers Grove IL 60516

SIEGEL, WILLIAM CARL, b Eau Claire, Wis, Sept 11, 32; m 62; c 5. RESOURCE ECONOMICS, RESOURCE LAW. *Educ:* Mich State Univ, BS, 54, MS, 57; Loyola Univ, La, LLB, 65, JD, 68. *Prof Exp:* Timber mgt asst, Sam Houston Nat Forest, US Forest Serv, 58, asst economist, Southern Forest Exp Sta, 58-60, assoc economist, 60-66, economist, 66-68, prin economist, 68-77, PROJ LEADER & CHIEF ECONOMIST, SOUTHERN FOREST EXP STA, US FOREST SERV, 77- *Concurrent Pos:* Univ teaching, 75- *Honors & Awards:* Super Serv Award, USDA, 85. *Mem:* Soc Am Foresters; Nat Tax Asn; Am Bar Asn; Forest Prod Res Soc. *Res:* Forestry economics, especially forest taxation, insurance, and credit with emphasis on the relationship between law and economics. *Mailing Add:* Southern Forest Exp Sta USFS 10210 Fed Bldg 701 Loyola Ave New Orleans LA 70113

SIEGELMAN, HAROLD WILLIAM, b Los Angeles, Calif, Feb 1, 20; m 47; c 2. PLANT BIOCHEMISTRY. *Educ:* Univ Calif, BS, 42; Univ Calif, Los Angeles, MS, 47, PhD, 51. *Prof Exp:* Horticulturist, Bur Plant Indust, Soils & Agr Eng, Plant Indust Sta, USDA, Wash, 51-53, Hort Crops Res Br, Agr Res Serv, 53-57, plant physiologist, Crops Res Div, 57-65; plant biochemist, 65-69, chmn dept biol, 69-74, SR PLANT BIOCHEMIST, BROOKHAVEN NAT LAB, 74- *Concurrent Pos:* Instr exp marine bot, Marine Biol Lab, Woods Hole, Mass, 68, instr-in-chg, 69-70. *Mem:* Am Soc Plant Physiologists; Am Chem Soc; Am Soc Biol Chemists; Biochem Soc; Am Acad Arts & Sci. *Res:* Effect of light on biochemistry of plants; algal viruses. *Mailing Add:* Dept Biol Bldg 463 Brookhaven Nat Lab Upton NY 11973

SIEGER, JOHN S(YLVESTER), b Pittsburgh, Pa, Oct 3, 25; m 51; c 4. CHEMICAL ENGINEERING. *Educ:* Carnegie Inst Technol, BS, 45; Columbia Univ, MS, 47, PhD(chem eng), 50. *Prof Exp:* Res engr, Pittsburgh Coke & Chem Co, 45-47; fel, Mellon Inst, 50-56; group leader, Pittsburgh Plate Glass Co, 56-59, head surfacing res dept, 59-64, sr res assoc, Glass Res Labs, PPG Indust Inc, 64-84; RETIRED. *Mem:* Am Chem Soc. *Res:* Research and development on the production and properties of flat glass. *Mailing Add:* 540 Landsdale Pl Pittsburgh PA 15228-1218

SIEGERT, ARNOLD JOHN FREDERICK, b Dresden, Ger, Jan 1, 11; nat US; m 44; c 1. STATISTICAL PHYSICS. *Educ:* Univ Leipzig, PhD(theoret physics), 34. *Prof Exp:* Lorentz Funds fel, Univ Leiden, Holland, 34-36; asst physics, Stanford Univ, 36-39; physicist, Tex Co, 39-42, Nat Geophys Co, 42, Stanolind Oil & Gas Co, Okla, 42 & radiation lab, Mass Inst Technol, 42-46; assoc prof physics, Syracuse Univ, 46-47; prof physics, Northwestern Univ, Evanston, 47-79; RETIRED. *Concurrent Pos:* Consult, Pan-Am Petrol Corp, 42-66, Rand Corp, 50-57, Lockheed Aircraft Corp, 58-61 & Argonne Nat Lab, 65-71; Guggenheim fel, Inst Adv Study, 53-54; NSF sr fels, Inst Theoret Physics, Amsterdam, Neth, 62-63 & Weizmann Inst, Israel, 63-64; vis prof, Univ Utrecht, Neth, 71-72 & 79-80, Kramers prof, 80. *Mem:* Fel Am Phys Soc. *Res:* Theoretical physics; quantum theory; random processes; exploration geophysics; gravity and magnetic methods; statistical mechanics. *Mailing Add:* 2347 Lake Ave Wilmette IL 60091

SIEGESMUND, KENNETH A, b Milwaukee, Wis, Nov 28, 32; m 59; c 4. ELECTRON MICROSCOPY, FORENSIC PATHOLOGY. *Educ:* Univ Wis, BS, 55, PhD(bot), 60. *Prof Exp:* Res assoc bot, 60-62, ASSOC PROF ANAT, MED COL WIS, 62- *Concurrent Pos:* Consult, Vet Admin Hosp, 62-78, Milwaukee County Hosp, 62-65 & Trinity Hosp, 78-85. *Mem:* Am Asn Anat; Electron Micros Soc Am; Neuroelec Soc; Am Acad Forensic Sci; AAAS. *Res:* Ultrastructure of the nervous system; applications of electron microscopy in diagnostic pathology, diagnosis and pathogenesis of occupational lung disease; forensic sci; immunodiagnostics. *Mailing Add:* Dept Anat & Cell Biol Med Col Wis 8701 Watertown Plank Rd PO Box 26509 Milwaukee WI 53226

SIEGFRIED, CLIFFORD ANTON, b Bismarck, ND, Oct 19, 47; m 70; c 3. AQUATIC ECOLOGY, ACIDIC DEPOSITION EFFECTS. *Educ:* Univ Calif, BS, 69, PhD(ecol), 74. *Prof Exp:* Post grad res scientist aquatic ecol, Univ Calif, Davis, 74-79; ASSOC SCIENTIST ENVIRON BIOL, BIOL SURV, NY STATE MUS, 79- *Concurrent Pos:* Instr, Life Sci Div, Yuba Col, 74; consult, Bir Bear Munic Water Dist, 76-79. *Mem:* Am Soc Limnol & Oceanog; NAm Lake Mgt Soc; Int Asn Ecol; AAAS. *Res:* Population dynamics, life history, production, and community structure of planktan and benthos communitios response of lake systems to acidification and effects of nutrient enrichment. *Mailing Add:* Biol Surv NY State Mus Albany NY 12230

SIEGFRIED, JOHN BARTON, b Philadelphia, Pa, Mar 19, 38; m; c 7. VISUAL PHYSIOLOGY, ELECTROPHYSIOLOGY. *Educ:* Univ Rochester, BA, 60; Brown Univ, MS, 62, PhD(physiol psychol), 67. *Prof Exp:* From asst prof to assoc prof psychol, Univ Houston, 66-73; PROF PHYSIOL OPTICS, PA COL OPTOM, 73- *Concurrent Pos:* Asst prof, dept neural sci, Grad Sch Biomed Sci, Univ Tex, Houston, 67-70; consult, US Army Human Eng Labs, Aberdeen Proving Ground, 67-; asst prof, dept ophthal, Baylor Col Med, 68-70; lectr, dept psychol, Univ Tex, Austin, 73; mem vision comt, Nat Acad Sci-Nat Res Coun, 73-85; adj assoc prof psychol, Lehigh Univ, 73-80; NSF res grant, 73, Nat Eye Inst, 80-88, & NIH res develop grant, 80; consult, Retina Dept, Wills Eye Hosp, Philadelphia, 76-80 & Ophthal Dept, St Christopher's Hosp for Children, Philadelphia, 78-86. *Mem:* Asn Res Vision & Ophthal; Int Soc Clin Electrophysiol Vision. *Res:* Processing of information by the human visual cortex as measured by the visual evoked cortical potential; the relation between electrophysiological measures and perception; color vision; electro-diagnostic procedures. *Mailing Add:* Dept Basic Sci Pa Col Optom 1200 W Godfrey Ave Philadelphia PA 19141

SIEGFRIED, ROBERT, b Columbus, Ohio, Jan 18, 21; m 46; c 4. HISTORY OF SCIENCE, CHEMISTRY. *Educ:* Marietta Col, BA, 42; Univ Wis, PhD(chem, hist sci), 52. *Prof Exp:* Asst prof sci, Col Gen Educ, Boston Univ, 52-54; asst prof chem, Univ Ark, 54-58; from asst prof to assoc prof div gen studies, Univ Ill, 58-63; assoc prof, 63-65, chmn dept, 64-73, PROF HIST SCI, UNIV WIS-MADISON, 65- *Concurrent Pos:* NSF fel, Royal Inst London, 67-68. *Mem:* AAAS; Hist Sci Soc. *Res:* History of chemistry, and the structure of matter, 1700-1850; chemical revolution. *Mailing Add:* 2206 W Lawn Ave Madison WI 53711

SIEGFRIED, ROBERT WAYNE, II, b Elmhurst, Ill, July 1, 50; m 78; c 2. ULTRASONICS, MATERIAL PROPERTIES. *Educ:* Calif Inst Technol, BS, 72; Mass Inst Technol, PhD(geophys), 77. *Prof Exp:* Sr scientist geophys, Corning Glass Works, 77-80; RES GEOPHYSICIST, ATLANTIC RICHFIELD CO, 80- *Mem:* Am Geophys Union; Soc Explor Geophysicist; Soc Petrol Engrs; Soc Prof Well Log Analysts; Soc Core Analysts; Sigma Xi. *Res:* Vapor phase deposition processes; effect of microstructure on the physical properties of materials; well log interpretation. *Mailing Add:* Six Creekwood Circle Richardson TX 75080

SIEGFRIED, WILLIAM, b Philadelphia, Pa, July 4, 25; m 49; c 2. ORGANIC CHEMISTRY. *Educ:* Bucknell Univ, BS, 50. *Prof Exp:* Chemist, Ohio Apex Div, FMC Corp, 50-52; chief chemist, Kindt-Collins Co, 52-62; dir res, Freeman Mfg Co, 62-65, Munray Prod Div, Fanner Mfg, 65-68 & Victrylite Candle Co, 68-70; chief chemist & prod mgr, Freeman Mfg Co, 70-79; managing dir res & develop, Blended Waxes Inc, 79-85; CHIEF CHEMIST, J F MCCAUGHIN CO, 85- *Res:* Developed high-temp sheet wax and the method of manufacturing it for use in tooling; developed instrumental color development methods to accurately color waxes for the manufacture of candles; developed waxes for use in investment casting. *Mailing Add:* 1409 Tanar Dr Valinda CA 91746

SIEGL, WALTER OTTO, b Rochester, NY, Mar 7, 42; m 69; c 1. ENVIRONMENTAL CHEMISTRY, ORGANIC CHEMISTRY. *Educ:* Ohio Wesleyan Univ, BA, 63; Emory Univ, MS, 66; Wayne State Univ, PhD(org chem), 69. *Prof Exp:* Fel chem, Stanford Univ, 69-72; RES SCIENTIST CHEM, FORD MOTOR CO, 72- *Mem:* Am Chem Soc. *Res:* Environmental analytical chemistry; chemistry of corrosion protective coatings; gas chromatography. *Mailing Add:* 11 Brookline Lane Dearborn MI 48120

SIEGLAFF, CHARLES LEWIS, b Waterloo, Iowa, Sept 30, 27; m 50; c 2. POLYMER CHEMISTRY, COLLOID CHEMISTRY. *Educ:* Univ Iowa, BS, 50, PhD(chem), 56; Univ Cincinnati, MS, 53. *Prof Exp:* Res chemist, Corn Prod Refining Co, 53 & Polymer Res Lab, Dow Chem Co, 55-60; res assoc, Diamond Shamrock Corp, 60-73, sr res assoc, 73-76, res fel, 76-84; vpres, Mitech, 84-86; PRES, MYSCI, 86- *Concurrent Pos:* Chmn, Gordon Res Conf Polymers, 74-75. *Mem:* Soc Rheol; Am Chem Soc; Japanese Soc Polymer Sci; Royal Soc Chem; NY Acad Sci. *Res:* Electrochemistry and transport phenomena; physical chemistry of polymer solutions; surface and polymer flow properties; colloid chemistry of latex systems; catalytic surface chemistry; rheology of polymers and dispersions. *Mailing Add:* 9768 Johnny Cake Ridge Mentor OH 44060

SIEGLER, PETER EMERY, b Budapest, Hungary, Feb 3, 24; m 57. INTERNAL MEDICINE, ALLERGY. *Educ:* Eotvos Lorand Univ, Budapest, MD, 51. *Prof Exp:* Instr path, Med Univ Budapest, 48-51, from instr to asst prof med, 51-56; instr, 59-61, assoc, 61-63, assoc prof, 63-68, dir clin pharmacol, 66-68, ASST PROF MED, HAHNEMANN MED COL & HOSP,. *Mem:* Am Soc Pharmacol & Exp Therapeut; Am Soc Clin Pharmacol & Therapeut; Am Fedn Clin Res; Am Acad Allergy; Am Col Allergists. *Res:* Clinical pharmacology; allergy. *Mailing Add:* 3837 Red Lion Rd Philadelphia PA 19114

SIEGLER, RICHARD E, BLOOD DISEASES, LEUKEMIA. *Educ:* Univ Chicago, MD, 55. *Prof Exp:* PROF PATH, CHARLES R DREW MED SCH, UNIV CALIF, LOS ANGELES, 78- *Mailing Add:* Dept Path Charles R Drew Med Sch Univ Calif 12020 S Wilmington Blvd Los Angeles CA 90059

SIEGMAN, A(NTHONY) E(DWARD), b Detroit, Mich, Nov 23, 31; m 56, 74; c 3. LASERS, ELECTRO-OPTICS. *Educ:* Harvard Univ, AB, 52; Univ Calif, Los Angeles, MS, 54; Stanford Univ, PhD(elec eng), 57. *Prof Exp:* Mem tech staff, Hughes Aircraft Co, 52-54; asst, Electronics Lab, 54-56, from asst prof to assoc prof elec eng, 56-86, dir, Edward L Ginzton Lab, 78-83, BURTON J & ANN M MCMURTRY PROF ENG, STANFORD UNIV, 86- *Concurrent Pos:* Vis prof appl physics, Harvard Univ, 65; consult, GTE Sylvania, United Technol & others; Guggenheim fel, Univ Zurich, 69-70; mem, Air Force Sci Adv Bd; chmn, Int Quantum Electronics Conf, 66, conf chmn, 68; co-dir, Far Eastern Laser Sch, Korea, 83, Winter Sch Lasers & Laser Optics, Taiwan, 84. *Honors & Awards:* W R G Baker Award, Inst Elec & Electronics Engrs, 72, J J Ebers Award, 77; R W Wood Prize, Optical Soc Am, 80, Frederic Ives Medal, 87; Benjamin W Lee Mem Lectr on Physics, Seoul, Korea, 81; Alexander von Humboldt Sr Scientist Award, 84; Quantum Electronics Award, Lasers & Electro-Optics Soc, Inst Elec & Electronic Engrs, 89. *Mem:* Nat Acad Sci; Nat Acad Eng; fel Am Phys Soc; fel Inst Elec & Electronics Engrs; fel Optical Soc Am; Am Soc Eng Educ; fel AAAS; Am Asn Univ Prof; Sigma Xi. *Res:* Lasers; quantum electronics; optics. *Mailing Add:* Ginzton Lab Stanford Univ Stanford CA 94305-4085

SIEGMAN, FRED STEPHEN, b Brooklyn, NY, Apr 15, 46. BIOCHEMISTRY. *Educ:* Brooklyn Col, BS, 67; Ind Univ, Bloomington, PhD(biochem), 75. *Prof Exp:* Asst biochemist, Evansville Ctr Med Educ, 73-76; appl specialist, Millipore Corp, 76-80, ultrafiltration prod mgr, 80-84, sales specialist, 84-87; Europ oper mgr, Intellegenetics Corp, 87-88; corp biotechnol, 88-90, SR SALES SPECIALIST, MILLIPORE CORP, 90- *Mem:* Parenteral Drug Asn. *Res:* The study of nucleic acid and protein synthesis and its controls in bacteriophage T4-infected E coli; metabolism of rat liver cells in in vitro cell suspensions, specifically the effect of alcohol on metabolism. *Mailing Add:* 1234 12th St No 6 Santa Monica CA 90401

SIEGMAN, MARION JOYCE, b Brooklyn, NY, Sept 7, 33. PHARMACOLOGY, PHYSIOLOGY. *Educ:* Tulane Univ, BA, 54; State Univ NY, PhD(pharmacol), 66. *Prof Exp:* Teaching asst pharmacol, State Univ NY Downstate Med Ctr, 61-66, res assoc, 66-67; from instr to assoc prof, 67-77, PROF PHYSIOL, JEFFERSON MED COL, 77- *Concurrent Pos:* NIH res grants, 68-; mem panel regulatory biol, NSF, 78-79; mem physiol study sect, NIH, 79-83. *Honors & Awards:* Lindback Award; Burlington Northern Found Award, 85. *Mem:* Biophys Soc; Soc Exp Med Biol; Am Physiol Soc; Soc Gen Physiol; Sigma Xi. *Res:* Mechanical properties of smooth muscle; regulation and energetics of contraction; excitation-contraction coupling; cation transport and metabolism of uterine smooth muscle; electron microscopy. *Mailing Add:* 1045 Lombard St Philadelphia PA 19147

SIEGMANN, WILLIAM LEWIS, b Pittsburgh, Pa, Sept 14, 43; m 65; c 7. UNDERWATER ACOUSTICS. *Educ:* Mass Inst Technol, BS, 64, MS, 67, PhD(appl math), 68. *Prof Exp:* Fel & res assoc mech, Johns Hopkins Univ, 68-70; from asst prof to assoc prof, 70-83, PROF MATH SCI, RENSSELAER POLYTECH INST, 83- *Concurrent Pos:* Prin investr, Off Naval Res, 77-, Unisys Corp, 85-87, Inst Naval Oceanog, 87- & NASA, 88-; vis mathematician, Naval Underwater Systs Ctr, 83. *Mem:* Fel Acoust Soc Am; Soc Indust & Appl Math; Am Phys Soc; Am Meteorol Soc; Am Asn Univ Prof. *Res:* Oceanic sound transmission and reception; atmospheric acoust propagation; asymptotic and numerical methods. *Mailing Add:* Dept Math Sci Rensselaer Polytech Inst Troy NY 12180-3590

SIEGMUND, DAVID O, b St Louis, Mo, Nov 15, 41; m 62; c 2. MATHEMATICAL STATISTICS. *Educ:* Southern Methodist Univ, BA, 63; Columbia Univ, PhD(statist), 66. *Prof Exp:* Asst prof statist, Columbia Univ, 66-67 & Stanford Univ, 67-69; assoc prof statist, Columbia Univ, 69-71, prof, 71-76; PROF STATIST, STANFORD UNIV, 76- *Concurrent Pos:* NSF fel, 71-72; Guggenheim fel, 74-75; Humboldt award, 80-81. *Mem:* Inst Math Statist; Berndulli Soc Math Statist & Probability. *Res:* Probability theory. *Mailing Add:* Dept Statist Stanford Univ Stanford CA 94305

SIEGMUND, OTTO HANNS, b Gross Neudorf, Ger, Aug 25, 20; nat US; m 50. VETERINARY MEDICINE. *Educ:* Mich State Col, DVM, 44. *Prof Exp:* Asst animal path, Univ Ill, 45-46; pharmacologist, Sterling-Winthrop Res Inst, 46-49; pathologist, Chem Corps, US Dept Army, 49-50; lectr vet therapeut, Univ Calif, 50-51; dir vet publ, animal sci res, Merck, Sharp & Dohme Res Labs, 51-83; RETIRED. *Mem:* Fel AAAS; Sigma Xi; Am Soc Mammal; Am Vet Med Asn; NY Acad Sci. *Res:* Veterinary pharmaceuticals and literature. *Mailing Add:* Prospect Hill Rd Royalston MA 01368

SIEGMUND, WALTER PAUL, b Bremen, Germany, Aug 26, 25; US citizen; m 50; c 3. FIBER OPTICS. *Educ:* Univ Rochester, NY, BA, 45, PhD, 52. *Prof Exp:* Assoc dir res, Am Optical Corp, Southbridge, Mass, 53-64, mgr, fiber optics dept, 64-66, mgr, fiber optics res & develop, 66-84; dir fiber optics technol, 84-85, DIR ADVAN TECHNOL, REICHERT FIBER OPTICS, DIV OF REICHERT-JUNG INC, SOUTHBRIDGE, MASS, 85- *Honors & Awards:* Karl Fairbanks Award, Soc Photo-Optical Instrumentation Engrs, 70; David Richardson Medal, Optical Soc Am, 77. *Mem:* Fel Optical Soc Am; fel Soc Photo-Optical Instrumentation Engrs. *Res:* Fiber optics including manufacturing processes and products; fiber optic components for image intensifiers; borescopic inspection instruments; medical endoscopes. *Mailing Add:* Cassidy Rd Pomfret Center CT 06259

SIEGRIST, JACOB C, b Oella, Md, June 11, 19; m 40. VETERINARY MEDICINE, DAIRY HUSBANDRY. *Educ:* Univ Md, BS, Cornell Univ, DVM, 50. *Prof Exp:* Pvt pract, Md, 50-51 & NJ 52-53; assoc prof vet path, Univ Md, 51-52; staff vet, Div, Schering Corp, 53-56, vet med dir, 56-62; vet dir, Inst Clin Med, Syntex Corp, 62-65, animal dir, Syntex Agribus, Inc, 65-71, vpres, Far East Opers, 71-75, VPRES INT ANIMAL HEALTH, SYNTEX, USA, INC, 75- *Mem:* Am Vet Med Asn; Indust Vet As (pres, 56); US Animal Health Asn; NY Acad Sci. *Res:* Ma control; reproduction physiology; endocrinology; uses of corticoids. *Mailing Add:* 1513 Country Club Rd Los Altos CA 94022

SIEGWARTH, JAMES DAVID, b Chehalis, Wash, June 22, 34; m 62; c 1. SOLID STATE PHYSICS. *Educ:* Univ Wash, 57, PhD(physics), 66. *Prof Exp:* Reactor physicist, Atomic Energy Div, Phillips Petrol Co, 57-60; res asst physics, Univ Wash, 61-66, sr res assoc ceramic eng, 66-67; PHYSICIST, NAT BUR STANDARDS, 67- *Mem:* Am Phys Soc. *Res:* Behavior of antiferromagnetic materials as a function of temperature using the Mossauer effect. *Mailing Add:* 2-1004 Nat Bur Standards 85 S 35th St Boulder CO 80302

SIEH, DAVID HENRY, b Columbus, Nebr, Aug 21, 47; div. CHROMATOGRAPHIC SEPARATIONS. *Educ:* Unvi Nebr-Lincoln, BA, 69, PhD(chem), 79. *Prof Exp:* Scientist, Frederick Cancer Res Facil, 78-81; res investr, 81-87, sr res investr, E R Squibb & Sons, Inc, 87-88, GROUP LEADER, BRISTOL-MYERS SQUIBB. *Concurrent Pos:* Mem, Anal Div, Am Chem Soc, Org Div, AAPS. *Mem:* Am Chem Soc; Am Asn Pharmaceut Scientists. *Res:* Trialkyltriazenes, including the carcinogenic and tumor inhibiting properties; develop analytical methods, emphasizing high-pressure liquid chromatography, thin-layer chromatography, GC, CZE and chiral chromatography. *Mailing Add:* Bristol-Myers Squibb One Squibb Dr New Brunswick NJ 08903-0191

SIEH, KERRY EDWARD, b Waterloo, Iowa, Oct 11, 50; c 3. GEOLOGY. *Educ:* Univ Calif, Riverside, AB, 72; Stanford Univ, PhD(geol), 77. *Prof Exp:* PROF GEOL, CALIF INST TECHNOL, 77- *Honors & Awards:* E B Burwell Jr Mem Award, Geol Soc Am, 80; Initiatives in Res Award, Nat Acad Sci, 82. *Mem:* Am Geophys Union; Geol Soc Am; Seismol Soc Am. *Res:* Historic and prehistoric behavior of earthquake faults, in particular the San Andreas fault of California using geomorphologic and stratigraphic methods; volcanic stratigraphy and processes. *Mailing Add:* Seismol Lab 252-21 Cal Inst Technol 1201 E Calif Blvd Pasadena CA 91125

SIEHR, DONALD JOSEPH, b Milwaukee, Wis, Nov 13, 28; m 56; c 2. FUNGAL PHYSIOLOGY, NATURAL PRODUCTS. *Educ:* Univ Wis, PhD(biochem), 57. *Prof Exp:* Res biochemist, Abbott Labs, 57-61; assoc prof, 61-69, PROF CHEM, UNIV MO-ROLLA, 69- *Concurrent Pos:* NIH sr fel, 68-69. *Mem:* AAAS; Am Chem Soc; Brit Biochem Soc; Mycol Soc Am; Brit Mycol Soc. *Res:* Differentiation in Basidiomycetes; plant growth substances; chemical transformations in biological systems; microbial chemistry and technology. *Mailing Add:* Dept Chem Univ Mo Rolla MO 65401

SIEKER, HERBERT OTTO, b Maplewood, Mo, Mar 20, 24; m 48; c 1. MEDICINE. *Educ:* Washington Univ, MD, 48; Am Bd Internal Med, dipl. *Prof Exp:* Intern med, 48-49, asst resident, 49-50, res fel, Sch Med, 50-51, sr asst resident, 53-54, instr & Life Ins Med Res Fund fel, 54-55, assoc, 55-56, from asst prof to assoc prof, 56-60, CHIEF PULMONARY-ALLERGY DIV, 65-79 PROF MED, SCHMED, DUKE UNIV, 61-,. *Concurrent Pos:* Chief pulmonary dis sect, Vet Admin Hosp, 55-56, asst chief med serv, 56-57, consult, 58, Durham & Fayetteville, 59-; consult, US Army Hosp, Ft Bragg, NC, 59- *Mem:* Asn Am Physicians; Soc Exp Biol & Med; Am Thoracic Soc; Am Soc Clin Invest; Am Fedn Clin Res. *Res:* Cardiorespiratory physiology and disease; allergic and immunologic disorders. *Mailing Add:* Pulmonary-Allergy Div Sch of Med Duke Univ PO Box 3822 Durham NC 27710

SIEKER, LARRY CHARLES, b Great Bend, Kans, Feb 22, 31; div; c 2. PROTEIN CRYSTALLOGRAPHY, TECHNIQUES OF CRYSTALLIZATION. *Educ:* Pac Lutheran Univ, BA, 54; Univ Wash, PhD, 81. *Prof Exp:* Technician biochem, Univ Wash, 59-62 & Univ Calif, San Diego, 62-64; physicist, Univ Wash, 64-71, res assoc biol struct, 71-81, res asst prof, 81-89, RES ASSOC PROF, UNIV WASH, 89- *Concurrent Pos:* Res assoc, Nat Ctr Sci Res, Marseille, France, 78-79; consult, protein crystallog on space flights, 87; vis prof, Dept Fundamental Res, Grenoble, France, 89- *Mem:* Am Crystallog Asn; AAAS. *Res:* Determination of structure and function of biological macromolecules by x-ray crystal analysis. *Mailing Add:* Dept Biol Struct SM-20 Univ Wash Seattle WA 98195

SIEKERT, ROBERT GEORGE, b Milwaukee, Wis, July 23, 24; m 51; c 3. NEUROLOGY. *Educ:* Northwestern Univ, BS, 45, MS, 47, MD, 48. *Prof Exp:* Instr anat, Sch Med, Univ Pa, 48-49; fel neurol, Mayo Grad Sch Med, 50 & 53-54, from instr to assoc prof, 55-73, PROF NEUROL, MAYO MED SCH, MAYO CLIN 69-, CONSULT,. *Concurrent Pos:* Head A sect neurol, Mayo Clin, 66-76, assoc bd govs, 73-80; mem bd trustees, Mayo Found, 74-81; ed-in-chief, Mayo Clin Proc, 82- *Mem:* AAAS; Am Neurol Asn; AMA; fel Am Acad Neurol. *Res:* Cerebrovascular disease; descriptions of transient ischemic episodes and investigation of therapeutic programs. *Mailing Add:* Dept Neurol Mayo Clin W8 200 First Ave SW Rochester MN 55901

SIEKEVITZ, PHILIP, b Philadelphia, Pa, Feb 25, 18; m 49; c 2. BIOCHEMISTRY. *Educ:* Philadelphia Col Pharm & Sci, BS, 42; Univ Calif, PhD(biochem), 49. *Hon Degrees:* DSc, Philadelphia Col Pharm & Sci, 71; PhD, Univ Stockholm, 74. *Prof Exp:* USPHS res fel biochem, Harvard Univ,

49-51; fel oncol, Univ Wis, 51-54; from asst prof to prof, 54-88, EMER PROF CELL BIOL, ROCKEFELLER UNIV, 88- *Mem:* Nat Acad Sci; Am Soc Biol Chemists; Am Soc Cell Biol (pres, 66-67); Am Soc Neurosci; NY Acad Sci (pres, 76). *Res:* Protein synthesis; oxidative phosphorylation; cell biology; neurobiology. *Mailing Add:* Rockefeller Univ York Ave & 66th St New York NY 10021

SIELKEN, ROBERT LEWIS, JR, b Little Rock, Ark, July 10, 44; m 68. MATHEMATICAL STATISTICS, OPERATIONS RESEARCH. *Educ:* DePauw Univ, BA, 66; Fla State Univ, MS, 68, PhD(statist), 71. *Prof Exp:* Asst prof, 71-76, ASSOC PROF STATIST, TEX A&M UNIV, 76- *Mem:* Inst Math Statist; Am Statist Asn; Biomet Soc; Opers Res Soc Am; Math Prog Soc. *Res:* Optimization theory; risk estimation; mathematical programming; stochastic approximation methods. *Mailing Add:* 3833 Texas Ave Bryan TX 77802

SIELOFF, RONALD F, b Detroit, Mich. PROCESS OPTIMIZATION, PROCESS DEVELOPMENT. *Educ:* Wayne State Univ, BS, 76; Ga Tech, PhD(org), 81. *Prof Exp:* Sr scientist process develop, Owens Corning Fiberglas, 82-86; advan process chemist coatings, 86-88, MGR RES & DEVELOP, G E PLASTICS, 89- *Honors & Awards:* Most Outstanding Sr Researcher, Am Chem Soc, 76. *Res:* Phenolic binder development; statistical process optimization; optical coatings development; coatings application development. *Mailing Add:* G E Plastics One Lexan Lane Mt Vernon IN 47620

SIEMANKOWSKI, FRANCIS THEODORE, b Buffalo, NY, Nov 12, 14; m 42; c 2. GEOLOGY, SCIENCE EDUCATION. *Educ:* State Col Teachers, BS, 39; Univ Buffalo, MEd, 50; State Univ NY Buffalo, EdD(sci educ), 70. *Prof Exp:* Teacher pub schs, NY, 39-41 & 45-49, prin & adult educ dir, 49-51, teacher, 51-54; assoc prof phys sci, 64-71, PROF GEOSCI & SCI EDUC, STATE UNIV NY BUFFALO, 71- *Concurrent Pos:* Consult, Teacher Corps, Peace Corps Spec Prog, Afghanistan, 71-74. *Mem:* AAAS; Nat Asn Geol Teachers; Geol Soc Am; Nat Asn Res Sci Teaching; Nat Sci Teachers Asn. *Res:* Individualizing the teaching of science in geology and other physical sciences. *Mailing Add:* 50 Cayuga Creek Rd Cheektowaga NY 14227

SIEMANKOWSKI, RAYMOND FRANCIS, biophysics, biochemistry, for more information see previous edition

SIEMANN, DIETMAR W, b Hanover, WGer, Jan 11, 50; Can citizen; m 72; c 1. CANCER RESEARCH, MEDICAL BIOPHYSICS. *Educ:* Univ Man, BSc, 72; Univ Toronto, MSc, 75, PhD(med biophysics), 77. *Prof Exp:* Res assoc, Univ Rochester Sch Med & Dent, 77-78, sr instr, 78-79, asst prof radiation oncol, 80-85, radiation biol & biophysics, 81-85, ASSOC PROF RADIATION ONCOL, BIOL & BIOPHYSICS, UNIV ROCHESTER SCH MED & DENT 85-, DIR, EXP THERAPEUT DIV, CANCER CTR, 88- *Concurrent Pos:* Prin investr grants, NIH, 82-; mem, Ad Hoc Rev Comt, Nat Cancer Inst, 82-, Rev Comt, Nat Cancer Inst Can, 87-, actg mem, Radiation Sensitizer/Radiation Protector Working Group, 83-; vchmn, Radiation Ther Oncol Group Tumor Biol Comt, 87-; assoc ed, Int J Radiation Oncol Biol Physics, 83-87, bd of ed, 86-89, sr ed, Biol, 89-; assoc dir, Univ Rochester Cancer Ctr, 88-; co-chair, 7th Chem Modifiers of Cancer Treatment Conf, 91; chair, 3rd Int Workshop on Tumor Hypoxia, 90; asst dir, Exp Therapeut Div, Univ Rochester Cancer Ctr, 84-88. *Honors & Awards:* Res Award, Radiation Res Soc, 90. *Mem:* Radiation Res Soc; Am Asn Cancer Res. *Res:* Response to antitumor agents assessed in cell cultures, multicell spheroids and in solid tumors and normal tissue in the laboratory as models for clinical cancer therapy; multi-modality therapies with chemotherapeutic agents, radiation, sensitizers and protectors. *Mailing Add:* Exp Therapeut Div Cancer Ctr Box 704 Univ Rochester 601 E Elmwood Ave Rochester NY 14642

SIEMANN, ROBERT HERMAN, b Englewood, NJ, Dec 3, 42; m 64; c 4. HIGH ENERGY PHYSICS. *Educ:* Brown Univ, ScB, 64; Cornell Univ, PhD(physics), 69. *Prof Exp:* Res assoc, Stanford Linear Accelerator Ctr, 69-72; assoc physicist, Brookhaven Nat Lab, 72-73; from asst prof to assoc prof, 73-84, PROF PHYSICS & DIR OPER, WILSON LAB, CORNELL UNIV, 84- *Mem:* AAAS; Am Physical Soc. *Res:* Accelerator design and construction. *Mailing Add:* Newman Lab Cornell Univ Ithaca NY 14853

SIEMENS, ALBERT JOHN, b Winnipeg, Man, Dec 7, 43; m 65; c 2. PHARMACOLOGY. *Educ:* Univ Man, BSc, 66, MSc, 69; Univ Toronto, PhD(pharmacol), 73. *Prof Exp:* Sr res scientist alcoholism & drug abuse, Res Inst Alcoholism, NY State Dept Ment Hyg, 73-80; asst dir, New Drug Develop, Pfizer, Inc, 80-; DIR, CLIN TRIALS, FAMILY HEALTH INT. *Concurrent Pos:* Adj asst prof biochem pharmacol, Sch Pharm, State Univ NY Buffalo, 74- *Mem:* Can Pharmaceut Asn; Res Soc Alcoholism; NY Acad Sci; Am Soc Pharmacol & Exp Therapeut. *Res:* Drug metabolism; drug interactions; pharmacology of marihuana; biochemical pharmacology of drug tolerance, dependence and addiction; alcoholism. *Mailing Add:* Clin Res Int Inc PO Box 13991 Research Triangle Park NC 27709

SIEMENS, JOHN CORNELIUS, b Shafter, Calif, Feb 22, 34; m 61; c 3. AGRICULTURAL ENGINEERING. *Educ:* Univ Calif, BS, 57; Univ Ill, MS, 58, PhD(soil mech), 63. *Prof Exp:* Instr agr eng, Univ Ill, 58-63; asst prof, Cornell Univ, 63-68; assoc prof, 68-76, PROF AGR ENG, UNIV ILL, URBANA, 76- *Mem:* Am Soc Agr Engrs. *Res:* Power and machinery area of agricultural engineering. *Mailing Add:* Dept Agr Eng Univ Ill 1304 W Pa Ave Urbana IL 61801

SIEMENS, PHILIP JOHN, b Elgin, Ill, Nov 13, 43. NUCLEAR PHYSICS. *Educ:* Mass Inst Technol, BSc, 65; Cornell Univ, PhD(physics), 70. *Prof Exp:* Actg Amanuensis, Niels Bohr Inst, 70-71; NATO fel, Univ Copenhagen, 71-72, univ fel, 72-73, lectr, 74-81; PROF PHYSICS, TEX A&M UNIV, 80- *Concurrent Pos:* Vis assoc prof, Univ Ill, 77; vis scientist, Lawrence Berkeley Lab, Univ Calif, 78-79. *Mem:* Fel Am Phys Soc. *Res:* Theoretical nuclear and many-body physics. *Mailing Add:* Ore State Univ 301 Wenigere Univ Corvallis OR 97331

SIEMER, EUGENE GLEN, b Cincinnati, Ohio, Jan 3, 26; m 49; c 3. AGRONOMY, PLANT MORPHOLOGY. *Educ:* Colo State Univ, BS, 50, MS, 61; Univ Ill, PhD(plant morphol), 64. *Prof Exp:* From instr to asst prof, 54-58, assoc prof, 64-77, PROF AGRON & SUPT MOUNTAIN MEADOW RES CTR, COLO STATE UNIV, 77- *Concurrent Pos:* Asst scientist, Univ Tenn-AEC Agr Res Lab, 68-69. *Mem:* Am Soc Agron; Am Forage & Grassland Coun; Sigma Xi. *Res:* Mountain meadow forage production and utilization; root vascular anatomy; progressive morphological development of maize, small grains, forage grasses and legumes; water management; climatology; information retrieval; radiobotany; plant responses to ground water levels. *Mailing Add:* Colo Agr Exp Sta PO Box 598 Gunnison CO 81230

SIEMERS-BLAY, CHARLES T, b Lodi, Calif, Aug 30, 44; div; c 2. SEDIMENTOLOGY. *Educ:* Ore State Univ, BS, 66; Ind Univ, MA, 68, PhD(geol), 71. *Prof Exp:* Asst prof geol, Ind Univ, NW Campus, 70-71; asst prof geol, Univ NMex, 71-75; sr res geologist, Cities Serv Co, 75-80; prof geol, Univ Wyo, 80-82; pres, Sedimentology Inc, Boulder, Colo, 82-90, chief scientist, RPI, 88-90; INT CONSULT, 90- *Concurrent Pos:* founder & pres, Alaska Res Assoc, Inc, 83-85; vis prof appl sedimentology, Wuhan Col Geol, Wuham & Beijing, People's Repub China, 88; lectr, Inst Technol Bandung, Indonesia, 89- *Honors & Awards:* A I Levorsen Award, Am Asn Petrol Geologists, 78. *Mem:* Soc Econ Paleontologists & Mineralogists; Am Asn Petrol Geologists; Sigma Xi; Int Asn Sedimentologists. *Res:* Sedimentology; stratigraphy; sedimentary petrology; paleoecology and interpretation of depositional paleoenvironments; genesis and hydrocarbon-bearing potential of ancient depositional systems; major work north slope and peninsular areas of Alaska, Rocky Mountain basins, Gulf Coast region, Columbia, Argentina and North Sea basin; character and orign of shorelines, Kauai, Hawaii; stratigraphy, depositional systems and petroleum potential of Indonesian petroliferous basins. *Mailing Add:* 1135 Pearl St Suite 1 Boulder CO 80302

SIEMIATYCKI, JACK, b Innsbruck, Austria, Dec 31, 46; Can citizen; m 82; c 2. PUBLIC HEALTH & EPIDEMIOLOGY, BIOSTATISTICS. *Educ:* McGill Univ, BSc, 67, MSc, 71, PhD(epidemiol), 76. *Prof Exp:* Res dir community health, Pointe St Charles Community Clin, 70-72; postdoctoral fel epidemiol, Int Agency Res Cancer, WHO, 77-78; from asst prof to assoc prof, 78-84, PROF EPIDEMIOL, ARMAND- FRAPPIER INST, UNIV QUE, CAN, 84- *Concurrent Pos:* Adj prof epidemiol & biostatist, McGill Univ, 79-, Sch Occup Health, 81-; consult, Int Joint Comn, Can-US, 82-88, Int Agency Res Cancer & Can Fed Health Dept; assoc ed, Am J Epidemiol, 89-, Int J Environ Health Res, 90- *Mem:* Soc Epidemiol Res; Am Pub Health Asn; Int Epidemiol Asn; Int Soc Environ Epidemiol; Can Pub Health Asn. *Res:* Epidemiologic research on role of environmental agents in causing human cancer; causes of insulin dependent diabetes; studies to advance the methodology of epidemiology and biostatistics. *Mailing Add:* Epidemiol Res Unit Armand-Frappier Inst 531 Blvd des Prairies Laval-des-Rapidos Laval PQ H7V 1B7 Can

SIEMS, NORMAN EDWARD, b St Louis, Mo, Feb 28, 44; m 66; c 2. NUCLEAR SCIENCE, ASTRONOMY. *Educ:* Rensselaer Polytechnic Inst, BS, 66; Johns Hopkins Univ, MS, 70; Cornell Univ, PhD(nuclear sci), 76. *Prof Exp:* Instr math, physics & reactor prin, US Naval Nuclear Power Sch, Bainbridge, Md, 66-70; instr, Quincy Col, 73-75, asst prof physics, 75-80, chmn dept, 77-80; from asst prof to assoc prof, 80-87, chmn dept, 84-87, PROF PHYSICS, JUNIATA COL, 87- *Concurrent Pos:* Instr, Malaysia, Ind Univ, 87-88. *Mem:* Am Asn Physics Teachers; Am Nuclear Sci Teachers Asn. *Res:* Investigation of the low-lying first excited state of Silver-110. *Mailing Add:* Dept Physics Juniata Col Huntingdon PA 16652

SIEMS, PETER LAURENCE, b London, Eng, Jan 26, 32; m 57; c 4. GEOLOGY. *Educ:* Univ London, BSc, 57; Colo Sch Mines, DSc(geol), 67. *Prof Exp:* Geologist, Anglo Am Corp S Africa, 57-61; from asst prof to assoc prof, 65-72, PROF GEOL, UNIV IDAHO, 72- *Concurrent Pos:* Vis prof, Fed Univ Bahia, Brazil, 70-71. *Mem:* Soc Econ Geol; Asn Explor Geochem; Am Inst Mining, Metall & Petrol Eng; Brit Inst Mining & Metall; Geol Soc Am. *Res:* Origin of mineral deposits; geochemical exploration for mineral deposits. *Mailing Add:* Dept of Geol Univ of Idaho Moscow ID 83843

SIEMSEN, JAN KARL, b Duisburg, Ger, May 24, 24; US citizen; m 55; c 4. NUCLEAR & INTERNAL MEDICINE. *Educ:* Univ Duesseldorf, MD, 48, Univ Basel, MD, 50; Am Bd Internal Med, dipl, 63; Am Bd Nuclear Med, dipl, 72. *Prof Exp:* Resident physician, Univ Basel, 50-52; med dir pulmonary dis, Park Sanitorium, Arosa, Switz, 52-54; resident physician, Laurel Heights Hosp, Shelton, Conn, 55-57; intern physician, Virginia Mason Hosp, Seattle, 59-60; resident physician, Vet Admin Hosp, Long Beach, Calif, 60-61, asst chief med & radioisotope serv, 62-65; assoc prof radiol & med, Sch Med, 65-75, assoc prof biomed chem, Sch Pharm, 68-75, PROF RADIOL & MED, SCH MED, UNIV SOUTHERN CALIF, 75- PROF BIOMED CHEM, SCH PHARM, 75-; DIR DEPT NUCLEAR MED, LOS ANGELES COUNTY-UNIV SOUTHERN CALIF MED CTR, 65- *Concurrent Pos:* Consult nuclear med, Vet Admin Hosp Long Beach, US Naval Hosp, Long Beach, Orthopaedic Hosp, Los Angeles & Intercommunity Hosp, Covina, 65- *Mem:* Fel Am Col Physicians; Am Col Nuclear Physicians; Asn Univ Radiologists; European Soc Nuclear Med; Radiol Soc NAm. *Res:* Development and evaluation of radiopharmaceuticals in medicine; in vivo kinetics studies of metabolic processes; modalities of radiotherapy with internally administered radionuclides; radiopharmacokinetics; image processing. *Mailing Add:* Dept of Nuclear Med LAC-USC Med Ctr Los Angeles CA 90033

SIENIEWICZ, DAVID JAMES, b Halifax, NS, Nov 15, 24; m 54; c 3. RADIOLOGY. *Educ:* Univ NB, BA, 45; Dalhousie Univ, MD, CM, 50; Royal Col Physicians & Surgeons Can, cert specialist diag & therapeut radiol, 54; Am Bd Radiol, dipl, 54. *Prof Exp:* Asst radiologist, Montreal Gen Hosp, 55-56, assoc radiologist, 56-58, radiologist-in-chief, 58-68, sr radiologist, 68-71; RADIOLOGIST, ST MICHAEL'S HOSP, TORONTO, 71-; ASSOC PROF, DEPT RADIOL, UNIV TORONTO, 71- *Concurrent Pos:* From asst prof to assoc prof radiol, McGill Univ, 55-71; vis fel, Stockholm, 68-69;

consult staff, Montreal Gen Hosp & St Michael's Hosp, Toronto, 71-; gov & mem bd gov, Toronto Inst Med Technol, 77-; chancellor, Am Col Radiol, 79-85. *Mem:* Fel Am Col Radiol (vpres, 85-86); Am Roentgen Ray Soc; Can Med Asn; Can Asn Radiologists (hon secy-treas, 63-65, pres, 75-76); Asn Univ Radiologists; Soc Thoracic Radiol; Radiol Soc NAm. *Res:* Diagnostic and therapeutic radiology. *Mailing Add:* Dept Radio 30 Bond St Toronto ON M5B 1W8 Can

SIERAKOWSKI, ROBERT L, b Vernon, Conn, Apr 11, 37; m 75; c 2. ENGINEERING MECHANICS, MATERIALS SCIENCE. *Educ:* Brown Univ, BSc, 58; Yale Univ, MS, 60, PhD(eng mech), 64. *Prof Exp:* Sr engr mech, Sikorsky Aircraft Div, United Technol Corp, 58-60; res asst, Yale Univ, 60-63; sr res scientist, United Technol Res Labs, 63-67; prof eng sci, Univ Fla, 67-83; CHMN, CIVIL ENG DEPT, OHIO STATE UNIV; CHMN, CIVIL ENG DEPT, OHIO STATE UNIV 83- *Concurrent Pos:* Adj asst prof eng mech, Rensselaer Polytech Inst, 64-67; Nat Res Coun fel, 72-73. *Mem:* Am Soc Mech Engrs; Am Inst Aeronaut & Astronaut; Soc Exp Stress Anal; Soc Advan Mat & Processing Eng; Am Soc Eng Educ. *Res:* Advanced structural composites; structural vibrations; biomechanics. *Mailing Add:* Dept Civil Eng 402 Hitchcock Rd Oh State Univ 2020 Neil Ave Columbus OH 43210-1275

SIEREN, DAVID JOSEPH, b Ashland, Wis, May 10, 41; m 72; c 1. TAXONOMIC BOTANY. *Educ:* Northland Col, AB, 63; Univ Ill, Urbana, MS, 65, PhD(bot), 70. *Prof Exp:* From asst prof to assoc prof, 69-83, chmn dept, 73-78, PROF BIOL, UNIV NC, WILMINGTON, 83-, DIR HERBARIUM, 69- *Mem:* Sigurd Olson Environ Inst; Am Asn Plant Taxonomists; Nature Conserv. *Res:* Vascular plant floristics. *Mailing Add:* Dept Biol Sci Univ NC Wilmington NC 28403-3297

SIERK, ARNOLD JOHN, b Batavia, NY, Nov 10, 46; m 68; c 3. NUCLEAR PHYSICS. *Educ:* Cornell Univ, BS, 68; Calif Inst Technol, PhD(nuclear physics), 73. *Prof Exp:* Fel theoret nuclear physics, Los Alamos Sci Lab, Univ Calif, 72-74; asst prof nuclear physics, Calif Inst Technol, 74-77; STAFF MEM, THEORET DIV, LOS ALAMOS NAT LAB, 77- *Concurrent Pos:* Vis staff mem, Los Alamos Sci Lab, Univ Calif, 74-77; fel, Alfred P Sloan Found, 75-77. *Mem:* Am Phys Soc; Sigma Xi. *Res:* Models of fission and heavy-ion fusion reactions; low-energy light-particle reactions; heavy-ion accelerators. *Mailing Add:* T-2 MS B243 Los Alamos Nat Lab PO Box 1663 Los Alamos NM 87545

SIERVOGEL, ROGER M, b Phoenix, Ariz, Dec 17, 44; m; c 2. GENETIC EPIDEMIOLOGY, CARDIOVASCULAR DISEASE. *Educ:* Ariz State Univ, BS, 67, MS, 68; Univ Ore, PhD, 71. *Prof Exp:* NIH fel human quant genetics, Sch Pub Health, Univ NC, 71-73, vis asst prof, Dept Biostatist, 73-74; res scientist human genetics, Fels Res Inst, 74-77, Fels asst prof, 77-78, Fels assoc prof, 79-84, FELS PROF COMMUNITY HEALTH, SCH MED, WRIGHT STATE UNIV, 85- *Concurrent Pos:* Prin investr grants, Nat Heart, Lung & Blood Inst, 76-77, 77-79 & 79-82, Nat Inst Child Health Human Dev, 76-94 & 90-95, Am Heart Asn, Miami Valley Chap, 77-78 & 78-79; mem, Epidemiol & Dis Control Study Sect, NIH, 84-88; exec comt, Human Biol Coun, 88-92; mem High Blood Pressure Res Coun & fel Epidemiol Coun, Am Heart Asn. *Mem:* Am Soc Human Genetics; Am Heart Asn; Human Biol Coun (secy-treas, 80-84); Soc Pediat Res. *Res:* Genetics of multifactorial traits, major gene effects, and genetic linkage in humans, especially genetic epidemiology of traits related to hypertension, body composition and growth. *Mailing Add:* Dept Community Health Div Human Biol Wright State Univ Sch of Med 1005 Xenia Ave Yellow Springs OH 45387-1698

SIESHOLTZ, HERBERT WILLIAM, industrial chemistry; deceased, see previous edition for last biography

SIESS, CHESTER P(AUL), b Alexandria, La, July 28, 16; m 41; c 1. STRUCTURAL ENGINEERING. *Educ:* La State Univ, BS, 36; Univ Ill, MS, 39, PhD(struct eng), 48. *Prof Exp:* staff mem, State Hwy Dept, La, 36-37; asst, Univ Ill, 37-39; testing engr, Dept Subways, Chicago, 39-41; engr, NY Cent Rwy, Ill, 41; spec res assoc, 41-45, from res asst prof to res assoc prof, 45-55, prof, 55-78, EMER PROF CIVIL ENG, UNIV ILL, URBANA, 78- *Concurrent Pos:* Mem adv comn reactor safeguards, Nuclear Regulatory Comn, 68-, chmn, 72. *Honors & Awards:* Wason Medal, Am Concrete Inst, 49, Turner Medal, 64; Award, Concrete Reinforcing Steel Inst, 56; Huber Award, Am Soc Civil Engrs, 56, Howard Award, 68, Reese Award, 70. *Mem:* Nat Acad Eng; hon mem Am Soc Civil Engrs; Int Asn Bridge & Struct Engrs; hon mem Am Concrete Inst (pres, 73-74). *Res:* Reinforced and prestressed concrete. *Mailing Add:* 3129 Newmark Lab Univ Ill 205 N Mathews St Urbana IL 61801-2397

SIESSER, WILLIAM GARY, b Parsons, Kans, Apr 25, 40; m 71; c 2. PALEONTOLOGY, STRATIGRAPHY-SEAIMENTATION. *Educ:* Univ Kans, BS, 62; La State Univ, MS, 67; Univ Cape Town, PhD(geol), 71. *Prof Exp:* Res officer marine geol, Univ Cape Town, 67-71, sr res officer marine geol, 71-78; assoc prof, 79-85, chmn, Geol Dept, 82-85, PROF GEOL, VANDERBILT UNIV, 85- *Concurrent Pos:* Vis prof, Cambridge Univ, UK, 74, Univ Melbourne, Australia, 78 & Univ Col London, UK, 85. *Mem:* Sigma Xi; Int Nannoplankton Asn; Soc Sedimentary Geol. *Res:* Biostratigraphy and paleoceanography using calcareous nannofossils. *Mailing Add:* Dept Geol Vanderbilt Univ Nashville TN 37235

SIEVER, LARRY JOSEPH, b Chicago, Ill, Sept 2, 47; m 89; c 1. PSYCHIATRY. *Educ:* Harvard Col, BA, 69; Stanford Med Sch, MD, 75. *Prof Exp:* Staff physician res, NIMH, 78-82; dir, Out-Patient Clin, 82-87, DIR, OUT-PATIENT PSYCHIAT DIV, BRONX VET ADMIN MED CTR, 87-; DIR, OUT-PATIENT PSYCHIAT DIV, MT SINAI SCH MED, 87-, PROF, 90- *Concurrent Pos:* Assoc prof, Mt Sinai Sch Med, 82-90. *Honors & Awards:* A G Bennet Award, Soc Biol Psychiat, 83. *Mem:* AAAS; Soc Biol Psychiat; Am Psychiat Asn; Am Col Neuropharmacol. *Res:* The investigation of neurotransmitter, hormone, and brain structural alterations in affective and personality disorders. *Mailing Add:* Dept Psychiat 116a Bronx Vet Admin Med Ctr Bronx NY 10468

SIEVER, RAYMOND, b Chicago, Ill, Sept 14, 23; m 45; c 2. SEDIMENTARY PETROLOGY, GEOCHEMISTRY. *Educ:* Univ Chicago, BS, 43, MS, 47, PhD(geol), 50. *Hon Degrees:* MA, Harvard Univ, 60. *Prof Exp:* Asst, Ill Geol Surv, 43-44, from asst geologist to geologist, 47-57; NSF sr fel & res assoc, 56-57, from asst prof to assoc prof, 57-65, chmn dept geol sci, 68-71 & 76-81, PROF GEOL, HARVARD UNIV, 65- *Concurrent Pos:* Res assoc, Oceanog Inst, Woods Hole, Mass, 57-70; vis scholar, Scripps Inst Oceanog, 81-82; Guggenheim fel, 81-82; fel, Japan Soc Prom Sci, Univ Tokyo, 81. *Honors & Awards:* Am Asn Petrol Geologists Award, 52; Soc Sedimentary Geol Award, 59, 90 & 91. *Mem:* Fel Am Acad Arts & Sci; fel Geol Soc Am; hon mem Soc Sedimentary Geol; Geochem Soc; Am Geophys Union. *Res:* Stratigraphy and sedimentation; origin of coal; silica in sediments; cementation of sandstones; marine sediments; plate tectonics and sediment formation; evolution of atmosphere and oceans. *Mailing Add:* Hoffman Lab Harvard Univ Cambridge MA 02138

SIEVERS, ALBERT JOHN, III, b Oakland, Calif, June 28, 33; m 59; c 4. SOLID STATE PHYSICS. *Educ:* Univ Calif, Berkeley, AB, 58, PhD(physics), 62. *Prof Exp:* Model maker microwave tubes, Varian Assocs, 51-54; res asst solid state physics, Univ Calif, Berkeley, 59-62; res assoc, 62-63, from instr to assoc prof, 63-71, PROF SOLID STATE PHYSICS, CORNELL UNIV, 71- *Concurrent Pos:* Consult, Lockheed Res & Develop, 66-67, Los Alamos Sci Lab, 68-69 & 78-87, Nat Acad Sci, 75 & Gen Motors Corp, 78-85, mem, Mat Res Coun, Advan Res Projs Agency, 69-; vis prof, Stanford Univ, 70, Univ Calif, Irvine, 71, Univ Canterbury, NZ, 76, Los Alamos Sci Lab, 77, Int Bus Mach, San Jose, 81, Sci Univ Tokyo, 84, Max Planck Inst, Fed Repub Ger, Stuttgart, 85; sr fel, NSF, 70-71; Erskine fel, Univ Canterbury, NZ, 75 & 76; vis scientist, NSF, 81; Humbolt sr scientist, 85. *Honors & Awards:* Frank Iskson Prize, Am Phys Soc, 88. *Mem:* Fel Am Phys Soc; fel Optical Soc Am. *Res:* Infrared spectroscopy of condensed matter in both time and frequency domain; surfaces superconductivity; semi-metals; semiconductors; lattice vibrations; far infrared spectroscopy of biological molecules; impurity modes; interfaces; selective surfaces; high Tc superconductor, pure and applied IR photonics. *Mailing Add:* 518 Clark Hall Dept of Physics Cornell Univ Ithaca NY 14853

SIEVERS, DENNIS MORLINE, b Fremont, Nebr, June 15, 44; m 64; c 1. ENVIRONMENTAL ENGINEERING, AGRICULTURAL ENGINEERING. *Educ:* Univ Nebr, BS, 67; Univ Mo-Columbia, MS, 69, PhD(sanit eng), 71. *Prof Exp:* Res asst agr eng, Univ Mo-Columbia, 67-69, Environ Protection Agency fel, 69-71, asst prof, 72-79, assoc prof agr eng, 79-90, PROF AGR ENG, UNIV MO-COLUMBIA, 90- *Mem:* Am Soc Agr Engrs; Water Pollution Control Fedn; Sigma Xi; Asn Ground Water Scientists & Engrs. *Res:* Water quality; aquatic ecology; waste treatment for agriculture. *Mailing Add:* Dept Agr Eng Univ Mo Columbia MO 65211

SIEVERS, GERALD LESTER, b Winona, Minn, July 15, 40; m 67. MATHEMATICAL STATISTICS. *Educ:* St Mary's Col, Minn, BA, 62; Univ Iowa, MS & PhD(statist), 67. *Prof Exp:* Assoc prof, 67-77, PROF MATH, WESTERN MICH UNIV, 77- *Mem:* Inst Math Statist; Am Statist Asn. *Res:* Nonparametric statistics; linear models. *Mailing Add:* Dept of Math Western Mich Univ Kalamazoo MI 49008

SIEVERS, ROBERT EUGENE, b Anthony, Kans, Mar 28, 35; m 61; c 2. ANALYTICAL CHEMISTRY, INORGANIC CHEMISTRY. *Educ:* Univ Tulsa, BChem, 56; Univ Ill, MS, 58, PhD(inorg chem), 60. *Prof Exp:* Res chemist, Monsanto Chem Co, Mo, 60; res chemist, Aerospace Res Labs, Wright-Patterson AFB, 60-63, group leader anal & inorg chem, 63-69, sr scientist & dir inorg & anal chem, 69-75; PROF CHEM & CHMN DEPT, UNIV COLO, BOULDER, 75- *Concurrent Pos:* Vis prof, Univ Tubingen, 68-69; adj prof, Wright State Univ, 69-75. *Honors & Awards:* Res & Develop Award, US Air Force, 62, Tech Achievement Award, 71. *Mem:* Am Chem Soc; Royal Soc Chem; Soc Environ Geochem & Health. *Res:* Trace analysis, environmental analytical chemistry; chromatography, highly volatile and soluble metal chelates, lanthanide Nuclear Magnetic Resonance shift reagents; inorganic stereochemistry, gas chromatography and mass spectrometry of metal chelates; fuel additives; trace analysis of anions. *Mailing Add:* Univ Colo Cires Campus Box 216 Boulder CO 80303

SIEVERS, SALLY RIEDEL, b Butte, Mont, Dec 23, 41. MATHEMATICAL STATISTICS, COMPUTER SCIENCES. *Educ:* Stanford Univ, BS, 63; Cornell Univ, PhD(math), 72. *Prof Exp:* Res assoc social psychol, Cornell Univ, 68-69; instr math, Ithaca Col, 69-72; res assoc math statist, Cornell Univ, 72-73, lectr math, 72-74; LECTR MATH, WELLS COL, 77- *Concurrent Pos:* Statist consult, Environ Protection Agency & Energy Dept Studies, Tech Empirics Corp, 78-80. *Mem:* Inst Math Statist; Am Statist Asn. *Res:* Theory of Ranking procedures; applied statistics. *Mailing Add:* Dept Math Wells Col Aurora NY 13026

SIEVERT, CARL FRANK, b Blue Island, Ill, Oct 25, 20; m 43; c 3. CHEMISTRY. *Educ:* Capital Univ, BS, 42; Univ Ill, PhD(biochem), 47. *Prof Exp:* Asst chem, Univ Ill, 42-44 & 46-47; from asst to assoc prof, Franklin & Marshall Col, 47-57; prof & head dept, Catawba Col, 57-59; PROF CHEM, CAPITAL UNIV, 59-, CHMN DEPT, 63- *Concurrent Pos:* At Armstrong Cork Co, Pa, 48. *Mem:* AAAS; Am Chem Soc. *Res:* Fatty acid metabolism; omega oxidation of fatty acids. *Mailing Add:* 3981 Winchester Rd Capital Univ Carroll OH 43112

SIEVERT, HERMAN WILLIAM, b Aurora, Ill, Sept 19, 28; m 53; c 2. BIOCHEMISTRY. *Educ:* NCent Col, Ill, BA, 50; Univ Wis, MS, 52, PhD(biochem), 58. *Prof Exp:* Biochemist, 58-70, head dept molecular biol, 70-74, mgr planning & admin, 74-77, mgr res qual assurance, 77-87, CLIN PROJ MGR, PHARM PROD, ABBOTT LABS, 87- *Mem:* AAAS; Am Chem Soc. *Mailing Add:* 233 E Washington Ave Lake Bluff IL 60044

SIEVERT, RICHARD CARL, b Brooklyn, NY, Nov 27, 37; m 59; c 3. PLANT PATHOLOGY. *Educ:* Cornell Univ, BS, 58; Univ Wis, PhD(plant path), 63. *Prof Exp:* Plant pathologist, US Army Biol Labs, 63-65; res plant pathologist, USDA, 65-78. *Concurrent Pos:* Asst prof agr biol, Univ Tenn, 65-74 & plant path, NC State Univ, 74-78. *Mem:* AAAS; Am Phytopath Soc; Am Inst Biol Sci; Sigma Xi. *Res:* Tobacco diseases; virology; nutrition of microorganisms; soil microbiology. *Mailing Add:* 102 Tranquil Circle Oxford NC 27565

SIEW, ERNEST L, b Nanking, Kiangsu, Dec 26, 40; US citizen; m 58; c 2. COMPUTER SCIENCES, ENVIRONMENTAL HEALTH. *Educ:* Columbia Univ, BA, 60; NY Univ, PhD(chem), 69. *Prof Exp:* Instr chem, Friends Acad, Long Island, 65-66; sr scientist, GAF Corp, 69-70; res scientist, Comput Dynetec Corp, 71-72; sr scientist, NY Univ Med Ctr, 72-73; res assoc chem, Univ Ky, 73-74; asst prof chem, Trenton State Col, 74-75; asst prof chem, State Univ NY, New Paltz, 75-76; instr chem, Roc Land Community Col, 76-77; res assoc chem, Adelphi Univ, 78-83; asst prof chem, Wagner Col, 83-84; ASST PROF CHEM, STATE UNIV NY, ALBANY, 84- *Concurrent Pos:* Res assoc chem, Rochester Univ, 74-75; res assoc radiochemistry, Albany Med Col, 85-86; adj assoc prof chem, Hofstra Univ & Nassau Community Col, 79-; Presidential fel, Univ Ky, 73; chmn, Acad-Indust Rel Comt, Am Chem Soc, 88- *Mem:* Am Chem Soc; Sigma Xi. *Res:* Spectroscopic investigation of cataract formation; ionizing radiation such as X-rays; microwaves; drugs such as steroids; metabolic disease; aging (senile cataract); light scanning method; differential scanning calorimetry. *Mailing Add:* Dept Chem State Univ NY Albany NY 12222

SIEWERT, CHARLES EDWARD, b Richmond, Va, Oct 20, 37. NUCLEAR ENGINEERING, ASTROPHYSICS. *Educ:* NC State Univ, BS, 60, MS, 62; Univ Mich, PhD(nuclear eng), 65. *Prof Exp:* Teaching asst physics, NC State Univ, 60-62 & Mid East Tech Univ, Ankara, 64-65; asst prof nuclear eng, 65-69, ASSOC PROF NUCLEAR ENG, NC STATE UNIV, 69- *Mem:* Am Nuclear Soc; Am Phys Soc. *Res:* Neutron transport theory and radiative transfer. *Mailing Add:* Dept of Nuclear Eng NC State Univ PO Box 8205 Raleigh NC 27695-8205

SIEWIOREK, DANIEL PAUL, b Cleveland, Ohio, June 2, 46; m 72; c 2. ELECTRICAL ENGINEERING, COMPUTER SCIENCE. *Educ:* Univ Mich, BSEE, 68; Stanford Univ, MSEE, 69, PhD(elec eng), 72. *Prof Exp:* assoc prof, 72-80, PROF ELEC ENG & COMPUT SCI, CARNEGIE-MELLON UNIV, 80- *Concurrent Pos:* Consult engr, Digital Equip Corp, 72-; assoc ed, Comput Systs, Asn Comput Mach, 72-78; mem bd dirs, Spec Interest Group Comput Archit, 75-79; consult, Naval Res Lab, 75-77, Res Triangle Inst, 78- & United Technol, 78- *Honors & Awards:* Frederick Emmos Terman Award, Am Soc Eng Educ,83; Eckert-Mauchly Award, Asn Comput Mach-Inst Elec & Electronic Engrs, 88. *Mem:* Fel Inst Elec & Electronic Engrs; Asn Comput Mach; Sigma Xi. *Res:* Computer architecture; multiprocessors; reliability; fault tolerant computing; design automation. *Mailing Add:* Sch Comput Sci Carnegie-Mellon Univ Pittsburgh PA 15213-3890

SIFFERMAN, THOMAS RAYMOND, b Chicago, Ill, July 28, 41; m 68; c 3. FLUID MECHANICS, ENVIRONMENTAL. *Educ:* Marquette Univ, BME, 64; Purdue Univ, MSME, 66, PhD(fluid mech), 70. *Prof Exp:* Co-op student, Allis-Chalmers, 61-64; res & teaching asst, Purdue Univ, 68-70; sr res scientist fluid flow, res & develop, Conoco, Inc, 70-82; RES ASSOC, DALLAS RES LAB, MOBIL R&D CORP, 82- *Concurrent Pos:* Vis assoc prof & consult petrol eng, Univ Tulsa, 81-82; activ leader, Explor & Prod Res, Cent Res Lab, Mobil Res & Develop Corp, Princeton, NJ, 87-88. *Mem:* Soc Petrol Engrs; Soc Rheology; Am Soc Mech Engrs; Sigma Xi. *Res:* Drilling fluids, especially hole cleaning, with emphasis on the rheology of non-Newtonian fluids including drilling muds, fracturing fluids, and heavy and waxy crudes; environmental waste management; 8 US patents; 22 publications. *Mailing Add:* Mobil R&D Corp PO Box 819047 Dallas TX 75381-9047

SIFFERT, ROBERT S, b NY, June 16, 18; m 41; c 2. ORTHOPEDIC SURGERY. *Educ:* NY Univ, BA, 39, MD, 43. *Prof Exp:* prof orthop & chmn dept, 66-86, BERNARD J LASKER DISTINGUISHED SERV PROF, MT SINAI SCH MED, 86- *Concurrent Pos:* Dir dept orthop, Mt Sinai Hosp, 60- & City Hosp at Elmhurst, 63-; mem bd, Care-Medico, 72. *Mem:* Am Orthop Asn; Am Acad Orthop Surg; Asn Bone & Joint Surg; Am Col Surgeons; NY Acad Med. *Res:* Clinical and laboratory research in problems relating to growth and deformity; community research. *Mailing Add:* 955 Fifth Ave New York NY 10021

SIFFORD, DEWEY H, b La Grange, Ark, Sept 9, 30; m 58; c 6. ORGANIC CHEMISTRY. *Educ:* Ark State Univ, BS, 52; Univ Okla, PhD(org chem), 62. *Prof Exp:* From asst prof to assoc prof, 61-65, actg chmn div, 68-69, prof chem, 65-83, chmn, div phys sci, 70-83, ACTG DEAN, ARK STATE UNIV, 80-, CHMN, DEPT CHEM, 83- *Mem:* Sigma Xi; Am Chem Soc. *Res:* Terpenes; venoms. *Mailing Add:* Dept Chem Box 429 Ark State Univ State University AR 72467

SIFNEOS, PETER E, b Greece, Oct 22, 20; nat US; div; c 3. MEDICINE. *Educ:* Sorbonne, cert, 40; Harvard Univ, MD, 46. *Prof Exp:* Teaching fel psychiat, Harvard Univ, 52-53, res fel ment health, 53-54, asst psychiat, 54-55, instr, 55-58, assoc, 58-64, asst prof, 64-68; assoc clin prof, 68-71, assoc prof, 71-74, PROF PSYCHIAT, HARVARD MED SCH, 74-; ASSOC DIR PSYCHIAT DEPT, BETH ISRAEL HOSP, 68- *Concurrent Pos:* Asst, Mass Gen Hosp, 53-55, from asst psychiatrist to assoc psychiatrist, 55-64, psychiatrist, 64-, dir inpatient & outpatient psychiat serv, 65-68; vis prof, Med Sch, Univ Oslo, 71-84; ed-in-chief, Psychother & Psychosom, 75-; vpres, Int Fedn Med Psychother, 73-88; vis prof psychiat, McGill Univ, 75-79. *Honors & Awards:* Dikemark H Froshaug Prize, 85. *Mem:* AAAS; AMA; Am Psychiat Asn; Am Psychosom Soc; Royal Soc Med. *Res:* Psychophysiological correlations of psychosomatic disorders; neurophysiological, neurochemical and psychological correlations of emotions; manipulative suicide attempts; short-term dynamic psychotherapy; teaching of psychiatry. *Mailing Add:* Beth Israel Hosp Psychiat Dept Harvard Med Sch 25 Shattuck St Boston MA 02115

SIFNIADES, STYLIANOS, b Piraeus, Greece, Dec 22, 35; US citizen; div; c 2. PHYSICAL CHEMISTRY, ORGANIC CHEMISTRY. *Educ:* Univ Athens, Dipl, 57; Univ BC, MSc, 62, PhD(chem), 65. *Prof Exp:* Sr res chemist, Allied-Signal Inc, 65-67, res group leader, 67-71, res assoc, 71-76, res tech supvr, 76-80, RES FEL RES & TECHNOL, ALLIED-SIGNAL INC, 80- *Mem:* Am Chem Soc. *Res:* Computer simulation of chemical processes; kinetic methods of optical resolution and asymmetric transformation; ion exchange processes; liquid membranes and extraction. *Mailing Add:* Allied-Signal Inc Box 1021R Morristown NJ 07962

SIFONTES, JOSE E, b Arecibo, PR, Oct 17, 26; US citizen; m 52; c 7. PEDIATRICS, PULMONARY. *Educ:* Syracuse Univ, MD, 48; Am Bd Pediat, dipl, 54. *Prof Exp:* Chief pediat, A Ruiz Soler Tuberc Hosp, 52-56; med officer in chg, Tuberc Res Ctr, USPHS, 56-58; assoc prof, Tuberc Res Ctr, USPHS, 58-66; dean, Sch Med, 66-71, chmn dept, 74-77, PROF PEDIAT, SCH MED, UNIV PR, SAN JUAN, 66- *Concurrent Pos:* Spec consult, USPHS, 58-66; consult & vis Prof, CARE proj, Honduras, 66; Nat Adv Coun, CDC 67-70; WHO consult, Med Educ, Surinam Proj, 72; mem adv comt, spec projs grant for med educ, USPHS, 70-71; mem comt, Int Child Health, Am Acad Pediat, 76-80; chmn, Sect Des of Chest, Am Acad Pediat, 64. *Mem:* Am Thoracic Soc; Am Acad Pediat; AMA; Am Pediat Soc; AAAS. *Res:* Pediatric pulmonary diseases tuberculosis; medical education. *Mailing Add:* 1658 Lilas San Juan PR 00927

SIFRE, RAMON ALBERTO, b Vega Alta, PR, May 15, 24; m 47; c 6. INTERNAL MEDICINE, GASTROENTEROLOGY. *Educ:* Univ Louisville, MD, 46; Univ PR, BS, 47; Am Bd Internal Med, dipl, 54 & 77; Am Bd Gastroenterol, dipl, 57, 77. *Prof Exp:* Intern, Grad Hosp, Univ Pa, 47-48, post grad student internal med, Grad Sch Med, 48-49, resident, Grad Hosp, 49-50, resident gastroenterol, 50-51; chief gastroenterol sect, Fitzsimons Army Hosp, 51-53; asst prof, 53-57, ASSOC PROF CLIN MED, MED SCH, UNIV PR, SAN JUAN, 57- *Concurrent Pos:* Asst attend physician, San Juan City Hosp, 54-57, assoc attend, 57-; asst attend physician, Presby Hosp, San Juan, 54-57, consult, 57-; consult, Mimiya Hosp, 54-; attend, Doctors' Hosp, 58-; chief gastroenterol sect, Univ Hosp, 59-62. *Mem:* Fel Am Col Physicians. *Mailing Add:* Cacique 2070 Ocean Park Santurce PR 00911

SIGAFOOS, ROBERT SUMNER, b Akron, Ohio, June 4, 20; m 74; c 3. PLANT ECOLOGY. *Educ:* Ohio State Univ, BS, 42, MSc, 43; Harvard Univ, MA, 48, PhD(biol), 51. *Prof Exp:* Assoc bot, Ohio State Univ, 42-43; botanist, Mil Geol Br, US Geol Surv, 48-57, botanist & hydrologist, Gen Hydrol Br, 57-66, res botanist, Water Resourses Div, 66-80; RETIRED. *Concurrent Pos:* Assoc prof lectr, George Washington Univ, 58-70, prof lectr, 70-76. *Mem:* Fel AAAS; fel Arctic Inst NAm. *Res:* Relationship of drainage basin vegetation to streamflow; effects of flooding upon flood-plain forests; tree growth in natural environments; botanical evidence of alpine glacier history; effects of high natural sulfur and carbon dioxide emissions upon wild plants. *Mailing Add:* 910 McDaniel Ct Herndon VA 22070

SIGAFUS, ROY EDWARD, b Warren, Ill, Nov 15, 20; m 43; c 4. AGRICULTURE. *Educ:* Univ Mass, BS, 48, MS, 49; Cornell Univ, PhD(crops), 51. *Prof Exp:* Teacher elem sch, Ill, 40-42; asst forage crops, Cornell Univ, 48-50; asst agronomist, Univ Ky, 50-52, asst prof crops, 52-53, from assoc prof to prof agron, 53-85; RETIRED. *Concurrent Pos:* Mem, Univ Ky-US Agency Int Develop contract team, Agr Univ Develop, Indonesia, 64-66 & Agr Res Sta Develop, Thailand, 66-69; mem comt forage crop variety eval, 69- *Res:* Forage crop variety evaluation. *Mailing Add:* 396 Bob-O-Link Dr Lexington KY 40503

SIGAI, ANDREW GARY, b Baltimore, Md, Dec 3, 44; m 69; c 1. HIGH TEMPERATURE CHEMISTRY. *Educ:* Rensselaer Polytech Inst, BS, 65, MS, 68, PhD(chem), 70. *Prof Exp:* Mem tech staff vapor phase crystal growth & characterization, RCA Labs, David Sarnoff Res Ctr, RCA Corp, 69-72; scientist photoreceptor technol, Joseph C Wilson Ctr Technol, Xerox Corp, 72-80; STAFF SCIENTIST, GTE LABS, 80- *Honors & Awards:* Leslie H Warner Tech Achievement Award, GTE Corp, 87. *Mem:* Am Chem Soc; Electrochem Soc; Am Vacuum Soc. *Res:* High temperature thermodynamics; vapor-phase crystal growth; xerography and photoreceptor technology; investigation of the evaporation properties of selenium alloys for xerographic applications; electroluminescence and lasers; III-V compound semiconductor technology; phosphor and phosphor coating research and development. *Mailing Add:* GTE Labs 40 Sylvan Rd Waltham MA 02154

SIGAL, NOLAN H, b Rochester, Pa, Dec 3, 49; m 71; c 3. B LYMPHOCYTES, MONOCLONAL ANTIBODIES. *Educ:* Princeton Univ, AB, 71; Univ Pa, PhD(immunol), 76, MD, 77. *Prof Exp:* asst prof, pediat & immunol Hosp Sick Children, Toronto, Ont Can, 80-83; SR DIR IMMUNOL RES LAB, MERCK, SHARP & DOHME, 83- *Mem:* Am Asn Immunologists; Clin Immunol Soc. *Mailing Add:* Dept Immunol Res Merck, Sharp & Dohme Res Lab PO Box 2000 Rahway NJ 07065

SIGAL, RICHARD FREDERICK, b Cleveland, Ohio, Mar 19, 43; m 79; c 1. EXPLORATION GEOPHYSICS, THEORETICAL PHYSICS. *Educ:* Case Inst Technol, BS, 65; Yeshiva Univ, MA, 67, PhD(physics), 71. *Prof Exp:* Adj asst prof physics, Hunter Col City Univ New York, 71-72; fel physics, Univ Alta, 72-74, res assoc, 74-75, vis asst prof physics, 75-76, res assoc, 76-77, instr, Col New Caledonia, 77-78; STAFF RES SCIENTIST, AMOCO PROD CO, 78- *Mem:* Am Phys Soc; Am Geophys Union; AAAS; Soc Exp Geophysicists. *Res:* The use of various electrical and electro-magnetic methods in geophysical exploration; estimation of rock parameters from geophysical methods for application to reservoir description. *Mailing Add:* Amoco Prod Co Box 3385 Tulsa OK 74102

SIGEL, BERNARD, b Wilno, Poland, May 14, 30; US citizen; m 56; c 5. SURGERY. *Educ:* Univ Tex, MD, 53. *Prof Exp:* Staff mem surg, Vet Admin Hosp, Coral Gables, Fla, 59-60; from asst prof to prof surg, Med Col Pa, 60-74; prof surg & dean, Abraham Lincoln Sch Med, Col Med, Univ Ill, 74-78; PROF & DIR SURG RES, MED COL PA, 89- *Concurrent Pos:* Mem coun thrombosis, Am Heart Asn; USPHS res career develop award, 63- *Mem:* Soc Univ Surgeons; Am Gastroenterol Asn; Am Col Surgeons; Am Asn Pathologists. *Mailing Add:* Dept Surg Med Col Pa 3300 Henry Ave Philadelphia PA 19129

SIGEL, CARL WILLIAM, b Skokie, Ill, Apr 10, 42; m 69; c 2. ORGANIC CHEMISTRY. *Educ:* Univ Ill, Urbana, BS, 63; Ind Univ, Bloomington, PhD(org chem), 67. *Prof Exp:* NIH fel, Univ Wis-Madison, 67-69; NIH fel, Univ Va, 69, res assoc cancer, 69-71; assoc head dept, med biochem, 85-86, DIV DIR, PHARMACOKINETICS & DRUG METAB, BURROUGHS-WELLCOME & CO, 86- *Concurrent Pos:* Adj assoc prof, Univ NC. *Mem:* Am Chem Soc; Am Asn Cancer Res. *Res:* Drug metabolism and pharmacokinetics, isolation, structural elucidation and synthesis of biologically active molecules. *Mailing Add:* Burroughs-Wellcome & Co 3030 Cornwallis Rd Rm 1109 Research Triangle Park NC 27709

SIGEL, M(OLA) MICHAEL, b Nieswiez, Poland, June 24, 20; nat US; m 41; c 5. VIROLOGY, IMMUNOLOGY. *Educ:* Univ Tex, AB, 41; Ohio State Univ, PhD(bact), 44; Am Bd Microbiol, dipl. *Prof Exp:* Officer in chg bacteriol, Serol & Virol Sect, US Army Serv Lab, 43-46; assoc virologist, Sch Med, Univ Pa, 46-50, asst prof, 50-53; chief reference diag & res unit, Virol & Rickettsial Sect, Commun Dis Ctr, USPHS, 53-55; assoc prof bact, Sch Med, Univ Miami, 55-58, prof microbiol & oncol, 58-78; prof microbiol & immunol & chmn dept, 78-90, prof ophthal, 84-90, DISTINGUISHED EMER PROF, SCH MED, UNIV SC, 90- *Concurrent Pos:* Vis prof, Univ Ala, 53; dir virus lab, Variety Children's Res Found, 55-60, res dir, Found, 60-70; consult, WHO, 56; vis prof, Univ WI; Res assoc, Lerner Marine Lab, Am Mus Natural Hist, Bimini; consult, Vet Admin Comt Infectious Dis; dir, Virol Lab, Sch Med, Univ Miami, 70-78, adj prof microbiol, 78-84; adj prof, Dept Epidemiol & Biostatist, Sch Pub Health, Univ SC, 80- *Mem:* Fel Am Acad Microbiol; Soc Exp Biol & Med; Am Asn Immunologists; Am Asn Cancer Res; fel NY Acad Sci; Sigma Xi; fel AAAS; Am Soc Microbiol; Am Asn Pathologists; Am Soc Cell Biol. *Res:* Myxoviruses and interferon; psittacosis; tumor virology; basic immunology; tumor immunology; phylogeny of immunity; regulation of immune responses by chemical agents; cytokines and epithelial cells in immunity; modes of action of cytokines. *Mailing Add:* Dept of Microbiol & Immunol Univ of SC Sch of Med Columbia SC 29208

SIGELL, LEONARD, b Portland, Ore, Dec 28, 38. PHARMACOLOGY. *Educ:* Ore State Univ, BS, 61; Univ Ore, PhD(pharmacol), 64. *Prof Exp:* Instr pharmacol, 64-66, instr clin pharmacol & exp med, 66-68, asst prof pharmacol & instr med, 68-72, assoc prof pharmacol, 72-81, DIR DRUG & POISON INFO CTR, COL MED, UNIV CINCINNATI, 72-, PROF PHARMACOL & ASSOC PROF EXP MED, 81- *Concurrent Pos:* Dir drug info, Cincinnati Gen Hosp, 66-72. *Mem:* Drug Info Asn; Am Asn Poison Control Ctrs. *Res:* Information science; cardiovascular and behavioral pharmacology; clinical pharmacology-drug epidemiology. *Mailing Add:* Drug & Poison Info Ctr 7701 Bridge Col Med Univ Cincinnati Cincinnati OH 45267

SIGG, ERNEST BEAT, physiology, for more information see previous edition

SIGGIA, ERIC DEAN, b Easton, Pa, Nov 24, 49. STATISTICAL MECHANICS, FLUID DYNAMICS. *Educ:* Harvard Univ, AB & AM, 71, PhD(physics), 72. *Prof Exp:* Asst prof physics, Univ Pa, 75-77; from asst prof to assoc prof, 77-85, PROF PHYSICS, CORNELL UNIV, 85- *Concurrent Pos:* Soc Fel jr fel, Harvard Univ, 71-75; Sloan Found grant, 81-83; Guggenheim Fel, 88-89. *Mem:* Am Phys Soc. *Res:* Theory of fluid turbulence and dynamics of phase transitions. *Mailing Add:* Dept Physics 109 Clark Hall Cornell Univ Ithaca NY 14853

SIGGIA, SIDNEY, b New York, NY, June 22, 20; m 44; c 2. ANALYTICAL CHEMISTRY. *Educ:* Queens Col, NY, BS, 42; Polytech Inst Brooklyn, MS, 43, PhD(org chem), 44. *Prof Exp:* From res analyst to mgr, Anal Dept, Gen Aniline & Film Corp, Pa, 44-58; dir anal res & serv, Res Lab, Olin Mathieson Chem Corp, 58-66; PROF CHEM, UNIV MASS, AMHERST, 66- *Honors & Awards:* Anachem Soc Award, 69; Fisher Award, Am Chem Soc, 75. *Mem:* AAAS; Sigma Xi; Am Chem Soc. *Res:* Organic and functional group analysis; analysis of mixtures, surface active agents and polymers; analytical separation; establishment of specifications; chemical kinetics; management of analytical chemical facilities. *Mailing Add:* 20 Courtland Dr Amherst MA 01002

SIGGINS, GEORGE ROBERT, b Miami, Okla, Dec 29, 37; m 78; c 1. NEUROSCIENCES, NEUROPHARMACOLOGY. *Educ:* Harvard Univ, AB, 60; Boston Univ, MA, 63, PhD(biol & physiol), 67. *Prof Exp:* Fel vascular physiol, Boston Univ, 67-68; fel pharmacol, NIH-NIMH, 68-70; res scientist, Lab Neuropharmacol, St Elizabeth's Hosp, 70-72, sect chief, 72-76, actg lab chief, 75-76; assoc dir neurobiol, A V Davis Ctr, Salk Inst, 76-84; MEM, SCRIPPS CLIN & RES FOUND, 84- *Honors & Awards:* A E Bennett Award, Soc Biol Psychiat, 71; A Cressy Morrison Award, NY Acad Sci, 71; Sr US Scientist Award, Alexander von Humboldt Found, 78; Bissendorf Lect Award, 88. *Mem:* Am Soc Pharmacol & Exp Therapy; Soc Neurosci. *Res:* Neurophysiology; psychopharmacology; autonomic physiology; microcirculation; neuronal cytochemistry; tissue cultures. *Mailing Add:* Dept Neuropharmacology, Res Inst Scripps Clin 10666 N Torrey Pines Rd La Jolla CA 92037

SIGGINS, JAMES ERNEST, b Salt Lake City, Utah, Oct 14, 28; m 58; c 2. MEDICINAL CHEMISTRY. *Educ:* Amherst Col, BA, 52; Univ Chicago, MS, 57, PhD(chem), 59. *Prof Exp:* Chemist, Res Ctr, Lever Bros Co, 52-54; res assoc synthetic org chem, Sterling-Winthrop Res Inst, 59-74; CHEMIST, NY STATE HEALTH LABS, 79- *Mem:* Am Chem Soc. *Res:* Heterocyclics; synthetic organic medicinals; radiopaque contrast agents. *Mailing Add:* 107 Van Dyke Pl Apt 10 Guilderland NY 12084-9699

SIGILLITO, VINCENT GEORGE, b Washington, DC, Feb 20, 37; m 58; c 2. APPLIED MATHEMATICS. *Educ:* Univ Md, College Park, BS, 58, MA, 62, PhD(math), 65. *Prof Exp:* Assoc chemist, Appl Physics Lab, Johns Hopkins Univ, 58-62, assoc mathematician, 62-65, sr staff mathematician, 65-76, PRIN STAFF MATHEMATICIAN, APPL PHYSICS LAB, JOHNS HOPKINS UNIV, 76-, GROUP SUPVR, INFO & SCI INFIRMARY GROUP, 81- *Concurrent Pos:* Instr eve col grad prog, NIH, 65-68; lectr, Johns Hopkins Univ, 66-; Parsons vis prof, Dept Math Sci, Johns Hopkins Univ, 78-79. *Res:* Neural networks. *Mailing Add:* Appl Physics Lab Johns Hopkins Univ Johns Hopkins Rd Laurel MD 20723

SIGLER, JOHN WILLIAM, b Ames, Iowa, Dec, 20, 46; m 76; c 3. ECOLOGICAL & ENVIRONMENTAL ASSESSMENTS. *Educ:* Utah State Univ, BS, 69, MS, 72; Univ Idaho, PhD(fisheries mgt), 80. *Prof Exp:* Res asst, Utah State Univ, 69-72; proj scientist environ protection team, Armament Lab, 72-73, Weapons Lab, US Air Force, 73-75; NSF res asst, Res Appl Nat Needs, Univ NM, 74-76, Univ Idaho, 76-81; consult biologist, W F Sigler & Assoc Inc, 81-86; MGR ENVIRON SCI GROUP SPECTRUM SCI & SOFTWARE, INC, 86- *Concurrent Pos:* Proj coordr & prin investr fisheries, Interstate Commerce Comn, Washington, DC, 84; proj mgr, Nat Res Surv, Energy Fuels Nuclear Inc, Ariz Strip, 85-86, Land Tenure Adjustment Environ Impact Statement, Barstow Resource Area, Calif, 86- *Mem:* Am Fisheries Soc; Ecol Soc Am; Am Inst Fisheries Res Biologists; Pac Fisheries Biologists; Sigma Xi. *Res:* Effects of chronic exposure of anadromous salmonids to turbidity, and its effect on growth and social behavior; toxic effects of unique Air Force wastes on aquatic life; utilization of adenosine triphosphate as an indicator of ecosystem condition in a large reservoir; production-biomass relationships in a large reservoir; recreation fisheries, management theory and applications; fishes of the Great Basin; reduction of feeding, growth and competitive capability in steelhead trout and chosalmon fry exposed to 25-200 NTU or chronic turbidity; utilization of adenosine triphosphate as a parameter in investigations of ecosystem dynamics and structure; bioassay and chemical analysis in support of environmental impact assessments and statements; utilization of chronic flow through and static bioassays in aquatic ecosystems for the determination of the toxic properties of unique Air Force waste products; the environmental effects of illuminating flare residue; environmental assessment for test firing of ammunition with standard and experimental propellants; environmental assessment for development and testing of the Close Air Support Missile; author of numerous publications. *Mailing Add:* 1780 N Research Park Way No 106 Logan UT 84321-1941

SIGLER, JULIUS ALFRED, JR, b Kissimmee, Fla, Dec 22, 40; m 65; c 3. SOLID STATE PHYSICS. *Educ:* Lynchburg Col, BS, 62; Univ Va, MS, 66, PhD(physics), 67. *Prof Exp:* Assoc prof, 67-71, assoc dean, 85-88, PROF PHYSICS & CHMN DEPT, LYNCHBURG COL, DIR GRAD SCI PROGS, 71- *Concurrent Pos:* Consult, Oak Ridge Assoc Univs, 73-74; vis prof physics, Ind State Univ & re-write ed, NSF Physics Technol Proj, 74-75; consult, Solar Tech Proj, NSF, 80. *Mem:* Am Asn Physics Teachers; Soc Physics Students; Nat Rwy Hist Soc. *Res:* Theoretical calculations of relative energies of various defects in quenched face centered cubic metals; variation of mechanical density of gold wires with tensile strains; development of physics curricula and teaching strategies; general curriculum development. *Mailing Add:* Dept Physics Lynchburg Col Lynchburg VA 24501

SIGLER, LAURENCE EDWARD, b Tulsa, Okla, Aug 26, 28; m 82; c 3. MATHEMATICS. *Educ:* Okla State Univ, ScB, 50; Columbia Univ, AM, 54, PhD(math), 63. *Prof Exp:* Instr math, Columbia Univ, 56-59; asst prof, Hunter Col, 61-65 & Hofstra Univ, 65-67; assoc prof, 67-72, PROF MATH, BUCKNELL UNIV, 72- *Mem:* Am Math Soc; Math Asn Am; Ital Math Union. *Res:* Asymptotic theory of ordinary differential equations; set theory; mathematical history. *Mailing Add:* Dept of Math Bucknell Univ Lewisburg PA 17837

SIGLER, MILES HAROLD, b Buffalo, NY, Feb 11, 29; m 54; c 3. NEPHROLOGY. *Educ:* Univ Rochester, BA, 51; Cornell Univ, MD, 55. *Prof Exp:* Intern med, State Univ NY Syracuse, 55-56; resident, Col Med, Thomas Jefferson Univ, 56-58; NIH fel nephrol, Hosp Univ Pa, 60-62; assoc med, Jefferson Med Col Hosp, 62-64; assoc med, 64-69, CHIEF DEPT NEPHROL, LANKENAU HOSP, 69- *Concurrent Pos:* Clin prof med, Col Med, Thomas Jefferson Univ, 72- *Mem:* Am Fedn Clin Res; Am Soc Nephrol; fel Am Col Physicians; Int Soc Nephrol; Am Heart Asn; Sigma Xi. *Res:* Renal mechanisms controlling fluid and electrolyte excretion during fasting; mechanism of the sodium diuresis of fasting; mechanism of the phosphateuresis of fasting; energy metabolism in uremia; slow continuous hemodialysis. *Mailing Add:* 416 Haywood Rd Merion Station PA 19066

SIGLER, PAUL BENJAMIN, b Richmond, Va, Feb 19, 34; m 58; c 5. BIOCHEMISTRY. *Educ:* Princeton Univ, AB, 55; Columbia Univ, MD, 59; Cambridge Univ, PhD(biochem), 68. *Prof Exp:* Intern & resident med, Col Physicians & Surgeons, Columbia Univ, 59-61; res assoc protein crystallog, Nat Inst Arthritis & Metab Dis, 61-63, mem staff, 63-64; fel protein crystallog, Lab Molecular Biol, Med Res Coun, Eng, 64-67; assoc prof, 67-74, prof biophys & theoret biol, 74-84, PROF BIOCHEM & MOLECULAR BIOL, UNIV CHICAGO, 84- *Mem:* AAAS; Am Chem Soc; Am Crystallog Soc. *Res:* X-ray diffraction; protein crystallography; membrane interactive proteins; genetic control; transfer ribonucleic acid; nucleic acid interactions of proteins. *Mailing Add:* Dept Cell Biol & Genetics Univ Chicago Pritzker Sch 920 E 58th St Chicago IL 60637

SIGLER, WILLIAM FRANKLIN, b LeRoy, Ill, Feb 17, 09; m 36; c 2. FISHERIES. *Educ:* Iowa State Univ, BS, 40, MS, 41, PhD(zool), 47. *Prof Exp:* Soil conservationist, Soil Conserv Serv, USDA, 35-37; consult, Cent Eng Co, Iowa, 40-41; res assoc fisheries, Iowa State Col, 41-42; asst prof fisheries & limnol, Utah State Univ, 47-50, prof wildlife sci & head dept, 50-74; pres & chmn bd dirs. W F Sigler & Assocs, Inc, 74-87; RETIRED. *Concurrent Pos:* Mem, Utah Water Pollution Control Bd, 57-65, chmn, 63-65; consult, US Surgeon Gen, 63-67; adv, Int Fish & Game Law Enforcement Asn. *Mem:* Fel AAAS; fel Int Acad Fishery Sci; Wildlife Soc; Am Fisheries Soc; Ecol Soc Am. *Res:* Basic stream productivity; fishery biology; wildlife law enforcement. *Mailing Add:* 309 E 200 S Logan UT 84321

SIGMAN, DAVID STEPHAN, b New York, NY, June 14, 39; m 63; c 2. BIOCHEMISTRY. *Educ:* Oberlin Col, AB, 60; Harvard Univ, AM, 62, PhD(chem), 65. *Prof Exp:* NIH res fel biochem, Sch Med, Harvard Univ, 65-67, instr, 67-68; from asst prof to assoc prof, 68-79, PROF BIOCHEM, SCH MED, UNIV CALIF, LOS ANGELES, 79- *Concurrent Pos:* Alfred P Sloan fel, 72-74; Josiah May fel, 75-76; mem, Molecular Biol Inst, 77- *Mem:* Am Soc Biochemists; Am Chem Soc; Sigma Xi. *Res:* Mechanism of enzyme action; role of metal ions in biological systems; structure of nucleic acids. *Mailing Add:* 1220 N Keater Ave Los Angeles CA 90049-1318

SIGMAR, DIETER JOSEPH, b Vienna, Austria, 1935; c 3. PLASMA PHYSICS, NUCLEAR ENGINEERING. *Educ:* Tech Univ Vienna, MS, 60, ScD(theoret physics), 65. *Prof Exp:* Asst prof theoret physics, Tech Univ, Vienna, Austria, 65-66; res staff plasma physics & thermonuclear fusion, Oak Ridge Nat Lab, Tenn, 66-70; res staff plasma physics, Res Lab Electronics, 70-72, from adj prof nuclear eng to assoc prof nuclear eng & aeronaut & astronaut, 76-84, SR RES SCIENTIST, PLASMA FUSION CTR, MASS INST TECHNOL, 89-; SR SCIENTIST, FUSION ENERGY DIV, OAIC RIDGE NAT LAB, 82- *Concurrent Pos:* Consult, Fusion Energy Div, Oak Ridge Nat Lab, 70-76 & Argonne Nat Lab, 74-76; vis fac mem, Los Alamos Sci Lab, 74 & 75; lectr plasma physics, Tech Univ, Vienna, 78, prof theoret physics. *Mem:* Fel Am Phys Soc; Am Nuclear Soc; Austrian Phys Soc. *Res:* Theoretical plasma physics; controlled thermonuclear fusion research; fudamental transport theory of magnetically confined fully ionized gases and its application to toroidal fusion reactors. *Mailing Add:* Plasma Fusion Ctr NW 16-243 MIT 167 Albany St Cambridge MA 02139

SIGMON, KERMIT NEAL, b Lincoln Co, NC, Apr 18, 36; m 60; c 1. NUMERICAL ANALYSIS, PARALLEL COMPUTING. *Educ:* Appalachian State Univ, BS, 58; Univ NC, Chapel Hill, MEd, 59; Univ Fla, PhD(math), 66. *Prof Exp:* Teacher, Charlotte-Mecklenburg Schs, NC, 59-63; PROF MATH, UNIV FLA, 66- *Concurrent Pos:* Ger Res Asn study grant, Hannover Tech Univ, 72-73; consult, Oak Ridge Nat Lab, 88-89. *Mem:* Am Math Soc; Math Asn Am; Sigma Xi; Soc Indust Appl Math. *Res:* Topological algebra; algebraic topology; numerical analysis; parallel computing. *Mailing Add:* Dept of Math Univ Fla 313 Walker Hall Gainesville FL 32611

SIGNELL, PETER STUART, b Lima, Ohio, June 29, 28; m 52; c 2. THEORETICAL PHYSICS, SCIENCE EDUCATION. *Educ:* Antioch Col, BS, 52; Univ Rochester, MS, 54, PhD, 58. *Prof Exp:* From instr to asst prof physics, Bucknell Univ, 57-59; from asst prof to assoc prof, Pa State Univ, 59-64; assoc prof, 64-65, PROF PHYSICS, MICH STATE UNIV, 65- *Res:* Nuclear forces; research and development for independent study instruction. *Mailing Add:* Dept Physics 106 Physics Astron Bldg Mich State Univ East Lansing MI 48824

SIGNER, ETHAN ROYAL, b Brooklyn, NY, Apr 3, 37; m 62, 79; c 3. PLANT AND MOLECULAR BIOLOGY, BOTANY-PHYTOPATHOLOGY. *Educ:* Yale Univ, BS, 58; Mass Inst Technol, PhD(biophys), 63. *Prof Exp:* NSF fel, Med Res Coun Lab Molecular Biol, Eng, 62-64; Am Cancer Soc fel, Pasteur Inst, Paris, 64-65, Jane Coffin Childs Mem Fund fel, 65-66; from asst prof to assoc prof microbiol, 66-72, PROF BIOL, MASS INST TECHNOL, 72- *Concurrent Pos:* Wellcome vis prof microbiol, Univ Calif, San Francisco, 82; vis prof genetics, Harvard Med Sch, 86-87; fel, Marion & Jasper Whiting Found, 88. *Mem:* AAAS; Am Soc Microbiol; NY Acad Sci; Intern Soc Plant Molecular Biol; Genetics Soc Am. *Res:* Genetics; microbiology; genetics of nodulation and nitrogen-fixation by root nodule bacteria of leguminous plants; genetic recombination in higher plants. *Mailing Add:* Dept of Biol Mass Inst of Technol Cambridge MA 02139

SIGNORINO, CHARLES ANTHONY, b Beaverdale, Pa, July 28, 32; m 54; c 8. ORGANIC CHEMISTRY, PHYSICAL CHEMISTRY. *Educ:* Pa State Univ, BS, 54; Univ Pa, MS, 56, PhD, 59; Westminster Theol Seminary, MAR, 78. *Prof Exp:* Asst instr chem, Univ Pa, 54-58; res chemist, Atlantic Refining Co, 58-62; assoc prof chem, Eastern Baptist Col, 62-68; dir tech serv, Colorcon, Inc, 66-68, vpres, 68-72, mem bd dirs, 72-83; PRES, EMERSON RESOURCES, 88- *Concurrent Pos:* Vis prof, Eastern Baptist Col, 68-69; consult, 84- *Mem:* Am Chem Soc; Am Pharmaceut Asn; Am Asn Pharmaceut Scientists. *Res:* New processes for monomer synthesis; polymerization kinetics; radiation induced polymerization; selective oxidation of hydrocarbons, especially olefins; patented lake manufacturing processes; patented special tablet coatings; specialty chemicals formulations. *Mailing Add:* 300 Lincoln Rd King of Prussia PA 19406

SIGURDSSON, HARALDUR, b Iceland, May 31, 39; div; c 2. PETROLOGY, VOLCANOLOGY. *Educ:* Queen's Univ, Belfast, BSc, 65; Durham Univ, Eng, PhD(geol), 70. *Prof Exp:* Res fel geol, Univ W Indes, Trinidad, 70-73; assoc prof, 74-80, PROF OCEANOG, UNIV RI, 80- *Mem:* Am Geophys Union; Geochem Soc. *Res:* Petrology of ocean ridge basalts; volcanic geology of Iceland and the Lesser Antilles Island Arc. *Mailing Add:* Grad Sch of Oceanog Univ of RI Narragansett RI 02882-1197

SIH, ANDREW, b New York, NY, Mar 10, 54; m 84; c 1. PREDATOR-PREY INTERACTIONS, EVOLUTION OF BEHAVIOR. *Educ:* State Univ NY, Stony Brook, BS, 74; Univ Calif, Santa Barbara, MS, 77, PhD(biol), 80. *Prof Exp:* Postdoctoral fel, Ohio State Univ, 80-81, Mich State Univ, 81-82 & Univ Calif, Berkeley, 82; from asst prof to assoc prof, 82-91, PROF, SCH BIOL SCI, UNIV KY, 91- *Concurrent Pos:* Prin investr, NSF, 85-94, panel mem, 91-; vis scientist, Univ Calif, Berkeley & Oxford Univ, 91. *Honors & Awards:* Murray F Buell Award, Ecol Soc Am, 80. *Mem:* Ecol Soc Am; Am Soc Naturalists; Soc Study Evolution; AAAS. *Res:* Evolution of behaviors, feeding, mating, predator avoidance, and life history traits that influence species interactions; mathematical modeling; focal organisms studied include amphibians, fish and aquatic insects. *Mailing Add:* Sch Biol Sci Univ Ky Lexington KY 40506-0225

SIH, CHARLES JOHN, b Shanghai, China; nat US; m 59. BIO-ORGANIC CHEMISTRY. *Educ:* Carroll Col, Mont, AB, 53; Mont State Col, MS, 55; Univ Wis, PhD(bact), 58. *Prof Exp:* Sr res microbiologist, Squibb Inst Med Res, 58-60; from assoc prof to prof, 60-77, F B POWER PROF PHARMACEUT CHEM, UNIV WIS-MADISON, 77- *Concurrent Pos:* Sci adv, Eastman Kodak. *Honors & Awards:* Ernest Volwiler Award, 77; Roussel Prize, 80. *Mem:* Am Soc Microbiol; Am Soc Biol Chemists; Acad Pharmaceut Sci; Am Chem Soc. *Res:* Chemical syntheses using enzymes; enzymatic mechanism of sterol side chain degradation; suicide inhibitors; natural products chemistry; natural products chemistry. *Mailing Add:* 6322 Landfall Dr Madison WI 53705

SIHAG, RAM K, b India, Jan 1, 50. BIOCHEMISTRY. *Educ:* P U Phandigarh, India, BS, 69; Hahisar Univ, 72; J-Nehru Univ, New Delhi, PhD(life sci), 77. *Prof Exp:* Postdoctoral fel, Univ Conn Health Ctr, 81-84, Pa State Univ, 84-85; instr, 85-89, RES ASSOC BIOCHEM, HARVARD MED SCH, 85-, RES ASST, MCLEAN HOSP, 85-, ASST PROF, DEPT PSYCHIAT, 91- *Mem:* Protein Soc; AAAS; Soc Cell Biol. *Mailing Add:* Ralph Lowell Labs McLean Hosp Harvard Med Sch 115 Mill St Belmont MA 02178

SIIROLA, JEFFREY JOHN, b Patuxent River, Md, July 17, 45; m 71; c 2. CHEMICAL PROCESS SYNTHESIS, CHEMICAL TECHNOLOGY ASSESSMENT. *Educ:* Univ Utah, BS, 67; Univ Wis-Madison, PhD, 70. *Prof Exp:* Res engr, Tenn Eastman Co, 72-74, sr res engr, 74-80, res assoc, Eastman Kodak Co, 80-88, SR RES ASSOC, EASTMAN CHEMICAL CO, 80- *Concurrent Pos:* Mem exec comt, Comput & Systs Technol Div, Am Inst Chem Engrs, 81-, prog chmn, 88-, trustee, Comput Aids Chem Eng Educ Corp, 83-, pres, 90-, ed bd, Indust & Eng Chem Res, 86-89. *Mem:* Am Inst Chem Engrs; Am Chem Soc; Sigma Xi; Am Asn Artificial Intel; Assoc Comput Mach. *Res:* Computer-aided chemical process design synthesis, analysis and optimization; symbolic programming; artificial intelligence; chemical technology assessment; chemical engineering education. *Mailing Add:* Eastman Chem Co PO Box 1972 Kingsport TN 37662

SIITERI, PENTTI KASPER, b Finland, Oct 20, 26; US citizen; m 50; c 5. BIOCHEMISTRY. *Educ:* Dartmouth Col, AB, 48; Univ NH, MS, 50; Columbia Univ, PhD(biochem), 63. *Prof Exp:* Res scientist, Lederle Labs, Am Cyanamid Co, 53-59; from assoc prof to prof obstet, gynec & biochem, Univ Tex Southwestern Med Br, Dallas, 62-73; PROF OBSTET & GYNEC, UNIV CALIF, SAN FRANCISCO, 73- *Concurrent Pos:* Mem, Endocrinol Sect, NIH, 69-75, chmn, 75-77; NIH res grant, 73- *Mem:* AAAS; Am Soc Biol Chem; Endocrine Soc; Soc Gynec Invest. *Res:* Mechanism of estrogen action; hormones in reproduction and cancer. *Mailing Add:* Dept Obstet & Gynec Univ Calif HSW 1656 San Francisco CA 94143

SIJ, JOHN WILLIAM, b St Louis, Mo, June 21, 43; m 65; c 2. PLANT PHYSIOLOGY. *Educ:* Eastern Ill Univ, BSEd, 65; Ohio State Univ, MS, 67, PhD(plant physiol), 71. *Prof Exp:* Res assoc, Evapotranspiration Lab, Kans State Univ, 71-72; from asst prof to assoc prof, 72-83, PROF, AGR RES & EXTEN CTR, TEX A&M UNIV, 83- *Honors & Awards:* Am Soybean Asn ICI Am Award, Soybean Researcher's Prog. *Mem:* Am Asn Plant Physiologists; Crop Sci Soc Am; Am Soc Agron; Sigma Xi. *Res:* Soybean physiology and management; alternate crop and cropping systems. *Mailing Add:* Res & Exten Ctr Tex A&M Univ Rte 7 Box 999 Beaumont TX 77713

SIKAND, RAJINDER S, b Barnala, India. PHYSIOLOGY, MEDICINE. *Educ:* King Edward Med Col, Lahore, Punjab, MB, BS, 46. *Prof Exp:* Res fel physiol, Sch Med, Yale Univ, 49-52; med officer, West Middlesex Hosp, UK, 53-54; med registr, King George V Hosp, Godalming, UK, 54-56; res fel physiol, Sch Med, Univ Md, Baltimore City, 61-63; res assoc, State Univ NY Buffalo, 63-64; res assoc, Max Planck Inst Med Res, Gottingen, 64-65; assoc clin prof internal med, Sch Med, Yale Univ, 65-75; CO-DIR, SECT PULMONARY DIS, DEPT MED, ST RAPHAEL'S HOSP, NEW HAVEN, CT, 75- *Mem:* Am Physiol Soc; fel Am Col Physicians; fel Am Col Chest Physicians; Am Thoracic Soc; Am Fedn Clin Res. *Res:* Cardiopulmonary physiology. *Mailing Add:* Pulmonary Unit St Raphael's Hosp 1450 Chapel St New Haven CT 06511

SIKARSKIE, DAVID L(AWRENCE), b Marquette, Mich, Aug 3, 37; div; c 3. ENGINEERING MECHANICS. *Educ:* Univ Pa, BS, 59; Columbia Univ, MS, 60, ScD(eng mech), 64. *Prof Exp:* Res asst solid mech, Columbia Univ, 62-63; mem tech staff, Ingersoll Rand Res Ctr, 63-66; from asst prof to assoc prof aerospace eng, Univ Mich, Ann Arbor, 66-72, prof aerospace eng, 72-79, chmn metall, mech & mats sci, 79-84; dean, 85-90, DIR SPEC PROJS, COL ENG, MICH TECH UNIV, 90- *Concurrent Pos:* Vis lectr, Princeton Univ, 65-66. *Mem:* Am Soc Mech Engrs; Soc Eng Sci; Sigma Xi; Nat Soc Prof Engrs; Am Asn Eng Educ. *Res:* Brittle fracture; rock mechanics; elasticity; nonlinear structural mechanics; biomechanics; composite material mechanics. *Mailing Add:* 25 Peepsock Houghton MI 49931

SIKDAR, DHIRENDRA N, b India, Nov 1, 30; m 61; c 2. PHYSICS, METEOROLOGY. *Educ:* Univ Calcutta, BSc, 49, MSc, 51; Univ Wis-Madison, PhD(meteorol), 69. *Prof Exp:* Asst Meteorol Dept, Govt India, 61-69; asst scientist, Space Sci & Eng Ctr, Univ Wis-Madison, 69-70, assoc scientist, 70-74; PROF ATMOS SCI, UNIV WIS-MILWAUKEE, 74- *Concurrent Pos:* NSF res grants, 70- *Mem:* Am Geophys Union; Sigma Xi; Am Meteorol Soc. *Res:* Satellite meteorology; severe storm circulations and energetics; Great Lake research; urban meteorology. *Mailing Add:* Dept Geol Sci Univ Wis PO Box 413 Milwaukee WI 53210

SIKDER, SANTOSH K, b Faridpur, Bangladesh, Jan 1, 49; m 86. IMMUNOCHEMISTRY, VIRAL IMMUNOLOGY. *Educ:* Dhaka Univ, Bangladesh, BS, 69, MS, 70; Jadapur Univ, Calcutta, India, PhD(org chem), 81. *Prof Exp:* Res & develop chemist pharmaceut labs, Universal Drug House, Calcutta, 71-74; jr res fel org chem, Jadapur Univ, Calcutta, 75-78, sr res fel, 78-81; postdoctoral res scientist microbiol, Columbia Univ, NY, 82-87; instr, 88-90, ASST PROF MED, CORNELL UNIV MED COL, NY, 90- *Mem:*

Am Asn Immunologists. *Res:* Regulation of human-immunodeficiency virus transcription; effect of different hormones-antihormones on the HIV-transactivation; contribution of different cellular factors and its interaction with viral genes and proteins particularly to explain the viral latency. *Mailing Add:* 528 Cumberland Ave Teaneck NJ 07666

SIKES, JAMES KLINGMAN, b Henderson, Tenn, Apr 12, 24; m 50; c 2. ANALYTICAL CHEMISTRY, PHYSIOLOGY. *Educ:* Abilene Christian Col, BS, 47; Tex Technol Col, MS, 51; Tex Tech Univ, MS, 67. *Prof Exp:* Supvr lab, Paymaster Oil Mills, Anderson Clayton Co, 47-54; chief chemist res, Plains Coop Oil Mill, 54-65; partner, Plains Lab, 65-73; dept head lab, Brookside Farms Lab Asn, 73-75; PRES, SCAN, INC, 75- *Concurrent Pos:* Mem res comt, Nat Cottonseed Prods Asn, 60-62. *Mem:* Am Oil Chemists Soc; Am Inst Chemists; AAAS; Am Soc Animal Sci; Asn Consult Chemists & Chem Engrs. *Res:* Cottonseed oil processing and refining; laboratory methodology; water quality; insecticides; herbicides; animal feeds; animal nutrition; soil and crop improvement. *Mailing Add:* Rte 2 PO Box 350-A15 Lubbock TX 79452

SIKINA, THOMAS, b Philadelphia, Pa, July 24, 50; c 1. ANTENNA ENGINEERING. *Educ:* Pa State Univ, BS, 72; Del Valley Col, BS, 77; Drexel Univ, MS, 85. *Prof Exp:* Antenna engr, RCA, 79-84; design specialist, ITT-Gilfillan, 84-86; CHIEF ENGR, CHU ASSOC INC, 86- *Mem:* Inst Elec & Electronic Engrs. *Mailing Add:* Raytheon Co Boston Post Rd Wayland MA 01778

SIKLOSI, MICHAEL PETER, b Akron, Ohio, Sept 8, 48; m 71. ORGANIC CHEMISTRY. *Educ:* Montclair State Col, BA, 70; Purdue Univ, MS, 72, PhD(chem), 77. *Prof Exp:* RESEARCHER CHEM, PROCTER & GAMBLE CO, 77- *Mem:* Am Chem Soc. *Res:* Peroxyacids; reactivity; decomposition mechanisms, reaction mechanisms; organometallic chemistry; reactions of allylic Grignard reagents; detergency. *Mailing Add:* 7299 Bobby Lane Cincinnati OH 45243-2003

SIKLOSSY, LAURENT, b Budapest, Hungary; US citizen. COMPUTER SCIENCE. *Educ:* Yale Univ, BA, 63; Harvard Univ, MA, 64; Carnegie-Mellon Univ, PhD(comput sci), 68. *Prof Exp:* Proj scientist comput sci, Carnegie-Mellon Univ, 65-68; asst prof info & comput sci, Univ Calif, Irvine, 68-70; asst prof comput sci, Univ Tex, Austin, 71-74, assoc prof, 74-78; dir grad studies, Dept Info Eng, 79-80, PROF, DEPT INFO ENG, UNIV ILL, CHICAGO, 78- *Concurrent Pos:* Regents grant, Univ Calif, Irvine, 69-70; consult, Hughes Aircraft Corp, Calif, 70; lectr & consult, US Dept State, Repub of Cameroun & Algeria, 70-71; vis titular prof, Univ Sao Paulo, 72; Univ Res Inst grant, Univ Tex, Austin, 70-75, NSF grants, 71-75; Arts & Sci Found grant, 72-75; Orgn Am States vis prof, Univ Tech Fed Santa Maria, Valparaiso, 74 & 75; Nat Acad Sci grant, 75; vis prof, Univ Paris, 76 & 81; consult, Jelmoli AG, Zurich, Switz, 77. *Mem:* Asn Comput Mach; Sigma Xi; Am Phys Soc. *Res:* Artificial intelligence; robotics; problem-solving; intelligent computer tutors; information structures; list processing; management information systems; use of computers in developing countries; speech processing by computer; author or coauthor of over 50 publications. *Mailing Add:* Postbus 71710 NL-1008 De Amsterdam Amsterdam Netherlands

SIKORA, JEROME PAUL, b Cleveland, Ohio, Apr 9, 47; m 71; c 2. STRUCTURAL MECHANICS, PHOTOMECHANICS. *Educ:* Univ Detroit, BS, 69. *Prof Exp:* PHYSICIST, D TAYLOR NAVAL SHIP RES & DEVELOP CTR, 69- *Mem:* Sigma Xi. *Res:* Optical methods such as holography, speckle, moire and photoelasticity to measure displacements, stresses, vibrations, and contour mapping of hard structures. *Mailing Add:* 14126 Chadwick Lane Rockville MD 20853

SIKORSKA, HANNA, b Warsaw, Poland, July 7, 50; Polish & Can citizen. MONOCLONAL ANTIBODIES, TUMOR IMMUNOLOGY. *Educ:* Univ Warsaw, BS & MS, 73; Polish Acad Sci, PhD(immunol), 78. *Prof Exp:* PhD fel immunol, Inst Rheumatology, Warsaw, 75-78, adj, 78-81; postdoctoral immunol, Inst Arnand-Frappier, Que, 81-82; postdoctoral thyroidology, Queen's Univ, Ont, 82; postdoctoral autoimmunity, McGill Univ, Que, 82-85; VPRES RES & DEVELOP, ROUGIER BIO-TECH LTD, MONTREAL, 85- *Mem:* Am Asn Immunologists; Can Soc Clin Res; Parenteral Drug Asn; Drug Info Asn. *Res:* Development of monoclonal antibody based radiopharmaceuticals for imaging of myocardial cell necrosis, tumors and occult infactions; monoclonal antibody conjugates for cancer therapy; anti-idiotypes as cancer vaccines; antibody delivery systems; marine toxins; pseudomonas aernginosa serotyping in cystic fibrosis. *Mailing Add:* Rougier Bio-Tech Ltd 8480 St Laurent Montreal PQ H2P 2M6 Can

SIKORSKA, MARIANNA, b Wilkolaz, Poland, Jan 20, 44. BIOCHEMISTRY. *Educ:* Polytech Univ, Warsaw, BChemE, 72; Polish Acad Sci, PhD(biochem), 78. *Prof Exp:* Res asst neurochem, Med Res Ctr, Polish Acad Sci, Warsaw, 69-78; RES ASSOC CELL PHYSIOL, DIV BIOL, NAT RES COUN, CAN, 78- *Res:* Role of nuclear protein kinesis in changes in chrometin structure. *Mailing Add:* Nat Res Coun Can Div Biol M-54 Montreal Rd Ottawa ON K1A 0R6 Can

SIKORSKI, JAMES ALAN, b Stevens Pt, Wis, Nov 9, 48; m 77; c 1. ENZYME INHIBITOR DESIGN & SYNTHESIS. *Educ:* Northeast La State Col, BS, 70; Purdue Univ, MS, 76, PhD(org chem), 81. *Prof Exp:* Res chemist II, Monsanto Agr Co, 76-78, sr res chemist, 78-82, res specialist, 82, res group leader, 82-87, SCI FEL, MONSANTO AGR CO, 87- *Concurrent Pos:* Instr org chem, St Louis Community Col, Florisant Valley, 77-78. *Mem:* Am Chem Soc; AAAS; Sigma Xi; Protein Soc. *Res:* Plant biochemistry; heterocycle synthesis; organophosphorus chemistry; rational design and synthesis of mechanism-based enzyme inhibitors; molecular modeling. *Mailing Add:* Monsanto Agr Co 800 N Lindbergh Blvd St Louis MO 63167

SIKOV, MELVIN RICHARD, b Detroit, Mich, July 8, 28; m 52; c 3. RADIATION BIOLOGY. *Educ:* Wayne State Univ, BS, 51; Univ Rochester, PhD(radiation biol), 55. *Prof Exp:* Res assoc radiation biol, Univ Rochester, 52-55; asst prof radiobiol, Col Med, Wayne State Univ, 55-61, assoc prof, 61-65; sr res scientist, 65-68, res assoc, 68-78, mgr develop toxicol, 78-81, SR STAFF SCIENTIST, PAC NORTHWEST LABS, BATTELLE MEM INST, 81- *Mem:* Radiation Res Soc; Health Physics Soc; Am Asn Pathologists; Soc Toxicol; Teratology Soc; fel Am Inst Ultrasound Med. *Res:* Age and environmental factors in radionuclide metabolism and toxicity; effects of radiation and chemicals on embryonic and postnatal development; relationships between teratogenic and oncogenic mechanisms. *Mailing Add:* Dept Biol & Chem P7-53 Battelle Northwest Battelle Blvd Richland WA 99352

SIKRI, ATAM P, b Sheikhupra, India, Mar 12, 39; US citizen; m 68; c 2. ENVIRONMENTAL ENGINEERING, ENVIRONMENTAL PROTECTION SPECIALIST. *Educ:* Univ Mich, BSE(chem eng) & BSE(metall eng), 63, MSE, 65; Univ Pa, PhD(chem eng), 70. *Prof Exp:* Res engr chem eng, DuPont Co & Cities Serv Co, 65-72; proj mgr chem eng, Johnson & Johnson, 72-74; sr engr chem eng, Hoffman-LaRoche Inc, 74-76; asst dir chem eng, Dept Energy, 76-82; sr proj engr petrol eng, US Corps Engrs, 82-84; sr nuclear engr, 84-86, sr proj engr chem eng, 86-91, TEAM LEADER ENVIRON ENG, US DEPT ENERGY, 91- *Concurrent Pos:* Teaching fel, Univ Pa, 66-68; lectr, Rutgers State Univ NJ, 70-71. *Mem:* Am Inst Chem Engrs; Am Chem Soc; Soc Petrol Engrs. *Res:* Rheological properties of polymeric solutions; extractive distillation of paraffins and olefins; pipelines calculations for gas-liquid systems; water accumulation model for natural gas-liquid wells. *Mailing Add:* 9431 Wooded Glen Ave Burke VA 22015-4231

SILAGI, SELMA, b Sept 5, 16; US citizen; m 36; c 2. GENETICS, CANCER. *Educ:* Hunter Col, AB, 36; Columbia Univ, MA, 38, PhD(genetics), 61. *Prof Exp:* Teacher pub schs, NY, 38-50; lectr biol, Queens Col, 57-59; NIH fel, 59-62; res assoc biochem genetics, Rockefeller Univ, 62-65; from asst prof to prof, genetics, 65-87, EMER PROF OF GENETICS, MED COL, CORNELL, UNIV, 87- *Concurrent Pos:* Guest investr, Rockefeller Univ & vis investr, Sloan-Kettering Inst Cancer Res, 65-66; fac res award, Am Cancer Soc, 70-75; mem, Cancer Spec Prog Adv Comt, Nat Cancer Inst, 73-75. *Mem:* AAAS; Genetics Soc Am; Am Soc Cell Biol; Harvey Soc; Tissue Cult Asn; Sigma Xi. *Res:* Cellular differentiation and gene action in mammalian cells in tissue culture; reversible suppression of malignancy by 5-bromodeoxyuridine; cancer immunology; cell hybridization and somatic cell genetics; mechanism of action of bomodeoxyuridine. *Mailing Add:* 3535 First Ave 11C San Diego CA 92103-4845

SILANDER, JOHN AUGUST, JR, b Highland Park, Ill, Mar 1, 45; m 71; c 3. PLANT ECOLOGY, EVOLUTION. *Educ:* Pomona Col, BA, 67; Univ Mich, MA, 69; Duke Univ, PhD(bot), 76. *Prof Exp:* Instr biol, Peace Corps Prog, Ghana, 69-71; asst prof, 76-83, ASSOC PROF BIOL SCI, UNIV CONN, 84- *Concurrent Pos:* Fulbright fel, Australian Nat Univ, 77-78. *Mem:* Ecol Soc Am; Soc Study Evolution; Am Soc Naturalists; Brit Ecol Soc. *Res:* Genetic, evolutionary, population and community aspects of plant ecology. *Mailing Add:* Ecol & Evolutionary Biol Univ of Conn Box U-42 Storrs CT 06268

SILBAR, RICHARD R(OBERT), b Milwaukee, Wis, Jan 19, 37; m 63. PHYSICS. *Educ:* Univ Mich, BS, 59, MS, 60, PhD(physics), 63. *Prof Exp:* Res assoc & instr physics, Johns Hopkins Univ, 63-65; res asst prof, Cath Univ Am, 65-67; group leader, 75-78, STAFF PHYSICIST, MEDIUM ENERGY NUCLEAR PHYSICS THEORY, LOS ALAMOS NAT LAB, 67- *Concurrent Pos:* Vis scientist, Swiss Inst Nuclear Res, 73-74; vis prof, State Univ NY, 76-77, Univ Mass, 85-86; longterm acad sci exchange, Inst Nuclear Res, Moscow, USSR, 78-; detaillee, Div Nuclear Physics, Dept Energy, 81-82, Div Adv Energy Proj, 91. *Mem:* Fel Am Phys Soc; Am Asn Artificial Intel. *Res:* Particle and nuclear physics theory; medium energy physics; beam optics; simulation. *Mailing Add:* Los Alamos Nat Lab T-5 MS-B283 Los Alamos NM 87545

SILBART, LAWRENCE K, b Chicago, Ill, Jan 22, 58. CANCER RESEARCH. *Educ:* Univ Mich, BS, 80, MS, 83, PhD(toxicol), 87. *Prof Exp:* Postdoctoral fel, Univ Mich, 87-89, res investr path, 89-91; ASST PROF PATH, UNIV CONN, 91-, DIR, CTR ENVIRON HEALTH, 91- *Mem:* Am Asn Immunologists; Am Asn Path; Soc Mucosal Immunol. *Mailing Add:* Univ Conn George White Bldg 3636 Horsebarn Rd Box U-39 Storrs CT 06269

SILBAUGH, STEVEN A, b Davenport, Iowa, Aug 27, 49; m 77; c 1. LUNG PHYSIOLOGY, PHARMACOLOGY. *Educ:* Univ NMex, PhD(biol), 80. *Prof Exp:* RES SCIENTIST, LILLY CORP CTR, 87- *Mem:* Am Physiol Soc; Ind Thoracic Soc; Sigma Xi. *Res:* Pulmonary physiology and pharmacology, aerosol science, development of novel agents to treat asthma. *Mailing Add:* 6906 N Delaware Indianapolis IN 46220

SILBER, HERBERT BRUCE, b New York, NY, Apr 3, 41; m 66; c 2. INORGANIC CHEMISTRY. *Educ:* Lehigh Univ, BS, 62, MS, 64; Univ Calif, Davis, PhD(inorg chem), 67. *Prof Exp:* Swedish Govt fel & Fulbright-Hays travel grant, Royal Inst Technol, Stockholm, 67-68 & Univ Md, College Park, 68-69; asst prof chem, Univ Md, Baltimore County, 69-75; prof chem, Univ Tex, San Antonio, 75-85; PROF CHEM, SAN JOSE STATE UNIV, 86. *Mem:* Am Chem Soc; Sigma Xi. *Res:* Lanthanide chemistry; metal ion complexation chemistry; solvation; relaxation kinetics; ultrasonic absorption. *Mailing Add:* Chem Dept San Jose State Univ San Jose CA 95192

SILBER, ROBERT, b Vienna, Austria, Jan 4, 31; US citizen; m 54; c 4. INTERNAL MEDICINE, HEMATOLOGY. *Educ:* NY Univ, BA, 50; State Univ NY, MD, 54. *Prof Exp:* From intern to resident med, Third Med Div, Bellevue Hosp, New York, 54-58; instr & USPHS spec fel biochem, Univ Wash, 60-62; asst prof, 62-66, ASSOC PROF MED, SCH MED, NY UNIV,

66-, DIR DIV HEMAT, 68- *Concurrent Pos:* Nat Cancer Inst clin fel, Col Med, Univ Utah, 56-57; Guggenheim Mem Found fel, 71-72. *Mem:* Am Soc Clin Invest; Am Soc Biol Chem; Am Fedn Clin Res; Soc Hemat; Harvey Soc. *Res:* Leukocyte enzymology; folic acid metabolism; disorders of the red cell membrane; biochemistry. *Mailing Add:* NY Univ Sch Med 550 First Ave New York NY 10016

SILBER, ROBERT, b Montgomery, WVa, Nov 8, 37; m 64; c 2. MATHEMATICS. *Educ:* Vanderbilt Univ, BA, 57; Univ Ala, MA, 63; Clemson Univ, PhD(math), 68. *Prof Exp:* Aerospace technologist, NASA, 58-63; asst prof math, 68-81, ASSOC PROF MATH, NC STATE UNIV, 81- *Mem:* Am Math Soc; Math Asn Am. *Res:* Optimization theory; functional analysis. *Mailing Add:* Dept of Math NC State Univ Box 8205 Raleigh NC 27695-8205

SILBERBERG, DONALD H, b Washington, DC, Mar 2, 34; m 59; c 2. NEUROLOGY. *Educ:* Univ Mich, MD, 58. *Hon Degrees:* MA, Univ Pa, 71. *Prof Exp:* Resident neurol, NIH, 59-61; Fulbright fel neurol & Neuro-ophthal, Nat Hosp, London, Eng, 61-62; USPHS spec fel neuro-ophthal, Sch Med, Wash Univ, 62-63; assoc neurol, 63-65, from asst prof to assoc prof neurol & ophthal, 65-71, vchmn neurol dept, 74-82, PROF NEUROL & OPHTHAL, SCH MED, UNIV PA, 71-, CHMN NEUROL DEPT, 82- *Concurrent Pos:* Consult, Philadelphia Vet Admin Hosp, 66- & Childrens Hosp Philadelphia, 67- *Mem:* Am Acad Neurol; Am Neurol Asn; Am Asn Neuropath; Am Soc Neurochem; Soc Neurosci. *Res:* Tissue culture studies of myelination; demyelinating diseases. *Mailing Add:* Dept Neurol Hosp the Univ Pa Philadelphia PA 19104

SILBERBERG, I(RWIN) HAROLD, b Austin, Tex, Feb 25, 26; m 49; c 4. CHEMICAL ENGINEERING. *Educ:* Univ Tex, BS, 47, MS, 51, PhD(chem eng), 58. *Prof Exp:* Sr res technologist, Field Res Lab, Socony Mobil Oil Co, 57-60; ASST DIR, UNIV DIV, TEX PETROL RES COMT, UNIV TEX, AUSTIN, 60- *Mem:* Am Chem Soc; Am Inst Chem Engrs; Am Soc Petrol Engrs; Am Inst Chemists; Nat Soc Prof Engrs. *Res:* Volumetric and phase behavior of fluids, particularly hydrocarbons; petroleum production and reservoir engineering; thermodynamics. *Mailing Add:* Petroleum Eng Dept Univ Tex CPE 3118 Austin TX 78712-1061

SILBERBERG, REIN, b Tallinn, Estonia, Jan 15, 32; US citizen; m 65; c 2. COSMIC RAY PHYSICS. *Educ:* Univ Calif, Berkeley, AB, 55, MA, 56, PhD(physics), 60. *Prof Exp:* Res asst physics, Univ Calif, 56-60; res assoc, Nat Acad Sci-Nat Res Coun, US Naval Res Lab, 60-62, actg chief scientist, Lab Cosmic Ray Physics, 81-83, res physicist, 62-90, dep br head, Gamma & Cosmic Ray, Astrophysics, 84-90; CONSULT, UNIV SPACE RES ASN, 90- *Concurrent Pos:* Assoc Dir, Int Sch Cosmic Ray Astrophysics, Erice, Italy, 78-86. *Mem:* Am Geophys Union; Radiation Res Soc. Fel Am Phys Soc; Am Astron Soc; Sigma Xi; Int Astron Union. *Res:* Cosmic ray effects on microelectronics; acceleration of cosmic rays in accretion disks of ultra-massive black holes; isotopic and elementary composition of cosmic rays; solar modulation and transformation of cosmic ray composition in space; spallation and fission reactions; neutrino astronomy; radiobiological effects of heavy cosmic ray nuclei. *Mailing Add:* Code 4154 US Naval Res Lab Washington DC 20375

SILBERBERG, RUTH, b Kassel, Germany, Mar 20, 06; nat US; m 33. PATHOLOGY. *Educ:* Breslau Univ, MD, 31. *Prof Exp:* Asst, Path Inst, Breslau Univ, 30-33; pathologist, Jewish Hosp, 33; vol res path, Dalhousie Univ, 34-36 & Wash Univ, 37-41; asst, NY Univ, 41-44; from instr to prof, 45-74, emer prof path, Sch Med, Wash Univ, 74-76; VIS SCIENTIST, HEBREW HADASSAH UNIV, 76- *Concurrent Pos:* Dean fel, NY Univ, 41-44; actg pathologist, Jewish Hosp, St Louis, 45-46; pathologist, Barnard Free Skin & Cancer Hosp, 47; sr pathologist, Hosp Div, City of St Louis, 47-59. *Mem:* Am Soc Exp Path; Soc Exp Biol & Med; Soc Develop Biol; Am Asn Path & Bact; Am Asn Cancer Res. *Res:* Experimental pathology; skeletal growth and aging; hormonal carcinogenesis; developmental potencies of the lymphocyte; pathogenesis of osteoarthrosis. *Mailing Add:* Dept of Path Box 1172 Jerusalem Israel

SILBERFELD, MICHEL, b Paris, France, Feb 15, 46; Can citizen; m; c 1. PSYCHIATRY, EPIDEMIOLOGY. *Educ:* McGill Univ, BSc, 66, MDCM, 70; Univ Toronto, MSc, 73. *Prof Exp:* ASST PROF EPIDEMIOL & PSYCHIAT, UNIV TORONTO, 76-; STAFF PSYCHIATRIST, WELLESLEY HOSP, 78- *Concurrent Pos:* Consult & res scientist psychiat & epidemiol, Addiction Res Found, 76-78; consult, Princess Margaret Hosp, 78-, mem staff, 79- *Res:* Social psychiatry. *Mailing Add:* Univ Toronto Toronto ON M5S 1A8 Can

SILBERG, STANLEY LOUIS, b Kansas City, Mo, Dec 27, 27; m 56; c 2. EPIDEMIOLOGY, MICROBIOLOGY. *Educ:* Univ Kans, AB, 51, MA, 52; Univ Minn, MPH, 59, PhD(epidemiol), 65. *Prof Exp:* Res asst mycol, Kans State Health Dept, 52-53 & Sch Med, Univ Kans, 53-54; res fel epidemiol, Univ Minn, 57-61; asst prof, Sch Med, Univ Mo-Columbia, 61-69; assoc prof biostatist & epidemiol, 69-72, actg chmn, 78-84, vchmn 81-84, PROF BIOSTATIST & EPIDEMIOL, SCH PUB HEALTH, HEALTH SCI CTR, UNIV OKLA, 72- *Concurrent Pos:* Mo Div Health grant, 63-65; USPHS grant, 65-; consult, USPHS Commun Dis Ctr, Ga, 66-76, US Pub Health Comn Corps, 57- *Honors & Awards:* Sigma Xi Res Soc. *Res:* Infectious disease epidemiology. *Mailing Add:* Univ Okla Health Sci Ctr Col Health Bldg PO Box 26901 Oklahoma City OK 73190

SILBERGELD, ELLEN K, b Washington, DC, 1945. TOXICOLOGY, NEUROSCIENCES. *Educ:* Vassar Col, AB, 67; Johns Hopkins Univ, PhD(eng), 72. *Prof Exp:* Fel neurosci, Johns Hopkins Sch Hyg, 72-75; staff fel neuropharm, Nat Inst Neurol & Commun Disorders & Stroke, NIH, 75-79, sect chief neurotoxicol, 79-81; chief scientist Toxicol Environ Defense Fund, 82-91; PROF TOXICOL, UNIV MD, 91- *Concurrent Pos:* Deleg, US & USSR Environ Health Exchange, 77-78; mem, US Dept Human Health Serv, 77-81; lectr continuing med, educ environ & occup med, 78-88; adv, Hyperkinesis & Diet, Nutrit Found, 75-78; bd mem Toxicology, Nat Acad Sci, 83-89, US Environ Protection Agency Sci Bd, 83-90, Sci Counr, Nat Inst Environ Health Sci, 90-; Baldwin Scholar, Col of Notre Dame; Fulbright Scholar. *Mem:* Soc Neurosci; Int Brain Res Orgn; Am Soc Pharmacol & Exp Therapeut; Am Soc Neurochem; Soc Occup & Environ Health; Am Pub Health Soc; Soc Toxicol. *Res:* Neurotoxicology and environmental toxicology; adverse effects of chemicals and drugs on nervous system function; lead poisoning; developmental neurosciences. *Mailing Add:* Toxicol Prog Univ MD Howard Hall 544 Baltimore MD 21201

SILBERGELD, MAE DRISCOLL, space physics, computer sciences; deceased, see previous edition for last biography

SILBERGELD, SAM, b Wengrow, Poland, Mar 1, 18; US citizen; m 52; c 4. PSYCHIATRY. *Educ:* Blackburn Col, AA, 38; Univ Chicago, BS, 39; Univ Ill, MS, 41, PhD(biochem), 43; Duke Univ, MD, 54. *Prof Exp:* Instr biochem, Mayo Found, 44-45; instr, Ill Neuropsychiat Inst, Univ Ill Med Sch, 45, asst prof chem, Chicago Undergrad Div, Univ Ill, 46-52; intern med, Cincinnati Gen Hosp, 54-55, resident internal med, 56; MED DIR, USPHS, 56- *Concurrent Pos:* Staff asst to dir, Div Biol Stand, Nat Inst Ment Health, NIH, 56-59, res grants specialist, Div Gen Med Sci, 59-61, chief gen clin res ctr br, Div Res Facil & Resources, 61-64, res grants specialist, Nat Inst Ment Health, 64 & Ment Health Career Develop Prog, 64-, res psychiatrist, Ment Health Study Ctr, 70-, actg dir, Div Intramural Res, Nat Inst Alcohol Abuse & Alcoholism, 76-77; USPHS ment health career develop award, 64-70; resident psychiat, Sch Med, Stanford Univ, 64-67, res assoc, 67-68, res psychiatrist, Lab Clin Sci, 68-70; physician's recognition award, AMA, 77-82. *Mem:* Am Psychiat Asn; NY Acad Sci; Am Chem Soc. *Res:* Clinical, community and psychobiochemical research, especially the metabolic and behavioral aspects of stress. *Mailing Add:* 10704 Clermont Ave Garrett Park MD 20896

SILBERGER, ALLAN JOSEPH, b York, Pa, Aug 24, 33; m 57; c 4. MATHEMATICS. *Educ:* Univ Rochester, AB, 55; Johns Hopkins Univ, MA, 62, PhD(math), 66. *Prof Exp:* Assoc math, Appl Physics Lab, Johns Hopkins Univ, 58-64, instr, Univ, 64-66; asst prof, Bowdoin Col, 66-71; mem, Inst Advan Study, 71-73; guest prof, Math Inst, Univ Bonn, WGer, 73-74; vis assoc prof, Univ Mass, 74-75; assoc prof, 75-80, PROF MATH, CLEVELAND STATE UNIV, 80- *Mem:* Am Math Soc; Math Asn Am. *Res:* Representation theory; group theory. *Mailing Add:* Dept Math Cleveland State Univ Cleveland OH 44115

SILBERGER, DONALD MORISON, b York, Pa, Feb 26, 30; m 80; c 5. MATHEMATICS. *Educ:* Harvard Univ, BA, 53; Univ Wash, MS, 61, PhD(math), 73. *Prof Exp:* High sch instr, Ohio, 56-58; NSF res asst, 62-63; instr math, Idaho State Univ, 63-65; assoc prof, Butler Univ, 65-67; lectr, Western Wash State Col, 67-68; from asst prof to assoc prof, Tougaloo Southern Christian Col, 68-74 & 76-77; spec asst prof math, Univ Colo, Boulder, 74-76; prof post grad math, Fed Univ Santa Catarina, Brazil, 77-82; assoc prof math, St Martin's Col, Olympia, Wash, 82-83; ASSOC PROF MATH, STATE UNIV NY, NEW PALTZ, 83- *Mem:* NY Acad Sci; Am Math Soc; Planetary Soc; Math Asn Am; Brazilian Soc Math. *Res:* Finite combinatorics; algebraic theory of semigroups; theory of regular finite simple hypergraphs; logic; universal terms; combinational algebra; sequence-valued functions of a sequence variable. *Mailing Add:* 44 Church St New Paltz NY 12561

SILBERGLEIT, ALLEN, b Springfield, Mass, Mar 8, 28; m 56; c 3. SURGERY, PHYSIOLOGY. *Educ:* Univ Mass, BA, 49, MS, 51; Univ Cincinnati, MD, 55; Wayne State Univ, PhD(physiol), 65; Am Bd Surg, dipl, 61; Bd Thoracic Surg, dipl, 65. *Prof Exp:* From intern to resident surg, Univ Minn, 55-60; med officer, USAF & chief surg serv, Sheppard AFB Hosp, Tex, 60-62; from instr to assoc prof, 62-70, PROF SURG & PHYSIOL, SCH MED, WAYNE STATE UNIV, 70- *Concurrent Pos:* Mich Heart Asn res grants; resident thoracic surg & clin investr, Allen Park Vet Admin Hosp, Detroit, 62-65; surgeon, Detroit Receiving Hosp, 62-, Harper Hosp, Detroit, 87- & dir surg, St Joseph Mercy Hosp, Pontiac, 66-; gov, Am Col Surgeons, 90- *Mem:* Fel Am Col Surgeons; Soc Thoracic Surg; Am Heart Asn; Am Physiol Soc; Sigma Xi; Am Trauma Soc. *Res:* Cardiopulmonary, vascular and gastrointestinal problems; surgical physiology; research methods; application of basic science in training of surgical residents. *Mailing Add:* St Joseph Mercy Hosp Pontiac MI 48341-2985

SILBERGLITT, RICHARD STEPHEN, b Brooklyn, NY, Mar 9, 42; m 66; c 2. SOLID STATE PHYSICS. *Educ:* Stevens Inst Technol, BS, 63; Univ Pa, MS, 64, PhD(physics), 68. *Prof Exp:* Res assoc theoret solid state physics, Univ Pa, 68; lectr physics, Univ Calif, Santa Barbara, 68-69; res assoc solid state physics, Brookhaven Nat Lab, 69-71; asst prog dir theoret physics, Physics Sect, NSF, 71, asst prog dir solid state & low temperature physics, Div Mat Res, 71-72, assoc prog dir solid state physics, 72-75; mem energy study staff, Nat Acad Sci, 76-80; mem staff, DHR Inc, 80-84, vpres, 84-87; sr scientist, Ques Tech, Inc, 87-90; CO SR SCIENTIST, TECH ASSESSMENT & TRANSFER, INC, 90- *Mem:* Am Phys Soc. *Res:* Effect of spin wave and spin-phonon interactions on dynamical properties; inelastic neutron scattering; unique properties of systems with layered structure of displaying soft modes; lattic instabilities in nonstoichiometric materials. *Mailing Add:* Tech Assessment & Transfer Inc Suite 212 133 Defense Hwy Annapolis MD 21401

SILBERHORN, GENE MICHAEL, b Lenawee Co, Mich, Apr 30, 38; m 60; c 2. MARINE BOTANY. *Educ:* Eastern Mich Univ, BS, 63; WVA Univ, MS, 65; Kent State Univ, PhD(bot), 70. *Prof Exp:* Teaching asst gen bot & gen biol, WVa Univ, 63-65, NSF teaching fel, 65; asst prof ecol & Plant ecol, Radford Col, 65-67; NSF trainee gen bot, Kent State Univ, 67-70; Killian fel & instr, Univ Alta, 70-72; sect head, Wetlands Res Sect, 72-80, assoc prof, 80-86 DEPT HEAD, WETLANDS ECOLOGY DEPT, VA INST MARINE SCI, 80-;. *Concurrent Pos:* ed, Wetlands J, Soc Wetlands Scientists, 84-87 & vpres, 88. *Mem:* Sigma Xi; Soc of Wetland Scientists. *Res:* Inventory and evaluation of tidal wetlands of Virginia; community structure of tidal freshwater marshes; monitoring of coastal habitats; reproductive life cycle of submerged aquatic vegetation. *Mailing Add:* Wetlands Ecol Dept Va Inst Marine Sci Gloucester Point VA 23062

SILBERLING, NORMAN JOHN, b Oakland, Calif, Nov 28, 28; m 78; c 2. STRATIGRAPHY, TECTONICS. *Educ:* Stanford Univ, BS, 50, MS, 53, PhD(geol), 57. *Prof Exp:* Geologist, US Geol Surv, 50-66; from assoc prof to prof geol, Stanford Univ, 66-75; GEOLOGIST, US GEOL SURV, 75- *Mem:* AAAS; Geol Soc Am; Am Geophys Union. *Res:* Pre-Tertiary stratigraphy and tectonics of western North America; paleontology and biostratigraphy of Triassic marine invertebrates. *Mailing Add:* US Geol Surv MS919, Box 25046 Denver Fed Ctr Denver CO 80225

SILBERMAN, EDWARD, b Minneapolis, Minn, Feb 8, 14; m 41; c 4. FLUID MECHANICS, WATER RESOURCES MANAGEMENT. *Educ:* Univ Minn, BCE, 35, MS, 36. *Prof Exp:* Water technician, State Planning Bd, Minn, 36; jr engr, flood control, Tenn Valley Authority, 37; jr engr construct, Minneapolis Dredging Co, 38; engr construct, US Civil Aeronaut Admin, 38-41 & 46; from res assoc to prof civil eng, 46-81, dir St Anthony Falls Hydraul Lab, 63-74, EMER PROF CIVIL ENG, UNIV MINN, MINNEAPOLIS, 82- *Concurrent Pos:* Comnr, Bassett Creek Water Mgt Comn, Hennepin County, Minn, 72-; chmn, Water Resources Planning & Mgt Div, Am Soc Civil Engrs, 81. *Mem:* Am Soc Civil Engrs; Soc Am Mil Engrs; Am Water Resources Asn (pres, 69); Int Asn Hydraul Res. *Res:* Water resources management; model studies of hydraulic and fluid flow phenomena, supercavitating flows; boundary layers; turbulence; air-water mixtures; flow losses in closed and open conduits; underwater acoustics. *Mailing Add:* 2325 Brookridge Ave Minneapolis MN 55422

SILBERMAN, ENRIQUE, b Buenos Aires, Arg, Dec 9, 21; m 49; c 2. MOLECULAR SPECTROSCOPY. *Educ:* Univ Buenos Aires, PhD(eng), 45. *Prof Exp:* Investr physics, Arg Atomic Energy Comn, 53-58, head dept, 58-63; prof, Univ Buenos Aires, 63-66; PROF PHYSICS, FISK UNIV, 66- *Concurrent Pos:* Guest prof, Univ Notre Dame, 63; consult, Arg Nat Coun Sci Res, 64; vis prof, Vanderbilt Univ, 67- *Mem:* AAAS; Am Asn Physics Teachers; Am Phys Soc; Arg Physics Asn. *Res:* Infrared and Raman spectroscopy; normal coordinates analysis; inorganic ions in solid solutions; vibrational determination of crystal structures; ferroelectrics; semiconductors. *Mailing Add:* Dept Physics Fisk Univ Nashville TN 37208

SILBERMAN, MILES LOUIS, b New York, NY, Sept 25, 40; m 61; c 1. GEOCHEMISTRY. *Educ:* City Univ New York, BS, 63; Univ Rochester, MS, 67, PhD(geol), 71. *Prof Exp:* Geologist, Pac Mineral Resources Br, 67-75, geologist, Alaskan Geol Br, 76-81, GEOLOGIST, MARINE GEOL, US GEOL SURV, 81- *Mem:* Geol Soc Am; Mineral Asn Can; Geochem Soc; Soc Econ Geologists. *Res:* Geochronology and geochemistry of igneous rocks and their associated ore deposits using chemical, isotopic and geol mapping techniques; exploration for precious and base metal deposits; regional and structural controls on localization of ore deposits. *Mailing Add:* 63 S Devinney St Golden CO 80401

SILBERMAN, ROBERT G, b New York, NY, Aug 20, 39; m 61; c 2. ORGANIC CHEMISTRY, CHEMICAL EDUCATION. *Educ:* Brooklyn Col, BS, 60; Cornell Univ, MS, 63, PhD(org chem), 65. *Prof Exp:* From asst prof to assoc prof, 65-82, PROF CHEM, STATE UNIV NY CORTLAND, 82- *Concurrent Pos:* Vis prof, Cornell Univ, 84 & 85, Univ Glasgow, Scotland; consult, Crown Restoration; mem chem & community implementation team, Chem Educ Div, Am Chem Soc, 88-90; consult, KSR Inc. *Mem:* Am Chem Soc; Nat Sci Teachers Asn; Sigma Xi. *Res:* Organic analysis; chemical education; laboratory programs; historic paint analysis; historic restoration of art objects. *Mailing Add:* Dept Chem State Univ of NY Cortland NY 13045

SILBERMAN, RONALD, b Jackson, Mich, July 31, 32; m 55; c 3. CLINICAL MICROBIOLOGY, IMMUNOLOGY. *Educ:* Temple Univ, BA, 58; Hahnemann Med Col, MS, 60; Univ Md, Baltimore City, PhD(microbiol), 70. *Prof Exp:* Asst prof microbiol & immunol, 69-74, asst prof, 71-74, ASSOC PROF PATH, SCH MED, MED CTR, LA STATE UNIV, SHREVEPORT, 74- *Concurrent Pos:* Chief microbiologist, Lab Serv, Vet Admin Med Ctr, Shreveport, 71-84; dir microbiol, Clin Lab, Med Ctr, La State Univ Hosp, Shreveport, 71-, asst dir, 72-; clin prof microbiol, Med Technol, La Tech Univ, Ruston, 81- *Mem:* Am Soc Microbiol; Asn Practrs in Infection Control. *Res:* Rickettsiology. *Mailing Add:* Dept Path Med Ctr La State Univ PO Box 33932 Shreveport LA 71130-3932

SILBERNAGEL, BERNARD GEORGE, b Wausau, Wis, Dec 13, 40; m 65; c 2. PHYSICS. *Educ:* Yale Univ, BS, 62; Univ Calif, San Diego, MS, 64, PhD(physics), 66. *Prof Exp:* Lectr physics, Univ Calif, Santa Barbara, 66-68, asst prof, 68-72; mem staff, 72-81, SR RES ASSOC, EXXON RES & ENG CO, 81- *Mem:* AAAS; Am Phys Soc; Am Chem Soc. *Res:* Magnetic resonance; solid state physics; superconductivity; magnetism. *Mailing Add:* Esso Res/Engr Co Clinton Township Rte 22 E Annandale NJ 08801

SILBERNAGEL, MATT JOSEPH, b Hague, NDak, May 13, 33; m 55; c 5. PLANT PATHOLOGY, PLANT BREEDING. *Educ:* Univ Wash, BS, 57; Wash State Univ, PhD(plant path), 61. *Prof Exp:* PLANT PATHOLOGIST, CROPS RES DIV, AGR RES SERV, USDA, 61- *Concurrent Pos:* AID consult, Brazil, 70, India-Pakistan, 77, Cent Int de Agr Trop Workshops, 75 & 81 & AID Title XII Tanzanian-Wash State Univ, Bean Collabr Res Support Prog, 80-92. *Mem:* Am Phytopath Soc; Am Soc Hort Sci. *Res:* Breeding common beans for disease resistance, environmental stress tolerance and seed quality; study of bean-disease-environment interactions; improvement of beans for direct mechanical harvesting. *Mailing Add:* Irrig Agr Res & Exten Ctr USDA Rte 2 Box 2953A Prosser WA 99350-9687

SILBERSCHATZ, ABRAHAM, b Haifa, Israel, May 1, 47; m 68; c 3. OPERATING SYSTEMS, DATABASE SYSTEMS. *Educ:* State Univ NY, Stony Brook, PhD(comput sci), 76. *Prof Exp:* From asst prof to assoc prof, 76-84, PROF COMPUTER SCI, UNIV TEX, AUSTIN, 84- *Mem:* Asn Comput Mach; Inst Elec & Electronic Engrs. *Res:* Operating systems; distributed systems; database systems. *Mailing Add:* Dept Comput Sci Univ Tex Austin TX 78712

SILBERSTEIN, EDWARD B, b Cincinnati, Ohio, Sept 3, 36; div; c 2. NUCLEAR MEDICINE, ONCOLOGY. *Educ:* Yale Univ, BS, 58; Harvard Univ, MD, 62; Am Bd Internal Med, cert, 69, recert, 80, cert hemat, 72, cert oncol, 81; Am Bd Nuclear Med, cert, 72. *Prof Exp:* Resident internal med, Cincinnati Gen Hosp, 63-64 & Univ Hosps of Cleveland, 66-67; trainee hemat, New Eng Med Ctr Hosps, 67-68; asst prof med, Univ Cincinnati, 70-75, assoc prof radiol, 72-77; assoc prof med, 75-78, PROF MED, UNIV CINCINNATI, 78-; ASSOC DIR NUCLEAR MED, RADIOISOTOPE LAB, UNIV CINCINNATI MED CTR, 68- *Concurrent Pos:* Am Cancer Soc res grants, 72 & 74; chmn adv panel radiopharmaceut, US Pharmacopoeia, 85 & 90 & mem, standards subcomt on radiopharmaceut 85-90; prof radiol, Univ Cincinnati, 77-; dir, Nuclear Med, Jewish Hosp, 76- *Honors & Awards:* Sigma Xi. *Mem:* AAAS; Am Col Physicians; Radiation Res Soc; Am Soc Hemat; fel Royal Soc Health; Am Soc Clin Oncol. *Res:* Effect of radiation on chromosomes; biological indicators of radiation damage; diagnosis of tumors with radiopharmaceuticals. *Mailing Add:* ML 577 Univ Cincinnati Med Ctr Cincinnati OH 45267

SILBERSTEIN, OTMAR OTTO, b Graz, Austria, Apr 18, 21; US citizen; m 47; c 2. FOOD SCIENCE. *Educ:* Mich State Univ, BS, 49, MS, 50; Cornell Univ, PhD(veg crops), 53. *Prof Exp:* Res asst, US Plant, Soil & Nutrit Lab, NY, 53-54; res chemist, Welch Grape Juice Co, 54-58; head dept food technol, Wallerstein Labs Div Baxter Labs, 58-63; dir res & develop, Gilroy Foods, Inc, 63-80; dir explor technol, McCormick & Co, Inc, 80-86; CONSULT, 86- *Mem:* Am Chem Soc; Inst Food Technologists; Am Asn Cereal Chemists. *Res:* Food enzymology and flavor chemistry; dehydration of vegetables; biotechnology; tissue culture. *Mailing Add:* 841 Sixth St Gilroy CA 95020

SILBERT, DAVID FREDERICK, b Cambridge, Mass. BIOCHEMISTRY. *Educ:* Harvard Univ, AB, 58, MD, 62. *Prof Exp:* Intern & resident med, Sch Med, Washington Univ, 62-64; res assoc microbial genetics, Nat Inst Arthritis & Metab Dis, 64-66; Am Cancer Soc fel biol chem, Sch Med, 66-68, from asst prof to assoc prof, 68-77, PROF BIOL CHEM, SCH MED, WASHINGTON UNIV, 77- *Concurrent Pos:* NIH res grant, 68-; res grant, Am Chem Soc, 75-; mem microbiol chem study sect, NIH, 75- *Mem:* Am Soc Biol Chem; Am Soc Microbiol; Am Chem Soc. *Res:* Biochemical genetics; membrane chemistry. *Mailing Add:* Dept Biol Chem Wash Univ Med Sch 4566 Scott Ave St Louis MO 63110

SILBERT, JEREMIAH ELI, BIOCHEMISTRY, METABOLISM OF PROTEOGLYCANS. *Educ:* Harvard Univ, MD, 57. *Prof Exp:* PROF MED, MED SCH, HARVARD UNIV, 81- *Mailing Add:* 106 Spooner Rd Brookline MA 02115

SILBERT, LEONARD STANTON, b Philadelphia, Pa, Dec 16, 20. PHYSICAL CHEMISTRY, ORGANIC CHEMISTRY. *Educ:* Philadelphia Col Pharm, BSc, 43; Univ Pittsburgh, PhD(chem), 53. *Prof Exp:* Jr prof asst, Eastern Regional Res Labs, USDA, 46-47, Nat Renderers Asn res fel, Eastern Utilization Res Br, 53-56, res chemist, Eastern Regional Res Ctr, 56-85; RETIRED. *Mem:* AAAS; Am Chem Soc; Am Oil Chemists Soc; Royal Soc Chem. *Res:* Synthesis of lipid compounds and their physical properties; structural determinations and applications of peroxide chemistry; organic thiocyanations; isopropenylation chemistry; free radical studies; food irradiation; carbanian. *Mailing Add:* Apt 105 Pastorius Bldg 7800 C Stenton Ave Philadelphia PA 19118

SILBEY, ROBERT JAMES, b Brooklyn, NY, Oct 19, 40; m 62; c 2. PHYSICAL CHEMISTRY. *Educ:* Brooklyn Col, BS, 61; Univ Chicago, PhD(chem), 65. *Prof Exp:* Nat Acad Sci-Nat Res Coun-Air Force Off Sci Res fel, Univ Wis, 65-66; from asst prof to assoc prof, 66-76, PROF CHEM, MASS INST TECHNOL, 77- *Concurrent Pos:* Sloan Found res fel, 68-70; Camille & Henry Dreyfus Found teacher scholar, 71-76; vis prof, Inst Theoret Physics, Univ Utrecht, 72-73 & 86; Guggenheim Found fel, 72-73. *Honors & Awards:* Alexander von Humboldt Sr Scientist Award, 89. *Mem:* Fel Am Phys Soc. *Res:* Quantum chemistry; theory of excited states of solids and molecules. *Mailing Add:* Dept Chem Mass Inst Technol Cambridge MA 02139

SILCOX, JOHN, b Saltash, Eng, May 26, 35; m 60; c 3. INELASTIC ELECTRON SCATTERING. *Educ:* Bristol Univ, BSc, 57; Cambridge Univ, PhD(physics), 61. *Prof Exp:* From asst prof to assoc prof eng physics, 61-70, actg chmn dept appl physics, 70-71, dir sch appl & eng physics, 71-74 & 79-83, PROF ENG PHYSICS, CORNELL UNIV, 70- *Concurrent Pos:* Guggenheim Found fel, 67-68; mem solid state sci comt, Nat Res Coun, Nat Acad Sci, 78-82, chmn, 87-88. *Mem:* Fel Am Phys Soc; Electron Micros Soc Am(pres, 79). *Res:* Transmission electron microscopy; defects in crystals; inelastic electron scattering. *Mailing Add:* Dept Eng Physics Cornell Univ 210 Clark Hall Ithaca NY 14853

SILCOX, WILLIAM HENRY, b Taft, Calif, July 21, 22; m 52; c 5. OFFSHORE DRILLING & OIL PRODUCTION SYSTEMS. *Educ:* Univ Calif, Berkeley, BS, 47. *Prof Exp:* Engr, Standard Oil Co Calif, 47-51, lead mech engr, 54-62; proj engr, USAF, Wright Field, 53-54; sr offshore engr, Offshore Technol & Planning Staff, Chevron Corp, 62-67, div offshore engr, 67-78, offshore eng mgr, 75-85; consult, 85-88; vpres technol, Wellstream Corp, 88-89; RETIRED. *Concurrent Pos:* Consult, NSF Deep Sea Drilling Proj, 67-85; mem, Nat Petrol Coun Task Group Arctic Oil & Gas Reserves, 78-79, Nat Res Coun Marine Bd, 80-83 & Ad Hoc Comt Int Standards, Am Petrol Inst, 83-85. *Honors & Awards:* Offshore Distinguished Achievement

Award for Individuals, 88. *Mem:* Nat Acad Eng; Soc Petrol Engrs. *Res:* Operation and development of offshore oil platforms; subsea drilling and production systems; arctic drilling and production systems for water depth to 6000 feet. *Mailing Add:* 14 Fernwood Dr San Francisco CA 94127

SILEN, WILLIAM, b San Francisco, Calif, Sept 13, 27; m 47; c 3. SURGERY. *Educ:* Univ Calif, Berkeley, BA, 46; Univ Calif, San Francisco, MD, 49; Am Bd Surg, dipl, 58. *Hon Degrees:* MA, Harvard Univ, 66. *Prof Exp:* Intern, Univ Calif Hosp, San Francisco, 49-50, asst resident gen surg, 50; ward surgeon, Travis AFB, 50-52; asst resident gen surg, Univ Calif Hosp, San Francisco, 52-56, chief resident, 56-57; from instr to asst prof surg, Sch Med, Univ Colo, 57-60; from asst prof to assoc prof, Sch Med, Univ Calif, San Francisco, 60-66; Johnson & Johnson prof surg, 75, PROF SURG, HARVARD MED SCH, 66-, DIR, ADVAN CLIN SURG, 66-; SURGEON-IN-CHIEF, BETH ISRAEL HOSP, BOSTON, 66- *Concurrent Pos:* Asst chief surg, Denver Vet Admin Hosp, 57-59, chief, 59-60; asst chief, San Francisco Gen Hosp, 60-61, chief, 61-66; chmn, Dept Surg, Beth Israel Hosp, 66-; consult surg, Children's Hosp Med Ctr, Boston, 68-; mem, Vet Admin Res Comt, 69-73; mem, Am Bd Surg, 70-73, sr mem, 73-; ed, Yr Bk Surg, 70-84; chmn, Comt Res, Am Gastroenterol Asn, 71-73, mem gov bd, 72-75; mem, Surg B Study Sect, NIH, 72-76; chmn, Workgroup on Eval of Ther, Nat Comn Digestive Dis, 77-78; co-ed Surg Alert, 84, consult ed, Gastroenterol, 86-91. *Mem:* Inst Med-Nat Acad Sci; Soc Clin Surgeons; Soc Univ Surgeons; fel Am Col Surgeons; Asn Acad Surg; Am Surg Asn; AMA; Soc Exp Biol & Med; Am Fedn Clin Res; Am Gastroenterol Asn (vpres, 76, pres-elect, 77, pres, 78). *Res:* Protective mechanisms in gastric and duodenal mucosa; physiologic effects of gastrointestinal operations. *Mailing Add:* Beth Israel Hosp 330 Brookline Ave Boston MA 02215

SILER, WILLIAM MACDOWELL, b Houston, Tex, Aug 5, 20; m 74; c 3. BIOMATHEMATICS, THEORETICAL BIOLOGY. *Educ:* Stevens Inst Technol, MS, 49; City Univ New York, PhD(biol), 72. *Prof Exp:* Proj engr, Exp Towing Tank, Stevens Inst Technol, 47-49; physicist radiol physics, Mem Sloan-Kettering Cancer Ctr, Mem Hosp, New York, 59-65; assoc prof & chmn med comput sci prog, Downstate Med Ctr, Brooklyn, 65-72; prof biomath, Univ Ala, Birmingham, 72-80, chmn dept, 72-75; dir clin comput, Caraway Med Ctr, Birmingham, Ala, 80-87; CONSULT, 87- *Concurrent Pos:* Mem, Study Sect Comput & Biomath Res, NIH, 69-73, Biostatist & Epidemiol Contract Rev Comt, Nat Cancer Inst, 81-85. *Honors & Awards:* Distinguished Data Processing Prof, Data Processing Mgt Asn, 84. *Mem:* Am Asn Physicists Med; Biomed Eng Soc; Sigma Xi; Inst Elec & Electronics Engrs. *Res:* Artificial intelligence and image processing; modelling biological systems; radiological physics; computers in cardiology. *Mailing Add:* Cardiol Dept Carraway Methodist Med Ctr Birmingham AL 35234

SILER-KHODR, THERESA M, b Pomona, Calif, June 17, 47; m 74; c 3. REPRODUCTIVE ENDOCRINOLOGY. *Educ:* Immaculate Heart Col, BA, 68; Univ Hawaii, PhD(biochem), 71. *Prof Exp:* Teaching asst biochem, Univ Hawaii, 68-71; scientist reproductive biol, Univ Calif San Diego, 72-74; asst prof obstet & gynec, Am Univ Beirut, 74-76; from asst prof to assoc prof, 76-86, PROF OBSTET & GYNEC, UNIV TEX, SAN ANTONIO, 86- *Concurrent Pos:* Prin investr, NIH, 81-84, core dir, 78- *Mem:* Am Endocrine Soc; Soc Gynec Invest; Am Fertil Soc; Sigma Xi; Soc Study Reproduction. *Res:* The nature and function of human chorionic gonadotropin releasing hormone in the human placenta. *Mailing Add:* 7703 Floyd Curl Dr San Antonio TX 78230

SILEVICH, MICHAEL B, b Boston, Mass, Oct 10, 42. IONOSPHERIC & SPACE PHYSICS. *Educ:* Northeastern Univ, BS, 65; MSEE, 66, PhD(elec eng), 71; Brandeis Univ, MA, 70. *Prof Exp:* PROF ELEC & COMPUT ENG, NORTHEASTERN UNIV, 81- *Concurrent Pos:* Travel grant, Fulbright, 73; dir, Ctr Electromagnetics Res, 83- *Mem:* Sr mem Inst Elec & Electronics Engrs; Am Geophys Union. *Res:* Ionospheric and space physics; microprocessor based design; electromagnetic theory; plasma science and engineering; space systems and electron devices; microwave devices. *Mailing Add:* Ctr Electromagnetics Res Northeastern Univ 235 Forsyth Bldg Boston MA 02115

SILEVITCH, MICHAEL B, ELECTROMAGNETISM. *Educ:* Northeastern Univ, BS, 65, MS, 66, PhD(elec eng), 71; Brandeis Univ, MA, 70. *Prof Exp:* PROF ELEC & COMPUT ENG, NORTH EASTERN UNIV, 81-, DIR, CTR ELECTRO-MAGNETICS RES, 83- *Mem:* Inst Elec & Electronical Engrs; Am Geophysical Union. *Res:* Space physics, particularly in the temporal and spatial structure of the Aurora Borealis and the propagation of magnetic storms; additional research interests include the kinetics of the photographic development process, the statistical mechanics of dense plasmas, properties of solid state plasma, electron and ion beam dynamics, and high power microwave devices. *Mailing Add:* Ctr for Electromagnetics Res Northeastern Univ 235 Forsyth Bldg Boston MA 02115

SILFLOW, CAROLYN DOROTHY, b Kendrick, Idaho, Jan 8, 50; m 81. CELL BIOLOGY. *Educ:* Pac Lutheran Univ, BS, 72; Univ Ga, PhD(bot), 77. *Prof Exp:* Fel, Dept Biol, Yale Univ, 77-81; asst prof cell & develop biol, 82-88, ASSOC PROF, DEPT GENETICS & CELL BIOL, UNIV MINN, 88- *Concurrent Pos:* Instr physiol, MBL, Woods Hole, Mass, 83. *Mem:* Am Soc Cell Biol; Soc Develop Biol; Int Soc Plant Molecular Biol. *Res:* Regulation of flagellar gene expression following removal of flagella in Chlamydomonas; tubulin gene structure and expression; molecular basis for microtubule heterogeneity; plant tubulin genes. *Mailing Add:* Dept Genetics & Cell Biol Univ Minn 250 Biosci Ctr St Paul MN 55108

SILFVAST, WILLIAM THOMAS, b Salt Lake City, Utah, June 7, 37; m 59; c 3. PHYSICS. *Educ:* Univ Utah, BS(math) & BS(physics), 61, PhD(physics), 65. *Prof Exp:* Res assoc physics, Univ Utah, 65-66; NATO fel, Oxford Univ, 66-67; MEM TECH STAFF, BELL TEL LABS, 67- *Concurrent Pos:* NATO fel. *Mem:* Am Phys Soc; Optical Soc Am. *Res:* Lasers and plasma physics, with specific interest in gaseous and metal vapor lasers; recombination lasers and laser-produced plasmas. *Mailing Add:* Bell Labs Rm 4C-424 Holmdel NJ 07733

SILHACEK, DONALD LE ROY, b Norfolk, Nebr, Nov 9, 37; m 60; c 2. ENDOCRINOLOGY, INSECT DEVELOPMENT. *Educ:* Univ Nebr, Lincoln, BSc, 58, MSc, 61; Univ Wis-Madison, PhD(biochem, entom), 66. *Prof Exp:* RES CHEMIST, INSECT ATTRACTANTS LAB, AGR RES SERV, USDA, 65- *Concurrent Pos:* Instr, Armstrong State Col, 66-68; asst prof, Univ Fla, 70-80, prof, 80- *Mem:* Am Chem Soc; Am Soc Zoologists. *Res:* Endogenous mechanisms controlling developmental protein metabolism in insects; hormonal mechanisms controlling development of insects. *Mailing Add:* Insect Attractants Lab USDA PO Box 14565 Gainesville FL 32604

SILHAVY, THOMAS J, b Wauseon, Ohio, Jan 13, 48; m; c 2. PROTEIN EXPORT. *Educ:* Harvard Univ, PhD(biochem), 75. *Hon Degrees:* DSc, Ferris State Col, 82. *Prof Exp:* Jane Coffin Childs Found fel, 75-77; Med Found Res fel, 78-79; instr advan bact genetics, Cold Spring Harbor Lab, 81-85; PROF MOLECULAR BIOL, PRINCETON UNIV, 84-, DIR GRAD STUDIES, MOLECULAR BIOL DEPT, 89- *Concurrent Pos:* Mem, Life Scis Peer Rev Comt, 85-, prog dir, Genetics Predoctoral Training Grant, NIH, 87-, co-dir, Life Scis Res Found, 89- *Honors & Awards:* Advan Technol Achievement Award, Litton, 82. *Res:* Genetic analysis in Escherichiacoli; molecular mechanisms of protein export and secretion; control of gene expression by transmember signal transduction. *Mailing Add:* Dept Molecular Biol Princeton Univ Princeton NJ 08544

SILJAK, DRAGOSLAV (D), b Beograd, Yugoslavia, Sept, 10, 33; m 67; c 2. SYSTEMS THEORY, CONTROL ENGINEERING. *Educ:* Univ Beograd, BSEG, 58, MSEE, 61, ScD(elec eng), 63. *Prof Exp:* Docent prof elec eng, Univ Beograd, 63-64; from assoc prof to prof, 64-84, B & M PROF ELEC ENG, UNIV SANTA CLARA, 84- *Concurrent Pos:* NASA res grant, Univ Santa Clara, 65-77; NSF grant, 68-69; Dept Energy contract, 77-80; fel Japan Soc Promotion Sci; Distinguished prof, Fulbright Found. *Mem:* Fel Inst Elec & Electronics Engrs; hon mem Serbian Acad Arts & Sci. *Res:* Concepts, characterizations and methodology in studies of dimensionality, uncertainty and structure of complex dynamic systems, with applications to electric power networks, large space structures, model ecosystems and economics; population biology; space vehicles. *Mailing Add:* Dept Elec Eng Univ Santa Clara Santa Clara CA 95053

SILK, JOHN KEVIN, b Cambridge, Mass, May 6, 38; m 60; c 2. X-RAY OPTICS, INFRARED INSTRUMENTATION. *Educ:* Harvard Univ, AB, 60; Mass Inst Technol, PhD(physics), 69. *Prof Exp:* Scientist asst, Raytheon Co, 60-64, consult, 64-68; sr staff scientist, Am Sci & Eng Co, 69-88; CHIEF SCIENTIST BLOCK ENG, CONTRAVES, 88- *Concurrent Pos:* Res asst plasma physics, Mass Inst Technol, 66-69. *Honors & Awards:* Skylab Achievement Award, NASA, 74 & Solar Physics Group Award, 75. *Mem:* Am Phys Soc; Int Soc Optical Eng. *Res:* X-ray telescope on skylab; x-ray imaging systems; x-ray diagnostics of laboratory fusion plasmas; x-ray lithography; nondestructive testing; infrared spectroscopy; IR scene simulation. *Mailing Add:* 992 Memorial Dr Cambridge MA 02138

SILK, JOSEPH IVOR, b London, Eng, Dec 3, 42; m 68; c 2. COSMOLOGY, PARTICLE ASTROPHYSICS. *Educ:* Cambridge Univ, MA, 63; Harvard Univ, PhD(astron), 68. *Prof Exp:* Res fel, Cambridge Univ, 68-69; res assoc, Princeton Univ, 69-70; from asst prof to assoc prof, 70-78, PROF ASTRON, UNIV CALIF, BERKELEY, 78- *Concurrent Pos:* Alfred P Sloan Found fel, 72-74; John Simon Guggenheim Found fel, 75-76; mem, Inst Advan Study, Princeton, 75-76; Miller prof, Miller Found Basic Res, 80-81; res assoc, Inst Astrophys, Paris, 82-83; prin investr, NASA, NSF, Dept Energy & Calspace. *Mem:* Am Astron Soc; Royal Astron Soc; Int Astron Union; fel AAAS; Am Physiol Soc. *Res:* Theoretical studies of cosmology, galaxy formation and star formation; interface of particle physics and astrophysics. *Mailing Add:* Dept Astron Univ Calif Berkeley CA 94720

SILK, MARGARET WENDY KUHN, b Baltimore, Md, Nov 16, 46; m 68; c 2. QUANTITATIVE BOTANY, AGRICULTURE. *Educ:* Harvard Univ, BA, 68; Univ Calif, Berkeley, PhD(bot), 75. *Prof Exp:* Teaching asst, Univ Calif, Berkeley, 70-75; res fel biol, Univ Pa, 75-76; ASST, ASSOC & RES SCIENTIST, AGR EXP STA & ASST, ASSOC & PROF QUANT PLANT SCI, UNIV CALIF, DAVIS, 76- *Concurrent Pos:* Vis Professorship Women, Nat Sci Found, Univ Calif, Berkeley, 86. *Mem:* Am Soc Plant Physiol; Bot Soc Am; Soc Develop Biol; Soc Math Biol. *Res:* Quantitative aspects of plant development, morphogenesis, physiology and plant-water-environment interactions. *Mailing Add:* 128 D St Davis CA 95616

SILL, ARTHUR DEWITT, b Akron, Ohio, Dec 1, 21; m 47; c 3. STRUCTURAL CHEMISTRY, ORGANIC CHEMISTRY. *Educ:* Ohio State Univ, BS, 48; Univ Cincinnati, MS, 61, PhD(org chem), 64. *Prof Exp:* Prin chemist, Battelle Mem Inst, 49-55; res asst org prod res, Merrell-Nat Labs Div, Richardson-Merrell, Inc, 55-64, proj leader org res dept, 64-71, org chemist, Anal Res Dept, 71-80, sr anal chemist, Merrell Dow Res Inst, 80-86; RETIRED. *Mem:* Am Chem Soc. *Res:* Identification of structure of organic substances, including impurities and metabolites, via spectroscopic methods, including mass spectrometry and nuclear magnetic resonance; organic synthesis; development of analytical methods for drug substances. *Mailing Add:* 22 Beckford Dr Greenhills OH 45218

SILL, CLAUDE WOODROW, b Layton, Utah, Oct 29, 18; m 55; c 3. ANALYTICAL CHEMISTRY. *Educ:* Univ Utah, AB, 39, MA, 41. *Prof Exp:* Researcher, Iowa State Col, 41-42; chemist, US Bur Mines, 42-51; chief anal chem br, AEC, 51-75; chief anal chem br, Health Serv Lab, Energy Res & Develop Admin, 75-77; sr scientist, Radiol & Environ Sci Lab, Dept Energy, 77-80; sr scientist, 80-82, sci specialist, 82-87, PRIN SCIENTIST, EG&G IDAHO, INC, 87- *Mem:* Sigma Xi; Health Physics Soc. *Res:* Synthesis of organo-uranium compounds: analytical methods. *Mailing Add:* 2751 S Blvd Idaho Falls ID 83404

SILL, LARRY R, b Fairmont, Minn, Sept 10, 37; m 59; c 2. MAGNETISM, LOW TEMPERATURE SOLID STATE PHYSICS. *Educ:* Carleton Col, BA, 59; Iowa State Univ, PhD(physics), 64. *Prof Exp:* From instr to prof, 64-78, actg head dept, 66-68, assoc dean, Col Lib Arts & Sci, 69-80, PROF PHYSICS, NORTHERN ILL UNIV, 78-; DIR, TECHNOL COMMERCIALIZATION CTR, 86- *Concurrent Pos:* Vis scientist, Argonne Nat Lab, 83-84. *Mem:* Am Phys Soc; Am Asn Physics Teachers; Sigma Xi; SUPA; Licensing Exec Soc. *Res:* Measurements of the low temperature transport and magnetic properties of rare earth compounds and alloys in bulk and in multilayered structures. *Mailing Add:* Dept Physics Northern Ill Univ De Kalb IL 60115

SILL, WILLIAM ROBERT, b Cleveland, Ohio, Oct 14, 37; m 63; c 2. GEOPHYSICS. *Educ:* Mich State Univ, BS, 60; Mass Inst Technol, MS, 63, PhD(geophys), 68. *Prof Exp:* Mem tech staff, Bellcomm Inc, Washington, DC, 67-72; assoc res prof, 72-81, RES PROF GEOPHYS, UNIV UTAH, 81-; PROF PHYSICS, MONTANA COL MINERAL SCI, 83- *Mem:* AAAS; Am Geophys Union. *Res:* Electrical properties of rocks and planetary interiors; solar wind-planetary interactions; lunar geophysics; mineral exploration. *Mailing Add:* Dept Physics & Geophys Eng Montana Col Mineral, Sci & Technol Butte MT 59701

SILLA, HARRY, b Jersey City, NJ, Dec 7, 29; m 57; c 3. CHEMICAL ENGINEERING. *Educ:* City Col New York, BChE, 54; Stevens Inst Technol, MS, 61, PhD(chem eng), 70. *Prof Exp:* Proj engr, Eng Ctr, Columbia Univ, 54 & 56-57; chem engr, Stevens Inst Technol, 57-59; proj leader, Aerochem Res Labs, Sybron Corp, 59-64; asst prof 64-78, PROF CHEM ENG, STEVENS INST TECHNOL, 78- *Concurrent Pos:* Mem, Solid Waste Task Force, NJ Energy Inst. *Mem:* Am Chem Soc; Am Inst Chem Engrs; Sigma Xi. *Res:* Combustion; process engineering; waste treatment. *Mailing Add:* Stevens Inst of Technol Dept Chem Eng Castle Point Sta Hoboken NJ 07030

SILLECK, CLARENCE FREDERICK, b Brooklyn, NY, Nov 12, 09; m 34; c 1. CHEMISTRY, PIGMENT DISPERSION. *Educ:* Polytech Inst Brooklyn, BS, 32, MS, 36. *Hon Degrees:* DSc, Polytech Inst NY, 79. *Prof Exp:* Tech dir, C J Osborn Co, 32-68, pres, 68-69; VPRES, C J OSBORN CHEM, INC, PENNSAUKEN, NJ, 69- *Concurrent Pos:* Trustee, Polytech Inst New York. *Mem:* AAAS; Am Chem Soc; fel Am Inst Chemists; NY Acad Sci; Fedn Socs Coatings Technol. *Res:* Synthetic resins and pigment dispersions for protective coatings; colloidal dispersion of carbon black in cellulose nitrate; dispersion of carbon black in alkyd resins; specialty finishes. *Mailing Add:* 186-20 Henley Rd Jamaica NY 11432

SILLIKER, JOHN HAROLD, b Ayer's Cliff, Que, June 20, 22; nat US; m 79; c 2. MICROBIOLOGY. *Educ:* Univ Southern Calif, AB, 47, MS, 48, PhD(bact), 50; Am Bd Med Microbiol, dipl. *Prof Exp:* USPHS fel bact, Hopkins Marine Sta, Stanford Univ, 50-51; asst prof, Sch Med, Univ Rochester, 51-52; bacteriologist, George W Gooch Labs, Ltd, Calif, 52-53; chief bacteriologist & assoc dir res, Swift & Co, Ill, 53-61; res assoc & consult, Dept Path, St James Hosp, Chicago Heights, Ill, 61-67; pres, Silliker Labs, 67-87, CONSULT MICROBIOLOGIST, SILLIKER LABS, 87- *Concurrent Pos:* Mem res coun, Inst Am Poultry Industs & Am Meat Inst Found, 53-61; lectr, Meat Hyg Training Ctr, USDA, 61-; mem, Comt on Salmonella Sampling & Methodology in Egg Prod, 64-; vis prof food sci, Univ Ill, 65; mem, Comt Salmonella, Nat Res Coun-Nat Acad Sci, 67-68; consult, Int Comn Microbiol Specifications for Foods, 70-74, mem, 74 & treas, 80; div lectr, Food Microbiol Div, Am Soc Microbiol, 78 & consult, 87-; mem, Subcomt Microbiol Criteria for Foods, Bat Res Coun-Nat Acad Sci, 80-84. *Honors & Awards:* Sci Award, Am Pub Health Asn, 63; Harold Barnum Indust Award, Int Asn Milk, Food & Environ Sanitarians, 87. *Mem:* Fel Am Pub Health Asn; fel Inst Food Technologists; fel Am Acad Microbiol; Soc Appl Microbiol; Royal Soc Health. *Res:* Fatty acid oxidation by bacteria; enteric and food bacteriology; microbiological methods for examination of foods, particularly Salmonella and other enteric organisms. *Mailing Add:* 1237 Primavera Dr Palm Springs CA 92262

SILLIMAN, RALPH PARKS, b Seattle, Wash, June 26, 13; m 37; c 2. FISH BIOLOGY. *Educ:* Univ Wash, BS, 36. *Prof Exp:* Tech asst Pac halibut res, Int Fish Comn, 36-37; aquatic biologist Pac sardine res, US Fish & Wildlife Serv, 38-45, aquatic biologist, Pac salmon res, 45-49, adminr anadromous & inland fishery res, 49-63, fish pop res, US Bur Commercial Fisheries, 64-69 & Nat Marine Fisheries Serv, 70-73; CONSULT FISHERY BIOLOGIST, 73- *Honors & Awards:* Bronze Med, US DC, 73. *Mem:* Am Fisheries Soc; Am Inst Fishery Res Biol (pres, 64-66); Am Inst Biol Sci; Sigma Xi. *Res:* Population dynamics; mathematical population models. *Mailing Add:* 4135 Baker NW Seattle WA 98107

SILLMAN, EMMANUEL I, b Philadelphia, Pa, Dec 7, 15. PARASITOLOGY, INVERTEBRATE ZOOLOGY. *Educ:* Univ Mich, PhD(zool), 54. *Prof Exp:* Lectr & asst prof zool, Ont Agr Col, 53-58; asst prof, Univ Man, 58-60; from assoc prof to prof biol, Duquesne Univ, 60-80; RETIRED. *Concurrent Pos:* Epidemiol of trop dis, US Army. *Mem:* AAAS; Am Soc Zoologists; Am Soc Parasitologists; Soc Syst Zool; Am Micros Soc; Nat Ctr Sci Educ. *Res:* Life histories of digenetic trematodes and warble flies; disease and parasites of fish; biology and ecology of protozoa, flatworms and gastropods. *Mailing Add:* 623 Burton Dr Pittsburgh PA 15235-4423

SILLS, MATTHEW A, b New York, NY, May 17, 55. BIOCHEMISTRY. *Educ:* Rutgers Univ, BS, 78; Univ Pa, PhD(pharmacol), 84. *Prof Exp:* Res assoc pharmacol, NIH, 84-86; sr scientist, 86-87, sr res scientist, 88-89, MGR, BIOCHEM PROFILING, CIBA-GEIGY PHARMACEUT, 89- *Concurrent Pos:* Pratt fel, Nat Inst Gen Med Sci, NIH, 84-86. *Mem:* Nat Soc Neurosci; Am Chem Soc; Am Soc Pharmacol & Exp Therapeut; AAAS. *Mailing Add:* Profiling Dept Ciba-Geigy Pharmaceut 556 Morris Ave Summit NJ 07901

SILSBEE, HENRY BRIGGS, b Washington, DC, Jan 15, 23; m 56. PHYSICS. *Educ:* Harvard Univ, BS, 43, MA, 47, PhD(physics), 51. *Prof Exp:* Asst, Carnegie Inst Dept Terrestrial Magnetism, 41; jr physicist, Nat Bur Standards, 43-44; jr scientist Manhattan proj, Los Alamos Sci Lab, 44-46; asst prof physics, Univ Calif, Berkeley, 51-58; assoc prof, Univ Wash, Seattle, 58-63; mem staff, Brookhaven Nat Lab, 63-64; assoc prof, 64-67, PROF PHYSICS, STATE UNIV NY, STONY BROOK, 67- *Res:* Molecular beams; solid state physics; low temperature physics. *Mailing Add:* Dept of Physics State Univ of NY Stony Brook NY 11794

SILSBEE, ROBERT HERMAN, b Washington, DC, Feb 24, 29; m 50; c 3. SOLID STATE PHYSICS, MAGNETIC RESONANCE. *Educ:* Harvard Univ, AB, 50, MA, 51, PhD(physics), 56. *Prof Exp:* Mem staff, Solid State Div, Oak Ridge Nat Lab, 56-57; from instr to assoc prof, 57-65, PROF PHYSICS, CORNELL UNIV, 65- *Concurrent Pos:* Sloan res fel, 58-60; NSF sr fel, 65-66; Guggenheim Found fel, 73-74; ed, Solid State Communs, 71- *Mem:* Am Phys Soc; AAAS. *Res:* Solid state physics; magnetic resonance; radiation damage; optical properties; electronic and spin transport in small conducting systems. *Mailing Add:* Dept Physics Cornell Univ Ithaca NY 14853

SILVA, ARMAND JOSEPH, b Waterbury, Conn, June 1, 31; m 54, 68; c 3. OCEAN & CIVIL ENGINEERING. *Educ:* Univ Conn, BSE, 54, MS, 56, PhD(civil eng), 65. *Prof Exp:* Instr civil eng, Univ Conn, 55-56; soils engr, Thompson & Lichtner Co, 56-58; from instr to prof civil eng, Worcester Polytech Inst, 58-76, head dept, 71-76; PROF OCEAN & CIVIL ENG, UNIV RI, 76-, DIR, MARINE GEOMECH RES GROUP, 81-, CHMN OCEAN ENG, 84- *Concurrent Pos:* Consult, 58-; pres & treas, Geotechnics, Inc, 56-67; NSF sci-fac fel, 62-63; grants & contracts, Worcester Polytech Inst, 66-67 & 71-72, NSF, 70, 79, 80-83, 87-90, Shell Develop Co, 71 & 85-, Off Naval Res, 72-82, Dept Energy, 74-87, US Geol Surv, 82-84, Bedford Inst Oceanog, 82-85 & SAIC/Corps Engrs, 90. *Mem:* Am Soc Civil Engrs; Int Soc Soil Mech & Found Engrs; Am Geophys Union; Marine Technol Soc; AAAS. *Res:* Ocean engineering; soil mechanics; geotechnical properties of ocean sediments; marine sediment processes; sediment coring technology; soil-structure interaction; marine geomechanics; creep behavior; ice scouring; dredge disposal. *Mailing Add:* Dept Ocean Eng Univ RI Narragansett RI 02882

SILVA, JAMES ANTHONY, b Kilauea, Hawaii, Sept 4, 30; m 67; c 2. SOIL FERTILITY. *Educ:* Univ Hawaii, BS, 51, MS, 59; Iowa State Univ, PhD(soil biochem), 64. *Prof Exp:* Asst-in-training sugar cane, Exp Sta, Hawaiian Sugar Planter's Asn, 51-53, asst agronomist, 53-59; asst soil biochem, Iowa State Univ, 59-64; asst prof soil sci, 64-70, assoc prof soil sci & sta statistician, 70-76, PROF SOIL SCI, COL TROP AGR, UNIV HAWAII, HONOLULU, 76- *Concurrent Pos:* NSF fel, 64-66; sabbatical, Cornell Univ, 71-72 & Ore State Univ, 82; prin investr int proj, Benchmark Soils Proj, AID, 76-84. *Mem:* Soil Sci Soc Am; Am Soc Agron; Int Soc Soil Sci. *Res:* Nutrient availability and uptake by plants, especially nitrogen, phosphorus, silicon and zinc; soil and plant tissue testing for nutrient requirements of tropical soils and crops, especially maize and sugarcane; transfer of agrotechnology based on the soil family of soil taxonomy; modeling growth of sugar cane. *Mailing Add:* Dept Soil Sci Univ Hawaii Honolulu HI 96822

SILVA, OMEGA LOGAN, b Dec 14, 36; US citizen; m 58, 82; c 1. INTERNAL MEDICINE, ENDOCRINOLOGY. *Educ:* Howard Univ, BS, 58, MD, 67; Am Bd Internal Med, dipl, 74, Am Bd Internal Med, Endocrinol & Metabolism, cert, 79. *Prof Exp:* Chemist, NIMH, 58-63; from intern to resident internal med, Vet Admin Hosp, 67-70, res assoc endocrinol, 71-74, clin investr, 74-77; from asst prof to assoc prof med, George Washington Univ, 74-85; adj assoc prof oncol, Howard Univ, 77-85, PROF ONCOL, HOWARD UNIV, 85- *Concurrent Pos:* Fel endocrinol, Vet Admin Hosp, Washington, DC, 70-71; Lucy Moten Travel fel for Study & Travel, Europe, 58. *Honors & Awards:* Merck Index Award, 57. *Mem:* Am Chem Soc; Am Med Women's Asn; Endocrine Soc; Am Diabetes Asn; fel Am Col Physicians. *Res:* Calcium metabolism; function of human calcitonin. *Mailing Add:* 354 N St SW Washington DC 20024

SILVA, PATRICIO, b Santiago, Chile, July 7, 39; c 3. NEPHROLOGY. *Educ:* Cath Univ Chile, BS, 59, MD, 64. *Prof Exp:* Resident, Clin Hosp, Univ Chile, 64-67; staff physician & consult nephrol, Regional Hosp Talca, Chile, 67-69; prof nutrit, Univ Chile, 68, prof pathophysiol, Cent Univ Talca, 69; res assoc pharmacol, Dartmouth Med Sch, 69-70; fel, Yale Univ Sch Med, 70-71, res assoc, 72; instr, 72-74, asst prof, 74-79, ASSOC PROF MED, HARVARD MED SCH, 79-; CHIEF, RENAL DIV, NEW ENGLAND REACONESS HOSP & JOSLIN DIABETES CTR, 90- *Concurrent Pos:* Fel, Nat Kidney Found, 70 & Conn Heart Asn, 71; clin asst, Boston City Hosp, 72-74; investr, Am Heart Asn, 77. *Mem:* Am Fedn Clin Res; Am Soc Nephrol; Am Physiol Soc; Am Soc Exp Biol & Med; AAAS; Am Soc Clin Invest. *Mailing Add:* Dept Med 110 Francis St Boston MA 02215

SILVA, PAUL CLAUDE, b San Diego, Calif, Oct 31, 22. BOTANY. *Educ:* Univ Southern Calif, BA, 46; Stanford Univ, MA, 48; Univ Calif, PhD(bot), 51. *Prof Exp:* Res fel bot, Univ Calif, 51-52; from instr to assoc prof, Univ Ill, 52-61, sr herbarium botanist, 61-67, RES BOTANIST, UNIV CALIF, BERKELEY, 67- *Concurrent Pos:* Guggenheim fel, 58-59; ed, Phycologia 61-69. *Honors & Awards:* Darbaker Award, Bot Soc Am, 58. *Mem:* Bot Soc Am; Am Soc Plant Taxonomists; Phycol Soc Am (secy, 54-57, vpres, 57, pres, 58); Int Asn Plant Taxon; Int Phycol Soc (pres, 65). *Res:* Morphology, taxonomy and ecology of marine algae. *Mailing Add:* Herbarium Univ of Calif Berkeley CA 94720

SILVA, RICARDO, b Hong Kong, Nov 20, 31; US citizen; m 63; c 4. ORGANIC CHEMISTRY. *Educ:* Univ Sydney, BS, 53; Univ Calif, Los Angeles, PhD(chem), 61. *Prof Exp:* NIH fel, Imp Col, Univ London, 61-62; from asst prof to assoc prof, 62-70, chmn dept, 73-83, PROF CHEM, CALIF STATE UNIV, NORTHRIDGE, 70- *Concurrent Pos:* NSF inst grants, 62-63 & 64-65, fel, 66-67; NIH res grant, 64-65; Calif Inst Technol Pres Fund grant,

72-73, Air Force Off Sci Res, 85. *Mem:* Am Chem Soc. *Res:* Nuclear magnetic resonance; polycyclic compounds; natural products; structure and reactivity; biogenetic origins; photochemical reactions of naturally occurring compounds; synthesis of indole alkaloids. *Mailing Add:* Dept of Chem Calif State Univ Northridge CA 91330

SILVA, ROBERT JOSEPH, b Oakland, Calif, Feb 16, 27; m 50; c 3. NUCLEAR CHEMISTRY. *Educ:* Univ Ore, BS, 51; Univ Calif, Berkeley, PhD(chem), 59. *Prof Exp:* Chem engr, Kaiser Aluminum Corp, Calif, 50-51; health physicist, Radiation Lab, Univ Calif, 51-54, lab technician, 54-57, res asst, 57-59; chemist, Oak Ridge Nat Lab, 59-66; nuclear chemist, Lawrence Radiation Lab, Univ Calif, 66-68; res scientist, Spec Training Div, Oak Ridge Assoc Univs, 68-70; mem staff, Oak Ridge Nat Lab, 70-77; STAFF SCIENTIST, LAWRENCE BERKELEY LAB, UNIV CALIF, 77- *Mem:* Am Phys Soc; Am Chem Soc; Sigma Xi. *Res:* Low energy nuclear reactions and scattering; nuclear spectroscopy; nuclear instruments; transuranium nuclear research; transuranium chemical research. *Mailing Add:* 111 Regent Pl Alamo CA 94507-1801

SILVA-HUTNER, MARGARITA, b Rio Piedras, PR, Nov 28, 15; m 56; c 1. MYCOLOGY, MICROBIOLOGY. *Educ:* Univ PR, BA, 36; Harvard Univ, AM, 45, PhD, 52; Am Bd Med Microbiol, dipl. *Prof Exp:* Instr mycol & dermat, Sch Trop Med, PR, 36-45, assoc, 45-49, asst mycol, Dept Dermat, Col Physicians & Surgeons, NY, 50-53, res assoc, 53-56, asst prof mycol & head mycol lab, 56-63, ASSOC PROF DERMAT, COL PHYSICIANS & SURGEONS, COLUMBIA UNIV, 63- *Concurrent Pos:* Teaching asst, Boston Univ, 47 & 49; consult, Communicable Dis Ctr, USPHS, 52-53 & Squibb Inst Med Res, 54-55; mem standards & exam comt pub health & med Lab mycol, Am Bd Med Microbiol; mem tech adv comt, Am Type Cult Collection; dir mycol lab, Presby Hosp, New York, 56-; consult, USPHS Hosp in Stapleton, Staten Island, NY, 63-79; mycologist dermat serv, Presby Hosp, New York, 76- *Mem:* AAAS; fel Am Acad Microbiol; Am Soc Microbiol; Med Mycol Soc of the Americas (pres, 78-79); hon mem Soc Brasileira Dermat; Sigma Xi. *Res:* Morphology, taxonomy and biology of pathogenic fungi. *Mailing Add:* Col Physicians-Surgeons 630 W 168th St New York NY 10032

SILVEIRA, AUGUSTINE, JR, b New Bedford, Mass, July 17, 34; m 60; c 2. ORGANIC CHEMISTRY. *Educ:* Southeastern Mass Univ, BS, 57; Univ Mass, PhD(chem), 62. *Hon Degrees:* ScD, Southeastern Mass Univ, 75. *Prof Exp:* Teaching fel chem, Univ Mass, 57-58, teaching assoc, 58-60, instr, 60-62; asst prof, Rutgers Univ, 62-63; assoc prof chem, 63-64, prof chem, 64-76, DISTINGUISHED TEACHING PROF, STATE UNIV NY COL OSWEGO, 76-, CHMN DEPT, 67-, EVALUATOR GRAD PROGS, 68- *Concurrent Pos:* Dir res, NIH res grant, 62-63; State Univ NY fac res fel, 64 & 67, res grant in aid, 64-65 & 67-69; dir, NSF Undergrad Sci Equip grant, 65-67 & 77-78, partic grant org mech, NSF Conf, Vt, 66; Am Coun Educ fel, Univ Calif, Irvine, 69-70, 76-77 & 91; indust consult, 70-; mem comn higher educ, Mid States Asn Cols & Sec Schs, 71-; dir, NSF res grants, 73-74, 77-78, 79-82, 85-88; consult, NY State Educ Dept, 75; vis prof, Univ Calif, Irvine, 76-77, 84, Calif State Univ, Long Beach, 76-77; res corp grant, State Univ NY, 77-78, fac grant, 78-79, fac grant undergrad instr, 77-79 & 80-82; grant, Eastern Col Sci Conf, 78; mem optom bd, State NY, 80-; State Univ NY fac exchange scholar, 81- *Honors & Awards:* President's Res Award, 83, Recipient of over 30 grants & awards. *Mem:* AAAS; Am Chem Soc; fel Am Inst Chemists; Sigma Xi. *Res:* Structure, synthesis and reactions involving organonitrogen and organometallic chemistry. *Mailing Add:* 2021 Benson Ave PO Box 98 Minetto NY 13115

SILVEIRA, MILTON ANTHONY, b Mattapoisett, Mass, May 4, 29; c 4. MECHANICAL ENGINEERING. *Educ:* Univ Vt, BSME, 51; Univ Va, MSAE, 60. *Hon Degrees:* Dr, Univ Vt, 77. *Prof Exp:* Res intern, Dynamic Loads Div, Langley Field, 51 & studies on helicopter vibration & dynamic probs, Vibration & Flutter Br, 55-61, actg head loads sect, Struct Br, Space Task Group, 61-63, from asst br chief to dep chief, Aerodyn Br, Spacecraft Res Div, 63-65, from tech mgr to prog mgr, Little Joe II Launch Vehicle, 64-65, head, Flight Performance & Dynamics Br, 65-67, asst to dir eng & develop for spec proj, 67-68, chief, Eng Analysis Off, 68-69, mgr, Space Shuttle Eng Off, 69-73, dep prog, Space Orbiter Proj, Johnson Spacecraft Ctr, 73-81, asst to dep adminr, 81-82, CHIEF ENGR, HQ, NASA, 83- *Honors & Awards:* Sustained Superior Performance Award, NASA, 65 & 69, Except Serv Medal, 69 & 81. *Mem:* Am Inst Aeronaut & Astronaut. *Res:* Engineering systems analysis; engineering of the space shuttle systems and directing in-house design and analysis efforts. *Mailing Add:* 7213 Evans Mill McLean VA 22101

SILVER, ALENE FREUDENHEIM, b New York, NY, Oct 31, 16; div; c 3. DEVELOPMENTAL BIOLOGY, DERMATOLOGY. *Educ:* Columbia Univ, BA, 38; Univ Ill, PhD(physiol), 47. *Prof Exp:* Res asst circulatory physiol, Col Med, Univ Ill, 42-46; res assoc, Michael Reese Hosp, Chicago, 46-47; res assoc wound healing, Johns Hopkins Hosp, Baltimore, 48-49; res assoc skin physiol, Brown Univ, 61-64, res assoc develop genetics, 65-70, res asst prof, 69-70; assoc prof biol, 70-75, PROF BIOL, RI COL, 76- *Concurrent Pos:* USPHS biomed sci support grant, 66-68; assoc mem, Inst Life Sci, Brown Univ, 70-75; sr investr biol & med, Brown Univ, 75- *Mem:* Int Pigment Cell Soc; Sigma Xi; Am Soc Cell Biol; Am Soc Zoologists; Soc Develop Biol. *Res:* Physiology of skin; developmental genetics of hair regeneration; phenotypic modulation of hair germ melanocytes during hair cycle; development of the eye in mutant and normal mice. *Mailing Add:* Dept Biol RI Col Mt Pleasant Ave Providence RI 02908

SILVER, ARNOLD HERBERT, b Brooklyn, NY, Sept 27, 31; m 52; c 5. SOLID STATE PHYSICS. *Educ:* Rensselaer Polytech Inst, BS, 52, MS, 54, PhD(physics), 58. *Prof Exp:* Asst physics, Rensselaer Polytech Inst, 52-55 & Brown Univ, 55-57; res engr, Sci Lab, Ford Motor Co, 57-62, sr res scientist, 62-64, prin res scientist assoc, 64-65, staff scientist, 65-69; dir, Electronics Res Lab, Aerospace Corp, 69-80; Sr scientist, 81-84, sr staff engr, TRW, Inc, 84-87; MGR, SUPERCONDUCTIVE ELECTRONICS RES, 87- *Honors & Awards:* Inst Elec & Electronics Engrs Microwave Theory & Techniques Distinguished Microwave Lectr, 88-89. *Mem:* AAAS; fel Am Phys Soc; Inst Elec & Electronics Engrs. *Res:* Nuclear and electron magnetic resonance; radio frequency spectroscopy; superconductivity; quantum effects in superconductors; electronic techniques and instrumentation; superconducting and cryogenic devices. *Mailing Add:* RI/2170 TRW Space & Technol Group One Space Park Redondo Beach CA 90278

SILVER, BARNARD STEWART, b Salt Lake City, Utah, Mar 9, 33; m 63; c 2. ENERGY ENGINEERING. *Educ:* Mass Inst Technol, BS, 57; Stanford Univ, MS, 58. *Prof Exp:* Engr, Aircraft Nuclear Propulsion Div, Gen Elec, 57; engr sugar mach, Silver Eng Works, 59-66, mgr sales, 66-71; chief engr, Union Sugar Div, Consolidated Foods Co, 71-74; dir du complexe, Sodesucre, Abidjan, Cote d'Ivoire, 74-76; supt eng & maint, Moses Lake Factory, U&I, Inc, 76-79; PRES, SILVER ENTERPRISES, 71-; PRES, SILVER ENERGY SYSTS CORP, 80- *Concurrent Pos:* Instr eng, Big Bend Community Col, 80-81. *Honors & Awards:* Decorated Chevalier, l'Orde Nat, Repub de Cote d'Ivoire, 76. *Mem:* Am Soc Mech Engrs; Asn Energy Engrs; Am Soc Sugar Beet Technologists; Sigma Xi. *Res:* Agronomic practices, production and utilization of Jerusalem artichokes in world climates; extraction of inulin and accompanying constituents from the Jerusalem Artichoke. *Mailing Add:* 4391 S Carol Jane Dr Salt Lake City UT 84124-3601

SILVER, DAVID MARTIN, b Chicago, Ill, Sept 25, 41; m 63; c 2. THEORETICAL CHEMISTRY. *Educ:* Ill Inst Technol, BS, 62; Johns Hopkins Univ, MA, 64; Iowa State Univ, PhD(chem), 68. *Prof Exp:* NSF fel chem, Harvard Univ, 68-69; vis scientist, Europ Ctr Atomic & Molecular Calculations, Orsay, France, 70; supvr, chem physics res, 77-83, CHEMIST, APPL PHYSICS LAB, JOHNS HOPKINS UNIV, 70-, PRIN STAFF, 76-, SUPVR, COMPUTATIONAL PHYSICS RES, 84- *Concurrent Pos:* Vis prof chem, Johns Hopkins Univ, 84-85. *Mem:* AAAS; Am Chem Soc; Am Phys Soc; Am Geophys Union. *Res:* Molecular physics; electron correlation and electronic structure of atoms and molecules; fluid mechanics; ophthalmology; molecular interactions; mathematical analysis; electromagnetic scattering. *Mailing Add:* Appl Physics Lab Johns Hopkins Univ Laurel MD 20723

SILVER, DONALD, b New York, NY, Oct 19, 29; m 58; c 4. CARDIOVASCULAR SURGERY, THORACIC SURGERY. *Educ:* Duke Univ, AB, 50, BS & MD, 55; Am Bd Surg, dipl, 65, Gen Vascular Surg, 83; Am Bd Thoracic Surg, dipl, 67. *Prof Exp:* Asst prof surg, Med Ctr, Duke Univ, 64-66, from assoc prof surg & dir vascular clin to prof, 66-75; PROF SURG & CHMN DEPT, UNIV MO MED CTR, 75- *Concurrent Pos:* Co-prin investr, NIH grant, 64-71, prin investr, 71-75 & 75-80; attend physician, Vet Admin Hosp, Durham, 65-75; consult, Watts Hosp, Durham, 65-75; consult, Harry S Truman Vet Admin Hosp, Columbia, Mo, 75; mem bd sci adv, Cancer Res Ctr, Columbia, Mo, 75; James IV Surg traveler. *Mem:* Sigma Xi; fel Am Col Surgeons; Int Cardiovasc Soc; Soc Univ Surgeons; Am Surg Asn; Soc Vascular Surg. *Res:* Thromboembolic phenomena with special emphasis on the vascular and fibrinolytic systems. *Mailing Add:* Dept Surg Univ Mo Med Ctr Columbia MO 65212

SILVER, EDWARD A, b Mt Vernon, NY, Aug 20, 48; m 71; c 1. MATHEMATICS, MATHEMATICS EDUCATION. *Educ:* Iona Col, BA, 70; Columbia Univ, MA, 73, MS & EDD(math educ), 77. *Prof Exp:* asst prof math, Northern Ill Univ, 77-79; asst prof, 79-81, ASSOC PROF MATH, SAN DIEGO STATE UNIV, 81- *Mem:* Math Asn Am; Am Educ Res Asn; Nat Coun Teachers Math; Cognitive Sci Soc. *Res:* Study of mathematical cognition, especially mathematical problem solving, using an eclectic approach that leans heavily on techniques drawn from artificial intelligence, cognitive psychology, and mathematics education. *Mailing Add:* 414 Manordale Rd Upper St Clair PA 15241

SILVER, EDWARD ALLAN, b Montreal, Que, June 13, 37; m 66; c 2. OPERATIONS MANAGEMENT, INDUSTRIAL ENGINEERING. *Educ:* McGill Univ, BEng, 59; Mass Inst Technol, ScD(opers res), 63. *Prof Exp:* Prof staff mem, Opers Res Group, Arthur D Little, Inc, Mass, 63-67; assoc prof bus admin, Boston Univ, 67-69; from assoc prof to prof mgt sci, Univ Waterloo, 69-81; PROF OPERS MGT, UNIV CALGARY, 81- *Concurrent Pos:* Lectr, Mass Inst Technol, 65-67; consult, Arthur D Little Inc, 63-67, Bell Can, US Army Inventory Res Off, Defence Res Bd Can, Can Ctr Remote Sensing, Can Gen Tower, Am Optical, Uniroyal, Standard Oil Ind & Nova; vis prof, Ecole Polytech Federale de Lausanne, Switz, 76-77 & 90, Stanford Univ, 86, Xian Jiotong Univ, China, 87, Exec Inst Advan Studies, Vienna, 90; Sem & In-house workshops, many orgn. *Honors & Awards:* Opers Res Div Award, Inst Indust Engrs, 86; Award Merit, Can Oper Res Soc, 90. *Mem:* Can Oper Res Soc (pres, 80-81); Opers Mgt Asn; Inst Mgt Sci; Am Production & Inventory Control Soc; Inst Indust Engrs; Int Soc Inventory Res (vpres, 90-94). *Res:* Applications of approximate solutions to quantitative models of complex problems in operations management and industrial engineering. *Mailing Add:* Fac Mgt Univ Calgary Calgary AB T2N 1N4 Can

SILVER, ELI ALFRED, b Worcester, Mass, June 3, 42; m 67, 87; c 2. OCEANOGRAPHY, GEOLOGY. *Educ:* Univ Calif, Berkeley, AB, 64; Scripps Inst Oceanog, Phd(oceanog), 69. *Prof Exp:* Res fel, Scripps Inst Oceanog, 69-70; res geologist, US Geol Surv, 70-73; from asst prof to assoc prof, 73- 79, chmn dept, 75-77, PROF EARTH SCI, UNIV CALIF, SANTA CRUZ, 79- *Concurrent Pos:* Chmn, Active Margins Panel, Ocean Drilling Prog, 80-82, chmn, Western Pac Panel, Ocean Drilling Prog, 83-85; assoc ed, J Geophys Res, 85- *Mem:* Fel Geol Soc Am; Am Geophys Union; Soc Explor Geophysicists; Seismol Soc Am; AAAS. *Res:* Structure and evolution of subduction zones and collision zones, using marine geophysical tools, focused on Indonesia, the Caribbean, the western coast of the United States, and Papua New Guinea. *Mailing Add:* Earth Sci Bldg Univ Calif Santa Cruz CA 95064

SILVER, ERNEST GERARD, b Munich, Ger, Dec 26, 29; US citizen; m 54; c 1. EXPERIMENTAL PHYSICS. *Educ:* Boston Univ, BA, 52; Harvard Univ Grad Sch, MS, 54; Oak Ridge Sch Reactor Technol, dipl, 55; Univ Tenn, Knoxville, PhD(physics), 65. *Prof Exp:* Staff physicist reactor physics & neutron physics, Oak Ridge Nat Lab, 55-74; staff mem, Inst Energy Anal, Oak Ridge Nat Lab, 74-75, exec officer, 75-77, asst mgr, Breeder Reactor Proj, 77-79, asst mgr, Nuclear Standards Mgt Ctr, 79-81. *Concurrent Pos:* asst ed, Nuclear Safety J, 81-84, ed-in-chief, 84- *Honors & Awards:* Fel Am Nuclear Soc. *Mem:* Am Nuclear Soc; Am Phys Soc; AAAS. *Res:* Energy policy; relation of energy use to national economy; neutron cross section for reactor application; time-dependent neutron diffusion. *Mailing Add:* 107 Lehigh Lane Oak Ridge TN 37830

SILVER, FRANCIS, b Martinsburg, WVa, Jan 4, 16; m 65. LOW LEVELS OF TOXICITY, SUBTLE DESTROYERS OF GREAT CIVILIZATIONS. *Educ:* Johns Hopkins Univ, BE, 37, Univ MD, BE, 53. *Prof Exp:* Lab asst & plant engr & prod mgr engs, Standard Lime & Stone Co, 37-43; plant engr & gen mgr, John W Bishop Co, 47-50; weight engr, Fairchild, Aircraft & Boeing Aircraft, 51-57; land survr, County Survr Lands, Berkeley, WVa, 58-88; CHEM ENGR, INDOOR & OUTDOOR AIR POLLUTION CONS, 42-, POLLUTION ENG & CONSULT PUBL, 86- *Concurrent Pos:* Consult air pollution, Am Acad Environ Med, 62- *Honors & Awards:* Jonathan Forman Award, Am Acad Environ Med, 82- *Mem:* Soc Clin Ecol; Am Soc Mech Eng; Am Acad Environ Med. *Res:* Difference in safety between oral ingestion and breathing for low water-solubility substances like mercury and pesticides; breathing pesticides, the core cause of mankind's problems; government proof of safety reliability on animal feeding experiments. *Mailing Add:* Environ Eng & Publ 1066 Nat Press Bldg Washington DC 20045

SILVER, FRANK MORRIS, b Eden, NC, Aug 30, 43; m 69; c 3. TECHNOLOGY ACQUISITIONS, POLYMER ALLOYS. *Educ:* Univ NC, Chapel Hill, BS, 65; Univ Wis-Madison, PhD(org chem), 70. *Prof Exp:* Sr res chemist, 70-74, res specialist, 74-78, group leader, 78-79, sr group leader, 79-81, MGR, MONSANTO CO, 81- *Mem:* Am Chem Soc. *Res:* Organic and polymer reaction mechanisms; solution and melt polymerization of novel polymers; organic and polymer synthesis and characterization; fiber spinning, characterization and end-use application of novel polymers; rubber-plastic blends; injection molding; thermoplastic elastomers; polymer alloys; nonwoven fabrics. *Mailing Add:* 84 Pinewood Hills Longmeadow MA 01106

SILVER, GARY LEE, b Columbus, Ga, Nov, 22, 36; m 68; c 1. ANALYTICAL CHEMISTRY. *Educ:* Mass Inst Technol, BS, 59; Univ NC, PhD(anal chem), 63. *Prof Exp:* Sr chemist, 63-80, FEL, MOUND LAB, EG&G INC, 80- *Mem:* Am Chem Soc. *Res:* Inorganic chemistry: lanthanides and actinides, particularly plutonium; radioactive materials and radioactive waste treatment; numerical methods. *Mailing Add:* Mound Lab EG&G Inc Miamisburg OH 45342

SILVER, GEORGE ALBERT, b Philadelphia, Pa, Dec 23, 13; m 37; c 3. MEDICINE, PUBLIC HEALTH. *Educ:* Univ Pa, BA, 34; Jefferson Med Col, MD, 38; Johns Hopkins Univ, MPH, 48. *Hon Degrees:* MA, Yale Univ, 70. *Prof Exp:* Asst demonstr bact, Jefferson Med Col, 39-42; asst prof pub health admin, Sch Hyg & Pub Health, Johns Hopkins Univ, 48-51; asst prof admin med, Col Physicians & Surgeons, Columbia Univ, 53-59; prof social med, Albert Einstein Col Med, 59-65; dep asst secy health, US Dept Health, Educ & Welfare, Washington, DC, 66-68, Health Exec Urban Coalition, Health Prog, 68-70; prof, 69-84, EMER PROF PUB HEALTH, SCH MED, YALE UNIV, 84- *Concurrent Pos:* Regional med officer, US Dept Agr Migrant Prog, 47-48; health officer, Eastern Health Dist, City Dept Health, Baltimore, Md, 48-51; chief dept social med, Montefiore Hosp, 51-65; consult, WHO, 68-mem tech bd, Milbank Mem Fund, 63-76. *Mem:* Sr mem Inst Med-Nat Acad Sci; Am Pub Health Asn; Sigma Xi; Fedn Am Scientists (secy, 80-89). *Res:* Social medicine; medical care organization and administration; health policy. *Mailing Add:* Dept Pub Health Sch Med Yale Univ 89 Trumbull New Haven CT 06510

SILVER, HENRY K, pediatrics; deceased, see previous edition for last biography

SILVER, HERBERT GRAHAM, b Somerset, Eng, Sept 10, 38, US citizen; wid. PHYSICAL CHEMISTRY, ELECTROCHEMISTRY. *Educ:* Univ London, BSc & ARCS, 60, PhD(phys chem), 63; Imp Col, Univ London, dipl, 63. *Prof Exp:* NIH grant & Fulbright scholar, Harvard Univ, 63-65; mem tech staff chem, Bell Tel Labs, Inc, 65-67 & Gen Tel & Electronics Labs, Inc, 67-72, eng specialist, GTE Sylvania Inc, 72-75; chief lamp engr & mgr linear lamps, Canrad-Hanovia, Inc Div, Canrad Precision Industs Inc, 75-77; mgr, Filter Performance & Contamination Labs, Pall Corp, 78-81; sr scientist, Advan Develop, Components Div Burndy Corp, 81-82; ENGR, DEFENSE ELECTRONICS DIV, UNISYS CORP, 83- *Mem:* AAAS; fel Royal Soc Chem; Sigma Xi; NY Acad Sci. *Res:* Gas discharges; materials research; electrochemistry; electroplating; development of high-efficiency gas discharge lamps for lighting, photochemical processes, medical therapeutics; filters and fluid clarification devices; adhesives and elastomers consultant; EMI/RFI elastomer gasketing consultant. *Mailing Add:* 14 Gay Dr Kings Point NY 11024

SILVER, HOWARD FINDLAY, b Denver, Colo, Sept 16, 30; m 61; c 3. CHEMICAL ENGINEERING. *Educ:* Colo Sch Mines, BS, 52; Univ Mich, MSc, 57, PhD(chem eng), 61. *Prof Exp:* Chem engr, E I du Pont de Nemours & Co, 52-53 & 55; res engr, Chevron Res Corp Div, Standard Oil Co Calif, 57-58 & 61-64; assoc prof chem eng, 64-68, PROF CHEM ENG, UNIV WYO, 68- *Concurrent Pos:* Prog mgr supporting res, Fossil Fuels & Advan Systs Dept, Elec Power Res Inst, 74-75, consult, Clean Liquid & Solid Fuels Dept, 75- *Mem:* Am Inst Chem Engrs; Am Chem Soc. *Res:* Conversion of coal and shale oil to synthetic liquid hydrocarbon products by means of hydrogenation and cracking reactions. *Mailing Add:* Dept Chem Eng Univ Wyo Laramie WY 82071

SILVER, HOWARD I(RA), b New York, NY, June 2, 39; m 62; c 2. ELECTRICAL ENGINEERING. *Educ:* City Col New York, BEE, 61; NY Univ, MEE, 64, PhD(elec eng), 68. *Prof Exp:* Eng trainee, Missile & Space Div, Gen Elec Co, 61-62; engr, Fed Labs, Int Tel & Tel Corp, 62-64; assoc engr, Sperry Gyroscope Co, 64-66; instr elec eng, NY Univ, 66-68; from asst prof to assoc prof, 68-76, chmn dept, 71-76, PROF ELEC ENG, FAIRLEIGH DICKINSON UNIV, 76- *Concurrent Pos:* Consult, Devenco Res Lab, 69, ATT Technol, 83-; instr & consult, Bell Tel Labs, 71-83. *Res:* Microprocessor system design; digital signal processing; logic design. *Mailing Add:* Dept Elec Eng Fairleigh Dickinson Univ Teaneck NJ 07666

SILVER, HULBERT KEYES BELFORD, b Montreal, Que, July 15, 41; m 66; c 3. MEDICINE. *Educ:* Bishop's Univ, BSc, 62; McGill Univ, MD & CM, 66, PhD(immunol), 74. *Prof Exp:* Asst res oncologist, Sch Med, Univ Calif, Los Angeles, 73-74, asst prof surg, 75-76; from asst prof to assoc prof, 76-84, PROF MED, UNIV BC, 84-; MED ONCOLOGIST, CANCER CONTROL AGENCY BC, 76- *Honors & Awards:* Medal Med, Royal Col Physicians & Surgeons Can, 75. *Mem:* Am Asn Cancer Res; Am Soc Clin Oncol; Int Soc Interferon Res; Royal Col Physicians & Surgeons Can; Am Col Physicians. *Res:* Clinical cancer treatment with chemotherapy and biological response modifiers; immune effects of biological response modifiers; immunodiagnosis. *Mailing Add:* BC Cancer Agency 600 W 10th Ave Vancouver BC V5Z 4E6 Can

SILVER, JACK, MEDICINE. *Prof Exp:* CHIEF DIV MOLECULAR MED & PROF MED, NORTH SHORE UNIV HOSP, CORNELL UNIV MED COL, 88- *Mailing Add:* Dept Med North Shore Univ Hosp Cornell Univ Med Col 350 Community Dr Manhasset NY 11030

SILVER, LAWRENCE, b New York, NY, Apr 15, 21; m 52; c 3. MEDICINE. *Educ:* Queens Col, BS, 41; Univ Idaho, MS, 42; NY Univ, MD, 50. *Prof Exp:* Intern, Beth Israel Hosp, New York, 50-51; asst resident internal med, Vet Admin Hosp, Bronx, NY, 51-52; fel, Mayo Found, Univ Minn, 52-53; asst physician & vis investr hypertension, Hosp, Rockefeller Inst, 53-54; assoc scientist, Med Res Ctr, Brookhaven Nat Lab, 56-58; assoc med dir, Dept Clin Invest, Chas Pfizer & Co, 58-59; res assoc hypertension, Brookhaven Nat Lab, 59-61; assoc med dir, Dept Clin Infest, Chas Pfizer & Co, 61-64; PHYSICIAN-IN-CHG, DEPT NUCLEAR MED, QUEENS HOSP CTR, 64- *Mem:* AAAS; Am Physiol Soc; Soc Exp Biol & Med. *Res:* Relationship of salt to hypertensive disease states; influence and mechanism of action of hormones on movement of salt and water in normals and hypertensives. *Mailing Add:* Nuclear Med Queens Hosp Ctr 82-68 164th St Jamaica NY 11432

SILVER, LEE MERRILL, b Philadelphia, Pa, Apr 27, 52; m 74; c 2. DEVELOPMENTAL GENETICS, MOLECULAR BIOLOGY. *Educ:* Univ Pa, BA & MS, 73; Harvard Univ, PhD(biophys), 78. *Prof Exp:* Res fel genetics, Sloan Kettering Cancer Inst, 77-79, assoc, 79-80; sr staff investr, Cold Spring Harbor Lab, 80-84; PROF, PRINCETON UNIV, 84- *Concurrent Pos:* Fel, Pop Coun, 77-78 & NIH, 78-79; asst prof genetics, Med Sch, Cornell Univ, 79-80 & State Univ NY Stony Brook, 80-; vis asst prof genetics, Albert Einstein Col Med, 80; ed, Mammalian Arome, 89- *Mem:* Am Soc Cell Biol; AAAS; Int Soc Differentiation; Genetics Soc Am. *Res:* Molecular embryology; molecular biology of spermatogenesis; the mouse T/t complex; chromosomal proteins; cell surface proteins. *Mailing Add:* Dept Molecular Biol Princeton Univ Princeton NJ 08544

SILVER, LEON THEODORE, b Monticello, NY, Apr 9, 25; m 47; c 2. PETROLOGY, GEOCHEMISTRY. *Educ:* Univ Colo, BS, 45; Univ NMex, MS, 48; Calif Inst Technol, PhD(geol & geochem), 55. *Prof Exp:* Field asst, jr geologist & asst geologist, US Geol Surv, Mineral Deposits Br, Colo & Ariz, 47-54; geologist, Astrogeol Br, US Geol Surv, 70-76; res geologist, Geochem Uranium, AEC Contract Res, Div Geol Sci, 52-55, from asst prof to prof geol, 55-83, W M KECK FOUND PROF RESOURCE GEOL, CALIF INST TECHNOL, 83- *Concurrent Pos:* Guggenheim fel, 64; mem subcomn geochronology, Int Union Geol Sci, 70-; consult, NASA, 71-; counr, Am Geol Soc, 74-76; chmn, Sci Eng & Pub Policy, Nat Acad Sci, 84-, counr, 89-92; mem, Gov Bd, Nat Res Coun, 89-92, chair, Comt Int Orgn & Progs, 90- *Honors & Awards:* Except Sci Achievement Medal, NASA, 71; Prof Excellence Award, Am Inst Prof Geologists, 72; Cert Spec Commendation, Geol Soc Am, 73. *Mem:* Nat Acad Sci; fel Geol Soc Am (vpres, 78, pres 79); fel Mineral Soc Am; Geochem Soc; Am Geophys Union; fel AAAS; Am Acad Arts & Sci. *Res:* Igneous and metamorphic petrology; geochemistry of uranium, thorium and lead; geochronology; regional geology of southwestern United States; tectonic history of North America; mineralogy and petrology of meteorites and lunar materials. *Mailing Add:* Div Geol & Planetary Sci Calif Inst Technol Pasadena CA 91125

SILVER, MALCOLM DAVID, b Adelaide, SAustralia, Apr 29, 33; m 57; c 3. PATHOLOGY. *Educ:* Univ Adelaide, MB, BS, 57, MD, 72; McGill Univ, MSc, 61, PhD(path), 63; Am Bd Path, dipl, 63; FRACP, 64; FRCP(C), 75. *Prof Exp:* Resident med officer, Royal Adelaide Hosp, Australia, 57-58; resident path, Royal Victoria Hosp, Montreal, 58-63; res fel exp path, John Curtin Sch Med Res, Australian Nat Univ, 63-65; from asst prof to prof, 65-79, CHMN, DEPT PATH, UNIV TORONTO, 85-; CHIEF PATH, TORONTO HOSP, 85- *Concurrent Pos:* Resident path, Path Inst, McGill Univ, 58-63; staff pathologist, Toronto Gen Hosp, 65-72, sr staff pathologist, 72-79; chmn, Dept Path, Univ Western Ont & chief path, Univ Hosp, London, Ont, 79-85. *Mem:* Am Asn Path & Bact; Am Heart Asn; Can Asn Path; Can Cardiovasc Soc; Int Acad Path. *Res:* Cardiovascular pathology. *Mailing Add:* Dept Path Banting Inst 100 College St Toronto ON M5G 1L5 Can

SILVER, MARC STAMM, b Philadelphia, Pa, Jan 5, 34; m 61; c 3. BIO-ORGANIC CHEMISTRY. *Educ:* Harvard Univ, AB, 55; Calif Inst Technol, PhD(chem, physics), 59. *Hon Degrees:* MA, Amherst Col, 69. *Prof Exp:* From instr to assoc prof, 58-69, PROF CHEM, AMHERST COL, 69- *Concurrent Pos:* NSF fels, Northwestern Univ, 61-62 & Weizmann Inst, 66-67; vis prof, Yale Univ, 77-78; res fel, Oxford Univ, 71 & Imp Col, 81-82, 86-87. *Mem:* AAAS; Am Soc Biol Chemists; Am Chem Soc. *Res:* Organic reaction mechanisms; mechanism of enzyme action. *Mailing Add:* Dept of Chem Amherst Col Amherst MA 01002

SILVER, MARSHALL LAWRENCE, b Los Angeles, Calif, Nov 26, 42; m 69; c 1. CIVIL ENGINEERING. *Educ:* Univ Colo, Boulder, BS, 65; Univ Calif, Berkeley, MS, 67, PhD(civil eng), 69. *Prof Exp:* Civil engr, John P Elliott-Consult Engr, Colo, 66-67; res engr, Inst Traffic & Transp Eng, Univ Calif, Berkeley, 67-69; PROF SOIL MECH, UNIV ILL, CHICAGO, 69- *Concurrent Pos:* Vis prof, Cent Univ, Caracas, 76 & Univ Tokyo, 77; Terzaghi fel, 81. *Mem:* Am Soc Civil Engrs; Am Railway Eng Asn; Am Soc Testing & Mat; Int Soc Soil Mech & Found Eng. *Res:* Dynamic behavior of soil materials; ground borne and industrial vibrations; earthquake engineering. *Mailing Add:* Dept Civil Eng Box 4348 Univ Ill Chicago IL 60680

SILVER, MARVIN, b New York, NY, Oct 22, 24; m 51; c 1. SOLID STATE PHYSICS. *Educ:* Rensselaer Polytech Inst, BEE, 45; NY Univ, MS, 51, PhD(physics), 59. *Prof Exp:* Electronic engr, Radio Corp, Am, 47-48 & Franklin Inst, 48-49; res physicist, Chatham Electronics Co, 52-55; instr, Hunter Col, 56-57; chief solid state br, US Res Off, NC, 58-67; assoc prof, 67-70, PROF PHYSICS, UNIV NC, CHAPEL HILL, 70- *Concurrent Pos:* From vis asst prof to adj prof, Univ NC, Chapel Hill, 59-67; consult, Kuthe Labs, 56-57. *Mem:* Am Phys Soc. *Res:* Electronic conductivity in organic solids; breakdown phenomena in gaseous electrical discharges. *Mailing Add:* Dept of Physics Univ of NC Chapel Hill NC 27514

SILVER, MARY WILCOX, b San Francisco, Calif, July 13, 41; c 2. BIOLOGICAL OCEANOGRAPHY, MARINE ECOLOGY. *Educ:* Univ Calif, Berkeley, AB, 63; Scripps Inst Oceanog, Univ Calif, San Diego, PhD(oceanog), 71. *Prof Exp:* Lectr marine biol, Moss Landing Marine Labs, Moss Landing, Calif, 70-71; asst prof marine biol, San Francisco State Univ, 71-72; from asst prof to assoc prof, 72-87, PROF MARINE SCI, UNIV CALIF, SANTA CRUZ, 87- *Mem:* Am Soc Limnol & Oceanog; Am Geophys Union; Phycol Soc Am; AAAS. *Res:* Biology of planktonic tunicates; ecology of suspended particulates (marine snow); pelagic detrital food webs; deep ocean particle flux. *Mailing Add:* Inst Marine Studies Univ Calif Santa Cruz CA 95064

SILVER, MELVIN JOEL, b Philadelphia, Pa, June 22, 20; m 55; c 1. BIOCHEMISTRY, PHARMACOLOGY. *Educ:* Temple Univ, AB, 41; Philadelphia Col Pharm, MSc, 43, DSc(bact), 53. *Prof Exp:* Res biochemist, 53-57, assoc prof pharmacol, 59-72, PROF PHARMACOL, JEFFERSON MED COL, THOMAS JEFFERSON UNIV, 72-, ASSOC, CARDEZA FOUND, 57-, SR MEM, 72- *Concurrent Pos:* Wellcome Trust res fel, Royal Col Surgeons Eng, 70-71. *Mem:* Am Soc Pharmacol & Exp Therapeut; Am Chem Soc; NY Acad Sci; Int Soc Thrombosis & Haemostasis; Sigma Xi. *Res:* Role of prostaglandin synthesis, phospholipids and platelets in hemostasis. *Mailing Add:* Dept Pharmacol Jefferson Med Col Cardeza Found 1015 Walnut St Philadelphia PA 19107

SILVER, MEYER, b New York, NY, Sept 12, 26; m 47; c 3. QUANTUM ELECTRONICS, LASERS. *Educ:* Brooklyn Col, BA, 49; Rensselaer Polytech Inst, MS, 57; Univ Notre Dame, PhD, 60. *Prof Exp:* Physicist, US Naval Ord Test Sta, 52-55, 60-62 & TRW Systs, Inc, Calif, 62-67; mem tech staff, Aerospace Corp, 67-70; chief laser systs sect, Martin-Marietta Corp, Fla, 70-71; div chief mil progs, Zenith Radio Res Corp, Calif, 71-72; dir advan eng, Appl Technol Div, Itek Corp, 72-75; mgr advan technol dept, Lockheed Missiles & Space Corp, 75-77; asst mgr, Optics Dept, TRW Defense, 77-80, MGR, OPTICAL COMPONENTS DEPT, SPACE SYSTS GROUP, TRW, INC, 80- *Mem:* Am Phys Soc; Sigma Xi; Res Soc Am; Optical Soc Am. *Res:* Acousto-optics-electronics; laser research; gas discharges; atomic physics; multiple photon processes. *Mailing Add:* 8611 Braeburn Dr Annandale VA 22003

SILVER, PAUL J, b Akron, Ohio, Dec 11, 51. CARDIOPULMONARY PHARMACOLOGY. *Educ:* Univ Akron, BS, 74, MS, 76; Univ Cincinnati, PhD(med physiol), 79. *Prof Exp:* Postdoctoral, Univ Tex Health Sci Ctr, 80-82; mgr, Cardiovasc Biochem Group, Wyeth Lab, 82-86; sect head, 86-87, DIR CARDIO-PHARMACOL, STERLING RES INST, 87- *Mem:* Biophys Soc; Am Soc Pharmacol & Exp Therapeut. *Mailing Add:* Sterling Res Inst 81 Columbia Turnpike Rensselaer NY 12144

SILVER, RICHARD N, b Bridgeport, Conn, July 18, 45; m 69; c 4. THEORETICAL SOLID STATE PHYSICS. *Educ:* Calif Inst Technol, BS, 66, PhD(theoret physics), 71. *Prof Exp:* Res assoc elem particle physics, Brown Univ, 71-72; res assoc solid state physics, Calif Inst Technol, 72-74; staff mem solid state physics, 74-79, neutron scattering, 79-86, STAFF MEM CONDENSED MATTER PHYSICS, 86- *Concurrent Pos:* IBM fel, Calif Inst Technol, 72-74. *Mem:* Am Phys Soc. *Res:* High excitation conditions in semiconductors; solid state lasers; statistical mechanics of phase transitions; neutron scattering experiments with pulsed spallation neutron sources; maximum entropy data analysis; quantum Monte Carlo. *Mailing Add:* MSB 262 Group T11 Los Alamos Nat Lab Los Alamos NM 87545

SILVER, RICHARD TOBIAS, b New York, NY, Jan 18, 29; m 63; c 1. MEDICINE. *Educ:* Cornell Univ, AB, 50, State Univ NY, MD, 53; Am Bd Internal Med, dipl, 62, cert med oncol, 73. *Prof Exp:* Intern med, NY Hosp-Cornell Med Ctr, 53-54, resident physician, 56-58; clin assoc, Gen Med Br, Nat Cancer Inst, 54-56; from instr to assoc prof, 58-73, PROF CLIN MED, MED COL, CORNELL UNIV, 73, CHIEF ONCOL SERV, NEW YORK HOSP-CORNELL MED CTR, 77- *Concurrent Pos:* Vis Fulbright lectr, res scholar & vis prof, Sch Med, Univ Bahia, 58-59; physician outpatients, New York Hosp, 58-62, from asst attend physician to assoc attend physician, 62-73, attend physician, 73-; hematologist & consult, Gracie Sq Gen Hosp, 59-; asst vis physician, 2nd Cornell Div, Bellevue Hosp, 62-68; consult leukemia & mem, Leukemia-Myeloma Task Force, NIH, 65-69, mem polycythemia study group, 67-; dir chemother serv, Div Hemat & Oncol, Cornell Med Ctr, 67-75; consult & attend hematologist, Englewood Hosp, NJ & Manhattan Eye, Ear & Throat Hosp; prin investr, Clin Chemother Prog, Cancer Control, 73-80; group vchmn, Cancer & Leukemia Group B, 76-; mem, Comt Pub Affairs, Am Soc Clin Oncol, 81. *Mem:* Am Soc Hemat; Harvey Soc; Am Fedn Clin Res; fel Am Col Physicians; Am Soc Clin Oncol. *Res:* Hematology and oncology, especially leukemia and oncology chemotherapy; over 130 publications and articles. *Mailing Add:* New York Hosp-Cornell Med Ctr 1440 York Ave New York NY 10021

SILVER, ROBERT, b Detroit, Mich, Aug 3, 21; m 47; c 2. PHYSICS. *Educ:* Wayne State Univ, BS, 48; Univ Calif, Berkeley, PhD(physics), 58. *Prof Exp:* Physicist, Lawrence Radiation Lab, Univ Calif, 51-57; sr res physicist, Gen Motors Res Labs, 57-66; assoc prof, 66-71, chmn dept, 74-82______ PROF PHYSICS & ASTRON, EASTERN MICH UNIV, 71-, chmn dept, 74-82. *Concurrent Pos:* Instr, Wayne State Univ, 59-61 & Univ Detroit, 63-66. *Mem:* Am Phys Soc; Am Asn Physics Teachers; Soc Sigma Xi. *Res:* Low energy particle scattering; nuclear reactors and neutron moderation; thermionic conversion; traffic dynamics; electrooptics; Mossbauer effect. *Mailing Add:* Three Buckingham Ct Ann Arbor MI 48104

SILVER, SIMON DAVID, b Detroit, Mich, June 22, 36; m 58; c 2. MOLECULAR BIOLOGY, MICROBIAL PHYSIOLOGY. *Educ:* Univ Mich, BA, 57; Mass Inst Technol, PhD(biophys), 62. *Prof Exp:* NSF fel, Med Res Coun-Microbial Genetics Res Unit, Hammersmith Hosp, London, 62-64; asst res biophysicist, Virus Lab, Univ Calif, Berkeley, 64-66; from asst prof to assoc prof, 66-76, PROF BIOL & MICROBIOL, WASHINGTON UNIV, 76- *Concurrent Pos:* Vis fel, Dept Biochem, John Curtin Sch Med Res, Australian Nat Univ, 74-75 & 79-80; ed-in-chief, J Bacteriol, 78-87; mem metabolic biol panel, NSF, 80-84. *Mem:* Am Soc Microbiol; Genetics Soc Am; Biophys Soc; Am Soc Biol Chemists; Soc Gen Microbiol. *Res:* Bacterial physiology; molecular genetics. *Mailing Add:* Univ Ill Col Med Box 6998 Chicago IL 60680

SILVER, SYLVIA, b Chicago, Ill, Dec 11, 42; m 63; c 2. MEDICAL TECHNOLOGY, MICROBIOLOGY. *Educ:* Drake Univ, BA, 65; Cath Univ Am, MTS, 75, DA(clin microbiol), 77. *Prof Exp:* Res asst biochem & vet physiol, Iowa State Univ, 66-68; res asst surg res, Harvard Med Sch, 68-69; instr med technol, Sch Med, Univ Md, 71-73; lectr, Montgomery Col, 75-78; asst prof, 78-84, ASSOC PROF PATH, SCH MED & HEALTH SCI, GEORGE WASH UNIV, 84-, DIR MED TECHNOL PROG, 78-, ASSOC PROF HEALTH CARE SCI, 88- *Concurrent Pos:* Consult, Dept Health & Human Serv, health professions, 79-81, Lister Hill Nat Ctr Biomed, Commun Nat Libr Med, 83-85; Wear vis prof, Wichita State Univ, 84-85. *Mem:* Am Soc Microbiol; Am Soc Clin Pathologists; Am Soc Med Technol; Am Soc Allied Health Professions; Sigma Xi; Int Soc AIDS Educ. *Res:* Isolation and identification of anaerobic bacteria; micro-methods in identification of bacteria; structure of hepatitis virus; clinical microbiology; laboratory medicine; computer assisted instruction; HIV prevention in university students. *Mailing Add:* George Wash Univ Med Ctr 2300 Eye St NW Washington DC 20037

SILVER, WARREN SEYMOUR, b New York, NY, Nov 14, 24; m 50; c 4. MICROBIOLOGY, SOILS & SOIL SCIENCES. *Educ:* Univ Md, BS, 49, MS, 50; Johns Hopkins Univ, PhD(biochem), 53. *Prof Exp:* Jr instr biol, Johns Hopkins Univ, 50-53; res scientist, Upjohn Co, 53-54; Nat Cancer Inst fel, Inst Microbiol, Rutgers Univ, 54-56; from asst prof to prof bact, Univ Fla, 56-67; prof life sci & chmn dept, Ind State Univ, 67-70; PROF BIOL, UNIV S FLA, 70- *Concurrent Pos:* Prog mgr, Competitive Res Grants Off, USDA, 79-80. *Honors & Awards:* Fel, Am Acad Microbiol. *Mem:* AAAS; Am Soc Microbiol; Brit Soc Gen Microbiol. *Res:* Bacterial nutrition; inorganic nitrogen metabolism by microorganisms; microbial ecology; respiratory enzymes; plant-microbe symbiosis; biological nitrogen fixation. *Mailing Add:* Dept Biol LIF 169 Univ S Fla Tampa FL 33620-5150

SILVERA, ISAAC F, b San Diego, Calif, Mar 25, 37; m 61; c 4. SOLID STATE PHYSICS. *Educ:* Univ Calif, Berkeley, AB, 59, PhD(physics), 65. *Prof Exp:* Franco-Am exchange fel, Lab Electrostatics & Physics of Metal, 65-66; mem tech staff, Sci Ctr, NAm Rockwell Corp, 66-71; prof exp physics, physics lab, Univ Amsterdam, 71-; AT LYMAN LAB PHYSICS, HARVARD UNIV. *Mem:* Am Phys Soc. *Res:* Far infrared and Raman spectroscopy of solids at low temperature; critical phenomena; light scattering in quantum solids and molecular beams; atomic hydrogen in the condensed phase; high pressure physics. *Mailing Add:* Lyman Lab Physics Harvard Univ Cambridge MA 02135

SILVERBERG, STEVEN GEORGE, b New York, NY, Nov 30, 38; m 68. SURGICAL PATHOLOGY. *Educ:* Brooklyn Col, AB, 58; Johns Hopkins Univ, MD, 62. *Prof Exp:* Intern med, Bellevue Hosp, 62-63; resident path, Yale Univ-Grace New Haven Hosp, 63-65; fel path, Mem Hosp for Cancer & Allied Dis, 65-66; from asst prof to assoc prof surg path, Med Col Va, 68-72; from assoc prof to prof path, Sch Med, Univ Colo, Denver, 72-81; PROF PATH & DIR ANAT PATH, GEORGE WASHINGTON UNIV, MED CTR, 81- *Concurrent Pos:* Exec dir, Colo Regional Cancer Ctr, 76-79; ed-in-chief, Int J Gynec Path, 85- *Mem:* Am Soc Clin Path; Int Acad Path; Soc Gynec Oncol; Am Soc Cytol; Int Soc Gynec Path; Arthur Purdy Stout Soc Surg Path (secy, 79-83). *Res:* Natural history and histogenesis of neoplasms; ultrastructure in surgical pathology; gynecologic and breast pathology; epidemiologic pathology. *Mailing Add:* Geo Washington Univ Med Ctr 2300 Eye St NW Washington DC 20037

SILVERBORG, SAVEL BENHARD, b Gardner, Mass, Jan 20, 13; m 43. FORESTRY. *Educ:* UNiv Idaho, BS, 36; Univ Minn, PhD(forest path), 48. *Prof Exp:* Asst, Univ Minn, 40-42; prin procurement inspector, US Dept Army Air Force, Ill, 42-43; forest pathologist, 47-60, assoc prof, 60-66, prof, 66-77, EMER PROF FOREST PATH & FOREST BOT, 77- *Mem:* AAAS; Soc Am Foresters; Am Phytopath Soc; Sigma Xi. *Res:* Diseases of Hevea brasiliensis; wood decay in buildings in New York; factors affecting the growth and survival of phytophthora palmivora; estimating cull in northern hardwoods; forest plantation diseases. *Mailing Add:* Dept Forest Bot & Path Col Environ Sci & Forestry Syracuse NY 13210-2788

SILVERMAN, ALBERT, b Boston, Mass, Oct 29, 19; m 41; c 2. NUCLEAR PHYSICS. *Educ:* Univ Calif, PhD(physics), 50. *Prof Exp:* Res assoc physics, 50-52, from asst prof to assoc prof, 52-60, PROF PHYSICS, CORNELL UNIV, 60-, MEM STAFF, LAB NUCLEAR STUDIES, 77- *Concurrent Pos:* Guggenheim & Fulbright fels, 59-60, 86-87. *Mem:* Fel Am Phys Soc. *Res:* High energy physics. *Mailing Add:* Lab Nuclear Studies Cornell Univ Ithaca NY 14853

SILVERMAN, ALBERT JACK, b Montreal, Que, Jan 27, 25; nat US; m 47; c 2. PSYCHIATRY, PSYCHOPHYSIOLOGY. *Educ:* McGill Univ, BSc, 47, MD & CM, 49; Am Bd Psychiat & Neurol, dipl, 55; Washington Psychoanal Inst, grad, 64. *Prof Exp:* Intern, Jewish Gen Hosp, Montreal, 49-50; resident psychiat, Med Ctr, Univ Colo, 50-53, instr, 53; instr, Sch Med, Duke Univ, 53-54, assoc, 54-56, from asst prof to assoc prof, 56-63; prof & chmn dept, Med Sch, Rutgers Univ, 63-70; chmn dept, 70-81, PROF PSYCHIAT, MED CTR, UNIV MICH, ANN ARBOR, 70- *Concurrent Pos:* Chief stress-fatigue sect, Aeromed Lab, Wright-Patterson AFB, 56-57; attend psychiatrist, Durham Vet Admin Hosp, 57-60, consult, 60-63; consult, Watts Hosp, 59-63, Lyons Vet Admin Hosp, 63-70, Fitkin Mem Hosp & Carrier Clin, 64-70, Ann Arbor Vet Admin Hosp, 70- & Dept of Defense, 70-; mem comt biol sci, Nat Inst Ment Health, 64-69, chmn, 68-69, mem res scientist develop comt, 70-74, chmn, 72-74, mem, small grants comt, 82-87; mem, Merit Rev Bd Behav Sci, Vet Admin, 75-78, chmn, 77-78; behav sci test comt, Nat Bd Med Examrs, 78-82, chmn, 84-87, mem, behav sci task force, 83-84 mem, comprehensive com, 87- *Mem:* Soc Biol Psychiat; Am Psychosom Soc (pres, 76-77); Am Psychiat Asn; Am Col Psychiat; Am Acad Psychoanal; Sigma Xi. *Res:* Psychosomatic medicine; psycho-endocrinology. *Mailing Add:* 19 Regent Dr Ann Arbor MI 48104-1738

SILVERMAN, ANN JUDITH, b Providence, RI, Nov 4, 46. NEUROENDOCRINOLOGY. *Educ:* Univ Calif, Los Angeles, BA, 67, PhD(zool), 70. *Prof Exp:* Fel, Dept Anat, Sch Med, Univ Rochester, 70-72, Dept Biol, Univ Calif, Los Angeles, 72-74; asst scientist, Wis Regional Primate Res Ctr, Univ Wis, 74-76; from asst prof to assoc prof, 76-85, PROF ANAT & CELL BIOL, COL PHYSICIANS & SURGEONS, COLUMBIA UNIV, 88- *Concurrent Pos:* Prin investr grant, Nat Inst Child Health & Human Develop, US PHS, 75-, co-prin investr, Nat Inst Arthritis, Metabolic & Digestive Dis, 77-; fel, Alfred P Sloan Found, 76; mem, Molecular Cell Neurobiol Panel, NSF, 81-84. *Mem:* AAAS; Soc Neurosci; Am Soc Anatomists. *Res:* Distribution and function of neurosecretory neurons related to reproductive function and stress. *Mailing Add:* Dept Anat Col Physicians & Surgeons Columbia Univ 630 W 168th St New York NY 10032

SILVERMAN, BENJAMIN DAVID, b New York, NY, Mar 14, 31; m 67; c 2. SOLID STATE PHYSICS. *Educ:* Brooklyn Col, BA, 53; Univ Rochester, MA, 55; Rutgers Univ, PhD, 59. *Prof Exp:* Prin res scientist, Raytheon Co, 59-66 & Electronics Res Ctr, NASA, 66-69; RES STAFF MEM, IBM CORP, 69- *Mem:* Fel Am Phys Soc. *Res:* Solid state theory; dielectrics; ferroelectrics; lattice dynamics; organic solids. *Mailing Add:* T J Watson Lab IBM Corp PO Box 218 Yorktown Heights NY 10598

SILVERMAN, BERNARD, b Richmond, Va, Aug 15, 22; m 57; c 2. ELECTRICAL ENGINEERING. *Educ:* Va Polytech Inst, BS, 42; Univ Ill, MS, 47, PhD(elec eng), 54. *Prof Exp:* Instr elec eng, Univ Ill, 47-53; engr, Gen Elec Co, 53-58; assoc prof elec eng, Syracuse Univ, 58-86; RETIRED. *Concurrent Pos:* Consult, NASA, 67 & IBM, 81. *Mem:* Sr mem Inst Elec & Electronics Engrs. *Res:* Nonlinear analysis; stability; communications; magnetic and dielectric devices; digital signal processing; reliability. *Mailing Add:* Dept Elec Eng Link Hall Syracuse Univ Syracuse NY 13244-1240

SILVERMAN, CHARLOTTE, b New York, NY, May 21, 13. EPIDEMIOLOGY, RADIATION. *Educ:* Brooklyn Col, BA, 33; Med Col Pa, MD, 38; Johns Hopkins Univ, MPH, 42, DrPH, 48; Am Bd Prev Med, dipl, 49. *Prof Exp:* Field analyst, US Children's Bur, DC, 42-43; field researcher, Tuberc Control Div, USPHS, 43-45; from asst dir to dir bur tuberc, Baltimore City Health Dept, 46-56; chief div epidemiol, State Dept Health, Md, 56-59 & Off Planning & Res, 59-62; consult, Nat Inst Mental Health, 62-64, asst chief social psychiat sect, 65-66, chief epidemiol studies br, 66-67, chief pop studies prog, Nat Ctr Radiol Health, 68-70, dep dir biol effects, Bur Radiol Health, 71-83, ASSOC DIR HUMAN STUDIES, OFF SCI & TECHNOL, CTR DEVICES & RADIOL HEALTH, FOOD & DRUG ADMIN, 83- *Concurrent Pos:* Lectr prev med, Sch Med, Johns Hopkins Univ, 47-52, lectr epidemiol, Sch Hyg & Pub Health, 54-56 & 66-85, asst prof, 55-64, assoc, 85- *Honors & Awards:* Award of Merit, Food & Drug Admin, 74. *Mem:* AAAS; fel Am Col Prev Med; fel Am Orthopsychiat Asn; Soc Epidemiol Res; fel Am Col Epidemiol; fel Am Pub Health Asn. *Res:* Epidemiology of chronic conditions, including ionizing and nonionizing radiation. *Mailing Add:* 4977 Battery Lane No 1001 Bethesda MD 20814-4927

SILVERMAN, DAVID J, b Summerville, SC, July 13, 43; m 66; c 2. MICROBIOLOGY, CELL BIOLOGY. *Educ:* Muhlenberg Col, BS, 65; Univ Tenn, Knoxville, MS, 67; WVa Univ, PhD(med microbiol), 71. *Prof Exp:* Res assoc microbiol, Dartmouth Med Sch, 71-73, instr, 73; asst prof, 73-81, ASSOC PROF MICROBIOL, 81-, ASSOC PROF PATH, SCH MED, UNIV MD, 87- *Concurrent Pos:* Nat Cancer Inst fel, Dartmouth Med Sch, 71-73. *Honors & Awards:* Eastman Kodak Med Photog Award, Am Soc Clin Pathologists. *Mem:* Am Soc Microbiol; Electron Micros Soc Am; Am Soc Trop Med Hyg; Am Soc Rickettsiol & Rickettsial Dis; Int Col Rickettsiologists. *Res:* Pathogenesis of rickettsial diseases; host-parasite interaction; electron microscopy. *Mailing Add:* Dept Microbiol & Immunol Univ MD Sch of Med Baltimore MD 21201

SILVERMAN, DAVID NORMAN, b South Bend, Ind, July 22, 42; m 66; c 2. PHARMACOLOGY, BIOPHYSICS. *Educ:* Mich State Univ, BS, 64; Columbia Univ, MA, 66, PhD(phys chem), 68. *Prof Exp:* NIH fel, Cornell Univ, 69-71; from asst prof to assoc prof pharmacol & biochem, 71-80, assoc dean, acad affairs & sponsored prog, 86-90, PROF PHARMACOL & BIOCHEM, COL MED, UNIV FLA, 80- *Concurrent Pos:* Fogarty fel, Umea Univ, Sweden, 78; mem adv bd, Molecular Pharmacol, 78-84. *Mem:* Am Soc Biol Chemist; Am Chem Soc; Am Soc Pharmacol & Exp Therapeut. *Res:* Stable isotopes in pharmacology and biochemistry; magnetic resonance; mass spectrometry; carbonic anhydrase. *Mailing Add:* Dept Pharmacol Univ Fla Col of Med Gainesville FL 32610

SILVERMAN, DENNIS JOSEPH, b Long Beach, Calif, Oct 7, 41; m 65; c 1. ELEMENTARY PARTICLE PHYSICS. *Educ:* Univ Calif, Los Angeles, BA, 63; Stanford Univ, MS, 64, PhD(physics), 68. *Prof Exp:* Res assoc, Princeton Univ, 68-69, instr, 69-70; res assoc, Univ Calif, San Diego, 70-71; from asst prof to assoc prof, 71-82, PROF PHYSICS, UNIV CALIF, IRVINE, 82- *Mem:* Am Phys Soc. *Res:* Quark models of elementary particle structure and interactions. *Mailing Add:* Dept of Physics Univ of Calif Irvine CA 92717

SILVERMAN, EDWARD, b Minneapolis, Minn, Nov 23, 17; m 61; c 2. MATHEMATICS. *Educ:* Univ Calif, AB, 38, MA, 39, PhD(math), 48. *Prof Exp:* Asst math, Univ Calif, 46-48; Off Naval Res fel, Inst Adv Study, 48-49; asst prof math, Kenyon Col, 49-51 & Mich State Univ, 54-57; mem staff, Sandia Corp, 51-54; assoc prof, 57-61, PROF MATH, PURDUE UNIV, LAFAYETTE, 61- *Mem:* Am Math Soc. *Res:* Multiple integral problems in the calculus of variations. *Mailing Add:* 1425 N Salisbury West Lafayette IN 47906

SILVERMAN, ELLEN-MARIE, b Milwaukee, Wis, Oct 12, 42; div; c 1. SPEECH PATHOLOGY, OTOLARYNGOLOGY. *Educ:* Univ Wis-Milwaukee, BS, 64; Univ Iowa, MA, 67, PhD(speech path), 70. *Prof Exp:* Speech clinician, Curative Workshop Milwaukee, 64-65; res assoc commun, Univ Ill, Urbana, 70-71; asst prof, Marquette Univ, 73-79; assoc prof speech path, Marquette Univ, 79-85; assoc clin prof otolaryngol, Med Col Wis, 79-83, speech pathologist, 85; SPEECH-LANG PATHOLOGIST, PVT PRACT, 85- *Concurrent Pos:* Nat Inst Dent Res fel, 69-71; asst prof otolaryngol, Med Col Wis, 75-79. *Mem:* Am Speech & Hearing Asn; Sigma Xi; Am Cleft Palate Asn; Am Speech & Hearing Asn; Sigma Xi. *Res:* Speech pathology, especially stuttering and voice; voice disorders associated with otolaryngology and counseling methods. *Mailing Add:* 5567 N Diversey Blvd Milwaukee WI 53217-5202

SILVERMAN, FRANKLIN HAROLD, b Providence, RI, Aug 16, 33; c 1. SPEECH PATHOLOGY, APPLIED STATISTICS. *Educ:* Emerson Col, BS, 60; Northwestern Univ, MA, 61; Univ Iowa, PhD(speech path), 66. *Prof Exp:* Res assoc stuttering, Univ Iowa, 65-68; asst prof speech path, Univ Ill, Urbana-Champaign, 68-71; assoc prof, 71-77, PROF SPEECH PATH, MARQUETTE UNIV, 77-; CLIN PROF REHABILITATION MED, MED COL WIS, 80- *Mem:* Fel Am Speech Language Hearing Asn; Sigma Xi. *Res:* Stuttering in elementary-school children; assessment of the impacts of stuttering therapy methods; research design in speech pathology and audiology; augmentative communication strategies for the severely communicatively impaired. *Mailing Add:* Col Speech Marquette Univ Milwaukee WI 53233

SILVERMAN, GERALD, b New York, NY, Aug 18, 25; m 63. MICROBIOLOGY, FOOD SCIENCE. *Educ:* Cornell Univ, BS, 50, PhD, 54. *Prof Exp:* Chemist, Gen Foods Corp, 54-58; assoc prof microbiol, Mass Inst Technol, 58-69; MICROBIOLOGIST & CHIEF, MICROBIOL BR SCI & ADVAN TECH, US ARMY NATICK RES & DEVELOP CTR, 69- *Concurrent Pos:* Adj prof, Univ RI, 69-; ed, J Food Serv Syst & J Food Safety; mem, Device Good Mfg Practices Adv Comt, Food & Drug Admin. *Mem:* Fel AAAS; Am Soc Microbiol; Inst Food Technologists; Soc Indust Microbiol. *Res:* Food safety; microbial toxins; thermal and irradiation sterilization; taxonomy; dehydration; public health; epidemiology. *Mailing Add:* US Army Natick Res Develop & Eng Ctr SSD Kansas St Natick MA 01760-5020

SILVERMAN, GORDON, b Brooklyn, NY, Mar 4, 34; m 57; c 3. ELECTRONICS. *Educ:* Columbia Col, AB, 55, BS, 56, MS, 57; Polytech Inst Brooklyn, PhD(syst sci), 72. *Prof Exp:* Sr engr, Int Tel & Tel Corp, 57-61; proj engr, Loral Electronics Corp, 61-64; affiliate, 64-80, SR RES ASSOC ELEC ENG, ROCKEFELLER UNIV, 80- *Concurrent Pos:* Mem adj fac, Fairleigh Dickinson Univ, 58- & Polytech Inst Brooklyn, 65-68. *Mem:* Inst Elec & Electronics Engrs; Sigma Xi. *Res:* Application of electronic instrumentation to the study of biological systems; analysis and simulation of human learning. *Mailing Add:* 70 Riverside Dr New York NY 10024

SILVERMAN, HAROLD, b Detroit, Mich, Jan 30, 50; m 81; c 2. BIOMINERALIZATION, SKELETAL MUSCLE FORM & FUNCTION. *Educ:* Univ Mich, BS, 72; Ohio Univ, MS, 74, PhD(zool), 77. *Prof Exp:* Asst muscle physiol, Muscular Dystrophy Asn, Univ Toronto, 77-80; asst prof biol, Pan Am Univ, 80-81; asst prof zool & physiol, 81-86, ASSOC PROF ZOOL & PHYSIOL, LA STATE UNIV, 86-, ASSOC DEAN, COL BASIC SCI. *Concurrent Pos:* NSF grants, 83- *Mem:* Am Asn Anat; Am Soc Zoologist. *Res:* Cellular and physiological processes of biominerazation of calcium concretions in freshwater mussels; skeletal muscle fiber response to neural overactivity biochemically and morphologically. *Mailing Add:* Zool Physiol La State Univ Baton Rouge LA 70803

SILVERMAN, HAROLD I, b Lawrence, Mass, Apr 27, 28; m 51; c 2. PHARMACY. *Educ:* Philadelphia Col Pharm & Sci, BSc, 51, MSc, 52, DSc, 56. *Prof Exp:* Instr pharmacog, Philadelphia Col Pharm & Sci, 52-56; asst prof pharm, Brooklyn Col Pharm & Long Island Univ, 56-59, assoc prof pharm & dir aerosol res lab, 59-64; vpres & sci dir, Knoll Pharmaceut Co, NJ, 64-68; chmn dept pharm, 68-73, prof pharm & assoc dean, Mass Col Pharm, 68-, dir Div Appl Sci, 73-; AT THOMPSON MED CO. *Concurrent Pos:* Sr scientist pharmaceut prod res & develop, Warner Lambert Res Inst, NJ, 58-61; lectr, New Eng Col Optom, 71- & Sch Med, Boston Univ, 71-; consult, Malmstrom Chem Corp, Warner Lambert Res Inst, J T Baker Chem Co, Gallard & Schlessinger Co, Thompson Med Co, Topps Chewing Gum, Inc, Cooper Labs, Inc, H V Shuster, Inc, Gillette Co, Corneal Sci, Inc, Arthur D Little, Inc, NJ Bd Pharm, Mass Bd Pharm, Boston Mus of Sci & R I Bd Optom; partic, US Pharmacopea, Nat Formulary Rev, 68-; mem human subjects comt, Peter Bent Brigham Hosp, Boston Univ Med Ctr, 74-; sci ed, Pharm Lett, 74-; contrib ed, Apothecary, 74-; grants from Smith, Kline & French, S B Penick, Baker Castor Oil Co & Pfeiffer Found. *Honors & Awards:* Newcomb Award, 56; Am Cyanamid Co Lederle res awards, 62, 63; Distinguished Serv Award,

Am Optom Asn, 74. *Mem:* Am Pharmaceut Asn; AAAS; Am Chem Soc; Soc Pharmacists in Indust; fel Soc Cosmetic Chemists. *Res:* Pharmaceutical chemistry and drug development; drug stabilization; methods of analysis; pharmacokinetics. *Mailing Add:* 45 Crest Rd Framingham MA 01701

SILVERMAN, HERBERT PHILIP, b Brooklyn, NY, Sept 8, 24; m 49; c 3. ELECTROCHEMISTRY. *Educ:* City Col New York, BS, 48; Stanford Univ, PhD(chem), 57. *Prof Exp:* Actg instr anal chem, Stanford Univ, 55; res chemist, Kaiser Aluminum & Chem Co, 55-58; sr scientist, Lockheed Missiles & Space Co, 58-61; group leader electrochem, Magna Corp, TRW, Inc, Anaheim, 61-62, dept mgr phys chem, 62-64, div mgr res & develop, 64-65, assoc mgr chem sci, TRW Systs, 65-66, mgr biosci, 66-78; group leader electrochem res, 78-79, HEAD MEMBRANE RES, OCCIDENTAL RES CORP, 79- *Concurrent Pos:* Lectr, Univ Southern Calif. *Mem:* AAAS; Am Chem Soc; Electrochem Soc; Sigma Xi. *Res:* Electrochemical processes; electrobiochemistry; photochemical influence on electrochemical reactions; biochemical influence on electrochemical reactions; nonaqueous electrochemistry; ion exchange membranes. *Mailing Add:* 1335 Meadow Lane Yellow Springs OH 45387

SILVERMAN, JACOB, b Brooklyn, NY, May 11, 23; m 51; c 2. PHYSICAL CHEMISTRY, SCIENCE ADMINISTRATION. *Educ:* Wayne State Univ, BSc, 43, PhD(phys chem), 49. *Prof Exp:* Lectr & res assoc chem, Univ Southern Calif, 49-51; radiol chemist, US Naval Radiol Defense Lab, 51-53; res chemist, Aerojet Gen Corp Div, Gen Tire & Rubber Co, 53-54; mgr chem & mat sci, Rocketdyne Div, NAm Rockwell, Inc, 54-75, dir energy systs, Rocketdyne Div, 75-78, dir, fossil energy systs, Energy Systs Group, Rockwell Int Corp, 78-84; RETIRED. *Mem:* AAAS; Am Chem Soc; Sigma Xi. *Res:* Technical management; research administration; fuel technology; physical chem. *Mailing Add:* 22963 Darien St Woodland Hills CA 91364

SILVERMAN, JEFFREY ALAN, CANCER RESEARCH, GROWTH REGULATION. *Educ:* Univ Pittsburgh, PhD(exp path), 84. *Prof Exp:* RES ASSOC, MED SCH, DARTMOUTH COL, 84- *Mailing Add:* 9000 Rockville Pike Bldg 37 Rm 3C25 Bethesda MD 20892

SILVERMAN, JERALD, b Brooklyn, NY, Mar 23, 42; m; c 2. NUTRITIONAL CARCINOGENESIS, ALTERNATIVES TO USE OF LABORATORY ANIMALS. *Educ:* Cornell Univ, BS, 64, DVM, 66; Am Col Lab Animal Med, dipl, 81. *Prof Exp:* Staff vet, Humane Soc NY, 66-67; pvt vet pract, Brooklyn, NY, 68-69 & Pearl River, NY, 70-75; res assoc prof path, NY Med Col, 78-85; dir, Res Animal Facil, Naylor Dana Inst Dis Prev, Am Health Found, 75-85; DIR, LAB ANIMAL CTR, OHIO STATE UNIV, 85-, ASSOC DIR, UNIV LAB ANIMAL RESOURCES & ASSOC PROF, VET PREVENTIVE MED, 85- *Concurrent Pos:* Assoc ed, Lab Animal, 79-, Lab Animal Sci, 82-86; consult, Revlon Health Care Group, Inc, 80-85, Avon Res, 82-89, Am Health Found, 85-87, Eugene Tech Int, 85-89. *Mem:* Am Vet Med Asn; Am Asn Lab Animal Sci; Am Soc Lab Animal Practitioners; Am Col Lab Animal Med; Am Asn Cancer Res; Vet Cancer Soc; Int Soc Vitamin & Nutrit Oncol. *Res:* Nutritional aspects of cancer prevention, especially dietary fat and vitamins; diseases of laboratory animals; alteratives to animal usage. *Mailing Add:* Lab Animal Ctr Ohio State Univ Columbus OH 43235-1969

SILVERMAN, JOSEPH, b New York, NY, Nov 5, 22; m 51; c 2. RADIATION CHEMISTRY, POLYMER CHEMISTRY. *Educ:* Brooklyn Col, BA, 44; Columbia Univ, AM, 48, PhD(chem), 51. *Prof Exp:* Staff phys chemist, Atomic Energy Div, H K Ferguson Co, NY, 51-52; res dir nuclear labs, Walter Kidde & Co, 52-55; vpres & lab dir, Radiation Appln, Inc, 55-58; assoc prof chem, State Univ NY, 58-59; assoc prof, 60-63, dir, Inst Phys Sci & Technol, 77-83, PROF CHEM & NUCLEAR ENG, UNIV MD, COLLEGE PARK, 63- *Concurrent Pos:* Guggenheim fel, 66-67; consult, Danish Atomic Energy Comn, 67-76; guest scientist, Atomic Energy Res Estab, Denmark; vis prof, Royal Mil Col Sci, Eng; ed, Int J Appl Radiation & Isotopes, 73-78; consult, Indust Res Inst, Japan, 73-; vis prof, Univ Tokyo, 74; vis distinguished prof nuclear energy, Univ Tokyo, 74; consult, UN Develop Prog, 76-83; consult, Int Atomic Energy Agency, 81. *Honors & Awards:* Radiation Indust Award, Am Nuclear Soc, 75. *Mem:* Am Chem Soc; fel Am Nuclear Soc; fel Am Phys Soc; fel Nordic Soc Radiol Chem & Technol. *Res:* Pure and applied polymer and radiation chemistry; radiation source technology. *Mailing Add:* Dept Chem Eng Univ Md College Park MD 20742-2115

SILVERMAN, LEONARD, ENGINEERING. *Educ:* Columbia Univ, BS, 62, MS, 63, PhD(elec eng), 66. *Prof Exp:* Actg asst prof elec eng & computer sci, Univ Calif, Berkeley, 66-68; from asst prof to assoc prof, Dept Elec Eng-Systs, 68-77, chmn, 82-84, PROF, DEPT ELEC ENG-SYSTS, UNIV SOUTHERN CALIF, 77-, DEAN ENG, 84- *Concurrent Pos:* Consult var aerospace & electronic co; dir, Bd Tandon Co, 88- *Honors & Awards:* Centennial Medal, Inst Elec & Electronics Engrs, 84. *Mem:* Nat Acad Eng; fel Inst Elec & Electronics Engrs; Soc Indust & Appl Math. *Res:* Theory and application of multivariable systems and control; author of more than 100 publications. *Mailing Add:* Sch Eng Univ Southern Calif Univ Par Los Angeles CA 90089-1450

SILVERMAN, MELVIN, b Montreal, Que, Jan 4, 40. MEDICINE. *Educ:* McGill Univ, BSc, 60, MD, CM, 64, FRCP(C), 69. *Prof Exp:* Assoc res scientist, Dept Med, NY Univ, 66-68; Med Res Coun Centennial fel, McGill Univ Med Clin Montreal Gen Hosp, 69-71, asst prof, Dept Med, McGill Univ, 70-71; from asst prof to assoc prof med, 71-81, assoc prof physics, 75-81, PROF MED & PHYSICS, UNIV TORONTO, 81- *Concurrent Pos:* Asst physician, Montreal Gen Hosp, 70-71; staff physician nephrol, Toronto Gen Hosp, 71-84; dir, Trihosp Nephrol Serv, Toronto Gen Hosp, Mt Sinai Hosp & Women's Col Hosp, 84-90, sr staff physician, Trihosp Nephrol Serv, 90-; dir, MRC Group Membrane Biol, Dept Med, Univ Toronto, 87-, subspecialty training prog, 90-; mem sci coun, Kidney Found Can, 77-85 & 78-84, Med Res Coun Can, 84-89; numerous res grants var orgn, 72- *Honors & Awards:* J Francis Mem Prize in Med, 64. *Mem:* Am Soc Clin Invest; Am Physiol Soc; Am Soc Nephrol; Am Biophys Soc; NY Acad Sci. *Res:* Membrane transport; renal microcirculation; renal disease mechanism. *Mailing Add:* Dept Med Univ Toronto Med Sci Bldg Rm 7226 Toronto ON M5S 1A8 Can

SILVERMAN, MEYER DAVID, b New York, NY, Jan 8, 15; m 40; c 1. PHYSICAL INORGANIC CHEMISTRY. *Educ:* Yale Univ, BChE, 34; George Washington Univ, MA, 42; Univ Tenn, PhD(phys chem), 50. *Prof Exp:* Sci aide, Food & Drug Admin, USDA, Washington, DC, 38, sci aide, Cotton Mkt Div, 39-41; jr chem engr, Edgewood Arsenal, Md, 41-42; res chem engr, Permutit Water Conditioning Corp, NY, 42-43; assoc chem engr, Oak Ridge Nat Lab, 43-46, chemist, 47-52, sr res chemist, 52-75, res staff mem chem eng, 75-82. *Concurrent Pos:* Fel, Oak Ridge Inst Nuclear Studies; consult chem eng, waste disposal (radioactive), coal conversion & recovery of metals from eastern oil shales, 82- *Mem:* Am Chem Soc; Sigma Xi. *Res:* Chemical potentials; ion exchange; reactor chemistry; radiation-induced corrosion; aerosol physics; nuclear reactor safety; coal conversion processes and plant equipment; high temperature energy storage; mineral recovery from eastern oil shales. *Mailing Add:* 397 East Dr Oak Ridge TN 37830

SILVERMAN, MICHAEL ROBERT, b Ft Collins, Colo, Oct 7, 43; m 64; c 2. MOLECULAR & MICROBIAL GENETICS. *Educ:* Univ Nebr, BS, 66, MS, 68; Univ Calif, San Diego, PhD(biol), 72. *Prof Exp:* USPHS trainee tumor virol, Med Sch, Univ Colo, 73-75; ASST RES BIOLOGIST MOLECULAR GENETICS, UNIV CALIF, SAN DIEGO, 75- *Mem:* Am Soc Microbiol. *Res:* Regulation of gene expression, particularly of genes which determine components of the flagellar organelle of Escherichia coli. *Mailing Add:* Agouron Inst 505 Coast Blvd S La Jolla CA 92037

SILVERMAN, MORRIS, b Brooklyn, NY, June 19, 26; m 55; c 2. BIOCHEMISTRY, BACTERIOLOGY. *Educ:* NY Univ, BA, 49; Univ Mich, MS, 51; Yale Univ, PhD(biochem), 60. *Prof Exp:* USPHS fel biochem, Pub Health Res Inst New York, 59-61; instr, 61-63, ASST PROF BIOCHEM, COL MED, STATE UNIV NY DOWNSTATE MED CTR, 63- *Mem:* AAAS; Am Chem Soc. *Res:* Carbohydrate metabolism; oxidative phosphorylation; oxidative enzymes; microbial metabolism. *Mailing Add:* Dept of Biochem State Univ NY Downstate Med Ctr Brooklyn NY 11203

SILVERMAN, MORRIS BERNARD, b Roxbury, Mass, June 28, 24; m 56. INORGANIC CHEMISTRY. *Educ:* Boston Univ, AB, 48; Univ Washington, Seattle, PhD(chem), 56. *Prof Exp:* Radiochemist, Tracerlab, Inc, 49-50; draftsman, Johnson Fare Box Co, 51; develop chemist, Armour & Co, 52-53; fel organometallics, Univ Wash, Seattle, 56-57; chemist, Calif Res Corp, Standard Oil Co Calif, 57-59; ASSOC PROF CHEM, PORTLAND STATE UNIV, 59- *Mem:* Am Chem Soc; Soc Cosmetic Chemists. *Mailing Add:* 7124 SW 5 ST Portland OR 97219-2230

SILVERMAN, MYRON SIMEON, b New York, NY, Aug 2, 15. BACTERIOLOGY, IMMUNOLOGY. *Educ:* Cornell Univ, BS, 37, MS, 38; Univ Calif, PhD, 50; Am Bd Microbiol, dipl. *Prof Exp:* Lab technician, Div Lab & Res, NY State Dept Health, 39-41; asst bact, Univ Calif, 48-50; supvry bacteriologist, US Naval Radiol Defense Lab, 50-62, head microbiol & immunol br, 62-69; res microbiologist, Naval Med Res Unit 1, Calif, 69-70; assoc dean, Grad Sch, 79-81, PROF ORAL BIOL & BACT, DENT RES CTR & SCH MED, UNIV NC, CHAPEL HILL, 70-, SPEC ASST TO DEAN, GRAD SCH, 81- *Concurrent Pos:* Nat Cancer Inst res fel, Guy's Hosp, Med Sch, Univ London, 60-61; Naval Radiol Defense Lab fel, Brookhaven Nat Lab, 68-69; res assoc, Univ Calif, Berkeley, 50-59, lectr, 58-68; mem comn radiation & infection & comn epidemiol surv, Armed Forces Epidemiol Bd. *Mem:* AAAS; Am Soc Microbiol; Radiation Res Soc; Soc Exp Biol & Med; Am Asn Immunol; Sigma Xi. *Res:* Induction of immune response; cellular interactions in the immune response; immunological aspects of oral disease. *Mailing Add:* 320 Ridgecrest Dr Chapel Hill NC 27514

SILVERMAN, NORMAN A, b Boston, Mass, Dec 19, 46. CARDIOTHORACIC SURGERY, INTRAOPERATIVE MIOCARDIAL PROTECTION. *Educ:* Boston Univ, MD, 71. *Prof Exp:* PROF SURG, COL MED, UNIV ILL, 80- *Mem:* Soc Univ Surgeons; Am Asn Thoracic Surg; Am Physiol Soc. *Res:* Adjudication of the efficacy of cardioplegia by postichemic assessment of ventricular function and energetics. *Mailing Add:* 11 Sycamore Lane Grosse Point MI 48230-1936

SILVERMAN, PAUL HYMAN, b Minneapolis, Minn, Oct 8, 24; m 45; c 2. PARASITOLOGY, IMMUNOLOGY. *Educ:* Roosevelt Univ, BS, 49; Northwestern Univ, MS, 51; Univ Liverpool, PhD(parasitol), 55. *Hon Degrees:* DSc, Univ Liverpool, 68. *Prof Exp:* Sr sci officer, dept parasitol, Moredun Inst, Edinburgh, Scotland, 56-59; head, dept immunoparasitol, Allen & Hansburys, Ltd, Ware, Eng, 60-62; prof zool, vet path & hyg, Univ Ill, Urbana, 63-72, chmn, dept zool, 64-65, head, 65-69; chmn, dept biol, Univ NMex, 72-73, actg vpres res & grad affairs, 73-74, vpres res & grad affairs, 74-77; provost res & grad studies, State Univ NY Albany, 77-79, pres res found, 79-80; pres, Univ Maine, 80-84; fel, Biol & Med Div, Lawrence Berkeley Lab, Univ Calif, 84-86, actg head, Biol & Med Div & actg dir, 86-87, dir, Donner Lab, 84-90, dir, Biotechnol Prog, 89-90; DIR, SCI AFFAIRS, BECKMAN INSTRUMENTS, INC, 90- *Concurrent Pos:* Consult-examr, Comn Cols & Univs, NCent Asn; prof & head, Natural Sci Div, Temple Buell Col, 70-71; adj prof, Univ Colo, Boulder, 70-72; consult comn malaria, Armed Forces Epidemiol Bd, 71-; mem bd dirs, NCent Asn Cols & Sec Schs, 71-, vchmn, Comn on Insts Higher Educ, 72-74 & 76-, chmn, 74-76 & mem bd dirs, Inhalation Toxicol Lab, Albuquerque, NMex; chmn acad bd, US Army Command & Gen Staff Col, Ft Levenworth, Ky, 84-; assoc dir, Lawrence Berkeley Lab, 86-90. *Mem:* Am Soc Microbologists; Am Asn Immunol; Am Soc Trop Med & Hyg; Am Soc Microbiologists; Royal Soc Trop Med & Hyg; NY Acad Sic; Sigma Xi. *Res:* Nature of host-parasite of relationship, particularly on immunological phenomena; in vitro culture of parasites. *Mailing Add:* Beckman Instruments Inc 2500 Harbor Blvd Fullerton CA 92634-3100

SILVERMAN, PHILIP MICHAEL, b Chicago, Ill, Oct 21, 42; div; c 2. BIOCHEMISTRY, MOLECULAR BIOLOGY. *Educ:* Univ Ill, Urbana, BS, 64; Univ Calif, Berkeley, PhD(biochem), 68. *Prof Exp:* From asst prof to prof molecular biol, Albert Einstein Col Med, 71-88; MEM & HEAD PROG

MOLECULAR & CELL BIOL, OKLA MED RES FOUND, 88- *Concurrent Pos:* Damon Runyon Mem Fund Cancer Res fel, Albert Einstein Col Med, 69-71; estab investr, Am Heart Asn, 75-80; Irma T Hirschl career scientist, 81-85; mem adv panel genetic biol, NSF, 81-86; Marjorie Nichlos chair med res, 88- *Mem:* Am Soc Biol Chemists; Am Soc Microbiol; Am Soc Cell Biol. *Res:* Biochemistry and genetics of bacterial conjugation; synthesis and function of membrane proteins. *Mailing Add:* Prog Molecular Cell Biol Okla Med Res Found 825 Northeast 13th St Oklahoma City OK 73104

SILVERMAN, RICHARD BRUCE, b Philadelphia, Pa, May 12, 46; m 83; c 3. ENZYME INHIBITION, ENZYME MECHANISMS. *Educ:* Pa State Univ, BS, 68; Harvard Univ, MA, 72, PhD(org chem), 74. *Prof Exp:* Fel biochem, Brandeis Univ, 74-76; from asst prof to assoc prof, 76-82, PROF CHEM, BIOCHEM, MOLECULAR & CELL BIOL, NORTHWESTERN UNIV, 86- *Concurrent Pos:* DuPont Young Fac fel, 76; prin investr grants, NIH, 77-; mem adv panels, NIH, 81-85, 87-, Res Career Develop Award, 82-87; Alfred P Sloan Found fel, 81-85; consult, Proctor & Gamble, 84, Abbott Labs, 87, Monsanto/Searle, 88-90. *Mem:* Am Chem Soc; fel AAAS; Am Soc Biochem & Molecular Biol; fel Am Inst Chemists. *Res:* Molecular mechanism of action, rational design and synthesis of medicinal agents that act as enzyme inactivators; elucidation of enzyme mechanisms. *Mailing Add:* Dept Chem Northwestern Univ Evanston IL 60208-3113

SILVERMAN, ROBERT, b Cleveland, Ohio, Oct 23, 28; m 48, 71; c 4. MATHEMATICS. *Educ:* Ohio State Univ, BSc, 51, MA, 54, PhD(math), 58. *Prof Exp:* Instr math, Ohio State Univ, 59; Nat Acad Sci-Nat Res Coun res assoc, Nat Bur Standards, 59-60; asst prof math, Syracuse Univ, 60-65; from assoc prof to prof, 68-80, EMER PROF MATH, WRIGHT STATE UNIV, 84- *Concurrent Pos:* Res assoc, Univ Western Ont, 80, 82. *Mem:* Math Asn Am; Am Math Soc; fel Nat Sci Found; Sigma Xi. *Res:* Combinatorial analysis; algebra. *Mailing Add:* Dept Math Wright State Univ PO Box 709 Yellow Springs OH 45387

SILVERMAN, ROBERT ELIOT, b Wilmington, Del, Feb 15, 50; m 77; c 2. ENDOCRINOLOGY. *Educ:* Univ Penn, BA, 72; Wash Univ Sch Med, MD, 78, PhD(biol chem), 78. *Prof Exp:* House officer, internal med, Yale-New Haven Hosp, 78-81; med staff fel, Nat Inst Health, 81-84; BR CHIEF DIABETES PROGS, NIDDK, NAT INST HEALTH, 84- *Mem:* Am Col Physicians; Endocrine Soc; Am Diabetes Assoc. *Res:* Management of and development of programs of support for biomedical research on diabetes mellitus and related topics. *Mailing Add:* Nat Inst of Diabetes and Digestive and Kidney Dis NIH Westwood Bldg Rm 626 Bethesda MD 20892

SILVERMAN, ROBERT HUGH, b Houston, Tex, Nov 24, 48; m 84; c 2. PATHOLOGY, INTERFERON. *Educ:* Mich State Univ, BSc, 70; Iowa State Univ, PhD(molecular biol), 77. *Prof Exp:* Fel, Roche Inst Molecular Biol, 77-79; mem sci staff, Nat Inst Med Res, Mill Hill, London, 79-80, Imperial Cancer Res Fund Labs, London, 80-81; PROF PATH, UNIFORMED SERV, UNIV HEALTH SCI, 82- *Concurrent Pos:* Vis scientist, German Cancer Res Ctr, Heidelberg, FRG. *Mem:* Int Soc Interferon Res; Am Asn Pathologists. *Res:* Biochemical mechanism of action of interferons, in particular the role of the unusual oligonucleotide series known as 2-5A. *Mailing Add:* Dept Path Uniformed Serv Univ Health Sci 4301 Jones Bridge Rd Bethesda MD 20814

SILVERMAN, SAM M, b New York. NY, Nov 16, 25; m 48, 66; c 5. GEOPHYSICS, HISTORY & PHILOSOPHY OF SCIENCE. *Educ:* City Col New York, BChE, 45; Ohio State Univ, PhD(phys chem), 52; Suffolk Univ Law Sch, JD, 82. *Prof Exp:* Res assoc phys chem, Ohio State Univ, 52-55; asst prof silicate chem, Univ Toledo, 55-57; chief, Polar Atmospheric Processes Br & dir, Geopole Observ, Air Force Cambridge Res Labs, 63-74, res physicist, 57-80; CONSULT, 80-; SR RES PHYSICIST, PHYSICS DEPT, BOSTON COL, 81- *Concurrent Pos:* Vis res assoc, Queens Univ, Belfast, 63-64; adv bd, Inst Space & Atmospheric Studies, Univ Sask, 65-69; abstractor & reviewer psychohist, Am J Psychother, 75-77; ed, Hist Geophys Newsletter; co-chmn, Interdivisional Comn Hist, Int Asn Geomagnetism & Aerorourg. *Mem:* Fel Am Phys Soc; Am Geophys Union; Sigma Xi; fel Explorers Club. *Res:* Polar cap upper atmosphere; aurora, airglow and related fields of upper atmosphere physics; law. *Mailing Add:* 18 Ingleside Rd Lexington MA 02173

SILVERMAN, SOL RICHARD, b New York, NY, Oct 2, 11; m 38; c 1. AUDIOLOGY, SPEECH PRODUCTION. *Educ:* Cornell Univ, BA, 33; Wash Univ, MS, 38, PhD(psychol), 42. *Hon Degrees:* DLitt, Gallaudet Col, 61; LHD, Hebrew Union Col, 62; LLD, Emerson Col, 66; DSc, Rochester Inst Technol, 89. *Prof Exp:* Instr audiol, 35-45, assoc prof educ, 45-48, dir, 47-72, prof audiol, 48-81, EMER DIR, CENT INST DEAF, WASH UNIV, 72-, EMER PROF AUDIOL, 81- *Concurrent Pos:* Dir, proj hearing & deafness, Cent Inst Deaf, Off Sci Res & Develop, 42-45; consult, probs deafness, Secy War, 44-48 & audiol, US Air Force, 51-53; mem, comt conserv hearing, Am Acad Ophthal & Otolaryngol, 57-76; mem, Nat Adv Coun Voc Rehab, 62-66; co-chmn, Int Cong Educ Deaf, Washington, DC, 63; chmn, Nat Adv Comt Educ Deaf, 66-70; mem, commun dis res training comt, Nat Inst Neurol Dis & Blindness, 67-71; mem nat adv group, Nat Tech Inst Deaf, 77-, chmn, 78-82; adj res scientist audiol, Inst Advan Study Commun Processes, Univ Fla, 82- *Honors & Awards:* Herbert Birkett Mem Lectr, McGill Univ, 49; J McKenzie Brown Mem Lectr, Children's Hosp, Los Angeles. *Mem:* Am Speech & Hearing Asn (pres, 53); Alexander Graham Bell Asn Deaf (pres, 57-60); Coun Educ Deaf (pres, 60-64); fel Acoust Soc Am; hon mem Asn Res Otolaryngol. *Res:* speech production; hearings aids; auditory tests. *Mailing Add:* 2510 NW 38th St Gainesville FL 32605

SILVERMAN, SOL ROBERT, b New York, NY, Nov 5, 18; m 51; c 3. ORGANIC GEOCHEMISTRY. *Educ:* NY Univ, BA, 40; Univ Chicago, MS & PhD(geol), 50. *Prof Exp:* Chemist, Chem Warfare Serv, US War Dept, 42-44 & Mat Lab, US Dept Navy, 46-47; geochemist, US Geol Surv, 50-51; res assoc, Calif Res Corp, 51-63, sr res assoc, Chevron Oil Field Res Co, 63-83; RETIRED. *Concurrent Pos:* Lectr petrol geol, Calif State Polytech Univ, 76- *Mem:* Fel AAAS; Am Chem Soc; Geochem Soc; Am Asn Petrol Geologists. *Res:* Distribution of stable isotopes in nature; petroleum geochemistry; biogeochemistry. *Mailing Add:* 1230 Casa del Rey La Habra CA 90631

SILVERMAN, WILLIAM BERNARD, b New York, NY, Mar 3, 32. PLANT PATHOLOGY. *Educ:* City Col New York, BSc, 53; Univ Minn, MSc, 56, PhD(plant path, bot), 58. *Prof Exp:* Asst plant path & bot, Univ Minn, 53-58; res assoc polyacetylenes, NY Bot Garden, 58-59; from asst prof to assoc prof, 59-74, PROF BIOL, COL ST THOMAS, 74- *Res:* Histology-cytology; microbiology. *Mailing Add:* Dept of Biol Col St Thomas 2115 Summit Ave St Paul MN 55105

SILVERN, LEONARD CHARLES, b New York, NY, May 20, 19; m 48, 69 & 85. ENGINEERING PSYCHOLOGY, GEOGRAPHIC INFORMATION SYSTEMS. *Educ:* Long Island Univ, BS, 46; Columbia Univ, MA, 48, EdD, 52. *Prof Exp:* Training supvr & coord, US Dept Navy, 39-49; training dir, Exec Dept, State Div Safety, NY, 49-55; resident eng psychologist, Lincoln Lab, Mass Inst Technol, Rand Corp, 55-56; head tech training eval & develop, Hughes Aircraft Co, 56-58, asst head sci educ, Corp Off, 58-60, dir educ & training res lab, Ground Systs Group, 60-62; dir human performance eng lab & consult eng psychologist to vpres tech, Norair Div, Northrop Corp, 62-64; prin scientist, Educ & Training Consult Co, 64-66, vpres behav systs, 66-68, pres, 68-80; PRES, SYSTS ENG LABS, 80- *Concurrent Pos:* Adj prof, Grad Sch, Univ Southern Calif, 57-65; reviewer, Comput Rev, 62-; vis prof, Univ Calif, Los Angeles, 63-72; consult, Electronic Industs Asn, 61-69, Univ Hawaii, 70-76, Hq, Air Training Command, USAF, Randolph AFB, Tex, 64-69, Centro Nacional de Productividad, Mexico City, 73-75, Ford Motor Co, 75-76, NS Dept Educ, Halifax, 75-79, Nat Training Systs, Inc, 76-81, Search, 76-, Nfld Pub Serv Comn, 78 & Legis Affairs Off, US Dept Agr, 80; dist opers officer, Los Angeles County Sheriff's Dept, Disaster Commun Serv, 73-75 & dist commun officer, 75-76; councilman, City of Sedona, Ariz, 88-92; treas, Sedona-Verde Valley Group, Sierra Club; chair, bd dirs, VOX POP, 83- *Mem:* Sr mem Inst Elec & Electronics Engrs; Am Psychol Asn; Soc Eng Psychol; Soc Wireless Pioneers; Am Radio Relay League; Sedona Westerners & Quarter Century Wireless Asn. *Res:* Human learning in engineering and physical science environments; computer-assisted instruction; systems engineering applied to complex social and political systems. *Mailing Add:* Systs Eng Labs Box 2085 Sedona AZ 86336-2085

SILVERNAIL, WALTER LAWRENCE, b St Louis, Mo, Sept 8, 21; m 44; c 3. INDUSTRIAL CHEMISTRY. *Educ:* Park Col, AB, 47; Univ Mo, AM, 49, PhD(chem), 54. *Prof Exp:* From assoc prof to prof chem, Ill Col, 49-56; assoc prof, Ferris Inst, 56-57; res chemist, Lindsay Chem Div, Am Potash & Chem Corp, 57-64, mgr tech serv, Kerr-McGee Chem Corp, West Chicago Plant, 64-73; CONSULT, 74- *Res:* Rare earth and thorium chemistry; ion exchange; glass polishing; chemical education. *Mailing Add:* 140 E Stimmel St Chicago IL 60185

SILVERS, J(OHN) P(HILLIP), b Chicago, Ill, Jan 23, 20; m 45; c 3. THERMODYNAMICS. *Educ:* Purdue Univ, BS, 44, MS, 46, PhD(heat transfer, thermodyn), 51. *Prof Exp:* Asst heat transfer & vibration analysis, Eng Exp Sta, Purdue Univ, 44-47, instr & admin asst to head dept thermodyn & heat transfer, 47-50; assoc scientist, Argonne Nat Lab, 50-55; assoc mgr dept appl res, Res & Adv Develop Div, Avco Corp, 55-57, tech asst to vpres res, 57-60, head adv prog undersea technol, 60-65, mgr marine technol dept, 65-66, mgr prog planning, Avco Space Systs Div, 66-68, asst dir eng, Avco Systs Div, 68-71; exec dir, Mass Sci & Technol Found, 72-77; ENERGY CONSULT, 77- *Concurrent Pos:* Sr scientist, res staff, Mass Inst Technol, 78-81; mgr thermal systs, Res & Develop Div, Dynatech, 81-85. *Mem:* Marine Technol Soc; Sigma Xi. *Res:* Nuclear reactor development; heat transfer; numerical analysis; marine technology; environmental science. *Mailing Add:* 327 Salem St Wilmington MA 01887

SILVERS, WILLYS KENT, b New York, NY, Jan 12, 29; m 56; c 2. GENETICS. *Educ:* Johns Hopkins Univ, BA, 50; Univ Chicago, PhD(zool), 54. *Prof Exp:* Assoc staff scientist, Jackson Lab, Bar Harbor, 57; assoc mem, Wistar Inst, Philadelphia, 57-65; assoc prof med genetics, 65-67, chmn genetics grad group, 80-90, PROF HUMAN GENETICS, SCH MED, UNIV PA, 67-, PROF PATH, 69- *Concurrent Pos:* USPHS fel, Brown Univ, 55-56 & Jackson Lab, Bar Harbor, Maine, 56; NIH career develop award, 64-71; mem allergy & immunol study sect, NIH, 62-66; assoc ed, J Exp Zool, 65-70 & 82-; mem primate res ctr adv comt, NIH, 68-71; sect ed, Immunogenetics & Transplantation, J Immunol, 73-77; mem comt cancer immunobiol, Nat Cancer Inst, 74-77; mem bd sci oversers, Jackson Lab, Bar Harbor, Maine, 81-89; chmn bd sci oversers, Jackson Lab, 87-89. *Mem:* AAAS; Genetics Soc Am; Am Genetic Asn (pres, 81-); Am Soc Human Genetics. *Res:* Mammalian genetics, with particular reference to the genetics of coat-color determinant and immunogenetics; biology and immunology of tissue transplantation; biology of skin. *Mailing Add:* Dept Human Genetics Univ Pa Sch Med Philadelphia PA 19104

SILVERSMITH, ERNEST FRANK, b Nuernberg, Ger, Oct 3, 30; nat US; m 53; c 4. ORGANIC CHEMISTRY. *Educ:* Harvard Univ, AB, 52; Univ Wis, PhD(chem), 55. *Prof Exp:* Res fel chem, Calif Inst Technol, 55-56; asst prof, Mt Holyoke Col, 56-58; res chemist, E I du Pont de Nemours & Co, 58-67; PROF CHEM, MORGAN STATE UNIV, 67- *Mem:* Am Chem Soc. *Res:* Photochemistry; spectroscopy; modern organic synthesis. *Mailing Add:* Dept of Chem Morgan State Univ Baltimore MD 21239

SILVERSTEIN, ABE, b Terre Haute, Ind, Sept 15, 08; m 50; c 3. MECHANICAL ENGINEERING. *Educ:* Rose Polytech Inst, BS, 29, MechEng, 34. *Hon Degrees:* ScD, Rose Polytech Inst, 59; DEng, Case Western Reserve Univ, 58; LHD, Yeshiva Univ, 60; DAS, Fenn Col, 64. *Prof Exp:* Aerodynamic res engr, Langley Res Ctr, Nat Adv Comt Aeronaut, Va, 29-40, head full scale wind tunnel, 40-43, chief engine installation res div, Lewis Flight Propulsion Lab, Ohio, 43-45, chief wind tunnel & flight res div, 45-49, chief res, 49-52, assoc dir, 52-58, dir off space flight progs, NASA, Washington, DC, 58-61, dir, Lewis Res Ctr, 61-70; dir environ planning, Repub Steel Corp, 70-78; RETIRED. *Concurrent Pos:* Consult mech eng, 76- *Mem:* Nat Acad Eng; fel Am Astronaut Soc; fel Am Inst Aeronaut & Astronaut; fel Royal Aeronaut Soc; Int Acad Astronaut. *Res:* Aerodynamic and propulsion aspects of aeronautical research; propulsion and power generation aspects of space; design and construction of facilities for space research. *Mailing Add:* 114 SW 52nd Terr Cape Coral FL 33914

SILVERSTEIN, ALEXANDER, neurology, psychiatry; deceased, see previous edition for last biography

SILVERSTEIN, ARTHUR M, b Aug 6, 28. IMMUNOLOGY, HISTORY OF MEDICINE. *Educ:* Ohio State Univ, AB, 48, MSc, 51; Rensselaer Polytech Inst, PhD(phys chem), 54. *Hon Degrees:* DSc, Univ Granada, Spain, 86. *Prof Exp:* Res asst, Dept Immunochem, Sloan-Kettering Inst Cancer Res, 48-49; biochemist, Dept Serol, NY State Dept Health Res Labs, 49-52, sr biochemist, 52-54; chief, Immunobiol Br, Armed Forces Inst Path, 54-64; assoc prof, 64-67, Odd Fellows res prof, 67-90, EMER PROF OPHTHALMIC IMMUNOL, JOHNS HOPKINS UNIV SCH MED, 91- *Concurrent Pos:* Res fel, Coun Res in Glaucoma & Allied Dis, 60; lectr, Int Cong Opthalmol, Kyoto, Japan, 78, Rome, Italy, 86, Ger Opthalmol Cong, 80; Dept Defense res & develop award, 64. *Honors & Awards:* Macy lectr, Harvard Univ Med Sch, 65; Ralph English Miller Mem lectr, Dartmouth Med Sch, 69; Jonas S Friedenwald Mem Award & lectr, Asn Vision Opthalmol, 73; Doyne Mem Medal & lectr, Oxford Ophthalmol Cong, 74. *Mem:* Sigma Xi; Am Asn Immunologists; Am Soc Exp Pathol; Brit Soc Immunol; Asn Res Vision & Ophthalmol; fel AAAS. *Res:* History of immunology; ocular immunology. *Mailing Add:* Inst Hist Med Sch Med Johns Hopkins Univ Baltimore MD 21205

SILVERSTEIN, CALVIN C(ARLTON), b Newark, NJ, Jan 31, 29; m 59; c 3. HEAT TRANSFER, ENERGY CONVERSION. *Educ:* Newark Col Eng, BS, 50; Princeton Univ, MSE, 51. *Prof Exp:* Res engr, Bendix Aviation Corp Res Labs, 52-54; asst proj engr, Martin Co, 55-57; proj eng, Bell Aircraft Corp, 57; prin mech engr, Cornell Aeronaut Lab, Inc, 57-61; res specialist, Atomics Int Div, N Am Aviation, Inc, 61-62; chief analyst & sr staff analyst, Hittman Assoc, Inc, 63-65; eng consult, 65-73; mgr prog develop, Westinghouse Elec Corp, 73-81; PRES, CCS ASSOCS, 81- *Mem:* Am Soc Mech Engrs. *Res:* Heat transfer; energy conversion; energy storage; power generation; heat pipe technology and applications; capillary-pumped heat transfer loops; hydrogen production and applications; fluidized beds; solar energy. *Mailing Add:* PO Box 563 Bethel Park PA 15102

SILVERSTEIN, EDWARD ALLEN, b Washington, DC, Aug 25, 30; m 56; c 3. NUCLEAR MEDICINE, NUCLEAR PHYSICS. *Educ:* Univ Chicago, BA, 50, MS, 53; Univ Wis, PhD(physics), 60. *Prof Exp:* Res asst nuclear physics, Univ Padua, 61-63; instr & res assoc, Univ Wis, 63-64; asst prof, Case Western Reserve Univ, 64-69; sr staff scientist, Bendix Aerospace Systs Div, 69-70; asst prof radiol, Med Col Wis, 70-75; radiation physicist & assoc dir sch nuclear med, Milwaukee County Gen Hosp, 70-75; group leader nuclear med physics, Sect Med Radiation Physics & asst prof, Rush Med Col, 75-81; PHYSICIST NUCLEAR MED, NORTHWESTERN MEM HOSP, 81-; asst prof, 81-90, ASSOC PROF, MED SCH, NORTHWESTERN UNIV, 90- *Mem:* Soc Nuclear Med; Am Asn Physicists Med. *Res:* Application of digital computer to nuclear medicine. *Mailing Add:* Dept Nuclear Med Northwestern Mem Hosp 250 E Superior St Chicago IL 60611

SILVERSTEIN, ELLIOT MORTON, b Chicago, Ill, Jan 2, 28; m 57; c 3. ELECTROOPTICS, INFRARED SYSTEMS. *Educ:* Univ Chicago, BA, 50, MS, 53, PhD(physics), 58. *Prof Exp:* Mem tech staff, Opers Anal Group, Hughes Aircraft Co, Calif, 58-61, group head, Surveyor Spacecraft Lab, 61-64, sr proj engr, Surveyor Lab, 64-65, sr staff physicist, 66-68; prin engr/scientist, Avionics Control & Info Systs Div, McDonnell Douglas Astronaut Co, Huntington Beach, 68-81; proj engr, Defense & Surveillance Opers, 81-88, PROJ ENGR, PROG OFF, DEFENSE SUPPORT PROG, AEROSPACE CORP, CALIF, 89- *Concurrent Pos:* Mem, Standards Comt, Optical Soc Am, 82-85. *Mem:* AAAS; Am Phys Soc; Optical Soc Am; Sigma Xi; Inst Elec & Electronic Engrs. *Res:* Theoretical and experimental studies in infrared and optical physics and in optical and electrooptical imaging, image processing, detection, communications and radar; design and analysis of spacecraft instrumentation. *Mailing Add:* 8004 El Manor Ave Los Angeles CA 90045-1434

SILVERSTEIN, EMANUEL, b New York, NY, Feb 14, 30; m 65; c 2. BIOCHEMISTRY, GENETICS. *Educ:* City Col New York, BS, 50; State Univ NY Downstate Med Ctr, MD, 54; Univ Minn, PhD(biochem), 63. *Prof Exp:* Intern med, Med Sch, Univ Minn, 54-55; intern path, Sch Med, Yale Univ, 55-56; res assoc exp path, Nat Inst Arthritis & Metab Dis, 56-58; fel med, Med Sch, Univ Minn, 58-59 & fel biochem 59-63; from asst prof to assoc prof med, 64-77, PROF MED & BIOCHEM, STATE UNIV NY DOWNSTATE MED CTR, 77- *Concurrent Pos:* Fel molecular biol, Mass Inst Technol, 63-64; vis scientist, Weizmann Inst Sci, 71 & 81; posdoc fel, Am Cancer Soc, 59-64. *Mem:* AAAS; Am Asn Path; Am Soc Biochem & Molecular Biol; Am Soc Microbiol; Genetics Soc Am; Am Soc Human Genetics. *Res:* Enzyme mechanism and regulation; medical and molecular genetics; protein synthesis; hormone receptors; cell differentiation; porphyrins and oxidative enzymes; chemical diagnosis and therapy of disease, neurochemistry; neurochemistry; molecular mechanism of behavior and brain function. *Mailing Add:* Dept Med State Univ NY Health Sci Ctr Brooklyn NY 11203-2098

SILVERSTEIN, HERBERT, b Philadelphia, Pa, Aug, 35; m 88; c 2. HEARING & BALANCE DISORDERS, OTOLOGY & NEUROTOLOGY. *Educ:* Dickenson Col, BSc, 57; Temple Univ, MD, 61, MSc, 63; Am Bd Otolaryngol, dipl, 67. *Prof Exp:* Intern, Philadelphia Gen Hosp, 61-62; resident surg, Philadelphia Vet Hosp, 62-63; resident otolaryngol, Mass Eye & Ear Infirmary, 63-66, asst, 68-71; attend otolaryngologist, Hosp Univ Pa, 71-74; assoc prof, Univ S Fla, 74-75; PRES & FOUNDER, EAR RES FOUND FLA, 79- *Concurrent Pos:* Asst prof otolaryngol, Univ Pa Med Sch, 71-73; asst clin prof surg, Univ Fla, Gainesville, 77; clin prof surg, Univ S Fla, 84; dir, Walker Biochem Lab, Mass Eye & Ear Infirmary, 68-71; assoc dir, Otology Res Lab, Presby Univ Pa Med Ctr, 71-72; consult, Philadelphia Vet Admin Hosp, 71-74, Childrens Hosp, 71. *Honors & Awards:* Charles Burr Award, 62; Res Award, Am Acad Otolaryngol, 66. *Mem:* Fel Am Acad Ophthal & Otolaryngol; AMA; fel Am Col Surgeons; fel Am Laryngol, Rhinological & Otol Soc. *Res:* Author of over 120 technical publications. *Mailing Add:* Ear Res Found 1921 Floyd St Sarasota FL 34239

SILVERSTEIN, MARTIN ELLIOT, b New York, NY, Sept 6, 22; m 62. MEDICAL SCIENCE, HEALTH SCIENCES. *Educ:* Columbia Univ, AB, 45; NY Med Col, MD, 48. *Prof Exp:* Asst res surg, Flower & Fifth Ave Hosps, 49-50, chief res, 51-52, instr bact & surg, NY Med Col, 53-57, assoc prof surg & assoc dean, 57-63; chmn exp surg, Menorah Inst Med Educ & Res, 63-66, chmn dept exp surg & exec dir, Menorah Med Ctr, 63-66, gen dir, 66; dir grad med educ, Bronx-Lebanon Hosp Ctr, 67-69; chief surg, Grand Canyon Hosp, 69-70; pres, Health Analysis Inc, 70-73; assoc prof surg & chief, Surg Trauma Sect, Col Med, Univ Ariz, 74-84, res prof surg, 84-85; CLIN PROF SURG & MIL MED, F EDWARD HEBERT SCH MED, UNIFORMED SERVS UNIV HEALTH SCI, 85- *Concurrent Pos:* Asst res, Metrop Hosp, New York, 50, chief res surg, 52-53, res investr, Col Burn Study, 52-54, mem staff, Res Unit, 54-55, vis surgeon, 55-63; asst surgeon & asst vis surgeon, Flower & Fifth Ave Hosps, 53-57, vis surgeon, 57-63; Dazian Found fel, 54-55; vis surgeon, Bird S Coler Hosp & Hebrew Home for Aged; trustee, Midwest Res Inst, 64-66, mem exec comt, Bd Trustees, 65-66; pres, Claudia Gips Found, 67-; dir med systs, Resource & Mgt Systs Corp, DC; NSF vis scientist, Auburn Univ; lectr, UNIVAC Int Exec Ctr, Italy, 69-70; Gov & dep secy gen, Int Coun Comput Commun, 74-; consult, US Arms Control & Disarmament Agency, 76; vis scholar, Ctr Strategic & Int Studies, 81-82; adj prof health care sci, Sch Med, George Washington Univ, 81-; sr fel sci & technol, Ctr Strategic & Int Studies, 83-, clin prof family & community int med, Sch Med, 84- *Mem:* Am Fedn Clin Res; Am Asn Surg Trauma; Harvey Soc; NY Acad Sci; fel Am Col Nuclear Med; fel Am Col Surgeons; fel Am Col Gastroenterol; Critical Care Soc; fel Am Col Emergency Physicians; fel Am Col Nuclear Med; fel NY Acad Med. *Res:* Control of hemodynamics and neurovascular syndromes in man; physiology of body water; burns; shock; metabolism of trauma; surgical physiology; curriculum design; man-machine systems; bioinstrumentation; operations research; computer applications; shock and hemodynamics; biological sensing; processing and telecommunications; emergency medical systems; societal impact of disasters; mitigation of mass casualties; disaster planning. *Mailing Add:* 7041 N Corrida de Venado Tucson AZ 85718

SILVERSTEIN, MARTIN L, b Philadelphia, Pa, May 22, 39. BOUNDARY THEORY, TIME CHANGE. *Educ:* Princeton Univ, PhD(math), 65. *Prof Exp:* Asst prof math, Rutgers Univ, 68-70; assoc prof, Univ Southern Calif, 70-77; PROF MATH, WASH UNIV, 77- *Mem:* Am Math Soc; Math Asn Am. *Mailing Add:* Wash Univ Box 1146 St Louis MO 63130

SILVERSTEIN, RICHARD, b Boston, Mass, Aug 9, 39; m 62; c 3. BIOCHEMISTRY. *Educ:* Brandeis Univ, BA, 60; Fla State Univ, PhD(chem), 65. *Prof Exp:* Staff scientist, Charles F Kettering Res Lab, 68-69; asst prof, 69-77, ASSOC PROF BIOCHEM, UNIV KANS MED CTR, 69- *Mem:* Am Soc Biol Chem; Am Chem Soc; NY Acad Sci; Sigma Xi. *Res:* Defense against endotoxin and tumor neurosis factor cachectin. *Mailing Add:* Dept Biochem Univ Kans Med Ctr Kansas City KS 66103

SILVERSTEIN, ROBERT MILTON, b Baltimore, Md, Mar 26, 16; m 43; c 1. NATURAL PRODUCTS CHEMISTRY. *Educ:* Univ Pa, BS, 37; NY Univ, PhD, 49. *Prof Exp:* Asst, NY Univ, 46-48; sr org chemist, Stanford Res Inst, 48-64, res fel, 64-69; prof, 69-87, EMER PROF CHEM, STATE UNIV NY COL FORESTRY, SYRACUSE UNIV, 87- *Concurrent Pos:* Co-ed, J Chem Ecol. *Honors & Awards:* Medal, Royal Swed Acad Agr, Int Soc Chem Ecol. *Mem:* Am Chem Soc; Sigma Xi; Entom Soc Am; Int Soc Chem Ecol. *Res:* Organic synthesis; mechanisms and isolation and structure elucidation of natural products; application of spectrometry to organic chemistry. *Mailing Add:* Dept Chem State Univ NY Col Forestry Syracuse NY 13210

SILVERSTEIN, SAMUEL CHARLES, b New York, NY, Feb 11, 37; m 67; c 2. CELL BIOLOGY, INFECTIOUS DISEASES. *Educ:* Dartmouth Col, BA, 58; Albert Einstein Col Med, MD, 63. *Prof Exp:* Intern, med, Univ Col Med Ctr, 63-64, postdoctoral, cell biol, Rockefeller Univ, 64-67, resident, Mass Gen Hosp, 67-68, assoc prof, cellular physiol & immunol, Rockefeller Univ, 72-83, PROF & CHMN, PHYSIOL & CELLULAR BIOPHYSICS, COLUMBIA UNIV COL PHYSICIANS & SURGEONS, 83- *Concurrent Pos:* Helen Hay Whitney fel, 64-67; established investr, Am Heart Asn, 72-77. *Honors & Awards:* The John Oliver La Gorce Medal, Nat Geog Soc, 67; Reticulendothelial Soc Award, 87. *Mem:* Am Soc Cell Biol; Am Soc Biol Chem & Molecular Biol; Am Physiol Soc; Am Soc Clin Investigation; Infect Dis Soc Am; Am Asn Immunol. *Res:* The roles of mononuclear leucocytes, and vascular endothelial cells in immunity inflammation, and host defense; the tools of cell and molecular biology, electrophysiology, and cellular immunology. *Mailing Add:* Dept Physiol Rm 11-511 Columbia Univ Col Physicians & Surgeons 630 W 168th St New York NY 10032

SILVERSTEIN, SAUL JAY, b Brooklyn, NY, Aug 23, 46; m 76; c 3. MOLECULAR BIOLOGY, GENE EXPRESSION. *Educ:* Cornell Univ, BS, 68; Univ Fla, PhD(microbiol), 71. *Prof Exp:* From asst prof to assoc prof, 74-87, PROF MICROBIOL, COLUMBIA Univ, 87- *Concurrent Pos:* Fel virol, Univ Chicago, 71; fel, Damon Runyon Walter-WinCholl Cancer Fund, 78; ed, Somatic Cell Genetics, 80-; mem, Exp Virol Study Sect, 82-86; vis prof, Japanese Nat Cancer Inst, 87. *Honors & Awards:* Career Develop Award, NIH. *Mem:* Harvey Soc; Am Soc Microbiol; Am Soc Virol. *Res:* Gene regulation, particularly applied to eukaryotic virus-host cell interactions. *Mailing Add:* Dept Microbiol Columbia Univ 360 W 168b St New York NY 10032

SILVERSTONE, HARRIS JULIAN, b New York, NY, Sept 18, 39; m 60; c 4. QUANTUM CHEMISTRY. *Educ:* Harvard Univ, AB, 60; Calif Inst Technol, PhD(chem), 64. *Prof Exp:* NSF fel, Yale Univ, 64; from asst prof to assoc prof, 65-71, PROF CHEM, JOHNS HOPKINS UNIV, 71- *Concurrent Pos:* Sloan Found fel, 69. *Mem:* Am Phys Soc; Am Chem Soc. *Res:* Application of quantum mechanics to chemistry. *Mailing Add:* Dept Chem Johns Hopkins Univ Baltimore MD 21218

SILVERT, WILLIAM LAWRENCE, b New York, NY, Dec 11, 37; Can citizen; m 68; c 2. MARINE ECOLOGY, BIOECONOMICS. *Educ:* Brown Univ, ScB, 58, PhD(physics), 65. *Prof Exp:* Res assoc physics, Mich State Univ, 64-66; asst prof, Case Western Reserve Univ, 66-69; asst prof physics & astron, Univ Kans, 69-72; assoc prof physics, 72-75, dir, E Coast Fisheries Mgt Proj, Inst Environ Studies, Dalhousie Univ, 75-78; res scientist, Marine Ecol Lab, 78-87, RES SCIENTIST, HABITAT ECOL DIV, BEDFORD INST OCEANOG, 88- *Concurrent Pos:* Lectr, Univ Mich, 65-66; consult, Bendix Corp, 66; Nat Acad Sci exchange scholar, Inst Physics Probs, Moscow, 66-67. *Mem:* Can Soc Theor Biol. *Res:* Theoretical marine ecology; systems analysis; resource management; bioeconomics; trophodynamics; ecosystem modeling. *Mailing Add:* 6113 Oakland Rd Halifax NS B3H 1P1 Can

SILVERTHORN, DEE UNGLAUB, b New Orleans, La, Dec 3, 48; m 72. INVERTEBRATE PHYSIOLOGY. *Educ:* Tulane Univ, BS, 70; Univ SC, PhD(marine sci), 73. *Prof Exp:* Res assoc biochem, Med Univ SC, 73-74, from instr to asst prof physiol, 74-77; res scientist physiol & biophys, Univ Tex Med Br, Galveston, 78-80; LECTR, DEPT ZOOL, UNIV TEX, AUSTIN, 86- *Concurrent Pos:* Vis asst prof biol, Univ Houston, 77-78. *Mem:* Am Soc Zoologists; Am Physiol Soc; Sigma Xi. *Res:* Endocrinology and physiology of thermal acclimation and osmoregulation in crustaceans; biochemistry and physiology of osmoregulation in crustaceans. *Mailing Add:* Dept Zool, Univ Tex Austin TX 78712-1064

SILVERTON, JAMES VINCENT, b Seaton Delaval, Eng, May 10, 34; m 64; c 2. PHYSICAL CHEMISTRY, CRYSTALLOGRAPHY. *Educ:* Glasgow Univ, BSc, 55, PhD(chem), 63. *Prof Exp:* Res assoc chem, Cornell Univ, 58-61; sr res fel solid state physics, UK Atomic Energy Authority, Eng, 61-62; asst lectr chem, Glasgow Univ, 62-63; asst prof, Georgetown Univ, 63-70; SCIENTIST, NIH, 70- *Concurrent Pos:* Petrol Res Fund starter grant, 63-65; prog dir, Nat Inst Dent Res training grant x-ray crystallog, 65-70. *Mem:* Am Chem Soc; Am Crystallog Asn; Royal Soc Chem; Sigma Xi. *Res:* Structures of complex inorganic and organic compounds by x-ray crystallographic techniques. *Mailing Add:* 8609 Hidden Hill Ln Potomac MD 20854

SILVESTER, JOHN ANDREW, b Kent, Eng, Apr 26, 50; m 80; c 2. PERFORMANCE MODELING, COMPUTER COMMUNICATION. *Educ:* Cambridge Univ, BA, 71, MA, 75; WVa Univ, MS, 73; Univ Calif, Los Angeles, PhD(comput sci), 80. *Prof Exp:* Mem staff comput sci, Univ Calif, Los Angeles, 73-78; ASSOC PROF COMPUT SCI, DEPT ELEC ENG SYST, UNIV SOUTHERN CALIF, 79- *Concurrent Pos:* Consult, 79-; chmn, Comsoc tech comt, comput commun, Inst Elec & Electronics Engrs. *Mem:* Inst Elec & Electronics Engrs; Asn Comput Mach. *Res:* Performance modelling of computer systems and networks, especially computer communications and multiple access techniques. *Mailing Add:* Dept Elec Eng Syst Univ Southern Calif Los Angeles CA 90089

SILVESTER, PETER PEET, b Jan 25, 35; Can citizen; m 58. ELECTRICAL ENGINEERING. *Educ:* Carnegie Inst Technol, BS, 56; Univ Toronto, MASc, 58; McGill Univ, PhD(elec eng), 64. *Prof Exp:* from lectr to assoc prof, 58-72, PROF ELEC ENG, McGILL UNIV, 72- *Concurrent Pos:* SERC sr fel, Imp Col, Univ London, 67-68 & 80-81; sr vis fel, Ctr Res & Develop, Gen Elec Co, 78-79; vis fel commoner, Trinity Col, Cambridge, 88-89. *Mem:* Soc Indust & Appl Math; fel Inst Elec & Electronics Engrs; NY Acad Sci. *Res:* Numerical analysis of electromagnetic field problems. *Mailing Add:* Dept Elec Eng McGill Univ 3480 University St Montreal PQ H3A 2A7 Can

SILVESTON, PETER LEWIS, b New York, NY, Mar 10, 31; c 3. REACTOR ENGINEERING, POLLUTION CONTROL. *Educ:* Mass Inst Technol, SB, 51, SM, 53; Munich Tech Univ, Dr Ing, 56. *Prof Exp:* Chem engr, Esso Res & Eng Co, 57-59; res engr, Res Div, Am Standard Corp, 59-61; asst prof chem eng, Univ BC, 61-63; assoc prof, 63-69, PROF CHEM ENG, UNIV WATERLOO, 69- *Mem:* Am Chem Soc; Chem Inst Can; Am Inst Chem Engrs. *Res:* Reactor design; kinetics; catalysis; waste treatment. *Mailing Add:* Dept Chem Eng Univ Waterloo Waterloo ON N2L 3G1 Can

SILVESTRI, ANTHONY JOHN, b Glassboro, NJ, Mar 14, 36; m 60; c 2. PETROLEUM CHEMISTRY. *Educ:* Villanova Univ, BS, 58; Pa State Univ, PhD(chem), 61. *Prof Exp:* Res chemist, Mobile Res & Develop Corp, 61-63, sr res chemist, 63-68, res assoc, 68-73, mgr, Anal & Spec Technol Group, 73-75, Catalysis Res Sect, Mobile Res & Develop Corp, 75-77, mgr Process Res & Develop Sect, 77-79, planning & eval, 79-80, Process Res & Tech Serv Div, 80-84, Prod Res & Tech Serv Div, 84-89, VPRES ENVIRON HEALTH & SAFETY, MOBILE RES & DEVELOP CORP, 89- *Mem:* Am Inst Chem Engrs; Am Chem Soc; Soc Automotive Engrs. *Res:* Heterogeneous catalysis and chemical kinetics. *Mailing Add:* Mobil Res & Develop Corp PO Box 1031 Princeton NJ 08543

SILVESTRI, GEORGE J, JR, b Jessup, Pa, Aug 3, 27; m 61; c 2. THERMODYNAMICS, POWER GENERATION CYCLE PERFORMANCE & ANALYSIS. *Educ:* Drexel Univ, BS, 53, MS, 56. *Prof Exp:* Develop engr, Westinghouse Elec Corp, 53-69, fel engr, 69-72 & 74, appl engr, 72-73; proj mgr, Elec Power Res Inst, 73-74; ADV ENGR, WESTINGHOUSE ELEC CORP, 74- *Concurrent Pos:* Chmn educ & res comt, Power Div, Am Soc Mech Engrs, 82-85 & 90-91, bd res & technol develop, 85-91, lectr, course on steam turbines, Power Div, 89-, perf test code course on testing, 91-, res comt properties steam. *Mem:* Am Soc Mech Engrs; Am Nuclear Soc. *Res:* Advanced power generation cycles; operation procedures to enhance steam turbine performance; low pressure turbine laboratory testing; improved steam property algorithms; steam turbine performance computer program. *Mailing Add:* 1840 Cheryl Dr Winter Park FL 32792

SILVETTE, HERBERT, b McKee's Rocks, Pa, Dec 23, 07; m 31. TOBACCO. *Educ:* Univ Va, BS & MS, 32, PhD(physiol), 34. *Prof Exp:* Instr biochem, Univ Va Med Sch, 28-30, instr physiol, 30-38, from asst prof to assoc prof pharmacol, 38-47; RETIRED. *Concurrent Pos:* Vis prof pharmacol, Meharry Med Col, 49-52, Univ Wash Col Med, 52-53 & Med Col Va, 53-75. *Mem:* Am Physiol Soc; Am Soc Pharmacol & Exp Therapeut; AMA. *Res:* Cultural pharmacology. *Mailing Add:* "Low Gear" Stanardsville VA 22973

SILVEY, J K GWYNN, limnology; deceased, see previous edition for last biography

SILVIDI, ANTHONY ALFRED, b Steubenville, Ohio, Jan 17, 20; m 47; c 4. BIOPHYSICS. *Educ:* Ohio Univ, BS, 43, MS, 45; Ohio State Univ, PhD(physics), 49. *Prof Exp:* Assoc prof physics & head dept, Col Steubenville, 49-51; res physicist, Cornell Aeronaut Lab, Inc, 51-52; from asst prof to assoc prof, 52-63, res assoc, 52-59, coord grad progs, 68-73, PROF PHYSICS, KENT STATE UNIV, 63- *Concurrent Pos:* Res physicist, Goodyear Aircraft Corp, 52-58; consult, Biochem Dept, Children's Hosp of Akron, Ohio, 70-; prin investr, DOE, NSF, AEC grants, 60-85. *Mem:* Am Phys Soc; Biophys Soc; Sigma Xi. *Res:* Nuclear magnetic resonance; biophysics; application of physical techniques for solutions of biological problems. *Mailing Add:* 311 Valley View Dr Kent OH 44240

SILVIUS, JOHN EDWARD, b Dover, Ohio, May 9, 47; m 69; c 2. PLANT PHYSIOLOGY, BOTANY. *Educ:* Malone Col, BA, 69; WVa Univ, PhD(plant physiol), 74. *Prof Exp:* Teacher biol, Dover Pub Schs, 69-71; vis lectr bot, Univ Ill, Champaign-Urbana, 74-75, res assoc agron, 75-76; plant physiologist, Sci & Educ Admin-Agr Res, USDA, 76-80; MEM FAC, SCI DEPT, CEDARVILLE COL, OHIO, 80- *Mem:* Am Soc Plant Physiologist; Am Soc Agron; Sigma Xi. *Res:* Physiological and biochemical mechanisms which regulate the photosynthetic production, partitioning and translocation of carbon assimilates in plants. *Mailing Add:* Cedarville Col Box 601 Biol Dept Cedarville OH 45314

SILZARS, ARIS, b Riga, Latvia, June 22, 40; m 65; c 2. ELECTRICAL ENGINEERING, PHYSICS. *Educ:* Reed Col, 63; Univ Utah, MA, 65, PhD(elec eng), 69. *Prof Exp:* Teaching asst, Univ Utah, 63-65; res asst, NASA, 65-68; mem tech staff, Watkins-Johnson Co, 69-73, sect head, EBS Devices, 73-74; mgr display devices, Tektronix Inc, 74-77, dir, Component Develop Group, 77-79, dir, Solid State Group, 79-90; PRES, LANVIDE ELECTRONIC COMPONENT, 90- *Concurrent Pos:* Consult, Dept Chem, Univ Utah, 68-70; adj assoc prof, Ore State Univ, 75- *Mem:* Sr mem Inst Elec & Electronics Engrs; Soc Info Display. *Res:* Semiconductor devices; hybrids; display devices. *Mailing Add:* 1515 Eastman Station Rd Landenberg PA 19550

SIMAAN, MARWAN, b July 23, 46; US citizen. ELECTRICAL ENGINEERING, GEOPHYSICS. *Educ:* Am Univ Beirut, BEE, 68; Univ Pittsburgh, MS, 70; Univ Ill, PhD(elec eng), 72. *Prof Exp:* Vis asst prof elec eng, Univ Ill, 72-74; res engr geophys, Shell Develop Co, 74-75; from assoc prof to prof, 76-89, BELL PA/BELL ATLANTIC PROF ELEC ENG, UNIV PITTSBURGH, 89- *Concurrent Pos:* Tech consult, Gulf Res & Develop Co, 79-85, Alcoa, 85-; assoc ed, J Optimization Theory & Applns, 86-; prin investr, govt & indust grants, 76-; co-ed, Multidimensional Systs & Signal Processing J, 89- *Mem:* Fel Inst Elec & Electronics Engrs; Am Asn Artificial Intelligence; Soc Explor Geophysicists; NY Acad Sci; Sigma Xi. *Res:* Optimization and optimal control; digital signal processing and geophysical applications; artificial intelligence and knowledge based engineering; manufacturing systems. *Mailing Add:* Dept of Elec Eng Univ of Pittsburgh Pittsburgh PA 15261

SIMANEK, EUGEN, b Prague, Czech, July 15, 33; m 55, 69; c 1. PHYSICS. *Educ:* Prague Tech Univ, MEE, 56; Czech Acad Sci, Cand Sci, 63. *Prof Exp:* Res physicist, Inst Physics, Czech Acad Sci, 56-68; theoret physicist, IBM Res Lab, Switz, 68-69; PROF PHYSICS, UNIV CALIF, RIVERSIDE, 69- *Concurrent Pos:* Vis assoc prof, Univ Calif, Los Angeles, 65-67; Energy Res & Develop Admin contract theory of superconductivity, 75; vis prof, Univ Neuchatel, Switz, 81; guest prof physics, Troisieme Cycle of French Switz, 81. *Res:* Theoretical solid state physics; quantum theory of metals; many body problem; critical phenomena; low temperature physics; superconductivity. *Mailing Add:* Dept Physics Univ Calif Riverside CA 92521

SIMANTEL, GERALD M, b Huron, SDak, Oct 12, 34; m 55; c 3. PLANT BREEDING. *Educ:* Ore State Univ, BS, 59; SDak State Univ, PhD(agron), 63. *Prof Exp:* plant breeder, res dept, 63-87, mgr, seed prod & develop, 87-90, SR PLANT BREEDER NAM, AMALGAMATED SUGAR CO, 90- *Mem:* Am Soc Agron; Crop Sci Soc Am; Am Soc Sugar Beet Technologists. *Res:* Development of more productive sugar beet varieties. *Mailing Add:* Hilleshog Mono-Hy Inc PO Box 1786 Nyssa OR 97913

SIMARD, ALBERT JOSEPH, b Hartford, Conn, July 11, 42. FOREST FIRE SCIENCE, SYSTEMS ANALYSIS. *Educ:* Univ Conn, BS, 63; Univ Calif, Berkeley, MSc, 68; Univ Wash, PhD(fire sci), 78. *Prof Exp:* Res scientist, Forest Fire Res Inst, Can Forest Serv, 67-79; PROJ LEADER, NCENT FOREST EXP STA, US FOREST SERV, 79- *Concurrent Pos:* Adj prof, Mich State Univ, 80- *Mem:* Soc Am Foresters. *Res:* Atmospheric relations to wildland fire; fire danger rating, fire management systems, fire weather, fire economics, fire ecology; information systems. *Mailing Add:* 482 E Jefferson Diamondale MI 48821

SIMARD, GERALD LIONEL, b Lewiston, Maine, May 11, 12; m 46; c 5. PHYSICAL CHEMISTRY. *Educ:* Bates Col, BS, 33; Mass Inst Technol, PhD(phys chem), 37. *Prof Exp:* Res chemist, Atlantic Refining Co, 37-38; indust fel, Battelle Mem Inst, 39, res engr, 39-43; group leader, Am Cyanamid Co, 43-53; sect leader, Schlumberger Well Surv Corp, Conn, 53-60, res mgr, 60-67; assoc prof, 67-77, EMER PROF CHEM ENG, UNIV MAINE, ORONO, 77- *Mem:* Am Chem Soc; Am Inst Chem Engrs; Sigma Xi. *Res:* Thermodynamics; surface chemistry; kinetics, catalysis; electrochemistry; instrumental analysis; pulp and paper technology; environmental chemistry. *Mailing Add:* PO Box 86 Winterport ME 04496

SIMARD, RENE, b Montreal, Que, Oct 4, 35; m 69; c 3. CELL BIOLOGY, MOLECULAR BIOLOGY. *Educ:* Univ Montreal, BA, 56, MD, 62; Univ Paris, DSc, 68, FRCP, 76. *Prof Exp:* Resident path, Mt Sinai Sch Med, 65; asst prof path, Univ Montreal, 68-69; from asst prof to assoc prof & dir cell biol, Sch Med, Univ Sherbrooke, 69-75; PROF PATH & DIR MONTREAL

CANCER INST, UNIV MONTREAL, 75-, VPRIN ACAD, 85- *Concurrent Pos:* Med Res Coun Can fel, Inst Cancer Res, Villejuif, France, 65-68 & scholar, Univ Sherbrooke, 68-; mem grants comt anat & path, Med Res Coun Can, 70-77; mem res adv group & grants comt, Nat Cancer Inst Can, 72-77; chmn, Quebec Health Res Coun, 75-78; pres, Med Res Coun, 78-81. *Mem:* AAAS; Am Soc Cell Biol; Inst Soc Cell Biol; Fr Soc Electron Micros; Can Soc Oncol (pres, 82-83). *Res:* Regulation of nucleic acids synthesis in eukaryotic cells; herpes viruses and cancer of the cervix. *Mailing Add:* Montreal Cancer Inst Univ Montreal 1560 E Sherbrook St Montreal PQ H2L 4M1 Can

SIMARD, RONALD E, b Aug 8, 39; Can citizen; m 66; c 2. MICROBIOLOGY. *Educ:* Univ Montreal, BSc, 62; McGill Univ, MSc, 65, PhD(microbiol), 70. *Prof Exp:* Microbiologist & indust scholar yeast res, J E Seagram & Sons, Inc, 62-65; microbiologist, Fed Dept Fisheries, 65-67; prof microbiol, Laval Univ, 70-75; prof microbiol, Univ Sherbrooke, 75-77; MEM FAC, FOOD SCI DEPT, UNIV LAVAL, 77- *Mem:* Can Soc Microbiol; Am Soc Microbiol; Can Inst Food Sci & Technol; Am Soc Enol. *Res:* Applied microbiology in fermentation; pollution; biological treatment of domestic wastes. *Mailing Add:* Food Dept Agr Laval Univ Ste-Foy PQ G1S 2K4 Can

SIMARD, THERESE GABRIELLE, b St-Lambert, Que, Mar 3, 28. ANATOMY. *Educ:* Univ Montreal, BA, 56, BSc, 62; Univ Mich, MSc, 64; Queen's Univ, Ont, PhD(anat), 66. *Prof Exp:* Prof anat, Univ Mich, 64 & Queen's Univ, Ont, 65-66; asst prof, 66-71, ASSOC PROF ANAT, UNIV MONTREAL, 71- *Honors & Awards:* Gold Medal, Int Soc Phys Med & Rehab, 72. *Mem:* Biofeedback Res Soc; Can Asn Anat; Am Asn Anat; Can Asn Phys Med & Rehab; Int Soc Electrophysiol Kinesiology (treas, 73-). *Res:* Electromyographic studies on the kinesiology of muscles and developmental method of studies; training of the neuromuscular action potential. *Mailing Add:* Dept Anat Univ Montreal Fac Med Montreal PQ H3T 1J4 Can

SIMBERLOFF, DANIEL S, b Easton, Pa, Apr 7, 42. ECOLOGY, MATHEMATICAL BIOLOGY. *Educ:* Harvard Univ, AB, 64, PhD(biol), 69. *Prof Exp:* From asst prof to assoc prof biol, 68-78, PROF BIOL, FLA STATE UNIV, 78- *Concurrent Pos:* NSF grants ecol, 69-82. *Honors & Awards:* Mercer Award, Ecol Soc Am, 71. *Mem:* Ecol Soc Am; Soc Study Evolution; Brit Ecol Soc; Japanese Soc Pop Ecol; Soc Syst Zool. *Res:* Biogeography; evolution. *Mailing Add:* Dept of Biol Sci Fla State Univ Tallahassee FL 32306

SIMCHOWITZ, LOUIS, RHEUMATOLOGY. *Educ:* New York Univ, MD, 70. *Prof Exp:* ASSOC PROF MED, WASHINGTON UNIV, 85- *Mailing Add:* Sch Med Washington Univ 660 S Euclid Ave St Louis MO 63110

SIMCO, BILL AL, b Mountainburg, Ark, July 14, 38; m 60. ICHTHYOLOGY. *Educ:* Col of Ozarks, BS, 60; Univ Kans, MA, 62, PhD(zool), 66. *Prof Exp:* Kettering intern biol, Kenyon Col, 65-66; from asst prof to assoc prof, 66-77, PROF BIOL, MEMPHIS STATE UNIV, 77- *Mem:* AAAS; Am Fisheries Soc; Sigma Xi; World Aquaculture Soc. *Res:* Studies on reproductive and stress physiology and factors limiting production of channel catfish in ponds and raceways; culture of catfish in recirculating raceways. *Mailing Add:* Dept of Biol Memphis State Univ Memphis TN 38152

SIME, DAVID GILBERT, b Glasgow, Scotland, July 4, 48. SOLAR CORONA, INTERPLANETARY MEDIUM. *Educ:* Univ Edinburgh, BSc Hons, 70; Univ Calif, San Diego, PhD(appl physics), 76. *Prof Exp:* Res asst, Univ Calif, San Diego, 70-76; res assoc, Swiss Fed Inst Technol, Zurich, 76-77; vis scientist, 77-78, staff scientist I, 78-80, staff scientist II, 80-85, SCIENTIST III, HIGH ALTITUDE OBSERV, BOULDER, 85- *Mem:* Am Geophys Union; Am Astron Soc. *Res:* Three dimensional structure of the solar corona and the interplanetary medium; instrument development and data processing, especially as applied to the sun. *Mailing Add:* High Altitude Observ NCAR PO Box 3000 Boulder CO 80307

SIME, RODNEY J, b Madison, Wis, July 3, 31; m 55; c 2. PHYSICAL CHEMISTRY. *Educ:* Univ Wis, BS, 55; Univ Wash, PhD(chem), 59. *Prof Exp:* From asst prof to assoc prof, 59-67, PROF CHEM, CALIF STATE UNIV, SACRAMENTO, 67- *Honors & Awards:* Alexander von Humboldt fel, Univ Tübingen, 64-66; guest prof, Swiss Fed Inst Technol, 74-75. *Res:* High temperature chemistry and vaporization processes of transition element halides; x-ray diffraction studies of crystal and molecular structure. *Mailing Add:* 609 Shangri Lane Sacramento CA 95825

SIME, RUTH LEWIN, b New York, NY, July 2, 39; m 68; c 2. PHYSICAL CHEMISTRY. *Educ:* Barnard Col, BA, 60; Radcliffe Col, MA, 61; Harvard Univ, PhD(chem), 65. *Prof Exp:* Asst prof chem, Calif State Col, Long Beach, 64-65, Sacramento State Col, 65-67 & Hunter Col, 67-68; INSTR CHEM, SACRAMENTO CITY COL, 68- *Mem:* AAAS; Am Crystallog Asn. *Res:* Biographical research of physicist Lise Meitner. *Mailing Add:* Sacramento City Col 3835 Freeport Blvd Sacramento CA 95822

SIMENSTAD, CHARLES ARTHUR, b Yakima, Wash, Feb 22, 47. MARINE ECOLOGY, FISHERIES BIOLOGY. *Educ:* Univ Wash, BS, 69, MS, 71. *Prof Exp:* FISHERIES BIOLOGIST MARINE BIOL, FISHERIES RES INST, UNIV WASH, 71- *Mem:* AAAS; Sigma Xi; Am Inst Fisheries Res Biologists; Ecol Soc Am. *Res:* Community and trophic ecology of nearshore marine communities; ecology of epibenthic zooplankton; food web structure and feeding ecology of marine fish assemblages; early marine life history of Pacific salmon. *Mailing Add:* Fisheries Res Inst W10 Univ of Wash Seattle WA 98195

SIMEON, GEORGE JOHN, b New York, NY, July 21, 34; m 65; c 3. JAPANESE SOCIETY & CULTURE, INDIGENOUS NAVIGATIONAL SYSTEMS. *Educ:* Univ Hawaii, BA, 62, MPH, 77; Univ Southern Calif, PhD 68. *Prof Exp:* Fel & researcher med anthrop, Org Am States, 68-69; researcher anthrop, US Fulbright Comn, 70-71 & Macquarie Univ, 72-74; researcher med anthrop, Nat Geog Soc & Wenner-Gren Found, 74; med data analyst, Dept Health Educ & Welfare, Kaiser Found Hosps, 75; researcher med anthrop, Cross-Cult res fel, Univ Hawaii & Indonesia Schs Pub Health, 76; researcher, Nat Museum Man, 79-80; fel, Nat Inst Alcohol Abuse & Alcoholism, Brown Univ, 80-81; PROF, HIROSAKI GAKUIN UNIV, JAPAN, 81- *Concurrent Pos:* Vis fel, Ohio State Univ, 79; consult, Public Health Social Sci. *Mem:* Am Pub Health Asn; Am Anthrop Asn. *Res:* Ethnomedicine and medical anthropology in reference to the acquisition and analysis of field data with a view towards developing an ethnomedical theory; traditional terrestrial and sea navigational systems; Japanese society & culture. *Mailing Add:* c/o Takeuchi PO Box 68 Waimea Kauai HI 96796-0068

SIMEONE, FIORINDO ANTHONY, b St Ambrose, Italy, Jan 20, 08; nat US; m 41; c 5. SURGERY. *Educ:* Brown Univ, AB, 29, ScM, 30; Harvard Univ, MD, 34; Am Bd Surg & Bd Thoracic Surg, dipl. *Hon Degrees:* ScD, Brown Univ, 54. *Prof Exp:* House officer surg, Mass Gen Hosp, 34-36, res surg, 38-39; asst surg, Harvard Med Sch, 38-40, asst genito-urinary surg, 40, instr, 40-41, assoc, 41-42, asst prof surg, 46-50; prof, Sch Med, Western Reserve Univ, 50-67; prof, 67-76, EMER PROF MED SCI, BROWN UNIV, 76- *Concurrent Pos:* Nat Res Coun fel & teaching fel physiol, Harvard Med Sch, 36-38; fel, Peter Bent Brigham Hosp, 39-41, surgeon-in-chief pro tempore, 59; asst, Mass Gen Hosp, 41-42, asst vis surgeon, 46-50, consult, 50-; consult, Mass Eye & Ear Infirmary & surg gen, US Army, 46-50; dir surg, Cleveland Metrop Gen Hosp, 50-67; surgeon-in-chief, Miriam Hosp, RI, 67-76, emer surgeon-in-chief, 76-; hon prof, Med Schs & hon dir prof units, St Bartholomew's & St Thomas Hosps, London, Eng, 56; lectr, Harvard Med Sch, 59; prof, Am Univ, Beirut & chief surg, Hosp, 60; mem subcomt cardiovasc syst & mem subcomt on shock, Nat Res Coun; mem subcomt metab in trauma, Adv Comt Metab, US Armed Forces & mem surg study sect, USPHS. *Honors & Awards:* Perpetual Student, St Bartholomew's Hosp, 56- *Mem:* AAAS; Am Cancer Soc; Soc Univ Surg; Soc Vascular Surg; AMA; Am Surg Asn; Am Col Surgeons. *Res:* Physiology of the autonomic nervous system and its effectors; physiology of the cardiovascular system; physiology of trauma. *Mailing Add:* Miriam Hosp 164 Summit Ave Providence RI 02911

SIMEONE, JOHN BABTISTA, b Providence, RI, Nov 20, 19; m 45. FOREST ENTOMOLOGY. *Educ:* Univ RI, BS, 42; Yale Univ, MF, 48; Cornell Univ, PhD(entom), 60. *Prof Exp:* Asst entom, 48-56, from asst prof to assoc prof, 56-64, chmn dept entom, 62-77, chmn dept environ & forest biol, 77-81, PROF FOREST ENTOM, STATE UNIV NY COL ENVIRON SCI & FORESTRY, SYRACUSE, 64- *Concurrent Pos:* Co-ed, J Chem Ecol, 75- *Mem:* AAAS; Entom Soc Am; Soc Am Foresters; Ecol Soc Am; NY Acad Sci; Sigma Xi. *Res:* Biology of insects causing deterioration of wood; chemical ecology; chemical ecology of forest insects. *Mailing Add:* State Univ NY Col Environ Sci & Forestry Syracuse NY 13210

SIMERAL, WILLIAM GOODRICH, b Portland, Ore, May 22, 26; m 49; c 4. MANAGEMENT. *Educ:* Franklin & Marshall Col, BS, 48; Univ Mich, MS, 50, PhD(physics), 53. *Prof Exp:* Res physicist, E I Du Pont de Nemours & Co, Inc, 53-56, res supvr, 56-57, sr res supvr, 57-64, res mgr, 64-66, asst dir res & develop, Plastics Dept, 66-68, dir Commercial Resins Div, 68-71, asst dir Cent Res Dept, 71-74, asst gen mgr, Plastics Dept, 74, vpres & gen mgr, 74-77, sr vpres, 77-81, dir, 77-87, exec vpres, 81-87; RETIRED. *Mem:* Am Phys Soc. *Res:* Physics of high polymers. *Mailing Add:* 800 Slashpine Ct Naples FL 33963

SIMERL, L(INTON) E(ARL), b Chillicothe, Ohio, Dec 2, 11; m 41; c 4. CHEMICAL ENGINEERING. *Educ:* Ohio State Univ, BChE, 35; Lawrence Col, MS, 37, PhD(chem eng), 39. *Prof Exp:* Develop engr, Mead Corp, Ohio, 39-41; supvr mat eng lab, Mfg Eng Dept, Marathon Corp, Wis, 46-53; chief develop sect, Res & Develop Dept, Film Div, Olin Industs, Inc, 53-56, dir res & develop, Film Div, Olin Mathieson Chem Corp, 56-62, dir packaging opers, Int Div, 62-65; vpres res & develop, Oxford Paper Co, 65-70; gen mgr, C H Dexter Div, Dexter Corp, Scotland, 71-73; CONSULT, 73- *Concurrent Pos:* Vol consult, Int Exec Serv Corps, 73- *Mem:* Am Chem Soc; Tech Asn Pulp & Paper Indust; Inst Food Technol. *Res:* Cellulose and lignin chemistry; protective packaging; design and construction of chemical plants; pulp; paper; cellophane; high polymers; chemical warfare agents. *Mailing Add:* 16 Town Line Hwy-N Bethlehem CT 06751

SIMHA, ROBERT, b Vienna, Austria, Aug 4, 12; nat US; m 41. MACROMOLECULAR SYSTEMS. *Educ:* Univ Vienna, PhD(physics), 35. *Hon Degrees:* Dr, Tech Univ Dresden, 87. *Prof Exp:* Res assoc, Univ Vienna, 35-38; vis fel, Columbia Univ, 39-40, res assoc, 40-41; lectr, Polytech Inst Brooklyn, 41-42; asst prof, Howard Univ, 42-45; lectr, Nat Bur Standards Grad Sch, 44-45, consult & coordr polymer res, 45-51, lectr, 47-48; prof chem eng, NY Univ, 51-59; prof chem, Univ Southern Calif, 59-67; PROF MACROMOLECULAR SCI, CASE WESTERN RESERVE UNIV, 68- *Concurrent Pos:* Lectr, Grad Div, Brooklyn Col, 40-42; vis prof, Univ Southern Calif, 58-59; chmn, First Winter Gordon Res Conf, 63; John F Kennedy Mem Found sr fel, Weizmann Inst, 66-67; sr vis res fel, Univ Manchester, 67-68; consult, UNIDO, India, 77-78, indust consult, 40-; guest prof, Tech Univ, Dresden, 85 & 89; vis res fel, Univ Stirling, 87. *Honors & Awards:* Cert Recognition, NASA, 88 & 89; Morrison Prize, NY Acad Sci, 48; US Dept Com Award, 48; Nat Bur Standards Award, 49; Bingham Medal, Soc Rheol, 73; High Polymer Physics Prize, Am Phys Soc, 81; Cert Recognition, NASA, 88-89. *Mem:* Fel AAAS; fel Am Phys Soc; Am Chem Soc; fel Am Inst Chemists; fel NY Acad Sci; Soc Rheology; Sigma Xi; fel, Wash Acad Sci. *Res:* Hydrodynamics of colloidal solutions; viscosity of liquids and macromolecular solutions; physical and thermodynamic properties of polymers; polymerization and depolymerization processes, including biological macromolecules. *Mailing Add:* Dept Macromolecular Sci Case Western Reserve Univ Cleveland OH 44106

SIMHAN, RAJ, b Visakhapatnam, India, Sept 18, 33; US citizen; m 72; c 2. GLASS & CERAMIC FIBERS, ELECTRIC MELTING OF GLASS. *Educ:* Andhra Univ, India, BS, 54; Banaras Hindu Univ, India, BS, 57, MS, 61; Sheffield Univ, UK, PhD(glass), 74. *Prof Exp:* Design engr furnace design,

Karrena Fuerungshan, Ger, 62-67; glass technologist, Consumers Glass, Can, 67-69; res assoc glass, Manville Corp, 73-87; MGR RES, GAF BLDG MAT CORP, 87- *Mem:* Am Ceramic Soc; Soc Glass Technol UK. *Res:* Surface studies of glass fiber surfaces; the sol-gel process of making ceramic fibers and coatings; effect of redox equilibrium on the fiberizing process; the electric melting of glass. *Mailing Add:* Gaf Bldg Mat Corp Fiber Glass Rd Nashville TN 37210

SIMIC, MICHAEL G, b N Becej, Yugoslavia, Jan 29, 32; m 68; c 1. ANTIOXIDANTS, FREE RADICAL PROCESSES. *Educ:* Univ Belgrade, dipl, 56; Durham Univ, Eng, PhD (radiation chem), 64. *Hon Degrees:* DSc, Univ Newcastle Tyne, 84. *Prof Exp:* Res Assoc radiation biol, zool dept, Univ Tex, 70-75; res staff food chem, US Army Res & Develop, Lab, Natrick, Mass, 75-80; proj leader, 80-86, GROUP LEADER BIOCHEM EFFECTS RADIATION, NAT BUR STANDARDS, 86- *Concurrent Pos:* Lectr mats sci, Univ Newcastle, Eng, 68-86; vis scientist pulse radiolysis, US Army Res & Develop Lab, Natrick, 70-86; res prof, Am Univ, 84-88. *Mem:* Am Chem Soc; Am Soc Photobiology; Biophys Soc; Radiation Res Soc; Soc Risk Anal; Oxygen Soc. *Res:* Mechanisms of electron transfer and free radical processes. *Mailing Add:* Nat Bur Standards Bldg 245 C205 Gaithersburg MD 20899

SIMILON, PHILIPPE LOUIS, b Liege, Belg, Oct 1, 53. TURBULENT TRANSPORT, WAVE PROPOGATION. *Educ:* Univ Liege, Belg, MA, 76; Princeton Univ, PhD(astrophys), 81. *Prof Exp:* Sr res assoc, Cornell Univ, 85-88; ASSOC PROF, YALE UNIV, 88- *Mem:* Fel Am Phys Soc. *Res:* Theoretical and numerical study of waves and turbulence in plasma, with applications to magnetic confinement fusion devices and to the solar atmosphere. *Mailing Add:* 303 Becton Ctr Yale Univ New Haven CT 06520

SIMINOFF, PAUL, b Brooklyn, NY, May 8, 23; m 51; c 3. VIROLOGY, IMMUNOBIOLOGY. *Educ:* Mich State Col, BS, 48; Univ Ill, MS, 49, PhD, 51. *Prof Exp:* Microbiologist, S B Penick & Co, NJ, 51-54 & Upjohn Co, Mich, 54-58; MICROBIOLOGIST, BRISTOL LABS, INC, 58- *Mem:* AAAS; Am Soc Microbiol; Sigma Xi. *Res:* Fermentation of antibiotics and vitamins; application of tissue culture to virus and cancer research; immunology and cancer; interferon induction; drug research in allergic and immune complex diseases. *Mailing Add:* 705 Sycamore Terr Dewitt NY 13214

SIMINOVITCH, LOUIS, b Montreal, Que, May 1, 20; m 44; c 3. BIOPHYSICS, MICROBIOLOGY. *Educ:* McGill Univ, BSc, 41, PhD(chem), 44. *Prof Exp:* Res phys chemist, Nat Res Coun Can, 44-47; Royal Soc Can fel biochem & microbiol, Pasteur Inst, Paris, 47-49, microbiol, Nat Ctr Sci Res, 49-53; Nat Cancer Inst Can fel, Connaught Med Res Lab, 53-56, from assoc prof to prof microbiol, 56-67, assoc prof med biophys, 58-60, chmn dept med cell biol, 69-72, assoc prof pediat, 72-78, PROF MED BIOPHYS, UNIV TORONTO, 60-, PROF MED GENETICS, 66-, UNIV PROF, 76-; DIR RES, SAMUEL LUNENFELD RES INST, MT SINAI HOSP, TORONTO, 83- *Concurrent Pos:* Fel, Can Royal Soc, 47-49 & Nat Cancer Inst Can, 53-56; head subdiv microbiol, div biol res, Ont Cancer Inst, Toronto, 57-63, head div, 63-69, sr scientist, 69-83; ed, Virol, 60-80, Bact Revs, 69-72 & J Molecular & Cellular Biol, 80-90; mem, panel sect, Nat Cancer Inst Can, 65-69, res adv group, 69-74, chmn, 70-72, virol & richettsiology study sect, NIH, 66-68, adv comt, 78-83, grants comts, cellular biol & genetics, Nat Res Coun Can, 66-69 & cancer, growth & differentiation, Med Res Coun Can, 67-70 & genetics, 71-74, health res & develop comt, Ont Coun Health, 66-82, chmn, 74, task force res grants rev comt, Prov Ont, 70-74, task force joint res rev comt, 73-79, Prov task force health res req, 74-76, & adv comt genetic serv, 76-82, gen coun comn genetic eng, United Church Can, 74-78 & adv bd, Ont Ment Health Found, 74-78; mem, working group human experimentation, Med Res Coun Can, 76-78, exec, 77-83, task force res Can, Sci Coun Can, 76-80 & comt sci & legal process, 78-81, adv comt Duke Univ Comprehensive Cancer Ctr, 77-82, sci adv comt, Connaught Res Inst, 80-84, res comt, Ont Cancer Treatment & Res Found, 81-86, adv bd, Gairdner Found, 83-, Health Res & Develop Coun Ont, 83-86, med planning comt, 83-87, med planning comt, Arthritis Soc, 83-87 & sci adv bd, Huntington's Soc Can, 84-89; chmn, subcomt study basic biol Can, Biol Coun Can, 68-69, task force genetic serv, Prov Ont, 74-76 & ad hoc comt guidelines handling recombinant DNA molecules & certain animal viruses, Med Res Coun Can, 75-77; geneticist-in-chief, Hosp Sick Children, Toronto, 70-85; mem bd dir, Mt Sinai Inst, Mt Sinai Hosp, Toronto, 75-82, Nat Cancer Inst Can, 75-85, pres, 82-84, Can Cancer Soc, 81-84, Can Weizmann Inst Sci, 72- & Ont Cancer Treatment & Res Found, 79-; nat corresp, comt genetic experimentation, Int Coun Sci Unions, 77-85; consult, Allelix, Inc, 82-87; mem fac res comt, Univ Toronto, 84-; mem, AIDS Study Steering Comt, Royal Soc Can, 87-88, Subcomt Task Force Coord Cancer Res, Univ Toronto, 87-89, Sci-Technol-Serv Subcomt, Sci Coun Can, Ottawa, 88-89; chmn, Subcomt Res, Royal Soc Study AIDS, 87-88, adv comt evolutionary biol, Can Inst Adv Res, 86-, external adv comt, Loeb Inst Med Res, 87-, sci adv comt, Rotman Res Inst Baycrest Ctr, 90-, sci adv comt, Montreal Cancer Inst, 90-; chmn res adv panel, Ont Cancer Treatment & Res Found, Toronto, 86-, mem exec comt of the bd, 91- *Honors & Awards:* Louis Rapkine mem lectr, Inst Pasteur, Paris, 64; Major G Seelig lectr, Washington Univ, St Louis, 66; Centennial Medal, Can, 67; Elizabeth Laird lectr, Univ Winnipeg, 76; Queen Elizabeth II Jubilee Silver Medal, 77; Flavelle Gold Medal, Royal Soc Can, 78; G Malcolm Brown mem lectr, Royal Col Physicians & Surgeons Can, 78; Jack Schultz mem lectr, Inst Cancer Res, Pa, 78; Officer of the Order of Can, 80; Izaak Walton Killam Mem Prize, 81; Wightman Award, Gairdner Found, 81; V W Scully mem lectr, Hamilton, Ont, 81; Maurice Grimes mem lectr, Can Cancer Soc, 81; Herzberg mem lectr, Carleton Univ, Ottawa, 85; Medal of Achievement Award, Inst Res Clin Montreal, 85. *Mem:* Am Asn Cancer Res; fel Royal Soc Can; Genetics Soc Can; Can Soc Cell Biol(pres-elect, 66, pres, 67); AAAS; fel Royal Soc London; Canadians for Health Res; Genetics Soc Am. *Res:* Differentiation in haemopoietic tissues in mice; biochemical and physiological genetics of bacteriophages; tumor viruses. *Mailing Add:* Samuel Lunenfeld Res Inst Mt Sinai Hosp 600 Univ Ave Toronto ON M5G 1X5 Can

SIMITSES, GEORGE JOHN, b Athens, Greece, July 31, 32; US citizen; m 60; c 3. ENGINEERING & STRUCTURAL MECHANICS. *Educ:* Ga Inst Technol, BS, 55, MS, 56; Stanford Univ, PhD(aeronaut & astronaut), 65. *Prof Exp:* Instr struct, Ga Inst Technol, 56-59, proj engr, Eng Exp Sta, 58-61, asst prof struct & design, 59-66, assoc prof spacecraft struct, 66-68, assoc prof eng sci & mech, 68-74, prof eng sci & mech, 74-86, prof aerospace eng, 86-89; PROF & HEAD AEROSPACE ENG & ENG MECH, UNIV CINCINNATI, 89- *Concurrent Pos:* Consult to numerous industs & co, 66- *Honors & Awards:* Sustained Res Award, Sigma Xi, 80. *Mem:* Soc Engr Sci; assoc fel Am Inst Aeronaut & Astronaut; fel Am Soc Mech Engrs; Am Acad Mech; Sigma Xi; Struct Stability Res Coun; Hellenic Soc. *Res:* Stability theory; optimization of structures; dynamic stability; creep buckling and rachetting; structural stability; mechanics of composite materials. *Mailing Add:* Dept Aerospace Eng & Eng Mech Univ Cincinnati Cincinnati OH 45221-0070

SIMIU, EMIL, b Bucharest, Romania, Apr 8, 34; US citizen; m 70; c 2. STRUCTURAL ENGINEERING. *Educ:* Inst Civil Eng, Bucharest, Dipl Ing, 56; Polytech Inst Brooklyn, MS, 68; Princeton Univ, PhD(civil eng), 71. *Prof Exp:* Design engr struct eng, Bucharest Design Inst, 56-62, Bechtel Corp, 63-65, Lev Zetlin & Assocs, 65-66 & Ammann & Whitney, Inc, 66-68; res asst civil eng, Princeton Univ, 68-71; res assoc, Nat Bur Standards, 71-73, res eng struct eng, 71-88; Fel, NAT INST STANDARDS & TECHNOL, 88- *Honors & Awards:* Gold Medal US Dept Com, 89. *Mem:* Am Soc Civil Engrs; Sigma Xi. *Res:* Dynamic loads on structures induced by wind, earthquake, and ocean waves; dynamic and fluidelastic response. *Mailing Add:* Ctr Bldg Technol Nat Bur Standards Gaithersburg MD 20899

SIMKIN, BENJAMIN, b Philadelphia, Pa, Apr 17, 21; m 47; c 2. ENDOCRINOLOGY. *Educ:* Univ Southern Calif, AB, 41, MD, 44. *Prof Exp:* Intern, Los Angeles County Hosp, 43-44; resident med, Cedars of Lebanon Hosp, 44-46; instr, 49-66, ASST CLIN PROF, SCH MED, UNIV SOUTHERN CALIF, 66-; ATTEND PHYSICIAN, CEDARS-SINAI MED CTR, 69- *Concurrent Pos:* Beaumont res fel med, Cedars of Lebanon Hosp, 46-47; fel, Michael Reese Hosp, 47-48; res fel, May Inst, Jewish Hosp, Cincinnati, 48-49; jr attend physician, Los Angeles County Hosp, 49-65, attend physician, 65-; asst adj, Cedars of Lebanon Hosp, 52-58, assoc attend physician, 58-60, attend physician & chief endocrine clin, 61-69. *Mem:* AAAS; Soc Exp Biol & Med; Endocrine Soc; Am Diabetes Asn; AMA; Sigma Xi. *Res:* Pituitary hormones; obesity; clinical endocrinology. *Mailing Add:* 6200 Wilshire Blvd Los Angeles CA 90048

SIMKIN, DONALD JULES, b Brooklyn, NY, Sept 5, 25; m 53; c 3. AEROSPACE ENGINEERING & TECHNOLOGY. *Educ:* Univ Calif, Berkeley, BS, 45, MS, 49,. *Prof Exp:* Engr, Shell Develop Co, 49-57; sr engr, Marquardt Corp, 57-58, supvr rocket propulsion, 58-60; dept head, Astropower Lab, Douglas Aircraft Co, 60-62; supvr spacecraft propulsion, Space & Info Div, Downey, Rockwell Int Corp, 62, chief, 62-66, mgr, 66-67, sr proj engr advan systs dept, 67-69, mgr mission & opers anal, space shuttle proj, 69-82, mgr independent res & develop, orbiter div, assoc chief proj engr, 84-87, proj mgr, adv eng, Orbiter Div, 88-90; ENGR, ENG DEPT, MCDONNELL DOUGLAS SPACE SYSTS, 90- *Concurrent Pos:* Lectr, Space Technol Series, Univ Calif, Los Angeles, 58-68; lectr, mgt, West Coast Univ, 83-84. *Honors & Awards:* Shuttle Flag Award. *Mem:* Assoc fel Am Inst Aeronaut & Astronaut; Am Chem Soc; Am Inst Chem Engrs; Combustion Inst; Am Soc Mech Engrs. *Res:* Spacecraft propulsion systems; thermodynamics of propellants; combustion phenomena; chemical kinetics; fluid dynamics; unit operations of chemical engineering; spacecraft design; space mission planning and operations. *Mailing Add:* Eng Dept McDonnell Douglas Space Systs 5301 Bolsa Ave Huntington Beach CA 92647

SIMKIN, SUSAN MARGUERITE, b Detroit, Mich, July 26, 40; m 61; c 2. ASTRONOMY. *Educ:* Earlham Col, BA, 62; Univ Wis, PhD(astron), 67. *Prof Exp:* Res assoc & lectr astron, Columbia Univ, 66-73; res assoc, Mich State Univ, 74-75, asst prof astron, 76-79, assoc prof physics & astron, 79-84, PROF PHYSICS & ASTRON, MICH STATE UNIV, 84- *Concurrent Pos:* NATO fel, Kapteyn Lab, 75-76; sr res fel, Mt Stromlo Observ, 76-80, vis fel, 80-81. *Mem:* AAAS; Int Astron Union; Am Astron Soc; Sigma Xi. *Res:* Astronomical photometry; spectroscopy; structure of galaxies; radio galaxies. *Mailing Add:* Dept Physics & Astron Mich State Univ East Lansing MI 48824

SIMKIN, THOMAS EDWARD, b Auburn, NY, Nov 11, 33; m 65; c 2. VOLCANOLOGY, GEOLOGY. *Educ:* Swarthmore Col, BS, 55; Princeton Univ, MSE, 60, PhD(geol), 65. *Prof Exp:* Indust engr, Proctor & Gamble Co, 55-56; hydrographer, US Coast & Geodetic Surv, 56-58; instr geol, State Univ NY Binghamton, 64-65; res assoc geophys sci, Univ Chicago, 65-67; supvr geol, Smithsonian Oceanog Sorting Ctr & res assoc, Petrol Div, Smithsonian Inst, 67-72, CUR PETROLOGY & VOLCANOLOGY, SMITHSONIAN INST, 72- *Concurrent Pos:* Secy for Americas Sci, Charles Darwin Found for Galapagos Isles, 70-90; dir, Global Volcanism Prog, Smithsonian, 84- *Mem:* Am Geophys Union; Int Asn Volcanology & Chem Earth's Interior. *Res:* Global volcanism and volcanology, particularly oceanic volcano evolution, calderas and contemporary volcanism; volcanology and petrology of Galapagos Islands and Scottish Tertiary Province. *Mailing Add:* NHB Stop 119 Smithsonian Inst Washington DC 20560

SIMKINS, CHARLES ABRAHAM, b Reading, Kans, Nov 3, 23; m 45; c 6. SOIL FERTILITY, AGRONOMY. *Educ:* Kans State Univ, BS(biol sci) & BS(agr), 48, MS, 50, PhD(soil), 58; Oak Ridge Inst Nuclear Studies, DRIP, 51. *Hon Degrees:* Dr, Univ Hungary, 73. *Prof Exp:* Res asst soils, Kans State Univ, 46-48, asst prof, 49; asst prof, Univ Idaho, 50-52; res asst, Kans State Univ, 52-53; assoc prof, Univ Minn, 53-58; chief soils, Develop & Res Corp, Iran, 58-63; soil scientist, Food & Agr Orgn, UN, Cyprus, 63-64; chief of party agr develop, Chile Proj, 64-70, PROF SOILS, UNIV MINN, ST PAUL, 70-, EXTEN SPECIALIST, AGR EXTEN, 74- *Concurrent Pos:* Consult, Govt Saudi Arabia, 60 & Govt Hungary, 71-72; proj mgr, UN Drug Fund, Lebanon, 73-74. *Mem:* Am Soc Agron; Soil Sci Soc Am. *Res:* Nutrition of wheat plant; pasture fertility research; potato fertility; land use. *Mailing Add:* USAID-MFAD-Uganda Proj 42 Nakosero Rd PO Box 7007 Kampala Uganda

SIMKINS, KARL LEROY, JR, b Aldine, NJ, July 2, 39; m 61; c 4. ANIMAL SCIENCE. *Educ:* Rutgers Univ, BS, 61; Univ Wis, MS, 62, PhD(animal nutrit & biochem), 65. *Prof Exp:* Res asst dairy sci, Univ Wis, 61-62, res asst dairy sci & biochem, 62-65; res nutritionist, 65-70, group leader, Clin Develop Lab, 70-77, mgr, animal indust develop, agr res ctr, 77-88, MGR, CLIN DEVELOP, AM CYANAMID CO, PRINCETON, NJ, 88- *Mem:* Am Soc Animal Sci; Am Dairy Sci Asn. *Res:* Appetite and growth regulation in domestic animals; development of recombinant bovine somatotropin in ruminants; efficacy and safety of growth regulators, antibacterials, anthelmintics, coccidiostats and pesticides; evaluation of compounds which alter rumen fermentation. *Mailing Add:* Agr Res Ctr Am Cyanamid Co PO Box 400 Princeton NJ 08543-0400

SIMKOVER, HAROLD GEORGE, b Montreal, Que, Mar 27, 23; nat US; m 47; c 2. ENTOMOLOGY. *Educ:* McGill Univ, BSc, 47; Univ Wis, MS, 48, PhD(entom), 51. *Prof Exp:* Jr entomologist & instr, Wash State Univ, 51-53; entomologist, Shell Develop Co, 53-63, patent agt, 63-66, sr field rep, Shell Chem Co, 66-69, sr field rep, Biol Sci Res Ctr, Shell Develop Co, Dublin, Calif, 69-84, sr field rep 84-86; RETIRED. *Mem:* AAAS; Entom Soc Am. *Res:* Economic entomology; pesticides. *Mailing Add:* 6321 Longcroft Dr Oakland CA 94611

SIMKOVICH, GEORGE, b Smithton, Pa, Apr 19, 28; m 63; c 3. MATERIAL SCIENCES, METALLURGY. *Educ:* Pa State Univ, BS, 52, MS, 55, PhD(metall), 59. *Prof Exp:* Res asst mineral prep, Pa State Univ, 52-55, asst, 55; res assoc metall, Yale Univ, 58-60; Nat Sci Found res fel, 60-61; res fel, Max Planck Inst Phys Chem, 61-62; scientist, Fundamental Res Lab, US Steel Corp, 62-64; assoc prof, 64-71, PROF METALL, PA STATE UNIV, UNIVERSITY PARK, 71- *Mem:* Am Inst Mining, Metall & Petrol Engrs; Am Soc Metals; Nat Asn Corrosion Engrs; Electrochem Soc. *Res:* Physical chemistry of metallurgy; physical chemistry of materials; high-temperature studies; oxidation; point defects in solids. *Mailing Add:* Dept Metall 206 Steidle Bldg Pa State Univ University Park PA 16802

SIMMANG, C(LIFFORD) M(AX), b San Antonio, Tex, Feb 14, 12; m 42; c 1. MECHANICAL ENGINEERING. *Educ:* Agr & Mech Col Tex, BS, 36, MS, 38; Univ Tex, PhD(mech eng), 52. *Prof Exp:* Jr engr petrol, Humble Oil Co, 36-37; from instr to prof mech eng, 38-77, head dept, 57-77, EMER PROF MECH ENG, TEX A&M UNIV, 77- *Honors & Awards:* Charles W Crawford Award, Eng, 79. *Mem:* Am Soc Mech Engrs; Soc Am Mil Engrs; Sigma Xi; Am Soc Eng Educ. *Res:* Heat transfer and coefficients in regenerative air heater; fluid flow, loss in tubing elbows and bends. *Mailing Add:* Dept Mech Eng Tex A&M Univ 401 North Ave E Bryan TX 77801

SIMMEL, EDWARD CLEMENS, b Berlin, Ger, Jan 30, 32; US citizen; m 83; c 3. AVIATION PSYCHOLOGY, BEHAVIORAL GENETICS. *Educ:* Univ Calif, Berkeley, AB, 55; Wash State Univ, PhD(exp psychol), 60. *Prof Exp:* Res trainee, Vet Admin Hosp, American Lake, Wash, 60; asst prof psychol, Western Wash Univ, 60-62 & Calif State Univ, Los Angeles, 62-65; from asst prof to prof psychol, 65-90, dir behav genetics lab, 77-84, EMER PROF PSYCHOL, MIAMI UNIV, 90- *Mem:* Am Psychol Soc; Behav Genetics Asn; Aerospace Med Asn; Sigma Xi. *Res:* Judgement and decision processes in aircrews; stress effects. *Mailing Add:* PO Box 759 Borrego Springs CA 92004

SIMMON, VINCENT FOWLER, b Los Angeles, Calif, Aug 9, 43; m; c 3. MICROBIOLOGY. *Educ:* Amherst Col, Mass, BA, 64; Univ Toledo, MS, 67; Brown Univ, RI, PhD(molecular biol), 72. *Prof Exp:* Res & develop chemist thermoplastics, Textileather Div, Gen Tire & Rubber Co, 64-65; fel microbiol, Stanford Univ, Stanford, Calif, 71-73; microbiologist, Stanford Res Inst, 73-75, mgr microbial genetics, 75-77, asst dir dept toxicol, SRI Int, 77-79; dir tech opers, Genex Corp, 79-80, vpres mkt, 80-81, vpres int develop, 82-83, sr vpres res & develop, 83-84; VPRES BIOTECHNOL & BIOMED RES, W R GRACE, COLUMBIA, MD, 84- *Concurrent Pos:* Vis asst prof, Dept Anesthesiol, Stanford Med Sch, 77-79; mem, Comt Chem & Environ Mutagens, Nat Res Coun, 79-81; bd dirs, Chem Inst Toxicol, 86- *Mem:* Am Environ Mutagen Soc; Am Soc Microbiol; AAAS; Sigma Xi; NY Acad Sci. *Res:* Biotechnology and biomedical research including genetic engineering for the production of chemicals, pharmaceuticals and therapeutic devices. *Mailing Add:* Genex Corp 12300 Washington Ave Rockville MD 20852

SIMMONDS, JAMES G, b Washington, DC, July 26, 35; m 88; c 2. APPLIED MECHANICS, SHELL THEORY. *Educ:* Mass Inst Technol, BS, 58, MS, 58, PhD (math), 65. *Prof Exp:* From asst prof appl math to prof appl math , 66-86, CHAIRED PROF APPL MATH, UNIV VA, 86- *Mem:* Soc Nat Philosophers; Math Asn Am; Am Soc Mech Engrs; AAAS. *Res:* Nonlinear theory of elastic shells; perturbation theory. *Mailing Add:* Dept Appl Math Univ Va Thornton Hill Charlottesville VA 22903

SIMMONDS, RICHARD CARROLL, b Baltimore, Md, Aug 22, 40. LABORATORY ANIMAL SCIENCE, RESEARCH ADMINISTRATION. *Educ:* Univ Ga, DVM, 64; Tex A&M Univ, MS, 70. *Prof Exp:* US Air Force, 64-85, res vet, Arctic Aeromed Lab, 64-67 & Arctic Med Res Lab, Alaska, 67-68; staff vet & resident, Lab Animal Med, Sch Aerospace Med, 68-70; area test dir, Lunar Quarantine Prog & staff vet, Air Force Detailee, Johnson Space Ctr, 70-73; staff vet & mgr, Joint US/USSR Biol Satellite Proj, Air Force Detailee, Ames Res Ctr, NASA, 73-76; dir, Dept Lab Animal Med, Sch Med, 76-82, dir, Instnl & Res Support, Uniformed Serv Univ Health Sci, 82-88; DIR, LAB ANIMAL MED, UNIV NEV, 88-, ASST DEAN, SCH MED, 89- *Honors & Awards:* Commendation Medal, US Air Force & US Army, 68; Meritorious Serv Medal, US Air Force, 74, 77; Meritorious Serv Medal, NASA, 76. *Mem:* Am Vet Med Asn; Am Asn Lab Animal Sci; Am Col Lab Animal Med; AAAS; Am Soc Mammal. *Res:* Biomedical effects of altered geophysical environments; laboratory animal science and mammalian thermal regulation. *Mailing Add:* Dir Lab Animal Med Univ Nev Reno NV 89557-0040

SIMMONDS, ROBERT T, b Hackensack, NJ, Aug 2, 32; m 57. PALEOBIOLOGY. *Educ:* Columbia Univ, BS, 54; Syracuse Univ, MS, 58; Univ Ill, PhD(geol), 61. *Prof Exp:* Instr geol, Denison Univ, 61; from assoc prof to prof earth sci, State Univ NY Col Oneonta, 61-87. *Mem:* AAAS; Nat Asn Geol Teachers. *Res:* Paleoecology, evolution, geotectonics. *Mailing Add:* 190 E Beach Rd Norland WA 98358

SIMMONDS, SIDNEY HERBERT, b Winnipeg, Man, Aug 29, 31; m 55; c 3. CIVIL ENGINEERING. *Educ:* Univ Alta, BSc, 54, MSc, 56; Univ Ill, PhD, 62. *Prof Exp:* Assoc prof, 57-70, PROF CIVIL ENG, UNIV ALTA, 70- *Concurrent Pos:* Consult structural engr; vis prof, Univ Tex, Austin, 77, Univ SC, 78, Royal Mil Col Can, 87. *Mem:* Am Concrete Inst; Am Soc Civil Engrs; Int Asn Shell Struct; Can Soc Civil Engrs. *Res:* Structural analysis and design. *Mailing Add:* Dept Civil Eng Univ Alberta Edmonton AB T6G 2G7 Can

SIMMONDS, SOFIA, b New York, July 31, 17; m 36. BIOCHEMISTRY. *Educ:* Columbia Univ, BA, 38; Cornell Univ, PhD(biochem), 42. *Prof Exp:* Asst biochem, Med Col, Cornell Univ, 41-42, res assoc, 42-45; instr physiol chem, Yale Univ, 45-46, from instr to asst prof microbiol, 46-50, from asst prof to assoc prof biochem & microbiol, 50-62, assoc prof biochem, 62-69, assoc prof, 69-75, dir undergrad studies, 73-85, prof molecular biophys & biochem, 75-88, assoc dean & dean undergrad studies, 88, lectr & dir undergrad studies, 90-91, EMER PROF, MOLECULAR BIOPHYSICS & BIOCHEM, YALE UNIV, 88- *Honors & Awards:* Garvan Medal, Am Chem Soc, 69. *Mem:* Am Soc Biol Chem; Am Chem Soc. *Res:* Amino acid metabolism; transmethylation in animals; amino acid and protein metabolism in micro-organisms. *Mailing Add:* Molecular Biophys & Biochem Sch Med Yale Univ PO Box 3333 New Haven CT 06510-8024

SIMMONS, ALAN J(AY), b New York, NY, Oct 14, 24; m 47; c 5. ELECTRICAL ENGINEERING. *Educ:* Harvard Univ, BS, 45; Mass Inst Technol, MS, 48; Univ Md, PhD(elec eng), 57. *Prof Exp:* Electronic scientist, US Naval Res Lab, 48-57; head, Microwave Dept, TRG, Inc, 57-71; group leader, Lincoln Lab, Mass Inst Technol, 71-87; CONSULT, 87- *Mem:* AAAS; fel Inst Elec & Electronics Engrs (pres, 86); Antennas Propagation Soc (pres, 86). *Res:* Microwave antennas, waveguides and components; electromagnetic theory; communication satellites. *Mailing Add:* PO Box 103 Winchester MA 01890-0103

SIMMONS, CHARLES EDWARD, b Oklahoma City, Okla, Apr 5, 27; m 53; c 2. PSYCHIATRY. *Educ:* Univ Okla, BS, 50, MD, 54. *Prof Exp:* Intern, Gorgas Hosp, Ancon, Panama, CZ, 54-55; resident psychiat, Griffith Mem Hosp, Norman, Okla, 55-57; resident, Parkland Hosp, Dallas, Tex, 57-58; from instr to asst prof, Univ Tex Southwest Med Sch Dallas, 58-60; mem staff, Chestnut Lodge, Md, 62-63; PROF PSYCHIAT, UNIV TEX MED SCH SAN ANTONIO, 67- *Concurrent Pos:* NIMH teaching fel psychiat, Univ Tex Southwest Med Sch Dallas, 57-58; consult, Wilford Hall Air Force Hosp, San Antonio, Tex, 67- *Mem:* AMA; Am Psychiat Asn; Am Psychoanal Asn. *Res:* Psychoanalysis. *Mailing Add:* PO Box 254 Port Mansfield TX 78598

SIMMONS, CHARLES FERDINAND, agronomy, soil science; deceased, see previous edition for last biography

SIMMONS, DANIEL HAROLD, b New York, NY, June 22, 19; m 42; c 2. PULMONARY DISEASE. *Educ:* Univ Calif, Los Angeles, BA, 41; Univ Southern Calif, MD, 48; Univ Minn, PhD(physiol), 53. *Prof Exp:* Instr math, Univ Calif, Los Angeles, 43; asst prof med, 53-55, from asst prof to assoc prof med & physiol, 55-65, prof physiol, 65-81, PROF MED, UNIV CALIF, LOS ANGELES, 65- *Concurrent Pos:* Nat Heart Inst trainee, Univ Minn, 51-53; sect chief, Vet Admin Ctr, Los Angeles, 53-61; dir res & assoc dir div med, Mt Sinai Hosp, 61-64, dir med res inst, Cedars-Sinai Med Ctr, 64-66. *Mem:* Am Physiol Soc; Am Fedn Clin Res; fel Am Col Chest Physicians; Am Thoracic Soc. *Res:* Mechanisms of clinically-related problems in respiration physiology. *Mailing Add:* Dept Med 32-176 Ctr Health Scis Univ Calif Sch Med Los Angeles CA 90024

SIMMONS, DANIEL L, b Provo, Utah, May 14, 55; c 3. MOLECULAR BIOLOGY. *Educ:* Brigham Young Univ, BS, 78, MS, 80; Univ Wis-Madison, PhD(oncol), 86. *Prof Exp:* Postdoctoral fel, Biol Labs, Harvard Univ, Cambridge, Mass, 86-89; ASST PROF CHEM, BRIGHAM YOUNG UNIV, 89- *Mem:* Sigma Xi; AAAS; Am Chem Soc. *Res:* Molecular mechanisms of transformation; immediate-early genes induced by Rous sarcoma virus. *Mailing Add:* 647 WIDB Brigham Young Univ Provo UT 84602

SIMMONS, DARYL MICHAEL, b Hudson, NY, May 14, 52. CHEMISTRY, DRUG INVESTIGATION. *Educ:* Col St Rose, BA, 79; Slippery Rock Univ, MS, 81. *Prof Exp:* Lab tech, Bender Hyg Labs, 75; prod tech, Sterling Org, 76-78; anal chemist, Stiefel Res Inst, 81-82; SR RES SCIENTIST, STERLING RES GROUP, 82- *Concurrent Pos:* Adj instr chem, biol & algebra, Columbia-Greene Commun Col, 82-; presenter, Am Asn Pharmaceut Scientists & Asn Off Anal Chemists, 89-; vis scientist, Pharmaceut Mfg Asn, 89-; judge, Pa Acad Sci. *Mem:* Pharmaceut Mfg Asn; Am Asn Pharmaceut Scientists. *Res:* Determination of chemical and physical properties of drugs corresponding to oral absorption and bioavailability; development of small animal models to evaluate absorption and bioavailability; development of a formulation for maximum systemic blood levels of investigation drugs; numerous publications. *Mailing Add:* RD 1 Box 20B West Coxsackie NY 12192

SIMMONS, DAVID J, b Jamaica Plain, Mass, Mar 10, 31; m 57; c 1. PHYSIOLOGY, ENDOCRINOLOGY. *Educ:* Boston Univ, BA, 54; Clark Univ, MA, 56; Univ Chicago, PhD(paleozool), 59. *Prof Exp:* Instr biol, Wright Jr Col, 59-60; res assoc physiol, Univ Chicago, 59-62; from asst physiologist to assoc physiologist, Radiol Physics Div, Argonne Nat Lab, 62-72; asst prof orthop surg, Sch Med, Wash Univ, 72-75, assoc prof res orthop surg, 75-86; assoc prof, 86-90, PROF, UNIV TEX MED BR, GALVESTON, 90- *Concurrent Pos:* NIH fel, Univ Chicago, 60-62. *Mem:* AAAS; Am Asn Anatomists; Int Soc Chronobiol; Soc Vert Paleont; Orthop Res Soc; Am Soc Bone & Mineral Res. *Res:* Skeletal development; collagen formation; cell population dynamics. *Mailing Add:* Dept Orthop Surg Three Adler Circle Galveston TX 77551

SIMMONS, DAVID RAE, b Oklahoma City, Okla, May 4, 40; m 61; c 3. MATHEMATICS. *Educ:* Centenary Col La, BS, 62; Univ Ark, MS, 66, PhD(math), 69. *Prof Exp:* Asst prof math, Centenary Col La, 69-74, actg chmn dept, 70-72; full prof math & chmn Dept Math & Comput Sci, La Col, 74-88. *Mem:* Am Math Soc; Asn Comput Mach; Math Asn Am. *Res:* Category theory; universal algebra, computer studies of semigroups of relations. *Mailing Add:* Dept Math & Comput Sci La Col 1140 Col Dr Pineville LA 71359

SIMMONS, DICK BEDFORD, b Houston, Tex, Dec 24, 37; m 59; c 3. COMPUTER & INFORMATION SCIENCES. *Educ:* Tex A&M Univ, BS, 59; Univ Pa, MS, 61, PhD(comput & info sci), 68. *Prof Exp:* Design engr, Radio Corp Am, 59-61; mem tech staff, Bell Tel Labs, 63-69, supvr comput lang & systs, 69-70; assoc prof, 71-81, PROF COMPUT SCI, TEX A&M UNIV, 81-, DIR COMPUT CTR, 72- *Concurrent Pos:* Consult, Tex A&M Univ, 70-71 & Datamaster Div, Am Chain & Cable Co, 72-; prin investr, Unitech Oper Systs Proj, 71-72 & NASA Automated Doc Study, 72. *Mem:* Asn Comput Mach; Inst Elec & Electronics Engrs; Asn Educ Data Systs; Am Soc Eng Educ. *Res:* Automatic documentation; computer languages; programmer productivity; computer architecture. *Mailing Add:* Dept Comput Sci Tex A&M Univ College Station TX 77843

SIMMONS, DONALD GLICK, b Waynesboro, Va, Sept 6, 38; m; c 2. VETERINARY MICROBIOLOGY. *Educ:* Bridgewater Col, BA, 62; Univ Ga, DVM, 67, MS, 69, PhD(vet microbiol), 71. *Prof Exp:* NIH spec fel vet med microbiol, Univ Ga, 67-71; from asst prof to prof vet med microbiol, NC State Univ, 71-80, prof microbiol, path & parasitol, 80-87; PROF & HEAD, DEPT VET SCI, PA STATE UNIV, 88- *Honors & Awards:* P P Levine Award, 80. *Mem:* Sigma Xi; Am Vet Med Asn; Am Col Vet Microbiologists; Am Asn Avian Pathologists. *Res:* Veterinary medical microbiology; avian medicine; respiratory and enteric viruses of poultry; respiratory bacteria of poultry. *Mailing Add:* 2400 Shingletown Rd State College PA 16801

SIMMONS, EMORY GUY, b Ind, Apr 12, 20. MYCOLOGY. *Educ:* Wabash Col, AB, 41; DePauw Univ, AM, 46; Univ Mich, PhD(bot), 50. *Hon Degrees:* DSc(microbiol) Kasetsant Univ, Thailand, 88. *Prof Exp:* Instr bact & bot, DePauw, 46-47; asst prof bot, Dartmouth Col, 50-53; mycologist, US Army Natick Labs, 53-58, head mycol lab, 58-74, prin investr, Develop Ctr Cult Collection of Fungi, 74-77; prof bot, 74-77, prof microbiol, Univ Mass, Amherst, 77-87; RETIRED. *Concurrent Pos:* Chmn adv comt fungi, Am Type Cult Collection; US rep, Expert Group on Fungus Taxon, Orgn Econ Coop & Develop; Secy Army res fel, Thailand, Indonesia, 68-69; adj prof, Univ RI, 72-74; mem exec bd, US Fedn Cult Collections, 74-76, pres, 76-78; pres & chmn bd, Second Int Mycol Cong, Inc, 75-78; mem adv comt cult collections, UN Environ Prog/UNESCO/Int Cell Res Orgn, 77- *Mem:* AAAS; Mycol Soc Am (secy-treas, 63-65, vpres, 66, pres, 68); Brit Mycol Soc; Int Asn Plant Taxonomists. *Res:* Taxonomic mycology; taxonomy of Fungi imperfecti; taxonomy and cultural characteristics of Ascomycetes. *Mailing Add:* 717 Thornwood Rd Crawfordsville IN 47933-2760

SIMMONS, ERIC LESLIE, b Santo Domingo, Dominican Repub, Feb 11, 17; nat US; m 43; c 4. BIOLOGY. *Educ:* Swarthmore Col, AB, 38; Ind Univ, PhD(zool), 44. *Prof Exp:* Res assoc, Metall Lab, 43-46, from instr to asst prof biol sci, 46-55, premed adv, 48-52, asst dean students, 52-54, assoc prof med, 55-78, PROF MED, SCH MED, UNIV CHICAGO, 78- *Mem:* Sigma Xi; assoc Am Soc Zoologists; Radiation Res Soc; Am Asn Lab Animal Sci; Int Soc Exp Hemat. *Res:* Radiobiology; effect of radiation on the hemopoietic system; cancer induction. *Mailing Add:* 106 S Bryan St Bloomington IN 47408

SIMMONS, FRANCIS BLAIR, b Los Angeles, Calif, Nov 15, 30; m 71; c 4. PHYSIOLOGY. *Educ:* Transylvania Col, AB, 52; Univ Louisville, MD, 56. *Prof Exp:* Intern med, Madigan Army Hosp, 56-57; res physiologist, Walter Reed Army Inst Res, 58-59; resident physician & res assoc otolaryngol, 59-62, assoc prof otolaryngol, 65-71, chief dept, 65-80, PROF OTOLARYNGOL, SCH MED, STANFORD UNIV, 71- *Res:* Auditory physiology and psychophysics; neurophysiology. *Mailing Add:* Stanford Univ Med Ctr Stanford CA 94305

SIMMONS, GARY ADAIR, b Dothan, Ala, Aug 7, 44; m 69; c 5. FOREST ENTOMOLOGY. *Educ:* Mich Tech Univ, BS, 66, MS, 68; Univ Mich, PhD(forestry), 72. *Prof Exp:* Res assoc, Univ Maine, Orono, 72-73, res asst prof, 73-74, asst prof forest entom & biomet, Dept Entom & Sch Forest Resources, 74-76; from asst prof to assoc prof entom, 76-84, PROF, DEPTS ENTOM & FORESTRY, MICH STATE UNIV, 84- *Concurrent Pos:* Prin investr, Dept Conserv, State Maine, 74-79, Thomson-Hayward Chem Co & Sumitomo Chem Co, Ltd, 75-76, Stauffer Chem Co, 76, Dept Natural Resources, Mich, 78-85, CANUSA, USDA, 78-84, Int Soc Arboricult, 82, H H & G Dow Found, 84-85 & R Gerstacher Found, 84-85; co-prin investr, Coop State Res Serv, USDA, 76-78, Forest Serv, 76-80, Animal & Plant Health Inspection Serv, 77-78, Mich Agr Exp Sta, 77, Off Educ, United Ser Orgn Health Educ & Welfare, 79-80 & Forest Serv, USDA, 79-81; leader, scientists' working group pop measurement, Can-US Spruce Budworms Prog, 79-84; assoc ed, NJ Appl Forestry, 83-87; chair entom sect, Am Registry Prof Entomologists, 83. *Mem:* Entom Soc Am; Soc Am Foresters. *Res:* Forest entomology, sampling, modeling; integrated pest management. *Mailing Add:* 243 Nat Sci-Entom Mich State Univ East Lansing MI 48824

SIMMONS, GARY WAYNE, b Parsons, WVa, June 17, 39; m 64; c 1. PHYSICAL CHEMISTRY, SURFACE CHEMISTRY. *Educ:* WVa Univ, BS, 61; Univ Va, PhD(chem), 67. *Prof Exp:* Fel physics, Ga Inst Technol, 66-67, res chemist, 67-70; asst prof, 70-74, assoc prof, 74-79, PROF CHEM, LEHIGH UNIV, 79- *Honors & Awards:* Melvin Romanoff Award, Nat Asn Corrosion Engrs, 74; Henry Marion Howe Medal, Am Soc Metals, 79. *Mem:* Am Chem Soc; Sigma Xi. *Res:* Fundamental properties of the solid-gas and solid-liquid interface and application of these properties to practical problems of catalysis, corrosion, stress corrosion cracking and corrosion fatigue; low energy electron diffraction, Auger electron spectroscopy, Mossbauer spectroscopy, x-ray photoelectron spectroscopy and electron microscopy. *Mailing Add:* 3660 Helen St Stafore Estates Bethlehem PA 18017

SIMMONS, GENE, b Dallas, Tex, May 15, 29; div; c 4. GEOPHYSICS. *Educ:* Tex A&M Univ, BS, 49; Southern Methodist Univ, MS, 58; Harvard Univ, PhD(geophys), 62. *Prof Exp:* Petrol engr, Humble Oil & Refining Co, 49-51; partner, Simmons Gravel Co, 53-62; asst prof geol, Southern Methodist Univ, 62-65; prof, 65-89, EMER PROF GEOPHYS, MASS INST TECHNOL, 89-; PRIN GEOPHYSICIST, HASER-RICHTER GEOSCI, INC, SALEM, NH, 89- *Concurrent Pos:* Mem var adv groups, NASA, 65-72, chief scientist, Manned Spacecraft Ctr, 69-71; mem, Comt C18, Am Soc Testing & Mat. *Honors & Awards:* Except Sci Achievement Medal, NASA, 71. *Mem:* Fel Am Geophys Union; Soc Explor Geophys; fel Geol Soc Am; Seismol Soc Am. *Res:* Thermal measurements of earth, moon and planets; physical properties of rocks at high pressures and temperatures; application of geophysical techniques to geological problems. *Mailing Add:* 180 N Policy St Salem NH 03079

SIMMONS, GEORGE ALLEN, b Birmingham, Ala, May 2, 26; m 51; c 3. CHEMISTRY GLASS TECHNOLOGY. *Educ:* Birmingham Southern Col, BS, 47; Ohio State Univ, MS, 49, PhD(chem), 52. *Prof Exp:* From asst prof to assoc prof chem, Birmingham Southern Col, 49-55; mgr res serv, Pittsburgh Plate Glass Co, 55-61, coordr melting, 61-62; spec proj chief, Owens Ill-Toledo, 62-66, tech dir new prod develop, 66-69, sr mkt mgr, 69-71; tech dir, Dominion Glass Co, Mississauga, Ont, Can, 71-76; sr vpres res & develop, Thatcher Glass Co, Elmira, NY, 76-81; tech dir, Ga Marble Co, Atlanta, 81-87; RETIRED. *Concurrent Pos:* Asst prof, Purdue Univ, 52-53; Du Pont fel, 52. *Mem:* Fel Am Inst Chemists; Am Chem Soc; fel Am Ceramic Soc. *Res:* Polymer composites; ultrafine grinding; mineral processing. *Mailing Add:* 4720 Norman Dr Kennesaw GA 30144

SIMMONS, GEORGE FINLAY, b Austin, Tex, Mar 3, 25; m 54; c 1. MATHEMATICS. *Educ:* Calif Inst Technol, BS, 46; Univ Chicago, MS, 48; Yale Univ, PhD(math), 57. *Prof Exp:* Instr math, Univ Col, Univ Chicago, 47-50, Univ Maine, 50-52 & Yale Univ, 52-56; asst prof, Univ RI, 56-58 & Williams Col, 58-62; from assoc prof to prof, 62-90, EMER PROF MATH, COLO COL, 90- *Mem:* Math Asn Am. *Res:* Topology; abstract algebra; analysis. *Mailing Add:* Dept of Math Colo Col Colorado Springs CO 80903

SIMMONS, GEORGE MATTHEW, JR, b Charleston, SC, Dec 25, 42; m 63; c 1. LIMNOLOGY. *Educ:* Appalachian State Univ, BS, 64; Va Polytech Inst & State Univ, PhD(zool), 68. *Prof Exp:* Asst prof biol & ecol, Va Commonwealth Univ, 68-71; from asst prof to assoc prof limnol, 71-78, assoc prof zool, 78-83, interim head, 88-89, PROF ZOOL, VA POLYTECHNIC INST & STATE UNIV, 83- *Mem:* Int Soc Theoret Appl Limnol; Am Soc Limnol & Oceanog; Explorer's Club. *Res:* Limnological studies of reservoir ecosystems; origin and role of freshwater benthic communities in antarctic lakes; productivity of tropical marine sand flats; Submarine ground water discharge and nutrient flux in coastal marine environments. *Mailing Add:* Dept Biol Va Polytechnic Inst & State Univ Blacksburg VA 24061-0406

SIMMONS, GUSTAVUS JAMES, b Ansted, WVa, Oct 27, 30; m 50; c 1. INFORMATION THEORY, CRYPTOGRAPHY. *Educ:* NMex Highlands Univ, BS, 55; Univ Okla, MS, 58; Univ NMex, PhD(math), 69. *Prof Exp:* Res assoc, Sandia Corp, 54-55, physicist, Nuclear Test Dept, 58-60; res scientist res staff, Lockheed Aircraft Corp, 55-56; sr group engr, Adv Electronics Div, McDonnell Aircraft Corp, 60-61; chief electronic engr, Electronic & Res Div, Fairbanks Morse & Co, 61; div supvr advan systs res, Sandia Corp, 62-70; dir res, Rolamite, 70-71; mgr, Math Dept, 71-87, SR RES FEL, SANDIA NAT LABS, 87- *Concurrent Pos:* Mem subcomt facil, Gov Tech Excellence Comt; ed, J Cryptology, Calif & Ars Combinatoria, Can. *Honors & Awards:* Nuclear Weapon Excellence Award, Dept Energy, 86; Ernest Orlando Lawrence Award, 86. *Mem:* Math Asn Am. *Res:* Digital message authentication theory; nuclear weapon system, especially the control and engineered use of such systems; combinatorial mathematics and graph theory; algorithms for high speed computation; cryptography. *Mailing Add:* Org 200 Sandia Nat Labs Bldg 807 Albuquerque NM 87185

SIMMONS, GUY HELD, JR, b Lafayette, Tenn, Oct 9, 36; m 59, 89; c 5. NUCLEAR MEDICINE. *Educ:* Western Ky Univ, BA, 61; Univ NC, MS, 64; Univ Cincinnati, PhD(nuclear eng), 72. *Prof Exp:* Instr physics, Western Ky Univ, 61; physicist radiation physics, USPHS, 61-72; ASSOC PROF RADIATION MED, UNIV KY, 72-; ASST CHIEF NUCLEAR MED, VET ADMIN HOSP, LEXINGTON, 72- *Concurrent Pos:* Ky Heart Asn grant, Univ Ky, 74-, Nat Cancer Inst grant, 75-; Vet Admin grant, 78; instr radiol, Med Ctr, Univ Cincinnati, 66-72; Food & Drug Admin grant, 88. *Mem:* Soc Nuclear Med; Am Asn Physicists Med; Am Col Radiol; Radiol Soc NAm. *Res:* Nuclear medicine instrumentation development and evaluation; digital image processing for clinical applications. *Mailing Add:* Nuclear Med Serv Vet Admin Hosp Lexington KY 40511

SIMMONS, HARRY DADY, JR, b Chicago, Ill, June 10, 38. MEDICINAL CHEMISTRY, ONCOLOGY. *Educ:* Univ Ill, BS, 60; Mass Inst Technol, PhD(chem), 66, State Univ NY, Downstate Med Ctr, MD, 77. *Prof Exp:* NIH fels inorg chem, Munich Inst Technol, 66-67; NIH fel chem, Brandeis Univ, 67-69; res chemist, Allied Chem Corp, 69-71; lab supvr, Kingsbrook Jewish Med Ctr, 71-74; med resident, Berkshire Med Ctr, 77-78, path resident, 78-80; res fel molecular biol, Albany Med Col, 80-81; res fel med eng, Polytech Inst, Troy, NY, 81-85; path resident, Mt Sinai Med Ctr, NY, 85-87; PATH FEL, MONTEFIORE MED CTR, BRONX, NY, 87- *Mem:* Am Chem Soc; NY Acad Sci; AMA; Am Meteorol Soc; Sigma Xi. *Res:* Synthetic organometallic chemistry of mercury, chromium, iron, group IV A metals; coordination compounds and complexes as catalysts; biomedical engineering; applications of scanning electron microscopy; anatomic pathology; educational programs for personal computer use. *Mailing Add:* 6255 Broadway Bronx NY 10471

SIMMONS, HOWARD ENSIGN, JR, b Norfolk, Va, June 17, 29; m 51; c 2. ORGANIC CHEMISTRY. *Educ:* Mass Inst Technol, BS, 51, PhD(org chem), 54. *Hon Degrees:* DSc, RPI, 87. *Prof Exp:* Res chemist, E I Du Pont de Nemours & Co, Inc, 54-59, res supvr, 59-70, assoc dir res, 70-74, dir res,

74-79, dir, 79-83, vpres, Cent Res & Develop Dept, 83-90, VPRES & SR SCI ADV, E I DU PONT DE NEMOURS & CO, INC, 90- *Concurrent Pos:* Sloan vis prof, Harvard Univ, 68; Kharasch vis prof, Univ Chicago, 78; adj prof, Univ Del, 70-; mem, Nat Sci Bd, 90. *Honors & Awards:* Chandler Medal, 91. *Mem:* Nat Acad Sci; fel AAAS; Am Chem Soc; Sigma Xi; Am Acad Arts & Sci; Nat Sci Bd. *Res:* Physical organic chemistry; reaction mechanisms; small ring chemistry; theoretical chemistry; chemistry of large molecules. *Mailing Add:* Cent Res & Develop Dept Exp Sta E I Du Pont de Nemours & Co Inc Wilmington DE 19898

SIMMONS, JAMES E, b Chicago, Ill, Sept 16, 25; m 48; c 2. EXPERIMENTAL NUCLEAR PHYSICS. *Educ:* Univ Calif, BS, 49, MA, 54, PhD(physics), 57; Univ Paris, dipl, 52. *Prof Exp:* PHYSICIST, LOS ALAMOS NAT LAB, UNIV CALIF, 57- *Concurrent Pos:* Vis scientist, Cen-Saclay Synchrotron, France, 80. *Mem:* Fel Am Phys Soc. *Res:* Low energy nuclear physics with neutron beams; medium energy experiments on spin dependence of nuclear forces; development of cryogenic instrumentation for nuclear physics; measurement of medium energy atomic beam properties by optical imaging methods. *Mailing Add:* Physics Div MS D456 Los Alamos Nat Lab Los Alamos NM 87545

SIMMONS, JAMES EDWIN, b Toledo, Ohio, July 13, 23; m 72; c 7. CHILD PSYCHIATRY. *Educ:* Toledo Univ, BS, 45; Ohio State Univ, MD, 47; Am Bd Psychiat & Neurol, dipl, 54, cert child psychiat, 60. *Prof Exp:* Intern, St Vincent's Hosp, Toledo, Ohio, 47-48; psychiat resident, Menninger Found, Kans, 48-51; from instr to Richter Prof psychiat, 53-88, coordr child psychiat serv, 62-74, actg chmn dept psychiat, 74-75, dir child psychiat serv, 75-88, RICHTER EMER PROF CHILD PSYCHIAT, SCH MED, IND UNIV, INDIANAPOLIS, 88- *Concurrent Pos:* Psychiatrist-dir, Child Guid Clin Marion County, Inc, Indianapolis, 53-57; fel child psychiat, Univ Louisville, 57-58; consult, La Rue D Carter Hosp, Indianapolis, 58; ed newslett, Am Asn Psychiat Serv Children, 70-74. *Mem:* Fel Am Acad Child Psychiat; fel Am Orthopsychiat Asn; fel Am Psychiat Asn; Am Acad Pediat; AMA. *Res:* Human neonatal behavior; follow-up of psychiatric hospitalization of children; teaching psychiatic examination of children to physicians. *Mailing Add:* 2950 Stauffer Row Indianapolis IN 46268

SIMMONS, JAMES QUIMBY, III, b Philadelphia, Pa, Apr 16, 25; m 54; c 4. PSYCHIATRY. *Educ:* Rutgers Univ, BS, 48; Bowman Gray Sch Med, Wake Forest Univ, MD, 52. *Prof Exp:* Intern, Walter Reed Army Med Ctr, 52-53; chief inpatient child psychiat, 62-68, assoc clin prof psychiat & assoc prog dir ment retardation, 68-72, PROF PSYCHIAT IN RESIDENCE & CHIEF MENT RETARDATION & CHILD PSYCHOL, NEUROPSYCHIAT INST, CTR FOR HEALTH SCI, UNIV CALIF, LOS ANGELES, 82- *Concurrent Pos:* Asst prof in residence psychiat, Neuropsychiat Inst, Ctr for Health Sci, Univ Calif, Los Angeles, 64-68; atten psychiatrist, Vet Admin Ctr, Los Angeles, 65-85. *Honors & Awards:* Distinguished Serv Medal: Army & Defense Superior Serv Medal, US Dept Defense. *Mem:* AAAS; fel Am Psychiat Asn; fel Am Acad Child & Adolescent Psychol; fel Am Col Psychol; Am Orthopsychiat Asn; Sigma Xi. *Res:* Behavior modification in schizophrenic and retarded children utilizing reinforcement principles; language in autistic and developmentally disabled children. *Mailing Add:* Dept Psych B 7 349 Npi Univ Calif 17201 Otsego St Encino CA 91316

SIMMONS, JAMES WOOD, b Chase City, Va, Sept 20, 16; m 41; c 3. MOLECULAR SPECTROSCOPY, SCIENCE EDUCATION. *Educ:* Hampden-Sydney Col, BS, 37; Va Polytech Inst, MS, 39; Duke Univ, PhD(physics), 48. *Prof Exp:* Instr physics, Va Polytech Inst, 39-41, asst prof, 46; from asst prof to assoc prof, Emory Univ, 48-59, chmn dept, 58-63, prof physics, 59-83; RETIRED. *Concurrent Pos:* Consult, Eng Exp Sta, Ga Inst Technol, 65-75. *Mem:* Am Phys Soc; Am Asn Physics Teachers. *Res:* Determination of molecular and nuclear properties by microwave spectroscopy; electronics instrumentation. *Mailing Add:* 2127 Spring Creek Rd Decatur GA 30033

SIMMONS, JEAN ELIZABETH MARGARET, b Cleveland, Ohio, Jan 20, 14; m 35; c 3. ORGANIC CHEMISTRY, BIOCHEMISTRY. *Educ:* Western Reserve Univ, BA, 33; Univ Chicago, PhD(org chem), 38. *Prof Exp:* From instr to prof chem & chmn dept, Barat Col, 38-58; prof chem, 59-84, chmn dept, 65-71 & 76-81, chmn div natural sci, 66-69, EMER PROF CHEM, UPSALA COL, 84-, ASST TO PRES, 69-73 & 78- *Concurrent Pos:* Instr, Univ Chicago, 38-41; coordr basic sci, Sch Nursing, Evangelical Hosp, 43-47; chmn comt study sci div of Upsala Col, Lutheran Church Am grant, 65-68; pres & trustee, Va Gildersleeve Int Fund Univ Women, Inc; delegate Int Fed Univ Women, Germany, 68, Japan, 74, Scotland, 77; pres, Grad Women in Sci, 70-71, Fedn Orgns Prof Women, 74-75; consult fund raising, Higher Educ, 72-; vis fel, Hist of Sci, Princeton Univ, 77; dir, officer & mem comt, NJ Educ Comput Ctr, 69-72; Int Fed deleg to UN conf, Austria, 79, Kenya, 81. *Honors & Awards:* Citation, Surgeon Gen US, 45; Award, Lindback Found, 64. *Mem:* Fel AAAS; Am Chem Soc; Am Asn Univ Women; Int Alliance Women; fel Sigma Xi; Int Fedn Univ Women. *Res:* Biuret reactions of polypeptides; respiratory pigments; protein chemistry; history of women in science. *Mailing Add:* 40 Balsam Lane Princeton NJ 08540

SIMMONS, JOE DENTON, b Elberton, Ga, Jan 14, 38; m 63; c 2. PHYSICAL CHEMISTRY, MOLECULAR SPECTROSCOPY. *Educ:* David Lipscomb Col, BA, 59; Vanderbilt Univ, PhD(chem), 63. *Prof Exp:* Nat Acad Sci-Nat Res Coun res fel spectros, 63-65, physicist, 65-78, sci adv, 79-80, DEP DIR, CTR BASIC STANDARDS, NAT BUR STANDARDS, 80- *Mem:* Am Phys Soc; AAAS; Optical Soc Am; Sigma Xi; Precision Measurement Asn; Am Soc Qual Control. *Res:* High resolution spectroscopy of polyatomic and diatomic molecules. *Mailing Add:* Ctr Basic Standards Bldg 221 Room B-160 Nat Bur Standards Washington DC 20234

SIMMONS, JOHN ARTHUR, b Santa Monica, Calif, Jan 25, 32; c 4. THEORETICAL MECHANICS, MATERIALS SCIENCE. *Educ:* Univ Calif, Berkeley, BA, 53, MA, 56, PhD(appl math), 62. *Prof Exp:* Res mathematician, Inst Eng Res, Univ Calif, Berkeley, 57-60, mathematician, Lawrence Radiation Lab, 60-61; fel, Miller Inst Basic Res Sci, 61-62; RES MATHEMATICIAN, NAT BUR STANDARDS, 62- *Concurrent Pos:* Com sci & te chnol fel, Staff of Congressman J W Symington, 72-73. *Mem:* Am Soch Mech Engrs; Am Phys Soc; Soc Indust & Appl Math; Am Inst Mech Eng; Am Soc Testing & Mat. *Res:* Dislocation theory; elastodynamics; plastic flow and fracture; acoustic emission and ultrasonics; residual and internal stresses; phase transformation kinetics. *Mailing Add:* Nat Inst Standards & Technol A167 Mat Bldg Gaithersburg MD 20899

SIMMONS, JOHN ROBERT, b Cokeville, Wyo, May 28, 28; m 52; c 4. BIOCHEMISTRY, GENETICS. *Educ:* Utah State Univ, BS, 55, MS, 56; Calif Inst Technol, PhD(biochem), 59. *Prof Exp:* USPHS fel biochem, Sch Med, Stanford Univ, 59-61; from asst prof to prof zool, 61-74, PROF BIOL, UTAH STATE UNIV, 74- *Res:* Proteins and nucleic acids in genetic function; genetic studies of plant tissues in vitro. *Mailing Add:* Dept Biol Utah State Univ Logan UT 84322-5305

SIMMONS, JOSEPH HABIB, b Marrakech, Morocco, Feb 19, 41; US citizen; m 62; c 2. SOLID STATE PHYSICS. *Educ:* Univ Md, College Park, BS, 62; John Carroll Univ, MS, 66; Cath Univ Am, PhD(physics), 69. *Prof Exp:* Res physicist, Lewis Res Ctr, NASA, 62-66; sr scientist solid state physics, Inorg Mat Div, Nat Bur Stand, 66-74; ADJ ASSOC PROF PHYSICS, CATH UNIV AM, 74- *Concurrent Pos:* Consult inorg mat div, Nat Bur Stand, 74- *Honors & Awards:* Superior Accomplishment Award, Nat Bur Stand, 71. *Mem:* Am Ceramic Soc; Am Phys Soc. *Res:* Phase transitions in glasses; thermodynamics and kinetics of liquid state; relaxation processes; optical fiber transmission lines. *Mailing Add:* Mat Sci & Eng Dept Univ Fla Gainesville FL 32611

SIMMONS, KENNETH ROGERS, reproductive physiology; deceased, see previous edition for last biography

SIMMONS, LEONARD MICAJAH, JR, b Hattiesburg, Miss, Nov 23, 37; m 59; c 3. THEORETICAL PHYSICS. *Educ:* Rice Univ, BA, 59; La State Univ, Baton Rouge, MS, 61; Cornell Univ, PhD(theoret phys), 65. *Prof Exp:* Res assoc physics, Univ Minn, 65-67 & Univ Wis-Madison, 67-69; asst prof, Univ Tex, Austin, 69-71; vis asst prof, Univ NH, 71-73; asst theoret div leader, Los Alamos Nat Lab, 74-76 & 83-86, assoc theoret div leader, 76-81 & 86-87, dep assoc dir physics & math, 81-83, STAFF MEM, LOS ALAMOS NAT LAB, 73- *Concurrent Pos:* Trustee, Aspen Ctr for Physics, 76-82, treas, 79-82 & pres, 85-88; vis prof physics, Wash Univ, 80-81; co-ed, Los Alamos Series Basic & Appl Sci, Univ Calif Press, 78-; NSF fel, 61; vpres, Santa Fe Inst, 86-88, exec vpres, 88-, trustee & sci bd, 90-; vis prof physics, Univ Ariz, 87-88; consult, Ariz Super Conducting Super Collider Project, 87, Ariz Ctr Study Complex Syst, 87. *Mem:* AAAS; Am Phys Soc; NY Acad Sci; Int Asn, Math & Physics. *Res:* Novel methods in field theory and for solution of nonlinear equations; coherent states; mathematical physics, properties of special functions; theory of elementary particles. *Mailing Add:* Santa Fe Inst 1120 Canyon Rd Santa Fe NM 87507

SIMMONS, NORMAN STANLEY, b New York, NY, May 28, 15; c 2. BIOLOGY, MEDICAL SCIENCES. *Educ:* City Univ New York, BS, 39; Harvard Univ, DMD, 39; Univ Rochester, PhD(exp path), 50. *Prof Exp:* EMER PROF HEALTH SCI, UNIV CALIF, LOS ANGELES, 50- *Concurrent Pos:* Fel & vis scientist, Harvard Univ; Career Res Award, NIH, 65-78. *Mem:* Biochem Soc; Biophys Soc; Inst Animal Dis Res. *Res:* Isolation and characterization of biological macromolecules; first isolation of pure and undegraded DNA; tobacco Mosai Virus first discovery of conformation dependent optically active electronic transitions in proteins and polypeptides; calcification of dental enamel, protein and apetile crystals. *Mailing Add:* Health Sci Univ Calif Los Angeles 886 Hilgard Ave Los Angeles CA 90024-3136

SIMMONS, PAUL C, b Jerome, Ariz, July 14, 32; m 52; c 3. METALLURGY, MATERIALS ENGINEERING. *Educ:* Univ Ariz, BS, 54, MS, 61, PhD(metall), 67. *Hon Degrees:* Prof Eng, Univ Ariz, 70. *Prof Exp:* Metallurgist, Guided Missile Mfg Div, Hughes Aircraft Co, 54-59, supvr metall eng, 59-63, process engr, 63-65, group head mat & processes, Missile Systs Div, 66-67 & Res & Develop Div, 67-68, sr tech staff asst mat & processes & prod effectiveness, 68-71, sect head prod anal & develop, 71-74, sr & chief scientist, asst labs mgr & prog mgr, Missile Develop Div, 74-91; CONSULT, 91- *Res:* High strength wire; precipitation hardening; fracture dynamics; short arc lamp technology; rocket motor and pressure vessel design; missile finish systems; long term storage of missiles; guidance unit technology; management principles; standardization; producibility. *Mailing Add:* 652 S Del Valle Tucson AZ 85711

SIMMONS, RALPH OLIVER, b Kensington, Kans, Feb 19, 28; m 52; c 4. PHYSICS. *Educ:* Univ Kans, BA, 50; Oxford Univ, BA, 53, MA, 57; Univ Ill, PhD(physics), 57. *Prof Exp:* Asst, Univ Ill, Urbana-Champaign, 54-55, res assoc, 57-59, from asst prof to assoc prof, 59-65, head physics dept, 70-86, PROF PHYSICS, UNIV ILL, URBANA-CHAMPAIGN, 65- *Concurrent Pos:* NSF sr fel, Ctr Study Nuclear Energy, Mol, Belg, 65; mem, Int Adv Bd, J of Physics C: Solid State Physics, 70-76; chmn, Div Solid State Physics, Am Phys Soc, 76-77, Off Phys Sci, Nat Res Coun, Nat Acad Sci, 78-81; mem assembly math & phys sci & Geophys Res Bd, Nat Res Coun, Nat Acad Sci, 78-81; consult, Argonne Nat Lab, 78-85; trustee, Argonne Univs Asn, 79-82; coun mem, Am Phys Soc, 88-; coun mem & chmn physics sect, AAAS, 85-86; vis scientist, L'Ecole Nonuale Supéneure, Paris, France, 85; chmn, adv comt Physics Today, Am Inst Physics, 88-89; Rhodes Scholar, 50-52. *Mem:* Fel AAAS; fel Am Phys Soc; Am Crystallog Asn; Am Asn Physics Teachers; Sigma Xi. *Res:* Lattice defects in solid helium, metals, semiconductors and ionic crystals; irradiation damage of solids; thermal, elastic and defect properties of noble gas crystals and other molecular solids; structure and dynamics of solids and fluids by neutron and x-ray scattering. *Mailing Add:* 431 Loomis Lab Physics Univ Ill at Urbana-Champaign Urbana IL 61801

SIMMONS, RICHARD LAWRENCE, b Boston, Mass, Feb 23, 34; m 58; c 2. SURGERY, IMMUNOLOGY. *Educ:* Harvard Univ, AB, 55; Boston Univ, MD, 59. *Prof Exp:* Asst surg, Columbia Univ, 63-64, instr, 64-68; from asst prof to prof surg & microbiol, Univ Minn, Minneapolis, 72-87; CHMN, DEPT SURG, UNIV PITTSBURGH, 87- *Concurrent Pos:* NIH fel, Columbia Univ, 60-61, Am Cancer Soc clin fel, 63-64; Markle Found scholar acad med, 69-75; consult, Vet Admin Hosp, Minneapolis, 71- *Honors & Awards:* Found Award, Am Asn Obstet & Gynec. *Mem:* Am Asn Immunol; Soc Univ Surg; Am Soc Nephrol; Transplantation Soc. *Res:* Transplantation biology; immunology of cancer. *Mailing Add:* Dept Surgery Univ Pittsburgh 497 Scaife Hall Pittsburgh PA 15261

SIMMONS, RICHARD PAUL, b Bridgeport, Conn, May 3, 31; m 59; c 2. MATERIALS SCIENCE ENGINEERING. *Educ:* Mass Inst Technol, BS, 53. *Hon Degrees:* DCom, Robert Morris Col, 87; LLD, Washington & Jefferson Col, 91. *Prof Exp:* Metallurgist, Allegheny Ludlum Steel Corp, 53-57; mgr processing, Titanium Metals Corp, 57-59; mgr qual control, Latrobe Steel Co, 59-62; asst gen mgr, Repub Steel Corp, 62-68; vpres mfg, Allegheny Ludlum Steel, 68-71, pres, 72-75, pres metals group, 76-80, pres & chief exec officer, 80-86, chmn & chief exec officer, 86-90, CHMN, ALLEGHENY LUDLUM CORP, 90- *Honors & Awards:* Ben Fairless Medal, Asn Inst Mech Engrs & Prof Engrs. *Mem:* Fel Am Soc Metals Int; Am Inst Mech Engrs & Prof Engrs. *Res:* Metallurgy; processing and quality control of steel. *Mailing Add:* Quaker Hollow Rd Sewickley PA 13154

SIMMONS, THOMAS CARL, b Williamsport, Pa, Nov 14, 20; m 52. ORGANIC CHEMISTRY. *Educ:* Pa State Univ, BS, 45, MS, 50, PhD(biochem), 52. *Prof Exp:* Asst biochem, Pa State Col, 45-52; org res chemist, 52-68, CHIEF, ORG CHEM SECT, US ARMY CHEM SYSTS LAB, 68- *Mem:* AAAS; Am Chem Soc; Sigma Xi; Am Defense Preparedness Asn; NY Acad Sci. *Res:* Physiological, fluorine, natural product and medicinal chemistry; chemistry of organo-phosphorus compounds and chemical warfare agents. *Mailing Add:* 2706 Bynum Hills Circle Bel Air MD 21014

SIMMONS, WILLIAM BRUCE, JR, b Bay City, Tex, Nov 5, 43; m 66; c 2. MINERALOGY, PETROLOGY. *Educ:* Duke Univ, BS, 66; Univ Ga, MS, 68; Univ Mich, Ann Arbor, PhD(mineral), 72. *Prof Exp:* Field geologist, Owens Ill Glass Co, 68; from instr to asst prof, 72-75, assoc prof, 75-79, PROF EARTH SCI, UNIV NEW ORLEANS, 79- *Mem:* AAAS; Mineral Soc Am; Geol Soc Am; Sigma Xi. *Res:* Mineralogy, petrology and geochemistry of pegmatite systems; mineralogy and petrology. *Mailing Add:* Dept Geol & Geophysics Univ New Orleans New Orleans LA 70148

SIMMONS, WILLIAM FREDERICK, b Philadelphia, Pa, Nov 14, 38. ENGINEERING, MECHANICS. *Educ:* Lehigh Univ, BS, 60; Pa State Univ, MS, 62; Johns Hopkins Univ, PhD(mech), 67. *Prof Exp:* Instr mech, Johns Hopkins Univ, 63-66, res asst, 66-67; res assoc, Mass Inst Technol, 67-69; asst scientist oceanog, Woods Hole Oceanog Inst, 69-71, assoc scientist, 71-74; mode exec officer, 74-77, polymode exec officer, Mass Inst Technol, 74-75, polymode exec scientist, 76-77; first global exp prog officer, World Meteorol Orgn, 77-80, consult, Global Atmospheric Res Prog Off, 81-91; CONSULT OCEANOG, 91- *Res:* Wave motions and nonlinear processes; thermo-mechanical coupling at the air-sea interface; theoretical and experimental studies of resonant interactions of internal waves; oceanic Mesoscale; deep dispersion, waste disposal assessment and models. *Mailing Add:* Box 412 Woods Hole MA 02543

SIMMONS, WILLIAM HOWARD, b Mansfield, Ohio, May 15, 47. PROTEIN-PEPTIDE CHEMISTRY. *Educ:* Wittenburg Univ, BA, 69; Bowling Green State Univ, MS, 73; Univ Ill, PhD(physiol), 79. *Prof Exp:* Teaching asst chem, Bowling Green State Univ, 71-75; res asst physiol chem, Ohio State Univ, 75-76; teaching asst physiol, Med Ctr, Univ Ill, 76-79, res assoc, 79-81; asst prof, 81-87, ASSOC PROF BIOCHEM, MED CTR, LOYOLA UNIV, CHICAGO, 87- *Concurrent Pos:* Lectr, Univ Ill Schs Dent & Pharm, 80-81; prin investr res grants, NIH, 81-87, Am Heart Asn, 87-90. *Mem:* Am Soc Biol Chemists; Soc Neurosci; Am Chem Soc; NY Acad Sci; AAAS; Sigma Xi; Protein Soc. *Res:* Mechanism of metabolism of vasoactive peptides in the lung; mechanism of metabolism of peptide neurotransmitters in the brain; purification of skin proteases; seasonal variation of pineal gland peptide. *Mailing Add:* Dept Molecular & Cellular Biochem Loyola Univ Chicago Stritch Sch Med 2160 S First Ave Maywood IL 60153

SIMMS, JOHN ALVIN, b Cleveland, Ohio, Apr 2, 31; m 55; c 2. ORGANIC POLYMER CHEMISTRY. *Educ:* NGa Col, BS, 51; Purdue Univ, MS, 53, PhD(org chem), 56. *Prof Exp:* SR RES FEL, FABRICS & FINISHES DEPT, E I DU PONT DE NEMOURS & CO, INC, 55- *Mem:* Am Chem Soc. *Res:* Synthesis of monomers a polymers; free radical polymerization; adhesives; elastoplastic film formers with superior photooxidative stability; isocyanate functional oligomers; group transfer polymerization. *Mailing Add:* Du Pont Exp Sta Bldg 328/109 Wilmington DE 19880-0328

SIMMS, NATHAN FRANK, JR, b Winston-Salem, NC, Oct 20, 32; m 59; c 3. PURE MATHEMATICS. *Educ:* NC Cent Univ, BS, 54, MS, 59; Lehigh Univ, PhD(math), 70. *Prof Exp:* Instr math, Fla A&M Univ, 59-60 & NC Cent Univ, 60-62; from asst prof to assoc prof math, Winston-Salem Univ, 64-72, prof, 72-; AT DEPT MATH & COMPUT SCI, NC STATE UNIV. *Concurrent Pos:* Consult, Regional Educ Lab Carolinas & Va, 70-72, Math Asn Am Minority Insts, 72-73 & Metric Educ Prog Winston-Salem/Forsyth Schs, 74-; NSF res grant, 73; dir, Div of Lib Arts & Sci. *Mem:* Math Asn Am; Sigma Xi; Am Math Soc. *Res:* Category theory and homological algebra; frobenius categories and spectral sequences. *Mailing Add:* 3924 Glen Oak Dr Winston-Salem NC 27105

SIMMS, PAUL C, b Jackson, Tenn, Nov 10, 32; m 59; c 2. NUCLEAR PHYSICS. *Educ:* NGa Col, BS, 53; Purdue Univ, PhD(physics), 58. *Prof Exp:* Res assoc physics, Columbia Univ, 59-60, asst prof, 60-64; assoc prof, 64-73, PROF PHYSICS, PURDUE UNIV, LAFAYETTE, 73- *Mem:* Am Phys Soc. *Res:* Nuclear structure. *Mailing Add:* Dept of Physics Purdue Univ Lafayette IN 47907

SIMNAD, MASSOUD T, b Teheran, Iran, Mar 11, 20; nat US; m 54; c 2. MATERIALS SCIENCE. *Educ:* Univ London, BS, 41; Cambridge Univ, PhD(phys chem), 45. *Prof Exp:* Res assoc, Imp Col, Univ London, 41-42 & Cambridge Univ, 45-48; Am Electrochem Soc Weston fel, Carnegie Inst Technol, 48, mem staff, Metals Res Lab, 49-56; head chem & metall II div, Gen Atomic Div, Gen Dynamics Corp, Calif, 56-60, asst chmn metall dept, 60-69; sr res adv, Gen Atomic Co, Inc, 69-73, sr tech adv, 73-81; CONSULT, 82- *Concurrent Pos:* Vis prof, Mass Inst Technol, 62-63; mem rev comt, Argonne Nat Lab, 74-80; vis lectr nuclear energy, Univ Calif, San Diego, 78-, adj prof, 82- *Honors & Awards:* Cert of Merit, Am Nuclear Soc, 65. *Mem:* Fel AAAS; Am Inst Aeronaut & Astronaut; fel Am Soc Metals; Mat Res Soc; fel Am Nuclear Soc; Electrochem Soc. *Res:* Nuclear reactor materials and fuels research and development; materials science and technology; energy conversion and utilization. *Mailing Add:* PO Box 8536 Rancho Santa Fe CA 92067

SIMON, ALBERT, b New York, NY, Dec 27, 24; m 72; c 3. PLASMA PHYSICS, CONTROLLED FUSION. *Educ:* City Col New York, BS, 47; Univ Rochester, PhD(physics), 50. *Prof Exp:* Physicist, Oak Ridge Nat Lab, 50-55, assoc dir neutron physics div, 55-61; head plasma physics div, Gen Atomic Div, Gen Dynamics Corp, 61-66; chmn dept mech & eng, 77-84, PROF MECH & ENG, UNIV ROCHESTER, 66-, PROF PHYSICS, 67- *Concurrent Pos:* Chmn, Div Plasma Physics, Physics Soc, 63-64; Guggenheim fel, 64-65; mem, Inst Advan Study, 74-75; sr vis fel, UK Sci Res Coun, Oxford Univ, 75; chmn, Nuclear Eng Div, Am Soc Eng Educ, 85-86. *Mem:* Fel Physics Soc; Am Soc Mech Engrs; AAAS; Am Soc Eng Educ. *Res:* Controlled thermonuclear reactor research. *Mailing Add:* Dept Mech Eng Univ Rochester Rochester NY 14627

SIMON, ALEXANDER, b New York, NY, Oct 13, 06; m 34; c 1. PSYCHIATRY. *Educ:* Columbia Univ, BA, 26, MD, 30; Am Bd Psychiat & Neurol, dipl, 38; Am Psychiat Asn, cert ment hosp adminr, 54. *Prof Exp:* Intern, St Joseph's Hosp, Paterson, NJ, 30-31; resident psychiat, St Elizabeth's Hosp, Washington, DC, 31-34; assoc neurol, Med Sch, George Washington Univ, 35-43; lectr, 43-45, from assoc prof to prof, 45-74, EMER PROF PSYCHIAT, SCH MED, UNIV CALIF, SAN FRANCISCO, 74- *Concurrent Pos:* Consult, Dept Army & Letterman Army Hosp, San Francisco, 46-, Vet Admin Hosp, Palo Alto, 49-58, Parks AFB, 53-56, Surgeon Gen, US Air Force, 53-59 & Travis AFB, 59-63; chmn dept psychiat, Univ Calif, San Francisco, 56-74; consult, Mayor's Comt, White House Conf on Aging, 59, Calif deleg, 61 & 71; consult, San Francisco Comt on Aging, USPHS, 60 & Gov Citizens Adv Comt on Aging; asst med supt, Langley Porter Neuropsychiat Inst, 43-56, med dir, 56-74; lectr, Sch Social Welfare, Univ Calif, Berkeley, 49-66; mem training study sect, Nat Inst Ment Health, 56-60 & ment health prog-proj comt, 61-65, chmn, 64-65, consult res utilization br, 63-65; mem Gov Interdept Comt Probs of Aging, 60-66; mem, Dept Ment Hyg Comt on Aging, 61-69, chmn, 64-69; mem adv comt housing for sr citizens, Housing & Home Finance Agency, Fed Housing Admin, 62-63; mem med adv bd, Nat Asn Prev Addiction to Narcotics, 64-69; trustee, Pac Inst Living, San Francisco, 64-70; mem med adv bd, Nat Aid to Visually Handicapped, 68-; comnr, Calif Comn on Aging, 76-79; psychiatrist to geriat serv unit, SE Community Ment Health, San Francisco, 74- *Honors & Awards:* Outstanding Civilian Serv Medal, US Army, 67; J Elliott Royer Award, 68; Outstanding Serv Award & Western Geront Soc Award, 77. *Mem:* AMA; fel Am Psychiat Asn; Am Psychopath Asn; Am Acad Neurol; Geront Soc. *Res:* Geriatric mental illness. *Mailing Add:* 1980 Vallejo St No 8 San Francisco CA 94123

SIMON, ALLAN LESTER, b Boston, Mass, Mar 18, 34; m 55; c 3. RADIOLOGY. *Educ:* Boston Univ, AB, 55; Tufts Univ, MD, 59; Yale Univ,73. *Prof Exp:* Intern med, Univ Md, 59-60; resident radiol, Beth Israel Hosp, Boston, 60-63; instr radiol, Yale-New Haven Med Ctr, 65-66; sr surgeon, NIH, 66-68; assoc prof radiol, Johns Hopkins Univ, 68-69; assoc prof, Univ Calif, San Diego, 69-72; prof diag radiol, Sch Med, Yale Univ, 72-75; CLIN PROF RADIOL, UNIV CALIF, SAN DIEGO, 75- *Concurrent Pos:* Nat Cancer Inst trainee radiol, 61-62; USPHS res fel cardiovasc radiol, Yale-New Haven Med Ctr, 63-65; consult, Nat Heart & Lung Inst & Clin Ctr, NIH, 68- *Mem:* Am Heart Asn; Am Col Radiol. *Res:* Diagnostic methods in cardiovascular disease; electronic processing of radiographic images. *Mailing Add:* Dept Radiol Alvarado Community Hosp San Diego CA 92120

SIMON, ANDREW L, b Kisujszallas, Hungary, Dec 1, 30; m 61; c 2. CIVIL ENGINEERING. *Educ:* Budapest Tech Univ, Dipl Ing, 54; Purdue Univ, PhD(fluid mech), 62. *Prof Exp:* Hydraul design engr, Eng Design Bur, State of Hungary, 54-56; stress analyst, Babcock & Wilcox Co, Ohio, 56-58; asst civil eng, Purdue Univ, 58-61; prof & head dept, WVa Inst Technol, 61-65; PROF CIVIL ENG & HEAD DEPT, UNIV AKRON, 65-, DIR, INST TECH ASSISTANCE, 77- *Concurrent Pos:* Lectr, Kanawha Valley Grad Ctr, WVa Univ, 63-64. *Mem:* Am Soc Civil Engrs; Am Soc Eng Educ. *Res:* Hydraulics; ground water seepage; two phase flows potential theory. *Mailing Add:* Dept of Civil Eng Univ of Akron Akron OH 44325

SIMON, ARTHUR BERNARD, b New York, NY, Jan 17, 27; m 51; c 2. MATHEMATICS. *Educ:* St Louis Univ, BS, 49; Univ Miami, MS, 54; Tulane Univ, PhD, 57. *Prof Exp:* Instr math, Yale Univ, 57-59; from asst prof to prof math, Northwestern Univ, 59-72; chmn dept, 72-80, PROF MATH, CALIF STATE UNIV, HAYWARD, 72- *Mem:* Am Math Soc. *Res:* Mathematical analysis; Banach algebras. *Mailing Add:* Dept Math Calif State Univ Hayward CA 94542

SIMON, BARRY MARTIN, b Brooklyn, NY, Apr 16, 46; m 71; c 5. MATHEMATICAL PHYSICS. *Educ:* Harvard Univ, BA, 66; Princeton Univ, PhD(physics), 70. *Prof Exp:* Instr math, Princeton Univ, 69-70, from asst prof to assoc prof math & physics, 70-76, prof, 76-81; prof math & theoret physics, 81-84, IBM PROF MATH & THEORET PHYSICS, CALIF INST TECHNOL, 84- *Concurrent Pos:* Sloan fel, Princeton Univ, 71-73; assoc ed, J Operator Theory, 78-, J Math Physics & J Statist Physics, 79-81 & Commun Math Physics & Duke Math J, 81-; Sherman B Fairchild distinguished vis scholar, Calif Inst Technol, 80-81. *Mem:* Am Math Soc; Am Phys Soc; Austrian Acad Sci. *Res:* Applications of rigorous mathematics to theoretical physics, especially to quantum physics; atomic and molecular physics; nonrelativistic quantum mechanics; quantum field theory; statistical mechanics. *Mailing Add:* Dept Math & Theoret Physics Mail Code 253-37 Calif Inst Technol Pasadena CA 91125

SIMON, CARL PAUL, b Chicago, Ill, Feb 7, 45; m 84; c 2. MATHEMATICAL ECONOMICS, DYNAMICAL SYSTEMS. *Educ:* Univ Chicago, BS, 66; Northwestern Univ, MS, 67, PhD(math), 70. *Prof Exp:* Instr math, Univ Calif, Berkeley, 70-72; asst prof math, 72-78, assoc prof math & econ, 78-88, PROF MATH & ECON, UNIV MICH, 88- *Concurrent Pos:* Vis asst prof math, Northwestern Univ, 75; assoc prof math & econ, Univ NC, 78-80. *Mem:* Am Math Soc; Soc Indust & Appl Math; Economet Soc; Sigma Xi. *Res:* Stability of dynamical systems; index of fixed points and singularities; mathematical economics; mathematical epidemiology. *Mailing Add:* Dept Math Univ Mich Ann Arbor MI 48109

SIMON, CHRISTINE MAE, b Memphis, Tenn, Oct 14, 49. EVOLUTIONARY BIOLOGY, SYSTEMATICS. *Educ:* Univ Fla, BS, 71, MS, 74; State Univ NY, Stony Brook, PhD(ecol & evolution), 79. *Prof Exp:* Res assoc evolutionary systs, Univ Chicago, 78-79; RES ASSOC, ZOOL DEPT, UNIV HAWAII, 80-, ASSOC ENTOMOLOGIST, BERNICE P BISHOP MUS, 81- *Concurrent Pos:* T Roosevelt Mem Fund grant, 76-77. *Mem:* Sigma Xi; Soc Study Evolution; Soc Syst Zool; Ecol Soc Am. *Res:* Geographic variation and its relationship to speciation; rates of evolution; interaction between ecology and genetics; macroevolution; phylogenetic systematics; history of evolutionary biology; periodical cicadas, Toxorhynchites mosquitos, Hawaiian Drosophila. *Mailing Add:* Dept Zool Univ Hawaii Edwardson Hall Honolulu HI 96822

SIMON, DAVID ZVI, b Montreal, Que, May 16, 29; m 48; c 4. MEDICINAL CHEMISTRY. *Educ:* Univ Montreal, BPh, 54, PhD(med chem), 65. *Prof Exp:* Nat Res Coun France fel, 66-67; asst prof, 67-71, ASSOC PROF MED CHEM, FAC PHARM, UNIV MONTREAL, 71- *Concurrent Pos:* Prof, Health Sci Centre & Res & Develop Inst, Ben-Gurion Univ, 74-75. *Res:* The role of the acetylenic bond in drug molecules. *Mailing Add:* Fac Pharm Univ Montreal Montreal PQ H3C 3J7 Can

SIMON, DOROTHY MARTIN, b Harwood, Mo, Sept 18, 19; m 46. PHYSICAL CHEMISTRY. *Educ:* Southwest Mo State Col, AB, 40; Univ Ill, PhD(chem), 45. *Hon Degrees:* DSc, Worcester Polytech Inst, 71; DE, Lehigh Univ, 78. *Prof Exp:* Asst chem, Univ Ill, 41-45; res chemist, E I du Pont de Nemours & Co, NY, 45-46; chemist, Clinton Lab, Tenn, 47; assoc chemist, Argonne Nat Lab, 48-49; aeronaut res scientist, Lewis Lab, Nat Adv Comt Aeronaut, 49-53, asst chief chem br, 54-55; Rockefeller fel, Cambridge Univ, 53-54; group leader combustion, Magnolia Petrol Co, Tex, 55-56; prin scientist & tech asst to pres res & advan develop div, 56-62, dir corp res, 62-64, vpres defense & indust prod group, 64-68, CORP VPRES & DIR RES, AVCO CORP, 68- *Concurrent Pos:* Fel, Univ Ill, 45; Marie Curie lectr, Pa State Univ, 62; dir, Econ Systs Corp, 66-72; mem comt sponsored res, Mass Inst Technol, 72-, Harvard Bd of Overseers Comt for Appl Res & NASA Space Systs & Technol Adv Comt, 78-; trustee, Worchester Polytech Inst, 73-; dir, Crown Zellerbach Corp & Conn Nat Bank, 78-; dir, Warner Lambert Co, 80 & The Charles Stark Draper Lab, 81; trustee, NEastern Univ, 80-; mem, Nat Mats Adv Bd, Nat Res Coun/Nat Acad Sci, 81; mem, President's Comt, Nat Medal Sci, 78-81; Chmn bd, Guggenheim Medal Award, 79-82. *Honors & Awards:* Rockefeller Pub Serv Award, 53; Achievement Award, Soc of Women Engrs, 56. *Mem:* AAAS; Am Chem Soc; fel Am Inst Aeronaut & Astronaut; Combustion Int; fel Am Inst Chemists. *Res:* Combustion; aerothermo chemistry; research management and strategic planning. *Mailing Add:* 222 Stagecoach Rd Chapel Hill NC 27514

SIMON, EDWARD, b Bradley Beach, NJ, Sept 24, 27; m 50; c 5. SOLID STATE SCIENCE. *Educ:* Rutgers Univ, BS, 50; Purdue Univ, MS, 52, PhD(physics), 55. *Prof Exp:* Res engr, Transitron, Inc, 54-58; vpres & founder, Solid State Prod, Inc, Unitrode Corp, 58-67, vpres, Unitrode Corp, 67-90; RETIRED. *Concurrent Pos:* Dir, Mass Microelectronics Ctr, 82-87, chmn bd, 87-89, emer chmn, 89- *Mem:* AAAS; Am Phys Soc; Electrochem Soc; Inst Elec & Electronics Engrs. *Res:* Semiconductor devices, design, research and development. *Mailing Add:* 72 Winding Pond Dr Londonderry NH 03053

SIMON, EDWARD HARVEY, b Elizabeth, NJ, June 25, 34; m 56; c 4. VIROLOGY. *Educ:* Rutgers Univ, BS, 56; Calif Inst Technol, PhD(biol), 60. *Prof Exp:* USPHS fel genetics, Carnegie Inst, 59-60; from asst prof to assoc prof, 60-70, PROF BIOL, PURDUE UNIV, LAFAYETTE, 70- *Concurrent Pos:* NSF sr res fel, Weizmann Inst, 66; vis prof, Hebrew Univ, Israel, 73 & Weizmann Inst, 79 & 85. *Mem:* AAAS; Am Soc Microbiol; Am Soc Virol; Am Asn Univ Professors; Int Soc Interferon Res. *Res:* Genetics of animal viruses; mode of action of interferon; isolation of interferon response mutants of cells and viruses. *Mailing Add:* Dept Biol Sci Purdue Univ Lafayette IN 47907

SIMON, ELLEN MCMURTRIE, b Norristown, Pa, Mar 29, 19; m 54; c 1. PROTOZOOLOGY, CRYOBIOLOGY. *Educ:* Ursinus Col, AB, 40; Univ Wis, MS, 52, PhD, 55. *Prof Exp:* Res assoc bact, Univ Wis, 55-58; res assoc zool, 63-64, asst prof, 64-76; res scientist genetics & develop, 77-86, ECOL, ETHNOL & EVOLUTION, UNIV ILL, URBANA, 87- *Concurrent Pos:* Vis asst prof microbiol, Univ Ill, 75 & 76, vis assoc prof, 77, 78, 79 & 85-86; mem exec bd, US Fedn for Cult Collections, 76-85; org lect, wet lab workshops, cyropreservation, 79, 80, 84, 88. *Mem:* Am Soc Microbiol; Soc Protozoologists. *Res:* Genetics; variants of Salmonella and Brucella; preservation of protozoa in liquid nitrogen; genetics and aging of ciliated protozoa. *Mailing Add:* Dept Ecol Ethnol & Evolution Univ Ill 515 Morrill Hall 505 S Goodwin Urbana IL 61801

SIMON, ERIC, b Egelsbach, Ger, Jan 8, 20; US citizen; m 51; c 3. CHEMICAL ENGINEERING, CHEMISTRY. *Educ:* City Col NY, BChE, 45. *Prof Exp:* Prof supvr, Harmon Color Works, 46-48; group leader res, Sun Chem Corp, 48-53; tech dir chem prod, Pigmentos y Oxidos, 54-74; OWNER, E S CONSULT ENG CO, 74- *Concurrent Pos:* Lectr, Cent Patronal de Nuevo Leon, 70-74. *Mem:* Asn Consult Chemists & Engrs; AAAS; Am Chem Soc; Inst Chem Eng Mex. *Res:* Development of appropriate technology for developing countries. *Mailing Add:* 6374 Creekbend Houston TX 77096-5661

SIMON, ERIC JACOB, b Wiesbaden, Ger, June 2, 24; US citizen; m 47; c 3. NEUROCHEMISTRY, PHARMACOLOGY. *Educ:* Case Inst Technol, BS, 44; Univ Chicago, MS, 47, PhD(org chem), 51. *Hon Degrees:* Dr, Univ Paris, 82. *Prof Exp:* Res assoc muscular dystrophy, Med Col, Cornell Univ, 53-59; from asst prof to prof exp med, 59-80, PROF PSYCHIAT & PHARMACOL, MED SCH, NY UNIV, 80- *Concurrent Pos:* Nat Found Infantile Paralysis fel biochem, Col Physicians & Surgeons, Columbia Univ, 51-53; lectr chem, City Col of New York, 53-59; chmn, Sect Biochem, NY Acad Sci, 84-86, gov-at-large, 86-89; mem, biomed review comt, Nat Inst Drug Abuse, 76-80, chairperson, 79-80; chmn, Int Narcotic Res Conf, 80-84; mem, Nat Adv Coun Drug Abuse, 88- *Honors & Awards:* Res Pacesetter Award, Nat Inst on Drug Abuse, 79; Louis & Bert Freedman Found Award, NY Acad Sci, 80; Nathan B Eddy Mem Prize, Comt on Probs Drug Dependence, 83. *Mem:* Fel AAAS; Am Chem Soc; Am Soc Pharmacol & Exp Therapeut; Am Soc Neurochem; Am Soc Biol Chem. *Res:* Study of opiate receptors and endogenous opioid peptides; metabolism of vitamin E; neurochemical studies in nerve cells in culture; receptor isolation and characterization; endogenous opioids. *Mailing Add:* Dept Psychiat 550 First Ave NY Univ Med Ctr New York NY 10016

SIMON, FREDERICK OTTO, b New York, NY, Dec 11, 39. GEOCHEMISTRY, ANALYTICAL CHEMISTRY. *Educ:* Am Univ, BS, 61, MS, 63; Univ Md, PhD(chem), 72. *Prof Exp:* Chemist, 61-73, proj leader, 73-75, supvr chemist anal chem & geochem, 75-79, RES CHEMIST, US GEOL SURV, 79- *Concurrent Pos:* Ed, Geochem Int J, Am Geol Inst, 72-76. *Mem:* Am Chem Soc; Geochem Soc; Geol Soc Am. *Res:* Inorganic geochemistry of coal; coal quality; application of analytical, inorganic and physical chemistry to problems in earth science; analysis of rarer elements in geologic materials. *Mailing Add:* 11813 Stuart Mill Rd Oakton VA 22124-1227

SIMON, FREDERICK TYLER, b Pittsburgh, Pa, May 9, 17; m 46; c 2. PHYSICAL CHEMISTRY, COLOR SCIENCE. *Educ:* Morris Harvey Col, BS, 55; Marshall Univ, MS, 57. *Prof Exp:* Asst colorist, Am Cyanamid Co, 38-40; head res lab, Philadelphia Qm Corps, US Army, 40-44; head spectros lab, Sidney Blumenthal & Co, 44-48; head qual control, Peerless Woolen Mills, 49-53; dir textiles, Good Housekeeping Inst, 53-54; res specialist, Union Carbide Corp, 54-68; J E SIRRINE PROF TEXTILE SCI, CLEMSON UNIV, 68- *Concurrent Pos:* Consult, Burlington Industs, Inc, 70-, Cherokee Finishing Corp, 70-, Sandoz Wander Inc, 71- & Diano Corp, 74-; deleg, Comt Colorimetry, Int Comn Illum, 71; chairholder, Color Mkt Group; mem, The Colour Group, Gt Brit. *Mem:* Am Chem Soc; Am Asn Textile Chem & Colorists; Intersoc Color Coun; Optical Soc Am; Sigma Xi. *Res:* Color science, including dyeing and coloration chemistry of dyes and pigments; polymer morphology, including microscopy, x-ray diffraction, crystallization kinetics, neutron scattering and general textile processing and fiber manufacture; computer color matching and industrial color toleracnces as well as fluorescent colorants. *Mailing Add:* PO Box 391 Clemson SC 29631

SIMON, GARY ALBERT, b Wilkes-Barre, Pa, Apr 24, 45. STATISTICS, APPLIED STATISTICS. *Educ:* Carnegie-Mellon Univ, BS, 66; Stanford Univ, PhD(statist), 72. *Prof Exp:* Asst prof statist, Princeton Univ, 71-75; ASSOC PROF, DEPT APPL MATH & STATIST, STATE UNIV NY STONY BROOK, 75- *Mem:* Am Statist Asn. *Res:* Analysis of categorical data and nonparametric statistics. *Mailing Add:* Dept Statist & Oper Res NY Univ New York NY 10003

SIMON, GEORGE WARREN, b Frankfurt, Ger, Apr 22, 34; US citizen; m 58; c 3. SOLAR PHYSICS, MAGNETOCONVECTION. *Educ:* Grinnell Col, AB, 55; Calif Inst Technol, MS, 61, PhD(physics), 63; Univ Utah, MBA, 76. *Hon Degrees:* DSc, Grinnell Col, 83. *Prof Exp:* Jr scientist, Atomic Power Div, Westinghouse Elec Corp, 55-56, assoc scientist, 58; mem tech staff radiation effects, Hughes Res Labs, 58-61 & Space Technol Labs, 61-63; dir comput lab, Nat Solar Observ, 69-76, dep dir observ, 75-76; chief solar res br, Air Force Geophys Lab, 76-83; space shuttle astronaut, 78-88; RES ASTROPHYSICIST, NAT SOLAR OBSERV, 63-; SR SCI, AIR FORCE PHILLIPS LAB, 83- *Concurrent Pos:* Physicist, Max Planck Inst Physics & Astrophys, 65-66 & 77; consult, NASA, 67-; vis assoc, Harvard Col Observ, 69-70; vis sci, Univ Cambridge, UK, 76, 82, 86 & 88; adj prof, Univ Calif Los Angeles, Univ Colo, Univ Ariz, 70- *Mem:* Am Astron Soc; Am Phys Soc; Int Astron Union; Am Geophys Union; fel Royal Astron Soc. *Res:* Magnetic and velocity fields and other inhomogeneities in the solar atmosphere; observations and kinemakical modeling of solar magnetoconvection; nuclear reactor physics; radiation effects in semiconductors; space physics. *Mailing Add:* 2308 Rancho Lane Alamogordo NM 88310

SIMON, GEZA, b Budapest, Hungary, Mar 6, 41; US citizen; m 65; c 2. CARDIOVASCULAR DISEASES, INTERNAL MEDICINE. *Educ:* NY Univ, BA, 64; State Univ NY Downstate Med Ctr, MD, 68; Mich State Univ, PhD(physiol), 74; Am Bd Internal Med, dipl, 73. *Prof Exp:* From intern to resident med, NY Univ-Bellevue Hosp, 68-70; med officer, US Army, 70-72; fel physiol & clin instr med, Mich State Univ, 72-74; NIH res fel med, Univ Mich Med Ctr, 74-75, res scientist, 75-76; asst prof, 76-82, ASSOC PROF MED, UNIV MINN, MINNEAPOLIS, 82- *Concurrent Pos:* Fel, Am Heart Asn. *Mem:* Soc Exp Biol Med; Am Fedn Clin Res; Int Soc Hypertension. *Res:* Human and experimental hypertension with emphasis on unidentified vasoactive humoral agents in volume-expanded hypertension. *Mailing Add:* Vet Admin Med Ctr 111J2 One Veterans Dr Minneapolis MN 55417

SIMON, HAROLD J, b Karlsruhe, Ger, Jan 22, 28; US citizen; m 49; c 3. MEDICINE. *Educ:* Harvard Med Sch, MD, 53; Rockefeller Inst, PhD(microbiol), 59. *Prof Exp:* Intern med, NY Hosp-Cornell Med Ctr, 53-54, asst resident, 54-56; asst prof med, Sch Med, Stanford Univ, 59-66; asst dean, Sch Med, 66-69, assoc dean educ & student affairs, 69-78, PROF MED & COMMUNITY MED, SCH MED, UNIV CALIF, SAN DIEGO, 66- *Concurrent Pos:* Arthritis Found fel, 59-61; USPHS grant, 59-78 & career develop award, 62-65; asst, Med Col, Cornell Univ, 54-56; asst physician, Hosp, Rockefeller Inst, 56-59; physician outpatients, NY Hosp-Cornell Med Ctr, 56-59; mem study sect commun dis, USPHS, 66-71; panelist & chmn sect III, Pac Intersci Cong, Japan, 66; convener, Panel Coccal Infections, Int Cong Chemother, Austria, 67; mem comt manned probs space flight, Nat Acad Sci, 68-; vis sr scholar, Div Int Health, Inst Med, Nat Acad Sci, 78-79. *Mem:* AAAS; Infectious Dis Soc Am; Am Col Physicians; Am Soc Clin Pharmacol & Therapeut; Am Fedn Clin Res; Sigma Xi. *Res:* Studies in host-parasite interactions; clinical pharmacology of antimicrobial agents; prevention of hospital-acquired infections; international medicine and epidemiology; medical education; student affairs. *Mailing Add:* Dept Comm Med Univ Calif San Diego La Jolla CA 92093

SIMON, HENRY JOHN, b London, Eng, May 4, 39; US citizen; m 60; c 4. SOLID STATE PHYSICS. *Educ:* Tufts Univ, BSEE, 60; Harvard Univ, MA, 65, PhD(appl physics), 69. *Prof Exp:* Sr res scientist, United Aircraft Res Lab, East Hartford, 69-70; asst prof physics, Worcester Polytech Inst, 70-72; asst prof, 72-81, PROF PHYSICS, UNIV TOLEDO, 81- *Concurrent Pos:* Vis prof physics, Univ Pa, 78-79; vis prof elec eng, Univ Md, 86-87. *Mem:* Am Phys Soc. *Res:* Interaction of intense light with matter; nonlinear optics; harmonic generation of light; surface plasmons. *Mailing Add:* Dept Physics & Astron Univ Toledo 2801 W Bancroft Toledo OH 43606

SIMON, HERBERT A, Milwaukee, Wis, June 15, 16; m 37; c 3. SIMULATION OF HUMAN THINKING. *Educ:* Univ Chicago, BA, 36, PhD(polit sci), 43. *Hon Degrees:* Honorary degrees from many American & European Univ. *Prof Exp:* Res asst, Int City Mgr Asn, 36-39; res dir pub admin, Univ Calif, Berkeley, 39-42; from asst prof to prof polit sci, Ill Inst Technol, 42-49; prof admin, Carnegie Inst Technol, 49-67; UNIV PROF COMPUT SCI & PSYCHOL, CARNEGIE-MELLON UNIV, 67- *Concurrent Pos:* Consult, Cowles Comns,46-60 & The Rand Corp, 52-70; chmn, Social Sci Res Coun, 61-65 & Nat Res Coun, 68-70; mem coun, Nat Acad Sci, 78-81 & 83-86; vis lectr, various univ, 59- *Honors & Awards:* Nobel Prize Econ, 78; Turing Award, Asn Comput Mach, 75; John von Neuman Theory Award, Inst Mgt Sci, 88. *Mem:* Asn Comput Mach; Am Psychol Asn; Am Soc Artificial Intel. *Res:* Computer simulation of human cognitive processes including problem solving, scientific discovery, learning, imagery; representation and testing of simulation models. *Mailing Add:* Dept Psychol Carnegie Mellon Univ Pittsburgh PA 15213

SIMON, HORST D, b Stadtsteinach, WGer, Aug 8, 53; US citizen; m; c 2. ALGORITHMS. *Educ:* Tech Univ, Berlin, WGer, dipl math, 78; Univ Calif, Berkeley, PhD(math), 82. *Prof Exp:* Asst prof, Dept Appl Math, State Univ NY, Stony Brook, 82-83; coordr & lectr, Boeing Computer Serv, 83-85, proj mgr, 85-86, mgr, Computational Math Group, 86-87, tech mkt & collabr, 87-89; MGR RES DEPT, NAT AERODYN SIMULATOR SYSTS DIV, AMES RES CTR, NASA, MOFFETT FIELD, CALIF, 89- *Concurrent Pos:* Managing ed, Int J High Speed Comput; ed, Proc Conf on Sci Applications of Connection Mach & J Matrix Analysis & Applications; proposal reviewer, NSF, Air Force Off Sci Res & Dept Energy; mem group & bd dirs, Spec Interest Group Numerical Math, Asn Comput Mach; mem, Activ Groups Linear Algebra & Supercomput, Soc Indust & Appl Math. *Honors & Awards:* Gordon Bell Award, 88. *Mem:* Soc Indust & Appl Math; Asn Comput Mach; Inst Elec & Electronic Engrs Computer Soc; Asn Appl Math & Mech. *Res:* Development of high performance algorithms for the wide variety of vector and parallel machines at National Aerodynamics Simulator. *Mailing Add:* Ames Res Ctr NASA MS T-045-1 Moffett Field CA 94305

SIMON, JACK AARON, b Champaign, Ill, June 17, 19. COAL GEOLOGY. *Educ:* Univ Ill, BA, 41, MS, 46. *Hon Degrees:* DSc, Northwestern Univ, 81. *Prof Exp:* Tech & res asst, Ill State Geol Surv, 37-42, from asst geologist to geologist, 45-67, prin geologist, 67-73, chief, 73-81, prin scientist, 81-83; RETIRED. *Concurrent Pos:* From assoc prof to prof metall & mining, Univ Ill, 67-77, 80-85, adj prof geol, 79-86. *Honors & Awards:* Gilbert H Cady Award, Geol Soc Am, 75; Percy W Nicholls Award, Am Inst Mining, Metall & Petrol Engrs. *Mem:* AAAS; Geol Soc Am; Soc Econ Geologists; Am Asn Petrol Geol; Am Inst Mining, Metall & Petrol Engrs; Am Inst Prof Geologists (vpres, 73). *Res:* Coal resources; coal mining geology; Pennsylvanian stratigraphy. *Mailing Add:* 502 W Oregon St Urbana IL 61801-4044

SIMON, JEROME BARNET, b Regina, Sask, Aug 21, 39; div; c 1. GASTROENTEROLOGY. *Educ:* Queens Univ, Can, MD, 62; Royal Col Physicians & Surgeons, FRCP(C), 67; Am Bd Internal Med, dipl, 69; Am Col Gastroenterol, FACG, 83; Am Col Physicians, FACP, 87. *Prof Exp:* House officer med, Montreal Gen Hosp, McGill Univ, 62-65, training resident gastroenterol, 65-66; fel, Liver Study Unit, Yale Univ, 66-69; from lectr to assoc prof, 69-88, PROF GASTROENTEROL, DEPT MED, QUEENS UNIV, 88- *Concurrent Pos:* Attend staff, Kingston Gen Hosp, Ont, 69-, head, Div Gastroenterol, 75-89; consult staff, Hotel Dieu & St Marys Hosp, 69-; mem, med adv bd, Can Liver Found, 72-75, vchmn & mem, bd dirs, 88-90; vis prof, Univ Alta, 72, Univ Toronto, 73, 77, 79 & 81, Dalhousie Univ, 75, Univ Conn & Mem Univ, 76, Univ Calgary, 77, McMaster Univ, 78, Wayne State Univ, 85, Univ Alberta, 89; mem, Prog Grants Comt, Med Res Coun Can, 75-76, Clin Invest Comt, 77-79; res comts, Can Asn Gastroenterol, 76-81, educ comt, 85-; admis comt, Am Asn Study Liver Dis, 78-80; fel & bursaries comt, Can Cancer Soc, Ont Div, 79-86, chmn, 82-84; postgrad course comt, Am Col Gastroenterol, 83-84; examr, Int Med, Royal Col Physicians & Surgeons, 85; counr, Can Soc Clin Invest, 86-, secy-treas, 90-; gov bd, Can Asn Study Liver, 85-, vpres, 87-88 & pres, 88-89. *Mem:* Am Fedn Clin Res; Am Asn Study Liver Dis; Am Gastroenterol Asn; Am Col Physicians; Can Med Asn; Am Col Gastroenterol. *Res:* Lipid metabolism in liver disease; clinical research gastroenterology and liver diseases. *Mailing Add:* Dept Med Queens Univ Kingston ON K7L 3J7 Can

SIMON, JIMMY L, b San Francisco, Calif, Dec 27, 30; m 53; c 2. MEDICINE. *Educ:* Univ Calif, Berkeley, AB, 52; Univ Calif, San Francisco, MD, 55. *Prof Exp:* Intern, Hosp, Univ Calif, 55-56; asst resident pediat, Grace-New Haven Hosp, 56-57; sr resident, Children's Hosp, Boston, Mass, 57-58; from instr to asst prof, Sch Med, Univ Okla, 60-64; dir, Kern County Gen Hosp, Bakersfield, Calif, 65-66; from assoc prof to prof pediat, Univ Tex Med Br Galveston & dep chmn dept, 66-74; PROF PEDIAT & CHMN DEPT, BOWMAN GRAY SCH MED, 74- *Mem:* Am Pediat Soc; Am Acad Pediat; AAS Pediat Asn; Am Bd Pediat. *Res:* Clinical pediatrics; ambulatory medical care. *Mailing Add:* Dept of Pediat Bowman Gray Sch of Med Winston-Salem NC 27103

SIMON, JOHN DOUGLAS, b Cincinnati, Ohio, Feb, 11, 57. PHYSICAL CHEMISTRY, CHEMICAL DYNAMICS. *Educ:* Williams Col, BA, 79; Harvard Univ, MA, 81, PhD(chem), 83. *Prof Exp:* Res assoc, Los Angeles, 83-85, asst prof chem, 85-88, ASSOC PROF CHEM, UNIV CALIF, SAN DIEGO, 88- *Mem:* Am Chem Soc; Am Phys Soc; AAAS. *Res:* Application of picosecond and subpicosecond spectroscopic techniques to chemical problems. *Mailing Add:* Dept Chem B-041 Univ Calif San Diego La Jolla CA 92093

SIMON, JOSEPH, b Pittsfield, Mass, Sept 17, 18; m 54; c 1. MICROBIOLOGY. *Educ:* Cornell Univ, BS, 40; Agr & Mech Col, Tex, MS, 42; Kans State Col, DVM, 46; Univ Wis, PhD(vet path, vet bact), 51. *Prof Exp:* From instr to asst prof vet sci, Univ Wis, 48-58; res assoc prof cancer res, Univ Fla, 58-60; prof hyg & vet res, 60-79, PROF PATH, UNIV ILL, URBANA, 60- *Mem:* AAAS; Am Vet Med Asn; Int Acad Path; Fedn Biol Soc; Am Col Vet Path; Sigma Xi. *Res:* Pathology and microbiology of bovine mastitis; reproductive diseases; carcinogenesis; enteric and pulmonary diseases of swine. *Mailing Add:* 1801 S Anderson Urbana IL 61801

SIMON, JOSEPH LESLIE, b Everett, Mass, Mar 16, 37. ZOOLOGY. *Educ:* Tufts Univ, BS, 58; Univ NH, MS, 60, PhD(zool), 63. *Prof Exp:* From instr to assoc prof, 63-78, PROF ZOOL, UNIV S FLA, 78 - *Concurrent Pos:* Fel, Systs Ecol Prog, Marine Biol Lab, Woods Hole, Mass, 65-67. *Mem:* AAAS; Am Soc Zool; Ecol Soc Am; Am Soc Limnol & Oceanog. *Res:* Systematics of and reproduction and development of Polychaetous Annelids; ecology of marine benthic communities. *Mailing Add:* Dept Biol Univ SFla 4202 Fowler Ave Tampa FL 33620

SIMON, JOSEPH MATTHEW, b Reading, Pa, July 14, 41; m 67. ANALYTICAL CHEMISTRY. *Educ:* Albright Col, BS, 63; Univ Pittsburgh, PhD(chem), 69. *Prof Exp:* Asst prof, 69-74, assoc prof, 74-81, PROF CHEM, POINT PARK COL, 81-, CHMN DEPT NATURAL SCI & TECHNOL, 74-, PRES, 86- *Mem:* AAAS; Am Chem Soc. *Res:* Medium effects and nonaqueous solutions; electrochemistry; polarography. *Mailing Add:* Point Park Col Blvd of Allies & Wood St Pittsburgh PA 15222

SIMON, LEE WILL, b Evanston, Ill, Feb 18, 40; m 66; c 3. ASTRONOMY. *Educ:* Northwestern Univ, Evanston, BA, 62, MS, 64, PhD(astron), 72. *Prof Exp:* Staff astronomer & prog supvr, Adler Planetarium, 69-76; planetarium dir, Morrison Planetarium, Calif Acad Sci, Golden Gate Park, San Francisco, 77-84. *Mem:* Am Astron Soc; Sigma Xi. *Res:* Stellar spectroscopy. *Mailing Add:* 245 San Marin Dr Novato CA 94945-1220

SIMON, MARK ROBERT, b New York, NY, July 26, 41; m 78. ANATOMY, PHYSICAL ANTHROPOLOGY. *Educ:* Hunter Col, City Univ NY, BA, 66, PhD(anthrop), 74. *Prof Exp:* Asst prof, 77-83, ASSOC PROF VET ANAT, UNIV ILL SCH VET MED, 83- *Mem:* Am Asn Vet Anatomists; Sigma Xi. *Res:* Experimental morphology; gross anatomy; craniofacial morphogenesis; mechanical and hormonal factors in cartilage growth. *Mailing Add:* 2001 S Lincoln Ave Univ of Ill Urbana IL 61801

SIMON, MARTHA NICHOLS, b New York, NY, Dec 31, 40; m 81; c 1. BIOCHEMISTRY, MOLECULAR BIOLOGY. *Educ:* Radcliffe, Col, AB, 62; Cornell Univ, PhD(chem), 68. *Prof Exp:* Res fel chem, Calif Inst Technol, 67-69; res assoc biochem, State Univ NY, Stony Brook, 69-72; sr res assoc biol, 72-75, asst biologist, 75-77, assoc Biologist, 77-84, BIOLOGY ASSOC , BROOKHEAVEN NAT LAB,85- *Concurrent Pos:* Spec fel, NIH, 71-72, co-investr, 75-77, investr, 77- *Mem:* AAAS; Biophys Soc. *Res:* Structure and funtion of nucleis acids; molecular details involved in DNA metabolism. *Mailing Add:* 145 Main St Setauket NY 11733

SIMON, MARVIN KENNETH, b New York, NY, Sept 10, 39; m 66; c 2. ELECTRICAL ENGINEERING. *Educ:* City Col New York, BEE, 60; Princeton Univ, MSEE, 61; NY Univ, PhD(elec eng), 66. *Prof Exp:* Mem tech staff commun, Bell Tel Labs, 61-63; instr elec eng, NY Univ, 63-66; mem tech staff commun, Bell Tel Labs, 66-68; mem tech staff commun, 68-80, SR RES ENGR, JET PROPULSION LAB, 80- *Concurrent Pos:* Instr elec eng, West Coast Univ, 69; consult, LinCom Corp, 73-76 & Axiomatix Corp, 76-; ed Communications, Inst Elec & Electronics Engrs, 74-76, chmn commun theory workshop, 75, chmn commun theory comt, 77-, fel grade, 78; lectr elec eng, Calif Inst Technol, 78, vis prof, 79. *Honors & Awards:* Tech Paper Award, NASA, 74. *Mem:* Sigma Xi; Fel Inst Elec & Electronics Engrs. *Res:* Digital communications as applied to space and satellite communication systems; modulation theory; synchronization techniques. *Mailing Add:* Jet Propulsion Lab 4800 Oak Grove Dr Mail Stop 161-228 Pasadena CA 91109

SIMON, MELVIN, b New York, NY, Feb 8, 37; m; c 3. BIOLOGY. *Educ:* City Col NY, BS, 59; Brandeis Univ, PhD(biochem), 63. *Prof Exp:* Postdoctoral fel, Princeton Univ, NJ, 64-65; from asst prof to prof, Dept Biol, Univ Calif, San Diego, La Jolla, 65-82; PROF, DIV BIOL, CALIF INST TECHNOL, PASADENA, 82- *Concurrent Pos:* Postdoctoral fel, USPHS, 63-64; John Simon Guggenheim mem fel, 78; Anne P & Benjamin F Biaggini prof biol sci, 87. *Honors & Awards:* Carter-Wallace Lectr, Princeton Univ, 88; Selman A Waksman Award Microbiol, Nat Acad Sci, 91. *Mem:* Nat Acad Sci; Am Acad Arts & Sci. *Res:* Author of various publications. *Mailing Add:* Div Biol Calif Inst Technol Pasadena CA 91125

SIMON, MICHAEL RICHARD, b New York, Oct 12, 43; m 70; c 2. IMMUNOLOGY, INTERNAL MEDICINE. *Educ:* State Univ NY, Binghamton, BA, 65; NY Univ, MD, 69; Stanford Univ, MA, 73; Am Bd Internal Med, dipl, 75; Am Bd Allergy & Immunol, dipl, 77, Am Bd Med Lab Immunol, dipl, 85, Am Bd Qual Assurance & Utilization, dipl, 90. *Prof Exp:* Intern internal med, Kings County Hosp-Downstate Med Ctr, Brooklyn, 69-70; gen med officer, Indian Hosp, USPHS, San Carlos, Ariz, 70-72; resident internal med, Affil Intern-Resident Phys Prog, Wayne State Univ, 73-75; fel allergy & immunol, Med Ctr, Univ Mich, 75-77; ASST PROF INTERNAL MED, SCH MED, WAYNE STATE UNIV, 77-; CO-CHIEF ALLERGY-IMMUNOL SECT, VET ADMIN MED CTR, ALLEN PARK, 77- *Concurrent Pos:* Allergy Found Am fel clin allergy & immunol res, 76-77; vis asst prof internal med, Div Clin Pharmacol, Stanford Univ Med Ctr, Stanford, Calif, 86. *Honors & Awards:* Cert of Spec Competance in Clin Immunol & Allergy, Royal Col Physicians & Surgeons Can, 85. *Mem:* Fel Am Col Physicians; Am Acad Allergy; Am Fedn Clin Res; Soc Exp Biol Med; fel Royal Col Physicians & Surgeons Can. *Res:* Immunology of infectious and granulomatous diseases; immunopharmacology. *Mailing Add:* Allergy & Immunol Sect 111F Vet Admin Hosp Allen Park MI 48101

SIMON, MICHAL, b Prague, Czech, Sept 28, 40; US citizen; m; c 1. ASTRONOMY. *Educ:* Harvard Univ, AB, 62; Cornell Univ, PhD(astrophys), 67. *Prof Exp:* Res fel astron, Calif Inst Technol, 67-69; from asst prof to assoc prof, 69-74, chmn dept earth & space sci, 80-83, PROF ASTRON, STATE UNIV NY STONY BROOK, 74- *Mem:* AAAS; Am Astron Soc; Int Astron Union. *Res:* Infrared astronomy; star formation. *Mailing Add:* Dept of Earth & Space Sci State Univ of New York Stony Brook NY 11794

SIMON, MYRON SYDNEY, b Burlington, Vt, Sept 23, 26; m 50; c 3. ORGANIC CHEMISTRY. *Educ:* Harvard Univ, AB, 46, AM, 48, PhD(chem), 49. *Prof Exp:* Res chemist, Polaroid Corp, 49-55, group leader, 55-59, from asst mgr to mgr, Org Chem Res Dept, 59-73, asst dir, Org Chem Res Div, 74-80, assoc dir, 81-88, res fel, 85-88; PRES, IMAGE-INATION ASSOCS, 89- *Mem:* Am Chem Soc. *Res:* Photographic materials; monomers; organic synthesis; diffusion transfer color photographic research; chemistry of dyes. *Mailing Add:* 20 Somerset Rd West Newton MA 02165-2722

SIMON, NANCY JANE, b Pittsburgh, Pa, June 1, 39. LOW TEMPERATURE PHYSICS. *Educ:* Univ Chicago, BS, 60; Radcliffe Col, AM, 61; Harvard Univ, PhD(physics), 68; Univ Colo, BA, 81. *Prof Exp:* PHYSICIST, INST MAT SCI & ENG, NAT BUR STANDARDS, 68- *Mem:* Sigma Xi. *Res:* Mossbauer effect; Kapitza coefficient; low temperature physics and engineering; cryogenic information retrieval; properties of structural materials at cryogenic temperatures; superconducting magnets for fusion energy systems. *Mailing Add:* Fracture & Deformation Div 430 Nat Bur Standards 325 Broadway Boulder CO 80303

SIMON, NORMAN M, b Chicago, Ill, Mar 30, 29; m 57; c 3. NEPHROLOGY, HYPERTENSION. *Educ:* Harvard Univ, BA, 50; Yale Univ, MS, 51; Northwestern Univ, MD, 55; Am Bd Internal Med, dipl, 63; Sub-Comt Nephrology, dipl, 76. *Prof Exp:* From instr to assoc prof med, 63-82, PROF CLIN MED, NORTHWESTERN UNIV, 82- *Concurrent Pos:* Nat Heart Inst res fel, Michael Reese Hosp, 56-57; fel med, Med Sch, Northwestern Univ, 61-63; assoc chief med, Passavant Mem Hosp, Chicago, 72-73; sr attending physician, Evanston Hosp, 79-, head, Div Nephrology, 88- *Mem:* AAAS; Int Soc Nephrology; Cent Soc Clin Res; fel Am Col Physicians; Am Soc Nephrology. *Res:* Renal hypertension; natural history of renal diseases; metabolism of uremia; chronic dialysis and renal transplantation. *Mailing Add:* Evanston Hosp 2650 Ridge Ave Evanston IL 60201

SIMON, NORMAN ROBERT, b New York, NY. ASTROPHYSICS. *Educ:* Syracuse Univ, BA, 59; City Col New York, MS, 64; Yeshiva Univ, PhD(physics), 68. *Prof Exp:* Fel astrophys, Nat Acad Sci-Nat Res Coun, 68-70; asst prof, 70-74, assoc prof, 74-79, PROF PHYSICS & ASTRON, UNIV NEBR, 79- *Mem:* Am Astron Soc; Int Astron Union. *Res:* Stellar interiors; pulsations of stars; stellar evolution; cepheid variable stars. *Mailing Add:* Behlen Lab of Physics & Astron Univ of Nebr Lincoln NE 68588-0111

SIMON, PHILIPP WILLIAM, b Sturgeon Bay, Wis, Mar 19, 50; m 73; c 2. PLANT BREEDING, HORTICULTURE. *Educ:* Carroll Col, BS, 72; Univ Wis, MS, 75, PhD(genetics), 77. *Prof Exp:* from asst prof to assoc prof, 80-85, PROF, DEPT HORT, UNIV WIS, 85-; RES PLANT GENETICIST, USDA AGR RES SERV, 78- *Honors & Awards:* Nat Food Processors Award, 81, 84. *Mem:* AAAS; Genetics Soc Am; Am Soc Hort Sci; Bot Soc Am; Am Genetics Asn. *Res:* Genetics and breeding of carrot, onion, garlic and cucumber; biochemical genetics and improvement of nutritional quality and culinary guality; gene mapping in carrot, onion, cucumber. *Mailing Add:* USDA ARS Dept Hort Univ Wis Madison WI 53706

SIMON, RALPH EMANUEL, b Passaic, NJ, Oct 20, 30; m 52; c 3. PHYSICS. *Educ:* Princeton Univ, BA, 52; Cornell Univ, PhD, 59. *Prof Exp:* Asst, Cornell Univ, 52-57; mem tech staff, Labs, RCA Corp, 59-68, dir, Conversion Devices Lab, David Sarnoff Res Ctr, 68-69, mgr advan technol, Electrooptics Oper, RCA Electronic Components, 69-70, mgr, Electrooptics Oper, 70-76, vpres, electro-optics devices div, RCA Electronics Components, 76-; EXEC VPRES & CHIEF OPER OFFICER, LYTEL INC, SOMERVILLE, NJ. *Mem:* Am Phys Soc; fel Inst Elec & Electronics Engrs. *Res:* Solid state physics. *Mailing Add:* 25 Dexter Dr N Basking Ridge NJ 07920

SIMON, RICHARD L, b Oakland, Calif, Dec 1, 50; m 85; c 2. MICRO-CLIMATOLOGY, FORENSIC METEOROLOGY. *Educ:* Univ Calif, Berkeley, BS, 73; San Jose State Univ, MS, 76. *Prof Exp:* Res asst, Ames Res Ctr, NASA, 74-75; meteorologist, Nat Environ Satellite Serv, Nat Oceanic & Atmospheric Admin, 76 & Pac Gas & Elect Co, 80-82; res assoc, San Jose State Univ, 77-78; meteorologist & pres, Global Weather Consults, Inc, 77-80; sr meteorologist, Am Energy Projs, Inc, 82-83; METEOROL CONSULT, 83- *Concurrent Pos:* Lectr, San Jose State Univ, 76 & 86 & San Jose Metrop Adult Educ Prog, 77-78. *Mem:* Am Meteorol Soc; Am Wind Energy Asn. *Res:* Wind energy meteorology; practical methodologies for the siting of wind turbines and assessment of their potential energy production. *Mailing Add:* 80 Alta Vista Ave Mill Valley CA 94941

SIMON, RICHARD MACY, b St Louis, Mo, July 27, 43. BIOMETRIC RESEARCH. *Educ:* Wash Univ, St Louis, BS, 65, PhD, 69. *Prof Exp:* Consult statist, Gen Elec Ctr Advan Studies, Santa Barbara, Calif, 67-69, Rand Corp, Santa Monica, Calif, 68-69; computer scientist, Div Computers Res & Technol, NIH, 69-71, head, Biostatist Info Systs Unit, Med Oncol, 71-75, head, Biostatist & Data Mgt Sect, Clin Oncol Prog, 76-78, CHIEF, BIOMET RES BR, CANCER THER EVAL PROG, DIV CANCER TREAT, NAT CANCER INST, NIH, BETHESDA, MD, 78- *Concurrent Pos:* Adv, Am Diabetes Asn; mem, Task Force Endpoints, AIDS studies, NIH, Proj Controlled Therapeut Trials, Int Union Control Cancer, 87-; mem, Bd Dirs, Soc Clin Trails, 86-; vis prof internal med, Univ Va, 89-90; reviewer, Am J Epidemiol, Can J Statist, J Chronic Dis, J Am Statist Asn, J Nat Cancer Inst. *Mem:* Fel Am Statist Asn; Am Asn Cancer Res; Am Soc Clin Oncol; Asn Comput Mach; Biomet Soc; Soc Clin Trials. *Mailing Add:* NIH Nat Cancer Inst Biomet Res Br Exec Plaza N Rm 739C 6130 Executive Blvd Rockville MD 20892

SIMON, ROBERT DAVID, b Chicago, Ill, June 14, 45; m 67; c 3. MICROBIOLOGY, ECOLOGY. *Educ:* Univ Chicago, BS, 67; Mich State Univ, PhD(bot), 71. *Prof Exp:* Assoc prof biol, Univ Rochester, 71-82; DEPT BIOL, STATE UNIV NY COL GENESEO, 82- *Mem:* Am Soc Microbiol. *Res:* Microbiology, ecology and biochemistry of the cyanabacteria. *Mailing Add:* Dept Biol State Univ NY Geneseo NY 14454

SIMON, ROBERT H, b Nashua, NH, July 10, 20; m 43; c 3. CHEMICAL ENGINEERING. *Educ:* Mass Inst Technol, SB & SM, 41; Ore State Univ, PhD(chem eng), 48. *Prof Exp:* Engr, Nat Defense Res Comt, Carnegie Inst Technol, 41-42; engr, Federated Metals Div, Am Smelting & Ref Co, 43-44; supvr nuclear fuel & waste processing, Knolls Atomic Power Lab, Gen Elec Co, 48-54, mgr naval nuclear reactor test loops, 55-59; proj engr, Gen Atomic Div, Gen Dynamics Corp, 59-61, mgr fuel fabrication eng, Gen Atomic Co, 62-63, proj mgr nuclear power reactors, 63-64, asst mgr eng, 64-66, asst mgr Wash rep power reactors, 67-69, proj mgr nuclear power reactors, Gulf Gen Atomic Co, 69-73, dir gas-cooled fast breeder reactor prog, 73-82, consult, gas-cooled reactors to utility group, 83-85; RETIRED. *Concurrent Pos:* Mem radioactive waste processing comt, Atomic Energy Comn, 51-53. *Mem:* Am Inst Chem Engrs; Am Nuclear Soc. *Res:* Gas-cooled reactors; nuclear reactor test facilites and components; radioactive waste processing and fission product recovery. *Mailing Add:* 1026 Santa Barbara San Diego CA 92107

SIMON, ROBERT H(ERBERT) M(ELVIN), b New York, NY, Nov 11, 24; m 50; c 4. MATERIALS SCIENCE ENGINEERING, POLYMER ENGINEERING. *Educ:* Univ Del, BChE, 48; Yale Univ, DEng(chem eng), 57. *Prof Exp:* Develop engr, Gen Elec Co, 48-52; sr engr, 56-66, specialist process technol, 66-68, group supvr, 68-74, sr technol specialist, 74-80, MONSANTO FEL, MONSANTO CO, 80- *Concurrent Pos:* Adj lectr chem eng, Yale Univ, 84-85. *Mem:* Am Inst Chem Engrs; Am Chem Soc. *Res:* Polymer process development polymer foams; polymer fabrication techniques; high pressure reactions. *Mailing Add:* 23 Caravelle Dr Longmeadow MA 01106

SIMON, ROBERT MICHAEL, b Philadelphia, Pa, Jan 12, 56; m 81; c 1. ORGANOSILICON & ORGANOLITHIUM CHEMISTRY. *Educ:* Ursinus Col, BS, 77; Mass Inst Technol, PhD(inorg chem), 82. *Prof Exp:* Mellon Found sci policy fel, 82-83, staff officer, 83-85, sr staff officer, 85-88, staff dir, bd chem sci & technol, Nat Acad Sci, 88-; SECY, ENERGY ADVISORY BD, US DEPT ENERGY AC-1, WASH, DC, CURRENTLY. *Mem:* Am Chem Soc; Am Inst Chem Engrs; Sigma Xi; AAAS. *Res:* Science policy issues related to the disciplines of chemistry, biochemistry and chemical engineering; synthetic inorganic-organometallic chemistry. *Mailing Add:* 6116 Quebec Pl Berwyn Heights MD 20740

SIMON, SANFORD RALPH, b New York, NY, Nov 6, 42; m 64; c 2. PATHOLOGY. *Educ:* Columbia Univ, AB, 63; Rockefeller Univ, PhD(biochem), 67. *Prof Exp:* Guest investr biochem, Rockefeller Univ, 67-69; from asst prof to assoc prof biochem, 69-87, PROF BIOCHEM & PATH, STATE UNIV NY STONY BROOK, 87- *Concurrent Pos:* Nat Heart, Lung & Blood Inst grant & Am Heart Asn grant-in-aid, 70-; Alfred P Sloan Found res fel, 72; Nat Heart, Lung & Blood Inst career develop award, 75-80. *Mem:* Am Soc Biochem & Molecular Biol. *Res:* Structure-function relationships of normal and modified hemoglobins and metalloproteins; role of protein conformation in mechanisms of enzyme action; applications of physical biochemistry in protein research; design and characterization of protease inhibitors; interactions of inflammatory cells with extracellular matrix. *Mailing Add:* Dept of Biochem State Univ of NY Stony Brook NY 11794-5215

SIMON, SELWYN, b Chicago, Ill, Mar 27, 25; m 51; c 2. MICROBIOLOGY. *Educ:* Purdue Univ, BS, 47, MS, 49; Ill Inst Technol, PhD, 56. *Prof Exp:* Res asst, Purdue Univ, 47-49; biochemist, 49-56, group leader, Food Lab, 56-62, asst mgr, 62-65, mgr food sci inst, films-packaging div, Union Carbide Corp, 65-85; SR TECHNICAL CONSULT, VISKASE CORP, 86- *Mem:* Am Soc Microbiol; Am Chem Soc; Am Meat Sci Asn; Sigma Xi. *Res:* Food technology; proteolytic enzymes; sausage and meat processing; texture; smoke process. *Mailing Add:* Viskase Corp 6830 W 65th St Chicago IL 60638

SIMON, SHERIDAN ALAN, b Buffalo, NY, Apr 20, 47; m 70. ASTROPHYSICS, WRITING. *Educ:* Univ Rochester, BS, 69, MA, 71, PhD(physics & astron), 78. *Prof Exp:* From asst prof to assoc prof physics, 74-89, PROF PHYSICS, GUILFORD COL, 89- *Concurrent Pos:* Kenan grant, Guilford Col, 76-78; Burroughs Wellcome Found grant, 78; res corp grant, 79-81; vis scientist, NASA, Langley, 81. *Honors & Awards:* Stoddard

Prize, Univ Rochester, 69. *Mem:* Am Physical Soc; Sigma Xi; Am Astron Soc; Am Asn Physics Teachers. *Res:* Stellar evolution; computer simulation of stellar evolution in the presence of arbitrary rotation; college level physics education; college level astronomy education; computer software for astronomy and physics education; popular articles; textbook in astronomy; design of fictional planets for science fiction writers. *Mailing Add:* Dept Physics Guilford Col Greensboro NC 27410

SIMON, SIDNEY, allergy, immunology, for more information see previous edition

SIMON, SIDNEY ARTHUR, b New York, NY, Feb 16, 43; div. PHYSICAL BIOLOGY. *Educ:* Ind Inst Technol, BS, 65; Ariz State Univ, MS, 68; Northwestern Univ, PhD(mat sci), 73. *Prof Exp:* Fel biol, Northwestern Univ, 73; fel physiol, Duke Univ, 73-74, assoc anesthesiol, 74-80, asst prof physiol, 74-80, anesthesiol, 80-88, ASSOC PROF ANESTHESIOL & PROF NEUROBIOL, DUKE UNIV, 88-, ASSOC PROF PHYSIOL, 80- *Mem:* Biophys Soc; Soc Gen Physiologists; Soc Neurosci. *Mailing Add:* Dept Neurobiol Duke Univ Durham NC 27710

SIMON, TERRENCE WILLIAM, b Cottonwood, Idaho, Aug 16, 46; m 71; c 1. MECHANICAL ENGINEERING. *Educ:* Wash State Univ, BS, 68; Univ Calif, Berkeley, MS, 71; Stanford Univ, PhD(mech eng), 80. *Prof Exp:* Engr, Nuclear Energy Div, Gen Elec Co, 68-74, Stearns Roger, Inc, 74-76; ASST PROF THERMOSCI, MECH ENG DEPT, UNIV MINN, 80- *Concurrent Pos:* Consult, Stanford Linear Accelerator Ctr, 77-78, Hewlett-Packard Inc, 78-80, Control Data Corp, 83. *Mem:* Am Soc Mech Engrs. *Res:* Experimental and analytical studies in the fundamentals of heat transfer and fluid mechanics. *Mailing Add:* Dept Mech Eng Univ Minn 111 Church St SE Minneapolis MN 55455

SIMON, WILBUR, b Geneva, Ill, July 5, 17; m 57; c 2. CHEMISTRY. *Educ:* Univ Ill, BS, 39; Univ Iowa, MS, 40, PhD(anal chem), 51. *Prof Exp:* Chem engr, Tenn Copper Co, 40-44; assoc chemist, Metall Lab, Univ Chicago, 44-45; jr res chemist, Carbide & Carbon Chem Corp, 45-49; head analyst & stand br, US Naval Radiol Defense Lab, 51-53; res chemist, Princeton Radiation Chem Lab Inc, 53-55 & Morton Chem Co Ill, 55-61; assoc res coordr, Universal Oil Prod Co, 61-64; OWNER & DIR, SIMON RES LAB, 64- *Mem:* AAAS; Am Chem Soc. *Res:* Analytical development; application of nuclear magnetic resonance to organic chemistry; nuclear magnetic resonance spectroscopy; disposal of solid and liquid wastes; removal of contaminants from sewage; environmental problems; overflow and backup of sewers; corrosion; metal cleaning; adhesives. *Mailing Add:* 816 Murray Ave Elgin IL 60120

SIMON, WILLIAM, b Pittsburgh, Pa, May 27, 29. PHYSICS, BIOMATHEMATICS. *Educ:* Carnegie Inst Technol, BS, 50; Harvard Univ, MA, 52, PhD(appl physics), 58. *Prof Exp:* Instrument sect head elec eng, Spencer Kennedy Lab, Boston, Mass, 53-57; sr systs engr, Nat Radio Co, Malden, 57-59; chief physicist, Image Instruments, Inc, Newton, 59-60; mem staff comput, Lincoln Lab, Mass Inst Technol, 61-64; res assoc, dept physiol, Harvard Med Sch, 64-68; assoc prof, 68-77, PROF BIOMATH, ROCHESTER SCH MED & DENT, 77-, PROF BIOPHYS, 82-, PROF MED INFORMATICS, 88- *Concurrent Pos:* Vis assoc prof elec eng, Mass Inst Technol, 74-75; consult physicist, Comstock & Westcott, Cambridge, Mass. *Mailing Add:* Dept Biophysics Box BPHY Rochester NY 14642

SIMONAITIS, ROMUALDAS, b Lithuania, June 10, 34; US citizen; m 63; c 2. CHEMICAL KINETICS, PHOTOCHEMISTRY. *Educ:* Ill Inst Technol, BS, 58; Univ Calif, Los Angeles, MS, 64; Univ Calif, Riverside, PhD(phys chem), 68. *Prof Exp:* Chemist, Douglas Aircraft Co, 62-63; actg instr chem, Univ Wis-Milwaukee, 63-64; fel, 68-70, res assoc, 70-80, SR RES ASSOC PHYS CHEM, IONOSPHERE RES LAB, PA STATE UNIV, 80- *Mem:* Am Chem Soc; Sigma Xi. *Res:* Photochemistry of atmospheric molecules; atom and free radical reactions important in planetary atmospheres. *Mailing Add:* 228 W Prospect Ave University Park PA 16802

SIMONDS, JOHN ORMSBEE, b Jamestown, NDak, Mar 11, 13; m 43; c 4. ECOLOGY. *Educ:* Mich State Univ, BS, 35; Harvard Univ, MLA, 39. *Hon Degrees:* DSc, Mich State Univ, 68. *Prof Exp:* Partner, Simonds & Simonds, Landscape Architects, 39-52; partner, Collins, Simonds & Simonds, Landscape Architects, 52-70; fac, Dept Archit, Carnegie-Mellon Univ, 55-67; partner, 70-82, EMER PARTNER, ENVIRON PLANNING & DESIGN PARTNERSHIP, 83-; CONSULT, COMPREHENSIVE LAND PLANNING, 83- *Concurrent Pos:* US consult, InterAm Housing & Planning Ctr, Bogota, 60 & 61; vis critic, Archit & Planning, Yale Univ, 61 & 62; lectr, 40 Univs in US & abroad, 61-; adv, US Bur Pub Roads, Dept Transp, 65-68; chmn, Urban Open Space Panel, White House Conf, 65; founding pres, Am Soc Landscape Architects Found, 66-68; mem, President's Task Force Environ, 68-70 & Fla Gov's Task Force on Resources & Environ, 68-71; consult, Chicago Cent Area Comt, 62 & Land & Nature Trust, Lexington, Ky, 87-89; hon corresp, Royal Town Planning Inst, Gt Brit. *Honors & Awards:* Am Soc Landscape Architects Medal, Am Soc Landscape Architects, 73. *Mem:* Am Soc Landscape Architects (vpres, 61-63, pres, 63-65); Royal Town Planning Inst; assoc mem Nat Acad Design. *Res:* Author and/or editor of five books and many articles and technical reports in field of environmental science, landscape architecture, community and regional planning, urban design, riverfront development, parks, recreation and growth management. *Mailing Add:* The Loft 17 Penhurst Rd Pittsburgh PA 15202

SIMONE, JOSEPH VINCENT, b Chicago, Ill, Sept 19, 35; m 60; c 3. HEMATOLOGY, PEDIATRICS. *Educ:* Loyola Univ, Chicago, MD, 60; Am Bd Internal Med & Am Bd Pediat, dipl, 67; Am Bd Pediat Hemat-Oncol, cert, 74. *Prof Exp:* Instr pediat, Col Med, Univ Ill, 66-67; from asst prof to assoc prof pediat, Univ Tenn Health Sci Ctr, 67-77; chief hemat & oncol, St Jude Children's Hosp, 73-77; prof pediat, Stanford Univ, 77-78; assoc dir clin res, 78-83, DIR CLIN RES, ST JUDE CHILDREN'S RES HOSP, 83- *Concurrent Pos:* Trainee pediat hemat, Col Med, Univ Ill, 63-66; chief hemat, St Jude Children's Hosp, 69-73. *Honors & Awards:* Richard & Hinda Rosenthal Award, 79. *Mem:* Fel Am Col Physicians; Am Soc Hemat; Soc Pediat Res; Am Asn Cancer Res; Am Pediat Soc; Am Soc Clin Oncol. *Res:* Cancer therapy. *Mailing Add:* 332 N Lauderdale St Memphis TN 38101

SIMONE, LEO DANIEL, b New York, NY, Oct 6, 35. PLANT MORPHOLOGY, BRYOLOGY. *Educ:* Manhattan Col, BS, 57; Columbia Univ, MA, 59, PhD(bot), 67. *Prof Exp:* Teaching asst bot, Columbia Univ, 60-64, preceptor, 64-66; ASSOC PROF BIOL, STATE UNIV NY POTSDAM, 66- *Concurrent Pos:* NY State grant-in-aid, 68-69. *Mem:* AAAS; Am Soc Plant Physiol; Am Inst Biol Sci; Am Bryol & Lichenological Soc; Bot Soc Am. *Res:* Apogamy and apospory in the Hepaticae; experimental plant morphology and aseptic culture of liverworts. *Mailing Add:* Dept of Biol State Univ NY Col Potsdam Potsdam NY 13676

SIMONELLI, ANTHONY PETER, b Bridgeport, Conn, June 28, 24; m 56; c 3. DRUG DELIVERY SYSTEMS, CHEMICAL KINETICS. *Educ:* Univ Conn, BA & BS, 55; Univ Wis, MS, 58, PhD(pharm), 60. *Prof Exp:* From asst prof to assoc prof pharm, Med Col Va, 60-64; from asst prof to assoc prof, Univ Mich, Ann Arbor, 64-72; PROF PHARMACEUT & CHMN SECT, SCH PHARM, UNIV CONN, 72- *Concurrent Pos:* Consult, Eli Lilly Labs, 69, Pfizer Labs, 73-83, Northeastern Univ, 75, Ortho Labs, 78, Ives Labs, 79-84 & SK&F Labs, 81-82, Lederle Labs, 84, Warner Lambert Labs, 88, Merck Labs, 87, Bristol Myers, 83, Sando* Labs, 85; chmn, Parmaceut Sect, Am Pharmaceut Asn Acad Parmaceut Sci, 69, Pharmaceut sci, AAAS, 83 & Pharmaceut Sect, Am Asn Cols Pharm, 85; vis prof, Univ Wis, 74, Univ Minn, 76 & Univ Assiut, Egypt, 83; mem, Sci Bd, Menley James Labs, 82-85. *Honors & Awards:* Ebert Prize, 67 & 69. *Mem:* Fel AAAS; Am Chem Soc; fel Am Pharmaceut Asn Acad Pharmaceut Sci; Soc Rheology; NAm Thermal Anal Soc; Am Asn Cols Pharm. *Res:* Biopharmaceutics; pharmacokinetics; biological calcification; amorphous-polymorphic drugs; solution drug interactions; transport phenomena; rheology; drug solubility and dissolution rates; biodegradable polymers as drug delivery systems; thermal analysis; drug formulations; solid state transformation kinetics; interfacial phenomena. *Mailing Add:* Sch Pharm Univ Conn U-92 Storrs CT 06268

SIMONEN, THOMAS CHARLES, b Munising, Mich, Aug 25, 38; m 64; c 2. PLASMA PHYSICS. *Educ:* Mich Technol Univ, BS, 60; Stanford Univ, MS, 64, PhD(elec eng), 66. *Prof Exp:* Mem tech staff elec eng, Hughes Aircraft Co, 60-62; res asst plasma physics, Garching, Ger, 67-68; res assoc plasma physics, Plasma Physics Lab, Princeton Univ, 68-70; exp physics prog leader, Lawrence Livermore Nat Lab, 70-88; DIR, DILL-D TOKAMAK PROG, GEN ATOMICS, 88- *Mem:* Fel Am Phys Soc. *Res:* Plasma confinement of magnetic fusion plasmas. *Mailing Add:* Gen Atomics 13-255 PO Box 85608 San Diego CA 92138

SIMONET, DONALD EDWARD, b Orlando, Fla, Mar 26, 46; m 67; c 1. ENTOMOLOGY, AGRICULTURE. *Educ:* Emory & Henry Col, BA, 67; Va Polytech Inst & State Univ, MS, 75, PhD(entomol), 78. *Prof Exp:* Asst prof entomol, Ohio Agr Res & Develop Ctr, 78-80; MOBAY VERO BEACH LAB. *Mem:* Entomol Soc Am; Int Orgn Biol Control; Entomol Soc Can. *Res:* Integrated pest management of insect pests of vegetables. *Mailing Add:* Mobay Vero Beach Lab PO Box 1508 Vero Beach FL 32960

SIMONI, ROBERT DARIO, b San Jose, Calif, Aug 18, 39; m 61; c 3. BIOCHEMISTRY. *Educ:* San Jose State Col, BA, 62; Univ Calif, Davis, PhD(biochem), 66. *Prof Exp:* Fel biochem, Johns Hopkins Univ, 66-70; from asst prof to assoc prof, 71-81, PROF BIOL SCI, STANFORD UNIV, 81- *Mem:* AAAS; Am Soc Biol Chemists. *Res:* Structure and function of biological membranes; mechanisms of solute transport; regulation of cholesterol metabolism. *Mailing Add:* Dept of Biol Sci Stanford Univ Stanford CA 94305

SIMONIAN, VARTKES HOVANES, pharmacognosy; deceased, see previous edition for last biography

SIMONIS, GEORGE JEROME, b Wisconsin Rapids, Wis, Nov 9, 46; m 68. MILLIMETER WAVE MATERIALS, INFRARED LASERS. *Educ:* Univ Wis-Platteville, BS, 68; Kans State Univ, PhD(physics), 73. *Prof Exp:* RES PHYSICIST LASER RES, HARRY DIAMOND LABS, DEPT ARMY, 72- *Mem:* Am Phys Soc; Optical Soc Am; Inst Elec & Electronics Engrs. *Res:* Infrared lasers; nonlinear optical interactions; diode lasers; near-millimeter waves. *Mailing Add:* 15230 Baughman Dr Silver Spring MD 20906

SIMONIS, JOHN CHARLES, b Marion, Ohio, July 24, 40; m 63; c 3. FLUID STRUCTURAL INTERACTION, REVERSE ENGINEERING OF PARTS FOR COMMERCIAL & MILITARY SYSTEMS. *Educ:* Case Inst Technol, BS, 62; Ga Inst Technol, MS, 64, PhD(eng mech), 68. *Prof Exp:* Teaching asst eng mech, Ga Inst Technol, 63-67; prin Engr, Babock & Wilcox, 67-80; instr math, San Antonio Community Col, 83-87; SR RES ENGR, SOUTHWEST RES INST, 80- *Concurrent Pos:* Adj prof mech engr, Univ Tex, San Antonio, 87- *Mem:* fel Am Soc Mech Engrs; Am Soc Mech Engrs. *Res:* Evaluate the effects of vibrating forces on structures, buildings and machines; developed unique technique to reverse engineer mechanical components in commercial or military systems. *Mailing Add:* 6220 Culebra Rd PO Drawer 28510 San Antonio TX 78228-0510

SIMONOFF, ROBERT, b Baltimore, Md, Feb 27, 20; m 43; c 2. ORGANIC CHEMISTRY. *Educ:* Univ Md, BS, 40, MS, 42, PhD(chem), 45. *Prof Exp:* Asst biochem, Sch Dent, Univ Md, 41-45, fel pharmacol, Sch Med, 49-50; org chemist, Gen Elec Co, 45-49; res chemist, Nat Aniline Div, Allied Chem & Dye Corp, 50-51; chief chemist, William H Rorer, Inc, 51-54; sr res chemist, L Sonneborn Sons, Inc, 54-60, asst supt, Sulfonate Dept, Sonneborn Chem & Refining Corp, 60, supt, 60-65, dir sulfonate res, Sonneborn Div, Witco Chem Co, Inc, Pa, 65-68, tech dir, Sonneborn Div, 68-80, sr div scientist, Sonneborn Div, Witco Chem Corp, 80-89; RETIRED. *Mem:* Am Chem Soc; fel Am Inst Chemists; Am Soc Testing & Mat. *Res:* Petroleum sulfonates; hydrogenations; synthetic medicinals; mineral oil; petrolatum. *Mailing Add:* 13-3 Tamaron Dr Waldwick NJ 07463-1129

SIMONS, BARBARA BLUESTEIN, b Boston, Mass, Jan 26, 41; div; c 3. DETERMINISTIC SCHEDULING THEORY, GRAPH THEORETIC ALGORITHMS. *Educ:* Univ Calif, Berkeley, PhD(computer sci), 91. *Prof Exp:* RES STAFF MEM, INT BUS MACH RES, 90- *Concurrent Pos:* Vis fac, Univ Calif, Santa Cruz, 84; mem, Grad Fel Eval Panel Comput Sci, Nat Res Coun, 85-87 & Res Initiation Awards Panel, NSF, 89; ed, Fault Tolerant Distributed Comput, Springer-Verlag Lect Notes Computer Sci, 90. *Mem:* Asn Comput Mach (secy, 90-92); AAAS. *Res:* Algorithms and graph theory, with an emphasis on scheduling and on optimization problems arising from parallelizing compilers; fault tolerant distributed computing. *Mailing Add:* Int Bus Mach Almaden Res Ctr K53-802 650 Harry Rd San Jose CA 95120-6099

SIMONS, DANIEL J, b Rochester, NY, Jan 17, 40; m 67. MICROBIOLOGY, INVERTEBRATE PHYSIOLOGY. *Educ:* Univ Rochester, AB, 61, MS, 63, PhD(biol), 68. *Prof Exp:* Res assoc biol, Univ Rochester, 68; sr res microbiologist, Biospherics, Inc, 68-69; from instr to assoc prof, 69-80, PROF ZOOL, MONTGOMERY COL, 80-, CHMN, DEPT BIOL, 80- *Mem:* AAAS; Am Soc Microbiol; Am Soc Zool. *Res:* Innovative learning experiences in teaching zoological sciences. *Mailing Add:* Dept Biol Montgomery Col Rockville MD 20850

SIMONS, DARYL B, b Payson, Utah, Feb 12, 18; m 44; c 2. HYDRAULIC ENGINEERING. *Educ:* Utah State Univ, BS, 47, MS, 48; Colo State Univ, PhD(civil eng), 57. *Prof Exp:* Design engr, McGraw Constructors & Engrs, 48; prof civil eng, Univ Wyo, 49-57; proj chief fluid mech res, US Geol Surv, 57-63; sect chief eng, 63-65, assoc dean res, 65-77, PROF CIVIL ENG, COLO STATE UNIV, 63- *Concurrent Pos:* Consult, Int Boundary Water Comn, US Bur Pub Rds & Corps Engrs. *Mem:* Am Soc Civil Engrs; Int Asn Hydraul Res. *Res:* Water resources; fluid mechanics and hydraulics, especially river mechanics, sediment transport, resistance of flow in alluvial channels and design of stable channels. *Mailing Add:* Simons & Assoc Inc 2821 Remington Ft Collins CO 80525

SIMONS, EDWARD LOUIS, b New York, NY, May 9, 21; m 43; c 2. ENVIRONMENTAL CHEMISTRY. *Educ:* City Col New York, BS, 41; NY Univ, MS, 43, PhD(chem), 45. *Prof Exp:* Asst chemist, NY Univ, 41-44; asst group leader, Kellex Corp, NJ, 44-45; res scientist, Carbide & Carbon Chem Corp, NY, 45-46; from instr to asst prof chem, Rutgers Univ, 46-51; res assoc, Gen Elec Res & Develop Ctr, Gen Elec Co, 51-68, mgr fuel cells prog, 68-69, inorg chemist, 69-71, mgr, Environ Info Ctr, 71-73, mgr Environ Protection Oper, 73-82; RETIRED. *Concurrent Pos:* Environ consult, 85- *Honors & Awards:* Award, Nat Asn Corrosion Engrs, 56. *Mem:* Am Chem Soc. *Res:* Fuel cells; corrosion; phase equilibria; thermogravimetry. *Mailing Add:* Two Woodside Dr Albany NY 12208-1145

SIMONS, ELIZABETH REIMAN, b Vienna, Austria, Sept 1, 29; nat US; m 51; c 2. BIOPHYSICAL CHEMISTRY. *Educ:* Cooper Union, BChE, 50; Yale Univ, MS, 51, PhD(phys chem), 54. *Prof Exp:* Asst, Biophys Div, Dept Physics, Yale Univ, 50-53; res chemist, Tech Opers, Inc, 53-54; instr chem, Wellesley Col, 54-57; from res asst to res assoc path, Children's Cancer Res Found, 57-63; res assoc biol chem, Harvard Med Sch, 63-65, lectr, 65-72; assoc prof biochem, 72-78, PROF BIOCHEM, SCH MED, BOSTON UNIV, 78- *Concurrent Pos:* Tutor, Harvard Univ, 71- *Mem:* Am Soc Biol Chemists; Am Chem Soc; Biophys Soc; Am Soc Hemat; Am Heart Asn; Soc Cell Biol. *Res:* Protein structure; hemostasis and platelet interactions; cell membrane biophysics; neutrophil biochemistry. *Mailing Add:* 117 Chestnut St West Newton MA 02165

SIMONS, ELWYN LAVERNE, b Lawrence, Kans, July 14, 30; m 72; c 3. VERTEBRATE PALEONTOLOGY, PRIMATOLOGY. *Educ:* Rice Univ, BS, 53; Princeton Univ, MA, 55, PhD, 56; Univ Col, Oxford, DPhil, 59; Yale Univ, MA, 67. *Prof Exp:* Lectr, Dept Geol, Princeton Univ, 59; asst prof zool, Univ Pa, 59-61; appointment fac head, Div Vert Paleont, Yale Peabody Mus, 60-77; prof paleont, Dept Geol & Geophys, Yale Univ, 67; prof anthrop & anat, 77-82, DIR, PRIMATE CTR, DUKE UNIV, 77-, JAMES B DUKE PROF BIOL ANTHROP & ANAT, 82- *Concurrent Pos:* Gen George Marshall scholar, Oxford Univ, 56-59, Boise Fund fel, 58-59; res assoc, Am Mus Natural Hist, 59-76; vis assoc prof geol & cur vert paleont, Yale Univ, 60-61; R C Hunt Mem fel anthrop, Viking Found, 65; mem, Grad Res Panel, NSF, 90- *Honors & Awards:* Anadale Mem Medal, Asiatic Soc Calcutta, 73; Alexander von Humboldt Sr Scientist Award, WGer, 75 & 76. *Mem:* Nat Acad Sci; Am Soc Zool; Am Asn Phys Anthrop; Brit Soc Study Human Biol; Am Soc Primatology; Int Primatological Soc; Soc Vert Paleont; Soc Study Evolution; Sigma Xi; AAAS. *Res:* Primatology; primate and human paleontology, early mammalian evolution and anatomy, particularly Pantodonta and related subungulates; primate husbandry and behavioral evolution of prosimians; fossil prosimians; paleocene of North America and Europe; earliest apes and monkeys of Africa; late Tertiary Old World apes, including human orgins; author of various publications. *Mailing Add:* Duke Primate Ctr 3705 Erwin Rd Durham NC 27705

SIMONS, FRANK, b Detroit, Mich, Nov 14, 17. GEOLOGY. *Educ:* Univ Calif, Los Angeles, BA, 40; Stanford Univ, PhD(geol), 51. *Prof Exp:* Res geologist, US Geol Surv, 41-88, consult res geologist, 88-89; RETIRED. *Mem:* Geol Soc Am; Soc Eng Geologists; AAAS. *Mailing Add:* 9005 W Second Ave Lakewood CO 80226

SIMONS, GENE R, b Staten Island, NY, July 19, 36; m 61; c 4. MANAGEMENT ENGINEERING. *Educ:* Rensselaer Polytech Inst, BS, 57, PhD(mgt sci), 69; Stevens Inst Technol, MS, 61. *Prof Exp:* Planning engr indust eng, Western Elec Co, 57-61; mem staff indust eng, Am Cyanamid Co, 61-62; asst prof indust eng, New Haven Col, 62-64, chmn, 64-66; instr, 66-69, asst prof, 69-70, curriculum chmn mgt eng, 71-77, ASSOC PROF MGT ENG, RENSSELAER POLYTECH INST, 70-, DIR INDUST & MGT ENG, 77- *Concurrent Pos:* Curric chmn mgt eng, Rensselaer Polytech Inst, 71-77; Pres, Gene Simons, Inc, 80-; vpres, Workshops, Menands, NY, 81-83. *Mem:* Inst Indust Engrs (vpres, 82-83); Proj Mgt Inst; Am Soc Eng Educ; Sigma Xi. *Res:* Management systems; maintenance planning and control. *Mailing Add:* Dept Decision Sci Rensselaer Poly Main Campus Troy NY 12180

SIMONS, HAROLD LEE, b New York, NY, Aug 25, 26; m 51; c 2. PHYSICAL CHEMISTRY. *Educ:* Princeton Univ, AB, 49; Yale Univ, MS, 51, PhD(chem), 53. *Prof Exp:* Res chemist, 53-70, group leader, 70-74, SR RES ASSOC, KENDALL CO, 75- *Concurrent Pos:* Lectr div continuing educ, Boston Univ, 62-65. *Mem:* Am Chem Soc; Soc Rheology; Am Phys Soc. *Res:* Characterization of polymers; physical properties of polymers. *Mailing Add:* 117 Chestnut St West Newton MA 02165

SIMONS, JOHN NORTON, b Lennox, SDak, Aug 13, 26; m 48; c 4. ENTOMOLOGY. *Educ:* Univ Calif, PhD(entom), 53. *Prof Exp:* Asst virologist, Everglades Exp Sta, Univ Fla, 53-59; sr entomologist, Stanford Res Inst, 59-70; sr scientist, Ciba-Geigy Corp, 70-77; PRES JMS FLOWER FARMS, INC, 77- *Concurrent Pos:* Vis prof entom, Univ Calif, Berkeley, 74-75. *Mem:* Entom Soc Am; Am Phytopath Soc. *Res:* Insect transmission of plant virus diseases; plant virus disease control. *Mailing Add:* 1105 25th Ave Vero Beach FL 32960

SIMONS, JOHN PETER, b Youngstown, Ohio, Apr 2, 45; m 68. PHYSICAL CHEMISTRY. *Educ:* Case Inst Technol, BS, 67; Univ Wis, PhD(chem), 70. *Prof Exp:* From asst prof to assoc prof, 71-79, PROF CHEM, UNIV UTAH, 79- *Concurrent Pos:* NSF fel, Mass Inst Technol, 70-71; Alfred P Sloan fel, Univ Utah, 73-77, Camille & Henry Dreyfus teaching fel, 75-80, David P Gardner fel, 79, 81; John Simon Gugenheim fel, 80-81. *Mem:* Am Chem Soc; Am Phys Soc; Sigma Xi; Asn Women Sci. *Res:* Quantum chemistry; statistical mechanics; molecular spectroscopy; negative molecular ions; solvation effects; molecular dynamics. *Mailing Add:* Dept Chem Univ Utah Salt Lake City UT 84112

SIMONS, MARR DIXON, b Murray, Utah, May 7, 25; m 50; c 2. PLANT PATHOLOGY. *Educ:* Iowa State Col, PhD(plant path, crop breeding), 52. *Prof Exp:* From asst prof to assoc prof bot & plant path, 52-59, PROF BOT & PLANT PATH, IOWA STATE UNIV, 60-; PLANT PATHOLOGIST, AGR RES SERV, USDA, 52- *Mem:* Am Phytopath Soc. *Res:* Diseases of oats; cereal rust diseases. *Mailing Add:* Dept Plant Path Iowa State Univ Ames IA 50011

SIMONS, MAYRANT, JR, b Charleston, SC, Aug 10, 36; m 59; c 3. ELECTRICAL ENGINEERING. *Educ:* Clemson Univ, BS, 58; Duke Univ, MS, 64, PhD(elec eng), 68. *Prof Exp:* Mem tech staff eng, Bell Tel Labs, NC, 61-66; SR ENGR, SEMICONDUCTOR RES CTR, RES TRIANGLE INST, 66- *Res:* Investigation of nuclear radiation effects on semiconductor materials, devices and circuits; semiconductors. *Mailing Add:* Res Triangle Inst Box 12194 Research Triangle Park NC 27709

SIMONS, ROBERT W, b Rockford, Ill, Jan 28, 45; m 74; c 2. CONTROL GENE EXPRESSION, MOBILE GENETIC ELEMENTS. *Educ:* Univ Ill, Urbana, BS, 72; Univ Calif, Irvine, PhD(molecular biol), 80. *Prof Exp:* Teaching asst genetics & biochem, Univ Calif, Irvine, 77-80; lectr genetics, Harvard Univ, 80-85; asst prof, 85-90, ASSOC PROF GENETICS, UNIV CALIF, LOS ANGELES, 90- *Concurrent Pos:* Damon Runyon Walter Winchel fel, Harvard Univ, 80-82, Leukemia Soc spec fel, 83-85; Am Cancer Soc fac researcher, Univ Calif, Los Angeles, 86-89; mem, Molecular Biol Inst, 85- *Mem:* Genetics Soc Am; AAAS; Am Soc Microbiol; Sigma Xi. *Res:* Molecular genetics of the control of gene expression with emphasis on mobile genetic elements and the determinants of RNA function and stability. *Mailing Add:* Dept Microbiol & Molecular Genetics 5304 Life Sci Bldg Univ Calif Los Angeles CA 90024

SIMONS, ROGER ALAN, b Detroit, Mich, May 11, 43; m 74; c 2. COMPUTER ALGORITHMS. *Educ:* Univ Calif, Los Angeles, AB, 64; Univ Calif, Berkeley, MA, 66, PhD(math), 72; Brown Univ, Providence, ScM, 83. *Prof Exp:* From instr to assoc prof math, Univ Wis-Green Bay, 70-81; PROF MATH & COMPUT SCI, RI COL, 81- *Concurrent Pos:* From vis asst prof to vis prof, 75-88, vis lectr, Univ Hawaii, 77, 85; consult, appln prog, US Fish & Wildlife Agency, 77, Woodbury Computer Assocs, Paramus, NJ, 83-84, subcomt children & families, RI State Legis, 86. *Mem:* Asn Comput Mach; Asn Symbolic Logic; Am Math Soc; Sigma Xi; Inst Elec & Electronics Engrs Comput Soc. *Res:* Parallel and sequential algorithms, complexity theory; boolean algebra and mathematical logic; applying mathematics to philosophical problems; theoretical aspects of computer design. *Mailing Add:* Dept Math & CS RI Col Providence RI 02908

SIMONS, ROGER MAYFIELD, b Portland, Ore, Oct 11, 26; m 54; c 2. APPLIED MATHEMATICS. *Educ:* Stanford Univ, BS, 49; Mass Inst Technol, PhD(math), 55. *Prof Exp:* Appl sci rep, IBM Corp, 55-59, staff mathematician, 59-60, mgr, Eng Comput Lab, 60-64, mgr eng & sci comput lab, 65-72, sr programmer, 72-90; LECTR, DEPT MATH & COMPUTER SCI, SAN JOSE STATE UNIV, 90- *Concurrent Pos:* Asst prof, Dept Math, San Jose State Col, 60. *Mem:* Asn Comput Mach; Sigma Xi. *Res:* Digital computer applications, simulation, and programming. *Mailing Add:* 20744 Scenic Vista Dr San Jose CA 95120

SIMONS, ROY KENNETH, b Kincheloe, WVa, Dec 26, 20; m 53; c 2. HORTICULTURE. *Educ:* Univ WVa, BS & MS, 47; Mich State Univ, PhD, 51. *Prof Exp:* Pomologist, Univ Del, 47-48; asst, Mich State Univ, 48-51; instr pomol, 51-53, asst prof, 53-58, assoc prof hort, 58-64, PROF HORT, UNIV ILL, URBANA, 64- *Concurrent Pos:* Ed, Fruit Varieties J, Am Pomol Soc. *Honors & Awards:* Stark Award, Am Soc Hort Sci, 67, Promotion Incentive Award, 77 & 78. *Mem:* Fel Am Soc Hort Sci; Rootstock Res Found; Int Dwarf Fruit Tree Asn. *Res:* Nutrition of deciduous fruit trees; soil moisture conservation to maintain optimum production of quality fruit; morphological and anatomical development of deciduous fruits; orchard cultural management practices in relation to dwarfing rootstocks; growth and development of fruit. *Mailing Add:* 1517 Alma Dr Champaign IL 61820

SIMONS, SAMUEL STONEY, JR, b Philadelphia, Pa, Sept 13, 45; m 70; c 2. MOLECULAR ENDOCRINOLOGY, STEROID HORMONES. *Educ:* Princeton Univ, AB, 67; Harvard Univ, MA, 69, PhD(chem), 72. *Prof Exp:* Fel molecular biol, Univ Calif, San Francisco, 72-75; staff fel, 75-78, sr staff fel, 78-80, RES CHEMIST ENDOCRINOL, NAT INST, DIABETES & DIGESTIVE & KIDNEY DIS, NIH, 80-, CHIEF, STEROID HORMONES SECT, 85- *Mem:* Am Chem Soc; Am Soc Biochem & Molecular Biol. *Res:* Mechanism of action of steroid hormones; steroid-receptor interactions; affinity labelling of steroid receptors; steroid control of gene transcription; glucocorticoid and antiglucorticoid steroids. *Mailing Add:* Nat Inst Diabetes & Digestive & Kidney Dis NIH Bldg 8 Rm B2A-07 Bethesda MD 20892

SIMONS, SANFORD L(AWRENCE), b New York, NY, Apr 10, 22; m 47; c 5. METALLURGY. *Educ:* Univ Mo, BS, 44. *Prof Exp:* Res engr, Battelle Mem Inst, 44; jr scientist, Manhattan Dist Proj, Los Alamos Sci Lab, 44-46; consult engr, Alldredge & Simons Labs, 46-48; asst, US Army rocket prog, Denver, 48-50; consult engr, 50-53; design engr, Heckethorn Mfg Co, 53-55; chief develop engr, Metron Instrument Co, 56-57; consult engr, 57-66; dir med eng, 66-72, CONSULT MED ENGR, SCH MED, UNIV COLO, 72-; PRES, SIENCO, INC, 72- *Mem:* Am Soc Metals; Am Soc Testing & Mat; Nat Soc Prof Engrs; Am Inst Mining, Metall & Petrol Engrs. *Res:* Metallurgy of copper-manganese systems and plutonium; development of rocketborne spectrographs; automatic perfusion apparatus; surgical prosthetics; medical and bio-engineering. *Mailing Add:* SIENCO 9188 S Turkey Creek Rd Morrison CO 80465

SIMONS, STEPHEN, b London, Eng, Aug 11. 38; m 63; c 1. MATHEMATICS. *Educ:* Cambridge Univ, BA, 59, PhD(math), 62. *Prof Exp:* Instr math, Univ BC, 62-63; res fel math, Peterhouse, Cambridge Univ, 63-64; asst prof, Univ BC, 64-65; from asst prof to assoc prof, 65-73, chmn dept math, 75-77, 88-89, PROF MATH, UNIV CALIF, SANTA BARBARA, 73- *Concurrent Pos:* NSF res grant, 65-76; mem bd dir, Calif Educ Comput Consortium, 74-75; mem bd trustees, Math Sci Res Inst, 88- *Mem:* Am Math Soc. *Res:* Functional analysis, linear and non-linear. *Mailing Add:* Dept Math Univ Calif Santa Barbara CA 93106

SIMONS, WILLIAM HADDOCK, b Vancouver, BC, Dec 2, 14; m 44; c 4. MATHEMATICS. *Educ:* Univ BC, BA, 35, MA, 37; Univ Calif, PhD(math), 47. *Prof Exp:* From asst prof to assoc prof math, Univ BC, 46-70; prof math, Ore State Univ, 70-82; RETIRED. *Concurrent Pos:* Lectr, Khaki Col Can, Eng; meteorologist, Can Meteorol Serv. *Mem:* Am Math Soc; Math Asn Am; Can Math Cong; London Math Soc. *Res:* Fourier coefficients of modular functions. *Mailing Add:* 3295 Chintimini Corvallis OR 97333

SIMONS, WILLIAM HARRIS, b Norwich, Conn, Apr 14, 38. MATHEMATICS. *Educ:* Carnegie-Mellon Univ, BS, 61, MS, 65, PhD(math), 69. *Prof Exp:* Asst prof, 69-74, assoc prof, 74-79, PROF MATH, WVA UNIV, 79- *Concurrent Pos:* Consult, US Bur Mines, 66-75 & US Dept Energy, 76-80. *Mem:* Soc Indust & Appl Math; Math Asn Am; Sigma Xi. *Res:* Disconjugacy of ordinary differential equations; magnetohydrodynamic generators and power plants; magnetic and electrostatic filtration; hydrogasification of coal to pipeline gas; biomechanics. *Mailing Add:* Dept Math WVa Univ Morgantown WV 26506

SIMONSEN, DAVID RAYMOND, b Clay Co, Nebr, July 29, 16; m 47; c 5. PHYSICAL CHEMISTRY. *Educ:* Dana Col, BA, 38; Univ Nebr, MA, 43, PhD(chem), 44. *Prof Exp:* Prin & instr high schs, Nebr, 38-41; teaching asst, Univ Nebr, 41-44; res chemist, Eastman Kodak Co, 44-52, asst supvr, 52-64, supvr, 64-81; RETIRED. *Concurrent Pos:* Chmn int comt, Contamination Control Soc, 76-78. *Mem:* Am Chem Soc; Instrument Soc Am; Inst Environ Sci (liaison vpres, 73-76); Sigma Xi. *Res:* Contamination control; instrumental gas analysis. *Mailing Add:* 106 White Oak Lane Greenwood SC 29646-9226

SIMONSEN, DONALD HOWARD, b Portland, Ore, June 12, 21; m 47; c 4. BIOCHEMISTRY. *Educ:* Reed Col, BA, 43; Ore State Col, MA, 45; Ind Univ, PhD(chem), 51. *Prof Exp:* Res assoc zool, Ind Univ, 47-50, fel, 51-52; res scientist virol, Upjohn Co, 52-56; from asst prof to assoc prof chem, 56-63, chmn dept phys sci & math, 60-6i, chmn dept chem, 61-64, assoc dean instr, 66-67, acad vpres, 67-69, actg pres, 69-70, PROF CHEM, CALIF STATE UNIV, LONG BEACH, 63- *Mem:* AAAS; Am Chem Soc; fel Am Inst Chem; Brit Biochem Soc; NY Acad Sci. *Res:* Growth factors for guinea pigs; tissue metabolism; salt and nitrogen metabolism in surgical patients; methods of blood volume determination; human plasma and plasma substitutes in treatment of clinical hypoproteinemia; biochemical genetics of Paramecia; microrespiration techniques; chemotherapy of virus disease. *Mailing Add:* 6350 El Paseo Ct Long Beach CA 90815

SIMONSEN, STANLEY HAROLD, b Missoula, Mont, Aug 25, 18; m 43; c 4. CHEMISTRY. *Educ:* Iowa State Teachers Col, AB, 40; Univ Ill, MS, 47, PhD(chem), 49. *Prof Exp:* Anal chemist, Vanadium Corp Am, 40-44; asst analytical chem, Univ Ill, 46-49; asst chem, 49-53, assoc prof, 53-63, assoc dir, Anal Chem Res Lab, 51-52, PROF CHEM, UNIV TEX, AUSTIN, 63- *Concurrent Pos:* Instr, Pa State Col, 43-44. *Mem:* Am Chem Soc. *Res:* Structural investigations with x-rays; x-ray diffraction; instrumental analytical methods; investigation of metallo-organic compounds useful in analytical chemistry. *Mailing Add:* Dept Chem WEL 3-148 Univ Tex Austin TX 78712

SIMONSON, GERALD HERMAN, b Albert Lea, Minn, Nov 19, 27; m 52; c 3. SOIL SCIENCE, AGRONOMY. *Educ:* Univ Minn, BS, 51, MS, 53; Iowa State Univ, PhD(soils), 60. *Prof Exp:* Asst soil scientist, NDak State Univ, 53-55; res assoc soil surv, Iowa State Univ, 55-60; asst prof soil classification, Mont State Col, 60-61; from asst prof to prof soil classification & surv, 61-71, PROF SOIL SCI, ORE STATE UNIV, 71- *Concurrent Pos:* Specialist, soil classification & erosion, W Africa, 88. *Mem:* Am Soc Agron; Sigma Xi; Int Soil Sci Soc. *Res:* Soil-landscape relationships in alluvial landscapes; nature of soil characteristics related to environment of formation; soil distribution patterns in Oregon; soil-natural vegetation relationships; andisols and spodosol classification on Oregon coast. *Mailing Add:* Dept of Soils Ore State Univ Corvallis OR 97331

SIMONSON, JOHN C, b Washington, DC, Nov 16, 60. GEOLOGY. *Educ:* Col William & Mary, BS, 82; Univ Tenn, MS, 85. *Prof Exp:* Explor geologist, Amoco Prod Co, 85-87; HYDRO-GEOLOGIST, ENVIRON RESOURCE MGT INC, 87- *Concurrent Pos:* Res fel, Oak Ridge Nat Lab, 85-87. *Mem:* Geol Soc Am. *Mailing Add:* 2124 S Hicks St Philadelphia PA 19145

SIMONSON, LLOYD GRANT, b San Jose, Calif, Dec 1, 43; m 68. MICROBIOLOGY, IMMUNOBIOLOGY. *Educ:* Western Ill Univ, BA, 66; Ill State Univ, MS, 68, PhD(microbiol), 74; Roosevelt Univ, MBA, 84. *Prof Exp:* Res microbiologist, 68-72; res microbiologist dent res & prin investr, 74-87; SCI DIR, HEAD SCI INVESTS DEPT, NAVAL DENTAL RES INST, 87-; ADJ PROF, NORTHWESTERN UNIV DENT SCH, 90- *Concurrent Pos:* Consult, Chicago Med Sch, 77-80; prog chmn, microbiol-immunol group Int Asn Dent Res-Am Asn Dent Res, Chicago Sect, 84-87, treas, 86-87, vpres, 87-88, pres, 88-89; grant reviewer, Med Res Coun, Can, NIH, Oral Biol & Med Study Sect. *Mem:* Am Soc Microbiol; Int Asn Dent Res; Am Asn Dent Res; AAAS; Sigma Xi. *Res:* Monoclonal antibodies; enzyme-linked immunoassay; affinity chromatography; development of immunodiagnostics; hybridoma technique; dental caries therapeutics; enzymology and molecular biology; cell mediated immunity and cancer; immobilized enzymes and intermolecular conjugation; bacterial adherence to surfaces and adherence-inhibition; surveys for microbial enzymes and fermentation; fluorescence immunoassay. *Mailing Add:* Naval Dent Res Inst Bldg 1-H NTC Great Lakes IL 60088-5259

SIMON WALHOUT, JUSTINE I, b Dec 11, 30; m 58; c 4. ORGANIC CHEMISTRY. *Educ:* Wheaton Col, Ill, BS, 52; Northwestern Univ, PhD(org chem), 56. *Prof Exp:* From asst prof to assoc prof chem, Rockford Col, 56-65, dept chmn, 57-59 & 87-; consult, Pierce Chem Co, Rockford Ill, 67-68; assoc adj prof, 81-82, assoc prof, 82-89, PROF CHEM, ROCKFORD COL, ILL, 89- *Concurrent Pos:* Mem GOB Testing Comt, Am Chem Soc, 87-91. *Mem:* Am Chem Soc; Sigma Xi; Am Asn Univ Women. *Mailing Add:* Dept Chem Rockford Col 5050 E State St Rockford IL 61108

SIMOPOULOS, ARTEMIS PANAGEOTIS, b Kampos-Avias, Greece, Apr 3, 33; US citizen; m 57; c 3. PEDIATRICS, ENDOCRINOLOGY. *Educ:* Barnard Col, Columbia Univ, BA, 52; Boston Univ, MD, 56; Am Bd Pediat, dipl, 64. *Prof Exp:* Spec lectr pediat, Sch Med, Ewha Woman's Univ, Seoul, Korea, 58-59; NIH fel hemat, 60-61; asst prof, Sch Med, George Washington Univ, 62-67; staff pediatrician, Nat Heart & Lung Inst, 68-71; prof assoc, Div Med Sci, Nat Acad Sci-Nat Res Coun, 71-74, actg exec secy, 74-75, exec secy, 75-76; co-chmn & exec secy, Joint Subcomt, Human Nutrit Res, Off Sci & Technol Policy, Exec Off of the President, 79-83; chief develop biol & nutrit br, Nat Inst Child Health & Human Develop, 77, vchmn & exec secy nutrit coord comt, Off of Dir, 77-78; chmn nutrit coord comt & spec asst for coordr nutrit res, Off of Dir, NIH, 78-86; dir, Div Nutrit Sci, Int Life Sci Inst Res Found, 87-88; dir, 89-90, PRES, CTR GENETICS, NUTRIT & HEALTH, INC, 90- *Concurrent Pos:* Mem acad staff pediat, Children's Hosp of DC, 62-67, mem assoc staff, Nursery Serv, 67-71; dir nurseries, George Washington Univ Hosp, 65-67; co-chmn perinatal comt, Working Party Biol Aspects Prev Ment Retardation for DC, 67-68; clin asst prof pediat, Sch Med, George Washington Univ, 67-71; mem bd, Capitol Head Start, Inc, Washington, DC, 68-70; consult, Endocrinol Br, Nat Heart & Lung Inst, 71-78; liaison, Div Med Sci, Nat Acad Sci-Nat Res Coun to Am Acad Pediat Comt Drugs, 71-76; mem res adv comt, Maternity Ctr Assoc, New York, 72-; exec dir bd, Maternal, Child & Family Health Res, 74-76; adv & mem, US deleg, Ad Hoc Comt Food & Nutrit Policy, UN Food & Agr Orgn, 78, mem, US deleg, Meeting on Infant & Young Child Feeding, WHO-UNICEF, 79, tech adv, US Govt Deleg, Meeting on the 3rd Draft Int Code Mkt Breast Milk Substitutes, WHO, 80; contrib ed, Nutrit Rev, 79-86; consult ed, Nutrit Res, 83-, Ann Internal Med, 84- & J AMA, 85-; mem, Expert Comt Nutrit, Int Life Sci Inst, 84-88; ed, World Rev Nutrit & Dietetics, 89- *Mem:* Fel Am Acad Pediat; Soc Pediat Res; Endocrine Soc; Am Pediat Soc; Am Inst Nutrit; Am Asn World Health (asst treas & vpres, 81-); Am Soc Clin Nutrit. *Res:* Genetics of endocrine diseases and growth problems in children; clinical nutrition; body weight, health, longevity and obesity; body weight standards; omega-3 fatty acids; genetic variation and nutrition; nutrition and fitness; antioxidant vitamins. *Mailing Add:* 4330 Klingle St NW Washington DC 20016

SIMOVICI, DAN, b Iassy, Romania, Feb 15, 43; US citizen; m 65; c 1. FORMAL LANGUAGES, AUTOMATA THEORY. *Educ:* Polytech Inst Iassy, Romania, MS, 65; Univ Iassy, MS, 70; Univ Bucharest, PhD(math), 74. *Prof Exp:* Asst prof comput sci, Univ Iassy, 71-78, assoc prof, 78-80; assoc prof comput sci, Univ Miami, 81-82; assoc prof, 82-84, PROF COMPUT SCI, UNIV MASS, 85- *Mem:* Am Math Soc; Asn Comput Mach; Comput Soc of Inst Elec & Electronics Engrs. *Res:* Theoretical computer science and related areas; formal languages; automata theory; switching theory; boolean algebras; database systems. *Mailing Add:* Dept Math & Comput Sci Univ Mass Boston MA 02125

SIMPKINS, JAMES W, b Port Clinton, Ohio, Sept 22, 48; m; c 3. PHARMACODYNAMICS. *Educ:* Univ Toledo, BS, 71, MS, 74; Mich State Univ, PhD(physiol), 77. *Prof Exp:* Teaching asst, Dept Biol, Univ Toledo, 71-74; technician, Dept Physiol, Med Col Ohio, Toledo, 72-74; instr, Sci Dept, Lansing Community Col, Mich, 75-77; res asst, Dept Physiol, Mich State Univ, 74-77; assoc chmn, Dept Pharmacodynamics, Univ Fla, 84-86, chmn, 86-88, asst dean res & grad studies, Col Pharm, 88-89, assoc dean, 89-91, from asst prof to assoc prof, 77-86, PROF, DEPT PHARMACODYNAMICS, UNIV FLA, 86- *Mem:* Endocrine Soc; Am Physiol Soc; Sigma Xi; AAAS; Soc Neurosci; Gerontol Soc Am; Am Asn Pharmaceut Scientists; Am Asn Col Pharm. *Res:* Mechanism of hypothalmic control on anterior pituitary function with particular interest in the role of catechol- and indolemines in the secretion of pituitary hormones during development, aging and the feedback of steroids and peptide hormones; pharmacology of opiate agonists and antagonists; pharmacology of brain-specific drug delivery system; numerous technical publications. *Mailing Add:* Dept Pharmacodynamics Col Pharm Univ Fla Box J-487 JHMEC Gainesville FL 32610

SIMPKINS, PETER G, b London, Eng, Nov 28, 34; m 59; c 3. AERONAUTICS, FLUID MECHANICS. *Educ:* Univ London, Dipl technol, 58, PhD(aeronaut), 64; Calif Inst Technol, MSc, 60. *Prof Exp:* Apprentice engr, Handley Page Co, Ltd, 53-57, mem res staff aerodyn, 57-58; res asst aeronaut, Imp Col, Univ London, 60-65; sr consult scientist, Avco Res & Develop Labs, 65-68; mem tech staff, Anal Mech Dept, 68-71 & Ocean Physics Res Dept, 71-74, mem tech staff, Mat Res Labs, Bell Labs, 74-83, DISTINGUISHED TECH STAFF, MAT PHYSICS RES, AT&T BELL LABS, 83- *Concurrent Pos:* Sr res fel, Southampton Univ, Eng, 73-74; consult, Eng Mech Div, NSF, 75-; mem, Space Lab 3 Rev Bd, NASA, 78-79, Mat Processing in Space Panel, 80-81; adj prof chem eng, Lehigh Univ, 90- *Mem:* Am Inst Physics; Am Phys Soc; Am Soc Mech Engr. *Res:* Fluid mechanics, convection, gas dynamics, heat transfer and wave propagation in solids. *Mailing Add:* AT&T Bell Labs Rm 1A121 600 Mountain Ave Murray Hill NJ 07974

SIMPLICIO, JON, b Bronx, NY, Sept 18, 42; m 69. BIOCHEMISTRY, INORGANIC CHEMISTRY. *Educ:* State Univ NY Stony Brook, BS, 64; State Univ NY Buffalo, PhD(chem), 69. *Prof Exp:* Res assoc biochem, Cornell Univ, 68-69; asst prof chem, Univ Miami, 70-77; mem staff, Allied Chem Corp, 77-82; MEM STAFF, CELANESE CORP, 82- *Mem:* Am Chem Soc. *Res:* Fast reaction mechanisms and kinetics of enzymes; kinetics of metalloporphyrins with nucelophiles and their interactions with micelles. *Mailing Add:* 20 Waterside Plaza New York NY 10010

SIMPSON, ANTONY MICHAEL, b Leamington, Eng, May 8, 41; m 69; c 2. SOLID STATE PHYSICS. *Educ:* Cambridge Univ, BA, 63; Dalhousie Univ, MSc, 65, PhD(physics), 69. *Prof Exp:* From asst prof to assoc prof, 68-89, PROF PHYSICS, DALHOUSIE UNIV, 89-, CHAIR, PHYSICS DEPT, 89- *Concurrent Pos:* Fel, Cambridge Univ, 71-72; vis sci, Univ BC, 83-84. *Mem:* Am Phys Soc; Can Asn Physicists; Can Acousts Asn. *Res:* Transport properties, ultrasonics and thermal expansion in solids. *Mailing Add:* Dept Physics Dalhousie Univ Halifax NS B3H 3J5 Can

SIMPSON, BERYL BRINTNALL, b Dallas, Tex, Apr 28, 42; div. BIOSYSTEMATICS. *Educ:* Radcliffe Col, BA, 64; Harvard Univ, MA & PhD(biol), 67. *Prof Exp:* Res botanist, Arnold Arboretum, Harvard Univ, 68, res fel bot, Gray Herbarium, 68-69, res assoc, 69-71; assoc cur dept bot, US Mus Natural Hist, Smithsonian Inst, 72-78; PROF BOT, UNIV TEX, AUSTIN, 78- *Mem:* Soc Study Evolution (pres, 86); Bot Soc Am; Am Soc Plant Taxon; fel AAAS. *Res:* Speciation problems of Andean plant genera; systematics of Andean genera; reproductive systems of angiosperms; pollination biology. *Mailing Add:* Dept Bot Univ Tex Austin TX 78713

SIMPSON, BILLY DOYLE, b Holdenville, Okla, Mar 12, 29; m 57; c 2. ORGANIC CHEMISTRY. *Educ:* Okla State Univ, BS, 51, MS, 53; Univ Kans, PhD(org chem), 57. *Prof Exp:* Res chemist, 57-66, adhesives develop & sales consult, 66-75, adhesive develop & lab coordr, 75-77, mkt segment mgr molded & extruded goods, 77-80, TECH SERV & DEVELOP MGR, PHILLIPS PETROL CO, 80- *Mem:* Am Chem Soc; Soc Petrol Engrs. *Res:* Nitrogen-containing compounds; petrochemicals; adhesives. *Mailing Add:* 2022 Hamlin Valley Dr Houston Bartlesville TX 77090-2019

SIMPSON, CHARLES FLOYD, veterinary pathology; deceased, see previous edition for last biography

SIMPSON, DALE R, b Wilmar, Calif, Dec 29, 30; m 58; c 2. PETROLOGY, MINERALOGY. *Educ:* Pa State Univ, BS, 56; Calif Inst Technol, MS, 58, PhD(geol), 60. *Prof Exp:* From asst prof to assoc prof geol, 60-66, PROF GEOL, LEHIGH UNIV. 66- *Mem:* Mineral Soc Am; fel Geol Soc Am. *Res:* Synthesis, stability and chemical variants of phosphate minerals; petrology of granitic pegmatites; energy storage using latent heat; effect of saline waters on igneous rocks. *Mailing Add:* Dept of Geol Lehigh Univ Bethlehem PA 18015

SIMPSON, DAVID ALEXANDER, b Englewood, NJ, Mar 2, 43. PHYSICAL ORGANIC CHEMISTRY. *Educ:* Allegheny Col, BS, 65; Univ Ill, MS, 68, PhD(org chem), 69. *Prof Exp:* Res chemist, 69-74; SR RES CHEMIST, ORG & PHYS ORG CHEM, HERCULES, INC, 74- *Mem:* Am Chem Soc; Inter-Am Photochem Soc; Sigma Xi. *Res:* Photochemistry; photopolymerization; photooxidation; photo crosslinking and photodegradation of polymers; laser induced chemical and photochemical reactions. *Mailing Add:* 1121 Flint Hill Rd Arundel Wilmington DE 19808

SIMPSON, DAVID GORDON, b Belfast, Northern Ireland, Jan 24, 20; US citizen; m 56; c 4. INTERNAL MEDICINE. *Educ:* Queen's Univ, Belfast, MB, BCh & BAO, 42, MD, 50. *Prof Exp:* Sr registr med, Northern Ireland Hosps Authority, 49-52; from resident to chief resident chest serv, Bellevue Hosp, Columbia Univ Div, New York, 52-56, instr med, Col Physicians & Surgeons, Columbia Univ, 56-64; assoc prof med & head div pulmonary dis, Sch Med, Univ Md, Baltimore, 64-76, assoc prof internal med, 76-84; RETIRED. *Concurrent Pos:* Consult, Vet Admin Hosp, Baltimore, 64-, Keswick Home, 64-, Md Gen Hosp, 64-, Mercy Hosp, 64- & Montebello State Hosp, 64- *Mem:* Am Thoracic Soc. *Res:* Pulmonary tuberculosis and other respiratory diseases. *Mailing Add:* 641 W University Pkwy Baltimore MD 21210

SIMPSON, DAVID PATTEN, b Eugene, Ore, Mar 20, 30; m 56; c 4. NEPHROLOGY. *Educ:* Harvard Univ, AB, 52; McGill Univ, MD, 57. *Prof Exp:* Intern, Grace-New Haven Community Hosp, 57-58; resident med, Scripps Clin & Res Found, 58-60; vis scholar, Col Physicians & Surgeons, Columbia Univ, 60-65; from asst prof to assoc prof med, Univ Wash, 65-74; PROF MED & DIR NEPHROL PROG, UNIV WIS CTR HEALTH SCI, 74- *Concurrent Pos:* USPHS res fel, 60-62; NY Heart Asn sr res fel, 62-65; chief nephrology, USPHS Hosp, 70-74. *Res:* Relationships between intermediary metabolism in the kidney and renal physiology; regulation of organ growth. *Mailing Add:* Rte 1 Box 1080 Lopez WA 98261

SIMPSON, EUGENE SIDNEY, b Schenectady, NY, July 14, 17. HYDROLOGY. *Educ:* City Col New York, BS, 44; Columbia Univ, MA, 49, PhD(geol), 60. *Prof Exp:* Geologist, US Geol Surv, 46-63; prof, 63-85, EMER PROF HYDROL, UNIV ARIZ, 85- *Mem:* Geol Soc Am; Am Soc Civil Eng; Am Geophys Union; Sigma Xi. *Res:* Dispersion of fluids in subsurface flow; ground water hydraulics and chemistry. *Mailing Add:* Dept Hydrol & Water Resources Univ Ariz Tucson AZ 85721

SIMPSON, EVAN RUTHERFORD, REPRODUCTIVE ENDOCRINOLOGY. *Educ:* Univ Edinburgh, Scotland, PhD(biochem), 67. *Prof Exp:* PROF OBSTET & GYNECOL & BIOCHEM, SOUTHWESTERN MED SCH, UNIV TEX, 82- *Res:* Regulation of steroid hormone biosynthesis. *Mailing Add:* Dept Biochem Southwestern Med Sch Univ Tex 5323 Harry Hines Blvd Dallas TX 75235

SIMPSON, EVERETT COY, b Maysville, NC, Feb 13, 25; m 51; c 4. ZOOLOGY, ENDOCRINOLOGY. *Educ:* Okla State Univ, BS, 50; Univ Ky, MS, 52, PhD(genetics), 60. *Prof Exp:* Res asst animal sci, Univ Ky, 51-52 & genetics, 56-60; asst prof biol, Memphis State Univ, 60-61; assoc prof, 61-64, assoc dir dept, 71, PROF BIOL, E CAROLINA UNIV, 64- *Concurrent Pos:* Res grants, 63-65 & 66-68; NSF inserv Inst Awards, 65-67; dir three inserv insts, 66, 67 & 68. *Mem:* AAAS; Am Genetic Asn. *Res:* Endocrinology, especially the physiology of reproduction. *Mailing Add:* Dept of Biol ECarolina Univ Greenville NC 27834

SIMPSON, FRANK, b Gatley, Eng, Dec 4, 41; Can & UK citizen; m 68; c 2. RESERVOIR GEOLOGY. *Educ:* Univ Edinburgh, BSc, 65; Jagiellonian Univ, Krakow, Dr Nat Sci(geol), 68. *Prof Exp:* Res geol, Sask Dept Mineral Resources, 69-74; assoc prof, 74-80, PROF GEOL, UNIV WINDSOR, 80- *Concurrent Pos:* Chmn dept, Univ Windsor, 82-88; hon prof geol, Univ St Andrews, Scotland, 83; pres, Coun Univ Dept Geol, Ont, 84-85. *Honors & Awards:* Ludwik Zejszner Science Award, Geol Soc Poland, 70. *Mem:* Can Soc Petrol Geologists; Geol Soc Poland; Soc Econ Paleontologists & Mineralogists; Asn Geoscientists Int Develop. *Res:* Sedimentology of siliciclastic deposits; cretaceous stratigraphy; petroleum geology; solution-generated collapse features; environmental management; geology of east and central Africa. *Mailing Add:* Dept Geol Univ Windsor 401 Sunset Ave Windsor ON N9B 3P4 Can

SIMPSON, FREDERICK JAMES, b Regina, Sask, June 8, 22; c 5. BACTERIOLOGY, SCIENCE ADMINISTRATION. *Educ:* Univ Alta, BSc, 44, MSc, 46; Univ Wis, PhD(bact), 52. *Prof Exp:* Jr res officer div appl biol, Nat Res Coun, Can, 46-48, asst res officer bact, Prairie Regional Lab, 52-57, from assoc res officer to sr res officer, 58-70, head physiol & biochem of bacteria, 58-70, asst dir, Atlantic Regional Lab, 70-73, dir Atlantic Regional Lab, 73-84; RETIRED. *Concurrent Pos:* Chmn, Atlantic Coun Provinces Coun Sci, 81-84. *Honors & Awards:* Queen's Silver Anniversary Medal. *Mem:* Aquaculture Asn Can; Can Soc Microbiol. *Res:* Metabolism of sugars; enzymic degradation of hemicelluloses and aromatic compounds; marine phycology. *Mailing Add:* 95 Empire St Apt 11 Bridgewater NS B4V 2L5 Can

SIMPSON, GEDDES WILSON, b Scranton, Pa, Aug 15, 08; m 33; c 4. ECONOMIC ENTOMOLOGY. *Educ:* Bucknell Univ, AB, 29; Cornell Univ, AM, 31, PhD(econ entom), 35. *Prof Exp:* Asst entom, State Univ NY Col Agr, Cornell Univ, 30-31; asst entomologist, Maine Exp Sta, 31-44, chg roguing serv, 38-46, chg Fla test plot, 39-59, assoc entomologist, 44-52, entomologist, 52-74, prof entom & chmn dept, Univ, 54-74, EMER PROF ENTOM, UNIV MAINE, ORONO, 74- *Concurrent Pos:* Ed-in-chief, Am Potato J, 75-88; asst to dir, Maine Agr Exp Sta, 76-88. *Mem:* Fel AAAS; Entom Soc Am; Potato Asn Am; Acadian Entom Soc. *Res:* Insect transmission of plant virus diseases; biology and control of aphids affecting potatoes. *Mailing Add:* 235 Fairground Rd Lewisburg PA 17837-1289

SIMPSON, HOWARD DOUGLAS, b Carrizozo, NMex, May 30, 37; m 67; c 1. CATALYSIS, HYDROTREATING. *Educ:* Univ NMex, BS, 59; Univ Tex, Austin, MS, 65, PhD(chem eng), 69. *Prof Exp:* Res engr, Unocal Corp, 71-76, sr res engr, 76-81, res assoc, 81-85, sr res assoc, 85-89, STAFF CONSULT, UNOCAL CORP, 89- *Mem:* Catalysis Soc; Am Chem Soc; Am Crystallog Asn; Sigma Xi. *Res:* New and improved hydrotreating catalysts for the petroleum industry. *Mailing Add:* 3772 Hamilton St Irvine CA 92714

SIMPSON, HOWARD EDWIN, b Grand Forks, NDak, June 27, 17; m 43; c 4. GEOLOGY. *Educ:* Univ NDak, BA, 40; Univ Ill, MS, 42; Yale Univ, PhD(geol), 53. *Prof Exp:* Geologist, Eng Geol Br, US Geol Surv, 47-60, geologist, Regional Geol Br, 60-66, geologist, Eng Geol Br, 66-77, geologist, Spec Proj Br, 77-81; CONSULT GEOL & QUAL ASSURANCE ENG, 81- *Mem:* Asn Eng Geol; Am Inst Prof Geologists; Geol Soc Am. *Res:* Application of geology to problems of urban development; geomorphology; Pleistocene geology. *Mailing Add:* 2020 Wash Ave Golden CO 80401-2361

SIMPSON, IAN ALEXANDER, b East Grinstead, Gt Brit, Mar 27, 48; m 77; c 2. DIABETES RESEARCH. *Educ:* Hull Univ, BSc, 71; Univ Col, London, PhD(biochem), 75. *Prof Exp:* Grad fel, Univ Col, London, 71-74; res assoc, Muscular Dystrophy Group, Gt Brit, Guys Hosp Med Sch, London, 74-77; res assoc, Physiologisch-Chemisches Inst, Univ Wuerzburg, WGer, 77-79; vis assoc, Cellular Metabolism & Obesity Sect, Nat Inst Arthritis, Diabetes & Digestive & Kidney Dis, NIH, 79-82, vis scientist, 82-85, VIS SCIENTIST & ASSOC CHIEF, EXP DIABETES METABOLISM & NUTRIT SECT, NIH, 85- *Concurrent Pos:* Res grants, Muscular Dystrophy Group Gt Brit, Guy's Hosp Med Sch, 74-77, Deutsche Forschungsgemeinschaft, Univ Wuerzburg, 77-79 & Am Diabetes Asn, NIH, 87-88 & 88-89. *Mem:* Am Diabetes Asn; Am Soc Biochem & Molecular Biol; Endocrine Soc. *Res:* Diabetes; biology; biochemistry. *Mailing Add:* Bldg 10 Rm 5N102 NIH Bethesda MD 20892

SIMPSON, JAMES EDWARD, b Chicago, Ill, July 6, 31; div; c 3. GRAPH THEORY, COMBINATORICS. *Educ:* Loyola Univ, Ill, BSEd, 53, MA, 56; Yale Univ, PhD(math), 61. *Prof Exp:* From instr to assoc prof math, Marquette Univ, 55-66; ASSOC PROF MATH, UNIV KY, 66- *Concurrent Pos:* NSF res grants, 63-66; vis Fulbright prof, Arya-Mehr Univ Technol, 68-70. *Mem:* Am Math Soc; Math Asn Am. *Res:* Functional analysis; spectral analysis of operators; graph theory; combinatorics; database management. *Mailing Add:* Dept of Math Univ of Ky Lexington KY 40506

SIMPSON, JAMES HENRY, b Haledon, NJ, Oct 13, 29; m 61; c 3. PHYSICS. *Educ:* Rutgers Univ, BS, 51, PhD(physics), 58. *Prof Exp:* Asst prof physics, Univ Del, 57-58; prin scientist, Singer Co, 58-80, mgr appl physics dept, Kearfott Div, 80-90; CONSULT, 90- *Concurrent Pos:* adj prof, Fordham Univ, 70- *Mem:* Am Phys Soc; Inst Elec & Electronic Eng; Sigma Xi. *Res:* Magnetic resonance; atomic physics; ring laser gyroscopes; optically-pumped, magnetic-resonance gyroscopes. *Mailing Add:* 112 Lily Pond Lane Katonah NY 10536

SIMPSON, JAMES R(USSELL), b Passaic, NJ, Mar 22, 11; m 35, 73; c 2. CIVIL ENGINEERING. *Educ:* Va Polytech Inst, BS, 34, MS, 42; Environ Engrs Intersoc, dipl, 56. *Prof Exp:* Indust hyg engr, USPHS, 35-36, sanit engr, 44-47; sanit engr, State Dept Health, Va, 36-37, pub health engr, 38-42; sanit engr, Fed Housing Admin, 43-44 & 47-56, chief sanit eng sect & spec asst tech studies prog, 56-59, chief standards & studies sect, 59-64, dep dir archit standards div, 64-67, dir off advan bldg technol, Dept Housing & Urban Develop, 67-70; HOUSING INDUST CONSULT, 70- *Mem:* Inst Elec & Electronics Engrs. *Res:* Building research. *Mailing Add:* 7721 Weber Ct Annandale VA 22003

SIMPSON, JOANNE, b Boston, Mass, Mar 23, 23; m 48, 65; c 3. METEOROLOGY. *Educ:* Univ Chicago, PhD, 49. *Hon Degrees:* DSc, State Univ NY, Albany, 91. *Prof Exp:* Instr meteorol, NY Univ, 43-44; instr, Univ Chicago, 44-45; instr physics & meteorol, Ill Inst Technol, 46-49, asst prof, 49-51; meteorologist, Woods Hole Oceanog Inst, 51-60; prof meteorol, Univ Calif, Los Angeles, 60-65; head exp br atmospheric physics & chem lab, Environ Sci Serv Admin, 65-71, dir exp meteorol lab, Nat Oceanic & Atmospheric Admin, 71-74; prof environ sci & mem ctr advan studies, Univ Va, 74-76, W W Corcoran prof, 76-81; head severe storms br, 79-88, CHIEF SCIENTIST, METEOROL & EARTH SCI DIRECTORATE, GODDARD SPACE FLIGHT CTR, NASA, 88-, SR FEL, 88- *Concurrent Pos:* Hon lectr, Imp Col, Univ London & Guggenheim fel, 54-55; counr, Am Meteorol Soc, 75-77 & 79-81; comnr, Sci Technol Activities, 82-87; chief scientist, Simpson Weather Assocs, 74-79; project scientist, Tropical Rainfall Measuring Mission, Goddard Space Flight Ctr, NASA, 86- *Honors & Awards:* Meisinger Award, Am Meteorol Soc, 62, Rossby Res Medal, 83; Silver Medal, Dept Com, 67, Gold Medal, 72; Vincent J Schaefer Award, Weather Modification Asn, 79; Except Sci Achievement Medal, NASA, 82. *Mem:* Nat Acad Eng; fel Am Meteorol Soc. *Res:* Convection in atmosphere; cumulus clouds; tropical meteorology; weather modification; satellite meteorology. *Mailing Add:* Code 912 Earth Sci Directorate Severe Storms Br Goddard Space Flight Ctr Greenbelt MD 20771

SIMPSON, JOHN ALEXANDER, b Portland, Ore, Nov 3, 16; m 46; c 2. PHYSICS. *Educ:* Reed Col, AB, 40; NY Univ, MS, 42, PhD(physics), 43. *Prof Exp:* Asst physics, NY Univ, 40-42, res assoc, Off Sci Res & Develop proj, 42-43; sci group leader, Manhattan Dist Proj, Metall Lab, 43-46, from instr to prof, 45-68, Edward L Ryerson distinguished serv prof, 68-74, ARTHUR H COMPTON DISTINGUISHED SERV PROF PHYSICS, UNIV CHICAGO & ENRICO FERMI INST NUCLEAR STUDIES, 74-, DIR, INST, 73-, EMER PROF PHYSICS, 87- *Concurrent Pos:* Estab lab astrophys & space res in Enrico Fermi Inst, Univ Chicago, 64; fel, Ctr Policy Study, 66-; sci consult, Argonne Nat Lab, 46-54; chmn comt biophys, Univ Chicago, 51-52; mem spec int comt, Int Geophys Yr, 54-60 & US nat comt & tech panel cosmic rays, 55-58; mem space sci bd, Nat Acad Sci, 58-66, consult, 66-; pres, Int Comn Cosmic Radiation, 64-67; mem astron missions bd, NASA. *Mem:* Nat Acad Sci; fel Am Phys Soc; fel Am Geophys Union; Am Astron Soc; Int Acad Astronaut. *Res:* Cosmic radiation origin; galactic, solar and magnetospheric acceleration of particles; experiments on nuclear composition, spectra, time variations of radiation with neutron monitors, satellite and space probes; interplanetary and solar magnetic fields deduced from particle propagation. *Mailing Add:* Enrico Fermi Inst Nuclear Studies Dept Physics Univ Chicago Chicago IL 60637

SIMPSON, JOHN AROL, b Toronto, Ont, Mar 30, 23; nat US; m 48; c 1. PHYSICS, SCIENCE ADMINISTRATION. *Educ:* Lehigh Univ, BS, 46, MS, 48, PhD(physics), 53. *Prof Exp:* Physicist, Nat Bur Standards, 48-62, chief electron physics sect, 62-69, dep chief optical physics div, 69-75, chief mech div, 75-78, dir, Ctr Mfg Eng, 81-91, DIR, MFG ENG LAB, NAT INST STANDARDS & TECHNOL, 91- *Concurrent Pos:* Asst, Lehigh Univ, 51-52. *Honors & Awards:* Doc Silver Medal & Gold Medal; Allen V Austin Measurement Sci Award; Am Machinist Award. *Mem:* Am Phys Soc; Nat Acad Eng; Am Phys Soc. *Res:* Metrology; optics. *Mailing Add:* 312 Riley St Falls Church VA 22046

SIMPSON, JOHN BARCLAY, b Oakland, Calif, June 8, 47; m; c 2. BEHAVIORAL PHYSIOLOGY. *Educ:* Univ Calif, Santa Barbara, BA, 69; Northwestern Univ, MA, 72, PhD(neurobiol & behav), 73. *Prof Exp:* Instr, Col Gen Studies, Univ Pa, 74-75; from asst prof to assoc prof, 75-82, PROF PSYCHOL, UNIV WASH, 82-, DIR, PHYSIOL & PSYCHOL PROG, 85- *Concurrent Pos:* Fel, Inst Neurol Sci, Univ Pa, 73-75; vis assoc prof physiol, Univ Calif, San Francisco, 76-80; vis prof, Howard Florey Inst Exp Physiol & Med, Univ Melbourne, Australia, 83. *Mem:* AAAS; Soc Neurosci; Soc Study Ingestive Behavior. *Res:* Neural involvement in body fluid regulation; neural control of ingestive behaviors. *Mailing Add:* Dept of Psychol NI-25 Univ of Wash Seattle WA 98195

SIMPSON, JOHN ERNEST, b Toledo, Ohio, Feb 10, 42; m 66; c 1. ORGANIC CHEMISTRY, ENOLOGY. *Educ:* Univ NMex, BS, 63, MS, 66, PhD(chem), 68. *Prof Exp:* Asst prof chem, Pomona Col, 67-68; assoc prof, 68-80, PROF CHEM, CALIF STATE POLYTECH UNIV, POMONA, 80- *Mem:* Am Chem Soc; AAAS; Sigma Xi; Am Soc Enologists. *Res:* Organic synthesis-crown ethers; bridged aromatics; C-13 labeled compounds; phenolics, isolation and identification in grapes and wines; nuclear magnetic resonance spectroscopy. *Mailing Add:* Chem Dept Calif State Polytech Univ 3801 W Temple Pomona CA 91768

SIMPSON, JOHN HAMILTON, b Montreal, Que, May 28, 15; m 53; c 2. THEORETICAL SOLID STATE PHYSICS, SOLID STATE ELECTRONICS. *Educ:* McGill Univ, BEng, 37; Bristol Univ, PhD(theoret physics), 50. *Prof Exp:* Test engr, Can Gen Elec Co, 37-38; res officer, Radio & Elec Eng Div, Nat Res Coun Can, 38-46, sr res officer, 48-61, prin res officer, Physics Div, 61-76, energy consult, 77-81; CONSULT PHOTOVOLTAICS, 81- *Concurrent Pos:* Lectr, Univ Ottawa & Carleton Univ; pres, Cosim Solar Res Ltd. *Mem:* Am Phys Soc; Can Asn Physicists; Inst Elec & Electronics Engrs. *Res:* Theory of optical-electrical properties of defects in ionic crystals; theory of semiconducting devices; theory of cooperative effects in dielectrics; direct current transmission line theory. *Mailing Add:* 2184 Braeside Ave Ottawa ON K1H 7J5 Can

SIMPSON, JOHN W(ISTAR), b Glenn Springs, SC, Sept 25, 14; m 48; c 4. ELECTRICAL ENGINEERING. *Educ:* US Naval Acad, BS, 37; Univ Pittsburgh, MS, 41. *Hon Degrees:* DSc, Seton Hill Col, 68. *Prof Exp:* Res engr, Westinghouse Elec Corp, 38-39, mgr, Navy & Marine Switchbd Sect, 39-46; mgr nuclear eng, Power Pile Div, Oak Ridge Nat Lab, 46-48; supv engr circuit breaker design, 48-49, asst eng mgr, Bettis Atomic Power Lab, 48-52, asst div mgr, 52-54, mgr pressurized water reactor proj, 54-55, div mgr, 55-58, vpres & gen mgr, 58-59, vpres & gen mgr, Atomic Power Div, 59-62, vpres eng & res, 62-63, vpres & gen mgr elec utility group, 63-69, PRES POWER SYSTS, WESTINGHOUSE ELEC CORP, 69- *Concurrent Pos:* Consult, US Navy, 45; mem US deleg, Int Conf Peaceful Uses Atomic Energy, 55 & 58. *Honors & Awards:* Edison Medal, 71. *Mem:* Nat Acad Eng; fel Inst Elec & Electronics Engrs; fel Am Soc Mech Engrs; Am Soc Naval Engrs; fel Am Nuclear Soc. *Res:* Design of nuclear reactors. *Mailing Add:* 36 E Beach Lagoon Dr Hilton Head SC 29928

SIMPSON, JOHN WAYNE, b Henryetta, Okla, Aug 17, 35; m 66. BIOCHEMISTRY. *Educ:* Phillips Univ, BA, 57; Rice Univ, MA, 59, PhD(biochem), 65. *Prof Exp:* Asst mem biochem, 67-71, ASSOC PROF BIOCHEM, UNIV TEX DENT SCI INST, HOUSTON, 71- *Concurrent Pos:* Fel, Rice Univ, 65-67. *Mem:* AAAS; Am Physiol Soc; Int Asn Dent Res. *Res:* Comparative aspects of free amino acid distribution; pathways of glucose degradation in invertebrates; intermediary metabolism in oral tissues. *Mailing Add:* Univ Tex Dent Sci Inst PO Box 20068 Houston TX 77025

SIMPSON, KENNETH L, b Los Angeles, Calif, June 24, 31; m 57; c 3. FOOD SCIENCE. *Educ:* Univ Calif, Davis, BS, 54, MS, 60, PhD(agr chem), 63. *Prof Exp:* NSF fel biochem, Unv Col, Wales, 63-64; from asst prof to assoc prof agr chem, 64-69, assoc prof food & resource chem, 69-72, PROF FOOD SCI & TECHNOL, UNIV RI, 72- *Concurrent Pos:* NSF res grants, sea grant; NIH grants; Sci Res Coun vis fel, Univ Liverpool, 71-72. *Honors & Awards:* Emil Racovitza Sci Medallion, Govt Romania, 76. *Mem:* Am Chem Soc; Inst Food Technol; World Mariculture Soc. *Res:* Chemistry and biochemistry of carotenoids in microorganism plants and animals; fish nutrition; utilization of fish processing waste; provitamin A analysis in fruit and vegetables; aquaculture of artemia-brine shrimp. *Mailing Add:* Dept Food Sci & Technol Nutrit Univ of RI Kingston RI 02881

SIMPSON, LARRY P, b Philadelphia, Pa, Oct 31, 40; m 68. CELL BIOLOGY, PROTOZOOLOGY. *Educ:* Princeton Univ, AB, 62; Rockefeller Univ, PhD(cell biol), 67. *Prof Exp:* Asst prof zool, 67-74, assoc prof cell biol, 74-75, PROF CELL BIOL, UNIV CALIF, LOS ANGELES, 75- *Concurrent Pos:* NATO fel sci, Univ Brussels, 67-68; Molecular Biol Inst. *Honors & Awards:* Hutner Award, Soc Protogeol, 80. *Mem:* Soc Protozool; Am Soc Cell Biol; Molecular Biol Inst. *Res:* Cell biology of parasitic protozoa, especially mitochondrial biogenesis. *Mailing Add:* Dept Biol Univ Calif 405 Hilgard Ave Los Angeles CA 90024

SIMPSON, LEONARD, b Hale Center, Tex, Apr 25, 32; m 64. INVERTEBRATE ZOOLOGY. *Educ:* Univ Calif, Berkeley, AB, 55, MA, 62, PhD(zool), 68. *Prof Exp:* Instr biol, Diablo Valley Col, 61-66; asst prof, 68-72, ASSOC PROF BIOL, PORTLAND STATE UNIV, 72- *Mem:* AAAS; Am Soc Zool; Sigma Xi; Western Soc Naturalists. *Res:* Neurosecretion and neuroendocrinology of Mollusca; reproductive endocrinology of invertebrates. *Mailing Add:* 4570 NW Columbia Ave Portland OR 97229

SIMPSON, LEONARD ANGUS, b Vancouver, BC, Aug 1, 39; m 67; c 3. MATERIALS SCIENCE, FRACTURE MECHANICS. *Educ:* Univ BC, BSc, 61, MSc, 63; Univ Wales, PhD(metall), 68. *Prof Exp:* Metallurgist, Res Lab, Gen Elec Co, NY, 63-64; Res officer, 68-83, MGR, MAT & MECH BRANCH, ATOMIC ENERGY CAN LTD, PINAWA, 83- *Concurrent Pos:* Vis scientist, Mat Res Lab, Brown Univ, 78-80; mem OECD/NEA Principle Working Group 3. *Mem:* Am Soc Testing & Mat. *Res:* Fracture mechanics of metals and nonmetals elastic-plastic fracture criteria; hydrogen embrittlement. *Mailing Add:* Whiteshell Lab AECL Res Pinawa MB R0E 1L0 Can

SIMPSON, MARGARET, b Hong Kong, Jan 19, 35; US citizen. INVERTEBRATE ZOOLOGY. *Educ:* Immaculate Heart Col, BA, 56; Cath Univ, MS, 59, PhD(zool), 61. *Prof Exp:* Technician sch med, Univ Southern Calif, 56-57; USPHS fel, 62-63; assoc prof biol, St Francis Col, Maine, 63-67; asst prof, Adelphi Univ, 67-73; assoc prof, 73-80, PROF BIOL, SWEET BRIAR COL, 80- *Concurrent Pos:* Danforth assoc, 78-83. *Mem:* Sigma Xi; Am Inst Biol Sci; Marine Biol Asn UK; Am Soc Zoologists. *Res:* Biology of polychaetes, specifically family Glyceridae, especially histology, embryology and venom glands. *Mailing Add:* Dept of Biol Sweet Briar Col Sweet Briar VA 24595

SIMPSON, MARION EMMA, b Odenton, Md, Feb 3, 27. PLANT PHYSIOLOGY, BOTANY. *Educ:* Univ Md, BS, 52, MS, 59, PhD(plant physiol), 62. *Prof Exp:* Clerk, 44-46, sci aid, 46-58, plant pathologist, 58-60, res plant pathologist, Plant Indust Sta, Cotton Div, 60-77, MEM STAFF, BELTSVILLE AGR RES CTR, USDA, 77- *Mem:* Am Soc Plant Physiologists; Am Phytopath Soc; Mycol Soc Am; Soc Indust Microbiol. *Res:* Field deterioration of cotton fiber and physiology of organisms involved; cellulase production by fungi and characterization of this enzyme. *Mailing Add:* 1546 Meyers Sta Rd Odenton MD 21113

SIMPSON, MELVIN VERNON, b New York, NY, July 5, 21; m 76; c 3. BIOCHEMISTRY. *Educ:* City Col New York, BS, 42; Univ Calif, PhD(biochem), 49. *Prof Exp:* Physicist, Philadelphia Navy Yard, 42-44; physicist, US Naval Ord Lab, 44-45; instr physiol med sch, Tufts Col, 49-51; from asst prof to assoc prof biochem sch med, Yale Univ, 52-62; Am Cancer Soc prof, Dartmouth Med Sch, 62-66; chmn dept biochem, 66-75, AM CANCER SOC PROF BIOCHEM, STATE UNIV NY STONY BROOK, 76- *Concurrent Pos:* USPHS fel, Wash Univ, 51-52; mem fel review panel, NIH, 66-70; mem adv comt nucleic acids & protein synthesis, Am Cancer Soc, 70-75; mem merit rev bd basic sci, US Vet Admin. *Honors & Awards:* Res Award, Union Carbide Co, 69 & 70. *Mem:* Am Soc Biol Chemists; Am Chem Soc; Am Soc Cell Biol; Biophys Soc. *Res:* Protein biosynthesis; ribosomes; mitochondria; DNA biosynthesis. *Mailing Add:* Dept of Biochem State Univ of NY Stony Brook NY 11794

SIMPSON, MURRAY, b New York, NY, July 27, 21; m 47; c 4. ELECTRONIC WARFARE SYSTEMS, ANTENNAS. *Educ:* City Col New York, BEE, 42; Polytech Univ NY, MEE, 52. *Prof Exp:* Engr, Int Tel & Tel Co, 42-44; engr, US Naval Res Lab, 44-46; sr engr, Raytheon Co, 46-48; proj engr, Fairchild Guided Missiles Co, 48-50; vpres & tech dir, Maxson Electronics Co, 50-62; pres, Sedco Systs Inc, 63-86; CONSULT, SIMPSON ASSOC, 86- *Honors & Awards:* Silver Medal, Electronic Warfare Soc, 81. *Mem:* Fel Inst Elec & Electronics Engrs; Air Force Asn. *Res:* Microwave electronic systems; development of advanced radar and electronic warfare systems; phased array antenna systems. *Mailing Add:* 466 Susan Ct West Hempstead NY 11552

SIMPSON, NANCY E, b Toronto, Ont, Oct 29, 24. HUMAN GENETICS. *Educ:* Univ Toronto, BPHE, 47, PhD, 59; Columbia Univ, MA, 51. *Prof Exp:* Res assoc human genetics, Hosp Sick Children, Toronto, 59-60; asst prof pharmacol, Univ Toronto, 61-65; from asst prof to assoc prof biol & pediat, 65-75, prof med genetics & chmn div, 75-80, prof, 80-89, EMER PROF MED GENETICS, DEPT PEDIAT, QUEEN'S UNIV, ONT, 89- *Concurrent Pos:* Res fel, Pop Genetics Res Unit, Oxford Univ, 60-61; Queen Elizabeth II scientist, 62-68; Med Res Coun Can fel & vis scientist, Galton Lab, Univ Col, Univ London, 72-73. *Mem:* Am Soc Human Genetics; Genetics Soc Can; Can Col Med Geneticists. *Res:* Diabetes; enzyme polymorphisms in man; human gene mapping; pharmacogenetics. *Mailing Add:* Dept of Pediat Queen's Univ Kingston ON K7L 3N6 Can

SIMPSON, OCLERIS C, b Normangee, Tex, Sept 10, 39; m; c 1. AGRICULTURAL RESEARCH. *Educ:* Prairie View A&M Univ, BS, 60; Iowa State, MS, 62; Univ Nebr, PhD(animal sci), 65. *Prof Exp:* Assoc prof, Ft Valley State Col, Ga, 65-69; res instr, Med Col Ga, 70-71; res assoc, Meharry Med Col, Nashville, Tenn, 72-74; res coordr, Ft Valley State Col, Ga, 74-78; res dir, Prairie View A&M Univ, 78-83, asst dir planning & eval, 83; DEAN RES & EXTEN, RES DIR & EXTEN ADMINR, LANGSTON UNIV, 83- *Concurrent Pos:* Mem US Invest Team China, 82, Joint Coun Food & Agr Sci, 1890 Land-Grant Col & Univs, 85-88, Int Sci & Educ Coun Tech Assistance Subcomt. *Res:* Animal science; biochemical studies on the metabolites of Vitamin A, C and TIBC; nutrition; numerous technical publications. *Mailing Add:* Langston Univ PO Box 730 Langston OK 73050

SIMPSON, RICHARD ALLAN, b Portsmouth, NH, June 25, 45. RADAR ASTRONOMY. *Educ:* Mass Inst Technol, BS, 67; Stanford Univ, MS, 69, PhD(elec eng), 73. *Prof Exp:* Res assoc, 73-76, SR RES ASSOC RADAR ASTRON, STANFORD UNIV, 76- *Concurrent Pos:* Vis res assoc, Arecibo Observ, 75-76 & 78. *Mem:* Am Astron Soc; Am Geophys Union; Inst Elec & Electronics Engrs; AAAS; Union Radio Scientifique Internationale. *Res:* Theoretical and experimental research on scattering of radio waves by planetary surfaces; inference of geophysical properties of surfaces from scattered waves. *Mailing Add:* 3326 Kipling St Palo Alto CA 94306-3012

SIMPSON, RICHARD S, b Pensacola, Fla, Sept 25, 35; m 59; c 3. ELECTRICAL ENGINEERING. *Educ:* Univ Fla, BSEE, 57, MSE, 58, PhD(elec eng), 61. *Prof Exp:* Res assoc elec eng, Univ Fla, 58-61; from asst prof to prof, Univ Ala, 61-69; prof elec eng, Univ Houston, 69-; SCIENTIST, TEXACO USA, HOUSTON RES CTR. *Mem:* Sr mem Inst Elec & Electronics Engrs; Am Soc Eng Educ. *Res:* Communication and telemetry systems; detection theory; video compression. *Mailing Add:* Texaco USA Houston Res Ctr Bellaire TX 77401

SIMPSON, ROBERT BLAKE, US citizen; c 1. MOLECULAR BIOLOGY. *Educ:* Univ Ill, Urbana, BS,69; Harvard Univ, Cambridge, MA, 72, PhD(biophysics), 79. *Prof Exp:* Sr fel, Dept Microbiol, Univ Wash, Seattle, 79-81; RES SCIENTIST & GROUP LEADER, MOLECULAR BIOL GROUP, PLANT CELL RES INST, ATLANTIC RICHFIELD CO, DUBLIN, CALIF, 81- *Mem:* Am Soc Microbiol; Plant Molecular Biol Asn; AAAS; Am Soc Plant Physiologists. *Res:* The mechanism of Ti-plasmid DNA transfer from Agrobacteria to plant cells; the introduction of useful genes into plants; regulation of gene expression in plants and resistance of plants to pathogens. *Mailing Add:* Plant Cell Res Inst Inc 6560 Trinity Ct Dublin CA 94568-2685

SIMPSON, ROBERT GENE, b Neodesha, Kans, Aug 13, 25; m 51; c 2. ENTOMOLOGY. *Educ:* Colo Agr & Mech Col, BS, 50, MS, 53; Kans State Univ, PhD(entom), 59. *Prof Exp:* Sales rep, Calif Spray-Chem Corp, 52-53; entomologist, State Dept Agr, Colo, 53-56; asst entom, Kans State Univ, 56-59; asst exten entomologist, Univ Nebr, 59-60; asst prof entom, 60-67, assoc prof, 67-79, PROF ENTOM, COLO STATE UNIV, 79- *Mem:* Entom Soc Am; AAAS. *Res:* Biological control; life history and control of field crops and horticultural insects. *Mailing Add:* 1317 Lory Ft Collins CO 80524

SIMPSON, ROBERT JOHN, b Newburgh, NY, Feb 2, 27; m 52; c 4. ANIMAL PHYSIOLOGY, ENDOCRINOLOGY. *Educ:* Houghton Col, BA, 50; Univ Ill, MS, 60, PhD(physiol), 63. *Prof Exp:* Biologist, Lederle Labs, Am Cyanamid Co, 52-58; teaching asst anat & physiol, Univ Ill, 58-62; asst prof biol & human & cell physiol, Muskingum Col, 62-65; asst prof human & animal physiol, 65-68, ASSOC PROF HUMAN & ANIMAL PHYSIOL & ENDOCRINOL, UNIV NORTHERN IOWA, 68- *Mem:* AAAS; Sigma Xi; Am Soc Zoologists; NY Acad Sci. *Res:* Mechanisms in calcification of epiphyseal cartilage; vitamin D actions; factors in calciphylaxis; calcium metabolism and hormones; effects of estrogen and progesterone on cartilage calcification; mineral metabolism interrelations; normal and pathological calcium deposition; lipoprotein metabolism. *Mailing Add:* Dept of Biol Univ of Northern Iowa Cedar Falls IA 50613

SIMPSON, ROBERT LEE, b San Francisco, Calif, Apr 3, 42; m 70; c 1. LIMNOLOGY, FRESHWATER WETLAND ECOLOGY. *Educ:* Fresno State Col, BA, 65, MA, 67; Cornell Univ, PhD(limnol), 71. *Prof Exp:* Teaching asst zool, Fresno State Col, 65-67; res asst limnol, Cornell Univ, 67-70; from asst prof to assoc prof, 70-79, chmn dept, 72-80, PROF BIOL, RIDER COL, 79-; DEAN, SCH SCI & MATH, WILLIAM PATTERSON COL. *Concurrent Pos:* Grants, Off Water Res & Technol, 75-78, 79-82, Environ Protection Agency, 76-77 & 78-81. *Mem:* Am Soc Limnol & Oceanog; Ecol Soc Am; Brit Ecol Soc; Estuarine Res Fedn; Am Inst Biol Sci. *Res:* Ecology of freshwater tidal and non-tidal wetlands including analysis of production, decomposition and nutrient cycling processes; impact of sewage and non-point source pollutants on freshwater wetlands. *Mailing Add:* Dean of Sch Sci & Math William Patterson Col Wayne NJ 07470

SIMPSON, ROBERT TODD, b Chicago, Ill, June 28, 38; m 63; c 4. BIOCHEMISTRY. *Educ:* Swarthmore Col, BA, 59; Harvard Univ, MD, 63, PhD(biol chem), 69. *Prof Exp:* Intern med, Peter Bent Brigham Hosp, Boston, 63-64; teaching asst biol chem, Harvard Med Sch, 65-69; sr surgeon, 69-77, CHIEF SECT DEVELOP BIOCHEM, LAB NUTRIT & ENDOCRINOL, NAT INST ARTHRITIS, METAB & DIGESTIVE DIS, NIH, 73-, CHIEF, LAB CELL & DEVELOP BIOL, 80-; CAPT, USPHS, 77- *Honors & Awards:* USPHS Commendation, 82. *Mem:* Am Soc Biol Chem; Am Soc Develop Biol. *Res:* Chromatin structure; histone-DNA interactions; chemical basis for gene regulation in eucaryotic cells. *Mailing Add:* Bldg Six Rm B1-28 Nat Insts Health Bethesda MD 20892

SIMPSON, ROBERT WAYNE, b Providence, RI, Dec 28, 28; m 54; c 4. VIROLOGY. *Educ:* Univ RI, BS, 51; Brown Univ, MS, 56; Rutgers Univ, PhD(virol), 58. *Prof Exp:* Asst virol, Inst Microbiol, Rutgers Univ, 55-58; mem res staff, Dept Virol, Pub Health Res Inst New York, Inc, 58-68; assoc prof, 68-75, PROF VIROL, WAKSMAN INST, RUTGERS UNIV, PISCATAWAY, 75- *Concurrent Pos:* Res asst prof path, Sch Med, NY Univ, 66-70; instr microbiol, Hunter Col, City Univ New York, 68; mem virol study sect, NIH, USPHS, 72-76, clin sci study sect, 79-80 & 83; mem Triage Rev, NIH, 82, NJ State Comn Cancer Res, 83-, trustee, Biomed Res Fund, New Life Found, 85- *Honors & Awards:* Minnie Rosen Award, Sch Med, Ross Univ, 84. *Mem:* AAAS; Sigma Xi; Am Soc Microbiol; Brit Soc Gen Microbiol; Am Soc Virol; Am Asn Univ Profs. *Res:* Arthritis-associated parvoviruses; antiviral agents. *Mailing Add:* Waksman Inst Rutgers Univ PO Box 759 Piscataway NJ 08855-0759

SIMPSON, ROGER LYNDON, b Roanoke, Va, Oct 25, 42; m 64; c 2. FLUID MECHANICS, HEAT TRANSFER. *Educ:* Univ Va, BME, 64; Stanford Univ, MSME, 65, PhD(mech eng), 68. *Prof Exp:* Develop engr, Atomic Power Equip Dept, Gen Elec Co, Calif, 68; from asst prof to assoc prof thermal & fluid sci, Southern Methodist Univ, 69-70, from assoc prof to prof civil & mech eng, 74-83; PROF AEROSPACE & OCEAN ENG, VA POLYTECH INST, 83- *Concurrent Pos:* Vis scientist, Max-Planck Inst fur Stroemungsforschung, W Ger, 75-76, NASA-Ames, 90, Univ Erlangen, Ger, 90. *Mem:* Fel Am Inst Aeronaut & Astronaut; fel Am Soc Mech Engrs; Sigma Xi. *Res:* Turbulent shear flows; structure of turbulent flows; laser anemometry; unsteady and separated flows; mass transfer; boundary layer control. *Mailing Add:* Dept Aerospace & Oceanog Eng Va Polytech Inst Blacksburg VA 24061

SIMPSON, RUSSELL BRUCE, b Jersey City, NJ, Feb 15, 42; m 66; c 2. VETERINARY MICROBIOLOGY. *Educ:* Tex A&M Univ, BS, 65, DVM, 66, MS, 74; Am Col Vet Microbiologists, dipl, 75. *Prof Exp:* Lab officer microbiol, Vet Div, Walter Reed Army Inst Res, 66-68 & US Navy Prev Med Univ, DaNang, 68-69; instr & asst prof vet microbiol, 69-76, ASSOC PROF VET MICROBIOL & PARASITOL, COL VET MED, TEX A&M UNIV, 76- *Mem:* Am Asn Equine Practr; Am Asn Vet Lab Diagnosticians; Asn Am Vet Med Cols; Am Vet Med Asn. *Mailing Add:* Dept Vet Microbiol Tex A&M Univ College Station TX 77843

SIMPSON, S(TEPHEN) H(ARBERT), JR, b Columbus, Tex, Mar 10, 07; m 59; c 2. ELECTRICAL ENGINEERING, COMMUNICATIONS. *Educ:* Agr & Mech Col Tex, BS, 28. *Prof Exp:* Mem staff, RCA Commun, Inc, 29-37, mgr, prog & radio- photo, 37-42, traffic engr, 46-53, asst vpres & dist mgr, 53; mgr commun res, Southwest Res Inst, 53-56, asst vpres, 56-58, asst to pres, 58-66; PRES, SOUTHWEST SOUND & ELECTRONICS INC, 66-, CHMN BD, 80- *Mem:* Inst Elec & Electronics Engrs; Sigma Xi; Audio Eng Soc. *Res:* International communications; special communications; program transmission; sound and closed circuit television hospital communications; industrial research adminstration; architectural acoustics and room equalization. *Mailing Add:* 2323 Loop 410 NW San Antonio TX 78230

SIMPSON, SIDNEY BURGESS, JR, b Russellville, Ark, Oct 8, 35; m 63; c 1. DEVELOPMENTAL BIOLOGY, NEUROBIOLOGY. *Educ:* Ark Polytech Col, BS, 57; Tulane Univ, MS, 62, PhD(zool), 63. *Prof Exp:* NIH fel, Case Western Reserve Univ, 64, sr instr anat, Sch Med, 64-65, asst prof, 65-71; from assoc prof to prof biol sci, dept biochem & molecular biol, Northwest Univ, 71-85; PROF & HEAD, DEPT BIOL SCI, UNIV ILL, CHICAGO, 85- *Concurrent Pos:* NIH career develop award, 66- *Honors & Awards:* Marcus Singer Medal, 86. *Mem:* AAAS; Soc Develop Biol; Soc Cell Biol; Soc Neuroscience. *Res:* Vertebrate and invertebrate regeneration. *Mailing Add:* Dept of Biol Sci MC 066 Univ Ill Chicago IL 60680

SIMPSON, STEPHEN G, b Allentown, Pa, Sept 8, 45; m 73; c 2. MATHEMATICS, MATHEMATICAL LOGIC. *Educ:* Lehigh Univ, BA & MS, 66; Mass Inst Technol, PhD(math), 71. *Prof Exp:* Gibbs instr, Yale Univ, 71-72; lectr, Univ Calif, Berkeley, 72-74; res fel, Oxford Univ, 74-75; asst prof, 75-77, assoc prof, 77-80, PROF MATH, PA STATE UNIV, 81- *Concurrent Pos:* NSF res grants, 71-74 & 75-; res fel, Sci Res Coun, UK, 74-75; res fel Alfred P Sloan Found, 80-82 & Deutsche Forschungsgemeinschaft, West Germany, 83-84; vis assoc prof, Univ Chicago, 78 & Univ Conn, 79-80; vis prof, Univ Paris, 81 & Univ Munich, 83-84. *Mem:* Am Math Soc; Asn for Symbolic Logic. *Res:* Foundation of mathematics; mathematical logic; combinatorics. *Mailing Add:* Dept Math Pa State Univ Univ Park PA 16802

SIMPSON, THOMAS A, b Adams, Mass, Oct 23, 25; m 54; c 2. MINING, GEOLOGY. *Educ:* Univ Mo-Rolla, BS, 51, EMines, 65; Univ Ala, MS, 59. *Prof Exp:* Geologist, US Geol Surv, 54-61; chief geologist, Econ Geol Div, Geol Surv Ala, 61-65, asst state geologist, planning sect, 65-75; actg head dept, Univ Ala, 78-79, assoc prof mineral eng, 75-86, res assoc, Mineral Resurces Inst, 76-86, emer adj assoc prof mineral eng, 86-90; RETIRED. *Concurrent Pos:* Res assoc, Mus Natural Hist, Univ Ala, 63-86, lectr univ, 65-75; mining geologist, Surinam, 64 & Venezuela, 67. *Mem:* Am Inst Mining, Metall & Petrol Engrs; fel Geol Soc Am; Asn Eng Geologists; Am Inst Prof Geologists; Am Asn Petrol Geol. *Res:* Economic geology; mining hydrology and hydrogeologic investigations; mine blasting studies; Southeast iron ore studies. *Mailing Add:* 72 Vestavia Hills Northport AL 35476

SIMPSON, TRACY L, b New York, NY, July 12, 37. BIOSILICIFICATION, SPONGE BIOLOGY. *Educ:* Brown Univ, AB, 59; Yale Univ, PhD(biol), 65. *Prof Exp:* From instr to asst prof, Tufts Univ, 64-67; assoc prof biol, 67-76, PROF BIOL, UNIV HARTFORD, 76- *Concurrent Pos:* Asst proj dir, Undergrad Equip Grant, NSF, 65-67 & prin investr res grant, 65-68, 68-70 & 71-73; vis assoc prof, Dartmouth Col, 74, Health Ctr, Univ Conn & Res Found grant, 74-76; prin investr, Int Silicon Symposium, 77-78; exchange scientist, US-France Sci Prog & res prof, Univ Claude Bernard, Lyon, 82-85. *Mem:* AAAS; Am Soc Zool; Am Micros Soc; NY Acad Sci; Sigma Xi; Microbiol Soc Am. *Res:* Cell biology of sponges; biology of silicon and silicification. *Mailing Add:* Dept Biol Univ Hartford West Hartford CT 06117

SIMPSON, WILBURN DWAIN, b Long Grove, Okla, Oct 4, 37; m 67; c 2. TELECOMMUNICATIONS, NUCLEAR PHYSICS. *Educ:* Univ Miss, BS, 59, MS, 61; Rice Univ, MA, 63, PhD(nuclear physics), 65. *Prof Exp:* Res assoc nuclear physics, Rice Univ, 65-67; asst physicist, Brookhaven Nat Lab, 67-69; vpres syst develop, Periphonics Corp, 69-80; vpres technol, Alta Technol Inc, 80-85; pres, Ayentka Consult Corp, 80-81; PRES, W D SIMPSON TECHNOL, INC, 85-; VPRES TECHNOL, SABER EQUIP CORP, 89- *Concurrent Pos:* Author, New Techniques in Software Proj Mgt, 87. *Mem:* AAAS; Am Phys Soc; NY Acad Sci; Inst Elec & Electronics Engrs; Asn Comput Mach. *Res:* Three-body systems in nuclear physics; intermediate energy nuclear physics; nuclear structure; nucleon-nucleon and nucleon-nucleus interactions; meson-nucleon interactions; computer controlled audio response; communication processors; electronic funds transfer terminals; computer controlled networks; automated fare collection; telecommunication devices; magnetic encoding. *Mailing Add:* 124 Catalpa Rd Wilton CT 06897

SIMPSON, WILLIAM ALBERT, b Pittsburgh, Pa, Oct 29, 34; m 58; c 1. FRACTALS, HISTORY OF MATHEMATICS. *Educ:* US Naval Acad, BS, 58; Univ Mich, MS, 64; Mich State Univ, PhD(math), 71. *Prof Exp:* Res asst, Inst Sci Technol, 65-66, asst prof math, Univ Mich, 71-73; asst prof, Off Institutional Res, 73-85, PROF MATH, LYMAN BRIGGS SCH, MICH STATE UNIV, 85- *Concurrent Pos:* Adj math instr, Univ Maryland, 61-63; adj asst prof, dept math, 73-74, adj assoc prof, mgt dept, 81-82, adj assoc prof, Dept Educ Admin, Mich State Univ, 82-85; Fulbright sr scholar, Univ Bath, England, 84. *Mem:* Math Asn Am; Asn Inst Res; AAAS. *Res:* Higher education, policy analysis, statistic studies, computer models; mathematics, non-linear systems & history of math. *Mailing Add:* 1410 Sherwood East Lansing MI 48823

SIMPSON, WILLIAM HENRY, b Woodbury, NJ, Mar 24, 42; m 69; c 2. SOLID STATE CHEMISTRY, PHOTOGRAPHIC CHEMISTRY. *Educ:* Col William & Mary, BS, 63; Univ Pa, PhD(phys chem), 67. *Prof Exp:* Fel org solid state, Franklin Inst Res Labs, Pa, 68-70, res chemist, 71-77; scientist, 77-86, SR SCIENTIST, POLAROID CORP, 86-; AT MEAD IMAGING. *Mem:* Am Chem Soc; Soc Imaging Sci & Technol. *Res:* Photochemistry; radiation chemistry of organic materials; photographic science; silver halide photographic emulsions. *Mailing Add:* Mead Imaging 3385 Newmark Dr Miamiburg OH 45342

SIMPSON, WILLIAM L, oncology, for more information see previous edition

SIMPSON, WILLIAM ROY, b Padernal, NMex, June 27, 24; m 47; c 3. PLANT PATHOLOGY. *Educ:* Univ Idaho, BS, 49, MS, 51. *Prof Exp:* Technician plant path, 47-49, from res asst to res assoc, 49-54, from jr plant pathologist to assoc plant pathologist, Br Exp Sta, 54-70, res plant pathologist, 54-80, RES PROF PLANT PATH, UNIV IDAHO, 70-, EXTEN PROF & EXTEN PLANT PATHOLOGIST, 80- *Mem:* Am Phytopath Soc; Sigma Xi. *Res:* Vegetable pathology, especially disease of tomatoes, beets, spinach, corn and onions; curly top virus disease of vegetable crops; stalk rot and head smut of corn; development of virus resistant tomato. *Mailing Add:* Res & Exten Ctr Univ Idaho PO Box 549 Parma ID 83660

SIMPSON, WILLIAM STEWART, b Edmonton, Alta, Apr 11, 24; US citizen; m 50; c 4. PSYCHIATRY, HOSPITAL ADMINISTRATION. *Educ:* Univ Alta, Edmonton, BS, 46, MD, 48. *Prof Exp:* Sect chief psychiat, C F Menninger Mem Hosp, 59-66; assoc dir psychiat, Menninger Sch Psychiat, Menninger Found, 66-68, dir field serv fund raising, 72-74; dir psychiat residency training & chief psychiat serv, Topeka Vet Admin Hosp, 74-77; sr psychiatrist & psychoanalyst, 77-84, assoc dir, med serv, Adult Outpatient Dept, 84-88, DIR, CTR SEXUAL HEALTH, MENNINGER FOUND, 86- *Concurrent Pos:* Mem fac, Menninger Sch Psychiat, 53-, mem exec comt & mgt comt, 54-77; first pres, Topeka Alcoholism Info Ctr, 63-67; assoc ed, Bull Menninger Clin, 64-70; clin dir psychiat & hosp admin, Topeka State Hosp, 54-59 & 68-72; mem bd dirs, Nat Coun Alcoholism, 68-77, vpres, 70-73, pres, 73-75; mem, Kans Citizens Adv Alcoholism, 73-77; lectr sex ther, Chinese Med Asn, Taipei, Taiwan, 85, Dept Psychol, Univ Warsaw, Poland, 87, Dept Psychiat, Tokai Univ Med Sch, Japan, 85. *Honors & Awards:* Silver Key Award, Nat Coun Alcoholism, 75. *Mem:* Fel Am Psychiat Asn; Am Psychoanal Asn; Am Asn Sex Educr, Counr & Therapists; Soc Sci Study Sex; Int Psychoanal Asn; Soc Sex Ther & Res. *Res:* Relationship between sex therapy and psychoanalysis. *Mailing Add:* 834 Buchanan Topeka KS 66606

SIMPSON, WILLIAM TRACY, b Berkeley, Calif, Dec 7, 20; m 44, 61; c 5. CHEMISTRY. *Educ:* Univ Calif, AB, 43, PhD(chem), 48. *Prof Exp:* Asst chem, Univ Calif, 46-48; instr, Univ Wash, 48-49, from asst prof to assoc prof, 49-57, prof, 57-64; prof, 65-77, chmn dept, 72-75, EMER PROF CHEM, UNIV ORE, 77- *Concurrent Pos:* Vis lectr, Univ Calif; vis prof, Fla State Univ; chmn, Gordon Conf Theoret Chem, 66; Fulbright vis prof, Lima, Peru, 71; assoc, Neurosci Res Prog, 63-65; assoc ed, J Chem Physics, 66-69. *Res:* Theoretical and experimental study of molecular electronic spectra; vacuum ultraviolet spectroscopy; electronic spectra of thin films and molecular crystals; theory of the index of refraction. *Mailing Add:* 1760 Royal Way San Luis Obispo CA 93405

SIMPSON-HERREN, LINDA, b Birmingham, Ala, July 7, 27; m 63; c 1. CHEMOTHERAPY. *Educ:* Univ Ala, BS, 48. *Prof Exp:* Instr physics, Univ Ala, 48; assoc physicist, 48-57, res physicist, 57-69, sr physicist, 69-71, SECT HEAD CELL & TISSUE KINETICS, SOUTHERN RES INST, 71- *Concurrent Pos:* Am ed, Cell & Tissue Kinetics. *Mem:* Am Asn Cancer Res; Cell Kinetics Soc (vpres, 77-78, pres, 78-79); AAAS; Sigma Xi; Health Physics Soc; Int Cell Cycle Soc. *Res:* Cell and tumor kinetics of experimental tumor systems to optimize scheduling of chemotherapy alone or in combination with surgery; radiation and drug effects; tumor and host heterogeneity; drug distribution. *Mailing Add:* Southern Res Inst PO Box 55305 Birmingham AL 35255-5305

SIMRALL, HARRY C(HARLES) F(LEMING), b Memphis, Tenn, Oct 16, 12; m 36. ELECTRICAL & MECHANICAL ENGINEERING. *Educ:* Miss State Univ, BS, 34 & 35; Univ Ill, MS, 39. *Prof Exp:* Instr elec eng, Miss State Univ, 34-35, instr drawing, 35-37, from instr to assoc prof elec eng, 37-44; engr, Cent Sta Eng, Indust Eng Dept, Westinghouse Elec Corp, 44-45; assoc prof elec eng, 45-47, prof elec eng & head dept, 47-57, dean col eng, 57-78, EMER DEAN COL ENG & EMER PROF ELEC ENG, MISS STATE UNIV, 78- *Mem:* Fel Inst Elec & Electronics Engrs; Am Soc Eng Educ; Illum Eng Soc; Am Soc Agr Engrs; Nat Soc Prof Engrs (vpres, 62-64, pres elect, 69-70, pres, 70-71). *Res:* Electric power. *Mailing Add:* 107 White Dr W Starkville MS 39759-2634

SIMRING, MARVIN, b Brooklyn, NY, June 16, 22; m 49; c 4. PERIODONTOLOGY, NUTRITION. *Educ:* Brooklyn Col, BA, 42; New York Univ Col Dent, DDS, 44; Am Bd Periodont, dipl, 55. *Prof Exp:* Captain, US Army, Dent Corps, 44-47; clin prof & dir periodont, NY Univ Col Dent, 47-49; assoc prof, Univ Fla, 79-82; VIS LECTR PERIODONT, NY UNIV, 79- *Concurrent Pos:* Dir & attend dentist, Dept Periodont, Jewish Hosp, Brooklyn, 53-79; consult, US First Army Hosp, New York, 62-69, Brooklyn Vet Hosp, 70-79, US Vet Hosp, Gainesville, Fla, 79-82; attend dentist, Shands Teaching Hosp, Fla, 79-82. *Honors & Awards:* Otto Loos Medal, Univ Frankfurt, W Germany, 80. *Mem:* Am Dent Asn; Am Acad Periodontol; fel Am Acad Oral Med; fel Am Col Dent; Am Asn Dent Res; fel Int Col Dent. *Res:* Relating periodontal health and disease to occlusion, splinting, appliances, nutrition, tooth movement, pulpal status, and root canal therapy. *Mailing Add:* 2501 NW 21st Ave Gainesville FL 32605

SIMS, ASA C, JR, b Asheville, NC, Sept 30, 19; m 43; c 2. PLANT PATHOLOGY. *Educ:* Hampton Inst, BS, 40; Ohio State Univ, MS, 54, PhD, 56; Harvard Univ, dipl educ mgt, 73. *Prof Exp:* Instr hort, Fla Agr & Mech Col, 48-49; asst prof, SC State Col, 49-52; asst bot, Ohio State Univ, 53-56; from assoc prof to prof biol, Southern Univ, Baton Rouge, 56-68; prof, 68-70, chmn dept biol, 68-69, chmn div sci, 69-70, DEAN ACAD AFFAIRS, SOUTHERN UNIV, NEW ORLEANS, 70- *Concurrent Pos:* Sci fac fel, Univ Minn, 65-66. *Mem:* Am Phytopath Soc; Bot Soc Am. *Res:* Fungus physiology; nature of disease; radioisotopes. *Mailing Add:* Dept Biol Univ New Orleans New Orleans LA 70126

SIMS, BENJAMIN TURNER, b Dec 11, 34; US citizen; m 64; c 1. MATHEMATICS. *Educ:* Univ Mo, AB, 56, MA, 59; Iowa State Univ, PhD(math), 62. *Prof Exp:* Asst prof math, San Jose State Univ, 62-64, Am Univ Beirut, 64-66 & San Jose State Univ, 66-67; assoc prof, 67-71, PROF MATH, EASTERN WASH UNIV, 71- *Mem:* Am Math Soc; Math Asn Am; Sigma Xi; Am Sci Affil. *Res:* Point-set topology. *Mailing Add:* Dept Math MS No 32 Eastern Wash Univ Cheney WA 99004-2415

SIMS, CHESTER THOMAS, b Winchester, Mass, Dec 14, 23; m 49, 80; c 2. METALLURGICAL ENGINEERING. *Educ:* Northeastern Univ, BS, 47; Ohio State Univ, MS, 51. *Prof Exp:* Res engr, Battelle Mem Inst, 47-50 & 52-55, asst div chief, 55-58; mat engr, Knolls Atomic Power Lab, Gen Elec Co, 58-60, metallurgist, Mat & Processes Lab, 60-63, proj leader, 64-65; mgr high-temperature mat, Gen Elec Co, Schenectady, NY, 66-68, mgr alloy & joining metall, 68-71, mgr advan mat, Gas Turbine Div, 71-81, mgr, Mat Info Serv Corp Eng, 81-86; PROF MAT ENG, RENSSELAER POLYTECH INST,

86- *Concurrent Pos:* Lectr, Rensselaer Polytech Inst, 63-64; consult, Pac Northwest Labs, Battelle Mem Inst, 64-; chmn gas turbine panel, Am Soc Mech Engrs-Am Soc Testing & Mat; dedicatee, Brazil-US Conf Refractory Metalls in Superalloys, 84; consult, 86- *Honors & Awards:* Distinguished Serv Award, Hudson-Mohawk Sect, Am Inst Mining, Metall & Petrol Engrs, 74; William Hunt Eisenman Award, Am Soc Metals, 76. *Mem:* Fel Am Soc Metals; Am Welding Soc; Am Inst Mining, Metall & Petrol Engrs. *Res:* Physical metallurgy of superalloys, refractory and scarce metals; elevated temperature oxidation and corrosion; high-temperature alloy development and ceramics for gas turbines; materials for high-temperature reactors. *Mailing Add:* PO Box 644 Lakeshore Dr Bolton Landing NY 12814

SIMS, ERNEST THEODORE, JR, horticulture, plant physiology; deceased, see previous edition for last biography

SIMS, ETHAN ALLEN HITCHCOCK, b Newport, RI, Apr 22, 16; m 39; c 3. MEDICINE, BIOCHEMISTRY. *Educ:* Harvard Univ, BS, 38; Columbia Univ, MD, 42. *Hon Degrees:* ScD, Univ Vt, 90. *Prof Exp:* House officer, New Haven Hosp, Conn, 42-44; instr med, Sch Med, Yale Univ, 47-50; from asst prof to prof, 50-85, EMER PROF MED, COL MED, UNIV VT, 85- *Concurrent Pos:* Brown res fel, New Haven Hosp, Conn, 46-47; Commonwealth fel, Sch Med, Case Western Reserve Univ, 64-65; assoc attend physician, Mary Fletcher & DeGoesbriand Mem Hosps, 50-64; dir metab unit, Dept Med, Univ Vt, 57-73; attend physician, Med Ctr, Hosp of Vt, 64-; vis prof med, Div Endocrinol, Sch Med, Tufts Univ, 74-75. *Honors & Awards:* Herman Award, Am Soc Clin Nutrit, 87. *Mem:* Fel Am Col Physicians; Am Fedn Clin Res; Endocrine Soc; Am Diabetes Asn; Am Soc Clin Nutrit; Sigma Xi. *Res:* Metabolic diseases; diabetes and obesity; software systems. *Mailing Add:* 51 Old Farm South Burlington VT 05403

SIMS, JAMES JOSEPH, b Woodland, Calif, June 13, 37; div; c 2. ORGANIC CHEMISTRY. *Educ:* Ariz State Univ, BS, 59; Univ Calif, Los Angeles, PhD(org chem), 63. *Prof Exp:* NSF fel, Swiss Fed Inst Technol, 63-64; lectr chem, 64-65, asst chemist, 65-70, assoc prof, 70-74, PROF PLANT PATH & CHEM, UNIV CALIF, RIVERSIDE, 74- *Mem:* Am Chem Soc; The Chem Soc. *Res:* The chemistry of natural products; isolation; structure proof; synthesis. *Mailing Add:* Dept Plant Path Univ Calif Riverside CA 92521

SIMS, JAMES R(EDDING), b Macon, Ga, July 2, 18; m 46; c 3. CIVIL ENGINEERING. *Educ:* Rice Inst, BS, 41; Univ Ill, MS, 50, PhD(eng), 56. *Prof Exp:* Instr civil eng, Rice Univ, 42-44 & 46-47, from asst prof to assoc prof, 47-58, chmn dept, 58-63, mgr campus bus affairs, 63- 69, vpres, 69-70, dir campus bus, 70-74, prof civil eng, 58-87, Herman & George R Brown chair civil eng, 74-87, EMER HERMAN & GEORGE R BROWN PROF CIVIL ENG, RICE UNIV, 87- *Concurrent Pos:* Consult, Humble Oil & Refining Co, Exxon, 53-62. *Mem:* Fel Am Soc Civil Engrs (vpres, 70-71, pres, 81-82). *Res:* Structural materials, particularly steel and concrete; structures in open sea subject to storm loading; protective construction subject to blast loading. *Mailing Add:* Dept Civil Eng Rice Univ PO Box 1892 Houston TX 77251

SIMS, JOHN DAVID, b Decatur, Ill, Dec 7, 39; c 2. SEDIMENTOLOGY, QUATERNARY GEOLOGY. *Educ:* Univ Ill, BS, 63; Univ Cincinnati, MS, 64; Northwestern Univ, PhD(geol), 67. *Prof Exp:* Res asst, Ill Geol Surv, 60-63; geologist, 67-82, SR RES GEOLOGIST, US GEOL SURV, 82- *Concurrent Pos:* Consult, Yugoslavia post earthquake study, Nat Sci Found, 79, People's Repub China, earthquake hazards, 87. *Honors & Awards:* Meritorious Award, Am Planning Asn, 80 & 81. *Mem:* Int Asn Sedimentologists; Soc Econ Paleont & Mineral; Am Geophys Union; Sigma Xi; Geol Soc Am; Am Quaternary Asn. *Res:* Late Cenozoic stratigraphic and sedimentologic studies of lacustrine sediments; earthquake-induced deformation of soft sediments; pleistocene paleolimnology and paleoclimatology; detailed geologic studies in and near the San Andreas fault zone. *Mailing Add:* US Geol Surv 345 Middlefield Rd MS 977 Menlo Park CA 94025

SIMS, JOHN LEONIDAS, b Sedalia, Ky, May 4, 30; m 51; c 4. AGRONOMY. *Educ:* Univ Ky, BS, 55, MS, 56; Iowa State Univ, PhD(soil microbiol), 60. *Prof Exp:* Asst soils, Univ Ky, 55-56 & Iowa State Univ, 56-60; asst prof agron, Univ Ark, 60-66; from asst prof to assoc prof, 66-70, PROF AGRON, UNIV KY, 75- *Concurrent Pos:* Vis assoc prof agron, La State Univ, 73-74. *Mem:* Am Soc Agron; Soil Sci Soc Am. *Res:* Soil plant relationships in the mineral nutrition of tobacco; microbial processes in soil as related to soil fertility and nitrogen fertilization of tobacco. *Mailing Add:* Dept of Agron Agr Sci Ctr Univ of Ky Lexington KY 40546-0091

SIMS, JOHN LEROY, b Houston, Tex, Sept 21, 12; m 42; c 3. CLINICAL MEDICINE. *Educ:* Rice Inst, BA, 33; Univ Tex, MD, 37. *Prof Exp:* From intern to resident med, Wis Gen Hosp, 37-42; from instr to assoc prof, 46-56, PROF MED, MED SCH, UNIV WIS-MADISON, 56-, COORDR, OUTREACH POSTGRAD MED EDUC, 83- *Concurrent Pos:* Mem staff, Wis Gen Hosp, 42- *Mem:* AAAS; fel AMA; Am Soc Internal Med; Am Heart Asn; fel Am Col Physicians. *Res:* Internal medicine; hepatic disease. *Mailing Add:* H6/516 Univ Hosp Univ of Wis Med Sch Madison WI 53792

SIMS, LESLIE BERL, b Royalton, Ill, Mar 21, 37; m 58; c 3. PHYSICAL CHEMISTRY. *Educ:* Southern Ill Univ, BA, 58; Univ Ill, MS, 61, PhD(chem), 67. *Prof Exp:* Asst prof chem, Mich State Univ, 64-67; asst prof, 67-70, assoc prof, 70-76, PROF CHEM, UNIV ARK, FAYETTEVILLE, 76-, CHMN DEPT, 79- *Mem:* Am Chem Soc; Sigma Xi. *Res:* Chemical kinetics; kinetic isotope effects; gas phase unimolecular reactions; theoretical kinetics; reaction dynamics; molecular vibrations. *Mailing Add:* Dept Chem 113 Cox Hall NC State Univ Box 8204 Raleigh NC 27695-8204

SIMS, PAUL KIBLER, b Newton, Ill, Sept 8, 18; m 40; c 2. GEOLOGY. *Educ:* Univ Ill, AB, 40, MS, 42; Princeton Univ, PhD(geol), 50. *Prof Exp:* Asst, Univ Ill, 40-42; spec asst geologist, State Geol Surv, Ill, 42-43; geologist, US Geol Surv, 43-44 & 46-61; dir, Minn Geol Surv, 61-73; GEOLOGIST, US GEOL SURV, 73- *Concurrent Pos:* Asst, Princeton Univ, 46-47; pres, Econ Geol Publ Co; co-ed, 75th Anniversary Vol, Econ Geol, 81. *Honors & Awards:* Thayer Lindsley Distinguished lectr, Soc Econ Geologists, 84-85, Ralph W Marsden Medal Award, 89; Goldich Medal, Inst Lake Superior Geol, 85. *Mem:* Geol Soc Am; Soc Econ Geol; Soc Econ Geol (pres, 75-76). *Res:* Early crustal evolution in Lake Superior and Midcontinent region; geology and magnetite iron ore deposits in New Jersey; geology and ore deposits of the Front Range, Colorado; precambrian geology and ore deposits of Minnesota, Wisconsin, and Michigan. *Mailing Add:* US Geol Surv Denver Fed Ctr Box 25046 MS 905 Denver CO 80225

SIMS, PETER JAY, IMMUNOCHEMISTRY, MEMBRANE BIOPHYSICS. *Educ:* Duke Univ, MD, 80, PhD(physiol & pharmacol), 80. *Prof Exp:* ASSOC MEM, OKLA MED RES FOUND, 85-; SCI DIR, OKLA BLOOD INST, 85-; ASSOC PROF MED, HEALTH SCI CTR, OKLA UNIV, 85- *Res:* Physical biochemistry. *Mailing Add:* Okla Med Res Found 825 NE 13th St Oklahoma City OK 73104

SIMS, PHILLIP LEON, b Mountain View, Okla, Apr 7, 40; m 62; c 2. RANGE SCIENCE. *Educ:* Okla State Univ, BS, 62, MS, 64; Utah State Univ, PhD(range sci), 67. *Prof Exp:* From asst prof to assoc prof range sci, Colo State Univ, 67-77; RES LEADER, US SOUTHERN GREAT PLAINS FIELD STA, SCI & EDUC ADMIN-AGR RES, USDA, 77- *Mem:* Soc Range Mgt; Am Soc Animal Sci; Brit Ecol Soc. *Res:* Range animal nutrition, management and improvements; dynamics of primary producer; grazing systems; secondary productivity of range ecosystems. *Mailing Add:* 2000 18th St Woodward OK 73801

SIMS, REX J, b Racine, Wis, July 1, 22; m 45; c 2. ORGANIC CHEMISTRY. *Educ:* Wabash Col, AB, 44; Northwestern Univ, PhD(org chem), 49. *Prof Exp:* Asst, Nat Defense Res Comt, Northwestern Univ, 43-45; res chemist, Swift & Co, Ill, 49-61; res scientist, Gen Foods Corp, 61-80, prin scientist, Tech Ctr, 80-88; RETIRED. *Honors & Awards:* Indust Achievement Award, Inst Food Technologists, 70. *Mem:* Am Chem Soc; Am Oil Chemists Soc. *Res:* Fats and oils; kojic acid and pyrones; fat oxidation; emulsification. *Mailing Add:* 51 Guyon St Pleasantville NY 10570

SIMS, ROBERT ALAN, b Colorado Springs, Colo, Nov 1, 36; m 59; c 1. INSTRUMENTATION, COMPUTER SCIENCE. *Educ:* Colo Sch Mines, Engr, 58, MS, 61; Univ Okla, PhD(chem eng), 68. *Prof Exp:* Asst prof eng, Wright State Univ, 68-69; asst prof instrumentation, 69-80, ASSOC PROF ELECTRONICS & INSTRUMENTATION, GRAD INST TECHNOL, UNIV ARK, LITTLE ROCK, 80- *Mem:* Instrument Soc Am; Sigma Xi. *Res:* Process control; computer simulation; instrumental applications of microcomputers & microcontrollers. *Mailing Add:* Univ Ark Etas 575 2801 S University Ave Little Rock AR 72204

SIMS, SAMUEL JOHN, b Los Angeles, Calif, Feb 20, 34; m 61; c 2. ECONOMIC GEOLOGY. *Educ:* Calif Inst Technol, BS, 55; Univ Tex, MA, 57; Stanford Univ, PhD(geol), 60. *Prof Exp:* Geologist, Soc des Mines de Fer de Mekambo, 60-62 Companhia Minas de Jangada, 62-64; geologist dept geol, Bethlehem Steel Corp, 64-78, sr geologist, 78-82, admnr, 82-85; CONSULT GEOLOGIST, 85- *Mem:* Geol Soc Am; Soc Econ Geol; Am Inst Mining, Metall & Petrol Eng. *Res:* Geological exploration for economic mineral deposits. *Mailing Add:* 768 Redfern Lane Bethlehem PA 18017

SIMS, WILLIAM LYNN, b Hazen, Ark, May 30, 24; m 51; c 3. VEGETABLE CROPS. *Educ:* Univ Wis, BS, 48, MS, 49, PhD, 54. *Prof Exp:* From asst prof to assoc prof hort, Tex Col Arts & Industs, 52-57; assoc agriculturist,Univ Calif, Davis, 57-63, agriculturist, 63-88; RETIRED. *Concurrent Pos:* Fulbright res scholar, NZ-US Educ Found, 74-75; assoc dir, Agr Develop Systs, Univ Calif/Egypt, US Agency Int Develop, 81-82; Int Exec Serv Corps, Dominican Republic & Morocco, 87 & 88. *Honors & Awards:* Carl Bittner Ext Award, 72; Hort Hon Award, Asn Portuguese Hort, 79. *Mem:* fel Am Soc Hort Sci; fel AAAS. *Res:* Agricultural extension; variety evaluation and growth regulations. *Mailing Add:* 823 Linden Lane Davis CA 95616

SIMSON, JO ANNE V, b Chicago, Ill, Nov 19, 36; div; c 3. CELL BIOLOGY, CYTOCHEMISTRY. *Educ:* Kalamazoo Col, BA, 59; Univ Mich, MS, 61; State Univ NY, PhD(anat), 68. *Prof Exp:* Instr anat, State Univ NY Upstate Med Ctr, 67-68; Nat Cancer Inst fel cell biol, Fels Res Inst, Sch Med, Temple Univ, 68-70; asst prof path, 70-75, from asst prof anat to assoc prof anat, 75-83, PROF ANAT, MED UNIV SC, 83- *Concurrent Pos:* Ed bd, Anat Rec, 75-85; Fogarty Int Fel, 87-88. *Mem:* Am Asn Cell Biol; Am Asn Anat; Histochem Soc (secy, 79-83, coun, 85-89). *Res:* Light and electron microscopic morphology and cytochemistry, especially of secretory processes; cellular ion localization and movement; membrane alterations during both exocytosis and endocytosis; cell membrane traffic. *Mailing Add:* Dept Anat Med Univ SC 171 Ashley Ave Charleston SC 29425

SINAI, JOHN JOSEPH, b Whiting, Ind, Oct 27, 30; m 57; c 5. PHYSICS. *Educ:* Miami Univ, BA, 53; Univ Ill, MS, 55; Purdue Univ, PhD(physics), 63. *Prof Exp:* Res assoc physics, Univ Chicago, 63-64; asst prof, 64-68, dir comput ctr, 66-70, assoc dean arts & sci, 70-74, actg chmn dept physics, 74-75, chmn dept, 75-78, assoc prof, 68-79, PROF PHYSICS, UNIV LOUISVILLE, 79- *Mem:* Am Phys Soc; Sigma Xi. *Res:* Vibrational spectra of disordered solids. *Mailing Add:* 60 Eastover Ct Louisville KY 40206

SINANOGLU, OKTAY, b Bari, Italy, Feb 25, 35; Turkish citizen; m 63; c 2. THEORETICAL CHEMISTRY. *Educ:* Univ Calif, Berkeley, BS, 56, PhD(theoret chem), 59; Mass Inst Technol, MS, 57. *Prof Exp:* Chemist, Lawrence Radiation Lab, Univ Calif, Berkeley, 59-60; from asst prof to assoc prof chem, 60-63, PROF CHEM, YALE UNIV, 63-, PROF MOLECULAR BIOPHYS, 65- *Concurrent Pos:* Sloan fel, 61-64; vis prof, Middle East Tech Univ, Ankara, Turkey, 62-64, consult prof, 64-; consult prof, Bogazici Univ, Turkey, 73-; consult, Rocket Oxidizers Res Prog, Adv Res Projs Agency, 64 & Turkish Sci & Tech Res Coun, 73-; ed of biol & phys scientists, NIH, 64;

mem Parr subcomt on theoret chem of Westheimer comt, Nat Acad Sci, 64-65; co-chmn elect, Gordon Res Conf physics & chem of biopolymers, 66; mem subpanel on atomic, molecular physics, Inst Defense Anal, 66; mem chem rev comt & consult, Argonne Univ Asn for Argonne Nat Lab, 67-70, consult, Nat Lab, 67-73. *Honors & Awards:* Turkish Sci Medal, 66; Alexander von Humboldt Award, 73. *Mem:* Am Inst Chem Engrs; Inst Am Chemists; Am Chem Soc; Am Phys Soc; Am Acad Arts & Sci. *Res:* Theoretical chemistry; quantum chemistry; theory of intermolecular forces; theory of solvent effects of biopolymer structure; biochemical reaction networks; many-electron theory of atoms and molecules. *Mailing Add:* Dept Chem Yale Univ PO Box 6666 New Haven CT 06511

SINCIUS, JOSEPH ANTHONY, b Amsterdam, NY, Nov 30, 26; m 49; c 2. PHOTOGRAPHIC CHEMISTRY. *Educ:* Union Univ, NY, BS, 46; Canisius Col, MS, 56; Mich State Univ, PhD, 60. *Prof Exp:* Chemist, Hooker Chem Soc, NY, 47-54, group leader, 54-56; asst, Mich State Univ, 56-57, 59; res chemist, E I Du Pont De Nemours & Co Inc, Parlin, NJ, 59-64, sr res chemist, 64-67, res supvr, 67-73, res assoc, 73-81, res fel, Photo Prod Dept, 81-86; RETIRED. *Concurrent Pos:* Consult, 86-89. *Mem:* Soc Photog Sci & Eng. *Res:* Kinetics and reaction mechanisms; radioactive tracer techniques; photo and photographic chemistry; image forming systems. *Mailing Add:* 199 Pinckney Rd Little Silver NJ 07739

SINCLAIR, A RICHARD, b Oklahoma City, Okla, Feb 5, 40; c 2. HYDRAULIC FRACTURING, WELL STIMULATION. *Educ:* Univ Okla, BS, 63, MS, 64. *Prof Exp:* Scientist, Ames Res Ctr, NASA, 64-67; sr res specialist petrol eng, Exxon Corp, 67-76; sr res assoc mech & petrol eng, Maurer Eng Inc, 76-81; PRES, SANTROL PRODS INC, 78- *Concurrent Pos:* Teacher, Halliburton Energy Inst, 76-; consult, Well Stimulation Inc, 81- *Honors & Awards:* Cedric Ferguson Medal, Soc Petrol Engrs, 71; Rossitor Raymond Award, Am Inst Mining, Metall & Petrol Engrs, 73. *Mem:* Soc Petrol Engrs. *Res:* Hydraulic fracturing; propping agents; specialty chemicals for oil field use. *Mailing Add:* Santrol Prods Inc Suite 1260 11757 Katy Freeway Houston TX 77079

SINCLAIR, ALASTAIR JAMES, b Hamilton, Ont, Aug 1, 35; m 64; c 2. EARTH SCIENCES, GENERAL. *Educ:* Univ Toronto, BASc, 57, MASc, 58; Univ BC, PhD(geol), 64. *Prof Exp:* Asst prof geol, Univ Wash, 62-64; from asst prof to assoc prof, 64-74, dept head, 85-90, PROF GEOL, UNIV BC, 74- *Concurrent Pos:* Consult, numerous mining co, 64-, Placer Dome Ltd, 78-86 & UN Seatrad Ctr, Malaysia, 83-85; mem, subcomt isotope studies & geochronology, Nat Res Coun Can, 66-69; exhange scientist, France, 72-73 & Brazil, 81, 82 & 87; hon mem Sci Tech Comn Geochem, Geol Soc Brazil, 82; Killam Sr fel, 90-91. *Honors & Awards:* Distinguished Serv Award, Mineral Deposits Div, Geol Asn Can, 90. *Mem:* Can Inst Mining & Metall; Asn Explor Geochemists; Soc Econ Geol; Geol Asn Can; hon mem Geol Soc Brazil. *Res:* Origin of mineral deposits; isotope geology related to mineral deposits; temperature of mineral deposition; zoning applied to mineral exploration; geostatistics; mineral exploration data analysis. *Mailing Add:* Dept Geol Sci Univ BC Vancouver BC V6T 2B4 Can

SINCLAIR, ANNETTE, b Hale, Mo, Aug 14, 16. MATHEMATICS. *Educ:* Cent Mo State Col, BS, 40; Univ Ill, AM, 45, PhD(math), 49. *Prof Exp:* Teacher pub schs, Mo, 34-42; actuarial clerk, Gen Am Life Ins Co, 42-44; asst, Univ Ill, 44-49; instr, Univ Tenn, 49-52; from asst prof to assoc prof math, Southern Ill Univ, 52-57; assoc prof math, Purdue Univ, West Lafayette, 57-86; RETIRED. *Concurrent Pos:* Vis assoc prof, Univ Okla, 66-68. *Mem:* Am Math Soc; Math Asn Am. *Res:* Theory of approximation by analytic functions. *Mailing Add:* 1857 Restful Dr Bradenton FL 34207

SINCLAIR, BRETT JASON, DATA NETWORK DESIGN, LASER COMMUNICATIONS. *Educ:* Pratt Inst, MA, 69; City Univ NY, BA, 72. *Prof Exp:* Res & develop laser commun, Capital Develop Group, 73-85; network analyst data commun, 83-85; CONSULT COMPUTER NETWORKS, SOS INC, 85- *Res:* Data communications network planning, design and tuning; design of non-mechanical laser scanning system for data communications and telecommunications. *Mailing Add:* SOS Inc 68 Spencer Rd Basking Ridge NJ 07920

SINCLAIR, CHARLES KENT, b Watertown, NY, Aug 9, 38; m 83. EXPERIMENTAL HIGH ENERGY PHYSICS, APPLIED PHYSICS & ACCELERATOR PHYSICS. *Educ:* Rensselaer Polytech Inst, BS, 60; Cornell Univ, PhD(Physics), 67. *Prof Exp:* Res assoc physics, Tufts Univ, 66-68, asst prof, 68-69; staff physicist, Stanford Linear Accelerator Ctr, 69-87; SR SCIENTIST, CEBAF, 87- *Mem:* Fel Am Phys Soc; Sigma Xi; Am Vacuum Soc. *Res:* Experimental tests of quantum electrodynamics; single and multiple meson photoproduction; proton compton effect; research with polarized high energy electrons and gamma rays; photocathode studies; microwave power production. *Mailing Add:* CEBAF 12000 Jefferson Ave Newport News VA 23606

SINCLAIR, CLARENCE BRUCE, botany, for more information see previous edition

SINCLAIR, D G, b Rochester, NY, Nov 2, 33; Can citizen; m 58; c 1. ACADEMIC ADMINISTRATION. *Educ:* Univ Toronto, DVM, 58, MSA, 60; Queen's Univ, PhD(physiol), 63. *Prof Exp:* USPHS fel, Columbia Univ, 62-63; fel physiol & Meres sr scholar med res, St John's Col, Cambridge Univ, 63-65; from asst prof to assoc prof, 66-72, prof physiol, 72, dean Fac Arts & Sci, 74-84, vprin, 84-88, DEAN FAC MED & VPRIN HEALTH SCI, QUEENS UNIV, ONT, 88- *Concurrent Pos:* Markle Scholar acad med, 66-71; Dir Med Res Coun, Ottawa, 83-84. *Mem:* Can Physiol Soc. *Res:* Gastrointestinal physiology. *Mailing Add:* Dean Med Fac Med Queen's Univ Kingston ON K7L 3N6 Can

SINCLAIR, DOUGLAS C, b Cambridge, Mass, July 13, 38; m 60; c 2. OPTICAL ENGINEERING. *Educ:* Mass Inst Technol, BS, 60; Univ Rochester, PhD(optics), 63. *Prof Exp:* Asst prof optics, Univ Rochester, 65-67; tech dir, Spectra-Physics, Calif, 67-69; from assoc prof to prof optics, Univ Rochester, 69-80; PRES, SINCLAIR OPTICS, NY, 80- *Concurrent Pos:* Lectr, Stanford Univ, 69; consult, Nat Acad Sci, 70; ed, Optical Eng, 72; ed, J Optical Soc Am, 76-78. *Honors & Awards:* Adolph Lomb Medal, Optical Soc Am, 68. *Mem:* AAAS; Optical Soc Am. *Res:* Design of optical systems; development of software for optical design. *Mailing Add:* 6780 Palmyra Rd Fairport NY 14450-3342

SINCLAIR, GEORGE, b Hamilton, Ont, Nov 5, 12; m 51; c 3. ELECTRICAL ENGINEERING. *Educ:* Univ Alta, BSc, 33, MSc, 35; Ohio State Univ, PhD(elec eng), 46. *Hon Degrees:* DSc, Ohio State Univ, 73. *Prof Exp:* From asst prof to prof elec eng, Univ Toronto, 47-78; pres, 51-73, CHMN BD, SINCLAIR RADIO LABS LTD, 73- *Concurrent Pos:* Chmn, Can Comn, Int Sci Radio Union, 51-61, int chmn subcomn, 54-60; Guggenheim fel, 58. *Honors & Awards:* McNaughton Medal, Inst Elec & Electronics Engrs, 75. *Mem:* Fel Inst Elec & Electronics Engrs; Eng Inst Can; fel AAAS; fel Royal Soc Can; fel Can Acad Eng. *Res:* Boundary-value problems of electromagnetic theory, particularly theory of slot antennas and applications of integral equation methods of solution. *Mailing Add:* Sinclair Radio Labs Ltd 85 Mary St Aurora ON L4G 3G9 Can

SINCLAIR, GEORGE M(ORTON), mechanics, metallurgy; deceased, see previous edition for last biography

SINCLAIR, GLENN BRUCE, b Auckland, NZ, Mar 7, 46; c 4. SOLID MECHANICS. *Educ:* Univ Auckland, BSc, 67, BE, 69; Calif Inst Technol, PhD(appl mech), 72. *Prof Exp:* Res scientist, Dept Sci & Indust Res, Appl Math Div, Wellington, 68-69; J Willard Gibbs instr appl sci & eng, Yale Univ, 72-74; lectr appl mech, Univ Auckland, 74-77; from asst prof to assoc prof, 77-82, PROF MECH ENG, 82-, HEAD DEPT, CARNEGIE-MELLON UNIV, 86- *Concurrent Pos:* Instr, Auckland Tech Inst, 64-66, 69, & 75-77; prof, Pratt & Whitney Aircraft Corp, Conn, 78, Fla, 79; vis prof, Dept Eng, Cambridge Univ, 81. *Mem:* Am Acad Mech. *Res:* Modelling and analysis of engineering problems in solid and structural mechanics particularly as related to fatigue and fracture mechanics and contact problems; development of both analytical and numerical methods for such problems. *Mailing Add:* Dept Mech Eng Carnegie-Mellon Univ Pittsburgh PA 15213

SINCLAIR, HENRY BEALL, b St Louis, Mo, Jan 11, 30; m 55; c 2. ORGANIC CHEMISTRY. *Educ:* Univ Calif, Berkeley, BS, 52; Mass Inst Technol, PhD(org chem), 59. *Prof Exp:* Res chemist, Mallinckrodt Chem Works, 52-54 & Procter & Gamble Co, 58-63; res chemist, 63-69, prin chemist, 69-78, RES CHEMIST, NORTHERN REGIONAL RES LAB, USDA, 78- *Mem:* Am Chem Soc; Royal Soc Chem; AAAS. *Res:* Amino acids; heterocyclics; opium alkaloids; carbohydrates and their derivatives. *Mailing Add:* 1019 W Teton Dr Peoria IL 61604

SINCLAIR, J CAMERON, b Butte, Mont, Dec 19, 18; m 45; c 4. NEUROPHYSIOLOGY & BIOMEDICAL ENGINEERING. *Educ:* Univ Calif, Berkeley, AB, 48; Univ Calif, Los Angeles, MA, 56; Iowa State Univ, PhD(physiol), 66. *Prof Exp:* Res asst biochem, White Mem Hosp, Los Angeles, 48-52; res asst surg, Univ Calif, Los Angeles, 54-56 & anat, 56-57; res assoc physiol, Univ Calif, San Francisco, 60-61; assoc prof biol, Gordon Col, 62-63; assoc prof, Buena Vista Col, 67-70; res scientist, Div Neuro-Pharmacol, 70-72, Educ Consult, Elem Math, NJ Neuropsychiat Inst, 73-88; RETIRED. *Concurrent Pos:* Res fel physiol, Univ Minn, 66-67; assoc prof psychol, Rider Col, 70-73; consult, Lift, Inc; life mem, Woman's pace; treas, Sch-Age Child Care Coalition. *Mem:* AAAS. *Res:* Instrumentation for cardiovascular physiology; neurophysiology of behavior and sensory integration; neurophysiological bases of mental health; diagnosis of mental illness; teaching math to children as a language is taught via computers. *Mailing Add:* 320 Genesee St Trenton NJ 08611-1912

SINCLAIR, JAMES BURTON, b Chicago, Ill, Dec 21, 27. PLANT PATHOLOGY. *Educ:* Lawrence Univ, BS, 51; Univ Wis-Madison, PhD(plant path), 55. *Prof Exp:* Res asst plant path, Univ Wis-Madison, 51-55, res assoc, 55-56; from asst prof to prof, La State Univ, Baton Rouge, 56-68, asst to chancellor, 66-68; PROF PLANT PATH, UNIV ILL, URBANA, 68- *Concurrent Pos:* Grants, Olin Mathieson Chem Corp, 60-66; Allied Chem Co, 64; Diamond Alkali Co, 66; E I Du Pont de Nemours & Co, Inc, 66-68; US Rubber Co, 66-68; partic, Adv Virol Sem, Univ Md, 63; secy, Cotton Dis Coun, 63-64, chmn, 65-66; La State Univ Found res grants, 65-67, grad sch for travel grant, 68; partic conf seed & soil treatment, Am Phytopath Soc, 65-67, chmn, 67, partic conf plant dis control & diag, 67; Conf Control Soil Fungi, Ariz, 68; Am Phytopath Soc for travel grant, 68; Agency Int Develop develop grant, 68-73; Int Cong Plant Path, London, 68, Minneapolis, 73 & Munich, 78; Indian Sci Cong, Kharapur, 70; Int Cong Plant Protection, Paris, 70; Sem Plant Protection Trop Food Crops, Ibadan, 71; Ford Found for travel grant, 71; campus coordr, Ill-Tehran Res Unit, 73-78; develop progs, Pakistan & Zambia, 80- *Honors & Awards:* Paul A Funk Award, 85; Soybean Res Recognition Award, ICI Americas/Am Soybean Asn, 83; Award of Distinguished Serv, US Dept Agr, 88; Res Award, Am Soybean Asn, 89. *Mem:* AAAS; Am Phytopath Soc; Am Inst Biol Sci. *Res:* Cotton seedling disease control and tolerance to fungicides; cytology and ultrastructure of Rhizoctonia; pathogenicity of Geotrichum; citrus fruit rots and viral diseases; seed and soil-borne diseases and microorganisms of soybean; soybean seed quality; biological control of plant diseases. *Mailing Add:* Dept Plant Path Univ Ill 1102 S Goodwin Ave Urbana IL 61801-4709

SINCLAIR, JAMES DOUGLAS, b Evanston, Ill, Nov 23, 45; m 67; c 3. RELIABILITY OF ELECTRONICS, ANALYTICAL CHEMISTRY. *Educ:* Purdue Univ, BS, 67; Univ Wis, PhD(inorg chem), 72. *Prof Exp:* HEAD MAT RELIABILITY & ELECTROCHEM RES, AT&T BELL LABS, 85- *Concurrent Pos:* Supvr, Contamination Res Group, 72-85. *Mem:* Am Chem Soc; Inst Elec & Electronic Engrs; Electrochem Soc; Nat Asn Corrosion

Engrs; Am Asn Aerosol Res. *Res:* Materials reliability of electronic devices; contamination research and control; interaction of corrosive particles and corrosive gases with electronic equipment; analytical techniques for measuring ionic contaminants on surfaces; clean room manufacturing technology; electrochemical corrosion processes. *Mailing Add:* 11 Mount Vernon Ave Summit NJ 07901

SINCLAIR, JOHN G, b Saskatoon, Sask, Sept 4, 40; m; c 3. PHARMACOLOGY, PAIN. *Educ:* Purdue Univ, PhD(neuropharmacol), 68. *Prof Exp:* PROF PHARMACOL, SCH PHARMACEUT, UNIV BC, 80- *Mem:* Int Asn Study Pain; Am Soc Pharmacol Therapeut; Neurosci Soc; Pharmacol Soc Can. *Res:* Neurophysiol and pharmacology of pain in animals. *Mailing Add:* Div Pharmacol Fac Pharmaceut Sci Univ BC Vancouver BC V6T 1W5 Can

SINCLAIR, JOHN HENRY, b Oakwood, Tex, Aug 14, 35; m 62; c 1. CELL BIOLOGY. *Educ:* Tex A&M Univ, BS, 58, MS, 59; Univ Chicago, PhD(zool), 66. *Prof Exp:* Res chemist, Coleman Instruments, Inc, 61-62; fel embryol, Carnegie Inst Technol, 66-68; from asst prof to assoc prof, 68-77, chmn dept, 73-77, PROF ZOOL, IND UNIV, BLOOMINGTON, 77- *Mem:* AAAS; Am Soc Cell Biol; Genetics Soc Am. *Res:* Chemical and physical characterization of nucleic acids of mitochondria and nucleoli; emphasis on zea mays; higher plants. *Mailing Add:* Dept Biol Ind Univ Rm 305 Bloomington IN 47405

SINCLAIR, KENNETH F(RANCIS), b Kentfield, Calif, Jan 26, 25; m 53; c 6. ENGINEERING PHYSICS. *Educ:* Univ Calif, BS, 50. *Prof Exp:* Jr investr, Lab Instrumentation Prog, US Naval Radiol Defense Labs, 50-54, investr, 54-56, head radiac eng prog, 56-58, head radiation instrumentation br, 58-67; res scientist, Hq, NASA, Moffett Field, 67-72; PRES, XETEX, INC, MOUNTAIN VIEW, 72- *Mem:* Sr mem Inst Elec & Electronics Engrs; Health Physics Soc; Am Soc Nondestructive Testing. *Res:* Remote sensing; systems analysis; nuclear instrumentation; nondestructive testing; data handling and processing; radiation detectors. *Mailing Add:* Xetex Inc 660 National Ave Mountain View CA 94043

SINCLAIR, MICHAEL MACKAY, b New Glasgow, NS, June 20, 44; m 70; c 1. FISHERIES MANAGEMENT, POPULATION BIOLOGY. *Educ:* Queen's Univ, Can, Bsc, 67; Southampton Univ, UK, MSc, 69; Univ Calif, San Diego, PhD(oceanog), 77. *Prof Exp:* Prof oceanog, Univ Que, Rimouski, 73-77, dir, MSc Oceanog Prog, 76-77; res scientist fisheries mgt, Bedford Inst Oceanog, 78-82, head, Pop Dynamics Sect, Marine Fish Div, 81-82; chief, Invert & Marine Plants Div, Halifax Lab, 82-88,; DIR, BIOL SCI BR, BEDFORD INST OCEANOG, 88- *Concurrent Pos:* Vis researcher, Sta Zoologique, Ville Franche-sur-Mer, 78. *Res:* Phytoplankton temporal and spatial distributions in relation to physical processes; population biology of marine animals. *Mailing Add:* PO Box 1006 Dartmouth NS B2Y 4A2 Can

SINCLAIR, NICHOLAS RODERICK, b Bradford-on-Avon, Eng, Apr 20, 36; Can citizen; m 61, 76; c 6. IMMUNOLOGY, ONCOLOGY. *Educ:* Dalhousie Univ, BSc, 57, MD, 62, PhD(biochem), 65. *Prof Exp:* Vis scientist immunol, Chester Beatty Res Inst, Eng, 65-67; asst prof immunol, 67-73, assoc prof microbiol & immunol, 73-77, assoc prof med, 77-80, actg chmn, Dept Bact & Immunol, 74-76, asst dean res, Fac Med, 78-80, PROF MICROBIOL & IMMUNOL, UNIV WESTERN ONT, 79-, PROF MED, 80-, CHMN, DEPT MICROBIOL & IMMUNOL, 81- *Concurrent Pos:* Med Res Coun Can fel, 64-67 & scholar, 67-72; assoc dir, Transplant Monitoring Lab, dept nephrol, Univ Hosp, 73- *Mem:* Can Soc Immunol (vpres, 81-); Am Asn Immunologists. *Res:* Immunobiology; tumour immunology. *Mailing Add:* Dept Microbiol & Immunol Univ Western Ont London ON N6A 5C1 Can

SINCLAIR, NORVAL A, b Sturgis, SDak, Sept 1, 35; m 58; c 3. MICROBIOLOGY. *Educ:* SDak State Univ, BS, 57, MS, 59; Wash State Univ, PhD(bact), 64. *Prof Exp:* Bacteriologist, Minn State Health Dept, 57-58; from instr to asst prof microbiol, Colo State Univ, 64-67; USPHS trainee, Hopkins Marine Sta, 67-68; from asst prof to assoc prof microbiol, 68-82, ASSOC HEAD MICROBIOL & IMMUNOL, UNIV ARIZ, 82- *Mem:* AAAS; Am Soc Microbiol. *Res:* Psychrophilic microbes; microbial physiology, growth and ecology; microbiology of groundwater. *Mailing Add:* Dept of Microbiol Univ of Ariz Tucson AZ 85721

SINCLAIR, PETER C, b Seattle, Wash, Feb 17, 29; m 60; c 4. ATMOSPHERIC PHYSICS. *Educ:* Univ Wash, BS, 52; Univ Calif, Los Angeles, MS, 58; Univ Ariz, PhD(atmospheric physics), 66. *Prof Exp:* Res asst meteorol, Univ Wash, 48-52, Univ Calif, Los Angeles, 57-60 & Univ Ariz, 60-65; asst prof, 65-68, ASSOC PROF ATMOSPHERIC SCI, COLO STATE UNIV, 69- *Concurrent Pos:* Mem bd dir, Waverly West Soaring Ranch, 68- *Mem:* AAAS; Am Meteorol Soc; Am Geophys Union. *Res:* Severe storms; atmospheric convection; airborne atmospheric instrumentation; weather modification. *Mailing Add:* Dept of Atmospheric Sci Colo State Univ Ft Collins CO 80523

SINCLAIR, PETER ROBERT, b Sydney, Australia, Jan 17, 42; m 71. PORPHYRIN BIO-SYNTHESIS. *Educ:* Univ Sydney, MS, 66; Univ Ky, PhD(biochem), 70. *Prof Exp:* RES BIOLOGIST, VET ADMIN, WHITE RIVER JUNCTION, VT, 77-; RES ASSOC PROF BIOCHEM, DARTMOUTH MED SCH, 77- *Mem:* Am Soc Biochem & Molecular Biol. *Res:* Regulation of heme and cytochrome P450 especially in hepatocyte cultures. *Mailing Add:* Vet Admin White River Junction VT 05001

SINCLAIR, RICHARD GLENN, II, b Parsons, Kans, Mar 10, 33; m 55; c 3. POLYMER CHEMISTRY. *Educ:* Univ Mo-Kansas City, BS, 60, PhD(polymer sci), 67. *Prof Exp:* Analyst, E I du Pont de Nemours & Co, 58-61; chemist, Chemagro Corp, 61-63; sr res chemist, 66-68, ASSOC SECT MGR, ORG & POLYMER CHEM SECT, BATTELLE MEM INST, 78- *Mem:* Am Chem Soc. *Res:* Discovery and development of novel high polymer systems; poly (lactic acid) controlled-release systems. *Mailing Add:* 985 Kenway Ct Columbus OH 43220

SINCLAIR, ROBERT, b Liverpool, Eng, Feb 15, 47; c 2. MATERIALS SCIENCE, ELECTRON MICROSCOPY. *Educ:* Cambridge Univ, BA, 68, PhD(mat sci), 72. *Prof Exp:* Res assoc, Univ Newcastle, Tyne, 71-73; res engr, Univ Calif, Berkeley, 73-76; from asst prof to assoc prof, 77-84, PROF MAT SCI, STANFORD UNIV, 84- *Concurrent Pos:* Res consult, Xerox Corp, 78-80, Raychem Corp, 84-88, Matsushita Elec, 87- & Kobe Develop Corp, 90-; Alfred P Sloan Found fel, 79. *Honors & Awards:* Robert Lansing Hardy Gold Medal, Am Inst Mining, Metall & Petrol Engrs, 76; Eli Franklin Burton Award, Electron Micros Soc Am, 77; Marcus E Grossman Award, Am Soc Metals, 82. *Mem:* Am Inst Mining, Metall & Petrol Engrs; Electron Micros Soc Am; Mat Res Soc. *Res:* Solid state phase transformations; high-resolution transmission electron microscopy; microstructure property relationships of materials; semiconductor interfaces. *Mailing Add:* Dept Mat Sci & Eng Stanford Univ Stanford CA 94305

SINCLAIR, ROLF MALCOLM, b New York, NY, Aug 15, 29; div; c 2. PHYSICS, ARCHAEOASTRONOMY. *Educ:* Calif Inst Technol, BS, 49; Rice Inst, MA, 51, PhD(exp physics), 54. *Prof Exp:* Res physicist, Westinghouse Res Labs, 53-56; res assoc nuclear physics, Physics Inst, Univ Hamburg, 56-57; res asst, Univ Paris, 57-58; mem res staff, Plasma Physics Lab, Princeton Univ, 58-69; PROG DIR, PHYSICS DIV, NSF, 69- *Concurrent Pos:* Res assoc, UK Atomic Energy Auth, 65-66, Culham Lab, Eng; mem, The Solstice Proj, 78-; NSF rep, US Solar Eclipse Expedition, India, 80; vis distinguished prof, NMex State Univ, 85; vis prof, Northern Arizona Univ, 86; vis scientist, Los Alamor Nat Lab, 88-89, consult, 90- *Mem:* Fel AAAS (secy physics sect, 72-); fel Am Phys Soc; Sigma Xi. *Res:* Atomic, plasma and nuclear physics; controlled thermonuclear power; administration of science; archaeoastronomy of US southwest and Meso America. *Mailing Add:* Physics Div Nat Sci Found Washington DC 20550

SINCLAIR, RONALD, b Dungannon, Northern Ireland, Oct 20, 31; div; c 3. CELL BIOLOGY, BIOCHEMISTRY. *Educ:* Queen's Univ, Belfast, BSc, 54, PhD(biochem), 58. *Prof Exp:* Res asst biochem, Queen's Univ, Belfast, 54-58; fel, Jackson Lab, 58-60; res fel chem biol unit, Univ Edinburgh, 60-65; ASSOC PROF BIOL, MCGILL UNIV, 65- *Mem:* AAAS; Tissue Cult Asn; Brit Biochem Soc; Brit Soc Cell Biol; Can Soc Cell Biol. *Res:* Cell growth. *Mailing Add:* Dept Biol McGill Univ 1205 Penfield Ave Montreal PQ H3A 1B1 Can

SINCLAIR, THOMAS FREDERICK, immunopathology, oncology, for more information see previous edition

SINCLAIR, THOMAS RUSSELL, b Indianapolis, Ind, Aug 4, 44; m 67; c 3. AGRONOMY, PLANT PHYSIOLOGY. *Educ:* Purdue Univ, Lafayette, BS, 66, MS, 68; Cornell Univ, PhD(field crop sci), 71. *Prof Exp:* Mem staff, Nat Sci Found Int Biol Prog, Duke Univ, 71-74; plant physiologist, Microclimate Proj, Sci & Educ Admin-Agr Res, 74-79, PLANT PHYSIOLOGIST, ENVIRON PHYSIOL UNIT, AGR RES SERV, USDA, 80- *Concurrent Pos:* Vis scientist, State Agr Univ, Wageningen, Neth, 73-74; mem staff, Dept Agron, Cornell Univ, 74-79; adj prof, Dept Agron & Fruit Crops, Univ Fla, 80- *Mem:* Am Soc Agron; Crop Sci Soc Am; Am Soc Plant Physiol. *Res:* Experimental and computer simulation research on crop productivity, especially soybean, by improving carbon dioxide assimilation rates, nitrogen fixation rates, seed growth characteristics, use of vegetative stands. *Mailing Add:* Agron Physiol Lab Univ Fla Bldg 164 Gainesville FL 32611

SINCLAIR, WARREN KEITH, b Dunedin, NZ, Mar 9, 24; nat US; m 48; c 2. BIOPHYSICS. *Educ:* Univ NZ, BSc, 44, MSc, 45; Univ London, PhD(physics), 50. *Prof Exp:* Physicist, Dept Sci & Indust Res, NZ Govt, 44-45; lectr radiol physics, Univ Otago & radiol physicist, Dunedin Pub Hosp, 45-47; teacher & lectr, Univ London & physicist, Royal Cancer Hosp, 47-54; prof physics, Univ Tex & chief physicist, MD Anderson Hosp & Tumor Inst, 54-60; sr biophysicist, Div Biol & Med, Argonne Nat Lab, 60-83, dir, 70-74, assoc lab dir, 74-81; prof radiobiol, 64-85, EMER PROF RADIOBIOL, UNIV CHICAGO, 85-; PRES, NAT COUN RADIATION PROTECTION & MEASUREMENTS, WHO, 77-81 & 81- *Concurrent Pos:* Consult, Humble Oil & Refining Co & Univ Tex MD Anderson Hosp & Tumor Inst; mem, Int Comn Radiation Protection, 77-; mem, US Nat Comt Pure & Appl Biophys & chmn, US Nat Comt Med Physics, 64-70; mem, Int Comn Radiation Units, 69-85; secy gen, Vth Int Cong Radiation Res, 74; US alt deleg, UN Sci Comn on the Effects of Atomic Radiation, 77- *Honors & Awards:* Curie lectr, 79; Failla lectr, Radiation Res Soc, 87; Coolidge Award, Am Asn Physicists Med, 86. *Mem:* Radiation Res Soc (pres, 78-79); Soc Nuclear Med; Asn Physicists in Med (pres, 60-61); Radiol Soc NAm; Brit Inst Radiol; Biophys Soc; Soc Risk Analysis; Health Physics Soc. *Res:* Radiation protection and radiological physics; radiation response in synchronized cell cultures; quantitative aspects of radiobiology; radiation risk estimation in humans. *Mailing Add:* Natl Coun on Radiation Prot Meas 7910 Woodmont Av Ste 800 Bethesda MD 20814-3045

SINCLAIR, WAYNE A, b Medford, Mass, Dec 15, 36; m 58; c 3. FOREST PATHOLOGY. *Educ:* Univ NH, BS, 58; Cornell Univ, PhD(plant path), 62. *Prof Exp:* From asst prof to assoc prof, 62-75, PROF PLANT PATH, CORNELL UNIV, 75- *Concurrent Pos:* Vis forest pathologist, Weyerhaeuser Co, Wash, 70. *Mem:* Am Phytopath Soc; Int Soc Arboricult. *Res:* Forest pathology; mycoplasmal diseases of trees. *Mailing Add:* Dept of Plant Path 334 Plant Sci Bldg Cornell Univ Ithaca NY 14853-5908

SINCLAIR, WILLIAM ROBERT, b Chicago, Ill, Apr 30, 24; m 58; c 4. PHYSICAL INORGANIC CHEMISTRY. *Educ:* Univ Chicago, PhD(chem), 52. *Prof Exp:* Instr chem, Univ Minn, 50-51; mem tech staff, Bell Tel Labs, Inc, 52-88; MARTIN GOFFMAN ASSOC, 88-; W ROBERT SINCLAIR ASSOC, 88- *Mem:* Am Chem Soc; Mat Res Soc. *Res:* Preparation and properties of glassy and crystalline inorganic films; high Tc superconductors. *Mailing Add:* 22 Manor Hill Rd Summit NJ 07901

SINCOVEC, RICHARD FRANK, b Pueblo, Colo, July 14, 42; c 2. COMPUTER SCIENCES, SOFTWARE SYSTEMS. *Educ:* Univ Colo, Boulder, BS, 64; Iowa State Univ, MS, 67, PhD(appl math), 68. *Prof Exp:* Instr math, Iowa State Univ, 64-68, jr mathematician, Ames Lab, 66-68; sr res mathematician, Exxon Prod Res Co, 68-70; asst prof comput sci & math, Kans State Univ, 70-74, from assoc prof to prof comput sci, 74-77; mgr numerical anal, Boeing Comput Serv Co, 77-80; PROF & CHMN, COMPUT SCI, UNIV COLO, 80- *Concurrent Pos:* Am Chem Soc Petrol Res Fund fel, 71-74; consult, Lawrence Livermore Lab, 71-88. *Mem:* Soc Indust & Appl Math; Asn Comput Mach. *Res:* Software engineering with Ada mathematical software components in Ada; Galerkin, finite element and collocation methods for solving ordinary and partial differential equations; numerical linear algebra; parallel computing. *Mailing Add:* Ames Res Ctr NASA R1ACS M-S 230-S Moffett Field CA 94035

SINDELAR, ROBERT D, b Chicago, Ill, Dec 3, 52; m 76; c 3. MEDICINAL CHEMISTRY, MOLECULAR MODELING. *Educ:* Millikin Univ, BA, 74; Univ Iowa, MS, 75, PhD(med chem), 80. *Prof Exp:* Fel chem, Univ Brit Columbia, 80-81; fel chem, State Univ NY, Buffalo, 81-83; asst prof med chem, 83-89, res asst prof, 85-89, ASSOC PROF MED CHEM & RES ASSOC PROF, RES INST PHARMACEUT SCI, UNIV MISS, 85- *Concurrent Pos:* Prin investr, Am Heart Asn, 85-90, First Chem Corp, 87-88, T-Cell Scis, 87-; comt mem, Nat Rho Chi Grad Scholarship Comt, 88-, Am Chem Soc, Div Med Chem, 85-; dir, Molecular Modeling Lab, Univ Miss Sch Pharm, 86-; vis res scientist, Ctr Molecular Design, Washington Univ, St Louis, Mo, 90. *Mem:* Am Chem Soc; Am Asn Col Pharm; Int Soc Heterocyclic Chem; Molecular Graphics Soc; Sigma Xi. *Res:* Complex biologically-active natural products serve as topographical models for drug design, molecular modeling, synthesis and pharmacological evaluation are utilized to study natural products with immunomodulatory, cardiovascular and antiviral activity. *Mailing Add:* Dept Med Chem Univ Miss Sch Pharm University MS 38677

SINDEN, JAMES WHAPLES, b Oak Park, Ill, Nov 12, 02; m 26; c 2. PLANT PATHOLOGY. *Educ:* Univ Kans, AB, 24; Cornell Univ, PhD(plant path), 37. *Prof Exp:* Asst instr plant path, Cornell Univ, 24-25, instr, 25-30; from asst prof to prof, Pa State Univ, 30-52; CONSULT, 52- *Concurrent Pos:* Spec investr, Trop Deterioration Lab, US Army, 45-46. *Mem:* AAAS; Am Phytopath Soc; Mycol Soc Am. *Res:* Physiology and pathology of the commercial mushroom; economic mycology. *Mailing Add:* Box 2281 State St Saxonburg PA 16056

SINDERMANN, CARL JAMES, b North Adams, Mass, Aug 28, 22; m 43; c 5. MARINE BIOLOGY, PARASITOLOGY. *Educ:* Univ Mass, BS, 49; Harvard Univ, AM, 51, PhD(biol), 53. *Prof Exp:* Teaching fel biol, Harvard Univ, 50; parasitologist, Biol Surv, State Dept Conserv, Mass, 50; instr biol, Brandeis Univ, 51-53, asst prof, 53-56; res biologist, Bur Commercial Fisheries, US Fish & Wildlife Serv, 54-59, chief Atlantic herring invests, 59-62, prog coordr, Atlantic herring progs, 62-63, dir biol lab, Md, 63-68, dir, Trop Atlantic Biol Lab, Fla, 68-71; dir, Middle Atlantic Coastal Fisheries Ctr, Nat Marine Fisheries Serv, Nat Oceanic & Atmospheric Admin, 71-76; dir, Sandy Hook Marine Lab, 76-85; MARINE BIOLOGIST, NAT OCEANIC ATMOSPHERIC ADMIN, OXFORD LAB, NAT MARINE FISHERIES SERV, 85- *Concurrent Pos:* Asst, Harvard Med Sch, 52; marine biologist, State Dept Sea & Shore Fisheries, Maine, 52-54; vis lectr, Georgetown Univ, 66-68; adj prof, Div Fisheries Sci, Rosenstiel Sch Marine & Atmospheric Sci, Univ Miami, 69-72; adj prof biol, Lehigh Univ, 73-; adj prof vet microbiol, NY State Vet Col, Cornell Univ, 75-; vis prof, Ont Vet Col, Guelph, 78-79; sci ed, Fishery Bull, US Dept Com, 80-83; vis prof, Univ Miami, 85-88, adj prof, 88- *Mem:* Soc Invert Path; Am Soc Parasitol; Nat Shellfisheries Asn; World Maricult Soc. *Res:* Parasites and diseases of marine organisms; immune responses of marine invertebrates; marine pollution, ecology. *Mailing Add:* NOAA Oxford Lab Nat Marine Fisheries Serv Oxford MD 21654

SINE, ROBERT C, b Waukegan, Ill, Nov 22, 36. OPERATOR THEORY, MARKOV. *Educ:* Univ Ill, PhD(math), 62. *Prof Exp:* PROF MATH, UNIV RI, 71- *Mem:* Am Math Soc. *Mailing Add:* Univ RI Dept Math Kingston RI 02881

SINENSKY, MICHAEL, b New York, NY, July, 2, 45; c 2. BIOCHEMISTRY. *Educ:* Columbia Col, NY, BS, 66; Harvard Univ, PhD(biochem & molecular biol), 72. *Prof Exp:* Jr fel biochem, Soc Fels, Harvard Univ, 72-74; asst prof chem, Univ Pa, 74-75; SR FEL, ELEANOR ROOSEVELT INST CANCER RES, 75-; prof, 80-85, ADJOINT PROF, BIOCHEM, BIOPHYS & GENETICS & PATH, UNIV COLO HEALTH SCI CTR, 85- *Mem:* Am Soc Biol Chemists. *Res:* Mechanism of regulation of cholesterol biosynthesis. *Mailing Add:* Eleanor Roosevelt Inst Cancer Res 1899 Gaylord St Denver CO 80206

SINES, GEORGE, JR, b Salem, Ohio, July 12, 23; m 56; c 3. METALLURGY, CERAMICS. *Educ:* Ohio State Univ, BME, 43; Univ Calif, Los Angeles, MS, 49, PhD(metall), 53. *Prof Exp:* Design & test engr, Krouse Testing Mach Co, Ohio, 43-44; asst instr physics, San Diego State Col, 46-47; lectr & asst engr, Univ Calif, Los Angeles, 47-53; asst prof metall, Inst Study Metals, Univ Chicago, 53-56; assoc prof eng, 56-62, PROF MAT, UNIV CALIF, LOS ANGELES, 62- *Concurrent Pos:* Fulbright res prof, Tokyo Inst Technol, 58-59; consult, Japanese Atomic Energy Res Inst, 58-59 & Douglas Aircraft Co, Inc, 59-71; NSF sr fel, Ctr Nuclear Studies, Mol, Belg, 65-66; assoc ed, J Eng Mat & Technol, 73-85. *Honors & Awards:* Templin Award, Am Soc Testing & Mat, 78. *Mem:* Am Soc Testing & Mat; Am Soc Metals; Am Ceramic Soc; Am Soc Mech Engrs. *Res:* Fracture of solids; interactions between crystal defects; diffusion in solids; mechanical properties of ceramics. *Mailing Add:* Dept Mat Sch Eng Univ Calif Los Angeles CA 90024

SINEX, FRANCIS MAROTT, b Indianapolis, Ind, Jan 11, 23; c 2. BIOCHEMISTRY. *Educ:* DePauw Univ, AB, 44; Ind Univ, MA, 45; Harvard Univ, PhD, 51. *Prof Exp:* Jr biochemist, Brookhaven Nat Labs, 50-51, biochemist & exec officer, Biochem Div, 51-57; chmn dept, 57-77, PROF BIOCHEM, SCH MED, BOSTON UNIV, 57-, HEAD SECT BIOMED GERONT, 78- *Mem:* Geront Soc (pres, 69-70); Am Soc Biol Chem; Alzheimer's Dis & Related Dis Asn. *Mailing Add:* Dept Biochem Sch Med Boston Univ 80 E Concord St Boston MA 02118

SINFELT, JOHN HENRY, b Munson, Pa, Feb 18, 31; m 56; c 1. CATALYSIS. *Educ:* Pa State Univ, BS, 51; Univ Ill, MS, 53, PhD(chem eng), 54. *Hon Degrees:* DSc, Univ Ill, 81. *Prof Exp:* Chem engr, 54-57, group leader, 57-62, res assoc, 62-68, sr res assoc, 68-72, sci adv, 72-79, SR SCI ADV, EXXON RES & ENG CO, 79- *Concurrent Pos:* Vis prof, Univ Minn, 69. *Honors & Awards:* Emmett Award, Catalysis Soc, 73; Lacey lectr, Calif Inst Technol, 73; Reilley lectr, Notre Dame Univ, 74; Petrol Chem Award, Am Chem Soc, 76; Dickson Prize Sci, Carnegie-Mellon Univ, 77; Gault lectr, Coun Europe Res Group Catalysis, 80; Chem Pioneer Award, Am Inst Chemists, 81; Welch lectr, Conf Chem Res, Robert A Welch Found, 81; Nat Medal Sci, 79; Gold Medal, Am Inst Chemists, 84; Murphree Award Indust & Eng Chem, Am Chem Soc, 86; Perkin Medal, Soc Chem Indust, 84; Am Phys Soc Int prize for New Materials, 78. *Mem:* Nat Acad Eng; Nat Acad Sci; Am Chem Soc; Catalysis Soc; Am Acad Arts & Sci. *Res:* Heterogeneous catalysis; surface science; petroleum chemistry. *Mailing Add:* Corp Res Sci Lab Exxon Res & Eng Co Clinton Twp Rte 22E Annandale NJ 08801

SING, CHARLES F, b Joliet, Ill, July 6, 36. HUMAN GENETICS, STATISTICS. *Educ:* Iowa State Univ, BS, 60; Kans State Univ, MS, 63; NC State Univ, PhD(statist genetics), 66. *Prof Exp:* Assoc prof, 72-76, PROF HUMAN GENETICS, MED SCH, UNIV MICH, ANN ARBOR, 76- *Mem:* Am Soc Human Genetics; Biometrics Soc; Genetics Soc; Sigma Xi. *Res:* Genetics of common diseases. *Mailing Add:* Dept Human Genetics Univ Mich Med Sch Ann Arbor MI 48104

SINGAL, DHARAM PARKASH, b Sonepat, India, May 5, 34; Can citizen; m 64; c 2. HUMAN HISTOCOMPATABILITY ANTIGENS. *Educ:* Univ Delhi, India, BSc, 52, MSc, 57; Wash State Univ, PhD(genetics), 68. *Prof Exp:* Asst fel immunol, Univ Calif, Los Angeles, 67-69; lectr path, 70-72, from asst prof to assoc prof, 72-78, PROF PATH & IMMUNOL, MCMASTER UNIV, 78-, SPEC PROF LAB MED & DIR HISTOCOMPATABILITY LAB, MED CTR, 70- *Concurrent Pos:* Consult, Can Red Cross Nat Lab, 77-84. *Mem:* Am Asn Immunologists; Int Transplantation Soc; Am Soc Histocompatibility & Immunogenetics; Can Soc Immunol. *Res:* Polymorphism; complexibility and characterization of human histocompatibility (HLA) antigens by serological, cellular and molecular techniques; the role of HLA antigens in disease and clinical transplantation; mechanism of enhanced allograft survival by blood transfusion; transplantation immunology. *Mailing Add:* Dept Path McMaster Univ 1200 Main St W Hamilton ON L8N 3Z5 Can

SINGAL, PAWAN KUMAR, b Oct 2, 46; m; c 2. CELL BIOLOGY, PHYSIOLOGY. *Educ:* Univ Alberta, Edmonton, Can, PhD(physiol), 74. *Prof Exp:* PROF CARDIOVASC PHYSIOL, DEPT PHYSIOL, UNIV MANITOBA, 77- *Mem:* Am Physiol Soc; Int Soc Heart Found. *Res:* oxygen radical injury in the heart; author of three books and ninety-five papers published. *Mailing Add:* Dept Physiol Univ Manitoba Winnipeg MB R3E 0W3 Can

SINGARAM, BAKTHAN, b Andrha Pradesh, India, May 20, 50; m 77; c 1. ORGANOBORANE CHEMISTRY, CHIRAL SYNTHESIS VIA ORGANOBORANES. *Educ:* Madras Univ, India, BSc, 69, MSc, 71, PhD(org chem), 77. *Prof Exp:* Res assoc, Purdue Univ, 77-80, Univ Col Swansea, Wales, UK, 80-82; asst res sci, Purdue Univ, 82-87, assoc res sci, 87-89; ASST PROF ORG CHEM, UNIV CALIF, SANTA CRUZ, 89- *Concurrent Pos:* Consult, Wallace Labs, Cranbury, NJ, 83- *Mem:* Am Chem Soc. *Res:* Organic chemistry; exploratory synthetic organic chemistry; asymmetric synthesis; development of synthetic methods and organometallic reagents for the synthesis of optically active organic compounds of biological and medicinal significance. *Mailing Add:* Dept Chem & Biochem Univ Calif Santa Cruz CA 95064

SINGER, ALAN G, b Berkeley, Calif, Sept 28, 40; m 63; c 3. VERTEBRATE PHEROMONES. *Educ:* Univ Calif, Berkeley, AB, 65; State Univ NY, PhD(chem), 74. *Prof Exp:* res asst, Kaiser Found Res Inst, 68-69; res assoc, Rockefeller Univ, 73-76, asst prof org chem, 76-87; assoc prof biochem, NY Osteop Med, 87-89; ASSOC MEM, MONEL CHEM CENSUS CTR, 89- *Concurrent Pos:* Asst prof chem, Kingsborough Community Col, 74-75. *Mem:* AAAS; NY Acad Sci; Sigma Xi; Am Chem Soc; Asn Chemoreception Sci. *Res:* Chemical identification of pheromones regulating hormone levels, reproductive behavior, and fertility in mammals; characterization and cloning of an aphrodisiac protein pheromone detected in the vomeronasal organ. *Mailing Add:* Monel Chem Census Ctr 3500 Market St Philadelphia PA 19104

SINGER, ALFRED, b Berlin, Germany, Dec 10, 46. IMMUNOLOGY. *Educ:* Columbia Univ, MD, 72. *Prof Exp:* SR INVESTR IMMUNOL, NAT CANCER INST, 78- *Mailing Add:* Nat Cancer Inst NIH Bldg 10 Rm 4B-17 Bethesda MD 20892

SINGER, ARNOLD J, b Paterson, NJ, Apr 7, 15; m 78; c 2. MEDICAL SCIENCES, GENERAL RESEARCH ADMINISTRATION. *Educ:* Rutgers Univ, BSc, 36, BS, 37; Cornell Univ, MS, 39, PhD(bact), 41. *Prof Exp:* Pharmacologist, State Dept Health, NJ, 40-43; dir res, Chatham Pharmaceut Inc, 46-48; dir res, Block Drug Co, 49-52; pres, Reed & Carnrick, 52-70, dir, Res Inst, 70-77, chmn bd, 70-80; VPRES, SIMBEC RES, 81- *Concurrent Pos:* Pres, Fesler Co, Inc, 61- *Mem:* AAAS; Am Chem Soc; NY Acad Sci. *Res:* Medicinal chemistry. *Mailing Add:* 204 Eagle Rock Ave Roseland NJ 07068-1718

SINGER, ARTHUR CHESTER, b Vienna, Austria, Aug 30, 36; US citizen; m 60; c 3. BIOSTATISTICS. *Educ:* Univ Ill, BS, 61, MS, 63, PhD(quantitative genetics), 67. *Prof Exp:* Asst prof genetics & statist, Ind State Univ, 67-70; PRIN STATISTICIAN, BIOSTATIST SECT, WYETH AYERST RES, 70- *Mem:* Biomet Soc. *Mailing Add:* Biostatist Sect Wyeth Ayerst Res Box 8299 Philadelphia PA 19101

SINGER, B, b San Francisco, Calif. MOLECULAR BIOLOGY, BIOCHEMISTRY. *Educ:* Univ Calif, Berkeley, BS, PhD. *Prof Exp:* Jr chemist, Shell Develop Co, 42-43; jr chemist biochem, Western Regional Res Lab, USDA, 43-46; from res assoc molecular biol to assoc res biochemist, 46-79, res prof, 79-85, EMER PROF BIOCHEM, UNIV CALIF, BERKELEY, 85-, PRIN INVESTR MOLECULAR BIOL, 69-, FAC SCIENTIST, DONNER LAB, 75- *Concurrent Pos:* Assoc ed, Cancer Res, 80 -; vis prof, Univ Queensland, Australia, 82, vis prof, Polish Acad Sci, Warsaw, 86; mem, chem pathol study sect, NIH, 81-86, mem, Sci Adv Bd, Nat Ctr Toxicol Res, Fed Drug Admin, 82-85. *Mem:* Am Assoc Cancer Res. *Res:* Relationship of nucleic acid structure to function; effect of modification by mutagens and/or carcinogens on biological activity of viruses, viral nucleic acids and mammalian cells; mechanism of repair of DNA damage; fidelity of replication over modified bases in DNA; in vitro and in vivo, modified base-containing oligonucleotide synthesis and characterization. *Mailing Add:* Donner Lab Univ Calif Berkeley CA 94720

SINGER, BARRY M, b New York, NY, Feb 15, 40; m 64; c 2. DEVICE PHYSICS, SUB-SYSTEMS RESEARCH. *Educ:* Univ Colo, Boulder, BS, 61; NY Univ, MS, 64; Polytech Inst NY, PhD(electrophysics), 68. *Prof Exp:* Sr engr, Raytheon Co, 60-63, sect engr, 63-65, sect head, 65-69; sr prog leader component & device res, 69-79, group dir, 79-82, group dir component & device res, 82-84, DEP DIR, PHILIPS LAB, 84-, DIR PHYSICS & MAT RES SECTOR, 84- *Concurrent Pos:* Consult, Vita Corp. *Mem:* Inst Elec & Electronics Engrs; Sigma Xi. *Mailing Add:* Philips Labs 345 Scarborough Rd Briarcliff Manor NY 10510

SINGER, BURTON HERBERT, b Chicago, Ill, June 12, 38; m 71; c 1. STATISTICS. *Educ:* Case Inst Technol, BS, 59, MS, 61; Stanford Univ, PhD(statist), 67. *Prof Exp:* From asst prof to assoc prof statist, 67-77, PROF MATH STATIST, COLUMBIA UNIV, 77- *Concurrent Pos:* Statist consult, Rand Corp, 71-, Union Carbide Corp, 71- & US AEC, 72-75; res assoc statist, Princeton Univ, 72-73. *Mem:* AAAS; Am Statist Asn; Psychomet Soc. *Res:* Discrimination and identification of mathematical models in the social sciences; designs for observational studies; inverse problems. *Mailing Add:* 193 Battery Rd New Canaan CT 06840

SINGER, DON B, b Woodward, Okla, May 18, 34; m 58; c 3. DEVELOPMENTAL PATHOLOGY. *Educ:* Baylor Col Med, Houston, Tex, MD, 59. *Hon Degrees:* MA, Brown Univ, 76. *Prof Exp:* Pathologist, Baylor Col Med & Tex Children's Hosp, Houston, 63-75; CHIEF PATH, WOMEN & INFANTS HOSP, PROVIDENCE, RI, 75-; PROF PATH, BROWN UNIV MED PROG, 75- *Mem:* Soc Pediat Path (treas, 85-); Int Acad Path; Am Asn Pathologists; Col Am Pathologists; Am Soc Clin Pathologists. *Res:* Clinical-pathologic relationships in fetal and neonatal diseases, particularly infectious diseases of neonates. *Mailing Add:* Dept Path Women & Infants Hosp 101 Dudley St Providence RI 02905

SINGER, DONALD ALLEN, b Ukiah, Calif, 43; m; c 2. APPLIED STATISTICS, MINERAL ECONOMICS. *Educ:* San Francisco State Univ, BA, 66; Pa State Univ, MS, 68, PhD(mineral & petrol), 71. *Prof Exp:* Syst analyst, Kennecott Copper Corp, 71-72, sr comput programmer, 72-73; GEOLOGIST, BR RESOURCE ANALYST, US GEOL SURV, 73- *Honors & Awards:* Meritorious Serv Award, US Dept Interior. *Mem:* Sigma Xi; Am Statist Asn; Int Asn Math Geol; Soc Econ Geologists; Japan Soc Mining Geologist. *Res:* Operational mineral resource classification; predicting the occurrence of mineral resources; modeling the search for mineral resources; mineral resource predictions for large regions. *Mailing Add:* US Geol Surv Mail Stop 984 345 Middlefield Rd Menlo Park CA 94025

SINGER, DONALD H, b New York, NY, Sept 27, 29; m 58; c 3. CARDIOLOGY. *Educ:* Cornell Univ, AB, 48; Stanford Univ, MA, 50; Northwestern Univ, MD, 54; Am Bd Internal Med, dipl, 63. *Prof Exp:* Intern, Michael Reese Hosp, Chicago, 54-55, fel, Cardiovasc Dis, Nat Heart Inst trainee, Cardiovasc Inst, 55-57; asst resident med, Beth Israel Hosp, Boston, Mass, 57-58; fel cardiol, Sch Med, Georgetown Univ, 60-61, chief med res & instr med, 61-62; res assoc, Col Physicians & Surgeons, Columbia Univ, 62-63, asst prof pharmacol, 63-68; assoc prof, 68-77, PROF MED & PHARMACOL, NORTHWESTERN UNIV MED SCH, 77-, DIR REINGOLD ECG CTR, 68-, CHESTER C & DEBORAH M COOLEY PROF CARDIOL, 89-, MEM FEINBERG CARDIOL RES INST, 89- *Concurrent Pos:* Fel, John Polachek Found Med Res, 63-66; estab investr, AMA, 67-72; mem res comt, Chicago Heart Asn, 68-75; attend physician, Passavant Mem Hosp, 69-73, Northwestern Mem Hosp, 73-; vis scientist, Univ Chicago, 78-79; dir cardiol clin, Northwestern Med Fac Found, 86-87; fel coun, circ Am Heart Asn, 78- *Mem:* Fel Am Col Physicians; Am Asn Univ Prof; Cardiac Muscle Soc; Am Physiol Soc; Electrophysiol Soc; AMA; Am Fed Clin Res; Am Soc Pharmacol & Exp Therapeut. *Res:* Cellular electrophysiology and electropharmacology (cellular mechanisms of cardiac arrhythmias and anti-arrhythmic agents, standard microelectrode, whole cell voltage clamp and patch clamp studies on human cardiac tissue/cells); autonomic regulation of heart rate/rhythm (normal and diseased hearts) using heart rate variability and power spectral analysis. *Mailing Add:* Northwestern Univ Med Sch Reingold ECG Ctr 320 E Superior St Chicago IL 60611

SINGER, EUGEN, b Levoca, Czech, Apr 1, 26; Can citizen; m 53; c 1. CHEMICAL ENGINEERING, INSTRUMENTATION. *Educ:* Prague Tech Univ, Dipl eng, 52, PhD(anal chem), 63. *Prof Exp:* Asst prof anal chem, Prague Tech Univ, 52-53; chief gas treatment div, Res Inst Inorganic Chem, Czech, 53-68; supvr monitoring & instrumentation develop, Ministry Environ Ont, 68-89; RETIRED. *Res:* Monitoring of ambient air pollutants, instrumentation development, acquisition, reduction and interpretation. *Mailing Add:* 43 Thornelisse Park Dr No 1520 Toronto ON M4H 1J4 Can

SINGER, GEORGE, b Belgrade, Yugoslavia, Apr 23, 37; US citizen; m 60; c 2. ACAROLOGY. *Educ:* City Col New York, BSc, 59; Univ Kans, MS, 62; Ore State Univ, PhD(entom), 65. *Prof Exp:* Asst prof zool, Univ Mont, 65-68; fel entom, McGill Univ, 69-71; proj assoc entom, Univ Wis-Madison, 71-75; RETIRED. *Res:* Biochemistry of chemoreception and repellency; sensory physiology; host parasite relationships; ecology and bionomics; systematics and evolution; pest control. *Mailing Add:* 4913 Goldfinch Dr Madison WI 53714

SINGER, HOWARD JOSEPH, b Annapolis, Md, Mar 2, 44; m 73; c 2. SPACE PHYSICS, MAGNETOSPHERIC PHYSICS. *Educ:* Univ Md, BS, 67; Boston Univ, MA, 72; Univ Calif, Los Angeles, MS, 75, PhD(geophysics & space physics), 80. *Prof Exp:* Instr astron, Wellesley Col, 69-70; mathematician, Fed Systs Div, IBM, 70-72; staff res assoc, Inst Geophysics & Planetary Physics, Univ Calif, Los Angeles, 72-74, res assoc, 74-80; res assoc, 80-81, ASST RES PROF, DEPT ASTRON, BOSTON UNIV, 82- *Concurrent Pos:* Dep sta sci leader, S Pole Sta Antarctica, Inst Geophysics & Planetary Physics, Univ Calif, Los Angeles, 72-73. *Mem:* Am Geophys Union; AAAS; Am Asn Physics Teachers; Sigma Xi. *Res:* Analysis of hydromagnetic waves in the earth's magnetosphere; solar wind interaction with the magnetosphere; management and analysis of digital data. *Mailing Add:* Air Force Geophys Lab PHG Hanscom AFB MA 01731

SINGER, IRA, b New York City, NY, May 28, 23. MEDICINE. *Educ:* Johns Hopkins Univ, AB, 47; Univ Chicago, MS, 49, PhD, 53; Am Bd Microbiol, dipl, 63. *Prof Exp:* Asst, Dept Bact & Parasitol, Univ Chicago, 49-53; res assoc, Rockefeller Inst Med Res, 54-56, assoc, 56-59; assoc prof microbiol, Dept Microbiol & Trop Med, Georgetown Univ, 59-63, prof microbiol, 63-65; assoc dir sci prog, Dept Postgrad Prog, 65-66, ASST DIR, DEPT UNDERGRAD EVAL & STANDARDS, AMA, 75- *Concurrent Pos:* Prof & actg chmn, Dept Microbiol, Georgetown Univ, 64-65. *Mem:* AAAS; Am Soc Parasitologists; Am Soc Trop Med & Hyg; Am Soc Exp Path; Am Soc Microbiol; NY Acad Sci; Soc Exp Biol & Med; Soc Protozollogists; AMA; World Med Asn. *Res:* Medicine; microbiology. *Mailing Add:* Dept Med Sch Serv Am Med Asn 535 N Dearborn St Chicago IL 60610

SINGER, IRWIN, b Brooklyn, NY. WATER METABOLISM RESEARCH, HOSPITAL ADMINISTRATION EMPLOYMENT. *Educ:* Cornell Univ, BA, 58; Albert Einstein Col, MD, 62. *Hon Degrees:* MA, Univ Pa, 73. *Prof Exp:* Intern & resident internal med, Bronx Munic Hosp Ctr, Albert Einstein Col Med, 62-64, sr resident internal med, 66-67; res assoc neurophysiol, Nat Inst Ment Health, NIH, 64-66; instr & fel med, Mass Gen Hosp, Harvard Med Sch, 67-69, asst prof med, 69-70; from asst prof to prof med, Sch Med, Univ Pa, 70-86; assoc dean vet affairs, 86-89, ASSOC DEAN ACAD AFFAIRS, VET ADMIN LAKESIDE MED CTR, 89-; PROF MED, NORTHWESTERN UNIV MED SCH, 86- *Concurrent Pos:* Investr, Marine Biol Lab, Woods Hole, Mass, 64-66, investr, 72-73; asst med, Mass Gen Hosp, Boston, Mass, 69-70; estab investr, Am Heart Asn, 69-74; chief, Renal-Electrolyte Sect, Med Serv, Philadelphia Vet Admin Med Ctr, 70-78, asst chief, Med Serv, 72-76 & chief, 77-82; staff physician, Hosp Univ Pa, 70-86, dir, Cont Ambulatory Peritoneal Dialysis Prog, Renal-Electrolyte Sect, 84-86; mem, Gov Adv Comt, Am Col Physicians, 76-86; vchmn, Dept Med, Sch Med, Univ Pa, 78-82, dir, Clin Res Prog, Renal-Electrolyte Sect, Dept Med, 83-86; vis prof med, Med Col Pa, 79-86; pres, John Morgan Soc, 80-81; chief staff, Vet Admin Lakeside Med Ctr, Chicago, 86- *Mem:* Soc Gen Physiologists; Am Physiol Soc; Am Soc Nephrol; Int Soc Nephrology; Am Soc Clin Invest; Cent Soc Clin Res; fel Am Col Physicians; fel NY Acad Sci. *Res:* Salt and water metabolism, particularly the latter, mostly writing; clinical nephrology (acid-base, electrolytes, salt & water) and some health services research. *Mailing Add:* Vet Admin Lakeside Med Ctr 333 E Huron St Chicago IL 60611

SINGER, IRWIN I, b New York, NY, Dec 20, 43; m 67; c 2. CELL BIOLOGY, VIROLOGY. *Educ:* City Col New York, BS, 65; NY Univ, MS, 67, PhD(biol), 70. *Prof Exp:* Instr biol, Mercy Col, 69-70; asst prof, St Louis Univ, 70-71; assoc investr electron micros, Inst Med Res, Bennington, 71-82; sr res fel, 82-85, SR INVESTR, MERCK SHARP & DOHME RES LABS, 85- *Mem:* Electron Micros Soc Am; Am Soc Cell Biol. *Res:* Ultrastructure and cytochemistry in regenerating Cnidaria; electron microscopy and pathobiology of the Parvoviruses; interactions between fibronectin, fibronectin receptors and the cytoskeleton at the cell attachment surface in vitro and during wound healing and rheumatoid arthritis in vivo; analysis of HMG-CoA reductase in hepatocytes and enterocytes treated with inhibitors of cholesterol synthesis; roles of laminin receptors and basement membranes in leukocytes during inflammation; intracellular localization of interleukin - 1B. *Mailing Add:* Merck Sharp & Dohme Res Labs Merck Co In 80A 54N PO Box 2000 Rahway NJ 07065

SINGER, ISADORE MANUAL, b Detroit, Mich, May 3, 24; m 44; c 3. PURE MATHEMATICS. *Educ:* Univ Mich, BS, 44; Univ Chicago, MS, 48, PhD(math), 50. *Hon Degrees:* DSc, Tulane Univ, 81; LLD, Univ Mich, 89 & Univ Ill Chicago, 90. *Prof Exp:* Moore instr math, Mass Inst Technol, 50-52; asst prof, Univ Calif, Los Angeles, 52-54; vis asst prof, Columbia Univ, 54-55; vis mem, Inst Advan Study, Princeton Univ, 56; prof math, Mass Inst Technol, 56-70, Norbert Wiener prof, 70-79; prof math, Univ Calif, Berkeley, 79-83, vis prof, 77-79, prof, 79-83, Miller Prof math, Univ Calif, Berkeley; John D MacArthur prof math, 83-87, INST PROF, MASS INST TECHNOL, 87- *Concurrent Pos:* Sloan fel, 59-62; Guggenheim fel, 68-69, 75-76; chmn, Comt Sci & Pub Policy, Nat Acad Sci, 73-78, counr & mem, Comn Math & Phys Sci, 78-81; vis prof math, Univ Calif Berkeley, 77-79; White House Sci Coun, 82-88; chair, Geom & Physics, Found France, 87-88. *Honors & Awards:* Bocher Prize, Am Math Soc, 69; Nat Medal Sci, 83. *Mem:* Nat Acad Sci; Am Acad Arts & Sci; Am Math Soc (vpres, 70-72); Am Phys Soc; Am Philos Soc. *Res:* Differential geometry; commutative Banach algebras; global analysis; author of various publications. *Mailing Add:* Bldg 2-174 Math Dept Mass Inst Technol Cambridge MA 02139

SINGER, JACK W, b New York, NY, Nov 09, 42; m 84; c 1. CELL BIOLOGY. *Educ:* Columbia Col, AB, 64; State Univ NY, MD, 68. *Prof Exp:* Intern & resident med,, Univ Chicago Hosp, 68-70; chief hemat lab div, Nat Ctr Dis Control, 70-72; fel hemat oncol, Univ Wash, 72-75; from asst prof to assoc prof med & oncol, 75-85, PROF MED & ONCOL, UNIV WASH, 86- *Concurrent Pos:* Chief med oncol, Vet Admin Med Ctr, 75-; mem, Fred Hutchinson Cancer Res Ctr. *Mem:* AAAS; Am Fedn Clin Res; Am Soc Hemat; Int Soc Exp Hemat; Am Soc Clin Oncol; Am Soc Clin Invest. *Res:* Control of cell proliferation in normal and leukemia hemopoiesis; cytokine regulation; human leukemia. *Mailing Add:* 1660 S Columbian Way Seattle WA 98108

SINGER, JEROME RALPH, b Cleveland, Ohio, Oct 16, 21; m 56; c 2. BIOENGINEERING, BIOPHYSICS. *Educ:* Univ Ill, BS, 51; Northwestern Univ, MS, 53; Univ Conn, PhD(physics), 55. *Prof Exp:* Engr, Van de Graaff Proj, Northwestern Univ, 51-53; instr physics, Univ Conn, 53-55; physicist, Solid State Div, US Naval Ord Lab, 55-56; chief staff physicist, Nat Sci Labs, Inc, Washington, DC, 56-57; assoc prof elec eng, 57-77, PROF ENG SCI, UNIV CALIF, BERKELEY, 77- *Concurrent Pos:* Vis lectr, Catholic Univ, 55-56 & George Washington Univ, 56-57; proj engr, Sperry Corp, 54; sci consult, Missile Systs Div, Lockheed Aircraft Corp, 57-; Telemeter Magnetics, Inc, 57-58; Aeroneutronics, Inc div, Ford Motor Co, 58- *Mem:* AAAS; Am Phys Soc; Optical Soc Am; sr mem Inst Elec & Electronics Eng; Brit Inst Physics. *Res:* Electronics; magnetic phenomena; quantum mechanical amplifiers; magnetic resonance; blood studies; flow and magnetic properties of blood; rheological properties of body fluids. *Mailing Add:* Electronic Res Labs Univ of Calif Berkeley CA 94720

SINGER, JOSHUA J, b Hartford, Conn, Feb 15, 42. MEMBRANE BIOPHYSICS, SMOOTH MUSCLE. *Educ:* Mass Inst Technol, SB, SM, EE(elec eng), 66; Harvard Univ, PhD(med sci physiol), 70. *Prof Exp:* Assoc prof, 74-86, PROF PHYSIOL, UNIV MASS MED SCH, 86- *Mem:* Biophysical Soc; Soc Neurosci; Soc Gen Physiologist; Am Physiol Soc; Inst Elec & Elect Engrs; Am Heart Asn. *Res:* Voltage, chemical, and mechanical control of ion channels in dissociated smooth muscle cells; relationship of membrane electrical events to cellular biochemical and physiological processes. *Mailing Add:* Dept Physiol Univ Mass Med Sch 55 Lake Ave N Worcester MA 01655

SINGER, KAY HIEMSTRA, LYMPHOID & EPITHELIAL INTERACTION, T-CELL MATURATION. *Educ:* Duke Univ, PhD(immunol), 77. *Prof Exp:* RES ASST PROF MED & MICROBIOL-IMMUNOL, DUKE UNIV, 79- *Res:* Immunology. *Mailing Add:* Duke Univ Med Ctr Box 2987 Durham NC 27710

SINGER, LAWRENCE ALAN, b Chicago, Ill, June 7, 36; m 67; c 2. ORGANIC CHEMISTRY. *Educ:* Northwestern Univ, BA, 58; Univ Calif, Los Angeles, PhD(org chem), 62. *Prof Exp:* Fel, Harvard Univ, 62-64; asst prof chem, Univ Chicago, 64-67; assoc prof, 67-71, PROF CHEM, UNIV SOUTHERN CALIF, 73- *Concurrent Pos:* A P Sloan fel, 70-72. *Mem:* Am Chem Soc; Royal Soc Chem. *Res:* Organic photochemistry; laser spectroscopic studies; organic free radicals; photoinduced electron transfer reactions. *Mailing Add:* Dept Chem Univ Southern Calif Los Angeles CA 90089-0744

SINGER, LEON, biochemistry; deceased, see previous edition for last biography

SINGER, LEONARD SIDNEY, b Middletown, Pa, Oct 9, 23; m 48; c 3. PHYSICAL CHEMISTRY. *Educ:* Pa State Col, BS(chem eng), 43; Univ Chicago, PhD(phys chem), 50. *Prof Exp:* Chem engr, Celanese Corp Am, 43-44; Du Pont fel chem, Cornell Univ, 50-51; phys chemist, US Naval Res Lab, 51-55; phys chemist, Union Carbide Corp, 55-75, corp res fel, Res Lab, Carbon Prod Div, 75-85; RETIRED. *Honors & Awards:* Charles E Pettinos Award, Am Carbon Soc, 77, George D Graffin Lectureship, 83. *Mem:* Am Chem Soc; Am Carbon Soc. *Res:* Magnetism; magnetic resonance; free radicals; carbon and graphite; carbon fibers. *Mailing Add:* 525 Race St Berea OH 44017-2220

SINGER, MARCUS, b Pittsburgh, Pa, Aug 28, 14; m 38; c 2. ANATOMY. *Educ:* Univ Pittsburgh, BS, 38; Harvard Univ, AM, 40, PhD(biol), 42. *Prof Exp:* Asst anat, Harvard Med Sch, 42-44, instr, 44-46, assoc, 46-48, asst prof, 48-51; from assoc prof to prof zool, Cornell Univ, 51-61; prof anat, 61-80, dir dept, 61-85, HENRY WILLSON PAYNE EMER PROF ANAT, SCH MED, CASE WESTERN UNIV, 85- *Concurrent Pos:* Guggenheim fel, 67; mem, Cell Biol Study Sect, NIH, 71-74 & Neurol B Study Sect, 76-79; mem, Int Brain Orgn, 78; Marcus Singer Int Symp Regeneration, 85- *Mem:* AAAS; Am Soc Zool; Asn Res Nerv & Ment Dis; Am Asn Anat; fel Am Acad Arts & Sci. *Res:* Regeneration; histochemistry; experimental morphology; nerve regeneration; neuroanatomy. *Mailing Add:* Dept Anat Case Western Reserve Univ Med Sch Cleveland OH 44106

SINGER, MAXINE FRANK, b New York, NY, Feb 15, 31; m 52; c 4. BIOCHEMISTRY. *Educ:* DSc, Wesleyan Univ, 77, Swarthmore Col, 78, Univ Md, 85, Cedar Crest Col, 86, City Univ New York, 88, Radcliffe Col & Williams Col, 90 & Franklin & Marshall Col, 91. *Prof Exp:* USPHS fel, NIH, 57, biochemist, Nat Inst Arthritis & Metab Dis, 58-74, head sect nucleic acid enzymol, 74-79, chief, Lab Biochem, 79-87, EMER SCIENTIST, NAT CANCER INST, 88-; PRES, CARNEGIE INST WASH, WASHINGTON, DC, 88- *Concurrent Pos:* Guest scientist, Weizmann Inst Sci, 71-72; bd of trustees, Wesleyan Univ, Conn, 72-75, & Whitehead Inst Biomed Res, 86-; fel, Yale Corp, Yale Univ, 75-; Dreyfus distinguished scholar, Swarthmore Col, 82; John Simon Guggenheim mem fel, 87; mem, Human Genome Orgn & Comt Sci, Eng & Pub Policy, Nat Acad Sci, 89- & Int Adv Bd Chulabborn Res Inst, Bangkok, 90-; dir, Whitehead Inst, 88- *Honors & Awards:* Superior Serv Award, HEW, 75; NIH Dirs Award, 77; G Burroughs Mider Lectr, NIH, 77. *Mem:* Inst Med-Nat Acad Sci; fel Am Acad Arts & Sci; Am Soc Biol Chem; Am Chem Soc; Am Soc Microbiol; AAAS. *Res:* Nucleic acid chemistry and metabolism; biochemistry of animal viruses; genome organization. *Mailing Add:* Carnegie Inst Wash 1530 P St NW Washington DC 20005

SINGER, MICHAEL, b New York, NY, Feb 25, 50. ALGEBRA, DIFFERENTIAL EQUATIONS. *Educ:* Univ Calif, Berkeley, PhD(math), 74. *Prof Exp:* PROF MATH, NC STATE UNIV, 85- *Mem:* Am Math Soc; Math Asn Am. *Mailing Add:* Dept Math NC State Univ Box 8205 Raleigh NC 27607

SINGER, PAUL A, b July 18, 49; m. CLINICAL RESEARCH. *Educ:* Univ Calif, Berkeley, BA, 71; Glasgow Univ, Scotland, PhD(biochem/immunol), 78. *Prof Exp:* Res asst, Dept Microbiol & Immunol, Los Angeles Med Sch, Univ Calif, 72-74; postdoctoral fel, Dept Immunogenetics, Max-Planck Inst Biol, 78-80; asst mem, Dept Immunol, Scripps Clin & Res Found, La Jolla, 83-89; consult, 89-90; ASSOC MEM, DEPT CLIN RES, WHITTIER INST DIABETES & ENDOCRINOL, LA JOLLA, 90- *Concurrent Pos:* Calif Lupus res award, 88, NIH res grant, 89. *Mem:* AAAS; Am Asn Immunologists; Brit Soc Immunol. *Res:* T cell receptor genes in autoimmune diabetes; molecular analyses of T cell receptor repertoire expression in murine models of SLE. *Mailing Add:* Dept Clin Res Whittier Inst 9894 Genesee Ave La Jolla CA 92037

SINGER, PHILIP C, b Brooklyn, NY, Sept 6, 42; m 65; c 4. ENVIRONMENTAL SCIENCES & ENGINEERING. *Educ:* Cooper Union, BCE, 63; Northwestern Univ, MS, 65; Harvard Univ, SM, 65, PhD, 69. *Prof Exp:* Asst prof civil eng, Univ Notre Dame, 69-73; assoc prof environ sci & eng, 73-78, PROF ENVIRON SCI & ENG, UNIV NC, 78-, DIR WATER RESOURCES ENG PROG, 79- *Concurrent Pos:* Guest prof, Swiss Fed Inst Water Res & Water Pollution Control, Dubendorf, Switz, 79; assoc ed, Environ Sci & Technol, J Am Chem Soc, 83-90; mem ed bd, Ozone Sci & Eng, J Int Ozone Asn, 86-; vis prof, Stanford Univ, 89-90; dipl, Am Acad Environ Engrs. *Mem:* Am Chem Soc; Am Water Works Asn; Water Pollution Control Fedn; Asn Environ Eng Prof; Int Ozone Asn. *Res:* Control of trihalomethane formation in drinking water; ozonation for water and wastewater treatment; chemical and physical processes for water quality management; aquatic chemistry; metal-organic interactions; chemistry and control of pollution by acidic mine drainage; treatment and environmental impact assessment of coal conversion wastewaters; chemistry of eutrophication. *Mailing Add:* Dept Environ Sci Eng Univ NC Pub Health Chapel Hill NC 27599-7400

SINGER, RICHARD ALAN, b New York, NY, May 21, 45; m 71. EXPERIMENTAL HIGH ENERGY PHYSICS. *Educ:* Mass Inst Technol, BS, 67, PhD(physics), 72. *Prof Exp:* Fel, 72-75, ASST PHYSICIST, ARGONNE NAT LAB, 75- *Mem:* Am Phys Soc; Sigma Xi. *Res:* Analysis of bubble chamber experiments; study of strong interactions at both intermediate and high energies; weak interaction experiments using both neutrino and antineutrino beams. *Mailing Add:* 474 Cassin Rd Naperville IL 60565

SINGER, ROBERT MARK, b Brooklyn, NY, Nov 2, 43. CELL BIOLOGY, ANATOMY. *Educ:* Long Island Univ, BA, 65; Hunter Col, MA, 67; Syracuse Univ, PhD(cell biol), 71. *Prof Exp:* Res fel cancer, Tufts Univ, 72-73, instr cancer, 73-74, asst prof cancer res, Dept Path, Sch Med, 75-78; mem staff, 78-80, CHMN DEPT ANAT, SCH DENT, FAIRLEIGH DICKINSON UNIV, 80- *Mem:* Am Soc Cell Biol. *Res:* Characterizing the role of tumor specific isoenzymes of alkaline phosphatase in cancer cells; glucocorticoid mediated alterations of isoenzyme profiles and the rate limiting effects of the cell cycle. *Mailing Add:* Sch Dent Fairleigh Dickinson Univ 140 Univ Plaza Dr Hackensack NJ 07601

SINGER, ROLF, b Schliersee, Ger, June 23, 06; nat US; c 1. MYCOLOGY. *Educ:* Univ Vienna, PhD(bot), 31; Acad Sci, USSR, Dr Biol, 40. *Hon Degrees:* DSc, Univ Lausanne, 77. *Prof Exp:* Asst cryptogams, Univ Barcelona, 34-35; sr sci specialist, Bot Inst, Acad Sci Leningrad, 35-40; res assoc, Farlow Herbarium, Harvard Univ, 41-44, asst cur, 44-48, actg dir, 46-48; prof extraordinary, Nat Univ Tucuman, 48-53; prof chem & head dept, Nebr Wesleyan Univ, 53-54; head dept bot, Inst M Lillo, Nat Univ Tucuman, 54-61; prof syst bot, Univ Buenos Aires, 61-68; vis res cur, 68-77, RES ASSOC, FIELD MUS NATURAL HIST, 77- *Concurrent Pos:* Oberlander Trust fel, 41-42; Guggenheim Mem Found fel, Subtrop Am, 42-43, 52-53; sci dir, Orgn for Flora Neotropica Inc, 54-68; mem sci mission, Caucasus Mts, Vienna Acad Sci, 28-29; Pyrenees, Coun Natural Sci, Spain, 34; Altai Mts, Acad Sci, USSR, 37; Brazil, Tierra del Fuego & Patagonia, Inst M Lillo, 50-66, Mex, 57-73; Yungas, NSF, 56; Peru, Philos Soc, 58; South Chile, 59-67; Columbia, 60, 68; Ecuador, NSF, 73; exchange scientist, Nat Acad Sci, 74; mem comt fungi, Int Bot Cong, 50-; Hon prof, Fed Univ Pernambuco, Univ Chile & Univ Buenos Aires; vis prof biol sci, Univ Ill, Chicago Circle, 68-77; researcher, Nat Coun Sci & Technol Develop, Nat Res Inst Amazonia, 77-; vis prof bot, Univ Vienna, 76 & 79. *Honors & Awards:* Distinguished Mycologist Award, Mycol Soc Am, 86. *Mem:* Mycol Soc Am; Mycol Soc France; Mycol Soc Mex; Mycol Soc Germany; Mycol Soc Austria. *Res:* Taxonomy of the Basidiomycetes; especially of the Agaricales, commercial growing of mushrooms; antibiotics and psychotropic substances derived from Basidiomycetes; forest botany and ecology of ectotrophic mycorrhiza; antarctic fungi. *Mailing Add:* Field Mus Natural Hist Roosevelt Rd & Lakeshore Dr Chicago IL 60605

SINGER, RONALD, b Cape Town, SAfrica, Aug 12, 24; m 50; c 4. ANATOMY, PHYSICAL ANTHROPOLOGY. *Educ:* Univ Cape Town, MB, ChB, 47, DSc(anat), 62. *Prof Exp:* Lectr anat, Univ Cape Town, 49-51, sr lectr, 51-60, assoc prof, 60-62; PROF ANAT & ANTHROP, UNIV CHICAGO, 62-, R R BENSLEY PROF BIOL & MED SCI, 73- *Concurrent Pos:* Rotary Found & Johns Hopkins Univ fels, 51-52; vis prof, Univ Ill, 59-60, Robert J Terry lectr, 64; mem sci group pop genetics primitive pop, WHO, 62; consult, Nat Found, 65-; subcomt phys anthrop, Int Anat Nomenclature Comt, 71- *Honors & Awards:* Cornwall & York Prize, 50. *Mem:* Fel AAAS; fel Royal Soc SAfrica; SAfrican Asn Adv Sci (vpres, 58-62); SAfrican Archaeol Soc (pres, 61-62); Int Quaternary Asn; Am Asn Phys Anthropol; Sigma Xi (pres, 68-69, 81-83). *Res:* Gross anatomy; paleoanthropology; genetics of African indigenous populations. *Mailing Add:* Dept Anat Univ Chicago 1025 E 57th St Chicago IL 60637-1839

SINGER, S(IEGFRIED) FRED, b Vienna, Austria, Sept 27, 24; nat US. GEOPHYSICS. *Educ:* Ohio State Univ, BEE, 43; Princeton Univ, AM, 44, PhD(physics), 48. *Hon Degrees:* DSc, Ohio State Univ, 70. *Prof Exp:* Instr physics, Princeton Univ, 43-44; mem staff electronic comput design, Naval Ord Lab, Md, 45-46; physicist, Appl Physics Lab, Johns Hopkins Univ, 46-50; sci liaison officer, Off Naval Attache, US Embassy, London, 50-53; from assoc prof to prof physics, Univ Md, 53-62; dir, Nat Weather Satellite Ctr, US Dept Com, 62-64; prof atmospheric sci & dean sch environ & planetary sci, Univ Miami, 64-67; dep asst secy sci progs, US Dept Interior, 67-70; dep asst adminr, Environ Protection Agency, 70-71; prof environ sci & mem Energy Policy Studies Ctr, Univ Va, 71-87; chief scientist, US Dept Transp, Washington, DC, 87-90. *Concurrent Pos:* Dir, Ctr Atmospheric & Space Physics, Univ Md, 59-62; vis researcher, Jet Propulsion Lab, Calif Inst Technol, 61-62; fed exec fel, Brookings Inst, 71; mem Comn IV, Int Sci Radio Union, 54-; tech panels rockets & cosmic rays, US Nat Comt, Int Geophys Year, 57-58; head sci eval group & sci consult, Select Comt Astronaut & Space Explor, US House Rep, 58; chmn subcomt basic res, Comt Sci & Technol, US Chamber Com, 60-62 & mem environ pollution adv bd, 66-67; mem spacecraft oceanog adv group, Naval Oceanog Off, 66-67; chmn adv comt environ effects supersonic transport, Dept Transp, 71; consult, Inst Defense Analysis, AEC & major industs; vchmn, Nat Adv Comt Oceans & Atmosphere, 81-87. *Honors & Awards:* President's Commendation Award, 58; 1st Astronaut Medal, Brit Interplanetary Soc, 62. *Mem:* Fel AAAS; fel Am Phys Soc; fel Am Inst Aeronaut & Astronaut; fel Brit Interplanetary Soc; fel Am Astronaut Soc. *Res:* Upper atmosphere, ionosphere currents, theory of magnetic storms, radiation belts; origin of meteorites, moon, solar system; environmental effects of pollution, especially on global climate; remote sensing, from earth satellites; pollution control techniques and economics; water and energy resources; effects of population growth. *Mailing Add:* Wash Inst 1015 18th St NW Suite 300 Washington DC 20036

SINGER, S(EYMOUR) J(ONATHAN), b New York, NY, May 23, 24; m 47; c 3. BIOLOGY. *Educ:* Columbia Univ, AB, 43, AM, 45; Polytech Inst Brooklyn, PhD(chem), 47, MA, 60. *Hon Degrees:* MA, Polytech Inst Brooklyn, 60. *Prof Exp:* Abbott fel, Calif Inst Technol, 47-48, USPHS fel, 48-50, sr res fel, 50-51; from asst prof to prof phys chem, Yale Univ, 51-61; PROF BIOL, UNIV CALIF, SAN DIEGO, 61- *Concurrent Pos:* Guggenheim fel, 59-60; mem adv panel molecular biol, NSF, 60-63; mem, Allergy & Immunol Study Sect, USPHS, 63-64; emer prof, Am Cancer Soc, 76- *Mem:* Nat Acad Sci; Am Acad Arts & Sci; Am Soc Cell Biol; Am Soc Biol Chemists. *Res:* Molecular biology; physical chemistry of proteins; immunochemistry; membrane biology; chemical cytology. *Mailing Add:* Dept Biol Univ Calif San Diego La Jolla CA 92037

SINGER, SAMUEL, b New York, NY, June 22, 27; m 57; c 3. MICROBIOLOGY. *Educ:* City Col New York, BS, 50; Univ Ky, MS, 52; NY Univ, PhD(protozool, physiol), 58. *Prof Exp:* Res microbiologist, Div Chemother, Burroughs Wellcome & Co, 52-62; sr res microbiologist, Dept Microbiol, Bioferm Div, Int Mineral & Chem Corp, Calif, 62-68; microbiologist, Brown & Williamson Tobacco Corp, Ky, 68-70; assoc prof, 70-75, PROF BIOL SCI, WESTERN ILL UNIV, 75- *Mem:* AAAS; Am Soc Microbiol; Soc Invert Path; Brit Soc Gen Microbiol; Sigma Xi; Soc Indust Microbiol. *Res:* Microbiol insecticides; fermentation of insecticidal bacillus; microbial nutrition; bacterial ecology. *Mailing Add:* Dept Biol Sci Western Ill Univ Macomb IL 61455

SINGER, SANFORD SANDY, b Brooklyn, NY, Sept 14, 40; m 71; c 2. BIOCHEMISTRY. *Educ:* Brooklyn Col, BS, 62; Univ Mich, Ann Arbor, MS, 64, PhD(biochem), 67. *Prof Exp:* Fel, Albert Einstein Col Med, 67 & 68; fel, Fels Res Inst, Med Sch, Temple Univ, 68-72; from asst prof to assoc prof, 72-84, PROF CHEM, UNIV DAYTON, 84- *Concurrent Pos:* Am Cancer Soc fel, Albert Einstein Col Med, 67-69. *Mem:* Sigma Xi; Am Soc Biochem & Molecular Biol; Endocrine Soc. *Res:* Studies of mechanism of action of adrenal glucocorticosteroids; studies of mechanism of carcinogenesis. *Mailing Add:* Dept Chem Univ Dayton 300 College Park Ave Dayton OH 45469

SINGER, SOLOMON ELIAS, chemistry, nuclear physics; deceased, see previous edition for last biography

SINGER, STANLEY, b North Adams, Mass, Oct 2, 25. CHEMISTRY, PHYSICS. *Educ:* Univ Calif, Los Angeles, BS, 46, PhD(chem), 50. *Prof Exp:* Res assoc chem, Univ Calif, Los Angeles, 50-51; chemist, US Naval Ord Test Sta, 51-52, res assoc, 53, head properties sect, 54-55, liquid propellants & combustion br, 55-58; head phys & inorg chem, Hughes Tool Co, 58-59; head propulsion physics dept & assoc head chem dept, Rocket Power, 59-64; dir physics, Dynamic Sci Corp, 64-65; DIR, ATHENEX RES ASSOC, 65- *Concurrent Pos:* Engr in residence, Eng Socs Comn Energy, 81-83; pres, Int Comt Ball Lightning Res. *Mem:* AAAS; Am Chem Soc; Sigma Xi; Am Inst Aeronaut & Astronaut; Combustion Inst. *Res:* Synthesis; polymers; combustion; rocket propellants; ion sources; arc-plasma chemistry; refractory metals; ion-molecule reactions; electric space propulsion; plasma-magnetic field interaction; fine particles; atmospheric electricity; lightning; atmospheric particles; superconductivity; synthetic fuels; environmental health hazards. *Mailing Add:* 381 S Meridith Ave Pasadena CA 91106

SINGER, SUSAN RUNDELL, b Schenectady, NY, July 1, 59; m 81; c 2. PLANT DEVELOPMENT, FLOWERING. *Educ:* Rensselaer Polytech Inst, BS, 81, MS, 82, PhD(biol), 85. *Prof Exp:* ASST PROF BIOL, CARLETON COL, 85- *Concurrent Pos:* Postdoc fel, Rensselaer Polytech Inst, 85. *Mem:* Soc Develop Biol; Bot Soc Am; Am Soc Plant Physiol; Int Soc Plant Molecular Biol; Int Working Group on Flowering. *Res:* Developmental regulation of early events leading to flowering; tissue culture with genetic and molecular approaches are utilized. *Mailing Add:* Dept Biol Carlton Col Northfield MN 55057

SINGER, THOMAS PETER, b Budapest, Hungary, July 10, 20; nat US; m 62; c 2. BIOCHEMISTRY. *Educ:* Univ Chicago, SB, 41, SM, 43, PhD(biochem), 44. *Prof Exp:* Asst med, Off Sci Res & Develop & Comt Med Res Projs, Chicago, 42-44, res assoc, Manhattan Proj, 44-46; asst prof agr biochem, Univ Minn, 46-47; asst prof biochem, Sch Med, Western Reserve Univ, 47-51; mem, Inst Enzyme Res, Univ Wis, 52-54; chief div enzyme res, Edsel B Ford Inst Med Res, Henry Ford Hosp, Detroit, Mich, 54-65; prof biochem, 65-68, prof biochem & biophys, 68-70, ADJ PROF BIOCHEM & BIOPHYS, SCH MED, UNIV CALIF, SAN FRANCISCO, 70-; HEAD MOLECULAR BIOL DIV, VET ADMIN HOSP, 65- *Concurrent Pos:* Guggenheim fel, Univ Paris & Cambridge Univ, 51-52; Guggenheim fel, 59; Orgn Am States fel, 71; Fulbright fel, 72; estab investr, Am Heart Asn, 54-59. *Mem:* Am Soc Biol Chemists; Am Chem Soc. *Res:* Flavin and flavoenzyme structure; enzyme regulation; mechanism of action and regulation of enzymes of the respiratory chain, nonheme iron-sulfur roles in oxidizing enzymes; mitochondrial biogenesis. *Mailing Add:* Dept Biochem & Biophys Rm 222 Univ Calif Sch Med 4150 Clement St San Francisco CA 94121-1563

SINGER, WILLIAM MERRILL, b New York, NY, Feb 26, 42. HOMOTOPY THEORY, HOMOLOGICAL ALGEBRA. *Educ:* Cornell Univ, BA, 63; Princeton Univ, MA, 65, PhD(math), 67. *Prof Exp:* Instr math, Mass Inst Technol, 67-69; asst prof, Boston Col, 69-74; assoc prof, 74-82, chmn, Dept Math, 84-87, PROF MATH, FORDHAM UNIV, 82- *Concurrent Pos:* NSF res grant, Boston Col, 71-72; NSF res grants, 73-74, 75-76, 76-77, 81-83, 83-85 & 85-87. *Mem:* Am Math Soc. *Res:* Algebraic topology; Steenrod algebra; Hopf algebras; homological algebra and its applications to homotopy theory; semi-simplicial methods; Adams spectral sequences; homotopy groups of spheres. *Mailing Add:* Dept of Math Fordham Univ Bronx NY 10458

SINGEWALD, MARTIN LOUIS, b Baltimore, Md, May 10, 09; m 33; c 3. INTERNAL MEDICINE. *Educ:* Johns Hopkins Univ, BE, 30, MD, 38. *Prof Exp:* From asst prof to assoc prof, 55-74, EMER ASSOC PROF MED, SCH MED, JOHNS HOPKINS UNIV, 74- *Concurrent Pos:* Mem coun arteriosclerosis, Am Heart Asn. *Mem:* Am Soc Internal Med; AMA; fel Am Col Physicians; Am Clin & Climatological Asn. *Res:* Coronary artery diseases. *Mailing Add:* 600 Brookwood Rd Baltimore MD 21229

SINGEWALD, QUENTIN DREYER, economic geology; deceased, see previous edition for last biography

SINGH, AJAIB, b Dholan Hithar, Pakistan, Jan 14, 35; m 64. PHYSICAL ORGANIC CHEMISTRY, POLYMER CHEMISTRY. *Educ:* Punjab Univ, India, BSc, 54, Hons, 56, MSc, 58; Univ Calif, Davis, PhD(phys org chem), 61. *Prof Exp:* Res fel chem, Harvard Univ, 61-62; res chemist, Am Cyanamid Co, 62-68, sr res chemist, Org Chem Div, 68-77, prin res chemist, Chem Res Div, 77-78; sr group leader, 78-82, TECH MGR POLYURETHANES, UNIROYAL CHEM CO, UNIROYAL INC, 82- *Concurrent Pos:* Vis res fel, Queen Mary Col, Univ London, 68-69. *Mem:* Am Chem Soc. *Res:* Polymer structure properties, degradation kinetics and mechanisms; specialty, polyurethane elastomers. *Mailing Add:* 58 Autumn Ridge Rd Huntington CT 06484-3631

SINGH, AJIT, b Indore, India, Oct 31, 32; m 55. POLYMER CHEMISTRY, PULP CHEMISTRY. *Educ:* Agra Univ, BSc, 50, MSc, 52; Univ Alta, PhD(radiation chem), 64. *Prof Exp:* Lectr chem, Holkar Col, Indore, 56-59; teaching asst, Univ Alta, 59-63; res assoc, Chem Div, Argonne Nat Lab, 63-65; asst res officer, 66-68, assoc res officer, 69-81, sr res officer, 81-87, SR SCIENTIST, WHITESHELL NUCLEAR RES ESTAB, ATOMIC ENERGY CAN LTD, 88- *Concurrent Pos:* Ed, Int J Radiation Physics & Chem, 75. *Mem:* Fel Chem Inst Can; Tech Asn Pulp & Paper Indust; Am Chem Soc; Am Soc Photobiol; Inter-Am Photochem Soc; Can Pulp Paper Asn. *Res:* Mechanisms of reactions in radiation chemistry, radiation biology, photochemistry and photobiology; physical and chemical properties of transient molecular species by the pulse radiolysis and flash photolysis techniques; industrial applications of radiation; role of free radicals in pulping; pulp bleaching; polymerization and polymer cross-linking. *Mailing Add:* Radiation Appln Res Br AECL Res Pinawa MB R0E 1L0 Can

SINGH, AMARJIT, b Ramdas, India, Nov 19, 24; m 53; c 3. MICROWAVE ELECTRON DEVICES & PLASMAS, SEMICONDUCTOR DEVICES. *Educ:* Punjab Univ, BSc, 44, MSc, 45; Harvard Univ, MEngSc, 47, PhD(electron physics), 49. *Hon Degrees:* DSc, Punjabi Univ, 75. *Prof Exp:* Lectr electronics, Univ Delhi, India, 49-53; sci officer microwave tubes, Nat Phys Lab, New Delhi, India, 53-57; asst dep dir electron tubes, Cent Electronics Eng Res Inst, India, 57-62; res engr microwave tubes, Univ Mich, Ann Arbor, 62-63; res engr semiconductor devices, Bell Telephone Labs, Murray Hill, NJ, 63; dir electron tubes & electronics, Cent Electronics Eng Res Inst, India, 63-84; nat proj coordr semiconductor & power electronics, United Nations Develop Prog, 84-87; VIS SCIENTIST ELECTRON TUBES, UNIV MD, COL PARK, 87- *Concurrent Pos:* Mem electronics comt, Govt India, 63-68, chmn Working Group Consumer Electronics & Instrumentation, 68-75, mem Univ Grants Comn, 75-78, mem Sci & Eng Res Coun, 78-82, chmn Res & Develop Comt, 78-84, mem gov body, Coun Sci & Indust Res, 74-76 & 80-82; vis scientist, Stanford Linear Accelerator Ctr, Palo Alto, 82, Plasma Fusion Lab, Mass Inst Technol, 86. *Mem:* Fel Inst Elec & Electronic Engrs; distinguished fel Inst Electronic & Telecommun Engrs; fel Indian Acad Sci. *Res:* Microwave electronics and microwave tubes; tuning of interdigital magnetrons, wide ranges obtained; depressed collectors for large-orbit gyrotrons, new scheme proposed and verified; depressed collectors for small-orbit gyrotrons of cavity as well as quasi-optical type; microwave plasmas: surface waves on plasma columns, launching efficiency studied; microwave semiconductors: millimeter wave source using surface plasmons. *Mailing Add:* 1201 T Energy Res Bldg Univ Md College Park MD 20742-3511

SINGH, AMREEK, b Etah, Uttar Pradesh, India, Apr 13, 35; Can citizen; m 77; c 1. HISTOLOGY, ULTRASTRUCTURAL PATHOLOGY. *Educ:* Agra Univ, India, BVSc & AH, 55; Univ Guelph, Can, MS, 68 & PhD(morphol), 71. *Prof Exp:* Instr path, Vet Col, Agra Univ, Mathura, India, 56-58, lectr histol, 58-61; asst prof histol & embryol, Col Vet Med Pant Nagar, India, 61-65; asst med histol, Sch Med, Geneva Univ, Switz, 71-74; from asst prof to assoc prof histol, Ont Vet Col, Guelph, Can, 74-85; PROF HISTOL, ATLANTIC VET COL, CHARLOTTETOWN, PEI, CAN, 85- *Concurrent Pos:* Vis fel, Commonwealth Sci & Indust Res Orgn, Australia, 78; Brit coun fel, Christie Hosp, Eng, 82; referee, Can J Comp Med; consult, Nat Dept Health & Welfare, Can, 77-85. *Mem:* AAAS; Am Asn Anatomists; Swiss Soc Cell & Molecular Biol; Electron Micros Soc Am; Soc Toxicol Can; World Asn Vet Anatomists. *Res:* Histology of normal and experimentally altered organ and tissues of animals; electron microscopy of liver, kidney and thyroid gland of animals exposed to known environmental pollutants in laboratory conditions. *Mailing Add:* Atlantic Vet Col Charlottetown PE C1A 4P3 Can

SINGH, ARJUN, b Gonda, India, Jan 27, 43; nat US; m 63; c 3. GENETICS. *Educ:* G V Pant Univ Agr & Technol, Pantnagar, India, BSc, 64; Univ Ill, Urbana, PhD(genetics), 69. *Prof Exp:* Res assoc, Sch Med, Case Western Reserve Univ, 69-70; asst specialist, Donner Lab, Univ Calif, Berkeley, 70-71; fel, Univ Rochester, 71-72, assoc, 73-74, instr, Sch Med, 74-76; res assoc, Univ Mass, 76-77; NIH trainee, Dept Biochem, Univ Wis-Madison, 77-78, assoc scientist, 78-80; sr scientist, Nabisco Brands Res Ctr, 80-81; scientist, 80-87, SCIENTIFIC MGR, GENENTECH, INC, 87- *Mem:* Genetics Soc Am; Am Soc Microbiol; AAAS. *Res:* Yeast genetics; genetic engineering; plant molecular biology; research administration. *Mailing Add:* Genentech Inc 460 Point San Bruno Blvd San Francisco CA 94080

SINGH, BALDEV, b Takhtupura, Punjab, India, Jan 9, 39; US citizen; m 67; c 2. PHARMACEUTICAL CHEMISTRY, GENERAL CHEMISTRY. *Educ:* D M Col Moga, Punjab, India, BSc, 59; Banaras Hindu Univ, India, MSc, 62; State Univ NY, PhD(med chem), 67. *Prof Exp:* Lab instr chem, D M Col Moga, Punjab, India, 59-60; lectr chem, S D Col Barnala, Punjab, India, 62-63; from assoc chemist to res chemist, 67-80, res fel, 80-90, FEL, STERLING RES GROUP, 90- *Mem:* Am Chem Soc. *Res:* Design and synthesis of organic compounds of biological interest with major emphasis on heterocyclic chemistry; fifty United States patents and published about 40 papers relating to chemotherapy and pulmonary and cardiovascular pharmacology. *Mailing Add:* Three Blue Mountain Trail East Greenbush NY 12061

SINGH, BALWANT, b Hasanpur, India, Jan 21, 34; US citizen; m 72. MICROBIOLOGY, EPIDEMIOLOGY. *Educ:* Aligarh Muslim Univ, India, BSc, 53; Agra Univ, BVSc & AH, 57; Univ Mo, MS, 62, PhD(microbiol), 65. *Prof Exp:* Vet officer, Uttar Pradesh State Animal Husb Dept, 57-61; res assoc epidemiol & microbiol, Univ Pittsburgh, 66-70, asst res prof, 70-76, res assoc prof epidemiol & microbiol, Grad Sch Pub Health, 76-86; CONSULT, HEALTH RES INT, SCI TECH INT, 86- *Mem:* Am Soc Microbiol; Am Pub Health Asn; Int Epidemiol Asn; Soc Epidemiol Res; Am Venereal Dis Asn. *Res:* Venereal diseases; public health microbiology; uterine cancer; reproductive health. *Mailing Add:* 5409 Guarino Rd Pittsburgh PA 15217

SINGH, BHAGIRATH, b Jaipur, India, Feb 8, 46; m 68; c 3. MOLECULAR IMMUNOLOGY, PEPTIDE SYNTHESIS. *Educ:* Rajasthan Univ, BSc, 63, MSc, 65; Agra Univ, PhD(med chem), 69. *Prof Exp:* Jr res fel chem, Cent Drug Res Inst, 66-69, sr res fel, 69-70; res fel org chem, Liverpool Univ, 70-73; res fel immunol, 73-77, from asst prof to assoc prof, 77-86, PROF IMMUNOL, UNIV ALTA, 86-, ACTG CHMN IMMUNOL, 88- *Concurrent Pos:* Prin investr, Med Res Coun Group Immunoregulation, 82-87; Heritage Scholar, Alta Heritage Found Med Res, 85-89, Heritage scientist, 89- *Mem:* Am Asn Immunologists; Can Soc Immunol; NY Acad Sci; Am Diabetes Asn; AAAS; Am Peptide Soc; Can Diabetes Asn. *Res:* Molecular immunology of proteins and synthetic peptides antigens; immunodomint sites on transplantation antigens; diabetes and autoimmunity; activation and antigen receptors on T lymphocytes; molecular biology of immune recognition; monoclonal antibodies; autoimmunity; diabetes. *Mailing Add:* Dept Immunol Univ Alta 845 Med Sci Bldg Edmonton AB T6G 2H7 Can

SINGH, BHARAT, b Calcutta, India, Feb 21, 39; m 64; c 2. PLANT PHYSIOLOGY, FOOD SCIENCE. *Educ:* Banaras Hindu Univ, BSc, 58; Ranchi Univ, India, MSc, 61; Univ BC, PhD(bot), 68. *Prof Exp:* Lectr bot, St Columba's Col, Hazaribagh, India, 61-64; res fel, Univ BC, 64-68, Med Res Coun Can fel, 68-69; fel, Utah State Univ, 69-70, vis asst prof food sci, 70-72; assoc prof, 72-75, PROF FOOD SCI, ALA A&M UNIV, 75- *Concurrent Pos:* Coordr, Peanut Collab Res Support Prog, Sudan, Carribean Countries (Food Technol Proj). *Honors & Awards:* Morrison-Evans Outstanding Scientist award, 80. *Mem:* Inst Food Technologists; Am Soc Agron; Crop Sci Soc Am; Am Asn Cereal Chemists; Am Peanut Res & Educ Soc. *Res:* Biochemical and functional characteristics of triticale; chemical regulation of growth and metabolism in plants; utilization of cellulosic biomass; toxicity of myristate; post harvest handling, storage and processing of peanuts. *Mailing Add:* Dept of Food Sci & Technol Ala A&M Univ Normal AL 35762

SINGH, CHANCHAL, b Singapore, April 4, 34; m 66; c 3. APPLIED MATHEMATICS. *Educ:* Punjab Univ, India, BA, 57; Univ Mich, Ann Arbor, MA, 65; Fla State Univ, MSc, 70, PhD(statistics), 72. *Prof Exp:* Asst prof math, State Univ NY, Potsdam, 66-69, asst prof math, St Lawrence Univ, Canton, NY, 72-76, assoc prof, 76-82, CUMMINGS PROF MATH, ST LAWRENCE UNIV, CANTON, NY, 82- *Concurrent Pos:* Chmn, Dept Math, St Lawrence Univ, Canton, NY; assoc ed, J Info Optimization Sci. *Honors & Awards:* Fulbright lectr, Univ Delhi, India, 88. *Mem:* Indian Genetics Soc; fel Am Dermatoglyphics Asn; Sigma Xi; Oper Res Soc Am; Int Math Prog Soc; Oper Res Soc India. *Res:* Optimization theory; quality theory and optimality conditions; finger prints as applied to genetics and physical anthropology; recreational mathematics. *Mailing Add:* Dept Math St Lawrence Univ Canton NY 13617

SINGH, DAULAT, b Dihwa, India, July 27, 39; m 54; c 3. SOIL CHEMISTRY, ANALYTICAL CHEMISTRY. *Educ:* Agra Univ, BSc, 57, MSc, 59; Univ Guelph, MSc, 65; Mich State Univ, PhD(soil chem), 69. *Prof Exp:* Res asst soil chem, Nat Sugar Inst, Kanpur, India, 60-63; res asst soil phosphorous, MacDonald Col, McGill Univ, 66; CHEMIST, ENVIRON, LAB DIV, MICH DEPT AGR, 69- *Mem:* Am Soc Agron; Int Soc Soil Sci; Indian Soil Sci Soc; Asn Off Anal Chem. *Res:* Mechanisms of soil-fertilizer interaction; soil phosphorous reaction products. *Mailing Add:* 2089 Ashland Okemos MI 48864

SINGH, DILBAGH, b Partabpura, India, Oct 15, 34; m 54; c 3. PLANT PATHOLOGY. *Educ:* Govt Col, Ludhiana, India, BSc, 56; Panjab Univ, India, BSc, 59, MSc, 61; Univ Wis, Madison, PhD(plant path), 68. *Prof Exp:* Lectr biol, Sikh Nat Col, India, 61-62; res asst plant path, Univ Wis, 62-67; chmn, Div Natural Sci, 77-80 & 83-89, PROF BIOL, BLACKBURN COL, 67-, CHMN, BIOL DEPT, 89- *Concurrent Pos:* Vis scientist, Univ Mass, 74-76. *Mem:* Bot Soc Am; Am Inst Biol Sci; Mycol Soc Am; Nat Geog Soc; Am Mus Natural Hist; Nat Asn Biol Teachers. *Res:* Vascular diseases of plants; effects of pathogenesis on the nitrogenous and carbohydrate contents of the xylem sap; seasonal variation in the xylem sap components; effects of water stress on growth and metabolism of wilt inducing fungi; effects of water stress on the flowering behavior of various ornamental plants, Begonia, Impatiens and Cloeus. *Mailing Add:* Dept Biol Blackburn Col 700 Col Ave Carlinville IL 62626

SINGH, GURDIAL, b Punjab, India, Aug 15, 34; m 60; c 3. POLYMER CHEMISTRY. *Educ:* Panjab Univ, India, BSc, 57, MSc, 59; Univ Cincinnati, PhD(chem), 64. *Prof Exp:* Demonstrator chem, Panjab Univ, India, 58, 59, jr fel, 59-60; teaching asst, Univ Cincinnati, 60-61, fel, 63-64; res assoc, Mass Inst Technol, 64-65; from res chemist to sr res chemist, 65-79, res assoc, 79-89, SR RES ASSOC, TEXTILE FIBERS DEPT, EXP STA, E I DU PONT DE NEMOURS & CO, 79- *Mem:* Am Chem Soc. *Res:* Synthetic polymer chemistry; organophosphorus chemistry; organic chemistry; organometallic chemistry. *Mailing Add:* EI Du Pont De Nemours & Co Exp Sta Wilmington DE 19898

SINGH, GURMUKH, b Dhoodt-Kalan, India, Mar 12, 50. PULMONARY PATHOLOGY. *Educ:* Univ Poona, MD, 71; Univ Pittsburgh, PhD(path), 78. *Prof Exp:* CHIEF LAB SERV, VET ADMIN MED CTR. *Mem:* Int Acad Path; Am Asn Pathologists. *Mailing Add:* Depth Path Vet Admin Med Ctr Pittsburgh PA 15240

SINGH, HAKAM, b Bagh, WPakistan, Feb 11, 28; m 50; c 4. POLYMER CHEMISTRY, SURFACE CHEMISTRY. *Educ:* Univ Delhi, BSc, 52, MSc, 54, PhD(chem), 59. *Prof Exp:* Lectr chem, Univ Delhi, 57-60; res assoc & lectr, Univ Southern Calif, 60-62; lectr, Univ Delhi, 62-63; reader, Punjabi Univ, 63-65; asst prof, IIT, New Delhi, 65-68; vis assoc prof, Univ Southern Calif, 68; res supr, 68-81, mgr, Polymer Labs, 81-87, vpres polymer res, 87-89, SR DIV VPRES POLYMER RES & CORP RES & DEVELOP, PROD RES & CHEM CORP, 89- *Concurrent Pos:* Nat fel polymer technol, Intra Sci Res Found, 70- *Mem:* NY Acad Sci; AAAS; Am Chem Soc. *Res:* Structural chemistry of clay minerals; electrochemistry of nerve impulse initiation; interactions of surfactants with metal ions; electrochemistry of crevice corrosion; design and development of protective coatings; development of high performing elastomers and their processing; synthesis of high temperature polysulfide polymers; development of cyanosiloxane sealants for aerospace applications; polythioether polymers , hybrid polymers, modified silicone polymers. *Mailing Add:* Res & Develop Lab Prod Res & Chem Corp 2820 Empire Ave Burbank CA 91504

SINGH, HARBHAJAN, b Delhi, India, July 12, 41; m 71; c 2. CHEMISTRY, BIOCHEMISTRY. *Educ:* Univ Delhi, BS, 61, MS, 63, PhD(chem), 66; Southern Ill Univ, MBA, 83. *Prof Exp:* Asst prof exp med, Sch Med, NY Univ, 66-77, adj assoc prof biochem, Sch Dent, 75-77; from res specialist to sr res specialist, 77-81, res group leader, 81-83, SR RES GROUP LEADER, DEPT ENVIRON SCI, MONSANTO AGR PROD CO, 83- *Concurrent Pos:* Coun Sci & Indust Res India sr res fel natural prod, Univ Delhi, 66-67; Med Res Coun Can fel, Univ Western Ont, 67-68; Hormel Inst fel, Univ Minn, 69; res biochemist, Lipid Metab Lab, Vet Admin Hosp, New York, 69-; adj asst prof biochem, Sch Dent, NY Univ, 72-74. *Mem:* Am Chem Soc; Int Soc Study Xenobiotics. *Res:* Chemistry and metabolism of myelin; bacterial lipid metabolism; spingolipids; chemistry of natural products; development of analytical methods; lipid metabolism; metabolic and environmental fate studies on pesticides; synthesis of labeled pesticides. *Mailing Add:* Monsanto Agr Co BB5K 700 Chesterfield Pkwy St Louis MO 63198

SINGH, HARPAL P, b India, Aug 16, 41; m 70; c 3. ENVIRONMENTAL HEALTH, TOXICOLOGY. *Educ:* Panjab Univ, India, BS, 60, MS, 62; Univ Tenn, Knoxville, PhD(biol), 70, MPH, 74. *Prof Exp:* Res asst zool, Punjab Agr Univ, Ludhiana, India, 62-64; asst prof biol, Bennett Col, Greensboro, NC, 69-70; assoc prof biol, Knox Col, Knoxville, 70-74; assoc prof & coordr allied health, 74-80, PROF BIOL & COORDR MED TECHNOL, SAVANNAH STATE COL, 80- *Concurrent Pos:* NIH grants, 77-; dir, MARC Prog, NIH, 84- *Mem:* Nat Sci Teachers Asn; Am Soc Allied Health Prof; AAAS. *Res:* Chemical and radiosensitivity of male germ cells. *Mailing Add:* Savannah State Col PO Box 20425 Savannah GA 31404

SINGH, HARWANT, b Amritsar, India, Nov 20, 34; m 55. FOOD IRRADIATION, BIOLOGICAL SYSTEMS. *Educ:* Agra Univ, BSc, 55; Vikram Univ, India, MSc, 59; Univ Alta, PhD(biochem), 62. *Prof Exp:* NIH res fel biochem, Univ Alta, 62-63; res assoc, Med Sch, Northwestern Univ, 63-65; asst res officer, 66-68, assoc res officer, 66-81, sr res officer, 81-87, SR SCIENTIST, WHITESHELL NUCLEAR RES ESTAB, ATOMIC ENERGY CAN LTD, 88- *Mem:* Chem Inst Can. *Res:* Structure of oligonucleotides and nucleic acids; mechanisms involved in protein synthesis; ribosome interactions; photo and radiation chemistry of ribosomes and their constituents; roles of free radicals in photo; radiation biology; food irradiation. *Mailing Add:* Radiation Appln Res Br AECL Res Pinawa MB R0E 1L0 Can

SINGH, INDER JIT, b India, Apr 28, 39. ANATOMY, DENTISTRY. *Educ:* Panjab Univ, India, BDS, 59; Univ Ore, PhD(anat), 69; Columbia Univ, DDS, 84. *Prof Exp:* House surgeon, Govt Dent Col & Hosp, India, 59-60; asst pedodont, Dent Sch, Univ Ore, 61-65; res assoc med psychol, Med Sch, 68-69; asst res scientist, Inst Dent Res, 69-71, assoc res scientist, 71-78, asst prof, 72-74, assoc prof, 74-79, PROF ANAT, COL DENT & GRAD SCH ARTS & SCI, NY UNIV, 79-, ASSOC CHMN DEPT ANAT, 78- *Concurrent Pos:* Fel, Guggenheim Dent Clin, New York, 60-61; NIH spec res fel, Lab Cellular Res, Inst Dent Res, 71 & 72; adj asst prof, Fordham Univ, 70-71; assoc prof, City Univ New York, 71- *Mem:* AAAS; Am Asn Anat; fel Geront Soc; NY Acad Sci; Int Asn Dent Res; Am Dent Asn; Acad Gen Dent; Sigma Xi. *Res:* Mammalian growth and development; experimental teratology; skeletal biology. *Mailing Add:* NY Univ Dent Ctr 342 E 24th St New York NY 10010

SINGH, INDERJIT, b Langrian, India, July 13, 43; US citizen; m 80. NEUROCHEMISTRY, NEUROBIOLOGY. *Educ:* Panjab Univ, India, BSc, 65, MSc, 67; Iowa State Univ, PhD(biochem), 74. *Prof Exp:* Res fel neurochem, Mass Gen Hosp, 75-76; res fel, 76-77, instr, 77-78, ASST PROF NEUROL, JOHNS HOPKINS SCH MED, 78- *Mem:* Am Soc Neurochem; Int Soc Neurochem; Am Soc Biol Chem. *Res:* Molecular mechanisms of the development of oligodendrocytes, myelinogenesis and the status of myelin in neuropathological disorders. *Mailing Add:* Dept Pediat & Cell Biol Med Univ SC 171 Ashley Ave Charleston SC 29425

SINGH, IQBAL, b Muzafaagarh, India, Feb 23, 26; nat US. ORTHOPEDIC SURGERY, IMMUNOLOGY. *Educ:* Panjab Univ, FSc, 42, MD, 49. *Prof Exp:* Instr physiol, Mission Med Col, India, 51; asst instr orthop, Univ Pa, 55-57; clin asst, Univ Calif, Los Angeles, 59-61; asst clin prof, Univ Calif, Irvine, 66-69; assoc prof orthop & chmn div, Dept Surg, Med Col Ohio, 69-76, assoc prof, 76-80, prof surg, 80-; AT HARRIMAN JONES MED CLIN. *Mem:* AMA; AAAS; Asn Clin Scientists; Can Orthop Asn; Am Asn Surg Trauma; Sigma Xi. *Res:* Cell mediated and humoral immunity; blocking factor; surgical extirpation supplemented with immunotherapy; horizontal and vertical transmission of immunity in patients; an animal model for human osteosarcoma. *Mailing Add:* Harriman Jones Med Clin 211 Cherry Ave Long Beach CA 90815

SINGH, JAG JEET, b Rohtak, India, May 20, 26; c 1. PHYSICS, ASTRONOMY. *Educ:* Panjab Univ, BS, MS, 48; Liverpool Univ, PhD(nuclear physics), 56. *Prof Exp:* Lectr physics, Panjab Educ Serv, India, 50-53; res fel nuclear physics, Univ Liverpool, 56-57; prof physics, Panjab Govt Col, 57-58; USAEC res fel nuclear physics, Univ Kans, 58-59; asst prof physics, Mem Univ Nfld, 59-60; assoc prof, WVa State Col, 60-62 & Col William & Mary, 62-64; staff scientist aerospace sci, 64-80, CHIEF SCIENTIST, INSTRUMENT RES DIV, LANGLEY RES CTR, NASA, 80- *Concurrent Pos:* Adv Gov Va, William & Mary Repr, Govs Adv Coun, Va Assoc Res Ctr, 62-63; lectr-consult, Langley Res Ctr, NASA, 62-64; adj prof physics & geophys sci, Old Dominion Univ & consult physicist, Col William & Mary, 74-; consult, Med Res Serv, Vet Admin Ctr, Hampton, 76-83. *Honors & Awards:* Appolo Achievement Award, NASA, 69, Technol Utilization Award, 77, 82, 86 & 87; Outstanding Performance Award, NASA, 86. *Mem:* Fel Brit Inst Physics; fel AAAS; assoc fel Am Inst Aeronaut & Astronaut; Am Phys Soc. *Res:* Materials science, environmental and optical physics relevant to NASA mission; Mossbauer spectroscopy and positron annihilation spectroscopy of structural alloys. *Mailing Add:* PO Box 325 Yorktown VA 23690

SINGH, JAGBIR, b Baraut, India, Jan 2, 40; m 68. PROBABILITY. *Educ:* Aligarh Muslim Univ, India, MS, 60; Fla State Univ, PhD(statist), 67. *Prof Exp:* Asst prof statist, J V Col, Baraut, India, 60-62; asst prof math, Ohio State Univ, 67-74; assoc prof, 74-78, PROF STATIST, TEMPLE UNIV, 78- *Concurrent Pos:* Sr prof, Indian Agr Res Sci Inst, New Delhi, 77-78. *Mem:* Inst Math Statist; Am Statist Asn; Biomet Soc; Int Statist Inst. *Res:* Paired comparison model building; estimation; applied probability; sampling. *Mailing Add:* Dept Statist Temple Univ Philadelphia PA 19122

SINGH, JARNAIL, b Amritsar, Punjab, India, Oct 26, 41; US citizen; m 68; c 2. TERATOLOGY, TOXICOLOGY. *Educ:* Panjab Univ, BS, 61; Punjab Agr Univ, MS, 64; Kans State Univ, PhD(cytogenetics), 68. *Prof Exp:* Assoc prof biol, 69-82, chmn, Div Math & Sci, 70-73, assoc dir Math Sci Inst, 83-85, PROF BIOL, STILLMAN COL, TUSCALOOSA, ALA, 82- *Concurrent Pos:* Prin investr, MBRS prog, Stillman Col, 73-, fac res & grant helper, 84-; Nat Res Serv fel, Univ Ala, 83-85; extramural assoc, NIH, 85. *Mem:* Behav Teratology Soc; Teratology Soc; Nat Minority Health Affairs Asn. *Res:* Teratological effects of air pollution gases; toxicological effect of air pollution gases during prenatal and postnatal development; teratogenicity of mycotoxins under protein deprived conditions. *Mailing Add:* Dept Math & Sci Stillman Col PO Drawer 1430 Tuscaloosa AL 35403

SINGH, JASWANT, b Gunna Ur, W Punjab, Sept 29, 37; m 69; c 1. PLANT PATHOLOGY. *Educ:* Khalsa Col, India, BSc, 58; Panjab Univ, India, MSc, 60; Univ Ill, Urbana, PhD(plant path), 66. *Prof Exp:* Demonstr chem, Khalsa Col, India, 58-61; res assoc & asst prof bot & plant path, Sci Res Inst, Ore State Univ, 66-68; assoc prof, 68-69, head dept, 77-82, PROF BIOL, MISS VALLEY STATE UNIV, 69-, HEAD DEPT, 83- *Mem:* AAAS. *Res:* Reproductive physiology; physiology and biochemistry of fungi; waste water treatment efficiencies. *Mailing Add:* 117 E Adams Greenwood MS 38930

SINGH, KANHAYA LAL, b Varanasi, India, Feb 15, 44; m 65; c 2. ANALYSIS, FUNCTIONAL ANALYSIS. *Educ:* Agra Univ, BSc, 62; Mem Univ, MA, 69; Tex A&M Univ, PhD(math), 80. *Prof Exp:* Fel, Lakehead Univ, 80-81; ASST PROF CALCULUS & DIFFERENTIAL EQUATIONS, UNIV MINN, DULUTH, 81- *Mem:* Am Math Soc; Math Asn Am. *Res:* Nonlinear functional analysis, fixed point theory, and approximation theory. *Mailing Add:* Dept Math Fayetteville State Univ Fayetteville NC 28301

SINGH, LAXMAN, b Meerut, India, Sept 23, 44; US citizen; m 66; c 2. AGRICULTURAL & FOOD CHEMISTRY, ANALYTICAL CHEMISTRY. *Educ:* Agra Univ, India, BS, 64; Indian Agr Res Inst, MS, 66; Miss State Univ, PhD(soil chem), 70. *Prof Exp:* Chief chemist, Runyon Testing Labs, Chicago, 70-73; MGR RES & DEVELOP, VITAMINS, INC, CHICAGO, 73- *Mem:* Am Chem Soc; Asn Vitamin Chemists. *Res:* Vitamins and vitamin D products; vegetable oil processing; molecular distillation; supercritical fluid extractions; liquid chromatography. *Mailing Add:* Vitamins Inc 809 W 58th St Chicago IL 60621

SINGH, MADAN GOPAL, b Batala, India, Mar 17, 46; Brit & French citizen; m 79; c 2. DECISION TECHNOLOGIES, SYSTEMS ENGINEERING. *Educ:* Univ Exeter, UK, BSc Hons, 69; Univ Cambridge, PhD(eng), 73; French State, Docteur es Sci, 78. *Hon Degrees:* MSc, Univ Manchester, 82. *Prof Exp:* Fel, St John's Col, Cambridge, 74-77; assoc prof eng, Univ Toulouse, 76-78; researcher, CNRS, France, 78-79; prof control eng, 79-87, head dept, 81-83 & 85-87, PROF INFO ENG, UNIV MANCHESTER INST SCI & TECHNOL, 87- *Concurrent Pos:* Co-ed-in-chief, J Large Scale Systs, 80-88 & J Info & Decision Technol, 88-; ed-in-chief, Encycl Systs & Control, 81-91; vchmn systs eng, Int Fedn Automatic Control, 81-84; hon prof, Beijing Univ Aeronaut, 88. *Honors & Awards:* Rank Zerox Lectr, 87. *Mem:* Fel Inst Elec & Electronic Engrs; fel Inst Elec Engrs. *Res:* Decision technologies for managerial decision making; complex systems theory. *Mailing Add:* Computation Dept UMIST Sackville St Manchester M6D 1QD England

SINGH, MADHO, b Mandha, India, Apr 25, 36; m; c 5. GENETICS, BIOMETRICS. *Educ:* Univ Rajasthan, BScAg, 54; Agra Univ, MScAg, 56; Univ Minn, PhD, 65. *Prof Exp:* Lectr animal genetics, SKN Col, Jobner, India, 56-60; res asst genetics, Univ Minn, 61-65; res assoc biol, Univ Chicago, 65-66; reader animal sci & head dept, Univ Udaipur, India, 66-68; res scientist zool, Univ Tex Austin, 68-69; from asst prof to assoc prof, 69-87, PROF BIOL, STATE UNIV NY COL, ONEONTA, 88- *Concurrent Pos:* Vis prof genetics, Sch Med, Univ Hawaii, 74-75; vis fel, Cornell Univ, 76; res awards, State Univ NY Res Found. *Mem:* NY Acad Sci; Genetics Soc Am; Biomet Soc; Soc Study Evolution; Am Soc Human Genetics; Sigma Xi. *Res:* Human genetics; biometrics; evolutionary biology; selection studies in mice; application of computers in genetic research; theoretical studies in population genetics. *Mailing Add:* Dept Biol State Univ NY Col Oneonta NY 13820-1380

SINGH, MAHENDRA PAL, b India, Sept 20, 41; m 65; c 3. STRUCTURAL ANALYSIS & DESIGN. *Educ:* Univ Roorkee, BE, 62, ME, 66; Univ Ill, PhD(civil eng), 72. *Prof Exp:* Asst engr civil eng, Western Railway, India, 63-68; sr engr structural design, Sargent & Lundy, Chicago, 72-76, supvr, 76-77; assoc prof, 77-82, PROF ENG MECH, VA POLYTECH INST & STATE UNIV, 82- *Concurrent Pos:* Consult, Sargent & Lundy, Chicago, 77-79, Woodward Clyde Consult, 79-80 & Stevenson & Assoc, Cleveland, 81-; engr, Lawrence Livermore Nat Lab, 79-80. *Mem:* Am Soc Civil Engrs; Earthquake Eng Res Inst; Indian Soc Earthquake Technol. *Res:* Structural engineering; structural reliability; soil dynamics; structural dynamics; earthquake and wind engineering. *Mailing Add:* 608 Piedmont St Blacksburg VA 24060

SINGH, MALATHY, b Vellore, India, June 15, 42; m; c 2. DEVELOPMENT OF THE LUNG, NEUROBIOLOGY. *Educ:* Univ Delhi, India, PhD(biochem), 70. *Prof Exp:* ASSOC RES SCIENTIST, DEPT PEDIAT, COLUMBIA UNIV, 78- *Mem:* Am Soc Biol Chemists; Am Inst Nutrit; Am Dietetic Asn. *Mailing Add:* Cornell Univ Med Ctr 525 E 68th St Rm PWC 253 New York NY 10021

SINGH, MANOHAR, b Punjab, India, Apr 13, 30; m 56; c 3. APPLIED MATHEMATICS, CONTINUUM MECHANICS. *Educ:* Panjab Univ, India, BA, 50, MA, 53; Brown Univ, MSc, 63, PhD(appl math), 65. *Prof Exp:* Lectr math, Govt Col, Panjab Univ, 53-61; asst prof, NC State Univ, 65-67; chmn dept, 78-81, PROF MATH, SIMON FRASER UNIV, 67- *Concurrent Pos:* Vis prof, Panjab Univ, 69-70. *Mem:* Can Math Cong; Can Appl Math Soc. *Mailing Add:* Dept Math & Statist Simon Fraser Univ Burnaby BC V5A 1S6 Can

SINGH, MANSA C, b Lyallpur, India, Oct 10, 28; m 63; c 3. ENGINEERING MECHANICS, STRUCTURAL ENGINEERING. *Educ:* Panjab Univ, India, BSc, 52; Univ Minn, MS, 56, PhD(struct & appl mech), 62. *Prof Exp:* Asst engr, Bhakra Dam Designs Directorate, 52-55; asst prof civil eng, Univ Kans, 61-63 & Punjab Eng Col, 63-64; asst prof eng mech, SDak State Univ, 64-68; ASSOC PROF MECH ENG, UNIV CALGARY, 68- *Mem:* AAAS; Am Soc Civil Engrs; Am Soc Eng Educ. *Res:* Viscoelastic behavior of surfaces of revolution under combined mechanical and thermal loads; thermoelastoplastic bending and stability of beam columns; application of group theory to problems of vibrations and wave propagation; impact of nonlinear viscoplastic and viscous rods; problem of notation in vector mechanics. *Mailing Add:* Mech Eng Dept Univ Calgary Calgary AB T2N 1N4 Can

SINGH, MRITYUNJAY, b Khujjhi, India, Mar 2, 57; m 74; c 4. PHASE EQUILIBRIA & PHASE TRANSFORMATION, ELECTRON MICROSCOPY. *Educ:* Gorakhpur Univ, India, BS, 77, MS, 80; Banaras Hindu Univ, PhD(metall eng), 83. *Prof Exp:* Proj scientist, Banaras Hindu Univ, India, 83-86; res assoc, La State Univ, 86-87; SR RES ASSOC, RENSSELAER POLYTECH INST, 87- *Mem:* Mat Res Soc; Electrochem Soc; Am Ceramic Soc. *Res:* Phase transformation and phase equilibria; crystal growth and characterization of semiconductors and optoelectronic materials; processing and characterizaion of high temperature advanced structural composites; composite materials. *Mailing Add:* 274A Mat Res Ctr Rensselaer Polytech Inst Troy NY 12180-3590

SINGH, PARAM INDAR, b Ferozepore, India, Nov 27, 46; US citizen; m 69; c 2. CARDIO-VASCULAR DEVICES. *Educ:* Univ Colo, BS, 68, MS, 70, PhD(aerospace eng sci), 74. *Prof Exp:* sr scientist, Avco Everrett Res Lab, 74-76, prin res scientist aerophysics, 76-82; prin staff scientist & vpres, 82-86, exec vpres, 86-90, CONSULT, ABIOMED, 90- *Concurrent Pos:* Prin investr, Permanent Cardiac Assistance Systs. *Mem:* AAAS; Am Heart Asn; Int Soc Artifical Internal Organs. *Res:* Applied physics; development of medical devices for cardio-vascular systems; biological fluid mechanics; laser effects and propagation; prosthetic calcification. *Mailing Add:* 40 N Hancock St Lexington MA 02173

SINGH, PRITHE PAUL, b Havialian, India, Sept 10, 30; m 59; c 2. NUCLEAR PHYSICS. *Educ:* Univ Agra, BSc, 51, MSc, 53; Univ BC, PhD(nuclear physics), 60. *Prof Exp:* Lectr physics, D C Jain Col, India, 53-54; res asst nuclear physics, Dept Atomic Energy, Govt of India, 54-55; Nat Res Coun Can fel, Atomic Energy Can, Ltd, 59-62; res assoc nuclear physics, Argonne Nat Lab, 62-64; from asst prof to assoc prof, 64-71, PROF PHYSICS, IND UNIV, BLOOMINGTON, 71- *Concurrent Pos:* Fac assoc, Argonne Nat Lab, 66-72; consult, US Naval Res Lab, 69-70; assoc dir res, Ind Univ Cyclotron Facil, 78-79, co-dir, 79-86. *Mem:* Fel Am Phys Soc; Sigma Xi. *Res:* Nuclear spectroscopy with neutrons and charged particles; inverse photodisintegration studies; nuclear reaction mechanism for alpha particle interaction with nuclei; statistical properties of nuclear cross sections; pion production; medium energy nuclear physics; heavy ion reactions; producer of radio series A Moment of Science. *Mailing Add:* Dept Physics Swain Hall W Ind Univ Bloomington IN 47405

SINGH, PRITHIPAL, b Amritsar, India, Apr 6, 39; m 63; c 2. ORGANIC CHEMISTRY. *Educ:* Khalsa Col, India, BS, 59; Banaras Hindu Univ, MS, 61; Toronto Univ, PhD(org chem), 67. *Prof Exp:* Teacher chem, Khalsa Col, Delhi Univ, 61-68 & Banaras Hindu Univ, 68-69; res chemist, 70-73, group leader, 73-74, sect mgr chem, 74-77, asst dir, 77-81, VPRES, SYVA CORP, 81- *Concurrent Pos:* Fel, Southampton Univ & Brit Coun travel grant, 69-70. *Mem:* Am Chem Soc; fel The Chem Soc; Am Asn Clin Chem; Interam Soc Photochemists; Am Asn Photochem & Photobiol. *Res:* Synthetic, structural organic chemistry; immunochemistry; photochemistry. *Mailing Add:* Chemtrak 484 Oakmead Pkwy Sunnyvale CA 94086-4708

SINGH, RABINDAR NATH, b Ludhiana, India, Apr 10, 31; m 57; c 3. AGRONOMY, SOILS. *Educ:* Panjab Univ, India, BSc, 55; Univ Tenn, MS, 59; Va Polytech Inst & State Univ, PhD(agron), 65. *Prof Exp:* Res asst soil fertil, Va Polytech Inst & State Univ, 63-65, fel & res assoc, 65-66; res assoc soil chem & fertil, WVA Univ 66-69, res assoc clay mineral, 69-71, asst prof soil fertil & clay mineral, 71-75, a assoc prof, 75-78 PROF AGRON & AGRONOMIST, WVA UNIV, 78- *Concurrent Pos:* Mem, Nat Task Force Sewage Sludge Crop Land, Coun Agr Sci & Technol, Environ Protection Agency. *Mem:* Am Soc Agron; Am Soc Soil Sci; Int Soc Soil Sci; Am Chem Soc. *Res:* Chemistry of soil phosphorus and micronutrients and their availability to crops; disposal of sewage sludge on agricultural land and mine soils; chemistry and mineralogy of coal overburden material; reclamation of strip mined land with industrial waste such as fly ash, balloon ash, kiln dust etc. *Mailing Add:* Dept Soil & Plant Sci WVa Univ Box 6108 Morgantown WV 26506

SINGH, RAGHBIR, b Punjab, India, Nov 1, 31; US citizen; m 66; c 1. AGRONOMY, PLANT PHYSIOLOGY. *Educ:* Panjab Univ, India, BSc, 52, MSc, 55; Univ Minn, PhD(agron), 64. *Prof Exp:* Res asst plant physiol, Ministry of Agr, India, 55-59; res fel, Univ Minn, 64-65; from asst prof to assoc prof biol, Chadron State Col, 65-67; PROF BIOL, BENEDICT COL, 67-, CHMN DIV SCI & MATH, 68- *Concurrent Pos:* NSF res grant, 66, undergrad equip, 67-69. *Res:* Nutrition, physiology and biochemistry of crop plants. *Mailing Add:* 1708 Pinewood Dr Columbia SC 29205

SINGH, RAJ KUMARI, b Agra, India, Feb 5, 48; Can citizen; m 68; c 3. NEURO-BIOLOGY, DISEASE MARKERS ON CELL SURFACES. *Educ:* Agra Univ, India, BSc, 64, MSc, 66; Liverpool Univ, Eng, PhD(biochem), 73. *Prof Exp:* Jr res fel path, Agra Univ, 67-69; Unichem fel med chem, Cent Drug Res Inst, 69-70; teaching fel biochem, 74-76, fel physiol, 80-82, RES ASSOC ZOOL, UNIV ALTA, 85- *Res:* Monoclonal antibodies; marker protein in brain cells; developmental biology and neurobiology; association of abnormal cell surface markers with diseases. *Mailing Add:* Dept Biol Vancouver Community Col 100 W 49th Ave Vancouver BC V5Y 2Z6 Can

SINGH, RAJENDRA, b Feb 13, 1950, nat US; m. ACOUSTICS, DIGITAL SIGNAL PROCESSING. *Educ:* Birla Inst Technol & Sci, India, BS Hons, 71; Univ Roorkee, India, MS, 73; Purdue Univ, PhD(mech eng), 75. *Prof Exp:* Grad inst res, Ray W Herrick Lab, Purdue Univ, 73-75, teaching asst, Mech Eng Dept, 74-75; adj lectr, Indust Eng & Opers Res Dept, 77-79; acoust dynamics engr, Carlyle Compressor Co, 75-77, sr acoust dynamics engr, 77-79; from asst prof to assoc prof, Mech Eng Dept, 79-87, DIR FLUID POWER LAB, OHIO STATE UNIV, 82-, PROF, MECH ENG DEPT, 87- *Concurrent Pos:* Mem, Computer Appln Tech Comt, Am Soc Heating, Refrig & Air Conditioning Engrs, chmn, subcomt on standards, 78-79, Comt Numerical Methods, Am Soc Mech Engrs, 83-, task group Impedance Tube Standards, 82-86; consult, over 20 orgn, 80-; grants & contracts, var corp and orgn, 80-93; key prof, Fluid Power Educ Found, 83-; bd gov, Nat Conf Fluid Power, 83-; chmn tech comt, Noise Control Methods, Inst Noise Control Eng, 90- *Mem:* Acoust Soc Am; Am Acad Mech; Am Soc Heating, Refrig & Air Conditioning Engrs; Am Soc Mech Engrs; Am Soc Eng Educ; Inst Noise Control Eng; Soc Exp Mech. *Res:* Acoustics, noise and vibration control; non linear dynamics; fluid power; digital signal processing; author of 1 book, 60 articles in archival journals and numerous technical papers. *Mailing Add:* Dept Mech Eng Ohio State Univ 206 W 18th Ave Columbus OH 43210-1107

SINGH, RAJINDER, b Adamke Cheema, WPakistan, Apr 1, 31; m 61; c 2. MATHEMATICAL STATISTICS. *Educ:* Panjab Univ, India, MA, 52; Univ Ill, Urbana, PhD(statist), 60. *Prof Exp:* Lectr statist, Panjab Univ, India, 52-56 & 60-62, reader, 62-64; asst prof math, Univ Ill, Urbana, 64-66; from asst prof to assoc prof, 66-75, PROF MATH, UNIV SASK, 75- *Res:* Estimation problems in statistics. *Mailing Add:* Dept Math Univ Saskatoon Saskatoon SK S7N 0W0 Can

SINGH, RAMA SHANKAR, b Varanasi, India, Sept 27, 38. SOLID STATE PHYSICS. *Educ:* Banaras Hindu Univ, India, BSc, 59, MSc, 61; Univ RI, PhD(elec eng), 71. *Prof Exp:* Resident res asst solid state physics, US Army Munition Command & Nat Res Coun/Nat Acad Sci, 71-72; from asst prof to assoc prof physics, Univ PR, Mayaguez, 72-78; staff mem, Lincoln Lab, Mass Inst Technol, Lexington, 78-80; sr process engr, Gen Elec Co, 80-84; SR PROF STAFF, MARTIN MARIETTA CO, 84- *Concurrent Pos:* Scientist I, PR Nuclear Ctr, Mayaguez, 72-78. *Mem:* Soc Photo-Optical Instrumentation Engrs; Am Phys Soc; sr mem Inst Elec & Electronics Engrs; Electrochem Soc. *Res:* Optical properties of materials from ultraviolet to far infrared; lattice dynamics and phase transition; Raman and Brillouin scattering; semiconductor memory devices; very-large-scale integrated circuits fabrication; radiation; hardening processing, testing and risk assessment of very-large-scale integration devices in radiation environments. *Mailing Add:* Electronics & Missile Div Martin Mariette PO Box 5837 MP 184 Orlando FL 32855

SINGH, RAMA SHANKAR, b Azamgarh, India, Mar 2, 45; m 76; c 2. POPULATION GENETICS, EVOLUTIONARY THEORY. *Educ:* Agra Univ, India, BSc, 65; Kanpur Univ, India, MSc, 67; Univ Calif, Davis, PhD(genetics), 72. *Prof Exp:* Lectr bot, Govt Agr Col Kanpur, 67-68; Ford Found fel biol, Univ Chicago, 72-73; res assoc, Harvard Univ, 73-75; from asst prof to assoc prof, 75-85, PROF BIOL, MCMASTER UNIV, 86- *Mem:* AAAS; Genetics Soc Am; Genetics Soc Can; Soc Study Evolution; Soc Am Naturalists. *Res:* Genetic variation and its role in adaptation and species formation; the role of sex in evolution; history of biology and philosophy. *Mailing Add:* Dept Biol McMaster Univ Hamilton ON L8S 4K1 Can

SINGH, RAMAN J, b Nabha, Panjab, India, Apr 16, 40; US citizen; m 69. INVERTEBRATE PALEONTOLOGY, TREPOSTOME BRYOZOANS. *Educ:* Panjab Univ, BSc, 59, MSc, 61; Univ Cincinnati, MS, 66, PhD(geol), 71. *Prof Exp:* Asst geologist, Panjab Govt, Chandigarh, 61-62; asst prof geol, Univ Tenn, Chattanooga, 69-71; from asst prof to assoc prof, 71-82, PROF GEOL, NORTHERN KY UNIV, 82- *Concurrent Pos:* Trustee, Behringer-Crawford Mus, Covington, Ky, 80-84; NDEA fel, Smithsonian Inst, 68-69. *Mem:* Paleont Res Inst; Nat Asn Geol Teachers; Sigma Xi. *Res:* Ultrastructure of trepostome bryozoans; teaching earth science. *Mailing Add:* Div Geol Northern Ky Univ Highland Heights KY 41076-1448

SINGH, RAMESHWAR, b Bihar, India, July 2, 37; m 54; c 4. HYDRAULICS, FLUID MECHANICS. *Educ:* Auburn Univ, BCE, 62, MS, 63; Stanford Univ, PhD(civil eng), 65. *Prof Exp:* Sectional officer design & construct, Irrig Dept, Govt Bihar, India, 56-60; asst prof civil eng, Univ BC, 65-67; assoc prof, 67-77, PROF CIVIL ENG, SAN JOSE STATE UNIV, 77- *Concurrent Pos:* Consult, Fraser River Flood Res, Can, 65-67 & Jennings, McDermitt & Heis Consult Firm, Calif, 68-70, Santa Clara Valley Water Dist, 70-90. *Mem:* Fel Am Soc Civil Engrs; Am Geophys Union; Soil Conserv Soc Am; Nat Soc Prof Engrs. *Res:* Hydraulics; hydrology; fluid mechanics with application of applied mathematics and computers. *Mailing Add:* Dept Civil Eng & Appl Mech San Jose State Univ One Washington Sq San Jose CA 95192

SINGH, RIPU DAMAN, b Patmau, Uttar Pradesh, India. PHYSICAL ANTHROPOLOGY, PRIMATOLOGY. *Educ:* Univ Lucknow, BA, 51, MA, 53; Univ Ore, MA, 69, PhD(anthrop), 71. *Prof Exp:* Asst prof anthrop, Univ Lucknow, 56-60; asst anthropologist, Anthrop Surv India, Govt India, 60-66; instr anthrop, Univ Ore, 66-70; from asst prof to assoc prof, 70-73, PROF ANTHROP, UNIV WINDSOR, 83- *Concurrent Pos:* Assoc anthrop. *Honors & Awards:* Birbal Sahni Award, 76. *Mem:* Am Asn Phys Anthrop; Can Asn Phys Anthrop; Ethnog & Folk Cult Soc India; Int Dermatoglyphic Asn; Am Dermatoglyphic Asn; Human Biol Coun. *Res:* Human population variations; dermatoglyphic variations and genetic patterns in caste populations of India and Canada; primate behavior and comparative primatology; human evolution. *Mailing Add:* Dept Sociol & Anthrop Univ Windsor Windsor ON N9B 3P4 Can

SINGH, RODERICK PATAUDI, b Georgetown, Guyana, Feb 26, 35; Can citizen; m 61; c 3. ANATOMY. *Educ:* Univ Western Ont, BA, 61, MSc, 63, PhD(anat), 66. *Prof Exp:* Instr anat, Med Sch, Wayne State Univ, 66-67; from lectr to asst prof, 67-72, ASSOC PROF ANAT, UNIV WESTERN ONT, 72- *Concurrent Pos:* Cytogeneticist, Dept Path, Univ Hosp, London, Ont, 70-; consult, Depts Pediat & Obstet & Gynec, St Joseph's Hosp, London, Ont. *Honors & Awards:* Award, Soc Obstet & Gynec Can, 68. *Mem:* Am Asn Anat; Can Asn Anat; Am Asn Phys Anthrop; Soc Study Human Biol. *Res:* Cytogenetics of human abortuses; embryology and morphology of the human ovary; mutagenesis. *Mailing Add:* Dept Anat Univ Western Ont London ON N6A 5C1 Can

SINGH, RUDRA PRASAD, b Sariya, India, Sept 1, 40; Can citizen; m 56; c 3. VIROLOGY. *Educ:* Agra Univ, India, BScAg, 59, MScAg, 61; NDak State Univ, Fargo, PhD(plant path), 66. *Prof Exp:* Sr res asst plant virol, Dept Agr, Uttar Pradesh, India, 61-62; fel, Nat Res Coun Can, 66-67; res scientist plant virol, 68-78, SR RES SCIENTIST, AGR CAN RES STA, 78- *Concurrent Pos:* Head, Potato Pest Mgt Sect, 85- *Mem:* Am Phytopath Soc; Potato Asn Am; Can Phytopath Soc; Indian Potato Asn; Europ Asn Potato Res. *Res:* Development of detection procedure for potato spindle tuber viroid; search for viroid resistance in tuber-bearing solanum species; development of serological methods for the detection of potato viruses; development of cDNA probes for potato virus detection in tubers; determination of the role of aphids in virus transmission; virus indicator plants. *Mailing Add:* Agr Can Res Sta PO Box 20280 Fredericton NB E3B 4Z7 Can

SINGH, SANKATHA PRASAD, b Varanasi, India, Jan 27, 37; m 59; c 2. MATHEMATICS. *Educ:* Agra Univ, BSc, 57; Benaras Hindu Univ, MSc, 59, PhD(math), 63. *Prof Exp:* Lectr math, Benaras Hindu Univ, 59-63; lectr, Univ Ill, 63-64; asst prof, Wayne State Univ, 64-65 & Univ Windsor, 65-67; assoc prof, 67-72, PROF MATH, MEM UNIV NFLD, 72- *Concurrent Pos:* Nat Res Coun grant, 65- & NATO res grants, 80; foreign ed, Indian J Math; dir,

NATO Adv Study Inst, 82-83, 84-85 & 90-91. *Mem:* Am Math Soc; Can Math Soc; fel Nat Acad Sci India; Indian Math Soc; fel Inst Math & Applns. *Res:* Transform calculus; approximation theory; fixed point theory (nonlinear functional analysis). *Mailing Add:* Dept Math Mem Univ Nfld St John's NF A1C 5S7 Can

SINGH, SANT PARKASH, b Anokh Singh Wala, India, Oct 2, 36; m 68. ENDOCRINOLOGY, NUTRITION. *Educ:* Panjab Univ, India, MBBS, 59; McGill Univ, MSc, 70; Am Bd Internal Med, dipl, 68; Am Bd Nuclear Med, dipl, 76; Am Bd Endocrinol & Metab, dipl, 77. *Prof Exp:* Intern med, Kingston Gen Hosp, Ont, 60-61; resident, Bergen Pines County Hosp, Paramus, NJ, 61-63; resident endocrinol, Philadelphia Gen Hosp, 63-64; resident med, Bergen Pines County Hosp, Paramus, NJ, 64-65; assoc endocrinologist, Brooklyn-Cumberland Med Ctr, 70-72, dir endocrinol sect, 72-73; asst prof med, State Univ NY Downstate Med Ctr, 71-73; assoc prof clin med, Northwestern Univ, Chicago, 73-74; assoc prof, 74-78, dir, div endocrinol & metab, 74-85, PROF MED & CHIEF OF ENDOCRINE-METAB DIV, CHICAGO MED SCH, 78-; ASSOC CHIEF OF STAFF & CHIEF ENDOCRINE-METAB SECT, VET ADMIN HOSP, NORTH CHICAGO, 73- *Concurrent Pos:* Fel, State Univ NY Downstate Med Ctr, 65-66; Med Res Coun Can fel, McGill Univ Clin Royal Victoria Hosp, Montreal, 66-70. *Mem:* Am Col Physicians; Endocrine Soc; Am Diabetes Asn; Am Fedn Clin Res; AMA. *Res:* Diabetes, growth thyroid pathophysiology; carbohydrate metabolism. *Mailing Add:* Chicago Med Sch North Chicago IL 60064

SINGH, SHIVA PUJAN, b Gonda, India, July 15, 47; US citizen; m 73; c 3. IMMUNOLOGY, MICROBIOLOGY. *Educ:* Pant Univ Agr & Technol, India, BSc, 69, MSc, 71; Auburn Univ, PhD(microbiol), 76. *Prof Exp:* Res assoc, Tuskegee Univ, 76; from asst prof to assoc prof, 76-86, coordr biomed res, 84-86, PROF BIOL & DIR SCI RES, ALA STATE UNIV, 86- *Concurrent Pos:* Fac fel, Argonne Nat Lab, 79; trainee, Auburn Univ, 80, 81 & 82; trainee & extramural assoc, NIH, 83; prog dir & prin investr, Minority Biomed Res Support Prog, 84-; prog dir, Minority Access to Res Careers Prog, 85-; orgnr & dir, Nat Workshop on Use of Monoclonal Antibodies, 87; consult, NIH, 87-; mem, Nat Minority Health Affairs Asn, 87- *Mem:* Indian Microbiologists Asn Am (pres, 87-88); Am Soc Microbiol. *Res:* Outer membrane proteins of gram-negative bacteria; preparation and use of monoclonal antibodies for detection and identification of microorganisms. *Mailing Add:* Dept Biol Ala State Univ 915 S Jackson St Montgomery AL 36101-0271

SINGH, SHOBHA, b Delhi, India, July 15, 28; m 46; c 5. LASERS, SOLID STATE PHYSICS. *Educ:* Univ Delhi, BSc, 49, MSc, 51; Johns Hopkins Univ, PhD(physics), 57. *Prof Exp:* Res asst physics, Nat Phys Lab, India, 51-53; asst prof, Wilson Col, 57-59; res officer, Atomic Energy Estab, India, 59-61 & Nat Res Coun Can, 61-64; mem tech staff physics, Bell Tel Labs, 64-89; PRIN SCIENTIST, POLAROID CORP, 90- *Concurrent Pos:* Fel, Johns Hopkins Univ, 57-59; Nat Res Coun Can fel, 61-63. *Mem:* Am Phys Soc; fel Optical Soc Am; Sigma Xi. *Res:* Spectra of solids, Raman spectra, laser induced non-linear phenomena in solids, electrochromics and three to five compound semiconductors. *Mailing Add:* Polaroid Corp 21 Osborn St Cambridge MA 02139

SINGH, SHYAM N, b Ballia, India, Jan 1, 52; m 75; c 3. FUNDAMENTAL & APPLIED COMBUSTION RESEARCH, EMISSION CONTROL & HEAT TRANSFER EQUIPMENT DESIGN. *Educ:* Banares Hindu Univ, India, BTech, 75; Pa State Univ, MS, 82. *Prof Exp:* Combustion develop engr, Midland Ross Corp, 79-81; res & develop engr, Agua-Chem Div, Cleaver-Brooks, 81-83, proj mgr prod heat recovery, Energy Systs Div, 83-85; dir res & eng, WB Combustion Inc, 85-87; mgr res eng, 87-88, res & develop, 88-89, MGR, NEW PROD DEVELOP, COMBUSTION DIV, ECLIPSE INC, 89- *Concurrent Pos:* Lectr, Dept Mech Eng, Univ Wis-Milwaukee, 87-; consult, Fredefort Malt Corp, 86-; prin investr, Ceramic Radiant Tube Tech, Gas Res Inst, 86- & combination of oil & gas fired burner, 87-, gas fired infrared technol, low emission burner. *Honors & Awards:* Ralph James Award, Am Soc Mech Eng, 91. *Mem:* Combustion Inst; Am Soc Mech Eng. *Res:* Combustion, heat transfer and pollution control areas; high temperature ceramic materials for conducting advanced combustion research to improve industrial burner use; solid fuel combustion. *Mailing Add:* 2113 Silverthorn Dr Rockford IL 61107

SINGH, SUKHJIT, b Ramidi, India, July 21, 41; US citizen; m. TOPOLOGY. *Educ:* Ariz State Univ, BA, 69; Pa State Univ, MA, 70, PhD(math), 73. *Prof Exp:* Asst, 70-73, Pa State Univ, 70-73, from asst prof to assoc prof math, 73-84; PROF MATH, SOUTHWEST TEX STATE UNIV, 85- *Mem:* Am Math Soc. *Res:* Decomposition spaces, shape theory and manifolds. *Mailing Add:* SW Tex State Univ Dept Math San Marcos TX 78666-4616

SINGH, SUMAN PRIYADARSHI NARAIN, b Ludhiana, India, June 23, 41; US citizen; m 75; c 1. FOSSIL ENERGY PROCESSES. *Educ:* Indian Inst Technol, Bombay, BTech, 64; Okla State Univ, MS, 67, PhD(chem eng), 73. *Prof Exp:* Process design engr natural gas liquids processing, Phillips Petrol Co, 66-69; engr petrol res, Exxon Res & Develop Labs, 74-75; from develop staff mem I to develop staff mem II coal processing, Oak Ridge Nat Lab, 76-81, task leader environ control technol, 81-82, prog mgr environ control tech, 82-86, PROG MGR, WASTE MGMT TECHNOL CTR, OAK RIDGE NAT LAB, 86- *Concurrent Pos:* NSF fel, Okla State Univ, 76; mem, Control Technol Task Group, Nat Acid Precipitation Assessment Prog. *Mem:* Am Inst Chem Engrs; Sigma Xi. *Res:* Development and assessment of environmental control processes, fossil energy liquefaction, gasification and beneficiation processes and petroleum technology. *Mailing Add:* 600 Fernwood Rd Knoxville TN 37923

SINGH, SURENDRA PAL, b Mawana, India, June 24, 53; m; c 1. QUANTUM & NONLINEAR OPTICS, LASERS. *Educ:* Banaras Hindu Univ, Varanasi, MS, 75; Univ Rochester, PhD(physics), 82. *Prof Exp:* Asst prof physics, 82-86, ASSOC PROF PHYSICS, UNIV ARK, 86- *Concurrent Pos:* CSIR Jr res fel, Banaros Hindu Univ, Varanasi, 75-76; Rhush Rhees fel, Univ Rochester, 76-79; vis fel, Joint Inst Lab Astrophysics, Univ Colo, Boulder, 89-90. *Mem:* Am Phys Soc; Optical Soc Am; India Phys Asn. *Res:* Coherence and fluctuations in light-matter interactions; more than 45 research articles in professional journals. *Mailing Add:* Dept Physics Univ Ark Fayetteville Fayetteville AR 72701

SINGH, SURINDER SHAH, b Kotshakir, India, Jan 5, 37; m 75; c 2. CLAY MINERALOGY, SOIL CHEMISTRY. *Educ:* Panjab Univ, India, BSc, 57, MSc, 60; Indian Agr Res Inst, New Delhi, PhD(soil sci), 63. *Prof Exp:* Res asst saline soils, Indian Dept Agr, Ludhiana, 59-60; fel soil sci, Univ Uppsala, 63-64 & soil res inst, Nat Res Coun Can, 64-65; res officer, Land Resource Ctr, Can Dept Agr, 65-67, res scientist, 67-85, proj leader, 87-90, SR RES SCIENTIST, LAND RESOURCE CTR, CAN DEPT AGR, 85- *Concurrent Pos:* Assoc Ed, Can J Soil Sci, 85- *Mem:* Int Asn Study Clays; Am Soc Agron; Can Soc Soil Sci; Int Soc Soil Sci. *Res:* Interactions between inorganic and organic soil components; isolation and characterization of the reactions responsible for controlling the distribution of components among the different phases of a soil; reactions of metals and inorganic pollutants; solubility products of sparingly soluble substances; acidic precipitation and aluminum mobility; pillared clay and clay transformations. *Mailing Add:* Land Resource Res Ctr Can Dept Agr Ottawa ON K1A 0C6 Can

SINGH, SURJIT, b Roorkee, India, Oct 9, 31; m 64; c 7. PHYSICAL CHEMISTRY, ENVIRONMENTAL SAFETY. *Educ:* Khalsa Col, Amritsar, India, BSc, 52; Panjab Univ, India, MSc, 55; St Louis Univ, PhD(chem), 63. *Prof Exp:* Instr chem, Hindu Col, Amritsar, 52-53; asst prof, Khalsa Col, India, 55-56 & Govt Col, Gurdaspur, 56-59; asst, St Louis Univ, 59-63; from asst prof to assoc prof, Waynesburg Col, 63-67; assoc prof, 67-80, PROF CHEM, STATE UNIV NY BUFFALO, 80- *Concurrent Pos:* Res assoc, Centre Neurochimie, Strasbourg, 78; pres & consult, Eagle Res Corp, Scipar, Inc. *Mem:* Am Chem Soc; Am Inst Chem Eng. *Res:* Hazardous waste management and resource recovery; nucleation phenomena; atmospheric chemistry; charge transfer spectra; fire safety of construction materials (thermodynamics and material properties); photochromism. *Mailing Add:* Dept Chem State Univ NY 1300 Elmwood Ave Buffalo NY 14222

SINGH, TEJA, b June 18, 28; Can citizen; c 2. FOREST HYDROLOGY, BIOMETRICS-BIOSTATISTICS. *Educ:* E Punjab Univ, India, BA, 49; Utah State Univ, Logan, MSc, 63. *Hon Degrees:* Doctorate, World Univ, 86. *Prof Exp:* Forestry training, Dehra Dun, India, 49-51; oper forestry, Himachal Pradesh & Punjab, India, 51-59; res asst, Utah State Univ, 60-62, Eastern Rockies Forest Conserv Bd, 63; tutor, Utah State Univ, 64; res officer, Can Dept Forestry, Calgary, 65-66; res scientist, Can Dept Environ & Fisheries, 67-77; res expert, UN Develop Porg, 77-79; RES SCIENTIST & BIOMETRICIAN, NORTHERN FORESTRY CTR, CAN FORESTRY SERV, AGR CAN, 79-; PRES, CAN RESOURCES DEVELOP & MGT LTD, 80- *Concurrent Pos:* Chief tech adv multidisciplinary hydrol res, Food & Agr Orgn Watershed Mgt & Coord Proj, Iran, 77-79; consult environ impact, Food & Agr Orgn & UNESCO, Rome, 80; Can ed, Nat Woodlands, 87-90; chmn, Alta Climat Asn, 89-90. *Mem:* Sigma Xi; Can Wildlife Fedn; Am Geophys Union; Soc Range Mgt; Can Inst Forestry; Soc Am Foresters; Ecol Soc Am; Soil Conserv Soc Am; Soc Int Develop; NZ Hydrol Soc. *Res:* Forestry ecosystem modeling; climatology; environmental quality, ecology, global warming, biometrics, computer simulation; energy from biomass; hydrologic research prairie provinces of Alberta, Saskatchewan & Manitoba; risk analysis; author of over 120 publications. *Mailing Add:* N Forest Res Ctr Forestry Can Edmonton AB T6H 3S5 Can

SINGH, TRILOCHAN (HARDEEP), b Vehari, Pakistan, Dec 31, 37; m 64; c 2. MECHANICAL ENGINEERING. *Educ:* Punjab Eng Col, BSc, 61; Univ Calif, Berkeley, MS, 66, PhD(mech eng), 70. *Prof Exp:* Asst engr, Oil & Natural Gas Comn, Dehradum, India, 60-64; lectr mech eng, Thapar Col Eng, Patiala, 64-65; res asst, Univ Calif, Berkeley, 65-70; asst prof, 70-76, NSF res initiation grant, 72-74, ASSOC PROF MECH ENG SCI, WAYNE STATE UNIV, 76- *Mem:* Soc Automotive Engrs; Combustion Inst; Air Pollution Control Asn. *Res:* Basic combustion studies; pollutant species formation; eliminations and control in different types of combustion systems. *Mailing Add:* Dept Mech Eng Wayne State Univ 2100 W Engineering Detroit MI 48202

SINGH, VIJAY P, b Agra, UP, India, July 15, 46; m 76; c 2. MATHEMATICAL MODELING, SYSTEM ANALYSIS. *Educ:* UP Agr Univ, BS, 67; Univ Guelph, MS, 70; Colo State Univ, PhD(hydrol), 74. *Prof Exp:* Engr irrig, Rockefeller Found, 67-68; res asst hydrol, Univ Guelph, 68-70; res asst, Colo State Univ, 70-74, res assoc, 74; asst prof, NMex Inst Mining & Technol, 74-77; assoc prof, George Washington Univ, 77-78; assoc prof civil eng, Miss State Univ, 78-81; PROF & DIR HYDROL & WATER RESOURCES, LA STATE UNIV, 81- *Concurrent Pos:* Vis scientist, Coun Sci & Indust Res, Govt India, 80-81; vis acad, Univ Wollongong, Australia, 82; sr res engr, US Army Engr Waterways Exp Sta, 82-85; ed, Indian Asn Hydrologists, India, 81-, Hydroelec Energy, China, 86-, Stochastic Hydrol & Hydraul, Agr Water Mgt, Irrig Sci, Natural Hazards & Water Mgt; consult, US govt agencies, int orgn, pvt co & Univs, 86-; res grants, NSF, US Geol Surv, US Dept Army, US Dept Agr, UN Educ, Sci, & Cultural Orgn & US Agency Int Develop. *Mem:* Am Geophys Union; Am Soc Civil Engrs; Int Asn Hydraul Res; fel Inst Engrs; fel Indian Asn Hydrologists; fel Am Water Res Asn. *Res:* Continuum mechanics; control theoretic and stochastic modeling of hydrologic processes, with particular regard to stream flow, sediment yield, irrigation, flood, frequency analysis, entropy theory, free boundary problems, and system applications; author of more than 300 research papers and 11 books. *Mailing Add:* Dept Civil Eng La State Univ Baton Rouge LA 70803

SINGH, VIJAY PAL, b New Delhi, India, July 25, 47; US citizen; m 72; c 2. ELECTROLUMINSCENT DISPLAY, SOLAR CELLS. *Educ:* Indian Inst Technol, Delhi, BTech, 68; Univ Minn, MS, 70, PhD(elec eng), 74. *Prof Exp:* Res asst prof, Inst Energy Conversion, Univ Del, 74-76; res engr, 76-80, sect head device res, 80-81, mgr mat & device res, Photon Power Inc, El Paso, Tex,

81-83; assoc prof, 83-90, PROF, ELEC ENG DEPT, UNIV TEX, EL PASO, 90- *Concurrent Pos:* Prin investr, NSF Grant, 89-92; secy El Paso sect Inst Elec & Electronic Engrs, 84-87; consult, Battelle, 86. *Mem:* Inst Elec & Electronics Engrs; Soc Info Display; Electrochem Soc. *Res:* Solar cells; thin film electroluminescent displays; superconductors; photo detectors. *Mailing Add:* Elec Eng Dept Univ Tex El Paso TX 79968

SINGH, VIJENDRA KUMAR, b Moradabad, India, Aug 15, 47; m 74. NEUROIMMUNOLOGY. *Educ:* Lucknow Univ, India, BSc, 64, MSc, 66; Univ BC, PhD(biochem), 72. *Prof Exp:* Res asst biochem, Coun Sci & Indust Res, India, 66-68; fel, Univ BC, 72-74, res assoc neurosci, 74-78, asst prof, dept path, 79-85; mem staff, 78-81, DIR RES, DIV IMMUNOL, CHILDREN'S HOSP, 81-; ASSOC PROF, DEVELOP CTR HANDICAPPED PERSONS & DEPT BIOL, UTAH STATE UNIV, LOGAN, 88- *Concurrent Pos:* Asst prof, basic & clin immunol, Med Univ SC, Charleston, 85-88. *Mem:* AAAS; Soc Neurosci; Am Asn Immunologists. *Res:* Structural-functional relationships between immune system and nervous system; immunologic mechanisms in neuropsychiatric disorders, especially Alzheimer's dementia and autism. *Mailing Add:* Develop Ctr Handicapped Persons & Dept Biol Utah State Univ Logan UT 84322

SINGH, VISHWA NATH, b July 6, 36; m; c 2. NUTRITION, BIOCHEMISTRY. *Educ:* Univ Delhi, India, PhD(biochem), 62. *Prof Exp:* CLIN RES SCIENTIST, HOFFMANN LA ROCHE, INC, 80- *Res:* Nutrition requirements of special population groups; diabetes; roles of vitamins in health and disease; carbohydrate and lipid metabolism. *Mailing Add:* Dept Vitamins & Clin Nutrit Hoffmann La Roche Inc 140 Kingsland St Bldg 76-412 Nutley NJ 07110

SINGHAL, AVINASH CHANDRA, b Aligarh, India, Nov 4, 41; m 67; c 3. SEISMIC ENGINEERING, STRUCTURAL ENGINEERING. *Educ:* St Andrews Univ, BSc, 59 & 60; Mass Inst Technol, SM, 61, CE, 62, ScD(civil eng), 64. *Prof Exp:* Prof civil eng, Laval Univ, Que, 65-69; asst prog mgr, TRW, 69-71; mgr systs eng, Gen Elec, 71-72 & Engrs India Ltd, 72-74; proj engr consult eng, Weidlinger Assoc, 74-77; assoc prof, 77-84, PROF CIVIL ENG & DIR EARTHQUAKE RES LAB, ARIZ STATE UNIV, 84- *Concurrent Pos:* Chmn Eng Mech Sub Task Comn, Am Soc Civil Eng, 68-70; fel Am Soc Civil Eng Comts, 69-73 & Tech Coun Lifeline Earthquake Eng, 78-; chmn sub comt, Am Soc Mech Eng, 82-86, 88 & 89; vis prof, Melbourne Univ, Australia, 84; res fel, Kobe Univ, Japan, 90. *Honors & Awards:* First Prize, Int Asn Shell Struct, 61; Henry Adams Medal, Inst Struct Eng, London, 71. *Mem:* Am Soc Civil Engrs; Earthquake Engr Res Inst. *Res:* Earthquake engineering; soil engineering; blast and vibrations; pipelines; author of 236 publications including six books. *Mailing Add:* Dept Civil Eng ECE 5306 Ariz State Univ Tempe AZ 85287-5306

SINGHAL, RADHEY LAL, b Gulaothi, India, July 12, 40; m 61; c 3. PHARMACOLOGY. *Educ:* Univ Lucknow, BSc, 57, MSc, 59, PhD(pharmacol), 61. *Prof Exp:* Res assoc pharmacol, Ind Univ, 62-64, instr, 64-65; from asst prof to assoc prof, 66-72, actg chmn dept, 74-76, PROF PHARMACOL, UNIV OTTAWA, 72-, CHMN DEPT, 76- *Concurrent Pos:* Med Res Coun Can scholar, Univ Ottawa, 67-72. *Mem:* Endocrine Soc; Int Soc Neurochem; Soc Toxicol; Am Soc Pharmacol & Exp Therapeut; Am Soc Biol Chem. *Res:* Endocrine and biochemical pharmacology; neuroendocrinological approaches to the study of brain function; environmental toxicology. *Mailing Add:* 22 Royal Hunt Ct Ottawa ON K1N 9M1 Can

SINGHAL, RAM P, b New Delhi, India, Aug 12, 39; US citizen; m 68; c 2. BIOCHEMISTRY, ANALYTICAL CHEMISTRY. *Educ:* Univ Lucknow, BS, 58, MS, 60; Univ Sci & Technol, France, Dipl, 64, PhD(biochem), 67. *Prof Exp:* Instr biochem, All-India Inst Med Sci, 60-62; researcher, Cancer Res Inst, Univ Lille, 62-67; scientist, Coun Sci & Indust Res, India, 67-68; USPHS fel, Wayne State Univ, 68-69; Univ Tenn-Oak Ridge Nat Lab fel, Oak Ridge Nat Lab, 70-71, res scientist, 72-74; vis scientist, Scripps Res Found, 74; from asst prof to assoc prof, 74-86 , PROF BIOCHEM, WICHITA STATE UNIV, 86- *Concurrent Pos:* Nat Ctr Sci Res researcher, Univ Sci & Technol, Univ Lille, 64-67; NSF exchange 64-67; NIH expert scientist, 82-83, WHO scientist, 86-87. *Honors & Awards:* Cancer Res Award, Am Cancer Soc, 75, Acad Res Enhancement Award, 85. *Mem:* AAAS; Am Chem Soc; Am Soc Biol Chemists & Molecular Biologists. *Res:* Novel methods in molecular biology; high-performance liquid chromatography, methods development, capillary electrophoresis; reaction to anticancer agent dirhodium tetraacetate with nucleic acids; significance of minor, modified components of RNAs and DNAs; chemical probe of DNA conformation; biotechnology; chemical probe of the nucleic acid structure-function relation; noval antiviral drug; reaction mechanism of boronate complexing; high-performance liquid chromatography; affinity chromotography. *Mailing Add:* Dept Chem Wichita State Univ Campus Box 51 Wichita KS 67208

SINGHAL, SHARWAN KUMAR, b Oct 8, 39; Can citizen; m 64; c 2. IMMUNOLOGY. *Educ:* McGill Univ, PhD(immunol), 68. *Prof Exp:* Med Res Coun Can fel tumor biol, Karolinska Inst, Sweden, 68-70; assoc prof, 70-79, PROF IMMUNOL, UNIV WESTERN ONT, 80- *Mem:* Am Asn Immunol; Can Soc Immunol; Scand Soc Immunol. *Res:* Regulation of the immune response at the cellular level. *Mailing Add:* Dept Microbiol & Immunol Univ Western Ont London ON N6A 5C1 Can

SINGHVI, SAMPAT MANAKCHAND, b Jodhpur, India, Oct 14, 47; m 71; c 2. BIOPHARMACEUTICS, PHARMACOKINETICS. *Educ:* BITS Pilani, Rajasthan, India, BPharm, 67; Philadelphia Col Pharm & Sci, MS, 70; State Univ NY, Buffalo, PhD(pharmaceut), 74; Rider Col, MBA, 79. *Prof Exp:* Res sci, Wyeth Labs, Am Home Prod, 69, 70; teaching asst chem, Philadelphia Col Pharm & Sci, 69-70; res asst pharmaceut, State Univ NY, Buffalo, 70-73; res investr, E R Squibb & Sons Inc, 74-78, sr res investr, 78-79, res group leader drug metab, 79-88, ASSOC DIR, WORLDWIDE REGULATORY AFFAIRS, BRISTOL-MYERS SQUIBB, 88- *Mem:* Am Soc Pharmacol & Exp Therapeut; Am Pharm Asn; Am Asn Pharmaceut Scientists; Int Soc Study Xenobiotics; Regulatory Affairs Prof Soc; Drug Info Asn. *Res:* Drug metabolism in animals; bioavailability and pharmacokinetics of various new drugs in animals and man. *Mailing Add:* Bristol-Myers Squibb PO Box 4000 Princeton NJ 08543-4000

SINGISER, ROBERT EUGENE, b Mechanicsburg, Pa, Aug 7, 30; m 54; c 2. RESEARCH ADMINISTRATION, PHARMACEUTICS. *Educ:* Temple Univ, BS, 52; Univ Fla, MS, 56; Univ Conn, PhD(pharm), 59. *Prof Exp:* Pharmaceut chemist, Merck & Co, Inc, 52-55; res pharmacist, 58-64, dept mgr, Pharmaceut Prod Res, 64-66, dir pharmaceut res & develop, 66-68, sci dir, Pharmaceut Prod Div, 68-70, vpres sci affairs, Pharmaceut Prod Div, 70-86, DIR NEW TECHNOL, PHARMACEUT PROD DIV, ABBOTT LABS, 86- *Mem:* Fel Am Found Pharmaceut Educ; Am Pharmaceut Asn; Am Soc Hosp Pharmacists; Int Pharmaceut Fedn; Am Asn Pharmaceut Scientists. *Res:* Ultrasonic emulsification; thermo-stable ointment bases; air-suspension tablet coating techniques; non-sterile, human pharmaceutical dosage forms; biopharmaceutics. *Mailing Add:* 326 S Hickory Haven Dr Gurnee IL 60031

SINGLER, ROBERT EDWARD, b Chicago, Ill, Mar 21, 41; m 70; c 3. ORGANIC CHEMISTRY. *Educ:* Loyola Univ, BS, 63; Southern Ill Univ, MA, 65; Univ Calif, Los Angeles, PhD(chem), 70. *Prof Exp:* RES CHEMIST POLYMER RES, US ARMY MAT & TECHNOL LAB, 70-, GROUP LEADER, POLYMER CHEM GROUP, POLYMER RES BR, 85- *Concurrent Pos:* Army res fel, Macromolecular Inst, Freiburg, Ger, 79-80. *Mem:* Am Chem Soc; Mat Res Soc. *Res:* Synthesis, characterization and development of cyclic phosphazenes and polyphosphazenes; fire resistant materials; rubber technology; thermal analysis of elastomeric materials. *Mailing Add:* Polymer Res Br Army Mat Technol Lab Watertown MA 02172-0001

SINGLETARY, JOHN BOON, b Houston, Tex, May 6, 28; m 55. NUCLEAR PHYSICS. *Educ:* Agr & Mech Col Tex, BS, 49, MS, 51; Northwestern Univ, PhD, 58. *Prof Exp:* Mem staff, Los Alamos Sci Lab, 57-59; res scientist, Lockheed Aircraft Co, 59-70; dept mgr, Braddock, Dunn & McDonald, 70-71; LAB SCIENTIST, HUGHES AIRCRAFT CO, 71- *Mem:* Am Phys Soc; Sigma Xi. *Res:* Neutron and low energy nuclear physics; environmental effects on spacecraft materials; radiation effects on electronic components; electromagnetic pulse environment and effects on components and systems. *Mailing Add:* 22316 Barbacoa Dr Saugus CA 91350

SINGLETARY, LILLIAN DARLINGTON, b Chicago, Ill. NUCLEAR PHYSICS. *Educ:* Northwestern Univ, BS, PhD(exp neutron & charged particle physics), 62. *Prof Exp:* Res scientist, Lockheed Res Lab, 62-69; EMP sect head, Lockheed Missiles & Space Co, 69-70; dept mgr, EMP & Appl Nuclear Technol Dept, Braddock, Dunn & McDonald, Inc, NMex, 70-71; SECT HEAD ADV TECHNOL, VULNERABILITY & HARDNESS LAB, TRW SYSTS GROUP, 71- *Concurrent Pos:* Consult, Los Alamos Sci Lab, 69-73. *Mem:* Inst Elec & Electronic Engrs; Am Phys Soc; Sigma Xi. *Res:* Analytical and test programs investigating electromagnetic pulse generated by nuclear weapon and system generated electromagnetic pulse generated by nuclear weapon photons incident on systems. *Mailing Add:* 32759 Seagate Dr No 106 Rancho Palos Verdes CA 90274

SINGLETARY, ROBERT LOMBARD, b Atlanta, Ga, Mar 21, 41; m 64; c 2. MARINE ECOLOGY, INVERTEBRATE ZOOLOGY. *Educ:* Univ NC, AB, 63; Univ RI, MS, 67; Univ Miami, PhD(marine biol), 70. *Prof Exp:* PROF BIOL, UNIV BRIDGEPORT, 70. *Mem:* Estuarine Res Fedn; Int Oceanog Found; Sigma Xi. *Res:* Biology of Echinoderms; benthic and intertidal ecology. *Mailing Add:* Dept Biol Univ Bridgeport Bridgeport CT 06601

SINGLETARY, THOMAS ALEXANDER, b Cairo, Ga, Sept 17, 37; m 65; c 1. ELECTRONICS, TECHNOLOGY. *Educ:* Ga Southern Col, BS, 59; Stout Univ, MS, 60; Univ Mo-Columbia, EdD(indust educ), 68. *Prof Exp:* Assoc prof, 60-77, INSTR ELECTRONICS TECHNOL, GA SOUTHERN COL, 77- *Concurrent Pos:* Systs consult, Statesboro Telephone Inc, 60-; TV consult, Westinghouse Elec Corp, 62-65; tech instr, Rockwell Mfg Co, 64-; reviewing consult, Delmar Publs, 72- *Mem:* Am Indust Arts Asn. *Res:* Electrofinishing technology as applied to metals; photographic media as applied to education. *Mailing Add:* Dept Tech Landrom Box 8044 Ga Southern Col Statesboro GA 30460

SINGLETON, ALAN HERBERT, b Punxsutawney, Pa, Nov 28, 36; m 58; c 4. CHEMICAL ENGINEERING. *Educ:* Univ Md, College Park, BS, 58; Lehigh Univ, MS, 62, PhD(chem eng), 68. *Prof Exp:* Res engr, US Naval Propellant Plant, 58-59; mgr explor eng, Air Prod & Chem, Inc, 59-68; section mgr, Res Dept, Bethlehem Steel Corp, 68-77; prog mgr, UCG Opers, 77-81, dept mgr synthetic fuels develop, Gulf Oil Corp Res, 81-85; PRES, ENERGY INT CORP, 85- *Concurrent Pos:* mem, Tex Energy & Natural Resources Adv Coun, 81-83; mem, vis comt, WVa Univ, 82-; mem, Adv Bd Pittsburgh Coal Conference, 81- *Mem:* Am Inst Chem Eng; Am Chem Soc; Sigma Xi. *Res:* Synthetic fuels process development; underground coal gasification process development and operations; organic chemical product and process development; catalysis; cryogenic refrigeration, liquefaction, and containment system development; synthesis gas conversion processes. *Mailing Add:* Energy Int Corp 135 William Pitt Way Pittsburgh PA 15238

SINGLETON, BERT, b New York, NY, July 13, 28; m 56; c 2. ANALYTICAL CHEMISTRY, ORGANIC CHEMISTRY. *Educ:* Cornell Univ, BChE, 50. *Prof Exp:* Res asst org fluorine chem, Cornell Univ, 51-53; process develop chemist, Merck Sharp & Dohme Res Labs, 55- 57, foreign projs chemist, 57-59, sr develop chemist, 59-65, sect head, Process Controls Res, 65-69, mgr process controls, 69-80, assoc dir anal res, 80-90; RETIRED. *Mem:* AAAS; Am Chem Soc; Sigma Xi. *Res:* Analytical methods; gas and liquid chromatography; ultraviolet and infrared spectrophotometry; microanalysis; automatic chemical and process control instrumentation; pharmaceuticals; steroids and vitamins; synthetic organic chemistry. *Mailing Add:* 443 Beechwood Pl Westfield NJ 07090

SINGLETON, CHLOE JOI, b Cleveland, Ohio, Dec 4, 43; m 84. MACROMOLECULAR SCIENCE. *Educ:* Case Inst Technol, BS, 67; Case Western Res Univ, MSE, 74, PhD(macromolecular sci), 75; Baldwin-Wallace Col, MBA, 81. *Prof Exp:* Physicist x-ray diffraction, B F Goodrich Tech Ctr, 67-69, res physicist, 69-70, physicist res & develop electron micros, 74-78, sr physicist thermal anal polymers, Res & Develop Ctr, 78-79, sr res & develop physicist/applns engr thermoplastic polyurethanes, 78-84; dir technol & mkt planning, Sherwin-Williams Co, 84-85, dir prof develop, 85-88; mkt res & commun mgr, Desoto Inc, 88-89, bus develop mgr, anal serv, 89-90; MKT RES MGR, WESTVACO CORP, 90- *Mem:* Am Chem Soc; Am Phys Soc; Soc Plastics Engrs; Soc Competitor Intel Prof. *Res:* Morphological characterization of plastic and rubber materials, using the tools of thermal analysis; electron microscopy and x-ray diffraction; applications engineering of thermoplastic polyurethanes; paints and coatings technology; market and business research. *Mailing Add:* WEstvaco Corp PO Box 70848 Charleston Heights SC 29415-0848

SINGLETON, DAVID MICHAEL, b Poole, Eng, Nov 3, 39; m 62; c 2. FREE RADICAL CHEMISTRY, CATALYSIS. *Educ:* Univ London, BSc, 60; McMaster Univ, PhD(org chem), 65. *Prof Exp:* Res assoc reductive reactions of chromous ion, with J K Kochi, Case Western Reserve Univ, 65-67; chemist, Petrol Chem Dept, Shell Develop Co, Calif, 67-72, hydroprocessing dept, 72-74, SR RES CHEMIST, SHELL DEVELOP CO, 74-, CHEM RES & APPL DEPT, 77- *Concurrent Pos:* Shell exchange scientist, Shell Res & Develop Co, Amsterdam, 75-76; adj prof, Univ Houston, 83-91. *Mem:* Am Chem Soc; Chem Inst Can; fel Royal Soc Chem; Southwest Catalysis Soc. *Res:* Free radical reactions; organic redox reactions by metal complexes; organometallic chemistry; catalysis of organic reactions by metal complexes; hydrocarbon chemistry; heterogeneous catalysis. *Mailing Add:* Westhollow Res Ctr Box 1380 Shell Develop Co Houston TX 77251-1380

SINGLETON, DONALD LEE, b Lyons, Kans, Mar 26, 44; m 66; c 2. PHYSICAL CHEMISTRY. *Educ:* Univ Calif, Davis, BS, 66; Northwestern Univ, MS, 70, PhD(phys chem), 71. *Prof Exp:* Vis scientist atmospheric chem, Nat Ctr Atmospheric Res, 70-72; fel, 72-74, RES OFFICER, NAT RES COUN CAN, 74- *Mem:* Am Chem Soc; Can Inst Chem. *Res:* Kinetics and mechanisms of atomic and free radical reactions; atmospheric chemistry; laser processing. *Mailing Add:* Inst Environ Chem Nat Res Coun Ottawa ON K1A 0R9 Can

SINGLETON, EDGAR BRYSON, b Warren, Ohio, June 17, 26; m 53; c 3. MOLECULAR PHYSICS. *Educ:* Ohio Univ, BS, 49, MS, 51; Ohio State Univ, PhD(physics), 58. *Prof Exp:* Instr & res assoc physics, Ohio State Univ, 58-59; from asst prof to assoc prof, 59-72, PROF PHYSICS, BOWLING GREEN STATE UNIV, 72- *Mem:* Optical Soc Am; Am Asn Physics Teachers; Sigma Xi. *Res:* Molecular physics and infrared spectroscopy in absorption and emission of radiation. *Mailing Add:* Dept Physics Bowling Green State Univ Bowling Green OH 43402

SINGLETON, GEORGE TERRELL, b Wichita Falls, Tex, Dec 16, 27; c 3. OTOLARYNGOLOGY. *Educ:* Midwestern Univ, BA & BS, 49; Baylor Univ, MD, 54; Am Bd Otolaryngol, dipl, 59. *Prof Exp:* Intern, Henry Ford Hosp, 54-55, asst resident, 55-57, sr resident, 57-58; assoc prof surg & head, div otolaryngol, 61-68, chief otolaryngol, 68-75, asst dean clin affairs, 70-77, PROF SURG, UNIV FLA, 68 - *Concurrent Pos:* NIH spec fel otolaryngol, Univ Chicago, 60-61; mem communicative dis res training comt, Nat Inst Neurol Dis & Stroke, 69-73; chief of staff, Shands Teaching Hosp & Clins, 72-76, actg hosp dir, 75-77. *Honors & Awards:* Am Acad Ophthal & Otolaryngol Award, 60; Harris P Mosher Mem Award, Am Laryngol, Rhinol & Otolaryngol Soc. *Mem:* Am Acad Ophthal & Otolaryngol; Am Laryngol, Rhinol & Otol Soc; Soc Univ Otolaryngol; Am Otol Soc. *Res:* Computer analysis of caloric induced nystagmus responses with known central peripheral and vestibular lesions and histopathology of human temporal bones and related clinical neuro-otology. *Mailing Add:* Dept Surg J-264 JHMHC Univ Fla Gainesville FL 32601

SINGLETON, HENRY E, b Haslet, Tex, Nov 27, 16. INDUSTRIAL & MANUFACTURING ENGINEERING. *Educ:* Mass Inst Technol, SB & SM, 40, ScD, 50. *Prof Exp:* Vpres, Litton Industs, 54-60; chief exec & chmn bd, 60-87, chmn, 87-91, CHMN, EXEC COMT BD DIRS, TELEDYNE CORP, 91- *Mem:* Nat Acad Eng. *Mailing Add:* Teledyne Corp 1901 Ave of the Stars Los Angeles CA 90067

SINGLETON, JACK HOWARD, b Rawtenstall, Eng, Sept 27, 26; m 54; c 2. PHYSICAL CHEMISTRY. *Educ:* Univ London, BSc, 47, dipl & PhD, 50. *Prof Exp:* Asst phys chem, Aberdeen Univ, 49-52; res assoc, Univ Wash, 52-55; engr, Res Labs, Westinghouse Elec Corp, 55-86; CONSULT, 86- *Mem:* Am Chem Soc; hon mem Am Vacuum Soc (pres,86). *Res:* Physical and chemi-sorption; catalysis on metals; ultrahigh vacuum. *Mailing Add:* 1184 St Vincent Dr Monroeville PA 15146

SINGLETON, JAMES L, b Coya, Chile, Dec 4, 20; US citizen; m 45; c 3. ELECTRICAL ENGINEERING. *Educ:* Univ Colo, BSEE, 49; US Air Force Inst Technol, MS, 52. *Prof Exp:* Instr, Tech Training Command, US Air Force, 49-50, supvr teletype mech course, 50-51, instr eng, US Naval Acad, 52-56, asst prof elec eng, US Air Force Acad, 56-60, res & develop officer, Electronics Systs Div, Systs Command, 60-64, res & develop staff officer, Hq, Pentagon, 64-67; engr analyst, Command & Control Systs, Anal Serv, Inc, 68-76; engr analyst, Electrospace Systs, Inc, 76-82; eng analyst, Mitre Corp, 82-; ENGR ANALYST, ELECTROSPACE SYSTS, INC. *Mem:* Sr mem Inst Elec & Electronics Engrs. *Res:* Engineering education; electronics; avionics; communications. *Mailing Add:* Electrospace Systs Inc Box 2668 Arlington VA 22202

SINGLETON, JOHN BYRNE, b Troy, NY, May 16, 30; m 52; c 6. ENGINEERING. *Educ:* Col of the Holy Cross, BS, 52; Univ RI, MS, 54. *Prof Exp:* Mem tech staff, Bell Tel Labs, 54-60, head, Integrated Circuits Dept, 60-70, Data & Digital Dept, 71-76, Digital Systs Dept, 77-84 & D-Channel Banks Dept, 84-; RETIRED. *Honors & Awards:* Centennial Medal, Inst Elec & Electronics Engrs. *Mem:* Sr mem Inst Elec & Electronics Engrs. *Res:* Design and development of digital transmission lines and terminals. *Mailing Add:* 1850 Sand Hill Rd No 28 Palo Alto CA 94304

SINGLETON, MARY CLYDE, b Enfield, NC, Mar 31, 12. ANATOMY, PHYSICAL THERAPY. *Educ:* Univ NC, Greensboro, BS,32; Duke Univ, MA, 60, PhD(anat), 64. *Prof Exp:* Clin supvr phys ther, Med Ctr, Duke Univ, 40-54; coordr ther, Georgia Warm Springs Found, 54-58; asst prof phys ther & instr anat, Univ NC, 62-66, asst prof anat, 66-73, assoc prof phys ther, 69-74, assoc prof anat, 73-77, prof phys ther, 74-81, prof anat, 77-81, EMER PROF, UNIV NC, CHAPEL HILL, 81- *Honors & Awards:* Golden Pen Award, Am Phys Ther Asn, 74, Lucy Blair Serv Award, 77; McMillian Lectureship, 77. *Mem:* AAAS; Am Asn Anat; Am Phys Ther Asn (pres, 50-52); Sigma Xi. *Res:* Neuroanatomy; gross anatomy; physical therapy. *Mailing Add:* 6212 Old NC 86 Chapel Hill NC 27516

SINGLETON, RICHARD COLLOM, b Schenectady, NY, Feb 21, 28; m 50; c 6. MATHEMATICAL STATISTICS, INFORMATION SCIENCE. *Educ:* Mass Inst Technol, BS & MS, 50; Stanford Univ, MBA, 52, MS, 59, PhD(math statist), 60. *Prof Exp:* Economist, Stanford Res Inst, 52-54, res engr, 54-56, systs analyst, 56-59, res math statistician, 59-63, sr res math statistician, Dept Math, 63-72, STAFF SCIENTIST, DEPT MATH STATIST, INFO, SERV & SYSTS DIV, SRI INT, 72-; ASSOC ED, INFO SCI, 68- *Mem:* Inst Math Statist; Opers Res Soc Am; sr mem Inst Elec & Electronics Eng. *Res:* Design of experiments in the physical and biological sciences; time series analysis; statistical inference; theory of error-correcting codes; economics; litigation support statistical analysis. *Mailing Add:* Dept Math Statist SRI Int Menlo Park CA 94025

SINGLETON, RIVERS, JR, b New Orleans, La, Sept 2, 39; m 63; c 3. MICROBIOL PHYSIOLOGY, SCIENCE & SOCIETY INTERACTION. *Educ:* Trinity Univ, BS, 61; Mich State Univ, MS, 63; Univ Kans, PhD(biochem), 69. *Prof Exp:* Dir, clin chem,chem div, First US Army Med Lab, 63-65; fel biochem, Case Western Reserve Univ, 69-72; res assoc biochem microbiol, Ames Res Ctr, NASA, 72-74; res asst prof microbiol, 74-88, res scientist, Sch Life & Health Sci, 88-89, ASSOC PROF, SCH LIFE & HEALTH SCI & DEPT ENGLISH, UNIV DEL, 89-, DIR, CTR SCI & CULT, 89- *Concurrent Pos:* Pre-doctoral fel, NIH, 66-69; post-doctoral fel, Am Cancer Soc, 70-72; res assoc, Nat Acad Sci, 72-74; Cong Sci Fel, Am Soc Microbiol, 88-89. *Mem:* Sigma Xi; Am Soc Microbiol; Asn Integrative Studies; AAAS. *Res:* The evolutionary aspects and chemical mechanisms whereby microorganisms adapt to and grow under extreme environmental conditions; interdisciplinary studies on science and society interaction. *Mailing Add:* Sch Life & Health Sci Univ Del Newark DE 19716

SINGLETON, ROBERT RICHMOND, b Brooklyn, NY, Mar 23, 13; m 38; c 2. MATHEMATICS. *Educ:* Dartmouth Col, AB, 34; Brown Univ, MSc, 35; Princeton Univ, PhD(math), 62. *Prof Exp:* Clerk, Metrop Life Ins Co, 36-37; res asst govt, Princeton Univ, 37-39, res assoc, 39-42; res asst, Merrill Flood & Assoc, 42-46; dir develop, Aero Serv Corp, 46-52; instrument develop consult, self-employed, 52-56; res consult, Gen Elec Co, 56-62; lectr, 62-76, adj prof, 76-79, EMER ADJ PROF MATH, WESLEYAN UNIV, 79- *Concurrent Pos:* Consult, Gov, WVa, 38-40. *Mem:* Math Asn Am. *Res:* Management science; graph theory. *Mailing Add:* PO Box 435 Portland CT 06480

SINGLETON, SAMUEL WINSTON, medicine, for more information see previous edition

SINGLETON, TOMMY CLARK, b Lufkin, Tex, Oct 5, 28; m 58; c 3. ORGANIC CHEMISTRY. *Educ:* Stephen F Austin State Col, BS, 49; Rice Univ, PhD(chem), 54. *Prof Exp:* Chemist, Naval Stores Sta, USDA, 55-56; sr process chemist, Res Dept, 56-69, process specialist, Process Technol Dept, 69-75, SR PROCESS SPECIALIST, PROCESS TECHNOL DEPT, MONSANTO CO, 75- *Mem:* Am Chem Soc. *Res:* Carbonylation chemistry; plant process problems; syntheses of vinyl monomers. *Mailing Add:* Monsanto Co 800 N Lindergh Blvd St Louis MO 63167

SINGLETON, VERNON LEROY, b Mill City, Ore, June 28, 23; m 47; c 3. NATURAL PRODUCTS CHEMISTRY, ENOLOGY. *Educ:* Purdue Univ, BSA, 47, MS, 49, PhD(agr biochem), 51. *Hon Degrees:* DSc, Univ Stellenbosch, 83. *Prof Exp:* grad asst biochem, Purdue Univ, 47-49; res chemist, Lederle Labs, Am Cyanamid Co, 51-54; assoc biochemist, Pineapple Res Inst, Univ Hawaii, 54-58, biochemist, 58; asst enologist, 58-63, assoc enologist, 63-66, assoc chemist, 66-69, lectr, 59-69, PROF ENOL & CHEMIST, AGR EXP STA, UNIV CALIF, DAVIS, 69- *Concurrent Pos:* Assoc prof, Univ Hawaii, 56-58; consult, Pineapple Res Inst, 59; abstr ed, Am Soc Enol, 59-73; sr res fel, Inst Denology & Viticulture, Univ Stellenbosch, SAfrica, 68-69; vis scientist Long Ashton Res Sta, Univ Bristol, UK, 75-76; vis prof, Univ Stellenbosch, 82, Lincoln Univ, NZ, 89. *Honors & Awards:* Andre Simon Lit Prize, 65; Biennial Wine Res Award, Soc Med Friends Wine, 77. *Mem:* Am Chem Soc; fel, Am Inst Chem; Am Soc Enol (treas, 69-73, 2nd vpres, 73, 1st vpres, 74, pres, 75-76); Inst Food Technologists; Phytochem Soc NAm; Sigma Xi; AAAS. *Res:* Chemistry of natural products, especially flavonoids, phenolic acids, tannins, mold products, and wines; food storage reactions; sensory analysis; biochemistry of fruits; chromatography. *Mailing Add:* Dept Viticulture & Enology Univ Calif Davis CA 95616-8749

SINGLETON, WAYNE LOUIS, b Oaktown, Ind, Feb 2, 44; m 69. ANIMAL SCIENCE, REPRODUCTIVE PHYSIOLOGY. *Educ:* Purdue Univ, Lafayette, BS, 66; SDak State Univ, MS, 68, PhD(animal sci), 70. *Prof Exp:* Asst prof, 70-76, ASSOC PROF ANIMAL SCI, COOP EXTEN SERV, PURDUE UNIV, LAFAYETTE, 76- *Mem:* Soc Study Reproduction; Sigma Xi. *Res:* Swine and beef cattle artificial insemination; reproductive efficiency of beef cattle and swine. *Mailing Add:* Dept Sci Purdue Univ West Lafayette IN 47907

SINGLEY, JOHN EDWARD, b Wildwood, NJ, July 31, 24; m 50; c 4. WATER CHEMISTRY. *Educ:* Ga Inst Technol, BS, 50, MS, 52; Univ Fla, PhD(water chem), 66. *Prof Exp:* Asst chem, Ga Inst Technol, 49-50; phys chemist, US Army Ord Rocket Res Ctr, Redstone Arsenal, Ala, 50-51; chemist & group leader, Res & Develop Dept, Tenn Corp, 51-58, supvr tech serv, 58-65; from instr to assoc prof, Ga State Col, 54-67; from assoc prof to prof, 67-90, EMER PROF WATER CHEM, UNIV FLA, 90-; VPRES, JAMES M MONTGOMERY, CONSULT ENGRS INC. *Concurrent Pos:* Consult, Panel on Public Water Supplies, Nat Acad Sci, 71-72 & President's Coun on Environ Qual, 74; sr staff consult, Environ Sci & Eng, Inc, Gainesville, 77-82, sr vpres res & develop, 82- *Honors & Awards:* Ambassadors Award, Am Water Works Asn, 73, Res Award, 85; Fuller Award, 76. *Mem:* Hon mem Am Water Works Asn; fel Am Inst Chem; Inter-Am Asn Sanit Eng; Sigma Xi; fel Inst Water & Environ Mgt UK. *Res:* Coagulation mechanisms; corrosion in potable water systems; color, iron and manganese in water supplies; water treatment processes. *Mailing Add:* 1719 NW 23 Blvd PHE Gainesville FL 32605

SINGLEY, MARK E(LDRIDGE), b Delano, Pa, Jan 25, 21; m 42; c 4. AGRICULTURAL ENGINEERING. *Educ:* Pa State Univ, BS, 42; Rutgers Univ, MS, 49. *Prof Exp:* From instr to prof, 47-82, prof II, 82-87, EMER PROF AGR ENG, RUTGERS UNIV, 87-; VPRES RES & DEVELOP, BEDMINSTER BIOCONVERSION CORP. *Concurrent Pos:* Pres, Agr Mus, State NJ, 85-89, trustee, 89- *Honors & Awards:* Massey-Ferguson Medal,Am Soc Agr Engrs, 87. *Mem:* Fel AAAS; fel Am Soc Agr Engrs; Sigma Xi. *Res:* Deep bed drying, pneumatic handling and characteristics of fibrous and granular farm crops; resource management; land use planning; composting of soild waste. *Mailing Add:* Dept Biol & Arg Eng Cook Col Rutgers State Univ PO Box 231 New Brunswick NJ 08903

SINGPURWALLA, NOZER DRABSHA, b Hubli, India, Apr 8, 39; m 69; c 1. OPERATIONS RESEARCH, STATISTICS. *Educ:* B V B Col Eng & Tech, India, BS, 59; Rutgers Univ, MS, 64; NY Univ, PhD(opers res). 68. *Prof Exp:* PROF OPERS RES, PROF STATIST & DIR, INST RELIABILITY & RISK ANALYSIS, GEORGE WASHINGTON UNIV, 69- *Concurrent Pos:* Vis prof statist, Stanford Univ, 78-79. *Honors & Awards:* Wilks Award for Contributions to Reliability & Life Testing Methodologies. *Mem:* Int Asn Statist in Phys Sci; fel Am Statist Asn; fel Inst Math Statist; Int Statist Inst. *Res:* Applications of statistics to reliability theory; development of statistical methodology. *Mailing Add:* Dept of Opers Res George Washington Univ Staughton Hall Rm 214B 707 32nd Washington DC 20052

SINGWI, KUNDAN SINGH, b Vdaipur, India, Mar 13, 19. CONDENSED MATTER THEORY. *Educ:* Allahabad Univ, BS, 38, MS, 40, DSc(physics), 49. *Prof Exp:* Chmn dept phys & astron, 79-82, PROF PHYSICS, NORTHWESTERN UNIV, 68- *Mailing Add:* Dept Physics & Astron Northwestern Univ Evanston IL 60208

SINHA, AKHOURI ACHYUTANAND, b Churamanpur, Bihar, India, Dec 17, 33; US citizen; m 79. ZOOLOGY, EMBRYOLOGY. *Educ:* Univ Allahabad, BS, 54; Patna Univ, MS, 56; Univ Mo, Columbia, PhD(zool), 65. *Prof Exp:* Lectr zool, Ranchi Univ, India, 56-61; fel anat, Univ Wis, 65; asst prof biol, Wis State Univ, Wis Claire, 65-67; sr scientist, Univ Minn, Minneapolis, 67-69, assoc prof zool & vet anat, 69-76, assoc prof genetics & cell biol, 76-81; RES PHYSIOLOGIST, RES SERV, VET ADMIN HOSP, 69-; PROF GENETICS & CELL BIOL, UNIV MINN, ST PAUL, 81- *Mem:* Am Asn Can Res; Am Asn Anat; Soc Study Reproduction; Am Soc Cell Biol; Indian Sci Cong Asn; Int Soc Differentiation. *Res:* Reproductive physiology; prostate biology and cancer of prostate including differentiation and regulation; tissue culture; immunocytochemistry; autoradiography; endocrinology. *Mailing Add:* Bldg 70, Res Serv 151 Veterans Affairs Med Ctr One Veterans Dr Minneapolis MN 55417

SINHA, AKHOURI SURESH CHANDRA, b Churamanpur, India, Mar 14, 38; m 68. ELECTRICAL ENGINEERING. *Educ:* Bihar Univ, BS, 57; Banaras Hindu Univ, BS, 61; Univ Mo-Columbia, MS, 66, PhD(elec eng), 69. *Prof Exp:* Asst prof elec eng, Ind Inst Technol, 69-77; assoc prof, 77-80, PROF ELEC ENG & CHMN DIV ENG, IND UNIV-PURDUE UNIV, 80- *Res:* Control systems; stability theory; optimal control theory. *Mailing Add:* Div Eng Ind Univ-Purdue Univ Indianapolis IN 46205

SINHA, AKHUARY KRISHNA, b Churamanpur, India, Jan 5, 41; m 70. GEOLOGY. *Educ:* Sci Col, Patna, BSc, 60; Patna Univ, MSc, 62; Univ Calif, Santa Barbara, PhD(geol), 69. *Prof Exp:* Lectr geol, Patna Univ, 63-65; Carnegie Inst fel, Dept Terrestrial Magnetism, Washington, DC, 69-71; asst prof, 71-76, ASSOC PROF GEOL, VA POLYTECH INST & STATE UNIV, 76- *Concurrent Pos:* NSF res grant, Va Polytech Inst & State Univ, 71- *Honors & Awards:* Cottrell Award, Res Corp, USA, 72. *Mem:* AAAS; Am Geophys Union; Geochem Soc; Geol Soc Am. *Res:* Common lead and strontium systematics, geochronology; trace element geochemistry; isotope geology; regional tectonics. *Mailing Add:* Dept Geol & Geochronology Va Polytech Inst & State Univ Blacksburg VA 24061

SINHA, ARABINDA KUMAR, b Kasiadanga, India, Mar 1, 33; m 62; c 1. PHYSIOLOGY, NEUROBIOLOGY. *Educ:* Calcutta Univ, BSc, 56, MSc, 61; Univ Calif, San Francisco, PhD(physiol), 69. *Prof Exp:* Demonstr & lectr physiol, Presidency Col, Calcutta, 61-62; actg asst prof, Univ Calif, Berkeley, 68; assoc physiol, Univ Calif, San Francisco, 68-69; asst prof, 72-78, ASSOC PROF PHYSIOL, ROBERT WOOD JOHNSON MED SCH, UNIV MED & DENT NJ, 78- *Concurrent Pos:* Mem grad fac, Rutgers Univ, 72- *Mem:* Am Physiol Soc; Soc Neurosci. *Res:* Cerebra circulation. *Mailing Add:* Robert Wood Johnson Med Sch Dept Physiol Univ of Med & Dent of NJ Piscataway NJ 08854-5035

SINHA, ASRU KUMAR, b Tamluk, India, Aug 10, 43; m 69; c 1. HORMONAL REGULATION, THROMBOTIC DISORDERS. *Educ:* City Col Calcutta, BSC, 61; Univ Col Sci, Calcutta, MSC, 63; Calcutta Univ, DSc, 70. *Prof Exp:* Res assoc microbiol, Miami Univ, Oxford, Ohio, 69-72, Med Ctr, Kans Univ, 72-74; res investr med, Univ Pa, 74-78; ASST PROF, THROMBOSIS CTR, TEMPLE UNIV, 78- *Mem:* Am Soc Cell Biol; AAAS; Am Soc Biol Chem. *Res:* Hormonal control of cellular behaviors, particularly by prostaglandins through cyclic nucleotides dependent and independent pathways and the hormonal memories. *Mailing Add:* Albert Einstein Col Med Platelet Res Dept Med Montefiore Med Ctr 111 E 210 St Bronx NY 10467

SINHA, BIDHU BHUSHAN PRASAD, b Mallehpur, India; Can citizen. BIOPHYSICS, HISTORY & PHILOSOPHY OF SCIENCE. *Educ:* Patna Univ, India, MSc, 61; Mem Univ Nfld, MSc, 67; Univ Mass, PhD(nuclear physics), 71. *Prof Exp:* Lectr physics, Patna Univ, 62-64; instr physics & sci, Eastport Cent Sch, Nfld, 66-68; lectr & instr physics, Univ Guelph, Ont, 72-73; dir, Inst Sci & Math, 74-75; prof physics & sci, Fanshawe Col Appl Arts & Technol, Ont, 78-79; DIR & PRES, INST SCI & MATH, 79- *Concurrent Pos:* Lectr tech physics, PATNA Inst Technol, India, 61-64; teaching fel, Physics Dept, Mem Univ Nfld, 64-66; dir sports, Eastport Sch, Nfld, 66-68; fel physics, Univ Toronto, 71-72; vis prof, World Open Univ, Calif & SD, 74-; vis student adminr, Grad Bus Sch, York Univ, 75-77. *Mem:* Am Phys Soc. *Res:* X-rays; molecular physics; nuclear accelerator physics; pioneering theoretical works in faster-than-light relativity; fundamental particles; parapsychophysics; pioneering quantum-relativity theory of life physics & species; faster-than-light physics--100% certainty-relation & cosmic-evolutionary-cycle of nothing-to-things-to-nothing as 4-dimensions-mix & break/up; relativity-theory based clean and cold real (4-D, D equals dimension) - nuclear energy as coupling of space (3-D) and time (1-D) - holes energy of (3/2)m C2-process (m equals any mass) (c equals light-velocity constant) (mistakeable in/name as cold fusion of 1989-public-news). *Mailing Add:* PO Box 892 Sta B London ON N6A 4Z3 Can

SINHA, BIRANDRA KUMAR, b Gaya, India, Jan 10, 45; US citizen; m 70. MEDICINAL CHEMISTRY, BIOPHYSICS. *Educ:* Ohio State Univ, PhD(med chem), 72. *Prof Exp:* Fel biochem, Ohio State Univ, 72-73; NIH fel molecular pharmacol, 74-75; sr investr med chem, Microbiol Assocs, 75-77; sr staff fel biophysics, Nat Inst Environ Health Sci, 77-82; SR INVESTR, NAT CANCER INST, 82- *Mem:* Sigma Xi; Am Asn Cancer Res; Am Soc Pharmacol & Exp Therapeutics. *Res:* Pharmacology of antitumor agents; mechanism of drug resistance; free radicals in toxicity. *Mailing Add:* Environ Biophys Dept Nat Inst Environ Health Sci PO Box 12233 Research Triangle Park NC 27709

SINHA, DIPEN N, b Mosaboni Mines, India, Mar 9, 51; US citizen; m 75; c 1. NONDESTRUCTING EVALUATION USING ULTRASOUND, SOLID STATE SENSORS & DEVICES. *Educ:* St Xavier's Col, BSc, 70; Indian Inst Technol, MSc, 72, DIIT, 73; Portland State Univ, PhD(physics), 80. *Prof Exp:* Staff mem, Rockwell Int, 83-86; postdoctoral fel physics, 80-83, STAFF PHYSICIST, LOS ALAMOS NAT LAB, 86- *Concurrent Pos:* Prin investr, Electronics Res Group, Los Alamos Nat Lab, 87-90, proj leader, 90-, chief scientist, 91- *Honors & Awards:* R&D 100 Award, 90. *Mem:* Am Phys Soc; Sigma Xi. *Res:* Development of a new nondestructive evaluation technique called acoustic resonance spectroscopy; vibrational characteristics of complex objects; effect of ultrasound on heat transfer in fluids. *Mailing Add:* Los Alamos Nat Lab MS D429 PO Box 1663 Los Alamos NM 87545

SINHA, HELEN LUELLA, b Man, Can, July 29, 36; m 63; c 2. QUALITY ASSURANCE, NURSING DOCUMENTATION. *Educ:* Univ Man, BN, 79 & MA, 83. *Prof Exp:* Instr eye, ear, nose, throat nursing, St Boniface Gen Hosp, 63-64; staff nurse, Palliative Care Unit, Munic Hosp, 81; res assoc qual assurance, Univ Man, 83-85; SPEC PROJ QUAL ASSURANCE DOC, ST BONIFACE GEN HOSP, 85- *Mem:* Can Asn Qual Assurance Prof; N Am Nursing Diag Asn; Can Nurses Asn. *Res:* Project to develop, evaluate and implement a new documentation system for nursing and a project to update and expand to all areas of nursing, the quality monitoring system in use in hospital. *Mailing Add:* Nursing Ser St Boniface Gen Hosp 409 Tache Ave Winnipeg MB R2H 2A6 Can

SINHA, INDRANAND, b Bihar, India, July 3, 31; m 52; c 3. MATHEMATICS. *Educ:* Benares Hindu Univ, BSc, 51, MSc, 53; Univ Wis, PhD(algebra), 62. *Prof Exp:* Lectr math, Univ Bihar, 53-63; asst prof, Mich State Univ, 63-65; assoc prof, Indian Inst Technol, Kanpur, 65-68; assoc prof, 68-71, PROF MATH, MICH STATE UNIV, 71- *Res:* Linear algebra, Grour-rings, linear groups. *Mailing Add:* Dept Math Mich State Univ East Lansing MI 48824

SINHA, KUMARES C, b Calcutta, India, July 12, 42; m 67; c 5. TRANSPORTATION ENGINEERING, URBAN SYSTEMS ENGINEERING. *Educ:* Jadavpur Univ, India, BCE, 61; Calcutta Univ, DTRP, 64; Univ Conn, MS, 66, PhD(civil eng), 68. *Prof Exp:* Jr lectr civil eng, Jadavpur Univ, India, 61-62; asst engr, Govt W Bengal, India, 62-64; res asst civil eng, Univ Conn, 64-68; asst prof civil eng, 68-72, assoc prof & dir urban transp prog, Marquette Univ, 72-74; assoc prof, 74-78, assoc dir, Ctr Pub Policy & Pub Admin, 78-80, PROF CIVIL ENG & HEAD TRANSP ENG, PURDUE UNIV, 78- *Concurrent Pos:* Systs eng consult, Southeast Wis Regional Planning Comn, 69-72; mem, Nat Transp Res Bd. *Honors & Awards:* Fred Burggraf Award, Nat Acad Sci, 72; Frank M Masters Award, Am Soc Civil Eng, 86. *Mem:* Am Pub Works Asn; Am Soc Eng Educ; Am Planning Asn; Am Inst Cert Planners; Am Road & Transp Builders Asn; fel Am Soc Civil Eng; fel Inst Transp Eng. *Res:* Transportation systems analysis; urban and regional planning and policy analysis; author or co-author of 200 technical publications. *Mailing Add:* Purdue Univ Sch Civil Eng West Lafayette IN 47907

SINHA, MAHENDRA KUMAR, b Kanpur, India, July 8, 31; m 57; c 2. SURFACE PHYSICS. *Educ:* Agra Univ, BSc, 49, MSc, 52; Pa State Univ, PhD(physics), 61. *Prof Exp:* Lectr physics, Christ Church Col, Kanpur, India, 52-57; res assoc, Pa State Univ, 61-62; sci officer, Atomic Energy Estab, Bombay, India, 62-64; fel physics, radio & elec eng div, Nat Res Coun Can, 64-66; from asst prof to assoc prof, 66-74, actg chmn dept, 77-78, PROF PHYSICS, N DAK STATE UNIV, 74- *Concurrent Pos:* Guest scientist, Max

Planck Inst Plasma Phys, Ger, 75-76. *Mem:* Am Phys Soc. *Res:* Field ion and field ion microscopy; sputtering of solids by medium energy gas ions; high voltage breakdown in vacuum and insulators; entrapment and surface damage of solid due to kiloelectron volt gas ions. *Mailing Add:* Dept Physics NDak State Univ Univ Station Fargo ND 58105

SINHA, NARESH KUMAR, b Gaya, India, July 25, 27; m 51; c 3. ELECTRICAL ENGINEERING. *Educ:* Benares Hindu Univ, BSc, 48; Manchester Univ, PhD(elec eng), 55. *Prof Exp:* From asst prof to assoc prof elec eng, Bihar Inst Technol, India, 50-61; asst res scientist, NY Univ, 61; assoc prof elec eng, Univ Tenn, 61-65; assoc prof, 65-71, chmn dept elec & comput eng, 82-88, PROF ELEC ENG, MCMASTER UNIV, 71-, DIR INSTRNL COMPUT, FAC ENG, 88- *Concurrent Pos:* Res contract, NASA, Ala, 64-65, Dept Commun, Ottawa, 73-76 & 77-78; Nat Res Coun Can res grant, 66- *Mem:* Sr mem Inst Elec & Electronics Engrs; Eng Inst Can; fel Inst Elec Engrs; Can Soc Elec Eng. *Res:* Optimum nonlinear filtering of random signals embedded in noise; adaptive and learning control systems; optimal control theory; sensitivity of systems to variations in parameters; application of microcomputer to process control. *Mailing Add:* Dept Elec Eng McMaster Univ 1280 Main St W Hamilton ON L8S 4L7 Can

SINHA, NAVIN KUMAR, b Patna, India, Oct 14, 45; m 71; c 2. MOLECULAR BIOLOGY. *Educ:* Patna Univ, BSc, 62, MSc, 64; Univ Minn, PhD(genetics), 72. *Prof Exp:* Res assoc biol, Mass Inst Technol, 72-73; fel biochem, Princeton Univ, 73-76; asst prof, 76-82, ASSOC PROF MICROBIOL, RUTGERS UNIV, 82- *Concurrent Pos:* NIH fel, Nat Cancer Inst, 74-76; prin investr, Nat Inst Gen Med Sci, 77-; vis scientist, Max-Planck Inst, Cologne, WGermany, 84-85; Alexander von Humboldt fel, 84-85. *Mem:* Am Soc Microbiol; Genetics Soc Am. *Res:* Mechanism of DNA replication; DNA protein interaction; molecular mechanisms of mutation. *Mailing Add:* Waksman Inst of Microbiol Rutgers Univ Piscataway NJ 08855

SINHA, OM PRAKASH, b Faizabad, India; US citizen; m 65; c 1. THEORETICAL SOLID STATE PHYSICS. *Educ:* Allahabad Univ, India, BS, 46, MS, 48; Wayne State Univ, MA, 65; Yeshiva Univ, NY, PhD(physics), 69. *Prof Exp:* Lectr physics, Agra Univ, India, 50-58; instr physics, Marygrove Col, Detroit, 63-65; res physicist, Energy Conversion Devices Inc, Troy, Mich, 69-71; fel, Simon Fraser Univ & Nat Res Coun Can, 71-72; from asst prof to assoc prof, Clark Col, 73-89; ASSOC PROF PHYSICS, CLARK-ATLANTA UNIV, 89- *Concurrent Pos:* Cottrell Res Grant, Res Corp, 75; NASA Grant, 77-79. *Mem:* Am Asn Physics Teachers; Am Phys Soc; Nat Inst Sci. *Res:* Physics of semiconductors; charge and mass transport in solids; transport and relaxation processes in liquids; chemical physics. *Mailing Add:* 2115 Coosawatte Dr NE Apt I-5 Atlanta GA 30319

SINHA, RAJ P, b Pahsara, India, Nov 11, 34; Can citizen; m 58; c 3. MICROBIAL GENETICS, BIOCHEMICAL GENETICS. *Educ:* Bihar Univ, India, BSc, 57; Univ Wyo, Laramie, MS, 62; Univ Man, Winnipeg, PhD(genetics), 67. *Prof Exp:* Agr supvr, Dept Agr, Govt Bihar, India, 57-60; fel, Dept Biol, Carleton Univ, Ottawa, Ont, 68-72; RES SCIENTIST MICROBIAL GENETICS, CAN AGR, FOOD RES INST, CENT EXP FARM, OTTAWA, 72- *Mem:* Am Soc Microbiol; Can Soc Food Technol; Can Soc Microbiol; Am Dairy Sci Asn; NY Acad Sci. *Res:* Genetic organization and function in lactic acid bacteria; genetic recombination and repair mechanism; DNA replication, phage-host relation, modifications and restrictions; plasmid genetics; recombinant DNA; genetic engineering. *Mailing Add:* Food Res Inst Cent Exp Farm Ottawa ON K1A 0C6 Can

SINHA, RAMESH CHANDRA, b Bareilly, India, Feb 10, 34; Can citizen; m 57; c 2. PLANT PATHOLOGY. *Educ:* Agr Univ, India, BSc, 53; Lucknow Univ, India, MSc, 56; London Univ, Eng, PhD(plant virol), 60, DSc(plant virol & mycoplasma), 74. *Prof Exp:* Exp officer plant virol, Rothamsted Exp Sta, Eng, 59-60; res assoc plant virol, Univ Ill, Urbana, 60-65; res scientist, 65-76, PRIN RES SCIENTIST PLANT VIROL & MYCOPLASMA, AGR CAN, OTTAWA, 76- *Mem:* Can Phytopath Soc; Am Phytopath Soc; Indian Phytopath Soc; Int Orgn Mycoplasmologists; fel Royal Soc Can. *Res:* Developed serological methods for rapid diagnosis of diseases in plants caused by non-helical, non-culturable mycoplasmas. *Mailing Add:* Plant Res Ctr Res Br Agr Can Ottawa ON K1A 0C6 Can

SINHA, RANENDRA NATH, b Calcutta, India, Jan 25, 30; Can citizen; m 63; c 2. INSECT ECOLOGY, ACAROLOGY. *Educ:* Univ Calcutta, BSc, 50; Univ Kans, PhD(entom, zool), 56. *Prof Exp:* Instr biol, St Xavier's Col, India, 51-52; res assoc zool, McGill Univ, 56-57; res scientist, 57-69, sr res scientist, 70-76, PRIN RES SCIENTIST, RES STA, CAN DEPT AGR, 76- *Concurrent Pos:* Nat Res Coun Can fel, 56-57; hon prof fac grad studies, Univ Manitoba, 61-; hon lectr entom, Kyoto Univ, 66-67. *Honors & Awards:* Gold Medal Award, Entom Soc Can, 85. *Mem:* Entom Soc Am; Japanese Soc Pop Ecol; Sigma Xi; Ecol Soc Am. *Res:* Ecology of stored grain and its products; ecosystem analysis by multivariate statistics; stored-product entomology and acarology; insect resistance to cereal varieties; arthropod-fungus interrelations. *Mailing Add:* Res Sta Can Dept Agr 195 Dafoe Rd Winnipeg MB R3T 2M9 Can

SINHA, SHOME NATH, b Dhanbad, India, Sept 15, 52. SYNTHESIS & PROCESSING OF MATERIALS BY DESIGN, MICROSTRUCTURE-PROPERTY-PROCESSING RELATIONSHIPS. *Educ:* BIT, Sindri, India, BScEngg, 73; Indian Inst Technol, Bombay, India, MScEngg, 77; Univ Utah, Salt Lake City, PhD(metall), 84. *Prof Exp:* Process metallurgist, Alloy Steels Plant, HSL, India, 74-79; postdoctoral appointee, Argonne Nat Lab, Ill, 85-88; ASST PROF METALL MAT PROCESSING MAJORS, UNIV ILL, CHICAGO, 88- *Concurrent Pos:* Adj asst prof, Univ Ill, Chicago, 86-87; fac assoc, Argonne Nat Lab, 89- *Mem:* Mat Res Soc; Am Ceramic Soc; Am Soc Metals. *Res:* Material processing of high temperature superconductors; high performance ceramic; metal composites; innovative synthesis; mechanical alloying; nanometer range particles; inductive plasma scintering; modeling and development of process by design. *Mailing Add:* CEMM Dept Mail Code 246 Chicago IL 60680

SINHA, SNEHESH KUMAR, b Banaras, India; Can citizen; m 56; c 2. APPLIED STATISTICS. *Educ:* Patna Univ, BA, 46, MA, 49 & 54; Univ London, MSc, 59, PhD(statist), 72; Univ Chicago, AM, 68. *Prof Exp:* Lectr math, Jamshedpur Coop Col, Univ Bihar, 53-57; asst prof math & statist, St Mary's Univ, NS, 59-61; from asst prof to assoc prof, 61-72, PROF STATIST, UNIV MAN, 72- *Mem:* Fel Royal Statist Soc; Int Statist Inst. *Res:* Life testing and reliability estimation; bayesian inference; properties of associated distributions in the presence of an outlier observation. *Mailing Add:* Dept Statist Univ Man Winnipeg MB R3T 2N2 Can

SINHA, SUNIL K, b Calcutta, India, Sept 13, 39; m 62; c 2. SOLID STATE PHYSICS. *Educ:* Cambridge Univ, BA, 60, PhD(physics), 64. *Prof Exp:* Vis scientist, Atomic Energy Estab, Trombay, India, 64-65; assoc, 65-66, from asst prof to prof physics, Iowa State Univ, 66-75; sr scientist, Argonne Nat Lab, 75-83; SR RES ASSOC & HEAD, CONDENSED MATTER GROUP, EXXON RES ENG CO, 83- *Concurrent Pos:* Vis fel, Japanese Soc Prom Sci, 77; adj prof physics, Northwestern Univ & Northern Ill Univ, 80-; Guggenheim fel, 82; acting group head, x-ray scattering, Brookhaven Nat Lab, 89-90. *Honors & Awards:* Doe-Bes Mat Sci Award, 82. *Mem:* Am Phys Soc; AAAS; Mat Res Soc; fel Am Phys Soc; Am Crystallog Asn. *Res:* Experimental investigation of magnetic structures and dynamics of magnetic system by neutron scattering; neutron and x-ray studies of fractal structure and dynamics, porous media; colloidal crystals, polymer conformations and phase transitions; synchrotron x-ray studies of surfaces, interfaces, thin films and surface phase transitions; high-temperature superconductivity. *Mailing Add:* CR-SL Exxon Res & Eng Co Rte 22E Annandale NJ 08801

SINHA, VINOD T(ARKESHWAR), b Patna, India, June 10, 41; m 66; c 2. CHEMICAL ENGINEERING. *Educ:* Bombay Univ, BChemEng, 64; Univ Alta, MSc, 64; Univ Calif, Davis, PhD(chem eng), 67. *Prof Exp:* Res chem engr, Am Cyanamid Co, 67-74, proj leader, 77-80, sr res chem engr, 74-82, prin res chem engr, 83-89, ASSOC RES FEL, AM CYANAMID CO, 90- *Mem:* Am Inst Chem Engrs; Am Chem Soc. *Res:* Fluid mechanics and hydrodynamic stability; heat and mass transfer; impact thermoplastics; extrusion, three phase fluidization, cost estimating; polymerization. *Mailing Add:* 17 Diamond Crest Lane Stamford CT 06903

SINHA, YAGYA NAND, b Muzaffarpur, Bihar, India, Oct 21, 36; m 58; c 4. ENDOCRINOLOGY. *Educ:* Bihar Univ, GBVC, 57; Mich State Univ, MS, 64, PhD(physiol), 67. *Prof Exp:* Vet asst surgeon, Bihar Govt, India, 57-59; res asst, Livestock Res Sta, Patna, India, 59-61; grad asst, Mich State Univ, 62-67; res assoc, Cornell Univ, 67-69; asst mem II, Scripps Clin & Res Found, 69-81, radiation safety officer, 82-87, SR MEM & DIR ANIMAL RES, WHITTIER INST DIABETES & ENDOCRINOL, SCRIPPS MEM HOSP, 82- *Concurrent Pos:* Res grants, Nat Inst Health. *Mem:* AAAS; Soc Exp Biol & Med; Endocrine Soc. *Res:* Endocrinology of prolactin and growth hormone. *Mailing Add:* Whittier Inst Diabetes & Endocrinol Scripps Mem Hosp 9894 Genesee Ave La Jolla CA 92037

SINIBALDI, RALPH MICHAEL, b Elmhurst, Ill, Nov 9, 47; div. HEAT SHOCK GENE EXPRESSION, DEVELOPMENTAL GENETICS. *Educ:* Univ Ill, Chicago, BS, 70, MS, 74, PhD(exp biol), 78. *Prof Exp:* Asst biol, 71-74 & 75-78, fel biochem, Med Ctr, 78-80 & res assoc, Univ Ill, Chicago, 80-81; trainee develop biol, Univ Chicago, 81-82; SR RES BIOLOGIST, SANDOZ CROP PROTECTION, 82. *Mem:* Am Soc Cell Biol; Genetics Soc Am; Am Soc Microbiol; Plant Molecular Biol Asn; AAAS. *Res:* Gene expression studies in corn; DNA mediated transformation of corn. *Mailing Add:* Dept Molecular Biol Zoecon Res Inst Sandoz Crop Protection Corp 975 Calif Ave Palo Alto CA 94304

SINIFF, DONALD BLAIR, b Bexley, Ohio, July 7, 35; m 59; c 3. ECOLOGY, BIOMETRY. *Educ:* Mich State Univ, BS, 57, MS, 58; Univ Minn, PhD(entom, fish & wildlife), 67. *Prof Exp:* Biometrician, Alaska Dept Fish & Game, 60-64; res fel, 64-67, assoc prof, 67-75, PROF ECOL, DEPT ECOL-BEHAV BIOL, UNIV MINN, MINNEAPOLIS, 75- *Concurrent Pos:* Prin investr, NSF Off Polar Prog grant, 67-; comnr, Marine Mammal Comn, 75-81. *Mem:* Wildlife Soc; Ecol Soc. *Res:* Vertebrate ecology; statistical and computer applications in field studies; population dynamics large mammals. *Mailing Add:* Dept Ecol & Behav Biol 108 Zool Bldg Univ Minn Minneapolis MN 55455

SINK, DAVID SCOTT, b Findlay, Ohio, July 10, 50; m 73. INDUSTRIAL ENGINEERING. *Educ:* Ohio State Univ, BS, 73, MS, 77, PhD(indust eng), 78. *Prof Exp:* Serv systs engr, Eastman Kodak Corp, 73-75; res assoc, Ohio State Univ, 75-77, instr indust eng, 77-78; ASST PROF INDUST ENG & MGT, OKLA STATE UNIV, 78- *Concurrent Pos:* Res & develop dir, Okla Productivity Ctr, 78-84, res, 84; prin investr, Okla State Univ, 81-84, Va Polytechnic Productivity Ctr, 84- *Honors & Awards:* Halliburton Award, 81; Dow Award, 82. *Mem:* Inst Indust Engrs; Am Soc Eng Educ; Acad Mgt. *Res:* Methods and techniques for measuring and improving productivity in all aspects of organizations. *Mailing Add:* Va Polytechnic Inst Productivity Ctr 567 Whittemore Blacksburg VA 24061-0118

SINK, DONALD WOODFIN, b Salisbury, NC, Nov 10, 37; m 60; c 3. INORGANIC CHEMISTRY. *Educ:* Catawba Col, AB, 59; Univ SC, PhD(inorg chem), 65. *Prof Exp:* Instr chem, Appalachian State Teachers Col, 60-61; asst prof, Lenoir-Rhyne Col, 65-67; asst prof, Northern Mich Univ, 67-68; from asst prof to assoc prof chem, Appalachian State Univ, 68-72, prof chem & sec educ, 73-75, asst dean, 75-86, assoc dean, 86-90, ACTG DEAN COL ARTS & SCI, APPALACHIAN STATE UNIV, 90-, PROF CHEM, 72- *Mem:* Am Chem Soc; Sigma Xi; Nat Teachers Asn. *Res:* Synthesis and study of cis-trans square planar isomers of palladium and platinum complexes; far infrared spectra of square planar complexes of palladium and platinum. *Mailing Add:* 201 SG Greer Hall Appalachian State Univ 210 Hillcrest Boone NC 28608

SINK, JOHN DAVIS, b Homer City, Pa, Dec 19, 34; m 64; c 2. BIOCHEMISTRY, BIOPHYSICS. *Educ:* Pa State Univ, BS, 56, MS, 60, PhD(biochem, animal sci), 62; Univ Pittsburgh, EdD, 86. *Prof Exp:* Admin officer, Pa Dept Agr, 62; asst prof animal sci, Pa State Univ, 62-66, assoc prof meat sci, 66-72, prof, 72-80; prof & chmn animal & vet sci, WVa Univ, 80-85; CHIEF EXEC OFFICER, PA STATE UNIV, 85- *Concurrent Pos:* NSF fel, 64-65; consult, Pa Dept Agr, 62-79; joint staff officer, USDA, 79-80. *Honors & Awards:* Darbaker Prize, Pa Acad Sci, 67. *Mem:* AAAS; Am Meat Sci Asn (pres, 74-75); Am Chem Soc; Biophys Soc; Am Soc Animal Sci; Am Asn Higher Ed; Soc Res Adminr; Am Asn Univ Adminr; Inst Food Technologists. *Res:* Lipid and steroid biochemistry; muscle biophysics and physiology; higher education; public policy. *Mailing Add:* Pa State Univ PO Box 519 Uniontown PA 15401

SINK, KENNETH C, JR, b Altoona, Pa, Oct 7, 37. CELL GENETICS. *Educ:* Pa State Univ, BS, 59, MS, 61, PhD(genetics, plant breeding), 63. *Prof Exp:* From asst prof to assoc prof, 63-75, PROF, MICH STATE UNIV, 75- *Mem:* Am Soc Hort Sci; Am Genetic Asn. *Res:* Cell and protoplast culture and fusion. *Mailing Add:* Dept Hort Mich State Univ 288 Plant & Soil Sci Bldg East Lansing MI 48824

SINKE, CARL, b Moline, Mich, Oct 15, 28; m 56; c 4. MATHEMATICS. *Educ:* Calvin Col, AB, 49; Purdue Univ, MS, 51, PhD(math), 54. *Prof Exp:* Asst prof, 56-64, chmn dept, 64-74, PROF MATH, CALVIN COL, 64- *Concurrent Pos:* Consult, Hq, Ord Weapons Command, Ill, 56-57 & Off Ord Res, NC, 57-58. *Mem:* Am Math Soc; Math Asn Am. *Res:* Analysis; asymptotic series; operations research; optimization problems. *Mailing Add:* 4511 36th St SE Grand Rapids MI 49508

SINKFORD, JEANNE C, b Washington, DC, Jan 30, 33; m 51; c 3. DENTISTRY, PHYSIOLOGY. *Educ:* Howard Univ, BS, 53, DDS, 58; Northwestern Univ, MS, 62, PhD(physiol), 63. *Hon Degrees:* DSc, Georgetown Univ, 78. *Prof Exp:* Res asst psychol, US Dept Health, Educ & Welfare, 53; instr dent, Col Dent, Howard Univ, 58-60; clin instr, Dent Sch, Northwestern Univ, 63-64; assoc prof & head, Dept Prosthodontics, 64-68, assoc dean, 67-74, PROF PROSTHODONTICS, HOWARD UNIV, 68-, DEAN, COL DENT, 75-, PROF, GRAD SCH ARTS & SCI, DEPT PHYSIOL, 76- *Concurrent Pos:* USPHS gen res & training grant, 65-; consult prosthodont, Freedmen's Hosp, Washington, DC, 64, res & prosthodont, Vet Admin Hosp, 65 & US Army grants, 65-; attend staff, Freedmen's Hosp, Howard Univ Hosp, 64-, Children's Hosp Nat Med Ctr, 75- & DC Gen Hosp, 75-; mem, numerous adv comts, govt agencies & nat socs, 65- *Mem:* Inst Med-Nat Acad Sci; Am Dent Asn; Int Asn Dent Res; Am Col Dent; Int Col Dent; Am Prosthodontic Soc; Sigma Xi; NY Acad Sci. *Res:* Endogenous anti-inflammatory substances; chemical healing agent; cyanoacrylates; gingival retraction agents; hereditary dental defects; oral endocrine effects; neuromuscular problems and temporomandibular joint. *Mailing Add:* Howard Univ Col Dent 600 W St NW Washington DC 20059

SINKOVICS, JOSEPH G, b Budapest, Hungary, June 17, 24; nat US; m 54; c 2. INFECTIOUS DISEASES, MEDICAL ONCOLOGY. *Educ:* Peter Pazmany Univ, Budapest, MD, 48; Am Bd Med Microbiol, dipl, 62; Am Bd Internal Med, dipl, 65, recert, 80, cert infectious dis, 72, cert med oncol, 77. *Prof Exp:* Adj prof virol, Eotvos Lorant Univ, Budapest, 49-53; sr investr, State Inst Pub Health, Hungary, 54-56; intern & resident internal med, Cook County Hosp, Chicago, 58-62; from asst prof to prof oncolhemat, M D Anderson Hosp, Univ Tex, Houston, 59-80; prof virol, Baylor Col Med, Tex, 80-83; PROF MED, DEPT MED & MED MICROBIOL, MED COL, UNIV SFLA, 83-; DIR, CANCER INST, ST JOSEPH'S HOSP, TAMPA, FLA, 83- *Concurrent Pos:* Rockefeller fel, Inst Microbiol, Rutgers Univ, 57; Am Cancer Soc fel, Univ Tex M D Anderson Hosp & Tumor Inst, 59, consult prof oncol, 80-; specialist in lab diag, Univ Budapest, 54; prin investr, USPHS res grants & Nat Cancer Inst grants & contracts, NIH, 62-78; vis prof virol, Baylor Col Med, Tex, 83-89; consult oncol, Bay Pines Vet Admin Hosp, St Petersburg, Fla, 83-90; mem, Nat Adv Coun Allergy Infectious Dis, NIH, Bethesda, 84-88. *Mem:* AMA; Am Asn Cancer Res; Am Soc Microbiol; Am Soc Clin Oncol; Infectious Dis Soc Am. *Res:* Virology; tumor immunology; immunotherapy of human tumors; cytotoxic lymphocytes; viral oncolysates; chemoimmunotherapy of sarcomas; established tissue cultures of human tumors; chemotherapy of human tumors; infectious diseases; septicemias in cancer patients. *Mailing Add:* Cancer Inst PO Box 4227 St Joseph's Hosp 3001 W Dr Martin Luther King Jr Blvd Tampa FL 33677-4227

SINKS, LUCIUS FREDERICK, b Newburyport, Mass, Mar 14, 31; m 56, 85; c 4. BIOPHYSICS, PEDIATRICS. *Educ:* Yale Univ, BS, 53; Jefferson Med Col, MD, 57; Ohio State Univ, MMSc, 63. *Prof Exp:* Assoc cancer res pediatrician, Roswell Park Mem Inst, 66-67, chief cancer res pediatrician, 67-76; prof pediat, Georgetown Univ, 76-81, chief, Div Pediat & Adolescent Oncol/Hemat, Vincent T Lombardi Cancer Res Ctr, 76-81; prof pediat, Tufts Univ, 81-87; CHIEF, CANCER CTR BR, DIV CANCER PREV & CONTROL, NAT CANCER INST. *Concurrent Pos:* Nat Cancer Inst spec fel, Cambridge, Eng, 64-66; Nat Cancer Inst grant, Roswell Park Mem Inst, 66-74; Nat Cancer Inst Advan Clin Oncol Training Prog grant, 68-78; from asst res prof to assoc res prof pediat, State Univ NY, Buffalo, 66-69, res prof pediat, 69-; chief, Div Pediat & Adolescent Oncol & Hemat, New Eng Med Ctr Hosps, 81- *Mem:* Am Asn Cancer Res; Soc Pediat Res; Am Soc Clin Oncol. *Res:* Pediatric oncology; cell biology. *Mailing Add:* Div Cancer Prev & Control Nat Cancer Inst Exec Plaza N Rm 308 Bethesda MD 20892-4200

SINKULA, ANTHONY ARTHUR, b Laona, Wis, Jan 2, 38; m 63; c 2. PHARMACEUTICAL & MEDICINAL CHEMISTRY. *Educ:* Univ Wis, BS, 59; Ohio State Univ, MS, 61, PhD(med chem), 63; Western Mich Univ, MBA, 66. *Prof Exp:* Res assoc pharm, Upjohn Co, 63-68, sr res scientist, 68-76, res head, 76-78, mgr, Res Prog Planning, 78-83, dir, Pharm Res & Drug Del Syst Res, 83-88, dir, Pharmaceut Res Lab, 88-90, DIR, RES PLANNING, UPJOHN CO, 90- *Concurrent Pos:* Fel, NIMH, 60-63. *Honors & Awards:* W E Upjohn Award. *Mem:* Am Pharmaceut Asn; Acad Pharmaceut Sci; Am Asn Pharmaceut Scientists. *Res:* Chemical modification of drugs; drug formulation research; pro-drug chemistry research; research administration. *Mailing Add:* The Upjohn Co 301 Henrietta St Kalamazoo MI 49001

SINNER, DONALD H, b Loup City, Nebr, Aug 26, 37. REGULATIONS & COMPLIANCE, PRODUCT AND PROCESS DEVELOPMENT. *Educ:* Univ Nebr, BsChE, 59; WVa Univ, MSChE, 68. *Prof Exp:* Develop engr chem & plastics, Union Carbide, 59-69, prod engr, Silicones Div, 69-75; dir res & develop, Chem Eng Res & Develop Ctr, Ecolabs, 75-84; asst vpres res & develop, Premier Indust Corp, 84-89; sr processing engr, Sigma Group M K Ferguson, 89-90; MGR REGULATORY AFFAIRS, GO JO INDUSTS INC, 90- *Concurrent Pos:* Chmn local sect, continuing educ comt, Am Inst Chem Eng. *Mem:* Am Inst Chem Engr; Soc Cosmetic Chemists. *Res:* Chemical engineering processes and developments; production processing, safety and compliance; pilot unit design, installation and operation; technical management and administration; engineering research & development, design engineering, cosmetics. *Mailing Add:* 6923 Donna Rae Dr Seven Hills OH 44131

SINNETT, CARL E(ARL), b Wilkinsburg, Pa, Aug 26, 22; m 48; c 3. PROCESS DESIGN & ECONOMIC EVALUATION, PETROLEUM REFINING & SYNTHETIC FUELS DEVELOPMENT. *Educ:* Carnegie Inst Technol, BS, 46, MS, 47. *Prof Exp:* Asst tech man, B F Goodrich Co, Ohio, 46; chem engr, Standard Oil Co, Ind, 47-51; proj engr, Econ & Comput Sci Div, Gulf Res & Develop Co, 51-68, sr proj engr, Corp Res & Chem Div, Gulf Sci & Technol Co, 68-78, sr proj engr, Chem & Minerals Div, Gulf Res & Develop Co, 78-83. *Mem:* Am Inst Chem Engrs. *Res:* Process design and economic evaluation of petroleum refining and energy conversion processes; developed process schemes and economics for producing various chemicals and chemical feedstocks from coal. *Mailing Add:* 43 Crystal Dr Oakmont PA 15139

SINNHUBER, RUSSELL OTTO, b Detroit, Mich, Apr 28, 17; m 42; c 3. FOOD SCIENCE, TOXICOLOGY. *Educ:* Mich State Univ, BS, 39; Ore State Univ, MS, 41. *Prof Exp:* From asst prof to assoc prof, 43-63, prof food sci & technol, Ore State Univ, 63-82; RETIRED. *Concurrent Pos:* Mem Agr Res Inst agr bd subcomt fish nutrit, Nat Acad Sci-Nat Res Coun. *Honors & Awards:* Conserv Serv Award, US Dept Interior, 66. *Mem:* Fel AAAS; Am Inst Nutrit; Am Oil Chem Soc; Inst Food Technol; Am Chem Soc; Am Asn Cancer Res. *Res:* Lipid chemistry and autoxidation; irradiation of seafoods; fish and shellfish nutrition; fatty acid metabolism; carcinogenesis; mycotoxins. *Mailing Add:* 5852 S Bay Rd Toledo OH 97391

SINNIS, JAMES CONSTANTINE, b Dover, NJ, May 31, 35; m 57; c 3. PLASMA PHYSICS. *Educ:* Stevens Inst Technol, ME, 57, MS, 59, PhD(physics), 63. *Prof Exp:* RES STAFF, PLASMA PHYSICS LAB, PRINCETON UNIV, 63- *Mem:* Am Phys Soc; AAAS. *Res:* Plasma physics research with special interest in development of fusion power via the magnetic confinement approach. *Mailing Add:* Plasma Physics Lab James Forrestal Campus B/Site NEWGUGG 316 Princeton Univ PO Box 451 Princeton NJ 08543

SINNOTT, GEORGE, b St Louis, Mo, Mar 13, 32; m 57; c 3. MOLECULAR PHYSICS. *Educ:* Univ Chicago, AB, 53, MS, 57; Washington Univ, PhD(physics), 64. *Prof Exp:* Physicist, Argonne Nat Lab, 56-58, Lockheed-Palo Alto Labs, Calif, 64-68 & Joint Inst Lab Astrophys, Univ Colo, Boulder, 68-71; chief, Fire Physics & Dynamics Prog, Nat Bur Standards, 72-74; sci policy asst to Congressman Charles Mosher & Timothy Wirth, 74-75; cong liaison officer, 75-78, assoc dir tech eval, 78-90, DIR INT & ACAD AFFAIRS, NAT ENG LAB, NAT INST STANDARDS & TECHNOL, 90- *Mem:* Am Phys Soc; Sigma Xi. *Mailing Add:* Admin Bldg Rm A505 Nat Inst Standards & Technol Gaithersburg MD 20899

SINNOTT, M(AURICE) J(OSEPH), b Detroit, Mich, Jan 19, 16; m 44; c 5. METALLURGICAL ENGINEERING. *Educ:* Univ Mich, BS, 38, MS, 41, ScD(metall eng), 46. *Prof Exp:* Plant metallurgist, Great Lakes Steel Corp, 38-40; res assoc, Eng Res Inst, Univ Mich, 40-43; sr develop engr, Goodyear Aircraft Corp, 43; instr chem & metall eng, Univ Mich, Ann Arbor, 44-46, from asst prof to assoc prof, 46-54, prof chem & metall eng, 54-84, assoc dean eng, 73-82; RETIRED. *Mem:* Am Soc Metals; Am Inst Mining, Metall & Petrol Engrs; Brit Inst Metal; Sigma Xi. *Res:* Metal physics; grain boundary phenomena; x-ray analysis; nucleation; growth. *Mailing Add:* 2115 Woodside Rd Ann Arbor MI 48104-4515

SINOR, LYLE TOLBOT, b Columbus, Ga, May 24, 57; m 79; c 2. BLOOD BANKING. *Educ:* Univ Kans, BS, 80, PhD(immunohemat), 83. *Prof Exp:* ASST DIR RES, COMMUNITY BLOOD CTR OF GREATER KANSAS CITY, MO, 83-, DIR SCI COMPUT & AUTOMATION, 85- *Concurrent Pos:* Consult, Immucor, Inc, 85- *Mem:* Am Asn Pathologists; assoc Am Soc Clin Pathologists; Am Asn Blood Banks; AAAS; NY Acad Sci; Int Soc Blood Transfusion. *Res:* Development of new clinical test procedures; study and characterization of blood group antigens. *Mailing Add:* Immucor Inc 3130 Gateway Dr PO Box 5625 Norcross GA 30091-5625

SINOTTE, LOUIS PAUL, b Haverhill, Mass, June 12, 27; m 55; c 5. PHARMACEUTICAL CHEMISTRY, ORGANIC CHEMISTRY. *Educ:* Mass Col Pharm, BS, 50; Purdue Univ, MS, 52, PhD(pharmaceut chem), 54. *Prof Exp:* Tech asst to dir qual control, 54-55, mgr pharmaceut control, 55-61, DIR QUAL CONTROL, MERCK SHARP & DOHME DIV, MERCK & CO, INC, 61- *Concurrent Pos:* Mem trustee adv comt, Rutgers Col Pharm. *Mem:* Am Chem Soc; Soc Cosmetic Chemists; Am Soc Qual Control; Am Pharmaceut Asn; Pharmaceut Mfrs Asn. *Res:* Pharmaceutical and medicinal research; quality control; manufacturing pharmacy. *Mailing Add:* Pharm Qual Reg Serv PO Box 1203 North Wales PA 19454-0203

SINSHEIMER, JOSEPH EUGENE, b New York, NY, Dec 30, 22; m 54; c 3. MEDICINAL CHEMISTRY, XENOBIOTIC METABOLISM. *Educ:* Univ Mich, BS, 48, MS, 50, PhD(pharmaceut chem), 53. *Hon Degrees:* Dr, Univ Ghent, Belgium, 74. *Prof Exp:* Assoc technologist org chem, Gen Foods Corp, 53-57, proj leader, 57; from asst prof to assoc prof pharmaceut chem, Univ RI, 57-60; assoc prof, 60-68, PROF MED & PHARMACEUT CHEM, COL PHARM, UNIV MICH, ANN ARBOR, 68-, PROF TOXICOL, SCH

PUB HEALTH, 80- *Concurrent Pos:* Consult, Labs Criminal Invest, Univ RI, 58-60; Am Found Pharmaceut Educ Pfeiffer Mem res fel, St Mary's Med Sch, London, 72; mem comt rev, US Pharmacopeia, 75-; fel, Belgium Nat Found Sci Res, Univ Ghent, 78-79. *Mem:* Am Chem Soc; Am Pharmaceut Asn; fel Am Asn Pharmaceut Sci; Am Asn Univ Professors; Environ Mutagen Soc; Soc Toxicol. *Res:* Analytical medicinal chemistry; drug metabolism; natural product and flavor chemistry; toxicology of aliphatic epoxides; genotoxicity of benzidine analogs. *Mailing Add:* Col of Pharm Univ of Mich Ann Arbor MI 48109-1065

SINSHEIMER, ROBERT LOUIS, b Washington, DC, Feb 5, 20; c 3. BIOCHEMISTRY, BIOPHYSICS. *Educ:* Mass Inst Technol, SB, 41, SM, 42, PhD(biophys), 48. *Hon Degrees:* DSc, St Olaf Col, 74, Northwestern Univ, 76. *Prof Exp:* Res assoc biol, Mass Inst Technol, 48-49; from assoc prof to prof biophys, Iowa State Col, 49-57; prof biophys, Calif Inst Technol, 57-77, chmn div biol, 68-77; chancellor, Univ Calif, Santa Cruz, 77-87; prof, 88-90, EMER PROF BIOL, UNIV CALIF, SANTA BARBARA, 90- *Concurrent Pos:* Chmn ed bd, Proc Nat Acad Sci, 72-80; mem bd sci adv, Jane Coffin Childs Fund Med Res, 73-82. *Honors & Awards:* Beijerinck Virol Medal, Royal Netherlands Acad Sci & Lett, 69. *Mem:* Nat Acad Sci; Nat Inst Med; Am Acad Arts & Sci; Am Soc Biol Chem; Biophys Soc (pres, 70-71). *Res:* Physical and chemical properties of nucleic acids; replication of nucleic acids; bacterial viruses. *Mailing Add:* Dept Biol Sci Univ of Calif Santa Barbara CA 93106

SINSKEY, ANTHONY J, b Highland, Ill, Apr 1, 40; m 69. FOOD SCIENCE, MICROBIOLOGY. *Educ:* Univ Ill, BSc, 62; Mass Inst Technol, ScD(food sci), 66. *Prof Exp:* Fel, Sch Pub Health, Harvard Univ, 66-67; from asst prof to assoc prof microbiol, 67-77, PROF APPL MICROBIOL, MASS INST TECHNOL, 77- *Honors & Awards:* Samuel Cate Prescott Award, 75. *Mem:* Fel Am Acad Microbiol; Am Chem Soc; Inst Food Technol; Brit Soc Appl Bact; Am Soc Microbiol. *Res:* Applied microbiology; single cell protein; recovery and characterization of microorganisms; genetics of industrial microorganism; mommolion cell culture. *Mailing Add:* Dept Biol Mass Inst Technol 77 Massachusetts Ave Cambridge MA 02139

SINSKI, JAMES THOMAS, b Milwaukee, Wis, June 23, 27. MEDICAL MYCOLOGY. *Educ:* Marquette Univ, BS, 47, MS, 52; Purdue Univ, PhD(mycol), 55; Am Bd Med Microbiol, dipl, 69. *Prof Exp:* Instr biol, Spring Hill Col, 55-57; head chem anal lab sugar res, Am Sugar Refinery, New Orleans, 57-58; chief mycol sect, Fort Detrick, Frederick, Md, 60-66; ASSOC PROF MED MYCOL, UNIV ARIZ, 66- *Concurrent Pos:* US Dept Health, Educ & Welfare trainee med mycol, Tulane Univ, 58-60; fac, Sino-Am Advan Workshop Med Mycol, Nanjing, China, 85. *Mem:* Am Soc Microbiol; Med Mycol Soc of the Americas; Int Soc Human & Animal Mycol; fel Am Acad Microbiol. *Res:* Coccidioidomycosis; dermatomycosis. *Mailing Add:* Dept Microbiol Univ Ariz Tucson AZ 85721

SINTES, JORGE LUIS, dentistry, nutritional biochemistry, for more information see previous edition

SINTON, JOHN MAYNARD, b Bozeman, Mont, Apr 12, 46; div; c 2. GEOLOGY, PETROLOGY. *Educ:* Univ Calif, Santa Barbara, AB, 69; Univ Ore, MS, 71; Univ Otago, NZ, PhD(geol), 76. *Prof Exp:* Teaching fel geol, Univ Otago, NZ, 71-75; fel mineral sci, Smithsonian Inst, 76-77; asst prof, 77-81, ASSOC PROF GEOL, UNIV HAWAII, 81- *Mem:* Am Geophys Union. *Res:* Ophiolites; the oceanic crust; Hawaiian volcanic and plutonic rocks. *Mailing Add:* Hawaii Inst Geophysics 2525 Correa Rd Honolulu HI 96822

SINTON, STEVEN WILLIAMS, SOLID STATE PHYSICS. *Educ:* Univ Colo, Boulder, BS, 76; Univ Calif, Berkeley, PhD(chem), 81. *Prof Exp:* SR RES CHEMIST, EXXON PROD RES CO, 81- *Mem:* Am Chem Soc. *Res:* Solid and liquid state nuclear magnetic resonance spectroscopy; multiphoton processes. *Mailing Add:* 750 Stierlin Rd Apt 98 Mountainview CA 94043

SINTON, WILLIAM MERZ, b Baltimore, Md, Apr 11, 25; m 60; c 3. ASTRONOMY. *Educ:* Johns Hopkins Univ, AB, 49, PhD(physics), 53. *Prof Exp:* Asst, Johns Hopkins Univ, 49-53, res staff, 53-54; res assoc, Harvard Univ, 54-56; astrophysicist, Smithsonian Inst, 56-57; astronr, Lowell Observ, Ariz, 57-66; prof astron, Univ Hawaii, 65-90, mem, Inst Astron, 67-90; CONSULT, 90- *Honors & Awards:* Lomb Medal, Optical Soc Am, 54. *Mem:* Optical Soc Am; Am Astron Soc; Int Astron Union; AAAS. *Res:* Infrared spectroscopy; temperatures of planets; infrared spectra of planets and stars; volcanism on Io. *Mailing Add:* 850 E David Dr Flagstaff AZ 86001

SIOMOS, KONSTADINOS, laser spectroscopy, laser technology, for more information see previous edition

SION, EDWARD MICHAEL, b Wichita, Kans, Jan 18, 46; m 68; c 2. STELLAR STRUCTURE & EVOLUTION, STELLAR SPECTROSCOPY. *Educ:* Univ Kans, BA, 68, MA, 69; Univ Pa, PhD(astrophys), 75. *Prof Exp:* Teaching asst, Ohio State Univ, 68-69; res asst, Univ Pa, 69-70, teaching fel, 71-75; asst prof astron, 75-83, assoc prof, 83-88, PROF ASTRON & ASTROPHYS, VILLANOVA UNIV, 88- *Concurrent Pos:* Prin investr, NSF, 78-, NASA, 80-; vis assoc prof physics, Ariz State Univ, 83-85; res assoc astrophys, Ctr Nat Res Sci, Toulouse, France, 90-91; vis scientist astrophys, Hubble Space Telescope Sci Inst, Baltimore, 91- *Mem:* Int Astron Union; Am Astron Soc; fel Royal Astron Soc; Sigma Xi; AAAS. *Res:* Late stages of stellar evolution, white dwarfs, cataclysmic variable stars, symbiotic stars. *Mailing Add:* Dept Astron & Astrophys Villanova PA 19085

SION, MAURICE, b Skopje, Yugoslavia, Oct 17, 28; Can citizen; m 57; c 3. MATHEMATICS. *Educ:* NY Univ, BA, 47, MS, 48; Univ Calif, PhD(math), 51. *Prof Exp:* Asst, Inst Math & Mech, NY Univ, 47-48; asst math, Univ Calif, Berkeley, 48-50, lectr, 50-51; mathematician, Nat Bur Standards, 51-52; mem, Inst Advan Study, 55-57; from asst prof to assoc prof, 60-64, head dept, 84-86 PROF MATH, UNIV BC, 64- *Concurrent Pos:* Instr, Univ Calif, Berkeley, 52-53, asst prof, 57-60; mem, Inst Advan Study, 62; Can Coun fel, Univ Florence & Univ Pisa, 70-71; vis prof, Univ of Strasbourg, France, 74-76, Univ Paris Six, France, 86-87. *Mem:* Am Math Soc. *Res:* Measure theory. *Mailing Add:* Dept Math Univ BC Vancouver BC V6T 1Y4 Can

SIOPES, THOMAS DAVID, b Bremerton, Wash, July 9, 39; m 62; c 2. AVIAN PHYSIOLOGY & BIOCHEMISTRY. *Educ:* Calif State Univ, Sacramento, BA, 64; Univ Calif, Davis, MS, 72, PhD(physiol), 78. *Prof Exp:* From asst prof to assoc prof, 78-89, PROF POULTRY SCI, NC STATE UNIV, 89- *Mem:* Poultry Sci Asn; Worlds Poultry Sci Asn; AAAS. *Res:* Environmental and reproductive physiology of birds: photo-periodism, biological rhythms and thermoregulation. *Mailing Add:* Dept Poultry Sci NC State Univ Raleigh NC 27650-7608

SIOUI, RICHARD HENRY, b Brooklyn, NY, Sept 25, 37; m 62; c 6. CHEMICAL ENGINEERING. *Educ:* Northeastern Univ, BS, 64; Univ Mass, Amherst, MS, 67, PhD(chem eng), 68; Sch Indust Mgt, Worcester Polytech Inst, dipl, 76; Dartmouth Col, Tuck Exec Prog dipl, 86. *Prof Exp:* Sr res engr, Grinding Wheel Div, 68-71, res supvr, 71-78, tech mgr, 78-81, Res Mgr 81-83, dir res, Diamond Prod Div, Norton Co, 83-87, DIR TECHNOL, SUPERABRASIVES DIV, 87- *Mem:* Am Inst Chem Engrs; Am Indian Sci & Eng Soc. *Res:* Development of new products which utilize the superabrasives--diamond and cubic boron nitride--for the grinding of hard materials much as ceramics and tool steels and processes for the manufacture of such products. *Mailing Add:* 18 Avery Heights Dr Holden MA 01520-1033

SIPE, DAVID MICHAEL, b Yokosuka, Japan, Apr 29, 59; US citizen; m 88. FLOW CYTOMETRY, FLUORESCENE. *Educ:* Stanford Univ, BS, 81; Carnegie Mellon Univ, PhD(biol sci & biochem), 90. *Prof Exp:* Immunochemist, Becton-Dickinson Immunocytometry Systs, 81-84; res asst biochem, Dept Biol Sci, Carnegie Mellon Univ, 84-90; POSTDOCTORAL FEL BIOCHEM, DEPT PATH, UNIV UTAH, 90- *Mem:* Am Soc Cell Biol; Sigma Xi. *Res:* Physiology and biochemistry of iron metabolism in yeast and mammalian cells; receptor-mediated endocytosis; flow cytometry; fluorescence. *Mailing Add:* Dept Path Sch Med Univ Utah 50 N Medical Dr Salt Lake City UT 84132

SIPE, HARRY CRAIG, b Flushing, Ohio, Nov 22, 17; m 46, 70. SCIENCE EDUCATION. *Educ:* Bethany Col, WVa, AB, 37; Univ Va, MA, 38; Peabody Col, PhD(sci educ), 52. *Prof Exp:* Teacher high sch, Va, 38-41; develop chemist, O'Sullivan Rubber Co, 41-42; instr physics, Bethany Col, WVa, 43-44; master, Woodberry Forest Sch, 44-46; asst prof, Mars Hill Col, 46-47; from asst prof to prof, Florence State Col, 47-54; phys scientist, Indian Springs Sch, 54-57; prof physics & sci ed, George Peabody Col, 57-67; prof, 67-82, EMER PROF TECH EDUC, STATE UNIV NY ALBANY, 82- *Concurrent Pos:* Vis prof, Ohio State Univ, 66-67. *Mem:* AAAS; Am Chem Soc; Nat Asn Res Sci Teaching (pres, 66); Nat Sci Teachers Asn; Am Asn Physics Teachers. *Res:* Theory of science education; cognitive development; teacher education. *Mailing Add:* PO Box 7 Glenmont NY 12077-0007

SIPE, HERBERT JAMES, JR, b Lewistown, Pa, Aug 17, 40. PHYSICAL CHEMISTRY. *Educ:* Juniata Col, BS, 62; Univ Wis-Madison, PhD(chem), 69. *Prof Exp:* from asst prof to assoc prof, 68-81, chmn dept, 76-82, PROF CHEM, HAMPDEN-SYDNEY COL, 81- *Concurrent Pos:* NSF-Undergrad Res Partic grants, 72, 73, 77 & 78; NSF-ISEP grants, 73 & 78; vis prof, Univ Ala, 80-81; NSF-RUI grant, 84; NSF-CSIP grant, 85, 87; sabbatical leave, Nat Inst Environ Health Sci, 87-88. *Mem:* AAAS; Am Chem Soc; Sigma Xi; Am Asn Univ Professors. *Res:* Electron paramagnetic resonance spectroscopy of organo-metallic compounds and free radical metabolites of xenobiotic compounds; endor spectroscopy. *Mailing Add:* Dept Chem Hampden-Sydney Col Hampden-Sydney VA 23943

SIPE, JERRY EUGENE, b Hickory, NC, Sept 19, 42; m 63; c 3. BIOCHEMISTRY, MICROBIOLOGY. *Educ:* Lenoir-Rhyne Col, BS, 64; Wake Forest Univ, PhD(biochem), 69 Univ Evansville, AAS, 88. *Prof Exp:* From instr biochem to asst prof, Bowman Gray Sch Med, Wake Forest Univ, 71-74; assoc prof biol, 74-87, PROF BIOL & CHEM ANDERSON UNIV, 88- *Concurrent Pos:* Dir teaching labs, Bowman Gray Sch Med, Wake Forest Univ, 69-74; Fulbright lectr fel biochem, Univ Nairobi, 81-83. *Mem:* AAAS; Am Soc Microbiol. *Res:* Nucleic acid methylation, especially ribosomal RNA methylation; methylation contributions to the growing cell. *Mailing Add:* Dept Biol Anderson Univ Anderson IN 46011

SIPERSTEIN, MARVIN DAVID, b Minneapolis, Minn, Sept 21, 25; m 52; c 3. BIOCHEMISTRY, INTERNAL MEDICINE. *Educ:* Univ Minn, BS, 46, MB, 47, MD, 48; Univ Calif, PhD(physiol), 53. *Prof Exp:* From asst prof to prof internal med, Univ Tex Health Sci Ctr Dallas, 64-73; PROF MED, UNIV CALIF, SAN FRANCISCO, 73-; CHIEF METAB SECT, VET ADMIN HOSP, 73- *Concurrent Pos:* NIH res career award, 61-73. *Honors & Awards:* Marchman Award, 59; Lilly Award, Am Diabetes Asn, 59. *Mem:* AAAS; Am Soc Biol Chem; Am Soc Clin Invest; Soc Exp Biol & Med; Asn Am Physicians (pres, 79-80). *Res:* Control of isoprene synthesis in normal and cancer cells; diabetes. *Mailing Add:* Vet Admin Hosp 111-F 4150 Clement St San Francisco CA 94121

SIPES, IVAN GLENN, b Tarentum, Pa, July 26, 42. PHARMACOLOGY, TOXICOLOGY. *Educ:* Univ Cincinnati, BS, 65; Univ Pittsburgh, PhD(pharmacol), 69. *Prof Exp:* Staff fel pharmacol, Nat Heart, Lung & Blood Inst, 69-71, sr staff fel, 71-73; from asst prof to assoc prof, 73-82, PROF & HEAD PHARMACOL & TOXICOL, COL MED, UNIV ARIZ, 82- *Concurrent Pos:* Spec lectr, George Wash Univ, 71-73; assoc ed, Life Sci J, 73-; fel Off Naval Res, 77-, Nat Inst Environ Health Sci, 77-79 & 78-83, Nat Cancer Inst, 77-80 & Int Union Pharmacol, Toxicol Sect; ed, Toxicol: Appl Pharmacol, NAS/NRC Comt Toxicol & Bd Environ Studies & Toxicol. *Mem:* Soc Toxicol; Am Soc Pharmacol & Exp Therapeut; Am Asn Cancer Res; Am Asn Study Liver Dis. *Res:* Role of biotransformation in drug or xenobiotic induced liver injury; organohalogen induced chemical carcinogenesis; pharmacokinetics of xenobiotics; disposition of polychlorinated biphenyls; halothane induced hepatitis. *Mailing Add:* Dept Pharmacol & Toxicol Univ Ariz Col Pharm 1703 E Mabel Tucson AZ 85721

SIPOS, FRANK, b Lucenec, Czech, July 13, 26; div; c 2. PEPTIDE CHEMISTRY. *Educ:* Charles Univ, Prague, Dr nat, 51; Czech Acad Sci, PhD(org chem), 56. *Prof Exp:* Teaching asst pharmaceut chem, Charles Univ, 48-51; scientist, Inst Org Chem & Biochem, Czech Acad Sci, 51-65; sr res chemist, Norwich Pharmacal Co, 66-70; sr res investr, Squibb Inst Med Res, 70-81, CHEM PROCESS TECHNICIAN, E R SQUIBB, 81- *Honors & Awards:* Czech State Prize, 63. *Mem:* Am Chem Soc. *Res:* Reaction mechanism; stereochemistry; synthesis of polypeptides. *Mailing Add:* 2325 Tamalpais Ave El Cerrito CA 94530

SIPOS, TIBOR, b Budapest, Hungary, May 13, 35; US citizen; m 59; c 3. BIOCHEMISTRY, MICROBIOLOGY. *Educ:* Lebanon Valley Col, BS, 64; Lehigh Univ, PhD(biochem), 68. *Prof Exp:* Sr res scientist enzymol, Wallerstein Co, 68-69; sr res scientist enzymol, 69-74, res assoc, Johnson & Johnson Res, 74-77, sr res assoc, 77-78, asst mgr, 78-79, mgr consumer dent care res, 80-87, DIR RES, JOHNSON & JOHNSON DENTAL CARE, CO, 87- *Concurrent Pos:* Adj asst prof med, Col Med & Dent, NJ Med Sch, 74- *Honors & Awards:* Philip B Hofmann Res Scientist Award, 78. *Mem:* AAAS; Am Chem Soc; Am Soc Microbiol; Int Asn Dent Res. *Res:* Application of enzymes in the health care field; mechanism of action of anticoagulants; prevention of dental caries, gingivitis and periodontitis; development of a synthetic anticaries vaccine. *Mailing Add:* 51 Bissell Rd Lebanon NJ 08833-9331

SIPOSS, GEORGE G, b Hungary, Apr 18, 31; m 57; c 4. MECHANICAL ENGINEERING. *Educ:* Can Inst Sci & Technol, dipl advan mech eng, 59; Pepperdine Univ, MBA, 76. *Prof Exp:* PRES & CHIEF EXEC OFFICER, AM ONNI MED INC, 85-; PRES, UNIVERSAL CONSULT DEVELOPMENTS, 89- *Concurrent Pos:* Consult & lectr, var co, univs & insts, 71- *Honors & Awards:* IR-100 Award, 80. *Mem:* Nat Asn Prof Engrs. *Res:* Open heart surgery disposable valves, filters, oxygenators. *Mailing Add:* Universal Consult Developments Inc 2940-H Grace Lane Costa Mesa CA 92626

SIPPEL, THEODORE OTTO, b West Englewood, NJ, Aug 19, 27; m 52; c 2. ANATOMY. *Educ:* Univ Rochester, AB, 48; Yale Univ, PhD(zool), 52. *Prof Exp:* From instr to asst prof biol, Johns Hopkins Univ, 52-57; sr instr anat, Sch Med, Western Reserve Univ, 57-59; from asst prof to assoc prof, 59-69, PROF ANAT, SCH MED, UNIV MICH, ANN ARBOR, 69- *Concurrent Pos:* Nat Eye Inst sr fel anat, Univ Dundee, 71-72; res assoc, Kresge Eye Inst, Mich, 64-65. *Mem:* Asn Res Vision & Ophthal; Am Asn Anat; Histochem Soc. *Res:* Metabolism of the lens; histochemistry. *Mailing Add:* Dept Anat 4643 Med Sci 2 Univ Mich 1301 Catherine Rd Ann Arbor MI 48109-0616

SIPPELL, WILLIAM LLOYD, forest entomology, for more information see previous edition

SIPRESS, JACK M, b Brooklyn, NY, Apr 9, 35; m 56; c 2. LIGHTWAVE COMMUNICATIONS, UNDERSEA COMMUNICATIONS. *Educ:* Polytech Inst Brooklyn, BEE, 56, MEE, 57, DrEE, 61. *Prof Exp:* sr res asst & teaching fel, Microwave Res Inst & Elec Eng Dept, Polytech Inst Brooklyn, 56-58; mem tech staff, AT&T Bell Labs, NJ, 58-61, supvr, Digital Transmission Studies Group, 61-69, head, T2 Digital Line Dept, 69-72, T4M Digital Line Dept, 72-76, Satellite Transmission Dept, 76-79, Lightwave Syst Develop Dept, 78-79, dir, Transmission Technol Lab, 80, DIR, UNDERSEA SYSTS LAB, 80- DIR, RES & DEVELOP & MFG, SUBMARINE SYSTS, AT&T, NJ, 89- *Honors & Awards:* Edwin Howard Armstrong Achievement Award, Inst Elec & Electronic Engrs Commun Soc, 88, Int Commun Award, 91. *Mem:* Fel Inst Elec & Electronics Engrs. *Res:* Light wave and undersea communications; twelve technical papers/talks related to design of active RC networks and digital transmission. *Mailing Add:* AT&T Holmdel NJ 07733

SIPSON, ROGER FREDRICK, b Buffalo, NY, Oct 7, 40; m 67. PHYSICS. *Educ:* Union Col, NY, BS, 62; Syracuse Univ, PhD(physics), 68. *Prof Exp:* Asst prof, 68-73, ASSOC PROF PHYSICS, MOORHEAD STATE UNIV, 73- *Mem:* Am Asn Physics Teachers. *Res:* Quantum theory of fields; mathematical physics. *Mailing Add:* Dept of Physics Moorhead State Univ 11 St S Moorhead MN 56560

SIQUIG, RICHARD ANTHONY, b Gilroy, Calif, Feb 2, 42; m 74. UNDERWATER ACOUSTICS. *Educ:* Calif Inst Technol, BS, 66; Univ Colo, MS, 71, PhD(astrophys), 74. *Prof Exp:* Vis asst prof astron, Ohio State Univ, 74-75; vis asst prof, Univ Wis-Madison, 75-76; res assoc, Hamburg Observ, 76-77; vis scientist, Nat Ctr Atmospheric Res, 78-79; res assoc, Univ Colo, 79; SR TECH ASSOC, OCEAN DATA SYSTS, INC, 80- *Mem:* Am Geophys Union. *Res:* Stability of stellar models during their evolution; solar variability and climate; planetary atmosphere cooling rates. *Mailing Add:* 54 Middle Canyon Rd Carmel Valley CA 93924

SIRAGANIAN, REUBEN PAUL, b Aleppo, Syria, Feb 7, 40; US citizen; m 70; c 2. IMMUNOLOGY. *Educ:* Am Univ Beirut, BS, 59; State Univ NY Downstate Med Ctr, MD, 62; Johns Hopkins Univ, PhD(immunol), 68. *Prof Exp:* Assoc mem immunol, Pub Health Res Inst New York, 68-73; HEAD SECT CLIN IMMUNOL, LAB IMMUNOL, NAT INST DENT RES, NIH, 73- *Concurrent Pos:* Res career develop award, NIH, 70-73, mem immunobiol study sect, 74-78; asst prof med, NY Univ, 71-73. *Mem:* Am Asn Immunol; Am Acad Allergy; AAAS; Am Soc Clin Invest. *Res:* Inflammation; immediate hypersensitivity reactions. *Mailing Add:* Lab Immunol Nat Inst Dent Res NIH Bethesda MD 20892

SIRBASKU, DAVID ANDREW, b St Paul, Minn, Nov 25, 41; m 64; c 2. BIOCHEMISTRY, CELL BIOLOGY. *Educ:* Col St Thomas, BS, 63; Univ Ill, PhD(biochem), 67. *Prof Exp:* NIH fel, Mass Inst Technol, 67-70; NIH spec fel, Univ Calif, San Diego, 70-72; from asst prof to assoc prof, 72-86, PROF BIOCHEM, UNIV TEX MED SCH, HOUSTON, 86- *Concurrent Pos:* Grant reviewer, NIH, Am Chem Soc, NSF & NATO; recipient, fac res award, Am Cancer Soc, 80-85. *Mem:* Am Soc Cell Biol; Endocrine Soc; Am Chem Soc; AAAS; Am Tissue Cult Asn; Am Asn Cancer Res. *Res:* Hormonal control of cell growth; growth promoting serum factors. *Mailing Add:* Dept Biochem Univ Tex Med Sch PO Box 20708 Houston TX 77225

SIRCAR, ANIL KUMER, b Calcutta, India, Jan 1, 28; m 52; c 3. POLYMER CHEMISTRY. *Educ:* Univ Dacca, BS, 48, MS, 49; Univ Calcutta, DPhil(polymer chem), 55. *Prof Exp:* Chemist, Sindri Fertilizers & Chem Ltd, India, 51-52; res asst polymer chem, Indian Asn Cultivation Sci, 52-57; NSF fel emulsion polymerization, Univ Minn, 58; res officer rubber chem, Indian Asn Cultivation Sci, 58-59; sr sci officer, Indian Rubber Mfrs Res Asn, 60; sect mgr rubber lab, Nat Rubber Mfrs Ltd, 60-65; Nat Acad Sci res assoc cotton fiber, Southern Regional Res Lab, USDA, 65-67; res chemist, 67-81, J Huber Corp, sr scientist, 81-83; anal scientist, Engelhardt Corp, 83-84; POLYMER RES SCIENTIST, CTR BASIC & APPL POLYMER RES, UNIV DAYTON, 85- *Concurrent Pos:* Mem bd examrs, Univ Calcutta, 62-65 & Indian Inst Technol, Kharagpur, India, 75. *Mem:* NAm Thermal Anal Soc; Am Chem Soc; Soc Plastics Engrs; SAMPE; Inst Confederation Thermal Anal. *Res:* Physical chemistry of polymers; rubber reinforcement, processing and compounding; thermal and thermomechanical analysis; cotton fiber; polyelectrolytes; polymer blends; material science; vibration clamping. *Mailing Add:* Res Inst Univ Dayton 300 College Park Dayton OH 45469-0001

SIRCAR, ILA, b April 1, 38; nat US; m 63; c 2. SYNTHETIC ORGANIC CHEMISTRY, MEDICINAL RESEARCH. *Educ:* Calcutta Univ, BS, 56, MS, 58, PhD(org chem), 64. *Prof Exp:* Res assoc, Dept Chem, Stevens Inst Technol, Hoboken, NJ, 74-75; res assoc, Dept Physiol, Sch Med, Rutgers Univ, 75-77; scientist, 77-80, SR RES ASSOC, RES DIV, WARNER-LAMBERT CO, 77- *Concurrent Pos:* NIH res fel, Dept Chem, Univ New Orleans, 65-66; Nat Res Coun-Nat Acad Sci fel, USDA, New Orleans, 66-67. *Mem:* Am Chem Soc. *Res:* Synthesis of sugar derivatives for photolabeling of biological receptor sites; cardiovascular agents; synthesis of heterocyclic compounds of medicinal interest, peptides; study of reactions and their mechanism. *Mailing Add:* 2800 Plymouth Rd Ann Arbor MI 48105

SIRCAR, JAGADISH CHANDRA, b Calcutta, India, Dec 1, 35; m 63; c 2. ORGANIC CHEMISTRY, MEDICINAL CHEMISTRY. *Educ:* Univ Calcutta, BSc, 56, MSc, 58, PhD(org chem), 64. *Prof Exp:* Lectr chem, WBengal Jr Educ Serv, 60-64 & Kalyani Agr Univ, India, 64-65; NIH fel org chem, La State Univ, New Orleans, 65-67; res chemist, USDA, New Orleans, 67-69; scientist, Warner-Lambert Co. 69-74, sr scientist, 74-79, res assoc, 79-82, sr res assoc, 82-90, ASSOC RES FEL, WARNER-LAMBERT CO, 90- *Concurrent Pos:* Nat Acad Sci-Nat Res Coun resident res assoc Naval Stores Lab, USDA, Olustee, Fla, 67-69. *Mem:* Am Chem Soc; NY Acad Sci; AAAS; Int Soc Heterocyclic Chem. *Res:* Synthetic organic chemistry; steroids; natural products; photochemistry; medicinal chemistry; reactions and reaction mechanisms; heterocyclic chemistry; chemotherapy. *Mailing Add:* Chem Dept 2800 Plymouth Rd Ann Arbor MI 48105

SIREK, ANNA, b Velke Senkvice, Slovak, Jan 12, 21; Can citizen; m 46; c 4. PHYSIOLOGY, SURGERY. *Educ:* Univ Bratislava, MD, 46; Univ Toronto, MA, 55, PhD, 60. *Prof Exp:* Res assoc, Banting & Best Dept Med Res, Univ Toronto, 54-60, lectr, 60-63, from asst prof to assoc prof, 63-72, dir, Dept Teaching Labs, 75-76, prof, 72-86, EMER PROF PHYSIO, UNIV TORONTO, 86- *Concurrent Pos:* Vis fel surg, Kronprinsessan Lovissas Barnsjukhus, Stockholm, Sweden, 47-50; fel, Hosp Sick Children, Toronto, Ont, 50-54; asst dir dept teaching labs, Univ Toronto, 69-75; vis prof, Dept Physiol, Sackler Sch Med, Univ Tel-Aviv, 78. *Honors & Awards:* Hoechst Centennial Medal, Frankfurt, Ger, 66. *Mem:* Can Fedn Biol Socs; Int Diabetes Fedn; Can Endocrine Soc; Can Diabetic Asn; Soc Exp Biol & Med. *Res:* Metabolic studies in animals deprived of endocrines by experimental surgery and the response of diabetic and Houssay animals to insulin and growth hormone. *Mailing Add:* Dept Physiol Med Sci Bldg Univ Toronto Toronto ON M5S 1A1 Can

SIREK, OTAKAR VICTOR, b Bratislava, Slovak, Dec 1, 21; Can citizen; m 46; c 4. PHYSIOLOGY. *Educ:* Univ Bratislava, MD, 46; Univ Toronto, MA, 51, PhD(physiol), 54. *Prof Exp:* Res assoc, Banting & Best Dept Med Res, 50-57, from asst prof to prof, 57-87, EMER PROF PHYSIOL, FAC MED, UNIV TORONTO, 87- *Concurrent Pos:* Vis fel biochem, Wenner Gren Inst, 47-50; asst biochemist, Hosp Sick Children, Toronto, 55-57; vis scientist, La Rabida Inst, Univ Chicago, 62-63; vis prof, Univ Calif, Los Angeles, 67 & Univ Tel-Aviv, 78. *Honors & Awards:* Hoechst Centennial Medal, Frankfurt, Ger, 66; C H Best Prize, Can Workshops Diabetes, 75. *Mem:* Am Diabetes Asn; Am Physiol Soc; Endocrine Soc; Int Diabetes Fedn; Can Fedn Biol Soc. *Res:* Diabetes mellitus in humans and experimental animals; effect of hormones on vascular connective tissue. *Mailing Add:* Dept Physiol Med Sci Blidg Univ Toronto Toronto ON M5S 1A8 Can

SIRIANNI, JOYCE E, b Niagara Falls, NY, Apr 27, 42. PHYSICAL & DENTAL ANTHROPOLOGY. *Educ:* State Univ NY Buffalo, BA, 65, MA, 67; Univ Wash, PhD(phys anthrop), 74. *Prof Exp:* Asst prof, 72-78, ASSOC PROF PHYS ANTHROP, STATE UNIV NY, BUFFALO, 78- *Mem:* Am Asn Phys Anthropologists; Int Primatol Soc; Am Asn Anatomists; AAAS; Am Asn Dental Res; Am Soc Primatologists; Can Asn Phys Anthropologists. *Res:* Study of normal craniofacial growth and development in old world monkeys; study of the normal range of dental variability seen in old world monkeys. *Mailing Add:* Dept Anthrop Spaulding-Ellicott SUNY Buffalo NY 14261

SIRICA, ALPHONSE EUGENE, b Waterbury, Conn, Jan 16, 44; m; c 2. LIVER CARCINOGENESIS, LIVER CELL CULTURE. *Educ:* St Michaels Col, BA, 65; Fordham Univ, MS, 68; Univ Conn Health Ctr, PhD(biomed sci), 76. *Prof Exp:* Res assoc cancer chemother, Microbiol Assocs Cancer Chemother Res Lab, 69-71; fel trainee chem carcinogenesis, McArdle Lab Cancer Res, 76-79; Asst prof anat & hepatic path, Med Sch, Univ Wis, 79-84; assoc prof, 84-90, PROF, DEPT PATH, MED COL VA, 90- *Concurrent Pos:* Prin investr, NIH Grant, 81-83, 81-84, 84-87, 88-89, 88-90, 91-; mem, Sci Adv Comt Carcinogenesis & Nutrit, Am Cancer Soc, 89-92. *Mem:* Am Asn Cancer Res; Am Soc Cell Biol; Tissue Culture Asn; AAAS; NY Acad Sci; Am Asn Pathologists; Asn Clin Scientists; Am Asn Study Liver Dis. *Res:* Pathobiology of hepatocarcinogenesis and cholangiocarcinogenesis;

pathology of intrahepatic biliary epithelium; bile ductular cell culture; regulation of biliary cell differentiation; growth neoplastic transformation of biliary cells in vivo and in culture. *Mailing Add:* Dept Path Med Col Va Va Commonwealth Univ Box 662 MCV Sta Richmond VA 23298-0662

SIRIGNANO, WILLIAM ALFONSO, b Bronx, NY, Apr 14, 38; m 77; c 3. COMBUSTION. *Educ:* Rensselaer Polytech Inst, BAEng, 59; Princeton Univ, MA, 62, PhD(aeronaut aerospace & mech sci), 64. *Prof Exp:* Mem res staff, Guggenheim Labs, Princeton Univ, 64-67, asst prof mech & aero eng, from assoc prof, to prof, 69; Ladd prof & head mech eng, Carnegie-Mellon Univ, 79-84; DEAN & PROF MECH ENG, SCH ENG, UNIV CALIF, IRVINE, 85- *Concurrent Pos:* Consult, indust & govt, 66-; lectr & consult, aeronaut res & develop, NATO Adv Group, 67, 75 & 80; chmn, Nat & Int Tech Conferences, Acad Adv Coun, Indust Res Inst, 85-88 & Combustion Sci Microgravity Disciplinary Working Group, 87-90; mem, Space Sci Applications Adv Comt, NASA, 85-90 & Comt Microgravity Res, Space Studies Bd, Nat Res Coun, 90-; assoc ed, Combustion Sci & Technol, 69-70; tech ed, J of Heat Transfer, 85-; United Aircraft Res fel, 73-74; treas int orgn & chmn, Eastern Sect, Combustion Inst. *Honors & Awards:* Pendray Aerospace Lit Award, 91. *Mem:* Fel Am Inst Aeronaut & Astronaut; Soc Indust & Appl Math; fel Am Soc Mech Engrs; Soc Automotive Engrs. *Res:* Theoretical, computational and some experimental studies of turbulent reacting flows, spray combustion, fuel-droplet heating and vaporization, ignition, combustion instability, and fire safety; contributed articles to national and international professional journals, also research monographs. *Mailing Add:* Sch Eng Univ Calif-Irvine Irvine CA 92717

SIRIWARDANE, UPALI, b Columbo, Sri Lanka, Mar 16, 50; m 78; c 1. BORON CHEMISTRY, ORGANOMETALLIC CHEMISTRY. *Educ:* Univ Sri Lanka, BSc, 76; Concordia Univ, Can, MSc, 81; Ohio State Univ, Columbus, PhD(chem), 85. *Prof Exp:* Asst analyst, Govt Analyst's Dept, Columbo, Sri Lanka, 76-77; res officer minerals technol, Ceylon Inst Sci & Indust Res, 77-79; res assoc organometallic chem, Southern Methodist Univ, 86-87, staff crystallogr, 87-89; ASST PROF CHEM, LA TECH UNIV, 89- *Mem:* Am Chem Soc; Sigma Xi. *Res:* Synthesis and characterization of air-sensitive organometallic compounds using vacuum line and inert atmosphere techniques; structure determination using x-ray crystallography and multi-nuclear nuclear magnetic resonance spectroscopy; computations and graphics using personal computers; glass blowing. *Mailing Add:* Dept Chem La Tech Univ Ruston LA 71272-0001

SIRKEN, MONROE GILBERT, b New York, NY, Jan 11, 21; m 54; c 2. MATHEMATICAL STATISTICS, APPLIED STATISTICS. *Educ:* Univ Calif, Los Angeles, BA, 46, MA, 47; Univ Wash, PhD(sociol & math statist), 50. *Prof Exp:* Soc Sci Res Coun fel, Univ Calif, Berkeley, 50-51; math statistician & soc sci analyst, US Bur Census, 51-53; actuary, Nat Off Vital Statist, 53-55, chief, Actuarial Anal & Surv Methods Br, Vital Statist Div, 55-63; dir, Div Health Records Statist, 63-67, dir, Off Statist Methods, 67-73, chief math statistician & statist adv, 73-76, assoc dir math statist, 76-80, ASSOC DIR RES & METHODOLOGY, NAT CTR HEALTH STATIST, 80- *Concurrent Pos:* Lectr, Dept Prev Med, Sch Med, Univ Wash, 50; ed collabr, Am Statist Asn, 60-72; adj prof, Sch Pub Health, Univ NC, 68-70, mem fac, 68-; vis prof, Sch Pub Health, Univ Calif, Berkeley, 71; adv consult, Nat Inst Neurol & Communicative Disorders & Stroke, 75- *Mem:* AAAS; fel Am Statist Asn; Int Statist Inst; Pop Asn Am; Am Pub Health Asn. *Res:* Investigation of errors of measurement in population data systems and design of efficient sample surveys. *Mailing Add:* Nat Ctr Health Statist 6525 Belcrest Rd Rm 915 Hyattsville MD 20782

SIRKIN, ALAN N, b Newark, NJ, June 3, 44; m 73; c 1. CONSTRUCTION RELATED ACTIVITIES. *Educ:* Univ Miami, BS, 67; Ga Inst Technol, MS, 69. *Prof Exp:* PRIN ENGR, ALAN SIRKIN CONSULT ENGRS, 69- *Concurrent Pos:* Pres construct, Sirkin Bldg Corp; partner real estate, Sirkin Enterprises; mem, Subsurface Drainage Task Force & Environ Qual Control Bd; reviewer, Urban Planning J. *Mem:* Am Soc Civil Engrs; Nat Asn Homebuilders. *Mailing Add:* One Lincoln Rd Bldg Suite 217 Miami Beach FL 33139

SIRKIN, LESLIE A, b Dover, Del, Sept 18, 33; m 59; c 3. GEOLOGY, PALYNOLOGY. *Educ:* Hamilton Col, BA, 54; Cornell Univ, MS, 57; NY Univ, PhD(geol), 65. *Prof Exp:* From asst prof to assoc prof, 62-72, chmn dept, 67-75, PROF EARTH SCI, ADELPHI UNIV, 72-, CHMN DEPT, 90- *Concurrent Pos:* Consult geol, 68- *Mem:* AAAS; fel Geol Soc Am; Am Asn Stratig Palynologists; Int Asn Quaternary Res. *Res:* Mesozoic, Cenozoic stratigraphy; environmental geology. *Mailing Add:* Dept of Earth Sci Adelphi Univ Garden City NY 11530

SIRLIN, ALBERTO, b Buenos Aires, Arg, Nov 25, 30; m 63; c 2. THEORETICAL PHYSICS. *Educ:* Univ Buenos Aires, Dr(phys & math sci), 53; Cornell Univ, PhD(physics), 58. *Prof Exp:* Res assoc physics, Columbia Univ, 57-59; from asst to assoc prof, 59-68, PROF PHYSICS, NEW YORK UNIV, 68- *Concurrent Pos:* Vis scientist, Europ Orgn Nuclear Res, 60-61 & 67-68; vis prof, Univ Buenos Aires, 62-63; NSF res grant, 70- *Mem:* Am Phys Soc. *Res:* Theoretical particle physics; weak and electromagnetic interactions; unified gauge theories; solitons. *Mailing Add:* Dept Physics NY Univ Four Washington Pl New York NY 10003

SIRLIN, JULIO LEO, b Buenos Aires, Argentina, Dec 18, 26; m 68; c 2. REPRODUCTIVE BIOLOGY, NEUROSCIENCES. *Educ:* Univ Buenos Aires, BSc, 50, DSc, 54. *Prof Exp:* Res fel genetics, Dept Biol, Univ Chile, 51-52; res assoc, Dept Animal Genetics, Univ Endinburgh, 53-59, mem staff, 60-67; assoc prof, 67-73, PROF, DEPT ANAT, MED COL, CORNELL UNIV, 73- *Mem:* Int Cell Res Orgn. *Res:* Fertilizing capacity of human spermatozoa: functional and structural correletes. *Mailing Add:* Dept Anat Cornell Univ Med Col 1300 York Ave New York NY 10021

SIROHI, RAJPAL SINGH, b Utter Pradesh, India, Apr 7, 43; m 72; c 1. OPTICAL METROLOGY. *Educ:* Agra Univ, India, BSc, 62, MSc, 64; Indian Inst Technol, Delhi, PhD(physics), 70. *Prof Exp:* Sci officer instrumentation, Indian Inst Sci, Bangalore, 65-66; lectr physics, Delhi, 70-71, asst prof mech eng, Madras, 71-79, head eng design, 83-85, PROF PHYSICS, INDIAN INST TECHNOL, MADRAS, 79-; ASSOC PROF PHYSICS, ROSE-HULMAN INST TECHNOL, TERRE HAUTE, IND, 85- *Concurrent Pos:* Prin investr, Coun Sci & Indust Res, New Delhi, 73-79, Defense Res Develop Orgn, Delhi, 79-85 & Dept Sci & Technol, New Delhi, 79-; Humboldt fel optics, Fed Inst Phys Technol, Brunswick, WGermany, 74-75 & 77; vis asst prof, Case Western Reserve Univ, 79-80; mem nat comt, Nat Phys Lab, New Delhi, 82-84; consult, Gen Optics, Ltd, Pondicherry, 84-85. *Mem:* Optical Soc Am; Optical Soc India; German Soc Appl Optics. *Res:* Development of techniques in speckle metrology; applications of speckle techniques and hologram interferometry to non-destructive testing; use of holographic elements for testing and measurement. *Mailing Add:* Eng Design Ctr Indian Inst Technol Madras 600036 India

SIROIS, DAVID LEON, b Skowhegan, Maine, Oct 15, 33; m 67; c 3. PLANT PHYSIOLOGY. *Educ:* Univ Maine, BS, 61, MS, 63; Iowa State Univ, PhD(plant physiol), 67. *Prof Exp:* From asst prof to prof plant physiologist, Boyce Thompson Inst Plant Res, 67-90; RETIRED. *Mem:* Plant Growth Regulator Soc Am; Weed Sci Soc Am; Sigma Xi. *Res:* Effects of exogenous chemical compounds on plant growth including herbicides and plant growth regulators with particular interest in chemicals with potential for enhancing crop yields. *Mailing Add:* Three Highgate NE Ithaca NY 14850

SIROIS, PIERRE, b Que, Can, Dec 12, 45; m 73; c 1. IMMUNOPHARMACOLOGY. *Educ:* Univ Laval, Que, BA, 67; Univ Sherbrooke, BSc, 71, MSc, 72, PhD(pharmacol), 75. *Prof Exp:* Fel, Royal Col Surgeons, Eng, 75-77 & Hosp Sick Children, Toronto, 77-78; asst prof, 78-83, assoc prof, 83-87, PROF PHARMACOL, UNIV SHERBROOKE, 87- *Concurrent Pos:* Instr, Regionale L'estrie, Sherbrooke, 69-71 & Col Sherbrooke, 72-74; fel, Imperial Col Sci & Technol, 76-77; MRC Scientist Award, 87-92. *Mem:* Brit Pharmacol Soc; Can Soc Clin Invest; Can Soc Immunologists; French-Can Asn Advan Sci. *Res:* Non-respiratory functions of lungs; mediators of hypersensitivity; leukotrienes. *Mailing Add:* Dept Pharmacol Fac Med Univ Sherbrooke Sherbrooke PQ J1H 5N4 Can

SIROTA, JONAS H, b New York, NY, July 26, 16. EXPERIMENTAL BIOLOGY. *Educ:* Brooklyn Col, BA, 36; NY Univ, MD, 40. *Prof Exp:* Asst prof, 58-76, prof, 76-, EMER PROF, STANFORD UNIV; PVT PRACT RENAL PHYSIOL. *Concurrent Pos:* Clin instr med & nephrology, Stanford Univ, 76- *Mem:* Am Physiol Soc; AMA; Am Col Physicians. *Mailing Add:* 60 N 13th St San Jose CA 95113

SIROTNAK, FRANCIS MICHAEL, b Throop, Pa, Aug 10, 29. MOLECULAR PHARMACOLOGY, GENETICS. *Educ:* Univ Scranton, BS, 50; Univ NH, MS, 52; Univ Md, PhD(microbiol), 54. *Prof Exp:* Asst microbiol, Agr Exp Sta, Univ NH, 51-52; asst, Univ Md, 52-54; bacteriologist, US Army Chem Corps, 56-57; from asst prof to assoc prof, 63-76, PROF BIOL, SLOAN-KETTERING DIV, MED COL, CORNELL UNIV, 76-, MEM, SLOAN KETTERING INST CANCER RES, 75-, LAB HEAD, 74- *Concurrent Pos:* Nat Cancer Inst career develop award, 66; res assoc, Sloan Kettering Inst Cancer Res, 57-59, assoc, 59-66, assoc mem, 66-75, sect head, 67-74; instr, Fairleigh-Dickinson Univ, 58-59; asst prof, Long Island Univ, 61- *Mem:* Am Soc Microbiol; Am Asn Cancer Res; NY Acad Sci. *Res:* Chemotherapy and drug resistance; biochemical genetics; biochemical control mechanisms; genetics of neoplastic transformation. *Mailing Add:* Sloan-Kettering Cancer Ctr 1275 York Ave New York NY 10021

SIROVICH, LAWRENCE, b Brooklyn, NY, Mar 1, 33; m 60; c 2. APPLIED MATHEMATICS. *Educ:* Johns Hopkins Univ, AB, 56, PhD, 60; Brown Univ, MA, 65. *Prof Exp:* Res scientist appl math, Courant Inst Math Sci, NY Univ, 62-63; from asst prof to assoc prof, 63-67, PROF APPL MATH, BROWN UNIV, 67- *Concurrent Pos:* Fulbright fel, Free Univ Brussels, 61-62; prof, Inst Henri Poincare, Univ Paris, 68-69; vis prof, Rockefeller Univ, 71-72, adj prof, 72-; Guggenheim fel, 78-79; ed, Appl Math Sci, Quart Appl Math & Bull Sci Math; assoc managing ed, Soc Indust & Appl Math J. *Mem:* AAAS; Am Phys Soc; Am Math Soc. *Res:* Applied mathematics; kinetic theory of gases; fluid dynamics; biophysics; asymptotic analysis. *Mailing Add:* 37 Riverside Dr New York NY 10023

SIRRIDGE, MARJORIE SPURRIER, b Kingman, Kans, Oct 6, 21; m 44; c 4. HEMATOLOGY, LABORATORY MEDICINE. *Educ:* Kans State Univ, BS, 42; Univ Kans, MD, 44. *Prof Exp:* Pvt pract hemat, 55-71; DOCENT & PROF INTERNAL MED & HEMAT, SCH MED, UNIV MO, KANSAS CITY, 71-, ASST DEAN CURRIC, 85- *Concurrent Pos:* Consult, Providence-St Margaret's Health Ctr, 60-, Children's Mercy Hosp, 83- *Mem:* Am Soc Hemat; AMA. *Res:* Antithrombin III; platelet function studies; monitoring heparin therapy; hypercoagulability; lupus anticoagulant. *Mailing Add:* Sch of Med Univ Mo 2411 Holmes Kansas City MO 64108

SIRY, JOSEPH WILLIAM, b New York, NY, Aug 7, 20; m 44; c 2. MATHEMATICS, PHYSICS. *Educ:* Rutgers Univ, BS, 41; Univ Md, MA, 47, PhD(math), 53. *Prof Exp:* Mem, Actuarial Div, Metrop Life Ins Co, 40-42; physicist, Naval Res Lab, 46-58, actg head theoret anal sect, Rocket-Sonde Res Br, 49-51, head, 51-53, lectr, 53, consult, 53-56, head theory & anal br, Proj Vanguard, 56-58, chmn, Vanguard Working Group on Orbits, 56-59; head theory & anal br, Vanguard Div, 58-59, head theory & planning staff, Beltsville Space Ctr, 59, CHIEF THEORY & ANAL STAFF & DIR TRACKING & DATA SYSTS DIRECTORATE, GODDARD SPACE FLIGHT CTR, NASA, 59- *Concurrent Pos:* Lectr grad math, Univ Md, 53-55; exec secy, Upper Atmosphere Rocket Res Panel, Spec Comt, Int Geophys Year, 54 & rep, US Nat Comt Spec Comt Meeting, Moscow, 58; lectr space technol, Univ Calif, 58; vis lectr, Univ Tex, 58; mem, Equatorial Range Comt, NASA, 59; vis lectr, NY Univ, 60; Am Geophys Union deleg, Gen Assembly Int Union Geodesy & Geophys, Helsinki, 60; mem comt high atmosphere, Int Asn Geomagnetism & Aeronomy. *Mem:* Am Math Soc; Am Astron Soc. *Res:* Chromatic polynomials; topology; compressible fluids; ionosphere; cosmic rays; orbit determination; flight mechanics; astrodynamics; upper atmosphere densities; space research, science and technology. *Mailing Add:* 4438 42 St NW Washington DC 20016

SIS, RAYMOND FRANCIS, b Munden, Kans, July 22, 31; m 53; c 5. VETERINARY ANATOMY. *Educ:* Kans State Univ, BS, 53, DVM & BS, 57; Iowa State Univ, MS, 62, PhD(vet anat), 65. *Prof Exp:* Asst vet clins, Iowa State Univ, 61-62, instr vet anat, 62-64, asst prof vet clin sci, 64-66; assoc prof, 66-68, head dept, 68-83, PROF VET ANAT, TEX A&M UNIV, 68- *Concurrent Pos:* Clin prof, Dent Inst, Univ Tex & prin investr biomed res; bd dir, Int Asn Aquatic Animal Med, 85-87. *Mem:* Am Vet Med Asn; Int Asn Aquatic Animal Med; Am Asn Vet Anat; World Asn Vet Anat. *Res:* Surgical anatomy; radiographic anatomy of the cat; feline anatomy and surgery; salivary glands of the cat; histology of marine fish and marine mammals. *Mailing Add:* Dept Vet Anat Tex A&M Univ College Station TX 77843

SISCOE, GEORGE L, b Lansing, Mich, June 13, 37; m 64. SPACE PHYSICS. *Educ:* Mass Inst Technol, BS, 60, PhD(physics), 64. *Prof Exp:* Res fel physics, Calif Inst Technol, 64-67; asst prof, Mass Inst Technol, 67-71; from assoc prof to prof meteorol, 71-77, PROF ATMOSPHERIC PHYSICS, UNIV CALIF, LOS ANGELES, 77- *Concurrent Pos:* Consult, McDonnell Douglas Corp, Ctr Space Res, Mass Inst Technol, TRW Inc, Jet Propulsion Lab & Boston Col; ed, J Geophys Res. *Mem:* Am Geophys Union; AAAS. *Res:* Space physics, especially the magnetosphere and the solar wind. *Mailing Add:* Dept Atmospheric Sci Univ Calif 405 Hilgard Ave Los Angeles CA 90024

SISCOVICK, DAVID STUART, b Baltimore, Md, Feb 19, 51; m 75; c 3. PREVENTIVE CARDIOLOGY, CLINICAL EPIDEMIOLOGY. *Educ:* Univ Pa, Philadelphia, BA, 71; Univ Md, MD, 76; Univ Wash, Seattle, MPH, 81. *Prof Exp:* Intern internal med, Univ Wash, Seattle, 76-77, resident internal med, 77-79, resident prev med & actg instr med, 79-81; clin asst prof epidemiol, Dept Epidemiol, Sch Pub Health & asst prof med, Dept Med, Sch Med, Univ NC, 81-88, res assoc, Health Serv Res Ctr, 82-88, educ health professions, Sch Med, 85-88; ASSOC PROF MED & EPIDEMIOL, HARBORVIEW MED CTR, 88- *Concurrent Pos:* Fels, US Teaching Sem Epidemiol & Prev Cardiovasc Dis, Am Heart Asn-Nat Heart, Lung & Blood Inst, 78 & Student Work Shop Epidemiol Methods, Soc Epidemiol Res, Nat Cancer Inst; teaching & res scholar, Am Col Physicians, 84-87; assoc ed, J Gen Internal Med, 85-; prev cardiol award, Nat Heart, Lung & Blood Inst, 86-90; reviewer, J Chronic Dis, Annals Internal Med & Health Serv Res; consult, Ctr Dis Control, Behav Epidemiol & Eval, Army Phys Fitness Res Inst, World Health Orgn, Community Health, Vet Admin Health Serv Res; assoc fel, Coun Epidemiol, Am Heart Asn, 86. *Mem:* Am Heart Asn; Soc Prev Cardiol; Am Fedn Clin Res; Soc Res & Educ Primary Internal Med; Am Col Physicians. *Res:* Cardiovascular epidemiology and preventive cardiology health services; physical activity, lipids, alcohol, hypertension, coronary heart disease and sudden cardiac death. *Mailing Add:* Div Gen Int Med Harborview Med Ctr ZA-60 325 Ninth Ave Seattle WA 27514

SISENWINE, SAMUEL FRED, b Philadelphia, Pa, Dec 30, 40; m 62; c 3. DRUG METABOLISM, ORGANIC CHEMISTRY. *Educ:* Philadelphia Col Pharm, BSc, 62; Univ Pa, PhD(chem), 66. *Prof Exp:* Sr chemist, 66-71, group leader, Metab Chem Sect, 71-78, mgr, Drug Diposition Sect, 78-85, corp radiation health safety officer, 71-85, assoc dir, 85-87, DIR, DRUG METAB DIV, WYETH AYERST RES, 87- *Mem:* Am Soc Pharmacol & Exp Therapeut. *Res:* Absorption, distribution, excretion and biotransformation; pharmacokinetics, species differences in drug metabolism. *Mailing Add:* Drug Metabolism Div Wyeth Ayerst Res CN 8000 Princeton NJ 08543-8000

SISK, DUDLEY BYRD, b Lexington, Ky, Jan 9, 38; m 61; c 2. VETERINARY PATHOLOGY. *Educ:* Auburn Univ, DVM, 62; Purdue Univ, MS, 67, PhD(vet physiol), 70. *Prof Exp:* Instr physiol, Purdue Univ, 65-70; asst prof vet physiol, Univ Mo-Columbia, 70-74; pathologist, Dept Vet Sci, Univ Ky, 74-80; PATHOLOGIST, VET DIAG LAB, 80- *Mem:* Am Vet Med Asn; Sigma Xi; Asn Am Vet Med Cols; Am Asn Vet Lab Diagnosticians. *Res:* Livestock diseases. *Mailing Add:* 423 E Washington St PO Box 1389 Thomasville GA 31792

SISKA, PETER EMIL, b Evergreen Park, Ill, Apr 11, 43; m 67; c 2. PHYSICAL CHEMISTRY, CHEMICAL PHYSICS. *Educ:* DePaul Univ, BS, 65; Harvard Univ, AM, 66, PhD(chem), 70. *Prof Exp:* Res assoc chem, James Franck Inst, Univ Chicago, 70-71; asst prof, 71-76, ASSOC PROF CHEM, UNIV PITTSBURGH, 76- *Concurrent Pos:* Alfred P Sloan Found res fel, 75. *Mem:* Am Chem Soc; Am Phys Soc. *Res:* Molecular dynamics of chemical reactions and energy transfer; intermolecular forces; crossed molecular beam studies; model calculations of collision dynamics and electronic structure. *Mailing Add:* Dept of Chem Univ of Pittsburgh Pittsburgh PA 15260

SISKEN, JESSE ERNEST, b Bridgeport, Conn, Dec 7, 30; m 54; c 3. CELL BIOLOGY. *Educ:* Syracuse Univ, AB, 52; Univ Conn, MS, 54; Columbia Univ, PhD(bot, cytochem), 57. *Prof Exp:* Asst bot, Univ Conn, 53-54; res asst bot & cytochem, Columbia Univ, 54-55; res assoc, City of Hope Med Ctr, 57-60, assoc res scientist, 60-66, scientist, 66-67; assoc prof, 67-75, PROF MICROBIOL & IMMUNOL, COL MED, UNIV KY, 75- *Mem:* Am Soc Cell Biol; Int Cell Cycle Soc (pres, 84-86); Soc Anal Cytol; NY Acad Sci; Am Asn Advan Sci; Sigma Xi; Cell Kinetics Soc. *Res:* Nucleic acid and protein synthesis in the mitotic cycle; population kinetics of proliferating cells; regulation of cell metabolism related to cell division; role of calcium in regulation of proliferation and cell division and cell metabolism. *Mailing Add:* Dept Microbiol & Immunol Univ Ky Col Med Lexington KY 40536

SISKIN, MILTON, b Cleveland, Tenn, May 14, 21; m 48. ORAL MEDICINE, ENDODONTICS. *Educ:* Univ Tenn, BA, 42, DDS, 45; Am Acad Oral Med, dipl; Am Bd Endodontics, dipl, 63. *Prof Exp:* From intern to resident, Walter G Zoller Mem Dent Clin, Billings Hosp, Univ Chicago, 46-47; from instr to assoc prof oral med & surg, 47-64, chief div oral med & surg & head dept oral med, 55-58, PROF ORAL MED & SURG, COL DENT, UNIV TENN, MEMPHIS, 64-, LECTR, GRAD SCH ORTHOD & DEPT GEN ANAT & EMBRYOL, 54- *Concurrent Pos:* Consult various orgns & hosps, 51-; Am Dent Asn del, Int Dent Cong, 62; asst ed, J Dent Med, 63; ed, Biol Human Dent Pulp, 73; ed endodontics sect, Clin Dent, 75 & Oral Surg, Oral Med & Oral Path, 75-; dir, Registry Periapical Lesions, Am Asn Endodontists, 65-66; mem rev bd, Am Asn Endodontists & Nat Med AV Ctr; mem bd, Am Bd Endodontics, 63-69 & 71-73; mem exec comt, Am Cancer Soc, 71-72, bd dirs, 71-72; dir, Am Col Stomatologic Surgeons; mem exec comt, Am Inst Oral Sci, 66-; mem, Coun Fed Dent Serv, Am Dent Asn, 71-; Am Dent Europe lectr, 62; lectr, Can Govt & Can Asn Endodontists, 70, Brit Dent Soc, 72, Royal Col Denmark, 74 & Int Dent Congress, 74. *Honors & Awards:* Hinman Medallion, 66. *Mem:* AAAS; fel Am Col Dent; fel Am Acad Oral Path; fel Am Asn Endodont; Am Acad Oral Med. *Res:* Oral diagnosis; pathology. *Mailing Add:* 1835 Union Ave Memphis TN 38104

SISKIND, GREGORY WILLIAM, b New York, NY, Mar 3, 34. IMMUNOLOGY. *Educ:* Cornell Univ, BA, 55; NY Univ, MD, 59. *Prof Exp:* From instr to asst prof med, Med Ctr, NY Univ, 65-69; assoc prof, 69-76, PROF MED, MED SCH, CORNELL UNIV, 76-, HEAD DIV ALLERGY & IMMUNOL, 69- *Concurrent Pos:* Res fel microbiol, Sch Med, Wash Univ, 61-62; fel biol, Harvard Univ, 62-64; fel med, Med Ctr, NY Univ, 64-65. *Mem:* AAAS; Am Asn Immunol; Transplantation Soc; Am Acad Allergy; Am Soc Clin Invest; Am Asn Phys. *Res:* Runting syndrome; immunologic tolerance; heterogeneity of antibody binding affinity and changes in antibody affinity during immunization; antigenic competition; idiopathic thrombocytopenic purpura; ontogeny of B-lymphocyte function; regulation of the immune response by auto-anti-idiotype antibody; IqD. *Mailing Add:* Dept Med Div Allergy & Immunol Cornell Univ Med Col 1300 York Ave New York NY 10021

SISLER, CHARLES CARLETON, b Oklahoma City, Okla, Jan 13, 22; m 57; c 3. CHEMICAL ENGINEERING. *Educ:* Mich State Univ, BS, 59, MS, 54. *Prof Exp:* Plant engr phosphates, 50-53, eng supvr, 53-64, eng mgr, 64-69, res mgr chem intermediates, 69-78, res mgr spec chem, 78-81, MGR PROCESS DEVELOP NUTRIT CHEM, MONSANTO CO, 81- *Mem:* Am Inst Chem Engrs; Am Chem Soc; Sci Res Soc Am. *Res:* Food and feed chemicals. *Mailing Add:* 762 Oak Valley Dr St Louis MO 63131

SISLER, EDWARD C, b Friendsville, Md, Jan 25, 30. PLANT PHYSIOLOGY. *Educ:* Univ Md, BS, 54, MS, 55; NC State Col, PhD, 58. *Prof Exp:* Res assoc, Brookhaven Nat Lab, 58-59; biochemist, Smithsonian Inst, 59-61; from asst prof to assoc prof, 61-85, PROF BIOCHEM, NC STATE UNIV, 85- *Concurrent Pos:* Vis prof, Univ Calif, Davis, 82, Hebrew Univ Jerusalem, Rehouot, Israel, 83-84. *Res:* Role of boron in plants; electron transport; citric acid and glyoxylate cycles in the purple sulfur bacteria; nucleotide phosphates levels as affected by visible radiation; alkaloid metabolism; ethylene action in plants. *Mailing Add:* Dept Biochem NC State Univ Raleigh NC 27695

SISLER, GEORGE C, b Winnipeg, Man, Dec 28, 23; m 48; c 2. PSYCHIATRY. *Educ:* Univ Man, MD, 46; FRCP(C), 55. *Prof Exp:* Resident, Winnipeg Psychopath Hosp, 47-49 & Norton Mem Infirmary, 50-51; resident neurol, Louisville Gen Hosp, 51-52; lectr, Univ Man, 52-54, head dept, 54-75, prof psychiat, 54-89, PSYCHIATRIST, HEALTH SCI CTR, UNIV MAN, 75-, SR SCHOLAR, FAC MED, 89- *Concurrent Pos:* Clin instr, Univ Louisville, 50-52; clin dir, Winnipeg Psychopath Hosp, 52-54; chief psychiat, Winnipeg Gen Hosp, 54-75 & psychiatrist, St Boniface Gen Hosp, 54-90; Sandoz traveling prof, 64. *Mem:* Fel Am Psychiat Asn; Can Med Asn; Can Psychiat Asn. *Res:* Psychopathology of organic brain damage; teaching and learning process in psychiatry. *Mailing Add:* Fac Med Univ Man Winnipeg MB R3E 0W3 Can

SISLER, HARRY HALL, b Ironton, Ohio, Mar 13, 17; m 40; c 4. CHLORAMINATION REACTIONS, HIGH ENERGY FUELS. *Educ:* Ohio State Univ, BSc, 36; Univ Ill, MSc, 37, PhD(chem), 39. *Hon Degrees:* DSc, Adam Mickewiec Univ, Poznan, Poland, 77. *Prof Exp:* Instr chem, Chicago City Col, 39-41; from instr to assoc prof chem, Univ Kans, 41-46; from asst prof to prof, Ohio State Univ, 46-56; prof & chmn, 56-68, dean arts & sci, 68-70, exec vpres, 70-73, dean, Grad Sch, 73-79, DISTINGUISHED SERV PROF CHEM, UNIV FLA, 79- *Concurrent Pos:* Chem consult, W R Grace & Co, 51-75, Koppers Co, 56-59, Batelle Mem Inst, 64-67, Tenn Valley Authority, 67, Martin Marietta, 74-77, Naval Ordinance Lab, Indian Head, Md, 74-77 & Mats Technol, Inc, 82-83; NSF lectr various univs, 55-80; mem chem adv panel, NSF, 59-62 & Oak Ridge Nat Lab, 62-66; Sloan vis prof chem, Harvard Univ, 62-63. *Honors & Awards:* James Flack Norris Award, Am Chem Soc, 79. *Mem:* Am Chem Soc. *Res:* Inorganic nitrogen and phosphorus chemistry; molecular addition compounds; hydrazine, chloramine and triazanium; salt chemistry. *Mailing Add:* Dept Chem Univ Fla Gainesville FL 32611

SISLER, HUGH DELANE, b Friendsville, Md, Nov 4, 22; m 50; c 3. PLANT PATHOLOGY. *Educ:* Univ Md, BS, 49, PhD(bot), 53. *Prof Exp:* Asst, 53-55, from asst prof to assoc prof, 55-64, chmn dept bot, 73-77, PROF PLANT PATH, UNIV MD, COLLEGE PARK, 64- *Concurrent Pos:* Ed, Phytopath, Am Phytopath Soc, 60-63; mem fel rev panel, NIH, 61-; NIH spec fel, State Univ Utrecht, Neth, 66- *Mem:* AAAS; fel Am Phytopath Soc; AAAS; Pesticide Sci Soc Japan. *Res:* Fungicidal action; fungus physiology; viruses. *Mailing Add:* Dept of Bot Univ of Md College Park MD 20742

SISODIA, CHATURBHUJ SINGH, b Delhi, India, Apr 2, 34; m 57; c 2. VETERINARY PHARMACOLOGY, TOXICOLOGY. *Educ:* Agra Univ, BVSc & AH, 58; Mich State Univ, MS, 60; Univ Minn, PhD(vet pharmacol), 64; Am Bd Vet Toxicol, dipl. *Prof Exp:* Res assoc vet pharmacol, Univ Minn, 64; assoc prof, Punjab Agr Univ, India, 65; prof, Col Vet Sci & Animal Husb, Mathura, India, 65-68; from asst prof to assoc prof, 68-75, dept head, 84-90, PROF VET PHYSIOL SCI, UNIV SASK, 75- *Mem:* Am Soc Vet Physiol & Pharmacol; Fel Am Acad Vet Comp Toxicol. *Res:* Pharmacokinetics, drug residues, poisonous plants and mycotoxins. *Mailing Add:* Western Col Vet Med Univ Sask Saskatoon SK S7N 0W0 Can

SISSENWINE, MICHAEL P, b Washington, DC, Feb 16, 47. BIOLOGICAL OCEANOGRAPHY, FISHERIES BIOLOGY. *Educ:* Univ Mass, BS, 69; Univ RI, PhD(oceanog), 75. *Prof Exp:* Res assoc oceanog, Univ RI, 73-75; OPERS RES ANALYST & CHIEF, FISHERIES SYSTS INVEST, NORTHEAST FISHERIES CTR, NAT MARINE FISHERIES SERV, 75-, DEP CHIEF, RESOURCE ASSESSMENT DIV, 80-, CHIEF, FISHERIES ECOL DIV, 85- *Concurrent Pos:* Consult, US Environ Protection Agency, 76-; US mem Demersal Fish Comt, Int Coun Explor Sea, Copenhagen, 77- *Mem:* Am Fisheries Soc; Int Estuarine Res Fedn. *Res:* Fish population dynamics; fisheries management systems; trophic interrelationships in marine ecosystems; biological systems models and simulations. *Mailing Add:* Northeast Fisheries Ctr Nat Marine Fisheries Serv Woods Hole MA 02543

SISSOM, LEIGHTON E(STEN), b Manchester, Tenn, Aug 26, 34; m 53; c 2. MECHANICAL ENGINEERING, ACCIDENT RECONSTRUCTION. *Educ:* Mid Tenn State Col, BS, 56; Tenn Polytech Inst, BSME, 62; Ga Inst Technol, MSME, 64, PhD(mech eng), 65. *Prof Exp:* Draftsman, Westinghouse Elec Corp, 53-57; mech designer, ARO, Inc, 57-58; instr eng sci, Tenn Polytech Inst, 58-61 & mech eng, 61-62; prof mech eng & chmn dept, Tenn Tech Univ, 65-79, dean eng, 79-88, prof mech eng, 88-89; PRES, SISSOM & ASSOCS, INC, 78- *Concurrent Pos:* Eng consult industs, ins co, law firms & govt agencies, 64-; secy-treas, vpres & pres, Tenn Tech Eng Develop Found, Inc, 70-; evaluator, Southern Asn Cols & Schs; chmn, Eng Deans Coun, 84-87; bd dir, Accreditation Bd Eng & Technol, 80-86, Am Soc Eng Educ, 84-87. *Honors & Awards:* Marlowe Award, Am Soc Eng Educ, 88, Scrievner Award, 89. *Mem:* Fel Am Soc Mech Engrs; fel Am Soc Eng Educ (pres, 91-92); Soc Automotive Engrs; Nat Soc Prof Engrs; Nat Acad Forensic Engrs. *Res:* Fluid handling; energy utilization; products liability; accident reconstruction; author or coauthor of over 100 publications, including 4 books. *Mailing Add:* Sisson & Assocs Inc 1151 Shipley Church Rd Cookeville TN 38501-7801

SISSOM, STANLEY LEWIS, b Italy, Tex, June 19, 32; m 60; c 2. INVERTEBRATE ZOOLOGY, AQUATIC ECOLOGY. *Educ:* NTex State Univ, BS, 54, MS, 59; Tex A&M Univ, PhD(zool), 67. *Prof Exp:* Instr biol, NTex State Univ, 59-61, Lamar State Col, 61-63 & Tex A&M Univ, 65-67; asst prof, 67-70, assoc prof, 70-77, PROF BIOL, SOUTHWEST TEX STATE UNIV, 77- *Mem:* Am Micros Soc; Crustacan Soc; Am Soc Zoologists. *Res:* Taxonomy and ecology of phyllopod crustaceans; ecology of temporary ponds; limnology. *Mailing Add:* Dept of Biol Southwest Tex State Univ San Marcos TX 78666

SISSON, DONALD VICTOR, b East Chain, Minn, Apr 18, 34; m 60; c 4. APPLIED STATISTICS. *Educ:* Gustavus Adolphus Col, BS, 56; Iowa State Univ, MS, 58, PhD(entom & statist), 62. *Prof Exp:* Asst prof appl statist, Utah State Univ, 59-60 & 62-65, assoc prof, 65-66; biol statistician, Abbott Labs, Ill, 66; from assoc prof toprof, appl statist & comput sci, 66-87, asst dean, Col Sci, 71-82 & 87-88, asst dean sci, 76-87, head, Appl Statist Dept, 82-87, PROF STATIST & EXP STA STATIST, UTAH STATE UNIV, 76- *Concurrent Pos:* Vis assoc prof statist, NC State Univ, 75-76; statist consult, Nepal, 87. *Mem:* Biomet Soc. *Res:* Biological statistics; design and analysis of experiments. *Mailing Add:* Agr Exp Sta UMC 4810 Utah State Univ Logan UT 84322

SISSON, GEORGE ALLEN, b Minneapolis, Minn, May 11, 20; m 44; c 3. OTOLARYNGOLOGY. *Educ:* Syracuse Univ, AB, 42, MD, 45. *Prof Exp:* Clin instr otolaryngol, State Univ NY Upstate Med Ctr, 51-54, from clin instr to clin asst prof surg, 51-68, from clin asst prof to clin prof otolaryngol, 55-68; PROF OTOLARYNGOL-HEAD & NECK SURG & CHMN DEPT, SCH MED, NORTHWESTERN UNIV, CHICAGO, 68- *Concurrent Pos:* Fel, Head & Neck Serv, Manhattan Eye, Ear & Throat Hosp, NY, 52-53; mem exec comt, Bd Dirs & Exam, Am Bd Otolaryngol; mem adv coun, Head & Neck Cancer Cadre, Nat Inst Neurol Dis & Stroke. *Mem:* Am Cancer Soc; Am Laryngol, Rhinol & Otol Soc; AMA; Am Col Surg; Am Acad Otolaryngol-Head & Neck Surg. *Res:* Cancer of the head and neck. *Mailing Add:* Otolaryngol-Head & Neck Surg Northwestern Univ Med Sch Chicago IL 60611

SISSON, GEORGE MAYNARD, b Boston, Mass, Feb 3, 22; m 52; c 3. PHYSIOLOGY, PHARMACOLOGY. *Educ:* Tufts Col, BS, 43; Univ Rochester, PhD(physiol), 52. *Prof Exp:* Asst, Univ Rochester, 48-49; jr physiologist, Brookhaven Nat Lab, 52; Nat Res Coun fel, Columbia Univ, 52-54; group leader pharmacol res, Am Cyanamid Co, 54-59; asst dir dept pharmacol, US Vitamin & Pharmaceut Corp, 59-61; dir pharmaceut prod info, Mead Johnson Res Ctr, 61-66, dir sci info & regulatory affairs, 66-77, dir drug regulatory affairs, 77-87; RETIRED. *Mem:* AAAS; Drug Info Asn; NY Acad Sci; Am Thoracic Soc; Am Fedn Clin Res. *Res:* Regulatory activities; pharmacodynamics; clinical pharmacology; toxicology; chronic obstructive lung disease; normal and pathological renal function; peripheral and cerebral vascular disease. *Mailing Add:* 4800 Ridge Knoll Dr Evansville IN 47710

SISSON, HARRIET E, b Duluth, Minn, July 2, 16; m 41; c 2. PHARMACY. *Educ:* Univ Minn, BS, 37, MS, 39; Univ Ore, PhD, 78. *Prof Exp:* Pharmacist, Univ Hosps, Univ Minn, 37-41; from instr to asst prof, Ore State Univ, 46-67, assoc prof pharm, 67-86; RETIRED. *Res:* Pharmaceutical chemistry; measurement of pharmacy aptitude. *Mailing Add:* 541 NW 18 Corvallis OR 97330

SISSON, JOSEPH A, b San Diego, Calif, Oct 24, 30; m 59; c 2. PATHOLOGY. *Educ:* San Diego Col, BA, 55; Wash Univ, MD, 60. *Prof Exp:* Res asst biochem, Scripps Clin, La Jolla, Calif, 55-56; intern path, Yale Univ New Haven Med Ctr, Conn, 60-61; resident, Albany Med Ctr, NY, 61-64; from instr to asst prof, Albany Med Col, 63-67; asst prof path, Creighton Univ, 68-69, chmn dept, 68-73, prof, 69-80; prof path, Eastern Va Med Sch, 80-83; RETIRED. *Concurrent Pos:* Fel, Albany New Med Col, 61-63; res grant, 65-67; asst attend pathologist, Albany Med Ctr Hosp, 65-67; attend pathologist, Vet Admin Hosp, Albany, NY, 66-67; dir path, Creighton Mem St Joseph's Hosp, 68-72; consult radiologist, US Vet Admin Hosp, Omaha, 68- *Mem:* Am Asn Path & Bact; Am Soc Exp Path. *Res:* Amino acid and lipid metabolism in pregnancy; biochemical aspects of atherosclerosis and thrombosis. *Mailing Add:* 6801 First St Lovett TX 79416

SISSON, RAY L, b Pueblo, Colo, Apr 24, 34; m 52; c 3. ELECTRICAL ENGINEERING. *Educ:* Univ Colo, BSEE, 60; Colo State Univ, MS, 66; Univ Northern Colo, EdD(voc educ), 73. *Prof Exp:* Asst res engr, Univ Colo Exp Sta, 60; asst engr, Parker & Assoc, Consult Engrs, 61-62; prof eng, 60-63, head dept, 63-69, head electronics & instrumentation, 68-73, dir elec area instr, 68-73, DEAN APPL SCI & ENG TECHNOL, UNIV SOUTHERN COLO, &#. *Concurrent Pos:* Consult, Exec Sys Prob Oriental Language, 79; chmn, Eng Technol Leadership Inst, 84-85; consult, NMex Highlands Univ, 85; consult, Mooreheat State Univ, 85; consult, SUNY, Alfred, 85; consult, SUNY, Farmingdale, 85; prog evaluate, Tech Adv Comt Accreditation Bd Eng Technol, Inst Elec & Electronic Engrs; chmn eng technol coun, Am Soc Eng Educ, 86-88, bd dir, 86-88; chmn, Eng Spectrum Task Force, Am Soc Eng Educ, 88. *Honors & Awards:* James H McGraw Award, Am Soc Eng Educ, 90. *Mem:* Am Soc Eng Educ; Inst Elec & Electronic Engrs. *Res:* Electrical networks; computer science; feedback control; antenna studies research. *Mailing Add:* Univ Southern Colo 2200 N Bonforte Blvd Pueblo CO 81001-4901

SISSON, THOMAS RANDOLPH CLINTON, pediatrics, obstetrics; deceased, see previous edition for last biography

SISTEK, VLADIMIR, b Prague, Czech, Sept 22, 31; m 54; c 3. ANATOMY, SURGERY. *Educ:* Charles Univ, Prague, MD, 56, PhD(surg), 67. *Prof Exp:* Resident surg, Regional Hosp, Most, Czech, 56-59; mem staff, Charles Univ, Prague, 59-63, asst prof surg, 63-68; asst prof anat, 69-73, ASSOC PROF ANAT, UNIV OTTAWA, 73- *Mem:* Can Asn Anat; Can Fedn Biol Socs. *Res:* Surgical anatomy; gastroenterology; medical education. *Mailing Add:* Dept Anat Fac Med Univ Ottawa 451 Smyth Rd Ottawa ON K1H 8M5 Can

SISTERSON, JANET M, b Edinburgh, Scotland, July 7, 40; m 65; c 2. NUCLEAR & MEDICAL PHYSICS. *Educ:* Univ Durham, Eng, BSc, 61; Univ London, DIC & PhD(physics), 65. *Prof Exp:* Basic grade physicist, London Hosp, 65-66; sr physicist, Chelsea Hosp Women, London, 66-68; res fel physics, Cambridge Electronic Accelerator, 68-73; res fel, 73-79, RES ASSOC PHYSICS, CYCLOTRON LAB, HARVARD UNIV, 73- *Mem:* Am Phys Soc; Am Asn Physicist Med; Am Nuclear Soc; Am Women Sci. *Res:* Medical applications of proton beams; proton activation analysis. *Mailing Add:* Cyclotron Lab Harvard Univ Cambridge MA 02138

SISTI, ANTHONY JOSEPH, organic chemistry; deceased, see previous edition for last biography

SISTO, FERNANDO, b Spain, Aug 2, 24; US citizen; m 46; c 3. AEROELASTICITY, TURBOMACHINERY. *Educ:* US Naval Acad, BS, 46; Mass Inst Technol, ScD, 52. *Hon Degrees:* MEng, Stevens Inst Technol, 62. *Prof Exp:* Chief propulsion div, Res Div, Curtiss-Wright Corp, 52-58; assoc prof, 58-59, head dept, 66-79, prof, 59-79, GEORGE M BOND PROF MECH ENG, STEVENS INST TECHNOL, 79- *Concurrent Pos:* Consult, Curtiss-Wright Corp, NJ, 58-60, Gen Elec Co, Ohio, 59-63, Gen Motors, Ind, 66-70, United Technol Corp, Conn, 73-79 & Westinghouse Corp, Philadelphia, 80-84; UNESCO-United Nations Develop Prog consult, Nat Aeronaut Lab, Bangalore, India, 78; lectr, Academic Sinica, China, 81; chair, Int Symposium Airbreathing Engines, Beijing, China, 81 & 85, Int Gas Turbine Conf, London, 82; rep, NSF workshop powerplant dynamics, Nat Aeronaut Lab, Bangalore, India, 88. *Honors & Awards:* Maury prize in physics, 45. *Mem:* Assoc fel Am Inst Aeronaut & Astronaut; fel Am Soc Mech Engrs; Am Soc Eng Educ; Am Physics Soc. *Res:* Aeroelasticity of turbomachines; turbomachinery theory; aerodynamics, unsteady flow; flight propulsion; applied mechanics; energy conversion; fluid dynamics; author of several publications. *Mailing Add:* Dept Mech Eng Stevens Inst Technol Castle Point Hoboken NJ 07030-5991

SISTROM, WILLIAM R, b Los Angeles, Calif, Feb 15, 27; m 52; c 4. MICROBIOLOGY. *Educ:* Harvard Univ, AB, 50; Univ Calif, Berkeley, PhD(microbiol), 54. *Prof Exp:* USPHS fel, 55-57; instr microbiol, Sch Med, NY Univ, 57-58; asst prof biol, Harvard Univ, 58-63; assoc prof, 63-70, chmn, Premedical Adv Comt, 74-91, PROF BIOL, UNIV ORE, 70- *Mem:* Am Soc Microbiol; Brit Soc Gen Microbiol. *Res:* Microbial physiology; bacterial photosynthesis. *Mailing Add:* Dept Biol Univ Ore Eugene OR 97403

SISTRUNK, THOMAS OLLOISE, inorganic chemistry, for more information see previous edition

SISTRUNK, WILLIAM ALLEN, b Mitchell, La, June 29, 19; m 45; c 5. FOOD TECHNOLOGY, BACTERIOLOGY. *Educ:* Southwestern La Inst, BS, 47; Ore State Col, MS, 49, PhD, 59. *Prof Exp:* Mkt specialist, USDA, 49-52; assoc horticulturist, 62-68, prof food sci, Univ Ark, Fayetteville, 68-88; RETIRED; assoc horticulturist, 62-68, PROF HORT FOOD SCI, UNIV ARK, FAYETTEVILLE, 68- *Mem:* Am Soc Hort Sci; Inst Food Technol. *Res:* Biochemistry; new horticultural varieties; plant nutrition effects on processing quality; objective tests for measuring quality of fruits and vegetables. *Mailing Add:* 1307 Success St Carthage TX 76796

SIT, WILLIAM YU, b Hong Kong, Feb 18, 44; m 70. ALGEBRA. *Educ:* Univ Hong Kong, BA, 67; Columbia Univ, MA, 69, PhD(math), 72; City Col City Univ NY, MS, 78. *Prof Exp:* Lectr, 71-72, instr, 72-73, asst prof, 73-80, ASSOC PROF MATH, CITY COL CITY UNIV NEW YORK, 80- *Mem:* Math Asn Am; Am Math Soc. *Res:* Differential algebra; invariants of differential dimension polynominals. *Mailing Add:* Dept of Math City Col of New York Convent Ave & W 138th St New York NY 10031

SITAR, DANIEL SAMUEL, b Thunder Bay, Ont, May 1, 44; m 68; c 2. CLINICAL PHARMACOLOGY, PHARMACOLOGY. *Educ:* Univ Man, BSc, 66, MSc, 68, PhD(pharmacol), 72. *Prof Exp:* Assoc fel, Univ Minn Sch Med, 71-73; lectr, Fac Med, McGill Univ, 73-75, asst prof med, 74-78, asst prof pharmacol, 75-78; teaching fel pharmacol, Fac Med, Univ Man, 68-71; res asst, Div Clin Pharmacol, Montreal Gen Hosp Res Inst, 73-78; from asst prof to assoc prof, 78-87, PROF MED & PHARMACOL, FAC MED, UNIV MAN, 87- *Concurrent Pos:* Res grant, Man Heart Found, 83-; Monat Scholar, McGill Univ, 75-78; sci staff, Health Sci Ctr, Univ Man & Deer Lodge Ctr, Winnipeg, 78-; Rh Inst grant, 80; Rosenstadt prof, Univ Toronto, 89-90. *Mem:* Am Soc Pharmacol & Exp Therapeut; Can Soc Clin Invest; Soc Toxicol Can; Pharmacol Soc Can; Am Soc Clin Pharmacol & Therapeut; fel Gerontol Soc Am,; Can Soc Clin Pharmacol. *Res:* Effects of development and disease on drug disposition and effect in man. *Mailing Add:* Dept Pharmacol & Therapeut Univ Man Winnipeg MB R3E 0W3 Can

SITARAMAN, YEGNASESHAN, b India, Oct 26, 36. ANALYSIS APPROXIMATION. *Educ:* Banaras Univ, MA, 50; Univ Kerala, PhD(math), 67. *Prof Exp:* Chmn, Math Dept & prof, math, Univ Kerala, 75-83; vis prof, Univ Toledo, 83-84; vis prof, math, Univ Louisvile, 84-88; PROF, MATH, KY WESLEYAN COL, 88- *Mem:* Am Math Soc; Math Asn Am; India Math Soc; Math Sci India. *Mailing Add:* 703 Princeton Pkwy #14 Owensboro KY 42301

SITARZ, ANNELIESE LOTTE, b Medellin, Columbia, Aug 31, 28; US citizen. PEDIATRIC HEMATOLOGY, PEDIATRIC ONCOLOGY. *Educ:* Bryn Mawr Col, BA, 50; Col Physicians & Surgeons, Columbia Univ, MD, 54. *Prof Exp:* Asst pediat, Col Physicians & Surgeons, 57-62, instr, 62-64, assoc, 64-68, asst prof, 68-74, assoc prof clin pediat, 74-83, PROF CLIN PEDIAT, COL PHYSICIANS & SURGEONS & CANCER RES CTR, COLUMBIA UNIV, 83-; ATTENDING PEDIATRICIAN, BABIES HOSP, CTR WOMEN & CHILDREN, 83- *Concurrent Pos:* Vis fel, Babies Hosp, Ctr Women & Children, NY, 57-59, asst pediatrician, 59-64, from asst attending pediatrician to assoc attending pediatrician, 64-83; consult, Overlook Hosp, Summit, NJ, 75- *Mem:* Fel Am Acad Pediat; Am Soc Hemat; Am Soc Clin Oncol; Am Asn Cancer Res; Int Soc Hemat; NY Acad Sci. *Res:* Cancer in children; efficacy of chemotherapy in pediatric solid tumors. *Mailing Add:* Babies Hosp Broadway & 167th St New York NY 10032

SITES, JACK WALTER, JR, b Clarksville, Tenn, Aug 6, 51; m 73; c 1. FISH & WILDLIFE SCIENCES. *Educ:* Austin Peay State Univ, BS, 73, MS, 75; Tex A&M Univ, PhD(vertebrate zool), 80. *Prof Exp:* Asst prof biol, zool & ecol, Dept Biol, Tex A&M Univ, 80-82; asst prof biol, zool & genetics, 82-86, ASSOC PROF ZOOL, DEPT ZOOL, BRIGHAM YOUNG UNIV, 86- *Concurrent Pos:* Res zoologist, The Nature Conservancy, 75-76. *Mem:* AAAS; Soc Study Evolution; Soc Syst Zool; Soc Conservation Biol; Am Soc Icthyologists & Herpetologists. *Res:* Mechanisms of chromosomal evolution and speciation, rates of speciation; genetic structure of natural populations of vertebrates; basic ecology; evolutionary biology; systematics. *Mailing Add:* 571 Widb Brigham Young Univ Provo UT 84602

SITES, JAMES RUSSELL, b Manhattan, Kans, Nov 18, 43; m 64; c 3. SOLID STATE ELECTRONICS, LOW TEMPERATURE PHYSICS. *Educ:* Duke Univ, BS, 65; Cornell Univ, MS, 68, PhD(physics), 69. *Prof Exp:* Programmer, Union Carbide Corp, 62-64; NSF fel, 65-69; fel physics, Los Alamos Sci Lab, 69-71; from instr to asst prof, 71-77, ASSOC PROF PHYSICS, COLO STATE UNIV, 77- *Mem:* Am Vacuum Soc; Am Phys Soc. *Res:* Investigation of electronic properties of compound semiconductor surfaces and interfaces; studies of thermal transport mechanisms in solid helium three; development of heterojunction solar cells. *Mailing Add:* Dept Physics Colo State Univ Ft Collins CO 80523

SITKOVSKY, MICHAIL V, b Pervomalsk, USSR, Sept 12, 47; US citizen; m 80. IMMUNOPHARMACOLOGY, MOLECULAR IMMUNOLOGY. *Educ:* Moscow State Univ, MSc, 70, PhD(biophys physiol), 73. *Prof Exp:* Res assoc biochem, Ctr Cancer Res, Mass Inst Technol, 81-83, res scientist, 83-84; SR INVESTR & HEAD, BIOCHEM & IMMUNOPHARMACOL UNIT, LAB IMMUNOL, NAT INST ALLERGY & INFECTIOUS DIS, NIH, 84- *Concurrent Pos:* Assoc ed, J Immunol Am Asn Immunologists, 91-93. *Mem:* Am Asn Immunologists. *Res:* Key proteins, enzymes, and messengers involved in the triggering and effector functions of cytotoxic T-lymphocytes; cytotoxicity and exocytosis; immunomodulating agents. *Mailing Add:* Lab Immunol Bldg 10 Rm 11N-311 Nat Inst Allergy & Infectious Dis-NIH Bethesda MD 20892

SITNEY, LAWRENCE RAYMOND, b Schenectady, NY, Oct 8, 23; m 56; c 2. PHYSICAL CHEMISTRY. *Educ:* Ohio State Univ, BS, 43, MS, 47, PhD(chem), 52. *Prof Exp:* Mem staff, Los Alamos Sci Lab, 52-59; staff scientist, Missiles & Space Div, Lockheed Aircraft Corp, Calif, 59-60; supvr adv nuclear systs appln, Rocketdyne Div, NAm Aviation, Inc, 60-61; tech mgr nuclear studies, Martin Co, Colo, 61-62; staff engr space power studies, 62-66, sr staff engr, 66-68, asst group dir, Advan Vehicle Systs Directorate, 68-70, assoc group dir, 70-75, dir, Advan Energy Systs, 75-81, sr engr plans & syst archit, Aerospace Corp, El Segundo, 82-85; staff mem, Bd Systs, Inc, Torrence, Calif, 85-89; RETIRED. *Mem:* Am Phys Soc. *Res:* Chemical thermodynamics; high temperature chemistry; heat of sublimation of graphite; space power; reusable space transportation systems; advanced energy conversion systems; solar thermal power systems. *Mailing Add:* 824 Los Lovatos Rd Sante Fe NM 87501-1280

SITRIN, MICHAEL DAVID, b Detroit, Mich, June 18, 48. GASTROENTEROLOGY, NUTRITION. *Educ:* Harvard Univ, MD, 74. *Prof Exp:* asst prof, 80-86, ASSOC PROF MED, UNIV CHICAGO, 87- *Res:* Vitamin D; calcium medabsorption. *Mailing Add:* Dept Med Univ Chicago Chicago IL 60637

SITRIN, ROBERT DAVID, b Utica, NY, July 24, 45; m 81; c 4. BIOCHEMISTRY, ANALYTICAL CHEMISTRY. *Educ:* Mass Inst Technol, BS, 67; Harvard Univ, MS, 68, PhD(chem), 72. *Prof Exp:* Fel chem, Woodard Res Inst, Switz, 72-73; assoc sr investr chem, 73-77, sr investr chem, 77-84, asst dir, protein biochem, Smith Kline & French Labs, 84-87; DIR BIOCHEM PROCESS RES & DEVELOP, MERCK SHARP & DOHME RES LABS, 87- *Mem:* Am Chem Soc; AAAS; Soc Induct Microbiol. *Res:* Amino glycoside antibiotics; natural product isolations; structure determination; natural products chemistry; high performance liquid chromatography; analytical biochemistry; instrumental analysis; laboratory automation antibiotic fermentation screen development; process development biopharmaceuticals; protein purification; process development. *Mailing Add:* Res Labs Merck Sharp & Dohme WP16-100 West Point PA 19486

SITTEL, CHESTER NACHAND, b Nashville, Tenn, Sept 2, 41; m 63; c 2. PROJECT MANAGEMENT. *Educ:* Vanderbilt Univ, BE, 63, MS, 66, PhD(chem eng), 69. *Prof Exp:* Res chem engr, Eastman Kodak Co, 69-72, sr res chem engr, 72-79, res assoc, 79, develop assoc, 79-81, res assoc, 81-88, sr develop assoc, 88-89, SR RES ASSOC, EASTMAN CHEM DIV, EASTMAN KODAK CO, 88- *Mem:* Am Inst Chem Engrs; Sigma Xi. *Res:* Development and implementation of process designs from laboratory-scale to commercialization. *Mailing Add:* 220 Brookfield Dr Kingsport TN 37663

SITTEL, KARL, b Frankfurt am Main, Ger, Oct 10, 16; nat US; m 53; c 1. APPLIED PHYSICS. *Educ:* Goethe Univ, Ger, PhD(physics), 42. *Prof Exp:* Asst, Max Planck Inst Biophys, Ger, 39-40; chief, Radiosonde Lab, Aerological Instruments, Marine Observ, 41-45; res & develop physicist, Aeromed Equip Lab, Naval Air Exp Sta, 47-50; sr staff physicist, Labs Res & Develop, Franklin Inst, 50-59; sr systs engr, Radio Corp Am, 59-63, leader systs physics, 63-67; consult, Environ Sci Lab, Valley Forge Space Tech Ctr, Gen Elec Co, 67-73, staff engr res & develop, 73-81; RETIRED. *Concurrent Pos:* Res assoc, Max Plank Inst Biophys, Ger, 46-47; consult, Jefferson Med Col, 53-56; grant, Woehler Found; mem, Coun Basic Sci, Am Heart Asn; consult, 68-85. *Mem:* Am Phys Soc; Sigma Xi. *Res:* Radiosonde development; viscoelasticity; advanced systems; biomedical technology; nuclear effects engineering. *Mailing Add:* 916 Denston Dr Ambler PA 19002

SITTENFIELD, MARCUS, b New York, NY, Jan 20, 19; m 45; c 2. CHEMICAL MARKET RESEARCH, ENVIRONMENTAL AUDITS. *Educ:* City Col NY, BChE, 38, MChE, 39. *Prof Exp:* Chem engr, Sherwood Refining Co, 40, Continental-Diamond Fiber Co, 40-42 & Chem Warfare Serv, 42-43; proj engr, Publicker Industs Inc, 43-47; CONSULT ENGR, MARCUS SITTENFIELD & ASSOCS, 47-; PRES, ROMAR CONSULTS, INC, 68- *Concurrent Pos:* Prin engr, Bechtel Corp, 58-59; consult, Off Res & Develop, Environ Protection Agency, 78-81. *Mem:* Am Inst Chem Engrs; Am Chem Soc (past chmn, Chem Mkt & Econ Div); fel Am Inst Chemists; Soc Prof Engrs. *Res:* Chemical marketing and market research; chemical plant location; environmental control; chemical markets in Latin America; chemical economics; quaternary ammonium compounds; synthetic tanning processes; destruction of chlorinated hydrocarbons as polychlorinated biphenyls. *Mailing Add:* 1015 Chestnut St Philadelphia PA 19107-4316

SITTERLY, CHARLOTTE MOORE, b Ercildoun, Pa, Sept 24, 98; m 37. ASTRONOMY, ASTROPHYSICS. *Educ:* Swarthmore Col, AB, 20; Univ Calif, PhD(astron), 31. *Hon Degrees:* DSc, Swarthmore Col, 62; Dr, Univ Kiel, 68; DSc, Univ Mich, 71. *Prof Exp:* Computer, Princeton Observ, 20-25 & 28-29, res asst, 31-36, res assoc, 36-45; physicist, Atomic Physics Div, Nat Bur Standards, 45-68, Off Standard Ref Data, 68-71 & US Naval Res Lab, 71-78; RETIRED. *Concurrent Pos:* Mem comt line spectra of elements, Nat Res Coun; mem comn standard wavelengths & spectral tables, Int Astron Union & comn fundamental spectros data, pres, 61-67; mem joint comn spectros, Int Coun Sci Unions, 50-58 & comt data sci & technol, 66-70; mem triple union comn spectros, Int Union Pure & Appl Physics, 60-65. *Honors & Awards:* Cannon Prize, Am Astron Soc, 37; Silver Medal, US Dept Com, 51, Gold Medal, 60; Fed Women's Award, 61; Cannon Centennial Medal, Wesley Col, 63; Nat Civil Serv League Career Serv Award, 66; William F Meggers Award, Optical Soc Am, 72. *Mem:* AAAS; fel Am Phys Soc; fel Optical Soc Am; Am Astron Soc (vpres, 58-60); hon mem Soc Appl Spectros. *Res:* Identification of lines in solar and sunspot spectra; analysis of atomic spectra; compilation of spectroscopic data derived from analyses of optical spectra; multiplet tables; atomic energy levels; solar spectrum. *Mailing Add:* 214 Penn Ave Chalfont PA 18914

SITTIG, DEAN FORREST, b Bellefonte, Pa, March 2, 61; m 89. MEDICAL INFORMATICS, EXPERT SYSTEMS. *Educ:* Pa State Univ, BS, 82, MS, 84; Univ Utah, PhD(med informatics), 88. *Prof Exp:* Assoc res scientist, 88-89, ASST PROF MED INFORMATICS, YALE UNIV, 89-, DIR EDUC MED INFORMATICS, YALE CTR MED INFORMATICS, 91- *Concurrent Pos:* Prin investr intelligent artifact & trend detection, Yale Univ, 89-; bd dirs, Soc Computers Critical Care, Pulmonary Med & Anesthesia, 90-; prin investr intel info synthesis for med monitors, NIH, 91-96. *Honors & Awards:* Martin Epstein Award, Symp Computer Appln Med Care, 87. *Mem:* Inst Elec & Electronic Engrs; Am Med Informatics Asn; Soc Technol Anesthesia; Soc Med Decision Making. *Res:* Real-time clinical decision support (expert/knowledge-based) systems for use in critical care, pulmonary medicine, and anesthesiology; real-time information synthesis, including artifact and trend detection, for the development of medical monitors. *Mailing Add:* 304 Alden Ave No 3 New Haven CT 06515-2114

SITTLER, EDWARD CHARLES, JR, b Freeport, NY, Oct 4, 47; m 71; c 2. ASTROPHYSICS. *Educ:* Hofstra Univ, BS, 72; Mass Inst Technol, PhD(physics), 78. *Prof Exp:* Fel astrophysics, Nat Res Coun, 78-79, ASTROPHYSICIST, GODDARD SPACE FLIGHT CTR, NASA, 80- *Mem:* Am Geophys Union; Am Phys Soc. *Res:* Planetary magnetospheres of Jupiter, Saturn, Uranus, and Neptune, including their respective novel satellites, Io, Titan, and Triton; transport of energy by electrons in the solar wind, which may contribute to understanding the mechanisms driving the Sun's coronal expansion; development of plasma analyzer systems for future space applications. *Mailing Add:* 1527 Elwyn Ave Crofton MD 21114

SITZ, THOMAS O, b Newport, RI, Dec 9, 44; m 64; c 1. BIOCHEMISTRY. *Educ:* Va Polytech Inst, BS, 67, PhD(biochem), 71. *Prof Exp:* Fel pharmacol, Baylor Col Med, 71-73, instr, 73-74, res assoc cell biol, 74-75; from asst prof to assoc prof chem, Old Dominion Univ, 75-82; ASSOC PROF BIOCHEM, VA TECH, 82- *Mem:* Am Chem Soc; Am Soc Microbiol; AAAS. *Res:* Secondary structure and methylation of 5.8S rRNA. *Mailing Add:* Dept Biol VA Polytech State Univ Blacksburg VA 24061-0406

SIU, CHI-HUNG, b Hong Kong, July 29, 47; m 76; c 3. DEVELOPMENTAL BIOLOGY. *Educ:* Int Christian Univ, Tokyo, BA, 69; Univ Chicago, PhD, 74. *Prof Exp:* Fel sci res, Scripps Clin & Res Found, 74-76; from asst prof to assoc prof, 76-90, PROF, BANTING & BEST DEPT, MED RES, UNIV TORONTO, CAN, 90- *Concurrent Pos:* MRC scholar, 78-83. *Mem:* Am Soc Cell Biol; Can Biochem Soc. *Res:* The role of plasma membrane macromolecules in specific cell to cell recognition and cell differentiation and the action of retinoids in cell differentiation. *Mailing Add:* C H Best Inst Univ Toronto Toronto ON M5G 1L6 Can

SIU, GERALD, BIOLOGY. *Educ:* Univ Md, BS, 80; Calif Inst Technol, PhD(chem biol), 86; Univ Calif, San Diego, MD, 89. *Prof Exp:* FEL, DEPT BIOL, UNIV CALIF, SAN DIEGO, 89- *Mem:* Am Asn Immunologists; Am Soc Microbiol; NY Acad Sci; AAAS. *Res:* Control of eukaryotic gene expression; molecular and cellular immunology; molecular evolution of multigene family; immunopathology; clinical immunology. *Mailing Add:* Dept Biol Univ Calif La Jolla CA 92093

SIU, TSUNPUI OSWALD, b Hong Kong, Dec 26, 45. CANCER PREVENTION & ETIOLOGY. *Educ:* Univ Calif, Los Angeles, BS, 69; Yale Univ, MS, 73; Harvard Univ, MS, 75, DSc(biostatist), 78. *Prof Exp:* Teaching fel physics, Yale Univ, 70-73, res asst theoret studies atomic struct, 70-73; analyst epidemiol data mgt, Sch Pub Health, Harvard Univ, 73-74, sr analyst, 74-76; asst prof community health & epidemiol, Queen's Univ, Ont, 76-81; SR SCIENTIST EPIDEMIOL, ALTA CANCER HOSP BD, 81-; assoc prof community health, Univ Calgary, 81-90; DEPT APPL MATH, CITY POLY TECH, HONG KONG, 90- *Concurrent Pos:* Lab instr biostatist, Sch Pub Health, Harvard Univ, 74-76; Nat Health Sci scholar, Health & Welfare Can, 78-81; expert lectr, World Bank Workshops Epidemiol Methods, China, 85; hon lectr community med, Univ Hong Kong, 85-86. *Mem:* Am Pub Health Asn; Can Pub Health Asn; Can Oncol Soc; Int Epidemiol Asn. *Res:* Cancer prevention with micro-nutrients; intelligent microcomputer systems in medicine; breast milk banking; epidemiology of Crohn's disease; epidemiology of cancer; folic acid on fragile x-syndrome; clinical trials. *Mailing Add:* Dept Appl Math City Poly Tech Hong Kong 83 Tat Chee Kowloon Tong Hong Kong

SIU, YUM-TONG, b Canton, China, May 6, 43; m 67; c 2. PURE MATHEMATICS. *Educ:* Univ Hong Kong, BA, 63; Univ Minn, MA, 64; Princeton Univ, PhD(math), 66; Yale Univ, MA, 70; Harvard Univ, MA, 82. *Prof Exp:* Asst prof math, Purdue Univ, 66-67 & Univ Notre Dame, 67-70; from assoc prof to prof math, Yale Univ, 70-78; prof math, Stanford Univ, 78-82; PROF MATH, HARVARD UNIV, 82- *Concurrent Pos:* Sloan fel, 71-73; Guggenheim fel, 86-87. *Mem:* AAAS; Am Math Soc. *Res:* Cohomology groups of coherent analytic sheaves on complex analytic spaces; extension of coherent analytic sheaves. *Mailing Add:* Dept Math Harvard Univ Cambridge MA 02138

SIURU, WILLIAM D, JR, b Detroit, Mich, Jan 29, 38; c 2. AUTOMOTIVE JOURNALISM, AEROSPACE JOURNALISM. *Educ:* Wayne State Univ, BSME, 60; Airforce Inst Technol, MSAE, 64; Ariz State Univ, PhD(mech eng), 75. *Prof Exp:* Chief, Launch Systs Br, Foreign Technol Div, 65-71, Technol Br, Air Force Rocket Propulsion Lab, 74-76; asst prof mech eng, US Mil Head, 76-79; comdr, Frank J Seilor Res Labs, 79-83; dir flight systs eng, Aeronaut Systs Div, 83-84; SR RES ASSOC, UNIV COLO, COLORADO SPRINGS, 86-; VPRES ENG, SPACE & AERONAUT SCI INC, 88- *Mem:* Soc Automotive Engrs; Am Inst Aeronaut & Astronaut; Soc Body Engrs. *Res:* Aircraft fire control systems; intelligent vehicle/highway systems; author of automotive and aviation technology. *Mailing Add:* 6341 Galway Dr Colorado Springs CO 80918

SIUTA, GERALD JOSEPH, b Yonkers, NY, Apr 6, 47; m 69; c 1. ORGANIC CHEMISTRY, MEDICINAL CHEMISTRY. *Educ:* Lehman Col, BA, 69; Fordham Univ, PhD(org chem), 74. *Prof Exp:* Res organic chemist, Med Res Div, 74-81, MGR, NEW PROD LICENSING, MED GROUP, LEDERLE LABS, AM CYANAMID CO, 81- *Concurrent Pos:* Lectr chem, Ladycliff Col, 76- *Mem:* Am Chem Soc; Sigma Xi; NY Acad Sci. *Res:* Synthetic organic chemistry; prostaglandins; complement inhibitors; antidiabetic agents; anti-atherosclerotic agents. *Mailing Add:* 14 Georgetown Oval New City NY 10956

SIVAK, ANDREW, b New Brunswick, NJ, May 31, 31; m 58; c 2. BIOCHEMISTRY, CELL BIOLOGY. *Educ:* Rutgers Univ, BS, 52, MS, 57, PhD(microbiol), 60. *Prof Exp:* USPHS fel, Univ Vienna, 60-61; biochemist Arthur D Little, Inc, 61-63; res dir, Bio-Dynamics, Inc, 63-64; res assoc, Med Ctr, NY Univ, 64-68, from asst prof to assoc prof environ med, 68-74; sr staff mem cell biol, Arthur D Little, Inc, 75-77, sect mgr, 75-87, vpres life sci, 77-89; PRES, HEALTH EFFECTS INST, 89-, EXEC DIR, ASBESTOS RES, 90- *Concurrent Pos:* Vis lectr, Harvard Sch Pub Health, 86- *Mem:* AAAS; Am Asn Cancer Res; Environ Mutagen Soc; Am Soc Cell Biol; Am Col Toxicol; Soc Toxicol. *Res:* Environmental toxicology; mechanisms of carcinogenesis and mutagenesis; cell membranes and control of cell division. *Mailing Add:* Health Effects Inst 141 Portland St Cambridge MA 02139

SIVAK, JACOB GERSHON, b Montreal, Que, June 22, 44; m 67; c 3. COMPARATIVE PHYSIOLOGY, PHYSIOLOGICAL OPTICS. *Educ:* Univ Montreal, LScO, 67; Ind Univ, MS, 70; Cornell Univ, PhD(physiol), 72, Pa Col Optometry, OD, 81. *Prof Exp:* From asst prof to assoc prof, 72-79, PROF OPTOM & BIOL, UNIV WATERLOO, 79-, DIR, SCH OPTOM, 84- *Concurrent Pos:* Res assoc, Mote Marine Lab, Fla; fel, Am Acad Optom, 71-; Lady Davis vis prof, Technion-Israel Inst Technol, 78-79 & I Taylor chmn biol, 82. *Honors & Awards:* Fry Award, Am Optom Found, 84. *Mem:* Am Acad Optom; Asn Res Vision & Ophthal. *Res:* Comparative anatomy and physiology of the eye with emphasis on refractive state and accomodative mechanisms; operations of the eye and refraction; optics of the crystalline lens and cataractogenesis. *Mailing Add:* Lab Comp Optom Univ Waterloo Sch Optom Waterloo ON N2L 3G1 Can

SIVASUBRAMANIAN, PAKKIRISAMY, b Andimadam, India, Aug 28, 39; m 67; c 3. DEVELOPMENTAL BIOLOGY. *Educ:* Annamalai Univ, India, BSc, 61; Univ Ill, MS, 71, PhD(physiol), 73. *Prof Exp:* Sci officer entom, Bhabha Atomic Res Ctr, India, 63-69; res fel insect develop, Biol Labs, Harvard Univ, 73-74; from asst prof to assoc prof, 75-83, PROF DEVELOP BIOL, UNIV NB, CAN, 83- *Mem:* Soc Develop Biol; Am Soc Zoologists. *Res:* Developmental neurobiology of insects; investigation of neuronal specificity and the mechanisms of establishment of neural networks in flies during metamorphosis by following the development of nerves in transplanted imaginal discs; insect neurochemistry. *Mailing Add:* Dept Biol Univ NB Fredericton NB E3B 6E1 Can

SIVAZLIAN, BOGHOS D, b Cairo, Egypt, Feb 11, 36; US citizen; m 63; c 2. OPERATIONS RESEARCH. *Educ:* Cairo Univ, BSc, 59; Case Western Reserve Univ, MS, 62, PhD(opers res), 66. *Prof Exp:* Sr mgt sci assoc opers res, B F Goodrich Co, 62-65; from asst prof to assoc prof, 66-72, PROF INDUST & SYSTS ENG, UNIV FLA, 72- *Concurrent Pos:* Consult, M&M Candies, Hackettstown, 61; consult scientist, Ft Belvoir, US Army, 68-69, White Sands Missile Range, 70-76 & Eglin AFB, 82-89; Nat Acad Sci exchange scholar, Poland, 73-74, Fulbright scholar to USSR, 80; res partic, Inst Energy Anal, Oak Ridge Assoc Univs, 77-78. *Mem:* Opers Res Soc Am. *Res:* Inventory and replacement theory; operations research; analysis of military systems; applied stochastic processes. *Mailing Add:* Dept Indust & Systs Eng Univ Fla Gainesville FL 32611

SIVCO, DEBORAH L, b Somerville, NJ, Dec 21, 57; m 81; c 2. MOLECULAR BEAM EPITAXY, III-V CRYSTAL GROWTH. *Educ:* Rutgers Univ, BA, 80; Stevens Inst Technol, BS, 88. *Prof Exp:* Lab technician, Laser Diode Labs, 80-81; MEM TECH STAFF, AT&T BELL LABS, 81- *Res:* Molecular beam epitaxial crystal growth of III-V semiconductor materials for optoelectronics and photonics. *Mailing Add:* AT&T Bell Labs Rm 1C-404 600 Mountain Ave Murray Hill NJ 07974-2070

SIVERS, DENNIS WAYNE, b Greeley, Colo, Jan 20, 44; m; c 2. HIGH ENERGY PHYSICS. *Educ:* Mass Inst Technol, BS, 66; Univ Calif, Berkeley, PhD(physics), 70. *Prof Exp:* Res assoc theoret physics, Argonne Nat Lab, 71-73 & Stanford Linear Accelerator, 73-75; vis scientist, Rutherford Lab, 75-76; asst physicist, 76-77, PHYSICIST THEORET PHYSICS, ARGONNE NAT LAB, 78- *Mem:* Fel Am Phys Soc. *Res:* Theory and phenomenology of high energy physics. *Mailing Add:* Argonne Nat lab 9700 S Cass Ave Argonne IL 60439

SIVIER, KENNETH R(OBERT), b Standish, Mich, Dec 10, 28; m 52; c 6. AEROSPACE ENGINEERING. *Educ:* Univ Mich, BSE(aeronaut eng) & BS(eng math), 51, PhD(aerospace eng), 67; Princeton Univ, MSE, 55. *Prof Exp:* Eng trainee, Naval Ord Lab, Aro, Inc, Tenn, 51-52; aeronaut eng, Aro, Inc, 52-53; res asst, Gas Dynamics Lab, Princeton Univ, 53-55; engr, McDonnell Aircraft Corp, 55-59, sr group engr, 59-62; res engr, Univ Mich, 62-67; ASSOC PROF AERONAUT & ASTRONAUT ENG, UNIV ILL, URBANA, 67- *Mem:* Am Inst Aeronaut & Astronaut. *Res:* Aerodynamic testing techniques and facilities; aerodynamic and aircraft design; aircraft flight mechanics; wind power; combustion of gases. *Mailing Add:* Aeronaut 101 Transp Bldg 104 S Mathews Urbana IL 61801

SIVINSKI, JACEK STEFAN, b Ashton, Nebr, June 23, 26; m 87; c 5. FOOD IRRADIATION, POSTHARVEST TECHNOLOGY. *Educ:* Iowa State Univ, BS, 57. *Prof Exp:* Sect Supvr, Facil Eng Div, Sandia Lab, 57-64, mem adv systs res staff, 64-66, mgr, Planetary Quarantine Dept, 66-74, mgr, Appl Biol & Isotope Utilization Dept, Sandia Labs, 74-81; DIR RADIATION TECHNOL PROGS, CH2M HILL, NMEX, 81- *Concurrent Pos:* Mem, Planetary Quarantine Adv Panel, NASA, 72-; mem, Comt Technol Transfer Develop Countries, Int Atomic Energy Agency; mem task force food irradiation, Coun Agr Sci & Technol, mem joint food irradiation technol & econ, UN Food & Agr Orgn/Int Atomic Energy Agency, mem Int Consult Group Food Irradiation; chmn, Food Irradiation Comt, Am Standard Testing & Mat, Food Irradiation Task Group, Agr Res Inst. *Mem:* Europ Soc Nuclear Methods Agr; Am Nuclear Soc; Inst Food Technologists; Eng Found; Int Asn Milk Food & Environ Sanitarians. *Res:* Systems analysis for low level aircraft penetration; desalination, particularly brackish waters; systems analysis for re-entry vehicle systems, lunar and planetary quarantine, space environments; radiation biology; radiation treatment of agricultural commodities; beneficial uses of nuclear byproducts; linear accelerator process development; isotope and linear accelerator facility and process technologies. *Mailing Add:* 12800 Comanche NE #26 Albuquerque NM 87111

SIVINSKI, JOHN A, b San Cloud, Minn, Nov 22, 38; m 62; c 3. PHYSICS. *Educ:* St Johns Univ, BS, 60; Vanderbilt Univ, MS, 64, PhD(physics), 65. *Prof Exp:* Proj mgr, Gen Electric Space Div, 65-72; PROG MGR, HUGHES DANBURY OPTICAL SYSTS, 72- *Mem:* Am Phys Soc. *Mailing Add:* Hughes Danbury Optical Systs 100 Wooster Heights Rd MS 845 Danbury CT 06810

SIVJEE, GULAMABAS GULAMHUSEN, b Zanzibar, Tanzania, Mar 11, 38; US citizen; m 59; c 3. AERONOMY. *Educ:* Univ London, BSc, 63; Johns Hopkins Univ, PhD(physics), 70. *Prof Exp:* Asst lectr physics, Makerere Col, Univ London, 63-65; res scientist physics, Stand Tel & Cables, Div Int Tel & Tel, 65-66; sr system analyst physics, Bendix Field Eng Corp, 70; fel physics, Johns Hopkins Univ, 71 & Space & Atmospheric Studies, Univ Sask, 71-72; asst prof, 72-76, ASSOC PROF GEOPHYS, GEOPHYS INST, UNIV ALASKA, 76- *Concurrent Pos:* Convener, Comt Airborne Studies Airglow,

Auroral & Magnetospheric Physics, aboard NASA's Convair 990 Jet Aircraft, 73- *Mem:* Am Geophys Union. *Res:* Energy, flux and pitch angle distribution of magnetospheric particles and their interactions with atmospheric constituents; UV spectroscopy of Venus and Jupiter; minor constituents, including gaseous pollutants in terrestrial atmosphere. *Mailing Add:* Dept Math & Sci Embry Riddle Aeron Univ Regional Airport Daytona Beach FL 32014

SIX, ERICH WALTHER, b Frankfurt, Ger, Sept 22, 26; m 57; c 1. BIOPHYSICS. *Educ:* Univ Frankfurt, Dr phil nat, 54. *Prof Exp:* Res fel radiobiol, Max Planck Inst Marine Biol, Ger, 54-56; res fel microbial genetics, Calif Inst Technol, 56-57; res assoc, Univ Southern Calif, 57, Max Planck Inst Biol, Ger, 58-59 & Univ Rochester, 59-60; from asst prof to assoc prof, 60-73, PROF MICROBIAL GENETICS, UNIV IOWA, 73- *Mem:* Am Soc Microbiol; Genetics Soc Am. *Res:* Microbial genetics; virology; molecular biology. *Mailing Add:* Dept of Microbiol Univ of Iowa Iowa City IA 52242

SIX, HOWARD R, b Princeton, WVa, Jan 5, 42; m 64; c 2. VACCINE DEVELOPMENT, IMMUNOCHEMISTRY. *Educ:* David Lipscomb Col, BA, 63; Vanderbilt Univ, PhD(microbiol), 72. *Prof Exp:* Post doctorate pharmacol, Wash Univ, 72-74; post doctorate molecular biol, Vanderbilt Univ, 74-75; from asst prof to assoc prof microbiol, Baylor Col Med, 75-88; VPRES RES & DEVELOP, CONNAUGHT LABS, INC, 89- *Concurrent Pos:* Adj prof, Univ Tex Med Sch, Houston, 76-79, adj assoc prof, 79-81; assoc ed, J Med Virol, 88- *Mem:* Am Asn Immunol; Infectious Dis Soc Am; Fedn Socs Exp Biol & Med; Am Soc Microbiol. *Res:* Development of vaccines for use in humans for the prevention of infectious diseases; definition of the pathogenic mechanisms producing disease during infection with microbial agents. *Mailing Add:* Two Bog Rd East Stroudsburg PA 18301

SIX, NORMAN FRANK, JR, b Tampa, Fla, July 24, 35; m 54; c 4. PHYSICS. *Educ:* Univ Fla, BS, 57, PhD(physics), 63; Univ Calif, Los Angeles, MS, 59. *Prof Exp:* Mem tech staff-physicist, Hughes Aircraft Co, Calif, 57-59; res asst astrophys, Univ Fla, 59-63; mgr geo-astrophys lab, Sci Res Labs, Brown Eng Co, Ala, 63-66; prof physics, Western Ky Univ, 66-74, prof physics & astron, 74-83, head dept, 66-83; asst to dir, Arecibo Observ, Cornell Univ, 83-86; vis scientist, Univ Space Res Asn, 86-88; asst dir, Space Sci Lab, 88-89, ASST ASSOC DIR SCI, NASA/MARSHALL SPACE FLIGHT CTR, 89- *Concurrent Pos:* Res assoc, Univ Fla, 64-66; consult, Brown Eng Co, Ala, 66- *Mem:* Am Astron Soc; Am Geophys Union. *Res:* Analytical and experimental studies in planetary radio emissions; lunar and solar physics; electromagnetic wave propagation; space environment; radio astronomy experiments from earth satellites and from the moon. *Mailing Add:* DS 01 NASA/Marshall Space Flight Ctr Huntsville AL 35812

SIZE, WILLIAM BACHTRUP, b Chicago, Ill, June 8, 43; m 68; c 3. GEOLOGY. *Educ:* Northern Ill Univ, BS, 65, MS, 67; Univ Ill, Urbana, PhD(geol), 71. *Prof Exp:* Asst prof geol, Eastern Ill Univ, 70-71; geologist, Hawaii Inst Geophys, 71-72; asst prof, 73-78, ASSOC PROF GEOL, 78-, DIR, GEOSCIENCES PROG, EMORY UNIV, 89- *Concurrent Pos:* Vis asst prof geol, Univ Hawaii, 71-72; corresp, Geodynamics Proj, Nat Acad Sci, 73-, Int Geol Correlation Prog, 77-; vis geologist, New Zealand Geol Surv, 90; Fulbright scholar, Norway. *Mem:* Sigma Xi; Geol Soc Am; Am Geophys Union. *Res:* Petrogenetic history of igneous rock textures; mechanics of igneous intrusions; origin of alkaline rocks; computer applications in petrology. *Mailing Add:* Geosciences Prog Emory Univ 1364 Clifton Rd NE Atlanta GA 30322

SIZEMORE, DOUGLAS REECE, b Detroit, Mich, Nov 2, 47; m 70; c 1. SOFTWARE SYSTEMS, STATISTICAL COMPUTING. *Educ:* Taylor Univ, BA, 69; Conservative Baptist Theol Sem, MA, 72; Univ Northern Colo, PhD(appl statist), 74. *Prof Exp:* Res fel, Univ Northern Colo, 73-74; assoc prof psychol, 74-82, assoc prof, 82-87, PROF COMPUT SCI, COVENANT COL, 87- *Concurrent Pos:* Consult, Chattem Inc, 77-, Tenn Valley Authority, 83- *Mem:* Asn Comput Mach; Am Statist Asn; AAAS; Data Processing Mgt Asn. *Res:* Regression methodology; software systems performance and measurement (software engineering); computer science education. *Mailing Add:* Scenic Hwy Covenant Col Lookout Mountain TN 37350

SIZEMORE, ROBERT CARLEN, b Lexington, Ky, Sept 30, 51; m 90; c 1. CELLULAR IMMUNOLOGY, IMMUNOREGULATION. *Educ:* Univ Ky, BS, 73, MS, 75; Univ Louisville, PhD(microbiol & immunol), 82. *Prof Exp:* Postdoctoral res assoc, Univ Miss Med Ctr, 82-84; DIR IMMUNOL, IMREG, INC, 84- *Concurrent Pos:* Adj asst prof, Tulane Univ Sch Med, 85- *Mem:* Am Asn Immunologists; Int Soc Develop & Comp Immunol; AAAS; Fedn Am Scientists; Am Soc Trop Med & Hyg. *Res:* Immunotherapeutics for acquired immune deficiency syndrome and other immune diseases; cellular immunology; immunoregulation; comparative immunology; immunoparasitology; immunomodulators; neuropeptides as links between the neuroendocrine and immune systems. *Mailing Add:* IMREG Inc 144 Elk Pl Suite 1400 New Orleans LA 70112

SIZEMORE, RONALD KELLY, b Farmville, Va, Feb 27, 47; m 84; c 2. MARINE MICROBIOLOGY. *Educ:* Wake Forest Univ, BS, 69; Univ SC, MS, 71; Univ Md, PhD(microbiol), 75. *Prof Exp:* Asst prof, Univ Houston, 75-81; assoc prof, 81-89, PROF BIOL & CHAIR, UNIV NC, WILMINGTON, 89- *Concurrent Pos:* Vis res scientist, Galveston Lab, Nat Marine Fisheries Serv, 75-81. *Mem:* Sigma Xi; Am Soc Microbiol; AAAS. *Res:* Ecology, taxonomy and molecular biology of marine bacteria with emphasis on the members of the genus Vibrio. *Mailing Add:* Dept Biol Univ NC Wilmington NC 28403-3297

SIZER, IRWIN WHITING, b Bridgewater, Mass, Apr 4, 10; m 35; c 1. BIOCHEMISTRY. *Educ:* Brown Univ, AB, 31; Rutgers Univ, PhD(physiol & biochem), 35. *Hon Degrees:* ScD, Brown Univ, 71. *Prof Exp:* Lab asst physiol, Rutgers Univ, 31-35; from instr to prof, 35-75, exec officer, 54-55, from actg head dept to head dept, 55-67, dean grad sch, 67-75, EMER PROF BIOCHEM, MASS INST TECHNOL, 75- *Concurrent Pos:* Mem comt physiol training, NIH, 48-64, chmn, 65, chmn comt gen res support, 65-69 & nat adv coun health res facilities, 70-72; consult, Johnson & Johnson Co, 49-77; trustee, Rutgers Univ, 62-71, mem bd gov, 68-71; mem, Corp Lesley Col, 62-88; trustee, Boston Mus Sci, 63-75; mem adv comt, Inst Microbiol, 63-68; consult probs Latin Am, Ford Found, 65-67; dir, Boston Fed Savings Bank, Lexington, 67; consult, Neurosci Res Prog, 67-75; trustee, Boston Biomed Res Inst, 68-; mem adv comt grad educ, Mass State Bd Higher Educ, 68-75; consult resource develop, Mass Inst Technol, 75-85; pres, Whitaker Health Sci Fund, Inc, 74-; trustee, Leobald Smith Res Inst, 85- *Honors & Awards:* Irwin Sizer Award, Mass Inst Technol, 75. *Mem:* Am Chem Soc; fel Am Inst Chem; Am Soc Biol Chem; fel Am Acad Arts & Sci. *Res:* Chemical stimulation of animals; spectroscopy of biological materials; enzyme kinetics; action of oxidases on proteins; x-ray photography of insects; intermediary metabolism of sulfur; enzymes of oxidation and transamination. *Mailing Add:* Rm E25-501 Mass Inst Technol Cambridge MA 02139

SIZER, WALTER SCOTT, b Providence, RI, Aug 15, 47; m 77. LINEAR ALGEBRA. *Educ:* Dartmouth Col, AB, 69; Univ Mass, MA, 72; Univ London, PhD(math), 76. *Prof Exp:* Vis lectr math, Univ Mass, Amherst, 76-77; vis asst prof, Southern Ill Univ, 77-80; asst prof, 80-85, assoc prof, 85-90, PROF MATH, MOORHEAD STATE UNIV, 90- *Concurrent Pos:* Vis lectr math, Mara Community Col, Kuantan, Malaysia, 86-87. *Mem:* Am Math Soc; Math Asn Am; Asn Women Math. *Res:* Similarity of matrices; representations of semigroup actions. *Mailing Add:* Dept Math Moorhead State Univ Moorhead MN 56560

SJOBERG, SIGURD A, ENGINEERING. *Prof Exp:* RETIRED. *Mem:* Nat Acad Eng. *Mailing Add:* 203 Pine Shadows Dr Seabrook TX 77586

SJOBLAD, ROY DAVID, b Worcester, Mass, Nov 22, 47; m 69; c 2. MICROBIOLOGY, BIOCHEMISTRY. *Educ:* Gordon Col, BS, 69; Univ Mass, MS, 71; Pa State Univ, PhD(agron), 76. *Prof Exp:* Res fel appl biol, Harvard Univ, 76-78; ASST PROF MICROBIOL, UNIV MD, 78- *Concurrent Pos:* Rockefeller Found fel, Harvard Univ, 76-77. *Mem:* Sigma Xi; Am Soc Microbiol; NY Acad Sci. *Res:* Microbial ecology; chemoreception in microorganisms; transformation of pesticides by microorganisms; fungal and algal enzymes. *Mailing Add:* Dept of Microbiol Univ of Md College Park MD 20742

SJODIN, RAYMOND ANDREW, b Salt Lake City, Utah, Oct 10, 27; m 54; c 2. BIOPHYSICS. *Educ:* Calif Inst Technol, BS, 51; Univ Calif, PhD(physiol), 55. *Prof Exp:* Asst physiol, Univ Calif, 51-55; res assoc biophys, Purdue Univ, 55-58; NIH fel, Univ Col, London, 58-59; res assoc physiol, Univ Uppsala, Sweden, 59-60; assoc prof, 60-66, PROF BIOPHYS, UNIV MD, BALTIMORE, 66-, INTERIM CHMN BIOPHYS, 88- *Mem:* Fel AAAS; Soc Gen Physiol; Am Physiol Soc; Biophys Soc. *Res:* Physical chemistry of membranes, nerve excitation and conduction of nerve impulses; ion fluxes across cell membranes in relation to electrical events and transport problems; cell electrofusion. *Mailing Add:* Dept Biophysics Univ Md Sch Med Howard Hall Baltimore MD 21201

SJOERDSMA, ALBERT, b Lansing, Ill, Aug 31, 24; m 50; c 4. EXPERIMENTAL MEDICINE, CLINICAL PHARMACOLOGY. *Educ:* Univ Chicago, BS, 45, PhD(pharmacol), 48, MD, 49; Am Bd Internal Med, dipl, 58. *Prof Exp:* Intern, Univ Hosp, Univ Mich, 49-50; resident, Cardiovasc Dept, Michael Reese Hosp, 51; resident med, USPHS Hosp, 51-53; clin investr, Nat Heart Inst, 53-58, chief exp therapeut br, 58-71; vpres & dir, Merrell Int Res Ctr, France, 71-76, sr vpres & dir, Merrell Nat Labs, Cincinnati & France, 76-78, vpres pharmaceut res & develop, Richardson-Merrell Inc, 78-81; vpres Pharmaceut Res, Dow Chem Co, 81-83; pres, Merrell Dow Res Inst, 83-89, EMER PRES, MARION MERRELL DOW RES INST, 89- *Concurrent Pos:* Nat Heart Inst res fel, Univ Chicago, 50-51; NIH fel, Malmo, Sweden, 59-60; spec lectr, George Washington Univ, 59-71; mem coun high blood pressure res, Am Heart Asn; hon chmn, Second World Conf, Clin Pharmacol & Therapeutics, 83. *Honors & Awards:* Theobold Smith Award, AAAS, 58; Harry Gold Award, Am Soc Pharmacol & Exp Therapeut, 77, Exp Therapeut Award, 90; Oscar B Hunter Award, Am Soc Clin Pharmacol & Therapeut, 81. *Mem:* AAAS; Am Soc Pharmacol & Exp Therapeut; Am Soc Clin Invest; Asn Am Physicians; Am Soc Clin Pharmacol & Therapeut; Am Chem Soc; Am Col Neuropsychopharmacol; Am Fedn Clin Res; Am Heart Asn; Am Med Asn. *Res:* Metabolism of biogenic amines; collagen metabolism; drug discovery. *Mailing Add:* Marrion Merrell Dow Res Inst 2110 E Galbraith Rd Cincinnati OH 45215-6300

SJOGREN, JON ARNE, b Karlstad, Wermland, Sweden, Oct 23, 51; US citizen. COMBINATORICS & FINITE MATHEMATICS. *Educ:* Ore State Univ, BS, 70; Univ Calif, Berkeley, PhD(math), 75; Duke Univ, MSEE, 86. *Prof Exp:* Mem tech staff, AT&T Bell Labs, 69-70; mathematician, Inst Hautes Etudes Sci, 75-76; res assoc math, Max-Planck Inst, Bonn, Ger, 76-78; lectr math, Fac Sci, Kuwait Univ, 78-80; instr, Ore State Univ, 80-81; asst prof math, Univ Portland, 81-84; eng consult, Res Triangle Inst, NC, 85-86; computer engr, Langley Res Ctr, NASA, 86-89; PROG MGR, MATH & INFO SCI, AIR FORCE OFF SCI RES, 89- *Mem:* Sr mem Inst Elec & Electronic Engrs; Am Math Soc; Soc Indust & Appl Math. *Res:* Combinatorial group theory and topology; enumerative combinatorics; matrix theory; applied functional analysis; signal processing; stochastic theory. *Mailing Add:* Air Force Off Sci Res NM Bldg 410 Bolling AFB Washington DC 20332-6448

SJOGREN, ROBERT ERIK, b Schenectady, NY, June 13, 31; m 72. MICROBIOLOGY, BIOCHEMISTRY. *Educ:* Cornell Univ, BS, 53; Univ Cincinnati, MS, 60, PhD(microbiol), 67. *Prof Exp:* Asst clin chemist, Ellis Hosp Lab, NY, 55-58; res assoc microbiol, Sterling Winthrop Res Inst, 60-64; ASSOC PROF MICROBIOL & BIOCHEM, UNIV VT, 67- *Concurrent Pos:* Hatch Act grant, Vt Agr Exp Sta, 68- *Mem:* Am Soc Microbiol; Sigma Xi. *Res:* Fungal metabolism; physical-chemical properties of protein; microbial ecology. *Mailing Add:* 96 Henry St Burlington VT 05401

SJOGREN, ROBERT W, JR, b Charlottesville, Va, Aug 19, 45; m 80; c 1. GASTROENTEROLOGY, HEPATOLOGY. *Educ:* Davidson Col, NC, BS, 67; Med Col Va, MD, 71. *Prof Exp:* Chief practice, Dept Med, Meddac, Ft McPherson, Ga, 75-76; chief teaching & practice, Dept Gastroenterol, Eisenhower Hosp, Ft Gordon, Ga, 78-80; STAFF PHYSICIAN PRACTICE, DEPT GASTROENTEROL, WALTER REED ARMY MED CTR, WASHINGTON, DC, 80-, ASST CHIEF RES, DEPT GASTROENTEROL, INST MED, 80- *Concurrent Pos:* Asst prof med teaching, Dept Med, Uniformed Serv Univ Health Sci, Bethesda, Md, 81- *Mem:* Fel Am Col Physicians; fel Am Col Gastroenterol; Am Soc Gastrointestinal Endoscopy; Am Motility Soc. *Res:* In vivo gastrointestinal motility in enteric infection and inflammation; human and animal studies; computer signal analysis; drug therapy. *Mailing Add:* Dept Gastroenterol Walter Reed Army Inst Res Washington DC 20307-5100

SJOLANDER, JOHN ROGERS, b LaCrosse, Wis, Aug 29, 24; m 45; c 3. ORGANIC CHEMISTRY. *Educ:* Univ Wis, BA, 46; Univ Minn, PhD(org chem), 50. *Prof Exp:* Res chemist, Merck & Co, Inc, 50-52; sr chemist, 52-54, proj coordr, 54-58, proj mgr, 58-61, tech & prod mgr, Film Dept, 61-63, tech dir, Film & Allied Prod Div, 63-69, dir corp tech planning & coord, 69-73, mgr agrichem proj, 73-77, TECH DIR, HOUSEHOLD & HARDWARE PRODS DIV, 3M CO, 77- *Mem:* Am Chem Soc. *Res:* Biaxially oriented plastic films; condensation polymers; food packaging; electrical insulating materials; research and development administration, appraisal and control; synthesis, screening and development of pesticides and growth regulators. *Mailing Add:* 3549 Mississippi Dr Minneapolis MN 55433

SJOLANDER, NEWELL OSCAR, industrial microbiology, for more information see previous edition

SJLUND, RICHARD DAVID, b Iron River, Mich, Dec 9, 39. PLANT TISSUE. *Educ:* Univ Wis-Milwaukee, BS, 63; Univ Calif, Davis, PhD(bot), 68. *Prof Exp:* Botanist, Ames Res Ctr, NASA, 63-64; ASSOC PROF BOT, UNIV IOWA, 73- *Concurrent Pos:* DAA fel, Ger, USA-USSR Res Exchange, Leningrad. *Mem:* Bot Soc Am; Am Soc Plant Physiol; Am Phytopathol Soc; AAAS. *Res:* Development of phleom sieve elements in plant tissue cultures; cell structures; protein synthesis; membrane transport; developing techniques for the isolation of phloem cells from tissue cultures; their use in forming phloem-specific monoclonal antibodies. *Mailing Add:* Dept Bot Univ Iowa Iowa City IA 52242

SJOSTRAND, FRITIOF S, b Stockholm, Sweden, Nov 5, 12; m 41, 55, 68; c 3. NEUROANATOMY. *Educ:* Karolinska Inst, Sweden, MD, 41, PhD, 45. *Hon Degrees:* PhD(biol), Univ Siena, Italy, 74 & NEastern Hill Univ, Shillong, India, 89. *Prof Exp:* Asst prof anat, Karolinska Inst, Sweden, 45; Swedish Med Res Coun fel biol, Mass Inst Technol, 47-48; assoc prof anat, Karolinska Inst, Sweden, 49-60; vis prof, 59-60, prof, 60-83, EMER PROF ZOOL, UNIV CALIF, LOS ANGELES, 83- *Concurrent Pos:* Mem & founder exec comt, Int Fedn Electron Micros Socs, 54-62 & Int Brain Res Orgn; ed & founder, J Ultrastruct Res, 57-89; prof & head dept histol, Karolinska Inst, Sweden, 60-62, NSF spec fel, 65-66; sr consult, Vet Admin Radioisotope Serv, 61, 65 & 66. *Honors & Awards:* Swedish Med Asn Award, 59; Anders Retzius Gold Medal, 67; Paul Ehrlich & Ludwig Darmstaedter Prize, 71. *Mem:* Fel Am Acad Arts & Sci; hon mem Soc Electron Micros Japan; hon mem Scand Electron Micros Soc. *Res:* Ultrastructure of cells as related to function; molecular structure and functional significance of cellular membranes; neuronal circuitry of the retina. *Mailing Add:* Dept of Biol Univ of Calif Los Angeles CA 90024-1606

SKADRON, GEORGE, b Vienna, Austria, July 1, 36; US citizen; m 65; c 1. PHYSICS, ASTROPHYSICS. *Educ:* Purdue Univ, BS, 57; Univ Rochester, MA, 60, PhD(physics), 65. *Prof Exp:* Res assoc physics, Univ Md, 65-67; Nat Acad Sci-Nat Res Coun res assoc, Res Labs, Environ Sci Serv Admin, 67-69; from asst prof to assoc prof, 69-75, prof physics, Drake Univ, 75-86; PROF & CHMN, PHYSICS DEPT, ILL STATE UNIV, 86- *Concurrent Pos:* Vis scientist, Max-Planck Inst Aeronomy, WGer, 75-76 & Max-Planck Inst Physics & Astrophys, 83. *Mem:* Am Phys Soc; Am Geophys Union; Union Concerned Scientists. *Res:* Cosmic radiation; plasma astrophysics; ionospheric physics; charged particle acceleration by shock waves in space. *Mailing Add:* Dept of Physics Ill State Univ Normal IL 60761

SKADRON, PETER, b Vienna, Austria, Jan 19, 34; US citizen; m 63; c 1. SOLID STATE PHYSICS. *Educ:* Purdue Univ, BS, 54, MS, 57, PhD(physics), 65. *Prof Exp:* Res physicist, Res & Develop Lab, Sprague Elec Co, 64-67; asst prof, 67-71, assoc prof, 71-80, PROF PHYSICS, BUTLER UNIV, 80- *Mem:* AAAS; Am Phys Soc. *Res:* Electrical transport properties in metals and semiconductors. *Mailing Add:* 6445 Park Central Dr W Indianapolis IN 46260

SKAFF, MICHAEL SAMUEL, b Boston, Mass, June 21, 36; m 64; c 3. MATHEMATICS. *Educ:* Univ Mich, BS, 58; Univ Ill, MS, 60; Univ Calif, Los Angeles, PhD(math), 68. *Prof Exp:* Comput engr, Douglas Aircraft Co, 62-63; sr staff mathematician, Hughes Aircraft Co, 63-68; PROF MATH, UNIV DETROIT, 68-, PRES, MICRO SCI, 84- *Concurrent Pos:* Adj prof, Lincoln Inst Land Policy, Harvard Univ, 68- & Int Asn Assessing Officers. *Mem:* Am Math Soc; Math Asn Am; Int Asn Assessing Officers. *Res:* Vector valued Orlicz spaces; computer simulation and modeling; calculus of variation and optimization; computer assited mass appraisal systems and computerized governmental software. *Mailing Add:* Dept Math Univ Detroit Detroit MI 48221

SKAGGS, LESTER S, b Trenton, Mo, Nov 21, 11; m 39; c 3. PHYSICS. *Educ:* Univ Mo, AB, 33, AM, 34; Univ Chicago, PhD(physics), 39. *Prof Exp:* Asst math, Univ Mo, 35; asst physics, Univ Chicago, 37-41; physicist, Michael Reese Hosp, 40-41, Carnegie Inst Technol, 41-43, Univ Mich, 43-44 & Univ Calif, 44-45; from asst prof to assoc prof, 48-56, prof med physics, 56-79, EMER PROF MED PHYSICS, UNIV CHICAGO, 79- *Concurrent Pos:* Physicist, Michael Reese Hosp, 45-49; sr scientist, King Faisal Specialist Hosp & Res Ctr, Riyadh, Saudi Arabia, 79-84. *Mem:* Fel Am Phys Soc; fel AAAS; fel Am Col Radiol; Am Asn Physicists Med; Radiol Soc NAm; fel Royal Soc Med. *Res:* High energy sources of radiation for therapy; dosimetry; neutron therapy. *Mailing Add:* Dept Radiation Oncol Univ Chicago 5841 S Maryland Ave Box 442 Chicago IL 60637

SKAGGS, RICHARD WAYNE, b Grayson, Ky, Aug 20, 42; m; c 2. AGRICULTURAL SYSTEMS. *Educ:* Univ Ky, BS, 64, MS, 66; Purdue Univ, PhD(agr eng), 70. *Prof Exp:* Res asst, Agr Eng Dept, Univ Ky, 62-64, grad res asst, 64-66; grad instr, Agr Eng Dept, Purdue Univ, 66-70; from asst prof to prof, 70-84, WILLIAM NEAL REYNOLDS PROF, BIOL & AGR ENG DEPT, NC STATE UNIV, 84- *Concurrent Pos:* Vchmn, Unsaturated Flow Comt, Am Soc Agr Engrs, 69-72, chmn, 72-75, mem, Benefits Agr Drainage Comt, 77-, vchmn, Drainage Res Comt, 77-79, chmn, 79-82, chmn & vchmn, Drainage Group, 83-87, mem, Res Comt, 84-86, Monograph Comt, 85-90, assoc ed, Trans Am Soc Agr Engrs, 85-87; mem, Tech Comt Subsurface Water, Am Geophys Union, 75-79; adv, Res Comt, Irrigation & Drainage Div, Am Soc Civil Engrs, 79-85; young researcher award, Am Soc Agr Engrs, 81; consult var firms, 81-90; vis prof, Ohio State Univ, Columbus, 86. *Honors & Awards:* Super Serv Award, USDA, 86 & 90; Hancor Soil & Water Eng Award, Am Soc Agr Engrs, 86. *Mem:* Nat Acad Eng; fel Am Soc Agr Engrs; Am Geophys Union; Soil Conserv Soc Am; Sigma Xi. *Res:* Author of various publications. *Mailing Add:* Dept Biol & Agr Eng NC State Univ Box 7625 Raleigh NC 27695-7625

SKAGGS, ROBERT L, b St Louis, Mo, Apr 2, 32; m 61; c 3. METALLURGICAL ENGINEERING. *Educ:* Mo Sch Mines, BS, 55; Iowa State Univ, MS, 58, PhD(metall), 67. *Prof Exp:* Develop engr, Pigments Dept, E I du Pont de Nemours & Co, 55-56; mat engr, Standard Oil Co Calif, 58-61; sr mat engr, Aeronaut Div, Honeywell Corp, 62-64; asst prof metall eng, Univ Ky, 67-69; assoc prof eng, 69-72, chmn dept, 72-77, PROF ENG, UNIV NEV, LAS VEGAS, 72- *Mem:* Am Soc Metals; Am Soc Eng Educ. *Res:* Metal joining; plastic deformation of alloys; corrosion. *Mailing Add:* Dept Mech Eng Univ Nev 4505 S Maryland Pkwy Las Vegas NV 89154

SKAGGS, SAMUEL ROBERT, b Philipsburg, Pa, June 23, 36; m 58; c 5. MATERIALS SCIENCE. *Educ:* NMex State Univ, BS, 58; Univ NMex, MS, 67, PhD(mat sci), 72. *Prof Exp:* Asst mech engr, Argonne Nat Lab, 58-60; staff mem, Los Alamos Sci Lab, 60-61 & 62-67; physicist, US Air Force Spec Weapons Ctr, 67-68; consult, US Air Force, 68-78; prog mgr fossil energy & mats, 82-86, PROG MGR ARMOR PROTECTIVE SYSTS, LOS ALAMOS NAT LAB, 86-, STAFF MEM, 60-67, 71- *Concurrent Pos:* sabbatical leave, Off Advan Res & Technol, Fossil Energy Div, US Dept Energy, 81-82. *Mem:* AAAS; Am Ceramic Soc. *Res:* High temperature properties and behavior of nonferrous metals and ceramics; ceramic, and CMC armors. *Mailing Add:* Los Alamos Nat Lab Box 1663/MS-K574 Los Alamos NM 87545

SKALA, JAMES HERBERT, b Oak Park, Ill, July 27, 29; m 49. ANALYTICAL BIOCHEMISTRY, NUTRITION. *Educ:* Beloit Col, BS, 50; Univ Minn, MS, 57, PhD(poultry sci), 61. *Prof Exp:* Food chemist tech serv div, Am Can Co, 50-53; biochemist, Peru Surv, Interdept Comt Nutrit Nat Defense, 59; asst prof poultry prod technol, Univ Wis-Madison, 60-69; chief anal biochem br, Chem Div, US Army Med Res & Nutrit Lab, Denver, 69-74; chief, Anal Biochem Sect, Letterman Army Inst Res, 74-76, chief, Biochem Div, Dept Nutrit, 76-81; RES CHEMIST, WESTERN HUMAN NUTRIT RES CTR, AGR RES SERV, USDA, 81- *Concurrent Pos:* Nutritionist, Uruguay Surv, Interdept Comt Nutrit Nat Defense, 62. *Mem:* Inst Food Technologists; Am Asn Clin Chem; Am Inst Nutrit; Poultry Sci Asn; Am Chem Soc. *Res:* Analytical biochemistry; clinical chemistry; human nutrition; especially biochemistry and dietary survey techniques; food chemistry; poultry meat and egg products. *Mailing Add:* PO Box 868 Winchester OR 97495-0868

SKALA, JOSEF PETR, b Prague Czech, Aug 15, 41; Can citizen; c 2. MOLECULAR ENDOCRINOLOGY, DEVELOPMENTAL MEDICINE. *Educ:* Charles Univ Prague, MD, 64; Univ BC, PhD(physiol), 73; FRCP(C), 77. *Prof Exp:* Resident pediat, As-Cheb Gen Hosp, Czech, 64-66; asst prof exp physiol, Charles Univ Prague, 66-68; vis scientist biochem, Univ Stockholm, Sweden, 68-69; fel, pediat, obstet & gynec, Univ BC, 69-72 & metabolism, Hammersmith Hosp, London, UK, 72-73; sr lectr, 73-74, from asst prof to assoc prof, 74-84, assoc obstet & gynec & assoc physiol, 78, PROF PEDIAT, UNIV BC, 84-, PRIN INVESTR, RES CTR, 85- *Honors & Awards:* Med Res Coun Scholar, 75-80. *Mem:* Am Soc Biol Chemists; Biochem Soc; Can Biochem Soc; Soc Pediat Res; Can Med Asn; Can Soc Endocrinol & Metab. *Res:* Molecular mechanisms of hormonal regulation; thermogenesis of brown adipos tissue; hormonal regulation in ontogenic development and cancer; bone marrow transplantation; oncology. *Mailing Add:* Res Ctr Univ BC 950 W 28th Ave Vancouver BC V5Z 4H4 Can

SKALAK, RICHARD, b New York, NY, Feb 5, 23; m 53; c 4. CIVIL ENGINEERING, FLUID MECHANICS. *Educ:* Columbia Univ, BS, 43, CE, 46, PhD(civil eng), 54. *Prof Exp:* Instr struct analysis & design, 46-54, from asst prof to assoc prof fluid mech, 54-64, prof fluid mech, columbia univ, 64-88; PROF BIOL ENG, UNIV CALIF, 88- *Concurrent Pos:* NSF fel, Cambridge Univ, 60-61; sr res fel, Gothenburg Univ, 67-68. *Mem:* AAAS; Am Soc Civil Engrs; Am Soc Eng Educ; Soc Mech Engrs; Soc Eng Sci; Biomed Eng Soc. *Res:* Surface waves, vibration and shock wave phenomena in liquids; fluid mechanics of biological systems; mechanics of blood flow. *Mailing Add:* Dept Appl Mech & Eng Sci/Bioeng Univ Calif San Diego La Jolla CA 92093-0412

SKALKA, ANNA MARIE, b New York, NY, July 2, 38; m 60; c 2. MOLECULAR BIOLOGY, VIROLOGY. *Educ:* Adelphi Univ, AB, 59; NY Univ, PhD(microbiol), 64. *Prof Exp:* Am Cancer Soc fel molecular biol, Carnegie Inst Genetics Res Unit, 64-66, fel, 66-69; asst mem, Dept Cell Biol, 69-71, assoc mem, 71-76, mem, 76-80, HEAD, LAB MOLECULAR & BIOCHEM GENETICS, ROCHE INST MOLECULAR BIOL, 80- *Concurrent Pos:* Vis prof, Dept Molecular Biol, Albert Einstein Col Med, 73-

& Rockefeller Univ, 75. *Mem:* AAAS; Am Soc Microbiol; Am Soc Biol Chem; Sigma Xi; Asn Women Sci. *Res:* Structure and function of DNA; host and viral functions in the synthesis of viral DNA and RNA; phage DNA as a vehicle for the amplification and study of eukaryotic genes; molecular biology of avian retroviruses. *Mailing Add:* Dept Molecular Genetics Hoffmann La Roche Inc Nutley NJ 07110

SKALKO, RICHARD G(ALLANT), b Providence, RI, Apr 10, 36; m 60, 85; c 3. DEVELOPMENTAL TOXICOLOGY, REPRODUCTIVE TOXICOLOGY. *Educ:* Providence Col, AB, 57; St Johns Univ, NY, MS, 59; Univ Fla, PhD(anat), 63. *Prof Exp:* From instr to asst prof anat, Cornell Univ Med Col, 63-67; from asst prof to assoc prof, La State Univ Med Ctr, New Orleans, 67-70; from assoc prof to prof anat & toxicol, Albany Med Ctr, 70-77; PROF ANAT & CHAIR DEPT, ETENN STATE UNIV COL MED, 77- *Concurrent Pos:* Dir embryol, NY Dept Health Birth Defects, Inst, 70-77; vis prof, Inst Toxicol & Develop Pharmacol, Berlin, 78; mem, Toxicol Study Sect, NIH, 84-89 & Human Embryol Study Sect, 90-94. *Mem:* Soc Toxicol; Teratology Soc; Am Asn Anatomists. *Res:* Embryology; toxicology; anatomy; cytology; developmental toxicology; reproductive toxicology. *Mailing Add:* Dept Anat ETenn State Univ PO Box 19960A Johnson City TN 37614

SKALNIK, J(OHN) G(ORDON), b Medford, Okla, May 30, 23; m 47; c 2. ELECTRICAL ENGINEERING. *Educ:* Okla Agr & Mech Col, BS, 44; Yale Univ, ME, 46, DEng, 55. *Prof Exp:* Instr elec eng, Yale Univ, 44-49, from asst prof to assoc prof, 49-65; chmn dept, 68-71, dean, Col Eng, 71-76, PROF ELEC ENG, UNIV CALIF, SANTA BARBARA, 65- *Mem:* Sr mem Inst Elec & Electronics Engrs. *Res:* Splitanode magnetrons; signal-to-noise ratio study; solid-state devices and circuits. *Mailing Add:* Dept Elec Eng & Comput Sci Univ Calif Santa Barbara CA 93106

SKALNY, JAN PETER, b Bratislava, Czech, Mar 19, 35; div; c 2. SILICATE CHEMISTRY. *Educ:* Univ Chem Tech, Prague, Eng Chem, 58; Acad Mining & Metall, Cracow, PhD(silicate chem), 65. *Prof Exp:* Technologist, Pragocement, Prague, 58-60; asst prof bldg mat, Slovak Tech Univ, Bratislava, 60-66; vis res worker cement chem, Cement & Concrete Asn Res Sta, Slough, Eng, 67; res fel cement & surface chem, Clarkson Col, 68-69; group leader, Tech Ctr, Am Cement Corp, 69-71, mgr prod develop, Pac Southwest Region, 71-72; res scientist, Martin Marietta Labs, 72-73, sr res scientist, 73-74, head cement dept, 74-78, assoc dir, 78-85; dir, Construct Mat Res, Res Div, W R Grace & Co, 85-91; CONSULT, 91- *Concurrent Pos:* Mem, Panel Waste Solidification, Comt Radioactive Waste Disposal, 76-77; Comt Status of US Cement & Concrete Res & Develop, 77-78, chmn, Comt Concrete Durability, Nat Acad Sci, 85-86. *Mem:* Fel Am Ceramic Soc (vpres, 86); Int Union Testing & Res Labs Mat & Structure; Am Soc Testing & Mat; Nat Acad Sci. *Res:* Building materials; cement production and hydration; admixture chemistry; research administration. *Mailing Add:* 5665 Vantage Point Rd Columbia MD 21044

SKALSKI, STANISLAUS, b Englewood, NJ, Feb 1, 34. SOLID STATE PHYSICS. *Educ:* Polytech Inst Brooklyn, BS, 58; Rutgers Univ, MS, 60, PhD(physics), 64. *Prof Exp:* Asst prof, 64-72, chmn dept, 72-78, ASSOC PROF PHYSICS, FORDHAM UNIV, 72- *Mem:* Am Phys Soc. *Res:* Ferromagnetism and superconductivity. *Mailing Add:* Nine Donnybrook Dr Demarest NJ 07627

SKAMENE, EMIL, b Aug 27, 41. CLINICAL IMMUNOLOGY. *Educ:* Charles Univ, Prague, MD, 64; Czech Acad Sci, PhD(immunol), 68; FRCP(C), 73, FACP, 74. *Prof Exp:* Postdoctoral res fel, Harvard Med Sch, 68-70; postdoctoral clin fel, McGill Univ, 70-74; res assoc, 74-84, SR SCIENTIST, MONTREAL GEN HOSP RES INST, 85-; PROF MED, MCGILL UNIV, 84- *Concurrent Pos:* Assoc dir, Div Clin Immunol & Allergy, Montreal Gen Hosp, 84-, sr physician, 87-; prof, McGill's Ctr Studies in Aging, 86-, McGill Ctr Human Genetics, 87- & McGill Inst Parasitol, 89-; dir, McGill Ctr Study Host Resistance, 88-; mem sci prog comt, Royal Col Physicians & Surgeons, Can, 89-91, res comt, 91-92, strategic plan task force, Can Inst Acad Med, 89-90; gov, Am Col Physicians, 91- *Honors & Awards:* Gold Medal, Royal Col Physicians & Surgeons, 80; Alexandre-Besredka Award, French-Ger Found Soc Immunol, 89; Cinader Award, Can Soc Immunol, 91. *Res:* Application of genetics to the study of macrophage responses to infection; genetic control of susceptibility to tuberculosis and leprosy in human populations; genetic basis for human disease; innovations for health care. *Mailing Add:* Dept Immunol Montreal Gen Hosp 1650 Cedar Ave Montreal PQ H3G 1A4 Can

SKANDALAKIS, JOHN ELIAS, b Molai, Sparta, Greece, Jan 20, 20; nat US; m 50; c 3. SURGICAL ANATOMY & TECHNIQUE. *Educ:* Nat Univ, Athens, MD, 46; Emory Univ, MS, 50, PhD, 62. *Hon Degrees:* LLD, Woodrow Wilson Col Law, 87. *Prof Exp:* Dir surg training prog, 57-72, chmn dept postgrad educ, 73-77, SR ATTEND SURGEON, PIEDMONT HOSP, 77-; PROF ANAT, SCH MED, EMORY UNIV, 63-, CHRIS CARLOS PROF SURG ANAT & TECHNIQUE, 77-, DIR CTR SURG ANAT & TECHNIQUE, 84- *Concurrent Pos:* Mem Bd Regents, Univ Syst Ga, regent at large, 81-, chmn, 83-84; pvt pract; distinguished prof, Emory Univ, 80. *Honors & Awards:* Rorer Award, Am J Gastroenterol, 65. *Mem:* AMA; fel Am Col Surg; Greek Surg Soc; Am Asn Anat; Am Asn Clin Anatomists; Soc Int Chir. *Res:* General and clinical surgery; surgical embryology; surgical anatomy and technique. *Mailing Add:* Emory Univ Sch Med 1462 Clifton Rd NE 303 Dental Bldg Atlanta GA 30322

SKAPERDAS, GEORGE T(HEODORE), b New York, NY, Jan 25, 14; m 45; c 2. FUEL TECHNOLOGY & PETROLEUM ENGINEERING. *Educ:* McGill Univ, BEng, 36; Mass Inst Technol, SM, 38, ScD(chem eng), 40. *Prof Exp:* Lab chemist, British-Am Oil Co, Can, 36; test engr, Aluminum Co Can, Ltd, Que, 38; asst, Mass Inst Technol, 38-40; chem & process engr, M W Kellogg Co, New York, 40-51, assoc dir chem eng, 51-67, mgr develop, 67-73, mgr process eng, Pullman Kellogg, 73-74, dir coal develop, 74-75, sr consult, 75-79; PRES, G T SKAPERDAS, P C, 80- *Concurrent Pos:* Adj prof, NY Univ, 47-53; consult engr, 79- *Mem:* Am Chem Soc; fel Am Inst Chem Engrs. *Res:* Gas absorption; heat transfer; corrosion; process development engineering; oxychlorination; coal gasification; hydrogen recovery; synthetic chemicals; air separation; economic evaluation; silicon manufacture. *Mailing Add:* 14 Wychview Dr Westfield NJ 07090-1821

SKARDA, R VENCIL, JR, b Los Angeles, Calif, May 22, 40; m 71; c 7. MATHEMATICS. *Educ:* Pomona Col, BA, 61; Calif Inst Technol, MS, 64, PhD(math), 66. *Prof Exp:* Asst prof, 65-71, ASSOC PROF MATH, BRIGHAM YOUNG UNIV, 71- *Mem:* Math Asn Am; Am Math Soc; London Math Soc. *Res:* Analytic number theory; functional analysis and functional iterations; combinatorics; inequalities in l-1-space; control theory. *Mailing Add:* Dept of Math Brigham Young Univ Provo UT 84602

SKARLOS, LEONIDAS, b Manchester, NH, Apr 11, 41; m 70. CHEMISTRY, MATHEMATICS. *Educ:* Univ Vt, BA, 64; Univ NH, MS, 66; Boston Col, PhD(chem), 69. *Prof Exp:* Sr chemist, 69-74, res chemist, Richmond Res Labs, 74-79, PROJ CHEMIST, PORT ARTHUR RES LABS, TEXACO INC, 79- *Res:* Developing methods of determining pollution resulting from coal gasification. *Mailing Add:* Texaco Res & Develop PO Box 1608 Port Arthur TX 77641

SKARSAUNE, SANDRA KAYE, b Burlington, Iowa, Apr 16, 43; m 66; c 1. CEREAL & ANALYTICAL CHEMISTRY. *Educ:* Cornell Univ, BS, 65; NDak State Univ, PhD(cereal chem), 69; Univ Mich, MBA, 87. *Prof Exp:* Asst prof cereals, NDak State Univ, 69-73; lab chief, Centro Indust Exp Para Exportacion, 73-75; GROUP LEADER CHEM, KELLOGG CO, 75-, INT QUAL MGR, 83- *Mem:* Am Asn Cereal Chemists; Inst Food Technol; Am Chem Soc. *Mailing Add:* 8190 Pennfield Rd 235 Porter St Battle Creek MI 49016

SKARSGARD, HARVEY MILTON, b Viscount, Sask, Feb 27, 29; m 59; c 3. PHYSICS. *Educ:* Univ Sask, MSc, 50; McGill Univ, PhD(physics), 55. *Prof Exp:* Seismic interpreter, Explor Dept, Imp Oil, Ltd, 51-53; Nat Res Coun fel nuclear physics, Atomic Energy Res Estab, Eng, 56-57; Nat Res Coun fel plasma physics, European Orgn Nuclear Res, Switz, 57-58; from asst prof to assoc prof, 58-69, PROF PHYSICS, UNIV SASK, 69- *Mem:* Am Phys Soc; Can Asn Physicists; Can Asn Univ Teachers. *Res:* Plasma physics; beam-plasma interactions; current-generated wave instabilities in toroidal geometry; turbulent heating. *Mailing Add:* Dept of Physics Univ of Sask Saskatoon SK S7N 0W0 Can

SKARSGARD, LLOYD DONALD, b Viscount, Sask, Aug 16, 33; m 60; c 4. RADIATION BIOLOGY, BIOPHYSICS. *Educ:* Univ Sask, BE, 55, MSc, 56; Univ Toronto, PhD(radiation physics), 60. *Prof Exp:* Res assoc biophys, Yale Univ, 60-62, asst prof, 62-67; assoc prof physics, McMaster Univ, 67-72; head biophysics dept, 72-78, HEAD, MED BIOPHYS UNIT, 78-, RES DIR, BC CANCER FOUND, 81- *Concurrent Pos:* Consult physicist, Hartford Hosp, Conn, 61-67; head, Batho Biomed Facil, Tri-Univ Meson Facil, Univ BC, & hon prof physics & path, 72- *Mem:* AAAS; Radiation Res Soc; Biophys Soc; Can Asn Physicists. *Res:* X-ray and gamma-ray spectra; radiobiology of pi-mesons and heavy ions; radiation damage and repair; radiosensitization of anoxic cells. *Mailing Add:* Dept Physics Univ BC 2075 Westbrook Pl Vancouver BC V6T 1W5 Can

SKARSTEDT, MARK T(EOFIL), b Washington, DC, Dec 5, 43; m 69; c 3. BIOCHEMISTRY, RESEARCH ADMINISTRATION. *Educ:* Univ Calif, Los Angeles, BSc, 65; Univ Miami, PhD(biochem), 71. *Prof Exp:* Fel biochem, Imp Col Sci & Technol, London, 71-73; instr med, State Univ NY, Downstate Med Ctr, 73-75; res scientist biochem, 75-77, RES SUPVR BIOCHEM, AMES DIV, MILES LABS, 77- *Mem:* Sigma Xi; AAAS; Am Diabetes Asn; Am Asn Clin Chem; Am Chem Soc. *Res:* Identification, purification and study of enzymes, development of their application to medical therapy and diagnosis and development of new formats for diagnostic tests. *Mailing Add:* 7839 W 9005 Pendleton IN 46064

SKARULIS, JOHN ANTHONY, b New Haven, Conn, Feb 18, 17; m 42; c 2. PHYSICAL CHEMISTRY. *Educ:* St John's Univ, NY, BS, 37, MS, 39; NY Univ, PhD(chem), 49. *Prof Exp:* Asst res chemist, Gen Chem Co, 40-42, res chemist, 44-45, supvr & explosive chemist, Gen Chem Defense Corp, 42-44; from instr to assoc prof, 45-54, PROF CHEM, ST JOHN'S UNIV, NY, 54- *Mem:* Am Chem Soc. *Res:* Phase rule studies; inorganic fluorine compounds. *Mailing Add:* 1 Harris Ct Bellmore NY 11710-3524

SKATRUD, THOMAS JOSEPH, b Manitowoc, Wis, Feb 27, 53. BIOENGINEERING. *Educ:* Univ Wis, Madison, BS, 75, PhD(biochem), 79. *Prof Exp:* Dir biochem, 79-80, VPRES, BIO-TECH RESOURCES, INC, 80- *Mem:* Am Chem Soc; Am Soc Plant Physiologists; AAAS. *Mailing Add:* Univ Foods Corp Tech Ctr 6143 N 60th St Milwaukee WI 53218-1606

SKAU, KENNETH ANTHONY, b Chicago, Ill, Apr 18, 47; m 72; c 2. PHARMACOLOGY. *Educ:* Ohio State Univ, BS, 70, PhD(pharm), 77. *Prof Exp:* Trainee pharmacol, May Grad Sch Med, 77-80, instr, 79-80; res asst prof pharmacol, Col Pharm, Univ Utah, 80-82; asst prof, 83-89, ASSOC PROF PHARMACOL, UNIV CINCINNATTI, 89- *Mem:* Soc Neurosci; AAAS; Sigma Xi; NY Acad Sci. *Res:* Pharmacology, biochemistry and neurobiology of acetylcholinesterase molecular forms and pathological conditions related to aberrations of these forms; mechanisms of muscle and nerve diseases; diabetic neuropathy. *Mailing Add:* Col Pharm Univ Cincinnatti Cincinnatti OH 45221

SKAUEN, DONALD M, b Newton, Mass, May 14, 16; m 42; c 2. PHARMACY. *Educ:* Mass Col Pharm, BS, 38, MS, 42; Purdue Univ, PhD(pharm), 49. *Prof Exp:* Asst, Mass Col Pharm, 38-40; chief pharmacist, Children's Med Ctr, Boston, 40-46; asst, Purdue Univ, 46-48; from asst prof to prof, 48-79, EMER PROF PHARM, UNIV CONN, 79- *Mem:* Am Soc Hosp Pharmacists; Am Pharmaceut Asn; Acad Pharmaceut Sci; AAAS. *Res:* Ultrasound and radioisotopes in pharmacy research; pharmaceutical research and development. *Mailing Add:* 16 Storrs Heights Rd Storrs CT 06268

SKAVARIL, RUSSELL VINCENT, b Omaha, Nebr, Dec 6, 36; m 60; c 4. GENETICS. *Educ:* Univ Omaha, BA, 58; Creighton Univ, MT, 59; Ohio State Univ, MSc, 60, PhD(zool), 64. *Prof Exp:* Assoc prof, 64-77, PROF GENETICS, STATIST & COMPUT APPLN, OHIO STATE UNIV, 77- *Mem:* Am Genetic Asn; Biomet Soc. *Res:* Use of computers in biology. *Mailing Add:* Dept of Genetics Ohio State Univ 484 W 12th Ave Columbus OH 43210

SKAVDAHL, R(ICHARD) E(ARL), b Detroit, Mich, Nov 24, 34; m 59; c 3. NUCLEAR ENGINEERING. *Educ:* Mass Inst Technol, SB, 56, ScD(nuclear eng), 62; Univ Mich, MSE, 57. *Prof Exp:* Res engr, Am Metal Prod Co, Mich, 58-60; sr engr, Gen Elec Co, Wash, 62-64, mgr fuel element design & eval, 64-65; mgr fuel element design & eval, Pac Northwest Labs, Battelle Mem Inst, 65-66; proj engr, Fast Ceramic Reactor Develop Prog, 66-69, proj engr demonstration plant develop, 69-70, mgr develop & test progs, 70-73, mgr, Clinch River Proj, 73-78, mgr, Boiling Water Reactor 4 Proj, 78-79, mgr, Boiling Water Reactor 4 & 5 Projs, 79-81, mgr, Nuclear Serv Mkt & Prod Planning, 83, mgr, Piping Improv Prog, 83-84, mgr, Waste Mgt Serv Oper, 84-85, SERV GEN MGR, ENG SERV, GEN ELEC CO, 85- *Mem:* Am Nuclear Soc. *Res:* Nuclear power reactor design, development and project management. *Mailing Add:* Gen Elec Co 175 Curtner Ave San Jose CA 95125

SKAVENSKI, ALEXANDER ANTHONY, b East Liverpool, Ohio, Jan 27, 43; m 67; c 2. PSYCHOPHYSIOLOGY, NEUROPHYSIOLOGY. *Educ:* Univ Md, BS, 65, PhD(psychol), 70. *Prof Exp:* Fel biomed eng, Johns Hopkins Univ, 70-72; from asst prof to assoc prof, 72-80, PROF PSYCHOL, NORTHEASTERN UNIV, 80- *Concurrent Pos:* Vis scholar, Univ Calif, Berkeley, 78-79; vis sr scientist, Eye Res Inst Retina Found, Boston. *Mem:* AAAS; Asn Res Vision & Ophthal; Soc Neurosci; Sigma Xi. *Res:* Eye movement control and the consequence of eye movement on vision and visual space perception. *Mailing Add:* Dept Psychol 125 NI Bldg Northeastern Univ 360 Huntington Ave Boston MA 02115

SKEAN, JAMES DAN, b Kenova, WVa, Feb 19, 32; m 55; c 4. MICROBIOLOGY. *Educ:* Berea Col, BS, 56; Univ Tenn, MS, 59, PhD(microbiol), 66. *Prof Exp:* Res asst dairying, Univ Tenn, 56-66; asst prof, 66-70, assoc prof, 70-80, PROF BIOL, WESTERN KY UNIV, 80- *Mem:* AAAS; Am Soc Microbiol. *Res:* Influence of psychophilic bacteria on quality of milk and dairy products; use of autogenous vaccines in control of staphylococcal bovine mastitis. *Mailing Add:* Dept of Biol Western Ky Univ Bowling Green KY 42101

SKEATH, J EDWARD, b Williamsport, Pa, June 12, 36; m 62; c 2. MATHEMATICS. *Educ:* Swarthmore Col, BA, 58; Univ Ill, MA, 60, PhD(math), 63. *Prof Exp:* Instr math, Cornell Univ, 63-65; asst prof, 65-71, from actg dean to dean men, 70-75, assoc prof, 71-78, PROF MATH, SWARTHMORE COL, 78-, CHMN MATH DEPT, 81- *Mem:* Am Math Soc; Math Asn Am. *Res:* Riemann surface theory; potential theory. *Mailing Add:* Dept of Math Swarthmore Col Swarthmore PA 19081

SKEEL, ROBERT DAVID, b Calgary, Alta, Can, Dec 17, 47. NUMERICAL ANALYSIS, SCIENTIFIC COMPUTING. *Educ:* Univ Alta, BSc, 69, PhD(comput sci), 74; Univ Toronto, MSc, 70. *Prof Exp:* From asst prof to assoc prof, 74-86, PROF COMPUT SCI, UNIV ILL, 86- *Concurrent Pos:* Vis res asst prof comput sci, Univ Ill, 73-74; res assoc, Univ Manchester, 80-81. *Mem:* Asn Comput Mach; Soc Indust & Appl Math. *Res:* Numerical methods for differential equations; computational molecular biophysics. *Mailing Add:* 2413 Digital Comput Lab 1304 W Springfield Ave Urbana IL 61801

SKEELES, JOHN KIRKPATRICK, b Alexandria, La, June 27, 45; m 67; c 4. VETERINARY MEDICINE, MICROBIOLOGY. *Educ:* Okla State Univ, BS, 67, DVM, 69; Univ Ga, MS, 77, PhD(microbiol), 78, Am Col Vet Microbiologists, dipl. *Prof Exp:* Vet, US Army, 69-75; vet med resident microbiol, Univ Ga, 75-78; PROF POULTRY DIS, UNIV ARK, 78- *Mem:* Am Vet Med Asn; Am Asn Avian Pathologists; Sigma Xi; Am Col Vet Microbiologists. *Res:* Microbial diseases of poultry. *Mailing Add:* Dept of Animal Sci Univ of Ark Fayetteville AR 72701

SKEEN, JAMES NORMAN, b Knoxville, Tenn, Feb 23, 42; m 66. FOREST ECOLOGY, TERRESTRIAL COMMUNITY ECOLOGY. *Educ:* Maryville Col, Tenn, BS, 64; Univ Ga, MS, 66, PhD(bot ecol), 69. *Prof Exp:* Asst prof biol, Mercer Univ, Atlanta, 69-70, actg chmn dept, 70-71; ecologist, Fernbank Sci Ctr, Atlanta, Ga, 72-85; ASSOC DIR, FERNBANK INC, 85- *Concurrent Pos:* Consult, Environ Sci Div, Oak Ridge Nat Lab, 76-77; adj assoc prof, Dept Biol, Emory Univ, 78-; sci adv bd, Marshall Forest (Nature Conserv), 80- *Mem:* Torrey Bot Club. *Res:* Community analysis and system maturity; regeneration dynamics; micrometeorology; selection and evaluation of biomass fuel species. *Mailing Add:* Fernbank Inc 1788 Ponce de Leon Ave Atlanta GA 30307

SKEEN, LESLIE CARLISLE, b Dearborn, Mich, Feb 28, 42; m 63; c 3. NEUROANATOMY, COMPARATIVE NEUROLOGY. *Educ:* Fla State Univ, PhD(psychobiol), 72. *Prof Exp:* Fel anat, Duke Univ Med Ctr, 73-76, asst prof med res, 76-77; ASST PROF NEUROSCI & PSYCHOL, UNIV DEL, 77- *Concurrent Pos:* Prin investr, Brain Res Lab, Univ Del, 77- *Mem:* AAAS; Am Asn Anatomists; Soc Neurosci; Sigma Xi. *Res:* Evolutionary, developmental, and structural aspects of the vertebrate olfactory system, and its contributions to complex behavioral patterns. *Mailing Add:* 515 Pershing Rd Hockessin DE 19707

SKEES, HUGH BENEDICT, b Elizabethtown, Ky, Sept 6, 27; m 56; c 7. APPLIED CHEMISTRY. *Educ:* St Louis Univ, BS, 54, MS, 63. *Prof Exp:* Chemist, Petrolite Corp, 56-62; proj engr explor res, Standard Register Co, 62-64, supvr, 64-67; tech dir, Wallace Bus Forms Inc, 67-70; appl res mgr, 70-80, tech dir res , 85-87 PRINTING PROD RES MGR, STANDARD REGISTER CO, 80-, TECH DIR, ADVAN GRAPH IC ARTS, 87- *Mem:* AAAS; Am Chem Soc. *Res:* Petroleum waxes and derivatives; application of waxes in packaging, polishes and carbon paper; business forms technology; printing; carbon paper; adhesives; coating technology; chemical and instrumental analysis; physical testing; instrument design; test development; carbonless paper and imaging technology; document security systems; electronic printing and imaging technology. *Mailing Add:* Standard Regist Co LTB PO Box 1167 Dayton OH 45401-1167

SKEGGS, LEONARD TUCKER, JR, b Fremont, Ohio, June 9, 18; m 41; c 3. BIOCHEMISTRY. *Educ:* Youngstown Univ, AB, 40; Western Reserve Univ, MS, 42, PhD(biochem), 48; Am Bd Clin Chem, dipl. *Hon Degrees:* DSc, Youngstown Univ, 60; LHD, Baldwin-Wallace Col, 80. *Prof Exp:* Chief, Biochem Sect & Hypertension Res Lab, Vet Admin Hosp, Cleveland, 47-68, dir, Hypertension Res Lab, 68-83, med investr hypertension, 76-82; res fel clin biochem, Case Western Reserve Univ, 48-49, from instr to sr instr biochem, 50-52, from asst prof to assoc prof, 52-69, prof, 69-83, emer prof, biochem, 83-; RETIRED. *Honors & Awards:* Flemming Award, 57; Van Slyke Medal, 63; Am Chem Soc Award, 66; Ames Award, 66; Middleton Award, 68; Stouffer Award, 68; Bendetti-Pichler Award Microchem, 71; John Scott Award, 72; Cleveland Award Artificial Organs, 78; Edward Longstreth Medal, Fraklin Inst, 80. *Mem:* Am Chem Soc; Am Soc Biol Chem; fel Am Asn Clin Chem; fel NY Acad Sci; Sigma Xi. *Res:* Hypertension; automatic chemical analysis; multiple automatic analysis. *Mailing Add:* 10212 Blair Lane Kirtland OH 44094

SKEHAN, JAMES WILLIAM, b Houlton, Maine, Apr 25, 23. GEOLOGY, TECTONICS. *Educ:* Boston Col, AB, 46, AM, 47; Weston Col, PhL, 47, STB, 54, STL, 55; Harvard Univ, AM, 51, PhD(geol), 53. *Hon Degrees:* DHumL, St Joseph's Col, 78. *Prof Exp:* Asst prof geophys, 56-61, from asst dir to assoc dir, Weston Observ, 56-72, actg dir, 73-74, chmn dept geol, 58-68, assoc prof geophys & geol, 62-68, chmn dept geol & geophys, 68-70, dir environ ctr, 70-72, PROF GEOPHYS & GEOL, BOSTON COL, 68-, DIR WESTON OBSERV, ENERGY RES CTR, 73- *Concurrent Pos:* Chmn, Eng Geol Div, Geol Soc Am, 75; dir & proj mgr, Narragansett Basin Coal Proj, 76- *Mem:* AAAS; Geol Soc Am; Nat Asn Geol Teachers (pres, 71-72); Am Geophys Union; Geol Soc London. *Res:* Geotectonics, origin and development of the earth's crust with special reference to the origin of mountains of Eastern North America and Western Europe; origin and evolution of metamorphic coal basins. *Mailing Add:* Western Observ 380 Concord Rd Weston MA 02193

SKEIST, IRVING, b Worcester, Mass, Apr 9, 15; m 39; c 4. POLYMER CHEMISTRY. *Educ:* Worcester Polytech Inst, BS, 35; Polytech Inst Brooklyn, MS, 43, PhD(polymer chem), 49. *Prof Exp:* Res chemist, Celanese Corp Am, 37-51; tech dir, Newark Paraffine Paper Co, 51-53 & Am Molding Powder, 53; mkt specialist, Gering Prod, Inc, 53-54; CONSULT & PRES, SKEIST LABS, INC, 54- *Mem:* Am Chem Soc; Soc Plastics Indust; Soc Plastics Engrs; Com Develop Asn; Chem Mkt Res Asn; Sigma Xi. *Res:* Epoxy resins; polymers; plastics; adhesives; coatings; fibers. *Mailing Add:* 32 Laurel Ave Summit NJ 07901-3464

SKELCEY, JAMES STANLEY, b Saginaw, Mich, Sept 26, 33; m 55; c 3. CERAMICS ENGINEERING. *Educ:* Univ Detroit, BS, 56; Mich State Univ, PhD(chem), 61. *Prof Exp:* Proj leader, 61-72, res specialist, Inorg Res & Semiplants, 72-79, RES LEADER, CERAMICS & ADVAN MAT RES, DOW CHEM CO, 79- *Mem:* Am Chem Soc; Sigma Xi. *Res:* Brine chemicals; inorganic fluids and polymers; flame retardant fillers; high temperature chemistry; ceramics. *Mailing Add:* 6015 Sturgeon Creek Pkwy Midland MI 48640

SKELL, PHILIP S, b New York, NY, Dec 30, 18; m 48; c 4. ORGANIC CHEMISTRY. *Educ:* City Col, BS, 38; Columbia Univ, MA, 41; Duke Univ, PhD(chem), 42. *Hon Degrees:* LLD, Lewis Col, 65. *Prof Exp:* Instr chem, City Col, 38-39; asst, North Regional Res Lab, USDA, Ill, 42-43; res assoc antibiotics, Univ Ill, 43-46; instr chem, Univ Chicago, 46-47; asst prof, Univ Portland, 47-52; from asst prof to prof, 52-74, Evan Pugh prof, 74-84, EVAN PUGH EMER PROF CHEM, PA STATE UNIV, 84- *Concurrent Pos:* Committeeman, Nat Res Coun; NSF Sr Scientist Award, 61; Guggenheim fel, 68; Alexander von Humboldt Found Sr Scientist award, 74-75. *Mem:* Nat Acad Sci; Am Chem Soc. *Res:* Free radicals; carbenes; methylenes; carbonium ions; nonmetal atomic chemistry; ground and excited states; transition metal atomic chemistry. *Mailing Add:* Chem Dept Pa State Univ 220 Whitmore Lab University Park PA 16802

SKELLEY, DEAN SUTHERLAND, b Melrose, Mass, Mar 27, 38; m 66; c 4. ENDOCRINOLOGY, CLINICAL CHEMISTRY. *Educ:* Bates Col, BS, 60; Ohio State Univ, MS, 65, PhD(physiol), 68. *Prof Exp:* Dir steroid lab, Col Vet Med, Ohio State Univ, 68-70; asst prof obstet & gynec & assoc dir reprod res lab, Baylor Col Med, 70-77; clin biochemist, dept path, Mem Hosp Syst, Houston, 77-83; dir opers, Severance Ref Lab, San Antonio, Tex, 83-84; vpres sci affairs, Cone Biotech Inc, Seguin, Tex, 85-86; vpres med prod develop, Biosysts Develop Co, San Antonio, Tex, 86-88; TECH DIR, NAT HEALTH LABS, INC, SAN ANTONIO, TEX, 89- *Concurrent Pos:* Consult biol diag prod, AMF, Inc, 73-85; consult, Ctr Dis Control, 74-76; consult radioimmunoassay, dept path, Mem Hosp Syst, Houston, 74-; exec ed, Ligand Rev, 79-82; pres, Tech & Prof Serv, Inc, 79- *Mem:* Coun Biol Ed; Am Asn Clin Chem; Am Med Writers Asn. *Res:* Radioimmunoassay of steroids, polypeptide hormones and pharmacological agents; competitive protein binding and radioreceptor assays; ligand assays; regulatory affairs; product development; chemiluminescent assays. *Mailing Add:* PO Box 160879 San Antonio TX 78280

SKELLEY, GEORGE CALVIN, JR, b Boise City, Okla, Jan 28, 37; m 58; c 2. ANIMAL SCIENCE. *Educ:* Panhandle Agr & Mech Col, BS, 58; Univ Ky, MS, 60, PhD(meats), 63. *Prof Exp:* Res asst meats, Univ Ky, 58-62; from asst prof to assoc prof, 62-72, PROF ANIMAL SCI, CLEMSON UNIV, 72- *Mem:* Am Meat Sci Asn; Sigma Xi; Am Soc Animal Sci; Inst Food Technol. *Res:* Evaluation of and effect of nutrition on beef and pork carcasses; studies on meat tenderness. *Mailing Add:* Dept Animal Sci Clemson Univ Clemson SC 29634-0361

SKELLY, DAVID W, b Buffalo, NY, Dec 9, 38; m 62; c 4. THIN FILM, VACUUM DEPOSITION. *Educ:* Canisius Col, BS, 60; Univ Notre Dame, PhD(phys chem), 65. *Prof Exp:* PHYS CHEMIST, GEN ELEC RES & DEVELOP LAB, 65- *Concurrent Pos:* Fel, Univ Notre Dame, 65; lectr, training courses in sputter deposition, vacuum technol; consult, thin film process develop & equipment deposition. *Honors & Awards:* Gen Electric Dushman Award. *Mem:* Am Vacuum Soc. *Res:* Metallizations for semiconductor devices; thin film deposition processes; sensor technology; liquid crystal display technology; electroluminescent displays; radiation effects in condensed matter; environmental coatings for high temperature applications. *Mailing Add:* Eight Hollywood Dr Burnt Hills NY 12027-9419

SKELLY, JEROME PHILIP, SR, b Vermillion Twp, Ill, Dec 15, 32; m 57; c 3. BIOCHEMISTRY, BIOPHARMACEUTICS. *Educ:* Wayne State Univ, BS, 64, MS, 66, PhD(chem), 69. *Prof Exp:* Res assoc coated abrasives, Mich Abrasive Co, 58-59, head lab, 59-63; res asst connective tissue res, Wayne State Univ, 63-68; chemist, Bur Med, Food & Drug Admin, 68-72, dir, Div Clin Res, Bur Drugs, 72-74, chmn bioavailability comt, 73-74; scholar, Sch Pharm, Univ Calif, 74-75; chief, Pharmacokinetics & Biopharmaceut Br, Food & Drug Admin, 75-79, dep dir, 79-83, dir, Div Biopharmaceut, 83-89, DEP DIR, OFF RES, CTR DRUG EVAL & RES, FOOD & DRUG ADMIN, 88-, ASSOC DIR, OFF GENERIC DRUGS, 90- *Concurrent Pos:* Sr exec serv & chmn, Pharmacokinetics, Pharmacodynamics & Drug Metab Sect, Am Asn Pharmaceut Scientists. *Mem:* Sigma Xi; Am Soc Clin Pharmacol & Therapeut; fel Am Asn Pharmaceut Scientists; Am Chem Soc; fel Am Col Clin Pharmacol; Controlled Release Soc. *Res:* Drug bioavailability, absorption, disposition, metabolism, elimination; drug dosage regimen; analysis of drug in physiological fluids; physical pharmacy; dosage form processing; development of in vitro systems to correlate and predict in vivo drug bioavailability and activity; dermato pharmacokinetics; controlled release drugs and specialized drug delivery systems. *Mailing Add:* Pharmacokinetics & Biopharmaceut 5600 Fishers Lane Rockville MD 20857

SKELLY, MICHAEL FRANCIS, b Washington, DC, Sept 22, 50; m 72; c 1. PHARMACOLOGY, BIOCHEMISTRY. *Educ:* Univ Va, Charlottesville, BA, 72; George Washington Univ, PhD(pharmacol), 81. *Prof Exp:* Lab technician, dept clin path, Univ Va, 70-72, lab specialist, dept pharmacol, 72-73; postdoctoral fel toxicol, dept environ health, Univ Cincinnati, 80-83, sr fel geriat, 83-85; cell biologist, Environ Health Res & Testing, Inc, Cincinnati, 86-89; STAFF SCIENTIST, BIOL RES FACULTY & FACILITY INC, 89-; DEVELOP BIOANAL CHEMIST & ASST TECH DIR, ANAL SERV DIV, HILL TOP BIOLABS INC, CINCINNATI, 89- *Mem:* Am Acad Clin Toxicol; Biomet Soc; Am Acad Vet Comp Toxicol. *Res:* Chemical carcinogenesis; diagnosis of Alzheimer's disease; biochemical correlates of aging; pulmonary toxicology; drug metabolism; general pharmacology; geriatric pharmacology and toxicology; analytical biochemistry. *Mailing Add:* 1051 Timber Trail Cincinnati OH 45224-1617

SKELLY, NORMAN EDWARD, b Minneapolis, Minn, Nov 27, 28; m 53; c 6. ANALYTICAL CHEMISTRY, PHYSICAL CHEMISTRY. *Educ:* Col St Thomas, BS, 51; Univ Iowa, MS, 53, PhD, 55. *Prof Exp:* assoc scientist, Dow Chem Co, 55-90; RETIRED. *Mem:* Am Chem Soc; Sigma Xi. *Res:* Liquid chromatography. *Mailing Add:* 2007 Sharon Ct Midland MI 48642

SKELTON, BOBBY JOE, b Clemson, SC, Feb 11, 35; m 56; c 4. HORTICULTURE, PLANT PHYSIOLOGY. *Educ:* Clemson Univ, BS, 57, MS, 60; Va Polytech Inst, PhD(plant physiol), 66. *Prof Exp:* From instr to assoc prof, 57-75, prof hort, 75-82, ASSOC VPRES & DEAN ADMIS & REGIST, CLEMSON UNIV, 82- *Mem:* Am Soc Plant Physiol; Am Soc Hort Sci; Am Asn Col Registrars & Admis Officers. *Res:* Mineral nutrition of horticultural crops; pomology. *Mailing Add:* 101 Sikes Hall Clemson Univ Clemson SC 29631

SKELTON, EARL FRANKLIN, b Hackensack, NJ, Apr 8, 40; m 62; c 2. SOLID STATE PHYSICS. *Educ:* Fairleigh Dickinson Univ, BS, 62; Rensselaer Polytech Inst, PhD(physics), 67. *Prof Exp:* Nat Acad Sci-Nat Res Coun res assoc solid state physics, Solid State Div, 67-68, res physicist, 68-76, HEAD, PHASE TRANSFORMATION SECT, US NAVAL RES LAB, 76- *Concurrent Pos:* Lectr, Prince George's Community Col, 68-74; assoc prof lectr, George Washington Univ, 74-79, prof lectr, 79-; lectr, Univ Md, 75-; liaison scientist, Off Naval Res, Tokyo, 78; vis scholar, Stanford Univ, 80-81; mem, Exec Comt, Stanford Synchrotron Radiation Lab Users' Orgn, 84-86. *Honors & Awards:* Yuri Gargaran Satellite Commun Medal, USSR, 79. *Mem:* Fel Am Phys Soc; Am Crystallog Asn; Am Asn Physics Teachers; Sigma Xi; Am Asn Univ Professors. *Res:* Theoretical and experimental investigation of response of materials to conditions of extreme pressure and temperature; high transition temperature superconductors; selected III-V and II-VI compounds; phase transformation toughening mechanisms in ceramics; conventional x-ray scattering techniques and synchrotron produced radiation for very rapid in situ measurements; phase transformation kinetics and ultra-high temperature studies. *Mailing Add:* US Naval Res Lab Code 4683 Overlook Ave SE Washington DC 20375-2221

SKELTON, MARILYN MAE, b Coffeyville, Kans, May 3, 36; m 58; c 3. FOOD SCIENCE, BIOCHEMISTRY. *Educ:* Kans State Univ, BS, 57, MS, 58; Univ Wyo, PhD(biochem), 70. *Prof Exp:* Instr res foods & nutrit, Dept Foods & Nutrit, Kans State Univ, 59-62; asst prof foods & nutrit, Div Home Econ, Univ Wyo, 62-69; teacher sci & math, Army Educ Ctr, US Army, Ger, 71-72, educ counsr couns & admin, 72-74; asst prof foods & nutrit, Kans State Univ, 75-77; ASSOC PROF HOTEL & RESTAURANT MGT & ACAD COORDR, WEEKEND COL, UNIV DENVER, 78- *Mem:* Inst Food Technologists; Int Food Serv Exec Asn. *Res:* Relationship of chemical composition and histology to meat tenderness; alkaline degradation of pectin; relationship of chemical and physical properties of fruits and vegetables to palatability; sensory evaluation of food service products. *Mailing Add:* Sch Hotel & Restaurant Mgt Univ Denver Denver CO 80208

SKELTON, ROBERT EUGENE, b Elberton, Ga, March 21, 38; m 71; c 4. ELECTRICAL ENGINEERING, AERONAUTICAL & ASTRONAUTICAL ENGINEERING. *Educ:* Clemson, Univ BS, 63; Univ Alabama, MS, 70; UNiv Calif, Los Angeles, PhD(mech), 76. *Prof Exp:* Engr Lockheed Missiles & Space Co, Huntsville, Ala, 63-65; engr consult, Sperry Rand Corp, Huntsville, Ala, 65-75; from asst prof to assoc prof, PROF AERONAUT & ASTRONAUT, PURDUE UNIV, 82- *Concurrent Pos:* Mem, Aeronaut & Space Engr Bd, Nat Res Coun, 83-88; prin investr, numerous res grants Nat Aeronaut & Space Admin, Air Force Off Sci Res, Nat Sci Found, 76-; vis prof, Australian Nat Univ, 84-85 & 87; fel Japan Soc Prom Sci, 86; Nat Res Coun ad hoc comt on Nat Aeronaut & Space Admin-Univ Relationships in Aero & Space Eng, 84-85; chmn, Automatic Control Soc, Int Elec & Elec Engr, Huntsville, Ala, 69. *Mem:* Sigma Xi; Inst Elec & Electron Engrs; assoc fel, Am Inst Aeronaut & Astronaut. *Res:* Dynamics & control of space vehicles. *Mailing Add:* 2601 Nottingham Pl Purdue Univ West Lafayette IN 47906

SKELTON, THOMAS EUGENE, b Six Mile, SC, Dec 15, 30; m 53; c 3. ENTOMOLOGY. *Educ:* Clemson Univ, BS, 53, MS, 56; Univ Ga, PhD(entom), 69. *Prof Exp:* Asst entomologist, 56-60, from asst prof to assoc prof entom, 69-76, PROF ENTOM, CLEMSON UNIV, 76- *Concurrent Pos:* Actg head, Dept Entom, Clemson Univ, 87-88. *Honors & Awards:* Distinguished Achievement Award, Entom Soc Am, 79. *Mem:* Entom Soc Am; Sigma Xi. *Res:* Economic entomology; insects affecting apples, peaches and vegetables. *Mailing Add:* Dept Entom Clemson Univ Clemson SC 29631

SKEWIS, JOHN DAVID, b Lancaster, Pa, Dec 18, 32; m 57; c 3. PHYSICAL CHEMISTRY. *Educ:* Pa State Univ, BA, 54; Lehigh Univ, MS, 57, PhD(chem), 59. *Prof Exp:* Asst chem, Lehigh Univ, 54-56; fel, Univ Southern Calif, 59-60; res chemist, Res Ctr, US Rubber Co, NJ, 60-68, sr res scientist, 68-72, mgr polymer physics res, 72-78, MGR CORP TIRE RES, RES CTR, UNIROYAL, INC, 78- *Mem:* Am Chem Soc. *Res:* Surface and colloid chemistry. *Mailing Add:* Apple Lane Roxbury CT 06783

SKIBINSKY, MORRIS, b New York, NY, Aug 3, 25; m 51; c 2. MOMENT SPACES, ADEQUATE SUBFIELDS. *Educ:* City Col, BS, 48; Univ NC, MA, 51, PhD(math statist), 54. *Prof Exp:* Asst prof math & statist, Purdue Univ, 54-55; vis asst prof math statist, Mich State Univ, 56; from asst prof to assoc prof math & statist, Purdue Univ, 57-62; vis assoc prof statist, Univ Minn, Minneapolis, 62-63; mathematician, Brookhaven Nat Lab, 63-68; PROF STATIST, UNIV MASS, AMHERST, 68- *Concurrent Pos:* Vis scholar, Univ Calif, 61-62; vis prof statist, Fla State Univ, 81-82. *Mem:* Math Asn Am; Inst Math Statist. *Res:* Probability; decision theory; theory of moment spaces; introduced (1967) and developed the concepts of "adequate subfields" and "canonical" (or "normalized") moments; current work on moment spaces for distributions on simplices. *Mailing Add:* Dept of Math & Statist Univ of Mass Amherst MA 01003

SKIBNIEWSKI, MIROSLAW JAN, b Warsaw, Poland, Sept 14, 57; US citizen; m 90. CONSTRUCTION AUTOMATION, TECHNOLOGY TRANSFER MANAGEMENT. *Educ:* Warsaw Tech Univ, MCE, 81; Carnegie Mellon Univ, MS, 83, PhD(civil eng), 86. *Prof Exp:* Staff engr, Pittsburgh Testing Lab, 81-82; res asst construct eng, Carnegie Mellon Univ, 82-86; asst prof, 86-90, ASSOC PROF CONSTRUCT ENG, PURDUE UNIV, 90- *Concurrent Pos:* NSF presidential young investr award, 86, prin investr, 87-; control group mem, Task Force Construct Robotics, Am Soc Civil Engrs, 87-89, Construct Res Coun, 87-; co-prin investr, Construct Indust Inst; comt appointee, Int Coun Bldg Res, Studies & Doc, 89-; lectr, Commonwealth Sci & Indust Res Orgn, Australia, 90; vis prof, Slovak Tech Univ, 90; co-chair, comt Int Coun Tall Bldg, 91- *Mem:* Am Soc Civil Engrs. *Res:* Technology transfer in the construction industry, with particular emphasis on construction automation, including application of robotics and expert systems in this domain; author of one book on robotics in civil engineering, two book chapters and approximately 60 technical papers in research journals and conference proceedings on construction engineering and management. *Mailing Add:* Div Construct Eng & Mgt Purdue Univ Civil Eng Bldg 1245 West Lafayette IN 47907

SKIDMORE, DUANE R(ICHARD), b Seattle, Wash, Mar 5, 27; m 62; c 4. CHEMICAL ENGINEERING, PHYSICAL CHEMISTRY. *Educ:* Univ NDak, BS, 49; Univ Ill, Urbana, MS, 51; St Louis Univ, PhL, 56; Fordham Univ, PhD(phys chem), 60. *Prof Exp:* AEC asst chem eng, Univ Ill, Urbana, 51; instr, Creighton Prep, Nebr, 60; Petrol Res Fund/AEC fel, Fordham Univ, 60; res chemist, E I du Pont de Nemours & Co, 61-64; from asst prof to assoc prof chem eng, Univ NDak, 64-72, actg dean, Col Eng, 68-69; prof, Sch Mines, WVa Univ, 72-78; PROF CHEM ENG, OHIO STATE UNIV, 78- *Concurrent Pos:* Consult coal utilization & environ qual control & sanit chem. *Mem:* AAAS; Am Chem Soc; Am Inst Chem Engrs. *Res:* Kinetics of gas-phase reactions; reactor design; coal utilization. *Mailing Add:* 960 Lynbrook Rd Indian Hills Worthington OH 43085

SKIDMORE, EDWARD LYMAN, b Delta, Utah, Jan 21, 33; m 53; c 6. WIND EROSION MODELING, SOIL CONSERVATION & AGRONOMY. *Educ:* Utah State Univ, BS, 58; Okla State Univ, PhD(soil sci), 63. *Prof Exp:* Assoc prof, 70-75, PROF AGRON, KANS STATE UNIV, 75-; RES SOIL SCIENTIST, WIND EROSION LAB, SCI & EDUC ADMIN-AGR RES, USDA, 63- *Concurrent Pos:* Consult, US Agenc Int Develop, 83; dir, Col on Soil Physics, Int Ctr Theoret Physics, Trieste, Italy, 83, 85, 87 & 89; assoc ed, Entom Sci Soc Am J, 84, 85; mem, bd dir, Am Soc Agron & Int Sci Soc Am, 84-87. *Mem:* AAAS; fel Am Soc Agron; fel Soil Sci Soc Am; fel Soil & Water Conserv Soc; Int Soil Tillage Res Orgn; World Asn Soil & Water Conserv. *Res:* Soil plant water relations; soil physics; wind erosion; agricultural micrometeorology; develop submodels to predict surface soil wetness; soil aggregate status; stochastic weather simulator. *Mailing Add:* USDA/ARS Wind Erosion Unit E Waters Hall Rm 148 Kans State Univ Manhattan KS 66506

SKIDMORE, WESLEY DEAN, b Pocatello, Idaho, Jan 18, 31; m 52; c 5. BIOCHEMISTRY. *Educ:* Univ Utah, BS, 53; George Washington Univ, MS, 58; Univ Calif, San Francisco, PhD(biochem), 65. *Prof Exp:* Prin investr radiobiol, Armed Forces Radiobiol Res Inst, 65-69, proj dir, 70-74; CHEMIST, CVM, FOOD & DRUG ADMIN, 74- *Mem:* Am Chem Soc. *Mailing Add:* 10721 Game ReserveRd 9 Gaithersburg MD 20879-3105

SKIFF, FREDERICK NORMAN, b Albany, NY, Jan 22, 57; m 81; c 3. PLASMA WAVES, PLASMA DIAGNOSTICS. *Educ:* Cornell Univ, BS, 79; Princeton Univ, MA, 81, PhD(physics), 85. *Prof Exp:* Res physicist, Ecole Polytech Fed Lausanne, Switz, 85-89; ASST PROF DEPT PHYSICS, UNIV MD, COLLEGE PARK, 89- *Concurrent Pos:* NSF presidential young investr, 90; Alfred P Sloan fel, 90-92. *Mem:* Am Phys Soc. *Res:* Plasma waves: excitation, propagation, absorption; both theory and experimental study of linear and non-linear plasma wave phenomena; chaos and non-liner dynamics in plasma physics; plasma diagnostics by laser induced fluorescence and by electromagnetic wave transmission. *Mailing Add:* Lab Plasma Res Univ Md College Park MD 20742

SKIFF, PETER DUANE, b Pittsburgh, Pa, Dec 16, 38; m 65. HISTORY & PHILOSOPHY OF SCIENCE. *Educ:* Univ Calif, Berkeley, AB, 59; Univ Houston, MS, 61; La State Univ, PhD(physics), 66. *Prof Exp:* Instr, La State Univ, 63-65; from asst prof to assoc prof, 65-75, PROF PHYSICS, BARD COL, 75- *Concurrent Pos:* Vis instr, Marist Col, 67-68; adj fac, Norwich Univ, 84-85. *Mem:* Am Phys Soc; Am Asn Physics Teachers; History Sci Soc; Archaeol Inst Am; Philos Sci Asn. *Res:* Foundations of quantum theory; quantum statistical mechanics; archaeometry; history of science; philosophy of science; quantum field theory. *Mailing Add:* Bard Col Annandale on Hudson NY 12504

SKILES, JAMES J(EAN), b St Louis, Mo, Oct 16, 28; m 48; c 3. ELECTRICAL ENGINEERING, ELECTRIC POWER SYSTEMS. *Educ:* Washington Univ, BSEE, 48; Univ Mo-Rolla, MSEE, 51; Univ Wis, PhD, 54. *Prof Exp:* Engr, Union Elec Co, Mo, 48-49; instr elec eng, Univ Mo-Rolla, 49-51; instr elec eng, Univ Wis-Madison, 51-54, from asst prof to assoc prof, 54-62, assoc chmn dept, 63-67, chmn dept, 67-72, dir, Univ-Indust Res Prog, 72-75, dir, Energy Res Ctr, 75-89, Wis Elec Utilities res found prof energy eng, 75-89, PROF ELEC ENG, UNIV WIS-MADISON, 62- *Concurrent Pos:* Consult, Allis-Chalmers Mfg Co, 56-62, Space Technol Labs, Inc, 60-63 & Astronaut Corp Am, Wis, 66-69. *Mem:* Am Soc Eng Educ; Inst Elec & Electronics Engrs. *Res:* Computer applications; power systems analysis; energy conservation and systems; real-time computer applications. *Mailing Add:* 8099 Coray Lane Verona WI 53593-9073

SKILLING, DARROLL DEAN, b Carson City, Mich, June 18, 31; m 51; c 3. PLANT PATHOLOGY, FORESTRY. *Educ:* Univ Mich, BS, 53, MFor, 54; Univ Minn, PhD(plant path), 68. *Prof Exp:* Res forester, Lake States Forest Exp Sta, 54-61, RES PLANT PATHOLOGIST, NCENT FOREST EXP STA, US FOREST SERV, 61- *Concurrent Pos:* Prof, Dept Plant Path, Univ Minn, 83- *Mem:* Am Phytopath Soc; Int Soc Plant Pathologists. *Res:* Epidemiology of conifer tree diseases; fungicide screening and control of foliage tree diseases; Scleroderris canker, Lophodermium needlecast, brown spot disease and Cylindrocladium root rot; development of disease rsistance in forest trees using tissue culture and somaclonal variation; epidemiology of conifer tree diseases; fungicide screening and control of conifer foliage pathogens scleroderris canker. *Mailing Add:* NCent Forest Exp Sta US Forest Serv Folwell Ave St Paul MN 55108

SKILLING, HUGH, b San Diego, Calif, Sept 2, 05; m 32; c 1. ELECTRICAL ENGINEERING. *Educ:* Stanford Univ, AB, 26, PhD(elec eng), 31; Mass Inst Technol, SM, 30. *Prof Exp:* Engr, Southern Calif Edison Co, 27-29; instr elec mach, 29, from instr to assoc prof elec eng, 31-42, actg head dept, 41-44, exec head dept, 44-64, actg dean eng, 44-46, PROF ELEC ENG, STANFORD UNIV, 42- *Concurrent Pos:* Consult, US Secy War, Bikini, 46, Dartmouth Col & Univ Hawaii, 57, Univ Alaska, 64 & Univ Wash, 66; vis prof, Cambridge Univ, 51-52 & 65; lectr, Coun Higher Sci Invests, Madrid, 52 & Univ Chile, 57. *Honors & Awards:* Nat Award & Teaching Medal, Inst Elec & Electronics Engrs, 65. *Mem:* AAAS; fel Inst Elec & Electronics Engrs. *Res:* Electric circuits; electric power transmission; transient electric currents; fundamentals of electric waves; preparation of doctoral students for engineering teaching. *Mailing Add:* Dept of Elec Eng Stanford Univ Stanford CA 94305

SKILLING, JOHN BOWER, b Los Angeles, Calif, Oct 8, 21; m 43; c 3. STRUCTURAL ENGINEERING. *Educ:* Univ Wash, BS, 47. *Prof Exp:* Design engr, W H Witt Co, 47-54; sr partner struct & civil eng, Skilling, Helle, Christiansen, Robertson, 54-82; CHMN, STRUCT & CIVIL ENG, SKILLING WARD MAGNUSSON BARKSHIRE INC, 82- *Concurrent Pos:* Mem adv comt, Am Inst Steel Construct, 67-68; mem bldg res adv bd, Nat Acad Eng, 65-; mem Seismic Design Comt, Nat Acad Eng & Nat Res Coun; Nat Res Coun Bldg Res Adv Bd. *Mem:* Nat Acad Eng; fel Am Soc Civil Engrs; Int Asn Bridge & Struct Eng; Int Asn Shell Struct; Am Concrete Inst; Am Inst Steel Construct. *Mailing Add:* 1420 Fifth Ave No 500 Seattle WA 98101-2341

SKILLMAN, ROBERT ALLEN, b Peoria, Ill, Aug 21, 41. FISHERIES MANAGEMENT, POPULATION ECOLOGY. *Educ:* Bradley Univ, BA, 63; Iowa State Univ, MS, 65; Univ Calif, Davis, PhD(zool), 69. *Prof Exp:* Fishery biologist, Nat Marine Fisheries Serv, Honolulu Lab, 69-79; tuna specialist, UN Food & Agr Orgn, 79-80; FISHERY BIOLOGIST, NAT MARINE FISHERIES SERV, HONOLULU LAB, 81- *Mem:* Am Fisheries Soc; Ecol Soc Am; Am Inst Fishery Res Biologists. *Res:* Quantitative analysis of the population dynamics of marine fishes, including but not limited to production model analysis and the estimation of population parameters for growth, mortality, and recruitment. *Mailing Add:* Nat Marine Fisheries Serv Lab 2570 Dole St Honolulu HI 96822-2396

SKILLMAN, THOMAS G, b Cincinnati, Ohio, Jan 7, 25; m 47; c 2. MEDICINE. *Educ:* Baldwin-Wallace Col, BS, 46; Univ Cincinnati, MD, 49. *Prof Exp:* Instr med, Univ Cincinnati, 54-57; asst prof, Ohio State Univ, 57-61; from assoc prof to prof, Creighton Univ, 61-67; prof, 67-82, Kurtz Prof Endocrinol, 74-82, EMER PROF MED, OHIO STATE UNIV, 82- *Mem:* Am Diabetes Asn; Am Fedn Clin Res. *Res:* Clinical diabetes. *Mailing Add:* Ohio State Col Med 370 W Ninth Ave Columbus OH 43210-1238

SKILLMAN, WILLIAM A, b Lakehurst, NJ, Jan 22, 28. ELECTRICAL ENGINEERING. *Educ:* Lehigh Univ, BS, 52; Univ Rochester, MS, 54. *Prof Exp:* CONSULT ENGR, WESTINGHOUSE CORP, 54- *Mem:* Fel Inst Elec & Electronics Engrs; Am Eng Soc. *Mailing Add:* Electronic Syst Group MS 1105 Westinghouse Elec Corp, Box 1693, Balto-Wash Intl Airport Baltimore MD 21203

SKINNER, BRIAN JOHN, b Wallaroo, SAustralia, Dec 15, 28; nat US; m 54; c 3. GEOCHEMISTRY, ECONOMIC GEOLOGY. *Educ:* Univ Adelaide, BSc, 50; Harvard Univ, AM, 52, PhD, 55. *Prof Exp:* Lectr crystallog, Univ Adelaide, 55-58; res geologist, US Geol Surv, 58-62, chief, Br Exp Geochem & Mineral, 62-66; prof geol, 66-72, chmn dept geol & geophys, 67-72, EUGENE HIGGINS PROF, YALE UNIV, 72- *Concurrent Pos:* Ed, Econ Geol, 70-; chmn comt mineral resources & the environ, Nat Acad Sci-Nat Res Coun, 73-75; chmn bd, Earth Sci & Resources, 88-90. *Honors & Awards:* McKinstry Mem lectr, Harvard Univ, 78; DuToit Mem lectr, SAfrica, 79; Medal, Soc Econ Geologists, 81. *Mem:* Fel Mineral Soc Am; Geochem Soc (pres, 73); Soc Econ Geologists; Geol Soc Am (pres, 85); Geol Asn Can; Geol Soc Australia. *Res:* Phase equilibria in systems containing sulfur; geochemistry of ore deposits. *Mailing Add:* Dept Geol & Geophys Yale Univ New Haven CT 06520

SKINNER, CHARLES GORDON, b Dallas, Tex, Apr 23, 23; m 44; c 2. ORGANIC CHEMISTRY, BIOCHEMISTRY. *Educ:* NTex State Univ, BS, 44, MS, 47; Univ Tex, PhD(org chem), 53. *Prof Exp:* Res chemist, Celanese Corp Am, 49-50; Lilly fel, Univ Tex, 53-54; res scientist, Clayton Found Biochem Inst, 55-64; asst dean basic sci, Tex Col Osteop Med, 75-79, prof biochem, 72-83; chmn dept, 69-74, PROF CHEM, NTEX STATE UNIV, 64-, CHMN DEPT BASIC HEALTH SCI, 79- *Concurrent Pos:* Consult eng, AID, Dallas. *Honors & Awards:* Daugherty Award, Am Chem Soc, 78. *Mem:* Am Chem Soc; Am Soc Biol Chem; Sigma Xi; Am Inst Chem. *Res:* Synthesis and biological activity of metabolite antagonists; vitamins; purine and pyrimidines; antitumor agents. *Mailing Add:* Box 5006 N T Sta Denton TX 76203

SKINNER, DALE DEAN, b Payette, Idaho, July 23, 31; m 56; c 4. UNDERWATER ACOUSTICS, ELECTRICAL ENGINEERING. *Educ:* Univ Idaho, BSEE, 53. *Prof Exp:* Jr engr, 53-54, intermediate res engr, Res Labs, 54-61, res engr, 61-63, fel engr, 63-69, mgr ultrasonic technol, Underwater Acoust, 69-73, FEL ENGR, OCEAN RES & ENG CTR, WESTINGHOUSE ELEC CORP, 74- *Mem:* Inst Elec & Electronics Engrs. *Res:* Underwater sound scattering; sonar system and transducer designs; use of ultrasonics for medical diagnostics and nondestructive testing. *Mailing Add:* 297 Riverdale Rd Severna Park MD 21146

SKINNER, DAVID BERNT, b Joliet, Ill, Apr 28, 35; m 56; c 4. SURGERY. *Educ:* Univ Rochester, BA, 58; Yale Univ, MD, 59; Am Bd Surg, cert, 66; Am Bd Thoracic Surg, cert, 66; FRCPS(E), 87. *Hon Degrees:* ScD, Univ Rochester, 80. *Prof Exp:* Intern, Mass Gen Hosp, Boston, 59-60, asst resident, 60-64, resident, 65; clin asst prof surg, Univ Tex Med Sch, San Antonio, 66-68; from asst prof to prof, Johns Hopkins Univ, 68-72; Dallas B Phemister Prof Surg & Chmn Dept, Pritzker Sch Med, Univ Chicago, 72-87; PRES, CHIEF EXEC OFFICER & ATTEND PHYSICIAN, NY HOSP, 87-; PROF SURG, MED COL, CORNELL UNIV, 87- *Concurrent Pos:* Am Cancer Soc fel, Harvard Med Sch, 65, teaching fel, 65; NIH res grants, Johns Hopkins Univ, 68-72, Markle Scholar, 69-74; NIH res grant, Univ Chicago, 72-; sr registr, Frenchay Hosp, Britol, Eng, 63-64; asst chief exp surg, US Air Force Sch Aerospace Med, San Antonio, 66-68; consult surg, Robert B Green Hosp, San Antonio, 66-68, Loch Raven Vet Admin Hosp, 68-72, Good Samaritan Hosp, 68-72, USPHS Hosp, Baltimore, 69-72 & US Naval Med Ctr, Bethesda, 70-72; asst & assoc ed, J Surg Res, 68-72, ed, 72-82 & Current Topics Surg Res, 69-71, assoc ed, Dis Esophagus, 86-; mem, President's Biomed Res Panel, 75-76; vis physician, Rockefeller Univ Hosp, 89- *Honors & Awards:* J Murray Beardsly Lectr, Brown Univ, 86; First Francis D Moore Lectr, Harvard Univ, 86; William Seybold Lectr, Univ Tex, 86; Edward Hallaran Benett Lectr, Univ Dublin, 86; D Hayes Agnew Centennial Lectr, Univ Pa, 87; Arnold Seligman Lectr, Mt Sinai Hosp, 87; First Joseph Bulkley Lectr, Little Co Mary Hosp, 87; Charles B Huggins Sci Lectr, Univ Acadia, 88; Dallas B Phemister Lectr, Univ Chicago, 88; Rosario P San Filippo Lectr, Lutheran Med Ctr, 88; McGraw Lectr, Detroit Surg Asn, 89; Karl Klassen Lectr, Ohio State Univ, 89. *Mem:* Inst Med-Nat Acad Sci; Soc Univ Surgeons (pres, 79); Soc Surg Chmn (pres, 80-82); Am Asn Thoracic Surg; Soc Vascular Surg; Am Surg Asn; AMA. *Res:* Esophageal and upper gastrointestinal physiology and disorders; pulmonary disorders; cardiovascular physiology and artificial circulation; author of numerous technical publications. *Mailing Add:* NY Hosp 525 E 68th St New York NY 10021

SKINNER, DOROTHY M, b Newton, Mass, May 22, 30; m 65. MOLECULAR BIOLOGY. *Educ:* Tufts Univ, BS, 52; Harvard Univ, PhD(biol), 58. *Prof Exp:* Asst dir admis, Jackson Col, Tufts Univ, 52-54; fel biochem, Yale Univ & Brandeis Univ, 58-62; asst prof physiol & biophys, Med Ctr, NY Univ, 62-66; fel, Oak Ridge Inst Nuclear Studies, 66-68; SR SCIENTIST, OAK RIDGE NAT LAB, 68- *Concurrent Pos:* Prof, Oak Ridge Grad Sch Biomed Sci, Univ Tenn, 68-83; mem molecular biol study sect, NIH, 72-76; assoc ed, Growth, 79-88; instr marine biol lab, Wood Hole, Mass, 71; adj prof biophys, E Tenn State Univ, 83-; vis prof women, NSF, 83, 85, mem selection comt panel, physiol processes, 86-89; chair elect, AAAS, chair, retiring chair, sect G, 83-86; governing bd, Crustacean Soc, 87- *Mem:* Soc Gen Physiol (treas, 73-75); fel AAAS; Am Soc Cell Biol; Am Soc Biol Chem; Soc Develop Biol; Sigma Xi; Crustacean Soc. *Res:* Macromolecular changes associated with growth and development in Crustacea; satellite DNAs, structure and functions. *Mailing Add:* Biol Div Oak Ridge Nat Lab PO Box 2009 Oak Ridge TN 37831-8077

SKINNER, G(EORGE) M(ACGILLIVRAY), b Buffalo, NY, Aug 26, 09; m 38; c 1. ENGINEERING PHYSICS. *Educ:* Univ Mich, BS, 33, MS, 34. *Prof Exp:* Res engr, Linde Air Prod Co, Union Carbide & Carbon Corp, 34-40, group leader, 40-48, sect head, 48-56, res supvr, Linde Div, Union Carbide Corp, 56-62, head, Develop Lab Div, 62-69, mgr admin, Linde Div Lab, Tarrytown Tech Ctr, Union Carbide Corp, 69-79; RETIRED. *Mem:* Am Welding Soc; Inst Elec & Electronics Engrs. *Res:* High temperature technique; high frequency dielectrics; metallurgy; method of oxyacetylene cutting; welding arc and gaseous conduction; fluid dynamics; magnetohydrodynamics; high intensity, high pressure arc research. *Mailing Add:* 4503 Bending Oak San Antonio TX 78249-1847

SKINNER, GEORGE T, b Dundee, Scotland, July 22, 23; US citizen; div; c 2. ATMOSPHERE DYNAMICS, COMPUTER HARDWARE SYSTEMS. *Educ:* St Andrews Univ, BS, 48; Calif Inst Technol, MS, 49, AE, 51, PhD(aeronaut), 55. *Prof Exp:* Asst res officer aerodyn, Nat Res Coun Can, 51; res engr, 58-63, PRIN AERONAUT ENGR, CALSPAN CORP, 63- *Mem:* Am Phys Soc; Am Soc Mech Engrs; Sigma Xi. *Res:* Atmospheric boundary layer flows; molecular beam research using shock tubes as gas source; radiation from collisionally excited molecules; contained airflow in automobile tires; space systems testing technology. *Mailing Add:* 108 Oak Park Dr Tullahoma TN 37388

SKINNER, GORDON BANNATYNE, b Winnipeg, Man, Jan 7, 26; nat US; m 52; c 3. PHYSICAL CHEMISTRY. *Educ:* Univ Man, BSc, 47, MSc, 49; Ohio State Univ, PhD(phys chem), 51. *Prof Exp:* Chemist, Monsanto Co, 51-64; assoc prof, 64-67, PROF CHEM, WRIGHT STATE UNIV, 67- *Mem:* Combustion Inst; Am Chem Soc; AAAS. *Res:* Thermodynamic studies of zirconium and titanium; kinetics of gas reactions at high temperatures; application of kinetics to problems in combustion and detonation; computer simulation of complex systems. *Mailing Add:* HCR 61 Box 77 37 Banks OR 97106-9700

SKINNER, H CATHERINE W, b Brooklyn, NY, Jan 25, 31; m 54; c 3. MINERALOGY, BIOINORGANIC CHEMISTRY. *Educ:* Mt Holyoke Col, BA, 52; Radcliffe Col, MA, 54; Univ Adelaide, PhD(mineral), 59. *Prof Exp:* Mineralogist crystallog, Harvard Med Sch, 54-55; mineralogist, Nat Inst Arthritis & Metab Dis, 61-65 & Nat Inst Dent Res, 65-66; res assoc molecular biophys & geol, sr res assoc & lectr surg, 72-75, assoc prof biochem in surg, 78-82, RES AFFIL GEOL & GEOPHYSICS & LECTR ORTHOP SURG, YALE UNIV, 82- *Concurrent Pos:* Mem insts & spec progs comt, Nat Inst Dent Res, 71-75; mem publ comt, Yale Univ Press, 74-76; Agassiz vis lectr biol, Harvsrd Univ, 76-77; assoc ed, Am Mineral & counr, Mineral Soc Am, 78-81; co-chmn panel geochemistry of fibrous materials related to health risks, Nat Acad Sci, 79-; Master, Jonathan Edwards Col, 77-82; pres, Conn Acad Arts & Sci, 85- *Mem:* Fel Mineral Soc Am; Am Crystallog Asn; Geol Soc Am; Mineral Asn Can; Am Soc Bone & Mineral Res. *Res:* Phase equilibria studies of calcium phosphates; crystal chemistry of the mineral portion of calcified tissues; teeth, bone and invertebrate hard tissues; mineral metabolism; sedimentary carbonate deposits; geochemistry; fibrous inorganic materials, asbestos materials and health. *Mailing Add:* Dept Geol Yale Univ New Haven CT 06511

SKINNER, HUBERT CLAYTON, b Tulsa, Okla, Oct 3, 29; m 58; c 3. GEOLOGY. *Educ:* Univ Okla, BS, 51, MS, 53, PhD(geol), 54. *Prof Exp:* Mus technician, Univ Okla, 51-52, asst geol, 52-53, instr, 53-54; from asst prof to assoc prof, 54-62, PROF GEOL, TULANE UNIV, 62- *Concurrent Pos:* Supvr paleo lab, La Div, Texaco, 54-57; ed, Tulane Studies Geol & Paleont, 62- *Mem:* Paleont Soc; Geol Soc Am; Am Asn Petrol Geologists; Brit Palaeont Asn. *Res:* Paleontology; stratigraphy; Cretaceous and Tertiary micropaleontology, paleoecology and stratigraphy of the Gulf Coast; history of geology. *Mailing Add:* Dept of Geol-24C Tulane Univ 6823 St Charles Ave New Orleans LA 70118

SKINNER, JAMES ERNEST, b Okmulgee, Okla, Apr 15, 40. NEUROSCIENCES. *Educ:* Pomona Col, BA, 62; Univ Calif, Los Angeles, MA, 64, PhD(physiol), 67. *Prof Exp:* Res physiologist, Univ Calif, Los Angeles & Brain Res Inst, 66-67, asst res physiologist, Ment Health Training Prog grant, 67-68; asst prof, 68-76, PROF NEUROL, BAYLOR COL MED, 76- *Concurrent Pos:* Mem, Basic Psychopharmacol/Neuropsychol Res Rev Comt, NIMH. *Mem:* Soc Neurosci; Am EEG Soc. *Res:* Brain mechanisms and behavior. *Mailing Add:* Neurophysiol Sect Dept Neurol Baylor Col of Med MS F 603 Houston TX 77030

SKINNER, JAMES LAURISTON, b Ithaca, NY, Aug 17, 53; m 86; c 1. STATISTICAL MECHANICS. *Educ:* Univ Calif, Santa Cruz, BA, 75; Harvard Univ, MA, 77, PhD(chem physics), 79. *Prof Exp:* NSF fel, Stanford Univ, 80-81; from asst prof to prof chem, Columbia Univ, 81-90; JOSEPH O HIRSCHFELDER PROF CHEM, UNIV WIS, 90- *Concurrent Pos:* Sloan fel, 84; Dreyfus teacher-scholar, 84; vis scientist, Inst Theoret Physics, Univ Calif, Santa Barbara, 87; vis prof physics, Univ Grenoble, 87. *Honors & Awards:* Presidential Young Investr Award, NSF, 84. *Mem:* Am Phys Soc; Am Chem Soc; AAAS. *Res:* Energy transport and relaxation; statistical mechanics; chemical reactions; theoretical chemistry. *Mailing Add:* Dept Chem Univ Wis Madison WI 53706

SKINNER, JAMES STANFORD, b Lucedale, Miss, Sept 22, 36; m 63, 77; c 2. PHYSIOLOGY. *Educ:* Univ Ill, Urbana, BS, 58, MS, 60, PhD(phys educ & physiol), 63. *Prof Exp:* Assoc physiol, Sch Med, George Washington Univ, 64, asst prof lectr, 64-65; res assoc cardiol, Sch Med, Univ Wash, Seattle, 65-66; asst prof, Lab Human Performance Res, Pa State Univ, University Park, 66-70; res assoc, Med Clin, Univ Freiburg, Ger, 70-71; assoc prof phys educ, Univ Montreal & res assoc, Inst Cardiol, 71-77; prof phys educ, Univ Western Ont, 77-82; PROF PHYS EDUC, ARIZ STATE UNIV, 82- *Concurrent Pos:* Dir, Exercise & Sport Res Inst, Ariz State Univ, 83- *Honors & Awards:* Citation Award, Am Col Sports Med, 86. *Mem:* Fel Am Col Sports Med (pres-elect, 78-79, pres, 79-80); Can Asn Sports Sci (secy, 76-78); Am Asn Health, Phys Educ & Recreation; fel Am Heart Asn; Am Acad Phys Educ (secy-treas, 90-92). *Res:* Physiology of exercise, especially pertaining to cardiovascular system; effects of increased physical activity on the course and severity of cardiovascular disorders; exercise, training and genetics. *Mailing Add:* 7868 E Granada Rd Scottsdale AZ 85257

SKINNER, JOHN TAYLOR, chemistry, for more information see previous edition

SKINNER, JOSEPH L, b Bartlesville, Okla, Dec 2, 31; m 54; c 2. CHEMICAL ENGINEERING. *Educ:* Okla Baptist Univ, AB, 53; Univ Okla, BS, 56, MS, 58, PhD(chem eng), 62. *Prof Exp:* Res engr, 62-64, sr res engr, 64-66, res group leader, 66-72, STAFF ENGR, CHEM RES, CONTINENTAL OIL CO, 72- *Res:* Aluminum alkyl chemistry; chemical reactor design; high pressure equipment; chemical kinetics; crystallization; sulfation-sulfonation reaction; process development studies; research in production of alcohols and olefins. *Mailing Add:* 1712 Dover Ponca City OK 74604

SKINNER, LINDSAY A, b Chicago, Ill, Mar 28, 38; m 62; c 3. APPLIED MATHEMATICS. *Educ:* Northwestern Univ, BS, 60, PhD, 63. *Prof Exp:* Mem res staff, Int Bus Mach Corp, Calif, 63-64; asst prof math, Purdue Univ, Lafayette, 64-69; ASSOC PROF MATH, UNIV WIS-MILWAUKEE, 69- *Mem:* Soc Indust & Appl Math. *Res:* Perturbation theory and asymptotic expansions. *Mailing Add:* Dept of Math Univ of Wis Milwaukee WI 53201

SKINNER, LOREN COURTLAND, II, b Borger, Tex, Aug 16, 40; m 65; c 3. PHYSICS, METALLURGY. *Educ:* Mass Inst Technol, SB, 62, SM, 64, PhD(metall), 65. *Prof Exp:* Res assoc metall, Mass Inst Technol, 65-66; physicist, Integrated Circuits Ctr, Motorola, Inc, 66-70; dept mgr res & develop, Microprod Div, Am Micro-Systs Inc, 70-72, M O S Mgr, Data Gen Semiconductor Div, 72-76; mem tech staff, Advan Microdevices, Inc, 76-78; mgr, Solid State Technol Ctr, 78-86, DIR RES, NAT SEMICONDUCTOR CORP, SANTA CLARA, 86- *Mem:* Am Inst Mining, Metall & Petrol Engrs; Am Phys Soc; Am Vacuum Soc; Electrochem Soc. *Res:* Metal oxide semiconductor integrated circuit processing; electronic materials research, including device behavior, thin film deposition and patterning; circuit application development. *Mailing Add:* 310 Donohoe St East Palo Alto CA 94303

SKINNER, MARGARET SHEPPARD, b Jamaica, NY, May 8, 38; m 69; c 2. PATHOLOGY. *Educ:* Emory Univ, MD, 62. *Prof Exp:* From instr to assoc prof path, Sch Med, Tulane Univ, 65-73; PATHOLOGIST, CEDARS OF LEBANON HOSP, MIAMI, 73- *Concurrent Pos:* NIH fel path, Tulane Univ, 65-68; asst vis pathologist, Charity Hosp, New Orleans, 65-68, vis pathologist, 68-73; consult staff, Lallie Kemp Charity Hosp, 72-73; clin assoc prof path, Univ Miami, 77- *Mem:* Am Asn Pathologists; Am Soc Microbiol; AMA. *Res:* Experimental cell pathology; surgical pathology. *Mailing Add:* 33 Grand Bay Circle Juno Beach FL 33406

SKINNER, MORRIS FREDRICK, geology, paleontology; deceased, see previous edition for last biography

SKINNER, NEWMAN SHELDON, JR, b Gadsden, Ala, Nov 13, 34; m 55; c 3. MEDICINE, PHYSIOLOGY. *Educ:* Auburn Univ, BS, 55; Med Col Ala, MD, 60. *Prof Exp:* Med intern, Med Ctr, Yale Univ, 60-61; res assoc cardiovasc physiol, Lab Cardiovasc Physiol, Nat Heart Inst, 61-63 & 64-66; asst prof physiol & med, Univ Tex, Southwestern Med Sch, 66-68; from assoc prof to prof med, Sch Med, Emory Univ, 68-72, dir, Div Clin Physiol, 69-72; prof physiol & med & chmn dept physiol, Bowman Gray Sch Med, Wake Forest Univ, 72-74, prof med & prof physiol, 74-88; RETIRED. *Concurrent Pos:* Res assoc med res, Med Col Ala, 63-64. *Mem:* Am Fedn Clin Res; Am Physiol Soc; Soc Exp Biol & Med; Microcirculatory Soc. *Res:* Regulation of peripheral blood flow, particularly skeletal muscle and adipose tissue blood flow; general cardiovascular physiology. *Mailing Add:* 2451 Boone Ave Winston-Salem NC 27103

SKINNER, (ORVILLE) RAY, theoretical physics; deceased, see previous edition for last biography

SKINNER, RICHARD EMERY, b Anderson, Ind, Feb 15, 34; m 54; c 2. ENGINEERING PHYSICS, MATHEMATICS. *Educ:* Reed Col, BA, 55; Calif Inst Technol, MS, 57. *Prof Exp:* Physicist, Atomics Int Div, NAm Aviation, Inc, 55 & 56-59, consult, 55-56; physicist, Radio Corp Am, 59-64; supvry physicist, Electromagnetic Res Inc, 64-65; tech dir, Manst Corp, 66-68; vpres & gen mgr, Parzen Res Div, Ovitron, 68-69; pres, Skinner Industs, Inc, 69-70, INDEPENDENT CONSULT, PRES & CHIEF ENGR, R E SKINNER & ASSOCS INC, 70- *Mem:* Am Phys Soc; Am Nuclear Soc; Nat Soc Prof Engrs. *Res:* Electronics; electrical, mechanical and structural engineering; land surveying and communications; plasma physics; nuclear reactor dynamics theory; computer science. *Mailing Add:* R E Skinner & Assocs Inc 1731 SE 55th Ave Portland OR 97215-3396

SKINNER, ROBERT DOWELL, b Waxahachie, Tex, Nov 23, 42; m 66; c 4. NEUROPHYSIOLOGY. *Educ:* Univ Tex, Arlington, BS, 65; Univ Tex Southwestern Med Sch, Dallas, PhD(biophysics), 69. *Prof Exp:* Instr, 70-71, from asst prof to assoc prof, 71-89, PROF ANAT, COL MED, UNIV ARK MED SCI, LITTLE ROCK, 89- *Concurrent Pos:* NIH fel, Harvard Med Sch, 69-70. *Mem:* Am Asn Anat; Soc Neurosci. *Res:* Locomotor systems; spinal cord physiology; brain stem physiology; non-linear system dynamics. *Mailing Add:* Dept of Anat Slot 510 Univ of Ark Med Sci Little Rock AR 72205

SKINNER, ROBERT L, b Salt Lake City, Utah, Nov 6, 30; m 53; c 4. ELECTRICAL ENGINEERING. *Educ:* Univ Utah, BS, 54, MS, 55. *Prof Exp:* Res dir, Ensco, Inc, 54-68; pres, Entec Inc, 68-; AT PRE ENGR TECHNOL. *Concurrent Pos:* Instr, Univ Utah, 55-56. *Mem:* Inst Elec & Electronics Engrs. *Res:* Specialized systems and instrumentation used in medical research. *Mailing Add:* 8456 S 1430 E Sandy UT 84070

SKINNER, WALTER SWART, b Middletown, NY, Oct 13, 21; wid; c 2. EARTH SCIENCE. *Educ:* Monmouth Col, Ill, BS, 43; Lehigh Univ, MS, 48. *Prof Exp:* Instr geol, Lehigh Univ, 46-48; deep well geologist, S Penn Oil Co, 48-54; staff geologist, Sun Oil Co, 54-59, dist geologist, 59-65; assoc prof phys sci, Duquesne Univ, 65-66, from assoc prof to prof earth sci, 66-86 & from actg chmn to chmn dept physics, 72-86; RETIRED. *Concurrent Pos:* Consult, Hard Rock Mining Co, 58-59. *Mem:* Am Asn Petrol Geol; Geol Soc Am; Am Inst Prof Geol; fel Explorers Club; NY Acad Sci. *Res:* Stratigraphy, structure and sedimentation of the New York, Pennsylvania and West Virginia areas; regional stratigraphy and sedimentation of eastern United States. *Mailing Add:* 17650 Heiser Rd Berlin Ctr OH 44401

SKINNER, WILFRED AUBREY, JR, organic chemistry; deceased, see previous edition for last biography

SKINNER, WILLIAM ROBERT, b Tampa, Fla, Jan 1, 30; m 57; c 4. PETROLOGY, STRUCTURAL GEOLOGY. *Educ:* Univ Tex, BS, 53; Columbia Univ, PhD(geol), 66. *Prof Exp:* Inspector construct, Pittsburgh Testing Lab, 57-59; asst prof, 66-71, ASSOC PROF GEOL, OBERLIN COL, 71-, CHMN DEPT, 74- *Mem:* Geol Soc Am. *Res:* Structure and petrogenesis of Precambrian metamorphic terranes; petrology and geochemistry of Alpine-type and stratiform igneous complexes. *Mailing Add:* Dept of Geol Oberlin Col Oberlin OH 44074

SKINNIDER, LEO F, b Paisley, Scotland, Oct 2, 29; Can citizen; c 8. PATHOLOGY, HEMATOLOGY. *Educ:* Univ Glasgow, MB, ChB, 51; FRCP(C), 68. *Prof Exp:* Assoc prof, 69-73, PROF PATH, UNIV SASK, SASKATOON, 73- *Concurrent Pos:* Consult, Can Tumor Reference Ctr, 72-; Nat Cancer Inst Can grant, 74-; deleg, Col Am Pathologists, 70-78. *Mem:* Am Soc Clin Pathologists; Can Asn Pathologists; Col Am Pathologists; Can Soc Hematol; Can Health Res. *Res:* Ultrastructure and agar-culture characteristics of cells of lymphomas and leukemias; effect of retinol and its analogies on the proliferation of lymphoid cells. *Mailing Add:* 341 Mt Allison Crt Saskatoon SK S7N 0W0 Can

SKIPP, BETTY ANN, b Chicago, Ill, May 7, 28; m 51; c 2. GEOLOGIC MAPPING, MICROPALEONTOLOGY. *Educ:* Northwestern Univ, BA, 49; Univ Colo, MS, 56, PhD(struct geol), 85. *Prof Exp:* Geologist, 52-64, RES GEOLOGIST, US GEOL SURV, 65- *Mem:* Fel Geol Soc Am; Am Asn Petrol Geologists; Soc Econ Paleontologists & Geologists; Paleont Res Inst. *Res:* Structure and stratigraphy of Proterozoic through Tertiary rocks of south-central Idaho and adjacent Montana using geologic mapping and geophysical surveys; the biostratigraphy of Carboniferous smaller calcareous foraminifera worldwide; Mississippian smaller calcareous foraminifers and algae. *Mailing Add:* 2035 Grape Ave Boulder CO 80304

SKIPSKI, VLAIDIMIR P(AVLOVICH), biochemistry; deceased, see previous edition for last biography

SKIRVIN, ROBERT MICHAEL, b Burlington, Wash, Oct 27, 47; m 73; c 1. PLANT BREEDING, PLANT PHYSIOLOGY. *Educ:* Southern Ill Univ, BS, 69, MS, 71; Purdue Univ, PhD(hort), 75. *Prof Exp:* Lab & field asst, Fruit Res Sta, 68-69; res asst grape physiol, Southern Ill Univ, 69-71; David Ross fel hort, Purdue Univ, 71-73, res asst, 73-74, from instr to asst prof, 74-76; asst prof hort, 76-81, assoc prof, 81-88, PROF HORT & FORESTRY, UNIV ILL, 88- *Concurrent Pos:* Vis prof, plant breeding, DISR, Lincoln, NZ. *Mem:* Am Soc Hort Sci; Am Pomol Soc; Int Asn Plant Tissue Cult; Sigma Xi. *Res:* Development of methods to utilize tissue culture techniques for the improvement of asexually propagated crops; breeding and genetic studies of small fruits with a particular interest in thornless blackberries. *Mailing Add:* Dept Hort Field Lab Univ Ill Urbana IL 61820

SKJEGSTAD, KENNETH, b Henning, Minn, Oct 18, 31. BOTANY. *Educ:* Moorhead State Col, BS, 53; Univ Calif, Los Angeles, PhD(bot), 60. *Prof Exp:* Assoc biol, Univ Calif, 58-60; from instr to assoc prof bot, Univ Minn, Minneapolis, 60-66; assoc prof, 66-69, PROF BIOL, MOORHEAD STATE UNIV, 69- *Mem:* AAAS; Bot Soc Am; Am Soc Plant Physiol. *Res:* Physiology of plant growth and development; plant genetics. *Mailing Add:* Dept of Biol Moorhead State Univ Moorhead MN 56563

SKJOLD, ARTHUR CHRISTOPHER, b Minneapolis, Minn, Dec 20, 43; m 72; c 1. BIOCHEMISTRY, MICROBIAL GENETICS. *Educ:* Macalester Col, BA, 66; Kans State Univ, PhD(biochem), 70. *Prof Exp:* Res assoc microbiol genetics, Albert Einstein Med Ctr, 70-73; asst dir res immunodiag, Kallestad Labs Inc, 73-77; sr res scientist, Urine Chem Lab, 77-85, SR STAFF SCIENTIST, DRY REAGENT CHEM LAB, AMES DIV, MILES LAB, 85- *Mem:* Am Asn Clin Chem. *Res:* Nucleic acids; microbial genetics; regulation of RNA and protein synthesis; immunochemistry; protein purification; urinalysis; regulation of bacterial growth. *Mailing Add:* Miles Diag Div Miles Inc PO Box 70 Elkhart IN 46515

SKJONSBY, HAROLD SAMUEL, b Sisseton, SDak, July 6, 37; m 61; c 3. HISTOLOGY, ANATOMY. *Educ:* Concordia Col, BA, 59; Univ NDak, MS, 62, PhD(anat), 64. *Prof Exp:* From asst prof to assoc prof, 64-74, prof, 74, ANAT SCI & CHMN DEPT, UNIV TEX DENT BR HOUSTON, 84- *Mem:* Int Asn Dent Res; Am Asn Dent Schs. *Res:* Histochemistry of tooth development; structure of bone and connective tissue. *Mailing Add:* Dept of Anat Sci UTHSC-Houston Dent Br PO Box 20068 Houston TX 77225

SKLANSKY, J(ACK), b Brooklyn, NY, Nov 15, 28; m 57; c 3. ELECTRICAL ENGINEERING, COMPUTER SCIENCE. *Educ:* City Col New York, BEE, 50; Purdue Univ, MSEE, 52; Columbia Univ, DSc(eng), 55. *Prof Exp:* Res engr, Electronics Res Labs, Columbia Univ, 54-55 & RCA Labs, NJ, 55-65; head systs res sect, Nat Cash Register Co, Ohio, 65-66; assoc prof elec eng, 66-69, chmn dept, 78-80, PROF ELEC ENG, INFO & COMUT SCI & RADIOL SCI, UNIV CALIF, IRVINE, 69- *Concurrent Pos:* Res grants, Nat Inst Gen Med Sci, NSF & Army Res Off. *Honors & Awards:* Fourth Ann Award, Pattern Recognition Soc. *Mem:* Asn Comput Mach; fel Inst Elec & Electronics Engrs; Comput Soc. *Res:* Automatic pattern classifiers; image processing by computer; medical imaging; biomedical engineering; digital system theory. *Mailing Add:* 16 Perkins Ct Irvine CA 92715-4043

SKLAR, LARRY A, BIOPHYSICS, IMMUNOLOGY. *Educ:* Stanford Univ, PhD(phys chem), 76. *Prof Exp:* ASSOC MEM, SCRIPPS CLIN & RES FOUND, 79- *Mailing Add:* Scripps Clin & Res Found Imm 12 10666 N Torrey Pines Rd La Jolla CA 92037

SKLAR, STANLEY, b Bronx, NY, July 12, 37; m 63; c 2. PHARMACEUTICS. *Educ:* Philadelphia Col Pharm & Sci, BS, 59; Purdue Univ, Lafayette, MS, 61; Univ Conn, PhD(pharm), 65. *Prof Exp:* Res pharmacist, Bristol Labs, 65-67; res pharmacist, Radnor, 67-72, unit supvr, 72-73, unit supvr, Pilot Plant, 73-80, QUAL ASSURANCE MGR, WYETH LABS, WEST CHESTER, 80- *Concurrent Pos:* Secy, Parenteral Drug Asn, 80-81. *Mem:* Am Pharmaceut Asn; Acad Pharmaceut Sci; Am Chem Soc. *Res:* Quality assurance and quality control functions, including parenteral dosage forms, fermentation production and a fine chemical facility. *Mailing Add:* Four Selwyn Dr Broomall PA 19008

SKLAREW, DEBORAH S, b New York, NY, Apr 6, 50; m 77; c 1. ORGANIC GEOCHEMISTRY, ANALYTICAL CHEMISTRY. *Educ:* City Col NY, BS, 70; Univ Calif, Berkeley, MS, 72; Univ Ariz, PhD(geosci), 78. *Prof Exp:* SR RES SCIENTIST, BATTELLE PAC NORTHWEST LABS, 78- *Mem:* Geochem Soc; Am Chem Soc. *Res:* Kerogen analysis in Precambrian and recent sedimentary rocks; characterization of fossil fuel effluents; sulfur and nitrogen gas analysis in oil shale retorts; organomercury analysis; PCB sorption studies. *Mailing Add:* 544 Franklin Richland WA 99352

SKLAREW, ROBERT J, b New York, NY, Nov 25, 41; m 70; c 2. CYTOKINETICS, MICROSCOPIC-IMAGING. *Educ:* Cornell Univ, Ithaca, NY, BA, 63; New York Univ, MS, 65, PhD(biol), 70. *Prof Exp:* Asst res scientist, dept med, 65-70, assoc res sci, 70-71, res sci, 71-73, sr res sci, 73-79, res asst prof, dept pathol, 79-86, dir, Cytokinetics & Imaging Lab, Goldwater Mem Hosp, New York Univ Med Ctr, 80-88; PROF MED & ANAT, NY MED COL, VALHALLA,88- *Concurrent Pos:* Res assoc, dept pathol, Lenox Hill Hosp, NY, 81-86; adj asst prof clin dermatol, New York Med Col, 85-86. *Mem:* Cell Kinetics Soc (secy, 83-86, pres, 87-88); AAAS; Tissue Cult Asn; Int Cell Cycle Soc; Am Soc Cell Biol; Soc Analysis Cytol. *Res:* Cell cycle kinetics of heteroploid subpopulations in solid tumors determined by microscopic imaging of autoradiographic labeling and DNA distribution patterns; perturbation of cytokinetics by hormones and cytotoxins in relation to the sensitivity spectrum of subpopulations as a basis for optimization of therapy. *Mailing Add:* Cancer Res Inst 100 Grasslands Rd Elmsford NY 10523

SKOBE, ZIEDOMIS, b Riga, Latvia, Apr 29, 41; US citizen; c 4. CELL BIOLOGY. *Educ:* Boston Univ, BA, 63, PhD(biol), 72; Clark Univ, MA, 67. *Prof Exp:* DEPT HEAD ELECTRON MICROSCOPY, FORSYTH DENT CTR, 72- *Concurrent Pos:* Lectr, Boston Univ, 75- *Mem:* Am Asn Dent Res; Int Asn Dent Res; Am Soc Cell Biol. *Res:* Structure of tooth enamel and the ultrastructure of the cells involved in amelogenesis in several mammals using both scanning and transmission electron microscopy. *Mailing Add:* 20 Prospect St Boston MA 02136

SKOCHDOPOLE, RICHARD E, b Ravenna, Nebr, Dec 11, 27; m 53; c 4. POLYMER BLENDS, PLASTIC FOAMS. *Educ:* Univ Nebr, BSc, 49; Iowa State Univ, PhD(phys chem), 54. *Prof Exp:* Fel, Ames Lab, AEC, 54-55; res chemist, Phys Res Lab, 56-63, sr res chemist, 63-64, assoc scientist, 64-68, dir converted plastics lab, 68, res mgr designed plastics res, 68-70, res mgr foam prod res, 70-73, assoc scientist foam prods res, 73-76, assoc scientist chem prod lab, 76-83, ASSOC DEVELOP SCIENTIST, ENG PLASTICS TECH SERV & DEVELOP, DOW CHEM CO, 83- *Mem:* Am Chem Soc; Sigma Xi; Soc Plastics Engrs. *Res:* Preparation and characterization of cellular plastics, particularly thermal properties; physical chemistry of polymers and polymer solutions; preparation and characterization of polymers; polymer blends. *Mailing Add:* 2525 Lambros Midland MI 48640

SKOFRONICK, JAMES GUST, b Merrill, Wis, Oct 11, 31; m 59; c 4. CHEMICAL PHYSICS, SURFACE PHYSICS. *Educ:* Univ Wis, BS, 59, MS, 61, PhD(nuclear physics), 64. *Prof Exp:* Res asst physics, Univ Wis, 59-64; from asst prof to assoc prof, 64-74, PROF PHYSICS, FLA STATE UNIV, 74- *Concurrent Pos:* Mem vis staff, Max Planck Inst Aerodyn, 79-90. *Mem:* Am Phys Soc; Europ Phys Soc; Sigma Xi. *Res:* Studies of the microscopic properties of neutral atoms and molecules by the use of colliding beam methods; studies of the properties of surfaces by colliding neutral atoms and molecules with surfaces; medical physics; teaching. *Mailing Add:* Dept Physics Fla State Univ Tallahassee FL 32306

SKOG, LAURENCE EDGAR, b Duluth, Minn, Apr 9, 43; m 68; c 1. PLANT TAXONOMY. *Educ:* Univ Minn, BA, 65; Univ Conn, MS, 68; Cornell Univ, PhD(bot), 72. *Prof Exp:* Fel, Royal Bot Garden, Edinburgh, 68-69; res asst bot, Cornell Univ, 69-72; asst ed, Flora NAm Prog, 72-73, assoc cur bot, 73-86, CUR BOT, 86-, CHMN BOT, SMITHSONIAN INST, 87- *Concurrent Pos:* Adj prof biol, George Mason Univ, 80-; res fel, Manchester Mus, Univ Manchester, Eng, 84. *Mem:* Bot Soc Am; Am Soc Plant Taxonomists; Int Asn Plant Taxon; Asn Trop Biol (secy-treas, 81-85); Orgn Flora Neotropica. *Res:* Taxonomy and floristics of neotropical Gesneriaceae, Coriariaceae; pollination biology; Flora of the Guianas. *Mailing Add:* Dept Bot Nat Mus Natural Hist Smithsonian Inst Washington DC 20560

SKOGEN, HAVEN SHERMAN, b Rochester, Minn, May 8, 27; m 49; c 1. ANALYTICAL & PETROLEUM CHEMISTRY. *Educ:* Iowa State Univ, BS, 50; Rutgers Univ, New Brunswick, MS, 54, PhD(ceramics), 55; Univ Chicago, MBA, 70. *Prof Exp:* Res engr, E I du Pont de Nemours & Co, 55-56; prof chem, Elmhurst Col, 56-57; asst chief engr, Stackpole Carbon Co, 58-62; asst plant mgr, Manatronics, Inc, 62-65; mgr, Allen-Bradley Co, 65-70; pres, Haven S Skogen & Assocs, 70-74; CHIEF CHEMIST, OCCIDENTAL OIL SHALE INC, 74- *Mem:* AAAS; Am Ceramic Soc; Nat Inst Ceramic Engrs; Am Soc Prof Engrs. *Res:* Shale oil production and control including environmental impact and economics. *Mailing Add:* 3152 Primrose Grand Junction CO 81506

SKOGERBOE, GAYLORD VINCENT, b Cresco, Iowa, Apr 1, 35; m 58; c 2. AGRICULTURAL ENGINEERING, CIVIL ENGINEERING. *Educ:* Univ Utah, BS, 58, MS, 59. *Prof Exp:* Hydraulic engr, Utah Water & Power Bd, 60-63; res proj engr hydraulics, Utah Water Res Lab, Utah State Univ, 63-68; prof irrigation & drainage, Colo State Univ, 68-84; PROF AGR & IRRIGATION ENG, UTAH STATE UNIV, 84- *Mem:* Am Soc Agr Engrs; Am Soc Civil Engrs; Int Soc Ecol Modelling; Int Water Resources Asn. *Res:* Interdisciplinary research involving physical and social scientists on topics related to agricultural development and environmental problems of irrigated agriculture. *Mailing Add:* Int Irrig Ctr Utah State Univ Logan UT 84322-4150

SKOGERBOE, RODNEY K, b Blue Earth, Minn, June 25, 31; m 58; c 4. ANALYTICAL CHEMISTRY. *Educ:* Mankato State Col, BA, 58; Mont State Univ, PhD(chem), 63. *Prof Exp:* Jr chemist, Ames Lab, US AEC, 58-60; asst prof chem, SDak Sch Mines & Technol, 63-64; res mgr, Cornell Univ, 64-69; from asst prof to assoc prof chem, 69-73, PROF CHEM & ATMOSPHERIC SCI, COLO STATE UNIV, 73-, CHMN, DEPT CHEM, 80- *Mem:* Am Chem Soc; Soc Appl Spectros (pres, 72). *Res:* Spectrochemical and radiochemical methods of trace analysis; application of statistics to chemical problems. *Mailing Add:* Dept of Chem Colo State Univ Ft Collins CO 80523

SKOGERSON, LAWRENCE EUGENE, b Ft Collins, Colo, Aug 19, 42; m 69; c 3. BIOCHEMISTRY. *Educ:* Grinnell Col, AB, 64; Univ Pittsburgh, PhD(biochem), 68. *Prof Exp:* Asst prof biochem, Col Physicians & Surgeons, Columbia Univ, 70-77; ASSOC PROF BIOCHEM, MED COL WIS, 77- *Concurrent Pos:* Am Cancer Soc fel, NIH, Bethesda, 68-70; USPHS res grant, Columbia Univ, 71-78, Am Cancer Soc fac res award, 72-77. *Mem:* Am Soc Biol Chemists; Harvey Soc. *Res:* Regulation of RNA and protein synthesis. *Mailing Add:* Ecogen Inc 2005 Cabot Blvd W Langhorne PA 19047-1810

SKOGLEY, CONRAD RICHARD, b Deer Lodge, Mont, Nov 9, 24; m 48; c 3. AGRONOMY, TURFGRASS MANAGEMENT. *Educ:* Univ RI, BS, 50, MS, 52; Rutgers Univ, PhD, 57. *Prof Exp:* Asst agronomist, Univ RI, 51-53; asst & res assoc, Rutgers Univ, 53-56, exten assoc, 56-57, asst exten specialist, 57-59, assoc exten specialist, 59-60; assoc prof agron, 60-70, prof plant & soil sci, 70-90, EMER PROF PLANT & SOIL SCI, UNIV RI, 90- *Mem:* Am Soc Agron; Crop Sci Soc Am; Int Turfgrass Soc. *Res:* Turfgrass management; establishment and maintenance of fine turf grasses; turfgrass breeding. *Mailing Add:* Ten Liberty Rd Slocum RI 02877

SKOGLEY, EARL O, b Mott, NDak, Mar 18, 33; m 55; c 3. SOIL FERTILITY, PLANT NUTRITION. *Educ:* NDak State Univ, BS, 55, MS, 57; NC State Univ, PhD(soil fertility), 62. *Prof Exp:* Instr soils, NC State Univ, 57-62; res assoc soil fertil, Cornell Univ, 62-63; from asst prof to assoc prof, 63-71, PROF SOIL FERTIL, MONT STATE UNIV, 71- *Concurrent Pos:* Res adv, USAID, Brazil, 67 & 68 & on contract with IRI Res Inst, Inc, NY; consult/team mem develop transmission environ report proposed twin 500 Kv power lines Mont, 77-78. *Mem:* Fel AAAS; Soil Sci Soc Am; Am Soc Agron; Int Soil Sci Soc. *Res:* Development of synthetic growth media for plant nutrition studies; investigation of plant nutrient relations in soils and plants; development of new soil test based on phytoavailability of nutrients. *Mailing Add:* 3535 Stucky Rd Bozeman MT 59715

SKOGLUND, WINTHROP CHARLES, b Lynn, Mass, Dec 7, 16; m 41; c 1. POULTRY HUSBANDRY. *Educ:* Univ NH, BS, 38; Pa State Col, MS, 40, PhD, 58. *Prof Exp:* Asst, Pa State Col, 38-40; instr poultry indust & asst res poultryman, Univ Del, 40-42, exten poultry specialist, 42-46, from asst prof to assoc prof poultry indust, 46-50, dir agr short course, 49-50; prof poultry husb & chmn dept, Univ NH, 50-81; RETIRED. *Mem:* Sigma Xi. *Res:* Nutrition; breeding; hatching; egg production. *Mailing Add:* 30 Bagdad Rd Durham NH 03824

SKOK, JOHN, b Rumania, Nov 18, 09; nat US; m 38; c 1. PLANT PHYSIOLOGY. *Educ:* Northern Ill State Teachers Col, BEd, 35; Univ Chicago, SM, 37, PhD(plant physiol), 41. *Prof Exp:* Asst plant physiol, Univ Chicago, 40-41; assoc veg crops, Univ Ill, 42-43, asst chief, 43-47, asst prof hort, Exp Sta, 47-50; plant physiologist, Argonne Nat Lab, 50-62; dean col lib arts & sci, prof, 65-75, EMER PROF BIOL SCI, NORTHERN ILL UNIV, 75- *Mem:* AAAS; Am Soc Plant Physiologists; Bot Soc Am; Am Soc Hort Sci. *Res:* Plant nutrition; photoperiodism; radiobiology. *Mailing Add:* 1200 Loren Dr De Kalb IL 60115-2105

SKOK, RICHARD ARNOLD, b St Paul, Minn, June 19, 28; c 2. FOREST ECONOMICS. *Educ:* Univ Minn, BS, 50, MF, 54, PhD, 60. *Prof Exp:* Asst prof forest econ, Sch Forestry, Mont State Univ, 58-59; from instr to assoc prof, Univ Minn, 59-65, asst dir sch forestry, 67-71, assoc dean forestry, 71-74, PROF FOREST ECON, UNIV MINN, ST PAUL, 65-, DEAN COL FORESTRY, 74- *Concurrent Pos:* Mem, Joint Coun Food & Agr Sci. *Mem:* Fel Soc Am Foresters; Forest Prods Res Soc. *Res:* Forest resource development-economics and policy. *Mailing Add:* 235 Natural Res Admin Bldg Univ Minn St Paul MN 55108

SKOLD, LAURENCE NELSON, b Haxtun, Colo, Apr 11, 17; m 40; c 1. AGRONOMY. *Educ:* Colo Agr & Mech Col, BS, 38; Kans State Col, MS, 40. *Prof Exp:* Asst, Kans State Col, 38-40; from asst agronomist to assoc agronomist, Ga Agr Exp Sta, 40-46; from asst prof to assoc prof agron, Univ Tenn, Knoxville, 47-56, prof & head dept, 56-60, agron adv, India Agr Prog, 61-66, prof plant & soil sci, 67-82; RETIRED. *Mem:* Am Soc Agron. *Res:* Field crops, culture and improvement. *Mailing Add:* 6704 Crystal Lake Dr Knoxville TN 37919

SKOLMEN, ROGER GODFREY, b San Francisco, Calif, Dec 30, 29. TREE PHYSIOLOGY, FOREST PRODUCTS. *Educ:* Univ Calif, BS, 57, MS, 58, PhD, 77. *Prof Exp:* Asst chem, Shell Develop Co, 55; res asst forestry, Univ Calif, 56-59; soil scientist, Pac Southwest Forest Exp Sta, US Forest Serv, 59-60, wood scientist, Hawaii Res Ctr, 61-72, res forester, Inst Pac Islands Forestry, 72-86; RETIRED. *Mem:* Forest Prod Res Soc; Soc Am Foresters; Plant Tissue Cult Asn. *Res:* Tissue culture propagation of trees; genetic tree improvement; eucalyptus biomass production; utilization of Hawaii-grown woods; tree, log, and wood quality; durability; preservation; seasoning; sawmilling; marketing; silviculture. *Mailing Add:* 403 Koko Isle Circle Honolulu HI 96825

SKOLNICK, HERBERT, b Brooklyn, NY, Jan 15, 19; m 48; c 1. GEOLOGY. *Educ:* Brooklyn Col, BS, 47; Univ Okla, MS, 49; Univ Iowa, PhD(geol), 52. *Prof Exp:* Sedimentologist, Gulf Res & Develop Co, 52-53, supvr geol lab, Western Gulf Oil Co, 53-56, div stratigrapher, 56-60, chief stratigrapher & paleontologist, Spanish Gulf Oil Co, 60-64, supvr stratig lab, Nigerian Gulf Oil Co, 64-67; supvr Stratig Lab, Houston Tech Serv Ctr, Gulf Res & Develop Co, 67-71, res assoc, 71-73, supvr Geol Sect, 73-77, sr res assoc, 77-80, adminr explor res function, 80-82; RETIRED. *Concurrent Pos:* Consult, environ groups concerned with air & water quality & hazardous waste & municipal dumps. *Mem:* Geol Soc Am; Am Asn Petrol Geologists; Soc Econ Paleontologists and Mineralogists; Sigma Xi. *Res:* Petrology, petrography and diagenesis of sedimentary rocks; evolution of subsurface fluid systems; plate tectonics; oceanography; geochemistry; stratigraphy and sedimentology; integration of listed disciplines as a tool for hydrocarbon exploration. *Mailing Add:* 109 S Ridge Dr Monroeville PA 15146

SKOLNICK, JEFFREY, b Brooklyn, NY, June 27, 53. POLYMER DYNAMICS, BIOPHYSICS. *Educ:* Wash Univ, BA, 75; Yale Univ, MPhil, 77, PhD(chem), 78. *Prof Exp:* Asst prof chem, La State Univ, 79-82; from asst prof to prof chem, Wash Univ, 82-89; FULL MEM, RES INST SCRIPPS CLIN, 89- *Concurrent Pos:* Vis prof, Wash Univ, 81; fel, Alfred P Sloan Found. *Mem:* Am Chem Soc; Am Phys Soc; Sigma Xi; NY Acad Sci; AAAS; Biophys Soc. *Res:* Statistical mechanics of polymer solutions, glasses and melts with an emphasis on local main chain dynamics; stress strain behavior of polymer glasses; theory of protein structure, folding kinetics and protein dynamics. *Mailing Add:* Dept Molecular Biol Res Inst Scripps Clin 10666 N Torrey Pines Rd La Jolla CA 92037

SKOLNICK, MALCOLM HARRIS, b Salt Lake City, Utah, Aug 11, 35; m 59; c 4. BIOMEDICAL ENGINEERING. *Educ:* Univ Utah, BS, 56; Cornell Univ, MS, 59, PhD(theoret physics), 63; Univ Houston, JD, 86. *Prof Exp:* Staff scientist elem sci study, Educ Develop Ctr, Mass, 62-63; mem, Sch Math, Inst Advan Study, Princeton Univ, 63-64; instr physics, Mass Inst Technol, 64-65; staff scientist elem sci study & dir Peace Corps Training & Support Serv, Educ Develop Ctr, Mass, 65-67; assoc prof physics, Health Sci Ctr, State Univ NY, Stony Brook, 67-70, dir health sci commun, 68-71, assoc prof path, 70-71; prof & chmn dept biomed commun, Univ Tex Med Sch, Houston, 71-83, PROF BIOPHYSICS, BIOMED SCI, 71-, DIR, NEUROPHYSIOL RES CTR, UNIV TEX, HEALTH & SCI CTR, 84- *Concurrent Pos:* Consult, Educ Develop Ctr, Mass, Comn Col Physics, US Peace Corps, NSF & Nat Sci Teachers Asn; chmn health care technol study sect, Nat Ctr Health Serv Res, 74-78. *Mem:* Inst Elec & Electronics Engrs; Soc Neurosci; Sigma Xi. *Res:* Neuron activiation and inhibition by external stimulation; electrostimulation effects on neuromodulation; simulation and modeling in biophysical systems; product liability; health care technology assessment. *Mailing Add:* Neurophysiol Res Ctr 1343 Moursund Houston TX 77030

SKOLNICK, MARK HENRY, b Temple, Tex, Jan 28, 46; m 70; c 2. GENETICS, POPULATION GENETICS. *Educ:* Univ Calif, Berkeley, BA, 68; Stanford Univ, PhD(genetics), 75. *Prof Exp:* Res asst prof dept biol, 74-77; asst res prof, 74-76, ASST PROF DEPT MED BIOPHYSICS & COMPUT, UNIV UTAH, 76-, ADJ ASST PROF DEPT BIOL, 78- *Concurrent Pos:* NIH res grant, 74-81, 76-80, 77-80 & 79-; Pub Health grant, 76-81; mem Int Union Sci Study of Pop, 76-; dir, Div Health, Utah State Dept Soc Serv, 77-; mem epidemiol comn, Nat Cancer Inst, NIH, 78- *Mem:* Am Soc Human Genetics. *Res:* Genetic epidemiological studies of Mormon genealogies; historical and genetic demography; computerized genealogical data bases, human chromosome mapping, relationship of HLA to disease; preventive medicine and screening for genetic diseases. *Mailing Add:* Dept Med Biophys Utah Sch Med 50 N Medical Dr Salt Lake City UT 84132

SKOLNICK, PHIL, b New York, NY, Feb 26, 47; m 85; c 1. PHARMACOLOGY. *Educ:* Long Island Univ, BSc, 68; George Washington Univ, PhD(pharmacol), 72. *Prof Exp:* Staff fel, NIH, 72-75, sr staff fel pharmacol, 75-77; pharmacologist, Nat Inst Alcohol Abuse, Alcoholism, Alcohol, Drug Abuse & Ment Health Admin, 77-78; sr investr & pharmacologist, 78-83, chief neurobiol, 83-87, CHIEF, LAB NEUROSCI, NIH, 87-; RES PROF PSYCHIAT, UNIFORMED SERV UNIV HEALTH SCI, 89- *Concurrent Pos:* Sr Exec Serv, Dept Health & Human Serv, 89- *Honors & Awards:* A E Bennett Award, 80; Mathilde Soloway Award, 83. *Mem:* Am Soc Pharmacol & Exp Therapeut; Int Soc Neurochem; Soc Biol Psychiat; Am Col Neuropsychopharmacol; Am Soc Neuroscience. *Res:* Neuropharmacology; neuroendocrinology; neurochemical correlates of behavior; neuroimmunology. *Mailing Add:* Lab Neurosci NIH Bldg 8 Rm 111 Bethesda MD 20892

SKOLNIK, HERMAN, b Harrisburg, Pa, Mar 22, 14; c 2. ORGANIC CHEMISTRY, INFORMATION SCIENCE. *Educ:* Pa State Univ, BS, 37; Univ Pa, MS, 41, PhD(org chem), 43. *Prof Exp:* Chemist, Roosevelt Oil Co, Mich, 37-38, testing labs, State Hwy Dept, Pa, 38-39 & Barrett Div, Allied Chem & Dye Corp, Pa, 39-42; res chemist, Hercules Inc, 42-53, mgr, Tech Info Div, 53-79; CONSULT, 79- *Concurrent Pos:* Grants, Univ Pa; ed, J Chem Doc, Terpene Chem-Chem Abstracts & J Chem Info & Computer Sci. *Honors & Awards:* Austin M Patterson Award Chem Doc, Am Chem Soc, 69, Chem Info Sci Award, 76; Terpene Perfum Award, Asn Engn & Tech, 86. *Mem:* Am Soc Info Sci; Am Chem Soc; Tech Asn Pulp & Paper Indust. *Res:* Chemical documentation; history of science; terpenes; heterocyclics; azeotropy. *Mailing Add:* 239 Waverly Rd Wilmington DE 19803

SKOLNIK, MERRILL I, b Baltimore, Md, Nov 6, 27; m 50; c 4. ELECTRICAL ENGINEERING. *Educ:* Johns Hopkins Univ, BE, 47, MSE, 49, DrEng, 51. *Prof Exp:* Res asst elec eng, Johns Hopkins Univ, 47-50, res assoc, Radiation Lab, 50-53; eng specialist, Sylvania Elec Prod Co, 54; staff mem, Lincoln Lab, Mass Inst Technol, 54-59; res mgr, Electronic Commun Inc, 59-64; staff mem, Inst Defense Anal, 64-65; SUPT RADAR DIV, US NAVAL RES LAB, 65- *Concurrent Pos:* Lectr, eve div, Northeastern Univ, 56-59, & Continuing Eng Educ, George Washington Univ , 73-; adj prof, Drexel Inst Technol, 60-66 & eve div, Johns Hopkins Univ, 70-; vis prof, Johns Hopkins Univ, 73-74; ed, Proc Inst Elec & Electronic Engrs, 86-89; Distinguished vis scientist, Jet Propulsion Lab, 90- *Honors & Awards:* Heinrich Hertz Premium, Brit Inst Electronic & Radio Engrs, 65; Harry Diamond Award, Inst Elect & Electronics Engrs, 83, Centennial Medal, 84; Harry Diamond Award, Inst Elec & Electronic Engrs, 83, Centennial Medal, 84. *Mem:* Nat Acad Eng; Fel Inst Elec & Electronics Engrs. *Res:* Radar; antennas; electronic systems; electronic countermeasures; electric arc discharges. *Mailing Add:* Code 5300 US Naval Res Lab Washington DC 20375

SKOLNIKOFF, EUGENE B, b Philadelphia, Pa, Aug 29, 28; m 57; c 3. SCIENCE TECH & INT AFFAIRS, SCIENCE & INT ORGANIZATIONS. *Educ:* Mass Inst Technol, SB & SM, 50 & PhD(polit sci), 65; Oxford Univ, BA & MA, 52. *Prof Exp:* Res asst elec eng, Uppsala Univ, Sweden, 50; indust liaison admin, Mass Inst Technol, 52-55; pvt & proj engr, US Army Security Agency, 55-57; systs analyst, Inst Defense Anal, 57-58; White House staff, Off Spec Asst to Pres, 58-63; dir, Ctr Int Studies, 72-87, PROF POL SCI, MASS INST TECHNOL, 65- *Concurrent Pos:* Consult, Sloan Found, Ford Found, Carnegie Corp, Resources for the Future, Agency Int Develop, Off Technol Assessment, Orgn Econ Coop & Develop, White House Off Sci & Tech Policy, 63-73; founder & pres, Sci & Pub Policy Studies Group, 68-73; vis res scholar, Carnegie Endorsement Int Peace, 69-70; counr class III, AAAS, 73-77; sr consult, White House Off Sci & Technol, 77-81; bd trustees, UN Res Inst Social Develop, 79-85; chmn bd, Ger Marshall Fund US, 80-86; mem adv comt, Nat Low Level Nuclear Waste, 80-87 & adv comt sci & technol, Dept State, 87-; chmn, Sci Policy Group, Nat Acad Sci, 82-84; mem, Comt Sci Eng & Pub Policy, AAAS, 84-89. *Honors & Awards:* Comdr Cross Decoration, Fed Repub Ger; Order of the Rising Sun, Golden Rays, Neck Ribbon, Japan. *Mem:* Fel Am Acad Arts & Sci; fel AAAS; Foreign Aid Soc; Am Polit Sci Asn; Sigma Xi; Coun Foreign Rels; Overseas Develop Coun. *Res:* Interaction of science, technology and public policy, with special emphasis on international scene and affairs, including national and comparative science policy. *Mailing Add:* Mass Inst Technol E38-762 Cambridge MA 02139

SKOMAL, EDWARD N, b Kansas City, Mo, Apr 15, 26; m 51, 87; c 3. RADIO SYSTEMS. *Educ:* Rice Univ, BA, 47, MA, 49. *Prof Exp:* Physicist, Socony-Mobil Field Res Lab, 49-51; supvry physicist, Nat Bur Standards, 51-56; adv develop engr, Sylvania Elec Microwave Physics Lab, 56-59; chief appl engr, Solid State Electronics Div, Motorola, Inc, 59-63; sr develop engr & physicist, Aerospace Corp, El Segundo, 63-67, staff scientist, 67-80, sr eng specialist, off chief engr, 80-84, dir Commun Dept, 84-86; RETIRED. *Concurrent Pos:* Mem, Presidential Joint Tech Adv Comt Electromagnetic Compatibility, 65-74; mem Comn, E, 72-; mem, US Nat Comt, Int Union Radio Sci, 84- *Honors & Awards:* Richar S Stoddart Award, 80. *Mem:* Am Phys Soc; fel Inst Elec & Electronic Engrs; Sigma Xi; Union Radio Scientists. *Res:* Electromagnetism; microwave physics and interaction with solids; radio wave propagation; electromagnetic interference processes; stochastic progresses; radio scattering; guided wave propagation; cryogenics; automatic vehicle locating systems. *Mailing Add:* 1831 Valle Vista Dr Redlands CA 92373

SKONER, PETER RAYMOND, b Johnstown, Pa, July 22, 57; m 80; c 3. SCIENCE & MATHEMATICS EDUCATION. *Educ:* Pa State Univ, BS, 79; Ind Univ Pa, MBA, 84; St Francis Col, MEd, 85. *Prof Exp:* Mining engr, Bethlehem Mines Corp, 79-82; ASST PROF PHYS SCI, ST FRANCIS COL, 84-, CHMN PHYSICS DEPT, 87- *Concurrent Pos:* Res fel, NASA & Am Soc Eng Educ, 85. *Mem:* Am Soc Eng Educ; Nat Sci Teachers Am; Nat Coun Teachers Math. *Res:* Science and mathematics education; study of retention of pre-engineering college students; use of computers in teaching statistics. *Mailing Add:* Physics Dept St Francis Col Loretto PA 15940

SKOOG, DOUGLAS ARVID, b Willmar, Minn, May 4, 18; m 42; c 2. CHEMISTRY. *Educ:* Ore State Col, BS, 40; Univ Ill, PhD(anal chem), 43. *Prof Exp:* Res chemist, Calif Res Corp, Standard Oil Co Calif, 43-47; from asst prof to prof, 47-76, assoc exec head dept chem, 61-76, EMER PROF CHEM, STANFORD UNIV, 76- *Mem:* Am Chem Soc; AAAS; Sigma Xi. *Res:* Instrumental analysis; spectrophotometry; organic reagents; complex ions. *Mailing Add:* Dept of Chem Stanford Univ Stanford CA 94305

SKOOG, FOLKE, b Fjaras,Sweden, July 15, 08; nat US; m 47; c 1. BIOCHEMISTRY. *Educ:* Calif Inst Technol, BS, 32, PhD(biol), 36. *Hon Degrees:* DPhil, Univ Lund, 56; DSc, Univ Ill, 80 & Univ Pisa, Italy, 91. *Prof Exp:* Teaching asst, Calif Inst Technol, 34-36; Nat Res Coun fel bot, Univ Calif, 36-37; instr & res assoc, Harvard Univ, 37-41; assoc prof bot, Johns Hopkins Univ, 41-44; biochemist, Off Qm Gen, Washington, DC, 43-44; tech rep, US Army Europ Theater Opers, Eng, Scand, Germany & Austria, 45-46; from assoc prof to prof, 47-49, EMER PROF BOT, UNIV-WIS MADISON, 79- *Concurrent Pos:* vis physiologist, Exp Sta, Pineapple Res Inst, Univ, Hawaii, 38; assoc physiologist, NIH, 43; lectr, Wash Univ, 46; vpres physiol sect, Int Bot Cong, Paris, 54, Edinburgh, 64, Leningrad, 75, mem comt algal cult, 54-; mem study sect genetics & morphol, NIH, 56-60, study sect cell biol, 61-64; mem, Panel Regulatory Biol, NSF, 56-60; mem, Surv Comn Sci & Technol Educ, Brazil, Nat Acad Sci, 60; mem, Adv Panel Appl Math, Phys Sci, Biol & Eng, NSF, 79-81; mem, adv comt, USSR & E Europe, Nat Acad Sci, 80-84. *Honors & Awards:* Nat Medal of Sci, 91; Hales Prize, 54, Barnes Life Mem Award, Am Soc Plant Physiol. *Mem:* Nat Acad Sci; Int Plant Growth Subst Asn (vpres, 76-79, pres, 79-82); foreign mem Swed Nat Acad Sci; Deutsch Akademie der Naturforscher Leopoldina; Am Acad Arts & Sci; foreign mem Acad Reg Sci Upsaliensis; Am Soc Plant Physiologists (vpres, 52-53, pres, 57-58); Am Soc Gen Physiologists (pres, 56-57); Soc Develop Biologists (pres, 70-71); Bot Soc Am. *Res:* Plant growth and development; cytokinins; plant tissue culture. *Mailing Add:* Dept Bot Univ Wis Madison WI 53706

SKOOG, IVAN HOOGLUND, b Kewanee, Ill, July 26, 28; m 55; c 2. ORGANIC CHEMISTRY. *Educ:* Univ Ill, BS, 50; Northwestern Univ, MS, 52, PhD, 55. *Prof Exp:* CHEMIST, MINN MINING & MFG CO, 54- *Mem:* Am Chem Soc; Soc Photog Sci & Eng. *Res:* Organic synthesis; photography. *Mailing Add:* 8573 Hidden Bay Trail N Lake Elmo MN 55042-9526

SKOOG, WILLIAM ARTHUR, b Culver City, Calif, Apr 10, 25; m 49; c 4. MEDICAL ONCOLOGY. *Educ:* Stanford Univ, AB, 46, MD, 49; Am Bd Internal Med, dipl, 57. *Prof Exp:* Intern med, Univ Hosp, Stanford Univ, 48-49, asst resident, 49-50; asst resident med, NY Hosp-Cornell Med Ctr & asst in med col, 50-51; sr resident, Wadsworth Vet Admin Hosp, Los Angeles, Calif, 51; jr res physician, Atomic Energy Proj, Univ Calif, Los Angeles, 54-55, from instr to asst prof med, Sch Med, 55-59, jr res physician, 55-56, asst res physician & co-dir metab res unit, Ctr for Health Sci, 56-59; asst clin prof med & assoc res physician oncol, Sch Med & assoc staff, Med Ctr, Univ Calif, San Francisco, 59-61; lectr, 61, co-dir, Health Sci Clin Res Ctr, 65-67, dir health sci clin res ctr, 67-72, assoc prof, 62-73, ASSOC CLIN PROF MED, SCH MED, UNIV CALIF, LOS ANGELES, 73- *Concurrent Pos:* Clin assoc hemat, Vet Admin Ctr, Los Angeles, 56-59; clin instr, Sch Med, Stanford Univ, 59-61; mem staff, Palo Alto-Stanford Hosp Ctr, 59-61; attend specialist, Wadsworth Vet Admin Hosp, Los Angeles, 62-68; vis physician, Harbor Gen Hosp, Torrance, 62-65, attend physician, 65-77; consult, Clin Lab, Univ Calif, Los Angeles Hosp, 63-68; mem affiliate consult staff, St John's Hosp, Santa Monica, 64-71, courtesy staff, 71-72; active staff, St Bernardine Hosp, San Bernardino, 72-; active staff San Bernardino Community Hosp, 72-; consult staff, Redlands Community Hosp, 72-; chief, Oncol Sect, San Bernardino County Med Ctr, 72-76. *Mem:* AMA; fel Am Col Physicians; Am Soc Clin Oncol; Am Fedn Clin Res; Western Soc Clin Res. *Res:* Hematology; hematologic malignancies, especially multiple myeloma; cancer chemotherapy. *Mailing Add:* 401 E Highland Ave Suite 552 San Bernardino CA 92404

SKOP, RICHARD ALLEN, b Baltimore, Md, Mar 12, 43; m 64; c 3. APPLIED MECHANICS. *Educ:* Wash Univ, St Louis, BA, 64; Univ Rochester, PhD(appl mech), 68. *Prof Exp:* Res engr appl mech, US Naval Res Lab, 67-71, head fluid mech sect, 71-78, head appl mech br, 78-86; CHMN, MARINE PHYSICS, UNIV MIAMI, 86- *Mem:* Am Soc Mech Engrs; Marine Technol Soc; Sigma Xi. *Res:* Fluid and structure interaction problems; wake dynamics; computational fluid dynamics. *Mailing Add:* 538 Heavirtree Hill Severna Park MD 21146

SKOPIK, STEVEN D, b Detroit, Mich, Dec 9, 40; m 60; c 3. PHYSIOLOGY. *Educ:* Defiance Col, BS, 62; Princeton Univ, MA, 64, PhD, 66. *Prof Exp:* From instr to lectr biol, Princeton Univ, 65-67; asst prof, 67-71, assoc chairperson dept biol sci, 73-76, ASSOC PROF BIOL, UNIV DEL, 71-, COODR PHYSIOL SECT, 76- *Mem:* AAAS; Entom Soc Am. *Res:* Role of biological clocks in the development of insects; role of clock systems in insect photoperiodism. *Mailing Add:* Sch Life & Health Sci & Physiol Univ Del Newark DE 19716

SKOPP, JOSEPH MICHAEL, b Long Beach, Calif, Nov 24, 49; m 77; c 3. SOIL PHYSICS. *Educ:* Univ Calif, Davis, BS, 71; Univ Ariz, MS, 75; Univ Wis, PhD(soils), 80. *Prof Exp:* ASSOC PROF SOIL PHYSICS, UNIV NEBR, 80- *Mem:* Soil Sci Soc Am; Am Geophys Union; AAAS. *Res:* Measurement and description of solute (plant nutrients, pollutants or microorganisms) movement in soils; physical processes, particularly oxygen and nutrient transport, limiting microbial activity in soil. *Mailing Add:* Dept Agron Unvi Nebr Lincoln NE 68583-0915

SKORCZ, JOSEPH ANTHONY, b Milwaukee, Wis, May 25, 36; m 60; c 3. INDUSTRIAL ORGANIC CHEMISTRY. *Educ:* Marquette Univ, BS, 57; Univ Wis, MS, 59; Brown Univ, PhD(org chem), 62. *Prof Exp:* Sr res chemist, Lakeside Labs Div Colgate-Palmolive Co, 62-67, admin asst to dir res, 67-70, dir chem mfg & environ control, 70-74, dir mfg, 74-75; asst vpres mfg, Merrell-National Labs Div, 75-76, dir chem synthesis opers, Richardson-Merrell, Inc, 76-81, TECH MGR, MERRELL DOW PHARMACEUT INC, 81-; AT DOW CHEM USA, MI. *Mem:* Am Chem Soc. *Res:* Production scale synthetic organic chemistry; pharmaceutical manufacturing technology. *Mailing Add:* Dow Chem USA Bldg 2020 Midland MI 48667-0001

SKORINKO, GEORGE, b Palmerton, Pa, Sept 25, 30; m 60; c 2. PHYSICS. *Educ:* Lehigh Univ, BS, 52; Boston Univ, MA, 53; Pa State Univ, PhD(physics), 60. *Prof Exp:* Physicist, Westinghouse Res Lab, 60-62; assoc prof, 62-74, PROF PHYSICS, BROOKLYN COL, 74- *Mem:* Optical Soc Am. *Res:* Infrared spectroscopy of molecules; extreme ultraviolet spectroscopy; solid state physics. *Mailing Add:* Bedford & Ave H Brooklyn Col Brooklyn NY 11210

SKOROPAD, WILLIAM PETER, b Alta, Can, May 23, 18; m 45; c 4. PLANT PATHOLOGY. *Educ:* Univ Alta, BSc & MSc, 50; Univ Wis, PhD(plant path), 55. *Prof Exp:* Agr res off, Fed Lab Plant Path, Alta, 52-59; from asst prof to prof plant path, Univ Alta, 59-85; RETIRED. *Mem:* Am Phytopath Soc; Can Phytopath Soc; Agr Inst Can. *Res:* Disease of rape crops, and foliage diseases of barley. *Mailing Add:* 7311 78th St Edmonton AB T6C 2M9 Can

SKORYNA, STANLEY C, b Warsaw, Poland, Sept 4, 20; Can citizen; m 70; c 3. GASTROENTEROLOGY, EXPERIMENTAL SURGERY. *Educ:* Univ Vienna, MD, 43, PhD(biol), 62; McGill Univ, MSc, 50. *Prof Exp:* Lectr, 55-59, asst prof, 59-62, ASSOC PROF SURG, MCGILL UNIV, 62-, DIR GASTROINTESTINAL RES LAB, 59-, DIR, RIDEAU INST, 69- *Concurrent Pos:* Res fel cancer, McGill Univ, 47-49; res fel, Nat Cancer Inst

Can, 49-51, sr res fel, 51-54; dir, Can Med Exped to Easter Island, 64-65; deleg, Biol Coun Can, 67; assoc prof biol, Univ de Montreal, 85- *Honors & Awards:* Gold Medal Surg, Royal Col Physicians & Surgeons, Can, 57. *Mem:* Am Asn Cancer Res; Am Gastroenterol Asn; Am Col Surgeons; Nutrit Soc Can. *Res:* Experimental carcinogenesis; pathophysiology of peptic ulcer; intestinal absorption of the metal ions, strontium and calcium; mucolytic action of amides; growth curve studies on Coelenterata. *Mailing Add:* Gastrointestinal Res Lab-McGill 740 Ave Dr Penfield Montreal ON H3A 1A4 Can

SKOSEY, JOHN LYLE, b Gillespie, Ill, Jan 19, 36; m 60; c 3. MEDICINE, PHYSIOLOGY. *Educ:* Univ Southern Ill, BA, 57; Univ Chicago, MD, 61, PhD(physiol), 64. *Prof Exp:* Intern, Univ Chicago Hosp, 61-62, jr asst resident med, Univ Chicago, 62-63; clin assoc, Endocrinol Br, Nat Cancer Inst, 63-65; sr asst resident med, Univ Chicago, 65-66, resident, 66-67, instr, 67-69, asst prof, 69-74, assoc prof, 74-78; assoc prof, 78-82, PROF, DEPT MED, UNIV ILL, 82- *Concurrent Pos:* USPHS trainee physiol, Univ Chicago, 62-63, USPHS clin trainee, 65-67; Arthritis Found fel, 68-71. *Mem:* Am Fedn Clin Res; Am Physiol Soc; Am Rheumatism Asn; Sigma Xi. *Res:* Cell physiology and pathophysiology. *Mailing Add:* Dept Med Univ Ill 840 S Wood St Chicago IL 60612

SKOTNICKI, JERAULD S, b Niagara Falls, NY, Jan 28, 51; m 74. MEDICINAL CHEMISTRY, HETEROCYCLIC CHEMISTRY. *Educ:* Col of the Holy Cross, AB, 73; Dartmouth Col, MA, 75; Princeton Univ, PhD(chem), 81. *Prof Exp:* Chemist, Lederle Labs, 74-77; res chemist, Du Pont, 81-82; sr chemist, Wyeth Labs, 82-86, res scientist, 86-88, PRIN SCIENTIST, WYETH-AYERST RES, 88- *Concurrent Pos:* Adj fac, Villanova Univ, 85-88 & 91- *Mem:* Am Chem Soc; AAAS; Inflammation Res Asn. *Res:* Design and synthesis of biologically important molecules including interleuken-1 inhibitors, phosphodiesterase inhibitors, prostaglandins, folic acid analogs and beta-lactam antibiotics; introduction and development of new synthetic methodologies. *Mailing Add:* Wyeth-Ayerst Res Princeton NJ 08543-8000

SKOUG, DAVID L, b Rice Lake, Wis, Dec 31, 37; m 61; c 2. MATHEMATICAL ANALYSIS. *Educ:* Wis State Univ, River Falls, BA, 60; Univ Minn, PhD(math), 66. *Prof Exp:* From asst prof to assoc prof, 66-75, chmn dept, 75-83, PROF MATH, UNIV NEBR, LINCOLN, 76- *Mem:* Am Math Soc; Math Asn Am; Korean Math Soc. *Res:* Mathematical research in the areas of integration in function space, Wiener space and integrals, Feynman integrals. *Mailing Add:* Math Dept 809 Oldfather Hall Univ Nebr Lincoln NE 68588-0323

SKOUGSTAD, MARVIN WILMER, analytical chemistry; deceased, see previous edition for last biography

SKOULTCHI, ARTHUR, b New York, NY, Aug 8, 40; m 65; c 2. CELL BIOLOGY. *Educ:* Princeton Univ, AB, 62; Yale Univ, MS, 65, PhD(molecular biophys & biochem), 69. *Prof Exp:* Fel biochem, Yale Univ, 69-70, spec fel, 72-73; fel biol, Mass Inst Technol, 70-72; PROF CELL BIOL, ALBERT EINSTEIN COL MED, 73- *Honors & Awards:* Fac res award, Am Cancer Soc, 74. *Mem:* Sigma Xi; Am Soc Cell Biol. *Res:* Mechanisms for controlling gene expression during differentiation of mammalian cells; somatic cell genetics of differentiation; messenger RNA biosynthesis in animal cells. *Mailing Add:* Dept Cell Biol Albert Einstein Col Med 1300 Morris Park Ave Bronx NY 10461

SKOULTCHI, MARTIN MILTON, b New York, NY, Oct 27, 33; m 60; c 3. ORGANIC CHEMISTRY, POLYMER CHEMISTRY. *Educ:* NY Univ, BA, 54, MS, 57, PhD(org chem), 60. *Prof Exp:* Assoc chemist organometallic chem, Res Div, Col Eng, NY Univ, 55-60; SR RES ASSOC, NAT STARCH & CHEM CORP, 60- *Mem:* AAAS; Am Chem Soc; Soc Photog Sci & Eng; NY Acad Sci. *Res:* Synthesis of speciality monomers and polymers; chemical reactions of and on polymers; photochemistry. *Mailing Add:* 6 Lilac Lane Somerset NJ 08873

SKOUTAKIS, VASILIOS A, b Skoura-Spartis, Greece, Sept 24, 43; US citizen; m 68. TOXICOLOGY. *Educ:* Univ NC, Charlotte, BS, 70, Chapel Hill, BS, 72; Univ Tenn Memphis, Dr Pharm, 74. *Prof Exp:* From asst prof to assoc prof, 74-83, dir, 77-83, PROF & DIR, DRUGS-CLIN PHARMACOL & TOXICOL, UNIV TENN, MEMPHIS, 83- *Concurrent Pos:* Pres, Clin Toxicol Consult, Inc, 79-; ed & publ, J Clin Toxicol Counsult, 79-; dir, Drug & Toxicol Info Ctr. *Mem:* Am Col Clin Pharm (treas, 83-86); Am Col Clin Pharmacol; Am Soc Hosp Phamacists; Am Asn Col Pharm. *Res:* Clinical drug research trials; developmental therapeutics; pharmcotherapeutics. *Mailing Add:* 7271 Deep Valley Dr Memphis TN 38138

SKOV, CHARLES E, b Kearney, Nebr, June 29, 33; m 54; c 3. SOLID STATE PHYSICS. *Educ:* Nebr State Col, Kearney, BA, 54; Univ Nebr, PhD(physics), 63. *Prof Exp:* Assoc prof, 63-73, PROF PHYSICS, MONMOUTH COL, ILL, 73- *Mem:* Am Phys Soc; Am Optical Soc; Am Asn Physics Teachers; Sigma Xi. *Res:* Electrical and optical properties of insulating crystals. *Mailing Add:* Dept of Physics Monmouth Col Monmouth IL 61462

SKOV, NIELS A, b Ribe, Denmark, Nov 6, 19; US citizen; m 53; c 2. PHYSICAL OCEANOGRAPHY, HISTORY OF SCIENCE. *Educ:* Technol Denmark, BS, 47; Ore State Univ, MS, 65, PhD(phys oceanog), 67. *Prof Exp:* PROF OCEANOG, EVERGREEN STATE COL, 72- *Mem:* AAAS. *Mailing Add:* Dept Oceanog Evergreen State Col 2700 Evergreen Pkwy NW SE3127 Olympia WA 98505-0002

SKOVE, MALCOLM JOHN, b Cleveland, Ohio, Mar 3, 31; m 56; c 2. SOLID STATE PHYSICS. *Educ:* Clemson Univ, BS, 56; Univ Va, PhD(physics), 60. *Prof Exp:* Asst prof physics, Ill State Univ, 60-61 & Univ PR, 61; from asst prof to assoc prof, 61-68, PROF PHYSICS, CLEMSON UNIV, 68- *Concurrent Pos:* Fulbright lectr, Haile Selassie Univ, 66-67; vis prof, Swiss Fed Inst Technol, 74-75; prog dir, NSF, 87-88. *Mem:* AAAS; Am Phys Soc; Am Asn Physics Teachers. *Res:* Effect of elastic strain on electrical properties of metals. *Mailing Add:* Dept of Physics Clemson Univ Clemson SC 29634-1911

SKOVLIN, JON MATTHEW, b Colfax, Wash, Oct 31, 30; m 52; c 4. RANGE SCIENCE, WILDLIFE BIOLOGY. *Educ:* Ore State Univ, BS, 52; Univ Idaho, MS, 59. *Prof Exp:* Range scientist, Res Br, US Forest Serv, Ore, 56-68; ecologist, UN Food & Agr Orgn, Nairobi, Kenya, 68-71; res biologist wildlife habitat res, Range & Wildlife Habitat Lab, 71-76, range scientist/proj leader, USDA Forest Serv, Ore, 76-81; range & animal scientist, Winrock Int, Kenya, 82-85; CONSULT, 85- *Concurrent Pos:* Consult, EAfrica natural resources & environ, 77, 78 & 80; prin investr, EAfrica rangeland invest, Kenya; cert rangeland consult, Soc Range Mgt; cert wildlife biologist, Wildlife Soc. *Mem:* Soc Range Mgt; Soc Am Foresters; EAfrican Wildlife Soc; Wildlife Soc; Int Soc Trop Foresters; Grassland Soc, Southern Africa. *Res:* Investigations of levels, seasons and systems of livestock grazing and interactions on related resources throughout semi-arid zones of western North America and arid zones of East Africa; over 50 scientific publications. *Mailing Add:* PO Box 2874 La Grande OR 97850

SKOVRONEK, HERBERT SAMUEL, b Brooklyn, NY, Apr 19, 36; m 60; c 2. ORGANIC CHEMISTRY, ENVIRONMENTAL ENGINEERING. *Educ:* Brooklyn Col, BS, 56; Pa State Univ, PhD(carbenes), 61. *Prof Exp:* From chemist to sr chemist, Texaco, Inc, 61-62; res chemist, Rayonier, Inc, 62-67; res chemist, J P Stevens & Co, Inc, 67, group leader, 67-71; res chemist, Indust Waste Technol Br, US Environ Protection Agency, 71-74, tech adv to dir, Indust Waste Treatment Res Lab, 74-75, tech adv to dir, Indust Environ Res Lab, 75-78; mgr environ control, Semet-Solvay Div, Allied Chem Corp, 78-79, environ specialist, 79-82; ADJ PROF, NJ INST TECHNOL, 82-; SR ENVIRON SCIENTIST, SCI APPLICATIONS INT CORP, 88- *Concurrent Pos:* Environ consult, Environ Serv, 82- *Mem:* Water Pollution Control Fedn; Am Chem Soc; Hazardous Mat Control Res Inst. *Res:* Pollution abatement, waste minimization technology for metal finishing, pharmaceuticals, paints and chemicals manufacturing; environmental impacts of industrial energy conservation practices; innovative use of canine olfaction in toxic pollutant detection; management and control of industrial, hazardous and waterborne wastes; hazard communication. *Mailing Add:* 88 Moraine Rd Morris Plains NJ 07950

SKOW, LOREN CURTIS, b Gainesville, Tex, Sept 4, 46; m 74. GENETICS, BIOCHEMISTRY. *Educ:* Abilene Christian Col, BSEd, 69, MS, 71; Tex A&M Univ, PhD(fisheries sci), 76. *Prof Exp:* Res assoc mammalian genetics, Oak Ridge Nat Lab, 78-79, staff scientist mammalian mutagenesis, 79-81; sr staff fel genetics, Nat Inst Environ Health Sci, 81-; AT DEPT VET ANAT, COL VET MED, TEX A & M UNIV. *Concurrent Pos:* NIH fel, Jackson Lab, 76-78; cystic fibrosis consult, NIH, 78- *Mem:* Genetics Soc Am; AAAS. *Res:* Comparative vertebrate genetics; gene mapping; organization of genes controlling rodent salivary secretions; genetics of rodent lens crystallins. *Mailing Add:* Dept Vet Anat-Col Vet Med Tex A & M Univ College Station TX 77843-4458

SKOWRONSKI, RAYMUND PAUL, b Detroit, Mich, Feb 7, 48; m 70. ADVANCED MATERIALS, COMPOSITES. *Educ:* Univ Mich, BS, 69; Tex A&M Univ, PhD(chem), 75. *Prof Exp:* Teaching asst chem, Tex A&M Univ, 69-70, instr, 70-72; MATERIALS SCIENTIST, ROCKWELL INT, 75- *Mem:* Am Chem Soc; Sigma Xi (pres, 83-84). *Res:* Materials Science; coatings; emissivity; emissivity coatings; spacecraft armor; atomic oxygen; narrow-band reflectors, heat pipes, survivability, radiation shielding. *Mailing Add:* Rockwell Int WB21 6633 Canoga Ave Canoga Park CA 91303

SKRABEK, EMANUEL ANDREW, b Baltimore, Md, Mar 3, 34; m 72; c 1. ENERGY CONVERSION, PHYSICAL CHEMISTRY. *Educ:* Univ Md, BS, 56; Univ Wis, MS, 58; Univ Pittsburgh, PhD(phys chem), 62. *Prof Exp:* Sr res scientist energy conversion, Martin-Marietta Corp, 62-69; sr scientist heat pipe mat, Dynatherm Corp, 69-72; MGR RES THERMOELEC, TELEDYNE ENERGY SYSTS, 72- *Mem:* Am Chem Soc; AAAS; Am Soc Testing & Mat; Sigma Xi. *Res:* Thermoelectric materials development and testing; heat pipe materials compatibility; wetting of surfaces; thermal conductivity of insulators; thermodynamic compatibility of materials at 1000 degrees centigrade and above. *Mailing Add:* 1510 Cranwell Rd Lutherville MD 21093-5836

SKRABLE, KENNETH WILLIAM, b Teaneck, NJ, Oct 10, 35; m 57; c 2. PHYSICS, RADIOLOGICAL HEALTH. *Educ:* Moravian Col, BS, 58; Vanderbilt Univ, MS, 63; Rutgers Univ, PhD(environ sci), 70. *Prof Exp:* Health physics supvr, Indust Reactor Lab, Inc, 59-63; radiation safety officer & lectr radiation sci, Rutgers Univ, 63-68; prof & chmn radiological sci dept, Lowell Technol Inst, 68-74; PROF RADIOLOGICAL SCI, UNIV LOWELL, 74- *Concurrent Pos:* Chmn, New Eng Consortium on Environ Protection, 71-72; consult, Yankee Atomic Elec Co, 75-76, US Nuclear Regulatory Comn, 80- *Mem:* Health Physics Soc; Am Nuclear Soc; Sigma Xi. *Res:* Air pollution, aerosols; naturally occurring and man-made radioactive aerosols; measurement and control of air pollutants and radioactivity; internal and external radiation dosimetry. *Mailing Add:* Dept Physics Univ Lowell One University Ave Lowell MA 01854

SKRAMSTAD, HAROLD KENNETH, b Tacoma, Wash, July 26, 08; m 40; c 4. PHYSICS. *Educ:* Univ Puget Sound, BS, 30; Univ Wash, Seattle, PhD(physics), 35. *Prof Exp:* Teacher high sch, Wash, 30-31; physicist, Nat Bur Stand, 35-46, chief, Guided Missiles Sect, 46-50, asst chief, Missile Develop Div, 50-53; chief, Missile Systs Div, US Naval Ord Lab, 53-54; asst chief systs, Data Processing Systs Div, Nat Bur Stand, 54-61; assoc tech dir, US Naval Ord Lab, 61-67; prof indust eng & sci dir comput ctr, 67-74, EMER PROF MGT SCI & EMER ADJ PROF MATH, UNIV MIAMI, 74- *Concurrent Pos:* Life mem, Simulation Coun; adj prof, Fla Inst Technol, 76-84. *Honors & Awards:* Reed Award, Am Inst Aeronaut & Astronaut, 47.

Mem: Assoc fel Am Inst Aeronaut & Astronaut; life sr mem Inst Elec & Electronics Engrs. *Res:* Primary ionization of gases; wind tunnel turbulence; boundary layer flow; development of guided missiles; aerodynamics; computers; simulators; automatic control. *Mailing Add:* 8045 S A1A Hwy Melbourne Beach FL 32951

SKRDLA, WILLIS HOWARD, b DeWitt, Nebr, Feb 22, 20; m 42; c 3. AGRONOMY. *Educ:* Univ Nebr, BSc, 41; Purdue Univ, PhD(agron), 49. *Prof Exp:* Asst, Purdue Univ, 46-49; assoc prof agron, Va Agr Exp Sta, Va Polytech Inst, 49-53; agronomist airport turfing, US Dept Air Force, 53-57; leader, Genetic Resources Unit, Int Ctr Res Semi-Arid Tropics, Hyderabad, India, 84-86; leader design team, Indo-US Proj, Plant Genetic Resources, New Delhi, 86; prof agron, 57-83, agronomist res leader & coordr, N Cent Regional Plant Sta, US Dept Agr, 57-83, EMER PROF AGRON, IOWA STATE UNIV, 83. *Mem:* Fel AAAS; Sigma Xi; Soc Econ Bot; Am Soc Agron. *Res:* Plant introduction; seed increase, distribution, evaluation and permanent storage of world collections of agronomic and horticultural crops; coordinate regional program in 13 states; field crop, forage and turf investigations. *Mailing Add:* 2136 N Duff Ave Ames IA 50010

SKRINAR, GARY STEPHEN, b Teaneck, NJ, Aug 28, 42; c 3. EXERCISE PHYSIOLOGY. *Educ:* Oklahoma City Univ, BA, 64; Univ Ill, MS, 65; Univ Pittsburgh, PhD(exercise phys motor learning), 78. *Prof Exp:* Instr phys educ, Midland Mich Pub Schs, 65-68, Oklahoma City Univ, 68-70 & Ore Sta Univ, 70-71; asst prof, Brookdale Community Col, 71-74; ASSOC PROF HEALTH SCI, BOSTON UNIV, 78- *Concurrent Pos:* Exercise tech, exercise leader, Univ Pittsburgh Cardiac Rehab Prog, 74-78; exercise prog dir, Am Col Sports Med, 78. *Mem:* Am Col Sports Med; Am Heart Asn. *Res:* Exercise physiology; cardiac rehabilitation; acquisition and maintenance of physical fitness. *Mailing Add:* Dept of Health Sci 635 Commonwealth Ave Boston MA 02215

SKRINDE, ROLF T, b Stanwood, Wash, Sept 1, 28; m 58; c 3. CIVIL & ENVIRONMENTAL ENGINEERING. *Educ:* Wash State Univ, SB & CE, 51; Mass Inst Technol, SM, 52, SanE, 56, PhD(sanit eng), 59. *Prof Exp:* Sanit eng adv, Ministry Health, Saudi Arabia & Thailand, 53-55; asst sanit eng, Mass Inst Technol, 55-58; asst prof civil eng, Wash State Univ, 58-60, assoc prof, 60-61; assoc prof, Wash Univ, 61-64, chmn dept civil eng, 64-65; dir res, SEATO Grad Sch Eng, Bangkok, 65-67; prof civil eng, Univ Mass, Amherst, 67-69; prof & chmn dept civil eng, Univ Iowa, 69-77; mem staff, Olympic Assocs Co, 77-84; FAC, CIVIL ENG DEPT, SEATTLE UNIV, 84- *Mem:* Am Soc Civil Eng; Water Pollution Control Fedn; Am Water Works Asn; Nat Soc Prof Engrs. *Res:* Corrosion control in potable water systems; stream pollution; waste water treatment; determination of public health effects of agricultural use on return irrigation waters; pesticide residues; reverse osmosis water treatment; metal plating waste treatment. *Mailing Add:* 16327 Inglewood Place NE Bothell WA 98011

SKRIVAN, J(OSEPH) F(RANCIS), b Baltimore, Md, Oct 25, 31; m 54; c 6. CHEMICAL ENGINEERING. *Educ:* Johns Hopkins Univ, BE, 53, MS, 56, DEng, 58. *Prof Exp:* Res engr, 58-63, sr res engr, 64-66, group leader eng res, 66-72, res mgr, 72-75, tech dir, 75-78, mgr mfg catalyst dept, 78-81, res dir, 81-86, DIR, ENG & OPERS, AM CYANAMID CO, 86- *Mem:* Am Inst Chem Engrs; Sigma Xi. *Res:* Kinetics; heat transfer; high temperature processing and plasma technology; process development; auto exhaust catalysts. *Mailing Add:* 154 Berrian Rd Stamford CT 06905

SKROCH, WALTER ARTHUR, b Arcadia, Wis, July 1, 37; m 63; c 2. WEED SCIENCE. *Educ:* Wis State Univ, River Falls, BS, 59; Univ Wis, Madison, MS, 61, PhD(hort), 65. *Prof Exp:* Assoc prof, 68-73, PROF HORT SCI, N C STATE UNIV, 73- *Mem:* Am Pomol Soc; Weed Sci Soc Am; Am Soc Hort Sci. *Res:* Herbicide activity, tree fruits, ornamental, Christmas trees, landscape and broadrange fumigation as preplant treatment of horticultural crops. *Mailing Add:* Dept Hort Sci N C State Univ 166 Kilgore Raleigh NC 27607

SKROMME, LAWRENCE H, b Roland, Iowa, Aug 26, 13; m 39; c 3. AGRICULTURAL ENGINEERING. *Educ:* Iowa State Univ, BSc, 37. *Prof Exp:* From draftsman to design engr, Goodyear Tire & Rubber Co, 37-41; from proj engr to asst chief engr, Harry Ferguson, Inc, 41-51; chief engr, Sperry New Holland, Sperry Rand Corp, 51-61, vpres eng, 61-79; CONSULT, AGR ENGR, 79- *Concurrent Pos:* Mem, Farm Resources & Facilities Res Adv Comt, USDA, 65; mem, Int Rels Comt & Nat Medal of Sci Comt, Engrs Joint Coun; mem, Gov Comt Preserv Agr Land; vpres, Farm & Home Found, Lancaster County, Pa; mem div eng, Nat Res Coun; power & mach rep, Int Comn Agr Eng; vpres, bd mem & pres, Agr Mach Sect, Comn Int Genie Rurale, 73-; consult agr eng, 79- *Honors & Awards:* John Deere Gold Medal, Am Soc Agr Engrs. *Mem:* Nat Acad Eng; fel Am Soc Agr Engrs (vpres, 52-55, pres, 59-60); Am Soc Eng Educ; Nat Soc Prof Engrs; Soc Automotive Engrs; NY Acad Sci. *Res:* Development of efficient farm machines to reduce labor and improve productivity. *Mailing Add:* 2144 Landis Valley Rd Lancaster PA 17601

SKRYPA, MICHAEL JOHN, b Woonsocket, RI, Sept 26, 27; m 53; c 3. COMMERCIALIZATION OF RESEARCH DISCOVERIES. *Educ:* Brown Univ, BSc, 50; Clark Univ, PhD(org chem), 54. *Prof Exp:* Res chemist, Solvay Process Div, Allied Chem Corp, 53-61, assoc res supvr, 61-62, res supvr, 62-63, mgr appl res, 63-67, mgr new polymer applns develop, Plastics Div, 67-72, res consult, Corp Res Ctr, 72-76, mgr com develop, Venture Mgt Div, 76-79, mgr mkt develop, New Ventures Group, Allied Corp, 79-83, consult, 83-86, MANAGING ASSOC, MISKCO ASSOC, 86- *Mem:* Soc Plastics Eng; Com Develop Asn; Am Chem Soc; Am Soc Metals. *Res:* Organic syntheses; polymerizations; physical chemistry of polymers; physical chemistry of metals; development of applications for chemicals, polymers and metals; direction and management of research organizations. *Mailing Add:* Three Village Rd Florham Park NJ 07932

SKRZYPEK, JOSEF, COMPUTATIONAL NEUROSCIENCE, COMPUTER VISION. *Educ:* Western New Eng Col, BS, 71; Univ Calif, Berkeley, MS, 74 & PhD(eng & comput sci), 79. *Prof Exp:* Res assoc electronic eng & comput sci, Univ Calif, Berkeley, 76-79, fel vision, 79-80; fel comput & vision, NY Univ, 80-81; res engr, Analogic Inc, 81-82; assoc prof electronics & comput sci, Northeastern Univ, 82-85; ASST PROF COMPUT SCI, UNIV CALIF, LOS ANGELES, 85- *Concurrent Pos:* Prin investr, Nat Sci Found, 83-85, Texas Instruments, 87-88, Hughes Electronics Res Lab, 88-89 & Army Res Off, 88-91; prof analog devices, Northeastern Univ, 83-85; coprin investr, Defense Adv Res Proj Agency, 86-88; consult, TRW, 87- *Mem:* Inst Elec & Electronics Engr; Am Asn Artificial Intel; Int Conf Neural Networks. *Res:* Computational aspects of phenomenon of vision in men and machines; analysis, modelling and synthesis of neural network computing architectures that underly vision. *Mailing Add:* Dept Comput Sci Univ Calif 3532 Bh 405 Hilgard Ave Los Angeles CA 90024

SKUBIC, PATRICK LOUIS, b Eveleth, Minn, Sept 9, 47. ELEMENTARY PARTICLE PHYSICS. *Educ:* SDak State Univ, BS, 69; Univ Mich, MS, 70, PhD(physics), 77. *Prof Exp:* Fel, Rutgers Univ, 77-80; ASST PROF, UNIV OKLA, 81- *Mem:* Am Phys Soc; Sigma Xi. *Res:* Experimental elementary particle physics; precise measurement of neutral hyperon polarization and magnetic moments in high energy neutral hyperon beams; evidence for production of particles containing heavy quarks in electron-positron collisions. *Mailing Add:* 1510 Melrose Dr Norman OK 73069

SKUCAS, JOVITAS, b Klaipeda, Lithuania, Sept 21, 36; US citizen; m 65; c 3. RADIOLOGY. *Educ:* Newark Col Eng, BS, 58, MS, 64; Hahnemann Med Col, MD, 68. *Prof Exp:* Intern med, St Vincent's Hosp, New York, 68-69, resident radiol, 69-72; instr, Univ Ind, 72-73; asst prof, 73-76, ASSOC PROF RADIOL, UNIV ROCHESTER, 76- *Concurrent Pos:* Vis radiologist, Isaac Gordon Ctr Digestive Dis, Genesee Hosp, NY, 75- *Mem:* Radiol Soc NAm; Inst Elec & Electronics Engrs; Asn Univ Radiologists; Roentgen Ray Soc; Am Gastroenterol Asn. *Res:* Application of engineering to radiological science; gastro-intestinal radiology. *Mailing Add:* PO Box 648 1156 Peck Rd Hilton NY 14468

SKUD, BERNARD EINAR, b Ironwood, Mich, Jan 31, 27; m 50; c 3. MARINE BIOLOGY, FISHERIES. *Educ:* Univ Mich, BS, 49, MS, 50; Princeton Univ, cert pub affairs, 68. *Prof Exp:* Asst freshwater fishes, Univ Mich, 49-50; fishery res biologist, Alaska salmon & herring, US Nat Marine Fisheries Serv, US Bur Commercial Fisheries, US Fish & Wildlife Serv, 50-56, supv fishery res biologist, 56-58, asst lab dir, Gulf of Mex fisheries, 58-61, lab dir, Biol Lab, Boothbay Harbor, 61-70, dir invests, Int Pac Halibut Comn, Seattle, 70-78,; chief, Div Permits & Regulations, US Nat Marine Fisheries Serv, Washington DC, 78-79, special sci asst to ctr dir, Narragansett, RI, 79-85; EXEC DIR, INT N PAC FISHERIES COMN, VANCOUVER, BC, 85- *Concurrent Pos:* Chmn, herring & pelagic fish subcomt, Int N Atlantic Fisheries Comn; Princeton fel Pub Affairs, 67-68; affil prof, Univ Wash, 71-78; adj prof, Univ RI, 80-; ed, J Northwest Atlantic Fishery Sci, 85-87; affil prof, Univ Western Wash, 87- *Mem:* Am Fisheries Soc; Am Soc Limnol & Oceanog; Am Soc Ichthyol & Herpet; Am Inst Fisheries Res Biol (pres, 81-83). *Res:* Pacific salmon; Pacific and Atlantic herring population dynamics; age and growth studies; biological oceanography; estuaries; off-shore lobsters; Pacific halibut; marine ecosystems; interspecific interactions. *Mailing Add:* Inter N Pacific Fish Comm 6640 N W Marine Dr Vancouver BC V6T 1X2 Can

SKUDRZYK, FRANK J, b Cieszyn, Poland, Jan 19, 43; m 85. ROCK MECHANICS, EXPLOSIVES AND BLASTING. *Educ:* Univ Mining & Metall, BS, 68, MS, 68, PhD(rock mech), 73. *Hon Degrees:* Dir Mines, Sec Level, Ministry Coal Mining, Poland, 75. *Prof Exp:* From asst prof to assoc prof mining eng, Univ Mining & Metall, 68-79; sr researcher rock mech, Univ Miss, Rolla, 79-82; head dept mining & geol eng, 83-84, PROF MINING ENG, UNIV ALASKA, FAIRBANKS, 82- *Concurrent Pos:* Prin investr var proj, Univ Alaska, Fairbanks, 82- *Mem:* Soc Exp Mech; Am Soc Testing & Mech; Soc Mining Eng; Soc Explosive Eng; Marine Technol Soc. *Res:* Experimental and theoretical rock mechanics with applications to design and stability of mining and civil structures in rock and frozen ground; rock fragmentation methods; design of lab and insitu testing equipment. *Mailing Add:* Dept Mining & Geol Eng Univ Alaska 205 Brooks Bldg Fairbanks AK 99701-1190

SKUJINS, JOHN JANIS, b Latvia, Apr 13, 26; US citizen; div; c 2. SOIL BIOCHEMISTRY, SOIL MICROBIOLOGY. *Educ:* Univ Calif, Berkeley, BA, 57, PhD(agr chem), 63. *Prof Exp:* Fel soil microbiol, Cornell Univ, 62-64; res biochemist, Univ Calif, Berkeley, 64-69; from assoc prof to prof, 69-89, EMER PROF BIOL & SOIL SCI, UTAH STATE UNIV, 89- *Concurrent Pos:* NSF, US Environ Protection Agency, USDA, study grants, soil microbiol, US Int Biol Prog; chmn, Int Symp Environ Biogeochem Inc, 73-91; consult, SAMDENE proj, Egypt, 74-79; teacher, Helsinki Univ, 77-78, Finland Agr Univ, Saltillo, Mex, 78, Swedish Univ Agr Sci, Uppsala, 84, Agr Univ, Riga, Latvi, 91; mem, Int Comt Microbial Ecol, 78-; assoc ed, Geomicrobiol J, 78-; ed, Arid Soil Res & Rehab, 85-; mem US Environ Protection Agency Rev Panel, 86; mem US Environ Protection Agency working group on Methods for Assessing Environmental Impacts of Microbial Prods; mem, US Dept Energy Rev Panel, 83; mem, Latvian Acad Sci, 90- *Mem:* Am Chem Soc; Am Soc Microbiol; Soil Sci Soc Am; AAAS; Int Soil Sci Soc; Am Soc Agron. *Res:* Ecology of arid lands; nitrogen fixation and cycling; soil enzymology; microbial ecology; enzymatic and microbial activities in adverse environmental conditions; ecology of forest soils. *Mailing Add:* Dept Biol Utah State Univ Logan UT 84322-5500

SKULAN, THOMAS WILLIAM, b Milwaukee, Wis, Feb 12, 32; m 52; c 3. PHARMACOLOGY, PHYSIOLOGY. *Educ:* Univ Wis-Madison, BS, 58, PhD(pharmacol), 62. *Prof Exp:* NIH fel, Univ Fla, 62-63; group leader diuretics, Sterling-Winthrop Res Inst, 66-81, head cardiovasc sect, 69-81; RETIRED. *Res:* Renal pharmacology, hypertension. *Mailing Add:* Park Guilderlan Apts Guilderlan Center NY 12085

SKULTETY, FRANCIS MILES, b Rochester, NY, June 6, 22; m 45; c 3. MEDICINE. *Educ:* Univ Rochester, BS, 44, MD, 46; Univ Iowa, PhD(anat), 58; Am Bd Neurol Surg, dipl, 54. *Prof Exp:* Intern, Worcester City Hosp, Mass, 46-47; asst resident neurol, Cushing Vet Admin Hosp, 49-50; sr resident neurosurg, Univ Iowa Hosps, 51-52, instr surg, 52-53, assoc, 53-54, from asst prof to prof, 54-66; interim dean, Col Med, Univ Nebr, Omaha, 78-79, Shakleford prof neurosurg & neuroanat, 66-87, assoc dean clin affairs, 74-78 & 79-87, prof & chmn dept neurosurg, 75-87, EMER PROF SURG & NEUROSURG, COL MED, UNIV NEBR, 87- *Concurrent Pos:* Fel neurosurg, Lahey Clin, 50-51; clin traineeship Nat Inst Neurol Dis & Stroke, Dept Physiol, Oxford Univ, 57-58. *Mem:* AMA; Am Asn Neurol Surg; fel Am Col Surg; Am Neurol Asn; Soc Neurosci. *Res:* Neuroanatomy; neurophysiology; neural regulation of intake. *Mailing Add:* 840 Crestridge Rd Omaha NE 68154

SKUMANICH, ANDREW P, b Wilkes-Barre, Pa, Oct 5, 29; m 55; c 3. ASTROPHYSICS. *Educ:* Pa State Univ, BS, 51; Princeton Univ, PhD(astrophys), 54. *Prof Exp:* Staff mem, Los Alamos Sci Lab, Univ Calif, 54-60; asst prof & res assoc physics & astron, Univ Rochester, 60-61; MEM SR STAFF, HIGH ALTITUDE OBSERV, NAT CTR ATMOSPHERIC RES, 61-; CONSULT, NASA, 74- *Concurrent Pos:* Consult, Los Alamos Sci Lab, 61-73; lectr, Univ Colo, 61-69, adj prof, 69-; vis scientist, Lab Stellar & Planetary Physics, 73-74. *Mem:* Am Phys Soc; Am Astron Soc; Int Astron Union. *Res:* Thermodynamics and hydrodynamics of high temperature plasma; visible, ultraviolet and x-ray spectroscopy; radiative processes; atomic and electronic collision phenomena; solar, atmospheric and chromospheric physics. *Mailing Add:* 400 13th St Boulder CO 80302

SKUP, DANIEL, b Chicago, Ill, Apr 1, 51; Can citizen; m 72; c 2. INTERFERON GENE EXPRESSION, PROTEASE INHIBITORS IN METASTATIC TUMOR CELLS. *Educ:* Moscow State Univ, BSc, 73, MSc, 74; McGill Univ, Montreal, PhD(biochem), 80. *Prof Exp:* Postdoctoral researcher, Inst Curie, Orsay, France, 80-82; from asst prof to assoc prof, 82-88, sci dir, 87-88, ASSOC PROF & DIR, MONTREAL CANCER INST, 88- *Concurrent Pos:* Consult, New Eng Nuclear, 79; Med Res Coun vis prof, Fac Med, Univ Ottawa, 83 & Mem Univ, St-Johns, 91; panel mem, Cell Biol & Metastasis Panel, Nat Cancer Inst Can, 84-87, Molecular Biol Comt, Med Res Coun Can, 87-89, Cancer Comt, Med Res Coun Can, Fels Panel, Nat Cancer Inst Can, 89- & Que Govt Mission Biotechnol Israel, 91. *Mem:* Am Soc Microbiologists; Int Soc Interferon Res. *Res:* Regulation of gene expression during cell differentiation and during tumor progression and metastasis; role of interferon as a modulator of mammalian development; role of protease inhibitors in metastatic tumor cells as a tool for molecular basis of tumor invasion. *Mailing Add:* Montreal Cancer Inst 1560 Sherbrooke St E Montreal PQ H2L 4M1 Can

SKUTCHES, CHARLES L, b Northampton, Pa, Nov 3, 41; m; c 2. EXPERIMENTAL BIOLOGY. *Educ:* Catawba Col, AB, 63; NC State Univ, MS, 66, PhD(nutrit biochem), 73. *Prof Exp:* NSF res asst, 63-65; biochemist, Smith, Kline & French Labs, Philadelphia, Pa, 66-69; NIH res fel, 69-73; res assoc, Dept Biochem & Biophys, Tex A&M Univ, College Station, 73-75; asst investr, 75-77, assoc investr, 77-87, SR SCIENTIST, LANKENAU MED RES CTR, WYNNEWOOD, PA, 87-, DIR SCI ADMIN, 91- *Concurrent Pos:* Mem, Animal Care & Use Comt, Lankenau Med Res Ctr, 79-83, chmn, 83-; consult, Renal Care Adv Panel, Abbott Labs, Ill, 85-86; adj assoc prof, Dept Med, Sch Med, Temple Univ, 89- *Mem:* Sigma Xi; Am Diabetes Asn; AAAS; Am Soc Biochem & Molecular Biol. *Res:* Author of numerous publications. *Mailing Add:* 1116 Med Sci Bldg Lankenau Med Res Ctr 100 Lancaster Ave W of City Line Wynnewood PA 19401

SKUTNIK, BOLESH JOSEPH, b Passaic, NJ, Aug 19, 41; m 67; c 2. PHYSICAL CHEMISTRY. *Educ:* Seton Hall Univ, BS, 62; Yale Univ, MS, 64, PhD(theoret phys chem), 67. *Prof Exp:* Res assoc phys chem, Brandeis Univ, 67-69; sr res scientist, Firestone Radiation Res Div, Firestone Tire & Rubber Co, 69-73; asst prof chem, Fairfield Univ, 73-79; res scientist, Ensign-Bickford Industs, Inc, 79-81, new prod mgr, 81-87, CHIEF SCIENTIST, ENSIGN-BICKFORD OPTICS CO, 87- & ENSIGN-BICKFORD COATINGS CO, 90- *Concurrent Pos:* Lectr, Brandeis Univ, 68-69; abstractor, Chem Abstr Serv, 69-85; consult, Acad Press, Inc, 74-76, J Wiley & Sons, 79 & Darworth Corp, 82; adj prof, Univ Hartford, 90- *Mem:* Am Phys Soc; Am Chem Soc; Am Ceramics Soc; Optical Soc Am; Mat Res Soc; Soc Photo-Optical Instrumentation Engrs; Soc Plastics Engrs. *Res:* Effects of radiation on matter, theoretical and applied; characterization and physical properties of irradiated polymers; electronic structure of atoms and molecules; inter- and intra-molecular energy transfer; theoretical atomic structure and spectroscopy; photochemistry; radiation chemistry; polymer science; specialty coatings; quantum chemistry; interaction of matter and energy; polymer processing; photo initiation of polymerization; specialty coatings for optics; solventless coatings; polymeric composites for conductivity; electron & UV radiation processing of materials; reinforced polymers; flush photolysis & kenetic studies of organic and bio-organic systems; water soluble vehicles for pigments and wood preservatives. *Mailing Add:* PO Box 261 West Simsbury CT 06092

SKY-PECK, HOWARD H, b London, Eng, July 24, 23; nat US; m 52; c 3. BIOCHEMISTRY. *Educ:* Univ Southern Calif, BS, 49, PhD, 56. *Hon Degrees:* MD, Royal Soc Med, Eng, 57. *Prof Exp:* Lab instr biochem, Univ Southern Calif, 50-52; res assoc, Presby Hosp, Chicago, 55; from instr to assoc prof, Col Med, 55-67, assoc prof, Grad Sch, 58-67, PROF BIOCHEM, GRAD SCH, UNIV ILL, 67- *Concurrent Pos:* Asst attend biochemist, Presby-St Lukes Hosp, 58-62, sr attend biochemist, 62-, dir, Clin Chem Lab, 67-75; mem biochem comt, Nat Cancer Chemother Serv Ctr, NIH, 60; actg chmn dept, Rush Med Col, 70-71, prof biochem & chmn dept, 71-80. *Mem:* AAAS; Am Chem Soc; Am Cancer Soc; Am Asn Clin Chem; Am Asn Cancer Res; Nat Acad Clin Biochem; Sigma Xi. *Res:* Amino acid metabolism in mouse brains; biochemical comparison between normal and neoplastic human cancer tissues and use of biochemical techniques in evaluation of chemotherapeutic agents in cancer. *Mailing Add:* Dept of Biochem Rush-Presby-St Luke's Med Ctr Chicago IL 60612

SKYPEK, DORA HELEN, b Noma, Fla, Sept 18, 15; m 44; c 2. MATHEMATICS EDUCATION, MATHEMATICS. *Educ:* Fla State Univ, BA, 37; Emory Univ, MA, 61; Univ Wis, PhD(math educ), 66. *Prof Exp:* Teacher math, Fla Pub High Schs, 37-44 & St John's Country Day Sch, Fla, 54-58; from asst prof to prof, 63-83, EMER PROF MATH & MATH EDUC, EMORY UNIV, 83- *Concurrent Pos:* Dir, Experienced Teacher Fel Prog, US Off Educ, 68-69; dir res proj, NSF, 74-76; co-dir women sci career workshop, NSF, 80; coordr working group women math, IV Int Cong Math Educ, 80; mem, Int Study Group, Psychol Math Educ. *Honors & Awards:* Thomas Jefferson Award, 82; Dora Helen Skypek Award, Asn Women Math Educ, 83. *Mem:* Nat Coun Teachers Math. *Res:* The teaching and learning of mathematics; related Piagetian theory; factors that support and inhibit women in science careers. *Mailing Add:* 2120 Trailmark Dr Decatur GA 30033

SLABY, FRANK J, ANATOMY. *Prof Exp:* ASSOC PROF ANAT, GEORGE WASHINGTON UNIV MED CTR, 81- *Mailing Add:* Dept Anat George Washington Univ Med Ctr 2300 Eye St NW Washington DC 20037

SLABY, HAROLD THEODORE, b Traverse City, Mich, Oct 11, 20; m 51; c 1. ALGEBRA. *Educ:* Wayne State Univ, AB, 46, MA, 48; Univ Wis, PhD(math), 53. *Prof Exp:* From instr to assoc prof, 53-89, EMER ASSOC PROF MATH, WAYNE STATE UNIV, 89- *Concurrent Pos:* Mem consult bur, Math Asn Am, 68-83. *Mem:* Am Math Soc; Math Asn Am; Sigma Xi. *Res:* Non associative algebra. *Mailing Add:* 2691 Burnham Rd Royal Oak MI 48073

SLABYJ, BOHDAN M, b Chernivci, Ukraine, Dec 3, 31; US citizen; m 63; c 3. FOOD SCIENCE. *Educ:* Univ Alta, BSc, 58, MSc, 60; Univ Wash, PhD(food sci), 68. *Prof Exp:* Instr bact, Univ Alta, 60-62; asst microbiologist, Univ Wash, 62-67; asst prof microbiol, Duquesne Univ, 67-69; fel, Albert Einstein Med Ctr, Univ Maine, 69-72, from asst prof to assoc prof, 72-85, PROF FOOD SCI, UNIV MAINE, 85- *Mem:* Am Soc Microbiol; Inst Food Technol; Soc Appl Bacteriol; Am Chem Soc. *Res:* Seafood processing, quality, and safety. *Mailing Add:* Dept Food Sci Univ Maine Orono ME 04469

SLACK, DERALD ALLEN, b Cedar City, Utah, Dec 22, 24; m 45; c 2. PHYTOPATHOLOGY, NEMATOLOGY. *Educ:* Utah State Agr Col, BS, 48, MS, 49; Univ Wis, PhD, 53. *Prof Exp:* Asst res & collabr, US Dept Agr, Utah State Agr Col, 46-47, asst, 47-49; asst, Univ Wis, 49-52; from asst prof to assoc prof, 52-60, PROF PLANT PATH, UNIV ARK, FAYETTEVILLE, 60-, HEAD DEPT, 64- *Concurrent Pos:* Mem & secy, Ark State Plant Bd, 64- *Honors & Awards:* Outstanding Plant Pathologist, Am Phytopath Soc. *Mem:* Am Phytopath Soc (secy, 78-80); Soc Nematol; Int Soc Plant Path. *Res:* Fruit diseases; plant parasitic nematodes. *Mailing Add:* Dept of Plant Path Univ of Ark Fayetteville AR 72701

SLACK, GLEN ALFRED, b Rochester, NY, Sept 29, 28; m 51; c 3. SOLID STATE PHYSICS. *Educ:* Rensselaer Polytech Inst, BS, 50; Cornell Univ, PhD, 56. *Prof Exp:* PHYSICIST, GEN ELEC RES & DEVELOP CTR, 56- *Concurrent Pos:* Guggenheim Mem fel, Oxford Univ, 66-67. *Mem:* Fel Am Phys Soc; Sigma Xi. *Res:* Thermal properties and heat transport in solids; preparation and chemistry of crystals; properties of semiconductors; ultrasonic phonon propagation in solids. *Mailing Add:* Gen Elec Res & Develop Ctr Bldg K-1 River Rd Schenectady NY 12345

SLACK, JIM MARSHALL, b Irving, Tex, Mar 11, 31; m 65; c 3. PHYSIOLOGY. *Educ:* Sam Houston State Col, BS, 52; Sam Houston State Univ, MA, 66; Tex A&M Univ, PhD, 71. *Prof Exp:* Res assoc biochem, Radiation Res Assocs, 62-65; asst prof biol, Hardin-Simmons Univ, 66-67; assoc prof, Howard Payne Col, 70-73; asst prof physiol, Auburn Univ, 73-79; HEAD DEPT PHYSIOL & CHEM, TEX CHIROPRACTIC COL, 79- *Concurrent Pos:* Dean basic sci, Tex Chiropractic Col, 88. *Mem:* Am Soc Zoologists; Sigma Xi. *Res:* Physiology of trauma and stress, with particular interest in the action of Prostaglandin and the Endogenous Opioids. *Mailing Add:* Dept Physiol & Chem Tex Chiropractic Col Pasadena TX 77505

SLACK, JOHN MADISON, b Polson, Mont, Mar 9, 14; m 40; c 1. MEDICAL BACTERIOLOGY. *Educ:* Univ Minn, AB, 36, MS, 37, PhD(bact), 40; Am Bd Microbiol, dipl, 62. *Prof Exp:* Asst bact, Univ Minn, 38-40; instr path & bact, Col Med, Univ Nebr, 40-42; bacteriologist, US Army Chem Warfare Labs, Ft Detrick, Md, 46; prof, 46-77, EMER PROF MICROBIOL, MED CTR, W VA UNIV, 77- *Mem:* AAAS; Am Soc Microbiol; fel Am Pub Health Asn. *Res:* Identification and classification of Actinomyces; fluorescent antibody techniques; nocardin; food poisoning. *Mailing Add:* Univ Village 12401 N 22nd St Apt B-209 Tampa FL 33612

SLACK, KEITH VOLLMER, b Louisville, Ky, May 20, 24; m 62; c 4. STREAM LIMNOLOGY, BIOLOGICAL WATER QUALITY. *Educ:* Univ Ky, BS, 49, MS, 50; Ind Univ, PhD(zool), 54. *Prof Exp:* Asst zool, Univ Ky, 49-50, State Lake & Stream Surv, Ind Univ, 50-52 & Ind Univ, 52-53; oceanogr, US Navy Oceanog, 53-60; RES LIMNOLOGIST, US GEOL SURV, 60- *Concurrent Pos:* Lectr limnol & aquatic ecol, Univ Ariz, Tucson, 65-71 & Geol Surv Nat Training Ctr, Denver; res adv ecol, Water Resources Div, US Geol Surv, 74-76 & 78-; mem, Interagency Working Group Biol & Microbiol Methods, 75-84. *Mem:* Am Inst Biol Sci; Ecol Soc Am; Am Soc Limnol & Oceanog; Int Asn Theoret & Appl Limnol; NAm Benthological Soc. *Res:* Interrelations between aquatic organisms and their environment; stream limnology; controls on benthic invertebrate community; distribution and abundance; aquatic biological methods. *Mailing Add:* 805 Gailen Ave Palo Alto CA 94303

SLACK, LEWIS, b Philadelphia, Pa, Apr 15, 24; m 48; c 3. NUCLEAR PHYSICS. *Educ:* Harvard Univ, SB, 44; Wash Univ, PhD(physics), 50. *Prof Exp:* Asst physics, Wash Univ, 46-50; physicist, US Naval Res Lab, 50-54; from assoc prof to prof physics, George Washington Univ, 54-62, actg head dept, 57-60; asst exec secy div phys sci, Nat Acad Sci-Nat Res Coun, 60-67, assoc dir, Am Inst Physics, 67-87. *Concurrent Pos:* Secy comt nuclear sci, Nat Res Coun, 62-67; secy, US Nat Comt for Int Union Pure & Appl Physics,

74-78; mem, Sci Manpower Comn, 68-87, pres, 74-76. *Mem:* AAAS; Am Phys Soc; Am Asn Physics Teachers. *Res:* Beta ray and gamma ray spectroscopy; angular correlation; science administration; science education for the public. *Mailing Add:* 27 Meadowbank Rd Old Greenwich CT 06870

SLACK, LYLE HOWARD, b Wellsville, NY, Jan 6, 37; m 54; c 4. CERAMICS. *Educ:* Alfred Univ, BS, 58, PhD(ceramic sci), 65. *Prof Exp:* Res engr, Hommel Co, Pa, 58-60 & Lexington Labs, Mass, 60-61; mem tech staff thin film electronics, Bell Tel Labs, 65-67; asst prof ceramic eng, Va Polytech Inst & State Univ, 67-71, assoc prof, 71-80; SR RES CHEMIST, E I DU PONT DE NEMOURS & CO, INC, NY, 80- *Concurrent Pos:* Consult, Naval Res Labs, 69-81. *Mem:* Fel Am Ceramic Soc; Nat Inst Ceramic Engrs; Am Soc Eng Educ; Electrochem Soc; Int Soc Hybrid Microelectronics. *Res:* Structure and electric properties of semiconducting glasses; structure and electronic conduction in oxide thin films; solar photovoltaic materials; solar thermal coatings; thick film resistors; thick film dielectrics. *Mailing Add:* 3203 Kammerer Dr Wilmington DE 19803

SLACK, NANCY G, b New York, NY, Aug 12, 30; m 51; c 3. PLANT ECOLOGY. *Educ:* Cornell Univ, BS, 52, MS, 54; State Univ NY, Albany, PhD(ecol), 71. *Prof Exp:* Demonstr bot & evolution, Bot Sch, Oxford Univ, Eng, 66-67; lectr bot, State Univ NY, Albany, 69; from asst prof to assoc prof biol, 71-81, chmn dept, 78-84, PROF BIOL, RUSSELL SAGE COL, TROY, NY, 81- *Concurrent Pos:* Consult, Environ Impact Studies, Environ-One Corp, 73-, Environ Assessment Study, Dunn Geosci, 74-, Environ Impact Studies, Environmed Inc, 75-; trustee & ecol consult land acquisition, The Nature Conserv, 73-, mem bd trustees, 89-; consult col sci progs, NY State Educ Dept, 75-76, 78, 82 & 83-90; Am Asn Univ Professors fel, 79-80; NSF travel award, 81; vis fel, Renssalaer Polytech Inst, 85; vis res scholar, Yale Univ, 90-91. *Honors & Awards:* Donald Richards Fund Award, NY Bot Garden, 74, Diamond Award, 75. *Mem:* Ecol Soc Am; Am Bryol & Lichenol Soc; Int Bryol Asn; AAAS; Sigma Xi; Hist Sci Soc. *Res:* Community ecology of bryophytes and vascular plants including Sphagnum bog ecology; species diversity and community structure in bryophytes; ecology of epiphytic bryophytes in North America; island biogeography and vegetation changes on islands due to human disturbance; bryophytes in relation to ecological niche theory; history of botany and ecology in US; biology of rare and endangered plant species. *Mailing Add:* Ridge Rd Scotia NY 12307

SLACK, NELSON HOSKING, b Burlington, Vt, Feb 7, 35; m 60; c 3. BIOSTATISTICS. *Educ:* Univ Vt, BS, 57; Rutgers Univ, MS, 63, PhD(reproductive phys & statist), 64. *Prof Exp:* Res asst dairy sci, Rutgers Univ, 63-64; assoc cancer res scientist biostatist, 64-77, dep dir clin trials, Nat Prostatic Cancer Proj, 77-84, MEM STAFF, BIOMATHEMATICS DEPT, ROSWELL PARK MEM INST, 84- *Concurrent Pos:* Res prof, Niagara Univ; asst res prof, State Univ NY Buffalo. *Mem:* AAAS; Am Statist Asn; Sigma Xi. *Res:* Cancer research. *Mailing Add:* Dept Biomath Roswell Park Mem Inst 666 Elm St Buffalo NY 14203

SLACK, STEVEN ALLEN, b Logan, Utah, May 6, 47; m 70; c 2. PLANT PATHOLOGY, PLANT VIROLOGY. *Educ:* Univ Ark, BSA, 69, MS, 71; Univ Calif, Davis, PhD(plant path), 74. *Prof Exp:* From asst prof to assoc prof, 75-85, PROF PLANT PATH, UNIV WIS-MADISON, 85- *Mem:* Am Phytopath Soc; Potato Asn Am; AAAS. *Res:* Characterization and serology of plant viruses, epidemiology and control of plant virus and bacterial diseases, and potato diseases. *Mailing Add:* Dept Plant Path 334 Plant Sci Cornell Univ Ithaca NY 19853

SLACK, WARNER VINCENT, b East Orange, NJ, June 10, 33; m 56; c 3. COMPUTER SCIENCE, PATIENT-COMPUTER DIALOGUE. *Educ:* Princeton Univ, AB, 55; Columbia Univ, MD, 59. *Prof Exp:* Intern med, Univ Wis Hosps, 59-60, resident, 60-61, instr, Univ, 65-66, from asst prof to assoc prof med & comput sci, 66-70; asst prof, 70-73, ASSOC PROF MED, HARVARD MED SCH, 73-, ASSOC PROF PSYCHIAT, 90-; CO-PRES, CTR CLIN COMPUT, 86- *Concurrent Pos:* Alumni Res Found fel, Univ Wis Hosps, 61 & 64-65, NIH spec res fel, 65-66; lectr, Univ Philippines, 63-64; co-dir, Div Comput Med, Beth Israel Hosp & Brigham & Women's Hosp, 80- *Res:* Application of computer techniques to clinical medicine, specifically the use of computers to interview and counsel patients regarding their medical problems and to help patients to make their own medical decisions; computers in psychiatry and psychotherapy; hospital-wide clinical computing systems. *Mailing Add:* Ctr Clin Comput Harvard Med Sch 350 Longwood Ave Boston MA 02115

SLADE, ARTHUR LAIRD, b Aiken, SC, Oct 21, 37; m 78; c 2. POLYMER CHEMISTRY, COMPUTER SCIENCE. *Educ:* Duke Univ, AB, 59; Univ NC, PhD(phys chem), 64. *Prof Exp:* Res chemist, Marshall Res & Develop Lab, E I Du Pond de Nemours & Co, 64-69, staff chemist, 69, res supvr, 69-72, sr financial analyst, 72-73, distrib mgr, 73-75, prin consult corp planning, 75-78, mgr planning & financial commun, 78-80, pub affairs mgr chemicals & pigments, 80-81, pub affairs mgr, textile fibers, 81-83, group mgr pub affairs, 83-85, PLANNING DIR EXTERNAL AFFAIRS, E I DU PONT DE NEMOURS & CO, 85- *Mem:* Sigma Xi; Am Chem Soc. *Res:* Electrolyte solutions; electrochemistry; polymer coatings; automotive specialty products; man made textile fibers; science communications; technical management. *Mailing Add:* External Affairs Dept E I du Pont de Nemours & Co Wilmington DE 19898

SLADE, BERNARD NEWTON, b Sioux City, Iowa, Dec 21, 23; m 46; c 2. ELECTRICAL ENGINEERING. *Educ:* Univ Wis, BS, 48; Stevens Inst Technol, MS, 54. *Prof Exp:* Advan develop engr, Tube Div, Radio Corp Am, NJ, 48-53, mgr advan develop, 53-55, res engr labs, 55-56; mgr semiconductor develop, IBM Corp, Hopewell, NY, 56-60, mgr components prod opers, 60-65, corp dir advan mfg tech, 65-66, corp dir mfg planning & controls, 66-69, corp dir mfg eng & technol, 69-81, adv, Mfg Technol, 81-84; consult, Arthur D Little, Inc, Cambridge, Mass, 84-86; SR MGR CONSULT, UNITED RESEARCH CO, MORRISTOWN, NJ, 86- *Concurrent Pos:* Adj instr, Pace Univ, 77-79; guest lectr, Harvard Grad Sch Bus, 84, Stanford Grad Sch Bus, 85, Cornell, Chicago Univ, Penn State Univ & Univ Penn. *Mem:* Sr mem Inst Elec & Electronics Engrs; Sigma Xi. *Res:* Development and design of solid state components, including transistors, diodes and other semiconductor devices; manufacturing productivity with emphasis on high technology. *Mailing Add:* 12 Merry Hill Rd Poughkeepsie NY 12603

SLADE, EDWARD COLIN, b Stockport, Eng, Mar 24, 35; US citizen; m 62; c 2. NATIONALLY RECOGNIZED AUTHORITY ON VALVES, FLUID FLOW COMPUTER APPLICATIONS. *Educ:* Stockport Col, Eng, BSME, 55. *Prof Exp:* Design engr, Standard Steel, 57-60, James M Montgomery, 60-61, Carnation Co, 61-66, Stone & Webster, 66-67 & Donald R Warren, 67-68; sr design engr, Stearns Roger Corp, 68-69 & 72-76; sr engr & qual assurance engr, Holmes & Narver, 69-71; staff engr, Bechtel, 72; sr design engr, Dravo Corp, 76-77; engr fluid systs, Martin Marietta Energy Systs, Dept Energy, US Oak Ridge Nat Lab, 77-86; PRES, CONTAINMENT, OAK RIDGE, TENN, 86-; CONSULT, 86- *Concurrent Pos:* Peer reviewer nuclear valves, Oak Ridge Nat Lab, 77-; secy & mem, Bd Human Resources, Oak Ridge, Tenn. *Honors & Awards:* Silver Medalist, Geneva, Switz, 80. *Mem:* Int Platform Asn. *Res:* Designed and built a valve that could have avoided the catastrophe of Three Mile Island; work and writing are directed at the complexities of flow and materials in this area; one US patent. *Mailing Add:* 203 Tusculum Dr Oak Ridge TN 37830

SLADE, H CLYDE, b Millestown, Nfld, July 2, 18; m 44; c 4. INTERNAL MEDICINE, PSYCHIATRY. *Educ:* Dalhousie Univ, MD & CM, 49; FRCPS(C), 49. *Prof Exp:* Assoc prof health care & epidemiol, Univ BC, 68-71, hon assoc prof psychiat, 68-84, assoc prof primary health care & dir div, 71-77, EMER PROF PSYCHIAT, UNIV BC, 84- *Concurrent Pos:* Assoc med, Vancouver Gen Hosp, 50-; consult, Shaughnessy Hosp, 52- *Mem:* Can Med Asn; Can Psychiat Asn. *Res:* Psychosomatic medicine; rheumatology. *Mailing Add:* 3070 W 44th Ave Vancouver BC V6N 3K6 Can

SLADE, JOEL S, b Brooklyn, NY, Jan 1, 47; m 78. HETEROCYCLIC CHEMISTRY, ASYMMETRIC SYNTHESIS. *Educ:* Lowell Tech Inst, BS, 68, MS, 74; Colo State Univ, PhD(chem), 79. *Prof Exp:* Fel org chem res, Univ Pa, 79-80; sr res chemist med chem, 80-85, PROCESS RES CHEMIST, CIBA-GEIGY CORP, 85- *Mem:* Am Chem Soc. *Res:* Preparation of biologically interesting molecules which have the potential to become new therapeutic agents. *Mailing Add:* 14 Theresa Dr Flanders NJ 07836

SLADE, LARRY MALCOM, b Durango, Colo, Feb 20, 36; m 62; c 5. HORSE HUSBANDRY, NUTRITION. *Educ:* Brigham Young Univ, BS, 62; Va Polytech Inst, MS, 65; Univ Calif, Davis, PhD(animal nutrit), 71. *Prof Exp:* Teacher high sch, Calif, 65-66; asst prof animal sci, Calif State Polytech Col, 70-72; assoc prof animal sci, Colo State Univ, 71-78; ASSOC PROF ANIMAL SCI, UTAH STATE UNIV, 78- *Concurrent Pos:* Subcomt Horse Nutrit, Nat Res Coun; consult. *Mem:* Am Soc Animal Sci; Equine Nutrit & Physiol Soc. *Res:* Nutrient requirements of horses; conformation and performance of horses. *Mailing Add:* Dept Animal Dairy & Vet Sci Utah State Univ Logan UT 84322

SLADE, MARTIN ALPHONSE, III, b Dunedin, Fla. CELESTIAL MECHANICS, RADIO ASTRONOMY. *Educ:* Mass Inst Technol, SB, 64, SM, 67, PhD(planetary sci), 71. *Prof Exp:* RESEARCHER, EARTH & LUNAR PHYSICS APPLN GROUP, JET PROPULSION LAB, CALIF INST TECHNOL, 71- *Concurrent Pos:* Mem, Lunar Sci Review Panel, Lunar Sci Inst, Houston, 75- *Mem:* Am Astron Soc; Am Geophys Union. *Res:* Rotational dynamics of the moon; very long baseline interferometry; analysis of lunar laser ranging data; testing gravitational theories; radar astronomy. *Mailing Add:* MS 238-420 Jet Propulsion Lab 4800 Oak Grove Dr Pasadena CA 91109

SLADE, NORMAN ANDREW, b Wichita, Kans, Oct 14, 43; m 64; c 3. POPULATION ECOLOGY. *Educ:* Kans State Univ, BS, 65; Utah State Univ, MS, 69, PhD(ecol), 72. *Prof Exp:* Res assoc statist ecol, San Diego State Univ, 71-72; from asst prof to assoc prof, 72-81, PROF SYST & ECOL, UNIV KANS, 81- *Concurrent Pos:* Vis scientist pop ecol, Unit Behav Syst, Nat Inst Mental Health, 76. *Mem:* Ecol Soc Am; Am Soc Mammalogists; Biometric Soc; AAAS; Am Soc Naturalists. *Res:* Mammalian population dynamics, interspecific competition and computer simulation models of ecological systems; biostatistics. *Mailing Add:* Mus of Natural Hist Dept of Syst & Ecol Univ of Kans Lawrence KS 66045-2454

SLADE, PAUL GRAHAM, b Blackpool, Eng, Dec, 1941; US citizen; m 65; c 2. CIRCUIT BREAKER DEVELOPMENT, SURFACE SCIENCE. *Educ:* Univ Wales, BS, 63, PhD (physics), 66; Univ Pittsburgh, MBA, 73. *Prof Exp:* Sr engr, Westinghouse Sci & Technol Ctr, 66-72, mgr, Power Interruption Res, 72-75, Power Interruption & Plasma Systs, 75-83 & Plasma & Nuclear Sci, 83-90, DIR INT TECHNOL DEVELOP, WESTINGHOUSE SCI & TECHNOL CTR, 90- *Concurrent Pos:* Mem, organizing comt, Holm Conf, Inst Elec & Electronic Engrs, 77-, ad com trans ed, Components Hybrids & Mfg Soc, 87- *Honors & Awards:* Ragnar Holm Sci Acievement Award, 85. *Mem:* Fel Inst Elec & Electronic Engrs; Inst Physics. *Res:* Science of electric contacts for power application; arcs and their interaction with electric contacts; contactor, molded case breaker, vacuum interruption and SF interruption phenomena; strategic planning. *Mailing Add:* Westinghouse Sci & Technol Ctr Pittsburgh PA 15235

SLADE, PHILIP EARL, JR, b Hatiesburg, Miss, Sept 2, 29; m 65; c 4. POLYMER CHEMISTRY. *Educ:* Miss Southern Col, BA, 51; Tulane Univ, MS, 53, PhD(chem), 55. *Prof Exp:* Res chemist, Chemstrand Corp, Ala, 55; assoc chem, George Washington Univ, 56-57; assoc prof, Miss Southern Col, 57-60; sr res chemist, Chemstrand Res Ctr, Inc, 60-67; group supvr, 67-74, SUPVR, NYLON TECH CTR, MONSANTO TEXTILES CO, 74- *Mem:* Am Chem Soc; Fiber Soc. *Res:* Polymer characterization; polymer solution properties; thermal analysis; fiber characterization; fiber finish analysis. *Mailing Add:* Tech Ctr Monsanto Textiles Co PO Box 12830 Pensacola FL 32575

SLADEK, CELIA DAVIS, b Denver, Colo, Mar 25, 44; m 70; c 3. MEDICAL NEUROENDOCRINOLOGY. *Educ:* Hastings Col, BA, 66; Northwestern Univ, Chicago, MS, 69, PhD(physiol), 70. *Prof Exp:* Asst prof physiol, Univ Ill Med Ctr, 70-73; res assoc prof neuroendocrinol, 74-76, from asst prof to assoc prof neurol & anat, 76-88, PROF NEUROL & NEUROBIOL, UNIV ROCHESTER MED SCH, 88- *Concurrent Pos:* Am Cancer Soc grant, Univ Ill Med Ctr, 71-72; Am Diabetes Asn grant, 74-75; Nat Inst Arthritis, Metab & Digestive Dis grant, 77-91, NIH res develop career award, 77-82 & Nat Heart Lung Blood Inst, 82-89; NSF equipment grant, 80-, Nat Inst Neurol Dis & Stroke, 91-96; ed, Brain Res Bulletin, 80-, Exp Neurol, 89- *Mem:* Soc Neurosci; Am Asn Anatomists; NY Acad Sci; Endocrine Soc; Am Physiol Soc. *Res:* Regulation of vasopressin and oxytocin secretion and gene expression; clinical abnormalities associated with inappropriate vasopressin secretion; hypertension; development and aging of the neurohypophyseal system. *Mailing Add:* Dept Neurobiol Univ of Rochester Med Sch Rochester NY 14642

SLADEK, JOHN RICHARD, JR, b Chicago, Ill, Feb 6, 43; m 70; c 3. NEUROSCIENCE, AGING & DEVELOPMENT. *Educ:* Carthage Col, BA, 65; Northwestern Univ, MS, 68; Univ Health Sci, PhD(anat), 71. *Prof Exp:* From asst prof to assoc prof, Anat Univ Rochester, 73-82, assoc prof, Ctr Brain Res, 79-82, PROF & CHAIR NEUROBIOL-ANAT, SCH MED, UNIV ROCHESTER, 82-, KILIAN & CAROLINE SCHMITT PROF, 87. *Concurrent Pos:* Consult, Nat Inst Aging, Nat Inst Neurol & Commun Dis & Stroke & NSF, 80; ed-in-chief, Exp Neurol, 88-; mem, Biol Neurosci Study Sect, NIMH, 80-84, Neurol B2 Study Sect, NIH, 85-89. *Mem:* Am Asn Anatomists; Histochem Soc; NY Acad Sci; Soc Neurosci. *Res:* Neuron interactions of aminergic and peptidergic neurons during development, aging and following transplantation; neuroendocrinology of the hypothalamo-neurohypophyseal system of vasopressin and oxytocin neurons; Parkinson's disease. *Mailing Add:* 6499 Lake Rd Bergen NY 14416

SLADEK, KARL JOSEF, III, chemical engineering, for more information see previous edition

SLADEK, NORMAN ELMER, b Montgomery, Minn, Aug 20, 39; m 64; c 3. PHARMACOLOGY. *Educ:* Univ Minn, Minneapolis, BS, 62, PhD(pharmacol), 66. *Prof Exp:* NIH fel, Univ Wis-Madison, 66-68; from asst prof to assoc prof, 68-79, PROF PHARMACOL, MED SCH, UNIV MINN, MINNEAPOLIS, 79- *Concurrent Pos:* NIH res career develop award, 72-77. *Mem:* AAAS; Am Asn Cancer Res; Am Soc Pharmacol & Exp Therapeut; Am Soc Microbiol. *Res:* Cancer chemotherapy; drug interactions; drug metabolism; carcinogenesis; immunopharmacology. *Mailing Add:* Dept Pharmacol 3-249 Millard Hall Univ of Minn Minneapolis MN 55455

SLADEK, RONALD JOHN, b Chicago, Ill, Sept 19, 26; m 53; c 6. SOLID STATE PHYSICS. *Educ:* Univ Chicago, PhD(physics), 54. *Prof Exp:* Res physicist, Westinghouse Elec Corp, 53-61; assoc prof, 61-66, actg head dept, 69-71, assoc dean sci, 74-87, PROF PHYSICS, PURDUE UNIV, 66- *Mem:* Fel Am Phys Soc; Sigma Xi. *Res:* Ultrasonic and electrical properties of solids, especially crystalline solids exhibiting a phase transition and superionic glasses; effects of low temperatures, stress, and magnetic fields thereon. *Mailing Add:* Dept Physics Purdue Univ West Lafayette IN 47907

SLAGA, THOMAS JOSEPH, b Smithfield, Ohio, Dec 15, 41; m 66; c 2. BIOCHEMICAL PHARMACOLOGY. *Educ:* Col Steubenville, BA, 64; Univ Ark, PhD(physiol biophys), 69. *Prof Exp:* Fel, McArdle Lab Cancer Res, Univ Wis-Madison, 68-71; res investr chem carcinogenesis, Pac Northwest Res Ctr, 71-73; asst mem, Fred Hutchinson Cancer Res Ctr, 73-76; staff mem, E Tenn Cancer Res Ctr, 76-78; sr staff mem cancer & toxicol, biol div, Oak Ridge Nat Lab, 76-; AT DEPT BIOCHEM, SCI PARK-RES DIV, UNIV TEX SYST CANCER CTR. *Concurrent Pos:* Asst prof pharmacol, Sch Med, Univ Wash, 74-76. *Mem:* Am Asn Cancer Res; Am Soc Invest Dermatol; Sigma Xi. *Res:* Mechanism of chemical carcinogenesis in both in vivo and in vitro; in particular, the early molecular events after the application of chemical carcinogens and tumor promoters, mechanism of action of the antitumor agents of the skin. *Mailing Add:* Dept Biochem Sci Park Res Div Univ Tex Syst Cancer Ctr PO Box 389 Smithville TX 78957

SLAGEL, DONALD E, b Louisville, Ky, Sept 30, 30; m 57; c 3. BIOCHEMISTRY, NEUROBIOLOGY. *Educ:* Univ Ky, BS, 54; Univ Wis, MS, 56, PhD(biochem, phys chem), 61. *Prof Exp:* ASSOC PROF SURG, UNIV KY, 64- *Concurrent Pos:* NIH fel neuropath, Univ Wis, 61-63 & neurobiol, Gothenburg Univ, 63-64. *Mem:* Am Chem Soc; Int Soc Neurochem; Am Asn Neuropath; Am Soc Neurochem; Sigma Xi; AAAS. *Res:* Chemistry and ultrastructure of the nervous system; use of athymic mouse-human tumor xenograft in experimental studies, including combined modality treatment studies; study of the athymic mouse-human brain tumor xenogroft genome for sequences related to cell transformation. *Mailing Add:* Dept of Surg Div of Neurosurg Univ of Ky Med Ctr Lexington KY 40536-0084

SLAGEL, ROBERT CLAYTON, b Sabetha, Kans, Jan 4, 37; m 61; c 2. ORGANIC CHEMISTRY. *Educ:* Western Mich Univ, BS, 58; Univ Ill, PhD(sesquiterpenoids), 62. *Prof Exp:* Res chemist, Archer Daniels Midland Co, 62-64, sr res chemist, 64-67; group leader basic res, ADM Chem Div, Ashland Oil & Refining Co, 67-68; group leader polymer synthesis, Calgon Corp, Subsid Merck & Co Inc, 68-69, sect leader specialty chem res, 69-71, mgr polymer res, 71-77, asst dir res & develop, 77-78, dir specialty chem res, 78-79; tech dir, 79-88, mgr mkt & com develop, 88-89, BUS MGR INKS & COATINGS GROUP, CHEM PROD DIV, UNION CAMP CORP, 89- *Mem:* Am Chem Soc; Tech Asn Pulp & Paper Indust. *Res:* Monomers and polymers, chiefly polyelectrolytes; organonitrogen chemistry; ozonization of organic compounds; carbenes, chiefly halocarbenes; rosin based resins; fatty acid derivatives; tall oil distillation. *Mailing Add:* Union Camp Corp 1600 Valley Rd Wayne NJ 07470

SLAGER, URSULA TRAUGOTT, b Frankfurt, Germany, Sept 15, 25; nat US; m 49. PATHOLOGY. *Educ:* Wellesley Col, BA, 48; Univ Md, MD, 52; Am Bd Path, dipl, 58. *Prof Exp:* Instr path, Sch Med, Univ Md, 54-55, assoc, 56-57; pathologist, Los Alamos Med Ctr, 57-59 & Res Dept, Martin Co, 59-60; assoc clin prof path, Univ Southern Calif, 61-67, assoc prof, 67-; pathologist, Rancho Los Amigos Med Ctr, 68-88; RETIRED. *Concurrent Pos:* Hitchcock fel neuropath, Univ Md, 53-55; assoc pathologist, Orange County Gen Hosp, 61-65, actg dir path serv, 65-67; mem staff, Univ Southern Calif Med Ctr, 67- *Mem:* AMA; Am Col Path; Int Acad Path; Am Soc Clin Path; Am Soc Neuropath. *Res:* Neuropathology; radiation damage. *Mailing Add:* 2800 Sunset Rd Bishop CA 93514

SLAGG, NORMAN, b New York, NY, Jan 8, 31; m 57; c 2. PHYSICAL CHEMISTRY, CHEMICAL KINETICS. *Educ:* Brooklyn Col, BS, 52; Polytech Inst Brooklyn, PhD(chem), 60. *Prof Exp:* Res chemist, Reaction Motors Div, Thiokol Chem Corp, 60-62; sr res engr, Lamp Div, Westinghouse Elec Corp, 62-67; instr phys chem, Fairleigh Dickinson Univ, 65-67; lectr, Rutgers Univ, 67-75; HEAD FAST REACTION SECT, EXPLOSIVE LAB, PICATINNY ARSENAL, 67- *Mem:* Am Chem Soc; Am Phys Soc; Sigma Xi. *Res:* Kinetics; mechanism of chemical reactions; photochemistry; reactions of molten salts with glasses and ceramics; shock tube techniques; time resolved spectroscopy; explosive phenomena. *Mailing Add:* 22 Marlton Dr Wayne NJ 07470

SLAGLE, JAMES R, b Brooklyn, NY, Mar 1, 34; m 58; c 5. COMPUTER SCIENCE, MATHEMATICS. *Educ:* St John's Univ, BS, 55; Mass Inst Technol, MS, 57, PhD(math), 61. *Prof Exp:* Staff mathematician, Lincoln Lab, Mass Inst Technol, 55-63; group leader, Lawrence Radiation Lab, Univ Calif, 63-67; chief, Heuristics Lab, NIH, 67-74; head, Comput Sci Lab, 74-80, spec asst, Navy Ctr Appl Res Artificial Intelligence, 80-84, DISTINGUISHED PROF COMPUT SCI, NAVAL RES LAB, 84- *Concurrent Pos:* Teacher elec eng, Mass Inst Technol, 62-63 & Univ Calif, Berkeley, 63-67; teacher comput sci, Johns Hopkins Univ, 67-73. *Mem:* Asn Comput Mach; fel Am Asn Artificial Intel; sr mem Inst Elec & Electronic Engrs. *Res:* Artificial intelligence; automatic pattern recognition; automatic theorem proving; automatic expert consultant systems. *Mailing Add:* 2117 W Hoyt Ave St Paul MN 55108

SLAGLE, WAYNE GREY, b Monkstown, Tex, Nov 23, 34; m 57. PARASITOLOGY. *Educ:* Tex A&M Univ, BS, 63, MS, 66, PhD(zool), 70. *Prof Exp:* Asst biol, Tex A&M Univ, 63-66, instr, 66-70; asst prof 70-77, ASSOC PROF BIOL, STEPHEN F AUSTIN STATE UNIV, 77- *Mem:* Am Soc Parasitol. *Res:* Biological control of helminth parasites which cause human disease. *Mailing Add:* Box 13003 2910 Dogwood Nacogdoches TX 75961

SLAKEY, LINDA LOUISE, b Oakland, Calif, Jan 2, 39. BIOCHEMISTRY. *Educ:* Siena Heights Col, BS, 62; Univ Mich, PhD(biochem), 67. *Prof Exp:* Instr chem, St Dominic's Col, 67-69; fel biochem, Univ Wis, 70-73; from asst prof to assoc prof, 73-87, PROF & HEAD BIOCHEM, UNIV MASS, 87- *Mem:* Tissue Culture Asn; Am Soc Biol Chemists; Am Soc Cell Biol. *Res:* lipid structure and metabolism; interaction of vascular endothelium with blood components; regulation of plasma membrane protein turnover; purinoceptors, extracellular nueleotide metabolism. *Mailing Add:* Dept Biochem Univ Mass Amherst MA 01003

SLAMA, FRANCIS J, b St Louis, Mo, Apr 17, 39; m 63; c 4. CHEMISTRY. *Educ:* St Louis Univ, AB, 62, PhD(chem), 69. *Prof Exp:* Chemist, Commercial Div, Calgon Corp, 62-66; res chemist, 69-75, RES SUPVR, AMOCO CHEM CORP, STANDARD OIL CO IND, 75- *Concurrent Pos:* Instr, Col of DuPage, 70-73; instr, Waybonsee Community Col, 74- *Mem:* Am Chem Soc. *Res:* Plastics; polymer structure and properties. *Mailing Add:* Amoco Chem Co Naperville IL 60566

SLAMECKA, VLADIMIR, b Brno, Czech, May 8, 28; US citizen; m 62; c 2. INFORMATION SCIENCE, COMPUTER SCIENCE. *Educ:* Columbia Univ, MS, 58, DLS, 62. *Prof Exp:* Chemist, Brookvale Brewery, 52-54; assoc ed, Mid-Europ Press, Inc, 56-57; head chem libr, Columbia Univ, 58-60, proj investr sci orgn, 60-62; mgr info systs design, Documentation, Inc, Md, 62-64; PROF INFO & COMPUT SCI, GA INST TECHNOL, 64-; CLIN PROF MED, EMORY UNIV, 80- *Concurrent Pos:* NSF grant, Sci Orgn Eastern Europe, 60-62; Fulbright prof, 63-64; consult, NSF, 65- & NIH, 70-; vchmn comt int sci & technol prog, Nat Acad Sci, 74-76; chmn, US Nat Comt FID, 74-78; vchmn US Nat Comt, UNESCO/Gen Info Prog, 78-81. *Honors & Awards:* Systs Res Found Award, 86. *Mem:* Fel AAAS; Am Soc Info Sci; Asn Comput Mach; Sigma Xi; NY Acad Sci. *Res:* Information science; national and international information systems. *Mailing Add:* Sch of Info & Comput Sci Ga Inst of Technol Atlanta GA 30332

SLANSKY, CYRIL M, b Albuquerque, NMex, July 8, 13; m 39; c 3. CHEMISTRY. *Educ:* Col of Idaho, BS, 36; Univ Calif, PhD(chem), 40. *Prof Exp:* Asst chem, Univ Calif, 37-39; res chemist, Dow Chem Co, Mich, 40-44 & Calif, 44-47; chemist, Hanford Works, Gen Elec Co, 47-52; chief, Works Lab, Am Cyanamid Co, 52-53; chem develop, Atomic Energy Div, Phillips Petrol Co, 53, sect head, 53-60, mgr, Chem Develop Br, 60-62, mem staff nuclear & chem tech, 62-66; mem staff nuclear & chem tech, 66-71; sr tech adv, Allied Chem Corp, 71-78, nuclear consult, Chem Progs, 78; NUCLEAR CONSULT, 78- *Concurrent Pos:* Mem radioactive waste mgt, Int Atomic Energy Agency, 69-71. *Mem:* AAAS; Am Chem Soc; Am Nuclear Soc; Am Inst Chem Eng. *Res:* Chemistry and technology of inorganic compounds and compounds from calcined dolomite; electrolytic production of magnesium; separations processes; nuclear fuel cycle; radioactive waste disposal; applications of nuclear heat. *Mailing Add:* 2815 Holly Pl Idaho Falls ID 83402-4631

SLANSKY, RICHARD CYRIL, b Oakland, Calif, Apr 3, 40; c 2. THEORETICAL HIGH ENERGY PHYSICS. *Educ:* Harvard Univ, BA, 62; Univ Calif, Berkeley, PhD(physics), 67. *Prof Exp:* Res fel physics, Calif Inst Technol, 67-69; from instr to asst prof, Yale Univ, 69-74; mem staff, 74-89, THEORET DIV LEADER, LOS ALAMOS SCI LAB, 89- *Mem:* Fel, Am Physical Soc. *Res:* Elementary particle physics. *Mailing Add:* T-DO MS B210 Los Alamos Nat Lab MS B285 Los Alamos NM 87545

SLAPIKOFF, SAUL ABRAHAM, b Bronx, NY, Nov 5, 31; m 75; c 2. BIOCHEMISTRY. *Educ:* Brooklyn Col, BA, 52; Tufts Univ, PhD(biochem), 64. *Prof Exp:* USPHS fel biochem, Sch Med, Stanford Univ, 64-66; asst prof, 66-72, ASSOC PROF BIOL, TUFTS UNIV, 72- *Concurrent Pos:* NRS fel genetic toxicol, 76-77; vis scientist, Mass Inst Technol, 77-78; vis prof , Environ Health Sect, Boston Univ Sch Pub Health, 85. *Mem:* AAAS; Am Chem Soc; Am Pub Health Asn; Sigma Xi; Soc Environ Toxicol & Chem. *Res:* Environmental toxicology. *Mailing Add:* Dept Biol Tufts Univ Medford MA 02155

SLATE, FLOYD OWEN, b Carroll Co, Ind, July 26, 20; m 39; c 3. APPLIED CHEMISTRY, CONCRETE. *Educ:* Purdue Univ, BS, 41, MS, 42, PhD(anal chem), 44. *Prof Exp:* Chemist, Purdue Univ, 41-44, chemist & asst prof hwy eng, 46-49; lab supvr, Manhattan Dist Proj, Columbia Univ, 44; asst chief chemist, Garfield Div, Houdaille-Hershey, Ill, 44-46; prof eng mat, Cornell Univ, 49-87; CONSULT, 87- *Concurrent Pos:* Adv, Int Coop Admin, Pakistan, 56; vpres res & develop & mem bd, Geotech & Resources, Inc, White Plains, NY, 59-63; consult, Pure Waters Prog, 69-75, and many others, 49-; prin investr, NSF, 59-; vis prof, Univ NSW, Australia, 81, Univ Witwatersrand, 83. *Honors & Awards:* Wason Medal, Am Concrete Inst, 57, 65, 74 & 84; Anderson Award, Am Concrete Inst, 83. *Mem:* Am Chem Soc; Am Soc Test & Mat; Am Concrete Inst; Am Inst Chem; Am Soc Civil Eng. *Res:* Low cost housing; concrete; engineering materials; soils; chemistry applied to engineering problems. *Mailing Add:* Sch Civil Eng Cornell Univ Ithaca NY 14853

SLATER, C STEWART, b Feb 24, 57; US citizen. CHEMICAL ENGINEERING. *Educ:* Rutgers Univ, BS, 79, MS, 82, MPh, 83, PhD(chem eng), 83. *Prof Exp:* Process develop engr, Proctor & Gamble Co, Cincinnati, Ohio, 79-81; teaching asst, dept chem & biochem eng, Rutgers Univ, 81-83; ASSOC PROF CHEM ENG, MANHATTAN COL, 83- *Concurrent Pos:* Consult, 86- *Honors & Awards:* Ralph R Teetor Award, 86; New Eng Educr Award, Am Soc Eng Educ, 87. *Mem:* Am Inst Chem Engrs; Sigma Xi; Am Chem Soc; NAm Membrane Soc; Am Soc Eng Educ. *Res:* Membrane process research, modeling and computer simulation; biochemical engineering purification processes; application of engineering processes to industrial and hazardous watewater renovation and reuse systems. *Mailing Add:* Dept Chem Eng Manhattan Col Riverdale NY 10471

SLATER, CARL DAVID, b Moundsville, WVa, Oct 26, 33; m 65; c 2. ORGANIC CHEMISTRY. *Educ:* WVa Univ, BS, 55; Ohio State Univ, PhD(org chem), 60. *Prof Exp:* Res assoc org chem, Mass Inst Technol, 60-61; proj chemist, Chem Div, Union Carbide Corp, 61-62; from asst prof to assoc prof org chem, NDak State Univ, 62-67; assoc prof org chem, Memphis State Univ, 67-80; prof & chmn dept phys sci, Northern Ky Univ, 80-87, prof, dept chem, 87-90; DEAN, COL ARTS & SCI, WASHBURN UNIV, 90- *Mem:* Am Chem Soc. *Res:* Mechanism of electrocyclic processes; synthesis and reactions of aminothiophene derivatives; kinetics and mechanism of reactions of halogen-containing nitro compounds; quantum chemical calculations of molecular properties; nuclear magnetic resonance shift correlations. *Mailing Add:* Col Arts & Sci Washburn Univ Topeka KS 66621

SLATER, DONALD CARLIN, b Pensacola, Fla, July 27, 45; m 75; c 1. EXPERIMENTAL PHYSICS. *Educ:* Stanford Univ, BS, 67; Mass Inst Technol, PhD(physics), 71. *Prof Exp:* Res assoc physics, Stanford Univ, 71-74 & Univ Va, 74-76; res scientist, KMS Fusion, Inc, 76-78, mgr, 79-85; MGR, ROCKETDYNE, 85- *Mem:* Am Phys Soc. *Res:* Experimental inertial confinement fusion; laser-plasma interaction; diagnostic instrumentation. *Mailing Add:* Rocketdyne 6633 Canoga Ave FA38 Canoga Park CA 91303

SLATER, EVE ELIZABETH, b W Orange, NJ, May 16, 45; m 81; c 2. CARDIOLOGY, ENDOCRINOLOGY. *Educ:* Vassar Col, AB, 67; Col Physicians & Surgeons, Columbia Univ, MD, 71. *Prof Exp:* Intern & resident med, Mass Gen Hosp, Boston, 71-73, fel cardiol, 73-75, chief resident med, 76; asst prof, Sch Med, Harvard Univ, 79-83; ASSOC PROF MED, COL PHYSICIANS & SURGEONS, COLUMBIA UNIV, 83-; EXEC DIR BIOCHEM & MOLECULAR BIOL, MERCK SHARP DOHME RES LABS, 83- *Concurrent Pos:* Estab investr, Am Heart Asn, 80 & mem, Coun High Blood Pressure; consult med, Mass Gen Hosp, Boston, 83- *Mem:* Fel Am Col Cardiol; Chilean Soc Cardiol; Am Soc Biol Chemists; Endocrine Soc; Am Heart Asn. *Res:* Biochemistry of the renin-angiotensin system; mechanism of insulin resistance. *Mailing Add:* Merck Sharp & Dohme Res Labs PO Box 2000 Rahway NJ 07065

SLATER, GEORGE E(DWARD), petroleum engineering, for more information see previous edition

SLATER, GEORGE P, b Findochty, Scotland, Mar 11, 32; m 56; c 2. GAS CHROMATOGRAPHY. *Educ:* Aberdeen Univ, BSc, 54; Univ Sask, MSc, 57; Queen's Univ, Belfast, PhD(chem), 61. *Prof Exp:* Anal chemist, Swift Canadian Co, 54-55; res asst chem, Univ Sask, 57-58; chemist, Polymer Corp, Can, 58; Nat Res Coun Can fel chem, Univ Sask, 61-62; SR RES OFFICER, NAT RES COUN CAN, 62- *Mem:* Chem Inst Can. *Res:* Water pollution; analysis of pulp mill effluent; gas chromotography-mass spectroscopy; volatile plant products affecting insect behavior; taste and odor problems in water supplies; biosynthesis of plant lipids. *Mailing Add:* Nat Res Coun Can Plant Biotechnol Inst Saskatoon SK S7N 0W9 Can

SLATER, GRANT GAY, b Rochester, NY, Jan 6, 18; m 48; c 2. NUTRITION, MEDICAL RESEARCH. *Educ:* Univ Miami, BS, 40, Univ Southern Calif, MS, 50, PhD(biochem), 54. *Prof Exp:* Lab dir, Chem Warefare Lab, Alaska, 43-44; biochemist, Res Inst, Cedars Lebanon Hosp, Los Angeles, 54-55; res physiol chemist, Univ Calif, Los Angeles, 55-58; res specialist, State Dept Mental Hyg, 58-61; biochemist, Vet Admin Ctr, 61-68; res biochemist, Gateways Hosp, 68-72; researcher I, Div Environ & Nutrit Sci, Sch Pub Health & Dept Psychiatry, Med Sch, Univ Calif, 72-82; CONSULT, 82- *Concurrent Pos:* Vis instr, Univ Southern Calif, 55-57, vis asst prof, 58-59; mem fac, dept psychiat, Univ Calif, Los Angeles, 58-88 & Brain Res Inst, 61-68; study grants, Anti-trypsin in Lung Dis from Air Pollution, Div Lung Dis, NIH, 74-75. *Mem:* Emer mem Am Chem Soc; emer mem Am Inst Chem; emer mem Am Physiol Soc; emer mem Endocrine Soc; emer mem Soc Neurosci; emer mem Electrophoresis Soc; emer mem Am Oil Chemists; Sigma Xi. *Res:* Basic methodological research on plasma proteins and studies on the relationship of plasma proteins in humans to lung and brain disease; nutritional studies relating dietary and plasma cholesterol in humans; design of medical equipment. *Mailing Add:* 986 Somera Rd Los Angeles CA 90077-2624

SLATER, JAMES ALEXANDER, b Belvidere, Ill, Jan 10, 20; m 43; c 4. ENTOMOLOGY, SYSTEMATICS. *Educ:* Univ Ill, BA, 42, MS, 47; Iowa State Col, PhD(entom), 50. *Prof Exp:* From instr to prof entom, Univ Conn, 54-88 head dept zool, entom & biochem, 64-67, head systs & evolutionary biol sect, 70-80; RETIRED. *Concurrent Pos:* Comnr, Conn Geol & Natural Hist Surv & state ornithologist, 63-72; panelist, Sci Div, NSF, 63-66; res assoc, Nat Insect Col Pretoria, SAfrica, 67-68; res assoc, Am Mus Natural Hist, 76- *Honors & Awards:* LD Howard Award, Entom Soc Am, 85. *Mem:* Entom Soc Am; Soc Syst Zool (pres, 81-83); Royal Entom Soc; SAfrican Entom Soc. *Res:* Systematics and bionomics of Hemiptera, Lygaeidae and Miridae. *Mailing Add:* 373 Bassettes Bridge Rd Mansfield Center CT 06250

SLATER, JAMES LOUIS, b Grand Rapids, Mich, Dec 2, 44; c 2. INORGANIC CHEMISTRY. *Educ:* Mich State Univ, BS, 67; Fla State Univ, PhD(inorg chem), 71. *Prof Exp:* Fel inorg chem, Ames Lab, US AEC, Iowa State Univ, 71-73; instr inorg chem, Univ Va, 73-74; asst prof inorg chem, 74-77, ASST PROF CHEM, COL STEUBENVILLE, 77- *Mem:* Am Chem Soc. *Res:* Chemical studies of the metal carbonyls. *Mailing Add:* Dept Chem Franciscan Univ Steubenville OH 43952

SLATER, JAMES MUNRO, b Salt Lake City, Utah, Jan 7, 29; m 48; c 5. RADIOTHERAPY. *Educ:* Univ Utah & Utah State, BS, 54; Sch Med, Loma Linda Univ, MD, 63. *Prof Exp:* Instr radiother, 67-68, asst clin prof radiol radiother, 68-70, from asst prof to assoc prof, 70-74, PROF RADIOL RADIOTHER, SCH MED, LOMA LINDA UNIV, 75- *Concurrent Pos:* Fel, White Mem Med Ctr, 67-68 & Univ Tex, M D Anderson Hosp & Tumor Inst, 68-69; dir radiation oncol, Sch Med, Loma Linda Univ, 70-, dir nuclear med, 75-, interim chmn dept radiol sci, 78-; mem prof educ comt, Am Cancer Soc, 75-76, chmn, 76-77, vpres, 78-79. *Honors & Awards:* Physician's Recognition Award, AMA, 69, 72 & 75. *Mem:* Am Soc Therapeut Radiologists; Am Soc Clin Oncol; AAAS; Am Cancer Soc; Am Radium Soc. *Res:* Cancer immunology, emphasizing the effect of ionizing radiation on the human immune system; computerized dosimetry for radiation therapy planning; treatment of malignant disease using ionizing irradiation. *Mailing Add:* Dept Radiation Sci 11234 Anderson St Loma Linda CA 92354

SLATER, KEITH, b Oldham, Eng, Dec 20, 35; m 59; c 3. TEXTILES. *Educ:* Univ Leeds, BSc, 56, MSc, 58, PhD(textiles), 65. *Prof Exp:* Asst master, Leeds Cent High Sch, Eng, 60-65; from asst prof to assoc prof, 65-75, PROF TEXTILES, UNIV GUELPH, 75-, ACAD DIR, PARIS SEMESTER, 90- *Concurrent Pos:* Nat Res Coun grant, Univ Guelph, 66-, Defence Res Bd grant, 68-; consult, Wool Bur Can, 69-70, Hart Chem Ltd, 69-71, Harding Carpets Ltd, 70-78, Johnson & Johnson, 80-82 & ISKA, 84-85; vis fel, Gonville & Caius Col, Cambridge, UK, 83-84; vis prof, eng dept, Cambridge Univ, UK, 83-84; vis fel, Clare Hall, Cambridge, UK, 90-91. *Mem:* Fel Inst Textile Sci (pres, 72-73); fel Textile Inst (vpres, 74-78); Textile Fedn Can (pres, 79-81). *Res:* Yarn irregularity; yarn hairiness; acoustic properties of textiles; comfort of textiles; textile drying behavior; progressive deterioration of textiles; design of surgical operating theater gowns; design of clothing for severe climates; social consequences of technological abuse. *Mailing Add:* Textile Sci Div Univ Guelph Guelph ON N1G 2W1 Can

SLATER, PETER JOHN, b Mt Vernon, NY, Sept 30, 46; m 70. MATHEMATICS, OPERATIONS RESEARCH. *Educ:* Iona Col, BS, 68; Univ Iowa, MS, 72, PhD(math), 73. *Prof Exp:* Asst prof math, Cleveland State Univ, 74; res assoc, Nat Bur Stand, 74-75; mathematician, Sandia Labs, 75-81; ASSOC PROF MATH, UNIV ALA, 81- *Mem:* Am Math Soc; Opers Res Soc Am. *Res:* The field of graph theory with particular emphasis on problems of N-connectivity, geodesics and facility location; network modelling of facilities. *Mailing Add:* Math Dept Univ Ala Box 1247 Huntsville AL 35899

SLATER, PHILIP NICHOLAS, b London, Eng, Feb 9, 32; US citizen; m 58; c 3. OPTICS, REMOTE SENSING. *Educ:* Univ London, BS, 55, PhD(appl optics) & dipl, Imp Col, 58. *Prof Exp:* Res physicist, Optics Res Sect, IIT Res Inst, 58-62, sect mgr, 62-66; PROF OPTICAL SCI, UNIV ARIZ, 66-, CHMN, REMOTE SENSING COMT, 76- *Mem:* Fel Optical Soc Am; Am Soc Photogram & Remote Sensing; fel Soc Photog Sci & Engrs; Int Soc Optical Engrs. *Res:* Remote sensing physics and sensor systems; absolute radiometriccalibration and atmospheric correction. *Mailing Add:* 1280 N Speedway Pl Tucson AZ 85715

SLATER, RICHARD CRAIG, b Jersey City, NJ, Nov 16, 46; m 73; c 1. PHYSICAL CHEMISTRY. *Educ:* Stevens Inst Technol, BS, 68; Columbia Univ, PhD(chem), 73. *Prof Exp:* Fel chem, Dept Chem, Columbia Univ, 73-76; RES SCIENTIST, AVCO EVERETT RES LAB, 76- *Mem:* Am Chem Soc; Am Phys Soc; Sigma Xi. *Res:* Laser applications in chemistry; chemical kinetics; vibrational energy transfer. *Mailing Add:* 37 Wyman St Waban MA 02168

SLATER, SCHUYLER G, b New Haven, Conn, Feb 22, 23. ORGANIC CHEMISTRY, NATURAL PRODUCTS CHEMISTRY. *Educ:* Univ Conn, BS, 44, MS, 47; Boston Univ, EdD(sci educ), 65. *Prof Exp:* Instr chem, Univ Conn, 46-49; instr sci, Boston Univ, 50-54 & Cent Conn State Col, 54-55; assoc prof, Univ Maine, 54-56; PROF CHEM, SALEM STATE COL, 56-, CHMN DEPT, 80- *Concurrent Pos:* Observer, Brit Open Univ, 71-72. *Mem:* Am Chem Soc. *Res:* Curriculum development. *Mailing Add:* PO Box 689 Charlestown RI 02813-0689

SLATER, WILLIAM E, b Springfield, Ohio, July 16, 31; m 58. EXPERIMENTAL HIGH ENERGY PHYSICS. *Prof Exp:* Fel, 60-62, from asst prof to assoc prof, 62-73, PROF PHYSICS, UNIV CALIF, LOS ANGELES, 73- *Mem:* Am Phys Soc. *Res:* Track chamber and counter experiments. *Mailing Add:* Dept Physics 3-174 Univ Calif 405 Hilgard Ave Los Angeles CA 90024

SLATES, HARRY LOVELL, b Canton, Ohio, Feb 7, 23; m 51; c 1. ORGANIC CHEMISTRY. *Educ:* Mt Union Col, BSc, 44; Ohio State Univ, MSc, 48. *Prof Exp:* Chemist qual control, Goodyear Tire & Rubber Co, 45; asst chem, Ohio State Univ, 47-48; sr res chemist, Merck & Co, Inc, 48-89; RETIRED. *Mem:* AAAS; Am Chem Soc; Am Inst Chem; Royal Soc Chem; Swiss Chem Soc. *Res:* Steroid sapogenins; organic synthesis and structure determination. *Mailing Add:* 601 S Chestnut St Westfield NJ 07090

SLATKIN, DANIEL NATHAN, b Montreal, Que, Aug 5, 34; US citizen. PATHOLOGY & RADIOBIOLOGY, BIOMATHEMATICS. *Educ:* McGill Univ, BSc, 55, MD, 59. *Prof Exp:* Intern, Mt Sinai Hosp, New York, 59-60; res assoc med, Brookhaven Nat Lab, 60-61; resident path, Montefiore Hosp, New York, 61-63 & neuropath, 63-64; resident pediat path, Presby Hosp, New York, 64-65; registr morbid anat, Hammersmith Hosp, London, Eng, 65-66; Anna Fuller Found fel biochem, Inst Sci Res Cancer, Villejuif, France, 66-67; assoc pathologist, McKellar Gen Hosp, Thunder Bay, Ont, 68-69; from instr to asst prof path, State Univ NY, Stony Brook, 69-83; cons path, Vet Admin Hosp, Northport, NY, 72-89; PATHOLOGIST & SCIENTIST, MED DEPT, BROOKHAVEN NAT LAB, 72-, CHIEF STAFF, CLIN RES CTR, 90- *Concurrent Pos:* Dir, Elex Anal Technol Corp, Upton, NY, 87-; assoc clin prof pediat & neurol surg, Albert Einstein Col Med, Bronx, NY, 88- *Mem:* Radiation Res Soc; NY Acad Sci; Am Asn Pathologists. *Res:* Neutron-capture therapy; malignant gliomas; radiobiology; lead toxicity; microbeam radiation therapy; biomathematics. *Mailing Add:* PO Box 701 Upton NY 11973

SLATKIN, MONTGOMERY (WILSON), b Toronto, Ont, June 29, 45; US citizen. EVOLUTIONARY BIOLOGY. *Educ:* Mass Inst Technol, SB, 66; Harvard Univ, PhD(appl math), 70. *Prof Exp:* Res assoc biol, Univ Chicago, 70-71, asst prof theoret biol, biophys & biol, 71-76, assoc prof, 76-77; assoc prof zool, 77-85, PROF INTEGRATIVE BIOL, UNIV WASH, 85- *Concurrent Pos:* Vis staff mem, Los Alamos Sci Labs, 70- *Res:* Mathematical population genetics and population ecology. *Mailing Add:* Dept Integrative Biol Univ of Calif Berkeley CA 94720

SLATON, JACK H(AMILTON), b Riverside, Ill, Mar 9, 25; m 54. DESIGN ELECTRONIC & UNDERWATER ACOUSTIC SYSTEMS. *Educ:* Ill Inst Technol, BS, 45; Calif Inst Technol, MS, 47; Univ Calif, Los Angeles, PhD(eng), 72. *Prof Exp:* Engr, Dayton Acme Co, Ohio, 45-46; asst prof eng res, Ord Res Lab, Pa State Col, 47-50; res scientist, Mil Physics Res Lab, Univ Tex, 50-51; res engr, NAm Aviation, Inc, 51-53; electronic engr, Naval Ocean Systs Ctr, 53-83; chief eng fel, Underseas Syst Div, Honeywell Inc, San Diego, 83-90; CHIEF ENG FEL, MARINE SYSTS, ALLIANT TECHSYSTS INC, SAN DIEGO, 90- *Honors & Awards:* L T E Thompson Award, 62; David Bushnell Award, Am Defense Preparedness Asn, 84. *Mem:* Sigma Xi. *Res:* Underwater acoustics; development of underwater ordnance; electronic circuit design. *Mailing Add:* 1659 Calle Candela La Jolla CA 92037

SLATOPOLSKY, EDUARDO, b Buenos Aires, Arg, Dec 12, 34; US citizen; m 59; c 3. VITAMIN D METABOLISM, RENAL OSTEODYSTROPHY. *Educ:* Nat Col Nicolas Avellaneda, BS, 52; Univ Buenos Aires, Arg, MD, 59. *Prof Exp:* Postdoctoral renal, USPHS, Renal Div, Dept Int Med, Wash Univ Sch Med, 63-65, instr med nephrol, 65-67, from asst prof to assoc prof med, Dept Nephrol, 67-75, DIR, CHROMALLOY AM KIDNEY CTR, WASH UNIV SCH MD, 67-, CO-DIR RENAL DIV, 72-, PROF MED, NEPHROL DEPT, 75-, JOSEPH FRIEDMAN PROF RENAL DIS MED, 91- *Concurrent Pos:* Adv mem, regional med prog, renal prog, 70-75; chmn, Transplantation Comt, Barne Hosp, 75-, fel comt, Kidney Found Eastern Mo & Metro-East, 78; adv comt mem, Artificial Kidney-Chronic Uremia Prog, NIH, 78-90; rep, Latin-Am Nephrol, 83-88; mem, Study Sect Gen Med, NIH, 84-88. *Honors & Awards:* Frederick C Bartter Award, 91. *Mem:* Am Fedn Clin Res; Int Soc Nephrol; Am Soc Nephrol; AAAS; Sigma Xi; Endocrine Soc. *Res:* Pathogenesis and treatment of secondary hyparathyroidism and bone disease in renal failure; studies are conducted at both levels: clinical, on patients maintained on chronic dialysis and on animals with experimentally induced renal failure; effects of calcitriol on PTH MRNA and the extra-renal production of calcitriol by macrophages are studies in great detail; vitro studies in primary culture of bovine parathyroid cells are used to understand the mechanisms that control the secretion of PTH. *Mailing Add:* Renal Div Sch Med Wash Univ One Barnes Hosp Plaza St Louis MO 63110

SLATTERY, CHARLES WILBUR, b LaJunta, Colo, Nov 18, 37; m 58; c 2. PHYSICAL CHEMISTRY. *Educ:* Union Col, Nebr, BA, 59; Univ Nebr, MS, 61, PhD(phys chem), 65. *Prof Exp:* From asst prof to assoc prof chem, Atlantic Union Col, 63-69; res assoc, Mass Inst Technol, 69-70; from asst prof to assoc prof, 70-78, PROF BIOCHEM, SCH MED, LOMA LINDA UNIV, 78-, PROF PEDIAT, 80-, DEPT CHMN, 83- *Concurrent Pos:* Prin investr, NIH Grants, 78-81 & 86-89. *Mem:* Sigma Xi; Am Soc Biochem & Molecular Biol; AAAS; Am Chem Soc; Am Heart Asn; Protein Soc. *Res:* Ultracentrifuge theory with computer application to the problem of sedimentation in multicomponent systems; physical chemistry of macromolecules, principally on the structure and interactions of the bovine and human caseins; enzyme complexes in blood coagulation. *Mailing Add:* Dept Biochem Loma Linda Univ Sch Med Loma Linda CA 92354

SLATTERY, JOHN C, b St Louis, Mo, July 20, 32; m 56; c 5. CHEMICAL ENGINEERING. *Educ:* Washington Univ, St Louis, BS, 54; Univ Wis, MS, 55, PhD(chem eng), 59. *Prof Exp:* From asst prof to assoc prof, 59-67, PROF CHEM ENG, NORTHWESTERN UNIV, EVANSTON, 67- *Mem:* Am Inst Chem Engrs; Soc Rheol; Am Chem Soc; Soc Natural Philos. *Res:* Interfacial phenomena; multiphase flows; fluid mechanics; continuum mechanics. *Mailing Add:* Dept Chem Eng Tex A&M Univ College Station TX 77843-3122

SLATTERY, LOUIS R, b Ft Leavenworth, Kans, Oct 16, 08; m 44; c 2. SURGERY. *Educ:* Columbia Univ, AB, 29, MD, 33; Am Bd Surg, dipl, 40. *Prof Exp:* PROF CLIN SURG, MED CTR, NY UNIV, 50- *Concurrent Pos:* Consult, Inst Rehab & Phys Med, 48- & St Francis Hosp, Port Jervis, 50-; vis surgeon, Bellevue Univ & Doctors Hosps, 50-; consult surgeon, Lenox Hill Hosp, 68 & Vet Admin Hosp, 68. *Mem:* AMA; Am Col Surg. *Mailing Add:* Dept Surg NY Univ Med Ctr New York NY 10016

SLATTERY, PAUL FRANCIS, b Hartford, Conn, July 21, 40; m 64; c 1. HIGH ENERGY PHYSICS. *Educ:* Univ Notre Dame, BS, 62; Yale Univ, MS, 63, PhD(physics), 67. *Prof Exp:* Atomic Energy Comn fel, 67-69, asst prof, 69-73, assoc prof, 73-78, PROF PHYSICS, UNIV ROCHESTER, 78- *Honors & Awards:* Leigh Page Mem Prize, 63. *Mem:* Am Phys Soc. *Res:* Experimental high energy physics; study of hadron induced reactions via electronic techniques. *Mailing Add:* Dept Physics & Astron Univ Rochester Rochester NY 14627

SLAUGHTER, CHARLES D, B Kansas City, Kans, Jan 23, 36; m 75; c 3. SYSTEMS DESIGN & SYSTEMS SCIENCE. *Educ:* Northern Ariz Univ, BS, 58. *Prof Exp:* Res asst astron, Kitt Peak Nat Observ, 59-65, res assoc, 65-66, sr res assoc, 67-75; sr sci programmer astron, NOAO, 76-80, chief programmer, 81-84; software systs engr, 85-86, mgr systs eng, 87-89, MGR DIV ENG, INDUST RES, PHOTOMETRICS, 90- *Concurrent Pos:* Pres, Uniforth Systs Inc, 84-91. *Mem:* Inst Elec & Electronic Engrs. *Res:* Imaging software for scientific CCD camera control systems, real-time process control and instrumentation control software. *Mailing Add:* Uniforth Systs Inc 2626 N Grannen Rd Tucson AZ 85745

SLAUGHTER, CHARLES WESLEY, b Baker, Ore, Oct 28, 41; m 62; c 3. FOREST HYDROLOGY. *Educ:* Wash State Univ, BS, 62; Colo State Univ, PhD(watershed mgt), 68. *Prof Exp:* Res hydrologist, US Army Cold Regions Res & Eng Lab, 68-76, PRIN WATERSHED SCIENTIST, INST NORTHERN FORESTRY, US FOREST SERV, 76- *Concurrent Pos:* Adj assoc prof, Univ Alaska, 69-76 & adj prof water resources, 76-; chmn res coord subcomt, Inter-Agency Tech Comt, Alaska, 69-; chmn, Arctic Directorate, US-MAB, 81-89; Fulbright res fel, Iceland, 85. *Mem:* Soc Am Foresters; Soil Conserv Soc Am; Am Geophys Union; Am Water Resources Asn. *Res:* Wildland and snow hydrology; watershed management; permafrost hydrology; sustainable development and conservation. *Mailing Add:* Inst Northern Forestry USDA Forest Serv 308 Tamana Dr Fairbanks AK 99775

SLAUGHTER, FRANK GILL, JR, b Jacksonville, Fla, May 15, 40; m 59; c 2. MATHEMATICS. *Educ:* Harvard Univ, BA, 61; Duke Univ, PhD(math), 66. *Prof Exp:* Asst prof, 66-70, ASSOC PROF MATH & STATIST, UNIV PITTSBURGH, 71- *Mem:* Am Math Soc; Math Asn Am. *Res:* General topology; generalizations of metric spaces. *Mailing Add:* Dept Math Univ Pittsburgh 4200 5th Ave Pittsburgh PA 15260

SLAUGHTER, GERALD M, b Ilion, NY, June 8, 28; m 53; c 2. WELDING & INSPECTION, CORROSION. *Educ:* Rensselaer Polytech Inst, BS, 49, MS, 51. *Prof Exp:* Res fel mat, Rensselaer Polytech Inst, 49-51; develop engr mat, Oak Ridge Nat Lab, Union Carbide, 51-63, group leader, 63-76, SECT MGR MAT, OAK RIDGE NAT LAB, MARTIN MARIETTA, 76- *Concurrent Pos:* chmn, Brazing & Soldering Comn, Am Welding Soc, 74-77, joint comt, Am Soc Testing Mat, Am Soc Mech Eng & Mat Properties Coun, 81-86; tech div bd, Am Soc Metals Int, 81, 84; bd trustees, Am Soc Metals Int, 84-87, Fedn Mat Soc, 87- *Honors & Awards:* Wasserman Award, Am Welding Soc, 78, McKay-Helm Award & Adams Lectr, 79. *Mem:* Fel Am Soc Mat Int; fel Am Soc Nondestructive Testing; fel Am Welding Soc; Fedn Mat Sci; Am Soc Testing Mat. *Res:* Energy producing and energy conserving concepts; clean fossil fuel; fusion; solar and advanced heat engines; nondestructive testing; metal processing; corrosion. *Mailing Add:* Oak Ridge Nat Lab PO Box 2008 Oak Ridge TN 37831-6157

SLAUGHTER, JOHN BROOKS, b Topeka, Kans, Mar 16, 34; m 56; c 2. COMPUTER SCIENCES. *Educ:* Kans State Univ, BS, 56; Univ Calif, Los Angeles, MS, 61; Univ Calif, San Diego, PhD(eng sci), 71. *Hon Degrees:* Numerous from US Univs, 81-89. *Prof Exp:* Engr simulation, Convair Div, Gen Dynamics Corp, 56-60; phys sci admin info systs, Naval Electronics Lab Ctr, 60-75; dir appl physics lab, Univ Wash, 75-77; asst dir, NSF, 77-79, dir, 80-82; acad vpres & provost, Wash State Univ, 79-80; chancellor, Univ Md, College Park, 82-88; PRES, OCCIDENTAL COL, 88- *Concurrent Pos:* Instr eng, Calif Western Univ, 61-63, UCLA, 63; lectr eng, San Diego State Univ, 64-66; fel, Naval Electronics Lab Ctr, 69; ed, J Comput & Elec Eng Pergamon Press, 72-; elected bd dirs, Am Cancer Soc, 74; mem, Nat Acad Eng, 76-79, Nat Medal Sci, 79-81, Nat Sci Bd, 80-83, Bd Dirs Md State Chamber of Com, 83-88, Prince George's Chamber of Com, 84-88, Prince George's County Econ Develop Corp, 84-88; chair, Inst Elec & Electronic Engrs, Minority Comt, 76-80, Gov Task Force on Team Pregnancy, Md, 84-85, Prince George's County Pub Schs Adv Coun, 85-86, Pres Comm of Nat Col Athletics Asn, 86-88. *Honors & Awards:* David Dodds Henry lectr, Univ Illinois, 85; Croft lectr, Univ Mo Columbia, 85; Louis Clark Vanaxem lectr, Princeton Univ, 83. *Mem:* Nat Acad Eng; fel AAAS; Inst Elec & Electronics Engrs. *Res:* Development of computer algorithms for system optimization and discrete signal processing with emphasis on application to ocean and environmental system problems; author of fifteen publications. *Mailing Add:* Occidental Col Pres Off 1600 Campus Rd Los Angeles CA 90041

SLAUGHTER, JOHN SIM, b Muskagee, Okla, Aug 2, 43; m 73; c 2. PSYCHOPHYSIOLOGY, EXPERIMENTAL PSYCHOLOGY. *Educ:* Lynchburg Col, BA, 67; Univ Denver, MA, 70, PhD(exp psychol), 71. *Prof Exp:* Instr psychol, Lynchburg Col, 67-68; res asst, Univ Denver, 68-71; instr, Fitzsimmons Army Hosp, 69-70; asst prof, 71-76, ASSOC PROF PSYCHOL, STATE UNIV NY COL, FREDONIA, 76- *Concurrent Pos:* Res grants, State Univ NY, 72-74. *Mem:* Am Psychol Asn; Soc Psychophysiol Res. *Res:* Role of peripheral autonomic responses in determining emotional development; computer applications for improving classsroom teachig including simulations, data collection and data analysis. *Mailing Add:* Dept Psychol State Univ NY Col Fredonia NY 14063

SLAUGHTER, LYNNARD J, b Pittsburgh, Pa, Apr 28, 38. COMPARATIVE PATHOLOGY. *Educ:* Tuskegee Univ, DVM, 65. *Prof Exp:* ASSOC PROF PATH, HOWARD UNIV, 81- *Mailing Add:* Dept Path Col Howard Univ Washington DC 20059

SLAUGHTER, MAYNARD, b Athens, Ohio, Jan 13, 34; m 53; c 5. CRYSTALLOGRAPHY. *Educ:* Ohio Univ, BS, 55; Univ Mo-Columbia, AM, 57; Univ Pittsburgh, PhD, 62. *Prof Exp:* Chem mineralogist, Gulf Res & Develop Co, 57-60; from asst prof to prof geochem, Univ Mo-Columbia, 60-69; PROF GEOCHEM, COLO SCH MINES, 69- *Concurrent Pos:* Chief scientist, Crystal Res Lab, Golden, Colo. *Mem:* Fel Mineral Soc Am; Am Crystallog Asn; Geol Soc Am; Clay Mineral Soc. *Res:* Crystalline structure of minerals; clay mineralogy; methods of rock analysis; theoretical mineralogy. *Mailing Add:* Dept of Chem & Geochem Colo Sch of Mines Golden CO 80401

SLAUGHTER, MILTON DEAN, b New Orleans, La, June 9, 44; m 67; c 3. THEORETICAL ELEMENTARY PARTICLE PHYSICS. *Educ:* Univ New Orleans, BS, 71, PhD(physics), 74. *Prof Exp:* Postdoctoral particle physics, Univ Md, College Park, 74-76; postdoctoral particle physics, Los Alamos Nat Lab, 76-77, staff mem detonation theory, 77-81, asst div leader, Theory Div, 81-87, staff mem particle physics, 87-89; CHMN & PROF PHYSICS, UNIV NEW ORLEANS, 89- *Concurrent Pos:* Vis assoc prof physics, Univ Md, College Park, 84-85; tech exec officer, Nat Soc Black Physicists, 84-90. *Mem:* Am Phys Soc. *Res:* Development of the algebraic approach to hadrons and glueballs; vector meson to pseudoscalar radiative and pionic transitions and mass splittings; theoretical description of the radiative decays of baryons, mesons and glueballs. *Mailing Add:* Dept Physics Univ New Orleans New Orleans LA 70148

SLAUNWHITE, WILSON ROY, JR, b Waltham, Mass, Sept 25, 19; m 42; c 4. BIOCHEMISTRY. *Educ:* Mass Inst Technol, BS & MS, 42, PhD(org chem), 48. *Prof Exp:* Res assoc, Radiation Lab, Mass Inst Technol, 42-45 & Naval Res Lab, 45-46; res fel med, Mass Gen Hosp, 48-52; Damon Runyon fel, Med Sch, Univ Utah, 52-53; sr cancer res scientist, Roswell Park Mem Inst, 53-55, assoc scientist cancer res, 55-60, prin cancer res scientist, 60-67; res dir, Med Found Buffalo, 67-69; dir endocrine labs, Children's Hosp, 69-73; assoc res prof pediat, 70-84, PROF BIOCHEM, STATE UNIV NY BUFFALO, 70-, ADJ ASSOC PROF MED TECHNOL, 85- *Concurrent Pos:* From asst res prof to assoc res prof biochem, State Univ NY Buffalo, 56-63, res prof, Roswell Park Grad Div, 63-70, chmn dept, 63-67; consult, Med Found Buffalo, 58-67; ed, Steroids, 64- & J Clin Endocrinol & Metab, Endocrine Soc, 68-70; consult, Roswell Park Mem Inst, 70- *Mem:* Endocrine Soc; Am Soc Biol Chem. *Res:* Steroid and thyroid endocrinology; glycoprotein structure. *Mailing Add:* Dept Biochem 81 Colony Ct Snyder NY 14226

SLAUTTERBACK, DAVID BUELL, b Indianapolis, Ind, July 15, 26; m 52; c 4. CELL BIOLOGY. *Educ:* Univ Mich, BS, 48, MS, 49; Cornell Univ, PhD(anat), 52. *Prof Exp:* Instr anat, Med Sch, NY Univ, 54-55; instr, Med Col, Cornell Univ, 55-59; from asst prof to assoc prof, 59-67, chmn dept, 67-83, PROF ANAT, MED SCH, UNIV WIS-MADISON, 67- *Concurrent Pos:* Am Cancer Soc res fel, Med Sch, NY Univ, 52-54; Nat Cancer Inst fel, 54-55. *Mem:* AAAS; Am Soc Cell Biol; Am Asn Anatomists. *Res:* Integral and peripheral membrane proteins; smooth endoplasmic reticulum; immunocytochemistry of muscle proteins. *Mailing Add:* Dept Anat Univ Wis 318 Med Sci Ctr Madison WI 53706

SLAVEN, ROBERT WALTER, b Salem, Mass, Mar 8, 48; m 69; c 2. ORGANIC CHEMISTRY, ANALYTICAL CHEMISTRY. *Educ:* Univ Lowell, BSc, 69; Univ Wis, PhD(org chem), 74. *Prof Exp:* Fel, Northeastern Univ, 74-76; res chemist org chem, Lorillard Res Ctr, 76-84; sr chemist, 84-85, MGR ANAL CHEM /D+C DIV, CIBA GEIGY, 85- *Mem:* Am Chem Soc; Sigma Xi. *Res:* Nuclear magnetic resonance; structure elucidation. *Mailing Add:* 16018 Malvern Hill Ave Baton Rough LA 70817-3125

SLAVIK, MILAN, b Kosice, Czech, May 18, 30; m; c 1. CLINICAL PHARMACOLOGY. *Educ:* Charles Univ, Prague, MD, 60; Czech Acad Sci, PhD(pharmacol), 69; Bd Internal Med, cert internal med, 63. *Prof Exp:* Instr Clin Internal Dis, Charles Univ, Prague, 59-60; intern & resident, Dept Internal Med, Inst Nat Health, Melnik, 60-63, internist, 63-66; asst prof, Charles Univ, 67-69; clin pharmacologist, Inst Pharmacol, Czech Acad Sci, 68-69; spec asst to chief, Cancer Ther Eval Br, Nat Cancer Inst, NIH, 71-73, from actg chief to chief, 73-76; vis asst prof, Georgetown Univ, 76-77, asst prof, 77-78; prof med & pharmacol, Univ NMex, 78-83; PROF MED, DIV CLIN ONCOL, UNIV KANS MED CTR & MED ONCOL SECT, VET ADMIN MED CTR, 83- *Concurrent Pos:* Chief, Invest Drug Br, Cancer Ther Eval Prog, DCT, Nat Cancer Inst, 74-76; consult physician, Med Oncol Br, NCI & Vet Admin, 73-75; consult fac mem, Inst Clin Toxicol, 73-; mem, chemother subcomt, NCI, 75-79, Ref Panel Am Hosp Formulary Serv, Am Soc Hosp Planning, 75-83, US Japan Coop Res Prog, 76-82, adv comt amino acid analog & depleting enzymes, NCI, 78-82, assoc mem, Southwest Oncol Group, 78-80, mem, 80-, Clin Res Ctr adv comt & chief clin pharmacol & exp therapeut, Cancer Res & Treatment Ctr, Univ NMex, 78-83, hemat & oncol vertical comt, 79-83; adj prof obstet & gynecol, Jefferson Med Col, Thomas Jefferson Univ, 79; consult physician, Vet Admin Med Ctr, 81-83; courtesy prof pharmaceut chem, Col Pharm, Univ Kans & staff physician, Med Serv, Vet Admin Med Ctr, 83-; mem res & develop comt, Vet Admin Med Ctr, 84-85, 89-, cancer res fel rev comt, Ladies Auxillary Vet Foreign War US, 85- *Mem:* Am Asn Cancer Res; Am Soc Clin Oncol; Am Soc Pharmacol & Exp Therapeut; Am Fedn Clin Res; Am Soc Clin Toxicol; Int Asn Study Lung Cancer; Czech Med Soc; Czech Soc Arts & Sci; Int Soc Chronobiol; Sigma Xi. *Res:* Clinical pharmacology and therapeutics; medical oncology; skin and connective tissue diseases. *Mailing Add:* Div Clin Oncol Dept Med Univ Kans Med Ctr 1010 N Kansas Blvd Wichita KS 67214

SLAVIK, NELSON SIGMAN, b St Louis, Mo, Feb 28, 48; m 70. BIOLOGICAL SAFETY, OCCUPATIONAL HEALTH & SAFETY. *Educ:* Kalamazoo Col, BA, 70; Univ Ill, Urbana, MS, 72, PhD(microbiol), 75. *Prof Exp:* Res assoc plant tissue cult, Univ Ill, Urbana, 75-77, biol safety officer, 77-85, asst prof, Occup Health & Safety, 81-85; PRES, ENVIRON HEALTH MGT SYSTS INC, 85- *Concurrent Pos:* Asst dir, Div Environ Health & Safety, Univ Ill, Urbana, 78-81. *Mem:* Sigma Xi. *Mailing Add:* Environ Health Mgt Systs Inc PO Drawer 6309 South Bend IN 46660

SLAVIN, BERNARD GEOFFREY, b San Francisco, Calif, Oct 18, 36; m 60; c 3. ANATOMY & ULTRASTRUCTURE, IMMUNOCYTOCHEMISTRY & MOLECULAR BIOLOGY. *Educ:* Univ Calif, Berkeley, BA, 59; Univ Calif, San Francisco, MA, 62, PhD(anat), 67. *Prof Exp:* Teaching fel anat, Univ Calif, 64-65, res asst, 65-66, lectr, 66-67; NIH fel, Yale Univ, 67-69; asst prof, 69-74, ASSOC PROF ANAT, UNIV SOUTHERN CALIF, 74- *Concurrent Pos:* NIH res grant, 75-78; vis prof, Hebrew Univ, 78-79; vis scientist, Cedars-Sinai Med Ctr, 90-91. *Mem:* AAAS; Am Asn Anatomists. *Res:* Hormonal influence on adipose tissue in vitro; histochemistry and electron microscopy of adipose cells; innervation of adipose tissue; immunocytochemistry of the endocrine pancreas with age. *Mailing Add:* Dept Anat & Cell Biol Univ Southern Calif Med Sch Los Angeles CA 90033

SLAVIN, JOANNE LOUISE, b Harvard, Ill, Apr 2, 52; m 79; c 2. DIETARY FIBER, SPORTS NUTRITION. *Educ:* Univ Wis-Madison, BS, 74, RD, 78, MS, 78, PhD(nutrit), 81. *Prof Exp:* ASSOC PROF NUTRIT, DEPT FOOD SCI & NUTRIT, UNIV MINN, 81- *Mem:* Am Dietetic Asn; Inst Food Technologists; Soc Nutrit Educ; Am Col Sports Med; Am Inst Nutrit. *Res:* Human nutrition; dietary fiber; nutrient bioavailability; sports nutrition; iron status of women athletes; carbohydrate metabolism; diet and cancer. *Mailing Add:* Dept Food Sci & Nutrit Univ Minn 1334 Eckles Ave St Paul MN 55108

SLAVIN, OVID, b Romania, Dec 20, 21; US citizen; m 45. DENTISTRY, BIOLOGY. *Educ:* Washington Univ, AB, 42, DDS, 45. *Prof Exp:* Intern, Guggenheim Dent Clin, NY, 45-46; instr pedodont, Sch Dent & Oral Surg, Columbia Univ, 52-58; asst prof, Seton Hall Col Med & Dent, 58-62, asst dir prototype prog handicapped children, 61-62; prof biomed eng, NY Inst Technol, 64-65, prof life sci & chmn dept, 65-67, dean admin, Long Island Ctr, 64-66; chief dentist, Brookdale Hosp Ctr, Brooklyn, NY, 67-71, asst dir, Comprehensive Child Care Prog, 68-71; regional dent consult, Maternal & Child Health Serv, Health Serv & Ment Health Admin, Dept Health, Educ & Welfare, Philadelphia, 71-77; assoc prof, Sch Dent, Temple Univ, 77-87; RETIRED. *Concurrent Pos:* Chief pedodont sect, Long Island Jewish Hosp, NY, 53-55; chief pedodont serv, Jewish Chronic Dis Hosp, NY, 53-62. *Mem:* Am Asn Pub Health Dentists; Am Acad Pedodont; Am Dent Asn; Am Soc Dent Children. *Res:* Dentistry for handicapped children; biomedical engineering, particularly its application to clinical procedures; pedodontics. *Mailing Add:* Six Bourne Hay Rd Sandwich MA 02563

SLAVIN, RAYMOND GRANAM, b Cleveland, Ohio, June 29, 30; m 53; c 4. INTERNAL MEDICINE, ALLERGY. *Educ:* Univ Mich, AB, 52; St Louis Univ, MD, 56; Northwestern Univ, Chicago, MS, 63. *Prof Exp:* Resident internal med, Sch Med, St Louis Univ, 59-61; asst internal med, Northwestern Univ, 64-65; from asst prof to assoc prof, 65-73, PROF INTERNAL MED, SCH MED, ST LOUIS UNIV, 73- *Concurrent Pos:* NIH Immunol Sci Study Sect grant, 85-89; bd chmn, Asthma & Allergy Found Am. *Honors & Awards:* John M Sheldon Lectr, 90. *Mem:* Fel Am Acad Allergy & Immunol (past pres); fel Am Col Physicians; Am Asn Immunologists. *Res:* Immunology, clinical allergy; allergic bronchopulmonary aspergillosis and other pulmonary hypersensitivity syndromes; delayed hypersensitivity-clinical states in which delayed hypersensitivity is suppressed; sinusitis and asthma. *Mailing Add:* 1402 S Grand St St Louis MO 63104

SLAVKIN, HAROLD CHARLES, b Chicago, Ill, Mar 20, 38. DEVELOPMENTAL BIOLOGY, CELL BIOLOGY. *Educ:* Univ Southern Calif, BA, 63, DDS, 65. *Prof Exp:* Fel, Dept Anat, Sch Med, Univ Calif, Los Angeles, 65-66; fel, Dept Biochem, Univ Southern Calif, 66-68, asst prof, Sch Dent, 68-70, fac, gerontol, Gerontol Inst, 69-70, assoc prof, 71-73, PROF BIOCHEM & NUTRIT, SCH DENT, UNIV SOUTHERN CALIF, 74- *Concurrent Pos:* Res Career Develop Award, US Pub Health Serv, 68-72; prin investr, Southern Calif State Dent Asn grants, 69-70, NIH, Prog Proj grants, 69-, Intercellular Commun grants, 72-80, Training grants, 69-85 & NIH grants, 81-84; vis scientist, Intramural Prog, Nat Inst Dent Res, NIH, 75-76; co-ed, J Craniogacial Genetics & Develop Biol, 80-, co-managing ed, Differentiation, 80-81; consult, NIH, 67-, NSF, 73-75, Med Res Coun, Australia, Can, & Gt Brit, 72-, Bd Sci Dirs, Nat Inst Dent Res, NIH, 76-80, Human Biol Series, US News & World Report, 81-85 & Oral Biol & Med Study Sect, NIH, 81-85. *Mem:* AAAS; Am Soc Cell Biol; Int Asn Dent Res; Int Soc Develop Biol; NY Acad Sci. *Res:* Epithelial-mesenchymal interactions during vertebrate epidermal organ development; induction of epithelial-specific gene products such as enamel proteins during tooth morphogenesis; immunogenetic studies of drug-induced craniofacial malformations in murine and human embryogenesis. *Mailing Add:* 815 Centinela Ave Santa Monica CA 90403-2313

SLAWSKY, ZAKA I, b Chevy Chase, Md, Apr 2, 10. PHYSICS. *Educ:* R P Inst, BS, 33; Calif Inst Technol, MS, 35; Univ Mich, PhD(physics), 38. *Prof Exp:* PROF PHYSICS, UNIV MD, 88- *Mem:* Nat Acad Sci; Am Phys Soc. *Mailing Add:* 4701 Willard Ave Apt 318 Chevy Chase MD 20815

SLAWSON, PETER (ROBERT), b Toronto, Ont, July 10, 39; m 62; c 3. AIR POLLUTION, MECHANICAL ENGINEERING. *Educ:* Univ Waterloo, BASc, 64, MASc, 66, PhD(mech eng), 67. *Prof Exp:* Res assoc air pollution, 67-69, asst prof mech eng, 69-74, assoc prof, 74-80, PROF MECH ENG, UNIV WATERLOO, 80- *Concurrent Pos:* Pres, Envirodyne Ltd, 74- *Res:* Diffusion; atmospheric dynamics as applied to the dispersal of air pollutants. *Mailing Add:* Dept Mech Eng Univ Waterloo Waterloo ON N2L 3G1 Can

SLAYDEN, SUZANNE WEEMS, b Heidelberg, Ger, Aug 11, 48; US citizen. ORGANIC CHEMISTRY. *Educ:* Univ Tenn, BS, 70, PhD(org chem), 76. *Prof Exp:* ASST PROF CHEM, GEORGE MASON UNIV, 76- *Mem:* Am Chem Soc. *Res:* Mechanisms of organoborate rearrangements. *Mailing Add:* Chem Dept George Mason Univ Fairfax VA 22030

SLAYMAKER, FRANK HARRIS, b Lincoln, Nebr, April 22, 14; m 49; c 4. ELECTRICAL ENGINEERING. *Educ:* Univ Nebr, BS, 41, MS, 46. *Prof Exp:* Prof elec engr, The Stomberg Carlson Co, 41-50, sr elec engr, sound equipment, 50-51, chief engr, 51-56, mgr appl physics Lab, 56-61; assoc dir, Stromberg-Carson Div, Gen Dynamics Corp, 61-62, exec asst vpres engr & res, 62-63; dir ASW prog, Elec Div, 63-65, res specialist, 65-70; scientist, Tropel Inc, 70-75; sr engr, Univ Rochester, 75-77, res assoc, Ctr Visual Sci, 77-79; scientist, Tropel Div, Coherent Radiation Inc, 79-85; RETIRED. *Mem:* Fel Inst Elec & Electronic Engrs; fel Acoustics Soc Am. *Res:* Research and development in fields of speech analysis-synthesis, transducer development, musical tone analysis-synthesis; optical intertrometers, optical transfer functions. *Mailing Add:* 134 Glen Haven Rd Rochester NY 14609-2057

SLAYMAKER, HERBERT OLAV, b Swansea, Wales, Jan 31, 39; Can citizen; m 67; c 4. MOUNTAIN ENVIRONMENTS. *Educ:* Cambridge Univ, BA, 61, PhD(geomorphol), 68; Harvard Univ, AM, 63. *Prof Exp:* Asst lectr geog, Univ Col, Wales, Aberystwyth, 64-66, lectr, 66-68; from asst prof to assoc prof, 68-81, head geog, 82-91, PROF GEOG, UNIV BC, 81- *Concurrent Pos:* Vis lectr, dept geog, Cambridge Univ, 66-68; vis prof, dept geog, Southern Ill Univ, 73; coordr interdisciplinary hydrol, Univ BC, 75-, pres fac asn, 78-79 & mem bd gov, 84-87; chmn, Int Geog Union Comn, 80-84. *Honors & Awards:* Sr Killam Res Fel, 87-88. *Mem:* Can Asn Geogrs; Can Quaternary Asn; Am Geophys Union; Int Asn Gerontol (treas, 89-). *Res:* Hydrology and geomorphology of drainage basins; sediment and solute budgets of British Columbia's mountain regions; natural rates of erosion and land use impacts. *Mailing Add:* 3330 W 21st Ave Vancouver BC V6S 1G7 Can

SLAYMAN, CAROLYN WALCH, b Portland, Maine, Mar 11, 37; m 59; c 2. GENETICS. *Educ:* Swarthmore Col, BA, 58; Rockefeller Univ, PhD(biochem genetics), 63. *Hon Degrees:* DSc, Bowdoin Col, 85. *Prof Exp:* Instr biol, Western Reserve Univ, 64-65, asst prof, 65-67; asst prof microbiol & physiol, 67-70, assoc prof, 70-72, assoc prof human genetics and physiol, 72-77, PROF HUMAN GENETICS & PHYSIOL, SCH MED, YALE UNIV, 77-, CHMN DEPT HUMAN GENETICS, 84- *Concurrent Pos:* NSF fel, Cambridge Univ, 63-64; consult, Adv Panel Genetic Biol, NSF, 74-77; assoc ed, Genetics, 77-82; mem bd overseers, Bowdoin Col, 77-89, bd trustees, 89-; chmn, Genetic Basis Dis Rev Comt, NIH, 83-85; mem, Nat Adv Gen Med Sci Coun, NIH, 89- *Res:* Genetic control of membrane transport. *Mailing Add:* Dept Human Genetics Sch Med Yale Univ New Haven CT 06520

SLAYMAN, CLIFFORD L, b Mt Vernon, Ohio, July 7, 36; m 59; c 2. PHYSIOLOGY. *Educ:* Kenyon Col, AB, 58; Rockefeller Univ, PhD(physiol), 63. *Hon Degrees:* DSc, Kenyon Col, 91. *Prof Exp:* NSF fel physiol, Cambridge Univ, 63-64; asst prof, Sch Med, Western Reserve Univ, 64-67; from asst prof to assoc prof, 67-80, PROF PHYSIOL, SCH, MED, YALE UNIV, 83- *Concurrent Pos:* NIH res grants, 65, 68, 73, 78, 81 & 86-91, res career develop award, 69; NSF res grant, 79-82; Dept Energy res grant, 85- *Mem:* Am Physiol Soc; Am Soc Microbiol; Soc Gen Physiologists; Am Soc Plant Physiologists; AAAS. *Res:* Membrane biophysics; transport, energy-coupling, and electrogenesis in microorganisms. *Mailing Add:* Dept Cellular & Molecular Physiol Yale Univ Sch Med New Haven CT 06510

SLEATOR, WILLIAM WARNER, JR, b Ann Arbor, Mich, Apr 5, 17; m 40; c 4. BIOPHYSICS, PHYSIOLOGY. *Educ:* Univ Mich, AB, 38, MS, 39, PhD(physics), 46. *Prof Exp:* Physicist, Ballistic Res Lab, Aberdeen Proving Ground, Md, 42-45; res assoc physics, Univ Minn, 46-49; from asst prof biophys to assoc prof physiol, Sch Med, Wash Univ, St Louis, 49-64, prof physiol & biophys, 64-69, actg chmn dept, 66-68; head dept physiol & biophys, 69-76, PROF PHYSIOL & BIOPHYSICS, UNIV ILL, URBANA, 69- *Concurrent Pos:* Mem, Physiol Study Sect, NIH, 76-80. *Mem:* Fel Am Phys Soc; Am Physiol Soc; Biophys Soc (secy, 62-67); Sigma Xi. *Res:* Scattering of elementary particles and light nuclei; light scattering and absorption by living muscle and muscle proteins; fundamental cellular processes in heart, skeletal and smooth muscle; ionic conductance changes during cardiac action potential; mechanism of contraction process and of coupling between excitation and contraction; ion flux measurements will tracers and ion-selective electrodes. *Mailing Add:* Dept of Physiol & Biophysics Univ of Ill 407 S Goodwin Urbana IL 61801

SLECHTA, ROBERT FRANK, b New York, NY, June 4, 28; m 53; c 1. REPRODUCTIVE PHYSIOLOGY. *Educ:* Clark Univ, AB, 49, MA, 51; Boston Univ, PhD(biol), 55. *Prof Exp:* Asst physiol, Worcester Found Exp Biol, 52-53; res assoc & instr, Tufts Univ, 55-58; from asst prof to assoc prof, 58-65, assoc dean, Grad Sch, 67-79, PROF BIOL, BOSTON UNIV, 65- *Mem:* AAAS; Soc Study Reproduction. *Res:* Reproductive and microcirculatory physiology. *Mailing Add:* Cummington St Boston Univ Boston MA 02215

SLEDD, MARVIN BANKS, mathematics, for more information see previous edition

SLEDD, WILLIAM T, b Murray, Ky, Aug 25, 35; div; c 2. MATHEMATICS. *Educ:* Murray State Col, BA, 56; Univ Ky, MA, 59, PhD(math), 61. *Prof Exp:* Asst math, Univ Ky, 56-60; from asst prof to assoc prof, 61-69, PROF MATH, MICH STATE UNIV, 69- *Mem:* Math Asn Am; Am Math Soc. *Res:* Summability theory; Fourier analysis. *Mailing Add:* Dept Math Mich State Univ East Lansing MI 48824-1027

SLEDGE, EUGENE BONDURANT, b Mobile, Ala, Nov 4, 23; m 52; c 2. BIOLOGY. *Educ:* Auburn Univ, BS, 49, MS, 55; Univ Fla, PhD(biol), 60. *Prof Exp:* Res asst, Auburn Univ, 53-55; asst, Univ Fla, 56-59; nematologist, Div Plant Indust, Fla State Dept Agr, 59-62; asst prof biol, Ala Col, 62-70; chmn dept, 70-72, PROF BIOL, UNIV MONTEVALLO, 70- *Mem:* Am Ornith Union; Wilson Ornith Soc. *Res:* Ornithology, particularly avian myology. *Mailing Add:* Dept of Biol Univ of Montevallo Montevallo AL 35115

SLEE, FREDERICK WATFORD, b Spokane, Wash, Mar 16, 37. NUCLEAR PHYSICS. *Educ:* Univ Wash, BS, 59, MS, 60, PhD(physics), 66. *Prof Exp:* ASSOC PROF PHYSICS, UNIV PUGET SOUND, 66-, CHMN PHYSICS, 80- *Mem:* Am Phys Soc; AAAS. *Res:* Electronic instrumentation applied to physics, geophysics and biophysics. *Mailing Add:* Dept Physics Univ Puget Sound 1500 N Warner Tacoma WA 98416

SLEEMAN, RICHARD ALEXANDER, b Bennington, Vt, Sept 15, 26; m 50. CHEMISTRY. *Educ:* Fordham Univ, BS, 49; NY Univ, MA, 51, EdD(phys sci), 55. *Prof Exp:* Instr phys sci, Vt State Teachers Col, Castleton, 49-54 & NY Univ, 54-55; asst prof chem, Kent State Univ, 56-60; PROF CHEM, NORTH ADAMS STATE COL, 60-, MEM FAC DEPT EDUC, 76- *Mem:* AAAS; Nat Asn Res Sci Teaching. *Res:* Physics; quantum mechanics and black body radiation. *Mailing Add:* Dept Educ North Adams State Col North Adams MA 01247

SLEEP, NORMAN H, b Kalamazoo, Mich, Feb 14, 45. GEOLOGY, GEOPHYSICS. *Educ:* Mich State Univ, BS, 67; Mass Inst Technol, MS, 69, PhD(geol), 73. *Prof Exp:* Asst prof, dept geol, Northwestern Univ, 73-79; PROF, DEPT GEOL & GEOPHYS, STANFORD UNIV, 79- *Mem:* Geol Soc Am. *Mailing Add:* Dept Geophys & Geol Stanford Univ Stanford CA 94305

SLEEPER, DAVID ALLANBROOK, b Exeter, NH, Feb 1, 22; m 49; c 1. BIOLOGY, ENTOMOLOGY. *Educ:* Univ NH, BS, 43; Cornell Univ, PhD(entom), 63. *Prof Exp:* Entomologist, Alaska Insect Proj, USDA, 48 & Arctic Health Res Ctr, USPHS, 49-54; instr biol, Cornell Univ, 61-63; asst prof, Elmira Col, 63-65; asst prof, 65-71, assoc prof, 71-81, prof biol, Hobart & William Smith Cols, 81-87; RETIRED. *Mem:* Ecol Soc Am; Soc Syst Zool; Animal Behav Soc. *Res:* Ecology, taxonomy, and physiology of biting diptera, especially blackflies of the family Simuliidae. *Mailing Add:* 142 Burleigh Dr Ithaca NY 14850-1744

SLEEPER, ELBERT LAUNEE, b Newton, Iowa, July 27, 27; m 49, 72, 77; c 4. ENTOMOLOGY. *Educ:* Ohio State Univ, BS, 50, MS, 51, PhD(entom), 56. *Prof Exp:* Asst, Ohio Biol Surv, Ohio State Univ, 53-56; from asst prof to assoc prof, 57-66, PROF ENTOM, CALIF STATE UNIV, LONG BEACH, 66- *Concurrent Pos:* Collabr, US Nat Park Serv, 57-80; consult lab nuclear med & radiation biol, Univ Calif, Los Angeles, 70-80. *Mem:* Ecol Soc Am; Am Entom Soc; Am Forestry Soc; Entom Soc Am. *Res:* Systematics of the Curculionoidea excluding the Scolytidae; dynamics of desert insect populations; systematics, zoogeography and ecological distribution of Curculionoidea of the new world. *Mailing Add:* Dept of Biol Calif State Univ Long Beach CA 90840

SLEETER, THOMAS DAVID, b Pasadena, Calif, Mar 31, 52; m 85; c 2. AQUACULTURE. *Educ:* Clark Univ, BA, 74; Harvard Univ, SM, 76, PhD(environ eng), 80. *Prof Exp:* Res scientist, Bermuda Biol Sta Res, 80-88; CURATOR, BERMUDA AQUARIUM & ZOO, 88- *Concurrent Pos:* Sci ed, Harvard Environ Law Rev, 77-78; consult, US Senate, 81-, Am Petrol Inst, 83-, NORAQUA A-S, 84-, US Environ Protection Agency, Nat Oceanic & Atmospheric Admin, Exxon, Brit Petrol Co, UNESCO & Int Maritime Orgn; adj prof, Nova Univ, 84- *Res:* Fate and effect of oil spills on marine life; national and regional oil spill contingency planning; environmental resource assessment for aquaculture; environmental engineering. *Mailing Add:* Bermuda Aquarium PO Box 141 Flatts Smiths Bermuda

SLEETH, BAILEY, b Linn, WVa, Nov 1, 00; m 33; c 2. PLANT PATHOLOGY. *Educ:* Univ WVa, BS, 27, MS, 28, PhD(plant path), 32. *Prof Exp:* Teacher high sch, WVa, 27-30; asst plant path, Univ WVa, 32-33, asst exten plant pathologist, 42; asst forest pathologist, Bur Plant Indust, US Dept Agr, 33-41, assoc pathologist, Rubber Plant Invests, 42-46, pathologist, Div Soil Mgt & Irrig, 46-51; plant pathologist, 51-66, EMER PLANT PATHOLOGIST, TEX AGR EXP STA, TEX A&M UNIV, 66-; agr consult & ed, Rio Farms Inc, 66-84; RETIRED. *Honors & Awards:* Arthur Potts Award, Rio Grande Valley Hort Soc. *Mem:* Am Phytopath Soc; Mycol Soc Am; Am Inst Biol Sci; Sigma Xi. *Res:* Forest tree diseases in eastern United States; guayale seedling diseases in California and Texas; citrus diseases. *Mailing Add:* 307 Nebr Ave Weslaco TX 78596

SLEETH, RHULE BAILEY, b Linn, WVa, Feb 6, 29; m 53; c 2. FOOD TECHNOLOGY, ANIMAL NUTRITION. *Educ:* WVa Univ, BS, 51; Univ Fla, MS, 53; Univ Mo, PhD(meat & food technol), 59. *Prof Exp:* Instr meat technol, Univ Mo, 55-59; food technologist, 59-63, sect head fresh meat develop, 63-64, asst mgr, Food Res Div, 64-68, asst dir, Food Res Div, 68-78, dir food res & develop, 78-80, VPRES FOODS RES & DEVELOP, ARMOUR & CO, 81- *Concurrent Pos:* Mem prog comt, Meat Indust Res Conf, 65-66, chmn, 67; chmn, Reciprocal Meat Conf, 68; contact, Europ Meeting Meat Res Workers. *Honors & Awards:* Signal Serv Award, Reciprocal Meat Conf, 71. *Mem:* Fel Inst Food Technologists; Am Soc Animal Sci; Am Meat Sci Asn (pres, 73-74); Soc Advan Food Serv Res. *Res:* Sausage and cured meat development; sterile and refrigerated canned meats; food service research; food chemistry; dairy, poultry and food oils; packaging research. *Mailing Add:* Armour Res Ctr 15101 N Scottsdale Rd Scottsdale AZ 85254

SLEEZER, PAUL DAVID, b Chicago, Ill, Jan 26, 36; m 63. ANALYTICAL CHEMISTRY. *Educ:* Univ Rochester, BS, 58; Univ Calif, Los Angeles, PhD(chem), 63. *Prof Exp:* Sr res chemist, Solvay Process Div, Allied Chem Corp, NY, 63-66; sr develop chemist, Bristol Labs, 66-72, DEPT HEAD ORG SYNTHESIS LABS, INDUST DIV, BRISTOL-MYERS CO, 72- *Mem:* Am Chem Soc. *Res:* Exploratory organic research; synthesis and process development; process analysis and control; organic reaction mechanisms; physical-organic; organometallics. *Mailing Add:* Bristol-Myers Co PO Box 4755 Syracuse NY 13221-4755

SLEICHER, CHARLES A, b Albany, NY, Aug 15, 24; m 53; c 2. CHEMICAL ENGINEERING. *Educ:* Brown Univ, ScB, 46; Mass Inst Technol, MS, 49; Univ Mich, PhD(chem eng), 55. *Prof Exp:* Res engr, Shell Develop Co, 55-59; NSF fel, Cambridge Univ, 59-60; assoc prof chem eng, 60-66, chmn dept, 77-89, PROF CHEM ENG, UNIV WASH, 66- *Concurrent Pos:* Grants, Am Chem Soc, 60-64, NSF, 61- & Chevron Res Found, 64-67; consult, Westinghouse-Hanford Co, 73-85. *Mem:* AAAS; Am Chem Soc; Am Inst Chem Engrs. *Res:* Turbulent diffusion, heat transfer; heat transfer with variable fluid properties; dispersion of toxins in the environment; aerosol deposition in the lung. *Mailing Add:* Dept Chem Eng BF-10 Univ Wash Seattle WA 98195

SLEIGHT, ARTHUR WILLIAM, b Ballston Spa, NY, Apr 1, 39; m 63; c 3. SOLID STATE CHEMISTRY, SUPERCONDUCTIVITY. *Educ:* Hamilton Col, BA, 60; Univ Conn, PhD(inorg chem), 63. *Prof Exp:* Fel crystallog, Univ Stockholm, 63-64; res chemist solid state chem, 64-79, res supvr, 79-81, res mgr, 81-85, RES LEADER, E I DU PONT DE NEMOURS & CO, 85- *Concurrent Pos:* Assoc ed, Mat Res Bull, 76- & Inorg Chem Rev, 79-; adj prof, Univ Del, 78-; mem panel, Sci Challenges Arising from Technol Needs, 78; chmn, Solid State Chem Subdiv, Am Chem Soc, 81; mem major mats facil comt, Nat Acad Sci, 84; mem, comt on interdisciplinary aspects of crystallogr, Nat Res Coun, 86-; mem, Panel on High-Temperature Superconductivity, Nat Acad Sci, 87; vis prof, Univ Rennes, France, 88. *Mem:* Am Chem Soc; Mats Res Soc. *Res:* Solid state chemistry, structure-property relationships for inorganic solids, especially oxides and sulfides, superconductivity, ionic conductivity, defects; crystal growth; crystallography; structural chemistry; heterogeneous catalysis; electrical, magnetic and optical properties. *Mailing Add:* Dept Chem Ore State Univ Corvalis OR 97331

SLEIGHT, STUART DUANE, b Lansing, Mich, Oct 19, 27; m 50; c 4. VETERINARY PATHOLOGY. *Educ:* Mich State Univ, DVM, 51, MS, 59, PhD(vet path), 61. *Prof Exp:* Vet, Columbus Vet Hosp, Wis, 51-58; from asst prof to assoc prof, 61-68, PROF PATH, MICH STATE UNIV, 68- *Mem:* Am Col Vet Path; Conf Res Workers Animal Dis; Am Vet Med Asn; NY Acad Sci; Soc Toxicol. *Res:* Studies emphasize assessment of the toxic & carcinogenic effects of exposure to important environmental chemicals, including N-nitrosamines & polyhalogenated aromatic hydrocarbons. *Mailing Add:* 522 E Fee Hall Dept Path Mich State Univ East Lansing MI 48824

SLEIGHT, THOMAS PERRY, b Glens Falls, NY, May 15, 43; m 66; c 4. COMPUTER SCIENCE. *Educ:* Ohio Univ, BS(chem) & BS(math), 65; State Univ NY Buffalo, PhD(chem), 69. *Prof Exp:* Sci Res Coun fel, Univ Leicester, 68-69; advan systs design suprv, 77-83, dir's asst, comput & info systs, 83-89, PRIN STAFF, APPL PHYSICS LAB, JOHNS HOPKINS UNIV, 69-, CRUISE MISSILES SYSTS ENGR, 89- *Mem:* Am Chem Soc; Int Elec & Electronic Engrs Comput Soc. *Res:* Advanced computer systems; software engineering; navy computer systems; real time software. *Mailing Add:* Appl Physics Lab John Hopkins Univ John Hopkins Rd Laurel MD 20707

SLEIN, MILTON WILBUR, b St Louis, Mo, Jan 26, 19; m 50; c 3. BIOCHEMISTRY. *Educ:* Washington Univ, AB, 40, MS, 43, PhD(biochem), 49. *Prof Exp:* Asst biochem, Sch Med, Washington Univ, 42-49; biochemist, US Army Biol Ctr, Ft Detrick, 49-71 & Edgewood Arsenal, 71-72; biochemist, Frederick Cancer Res Ctr, 72-75; RETIRED. *Mem:* Am Soc Biol Chemists. *Res:* Enzymology; microbiological biochemistry. *Mailing Add:* 7065 Catalpa Rd Frederick MD 21701

SLEISENGER, MARVIN HERBERT, b Pittsburgh, Pa, June 3, 24; m 48; c 1. MEDICINE. *Educ:* Harvard Med Sch, MD, 47; Am Bd Internal Med, dipl, 54. *Prof Exp:* Intern med, Beth Israel Hosp, Boston, Mass, 47-48, chief resident, 49-50; intern, Beth Israel Hosp, Newark, NJ, 48-49; resident gastroenterol, Hosp Univ Pa, 50-51; instr med, Med Col, Cornell Univ, 52-53, from asst prof to assoc prof clin med, 54-65, prof med, 65-68; PROF MED & VCHMN DEPT, MED CTR, UNIV CALIF, SAN FRANCISCO, 68- *Concurrent Pos:* Fel gastroenterol, Hosp, Univ Pa, 50-51; fel med, Med Col, Cornell Univ, 51-52; assoc, Harvard Med Sch, 49-50; instr, Sch Med, Tufts Univ, 49-50; asst physician, out-patient clin, New York Hosp, 51-54, physician, 54-56, chief gastrointestinal clin, 54-68, from asst attend physician to attend physician, 56-68; consult, Rockefeller Inst Hosp, New York & aerospace prog, US Air Force, 64-; chief med serv, Ft Miley Vet Admin Hosp, San Francisco, Calif, 68- *Mem:* AAAS; Am Soc Clin Invest; Harvey Soc; Am Gastroenterol Asn; Am Fedn Clin Res. *Res:* Intestinal absorption; esophageal motility; fractionation of gastric juice mucoproteins; experimental ulcerative colitis; measurement of enzymes. *Mailing Add:* Dept Med Univ Calif Box 0120 Rm 997 San Francisco CA 94143

SLEMMONS, DAVID BURTON, b Alameda, Calif, Dec 31, 22; m 46; c 2. GEOLOGY, GEOPHYSICS. *Educ:* Univ Calif, BS, 47, PhD(geol), 53. *Prof Exp:* From asst prof to assoc prof, Univ Nev, 51-63, dir seismog sta, 52-64, chmn dept geol & geog, 66-70, prof geol & geophys, 63-89, dir ctr neotectonic studies, 85-89, EMER PROF GEOL & GEOPHYS, UNIV NEV, 89- *Concurrent Pos:* Prog dir geophys, Earth Sci Div, NSF, 70-71; del, 2nd & 3rd US-Japan Conf Earthquake Prediction; dir, Seismog Soc Am; mem, Earthquake Eng Res Inst. *Honors & Awards:* G K Gilbert Award, Carnegie Inst. *Mem:* Geol Soc Am; Seismol Soc Am; Soc Econ Geol; Am Inst Mining, Metall & Petrol Eng; Asn Eng Geol. *Res:* Geology, geomorphology, neotectonics, seismology and volcanology; petrography; universal stage determination of plagioclase; surface faulting; seismicity, faulting mechanics, earthquake hazards, seismic risk for engineering structures, seismic potential of faults; seismic safety of dams and nuclear reactors. *Mailing Add:* 2905 Autumn Haze Lane Las Vegas NV 89117

SLEMON, GORDON R(ICHARD), b Bowmanville, Ont, Aug 15, 24; m 49; c 4. ELECTRICAL ENGINEERING. *Educ:* Univ Toronto, BASc, 46, MASc, 48; Univ London, DIC & PhD(eng), 52, DSc, 68. *Prof Exp:* Lectr, Imp Col, Univ London, 49-53; asst prof elec eng, NS Tech Col, 53-55; head dept, 66-76, dean, fac appl sci & eng, 79-86, PROF, ELEC ENG, UNIV TORONTO, 55- *Concurrent Pos:* Pres, Elec Eng Consociates, 76-79; chmn bd, Innovation Found, 80- & Microelectronics Develop Ctr, 83-87. *Honors & Awards:* Western Elec Award, Am Soc Eng Educ, 65; Centennial Medal, Can, 67; Gold Medal, Yugoslav Union Elec Power Indust & Yugoslav Union Nikola Tesla Socs; Nikola Tesla Award, Inst Elec & Electronic Engrs, 90. *Mem:* Am Soc Eng Educ; fel Inst Elec & Electronics Engrs; fel Inst Elec Engrs UK; fel Eng Inst Can; Can Soc Elec Eng. *Res:* Power systems; electric propulsion; rotating machines; magnetics. *Mailing Add:* Dept Elec Eng Univ Toronto Toronto ON M5S 1A4 Can

SLEPECKY, RALPH ANDREW, b Nanticoke, Pa, Oct 8, 24; m 67; c 2. MICROBIOLOGY. *Educ:* Franklin & Marshall Col, BS, 48; Pa State Univ, MS, 50; Univ Md, PhD(bact), 53. *Prof Exp:* Asst bact, Pa State Univ, 48-50 & Univ Md, 50-53; asst prof biol, Franklin & Marshall Col, 53-56; res scientist, Univ Tex, 56-58; asst prof biol, Northwestern Univ, 58-63; assoc prof microbiol, 63-68, adminr biol res labs, 64-70, PROF MICROBIOL, SYRACUSE UNIV, 68- *Concurrent Pos:* Instr, Montgomery Jr Col, 52; bacteriologist, Stand Brands, Inc, 52; Found Microbiol lectr, 71-72; vis scientist, NIH, 76-77. *Mem:* AAAS; Am Soc Microbiol; Am Chem Soc. *Res:* Morphogenesis differentiation and heat resistance of bacterial spores. *Mailing Add:* 100 Dorset Rd Syracuse NY 13210

SLEPER, DAVID ALLEN, b Buffalo Ctr, Iowa, Aug 25, 45; m 65; c 2. PLANT BREEDING. *Educ:* Iowa State Univ, BS, 67, MS, 69; Univ Wis, PhD(plant breeding, genetics), 73. *Prof Exp:* Asst prof agron, Univ Fla, 73-74; from asst prof to assoc prof, 74-84, PROF AGRON, UNIV MO, 84- *Honors & Awards:* Merit Cert Am Forage & Grassland Coun, 82. *Mem:* Fel Am Soc Agron; AAAS; fel Crop Sci Soc Am; Am Forage & Grass Coun; Sigma Xi. *Res:* Investigations on the breeding and genetics of Festuca arundinacea and Dactylis glomerata. *Mailing Add:* Dept Agron Rm 209 Waters Hall Univ of Mo Columbia MO 65211

SLEPETYS, RICHARD ALGIMANTAS, b Ukmerge, Lithuania, Apr 4, 28; US citizen; m 54; c 1. PHYSICAL CHEMISTRY, PIGMENT TECHNOLOGY. *Educ:* Univ Detroit, BChE, 54; Newark Col Eng, MSChE, 57; Rutgers Univ, PhD(chem), 67. *Prof Exp:* Jr technologist, NL Industs, Inc, 54-59, technologist, 59-66, sr technologist, 66-72, res & develop sect mgr, 72-78; res group leader, Englehard Minerals & Chem Corp, 78-80, res group leader, 80-89, res assoc, 89, SR RES ASSOC, ENGLEHARD CORP, 89- *Mem:* Am Chem Soc; Am Inst Chem; Tech Asn Pulp & Paper Indust; Clay Minerals Soc; Mineral Soc; Am Soc Testing & Mats. *Res:* Titanium dioxide pigment technology; lattice energies; crystal structure; kaolin pigments and extenders; light scattering. *Mailing Add:* Engelhard Corp Pigments & Additives Div Menlo Park Edison NJ 08818

SLEPIAN, DAVID, b Pittsburgh, Pa, June 30, 23; m 50; c 3. MATHEMATICS, ELECTRICAL ENGINEERING. *Educ:* Harvard Univ, PhD(physics), 49. *Prof Exp:* Parker fel, Harvard Univ, 49-50; res mathematician, Bell Tel Labs Inc, 50-82; prof elec eng, Univ Hawaii, 70-80; RETIRED. *Concurrent Pos:* Vis Mackay prof, Univ Calif, Berkeley, 58-59, regent's lectr, 77. *Honors & Awards:* Von Neumann Lectr, Soc Indust & Appl Math, 82. *Mem:* Nat Acad Sci; Nat Acad Eng; AAAS; Inst Elec & Electronic Engrs; AAAS. *Res:* Communication theory; applied mathematics. *Mailing Add:* 212 Summit Ave Summit NJ 07901

SLEPIAN, PAUL, b Boston, Mass, Mar 26, 23; div; c 2. MATHEMATICS. *Educ:* Mass Inst Technol, SB, 50; Brown Univ, PhD(math), 56. *Prof Exp:* Instr math, Brown Univ, 54-56; mathematician, Ramo-Wooldridge Corp, Calif, 56 & Hughes Aircraft Co, 56-60; assoc prof math, Univ Ariz, 60-62; from assoc prof to prof, Rensselaer Polytech Inst, 62-69; chmn dept, Bucknell Univ, 69-70; PROF MATH, HOWARD UNIV, 70- *Concurrent Pos:* Lectr, Univ Southern Calif, 56-60; vis staff mem, Los Alamos Sci Lab, Los Alamos, NMex, 76, 78 & 79; instr, Lorton Prison Col Prog, Lorton, Va, 78- *Mem:* Am Math Soc; Soc Indust & Appl Math; Math Asn Am. *Res:* Surface area; applications of mathematics to circuit theory; general topology. *Mailing Add:* 2613 Guilford Ave Baltimore MD 21218

SLESNICK, IRWIN LEONARD, b Canton, Ohio, Aug 5, 26; m 47; c 5. SCIENCE EDUCATION. *Educ:* Bowling Green State Univ, BA & BS, 49; Univ Mich, MS, 53; Ohio State Univ, PhD, 62. *Prof Exp:* Instr high sch, Ohio, 49-55; instr unified sci, Ohio State Univ, 56-63; assoc prof, 63-66, coordr sci educ, 71-73, coordr sci educ, 80-85, PROF BIOL, WESTERN WASH UNIV, 66- *Concurrent Pos:* Consult, AID, India, 66-67, sci educ adv, 67-70; consult, UNESCO, 71-; consult, Biol Sci Curric Study, Scott, Foresman & Co, 72-76, ed adv, 77- *Mem:* Nat Sci Teachers Asn. *Res:* Population education; curriculum development in interdisciplinary studies. *Mailing Add:* 518 Highland Dr Bellingham WA 98225

SLESNICK, WILLIAM ELLIS, b Oklahoma City, Okla, Feb 24, 25. MATHEMATICS. *Educ:* US Naval Acad, BS, 45; Univ Okla, BA, 48; Oxford Univ, BA, 50, MA, 54; Harvard Univ, AM, 52. *Hon Degrees:* AM, Dartmouth Col, 72. *Prof Exp:* Teacher, St Paul's Sch, NH, 52-62; from asst prof to assoc prof, 62-71, vis instr, 58-59, asst dir educ uses, Kiewit Comput Ctr, 66-69, PROF MATH, DARTMOUTH COL, 71- *Concurrent Pos:* Mem, advan placement exam comt, Col Entrance Exam Bd, 67-71; Nat Humanities Fac, 72-; Rhodes Scholar, 48. *Mem:* Nat Coun Teachers of Math; Math Asn Am. *Mailing Add:* Dept Math & Computer Sci Dartmouth Col Hanover NH 03755

SLESSOR, KEITH NORMAN, b Comox, BC, Nov 4, 38; m 60; c 3. INSECT SEMIOCHEMICALS. *Educ:* Univ BC, BSc, 60, PhD(org chem), 64. *Prof Exp:* Nat Res Coun Can fels, Royal Free Hosp Med Sch, Univ London, 64-65 & Inst Org Chem, Univ Stockholm, 65-66; from asst to assoc prof, 66-82, PROF ORG BIOCHEM, SIMON FRASER UNIV, 82- *Concurrent Pos:* Nat Res Coun Can Sr Fel, Can Forest Serv, Sault Ste Marie, Can, 79-80. *Mem:* Am Chem Soc; Entom Soc Am. *Res:* Pheromone determination, structure and synthesis; the isolation, identification, synthesis and application of insect semiochemicals; forest lepidopteran pests - such as defoliators, cone and seed pests, wood boring beetles - responsible for the destruction of commercial timber, and social insects, particularly the honey bee. *Mailing Add:* Dept Chem Simon Fraser Univ Burnaby BC V5A 1S6 Can

SLETTEBAK, ARNE, b Danzig, Aug 8, 25; nat US; m 49; c 2. ASTROPHYSICS. *Educ:* Univ Chicago, SB, 45, PhD(astron), 49. *Prof Exp:* Asst, Yerkes Observ, Univ Chicago, 45-49; from instr to assoc prof, 49-59, dir, Perkins Observ, 59-78, chmn dept, 62-78, PROF ASTRON OHIO STATE UNIV, 59- *Concurrent Pos:* Fulbright res fel, Hamburg Observ, Ger, 55-56; mem bd dirs, Asn Univs for Res Astron, 61-79, chmn sci comt, 70-73, steering comt earth sci curric proj, 65-68; comt astron, Nat Res Coun; adv, Off Naval Res, 63-66; mem, Adv Panel Astron, NSF, 68-71; Fulbright lectr & guest prof, Univ Vienna Observ, Austria, 74-75 & 81. *Mem:* Am Astron Soc; Int Astron Union (pres, 76-79). *Res:* Stellar rotation; spectroscopic investigations of normal and peculiar stars. *Mailing Add:* Dept Astron Ohio State Univ 174 W 18 Ave Columbus OH 43210

SLEZAK, FRANK BIER, b Manhasset, NY, Nov 19, 28; m 51; c 2. ORGANIC CHEMISTRY. *Educ:* Antioch Col, BS, 51; Okla State Univ, MS, 53, PhD(org chem), 55. *Prof Exp:* Org res chemist, Diamond Alkali Co, 54-59, group leader condensation polymers, 59-63; group leader chem res & develop, Union Camp Corp, NJ, 63-69; assoc prof, 69-74, chmn dept biol & chem, 71-77, PROF CHEM, MERCER COUNTY COMMUNITY COL, 74- *Mem:* AAAS; Am Chem Soc. *Res:* Synthetic organic chemistry related to biological activity; heterocyclics; condensation polymerization. *Mailing Add:* 9 Pine Knoll Dr Lawrenceville NJ 08648-3142

SLEZAK, JANE ANN, b Amsterdam, NY. BIOMEDICAL ENGINEERING. *Educ:* State Univ NY, Albany, BS, MS; Rensselaer Polytech Inst, PhD(chem). *Prof Exp:* Res assoc chem, Univ Pittsburgh & Syracuse Univ; lectr chem, State Univ NY; asst prof chem, Schenectady Community Col; RES ASSOC BIOMED ENG, RENSSELAER POLYTECH INST, 79-; CHMN, CHEM DEPT, FULTON MONTGOMERY COMMUNITY COL. *Concurrent Pos:* Fel, AEC, NSF, Gen Elec, & NIH. *Mem:* Am Phys Soc; Sigma Xi; Am Chem Soc. *Res:* Materials properties of biological and calcified tissue using spectroscopy, magnetic resonance and physical chemistry; analyses of composite behavior and methods of failure of biological specimens. *Mailing Add:* 191 Church St Amsterdam NY 12010

SLICHTER, CHARLES PENCE, b Ithaca, NY, Jan 21, 24; m 80; c 6. SOLID STATE PHYSICS. *Educ:* Harvard Univ, AB, 45, AM, 47, PhD(physics), 49. *Prof Exp:* From instr to assoc prof, 49-55, PROF PHYSICS, UNIV ILL, URBANA-CHAMPAIGN, 55-, CTR ADVAN STUDY, 68- *Concurrent Pos:* Morris Loeb lectr, Harvard Univ, 61; mem sci adv comt, Off of the President, 65-69, comt nat medal sci, 69-74; Harvard Corp, 70-; dir, Polaroid Corp; former trustee & mem corp, Woods Hole Oceanog Inst. *Honors & Awards:* Langmuir Prize, Am Phys Soc, 69. *Mem:* Nat Acad Sci; fel AAAS; Am Acad Arts & Sci; Am Philos Soc; fel Am Phys Soc. *Res:* Nuclear magnetic resonance in solids. *Mailing Add:* Dept Physics Univ Ill Urbana IL 61801

SLICHTER, WILLIAM PENCE, physical chemistry; deceased, see previous edition for last biography

SLIDER, H C (SLIP), b Paden City, WVa, June 26, 24; m 46; c 2. PETROLEUM ENGINEERING. *Educ:* Ohio State Univ, BEM & MS, 49. *Prof Exp:* Exploitation engr, Shell Oil Co, 49-50, reservoir engr, 51-52, div reservoir engr, 53-56; prof, Ohio State Univ, 56-83, emer prof petrol eng, 83-90; RETIRED. *Concurrent Pos:* Consult, Shell Oil Co, 56, Humble Oil Co, 57 & 65, Esso Prod Res, Jersey Prod Res & Carter Prod Res Co, 58-65, Ins Co Am, 62-63 & 65-66, Texaco Inc, 66-, Schlumberger, 69 & 72, Petroleos del Peru, 70-71, Off Technol Assessment, US Cong, 76, Japan Petrol Develop Co, 77, Caltex, Indonesia, 79, Aramco, Saudi Arabia, Texaco, Angola & Nigeria, & Texaco Prod Serv, Eng, 80, Occidental of Peru, 83, Union Tex Petrol, Sun Oil & TRINMAR, Trinidad, 84, US Bur Land Mgt, 85, PT Indrillco Sakti, Indonesia, 88, China Nat Petrol, Peoples Republic China, 90, PT Loka Datamus Induh, Indonesia, 90, Ohio Oil & Gas Asn, 89, Quakerstate, 89-; vis prof, Univ Indust de Santander, Colombia, 77 & Tech Univ Clausthal, Ger, 78; distinguished lectr, Soc Petroleum Eng, 78-79. *Mem:* Am Inst Mining, Metall & Petrol Engrs; Am Inst Chem Engrs; Am Soc Eng Educ; Am Arbit Asn. *Res:* Petroleum production reservoir engineering. *Mailing Add:* Ohio State Univ Dept Chem Eng 140 W 19th Ave Columbus OH 43210

SLIEMERS, FRANCIS ANTHONY, JR, b Lima, Ohio, June 28, 29; m 52; c 8. POLYMER CHEMISTRY, PLASMA POLYMERIZATION. *Educ:* Univ Notre Dame, BS, 51; Ohio State Univ, MSc, 54. *Prof Exp:* Res assoc chem, Ohio State Univ, 54-56; prin chemist, Battelle Mem Inst, 56-62, sr res chemist, 62-71 & 73-79, assoc chief, 71-73, assoc sect mgr, 79-82, proj mgr, 82-87; RETIRED. *Concurrent Pos:* Consult, FAS, 87- *Mem:* Am Chem Soc. *Res:* Physical characterization of polymers; plasma polymerization. *Mailing Add:* 1514 Bolingbrook Dr Columbus OH 43228-9589

SLIEPCEVICH, CEDOMIR M, b Anaconda, Mont, Oct 4, 20; m 55. CHEMICAL ENGINEERING. *Educ:* Univ Mich, BS, 41, MS, 42, PhD(chem eng), 48. *Prof Exp:* Asst, Univ Mich, 41-46, from instr to assoc prof chem & metall eng, 46-55; prof chem eng, Univ Okla, 55-63, chmn, Sch Chem Eng, 55-59, assoc dean, Col Eng, 56-63, chmn, Sch Gen Eng, 58-63, res prof eng, 63-91, EMER PROF ENG, UNIV OKLA, 91- *Concurrent Pos:* Consult, 43-; sr chem engr, Monsanto Chem Co, 52-53; dir res & eng, Constock Liquid Methane Corp, NY, 55-60; mem adv panel to engr sec, NSF & numerous comts, 61-64; dir, Constock-Pritchard Corp, 61-63, Autoclave Engrs, Inc, 61- & Repub Geothermal, Inc, 74-75 & E-C Corp, 66-78; mem, adv comt to US Coast Guard on Transp of Hazardous Mat, Nat Acad Sci, 64-76; dir, Engr Col Res Coun, Am Soc Engr Educ, 65-68; OK rep, Southern Interstate Nuclear Bd, 65-68; pres, Univ Engrs, Inc, 65-78 & Univ Technologiists, Inc, 77-; mem-at-large, Div Chem & Chem Technol, Nat Acad Sci-Nat Res Coun, 66-69; mem, Liquid Natural Gas Task Force Nat Gas Surv, Fed Power Comn, 72-73; mem, adv panel on energy, US Off Sci & Technol, 72-73; mem, comt processing & utilization of fossil fuels & chmn, coal liquefaction ad hoc comt, Nat Res Coun, 75-77, mem hydrogen panel, comt advan energy storage systs, 77-79, mem, comt demilitarizing chem munitions & agents, 83-84, mem, Nat Transp Bd Pipelines, Land Use & Pub Safety, 87-88. *Honors & Awards:* McGraw Award, Am Soc Eng Educ, 58, Westinghouse Award, 64; Int Ipatieff Award, Am Chem Soc, 59; Sigma Xi Lectr, 62; Peter C Reilly Lectr, Notre Dame, 72; Donald L Katz Lectr, Univ Mich, 76; Walker Award, Am Inst Chem Engrs, 78; Gas Indust Res Award, Sprague Schlumberger-AGA Operating Sect, 86. *Mem:* Nat Acad Eng; fel AAAS; Am Chem Soc; Am Soc Eng Educ; fel Am Inst Chem Engrs. *Res:* High pressure equipment design; chemical reaction kinetics; process control and system identification; energy scattering; cryogenics; thermodynamics; flame dynamics; liquefaction; ocean transport and storage of natural gas; fundamental behavior of flames and combustion; desalination. *Mailing Add:* Rte 1 Box 41-B1 Washington OK 73093

SLIFE, CHARLES W, b Urbana, Ill, Dec 10, 49; m 71; c 3. CELL BIOLOGY, EXTRACELLULAR MATRIX. *Educ:* Univ Ill, BS, 73; Univ Wis, PhD(biochem), 78. *Prof Exp:* Res fel biochem, Johns Hopkins Univ, 78-82; ASST PROF BIOCHEM, SCH MED, EMORY UNIV, 82- *Concurrent Pos:* Prin investr, NIH, 83- *Mem:* Am Soc Biochem & Molecular Biol; Am Soc Cell Biol; AAAS. *Res:* The molecular mechanisms of cell-extracellular matrix interactions and their role in regulating cell growth and differentiation and tissue formation and repair. *Mailing Add:* Dept Biochem Gillette Res Inst 401 Prof Dr Gaithersburg MD 20879

SLIFE, FRED WARREN, b Milford, Ill, Nov 6, 23; m 47; c 4. AGRONOMY. *Educ:* Univ Ill, BS, 47, MS, 48, PhD, 52. *Prof Exp:* From instr to assoc prof, 47-60, prof agron, 60-77, PROF CROP PROD, UNIV ILL, URBANA, 77- *Mem:* Am Soc Agron; Weed Sci Soc Am. *Res:* Weed control, especially penetration and translocation; metabolism of herbicides in plants. *Mailing Add:* Dept Agron Univ Ill Urbana IL 61801

SLIFKIN, LAWRENCE, b Bluefield, WVa, Sept 29, 25; m 48; c 4. SOLID STATE PHYSICS, PHOTOGRAPHIC SCIENCE. *Educ:* NY Univ, BA, 47; Princeton Univ, PhD(phys chem), 50. *Prof Exp:* Res assoc physics, Univ Ill, 50-52, res asst prof, 52-54; asst prof, Univ Minn, 54-55; from asst prof to assoc prof, 55-63, Bowman Gray prof undergrad teaching, 79-82, PROF PHYSICS, UNIV NC, CHAPEL HILL, 63-, ALUMNI DISTINGUISHED PROF, 83- *Concurrent Pos:* NSF sr fel, Clarendon Lab, Oxford Univ, 62-63; foreign collabr, Ctr Nuclear Studies, Saclay, France, 75-76. *Honors & Awards:* Jesse Beams Award, Am Phys Soc, 77. *Mem:* Fel Am Phys Soc; Am Asn Physics Teachers; Mat Res Soc; fel Soc Photog Scientists & Engrs. *Res:* Diffusion in solids; defects in ionic crystals; the photographic process. *Mailing Add:* Dept Physics & Astron Univ NC Chapel Hill NC 27599-3255

SLIFKIN, MALCOLM, b Newark, NJ, Nov 9, 33; m 66; c 2. MEDICAL MICROBIOLOGY, VIROLOGY. *Educ:* Furman Univ, BS, 55; Univ NC, MSPH, 56, MS, 59; Rutgers Univ, PhD(serol), 62; Am Bd Med Microbiol, dipl. *Prof Exp:* Instr parasitol, Sch Med, Yale Univ, 6264; clin asst prof path, Sch Med, Univ Pittsburgh, 75-90; microbiologist, 65-71, HEAD SECT MICROBIOL, DEPT LAB MED, ALLEGHENY GEN HOSP, 71-; PROF, PATH & LAB MED, MED COL PA, ALLEGHENY CAMPUS, 90- *Concurrent Pos:* Fel microbiol, Sch Med, Yale Univ, 62-64; adj asst prof, Pa State Univ, 65-71, adj assoc prof, 71- *Mem:* Am Soc Microbiol; fel Am Acad Microbiol; NY Acad Sci; Sigma Xi. *Res:* Oncogenic simian adenoviruses; tissue culture of chemically induced liver cancer and preneoplastic liver; microcolony varients of Staphylococcus aureus; rapid niacin tests for Mycobacterium tuberculosis; applied clinical microbiology; choriogonadotropin-like antigens in bacteria and cancer cells; rapid identification of bacteria; rapid detection of bacteria from clinical specimens; development of new parasitological staining methods; diagnostic use of lectins in microbiology. *Mailing Add:* Microbiol Sect Dept Lab Med Allegheny Gen Hosp Pittsburgh PA 15212

SLIFKIN, SAM CHARLES, organic chemistry, for more information see previous edition

SLIGAR, STEPHEN GARY, b Inglewood, Calif, Mar 19, 48; m 87; c 2. BIOCHEMISTRY. *Educ:* Drexel Univ, BS, 70; Univ Ill, Urbana, MS, 71, PhD(physics, biochem), 75. *Prof Exp:* Physicist, Naval Air Propulsion Test Ctr, Aeronaut Engine Lab, Philadelphia, 66-69; res student physics, Drexel Univ, 68-70; resident, Cent States Univ Honors Prog, Argonne Nat Lab, 70, guest assoc molecular biol, Div Biol & Med Res, 71-73; res asst, Univ Ill, Urbana, 72-75, res assoc biochem, 75-77; asst prof, dept molecular biophys & biochem, Yale Univ, 77-82; PROF DEPT CHEM & BIOCHEM, UNIV ILL, URBANA, 82- *Concurrent Pos:* Fulbright Res Scholar, Paris, France, 89-90. *Mem:* Biophys Soc; Am Soc Biol Chemists; Am Chem Soc; Am Phys Soc; AAAS. *Res:* Physical biochemistry, mechanisms of energy transfer and oxygenation reactions; thermodynamics of regulation and control in biological catalysis; kinetic and equilibrium description of multi-protein systems and complexes; biochemical pharmacology; genetic engineering; protein structure - function. *Mailing Add:* Dept Biochem Univ Ill 1209 W California St Urbana IL 61801

SLIGER, WILBURN ANDREW, b Oklahoma City, Okla, Jan 21, 40; m 58; c 3. FISH BIOLOGY, FRESH WATER BIOLOGY. *Educ:* Cent State Univ, Okla, BS, 64; Okla State Univ, Stillwater, MS, 67, PhD(zool), 75. *Prof Exp:* From Instr to assoc prof, 66-84, chmn dept, 85, PROF BIOL, UNIV TENN, MARTIN, 84- *Concurrent Pos:* Vis prof biol, Murray State Univ, Ky, 76-77. *Mem:* Am Fisheries Soc; Sigma Xi (secy, 75-79). *Res:* Effects of water pollutants on the physiology of fish. *Mailing Add:* Dept Biol Univ Tenn Martin TN 38238

SLIKER, ALAN, b Cleveland, Ohio, June 7, 27; m 56; c 3. WOOD TECHNOLOGY. *Educ:* Duke Univ, BS, 51, MF, 52; NY Col Forestry, Syracuse Univ, PhD, 58. *Prof Exp:* Packaging engr, US Forest Prod Lab, Wis, 52-53; from instr to assoc prof forest prod, Mich State Univ, 55- 74, prof forestry, 74-91; RETIRED. *Mem:* Forest Prod Res Soc; Soc Am Foresters; Sigma Xi; Soc Wood Sci & Technol; Soc Exp Mech. *Res:* Mechanical properties of wood, particularly those properties concerned with structural use of wood. *Mailing Add:* 1800 Lindbergh Dr Lansing MI 48910

SLIKER, TODD RICHARD, b Rochester, NY, Feb 9, 36; m 63; c 2. SOLID STATE PHYSICS. *Educ:* Univ Wis, BS, 55; Cornell Univ, PhD(physics), 62; Harvard Univ, MBA, 70; Univ Denver, JD, 82. *Prof Exp:* Res assoc physics, Cornell Univ, 62; sr staff physicist, Electronic Res Div, Clevite Corp, 62-65, head appl physics sect, 65-68; asst to pres, Granville-Phillips Co, 70; vpres & gen mgr, McDowell Electronics, Inc, 70-71; pres, CA Compton Inc, 71-77; chief acct, C & S Inc, 77-80, vpres finance, 80-82. *Concurrent Pos:* Real estate mgr, 72- *Res:* Nuclear magnetic and electron paramagnetic resonance of solids; photolysis of silver chloride; linear electro-optic effects and devices; low frequency piezoelectric tuning fork filters; high frequency resonators and acoustic delay lines. *Mailing Add:* 1658 Bear Mountain Dr Boulder CO 80303

SLIKKER, WILLIAM, JR, DEVELOPMENTAL TOXICOLOGY, METABOLISM. *Educ:* Univ Calif, Davis, PhD(pharmacol & toxicol), 78. *Prof Exp:* CHIEF PHARMACODYNAMICS, NAT CTR TOXICOL RES, 78- *Res:* Neurotoxicology. *Mailing Add:* Div Reprod & Develop Toxicol HFT- 132 Bldg 53D Nat Ctr Toxicol Res Jefferson AR 72079

SLILATY, STEVE N, b Feb 11, 52; US citizen. BIOCHEMISTRY, MOLECULAR BIOLOGY. *Educ:* Cornell Univ, BS, 76; Univ Ariz, PhD(cell & develop biol), 83. *Prof Exp:* Res assoc, Univ Ariz, 83-87; coun scientist, Protein Eng, Enzyme Mechanisms & Molecular Biol, Nat Res Coun, Montreal, 87-91; FOUNDER, PRES & CHIEF EXEC OFFICER, QUANTUM BIOTECHNOLOGIES INC, MONTREAL, 91- *Concurrent Pos:* Fel cancer biol, Univ Ariz Med Ctr, 84-86. *Mem:* AAAS. *Res:* Determination of the chemical events at the level of bond rearrangement responsible for cleavage and, therefore, inactivation of the repressor protein LexA. *Mailing Add:* Quantum Biotechnologies Inc 6100 Royalmount Ave Montreal PQ H4P 2R2 Can

SLINEY, DAVID H, b Washington, DC, Feb 21, 41; m 66; c 3. HEALTH PHYSICS. *Educ:* Va Polytech Inst, BS, 63; Emory Univ, MS, 65; Univ London, PhD, 91. *Prof Exp:* CHIEF LASER BR, US ARMY ENVIRON HYG AGENCY, 65- *Concurrent Pos:* Consult, WHO, 65-, UNESCO, 85-; mem Army-Indust Comt Laser Safety, 67; consult laser hazards, NASA, 67-; US Coast & Geod Surv, 68-; mem comt laser safety, Am Nat Standards Inst, 68-; ed, Health Physics Jour, 76-87, Lasers Life Sci, 85-, Lasers Surg Med, 88-; Hayes-Fulbright fel, Yugoslavia, 76; US chief deleg, Comt TC76, Lasers of Int Electrotech Comm, 77-; mem, Nat Acad Sci/Nat Res Coun Panel, Health Aspects Video Viewing, 80-82, Food & Drug Admin Tech Electronic Prod Radiation Safety Comt, 81-84 & Int Non-Ionizing Radiation Comt, 80-; chmn, Comt Phys Agents, Am Conf Govt Indust Hygienists; consult, Laser Technol United Nations Educ & Sci Cult, 86- *Mem:* Optical Soc Am; Health Physics Soc; Am Conf Govt Indust Hygienists; Sigma Xi; Asn Res Vision & Ophthalmol; Am Soc Laser Med & Surg; Soc Photo-optical Instrumentation Engr. *Res:* Criteria for laser hazard analysis; standards for laser exposure; optical radiation hazards; non-ionizing radiation; medical application lasers. *Mailing Add:* US Army Environ Hyg Agency Aberdeen Proving Ground MD 21010-5422

SLINGERLAND, RUDY LYNN, b Troy, Pa, Apr 7, 47; m 84; c 2. GEOLOGY, SEDIMENTOLOGY. *Educ:* Dickinson Col, BS, 69; Pa State Univ, MS, 73, PhD(geol), 77. *Prof Exp:* From asst prof to assoc prof, 83-88, PROF GEOL, PA STATE UNIV, 89- *Concurrent Pos:* Consult sedimentary basin anal. *Mem:* Geol Soc Am; Am Geophys Union; Sigma Xi; Soc Econ Paleontologists & Mineralogists; Int Asn Sedimentologists. *Res:* Sedimentology; coal geology; coastal geology; fluvial geomorphology. *Mailing Add:* 303 Deike Bldg University Park PA 16802

SLINKARD, ALFRED EUGENE, b Rockford, Wash, Apr 5, 31; m 51; c 4. PLANT BREEDING. *Educ:* Wash State Univ, BS, 52, MS, 54; Univ Minn, PhD(plant genetics), 57. *Prof Exp:* Asst prof agron & asst agronomist, Univ Idaho, 57-66, assoc prof agron & assoc agronomist, 66-72; SR RES SCIENTIST, UNIV SASK, 72- *Concurrent Pos:* Ed, Can J Plant Sci, 83-85; consult, numerous foreign countries. *Honors & Awards:* Outstanding Res Award, Can Soc Agron, 85. *Mem:* AAAS; Crop Sci Soc Am; Am Soc Agron; fel Agr Inst Can; Genetics Soc Can; Can Soc Agron; Sigma Xi; hon mem Can Seed Growers Asn, 91. *Res:* Genetics and breeding of pea and lentil; establishment of lentil as a commercial crop in Western Canada, including development of a package of agronomic practices and the licensing of four cultivars, Laird, Eston, Rose and Indianhead licensed a pea cultivar Bellevue. *Mailing Add:* Dept Crop Sci Univ Sask Saskatoon SK S7N 0W0 Can

SLINKARD, WILLIAM EARL, b Omaha, Nebr, May 14, 43; m 65; c 2. INORGANIC CHEMISTRY. *Educ:* Trinity Univ, Tex, BS, 65; Ohio State Univ, PhD(inorg chem), 69; Corpus Christi State Univ, Tex, BS, 86. *Prof Exp:* STAFF CHEMIST, HOECHST CELANESE CHEM CO, 69- *Mem:* Am Chem Soc; Catalysis Soc. *Res:* catalytic vapor phase oxidation of hydrocarbons, methanol synthesis, steam reforming of natural gas, methanol homologation reactions and fuel applications of methanol. *Mailing Add:* Tech Ctr Hoechst Celanese Chem Group Box 9077 Corpus Christi TX 78469

SLIVA, PHILIP OSCAR, b Yonkers, NY, Apr 22, 38; m 61; c 2. SOLID STATE PHYSICS. *Educ:* Clarkson Col Technol, BS, 60; Purdue Univ, PhD(solid state physics), 67. *Prof Exp:* Assoc scientist, Solid State Res Br, Xerox Corp, 67-68, scientist, Exp Physics Br, 68-73, sr scientist, Explor Photoconductor Physics Area, 73-76, mgr, Photoconductor Characterization area, 76-90, VPRES SUPPLIES DEVELOP & MFG, XEROX CORP, 90- *Mem:* Am Phys Soc. *Res:* Acoustoelectric effects in III-IV semiconductors; solid state switching and memory devices in amorphous and polycrystalline materials; charge generation and transport in photoconductors. *Mailing Add:* 2400 Turk Hill Rd Victor NY 14564

SLIVINSKY, CHARLES R, b St Clair, Pa, May 20, 41; m 63; c 2. ELECTRICAL ENGINEERING, COMPUTER SCIENCE. *Educ:* Princeton Univ, BSE, 63; Univ Ariz, MS, 66, PhD(elec eng), 69. *Prof Exp:* From asst prof to assoc prof, 68-77, PROF ELEC ENG, UNIV MO-COLUMBIA, 77- *Concurrent Pos:* NSF sci equip grant, 69-71, res initiation grant, 71-72, solid state power control stability anal grant, 72-74, res equip grant, 77-79; consult, Air Force Flight Dynamics Lab, 75-; US Air Force Off Sci Res grant, 76-79; NSF res Eqpt grants, 80, sci res grant, 81, Power Systs Res grants, 82-90. *Mem:* AAAS; Inst Elec & Electronics Engrs; Am Soc Eng Educ; Sigma Xi. *Res:* Automatic control; computer control and signal processing; electrical power systems. *Mailing Add:* Dept Elec Eng Univ Mo Columbia MO 65211

SLIVINSKY, SANDRA HARRIET, b New York, NY; c 1. HIGH TEMPERATURE TECHNOLOGY, LASER PROCESSING. *Educ:* Alfred Univ, BA, 62; Pa State Univ, MS(physics), 66; Univ Calif, Davis-Livermore, MS(appl sci), 73; Stanford Univ, PhD(mat sci), 83. *Prof Exp:* Res assoc, Dikewood Corp, 66-68; physicist, Lawrence Livermore Lab, 68-73; sr res engr, Lockheed Missiles & Space Co, 73-76; sr engr, Gen Elec Co, 76-85, mat specialist, Watkins Johnson, 85-86; sr scientist, United Technol-CSD, 86-90; PRIN INVESTR, ASTRONAUT LAB/UNIV DAYTON, EDWARDS AFB, 90- *Mem:* Am Physics Soc; Soc Women Engrs; Inst Elec & Electronics Engrs; Am Ceramic Soc; Sigma Xi; Soc Advan Mat & Process Eng. *Res:* Development of a laser processing methods for skiving plastics over metal substrates; nitrogen-nitride equilibria in molten tin alloys, particularly for uranium and thorium metals using a modified sieverts apparatus. *Mailing Add:* 1529 Elmar Way San Jose CA 95129

SLOAN, ALAN DAVID, b New York, NY, July 5, 45; m 69. APPLIED MATHEMATICS. *Educ:* Mass Inst Technol, BS, 67; Cornell Univ, PhD(math), 71. *Prof Exp:* Fel, Carnegie-Mellon Univ, 71-72; vis asst prof physics, Princeton Univ, 74-75; from instr to asst prof, 72-77, ASSOC PROF MATH, GA INST TECHNOL, 77- *Mem:* Am Math Soc; Soc Indust & Appl Math; Math Asn Am; Sigma Xi. *Res:* Applications of functional analysis and nonstandard analysis to mathematical physics especially to the area of quantum theory. *Mailing Add:* Sch of Math Ga Inst of Technol Atlanta GA 30332

SLOAN, DONALD LEROY, JR, biochemistry; deceased, see previous edition for last biography

SLOAN, FRANK A, b Greensboro, NC, Aug 15, 42; m; c 2. HEALTH POLICY. *Educ:* Oberlin Col, AB, 64; Harvard Univ, PhD(econ), 69. *Prof Exp:* Res assoc, Dept Econ, Rand Corp, 68-71; asst prof, Dept Econ, Col Bus Admin, Univ Fla, 71-73, asst prof, Dept Community Health & Family Med, 72-73, res assoc, Health Systs Div, 72-76, assoc prof, Dept Econ, Community Health & Family Med, 73-76; prof econ, Vanderbilt Univ, 76-84, chmn, Dept Econ & Bus Admin, 86-89, Alexander Heard distinguished serv prof, 90-91, SR RES FEL & DIR, HEALTH POLICY CTR, INST PUB POLICY STUDIES, VANDERBILT UNIV, 76-, CENTENNIAL PROF ECON, 84- *Concurrent Pos:* Prin investr, HEW, 72-78, Nat Ctr Health Serv Res, 75-90, Social Security Admin, 75-77, Health Care Financing Admin, Dept Health & Human Serv, 78-, Robert Wood Johnson Found, 85-89; mem, Nat Adv Allergy & Infectious Dis Coun, HEW, 71-74, Adv Comt Cost & Financing Grad Med Educ Nat Acad Sci, 77, Health Resources Admin Nursing Res & Educ Adv Comt, 77-78, Nat Coun Health Care Technol, Dept Health & Human Serv, 79-81, Coun Res & Develop, Am Hosp Asn, 82-84 & Study Sect, Health Serv Res Rev, 85-89; consult, numerous govt agencies & indust, 83-91; res assoc, Nat Bur Econ Res, Inc, 88. *Mem:* Inst Med-Nat Acad Sci; Am Econ Asn. *Res:* Health care law and economics; author of numerous technical publications. *Mailing Add:* Health Policy Ctr Vanderbilt Univ Box 1503 Sta B Nashville TN 37235

SLOAN, GILBERT JACOB, b Elizabeth, NJ, July 25, 28; m 57; c 2. PHYSICAL ORGANIC CHEMISTRY. *Educ:* Mich Col Mining & Technol, BS, 48; Univ Mich, PhD(chem), 54. *Prof Exp:* Res chemist, E I Dupont de Nemours & Co Inc, 53-73, supvr, 73-80, res mgr cent res & develop dept, 80-85, sr res fel, polymer prod dept, 85-89, SR RES FEL, DUPONT FIBERS, 89- *Concurrent Pos:* Assoc ed, J Crystal Growth, 67-84. *Mem:* Am Chem Soc; Sigma Xi. *Res:* Stable free radicals; purification of organic compounds; crystal growth. *Mailing Add:* Du Pont Fibers Wilmington DE 19880-0302

SLOAN, HERBERT, b Clarksburg, WVa, Oct 10, 14; m 43; c 5. THORACIC SURGERY. *Educ:* Washington & Lee Univ, AB, 36; Johns Hopkins Univ, MD, 40. *Prof Exp:* Assoc prof, Univ Mich, Ann Arbor, 53-62, prof surg, 62, head sect thoracic surg, 70-86; RETIRED. *Concurrent Pos:* Assoc ed, Ann Thoracic Surg, 64-69, ed, 69-84; mem, Am Bd Thoracic Surg, 66-71, vchmn, 71-72, chmn credentials comt, 69-, vchmn nominating comt, 71-72, secy, 73-86; mem, Residency Rev Comt Thoracic Surg, 72-73; mem, Adv Group Cardiac Surg, Vet Admin, 71-81; mem, Surg Study Group, Inter-Soc Comn Heart Dis Resources, 71-72, Cardiac Surg Rev Panel, 73-; liaison mem, Am Bd Thoracic Surg to Inter-Soc Comn Heart Dis Resources, 73; mem, Adv Comt, Second Henry Ford Hosp Int Symp Cardiac Surg, 73-75, assoc ed, Proc Symp, 75; prog chmn, Adv Comt Thoracic Surg, Am Col Surgeons, 75-78. *Mem:* Am Surg Asn; Am Col Surgeons; Am Heart Asn; Soc Thoracic Surgeons (vpres, 73-74, pres, 74-75); Am Asn Thoracic Surg (vpres, 78-79); Sigma Xi. *Res:* Cardiac surgery. *Mailing Add:* 48109 Sect Thoracic Surgery Ann Arbor MI 48105

SLOAN, MARTIN FRANK, b St Louis, Mo, Oct 9, 34; m 60; c 2. INDUSTRIAL ORGANIC CHEMISTRY. *Educ:* Wash Univ, BA, 56; Univ Wis, PhD, 60. *Prof Exp:* Res chemist, Hercules Inc, 60-69, res supvr, 69-74, supvr mkt develop, 74-76, develop mgr, 76-80, res assoc, 80-87, MGR PATENT COORD, HERCULES INC, 88- *Mem:* Am Chem Soc. *Mailing Add:* 2203 Pennington Dr Wilmington DE 19810

SLOAN, NORMAN F, forest entomology, wildlife ecology; deceased, see previous edition for last biography

SLOAN, NORMAN GRADY, b Oklahoma City, Okla, May 8, 37; m 74; c 4. ENGINEERING. *Educ:* Univ Okla, BS, 60; Univ Tulsa, MS, 71. *Prof Exp:* Mech engr air conditioning design, US Corps Engrs, 60-61; proj engr natural gas processing plants, Dresser Eng Co, 61-91; PROJ MGR, PRO-QUIP, 91- *Mem:* Am Soc Mech Engrs. *Res:* Design of natural gas processing plants. *Mailing Add:* 11434 S Yale Tulsa OK 74137

SLOAN, ROBERT DYE, b Clarksburg, WVa, Feb 17, 18; m 46; c 2. MEDICINE. *Educ:* Washington & Lee Univ, AB, 39; Johns Hopkins Univ, MD, 43. *Prof Exp:* From instr to assoc prof radiol, Johns Hopkins Univ, 48-55; prof radiol & chmn dept, Med Ctr, Univ Miss, 55-82; RETIRED. *Concurrent Pos:* Consult, Vet Admin Hosp, 55- *Mem:* Am Roentgen Ray Soc; Radiol Soc NAm. *Res:* Diagnostic radiology; intestinal obstruction. *Mailing Add:* PO Box 1225 Raton NM 87740

SLOAN, ROBERT EVAN, b Champaign, Ill, July 17, 29; m 53; c 2. STRATIGRAPHY, PALEONTOLOGY. *Educ:* Univ Chicago, PhB, 48, SB, 50, SM, 52, PhD(geol), 53. *Prof Exp:* Asst prof, 53-63, assoc prof, 64-71, PROF GEOL, UNIV MINN, MINNEAPOLIS, 71- *Mem:* Soc Econ Paleont & Mineral; Soc Vert Paleont; Soc Study Evolution; Geol Soc Am; Soc Syst Zool. *Res:* Mesozoic and Paleocene mammals; Multituberculata; terrestrial vertebrate paleoecology; Cretaceous and Paleocene stratigraphy of western North America. *Mailing Add:* Dept Geol, Geophysics & Earth Sci Univ Minn Minneapolis MN 55455-0214

SLOAN, ROBERT W, b Rankin, Ill, July 18, 24; m 49; c 2. MATHEMATICS. *Educ:* US Naval Acad, BS, 46; Univ Ill, MS, 51, PhD(math), 55. *Prof Exp:* Asst prof math, Univ NH, 55-56; asst prof, Carleton Col, 56-59; prof, State Univ NY Col Oswego, 59-65; chmn dept, 65-76, PROF MATH, ALFRED UNIV, 65- *Mem:* Math Asn Am. *Res:* Analysis and numerical analysis. *Mailing Add:* Dept of Math Alfred Univ Alfred NY 14802

SLOAN, WILLIAM COOPER, b Asheville, NC, Aug 26, 27; m 51; c 2. BIOLOGY. *Educ:* Univ Fla, BS, 52, MS, 54, PhD(biol), 58. *Prof Exp:* Asst biol, Univ Fla, 52-54 & 55-58; entom, Univ Minn, 54-55; instr biol, Vanderbilt Univ, 58-60; NIH fel, Univ Calif, 60-61; asst prof zool, San Diego State Univ, 61-65, assoc prof, 65-68, prof biol, 68-90, assoc chair, biol dept, 84-90; RETIRED. *Mem:* AAAS; Sigma Xi. *Res:* Comparative physiology; nitrogen metabolism and excretion in invertebrates. *Mailing Add:* 3162 Monroe Ave No 8 San Diego CA 92116

SLOANE, CHRISTINE SCHEID, b Washington, DC, May 1, 45; m 69; c 2. ATMOSPHERIC CHEMISTRY & PHYSICS. *Educ:* Col William & Mary, BS, 67; Mass Inst Technol, PhD(chem physics), 71. *Prof Exp:* Res assoc phys chem, Univ Calif, Berkeley, 72-73; asst prof chem, Oakland Univ, 74-78; assoc sr res scientist, 78-80, staff res scientist, 80-85, SR STAFF RES SCIENTIST ENVIRON SCI, GEN MOTORS RES LABS, 85- *Honors & Awards:* John C Campbell Award. *Mem:* Am Chem Soc; Am Asn Aerosol Res; AAAS; Am Meteorol Soc; Air & Waste Mgt Asn. *Res:* Emissions modeling; toxic emission control; visibility; atmospheric optics. *Mailing Add:* Gen Motors Res Labs Warren MI 48090-9055

SLOANE, HOWARD J, b New York, NY, May 9, 31; m 57; c 2. SPECTROCHEMISTRY. *Educ:* Trinity Col, BS, 53; Wesleyan Univ, MA, 55. *Prof Exp:* Infrared spectroscopist, Dow Chem Co, 55-60; chief chemist, Beckman Instruments, Inc, 60-67; dir appln res, Cary Instruments Div, Varian Assocs, 67-72; sr scientist, Beckman Instruments, Inc, 72-74, mgr appln res, Sci Instruments Div, 74-77; PRES, SAVANT, SLOANE AV ANALYSIS & TRAINING, 77- *Concurrent Pos:* Lectr, Raman Inst & Workshop, Univ Md, 68-72 & absorption spectros, Ariz State Univ, 60-, Am Chem Soc, 83-, Finnigan MAT Inst, 79-86. *Mem:* Am Chem Soc; Soc Appl Spectros; Coblentz Soc (treas, 72-83); NY Acad Sci. *Res:* Instrumentation and applications of spectroscopy, especially vibrational and atomic. *Mailing Add:* Savant Sloane AV Anal Training PO Box 3670 Fullerton CA 92634-3670

SLOANE, NATHAN HOWARD, b Boston, Mass, Sept 15, 17; m 46; c 2. BIOCHEMISTRY. *Educ:* Mass Col Pharm, BS, 39; Mass Inst Technol, MPH, 43; Harvard Univ, PhD, 50. *Prof Exp:* Chemist, Lederle Labs, Am Cyanamid Co, 43-45, group leader biochem, 51-56; biochemist, Ciba Pharmaceut Prod, Inc, 49-51; biochemist, Nat Drug Co, Vick Chem Co, 56-58; biochemist & sr fel, Mellon Inst, 58-64; prof biochem, Col Med, Univ Tenn, Memphis, 64-86; VIS PROF, SCH PUB HEALTH, HARVARD UNIV, 86- *Res:* Anti-carcogenic proteins in urine and plasma; chemical carcinogenesis. *Mailing Add:* 1842 Brookside Dr Memphis TN 38138-2547

SLOANE, NEIL JAMES ALEXANDER, b Beaumaris, Wales, Oct 10, 39. MATHEMATICS, ELECTRICAL ENGINEERING. *Educ:* Univ Melbourne, BEE, 59, BA, 60; Cornell Univ, MS, 64, PhD(elec eng), 67. *Prof Exp:* Asst prof elec eng, Cornell Univ, 67-69; MEM TECH STAFF, MATH DEPT, BELL TEL LABS, 69- *Honors & Awards:* Chauvenet Prize, Math Asn Am, 79. *Mem:* Am Math Soc; Math Asn Am; fel Inst Elec & Electronics Engrs. *Res:* Coding theory; communication theory; combinatorial mathematics, graph theory. *Mailing Add:* Rm 2C-376 Bell Tel Labs Murray Hill NJ 07974

SLOANE, ROBERT BRUCE, b Harrogate, Eng, Mar 28, 23; US citizen; m 46; c 6. PSYCHIATRY. *Educ:* Univ London, MB, BS, 45, MD, 50, dipl psychol med, 51; FRCP(C), 74, cert psychiat, RCPC, 56; Am Bd Psychiat & Neurol, dipl, 63; FRCP, 74. *Prof Exp:* House physician, London Hosp, 45-46; asst registr, Nat Hosp Nerv Dis, London, 49; registr neurol, Guy's Hosp Med Sch, 50; registr & sr registr, Maudsley Hosp, 50-52, sr registr, 54-55; resident & chief resident, Mass Gen Hosp, Boston, 53, Milton res fel, 53-54; asst psychiatrist, Allan Mem Inst, McGill Univ, 55-57; prof psychiat & head dept, Queen's Univ, Ont, 57-64; prof & chmn dept, Health Sci Ctr, Temple Univ, 64-72; prof & chmn, Dept Psychiat, 72-, FRANZ ALEXANDER PROF PSYCHIAT, MED SCH, 81-, EXEC DIR, PACIFIC GER EDUC CTR, UNIV SOUTHERN CALIF, 83- *Honors & Awards:* Res Award, Soc Psychotherapy, 80. *Mem:* Am Psychiat Asn; Am Col Neuropsychopharmacology. *Res:* Research and writing in psychotherapy, depression and aging. *Mailing Add:* USC Dept of Psychiat LAC-USC Med Ctr Los Angeles CA 90033

SLOANE, THOMPSON MILTON, b Baltimore, Md, Aug 30, 45. COMBUSTION CHEMISTRY. *Educ:* Univ Ariz, BS, 67; Mass Inst Technol, PhD(phys chem), 72. *Prof Exp:* Chemist, Lawrence Berkeley Lab, Univ Calif, 72-73; from assoc sr res chemist to sr res chemist res labs, 73-82, asst dept head, 84-87, PRIN RES SCIENTIST, PHYS CHEM DEPT, GEN MOTORS CORP, 87- *Mem:* Am Chem Soc; Sigma Xi; Combustion Inst. *Res:* combustion chemistry. *Mailing Add:* Phys Chem Dept Res Labs Gen Motors Tech Ctr Warren MI 48090-9055

SLOAT, BARBARA FURIN, b Jan 20, 42; m 68; c 2. YEAST CELL MORPHOGENESIS, GENDER & SCIENCE. *Educ:* Univ Mich, PhD(zool), 68. *Prof Exp:* ASST RES SCIENTIST CELL BIOL, DIV BIOL SCI, UNIV MICH, 77- *Concurrent Pos:* Dir, Women in Sci Prog, Univ Mich, 80-84, assoc dir, honors prog, 86-87; lectr, Residential Col, 84- *Mem:* Am Soc Cell Biol; NY Acad Sci; AAAS; Sigma Xi; Am Women Sci (pres elect, 90). *Res:* Cellular morphogenesis in yeast; liposomal hydrolase characterization and activity in developing brain and liver of rat; factors that influence girls and women to choose and remain in scientific majors and careers. *Mailing Add:* Univ Mich 2010 Hall Ave Ann Arbor MI 48104

SLOAT, CHARLES ALLEN, b Cashtown, Pa, Dec 12, 98; m 46. CHEMISTRY. *Educ:* Gettysburg Col, BS, 23; Haverford Col, AM, 24; Princeton Univ, PhD(chem), 30. *Prof Exp:* Teaching asst chem, Haverford Col, 23-24 & Princeton Univ, 24-27; from asst prof to prof, 27-68, EMER PROF CHEM, GETTYSBURG COL, 68- *Concurrent Pos:* Capt, Chem Warfare Serv, US Army, 42-45, major, 45-46. *Mem:* AAAS; Am Chem Soc; Am Soc Metals. *Res:* Crystals; phenoména due to forces at crystal faces as studied by mutual orientation; adsorption of solutes by crystals in relation to compatability of space lattice. *Mailing Add:* 29 W Broadway Gettysburg PA 17325

SLOBIN, LAWRENCE I, b New York, NY, June 20, 38; m 64; c 1. IMMUNOLOGY. *Educ:* Queens Col, City Univ New York, BS, 59; Univ Calif, Berkeley, PhD(biochem), 64. *Prof Exp:* USPHS Teaching fel, Weizmann Inst Sci, 64-65; USPHS teaching fel, Univ Calif, San Diego, 65-66, Ann Cancer Soc fel biochem, 66-67; asst prof microbiol, Cornell Univ, 67-73; sr res assoc biochem, State Univ Leiden, Neth, 74-77; assoc prof, 77-83, PROF BIOCHEM, SCH MED, UNIV MISS, 83- *Mem:* Am Soc Biol Chemists; AAAS; Am Soc Microbiol; Am Soc Cell Biol. *Res:* Regulation of protein synthesis in mammaliam cells; structure and function of protein synthesis factors. *Mailing Add:* Dept Biochem Sch Med Univ Miss 2500 N State St Jackson MS 39216

SLOBODA, ADOLPH EDWARD, b New York, NY, Jan 17, 28; m 54; c 4. CELL PHYSIOLOGY, PHARMACOLOGY. *Educ:* Champlain Col, AB, 52; NY Univ, MS, 57, PhD(cytophysiol), 61. *Prof Exp:* Biologist, 52-56, biostatistician, 57, res biologist, 58-65, SR RES BIOLOGIST & GROUP LEADER INFLAMMATION & IMMUNOSUPPRESSION, LEDERLE LABS, AM CYANAMID CO, 66- *Mem:* Am Soc Pharmacol & Exp Therapeut; NY Acad Sci. *Res:* Drug effects on various aspects of inflammation and the immune system; relationship of immunosuppression and cancer. *Mailing Add:* Oncol & Immunol Res Lederle Labs Am Cyanamid Co Pearl River NY 10965

SLOBODA, ROGER D, b Troy, NY, May 18, 48; m 70; c 2. BIOLOGY, BIOCHEMISTRY. *Educ:* State Univ NY, BS, 70; Rensselaer Polytech Inst, PhD(develop biol), 74. *Prof Exp:* Fel biol, Yale Univ, 74-77; from asst prof to assoc prof, 77-88, PROF BIOL, DARTMOUTH COL, 88- *Concurrent Pos:* Instr, Physiol Course, Marine Biol Lab, 81-83, bd trustees, 90-93; investr, Palmer Sta, Antarctica, 85. *Mem:* Am Soc Cell Biol; AAAS; Sigma Xi. *Res:* Biochemistry of assembly and function of microtubules and associated proteins with specific interests in their roles in cell division and intracellular particle motility. *Mailing Add:* Dept Biol Sci Dartmouth Col Hanover NH 03755

SLOBODCHIKOFF, CONSTANTINE NICHOLAS, b Shanghai, China, Apr 23, 44; US citizen; m 71; c 2. EVOLUTIONARY BIOLOGY, BEHAVIORAL ECOLOGY. *Educ:* Univ Calif, Berkeley, BS, 66, PhD, 71. *Prof Exp:* From asst prof to assoc prof, 71-82, PROF BIOL, NORTHERN ARIZ UNIV, 82- *Concurrent Pos:* Fulbrigt fel, Kenya, 83; vis prof, Kenyatta Univ, 83. *Mem:* Ecol Soc Am; Soc Study Evolution; AAAS; Animal Behavior Soc. *Res:* Ecological, genetic, and behavioral factors contributing to the development and maintenance of social systems; communication in animal systems; development of animal "language" systems. *Mailing Add:* Dept Biol Sci Northern Ariz Univ Flagstaff AZ 86011

SLOBODKIN, LAWRENCE BASIL, b New York, NY, June 22, 28; m 52; c 3. ECOLOGY. *Educ:* Bethany Col, WVa, BS, 47; Yale Univ, PhD(zool), 51. *Prof Exp:* Chief invests, US Fish & Wildlife Serv, 51-52, fisheries res biologist, 52-53; vis investr, Univ Mich, Ann Arbor, 53-54, instr zool, 53-57, from asst prof to prof, 57-68; chmn prog ecol & evolution, 69-74, PROF BIOL, STATE

UNIV NY STONY BROOK, 68- *Concurrent Pos:* Guggenheim fel, 61, 74, fel, Woodrow Wilson Inst, 90; vis prof, Tel Aviv Univ, 65-66, Tsukua Univ, Japan, 89; distinguished vis scientist, Smithsonian Inst, 74-75. *Honors & Awards:* Russel Award, 61. *Mem:* Soc Gen Syst Res (pres, 67); Am Acad Arts & Sci; Am Soc Nat (pres, 85); Ecol Soc Am. *Res:* Theoretical and experimental population ecology; evolutionary strategy; ecological planning and decision making with reference to environmental management; biology of hydra aquatic toxicology; theory of simplicity. *Mailing Add:* Dept of Ecol & Evolution State Univ of NY Stony Brook NY 11794

SLOBODRIAN, RODOLFO JOSE, b Buenos Aires, Arg, Jan 1, 30; m 59; c 3. PHYSICS. *Educ:* Univ Buenos Aires, Bachelor, 48, LicSc, 53, DSc (physics), 55. *Prof Exp:* Investr nuclear physics, Arg Nat AEC, 53-63; prof physics, Nat Univ La Plata, 58-63; physicist, Lawrence Radiation Lab, Univ Calif, 63-68; chmn dept, 85-88, PROF PHYSICS, LAVAL UNIV, 68- *Concurrent Pos:* Asst to chair theoret physics, Univ Buenos Aires, 54-55, head lab spec physics, 57-58; Arg Nat AEC study mission, Radiation Lab, Univ Calif, 55-57. *Mem:* Am Phys Soc; NY Acad Sci; Can Asn Physicists. *Res:* Nuclear reactions; nucleon-nucleon interactions, final state interactions, polarization phenomena, multibody channels and nucleon transfer reactions; space physics. *Mailing Add:* Dept of Physics Laval Univ Quebec PQ G1K 7P4 Can

SLOCOMBE, JOSEPH OWEN DOUGLAS, b Port-of-Spain, Trinidad, July 27, 31; Can citizen; m 63; c 3. PARASITOLOGY, VETERINARY MEDICINE. *Educ:* Univ West Indies, dipl, 55; Univ Toronto, DVM, 61; Cornell Univ, PhD(parasitol), 69. *Prof Exp:* Asst to plant pathologist, Cent Exp Sta, Govt Trinidad & Tobago, 55-56, vet, Tobago, 61-65; asst parasitol, NY State Vet Col, Cornell Univ, 65-69; from asst prof to assoc prof, 69-76, PROF PARASITOL, ONT VET COL, UNIV GUELPH, 76- *Concurrent Pos:* Ont Racing Comn grant, Univ Guelph, 70-78; Nat Res Coun Can grant, 71-89; res grants, Ont Ministry Agr & Food, 74-78, E P Taylor res fund, 76-78, Can Vet res fund, 77-79 & Can Dept Agr, 78-79. *Mem:* Am Soc Parasitol; Am Asn Vet Parasitologists; Am Asn Equine Practrs; Am Heartworm Soc; Am Soc Parasitol; Am Vet Med Asn; Can Asn Advan Vet Parasitol; Can Vet Med Asn; Conf Res Workers Animal Dis; World Asn Advan Vet Parasitol. *Res:* Strongyles in horses; heartworm in dogs; Bovine parasitism; surveillance for parasitisms in domestic animals. *Mailing Add:* Dept Path Ont Vet Col Univ Guelph Guelph ON N1G 2W1 Can

SLOCOMBE, ROBERT JACKSON, b Peabody, Kans, May 22, 17; div; c 2. ORGANIC POLYMER CHEMISTRY. *Educ:* Univ Kans, AB, 39, MA, 41, PhD(org chem), 43; Univ Mo, St Louis, AB, 86. *Prof Exp:* Asst instr chem, Univ Kans, 40-43; res chemist, Monsanto Co, 43-45, res group leader, Ala, 45-50, Ohio, 50-61, Mo, 61-69, sr res specialist, 69-79; CONSULT, 79- *Mem:* AAAS; Am Chem Soc; Am Inst Chem Eng. *Res:* Preparation and production of isocyanates; reactions of phosgene; synthesis and stabilization of high polymers; multicomponent copolymerization; protein fractionation; reactive copolymers; epoxy matrix resins in composites; electrostatic printing inks; photopolymers; cellular plastics; bio-medical polymer systems; light activated printing inks. *Mailing Add:* 7825 Stanford Ave St Louis MO 63130

SLOCUM, DONALD WARREN, b Rochester, NY; m 90; c 2. MATERIALS CHEMISTRY, MACROCYCLES. *Educ:* NY Univ, PhD(chem), 63. *Prof Exp:* From asst prof to prof chem, Southern Ill Univ, 65-81, adj prof, 81-84; sr scientist, Gulf Res & Develop Co, Pittsburgh, 79-82; prog leader, Div Educ Progs, Argonne Nat Lab, 85-90; HEAD, DEPT CHEM, WESTERN KY UNIV, BOWLING GREEN, 90- *Concurrent Pos:* Vis prof, Dept Chem, Univ Ill, 70, Dept Inorg Chem, Univ Bristol, UK, 72 & Dept Chem, Univ Cincinnati, 76; vis consult, Hooker Res Labs, Grand Island, NY, 75; vis scientist, Pittsburgh Energy Res Ctr, US Energy Res & Develop Admin, 77; fac res partic, Chem Eng Div, Argonne Nat Lab, 78 & 79; vis lectr, Carnegie-Mellon Univ & Univ Pittsburgh, 83-84. *Mem:* Am Chem Soc; Chem Soc Gt Brit; Catalysis Soc; Org Reactions Catalysis Soc. *Res:* Organic, organometallic and inorganic synthesis and mechanism; stereochemistry; heterocyclic chemistry; homogeneous and heterogeneous catalysis; synthetic fuels, fine chemicals, polymers; chelating and encapsulating agents; chemistry of coal; supercritical media; separation and spectrascopic techniques; zeolites macrocycles. *Mailing Add:* Dept Chem Western Ky Univ Bowling Green KY 42101

SLOCUM, HARRY KIM, b Buffalo, NY, June 8, 47; m 74; c 3. BIOCHEMISTRY, BIOLOGY. *Educ:* State Univ NY, Buffalo, BA, 69, PhD(biochem), 74. *Prof Exp:* Res assoc immunol, Scripps Clin & Res Found, 74-76; cancer res scientist I, 76-77, SCIENTIST II PHARMACOL, ROSWELL PARK MEM INST, 77- *Mem:* Tissue Cult Asn; Am Asn Cancer Res; Am Soc Clin Oncol. *Res:* Characterization of cells comprising human solid tumors, including determinants of drug action, cellular interactions and heterogeneity in drug response. *Mailing Add:* Grace Cancer Drug Ctr Roswell Park Cancer Inst Buffalo NY 14263

SLOCUM, RICHARD WILLIAM, b Bryn Mawr, Pa, May 16, 34; c 5. ENGINEERING. *Educ:* Mass Inst Technol, BS, 55, PhD(nuclear physics), 59. *Prof Exp:* Scientist, Raytheon Co, 58-61; sect head, Aerospace Corp, 61-69; spec asst to dir, Advan Res Proj Agency, US Govt, 69-71; DIR RES & DEVELOP, AIR LOGISTICS CORP, 72- *Concurrent Pos:* Instr, Calif State Univ, Long Beach, 62-64, Marymount Col, 62-65, Univ Calif, Los Angeles, 63, Univ Southern Calif, 63-64. *Mem:* Soc Naval Architects & Engrs; Inst Elec & Electronics Engrs; Am Phys Soc; Arctic Inst NAm. *Res:* Arctic operations; ships and platforms; undersea vehicles and systems; air cushion vehicles; meteorological and surveillance satellites and systems. *Mailing Add:* 2138 Kinnelon Canyon Rd Pasadena CA 91107-1032

SLOCUM, ROBERT EARLE, b El Reno, Okla, Nov 28, 38; m 67. PHYSICS. *Educ:* Univ Okla, BS, 60, MEP, 63; Univ Tex, Austin, PhD(physics), 69. *Prof Exp:* Res engr, Tex Instruments Inc, 60-63; res engr, Boeing Co, 63-64; mem tech staff atomic physics, Equip Res & Develop Lab, Tex Instruments Inc, 66-73, mgr, Advan Magnetics Progs, 74-82; PRES, POLATOMIC INC, RICHARDSON, TEX, 82- *Mem:* AAAS; Am Phys Soc; Optical Soc Am. *Res:* Optical pumping with application to magnetometers for space and geophysical applications; thin film Hertzian polarizers; polarized optical systems. *Mailing Add:* 307 Arborcrest Richardson TX 74080

SLOCUM, ROBERT RICHARD, b Traverse City, Mich, June 21, 31; m 60; c 2. SOLID STATE PHYSICS, RADIOLOGICAL PHYSICS. *Educ:* Berea Col, AB, 52; Mich State Univ, MS, 56; Col William & Mary, PhD(physics), 69. *Prof Exp:* Instr physics, Colgate Univ, 56-57, 58-59; res physicist, Airborne Instruments Labs, NY, 59-60; from asst prof to assoc prof physics, Old Dom Col, 60-69; ASSOC PROF PHYSICS, CENT MICH UNIV, 69- *Concurrent Pos:* NSF sci fac fel, 66-67. *Mem:* Am Asn Physics Teachers; Am Phys Soc. *Res:* Semiconductor devices; nuclear magnetic resonance in metals. *Mailing Add:* Dept of Physics Cent Mich Univ Mt Pleasant MI 48859

SLODKI, MOREY ELI, b Chicago, Ill, June 16, 28; m 67; c 2. BIOCHEMISTRY. *Educ:* Univ Ill, BS, 48; Univ Iowa, PhD, 55. *Prof Exp:* LEAD SCIENTIST MICROBIOL PROPERTIES RES, NORTHERN REGIONAL RES CTR, AGR RES SERV, USDA, 55- *Mem:* Am Chem Soc; Am Soc Microbiol; Soc Indust Microbiol. *Res:* Microbial enzymes; biological nitrogen fixation; exocellular microbial polysaccharides. *Mailing Add:* Northern Regional Res Ctr Peoria IL 61614

SLODOWSKI, THOMAS R, b Jersey City, NJ, Dec 21, 26; m 52; c 2. EXPLORATION GEOLOGY. *Educ:* Calif Inst Technol, BS, 53; Princeton Univ, PhD(geol), 56. *Prof Exp:* Geologist, Am Overseas Petrol Ltd, 56-70; geologist, Explor Dept, Standard Oil Co Calif, 70-74; geologist, Geothermal Div, Union Oil Co Calif, 74-80; SR STAFF EXPLORATIONIST, TEX EASTERN CORP, 80- *Mem:* Fel Geol Soc Am; Am Asn Petrol Geol. *Res:* Regional and field geology. *Mailing Add:* 2660 Marilee Houston TX 77057

SLOGER, CHARLES, b Albany, NY, Dec 22, 38; m 67; c 2. PLANT PHYSIOLOGY. *Educ:* State Univ NY Albany, BS, 61, MS, 63; Univ Fla, PhD(bot), 68. *Prof Exp:* PLANT PHYSIOLOGIST, NITROGEN FIXATION & SOYBEAN GENETICS LAB, PLANT SCI INST, BELTSVILLE AGR RES CTR, USDA, 68- *Mem:* Am Soc Plant Physiol; Am Soc Microbiol. *Res:* Physiology of symbiotic nitrogen fixation; mineral nutrition. *Mailing Add:* Agr Res Serv NE Region USDA Beltsville Agr Res Ctr-West Beltsville MD 20705

SLOMA, LEONARD VINCENT, b Chicago, Ill, June 28, 20; m 46; c 2. ENGINEERING PHYSICS. *Educ:* Northwestern Univ, BS, 42, PhD, 51; Mass Inst Technol, SM, 48. *Prof Exp:* Consult, Arthur D Little, Inc, 46-48; instr mech eng, Northwestern Univ, 48-51; res engr, Autonetics Div, NAm Aviation, Inc, 51-53, supvr systs anal, 53-55, staff specialist, 55-56; scientist, 56-57, sect mgr, 57-59, assoc dir & head physics & electronics dept, 59-84, SR TECH COUN, RES CTR, BORG-WARNER CORP, 84- *Mem:* Sigma Xi; Soc Photo-optical Instrumentation Engrs. *Res:* Applied and fluid mechanics; heat transfer; refrigeration; air conditioning; acoustics; electronics; solid state circuitry; solid state physics; electrochemical physics; servo controls; systems analysis. *Mailing Add:* 6221 N Kirkwood Ave Chicago IL 60646

SLOMIANY, AMALIA, DIGESTIVE DISEASE. *Educ:* NY Med Col, PhD(biochem), 73. *Prof Exp:* PROF MED, NY MED COL, 81- *Res:* Glycocongujates; glycoprotein and glycolipids. *Mailing Add:* Dept Med Metropolitan Hosp NY Med Col 110 Bergen St Newark NJ 07103-2425

SLOMP, GEORGE, b Grand Rapids, Mich, Feb 15, 22; m 51; c 4. PHYSICAL CHEMISTRY. *Educ:* Calvin Col, AB, 42; Ohio State Univ, PhD(chem), 49. *Prof Exp:* Lab asst chem, Calvin Col, 41-42; asst inorg chem, Ohio State Univ, 42-43, org chem, 46-49; asst org synthesis, Am Petrol Inst, 43-44; res chemist, Am Oil Co, Tex, 44-46; SR SCIENTIST, UPJOHN CO, 49- *Res:* Synthesis and reactions of steroidal hormones and natural products; reaction mechanisms; ozonolysis; catalytic hydrogenation; Grignard reactions; nuclear magnetic resonance spectroscopy; molecular structure determination; computer programming. *Mailing Add:* Phys & Anal Chem Upjohn Co Kalamazoo MI 49001-0199

SLONCZEWSKI, JOAN LYN, b Hyde Park, NY, Aug 14, 56; m; c 2. BACTERIAL PHYSIOLOGY. *Educ:* Yale Univ, PhD(molecular biophysics-biochem), 82. *Prof Exp:* Fel NIH Grant, Univ Pa, 82-84; ASST PROF BIOL, KENYON COL, 84- *Concurrent Pos:* Vis asst prof molecular biol, Princeton. *Honors & Awards:* Young Investr Award, Am Soc Microbiol, 87. *Mem:* Am Soc Microbiol; Am Soc Cell Biol; AAAS. *Res:* Bacterial pH regulation and gene expression. *Mailing Add:* Dept Biol Kenyon Col Gambier OH 43022

SLONCZEWSKI, JOHN CASIMIR, b New York, NY, July 26, 29; m 55; c 3. SOLID STATE PHYSICS. *Educ:* Worcester Polytech Inst, BS, 50; Rutgers Univ, PhD(physics), 55. *Prof Exp:* RES STAFF MEM, WATSON RES CTR, IBM CORP, 55- *Mem:* Fel Am Phys Soc. *Res:* Theories of ferromagnetism; magnetic domain phenomena; structural phase transitions and electron-lattice interactions. *Mailing Add:* 161 Allison Rd Katonah NY 10536

SLONECKER, CHARLES EDWARD, b Gig Harbor, Wash, Nov 30, 38; m 61; c 3. ANATOMY. *Educ:* Univ Wash, DDS, 65, PhD(biol struct), 67. *Prof Exp:* Sci asst path, Int Path, Univ Bern, 67-68; from asst prof to assoc prof anat, Univ BC, 68-76, head dept, 81-92, dir ceremonies & spec events, 90-95, PROF ANAT, UNIV BC, 76- *Concurrent Pos:* Nat Inst Allergy & Infectious Dis fels, Univ Wash, 65-67; Swiss Nat Fund grant, 67-69; BC Med Res Found grant, 68-69; Med Res Coun grant, 70-75; res grants, G&F Heighway Fund, 75- & Muscular Dystrophy of Can, 78-80. *Honors & Awards:* Centennial Gold Medal, Am Asn Anatomists, 87. *Mem:* Am Asn Anatomists; Can Asn Anatomists. *Res:* Lymphocytic tissue morphology and physiology; cellular immunology; radiobiology; hematology and radioautography. *Mailing Add:* Dept Anat Fac Med Univ BC Vancouver BC V6T 1Z2 Can

SLONIM, ARNOLD ROBERT, b Springfield, Mass, Feb 15, 26; m 51, 84; c 3. BIOTECHNOLOGY, SCIENCE ADMINISTRATION. *Educ:* Tufts Col, BS, 47; Boston Univ, AM, 48; Johns Hopkins Univ, PhD(biol), 53. *Prof Exp:* Res asst nutrit, Sterling-Winthrop Res Inst, 48-49; res asst, NIH grantee pharmacol, Sch Med, George Washington Univ, 49-50; res asst & jr instr biol, Johns Hopkins Univ, 50-53; res assoc chemother, Children's Cancer Res Found, Harvard Med Sch, 53-54; head chem lab, Lynn Hosp, Mass, 55-56; res scientist (chief appl ecol, supvry res biologist, physiologist, biochemist, phys sci admin, mgr biotechnol), Aerospace Med Res Lab, Wright-Patterson AFB, Ohio, 56-86; CONSULT & PRES, ARSLO ASSOCS, DAYTON & COLUMBUS, OHIO, 87- *Concurrent Pos:* Lectr, Mass Sch Physiother, 55-56; mem, Int Bioastronaut Comt, Int Astronaut Fedn, 67-70 & 84-; mem, comt biol handbooks, Fedn Am Socs Exp Biol, 67-71; mem, environ carcinogens prog, Int Agency Res Cancer/WHO, 81-; legal expert, Environ Pollution, Denver, Co, 89-90. *Mem:* Sigma Xi; NY Acad Sci; Aerospace Med Asn; Am Soc Biochem & Molecular Biol; Am Physiol Soc; Int Acad Aviation & Space Med. *Res:* Aerospace physiology and biochemistry; life support system requirements; environmental pollution (esp potable water standards, aquatic toxicity); biomechanical/biodynamic stress (esp acceleration, vibration); biotechnology and environmental health consultant. *Mailing Add:* 630 Cranfield Pl Columbus OH 43213-3407

SLONIM, JACOB, b Israel, June 25, 45; Can citizen; m 66; c 3. NAMING IN DISTRIBUTED SYSTEMS, LANGUAGES FOR DISTRIBUTED SYSTEMS. *Educ:* Univ Western Ont, BS, 72, MS, 73; Univ Kans State, PhD(computer sci), 78. *Prof Exp:* Asst prof, distrib data base mgt software, Kans State Univ, 77-78; researcher distrib data base mgt software, NDX Corp, 79-82, Geac Computers, 82-88; prof, distrib data base mgt software, Univ Western Ont, 86-89; HEAD RES, IBM CAN CTR ADVAN SCI, 89-, ARCHITECT, DATABASES, 89-; PROF DISTRIB DATA BASE MGT SOFTWARE, UNIV TORONTO, 90- *Concurrent Pos:* Asst prof, Kans State Univ, 78-79; exec vpres, Res & Develop Dept, NDX Corp, 79-82; software dir, Res & Develop Dept, Geac Computers Coun, 82-88; prof, Univ Western Ont, 86-89; head res, IBM, Can, 90; adj prof, Univ Waterloo, 91. *Honors & Awards:* Kaplan Award, Pres Israel for Contribution to Israel Computer Indust, 80. *Mem:* Asn Comput Mach; sr mem Inst Elec & Electronics Engrs. *Res:* Creating a complete application level architecture and a set of tools for building reliable distributed systems. *Mailing Add:* 230 Scarbrough Rd Toronto ON M4E 3H6 Can

SLONKA, GERALD FRANCIS, parasitology, for more information see previous edition

SLOOP, CHARLES HENRY, PHYSIOLOGY, MICROCIRCULATION. *Educ:* Wake Forest Univ, PhD(physiol), 73. *Prof Exp:* ASSOC PROF PHYSIOL, LA STATE UNIV MED CTR, 76- *Res:* Lipoproteins. *Mailing Add:* La State Univ Med Ctr 1542 Tulane Ave New Orleans LA 70112

SLOOPE, BILLY WARREN, b Clifton Forge, Va, Jan 4, 24; m 51; c 2. PHYSICS, THIN FILMS. *Educ:* Univ Richmond, BS, 49; Univ Va, MS, 51, PhD(physics), 53. *Prof Exp:* Asst prof physics, Clemson Col, 53-55; from asst prof to assoc prof, Univ Richmond, 55-61, adj assoc prof, 61-68; head dept physics,Va Commonwealth Univ, 68-79, prof, 68-88; RETIRED. *Concurrent Pos:* Sr res physicist, Va Inst Sci Res, 56-68, head, Physics Div, 61-68; Horsley Res Award, Va Acad Sci, 61. *Mem:* Am Vacuum Soc; Am Asn Physics Teachers. *Res:* Epitaxial thin films and their properties. *Mailing Add:* 8718 Avalon Dr Richmond VA 23229

SLOSS, JAMES M, b Birmingham, Ala, Feb 14, 31. PARTIAL DIFFERENTIAL EQUATIONS, NUMERICAL ANALYSIS. *Educ:* Pamona Col, BA, 53; Univ Calif, Berkeley, PhD(math), 62. *Prof Exp:* Actg asst prof, 61-62, from asst prof to assoc prof, 63-73, PROF MATH, UNIV CALIF, SANTA BARBARA, 74- *Mem:* Am Math Soc; Soc Indust & Appl Math. *Mailing Add:* Univ Calif Santa Barbara CA 93105

SLOSS, PETER WILLIAM, b Butte, Mont, May 11, 42; m 66; c 3. GEOLOGY, OCEANOGRAPHY. *Educ:* Northwestern Univ, BS, 64; Univ Chicago, MS, 66; Rice Univ, PhD(geol), 72. *Prof Exp:* Sr res technician meteorol, Univ Chicago, 67-68; assoc researcher, Inst Storm Res, Houston, 68-69; vis asst prof geol, Mich State Univ, 72-73; phys scientist oceanog, 73-78, GEOLOGIST, NAT OCEANIC & ATMOSPHERIC ADMIN, NAT GEOPHYSICAL DATA CTR, 78-, CHIEF DATA SYSTEMS & PROD, MARINE GEOL & GEOPHYSICS DIV, 81- *Mem:* Am Meteorol Soc; Am Geophys Union; Sigma Xi. *Res:* Interdisciplinary studies spanning geology, meteorology and oceanography; management of data from such studies; tsunamis; computer graphics. *Mailing Add:* Nat Oceanic & Atmospheric Admin Code E/GC3 325 Broadway Boulder CO 80303

SLOSSER, JEFFREY ERIC, b Winslow, Ariz, Dec 1, 43; m 68; c 2. ENTOMOLOGY. *Educ:* Ariz State Univ, BS, 66; Univ Ariz, MS, 68, PhD(entom), 71. *Prof Exp:* Res assoc entom, Univ Ariz, 68-70 & Univ Ark, 72-75; asst prof, 75-79, assoc prof, 79-86 PROF ENTOM, TEX A&M UNIV, 86- *Concurrent Pos:* Ed, Southwestern Entomologist, 86-90; pres elect, Southwestern Entom Soc, 91. *Mem:* Entom Soc Am; Sigma Xi. *Res:* Integrated control and population dynamics of cotton and wheat insect pests in the rolling plains of Texas. *Mailing Add:* Tex Agr Exp Sta PO Box 1658 Vernon TX 76384

SLOSSON, JAMES E, b Van Nuys, Calif, Apr 12, 23; m 47; c 2. HYDROLOGY AND WATER RESOURCES. *Educ:* Univ Southern Calif, BA, 49, MS, 50, PhD(geol), 58. *Prof Exp:* Prof geol, Los Angeles Valley Col, 50-73 & 75-84; CONSULT ENG GEOLOGIST, 60- *Concurrent Pos:* Geologist, US Geol Surv, 49-50; res geologist, Gulf Oil Corp, 52-56; NSF grant mineralogy & geol, Univ Ill, 57; consult eng geologist for various projs, 58-73; chief eng geol, Slosson & Assoc, 60-73 & 75-; mem, Eng Geologists Qual Bd, City of Los Angeles, 61-76, County of Los Angeles, 66-68 & 81-, chmn, Eng Geol Rev & Appeals Bd, 72; mem, Gov Earthquake Coun, 73-74, Am Asn State Geologists, 73-75 & Nat Acad Sci panel on mudslides, 74; state geologist, Calif Div Mines & Geol, 73-75; comnr, Seismic Safety Comn, 75-78; lectr, Environ Mgt Inst, Sch Pub Admin, Univ Southern Calif, 74-; mem, Adv Comt for Socioecon & Polit Consequences of Earthquake Prediction, Univ Colo, NSF Study, 75-76; mem, Bd Registration Geologists & Geophys, 78-85; mem, Calif Earthquake Prediction Eval Coun, 75-; guest lectr, Harvard Univ Grad Sch, Calif State Northridge, Occidental Col, Univ Nev & Univ Calif, Los Angeles, Berkeley, Irvine & Davis; mem Nat Res Coun Comt Ground Failure Hazards, 86-; chair, FEMA/Colo Pub Safety Comt, Landslide Hazard Mitigation Proj, 86-; coord, ASCR/Off Emergency Serv, Disaster Preparedness Comt, 83- *Honors & Awards:* Ichard Jahns distinguished lectr, GSA, 89. *Mem:* Am Asn Petrol Geol; fel Earthquake Eng Res Inst; Am Soc Civil Eng; Asn Eng Geol; fel Geol Soc Am; Sigma Xi; Am Geophys Union; Seismol Soc Am. *Res:* Engineering geology; seismic research; authored over 120 papers on geol practices. *Mailing Add:* 15373 Valley Vista Blvd Sherman Oaks CA 91403

SLOTA, PETER JOHN, JR, b Cleveland, Ohio, July 11, 24; m 59. CHEMISTRY. *Educ:* Hiram Col, BA, 46; Temple Univ, MA, 49, PhD(chem), 54. *Prof Exp:* Instr chem, Hiram Col, 46-47, instr night sch, Drexel Inst, 51-53; res assoc chem, Univ Southern Calif, 54-57; Alexander von Humboldt stipend, Res Inorg Chem, Univ Munich, 57-59; res chemist & head organometallics br, US Naval Ord Lab, 59-70; assoc dir archaeol res unit, Dry Lands Res Inst, 72-74, ASSOC RES CHEMIST, RADIOCARBON DATING & RES LAB, UNIV CALIF, RIVERSIDE, 74- *Res:* Organometallic and phosphorus chemistry; radiocarbon-14; chemistry of the environment; archaeological dating and ecological research. *Mailing Add:* 2903 Ivy St Riverside CA 92506

SLOTE, LAWRENCE, b New York, NY, May 23, 24; m 60; c 1. OCCUPATIONAL SAFETY, HUMAN FACTORS. *Educ:* Univ RI, BSME, 47; NY Univ, MME, 48, EngScD(civil eng), 61. *Prof Exp:* Sr res scientist, Res Div, Sch Eng & Sci, NY Univ, 48-63; environ engr, Pollution Control & Life Support, Advan Civil Systs, Grumman Aerospace Corp, 63-73; sr environ engr & scientist, Bradford Comput & Systs, Inc, 73-74; assoc prof allied health sci, York Col, City Univ New York, 74-75; dir res, Ctr for Safety, 75-78, PROF OCCUP HEALTH & SAFETY, NY UNIV, 75-, CHMN DEPT, 78-, DIR, CTR FOR SAFETY, 78- *Concurrent Pos:* Adj prof, Dept Indust Eng, Sch Eng & Sci, NY Univ, 61-67 & Dept Grad Mgt Eng, C W Post Col, Long Island Univ, 65-75; mem, US Nat Comt Eng in Med & Biol, Nat Acad Eng, 66-69; mem bd adminrs, North Shore Jr Sci Mus, 70-71. *Mem:* Syst Safety Soc; Sigma Xi; fel NY Acad Sci; assoc fel NY Acad Med. *Res:* Environmental control of closed ecological systems; biological waste treatment systems; biomedical and biomechanical engineering; cryptobioclimatology; mathematical modeling of environmental pollution; immobilized enzymes and wastewater treatment; human factors; safety sciences. *Mailing Add:* 3749 Starr King Circle Palo Alto CA 94306

SLOTKIN, THEODORE ALAN, b Brooklyn, NY, Feb 17, 47; m 67. PHARMACOLOGY. *Educ:* Brooklyn Col, BS, 67; Univ Rochester, PhD(pharmacol), 70. *Prof Exp:* NIMH trainee biochem, 70-71, asst prof, 71-75, assoc prof, 75-79, PROF PHARMACOL, DUKE UNIV, 79- *Mem:* Am Soc Pharmacol & Exp Therapeut; Am Soc Neurochem; Soc Neurosci. *Res:* Neuropharmacology; neurochemistry; developmental neurobiology; drug abuse. *Mailing Add:* Dept Pharmacol Duke Univ 405 Nanaline H Duke Durham NC 27710

SLOTNICK, DANIEL LEONID, computer sciences, hardware systems; deceased, see previous edition for last biography

SLOTNICK, HERBERT, b Malden, Mass, Oct 6, 28; m 53; c 2. CHEMICAL ENGINEERING, PHYSICAL CHEMISTRY. *Educ:* Northeastern Univ, BS, 51; Worcester Polytech Inst, MS, 53; Mass Inst Technol, SM, 55; Univ Conn, PhD(eng), 64. *Prof Exp:* Proj chemist, Pratt & Whitney Aircraft, United Aircraft Corp, 54-67; PROF CHEM, CENT CONN STATE COL, 67- *Concurrent Pos:* Res partic, NSF Acad Year Exten, Cent Conn State Col, 70-72. *Mem:* Am Chem Soc; Am Inst Chem Engrs; Sigma Xi. *Res:* Active nitrogen reactions; air-water pollution; materials; high vacuum, inert gas, liquid metal technology; materials related to biomedical science. *Mailing Add:* Chem Dept Cent Conn State Univ New Britain CT 06050

SLOTNICK, VICTOR BERNARD, b Chicago, Ill, Oct 27, 31; m 60; c 3. MICROBIOLOGY, MEDICINE. *Educ:* Roosevelt Univ, BS, 53; Univ Chicago, MS, 55; Hahnemann Med Col, PhD(microbiol), 60; Jefferson Med Col, MD, 65; Am Bd Family Pract, dipl, 77. *Prof Exp:* Assoc microbiol, Univ Chicago, 54; res assoc, Virus Lab, Ill State Dept Pub Health, 54-55; assoc microbiol, Hahnemann Med Col, 57-60; sr virologist, Merck Inst, 60-61; intern, Albert Einstein Med Ctr, 65-66; asst dir, 66-68, assoc dir, 68-75, dir clin res, McNeil Labs Inc, 75-81, CLIN RES FEL, MCNEIL PHARMACEUT, 81- *Concurrent Pos:* NIH res fel, Jefferson Med Col, 64-65, clin instr, Dept Family Med, 75- *Mem:* Am Soc Microbiol; AMA; NY Acad Sci; Am Col Neuropsychopharmacol. *Res:* Virology; immunology; antiviral chemotherapy; cytology; biochemistry; experimental teratology; psychopharmacology. *Mailing Add:* 312 Melrose Ave Melrose Station PA 19066

SLOTSKY, MYRON NORTON, b Portland, Maine, Apr 30, 35; m 62; c 2. INDUSTRIAL PHARMACY. *Educ:* Mass Col Pharm, BS, 57, MS, 59, PhD(pharm), 67. *Prof Exp:* Chemist, Res Dept, Gillette Safety Razor Co, 59-62, sr cosmetic chemist, Toiletries Develop Sect, 62-64, chem & biol sect, 64; sr pharmaceut chemist, Colgate-Palmolive Co, 66-68; dir prod develop, 68-75, spec develop proj analyst, 75-80, mgr pilot opers, 80-86, MGR CLIN SUPPLIES, MARION LABS, INC, 86- *Mem:* Am Chem Soc; Am Pharmaceut Asn; Acad Pharmaceut Sci; Am Asn Pharmaceut Scientists. *Res:* Tablet and capsule formulation, suspension and solution technology; emulsion and aerosol technology; problems related to stabilization, preservation and production scale-up of cosmetics and pharmaceuticals; dandruff; acne; processes development; all areas of clinical manufacturing and packaging. *Mailing Add:* Marion Merrell Dow Inc PO Box 9627 Park A Kansas City MO 64134-0627

SLOTTA, LARRY STEWART, b Billings, Mont, Aug 20, 34; m 58; c 3. CIVIL ENGINEERING, HYDRAULICS. *Educ:* Univ Wyo, BS, 56, MS, 59; Univ Wis, PhD(civil eng), 62. *Prof Exp:* instr gen eng, Univ Wyo, 57-58; instr civil eng, Univ Wis, 58-61, res assoc, Dept Meteorol, 61-62; from asst prof to prof civil eng, 62-85, dir ocean eng prog, 71-76, EMER PROF CIVIL ENG, ORE STATE UNIV, 86-; DIR ENG, TEX A&M UNIV, GALVESTON, 89- *Concurrent Pos:* Fulbright grant grad study, Tech Univ, Netherlands, 56; vis prof, Dartmouth Col, 66, Kyoto Univ, Japan, 69, Delft Univ, Netherlands, 76 & Univ Melbourne, Australia, 77; pres, Slotta Eng Assoc, Inc; mem, Comt Bioeng & Human Factors, Am Soc Civil Engrs, 72, Task Comt Environ Effects Hydraul Structures, 73-76, chmn, Comt Tidal Hydraul, 73-76, mem, Tech Activities Comt on Ocean Eng, 74-76, Waterways, Port, Coastal & Ocean Eng Div Task Comt on Ocean Eng, 78-90, Tech Comt Ocean Eng, 85-90, chmn, 86-90; instr, Linn-Benton Community Col, 88; consult numerous companies, 57-91. *Mem:* Am Soc Eng Educ; Am Soc Civil Engrs; Int Asn Hydraul Res; Am Soc Mech Engrs; AAAS; Marine Technol Soc; Am Geophys Union; Sigma Xi; Nat Well Water Asn; Asn Ground Water Scientists & Engrs. *Res:* Fluid mechanics; systems engineering; computer applications; dredge spoil distribution and estuarine effects; ocean engineering: coastal hydraulics, dredging equipment and processes, harbor and related infrastructure design, environmental field studies; applied hydraulics: water resources engineering, hydropower site assessment, hydraulic modeling and computational fluid dynamics; environmental engineering: hazardous waste management, site investigation and assessment, remediation plans; author of numerous publications. *Mailing Add:* Dir Eng Tex A&M Univ PO Box 1675 Galveston TX 77553

SLOTTER, RICHARD ARDEN, b Souderton, Pa, Mar 3, 32; m 54; c 3. INORGANIC CHEMISTRY. *Educ:* Bluffton Col, BS, 54; Univ Mich, MS, 57, PhD(chem), 60. *Prof Exp:* From asst prof to assoc prof chem, Bluffton Col, 58-64; from assoc prof to prof, ROBERT COL, Istanbul, 64-71, chmn dept, 66-71; vis assoc prof, Bucknell Univ, 71-72; asst chmn dept chem, Northwestern Univ, Evanston, 72-86, lectr, 72-86; DEAN ACAD AFFAIRS, BLUFFTON COL, 86- *Mem:* AAAS; Am Chem Soc; Sigma Xi. *Res:* Polarography; coordination complexes; electrochemical kinetics. *Mailing Add:* 175 Sunset Dr Bluffton OH 45817-1113

SLOTTERBECK-BAKER, OBERTA ANN, b Cincinnati, Ohio, July 3, 36; m 70; c 1. SYMBOLIC ALGEBRA, PARALLEL COMPUTING. *Educ:* Ohio State Univ, BS, 58; Univ Tex, Austin, MA, 66, PhD(math), 69. *Prof Exp:* Teacher, Columbus Pub Schs, Ohio, 58-60 & Union County Regional Schs, NJ, 60-64; asst math, Univ Tex, Austin, 65-69; res fel, Univ Fla, 69-70, asst prof, 70-74; from asst prof to assoc prof, 74-82, PROF MATH & COMPUTER SCI, HIRAM COL, 83- *Mem:* Am Math Soc; Asn Comput Mach; Math Asn Am. *Res:* Symbolic/algebraic computing algorithms; Macsyma development work; parallel algorithms. *Mailing Add:* Dept Math Sci Hiram Col Hiram OH 44234

SLOVACEK, RUDOLF EDWARD, b Bloomington, Ind, Jan 4, 48; div; c 3. BIOCHEMISTRY, BIOENERGETICS. *Educ:* Univ Rochester, BA, 70, MS, 72, PhD(biol), 75. *Prof Exp:* Res assoc marine biol, Nat Res Coun, 75-76; res assoc, Brookhaven Nat Lab, 76-78, asst biophysicist biol, 78-80; sr scientist biol, Corning Glass Works, 80-86; res assoc, 86-87, mgr, Phys Measurement Systs, 88-89, SR STAFF SCIENTIST, CIBA CORNING DIAGNOSTICS CORP, 90- *Mem:* Electro Chem Soc; AAAS; NY Acad Sci; Int Soc Optical Eng. *Res:* Physical and chemical studies of energy conversion process and its regulatory mechanisms in photosynthesis; application of fiber optics and optical techniques to immunodiagnostic and clinical chemistry measurements in medicine. *Mailing Add:* 60 King St Norfolk MA 02056

SLOVIN, SUSAN FAITH, b Feb 5, 53; m 90. CELL-MEDIATED IMMUNOLOGY, IMMUNOTHERAPY. *Educ:* Columbia Univ, MA, 76, MPhil, 77, PhD(pathobiol), 78; Jefferson Med Col, MD, 90. *Prof Exp:* RES ASST PROF MED & MICROBIOL, THOMAS JEFFERSON UNIV SCH MED, 84- *Concurrent Pos:* Adj asst prof microbiol & immunol, NY Med Col, Valhalla. *Mem:* Sigma Xi; Am Asn Immunologists; NY Acad Sci; Am Asn Cancer Res; Soc Biol Ther. *Res:* Tumor immunology. *Mailing Add:* 150 E 85th St New York NY 10028

SLOVITER, HENRY ALLAN, b Philadelphia, Pa, June 16, 14. PHYSIOLOGICAL CHEMISTRY. *Educ:* Temple Univ, AB, 35, AM, 36; Univ Pa, PhD(org chem), 42, MD, 49. *Prof Exp:* Chemist, US Navy Yard, Philadelphia, 36-45; chemist, Harrison Dept Surg Res, Sch Med, Univ Pa, 45-49, intern, Hosp, 49-50; res fel, Nat Inst Med Res, London, 50-52; asst prof physiol chem & res asst prof surg, 52-57, res assoc prof neurosurg, 57-66, res prof neurosurg, 66-75, prof biochem, 66-75, PROF SURG RES & PROF BIOCHEM & BIOPHYSICS, SCH MED, UNIV PA, 75- *Concurrent Pos:* Vis scientist, Univ Tokyo, 63; US-USSR health exchange scientist, Sechenov Inst Physiol, Moscow, 65, Inst Biol & Med Chem Moscow, 71; US-India exchange scientist, Christian Med Col, Vellore, India, 67; proj officer award to Inst for Biol Res, Fogarty Int Ctr, NIH, Belgrade, Yugoslavia, 72-75, fel, Med Res Coun Exp Haemat Unit, St Mary's Hosp Med Sch, London, 78; vis scientist, Tokyo Metrop Inst Med Sci, 84. *Honors & Awards:* Glycerine Res Award, 54. *Mem:* AAAS; Am Physiol Soc; Am Soc Biol Chem; Int Soc Neurochem; Int Soc Blood Transfusion; hon mem Belgian Soc Anesthesiol. *Res:* Brain metabolism; erythrocyte lipids and metabolism; erythrocyte substitutes. *Mailing Add:* Sch of Med Univ of Pa Philadelphia PA 19104

SLOVITER, ROBERT SETH, b Nov 29, 50; m; c 2. ELECTROPHYSIOLOGY, NEUROANATOMY. *Educ:* Penn State Univ, PhD(pharmacol), 78. *Prof Exp:* ASSOC PROF PHARMACOL & NEUROL, COL PHYSICIANS & SURGEONS, COLUMBIA UNIV, 88-; DIR, NEUROL RES CTR, HELEN HAYES HOSP, 86- *Mailing Add:* Neurol Res Ctr Helen Hayes Hosp NY State Dept Health 9W West Haverstraw NY 10993

SLOWEY, JACK WILLIAM, b Wauwatosa, Wis, Mar 19, 32; m 52; c 4. AERONOMY. *Educ:* Univ Wis, BS, 55, MS, 56. *Prof Exp:* Physicist, 56-59, ASTRONOMER, SATELLITE TRACKING PROG, ASTROPHYS OBSERV, SMITHSONIAN INST, 59- *Concurrent Pos:* Lectr, Harvard Col, 57-60 & Boston Univ, 68-69; consult, IBM Corp, 62-75 & Boston Col, 75- *Mem:* Am Geophys Union; Am Astron Soc. *Res:* Artificial earth satellite orbits; structure and variations of earth's upper atmosphere. *Mailing Add:* Smithsonian Astrophys Observ 60 Garden St Cambridge MA 02138

SLOWIK, JOHN HENRY, b Hastings, Nebr, Sept 12, 45; m 72; c 2. SOLID STATE PHYSICS. *Educ:* Manhattan Col, BS, 67; Univ Ill, MS, 71, PhD(solid state physics), 73. *Prof Exp:* Nuclear physics, Johns Hopkins Univ, 67; asst physics, Univ Ill, 68-73; assoc scientist, 73-74, SCIENTIST SOLID STATE PHYSICS, XEROX CORP, 75- *Mem:* Am Inst Physics; Sigma Xi; Am Phys Soc. *Res:* Extreme-ultraviolet spectroscopy; charge transport in disordered solids; interfacial charge transport; transient electronics. *Mailing Add:* 35 Coachman Dr Penfield NY 14526

SLOWINSKI, EMIL J, JR, b Newark, NJ, Oct 12, 22; m 51; c 5. PHYSICAL CHEMISTRY. *Educ:* Mass State Col, BS, 46; Mass Inst Technol, PhD(phys chem), 49. *Prof Exp:* Instr chem, Swarthmore Col, 49-52; from instr to assoc prof, Univ Conn, 53-64; prof chem, Macalester Col, 64-88; RETIRED. *Concurrent Pos:* Indust fel, Monsanto Chem Co, 53-54; NSF fel, 60-61; Nat Acad Sci exchange prof, 68-69. *Mem:* Am Chem Soc. *Res:* Mathematical preparation for general chemistry; qualitative analysis and the properties of ions in aqueous solution; chemical principles; chemical principles in the laboratory. *Mailing Add:* 806 Bachelor Ave St Paul MN 55118

SLOYAN, MARY STEPHANIE, b New York, NY, Apr 18, 18. MATHEMATICS. *Educ:* Georgian Court Col, AB, 45; Catholic Univ, MA, 49, PhD(math), 52. *Prof Exp:* From asst prof to assoc prof math, 52-59, pres, 68-74, PROF MATH, GEORGIAN COURT COL, 59- *Concurrent Pos:* Vis lectr math, Cath Univ Am, 60-82; bd gov, Math Asn Am, 88-91, Comt Sections, 88-94. *Mem:* Am Math Soc; Math Asn Am; Sigma Xi. *Res:* Metric geometry; application of complex variables to geometry. *Mailing Add:* Dept Math Georgian Court Col Lakewood NJ 08701

SLOYER, CLIFFORD W, JR, b Easton, Pa, Apr 30, 34; m 62; c 4. TOPOLOGY. *Educ:* Lehigh Univ, BA, 56, MS, 58, PhD(math), 64. *Prof Exp:* Instr math, Lehigh Univ, 58-64; from asst prof to assoc prof, 64-74, asst chmn dept, 69-78, PROF MATH, UNIV DEL, 74- *Concurrent Pos:* Vis prof, Grad Lib Studies Prog, Wesleyan Univ, Middleton, CT, 64-; vis lectr pub schs, Pa, 65-66 & Del, 66-67. *Mem:* Math Asn Am; Nat Coun Teachers Math; Sch Sci & Math Assoc. *Res:* Several complex variables; continuation of meromorphic functions on complex analytic manifolds; secondary programs for gifted students. *Mailing Add:* Dept of Math Univ of Del Newark DE 19711

SLUDER, EARL RAY, b Newland, NC, Nov 9, 30; m 57; c 3. FORESTRY, GENETICS. *Educ:* NC State Univ, BS, 56, MS, 60, PhD(forestry, genetics), 70. *Prof Exp:* Forester, Container Corp Am, 56-57; res forester timber mgt, 57-66, RES FORESTER TREE IMPROV, SOUTHEASTERN FOREST EXP STA, USDA FOREST SERV, 66- *Mem:* Soc Am Foresters. *Res:* Genetic improvement of the southern yellow pines; tree breeding southern pines. *Mailing Add:* 742 Forest Lake Dr N Macon GA 31210

SLUDER, GREENFIELD, CELL BIOLOGY. *Educ:* Univ Pa, PhD(biol), 76. *Prof Exp:* SR STAFF SCIENTIST, WORCESTER FOUND EXP BIOL, 81- *Res:* Biophysical analysis of spinal assembly; analysis of centrosome formation & reproduction. *Mailing Add:* Worcester Found Exp Biol 222 Maple Ave Shrewsbury MA 01545

SLUSARCHYK, WILLIAM ALLEN, b Port Jefferson, NY, June 6, 40; m 68. ORGANIC CHEMISTRY, PHARMACEUTICAL CHEMISTRY. *Educ:* Brown Univ, BS, 61; Pa State Univ, PhD(org chem), 65. *Prof Exp:* RES FEL, BRISTOL-MYERS SQUIBB PHARM RES INST, PRINCETON, 65- *Mem:* Am Chem Soc; NY Acad Sci; Am Soc Microbiol. *Res:* Antivirals semi-synthetic antibiotics, monobactams, penicillins and cephalosporins; isolation, synthesis and structural elucidation of antibiotics. *Mailing Add:* 19 Richmond Dr Skillman NJ 08558

SLUSARCZUK, GEORGE MARCELIUS JAREMIAS, b Stanyslaviv, Ukraine, Jan 14, 32; US citizen; m 64; c 1. ORGANIC CHEMISTRY, ENVIRONMENTAL ANALYSIS. *Educ:* Wayne State Univ, BS, 60, MS, 62; Univ Pa, PhD(org chem), 67. *Prof Exp:* Chemist, Res & Develop Ctr, Gen Elec Corp, 62-64, mem staff res & develop, 67-76; spectroscopist, Univ Pa, 65-67; SR RES ASSOC, CORP RES CTR, INT PAPER CORP, 76- *Mem:* Am Chem Soc; Am Indust Hyg Asn; Shevchenko Sci Soc. *Res:* Environmental organic trace pollutants; pollutants of the working place; trace elemental analysis. *Mailing Add:* Corp Res Ctr Int Paper Corp Long Meadow Rd Tuxedo Park NY 10987

SLUSAREK, LIDIA, b Poland; US citizen; m; c 1. ANALYTICAL CHEMISTRY. *Educ:* Polytech Inst, Poland, MS, 69; Columbia Univ, PhD(chem), 76. *Prof Exp:* Res asst med chem, Albert Einstein Col Med, 69-71; res investr anal chem, Squibb Inst Med Res, 76-78; sect head mat control, E R Squibb & Sons, 78-79; group leader, Indust Lab Div, 79-80, sect supvr, 80-88, DIR, ANAL LABS, EASTMAN KODAK, 88- *Mem:* Am Chem Soc. *Res:* Development assays for drugs and impurities in formulations, bulk materials and body fluids; electrochemical analysis; chemical and physical analysis of photographic components. *Mailing Add:* 270 Panorama Trail Rochester NY 14625

SLUSHER, RICHART ELLIOTT, b Higginsville, Mo, May 20, 38; m 61; c 3. PHYSICS. *Educ:* Univ Mo-Rolla, BS, 60; Univ Calif, Berkeley, PhD(physics), 66. *Prof Exp:* MEM TECH STAFF, BELL TEL LABS, 65- *Honors & Awards:* Einstein Prize for Laser Sci, 89. *Mem:* Fel Am Phys Soc; fel Optical Soc Am. *Res:* Laser scattering from plasmas, solids and liquids; nonlinear optics of resonant coherent pulses; nuclear double resonance; astrophysics; quantum optics. *Mailing Add:* Bell Tel Labs Rm 1D-368 600 Mountain Ave Murray Hill NJ 07974

SLUSKY, SUSAN E G, b New York, NY, Dec 6, 49; m 71; c 3. SOLID STATE PHYSICS, APPLIED PHYSICS. *Educ:* Brown Univ, AB, 71; Univ Pa, MS, 72; Princeton Univ, PhD(physics), 78. *Prof Exp:* MEM TECH STAFF PHYSICS, BELL LABS, 78- *Mem:* Am Phys Soc. *Res:* Magnetic materials; superconducting electronics; III-V semiconductors. *Mailing Add:* AT&T Bell Labs Rm 7E408 600 Mountain Ave Murray Hill NJ 07974

SLUSS, ROBERT REGINALD, b Louisville, Ohio, July 18, 28; m 71; c 1. ENTOMOLOGY, ECOLOGY. *Educ:* Colo Col, BS, 53; Colo State Univ, MS, 55; Univ Calif, Berkeley, PhD(entom), 66. *Prof Exp:* Microbiologist, Pink Bollworm Res Ctr, USDA, 55-59; res assoc entom, Gill Tract, Univ Calif, 59-66; from asst to assoc biol, San Jose State Col, 66-69; assoc prof, State Univ NY Col Old Westbury, 69-70; PROF BIOL, EVERGREEN STATE COL, 70- *Mem:* Ecol Soc Am. *Res:* Insect pathology; insect population ecology; insect physiology; natural history. *Mailing Add:* Dept of Biol Evergreen State Col Olympia WA 98505

SLUSSER, M(ARION) L(ILES), b Memphis, Tenn, May 9, 19; m 46; c 3. CHEMICAL ENGINEERING. *Educ:* Univ Okla, BS, 48; Univ Colo, MS, 49. *Prof Exp:* Res engr, Field Res Lab, Mobil Res & Develop Corp, 49-55, sr res engr, 55-67, engr assoc, 67-82; PRES, PROD DIAGNOSTICS, INC, 82- *Concurrent Pos:* Consult, 82-84. *Mem:* Sigma Xi; Soc Petrol Engrs. *Res:* Well completion and oil production problems; oil well stimulation; thermal recovery methods for recovery of oil from oil shale by in-place methods; oil well stimulation by fracturing and acidizing. *Mailing Add:* 3804 Glenbrook Arlington TX 76015

SLUTSKY, ARTHUR, b Toronto, Ont, Dec 31, 48; m 71; c 2. ADULT RRESPIRATORY DISTRESS SYNDROME, MECHANISMS OF LUNG INJURY. *Educ:* Univ Toronto, BASc, 70, MASc, 72; McMaster Univ, MD, 76. *Prof Exp:* From instr to asst prof med, Harvard Univ, 80-84; assoc prof med, 84-88, PROF MED, SURG & BIOMED ENG, UNIV TORONTO, 88- *Concurrent Pos:* Mem, Adv Comt & Anesthesia Devices, Food & Drug Admin, 83-88, consult, Adv Comt Respiratory & Anesthesia Devices, 88-; scientist A, Med Res Coun Can, 85-90; counr, Can Soc Clin Invest, 86-90; mem, Steering Comt Crit Care Med, Am Col Chest Physicians, 87-89. *Mem:* Am Col Chest Physicians; Am Physiol Soc; AAAS; Am Soc Clin Invest. *Res:* Biology and pathophysiology of acute lung injury. *Mailing Add:* 600 University Ave Suite 656A Toronto ON M5G 1X5 Can

SLUTSKY, HERBERT L, b Chicago, Ill, Nov 6, 25; m 55; c 2. EPIDEMIOLOGY. *Educ:* Univ Ill, BS, 50, MS, 51, PhD(geog, physiol), 59. *Prof Exp:* Vis lectr geog, Univ Ill, 58; assoc prof, 59-67, PROF GEOG, ROOSEVELT UNIV, 67-, HEAD DEPT, 59- *Mem:* Am Geog Soc; Asn Am Geog; Int Soc Biometeorol; Am Pub Health Asn. *Res:* Physiology; geographical distribution of disease; ecological studies of protein, malnutrition and kwashiorkor; pediatric lead poisoning; tubercullosis; salmenella. *Mailing Add:* Dept Geog Roosevelt Univ 430 S Michigan Ave Chicago IL 60605

SLUTSKY, LEON JUDAH, b New York, NY, Oct 9, 32. PHYSICAL CHEMISTRY. *Educ:* Cornell Univ, BA, 53; Mass Inst Technol, PhD, 57. *Prof Exp:* Instr chem, Univ Tex, 57-59, asst prof, 59-61; asst prof, 61-69, PROF CHEM, UNIV WASH, 69- *Mem:* Am Phys Soc. *Res:* Lattice dynamics; mechanical properties of solids; surface chemistry. *Mailing Add:* Dept of Chem Univ of Wash Seattle WA 98195

SLUTZ, RALPH JEFFERY, b Cleveland, Ohio, May 18, 17; m 46; c 4. NUMERICAL METHODS. *Educ:* Mass Inst Technol, BS & MS, 39; Princeton Univ, PhD(theoret physics), 46. *Prof Exp:* Asst, Mass Inst Technol, 37-38; asst, Princeton Univ, 39-42, instr, 41-42; tech aide, Nat Defense Res Comt, 42-45; comput design engr, Inst Adv Study, 46-48; physicist, Nat Bur Standards, 48-49, asst chief electronic comput sect, 49-53, consult comput & math, 53-54, asst chief cent radio propagation lab, 54, chief radio propagation physics div, 54-60; guest worker magnetosphere res, Max-Planck Inst Physics & Astrophys, 60-61; sr scientist & consult upper atmosphere & space physics div, Nat Bur Standards, 61-65; sr scientist, Space Disturbances Lab, Environ Sci Serv Admin, Nat Oceanic & Atmospheric Admin, 65-69, actg dir, Space Environ Lab, 69-70, chief numerical anal & comput techniques group, 70-73, sr scientist, Environ Res Labs, 73-80; sr res assoc, Univ Colo Coop Inst Res Environ Sci, 80-90; GUEST RESEARCHER, NAT OCEANIC & ATMOSPHERIC ADMIN, 90- *Concurrent Pos:* Mem, US nat comt, Int Sci Radio Union, 56-59; consult, President's Sci Adv Comt, 64-67; mem study group IV, US preparatory comt, Int Radio Consult Comt; vis lectr, Univ Col, 70-72, adj prof elec eng, 72- *Honors & Awards:* Gold Medal, Dept Commerce, 52. *Mem:* Am Phys Soc; Inst Elec & Electronic Eng; Am Geophys Union; Int Asn Geomag & Aeronomy; AAAS. *Res:* Climate analysis; numerical forecasting; computer techniques. *Mailing Add:* 745 Mapleton Ave Boulder CO 80304

SLUTZKY, GALE DAVID, b Omaha, Nebr, Mar 13, 52; m 72, 83; c 2. MATERIALS HANDLING, ROBOTICS. *Educ:* Univ Wyo, BA, 76; Univ Tenn, MA, 81. *Prof Exp:* Res asst image processing elec eng, Univ Tenn, 76-83; asst dir, 83-87, ASSOC DIR ROBOTICS, CINCINNATI CTR ROBOTICS RES, UNIV CINCINNATI, 87-; CTR MGR MAT HANDLING PACKAGING, CTR AUTOMATED PACKAGING & MAT HANDLING, IAMS, 90- *Concurrent Pos:* Consult, Nichols Res Corp, 83-, anthropology, Univ Cincinnati, 86. *Mem:* Soc Mfg Engrs; Inst Indust Engrs; Am Soc Mech Engrs; Sigma Xi. *Res:* Pragmatic solutions of day-to-day problems encountered in packaging and materials handling; order picking; automated palletizing; load forming; simulation; warehousing; environmental packaging; robotics; machine vision applications. *Mailing Add:* 7752 Montgomery Rd No 81 Cincinnati OH 45236

SLY, PETER G, b Sidcup, Eng, Feb 11, 39; m 64; c 2. LAKE & MARINE GEOLOGY, SCIENCE POLICY. *Educ:* Univ London, BSc, 60; Univ Liverpool, PhD(geol), 67, DSc, 91. *Prof Exp:* Head process res, 72-79, res scientist, 79-88, DIR, SCI PROG, CAN CTR INLAND WATERS, 88- *Concurrent Pos:* Tech adv, Geonautics Ltd, 78-80. *Mem:* Can Asn Environ Analysis Lab; Int Asn Great Lakes Res; Geol Asn Can; Sigma Xi. *Res:* Marine geology in United Kingdom; geology and benthic fauna of large lake sediments, fish habitat and substrate analyses; develoment and testing of equipment; contaminant effects; application of submersible and diver usable techniques; foundations for offshore structures; science policy. *Mailing Add:* Rawson Acad Aquatic Sci Suite 404 One Nicholas St Ottawa ON K1N 7B7 Can

SLY, RIDGE MICHAEL, b Seattle, Wash, Nov 3, 33; m 57; c 2. PEDIATRIC ALLERGY & IMMUNOLOGY. *Educ:* Kenyon Col, AB, 56; Washington Univ, MD, 60; Am Bd Pediat, dipl, 65, cert allergy, 67; Am Bd Allergy & Immunol, cert 72, recert, 77 & 87. *Prof Exp:* NIH fel pediat allergy & immunol, Med Sch, Univ Calif, Los Angeles, 65-67; from asst prof to prof pediat, La State Univ Sch Med, New Orleans, 67-78; PROF PEDIAT, SCH MED & HEALTH SCI, GEORGE WASHINGTON UNIV & DIR ALLERGY & IMMUNOL, CHILDREN'S NAT MED CTR, 78- *Concurrent Pos:* Vis physician, Charity Hosp, New Orleans, 67-78. *Honors & Awards:* Peshkin Mem Award, Asn Care Asthma, 83. *Mem:* Am Acad Allergy; Am Acad Pediat; Am Col Allergists; Am Thoracic Soc; Asn for the Care of Asthma (pres, 80-81). *Res:* Pulmonary physiology and pharmacology of asthma; exercise induced asthma; mortality from asthma. *Mailing Add:* Children's Nat Med Ctr 111 Michigan Ave NW Washington DC 20010-2970

SLY, WILLIAM GLENN, b Arcara, Calif, June 15, 22. PHYSICAL CHEMISTRY. *Educ:* San Diego State Col, BS, 51; Calif Inst Technol, PhD(chem), 55. *Prof Exp:* NSF fel, Calif Inst Technol, 55-56, Hale fel, 56-57; fel, Mass Inst Technol, 57-58; from asst prof to assoc prof, 58-66, PROF CHEM, HARVEY MUDD COL, 66- *Concurrent Pos:* Res assoc, Calif Inst Technol, 59-61, sr res fel, 61-62; NSF sr fel, Swiss Fed Inst Technol, 65-66; guest prof, Ore State Univ, Corvallis, 72-73; vis scholar, Univ Calif, San Diego, 80. *Mem:* Am Chem Soc; Am Crystallog Asn; Inst Elec & Electronics Eng; Am Inst Physics. *Res:* Molecular structure; x-ray crystallography; application of high speed computers to structural analysis; metal complexes; organic molecules. *Mailing Add:* Dept Chem Harvey Mudd Col Claremont CA 91711

SLY, WILLIAM S, b East St Louis, Ill, Oct 19, 32; m 60; c 7. BIOCHEMISTRY, GENETICS. *Educ:* St Louis Univ, MD, 57. *Prof Exp:* Intern & asst resident, Ward Med Barnes Hosp, St Louis, Mo, 57-59; clin assoc, Nat Heart Inst, NIH, Bethesda, Md, 59-63, res biochemist, 59-63; dir, Div Med Genetics, Dept Med & Pediat, Sch Med, Wash Univ, 64-84, from asst prof to prof med, 64-78, from asst prof to prof pediat, 67-78, prof pediat, med & genetics, 78-84; PROF BIOCHEM & CHMN , E A DOISY DEPT BIOCHEM & PROF PEDIAT, SCH MED, ST LOUIS UNIV, 84- *Concurrent Pos:* Vis physician, Nat Heart Inst, Bethesda, Md, 61-63 & Pediat Genetics Clin, Univ Wis-Madison, 63-64; Am Cancer Soc fel, Lab Enzymol, Nat Ctr Sci Res, Gif-sur-Yvette, France, 63 & Dept Biochem & Genetics, Univ Wis, 63-64; attend physician, St Louis County Hosp, Mo, 64-84; asst physician, Barnes Hosp, St Louis, Mo, 64-84 & St Louis Children's Hosp, Mo, 67-84; consult genetics, Homer G Philips Hosp, St Louis, Mo, 69-81; mem, Steering Comt, Human Cell Biol Prog, 71-73 & Comt Genetic Coun, Am Soc Human Genetics, 72-76; Genetics Study Sect, Div Res Grants, NIH, 71-75; traveling fel award, Royal Soc Med, 73; sabbatical, Genetics Lab, Dept Biochem, Oxford Univ, Eng, 73-74, & Dept Biochem, Stanford Univ, Calif, 81-82; active staff, Cardinal Glennon Children's Hosp, St Louis, Mo, 84-; med adv bd, Howard Hughes Med Inst, 89- *Honors & Awards:* Merit Award, NIH, 88; Passaro Found Award, 91. *Mem:* Nat Acad Sci; Soc Pediat Res; Am Soc Human Genetics; Am Soc Clin Invest; Am Chem Soc; AMA; Sigma Xi; AAAS; Genetics Soc Am; Am Soc Microbiol. *Res:* Biochemical regulation; enveloped viruses as membrane probes in human diseases; lysosomal enzyme replacement in storage diseases; somatic cell genetics. *Mailing Add:* E A Doisy Dept Biochem St Louis Univ Sch Med 1402 S Grand Blvd St Louis MO 63104

SLYE, JOHN MARSHALL, b Boulder, Colo, Nov 27, 23. PURE MATHEMATICS. *Educ:* Calif Inst Technol, BS, 45; Univ Tex, PhD(pure math), 53. *Prof Exp:* Technician physics cyclotron oper, Los Alamos Sci Lab, Calif, 46-48; instr pure math, Univ Tex, 50-53; from instr to assoc prof math, Univ Minn, Minneapolis, 53-69; ASSOC PROF MATH, UNIV HOUSTON, 69- *Mem:* AAAS; Am Math Soc; Am Phys Soc. *Res:* Two dimensional spaces; point set theory. *Mailing Add:* Dept Math Univ Houston 4800 Calhoun Rd Houston TX 77204

SLYH, JOHN A(LLEN), ceramics engineering; deceased, see previous edition for last biography

SLYSH, ROMAN STEPHAN, b Ukraine, June 11, 26; US citizen; m 54; c 4. POLYMER ANALYSIS. *Educ:* Williams Col, BA, 53; Union Col, MS, 55; Pa State Univ, PhD(fuel sci, chem), 60. *Prof Exp:* Res asst chem, Pa State Univ, 55-59; sr scientist, 70-72, res assoc, 72-86, PROJ MGR DEVELOP ENG, AMP, INC, 86-; sr scientist, 70-72, res assoc, 72-86, PROJ DEVELOP ENG, AMP, INC, 86- *Mem:* Am Chem Soc. *Res:* Polymer modification; crosslinking; characterization and compounding; polymer composition; adhesives and sealants; adhesion to metals and plastics. *Mailing Add:* Advan Develop Lab PO Box 3608 AMP Inc 97-03 Harrisburg PA 17105-3608

SLYTER, ARTHUR LOWELL, b Havre, Mont, Oct 23, 41; m 64; c 2. REPRODUCTIVE PHYSIOLOGY, ANIMAL SCIENCE. *Educ:* Kans State Univ, BS, 64, PhD(reproductive physiol), 69; Univ Nebr, Lincoln, MS, 66. *Prof Exp:* PROF ANIMAL SCI & LIVESTOCK RES, S DAK STATE UNIV, 70- *Mem:* Am Soc Animal Sci; Soc Study Reproduction. *Res:* Reproductive physiology and efficiency in beef cattle and sheep. *Mailing Add:* Dept Animal Sci SDak State Univ Brookings SD 57007

SLYTER, LEONARD L, b Fontana, Kans, Nov 13, 33; m 48; c 2. MICROBIOLOGY, NUTRITION. *Educ:* Kans State Univ, BS, 55; Univ Mo, MS, 49; NC State Univ, PhD(animal nutrit), 63. *Prof Exp:* Res assoc bacteriol, Univ Ill, 62-64; res chemist animal sci, Res Div, USDA, 64-72, nutrit microbiol lab, Nutrit Inst, Agr Res Serv, 72-75, res chemist feed energy conserve lab, Animal Phys Genetics Inst, 75-78, RES MICROBIOL, RUMINAL NUTRIT LAB, ANIMAL SCI INST, AGR RES SERV, USDA, 79- *Mem:* Am Soc Microbiol; Am Soc Animal Sci. *Res:* Nutritional requirements, ecology and biochemical processes of microorganisms, particularly those involving ruminal bacteria and protozoa, pure and mixed cultures and continuous culture techniques. *Mailing Add:* 117 Periwinkle Ct Greenbelt MD 20770

SLYWKA, GERALD WILLIAM ALEXANDER, b Hafford, Sask, Apr 23, 39; m 65; c 4. ANALYTICAL CHEMISTRY, TOXICOLOGY. *Educ:* Univ Sask, BSP, 61, MSc, 63; Univ Alta, PhD(pharmaceut chem), 69; Univ Tenn, BS, 76; Cent Mich Univ, MA, 78. *Prof Exp:* Res chemist, Food & Drug Directorate, Ottawa, Ont, 63-64; lectr pharm, Univ Sask, 64-65; toxicologist, Crime Detection Lab, Royal Can Mounted Police, Regina, Sask, 69-71, head toxicol sect, Vancouver, BC, 71-72; asst prof med chem & head anal sect, Col Pharm, Univ Tenn Ctr Health Sci, Memphis, 72-75; assoc prof, 75-84, PROF MED CHEM, SCH PHARM, FERRIS STATE COL, 84- *Mem:* Int Asn Forensic Toxicologists; Am Chem Soc; Am Asn Poison Control Ctrs. *Res:* Bioavailability; instrumentation; drug metabolism; drug abuse. *Mailing Add:* 7630 Crestview Dr Reed City MI 49677

SMAGORINSKY, JOSEPH, b New York, NY, Jan 29, 24; m 48; c 5. DYNAMIC METEOROLOGY. *Educ:* NY Univ, BS, 47, MS, 48, PhD(meteorol), 53. *Hon Degrees:* DSc, Univ Munich, 72. *Prof Exp:* Asst & instr meteorol, NY Univ, 46-48; res meteorologist, US Weather Bur, 48-50; res meteorologist, Inst Adv Study, 50-53; head numerical weather prediction unit, Nat Weather Serv, Nat Oceanic & Atmospheric admin, 53-54, chief comput sect, Joint Numerical Weather Prediction Unit, 54-55, chief gen circulation res lab, 55-63, dir geophys fluid dynamics lab & dep dir meteorol res, 64-65, actg dir inst atmospheric sci, 65-66, dir geophys fluid dynamics lab, 65-83, CONSULT, NAT OCEANIC & ATMOSPHERIC ADMIN, 83-84. *Concurrent Pos:* Mem comt atmospheric sci, panel weather & climate modification, Nat Acad Sci, 63, interdept comt panel comput tech, 64, panel on pollution, Presidential Sci Adv Comt, 65; vis prof, 68-83, vis sr fel, Princeton Univ, 83-; officer, Int Joint Organizing Comt Global Atmospheric Res Prog, 68-80, chmn, Int Joint Sci Comt World Climate Res Prog, 80-81; vchmn, Nat Acad Sci-US Comt Global Atmospheric Res Prog, 67-73 & 80-87, officer, 74-77, mem, Climate Bd, 77-87, chmn, Climate Res Comt, 81-87; Brittingham vis prof, Univ Wis, 86. *Honors & Awards:* Gold Medal, US Dept Commerce, 66; Environ Sci Serv Admin Award, 70; Meisinger Award, Am Meteorol Soc, 67, Carl-Gustaf Rossby Res Medal, 72, Cleveland Abbe Award, 80, Charles Franklin Brooks Award, 91; Buys Ballot Medal, Royal Netherlands Acad Arts & Sci, 73; Int Meteorol Orgn Prize, World Meteorol Orgn, 74; Symons Mem Award, Royal Meteorol Soc, 81. *Mem:* Fel Am Meteorol Soc; hon mem Royal Meteorol Soc; fel Am Acad Arts & Sci, 86; Sigma Xi. *Res:* Geophysical fluid dynamics and thermodynamics; geophysical applications of high speed computers; atmospheric general circulation and theory of climate; atmospheric predictability. *Mailing Add:* 21 Duffield Pl Princeton NJ 08540-2605

SMAIL, JAMES RICHARD, b Youngstown, Ohio, Dec 6, 34; m 58; c 2. EMBRYOLOGY, MARINE BIOLOGY. *Educ:* Oberlin Col, AB, 57; Univ Ill, PhD(zool), 65. *Prof Exp:* Instr, 63-65, asst prof, 65-72, ASSOC PROF BIOL, MACALESTER COL, 72- *Mem:* AAAS; Am Inst Biol Sci; Int Oceanog Found. *Res:* Mechanism of hatching in birds, especially the role of the musculus complexus in the hatching process; changes in surface ultrastructure of sea urchin eggs during cleavage; ecology of coral. *Mailing Add:* Dept Biol Macalester Col 1600 Grand Ave St Paul MN 55105

SMALE, STEPHEN, b Flint, Mich, July 15, 30; m 54; c 2. MATHEMATICS. *Educ:* Univ Mich, BS, 52, MS, 53, PhD(math), 56. *Hon Degrees:* DSc, Univ Warwick, Eng, 74, Queens Univ, 87. *Prof Exp:* Instr math, Univ Chicago, 56-58; mem Inst Adv Study, 58-60; prof, Columbia Univ, 61-64; assoc prof, 60-61, PROF MATH, DEPT MATH, UNIV CALIF, BERKELEY, 64- *Concurrent Pos:* Alfred P Sloan res fel, 60-62; res prof, Miller Inst Basic Res Sci, Berkeley, 67-68, 79-80 & 90; colloquium lectr, Am Math Soc, 72; fac res lectr, Univ Calif, Berkeley, 83. *Honors & Awards:* Veblen Prize for Geom, Am Math Soc, 65; Fields Medal, Int Math Union , 66; Chauvenet Prize, Math Asn Am, 88; Von Neumann Award, Soc Indust & Appl Math, 89. *Mem:* Nat Acad Sci; Am Acad Arts & Sci; Int Union Math; Am Math Soc; fel Econometric Soc. *Res:* Differential topology; global analysis. *Mailing Add:* Dept Math Univ Calif Berkeley CA 94720

SMALL, ARNOLD MCCOLLUM, JR, b Springfield, Mo, Sept 16, 29; div; c 5. PSYCHOACOUSTICS. *Educ:* San Diego State Univ, BA, 51; Univ Wis, MS, 53, PhD(psychol), 54. *Prof Exp:* Res assoc, Mass Inst Technol, 51; asst psychol, Univ Wis-Madison, 51-54; asst prof psychol & dir bioelec lab, Lehigh Univ, 54-58; from asst prof to assoc prof, 58-64, PROF SPEECH PATH, AUDIOL & PSYCHOL, 64-, DIR DIV PHYS ED, UNIV IOWA, 86- *Concurrent Pos:* NIH fel, 54 & res grants, 56-58 & 68-73; NSF Res grants, 54-56, 60-66 & 77-80; Off Naval Res res grant, 54-58; vis scholar, Stanford Univ, 76-77, Univ Sydney, Australia, 89; consult, Vet Admin, 63-, Radio Corp Am, 66 & Cent Inst Deaf, 78-; assoc ed, J Speech & Hearing Res, 58-64 & J Acoust Soc Am, 70-77; ed, Am Speech & Hearing Asn Reports, 84-90. *Mem:* Fel AAAS; fel Am Speech & Hearing Asn; fel Acoust Soc Am; sr mem Inst Elec & Electronics Engrs; Psychonomic Soc. *Res:* Psycho-acoustics and physiological acoustics; psychological and physiological aspects of sensory processes; audition; computer science. *Mailing Add:* Dept Psychol Univ Iowa 127c SHC Iowa City IA 52242

SMALL, DONALD BRIDGHAM, b Philadelphia, Pa, May 25, 35; m 60; c 3. MATHEMATICS. *Educ:* Middlebury Col, BA, 57; Univ Kans, MA, 59; Univ Conn, PhD(math), 68. *Prof Exp:* Instr math, Univ Conn, 60-67; asst prof, Eastern Conn State Col, 67-68; asst prof, 68-74, chmn div nat sci, 76-80, ASSOC PROF MATH, COLBY COL, 74- *Concurrent Pos:* Dir, Maine High Sch Lect Prog; chmn, northeastern sect, Math Asn Am, 77-79, Calculus Articulation Panel, 82-86, gov, 82-85 & 88-91; vis prof, Harvey Mudd Col, 85-86, Claremont McKenna Col, 86, Pomona Col, 87. *Honors & Awards:* Cert Meritorious Serv, Math Asn Am. *Mem:* Am Math Soc; Math Asn Am. *Res:* Graph theory; curriculum development (computer algebra systems in calculus). *Mailing Add:* Dept Math Colby Col Waterville ME 04901

SMALL, ERNEST, b Ottawa, Ont, Mar 10, 40. BIOSYSTEMATICS. *Educ:* Carleton Univ, BA, 63, BSc, 65, MSc, 66; Univ Calif, Los Angeles, PhD(bot), 69. *Prof Exp:* RES BIOLOGIST, ECONOMIC PLANT SECT, CAN DEPT AGR, 69- *Honors & Awards:* G M Cooley Award, Am Soc Plant Taxon, 74. *Mem:* Am Soc Plant Taxon; Int Asn Plant Taxon; Linnean Soc London. *Res:* Systematics of cultivated plants. *Mailing Add:* 12 Gervin Nepean ON K2G 0J8 Can

SMALL, ERWIN, b Boston, Mass, Nov 28, 24. VETERINARY MEDICINE. *Educ:* Univ Ill, BS, 55, DVM, 57, MS, 65. *Prof Exp:* Intern, Angell Mem Hosp, Boston, Mass, 57-58; from instr to assoc prof, 58-67, head small animal med, Col Vet Med, 70-80, interim dept chair, 87-88, PROF VET CLIN MED, UNIV ILL, URBANA, 67-, ASSOC DEAN, ALUMNI & PUB AFFAIRS, 78-, DIR VET MED TEACHING HOSP, 88- *Concurrent Pos:* Fel, Nat Heart Inst, 65 & Morris Animal Found, 67-68. *Honors & Awards:* AAHA Award, 83. *Mem:* NY Acad Sci; Sigma Xi; Am Col Vet Dermat (pres, 79-82); Am Acad Vet Allergy (pres, 86-88); Nat Acad Pract Vet Med. *Res:* Serodiagnostic and immunologic approaches to the study of hemotrophic parasites. *Mailing Add:* Col Vet Med Univ Ill 1008 W Hazelwood Urbana IL 61801-4795

SMALL, EUGENE BEACH, b Reed City, Mich, Jan 7, 31; m 65; c 4. PROTISTOLOGY, MICROBIAL ECOLOGY. *Educ:* Wayne State Univ, BS, 53, MS, 56; Univ Calif, Los Angeles, PhD(zool), 64. *Prof Exp:* Asst prof zool, Univ Ill, 64-70; ASSOC PROF ZOOL, UNIV MD, COLLEGE PARK, 70- *Mem:* Soc Protozoologists (treas, 62-64); Am Micros Soc; AAAS; Soc Evolutionary Protistologists; Sigma Xi. *Res:* Protistan organisms: their morphology, morphogenesis, ecology, and evolutionary history; development of electron microscopical techniques applicable to the study of unicellular organisms; systematics of the phylum ciliophora. *Mailing Add:* Dept Zool Univ Md College Park MD 20742

SMALL, GARY D, b Atkinson, Nebr, Oct 17, 37; m 64; c 2. BIOCHEMISTRY. *Educ:* Nebr State Col, BS, 59; Western Reserve Univ, PhD(biochem), 65. *Prof Exp:* Nat Cancer Inst fel Dept Biochem, Univ Wash, 65-67; from asst prof to assoc prof, 67-76, PROF BIOCHEM, UNIV SDAK, 76- *Concurrent Pos:* Vis biochemist, Nat Cancer Inst fel, Brookhaven Nat Lab, 74-75. *Mem:* AAAS; Sigma Xi; Am Soc Biol Chemists; Am Soc Photobiol. *Res:* Enzymology of nucleases; effects of ultraviolet radiation on nucleic acids; biological repair of nucleic acid damage. *Mailing Add:* Dept of Biochem Univ of SDak Vermillion SD 57069

SMALL, HAMISH, b Antrim, Northern Ireland, Oct 5, 29; US citizen; m 54; c 2. PHYSICAL CHEMISTRY, ANALYTICAL CHEMISTRY. *Educ:* Queen's Univ, Belfast, BSc, 49, MSc, 52. *Prof Exp:* Chemist phys chem, Atomic Energy Res Estab, Eng, 49-55; chemist, 55-62, sr res chemist, 62-63, assoc scientist, 63-74, RES SCIENTIST PHYS CHEM, DOW CHEM CO, 74- *Honors & Awards:* Appl Anal Chem Award, Soc Anal Chemists Pittsburgh, 77; Albert F Sperry Medal, Instrument Soc Am, 78; A O Beckman Award, Instrument Soc Am, 83; Herbert H Dow Gold Medal, 83; Stephen Dal Nogare Chromatography Award, 84. *Mem:* Sigma Xi; Am Chem Soc. *Res:* Separation science; liquid chromatography; hydrodynamic chromatography; ion chromatography. *Mailing Add:* 4176 Oxford Dr Leland MI 49654

SMALL, HENRY GILBERT, b Chicago, Ill, June 17, 41; m 71; c 1. CITATION ANALYSIS & BIBLIOMETRICS, INFORMATION SCIENCE. *Educ:* Univ Ill, BA, 63; Univ Wis, MA, 66, PhD(hist sci & chem), 71. *Prof Exp:* Res assoc, Ctr Hist & Philos Physics, Am Inst Physics, 69-70, actg dir, 71-72; sr res scientist, 72-77, dir contract res, 77-80, DIR CORP RES, INST SCI INFO, 80- *Concurrent Pos:* Sr fel, dept hist & sociol sci, Univ Pa, 74-79; coun mem, Soc Social Studies Sci, 79-81; mem sci & arts comt, Franklin Inst, 85- *Mem:* Am Soc Info Sci; Soc Social Study Sci; Hist Sci Soc; fel AAAS. *Res:* Bibliometrics and especially citation and co-citation analysis, to study the structure and development of science; application of statistical methods, clustering and scaling to bibliometric data for mapping scientific fields; constructing science indicators and information retrieval. *Mailing Add:* Inst Sci Info 3501 Market St Philadelphia PA 19104

SMALL, IVER FRANCIS, b Sask, Can, Sept 19, 23; US citizen; m 54; c 4. PSYCHIATRY. *Educ:* Univ Sask, BA, 51; Univ Man, MD, 54; Univ Mich, MS, 60. *Prof Exp:* Asst prof psychiat, Med Sch & dir inserv psychiat, Hosp, Univ Ore, 60-62; dir inpatient serv psychiat, Malcolm Bliss Ment Health Ctr, 62-65; assoc prof, 65-69, PROF PSYCHIAT, SCH MED, IND UNIV, INDIANAPOLIS, 69-; ASST SUPT PSYCHIAT, LARUE D CARTER MEM HOSP, 65- *Concurrent Pos:* Asst prof psychiat, Sch Med, Washington Univ, 62-65, vis physician, Unit I, St Louis City Hosp, 6265. *Mem:* AMA; Soc Biol Psychiat (secy-treas, 70-); Am Psychiat Asn; Sigma Xi. *Res:* Work in clinical psychiatry; follow-up studies; the convulsive therapies; and the neuropsychology of mental illness. *Mailing Add:* Larue D Carter Mem Hosp 1315 W Tenth St Indianapolis IN 46202

SMALL, JAMES GRAYDON, b Seattle, Wash, Feb 10, 45; m 75. QUANTUM OPTICS, MEDICAL ULTRASOUND. *Educ:* Mass Inst Technol, BS, 67, PhD(physics), 74. *Prof Exp:* Lectr, Univ Ariz, 74-75, asst prof optical sci, 75-80, adj asst prof surg, 79-80; assoc prof physics & assoc dir, Inst Modern Optics, Univ NMex, 80-83; physics div leader, Tetra Corp, 83-84; SR SCIENTIST, HUGHES AIRCRAFT, 85- *Concurrent Pos:* Adj prof physics, Univ NMex, 88- *Mem:* Optical Soc Am. *Res:* Applications of stable lasers; special relativity. *Mailing Add:* Assoc Prof Physics Univ NMex Albuquerque NM 87131

SMALL, JOYCE G, b Edmonton, Alta, June 12, 31; US citizen; m 54; c 4. PSYCHIATRY. *Educ:* Univ Sask, BA, 51; Univ Man, MD, 56; Univ Mich, MS, 59; Am Bd Psychiat & Neurol, dipl, 61. *Prof Exp:* Intern, Winnipeg Gen Hosp, Man, 5556; resident psychiat, Ypsilanti State Hosp, Mich, 56-59; instr, Neuropsychiat Inst, Univ Mich, 59-60; from instr to asst prof, Sch Med, Univ Ore, 60-62; clin dir, Malcolm Bliss Ment Health Ctr, 62-65; asst prof psychiat, Sch Med, Washington Univ, 62-65; assoc prof, 65-69, PROF PSYCHIAT, SCH MED, IND UNIV, INDIANAPOLIS, 69-; CLIN DIR, LARUE D CARTER MEM HOSP, 65- *Concurrent Pos:* Teaching fel biochem, Univ Man, 55-56; res assoc neurol & psychiat consult, Crippled Children's Div, Med Sch, Univ Ore, 60-62; vis physician, St Louis City Hosps, 62-65; attend staff, Vet Admin Hosp, Indianapolis, 65-69, Univ Hosp Indianapolis, 74- & Wisford Mem Hosp, 79-; assoc mem, Inst Psychiat Res, 74- *Mem:* AAAS; fel Am Psychiat Asn; NY Acad Sci; Am Electroencephalog Soc; Soc Biol Psychiat. *Res:* Clinical psychiatry; electroencephalography; neurophysiology; psychopharmacology. *Mailing Add:* Larue D Carter Mem Hosp Indianapolis IN 46202

SMALL, LANCE W, b New York, NY, Apr 16, 41; m 65; c 2. NONCOMMUTATIVE RING THEORY. *Educ:* Univ Chicago, BS & MS, 62 & PhD(math), 65. *Prof Exp:* Asst prof math, Univ Calif Berkeley, 65-69; assoc prof math, Univ Southern Calif, 69-70; assoc prof, 70-74, PROF MATH, UNIV CALIF SAN DIEGO, 74- *Concurrent Pos:* Sr vis fel, Univ Leeds, Eng, 67-68, 72-73 & 84, hon vis fel, 78-; fel, Inst Advan Studies, Jerusalem, Israel, 77-78; mem coun, Am Math Soc, 83-87. *Mem:* Am Math Soc. *Res:* Noncommutative rings, particularly rings with polynomial identity and noetherian rings. *Mailing Add:* Dept Math Univ Calif San Diego La Jolla CA 92093

SMALL, LAVERNE DOREYN, b Black Earth, Wis, Dec 22, 16; m 38; c 2. MEDICINAL CHEMISTRY. *Educ:* Univ Minn, BS, 38, MS, 43, PhD(pharmaceut chem), 45. *Prof Exp:* Pharmacist, Walgreen Drug Co, Minn, 38-40; pharmacist, Johnson Co, Minn, 40-42; res chemist, Sterling-Winthrop Res Inst, 45-48; assoc prof pharmaceut chem, 48-54, prof & chmn pharm & pharmaceut chem, 54-73, prof med chem & pharmacog, 73-80, prof, 80-83, EMER PROF BIOMED CHEM, COL PHARM, UNIV NEBR, 83- *Mem:* Am Chem Soc; Am Pharmaceut Asn; Sigma Xi. *Res:* Synthesis and testing of compounds related to quinidine as cardiac antiarrhythmic agents; synthesis and testing of compounds related to allicin as antibacterial and antifungal agents. *Mailing Add:* 128 N 13th St Apt 705 Lincoln NE 68508-1501

SMALL, LAWRENCE FREDERICK, b St Louis, Mo, Feb 16, 34; m 63; c 3. BIOLOGICAL OCEANOGRAPHY. *Educ:* Univ Mo, AB, 55; Iowa State Univ, MS, 59, PhD(zool), 61. *Prof Exp:* Instr limnol, Iowa State Univ, 60-61; from asst prof to assoc prof, 61-72, PROF BIOL OCEANOG, ORE STATE UNIV, 72- *Concurrent Pos:* NSF fels, 61-64 & 70-72, res grant, 73-90; AEC fel, 63-66; USPHS training grant, 63-67, fel, 64-67; EPA res grant, 77-79, DOE grant, 81-91, Off Naval Res grant, 87-89; Int AEA spec serv res award, Monaco, 70-71, 77-78, 88 & 90; Nat Acad Sci res grant, Yugoslavia, 72; res grant Sea Grant, Nat Oceanic & Atmospheric Asn, 72 & 85-86; res contract, US Army Corps of Engrs, 74, Nat Aeronaut & Space Admin, 83. *Mem:* AAAS; Am Soc Limnol & Oceanog; Sigma Xi; Phycological Soc; Oceanog Soc. *Res:* Phytoplankton and zooplankton ecology and physiology; energy and material transfer in lower marine trophic levels. *Mailing Add:* Col Oceanog Ore State Univ Oceanog Admin Bldg 104 Corvallis OR 97331-5503

SMALL, PARKER ADAMS, JR, b Cincinnati, Ohio, July 5, 32; m 56; c 3. IMMUNOLOGY, MEDICINE. *Educ:* Univ Cincinnati, MD, 57; Tufts Univ, BS, 86. *Prof Exp:* Intern med, Univ Pa Hosp, Philadelphia, 57-58; res assoc immunol, USPHS, 58-60, surgeon, NIMH, 61-64, sr surgeon, Sect Phys Chem, 64-66; chmn dept, 66-76, PROF IMMUNOL & MED MICROBIOL, COL MED, UNIV FLA, 66-, PROF PEDIATRICS, 79- *Concurrent Pos:* USPHS res fel, 55 & spec fel, 60; res fel, Wright-Fleming Inst Microbiol, St Mary's Hosp Med Sch, 60-61; vis prof immunol, Univ Lausanne, 72, Univ Lagos, Nigeria, 82 & Al Hada Hosp, Saudi Arabia, 83; vis scholar, Asn Am Med Cols, 73; consult, Med Scientist Training Comt, Nat Inst Gen Med Sci & WHO; mem, bd dir, Biol Sci Curriculum Studies, 84-90, exec comt, 87-90; mem, ed adv comt, Nat Fund Med Educ, 84-87, Nat Bd Med Examiners Study comt to rev part I & II, 83-85, Nat Vaccine Adv Comt, 88-91; secy-treas, City of Oakland, Md, 64-65, mayor, 65-66; chmn, Citizens Pub Sch, Gainesville, Fla, 69-70; ed, Sec Innunol Systs, 71, Patient Oriented Prob Solving Syst Immunol, 82 & Pharmacol, 85; consult ed, ed bd, Microbios & Cytobios. *Mem:* AAAS; Fedn Am Scientists; Am Asn Immunologists; Sigma Xi; Physicians for Soc Responsibility; Am Soc Med; Am Med Asn. *Res:* Host defense against influenza; medical education. *Mailing Add:* Dept Immunol & Med Microbiol Univ Fla Col Med Box J266 Gainesville FL 32610

SMALL, ROBERT JAMES, b Philadelphia, Pa, Nov 23, 38; m 63; c 2. ORGANIC CHEMISTRY, PHOTOCHEMISTRY. *Educ:* Norwich Univ, BS, 61; Tex Tech Univ, MS, 64; Univ Ariz, PhD(org chem), 71. *Prof Exp:* Fel org chem, Univ Ky, 72-73; res chemist, Celanese Chem Co, 73-74; sr res chemist org chem, Ashland Chem Co, 74-80; proj leader, Ciba-Geigy Corp, 80-85; res dir, Wesley Industs, 86-90; SR RESEARCHER, FIRST CHEMICAL CORP, 90- *Concurrent Pos:* Co-owner, Data Phase. *Mem:* Am Chem Soc; Royal Soc Chemists; Am Inst Chemists; Sigma Xi. *Res:* Heterogenous and homogenous catalytic oxidation of olefins and heterogenous oxidative dehydrogenation of aliphatic systems; process development of reductive methylation of amines and hydrazine derivatives; photochemistry of oximes; paper dyes; hydroxamethylation. *Mailing Add:* 354 Teakwood Dr Satsuma AL 36572

SMALL, S(AUL) MOUCHLY, b New York, NY, Oct 11, 13; m 37; c 4. PSYCHIATRY. *Educ:* City Col New York, BS, 33; Cornell Univ, MD, 37; Am Bd Psychiat & Neurol, dipl. *Prof Exp:* Asst psychiat, Med Col & asst res psychiat, Inst Human Rels, Yale Univ, 38-39; from asst resident to resident psychiatrist, Payne Whitney Clin, Cornell Univ, 39-43, instr psychiat, 40-43; asst med dir & psychiat consult, Nat Hosp Speech Dis, 43-47; lectr psychiat, Columbia Univ, 48-51; prof psychiat & chmn dept Sch Med, State Univ NY, Buffalo, 51-78; head psychiat, Buffalo Gen Hosp, 63-78; EMER PROF PSYCHIAT, SCH MED, STATE UNIV NY, BUFFALO, 78-, EMER DIR, AM BD PSYCHIAT & NEUROL, 88- *Concurrent Pos:* Consult, Vassar Col, 43-47; psychiatrist, Rehab Ctr, NY Hosp, 44-46; from adj attend psychiatrist to assoc attend psychiatrist, Mt Sinai Hosp, 46-51, Clarence P Oberndorf vis psychiatrist, 66; consult to Surgeon Gen, US Army, 47-; consult, Buffalo Gen Hosp, 51-; dir psychiat, E J Meyer Mem Hosp, Buffalo, 51-68 & 74-78 med coordr, Dept Psychiat, 68-74; consult & lectr, US Vet Admin, chief consult & mem, Dean's Comt, Vet Admin Hosp, Buffalo, 52-; examr, US Info Agency, 59-62; chmn, Part III Comt, Patient Mgt Problems, Nat Bd Med Examrs, 60-; univ chief, dept Psychiat, Buffalo Children's Hosp, 68-; chmn, Sci Adv Coun, Muscular Dystrophy Asn, Inc; mem corp, Am Col Psychiatrists, Bd Regents; dir, Am Bd Psychiat & Neurol, mem, Part I Exam Comt; pres, Muscular Dystrophy Asn Inc, 80-89. *Honors & Awards:* Stockton Kimbell Fac Award, 65. *Mem:* AMA; Am Psychoanal Asn; Asn Am Med Cols; Am Psychosom Soc; fel Am Psychiat Asn. *Res:* Anorexia nervosa; psychology of chemical warfare; unconscious determination of vocational choice; stuttering; physiological validation of psychoanalytic theory; psychodynamic factors in surgery; psychopathology of alcoholism; hyperbaric oxygen effect on cognition and behavior in the aged; continuing medical education for psychiatrists; evaluation, self-assessment. *Mailing Add:* Erie County Med Ctr 462 Grider St Buffalo NY 14215

SMALL, TIMOTHY MICHAEL, b Muncie, Ind, Sept 29, 40; m 75; c 4. RESEARCH ADMINISTRATION, TECHNICAL MANAGEMENT. *Educ:* Ind Univ, BS, 63, MS, 64, PhD(physics), 68. *Prof Exp:* Res physicist, US Army Nuclear Effects Lab, Edgewood Arsenal, 68-70; actg mgr proj eng, Gen Elec Mgt & Tech Serv Dept, NASA Miss Test Facil, 70-71; res physicist, 71-78, develop proj officer, 78-81, tech asst, Mobility Equip Res & Develop Command, US Army, 81-83; tech base adminr, US Army Materiel Command, 83-86; DEP DIR RES & ANALYSIS, US ARMY FOREIGN SCI & TECHNOL CTR, 86- *Mem:* Am Phys Soc; AAAS; Sigma Xi. *Res:* Applications of nuclear physics to detection of objects. *Mailing Add:* RR Two Box Nine Scottsville VA 24590

SMALL, WILLIAM ANDREW, b Cobleskill, NY, Oct 16, 14; m 39; c 1. MATHEMATICS. *Educ:* US Naval Acad, BS, 36; Univ Rochester, AB, 50, AM, 52, PhD(math), 58. *Prof Exp:* Instr math, DeVeaux Sch, NY, 45-48; instr Univ Rochester, 51-55; asst prof, Alfred Univ, 55-56; asst prof, Grinnell Col, 56-58, assoc prof & chmn dept, 58-60; prof, Tenn Polytech Inst, 60-62; chmn dept, 62-78, prof math, 62-85, EMER PROF, STATE UNIV NY COL GENESEO, 85- *Concurrent Pos:* Fulbright-Hays lectr, Univ Aleppo, 64-65; assoc ed, Philosophia Mathematica. *Mem:* Math Asn Am. *Res:* Mathematical theory of probability; philosophy and history of mathematics; mathematical statistics. *Mailing Add:* 28 Court St Geneseo NY 14454

SMALLEY, ALFRED EVANS, b Chester, Pa, Feb 29, 28; c 2. ECOLOGY, ORNITHOLOGY. *Educ:* Pa State Univ, BS, 50, MS, 52; Univ Ga, PhD(zool), 59. *Prof Exp:* Instr biol, Univ Ky, 58-59; from instr to assoc prof, 59-75, PROF BIOL, TULANE UNIV LA, 75- *Mem:* Soc Syst Zool; Am Ornith Union; Crustacean Soc. *Res:* Ecology of aquatic ecosystems; marine invertebrate zoology; taxonomy of freshwater decapod crustacea. *Mailing Add:* Dept Biol Tulane Univ New Orleans LA 70118

SMALLEY, ARNOLD WINFRED, b Shreveport, La, Aug 2, 33; m 66. ORGANIC CHEMISTRY. *Educ:* Wiley Col, BS, 59; Univ Kans, MS, 62; Univ Mass, PhD(org chem), 65. *Prof Exp:* ASSOC PROF CHEM, SOUTHERN UNIV, BATON ROUGE, 65- *Mem:* Am Chem Soc. *Res:* Stabilities of metallocenyl substituted cations; electrophilic substitution reactions of metallocenes; quaternary phosphonium hydroxide decompositions; organic pollutants in municipal water supplies. *Mailing Add:* Dept Chem Southern Univ Box 9261 Baton Rouge LA 70813-9261

SMALLEY, EUGENE BYRON, b Los Angeles, Calif, July 11, 26; m 78; c 5. PLANT PATHOLOGY. *Educ:* Univ Calif, Los Angeles, BS, 49, Univ Calif, MS, 53, PhD(plant path), 57. *Prof Exp:* Asst plant path, Univ Calif, 53-56; asst prof plant path & forestry, 57-64, assoc prof plant path, 64-69, PROF PLANT PATH, UNIV WIS-MADISON, 69- *Mem:* Am Phytopath Soc; AAAS. *Res:* Forest pathology; vascular wilts of woody plants; diseases of garlic; mycotoxins; dutch elm disease; insect borne pathogens of trees. *Mailing Add:* Dept Plant Path Univ Wis 1630 Linden Dr Madison WI 53706

SMALLEY, GLENDON WILLIAM, b Bridgeton, NJ, Jan 23, 28; m 54; c 2. FOREST SOILS, SILVICULTURE. *Educ:* Mich State Univ, BS, 52, MS, 56; Univ Tenn, PhD, 75. *Prof Exp:* Forester, Sam Houston Nat Forest, Southern Region, US Forest Serv, 53-55, forester & asst dist ranger, Ouachita Nat Forest, 55-56, res forester, Southern Forest Exp Sta, Birmingham Res Ctr, 56-63, Silvicult Lab, 63-74, RES SOIL SCIENTIST SILVICULT LAB, SOUTHERN FOREST EXP STA, USDA FOREST SERV, SEWANEE, TENN, 74- *Mem:* Soc Am Foresters; Soil Sci Soc Am; Am Soc Agron; Ecol Soc Am. *Res:* Detailed planning, conducting, supervising and evaluating fundamental and applied research in forest soils for Cumberland Plateau and Highland Rim regions of Tennessee and Alabama. *Mailing Add:* Glenmary Farm Rabbitrun Lane Rte 1 Box 544 Sewanee TN 37375

SMALLEY, HARRY EDWIN, b Brooklyn, NY, Oct 23, 24; m 61; c 2. VETERINARY TOXICOLOGY. *Educ:* Trinity Univ, BS, 51, MS, 55; Tex A&M Univ, DVM, 59; Am Bd Vet Toxicol, dipl. *Prof Exp:* Biologist, Tex State Dept Health, Austin, 51-52; sanitarian, San Antonio Health Dept, 52-53; parasitologist, Grad Sch Med, Baylor Univ, 53-55; vet agr res serv, USDA, DC, 59-61, inspector chg, 61-63; res vet pharmacol div, US Food &

Drug Admin, 64-66; vet toxicol & entom res lab, 66-73, RES LEADER VET TOXICOL RES GROUP, AGR RES SERV, USDA, 73-, LAB DIR VET TOXICOL & ENTOM RES LAB, 74- *Concurrent Pos:* Area consult, Vector Control Surv, WHO-Pan-Am Sanit Bur, 52-53; panel mem, US Civil Serv Bd Exam Lab Animal Officers, 64-; tech adv agr res serv, USDA, 74-; consult ed, J Environ Qual, 75- *Mem:* Am Bd Vet Toxicol (secy-treas, pres, 73-76); AAAS; Am Col Vet Toxicol; Soc Toxicol; Am Vet Med Asn. *Res:* Action of pesticides and drugs on reproduction and teratology; comparative toxicology; ectoparasite pathology; disease vectors identification and control; research administration; toxicology of insect growth regulators. *Mailing Add:* 1114 Berkley Dr College Station TX 77840

SMALLEY, KATHERINE N, b Chicago, Ill, Oct 24, 35; m 59; c 2. ANIMAL PHYSIOLOGY, ENDOCRINOLOGY. *Educ:* Rockford Col, BA, 56; Univ Iowa, MS, 60, PhD(zool), 63. *Prof Exp:* Instr zool, Univ Iowa, 63-64; From asst prof to assoc prof, 66-82, PROF BIOL, EMPORIA STATE UNIV, 82- *Concurrent Pos:* Prin investr res grant, 64, 66, assoc prin investr res grant, NIH, 66-72; vis assoc prof, Univ Mich, 82. *Mem:* AAAS; Am Soc Zoologists; Sigma Xi. *Res:* Endocrinology of reproduction in amphibians, with emphasis on androgens and estrogens in females. *Mailing Add:* 1527 Rural Emporia KS 66801

SMALLEY, LARRY L, b Grand Island, Nebr, Aug 7, 37; m 57; c 3. GENERAL RELATIVITY, GRAVITATION. *Educ:* Univ Nebr, BS, 59, MS, 64, PhD, 67; Univ Ala Huntsville, BA, 87. *Prof Exp:* Instr physics, US Naval Nuclear Power Sch, Conn, 61-62; from asst prof to assoc prof, 67-80, chmn dept, 73-85, PROF PHYSICS, UNIV ALA, HUNTSVILLE, 80- *Concurrent Pos:* Asst reactor engr, Hallam Nuclear Power Fac, Nebr, 65; res physicist, Phys Sci Lab, Army Missile Command, Redstone Arsenel, 68; Nat Acad Sci/Nat Res Coun sr fel, NASA, 74-75; Humboldt fel, Univ Colgne, 76-77 & 80; space scientist, Space Sci Lab, Marshall Space Flight Ctr, 79-89; consult, Teledyne Brown Eng, Huntsville, Ala, 89-; mem, Strategic Defense Comn, Huntsville, Ala, 90. *Mem:* Am Phys Soc; Sigma Xi. *Res:* Theoretical physics, especially gravitational physics and discrete spacetime; fluid dynamics; impact machines. *Mailing Add:* Dept Physics Univ Ala Huntsville Huntsville AL 35899

SMALLEY, RALPH RAY, b Starkey, NY, Aug 26, 19; m 46; c 3. ORNAMENTAL HORTICULTURE, AGRONOMY. *Educ:* Cornell Univ, BS, 50, MS, 51; Univ FLa, PhD(soil fertil & hort corn), 61. *Prof Exp:* Asst prof soils & crops, State Univ NY Agr & Tech Col Farmingdale, 51-58; res asst turfgrass, Univ Fla, 58-61, turf technologist, 61-62; prof soils, field crops & turfgrass, State Univ NY Agr & Tech Col Cobleskill, 62-83; RETIRED. *Concurrent Pos:* Vis prof turfgrass, Cornell Univ, 73-74; Teacher fel, Nat Asn Cols & Teachers Agr, 82; consult soils & ornamental hort, 83- *Mem:* Am Soc Agron; Coun Agr Sci & Technol; fel Nat Asn Cols & Teachers Agr; Sigma Xi. *Res:* Effect of amendments on the physical and chemical properties of soil; turfgrass production and management; fine Bermudagrass response to physical soil amendments; turfgrass tillering; turfgrass irrigation. *Mailing Add:* RFD 1 Box 57H2 Howes Cave NY 12092

SMALLEY, RICHARD ERRETT, b Akron, Ohio, June 6, 43; m 68, 80; c 1. CHEMICAL PHYSICS. *Educ:* Univ Mich, BS, 65; Princeton Univ, MA, 71, PhD(chem), 73. *Prof Exp:* Res chemist, Shell Chem Co, 65-69; res assoc chem, Univ Chicago, 73-76; from asst prof to prof, 76-81, GENE & NORMAN HACKERMAN PROF CHEM, RICE UNIV, 82-, PROF PHYSICS, 90- *Concurrent Pos:* Harold W Dodds fel, Princeton Univ, 73 & Alfred P Sloan fel, 78-80; mem, Steering Comt, Rice Quantum Inst, 79-, chmn Inst, 86-; mem vis comt, Brookhaven Nat Lab, 83-84, Comt Atomic, Molecular & Optical Sci, Nat Res Coun, 88-91; chmn, Gordon Conf Metal & Semiconductor Clusters, 87. *Honors & Awards:* Irving Langmuir Prize Chem Physics, 91. *Mem:* Nat Acad Sci; fel Am Phys Soc; Am Chem Soc; Mat Res Soc; Sigma Xi; AAAS; Am Inst Physics. *Res:* Spectroscopic study of the unperturbed gas-phase structure and elementary chemical and photophysical processes of polyatomic molecules, radicals, and ions, including simple clusters of these with each other and with atoms; cluster structure and surface chemistry. *Mailing Add:* Dept of Chem Rice Univ Box 1892 Houston TX 77251

SMALLEY, ROBERT GORDON, b Chicago, Ill, June 1, 21; m 46; c 2. GEOLOGY, GEOCHEMISTRY. *Educ:* Univ Chicago, SB, 42, MS, 43, PhD, 48. *Prof Exp:* Geologist, US Geol Surv, 43-44; geologist, Stand Oil Co, Calif, 48-49, res geologist, La Habra Lab, Calif Res Corp, 49-59, sr res geologist, 59-66, SR RES ASSOC GEOCHEM, CHEVRON RES CO, 66- *Mem:* Geol Soc Am; Geochem Soc; Soc Appl Spectros. *Res:* Geochemistry of sediments and sedimentary rocks, carbonates and natural waters; petrology; igneous and metamorphic rocks; economic and petroleum geology. *Mailing Add:* PO Box 180 Lake Arrowhead CA 92352-0180

SMALLWOOD, CHARLES, JR, b Philadelphia, Pa, May 20, 20; m 44; c 3. SANITARY ENGINEERING. *Educ:* Case Western Reserve Univ, BS, 42; Harvard Univ, SM, 48. *Prof Exp:* Jr engr, Utilities Installation, US Eng Dept, Mich, 42-43; jr engr, Havens & Emerson, Consult Engrs, 46; asst, Harvard Univ, 46-48; from asst prof to assoc prof civil eng, NC State Univ, 50-58, grad adminr, 58-77, prof, 58-85, EMER PROF CIVIL ENG, NC STATE UNIV, 85- *Concurrent Pos:* Consult, Charles T Main, Inc, NC, 50- & J Harwood Beebe, SC, 56- *Mem:* Am Soc Civil Engrs; Am Water Works Asn; Water Pollution Control Fedn. *Res:* Biological treatment of wastes; hydrology; hydraulics; analysis of industrial wastes; radioactive wastes. *Mailing Add:* Dept Civil Eng NC State Univ Box 7908 Raleigh NC 27695-7908

SMALLWOOD, JAMES EDGAR, b Dallas, Tex, Oct 26, 45; m 67, 90; c 2. RADIOGRAPHIC ANATOMY, XERORADIOGRAPHY. *Educ:* Tex A&M Univ, DVM, 69, MS, 72. *Prof Exp:* From instr to assoc prof vet anat, Tex A&M Univ, 69-81; prof, 81-89, DISTINGUISHED PROF VET ANAT, NC STATE UNIV, 89- *Mem:* Am Vet Med Asn; Am Asn Vet Anatomists; World Asn Vet Anatomists. *Res:* Radiographic anatomy of domestic mammals and birds; skeletal development in the horse using xeroradiography. *Mailing Add:* APR Dept NC State Univ Col Vet Med 4700 Hillsborough Raleigh NC 27606

SMALLWOOD, RICHARD DALE, b Portsmouth, Ohio, Oct 9, 35; m 59; c 3. OPERATIONS RESEARCH, SYSTEMS ANALYSIS. *Educ:* Mass Inst Technol, SB, 57, SM, 58, ScD(elec eng), 62. *Prof Exp:* Lectr opers res, Mass Inst Technol, 62-64; asst prof eng-econ systs, Stanford Univ, 64-67, assoc prof, 67-73; res scientist anal res, Xerox Palo Alto Res Ctr, 73-79; PRES, APPL DECISION ANAL, INC, 79- *Concurrent Pos:* Consult prof, Dept Eng-Econ Systs, Stanford Univ, 73-; consult var govt & indust orgn. *Mem:* Inst Mgt Sci; Opers Res Soc Am; Inst Elec & Electronics Engrs. *Res:* Decision analysis; market analysis systems; modeling; man-machine systems; analysis of health care systems. *Mailing Add:* Appl Decision Analysis Inc 3000 Sand Hill Rd Bldg 4 Suite 255 Menlo Park CA 94025

SMALTZ, JACOB JAY, engineering computing; deceased, see previous edition for last biography

SMARANDACHE, FLORENTIN, b Balcesti, Vilcea, Romania, Dec 10, 54; US citizen; m 77; c 2. NUMERICAL FUNCTIONS IN NUMBER THEORY, PROPOSED PROBLEMS OF MATHEMATICS. *Educ:* Univ Craiova, Romania, MSc(math) & MSc(computer sci), 79. *Prof Exp:* Mathematician, IUG, Graiova, 79-81; prof math, Col Balcesti, 81-82, Sefrou, Morocco, 82-84, Nicolae Balcescu Col, Craiova, 84-85 & Sch Dragotesti, 85-86; pvt tutoring & math lessons for students, Craiova, 86-88 & pol refugee camp, Istanbul & Ankara, Turkey, 88-90; SOFTWARE ENGR/ COMPUTER SCIENTIST, HONEYWELL COM FLIGHT SYST GROUP, HONEYWELL, INC, PHOENIX, ARIZ, 90- *Concurrent Pos:* Coop prof, Lycde Sidi el Hassan Lyoussi, Sefrou, Morocco, 82-84; reviewer, number theory, Zentralblatt für Mathematik, Berlin, Ger, 85- *Mem:* Math Asn Am; Soc Statist Math Romania. *Res:* Function in the number theory called Smarandache Function and problems concerning it; propose original problems of general mathematics for journals and international competitions and olympiads for college students. *Mailing Add:* PO Box 42561 Phoenix AZ 85080

SMARDON, RICHARD CLAY, b Burlington, Vt, May 13, 48; m 72; c 2. ENVIRONMENTAL PLANNING, COASTAL ZONE PLANNING. *Educ:* Univ Mass, Amherst, BS, 70, MLA, 73; Univ Calif, Berkeley, PhD(environ planning), 82. *Prof Exp:* Assoc planner, Exec Off Environ Affairs, 73-75; environ impact assessment specialist, USDA Exten Serv, Oregon State Univ, 75-76; landscape architect, Pac Forest & Range Exp Sta, 77; res landscape architect, Dept Landscape Archit, Univ Calif, Berkeley, 77-79; assoc, Grad Prog Environ Sci, 80-82, SR RES ASSOC, COL ENVIRON SCI & FORESTRY, STATE UNIV NY, SYRACUSE, 82-, DIR INST ENVIRON POLICY & PLANNING, 86- *Concurrent Pos:* Adj asst prof, Dept Forestry & Wildlife Mgt, Univ Mass, 74-75; environ planner, US Geol Surv, Syracuse, NY, 80-82; consult, Ecology Compliance Ltd, Syracuse, NY, 81-; co-dir, Great Lakes Res Comm, 86-. mem, Great Lakes Adv Coun, NY; co-owner & vpres, Integrated Site, Inc, Syracuse, NY, 90- *Mem:* AAAS; Int Asn Impact Assessment; Landscape Res Group; Coastal Soc. *Res:* Landscape perception and visual resource management, integrated environmental planning for developing countries, coastal zone planning; numerious articles published in various journals. *Mailing Add:* Inst Environ Policy & Planning Col Environ Sci & Forestry Syracuse NY 13210

SMARDZEWSKI, RICHARD ROMAN, b Nanticoke, Pa, July 4, 42. ANALYTICAL CHEMISTRY. *Educ:* King's Col, Pa, BS, 64; Iowa State Univ, Ames, PhD(inorg chem), 69. *Prof Exp:* Sci Res Coun fel inorg chem, Univ Leicester, Eng, 69-70; NSF fel phys chem, Univ Va, Charlottesville, 71-72; Nat Res Coun res assoc, 72-74, res chemist inorg chem, 74-79; CHIEF SCIENTIST, US ARMY CHEM RES DEVELOP & ENG CTR, ABERDEEN PROVING GROUND, 79- *Concurrent Pos:* Adv, Nat Res Coun, Washington, DC. *Mem:* Am Chem Soc; Am Inst Chemists; Sigma Xi; Am Vacuum Soc. *Res:* Analytical chemistry. *Mailing Add:* 813 Maxwell Pl Bel Air MD 21014-3293

SMARR, LARRY LEE, b Columbia, Mo, Oct 16, 48; m 73; c 2. BLACK HOLES, RADIO JETS. *Educ:* Univ Mo, BA & MS, 70; Stanford Univ, MS, 72; Univ Tex, Austin, PhD(physics), 75. *Prof Exp:* Lectr astrophysics, Dept Astrophysical Sci, Princeton Univ, 74-75, res assoc, Observ, 75-76; res affil, Dept Physics, Yale Univ, 78-79; jr fel physics, Dept Physics & Astron, Harvard Soc Fellows, 76-79; from asst prof to assoc prof, 79-85, PROF ASTROPHYSICS, DEPT ASTRON & PHYSICS, UNIV ILL, 85-, DIR, NAT CTR SUPERCOMPUT APPLNS, 85- *Concurrent Pos:* Physicist, B Div, Lawrence Livermore Nat Lab, 76-79, consult, 76-; vis fel, Cambridge Univ, 78; assoc ed, J Comput Physic, 77-80; consult, Smithsonian Astrophys Observ, 79-81, Los Alamos Nat Lab, 83-; co-dir, Ill Alliance to Prevent Nuclear War, 81-84; Harlow Shapley Lectr, Am Astron Soc, 81-84; assoc ed, Computational Physics, 77-80; mem, NSF subcomt on computational Theoret Res, 81; Max-Planck Inst fel, 82-83; Nat Res coun, Comn Phys Sci Math & Resources, 87-; govt Univ Indust Res Roundtable, Nat Acad Sci, 87-; Alfred P Sloan Res fel, 80-84; Nat Sci Found fel, 70-73; Woodrow Wilson fel, 70. *Honors & Awards:* Lane Scholar, Univ Tex, 73-74. *Mem:* Am Phys Soc; Am Astron Soc; Int Soc Gen Relativity & Gravitation; AAAS. *Res:* Relativistic astrophysics; radio galaxies; numerical relativity; numerical hydrodynamics. *Mailing Add:* Astron Bldg Rm 341 Univ Ill 1011 W Springfield Urbana IL 61801

SMART, BRUCE EDMUND, b Philadelphia, Pa, Oct 9, 45; m 69. PHYSICAL ORGANIC CHEMISTRY. *Educ:* Univ Mo-Kansas City, BS(chem) & BS(math), 67; Univ Calif, Berkeley, PhD(chem), 70. *Prof Exp:* Staff scientist, 70-76, res supvr, 77-81, RES MGR, E I DU PONT DE NEMOURS & CO, INC, 82- *Mem:* Am Chem Soc (secy-treas, Div Fluorine Chem, 78-79); Sigma Xi; AAAS. *Res:* Organofluorine chemistry; small ring systems; carbonium ion chemistry; molecular rearrangements; thermochemistry; polymer chemistry. *Mailing Add:* Cent Res Dept Exp Sta E I du Pont de Nemours & Co Wilmington DE 19880

SMART, G N RUSSELL, b Montreal, Que, May 28, 21; m 46; c 3. ORGANIC CHEMISTRY. *Educ:* McGill Univ, BSc, 42, PhD(org chem), 45. *Prof Exp:* Asst, McGill Univ, 42-44, lectr chem, 44-45; Nat Res Coun Can fel, Univ Toronto, 45-46, res fel, Iowa State Univ, 46-47; from asst prof to prof, 47-78, head dept, 62-78, sr prof, 78-87, EMER PROF CHEM, MUHLENBERG COL, 87- *Concurrent Pos:* Indust consult; exec dir, Tuition Exchange, Inc, 72- *Honors & Awards:* Lindback Distinguished Teaching Award, 61. *Mem:* Sigma Xi; Am Asn Univ Prof; Am Chem Soc. *Res:* Stereochemistry; conformational analysis; organometallic and organosilicon chemistry; explosives. *Mailing Add:* 2219 Gordon St Allentown PA 18104

SMART, GROVER CLEVELAND, JR, b Stuart, Va, Nov 6, 29; m 57; c 2. AGRICULTURE, NEMATOLOGY. *Educ:* Univ Va, BA, 52, MA, 57; Univ Wis, PhD(plant path), 60. *Prof Exp:* Asst biol, Univ Va, 56-57; res asst plant path, Univ Wis, 57-60; asst prof plant path & physiol, Tidewater Res Sta, Va, Agr Exp Sta, Va Polytech Inst, 60-64; from asst prof to assoc prof, 64-73, asst chmn, 76-79, actg chmn, 79-80, asst chmn, Dept Entomol & Nematol, 80-81, PROF NEMATOL, INST FOOD & AGR SCI, UNIV FLA, 73 - *Concurrent Pos:* Co-ed, Nematology News Lett, 66-68, ed-in-chief, J Nematology, 72-74; chmn, Honors & Awards comt, Soc Nemotologists, 80-82, Archives comt, 83-88. *Mem:* Soc Nematologists; Sigma Xi; Soc Europ Nematol; Orgn Nematol Trop Am. *Res:* Control of insects using entomogenous nematodes; morphology of nematodes; biological control of plant parasitic nematodes. *Mailing Add:* Dept Entomol & Nematol Univ Fla Inst Food & Agr Sci Gainesville FL 32611-0740

SMART, JAMES BLAIR, b Des Moines, Iowa, Oct 6, 36; m 63; c 3. PHYSICAL INORGANIC CHEMISTRY, CLINICAL CHEMISTRY. *Educ:* Carroll Col (Mont), BA, 59; Univ Detroit, MS, 62; Wayne State Univ, PhD(inorg chem), 66. *Prof Exp:* Res assoc, Mich State Univ, 66-67; asst prof chem, Xavier Univ, 67-74; dir mfg, Nuclear Diag, Inc, 74-78, dir tech opers, 78-81, dir prod develop, 81-85; RES SCIENTIST, UNIV MICH MED SCH, 85- *Concurrent Pos:* Res Corp grant, 68-70. *Mem:* Am Chem Soc; Royal Soc Chem; Am Soc Qual Control; Soc Appl Spectros; Am Asn Clin Chemists. *Res:* Spectroscopic investigations of the effects on chemical and physical properties of through-space interactions between pi-systems and sigma-bonded organometallic compounds; applications of magnetic resonance spectroscopy; radioimmunoassay. *Mailing Add:* 4834 Whitman Circle Ann Arbor MI 48103-9439

SMART, JAMES CONRAD, inorganic chemistry, for more information see previous edition

SMART, JAMES SAMUEL, b New Bloomfield, Mo, Aug 31, 19; m 42. PHYSICS. *Educ:* Westminster Col, Mo, AB, 39; La State Univ, MS, 41; Univ Minn, PhD(physics), 48. *Prof Exp:* Instr aviation cadet eng, US Army Air Force, 41-43; physicist, Bur Ships, US Dept Navy, 43-46; asst physics, Univ Minn, 46-48; physicist, US Naval Ord Lab, 48-55; sci liaison officer, US Off Naval Res, Eng, 55-57, Wash, DC, 58-60; vis physicist, Brookhaven Nat Lab, 57-58; mem sr staff, Res Ctr, Int Business Mach Corp, 60-80; RETIRED. *Concurrent Pos:* Consult, US Off Naval Res, 58; mem, Nat Res Coun, 63-66; secy, Int Cong Magnetism, 67. *Mem:* Am Phys Soc; Am Asn Physics Teachers; Brit Inst Physics; Am Geophys Union. *Res:* Origin of chemical elements; magnetism; solid state physics; hydrology. *Mailing Add:* 71 Mt Airy Rd Croton-on-Hudson NY 10520-2126

SMART, JOHN RODERICK, b Laramie, Wyo, Sept 16, 34; m 59; c 5. NUMBER THEORY. *Educ:* San Jose State Col, AB, 56; Mich State Univ, MS, 58, PhD(math), 61. *Prof Exp:* From asst prof to assoc prof, 62-71, PROF MATH, UNIV WIS-MADISON, 71- *Concurrent Pos:* NSF fel, Courant Inst Math Sci, NY Univ, 61-62 & Glasgow Univ, 65-66. *Mem:* Am Math Soc; Math Asn Am. *Res:* Analytic number theory; automorphic and modular functions; discontinuous groups. *Mailing Add:* Dept Math Univ Wis 480 Lincoln Dr Madison WI 53706

SMART, KATHRYN MARILYN, microbiology, virology, for more information see previous edition

SMART, LEWIS ISAAC, b Nowata, Okla, Apr 1, 36; m 56; c 2. ANIMAL NUTRITION. *Educ:* Okla State Univ, BS, 60; Univ Ill, Urbana, MS, 62; Kans State Univ, PhD, 70. *Prof Exp:* Asst, Univ Ill, Urbana, 60-62; asst prof, Southern State Col Ark, 62-67; mem staff, Kans State Univ, 67-69; asst prof, 69-71, assoc prof, 71-77, PROF ANIMAL SCI, LA STATE UNIV, BATON ROUGE, 77- *Mem:* Am Soc Animal Sci. *Res:* Beef cattle nutrition and basic nutrition as related to animals. *Mailing Add:* 500 Chris St Natchitoches LA 71457

SMART, WESLEY MITCHELL, b San Francisco, Calif, Dec 12, 38; div; c 3. HIGH ENERGY PHYSICS. *Educ:* Univ Calif, Berkeley, BA, 61, MA, 65, PhD(high energy physics), 67. *Prof Exp:* Technician bubble chambers/high energy physics, Lawrence Berkeley Lab, Univ of Calif, 56-67; physicist bubble chambers/high energy physics, Stanford Linear Accelerator Ctr, Stanford Univ, 67-71; PHYSICIST BUBBLE CHAMBERS/HIGH ENERGY PHYSICS, FERMI NAT ACCELERATOR LAB, 71- *Mem:* Am Phys Soc. *Res:* Design, construction, and operation of the Fermi Lab bubble chamber and high energy physics experiments using bubble chambers. *Mailing Add:* Fermi Nat Accelerator Lab PO Box 500 Batavia IL 60510

SMART, WILLIAM DONALD, b Waukegan, Ill, Jan 26, 27; m 53; c 5. ORGANIC CHEMISTRY. *Educ:* Northwestern Univ, BS, 51; Univ Ill, MS, 53; Univ Chicago, MBA, 64. *Prof Exp:* Chemist, Abbott Labs, North Chicago, 53-60, group leader chem develop, 60-63, sect head, 63-64, mgr, 64-67, dir chem mfg, 67-69, vpres hosp equip mfg, 69-72, vpres, Mkt Hosp Prod Div, Abbott Labs, 72-75, vpres & gen mgr, Agr & Vet Prod Div, 75-76, exec vpres, Ross Labs Div, 76-80, pres, Ross Labs Div & vpres, Abbott Labs, 80-87; RETIRED. *Mem:* Am Chem Soc. *Mailing Add:* 3901 W Madura Rd Gulf Breeze FL 32561

SMAT, ROBERT JOSEPH, b Chicago, Ill, June 24, 38; div; c 2. ORGANIC CHEMISTRY. *Educ:* St Joseph's Col, Ind, BS, 60; Iowa State Univ, MS, 62; Ill Inst Technol, PhD(org chem), 66. *Prof Exp:* Res chemist, Chems, Dyes & Pigments Dept, Org Chem Div, Jackson Lab, E I Dupont de Nemours & Co, Inc, 66-72, patent chemist, 72-78, sr patent chemist, Elastomer Chem Dept, 78-80, patents consult, Polymer Prod Dept, 80-89, PATENT ASSOC, ELECTRONICS DEPT, E I DUPONT DE NEMOURS & CO, INC, 89- *Mem:* Am Chem Soc; The Chem Soc; Sigma Xi. *Res:* Synthesis and stereochemistry of small ring heterocyclic compounds; textile dyes and polymer finishes; thermoplastic film-forming polymers. *Mailing Add:* Electronics Dept Barley Mill Plaza 30-1236 E I du Pont de Nemours & Co Inc Wilmington DE 19880-0030

SMATHERS, GARRETT ARTHUR, b Canton, NC, Mar 15, 26; m 56; c 2. PLANT ECOLOGY, SCIENCE ADMINISTRATION. *Educ:* Univ NC, Asheville, dipl, 50; Furman Univ, BS, 52; Western Carolina Univ, MA, 55; Univ Hawaii, PhD(plant ecol), 72. *Prof Exp:* Chemist, Taylor Colquitt Co, 52-53, Am Enka Corp, 54-55; high sch teacher sci, Waynesville, NC, 53-54, 55-59, actg prin, 55-59; supvry park naturalist & res, Nat Park Serv, 59-66; res asst plant ecol, Univ Hawaii, 66-67; res biologist, Hawaii Volcanoes Nat Park, Nat Park Serv Coop Park Studies Unit, western Carolina Univ, 67-70, instr, Mather & Albright Training Ctrs, 70-71, regional chief scientist, Univ Wash, Pac Northwest Region, 70-73, chief scientist, Res Sci Admin, Nat Park Serv Sci Ctr, 73-77, adj prof biol, 80-83, sr res scientist, 77-83; RETIRED. *Concurrent Pos:* Mem, Adv Coun, Dept Forestry, Miss State Univ, 74-; prof B-1 coordr, Mem Org Preserves US-USSR Bilateral Agreement, Nat Park Serv, 75-; mem, NC Forestry Coun, 80; mem, NC environ mgt comn, 83; adj prof, Univ NC, Asheville, 83-; mem, NC Environ Mgt Comn, 85-89. *Mem:* Am Ecol Soc; Am Inst Biol Sci. *Res:* Study of the invasion, succession and recovery of vegetation on volcanic substrates; preparation of ecological atlases of present and proposed national parks; phytogeography of southern Appalachians; watershed drinking water supply, western North Carolina; environmental management and planning; environmental sciences; hydrology and water resources. *Mailing Add:* Six Briarknoll Ct Asheville NC 28803

SMATHERS, JAMES BURTON, b Prairie du Chieu, Wis, Aug 26, 35; m 57; c 4. MEDICAL PHYSICS, RADIATION SAFETY. *Educ:* NC State Col, BNE, 57, MS, 59; Univ Md, PhD(nuclear eng), 67. *Prof Exp:* Res engr, Atomics Int, 59; sect chief, Walter Reed Army Inst Res, 61-67; prof nuclear eng, Tex A&M Univ, 67-80, prof bioeng, 76-80; PROF RADIATION ONCOL, DEPT RADIATION ONCOL, UNIV CALIF, LOS ANGELES, 80- *Concurrent Pos:* Consult, Nat Cancer Inst, 82- *Mem:* Am Asn Physicists Med; Am Nuclear Soc; Health Physics Soc; Am Col Med Physics; Am Col Radiol; Am Soc Therapeut Radiation Oncol. *Res:* Applications of radiation to medicine and biology. *Mailing Add:* 18229 Minnehaha St Northridge CA 91326

SMATRESK, NEAL JOSEPH, b Worcester, Mass, July 9, 51; m 78; c 2. RESPIRATION PHYSIOLOGY, SENSORY PHYSIOLOGY. *Educ:* Gettysburg Col, BA, 73; State Univ NY, Buffalo, MA, 78; Univ Tex, Austin, PhD(zool), 80. *Prof Exp:* Trainee physiol, Univ Pa, 80-82; asst prof, 82-88, ASSOC PROF BIOL, UNIV TEX, ARLINGTON, 88- *Mem:* Am Physiol Soc; Am Soc Zool; Am Fisheries Soc; AAAS; Sigma Xi. *Res:* Control and coordination of respiration and heart rate in lower vertebrates; oxygen sensitive chemoreceptors; neural control of ventilation in fish and bimodal breathers. *Mailing Add:* Dept Biol Univ Tex-Arlington Box 19498 Arlington TX 76019

SMAY, TERRY A, b Oakland, Iowa, Aug 30, 35; m 54; c 3. ELECTRICAL ENGINEERING. *Educ:* Iowa State Univ, BS, 57, MS, 59, PhD(elec eng), 62. *Prof Exp:* Elec engr, Remington Rand Univac Div, Sperry Rand Corp, 57-58; from instr to asst prof elec eng, Iowa State Univ, 58-62; sr res scientist, Res Div, Control Data Corp, 62-65, supvr govt systs div, 65-66, dept mgr thin film memory develop, 66-70; assoc prof, 70-76, PROF ELEC ENG, IOWA STATE UNIV, 76- *Concurrent Pos:* Lectr, Univ Minn, 62-64. *Mem:* Inst Elec & Electronics Engrs. *Res:* Magnetic film memory development; high speed memory components. *Mailing Add:* Dept Elec Eng Iowa State Univ 213 Coover Ames IA 50011

SMAYDA, THEODORE JOHN, b Peckville, Pa, Aug 28, 31; m 56; c 2. BIOLOGICAL OCEANOGRAPHY. *Educ:* Tufts Univ, BS, 53; Univ RI, MS, 55; Univ Oslo, Dr Philos, 67. *Prof Exp:* Asst biol oceanogr, 59-61, asst prof biol oceanog, 61-66, assoc prof oceanog, 66-70, PROF OCEANOG & BOT, GRAD SCH OCEANOG, UNIV RI, 70- *Concurrent Pos:* Mem working panel phytoplankton methods, comt oceanog, Nat Acad Sci, 65-69; adv comt algae, Smithsonian Oceanog Sorting Ctr, 71-74; mem SCOR Working Group 33, phytoplankton. *Mem:* Fel AAAS; Am Soc Limnol & Oceanog; Phycol Soc Am; Int Phycol Soc; Marine Biol Asn; Plankton Soc Japan. *Res:* Ecology and physiology of marine phytoplankton; estuarine and coastal ecology; tropical and upwelling ecology; red tides, noxious blooms. *Mailing Add:* Dept Bot Univ RI Kingston RI 02881

SMEACH, STEPHEN CHARLES, b Hanover, Pa, Feb 25, 45; m 65; c 2. BIOMATHEMATICS. *Educ:* Univ Del, BS, 67; NC State Univ, MA, 70, PhD(biomath), 73. *Prof Exp:* NIH fel biomath, Biomath Prog, NC State Univ, 67-73; asst prof math, Univ South Fla, 73-80; WITH DEPT SCI EVAL, G D SEARLE & CO, 80- *Mem:* Soc Math Biol; Biomet Soc; Am Statist Asn; Sigma Xi. *Res:* Mathematical modeling of biological phenomenon, specifically statistical and stochastic approaches to include time series studies of population density behavior and stochastic models for drug concentration changes in the blood plasma. *Mailing Add:* Dept Sci Eval G D Searle & Co PO Box 1045 Skokie IL 60076

SMEAL, PAUL LESTER, b Clearfield, Pa, June 11, 32; m 54; c 3. ORNAMENTAL HORTICULTURE, NURSERY PRODUCTION. *Educ:* Pa State Univ, BS, 54; Univ Md, MS, 58, PhD(hort), 61. *Prof Exp:* Res asst hort, Univ Md, 59-60; from asst prof to assoc prof, 60-67, PROF HORT, VA POLYTECH INST & STATE UNIV, 67- *Concurrent Pos:* Instr, Flower

Show Sch, Nat Coun State Garden Clubs; secy-treas, Southern Region, Am Soc Hort Sci. *Honors & Awards:* Carl S Brittner Exten Award, Am Soc Hort Sci; Nursery Exten Award, Am Asn Nurserymen. *Mem:* Fel Am Soc Hort Sci; Int Plant Propagators Soc. *Res:* Ornamentals; plant propagation; nursery management; production of nursery stock, greenhouse flowers and bedding plants produced by nurserymen and flower growers; work with county extension personnel, nurserymen and professional grounds management personnel. *Mailing Add:* Dept Hort Va Polytech Inst & State Univ Blacksburg VA 24061-0327

SMEBY, ROBERT RUDOLPH, b Chicago, Ill, Dec 24, 26; m 50; c 4. BIOCHEMISTRY. *Educ:* Univ Ill, BS, 50; Univ Wis, MS, 52, PhD(biochem), 54. *Prof Exp:* Biochemist, R J Reynolds Tobacco Co, 54-56; biochemist, Miles-Ames Res Lab, Miles Labs, Inc, 56-59; BIOCHEMIST CLEVELAND CLIN, 59- *Concurrent Pos:* Adj prof, John Carroll Univ, 65-70 & Cleveland State Univ, 70-88; pres, bd trustees, Scientists Ctr Animal Welfare, 91. *Mem:* Fel AAAS; Am Chem Soc; Am Physiol Soc; fel Am Heart Asn; fel Am Inst Chemists. *Res:* Synthesis of peptides; isolation of substances from natural products with biological activity. *Mailing Add:* Res Inst 9500 Euclid Ave Cleveland OH 44195

SMECK, NEIL EDWARD, b Lancaster, Ohio, July 9, 41; m 65; c 3. AGRONOMY. *Educ:* Ohio State Univ, BS, 63, MS, 66; Univ Ill, PhD(agron), 70. *Prof Exp:* Soil scientist soil surv, Soil Conserv Serv, 63-66; asst prof agron, NDak State Univ, 69-71; ASSOC PROF AGRON, OHIO STATE UNIV-OHIO AGR RES & DEVELOP CTR, 71- *Mem:* Am Soc Agron; Soil Sci Soc Am; Soil Conserv Soc Am. *Res:* Soil genesis, morphology, and classification; sediment chemistry and mineralogy; weathering strip mine spoils. *Mailing Add:* Dept Agron Ohio State Univ Kottman Hall Columbus OH 43210

SMEDES, HARRY WYNN, b Spokane, Wash, Sept 11, 26; m 45; c 4. GEOLOGY, RESOURCE MANAGEMENT. *Educ:* Univ Wash, BS, 48, PhD(geol), 59. *Prof Exp:* Instr geol, Kans State Col, 51-53; geologist, Mineral Deposits Br, US Geol Surv, 53-61, Northern Rocky Mountains Br, 61-69 & Rocky Mountain Br Environ Geol, Colo, 69-75, staff geologist, Dept Interior Resource & Land Invest, 72-76, proj chief & coordr, 76-80; AT OFF WASTE ISOLATION, DEPT ENERGY, 80- *Concurrent Pos:* Mem, Boulder Batholith Proj, 53-72; chief area pub unit, US Geol Surv, 64-66, chief Absaroka volcanics proj, Yellowstone Nat Park, 66-70, chief geol mapping res proj, Rocky Mountain Br Environ Geol & remote sensing studies for Landsat & Skylab Prog, 69-75; mem geol verification team for US-Russian Peaceful Nuclear Explosives Treaty, 76- *Honors & Awards:* AIL Award, Am Soc Photogram, 75. *Mem:* Fel Geol Soc Am. *Res:* Mapping techniques; automated cartography. *Mailing Add:* Two Cool Brook Irvine CA 92715

SMEDFJELD, JOHN B, b Oslo, Norway, Apr 18, 35; US citizen; m 58; c 2. AEROELASTICITY, STRUCTURAL DYNAMICS. *Educ:* Pratt Inst, BME, 55, MS, 59. *Prof Exp:* Res engr, Grumman Aerospace Corp, 55-57, dynamic anal engr, 57-63, adv develop group leader dynamics, 63-65, aeroelasticity methods group leader, 65-72, head dynamic structural methods group, 72-77, TECH SPECIALIST STRUCT MECH, GRUMMAN AIRCRAFT SYST DIV, BETHPAGE, 77- *Mem:* Am Inst Aeronaut & Astronaut. *Res:* Gust response analysis; flutter analysis; unsteady aerodynamics; applied leads; structural optimization. *Mailing Add:* Nine Darrell St East Northport NY 11731

SMEDLEY, WILLIAM MICHAEL, b Chicago, Ill, Aug 2, 16; m 43; c 1. ORGANIC CHEMISTRY. *Educ:* Northwestern Univ, BS, 38, MS, 40. *Prof Exp:* Instr chem & physics, 40-47, from asst prof to assoc prof chem, 48-59, PROF CHEM, US NAVAL ACAD, 60- *Concurrent Pos:* Asst prof, Univ Md, 56-58, NSF fac fel, 64-65; dir res & vpres, Appl Sci & Chem Corp, Md & Am Chem Co; dir & vpres, Everett Factories, Mass. *Mem:* AAAS; Am Chem Soc. *Res:* Chlorination dioxanes; 2, 5-diphenyl dioxane, 2, 5-dichlorodioxane and derivatives; heterocyclics; derivatives of quinoline-quinone; chemistry of explosives. *Mailing Add:* 5017 Riverdale Rd Riverdale MD 20737-1915

SMEDSKJAER, LARS CHRISTIAN, b Copenhagen, Denmark, Oct 3, 44; m 75. EXPERIMENTAL PHYSICS, METALLURGY. *Educ:* Tech Univ Denmark, Cand Polyt, 69, PhD(exp physics), 72. *Prof Exp:* Amanvensis physics, Tech Univ Denmark, 72-74, assoc prof, 74; PHYSICIST METALL, ARGONNE NAT LAB, 74- *Mem:* Danish Soc Engrs. *Res:* Metal physics; positron physics. *Mailing Add:* 2726 63rd St Woodridge IL 60517

SMEINS, FRED E, b Luverne, Minn, Feb 14, 41; m 60; c 2. PLANT ECOLOGY. *Educ:* Augustana Col, SDak, 63; Univ Sask, MA, 65, PhD(plant ecol), 67. *Prof Exp:* Asst prof biol, Univ NDak, 67-69; from asst prof to assoc prof, 69-80, PROF RANGE SCI, TEX A&M UNIV, 80- *Res:* Ecology of wetland and grassland vegetation. *Mailing Add:* 2005 Nueces Dr College Station TX 77840-4844

SMELLIE, ROBERT HENDERSON, JR, b Glasgow, Scotland, June 2, 20; US citizen; m 45; c 3. PHYSICAL CHEMISTRY. *Educ:* Trinity Col, Conn, BS, 42, MS, 44; Columbia Univ, PhD(phys chem), 51. *Prof Exp:* Instr chem, Trinity Col, Conn, 43-44; anal foreman & supvr, Tenn Eastman Corp, 44-46; asst chem, Columbia Univ, 47-48; from instr to prof, 48-64, chmn dept, 63-71, SCOVILL PROF CHEM, TRINITY COL, CONN, 64- *Mem:* Am Chem Soc. *Res:* Kinetics of nitrile reactions; analytical chemistry of uranium; chemistry and electrokinetics of sulfur sols; flocculation of suspensions; combustion chemistry. *Mailing Add:* 69 Montclair Dr West Hartford CT 06107-2447

SMELT, RONALD, aerodynamics, for more information see previous edition

SMELTZER, DALE GARDNER, research systems, agricultural development; deceased, see previous edition for last biography

SMELTZER, RICHARD HOMER, b Sapulpa, Okla, Aug 14, 40; m 63; c 3. PLANT MORPHOGENESIS. *Educ:* Okla State Univ, BS, 62; Stephen F Austin State Univ, MF, 70; Lawrence Univ, MS, 72, PhD(paper chem), 75. *Prof Exp:* Inspector lumber, Southern Pine Inspection Bur, 62-65; asst qual control supt, Temple Indust, Inc, 65-68; RES ASSOC PLANT MORPHOGENESIS, INT PAPER CO, 75- *Mem:* Int Plant Propagation Soc; Int Asn Plant Tissue Cult. *Res:* Asexual propagation of trees through tissue culture and related biochemistry; evaluation of vegetative propagules. *Mailing Add:* Int Paper Co Rte 1 PO Box 471 Bainbridge GA 31717

SMELTZER, WALTER WILLIAM, b Moose Jaw, Sask, Dec 4, 24; m 60. METALLURGY, MATERIALS SCIENCE. *Educ:* Queen's Univ, Ont, BSc, 48; Univ Toronto, PhD(phys chem), 53. *Hon Degrees:* Dr, Univ Dijon, 81. *Prof Exp:* Res chemist, Nat Res Coun Can, 48-50; res chem engr, Aluminum Co, Ltd, 53-55; res metall engr, Metals Res Lab, Carnegie Inst Technol, 56-59; from asst prof to prof, 59-91, EMER PROF METALL, MCMASTER UNIV, 91- *Concurrent Pos:* Sr fel, Brit Res Coun, 76 & NATO, 79-80. *Honors & Awards:* Centennial Medal for Serv to the Nation, Govt Can, 68,; Albert Saveur Achievement Award, Am Soc Metals, 86. *Mem:* Nat Asn Corrosion Engrs; Electrochem Soc; fel Am Soc Metals; fel Royal Soc Can. *Res:* Adsorption and oxidation kinetics of metals; thermodynamic properties of solids; lattice defect structures of solid metal oxides and their influence on mass and thermal transport properties. *Mailing Add:* Dept Mat Sci & Eng McMaster Univ Hamilton ON L8L 4K1 Can

SMERAGE, GLEN H, b Topsfield, Mass, May 3, 37; m 60; c 2. SYSTEMS ENGINEERING. *Educ:* Worcester Polytech Inst, BS, 59; San Jose State Col, MS, 63; Stanford Univ, PhD(elec eng), 67. *Prof Exp:* Res asst & test engr, Instrumentation Labs, Mass Inst Technol, 59-60; jr engr, Western Develop Labs, Philco Corp, 60-61; engr, Electronic Defense Labs, Sylvania Electronics Systs, 62-67; asst prof elec eng, Utah State Univ, 67-76; ASSOC PROF AGR ENG, UNIV FLA, 76- *Mem:* Inst Elec & Electronics Engrs. *Res:* Modeling and analysis of systems; current emphasis on human social systems and their interaction with the physical and biological environment. *Mailing Add:* 2104 NW 15 Ave Gainesville FL 32605

SMERDON, ERNEST THOMAS, b Ritchey, Mo, Jan 19, 30; m 51; c 3. CIVIL & AGRICULTURAL ENGINEERING. *Educ:* Univ Mo, BS, 51, MS, 56, PhD(agr eng), 59. *Prof Exp:* Res engr, Univ Mo, 56-57, instr civil eng, 57-58, instr agr eng, 58-59; from assoc prof to prof, Tex A&M Univ, 59-68, prof civil eng & dir, Water Res Inst, 64-68; prof agr eng & chmn dept, Univ Fla, 68-74, asst dean res, 74-76; vchancellor acad affairs, Univ Tex Syst, Univ Tex, Austin, 76-82, prof civil eng & dir, Ctr Res Water Resources, 82-87; DEAN, COL ENG & MINES, UNIV ARIZ, TUCSON, 88- *Concurrent Pos:* Inst Int Educ & Ohio State Univ consult, Punjab Agr Univ, India, 65; consult govt of SVietnam, Guyana & El Salvador, 71, Bahamas, Peru & Brazil, 76 & USAID, Pakistan, 81; chmn, Univ Coun Water Resources, 72-74. *Mem:* Fel AAAS; Am Soc Agr Engrs; Am Soc Civil Engrs; Am Geophys Union; Am Asn Higher Educ. *Res:* Water resources development and irrigation; energy use and conservation; research administration. *Mailing Add:* Civil Eng Bldg 72 Tucson AZ 85721

SMERDON, MICHAEL JOHN, MOLECULAR BIOLOGY. *Educ:* Ore State Univ, PhD(biochem & biophysics), 76. *Prof Exp:* ASSOC PROF BIOCHEM & BIOPHYS, WASH STATE UNIV, 84-, ASSOC IN GENETICS & CELL BIOL, 80- *Res:* Repair of DNA damage by carcinogens; structure and function of DNA in human cells. *Mailing Add:* Biochem & Biophysics Prog Wash State Univ Pullman WA 99164-4660

SMERIGLIO, ALFRED JOHN, b Port Chester, NY, May 17, 37; m 65. BIOLOGY, COMPARATIVE ANATOMY. *Educ:* NY Univ, BS, 59, MA, 60, EdD(biol), 64. *Prof Exp:* From instr to asst prof biol, NY Univ, 59-66; asst prof, Jersey City State Col, 66-67; from asst prof to assoc prof, 67-70, PROF BIOL, NASSAU COMMUNITY COL, 75-, CHMN DEPT ALLIED HEALTH SCI, 70- *Concurrent Pos:* Asst res scientist biophys res lab, NY Univ, 65- *Mem:* AAAS; Nat Sci Teachers Asn. *Res:* Mammalian biology; physiological patterns of behavior in the Albino rat; piezoelectric properties of mineralized tissue. *Mailing Add:* Dept of Allied Health Sci Nassau Community Col Garden City NY 11530

SMETANA, ALES, b Hradec Kralove, Czech, Apr 4, 31; m 57; c 3. ENTOMOLOGY, ZOOGEOGRAPHY. *Educ:* Charles Univ, Prague, MD, 56; Czech Acad Sci, CSc(systs Anoplura), 60. *Prof Exp:* Res scientist, Inst Parasitol, Czech Acad Sci, 56-70; res scientist, Nat Mus, Prague, 70-71; RES SCIENTIST, BIOSYSTS RES INST, CAN DEPT AGR, 71- *Concurrent Pos:* Nat Res Coun Can fel, Entom Res Inst, Ottawa, 67-69. *Mem:* NY Acad Sci; Entom Soc Can; Entom Soc Am. *Res:* Systematics of the insect order Coleoptera, especially aquatic Coleoptera and the family Staphylinidae; zoogeography, particularly the holarctic distribution of Coleoptera. *Mailing Add:* Biosysts Res Centre Ottawa ON K1A 0C6 Can

SMETANA, FREDERICK OTTO, b Philadelphia, Pa, Nov 29, 28; m 52; c 4. MECHANICAL ENGINEERING, AEROSPACE ENGINEERING. *Educ:* NC State Col, BME, 50, MSME, 53; Univ Southern Calif, PhD(eng), 61. *Prof Exp:* Vpres, Philcord Corp, NC, 50-51; flight test analyst, Douglas Aircraft Co, Calif, 51-52; teaching asst mech eng, NC State Col, 52-53; res scientist, Eng Ctr, Univ Southern Calif, 55-62; assoc prof 62-65, PROF MECH ENG,NC STATE UNIV, 65- *Concurrent Pos:* Asst dir, NC Sci & Technol Res Ctr, 66-84; past consult, Pneumafil Corp, Litton Systs, Inc, Corning Glass Works, Waste King Corp & Servomechanisms, Inc; consult, US Army Armament Res & Develop Command, 76-79, US Army Res Off, 84-85. *Mem:* Am Inst Aeronaut & Astronaut. *Res:* Vehicle design; air data instrumentation; dynamic response flight testing; flight data systems identification; Rankine cycle solar electric power generation. *Mailing Add:* Dept Mech & Aerospace Eng NC State Univ Raleigh NC 27695-7910

SMETHIE, WILLIAM MASSIE, JR, b Rocky Mount, NC, Mar 29, 45; m 69; c 2. CHEMICAL & PHYSICAL OCEANOGRAPHY. *Educ:* Wofford Col, BS, 67; San Jose State Univ, MA, 73; Univ Wash, PhD(oceanog), 79. *Prof Exp:* Sr oceanographer, Univ Wash, 78-79; post doctoral scientist, Lamont-Doherty Geol Observ, 79-80, assoc res scientist, 80-87, res scientist, 87-90, SR RES sciENTIST, LAMONT-DOHERTY GEOL OBSERV, 90- *Mem:* Am Geophys Union; AAAS; Oceanog Soc. *Res:* Investigation of mixing, circulation, and water mass formation in the ocean and other marine systems from the distribution of naturally occurring and man-made substances. *Mailing Add:* Lamont-Doherty Geol Observ Palisades NY 10964

SMIBERT, ROBERT MERRALL, II, b New Haven, Conn, Dec 9, 30; m 61; c 2. MICROBIOLOGY. *Educ:* Univ Conn, BA, 52; Univ Md, MS, 57, PhD(microbiol), 59. *Prof Exp:* Instr microbiol, Sch Med, Temple Univ, 59-60; assoc prof vet sci, 60-65, prof microbiol, 68-78, prof bact, Anaerobe Lab, 78-80, PROF MICROBIOL, DEPT ANAEROBIC MICROBIOL, VA POLYTECH INST & STATE UNIV, 80- *Mem:* AAAS; Am Soc Microbiol; Soc Gen Microbiol; Am Venereal Dis Asn. *Res:* Microbial physiology and nutrition; taxonomy of bacteria; Mycoplasma; vibrios; campylobacter; leptospirosis; Treponemas; Borrelias. *Mailing Add:* 904 Buchanan Dr Blacksburg VA 24060

SMID, JOHANNES, b Amsterdam, Netherlands, Jan 18, 31; m 56; c 4. PHYSICAL CHEMISTRY, POLYMER CHEMISTRY. *Educ:* Free Univ, Amsterdam, BSc, 52, MSc, 54; State Univ NY, PhD(phys chem), 57. *Prof Exp:* Res assoc polymer chem, 59-63, from asst prof to assoc prof, 63-70, PROF POLYMER CHEM, STATE UNIV NY COL ENVIRON SCI & FORESTRY, 70- *Concurrent Pos:* Vis prof, Univ Nijmegen, 69-70, Louis Pasteur Univ, Strasbourg, 77-78, Univ Twente, Neth, 85-86. *Mem:* AAAS; Am Chem Soc. *Res:* Solvent-solute interactions; ion pair structures; ion-binding to macromolecules; polymer electrolytes; hydrogels; radiopaque polymers. *Mailing Add:* Dept Chem Col Environ Sci & Forestry State Univ NY Syracuse NY 13210-2786

SMID, ROBERT JOHN, b US. CERAMICS ENGINEERING. *Educ:* Univ Ill, BS, 62, MS, 65, PhD(ceramics eng), 68. *Prof Exp:* Sr engr, 68-74, prin engr, 74-80, ADV SCIENTIST, BETTIS ATOMIC POWER LAB, WESTINGHOUSE ELEC CORP, 80- *Mem:* Am Ceramic Soc. *Res:* Advanced naval nuclear fuel systems. *Mailing Add:* 1713 Ridge Rd Library PA 15129

SMIDT, FRED AUGUST, JR, b Sioux City, Iowa, July 19, 32; m 56; c 3. RESEARCH ADMINISTRATION, MATERIALS SCIENCE. *Educ:* Univ Nebr, BSc, 54; Iowa State Univ, PhD(phys chem), 62. *Prof Exp:* Sr engr, Hanford Lab, Gen Elec Co, 62-65; sr res scientist, Pac Northwest Labs, Battelle Mem Inst, 65-69; res metallurgist, Reactor Mat Br, Metall Div, Naval Res Lab, 69-71, sect head, 71-77; prog monitor, Reactor Res & Technol Div, US Dept Energy, 77-78; spec prog coordr, Mat Sci & Components Directorate, Naval Res Lab, 80-82, sect head, Reactor Mat Br, Metall Div, 78-82, staff specialist, Off Secy Defense (DDR&E), 89-90, HEAD, SURFACE MODIFICATION BR, NAVAL RES LAB, 82- *Honors & Awards:* Dudley Medal, Am Soc Testing & Mat, 79. *Mem:* Fel Am Soc Mat; Am Inst Mining, Metall & Petrol Engrs; Sigma Xi; Mat Res Soc. *Res:* Ion implantation for materials processing; irradiation damage to metals; electron microscopy; fast breeder and controlled thermonuclear reactor materials; ion beam assisted deposition of thin films; coatings for spacecraft; wear resistant coatings; pulsed laser deposition of thin films. *Mailing Add:* Surface Modification Br Naval Res Lab Code 4670 Washington DC 20375-5000

SMIKA, DARRYL EUGENE, b Hill City, Kans, July 1, 33; m 56, 87; c 4. SOIL CHEMISTRY. *Educ:* Kans State Univ, BS & MS, 56, PhD(soil chem), 69. *Prof Exp:* Soil conservationist, Soil Conserv Serv, USDA, Kans 56-57, soil scientist, Soil & Water Conserv Res Div, Agr Res Serv NDak, 57-61, res soil scientist, Exp Sta, Nebr, 61-73, res soil scientist, Cent Great Plains Res Sta, Soil & Water Conserv Res Div, Agr Res Serv, 73-88; RETIRED. *Mem:* Fel Am Inst Chemists; fel Soil Conserv Soc Am; Am Soc Agron; Soil Sci Soc Am; Can Soil Sci Soc. *Res:* Soil moisture and fertility under dryland conditions with present emphasis on cropping systems with no tillage. *Mailing Add:* 9901 N Country Rd 17 Ft Collins CO 80524

SMILEN, LOWELL I, b New York, NY, Apr 18, 31; m 63; c 2. ELECTRICAL ENGINEERING. *Educ:* Cooper Union, BSEE, 52; Univ Calif, Los Angeles, MS, 56; Polytech Univ NY, PhD(elec eng), 62. *Prof Exp:* Res engr, Hughes Aircraft Co, Calif, 52-56; sr res assoc, Polytech Inst Brooklyn, 56-62, asst prof electrophys, 62-64; res sect head, Sperry Gyroscope Co, 64-67; asst chief engr res, Loral Electronic Systs, Bronx, 67-70; leader advan microwave technol, Missile & Surface Radar Div, RCA Corp, NJ, 70-72; vpres eng, Laser Link Corp, 72-74; vpres eng, Almac/Stroum Electronics Corp, Div KDM Electronics Corp, 74-85, vpres eng, Almac Corp, Div Lex Electronics, 85-90; TECH CONSULT, 90- *Concurrent Pos:* Adj prof, Polytech Inst Brooklyn; mem eng sci dept, Hofstra Univ, 65; bd mem, Northcon Electronic Conv. *Honors & Awards:* Hughes Coop Fel, 52; Jack Chase Northcon Award, 90. *Mem:* Sr mem Inst Elec & Electronic Engrs; Asn Comput Mach; Am Electronics Asn. *Res:* Network theory and synthesis; microwave components and systems; antenna feed systems; solid state phased array antennas and radars; applied information theory and coding; applications of microprocessors and microcomputers; microcomputer systems design. *Mailing Add:* 13455 NE 27th Pl Bellevue WA 98005

SMILES, KENNETH ALBERT, b Elizabeth, NJ, Aug 5, 44; m 66; c 2. DERMATOLOGY, ENVIRONMENTAL PHYSIOLOGY. *Educ:* Denison Univ, BS, 66; Ind Univ, PhD(physiol), 70. *Prof Exp:* Res assoc physiol, Ind Univ, 70; res physiologist, Aerospace Med Res Lab, US Air Force, 70-72, actg chief environ physiol br, 72-74; group leader dermat group antiperspirant res, Carter Prods Res, Div Carter-Wallace, Inc, 74-76; sr med res assoc, 76-80, asst dir, 80-83, ASSOC DIR DERMAT, SCHERING CORP, 84- *Honors & Awards:* Sci Achievement Award, Systs Command, US Air Force, 72. *Mem:* Am Acad Dermat; Soc Invest Dermat; AAAS. *Res:* Mechanisms of anhydrosis and dermatological testing. *Mailing Add:* Oclassen Pharmaceut Inc 100 Pelican Way San Rafael CA 94901

SMILEY, HARRY M, b Cynthiana, Ky, Oct 6, 33; m 56; c 3. PHYSICAL CHEMISTRY. *Educ:* Eastern Ky State Col, BS, 55; Univ Ky, MS, 57, PhD(chem), 60. *Prof Exp:* Asst chem, Univ Ky, 55-60; res chemist, Union Carbide Corp, 60-67; assoc prof, 67-70, PROF CHEM & CHMN DEPT, EASTERN KY UNIV, 70- *Mem:* Am Chem Soc. *Res:* Physical and thermodynamic properties of non-ideal solutions, especially activity coefficients and heats of mixing. *Mailing Add:* 2083 Greentree Dr Richmond KY 40475-9625

SMILEY, JAMES DONALD, b Lubbock, Tex, Dec 6, 30; m 57; c 5. IMMUNOLOGY, MEDICINE. *Educ:* Tex Tech Col, BS, 52; Johns Hopkins Univ, MD, 56. *Prof Exp:* Intern & resident med, Columbia-Presby Hosp, 56-58; res assoc biochem, Nat Inst Arthritis & Metab Dis, 58-60; from instr to assoc prof, Univ Tex Health Sci Ctr, 60-70, PROF MED, UNIV TEX SOUTHWESTERN MED CTR, DALLAS, 70- *Concurrent Pos:* USPHS spec res fel, 60-63; USPHS career develop award, 68-73; sr investr, Arthritis Found, 63-68, Russell Cecil fel, 68; mem, Rheumatol Comt, Am Bd Internal Med, 70-73; assoc ed, J Clin Invest, 72-77. *Mem:* Am Rheumatism Asn; Am Soc Clin Invest; Am Asn Immunologists; Nat Soc Clin Rheumatologists; Am Chem Soc. *Res:* Connective tissue biochemistry; clinical immunology related to research in rheumatoid arthritis. *Mailing Add:* Presbyterian Hosp 8200 Walnut Hill Lane Dallas TX 75231

SMILEY, JAMES RICHARD, b Montreal, Que, June 22, 51; c 1. MOLECULAR BIOLOGY. *Educ:* McGill Univ, BSc, 72; McMaster Univ, PhD(biol), 77. *Prof Exp:* Fel biol, Yale Univ, 77-78, fel virol, 78-79; ASST PROF PATH, MCMASTER UNIV, 79- *Res:* Control of the expression of herpes simplex viral genes; mechanism of viral DNA replication. *Mailing Add:* Dept Path McMaster Univ Fac Med Hamilton ON L8N 4L8 Can

SMILEY, JAMES WATSON, b Charleston, WVa, Feb 17, 40; m 61; c 2. PHYSIOLOGY. *Educ:* Univ SC, BS, 62, MS, 65, PhD, 68. *Prof Exp:* Instr biol, Univ SC, 65-68; asst prof, Northeast La Univ, 68-71; assoc prof, 71-77, PROF BIOL, COL CHARLESTON, 77-, CHMN DEPT, 78- *Mem:* AAAS. *Res:* Endocrine control of molting in Crustacea. *Mailing Add:* Dept of Biol Col of Charleston Charleston SC 29424

SMILEY, JONES HAZELWOOD, b Casey Co, Ky, Apr 23, 33; m 53; c 1. AGRONOMY, PLANT PATHOLOGY. *Educ:* Univ Ky, BS, 59, MS, 60; Univ Wis-Madison, PhD(genetics), 63. *Prof Exp:* Res asst plant path, Univ Ky, 58-60; res asst genetics, Univ Wis-Madison, 60-63; asst prof agron, 63-68, assoc prof, 68-72, prof agron & plant path, 72-80, EXT PROF, DEPT AGRON, UNIV KY, 80- *Mem:* Am Soc Agron. *Res:* Tobacco breeding, management and diseases. *Mailing Add:* Dept Agron Rm 212C Agr Sci Ctr Univ Ky Lexington KY 40506

SMILEY, RICHARD WAYNE, b Paso Robles, Calif, Aug 17, 43; m 67; c 1. SOIL BORNE PLANT PATHOGENS, DISEASE CONTROL. *Educ:* Calif State Polytech Univ, BS, 65; Wash State Univ, MS, 69, PhD(plant path), 72. *Prof Exp:* Soil scientist fertility res, USDA Agr Res Serv, 66-69; res asst root dis, Dept Plant Path, Wash State Univ, 69-72; NATO fel, Commonwealth Sci & Indust Res Orgn, Soils Div, Adelaide, SAustralia, 72, vis res scientist soil microbiol, 73; from asst prof to assoc prof plant path, Cornell Univ, 73-85; PROF PLANT PATH, ORE STATE UNIV, 85- *Concurrent Pos:* Vis scientist, Victoria Dept Agr, Melbourne, Australia, 81-82; ed-in-chief, Am Phytopath Soc Press, 87-91; supt, Columbia Basin Agr Res Ctr, 85- *Mem:* Am Phytopath Soc; Am Soc Agron; Int Soc Plant Path; Int Turfgrass Soc. *Res:* Disease control investigations on cereal grains and turfgrasses, with emphasis on cultural or integrated chemical-biological control strategies for diseases caused by soilborne plant pathogens. *Mailing Add:* Columbia Basin Agr Res Ctr PO Box 370 Pendleton OR 97801

SMILEY, ROBERT ARTHUR, b Cleveland, Ohio, Mar 14, 25; m 49; c 10. ORGANIC CHEMISTRY. *Educ:* Case Inst Technol, BS, 50; Purdue Univ, PhD(org chem), 54. *Prof Exp:* From res chemist to sr res chemist, Explosives Dept, 54-69, tech asst, 69-70, mgr polymer intermediates dept, 70-74, res assoc, 74-81, RES FEL, PETROCHEM DEPT, E I DU PONT DE NEMOURS & CO, INC, 81- *Mem:* Am Chem Soc; Catalysis Soc. *Res:* Preparation and reactions of aliphatic nitro compounds; reactions of nitric acid and nitrogen oxides with organic compounds; heterogeneous catalysis; preparation and reactions of nitriles. *Mailing Add:* Exp Sta Bldg 336 1103 Norbee Dr Wilmington DE 19803-4123

SMILEY, TERAH LEROY, b Clay Co, Kans, Aug 21, 14; c 5. GEOCHRONOLOGY. *Educ:* Univ Ariz, BA, 46, MA, 49. *Prof Exp:* Ranger naturalist & actg custodian, Nat Park Serv, 39-41; asst dendrochronologist, 46-51, asst archaeol, 51-54, geochronology labs, 56-67, prof geochronology & head dept, 67-70, assoc head, dept geosci & chief res labs, 70-74, prof, 74-84, EMER PROF GEOSCIENCE, UNIV ARIZ, 84- *Concurrent Pos:* Res fel, Clare Col, Cambridge Univ, 69; vis prof geol inst, Univ Uppsala, Sweden, 69-70. *Mem:* Fel Geol Soc Am; Am Meteorol Soc; Am Quaternary Asn; fel AAAS. *Res:* Arid lands studies, paleoclimatology of Southwest United States. *Mailing Add:* Dept Geoscience Univ Ariz Tucson AZ 85721

SMILEY, VERN NEWTON, b Goshen, Ind, Sept 7, 30; m 56. ATMOSPHERIC PHYSICS. *Educ:* Univ Wis, BS, 55, MS, 56; Univ Colo, PhD(physics), 59. *Prof Exp:* Sr engr, Gen Dynamics/Convair, 59-61; resident res assoc, US Navy Electronics Lab Ctr, 61-62, res physicist, 62-71; prof physics, Univ Nev, Reno, 71-80, res prof atmospheric optics, Desert Res Inst, 71-80; sect head, energy measurement group, EG&G, Inc, 80-86; prog mgr, Off Naval Res, Pasadena, 86-88, San Diego, 88-90; TECH STAFF, NAVAL OCEAN SYSTS CTR, 90- *Concurrent Pos:* Consult, Gen Dynamics/Convair, 61-63; US Navy res fel & vis scientist, York, Eng, 66-67; lectr appl physics & info sci dept, Univ Calif, San Diego, 71; liaison scientist, Off Naval Res, London, 77-79. *Mem:* Optical Soc Am; Sigma Xi. *Res:* Multi-layer thin films; infrared radiometers; gas phase lasers; scanning active interferometers; laser amplifiers; thin film lasers; air pollution; atmospheric remote sensing with optical instrumentation; solar energy research; electro-optics and fiber optics research and development; high power lasers; adaptive optics. *Mailing Add:* Naval Ocean Systems Ctr Code 808 271 Catalina Blvd San Diego CA 92152

SMILLIE, LAWRENCE BRUCE, b Galt, Ont, July 5, 28; m 56; c 4. BIOCHEMISTRY. *Educ:* McMaster Univ, BSc, 50; Univ Toronto, MA, 52, PhD(biochem), 55. *Prof Exp:* Asst prof biochem, Univ Alta, 55-57; Nat Acad Sci-Nat Res Coun Donner fel med res, Univ Wash, 57-58; assoc prof, 58-67, PROF BIOCHEM, UNIV ALTA, 67- Prof Exp: Vis scientist, Lab Molecular Biol, Cambridge Univ, 63-64; vis scientist, Dept Biochem, Univ Birmingham, 71-72 & Ludwig Inst Cancer Res, MRC Ctr, Cambridge Univ, 84-85. *Mem:* Can Biochem Soc; Am Soc Biol Chemists; Brit Biochem Soc; fel Royal Soc Can. *Res:* Chemistry and functional role of the proteins of muscle and contractile systems; molecular cloning; DNA sequencing and oligonucleotide directed mutagenesis. *Mailing Add:* Dept Biochem Univ Alta Edmonton AB T6G 2H7 Can

SMILOWITZ, BERNARD, b Philadelphia, Pa. APPLIED PHYSICS, ELECTRICAL ENGINEERING. *Educ:* Temple Univ, AB, 55; Univ Pittsburgh, PhD(physics), 63. *Prof Exp:* Instr physics, Temple Univ, 63-66; res scientist, AIL Div, Cutler Hammer Inc, 66-77, sr res scientist appl physics, AIL Div, Eaton Corp, 77-83; vpres & dir res, General Microwave Corp, 83-86; ASSOC PROF ENG, HOFSTRA UNIV, 86- *Mem:* Am Phys Soc; Math Asn Am; Sigma Xi; Inst Elec & Electronics Engrs; NY Acad Sci. *Res:* Applied physics and material science research associated with the development of microwave solid state devices and components. *Mailing Add:* 53 Brandy Lane Lake Grove NY 11755

SMILOWITZ, HENRY MARTIN, b Brooklyn, NY, Sept 25, 46; m; c 3. NEUROBIOLOGY, CELL BIOLOGY. *Educ:* Reed Col, AB, 68; Mass Inst Technol, PhD(biochem), 72. *Prof Exp:* Fel microbiol, Med Sch, Tufts Univ, 72-73, neurobiol, Harvard Med Sch, 73-76; asst prof, 76-82, ASSOC PROF PHARMACOL, HEALTH CTR, UNIV CONN, 82- *Mem:* Am Soc Cell Biol; Soc Neurosci; NY Acad Sci; Int Soc Neurosci; AAAS; Am Soc Pharmacol & Exp Therapeut. *Res:* Regulation and development of neuromuscular junction function post-synaptic mechanisms; acetylcholine receptor and esterase: development, intracellular transport and localization; phosphorylation; role of calcium in cellular regulation; studies on voltage sensitive calcium channel of skeletal and cardiac muscle. *Mailing Add:* Dept Pharmacol Health Ctr Univ Conn Farmington CT 06032

SMILOWITZ, ZANE, b New York, NY, Sept 13, 33; m 59; c 3. ENTOMOLOGY. *Educ:* Univ Ga, BS, 61; Cornell Univ, MS, 65, PhD(entom), 67. *Prof Exp:* Entomologist, Trubeck Labs, 61; teaching & res asst, Cornell Univ, 61-67; PROF ENTOM, PA STATE UNIV, 67- *Mem:* AAAS; Entom Soc Am; Can Entom Soc; Sigma Xi; Potato Asn Am. *Res:* Research and development of integrated pest management systems for potato pests; insecticide resistance management; host plant interactions; plant responses to interacting biotic and abiotic stresses. *Mailing Add:* Dept Entom Pa State Univ 501 Agr Sci & Indust Bldg University Park PA 16802

SMIT, CHRISTIAN JACOBUS BESTER, b Piet Retief, SAfrica, Jan 10, 27; m 52, 68; c 4. FOOD SCIENCE, AGRICULTURAL CHEMISTRY. *Educ:* Univ Pretoria, BS, 47, HED, 48; Univ Calif, Berkeley, PhD, 53. *Prof Exp:* Tech asst chem, SAfrican Dept Agr, 44-46; asst prof officer, Agr Res Inst, Univ Pretoria, 48-49; prof officer, Fruit & Food Technol Res Inst, SAfrica, 53-58, first prof officer & chief food technol sect, 58-60; prof food sci & head dept, Stellenbosch, 60-63; res food scientist, Sunkist Growers Inc, 63-68; head dept & chmn div, 73-80, actg dean, Col Agr, 80-81, PROF FOOD SCI, UNIV GA, 68-, ASSOC DEAN & DIR RESIDENT INSTRUCTION, COL AGR, 81- *Concurrent Pos:* Consult, Nat Nutrit Res Inst, SAfrica, 60-63. *Mem:* Inst Food Technol; hon mem SAfrican Asn Food Sci & Technol. *Res:* Fruit and vegetable chemistry and processing; occurrence, manufacture and use of pectic substances. *Mailing Add:* Rm 102 Conner Hall Col Agr Univ Ga Athens GA 30602

SMIT, DAVID ERNST, b Beloit, Wis, Sept 1, 42; m 67; c 3. SEDIMENTARY GEOLOGY, STRATIGRAPHY. *Educ:* Augustana Col, AB, 64; Univ Iowa, MS, 67, PhD(geol), 71. *Prof Exp:* Teaching asst geol, Univ Iowa, 65-67, res asst, 67-70; from instr to asst prof, Univ Wis-Stevens Point, 70-74; asst prof geol, Wichita State Univ, 74-78, res assoc, 78-80; sr geologist, Energy Reserves Group, 80-82; RESOURCES INC, 79-;; CONSULT GEOLOGIST, 82. *Concurrent Pos:* Res assoc, Univ Iowa, 71; pres, Quad Resources Inc, 79-; co-dir NURE prog, Wichita State Univ, 79-80; consult geologist, 82- *Mem:* Geol Soc Am; Soc Econ Paleontologists & Mineralogists; Int Asn Sedimentologists; Int Geol Cong. *Res:* Depositional environments and diagenesis of shallow marine-platform sedimentary rocks; comparative sedimentology of recent limestones and the equivalent ancient rocks; location and mode of occurence of uranium in sedimentary rocks. *Mailing Add:* 221 Timber Ridge Court Edmond OK 73034

SMIT, JAN, b Midwoud, Netherlands, Aug 30, 21; m 48; c 2. SOLID STATE PHYSICS. *Educ:* Delft Univ Technol, Ingenieur, 48; State Univ Leiden, PhD(physics), 56. *Prof Exp:* Engr, Philips, Endhoven, Netherlands, 41-45, physicist, 48-63; prof solid state magnetism, 63-74, PROF MAT SCI, UNIV SOUTHERN CALIF, 74- *Concurrent Pos:* Consult, Ampex Corp, 64- *Mem:* Am Phys Soc. *Mailing Add:* 3855 Pirate Dr Palos Verde Pensla CA 90274

SMITH, A(LBERT) LEE, b Omaha, Nebr, Apr 11, 24; m 48; c 6. PHYSICAL CHEMISTRY, THERMODYNAMICS & MATERIAL PROPERTIES. *Educ:* Iowa State Univ, BS, 46; Ohio State Univ, PhD(phys chem), 50. *Prof Exp:* Res assoc, Res Found, Ohio State Univ, 50-51; supvr spectros lab, 51-69, mgr anal dept, 69-80, EMER SCIENTIST, DOW CORNING CORP, 89- *Honors & Awards:* Williams-Wright Award, Indust Molecular Spectros, 87. *Mem:* Am Chem Soc; Soc Appl Spectros. *Res:* Infrared spectra of organosilicon compounds. *Mailing Add:* 400 Rollcrest Ct Midland MI 48640

SMITH, A(POLLO) M(ILTON) O(LIN), b Columbia, Mo, July 2, 11; m 43; c 3. AERONAUTICAL ENGINEERING. *Educ:* Calif Inst Technol, MS, 38. *Hon Degrees:* DSc, Univ Colo, 75. *Prof Exp:* Aerodynamicist, Douglas Aircraft Co, 38-42; chief engr rocket propulsion, Aerojet Eng Corp, 42-44; asst chief aerodynamicist, Douglas Aircraft Co, Long Beach, McDonnell Douglas Corp, 44-48, supvr design res, 48-54, supvr aerodyn res, 54-69, chief aerodyn eng res, 69-75; consult, 75-85; prof, Calif State Univ, Long Beach, 85-86; RETIRED. *Concurrent Pos:* Mem subcomt internal flow, Nat Adv Comt Aeronaut, 48-51; mem, US Naval Tech Mission, Europe, 45; lectr, Univ Calif, Los Angeles, 54-58; Am ed, J Comput Methods Appl Mech & Eng, 72-77; consult, McDonnell Douglas Corp, 75-85, Dynamics Technol, Inc, 77-85 & Bolt, Beranek & Newman, 78-82; lectr, Peking Inst Aeronaut & Astronaut, 79; mem, Aeronaut Adv Comn, NASA, 80-83; Adj prof, UCLA, 75-80. *Honors & Awards:* Goddard Award, Am Inst Aeronaut & Astronaut, 54; co-winner Casey Baldwin Award, Can Aeronaut & Space Inst, 71; Wright Bros lectr, Am Inst Aeronaut & Astronaut, 74; Fluids Eng Award, Am Soc Mech Engrs, 85. *Mem:* Nat Acad Eng; hon fel Am Inst Aeronaut & Astronaut; Am Soc Mech Engrs; Am Phys Soc. *Res:* Aerodynamic and applied mechanics, especially boundary layer and heat transfer; inviscid flow theory. *Mailing Add:* 2245 Ashbourne Dr San Marino CA 91108

SMITH, A MASON, MICROBIOLOGY, IMMUNOLOGY. *Educ:* NC State Univ, PhD(immunol), 70. *Prof Exp:* ASSOC PROF MICROBIOL & IMMUNOL, SCH MED, E CAROLINA UNIV, 74- *Mailing Add:* Sch Med E Carolina Univ Brody Bldg Greenville NC 27858

SMITH, A(LLEN) N(ATHAN), b New Orleans, La, Oct 24, 21. CHEMICAL ENGINEERING. *Educ:* Tulane Univ, BS, 41; Ga Inst Technol, MS, 43; Ore State Univ, PhD(chem eng), 48. *Prof Exp:* Staff mem, J E Sirrine & Co, Engrs, SC, 41; chem engr, Aberdeen Proving Ground, Md, 41-42; technologist, Shell Oil Co, 43-45; chem engr, Union Oil Co, 45-56; assoc prof chem eng, Univ Louisville, 48-52; from assoc prof to prof, 52-63, CHMN DEPT CHEM ENG, SAN JOSE STATE UNIV, 63- *Mem:* Am Soc Eng Educ; fel Am Inst Chem Engrs; Sigma Xi. *Res:* Dialysis; fluid flow; heat and mass transfer. *Mailing Add:* Dept of Chem Eng San Jose State Univ San Jose CA 95192

SMITH, AARON, b Boston, Mass, Nov 3, 30. CLINICAL PSYCHOLOGY, HEALTH SERVICE RESEARCH. *Educ:* Brown Univ, AB, 52; Univ Ill, PhD(clin psychol), 58. *Prof Exp:* Res psychologist, Philadelphia State Hosp, 58-62; dir res, Haverford State Hosp, 62-73, asst hosp dir, 73-75; DIR RES, VET ADMIN MED CTR, RENO, 75-; RES ASSOC PROF, SCH MED, UNIV NEV, 75- *Concurrent Pos:* Co-dir, Northeast Psychol Clin, 59-75; exec dir, Sierra Biomed Res Corp, 89- *Mem:* Am Psychol Asn; Geront Soc Am. *Res:* Developing techniques for assessing the outcome of various health treatment programs and predicting these outcomes from patient variables. *Mailing Add:* Vet Admin Med Ctr 151 1000 Locust St Reno NV 89520

SMITH, ALAN B, b Chicago, Ill, Dec 19, 24; m 43; c 1. NUCLEAR PHYSICS. *Educ:* Beloit Col, BA, 49; Ind Univ, MS, 50, PhD(nuclear physics), 53. *Prof Exp:* Assoc physicist, Argonne Nat Lab, 53-58, head, Appl Nuclear Physics Sect, Appl Physics Div, 58-61, SR PHYSICIST, APPL NUCLEAR PHYSICS SECT, APPL PHYSICS DIV, ARGONNE NAT LAB, 61- *Mem:* Fel Am Phys Soc. *Res:* Neutron and fission physics. *Mailing Add:* Stonehedge Lincoln MA 01773

SMITH, ALAN BRADFORD, b Karuizawa, Japan, July 28, 32; US citizen; m 57; c 2. MAGNETIC DATA STORAGE, MAGNETIC MEASUREMENTS. *Educ:* Swarthmore Col, BS, 53; Rensselaer Polytech Inst, MEE, 59; Harvard Univ, MA, 60, PhD(appl physics), 66. *Prof Exp:* Engr, Sprague Elec Co, 53-59; res asst appl physics, Harvard Univ, 61-65; mem res staff, Sperry Rand Res Ctr, 65-83; CONSULT ENGR, DIGITAL EQUIPMENT CORP, 83- *Concurrent Pos:* Coordr, courses in Bubble Domain Memory Technol, UCLA, 75, 77, 79-80; ed-in-chief, Transactions on Magnetics, Inst Elec & Electronics Engrs, 79-81. *Mem:* Am Phys Soc; sr mem Inst Elec & Electronics Engrs; Magnetics Soc (pres, 85-86). *Res:* Thin-film magnetic recording heads; magneto-optics and magnetic measurements; magnetic bubble domain memory devices; ferromagnetic resonance; magnetoacoustic interactions in solids. *Mailing Add:* Stonehedge Lincoln MA 01773

SMITH, ALAN JAY, b New York, NY, Apr 10, 49. PERFORMANCE ANALYSIS. *Educ:* Mass Inst Technol, BS, 71; Stanford Univ, MS, 73, PhD(comput sci), 74. *Prof Exp:* From asst prof to assoc prof, 74-86, PROF COMPUT SCI, UNIV CALIF, BERKELEY, 86- *Concurrent Pos:* Vchmn, elec eng & comput sci dept, Univ Calif, Berkeley, 82-84; mem, Comput Measurement Group. *Mem:* Asn Comput Mach; Sigma Xi; Inst Elec & Electronics Engrs; Soc Indust & Appl Math. *Res:* Computer system performance; operating systems. *Mailing Add:* Computer Sci Div Univ Calif Berkeley CA 94720

SMITH, ALAN JERRARD, b London, Eng, May 8, 29; m 58; c 2. PHARMACOLOGY. *Educ:* Univ London, BSc, 50, PhD(chem), 56. *Prof Exp:* Asst, Univ Col, Exeter, Eng, 53-54; res chemist, Admiralty Mat Lab, Eng, 54-57; group leader res, Rohm and Haas Co, 57-63, regulatory liaison, 63-64; asst mgr anal develop, 64-70, mgr qual control tech serv, 70-78, mgr, 78-79, ASST DIR, CORP QUAL ASSURANCE, AYERST LABS INC, 79-; DIR QUAL AFFAIRS, WHITEHALL LAB. *Concurrent Pos:* Mem, Comt Quality Systs, Am Soc Testing & Mat, 78- *Mem:* Am Soc Qual Control; Am Chem Soc. *Res:* Pharmaceutical quality assurance and control; stability of pharmaceuticals; analytical methods for pharmaceuticals; ion exchange; redox polymers; chemistry of transition metals. *Mailing Add:* Qual Affairs Whitehall Lab 685 Third Ave New York NY 10017-4676

SMITH, ALAN LYLE, b Bartley, Nebr, June 27, 41; m 61; c 2. PLANT ECOLOGY, WETLANDS MITIGATION. *Educ:* Kearney State Col, BS, 64; Univ Nebr, MS, 68; Tex A&M Univ, PhD(ecol), 71. *Prof Exp:* Instr chem, Cozad High Sch, Nebr, 64-65; range scientist veg control, Tex Trans Inst, 72; plant ecologist, Dames & Moore, 72-80; pres, ALS Consults, Inc, Houston, 81-90; MGR ENVIRON SERV, GULF COAST REGION, MAXIM ENGRS, INC, 90- *Concurrent Pos:* Proj ecologist, Prep Environ Report, Seadock Inc, 73-74; ecol consult, Environ Report LOOP Inc, 73-75; proj mgr, Site Selection Petrochem Facil Int Consortium, 74, Seagrass Transplant Prog, Lower Laguna Madre, Tex, 84. *Mem:* Nat Asn Environ Prof. *Res:* Impact of crude oil on coastal marshes and problems associated with establishing vegetation on problem soils, especially related to strip mining reclamation; wetland mitigation; transplanting seagrasses. *Mailing Add:* 5418 Arncliffe Houston TX 77088

SMITH, ALAN PAUL, b Morristown, NJ, Mar 31, 45. PLANT ECOLOGY. *Educ:* Earlham Col, BA, 67; Duke Univ, MA, 70, PhD(bot), 74. *Prof Exp:* Asst prof biol, Univ Pa, Philadelphia, 74-80; assoc prof biol, Univ Miami, Coral Gables, 82-; BIOLOGIST ECOL, SMITHSONIAN TROP RES INST, BALBOA, PANAMA, 74-, ASST DIR, 88- *Mem:* Sigma Xi; Torrey Bot Club; Ecol Soc Am; Asn Trop Biol. *Res:* Ecological and evolutionary significance of latitudinal gradients in plant form; plant ecology of tropical alpine zones of the world; ecophysiology of tropical forest plants. *Mailing Add:* Smithsonian Trop Res Inst APO Miami FL 34002

SMITH, ALAN REID, b Sacramento, Calif, July 14, 43; m 66; c 2. BOTANY. *Educ:* Kans State Univ, BS, 65; Iowa State Univ, PhD(bot), 69. *Prof Exp:* Asst res, 69-77, assoc res, 77-83, RES BOTANIST, UNIV CALIF, BERKELEY, 83- *Concurrent Pos:* Ed, Pteridologia, 78-84, Am Fern J, 85-89; vpres, Am Fern Soc, 90- *Mem:* Am Soc Plant Taxon; Am Fern Soc; Brit Pteridological Soc. *Res:* Taxonomy of ferns; Thelypteris; pteridophytes of Chiapas, Mexico, Venezuela. *Mailing Add:* Univ Herbarium Univ of Calif Berkeley CA 94720

SMITH, ALBERT A, JR, b Yonkers, NY, Dec 2, 35. ELECTRICAL MAGNETIC. *Educ:* Milwaukee Sch Eng, BSEE, 61, NY Univ, MSEE, 64. *Prof Exp:* Staff engr, W Tech Div, Adler, 61-64; SR ENGR, IBM CORP, 64- *Concurrent Pos:* Chmn tech comt electromagnetic environ, Inst Elec & Electronics Engrs, 81. *Mem:* Inst Elec & Electronics Engrs. *Mailing Add:* IBM Corp 69ra/170 Neighborhood Rd Kingston NY 12401

SMITH, ALBERT CARL, b Los Angeles, Calif, Sept 13, 34; m 67; c 2. PATHOBIOLOGY, PATHOLOGY. *Educ:* Univ Calif, Los Angeles, BA, 56; Univ Calif, Irvine, PhD(biol sci), 67; Univ Hawaii, MD, 75, Am Bd Path, cert path, 79. *Prof Exp:* Asst to sr scientists, Univ Hawaii, 59; res scientist, Calif Dept Fish & Game, 61-63, 65-66 & 70-71; lab technician, Allergan Pharmaceut, 63; lab technician, Orange County Gen Hosp, 64; lab technician, Univ Calif, Irvine, 64-65, res asst organismic biol, 66, res assoc & instr pop & environ biol, 66-67; from asst prof to assoc prof biol, Univ Hawaii, Hilo, 67-73, mem grad fac zool, Univ Hawaii, Honolulu, 68-77; mem staff, Oceanic Inst, Hawaii, 77-79; dir, Med Labs Hawaii, Inc, 79-80; asst prof path, Col Med & Vet Med, Univ Fla, Gainesville, 80-85; chief clin lab, 80-85, CHIEF CLIN PATH, DIR, PATH SPEC STUDIES & CHIEF, CHEM SECT, CLIN LAB, VET ADMIN MED CTR, BAY PINES, FL, 85- *Concurrent Pos:* Res assoc, dept pop & environ & pop environ biol, Univ Calif, Irvine, 67-; res grants, Bur Nat Marine Fisheries Serv, US Dept Comm, 68-69; res grant Univ Hawaii, 68-72, sea grant, 71-73; consult dir, Genetics Lab, Calif State Fisheries Lab, Long Beach, 69-; prog dir path, Aquatic Sci, Inc, Fla, 69-70, consult, 70-; Marine Biol Consults, Inc, Calif, 70-; grant Am Found Oceanog, 69-77, 77 & 78; Calif Dept Fish & Game, 70-71; chief consult, Hawaii BioMarine, 70-; res assoc & consult, Oceanic Inst, 71-; sr sci staff consult pathobiol, Pan Pac Inst Ocean Sci, 73-; grants, US Energy Res & Develop Admin, 71-72, NIH, 72-73, NSF, 72-74 & Puerto Rico Undersea Lab, 73, USAID, 76- Rockefeller Found, 78-79 & Engelhard Found, 78-79; res fel path, Univ Hawaii, 75-76; clin path residency, St Francis Hosp, Honolulu, 77-78 & Queens Med Ctr, 78-79; marine affairs coordr, State Hawaii, 78; consult, Astromarine, Kuai, Hawaii, Hawaiian Elec Comp, Honolulu, State Univ Syst Fla, 81, Sea Grant, 80, clin path, Sunland Ctr, Gainesville, Fla, Med-Marine Res, 83-87, Elka, 88, Exxon, 89; res fac, Univ Fla, 81-85. *Mem:* AAAS; Soc Invert Path; NY Acad Sci; hon mem Am Longevity Soc; hon mem Int Soc Aquatic Med; Col Am Pathologists; Int Soc Develop & Comp Immunologists. *Res:* Evolution; experimental taxonomy; electrophoretic technique; diving and deep sea biology; pathobiology; proteins of the eye lens; immunobiology and serology; chemical phylogenetics; marine biomedicine; psychobiology, comparative hematology. *Mailing Add:* 15928 Redington Dr Redington Beach FL 33705

SMITH, ALBERT CHARLES, b Springfield, Mass, Apr 5, 06; m 35, 66; c 2. BOTANY. *Educ:* Columbia Univ, AB, 26, PhD(bot), 33. *Prof Exp:* From asst cur to assoc cur, NY Bot Garden, 28-40; cur herbarium, Arnold Arboretum, Harvard Univ, 40-48, ed j, 41-48, cur div phanerogams, Dept Bot, Smithsonian Inst, 48-56; prog dir syst biol, NSF, 56-58; dir, Mus Natural Hist, US Nat Mus, 58-62, asst secy, Smithsonian Inst, 62-63; dir res & prof bot, Univ Hawaii, 63-65, Wilder prof, 65-70; Torrey prof bot, Univ Mass, Amherst, 70-76; ED CONSULT, PAC TROP BOT GARDEN, 77- *Concurrent Pos:* Mem bot expeds, Colombia, Peru, Brazil, Fiji, Brit Guiana & W Indies, 26-69; fel, Bishop Mus, Yale Univ, 33-34; Guggenheim fel, 46-47; Ed, Brittonia, 35-40, J Arnold Arboretum, 41-48, Sargentia, 42-48 & Allertonia, 77-88. *Honors & Awards:* Robert Allerton Award, 79. *Mem:* Nat Acad Sci; fel Am Acad Arts & Sci; Linnean Soc London; Asn Trop Biol (pres, 67-68); Int Asn Plant Taxon (vpres, 59-64). *Res:* Taxonomy and phytogeography of flowering plants, especially of tropical America and southwest Pacific. *Mailing Add:* Dept of Bot Univ of Hawaii Honolulu HI 96822

SMITH, ALBERT ERNEST, JR, b Ransom, Kans, Dec 4, 38; m 60; c 2. PLANT PHYSIOLOGY. *Educ:* Ft Hays Kans State Col, BSc, 64; Tex A&M Univ, PhD(range sci), 69. *Prof Exp:* Assoc prof, 69-80, PROF AGRON, UNIV GA, 80- *Honors & Awards:* Gamma Sigma Delta Distinguished Jr Fac Award, 78; Trans Am Soc Ag Eng Outstanding Paper Award, 79. *Mem:* Am Soc Plant Physiol; Am Soc Agron; Weed Sci Soc Am; Am Soc Range Mgt. *Res:* Plant physiology and biochemistry of the modes of actions of herbicides; teaching agronomy 834, advanced chemical weed control. *Mailing Add:* Dept Agron & Weed Sci Univ Ga Exp Sta Griffin GA 30223

SMITH, ALBERT ERNEST, b Windham, Vt, Nov 1, 27; m 50; c 2. PHYSICS. *Educ:* Atlantic Union Col, BA, 49; Mich State Univ, MS, 51, PhD(physics), 54. *Prof Exp:* Instr physics, Mich State Univ, 53-54; asst prof, Union Col, Nebr, 54-57; res physicist, Radio Corp Am, NJ, 57-59; prof, Atlantic Union Col, 59-69, dean, 67-69; assoc dir phys sci, Tech Opers Inc, 69-71; PROF PHYSICS, LOMA LINDA UNIV, LA SIERRA CAMPUS, 71- *Mem:* Optical Soc Am. *Res:* Physical optics; image theory; photographic methods; optical instrumentation. *Mailing Add:* 72 San Mateo Rd Berkeley CA 94707

SMITH, ALBERT GOODIN, b Charleston, Mo, Aug 26, 24; m 53. PATHOLOGY. *Educ:* Washington Univ, MD, 47. *Prof Exp:* Intern, St Luke's Hosp, St Louis, 47-48; asst resident & resident path, Hosp, Univ Ark, 48-50; vol asst surg path, Col Physicians & Surgeons, Columbia Univ, 50; asst resident, resident & instr path, Sch Med, Duke Univ, 5051, assoc, 52-55, from asst prof to assoc prof, 55-66; prof path & dep chmn dept, Col Med, Univ Tenn & dir labs, City Memphis Hosps, 66-70; PROF PATH & HEAD DEPT, SCH MED, LA STATE UNIV, SHREVEPORT, 70- *Concurrent Pos:* Chief lab serv, Vet Admin Hosp, Shreveport, La, 70-; chief path serv, Confederate Mem Med Ctr, Shreveport, 71- *Mem:* AAAS; Am Soc Clin Path; Am Soc Exp Path; AMA; Am Asn Pathologists & Bacteriologists. *Res:* Surgical pathology; tissue culture; nucleic acid tissue effects; teratology. *Mailing Add:* PO Box 4159 Centenary Shreveport LA 71134-0159

SMITH, ALBERT MATTHEWS, b Bangor, Maine, Dec 25, 27; m 50; c 2. ANIMAL NUTRITION. *Educ:* Univ Maine, BS, 52; Cornell Univ, MS, 54, PhD(animal nutrit), 56. *Prof Exp:* From instr to asst prof animal husb, Cornell Univ, 55-57; from asst prof to assoc prof, 57-64, chmn dept animal sci, 63-79, PROF ANIMAL NUTRIT, UNIV VT, 64-, ANIMAL NUTRITIONIST, 61-, ASSOC DEAN, COL AGR & ASSOC DIR, VT STATE AGR EXP STA, 75- *Concurrent Pos:* Consult, Rep Korea, AID, 73. *Mem:* Am Dairy Sci Asn; Am Soc Animal Sci; Sigma Xi; fel AAAS. *Res:* Dairy cattle nutrition; forage evaluation, especially role of forages in summer and winter feeding regimes; mineral metabolism. *Mailing Add:* 204 Carrigan Hall Univ Vt Burlington VT 05401

SMITH, ALDEN ERNEST, b Lockport, NY, Apr 25, 23. SCIENCE EDUCATION, AQUATIC BIOLOGY. *Educ:* Univ Colo, Boulder, BA, 50; Univ Buffalo, EdM, 60; Syracuse Univ, MS, 64; State Univ NY Buffalo, EdD(sci educ), 71. *Prof Exp:* Lab asst bot, Brookhaven Nat Lab, 54-56; teacher jr high sch, Lockport Bd Educ, NY, 56-59, high sch, 59-65; from assoc prof to prof biol, State Univ NY Col Buffalo, 65-88; RETIRED. *Mem:* AAAS; Am Inst Biol Sci; Nat Asn Biol Teachers; Nat Sci Teachers Asn. *Res:* Methods of teaching biology at the college level; aquatic plants; biology of organisms in fresh water environments; wild and cultivated poisonous plants. *Mailing Add:* 7047 Old English Rd Lockport NY 14094

SMITH, ALEXANDER GOUDY, b Clarksburg, WVa, Aug 12, 19; m 42; c 2. ASTROPHYSICS,PHOTOGRAPHIC RESEARCH. *Educ:* Mass Inst Technol, SB, 43; Duke Univ, PhD(physics), 49. *Prof Exp:* Mem staff Radiation Lab, Mass Inst Technol, 42-46; instr & asst Duke Univ, 46-48; from asst prof to prof physics, 48-56, asst dean grad sch, 61-69, chmn dept astron, 62-71, actg dean grad sch, 71-73, distinguished alumni assoc prof, 81-83, PROF ASTRON & PHYSICS, UNIV FLA, 56-, DISTINGUISHED SERV PROF ASTRON, 82- *Concurrent Pos:* Consult, US Air Force, 54-65; mem bd dirs, Assoc Univ for Res in Astron, 60-63, consult, 64-69; mem users' comt, Nat Radio Astron Observ, 66-78, vis comt, 68-71; mem comt astron, Nat Res Coun, 66-69, chmn, 68-69; adv panel astron, NSF, 69-72; ed, Am Astron Soc Photo Bull, 75-87; trustee SE Univ Res Asn, 82- *Honors & Awards:* Medal, Fla Acad Sci, 65. *Mem:* Fel AAAS; fel Am Phys Soc; fel Optical Soc Am; Am Astron Soc; Int Astron Union; Soc Photog Scientists & Engrs; fel Royal Micros Soc. *Res:* Magnetron design; microwave molecular spectroscopy; atmospheric optics; hypersensitization of photographic materials for research in astrophysics; planetary radio astronomy; optical and radio variations of quasars. *Mailing Add:* Dept Astron Univ Fla 211 Space Sci Bldg Gainesville FL 32611

SMITH, ALICE LORRAINE, b Trinity, Tex. PATHOLOGY, CYTOLOGY. *Educ:* Univ Tex, BA, 40, MD, 46; Am Bd Path, dipl, 51. *Prof Exp:* Asst prof path, Univ Tex Southwestern Med Sch, 50-54; asst pathologist, Univ Hosp, Baylor Univ, 54-55; from asst prof to assoc prof path, Col Med, 55-57, prof, Res Inst & assoc prof, Col Dent, 57-61; assoc prof, 62-76, PROF PATH, UNIV TEX SOUTHWESTERN MED SCI CTR, DALLAS, 76-; DIR, DIV DIAG CYTOL & DIR, SCH CYTOTECHNOL, PARKLAND MEM HOSP, 62- *Concurrent Pos:* Pathologist, Wadley Res Inst & Blood Bank, 57-61; clin assoc, Univ Tex Health Sci Ctr, Dallas, 58-62. *Honors & Awards:* Commissioner's Spec Citation, Food & Drug Admin. *Mem:* Fel Am Col Physicians; fel Am Soc Clin Path; fel Am Col Path; AMA. *Mailing Add:* Dept Path Univ Tex Southwestern Med Ctr Dallas TX 75235

SMITH, ALLAN EDWARD, b Hull, Eng, June 2, 37; m 70. AGRICULTURE, ORGANIC CHEMISTRY. *Educ:* Univ Liverpool, BSc, 59, PhD(org chem), 63. *Hon Degrees:* DSc, Univ Liverpool, 82. *Prof Exp:* Fel radiol sci, Johns Hopkins Hosp, 63-65; res chemist, Agr Div, Imp Chem Industs, 65-67; RES CHEMIST, CAN DEPT AGR, 67- *Mem:* Europ Weed Res Soc; Chem Inst Can; Am Chem Soc; Weed Sci Soc Am. *Res:* Fate of herbicides after application, their biological and chemical degradation and metabolism in soil. *Mailing Add:* Can Dept Agr Box 440 Regina SK S4P 3A2 Can

SMITH, ALLAN LASLETT, b Newark, NJ, June 21, 38; m 60; c 2. PHYSICAL CHEMISTRY. *Educ:* Harvard Univ, BA, 60; Mass Inst Technol, PhD(phys chem), 65. *Prof Exp:* Nat Acad Sci-Nat Res Coun fel, Nat Bur Standards, 65-66; from asst prof to assoc prof chem, Yale Univ, 66-74; head dept math, Daycroft Sch, Rock Ridge, Conn, 75; assoc prof, 75-82, PROF CHEM, DREXEL UNIV, 83- *Concurrent Pos:* Alfred P Sloan Found fel, 70; NATO sr fel, Phys Chem Lab, Oxford Univ, 71. *Mem:* Am Phys Soc; Am Chem Soc; Inst Elec & Electronic Engrs Comput Sci; AAAS. *Res:* Laser fluroescence and flash photolysis of transient species; photochemical kinetics; computers in chemical education; forty publications in gas phase molecular spectroscopy. *Mailing Add:* Dept Chem Drexel Univ Philadelphia PA 19104

SMITH, ALLEN ANDERSON, b Boston, Mass; m 74; c 2. DEVELOPMENTAL BIOLOGY, HISTOCHEMISTRY. *Educ:* Brown Univ, AB, 61; Univ Ore, PhD(anat), 69. *Prof Exp:* Instr anat, Hahnemann Med Col, 69-70; instr zool, Tel Aviv Univ, 70-71; res asst, Temple Univ, 71-74; asst prof, 74-78, ASSOC PROF BIOL, WIDENER UNIV, 78- *Concurrent Pos:* Vis scientist, Dept Org Chem, Weizmann Inst, 81. *Mem:* Am Chem Soc; Soc Develop Biol; Histochem Soc. *Res:* Sweat glands; epitheliomesenchymal interactions; pharmacology of rotaxanes. *Mailing Add:* Div Sci Widener Univ Chester PA 19013

SMITH, ALLIE MAITLAND, b Lumberton, NC, June 9, 34; m 57; c 3. HEAT TRANSFER, OPTICS. *Educ:* NC State Univ, BSME, 56, MS, 61, PhD, 66. *Prof Exp:* Assoc engr, Martin Co, 56-57; develop engr, Western Elec Co, 57-58; instr eng, NC State Col, 58-60; mem tech staff, Bell Tel Labs, 60-62; res engr, Res Triangle Inst, 62-66; supvr res, Aro, Inc, 66-79; PROF MECH ENG & DEAN, SCH ENG, UNIV MISS, 79- *Concurrent Pos:* Asst prof mech eng, Exten Div, NC State Col, 61-62; part-time assoc prof aerospace eng, Space Inst, Univ Tenn, 67-79; chmn, Thermophysics Tech Comt, Am Inst Aeronaut & Astronaut, 75-77, 10th Thermophysics Conf, 75, Terrestrial Energy Systs Comt, 77-81, 17th Aerospace Sci Meetings, 79; assoc ed, Am Inst Aeronaut & Astronaut J, J Thermophysics & Heat Transfers; ed, Radiative Transfer & Thermal Control, Thermophysics of Spacecraft & Outer Planet Entry Probes. *Honors & Awards:* Space Shuttle Flag Plaque Award, Hermann Oberth Award & Thermophysics Award, Am Inst Aeronaut & Astronaut. *Mem:* Fel Am Inst Aeronaut & Astronaut. *Res:* Radiative characteristics of surfaces and solidified gases; effects of space environment on thermal control materials; space simulation; radiation gas dynamics; solid state diffusion; heat transfer; fluid mechanics; cryogenics; vacuum. *Mailing Add:* Sch Eng Univ Miss University MS 38677

SMITH, ALTON HUTCHISON, b Long Beach, Calif, July 28, 30; m; c 1. TOPOLOGY. *Educ:* Pepperdine Col, BA, 51; Univ Southern Calif, MA, 52, PhD(math), 56. *Prof Exp:* Res mathematician, Ramo-Wooldridge Corp, 56-57; from asst prof to assoc prof 57-65, PROF MATH, CALIF STATE UNIV, LONG BEACH, 65- *Concurrent Pos:* Consult, Ramo-Wooldridge Corp, 57-58. *Mem:* Am Math Soc; Math Asn Am. *Res:* Algebraic topology; spaces with operators. *Mailing Add:* Dept Math Calif State Univ 1250 Bellflower Blvd Long Beach CA 90840

SMITH, ALVIN WINFRED, b Kooskia, Idaho, Sept 25, 33; m 58; c 4. MARINE VIROLOGY. *Educ:* Wash State Univ, BA, 55, DVM, 57; Tex A&M Univ, MS, 67; Univ Calif, Berkeley, PhD(comp path), 75. *Prof Exp:* Chief, Res Animal Br, Sch Aerospace Med, 67-69, Res Animal Div, Naval Biosci Lab, 69-78, virol, Naval Ocean Syst Ctr, 78-80; res veterinarian, 80-81, DIR RES, SCH VET MED, ORE STATE UNIV, 81- *Concurrent Pos:* Chief, Marine Mammal Res Div, Naval Biosci Lab, 74-78, Virol Sect, San Diego Zoo, 78-80. *Mem:* Am Vet Med Asn; Am Col Lab Animal Med; Int Asn Aquatic Animal Med. *Res:* Mechanisms of transmission and survival of infectious disease agents in nature. *Mailing Add:* Sch Vet Med Ore State Univ Corvallis OR 97331

SMITH, AMELIA LILLIAN, b Philadelphia, Pa, Mar 25, 24; wid; c 2. PHYSIOLOGY. *Educ:* Ursinus Col, BS, 48; Rutgers Univ, MS, 62, PhD(physiol), 72. *Prof Exp:* Res assoc physiol, Merck Inst Therapeut Res, 50-54; instr radiation physics, Lyons Inst, 54-58; biophysicist, Rutgers Univ, NB, 63-64; assoc prof, 66-76, prof physiol, 76-80, PROF BIOL SCI, KEAN COL NJ, 80- *Concurrent Pos:* Res Pharmacol, UMDNJ, 82. *Res:* Role of Cahtpase and alkaline protease in cardiac contraction band formation; electron-microscopy, electrolytic and biochemical assays including acid hydrolases as well as isolation and characterization of acid, neutral and alkaline proteases and study of effects of stress models on these enzymes; radiation science. *Mailing Add:* Dept of Biol Kean Col of NJ Union NJ 07083

SMITH, ANDERSON DODD, b Richmond, Va, May 3, 44; m 66; c 2. GERONTOLOGY, EXPERIMENTAL PSYCHOLOGY. *Educ:* Washington & Lee Univ, BA, 66; Univ Va, MA, 69, PhD(exp psychol), 70. *Prof Exp:* From asst prof to prof psychol, 70-84, DIR, GA INST TECHNOL, 85- *Concurrent Pos:* NIH res grant, Nat Inst Aging, 72-, NIMH grant, 81; ed psychol sci, J Gerontol, 81-84; affil scientist, Yerkes Regional Primate Ctr, 81-90; adj prof, Ga State Univ, 84-, Univ Ga, 89- *Honors & Awards:* Monie Ferst Res Award, 84. *Mem:* Sigma Xi; fel Am Psychol Asn; Psychonomic Soc; fel Gerontol Soc; fel Am Psychol Soc. *Res:* Experimental psychology of human memory; age-related differences in encoding, storage and retrieval processes. *Mailing Add:* Sch Psychol Ga Inst Technol Atlanta GA 30332

SMITH, ANDREW GEORGE, b Williamsport, Pa, July 11, 18; m 45; c 2. MICROBIOLOGY, MEDICAL MYCOLOGY. *Educ:* Pa State Univ, BS, 40; Univ Pa, MS, 47, PhD, 50; Am Bd Med Microbiol, dipl, 71, 83. *Prof Exp:* From asst prof to assoc prof microbiol, Sch Med, Univ Md, 50-66; dir bact prod div, BBL Div, BioQuest, Md, 66-69; from assoc prof to prof microbiol, Sch Med, Univ Vt, 69-72; from assoc prof to prof path, Sch Med, Univ Md, 72-83, dir, Microbiol Lab Hosp, 72-83, res prof med technol, 83-85, assoc prof med dermatol, 77-85; RETIRED. *Concurrent Pos:* Consult, Vet Admin Hosp, Baltimore, 62-69 & 73-84. *Honors & Awards:* Lederle Med Fac Award, 55; Barnett L Cohem Award, Am Soc Microbiol, 76. *Mem:* Fel Am Acad Microbiol; Am Soc Microbiol; Med Mycol Soc Ams. *Res:* Bacterial cytology; applied and clinical microbiology; medical mycology. *Mailing Add:* 4025 Sont Hill Dr Ellicott City MD 21043

SMITH, ANDREW PHILIP, membrane biochemistry, neurochemistry, for more information see previous edition

SMITH, ANDREW THOMAS, b Glendale, Calif, Mar 14, 46; m; c 2. BEHAVIORAL ECOLOGY, POPULATION BIOLOGY. *Educ:* Univ Calif, Berkeley, AB, 68; Univ Calif, Los Angeles, PhD(biol), 73. *Prof Exp:* Lectr zool, Univ Alta, 73-74; asst prof biol, Univ Miami, 74-78; asst prof, 78-83, ASSOC PROF ZOOL, ARIZ STATE UNIV, 83- *Concurrent Pos:* Hon consult, Int Union Conserv Nature, 78-; prin investr, Nat Geog Soc, 84-85, Nat Acad Sci, 85; prin investr, NSF, 90-92. *Mem:* AAAS; Am Soc Mammalogists; Ecol Soc Am; Soc Study of Evolution; Soc Conserv Biol; Animal Behav Soc. *Res:* Conservation ecology; population biology; dispersal; biogeography; mammalogy; reproductive strategies; behavioral ecology. *Mailing Add:* Dept of Zool Ariz State Univ Tempe AZ 85287

SMITH, ANN, b London, Eng, June 3, 46. HEMEIRON METABOLISM, RECEPTOR TRANSPORT. *Educ:* Univ London, PhD(biochem), 74. *Prof Exp:* Asst prof, 83-86, ASSOC PROF BIOCHEM, MED CTR, LA STATE UNIV, 86- *Mem:* Am Soc Biol Chemists; Biochem Soc Eng; Sigma Xi; AAAS. *Res:* Structure and function of hemopexin; liver cell functions; prothyrin phototherapy. *Mailing Add:* Dept Biochem & Molecular Biol La State Univ Med Ctr 1091 Perdido St New Orleans LA 70112

SMITH, ANTHONY JAMES, b Kansas City, Mo, Aug 19, 18; m 43; c 4. ELECTROCHEMISTRY. *Educ:* Univ Mo, AB, 42. *Prof Exp:* Res chemist, Nat Fertilizer Develop Ctr, Tenn Valley Authority, 42-81; RETIRED. *Mem:* Am Chem Soc; Int Asn Hydrogen Energy; Nat Mgt Asn. *Res:* Electrolytic production of hydrogen; corrosion; microwave dielectric properties, density, pH, specific gravity, vapor pressure, viscosity of phosphatic solutions; electrolytic production of potassium phosphates; purification of phosphoric acid. *Mailing Add:* 710 Prospect St Florence AL 35630

SMITH, ARCHIBALD WILLIAM, b Edmonton, Alta, Jan 6, 30; m 53; c 3. LASERS. *Educ:* Univ Alta, BSc, 52, MSc, 53; Univ Toronto, PhD(physics), 55. *Prof Exp:* Staff mem, Defense Res Bd, Can, 56-61; staff mem, Thomas J Watson Res Ctr, IBM Corp, 62-76, sr adv engr, 77-80; with Discovision, 80-81; ADV ENGR, STORAGE TECHNOL CORP, 81- *Mem:* sr mem Inst Elec & Electronics Eng. *Res:* Laser and semiconductor physics. *Mailing Add:* Storage Technol Corp 2270 S 88 St Louisville CO 80028-4257

SMITH, ARLO IRVING, b Ft Smith, Ark, July 23, 11; m 37; c 3. BIOLOGY. *Educ:* Hendrix Col, AB, 32; Northwestern Univ, MS, 35; Univ Wash, PhD(bot), 38. *Prof Exp:* Prof biol, McMurry Col, 38-39; instr, Tex Tech Col, 39-42, asst prof bot, 45-46; from assoc prof to prof biol, 46-77, EMER PROF BIOL, SOUTHWESTERN AT MEMPHIS, 77- *Concurrent Pos:* Dir, Southwestern Arboretum, 55-77. *Mem:* Fel AAAS; Am Inst Biol Sci; Ecol Soc Am; Bot Soc Am. *Res:* Systematic botany; ecology; science education; wild flowers of south central United States, including some common trees, vines, shrubs and ferns. *Mailing Add:* 1914 Poplar No 302 Memphis TN 38104

SMITH, ARTHUR CLARKE, b Bartlesville, Okla, Sept 23, 29; m 55; c 3. SOLID STATE PHYSICS. *Educ:* Univ Kans, BS, 51; Harvard Univ, MA, 54, PhD(appl physics), 58. *Prof Exp:* Res fel & instr appl physics, Harvard Univ, 58-59; from asst prof to assoc prof, 59-68, PROF ELEC ENG, MASS INST TECHNOL, 68- *Mem:* Am Asn Physics Teachers; Am Phys Soc. *Mailing Add:* 51 Follen Rd Lexington MA 02173

SMITH, ARTHUR GERALD, b Newton, Kans, Jan 12, 29; m 49; c 1. ELECTROCHEMISTRY, ACCELERATED CORROSION. *Educ:* Phillips Univ, AB, 50; Iowa State Univ, MS, 53. *Prof Exp:* Chemist, Standard Oil Co, 53-58; res scientist, 58-68, sr res scientist, 68-76, PRIN RES SCIENTIST ASSOC, FORD MOTOR CO, 76- *Mem:* Am Chem Soc. *Res:* Paint adhesion failure mechanism studies; development of novel paints and paint application techniques; corrosion studies of single and multimetal systems; development of accelerated corrosion tests. *Mailing Add:* 3404 Washington Midland MI 48640

SMITH, ARTHUR HAMILTON, b Santa Barbara, Calif, Mar 28, 16; m 39; c 2. PHYSIOLOGY. *Educ:* Univ Calif, AB, 38, PhD(comp physiol), 48. *Prof Exp:* Asst animal husb, Univ Calif, Davis, 37-41 & 46-47, sr biochemist, 48; physiologist, Radiation Lab, Univ Calif, Berkeley, 48, NRC-AEC fel med sci, 48-50; lectr poultry husb, 50-51, asst prof, 51-55, assoc prof & assoc physiologist, Agr Exp Sta, 55-62, prof poultry husb & physiologist, 62-64, prof physiol & physiologist, 64-86, EMER PROF PHYSIOL & EMER PHYSIOLOGIST, AGR EXP STA, UNIV CALIF, DAVIS, 86- *Concurrent Pos:* Mem Comn Gravitational Physiol, Int Union Physiol Sci, 73; secy-treas, Galileo Found, 87- *Mem:* Aerospace Med Soc; Soc Exp Biol & Med; Am Phys Soc; Biophys Soc; Undersea Med Soc; Hist Sci Soc. *Res:* Environmental physiology; gravitational physiology. *Mailing Add:* Dept of Animal Physiol Univ of Calif Davis CA 95616-8519

SMITH, ARTHUR JOHN STEWART, b Victoria, BC, June 28, 38; m 66; c 2. EXPERIMENTAL HIGH ENERGY PHYSICS. *Educ:* Univ BC, BA, 59, MSc, 61; Princeton Univ, PhD(physics), 66. *Prof Exp:* Volkswagen Found fel physics, Deutsches Elektronen-Synchrotron, Hamburg, WGer, 66-67; from instr to assoc prof, Princeton Univ, 67-78, assoc chmn, 80-83, PROF PHYSICS, PRINCETON UNIV, 78-, CHMN, 90- *Concurrent Pos:* Vis scientist, Brookhaven Nat Lab & Fern Lab, SSC Lab. *Mem:* Fel Am Phys Soc. *Res:* Experimental high energy particle physics; electromagnetic and weak interactions. *Mailing Add:* Joseph Henry Labs Princeton Univ Princeton NJ 08544

SMITH, ARTHUR R, b Pittsburgh, Pa, Feb 9, 31; m 53; c 3. ECONOMIC GEOLOGY. *Educ:* Pa State Univ, BS, 52; Univ Calif, Berkeley, MS, 58, MBA, 70. *Prof Exp:* Explor geologist, Phelps Dodge Corp, 58-63; geologist, Calif Div Mines & Geol, 63-70; SR RES GEOLOGIST, MINERAL EXPLOR & DEVELOP DEPT, UTAH INT INC, SAN FRANCISCO, 70- *Mem:* Am Inst Mining, Metall & Petrol Eng. *Res:* Regional geology; mineral economic studies; geochemical exploration methods. *Mailing Add:* Dept Earth Sci West Chester Univ West Chester PA 19383

SMITH, B(LANCHARD) D(RAKE), JR, b New Orleans, La, Aug 22, 25; m 45; c 5. ELECTRICAL ENGINEERING. *Educ:* Ga Inst Technol, BS, 45; Mass Inst Technol, MS, 48. *Prof Exp:* Asst elec eng, Mass Inst Technol, 46-48; from engr to mgr transp systs ctr, Melpar, Inc, Westinghouse Air Brake Co, 48-68; vpres & tech dir, Appl Systs Technol, Inc, 68-78; consult, 78-80;

chief scientist, Melpar Div, E-Systs, 80-89; CONSULT & CHIEF SCIENTIST, ST RES, 89- *Honors & Awards:* Thompson Award, Inst Elec & Electronic Engrs, 55. *Mem:* Inst Elec & Electronic Engrs; Sigma Xi. *Res:* Electronic systems. *Mailing Add:* 2509 Ryegate Lane Alexandria VA 22308

SMITH, BARBARA D, b Boston, Mass, Mar 17, 43; m; c 3. CONNECTIVE TISSUE, GENE EXPRESSION. *Educ:* Boston Univ, PhD(biochem), 70. *Prof Exp:* From asst prof to assoc prof, 76-91, PROF BIOCHEM, SCH MED, BOSTON UNIV, 91-; RES CHEMIST, VET ADMIN MED CTR, BOSTON, MASS, 76- *Concurrent Pos:* Res, prin investr, 76- *Mem:* Am Soc Cell Biol; Am Soc Biochem & Molecular Biol. *Mailing Add:* 20 Harrison St Brookline MA 02146-6958

SMITH, BENJAMIN WILLIAMS, b Falls Church, Va, Aug 9, 18; m 40; c 5. BIOCHEMISTRY. *Educ:* Va Polytech Inst, BS, 40; George Washington Univ, MS, 47, PhD(biochem), 51. *Prof Exp:* From instr to assoc prof, 49-69, PROF BIOCHEM, MED SCH, GEORGE WASHINGTON UNIV, 69- *Mem:* AAAS; Asn Am Med Cols. *Res:* Enzymes; amylase; carbohydrate metabolism. *Mailing Add:* 2300 Eye St NW Rm 543 2300 Eye St NW Washington DC 20037

SMITH, BERNARD, b New York, NY, Aug 11, 27. PHYSICS, OPERATIONS RESEARCH. *Educ:* City Col New York, BS, 48; Columbia Univ, AM, 51, PhD(physics), 54. *Prof Exp:* Lectr elec eng & physics, City Col New York, 48-54; mem tech staff, Bel Tel Labs, Inc, 54-59; staff consult, Gen Tel & Electronics Labs, Inc, 59-61, mgr, 61-63, sr scientist & mgr, 63-70; chief scientist, Marcom Inc, 70-71, vpres & chief scientist, 71-84; dir telecomm, City NY Off Telecomm, 84-88; RETIRED. *Mem:* Asn Comput Mach; Am Phys Soc; sr mem Inst Elec & Electronics Engrs; Soc Indust & Appl Math; Inst Elec & Electronic Engrs Commun Soc; Inst Elec & Electronic Engrs Computer Soc. *Res:* Cryophysics; telecommunication systems; statistical communication theory; operations research; computer science. *Mailing Add:* 98-05 63 Rd Rego Park NY 11374

SMITH, BERTRAM BRYAN, JR, b Fort Jackson, SC, Sept 20, 42; m 72. SCIENCE POLICY. *Educ:* Univ Ala, BS, 64; Purdue Univ, PhD(chem), 70. *Prof Exp:* Gen phys scientist, Foreign Sci Technol Ctr, 70-79, GEN PHYS SCIENTIST, OFF OF DEP CHIEF OF STAFF FOR INTELL, DEPT ARMY, 79- *Mem:* Am Chem Soc; Am Phys Soc; AAAS. *Res:* Liquid theory; scattering theory. *Mailing Add:* 9543 Hunt Square Ct Springfield VA 22153

SMITH, BETTY F, b Magnolia, Ark, June 29, 30. TEXTILE CHEMISTRY, CARBOHYDRATE CHEMISTRY. *Educ:* Univ Ark, BS, 51; Univ Tenn, MS, 57; Univ Minn, PhD(textile), 60, PhD(biochem), 65. *Prof Exp:* Home agent home econ, Ark Agr Exten Serv, 51-56; assoc prof textiles, Cornell Univ, 65-70, chmn, Dept Textiles & Clothing, 68-69; PROF & HEAD DEPT, DEPT TEXTILES & CONSUMER ECON, UNIV MD, 70- *Mem:* Am Chem Soc; Am Asn Textile Chemists & Colorists; fel Textile Inst. *Res:* Flammability of polyester cotton blends and flammability test methods; performance properties of textile materials; chemical finishing of textiles. *Mailing Add:* 9216 St Andrews Pl College Park MD 20742

SMITH, BILL ROSS, b Stamford, Tex, Sept 22, 41. SOIL SCIENCE. *Educ:* Tex Tech Univ, BS, 64; Univ Ariz, MS, 66; NC State Univ, PhD(soil sci), 70. *Prof Exp:* Soil scientist, Soil Conserv Serv, USDA, 63 & Wake County Health Dept, NC, 70-73; from asst prof to assoc prof, 73-88, PROF AGRON & SOILS, CLEMSON UNIV, 88- *Mem:* Am Soc Agron; Soil Sci Soc Am; Soil & Water Conserv Soc. *Res:* Soil genesis and classification; evaluation of soils for different kinds of land use; soil mineralogy. *Mailing Add:* Dept Agron & Soils Clemson Univ Clemson SC 29634-0359

SMITH, BOB L(EE), b Topeka, Kans, Jan 26, 26; m; c 2. CIVIL ENGINEERING. *Educ:* Kans State Univ, BS, 48, MS, 53; Purdue Univ, PhD, 64. *Prof Exp:* From instr to assoc prof, 48-65, PROF CIVIL ENG, KANS STATE UNIV, 65- *Concurrent Pos:* Mem low volume roads & oper effects of geometrics comts, Transp Res Bd, Nat Acad Sci-Nat Res Coun, geometric design comt, Am Soc Civil Engrs. *Mem:* Am Soc Civil Engrs; Sigma Xi; Inst Transp Engrs. *Res:* Trip generation and distribution; economic analysis as related to transportation systems; traffic engineering; geometric design of highways; traffic assignment; highway safety design; accident reconstruction; expert systems. *Mailing Add:* 737 Midland Manhattan KS 66502

SMITH, BRAD KELLER, b Santa Monica, Calif, Apr 2, 55; m 82. TECTONOPHYSICS. *Educ:* Univ Wash, BS, 77; Univ Calif, Berkeley, MA, 79, PhD(geol), 82. *Prof Exp:* Teaching fel crystallog, Swiss Fed Polytech, 82-83; ASST PROF GEOL, ARIZ STATE UNIV, 84- *Mem:* Mineral Soc Am; Am Geophys Union. *Res:* High pressure-temperature deformation of silicate minerals; defect microstructures caused by deformation; high-resolution transmission electron microscope imaging of dislocation structures in minerals. *Mailing Add:* Dept Earth Sci Pacific Lutheran Univ Dacoma WA 78447

SMITH, BRADFORD ADELBERT, b Cambridge, Mass, Sept 22, 31; m 54; c 4. ASTRONOMY. *Educ:* Northeastern Univ, BS, 54. *Prof Exp:* Res engr, Williamson Develop Co, 54-55; assoc astronr, Res Ctr, NMex State Univ, 57-64, dir observ, 64-69, dir planetary progs, 69-74; assoc prof lunar & planetary lab & assoc astronomer, Steward Observ, Univ Ariz, 74-88. *Mem:* Am Astron Soc; Int Astron Union. *Res:* Planetary and lunar astronomy; image aberration electromechanical optical servo systems. *Mailing Add:* 82-6012 Puuhonua Napoopoo HI 96704

SMITH, BRADLEY EDGERTON, b Cedar-Vale, Kans, Jan 4, 33; m 53; c 2. ANESTHESIOLOGY. *Educ:* Tulsa Univ, BSc, 54; Okla Univ, MD, 57. *Prof Exp:* Res fel obstet anesthesiol, Columbia Univ, 60-61; instr anesthesiol, Yale Univ, 62-63; assoc prof, Univ Miami, 63-69; PROF ANESTHESIOL & CHMN DEPT, VANDERBILT UNIV, 69- *Concurrent Pos:* Consult, FDA, 68-74 & 75-76, mem adv coun anesthetic & respiratory drugs, 70-72; assoc examr, Nat Bd Respiratory Ther, 69- & Am Bd Anesthesiologists, 77-; consult, Vet Admin, 69-; mem comt anesthetic toxicity, Nat Res Coun-Nat Acad Sci, 72-74; fac Sen, Vanderbilt Univ, 72-76. *Mem:* Am Col Chest Physicians; Asn Univ Anesthetists; Am Soc Anesthesiologists; assoc fel Am Col Obstet & Gynec; Soc Obstet Anesthesia & Perinatology (pres, 70-71). *Res:* Obstetric anesthesia; anesthetic toxicity; developmental pharmacology and teratology; perinatal physiology; resuscitation of the newborn. *Mailing Add:* Dept Anesthesiol Vanderbilt Univ Sch Med 2301 TVC Nashville TN 37232-2125

SMITH, BRADLEY RICHARD, b Provo, Utah, Feb 26, 56; m 83; c 3. SCIENTIFIC COMPUTER-AIDED VISUALIZATION. *Educ:* Univ Utah, BUS, 80; Johns Hopkins Univ, MA, 83; Duke Univ, PhD(anat), 88. *Prof Exp:* PRES, BIOIMAGE, 83-; INSTR ANAT, DURHAM ARTS COUN, 87-; RES ASSOC, DEPT RADIOL, MED CTR, DUKE UNIV, 89- *Mem:* Am Soc Zoologists; Asn Med Illusr. *Res:* Magnetic resonance microscopy of developing cardiovascular system; computer-aided reconstruction of 3-dimensional data sets; computer visualization and volume rendering of medical imaging. *Mailing Add:* 17 Prentiss Pl Durham NC 27707-3974

SMITH, BRIAN RICHARD, b Glen Cove, NY, May 7, 52. HEMATOLOGY, BONE MARROW TRANSPLANTATION. *Educ:* Princeton Univ, AB, 72; Harvard Univ, MD, 76. *Prof Exp:* Asst prof med, Med Sch, Harvard Univ, 85-89; ASSOC PROF MED, MED SCH, YALE UNIV, 89- *Mem:* Am Asn Immunol; Am Soc Hemat; fel Am Col Physicians. *Mailing Add:* Dept Lab Med Yale Med Sch 333 Cedar St New Haven CT 06510

SMITH, BRIAN THOMAS, b Toronto, Ont, Apr 20, 42; m 65; c 2. NUMERICAL ANALYSIS, NUMERICAL SOFTWARE. *Educ:* Univ Toronto, BS, 65, MS, 67, PhD(comput sci), 69. *Prof Exp:* Res asst appl math, Swiss Fed Inst Technol, 69; asst scientist comput sci, Argonne Nat Lab, 70-75, scientist, 76-89; PROF, DEPT COMPUT SCI, FARRIS ENG CTR, UNIV NMEX, 89- *Concurrent Pos:* Mem numerical software work group 2.5, Int Fedn Info Processing; mem, Fortran Standards Comt, Am Nat Standards Inst; consult, Numerical Algorithms Group, Inc; mem, Lang Working Group, Dept Energy. *Mem:* Soc Indust & Appl Math; Asn Comput Mach. *Res:* Numerical software; computational aspects related to study of nonassociative algebras; automated reasoning; proving claims about programs; Fortran standardization. *Mailing Add:* Dept Computer Sci Farris Eng Ctr Univ of NMex Albuquerque NM 87131

SMITH, BRUCE BARTON, b Poplar Bluff, Mo, Sept 28, 41; m 63; c 2. PLANT MORPHOLOGY. *Educ:* Ark State Univ, BS, 63; Univ Miss, MS, 66; Univ SC, PhD(biol), 71. *Prof Exp:* Instr biol, Parsons Col, 66-67; asst prof, Atlantic Christian Col, 67-68; instr, Univ SC, 69-71; asst prof, 71-74, assoc prof, 74-81, PROF BIOL, YORK COL PA, 81-, CHMN DEPT, 81- *Concurrent Pos:* Sigma Xi res grant-in-aid, York Col Pa, 71- *Mem:* Bot Soc Am; Am Inst Biol Sci. *Res:* Angiosperm embryology and its use in phylogenetic studies of flowering plants. *Mailing Add:* Dept of Biol York Col of Pa Country Club Rd York PA 17403

SMITH, BRUCE H, b New York, NY, Feb 16, 19; m 43; c 4. MEDICINE. *Educ:* Syracuse Univ, AB, 40, MD, 43. *Prof Exp:* Med Corps, US Navy, 43-71, resident path, US Naval Hosp, Brooklyn, NY, 45-47; resident, Long Island Col Hosp, 47-49, dir labs, US Naval Hosp, Mare Island, Calif, 50-55, Philadelphia, 55-63, dep dir, Armed Forces Inst Path, 63-67, dir, 67-71; PROF PATH, SCH MED, GEORGE WASHINGTON UNIV, 71- *Concurrent Pos:* Fel path, Harvard Univ, 49-50; vis prof, Sch Med, Temple Univ, 57-64; clin prof, Georgetown Univ, 67-71. *Mem:* Fel Am Col Physicians; Int Acad Path; NY Acad Sci; AAAS; Am Asn Pathologists; Col Am Pathol. *Res:* Pathology. *Mailing Add:* Dept Path Vet Admin Med Ctr 50 Irving St NW Washington DC 20422

SMITH, BRUCE NEPHI, b Logan, Utah, Apr 3, 34; m 59; c 6. PLANT PHYSIOLOGY. *Educ:* Univ Utah, BS, 59, MS, 62; Univ Wash, PhD(bot), 64. *Prof Exp:* Asst bot, Univ Utah, 58-60; asst, Univ Wash, 60-62, actg instr, 62-63, asst, 63-64; res fel plant physiol, Univ Calif, Los Angeles, 64-65; res fel geochem, Calif Inst Technol, 65-68; asst prof bot, Univ Tex, Austin, 68-74; assoc prof, bot, Brigham Young Univ, 74-79, chmn bot & range sci, 76-79, dean col Biol & Agr, 82-88, PROF BOT, BRIGHAM YOUNG UNIV, 79- *Concurrent Pos:* Guest prof, Tech Univ Munich, 89; Orgn, Econ Coop & Develop fel, 89. *Mem:* AAAS; Am Soc Plant Physiol; Bot Soc Am; Geochem Soc; Soc Environ Geochem & Health; Am Asn Univ Prof; Sigma Xi. *Res:* Carbon, hydrogen, oxygen and nitrogen cycles followed by fractionation of natural abundance ratios of the stable isotopes; plant volatiles; trace metals in plants; plant metabolism and growth. *Mailing Add:* Dept Bot & Range Sci Brigham Young Univ Provo UT 84602

SMITH, BRYCE EVERTON, b Lotumbe, Zaire, Oct 21, 30; US citizen; m 52; c 3. ECOLOGY. *Educ:* Univ Mich, BSF, 52, AM, 57; Univ Wis, PhD(bot), 65. *Prof Exp:* Forester, Bowaters Southern Paper Corp, 54-55; teacher, High Schs, 57-60; asst prof biol sci, Western Ill Univ, 65-67; assoc prof biol, Eastern Conn State Col, 67-70; chmn dept biol sci, 70-76, assoc prof biol, 70-76, PROF BIOL, LAKE SUPERIOR STATE UNIV, 76- *Mem:* AAAS; Am Inst Biol Sci; Ecol Soc Am; Sigma Xi. *Res:* Interrelationships of higher plants in forest communities. *Mailing Add:* Biol & Chem Dept Lake Superior State Col Sault Ste Marie MI 49783

SMITH, BUFORD DON, b Omega, Okla, Feb 18, 25; m 47; c 2. CHEMICAL ENGINEERING. *Educ:* Okla State Univ, BS, 50, MS, 51; Univ Mich, PhD(chem eng), 54. *Prof Exp:* Chem engr, Humble Oil & Refining Co, 54-58; assoc prof chem eng, Purdue Univ, 58-65; PROF CHEM ENG & DIR THERMODYN RES LAB, WASH UNIV, 65- *Concurrent Pos:* Consult, Allison Div, Gen Motors Corp, 63, Sun Oil Co, 65 & Monsanto Co & Am Oil Co, 66. *Mem:* Am Inst Chem Engrs; Am Chem Soc. *Res:* Design of vapor-liquid separation processes; thermodynamics of liquid mixtures. *Mailing Add:* 94 Lake Forest St Louis MO 63117

SMITH, BURTON JORDAN, b Chapel Hill, NC, Mar 21, 41; m 66 66; c 2. COMPUTER ARCHITECTURE. *Educ:* Univ NMex, BSEE, 67; Mass Inst Technol, MSEE, 68, EE, 69, ScD(elec eng), 72. *Prof Exp:* Teaching asst elec eng, Mass Inst Technol, 67-70, instr, 70-72; from asst prof to assoc prof elec eng, Univ Colo, Denver, 72-79; vpres, Res & Develop, Denelcor, 79-85; fel, supercomputing Res Ctr, 85-88; CHMN & CHIEF SCIENTIST, TERA COMPUTER CO, 88- *Concurrent Pos:* Consult, Hendrix Electronics, Inc, 67-72 & Denelcor, Inc, 74-; dir, Sci Electronics Corp, 68-70. *Honors & Awards:* Eckert-Mauchly Award, Inst Elec & Electronic Engrs-Asn Comput Mach, 91. *Mem:* Inst Elec & Electronic Engrs; Asn Comput Mach. *Res:* Architecture of parellel computers; interface between hardware and software. *Mailing Add:* Tera Computer Co 400 N 34th St Suite 300 Seattle WA 98103

SMITH, BYRON COLMAN, b Crawfordsville, Ind, Apr 14, 24; m 71. COMPARATIVE ANATOMY. *Educ:* Ind State Univ, BS, 48; DePauw Univ, MA, 52; Univ Ga, PhD(zool), 58. *Prof Exp:* Asst prof zool, Univ SC, 58-64; assoc prof, 64-71, PROF BIOL, UNIV SOUTHERN MISS, 71- *Mem:* AAAS. *Res:* Acarology; ecology of the desert spider mite; Tetranychus desertorium banks on cotton; micro-fauna population of soils in the Sand Hill region of South Carolina and areas of the Piedmont and Coastal Plains. *Mailing Add:* Dept Biol Box 8444 Univ Southern Miss Hattiesburg MS 39401

SMITH, C(HARLES) WILLIAM, b Va, Jan 1, 26; m 50; c 2. ENGINEERING MECHANICS. *Educ:* VaPolytech Inst, BS, 46, MS, 49. *Prof Exp:* From instr to prof, 47-81, ALUMNI DISTINGUISHED PROF ENG SCI & MECH, VA POLYTECH INST & STATE UNIV, 81- *Concurrent Pos:* Instr, Exten, Univ Va, 57-58; lectr grad eng training progs, Western Elec Co & Gen Elec Co, 63-64; consult, Brunswick Corp, 65, Masonite Corp, 70, Polysci Corp, 71, US Army Missile Command, 71-72 & Kollmorgen Corp, 72-74; proj dir, NASA grants, 71-75; prin investr, NSF, 73-86, proj dir, 73-; proj dir, Delft Univ Technol, 75-76, Flight Dynamics Lab, US Air Force, 75-77, 86- & Oak Ridge Nat Lab, 75-78; ed, Fracture Mechanics, 78; bd ed, Theoret & Appl Fracture Mechs, 82- *Honors & Awards:* M M Frocht Award, Soc Exp Stress Analysis, 83; Except Sci Achievement Award, NASA, 86. *Mem:* Int Asn Struct Mech in Reactor Technol; fel Soc Exp Stress Anal; Am Soc Testing & Mat; Am Soc Mech Engrs; Soc Eng Sci. *Res:* Theoretical and experimental continuum solid mechanics, especially fracture mechanics and experimental stress analysis. *Mailing Add:* Dept Eng Sci & Mech Va Polytech Inst & State Univ Blacksburg VA 24061

SMITH, CALVIN ALBERT, b Troy, NH, Mar 11, 35; m 57; c 2. BOTANY. *Educ:* Wheaton Col, Ill, BS, 57; Miami Univ, MA, 60; Rutgers Univ, PhD(bot), 63. *Prof Exp:* From asst prof to assoc prof, 63-75, PROF BIOL, BALDWIN-WALLACE COL, 75- *Mem:* Bot Soc Am; Am Inst Biol Sci. *Res:* Shoot apices in the family Moraceae. *Mailing Add:* Dept Biol Baldwin-Wallace Col Berea OH 44017

SMITH, CAREY DANIEL, b Kenedy, Tex, July 10, 32; m 54; c 4. UNDERSEA WARFARE TECHNOLOGY. *Educ:* Univ Tex, Austin, BA & BS, 59. *Prof Exp:* Res physicist, Appl Res Labs, Univ Tex, Austin, 59-64; supvry engr, Sonar Signal Processing Sect Head, Bur Ships, Dept Navy, 64-66, supvry physicist & tech dir, Sonar Technol Br, Naval Ships Systs Command, 66-69, supvry physicist & dir, 69-74, supvry physicist & dir, Sonal Technol Off, Naval Sea Systs Command, 74-79 & Undersea Warfare Tech Off, 79-87; CONSULT, UNDERSEA WARFARE TECHNOL MGT, 87- *Concurrent Pos:* Sonar foreign liaison officer, Naval Ship-Sea Systs Command, 66-79; ASW foreign liaison officer, Naval Sea Systs Command, 79-87, dir, Combat Systs Technol Off, 79-81; chmn, Sonar Panel, Tech Coop Prog, Dept Defense, 83-87; sr navy tech adv, Undersea Warfare Systs Div, Am Defense Prepareness Asn, 80-87. *Honors & Awards:* Chevalier Award, Pres of France, 81. *Mem:* Fel Acoust Soc Am. *Res:* Design and field test a three color display that significantly increased the dynamic range of high resolution object location sonars; ocean environmental acoustics. *Mailing Add:* 1638 Dineen Dr McLean VA 22101

SMITH, CARL CLINTON, b Lima, Ohio, July 12, 14; m; c 3. DRUG METABOLISM, PRIMATES AS ANIMAL MODELS. *Educ:* DePauw Univ, AB, 36; Univ Cincinnati, MS, 37, PhD(biochem), 40. *Prof Exp:* EMER PROF ENVIRON HEALTH, COL MED, UNIV CINCINNATI, 51- *Concurrent Pos:* Consult, toxicol. *Mem:* Soc Toxicol; Am Soc Pharmacol Exp Therapeut; Am Chem Soc; Am Col Toxicol; fel AAAS. *Res:* Toxicology; drug metabolism. *Mailing Add:* Dept Environ Health Col Med Univ Cincinnati 3223 Eden Ave Cincinnati OH 45267

SMITH, CARL HOFLAND, b Minneapolis, Minn, June 6, 42; m 66; c 2. MAGNETIC MATERIALS, AMORPHOUS ALLOYS. *Educ:* Hamilton Col, AB, 64; Univ Minn, MA, 69 & PhD(physics), 71. *Prof Exp:* Vis asst prof physics, Macalester Col, St Paul, Minn, 69-71; entrepreneur, Portstar Industs, Nyack, NY, 71-72; chief scientist, Auto Res Corp, subsid of Bijur Lubricating Co, Oakland, NJ, 72-79; sr develop assoc, Metglas Prods, 79-83, sr res assoc, Corp Res, 83-85, SUPVR MAGNETIC ALLOYS RES, ALLIED-SIGNAL RES & TECHNOL, MORRISTOWN, NJ, 85- *Mem:* Sigma Xi; Am Phys Soc; sr mem Inst Elec & Electronics Engrs. *Res:* Measurement and study of magnetic properties of rapidly quenched ferromagnetic amorphous alloys and their application to high-frequency and pulse-power systems. *Mailing Add:* Allied-Signal Inc PO Box 1021R Morristown NJ 07962-1021

SMITH, CARL HUGH, b New York, NY, Nov 18, 34; m 69; c 2. PERINATAL RESEARCH, CLINICAL CHEMISTRY. *Educ:* Swarthmore Col, BA, 55; Yale Univ, MD, 59. *Prof Exp:* From instr to asst prof path, 65-72, from asst prof to assoc prof pediat & path, 72-82, PROF PEDIAT & PATH, WASH UNIV SCH MED, 82-, DIR CLIN LABS, 88- *Concurrent Pos:* Mem, Human Embryol & Develop Study Sect, NIH, 77-79. *Honors & Awards:* Borden Res Award, 59. *Mem:* Perinatal Res Soc; Soc Pediat Res; Am Physiol Soc; Soc Gynecol Invest; Am Asn Clin Chemists. *Res:* Transfer of amino acids, calcium and glucose by placenta, structure and function of its plasma membranes. *Mailing Add:* Dept Pediat Wash Univ Children's Hosp 400 S Kingshighway Blvd St Louis MO 63110

SMITH, CARL WALTER, b Salem, Mass, Dec 15, 37; m 61. PHYSICS. *Educ:* Earlham Col, BA, 60; Brown Univ, ScM, 63, PhD(physics), 66. *Prof Exp:* PHYSICIST, SANDIA LABS, 76- *Mem:* Am Geophys Union. *Res:* Mechanical wave propagation. *Mailing Add:* Org 9311 Sandia Nat Labs Box 5800 Albuquerque NM 87185

SMITH, CARL WALTER, JR, b Lamont, Okla, Mar 20, 27; m 55; c 2. NUCLEAR MEDICINE, ENDOCRINOLOGY. *Educ:* Univ Okla, BA & MD, 53; Am Bd Nuclear Med, dipl; Am Bd Internal Med, dipl. *Prof Exp:* Asst prof med, 59-63, asst prof med & radiol, 63-65, dir outpatient clins, 60-65, assoc prof med, prof radiol sci, dir Div Nuclear Med, Col Med, Univ Okla Health Sci Ctr, 65-82; ASST CHIEF, NUCLEAR MED, VA MED CTR, 82- *Concurrent Pos:* Mem, Okla Rad Adv Comt, 74. *Mem:* Am Soc Nuclear Med; Endocrine Soc; Am Col Nuclear Physicians; AMA. *Res:* Endocrinological diseases; nuclear medicine, including methodology, development of radiopharmaceuticals and clinical investigation related to nuclear medicine. *Mailing Add:* VAMC 115 921 NE 13th St Oklahoma City OK 73104

SMITH, CAROLYN JEAN, b Fitzgerald, Ga; m 75. CHEMISTRY. *Educ:* Mercer Univ, AB, 59; Emory Univ, PhD(org chem), 62. *Prof Exp:* Teaching asst chem, Emory Univ, 59-60; res chemist, E I du Pont de Nemours & Co, Inc, 62-71; asst prof, Lincoln Univ, 72-73; instr chem, Del Tech & Community Col, 73-75; lectr, Wilmington Col, 75-76; assoc prof, Cheyney State Col, 76-82; assoc prof chem, 83-90, PROF CHEM & PHYSICS, DEL COUNTY COMMUNITY COL, 91- *Concurrent Pos:* Instr, Oxford Col, Emory Univ, 62; vis assoc prof, Lincoln Univ, 77; contrib ed, World Bk Encycl, 83- *Mem:* Am Chem Soc; Am Asn Univ Professors. *Mailing Add:* Del County Community Col Media PA 19063

SMITH, CARROLL N, b Menlo, Iowa, Nov 5, 09; m 37; c 1. MEDICAL ENTOMOLOGY. *Educ:* George Washington Univ, AB, 32, MA, 34, PhD(med entom), 41. *Prof Exp:* Jr entomologist, USDA, Washington, DC, 35-37, asst entomologist, Mass, 37-41, assoc entomologist, Ga, 41-46, entomologist, Fla, 46-63, dir insect attractants behav & basic biol res lab, 63-69; consult-dir, Res Unit Genetic Control Mosquitoes, WHO, New Delhi, 70; ED, annual rev entom, Annual Rev Inc, 71-77; RETIRED. *Concurrent Pos:* Courtesy prof entom, Univ Fla, Gainesville, 63-69; mem, WHO Expert Panel Insecticides & Food & Agr Orgn expert panel on tick-borne dis of livestock; assoc mem, Rickettsial Dis Comn & Malaria Comn Armed Forces Epidemiol Bd; consult, S C Johnson & Son, 74-75. *Honors & Awards:* Medal of Honor, Am Mosquito Control Asn, 76. *Mem:* Hon mem Entom Soc Am (pres, 64); Am Mosquito Control Asn. *Res:* Biology, behavior and control of arthropods affecting man and animals. *Mailing Add:* 317 NW 32nd St Gainesville FL 32607

SMITH, CARROLL WARD, b Abilene, Tex, Dec 24, 27; m 55; c 4. BIOCHEMISTRY, ENVIRONMENTAL HEALTH. *Educ:* Univ Okla, BS, 58, MS, 59, PhD(environ health), 68. *Prof Exp:* Res chemist, Samuel Roberts Noble Found, Okla, 59-64; res biochemist, Civil Aeromed Res Inst, Fed Aviation Agency, Oklahoma City, 64-65; asst prof chem, 68-74, assoc prof, 74-81, PROF CHEM, HARDING COL, 81-, RES ASSOC PHYSIOL OF EXERCISE, 68- *Mem:* Am Chem Soc; Nat Speleol Soc. *Res:* Biochemistry and physiology of exercise with emphasis on preventive and rehabilitative medicine. *Mailing Add:* Dept Phys Sci Harding Univ Searcy AR 72143

SMITH, CASSANDRA LYNN, b New York, NY, May 25, 47. GENOMICS, CHROMOSOME STRUCTURE. *Educ:* WVa Univ, BA, 67, MS, 70; Tex A&M Univ, PhD(genetics), 74. *Prof Exp:* Fel, Dept Genetics, Pub Health Res Inst of City of NY, Inc, 74-78; res assoc, Chem Dept, Columbia Univ, 78-81, assoc res scientist, Dept Genetics & Develop, 81-87, asst prof, Dept Microbiol & Dept Psychiat, 87-89; sr scientist, Human Genome Ctr, 89-91, SR SCIENTIST CHEM BIODYNAMICS, LAWRENCE BERKELEY LAB, 91-; ASSOC PROF RESIDENCE, DEPT MOLECULAR & CELL BIOL, UNIV CALIF, BERKELEY, 88- *Concurrent Pos:* Consult, Pharmacia-Lkb, Sweden, 84-89, FMC Corp, 86-89, Promega, 90 & Bolhringer-Mannheim Gmbh Ger, 90-; managing ed, Int J Human Genome, 89-; exec ed, Gene Anali Tech & Applications, 89- *Mem:* Am Soc Microbiol; Genetics Soc Am; Am Soc Biochem & Molecular Biol; AAAS; Int Human Genome Orgn; Harvey Soc. *Res:* Developing techniques that allow molecular characterization of whole chromosomes; structure and function of chromosomes in both prokaryotes and eukaryotes. *Mailing Add:* Dept Molecular & Cell Biol Univ Calif 529 Stanley Berkeley CA 94720

SMITH, CATHERINE AGNES, b St Louis, Mo, Jan 5, 14. OTOLOGY, ANATOMY. *Educ:* Washington Univ, PhD(anat), 51. *Prof Exp:* Asst otolaryngol, Med Sch, Washington Univ, 48-54, res assoc clin otolaryngol, 54-59, from res asst prof to res prof otolaryngol, 58-69; prof, 69-79, EMER PROF OTOLARYNGOL, MED SCH, UNIV ORE, 79- *Concurrent Pos:* Inst, Washington Univ, 53-54; res assoc, Cent Inst Deaf, 54-62, res collabr, 62-69. *Honors & Awards:* Award Merit, Am Otol Soc, 75; Shambaugh Prize in Otology, 77; Award Merit, Asn Res Otologists, 81. *Mem:* Am Asn Anatomists; Am Otol Soc; Am Soc Cell Biol; Col Otorhinolaryngol Amicitiae Sacrum. *Res:* Ultrastructure and histology of the ear; neurophysiology of the inner ear. *Mailing Add:* 16200 S Pacific Hwy No 34 Lake Oswego OR 97034

SMITH, CECIL RANDOLPH, JR, b Denver, Colo, May 31, 24; m 54; c 3. NATURAL PRODUCTS CHEMISTRY, LIPID CHEMISTRY. *Educ:* Univ Colo, BA, 46, MS, 48; Wayne State Univ, PhD(org chem), 55. *Prof Exp:* Asst chem, Univ Colo, 46-47; org chemist, US Bur Mines, Wyo, 47-51, Julius Hyman & Co, Colo, 51-52 & Northern Regional Res Lab, USDA, 56-62; asst prof chem, Western Mich Univ, 62-63; org chemist, Northern Regional Res Ctr, 63-85, res leader, USDA, 73-85; vis scientist, Inst Chem Natural Substances, Nat Ctr Sci Res, Gif-sur-Yvette, France, 85; asst dir, Cancer Res Inst, Aris State Univ, 86-88; COLLABR, WESTERN COTTON RES CTR, USDA, 88- *Concurrent Pos:* Res fel, Nat Heart Inst, Glasgow, 55-56. *Honors & Awards:* Alton E Bailey Award, Am Oil Chemists Soc, 84. *Mem:* Am Chem Soc; Am Oil Chem Soc; Sigma Xi; Am Soc Pharmacognosy. *Res:* Shale oil; alkaloids; fatty acids; natural products; medicinal chemistry; detection, isolation and characterization of biologically active natural products, especially those useful for control of cancer and insect pests. *Mailing Add:* 514 E Colgate Dr Tempe AZ 85283-1906

SMITH, CEDRIC MARTIN, b Stillwater, Okla, Feb 1, 27; div; c 3. ADDICTION MEDICINE, NEUROPHARMACOLOGY. *Educ:* Okla State Univ, BS, 49; Univ Ill, BS, 50, MS & MD, 53. *Prof Exp:* Asst pharmacol, Col Med, Univ Ill, 55-58; intern, Philadelphia Gen Hosp, 53-54; from instr to prof pharmacol, Col Med, Univ Ill, 54-66, actg head dept, 65-66; chmn dept, 66-73, PROF PHARMACOL & THERAPEUT, SCH MED & DENT, STATE UNIV NY BUFFALO, 66- *Concurrent Pos:* USPHS spec fel, Univ Göttingen, 61-62; staff scientist, Inst Defense Analysis, 64-65; mem, grants rev study sect pharmacol, NIH, 65-68; founding dir, NY State Res Inst Alcoholism, 70-79, sr assoc res scientist, 79-; mem adv comt drug abuse, NY State Dept Health & Div Substance Abuse, 79-; mem med staff, Erie County Med Ctr, 76-; preceptor, Family Pract Ctr, 78-83; spec lectr, Japan Soc Neuropsychopharmacol, 85; mem, Res Rev Comt Nat Inst Drug Abuse, 85-88; vis prof, Univ Ky Col Med; cert, Am Soc Addiction Med. *Mem:* AAAS; Am Soc Pharmacol & Exp Therapeut; Am Soc Clin Pharmacol & Therapeut; Am Soc Addiction Med; Int Brain Res Orgn; Am Col Clin Pharmacol; AMA; Col Int Neuropsychopharmacol; Sigma Xi; Res Soc Alcoholism; Asn Chemoreception Sci. *Res:* Addiction medicine; neuropharmacology of muscle sensory receptors; psychotropic drugs; non-medical drug use; alcohol and intoxication; drug-alcohol interactions; medical education. *Mailing Add:* Dept Pharmacol & Therapeut 102 Farber Hall State Univ NY Buffalo NY 14214

SMITH, CHARLES ALLEN, b Lexington, Ky, Aug 4, 44; m 75; c 2. MOLECULAR BIOLOGY. *Educ:* Mass Inst Technol, SB, 66; Calif Inst Technol, PhD(biophys), 71. *Prof Exp:* Fel, 72-75, SR RES ASSOC BIOPHYS, STANFORD UNIV, 75- *Mem:* Sigma Xi. *Res:* Mechanisms for replication, repair, and function of eukaryotic DNA; organization of DNA in chromosomes. *Mailing Add:* Dept Biol Stanford Univ Stanford CA 94305

SMITH, CHARLES ALOYSIUS, b Minneapolis, Minn, Aug 18, 39; m 70; c 1. ANALYTICAL CHEMISTRY. *Educ:* Col St Thomas, BS, 61; Kans State Univ, PhD(chem), 66. *Prof Exp:* Chemist, McDonnell-Douglas Corp, Santa Monica, 65-75; chemist, Dept Entom, Univ Calif, Riverside, 75-78; mgr qual assurance methods, McGaw Labs, Irvine, Ca, 78-80; CHEMIST, MCDONNEL-DOUGLAS CORP, HUNTINGTON BEACH, CA, 80- *Mem:* Am Chem Soc. *Res:* Analytical chemical methods development and applications. *Mailing Add:* 17669 San Vicente Fountain Valley CA 92708-1699

SMITH, CHARLES BRUCE, b Dec 23, 36. NEUROPHARMACOLOGY, NEUROCHEMISTRY. *Educ:* Harvard Univ, MD, 65, PhD(pharmacol), 66. *Prof Exp:* dir neural & behav sci prog, 81-87, PROF PHARMACOL, SCH MED, UNIV MICH, 76- *Mem:* Am Soc Pharmacol & Exp Therapeut; Am Col Neuropsycholpharmacol. *Res:* Neuroreceptors. *Mailing Add:* 3625 Daleview Dr Ann Arbor MI 48105

SMITH, CHARLES E, b Omaha, Nebr, Nov 16, 17; m 41; c 2. PSYCHIATRY. *Educ:* George Washington Univ, AB, 39, MD, 41. *Prof Exp:* Intern, USPHS Hosp, Baltimore, Md, 41-42; staff psychiatrist, Vet Admin Hosp, Northport, NY, 45-49; chief med officer, Fed Correction Inst, Ky, 49-50; resident psychiatrist, USPHS Hosp, Staten Island, NY, 50-51; chief psychiat serv, Med Ctr Fed Prisoners, Mo, 51-55; asst med dir, Fed Bur Prisons, 56-62, med dir, 62-66; chief serv, West Side Div, St Elizabeth's Hosp, DC, 66-67; dir Ment Health Serv, NC Dept Corrections, 74-82, consult psychiatrist, Dorothea Dix Hosp, Raleigh, NC, 83-86; assoc prof, 67-74, prof, 74-86, EMER PROF PSYCHIAT, SCH MED, UNIV NC, CHAPEL HILL, 86-, CONSULT PSYCHIATRIST, PEDIAT NEUROL CLIN, UNIV NC HOSPS, 86- *Concurrent Pos:* Mem, Bd Dirs, Washington DC Area Coun Alcoholism, 61-67; mem, Prof Coun, Nat Coun Crime & Delinquency, 65-70; consult, NC Dept Corrections, 67-82. *Mem:* fel Am Psychiat Asn; fel Am Orthopsychiat Asn. *Res:* Legal aspects of psychiatry; correctional treatment of the mentally ill offender. *Mailing Add:* Dept Psychiat Univ NC Sch Med Chapel Hill NC 27514

SMITH, CHARLES EDWARD, b Clayton, Ala, June 8, 34; m 60; c 3. ELECTRICAL ENGINEERING. *Educ:* Auburn Univ, BEE, 59, MS, 63, PhD(elec eng), 68. *Prof Exp:* Res engr, Auburn Res Found, 59-68; assoc prof, 68-76, PROF ELEC ENG, UNIV MISS, 77-, CHMN DEPT, 75- *Concurrent Pos:* Actg chmn, Dept Comput & Info Sci, Univ Miss, 85-87, assoc dean grad studies, Sch Eng, 89- *Mem:* Inst Elec & Electronics Engrs; Am Soc Eng Educ; Sigma Xi. *Res:* Antennas; microwave circuits; communication systems; microwave measurements; computer-aided design. *Mailing Add:* Dept Elec Eng Univ Miss University MS 38677

SMITH, CHARLES EDWARD, JR, b Sharpsburg, Ky, Oct 26, 27; m 49. LIMNOLOGY, PHYCOLOGY. *Educ:* Eastern Ky Univ, BS, 54; Univ Ky, MS, 56; Univ Louisville, PhD(biol), 63. *Prof Exp:* Mat testing engr, Dept Hwy, Frankfort, Ky, 49-50, off engr, 50-51; asst zool, Univ Ky, 54-56; teacher, High Sch, Ky, 56-61; res asst algal physiol, Potamological Inst, Univ Louisville, 61-63; from asst prof to assoc prof biol, 63-71, admin asst, 67-68, assoc dir, 68-69, dir off res, 68-83, PROF BIOL, BALL STATE UNIV, 71- *Mem:* AAAS; Am Soc Limnol & Oceanog; Am Phycol Soc; Int Asn Theoret & Appl Limnol. *Res:* Physiology and ecology of phytoplankton, especially members of the cyanophyta; algal physiology; tissue culture propagation of orchids; use of computers in the teaching of biology. *Mailing Add:* 1005 Bittersweet Ln Muncie IN 47304

SMITH, CHARLES EUGENE, b Atlanta, Ga, June 22, 50; m 75; c 1. NEUROBIOLOGY. *Educ:* Mass Inst Technol, BS, 72; Univ Chicago, MS, 73, PhD(biophysics), 79. *Prof Exp:* Fel, Med Univ SC, 79-80, asst prof biomet, 80-89; ASSOC PROF BIOMATH, DEPT STATIST, NC STATE UNIV, 89- *Concurrent Pos:* Co-dir, Cardiomet Scientist Training Prog, 81-82. *Mem:* Acoust Soc Am; Inst Elec & Electronics Engrs; Biomet Soc; Am Statist Asn. *Res:* Applied stochastic processes. *Mailing Add:* Dept Statist NC State Univ Box 8203 Raleigh NC 27695-8203

SMITH, CHARLES FRANCIS, JR, b Casper, Wyo, Aug 8, 36; m 60; c 3. RADIOCHEMISTRY, ENVIRONMENTAL CHEMISTRY. *Educ:* Purdue Univ, BS, 58, MS, 61; Univ Calif, Berkeley, PhD(nuclear chem), 65. *Prof Exp:* CHEMIST, LAWRENCE LIVERMORE NAT LAB, UNIV CALIF, 65-, PROJ LEADER, 74- *Concurrent Pos:* Staff mem anal geochem, Nat Uranium Resources Eval Prog; consult, containment underground nuclear explosions, Defense Nuclear Agency, 88-, chmn, chem & radiochem adv team, 88- *Res:* Radiochemistry and chemistry of gaseous products and fission product gases; analytical geochemistry; neutron activation analyses. *Mailing Add:* Lawrence Livermore Nat Lab L-232 Box 808 Livermore CA 94551

SMITH, CHARLES G, b Chicago, Ill,. Oct 26, 27; m 50; c 4. CANCER RESEARCH, DRUG DISCOVERY. *Educ:* Ill Inst Technol, BS, 50; Purdue Univ, MS, 52; Univ Wis, PhD(biochem), 54. *Prof Exp:* Mgr biochem, Upjohn Co, 54-67; vpres res & develop & pres Squibb Inst, ER Squibb & Sons, 67-75; vpres res & develop, Revlon Health Care Group, 75-86; CONSULT, 86- *Mem:* Am Chem Soc; Fedn Soc Biol Chem; Am Asn Cancer Res; AAAS. *Res:* New drug discovery and development; cancer, cardiovascular disease, anihypertensive agents, antihypersensitivity diseases. *Mailing Add:* PO Box 8002 Rancho Santa Fe CA 92067-8002

SMITH, CHARLES HADDON, b Dartmouth, NS, Sept 3, 26; m 49; c 4. GEOLOGY. *Educ:* Dalhousie Univ, BSc, 46, MSc, 48; Yale Univ, MS, 51, PhD(geol), 52. *Prof Exp:* Instr eng, Dalhousie Univ, 46-48; geologist, Cerro de Pasco Copper Corp, Peru, 49; geologist, Geol Surv Can, 51-64, chief petrol sci div, 64-67 & crustal geol div, 67-68; sci adv, Sci Coun Can, 68-70; dir planning, 70-71, asst dep minister sci & technol, 72-75, sr asst dep minister, Can Dept Energy, Mines & Resources, 75-82; PRES, CHARLES H SMITH CONSULT, 82- *Concurrent Pos:* Dep secy gen, Int Upper Mantle Comt, chmn, Can Upper Mantle Comn; Sci Adv, Candian Commn for UNESCO; exec dir, Can Nat Comm for World Energy Conf; Pres, Canadian Geoscience Coun; dir, 14th World Energy Cong, Montreal, 89; distinguished lectr award, Soc Econ Geologists. *Mem:* Can Inst Mining & Metall (vpres); Mineral Soc Am; Soc Econ Geol (vpres); Geol Asn Can; fel Royal Soc Can (Foreign Secy, 86). *Res:* Petrology and economic geology; study of ultrabasic rocks; history of geoscience research. *Mailing Add:* 2056 Thistle Crescent Ottawa ON K1H 5P5 Can

SMITH, CHARLES HOOPER, b Winnfield, La, July 24, 17; m 45; c 3. CHEMISTRY. *Educ:* La Polytech Inst, BS, 38; La State Univ, MS, 40, PhD(phys chem), 47. *Prof Exp:* Asst prof chem, La Polytech Inst, 40-42; chemist, US Rubber Co, Mich, 42-45; asst & Am Chem Soc fel chem, La State Univ, 47-48; from assoc prof to prof chem, La Tech Univ, 48-80, head dept chem, 54-78; fel, Univ Col NWales, 80-81; RETIRED. *Mem:* Am Chem Soc. *Res:* Spectra of deuterated toluenes and deuterated formamide; analysis of blood. *Mailing Add:* 1600 Cooktown Rd Ruston LA 71270

SMITH, CHARLES IRVEL, b Baltimore, Md, Aug 22, 23; m 50; c 3. MEDICINAL CHEMISTRY. *Educ:* Univ Md, BS, 44, PhD(pharm chem), 50. *Prof Exp:* Asst, Dent Sch, Univ Md, 46-50; from sr res asst to instr physiol chem, Johns Hopkins Univ, 50-52; sr res scientist, Squibb Inst Med Res, 52-60; assoc prof med chem, 60-74, chmn dept, 75-82, PROF MED CHEM, COL PHARM, UNIV RI, 74- *Mem:* Fel AAAS; Am Chem Soc; Sigma Xi. *Res:* Drug Assay; radiopharmaceuticals; drug metabolism; drug design and synthesis; medicinal chemistry on enzyme inhibitors, antispasmodics, anticonvulsants, narcotic agents, antagonists, and antimalarials. *Mailing Add:* 20 Nichols Rd Kingston RI 02881-1804

SMITH, CHARLES ISAAC, b Hearne, Tex, Feb 9, 31. GEOLOGY. *Educ:* Baylor Univ, BS, 52; La State Univ, MS, 55; Univ Mich, PhD, 66. *Prof Exp:* Geologist, Shell Develop Co, 55-65; from asst prof to assoc prof, 65-72, prof geol & mineral, Univ Mich, Ann Arbor, 72-77, chmn dept, 71-77; PROF GEOL, UNIV TEX, ARLINGTON, 77-, CHMN DEPT, 77- *Mem:* Geol Soc Am; Am Asn Petrol Geol. *Res:* Stratigraphy; sedimentation. *Mailing Add:* Dept of Geol Univ of Tex Arlington TX 76019

SMITH, CHARLES JAMES, b Buffalo, NY, Sept 29, 25. NEUROPSYCHOLOGY. *Educ:* Univ Buffalo, BA, 48; McGill Univ, MA, 51, PhD, 54. *Prof Exp:* Instr psychol, Univ Mich, 53-61; ASSOC PROF PSYCHOL, STATE UNIV NY BUFFALO, 61- *Concurrent Pos:* Vis res fel physiol, John Curtin Sch Med Res, Australian Nat Univ, 67-68. *Mem:* Soc Neurosci; Int Brain Res Orgn; Sigma Xi. *Res:* Brain function; psychophysiology of vision. *Mailing Add:* Dept of Psychol State Univ of NY Buffalo NY 14260

SMITH, CHARLES LEA, b Alto, Tex, Mar 8, 18; m 46; c 2. CHEMISTRY. *Educ:* Stephen F Austin State Col, BA, 38. *Prof Exp:* Teacher, High Sch, 38-41; chemist, Trojan Powder Co, Pa, 42-44 & Standard Oil Co, NJ, 44-46; mat engr, Naval Air Exp Sta, Pa, 46-47; res engr, Battelle Mem Inst, 47-50, prin chemist, Battelle Develop Corp, 50-57; mgr proj develop, Southern Res Inst, 57-60; res adminr, Wyeth Labs, Inc, 60-87; RETIRED. *Mem:* Am Chem Soc. *Res:* Alkyd resins; lacquers; drying oils; polyhydric alcohols; preservation and protection of materials; rubber and rubber-like materials; leather. *Mailing Add:* 536 Weadley Rd Strafford Wayne PA 19087

SMITH, CHARLES O(LIVER), b Clinton, Mass, May 28, 20; m 46; c 7. METALLURGY. *Educ:* Worcester Polytech Inst, BS, 41; Mass Inst Technol, MS, 47, ScD, 51. *Prof Exp:* Engr, Blake Mfg Co, 40-43; instr mech eng, Worcester Polytech Inst, 41-43; instr metall, Mass Inst Technol, 46-47, from instr mech eng to asst prof, 47-51; res engr, Mech Testing Div, Res Labs, Aluminum Co Am, 51-54; eng consult, E I du Pont de Nemours & Co, 55; lectr reactor mat, Oak Ridge Nat Lab, 55-65; prof eng, Univ Detroit, 65-76, chmn dept, 65-68; prof eng, Univ Nebr, 76-81; prof eng, Rose-Hulman Inst Technol, 81-86; RETIRED. *Honors & Awards:* Fred Merryfield Design Award, 81. *Mem:* Am Soc Metals; fel Am Soc Eng Educ; Sigma Xi; fel Am Soc Mech Engrs. *Res:* Materials and their application with special reference to design; author of over 180 publications in engineering. *Mailing Add:* 1920 College Ave Terre Haute IN 47803

SMITH, CHARLES R, b Campti, La, Sept 11, 36; m 59; c 2. MATHEMATICS. *Educ:* Northwestern State Univ, BS, 57; Okla State Univ, MS, 64, EdD(math), 66. *Prof Exp:* Teacher, La, 59-63; asst prof, 66-69, ASSOC PROF MATH, NORTHEAST LA UNIV, 69- *Concurrent Pos:* NSF sci faculty fel, Univ Wash, 70-71. *Mem:* Math Asn Am. *Res:* Convexity; combinatorial geometry; functional analysis. *Mailing Add:* Dept Math Northeast La Univ 700 Univ Ave Monroe LA 71209

SMITH, CHARLES RAY, b Fayetteville, Tenn, May 15, 33; m 58; c 2. THEORETICAL PHYSICS. *Educ:* Vanderbilt Univ, BA, 55, MS, 62; Univ Colo, PhD(physics), 67. *Prof Exp:* Teaching asst physics, Univ Wis, 57-58; from physicist to aero-res engr, Redstone Arsenal, Ala, 58-61; physicist, Nat Bur Stand, Colo, 61-64; from asst prof to assoc prof physics, Univ Wyo, 64-77. *Mem:* Am Asn Physics Teachers. *Res:* Many-body physics; quantum electrodynamics; laser-plasma interactions. *Mailing Add:* DASB-H-YP PO Box 1500 Huntsville AL 35807-3801

SMITH, CHARLES ROGER, veterinary physiology, for more information see previous edition

SMITH, CHARLES SYDNEY, JR, b Lorain, Ohio, Apr 29, 16; m 40; c 3. PHYSICS. *Educ:* Case Inst Technol, BS, 37; Mass Inst Technol, ScD(physics), 40. *Prof Exp:* Instr physics, Univ Pittsburgh, 40-42; from instr to prof, Case Inst Technol, 42-68; distinguished prof & dir, Mat Res Ctr, 68-81, EMER PROF PHYSICS, UNIV NC, CHAPEL HILL, 81- *Concurrent Pos:* Bell Tel Labs, 52-53; consult, Union Carbide Corp. *Mem:* Fel Am Phys Soc. *Res:* Solid state physics. *Mailing Add:* Dept of Physics & Astron CB #3255 Phillips Hall Univ of NC Chapel Hill NC 27514-3253

SMITH, CHARLES WELSTEAD, b Asheville, NC, Aug 21, 27; m 50; c 4. PHYSIOLOGY. *Educ:* Wheaton Col, BS, 48; Univ Mich, MS, 49, MS, 53, PhD(physiol), 55. *Prof Exp:* Instr physiol, Sch Med, Univ Mich, 55-56; from instr to assoc prof, NJ Col Med & Dent, 56-64; from assoc prof to prof, 64-88, EMER PROF PHYSIOL, COL MED, OHIO STATE UNIV, 88- *Mem:* Am Physiol Soc. *Res:* Respiration; oxygen toxicity; blood gases; cardiac output; pulmonary blood flow; coronary blood flow. *Mailing Add:* Dept of Physiol Ohio State Univ Col of Med Columbus OH 43210

SMITH, CHARLES WILLIAM, JR, b Greensburg, Pa, May 13, 40; m 64; c 2. LOW TEMPERATURE PHYSICS. *Educ:* Allegheny Col, BS, 62; Ohio Univ, PhD(physics), 68. *Prof Exp:* From asst prof to assoc prof, 68-80, COOP ASSOC PROF ENG, UNIV MAINE, ORONO, 77-, PROF PHYSICS, 80-, DEPT CHMN, 86- *Mem:* Sigma Xi; Am Phys Soc; Am Asn Physics Teachers. *Res:* Low temperature condensed matter physics; superconductivity; liquid helium. *Mailing Add:* Dept Physics & Astron Univ Maine Orono ME 04469

SMITH, CHARLINE GALLOWAY, b Louisiana, Mo, Apr 9, 25; div; c 2. PHYSICAL ANTHROPOLOGY. *Educ:* Univ Utah, BS, 65, PhD(anthrop), 70. *Prof Exp:* Staff nurse surg, Barnes Hosp, St Louis, Mo, 46-47; staff nurse, Am Hosp, Chicago, 48-49; staff nurse, Lutheran Hosp, Los Angeles, 49-54; specialist thoracic intensive care, LDS Hosp, Salt Lake City, 62-63; instr anthrop, Div Continuing Educ, Univ Utah, 66-68, assoc ed, 66-70, teaching asst, 70; from asst prof to assoc prof phys anthrop, 70-91, EMER PROF, UNIV MONT, 91- *Mem:* AAAS; Am Asn Phys Anthrop; Am Anthrop Asn; Am Diabetes Asn; Am Ethnol Soc. *Res:* Diabetes Mellitus among American Indians; medical ethnobotany; sex ratios among primitive groups; cerebral dominance and handedness. *Mailing Add:* 638 Montana Ave Missoula MT 59801

SMITH, CHARLOTTE DAMRON, b Columbus, Ohio, Nov 13, 19; m 57. BIOCHEMISTRY. *Educ:* Wellesley Col, BA, 40; Rutgers Univ, MS, 42; George Washington Univ, PhD(biochem), 51. *Prof Exp:* Asst chem, Rutgers Univ, 40-42; res chemist, E I du Pont de Nemours & Co, 42-44; asst biochem, George Washington Univ, 47-51; res fel, Nat Cancer Inst, 51-54; res assoc environ med, Sch Hyg, Johns Hopkins Univ, 53-55; assoc sci info exchange, Smithsonian Inst, 55-81; RETIRED. *Mem:* Sigma Xi. *Res:* Metabolism of ascorbic acid in the guinea pig and of carcinogen 2-acetylamino fluorene in the rat; mode of action of chromium compounds in causing human lung cancer. *Mailing Add:* 3708 Manor Rd Chevy Chase MD 20815

SMITH, CHESTER MARTIN, JR, b Randolph, Vt, Sept 8, 35; m 58; c 2. COMPUTER SCIENCE. *Educ:* Univ Vt, BA, 57; Pa State Univ, MS, 59, PhD(mineral), 64. *Prof Exp:* Res asst comput sci, 61-62, ASST PROF COMPUT SCI, PA STATE UNIV, 63- *Concurrent Pos:* Chmn, Share Inc, 67-69, mgr, 69-70, dir, 70-71, secy, 71-72; chmn, Fortran Data Base Comt, Conf Data Systs Lang, 74-79. *Mem:* Asn Comput Mach. *Res:* Programming languages; compiler construction; information retrieval; data base management and standards; historical place name data bases; micro-computers. *Mailing Add:* Dept Comput Sci Pa State Univ University Park PA 16802

SMITH, CHRISTINE H, FOOD & DRUG INTERACTION. *Educ:* Univ Southern Calif, PhD(pharmacol). *Prof Exp:* ASSOC PROF FOOD & DRUG INTERACTION, CALIF STATE UNIV, NORTHRIDGE. *Mailing Add:* 9315 Wystone Ave Northridge CA 91324

SMITH, CHRISTOPHER CARLISLE, b Boston, Mass, June 18, 38; m 60; c 3. EVOLUTIONARY ECOLOGY. *Educ:* Univ Colo, BA, 60; Univ Wash, MA, 63, PhD(ecol), 65. *Prof Exp:* Asst prof biol, Fisk Univ, 65-67; res assoc ecol, Smithsonian Trop Res Inst, 67-68; asst prof zool, Univ Mo-Columbia, 68-70; assoc prof, 70-81, PROF ZOOL, KANS STATE UNIV, 81- *Mem:* AAAS; Ecol Soc Am; Soc Study Evolution; Am Soc Mammal; Am Soc Naturalists. *Res:* Relationship between mammalian social organization and ecology; relationship between animals and the fruiting pattern in forest trees; ecology of wind pollination. *Mailing Add:* Div of Biol Kans State Univ Manhattan KS 66506

SMITH, CLAIBOURNE DAVIS, b Memphis, Tenn, Jan 6, 38; m 59; c 2. ORGANIC CHEMISTRY. *Educ:* Univ Denver, BS, 59, MS, 61; Univ Ore, PhD(org chem), 64. *Prof Exp:* Res asst org chem, Denver Res Inst, Univ Denver, 59-61; res chemist, Cent Res Dept, 64-71, tech prog mgr, Fabrics & Finishes Dept, 71-73, sales mgr, 73-74, TECH MGR, INDUST PROD DIV, FABRIC & FINISHES DEPT, E I DU PONT DE NEMOURS & CO, INC, 74- *Mem:* Am Chem Soc. *Res:* Organic synthesis of polynitrafluoro aromatics; polymeric binders and thermal stable organic polymers; non-benzoid aromatic hydrocarbons and strained small ring compounds. *Mailing Add:* 901 Barnstable Ct RD 2 Hockessin DE 19707-9611

SMITH, CLAIRE LEROY, b Atlantic, Iowa, May 1, 23; m 53; c 4. MICROBIOLOGY. *Educ:* Univ Omaha, BS, 53; Univ Iowa, MS, 55. *Prof Exp:* Bacteriologist, Grain Processing Corp, 55-88; RETIRED. *Mem:* Soc Indust Microbiol; Am Soc Microbiol. *Res:* Industrial fermentations; brewing; vitamins; amino acids; antibiotics; enzymes. *Mailing Add:* 115 Lord Ave Muscatine IA 52761

SMITH, CLARENCE LAVETT, b Hamburg, NY, Dec 19, 27; m 54; c 2. ZOOLOGY. *Educ:* Cornell Univ, BS, 49; Tulane Univ, MS, 51; Univ Mich, PhD, 59. *Prof Exp:* Chmn, 75-82, CUR DEPT ICHTHYOL, AM MUS NATURAL HIST, 62- *Concurrent Pos:* Vis prof, Univ Okla, 69, Ohio State Univ, 59, 63 & 71 & Univ Mich, 76, 78 & 80; assoc prof, Col Guam, 60-61 & Univ Hawaii, 61-62; scientist, Aquanaut Proj, Tektite II, 70; adj prof, City Col New York, 70- & Rutgers Univ, 82- *Mem:* Ecol Soc Am; Am Soc Ichthyol & Herpet; Am Fisheries Soc; Am Soc Limnol & Oceanog. *Res:* Ichthyology; taxonomy, ecology, morphology and distribution of recent fishes; ecology of coral reef fishes and their larvae; freshwater fishes of New York State. *Mailing Add:* Am Mus Nat Hist Dept Ichthyol Central Park W at 79th New York NY 10024

SMITH, CLAY TAYLOR, b Omaha, Nebr, June 30, 17; m 40; c 2. GEOLOGY. *Educ:* Calif Inst Technol, BS, 38, MS, 40, PhD(geo), 43. *Prof Exp:* Recorder, US Geol Surv, 38, 39, jr geologist, 40-42; geol field engr, Consol Mining & Smelting Co Can, 43; asst geologist, Union Mines Develop Corp, NY, 43-46; field geologist, US Vanadium Corp, 46-47; asst prof eng, 47, from assoc prof to asst prof geol, 47-56, head dept, 52-66, dean student & admissions, 67-68, prof geol, 56-87, EMER PROF, NMEX INST MINING & TECHNOL, 87- *Concurrent Pos:* Consult raw mat resource eval, 50-; expert witness court cases involving raw material resources, 71-; dir alumni rels & ann giving, Nmex Inst Minig & Technol. *Mem:* AAAS; Soc Econ Geol; Geol Soc Am; Nat Asn Geol Teachers; Am Inst Prof Geologists; Sisma Xi. *Res:* Secondary earth science education; raw material resources of New Mexico; low angle faulting along the Rio Grande rift; geology of chromite deposits; origin of sedimentary type uranium ores; geology of ferroalloy elements. *Mailing Add:* Dept Geosci NMex Inst Mining & Technol Socorro NM 87801

SMITH, CLAYTON ALBERT, JR, b Champaign, Ill, June 10, 34. POSITIONAL ASTRONOMY. *Educ:* Univ Chicago, BA, 56, BS, 57; Georgetown Univ, PhD(astron), 69. *Prof Exp:* Chief, Cataloging Br, 80-84, ASTRONOMER ASTROMETRY, NAVAL OBSERV, 59-, CHIEF, ANAL DIV, 85- *Concurrent Pos:* Resident dir, Yale-Columbia Southern Observ, Arg, 68-70. *Mem:* Am Astron Soc; Astron Soc Pac; Sigma Xi; AAAS; Int Astron Union. *Res:* Formulation of catalogs of differential and fundamental systems of stellar positions and proper motions. *Mailing Add:* 6625 Harlan Pl NW Washington DC 20012-2138

SMITH, CLIFFORD JAMES, b Brooklyn, NY, Oct 30, 38; m 59; c 4. ANIMAL PHYSIOLOGY. *Educ:* Cornell Univ, BS, 60; Univ Md, PhD(physiol, biochem), 64. *Prof Exp:* Res asst physiol, Univ Md, 60-64; NIH fel anat, Univ Vt, 64-65; from asst prof to assoc prof, 65-75, chmn dept, 70-75, PROF BIOL, UNIV TOLEDO, 75- *Mem:* Am Soc Zool; Poultry Sci Asn. *Res:* Physiological control of hunger and appetite. *Mailing Add:* Dept Biol Univ Toledo Col Arts & Sci Toledo OH 43606

SMITH, CLOYD VIRGIL, JR, b Seminole, Okla, Dec 2, 36; m 60; c 3. SOLID MECHANICS. *Educ:* Ga Inst Technol, BCE, 58; Stanford Univ, MSCE, 59; Mass Inst Technol, ScD(civil eng), 62. *Prof Exp:* Res engr, Jet Propulsion Lab, Calif Inst Technol, 63-64; ASSOC PROF AEROSPACE ENG, GA INST TECHNOL, 64- *Mem:* Am Inst Aeronaut & Astronaut. *Res:* Elastic stability; matrix methods of structural analysis; nonlinear elasticity; structural dynamics. *Mailing Add:* 2949 Green Oak Circle Atlanta GA 30345

SMITH, CLYDE F, b Riverdale, Idaho, Aug 10, 13; m 36; c 3. ENTOMOLOGY. *Educ:* Utah State Agr Col, BS, 35, MS, 37; Ohio State Univ, PhD(entom), 39. *Prof Exp:* Asst entom, Utah State Agr Col, 34-36 & 37; asst, Ohio State Univ, 36-38, asst, Exten, 38 & 39; from asst entomologist to res prof entom, 39-50, head dept, 50-64, prof, 64-78, EMER PROF ENTOM, NC STATE UNIV, 78- *Mem:* AAAS; Entom Soc Am; Soc Syst Zool. *Res:* Biology; entomology, ecology and control of fruit insects; taxonomy of Aphididae (Homoptera) and Aphidiinae (Hymenoptera). *Mailing Add:* 2716 Rosedale Ave Raleigh NC 27607

SMITH, CLYDE KONRAD, b Sturgeon Bay, Wis, Dec 9, 25; m 47. VETERINARY MICROBIOLOGY. *Educ:* Mich State Univ, BS, 47, DVM, 51, MS, 53; Univ Notre Dame, PhD, 66; Am Col Vet Microbiologists, dipl. *Prof Exp:* Instr bact, Mich State Univ, 51, asst prof, 56-66; from assoc prof to prof vet sci, Ohio Agr Res & Develop Ctr, 66-86; RETIRED. *Mem:* AAAS; Am Soc Microbiol; Am Vet Med Asn; Am Asn Bovine Practr; Conf Res Workers Animal Dis. *Res:* Ruminant nutrition and physiology and germfree ruminants; respiratory and enteric diseases of sheep, feeder calves and dairy calves. *Mailing Add:* 3743 Bayshore Dr Sturgeon Bay WI 54235

SMITH, COLIN MCPHERSON, b Edinburgh, Scotland, Mar 14, 27; Can citizen; m 62; c 4. PSYCHIATRY, GERIATRIC MEDICINE. *Educ:* Univ Glasgow, MB, ChB, 49, MD, 59; DPM & RCP(I), 53; FRCP(C), 56; Univ Sask, MD, 62; FRCPsychiat, 72. *Prof Exp:* Res asst psychol, Univ London,

58-59; PVT PRACT PSYCHIAT, 87-; CLIN PROF PSYCHIAT & MED, UNIV SASK, 71- *Concurrent Pos:* Mem, Alcohol Comn Sask, 68-80; psychiat, Regina Gen Hosp, 74-, Wascana Hosp, 77-, Plains Health Ctr, 78- & Pasqua Hosp, 78-; consult, Royal Univ Hosp, 75- & Staff Plains Health Ctr, 81-; examr psychiat, Royal Col Physicians & Surgeons Can, 65-78, consult, 87- *Honors & Awards:* Ment Health Res Award, Can Ment Health Asn, 62. *Mem:* Can Psychiat Asn (pres, 74-75); fel Am Geriat Soc; fel Am Psychiat Asn; Can Asn Geront (treas, 83-87); fel Am Psychiat Asn. *Res:* Geriatric psychiatry; social psychiatry, alcoholism; gerontology; research design; author over 100 scientific articles and books. *Mailing Add:* 4437 Castle Rd Regina SK S4S 4W4 Can

SMITH, COLLEEN MARY, b Minneapolis, Minn, Oct 4, 43. BIOCHEMISTRY. *Educ:* Univ Minn, BA, 63; Univ Utah, PhD(biochem), 69. *Prof Exp:* Instr, 72-74, ASST PROF BIOCHEM, SCH MED, TEMPLE UNIV, 74- *Concurrent Pos:* NIH fel, Johnson Res Found, Univ Pa, 70-72. *Mem:* AAAS; Am Chem Soc; Biophys Soc; Fedn Am Soc Exp Biol. *Res:* Metabolic regulation; co-factor biosynthesis; enzymology. *Mailing Add:* Dept Biochem Sch Med Temple Univ 3400 N Broad St Philadelphia PA 19140

SMITH, CONSTANCE META, b Kingston NY, Dec 31, 49. AGRICULTURAL FUNGICIDES. *Educ:* Vassar Col, BA, 71; Northwestern Univ, MS, 73; Cornell Univ, PhD(plant path), 79. *Prof Exp:* RES BIOLOGIST, E I DU PONT DE NEMOURS, 79- *Concurrent Pos:* Mem, Fungicide Resistance Action Comt. *Mem:* Am Phytopath Soc. *Res:* Discovery and development of agricultural fungicides. *Mailing Add:* Rodney Pl 1404 Penn Ave Wilmington DE 19806

SMITH, CORNELIA MARSCHALL, b Llano, Tex, Oct 15, 95; wid. MORPHOLOGY. *Educ:* Baylor Univ, BA, 18; Univ Chicago, MA, 23; Johns Hopkins Univ, PhD(biol), 28. *Prof Exp:* From instr to asst prof bot, Baylor Univ, 28-35; prof biol & head dept, Stetson Univ, 35-40; prof, 40-67, chmn dept, 43-67, dir, Strecker Mus, 45-67, Piper prof, 67-80, EMER PROF BIOL, BAYLOR UNIV, 80- *Concurrent Pos:* Secy, Tex Bd Exam Basic Sci, 49-67. *Mem:* AAAS; Bot Soc Am; Modern Lang Asn Am; Sigma Xi. *Res:* Morphology of Dionaea; toxicity of Polistes venom; electrophoretic comparison of six species of Yucca and Hesperaloe. *Mailing Add:* 801 James Waco TX 76706

SMITH, CRAIG LA SALLE, b Miami Beach, Fla, Mar 29, 43; m 67. ORGANIC CHEMISTRY, MARINE SCIENCE. *Educ:* Johns Hopkins Univ, BA, 64; Univ Fla, PhD(chem), 68. *Prof Exp:* Fel, Ga Inst Technol, 67-70; ASSOC MARINE SCIENTIST, VA INST MARINE SCI, 70-; ASST PROF MARINE SCI, COL WILLIAM & MARY, 70-; ASST PROF MARINE SCI, UNIV VA, 70- *Mem:* Am Chem Soc. *Res:* Heterocyclic organic chemistry; carbanion chemistry; mass spectroscopy; oil pollution; organic geochemistry. *Mailing Add:* Dept Marine Sci Col William & Mary Gloucester Point VA 23062

SMITH, CURTIS ALAN, b Long Beach, Calif, Sept 8, 48. PHYSIOLOGY. *Educ:* Calif State Col, Fullerton, BA, 70; Univ Calif, San Francisco, PhD(physiol), 78. *Prof Exp:* FEL PHYSIOL, DEPT PREV MED, UNIV WIS-MADISON, 78- *Mem:* Assoc Am Physiol Soc. *Res:* Control of breathing. *Mailing Add:* Dept Prev Med Univ Wis 504 N Walnut St Madison WI 53705

SMITH, CURTIS GRIFFIN, b Milwaukee, Wis, Nov 14, 23; m 47; c 3. PHYSIOLOGY. *Educ:* Univ Chicago, AB, 47, PhD, 54. *Prof Exp:* Instr biophys, Sch Med, Univ Calif, Los Angeles, 54-55; from instr to assoc prof physiol, 55-69, PROF PHYSIOL, MT HOLYOKE COL, 69- *Concurrent Pos:* Vis prof, Univ EAnglia, 69-70. *Mem:* Am Soc Biol Chem; Am Chem Soc; Brit Soc Gen Microbiol; Sigma Xi; Soc Neurosci. *Res:* Biochemical genetics; molecular biophysics; biochemistry and physiology of neurotransmitters; neurophysiology. *Mailing Add:* Dept of Biol Sci Mt Holyoke Col South Hadley MA 01075

SMITH, CURTIS PAGE, b Long Prairie, Minn, Dec 25, 38; m 60; c 2. ORGANIC CHEMISTRY. *Educ:* Univ Mich, BSCh, 61; State Univ NY, PhD(chem), 67. *Prof Exp:* Fel chem, State Univ NY Stony Brook, 66-68; sr res chemist, Olin Res Ctr, 68-69; staff scientist, D S Gilmore Res Lab, The Upjohn Co, 69-85, STAFF SCIENTIST, UPJOHN CHEM DIV, DOW CHEM USA, 85- *Mem:* Am Chem Soc; Sigma Xi. *Res:* Organo phosphorus chemistry; mechanism of organic reactions; biomedical applications of polyurethanes. *Mailing Add:* 254 Northrolling Acres Rd Cheshire CT 06410-2150

SMITH, CURTIS R, b Mineola, Tex, Nov 12, 36; m 58; c 3. AUDIOLOGY. *Educ:* Univ Southern Miss, BS, 60, MS, 61, PhD(audiol), 65. *Prof Exp:* Chief audiol, Brooke Gen Hosp, Ft Sam Houston, Tex, 65-66; dir grad training audiol, Our Lady of the Lake Col, 66-69; ASSOC PROF SPEECH COMMUN, AUBURN UNIV & DIR SPEECH & HEARING CLIN, 69- *Concurrent Pos:* Chief clin audiologist, Harry Jersig Speech & Hearing Ctr, 66-69. *Mem:* Am Speech & Hearing Asn. *Res:* Hearing mechanism and vestibular system. *Mailing Add:* Dept Speech Commun Path-Audiol Auburn Univ Auburn AL 36849

SMITH, CURTIS WILLIAM, b Omaha, Ill, Jan 14, 18; m 42; c 5. ORGANIC CHEMISTRY. *Educ:* Southern Ill Univ, BEd, 40; Univ Ill, PhD(org chem), 43. *Prof Exp:* Chemist, Res & Develop, Shell Develop Co, 43-52, mem staff mgt, 52-65, mgr, Ind Chem Div, Shell Chem Co, 65-71, asst to pres, Shell Develop Co, 71-77, sr consult, Chem Indust Regulations, Shell Oil Co, 77-83; RETIRED. *Mem:* Sigma Xi; Am Chem Soc. *Res:* Develop synthesis of tryptophan using quaternary ammonium akylation; penicillin; chemistry of acrolein. *Mailing Add:* 163 Stoney Creek Houston TX 77024

SMITH, CYRIL BEVERLEY, b Winnipeg, Man, Feb 21, 21; m 52. PLANT NUTRITION. *Educ:* Univ Man, BSA, 42, MSc, 45; Pa State Univ, PhD(hort), 50. *Prof Exp:* From instr to assoc prof, 47-65, PROF PLANT NUTRIT, PA STATE UNIV, UNIVERSITY PARK, 65- *Mem:* Am Soc Hort Sci; Am Soc Plant Physiol; Am Chem Soc; Sigma Xi. *Res:* Use of plant analysis in studying nutritional status of plants. *Mailing Add:* 232 Belle Ave Boalsburg PA 16827

SMITH, CYRIL STANLEY, b Birmingham, Eng, Oct 4, 03; US citizen; m 31; c 2. SOLID STATE PHYSICS. *Educ:* Univ Birmingham, BSc, 24; Mass Inst Technol, DSc(metall), 26. *Hon Degrees:* DLitt, Case Inst Technol, 65; ScD, Univ Pa, 74, Univ Mass, 79, Lehigh Univ, 82. *Prof Exp:* Res metallurgist, Am Brass Co, 27-42; assoc div leader, Los Alamos Sci Lab, 43-46; dir, Inst Study Metals, Univ Chicago, 46-61; inst prof, 61-69, EMER INST PROF, MASS INST TECHNOL, 69- *Concurrent Pos:* Mem gen adv comt, US AEC, 46-52; Guggenheim Found fels, 55-56 & 78-79; mem, President's Sci Adv Comt, 59; mem coun, Smithsonian Inst, 66-76. *Honors & Awards:* Gold Medal, Am Soc Metals, 61; Douglas Gold Medal, Am Inst Mining & Metall Engrs, 63; Leonardo Medal, Soc Hist Tech, 66; Platinum Medal, Inst Metals, London, 70; Dexter Award, Am Chem Soc, 81; Pomerance Medal, Am Inst Archaeol, 82. *Mem:* Nat Acad Sci; Am Philos Soc; Am Soc Metals; Hist Technol Soc (pres, 63-65); Metall Soc; His Sci Soc; Am Acad Arts & Sci; Akademie der Wissenschaften, Göttingen; Indian Inst Metals; Jap Inst Metals. *Res:* General theory of structure as hierarchy; history of technology and science and their relations to art; author or coauthor of numerous publications. *Mailing Add:* 31 Madison St Cambridge MA 02138

SMITH, DALE, b Fairmont, Nebr, Apr 13, 15; m 40; c 2. AGRONOMY, PLANT PHYSIOLOGY. *Educ:* Univ Nebr, BSc, 38; Univ Wis, MSc, 40, PhD(agron, plant physiol), 47. *Prof Exp:* From asst prof to assoc prof, 46-56, prof agron, 56-77, EMER PROF AGRON, UNIV WIS-MADISON, 77- *Concurrent Pos:* NATO fel, Eng, 64; Haight travel award, Asia, Australia & NZ, 70; mem, Acad Guest, Swiss Inst Technol, Zurich, 79; adj prof plant sci, Univ Ariz, Tucson, 79-82; hon mem, NAm Alfalfa Improv Conf. *Honors & Awards:* Crop Sci Award, Am Soc Agron, 63; Merit Cert, Am Forage & Grassland Coun, 65, Medallion Award, 75, Distinguished Grasslander, 82. *Mem:* Fel AAAS; fel Am Soc Agron. *Res:* Forage management; chemical composition; growth responses, cold hardiness and food reserves in forage plants. *Mailing Add:* PO Box 1400 Sun City AZ 85372-1400

SMITH, DALE METZ, b Portland, Ind, Dec 23, 28; m 50; c 2. SYSTEMATIC BOTANY. *Educ:* Ind Univ, BS, 50, PhD(bot), 57; Purdue Univ, MS, 52. *Prof Exp:* Instr bot, Univ Ariz, 52-53; from instr to assoc prof, Univ Ky, 55-61; assoc prof, Univ Ill, 61-64; assoc prof, 64-71, prof, 71-88, EMER PROF BOT, UNIV CALIF, SANTA BARBARA, 88- *Concurrent Pos:* Chmn dept biol sci, Univ Calif, Santa Barbara, 79-81. *Honors & Awards:* Cooley Award, Am Soc Plant Taxon, 63. *Mem:* AAAS; Bot Soc Am; Am Soc Plant Taxon; Am Fern Soc; fel Linnean Soc London. *Res:* Biosystematics of Phlox; cytotaxonomy and chemotaxonomy. *Mailing Add:* HC-65 Box 100-BB Windsor KY 42565

SMITH, DALLAS GLEN, JR, b Gainesboro, Tenn, June 25, 40; m 61; c 3. ENGINEERING MECHANICS. *Educ:* Tenn Polytech Inst, BS, 63; Tenn Technol Univ, MS, 66; Va Polytech Inst, PhD(eng mech), 69. *Prof Exp:* Bridge design engr, Bridge Div, Tenn Dept Hwy, 63-65; asst prof eng mech, Va Polytech Inst, 69-70; asst prof eng sci, 70-74, assoc prof, 74-79, prof eng sci & mech, 79-82, PROF CIVIL ENG, TENN TECHNOL UNIV, 82- *Concurrent Pos:* Consult, US Army Missile Command, Redstone Arsenal, 71- *Mem:* Am Soc Testing & Mat; Soc Exp Stress Analysis; Am Soc Eng Educ. *Res:* Brittle fracture mechanics; fiber-reinforced composite materials; experimental mechanics. *Mailing Add:* Dept Civil Eng Tenn Technol Univ Cookeville TN 38505

SMITH, DANIEL JAMES, b Rochester, NY, June 17, 44; m 67; c 3. IMMUNOLOGY, ORAL BIOLOGY. *Educ:* Houghton Col, BS, 66; NY Med Col, PhD(immunol), 72. *Prof Exp:* Staff assoc, Forsyth Dent Ctr, 72-74, asst mem staff immunol, 74-76, assoc mem staff, 79-84; clin instr, 76-79, asst clin prof, 79-87, ASSOC CLIN PROF, HARVARD SCH DENT MED, HARVARD UNIV, 87-; SR MEM STAFF, FORSYTH DENT CTR, 84- *Mem:* Am Asn Immunologists; Int Asn Dent Res. *Res:* Nature and function of secretory immune system, role of immunity in diseases of oral cavity, oral microbiology, and ontogeny of the secretory immune system. *Mailing Add:* Dept Immunol Forsyth Dental Ctr 140 Fenway Boston MA 02115

SMITH, DANIEL JOHN, b Horicon, Wis, Apr 19, 46. BIOCHEMISTRY, ORGANIC CHEMISTRY. *Educ:* Wis State Univ, BS, 68; Univ Calif, Berkeley, PhD(org chem), 74. *Prof Exp:* Fel biochem, Univ Calif, Los Angeles, 74-76; vis scientist, Syva Res Inst, 76-77; ASST PROF BIOCHEM, UNIV AKRON, 77- *Mem:* Am Chem Soc; Control Release Soc. *Res:* Mechanism of photophosphorylation and oxidative phosphorylation; mechanism of control release organotin compounds as molluscicides. *Mailing Add:* 6482 Dresher Trail Stow OH 44224

SMITH, DANIEL MONTAGUE, b Gainesville, Tex, Oct 17, 32; m 53; c 3. PHYSICS. *Educ:* NTex State Col, BA, 52, MS, 53; Univ Tex, PhD(physics), 59. *Prof Exp:* Lab asst, NTex State Col, 50-52; asst, Univ Tex, 55-58, scientist, 57-58; physicist, Oak Ridge Nat Lab, 58-61; mem tech staff, Tex Instruments Inc, Tex, 61-75; mgr device technol, Nitron Div, McDonnell Douglas, 75-77; eng mgr, Nat Semiconductor Corp, 77-79; ENG MGR, MOTOROLA SEMICONDUCTOR PROD SECTOR, 79- *Mem:* Am Phys Soc; Inst Elec & Electronics Engrs. *Res:* Medium energy experimental nuclear physics; semiconductor device design and development. *Mailing Add:* 10301 Parkfield Austin TX 78758

SMITH, DARRELL WAYNE, b Long Beach, Calif, July 31, 37; m 58; c 3. MATERIALS SCIENCE, PHYSICAL METALLURGY. *Educ:* Mich Technol Univ, BS, 59; Case Western Reserve Univ, MS, 65, PhD(phys metall), 69. *Prof Exp:* Metallurgist, Babcock & Wilcox Co, 59-62 & Gen Elec Co, 62-68; res assoc, Case Western Reserve Univ, 68-69; res metallurgist, Gen

Elec Co, 69-70; asst prof, 70-75, assoc prof, 75-81, PROF METALL ENG, MICH TECHNOL UNIV, 81- *Mem:* Am Soc Metals; Am Powder Metall Inst; Am Soc Testing Mat. *Res:* Physical properties of consolidated powders, powder metallurgy; powder metallurgy. *Mailing Add:* Dept Metall Eng Mich Technol Univ Houghton MI 49931

SMITH, DARRYL LYLE, b Minneapolis, Minn, July 30, 46. SOLID STATE PHYSICS. *Educ:* St Mary's Col, Minn, BA, 68; Univ Ill, MS, 71, PhD(physics), 74. *Prof Exp:* INSTR APPL PHYSICS, CALIF INST TECHNOL, 74- *Mem:* Am Phys Soc; Sigma Xi. *Res:* Optical properties of solids. *Mailing Add:* 3000 Trinity Dr #58 Los Alamos CA 87544

SMITH, DARWIN WALDRON, b Los Angeles, Calif, Mar 25, 31; m 52; c 3. PHYSICAL CHEMISTRY. *Educ:* Univ Calif, Los Angeles, BS, 53; Calif Inst Technol, PhD(chem), 59. *Prof Exp:* NSF fel chem, Math Inst, Oxford Univ, 59; from asst prof to assoc prof chem, Univ Fla, 60-68; ASSOC PROF CHEM, UNIV GA, 68- *Mem:* Am Phys Soc; Am Chem Soc. *Res:* Quantum chemistry; theory of molecular structure. *Mailing Add:* Dept of Chem Univ of Ga Athens GA 30602

SMITH, DAVID, b Fall River, Mass, Nov 7, 39; m 67; c 8. PHYSICAL CHEMISTRY. *Educ:* Providence Col, BS, 61; Mass Inst Technol, PhD(chem), 65. *Prof Exp:* Instr chem, Brooklyn Col, 65-68; from asst prof to assoc prof, 68-82, PROF CHEM, PA STATE UNIV, HAZLETON, 82- *Mem:* Am Chem Soc; Am Phys Soc. *Res:* Low temperature heat capacities; hindered rotation in solids. *Mailing Add:* Dept Chem Pa State Univ Hazleton PA 18201

SMITH, DAVID ALEXANDER, b New York, NY, Jan 6, 38; m 58; c 4. NUMERICAL ANALYSIS, MATHEMATICS EDUCATION. *Educ:* Trinity Col, Conn, BS, 58; Yale Univ, PhD(math), 63. *Prof Exp:* Asst prof math, 62-68, dir grad studies, 68-71, dir undergrad studies, 82-84, ASSOC PROF MATH, DUKE UNIV, 68- *Concurrent Pos:* Vis assoc prof, Case Western Reserve Univ, 75-76; series ed math, Proj Conduit, 75-; calculus proj adv, DC Heath & Co, 76-79; assoc ed, Math Mag, 81-85 & Col Math J, 86-89; vis prof, Benedict Col, 84-86; United Negro Col Fund scholar-at-large, 84-85; chair, Comt on Computers in Math Educ, 87- *Mem:* Soc Indust & Appl Math; AAAS; Am Math Soc; Math Asn Am. *Res:* Abstract algebra; arithmetic functions; algorithmic algebra; combinatorial theory; numerical analysis; uses of computers in mathematics; application in social and biological sciences. *Mailing Add:* Dept Math Duke Univ Durham NC 27706

SMITH, DAVID ALLEN, b Osceola, Nebr, Feb 25, 33; div; c 3. MOLECULAR BIOLOGY, RADIOBIOLOGY. *Educ:* Dana Col, BA, 59; Univ Southern Calif, PhD(biochem), 64. *Prof Exp:* Biochemist, Biomed Res Group, Los Alamos Sci Lab, Univ Calif, 64-77; molecular biologist, Off Health & Environ Res, 77-88, actg dep dir, 79-88, DIR, HEALTH EFFECTS RES DIV, DEPT ENERGY, 88- *Concurrent Pos:* Exec dir, Health & Environ Res Adv Comt, Dept Energy, 83-88. *Mem:* AAAS; Am Chem Soc; Am Soc Biochem & Molecular Biol; Environ Mutagenesis Soc. *Res:* Molecular biology; molecular genetics; environmental mutagenesis; nucleic acids and nucleic acid enzymology; biotechnology; DNA repair; radiobiology. *Mailing Add:* Off Health & Environ Res ER-72 GTN Dept of Energy Washington DC 20545

SMITH, DAVID ALLEN, b Osceola, Nebr, Feb 25, 33; div; c 3. RADIOBIOLOGY, TOXICOLOGY. *Educ:* Dana Col, BA, 59; Univ Southern Calif, PhD(biochem), 64. *Prof Exp:* Biochemist, Los Alamos Nat Lab, 64-81; molecular biologist, Dept Energy, 77-88; dep dir, Health Effects & Res, 79-88, DIR, HEALTH EFFECTS & LIFE SCI RES, DEPT ENERGY, 88- *Mem:* Am Chem Soc; Am Soc Biochem & Molecular Biol; AAAS. *Res:* Nucleic acid enzymology; physical properties of nucleic acids; radiobiology. *Mailing Add:* Health Effects & Life Sci Res Div Dept Energy Washington DC 20586

SMITH, DAVID BEACH, b Newton, NJ, Dec 3, 11; m 42; c 4. SYSTEMS ENGINEERING, ELECTRICAL ENGINEERING. *Educ:* Mass Inst Technol, BS, 33, MS, 34. *Prof Exp:* Engr, Philco Corp, 34-39, dir res, 39-45, vpres res, 45-46, vpres res & eng, 46-58, vpres tech affairs, 58-61, vpres res & eng, Philco-Ford, 61-64; prof systs eng, Moore Sch, Univ Pa, 64-67; pres, HRB Singer, Inc, 67-69; vis lectr, Univ Pa, 69-72; prof & dir grad prog eng mgt, 72-78, lectr, 78-90, EMER PROF, DREXEL UNIV, 78- *Concurrent Pos:* Mem, Nat TV Syst Comt, 40-41; mem bd dirs, Philco Corp, 46-56; vchmn, Second Nat TV Syst Comt, 50-53; mem adv comt, Signal Corp, US Army, 53-56; mem bd dirs, Narco Sci, 65-82; mem, Sci & Arts Comt, Franklin Inst, 68-, chmn, 81. *Mem:* AAAS; Am Soc Eng Educ; fel Inst Elec & Electronics Engrs; Sigma Xi. *Res:* Research and development management. *Mailing Add:* 1400 Waverly Rd-B-129 Gladwyne PA 19035

SMITH, DAVID BURRARD, b Kidderminster, Eng, Dec 6, 16; Can citizen; m 42; c 3. BIOCHEMISTRY. *Educ:* Univ BC, BA, 39, MA, 41; Univ Toronto, PhD(biochem), 50. *Prof Exp:* Res officer, Div Biosci, Nat Res Coun Can, 50-66; prof, 66-82, EMER PROF BIOCHEM, UNIV WESTERN ONT, 82- *Mem:* AAAS; Can Soc Cell Biol; Am Soc Biol Chem; Can Biochem Soc. *Res:* Physical chemistry of proteins, especially hemoglobin; amino acid sequence and oxygen equilibrium of hemoglobin; hemoglobin haptoglobin complex; cross-linking proteins with bifunctional reagents. *Mailing Add:* Dept Biochem Univ Western Ont London ON N6A 5C1 Can

SMITH, DAVID CLEMENT, IV, b Midland, Tex, Apr 26, 51. PHYSICAL OCEANOGRAPHY, OCEAN NUMERICAL MODELS. *Educ:* Univ Houston, BS, 73; Tex A&M Univ, MS, 75, PhD(oceanog), 80. *Prof Exp:* Assoc, Mesoscale Air Sea Interaction Group, Fla State Univ, 80-81; asst res scientist, Dept Oceanog, Tex A&M Univ, 83-85; adj res prof, Naval Postgrad Sch, 81-83, asst prof, Dept Oceanog, 85-90; OCEANOG, POLAR SCI CTR, APPL PHYSICS LAB, UNIV WASH, 90- *Mem:* Am Meteorol Soc; Am Geophys Union. *Res:* Physical oceanographic research in ocean mesoscale dynamics through the use of regional ocean numerical models. *Mailing Add:* Polar Sci Ctr/Appl Physics Lab Univ Wash 1013 NE 40th St Seattle WA 98105

SMITH, DAVID EDMUND, b Brentford, Eng, Nov 3, 34; m 61; c 4. SATELLITE GEODESY, CELESTIAL MECHANICS. *Educ:* Univ Durham, Eng, BSc, 58; Univ London, MSc, 62, PhD(satellite geod), 66. *Prof Exp:* Sci officer math, Radio & Space Res Sta, 58-68; sr scientist geod, EG&G, Wolf Res & Develop Corp, 68-69; staff scientist geophys, 69-71, head geodynamics br, 71-87, ASSOC CHIEF, LAB TERRESTRIAL PHYSICS, NASA GODDARD SPACE FLIGHT CTR, 88- *Concurrent Pos:* Mem working group 1 satellite geod & geodynamics, Comt Space Res, 67-; mem comn satellite geod, Joint Comt Space Res & Int Union Geod & Geophys, 71-; mem study group fundamental geod constants, Int Asn Geod, 74-87, mem study group ref systs geod & geodynamics, 76-, mem study group on parameters of common relevance to astron, geodesy & geodynamics, 87-; mem working group measurement earth rotation, Int Astron Union, 78-; proj scientist, Crustal Dynamics Proj, NASA, 80-. prin investr radar altimeter, mem radio sci team, NASA Mars Observer Mission; proj scientist, Lageos Satellite, 76-80. *Honors & Awards:* Except Sci Achievement Medal, NASA, 74; John C Lindsay Mem Award, Goddard Space Flight Ctr, 78. *Mem:* Royal Astron Soc; fel Am Geophys Union (pres, geodesy sect, 88-). *Res:* Determination of shape and size of earth, its tectonics, gravity field, internal structure, rotations and tides; motion of artificial satellites and their perturbation. *Mailing Add:* Lab Terrestrial Physics Code 620 Goddard Space Flight Ctr Greenbelt MD 20771

SMITH, DAVID ENGLISH, b San Francisco, Calif, June 9, 20; m 48; c 3. PATHOLOGY. *Educ:* Cent Col, Mo, AB, 41; Washington Univ, MD, 44; Am Bd Path, dipl, 50. *Prof Exp:* Intern & resident path, Barnes Hosp, St Louis, Mo, 44-46; from instr to assoc prof path, Sch Med, Washington Univ, 48-55, asst head dept, 53-54; prof, Sch Med, Univ Va, 55-73, chmn dept, 58-73, dir cancer ctr, 72-73; prof path, Northwestern Univ, Evanston, 74-75; prof path, Univ Pa, Philadelphia, 76-80; PROF PATH & ASSOC DEAN, TULANE UNIV, 80- *Concurrent Pos:* Mem, Exec Comt, Nat Bd Med Examrs, 70-80, vpres & dir undergrad div, 75-80, secy, 77-80; trustee, Am Bd Path, 66-73; assoc dir, Am Bd Med Specialists, 74-75. *Mem:* Am Soc Clin Path; Am Asn Pathologists & Bacteriologists; AMA; Am Acad Neurol; Int Acad Path (pres, 64-65). *Res:* Neuropathology; quantitative histochemistry; evaluation of medical education. *Mailing Add:* 59 Colony Park Circle Galveston TX 77551-1739

SMITH, DAVID FLETCHER, b Sewickley, Pa, Jan 30, 46; m 67; c 2. COMMERCIAL DIAGNOSTIC ASSAYS, BIOTECHNOLOGY. *Educ:* Tex Lutheran Col, BS, 67; Univ Tex, Houston, MS, 69, PhD(biochem), 72. *Prof Exp:* Postdoctoral fel biochem, Dept Chem, Fla State Univ, Tallahassee, 72-73; res assoc membrane biol, Am Red Cross Blood Res Lab, Bethesda, Md, 73-76; sr staff fel biochem, Lab Biochem Pharm, Nat Inst Arthritis, Diabetes & Digestive & Kidney Dis, NIH, 77-80; from asst prof to assoc prof, Va Polytech Inst & State Univ, 80-89; BIOCHEMIST, UNIV GA, ATHENS, 89- *Concurrent Pos:* Vis prof, Biochem Inst, Univ Freiburg, Ger, 75-76; prin investr, NIH, 82-, NSF, 84- & NIH small bus innovative res grant award, 89-90; mem, Pathobiochem Study Sect, NIH, 89-; adj res scientist, Complex Carbohydrate Res Ctr, Univ Ga, Athens, 89-; vpres & dir, ELA Technol, Inc, Athens, 89-90; chief sci officer, Sealite Sci, Inc, Atlanta, Ga, 91- *Mem:* Am Soc Biochem & Molecular Biol; Soc Complex Carbohydrates; Sigma Xi. *Res:* Structure and function of oligosaccharides on cell surface glyco conjugates, glycolipids and glycoproteins; methods to exploit recombinant bioluminescent proteins as reagents for diagnostic assays. *Mailing Add:* Dept Biochem Univ Ga Athens GA 30602

SMITH, DAVID HARRISON, JR, b Hibbing, Minn, Mar 7, 26; m 83; c 2. GERMPLASM RESOURCES, PLANT BREEDING. *Educ:* Hamline Univ, BS, 50; Univ Minn, MS, 55; Mich State Univ, PhD(crop sci), 63. *Prof Exp:* Instr, High Sch, 55-56; instr bot & zool, Brainerd Jr Col, 56-59; asst prof crop sci, Mich State Univ, 63-65; res geneticist, Plant Sci Res Div, 65-80, CUR, USDA SMALL GRAINS COLLECTION, 80- *Mem:* Am Soc Agron; Crop Sci Soc Am; Am Genetic Asn. *Res:* Isolation and incorporation of resistance to the Cereal Leaf Beetle into adapted wheat, barley and oat germ-plasm; collection, maintenance and distribution of cereal germplasm. *Mailing Add:* USDA ARS BARC-W B046 Beltsville MD 20285

SMITH, DAVID HIBBARD, b Springfield, Mo, July 29, 41; m 66; c 2. ORGANIC CHEMISTRY. *Educ:* Univ Notre Dame, BS, 63; Univ Mo-Columbia, PhD(org chem), 71. *Prof Exp:* From asst prof to assoc prof, 70-83, PROF CHEM, DOANE COL, 83-, CHMN, NAT SCI DIV, 79- *Concurrent Pos:* Mem, Environ Control Coun, State Nebr, 74-79; res assoc, Univ Va, 82-83; vis prof, Univ Nebr, Lincoln, 87 & 88. *Mem:* Am Chem Soc. *Res:* Nitrosamines; azirdines; allelopathic chemicals. *Mailing Add:* 1175 Driftwood Dr Crete NE 68333-1726

SMITH, DAVID HUSTON, b Seattle, Wash, July 14, 37; m 58; c 2. CHEMISTRY. *Educ:* Whitman Col, BA, 59; Cornell Univ, MS, 62; Univ Tenn, Knoxville, PhD(anal chem), 70. *Prof Exp:* CHEMIST, OAK RIDGE NAT LAB, UNION CARBIDE NUCLEAR CO, 62- *Res:* Applications of mass spectrometry to safeguards; mass spectrometric research and development; design and development of computer programs for mass spectrometric data. *Mailing Add:* 104 Wilderness Lane Oak Ridge TN 37831

SMITH, DAVID I, b Cooperstown, NY, May 22, 54; m 81; c 3. MOLECULAR GENETICS OF LUNG & RENAL CANCER, PHYSICAL MAP OF CHROMOSOME THREE. *Educ:* Univ Wis-Madison, BS, 74, PhD(biochem), 78. *Prof Exp:* Postdoctoral researcher molecular biol, Albert Einstein Col Med, 78-80; sr res scientist, Enzo Biochem, NY, 80-81; postdoctoral fel biol chem, Univ Calif, Irvine, 81-85; asst prof, 85-90, ASSOC PROF MOLECULAR BIOL & GENETICS, WAYNE STATE UNIV, 90- *Concurrent Pos:* Mem Mammalian Genetics Study Sect, NIH, 90-; Basil O'Connor starter res grant, March of Dimes Birth Defects Found, 88. *Mem:* AAAS; Am Soc Human Genetics. *Res:* Constructing a precise physical map for human chromosome 3 to facilitate the isolation of chromosome 3 genes associated with specific diseases including small cell lung cancer, renal cell carcinoma and Von Hippel Lindau disease. *Mailing Add:* 3136 Scott Hall Wayne State Univ 540 E Canfield Detroit MI 48201

SMITH, DAVID JOHN, b Melbourne, Australia, Oct 10, 48; m 71; c 2. HIGH RESOLUTION ELECTRON MICROSCOPY, STRUCTURE OF MATERIALS. *Educ:* Univ Melbourne, BSc, 70, PhD(physics), 78, DSc, 88. *Prof Exp:* Staff demonstr, Sch Physics, Univ Melbourne, 75, vis scientist, 84; res asst, Cavendish Lab, Cambridge Univ, 76-80, sr res asst & actg dir, High Resolution Electron Microscope, 78-80, sr res asst & dir, 80-84; assoc prof, 84-87, PROF, CTR SOLID STATE SCI & DEPT PHYSICS, ARIZ STATE UNIV, 87-, DIR, NAT FACIL HIGH RESOLUTION ELECTRON MICROS, 91- *Honors & Awards:* Charles Vernon Boys Prize, Inst Physics, UK, 85. *Mem:* Royal Micros Soc; fel Inst Physics UK; Electron Micros Soc Am; Mat Res Soc. *Res:* High resolution electron microscopy, instrumentation and applications; characterization of surfaces of metals, oxides and semiconductors; the atomic structure of small and extended defects in solids. *Mailing Add:* Ctr Solid State Sci Ariz State Univ Tempe AZ 85287

SMITH, DAVID JOSEPH, b Parkersburg, WVa, Dec 17, 43; m 64; c 2. PHARMACOLOGY. *Educ:* Bethany Col, BS, 65; WVa Univ, PhD(pharmacol), 69. *Prof Exp:* Fel, Univ Iowa, 69-71; asst prof, 71-74, assoc prof, 71-81, PROF ANESTHESIOL & PHARMACOL, WVA UNIV, 81-, DIR, ANESTHESIOL RES LAB, 74- *Concurrent Pos:* Chmn, WVa Affil, Am Heart Asn, 83-85; mem res comt, WVa Heart Asn. *Mem:* Am Soc Pharmacol & Exp Therapeut; Am Soc Anesthesiologists; AAAS; Sigma Xi; Soc Neurosci. *Res:* Neuropharmacological research of drug action on neurochemical processes of neurotransmitter metabolism; metabolic changes in transmitter metabolism are correlated with alterations in central nervous system function, particularly nociceptive behavior. *Mailing Add:* Clin Invest Dept Wm Beaumont Army Med Ctr Ft Bliss El Paso TX 79920-5001

SMITH, DAVID LEE, b June 7, 44; US citizen; m 67; c 2. CHROMATOGRAPHY, MASS SPECTROMETRY. *Educ:* Univ Kans, BS, 66, PhD(chem), 69. *Prof Exp:* Instr chem, Univ Utah, 69-76, from res asst prof to res assoc prof med chem, 76-84; assoc prof, 84-88, PROF MED CHEM, PURDUE UNIV, 89- *Concurrent Pos:* NATO sr res scientist fel, 74. *Mem:* Am Chem Soc; Am Soc Mass Spectros; Am Asn Col Pharm. *Res:* Bio-medical applications of mass spectrometry and chromatography. *Mailing Add:* Dept Med Chem Purdue Univ West Lafayette IN 47907

SMITH, DAVID MARSHALL, b Gary, Ind, July 22, 43. PATHOLOGY, IMMUNOLOGY. *Educ:* Colo State Univ, DVM, 68; Mass Inst Technol, PhD(biochem & path), 74. *Prof Exp:* Intern vet med, Angell Mem Animal Hosp, 68-69; res fel path, Harvard Med Sch, 69-70 & Mass Inst Technol, 70-73; veterinarian, Mass Inst Technol, 73-74; asst group leader, 74-88, EXP PATHOLOGIST, LOS ALAMOS SCI LAB, UNIV CALIF, 74-, PROG MGR, 88- *Concurrent Pos:* Res affil, Forsyth Dent Ctr, 72-74. *Mem:* Am Vet Med Asn; Am Acad Clin Toxicol; Reticuloendothelial Soc; Hamster Soc; Am Animal Hosp Asn. *Res:* Immunology of carcinogenesis; lung cancer; radiation-induced carcinogenesis; toxicology. *Mailing Add:* 1066 Encantado Dr Santa Fe NM 87501

SMITH, DAVID MARTYN, b Bryan, Tex, Mar 10, 21; m 51; c 2. SILVICULTURE, FOREST ECOLOGY. *Educ:* Univ RI, BS, 41; Yale Univ, MF, 46, PhD, 50. *Hon Degrees:* DSc, Bates Col, 86. *Prof Exp:* Instr silvicult, Yale Univ, 46-47 & 48-51, from asst prof to prof, 51-67, asst dean sch, 53-58, Morris K Jesup prof, 67-90, EMER PROF SILVICULT, SCH FORESTRY & ENVIRON STUDIES, YALE UNIV, 90- *Concurrent Pos:* Vis prof, Univ Munich, 81; hon mem, Acad Forest Sci, Mex. *Honors & Awards:* Distinguished Serv Award, Am Forestry Asn, 90. *Mem:* Ecol Soc Am; fel Soc Am Foresters. *Res:* Silviculture; regeneration and manipulation of forest vegetation, especially stratified mixtures. *Mailing Add:* Sch Forestry & Environ Studies Yale Univ New Haven CT 06520

SMITH, DAVID PHILIP, b Minneapolis, Minn, July 21, 35; m 52; c 4. MAGNETIC RECORDING, TRIBOLOGY. *Educ:* Univ Minn, BS, 62, MS, 66, PhD(elec eng), 72. *Prof Exp:* Physicist, Cent Res Labs, 62-66, res physicist, 66-69, sr res physicist, 70-71, res specialist, Cent Res Labs, 71-72, supvr, 72-74, mgr, 74; mkt mgr anal syst, 3M Co, St Paul, Minn, 74-79; sr res specialist, 3M Data Cartridge Lab, 79-88; DIV SCIENTIST, 3M DATA STORAGE PROD, 88- *Concurrent Pos:* Mem, Res Comt Tribology, Am Soc Mech Engrs, 89- *Honors & Awards:* Wayne B Nottingham Prize, Annual Phys Electronics Conf, 71. *Mem:* Am Soc Mech Engrs. *Res:* Tribology of magnetic recording devices; surface science; solid state physics; ion surface interactions. *Mailing Add:* 3M Ctr Bldg 236-GD-18 St Paul MN 55144

SMITH, DAVID R(ICHARD), b London, Eng, July 29, 36; m 61; c 3. ELECTRICAL ENGINEERING. *Educ:* Univ London, BS, 57; Univ Wis, PhD(elec eng), 61. *Prof Exp:* Res fel, Nat Phys Lab, Teddington, Eng, 61-63; asst prof bioeng, Case Inst Technol, 63-66; assoc prof elec sci, 66-71, PROF COMPUT SCI, STATE UNIV NY STONY BROOK, 72- *Mem:* Inst Elec & Electronics Engrs. *Res:* Computer architecture and digital systems design. *Mailing Add:* Dept Comput Sci State Univ NY Stony Brook NY 11794

SMITH, DAVID REEDER, b Murray, Utah, Nov 1, 38; m 63; c 3. SOLID STATE PHYSICS. *Educ:* Univ Utah, BS, 63; Purdue Univ, West Lafayette, MS, 66, PhD(physics), 69. *Prof Exp:* From asst prof to assoc prof physics, SDak Sch Mines & Technol, 78-87; RES PHYSICIST, US DEPT COM, NAT INST STANDARDS & TECHNOL, BOULDER CO, 87- *Mem:* AAAS; Sigma Xi; Am Asn Physics Teachers. *Res:* Low temperature solid state physics; cryogenics; heat transfer. *Mailing Add:* 1013 Alsace Way Lafayette CO 80026

SMITH, DAVID ROLLINS, b Rockford, Ill, July 27, 37; m 67; c 2. SYSTEMATICS, ENTOMOLOGY. *Educ:* Ore State Univ, BA, 60, PhD(entom), 67. *Prof Exp:* RES ENTOMOLOGIST, SYST ENTOM LAB, AGR RES SERV, USDA, 65- *Concurrent Pos:* Ed, Proc Entom Soc Wash, 80-83, pres, Entom Soc Wash, 91. *Mem:* Entom Soc Am; Soc Syst Zool; Am Entom Soc; Japanese Entom Soc; Am Registry Prof Entomologists. *Res:* Systematics of sawflies and ants. *Mailing Add:* Syst Entom Lab C/O US Nat Museum Washington DC 20560

SMITH, DAVID S, b Ipswich, Mass, June 29, 21; m 47; c 3. PEDIATRICS. *Educ:* Dartmouth Col, AB, 42; Univ Pa, MD, 44; Am Bd Pediat, dipl, 50. *Prof Exp:* PROF PEDIAT, SCH MED, TEMPLE UNIV, 68- *Concurrent Pos:* Consult, Nazareth Hosp, Philadelphia, 66-; prin investr, Clin Res Ctr, 76-; dir inpatient serv, St Christopher's Hosp Children, 66-76, actg chmn, Dept Pediat, 76- *Honors & Awards:* Arthur Dannenberg MD lectr, Albert Einstein Med Ctr, 79. *Mem:* Am Acad Pediat; AMA. *Res:* Infectious disease. *Mailing Add:* 4012 Primrose Rd Temple Univ Philadelphia PA 19114

SMITH, DAVID SPENCER, b London, Eng, Apr 10, 34; m 64. CELL BIOLOGY. *Educ:* Cambridge Univ, BA, 55, MA & PhD(zool), 58. *Prof Exp:* Fel cell biol, Rockefeller Univ, 58-59, res assoc, 59-61; res fel zool, Cambridge Univ, 61-63; asst prof biol, Univ Va, 63-66; assoc prof med, anat & biol, 66-70, PROF MED & PHARMACOL, UNIV MIAMI, 70- *Concurrent Pos:* USPHS fel, 59; NSF res grant, 63-66; external dir res, Int Ctr Insect Physiol & Ecol, 70, dir res, Nairobi, Kenya; NIH res grants, 71-; Hope prof entom (zool), Univ Oxford, 80. *Mem:* Am Soc Cell Biol; Royal Entom Soc London. *Res:* Electron microscopic studies on vertebrate and invertebrate animal tissues, especially muscle fibers and central and peripheral nervous systems; studies on structure and function of cellular membranes. *Mailing Add:* 3000 Seminole Miami FL 33133

SMITH, DAVID VARLEY, b Memphis, Tenn, Apr 21, 43; m 65; c 3. NEUROSCIENCE, PSYCHOLOGY. *Educ:* Univ Tenn, BS, 65, MA, 67; Univ Pittsburgh, PhD(psychobiol), 69. *Prof Exp:* Res assoc, Rockefeller UniY, 69-71; asst prof, 71-75, assoc prof, 75-80, PROF PSYCHOL, UNIV WYO, 80-; OTOLARYNGOL & MAXILLOFACIAL SURGEON, UNIV CINN, COL MED, 85- *Concurrent Pos:* Adj asst prof psychol, Hunter Col, City Univ New York, 70-71; Nat Inst Neurol Commun Dis & Stroke res career develop award, 76-82; asst prog dir, NSF, 77-78. *Mem:* AAAS; Soc Neurosci; Asn Chemoreceptive Sci. *Res:* Gustatory physiology and behavior; taste quality coding; neurophysiology. *Mailing Add:* Otolaryngol & Maxillofacial Surg Univ Cinn Col Med 231 Bethesda Ave Cincinnati OH 45267-0528

SMITH, DAVID WALDO EDWARD, b Fargo, NDak, Apr 3, 34; m 60. PATHOLOGY, MOLECULAR BIOLOGY. *Educ:* Swarthmore Col, BA, 56; Yale Univ, MD, 60. *Prof Exp:* From intern to asst resident, Yale-New Haven Med Ctr, 60-62; res assoc, Molecular Biol Lab, Nat Inst Arthritis & Metab Dis, 62-64, investr, Lab Exp Path, 64-67; assoc prof path & microbiol, Ind Univ, Bloomington, 67-69; PROF PATH, MED SCH, 69- , PROF, BUEHLER CTR AGING, NORTHWESTERN UNIV, CHICAGO, 88- *Concurrent Pos:* Res fel, Yale-New Haven Med Ctr, 60-62; res career develop award, Nat Inst Gen Med Sci, 68-69; mem, Pathological Chem Study, NIH, 75-79, sabbatical leave, 86-87. *Mem:* AAAS; Am Asn Pathologist; Am Soc Hematol; Am Soc Biol Chemists; Geront Soc Am. *Res:* Transfer RNA; genetic control of protein synthesis; hemoglobin synthesis; gender and longevity; gerontology. *Mailing Add:* Dept Pathol Northwestern Univ Med Sch 303 E Chicago Ave Chicago IL 60611

SMITH, DAVID WARREN, b Garden Prairie, Ill, Jan 28, 39; m 62; c 3. ANALYTICAL CHEMISTRY. *Educ:* Northern Ill Univ, BS, 61; Iowa State Univ, PhD(anal chem), 68. *Prof Exp:* Teacher, High Sch, Ill, 61-63; sr chemist, Mallinckrodt, Inc, 68-70, sr res assoc, 70-73, group leader, 73-78, mgr res & develop, 78-79; MGR, CHEM SERV, ETHYL PETROL ADDITIVES INC, 79- *Mem:* Am Chem Soc; Soc Tribiologist & Lubrication Engrs; Soc Automotive Engrs. *Res:* Analytical chemistry of alkaloids; analytical chromatography; process research and development of natural products. *Mailing Add:* Ethyl Petroleum Additive Div 1530 S Second St St Louis MO 63104-3896

SMITH, DAVID WILLIAM, b Wisconsin Rapids, Wis, June 16, 38; m 77; c 2. FORESTRY. *Educ:* Iowa State Univ, BS, 60, MS, 68, PhD(forest biol), 70. *Prof Exp:* Exten forester, Iowa State Univ & Coop Exten Serv, Iowa, 66-67; asst prof forest technol & chmn, Glenville State Col, 70-72; asst prof forestry, 72-78, assoc prof forest soils & silvicult & chmn forest biol sect, 78-85, PROF SILVICULT & FOREST SOILS & ASST DIR, SCH FORESTRY & WILDLIFE RESOURCES, VA POLYTECH INST & STATE UNIV, 85- *Mem:* Soil Sci Soc Am; Soc Am Foresters. *Res:* Nutrient cycling in forest systems, specifically the effects of silvicultural practices on site productivity through changes in soil physical and chemical properties; hardwood regeneration. *Mailing Add:* Sch Forestry & Wildlife Resources Va Polytech Inst & Sta Univ Blacksburg VA 24061-0324

SMITH, DAVID WILLIAM, b Edson, Alta, Jan 18, 33; m 56; c 3. PLANT ECOLOGY. *Educ:* Univ Alta, BSc, 56, MSc, 59; Univ Toronto, PhD(ecol), 67. *Prof Exp:* Res officer host, Res Br, Can Dept Agr, 58-67; asst prof ecol, 67-72, asst prof, 72-77, ASSOC PROF BOT, UNIV GUELPH, 77- *Concurrent Pos:* Grants, Ont Dept Lands & Forests, Univ Guelph, 70-73 & Nat Res Coun Can, 71-80; Indian & Northern Affairs contract, 77-82. *Mem:* Can Bot Asn; Ecol Soc Am. *Res:* Vegetation dynamics and productivity of natural systems. *Mailing Add:* Dept Bot Univ Guelph Guelph ON N1G 2W1 Can

SMITH, DAVID WILLIAM, b Dayton, Ohio, March 17, 48; m 69; c 2. MICROBIAL ECOLOGY, PHYSIOLOGICAL ECOLOGY. *Educ:* Univ Calif, San Diego, BA, 69; Ind Univ, MA, 71; Univ Wis, PhD(bacteriol), 72. *Prof Exp:* Fel, Dept Bacteriol, Univ Calif, Los Angeles, 73-74; asst prof, 75-81, ASSOC PROF MICROBIOL, UNIV DEL, 81- *Concurrent Pos:* Prin investr res grants, Sea Grant Off, Nat Oceanic & Atmospheric Admin, 77- *Mem:* Am Soc Microbiol; AAAS; Fedn Am Scientists; Soc Gen Microbiol. *Res:* Sulfur and nitrogen cycle activities in salt marsh sediments as related to physical and chemical factors. *Mailing Add:* Dept Micro Biol Univ Del Newark DE 19711

SMITH, DAVID YOUNG, b Schenectady, NY, July 24, 34; m 63; c 2. SOLID STATE PHYSICS, OPTICAL PHYSICS. *Educ:* Rensselaer Polytech Inst, BS, 56; Univ Rochester, PhD(physics), 62. *Prof Exp:* Asst physics, Univ Rochester, 60-62; res assoc, Univ Ill, 62-63, res asst prof, 63-66; NSF fel, Physics Inst, Univ Stuttgart, 66-67; physicist, Solid State Sci Div, Argonne Nat Lab, 67-85, asst div dir, 74-79; PROF PHYSICS, UNIV VT, 86-, CHMN DEPT, 86- *Concurrent Pos:* Vis assoc prof, Mich State Univ, 71-72; Ger Acad Exchange Serv res fel, Physics Inst, Univ Stuttgart, 75-76; guest prof, Physics Inst, Univ Stuttgart & Max Planck Inst, Stuttgart, 79-80. *Mem:* Am Phys Soc; Sigma Xi. *Res:* Theoretical solid state physics especially the electronic states and optical properties of pure crystals and of defects; x-ray optics. *Mailing Add:* Cook Physical Sci Bldg Univ Vt Burlington VT 05405

SMITH, DEAN FRANCIS, b Los Angeles, Calif, July 25, 42; m 67; c 2. PLASMA PHYSICS. *Educ:* Mass Inst Technol, BS, 64; Stanford Univ, MS, 66, PhD(astrophys), 69. *Prof Exp:* Vis scientist, High Altitude Observ, Nat Ctr Atmospheric Res, 70-72, scientist, 72-78; sr res assoc, dept astro-geophysics, Univ Colo, 78-81; SR ASSOC, BERKELEY RES ASSOCS, 81- *Concurrent Pos:* US-USSR, Cultural Exchange fel, Sternberg Astron Inst, Moscow, 69-70; lectr, Univ Colo, 71-83; Nat Res Coun sr assoc, Nat Oceanic Atmospheric Admin Lab, 90-91. *Mem:* Int Astron Union; Am Phys Soc; Am Astron Soc; Astron Soc Australia; Int Union Radio Sci; Int Asn Geomagnetism & Aeronomy. *Res:* Plasma astrophysics; theory of solar radio bursts; theory of flares and particle acceleration on the sun; theory of reconnection; theory of pulsar and planetary magnetospheres. *Mailing Add:* Berkeley Res Assocs 290 Green Rock Dr Boulder CO 80302

SMITH, DEAN HARLEY, b Dayton, Wash, May 4, 22; m 45; c 2. VETERINARY MEDICINE, CLINICAL PATHOLOGY. *Educ:* Wash State Univ, BS, 44, DVM, 49; Ore State Univ, MS, 59. *Prof Exp:* Vet, Button Vet Hosp, Tacoma, Wash, 49-50 & Dayton, 50-52 & 54-56; res asst, Ore State Univ, 56-59, from asst prof to prof vet med, 59-76; supvr fed-state progs, Ore Dept Agr, 76-85. *Concurrent Pos:* Fulbright lectr, Col Vet Med, Cairo Univ, 65-66. *Mem:* Am Vet Med Asn; US Livestock Sanit Asn; Am Pub Health Asn. *Res:* Animal disease diagnosis with special emphasis on sheep and cattle. *Mailing Add:* PO Box 237 Dayton WA 99328

SMITH, DEAN ORREN, b Colorado Springs, Colo, May 28, 44; m 65; c 2. NEUROPHYSIOLOGY. *Educ:* Harvard Univ, BA, 67; Stanford Univ, AM, 69, PhD(biol sci), 71. *Prof Exp:* Fel physiol, Univ Götenborg, 71-72; fel, Tech Univ München,72-74, scholar, 74-75; actg asst prof biol, Univ Calif, Los Angeles, 75-76; from asst prof to assoc prof, 76-83, PROF PHYSIOL, UNIV WIS-MADISON, 83-, ASSOC DEAN GRAD SCH, 84- *Concurrent Pos:* Helen Hay Whitney Found fels, Univ Götenborg, 71-72 & Tech Univ, M06nchen, 72-74; A P Sloan res fel, Univ Wis-Madison, 78-80, res career develop award, 79-83, Romnes fel, 83-, mem aging rev comt, 85- *Mem:* Soc Neurosci; Am Physiol Soc; fel Geront Soc Am. *Res:* Integration in the nervous system at the level of the axon and the synapse; changes in synaptic mechanisms during aging. *Mailing Add:* Dept Physiol Univ Wis 1300 University Ave Madison WI 53706R

SMITH, DEANE KINGSLEY, JR, b Berkeley, Calif, Nov 8, 30; m 57; c 5. MINERALOGY, CRYSTALLOGRAPHY. *Educ:* Calif Inst Technol, BS, 52; Univ Minn, PhD(geol), 56. *Prof Exp:* Instr field geol, Univ Minn, 56; Portland Cement Asn fel, Nat Bur Standards, 56-60; chemist, Lawrence Livermore Lab, Univ Calif, 60-68; assoc prof mineral, 68-71, PROF MINERAL, PA STATE UNIV, UNIVERSITY PARK, 71- *Concurrent Pos:* Chmn, JCPDS-Int Ctr Diffraction Data, 78-82, 86-90. *Mem:* Fel Geol Soc Am; Am Soc Testing & Mat; fel Mineral Soc Am; Am Crystallog Asn (secy, 76-78); Mineral Asn Can. *Res:* Defects in crystals and crystal structures of inorganics and minerals; uranium and nuclear waste management; applications of powder x-ray diffractometry. *Mailing Add:* Dept Geosci Pa State Univ 239 Deike Bldg University Park PA 16802

SMITH, DELMONT K, b Pocatello, Idaho, June 9, 27; m 46; c 5. ORGANIC CHEMISTRY. *Educ:* Utah State Univ, BS, 49, MS, 55; Purdue Univ, PhD(org chem), 54. *Prof Exp:* Res chemist, Rayonier, Inc, 54-56, sect supvr, 56-59, asst res mgr, 59, div mgr, 59-61; res supvr, Chicopee Mfg Co, 61-66, dir woven prod res, 66-70, dir technol planning, 70-84; CONSULT, 85- *Mem:* Am Chem Soc; Tech Asn Pulp & Paper Indust; Am Asn Textile Chemists & Colorists; Int Nonwovens & Disposables Asn. *Res:* Organic halogen compounds; high polymers; cellulose and cellulose derivatives; textile technology; nonwoven technology. *Mailing Add:* 3112 E Hampton Ave Mesa AZ 85204

SMITH, DENISE MYRTLE, b Chester, Pa, Sept 15, 55. PROTEIN CHEMISTRY & FUNCTIONALITY, MEAT SCIENCE. *Educ:* Va Polytech Inst & State Univ, BS, 77; Ore State Univ, MS, 79; Wash State Univ, PhD(food sci), 84. *Prof Exp:* Asst prof, 85-90, ASSOC PROF FOOD SCI, FOOD SCI & HUMAN NUTRIT, MICH STATE UNIV, 90- *Concurrent Pos:* Assoc ed, Poultry Sci, Poultry Sci Asn, 89- *Mem:* Inst Food Technologists; Am Meat Sci Asn; Poultry Sci Asn. *Res:* Meat, milk and egg protein chemistry, functionality and rheology; improved technologies for the processing of meat and egg products; development of ELISAs as indicators of processing adequacy. *Mailing Add:* Dept Food Sci & Human Nutrit Mich State Univ East Lansing MI 48824-1224

SMITH, DENNIS CLIFFORD, b Lincoln, Eng, Mar 24, 28; m 55; c 6. BIOMATERIALS. *Educ:* Univ London, BSc, 50, MSc, 53, DSc, 79; Univ Manchester, PhD(chem), 57. *Hon Degrees:* DSc, Univ London, 79. *Prof Exp:* Asst lectr dent mat, Univ Manchester, 52-69, reader, 69; PROF BIOMAT & DIR CTR FOR BIOMAT, UNIV TORONTO, 69- *Concurrent Pos:* Vis assoc prof, Northwestern Univ, 60-61. *Honors & Awards:* Wilmer Souder Award, Int Asn Dent Res, 76; Clemson Award, Soc Biomat, 76; Nakabayashi Mem Award, Pierre Fouchard Soc, Japan, 82. *Mem:* Int Asn Dent Res; Soc Biomat; Can Soc Biomat; Adhesion Soc; hon fel Int Col Dentists. *Res:* Polymer chemistry; tissue reaction to materials; physical properties of materials. *Mailing Add:* Fac Dent-Ctr Bio Mat Univ Toronto 124 Edward St Toronto ON M5G 1G6 Can

SMITH, DENNIS EUGENE, b Blue Earth, Minn, May 26, 49; m 70; c 3. IMMUNOASSAY, CELL BIOLOGY. *Educ:* Univ Minn, BS, 77, PhD(pathobiol), 82. *Prof Exp:* res scientist immunol, Ames Div, 82-86, RES GROUP LEADER, BAXTER DADE DIV, MILES LABS, 86- *Mem:* Am Asn Clin Chem; Am Soc Cell Biol. *Res:* Development of non-isotopic immunoassays for drugs and proteins; protein chemistry; monoclonal antibodies. *Mailing Add:* Baxter Dade Div MS420 PO Box 520672 Miami FL 33152

SMITH, DENNIS MATTHEW, b Chicago, Ill, Mar 18, 52; m 80; c 1. PULMONARY PATHOBIOLOGY, BIOLOGY EDUCATION. *Educ:* Loyola Univ Chicago, BS, 74, PhD(anat), 79. *Prof Exp:* Asst prof 80-86, chmn, 88-90, ASSOC PROF BIOL SCI, WELLESLEY COL, 86- *Concurrent Pos:* Consult, var indust firms, 89- *Mem:* Am Soc Cell Biol; Am Asn Anatomists; Sigma Xi; AAAS. *Res:* Pathobiology of the mammalian distal lung, especially electron microscopy; injurious effects of chronic beta-adrenergic blockade on the distal lung; culture of lung cells and their interactions. *Mailing Add:* Dept Biol Sci Wellesley Col Wellesley MA 02181

SMITH, DENNISON A, b Newton, Mass, June 19, 43; m 69. NEUROSCIENCES. *Educ:* Colgate Univ, AB, 65; Univ Mass, MS, 67, PhD(psychol), 70. *Prof Exp:* From asst prof to assoc prof psychol, 69-87, chmn psychobiol prog, 76-87, CHMN PSYCHOL DEPT, OBERLIN COL, 78-, PROF PSYCHOL & NEUROSCI, 87- *Concurrent Pos:* Chmn, Neurosci Prog, Oberlin Col, 88- *Mem:* AAAS; Soc Neurosci; Sigma Xi. *Res:* Neuropharmacology. *Mailing Add:* Neurosci Prog Sperry Hall Oberlin Col Oberlin OH 44074

SMITH, DIANE ELIZABETH, b New York, NY, Nov 15, 37. NEUROANATOMY,IMMUNOCYTOCHEMISTRY. *Educ:* Bucknell Univ, BS, 59; Am Univ, MS, 65; Univ Pa, PhD(anat), 68. *Prof Exp:* Res biologist, Clin Neuropath Sect, Surg Neurol Br, Nat Inst Neurol Dis & Blindness, 62-65; asst prof anat, Daniel Baugh Inst Anat, Jefferson Med Col, Thomas Jefferson Univ, 69-75, assoc prof, 75; assoc prof, 75-80, PROF ANAT, LA STATE UNIV SCH MED, 80- *Concurrent Pos:* NIH res fel anat, Harvard Med Sch, 68-69; mem sci progs & adv comn, NIH & NINCDS, 84-86; mem educ affairs comt, Am Asn of Anatomists, 85-88; vis prof, Neurobiol, Cornell Univ Med Col, 85. *Mem:* AAAS; Sigma Xi; Am Asn Anatomists; Int Soc for Dev Neurosci. *Res:* Localization of immunocytochemically labeled neurotransmitters and neuromodulators in neurological mutant mice; alteration in neuronal circuitry as a result of genetic compromise. *Mailing Add:* Dept of Anat La State Univ Med 1901 Perdido St New Orleans LA 70112

SMITH, DON WILEY, b Weinert, Tex, Nov 11, 36; m 58; c 4. AGRONOMY, PLANT PHYSIOLOGY. *Educ:* Tex Tech Col, BS, 58; Univ Wis, MS, 60, PhD(agron), 63. *Prof Exp:* Res assoc agron, Univ Wis, 62-63; asst prof bot, Colo State Univ, 65-67; ASSOC PROF BIOL, UNIV NTEX, 67- *Concurrent Pos:* Consult on botany, pesticide safety & wildflower culture. *Mem:* Am Soc Plant Physiol; Weed Sci Soc Am; Bot Soc Am. *Res:* Mechanism of action of boron; gene transplants in tomato; effects of microwaves on plant cells; physiology of wildflower. *Mailing Add:* Dept of Biol Sci Univ of N Tex PO Box 5218 Denton TX 76203-5218

SMITH, DONALD ALAN, b Toronto, Ont, Aug 29, 30; Can citizen; m 53; c 5. VERTEBRATE ZOOLOGY. *Educ:* Univ Toronto, BA, 52, MA, 53, PhD(exp biol), 57. *Prof Exp:* Lectr, 57-58, asst prof, 58-63, ASSOC PROF BIOL, CARLETON UNIV, 63-, CUR, CARLETON UNIV MUS ZOOL, 73- *Concurrent Pos:* Vis prof, Makerere Univ, Uganda, 66-67; asst to ed, Can Field-Naturalist, 72-81. *Mem:* Am Soc Mammal; Can Soc Zool; Soc Preserv Nat Hist Collections. *Res:* Ecology, distribution, taxonomy, conservation, behavior and environmental physiology of vertebrates, especially rodents, bats and insectivores; reproductive biology; ectoparasites of mammals, especially fleas. *Mailing Add:* Dept Biol Carleton Univ Ottawa ON K1S 5B6 Can

SMITH, DONALD ARTHUR, b Can, Feb 2, 26; nat US; m 49; c 4. POLYMER CHEMISTRY. *Educ:* Univ BC, BA, 48; Univ Toronto, PhD(chem), 51. *Prof Exp:* Res assoc, Kodak Park Works, 51-77, sr lab head, 68-77, ASST DIR CHEM DIV, RES LABS, EASTMAN KODAK CO, 77- *Mem:* Am Chem Soc. *Res:* Sterochemistry; high polymers; synthesis of hydrophilic monomers and polymers. *Mailing Add:* 65 Pinegrove Ave Rochester NY 14617-2601

SMITH, DONALD EUGENE, b Tunkhannock, Pa, Jan 26, 34; m 58; c 3. REPRODUCTIVE PHYSIOLOGY, ENDOCRINOLOGY. *Educ:* Bloomsburg State Col, BScEd, 55; Ohio State Univ, MSc, 58, PhD(physiol, zool), 62. *Prof Exp:* Pub sch teacher, Pa, 55-56; asst zool, Ohio State Univ, 56-59, instr, 59-60; from instr to asst prof, Ohio Wesleyan Univ, 60-66; assoc prof, 67-72, PROF ZOOL, NC STATE UNIV, 72- *Concurrent Pos:* NSF res partic, Univ Ill, 65, vis asst prof, 66-67; vis prof, Duke Univ Med Ctr, 81. *Mem:* AAAS; Am Soc Zool; Sigma Xi; Soc for Study of Reproduction. *Res:* Mechanisms of hormone action; effects of estrogen and progesterone on uterine glucose metabolism; effects of metals on uterine steroid hormone receptors. *Mailing Add:* Dept Zool Box 7617 NC State Univ Raleigh NC 27695

SMITH, DONALD EUGENE, b Alice, Tex, Sept 29, 44; m 65; c 1. PHYSICAL CHEMISTRY, MATERIALS SCIENCE. *Educ:* Okla State Univ, BS, 66, PhD(phys chem), 71. *Prof Exp:* Res engr, Inland Steel Co, 70-73, sr res engr, 73-75, supv res engr, 75-78, gen supv res engr steel res, 78; PROD MGR, BELL LABS, 78- *Mem:* Nat Coil Coaters Asn; Am Soc Testing & Mat; Nat Asn Corrosion Engrs. *Res:* Corrosion; organic coatings for metals; surface structure and characterization; vibrational spectroscopy. *Mailing Add:* Bell Labs Rm 1M525 101 Crawfords Corner Rd PO Box 3030 Holmdel NJ 07733

SMITH, DONALD FOSS, b Athens, Tenn, Feb 14, 13; m 40; c 2. PHYSICAL CHEMISTRY. *Educ:* Univ Chattanooga, BS, 34; Univ Tenn, MS, 36; Univ Va, PhD(chem), 39. *Prof Exp:* Asst prof chem, Judson Col, 39-40 & The Citadel, 40-43; assoc explosives chemist, US Bur Mines, 43-44; asst prof chem, Pa Col Women, 44-45; from asst prof to assoc prof, Univ Vt, 45-51; from assoc prof to prof chem, Univ Ala, Tuscaloosa, 51-83, chmn dept, 81-83; RETIRED. *Mem:* Am Chem Soc; Am Inst Chem. *Res:* Heat capacity determinations at high temperatures; solubility determinations; cryoscopic determinations in fused salt systems; conductance in fused salts; heat capacities at low temperatures. *Mailing Add:* 25 Ridgeland Tuscaloosa AL 35406

SMITH, DONALD FREDERICK, b Picton, Ont, Nov 25, 49; m 74; c 3. VETERINARY SURGERY. *Educ:* Univ Guelph, DVM, 74; Am Col Vet Surgeons, dipl. *Prof Exp:* Intern vet med, Univ Pa, 74-75, resident vet surg, 75-77; assoc prof surg, Univ Wis, Madison,83-86; asst prof large animal surg, NY State Col Vet Met, 77-82, prof & chmn, Dept Clin Sci, 87-90; ASSOC DEAN VET MED, COL VET MED, CORNELL UNIV, 90- *Mem:* Comp Gastroenterol Soc. *Res:* Surgery of the borine gastrointestinal; surgery of the gastrointestinal tract; metabolic alkalosis. *Mailing Add:* Dept Vet Med Col Vet Med Cornell Univ Ithaca NY 14853-6401

SMITH, DONALD LARNED, b White Plains, NY, June 8, 40; m 67; c 1. NUCLEAR PHYSICS. *Educ:* Ga Inst Technol, BS, 62; Mass Inst Technol, PhD(physics), 67. *Prof Exp:* Asst physicist, 69-73, PHYSICIST, APPL PHYSICS DIV, ARGONNE NAT LAB, 73- *Mem:* Am Phys Soc; Am Nuclear Soc. *Res:* Nuclear measurement techniques; gamma-ray spectroscopy and associated correlations; radiation interaction with matter; neutron cross sections and neutron scattering phenomena. *Mailing Add:* Argonne Nat Lab Appl Physics Div 9700 S Cass Ave Bldg 314 Argonne IL 60439

SMITH, DONALD RAY, b Seminole, Okla, Jan 23, 39; m 64; c 2. MATHEMATICS. *Educ:* Auburn Univ, BS, 61; Stanford Univ, PhD(math), 65. *Prof Exp:* Vis mem, Courant Inst Math Sci, NY Univ, 65-66; asst prof, 66-71, assoc prof, 71-80, PROF MATH, UNIV CALIF, SAN DIEGO, 80- *Concurrent Pos:* NSF res grant, 67-69. *Mem:* Am Math Soc; Soc Indust & Appl Math. *Res:* Ordinary and partial differential equations. *Mailing Add:* Dept Math C-012 Univ Calif San Diego La Jolla CA 92093

SMITH, DONALD REED, b Hamilton, Ont, Sept 3, 36; m 60; c 3. PHYSICAL CHEMISTRY. *Educ:* McMaster Univ, BSc, 58; Univ Leeds, PhD(radiation chem), 61. *Prof Exp:* Demonstr phys chem, Univ Leeds, 58-61; from asst res off to assoc res off, Chalk River Nuclar Lab, Atomic Energy Can Ltd, 61-70, sr res off, 70-89, head, Phys Chem Br, 69-89; DIR STRAT DEV, SPACE STA PROG, 89- *Concurrent Pos:* Emmanuel Col vis fel, Univ Cambridge, Eng, 75-76; dir nuclear reactor & prof chem, McMaster Univ, 82-; pres, Nuclear Activation Serv Ltd, 84- *Mem:* Fel Chem Inst Can. *Res:* Electron spin resonance and laser magnetic resonance spectroscopy; radiation chemistry; isotope separation. *Mailing Add:* Strat Dev Space Sta Prog Montreal Rd Bldg R-88 Hamilton ON K1A 0R6 Can

SMITH, DONALD ROSS, b Indianapolis, Ind, Jan 24, 40; m 63; c 2. POLYMER CHEMISTRY, ORGANIC CHEMISTRY. *Educ:* Tufts Univ, BS, 62; Northeastern Univ, PhD(polymer chem), 72. *Prof Exp:* Sr chemist, 71-78, RES ASSOC, DENNISON MFG CO, 79- *Mem:* Soc Glass Decorators; Royal Soc Arts; Am Chem Soc. *Res:* Thermosetting label systems comprising release, protective lacquers, inks and adhesive lacquers; water repellant coatings for paper. *Mailing Add:* 22 Winfield Rd Hingham MA 02043

SMITH, DONALD STANLEY, b New Westminster, BC, Dec 23, 26; US citizen; m 48; c 1. MECHANICAL ENGINEERING. *Educ:* Univ Calif, Berkeley, BS, 50, MS, 66, PhD(mech eng), 69. *Prof Exp:* Engr, Procter & Gamble Co, 50-54; mgt consult, McKinsey & Co, 54-58; asst vpres eng, Hallamore Electronics Co, 58-59; independent consult, 59-60; asst vpres eng, Aircraft Div, Hughes Tool Co, 60-64; res engr, Univ Calif, 64-69; from assoc prof to prof, Calif State Univ, Chico, 69-88, chmn, Dept Mech Eng, 72-88; RETIRED. *Mem:* Combustion Inst. *Res:* Engine generated air pollution. *Mailing Add:* Two Canterbury Circle Chico CA 95926

SMITH, DONALD W(ANAMAKER), b Bethlehem, Pa, Aug 30, 23; m 46; c 2. ENVIRONMENTAL ENGINEERING. *Educ:* US Naval Acad, BS, 45. *Prof Exp:* Engr chem, E I Du Pont De Nemours & Co Inc, 47-56, tech supt, 56-59, staff mem, Develop Dept, 61-76, admin asst, Off Environ Affairs, 76-86; RETIRED. *Concurrent Pos:* Consult, 86- *Mem:* Am Inst Chem Engrs. *Res:* Corporate environmental management. *Mailing Add:* Coffee Run Apts C4I 614 Loveville Rd Hockessin DE 19707-0369

SMITH, DONALD WARD, b Flint, Mich, Jan 23, 26; m 45; c 5. MICROBIOLOGY, IMMUNOLOGY. *Educ:* Mich Col Mining & Technol, BS, 48; Univ Mich, MS, 50, PhD(bact), 51. *Prof Exp:* Instr bact, Univ Mich, 52-54; from asst prof to assoc prof, 54-65, PROF MED MICROBIOL, UNIV WIS-MADISON, 65- *Concurrent Pos:* USPHS fel, Univ Mich, 51-52; consult, Tuberc Div Commun Dis, WHO; mem tuberc panel, US-Japan Coop Med Sci Prog, NIH, 65-69; mem, Tuberc-Leprosy Spec Study Sect, Nat Inst Allergy & Infectious Dis, 67-69. *Mailing Add:* Dept Med Microbiol Univ Wis Madison 1300 University Ave Madison WI 53706

SMITH, DONN LEROY, b Denver, Colo, Nov 1, 15; m 37; c 2. CLINICAL PHARMACOLOGY. *Educ:* Univ Denver, AB, 39, MS, 41; Univ Colo, PhD(physiol, pharmacol), 48, MD, 58. *Prof Exp:* Asst prof physiol, Univ Denver, 48-50; from asst prof to assoc prof pharmacol, Med Sch, Univ Colo, 50-60, assoc dean, 60-63; dean, Sch Med & prof physiol, Univ Louisville, 63-69; dir, Med Ctr & dean, Col Med, 69-76, prof, 76-85, EMER PROF PHARMACOL & THERAPEUT & DEAN, UNIV S FLA, 85- *Mem:* AAAS; AMA; Am Soc Pharmacol & Exp Therapeut; Soc Exp Biol & Med; fel Am Col Clin Pharmacol. *Res:* Analgesia; traumatic shock; experimental hypertension. *Mailing Add:* 5212 E 127th Ave Tampa FL 33617

SMITH, DORIAN GLEN WHITNEY, b London, Eng, Oct 11, 34; m 59; c 3. GEOLOGY, MINERALOGY. *Educ:* Univ London, BSc, 59; Univ Alta, MSc, 60; Cambridge Univ, PhD(petrol), 63; Oxford Univ, MA, 64. *Prof Exp:* Demonstr mineral, Oxford Univ, 63-66; from asst prof to assoc prof geol, 66-74, PROF GEOL & CUR MINERALS & METEORITES, UNIV ALTA, 74- *Concurrent Pos:* Nuffield Found Travel Award mineral & petrol, Cambridge Univ, 71; counr, Int Mineral Asn, 82-; consult, UN, 82-84. *Mem:* Fel Geol Soc London; Mineral Soc London; Mineral Asn Can (pres, 78-79); Mineral Soc Am; Geochem Soc; Geol Asn Can. *Res:* High temperature thermal metamorphism; electron microprobe applications in mineralogy and petrology; study of bonding in minerals by soft x-ray spectroscopy; energy dispersive electron microprobe analysis; clay mineralogy; meteorites; mineral data base management. *Mailing Add:* Dept Geol Univ Alta Edmonton AB T6G 2M7 Can

SMITH, DOROTHY GORDON, b Barbados, BWI, Mar 5, 18; US citizen. MICROBIOLOGY. *Educ:* Queen's Univ, BA, 40; Rutgers Univ, PhD(microbiol), 47; Am Bd Microbiol, dipl. *Prof Exp:* Technician epidemiol, Meningitis Comn, Johns Hopkins Univ, 41-43; asst chemother, Merck Inst, 43-46; bacteriologist, Biol Labs, US Army, 47-62, microbiologist & biol sci adminr, Ft Detrick, 62-70; Infection Surveillance Officer, Frederick Mem Hosp, 71-76; RETIRED. *Concurrent Pos:* Mem, Antibiotic Utilization Comt, Frederick Mem Hosp. *Mem:* Fel AAAS; Am Soc Microbiol; Am Acad Microbiol; Soc Exp Biol & Med; Am Asn Contamination Control. *Res:* Arbovirus relationships; virus vaccines; immunology; environmental balance and imbalance; ecological control; infection control in the community hospital; viral diseases. *Mailing Add:* 506 Fairview Ave Frederick MD 21701

SMITH, DOUGLAS, b St Joseph, Mo, 1940; m 66. GEOLOGY. *Educ:* Calif Inst Technol, BS, 62, PhD(geol), 69; Harvard Univ, Am, 63. *Prof Exp:* Fel, Geophys Lab, Carnegie Inst, Washington, 68-71; from asst prof to assoc prof, 71-82, PROF GEOL, UNIV TEX, AUSTIN, 82- *Mem:* Mineral Soc Am; Geol Soc Am; Am Geophys Union. *Res:* Igneous and metamorphic petrology; experimental studies of phase equilibria; physical conditions and chemistry of rock-forming processes. *Mailing Add:* Dept Geol Univ Tex Austin TX 78712

SMITH, DOUGLAS ALAN, b Hempstead, NY, Jan 16, 59; m; c 2. MOLECULAR MODELING. *Educ:* Univ Scranton, BS, 79; Carnegie-Mellon Univ, MS, 84, PhD(chem), 85. *Prof Exp:* Res asst prof org chem, Dept Chem, Univ SC, 85-87; ASST PROF ORG CHEM, DEPT CHEM, UNIV TOLEDO, 87-, MEM CTR DRUG DESIGN & DEVELOP, 88- *Concurrent Pos:* Consult, Afton Chem, Inc, 84-88, Chomerics, Inc, subsid W R Grace, Inc, 90-91, MacMillan, Sobanski & Todd, 90; instr org chem, dept chem, Univ SC, 86; postdoctoral fel Nat Res Serv Award, NIH, 87; vis scientist, Mat Lab, Polymer Br, Wright-Patterson AFB, 89-91; mem steering comt, High Temperature Polymer & Molecular Modeling Focus Group, Ohio Aerospace Inst, 90- *Mem:* Am Chem Soc; Sigma Xi. *Res:* Synthetic, physical and computational organic chemistry; synthetic and computational studies of pericyclic reactions; remote asymmetric induction; nitrenium ions and nonlinear optical properties of organic molecules. *Mailing Add:* Dept Chem Univ Toledo Toledo OH 43606-3390

SMITH, DOUGLAS CALVIN, b Kokomo, Ind, Aug 5, 49. PSYCHOBIOLOGY, NEUROPHYSIOLOGY. *Educ:* Tex A&M Univ, BS, 71, MS, 73; Kans State Univ, PhD(psychol), 77. *Prof Exp:* Fel physiol & psychol, Univ Ill, 77-79; ASST PROF PSYCHOBIOL, SOUTHERN ILL UNIV, 79- *Concurrent Pos:* Lectr, Med Sch, Univ Ill, 77-79; prin investr, Visual Suppression NSF grant, 80-82. *Mem:* Soc Neurosci; Asn Res Vision & Opthal. *Res:* Electrophysiological and behavioral investigation of the effects of abnormalities in the development of the visual system of mammals with binocular vision, as well as the neural basis of memory. *Mailing Add:* Dept Psychol Southern Ill Univ Carbondale IL 62901

SMITH, DOUGLAS D, MATHEMATICAL SCIENCES. *Educ:* Pa State Univ, PhD(math), 71. *Prof Exp:* CHMN, MATH SCI, UNIV NC, 83- *Mem:* Am Math Soc; Math Asn Am. *Mailing Add:* Dept Math Univ NC 601 S Col Rd Wilmington NC 28406

SMITH, DOUGLAS LEE, b St Louis, Mo, Sept 22, 43; m 65; c 2. GEOLOGY. *Educ:* Univ Ill, BS, 65; Univ Minn, PhD(geophys), 72. *Prof Exp:* From asst prof to assoc prof, 72-80, PROF GEOL, UNIV FLA, 80- *Mem:* Am Geophys Union; Soc Explor Geophysicists; Geol Soc Am. *Res:* Geothermal conditions and their implications for energy resources; tectonic conditions; nature of earth's crust. *Mailing Add:* Dept Geol Univ Fla Gainesville FL 32611

SMITH, DOUGLAS LEE, b Staten Island, NY, Nov 16, 37; m 59; c 3. X-RAY CRYSTALLOGRAPHY. *Educ:* Dartmouth Col, AB, 58; Univ Wis, PhD(phys chem), 62. *Prof Exp:* Res chemist, Sandia Corp, NMex, 62-65; sr res chemist, 65-70, RES ASSOC, EASTMAN KODAK CO, 70- *Concurrent Pos:* Vis scientist, MIT, 86. *Honors & Awards:* Journal Award-Sci, Soc Photog Scientists & Engr, 75. *Mem:* AAAS; Am Crystallog Asn; Am Chem Soc. *Res:* X-ray crystal structure studies of biological macromolecules. *Mailing Add:* Eastman Kodak Res Labs Eastman Park Bldg 82 Rochester NY 14650-2158

SMITH, DOUGLAS LEE, b San Diego, Calif, June 22, 30; m 51; c 6. OCCUPATIONAL HEALTH, TOXICOLOGY. *Educ:* Univ Utah, BS, 51, PhD(pharmacog & pharmacol), 56. *Prof Exp:* Supvr pharmacol, Aerospace Med Lab, Wright Patterson AFB, 58-59; from instr to assoc prof physiol, US Air Force Acad, 59-66; criteria mgr health standards develop, 71-74, asst br chief, 74-76, toxicologist, Western Area Lab, 76, sr review pharmacologist, 76-80, br chief, Priorities & Res Analysis, 80, SR SCI ADVR TO DIR, NAT INST OCCUP SAFETY & HEALTH, 81- *Concurrent Pos:* Res assoc neurophysiol, Dept Sci Res, US Air Force Acad, 63; Nat Inst Occup Safety & Health rep occup health to Environ Protection Agency for Dept Health, Educ & Welfare, 72-74; liaison rep, Dept Labor, 74- *Honors & Awards:* Achievement Award, Nat Inst Occup Safety & Health, USPHS, 76. *Mem:* Am Conf Govt Indust Hygienists; Am Indust Hyg Asn; Comn Officers Asn

Pub Health Serv; Asn Mil Surgeons, US. *Res:* Toxicological research; industrial hygiene evaluation; criteria development for occupational health standards; evaluations for scientific merit. *Mailing Add:* NIOSH-Off Dir 5600 Fishers Lane Rm 8A-53 Rockville MD 20857

SMITH, DOUGLAS ROANE, b St Louis, Mo, Nov 8, 30; m 53; c 3. BOTANY. *Educ:* Ill State Univ, BS, 53; Univ Ill, Urbana, MS, 57; Wash State Univ, PhD(bot), 69. *Prof Exp:* Instr biol, Lincoln Col, 57-60; field rep, Hosp Labs, Aloe Sci, 60-61; instr biol, Millikin Univ, 61-63; asst prof, Col Guam, 63-65; assoc bot, Miami Univ, 65-67; res asst, Wash State Univ, 67-68; assoc prof, 68-72, PROF BOT, UNIV GUAM, 72- *Concurrent Pos:* Consult, Environ & Energy Res, 78-88; dir, Guam Energy Off, 83-85. *Mem:* AAAS; Am Bryol & Lichenological Soc; Bot Soc Am; Am Inst Biol Sci; Int Asn Plant Taxon; Sigma Xi. *Res:* Phytogeography of mosses of Hawaiian Islands and Micronesia; solar energy conversion of sea water to drinking water; bryology of the western Pacific; plant taxonomy. *Mailing Add:* PO Box 1784 Agana GU 96910

SMITH, DOUGLAS STEWART, b Fargo, NDak, Nov 26, 24; m 49; c 5. CLINICAL STUDIES. *Educ:* NDak State Univ, BS, 49; Mass Inst Technol, PhD, 52. *Prof Exp:* Res chemist, G D Searle & Co, 52-55; dir, res & develop, J B Williams Co, 55-58; tech dir, Vick Mfg Div, 58-61, dir, explor res, Vick Div Res& Develop, 61-64, assoc dir, Cent Sci Servs Dept, 64-69, spec proj mgr, 69-71, asst dir, develop, 71-72, dir, drug (colds) prod, 72-82, AREA DIR, DEVELOP & COM PROD, RICHARDSON VICKS, INC, 82- *Mem:* AAAS; Am Chem Soc. *Res:* Pharmaceutical development and clinical evaluation. *Mailing Add:* 42 Tory Hill Lane Norwalk CT 06853

SMITH, DOUGLAS WEMP, b Los Angeles, Calif, July 13, 38; m 75; c 3. MOLECULAR BIOLOGY, GENETICS. *Educ:* Stanford Univ, BS, 60, PhD(biophys), 67; Univ Ill, Urbana, MS, 62. *Prof Exp:* NIH fel, Max Planck Inst Virus Res, Tübingen, Ger, 67-69; from asst prof to assoc prof, 69-82, PROF BIOL, UNIV CALIF, SAN DIEGO, 83- *Concurrent Pos:* Acad res grant, Univ Calif, 70-71, Cancer Res Coord Comt grants, 70-72, 74-75, 78-79 & 81-82; Am Cancer Soc grants, 70-75; NIH grants, 76-81 & 82-86 & 86-89. *Mem:* AAAS; Biophys Soc; Am Soc Biochem Molecular Biol; NY Acad Sci; Am Soc Microbiol; Protein Soc. *Res:* Biochemistry; microbiology; recombinant DNA research; DNA replication and repair in prokaryotes; structure and function of bacterial origins; atomic physics; optical pumping and hyperfine structure; use of computers in molecular biology. *Mailing Add:* Dept Biol B-022 Univ Calif San Diego La Jolla CA 92093

SMITH, DUDLEY TEMPLETON, b Washington, DC, June 8, 40; m 65; c 2. WEED SCIENCE, RESEARCH MANAGEMENT & ADMINISTRATION. *Educ:* Univ Md, College Park, BS, 63, MS, 65; Mich State Univ, PhD(crop sci), 68; Univ Houston, MBA, 82. *Prof Exp:* Asst prof crop sci, 68-72, assoc prof, 72-73, asst dir, 73-79, ASSOC DIR, TEX AGR EXP STA, TEX A&M UNIV, 79- *Concurrent Pos:* Chmn, Bd Dirs Title XII Sorghum/Millet Consortium & Peanut CRSP. *Mem:* Am Soc Agron. *Res:* Herbicide behavior, residues and movement in soil and water; weed control in crops; growth, phenology and competition of perennial and annual weeds; international agricultural research; planning and management of in state and federal activities. *Mailing Add:* Tex Agr Exp Sta Tex A&M Univ College Station TX 77843

SMITH, DURWARD A, b Raymond, Wash, Jan 4, 47. FOOD SCIENCE, ENGINEERING. *Educ:* Univ Wash, BA, 70; Univ Idaho, BS, 72; La State Univ, MS, 73, PhD(food sci), 76; Jones Law Inst, JD, 84. *Prof Exp:* From asst prof to assoc prof food sci, Dept Hort, Auburn Univ, 83-90; ASSOC PROF, UNIV NEBR, 90- *Concurrent Pos:* Attorney, 85-; consult regulatory compliance food indust. *Mem:* Inst Food Technologists; Am Hort Soc. *Res:* Food engineering and science; high temperature, short time processing and its effect on product quality; nutritive value, biological and cellular integrity; improved efficiency peeling, shelling and skinning systems. *Mailing Add:* Univ Nebr 253 Food Industries Complex Lincoln NE 68583-0919

SMITH, DWIGHT GLENN, b Binghampton, NY, Apr 15, 43; m 68; c 2. POPULATION ECOLOGY. *Educ:* Elizabethtown Col, BS, 66; Brigham Young Univ, MS, 68, PhD(zool), 71. *Prof Exp:* from asst prof to assoc prof, 70-81, PROF BIOL, SOUTHERN CONN STATE UNIV, 82- *Concurrent Pos:* Ecol consult, Environ Pop Educ Asn, 74-; mem sci staff referee Condor, Cooper Ornith Soc, 75- *Mem:* Am Ornithologists Union; Cooper Ornith Soc. *Res:* Investigations of vertebrate predator and prey relationships with emphasis on mathematical models of habitat partitioning; habitat evaluation. *Mailing Add:* Dept Biol Southern Conn State Univ 501 Crescent St New Haven CT 06515

SMITH, DWIGHT MORRELL, b Hudson, NY, Oct 10, 31; m 55; c 3. PHYSICAL CHEMISTRY, ANALYTICAL CHEMISTRY. *Educ:* Cent Col, Iowa, BA, 53; Pa State Univ, PhD(chem), 57. *Hon Degrees:* ScD, Central Col, 86; DLitt, Univ Denver, 90. *Prof Exp:* Instr chem, Calif Inst Technol, 57-59; sr chemist, Texaco, Inc, 59-61; asst prof chem, Wesleyan Univ, 61-66; from assoc prof to prof, Hope Col, 66-72; chancellor, 84-89, PROF CHEM & CHMN DEPT, UNIV DENVER, 72- *Concurrent Pos:* NSF fac fel, Scripps Inst Oceanog, 71-72. *Mem:* Catalysis Soc; Am Chem Soc; AAAS. *Res:* Catalysis; infrared spectroscopy; kinetics; electrochemistry; surface chemistry. *Mailing Add:* Dept Chem Univ Denver Denver CO 80208

SMITH, DWIGHT RAYMOND, b Sanders, Idaho, July 28, 21; m 44; c 2. FISH & WILDLIFE SCIENCES. *Educ:* Univ Idaho, BS, 49, MS, 51; Utah State Univ, PhD(ecol), 71. *Prof Exp:* Res biologist, Idaho Fish & Game Dept, 50-52, area big game mgr, 53-56; range scientist, US Forest Serv, 56-61, wildlife res biologist, 62-65; from asst prof to prof wildlife biol, 65-83, EMER PROF, COLO STATE UNIV, 84- *Mem:* Sigma Xi. *Res:* Large terrestrial ungulates and relationships to habitat; inventory procedures and ecological concepts related to wildlife planning; environmental law as a tool of natural resources management. *Mailing Add:* Colo State Univ 2211 W Mulberry #213 Ft Collins CO 80521

SMITH, E(ASTMAN), b Springfield, Mass, Apr 2, 97; m 33. OPHTHALMOLOGY INSTRUMENTATION, VIOLIN TONE. *Educ:* Mass Inst Technol, BS, 22, MS, 31, ScD, 34. *Prof Exp:* Machinist, Wright Aero Corp, 22-23; in chg prod schedule control, Gilbert Clock Co, 23; in chg res, Mack Trucks, 23-25; tech writer, 25-26; engr, Advert Dept, Johns-Manville Corp, 26-27; instr & lectr sci, NY Univ, 27-29; from asst prof to assoc prof physics, Newark Col Eng, 33-39; consult engr, Shaw-Porter Automatic Transmission, 39-41; optical & mech engr, Pioneer Div, Bendix Aviation Corp, 41-42; consult engr, Perfex Corp, 42-43; dir res & develop, Milwaukee Gas Specialty Co, 43-44; res engr, Woods Hole Oceanog Inst, 44-45; res assoc prof mech eng, Univ Mo, 45-63; DIR, OPTONE INSTRUMENTS, 63- *Mem:* Fel AAAS; Am Soc Mech Engrs; emer mem Optical Soc Am. *Res:* Mechanical engineering; vibration measurement; sound control in musical instrument structures; eyesight instruments. *Mailing Add:* Optone Instruments Cranfield Circle RR 4 Box 460 Mountain Home AR 72653

SMITH, EARL W, b Chicago, Ill, Oct 14, 40. THEORETICAL PHYSICS. *Educ:* Univ Fla, BS, 62, PhD(physics), 66. *Prof Exp:* Proj leader, Nat Bur Standards, Boulder, Colo, 67-83; CONSULT, BALL AEROSPACE, 83- *Mem:* Am Phys Soc; Soc Photo-Optical Instrumentation Engrs. *Mailing Add:* Ball Aerospace PO Box 1062 Boulder CO 80306

SMITH, EDDIE CAROL, b Lexington, KY, Apr 13, 37; m 61; c 2. BIOCHEMISTRY. *Educ:* Univ Ky, BS, 59; Iowa State Univ, PhD(biochem), 63. *Prof Exp:* Asst biochem, Iowa State Univ, 59-63, NIH res assoc, 63-64; scholar, Univ Calif, Los Angeles, 64-65; from asst prof to assoc prof, 65-74, prof biochem, 74-80, DAVID ROSS BOYD PROF CHEM, UNIV OKLA, 80-, ASSOC DEAN, GRAD COL, 81- *Concurrent Pos:* Am Cancer Soc grant, 66-67; NSF res grant, 69-74. *Mem:* Am Chem Soc; Am Soc Biol Chemists. *Res:* Enzymic studies of alcoholic animals; regulation of metabolism; metabolic role of plant peroxidases. *Mailing Add:* Grad Col Univ Okla Buchanan Hall 314 Norman OK 73019-5708

SMITH, EDGAR CLARENCE, JR, b Los Angeles, Calif, July 20, 26; m 48; c 3. COMPUTER SCIENCE. *Educ:* Stanford Univ, BS, 49, MS, 50; Brown Univ, PhD(math), 55. *Prof Exp:* Instr math, Univ Ore, 53-54; asst prof, Univ Utah, 54-55; appl sci rep, IBM Corp, 55-58, univ rep, 58-60, mgr univ prog, 60-61, systs anal mgr, 61-65, large sci acct support, 66-68, mgr sci mkt, 68-71, mgr prog develop, 71-74, systs & prog consult, 74-78, product mgr, 78-86, exec briefing mgr, 87-89, CONSULT, IBM CORP, 89- *Concurrent Pos:* Teacher, Exten Div, Univ Calif, 57-58. *Mem:* Am Math Soc; Asn Comput Mach. *Res:* Boolean algebra; numerical analysis; applications of digital computers. *Mailing Add:* 269 Shelter Rock Rd Stamford CT 06903

SMITH, EDGAR DUMONT, organic chemistry; deceased, see previous edition for last biography

SMITH, EDGAR EUGENE, b Hollandale, Miss, Aug 6, 34; m 55; c 4. BIOCHEMISTRY. *Educ:* Tougaloo Col, BS, 55; Purdue Univ, MS, 57, PhD(biochem), 60. *Hon Degrees:* DSc, Morehouse, Sch Med, 89. *Prof Exp:* Res fel surg-biochem, Harvard Med Sch, 59-61, res assoc, 61-68; asst prof biochem & surg, Sch Med, Boston Univ, 68-71, assoc prof biochem & surg & asst dean student affairs, 71-74; assoc prof biochem & provost acad affairs, Univ Mass Med Sch, Worcester, 74-83-; VPRES ACAD AFFAIRS UNIV MASS SYST, 83- *Concurrent Pos:* NIH res grant, 66-69; assoc surg res, Beth Israel Hosp, Boston, 59-68; Robert Wood Johnson Health Policy Fel, 77-78; career develop award, Nat Cancer Inst, NIH. *Mem:* AAAS; Am Soc Biol Chemists; Am Chem Soc; NY Acad Sci; fel Am Inst Chemists. *Res:* Usefulness of certain enzymes in the development of new techniques for cancer diagnosis and prognosis; pyrimidine biosynthesis in normal and neoplastic human tissue; biochemistry of cell division; sickle cell anemia. *Mailing Add:* Off Pres Univ Mass Syst 250 Stuart St Boston MA 02116

SMITH, EDGAR FITZHUGH, b Rattan, Tex, Dec 13, 19; m 45; c 2. ANIMAL HUSBANDRY. *Educ:* Agr & Mech Col, Tex, BS, 41, PhD(range mgt), 56; Kans State Univ, MS, 47. *Prof Exp:* Asst prof animal husb, Ark State Col, 47-48; from asst prof to prof animal sci & indust, Kans State Univ, 48-85, animal scientist, Agr Exten Serv, 70-85; RETIRED. *Mem:* Am Soc Animal Sci; Soc Range Mgt. *Res:* Beef cattle production and grazing. *Mailing Add:* 20 Vista Lane Manhattan KS 66502

SMITH, EDITH LUCILE, b Jackson, Miss, Sept 9, 13. BIOCHEMISTRY, MICROBIOLOGY. *Educ:* Tulane Univ, BS, 35, MS, 37; Univ Rochester, PhD(biochem), 50. *Hon Degrees:* DSc, Tulane Univ, 84. *Prof Exp:* Lab asst chem, Newcomb Col, Tulane Univ, 35-36, from lab asst to assoc instr biochem, Tulane Univ, 36-47; asst prof biophys, Univ Pa, 55-58; from assoc prof to prof, 58-78, EMER PROF BIOCHEM, DARTMOUTH MED SCH, 78- *Concurrent Pos:* Fel biophys, Univ Pa, 50-54; Brit & Am Cancer Socs exchange fel, Cambridge Univ, 54-55. *Mem:* AAAS; Am Soc Biol Chemists; Am Soc Microbiol; Am Chem Soc. *Res:* Oxidative enzymes, particularly cytochrome pigments; respiratory chain systems of mammalian tissues and microorganisms; oxidative enzyme systems of photosynthetic bacteria. *Mailing Add:* Dept Biochem Dartmouth Med Sch Hanover NH 03755

SMITH, EDWARD, b Liberty, NY, Aug 26, 34; m 65; c 3. ANALYTICAL CHEMISTRY. *Educ:* Long Island Univ, BSPharm, 55; Univ Mich, MS, 58, PhD(pharmaceut chem), 62. *Prof Exp:* Anal chemist, US Food & Drug Admin, 62-65, res chemist, Div Pharmaceut Chem, Bur Sci, 65-70, sr res chemist, Div Drug Chem, Bur Drugs, 70-85; SR RES CHEMIST, BIOPHARMACEUT RES BR, FDA, 85- *Mem:* Am Pharmaceut Asn; Am Chem Soc; Acad Pharmaceut Sci; fel Asn Off Analytical Chem; Am Asn Pharmaceut Scientists. *Res:* Analysis of pharmaceuticals and their active constitutents and possible degradation products using chromatographic techniques; electrometric methods and nuclear chemical techniques; structure proof using spectrophotometric techniques; analysis of drugs in biological fluid samples; sepn stereiosmers (chiral sepns). *Mailing Add:* 14203 Castaway Dr Rockville MD 20853

SMITH, EDWARD HOLMAN, b Abbeville, SC, Sept 2, 15; m 47; c 4. ECONOMIC ENTOMOLOGY. *Educ:* Clemson Univ, BS, 38; Cornell Univ, MS, 40, PhD, 47. *Prof Exp:* Asst prof entom, Exp Sta, State Univ NY Col Agr, Cornell Univ, 47-50, from assoc prof to prof, 55-64; head dept, NC State Univ, 64-67; prof entom & dir coop exten, Cornell Univ, 67-72, chmn dept entom, 72-81. *Mem:* Entom Soc Am; AAAS; Sigma Xi. *Res:* Fruit insects; insect biology and control; mode of ovicidal action. *Mailing Add:* Col of Agr & Life Sci Cornell Univ 162 Comstock Hall Ithaca NY 14850

SMITH, EDWARD J(OSEPH), b New York, NY, Dec 12, 20; m 54; c 2. ELECTRICAL ENGINEERING. *Educ:* Cooper Union, BEE, 45: Polytech Inst Brooklyn, MEE, 48, PhD(elec eng), 51. *Prof Exp:* Res engr, Remington Rand Co, Conn, 45-47; instr elec eng, NY Univ, 47-48; res assoc, 50-53, res assoc prof, 53-57, assoc prof, 57-59, dir, Comput Ctr, 59-63, head, dept elec eng, 67-71 & dept elec eng & comput sci, 78-81, prof, 59-86, EMER PROF ELEC ENG, POLYTECH INST BROOKLYN, 86- *Concurrent Pos:* Vis prof, Eindhoven Technol Univ, 63-64. *Mem:* Asn Comput Mach; Inst Elec & Electronics Engrs; Am Soc Eng Educ; NY Acad Sci; AAAS. *Res:* Computers; logic design; computer architecture; switching and automata theory; nonlinear magnetics. *Mailing Add:* Dept Elec Eng & Comput Sci Polytech Inst NY Brooklyn NY 11201

SMITH, EDWARD JOHN, b Dravosburg, Pa, Sept 21, 27; m 53; c 4. PHYSICS, SPACE MAGNETISM. *Educ:* Univ Calif, Los Angeles, BA, 51, MS, 52, PhD(physics), 60. *Prof Exp:* Res geophysicist, Inst Geophys, Univ Calif, Los Angeles, 55-59; mem tech staff, Space Tech Labs, 59-61; MEM TECH STAFF, JET PROPULSION LAB, 61- *Honors & Awards:* Medal Exceptional Sci Achievement, NASA. *Mem:* AAAS; Sigma Xi; Int Sci Radio Union; Am Geophys Union; Am Astron Soc. *Res:* Planetary magnetism; space physics; interplanetary physics; wave-particle interactions in plasmas; propagation of electromagnetic waves; solar-terrestrial relations. *Mailing Add:* 2536 Boulder Rd Altadena CA 91001

SMITH, EDWARD LEE, b Apache, Okla, June 6, 32; m 58; c 2. PLANT GENETICS, FIELD CROPS. *Educ:* Okla State Univ, BS, 54, MS, 59; Univ Minn, PhD(plant genetics), 62. *Prof Exp:* Instr agron, Okla State Univ, 57-58; asst prof, Univ Tenn, 62-63; asst prof, Okla State Univ-Ethiopian Contract, 63-65; asst prof, Univ Ill, 65-66; assoc prof, 66-71, PROF AGRON, OKLA STATE UNIV, 71- *Mem:* AAAS; Crop Sci Soc Am; Am Soc Agron; Genetics Soc Am. *Res:* Wheat breeding and genetics; milling and baking quality in wheat; disease and insect resistance in small grains; heterosis, cytoplasmic male sterility and fertility restoration in wheat. *Mailing Add:* Dept of Agron Okla State Univ Stillwater OK 74078

SMITH, EDWARD M(ANSON), b Sharpsburg, Ga, Feb 16, 25; m 46; c 4. AGRICULTURAL ENGINEERING. *Educ:* Univ Ga, BS, 49; Kans State Univ, MS, 50. *Prof Exp:* Asst prof, Southwest Tex State Col, 50-52; assoc prof agr eng & assoc agr engr, USDA & Okla State Univ, 52-57; ASSOC PROF AGR ENG & ASSOC AGR ENGR, EXP STA, UNIV KY, 57- *Mem:* Am Soc Agr Engrs. *Res:* Farm machinery. *Mailing Add:* Dept of Agr Eng Univ of Ky Lexington KY 40506

SMITH, EDWARD RUSSELL, b Knoxville, Tenn, May 18, 44; m 69; c 5. NUTRITION. *Educ:* Univ Louisville, AB, 66, PhD(biochem), 71. *Prof Exp:* Instr biochem, obstet & gynec, Col Med, Univ Nebr, Omaha, 71, res asst prof, 71-75; asst prof, 75-80, ASSOC PROF OBSTET & GYNEC, UNIV TEX MED BR GALVESTON, 80- *Mem:* AAAS; Am Chem Soc; Sigma Xi; Geront Soc; Soc Study Reproduction. *Res:* Cause of reproductive senescence; regulation of protein synthesis and degradation; hormone action upon target tissues. *Mailing Add:* Obstet/Gynec C02 Univ Tex Galveston TX 77550

SMITH, EDWARD S, b New York, NY, Dec 4, 24. METALLURGICAL ENGINEERING. *Educ:* Va Tech Inst, BS, 48; Univ Pittsburgh, MS, 67. *Prof Exp:* SR MAT ENGR, PRATT & WHITNEY, 80- *Mem:* Fel Am Soc Metals; Am Soc Mfg Engrs; Am Inst Mining Metall & Petrol Engrs. *Mailing Add:* 3181 Medinah Circle Lake Worth FL 33467

SMITH, EDWIN BURNELL, b Wellington, Kans, Dec 1, 36; m 58; c 3. PLANT TAXONOMY, BIOSYSTEMATICS. *Educ:* Univ Kans, BS, 61, MA, 63, PhD(bot), 65. *Prof Exp:* Asst prof bot, Rutgers Univ, 65-66; vis cytologist, Brookhaven Nat Lab, 66; from asst prof to assoc prof, 66-76, chmn, Dept Bot & Bact, 78-81, PROF BOT, UNIV ARK, FAYETTEVILLE, 76- *Concurrent Pos:* Consult, Brookhaven Nat Lab, 66. *Mem:* Bot Soc Am; Am Soc Plant Taxon; Int Asn Plant Taxon. *Res:* Flora of Arkansas; biosystematics of flowering plants, especially Compositae; taxonomy of Coreopsis and Coreocarpus. *Mailing Add:* Dept Bot & Microbiol Univ Ark Fayetteville AR 72701

SMITH, EDWIN E(ARLE), chemical engineering, for more information see previous edition

SMITH, EDWIN LAMAR, JR, b San Marcos, Tex, Aug 13, 36; m 66; c 2. RANGE MANAGEMENT. *Educ:* Colo State Univ, BS, 58, MS, 64, PhD(soil sci), 66. *Prof Exp:* Instr range mgt, Colo State Univ, 61-64, instr forestry, 65-66; range adv, Brazil Contract, 66-69, from asst prof to assoc prof watershed mgt, 69-72, chief of party agr, Brazil Contract, 72-73, ASSOC PROF RANGE MGT, SCH RENEWABLE NATURAL RESOURCES, UNIV ARIZ, 74-, CHMN, DIV RANGE MGT, 80- *Mem:* AAAS; Soc Range Mgt. *Res:* Range ecology; range inventory, monitoring and soil vegetation relationships. *Mailing Add:* Sch Renewable Natural Resources Biol Sci E Rm 325 Univ Ariz Tucson AZ 85721

SMITH, EDWIN LEE, b Shelton, Nebr, Aug 12, 07; m 33; c 2. PHYSIOLOGY. *Educ:* Univ Nebr, BS, 35, MS, 38; Univ Chicago, PhD(physiol), 41. *Prof Exp:* Instr physiol, Col Med, Univ Ill, 41-43; asst prof, Med Col Va, 43-47; PROF PHYSIOL, UNIV TEX DENT BR HOUSTON, 47- *Mem:* AAAS; Am Physiol Soc; Soc Exp Biol & Med; Int Asn Dent Res; Sigma Xi. *Res:* Bioassay; pharmacology and physiology of circulation; renal physiology; maximum capacity of the vascular system; experimental renal hypertension; digitalis assay; barbiturates. *Mailing Add:* 4322 Briarbend Houston TX 77035

SMITH, EDWIN MARK, b Grand Rapids, Mich, Apr 10, 27; m 40; c 2. MEDICINE. *Educ:* Univ Mich, BS, 50, MD, 53. *Prof Exp:* Intern, Univ Hosp, Univ Mich, Ann Arbor, 53-54, resident phys med & rehab, 54-57, res assoc, Univ, 57-59, from asst prof to prof phys med & rehab, Sch Med, 59-78; PVT PRACT, FLINT, MICH, 78- *Mem:* AMA; Asn Electromyog & Electrodiag; Cong Rehab Med; Am Acad Phys Med & Rehab. *Res:* Physical medicine and rehabilitation; mechanics of deformity formation in rheumatoid arthritis; design and development of orthetic devices. *Mailing Add:* G5067 W Bristol Rd Flint MI 48507

SMITH, EILEEN PATRICIA, b Trenton, NJ, Mar 20, 41. PHYSICAL ORGANIC CHEMISTRY. *Educ:* Univ Pa, BSChem, 62, PhD(org chem), 67. *Prof Exp:* Instr chem, Mercer County Community Col, 67-68; asst prof, 68-72, ASSOC PROF CHEM, TRENTON STATE COL, 72- *Mem:* Sigma Xi; NY Acad Sci; Am Chem Soc; Am Inst Chemists. *Res:* Mass spectral studies; organic laboratory experiments; liquid crystal studies. *Mailing Add:* 625 Paxson Ave Trenton NJ 08619

SMITH, ELBERT GEORGE, b Eugene, Ore, July 18, 13. CHEMISTRY. *Educ:* Ore State Col, BA, 36; Iowa State Col, PhD(physiol & nutrit chem), Iowa State Col, 43. *Prof Exp:* From asst to instr chem, Iowa State Col, 36-43; asst prof, Hamline Univ, 43-46 & Univ Denver, 46-47; from asst prof to assoc prof, Univ Hawaii, 47-58; from assoc prof to prof, 58-78, EMER PROF CHEM, MILLS COL, 78- *Concurrent Pos:* Staff mem surv chem notation systs, Nat Res Coun, 61-64, mem comt mod methods handling chem info, 64-70. *Mem:* Fel AAAS; Chem Notation Asn (pres, 72); Am Chem Soc. *Res:* Nutritional biochemistry; chemical structure information retrieval; Wiswesser notation. *Mailing Add:* 6360 Melville Dr Oakland CA 94611

SMITH, ELDON RAYMOND, b Halifax, Can, May 21, 39; m 64; c 2. CARDIOLOGY. *Educ:* Dalhousie Univ, MD, 67; FRCP(C), 72. *Prof Exp:* From lectr to assoc prof med, Dalhousie Univ, 73-80; head, Cardiol Div, Foothills Gen Hosp, 80-85; prof med/physiol & head, Cardiol Div, 80-85, prof med & chmn dept, 85-90, ASSOC DEAN CLIN AFFAIRS, UNIV CALGARY, 90- *Concurrent Pos:* Examr cardiol, Royal Col Physicians Can, 77-80, 85-; sci officer, Med Res Coun Can, 82-83; chmn, Sci Rev Comt, Can Heart Found, 84-88; dir med, Foothills Gen Hosp, 85-90; mem specialties comt, Royal Col Physicians & Surgeons, Can, 88- *Honors & Awards:* Nat Res Award, Can Cardiovasc Soc, 73. *Mem:* Am Fedn Clin Res; Can Soc Clin Invest; Can Cardiovasc Soc (pres, 90-); Royal Col Physicians & Surgeons Can; fel Am Heart Asn; NY Acad Sci; Am Col Cardiol. *Res:* Cardiovascular physiology with specific interests in ventricular septal mechanics; pericardial function and control of venous capacitance; clinical studies relating to sudden cardiac death. *Mailing Add:* Dept Med Foothills Hosp 1403 29th St Northwest Calgary AB T2N 2T9 Can

SMITH, ELIZABETH KNAPP, b Coraopolis, Pa, Dec 15, 17; m 51. CLINICAL BIOCHEMISTRY, PEDIATRIC ENDOCRINOLOGY. *Educ:* Fla State Col Women, BS, 38; Univ Mich, MS, 39; Univ Iowa, PhD(biochem), 43. *Prof Exp:* Asst pediat, Univ Iowa, 39-43, res assoc, 44-47, res asst prof, 47-50; asst, Rackham Arthritis Res Unit, Univ Mich, 43-44; from asst prof to assoc prof obstet & gynec, Sch Med, Univ Wash, 50-58, res assoc prof pediat, 58-83, res assoc prof, Lab Med, 71-83, emer prof, 83-; RETIRED. *Concurrent Pos:* Clin chemist, Children's Hosp Med Ctr, 58-83. *Mem:* AAAS; Am Chem Soc; Endocrine Soc; Am Asn Clin Chem; Lawson Wilkins Pediat Endocrine Soc. *Res:* Endocrine and metabolic disorders in children; metabolism of adrenocortical hormones in infancy and childhood. *Mailing Add:* Dept Lab Children's Hosp Med Ctr 4800 Sand Pt Way NE Seattle WA 98105

SMITH, ELIZABETH MELVA, b Regina, Sask, Nov 4, 43. STEROID CHEMISTRY, SYNTHETIC ORGANIC CHEMISTRY. *Educ:* Univ Sask, Saskatoon, BSP, 65, PhD(pharmaceut chem), 69. *Prof Exp:* Fel chem, La State Univ, New Orleans, 69-70 & Wayne State Univ, Detroit, 70; res assoc, Dept Chem, Univ Ala, 70-72; fel, Schering Corp, 73, sr scientist, 74-77, from prin scientist chem to sr prin scientist chem, 77-89; CONSULT, 89- *Concurrent Pos:* Med Res Coun Can fel, 69 & 70. *Mem:* Am Chem Soc. *Res:* Steroid synthesis; synthesis and chemistry of heterocyclic compounds; amino acid chemistry. *Mailing Add:* Schering Corp 60 Orange St Bloomfield NJ 07003

SMITH, ELMER ROBERT, b Adams, Wis, Nov 14, 23; m 57; c 2. CHEMICAL ENGINEERING, BIOPHYSICS. *Educ:* Univ Wis, BS, 44. *Prof Exp:* Asst biophys, Univ Calif, 51-54 & Sloan-Kettering Inst, 54-56; instr, Med Col, Cornell Univ, 56; sr scientist, Bettis Atomic Power Lab, Westinghouse Elec Corp, 57-61; supvr radiochem, Hazleton Nuclear Sci Corp, 61-64; mgr chem & tech servs, Environ Systs Div, Nus Corp, 64-72, staff consult, 72-73, prin engr, 73-75, consult engr, 75-78, exec engr, 78-82; RETIRED. *Mem:* AAAS; Am Chem Soc; Sigma Xi. *Res:* Industry; air pollution; nuclear reactor safeguards and siting; radiochemistry; analytical methods in chemistry and radiochemistry; meteorology; reactor chemistry; environmental monitoring; low and high level radioactive solid waste disposal; environmental impact assessment; project management; pollution control; risk assessment. *Mailing Add:* 11206 Healy St Silver Spring MD 20902

SMITH, ELSKE VAN PANHUYS, b Monte Carlo, Monaco, Nov 9, 29; nat US; wid; c 2. ASTRONOMY. *Educ:* Radcliffe Col, BA, 50, MA, 51, PhD(astron), 56. *Prof Exp:* Harvard res fel solar physics, Sacramento Peak Observ, 55-62; vis fel, Joint Inst Lab Astrophys, Colo, 62-63; assoc prof astron, Univ Md, Col Park, 63-75, asst provost, Div Math & Phys Sci & Eng, 73-78, actg dir, astron prog, 75, prof, 75-80, , asst vchancellor acad affairs, 78-8; DEAN, COL HUMANITIES & SCI & PROF PHYSICS, VA COMMONWEALTH UNIV, RICHMOND, 80- *Concurrent Pos:* Res assoc, Lowell Observ, 56-57; consult, Goddard Space Flight Ctr, NASA, 63-65; counr, Am Astron Soc, 77-80; chmn, US Nat Comt, Int Astron Union, 78-80. *Mem:* Fel AAAS; Int Astron Union; Am Astron Soc. *Res:* Active regions on the sun, especially flares and plages; solar chromosphere; interstellar polarization; solar physics. *Mailing Add:* Col Humanities & Sci Va Commonwealth Univ Richmond VA 23284-2019

SMITH, EMIL L, b New York, NY, July 5, 11; m 34; c 2. BIOCHEMISTRY, BIOPHYSICS. *Educ:* Columbia Univ, BS, 31, PhD(biophys), 37. *Prof Exp:* Asst zool, Columbia Univ, 31-34, asst biophys, 34-36, instr, 36-38; res assoc, Rockefeller Inst, 40-42; biophysicist, Biol Lab, E R Squibb & Sons, 42-46; assoc res prof biochem & physiol, Col Med, Univ Utah, 46-47, from assoc prof to prof biochem & from assoc res prof to res prof med, 47-63; prof biol chem & chmn dept, 63-79, EMER PROF, SCH MED, UNIV CALIF, LOS ANGELES, 79- *Concurrent Pos:* Guggenheim fel, Cambridge Univ, 38-40; hon fel, Yale Univ, 40; mem, Panel Comt Growth, Nat Res Coun, 49-53; mem, Sect Arthritis & Metab, USPHS, 49-50, biochem, 50-54, Adv Comt Biochem, US Off Naval Res, 57-60, US Nat Comt Biochem, 58-62, chmn, 59-62; Reynolds lectr, Univ Utah, 58, Bloor lectr, Univ Rochester, 59,Hanna lectr, Western Reserve Univ, 66 & Alexander Agassiz lectr, Harvard Univ, 68; mem, Comt Int Orgn & Prog, Nat Acad Sci, 62-72, chmn, 64-68; mem, Bd Trustees, Calif Found Biochem Res, 64-, Adv Coun, Life Ins Med Res Fund, 66-70 & Sci Adv Panel, Ciba Found, 67-79; vis prof, Col France, 68; mem, vis comt, Dept Biol Chem, Harvard Med Sch, 68-71; mem, Comt Scholarly Commun with People's Repub China, 70-76, chmn, 72-75; mem, Bd Int Sci Exchange, Nat Res Coun-Nat Acad Sci, 73-77 & Comn Int Relations, 78-82; prog biomat & biotech, UNESCO, 79-87; exec coun, Protein Soc, 86- *Honors & Awards:* Annual Lectr & Medalist, Ciba Found, 68; Margaret J Hastings Distinguished lectr, Res Inst Scripps Clinic, 80; Abraham White Mem lect, 82; Stein-Moore Award, Protein Soc, 87. *Mem:* Nat Acad Sci; Am Chem Soc; Am Acad Art & Sci; Am Philos Soc; Am Soc Biol Chemists; foreign mem Acad Sci USSR. *Res:* Chemistry of proteins; milk proteins; amino acids; proteolytic enzymes; peptides; enzymology; histones; cytochromes; dehydrogenases; biochemical evolution. *Mailing Add:* Dept Biol Chem Univ Calif Sch Med Los Angeles CA 90024-1737

SMITH, EMIL RICHARD, b Bridgewater, Mass, July 25, 31; m 56; c 5. PHARMACOLOGY. *Educ:* Northeastern Univ, BS, 54; Tufts Univ, MS, 56, PhD(pharmacol), 58. *Prof Exp:* Assoc res pharmacologist, Sterling-Winthrop Res Inst, 60-62; res pharmacologist, Mason Res Inst, Worcester, Mass, 62-67; sect head, gen pharmacol & toxicol, Res Labs, Astra Pharmaceut Prod, Inc, 67-72; sect head, chem carcinogenesis, Mason Res Inst, 72-75; ASSOC PROF PHARMACOL, UNIV MASS MED SCH, WORCESTER, 75- *Concurrent Pos:* USPHS res fel pharmacol, Sch Med, Univ Buffalo, 58-60; lectr, Albany Med Col, 61-62, res pharmacologist, St Vincent Hosp, 63-; asst prof, Sch Med, Tufts Univ, 69-73; lectr pharmacol, Sch Med, Univ Mass, 74-75. *Mem:* Am Soc Pharmacol & Exp Therapeut; Soc Toxicol. *Res:* Cardiovascular and autonomic pharmacology; toxicology. *Mailing Add:* Dept of Pharmacol Univ of Mass Med Sch Worcester MA 01655

SMITH, EMMA BREEDLOVE, b Whitmell, Va, Dec 24, 31; m 55; c 2. MATHEMATICS EDUCATION, COMPUTER ASSISTED INSTRUCTION. *Educ:* Va State Univ, BS, 54, MS, 55; Ind Univ, PhD(math educ), 73. *Prof Exp:* Instr math, SC State Col, 55-58, Florissant Valley Community Col, 74-75; instr, 58-70, PROF MATH, VA STATE UNIV, 75- *Concurrent Pos:* Instr, NSF Inst Math & Sci Teachers, 58-70, dir tests, Va Conf Math & Sci Teachers, 60-70; asst instr, math dept, Univ Pa, 61-62, Ind Univ, 71-72; supvr student teachers, Ind Univ Sch Educ, 72-73; dir math lab, Univ Middle Sch, 72-73. *Mem:* Math Asn Am; Nat Tech Asn; AAAS; Nat Coun Teachers Math. *Res:* Mathematics; mathematics education; computer assisted instruction; factors influencing achievement and attitudes toward mathematics; women and minorities overcoming barriers to success in mathematics and science. *Mailing Add:* Math Dept Va State Univ Box 68 Petersburg VA 23803

SMITH, ERIC HOWARD, b Cincinnati, Ohio, July 4, 43. SYSTEMATIC ENTOMOLOGY. *Educ:* Miami Univ, Ohio, BA, 66; Purdue Univ, MS, 70; Ohio State Univ, PhD(entom), 73. *Prof Exp:* collection mgr, Div Insects, Field Mus Natural Hist, 75-80; tech dir, Orkin Nat Serv Dept, 81-89. *Mem:* Coleopterists Soc; Entom Soc Am; Soc Syst Zool. *Res:* Primarily the systematics, but also all other aspects of the Chrysomelidae (Insecta: Coleoptera) with emphasis on the subfamily Alticinae or flea beetles. *Mailing Add:* 1668 Olde Oak Dr Lithia Springs GA 30057

SMITH, ERIC MORGAN, b Lafayette, Ind, Feb 13, 53; m 79; c 1. NEUROIMMUNOLOGY, PSYCHONEUROIMMUNOLOGY. *Educ:* Syracuse Univ, BS, 75; Baylor Col, PhD(virol), 80. *Prof Exp:* Postdoctoral fel virol & immunol, Univ Tex Med Br, 79-81, from asst prof to assoc prof microbiol, 82-90, assoc prof psychiat, 86-90, PROF PSYCHIAT & MICROBIOL, UNIV TEX MED BR, 90- *Mem:* Am Asn Immunologists; Soc Neurosci; Am Soc Microbiol; Sigma Xi; AAAS; Int Soc Neuroimmunomodulation. *Res:* Characterizing the production and action of neuropeptides in the immune system. *Mailing Add:* Dept Psychiat D29 Univ Tex Med Br Galveston TX 77550

SMITH, ERLA RING, b Colma, Calif, Feb 18, 38; m 60; c 2. NEUROENDOCRINOLOGY, ANATOMY. *Educ:* Univ Wash, BA, 59, PhD(biol struct), 65. *Prof Exp:* Teaching asst biol struct, Univ Wash, 59-64; from res assoc to sr res assoc, 68-82, actg assoc prof, 83-86, SR RES ASSOC NEUROENDOCRINOL, STANFORD UNIV, 87- *Concurrent Pos:* Nat Inst Arthritis & Metab Dis fel neuroendocrinol, Stanford Univ, 65-68. *Mem:* AAAS; Endocrine Soc; Am Physiol Soc; Soc Neurosci. *Res:* Reproductive physiology; endocrinology; sex behavior. *Mailing Add:* Dept Molecular & Cellular Physiol Stanford Univ Sch Med, Beckman Ctr Stanford CA 94305-5426

SMITH, ERNEST KETCHAM, b Peking, China, May 31, 22; US citizen; m 50; c 3. RADIO PHYSICS, TELECOMMUNICATIONS. *Educ:* Swarthmore Col, BA, 44; Cornell Univ, MS, 51, PhD(radio wave propagation), 56. *Prof Exp:* Asst radio engr, Mutual Broadcasting Syst, 46-47, chief plans & allocations div, 47-49; res asst, Cornell Univ, 50-51 & 52-54; proj leader, Nat Bur Stand, 51-52; proj leader, Nat Bur Stand, 54-57, asst chief ionosphere res sect, Boulder Labs, 57, chief sect, 57-60, chief ionosphere res & propagation div, 60-62, chief upper atmosphere & space physics div, 62-65, chief aeronomy div, 65; dir aeronomy lab, Inst Telecommun Sci & Aeronomy, Environ Sci Serv Admin, 65-67, actg dir, Inst Telecommun Sci, 67-68, actg dir off univ rels, Res Labs, 68-70; assoc dir, Inst Telecommun Sci, 70-72, Consult to dir, Inst Telecommun Sci, Off Telecommun, US Dept Com, 72-76; mem tech staff, Calif Inst Technol, Jet Propulsion Lab, 76-87; PROF ADJ, UNIV COLO, BOULDER, 87- *Concurrent Pos:* Int vchmn study group six, Int Telecommun Union, Consultative Comt Int Radio, Dept State, 59-70, chmn US study group six, US Nat Comt, 70-76; vis prof, Colo State Univ, 63, affil prof, 64-69; assoc, Harvard Col Observ, 66-75; adj prof, Univ Colo, Boulder, 69-78; mem-at-large, US Nat Comt, Int Union Radio Sci, mem, Comn C, E, F & G. *Mem:* Fel AAAS; fel Inst Elec & Electronics Engrs; Sigma Xi; Int Union Radio Sci; Am Geophys Union. *Res:* Sporadic-E region of the ionosphere; radio scattering from the ionospheric F-region; radio refractive index of the nonionized atmosphere; very high frequency propagation via the ionosphere; natural noise; earthspace propagation. *Mailing Add:* 5159 Idylwild Tr Boulder CO 80301-3618

SMITH, ERNEST LEE, JR, b Nashville, Tenn, Aug 3, 34; m 60; c 2. OPERATIONS RESEARCH. *Educ:* Vanderbilt Univ, BA, 58. *Prof Exp:* Mathematician, US Army, 58-60 & Defense Atomic Support Agency, 60-63; br chief anal syst, 63-65, div chief appl prog div, 65-70, dep chief oper, 71-74, chief, Syst Planning & Eng Off, Nat Mil Command Syst Support Ctr, 74-76; chief plans div, 76-79, tech adv to dep dir, plans prog & mgt, 79-80, CHIEF, HARDWARE SYSTS DIV, COMMAND & CONTROL TECH CTR, 80- *Concurrent Pos:* Rep, Defense Commun Agency-Advan Airborne Command Post Software Develop Team, 73. *Mem:* Asn Comput Mach. *Res:* CCTC ADP/Communication program planning, budgeting and contracting; directing studies, analyses and engineering necessary to examine alternative approaches and to insure new technology is being planned to future ADP/ Communication capabilities. *Mailing Add:* Huntington Col Campus Box 84 1500 E Fairview Ave Montgomery AL 36194

SMITH, ERVIN PAUL, animal science, for more information see previous edition

SMITH, EUCLID O'NEAL, b Jackson, Miss, May 27, 47; m. ZOOLOGY. *Educ:* Miss State Univ, BA, 69; Univ Ga, MA, 72; Ohio State Univ, PhD(anthrop), 77. *Prof Exp:* Asst prof anthrop, Emory Univ, 76-83; asst res prof, 77-84, ASSOC RES PROF, YERKES REGIONAL PRIMATE RES CTR, 84-; ASSOC PROF ANTHROP, EMORY UNIV, 83- *Concurrent Pos:* Assoc ed, Am J Primatology, 84-, book rev ed, 88-90. *Mem:* Am Anthrop Asn; Am Asn Phys Anthropologists; Am Soc Primatologists; Animal Behav Soc; Int Primatol Soc; AAAS. *Res:* Primate social behavior; behavioral ecology; developmental sociobiology. *Mailing Add:* Dept Anthrop Emory Univ Atlanta GA 30322

SMITH, EUGENE I(RWIN), b Buffalo, NY, Mar 4, 44; m 73. IGNEOUS PETROLOGY, STRUCTURAL GEOLOGY. *Educ:* Wayne State Univ, BS, 65; Univ NMex, MS, 68, PhD(geol), 70. *Prof Exp:* Geologist, US Geol Surv, Ctr Astrogeol, 66-68; res assoc geol, Univ NMex, 70-72; asst prof, Univ Wis-Parkside, 72-76, ASSOC PROF GEOL, 76-80; assoc prof, 80-88, PROF GEOL, UNIV NEV, LAS VEGAS, 88- *Concurrent Pos:* Vis assoc prof geol, Univ Nev, Las Vegas, 78-79. *Mem:* Geol Soc Am; Am Geophys Union; AAAS. *Res:* Geological, petrographic and geochemical study of volcanic rocks formed during regional extension. *Mailing Add:* Dept Geosci Univ Nev Las Vegas NV 89154

SMITH, EUGENE JOSEPH, b New York, NY, Jan 26, 29; m 56; c 3. BIOCHEMISTRY. *Educ:* Queens Col, NY, BS, 51; Univ Conn, MS, 55; Duke Univ, PhD(biochem), 59. *Prof Exp:* Arthritis & Rheumatism Found fel, 59-61; asst prof biochem, Schs Med & Dent, Georgetown Univ, 61-68; res chemist, Food & Drug Admin, 68-69; RES CHEMIST, INST GENETICS & PHYSIOL, SCI & EDUC ADMIN-AGR RES, USDA, 69- *Concurrent Pos:* Consult, Walter Reed Armed Forces Inst Dent Res, 63-67. *Mem:* AAAS; Am Soc Biochemists. *Res:* Microbial metabolism and polysaccharides; nucleotides and amino-sugars; nucleic acid metabolism; avian tumor-virus research. *Mailing Add:* USDA Regional Poultry Res Lab 3606 E Mt Hope Rd East Lansing MI 48823

SMITH, EUGENE WILLIAM, botany, microbiology, for more information see previous edition

SMITH, F(REDERICK) DOW(SWELL), b Winnipeg, Man, Jan 2, 21; nat US; m 49; c 4. OPTICS. *Educ:* Queen's Univ, Ont, BA, 47, MA, 48; Univ Rochester, PhD(optics), 51. *Prof Exp:* Asst optics, Univ Rochester, 48-51; instr physics, Phys Res Lab, Boston Univ, 51-52, from asst prof to assoc prof, 52-55, chmn dept physics, 53-58, dir lab, 55-58; mgr advan tech div, Itek Corp, 58-67, vpres & corp scientist, 67-74; pres, New Eng Col Optom, 79-86; CONSULT, 75- *Concurrent Pos:* Asst, Res Coun, Ont, 47-48 & Bausch & Lomb Optical Co, NY, 49-50; mem vision comt, Nat Res Coun-US Armed Forces, 57-, chmn, 78; mem US comt, Int Comn Optics, 58-61 & 67-, vpres, 75-81. *Honors & Awards:* Goddard Award for Excellence in Aerospace Photog. *Mem:* Fel AAAS; fel Optical Soc Am (pres, 74, treas, 79-); fel Am Acad Optom. *Res:* Physical and geometrical optics; interferometry; aerial photography; physiological and ophthalmic optics. *Mailing Add:* RR Two Box 505 Rumney NH 03266

SMITH, F HARRELL, b Auburn, WVa, June 28, 18; m 46; c 4. ANIMAL SCIENCE, AGRICULTURAL MECHANICS. *Educ:* WVa Univ, BS, 42, MS, 49; Va Polytech Inst, 50; Pa State Univ, EdD(agr educ & mech), 58. *Prof Exp:* High sch teacher, 42; teacher agr, Potomac State Col, WVa Univ, 46-54, head dept, 54-60; PROF AGR & HEAD DEPT, UNIV MD EASTERN SHORE, 60- *Res:* Dairy and animal husbandry; herdsmanship and farm management; agricultural uses for Loblolly pine bark. *Mailing Add:* Dept Agr Univ Md, Eastern Shore Princess Anne MD 21853

SMITH, FELIX TEISSEIRE, b San Francisco, Calif, Aug 19, 20. ATOMIC PHYSICS. *Educ:* Williams Col, Mass, BA, 42; Harvard Univ, LLB, 49, MS, 53, PhD(chem), 56. *Prof Exp:* PHYSICIST, SRI INT, 56-, DIR MOLECULAR PHYSICS LAB, 74- *Concurrent Pos:* Mem comt atomic & molecular physics, Nat Acad Sci-Nat Res Coun, 71; chmn comt atomic & molecular physics, Nat Acad Sci, 73-75; chmn, Int Conf Physics Electronic & Atomic Collisions, 75-77. *Mem:* Fel Am Phys Soc; Am Chem Soc; Brit Inst Physics. *Res:* Quantum and semiclassical collision theory of electrons, atoms, ions and small molecules; differential scattering and collision spectroscopy; three-body processes. *Mailing Add:* 1030 Palo Alto Ave Palo Alto CA 94301-2224

SMITH, FLOYD W, b Limon, Colo, May 31, 20; m 50; c 3. SOIL FERTILITY, SOIL CHEMISTRY. *Educ:* Kans State Univ, BS, 42; Mich State Univ, MS, 46, PhD(soil sci), 49. *Prof Exp:* From asst prof to assoc prof, 46-50, actg head dept agron, 64-65, assoc dir exp sta, 65, PROF SOIL SCI, KANS STATE UNIV, 50-, DIR, KANS AGR EXP STA, 65- *Concurrent Pos:* Guest lectr, Mich State Univ, 63. *Mem:* Am Soc Agron; Soil Sci Soc Am; Crop Sci Soc Am; Soil Conserv Soc Am; fel Am Inst Chem. *Res:* Fertilizer research with corn, grain sorghum and wheat; productivity indexes for principal soil types in Kansas. *Mailing Add:* Kans Water Res Inst 14 Waters Hall Kans State Univ Manhattan KS 66506

SMITH, FRANCIS MARION, b Columbus, Kans, Nov 16, 23; m 45; c 3. NUCLEAR CHEMISTRY. *Educ:* Kans State Univ, BS, 44, MS, 48. *Prof Exp:* Instr chem, Kans State Univ, 45-49; chemist anal res & asphalt rheol, Stand Oil Co, Ind, 49-56; chemist emission spectros, Hanford Labs, Gen Elec Co, 56-65; res scientist, Battelle-Northwest Labs, 65-70; advan scientist, Westinghouse Hanford Co, 70-80, sr scientist, 80-89; RETIRED. *Res:* Emission spectroscopy; radiometallurgy; nuclear safeguards; calorimetry; hazardous materials shipping. *Mailing Add:* 9013 Franklin Rd Pasco WA 99301

SMITH, FRANCIS WHITE, b Capetown, SAfrica, July 20, 31; m 56; c 3. ANALYTICAL CHEMISTRY. *Educ:* Univ Cape Town, 52, Hons, 54, PhD(phys anal chem), 67. *Prof Exp:* Chemist, Metal Box Co, SAfrica, 52-54 & Schweppes Ltd, Eng, 55-56; res chemist, B F Goodrich Res Ctr, Ohio, 56-59; chemist, Geol Surv Dept, Uganda, 59-66; from asst prof to assoc prof, 67-80, PROF CHEM, YOUNGSTOWN STATE UNIV, 80- *Mem:* Am Chem Soc. *Res:* Ion-exchange; high-performance liquid chromatography; spectrographic analysis. *Mailing Add:* Dept of Chem Youngstown State Univ Youngstown OH 44555

SMITH, FRANCIS XAVIER, b Chelsea, Mass, Aug 28, 45; m 71; c 3. ORGANIC CHEMISTRY. *Educ:* Lowell Technol Inst, BS, 67, MS, 69; Tufts Univ, PhD(chem), 72. *Prof Exp:* Postdoctoral assoc org chem, Univ Va, 72-74; ASSOC PROF CHEM, KINGS COL, 74- *Concurrent Pos:* Consult analytical chem. *Mem:* Am Chem Soc; Sigma Xi; Am Inst Chemists. *Res:* Synthesis of organic compounds; chemistry of nitrogen heterocycles; analytical methods. *Mailing Add:* Dept Chem Kings Col Wilkes-Barre PA 18711

SMITH, FRANK A, b New York, NY, Jan 19, 37; m 65. MATHEMATICS. *Educ:* Brooklyn Col, BA, 58; Purdue Univ, MS, 60, PhD(math), 65. *Prof Exp:* Asst prof math, Ohio State Univ, 64-66 & Univ Fla, 66-67; lectr, Univ Leicester, 67-68; fel, Carnegie-Mellon Univ, 68-69; assoc prof, 69-84, PROF MATH, KENT STATE UNIV, 84- *Concurrent Pos:* Vis prof, Pitzer Col, 80-81. *Mem:* NY Acad Sci; Am Math Soc; Math Asn Am. *Res:* Structure of ordered semigroups and semirings and extensions of such orders; embeddings of subspaces of topological spaces with certain extension properties. *Mailing Add:* Dept Math Kent State Univ Kent OH 44242

SMITH, FRANK ACKROYD, b Winnipeg, Man, Feb 14, 19; m 44; c 2. PHYSIOLOGICAL CHEMISTRY. *Educ:* Ohio State Univ, BA, 40, MSc, 41, PhD(physiol chem), 44; Am Bd Clin Chem, dipl. *Prof Exp:* Assoc scientist, Atomic Energy Proj, 44, from instr to asst prof toxicol, Sch Med & Dent, 46-58, from asst prof to assoc prof radiation biol & toxicol, 58-83, assoc prof toxicol in radiation biol & biophys, 83-85, EMER PROF TOXICOL IN RADIATION BIOL & BIOPHYS, SCH MED & DENT, UNIV ROCHESTER, 85- *Concurrent Pos:* Mem fluoride panel, Comt Biol Effects of Air Pollutants, Nat Acad Sci-Nat Res Coun; dent study sect, Div Res Grants, NIH, 69-72; ad hoc reviewer criteria doc, Nat Inst Occup Safety & Health, HEW, EPA, WHO (fluoride), Conn Dept Health Serv. *Honors & Awards:* Adolph G Kammer Merit in Authorship Award, Am Occup Med Asn, 78. *Mem:* Am Chem Soc; Am Indust Hyg Asn; Soc Toxicol; Am Soc Pharmacol & Exp Therapeut; Sigma Xi. *Res:* Absorption, distribution and excretion of toxic materials, especially fluorides; mechanism of action of toxic agents; clinical chemistry as criterion of toxicity in industrial hygiene. *Mailing Add:* Dept Biophys Univ Rochester Sch Med & Dent Rochester NY 14642

SMITH, FRANK E, b Edmonton, Alta, Aug 5, 36; m 63; c 2. ONCOLOGY, CHEMOTHERAPY. *Educ:* Univ Alta, MD, 60. *Prof Exp:* Intern, Edmonton Gen Hosp, Alta, 60-61; resident med, 61-64, instr pharmacol & med, 66-69, asst prof, 69-76, ASSOC, PROF PHARMACOL & MED, BAYLOR COL MED, 76- *Concurrent Pos:* Fel cancer chemother, Baylor Col Med, 64-66; attend physician & consult chemother, Vet Admin Hosp; asst, Ben Taub Hosp; assoc physician, Methodist Hosp. *Mem:* Am Col Clin Pharmacol & Chemother; Am Soc Clin Oncol; Am Col Physicians. *Res:* Internal medicine; clinical oncology and pharmacology; cancer chemotherapy. *Mailing Add:* 1200 Moursand Ave Suite 865E Houston TX 77030

SMITH, FRANK HOUSTON, b Cornelius, NC, May 18, 03. PHYSIOLOGICAL EFFECTS ON ANIMALS. *Educ:* Davidson Col, BS, 26; NC State Univ, MS, 31. *Prof Exp:* Prof, dept animal sci, NC State Univ, 61-73; RETIRED. *Mem:* Emer mem Am Chem Soc; emer mem Am Oil Chemists Soc. *Res:* Toxic principle of cotton seeds. *Mailing Add:* 2506 Stafford Ave Raleigh NC 27607

SMITH, FRANK ROYLANCE, b London, Eng, Mar 20, 32; m 61; c 2. PHYSICAL CHEMISTRY, ELECTROCHEMISTRY. *Educ:* Univ London, BSc, 56, PhD(chem), 64. *Prof Exp:* Lab asst, Distillers Co, 51-55; sr chemist, Mullard Res Labs, 60-64; fel, La State Univ, 64-65; from asst prof to assoc prof chem, 65-74, PROF CHEM, MEM UNIV NFLD, 74- *Concurrent Pos:* Consult, Bell Northern Res Ltd, Ottawa, 77-79 & Instrumar, St John's, Nfld, 80, Canpolar, 87-88; vis, Phys Chem Dept, Univ Cambridge, 74-75. *Mem:* The Chem Soc; Chem Inst Can; Can Asn Physicists. *Res:* Electrochemical kinetics of electrolytic hydrogen evolution; redox reactions; solar photoelectrolysis of water with semiconductor electrodes; lead-acid batteries; metal deposition; diffusion of hydrogen isotopes through metals; catalysis and electrocatalysis. *Mailing Add:* Dept Chem Mem Univ Nfld St John's NF A1B 3X7 Can

SMITH, FRANK W(ILLIAM), b Philadelphia, Pa, Dec 1, 19; m 49; c 2. CHEMICAL ENGINEERING. *Educ:* Villanova Col, BChE, 41; Ill Inst Technol, MS, 42; Mass Inst Technol, ScD, 49. *Prof Exp:* Asst chem eng, Ill Inst Technol, 41-42; res assoc combustion, Mass Inst Technol, 42-49 & US Bur Mines, 49-55; asst tech dir, Abex Corp, 55-60, dir chem res, 60-63; DIR RES & ENG, MINE SAFETY APPLIANCES CO, 63-, VPRES, 67- *Concurrent Pos:* Mem Gov Sci Adv Comn. *Mem:* Am Chem Soc; Soc Automotive Engrs; Am Inst Chem Engrs; Am Inst Chemists; Am Inst Mining, Metall & Petrol Engrs. *Res:* Coal chemistry; combustion; polymers; sintered metals; safety equipment; process instrumentation; research planning and management. *Mailing Add:* 61A Hummingbird Dr Merrimack NH 03054-2759

SMITH, FRED GEORGE, JR, b Calif, Jan 1, 28; m 51; c 3. PEDIATRICS, NEPHROLOGY. *Educ:* Univ Calif, Los Angeles, BS, 51, MD, 55. *Prof Exp:* Intern pediat, Ctr Health Sci, Univ Calif, Los Angeles, 55-56; resident, Univ Minn Hosps, 56-57; chief resident, Ctr Health Sci, Univ Calif, Los Angeles, 57-58, from asst prof to prof, Sch Med, 60-73; prof pediat, Col Med, Univ Iowa, 73-89, chmn dept, 73-86; VPRES, AM BD PEDIAT, 89- *Concurrent Pos:* USPHS fel, St Mary's Hosp Med Sch, London, Eng, 68-69. *Mem:* Soc Pediat Res; Am Soc Pediat Nephrology; Am Pediat Soc. *Res:* Developmental and fetal renal physiology. *Mailing Add:* Am Bd Pediat 111 Silver Cedar Ct Chapel Hill NC 27514-1651

SMITH, FRED R, JR, b Pittsburgh, PA, March 23, 40; m 68; c 4. COMPUTER SCIENCE, TEACHING FACULTY COMPUTER USE. *Educ:* Duquesne Univ, BS, 62; WVa Univ, PhD(physics), 73. *Prof Exp:* Physicist G5-7, Nat Bur Standards, 62-64; res asst, Univ Md, 66-67; teaching asst, WVa Univ, 67-72; INSTR, PHYSICS, COASTAL CAROLINA COMMUNITY COL, 74- *Concurrent Pos:* Lab asst, Waynesburg Col, 69-71. *Mem:* Am Physical Soc; Math Assoc Am. *Mailing Add:* 521 University Dr Jacksonville NC 28546

SMITH, FREDERICK ADAIR, JR, b Trinity, Tex, Dec 8, 21; m 49; c 2. THEORETICAL MECHANICS, APPLIED MECHANICS. *Educ:* US Mil Acad, BS, 44; Johns Hopkins Univ, MS, 49; George Washington Univ, MBA, 63; Univ Ill, PhD(theoret & appl mech), 68. *Hon Degrees:* LHD, Mt St Mary Col, 83. *Prof Exp:* US Army, 44-, asst prof physics, US Mil Acad, 49-52, instr, Army Command & Gen Staff Col, 55-58, gen staff officer, 59-62, infantry comdr, Seventh Army, Europe, 63-65, prof mech, US Mil Acad, 65-74, head dept, 69-74, dean acad bd, US Mil Acad, 74-85, trustee, Asn Grad, US Army, 75-86; RETIRED. *Concurrent Pos:* Trustee, Asn Grad, US Mil Acad. *Mem:* Am Acad Mech. *Res:* Elasticity; materials science. *Mailing Add:* 606 Balfour Dr San Antonio TX 78239

SMITH, FREDERICK ALBERT, b Janesville, Wis, Sept 11, 11; m 35; c 2. PHYSICAL CHEMISTRY, ORGANIC CHEMISTRY. *Educ:* Univ Wis, BS, 34. *Prof Exp:* Control chemist, Nat Aniline Div, Allied Chem & Dye Corp, 34-39; res chemist, Linde Air Prod Co Div, Union Carbide Corp, 39-49, sales develop, 49-55, chemist, Silicones Div, 55-65, sr scientist, 65-76; RETIRED. *Concurrent Pos:* Consult, Lawrence Livermore Lab, Univ Calif, 76-81. *Mem:* Am Chem Soc. *Res:* Chemistry of high polymers; metal-organic compounds; silicones; silicone elastomers and resins. *Mailing Add:* 60 Stonecrest Rd Ridgefield CT 06877

SMITH, FREDERICK EDWARD, b Springfield, Mass, July 23, 20; m 45; c 3. THEORETICAL ECOLOGY, LANDSCAPE ECOLOGY. *Educ:* Univ Mass, BS, 41; Yale Univ, PhD(zool), 50. *Prof Exp:* From instr to prof zool, Univ Mich, 50-66, prof natural resources, 66-69; prof advan environ studies in resources & ecol, Grad Sch Design, Harvard Univ, 69-82, chmn, Landscape Archit, 81-82; RETIRED. *Concurrent Pos:* Mem, Nat Sci Bd, 68-74. *Mem:* Ecol Soc Am (pres, 73-74). *Res:* Form of population growth and population interactions; community studies; ecosystem science; landscape ecology; computer modeling. *Mailing Add:* 122 Gardiner Rd Woods Hole MA 02543

SMITH, FREDERICK GEORGE, b Oak Park, Ill, Aug 16, 17; m 43; c 1. BIOCHEMISTRY. *Educ:* Univ Chicago, BS, 39; Univ Wis, MS, 41, PhD(biochem), 43. *Prof Exp:* Asst biochem & plant path, Univ Wis, 39-43, res assoc, 43-44; asst prof chem, State Univ NY Col Agr, Cornell Univ, 44-47; res assoc biochem, Univ Rochester, 47-48; assoc prof bot, 48-56, head dept bot & plant path, 64-79, PROF BOT & BIOCHEM, IOWA STATE UNIV, 56- *Mem:* AAAS; Am Soc Biol Chemists; Am Soc Plant Physiol; Sigma Xi. *Res:* Fungus physiology; biochemistry of plant disease resistance; respiratory enzymes. *Mailing Add:* 2216 State Ave RR 5 Ames IA 50010

SMITH, FREDERICK GEORGE WALTON, oceanography; deceased, see previous edition for last biography

SMITH, FREDERICK PAUL, b Pittsburgh, Pa, June 14, 51; m 82; c 2. FORENSIC DRUG TESTING, CRIMINALISTICS. *Educ:* Antioch Col, BA, 74; Univ Pittsburgh, MS, 76, PhD(anal chem), 78. *Prof Exp:* Res asst, Western Psychiat Inst & Clin, Univ Pittsburgh, 74-75; criminalist I, Pittsburgh & Allegheny County Crime Lab, 75-76; res scientist, Off Chief Med Examr City New York, 79; asst prof, 79-82, ASSOC PROF FORENSIC SCI, UNIV

ALA, BIRMINGHAM, 82- *Concurrent Pos:* Sci dir, AccuTex Anal Labs Inc, 89-; sr appointment res fac, US Naval Res Lab, Am Soc Eng Educ, 89; Fulbright res scholar, Univ Strathclyde, Glasgow, Scotland, 90. *Mem:* Am Acad Forensic Sci; Am Chem Soc; Forensic Sci Soc; Sigma Xi. *Res:* Detection of drugs in hair and its application to legal questions; fire-related phenomena and their legal applications. *Mailing Add:* Univ Ala 101 MCJB Birmingham AL 35209

SMITH, FREDERICK T(UCKER), b Waltham, Mass, Nov 24, 20; m 49; c 1. SYSTEMS ENGINEERING, SOFTWARE ENGINEERING. *Educ:* Tufts Univ, BS, 43; Mass Inst Technol, MS, 48; Univ Calif, Los Angeles, PhD, 65. *Prof Exp:* Flight test engr, Instrumentation Lab, Mass Inst Technol, 48-51; systs engr, NAm Aviation, Inc, 51-54; res engr, Rand Corp, 54-65; SR STAFF ENGR, SYSTS ENG, SINGER-LIBRASCOPE, 65- *Res:* Anti-submarine warfare systems; computer science; applied mathematics. *Mailing Add:* Librascope Corp 833 Sonora Ave Glendale CA 91201

SMITH, FREDERICK W(ILSON), b Lansdowne, Pa, Mar 15, 17; m 42; c 4. CHEMICAL ENGINEERING. *Educ:* Univ Mich, BSE, 38, MSE, 39. *Prof Exp:* Chem engr, E I du Pont de Nemours & Co, Del & WVa, 39-44; from head sect to supvr, Chem Eng Dept, BASF Wyandotte Corp, 44-58, mgr chem eng res & semicommercial chem, 58-62, res staff consult, Res Div, 62-79; RETIRED. *Mem:* Am Chem Soc; Am Inst Chem Engrs. *Res:* Organic and inorganic synthesis; polyethylene; synthetic detergents; sodium carboxymethylcellulose; pilot plant research and semicommercial chemicals production supervision; research project and economic evaluations. *Mailing Add:* 7814 Park Ave Allen Park MI 48101-1714

SMITH, FREDERICK WILLIAM, b Albany, NY, Aug 2, 42; m 65; c 2. EXPERIMENTAL SOLID STATE PHYSICS. *Educ:* Lehigh Univ, BA, 64; Brown Univ, PhD(physics), 69. *Prof Exp:* Res fel, Rutgers Univ, New Brunswick, 68-70; asst prof, 70-77, assoc prof, 77-81, PROF PHYSICS, CITY COL NEW YORK, 81- *Concurrent Pos:* Alexander von Humboldt fel, Max-Planck Inst, 77-78. *Mem:* Am Phys Soc; Mat Res Soc. *Res:* surface reactions on semiconductors; epitaxial growth of thin films; amorphous semiconductor films. *Mailing Add:* Dept of Physics City Col New York Convent Ave and 138 St New York NY 10031

SMITH, FREDERICK WILLIAMS, b Mooresville, Ala, Sept 6, 22; m 46; c 2. SURGERY. *Educ:* Vanderbilt Univ, BA, 42, MD, 44; Am Bd Surg, dipl. *Prof Exp:* Intern surg, Duke Univ Hosp, 44-45; from jr resident to sr chief resident, Jefferson Hillman Hosp, 47-50; from instr to assoc prof surg, Med Col, Univ Ala, Birmingham, 64-73; mem clin fac, 73-81, ASSOC PROF SURG, SCH PRIMARY MED CARE, UNIV ALA, HUNTSVILLE, 81-; PATHOLOGIST, JEFFERSON HILLMAN HOSP, 47- *Concurrent Pos:* Consult, Tuberc Sanitorium, Flint, Ala, 50-52; chief surg, Huntsville Hosp, 62-63, Crestwood Hosp, 66. *Mem:* Fel Am Col Surgeons. *Mailing Add:* 205 St Clair Ave Huntsville AL 35801

SMITH, FREDERICK WILLIS, b Seattle, Wash, Apr 28, 38; m 60; c 2. MECHANICAL ENGINEERING. *Educ:* Univ Wash, BS, 61, MS, 63, PhD(mech eng), 66. *Prof Exp:* From asst prof to assoc prof mech eng, Colo State Univ, 65-77; MGR, SEMICONDUCTOR PROD DIV, MOTOROLA INC, 77- *Concurrent Pos:* Consult fracture mech. *Mem:* Am Soc Mech Engrs. *Res:* Fracture mechanics; elasticity; snow mechanics. *Mailing Add:* 1216 Parkwood Dr Ft Collins CO 80525

SMITH, G(EORGE) V, b Clarksburg, WVa, Apr 7, 16; m 49; c 1. METALLURGY. *Educ:* Carnegie Inst Technol, BS, 37, ScD(metall), 41. *Prof Exp:* Asst metall, Metals Res Lab, Carnegie Inst Technol, 37-39; metallurgist, Res Lab, US Steel Corp, 41-55; Francis Norwood Bard prof metall eng, Cornell Univ, 55-70; CONSULT ENGR, 70- *Concurrent Pos:* Adj prof, Polytech Inst Brooklyn, 48-55; asst dir, Sch Chem & Metall Eng, Cornell Univ, 57-62; consult, US Steel Corp, Socony Mobil Oil Co, Babcock & Wilcox Co, Gulf Gen Atomic & Metal Properties Coun, Atomic Energy Comn. *Honors & Awards:* Award, Am Soc Testing & Mat; G Hall Taylor Medal, Am Soc Mech Engrs. *Mem:* Am Soc Metals; fel Am Soc Testing & Mat; Am Inst Mining, Metall & Petrol Engrs; Am Soc Mech Engrs. *Res:* Plastic deformation of metals; elevated temperature properties. *Mailing Add:* 104 Berkshire Rd Ithaca NY 14850

SMITH, GAIL PRESTON, b Unionville, Pa, Jan 25, 15; m 37; c 2. APPLIED PHYSICS. *Educ:* Geneva Col, BS, 34; Syracuse Univ, MA, 36; Univ Mich, PhD(physics), 41. *Prof Exp:* Instr physics, Battle Creek Col, 36; instr math & physics, Geneva Col, 36-37; asst physics, Univ Mich, 37-41; res physicist, Corning Glass Works, 41-50, sr res assoc, 50-80, mgr gen prod develop, 61-66, mgr int res, 66-70, dir tech staff serv, 70-78, dir int res, 78-80; CONSULT, 80- *Mem:* Fel AAAS; fel Brit Inst Physics; fel Am Ceramic Soc; fel Soc Glass Technol; Am Phys Soc; Europ Phys Soc. *Res:* Beta-ray spectroscopy; density and expansivity of glasses; dielectric properties of glasses and glass electronic components; optical properties and applications of glasses and coatings; structural application of glasses and glass-ceramics. *Mailing Add:* 75 Caton Rd Corning NY 14830-3743

SMITH, GALE EUGENE, b Van Wert, Ohio, Feb 11, 33. PHOTOGRAPHIC CHEMISTRY, GRAPHIC ARTS. *Educ:* Bowling Green State Univ, BA, 55; Mich State Univ, PhD, 63. *Prof Exp:* Res asst phys chem, Mich State Univ, 60-63; RES CHEMIST, RES LABS, EASTMAN KODAK CO, 63- *Mem:* Am Chem Soc; Soc Photog Sci & Eng; Tech Asn Graphic Arts. *Res:* Offset lithography; electrochemistry of photographic developers; reaction mechanisms in photographic systems. *Mailing Add:* 299 Seneca Park Ave Rochester NY 14617

SMITH, GARDNER WATKINS, b Boston, Mass, July 2, 31; m 58; c 3. SURGERY. *Educ:* Princeton Univ, AB, 69; Harvard Med Sch, MD, 56; Am Bd Surg, dipl, 64; Am Bd Thoracic Surg, dipl, 65. *Prof Exp:* From instr to assoc prof surg, Sch Med, Univ Va, 63-70; PROF SURG, SCH MED, JOHNS HOPKINS UNIV, 70-; PROF SURG, SCH MED, UNIV MD, BALTIMORE CITY, 70-; surgeon-in-chief, Baltimore City Hosps, 70-78; dep dir, Dept Surg, Johns Hopkins Hosp, 78-85; CHMN, SECT OF SURG SCI, FRANCIS SCOTT KEY MED CTR, 85- *Concurrent Pos:* Fel surg, Sch Med, Johns Hopkins Univ, 57-58; consult, Vet Admin Hosps, Salem, Va, 68-70 & Baltimore, 71- & Greater Baltimore Med Ctr, 71- *Mem:* Am Col Surgeons; Am Surg Asn; Soc Univ Surgeons; Soc Vascular Surg; Soc Surg Alimentary Tract. *Res:* Physiology of portal hypertension; clinical research in gastrointestinal and vascular surgery. *Mailing Add:* Francis Scott Key Med Ctr 4940 Eastern Ave Baltimore MD 21224-2780

SMITH, GARMOND STANLEY, b Wayne, WVa, July 9, 32; m 53; c 6. ANIMAL NUTRITION, TOXICOLOGY. *Educ:* WVa Univ, BS, 53, MS, 57, PhD(agr biochem), 59. *Prof Exp:* Asst agr biochem, WVa Univ, 57-59; res assoc animal sci, Univ Ill, 59-60, asst prof, 60-65; instr biol, sci & relig, Lincoln Christian Col, 63-65; PROF ANIMAL NUTRIT, NMEX STATE UNIV, 68- *Concurrent Pos:* Invited res reports, Germany, 79, Austria & SAfrica, 81, Japan, 83, 88 & Can, 84; consult, SAfrica, Int Atomic Energy Agency & Food & Agr Orgn, Vienna, 81; vis prof toxicol, Univ Kansas Med Ctr, 82; vis prof food sci, Kyoto Univ, Japan, 88. *Mem:* Fel AAAS; Am Inst Nutrit; Am Soc Animal Sci; Am Inst Biol Sci; Coun Agr Sci Technol; Am Inst Chemists. *Res:* Nonprotein nitrogen, ruminants; metabolic changes in starvation and refeeding; vitamin A nutrition; nitrate toxicity; potassium-40 as an index of lean body mass; silica in animal metabolism; recycling of nutrients in agricultural and municipal wastes; improving animal tolerances of toxicants in forage and feeds; xenobiotics metabolism. *Mailing Add:* Dept Animal Sci NMex State Univ Las Cruces NM 88003

SMITH, GARRY AUSTIN, b Alta, Can, Sept 25, 40; m 63; c 2. PLANT GENETICS. *Educ:* NMex State Univ, BS, 64, MS, 66; Ore State Univ, PhD(genetics), 68. *Prof Exp:* Res asst, NMex State Univ, 64-66, Ore State Univ, 66-68; res geneticist, USDA Agr Res Serv, Canal Pt, Fla, 68-69; supvr res geneticist, sci & educ admin, Crops Res Lab, 69-88, RES LEADER, USDA AGR RES SERV, USDA N CROPS SCI LAB, 88- *Concurrent Pos:* Acad fac affil, Colo State Univ, 69-88, grad fac, 70; assoc ed, Crop Sci, Crops Sci Soc Am, 77-80; adj prof, NDak State Univ. *Mem:* Am Genetic Asn; fel Am Soc Agron; Am Soc Sugar Beet Technologists; fel Crop Sci Soc Am. *Res:* Quantitative plant genetics; inheritance of disease resistance. *Mailing Add:* US Dept Agr Agr Res Serv N Crop Sci Lab Univ Sta Box 5677 Fargo ND 58105-5677

SMITH, GARY CHESTER, b Ft Cobb, Okla, Oct 25, 38; m 65; c 6. MEAT SCIENCE, FOOD SCIENCE & TECHNOLOGY. *Educ:* Calif State Univ, Fresno, BS, 60; Wash State Univ, MS, 62; Tex A&M Univ, PhD(meat sci), 68. *Prof Exp:* Mgt trainee, Armour & Co, Wash, 62; instr animal sci, Wash State Univ, 62-65; from asst prof to prof meat sci, 68-82, HEAD DEPT ANIMAL SCI, TEX A&M UNIV, 82- *Concurrent Pos:* Mem, Comts Off Technol Assessment, Nat Res Coun-Nat Acad Sci & USDA. *Honors & Awards:* Distinguished Res Award, Am Soc Animal Sci, 74; Distinguished Res Award, Nat Livestock Mkt Asn, 79; Distinguished Res Award, Am Meat Sci Asn, 82. *Mem:* Am Soc Animal Sci; Am Meat Sci Asn (pres, 76-77); Inst Food Technologists; Am Dairy Sci Asn; Coun Agr Sci & Technol; Int Asn Milk, Environ & Food Sanitarians; Am Regist Prof Am Scientists. *Res:* Meat packaging; chemical, physical and histological muscle properties as related to palatability; quantitative and qualitative evaluation of beef, pork, lamb and goat carcasses; growth and development of meat animals. *Mailing Add:* 2219 Apache Ct Ft Collins CO 80525-1828

SMITH, GARY EUGENE, b Louisville, Ky, Oct 27, 32; m 55; c 2. POLYMER CHEMISTRY. *Educ:* Univ Ky, BS, 55. *Prof Exp:* Sr chemist, Shell Chem Co, 55-60; tech dir, Mangolia Plastics, Inc, 60-68; prod mgr, Ciba Prod Co, 68-70; dir res & develop, Chem Dynamics, Inc, 70-76; lab dir, Resinoid Eng, Inc, 76-81; prod mgr, Wilson Sporting Goods, 81-87; DIR TECHNOL, NEWPORT ADHESIVES & COMPOSITES, 87- *Concurrent Pos:* Chemist, Army Chem Ctr, US Army, 56-58. *Mem:* Soc Plastics Engrs. *Res:* Epoxy resin formulations and curing agents; reinforced polymer products. *Mailing Add:* Newport Adhesives & Composites 3401 Fordham Ave Santa Ana CA 92704

SMITH, GARY JOSEPH, METAGENESIS, CHEMICAL CARCINOGENESIS. *Educ:* NC State Univ, PhD(genetics), 75. *Prof Exp:* ASST PROF, UNIV NC, 81- *Mailing Add:* Dept Path Univ NC Bldg CB 7525 Chapel Hill NC 27599

SMITH, GARY KEITH, b Easton, Pa, Aug 31, 52; m 81. CELL BIOLOGY, REDUCED FOLATE TRANSPORTER. *Educ:* Lebanon Valley Col, BS, 74; Lehigh Univ, MS, 77, PhD(chem), 78. *Prof Exp:* Postdoctroal fel, Pa State Univ, 78-81; SR BIOCHEMIST, WELLCOME RES LABS, 81- *Concurrent Pos:* NIH postdoctoral fel, Pa State Univ, 79. *Mem:* Am Soc Biochem & Molecular Biol; Am Chem SOc. *Res:* biochemistry of pterins and folates and their relevance to cancer chemotherapy; mechanics of cell death induced by toxins and drugs; enzyme-antibody conjugates in therapy. *Mailing Add:* Div Cell Biol Research Triangle Park NC 27709

SMITH, GARY LEE, b Rock Springs, Wyo, May 27, 47; m 67; c 1. VIROLOGY. *Educ:* Univ Wyo, BS, 69; Kans State Univ, PhD(microbiol), 72. *Prof Exp:* Fel oncol, Leukemia Soc Am, Univ Wis, 73-74; asst prof, Univ Nebr, 74-77, assoc prof microbiol, 77-87; RES SCIENTIST, ELI LILLY & CO, 87- *Mem:* Am Soc Cell Biol; Sigma Xi; Am Soc Microbiol; AAAS. *Res:* Control of cellular proliferation by growth factors; endocrinology of cellular growth and development. *Mailing Add:* Lilly Res Labs Div Eli Lilly & Co Indianapolis IN 46285-3313

SMITH, GARY LEROY, b Mitchell, SDak, Nov 30, 35; m 59; c 4. NUCLEAR PHYSICS, SYSTEMS ANALYSIS. *Educ:* Univ Calif, Davis, BS, 63, MA, 65, PhD(physics), 69. *Prof Exp:* Nat Res Coun Res assoc, US Naval Res Lab, 69-70; sr staff physicist, Johns Hopkins Univ, appl physics lab, 70-75, asst group supvr, 75-76, asst acoustics prog mgr, 75-77, acoustics prog mgr, 77-79, asst dept supv, 79-84, sec prog mgr, fleet ballistic missle submarine, 81-88, assoc dept supv, 84-88; from dep asst dir to asst dir, Johns Hopkins Univ, Res Explor Develop, 88-90, asst dir, Res & Progs, 90-91, ASSOC DIR, JOHNS HOPKINS UNIV, 91- *Mem:* Am Phys Soc; AAAS. *Res:* Resonance-neutron capture gamma-ray spectroscopy on rare-earth nuclei; underwater acoustics. *Mailing Add:* Johns Hopkins Appl Physics Lab Johns Hopkins Rd Laurel MD 20723-6099

SMITH, GARY RICHARD, b Palo Alto, Calif, Nov 15, 48; m 70; c 1. THEORETICAL PLASMA PHYSICS, ATMOSPHERIC DYNAMICS. *Educ:* Oberlin Col, BA, 70; Univ Calif, Berkeley, PhD(physics), 77. *Prof Exp:* PHYSICIST, LAWRENCE LIVERMORE NAT LAB, 77- *Concurrent Pos:* Lectr, Univ Calif, Davis, 80-81. *Mem:* Am Phys Soc; Sigma Xi; Am Geophys Union. *Res:* Theoretical modeling of plasma heating and current drive in tokamak controlled-fusion devices; physics of earth's atmosphere. *Mailing Add:* Lawrence Livermore Nat Lab L-630 PO Box 5511 Livermore CA 94550

SMITH, GASTON, b Poplarville, Miss, Apr 7, 27; m 50; c 2. MATHEMATICS. *Educ:* Univ Southern Miss, BS, 49; Univ Ala, MA, 55 & 57, PhD(math), 63. *Prof Exp:* Instr high sch, Miss, 51-52; instr math, Sunflower Jr Col, 52-56; asst prof, Univ Southern Miss, 57-60; instr, Univ Ala, 60-63; prof, Univ Southern Miss, 63-67; PROF & CHMN DEPT MATH, WILLIAM CAREY COL, 67- *Mem:* Am Math Soc; Math Asn Am; Can Math Cong. *Res:* Summability. *Mailing Add:* Dept of Math William Carey Col Hattiesburg MS 39401

SMITH, GENE E, b Fulton Co, Ohio, June 6, 36; m 58; c 5. MECHANICAL ENGINEERING. *Educ:* Univ Mich, BSME, 59, MSME, 60, PhD(mech eng), 63. *Prof Exp:* From asst prof to assoc prof, 63-78, PROF MECH ENG, UNIV MICH, ANN ARBOR, 78- *Concurrent Pos:* Develop engr, Gen Motors Corp, 65-67. *Mem:* Soc Automotive Engrs; Am Inst Chem Engrs; Am Soc Eng Educ; Am Soc Mech Engrs. *Res:* Thermodynamics; heat transfer; phase equilibrium at low temperatures; direct energy conversion. *Mailing Add:* 2420 Bunker Hill Ann Arbor MI 48105

SMITH, GEOFFREY W, b Boston, Mass, Sept 29, 39; m 65; c 1. GEOLOGY. *Educ:* Tufts Univ, BS, 61; Univ Maine, MS, 64; Ohio State Univ, PhD(geol), 69. *Prof Exp:* Instr geol, Colby Col, 68-69; asst prof, 69-74, chmn dept, 74-80, ASSOC PROF GEOL, OHIO UNIV, 74- *Mem:* Am Quaternary Asn; Nat Asn Geol Teachers; AAAS; Geol Soc Am; Soc Econ Paleont & Mineral. *Res:* Glacial geology; geomorphology; quaternary stratigraphy; deglaciation studies and quaternary stratigraphy in Maine, British Columbia and Southeastern Ohio. *Mailing Add:* Dept Geol Ohio Univ Athens OH 45701

SMITH, GEORGE BYRON, b Pittsburgh, Pa, Apr 18, 33; m 55; c 6. PHYSICAL CHEMISTRY. *Educ:* Univ Pittsburgh, BS, 54, PhD(phys chem), 59. *Prof Exp:* Sr chemist, 59-67, SECT LEADER, ANAL & PHYS RES DEPT, MERCK, SHARP & DOHME RES LABS, 67- *Res:* Physical analytical chemistry; chemical kinetics. *Mailing Add:* 100 George Ave Edison NJ 08820-3113

SMITH, GEORGE C, b West Unity, Ohio, May 23, 35. PHYSICS, COMPUTERS. *Educ:* Cornell Univ, AB, 57, MS, 62, PhD(eng physics), 65; Univ NMex, JD(law), 73. *Prof Exp:* Physicist solid state, Sandia Labs, 66-70; sr scientist computing, Opers Res Inc, 75-76; physicist systs, Lawrence Livermore Nat Lab, 77-91. *Concurrent Pos:* Mem, Telluride Asn, 55-56; foreign scientist, Alexander von Humboldt Found, Ger, 65-66; consult, Opers Res Inc, 76-77; Arms Control fel, Stanford Univ, 84-85. *Mem:* Am Phys Soc; Inst Elec & Electronic Engrs; Am Inst Aeronaut & Astronaut; Sigma Xi; Int Soc Optical Eng. *Res:* Systems optimization; computer modeling; strategic studies; foreign technologies; advanced energy concepts; aerospace; weapons. *Mailing Add:* Lawrence Livermore Nat Lab Mail Code L-389 PO Box 808 Livermore CA 94550

SMITH, GEORGE C(UNNINGHAM), b Pittsburgh, Pa, Feb 16, 26; m 53; c 3. CHEMICAL ENGINEERING, ENVIRONMENTAL SCIENCE. *Educ:* Univ Pittsburgh, BSChE, 48, MS, 50; Carnegie Inst Technol, PhD(chem eng), 56. *Prof Exp:* Process engr, Gen Elec Co, 50-52; res engr, E I du Pont de Nemours & Co, 56-60; sr res engr, Jones & Laughling Steel Corp, 60-63, res assoc process metall, 63-69, staff engr, 69-70, tech coordr environ control, 70-84; CONSULT, WASTE WATER TREATMENT & SOLID WASTE MGT, 84- *Mem:* Am Inst Chem Engrs; Am Chem Soc; Am Inst Mining, Metall & Petrol Engrs; Air Pollution Control Asn. *Res:* Waste water treatment and air cleaning in iron and steel industry; solid waste and toxic materials management; steelmaking processes, especially fluid mechanics of basic oxygen processes; fluid mechanics. *Mailing Add:* 866 Foxland Dr Pittsburgh PA 15243

SMITH, GEORGE DAVID, b Youngstown, Ohio, Aug 24, 41; div; c 2. X-RAY CRYSTALLOGRAPHY, PHYSICAL CHEMISTRY. *Educ:* Westminster Col, Pa, BS, 63; Ohio Univ, PhD(chem), 68. *Prof Exp:* Res assoc, Mont State Univ, 68-74; SR RES SCIENTIST, MED FOUND BUFFALO, 82; ASST RES PROF, ROSWELL PARK DIV, GRAD SCH STATE UNIV NY, BUFFALO, 82- *Mem:* Am Chem Soc; Am Crystallog Asn; Sigma Xi; AAAS. *Res:* X-ray crystal structures of ionophores, antibiotics, and polypeptides; correlation of structure to function; protein crystallography; crystallographic studies of human insulin. *Mailing Add:* Med Found of Buffalo 73 High St Buffalo NY 14203

SMITH, GEORGE ELWOOD, b White Plains, NY, May 10, 30; wid; c 3. SOLID STATE ELECTRONICS. *Educ:* Univ Pa, BS, 55; Univ Chicago, MS, 56, PhD(physics), 59. *Prof Exp:* Mem staff, 59-64, HEAD MOS DEVICE DEPT, BELL LABS, 64- *Honors & Awards:* Ballantine Medal, Franklin Inst, 73; Liebmann Award, Inst Elec & Electronics Engrs, 74. *Mem:* Nat Acad Eng; fel Inst Elec & Electronics Engrs; fel Am Phys Soc. *Res:* Band structure of semimetals; thermoelectric effects; electronic transport phenomena; optical properties of semiconductors; optoelectronic devices; electrical conduction in metal oxides; semiconductor devices; charge coupled devices; integrated circuits. *Mailing Add:* PO Box 787 Barnegat NJ 08005-0787

SMITH, GEORGE FOSTER, b Franklin, Ind, May 9, 22; m 50; c 3. MICROELECTRONICS, LASERS. *Educ:* Calif Inst Technol, BS, 44, MS, 48, PhD(physics), 52. *Prof Exp:* Asst, Calif Inst Technol, 47-50, res assoc, 48-50; res physicist, Hughes Aircraft Co, 52-57, dept mgr, 57-62, assoc dir, 62-69, vpres, 65-81, dir res labs, 69-87, sr vpres, 81-87; RETIRED. *Concurrent Pos:* Res engr, Eng Res Assocs, 46-48; adj assoc prof, Univ Southern Calif, 60-62; consult, Army Sci Adv Panel, 75-78; mem policy bd, Hughes Aircraft Co, 65-87. *Honors & Awards:* Frederick Philips Award, Inst Elec & Electronic Engrs, 88. *Mem:* AAAS; fel Am Phys Soc; fel Inst Elec & Electronic Engrs; Sigma Xi. *Res:* Research management; electron devices; lasers; microelectronics; displays; physical electronics. *Mailing Add:* 6423 Riggs Pl Los Angeles CA 90045

SMITH, GEORGE IRVING, b Waterville, Maine, May 20, 27; m 74; c 5. QUATERNARY GEOLOGY. *Educ:* Colby Col, AB, 49; Calif Inst Technol, MS, 51, PhD(geol), 56. *Prof Exp:* Mem staff geol, Occidental Col, 51-52; geologist, 52-66, chief light metals & indust minerals br, 66-69, geologist, 69-78, coordr climate prog, 78-81, GEOLOGIST, US GEOL SURV, 81- *Concurrent Pos:* Fulbright scholar grant, Australia, 81. *Honors & Awards:* Meritorious Serv Award, Dept Int, 83. *Mem:* Geol Soc Am; Geochem Soc; Mineral Soc Am; Soc Econ Geol; Sigma Xi; Am Quaternary Asn. *Res:* Structure and stratigraphy of Mojave Desert area; Quaternary deposits and climates; evaporite deposits; volcanic petrology. *Mailing Add:* US Geol Surv 345 Middlefield Rd Menlo Park CA 94025

SMITH, GEORGE LEONARD, JR, b State College, Pa, Sept 6, 35; m; c 2. INDUSTRIAL ENGINEERING. *Educ:* Pa State Univ, BS, 57; Lehigh Univ, MS, 58 & 67; Okla State Univ, PhD(indust eng), 69. *Prof Exp:* Grad asst indust eng, Lehigh Univ, 57-58; instr, Prod Tech, Pa State Univ, York Campus, 58-59 & indust eng, Lehigh Univ, 59-67; grad res asst, Okla State Univ, 67-68; PROF & CHMN INDUST ENG, OHIO STATE UNIV, 68- *Concurrent Pos:* Labor arbitrator, Fed Mediation & Conciliation Serv, 70-; consult, Amalgamated Meat Cutters & Butcher Workmen, 78-; dir, Ergonomics Div, Am Inst Indust Engrs, 76-78; ed, Human Factors, 80- *Mem:* fel Inst Indust Eng; fel Human Factors Soc; Am Soc Eng Educ. *Res:* Person and machine systems analysis and design, human performance, design methods with particular emphasis on design of work and workspaces. *Mailing Add:* Dept Indust & Systs Eng Ohio State Univ Columbus OH 43210

SMITH, GEORGE PEDRO, b Norfolk, Va, Oct 26, 23; m 45; c 3. PHYSICAL CHEMISTRY. *Educ:* Univ Va, BS, 44, PhD(chem), 50. *Prof Exp:* GROUP LEADER, OAK RIDGE NAT LAB, 50- *Concurrent Pos:* Lectr, Univ Tenn, Knoxville, 52-63, prof, 64-78, adj prof, 81-; prof, Tech Univ Denmark, 72-73; ed, Advan in Molten Salt Chem, 71-76; lectr, Norwegian Inst Technol, 78. *Mem:* AAAS; Am Chem Soc; Electrochem Soc; Sigma Xi. *Res:* Molten salt chemistry; applied spectroscopy and photochemistry. *Mailing Add:* 7925 Chesterfield Dr Knoxville TN 37909-2916

SMITH, GEORGE THOMAS, b Evansville, Ind, Oct 19, 31; m 64; c 5. PATHOLOGY. *Educ:* Univ Md, BS, 52, MD, 56; Vienna Acad Med, cert, 58; Mass Inst Technol, MS, 76. *Prof Exp:* Intern, Royal Victoria Hosp, Montreal, 56-57; resident path, Peter Bent Brigham Hosp, 59-60, sr asst resident, 61-62; asst resident, Children's Hosp, Boston, 62; chief resident, Peter Bent Brigham Hosp, 63, assoc pathologist, Hosp & Harvard Med Sch, 63-64; res prof path, Desert Res Inst, 65-70; actg dean sch med, 67-70, dean sch med sci & prof path, Univ Nev, Reno, 70-77, dir labs environ pathophysiol, 65-77; assoc chief staff & educ, Boston Vet Admin Hosp, 77-78; PROF PATH & ASSOC DEAN, UNIV ALA, 79-; DIR, SOUTHEASTERN REGIONAL MED EDUC CTR, 79- *Concurrent Pos:* Nat Heart Inst res trainee, Peter Bent Brigham Hosp, 61-62; res fel path, Harvard Med Sch, 60-61; fel, Free Hosp Women, 62-63; trainee, Congenital Heart Dis Training & Res Ctr, Chicago, 62; NIH career develop award cardiovasc path, 64; mem, Regional Adv Bd Heart Dis, Cancer & Stroke, Utah, 65 & Nev Heart Dis, Cancer & Stroke Adv Bd, 66; prof path, Sch Med, Tufts Univ & Boston Univ, 77-78. *Mem:* AAAS; AMA; Am Fedn Clin Res; Am Soc Exp Path; Int Acad Path. *Res:* Cardiovascular and pulmonary disease. *Mailing Add:* Southeast Reg Med Educ Ctr Vet Admin 930 S 20th St Birmingham AL 35205

SMITH, GEORGE WOLFRAM, b Des Plaines, Ill, Sept 19, 32; m 56; c 2. LIQUID CRYSTALS, CHEMICAL PHYSICS. *Educ:* Knox Col, BA, 54; Rice Univ, MA, 56, PhD(physics), 58. *Prof Exp:* Welch Found fel physics, Rice Univ, 58-59; sr res physicist, 59-76, dept res scientist, 76-81, sr staff res scientist, 81-87, PRINCIPAL RES SCIENTIST, GEN MOTORS RES LAB, 87- *Concurrent Pos:* Instr, Lawrence Inst Technol, 63-65; lectr, Cranbrook Inst Sci, 63-87; coun, Gordon Res Conf, 78; co-ed, Particulate Carbon, Formation During Combustion, 81; sci adv comt, Cronbrook Inst Sci, 89-; chmn, comt applications physics, Am Phys Soc, 91. *Honors & Awards:* Cambell Award, 80; McCuen Award, 85. *Mem:* Fel Am Phys Soc; Sigma Xi; Am Carbon Soc; Combustion Inst. *Res:* Low temperature physics; nuclear magnetic resonance; molecular structures and motions in solid and liquid states; internal friction; liquid crystals; thermomagnetic gas torque; physics of carbon; phase transformations in solids; 80 publications and 10 patents. *Mailing Add:* Dept Physics Gen Motors Res Labs 30500 Mound Rd Warren MI 48090-9055

SMITH, GERALD A, b Akron, Ohio, Jan 8, 36; m 58; c 2. EXPERIMENTAL HIGH ENERGY PHYSICS. *Educ:* Miami Univ, BA, 57; Yale Univ, MS, 58, PhD(physics), 61. *Prof Exp:* Physicist, lectr & asst prof, Lawrence Radiation Lab, Univ Calif, Berkeley 61-67; prof physics, Mich State Univ, 67-82; prof physics & head dept, 83-88, PROF PHYSICS & DIR, LAB ELEM PARTICLES SCI, PA STATE UNIV, 88- *Concurrent Pos:* From lectr to asst prof, Univ Calif, Berkeley, 63-67; consult, Argonne Nat Lab, 68-72, Argonne Univs Asn, 71-73 & NSF, 73-76; mem & chmn, Fermilab User's Orgn, 71-74; prin investr, Nat Sci Found grants, 71-; mem, Bd Trustees, Argonne Univs Asn, 76-78; assoc lab dir high energy physics, Argonne Nat Lab, 78; mem & chmn, High Energy Discussion Group Exec Comt, Brookhaven Nat Lab, 81-85; mem, EPAC, Stanford Linear Accelerator Ctr, 82-84; mem, PSCC Comt, Europ Orgn Nuclear Res, 84; mem, Mat Res Lab Adv Comt, Pa State

Univ, 84-88; consult, Rand Corp, 87-; prin investr, USAF Off Sci Res grants, 87-, Jet Propulsiol Lab contract, 88- & Dept Energy grant, 90- *Mem:* Fel Am Phys Soc; AAAS; Sigma Xi. *Res:* High energy particle physics; electronic detectors; analysis; properties of hadronic states of matter. *Mailing Add:* 303 Osmond Lab Pa State Univ University Park PA 16802

SMITH, GERALD DUANE, b Cass City, Mich, Aug 31, 42; m 67; c 1. PHYSICAL CHEMISTRY, HEALTH PHYSICS. *Educ:* Huntington Col, BS, 64; Purdue Univ, PhD(health physics), 72. *Prof Exp:* AEC fel trainee, Battelle Northwest Labs, 64-65; instr chem, Owosso Col, 65-67; PROF CHEM, HUNTINGTON COL, 67-, VPRES & DEAN, 82- *Concurrent Pos:* Consult, Ind Radiation Emergency Response Team, 74-82. *Mem:* Am Chem Soc; Sigma Xi; Am Asn Physics Teachers. *Res:* Radiation dosimetry; crystal growth; reaction kinetics. *Mailing Add:* Dept of Chem Huntington Col Huntington IN 46750-1299

SMITH, GERALD FLOYD, b Louisville, Ky, Jan 4, 42; m 65; c 2. BLOOD COAGULATION, PHARMACEUTICAL RESEARCH. *Educ:* Univ Louisville, BS, 63, MS, 65, PhD(chem), 68; Indiana Univ, JD, 86. *Prof Exp:* NIH training grant molecular path, Sch Med, Univ Louisville, 68-71; Eli Lilly sr res fel, Eli Lilly Res Clin, 71-72, SR SCIENTIST, ELI LILLY RES LABS, 73- *Mem:* Sigma Xi; AAAS; Am Chem Soc; Am Heart Asn; Int Soc Thrombosis & Haemostasis. *Res:* Fibrinogen-fibrin chemistry; blood coagulation-thrombosis; inflammation; proteases-inhibitors; warfarin; heparin; therapeutic agents; vitamin K; metastasis; fibrinolysis, thrombolytics, plasminogen activators. *Mailing Add:* Dept M304 Bldg B8 Rm 463 Eli Lilly Res Labs 307 E McCarty St Indianapolis IN 46285

SMITH, GERALD FRANCIS, b Buffalo, NY, Oct 17, 28; m 56. APPLIED MATHEMATICS, MECHANICS. *Educ:* Univ Buffalo, BS, 52; Brown Univ, PhD(appl math), 56. *Prof Exp:* Res assoc appl math, Brown Univ, 56; mathematician, Calif Res Crop Div, Stand Oil Co Calif, 56-58; asst prof mech, Lehigh Univ, 58-60; asst prof eng & appl sci, Yale Univ, 60-64; assoc prof math, Univ Wis, Milwaukee, 64-65; prof, 65-80, DIR, CTR APPLN MATH, LEHIGH UNIV, 80- *Concurrent Pos:* NSF study grants, 61-64 & 66-68. *Mem:* Am Math Soc. *Res:* Theory of invariants and continuum mechanics. *Mailing Add:* Dept Mech Eng Packard Lab Bldg 19 Lehigh Univ Bethlehem PA 18015

SMITH, GERALD LYNN, b Seminole, Okla, Feb 3, 34; m 65. ELECTRICAL ENGINEERING, MATHEMATICS. *Educ:* Univ Okla, BSME & BSEE, 57, MSEE, 59, PhD(plasma dynamics), 66. *Prof Exp:* Assoc res engr, Cities Serv Res Co, 57-58; design engr, Douglas Aircraft Co, 59-60; instr elec eng, Univ Okla, 60-63; instr col eng, Okla State Univ, 63-64; head dept, Univ Tulsa, 64-70, assoc prof elec eng, 64-78; SR DESIGN ENG, ENGR ELEC DESIGN SECT, MCDONNELL DOUGLAS CORP, 85- *Mem:* AAAS; Am Soc Mech Engrs; Am Inst Mining, Metall & Petrol Engrs; Am Soc Eng Educ; Inst Elec & Electronics Engrs (secy, 67-68). *Res:* Electrodynamics; plasmadynamics; digital electronics; systems analysis and design. *Mailing Add:* Engr Elec Design Dept McDonnell-Douglas Corp 2000 N Memorial Dr Tulsa OK 74115

SMITH, GERALD M(AX), b Osborne, Kans, Jan 2, 20; m 41; c 3. ENGINEERING. *Educ:* Kans State Univ, BS, 48, MS, 51. *Prof Exp:* Res engr, Kans State Univ, 48-50, asst prof appl mech, 50-53; from assoc prof to prof, 53-83, chmn dept, 72-83, emer prof eng mech, Univ Nebr-Lincoln, 85-; RETIRED. *Honors & Awards:* Wasson Medal, Am Concrete Inst, 53 & 65. *Mem:* Am Soc Eng Educ. *Res:* Mechanics of deformable bodies; vibrations; dynamics computers and numerical methods. *Mailing Add:* 2431 S 58th St Lincoln NE 68506

SMITH, GERALD RALPH, b Vandalia, Ill, Feb 19, 44. MOLECULAR BIOLOGY. *Educ:* Cornell Univ, BS, 66; Mass Inst Technol, PhD(biol), 70. *Prof Exp:* Fel, Dept Biochem, Univ Calif, Berkeley, 70-72 & Dept Molecular Biol, Univ Geneva, 72-75; from asst prof to assoc prof molecular biol, Univ Ore, 75-81; RES SCIENTIST, FRED HUTCHINSON CANCER RES CTR, 81- *Concurrent Pos:* Helen Hay Whitney Found fel, 70-73; Swiss NSF int fel, 73-74. *Mem:* Am Soc Microbiol; AAAS. *Res:* Regulation of gene expression in bacteria and bacteriophage; molecular mechanisms of genetic recombination. *Mailing Add:* Fred Hutchinson Cancer Ctr 1124 Columbia St Seattle WA 98104

SMITH, GERALD RAY, b Los Angeles, Calif, Mar 20, 35; m 55; c 3. ZOOLOGY. *Educ:* Univ Utah, BS, 57, MS, 59; Univ Mich, PhD(zool), 65. *Prof Exp:* Asst prof zool & geol, Univ, 69-72, assoc cur, Mus Paleont, 69-72, assoc prof, Univ, 72-81, dir, Mus Paleont, 74-81, PROF ZOOL & GEOL, UNIV MICH, ANN ARBOR, 81-, CUR FISHES, MUS PALEONT, 69-, CUR LOWER VERT PALEONT, 72- *Mem:* AAAS; Soc Study Evolution; Soc Syst Zool; Am Soc Ichthyologists & Herpetologists (pres, 91); Soc Vertebrate Paleont. *Res:* Evolution of North American freshwater fishes. *Mailing Add:* Mus Zool Univ Mich Ann Arbor MI 48109

SMITH, GERALD RAY, b Prattville, Ala, May 12, 52; m 77. LEGUME BREEDING, FORAGE MANAGEMENT. *Educ:* Auburn Univ, BS, 75, MS, 77; Miss State Univ, PhD(agron), 81. *Prof Exp:* Plant breeder, Northrup King Co, 77-78; ASST PROF, TEX A&M UNIV, 81- *Mem:* Am Soc Agron; Crop Sci Soc Am. *Res:* Plant genetic control of legumes-rhizobia dinitrogen fixation; improvement of forage legumes through breeding and selection for increased dinitrogen fixation; pest resistance and improved reseeding. *Mailing Add:* Tex Agr Exp Sta PO Box E Overton TX 75684

SMITH, GERALD WAVERN, b Des Moines, Iowa, Dec 1, 29; m 58; c 1. ENGINEERING ECONOMICS, INDUSTRIAL ENGINEERING. *Educ:* Iowa State Univ, BS, 52, MS, 58, PhD(eng), 61. *Prof Exp:* From instr to assoc prof, Iowa State Univ, 56-67, Alcoa prof, 68-71, prof indust eng, 67-88, EMER PROF, IOWA STATE UNIV, 88- *Honors & Awards:* Wellington Award, Eng Econ, 86. *Mem:* Am Inst Indust Engrs; Am Soc Eng Educ. *Res:* Engineering economy; engineering valuation; management of capital expenditures by public and private organizations; capital expenditures by public and private organizations; capital expenditure decisions for public utilities. *Mailing Add:* 2808 Arbor St Ames IA 50011

SMITH, GERARD PETER, b Philadelphia, Pa, Mar 24, 35; m 62; c 4. NUTRITION. *Educ:* St Joseph's Univ, Pa, BS, 56; Univ Pa, MD, 60. *Prof Exp:* Assoc physiol, Sch Med, Univ Pa, 64-65, asst prof, 65-68; from asst prof to assoc prof, 68-73, PROF PSYCHIAT, MED COL, CORNELL UNIV, 73- *Concurrent Pos:* Head, Div Behav Sci, Cornell Univ, 69-; dir, Edward W Bourne Behav Res Lab, 69-, Eating Disorders Inst, New York, 84-88; NIMH career scientist award. *Honors & Awards:* Curt Richter Lecture, 76; Leon Lectr, 90. *Mem:* Am Physiol Soc; Endocrine Soc; Soc Neurosci; Asn Res Nerv & Ment Dis; Soc Biol Psychiat; Soc Study Ingestive Behav. *Res:* Behavioral neuroscience of eating and its disorders. *Mailing Add:* Dept Psychiat NY Hosp Cornell Med Ctr 21 Bloomingdale Rd White Plains NY 10605

SMITH, GERARD VINTON, b Delano, Calif, Oct 14, 31; m 56; c 3. PHYSICAL ORGANIC CHEMISTRY. *Educ:* Col of the Pac, BA, 53, MS, 56; Univ Ark, PhD(phys org chem), 59. *Prof Exp:* Res assoc, Northwestern Univ, 59-60; instr chem, 60-61; asst prof, Ill Inst Technol, 61-66; assoc prof, 66-73, PROF CHEM, SOUTHERN ILL UNIV, 73-, DIR MOLECULAR SCI PROG, 78- *Concurrent Pos:* Mem, Int Cong Catalysis; chmn, Gordon Res Conf, 78; chmn, Org Reactions Catalysis Soc, 78-80, bd, 89- *Mem:* AAAS; Am Chem Soc; Catalysis Soc (treas, 76-89); Org Reactions Catalysis Soc. *Res:* Mechanisms of heterogeneous catalysis; hydrogenation and exchange; oxidation; stereochemistry; asymmetric induction; hydrodesulfurization; substituent effects; nuclear magnetic resonance; mass spectrometry; gas-liquid chromatography; coal conversion processes. *Mailing Add:* Molecular Sci Prog Southern Ill Univ Carbondale IL 62901-4406

SMITH, GERRIT JOSEPH, b Syracuse, NY, Dec 18, 38. THEORETICAL PHYSICS, SCIENCE PHILOSOPHY. *Educ:* LeMoyne Col, BS, 60, Boston Col, MS, 62; Syracuse Univ, PhD(physics), 71. *Prof Exp:* Instr physics, Regis High Sch, New York, 70-71; res, Syracuse Univ, 71-72; asst prof, 72-80, ASSOC PROF PHILOS SCI, FORDHAM UNIV, 80- *Mem:* Am Phys Soc. *Res:* Measurability analysis of the gravitational field as part of the quantization of general relativity; relativistic conservation laws and the dimensionality of space-time; role of relativity principles in theory change. *Mailing Add:* 2505 Lorillard Pl No 4J Bronx NY 10458

SMITH, GILBERT EDWIN, b Nelsonville, Ohio, Oct 26, 22; m 51. GEOLOGY. *Educ:* Ohio Univ, BS, 50; WVa Univ, MS, 51. *Prof Exp:* Geologist, Ohio Div Geol Surv, 51-53, coal geologist, 53-56; geologist, Aluminum Co Am, 56-61; consult, 61-63; coal geologist, Ky Geol Surv, 63-66, geologist & head coal sect, 66-78; geologist & head coal sect, 78-80, ASSOC DIR, INST MINING & MINERALS RES, UNIV KY, 78- *Mem:* Am Inst Mining Metall & Petrol Engrs; Fel Geol Soc Am; Am Asn Stratig Palynologists; Mine Inspectors Inst Am. *Res:* Coal geology; mapping; resources; Pennsylvanian stratigraphy; reclamation; mining geology. *Mailing Add:* 1710 Blue Licks Rd Lexington KY 40504

SMITH, GILBERT HOWLETT, b Cornwall, NY, July 25, 38; m 61; c 5. CELL BIOLOGY, MOLECULAR BIOLOGY. *Educ:* Hartwick Col, AB, 59; Brown Univ, ScM, 63, PhD(biol), 65. *Prof Exp:* Staff fel, 65-67, head, Ultrastruct Res Sect, 67-70, sr staff scientist, Lab Biol, 70-75, sr staff scientist cancer res, lab molecularbiol, Nat Cancer Inst, NIH, 76-85; SR INVESTR ONCOGENETICS, LAB TUMOR IMMUNOL & BIOL, NAT CANCER INST, 85- *Mem:* Am Asn Cancer Res; Am Soc Cell Biol; Am Soc Microbiol; Sigma Xi. *Res:* Genetic, molecular and cellular mechanisms by which mammary epithelial cells functionally differentiate and the relationship of these mechanisms to malignant transformation of mammary cells by various carcinogenic stimuli; identification and characterization of mammary specific epithelial stem cells. *Mailing Add:* Lab Tumor Immunol & Biol Bldg Ten Rm 8B07 Nat Cancer Inst Bethesda MD 20892

SMITH, GLENN EDWARD, b Charleston, WVa, Mar 26, 23; m 54; c 4. PLANT PATHOLOGY. *Educ:* Morris Harvey Col, BS, 52; Ohio State Univ, MS, 54, PhD(bot & plant path), 60. *Prof Exp:* Asst prof bot & plant path, Ohio State Univ, 57-67; prof biology, Univ Charleston, 67- *Mem:* Am Phytopath Soc; Soc Nematol. *Res:* Phytonematology. *Mailing Add:* Dept Natural Sci Univ Charleston 2300 MacCorkle S East Charleston WV 25304

SMITH, GLENN S, b Huron, SDak, Aug 4, 52. DRILLING EXPLORATION. *Educ:* Colo State Univ, BS, 74; Ga Inst Technol, MSEE, 75. *Prof Exp:* Sr develop engr, Schlumberger Well Serv, Tex, 75-81; PRES, SMITH ENERGY CORP, 81- *Mem:* Inst Elec & Electronics Engrs. *Mailing Add:* Smith Energy Corp PO Box 907 Greeley CO 80631

SMITH, GLENN SANBORN, b Antler, NDak, Dec 21, 07; m 30; c 3. PLANT BREEDING. *Educ:* NDak Agr Col, BS, 29; Kans State Univ, MS, 31; Univ Minn, PhD(plant breeding, genetics), 47. *Hon Degrees:* DSc, ND State Univ, 90. *Prof Exp:* Jr agronomist, Bur Plant Indust, USDA, 29-35, asst agronomist, 35-42, from assoc agronomist to agronomist, Bur Plant Indust, Soils & Agr Eng, 42-47; assoc dean sch agr & assoc dir exp sta, 47-51, chief div plant indust, Exp Sta, 51-54, dean grad sch, 54-73, fac lectr, 65, prof, 47-78, EMER PROF AGRON, NDAK STATE UNIV, 78- *Concurrent Pos:* Consult wheat breeding, Ministry Agr, Repub Uruguay, 77-78 & agron curriculum, Fac Agr, Univ Repub, Uruguay, 79; Agron Deleg, China, 83. *Mem:* Sigma Xi; Crop Sci Soc Am; Fel Am Soc Agron. *Res:* Durum, hard red spring wheat, and oat breeding and genetics; pathology and quality problems. *Mailing Add:* 1115 N 14th St Fargo ND 58102

SMITH, GORDON MEADE, b Alva, Okla, June 21, 30; m 52; c 5. PHYSICAL CHEMISTRY. *Educ:* Okla State Univ, BS, 53, MS, 55; Univ Fla, PhD(chem), 58. *Prof Exp:* Asst chem, Univ Fla, 56-58; MEM STAFF, LOS ALAMOS SCI LAB, 58- *Res:* Ionospheric and atmospheric chemistry; active modification of the ionosphere; atmospheric nuclear weapon phenomenology. *Mailing Add:* 415 Estante Way Los Alamos NM 87544

SMITH, GORDON STUART, x-ray crystallography, for more information see previous edition

SMITH, GRAHAM MONRO, b Bayshore, NY, Nov 11, 47; m 71; c 2. COMPUTER GRAPHICS. *Educ:* Adelphi Univ, BA, 69; Univ Buffalo, PhD(chem), 74. *Prof Exp:* Fel theoret chem, Princeton Univ, 74-75; fel theoret chem, Univ Calif, Santa Cruz, 75-76; SR INVESTR THEORET CHEM, MERCK SHARP & DOHME RES LABS, 76- *Mem:* Am Chem Soc; AAAS; Asn Comput Mach. *Res:* Developing computational tools, including real time computing graphics, and applying them to structural problems in organic and medicinal chemistry. *Mailing Add:* 1210 Knollbrook Dr Lansdale PA 19446

SMITH, GRAHAME J C, b Kapuda, SAustralia, Feb 16, 42. INSECT ECOLOGY. *Educ:* Univ Adelaide, BS, 62; Cornell Univ, MS, 65, PhD(insect ecol), 67. *Prof Exp:* From instr to asst prof ecol, Brown Univ, 67-72; asst dean, Col Lib Arts, 75-76, ASSOC PROF ECOL, BOSTON UNIV, 72- *Mem:* Ecol Soc Am. *Res:* Behavior of insect parasitoids on different host species and factors which affect this behavior, particularly interactions with other individuals of the same parasitoid species. *Mailing Add:* 119 Cambridge Turnpike Lincoln MA 01773-1904

SMITH, GRANT GILL, b Fielding, Utah, Sept 25, 21; m 46; c 6. ORGANIC CHEMISTRY. *Educ:* Univ Utah, BA, 43; Univ Minn, PhD(org chem), 49. *Prof Exp:* Asst chem, Univ Minn, 43-44, 46-48, actg instr, 48-49; from instr to assoc prof, Wash State Univ, 49-61; assoc prof, 61-63, faculty honors lectr, 67, PROF CHEM, UTAH STATE UNIV, 63- *Concurrent Pos:* Researcher, Univ London, 57-58; NIH sr fel, Stanford Univ, 69-70. *Honors & Awards:* Utah Award, Am Chem Soc, 77. *Mem:* AAAS; Am Chem Soc; Royal Soc Chem; Int Soc for Study Origin Life; Am Soc Mass Spectros. *Res:* Physical organic chemistry; mechanisms of gas phase reactions; proximity effects in organic reactions; organic mass spectroscopy; organic geochemistry; chemometrics; biotechnology. *Mailing Add:* Dept Chem 03 - Biochem Utah State Univ Logan UT 84322-0300

SMITH, GRANT WARREN, II, b Kansas City, Mo, Jan 21, 41; m 62; c 1. ETHNOPHARMACOLOGY. *Educ:* Grinnell Col, BA, 62; Cornell Univ, PhD(chem), 66. *Prof Exp:* Asst chem, Grinnell Col, 62; teaching asst, Cornell Univ, 62-63, asst prof, 66-68; head dept chem, Univ Alaska, Fairbanks, 68-73, actg head dept gen sci, 72-73, assoc prof chem, 68-77, prof 77-78; dean, Sch Sci & Technol & prof chem, Univ Houston, Clear Lake City, 78-84; vpres acad affairs, 84-86, PRES & PROF CHEM, SOUTHEASTERN LA UNIV, 86- *Concurrent Pos:* DuPont fel, Cornell Univ, 67; vis prof, Cornell Univ, 73-74; Am Coun Educ fel, Acad Admin Internship Prog, 73-74; pres, Statewide Assembly, Univ Alaska Syst, 76-77. *Mem:* AAAS; Am Chem Soc; Am Asn Higher Educ; Sigma Xi; Soc Econ Bot; Am Asn Univ Adminr; Am Soc Pharmacog. *Res:* Organic photochemistry of unsaturated molecules and arctic water pollutants; ethnopharmacology and chemistry of arctic natural products. *Mailing Add:* Southestern La Univ PO Box 784-Univ Sta Hammond LA 70402

SMITH, H VERNON, JR, Univ Tex, Austin, BES, 64; Univ Ill, MS, 65; Univ Wis-Madison, PhD(physics), 71; m 64; c 1. ION SOURCE PHYSICS, ACCELERATOR PHYSICS. *Educ:* Univ Tex, Austin, BES, 64; Univ Ill, Urbana, MS, 65; Univ Wis-Madison, PhD(physics), 71. *Prof Exp:* Asst prof physics, Prairie View A&M Univ, 70-71; postdoctoral, Rice Univ, 71; asst scientist physics & nuclear eng, Univ Wis-Madison, 71-78; STAFF MEM ACCELERATOR TECHNOL, LOS ALAMOS NAT LAB, 78- *Mem:* Am Phys Soc; Nuclear & Plasma Sci Soc. *Res:* Ion source; diagnostics of ion source plasmas; diagnostics of particle beams extracted from ion sources. *Mailing Add:* Los Alamos Nat Lab PO Box 1663 Mail Stop H818 Los Alamos NM 87545

SMITH, HADLEY J(AMES), b Detroit, Mich, May 5, 18; m 52; c 6. ENGINEERING MECHANICS. *Educ:* Univ Mich, BS, 40, PhD(eng mech), 57. *Prof Exp:* Jr engr, Res Lab, Detroit Edison Co, 40-41; prod engr, Com Res Labs, Inc, 46-51; res engr, Res Inst, 52-55, from instr to assoc prof, 55-62, PROF ENG MECH, UNIV MICH, ANN ARBOR, 62- *Concurrent Pos:* Fac res fel, Rackham Sch Grad Studies, 58; NSF fel, Harvard Univ, 59; guest scientist, Los Alamos Sci Lab, 74. *Mem:* Am Phys Soc; Am Soc Mech Engrs. *Res:* Mathematical physics; hydraulics; hydrodynamics; thermodynamics; kinetic theory; statistical mechanics; heat and mass transfer. *Mailing Add:* 2001 Hall Ave Ann Arbor MI 48104

SMITH, HAL LESLIE, b Cedar Rapids, Iowa, Mar 18, 47; m 70. MATHEMATICS. *Educ:* Univ Iowa, BA, 69, PhD(math), 76. *Prof Exp:* Instr math, Univ Utah, 76-79; from asst prof to assoc prof, 79-87, PROF, ARIZ STATE UNIV, 87- *Mem:* Am Math Soc; Soc Indust & Appl Math; AAAS. *Res:* Differential equations; biomathematics. *Mailing Add:* Dept Math Ariz State Univ Tempe AZ 85287

SMITH, HAMILTON OTHANEL, b New York, NY, Aug 23, 31; m 57; c 5. MICROBIAL GENETICS. *Educ:* Univ Calif, Berkeley, AB, 52; Johns Hopkins Univ, MD, 56. *Prof Exp:* Intern, Barnes Hosp, St Louis, 56-57; res, Henry Ford Hosp, Detroit, 59-62; USPHS res fel microbial genetics, Univ Mich, 62-64, res assoc, 64-67; from asst prof to assoc prof microbiol, 67-73, prof, 73-81, PROF MOLECULAR BIOL & GENETICS, SCH MED, JOHNS HOPKINS UNIV, 81- *Concurrent Pos:* Guggenheim fel, 75-76; ed, Gene, 76- *Honors & Awards:* Nobel Prize Med, 78. *Mem:* Nat Acad Sci; Am Soc Microbiol; Am Soc Biol Chemists; AAAS; fel Am Acad Arts & Sci. *Res:* Genetic recombination; biochemistry of DNA recombination and DNA methylation and restriction. *Mailing Add:* Dept Microbiol Sch Med Johns Hopkins Univ Baltimore MD 21205

SMITH, HARDING EUGENE, b San Jose, Calif, May 10, 47; m 89. OBSERVATIONAL COSMOLOGY, ACTIVE GALAXIES & QUASARS. *Educ:* Calif Inst Technol, BS, 69; Univ Calif, Berkeley, MA, 72, PhD(astron), 74. *Prof Exp:* Asst res physicist, 74-78, from asst prof to assoc prof, 78-86, PROF PHYSICS, UNIV CALIF, SAN DIEGO, 86- *Concurrent Pos:* Consult, Cosmos Telecourse, Univ Calif, San Diego Sci Teacher Insts & other sci educ; chair, Keck Telescope Sci Steering Comt, 84-86; Morrison fel, Lick Observ, Univ Calif, Santa Cruz, 88; vis scholar, Inst Astron, Cambridge, 89; vis assoc physics, Calif Inst Technol, 89- *Mem:* Am Astron Soc; Int Astron Union; Astron Soc Pac; AAAS. *Res:* Nature of quasar emission lines and their relation to the formation of galaxies; chemical and physical evolution of galaxies; physics of active galactic nuclei. *Mailing Add:* Ctr Astrophys Univ Calif San Diego La Jolla CA 92093-0111

SMITH, HARLAN EUGENE, b Farmington, Ark, Oct 27, 20; m 48; c 2. PLANT PATHOLOGY. *Educ:* Univ Ark, BS, 49; Univ Wis, PhD(plant path, bot), 52. *Prof Exp:* Lab asst bact, Univ Ark, 47-48; asst, Univ Wis, 48-52; exten plant pathologist, Univ Ark, 52-55; exten plant pathologist, Tex A&M Univ, 55-62; plant pathologist, fed exten serv, USDA, 62-85, coordr, plant health exten prog, 72-85; RETIRED. *Concurrent Pos:* Founder, Smith's Plant Health Co, 85-; independent consult, tree, turf, plant health prob, Northern Va, 84- *Mem:* Am Phytopath Soc; Soc Nematol; Int Soc Arboricult. *Res:* Planning and developing extension youth and adult education programs in science and arts of plant pathology, especially plant disease control; air pollution injury to plants. *Mailing Add:* 1900 Country Club Dr Titusville FL 32780-5314

SMITH, HARLAN J, b Wheeling, WVa, Aug 25, 24; m 50; c 4. ASTRONOMY. *Educ:* Harvard Univ, BA, 49, MA, 51, PhD(astron), 55. *Hon Degrees:* Dr, Nicholas Copernicus Univ, Torun, Poland, 73, Denison Univ, Granville, Ohio, 83. *Prof Exp:* Instr astron, Observ, Yale Univ, 53-57, from asst prof to assoc prof, 57-63; chmn dept, 63-78, dir McDonald Observ, 63-89, PROF ASTRON, UNIV TEX, AUSTIN, 63- *Concurrent Pos:* Mem space sci bd, Nat Res Coun, 77-80, chmn comt astron & astrophysics, space sci bd; Astron Union & Int Sci Radio Union; co-ed, Astron J, Am Astron Soc, 58-63; chmn bd dirs, Asn Univs Res Astron, 80-82. *Mem:* Am Astron Soc (actg secy, 61-62, vpres, 77-80); Am Geophys Union; Royal Astron Soc. *Res:* Variable stars; planets; quasars. *Mailing Add:* McDonald Observ, Astronomy Dept Univ Tex-Austin Austin TX 78712

SMITH, HARLAN MILLARD, b Iowa City, Iowa, Sept 2, 21; m 54; c 3. PHYSICAL CHEMISTRY. *Educ:* Carroll Col, BA, 42; Univ Chicago, PhD(phys chem), 49. *Prof Exp:* From res chemist to head, fertilizer res sect, Exxon Res & Eng Co, Exxon Chem Co, 49-65, proj develop adv, 65-79, sr res assoc, 79-85; RETIRED. *Concurrent Pos:* Consult, 85-87. *Mem:* Am Chem Soc. *Res:* Raman spectra of aqueous sulfuric acid solutions; lubricating oil additives; industrial lubricants; fertilizers; pesticides. *Mailing Add:* 66 Cray Terr Fanwood NJ 07023

SMITH, HAROLD CARTER, b Statesboro, Ga, Aug 13, 20; m 49; c 2. BIOCHEMISTRY. *Educ:* Ga Southern Col, BS, 60; Univ NC, PhD(biochem), 64. *Prof Exp:* Dir labs, Evans County Heart Res Proj, Claxton, Ga, 59-60; res fel, McArdle Lab Cancer Res, Sch Med, Univ Wis, 64-67; dir biochem sect, Surg Biol Lab & asst prof biochem, Sch Med, Univ NC, Chapel Hill, 67-69; chemist, Res Dept, R J Reynolds Tobacco Co, 69-70; asst prof, 70-82, EMER LECTR, SURG & BIOCHEM, SCH MED, UNIV NC, CHAPEL HILL, 82- *Mem:* AAAS; Am Chem Soc; NY Acad Sci. *Res:* Chemical carcinogenesis; enzymes and steroids in breast and thyroid cancer; iodoamino acids; biochemistry of wound healing. *Mailing Add:* 744 Tinkerbell Rd Chapel Hill NC 27514-3015

SMITH, HAROLD GLENN, b Lafayette, La, July 3, 27; m 50; c 3. LATTICE DYNAMICS, SUPERCONDUCTIVITY. *Educ:* Univ Southwestern La, BS, 49; Tulane Univ, MS, 51; Iowa State Univ, PhD(physics), 57. *Prof Exp:* Jr res assoc physics, Iowa State Univ, 51-54, asst, 54-57; PHYSICIST, OAK RIDGE NAT LAB, 57- *Concurrent Pos:* Guest scientist, AERE, Harwell Eng, 61-62, Laue-Langeuin Inst, Genoble, France, 74-75. *Mem:* Fel Am Phys Soc; Sigma Xi; Am Crystallog Asn. *Res:* Neutron diffraction; x-ray crystallography; atomic, molecular, and solid state physics; martensitic transformations. *Mailing Add:* 103 Walton Lane Oak Ridge TN 37830

SMITH, HAROLD HILL, b Arlington, NJ, Apr 24, 10; m 39; c 4. GENETICS. *Educ:* Rutgers Univ, BS, 31; Harvard Univ, AM, 34, PhD(genetics), 36. *Prof Exp:* Asst, Dept Genetics, Carnegie Inst, 31-32; asst bot, Harvard Univ, 34-35; asst geneticist, Bur Plant Indust, USDA, 35-43; from assoc prof to prof plant genetics, Cornell Univ, 46-57; sr geneticist, Brookhaven Nat Lab, 55-78; RETIRED. *Concurrent Pos:* Consult, Chem Corps, US Dept Army, Md, 47-50; Guggenheim fel, 52; Fulbright lectr, Amsterdam, 53; sr scientist, Int Atomic Energy Agency, Vienna, 58-59; vis prof, Univ Calif, Berkeley, 66 & Univ Buenos Aires, 66; hon res assoc, Univ Col, London, 66; acad sci exchangee, Romania, 70; adj prof, NY Univ, 77-; consult, Brookhaven Nat Lab, 78. *Mem:* AAAS; Genetics Soc Am; Tissue Cult Asn; Soc Develop Biol; Am Genetic Asn (pres, 76). *Res:* Plant cytogenetics; experimental evolution; mutagenesis; radiation genetics; plant tumors; genetic control of differentiation; plant cell genetics. *Mailing Add:* Tower Hill Rd PO Box 278 Shoreham NY 11786

SMITH, HAROLD LINWOOD, b Richmond, Va, Dec 7, 27; m 49; c 2. PHYSICAL PHARMACY, BIOPHARMACEUTICS & PHARMACOKINETICS. *Educ:* Med Col Va, BS, 56, PhD(pharm chem), 62. *Prof Exp:* Lederle fel, Univ Mich, 61-63; chemist, Lederle Labs, Am Cyanamid, 63-64, proj leader, 64-68, group leader, 68; asst prof, 68-76, ASSOC PROF PHARM, MED COL VA, VA COMMONWEALTH UNIV, 76- *Concurrent Pos:* Dir opers, Bioclid Inc, 89- *Honors & Awards:* Lunsford Richardson Pharm Award, Merrell-Nat Labs, Richardson-Merrell Inc, 60. *Mem:* Am Pharmaceut Asn; Acad Pharmaceut Sci. *Res:* Rheology of pharmaceutical and biological systems; protein binding; solution theory; solid dissolution rate studies; pharmaceutic dosage form studies. *Mailing Add:* Med Col Va Sch Pharm Va Commonwealth Univ Richmond VA 23295

SMITH, HAROLD W(OOD), b Brookfield, Mo, Feb 8, 23; m 42; c 1. ELECTRICAL ENGINEERING. *Educ:* Univ Tex, BS, 44, MS, 49, PhD(elec eng), 54. *Prof Exp:* Asst prof, 46-58, PROF ELEC ENG, UNIV TEX, AUSTIN, 58-, DIR GEOMAGNETICS LAB, 66- *Mem:* Inst Elec & Electronics Engrs; Am Geophys Union. *Res:* Geomagnetics; electrical geoscience; information science. *Mailing Add:* Dept Elec Eng Univ Tex Austin TX 78712

SMITH, HAROLD WILLIAM, b Toronto, Ont, Aug 15, 28; m 50; c 2. CONTROL ENGINEERING. *Educ:* Univ Toronto, BASc, 50; Mass Inst Technol, ScD(instrumentation), 61. *Prof Exp:* Assoc prof, 66-69, PROF ELEC ENG, UNIV TORONTO, 69- *Concurrent Pos:* Mem assoc comt automatic control, Nat Res Coun Can, 64-, chmn, 70-; consult, Falconbridge Nickel Mines Ltd, 67-; mem reactor control comt, Atomic Energy Control Bd Can, 72- *Res:* Modeling and control of industrial processes; multivariable systems; metallurgical applications. *Mailing Add:* Dept Elec Eng Univ Toronto 10 Kings Col Rd Toronto ON M5S 1A4 Can

SMITH, HARRY ANDREW, b Grand Rapids, Mich, Aug 29, 33; m 56; c 3. ORGANIC CHEMISTRY, POLYMER CHEMISTRY. *Educ:* Univ Mich, BS, 55, MS, 57, PhD(org chem), 60. *Prof Exp:* Chemist, Polymer Res Lab, 60-62, res chemist, 62-66, Sci Proj Lab, 66-68, sr res chemist, Chem Lab, 68-71 & Org Chem Prod Res Lab, 71-74, res specialist II, Org Chem Res Lab, 74-76, res assoc, Designed Polymers & Chem Lab, 76-84, res assoc, Saran & Convented Prod, 84-87, RES ASSOC CONSUMER PROD LAB, DOW CHEM CO, 87- *Concurrent Pos:* fel, NSF, 56, 57. *Mem:* AAAS; Am Chem Soc; Sigma Xi; Soc Cosmetic Chemists; Cosmetic Toiletry & Fragrance Asn. *Res:* Condensation and ring opening polymerization, including preparation and characterization of phenylene sulfide polymers, carbonyl polymers, carbonyl-epoxide copolymers, phenolic resins, solvents and solvency; formulated consumer products. *Mailing Add:* 4608 James Dr Midland MI 48640

SMITH, HARRY FRANCIS, b Sioux City, Iowa, Mar 29, 41; m 65, 86; c 6. RING THEORY. *Educ:* Univ Calif, Berkeley, BA, 67; Univ Iowa, MS, 69, PhD(math), 72. *Prof Exp:* Asst prof math, Univ Iowa, 72-73311 asst prof math, Madison Col, 73-77; from asst prof to assoc prof, 77-85, prof math, Iowa State Univ, 85-87; LECTR, UNIV NEW ENG, ARMIDALE, NSW, AUSTRALIA, 88- *Concurrent Pos:* Nat Acad Sci exchange scientist, Steklov Inst Math-Acad Sci USSR & Moscow State Pedag Inst, 82-83; Fulbright lectr, Univ Philippines & Ateneo Univ, Manila, 85-86; vis sr lectr, Univ Papua New Guinea, 87. *Mem:* Am Math Soc; Australian Math Soc. *Res:* Nonassociative algebra, alternative rings and their generalizations. *Mailing Add:* Dept Math Univ New Eng Armidale 2351 Australia

SMITH, HARRY JOHN, b Arundel, Quebec, Sept 3, 27; wid; c 2. VETERINARY PARASITOLOGY, MEDICAL ENTOMOLOGY. *Educ:* McGill Univ, BSc, 54; Ont Vet Col, DVM, 58; Univ Toronto, MVSc, 60. *Prof Exp:* Res officer parasitol, Agr Can, 58-62, vet, 62-65, res scientist, 65-90; RETIRED. *Concurrent Pos:* Hon lectr, Mt Allison Univ, 72-77; vis prof, AVC, 91. *Mem:* Am Asn Vet Parasitologist; World Asn Advan Vet Parasitol; Wildlife Dis Asn; Can Asn Advan Vet Parisitol. *Res:* Epidemiology and ecology of livestock gastrointestinal helminths with special emphasis on parasitic gastroenteritis; ecology and diagnosis of trichinosis. *Mailing Add:* Nine Weldon St PO Box 184 Sackville NB E0A 3C0 Can

SMITH, HARRY LOGAN, JR, b Philadelphia, Pa, June 4, 30; m 53; c 5. MICROBIOLOGY. *Educ:* Temple Univ, AB, 52; Jefferson Med Col, MS, 54, PhD, 57. *Prof Exp:* Asst, 53-57, from instr to assoc prof, 57-74, PROF MICROBIOL, JEFFERSON MED COL, 74- *Concurrent Pos:* NIH res career develop award, 62-66; consult cholera, WHO, 66. *Mem:* Am Soc Microbiol. *Res:* Vibrios; cholera. *Mailing Add:* Jefferson Med Col 1020 Locust St Philadelphia PA 19107

SMITH, HARVEY ALVIN, b Easton, Pa, Jan 30, 32; m 55; c 3. MATHEMATICS, PHYSICS. *Educ:* Lehigh Univ, BS, 52; Univ Pa, MS, 55, AM, 58, PhD(math), 64. *Prof Exp:* Physicist, Opers Res Div, Fire Control Instrument Group, Frankford Arsenal, 52-54; engr, Radio Corp Am, 54-57; sr systs analyst, Remington Rand Univac Div, Sperry Rand Corp, 57-58; mem tech staff, Auerbach Electronics Corp, 58-59; instr math, Drexel Inst, 59-60, asst prof, 60-64; NSF sci fel, Univ Pa, 64-65; mem tech staff weapons systs eval group, Inst Defense Analysis, DC, 65-66; from assoc prof to prof math, Oakland Univ, 66-77; chmn dept, 77-82, PROF MATH, ARIZ STATE UNIV, 82- *Concurrent Pos:* Consult, Ford Found proj measurement of delinquency, Dept Sociol, Univ Pa, 63-64, US Army Security Agency, 67-68 & Inst Defense Analysis, 67-69; consult, Exec Off of President, 68-73, dep chief systs eval, Off Emergency Preparedness; consult, US Arms Control & Disarmament Agency, 73-79 & Los Alamos Nat Lab, 80- *Honors & Awards:* Exec Off of President Meritorious Serv Award. *Mem:* Am Math Soc; Soc Indust & Appl Math; Sigma Xi. *Res:* Functional analysis; applied mathematics; systems analysis; operations research; representations of locally compact groups; strategic policy studies; twisted group algebras; integral operators. *Mailing Add:* Dept Math Ariz State Univ Tempe AZ 85287

SMITH, HASTINGS ALEXANDER, JR, b Lexington, Ky, Apr 20, 43; m 65; c 2. NUCLEAR PHYSICS. *Educ:* Purdue Univ, BS, 65, MS, 67, PhD(nuclear physics), 70. *Prof Exp:* Appointee nuclear physics, Los Alamos Sci Lab, Univ Calif, 70-72; from asst prof to assoc prof physics, Ind Univ, 72-78; STAFF SCIENTIST, LOS ALAMOS NAT LAB, UNIV CALIF, 78- *Mem:* Am Phys Soc; AAAS; Am Asn Physics Teachers. *Res:* Intermediate and low-energy nuclear physics; gamma-ray and beta-ray spectroscopy; nuclear reactions; nuclei far from stability; non-destructive assay of special nuclear materials. *Mailing Add:* 760 Los Pueblos Los Alamos NM 87544

SMITH, HAYWOOD CLARK, JR, b Raleigh, NC, Oct 11, 45; m 69; c 2. DYNAMICAL ASTRONOMY, STATISTICAL ASTRONOMY. *Educ:* Univ NC, Chapel Hill, AB, 67; Univ Va, MA, 69, PhD(astron), 72. *Prof Exp:* Vis asst prof, 72-78, asst prof, 78-79, ASSOC PROF ASTRON, UNIV FLA, 79- *Mem:* Int Astron Union; Am Astron Soc. *Res:* Dynamical evolution of clusters of stars and galaxies; statistical astronomy; especially absolute magnitude calibration using astrometric data. *Mailing Add:* Dept Astron 211 Space Sci Res Bldg Univ Fla Gainesville FL 32611

SMITH, HELENE SHEILA, b Philadelphia, Pa, Feb 13, 41; m 62; c 1. MOLECULAR BIOLOGY. *Educ:* Univ Pa, BS, 62, PhD(microbiol), 67. *Prof Exp:* Asst res prof virol, Univ Calif, Berkeley, 71-75, assoc res prof, 75-77; staff researcher biol, Donner Lab, Lawrence Berkeley Lab, 77-82; ASST DIR, PERALTA CANCER RES INST, 80- *Concurrent Pos:* Mem, Grad Group Genetics, Univ Calif, Berkeley, 79-82, Cell Biol Panel, NSF, 80-84; adj assoc prof, Univ Calif, San Francisco, 84- *Mem:* Am Asn Cancer Res; Am Asn Cell Biol; Tissue Cult Asn; Soc Anal Cytol. *Res:* Biology of human mammary epithelial cells in culture, and the use of these cells to study radiation induced survival and carcinogenesis, chemotherapeutic drug sensitivity, and tumor heterogeneity. *Mailing Add:* Geraldine Bruch Cancer Res Ctr 2330 Clay St San Francisco CA 94115

SMITH, HENRY I, ELECTRICAL ENGINEERING. *Prof Exp:* Engr, Lincoln Lab, 68-80, mgr, 77-80, PROF ELEC ENG, MASS INST TECHNOL, 80- *Concurrent Pos:* Adj prof, Submicron Struct Lab, Mass Inst Technol, 77-80; vis scientist, Univ Col, London, 72, Thompson CSF, Paris, 74, Norweg Inst Technol, Trondheim, Norway, 76, Nippon Tel & Tel Corp, Atsugi, Japan, 90 & Univ Glasgow, 90. *Mem:* Nat Acad Eng; Am Phys Soc; Am Vacuum Soc; Mat Res Soc; Sigma Xi; fel Inst Elec & Electronics Engrs. *Res:* Submicron structures; nanofabrication; methods for preparing semiconductor-on-insulator films; electronic devices; quantum effects in sub-100 nm structures. *Mailing Add:* Mass Inst Technol 77 Massachusetts Ave Rm 39-427 Cambridge MA 02139

SMITH, HERBERT L, b Mayport, Pa, June 28, 29; m 52; c 4. INORGANIC CHEMISTRY. *Educ:* Univ Pittsburgh, BSEd, 53, MLitt, 57, PhD(inorg chem), 65. *Prof Exp:* Jr fel glass sci, Mellon Inst, 54-65; chmn dept, 71-79, PROF CHEM, SLIPPERY ROCK STATE COL, 65- *Mem:* Am Chem Soc; Sigma Xi. *Res:* Effects of irradiation on glasses and crystals; effects of electrolytes on the circular dichroism of coordination compounds. *Mailing Add:* Dept Chem Slippery Rock Univ Slippery Rock PA 16057-1326

SMITH, HOBART MUIR, b Stanwood, Iowa, Sept 26, 12; wid; c 2. VERTEBRATE ZOOLOGY. *Educ:* Kans State Col, BS, 32; Univ Kans, AM, 33, PhD(zool), 36. *Prof Exp:* Nat Res Coun fel biol, Univ Mich, 36-37; asst, Chicago Acad Sci, 37-38 & Chicago Mus Natural Hist, 38; Bacon traveling scholar, Smithsonian Inst, 38-41; instr zool, Univ Rochester, 41-45; asst prof comp anat, Univ Kans, 45-46; assoc prof, Agr & Mech Col, Tex, 46-47; from asst prof to prof comp anat & herpet, Univ Ill, Urbana, 47-68; chmn dept environ, pop & organismic biol, 70-74 & 78-79, prof, 68-83, EMER PROF ENVIRON, POP & ORGANISMIC BIOL, UNIV COLO, BOULDER, 83- *Mem:* Am Soc Ichthyologists & Herpetologists (vpres, 37); Soc Study Amphibians & Reptiles; Soc Syst Zool (pres, 65); Herpetologists League (pres, 47-59). *Res:* Herpetology; principles of taxonomy; zoogeography; comparative anatomy. *Mailing Add:* Dept of Environ Pop & Org Biol Univ of Colo Boulder CO 80309-0334

SMITH, HOMER ALVIN, JR, b Houston, Tex, Feb 23, 32; m 59; c 2. ORGANOMETALLIC CHEMISTRY, MEDICINAL CHEMISTRY. *Educ:* Rice Univ, BA, 53; Okla State Univ, PhD(chem), 61. *Prof Exp:* Asst prof chem, Tarkio Col, 61-64; from assoc prof to prof chem, Hampden-Sydney Col, 64-85, chmn dept, 66-68, 74-76 & 80-81; PROF CHEM & CHMN DEPT, MILLIKIN UNIV, 85- *Concurrent Pos:* NSF sci fac fel, Duke Univ, 68-69 & Ind Univ, 76-78. *Mem:* Am Chem Soc; Royal Soc Chem; Int Union Pure & Appl Chem. *Res:* syntheses involving organometallic intermediates and strong base systems; synthesis of potential medicinals. *Mailing Add:* Dept Chem Millikin Univ Decatur IL 62522

SMITH, HORACE VERNON, JR, b Rockford, Ill, July 23, 42; m 64; c 1. ACCELERATOR PHYSICS. *Educ:* Univ Tex, Austin, BES, 64; Univ Ill, Urbana, MS, 65; Univ Wis-Madison, PhD(physics), 71. *Prof Exp:* Asst prof physics, Prairie View Agr & Mech Col, 70-71; res assoc, 71-74, asst scientist nuclear eng & physics, Univ Wis-Madison, 74-78; STAFF MEM, LOS ALAMOS NAT LAB, 78- *Concurrent Pos:* Consult. *Mem:* Am Phys Soc; Inst Elec & Electronics Engrs. *Res:* Ion source development. *Mailing Add:* Los Alamos Nat Lab AT-10 MSH 818 PO Box 1663 Los Alamos NM 87545

SMITH, HOWARD DUANE, b Fillmore, Utah, June 25, 41; m 61; c 2. MAMMALOGY, ECOLOGY. *Educ:* Brigham Young Univ, BS, 63, MS, 66; Univ Ill, Urbana, PhD(vert ecol), 69. *Prof Exp:* Instr biol, Univ Ill, 68-69; from asst prof to assoc prof, 69-80, PROF ZOOL, BRIGHAM YOUNG UNIV, 81- *Concurrent Pos:* Collabr, Intermountain Forest & Range Exp Sta, US Forest Serv, 65-; NSF fels, Brigham Young Univ, 69-; consult, Wilderness Assocs, 74-; mem, Am Mus Natural Hist; consult terrestrial wildlife. *Mem:* AAAS; Am Soc Mammal; Wildlife Soc; Ecol Soc Am. *Res:* Small mammal populations; demography; bioenergetics; environmental impact studies on man; environmental impact of coal generating power plants or pesticide applications on the biota; wildlife biology. *Mailing Add:* 163 Widtsoe Bldg Brigham Young Univ Provo UT 84602

SMITH, HOWARD E, b San Francisco, Calif, Aug 1, 25; m 60; c 3. ORGANIC CHEMISTRY. *Educ:* Univ Calif, BS, 51; Stanford Univ, MS, 54, PhD(chem), 57. *Prof Exp:* Asst res chemist, Calif Res Corp, Stand Oil Co, Calif, 51-52; res assoc, Stanford Univ, 56; USPHS fel, Wayne State Univ, 56-58; fel, Swiss Fed Inst Technol, Zurich, 58-59; from asst prof to assoc prof chem, 59-71, PROF CHEM, VANDERBILT UNIV, 71- *Mem:* AAAS; Am Chem Soc; The Chem Soc. *Res:* Natural products; stereochemistry. *Mailing Add:* Dept Chem Vanderbilt Univ Nashville TN 37235

SMITH, HOWARD EDWIN, b Dayton, Ohio, Nov 9, 23; m 46; c 2. MECHANICAL ENGINEERING, AEROSPACE ENGINEERING. *Educ:* Univ Dayton, BME, 51; Univ Cincinnati, MS, 61, PhD, 69. *Prof Exp:* Tool designer, Master Elec Co, 41-43, 46-47, process engr, 51-57; from instr to prof mech eng, Univ Dayton, 57-85, chmn dept, 66-85; RETIRED. *Concurrent Pos:* Consult, Aerospace Res Labs, Wright Patterson AFB, 64-66. *Mem:* Am Soc Mech Engrs; Am Inst Aeronaut & Astronaut; Am Soc Eng Educ. *Res:* Experimental and theoretical analysis of the flow field and heat transfer downstream of a rearward-facing step in supersonic flow of air. *Mailing Add:* 4487 Lotz Rd Dayton OH 45429

SMITH, HOWARD JOHN TREWEEK, b Hornchurch, Eng, June 21, 37; m 63; c 2. LOW TEMPERATURE PHYSICS. *Educ:* Univ London, BSc, 58, PhD(physics), 61. *Prof Exp:* Res physicist, Petrocarbon Develop Ltd, Eng, 61-64; res physicist, Ferranti Electronics Ltd, Ont, 64; from asst prof to assoc prof, 64-85, PROF PHYSICS, UNIV WATERLOO, 85- *Res:* Superconductivity tunneling; far infrared spectroscopy; low temperature heat engines. *Mailing Add:* Dept Physics Univ Waterloo 298 Criagleith Waterloo ON N2L 3G1 Can

SMITH, HOWARD LEROY, b Eldorado, Kans, Nov 12, 24; m 47; c 5. ORGANIC CHEMISTRY. *Educ:* Univ Calif, BS, 48; Mass Inst Technol, PhD, 51. *Prof Exp:* Org chemist, Jackson Lab, 51-55, div head, 55-59, asst lab dir, 59-60, res dir, 60-65, asst gen supt process dept, Chambers Works, 65-66, supt miscellaneous intermediates area, 66-67, asst gen supt dyes & chem dept, 67-68, asst works mgr, 68-70, works mgr, 70-71, dir mfg serv, Org Chem Dept, 71-75, dir mfg, 75-76, DIR, EQUIP & MAGNETIC PROD, PHOTO PROD DEPT, E I DU PONT DE NEMOURS & CO, INC, 76- *Mem:* Am Chem Soc. *Res:* Dyes and textile chemicals; fluorocarbon chemistry; petroleum chemicals. *Mailing Add:* RD 1 Box 272 Landenberg PA 19350-9801

SMITH, HOWARD WESLEY, b New York, NY, Nov 24, 29. AEROSPACE ENGINEERING. *Educ:* Wichita State Univ, BS, 51, MS, 58; Okla State Univ, PhD(aerospace eng), 68. *Prof Exp:* Jr engr, Boeing Co, Kans, 50-51, stress analyst, 52-55, struct engr, 56-58, group supvr, 59-63, struct res mgr, 65-68, mem hq staff, Seattle, Wash, 69-70; PROF AEROSPACE STRUCT, UNIV KANS, 70- *Concurrent Pos:* Geront fel, Univ Kans, 80. *Honors & Awards:* Tasker Howard Bliss Medal, Soc Am Mil Engrs, 74; Space Shuttle Plaque, Am Inst Aeronaut & Astronaut, 84. *Mem:* Am Inst Aeronaut & Astronaut; Soc Exp Stress Analysis; Am Soc Eng Educ; Soc Am Military Engrs; Soc Advan Mat & Process Eng. *Res:* Aircraft loads; stresses; materials; composites; crashworthiness; cost models; optimization; teaching methods; biomechanics of bone. *Mailing Add:* Dept of Aerospace Eng 2003 Learned Hall Univ of Kans Lawrence KS 66045

SMITH, HUGO DUNLAP, b Natick, Mass, Nov 28, 23; m 49; c 5. PEDIATRICS. *Educ:* Yale Univ, BS, 44; Harvard Univ, MD, 47; Am Bd Pediat, dipl, 52. *Prof Exp:* Intern, Children's Hosp, Boston, 47-48; resident, Hosp Univ Pa, 48-49; resident & chief resident, Children's Hosp, Cincinnati, 49-52, asst chief staff & med dir clins, 54-69; PROF PEDIAT & ASSOC DEAN CURRIC, SCH MED, TEMPLE UNIV, 70- *Concurrent Pos:* From instr to prof pediat, Univ Cincinnati, 50-69; consult, US Air Force Hosp, Wright Patterson AFB, 58-66; ed, Am J Dis Children, 63-72. *Mem:* Am Pediat Soc; Am Diabetes Asn; Am Acad Pediat; Ambulatory Pediat Asn. *Res:* Medical education, curricula and teaching methodology; clinical and emotional aspects of juvenile diabetes mellitus. *Mailing Add:* 526 E Evergreen Ave Temple Univ Sch Med 3400 N Broad St Philadelphia PA 19118

SMITH, IAN CORMACK PALMER, b Winnipeg, Man, Sept 23, 39; m 65; c 4. MEDICAL PHYSICS. *Educ:* Univ Manitoba, BSc, 61, MSc, 62; Cambridge Univ, PhD(theoret chem), 65. *Hon Degrees:* FilDr, Stockholm, 86; DSc, Winnipeg, 90. *Prof Exp:* NATO fel, Stanford Univ, 65-66; mem res staff, Bell Tel Labs, 66-67; res officer, 67-87, DIR GEN, NAT RES COUN CAN, 87- *Concurrent Pos:* Consult, Bell Tel Labs, 68-70, CPC Int, 75-85 & Smith Kline & French, 79-82; adj prof chem, Carleton Univ, 73-90; adj prof biophys, Univ Ill, Chicago, 75-80; adj prof chem & biochem, Univ Ottawa, 77-; allied scientist, Ottawa Civic Hosp, 85-, Ottawa Gen Hosp, 87- & Ont Cancer Found, 88- *Honors & Awards:* Merck, Sharp & Dohme Award, Chem Inst Can, Labatt Award; Ayerst Award, Can Biochem Soc; Barringer Award, Can Spectros Soc, Herzberg Award; Organon Teknika Award, Can Soc Clin Chem. *Mem:* AAAS; fel Chem Inst Can; Am Chem Soc; Biophys Soc; Can Biochem Soc; fel Royal Soc Can. *Res:* Nuclear magnetic resonance, imaging and spectroscopy; optical spectroscopy; infrared spectroscopy application of these techniques to problems in molecular biology and medicine, especially biological membranes and cancer. *Mailing Add:* 14 Kindle Ct Ottawa ON K1J 6E2 Can

SMITH, IAN MACLEAN, b Glasgow, Scotland, May 21, 22; nat US; m 48; c 5. INFECTIOUS DISEASES, GERIATRIC INTERNAL MEDICINE. *Educ:* Glasgow Univ, MB, ChB, 44, MD, 57; FRCPG, 49; FRCPath, 76. *Prof Exp:* House physician internal med, Stobhill Hosp, Scotland, 44-45; clin asst, Royal infirmary, 47; registr path, Post-grad Med Sch, London, 47-48, house physician internal med, 48; tutor med path, Royal Hosp & Univ Sheffield, 48-49; fel internal med, Johns Hopkins Hosp, 49-51; asst resident, Wash Univ & Barnes Hosp, 51-53; asst prof, Rockefeller Inst & asst physician, Hosp, 53-55; chief infectious dis lab, Univ Hosps, Univ Iowa, 55-74, from asst prof to prof internal med, Col Med, 55-76; prof & chmn dept, Col Med, East Tenn State Univ, 76-78; PROF INTERNAL MED, COL MED, UNIV IOWA, 78- *Concurrent Pos:* Consult, Iowa State Dept Health, 58-76. *Mem:* Am Geriatric Soc. *Res:* Epidemiology and treatment of infectious diseases; epidemiology of elderly patients in acute care hospital. *Mailing Add:* Dept Med E419 GH Univ Iowa Hosp & Clin Iowa City IA 52242

SMITH, IEUAN TREVOR, b Bromley, Eng, Jan 11, 33; m 58; c 3. CHEMISTRY. *Educ:* Univ London, BSc, 54, MSc, 60. *Prof Exp:* Res chemist, Cray Valley Prod Ltd, Eng, 57-61; sr res officer, Paint Res Sta, 61-64; tech mgr, Epoxlite Ltd, 64-66; sr res chemist, Toni Co, 66-67, res supvr, 67-69, res supvr, 69-77, prin res assoc, 78-80, group leader, Gillette Res Inst, Gillette Co, 80-87; FREELANCE SCI & MED WRITER & ED, 87- *Res:* Properties and structure of polymers; polyelectrolyte behavior; surface and colloid chemistry; infrared spectroscopy, especially of surface species and proteins; keratin fibers, structure and properties; health, fitness, science and technology writing. *Mailing Add:* 11203 Lund Pl Kensington MD 20895

SMITH, ISAAC LITTON, b Russellville, Ala. ANALYTICAL CHEMISTRY. *Educ:* Florence State Univ, Ala, 61; Univ Ala, PhD(anal chem), 74. *Prof Exp:* Chemist, Reynolds Metals Co, 68-70; ANAL CHEMIST, ETHYL CORP, 74- *Concurrent Pos:* Mem res comt, Water Pollution Control Fedn, 74. *Mem:* Am Chem Soc; Water Pollution Control Fedn; Sigma Xi. *Res:* Analytical methods development; investigation of plant production problems; development and design of continuous monitor instrumentation. *Mailing Add:* Ethyl Corp PO Box 341 Baton Rouge LA 70821

SMITH, ISSAR, b New York, NY, Dec 4, 33; m 55; c 2. MOLECULAR BIOLOGY. *Educ:* City Col New York, BA, 55; Columbia Univ, MA, 57, PhD(biol), 61. *Prof Exp:* Fel, Sloan-Kettering Inst Cancer Res, 61-62; fel microbiol, Sch Med, NY Univ, 62-63; fel molecular biol, Albert Einstein Col Med, 63-64, res asst prof path, 64-67; assoc microbiol, 67-74, assoc mem, 74-79, MEM, PUB HEALTH RES INST CITY NEW YORK, 79-; RES PROF, SCH MED, NY UNIV, 79- *Concurrent Pos:* USPHS fel, 61-62, trainee, 62-63; Am Cancer Soc fel, 63-64, res grant, 65-; NIH career develop award, 71-, res grant, 72- *Mem:* AAAS; Am Soc Microbiol. *Res:* Genetics and physiology of ribosomes; evolutionary interrelationships between bacteria; nucleic acids; prokaryote differentiation and regulation. *Mailing Add:* Dept Microbiol NY Univ Sch Med 550 First Ave New York NY 10016

SMITH, J C, b Hudson, NC, Apr 19, 33; m 55; c 2. STRUCTURAL ENGINEERING. *Educ:* NC State Univ, BCE, 55, MS, 60; Purdue Univ, PhD(civil eng), 66. *Prof Exp:* Teaching asst, 55-56 & 58-60, from instr to asst prof, 60-74, ASSOC PROF CIVIL ENG, NC STATE UNIV, 74- *Mem:* Fel Am Soc Civil Engrs. *Res:* Structural analysis and design; closed form solutions of bridge floor systems; dynamic analysis of grids; numerical methods in structural engineering. *Mailing Add:* Dept of Civil Eng NC State Univ Box 7908 Raleigh NC 27695-7908

SMITH, J DUNGAN, b Attleboro, Mass, May 24, 39; m 59; c 3. PHYSICAL OCEANOGRAPHY, GEOLOGICAL OCEANOGRAPHY. *Educ:* Brown Univ, BA, 62, MS, 63; Univ Chicago, PhD(geophys), 68. *Prof Exp:* From actg asst prof to assoc prof, 67-77, PROF, DEPT OCEANOG & GEOPHYS PROG & ADJ PROF GEOL SCI, UNIV WASH, 77-, CHMN GEOPHYS PROG, 80- *Mem:* AAAS; Am Geophys Union; Sigma Xi; Int Asn Hydraul Res. *Res:* Coastal oceanography; mechanics of turbulent boundary layers; erosion and sediment transport; fluvial geomorphology; geophysical fluid mechanics. *Mailing Add:* Dept Geol Sci-AJ-20 Univ Wash Seattle WA 98195

SMITH, J RICHARD, b Utah, July 18, 24. NUCLEAR PHYSICS. *Educ:* Brigham Young Univ, BA, 49; Rice Univ, MA, 51, PhD(physics), 53. *Prof Exp:* Consult, IAEA South Korea, 85-88; SCIENTIFIC SPECIALIST, EG&G, IDAHO, 88- *Mem:* Am Phys Soc; Am Nuclear Soc. *Res:* Measurement of neutron multiplication. *Mailing Add:* 854 Claire View Lane Idaho Falls ID 83402

SMITH, JACK, b Morristown, NJ, Nov 28, 27; m 54; c 3. ELECTROMAGNETICS, ATMOSPHERIC SENSING. *Educ:* Univ Ariz, BS, 52, MS, 58, PhD(elec eng), 64. *Prof Exp:* Test engr, Gen Elec Co, 52-53, engr, 53-56; instr elec eng, Univ Ariz, 57-64, asst prof, 64, res assoc, Appl Res Lab, 57-64; assoc prof, 64-73, dean, Col Eng, 76-82, PROF ELEC ENG, UNIV TEX, EL PASO, 73-, SCHELLENGER PROF ELEC ENG, 82- *Concurrent Pos:* NSF sci fac fel, 61-62; consult, Atmospheric Sci Lab, US Army, 71- *Mem:* Inst Elec & Electronics Engrs; Am Geophys Union; Am Soc Eng Educ. *Res:* Atmospheric effects on high frequency electromagnetic wave propagation; lightning; upper atmosphere and ionospheric ionization characteristics and variations. *Mailing Add:* Dept of Elec Eng Univ of Tex El Paso TX 79968

SMITH, JACK CARLTON, b Kansas City, Mo, May 8, 13; m 57. POLYMER PHYSICS. *Educ:* Ohio State Univ, BEngPhysics, 35, MSc, 36; Calif Inst Technol, PhD(physics), 42. *Prof Exp:* Asst physics, Ohio State Univ, 36-37; teaching scholar, Calif Inst Technol, 37-39; physicist, US Dept Navy, 40-44; res physicist, Los Alamos Sci Lab, Univ Calif, 44-45; pioneering res div, Textile Fibers Dept, E I du Pont de Nemours & Co, 46-54; physicist, polymers div, Nat Bur Standards, 54-81; RETIRED. *Concurrent Pos:* Guest worker, polymers div, Nat Bur Standards, 81- *Mem:* Am Phys Soc. *Res:* High polymer physics; physical properties of textile yarns; mechanical properties of composite materials. *Mailing Add:* 3708 Manor Rd Apt 3 Chevy Chase MD 20815

SMITH, JACK HOWARD, b Middletown, NY, Nov 8, 21; m 49; c 3. THEORETICAL PHYSICS. *Educ:* Cornell Univ, AB, 43, PhD(theoret physics), 51. *Prof Exp:* Asst, Cornell Univ, 43-44; jr scientist, Theoret Div, Los Alamos Sci Lab, 44-46; asst, Cornell Univ, 46-49; mem staff, Theoret Div, Los Alamos Sci Lab, 49-51; res assoc theoret physics, Knolls Atomic Power Lab, Gen Elec Co, 51-63; PROF PHYSICS, STATE UNIV NY ALBANY, 63- *Mem:* AAAS; Am Asn Physics Teachers; Acoust Soc Am; Am Phys Soc. *Res:* Electromagnetic scattering; nuclear scattering of high energy electrons; neutron diffusion; reactor physics; reactor shielding; musical acoustics. *Mailing Add:* 1030 Atateka Rd Schenectady NY 12309

SMITH, JACK LOUIS, b Huntington, WVa, July 15, 34; m 61; c 2. BIOCHEMISTRY, NUTRITION. *Educ:* Univ Cincinnati, BS, 56, PhD(biochem), 62. *Prof Exp:* NIH trainee animal nutrit, Univ Ill, 61-63; Muscular Dystrophy Asn Am fel, Stanford Res Inst, 64; from asst prof to assoc prof biochem, Sch Med, Tulane Univ, 65-74, from asst prof to assoc prof biochem-nutrit, Sch Pub Health & Trop Med, 68-74, adj prof biochem, Sch Pub Health & Trop Med, 74-79; SWANSON ASSOC PROF BIOCHEM, 74- *Concurrent Pos:* Prin investr biochem, Touro Res Inst, 69-74. *Mem:* AAAS; Am Chem Soc; Am Inst Nutrit; Am Soc Clin Nutrit; Am Bd Nutrit; Am Dietetic Asn. *Res:* Nutritional biochemistry; nutritional assessment. *Mailing Add:* Dept Nutrit & Dietetics Univ Del Alison Hall Newark DE 19716

SMITH, JACK R(EGINALD), b Carrington, NDak, June 16, 35; m 59; c 3. ELECTRICAL ENGINEERING. *Educ:* Univ Southern Calif, BS, 58, MS, 60, PhD(elec eng), 64. *Prof Exp:* Mem tech staff, Hughes Aircraft Co, 58-59; res engr, Jet Propulsion Lab, 61; res assoc elec eng, Univ Southern Calif, 63-64; from asst prof to assoc prof, 64-70, PROF ELEC ENG, UNIV FLA, 70- *Mem:* Inst Elec & Electronics Engrs; Int Fedn Med Electronics & Biol Eng. *Res:* Biomedical engineering. *Mailing Add:* Dept of Elec Eng Univ of Fla Gainesville FL 32611

SMITH, JACKSON BRUCE, b Mt Holly, NJ, Mar 2, 38; m 63; c 2. INTERNAL MEDICINE, IMMUNOLOGY. *Educ:* Wake Forest Col, BS, 60, MD, 65. *Prof Exp:* Clin res fel, Inst Cancer Res, 67-69; fel, Univ Col, London, Eng, 72-74; clin assoc med, Univ Pa, 75-78; res physician, Inst Cancer Res, 74-81; assoc prof, 81-85, PROF MED & MICROBIOL, JEFFERSON MED COL, 85- *Concurrent Pos:* Clin assoc med, Pa Hosp, 76-81; Am Cancer Soc, grant, 78-80; adj asst prof med, Univ Pa Sch Med, 78-81; grant, NIH, 81-85 & 89-92. *Mem:* AAAS; Am Col Physicians; Am Fedn Clin Res; Am Asn Immunologists; Am Asn Cancer Res. *Res:* Immune system regulatory mechanisms in normal individuals and in patients and laboratory animals with lymphoproliferative and autoimmune disorders; immunological mechanisms of repeated pregnancy loss; maternal-fetal immunology. *Mailing Add:* Div Rheumatol Jefferson Med Col 1015 Walnut St Rm 613 Philadelphia PA 19107

SMITH, JAMES ALAN, b Detroit, Mich, Nov 19, 42; m 65; c 2. REMOTE SENSING, SCENE RADIATION MODELING. *Educ:* Univ Mich, BS, 63, MS, 65, PhD(physics), 70. *Prof Exp:* Res asst, Willow Run Labs, Univ Mich, 64-66, from res asst to assoc, Dept Physics, 66-70; asst prof remote sensing, Dept Earth Resources, 70-74, assoc prof, 74-78, prof remote sensing & comput appl, Dept Forestry, Colo State Univ, 78-85; head, Biospheric Sci Br, 85-90, assoc chief, Sci Info Systs Ctr, 90, STAFF SCIENTIST, LAB TERRESTRIAL PHYSICS, NASA GSFC, 90- *Concurrent Pos:* Assoc dir, Comput Ctr, Colo State Univ, 74-76; consult, numerous Fed Agencies & Indust; prin investr, NASA, Army Res Off, US Forest Serv, US Geol Surv, Corp Engrs, & US Fish & Wildlife Serv; assoc ed, Inst Elec & Electronic Engrs, Trans Geosci & Rem Sens, 83-91, ed, 91- *Mem:* Sr mem Inst Elec & Electronic Engrs; Geosci & Remote Sensing Soc; Am Geophys Union. *Res:* Modeling of optical reflective and thermal radiation patterns from earth surface features and the application of such models to remote sensing; large scale ecosystem analysis and modeling. *Mailing Add:* 11808 Bright Passage Columbia MD 21044-4139

SMITH, JAMES ALLBEE, b Detroit, Mich, Oct 20, 37; m 58; c 2. ANALYTICAL CHEMISTRY. *Educ:* Univ Mich, Ann Arbor, BS, 59; Ohio State Univ, PhD(org chem), 64. *Prof Exp:* Res chemist, Res & Develop Dept, Union Carbide Corp, 64-67, proj scientist catalysis, 67-68; res assoc coordr chem, Case Western Reserve Univ, 68-70; sr res chemist, Eng Develop Ctr, C E Lummus Co, 70-73; chemist, WVa Dept Agr, 73-78, asst dir, 78-81; MGR, TECH TESTING LABS, 81- *Mem:* Am Chem Soc; Asn Off Anal Chemists. *Res:* Development of methodology for residue analysis. *Mailing Add:* 1207 Larchwood Rd Charleston WV 25314-1232

SMITH, JAMES CECIL, b Little Orleans, Md, Jan 17, 34; m 61; c 3. PHYSIOLOGY, BIOCHEMISTRY. *Educ:* Univ Md, BS, 56, MS, 59, PhD(animal nutrit), 64. *Prof Exp:* Health serv officer, NIH, 59-61; biol chemist, Univ Calif, Los Angeles, 64-65; res physiologist, Vet Admin Hosp, Long Beach, Calif, 65-66; res biochemist, Vet Admin Hosp, Washington, DC, 66-77, chief, Trace Element Res Lab, 71-77; RES LEADER, VITAMIN & MINERAL NUTRIT LAB, USDA, 77- *Honors & Awards:* Klaus Schwarz Award, 82. *Mem:* Am Inst Nutrit. *Res:* Trace element metabolism such as zinc and copper. *Mailing Add:* Human Nutrit Res Ctr Vitamin & Mineral Nurtrit Lab US Dept Agr Rm 117 Bldg 307 BARC-E Beltsville MD 20705

SMITH, JAMES CLARENCE, JR, b Martinsville, Va, Aug 16, 39; m 62; c 2. MATHEMATICS. *Educ:* Davidson Col, BS, 61; Col William & Mary, MS, 64; Duke Univ, PhD(math), 67. *Prof Exp:* Aerospace technologist, Langley Res Ctr, NASA, 61-67; asst prof, 67-71, ASSOC PROF MATH, VA POLYTECH INST & STATE UNIV, 71- *Concurrent Pos:* Consult, Langley Res Ctr, NASA, 68 & 69. *Mem:* Am Math Soc; Math Asn Am. *Res:* Topology; dimension theory. *Mailing Add:* Dept Math Va Polytech Inst & State Univ Blacksburg VA 24061

SMITH, JAMES DAVID BLACKHALL, b Peterhead, Scotland, Apr 25, 40; c 2. POLYMER CHEMISTRY. *Educ:* Aberdeen Univ, BSc, 62, PhD(polymer chem), 65. *Prof Exp:* NSF res grant, Polymer Res Ctr, State Univ NY, 65-66; sr chemist, Laporte Industs Ltd, Luton, Eng, 67-68; mgr instalation systs, Res & Develop Ctr, 68-82, Polymer & Composite Res Dept, 82-87, CONSULT SCIENTIST, MATS TECHNOL DIV, WESTINHOUSE ELEC CORP, 87- *Mem:* Am Chem Soc; Royal Soc Chem; Inst Elec & Electronics Engrs. *Res:* Polymerization kinetics; electroinitiated polymerization reactions; polyester and epoxy resin technology; flame retardants; insulation and dielectric properties of polymers. *Mailing Add:* Insulation Dept Westinghouse Elec Corp R&D Ctr Pittsburgh PA 15235

SMITH, JAMES DONALDSON, b Wilkesboro, Pa, Aug 23, 22. GEOLOGY & MINING. *Educ:* Pa State Univ, BS, 49. *Prof Exp:* Adj fac teacher, City Col Beaver County, Pa, 83-88; GEOL & MINING CONSULT, 88- *Mem:* Fel Geol Soc Am; Am Inst Prof Geologists; Asn Eng Geologists; Assoc Inst Mining Engrs. *Mailing Add:* 111 Crest Dr RD 2 Beaver PA 15009

SMITH, JAMES DOUGLAS, b Paullina, Iowa, Dec 14, 27; m 55; c 2. DEVELOPMENTAL GENETICS, REGULATION OF SEED DEVELOPMENT. *Educ:* Iowa State Univ, BS, 50, MS, 56, PhD(genetics), 60. *Prof Exp:* Plant breeder, United-Hagie Hybrids, Inc, 53, consult, 53-59; from asst prof to assoc prof, 59-70, chmn fac, 62-80, PROF GENETICS, TEX A&M UNIV, 70- *Mem:* Am Genetics Asn; Genetics Soc Can; Genetics Soc Am; Bot Soc Am; Am Soc Plant Physiol. *Res:* Genetic, metabolic and environmental regulation of seed development; genetic dissection of the biosynthesis and functions of abscisic acid and gibberellins in developing seed of Zoa mays; metabolic effects resulting from modulation of water, temperature, phytohormones, carbohydrates and amino acids on seed grown in vitro. *Mailing Add:* Dept Soil & Crop Scis Tex A&M Univ Col Agr College Station TX 77843-2474

SMITH, JAMES DOYLE, b Charlottesville, Va, Jan 27, 21; m 44; c 2. ORGANIC CHEMISTRY. *Educ:* Univ Va, BS, 42, MS, 44, PhD(chem), 46. *Prof Exp:* Assoc, Med Col Va, Va Commonwealth Univ, 46-48, asst prof chem, 48-51, assoc prof, 51-62, prof, 62-81, actg chmn, 61, chmn dept, 62-74, actg chmn, Dept Pharmaceut Chem, 76-77; RETIRED. *Mem:* AAAS; Am Chem Soc; Am Pharmaceut Asn. *Res:* Synthetic organic and medicinal chemistry; amino acids; antimalarials; anti-tumor agents. *Mailing Add:* Sch Pharm Med Col Va Va Commonwealth Univ Box 540 Richmond VA 23298

SMITH, JAMES EARL, b May 28, 49; c 3. MECHANICAL & AEROSPACE ENGINEERING. *Educ:* WVa Univ, BSAE, 72, MSAE, 74, PhD, 84. *Prof Exp:* Staff engr, Morgantown Energy Res Ctr, Dept of Energy, 74-76; instr, dept aerospace eng, 76-78, adj asst prof, dept gen eng, 78-84, asst res prof, Col Eng, 84-85, ASST RES PROF, DEPT MECH & AEROSPACE ENG, WVA UNIV, 85- *Mem:* Sigma Xi; Am Inst Aeronaut & Astronaut; Am Soc Mech Engrs; Soc Automotive Engrs; Am Soc Eng Educ. *Res:* New concept of movement to engine design. *Mailing Add:* Dept Mech & Aerospace Eng Col Eng WVa Univ Morgantown WV 26506

SMITH, JAMES EDWARD, b Atlanta, Ga, Apr 7, 35; m 78; c 2. GEOCHEMISTRY, GEOPHYSICS. *Educ:* Univ of the South, BS, 58; Rochester Univ, PhD(statist, mech), 65. *Prof Exp:* Phys chemist, Union Carbide Res Inst, NY, 63-64; NIH fel chem, Ill Inst Technol, 64-65; phys chemist, Chem Lab, Phillips Petrol Co, 65-88; PHYS CHEMIST, OIL & GAS SEARCH SERV INC, 88- *Mem:* Soc Explor Geophysicists; Soc Petrol Eng; Sigma Xi (secy, 74-75, vpres, 75-76, pres, 76-77); Asn Petrol Geochem Explorationist. *Res:* Relative abundance and isotopic composition of alkanes in sediments; petroleum migration, accumulation and dissipation in the earth; petroleum geochemistry; shale compaction over geologic time and geopressures; transport phenomena in the earth; salinity changes accompanying shale compaction; theoretical interpretation of electric logs; synthetic seismograms; petroleum prospecting and economics; well logging. *Mailing Add:* 1209 Harris Dr Bartlesville OK 74006

SMITH, JAMES EDWARD, JR, b Cincinnati, Ohio, Dec 5, 41; m 69; c 4. TECHNOLOGY TRANSFER. *Educ:* Univ Cincinnati, BS, 63, MS, 66; Wash Univ, DSc, 69. *Prof Exp:* Res sanitary engr, Advan Waste Treatment Lab, 68-71, sanitary engr, 71-76, head municipal technol transfer staff, 76-77, ENVIRON ENGR, CTR ENVIRON INFO, US ENVIRON PROTECTION AGENCY, 77- *Concurrent Pos:* Consult, sludge treatment & disposal for indust residues, Arg govt, 81 & 85; consult, Environ Hazards & Food Protection Unit, Environ Health Div, WHO, Switz, 84-85. *Mem:* Am Soc Civil Engrs; Water Pollution Control Fedn; Int Asn Water Pollution Res. *Res:* Land application of sludge; sludge dewatering; drinking water treatment; definition of the effects and control of non-ionizing radiation; 60 publications. *Mailing Add:* 5821 Marlborough Dr Cincinnati OH 45230

SMITH, JAMES EDWARD, behavioral pharmacology, behavioral neurochemistry, for more information see previous edition

SMITH, JAMES ELDON, b Ft Wayne, Ind, June 9, 28; m 51; c 5. MICROBIOLOGY. *Educ:* DePauw Univ, BA, 50; Purdue Univ, MS, 52, PhD, 55. *Prof Exp:* Fel, Am Cancer Soc, Purdue Univ, 55-56, Life Ins Med Res Fund, 56-57; from res asst prof to asst prof bact, 57-70, ASSOC PROF MICROBIOL, SYRACUSE UNIV, 70- *Mem:* Am Soc Microbiol; Electron Micros Soc Am. *Res:* Virology and microbial genetics; physiology of virus reproduction; mycology and plant pathology; germfree animal physiology. *Mailing Add:* Dept Biol Syracuse Univ Syracuse NY 13244

SMITH, JAMES F, b Syracuse, NY, Apr 26, 30. MATHEMATICAL ANALYSIS. *Educ:* Bellarmine Col, NY, AB, 54; Cath Univ Am, MS, 57, PhD(math), 59; Woodstock Col, Md, STL. *Prof Exp:* From instr to assoc prof, 64-74, PROF MATH, LE MOYNE COL, NY, 74- *Mem:* Am Math Soc; Math Asn Am. *Res:* Banach algebras; Hilbert space; structure and spectral theory. *Mailing Add:* Dept Comput Sci Le Moyne Col Syracuse NY 13214

SMITH, JAMES G(ILBERT), b Benton, Ill, May 1, 30; m 55; c 1. ELECTRICAL ENGINEERING. *Educ:* Univ Mo, BSEE, 57, MSEE, 59, PhD(eng physics), 67. *Prof Exp:* Instr elec eng, Sch Mines, Univ Mo-Rolla, 57-59 & 61-66, asst prof, 59-61; asst prof, Sch Technol, 66-69, assoc prof, 69-72, chmn, Dept Elec Sci & Systs Eng, 71-80, PROF ELEC ENG, COL ENG & TECHNOL, SOUTHERN ILL UNIV, 72- *Concurrent Pos:* Fac Fel, NSF, 62-63. *Mem:* Am Soc Eng Educ; Inst Elec & Electronics Engrs; AAAS. *Res:* Electromagnetics and antennas, lightning and electrical properties of materials; electrophoresis and electroosmosis, physical and chemical properties of coal. *Mailing Add:* Dept Elec Eng Southern Ill Univ Carbondale IL 62901

SMITH, JAMES GRAHAM, organic chemistry; deceased, see previous edition for last biography

SMITH, JAMES H, b Oneida, Tenn, Feb 28, 34; c 2. PHYSICS, MATHEMATICS. *Educ:* Eastern Ky Univ, BS, 60; Univ Tenn, MS, 71. *Prof Exp:* Technician electronics, US Navy Airforce, 52-55; technician commun, Am Tel & Tel Co, 55-58; instr physics, Eastern Ky Univ, 60-61; assoc physicist phys tests, 62-68, PHYSICIST NONDESTRUCTIVE TESTING, NUCLEAR DIV, MARTIN MARIETTA ENERGY SYSTS, INC, OAK RIDGE NAT LAB, 68- *Concurrent Pos:* Instr, Univ Tenn; mem bd dirs, Am Soc Nondestructive Testing, 83-85. *Honors & Awards:* Achievement Award, Am Soc Nondestructive Testing, 76. *Mem:* Fel Am Soc Nondestructive Testing; Am Welding Soc; Soc Exp Stress Anal. *Res:* Nondestructive testing, specifically ultrasonics and eddy currents. *Mailing Add:* Oak Ridge Nat Lab PO Box 2008 Oak Ridge TN 37831

SMITH, JAMES HAMMOND, b Colorado Springs, Colo, Feb 2, 25; m 50, 55; c 4. NUCLEAR PHYSICS. *Educ:* Stanford Univ, AB, 45; Harvard Univ, AM, 47, PhD(physics), 52. *Prof Exp:* Jr lab technician, Oak Ridge Nat Lab, 50-51; instr physics, 51, from asst prof to assoc prof, 53-60, assoc head dept, 72-80, PROF PHYSICS, UNIV ILL, URBANA, 60- *Concurrent Pos:* Guggenheim fel, 66. *Mem:* Fel Am Phys Soc. *Res:* Photonuclear reactions; K meson decays; high energy nuclear reactions. *Mailing Add:* Dept Physics Loomis Lab 1110 W Green St Univ Ill Urbana IL 61801

SMITH, JAMES HART, b North Plainfield, NJ, Jan 20, 42. ENVIRONMENTAL CHEMISTRY. *Educ:* Yale Univ, BSci, 63; Univ Calif, Berkeley, PhD(chem), 67. *Prof Exp:* Fel, Calif Inst Technol, 67-70; DIR, DEPT PHYS CHEM, SRI INT, 70- *Mem:* Am Chem Soc. *Res:* Prediction of the environmental fate of chemicals; collection and analysis of environmental samples; measurement of physical properties of chemicals. *Mailing Add:* 30 Medway Rd Woodside CA 94062-2613

SMITH, JAMES JOHN, b St Paul, Minn, Jan 28, 14; div; c 4. PHYSIOLOGY. *Educ:* St Louis Univ, BS, 35, MD, 37; Northwestern Univ, MS, 40, PhD(physiol), 46. *Prof Exp:* Intern, St Paul Ramsey Med Ctr, St Paul, Minn, 37-38; asst path, Cook County Hosp, Chicago, Ill, 38-39; assoc prof physiol & dean, Sch Med, Loyola Univ Chicago, 46-50; chief, Med Educ Div, Cent Off, US Vet Admin, DC, 50-52; chmn dept, 52-78, PROF PHYSIOL, MED COL WIS, 52-, PROF MED, 78-; DIR, HUMAN PERFORMANCE LAB, ZABLOCKI VET ADMIN CTR, MILWAUKEE, WIS, 82- *Concurrent Pos:* Fulbright res prof, Heidelberg, 59-60; fel, Cardiovas Sect, Am Physiol Soc, 64; nat bd consults to surgeon gen, USAF, 52-54. *Mem:* Soc Exp Biol & Med; Am Physiol Soc; Sigma Xi; Am Heart Asn; Geront Soc Am. *Res:* Cardiovascular physiology, particularly effect of aging and circulatory disease on autonomic response of circulatory system to non-exercise and exercise stress; cardiovascular evaluation; peripheral circulation; circulatory control, aging, stress response. *Mailing Add:* Dept Physiol Med Col Wis Milwaukee WI 53226

SMITH, JAMES L, b Lackawanna, NY, Aug 5, 29; m 54; c 4. MATHEMATICS, GEOMETRY. *Educ:* Univ Louisville, BA, 51; Univ Pittsburgh, MS, 55; Okla State Univ, EdD(found in geom), 63. *Prof Exp:* Asst math, Univ Pittsburgh, 54-56; instr, Westminster Col, Pa, 56-61; asst, Okla State Univ, 61-63; assoc prof, 63-75, chmn dept, 65-71 & 76-82, PROF MATH, MUSKINGUM COL, 75- *Concurrent Pos:* Vis lectr, NSF summer insts, Southwestern State Col, Okla, 63 & Northeast Mo State Col, 64, dir & instr, Teacher Oriented Insts, 65-66; NSF sci fac fel, Wash State Univ, 67-68; vis assoc prof, Univ NH, 71-72; assoc dir, NSF Pre-Col Teacher Develop Proj, 77-78, dir, NSF CAUSE Comput Lit Proj, 77-81; Fulbright prof math, Univ Malawi, Africa, 79-80; vis prof, Ohio State Univ, 87-88. *Mem:* Math Asn Am; Am Math Soc; Nat Coun Teachers Math. *Res:* Function approach to geometry; technology applications to geometry. *Mailing Add:* Dept Math & Comput Sci Muskingum Col New Concord OH 43762-1199

SMITH, JAMES LAWRENCE, b Detroit, Mich, Sept 3, 43; m 65; c 2. MAGNETISM, SUPERCONDUCTIVITY. *Educ:* Wayne State Univ, BS, 65; Brown Univ, PhD(physics), 74. *Prof Exp:* Staff mem physics, 73-82, fel, Los Alamos, 82-86, dir, Ctr Mat Sci, 86-87, FEL, LOS ALAMOS NAT LAB, UNIV CALIF, 87- *Honors & Awards:* E O Lawrence Award, 86; Int Prize for New Mat, Am Phys Soc, 90. *Mem:* Fel Am Phys Soc; AAAS; Mat Res Soc. *Res:* Study of electronic behavior of actinides; occurence of superconductivity and magnetism in transition metals; high temperature superconductivity. *Mailing Add:* MS K763 Los Alamos Nat Lab PO Box 1663 Los Alamos NM 87545

SMITH, JAMES LEE, b Clinton Co, Ind, Dec 23, 28. MICROBIOLOGY. *Educ:* Ind Univ, AB, 52, MA, 54, PhD(bact), 62. *Prof Exp:* Med bacteriologist, US Army, Ft Detrick, Md, 54-59; USPHS fel microbiol, Univ Chicago, 61-63; RES MICROBIOLOGIST, EASTERN MKT & NUTRIT RES DIV, USDA, 63- *Mem:* AAAS; Am Soc Microbiol; Am Chem Soc; Brit Soc Gen Microbiol. *Res:* Bacterial physiology and nutrition. *Mailing Add:* 507 Oak St North Wales PA 19454-3020

SMITH, JAMES LEE, b Thayer, Kans, Feb 27, 35; m 68; c 2. BIOLOGY. *Educ:* San Francisco State Col, BA, 58; Univ Calif, Berkeley, PhD(bot), 63. *Prof Exp:* Asst prof biol, Chico State Col, 63-64 & Univ Colo, Boulder, 64-69; asst prof, 69-74, ASSOC PROF BIOL, CENT MO STATE UNIV, 74- *Mem:* AAAS; Bot Soc Am; Am Bryol & Lichenological Soc; Sigma Xi. *Res:* Phytoplankton ecology. *Mailing Add:* Dept of Biol Cent Mo State Univ Warrensburg MO 64093

SMITH, JAMES LEE, b Peoria, Ill, Feb 16, 37; m 87; c 4. EROSION, IRRIGATION & MACHINE DEVELOPMENT. *Educ:* Univ Ill, BS, 61, MS, 64; Univ Minn, PhD(agr eng), 71. *Prof Exp:* Proj engr, Air Force Weapons Lab, Kirkland AFB, 63-64; instr, Univ Utah, Salt Lake City, 64-67; proj engr, Int Harvester Co, Hinsdale, 67-69; NSF trainee, Univ Minn, St Paul, 69-71; from assoc prof to prof, Colo State Univ, Ft Collins, 71-81; HEAD PROF, UNIV WYO, LARAMIE, 81- *Concurrent Pos:* Lectr, Ill Inst Technol, 67-69; Fulbright lectr & vis prof, Univ Nairobi, Kenya, 87-88. *Mem:* Am Soc Agr Engrs; Am Soc Eng Educ. *Res:* Erosion, irrigation and machine development; problems relating to soil and water conservation, erosion control and development of farm machinery industry to serve small farms. *Mailing Add:* Agr Eng Dept Univ Wyo Laramie WY 82071

SMITH, JAMES LEWIS, b Tacoma, Wash, Feb 19, 49; m 84. ANALYTICAL CHEMISTRY. *Educ:* Univ Puget Sound, BSc, 71, MSc, 74; Univ BC, PhD(chem), 80. *Prof Exp:* Sr anal chemist, Seattle Trace Organics Monitoring Lab, 80-81; res chemist analytical chem, Nat Oceanic & Atmospheric Admin, 81; asst prof chem, 81-86, ASSOC PROF CHEM, NMEX INST MINING & TECHNOL, 86- *Concurrent Pos:* Res collabr, Environ Sci Res Group, Los Alamos Nat Lab, 86; res scientist, NMex Petrol Recovery Res Ctr, 87- *Mem:* Am Chem Soc; Sigma Xi. *Res:* Analytical chemistry method development as applied to environmental and petroleum chemistry; applications of nuclear magnetic resonance to study nucleic acids and fluid flow. *Mailing Add:* Chem Dept NMex Inst Mining & Technol Socorro NM 87801

SMITH, JAMES LYNN, b Columbia, Miss, Sept 15, 40; m 65. SOLID STATE PHYSICS, OPTICAL PHYSICS. *Educ:* Univ Southern Miss, BS, 62; Auburn Univ, MS, 65, PhD(physics), 68. *Prof Exp:* RES PHYSICIST, PHYS SCI LAB, US ARMY MISSILE COMMAND, 68- *Mem:* Am Optical Soc; Am Phys Soc. Sigma Xi. *Res:* Ultra high vacuum techniques; Hall effect in thin metal films; dielectric breakdown in evaporated films; gallium arsenide injection lasers; laser material degradation; optical data processing; laser radar. *Mailing Add:* 426 High Sch Dr Grande Prairie TX 75050

SMITH, JAMES PAYNE, JR, b Oklahoma City, Okla, Apr 13, 41. PLANT TAXONOMY, AGROSTOLOGY. *Educ:* Tulsa Univ, BA & BS, 63; Iowa State Univ, PhD(bot), 68. *Prof Exp:* Chmn, biol sci, 78-83, PROF BOT, HUMBOLDT STATE UNIV, 78-, DEAN, COL OF SCI, 84- *Mem:* Am Soc Plant Taxon; Bot Soc Am; Soc Study Evolution; Soc Econ Bot; Int Asn Plant Taxonomists. *Res:* Taxonomy of flowering plants; flowering plants of northern California; grasses of the US, especially California; rare and endangered plants of California. *Mailing Add:* Off Dean Col Sci Humboldt State Univ Arcata CA 95521

SMITH, JAMES R, b Springfield, Mo, Apr 28, 41; m 75; c 2. CELL BIOLOGY. *Educ:* Univ Mo, BS, 63; Univ Ariz, BS, 66; Yale Univ, MPh, 68, PhD(molecular biophys), 70. *Prof Exp:* Teaching asst, Biophys Lab, Yale Univ, 68-69; res assoc microbiol, Stanford Univ, 70-72; res physiologist, Vet Admin Hosp, Martinez, Calif, 72-75; assoc scientist, 75-80, sr scientist, Walton Jones Cell Sci Ctr, 80-83; assoc prof cell genetics, 83-87, PROF CELL GENETICS, DIV MOLECULAR VIROLOGY, BAYLOR COL MED, 87- *Concurrent Pos:* Vis asst res physiologist, Univ Calif, Berkeley, 72-75; adj prof, Dept Biol Sci, State Univ NY Plattsburgh & Biol Dept, North Country Community Col, NY, Saranac Lake; adj assoc prof, Dept Molecular Pharmacol, Univ RI, Kingston & Dept Med Microbiol, Univ Vt, Burlington; adj affil assoc prof life sci, Worcester Polytech Inst, Mass; co-dir, Roy M & Phyllis Gough Huffington Ctr Aging, 86-; mem, Nat Adv Coun Aging, Nat Inst Health. *Mem:* Tissue Cult Asn; Am Soc Cell Biol; fel Gerontol Soc Am; AAAS. *Res:* Cellular aging; control of cell proliferation. *Mailing Add:* Div Molecular Virol Baylor Col Med Tex Med Ctr One Baylor Plaza Houston TX 77030

SMITH, JAMES REAVES, b Columbia, SC, June 6, 42; m 64; c 2. ALGEBRA. *Educ:* Univ SC, BS, 63, PhD(math), 68. *Prof Exp:* Asst prof, 68-72, assoc prof, 72-78, PROF MATH, APPALACHIAN STATE UNIV, 78- *Mem:* Am Math Soc; Math Asn Am; Nat Coun Teachers Math. *Res:* Study of regular modules, those whose submodules are all pure; also projective simple modules. *Mailing Add:* Dept Math Sci Appalachian State Univ Boone NC 28608

SMITH, JAMES ROSS, b Kingsport, Tenn, Nov 15, 43; m 63; c 2. QUALITY CONTROL, OPERATIONS RESEARCH. *Educ:* Va Polytech Inst & State Univ, BS, 65, MS, 67, PhD(indust eng & opers res), 71. *Prof Exp:* Indust engr, Holston Defense Corp, 66-69; grad asst indust eng, Va Polytech Inst & State Univ, 69-71; asst prof, 71-74, ASSOC PROF INDUST ENG, TENN TECHNOL UNIV, 74- *Mem:* Am Inst Indust Engrs; Am Soc Qual Control. *Res:* Applied statistics and applied operations research. *Mailing Add:* Dept Indust Eng Tenn Technol Univ Cookville TN 38505

SMITH, JAMES S(TERRETT), b Pittsburgh, Pa, Aug 21, 17; m 43; c 5. METALLURGY, CHEMISTRY. *Educ:* Yale Univ, BS, 40, PhD(phys chem), 43. *Prof Exp:* Chemist, Off Sci Res & Develop, Yale Univ, 42; res chemist, Manhattan Proj, Columbia Univ, 43-45 & Carbide & Carbon Chems Corp, NY, 45; res chemist, E I du Pont de Nemours & Co, 45-50, res supvr, 51; sect head, 52-54, res mgr, 54-65, eng mgr metall, 65-77, CONSULT METALL, GTE SYLVANIA, 77- *Mem:* Am Chem Soc; Sigma Xi; Am Phys Soc. *Res:* Physical and process powder metallurgy of ceramics, molybdenum and wolfram. *Mailing Add:* RD 4 Box 11 Towanda PA 18848

SMITH, JAMES STANLEY, b Ithaca, NY, Apr 7, 39. ENVIRONMENTAL ANALYTICAL CHEMISTRY. *Educ:* Williams Col, AB, 60; Iowa State Univ, PhD(org chem), 64. *Prof Exp:* Fel org chem, Univ Ill, 64-65; asst prof, Eastern Mich Univ, 66-68; fel mass spectros, Cornell Univ, 68-69; supvr anal chem, Allied Corp, 69-81; DIR, ANAL LAB, ROY F WESTON, INC, 81- *Concurrent Pos:* Vis lectr org chem, Univ Ill, 66-69. *Mem:* Am Soc Testing Mat; Am Soc Mass Spectrometry; Am Chem Soc. *Res:* Analysis of drinking water, waste water, air and hazardous chemical waste; development of methodologies to determine environmental pollutants. *Mailing Add:* 7A Grace's Dr Coatesville PA 19320-1205

SMITH, JAMES THOMAS, b Springfield, Ohio, Nov 8, 39; m 63; c 1. MATHEMATICS. *Educ:* Harvard Univ, BA, 61; San Francisco State Col, MA, 64; Stanford Univ, MS, 66; Univ Sask, PhD(math), 70. *Prof Exp:* Mathematician, US Naval Radiological Defense Lab, 62-67; instr math, San Francisco State Col, 66-67 & Univ Sask, Regina Campus, 68; assoc prof, 69-77, PROF MATH, SAN FRANCISCO STATE UNIV, 77- *Concurrent Pos:* chmn dept, San Francisco State Univ 75-82; software engr, Blaise Comput Inc, 84-85. *Mem:* Inst Elec & Electronic Engrs; German Math Asn; Math Asn Am; Asn Comput Mach. *Res:* Foundations of geometry; microcomputer systems software. *Mailing Add:* Dept of Math San Francisco State Univ San Francisco CA 94132

SMITH, JAMES W(ILMER), b Kamloops, BC, June 13, 31; m 58; c 3. CHEMICAL ENGINEERING. *Educ:* Univ BC, BASc, 54, MASc, 55; PhD(chem Univ London, PhD(chem eng), 60, CIH, 81. *Prof Exp:* Process engr, Du Pont Can, Ont, 55-57; fel chem eng, Univ BC, 61-62; from asst prof to assoc prof, 62-70, assoc chmn chem eng, 75-81, PROF CHEM ENG, UNIV TORONTO, 70-, CHMN, 85-; pres, 75-85, CHMN CHEM ENG, RES CONSULTS LTD, 85- *Mem:* Chem Inst Can; Can Soc Chem Eng; Air Pollution Control Asn; Am Indust Hygiene Asn. *Res:* Heat transfer, fluid flow and chemical reaction; properties of particulate systems; industrial hygiene. *Mailing Add:* Dept Chem Eng Univ Toronto Toronto ON M5S 1A4 Can

SMITH, JAMES WARREN, b Logan, Utah, July 5, 34; m 58; c 2. CLINICAL PATHOLOGY, PARASITOLOGY. *Educ:* Univ Iowa, BA, 56, MD, 59. *Prof Exp:* Resident path, Univ Iowa Hosp, 60-65; pathologist, US Naval Hosp, Chelsea, Mass, 65-67; asst prof, Med Col, Univ Vt, 67-70; PROF PATH, IND UNIV MED CTR, INDIANAPOLIS, 70- *Mem:* Am Soc Clin Path; Col Am Path; Am Soc Microbiol; AMA; Infectious Dis Soc Am; Am Soc Trop Med & Hyg. *Res:* Pneumocystis carinii; clinical microbiology. *Mailing Add:* Ind Univ Med Ctr Indianapolis IN 46202-5250

SMITH, JAMES WILLIE, JR, b Jackson, Miss, Mar 17, 44; m 69. ENTOMOLOGY. *Educ:* Miss State Univ, BS, 66; Univ Calif, Riverside, PhD(entom), 70. *Prof Exp:* Res asst entom, Univ Calif, Riverside, 66-69; asst prof, 70-74, ASSOC PROF ENTOM, TEX A&M UNIV, 74- *Mem:* AAAS; Entom Soc Am. *Res:* Insect pest management of field crops; population ecology; resistant plant varieties; biological control. *Mailing Add:* Dept Entom Tex A&M Univ Col Agr College Station TX 77843-2475

SMITH, JAMES WINFRED, b Greenwood, Miss, Jan 27, 43; m 66; c 3. ENTOMOLOGY, ECOLOGY. *Educ:* Miss State Univ, BS, 65; La State Univ, MS, 67, PhD(entomol), 70. *Prof Exp:* Asst prof biol, Motlow State Community Col, 70-71; RES ENTOMOLOGIST & RES LEADER, BIOENVIRON INSECT LAB, ARS, USDA, 71-, RES LEADER, BOLL WEEVIL RES UNIT, 87- *Concurrent Pos:* Assoc, Dept Entomol, Miss State Univ, 74-; Delta Coun res award, 85-86. *Mem:* Entomol Soc Am; Acarological Soc Am; Sigma Xi. *Res:* Field research on the ecology, population dynamics, and control of insect pests by bioenvironmental methods. *Mailing Add:* Boll Weevil Res Unit USDA Agr Res Serv PO Box 5367 Mississippi State MS 39762

SMITH, JAN D, b Pretoria, SAfrica, Feb 6, 39; m 62; c 3. ANESTHESIOLOGY, INTERNAL MEDICINE. *Educ:* Univ Pretoria, MB ChB, 62; Royal Col Physicians UK, MRCP, 73; Am Bd Anesthesiol, dipl, 69; Am Bd Internal Med, dipl, 80, Am Bd Internal Med (pulmonary disorder), dipl, 82. *Prof Exp:* Resident anesthesiol, Sch Med, Harvard Univ, Peter Bent Brigham Hosp, 64-66; fel, Sch Med, Univ Pittsburgh, 66-69, resident, 71; resident int med, Groote Schuur Hosp, Univ Cape Town, 70-71; asst prof int med, Sch Med, Univ Iowa, 74-76; from asst prof to assoc prof anesthesiol & internal med, Health Sci Ctr, Univ Tex, 76-83; prof anesthesiol & internal med, Univ Neb, 83-85; PROF ANESTHESIOL, NE OHIO COL MED, 85- *Mem:* Am Fedn Clin Res; Am Col Chest Physicians; Am Col Physicians; Am Soc Anesthesiol; Am Thoracic Soc; Soc Critical Care Med. *Res:* Pathogenesis of Shock Lung; carbon dioxide physiology. *Mailing Add:* 1323 Presby Univ Hosp Pittsburgh PA 15213

SMITH, JAN G, b Yoe, Pa, Sept 25, 38; m 60; c 1. GEOLOGY. *Educ:* Pa State Univ, BS, 60, MS, 61; Univ Tasmania, PhD(geol), 64. *Prof Exp:* Geologist, Continental Oil Co Australia, Ltd, 64-66; res geologist, Gulf Res & Develop Co, 66-68, supv geologist, 68-88; AGENT, STATEFARM CO, 88- *Mem:* Geol Soc Am; Am Asn Petrol Geologists; Am Geophys Union. *Res:* Tectonics. *Mailing Add:* Statefarm 611 Jackson Richmond TX 77469

SMITH, JAY HAMILTON, b Rexburg, Idaho, June 5, 27; m 49; c 5. SOIL MICROBIOLOGY. *Educ:* Brigham Young Univ, BS, 51; Utah State Univ, MS, 53; Cornell Univ, PhD(soil microbiol), 55. *Prof Exp:* Soil microbiologist, Soil & Water Conserv Res Div, US Dept Agr, SC, 55-58, soil scientist, Soils Lab, Agr Res Serv, Md, 58-64, soil scientist, Snake River Conserv Res Ctr, Agr Res Serv, USDA, 64-87; RETIRED. *Concurrent Pos:* Soil scientist, Clemson Col, 55-58; prof, Utah State Univ, 64- & Univ Idaho, 68- *Mem:* Int Humic Substances Soc; Soil Sci Soc Am; Sigma Xi; Am Soc Agron. *Res:* Instrumentation of soil organic matter and nitrogen; chemistry of soil nitrogen; microbiology of irrigation water and drainage; land disposal of food processing wastes. *Mailing Add:* 3787 N 3575 E Kimberly ID 83341

SMITH, JEAN BLAIR, b Detroit, Mich, Sept 23, 42; m 67; c 2. CLINICAL CHEMISTRY. *Educ:* WVa Univ, BA, 65; Univ Kans, PhD(anal chem), 68. *Prof Exp:* Staff electrochemist, Artificial Eye Proj, Univ Utah, 72-76, res instr, Bioeng Dept, 76-77; res chemist, Vet Admin Hosp, 77-81; RES ASST PROF, DEPT PATH, MED CTR, UNIV UTAH, 81- *Mem:* Am Asn Clin Chemists. *Res:* Role of prostaglandins in platelet aggregation; ion transport mechanisms of erythrocytes. *Mailing Add:* 216 Spring Valley Rd West Lafayette IN 47906

SMITH, JEAN E, b Buffalo, NY, Jan 15, 32; m 65. ANIMAL ECOLOGY. *Educ:* Ohio Univ, BS, 54, MS, 56; Univ Wyo, PhD(zool), 62. *Prof Exp:* Instr biol, Alderson-Broaddus Col, 56-57; asst prof, Adams State Col, 62-66; assoc prof, Metrop State Col, 66-69; assoc prof, 69-80, chmn dept, 78-86, PROF BIOL SCI, CARROLL COL, MONT, 80- *Mem:* AAAS; Am Ornith Union; Cooper Ornith Soc. *Res:* Vertebrate ecology; animal behavior. *Mailing Add:* Dept of Biol Carroll Col Helena MT 59601

SMITH, JEFFREY DREW, b Wearhead, Durham, Eng, Aug 2, 22; m 50; c 2. PLANT PATHOLOGY, MYCOLOGY. *Educ:* Univ Durham, BSc, 46, MSc, 57. *Prof Exp:* Demonstr agr bot, King's Col, Univ Durham, 46-48, jr lectr plant path, 48-51; res plant pathologist, Sports Turf Res Inst, Eng, 51-58; exten plant pathologist, NScotland Col Agr, Aberdeen, 58-60; prin sci officer, Res Div, NZ Dept Agr, 60-64; res scientist, res br, Can Dept Agr, 65-; RES SCIENTIST, UNIV SASK, 85- *Concurrent Pos:* NZ Dept Agr overseas res grant, Rothamsted Exp Sta, Eng, 62; vis scientist, Ore State Univ, 70 & Norweg Plant Protection Inst, 74-75; res scientist, Can Int Develop Agency, Kenya, 81-82. *Mem:* Can Phytopath Soc; Brit Inst Biol; Brit Mycol Soc. *Res:* Epidemiology and control of diseases of forage, range and turf grasses; psychophilic fungi and mycotoxicology. *Mailing Add:* 306 Egbert Ave Saskatoon SK S7N 1X1 Can

SMITH, JEROME ALLAN, b Lansing, Mich, Apr 17, 40; m 62; c 2. AERONAUTICAL ENGINEERING, FLUID MECHANICS. *Educ:* Univ Mich, BSE, 62, Calif Inst Technol, MS, 63, PhD, 67. *Prof Exp:* From asst prof to prof aerospace & mech sci, Princeton Univ, 67-79; tech dir, Off Naval Res, 79-83; VPRES MISSILE SYST, MARTIN MARIETTA CORP, 87- *Concurrent Pos:* Mem, Lab Adv Bd Surface Weapons, USN, 73-78, chmn, 75-78; consult, McDonnell Douglas Res Labs, 74-79. *Mem:* Am Phys Soc; AAAS; Sigma Xi. *Res:* Experimental investigation of high speed flow-shock tubes and hypersonic wind tunnels. *Mailing Add:* 8629 Summerville Pl Orlando FL 32819-3850

SMITH, JEROME H, b Omaha, Nebr, Oct 9, 36; m; c 3. INFECTIOUS & TROPICAL DISEASES, PARASITOLOGY. *Educ:* Univ Nebr, BS, 61, MS, 62, MD, 63; Harvard Sch Pub Health, MSc, 69. *Prof Exp:* Prof path & lab med, Tex A&M Univ, 84-89; PROF PATH, UNIV TEX MED BR, GALVESTON, 76-84, 90- *Mem:* Am Soc Parasitologists; Am Soc Clin Path; Am Soc Trop Med & Hyg; US-Can Acad Path. *Res:* Schistosomiasis; sarcocystosis; infectious diseases; pathology. *Mailing Add:* 2011 Town Hill Rd Houston TX 77062

SMITH, JEROME PAUL, b Ft Wayne, Ind, Apr 18, 46. ANALYTICAL CHEMISTRY, OCCUPATIONAL HEALTH. *Educ:* Col St Thomas, BA, 69; Univ Colo, PhD(anal chem), 73. *Prof Exp:* Asst res chemist, Univ Calif, Riverside, 73-77; AEROSOL RES SPECIALIST, NAT INST OCCUP SAFETY & HEALTH, 78- *Res:* Analytical chemistry applied to the measurement of contaminants in the workplace; health assessment for contaminants in the workplace. *Mailing Add:* MCRB N10SH 4676 Columbia Pkwy Columbia Pkwy OH 45226-1922

SMITH, JERRY HOWARD, b Mobile, Ala, Jan 28, 44. ORGANIC CHEMISTRY, BIO-ORGANIC CHEMISTRY. *Educ:* Auburn Univ, BS, 66; Emory Univ, PhD(org chem), 70. *Prof Exp:* Fel, Univ Chicago, 70-72, NIH fel, 71-72; asst prof chem, Marquette Univ, 72-76; RES CHEMIST, ICI AMERICAS, 77- *Mem:* Am Chem Soc. *Res:* Organic chemistry. *Mailing Add:* Chem Technol Group ICI Americas Inc Wilmington DE 19897

SMITH, JERRY JOSEPH, b Oblong, Ill, Feb 8, 39; m 55; c 4. PHYSICAL INORGANIC CHEMISTRY. *Educ:* Univ Ill, BS, 61; Univ Calif, Berkeley, PhD(inorg chem), 65. *Prof Exp:* Instr inorg chem & sr res assoc, Univ Wash, 66-70; asst prof chem, Drexel Univ, 70-76; chemist, Chicago Off Naval Res, 76-78; CHEMIST, ARLINGTON OFF NAVAL RES, 78- *Mem:* Electrochem Soc; Am Chem Soc. *Res:* Electrochemistry; surface science; interfacial processes. *Mailing Add:* Off Naval Res Code 472 Arlington VA 22217

SMITH, JERRY MORGAN, b Winchester, Va, Mar 13, 34; m 57; c 2. PHARMACOLOGY, TOXICOLOGY. *Educ:* The Citadel, BS, 56; Med Col SC, MS, 59; Univ Kans, PhD(pharmacol), 64. *Prof Exp:* Fel pharmacol, Emory Univ, 63-65; NIH fel, 64-65; res pharmacologist, Lederle Labs, Am Cyanamid Co, 65-69; sect head toxicol, Wellcome Res Labs, Burroughs Wellcome & Co, 69-70; dir toxicity, Biodynamics, Inc, 70-73; dir toxicol, Rohm & Haas Co, 73-90; RES ANALYST, UNIV DEL, 90- *Res:* Renal and biochemical pharmacology; teratology; toxicology, especially pharmaceutical, pesticide, food and chemical. *Mailing Add:* Grad Col Marine Studies Univ Del Robinson Hall Newark DE 19716

SMITH, JERRY WARREN, b Welch, WVa, Oct 8, 42; m 67; c 1. VIROLOGY, IMMUNOLOGY. *Educ:* Marshall Univ, BS, 64; Ohio Univ, MS, 66; Univ Iowa, PhD(microbiol), 70. *Prof Exp:* NIH fel, Baylor Col Med, 70-72; asst prof, 72-77, assoc prof microbiol, LA State Univ Med Ctr, New Orleans, 77-; AT RES & DEVELOP DIV, DIFCO LABS, INC. *Mem:* Am Soc Microbiol. *Res:* Immunological interactions between viruses and hosts; relationship between herpes viruses and human cancers. *Mailing Add:* Res & Devel Div, Difco Labs 1180 Ellsworth Rd Ann Arbor MI 48108

SMITH, JESSE GRAHAM, JR, b Winston-Salem, NC, Nov 22, 28; m 50; c 3. DERMATOLOGY, GERONTOLOGY. *Educ:* Duke Univ, MD, 51. *Prof Exp:* From asst resident to resident dermat, Duke Univ Hosp, 54-56 & Jackson Mem Hosp, 56-57; from instr to asst prof, Univ Miami, 57-60; from assoc prof to prof, Sch Med, Duke Univ, 60-67; actg chmn, dept path, 73-75, PROF DERMAT & MED & CHMN DEPT DERMAT, MED COL GA, 67- *Concurrent Pos:* Nat Inst Arthritis & Metab Dis fel, 57-60; mem, Gen Med A Study Sect, NIH, 64-69, chmn, 68-69, mem, Dermat Training Grants Comt, 69-73 & Adv Coun, Nat Inst Arthritis, Metab & Digestive Dis, 75-79; chief staff, Eugene Talmadge Mem Hosp, 70-72; mem & dir, Am Bd Dermat, 74-83, pres, 80-81; ed, Jour Am Acad Dermatol, 78-88; chmn, Sect Dermat, South Med Asn, 73-74, & AMA, 81-85. *Honors & Awards:* Am Dermat Asn Award, 59 & 60; Clyde L Cummer Gold Award, Am Acad Dermat, 63. *Mem:* Soc Invest Dermat (pres, 78-79); AMA; Am Dermat Asn (secy, 76-81, pres, 81-82); fel Am Col Physicians; Am Fedn Clin Res; Asn Professors Dermat (pres, 84-86); Am Acad Dermat (pres, 89-90); Soc Investigative Dermat (pres, 79-80). *Res:* Aging of the skin. *Mailing Add:* Dept of Dermat Med Col of Ga Augusta GA 30912

SMITH, JOE K, b Burlington, Ky, Feb 5, 30; m 60. MATHEMATICS. *Educ:* Eastern Ky State Col, BS, 52; Fla State Univ, MS, 57, EdD(math ed), 67. *Prof Exp:* Instr math, Cent Fla Jr Col, 58-60; instr, Miami Dade Jr Col, 60-63, assoc prof, 64-65; teacher high sch, Fla, 65-66; asst prof, Western Ky Univ, 66-69, assoc prof, 69-71; ASSOC PROF MATH, NORTHERN KY UNIV COL, 71- *Mem:* Math Asn Am; Am Meteorol Soc; Nat Coun Teachers Math. *Res:* Mathematics education. *Mailing Add:* 214 Buckingham Dr Florence KY 41042

SMITH, JOE M(AUK), b Sterling, Colo, Feb 14, 16; m 43; c 2. CHEMICAL ENGINEERING. *Educ:* Calif Inst Technol, BS, 37; Mass Inst Technol, ScD(chem eng), 43. *Prof Exp:* Design engr, Tex Co, NY, 37-38; res engr, Stand Oil Co, Calif, 38-41; asst, Nat Defense Res Comt, Mass Inst Technol, 42-43, instr chem eng, 43; asst prof, Univ Md, 43-44; proj engr, Publicker Com Alcohol Co, Pa, 44-45; from asst prof to prof chem eng, Purdue Univ, 45-57, asst dir eng exp sta, 54-57; dean col technol, Univ NH, 57; prof chem eng & chmn dept, Northwestern Univ, 57-60, Walter P Murphy distinguished prof, 59-61; PROF CHEM ENG & CHMN DEPT, UNIV CALIF, DAVIS, 61- *Concurrent Pos:* Res Corp grant, 46-48; Guggenheim fel & Fulbright res scholar, Delft Univ Technol, 53-54; Ford Found scholar, Argentina, 61; hon prof, Univ Buenos Aires, 63-; Fulbright awards, Argentina, 63, 66, Spain, 65 & Brazil, 90. *Honors & Awards:* William H Walker Award, Am Inst Chem Engrs, 60, R H Wilhelm Award, 77, W K Lewis Award, 83. *Mem:* Nat Acad Eng; Am Chem Soc; Am Inst Chem Engrs. *Res:* Interaction of physical and chemical processes in heterogeneous reactions; heat transfer combined with chemical reaction; applied chemical kinetics and reactor design. *Mailing Add:* Dept Eng Univ Calif Davis CA 95616

SMITH, JOE NELSON, JR, b Washington, DC, June 24, 32; m 52; c 3. SURFACE PHYSICS, SOLID STATE PHYSICS. *Educ:* Calif Inst Technol, BS, 58, MS, 59; Nat Univ Leiden, PhD, 70. *Prof Exp:* Staff mem atomic physics, Gen Atomic Div, Gen Dynamics Corp, 59-69; vis scientist, FOM Inst Atomic & Molecular Physics, Neth, 69-70; sr staff physicist, Gulf Radiation Technol Div, Gulf Energy & Environ Systs Co, 70-72; sr staff physicist, IRT Corp, 72-74; staff surface physicist, Gen Atomic Co, 74-83; Staff scientist, Ga Technol, Inc, 83-88; STAFF SCIENTIST, GEN ATOMICS, 88- *Mem:* Am Phys Soc; Am Vacuum Soc; Am Nuclear Soc. *Res:* Experimental research in particle-surface interactions including momentum and energy transfer, chemical reaction and catalysis, surface ionization; sputtering, radiation damage in surface region of solids, secondary ion mass spectrometry,auger electron spectroscopy, thin film phenomena and high voltage vacuum breakdown; thermionic energy conversion. *Mailing Add:* 4688 Sun Valley Rd Del Mar CA 92014

SMITH, JOHN, b Selkirk, Scotland, May 4, 38; m 64; c 1. ELEMENTARY PARTICLE PHYSICS. *Educ:* Univ Edinburgh, BS, 60, MS, 61, PhD, 63. *Prof Exp:* Joint Inst Nuclear Res, Dubna, Russia, 63; NATO res fel, Niels Bohr Inst, Copenhagen, Denmark, 64-65; Rothman res fel, Univ Adelaide, 66-67; res assoc, 67-69, asst prof, 69-74, assoc prof, 74-78, PROF PHYSICS, INST THEORET PHYSICS, STATE UNIV NY STONY BROOK, 78- *Mem:* Am Inst Physics. *Res:* Elementary particle physics. *Mailing Add:* Inst Theoret Phys State Univ NY Stony Brook Stony Brook NY 11794-3840

SMITH, JOHN BRYAN, b June 17, 42; Brit citizen; m 67; c 2. HEMATOLOGY, THROMBOSIS. *Educ:* London Univ, PhD(biochem), 71. *Prof Exp:* Prof pharmacol, Thomas Jefferson Univ, 71-82; PROF PHARMACOL, TEMPLE UNIV MED SCH, 82-, CHMN, DEPT PHARMACOL, 86- *Concurrent Pos:* Mem, Prog Comt, Am Soc Pharmacol & Exp Therapeut, 87-93; vpres, Mid-Atlantic Pharmacol Asn, 89-91, pres, 91- *Mem:* AAAS; Am Soc Pharmacol & Exp Therapeut; Sigma Xi. *Res:* Involvement of blood platelets in hemostasis and thrombosis. *Mailing Add:* Dept Pharmacol Temple Univ Med Sch Philadelphia PA 19140

SMITH, JOHN COLE, b Anniston, Ala, Jan 26, 35; m 59; c 2. ENTOMOLOGY. *Educ:* Auburn Univ, BS, 57, MS, 61; La State Univ, PhD(entom), 65. *Prof Exp:* Asst prof entom, 65-71, ASSOC PROF ENTOM, VA POLYTECH INST & STATE UNIV, 71- *Mem:* Entom Soc Am; Am Peanut Res & Educ Soc. *Res:* Research on insects affecting peanuts and soybeans, primarily through chemical and biological control with major emphasis on resistant plant lines. *Mailing Add:* Holland Sta 400 Kingsdale Rd Suffolk VA 23437

SMITH, JOHN EDGAR, b West Alexander, Pa, June 14, 39. NUTRITIONAL BIOCHEMISTRY. *Educ:* WLiberty State Col, BS, 61; WVa Univ, MS, 65; Univ Nebr, Lincoln, PhD(nutrit), 70. *Prof Exp:* Teacher high sch, Ohio, 61-62; trainee med, Columbia Univ, 69-72, res assoc, 72-80; MEM FAC NUTRIT PROG, COL HUMAN DEVELOP, PA STATE UNIV, 80- *Mem:* AAAS; Am Chem Soc; Am Soc Cell Biol; Am Inst Nutrit. *Res:* Fat-soluble vitamin transport in blood. *Mailing Add:* 308 Toftrees Ave Apt 232 State Col PA 16803-2024

SMITH, JOHN ELVANS, b Washington, DC, Sept 6, 29; m 62; c 3. ANALYTICAL CHEMISTRY. *Educ:* Univ Colo, BA, 52, PhD(anal chem), 60. *Prof Exp:* Res chemist, US Naval Res Lab, DC, 60-62; asst prof chem, Pueblo Col, 62-64, assoc prof, Southern Colo State Col, 64-70, PROF CHEM, UNIV SOUTHERN COLO, 70- *Mem:* Am Chem Soc; Nat Educ Asn. *Res:* Energy conservation; gas chromatography; electroanalytical chemistry; spectrographic and spectrophotometric analysis. *Mailing Add:* Dept Chem Univ Southern Colo Pueblo CO 81001

SMITH, JOHN ERNEST, JR, solid state physics, for more information see previous edition

SMITH, JOHN F(RANCIS), b Kansas City, Kans, May 9, 23; m 47; c 2. METALLURGY, PHYSICAL CHEMISTRY. *Educ:* Univ Mo-Kansas City, BA, 48; Iowa State Univ, PhD(phys chem), 53. *Prof Exp:* Assoc prof, 54-63, chmn dept, 66-70, div chief, Inst Atomic Res, 66-70, chemist & metallurgist, 53-66, PROF MAT SCI & ENG, IOWA STATE UNIV, 63-, SR METALLURGIST & SECT CHIEF, AMES LAB, DEPT ENERGY, 70- *Concurrent Pos:* Mem, World Metall Cong, Chicago, 57; consult, Tex Instruments, 58-63, Argonne Nat Lab, 64-70, Iowa Hwy Comn, 73, Los Alamos Nat Lab, 83-86 & Nat Bur Standards, 88- *Mem:* Fel Am Inst Chemists; fel Am Soc Metals; Am Crystallog Asn; Metall Soc; Am Inst Mining, Metall & Petrol Engrs. *Res:* Crystal structures and thermodynamics of alloys and intermetallic compounds; relationships between energetics, crystal structures, and physical properties; nondestructive evaluation. *Mailing Add:* Dept Mat Sci & Eng 122 Wilhelm Hall Iowa State Univ Ames IA 50011

SMITH, JOHN HENRY, b Gilman, Iowa, July 18, 04; m 36. STATISTICS. *Educ:* Iowa State Teachers Col, BA, 35; Univ Chicago, MBA, 39, PhD(bus), 41. *Prof Exp:* Instr statist, Univ Chicago, 40-42; statistician, US Bur Labor Statist, 42-47; prof statist, 47-73, EMER PROF STATIST, AM UNIV, 73- *Mailing Add:* Realife Coop Owatonna MN 55060

SMITH, JOHN HENRY, b Rome, NY, July 3, 37. MECHANICAL PROPERTIES, FRACTURE MECHANICS. *Educ:* Lafayette Col, AB & BS, 58; Univ Mo, MS, 59; Mass Inst Technol, ScD(metall), 64. *Prof Exp:* Aerospace scientist, NASA Lewis Res Ctr, 64-66; metallurgist, Res Ctr, US Steel Corp, 66-74; metallurgist, 74-86, DEP CHIEF, METALL DIV, NAT BUR STANDARDS, 86- *Concurrent Pos:* Res assoc , dept metall, Mass Inst Technol, 66; Dept Commerce Sci & Technol fel, 80-81. *Mem:* Am Soc Testing & Mat; Am Soc Metals; Am Inst Mining, Metall & Petrol Engrs; Am Welding Soc; Am Soc Mech Engrs; Int Inst Welding. *Res:* Mechanical properties of materials; non-destructive testing; structural integrity; physical metallurgy; welding; pressure vessels and piping materials; structural analysis and fracture mechanics. *Mailing Add:* B-261 Mat Bldg 223 Nat Inst Standards & Technol Gaithersburg MD 20899

SMITH, JOHN HOWARD, b Ithaca, NY, Jan 21, 37. MATHEMATICS. *Educ:* Cornell Univ, AB, 58; Mass Inst Technol, PhD(math), 63. *Prof Exp:* Instr math, Univ Mich, 63-65, asst prof, 65-66; vis lectr, Mass Inst Technol, 66-67; from asst prof to assoc prof, 67-82, PROF MATH, BOSTON COL, 82- *Mem:* Am Math Soc; Math Asn Am. *Res:* Algebraic number theory; linear algebra; combinatorics. *Mailing Add:* Dept Math Boston Col Chestnut Hill MA 02167

SMITH, JOHN LESLIE, JR, b Waco, Tex, Dec 2, 24; m 64; c 2. MEDICINE, DERMATOPATHOLOGY. *Educ:* Tulane Univ, MD, 48; Am Bd Path, dipl, 55, cert clin path, 56 & cert dermatopath, 75. *Prof Exp:* Asst pathologist, Armed Forces Inst Path, 55-57; from asst pathologist to assoc pathologist, Univ Tex M D Anderson Hosp & Tumor Inst Houston, 57-74; from asst prof to prof path, 57-80, Univ Tex Grad Sch Biomed Sci, 74-80, Ashbel Smith prof, 80-86, ASHBEL SMITH EMER PROF PATH, UNIV TEX M D ANDERSON CANCER CTR, 86-, PATHOLOGIST, 74- *Mem:* Int Acad Path; Am Soc Clin Path; Am Acad Dermat; Am Soc Dermatopath; AMA; Col Am Pathologists. *Res:* Pathology; neoplastic diseases. *Mailing Add:* Dept of Path M D Anderson Cancer Ctr Houston TX 77030

SMITH, JOHN M, b Indianapolis, Ind, May 20, 22; m 48; c 3. ECONOMIC GEOLOGY, CLAY MINERALOGY. *Educ:* Ind Univ, BS, 52, MA, 54. *Prof Exp:* Geologist, Ind Geol Surv, 54-57; geologist, 57-67, CHIEF GEOLOGIST & DIR, GA KAOLIN CO INC, 67- *Concurrent Pos:* Mem, State Bd Regist Prof Geologists, Ga, 75-81. *Mem:* Fel Geol Soc Am; Clay Minerals Soc; Am Inst Mining, Metall & Petrol Engrs. *Res:* Evaluation of United States and foreign nonmetallic mineral deposits; direction of exploration and development of nonmetallics; research and development of clay minerals, especially kaolinite. *Mailing Add:* 624 Old Club Rd S Macon GA 31210

SMITH, JOHN MELVIN, b Washington, DC, Apr 16, 37; m 59; c 4. MATHEMATICS EDUCATION. *Educ:* Univ Richmond, BS, 59; Univ Md, MA, 61, PhD(math, educ), 70. *Prof Exp:* Instr math, Georgetown Univ, 62-66; from instr to assoc prof, 66-75, PROF MATH EDUC, GEORGE MASON UNIV, 75- *Concurrent Pos:* Mathematician, Nat Bur Standards, 57- *Mem:* Am Math Soc; Math Asn Am; Nat Coun Teachers Math. *Res:* Matrix theory; computer applications; elementary school mathematics. *Mailing Add:* Dept Math George Mason Univ Fairfax VA 22030-4444

SMITH, JOHN ROBERT, b Salt Lake City, Utah, Oct 1, 40; m 62; c 2. THEORETICAL SOLID STATE PHYSICS. *Educ:* Toledo Univ, BS, 62; Ohio State Univ, PhD(physics), 68. *Prof Exp:* Aerospace engr surface physics, Lewis Res Ctr, NASA, 65-68, fel solid state theory, Univ Calif, San Diego, 70-72; sr res physicist & head, Surface & Interface Physics Group, 72-80, sr staff scientist & head, Solid State Physics Group, 80-86, PRIN RES SCIENTIST, GEN MOTORS, 86- *Concurrent Pos:* Air Force Off Sci Res & Nat Res Coun fel, Univ Calif, 70-72; adj prof, physics dept, Univ Mich, 83- *Honors & Awards:* David J Adler Award, Am Phys Soc, 91. *Mem:* Am Vacuum Soc; Sigma Xi; Fel Am Phys Soc. *Res:* Theory of solid surfaces, electronic properties, magnetic properties and chemisorption; adhesion, metal contact electronic structure; defects and universal features of bonding in solids. *Mailing Add:* Physics Dept Gen Motors Res Warren MI 48090-9055

SMITH, JOHN ROBERT, b Los Angeles, Calif, Apr 17, 48; c 2. PHARMACOLOGY. *Educ:* Loyola Univ, Calif, BS, 70; Ore State Univ, MS, 74, PhD(pharmacol), 76. *Prof Exp:* Fel pharmacol, Sch Med, Univ Wash, 75-77; asst prof, Sch Dent Med, Southern Ill Univ, 77-79; asst prof & sr researcher comp pharmacol, Marine Sci Ctr, Ore State Univ, 79-; ASSOC PROF, DEPT PHYSIOL & PHARMACOL, SCH DENT, ORE HEALTH SCI UNIV. *Mem:* Am Soc Pharmacol & Exp Therapeut; Int Asn Dent Res; PANWAT. *Res:* Comparative neuropharmacology and toxicology. *Mailing Add:* Dept Physiol & Pharmacol Dent Sch Ore Health Sci Univ 611 SW Campus Dr Portland OR 97201

SMITH, JOHN THURMOND, b Mo, May 29, 25; wid. BIOCHEMISTRY. *Educ:* Culver-Stockton Col, BA, 51; Univ Mo, MS, 53, PhD(agr chem), 55. *Prof Exp:* Asst agr chem, Univ Mo, 52-55, res assoc biochem, 55; res assoc biochem, 56-58, from instr to prof, 58-90, EMER PROF NUTRIT, UNIV TENN, KNOXVILLE, 90- *Mem:* Am Inst Nutrit; Am Chem Soc. *Res:* Chemistry of sulfur compounds and sulphatases; dietary sulfur and xenobiotic metabolism; enzyme chemistry; food phosphates; importance of inorganic sulfate, enzymes, metabolic obesity. *Mailing Add:* Col Human Ecol Univ Tenn Knoxville TN 37996-1900

SMITH, JOHN W(ARREN), b De Soto, Mo, Nov 18, 43; m 64; c 3. ENVIRONMENTAL ENGINEERING, CIVIL ENGINEERING. *Educ:* Mo Sch Mines & Metall, BS, 65; Univ Mo-Rolla, MS, 67, PhD(civil eng), 68. *Prof Exp:* Instr civil eng & res asst environ eng, Univ Mo-Rolla, 65-68; proj engr, Esso Res & Eng Co, 68-70; from asst prof to assoc prof environ eng, 70-75, actg dir, Ctr Alluvial Valley Studies, 77-79, PROF CIVIL ENG, MEMPHIS STATE UNIV, 75- *Concurrent Pos:* Mem bd consults, Ryckman, Edgerly, Tomlinson & Assocs, 71-72; actg dep city engr, City of Memphis, 73-74. *Mem:* Water Pollution Control Fedn; Am Water Works Asn. *Res:*

Industrial and municipal waste treatment; water resources; biological wastewater treatment using fixed film reactors, hazardous waste analysis and management, water reuse; kinetics of fixed film reactors and low energy treatment systems. *Mailing Add:* Dept Civil Eng Memphis State Univ Memphis TN 38152

SMITH, JOHN WOLFGANG, b Vienna, Austria, Feb 18, 30; US citizen. MATHEMATICS. *Educ:* Cornell Univ, AB, 48; Purdue Univ, MS, 50; Columbia Univ, PhD(math), 57. *Prof Exp:* Aerodynamicist, Bell Aircraft Corp, 50-53; C L E Moore instr math, Mass Inst Technol, 58-61; asst prof, Univ Calif, Los Angeles, 61-64; assoc prof, 64-67, PROF MATH, ORE STATE UNIV, 67- *Res:* Differential geometry; algebraic topology. *Mailing Add:* Dept of Math Ore State Univ Corvallis OR 97331

SMITH, JONATHAN JEREMY BERKELEY, b Leicester, Eng, Dec 26, 40; m 67, 81; c 2. SENSORY PHYSIOLOGY. *Educ:* Cambridge Univ, BA, 62, PhD(zool), 65, MA, 66. *Prof Exp:* From asst prof to assoc prof, 65-82, vdean arts & sci, 83-88, PROF ZOOL, UNIV TORONTO, 82- *Mem:* Brit Soc Exp Biol; Can Soc Zool; Asn Chemoreception Scis. *Res:* Sensory physiology; chemoreception and feeding in insects, particularly blood-feeders; hearing mechanisms in lower vertebrates. *Mailing Add:* Dept Zool Univ Toronto Toronto ON M5S 1A1 Can

SMITH, JOSEF RILEY, b Council Bluffs, Iowa, Oct 1, 26; m 73; c 7. PULMONARY PHYSIOLOGY. *Educ:* Northwestern Univ, MB, 50, MD, 51; Marquette Univ, MSc, 64; Am Bd Internal Med, cert, 57. *Prof Exp:* Instr, Dept Med, Univ Miss, Jackson, 56-59; asst prof, Dept Med, Marquette Univ, Milwaukee, 59-63; from assoc prof to prof internal med, Univ Mich, Ann Arbor, 63-72; prof, Col Med, Northeast Ohio Univ, 77-79; PVT PRACT INTERNAL MED, 87- *Concurrent Pos:* Nat Tuberc Asn fel, Univ Miss Med Sch, 57-59; Vet Admin clin investr cardiol, Vet Admin Hosp, Wood, Wis, 59-62; NIH fel biomed eng, Drexel Inst Technol & Marquette Univ, 62-63; counr, Dept Internal Med, Univ Mich Med Sch, 69-72; mem, Ad Hoc Comt Biomed Eng, Am Col Physicians, 69-71; controller, Mahoning County Tuberc Clin, 73-79; med dir respiratory ther, St Joseph Hosp, Tucson, Ariz, 86-91. *Mem:* Fel Am Col Physicians; fel Sigma Xi; Am Thoracic Soc. *Res:* Pulmonary physiology with particular interest in gas transfer. *Mailing Add:* 5551 E Hampton Suite 109 Tucson AZ 85712

SMITH, JOSEPH COLLINS, b Knoxville, Tenn, Oct 26, 28; m 53; c 1. ENGINEERING, ASTRONOMY. *Educ:* US Naval Acad, BS, 53; George Washington Univ, MA, 64. *Prof Exp:* Mem staff opers of Comdr Submarines Pac, 64-66; commanding officer submarines, USS Odax (55-484), 66-68; mem staff opers & plans submarines of Chief Naval Opers, 68-71; asst chief staff opers of Comdr Antisubmarine Warfare Forces, US Sixth Fleet, 71-74; prof & head dept naval sci, Iowa State Univ, 74-76; supt-in-chg time & navig, US Naval Observ, 76-79; consult, 79-82; dir, Space Studies & spec asst to pres, Nat Defense Univ, 82-85; VPRES, AM PRIDE, INC, 85- *Concurrent Pos:* Mem bd, Coast Guard Adv Bd, 76-; mem found, Nat Defense Univ; consult, 85- *Mem:* Naval Inst; Inst Navig. *Res:* Astrometric instrumentation; timing; fundamental positioning; global navigational equipment; systems improvements; radio interferometry; Strategic Defense Initiative; satellite systems. *Mailing Add:* 2711 Churchcreek Lane Poplar Pt Edgewater MD 21037

SMITH, JOSEPH DONALD, b New Brunswick, NJ, Sept 18, 43. BIOCHEMISTRY. *Educ:* Columbia Univ, AB, 65; Univ Chicago, PhD(biochem), 69. *Prof Exp:* Fel molecular biol, Albert Einstein Col Med, 70-74; res scientist neurosci, NY State Psychiat Inst, 74-75; asst prof chem, Miami Univ, 75-82; assoc prof , 82-89, PROF CHEM, SOUTHEASTERN MASS UNIV, 89-, CHAIRPERSON, CHEM DEPT, 90- *Concurrent Pos:* Asst prof biochem, Sch Med, Wright State Univ, 75-77. *Mem:* Am Chem Soc; AAAS; Am Soc Microbiol; NY Acad Sci; Am Soc Biochem & Molecular Biol; Protein Soc. *Mailing Add:* Dept Chem Southeastern Mass Univ North Dartmouth MA 02747

SMITH, JOSEPH EMMITT, b Big Spring, Tex, Jan 24, 38; m 60; c 2. CLINICAL PATHOLOGY, PHYSIOLOGY. *Educ:* Tex A&M Univ, BS, 59, DVM, 61; Univ Calif, Davis, PhD(comp path), 64. *Prof Exp:* Lab asst vet anat, Tex A&M Univ, 58-59, NSF res trainee reproduction physiol, 59-61; USPHS trainee metab & hemat dis, Univ Calif, Davis, 61-64; assoc res scientist, City of Hope Med Ctr, Duarte, Calif, 64-66; from asst prof to assoc prof path, Okla State Univ, 66-69; PROF PATH, COL VET MED, KANS STATE UNIV, 69-, HEAD DEPT PATH, 87- *Concurrent Pos:* USPHS res grant, 65-; career develop awardee, NIH. *Mem:* Am Vet Med Asn; Am Asn Clin Chemists; Am Soc Vet Clin Path; Soc Exp Biol & Med; Am Col Vet Path; Am Asn Hemat. *Res:* Inherited metabolic errors of animals which serve as models of human disorders, particularly those of erythrocyte metabolism; clinical enzymology. *Mailing Add:* Col of Vet Med Kans State Univ Manhattan KS 66506-5605

SMITH, JOSEPH H, JR, b Moscow, Pa, July 9, 25. ELECTRICAL ENGINEERING. *Educ:* Rensselaer Polytech Inst, BEE, 45, MEE, 54. *Prof Exp:* From instr to asst prof, 48-59, asst dept head, 59-60, exec officer dept elec eng, 60-69, ASSOC PROF ELEC ENG, RENSSELAER POLYTECH INST, 60-, ASST DEAN ENG FOR PRE-ENG CURRIC, 69- *Mem:* AAAS; Am Soc Eng Educ. *Res:* Electromechanical energy conversion, especially recent developments in speed control of alternating current saturistor motors; transmission or transfer of power in a multimachine system. *Mailing Add:* Dept Elec Comp & Syst Eng Rensselaer Polytech Inst Troy NY 12180

SMITH, JOSEPH HAROLD, b Fielding, Utah, Oct 5, 14; m 40; c 5. PHYSICAL CHEMISTRY, INORGANIC CHEMISTRY. *Educ:* Univ Utah, BS, 36, MA, 38; Univ Wis, PhD(phys chem), 41. *Prof Exp:* Asst chem, Univ Utah, 36-38; asst, Univ Wis, 38-41; instr, Univ Ill, 41-43; from assoc prof to prof chem, Univ Mass, Amherst, 43-77; RETIRED. *Concurrent Pos:* Consult, Chicopee Mfg Corp, 51-; indust consult, Johnson & Johnson. *Mem:* Am Chem Soc. *Res:* Chemical kinetics; coordination complexes; spectrophotometrics of complex ions; gas phase kinetics of nitrogen oxides; carbon dioxide absorption rates in alkaline solutions. *Mailing Add:* 84 Stoney Hill Rd Amherst MA 01002

SMITH, JOSEPH JAMES, b New York, NY, Apr 6, 21; m 46; c 7. CREATIVITY, INNOVATION. *Educ:* Fordham Univ BS, 43. *Hon Degrees:* DS, WVa State Col, 75; DS, Marshall Univ, 76. *Prof Exp:* Res chemist, Bakelite Co Div, Union Carbide & Carbon Corp, 43-51, proj leader, 51-53, group leader, 53-56, sect head, Union Carbide Plastics Co, 56-64, asst dir, 64-66, tech mgr polyolefins & asst dir res & develop, Chem Div, 66-67, dir chem & plastics div, 67-82; CONSULT, CREATIVITY, INNOVATION, R & D ORGN, J J SMITH CONSULT, 82- *Honors & Awards:* Medal, Am Inst Chemists, 43. *Mem:* Am Chem Soc. *Res:* Surface chemistry; corrosion and protective coatings; infrared and ultraviolet spectrophotometry; chemistry of organic high polymers; catalysis; organometallic compounds; polyolefins; electron irradiation; program management of research and development organizations; creativity and innovation. *Mailing Add:* 1580 Virginia St E Charleston WV 25311

SMITH, JOSEPH JAY, b Cementon, NY, Feb 13, 15; m 62; c 3. OBSTETRICS & GYNECOLOGY. *Educ:* Cornell Univ, AB, 36; Long Island Col Med, Cornell Univ, MD, 41. *Prof Exp:* Intern, Montefiore Hosp, New York, 41-42 & Fordham Hosp, 42-43, resident path, 45-46; intern obstet & gynec, Long Island Col Hosp, 46-47; from asst resident to resident, Greenpoint Hosp, Brooklyn, 47-50; clin instr, Col Med, State Univ NY Downstate Med Ctr, 50-54; from clin asst prof to clin assoc prof, 55-66, dir dept gynec & obstet, Lincoln Hosp, 63-70, assoc prof, 66-73, PROF GYNEC & OBSTET, ALBERT EINSTEIN COL MED, 73- *Concurrent Pos:* Attend obstetrician, Bronx Munic Hosp Ctr, 57- *Mem:* Fel Am Col Surgeons; fel Am Col Obstet & Gynec. *Res:* Resident and student training in obstetrics and gynecology. *Mailing Add:* Dept Obstet-Gynec Albert Einstein Col Med Bronx NY 10461

SMITH, JOSEPH LECONTE, JR, b Macon, Ga, Sept 4, 29; m 53; c 3. MECHANICAL & CRYOGENIC ENGINEERING. *Educ:* Ga Inst Technol, BME, 52, MS, 53; Mass Inst Technol, DSc(mech eng), 59. *Prof Exp:* Res asst, 55-56, from instr to assoc prof, 56-69, PROF MECH ENG, MASS INST TECHNOL, 69-, PROF IN CHG CRYOGENIC ENG LAB, 64- *Concurrent Pos:* Consult, Los Alamos Sci Lab, 58-66, Cambridge Electron Accelerator, Harvard Univ, 59-65, Arthur D Little, Inc, 69-71 & Westinghouse Res Lab, 70- *Mem:* Nat Acad Eng; Am Soc Mech Engrs; Inst Elec & Electronics Engrs; Sigma Xi. *Res:* Thermodynamics; heat transfer; fluid mechanics. *Mailing Add:* Dept Mech Eng Bldg 41-204 Mass Inst Technol Cambridge MA 02139

SMITH, JOSEPH PATRICK, b Lackawanna, NY, July 9, 51; m 79; c 1. SPECTROSCOPY, SURFACE & COLLOID SCIENCE. *Educ:* Univ Rochester, BS, 72; Univ Calif, Berkeley, PhD(phys chem), 78. *Prof Exp:* NSF Nat Needs Res fel, Dept Biochem, Univ Wis-Madison, 78-79; fel, Chem Div, Argonne Nat Lab, 79-81; sr res chemist, Long Range Res Div, 81-86, res specialist, Reservoir Div, 86-90, RES SPECIALIST, PROD OPERS DIV, EXXON PROD RES CO, 90- *Mem:* Am Chem Soc; Soc Petrol Eng. *Res:* Environmental science: modeling of offshore and coastal discharges; physico-chemical aspects of petroleum production (wettability, surfactant flooding); spectroscopy and nitrogen fixation. *Mailing Add:* Exxon Prod Res Co PO Box 2189 Houston TX 77252-2189

SMITH, JOSEPH VICTOR, b Eng, July 30, 28; nat US; m 51; c 2. MINERALOGY. *Educ:* Cambridge Univ, BA, 48, MA & PhD(physics), 51. *Prof Exp:* Fel crystallog, Geophys Lab, Carnegie Inst, 51-54; demonstr mineral & petrol, Cambridge Univ, 54-56; from asst prof to assoc prof mineral, Pa State Univ, 56-60; prof mineral & crystallog, 60-76, LOUIS BLOCK PROF PHYS SCI, UNIV CHICAGO, 76-, EXEC DIR, CONSORTIUM ADVAN RADIATION SOURCES, 89- *Concurrent Pos:* Consult, Linde Div, Union Carbide Corp, 56-87; ed, X-ray Powder Data File, Am Soc Testing & Mat, 58-68. *Honors & Awards:* Mineral Soc Am Award, 61; Murchison Medal, 80; Roebling Medal, 81. *Mem:* Nat Acad Sci; fel Geol Soc Am; fel Royal Soc; fel Am Geophys Union; fel Am Acad Arts & Sci; fel Mineral Soc Am (pres, 72-73); hon fel Mineral Soc Great Britain; fel Meteoritical Soc; fel AAAS. *Res:* Mineralogy applied to petrology, geochemistry and industrial chemistry. *Mailing Add:* Dept of Geophys Sci Univ of Chicago Chicago IL 60637

SMITH, JOSEPHINE REIST, b Altoona, Pa, Nov 26, 29; m 51; c 3. MEDICAL MICROBIOLOGY. *Educ:* Pa State Univ, BS, 52; Temple Univ, MS, 71, PhD(pharm chem), 76. *Prof Exp:* Technician immunol, Protein Found, Dept Phys Chem, Harvard Univ, 53-54; res asst, Dept Biol, Haverford Col, 64-69; INSTR MICROBIOL, DEPT BIOL, MONTGOMERY COUNTY COMMUNITY COL, 75- *Mem:* Am Soc Microbiol; Sigma Xi; AAAS. *Res:* Application of pharmacokinetics to the study of immune responses. *Mailing Add:* 12 Forest Rd Wayne PA 19087

SMITH, JUDITH TERRY, b New York, NY, Mar 4, 40; m 69; c 3. PALEONTOLOGY. *Educ:* Barnard Col, Columbia Univ, AB, 62; Stanford Univ, MS, 64, PhD(geol), 68. *Prof Exp:* Curatorial asst, 68-69, RES ASSOC PALEONT, DEPT GEOL, STANFORD UNIV, 69-; GEOLOGIST, US GEOL SURV, 70 & 72- *Concurrent Pos:* Res assoc, US Geol Surv, 72-73. *Mem:* Geol Soc Am; Paleont Soc; Paleont Res Inst. *Res:* Cenozoic molluscan paleontology; biostratigraphy and paleoecology, especially zonation and distribution problems involving pectinids; Tethyan and related faunas. *Mailing Add:* 1527 Byron St Palo Alto CA 94301

SMITH, JULIAN CLEVELAND, b Westmount, Que, Mar 10, 19; US citizen; m 46; c 3. CHEMICAL ENGINEERING. *Educ:* Cornell Univ, BChem, 41, ChemE, 42. *Prof Exp:* Engr chem eng, E I du Pont de Nemours & Co, 42-46; from asst prof to prof, 46-86, assoc dir, 73-75, dir, 75-83, EMER PROF CHEM ENG, CORNELL UNIV, 86- *Concurrent Pos:* Prof engr, NY State Educ Dept, 57-; consult, E I du Pont de Nemours & Co, 55-89, US Army Corps Engrs, 57-65, Atlantic Richfield Hanford Co, 71-77, Rockwell Int Co, 77-87, Westinghouse Hanford Co, 87-88, Am Cyanamid, 89- *Honors & Awards:* Julian C Smith lectr chem eng, Cornell Univ, 88. *Mem:* Fel Am Inst Chem Engrs; Am Chem Soc. *Res:* Mixing of liquids and pastes; flow of granular solids; centrifugal separation. *Mailing Add:* Sch Chem Eng Cornell Univ Ithaca NY 14853

SMITH, KAREN ANN, b Idaho Falls, Idaho, Aug 30, 58; m 89; c 4. ANALYTICAL CHEMISTRY. *Educ:* Pa State Univ, BS, 78; Univ Ill, PhD(phys chem), 84. *Prof Exp:* res chemist, Res & Develop Div, Colgate-Palmolive Co, 84-90; NUCLEAR MAGNETIC RESONANCE SPECTROSCOPIST, IOWA STATE UNIV, 91- *Mem:* Am Chem Soc; Am Inst Chemists; NY Acad Sci; AAAS. *Res:* Solid-state and solution nuclear magnetic resonance. *Mailing Add:* RR 4 Box 101B Boone IA 50036-9308

SMITH, KATHLEEN, b Fayetteville, Ark, Oct 9, 22. PSYCHIATRY. *Educ:* Univ Ark, BS, 44; Wash Univ, MD, 49; Am Bd Psychiat & Neurol, dipl, 56. *Prof Exp:* Intern, St Louis City Hosp, 49-50; resident psychiat, Barnes & McMillan Hosps, 50-52 & Malcolm Bliss Psychiat Hosp, 52-53; from instr to assoc prof, Sch Med, Washington Univ, 53-72, prof psychiat, 72-; RETIRED. *Concurrent Pos:* USPHS fel, 52-53; physician, Malcolm Bliss Ment Health Ctr, 53-, dir inpatient serv, 57-60, dir training, 60-64, med supt, 64-84; vis physician, St Louis City Hosp, 56-, consult, Nurses Infirmary, 55-57. *Mem:* AMA; Am Psychiat Asn. *Res:* Odor of schizophrenic sweat; gas chromatography; multihospital studies of tranquilizers and antidepressants. *Mailing Add:* APA Rte 1 Box 312 DuQuoin IL 62832

SMITH, KEITH DAVIS, b Portsmouth, Ohio, Dec 14, 30; m 58; c 3. ENDOCRINOLOGY, INTERNAL MEDICINE. *Educ:* Pa State Univ, BS, 52; Sch Med, Univ Pittsburgh, MD, 59. *Prof Exp:* Chief div endocrinol, Mercy Hosp, Pittsburgh, 64-67; asst prof med, Sch Med, Temple Univ, 68-71; assoc prof reproductive med, 71-75, PROF REPRODUCTIVE MED, UNIV TEX MED SCH, HOUSTON, 75- *Concurrent Pos:* USPHS fel, Albert Einstein Med Ctr, Philadelphia, 62-64, assoc mem div endocrinol, 68-71; clin instr med, Sch Med, Univ Pittsburgh, 65-67; prin investr contract endocrine changes vasectomized men, NIH, 72-79, co-investr prog proj multidisciplinary approach control male reproduction, 74-83; co-investr grant develop contraceptive agents human male, Ford Found, 74-78. *Mem:* Am Fertil Soc; Am Psychosomatic Soc; Endocrine Soc; Soc Study Reproduction; Am Soc Andrology. *Res:* Ovarian and testicular function; control by the pituitary and hypothalamus; effects of various physical, emotional and chemical stimuli on the hypothalamo-pituitary-gonadal axis. *Mailing Add:* Tex Inst Reprod Med & Endo 7800 Fannin St No 500 Houston TX 77054

SMITH, KEITH JAMES, b Cedar Rapids, Iowa, Mar 16, 37; c 2. ANIMAL NUTRITION. *Educ:* Iowa State Univ, BS, 59, PhD(ruminant nutrit), 63. *Prof Exp:* Asst prof basic nutrit, Iowa State Univ, 63; asst dir res & educ, Nat Cotton Seed Prod Asn, Inc, 66-75; animal nutritionist, 75-77, STAFF VPRES, RES & UTILIZATION, AM SOYBEAN ASN, 77- *Mem:* Am Inst Nutrit; Am Soc Animal Sci; Am Asn Food Technologists; Am Oil Chemists Soc. *Res:* Oilseed protein feed product utilization. *Mailing Add:* Am Soybean Asn Box 27300 777 Craig Rd St Louis MO 63141

SMITH, KELLY L, b Eugene, Ore, Mar 10, 51; m 74. CONTROLLED RELEASE, MEMBRANE TRANSPORT. *Educ:* Stanford Univ, BS & MS, 74. *Prof Exp:* Chem engr, Alza Corp, 74-78; DIR CONTROLLED RELEASE DIV, BEND RES INC, 78-, SECY, 88- *Concurrent Pos:* Mem bd dirs, Bend Res Inc, 81-; dir, Consep Membranes Inc, 84-, secy, 87-; mem bd gov, Controlled Release Soc, 87-90. *Mem:* Am Chem Soc; Am Inst Chem Engrs; Controlled Release Soc. *Res:* Controlled release theory; development of controlled release technologies and products in pharmaceuticals, animal health and agricultural fields; membrane transport and separation. *Mailing Add:* Bend Res Inc 64550 Res Rd Bend OR 97701

SMITH, KENDALL A, b Akron, Ohio, Feb 28, 42. T-CELLS. *Educ:* Ohio State Univ, MD, 68. *Prof Exp:* From asst prof to assoc prof, 74-82, PROF MED, DARTMOUTH MED SCH, 82- *Mem:* Am Soc Clin Invest; Am Asn Immunologists; Am Soc Hemat; Am Soc Clin Oncol; Am Fedn Cancer Res. *Mailing Add:* MD Ohio State Univ 11 Pleasant St Hanover NH 03755

SMITH, KENDALL O, b Wilson, NC, Sept 5, 28; m 49; c 1. MICROBIOLOGY, VIROLOGY. *Educ:* George Washington Univ, BA, 51; Univ NC, MS, 57, PhD(microbiol), 59. *Prof Exp:* Fel biophys, Univ NC, 59-60; from instr to assoc prof virol, Col Med, Baylor Univ, 60-65; res microbiologist, Div Biol Stand, NIH, 65-69; PROF MICROBIOL, MED SCH, UNIV TEX, SAN ANTONIO, 69- *Concurrent Pos:* Res fel, NIH, Baylor Univ, 61-62, res career develop award, 63. *Mem:* Am Soc Microbiol; Am Asn Immunol. *Res:* Correlation of physical and biological properties of viruses; viruses associated with chronic, degenerative diseases of man; chemotherapy of herpes infections; detection of viral antigens and antibodies by enzyme linked immunosorbent assay; reliable medical diagnostic technology for developing countries. *Mailing Add:* Dept of Microbiol Univ of Tex Med Sch San Antonio TX 78284

SMITH, KENDRIC CHARLES, b Oakwood, Ill, Oct 13, 26; m 55; c 2. PHOTOBIOLOGY, RADIATION BIOLOGY. *Educ:* Stanford Univ, BS, 47; Univ Calif, PhD(biochem), 52. *Prof Exp:* Res asst radiol, Med Sch, Univ Calif, 54-56; res assoc, 56-62, from asst prof to assoc prof, 62-73; prof radiol, 73-88, PROF RADIATION ONCOL, SCH MED, STANFORD UNIV, 88- *Concurrent Pos:* USPHS fel, Univ Calif, 52-54; USPHS career develop award, Stanford Univ, 66-71; mem, Comt Photobiol, Nat Acad Sci-Nat Res Coun, 64-74, chmn, 70-74; exec ed, Photochem & Photobiol, 66-72 & Photochem Photobiol Rev, 76-83; mem, US Nat Comt, Int Union Biol Sci, 67-73 & secy, 71-73; mem, Radiation Bio-effects & epidemiol Adv Comt, Food & Drug Admin, 74-75. *Honors & Awards:* Finsen Medal in Photobiol, 84. *Mem:* Am Inst Biol Sci(v pres, 81-82, pres, 82-83, past pres, 83-84); Am Soc Biol Chemists; Am Soc Microbiol; Brit Photobiology Soc; Am Soc Photobiology (pres, 72-74); Radiation Res Soc. *Res:* photochemical and radiation chemical reactions of nucleic acids; genetic control and biochemical mechanisms for repair of radiation damage; the molecular mechanisms of mutagenesis. *Mailing Add:* 927 Mears Ct Stanford CA 94305-1041

SMITH, KENNAN TAYLOR, b Green Bay, Wis, July 17, 26; m 48; c 2. MATHEMATICS. *Educ:* Bowling Green State Univ, BA, 47; Harvard Univ, MA, 48; Univ Wis, PhD(math), 51. *Prof Exp:* Asst prof math, Univ Kans, 52-59; from assoc prof to prof, Univ Wis-Madison, 59-68; PROF MATH, ORE STATE UNIV, 68- *Concurrent Pos:* Fulbright scholar, France, 51-52. *Mem:* Am Math Soc. *Res:* Linear topological spaces; Hilbert spaces. *Mailing Add:* Dept of Math Ore State Univ Corvallis OR 97331

SMITH, KENNETH A, b Winthrop, Mass, Nov 28, 36; m; c 4. MECHANICAL ENGINEERING. *Educ:* Mass Inst Technol, SB, 58, SM, 59, ScD(mech eng), 62. *Prof Exp:* Dir, Whitaker Col Health Sci & Technol, 89-91; from asst prof to assoc prof chem eng, Mass Inst Technol, 61-71, actg head, Dept Chem Eng, 76-77, Joseph R Mares prof chem eng, 78-81, assoc provost, 80-91, vpres res, 81-91, PROF CHEM ENG, MASS INST TECHNOL, 71-, EDWIN R GILLILAND PROF, 89- *Concurrent Pos:* Consult for major int petrol & chem co; postdoctoral fel, Cavendish Lab, Univ Cambridge, 64-65. *Honors & Awards:* Prof Progress Award, Am Inst Chem Engrs, 81. *Mem:* Nat Acad Eng; Am Chem Soc; Am Inst Chem Engrs; AAAS; Sigma Xi. *Res:* Fluid mechanics; heat and mass transfer; polymer characterization; desalination; liquefied natural gas; biomedical engineering; author of 78 publications. *Mailing Add:* Off Vpres Res Dept Chem Eng Bldg 66-405 Mass Inst Technol Cambridge MA 02139

SMITH, KENNETH CARLESS, b Toronto, Ont, May 8, 32. MAN-MACHINE INTERFACE, SPECIAL PURPOSE PROCESSORS. *Educ:* Univ Toronto, BASc, 54, MASc, 56, PhD(physics), 60. *Prof Exp:* Transmission engr tel, Can Nat Tel, 54-55; res engr digital electronics, Univ Toronto, 56-58, asst prof electronics, 60-61; res asst prof comput design, Dept Elec Eng, Univ Ill, 61-64, assoc prof, 64-65; assoc prof, 65-70, chmn dept, 76-81, PROF ELEC ENG & COMPUT SCI, DEPT ELEC ENG, UNIV TORONTO, 70-, PROF INFO SCI , FAC LIBR SCI, 81- *Concurrent Pos:* Res engr, Comput Ctr, Univ Toronto, 60-61; chief engr & prin investr, Digital Comput Lab, Univ Ill, 61-65, eng consult, Training Res Lab, 62-64, consult, Pattern Processor Proj, 65-70; dir & founder, Elec Eng Consociates, Ltd, 68-87, pres, 74-76; founding mem, Comput Systs Res Inst, Univ Toronto, 68-; dir, several small US & Can co, 68-; vpres, Owl Instruments, Ltd, Med Instrument Mfg, 70-80; assoc, Inst Biomed Eng, Univ Toronto, 74-; mem exec comt, awards chmn, Inst Elec & Electronics Engrs, Solid State Circuits Conf, 75-; chmn, pub coun, Can Soc Elec Eng, 83-; mem bd dir, Can Soc Prof Eng, 85- *Mem:* Fel Inst Elec & Electronics Engrs; Can Soc Elec Eng; Can Info Process Soc; Can Soc Prof Engrs. *Res:* Linear and digital circuits and systems including multivalued logic, parallel and other special-purpose processors, the man-machine interface and input and output systems with application in industry, education, medicine and music. *Mailing Add:* Dept Elec Eng Univ Toronto 10 King's Col Rd Toronto ON M5S 1A4 Can

SMITH, KENNETH EDWARD, b Milwaukee, Wis, Dec 29, 43; m 66; c 2. ANALYTICAL CHEMISTRY, SPECTROSCOPY. *Educ:* Univ Wis-Milwaukee, BS, 66; Univ Iowa, PhD(anal chem), 71. *Prof Exp:* NSF fel, Kans State Univ, 70-72; asst prof chem, Eastern Ill Univ, 72-73; assoc chemist, Wildlife Res Sect, Ill Natural Hist Surv, 73-80; mgr anal chem, Inst Gas Technol, 80-83; mgr anal chem, Swift Adhesives, 83-92, mgr aqueous adhesives, 87-90; SR DIR LAB SERV, NSF INT, ANN ARBOR, MICH, 90- *Concurrent Pos:* Instr, Kans State Univ. *Mem:* Am Chem Soc; Soc Appl Spectros; Am Soc Testing & Mat. *Res:* Analytical atomic spectroscopy; non-flame absorption systems; low-pressure plasmas in spectroscopy; temperature effects in plasmas; environmental metal analysis as contaminants. *Mailing Add:* 4375 Plymouth Rd Ann Arbor MI 48105

SMITH, KENNETH JUDSON, JR, b Raleigh, NC, Sept 4, 30; m 54; c 2. PHYSICAL CHEMISTRY, POLYMER PHYSICS. *Educ:* ECarolina Col, AB, 57; Duke Univ, MA, 59, PhD(phys chem), 62. *Prof Exp:* From res chemist to sr res chemist, Chemstrand Res Ctr, Inc, 61-68; from asst prof to prof chem, Col Environ Sci & Forestry, State Univ NY, 68-84, asst dir, polymer res ctr, 71-82, chmn dept, 72-84. *Concurrent Pos:* Dir org mat sci prog, Col Environ Sci & Forestry, State Univ NY, 71-74. *Mem:* AAAS. *Res:* Mechanical behavior of polymers; rubber elasticity; statistical mechanics of rubber networks at large deformations; stress induced crystallization in polymer networks; elastic behavior of composite and interpenetrating networks; thermoelastic behavior of polyelectrolyte networks. *Mailing Add:* 108 Scottholm Blvd Syracuse NY 13224

SMITH, KENNETH LARRY, b Minerva, Ohio, Apr 30, 41; m 78; c 2. DAIRY SCIENCE, MASTITIS. *Educ:* NMex State Univ, BS, 64; Ohio State Univ, MS, 66, PhD(dairy sci), 70. *Prof Exp:* Res assoc dairy sci, 64-70, from asst prof to assoc prof, 70-84, PROF DAIRY SCI, OHIO AGR RES & DEVELOP CTR, OHIO STATE UNIV, 84- *Concurrent Pos:* Vis scientist, Commonwealth Sci & Indust Res Orgn, Armindale Australia, 81. *Honors & Awards:* West Agro Award, Am Dairy Sci Asn, 84. *Mem:* Am Dairy Sci Asn; Nat Mastitis Coun. *Res:* Specific and nonspecific resistance to infection of the bovine mammary gland; diagnosis, therapy and control of bovine mashh. *Mailing Add:* Dept Dairy Sci Ohio Agr Res & Develop Ctr Wooster OH 44691

SMITH, KENNETH LEROY, b Holly, Colo, Oct 15, 31; m 60. DAIRY MICROBIOLOGY. *Educ:* Univ Mo, BS, 53, MS, 56, PhD(dairy microbiol), 60. *Prof Exp:* Instr dairy mfg, Univ Mo, 57-60, asst prof, 60; from asst prof to assoc prof dairy sci, 60-77, MICROBIOLOGIST, INST FOOD & AGR SERV, UNIV FLA, 66-, ASSOC PROF MICROBIOL & CELL SCI, 77- *Mem:* AAAS; Am Soc Microbiol; Am Dairy Sci Asn. *Res:* Application of microbial physiology and taxonomy in dairy science, especially in mastitis, the microbial flora and antibiotic residues in dairy products. *Mailing Add:* 1029 NW 36th Terr Gainesville FL 32605

SMITH, KENNETH MCGREGOR, b Wheeling, WVa, Mar 24, 23; m 76; c 3. PHYSICAL CHEMISTRY. *Educ:* Ohio Wesleyan Univ, BA, 47; Ohio State Univ, MSc, 50, PhD(chem), 55. *Prof Exp:* Res chemist, 55-61, tech serv rep, 61-68, tech specialist, 68-76, TECH ASSOC, E I dU PONT DE

NEMOURS & CO, INC, 76- *Concurrent Pos:* Chmn comt photog processing, Nat Standards Inst; US deleg, Int Standards Orgn. *Mem:* Am Chem Soc; Soc Photog Scientists & Engrs. *Res:* Chemical kinetics; photographic chemistry; photographic pollution abatement. *Mailing Add:* 102 W Park Pl Newark DE 19711-4569

SMITH, KENNETH RUPERT, JR, b St Louis, Mo, Sept 23, 32; m 56; c 7. NEUROSURGERY. *Educ:* Washington Univ, MD, 57. *Prof Exp:* Trainee anat, Washington Univ, 59-60 & 62, instr neurosurg, 63-66, instr anat, 64-66; from asst prof to assoc prof neurosurg, 6671, PROF NEUROSURG, SCH MED, ST LOUIS UNIV, 71-, CHMN SECT, 68- *Concurrent Pos:* Nat Inst Neurol Dis & Blindness spec fel, Washington Univ, 64-65 & Oxford Univ, 65-66. *Mem:* AAAS; AMA; Cong Neurol Surg (vpres, 75); Am Asn Anatomists; Am Asn Neurol Surg; Soc Neurol Surg. *Res:* Electron microscopy of the nervous system; neurophysiology of sensory receptors; clinical evaluation of cerebral circulation. *Mailing Add:* Dept Surg 3635 Vista & Grand St Louis MO 63110-0250

SMITH, KENNETH THOMAS, b Jan 4, 49; m 70; c 3. TRACE MINERAL METABOLISM, CALCIUM METABOLISM. *Educ:* Rutgers Univ, PhD(nutrit), 79. *Prof Exp:* SECT HEAD, NUTRIT RES, MIAMI VALLEY LABS, PROCTER & GAMBLE CO, 85- *Mem:* Sigma Xi; Inst Food Technologists; Am Inst Nutrit; Fel Am Col Nutrit. *Mailing Add:* Miami Valley Labs Procter & Gamble Co PO Box 39175 Cincinnati OH 45247

SMITH, KENT FARRELL, b Fish Haven, Idaho, June 26, 35; m 57; c 5. INTEGRATED CIRCUITS. *Educ:* Utah State Univ, BS, 57, MS, 58; Univ Utah, PhD(elec eng), 82. *Prof Exp:* Res engr, Stanford Res Inst, 59-61; electronic specialist, Phillips Petrol Co, 61-66; tech dir, Res & Develop Ctr, Gen Instrument, 66-72; scientist, Res Inst, 72-78, ASSOC PROF COMPUT SCI, UNIV UTAH, 78-, ACTG CHMN DEPT, 85- *Concurrent Pos:* VPres, LSI Testing, 67-72; consult, Gen Instrument Corp, 72- *Res:* Design methodology for the implementation of very large scale integrated circuits, using path programmable logic. *Mailing Add:* Dept Comput Sci Univ Utah 3190 Merrill Eng Salt Lake City UT 84112

SMITH, KEVIN MALCOLM, b Birmingham, Eng, Mar 15, 42; m 65; c 2. ORGANIC CHEMISTRY, BIOLOGICAL CHEMISTRY. *Educ:* Univ Liverpool, Eng, BSc, 64, PhD(chem), 67, DSc, 77; FRIC, 77. *Prof Exp:* Fel chem, Harvard Univ, 67-69; lectr chem, Univ Liverpool, Eng, 69-77; PROF CHEM, UNIV CALIF, DAVIS, 77-, CHMN CHEM, 90- *Concurrent Pos:* Fulbright travel grant, 67-69; org chem consult, Palmer Res Labs, 74-77; consult & sci adv bd, Aquanautics Corp, 85-; sci adv bd, Quadralogic Technol, 89- *Honors & Awards:* Leverhulme Prize, Soc Chem Indust, Eng, 64; Parke-Davis Prize, Parke-Davis & Co, 67; Corday-Morgan Medal & Prize, Chem Soc, 78; Potts Medal, Univ Liverpool, 88. *Mem:* Am Chem Soc; Royal Soc Chem; Sigma Xi. *Res:* Chemistry, biochemistry and spectroscopy of porphyrins, chlorophylls, bile pigments and their diverse metal complexes. *Mailing Add:* Dept of Chem Univ of Calif Davis CA 95616

SMITH, KIMBERLY GRAY, b Manchester, Conn, July 19, 48; m 72; c 1. VERTEBRATE ECOLOGY, COMMUNITY ECOLOGY. *Educ:* Tufts Univ, BS, 71; Univ Ark, MS, 75; Utah State Univ, PhD(biol & ecol) 82. *Prof Exp:* Res assoc ecol, Manomet Bird Observ, 77-80; res asst ecol, Utah State Univ, 78-80; res ecologist, Bodega Marine Lab, Univ Calif, Berkeley, 80-81; vis asst prof, 81-85, asst prof, 85-87, ASSOC PROF ZOOL, UNIV ARK, 87- *Mem:* AAAS; Am Ornithologists Union; Cooper Ornith Soc; Ecol Soc Am; Wildlife Soc; Wilson Ornith Soc; Sigma Xi. *Res:* Vertebrate ecology; habitat selection; community structure; reproductive ecology; role of food supplies. *Mailing Add:* Dept Biol Sci Univ Ark Fayetteville AR 72701

SMITH, KIRBY CAMPBELL, b Dallas, Tex, Feb 7, 40. MATHEMATICS. *Educ:* Southern Methodist Univ, BA, 62; Univ Wis, MS, 64, PhD(algebra), 69. *Prof Exp:* Asst prof math, Univ Miss, 68-70; asst prof math, Univ Okla, 70-75; ASSOC PROF MATH, TEX A&M UNIV, 75- *Mem:* Math Asn Am; Am Math Soc. *Res:* Noncommutative ring theory. *Mailing Add:* Dept Math Tex A&M Univ College Station TX 77843

SMITH, L DENNIS, b Municie, Ind. EMBRYOLOGY, BIOLOGY. *Educ:* Indiana Univ, BA, 59, PhD(expl biol), 63. *Prof Exp:* Prof biol, Purdue Univ, 73-87, from assoc head to head, Dept Biol Sci, 79-87; dean, Sch Biol Sci, 87-90, PROF, DEPT DEVELOP & CELL BIOL, UNIV CALIF, IRVINE, 87-, EXEC VCHANCELLOR, 90- *Concurrent Pos:* Asst embryologist, Argonne Nat Lab, 64-67, assoc biologist, 67-69; assoc prof, Purdue Univ, 69-73; res career develop award, Indiana Univ, 70-75; mem, NIH, 71-75; instr embryol, Woods Hole Marine Biol Lab, 72, 73, 74 & 88-89; mem, NASA, Space Biol Peer Rev Panel, Am Inst Biol Sci, 80-85 & Comt Space Biol & Med, Space Studies Bd, 84-, chmn, 86-; Guggenheim fel, 87; mem, bd sci counselors, 89- *Mem:* Soc Develop Biol; Int Soc Develop Biol; AAAS; Am Soc Biochem & Molecular Biol. *Res:* Regulation of protein synthesis during cogenesis and oocyte maturation with emphasis on recruitment of MRNA for translation; steroid interactions with the amphibian oocyte in the induction of oocyte maturation with emphasis on the mechanism(s) involved in cell cycle regulation; role of cytoplasmic germ plasm in leading to the formation, migration, and differentiation of primordial germ cells. *Mailing Add:* 509 Admin Univ Calif Irvine CA 92717

SMITH, L(EROY) H(ARRINGTON), JR, b Baltimore, Md, Nov 3, 28; m 51; c 3. MECHANICAL ENGINEERING. *Educ:* Johns Hopkins Univ, BE, 49, MS, 51, DEng, 54. *Prof Exp:* Compressor aerodynamicist gas turbines, Flight Propulsion Div, 54-61, supvr turbomach develop, 61-67, mgr compressor aerodyn develop unit, 67-69, mgr compressor & fan design tech opers, 69-71, mgr adv fan & compressor aerodyn, 71-74, mgr adv turbomach aerodyn, 75-80, MGR TURBOMACH AERO TECHNOL, GEN ELEC AIRCRAFT ENGINES, 81- *Honors & Awards:* Gas Turbine Award, Am Soc Mech Engrs, 81 & 87; Charles P Stienmetz Award, Gen Elec Co, 87; R Tom Sawyer Award, Am Soc Mech Engrs, 87. *Mem:* Fel Am Soc Mech Engrs; Nat Acad Eng. *Res:* Fluid mechanics; turbomachinery. *Mailing Add:* Gen Elec Aircraft Engines Mail Drop A322 Cincinnati OH 45215-6301

SMITH, LARRY, b Hughes Springs, Tex, June 26, 44; m 66; c 2. COLLOID CHEMISTRY, MANAGEMENT OF TECHNOLOGY APPLICATION. *Educ:* NTex State Univ, BA, 68, PhD(phys chem, org chem), 70; Southern Methodist Univ, MBA, 87. *Prof Exp:* Res chemist, Corp Res, Am Hoechst Corp, 71-72; res chemist, 73-82, mgr, Adv Recovery Processes, Sun Oil Co, 82-86; DIR RES ADMIN, SOUTHERN METHODIST UNIV, 88- *Concurrent Pos:* Res assoc, Univ Tex, Dallas, 72-73. *Mem:* Am Chem Soc; Soc Petrol Engrs; Soc Res Admin. *Res:* Surfactant, polymer and caustic flooding as methods of enhancing oil recovery; project management and the process of technology transfer from research and development lab to field pilots and to field-wide application; management of technology transfer. *Mailing Add:* 604 Stardust Lane Richardson TX 75080

SMITH, LARRY, b New York, NY, May 13, 42; m 64. MATHEMATICS. *Educ:* Brooklyn Col, BS, 62; Yale Univ, PhD(math), 66. *Prof Exp:* Actg instr math, Yale Univ, 65-66; instr, Princeton Univ, 66-68, asst prof, 68-69; assoc prof math, Univ Va, 69-77; INSTR MATH, CALIF STATE UNIV, LONG BEACH & CHAPMAN COL, 77- *Concurrent Pos:* Air Force Off Sci Res fel math, Inst Advan Sci Study, France, 68-69. *Mem:* Am Math Soc. *Res:* Algebraic topology. *Mailing Add:* 19622 Canberra Lane Huntington Beach CA 92646

SMITH, LARRY DEAN, b Tonkawa, Okla, Apr 1, 39; m 61; c 3. ENZYMOLOGY, CELL CULTURE. *Educ:* Okla State Univ, BS, 62, PhD(biochem), 67. *Prof Exp:* Assoc chemist, S R Noble Found, 67-71; assoc prof chem, Southwestern Col, 71-88; DIR, H L SNYDER MEM RES FOUND, 73- *Concurrent Pos:* Trainee endocrinol, Univ Calif, Riverside, 86-87. *Mem:* Am Asn Cancer Res; Am Asn Clin Chem; Am Chem Soc; Sigma Xi; AAAS. *Res:* Mechanism of neoplastic transformation as well as from neoplasia to metastasis; clinical assay procedures for detection of malignancy. *Mailing Add:* H L Snyder Memorial Res Found Box 745 Winfield KS 67156

SMITH, LAURA LEE WEISBRODT, b Georgetown, Ohio, July 16, 03; c 2. FOOD CHEMISTRY. *Educ:* Miami Univ, BS, 25; Iowa State Col, MS, 27; Univ Calif, PhD(nutrit), 30. *Prof Exp:* Asst, Iowa State Col, 25-27; Smith Incubator Co fel, Cornell Univ, 30-31, instr home econ, 37-42; teacher pub schs, NY, 42-45; consult, Inter-Am Inst Agr Sci, 47-55; from asst prof to prof food chem, Sch Hotel Admin, 55-72, EMER PROF FOOD CHEM, CORNELL UNIV, 72- *Mem:* Am Chem Soc; fel Am Inst Chemists; NY Acad Sci. *Res:* Use of modified starches in ready food programs; evaluation of frozen sauces; methods of detecting breakdown of fats; nutrition; science education. *Mailing Add:* 1707 Slaterville Rd Ithaca NY 14850

SMITH, LAWRENCE HUBERT, b Jackson, Mich, Apr 2, 30; m 50; c 5. PLANT PHYSIOLOGY. *Educ:* Mich State Univ, PhD(crops), 59. *Prof Exp:* Instr pub schs, Mich, 55-57; asst, Mich State Univ, 57-59; from asst prof to assoc prof, 59-68, PROF AGRON & PLANT GENETICS, UNIV MINN, ST PAUL, 68- *Mem:* Am Soc Agron; Am Soc Plant Physiol. *Res:* Physiological genetics; crop physiology. *Mailing Add:* Dept Agron 411 Borlang Hall Univ Minn 1991 Buford Cir St Paul MN 55108

SMITH, LAWTON HARCOURT, b Poughkeepsie, NY, Nov 15, 24; m 46; c 1. RADIOBIOLOGY. *Educ:* Univ Conn, BA, 50; Syracuse Univ, MS, 53, PhD(zool), 54. *Prof Exp:* BIOLOGIST, OAK RIDGE NAT LAB, 54- *Concurrent Pos:* NSF sr fel, Neth, 60. *Mem:* Radiation Res Soc; Am Physiol Soc; Soc Exp Biol & Med. *Res:* Radiation injury, protection and recovery in mammals. *Mailing Add:* 3436 Dolphin St Destin FL 32541

SMITH, LEHI TINGEN, b Oakley, Idaho, Nov 29, 27; m 56; c 4. MATHEMATICS. *Educ:* Ariz State Univ, BS, 48, MA, 55; Stanford Univ, EdD, 59. *Prof Exp:* Teacher high schs, Ariz, 53-54 & 55-57; asst math, Brigham Young Univ, 54; from asst prof to assoc prof, 59-70, PROF MATH, ARIZ STATE UNIV, 70- *Mem:* Math Asn Am; Nat Coun Teachers Math. *Res:* Cultural influence on mathematical perceptions; design and appraisal of improved curricula for preparation of prospective mathematics teachers. *Mailing Add:* Dept of Math Ariz State Univ Tempe AZ 85281

SMITH, LELAND LEROY, b Bradenton, Fla, May 14, 26; m 53; c 4. ORGANIC CHEMISTRY, BIOCHEMISTRY. *Educ:* Univ Tex, BA, 46, MA, 48, PhD(chem), 50. *Prof Exp:* Res assoc chem, Columbia Univ, 50-51; res chemist, Southwest Found Res & Educ, 52-54; group leader, Lederle Labs, Inc, Am Cyanamid Co, 54-60 & Wyeth Labs, Inc, 60-64; assoc prof biochem, 64-68, PROF BIOCHEM, UNIV TEX MED BR GALVESTON, 68- *Concurrent Pos:* Vis scientist, Worcester Found Exp Biol, 51-52 & Oak Ridge Inst Nuclear Studies, 53. *Mem:* Am Chem Soc; Am Soc Biol Chemists. *Res:* Steroid chemistry and biochemistry; steroid analysis, synthesis and biosynthesis; microbiological transformations of steroids; natural products chemistry; oxygen biochemistry. *Mailing Add:* Univ Tex Med Br Galveston TX 77550

SMITH, LEO ANTHONY, b Waycross, Ga, May 22, 40; m 62; c 3. INDUSTRIAL ENGINEERING. *Educ:* Ga Inst Technol, BS, 62, MS, 64; Purdue Univ, PhD(indust eng), 69. *Prof Exp:* Asst prof, 69-73, ASSOC PROF INDUST ENG, AUBURN UNIV, 73- *Mem:* Am Inst Indust Engrs; Human Factors Soc; Ergonomics Soc; Am Indust Hyg Asn. *Res:* Evaluation of physiological and psychological aspects of human performance in man-machine systems; design of man-machine systems. *Mailing Add:* Dept Indust Eng Auburn Univ Main Campus Auburn AL 36849

SMITH, LEONARD CHARLES, biochemistry; deceased, see previous edition for last biography

SMITH, LEROY H, JR, AERO TECHNOLOGY. *Educ:* Johns Hopkins Univ, BE, 49, MS, 51, DEng, 54. *Prof Exp:* MGR TURBO MACH, AERO TECHNOL, GEN ELEC, 83- *Concurrent Pos:* Chmn, Aircraft Engine's Fan & Compressor Aerodyn Design Bd, Gen Elec. *Honors & Awards:* R Tom Sawyer Award, Am Soc Mech Engrs, 87. *Mem:* Nat Acad Eng; fel Am Soc Mech Engrs. *Res:* Author of various publications; granted several patents. *Mailing Add:* GE Aircraft Engines One Neumann Way PO Box 156301 MD A-322 Cincinnati OH 45215-6301

SMITH, LESLIE E, b New York, NY, Jan 6, 41; m 63; c 2. PHYSICAL CHEMISTRY. *Educ:* Case Inst Technol, BSc, 62; Cath Univ Am, PhD(chem), 70. *Prof Exp:* Phys chemist, Polymers Div, Nat Bur Standards, 64-66 & 69-74, chief polymer stability & standards, 74-82; CHIEF, POLYMERS DIV, NAT INST STANDARDS & TECHNOL, 82- *Concurrent Pos:* ed, Polymer Commun, 84- *Mem:* AAAS; Am Chem Soc. *Res:* Interfacial phenomena; optical properties of surfaces; ellipsometry; adsorption of polymers; polymer decomposition; diffusion and transport through polymers. *Mailing Add:* Polymers Div Nat Inst Standards & Technol Washington DC 20899

SMITH, LESLIE GARRETT, b Rotherham, Eng, Nov 14, 27; nat US; m 55; c 2. IONOSPHERE, ATMOSPHERIC ELECTRICITY. *Educ:* Cambridge Univ, BA, 48, PhD(physics), 51. *Prof Exp:* Res assoc meteorol, Univ Chicago, 51-52; res geophysicist, Univ Calif, Los Angeles, 52-53; physicist geophys res directorate, Air Force Cambridge Res Ctr, 53-58; dir space sci lab, Tech Div, GCA Corp, 58-72; prin res scientist, Dept Elec Eng, 72-75, PROF ELEC ENG, UNIV ILL, URBANA-CHAMPAIGN, 75- *Concurrent Pos:* UNESCO consult to India, 81. *Honors & Awards:* Darton Prize, Royal Meteorol Soc, 54. *Mem:* Am Geophys Union. *Res:* Atmospheric physics; atmospheric electricity; ionospheric physics; sounding rockets. *Mailing Add:* 2305 Southmoor Dr Champaign IL 61821-5812

SMITH, LEVERETT RALPH, b Oakland, Calif, Feb 24, 49; m 78. SCIENCE EDUCATION. *Educ:* Univ Calif, Santa Cruz, BA, 70; Cornell Univ, MS, 74, PhD(chem), 76. *Prof Exp:* Teacher chem & math, Techiman Sec Sch, Ghana, 70-72; fel chem, Dept Entomol, Cornell Univ, 76-77; assoc fel chem, Col Environ Sci & Forestry, State Univ NY, 77-78; asst prof chem, Oberlin Col, 78-81; chem ed, Acad Press, 81-83; ENVIRON LAB MGR, KENNEDY/JENKS/CHILTON, 83- *Mem:* Am Chem Soc; Sigma Xi. *Res:* Exploratory organic chemistry; chemical ecology; environmental science; analytical chemistry. *Mailing Add:* 622 Clayton Ave El Cerrito CA 94530

SMITH, LEVERING, b Joplin, Mo, Mar 5, 10; m 33. ORDNANCE. *Educ:* US Naval Acad, BS, 32. *Hon Degrees:* NMex State Univ, LLD, 61. *Prof Exp:* Sect head, Res Div, Bur Ord, US Navy, 44-47, dep head explosives dept, Ord Test Sta, 47-49, head, Rockets & Explosives Dept, 49-51; assoc tech dir, 51-54, commanding officer, Ord Missile Test Facil, 54-56, br head & dep tech dir, Polaris Proj, 56-57, tech dir, 57-65, dir, 65-71, dir strategic systs projs, 71-77; Vice Admiral USN, 77; RETIRED. *Concurrent Pos:* Nat security consult, 77-; mem, President's Comm on Strategic Forces, 83. *Honors & Awards:* Hickman Award, Rocket Soc, Am Inst Aeronaut & Astronaut, 56; US Navy League Parsons Award, 61; Gold Medal, Am Soc Naval Engrs, 61; Conrad Award, 66; Knight Comdr, Order Brit Empire, 72; James Forrestal Award, Nat Security Indust Asn, 78. *Mem:* Nat Acad Engrs; fel Am Inst Aeronaut & Astronaut; Am Soc Naval Engrs. *Res:* Technology of propellants and high explosives; interior ballistics of rockets; statistical analysis; operations research. *Mailing Add:* 3306 Curlew St San Diego CA 92103-5541

SMITH, LEWIS DENNIS, b Muncie, Ind, Jan 18, 38; m 61; c 2. DEVELOPMENTAL BIOLOGY. *Educ:* Ind Univ, Bloomington, AB, 59, PhD(exp embryol), 63. *Prof Exp:* Res assoc, Ind Univ, 63-64; asst embryologist, Argonne Nat Lab, 64-67, assoc biologist, 67-69; assoc prof, Purdue Univ, 69-73, assoc head, 79-80, prof biol & develop biol, 73-, head biol, 81-; dean, Sch Biol, EXEC VCHANCELLOR, UNIV CALIF, 90- *Concurrent Pos:* Instr embryol, Woods Hole Marine Biol Lab, 72, 73 & 74; assoc ed, Develop Biol, 74-85, Wilhelm Roax's Archiv Develop Biol, 75-78; mem, Cell Biol Study Sect, NIH, 71-75, chmn, 77-79; mem, Am Inst Biol Sci, Space Biol Panel, NASA & Am Inst Biol Sci, 80-; mem, Comt Space Biol Med, Space Sci Bd. *Mem:* Soc Develop Biol; Int Soc Develop Biol; AAAS. *Res:* Mechanism of steroid action in induction of oocyte maturation; regulation of cell cycle; regulation of RNA and protein synthesis during oogenesis and oocyte maturation; role of germinal plasm in formation and migration of primordial germ cells. *Mailing Add:* Exec VChancellor Univ Calif 509 Adm Irvine CA 92717

SMITH, LEWIS OLIVER, JR, b Eckley, Colo, Nov 20, 22; m; c 2. ORGANIC CHEMISTRY. *Educ:* Grove City Col, BS, 44; Univ Rochester, PhD(org chem), 47. *Prof Exp:* Asst chem, Univ Rochester, 44-46; head dept, Polytech Inst PR, 47-48; instr, Wilson Col, 48-50, asst prof, 50-52; from asst prof to prof chem, Valparaiso Univ, 52-88, chmn dept, 59-61 & 63-65; RETIRED. *Mem:* Am Chem Soc. *Res:* Yield studies in organic chemical reactions; molecular analogies; synthesis and comparison of properties; matrix catalysis. *Mailing Add:* 2702 Maplewood Ave Valparaiso IN 46383

SMITH, LEWIS TAYLOR, b Seal Beach, Calif, Nov 2, 25; m 50; c 2. RADIATION PHYSICS. *Educ:* Univ Calif, Los Angeles, BS, 57, MS, 59, PhD(physics), 66. *Prof Exp:* Mem tech staff physics, Ground Systs Div, Hughes Aircraft Co, 65-71; scientist, Aerospace Div, Martin Marietta Corp, 72-74; sr physicist, Space Div, Gen Elec Co, 74-76; sr scientist physics, Electronics Div, Northrop Corp, 76-80; MEM TECH STAFF, LITTON SYSTS, 80- *Mem:* Am Phys Soc; IEEE. *Res:* Effects of nuclear weapon radiation upon materials and electronics, parts, circuits and systems. *Mailing Add:* 118 Cheyenne Trail Huntsville AL 35806

SMITH, LEWIS WILBERT, b York, Pa, Jan 13, 37; m 63; c 2. RUMINANT NUTRITION, FORAGE BIOCHEMISTRY. *Educ:* Univ Md, BS, 59, MS, 61, PhD(animal sci), 68. *Prof Exp:* health serv officer, radiochem, Comn Corps, Pub Health Serv, RA Taft Sanit Eng Ctr, Ohio, 62-64; dairy scientist, Dairy Cattle Res Br, 65-72, res animal scientist, Agr Environ Qual Inst, 73-75, lab chief, Feed Energy Conserv Lab, Animal Physiol & Genetics Inst, 75-79, dir Animal Sci Inst, 79-88, NAT PROG LEADER ANIMAL NUTRIT, AGR RES SERV, USDA, BELTSVILLE, MD, 88- *Concurrent Pos:* Mem, Animal Waste Mgt Comt, 70-73, chmn, 72-73, Regulatory Agency Comt, 76-, mem, Subcomt Environ Qual, 78-81, Prog Comt Animal Waste Mgt, Am Soc Animal Sci, 82-83; Grad asst animal nutrit, Dept Animal Sci, Univ Md, 58-59. *Mem:* Am Dairy Sci Asn; Am Soc Animal Sci; Am Registry Prof Animal Scientist. *Res:* Ruminant nutrition and biochemistry; the improvement of production efficiency, quality of animal products and profitability; analytical chemistry; biochemistry of forages; digestive kinetics of structural carbohydrates in forages. *Mailing Add:* 406 Neale Court Silver Spring MD 20901

SMITH, LLOYD HOLLINGWORTH, JR, b Easley, SC, Mar 27, 24; m 54; c 6. MEDICINE. *Educ:* Washington & Lee Univ, AB, 44; Harvard Med Sch, MD, 48; Am Bd Internal Med, dipl, 59. *Prof Exp:* Intern & asst resident med, Mass Gen Hosp, 48-50; chief med resident, 56, chief endocrine & metab unit, 58-63; vis investr, Oxford Univ, 63-64; chmn dept, 64-85, PROF MED, SCH MED, UNIV CALIF, SAN FRANCISCO, 64-, ASSOC DEAN, 85- *Concurrent Pos:* Harvard Soc Fels fel biochem, Harvard Univ & Pub Health Res Inst New York, Inc, 52-54; USPHS res fel, Karolinska Inst, Sweden, 54-55; res fel biochem, Huntington Labs, 57-58; counr, Nat Heart Inst; mem, President's Sci Adv Comt, 70-74; mem, Bd Overseers, Harvard Univ, 74-80. *Mem:* Inst Med-Nat Acad Sci; Am Soc Clin Invest; Endocrine Soc; Am Fedn Clin Res; Am Soc Biol Chemists; Asn Am Physicians. *Res:* Medical research, particularly areas of inherited metabolic diseases. *Mailing Add:* San Francisco Med Ctr Sch Med Univ Calif San Francisco CA 94143

SMITH, LLOYD MUIR, b Calgary, Alta, Feb 20, 17; nat US; m 48; c 2. FOOD CHEMISTRY. *Educ:* Univ Alta, BSc, 43, MSc, 49; Univ Calif, PhD(agr chem), 53. *Prof Exp:* Instr dairy indust, Univ Alta, 46-49, lectr, 49-50; asst dairy indust, Univ Calif, Davis, 50-52; asst prof dairying, Univ Alta, 52-54; from asst prof dairy indust to prof food sci & technol, 54-87, chemist, 67-87, EMER PROF FOOD SCI & TECHNOL, UNIV CALIF, DAVIS, 87- *Concurrent Pos:* Sr Fulbright res scholar, Dept Sci & Indust Res, NZ, 61-62; vis prof, dept food sci, Rutgers Univ, 75-76; Fulbright lectr, dept foods & exp nutrit, Univ Sao Paulo, Sao Paulo, Brazil, 86. *Honors & Awards:* Award of Merit, Am Oil Chem Soc, 83. *Mem:* Am Oil Chem Soc; Am Dairy Sci Asn; Nutrition Today Soc; Inst Food Technologists. *Res:* Chemistry of fats and other lipids; composition, structure and deterioration of lipids in foods; technology of edible fats, oils and emulsions; chemistry of milk. *Mailing Add:* Dept Food Sci & Technol 110 FS & TB Univ Calif Davis CA 95616

SMITH, LLOYD P, physics, engineering; deceased, see previous edition for last biography

SMITH, LORRAINE CATHERINE, b Toronto, Ont, May 8, 31; m 53; c 5. SCIENTIFIC EDITING, TOXICOLOGICAL DATA EVALUATION. *Educ:* Univ Toronto, BA, 53, MA, 54; Univ Ottawa, PhD(biol), 63. *Prof Exp:* Res asst physiol, Univ Toronto, 54-55; defence res sci officer, Defence Res Med Labs, Defence Res Bd Can, 55-57; res assoc physiol biol, Univ Ottawa, 63-65; res assoc zool biol, Carleton Univ, 70-75, sessional lectr, 73, res fel, 75-77; asst ed, Can J Fisheries & Aquatic Sci, Dept Fisheries & Oceans, Can, 76-86; SCI EVALUATOR, HEALTH PROTECTION BR, HEALTH & WELFARE DEPT, CAN, 86- *Concurrent Pos:* Ed, Canadian Field-Naturalist, 72-81; contractor, Nat Mus Natural Sci, Can, 75-77; mem working group biosphere reserves, Can/Man & Biosphere Proj, UNESCO, 82-89. *Mem:* Can Soc Zoologists (secy, 85-); Soc Toxicol Can; Coun Biol Ed. *Res:* Analysis and evaluation of toxicological scientific data; physiology and ecology of small mammals, especially population dynamics and reproductive biology; scientific editing and communications; general natural history. *Mailing Add:* 2144 Huntley Rd RR 3 Stittsville ON K2S 1B8 Can

SMITH, LOUIS C, b Hobbs, NMex, Nov 24, 37; div; c 3. BIOCHEMISTRY. *Educ:* Abilene Christian Col, BS, 59; Univ Tex, Austin, PhD(biochem), 63. *Prof Exp:* Res assoc biochem, Mass Inst Technol, 63-66, instr, 64-66; asst prof, 66-74, assoc prof biochem, 74-77, assoc prof, 77-79, PROF EXP MED, BIOCHEM & CELL BIOL, BAYLOR COL MED, 79- *Concurrent Pos:* NIH fel, Mass Inst Technol, 63-65; estab investr, Am Heart Asn, 72-77; mem study sect, Nat Heart, Lung & Blood Inst, NIH, 75-87. *Mem:* AAAS; Am Chem Soc; Am Soc Cell Biol; NY Acad Sci; Am Soc Biol Chemists; Soc Anal Cytol; Inst Elec & Electronic Engrs; Int Atherosclerosis Soc. *Res:* Lipoprotein structure, metabolism and transport of lipids and xenobiotics; digital fluorescence imaging microscopy. *Mailing Add:* Dept of Med Baylor Col Med Alkek A601 Houston TX 77030-2797

SMITH, LOUIS CHARLES, b Rochester, NY, Sept 16, 18; m 41; c 2. PHYSICAL ORGANIC CHEMISTRY. *Educ:* Univ Rochester, BS, 40; Columbia Univ, PhD(chem), 44. *Prof Exp:* Instr chem, Univ Vt, 43-44; sr res assoc, Nat Defense Res Comt, Carnegie Inst Technol, 44-45; res chemist, Gen Elec Co, 45-46; sect chief, US Naval Ord Lab, 46-49; group leader, Los Alamos Sci Lab, 49-74, assoc group leader 74-79; RETIRED. *Concurrent Pos:* Consult, 86- *Mem:* Am Chem Soc; Am Phys Soc; Sigma Xi. *Res:* Development of military high explosives; properties of military explosives and propellants. *Mailing Add:* 27 Allegheny Irvine CA 92720

SMITH, LOUIS DE SPAIN, b Odessa, Wash, Oct 12, 10; m 36; c 3. MICROBIOLOGY. *Educ:* Univ Idaho, BS, 32, MS, 35; Univ Wash, PhD(microbiol), 48. *Hon Degrees:* ScD, Univ Idaho, 71, Mont State Univ, 76. *Prof Exp:* Cytologist inst cancer res, Univ Pa, 35-40; bacteriologist biochem res found, Franklin Inst, 40-42 & 47-49; from assoc prof to prof bact, Vet Res Lab, Exp Sta, Mont State Univ, 50-67; prof microbiol, Va Polytech Inst & State Univ, 67-76; RETIRED. *Concurrent Pos:* Head dept bot & bact, Mont State Univ, 57-64, dean grad sch, 64-67. *Mem:* AAAS; hon mem Am Soc Microbiol; Am Acad Microbiol. *Res:* Pathogenic anaerobic bacteria. *Mailing Add:* NE 25th St E Wenatchee WA 98802

SMITH, LOUIS LIVINGSTON, b College Place, Wash, May 19, 25; c 1. SURGERY. *Educ:* Walla Walla Col, BA, 48; Loma Linda Univ, MD, 48. *Prof Exp:* From intern to sr resident, Los Angeles County Gen Hosp, 48-57; from instr to assoc prof surg, Loma Linda Univ, 56-69, chief, 71-90, EMER CHIEF, PERIPHERAL VASCULAR SURG SERV, MED CTR, 90-, DIR SURG RES LAB, 59-, PROF SURG, LOMA LINDA UNIV, 69- *Concurrent Pos:* NIH res fel, Peter Bent Brigham Hosp & Harvard Univ, 57-59; Harvey Cushing res fel, 58-59. *Mem:* AMA; fel Am Col Surgeons; Am Surg Asn; Int Cardiovasc Soc; Int Soc Surg; Soc Vasc Surg. *Res:* Surgical metabolism; shock; surgical trauma. *Mailing Add:* Dept of Surg Loma Linda Univ Loma Linda CA 92350

SMITH, LOWELL R, b Minneapolis, Minn, Nov 21, 33; m 60; c 2. ORGANIC CHEMISTRY. *Educ:* Univ Minn, BA, 55, PhD(org chem), 60. *Prof Exp:* Chemist, Russell-Miller Milling Co, 55-56; sr res chemist, Monsanto Co, 60-63, sr res specialist org chem, 63-75, sci fel, 76-90, SR SCI FEL, MONSANTO CO, 90- *Mem:* Am Chem Soc. *Res:* Indole, organophosphorus and heterocyclic chemistry; chemistry of oxalyl chloride; reaction mechanisms; chemical processing; amino acids; isocyanates. *Mailing Add:* Monsanto Co T-408 800 N Lindbergh Blvd St Louis MO 63166

SMITH, LOWELL SCOTT, b Akron, Ohio, July 20, 50. SOLID STATE PHYSICS. *Educ:* Univ Rochester, BS, 72; Univ Pa, PhD(physics), 76. *Prof Exp:* PHYSICIST, GEN ELEC CORP RES & DEVELOP, 76- *Mem:* Am Phys Soc; Inst Elec & Electronics Engrs. *Res:* Medical ultrasonic imaging; transducer design nuclear magnetic resonance imaging and spectroscopy; physical acoustics. *Mailing Add:* Gen Elec Corp Res & Develop Div PO Box 8 Schenectady NY 12301

SMITH, LUCIAN ANDERSON, b Mayfield, Ky, Nov 17, 10; m 37; c 5. MEDICINE. *Educ:* Wabash Col, AB, 30; Rush Med Col, MD, 35; Am Bd Internal Med, dipl, 43; Am Bd Gastroenterol, dipl, 43. *Prof Exp:* Intern, Presby Hosp, Chicago, 34-35; first asst internal med & neurol, 39-43, head sect internal med, 47-70, assoc prof, 47-77, SR CONSULT, SECT MED, MAYO CLIN, 70-, EMER PROF MED, MAYO GRAD SCH MED, UNIV MINN, 77- *Concurrent Pos:* Fel internal med, Mayo Clin, 36-39. *Mem:* Am Gastroenterol Asn; AMA; fel Am Col Physicians; Sigma Xi. *Mailing Add:* Mayo Clin Emer Staff 200 First St SW Rochester MN 55905

SMITH, LUTHER MICHAEL, b Clayton, Ga, Jan 6, 48; m 68; c 2. ENGINEERING DEVELOPMENT, COCHLEAR IMPLANT DESIGN. *Educ:* Univ Colo, BS, 74; Colo State Univ, MS, 79. *Prof Exp:* Comn officer, US Publ Health Serv Corps, 74-75; sr electronic specialist, Colo State Univ, 76-77, res asst, 78-79; design engr, Varian/Diasonics Ultrasound, 79-84; proj engr, Symbion Inc, 84-89; PROG DIR, SINAI SAMARITAN MED CTR, 89- *Concurrent Pos:* Consult, 88- *Mem:* Am Soc Artificial Internal Organs. *Res:* Totally implantable mechanical artificial heart and ventricular assistant device; design and patient testing of neural stimulation devices and patient test equipment for stimulation of the cochlea of deaf patients; cochlear implant hearing prosthesis design; author of various publications. *Mailing Add:* 5785 S Timberlane Rd New Berlin WI 53146

SMITH, LUTHER W, b Greenfield, Mass, Apr 26, 32. PHYSICS, MATHEMATICS. *Educ:* Univ Mass, BA, 53; Univ Kans, MS, 56. *Prof Exp:* Res physicist, Res Ctr, 58-82; optical engr, AO Instrument Group, Warner-Lambert, 82-; OPTICAL MGR, AM OPTICAL. *Mem:* Optical Soc Am. *Res:* Diffraction theory of image formation; phase microscopy; surface guided waves. *Mailing Add:* Eight Meadow Lane Dudley MA 01570

SMITH, LYLE W, b Normal, Ill, Feb 20, 20; m 42; c 2. PHYSICS. *Educ:* Univ Ill, BS, 42, MS, 43, PhD(physics), 48. *Prof Exp:* Asst physics, Univ Ill, 43-44 & 46-47, asst chem, 44-46, assoc physics, 47-48; physicist, 48-61, SR PHYSICIST, BROOKHAVEN NAT LAB, 61- *Mem:* AAAS; fel Am Phys Soc; NY Acad Sci; Sigma Xi. *Res:* High energy particle accelerators; high energy elementary particle physics. *Mailing Add:* Brookhaven Nat Lab Upton Long Island NY 11973

SMITH, LYNWOOD S, b Snohomish, Wash, Nov 15, 28; m 51; c 3. FISH PHYSIOLOGY. *Educ:* Univ Wash, BS, 52, MS, 55, PhD(zool), 62. *Prof Exp:* Teacher high sch, Wash, 52-53; instr biol & zool, Olympic Col, 55-60; asst prof zool, Univ Victoria, 62-65; from asst prof to assoc prof, Col Fisheries, 65-74, PROF FISHERIES, COL FISHERIES, UNIV WASH, 74-, ASSOC DIR INSTR, SCH FISHERIES, 86- *Mem:* Am Fisheries Soc; Am Asn Univ Prof. *Res:* Osmoregulation and blood circulation, effects of pollutants, general environmental physiology and functional anatomy in teleost fish; effects of stress and exercise on salmonids particularly during smolting; digestion physiology in fish; physiological effects of oil on fish. *Mailing Add:* Sch Fisheries Univ Wash Seattle WA 98195

SMITH, M SUSAN, b Detroit, Mich, July 18, 42. NEUROENDOCRINOLOGY, REPRODUCTIVE PHYSIOLOGY. *Educ:* NTex State Univ, BA, 64; Fla State Univ, MS, 69; Univ Ga, PhD(physiol), 72. *Prof Exp:* Fel physiol, Emory Univ, 71-73, instr, 73-74; from asst prof to assoc prof ,Med Sch Univ Mass, 74-79; res assoc prof, 79-81, assoc prof, 82-87, PROF PHYSIOL, SCH MED, UNIV PITTSBURGH, 87- *Concurrent Pos:* Vis lectr, Med Sch, New York Univ, 78; NIH Career Develop Award, Nat Inst Child Health & Human Develop, 78, Pop Res Comt, 87-; mem Biochem Endocrinol Study Section, NIH, 82-86. *Mem:* Endocrine Soc; Soc Study Reprod; Am Physiol Soc; AAAS; Soc for Neuroscience; Int Soc Neuroendocrinol. *Res:* Regulation of pituitary gonadotropin and prolactin secretion by the hypothalmic-pituitary-ovarian axis. *Mailing Add:* Dept Physiol Sch Med Univ Pittsburgh Pittsburgh PA 15261

SMITH, MALCOLM (KINMONTH), b Morristown, NJ, Dec 19, 19; m 60; c 3. PHYSICS, INSTRUMENTATION. *Educ:* Haverford Col, BS, 41; Columbia Univ, MA, 54. *Prof Exp:* Teacher, Gow Sch, NY, 48-52 & Putney Sch, Vt, 52-56; engr, Woods Hole Oceanog Inst, 56-58; ed, Phys Sci Study Comt, 58-60; exec officer physics, Sci Teaching Ctr, Mass Inst Technol, 60-66; assoc prof, Lowell Technol Inst, 66-72, prof physics & appl physics, 72-85, EMER PROF, UNIV LOWELL, 85- *Concurrent Pos:* Consult, Educ Develop Ctr, Newton, Mass, 57-68, Inst Serv to Educ, DC, 67-69 & Tech Educ Res Ctr, Cambridge, Mass, 70-80. *Mem:* Inst Elec & Electronics Eng; Am Asn Physics Teachers. *Mailing Add:* 2035 Berry Roberts Dr Sun City Center FL 33573

SMITH, MALCOLM CRAWFORD, JR, b Kingsville, Tex, Jan 2, 36; m 61; c 2. FOOD SYSTEMS, NUTRITION. *Educ:* Tex A&M Univ, DVM, 59; Purdue Univ, MS, 65. *Prof Exp:* Base vet, US Air Force, Holloman AFB, NMex, 59-62; chief vet serv, Sidi Slimane Air Base, Morocco, 62-63, Toul Rosieres Air Base, France, 63-65; Food Systs Consult, 83- *Mem:* Asn Military Surg US; Inst Food Technol. *Res:* Food technology; nutrition; aerospace life support systems; public health; marine biology. *Mailing Add:* Consult 3410 Miramar Dr La Porte TX 77571

SMITH, MANIS JAMES, JR, b Memphis, Tenn, Sept 26, 40; div. CARDIOVASCULAR ENDOCRINOLOGY. *Educ:* Memphis State Univ, BS, 62; Palmer Col Chiropractic, DC, 65; La State Univ Med Ctr, PhD(physiol), 74. *Prof Exp:* Fel, 75-77, ASST PROF, DEPT PHYSIOL & BIOPHYS, UNIV MISS MED CTR, 77- *Concurrent Pos:* Prin investr, Miss Heart Asn, 78-85. *Mem:* AAAS. *Res:* Control of adrenal steroidogenesis; radioimmunoassay development. *Mailing Add:* Dept Physiol & Biophys Univ Miss Med Ctr Jackson MS 39216

SMITH, MARCIA SUE, b Greenfield, Mass, Feb 22, 51. SCIENCE POLICY. *Educ:* Syracuse Univ, BA, 72. *Prof Exp:* Corresp & admin asst space, Am Inst Aeronaut & Astronaut, 73-75; analyst aerospace & energy technol, Sci Policy Res Div, Cong Res Serv, Libr Cong, 75-80, specialist, aerospace & energy systs, 80-85; exec dir, Nat Comn Space, 85-86; SPECIALIST, AEROSPACE POLICY, CONG RES SERV, 86- *Concurrent Pos:* Distinguished lectr, Am Inst Aeronaut & Astronaut, 83-; chmn, US Task Force Int Space Yr, Int Astron Fedn. *Mem:* fel Am Inst Aeronaut & Astronaut; fel Brit Interplanetary Soc; Sigma Xi; NY Acad Sci; Am Astronaut Soc (pres, 85-86); Int Acad Astronaut; Int Inst Space Law; Women in Aerospace (pres, 87). *Res:* Space science. *Mailing Add:* 6015 N Ninth St Arlington VA 22205

SMITH, MARIAN JOSE, b Hoboken, NJ, Oct 24, 15. BIOCHEMISTRY. *Educ:* Col St Elizabeth, AB, 36; Fordham Univ, MS, 54, PhD(biochem), 60. *Prof Exp:* Chemist, Reed & Carnrick, 37-45; teacher parochial schs, 45-47; from instr to assoc prof, 48-69, chmn chem dept, 76-79 & 82-87, PROF CHEM, COL ST ELIZABETH, 69- *Concurrent Pos:* AEC equip grant, 61-62; NSF travel grant, Int Symp, Goettingen, Ger, 61; Ciba Corp res grant, Univ Glasgow, 62; USPHS grant, 65-66; NSF res partic grant, 68; NSF grant for sci educ workshops, 80 & 81. *Mem:* Am Chem Soc; fel Am Inst Chemists. *Res:* Activity of enzymes related to nucleic acid metabolism in tumor-bearing rats; biochemistry aging. *Mailing Add:* Dept Chem Col of St Elizabeth Convent Station NJ 07961

SMITH, MARIANNE (RUTH) FREUNDLICH, b Karlsruhe, Ger, June 11, 22; US citizen; c 4. MATHEMATICS. *Educ:* Queens Col, BS, 43; Univ Ill, MS, 44, PhD(math), 47. *Prof Exp:* Asst physics, Univ Ill, 43-44, asst math, 46-47; lectr, Univ Calif, Berkeley, 47-49, instr, 49-50; vis prof physics, Univ Pittsburgh, 51-52; lectr math, Univ Calif, Berkeley, 57-58; from asst prof to assoc prof, Calif State Col Hayward, prof, 73-85, EMER PROF MATH, CALIF STATE UNIV, 85- *Mem:* Am Math Soc; Math Asn Am; Am Asn Univ Prof. *Res:* Normed rings; duality theorems; functional analyses, differential equations. *Mailing Add:* Dept Math Calif State Univ Hayward CA 94542

SMITH, MARION BUSH, JR, b Ferriday, La, Feb 25, 29. MATHEMATICS. *Educ:* La State Univ, BS, 49, MS, 51; Univ NC, PhD(math), 57. *Prof Exp:* Instr math, Univ NC, 53-57; asst prof, Fla State Univ, 57-58 & Univ Utah, 58-61; vis assoc prof, Univ Wis-Madison, 61-63, assoc prof & chmn dept exten div, 63-64; assoc prof math, Univ Wis-Baraboo/Sauk County Campus, 64-72, vchancellor ctr syst, 64-68, chmn dept math, 68-72, 74-77; PROF MATH, CALIF, STATE COL, BAKERSFIELD, 72- *Mem:* Am Math Soc; Math Asn Am. *Res:* History of mathematics; abstract topological spaces; foundations of geometry. *Mailing Add:* 3333 El Encanto Ct No 54 Bakersfield CA 93301

SMITH, MARION EDMONDS, b Susanville, Calif, July 13, 26; m 55; c 2. BIOCHEMISTRY. *Educ:* Univ Calif, BA, 52, MA, 54, PhD(biochem), 56. *Prof Exp:* Asst, Univ Calif, 53-56; res assoc med, Sch Med, Stanford Univ, 56-63, instr, 63-71, sr scientist med, 71-74; NEUROCHEMIST, VET ADMIN HOSP, PALO ALTO, 62-; adj prof neurol, 74-83, PROF NEUROL, VET ADMIN HOSP, PALO ALTO, 83- *Concurrent Pos:* Bd sci coun, NIH, 73-77 neurol Disorder Prog Proj Rev Comt, NINCDS, 80-84; nat Mult Sclerosic Fundamental Sci Adv Bd, 81-86; Neurobiology Merit Rev, Vet Admin, 83-86; Prin Investr, NIH, 62- *Honors & Awards:* Javits Investr Award, 88. *Mem:* Am Soc Neurochemistry; Int Soc Neurochem; Am Soc Neurochem (secy, 81-87); Soc Neurosci; Am Soc Biol Chemists. *Res:* Lipid and protein metabolism in nervous system; biochemistry of demyelinating diseases; myelin metabolism and function; astrocyte activation; myelin phagocytosis. *Mailing Add:* Dept Neurol 127A Vet Admin Hosp 3801 Miranda Ave Palo Alto CA 94304

SMITH, MARION L(EROY), b Sharon, Pa, June 3, 23; m 44; c 3. MECHANICAL ENGINEERING. *Educ:* La State Univ, BS, 44; Ohio State Univ, MS, 48. *Prof Exp:* From instr to assoc prof, 47-58, prof mech eng & assoc dean, Col Eng, 58-84, EMER ASSOC DEAN, OHIO STATE UNIV, 84- *Concurrent Pos:* Mem staff year-in-indust prog, Eng Dept, E I du Pont de Nemours & Co, 56-57; mem, State Bd Regist Prof Engrs & Surveyors, 75-84, chmn, 81-84; mem, Uniform Exam Comt, Nat Coun Eng Examrs, 76, chmn, 81-83. *Mem:* fel Am Soc Mech Engrs; fel Am Soc Eng Educ; Soc Automotive Eng; Nat Soc Prof Engrs; Nat Coun Eng Examrs. *Res:* Fundamentals of fuels and combustion; mechanism of oxidation of hydrocarbons, particularly pre-combustion reactions; dual fuel diesel engines; characteristics of miniature engine generator sets; methods of heat dissipation from aircraft compartments; gas turbines. *Mailing Add:* 4135 Rowanne Ct Columbus OH 43214

SMITH, MARK ANDREW, b East Brady, Pa, Nov 24, 47; m 69; c 2. MATHEMATICS, FUNCTIONAL ANALYSIS. *Educ:* Indiana Univ Pa, BS, 69; Univ Ill, Urbana, MS, 70, PhD(math), 75. *Prof Exp:* Asst prof, Lake Forest Col, 75-77; from asst prof to assoc prof, 77-84, PROF MATH, MIAMI UNIV, 84- *Mem:* Am Math Soc; Sigma Xi. *Res:* Geometry of Banach spaces. *Mailing Add:* Dept Math Miami Univ Oxford OH 45056

SMITH, MARK K, b New York, NY, Feb 14, 28. ENGINEERING ADMINISTRATION. *Educ:* Mass Inst Technol, BS, 51, PhD, 54. *Prof Exp:* Vpres, Geophys Serv, Inc, 62-67, pres, 67-69; vpres, Tex Instruments, 69-73; CONSULT, 73- *Mem:* Nat Acad Eng. *Mailing Add:* PO Box 189 Main St Norwich VT 05055

SMITH, MARK STEPHEN, US citizen. EMERGENCY MEDICINE. *Educ:* Swarthmore Col, BA, 68; Stanford Univ, MS, 71; Yale Univ, MD, 77. *Prof Exp:* Asst prof & actg chmn, 84-85, assoc prof, 85-89, PROF EMERGENCY MED, GEORGE WASHINGTON UNIV MED CTR, 89-, CHMN, 85- *Mem:* Am Col Emergency Physicians; Soc Acad Emergency Med; Am Col Physician Execs; Math Asn Am; Sigma Xi. *Res:* Application of computers to emergency medicine; clinical decision making in emergency medicine; disaster medicine. *Mailing Add:* 9005 Jones Mill Rd Chevy Chase MD 20815

SMITH, MARTHA KATHLEEN, b Detroit, Mich, Mar 14, 44. MATHEMATICS. *Educ:* Univ Mich, Ann Arbor, BA, 65; Univ Chicago, MS, 67, PhD(math), 70. *Prof Exp:* G C Evans instr math, Rice Univ, 70-72; asst prof, Wash Univ, 72-73; from asst prof to assoc prof, 73-85, PROF MATH, UNIV TEX, AUSTIN, 85- *Mem:* Am Math Soc; Math Asn Am; Asn Women Math. *Res:* Ring theory. *Mailing Add:* Dept Math Univ Tex Austin TX 78712

SMITH, MARTIN BRISTOW, b Owatonna, Minn, Feb 18, 16; m 45; c 2. PHYSICAL CHEMISTRY. *Educ:* Univ Chicago, BS, 36, PhD(phys chem), 42. *Prof Exp:* Jr chemist, Universal Oil Prod, Ill, 36-38; rubber chemist, US Rubber Co, Mich, 42-46; chemist, Consol Vultee Co, 46-47; phys chemist, Ethyl Corp, 47-62, res assoc, 62-77; RETIRED. *Mem:* Fel Am Chem Soc. *Res:* Micro-calorimetry; heats of mixing of strong electrolytes; physical property measurements; thermody- namic calculations; monomer-dimer equilibria of aluminum alkyls; heats of formation of aluminum alkyls and related compounds. *Mailing Add:* 1742 Carl Ave Baton Rouge LA 70808-2904

SMITH, MARTYN THOMAS, b Lincoln, UK, Aug 17, 55; m 79; c 2. CARCINOGENESIS, RISK ASSESSMENT. *Educ:* Queen Elizabeth Col, Univ London, BS, 77; Med Col St Bartholomews Hosp, PhD (biochem), 80. *Prof Exp:* Asst forensic med, Karolinska Inst, 80-81; teaching fel, Toxicol Pharmacol, Sch Pharmacy, Univ London, 81-82; asst prof toxicol, 82-87, ASSOC PROF TOXICOL, UNIV CALIF, BERKELEY, 87- *Concurrent Pos:* Assoc dir, Health Effects Component, UC Toxic Substances Prog, 85-; prin invest, toxicol, Health Risk Assoc, Superfund Prog Proj, NIEHS, 87-; staff scientist, Lawrence Berkeley Lab, 87- *Mem:* Genetic & Environ Toxicol Asn; Am Asn Cancer Res; Am Asn Adv Sci; Soc Toxicol; Soc Free Radical Res. *Res:* Mechanisms by which toxic chemicals damage cells and alter the genetic material to produce disease, such as cancer; specific chemicals of interest are benzene, bipyridyl herbicides, MPTP and quinones. *Mailing Add:* Environ Health Sci Univ Calif Berkeley 322 Warren Hall Berkeley CA 94720

SMITH, MARVIN ARTELL, b Ogden, Utah, Apr 8, 36; m 60; c 8. NUCLEIC ACIDS, PLANT MOLECULAR BIOLOGY. *Educ:* Utah State Univ, BS, 60; Univ Wis-Madison, MS, 62, PhD(biochem), 64. *Prof Exp:* Res fel, Med Ctr, NY Univ, 64-66; from asst prof to assoc prof, 66-74, PROF CHEM & BIOCHEM, BRIGHAM YOUNG UNIV, PROVO, UTAH, 74- *Concurrent Pos:* Fel, Warf Res, Univ Wis, 60-62, NIH, 62-64, & 64-66, spec fel, 72-73; vis scientist chem & biochem, Union Carbide, NY, 64, Univ Calif, Davis, 72-73, plant genetics, Weizmann Inst, Israel, 80; Donald F Jones res fel, 72-73; vis prof chem & biochem, Kuwait Univ, 78-80. *Mem:* Am Soc Biol Chemists; Int Soc Plant Molecular Biol; AAAS; Sigma Xi. *Res:* Developmental biochemistry and gene expression, particularly in higher plants; coordination of nuclear and organelle genetic and enzymic systems. *Mailing Add:* Grad Sect Biochem Brigham Young Univ Provo UT 84602

SMITH, MARY ANN HARVEY, b Camden, Ark, Jan 29, 40; m 71. NUTRITION, MENTAL RETARDATION. *Educ:* Henderson State Col, BSE, 60; Univ Tenn, MS, 62, PhD(food sci, nutrit), 65. *Prof Exp:* Instr food sci, Univ Ala, 61-62; therapeut dietitian, Ft Sanders Hosp, 64-65; asst prof foods & nutrit, Mid Tenn State Univ, 65-67; from asst prof to assoc prof nutrit, Child Develop Ctr, Univ Tenn Med Units, Memphis, 67-77, assoc dir, 84-87, chief dept, 67-82, PROF NUTRIT, CHILD DEVELOP CTR, UNIV TENN MED UNITS, MEMPHIS, 77-, PROF & DIR, DIV CLIN NUTRIT, 86-; PROF & COORDR, CLIN NUTRIT PROG, MEMPHIS STATE UNIV, 86- *Concurrent Pos:* Adj prof nutrit, Col Home Econ, Univ Tenn, 67-87, Sch Home Econ, Univ Ark, 70-87 & Dept Home Econ, Col Educ, Univ Miss, 79-87. *Mem:* Am Dietetic Asn; Soc Nutrit Educ; Am Asn Ment Deficiency; Nutrit Today Soc. *Res:* Nutrition as related to mental retardation, including nutritional status of the retarded, feeding techniques and management of inborn errors of metabolism; imprimentation of national dietary guidelines; day care-nutrition guidelines for children with special needs. *Mailing Add:* Div Clin Nutrit 711 Jefferson Ave Memphis TN 38105

SMITH, MARY BUNTING, b Brooklyn, NY, July 10, 10; m 37, 79; c 4. MICROBIOLOGY. *Educ:* Vassar Col, AB, 31; Univ Wis, AM, 32, PhD, 34. *Prof Exp:* Asst agr bact & chem, Univ Wis, 33-35; asst biol, Bennington Col, 36-37; instr physiol & hyg, Goucher Col, 37-38; asst bact, Yale Univ, 38-40; lectr bot, Wellesley Col, 46-47; lectr microbiol, Yale Univ, 48-55; prof bact & dean, Douglass Col, Rutgers Univ, 55-60; pres, Radcliffe Col, 60-72; asst to pres, Princeton Univ, 72-75; Consult, 75- *Concurrent Pos:* Comnr, Atomic Energy Comn, 64-65; mem nat sci bd, NSF, 65-70. *Mem:* Am Soc Microbiol. *Res:* Bacteriology; bacteriostatic action of dyes; role of carotenoids in bacteria and green plants; microbial genetics; color inheritance in Serratia. *Mailing Add:* 135 McCullum Rd New Boston NH 03070

SMITH, MAURICE JOHN VERNON, b London, Eng, Aug 11, 29; nat US; c 5. UROLOGY. *Educ:* Cambridge Univ, BA, 55, MB, BChir, 56; Columbia Univ, PhD(anat), 65. *Prof Exp:* Intern & sr house officer, St Thomas Hosp, London, 56-58; resident gen surg, Univ Hosp, Saskatoon, Sask, 58-59; resident urol, Col Physicians & Surgeons, Columbia Univ, 59-60, asst anat, 60-64, assoc urol, 64-66; resident, Bowman Gray Sch Med, 66-68; from asst prof to assoc prof, 68-75, PROF UROL, MED COL VA, VA COMMONWEALTH UNIV, 75- *Concurrent Pos:* Consult, McGuire Vet Hosp, Richmond, Va, 69- & Portsmouth Naval Hosp, 69- *Mem:* Am Urol Asn; Soc Univ Urologists; fel Am Col Surgeons. *Res:* Renal calculi; renal and prostatic cancer. *Mailing Add:* Div Urol Med Col Va Box 118 Richmond VA 23298

SMITH, MAURICE VERNON, b Toronto, Ont, June 4, 20; m 48; c 4. APICULTURE. *Educ:* Univ Toronto, BSA, 42, MSA, 54; Cornell Univ, PhD, 57. *Prof Exp:* From asst to assoc prof environ biol, Univ Guelph, 49-83; RETIRED. *Res:* Honeybee behavior and pollination; bee breeding; insect photography. *Mailing Add:* Three Young St Guelph ON N1G 1M1 Can

SMITH, MAYNARD E, b Boston, Mass, Nov 29, 16; m 41; c 2. ANALYTICAL CHEMISTRY. *Educ:* Mass Inst Technol, PhD(chem), 49. *Prof Exp:* Asst chem, Mass Inst Technol, 41-49; anal chemist, Los Alamos Nat Lab, 49-72, sect leader, 72-81; RETIRED. *Mem:* Am Chem Soc; Am Inst Chemists. *Res:* Gases in metals; mass spectrometry; gas chromatography. *Mailing Add:* 75 Mesa Verde Dr Los Alamos NM 87544

SMITH, MELVIN I, b New York, NY, July 21, 24; m 46; c 4. MECHANICAL & CHEMICAL ENGINEERING. *Educ:* City Col New York, BChE, 44; Columbia Univ, MS, 47, EngScD, 55. *Prof Exp:* Res technologist, Res Dept, Socony Mobil Oil Co, Inc, 47-58, supv technologist, 58-72, mgr lubricants, Tech Serv Dept, Mobil Oil Corp, 72-84; RETIRED. *Mem:* Am Soc Lubrication Engrs; Am Soc Mech Engrs. *Res:* Application problems and formulation of lubricating, hydraulic and metal processing fluids. *Mailing Add:* 75 Mallard Dr Avon CT 06001

SMITH, MEREDITH FORD, b Milwaukee, Wis; m; c 2. INTERNATIONAL NUTRITION, COMMUNITY NUTRITION. *Educ:* Trinity Univ, Tex, BS, 70; Va Polytech Inst, PhD(human nutrit), 78. *Prof Exp:* Home economist, Consumer & Food Econ Inst, USDA, 71-75; teaching asst foods & nutrit, Va Polytech Inst & State Univ, 74-77; instr, Human Develop & Consumer Sci Dept, Univ Houston, 77-78, asst prof, 78-81; ASSOC PROF COMMUNITY NUTRIT & INT NUTRIT, DEPT FOODS & NUTRIT, KANS STATE UNIV, 81- *Concurrent Pos:* Food serv proj assoc, Coord Vocational Acad Educ Workshop, Univ Houston, 79; nutrit consult, Tex Dept Human Resources, 78-79 & Cath Univ, Santiago, 79, Off Int Coop & Develop, USDA, 80; tech advisor, Off Nutrit, USAID, 80, eval PL 480 title foods, Honduras, 86; prin investr, Analysis of Indigenous Food Patterns in Low-Income Families, USDA; dir, Human Ecol proj, Nat Univ Asancion, Paraguay, 88-; fel, March Dimes. *Mem:* Soc Nutrit Educ; Am Pub Health Asn; Am Home Econ Asn; Sigma Xi. *Res:* Nutritional assessment of at-risk populations in US and developing countries; factors influencing nutritional states and food behavior; evaluation of domestic and international community nutrition programs. *Mailing Add:* 501 Sunset Manhattan KS 66502

SMITH, MICHAEL, b London, Eng, Dec 18, 32; m 61. PHYSICAL CHEMISTRY. *Educ:* Univ Southampton, BSc, 54, PhD(phys chem), 58. *Prof Exp:* Asst lectr radiation chem, Kings Col, Univ Durham, 57-60; sr chemist, Am Cyanamid Co, 60-62; assoc scientist, Xerox Corp, 62-63, scientist, 63-65, sr scientist, 65-67, systs photoconductors, 67-69, lab mgr xerography, 69-71, process sect mgr, Webster, 71-73, mgr, Xerographic Technol Dept, 73-75, vpres, Info Technol Group, 74-76, group dir, Rank Xerox Eng, 76-78, vpres, Reprod Technol Group, 78-81, vpres & gen mgr, Supplies & Mat Bus Unit, 81-89, sr vpres, Reproduction/IOT Unit, 89-90, CORP OFFICER, XEROX CORP, 86-, DIR, RANK XEROX, 90- *Concurrent Pos:* Atomic Energy Res Estab fel, 57-60. *Res:* Radiation chemistry; organic semiconductors; xerography; imaging technologies and systems. *Mailing Add:* Seven Duxbury Way Rochester NY 14618

SMITH, MICHAEL, b Blackpool, Eng, Apr 26, 32; Can citizen; m 60; c 3. BIOCHEMISTRY, BIOTECHNOLOGY. *Educ:* Univ Man, BSc, 53, PhD(chem), 56. *Prof Exp:* Fel, BC Res Coun, 56-60; res assoc, Inst Enzyme Res, Univ Wis, 60-61; head chem sect, Vancouver Lab, Fisheries Res Bd Can, 61-66; assoc prof, 66-70, PROF BIOCHEM, UNIV BC, 70- *Concurrent Pos:* Med res assoc, Med Res Coun Can, 66- *Honors & Awards:* Gairdner Found Int Award, 86. *Mem:* Fel Chem Inst Can; fel Royal Soc Chem; Sigma Xi. *Res:* Nucleic acid and nucleotide chemistry and biochemistry using in-vitro mutagenesis to study gene expression. *Mailing Add:* Biotechnol Lab Univ BC 2075 Westbrook Mall Vancouver BC V6T 1W5 Can

SMITH, MICHAEL A, b Boston, Mass, Nov 08, 44; m; c 1. PETROLEUM GEOLOGY, REMOTE SENSING. *Educ:* Univ Mich, BS, 66; Univ Kans, MS, 69; Univ Tex, PhD(geol), 75. *Prof Exp:* Marine geologist, US Geol Surv, 66-68, petrol geologist & geochemist, 75-81; res geochemist, Getty Oil Co, 81-83, supvr basin eval & struct geol, 83-84; RES ASSOC, TEXACO INC, 84- *Concurrent Pos:* Mem, IGCP Proj 219, 86-91; adj prof, Dept Geol, Emory Univ, Atlanta, 88-89. *Mem:* Fel Geol Soc Am; Am Asn Petrol Geologists; Am Geophys Union; Am Asn Stragtig Palynologists; Soc Org Petrol. *Res:* Geological and depositional facies controls on worldwide petroleum occurrence; international basin geology; remote sensing using satellite data interpretation and image processing; biostratigraphy, paleogeography and paleoecology; organic petrology and regional source-rock geochemistry. *Mailing Add:* 1123 Shillington Dr Katy TX 77450

SMITH, MICHAEL CLAUDE, b Winston-Salem, NC, May 31, 49; m 70; c 4. INDUSTRIAL & MANUFACTURING ENGINEERING. *Educ:* Univ Tenn, Knoxville, BSIE, 71, MSIE, 74; Univ Mo-Columbia, PhD(indust eng), 77. *Prof Exp:* Indust engr, Buckeye Cellulose Corp, 71-72, St Mary's Med Ctr, Knoxville, Tenn, 72-74; asst prof, Ore State Univ, 77-79; asst prof indust eng, Univ Mo-Columbia, 79-82; SR ENGR, SCI APPLN INT CORP, 83- *Concurrent Pos:* Consult, Pharm Serv, Vet Admin, 78-, Univ Mo Hosp & Clin, 79-82; prin investr, Directorate Mgt Sci, Air Force Logistic Command, 81-82. *Mem:* Sr mem Am Inst Indust Eng; Operations Res Soc Am. *Res:* Concurrent engineering in design and manufacturing; organizational assessment and strategic planning; information systems planning. *Mailing Add:* Advan Concepts Div Sci Appln Int Corp 1710 Goodridge Dr McLean VA 22102

SMITH, MICHAEL HOWARD, b San Pedro, Calif, Aug 30, 38; m 58; c 2. ECOLOGY, EVOLUTION. *Educ:* San Diego State Col, AB, 60, MA, 62; Univ Fla, PhD(zool), 66. *Prof Exp:* Res assoc zool, Inst Ecol, Univ Ga, 66-67, asst prof, 67-71; assoc prof, Univ Tex, 70-71; assoc prof, 70-77, PROF ZOOL, UNIV GA, 77-; DIR SAVANNAH RIVER ECOLOGY LAB, 73-

Concurrent Pos: NSF res grant, 66-68; AEC Comn grants, 67-; dir, Savannah River Ecol Lab, Aiken, SC, 73- *Mem:* Am Soc Mammal; Soc Study Evolution; Ecol Soc Am; Am Soc Ichthyologists & Herpetologists; Am Soc Naturalists; Soc Syst Zool; Wildlife Soc; Am Fisheries Soc; Soc Pop Ecol. *Res:* Behavior; various biological aspects of vertebrates as they relate to the study of speciation and population ecology. *Mailing Add:* Savannah River Ecol Lab Drawer E Aiken SC 29802

SMITH, MICHAEL JAMES, b East St Louis, Ill, Feb 18, 45; m 67; c 2. ENVIRONMENTAL CHEMISTRY, HAZARDOUS & NUCLEAR WASTE DISPOSAL & TREATMENT. *Educ:* Southern Ill Univ, BA, 67; Univ Mo, Columbia, MA, 69, PhD(environ chem), 72; Stanford Univ, MBA, 89. *Prof Exp:* From asst prof to assoc prof anal & environ chem, Wright State Univ, 72-77, assoc dir to dir, Brehm Lab, 74-77; mgr, Chem Sci Lab, Rockwell Hanford Opers, 77-79, mgr engineered barriers proj, 79-80, mgr, Eng Dept, 80-84, prin mgr res, 84-86; dir, Western Eng Div, Albuquerque, 86-89, VPRES ANALYTICAL SERV, IT CORP, KNOXVILLE, TENN, 90- *Concurrent Pos:* Consult, Monsanto Res Corp, Am Nuclear Soc; Nat Mgt NSF & Inst Environ Educ, 74; mem proposal rev panel, NSF Off Exp Prog, 74-77; consult, Miami Conservancy Dist, 75-76; consult, Atlantic Richfield Hanford Co, 76-77; mem nuclear safety rev group, Dept Energy, 89- *Honors & Awards:* Am Bicentennial Comn, 75. *Mem:* Am Chem Soc; AAAS; Am Nuclear Soc. *Res:* Environmental engineering (air, water and hazardous waste systems) including hazardous chemical and nuclear waste disposal; coal acid mine drainage treatment; heavy metal contamination studies, pesticides analyses, air toxics and water and hazardous waste treatment; RI/FS and remediation projects; environmental chemistry. *Mailing Add:* 12220 Ansley Ct Knoxville TN 37922

SMITH, MICHAEL JAMES, b Madison, Wis, May 12, 45; m 68; c 2. HUMAN FACTORS ENGINEERING, ERGONOMICS. *Educ:* Univ Wis-Madison, BA, 68, MA, 70, PhD(indust psychol), 73. *Prof Exp:* Res analyst, Wis Dept Indust & Labor, 71-74; res psychologist, Nat Inst Occup Safety & Health, 74-84; PROF TEACHING & RES, DEPT INDUST ENG, UNIV WIS, 84- *Concurrent Pos:* Mem, World Health Orgn Panel on Psychosocial Hazards, 82-84, Am Psychol Asn Panel on Job Design & Stress, 90, Canadian Adv Panel on Cumulative Trauma Disorders, 91-; Sci adv, Workplace Health Fund, 88-; ed, Int J Human-Computer Interaction, 89-; bd mem, Int Comn on Human Aspects of Comput, 90- *Honors & Awards:* Super Serv Award, US Pub Health Serv, 80. *Mem:* Human Factors Soc; Asn Comput Math; Am Soc Testing & Mat; Inst Indust Engrs. *Res:* Relationship between the design of work and human responses to that design; ergonomics; job stress; organizational structure; safety and health exposures. *Mailing Add:* Dept Indust Eng Univ Wis 1513 University Ave Madison WI 53706

SMITH, MICHAEL JOSEPH, b Bay City, Mich, Jan 20, 39; m 78; c 2. DEVELOPMENTAL & MOLECULAR BIOLOGY. *Educ:* St Mary's Col, Calif, BS, 63; Univ BC, PhD(zool), 69. *Prof Exp:* Asst prof zool, Univ Nebr, Lincoln, 69-71; res fel biol, Calif Inst Technol, 71-73, sr res fel, 73-76; from asst prof to assoc prof, 76-87, PROF BIOL, SIMON FRASER UNIV, 87- *Concurrent Pos:* Spec res fel, USPHS, Calif Inst Technol, 71-73; res grant, Nat Res Coun Can, Simon Fraser Univ, 77-88 & Brit Columbia Health Care Res Found grant, 78-82. *Mem:* Can Soc Cell Biologist; Soc Develop Biol. *Res:* Molecular biology of eukaryote development; phylogeny and evolution of genomic and mitochondrial DNA sequence. *Mailing Add:* Dept Biol Sci Simon Fraser Univ Burnaby BC V5A 1S6 Can

SMITH, MICHAEL KAVANAGH, b San Francisco, Calif, May 20, 49; m 76; c 2. PROGRAM VERIFICATION, FORMAL SEMANTICS OF PROGRAMMING LANGUAGES. *Educ:* Princeton Univ, BSE, 71; Univ Tex, PhD(computer sci), 81. *Prof Exp:* Res assoc, Univ Tex, 82-85; vpres, 83-85, pres, 85-87, FOUNDER & DIR, COMPUTATIONAL LOGIC INC, 83-, EXEC VPRES, 87- *Mem:* Asn Comput Mach; Inst Elec & Electronics Engrs; AAAS. *Res:* Formal semantics of programming languages, program verification, and mechanical theorem proving, especially as applied to the Ada programming lanaguage. *Mailing Add:* 5609 Wagon Train Cove Austin TX 78749

SMITH, MICHAEL LEW, b Ashland, Kans. ENDOCRINOLOGY, HUMAN PHYSIOLOGY. *Educ:* Emporia Kans State Univ, BA, 70; Purdue Univ, PhD(chem), 74. *Prof Exp:* Biochemist, Madigan Army Med Ctr, 75-77, chief, Res Lab, 77-80; asst chief, Dept Clin Res, William Beaumont Army Med Ctr, 80-86; PROF, DEPT SOCIAL, SW TEX STATE UNIV, 86- *Mem:* Endocrine Soc; Am Asn Clin Chemists. *Res:* Functions of prolactin in male mammals. *Mailing Add:* Dept Social & Anthrop SW Tex State Univ San Marcos TX 78666

SMITH, MICHAEL R, b Portland, Ore, June 15, 45. BACTERIAL PHYSIOLOGY, METHANOGENESIS. *Educ:* Calif State Univ, Long Beach, BS, 67; Univ Calif, Los Angeles, PhD(microbiol), 76. *Prof Exp:* Fel microbiol, Dept Microbiol, Univ Calif, Los Angeles, 70-76, postdoctoral scholar microbiol, Sch Pub Health, 76-79, res assoc, 79-80; RES MICROBIOLOGIST, WESTERN REGIONAL RES CTR, AGR RES SERV, USDA, 80- *Mem:* Am Soc Microbiol; AAAS; Sigma Xi. *Res:* Bacterial methanogenesis; bacterial spore germination; agricultural research; methane production; anaerobic microbiology. *Mailing Add:* Western Regional Res Ctr USDA 800 Buchanan St Albany CA 94710

SMITH, MILTON LOUIS, b Childress, Tex, May 30, 39; m 66; c 2. INDUSTRIAL ENGINEERING. *Educ:* Tex Tech Univ, BS, 61, MS, 66, PhD(indust eng), 68. *Prof Exp:* Asst, 65-68, assoc prof, 68-78, PROF INDUST ENG, TEX TECH UNIV, 78- *Mem:* Am Inst Indust Engrs; Opers Res Soc Am. *Res:* Job sequencing; hail damage to solar collectors/reflectors; risk and lacerative hazard from fractured glass; modeling of flexible manufacturing systems. *Mailing Add:* Dept of Indust Eng Tex Tech Univ Lubbock TX 79409

SMITH, MILTON REYNOLDS, b Chicago, Ill, Mar 15, 34; m 55; c 1. BIOCHEMISTRY, PHYSIOLOGY. *Educ:* Knox Col, BA, 56; Univ Ariz, MS, 58, PhD(biochem), 63. *Prof Exp:* Res assoc biochem, Univ Ariz, 58-63; sr biochemist, Eli Lilly & Co, 63-80; TECH DIR, HEPAR INDUSTS, 80- *Concurrent Pos:* Instr, Ind Cent Col, 67-71; career consult, Knox Col, 72. *Mem:* AAAS; Am Chem Soc; Am Inst Chemists; NY Acad Sci. *Res:* Isolation research in the fields of lipids, proteins and enzymes; protein chemistry and endocrinology. *Mailing Add:* 7010 Parkshore Ct Middleton WI 53562-3702

SMITH, MORRIS WADE, b Baytown, Tex, Aug 1, 38; m 62; c 2. HORTICULTURE, PLANT PHYSIOLOGY. *Educ:* Tex Technol Univ, BS, 64; Tex A&M Univ, MS, 72, PhD(hort), 77. *Prof Exp:* Technician II, Dept Hort, Tex A&M Univ, 65-78; res horticulturist, Sci & Educ Admin-Agr Res, Coastal Plain Exp Sta, 78-84, RES HORTICULTURIST, SCI & EDUC ADMIN-AGR RES, USDA, 84- *Mem:* Am Soc Hort Sci. *Res:* Nutrition concerning container-grown ornamental plants; irrigation; herbicide and plant growth regulator research. *Mailing Add:* Sci & Educ Admin-Agr Res USDA 2423 Michael Dr Tifton GA 31794

SMITH, NAT E, b Bartow, Fla, Nov 29, 22; m 53; c 7. MEDICINE. *Educ:* Erskine Col, AB, 43; Med Col Ga, MD, 49. *Prof Exp:* Intern, Gorgas Hosp, Ancon, CZ, 49-50; med resident, DC Gen Hosp & George Washington Univ, Hosp, 52-54; staff physician, Vet Admin, 55-57; from instr to assoc prof med, Col Med, Univ Ill, 57-69, assoc dean, 62-75, prof med, 69-75; dean sch med, Mercer Univ, 74-76; ASSOC DEAN & PROF MED, BOWMAN GRAY SCH MED, WAKE FOREST UNIV, 76- *Mem:* Am Rheumatism Asn. *Res:* Rheumatic diseases; methods and evaluation of medical education. *Mailing Add:* Bowman Gray Sch Med Wake Forest Univ Winston-Salem NC 27103

SMITH, NATHAN ELBERT, b Avon, NY, Apr 30, 36; m 58; c 2. ANIMAL SCIENCE. *Educ:* Cornell Univ, BS, 67; Univ Calif, Davis, PhD(nutrit), 70. *Prof Exp:* Res nutritionist, Univ Calif, Davis, 70-71; asst prof animal sci, Cornell Univ, 71-74; ASST PROF ANIMAL SCI, UNIV CALIF, DAVIS, 74- *Mem:* Am Dairy Sci Asn; Am Soc Animal Sci. *Res:* Animal nutrition with emphasis on dairy nutrition, animal metabolism, mathematical and computer modeling of animal metabolism and animal production systems. *Mailing Add:* Purina Mills Inc PO Box 66812 St Louis MO 63166-6812

SMITH, NATHAN JAMES, b Cuba City, Wis, Oct 12, 21; m 46; c 3. PEDIATRICS. *Educ:* Univ Wis, BA, 43, MD, 45. *Prof Exp:* Fulbright scholar, Univ Paris, 50; instr pediat, Sch Med, Temple Univ, 51-53; from asst prof to assoc prof, Univ Calif, Los Angeles, 54-56; prof & chmn dept, Sch Med, Univ Wis, 57-65; PROF PEDIAT, SCH MED, UNIV WASH, 65-; PEDIATRICIAN-IN-CHIEF, KING COUNTY HOSP, 65- *Concurrent Pos:* Mem hemat training grant comt, NIH, 60-; spec asst nutrit progs to Secy Health, Educ & Welfare, 70-71. *Mem:* Am Soc Hemat; Soc Pediat Res; Am Pediat Soc; Soc Exp Biol & Med; Am Acad Pediat. *Res:* Nutrition as applied to pediatrics; sports and nutrition; sports and children. *Mailing Add:* Dept Pediat Univ Wash Rd 20 Seattle WA 98104

SMITH, NATHAN LEWIS, III, b Baltimore, Md, May 12, 43; m 65; c 2. ENZYMOLOGY, IMMUNOCHEMISTRY. *Educ:* Univ Miami, BS, 66; Univ Calif, Irvine, PhD(biol), 72. *Prof Exp:* Jr specialist biochem, Univ Calif, Irvine, 71-72, asst res biologist, 72-74; group leader diag, Nelson Res & Develop Co, 73-74; STAFF IMMUNOLOGIST, CORDIS CORP, 75- *Mem:* Am Chem Soc. *Res:* Design and development of clinical diagnostic test procedures especially enzyme tagged immunoassays. *Mailing Add:* Cytosignet Inc PO Box 219 North Andover MA 01845-0219

SMITH, NATHAN MCKAY, b Wendell, Idaho, Apr 22, 35; m 53; c 5. ZOOLOGY. *Educ:* Eastern Ore Col, BS, 61; Ore State Univ, MS, 64; Brigham Young Univ, MLS, 69, PhD(zool), 72. *Prof Exp:* Teacher, The Dalles, Ore, 61-63 & La Grande, 65-66; PROF LIBR INFO SCI, BRIGHAM YOUNG UNIV, 70- *Mem:* Soc Study Amphibians & Reptiles. *Res:* Reptilian taxonomy, comparative anatomy and natural history. *Mailing Add:* 5042 Harold B Lee Libr Brigham Young Univ Provo UT 84602

SMITH, NEAL A(USTIN), electrical engineering; deceased, see previous edition for last biography

SMITH, NED ALLAN, b Philadelphia, Pa, Feb 21, 40; m 63; c 2. COASTAL BIOLOGY. *Educ:* Juniata Col, BS, 62; Univ Pittsburgh, MS, 65, PhD(biol), 67. *Prof Exp:* USPHS postdoctoral fel, Tufts Univ, Medford, 67-69; asst prof zool, Univ NC, Chapel Hill, 69-75; educ & exhibs consult, 75, dir, NC Aquarium Pine Knoll Shores, 75-89, DIR, NC AQUARIUMS, NC OFF MARINE AFFAIRS, 89- *Mem:* AAAS; Am Asn Zool Parks & Aquariums. *Res:* Neurophysiological control of spontaneously active invertebrate muscle. *Mailing Add:* 417 N Blount St Raleigh NC 27601

SMITH, NED PHILIP, b Beaver Dam, Wis, Dec 3, 42. PHYSICAL OCEANOGRAPHY. *Educ:* Univ Wis-Madison, BS, 65, MS, 67, PhD(limnol & oceanog), 72. *Prof Exp:* Asst prof phys oceanog, Univ Tex, Austin, 72-77; assoc res scientist, 77-82, SR RES SCIENTIST, HARBOR BR FOUND, INC, FT PIERCE, FLA, 82-, SR SCI, HARBOR BR OCEANOG INST. *Mem:* Am Meteorol Soc; Oceanog Soc; Am Geophys Union. *Res:* Descriptive physical oceanography, including continental shelf circulation and intracoastal tides; heat budget of estuaries. *Mailing Add:* Harbor Br Oceanog Inst 5600 Old Dixie Hwy Ft Pierce FL 33946

SMITH, NELSON S(TUART), JR, b Weirton, WVa, Aug 9, 29; m 51; c 3. ELECTRICAL ENGINEERING. *Educ:* WVa Univ, BSEE, 56, MSEE, 58; Univ Pittsburgh, DSc(electrets), 62. *Prof Exp:* From instr to assoc prof, 56-70, PROF ELEC ENG, W VA UNIV, 70- *Concurrent Pos:* NSF res grant, 63-66; elec engr, Morgantown Energy Technol Ctr, 65- *Mem:* Am Soc Eng Educ; Inst Elec & Electronics Engrs. *Res:* Dielectric absorption currents; electrets; electrogasdynamics; solid state gas sensors. *Mailing Add:* Dept of Elec Eng WVa Univ Morgantown WV 26506-6101

SMITH, NEVILLE VINCENT, b Leeds, Eng, Apr 21, 42; m 70; c 2. SURFACE SCIENCE. *Educ:* Queens' Col, Cambridge Univ, BA, 63, MA & PhD(physics), 67. *Prof Exp:* Res assoc physics, Stanford Univ, 66-69; MEM STAFF, BELL LABS, 69- *Concurrent Pos:* Res head, Condensed State Physics Dept, Bell Labs, 78-81. *Honors & Awards:* Davisson-Germer Prize, Am Phys Soc, 91. *Mem:* Fel Am Phys Soc. *Res:* Optical properties and band structures of solids; photoemission and electronic structure of solids and surfaces; synchrotron radiation spectroscopies; inverse photoemission; spin-polarized electron spectroscopy. *Mailing Add:* Bell Labs 600 Mountain Ave Murray Hill NJ 07974

SMITH, NORMAN B, b Springfield, Ill, May 27, 27; m 51; c 4. DEXTROSE CRYSTALLIZATION, FLUE GAS DESULFURIZATION. *Educ:* Univ Ill, BS, 51, BS, 52. *Prof Exp:* Asst supt fermentation, Pabst Brewing Co, Peoria, Ill, 52-55; mgr fermentation, Grain Processing Corp, Muscatine, Iowa, 55-59, res engr, 59-66; chief chem engr, Stanley Consults Inc, 66-87; CHIEF CHEM ENGR & CHIEF EXEC OFFICER, BIOTEK ASSOC INT, 87- *Mem:* Am Chem Soc; Nat Soc Prof Engrs; Am Standard Testing Methods; Soc Indust Microbiologists. *Res:* Enzyme purification/utilization for starch hydrolysis; awarded US and foreign patents in field of dextrose crystallization and production of corn starch/ethyl alcohol; author on subjects of flue gas desulfurization, fuel alcohol production, energy conservation, quality control procedures. *Mailing Add:* PO Box 898 Muscatine IA 52761

SMITH, NORMAN CUTLER, b Paterson, NJ, Mar 18, 15; m 42; c 2. PHOTOGEOLOGY, SCIENTIFIC EDITING & WRITING. *Educ:* Wash & Lee Univ, AB, 37. *Prof Exp:* Field geologist, Standard Oil Co, Venezuela, 38-40; teaching fel, Harvard Univ Grad Sch Geol, 40-42; geologist, Humble Oil & Refining Co, 46-49; independent consult geologist, 72-79; mgr data anal & reporting, Law Eng Testing Co, 79-82; VPRES, OIL MINING CORP, 87-, DIR, SR ACAD INTERGENERATIONAL LEARNING, 89- *Concurrent Pos:* Mem, Nat Comt Geol, 63-72; instr, Geol Col Srs, NC Univ, Asheville, 88-, Adult Educ Brevard Col, 88-89, Adult Educ Blue Ridge Col, 88-89; Founder, pres & chmn bd, Coun Sci Socs, Dallas-Ft Worth, 57-60; bd dirs, Tulsa Found, 74-78, Tulsa Sci Ctr, 76-78; exec dir, Am Asn Petrol Geologists, 62-72. *Mem:* Fel Geol Soc Am; fel AAAS; emer mem Am Asn Petrol Geologists. *Res:* Articles and chapters in books on exploration geology; more than 150 proprietary reports on photogeological reconnaissance and petroleum exploration for more than 50 oil companies and engineering firms; manuals for report preparation and undergraduate counseling. *Mailing Add:* 105 Windward Dr Asheville NC 28803

SMITH, NORMAN DWIGHT, b Natural Bridge, NY, Jan 26, 41; m 67; c 2. SEDIMENTOLOGY. *Educ:* St Lawrence Univ, BS, 62; Brown Univ, MS, 64, PhD(geol), 67. *Prof Exp:* From asst prof to assoc prof, 67-78, PROF GEOL, UNIV ILL, CHICAGO, 78-, DEPT HEAD, 88- *Concurrent Pos:* Vis assoc prof geol, Univ Alta, 74-75; consult geologist, Anglo-Am Corp, SAfrica, 78-80 & Coun Sci & Indust Res, SAfrica, 82; Fulbright fel, 82; ed, J Sedimentary Petrol, 83-88; fel, Coun Sci & Indust Res, SAfrica, 89. *Mem:* Soc Econ Paleontologists & Mineralogists; Geol Soc Am; Int Asn Sedimentologists; Nat Asn Geol Teachers; Am Asn Petrol Geologists. *Res:* Fluvial sedimentology; stratigraphy and sedimentology of clastic rocks; limnology and lacustrine sedimentology; origin of placers; glacimarine sedimentation. *Mailing Add:* Dept Geol Sci Univ Ill Chicago Chicago IL 60680

SMITH, NORMAN OBED, b Winnipeg, Man, Jan 23, 14; nat US; m 44; c 3. PHYSICAL CHEMISTRY. *Educ:* Univ Man, BSc, 35, MSc, 36; NY Univ, PhD(phys chem), 39. *Prof Exp:* Asst chem, Univ Man, 39-40, lectr, 40-46, from asst prof to assoc prof, 46-50; from assoc prof to prof, 50-84, chmn dept, 74-78, EMER PROF CHEM, FORDHAM UNIV, 84- *Concurrent Pos:* Res assoc, Air Force contract, 50-51. *Mem:* Sr mem Am Chem Soc; fel Chem Inst Can. *Res:* Heterogeneous equilibria; hydrates; solid solutions; clathrates; solubility of gases; author or co-author of more than 50 publications. *Mailing Add:* Dept Chem Fordham Univ Bronx NY 10458

SMITH, NORMAN SHERRILL, b Roseburg, Ore, May 22, 32; m 57; c 2. WILDLIFE ECOLOGY. *Educ:* Ore State Univ, BSc, 58; Univ Mont, MSc, 62; Wash State Univ, PhD(zool), 69. *Prof Exp:* Lab technician II, Hopland Field Sta, Univ Calif, Davis, 60-62; res biologist, EAfrica Agr Forest Res Orgn, Kenya, 62-64; ASST UNIT LEADER WILDLIFE RES, ARIZ COOP WILDLIFE RES UNIT, UNIV ARIZ, 68- *Concurrent Pos:* Fulbright Sr res scholar, 62-63; Rockefeller Res grant, 63-64. *Mem:* Wildlife Soc; Am Soc Mammal. *Res:* Ecology of big game animals; reproduction physiology of wild mammals; metabolism and water requirements of game mammals. *Mailing Add:* Dept Wildlife 214 Biol Sci Bldg E Univ Ariz Tucson AZ 85721

SMITH, NORMAN TY, b Ft Madison, Iowa, May 5, 32; m 58; c 1. ANESTHESIOLOGY, PHARMACOLOGY. *Educ:* Harvard Med Sch, MD, 57; Am Bd Anesthesiol, dipl, 63. *Prof Exp:* Intern pediat, Children's Med Ctr, Boston, Mass, 57-58; resident anesthesia, Mass Gen Hosp, 58-60; from instr to asst prof, Sch Med, Stanford Univ, 62-72; assoc prof, 72-74, vchmn dept, 72-77, PROF, DEPT ANESTHESIA, UNIV CALIF, SAN DIEGO, 74- *Concurrent Pos:* NIH career develop award, 66-71; actg chief anesthesia, Vet Admin Hosp, Palo Alto, 62-64; vis scientist, Univ Wash, 67; vis prof, Inst Med Physics, Utrecht, Holland, 70-71; chief anesthesia, Vet Admin Hosp, San Diego, 72-77; dir anesthesia res, Univ Calif & US Naval Hosps, 75-77; ed, J Clin Monitoring, 84-; pres, Soc Technol in Anesthesia, 88- *Honors & Awards:* Detur Award, 53. *Mem:* AAAS; fel Am Col Anesthesiol; Ballistocardiographic Res Soc (pres, 73-76); Am Heart Asn; Am Soc Pharmacol & Exp Therapeut. *Res:* Cardiovascular pharmacology and physiology; ballistocardiography; control systems theory; control systems; analog and digital computation; on-line data processing; multiple drug interaction. *Mailing Add:* Dept Anesthesia Univ Calif San Diego VA Med Ctr San Diego CA 92161

SMITH, OLIN DAIL, b Tonkawa, Okla, Dec 15, 31; m 51; c 3. PLANT BREEDING. *Educ:* Okla State Univ, BS, 54, MS, 61; Univ Minn, PhD(agron), 69. *Prof Exp:* Supt, Wheatland Conserv, Exp Sta, Okla State Univ, 57-58, univ instr agron, 58-62, asst secy-mgr, Okla Crop Improv Asn, 62-65; res fel, Univ Minn, 65-70; asst prof agron, 70-75, assoc prof, 75-82, PROF SOIL & CROP SCI, TEX A&M UNIV, 82- *Mem:* Am Soc Agron; Crop Sci Soc Am; fel Am Peanut Res & Educ Soc (pres, 72-73). *Res:* Plant breeding and genetics; inheritance by Arachis Hypogaea; peanut variety improvement; breeding peanuts for disease and insect resistance. *Mailing Add:* Dept Soil & Crop Sci Tex A&M Univ College Station TX 77843

SMITH, OLIVER HUGH, molecular genetics; deceased, see previous edition for last biography

SMITH, OMAR EWING, JR, b Memphis, Tenn, Oct 21, 31; m 55; c 3. ECONOMIC ENTOMOLOGY, MEDICAL ENTOMOLOGY. *Educ:* Memphis State Univ, BS, 54; Iowa State Univ, MS, 58, PhD, 61. *Prof Exp:* Instr entom, Iowa State Univ, 60-61; from asst prof to assoc prof, 61-77, PROF BIOL, MEMPHIS STATE UNIV, 77- *Concurrent Pos:* Univ grant, 65-67. *Mem:* Entom Soc Am. *Res:* Mosquito and arthropod research; agricultural pests and insecticides; taxonomy of insects; extension entomology and related fields. *Mailing Add:* Dept of Biol Memphis State Univ Memphis TN 38152

SMITH, ORA, b Freeburg, Ill, Apr 13, 00; m 27; c 2. PLANT & FOOD SCIENCE. *Educ:* Univ Ill, BS, 23; Iowa State Univ, MS, 24; Univ Calif, PhD(plant physiol), 29. *Prof Exp:* Instr hort, Iowa State Univ, 24-27; res asst vegetable crops, Univ Calif, 27-29; asst prof hort, Okla State Univ, 29-30; from asst prof to prof, 30-67, EMER PROF VEGETABLE CROPS, CORNELL UNIV, 67- *Concurrent Pos:* Hort rep, Farm Credit Admin, 34-35; collabr, USDA, 31-39; consult indust waste comt, Nat Tech Task, 52-56; fel, Cornell Univ, 38-39; consult, US Army Quartermaster Gen Off, 44-45; prof, Interam Inst Agr Sci, 46-47; consult, Standard Oil Develop Co, 47-49; res dir, Potato Chip Inst Int, 49-76. *Mem:* Potato Asn Am (pres, 38-39); Inst Food Technol; Europ Asn Potato Res; Can Inst Food Sci & Technol; Am Oil Chem Soc. *Res:* Quality improvement of all forms of processed potatoes, including color, flavor, odor, texture and maintenance of high quality in storage. *Mailing Add:* 1707 Slaterville Rd Ithaca NY 14850

SMITH, ORA E, international technology, for more information see previous edition

SMITH, ORA KINGSLEY, b Orange, NJ, Mar 9, 27; m 53; c 4. PHYSIOLOGY, MEDICINE. *Educ:* Wellesley Col, BA, 49; Yale Univ, MD, 53. *Prof Exp:* Intern, Yale-New Haven Hosp, 53-54; from res asst to res assoc physiol, Sch Med, Yale Univ, 56-67, res assoc med, 67-70, epidemiol & pub health, 70-89; RETIRED. *Concurrent Pos:* Fel med, Georgetown Univ, 54-56; asst fel, John B Pierce Found, 70-72, assoc fel, 72-89. *Mem:* Endocrine Soc; Am Diabetic Asn; Brit Diabetic Asn; Am Physiol Soc; Sigma Xi. *Res:* Endocrinology and metabolism as involved in environmental adaptation. *Mailing Add:* Ten Gracie Sq-10A New York NY 10028

SMITH, ORRIN ERNEST, b Albany, Ore, Nov 20, 35; m 56; c 2. HORTICULTURE, PLANT PHYSIOLOGY. *Educ:* Ore State Univ, BS, 57; Univ Calif, Davis, PhD(plant physiol), 62. *Prof Exp:* Res plant physiologist cotton br, Crops Res Div, Agr Res Serv, USDA, 62-66; asst plant physiologist, Univ Calif, Riverside, 66-70, assoc prof plant physiol, assoc plant physiologist & vchmn dept plant sci, 70-75; chmn, Dept Hort, Wash State Univ, 75-80; assoc dean & dir resident instr, 80-83, ASSOC DEAN & DIR EXTENSION SERV, ORE STATE UNIV, 83- *Concurrent Pos:* Fulbright sr res scholarship, 72; review ed, Am Soc Hort Sci, 74-78. *Mem:* Fel Am Soc Hort Sci (vpres, 82-83); Am Soc Plant Physiol. *Res:* Hormonal regulation of plant growth and development; physiology of dormancy and germination; seed vigor. *Mailing Add:* Col Agr Sci Ore State Univ Corvallis OR 97331

SMITH, ORVILLE AUVERNE, b Nogales, Ariz, June 16, 27; m 53; c 2. NEUROPHYSIOLOGY. *Educ:* Univ Ariz, BA, 49; Mich State Univ, MA, 50, PhD, 53. *Prof Exp:* Instr psychol, Mich State Univ, 51-54; fel neuroanat, Univ Pa, 54-56; fel neurophysiol, Sch Med, Univ Wash, 56-58; from instr to asst prof anat & physiol, 58-59, from asst prof to assoc prof physiol & biophys, 62-67, from asst dir to assoc dir ctr, 62-71, DIR REGIONAL PRIMATE RES CTR, SCH MED, UNIV WASH, 71-, PROF PHYSIOL & BIOPHYS, 67- *Mem:* Am Physiol Soc; Am Asn Anatomists; Am Soc Primatologists (pres, 77-80); Pavlovian Soc NAm (pres, 78-79). *Res:* Physiological basis of behavior; cardiovascular control; neuroanatomy. *Mailing Add:* Dept Physiol & Biophys Univ Wash Sch Med HSB 1-421 SJ-50 Seattle WA 98195

SMITH, OTTO J(OSEPH) M(ITCHELL), b Urbana, Ill, Aug 6, 17; m 41; c 4. LARGE SYSTEMS, SOLAR ENERGY. *Educ:* Univ Okla, BS, 38; Stanford Univ, PhD(elec eng), 41. *Prof Exp:* Asst, Stanford Univ, 38-41; instr power & high voltage, Tufts Col, 41-43; asst prof commun, Univ Denver, 43-44; res engr, Electronics Dept, Westinghouse Elec Corp, Pa, 44-45; chief elec engr, Summit Corp, 45-47; from lectr to prof, 47-88, EMER PROF CONTROL & LARGE SYSTS, UNIV CALIF, BERKELEY, 88- *Concurrent Pos:* Vis prof, Inst Tech Aeronaut, Brazil, 54-56; Guggenheim fel, Polytech Univ, Darmstadt, Ger, 60; vis lectr, Kiev Inst Electrotechnology, Polytech Mus & Inst Electromechanics, USSR, 60; deleg, Cong Int Fedn Automatic Control, Moscow, 60; vis sr res fel, Monash Univ, Australia, 66-67; vis lectr, Fed Sch Eng, Itajuba, Brazil, 71, prof, 74; mem rev comt, Solar Thermal Test Facil, Users Asn, 77-78; NSF appointee, Acad Econ Studies & Inst Power Designs, Romania, 73; vis prof, Tech Univ Eindhoven, Neth, 74. *Mem:* Fel AAAS; fel Inst Elec & Electronics Engrs; Am Soc Eng Educ; Am Wind Energy Asn; Int Solar Energy Soc. *Res:* Electronics; feedback systems; servomechanisms; statistical and nonlinear synthesis; economic analogs; cybernetics; optimal economic planning; construction of digital computer programs for optimizing the use of government resources for maximum economic growth; solar-thermal-electric power system design; wind-electric systems. *Mailing Add:* 612 Euclid Ave Berkeley CA 94708

SMITH, P GENE, electrical engineering, communications, for more information see previous edition

SMITH, P SCOTT, b Richmond, Va, Dec 22, 22. PHYSICS. *Educ:* Cornell Univ, PhD, 51. *Prof Exp:* Asst physics, US Naval Res Lab, 43-45 & Cornell Univ, 46-50; asst prof, Kans State Col, 51-53; assoc prof, 53-71, PROF PHYSICS, EASTERN ILL UNIV, 71- *Mem:* Am Phys Soc; Am Asn Physics Teachers. *Res:* Nuclear, space and radiological physics; quantification of nuclear shell structure; rocketry and celestial mechanics; nuclear weapons. *Mailing Add:* Dept of Physics Eastern Ill Univ Charleston IL 61920

SMITH, PATRICIA ANNE, b Rockwood, Tenn, Sept 1, 35. DEVELOPMENTAL GENETICS. *Educ:* Carson-Newman Col, BS, 58; Northwestern Univ, MS, 65, PhD(genetics), 66. *Prof Exp:* Res asst genetics, Oak Ridge Nat Lab, 57-59; teacher jr high sch, Tenn, 59-61; res asst biochem, NMex Highlands Univ, 61; res assoc develop genetics, Northwestern Univ, 66-68; asst prof, 68-71, assoc prof, 71-79, PROF GENETICS, NORTHEASTERN ILL UNIV, 79- *Concurrent Pos:* Lectr eve div, Northwestern Univ, 65-72, NIH fel, 66-67. *Mem:* AAAS; Genetics Soc Am; Am Soc Zoologists. *Res:* Control of cell division and differentiation in the ovaries of Drosophila melanogaster; genetic control of synaptonemal complex formation. *Mailing Add:* Dept Biol Northeastern Ill Univ Bryn Mawr & St Louis Chicago IL 60625

SMITH, PATRICIA LEE, b Houston, Tex, Sept 16, 46. STATISTICS. *Educ:* Southwestern Univ, BA, 68; Purdue Univ, MS, 70; Tex A&M Univ, PhD(statist), 75. *Prof Exp:* Instr math, Purdue Univ, 70-72; asst prof statist, Tex A&M Univ, 76-80 & Old Dominion Univ, 80-82; site mgr comput opers, Comput Dynamics, Inc, 82-85; ASSOC RES MATHEMATICIAN, SHELL DEVELOP CO, 85- *Mem:* Am Statist Asn. *Res:* Linear models with emphasis on applications of splines to design of experiments, regression and time series. *Mailing Add:* PO Box 1380 Houston TX 77001

SMITH, PAUL AIKEN, b Bucaramanga, Colombia, Jan 12, 34; US citizen; m 58; c 2. PHYSICS, ASTRONOMY. *Educ:* Park Col, AB, 56; Wash Univ, AM, 59; Tufts Univ, PhD(physics), 64. *Prof Exp:* PROF PHYSICS, COE COL, CEDAR RAPIDS, IOWA, 64- *Res:* Elementary particle physics; astronomy; computers. *Mailing Add:* 1024 Maplewood Dr NE Cedar Rapids IA 52402

SMITH, PAUL CLAY, b Gray Hawk, Ky, Apr 3, 34; m 57; c 3. VETERINARY VIROLOGY, VETERINARY PATHOLOGY. *Educ:* Auburn Univ, DVM, 59; Ohio State Univ, MS, 66; Iowa State Univ, PhD, 77. *Prof Exp:* Chief rabies diag, SEATO Med Res Labs, Walter Reed Army Inst Res, 66-69, asst chief diag lab, 69-70; vet med officer virol invests, Nat Animal Dis Lab, USDA, 70-77; mem staff, Col Vet Med, Univ Tenn, 77-80; HEAD, DEPT MICROBIOL, SCH VET MED, AUBURN UNIV, 80- *Mem:* Wildlife Dis Asn; Conf Res Workers Animal Dis. *Res:* Bovine herpes viruses and respiratory disease of cattle; pseudorabies in swine; veterinary immunology. *Mailing Add:* Dept Microbiol Greene Hall Auburn Univ Auburn AL 36849

SMITH, PAUL DENNIS, b Baltimore, Md, Nov 14, 42; c 2. ENVIRONMENTAL MUTAGENESIS, MOLECULAR GENETICS. *Educ:* Loyola Col, BS, 64; Univ NC, PhD(zool), 68. *Prof Exp:* Fel genetics, Univ Conn, 68-70; from asst prof to assoc prof genetics, dept biol, Emory Univ, 70-84; prof biol & chmn dept, Southern Methodist Univ, 84-88; CHMN BIOL SCI, WAYNE STATE UNIV, 88- *Concurrent Pos:* NIH res grant, 71-, res career develop award, Gen Med Sci, 74-79. *Mem:* Environ Mutagen Soc; AAAS; Genetics Soc Am; Am Soc Microbiol. *Res:* DNA repair and mutagenesis in Drosophila. *Mailing Add:* 309 Nat Sci Bldg Wayne State Univ Detroit MI 48202

SMITH, PAUL E, JR, b Elizabeth, NJ, May 16, 23; m 47; c 3. ELECTRICAL ENGINEERING. *Educ:* Rensselaer Polytech Inst, BEE, 47; Mass Inst Technol, EE, 53. *Prof Exp:* Instr elec eng, Mass Inst Technol, 50-53, asst prof, 53-57; engr, Conval Corp, 57-58; chief engr, Feedback Controls, Inc, Mass, 58-68; sr eng specialist, GTE Sylvania Electronics Systs-East Div, 68-71; sr engr, Ikor, Inc, Burlington, 71-77, New Ikor Inc, Omniwave Electronics Corp, Gloucester, 77-80; ENGR, WINCOM CORP, 80- *Mem:* Inst Elec & Electronics Engrs. *Res:* Process control; servomechanics. *Mailing Add:* Sheridan Engineering Corp 51 Canal St Rear PO Box 49 Salem MA 01970

SMITH, PAUL EDWARD, b Emmetsburg, Iowa, May 24, 33; m 57; c 5. ECOLOGY, FISH BIOLOGY. *Educ:* Northern Iowa Univ, BA, 56; Univ Iowa, PhD(zool), 62. *Prof Exp:* Sverdrup fel plankton behav, Scripps Inst, Univ Calif, 62-63; RES BIOLOGIST & GROUP LEADER, LA JOLLA LAB, SOUTHWEST FISHERY SCI CTR, NAT MARINE FISHERIES SERV, 63- *Concurrent Pos:* Adj prof oceanog, Scripps Inst Oceanog, Univ Calif. *Honors & Awards:* Gold Medal, Dept Com, 90. *Mem:* Ecol Soc Am; Am Soc Limnol & Oceanog. *Res:* Zooplankton ecology and fish larva biology in marine environments, especially as influenced by temporal and spatial variations in small scale distribution; development of field and statistical methods for economical and concise descriptions of small scale distribution. *Mailing Add:* Southwest Fishery Sci Ctr PO Box 271 La Jolla CA 92037-0271

SMITH, PAUL FRANCIS, b Brookville, Pa, Apr 3, 27; m 51; c 4. MICROBIOLOGY. *Educ:* Pa State Univ, BS, 49; Univ Pa, MS, 50, PhD(bact), 51. *Prof Exp:* Asst instr bact, Univ Pa, 49-51; res microbiologist, Merck & Co, Inc, NJ, 51-52; from instr med microbiol to asst prof microbiol, Sch Med, Univ Pa, 52-61, res microbiologist, Inst Coop Res, 52-58; PROF & CHMN DEPT, SCH MED, UNIV SDAK, 61- *Concurrent Pos:* Vis scientist, Univ Utrecht, 70; treas, Int Orgn Mycoplasmology, 81-85; group rep, Am Soc Microbiol, 86-88. *Honors & Awards:* Lederle Med Fac Award, 60; Klieneberger-Nobel Award, 90. *Mem:* AAAS; fel Am Soc Microbiol; Am Chem Soc; Int Orgn Mycoplasmology. *Res:* Biochemistry and physiology of mycoplasmas; lipid biochemistry; structural characterization of glycolipids and lipoglycans from mycoplasmas and the relationship of those structural features to mycoplasma-host cell interactions such as adherence, pathogenicity, immune response. *Mailing Add:* Dept Microbiol Univ SDak Vermillion SD 57069

SMITH, PAUL FREDERICK, b Copeland, Kans, Dec 17, 16; m 40; c 2. PHYSIOLOGY, HORTICULTURE. *Educ:* Univ Okla, BS, 38, MS, 40; Univ Calif, PhD(physiol), 44. *Prof Exp:* Asst bot, Univ Okla, 38-39; asst plant physiol, Univ Calif, 40-42; asst physiologist, 43-47, assoc physiologist, 48-49, physiologist, 50-55, sr physiologist, 56-58, prin physiologist, 59-71, head physiologist, 71-75, AGR RES SERV, USDA, WORLD CITRUS CONSULT, 75- *Mem:* Bot Soc Am; fel Am Soc Hort Sci; Am Inst Biol Scientists; Am Soc Plant Physiol. *Res:* Pollen tube growth; bud dormancy and inhibition; vegetative propagation; mineral nutrition of citrus; citrus culture. *Mailing Add:* 2695 Ashville St Orlando FL 32818

SMITH, PAUL GORDON, b Fair Oaks, Calif, Mar 19, 15; m 37; c 3. VEGETABLE CROPS. *Educ:* Univ Calif, BS, 37; Univ Wis, PhD(plant path), 41. *Prof Exp:* Asst plant path & agron, Univ Wis, 37-43; instr, 41-44, asst prof truck crops, 44-52, assoc prof veg crops, 52-59, from asst olericulturist to assoc olericulturist, 44-59, prof, 59-80, olericulturist, Exp Sta, 59-80, EMER PROF VEG CROPS, UNIV CALIF, DAVIS, 80- *Concurrent Pos:* Specialist, Foreign Econ Admin, 42-44. *Mem:* Am Soc Hort Sci; Am Phytopath Soc; Soc Econ Botanists; Am Genetics Asn. *Res:* Breeding for disease resistance in vegetable crops. *Mailing Add:* 1013 Cabot Davis CA 95616

SMITH, PAUL HOWARD, b Akron, Ohio, Sept 26, 24; m 50; c 2. BACTERIOLOGY. *Educ:* Va Polytech Inst, BS, 53, MS, 55; Univ Calif, PhD(microbiol), 59. *Prof Exp:* Res bacteriologist, Univ Calif, 57-59; from asst prof to assoc prof biol sci, Univ Fla, 59-69; vis prof civil eng, Stanford Univ, 69-70; prof microbiol & cell sci, Univ Fla, 70-91, chairperson dept, 71-90, microbiologist, 74-90, EMER PROF MICROBIOLOGIST, UNIV FLA, 90- *Res:* Microbiology of domestic sewage sludge; rumen fermentation; methane evolution; degradation of cellulose. *Mailing Add:* 2120 SW 44th Ave Gainesville FL 32608

SMITH, PAUL JOHN, b Philadelphia, Pa, Jan 6, 43; m 70. STATISTICS, MATHEMATICS. *Educ:* Drexel Univ, BS, 65; Case Western Reserve Univ, MS, 67, PhD(math), 69. *Prof Exp:* Asst prof math, Wayne State Univ, 69, Ind Univ, 69-71; asst prof, 71-76, ASSOC PROF MATH, UNIV MD, 76- *Concurrent Pos:* Statist consult, US Selective Serv Syst, 73, Sci Educ Systs, Inc, 73-74, US Consumer Prod Safety Comn, 76-77, Nat Inst Mental Health, 77-80, Nat Cancer Inst, 85-86. *Mem:* Inst Math Statist; Am Statist Asn; Am Soc Qual Control. *Res:* Nonparametric, robust and multivariate statistical inference. *Mailing Add:* Dept Math Univ Md College Park MD 20742

SMITH, PAUL KENT, physical chemistry; deceased, see previous edition for last biography

SMITH, PAUL L, b London, Eng, Oct 12, 50; m 72. JURASSIC BIOCHRONOLOGY. *Educ:* Univ London, BSc, 72; Portland State Univ, MS, 76; McMaster Univ, PhD(geol), 81. *Prof Exp:* ASST PROF GEOL, UNIV BC, 80- *Concurrent Pos:* Mem Jurassic subcommission, Int Union Geol Sci. *Mem:* Geologists Asn London; Paleontol Asn; fel Geol Asn Can; Paleontol Res Inst; Paleontol Soc. *Res:* Stratigraphy biochronology and basin analysis of the Jurassic of western North America. *Mailing Add:* Dept Geol Sci Univ BC 6339 Stores Rd Vancouver BC V6T 1Z4 Can

SMITH, PAUL LETTON, JR, b Columbia, Mo, Dec 16, 32; m 54; c 5. ATMOSPHERIC PHYSICS, ELECTRICAL ENGINEERING. *Educ:* Carnegie Inst Technol, BS, 55, MS, 57, PhD(elec eng), 60. *Prof Exp:* From instr to asst prof elec eng, Carnegie Inst Technol, 55-63; sr engr, Midwest Res Inst, 63-66; res engr & assoc prof meteorol, 66-68, assoc prof meteorol & elec eng, 68-73, head eng group, Inst Atmospheric Sci, 68-75, sr scientist & head data acquistition & anal group, 76-81, RES PROF METEOROL & ELEC ENG, SDAK SCH MINES & TECHNOL, 73-, DIR, INST ATMOSPHERIC SCI, 81- *Concurrent Pos:* NSF fel meteorol, McGill Univ, 64, vis prof, 69-70; chief scientist, Hq, US Air Force Air Weather Serv, 74-75; actg dir, Inst Atmospheric Sci, 76-77; Vis scientist, Alta Res Coun, 84-85; Fulbright Lectr, Univ Helsinki, 86; fel, Am Meteorol Soc. *Mem:* Am Meteorol Soc; Am Soc Eng Educ; Inst Elec & Electronics Engrs; Sigma Xi; Weather Modification Asn. *Res:* Radar meteorology; cloud and precipitation physics; weather modification; remote sensing; meteorological instrumentation; electrostatic precipitation; propagation and scattering of electromagnetic and acoustic waves in the atmosphere. *Mailing Add:* Inst Atmospheric Sci SDak Sch of Mines & Technol Rapid City SD 57701-3995

SMITH, PAUL VERGON, JR, b Lima, Ohio, Apr 25, 21; m 45; c 4. CHEMISTRY. *Educ:* Miami Univ, AB, 42; Univ Ill, MS, 43, PhD(org chem), 45. *Prof Exp:* Instr, Miami Univ, 42; asst, Univ Ill, 42-43; from asst to res chemist, Off Rubber Reserve, 43-46; res chemist & group leader, Esso Res & Eng Co, 46-54, sci liaison, Esso Res Ltd, Eng, 55-57, asst dir, Chem Res Div, NJ, 57-60, Cent Basic Res Lab, 60-66, Res Dept, Esso Petrol Co Ltd, Eng, 66-67, dir Chem Dept, Esso Res SA, Belg, 67-71, head educ & sci rels, Esso Res & Eng Co, 71-73, mgr educ & sci rels, 73-78, pub affairs, 78-81, mgr educ & prof soc rels, Pub Affairs, Exxon Res & Eng Co, 81-86; RETIRED. *Concurrent Pos:* Mem adv bd, Petrol Res Fund, 65-66, CACHE, Inc, 80-86; dir, Am Chem Soc, 78-86, chmn bd, 84-86; bd dirs & treas, Jr Eng Tech Soc, Inc, 80-89; bd dirs, Centcom, Ltd, 84-86. *Honors & Awards:* President's Award, Am Asn Petrol Geologists, 55. *Mem:* AAAS; Am Soc Eng Educ (dir, 78-86, vpres, 80-86); Am Inst Chem Eng; Am Chem Soc; NY Acad Sci. *Res:* Synthetic rubber; detergents; physical separation of organic compounds; oil additives and synthetic lubricants; geochemistry; origin of petroleum; oxo process; plasticizers; polypropylene. *Mailing Add:* 713 Pineside Lane Naples FL 33963-8523

SMITH, PAULA BETH, b Malone, NY, Apr 7, 56; m 83; c 1. DEVELOPMENTAL IMMUNOLOGY, COMPARATIVE IMMUNOLOGY. *Educ:* Rochester Inst Technol, BS, 78; Pa State Univ, PhD(biol), 84. *Prof Exp:* Postdoctoral res assoc, 85-88, instr, 88-91, ASST PROF ANAT, MED CTR, UNIV NEBR, 91- *Mem:* Am Soc Zoologists; Int Soc Develop & Comp Immunol. *Res:* Development of the hemopoietic and immune systems and uses Xenopus laevis, the African clawed frog, as an animal model; commitment of embryonic mesoderm to hemopoiesis; colonization of hematopoietic microenvironments such as the liver and thymus; peripherization and function of mature hemopoietic cells that reach the spleen and other peripheral lymphoid organs. *Mailing Add:* Anat Dept Univ Nebr Med Ctr 600 S 42nd St Omaha NE 68198

SMITH, PERCY LEIGHTON, b Dunbar, WVa, July 23, 19; m 39; c 2. CHEMISTRY. *Educ:* Morris Harvey Col, BSc, 50. *Prof Exp:* Lab asst, Union Carbide Corp, 39-47, res chemist, 47-66, group leader, 66-74, assoc dir res & develop, 74-79; RETIRED. *Mem:* Am Chem Soc. *Res:* Analytical and organophosphorus chemistry; epoxy and polyester resins; polyurethanes. *Mailing Add:* 1508 Meyers Ave Dunbar WV 25064

SMITH, PERRIN GARY, b Bowie, Tex, Sept 3, 12; m 37; c 1. CHEMICAL ENGINEERING. *Educ:* Austin Col, BS, 34; Northwestern Univ, PhD(org chem), 38. *Prof Exp:* Abbott Labs fel, Northwestern Univ, 38-40; chemist res dept, Sharples Chem Inc, Mich, 40-44, chemist develop dept, Pa, 44-54, asst to pres, 54-56; asst to gen mgr, Indust Div, Pennwalt Corp, Philadelphia, 56-65, proj evaluator, Eng Dept, King of Prussia, 65-77; RETIRED. *Mem:* Am Chem Soc. *Res:* Amines; rubber chemicals; pharmaceutical intermediates; sulfer chemicals; process development. *Mailing Add:* 352 Black Horse Rd Chester Springs PA 19425

SMITH, PETER, b Sale, Eng, Sept 7, 24; m 51; c 4. PHYSICAL CHEMISTRY. *Educ:* Cambridge Univ, BA, 46, MA, 49, PhD(phys chem), 53. *Prof Exp:* Jr sci officer, Chem Dept, Royal Aircraft Estab, Eng, 43-46; demonstr phys & inorg chem, Univ Leeds, 50-51; asst prof chem, Purdue Univ, 54-59; from asst prof to assoc prof, 59-70, PROF CHEM, DUKE UNIV, 70- *Concurrent Pos:* Fulbright Res fel, Harvard Univ, 51-54. *Mem:* Am Chem Soc; Am Phys Soc; Royal Soc Chem. *Res:* Application of electron paramagnetic resonance spectroscopy to kinetic and structural problems; chemical kinetics, especially solution processes; biophysical chemistry. *Mailing Add:* Dept Chem Paul M Gross Chem Lab Duke Univ Durham NC 27706

SMITH, PETER A(LAN) S(OMERVAIL), b Erskine Hill, Eng, Apr 16, 20; nat US; m 52; c 2. HETEROCYCLIC CHEMISTRY. *Educ:* Univ Calif, BSc, 41; Univ Mich, PhD(inorg chem), 44. *Prof Exp:* Res assoc chem, 44-45; from instr to prof chem, 45-90, EMER PROF, UNIV MICH, ANN ARBOR, 90- *Concurrent Pos:* Fulbright res scholar, Univ Auckland, 51; mem, Am Chem Soc Adv Comt, Chem Corps, US Army, 55-60, mem sch & training comt, Chem Corps Adv Coun, 58-64; consult, Parke, Davis/Warner-Lambert Co, 57-; chmn comt nomenclature, Div Chem & Chem Technol, Nat Acad Sci-Nat Res Coun, 60-68; guest res prof, Inst Org Chem, Stuttgart Tech Univ, 61-62; bk rev ed, J Am Chem Soc, 70-91; Comn Nomenclature of Org Chem, Int Union Pure & Appl Chem, 84-, chmn, 88-, Am Chem Soc, chmn, 90- *Mem:* Am Chem Soc. *Res:* Heterocyclic chemistry; carbenes and nitrenes; organic nitrogen compounds; azides; hydrazines; hydroxylamines; coal asphaltenes. *Mailing Add:* Dept Chem Univ Mich Ann Arbor MI 48109

SMITH, PETER BLAISE, CELL DIFFERENTIATION, MEMBRANE BIOCHEMISTRY. *Educ:* Univ Tenn, PhD(microbiol), 73. *Prof Exp:* ASSOC PROF BIOCHEM, BOWMAN GRAY SCH MED, WAKE FOREST UNIV, 76- *Mailing Add:* Dept Biochem Bowman Gray Sch Med Wake Forest Univ 300 S Hawthorne Rd Winston Salem NC 27103

SMITH, PETER BYRD, medical bacteriology, for more information see previous edition

SMITH, PETER DAVID, b Providence, RI, Oct 25, 38; c 3. MATHEMATICS, COMPUTER SCIENCE. *Educ:* Col Holy Cross, AB, 60; Naval Postgrad Sch, MS, 64; Univ Wis, PhD(math), 68; Mich State Univ, MS, 75. *Prof Exp:* Instr math & comput sci, US Naval Postgrad Sch, 62-64, Xavier Univ La, 68-69; assoc prof, 69-88, PROF MATH & COMPUT SCI, ST MARY'S COL, IND, 88- *Concurrent Pos:* Consult, St Mary's Bus Off, 81. *Mem:* Asn Educ Data Systs; Asn Comput Mach; Asn Small Comput Users in Educ. *Mailing Add:* St Mary's Col Notre Dame IN 46556

SMITH, PETER LLOYD, b Victoria, BC, Apr 28, 44; m 68. ATOMIC & MOLECULAR SPECTROSCOPY, ULTRAVIOLET INSTRUMENTATION. *Educ:* Univ BC, BS, 65; Calif Inst Technol, PhD(physics), 72. *Prof Exp:* Res fel physics, Calif Inst Technol, 72; asst prof, Harvey Mudd Col, 72-73; res fel, 73-75, RES ASSOC PHYSICS, CTR ASTROPHYS, HARVARD COL OBSERV, 75- *Mem:* Am Phys Soc; Optical Soc Am; Am Astron Soc; Int Astron Union. *Res:* Visible and visible ultraviolet spectroscopy of atoms, ions and molecules; astrophysical applications include interstellar clouds, the sun, comets and plasmas; atmospheric chemistry; transitions probabilities; photon and electron cross sections for allowed and forbidden transitions; ultraviolet spectroscopic instrumentation for astronomy. *Mailing Add:* Harvard Smithsonian Ctr Astrophysics 60 Garden St P-246 Cambridge MA 02138

SMITH, PETER WILLIAM E, b London, Eng, Nov 3, 37; Can citizen; m; c 2. PHYSICS, ELECTRICAL ENGINEERING. *Educ:* McGill Univ, BSc, 58, MSc, 61, PhD(physics), 64. *Prof Exp:* Engr, Can Marconi Co, 58-59; mem tech staff laser res, Bell Tel Labs, 63-83, dist res mgr, guided wave & opto-electronics res, 84-89, DIV MGR, PHOTONIC SCI & TECHNOL RES, BELLCORE, 89- *Concurrent Pos:* Vis Mackay lectr, Univ Calif, Berkeley, 70-71; vis scientist, Lab d'Optique Quantique, Ecole Polytech, Palaiseau, France, 78-79. *Honors & Awards:* Sr Scientist Award, NATO, 79; Quantum Electronics Award, Inst Elec & Electronics Engrs, 86. *Mem:* Am Phys Soc; fel Optical Soc Am; fel Inst Elec & Electronics Engrs; Can Asn Phys. *Res:* Quantum electronics; atomic physics; gas lasers and laser devices. *Mailing Add:* Bellcore Red Bank NJ 07701-7020

SMITH, PHILIP EDWARD, b Johnson Co, Ill, Dec 25, 16; m 42; c 4. PARASITOLOGY. *Educ:* Southern Ill Univ, BEd, 40; Univ Ill, MS, 42; Johns Hopkins Univ, ScD(parasitol), 49. *Prof Exp:* Mus technician, Southern Ill Univ, 36-39, asst zool, 39-40; asst zool, Univ Ill, 40-42; asst parasitol, Sch Hyg & Pub Health, Johns Hopkins Univ, 47-49; asst prof zool, Okla Agr & Mech Col, 49-52; from asst prof to prof prev med & pub health, 52-70, assoc dean grad col med, Med Ctr, 56-70, assoc dean student affairs, Sch Med, 60-70, prof parasitol, Sch Health, 70-72, dean, Sch Health Related Professions, 70-72, dean, Col Allied Health Professions, 72-73, actg dean, Col Health & Allied Health Professions, 73-74, dean, Col Health, 74-81, dean, Col Allied Health, 81-82, REGENTS PROF & EMER DEAN, UNIV OKLA HEALTH SCI CTR, 82- *Mem:* Am Soc Parasitol; Am Soc Trop Med & Hyg; Micros Soc Am. *Res:* Morphology, life history and host-parasite relations of nematodes. *Mailing Add:* 241 NW 32nd St Oklahoma City OK 73118-8610

SMITH, PHILIP LEES, b Atlanta, Ga, Mar 3, 45; m 81; c 2. SLEEP DISORDERS. *Educ:* Harvard Univ, BA, 67; Tulane Univ Med Sch, MD, 72. *Prof Exp:* Res asst, Thorndike Mem Lab, Harvard Univ, 67-68; from instr to asst prof med, 77-85, INSTR ANESTHESIOL, JOHNS HOPKINS UNIV, 77-, ASSOC PROF MED, 85- *Concurrent Pos:* Dir, ICU, Francis Scott Key Med Ctr, 77-85, dir med ctr, 81-; dir, Johns Hopkins Sleep Disorders Ctr, 81- *Mem:* Am Col Chest Physicians; Am Thoracic Soc; Am Heart Asn; Asn Psychol Study Sleep; Geront Soc Am; Clin Sleep Society. *Mailing Add:* Francis Scott Key Med Ctr 4940 Eastern Ave Baltimore MD 21224

SMITH, PHILIP WESLEY, b Gainesville, Fla, Nov 28, 46; m 69; c 2. SPLINES & APPROXIMATION THEORY, NUMERICAL ANALYSIS. *Educ:* Univ Va, BA, 68; Purdue Univ, MS, 70, PhD(math), 72. *Prof Exp:* From asst prof to assoc prof math, Tex A&M Univ, 72-78; vis res prof math, Univ Alta, 78-79; vis scientist, TJ Watson IBM Res Ctr, 79-80; prof math, Old Dominion Univ, 80-85; MGR MATH SOFTWARE DESIGN, INT MATH & STATIST LIBR, 85- *Concurrent Pos:* Co-prin investr, Army Res Off, 74-82; consult, White Sands Missile Range, 75-76 & Gen Motors Res, 84-85. *Mem:* Soc Indust & Appln Math; Am Math Soc. *Res:* Numerical analysis applied to problems in approximation theory; shape perserving approximations and nonlinear eigenvalue problems. *Mailing Add:* 2500 Park W Tower One 2500 City W Blvd Houston TX 77042-3020

SMITH, PHILLIP DOYLE, b Berkeley, Calif, Apr 16, 46. CELLULAR IMMUNOLOGY, GASTROENTEROLOGY. *Educ:* Univ Calif, Berkeley, BA, 68; Univ Rochester, MD, 73. *Prof Exp:* Res fel cell biol, Cancer Res Inst, Univ Vienna, 71-72; intern & resident internal med, Vanderbilt, 73-76; res fel gastroenterol, Univ Colo, 76-79; guest investr parasite immunol, 80-82, SR STAFF RES FEL CELL IMMUNOL, NIH, 82- *Mem:* Am Asn Immunologists; Am Fedn Clin Res; Am Gastroenterol Asn; Am Col Physicians; Am Soc Trop Med. *Res:* Cellular immune responses to gastrointestinal infections. *Mailing Add:* NIH Bldg 30 Rm 322 9000 Rockville Pike Bethesda MD 20892

SMITH, PHILLIP J, b Muncie, Ind, Oct 2, 38; m 60; c 3. SYNOPTIC METEOROLOGY, ATMOSPHERIC ENERGETICS. *Educ:* Ball State Univ, BS, 60; Univ Wis, MS, 64, PhD(meteorol), 67. *Prof Exp:* Res asst meteorol, Univ Wis, 63-67; from asst prof to assoc prof, 67-81, PROF METEOROL, PURDUE UNIV, WEST LAFAYETTE, 81- *Concurrent Pos:* Prin investr, NSF & NASA grants, 69-; sr fel, Nat Ctr Atmospheric Res, 72-73, assoc dept head, 88- *Mem:* Fel Am Meteorol Soc; Am Geophys Union; Sigma Xi. *Res:* The energetics & dynamics of synoptic scale systems; Development of extratropical cyclones. *Mailing Add:* Dept of Earth & Atmospheric Scis Purdue Univ West Lafayette IN 47907

SMITH, PIERRE FRANK, b North Tonawanda, NY, Aug 17, 20. PHARMACY, MEDICINAL CHEMISTRY. *Educ:* Univ Buffalo, BS, 41; Univ Md, PhD(pharmaceut chem), 47. *Prof Exp:* Asst chem, Univ Md, 41-44; asst prof pharmaceut chem, Western Reserve Univ, 47-49; assoc prof, Rutgers Univ, 49-57; prof & chmn dept, Col Pharm, Univ RI, 57-65; chmn dept, 65-77, PROF, COL PHARM, NORTHEASTERN UNIV, 65- *Mem:* Am Chem Soc; Am Pharmaceut Asn; Am Inst Hist Pharm; Am Acad Pharmaceut Sci. *Res:* Organic synthesis; analytical chemistry; synthesis and study of tranquilizers and central nervous systems depressants; physical pharmacy. *Mailing Add:* 56 Marion St Natick MA 01760-3644

SMITH, QUENTON TERRILL, b Ames, Iowa, Jan 6, 29; m 88. ORAL BIOCHEMISTRY, ORAL BIOLOGY. *Educ:* Iowa State Univ, BS, 51, MS, 52; Univ Minn, PhD(physiol chem), 59. *Prof Exp:* Res assoc dermat, 59-69, from lectr to asst prof biochem, 59-69, assoc prof oral biol & biochem, 69-73, prof oral biol, 73-88, PROF ORAL SCI, SCH DENT, UNIV MINN, MINNEAPOLIS, 88- *Concurrent Pos:* USPHS spec fel, Argonne Nat Lab, 68-69; adj prof biochem & med sci, Univ Minn, Minneapolis, 73- *Mem:* Am Soc Biochem & Molecular Biol; AAAS; Soc Exp Biol & Med; Int Asn Dent Res. *Res:* Biochemistry of oral fluids, especially the relationship of salivary and gingival crevicular fluid composition to oral and systemic disease; connective tissue biochemistry. *Mailing Add:* 17-252 Moos Health Sci Tower Univ Minn Minneapolis MN 55455

SMITH, R JAY, b Flint, Mich, July 26, 22. ZOOLOGY. *Educ:* Alma Col, AB, 47; Univ Mich, MS, 49, PhD(zool, parasitol), 53. *Prof Exp:* Instr biol, Marquette Univ, 48-49; asst prof, Dubuque Univ, 53-55 & Col William & Mary, 55-56; asst prof zool, Univ Miami, 56-57; asst prof, 57-80, ASSOC PROF BIOL, UNIV DETROIT, 80- *Concurrent Pos:* USPHS fel parasitol, Univ PR, 59. *Mem:* Am Soc Parasitologists; Am Micros Soc; Am Soc Zoologists. *Res:* Invertebrate zoology; helminthology; trematodes as hosts to ancylid snails; water pollution biology; indicators of pollution, especially coliforms and invertebrates. *Mailing Add:* Dept Biol Univ Detroit 4001 W McNicholas Rd Detroit MI 48221

SMITH, R L, b Rigby, Idaho, Jan 13, 24; m 44; c 5. SOIL CHEMISTRY. *Educ:* Utah State Univ, BS, 51, MS, 52; Univ Calif, Los Angeles, PhD(plant sci), 55. *Prof Exp:* Asst soils, Utah State Univ, 51-52; asst plant sci, Univ Calif, Los Angeles, 52-54; asst pomologist, Univ Calif, 54-55; asst prof soils & asst soil chemist, 55-59, assoc prof & assoc soil chemist, 59-64, prof soils & soil chemist, 73-80, head Dept Soil Sci & Biometeorol, 73-80, PROF SOIL SCI & BIOMETEOROL, UTAH STATE UNIV, 80- *Mem:* Am Soc Agron; Soil Sci Soc Am; Am Soc Plant Physiol; Am Soc Hort Sci. *Res:* Soil chemistry of micro-nutrient elements in soils; lime-induced chlorosis; nitrogen interchanges in soil; soil and plant root relationships. *Mailing Add:* 565 E 400 S Logan UT 84321

SMITH, R LOWELL, b Toledo, Ohio, June 24, 40; m 63; c 2. ACCELERATED LIFE TESTING, NUMERICAL DATA PROCESSING. *Educ:* Univ Toledo, BS, 62; Case Western Reserve Univ, PhD(physics), 70. *Prof Exp:* Res fel, Case Western Reserve Univ, 70-71; staff scientist, Ultrascan Co, 71-72; sr res assoc, Horizons Res Inc, 72-77; SR SCIENTIST, TEX RES INST, AUSTIN, 77- *Mem:* Acoust Soc Am; Inst Elec & Electronics Engrs; Am Soc Nondestructive Testing. *Res:* Electrographics; electroacoustics; viscoelasticity; experimental design; reliability analysis; nondestructive testing; mathematical modeling; impedance spectroscopy; statistical inference; regression analysis; computer controlled data acquisition; instrumentation development; project management; technical writing. *Mailing Add:* Tex Res Inst 9063 Bee Caves Rd Austin TX 78733-6201

SMITH, RALPH CARLISLE, b West New York, NJ, May 24, 10; m 54. EDUCATION ADMINISTRATION, HISTORY. *Educ:* Rensselaer Polytech Inst, ChE, 31; George Washington Univ, JD, 39; Univ NMex, PhD, 62. *Prof Exp:* Chemist org, E I du Pont de Nemours & Co, 31-36; patent examr, US Patent Off, 36-38; chem engr org, Colgate Palmolive Co, 39-42; Lt Col nuclear, US Army Engrs, Manhattan Proj, 42-47; asst dir, Los Alamos Sci Lab, 47-57; asst to pres, Nuclear Div, ACF Industs, 57-60; dean & pres, NMex Highlands Univ, 61-71, grad dean, 71-77; ADJ PROF, COLUMBIA COL, 78- *Mem:* Am Inst Chemists; Am Inst Chem Engrs; Am Nuclear Soc; Am Phys Soc. *Mailing Add:* 55 Terra Vista Ave 2 San Francisco CA 94115

SMITH, RALPH E(DWARD), b Porterdale, Ga, May 6, 23; m 48; c 3. AGRICULTURAL ENGINEERING. *Educ:* Univ Ga, BS, 48, MS, 61; Okla State Univ, PhD(agr eng), 66. *Prof Exp:* Instr agr, Polk County Bd Educ, Ga, 48-54; instr physics & agr eng, Abraham Baldwin Agr Col, 54-56; from instr to assoc prof agr eng, Univ Ga, 56-89; RETIRED. *Concurrent Pos:* Vis prof, Okla State Univ, 65-66. *Mem:* Am Soc Agr Engrs; Am Soc Eng Educ; Am Soc Heating, Refrig & Air-Conditioning Engrs; Sigma Xi. *Res:* Environmental control engineering for livestock; farm electrification engineering; process engineering for agricultural engineering; processes of transient conduction heat transfer. *Mailing Add:* PO Box 1223 Athens GA 30603

SMITH, RALPH E, b Yuma, Colo, May 10, 40; m 88; c 1. VIROLOGY, ONCOLOGY. *Educ:* Colo State Univ, BS, 61; Univ Colo, Denver, PhD(microbiol), 68. *Prof Exp:* Teaching fel, dept microbiol & immunol, Duke Univ Med Ctr, Durham, NC, 68-70, from asst prof to prof viral oncol, 70-82; prof vet virol & head, dept microbiol, 83-87, interim vpres res, 89-90, ASSOC VPRES RES & PROF MICROBIOL, COLO STATE UNIV, FT COLLINS, 87- *Concurrent Pos:* Prin investr, Nat Cancer Inst grant, NIH, 70-; consult, Bellco Glass, Vineland, NJ, 76-80, Procter & Gamble Co, 82-87 & Schering Plough Corp, Bloomfield, NJ, 87-; Eleanor Roosevelt fel, Wellcome Res Found, Beckenham, Kent, Eng, 78-79. *Mem:* Am Soc Microbiol; NY Acad Sci; Am Asn Immunologists; Am Soc Virol; Am Soc Clin Path; Am Asn Avian Pathologists. *Res:* Biology of avian retroviruses; characterization of avian osteopetrosis and associated diseases; description of widespread disease in chicken's body caused by RAV-7 avian retrovirus; cell and tissue culture techniques; molecular cloning of avian retroviruses; characterization of lung tumors caused by an avian retrovirus. *Mailing Add:* Off Vpres Res Colo State Univ Ft Collins CO 80523

SMITH, RALPH EMERSON, b Beckley, WVa, Jan 13, 16; m 50; c 2. GROUNDWATER GEOLOGY. *Educ:* Tex Christian Univ, AB, 37, MS, 39. *Prof Exp:* Asst, Tex Christian Univ, 37-39 & Univ Okla, 39-42; assoc prof geol & geog, Drury Col, 46-47; geologist, Geol Div, Fuels Br, US Geol Surv, 47-48, geologist, Water Res Div, Ground Water Br, 49-75, geologist, Pub Lands Hydrol Prog, 64-75; consult, 75-83; RETIRED. *Concurrent Pos:* Supvr water well drilling, Indonesia, 80. *Res:* Water supply for stock and camp sites; relation of geology, weather and vegetation to water on public domains in the western mountain states; area studies; evaluate the possibility for additional ground water. *Mailing Add:* 8281 Chase Way Arvada CO 80003

SMITH, RALPH G, b St John, NB, Jan 11, 20; nat; m 42; c 6. INDUSTRIAL HYGIENE, ANALYTICAL CHEMISTRY. *Educ:* Wayne State Univ, BS, 42, MS, 49, PhD(chem), 53. *Prof Exp:* Chemist, Rotary Elec Steel Co, Mich, 40-42; assoc indust hygienist & chief chemist, Bur Indust Hyg, Detroit Dept Health, Mich, 46-55; assoc prof indust med & hyg, Sch Med, Wayne State Univ, 55-63, prof occup & environ health, 63-70; PROF ENVIRON & INDUST HEALTH, SCH PUB HEALTH, UNIV MICH, ANN ARBOR, 70- *Mem:* AAAS; Am Conf Govt Indust Hygienists; Am Indust Hyg Asn; Air Pollution Control Asn; Am Chem Soc. *Res:* Chemistry and toxicology; air analysis; analysis of biological samples for toxic substances; analytical chemistry of beryllium, ozone, mercury and lead; toxicity of air pollutants; mercury and chlorine. *Mailing Add:* 24711 Tudor Lane Franklin MI 48025

SMITH, RALPH GRAFTON, b Oxford Co, Ont, Mar 15, 00; nat US; m 74; c 4. PHARMACOLOGY. *Educ:* Ga Inst Technol, BEE, 63, PhD(electrical eng), 69; Mass Inst Technol, MSEE, 64. *Prof Exp:* Asst biochem, Univ Toronto, 21-22 & Connaught Labs, 23-25; intern, Toronto Gen Hosp, 25-26; Nat Res Coun fels, Washington Univ, 26-27 & Univ Chicago, 27-28; from instr to assoc prof pharmacol, Univ Mich, 28-43, asst secy med sch, 42-43; prof pharmacol, Tulane Univ, 43-50; from chief new drug br to dir div new drugs, Bur Med, US Food & Drug Admin, 50-66, actg dir, Bur Med, 62-64, asst to dir bur med for Nat Acad Liaison, 66-68, dir off med support, 68-70; consult div med sci, Nat Acad Sci-Nat Res Coun, 70-73; RETIRED. *Concurrent Pos:* Ed pharmacol sect, Biol Abstr, 40-67. *Mem:* Inst Elec & Electronics Engrs Antennas & Propagat Soc; Am Soc Pharmacol & Exp Therapeut; Inst Elec & Electronics Engrs Microwave Theory & Techniques Soc; Pan Am Med Asn; Sigma Xi. *Res:* Alveolar airarterial blood equilibrium; respiratory stimulants; cyanide poisoning and sulfur metabolism; diffusible calcium of blood serum; ergot; antimony metabolism; new drugs. *Mailing Add:* 1026 Noyes Dr Silver Spring MD 20910

SMITH, RALPH INGRAM, b Cambridge, Mass, July 3, 16; m 40; c 5. ZOOLOGY. *Educ:* Harvard Univ, BA, 38, MA, 40, PhD(zool), 42. *Prof Exp:* From instr to prof, 46-87, EMER PROF ZOOL, UNIV CALIF, BERKELEY, 87- *Concurrent Pos:* Fulbright lectr, Univ Glasgow, 53-54; Fulbright vis prof, Univ Turku, 61-62; Guggenheim fel, Univ Newcastle, Upon Tyne, 68-69. *Mem:* Soc Exp Biol; corresp mem Finnish Zool Bot Soc; Am Soc Zoologists; foreign mem Finnish Acad Sci & Letters. *Res:* Invertebrate zoology and comparative physiology. *Mailing Add:* Dept Integrative Biol Univ of Calif Berkeley CA 94720

SMITH, RALPH J(UDSON), b Herman, Nebr, June 5, 16; m 38; c 4. ELECTRICAL ENGINEERING. *Educ:* Univ Calif, BS, 38, MS, 40, EE, 42; Stanford Univ, PhD(elec eng), 45. *Prof Exp:* Jr engr, Stand Oil Co, Calif, 38-40; instr eng, San Jose State Col, 40-42, head dept, 45-52, chmn div eng, math & aeronaut, 52-57; adv electronics, Repub of Philippines, 57-58; instr, 42-45, prof, 58-81, EMER PROF ELEC ENG, STANFORD UNIV, 81- *Concurrent Pos:* Consult, State Dept Educ, Calif, 60. *Mem:* Fel Am Soc Eng Educ; fel Inst Elec & Electronics Engrs. *Res:* Author of three books. *Mailing Add:* Dept Elec Eng Stanford Univ Stanford CA 94305

SMITH, RAOUL NORMAND, b West Warwick, RI, May 15, 38; m 66; c 2. NATURAL LANGUAGE PROCESSING, KNOWLEDGE REPRESENTATION & EXPERT SYSTEMS. *Educ:* Brown Univ, AB, 63, AM, 64, PhD(computational ling), 68. *Prof Exp:* From instr to asst prof ling, Northwestern Univ, 67-73, assoc prof computational & math ling, 73-81; prin mem tech staff, comput sci lab, Gen Tel & Electronics Labs, Inc, 81-83; PROF COMPUT SCI, NORTHEASTERN UNIV, 83- *Concurrent Pos:* Orin investr, Am Coun Learned Socs Grant, 74, Am Philos Soc Grant, 74, Nat Endowment Humanities Res Grants, 75 & 76-77 & Dig Equip Corp Grant, 85; vis prof, Univ Maine, Orono, 78, & Jilin Univ Technol, Changchun, People's Repub China, 85; mem, Nat Acad Sci, Comn Microcomputers Develop Countries, 86; chmn bd, Cognitive Computers, Inc, 85-87; consult to various orgns. *Mem:* Asn Comput Mach; Am Asn Artificial Intel; Asn Computational Ling; Inst Elec & Electronics Engrs Comput Soc; Sigma Xi. *Res:* natural language interfaces and expert systems. *Mailing Add:* Col Comput Sci Northeastern Univ Boston MA 02115

SMITH, RAPHAEL FORD, b Wilson, NC, Jan 22, 33; m 58; c 4. MEDICINE, CARDIOLOGY. *Educ:* Vanderbilt Univ, BA, 55; Harvard Med Sch, MD, 60; FACP, 68; FACC, 69. *Prof Exp:* Intern asst resident, Mass Gen Hosp, 60-62, resident, 65-66; res asst aviation med, US Naval Sch Aviation Med, 62-65; chief, Cardiol Br, Naval Aerospace Med Inst, 66-69; from asst prof to assoc prof, 69-82, PROF MED, SCH MED, VANDERBILT UNIV, 74-, SCH ENG, 76-; CHIEF, CARDIOL SECT, NASHVILLE VET ADMIN HOSP, 75- *Concurrent Pos:* Consult, US Naval Hosp, 67-69 & NIH, Specialized Ctrs Res, 74-77. *Honors & Awards:* Skylab Achievement Award, NASA, 74. *Mem:* Am Heart Asn; Southern Soc Clin Invest. *Res:* Cardiac electrophysiology; aerospace medical research. *Mailing Add:* 1310 24th Ave S Nashville TN 37203

SMITH, RAY FRED, b Los Angeles, Calif, Jan 20, 19; m 40; c 3. ENTOMOLOGY. *Educ:* Univ Calif, BS, 40, MS, 41, PhD(entom), 46. *Hon Degrees:* DAgrSc, Landbouwhogesch, Wageningen, 76. *Prof Exp:* Field entomologist, Balfour-Guthrie Investment Co, 40; field & lab asst entom, 40-45, assoc, Exp Sta, 45-46, from instr & jr entomologist to prof 7 entomologist, 46-83, exec dir, consortium Int Corp Protection, 79-85, EMER PROF ENTOM, UNIV CALIF, BERKELEY, 83- *Concurrent Pos:* Guggenheim fel, 50; consult, Food & Agr Orgn, UN. *Honors & Awards:* C W Woodworth Award, 71; Hon Award, Consortium Integrated Pest Mgt, 85. *Mem:* Nat Acad Sci; fel & hon mem Entom Soc Am (pres, 76); fel Entom Soc Can; fel AAAS; fel Am Acad Arts & Sci. *Mailing Add:* 3092 Hedaro Ct Lafayette CA 94549

SMITH, RAYMOND CALVIN, b Glendale, Calif, Nov 17, 34; m 56; c 2. PHYSICAL OCEANOGRAPHY. *Educ:* Mass Inst Technol, SB, 56; Stanford Univ, PhD(physics), 61. *Prof Exp:* Res fel, Cambridge Electron Accelerator, Harvard Univ, 61-63; res oceanogr, Univ Calif, San Diego, 63-80; AT DEPT OF GEOG, UNIV CALIF, SANTA BARBARA. *Mem:* AAAS; Optical Soc Am; Am Geophys Union; Am Soc Limnol & Oceanog. *Res:* Environmental optics; primary productivity; remote sensing; ecology of southern ocean. *Mailing Add:* Dept of Geog Univ Calif Santa Barbara CA 93106

SMITH, RAYMOND JAMES, b Manchester, NH, July 16, 24. CIVIL ENGINEERING, GEOLOGY. *Educ:* Calif Inst Technol, BS, 45, MS, 48; Princeton Univ, MA, 50, PhD, 51. *Prof Exp:* Investr for Princeton, Caribbean, 48-54; asst prof, La State Univ, 54-57; res engr & geologist, NY Explor Co, 57-61; civil engr-geologist, US Naval Civil Eng Lab, Calif, 61-68; prof oceanog, Naval Postgrad Sch, 68-71; CONSULT CIVIL ENGR & GEOLOGIST, 71- *Mem:* Geol Soc Am; Am Soc Civil Engrs; Soc Econ Geologists; Am Geophys Union. *Res:* Application of geology to civil engineering. *Mailing Add:* 791 Via Ondulando Ventura CA 93003

SMITH, RAYMOND V(IRGIL), b Esbon, Kans, Nov 17, 19; m 48; c 5. MECHANICAL ENGINEERING. *Educ:* Univ Colo, BS, 48, MS, 51; Univ Utah, MS, 57; Oxford Univ, DPhil, 68. *Prof Exp:* Design engr, Boeing Airplane Co, 41-44 & 48; instr mech eng, Colo Sch Mines, 49-52; res engr, Sandia Corp, 52-53; asst prof mech eng, NMex State Univ, 53-54; assoc prof, Univ Utah, 54-57 & Colo State Univ, 57-61; mech engr, Cryogenic Eng Lab,

Nat Bur Standards, 58-71; prof mech eng, Wichita State Univ, 71-84; CONSULT, 84- *Concurrent Pos:* With UK Atomic Energy Res Estab, 66-67, 78 & 81. *Honors & Awards:* NBS Distinguished Auth, NASA Tech Utilization. *Mem:* Am Soc Mech Engrs; Am Soc Eng Educ; India Inst Sci. *Res:* Fuel combustion mechanism; two-phase flow cryogenic studies of heat transfer, thermodynamics and fluid mechanics; technology assessement; rehabitation engineering. *Mailing Add:* 5 Crestview Lakes Wichita KS 67220

SMITH, REGINALD BRIAN, b Warrington, Eng, Feb 7, 31; US citizen; m 63; c 2. RESPIRATORY CARE. *Educ:* Univ London, BS & MB, 56. *Prof Exp:* Clin instr, Univ Pittsburgh, 65-69, from asst prof to prof anesthesiol, 69-78; PROF ANESTHESIOL, CHMN DEPT & DIR ANESTHESIOL RESIDENCY, UNIV TEX HEALTH SCI CTR, SAN ANTONIO, 78-, CHIEF ANESTHESIOLOGIST, TEACHING HOSP, 78- *Concurrent Pos:* Dir, anesthesiol dept, Eye & Ear Hosp, Pittsburgh, 71-76; vchmn anesthesiol dept, Univ Pittsburgh, 73-77, actg chmn, 77-78; ed, Int Ophthal Clin, 73 & Int Anesthesiol Clin, 83; anesthesiologist-in-chief, Presbyterian Univ Hosp, Pittsburgh, 76-78. *Mem:* Royal Col Surgeons Eng; fel Am Col Anesthesiol; fel Am Col Chest Physicians; fel Am Col Physicians; Am Soc Anesthesiol. *Res:* Effects of anesthetic agents in the eye (intraocular pressure, oculocardiac reflex, drug interaction); anesthetic techniques used on endoscopy; techniques of artificial ventilation (high frequency ventilation, transtracheal ventilation, apneic ventilation). *Mailing Add:* Dept Anesthesiol Univ Tex Health Sci Ctr 7703 Floyd Curl Dr San Antonio TX 78284

SMITH, REID GARFIELD, b Toronto, Ont, Oct 4, 46; m 81; c 3. NOWLEDGE-BASED SYSTEM DESIGN, MACHINE LEARNING. *Educ:* Carleton Univ, BEng, 68, MEng, 69; Stanford Univ, PhD(elec eng), 79. *Prof Exp:* Defense sci officer, Defense Res Estab Atlantic, 69-81; PROG LEADER, SCHLUMBERGER-DOLL RES CTR, 81- *Concurrent Pos:* Lectr comput sci, Dalhousie Univ, 81. *Mem:* Asn Comput Mach; Inst Elec & Electronics Engrs; Am Asn Artificial Intel; Can Soc Comput Studies Intel; AAAS. *Res:* Knowledge-based system design, concentrating on knowledge acquisition via interactive machine learning and construction of KBS substrates for representation of domain knowledge, control and user interfaces. *Mailing Add:* Schlamberger Lab Comput Sci 8311 N RR b20 PO Box 200015 Austin TX 78720

SMITH, REX L, b Beaver, Utah, June 7, 29; c 4. PLANT MOLECULAR GENETICS, PLANT BREEDING. *Educ:* Utah State Univ, BS, 63; Iowa State Univ, PhD(plant breeding & genetics), 67. *Prof Exp:* From asst prof to assoc prof, 67-78, PROF AGRON, UNIV FLA, 78- *Mem:* AAAS; Am Soc Agron; Int Soc Plant Molecular Biol; Am Genetic Asn. *Res:* Molecular genetics; genetics and plant breeding. *Mailing Add:* Dept Agron Univ Fla Gainesville FL 32611

SMITH, RICHARD A, b Norwalk, Conn, Oct 13, 32. FAMILY MEDICINE. *Educ:* Howard Univ, BS, 53, MD, 57; Columbia Univ, MPH, 60; Am Bd Prev Med, dipl, 67. *Prof Exp:* Intern, USPHS Hosp, Seattle, 57-58; resident, Los Angeles City Health Dept, 58-59; epidemiologist, Wash State Health Dept, 60-61; sr Peace Corps physician, Lagos, Nigeria, 61-63; asst prof, Dept Prev Med, Howard Univ, 63-68; assoc prof & dir, Medex Prog, Sch Pub Health & Community Med, Univ Wash, 68-72; ADJ PROF FAMILY PRACT & COMMUNITY HEALTH & DIR, MEDEX GROUP, JOHN A BURNS SCH MED, UNIV HAWAII, 72- *Concurrent Pos:* Africa regional med officer, Med Prog Div, Peace Corps, 63-64, dep dir, 64-65; exec mgt trainee, Off Surgeon Gen, 65-66, spec asst dir, Off Int Health, 66, chief, Off Planning, 67, dep dir, Off Int Health, 67-68; clin asst prof, Dept Community & Int Health, Sch Med, Georgetown Univ, 67-68; adv, US deleg WHO, 67 & 70, mem, Int Task Force World Health Manpower, 70, consult, 77-; mem, Nat Adv Allied Health Prof Coun, NIH, 71 & bd dirs, Am Inst Res, 81- *Honors & Awards:* William A Jump Award, HEW, 68; Gerard B Lambert Award, 71; Rockefeller Pub Serv Award, 81. *Mem:* Inst Med-Nat Acad Sci; Am Pub Health Asn; Am Soc Trop Med & Hyg; fel Am Col Prev Med. *Res:* Family practice and community health care; author of 25 technical publications. *Mailing Add:* Medex Group Univ Hawaii 1833 Kalakaua Ave Suite 700 Honolulu HI 96815

SMITH, RICHARD ALAN, b Moscow, Idaho, Aug 6, 40; m 67; c 2. BIOCHEMISTRY. *Educ:* Whitman Col, BA, 62; Univ Minn, St Paul, PhD(biochem), 67. *Prof Exp:* Res assoc biochem, Univ Hawaii, 67-69; from asst prof to assoc prof, 69-89, PROF CHEM, STATE UNIV COL ARTS & SCI, 89-, CHMN DEPT, 84- *Concurrent Pos:* Vis prof, Univ Rochester, 83-84. *Mem:* AAAS; Am Chem Soc; Sigma Xi; Am Soc Biochem & Molecular Biol. *Res:* Structure-function relationships in enzymes; amine oxidases. *Mailing Add:* Dept Chem State Univ Col Arts & Sci Geneseo NY 14454

SMITH, RICHARD ANDREW, b Buffalo, NY, Aug 9, 50; m 74. HEMATOLOGY, ANATOMY. *Educ:* Canisius Col, BA, 72; State Univ NY, Buffalo, MA, 76, PhD(anat), 78. *Prof Exp:* Grad asst, 74-77, CLIN INSTR ANAT, STATE UNIV NY, BUFFALO, 78- *Mem:* AAAS; Am Soc Zoologists; Sigma Xi. *Res:* Morphological hematology; origin of the hemopoietic stem cell in mammalian embryos; endocrine control of hemopoiesis. *Mailing Add:* Dept Path Millard Fillmore Hosp Buffalo NY 14209

SMITH, RICHARD AVERY, b Long Beach, Calif, July 22, 24; m 47; c 4. EARTH SCIENCES & SCIENCE EDUCATION. *Educ:* Stanford Univ, BS, 49, EdD(teacher educ), 56; Univ Northern Colo, MA, 50. *Prof Exp:* Instr, Menlo Sch & Col, 49-50; teacher high sch, Calif, 51-55; from asst prof to assoc prof, 55-64, chmn, Dept Natural Sci, 68-82, assoc dean sch sci, 83-87 PROF PHYS SCI & SCI EDUC, SAN JOSE STATE UNIV, 64- *Concurrent Pos:* Sci consult, Peace Corps, Philippines, 64-66; chmn adv comt sci & educ, Calif State Bd Ed, 66-71, Pacific Educ Projs, 80- *Mem:* Fel AAAS; Nat Sci Teachers Asn; Nat Asn Res Sci Teaching; Nat Asn Geol Teachers. *Res:* Improvement of science teaching. *Mailing Add:* Dept Geol San Jose State Univ San Jose CA 95192-0102

SMITH, RICHARD BARRIE, b Vernon, BC, Apr 18, 34; m 59; c 2. FOREST PATHOLOGY, FOREST ECOLOGY. *Educ:* Univ BC, BSF, 57, PhD(forest ecol), 63; Yale Univ, MF, 58. *Prof Exp:* Res officer forest path, Can Dept Agr, 59-63; res officer, 63-65, res scientist II environ forestry & forest path, 65-85, RES SCIENTIST III ENVIRON FORESTRY, FORESTRY CAN, PAC FORESTRY CTR CAN, 85- *Concurrent Pos:* Int Joint Comn, 71. *Mem:* Can Inst Forestry; Can Phytopath Soc; Int Mountain Soc; Can Soc Soil Sci; Soil & Water Conserv Soc Am. *Res:* Edaphotopes of forest ecosystems; impact and biology of dwarf mistletoes on western North American conifers; environmental impact of forest management practices; forest productivity and natural revegetation on landslides. *Mailing Add:* Pac Forestry Centre 506 W Burnside Rd Victoria BC V8Z 1M5 Can

SMITH, RICHARD CARPER, b Jacksonville, Fla, May 9, 38; m 66; c 2. OPTICS. *Educ:* Davidson Col, BS, 60; Lehigh Univ, MS, 62, PhD(physics), 66. *Prof Exp:* Res physicist, Nat Security Agency, 67-68; from asst prof to assoc prof, 68-88, PROF PHYSICS, UNIV WFLA, 88- *Mem:* Am Asn Physics Teachers. *Res:* Use of coherent optical processing systems for image storage and enhancement. *Mailing Add:* Dept of Physics Univ WFla Pensacola FL 32514

SMITH, RICHARD CECIL, b Sydney, Australia, Feb 21, 40; US citizen; m 63; c 2. APPLIED NUCLEAR PHYSICS, ADAPTIVE OPTICS. *Educ:* Princeton Univ, AB, 62; Univ Md, Coll Park, PhD(physics), 70. *Prof Exp:* Res asst high energy physics, dept physics & astron, Univ Md, 65-70; sr physicist, Westinghous Elec Corp, 70-78, fel physicist, Res & Develop Ctr, 78-87; sr scientist prog mgr, Kaman Instrument Corp, 87-90, SR SCIENTIST, KAMAN SCI CORP, 90- *Concurrent Pos:* Mem, Westinghouse Res & Develop Planning, 78; dep prin investr, Starlab Wavefront Control Exp, 90-91. *Mem:* Am Phys Soc; Am Nuclear Soc; Int Soc Optical Eng. *Res:* Electron beam technology; gas discharge excitation; elementary particles; applied nuclear physics; neutron activation; DFN uranium exploration; oil well logging instrumentation; neutron generators; nuclear power instrumentation; Monte Carlo neutron transport codes; nuclear reactor design methods; proximity sensors; optical image processing; adaptive optics. *Mailing Add:* Kaman Sci Corp PO Box 7463 Colorado Springs CO 80933

SMITH, RICHARD CHANDLER, b St Paul, Minn, Sept 10, 13; m 44; c 1. FOREST ECONOMICS. *Educ:* Univ Minn, BS, 37; Duke Univ, MF, 47, DF, 50. *Prof Exp:* Field asst & jr forester, Forest Serv, USDA, 33-39; forester, Am Creosoting Co, 40-42; prof, 47-82, EMER PROF FORESTRY, UNIV MO-COLUMBIA, 82- *Concurrent Pos:* Res forester, Forest Serv, USDA, 62-63. *Honors & Awards:* Forest Conserv Award, Mo Conserv Fedn, 71. *Mem:* Fel Soc Am Foresters. *Res:* Forest economics and management; economics of timber production and multiple-use forestry. *Mailing Add:* Sch Natural Resources Univ Mo Columbia MO 65211

SMITH, RICHARD CLARK, b Salem, Ind, Apr 17, 27; m 57; c 1. PLANT PHYSIOLOGY. *Educ:* Vanderbilt Univ, AB, 49; Duke Univ, AM, 52, PhD(bot), 57. *Prof Exp:* Instr bot, Univ Tenn, 56-58 & Miami Univ, 58-59; from instr to asst prof, Rutgers Univ, 59-62; plant physiologist, Univ Calif, Davis, 62-64; from asst prof to assoc prof bot, 64-77, PROF BOT, UNIV FLA, 77- *Concurrent Pos:* Consult, Univ Ill, 72. *Mem:* AAAS; Bot Soc Am; Am Soc Plant Physiol; Scand Soc Plant Physiol. *Res:* Absorption and translocation of mineral ions in plants; metabolic activities of roots; plant growth. *Mailing Add:* Dept Bot-220 Bartram Hall Univ Fla Gainesville FL 32611

SMITH, RICHARD DALE, b Lawrence, Mass, July 1, 49; m 85; c 1. CERAMIC ENGINEERING, BIOCHEM. *Educ:* Lowell Technol Inst, BS, 71; Univ Utah, PhD(physical chem), 75. *Prof Exp:* Res scientist, 76-78, sr res scientist, 78-83, staff scientist, 83-88, SR STAFF SCIENTIST, BATTELLE, PAC LAB, 88- *Mailing Add:* Box 999 Battelle Northwest Richland WA 99352

SMITH, RICHARD DEAN, chemical engineering; deceased, see previous edition for last biography

SMITH, RICHARD ELBRIDGE, b Keene, NH, May 30, 32; m 63; c 2. GEOLOGY, GEOCHEMISTRY. *Educ:* Univ NH, BA, 59; Univ Ill, Urbana, MS, 60; Pa State Univ, PhD(petrol), 66. *Prof Exp:* Res oceanogr, Ocean Sci Dept, US Naval Oceanog Off, 66-70, actg head, Marine Chem Br, Res & Develop Dept, 70-75; CHIEF GEOLOGIST, OFF STRATEGIC PETROL RESERVE, DEPT ENERGY, 75- *Mem:* Geol Soc Am; Soc Econ Paleontologists & Mineralogists. *Res:* Geochemistry trace metals of coastal marine sediments; crude oil storage in salt domes. *Mailing Add:* 7829 Willowbrook Rd Fairfax Station VA 22039

SMITH, RICHARD FREDERICK, b Lockport, NY, Jan 31, 29; m 51; c 3. ORGANIC CHEMISTRY. *Educ:* Allegheny Col, BS, 50; Univ Rochester, PhD(chem), 54. *Prof Exp:* Res chemist, Monsanto Chem Co, 53-55; res assoc, Sterling-Winthrop Res Inst, 55-57; from assoc prof to prof chem, State Univ NY Albany, 57-65; chmn dept chem, 65-68, prof 65-74, DISTINGUISHED TEACHING PROF, STATE UNIV NY COL GENESEO, 74- *Concurrent Pos:* NSF fac fel, Univ Calif, Los Angeles, 62-63; vis prof, Dartmouth Col, 81, 83 & 87 Wesleyan, 85-86. *Mem:* Am Chem Soc; Royal Soc Chem. *Res:* Heterocycles; amine-imides; organic hydrazine derivatives. *Mailing Add:* Dept of Chem State Univ of NY Col Geneseo NY 14454

SMITH, RICHARD G(RANT), b Flint, Mich, Jan 19, 37; m 65; c 3. APPLIED PHYSICS, ELECTRICAL ENGINEERING. *Educ:* Stanford Univ, BS, 58, MS, 59, PhD(elec eng, appl physics), 63. *Prof Exp:* Mem tech staff, Bell Tel Labs, 63-68, supvr, 68-82, dept head, 82-87, DIR, AT&T BELL LABS, 87- *Concurrent Pos:* Chmn, Conf Laser & Electro-optical Systs, 80-; pres, Inst Elec & Electronic Engrs Lasers & Electro-optics Soc, 81. *Honors & Awards:* Centennial Award, Inst Elec & Electronics Engrs. *Mem:* Am Phys Soc; fel Inst Elec & Electronics Engrs; fel Optical Soc Am. *Res:* Quantum theory of nonlinear effects; nonlinear optics; lasers; optical fiber communications. *Mailing Add:* AT&T Bell Labs Rte 222 Breinigsville PA 18031

SMITH, RICHARD HARDING, b Philadelphia, Pa, Apr 27, 50. IMMUNOCHEMISTRY, RECEPTOR BIOCHEMISTRY. *Educ:* Millersville Univ, BS, 78; Wesleyan Univ, Middletown, PdD(biochem), 83. *Prof Exp:* Postdoctoral fel, Univ Mich Med Sch, 83-86, res assoc, 86-87; SR SCIENTIST, BIOQUANT, INC, ANN ARBOR, 86- *Concurrent Pos:* Mem, Inst Rev Bd, Lancaster Osteop Hosp, 77-78; consult, Defined Healthcare Res, Inc, NY, 90-91. *Mem:* AAAS; Am Asn Immunologists. *Res:* Antibody-based analytical systems; chemical modification of antibodies and analytes; biosensors for alternative site monitoring; use of membranes in immunochromatography. *Mailing Add:* 1919 Green Rd Ann Arbor MI 48105

SMITH, RICHARD HARRISON, b Ellenville, NY, Nov 28, 20; m 46; c 2. FOREST ENTOMOLOGY. *Educ:* State Univ NY, BS, 42, MS, 47; Univ Calif, PhD(entom), 61. *Prof Exp:* Forest entomologist, Bur Entom & Plant Quarantine, 46-52, FOREST ENTOMOLOGIST, FOREST SERV, USDA, 53- *Mem:* Entom Soc Am; Am Inst Biol Scientists. *Res:* Biology and control of Lyctus and Dendroctonus terebrans; resistance of pines to bark beetles and the pine reproduction weevil; variation, distribution and gentics of monoterpenes of pine xylem resin; forest insect research; direct control of bark beetles; residual insecticides for bark beetles. *Mailing Add:* Box 245 Berkeley CA 94701

SMITH, RICHARD JAMES, b Lansing, Mich, Jan 31, 47; m 70; c 4. SOLID STATE PHYSICS, SURFACE PHYSICS. *Educ:* St Mary's Col, Minn, BA, 69; Iowa State Univ, PhD(solid state physics), 75. *Prof Exp:* Res assoc physics, Mont State Univ, 75-77; asst scientist, Brookhaven Nat Lab, 77-80; MEM FAC PHYS DEPT, MONT STATE UNIV, 80- *Concurrent Pos:* Sabbatical leave, FOM Inst, Amsterdam, Neth, 87-88. *Mem:* Am Phys Soc; Am Vacuum Soc; Mat Res Soc; Am Asn Physics Teachers. *Res:* Electronic and structural studies of surfaces of condensed matter; ion-solid interactions. *Mailing Add:* Phys Dept Mont State Univ Bozeman MT 59717

SMITH, RICHARD JAY, b Brooklyn, NY, Aug 10, 48; m 70; c 3. CRANIOFACIAL BIOLOGY, HOMINOID EVOLUTION. *Educ:* Brooklyn Col, BA, 69; Tufts Univ, MS & DMD, 73; Yale Univ, PhD(anthrop), 80. *Prof Exp:* Resident orthod, Health Ctr, Univ Conn, 73-76, asst clin prof, 76-79; from asst prof to assoc prof orthod, Dent Sch, Univ Md, 79-84; prof & chmn, Dept Orthod, 84-91, prof biomed sci, 87-91, PROF ANTHROP, WASH UNIV SCH DENT MED, 91- *Concurrent Pos:* Dir, Postgrad Prog, Dept Orthod, Dent Sch, Univ Md, 80-; vis assoc prof, Dept Cell Biol, Med Sch, Johns Hopkins Univ, 81-; ed-in-chief, J Baltimore Col Dent Surg, 81-; assoc dean, 87, adj prof anthrop, 85-91, dean, Sch Dent Med, Wash Univ, 89-91. *Mem:* Am Asn Orthodonists; Int Asn Dent Res; Am Asn Phys Anthropologists; Am Dent Asn; Soc Study Evolution. *Res:* Functional morphology of craniofacial variation in mammals, particularly primates; biomechanical modeling correlates; paleontology; allometry. *Mailing Add:* Dept Anthrop Wash Univ McMillian Hall St Louis MO 63130

SMITH, RICHARD LAWRENCE, phycology; deceased, see previous edition for last biography

SMITH, RICHARD LLOYD, b Binghamton, NY, Oct 15, 45; m 68; c 1. ASTROPHYSICS. *Educ:* Rensselaer Polytech Inst, BS, 67; Mass Inst Technol, PhD(physics), 71. *Prof Exp:* Res fel physics, Calif Inst Technol, 71-72; asst prof physics, Rensselaer Polytech Inst, 72-77; staff mem, Syst Sci Div, Comput Sci Corp, 77-85; STAFF MEM, ANALYTICAL SCI CORP, 85- *Mem:* Am Astron Soc; Sigma Xi; Am Astronaut Soc. *Res:* Orbit determination. *Mailing Add:* 505 Burnt Mill Ave Silver Spring MD 20901

SMITH, RICHARD MERRILL, b South Bend, Ind, Nov 3, 42; m 64; c 2. PHYSIOLOGY. *Educ:* Ind Univ, Bloomington, AB, 64, PhD(physiol), 69. *Prof Exp:* Rockefeller Found vis prof physiol, Mahidol Univ, Thailand, 69-71; from asst prof to assoc prof, 71-87, PROF PHYSIOL, SCH MED, UNIV HAWAII, 87- *Concurrent Pos:* NIH pulmonary fac training award, 77-82. *Res:* Lung defense mechanisms in O2 toxicity. *Mailing Add:* Dept Physiol Univ Hawaii Sch Med 1960 East-West Rd Honolulu HI 96822

SMITH, RICHARD NEILSON, b Springfield, Mass, May 20, 18; m 37. PHYSICAL CHEMISTRY, ELECTROCHEMISTRY. *Educ:* Univ Mass, BS, 41; Univ Del, MS, 50, PhD(phys chem), 56. *Prof Exp:* Sr res chemist, Gen Chem Co Div, Allied Chem Corp, 41-55; sect mgr, Am Mach & Foundry Co, 55-71; mgr, Spec Proj Lab, Sybron Corp, 71-78; sr chemist, Southern Res Inst, 78-83; RETIRED. *Mem:* Am Chem Soc; Sigma Xi (secy, Sci Res Soc Am, 63). *Res:* Permselective membranes; electrodialysis processes; process and equipment development; specialty and heavy chemicals; plastics; film forming and processing. *Mailing Add:* 3545 Stonehenge Pl Irondale Birmingham AL 35210

SMITH, RICHARD PAUL, b Omaha, Nebr, Jan 31, 43; m 63; c 3. SUPERCONDUCTING MAGNET TECHNOLOGY, EXPERIMENTAL HIGH ENERGY PHYSICS. *Educ:* Univ Nebr, BS, 65; Syracuse Univ, MS, 67, PhD(physics), 72. *Prof Exp:* Researcher physics, 72-79, group leader & prin investr superconducting magnets, Argonne Nat Lab, 79-83; head, dept cryogen, Fermi Nat Acclerator Lab, 83-86; MEM, E740, E771 COLLABORATIONS, 87- *Concurrent Pos:* Consult superconducting magnets. *Mem:* Am Phys Soc; Sigma Xi. *Res:* High energy particle physics, weak interactions; computer controlled film scanning; superconducting magnet design. *Mailing Add:* Fermi Nat Accelerator Lab MS 357 PO Box 500 Batavia IL 60510

SMITH, RICHARD PEARSON, b Garland, Utah, Mar 4, 26. PHYSICAL CHEMISTRY. *Educ:* Univ Utah, BA, 48, PhD(phys chem), 51. *Prof Exp:* Jr fel, Harvard Univ, 51-53; from asst prof to assoc prof chem, Univ Utah, 53-61; mem staff, Exxon Res & Eng Co, 61-82; RETIRED. *Res:* Electronic structure of molecules; polarizabilities; dipole moments; substituent effects on rates and equilibria; polymer structure and property relationships; computer applications in chemistry. *Mailing Add:* 958 Willow Grove Rd Westfield NJ 07090

SMITH, RICHARD R, b Mendota, Ill, Aug 7, 36; m 59; c 4. PLANT BREEDING, PLANT GENETICS. *Educ:* Univ Ill, BSAgr, 62, MS, 63; Iowa State Univ, PhD(plant breeding), 66. *Prof Exp:* Asst prof agron, 66-72, ASSOC PROF AGRON, UNIV WIS-MADISON, 72-, RES GENETICIST, AGR RES SERV, USDA, 66- *Mem:* Crop Sci Soc Am; Am Soc Agron. *Res:* Quantitative genetics; statistics; agronomy; plant pathology. *Mailing Add:* 1575 Linden Dr Madison WI 53706

SMITH, RICHARD S, JR, b Somerville, NJ, Mar 2, 30; m 56; c 4. PLANT PATHOLOGY. *Educ:* Utah State Univ, BS, 58; Univ Calif, Berkeley, PhD(plant path), 63. *Prof Exp:* Lab technician plant path, Univ Calif, Berkeley, 60-61; plant pathologist, Pac Southwest Forest & Range Exp Sta, USDA, 61-76; supvry plant pathologist, Pac Southwest Region, Forest Serv, USDA, San Francisco, Calif, 76-87, SUPVRY PLANT PATHOLOGIST, FOREST INSECT & DIS RES, USDA, WASHINGTON, DC, 87- *Mem:* Am Phytopath Soc. *Res:* Epidemiology; control of diseases of forest tree seedlings; root diseases of mature western forest trees; management of forest disease problems; diseases of semi-tropical forest trees. *Mailing Add:* Forest Insect & Dis Res USDA PO Box 96090 Washington DC 20090-6090

SMITH, RICHARD SCOTT, b San Francisco, Calif, Aug 7, 39; m 79; c 2. IMMUNOLOGY, IMMUNOCHEMISTRY. *Educ:* Northwestern Univ, BA, 62; Ariz State Univ, MS, 65, PhD(immunol), 67. *Prof Exp:* Res scientist immunol, Children's Asthma Res Inst & Hosp, 69-73; sr res scientist, Hyland Labs, 74-76; mgr clin res & develop immunol, Becton Dickinson Immunodiag, 76-; AT BIOTECHNOL CTR, JOHNSON & JOHNSON. *Concurrent Pos:* Res fel, Scripps Clin & Res Found, 67-69. *Mem:* Am Asn Immunologists; Am Acad Allergy. *Res:* Development of new protein radioimmunoassays for hormones and tumor antigens. *Mailing Add:* Johnson & Johnson Biotechnol Ctr 4245 Sorrento Valley Blvd San Diego CA 92121

SMITH, RICHARD SIDNEY, b Eng, May 22, 33; Can citizen; m 59; c 4. PHYSIOLOGY. *Educ:* Univ Alta, BSc, 59, MD, 61; Karolinska Inst, MD, 64. *Prof Exp:* Docent neurophysiol, Karolinska Inst, Sweden, 64-66; from asst prof to assoc prof surg, 66-75, PROF SURG & MED RES COUN CAN ASSOC, UNIV ALTA, 75- *Concurrent Pos:* Med Res Coun Can scholar, Univ Alta, 66-69. *Mem:* Can Physiol Soc; Neurosci Soc. *Res:* Neuromuscular physiology; function of sense organs in muscle; axoplasmic transport. *Mailing Add:* Neurophysiol Lab Dept Surg Univ Alta Edmonton AB T6G 2G3 Can

SMITH, RICHARD THOMAS, b Oklahoma City, Okla, Apr 15, 24; m 46; c 5. PEDIATRICS, PATHOLOGY. *Educ:* Univ Tex, BA, 44; Tulane Univ, MD, 50; Am Bd Pediat, dipl; Am Bd Allergy & Immunol, dipl. *Prof Exp:* Intern pediat, Univ Minn Hosps, 50-51, resident, Med Sch, 51-52, asst prof pediat, 55-56; assoc prof, Southwestern Med Sch, Univ Tex, 57-58; prof pediat & head dept, Col Med & chief pediat, Hosp & Clin, 58-67, vpres advan, 84-88, PROF PATH & CHMN DEPT, COL MED,UNIV FLA, 67-, C A STETSON PROF EXP MED, 81- *Concurrent Pos:* Nat Res Coun fel med sci, Med Sch, Univ Minn, 52-53, Helen Hay Whitney Found res fel rheumatic fever & allied dis, 53-55; sr investr, Arthritis & Rheumatism Found, 55-58. *Mem:* AAAS; Am Pediat Soc; Soc Pediat Res; Soc Exp Biol & Med; Am Soc Path. *Res:* General and clinical immunology; tumor immunobiology. *Mailing Add:* Dept Path Univ Fla Col Med Box J-215 Gainesville FL 32610

SMITH, RICHARD THOMAS, b Allentown, Pa, June 15, 25; m 56; c 2. NONDESTRUCTIVE EVALUATION, OCEAN ENGINEERING. *Educ:* Lehigh Univ, BS, 46, MS, 47; Ill Inst Technol, PhD(elec eng), 55. *Prof Exp:* Engr, Radio Corp Am, 47; instr elec eng, Lehigh Univ, 47-50; asst, Ill Inst Technol, 50-52; design engr, Gen Elec Co, 52-55, eng analyst, Analytical Eng Sect, 55-58; assoc prof elec eng, Univ Tex, 58-61; Westinghouse prof, Va Polytech Inst, 61; proj dir, Tracor, Inc, 62-64; asst dir, Dept Electronics & Elec Eng, Southwest Res Inst, 64-65, dir & vpres, Dept Instrumentation Res, 65-66; Okla Gas & Elec Prof elec eng, Univ Okla, 66-68; prof elec mach, Renssalaer Polytech Inst, 68-70; Alcoa Found Distinguished Prof elec eng, Univ Mo-Rolla, 70-73; inst engr, Southwest Res Inst, 73-83; prof, Univ Tex, 83-87; CONSULT ENGR, 87- *Concurrent Pos:* Consult, Jack & Heintz & Lear Inc, 59, Southwest Res Inst, 61 & 70-73 & Gen Elec Co, 69-70; ed, Trans on Power Apparatus & Systs, Inst Elec & Electronics Engrs, 66-68; NSF fel, Univ Colo, 70; adv, Int Electrotech Comn; adj prof, Univ Tex, 73-75 & St Mary's Univ, 77-81. *Mem:* Sr mem Inst Elec & Electronics Engrs; fel Brit Inst Engrs; assoc fel Am Inst Aeronaut & Astronaut. *Res:* Electrical machines and power systems; numerous transactions publications; author of one book. *Mailing Add:* 402 Yosemite Dr San Antonio TX 78232

SMITH, ROBERT, b Dublin, Ireland, Apr 2, 21; m 47; c 4. FAMILY MEDICINE. *Educ:* Univ Dublin, BA, 44, MB, BCh & BAO, 45, MA, 54, MD, 57. *Prof Exp:* Asst prof physiol, Trinity Col, Dublin, 44-46; physician, Royal Army Med Corps, 46-48; practitioner family med, Nat Health Serv Brit, 48-63; sr lectr, Guy's Hosp Med Sch, Univ London, 63-68; assoc prof prev med, 68-70, prof family med & chmn dept, Univ NC, Chapel Hill, 70-75; PROF FAMILY MED & DIR DEPT, UNIV CINCINNATI, 75- *Concurrent Pos:* Vchmn, NC Regional Med Prog, 70-72; Wander lectr, Royal Soc Med, 70; vpres, NC Health Coun, 73-74, pres elect, 74-75; mem, US Pharmacopeia Adv Panel Family Pract. *Honors & Awards:* Hawthorne Prize, Brit Med Asn, 58; Int Prize, Royal Col Gen Practitioners, 59. *Mem:* Fel Royal Col Gen Practitioners; Am Acad Family Physicians; Royal Soc Med; Soc Teachers Family Med. *Res:* Pain threshold and its relationship to patient behavior; role of family in health and disease; behavioral and social factors and their interaction with pathophysiology of disease. *Mailing Add:* Dept Family Med Univ Cincinnati Cincinnati OH 45267

SMITH, ROBERT ALAN, b Glendale, Calif, Oct 30, 39; m 61; c 2. BORON CHEMISTRY. *Educ:* Univ Calif, Los Angeles, BS, 62; State Univ NY Buffalo, PhD(org chem), 68. *Prof Exp:* Sr res chemist petrochem, Atlantic Richfield Co, 68-69; fel, Calif Inst Technol, 69-70; sr res chemist boron chem, 70-90, SR SCIENTIST, US BORAX RES CORP, SUBSID RIO TINTO ZINC, LTD, 90- *Mem:* Am Chem Soc; Sigma Xi. *Res:* Inorganic and organic borate chemistry. *Mailing Add:* 21260 Trial Ridge Yorba Linda CA 92686-7806

SMITH, ROBERT BAER, b Logan, Utah, Oct 6, 38; m 60; c 2. GEOPHYSICS, GEOLOGY. *Educ:* Utah State Univ, BS, 60, MS, 65; Univ Utah, PhD(geophys), 67. *Prof Exp:* Am exchange scientist, Brit Antarctic Surv, 62-63; res asst seismol, 64-67, from asst prof to assoc prof geophys, 67-76, PROF GEOL & GEOPHYS, UNIV UTAH, 76-, DIR SEISMOG STAS, 80- *Concurrent Pos:* Consult geophys surv groups, 67- *Honors & Awards:* Antarctic Serv Medal, 63. *Mem:* AAAS; Soc Explor Geophys; Am Geophys Union; Geol Soc Am; Seismol Soc Am. *Res:* Earthquake seismology; micro-earthquakes; focal mechanisms; computer graphics; earthquake prediction; crust-mantle refraction studies; seismic profiling; geophysics as applied to regional tectonics; exploration seismology. *Mailing Add:* Dept of Geophys Univ of Utah Salt Lake City UT 84112

SMITH, ROBERT BRUCE, b Philadelphia, Pa, July 8, 37; m 59; c 3. ORGANIC CHEMISTRY, ACADEMIC ADMINISTRATION. *Educ:* Wheaton Col, Ill, BS, 58; Univ Calif, Berkeley, PhD(org chem), 62. *Prof Exp:* Asst prof chem, Nev Southern Univ, 61-66, assoc prof chem & chmn dept phys sci, 66-67, chmn dept chem, 67-68; PROF CHEM & DEAN COL SCI MATH & ENG, UNIV NEV, LAS VEGAS, 68- *Mem:* AAAS; Am Chem Soc. *Res:* Humanistic strategies in science teaching. *Mailing Add:* Col Sci Math & Eng Univ Nev 4505 Maryland Pkwy Las Vegas NV 89154

SMITH, ROBERT C, b Chicago, Ill, Sept 15, 32; m 57; c 3. BIOCHEMISTRY. *Educ:* Elmhurst Col, BS, 54; Univ Ill, MS, 58, PhD(biochem), 60. *Prof Exp:* USPHS res fel, 59-61; from asst prof to assoc prof, 61-68, alumni assoc prof, 68-69, ALUMNI PROF ANIMAL SCI, AUBURN UNIV, 69- *Mem:* NY Acad Sci; Am Soc Biol Chemists. *Res:* Nucleic acids; nucleotides; uric acid; 3-ribosyluric acid. *Mailing Add:* Dept Animal & Dairy Sci Auburn Univ Auburn AL 36849-5415

SMITH, ROBERT CLINTON, b St Thomas, Ont, Mar 11, 32; m 55; c 2. THEORETICAL PHYSICS. *Educ:* Univ Western Ont, BSc, 54; McGill Univ, MSc, 56, PhD(theoret physics), 60. *Prof Exp:* Lectr, Univ Ottawa, 58-60, asst prof, 60-66, secy, Fac Sci & Eng, 78-86, chmn dept physics, 86-91, ASSOC PROF PHYSICS, UNIV OTTAWA, 66- *Mem:* Am Asn Physics Teachers; Can Asn Physicists; Can Asn Univ Teachers. *Mailing Add:* Dept Physics Univ Ottawa Ottawa ON K1N 6N5 Can

SMITH, ROBERT EARL, b Indianapolis, Ind, Sept 13, 23; m 47, 88; c 4. AERONOMY. *Educ:* Fla State Univ, BS, 59, MS, 60; Univ Mich, PhD(atmospheric sci), 74. *Prof Exp:* USAF 43-63; dep chief, Atmospheric Sci Div, George C Marshall Space Flight Ctr 63-86; sr res engr, Comput Sci Corp 87-89; CHIEF, SPACE SCI & APPLN DIV, FWG ASSOC, INC, 89- *Mem:* Am Meteorol Soc; Am Inst of Aeronaut & Astronaut. *Res:* Temperature and dynamic structure of the upper atmosphere from 6300 angstrom units; atomic oxygen airglow emissions; natural space environment for the Space Station; natural environment for the National Aerospace Plane. *Mailing Add:* FWG Assoc Inc 7501 S Memorial Pkwy Huntsville AL 35802

SMITH, ROBERT EDWARD, b Winnipeg, Man, Oct 4, 29; m 56; c 4. PEDOLOGY. *Educ:* Univ Man, BA, 52, 55. *Prof Exp:* Pedologist, Man Dept Agr, 56-63; pedologist, Land Resource Res Inst, Can Dept Agr, 63-68, head, Pedology Sect, 68-, mem staff, Man Soul Surv Unit, 73-; HEAD SOIL SURV, UNIV MAN, WINNIPEG. *Concurrent Pos:* Adj prof soil sci, Univ Man, 74- *Mem:* Agr Inst Can; Can Soc Soil Sci; Int Soc Soil Sci. *Res:* Soil characterization, genesis and classification. *Mailing Add:* Rm 360 Ellis Bldg Univ Man Winnipeg MB R3T 2N2 Can

SMITH, ROBERT ELIJAH, b Pittsburgh, Pa, Aug 14, 11; m 38; c 4. COMPUTER SCIENCE, STATISTICS. *Educ:* State Univ Iowa, BA, 34; Univ Pittsburgh, PhD(statist, educ), 51. *Prof Exp:* Teacher, Iowa, Pa & NJ, 34-44; prof math & head dept, Duquesne Univ, 44-57; analyst computer, Univac Div, Sperry Rand Corp, Minn, 57-58; prin consult, Control Data Corp, Minneapolis, 58-80; RETIRED. *Mem:* Am Math Soc. *Res:* Computer programming projects. *Mailing Add:* 6912 Creston Rd Edina MN 55435

SMITH, ROBERT ELPHIN, b Pasadena, Calif, Sept 26, 29; m 59; c 2. PHYSIOLOGY, BIOMEDICAL ENGINEERING. *Educ:* Calif Inst Technol, BS, 51; Univ Wash, PhD(physiol, biophys), 62. *Prof Exp:* Civil engr, CZ Govt, 51-52; design engr, Lockheed Aircraft Corp, 52-53, tech writer, 53-54; civil engr, Daniel, Mann, Johnson & Mendenhall, 54; res asst physiol, Univ Wash, 57-58; asst prof physiol & biophys, Univ Ky, 62-68; asst prof human physiol, 68-70, ASSOC PROF HUMAN PHYSIOL, SCH MED, UNIV CALIF, DAVIS, 70- *Concurrent Pos:* US Air Force contract grant, 64-67; NASA grant, 64-68; NIH fel, 64-66, grant, 67-69. *Mem:* AAAS; Biophys Soc; Inst Elec & Electronics Eng; Arctic Inst NAm; Int Primatol Soc. *Res:* Physiological control systems; temperature regulation; circadian rhythms; menstrual cycles; peripheral circulation. *Mailing Add:* Dept of Human Physiol Univ of Calif Sch of Med Davis CA 95616

SMITH, ROBERT EMERY, b Jacksonville, Fla, Feb 2, 42; m 63; c 2. ENGINEERING PHYSICS. *Educ:* Duke Univ, BS, 63; Washington Univ, MA, 65, PhD(physics), 69. *Prof Exp:* Res scientist, Carbon Prod Div, Union Carbide Corp, 69-89; RES SCIENTIST, UCAR CARBON CO, 89- *Res:* Mathematical models of industrial products and processes, principally by finite element analysis; stress and heat transfer models; physical properties of graphite. *Mailing Add:* UCAR Carbon Co 12900 Snow Rd Parma OH 44130

SMITH, ROBERT EWING, b Montreal, Que, Sept 20, 34; m 59; c 4. NUTRITION. *Educ:* McGill Univ, BSc, 55, MSc, 57; Univ Ill, PhD(animal sci), 63. *Prof Exp:* Res scientist, Animal Res Inst, Can Dept Agr, 57-60 & 63-67; mgr poultry res, Quaker Oats Co, 67-69, mgr nutrit res, 69-73, dir qual assurance, 73-77, vpres foods res & develop, 77-79; VPRES RES & DEVELOP, SWIFT CO, 79- *Concurrent Pos:* Indust liaison, Food & Nutrit Bd, Am Acad Pediat; panel chmn nutrit comt, Am Corn Millers Fedn. *Mem:* Poultry Sci Asn; Am Inst Nutrit; Nutrit Soc Can; Am Acad Pediat; Soc Nutrit Educ. *Res:* Nutritional quality of human and pet foods; proteins and amino acid requirements and interrelationships. *Mailing Add:* Res & Develop Nabisco Brands Inc PO Box 1944 East Hanover NJ 07936-1944

SMITH, ROBERT FRANCIS, b Independence, Mo, May 4, 43; m 76; c 2. BEHAVIOR-ETHOLOGY. *Educ:* Univ Mo, BA, 73, MA, 76; Univ Kans, PhD(exp psychol), 84. *Prof Exp:* Res assoc, Dept Otorhinolaryngology, Kans Univ Med Ctr, 73-78; CO-PRIN INVESTR, BEHAV RADIOL LABS, KANS CITY VET ADMIN MED CTR, 78- *Concurrent Pos:* Chmn subcomt, ANSI Working Group on Biorhythms, 83-; consult, West Assocs, Rosemead, Calif, 84-85, Midwest Res Inst, 85-; referee bioelectromagnetics, J Bioelectro Magnetics Soc, 88- *Mem:* Bioelectromagnetics Soc. *Res:* Experimental study of effects of low-frequency (DC-100 Hz) electromagnetic fields on mammals; evaluation of electromagnetic effects on learning, locomotor activity, agonistic behavior, body temperature (telemetry), metabolism, development, and brain function. *Mailing Add:* Behav Radiol Labs Res Serv 151 KCVA Med Ctr 4801 Linwood Blvd Kansas City MO 64128

SMITH, ROBERT JAMES, b Brooklyn, NY, May 30, 44; m 70; c 1. ARTHRITIC DISEASES. *Educ:* St John's Univ, BS, 66; Univ Md, MS, 70, PhD(physiol), 71. *Prof Exp:* Instr physiol, Univ Md, 71-72; fel pharmacol, NIH, Sch Med, Tulane Univ, 72-74; sr scientist arthritic dis, Schering Corp, 74-78; from res scientist to sr res scientist, 78-88, SR SCIENTIST, ARTHRITIC DIS, UPJOHN CO, 88- *Concurrent Pos:* Field ed, CRC Press, 80-; bd mem, Inflammation Res Asn. *Mem:* Am Soc Pharmacol & Exp Therapeut; Am Asn Immunologists; Am Rheumatism Asn; Am Soc Hemat; NY Acad Sci; AAAS. *Res:* Hypersensitivity disease; discovery and development of therapeutic agents for treatment of arthritic disease; cellular components of inflammatory joint disease. *Mailing Add:* Dept Hypersensitivity Dis Res Upjohn Co 301 Henrietta St Kalamazoo MI 49001

SMITH, ROBERT JOHNSON, b Blodgett, Mo, July 23, 16; m 48; c 3. ORGANIC CHEMISTRY. *Educ:* Southeast Mo State Col, BS, 36; Univ Iowa, PhD(chem), 50. *Prof Exp:* Instr high schs, Mo, 36-42; from asst prof to prof chem, Southeast Mo State Col, 46-55; asst, Univ Iowa, 49-50; assoc prof, Eastern Ill Univ, 55-60, prof chem, 60-84; RETIRED. *Mem:* Am Chem Soc. *Res:* Mechanism and rate of addition of bromine to olefins in carbon tetrachloride solution; preparation and reactions of mercurials derived from olefins. *Mailing Add:* 1514 Second St Charleston IL 61920

SMITH, ROBERT KINGSTON, b Melrose, Mass, May 15, 24; m 78; c 1. INORGANIC CHEMISTRY. *Educ:* Univ Mass, BS & MS, 50; Univ Wyo, PhD(chem), 66. *Prof Exp:* Tech asst food technol, Mass Inst Technol, 54-56; instr chem, Univ Wyo, 56-63, asst, 63-66; asst prof chem, 66-68, asst dean col arts & sci, 68-70, actg dean, 70-71, ASSOC PROF CHEM, YOUNGSTOWN STATE UNIV, 68-, ASST DEAN COL ARTS & SCI, 71- *Mem:* Am Chem Soc. *Res:* Complexes of group VA elements, especially those of the antimony halides. *Mailing Add:* Off of the Dean Col Arts & Sci Youngstown State Univ Youngstown OH 44555

SMITH, ROBERT L, b New York, NY, Mar 29, 41; m 68; c 3. NEUROSCIENCE, AUDITORY NEUROPHYSIOLOGY. *Educ:* City Col New York, BEE, 62; New York Univ, MSEE, 66; Syracuse Univ, PhD(neurosci), 73. *Prof Exp:* Develop engr, Wheeler Lab, Great Neck, NY, 62-64; lectr elec eng, City Col New York, 64-66; instr elec eng, 70-74, from asst prof to assoc prof sensory res, 74-85, PROF NEUROSCI, SYRACUSE UNIV, 85- *Concurrent Pos:* Fel res career develop, NIH, Syracuse Univ, 79-84; assoc ed, J Acoust Soc Am, 86-89. *Mem:* Fel Acoust Soc Am; Sigma Xi; Asn Res Otolaryngol. *Res:* Neurophysiology and neural coding in the auditory nervous system; single unit recording from the cochlea, auditory nerve and cochlear nucleus; mathematical modeling of the results and systems analysis of the auditory system. *Mailing Add:* 3 Haverhill Pl Syracuse NY 13214

SMITH, ROBERT LAWRENCE, b Albemarle, NC, Aug 29, 39; m 62; c 2. ORGANIC CHEMISTRY, MEDICINAL CHEMISTRY. *Educ:* Univ NC, BS, 61; Univ Maine, MS, 63, PhD(org chem), 65. *Prof Exp:* NIH res fel org chem, Univ NC, 65-67; sr res chemist, 67-74, res fel, 74-75, asst dir, 75-77, assoc dir, 77-79, dir, 79-84, SR DIR, MERCK SHARP & DOHME RES LABS, 84- *Mem:* AAAS; Am Chem Soc; Asn Res Vision & Ophthal. *Res:* General organic synthesis; synthesis of medicinals; development of orally active HMG-CoA reductase inhibitors and topically effective carbonic anhydrase inhibitors. *Mailing Add:* Merck Sharp & Dohme Res Labs W 26 410 West Point PA 19486

SMITH, ROBERT LEE, b Schaller, Iowa, Oct 31, 23; m 47; c 3. CIVIL ENGINEERING. *Educ:* Univ Iowa, BS, 47, MS, 48. *Prof Exp:* Asst prof hydraul, Univ Kans, 48-52; exec dir, Iowa Natural Resources Coun, 52-55; exec secy & chief engr, Kans Water Resources Bd, 55-62; Parker prof water resources, Univ Kans, 62-66, prof civil eng, chmn dept & dir water resources inst, 66-72, Deane Ackers prof civil eng, 70-88; RETIRED. *Concurrent Pos:* Chmn, Interstate Conf Water Probs, 61 & Inter-Agency Comt Water Resources Res, 66-; tech asst, Off Sci & Technol, Exec Off Pres, 66-; mem ex officio, US Nat Comt, Int Hydrol Decade, 66-, mem, 68-71; water resources consult, Black & Veatch Consult Engrs, 68- *Honors & Awards:* Centennial Plaque, US Geol Surv, 80; Julian Hinds Award, Am Soc Civ Engrs, 88. *Mem:* Nat Acad Eng; fel Am Soc Civil Engrs; Nat Soc Prof Engrs; Am Water Works Asn; Am Geophys Union; fel AAAS; Am Soc Eng Educ. *Res:* Hydrology; water resources planning; Midwestern water problems; water policy. *Mailing Add:* 4001 Vintage Ct Lawrence KS 66047

SMITH, ROBERT LELAND, b Sacramento, Calif, June 30, 20; m 52; c 3. VOLCANOLOGY, RARE METAL GEOCHEMISTRY. *Educ:* Univ Nev, BS, 42. *Hon Degrees:* DSc, Univ Lancaster, UK, 89. *Prof Exp:* From jr geologist to prin geologist, US Geol Surv, 43-60, chief field geochem & petrol br, 60-66, res geologist, 66-82, SR RES GEOLOGIST, US GEOL SURV, 82- *Concurrent Pos:* Mem earth sci div, Nat Res Coun, 62-65; mem, US-Japan Sci Coop Volcano Res, 63-65; mem preliminary exam team, First Lunar Samples, Apollo 11 & 12, NASA, 69. *Honors & Awards:* Distinguished Serv Medal, US Dept Interior, 83; First Thorarinson Medal, Int Asn Volcanology & Chem Earth's Interior, 87. *Mem:* Sr fel Geol Soc Am; fel Mineral Soc Am; Geochem Soc; Am Ornithologists Union; fel AAAS. *Res:* Mineralogy;

petrology; geochemistry; pyroclastic rocks; volcanic glasses; rhyolitic volcanism; volcano tectonics, calderas and eruption cycles; geology of the Valles Mountains of New Mexico; geothermal resources; rare metals in igneous rocks; volcano hazards; bolivan tin belt. *Mailing Add:* 11064 Fair Oaks Blvd Fair Oaks CA 95628-5944

SMITH, ROBERT LEO, b Brookville, Pa, Mar 23, 25; m 52; c 4. WILDLIFE MANAGEMENT, ECOLOGY. *Educ:* Pa State Univ, BS, 49, MS, 54; Cornell Univ, PhD(wildlife mgt, ecol, soils), 56. *Prof Exp:* Instr agr, Jefferson County Bd Educ, Pa, 49-50; asst, Cornell Univ, 54-56; asst prof biol, State Univ NY Col Plattsburgh, 56-58; PROF WILDLIFE MGT, WVA UNIV, 58-, WILDLIFE ECOLOGIST, 74- *Concurrent Pos:* Mem task forces, Nat Res Coun, 81. *Mem:* Wildlife Soc; Ecol Soc Am; Cooper Ornith Soc; Am Soc Mammalogists; Am Ornithologists Union; Am Inst Biol Sci. *Res:* Habitat selection; structure of forest bird communities; ecology of highly disturbed lands; succession. *Mailing Add:* Div Forestry WVa Univ PO Box 6125 Morgantown WV 26506-6125

SMITH, ROBERT LEONARD, b New Orleans, La, Jan 19, 44; m 66; c 1. INDUSTRIAL HYGIENE CHEMISTRY, ENVIRONMENTAL HEALTH. *Educ:* La State Univ, New Orleans, BS, 65, PhD(chem), 70. *Prof Exp:* Res chemist, Res & Develop Lab, 70-77, environ chemist, 77-79, supvr, Toxicol & Indust Hygiene Lab, 79-83, mgr, Corp Regulatory Affairs, 83-86, MGR, CORP TOXICOL, ETHYL CORP, 87-, DIR, CORP TOXICOL & REGULATORY AFFAIRS, 88- *Mem:* Am Chem Soc. *Res:* Absorption spectroscopy; photochemistry. *Mailing Add:* 451 Florida St Baton Rouge LA 70801

SMITH, ROBERT LEWIS, b Ranger, Tex, June 22, 38; m 60; c 4. BIOCHEMISTRY. *Educ:* Abilene Christian Col, BS, 61; Univ Tenn, Memphis, MS, 62, PhD(biochem), 66. *Prof Exp:* Vis lectr chem, Queens Col, NC, 62-63; res assoc protein chem & enzym, Biol Dept, Brookhaven Nat Lab, 66-68; asst prof biochem, 68-71, ASSOC PROF BIOCHEM, LA STATE UNIV, SHREVEPORT, 71- *Concurrent Pos:* Res chemist, Vet Admin Hosp, Shreveport, La, 69-77. *Mem:* Sigma Xi; Am Chem Soc. *Res:* Protein chemistry; plasma proteins; blood coagulation; membrane transport of amino acids and peptides. *Mailing Add:* Dept Biochem Sch Med La State Univ Box 33932 Shreveport LA 71130

SMITH, ROBERT LLOYD, b Chicago, Ill, Dec 10, 35; m 86; c 3. PHYSICAL OCEANOGRAPHY. *Educ:* Reed Col, BA, 57; Univ Ore, MA, 59; Ore State Univ, PhD(oceanog), 64. *Prof Exp:* From instr to assoc prof, 62-75, PROF PHYS OCEANOG, ORE STATE UNIV, 75- *Concurrent Pos:* NATO fel, Nat Inst Oceanog, Eng, 65-66; sci officer, Off Naval Res, 69-71; vis prof, Inst Meerekunde, Univ Kiel, Ger, 79; vis scientist, Commonwealth Sci & Indust Res Orgn Marine Lab, Hobart, Australia, 88; mem, outer continental shelf adv comt, Minerals Mgt Serv, Dept Interior, 86-; ed, Progress in Oceanog, 85- *Mem:* Fel AAAS; Am Geophys Union; Oceanog Soc; Am Meteorol Soc. *Res:* General physical oceanography, currents, upwelling, coastal oceanography, underwater sound. *Mailing Add:* Col Oceanog Ore State Univ Corvallis OR 97331-5503

SMITH, ROBERT LLOYD, b Spirit Lake, Iowa, July 25, 41; div; c 1. BEHAVIOR-ETHOLOGY, ENTOMOLOGY. *Educ:* NMex State Univ, BS, 68, MS,71; Ariz State Univ, PhD(zool),75. *Prof Exp:* Entomologist, USDA, Agr Res Serv, Western Cotton Res Lab, 75-77; asst prof, Dept Entom, 77-83, ASSOC PROF, DEPT ENTOM, UNIV ARIZ, 83- *Mem:* AAAS; Am Inst Biol Sci; Am Soc Naturalists; Animal Behav Soc; Entomol Soc Am. *Res:* Insect behavior; evolutionary biology; aquatic entomology; urban entomology; reproductive biology. *Mailing Add:* Dept Entomol Univ Ariz Tucson AZ 85721

SMITH, ROBERT OWENS, b Elizabethton, Tenn, May 13, 37; m 61; c 2. EXPERIMENTAL SOLID STATE PHYSICS, ENVIRONMENTAL PHYSICS. *Educ:* Univ Colo, BS, 62; Rutgers Univ, MS, 64, PhD(physics), 69. *Prof Exp:* Mem tech staff explor develop, Bell Labs, 62-64; res fel solid state physics, Rutgers Univ, 64-69; chmn dept, 74-84, PROF PHYSICS, MONMOUTH COL, 69- *Mem:* Am Phys Soc; Am Asn Physics Teachers. *Res:* Solar energy conversion. *Mailing Add:* 72 Meyers Mill Rd Colts Neck NJ 07722

SMITH, ROBERT PAUL, b St Lucas, Iowa, Sept 9, 42; m 69; c 2. MATHEMATICS, STATISTICS. *Educ:* Loras Col, BS, 64; Univ Ariz, MA, 66, PhD(math), 71. *Prof Exp:* ASST PROF MATH, ARK STATE UNIV, 69- *Mem:* Inst Math Statist. *Res:* Statistical inference and hypothesis testing for continuous time parameter stochastic processes. *Mailing Add:* Dept Math Box 70 State Univ Ark State University AR 72467

SMITH, ROBERT SEFTON, b Baltimore, Md, Aug 16, 41; m; c 3. COMPUTER-AIDED INSTRUCTION IN MATHEMATICS. *Educ:* Morgan State Col, BS, 63; Pa State Univ, MA, 67, PhD(math), 69. *Prof Exp:* asst prof math, 69-77, ASSOC PROF MATH & STATIST, MIAMI UNIV, 77- *Mem:* Math Asn Am. *Res:* Computer-aided instruction in mathematics; lattice theory. *Mailing Add:* Dept Math & Statist Miami Univ Oxford OH 45056-1641

SMITH, ROBERT VICTOR, b Brooklyn, NY, Feb 16, 42; m 66; c 2. PHARMACEUTICAL CHEMISTRY, ANALYTICAL CHEMISTRY. *Educ:* St John's Univ, NY, BS, 63; Univ Mich, Ann Arbor, MS, 64, PhD(pharmaceut chem), 68. *Prof Exp:* From asst prof to assoc prof med chem, Col Pharm, Univ Iowa, 68-74; assoc prof & asst dir, 74-77, assoc dir, 77-78, PROF, DRUG DYNAMICS INST, COL PHARM, UNIV TEX, AUSTIN, 77-, DIR, 78- *Concurrent Pos:* NIH award, 74, 76 & 79; mem rev comt, US Pharmacopoeia, 75-80. *Mem:* Am Chem Soc; fel Acad Pharmaceut Sci. *Res:* Drug metabolism; analysis of drugs alone and in dosage forms; analysis of drugs in biological fluids. *Mailing Add:* Dean - Col Pharm Wash State Univ Pullman WA 99164-6510

SMITH, ROBERT W, b Detroit, Mich, May 4, 09. ATOMIC PHYSICS. *Educ:* Univ Tenn, BS, 29; Univ Mich, PhD(physics), 33. *Prof Exp:* Chmn, Mat Sci Dept, Gen Motors Inst, 66-74; RETIRED. *Mem:* Fel Am Soc Metals Int; Am Soc Testing & Mat; Soc Automotive Engrs. *Mailing Add:* 1608 Kensington Flint MI 48503

SMITH, ROBERT WILLIAM, b Ft Worth, Tex, Apr 14, 39; m 58; c 2. MICROBIOLOGY, BIOCHEMISTRY. *Educ:* N Tex State Univ, BA, 60; Okla State Univ, PhD(microbiol), 65. *Prof Exp:* Fel, 64-65, res assoc, 65-67, asst prof microbiol, 67-74, EXEC OFFICER, DEPT BIOL SCI, PURDUE UNIV, WEST LAFAYETTE, 74- *Mem:* Am Soc Microbiol. *Res:* Microbial physiology; structure genetics; protein chemistry; nature of self-associating protein systems; thermophily. *Mailing Add:* 7200 N County Rd 75 E West Lafayette IN 47906

SMITH, ROBERT WILLIAM, b Chelsea, Mass, Mar 21, 43; m 68; c 2. ECOLOGY, STATISTICS. *Educ:* Univ Calif, Berkeley, BA, 65; Univ Wash, BS, 69; Univ Southern Calif, PhD(biol), 76. *Prof Exp:* Instr ecol, Univ Southern Calif, 76-77; CONSULTS DATA ANALYSIS, SOUTHERN CALIF EDISON, SCI APPL, INC, WOODWARD-CLYDE CONSULTS, INC, HARBORS ENVIRON PROJ, INST MARINE & COASTAL STUDIES, UNIV SOUTHERN CALIF, LA CO SANITATION DIST, 77- & LOCKHEED AIRCRAFT SERV, MARINE BIOL CONSULTS, 78-; PRES, ECOANALYSIS INC. *Mem:* AAAS; Am Soc Naturalists; Ecol Soc Am; Am Statist Asn. *Res:* Development of analytical techniques for ecological-survey data; development of computer software for data analysis. *Mailing Add:* 221 E Matilija Suite A Ojai CA 93023

SMITH, ROBERTA HAWKINS, b Tulare, Calif, May 3, 45; m 69; c 2. PLANT PHYSIOLOGY, PLANT SCIENCE. *Educ:* Univ Calif, Riverside, BA, 67, MS, 68, PhD(plant sci & physiol), 70. *Prof Exp:* Asst prof biol, Sam Houston State Univ, 73-74; from asst prof to assoc prof plant sci, 74-83, PROF SOIL & CROP SCI, TEX A&M UNIV, 83-, EUGENE BUTLER PROF AGR BIOTECHNOL, 86- *Concurrent Pos:* Nat corresp, Int Asn Plant Tissue Cult, 82-86; chmn, Plant Div, Tissue Cult Asn, mem exec bd; chmn, fac plant physiol, Tex A&M Univ, 87-89, L-7, Crop Sci, 91; gov bd, Int Crops Res Inst for the Semi-Arid Tropics, 89-92. *Mem:* Am Soc Plant Physiol; Am Soc Plant Physiol; Int Asn Plant Tissue Cult; Tissue Cult Asn; Crop Sci Soc Am. *Res:* Plant morphology; plant tissue culture; crop improvement. *Mailing Add:* Dept Soil & Crop Sci Tex A&M Univ College Station TX 77843

SMITH, ROBERTS ANGUS, b Vancouver, BC, Dec 22, 28; nat US; m 53; c 4. BIOCHEMISTRY. *Educ:* Univ BC, BSA, 52, MSc, 53; Univ Ill, PhD(biochem), 57. *Prof Exp:* Instr chem, Univ Ill, 57-58; from asst prof to prof, 58-87, EMER PROF CHEM, UNIV CALIF, LOS ANGELES, 87- *Concurrent Pos:* Dir, ICN Pharmaceut Inc, 61-; Guggenheim fel, Cambridge Univ, 63; pres, VIRATEK Inc, 80- *Mem:* Am Chem Soc; Am Soc Biol Chemists; Am Asn Cancer Res. *Res:* Biological phosphoryl transfer reactions; chromosomal protein modification; antiviral and anticancer agents. *Mailing Add:* Dept Chem Univ Calif 405 Hilgard Ave Los Angeles CA 90024

SMITH, RODERICK MACDOWELL, b Boston, Mass, Mar 15, 44; m 66; c 3. FISHERIES. *Educ:* Earlham Col, BA, 65; Univ Mass, MS, 69, PhD(fisheries), 72. *Prof Exp:* Consult, Mass Div Fisheries & Game, 71; asst prof marine sci, Stockton State Col, 71-74; asst prof zool, 74-81, ADJ ASST PROF ZOOL, UNIV NH, 81- *Concurrent Pos:* Mem, Acad Adv Coun Study Comn, 73; consult biologist, Wetlands Inst, Lehigh Univ, 73-74. *Mem:* Am Fisheries Soc. *Res:* The reestablishing of anadromous fish runs in New England coastal plain rivers, including Pacific salmon introductions; biocide effects on larval marine fishes. *Mailing Add:* 21 Dana Ave Kittery ME 03904

SMITH, RODGER CHAPMAN, b South Hadley, Mass, July 18, 15; m 40; c 3. INORGANIC CHEMISTRY. *Educ:* Univ Mass, BS, 38. *Prof Exp:* Asst head fertilizer res, Eastern States Farmers' Exchange, Inc, 46-54, head fertilizer res, 55-62; mgr agr technol serv, Southwest Potash Corp, NY, 62-66, mgr mkt develop, 66-71; dir mkt develop, Amax Chem Corp, 71-83; CONSULT, CHEM MKTG & TECHNOL, 83- *Mem:* Am Chem Soc; Am Soc Hort Sci; Am Inst Chemists; Am Soc Agron. *Res:* Administration of market development for potash, phosphate rock and heavy chemicals in the United States and other countries; coordination with industry and governmental agencies; processes for mixed fertilizer granulation; global consulting on marketing and technology investigations for government and industries. *Mailing Add:* 1206 W Camino Del Pato Green Valley AZ 85614

SMITH, ROGER ALAN, b Pomona, Calif, July 16, 47; m 69; c 1. THEORETICAL PHYSICS. *Educ:* Oberlin Col, BA, 68; Stanford Univ, MS, 69, PhD(physics), 73. *Prof Exp:* Res assoc physics, Univ Ill, Urbana-Champaign, 73-75; Res assoc physics, State Univ NY Stony Brook, 75-77, lectr, Inst Theoret Physics, 77-79, res assoc physics, 79-81; asst prof, 81-84, ASSOC PROF PHYSICS, TEXAS A&M UNIV, 84- *Mem:* Am Phys Soc. *Res:* Properties of the nucleon-nucleon interaction; techniques for many-body calculations; structure of neutron stars. *Mailing Add:* Dept of Physics Texas A&M Univ College Station TX 77843-4242

SMITH, ROGER BRUCE, b New Bethlehem, Pa, Sept 12, 47; m 67; c 3. TOXICOLOGY. *Educ:* Philadelphia Col Pharm & Sci, BS, 70, MS, 73, PhD(pharmacol), 79; Am Bd Toxicol, dipl. *Prof Exp:* Res scientist, McNeil Pharmaceut, Johnson & Johnson, 77-79, sr scientist, 79-82, prin scientist, 77-87; RES FEL, R W JOHNSON PHARMACEUT RES INST, JOHNSON & JOHNSON, 87- *Concurrent Pos:* Adj assoc prof, Phil Col Pharm & Sci, 86. *Mem:* Soc Toxicol;; Sigma Xi; Am Col Toxicol; Dip Am Bd Toxicol. *Res:* Assessing appropriateness of existing animal models for toxicity studies and establishing reliable new predictive models where routine systems have failed; preclinical toxicity assessment of pharmacuticals. *Mailing Add:* Drug Safety Evaluation R W Johnson Pharmaceut Res Inst Mckean Rd Spring House PA 19477

SMITH, ROGER DEAN, b New York, NY, Oct 6, 32; m 57; c 4. PATHOLOGY, VIROLOGY. *Educ:* Cornell Univ, AB, 54; NY Med Col, MD, 58. *Prof Exp:* Resident surg, Detroit Receiving Hosp, Mich, 59-60; instr path, Col Med, Univ Ill, Chicago, 62-66, from asst prof to assoc prof, 66-72, asst dean col med, 70-72; prof path & dir dept, 72-90 EMER PROF, UNIV CINCINNATI, 90- *Concurrent Pos:* USPHS grant, 62-65; Nat Cancer Inst spec fel, 65-66; resident, Presby-St Luke's Hosp, Chicago, 62-66, consult; co-investr, Ill Div, Am Cancer Soc grant, 65-66; asst attend pathologist, Res & Educ Hosp, 66-; consult, Vet Admin Hosp, Chicago; prin investr, NIH grants, 67-72 & 76-79; Assoc Path Chair, Col Am Pathologists. *Mem:* Int Acad Path; Am Asn Path; Soc Exp Biol & Med; Am Soc Nephrology; Am Soc Clin Path. *Res:* Virus pathology and virus-cell relationships; renal and virus pathology; experimental renal disease; persistent virus infections and possible relation to glomerulonephritis. *Mailing Add:* Dept of Path Univ of Cincinnati Col of Med Cincinnati OH 45229

SMITH, ROGER ELTON, b Stillwater, Okla, Apr 16, 41. CIVIL ENGINEERING, HYDROLOGY. *Educ:* Tex Tech Univ, BSc, 63; Stanford Univ, MSc, 64; Colo State Univ, PhD(civil eng), 70. *Prof Exp:* Design engr, Metcalf & Eddy Engrs, Calif, 64-65; engr, Peace Corps, Pakistan, 65-67; res hydraul engr, Southwest Watershed Res Ctr, Agr Res Serv, 70-76, RES HYDRAULIC ENGR, FED RES, SCI EDUC ADMIN, USDA, 76- *Mem:* Am Soc Civil Engrs; Am Geophys Union. *Res:* Soil infiltration from rainfall; watershed response in relation to physical features; stochastic rainfall models; hydraulics of alluvial streams, including measuring techniques and unsteady flow phenomena. *Mailing Add:* Fed Bldg 301 S Howes PO Box E Ft Collins CO 80522

SMITH, ROGER FRANCIS COOPER, b Kapunda, South Australia, Mar 6, 40; m 73. ZOOLOGY, ECOLOGY. *Educ:* Univ Adelaide, BSc, 62; Australian Nat Univ, MSc, 66; Univ Alta, PhD(zool), 73. *Prof Exp:* Exp officer, Commonwealth Sci & Insust Res Orgn, Div Wildlife Res, 63-64; res asst physiol, Dept Zool, Australian Nat Univ, 66, res assoc, Dept Zool, Univ Alta, 72-73; asst prof, 73-79, ASSOC PROF, DEPT ZOOL, BRANDON UNIV, 79- *Mem:* Brit Ecol Soc; Australian & NZ AAS; Wildlife Soc; Can Soc Zoologists; Am Mammal Soc. *Mailing Add:* Dept Zool Brandon Univ Brandon MB R7A 6A9 Can

SMITH, ROGER M, b Winnipeg, Man, Sept 12, 18; m 47; c 4. RESEARCH ADMINISTRATION. *Educ:* Univ Man, BSc, 40. *Prof Exp:* Sr supvr reactor opers, Chalk River Nuclear Labs, Atomic Energy Can Ltd, 46-52, supt prod planning & control, 53-58; dir div safeguards, Int Atomic Energy Agency, 58-60; mgr admin div, Whiteshell Nuclear Res Estab, 60-82; dir, safeguard develop, Atomic Energy Can Ltd, 76-83; CONSULT, 83- *Honors & Awards:* Distinguished Serv Award, Inst Nuclear Mat Mgt. *Mem:* Can Asn Physicists; Inst Nuclear Mat Mgt. *Res:* Technical administration and planning; development of safeguard techniques. *Mailing Add:* Seven McWilliams Pl Pinawa MB R0E 1L0 Can

SMITH, ROGER POWELL, b Hokuchin, Korea, July 16, 32; US citizen; m 56; c 3. TOXICOLOGY. *Educ:* Purdue Univ, BS, 53, MS, 55, PhD(pharmaceut chem), 57. *Hon Degrees:* MA, Dartmouth Col, 75. *Prof Exp:* From instr to assoc prof, 60-73, chmn dept, 76-87, PROF PHARMACOL & TOXICOL, DARTMOUTH MED SCH, 76- *Concurrent Pos:* Consult, Vet Admin Ctr, White River Junction, Vt, 65-; assoc staff mem, Mary Hitchcock Mem Hosp, 68-; mem toxicol study sect, NIH, 68-72; USPHS career develop award, 66-71; assoc ed, Toxicol & Appl Pharmacol, 72-78; mem pharmacol toxicol prog comt, Nat Inst Gen Med Sci, 72-79; mem comt med & biol effects environ pollutants, Div Med Sci, Nat Res Coun, 75-77; adj prof, Vt Law Sch 81-87, Dartmouth Col, 84-; mem Enivon Health Sci Rev Comt, 85-87; mem Toxicol Info Prof Comt, Nat Res Coun, 86-88. *Mem:* Fel AAAS; fel Acad Toxicol Sci; Soc Toxicol; Am Soc Pharmacol & Exp Therapeut; Sigma Xi. *Res:* Experimental toxicology; red cell metabolism; nitric oxide vasodilator drugs; abnormal blood pigments. *Mailing Add:* Dept Pharmacol & Toxicol Dartmouth Med Sch Hanover NH 03756

SMITH, ROLF C, JR, INNOVATION, CREATIVE PROBLEMSOLVING TECHNIQUES. *Educ:* Tex A&M Univ, BA, MS. *Prof Exp:* Commun electronics officer & dir, Off Innovation, USAF, 63-87; PRES, OFF STRATEGIC INNOVATION, 87-; DIR, EXXON'S INNOVATION NETWORK, HOUSTON, 87- *Concurrent Pos:* Adj lectr, Ctr Creative Leadership, Greensboro, NC; adj prof, Univ Houston; headmaster, Sch Innovators, 89-; consult, Ford, DuPont, Union Carbide, E-Systs, Waste Mgt Corp & Small Bus Admin; intel officer, NATO, Ger. *Res:* Creative problem solving techniques; quality and continuous improvement facilitation; artificial intelligence. *Mailing Add:* Exxon Co PO Box 2180 Houston TX 77252-2180

SMITH, RONALD E, b Beeville, Tex, Mar 30, 36; m 57; c 3. PHYSICS. *Educ:* Tex A&M Univ, BS, 58, MS, 59, PhD(physics), 66. *Prof Exp:* Instr physics, Tex A&M Univ, 66-70; assoc prof, 66-70, PROF PHYSICS, NORTHEAST LA UNIV, 70-, HEAD DEPT, 77- *Mem:* Am Phys Soc; Am Asn Physics Teachers. *Res:* Nuclear magnetic resonance; electron spin resonance; solid state. *Mailing Add:* Dept of Physics Northeast La Univ Monroe LA 71209

SMITH, RONALD GENE, b Woodland, Wash, Jan 24, 44; m 70; c 2. ORGANIC CHEMISTRY, ANALYTICAL PHARMACOLOGY. *Educ:* Whitworth Col, BS, 66; Purdue Univ, PhD(org chem), 72. *Prof Exp:* Res assoc org chem, Ore Grad Ctr, 72-74, instr, 74-77; from asst prof to prof org chem, Univ Tex M D Anderson Hosp & Tumor Inst, Houston, 77-83; SR RES SPECIALIST, MONSANTO AGR CO, 83- *Concurrent Pos:* Asst prof, Grad Sch Biomed Sci, Univ Tex Health Sci Ctr, 78-80, fac mem, 80- *Mem:* Am Asn Cancer Res; Am Chem Soc; Am Soc Mass Spectrometry. *Res:* Application of mass spectrometry to the pharmacology of antitumor agents; identification of drug metabolites; pharmacokinetics; mechanism of drug action. *Mailing Add:* Monsanto Agr Co 800 N Lindbergh Blvd St Louis MO 63166

SMITH, RONALD W, b June 15, 36; m 60; c 3. PHYSICAL CHEMISTRY, CHEMICAL ENGINEERING. *Educ:* Pa State Univ, BS, 58; Univ Del, PhD, 65. *Prof Exp:* Res chemist, Hercules, Inc, 64-69, tech develop rep, 69-70, develop supvr, 70-72, tech mgr, 72-75; asst gen mgr, Haveg Industs, Inc, 75-80; PROD MGR, HERCULES INC, 81- *Mem:* Am Chem Soc; Am Inst Chem Engrs. *Res:* Surface chemistry; environmental science, pollution control. *Mailing Add:* 27D Walnut Hill Rd Hockessin DE 19707-9609

SMITH, ROSE MARIE, b Beaumont, Tex, Mar 3, 34; m 55; c 3. MATHEMATICS. *Educ:* Lamar Univ, BS, 55; Tex Woman's Univ, MA, 68; Okla State Univ, EdD, 75. *Prof Exp:* Teacher math & music, Grapevine Pub Schs, Grapevine, Tex, 55-66; PROF MATH, TEX WOMAN'S UNIV, 66-, CHMN, DEPT MATH, COMPUT SCI & PHYSICS, 86- *Mem:* Nat Coun Teachers Math; Math Asn Am. *Res:* Mathematical education. *Mailing Add:* 2106 Bell Ave Denton TX 76201-2011

SMITH, ROSS W, b Turlock, Calif, Dec 11, 27; m 55; c 3. MINERAL ENGINEERING, SURFACE CHEMISTRY. *Educ:* Univ Nev, BS, 50; Mass Inst Technol, SM, 55; Stanford Univ, PhD(mineral eng), 69. *Prof Exp:* Jr mining engr, Consol Coppermines Corp, Nev, 50; res asst metall, Mass Inst Technol, 53-55; assoc mfg process engr, Portland Cement Asn, 55-57; proj engr, Res Found, Colo Sch Mines, 58-60; assoc prof metall, SDak Sch Mines & Technol, 60-66; actg instr mineral eng, Stanford Univ, 66-68; assoc prof metall, 68-69, PROF METALL & CHMN DEPT CHEM & METALL ENG, UNIV NEV, RENO, 69- *Concurrent Pos:* Dept Health, Educ & Welfare grant, 64-66. *Mem:* Am Inst Mining, Metall & Petrol Engrs; Am Chem Soc; Am Inst Chem Engrs; Sigma Xi. *Res:* Comminution; pipeline flow of liquid-solid slurries; flotation; surface chemistry. *Mailing Add:* Dept of Chem & Metall Eng Univ of Nev Reno NV 89507

SMITH, ROY E, b Chippewa Falls, Wis, Apr 28, 26; m 47; c 3. SCIENCE EDUCATION, PHYSICS. *Educ:* Wis State Univ-Eau Claire, BS, 50; Univ Wis, MS, 57; Ohio State Univ, PhD(sci educ), 66. *Prof Exp:* Jr high sch teacher, Ill, 50-51; pub sch teacher, Wis, 51-56; from asst prof to assoc prof, 57-65, PROF PHYSICS & HEAD DEPT, UNIV WIS-PLATTEVILLE, 66- *Concurrent Pos:* Consult, Wis Mold & Tool Co, 57-63. *Mem:* Am Asn Physics Teachers; Nat Sci Teachers Asn. *Res:* Use of unit operators and dimensional methods as a vehicle for teaching fundamental quantitative physical science. *Mailing Add:* Dept Physics Univ Wis Platteville WI 53818

SMITH, ROY JEFFERSON, JR, b Covington, La, Nov 25, 29; m 52; c 2. WEED SCIENCE. *Educ:* Miss State Univ, BS, 51, MS, 52; Univ Ill, PhD(weed sci), 55. *Prof Exp:* SUPVRY RES AGRONOMIST WEED SCI, AGR RES SERV, USDA, 55- *Concurrent Pos:* Mem grad staff, Univ Ark, 55-; adv, Rockefeller Found, Int Rice Res Inst, Philippines, 64; adv, Inst Tech Interchange, East-West Ctr, 67-69; coop scientist, US Dept Army, 68-70; coop scientist for pest mgt proj in Pakistan & for biol control weeds with Univ Ark, 73-88. *Honors & Awards:* Super Serv Award, USDA, 67; Outstanding Researcher Award, Weed Sci Soc Am, 82. *Mem:* Fel Weed Sci Soc Am; Int Weed Sci Soc; Sigma Xi; Am Soc Agron. *Res:* Biology and interference of weeds in agronomic crops; integrated weed management systems for rice; biological control of weeds with plant pathogens; rice germplasm tolerance to herbicides; alleopathy of rice germplasm. *Mailing Add:* PO Box 351 Stuttgart AR 72160-0351

SMITH, ROY MARTIN, b Alamo, Tenn, Oct 21, 27; m 48; c 2. DENTISTRY, ORAL PATHOLOGY. *Educ:* Univ Tenn, DDS, 51, MS, 63. *Prof Exp:* Pvt pract, 53-58; asst prof oral med & surg, 58-62, assoc prof oral diag & chmn dept, 62-64, assoc prof path, 63-73, PROF ORAL DIAG, COL DENT, UNIV TENN, MEMPHIS, 64-, ASST DEAN ACAD AFFAIRS, 83- *Concurrent Pos:* Consult, Vet Admin Hosp, Memphis, Tenn, Methodist Hosp, Memphis, Tenn. *Mem:* Am Dent Asn; Am Acad Oral Path; Am Col Dent; Am Asn Dent Schs. *Res:* Transplantation of intraoral tissues; oral carcinogenesis; pharmacologic effects of eugenol. *Mailing Add:* 6473 Heather Dr Memphis TN 38119

SMITH, RUFUS ALBERT, JR, b Shreveport, La, Jan 29, 32; m 60; c 2. HORTICULTURE. *Educ:* La State Univ, BS, 56, MS, 58; Wash State Univ, PhD(hort), 67. *Prof Exp:* Instr hort, Western Ill Univ, 61-63; asst prof, Ore State Univ, 67-73; assoc prof, 73-79, PROF HORT, UNIV TENN, MARTIN, 81- *Mem:* Am Soc Hort Sci; Am Hort Soc. *Mailing Add:* Sch Agr Univ Tenn Martin TN 38238

SMITH, RUSSELL AUBREY, b Little Rock, Ark, June 8, 36; m 60; c 3. MECHANICAL ENGINEERING, MOTOR VEHICLE SAFETY. *Educ:* Rice Univ, BA & BSME, 58; Cath Univ Am, MME, 64, PhD(mech eng), 69. *Prof Exp:* Mem fac, Mech Eng Dept, Cath Univ Am, 66-76, chmn dept, 73-76; mem, Accident Invest Div, Nat Hwy Safety Admin, 76-82; FAC MECH ENGR, US NAVAL ACAD, 82- *Concurrent Pos:* Consult, Forensic Technol Int. *Mem:* Am Soc Mech Engrs; Am Soc Eng Educ; Soc Automotive Engrs. *Res:* Engineering mechanics; vehicle collision mechanics. *Mailing Add:* Dept Mech Eng US Naval Acad Annapolis MD 21402

SMITH, RUSSELL D, b Mexico, Mo, Feb 7, 50; m 72; c 2. CERAMICS ENGINEERING. *Educ:* Univ Mo, Rolla, BS, 72 & MS, 73. *Prof Exp:* Res engr, Dow Chem, 73-77; prod develop mgr, Standard Oil Eng Mats, 80-84; tech mgr, Standard Oil Eng Mats, 84-89; sr res engr, 77-80, GEN MGR, CARBORUNDUM, 89- *Mem:* Am Ceramic Soc; Nat Inst Ceramic Engrs. *Res:* Insulation; filtration; reinforcement; electronic ceramics. *Mailing Add:* 107 Timberlink Grand Island NY 14072

SMITH, RUSSELL LAMAR, b Oconee, SC, Jan 25, 59. PHOTOSYNTHESIS, NITROGEN-FIXATION. *Educ:* Clemson Univ, BS, 80; Univ Tex, MA, 84, PhD(biol sci), 86. *Prof Exp:* Vis scientist, Ctr Nuclear Studies, Grenoble, France, 86-87; postdoctoral, Biotechnol Inst, Penn State Univ, 87-89; postdoctoral, Memphis State Univ, 89-90; ASST PROF BIOCHEM, DEPT CHEM, UNIV TEX, ARLINGTON, 90- *Mem:* AAAS; Am Soc Microbiol; Am Chem Soc; Am Soc Plant Physiologists; Sigma Xi. *Res:* Enzymology and molecular biology of photosynthetic prokaryotes; inorganic nitrogen utilization in cyanobacteria. *Mailing Add:* Dept Chem Univ Tex PO Box 19065 Arlington TX 76019-0065

SMITH, SAM CORRY, b Enid, Okla, July 3, 22; m 45; c 3. NUTRITION. *Educ:* Univ Okla, BS, 47, MS, 48; Univ Wis, PhD(biochem), 51. *Prof Exp:* Spec instr chem, Univ Okla, 47-48, asst prof biochem, Sch Med, 51-54, assoc prof, 54-55; secy, Williams-Waterman Fund, Res Corp, NY, 55-67, assoc dir grants, 57-65, dir, 65-68, chmn, Williams-Waterman Prog Comt & chmn adv comt grants, 67-75, vpres grants, 68-75; exec dir, M J Murdock Charitable Trust, 75-88; CONSULT, 88- *Concurrent Pos:* Pub trustee, Nutrit Found, 76-84 & Int Life Sci Inst-Nut Found, 84-87. *Mem:* AAAS; Am Chem Soc. *Res:* Human nutrition; amino acid and vitamin metabolism; international nutrition. *Mailing Add:* 5204 DuBois Dr Vancouver WA 98661-6617

SMITH, SAMUEL, b Bronx, NY, Sept 13, 27; m 51; c 5. POLYMER CHEMISTRY. *Educ:* City Col New York, BS, 48; Univ Mich, MS, 49. *Prof Exp:* Chemist, Inst Paper Chem, 49-51; sr chemist, Cent Res, 51-57, sr chemist, Chem Div, 57-61, supvr polymer res, 61-67, res assoc, 67-75, corp scientist, Cent Res Labs, 75-83, CORP SCIENTIST, SPEC FILM LAB, MINN MINING & MFG CO, 83- *Honors & Awards:* Henry Millson Award, Am Asn Textile Chemists & Colorists, 80; Am Chem Soc Award for Creative Invention, 88. *Mem:* AAAS; Am Chem Soc. *Res:* Ring-opening polymerization; elastomeric resins; surface chemistry relating to adhesion and desorption processes; fluorochemical polymers and textile finishes. *Mailing Add:* 3100 Shorewood Lane St Paul MN 55113

SMITH, SAMUEL COOPER, b Lock Haven, Pa, Sept 21, 34; m 55; c 2. BIOCHEMISTRY. *Educ:* Pa State Univ, BS, 55, MS, 59, PhD(biochem), 62. *Prof Exp:* From asst prof to assoc prof, 61-74, PROF ANIMAL SCI & BIOCHEM, UNIV NH, 74- *Mem:* AAAS; Am Chem Soc; Am Oil Chem Soc; Sigma Xi. *Res:* Lipid biochemistry in tissue culture systems. *Mailing Add:* Animal Sci 407 Kendall Hall Univ of NH Durham NH 03824

SMITH, SAMUEL H, b Salinas, Calif, Feb 4, 40; m 60; c 2. PLANT PATHOLOGY, PLANT VIROLOGY. *Educ:* Univ Calif, Berkeley, BS, 61, PhD(plant path), 64. *Prof Exp:* NATO fel plant path, Glasshouse Crops Res Inst, Eng, 64-65; asst prof, Univ Calif, Berkeley, 65-69; assoc prof, Fruit Res Lab, 69-71 & Buckhout Lab, 71-74, head dept, 76-81, PROF PLANT PATH, BUCKOUT LAB, PA STATE UNIV, UNIVERSITY PARK, 74-, HEAD DEPT, 76-, DEAN, COL AGR & DIR PA AGR EXP STA & PA COOP EXTEN SERV, 81- *Mem:* Am Phytopath Soc; Brit Asn Appl Biol. *Mailing Add:* Pres Off Wash State Univ Pullman WA 99164-1048

SMITH, SAMUEL JOSEPH, b Montgomery, Ala, July 19, 39. SOIL CHEMISTRY, WATER CHEMISTRY. *Educ:* Auburn Univ, BS, 61; Iowa State Univ, PhD(soil chem), 67. *Prof Exp:* Res assoc, Iowa State Univ, 67-69; SOIL SCIENTIST RES, AGR RES SERV, USDA, 69- *Honors & Awards:* Super Serv Award, USDA, 81, Distinguished Serv Award, 84. *Mem:* Am Soc Agron; Soil Sci Soc Am; Am Inst Chemists. *Res:* Behavior and fate of agricultural chemicals in the environment. *Mailing Add:* Agr Res Serv USDA PO Box 1430 Durant OK 74702

SMITH, SELWYN MICHAEL, b Sydney, Australia, Aug 12, 42; Can citizen; c 2. FORENSIC PSYCHIATRY. *Educ:* Sydney Univ, MB & BS, 66; London Univ, DPM, 69; Univ Birmingham, MEng, 74; FRCP(C), 76; Am Bd Psychiat & Neurol, dipl, 78; Am Bd Forensic Psychol, dipl, 80. *Prof Exp:* House physician & surgeon, Sydney Hosp, 67; registrar psychiat, All Saints Hosp, Birmingham, Eng, 68-70; hon res fel psychiat, United Birmingham Hosp, 70-72, hon sr registrar, 71-75; dir forensic psychiat, Royal Ottawa Hosp, 75-80; prof psychiat, Univ Ottawa, 80-86; PVT PRACT, 86- *Concurrent Pos:* Consult psychiat, Ottawa Gen Hosp & Brockville Hosp, 76-; vis prof, dept psychol, Carleton Univ, 77-; chmn, forensic serv comt, Region Ottawa-Carleton & Eastern Ont, 77-; mem, fed & prov subcomt environ health, Task Force Acceptable Lead Level Blood, Health & Welfare, Can, 77-, secure serv adv comt adolescent serv, Ont Ministry Community & Social Serv, 80-, sci prog comt, Second World Cong Prison Health Care, Fed Govt Can, 83-; psychiatrist-in-chief, Royal Ottawa Hosp, 78-; intern, Acad Law & Ment Health. *Honors & Awards:* Cloake Medal, 73; Bronze Medal, Royal Col Psychiatrists, 74. *Mem:* Fel Royal Col Psychiatrists; Am Acad Psychiat & Law (pres elect); Am Col Psychiatrists; Can Psychiat Asn; Can Asn Treatment Offenders; fel Royal Australian & Col Psychiatrists NZ; fel Am Psychiat ASn. *Res:* Child abuse; forensic and legal issues. *Mailing Add:* 155 Queen St Ottawa ON K1P 6L1 Can

SMITH, SHARON LOUISE, b Denver, Colo, June 14, 45. OCEANOGRAPHY. *Educ:* Colo Col, BA, 67; Univ Auckland, NZ, MSc, 69; Duke Univ, PhD(zool), 75. *Hon Degrees:* DSc, Colo Col, 89. *Prof Exp:* Biologist, Raytheon Corp, 69-71; fel, Dalhousie Univ, 75-78; from asst oceanogr to oceanogr, 78-85, OCEANOGR WITH TENURE, BROOKHAVEN NAT LAB, 85-, DEP DIV HEAD, 90- *Concurrent Pos:* Adj asst prof, Univ Wash, 80-85, State Univ NY, Stony Brook, 79-; ed bd, Am Soc Limnol & Oceanog; assoc prog mgr, NSF, 88-89, mem Rev & Oversight Comt; mem, Steering Comt, Global Ocean Flux Study, Global Ecosyst Dynamics, Artic Syst Sci chair, Indian Ocean Study, US-JGOFS. *Mem:* Oceanogr Soc; Am Soc Limnol & Oceanog; Am Geophys Union. *Res:* Ecology of zooplankton, herbivorous crustaceans; food chain dynamics, biogeochemical cycling in productive areas of oceans; secondary production; life history strategies; animal/plant interactions. *Mailing Add:* Brookhaven Nat Lab Bldg 318 Upton NY 11973

SMITH, SHARRON WILLIAMS, b Ashland, Ky, Apr 3, 41; m 64; c 2. BIOCHEMISTRY. *Educ:* Transylvania Col, BA, 63; Univ Ky, PhD(biochem), 74. *Prof Exp:* Chemist, Charles Pfizer Pharmaceut Co, 63 & Procter & Gamble, 63-64; teacher sci, Lexington Pub Schs, Ky, 64-67; chemist biol membranes, Lab Cell Biol, Nat Heart & Lung Inst, 74-75; asst prof, 75-81, assoc prof, 81-87, chmn dept, 82-86, PROF CHEM, HOOD COL, 87- *Concurrent Pos:* Actg dean, Hood Grad Sch, 89-90. *Mem:* AAAS; Am Chem Soc. *Res:* Membrane biochemistry, proteins, phospholipids, phosphonoglycans. *Mailing Add:* Dept Chem Hood Col Frederick MD 21701

SMITH, SHELBY DEAN, b Macomb, Ill, Nov 25, 23; m 45; c 2. MATHEMATICS. *Educ:* Western Ill Univ, BS & MS, 50; Univ Ill, PhD, 66. *Prof Exp:* High sch teacher, Ill, 50-54; from asst prof to prof, 56-87, EMER PROF MATH SCI, BALL STATE UNIV, 87- *Res:* Mathematics education. *Mailing Add:* 2009 N Winthrop Rd Ball State Univ Muncie IN 47304

SMITH, SHELDON MAGILL, b St Paul, Minn, Apr 19, 31; div; c 3. OPTICAL PHYSICS. *Educ:* Univ Calif, Berkeley, BS, 53, Davis, MA, 62. *Prof Exp:* Res scientist solar physics, 63-72, res scientist planetary atmosphere, 74-80, res scientist space sci, Astrophy Br, Ames Res Ctr, NASA, 80-89; SR STAFF SCIENTIST, STERLING SOFTWARE, 90- *Concurrent Pos:* Prin investr, airborn eclipse exped NASA, 63, 65, 66, 70 & 75; mem tech comt, Int Comn Illum, 85, Am Soc Testing & Mat, 89-90. *Mem:* Optical Soc Am; Am Astron Soc; Int Soc Optical Eng; AAAS. *Res:* Radial-gradient filter technique for photograpy of the solar corona applied to telescopes on aircraft platforms; neutral-sheet theory of the magnetic field configuration of solar streamers; infrared reflectance of the Venus clouds; reflectance of IR-black coatings and BRDF measurement; reflecting layer model of the far-infrared reflectance of coatings; scattering by very rough surfaces. *Mailing Add:* Ames Res Ctr NASA Moffett Field CA 94035

SMITH, SIDNEY R, JR, b New Orleans, La, Oct 5, 35; m 58; c 4. ZOOLOGY, BIOCHEMISTRY. *Educ:* Univ Conn, BA, 57; Howard Univ, MS, 59, PhD, 63. *Prof Exp:* Fel endocrinol, Univ Wis, 63-64; asst prof biol, Morehouse Col, 64-66; NIH fel biochem, Univ Conn, 66-68; sr res scientist, 68-71, head sect immunol, 71-83, ASSOC DIR, ALLERGY & INFLAMMATION, SCHERING CORP, 83- *Concurrent Pos:* Adj assoc prof, Fairleigh Dickinson Univ, 74-83. *Mem:* AAAS; Am Asn Immunol; Am Inst Chem; NY Acad Sci; Int Soc Immunopharmacol. *Res:* Embryology; endocrinology; immunology. *Mailing Add:* 700 Spring Ave Ridgewood NJ 07451

SMITH, SIDNEY RUVEN, b Hamilton, Ont, Aug 25, 20; m 53; c 6. PHYSICAL CHEMISTRY. *Educ:* McMaster Univ, BSc, 42, MSc, 43; Ohio State Univ, PhD(phys chem), 52. *Prof Exp:* Jr res chemist, Nat Res Coun Can, 43-46; phys chemist, US Naval Ord Test Sta, Calif, 52-60; from asst prof to assoc prof, 60-74, PROF CHEM, UNIV CONN, 74- *Mem:* Am Chem Soc; Am Phys Soc. *Res:* Mass spectrometry; kinetics; stable isotopes. *Mailing Add:* 31 Lynwood Dr Storrs CT 06268

SMITH, SIDNEY TAYLOR, b Montezuma, Ga, May 27, 18; m 49; c 1. ELECTRON TUBES. *Educ:* Ga Inst Technol, BS, 39; Yale Univ, DEng, 42. *Prof Exp:* Electronic scientist, Naval Res Lab, 42-51, Hughes Res Labs, 51-56; head, Electron Tubes Br, 56-74, ELECTRONIC CONSULT, NAVAL RES LAB, 74- *Concurrent Pos:* Mem adv group electron tube, Dept Defense, 50-74. *Mem:* Fel Inst Elec & Electronics Engrs; Am Phys Soc. *Res:* Published 20 technical papers on electron tubes. *Mailing Add:* 4514 Bee St Alexandria VA 22310

SMITH, SPENCER B, b Ottawa, Ont, Jan 31, 27; m 54. OPERATIONS RESEARCH. *Educ:* McGill Univ, BE, 49; Columbia Univ, MS, 50, EngScD(indust eng), 58. *Prof Exp:* Instr indust eng, Columbia Univ, 50-58; admin engr, Mergenthaler Linotype Co, 53-58; mgr opers res, Semiconductor Div, Raytheon Co, 58-61 & Montgomery Ward & Co, 61-66; assoc prof, 66-71, actg chmn dept indust eng, 70-71, prof & chmn dept indust eng, 71-77, PROF MGT SCI OF STUART SCH BUS ADMIN, ILL INST TECHNOL, 77- *Concurrent Pos:* Consult, UN, 57, Chicago Mercantile Exchange, 74 & Inst Gas Technol, 76; Harris Trust & Savings Bank res grant, 68-70; Ill Law Enforcement Comn res grant, 72-73; res grant, Am Prod & Inventory Control Soc, 80; res grant, US Army Corps Engrs, 81. *Mem:* Am Inst Indust Engrs; Am Statist Asn; Am Soc Mech Engrs; Opers Res Soc Am; Inst Mgt Sci; Am Prod & Inventory Control Soc; Soc Mfg Engrs. *Res:* Mathematical programming; inventory theory; forecasting; production planning; simulation; information systems; law enforcement; planning of energy production and distribution systems; philanthropy. *Mailing Add:* Stuart Sch of Bus Admin Ill Inst of Technol Chicago IL 60616

SMITH, SPURGEON EUGENE, b San Marcos, Tex, July 17, 25; m 48; c 2. TOPOLOGY. *Educ:* Southwest Tex State Col, BS, 46. *Prof Exp:* Res mathematician, Defense Res Lab, Tex, 51-57; vpres & dir res, Textran Corp, 57-62; prin scientist & dir, Tracor Inc, 62-69, vpres & dir advan res sci & systs group, 69-89; RETIRED. *Concurrent Pos:* consult, cove theory. *Mem:* Am Math Soc; Acoust Soc Am. *Res:* Functions of a complex variable; probability; decision theory; complex group decision maps; data compaction. *Mailing Add:* 1305 Bradwood Rd Austin TX 78722

SMITH, STAMFORD DENNIS, b San Jose, Calif, Feb 27, 39; m 63; c 2. ENTOMOLOGY, HYDROBIOLOGY. *Educ:* San Jose State Col, BA, 61; Univ Idaho, MS, 64, PhD(entom), 67. *Prof Exp:* Asst prof biol, Kans State Col Pittsburg, 66-68; asst prof, 68-72, assoc prof, 72-80, PROF BIOL, CENT WASH UNIV, 80- *Mem:* Entom Soc Am; Soc Syst Zool. *Res:* Ecology and systematics of Trichoptera; biology of aquatic insects. *Mailing Add:* Dept Biol Sci Cent Wash Univ Ellensburg WA 98926

SMITH, STANFORD LEE, b Detroit, Mich, June 3, 35; m 58, 77, 83; c 2. ORGANIC CHEMISTRY. *Educ:* Albion Col, BA, 57; Iowa State Univ, PhD(org chem), 61. *Prof Exp:* Res assoc & instr chem, Iowa State Univ, 61-62; from asst prof to assoc prof chem, 62-84, PROF CHEM & RADIOL, UNIV KY, 84-, DIR, NMR SPECTROS CTR, 87- *Concurrent Pos:* UN consult, Cent Testing Lab, Pakistan, 75; fac consult, Varian Workshops, 81-82 & 84; dir instrumentation, magnetic Resonance Imaging & Spectros Ctr, Univ Ky, 85-90. *Mem:* AAAS; Am Chem Soc; Am Asn Univ Professors; Soc Magnetic Resonance Med; Int Soc Magnetic Resonance; Am Asn Physicist Med. *Res:* High resolution nuclear magnetic resonance spectroscopy; biochemical structure studies; molecular structure and associations; magnetic resonance imaging. *Mailing Add:* Dept Chem Univ Ky Lexington KY 40506-0055

SMITH, STANLEY GALEN, b Laramie, Wyo, Mar 25, 26; m 50; c 3. SYSTEMATIC BOTANY, AQUATIC ECOLOGY. *Educ:* Univ Calif, Berkeley, BA, 49, MS, 51, PhD(bot), 61. *Prof Exp:* Asst prof bot, Iowa State Univ, 60-65; assoc prof, 65-76, PROF BIOL, UNIV WIS-WHITEWATER, 76- *Concurrent Pos:* Univ res grants, 66-68, Wis Dept Natural Resources, 67-68 & US Off Water Resources, 68-69. *Mem:* Ecol Soc Am; Asn Aquatic Vascular Plant Biologists; Sigma Xi; Soc Wetland Scientists. *Res:* Scirpus lacustris complex and Typha biosystematics, ecology, floras, wetland vegetation ecology rivers, lakes, marshes, ferns, conservation, especially freshwater wetlands. *Mailing Add:* Dept Biol Univ Wis 800 W Main Whitewater WI 53190

SMITH, STANLEY GLEN, b Glendale, Calif, June 20, 31; m 65. ORGANIC CHEMISTRY. *Educ:* Univ Calif, Berkeley, BS, 53; Univ Calif, Los Angeles, PhD(chem), 59. *Prof Exp:* From instr to assoc prof, 60-73, PROF CHEM, UNIV ILL, URBANA, 73- *Concurrent Pos:* Sloan fel, 64-66. *Mem:* Am Chem Soc. *Res:* Physical organic chemistry; reaction kinetics; computer-based teaching. *Mailing Add:* 254 Roger Adam Lab 1209 California Urbana IL 61801-3731

SMITH, STEPHEN ALLEN, b Marietta, Ohio, Sept 7, 42; m 75. OPERATIONS RESEARCH. *Educ:* Univ Cincinnati, BS, 65; Stevens Inst Technol, MS, 67; Stanford Univ, PhD(eng econ), 72. *Prof Exp:* Mem tech staff opers res, Bell Tel Labs, 65-68; res scientist opers res, Xerox Palo Alto Res Ctr, 72-; AT CTR INFO SYST RES, MASS INST TECHNOL. *Mem:* Inst Mgt Sci; Inst Elec & Electronics Engrs. *Res:* Inventory control, queueing theory, and office systems. *Mailing Add:* 11 Colonial Dr Westford MA 01886

SMITH, STEPHEN D, b Philadelphia, Pa, Jan 15, 39; m 73; c 3. ANATOMY, EMBRYOLOGY. *Educ:* Wesleyan Univ, AB, 61; Tulane Univ, PhD(anat), 65. *Prof Exp:* Instr anat & ophthal, Tulane Univ, 64-65; from instr to assoc prof, 65-88, PROF ANAT, UNIV KY, 89- *Mem:* AAAS; Am Asn Anatomists; Soc Develop Biol; Am Soc Zoologists; NY Acad Sci; Int Soc Bioelec; Bioelectrochem Soc; Bioelec Rep Growth Soc. *Res:* Control of regeneration and differentiation-growth-limiting mechanisms; effects of physical stress and magnetic-electrical fields development. *Mailing Add:* Dept Anat Chandler Med Ctr Univ Ky Lexington KY 40536-0084

SMITH, STEPHEN D, b Houston, Tex, June 11, 48. GROUP THEORY, REPRESENTATION THEORY. *Educ:* Mass Inst Technol, BS, 70; Oxford Univ, PhD(math), 73. *Prof Exp:* Res instr, 73-75, from asst prof to assoc prof, 75-84, PROF MATH, UNIV ILL, 84- *Mem:* Am Math Soc. *Res:* Classification of finite simple groups, work groups & geometrics. *Mailing Add:* Univ Ill Box 4348 Chicago IL 60680

SMITH, STEPHEN JUDSON, b Fairfield, Iowa, June 14, 24; m 51; c 5. LASER SPECTROSCOPY. *Educ:* Kalamazoo Col, BA, 49; Harvard Univ, MA, 50, PhD(physics), 54. *Prof Exp:* Asst physics, Harvard Univ, 53-54; physicist, Nat Bur Standards, 54-86; RETIRED. *Concurrent Pos:* Fel, Joint Inst Lab Astrophys, Univ Colo, Boulder, 62-, lectr, Univ, 62-66, adj prof, 66-; Dept of Com sci & technol fel, 75-76; NSF prog dir, 75-76, 89-90; Alexander von Humboldt Found sr US scientist award, Univ Munich, 78-79. *Mem:* Am Phys Soc. *Res:* Multi-photon ionization including angular distributions; effects of laser field fluctuations on nonlinear atoms absorption. *Mailing Add:* Joint Inst for Lab Astrophys Univ Colo Boulder CO 80309-0440

SMITH, STEPHEN ROGER, b Fayette, Ala, Nov 21, 39; m 66; c 2. PHYSICS. *Educ:* Mass Inst Technol, SB, 62, PhD(physics), 69. *Prof Exp:* Instr physics, Princeton Univ, 69-72; from asst prof physics to assoc prof, 72-79; sr staff scientist, 79-80, dir res & eng, 80-84, div dir res, fisher controls, 84-85, GEN MGR, EMR PHOTOELECTRIC, PRINCETON, NJ, 85- *Mem:* Am Phys Soc; Am Asn Physics Teachers. *Res:* Quantum optics; photodetectors. *Mailing Add:* EMR Photoelec PO Box 44 Princeton NJ 08542-0044

SMITH, STEVEN JOEL, b Everett, Mass, Aug 4, 40; m 64; c 2. PHARMACOLOGY, BIOCHEMISTRY. *Educ:* Univ Mass, BA, 62; Baylor Col Med, MS, 64, PhD(pharmacol), 69. *Prof Exp:* Instr, Pa State Univ, 69-70, asst prof pharmacol, Hershey Med Ctr, 70-81; SR SCI, DEPT DRUGS, AM MED ASN, 81- *Mem:* AAAS; Sigma Xi; Soc Neurosci; Am Soc Pharmacol & Exp Therapeut. *Res:* Effects of xenobiotics and endogenous compounds on nucleolar RNA synthesis and processing of 45S RNA in normal and neoplastic tissues; nuclear RNA of brain and liver; isolation of subcellular organelles. *Mailing Add:* Dept Drugs Am Med Asn 515 N State St Chicago IL 60610

SMITH, STEVEN PATRICK DECLAND, b Tampa, Fla, July 12, 39; m 63; c 2. AEROSPACE ENGINEERING, NUCLEAR PHYSICS. *Educ:* Univ Fla, BSME, 62, MSE, 63, PhD(nuclear sci), 67. *Prof Exp:* RES AEROSPACE ENGR, US ARMY MISSILE RES & DEVELOP COMMAND, 67- *Mailing Add:* 827 Tannahill Dr S E Huntsville AL 35802

SMITH, STEVEN SIDNEY, b Idaho Falls, Idaho, Feb 11, 46; m 74. CELL BIOLOGY, TUMOR BIOLOGY. *Educ:* Univ Idaho, BS, 68; Univ Calif, Los Angeles, PhD(molecular biol), 74. *Prof Exp:* Lectr molecular biol, Univ Bern, Switz, 74-77; res assoc cell biol, Scripps Clin & Res Found, 78-81; asst res scientist, 82-87, ASSOC RES RES SCIENTIST, CITY HOPE NAT MED CTR, 87- *Concurrent Pos:* Consult, Molecular Biosystems Inc, 81-84; prin investr, Inst Gen Med Sci, NIH & Coun Tobacco Res USA, Inc, 83- & March of Dimes, 88- *Mem:* Am Crystallog Asn; Am Asn Cancer Res; Am Soc Cell Biol. *Res:* Mechanisms by which somatically inheritable patterns of gene expression are maintained and altered; the role of DNA methylation patterns in gene expression, chromosome damage and carcinogenesis. *Mailing Add:* City Hope Nat Med Ctr 1500 E Duarte Rd Duarte CA 91010

SMITH, STEWART EDWARD, b Baltimore, Md, Oct 5, 37; m 62; c 2. COAL SCIENCE, CHEMICAL KINETICS. *Educ:* Howard Univ, Wash, BS, 60; Ohio State Univ, PhD(chem), 69. *Prof Exp:* Teaching asst & phys chem, Ohio State Univ, 64-69; chemist, Sun Oil Co, 69-71; chemist, E I Du Pont de Nemours & Co, 63-64 & 72-74 tech serv rep, 74-78; chemist, Exxon Res & Eng Co, 78-81, group head, 81-82, coordr, 82-84, chemist, 84-86; ADV ENGR, WESTINGHOUSE, 86- *Mem:* Am Chem Soc; AAAS; Sigma Xi. *Res:* Gas-phase hydrocarbon oxidation kinetics; heterogeneons catalysis; polymer chemistry; coal science including coal characterization, liquefaction and combustion; surfactants; water chemistry. *Mailing Add:* 125 Amberwood Ct Bethel Park PA 15102

SMITH, STEWART W, b Minneapolis, Minn, Sept 15, 32; m 56; c 3. GEOPHYSICS. *Educ:* Mass Inst Technol, SB, 54; Calif Inst Technol, MS, 58, PhD(geophys), 61. *Prof Exp:* Seismologist, Shell Oil Co, 54-57; from asst prof to assoc prof geophys, Calif Inst Technol, 61-70; chmn geophys prog, 70-80, pres, Inc Res Insts Seismol, 85- 89, PROF GEOPHYS, UNIV WASH, 70- *Mem:* Am Geophys Union; Seismol Soc Am; Earthquake Eng Res Inst. *Res:* Seismology; free oscillations of the earth; instrumentation for long period seismic waves; elastic strain accumulation in the earth's crust; earthquake risk assessment. *Mailing Add:* Geophys Prog AK-50 Univ Wash Seattle WA 98195

SMITH, STUART, b Durham, Eng, Oct 12, 40. BIOCHEMISTRY, ENZYMOLOGY. *Educ:* Univ Birmingham, Eng, PhD(biochem), 65, DSc, 80. *Prof Exp:* SR RES BIOCHEMIST, CHILDREN'S HOSP MED CTR, OAKLAND, 80- *Mem:* Biochem Soc; AAAS; Am Soc Exp Biologists. *Mailing Add:* Children's Hosp Oakland Res Inst 747 52nd St Oakland CA 94609

SMITH, STUART D, b Montreal, Que, Jan 9, 41; m 63; c 3. OCEANOGRAPHY, ATMOSPHERIC CHEMISTRY & PHYSICS. *Educ:* McGill Univ, BEng, 62; Univ BC, PhD(oceanog & physics), 66. *Prof Exp:* Sci officer oceanog, 62-66, RES SCIENTIST OCEANOG, ATLANTIC OCEANOG LAB, BEDFORD INST, 66- *Mem:* Can Meteorol & Oceanog Soc (pres, 85-86 & 87-88); Am Geophys Union; Am Meteorol Soc. *Res:* Wind stress; heat flux; evaporation; evaporation, wind stress and boundary-layer turbulence over the open ocean and over drifting sea ice; surface wave generation; dynamics of iceberg drift; carbon dioxide exchange at sea surface. *Mailing Add:* Ocean Circulation Div Bedford Inst Oceanog PO Box 1006 Dartmouth NS B2Y 4A2 Can

SMITH, SUSAN MAY, b Winnipeg, Man, Jan 14, 42. ECOLOGY, ANIMAL BEHAVIOR. *Educ:* Univ BC, BSc, 63, MSc, 65; Univ Wash, PhD(zool), 69. *Prof Exp:* Asst prof biol, Wellesley Col, 69-73; mem fac, Dept Biol, Univ Costa Rica, 73-77; asst prof biol, Adelphi Univ, 77-79; ASST PROF BIOL, MT HOLYOKE COL, 79- *Mem:* AAAS; Asn Study Animal Behav; Cooper Ornith Soc; Wilson Ornith Soc; Sigma Xi. *Res:* Territoriality, social dominance and population regulation; animal communication; behavior of predators and the reactions of their prey; interspecific competition and niche overlap. *Mailing Add:* Dept of Biol Sci Mt Holyoke Col South Hadley MA 01075

SMITH, SUSAN T, b Detroit, Mich, Nov 22, 37; m 64; c 2. BIOCHEMISTRY, CLINICAL CHEMISTRY. *Educ:* Univ Mich, BS, 59; Duke Univ, PhD(biochem), 67. *Prof Exp:* Asst prof chem, ECarolina Univ, 67-69; teaching supvr med lab asst prog, Beaufort County Tech Inst, 69-71; CHAIRPERSON DEPT MED TECHNOL, SCH ALLIED HEALTH & SOCIAL PROFESSIONS, ECAROLINA UNIV, 72- *Mem:* AAAS; Am Soc Med Technol. *Res:* Mechanism of action of flavoproteins, especially xanthine oxidase and related enzymes. *Mailing Add:* Sch Allied Hlth & Soc Profsns ECarolina Univ Greenville NC 27834

SMITH, TERENCE E, b Penarth, UK, Mar 11, 36; m 62; c 2. PETROLOGY, GEOCHEMISTRY. *Educ:* Univ Wales, BSc, 59, PhD(geol), 63. *Prof Exp:* Sci off, Geol Surv Gt Brit, 62-65; lectr geol, Sunderland Tech Col, Eng, 65-67 & Univ WI, 67-69; from asst prof to assoc prof, 69-76, PROF GEOL, UNIV WINDSOR, 76- *Res:* Metamorphic petrology and structural geology of the Scottish Highlands; clastic sedimentation and structure in British Lower Paleozoic and West Indian Tertiary sediments; petrology and geochemistry of Nova Scotia granitic batholith; coast complex of British Columbia, Pennsula Rouges batholith of Southern California and Tertiary volcanoes in Jamaica. *Mailing Add:* Dept of Geol Univ of Windsor Windsor ON N9B 3P4 Can

SMITH, TERRY DOUGLAS, b Bethel Springs, Tenn, Nov 20, 42. MEDICINAL CHEMISTRY. *Educ:* Univ Tenn, Memphis, BS, 64; Univ Mich, MS, 65, PhD(med chem), 68. *Prof Exp:* Res investr radiopharmaceut, E R Squibb & Sons, Inc, NJ, 69-70; asst prof pharmaceut, radiol & nuclear med, Col Pharm, Univ Tenn, Memphis, 70-72; assoc chemist, Brookhaven Nat Lab, 73-75; sr radio pharm chemist, Mallinckrodt, Inc, 75-80; dir, res & develop nuclear div, Syncor Int Corp, 80-83, dir biomed group, 83-85; DIR RES, BERLEX LABS INC, 90- *Res:* Design and preparation of radiolabeled compounds for diagnosis of selected pathological conditions by external body scanning techniques. *Mailing Add:* 26 Millbrook Ct Danville CA 94526

SMITH, TERRY EDWARD, b Evansville, Ind, Aug 23, 40; m 62; c 3. POLYMER CHEMISTRY. *Educ:* David Lipscomb Col, BA, 62; Ga Inst Technol, PhD(phys chem), 67. *Prof Exp:* Res chemist, Am Cyanamid, Co, 67-72; res specialist, 72-76, group leader, 76-84, SECT MGR, GAF CHEMICALS CORP, 85- *Mem:* Am Chem Soc. *Res:* Polymer solutions and blends; light scattering; polymer characterization; polymer synthesis. *Mailing Add:* 143 Lake Rd Morristown NJ 07960

SMITH, THEODORE BEATON, b Columbus, Ohio, Feb 14, 18; m 47. METEOROLOGY. *Educ:* Ohio State Univ, BA, 38; Calif Inst Technol, MS, 40 & 42, PhD(meteorol), 49. *Prof Exp:* Instr meteorol, Calif Inst Technol, 42-44 & 47-48; res meteorologist, Am Inst Aerologic Res, 48-55; res meteorologist, Meteorol Res, Inc, 55-70, vpres res, 70-78, pres, 78-87; RETIRED. *Mem:* AAAS; fel Am Meteorol Soc. *Res:* Cloud physics; turbulent diffusion. *Mailing Add:* 1491 Linda Vista Ave Pasadena CA 91103

SMITH, THEODORE CRAIG, b Mansfield, Ohio, Sept 18, 30; m 52, 80; c 4. ANESTHESIOLOGY, PHARMACOLOGY. *Educ:* Ohio Wesleyan Univ, BA, 52; Univ Wis, MS, 60; Univ Cincinnati, MD, 56; Univ Pa, BBA, 78. *Prof Exp:* Intern & resident, Univ Wis, Hosps, 56-60; from asst prof to assoc prof anesthesia, Univ Pa, 62-72, prof, 72-82; mem fac, Dept Anesthesiol & prof anesthesiol & pharmacol, 80-87, PROF ANESTHESIOL, STRITCH SCH MED, LOYOLA UNIV CHICAGO, 87-; CHIEF ANESTHESIOL, E A HINES JR VET AFFAIRS HOSP, HINES, ILL, 87- *Concurrent Pos:* Chief anesthesiol, Vet Admin Hosp, Philadelphia, 78-80. *Mem:* Am Physiol Soc; Asn Univ Anesthetists; Am Soc Anesthesiol. *Res:* Respiratory physiology and pharmacology and their applications to anesthesiology. *Mailing Add:* 350 Fairbank Rd Riverside IL 60546

SMITH, THEODORE G, b Baltimore, Md, Aug 12, 34; m 65; c 2. CHEMICAL ENGINEERING. *Educ:* Johns Hopkins Univ, BEngSci, 56, MS, 58; Washington Univ, DSc(chem eng), 60. *Prof Exp:* Chem engr, Res Div, E I Du Pont de Nemours & Co, Del, 60-62, WVa, 62-63; from asst prof to assoc prof, 63-68, PROF CHEM ENG, UNIV MD, COLLEGE PARK, 71- *Mem:* AAAS; Am Inst Chem Engrs; Am Chem Soc. *Res:* Polymer plastics; fractionation, crystallization and solubility; diffusion through polymers; large scale chromatography; control of chemical processes; rheology; reactor design; kinetics. *Mailing Add:* Dept of Chem Eng Univ of Md College Park MD 20742

SMITH, THEODORE ISAAC JOGUES, b Brooklyn, NY, Jan 13, 45. AQUACULTURE, FISHERIES REHABILITATION. *Educ:* Cornell Univ, BS, 66; C W Post Col, MS, 68; Univ Miami, PhD(marine sci), 73. *Prof Exp:* SR MARINE SCIENTIST AQUACULT, SC WILDLIFE & MARINE RESOURCES DEPT, 73- *Mem:* World Aquacult Soc; Southeastern Estuarine Res Soc; Gulf & Caribbean Fisheries Inst; Am Fisheries Soc. *Res:* Determination of biological requirements for commercially important species; development of applicable techniques for use in mariculture; technical and advisory services for mariculture and related industries. *Mailing Add:* SC Wildlife & Marine Resources Dept 217 Fort Johnson Rd Charleston SC 29412

SMITH, THOMAS CALDWELL, b Charleston, WVa, Feb 20, 41; m 65; c 2. PHYSIOLOGY, BIOPHYSICS. *Educ:* Univ Richmond, BS, 63, MS, 65; Med Col Va, PhD(physiol), 69. *Prof Exp:* Instr physiol, Med Col Va, 68-69; from asst prof to assoc prof, 69-83, PROF PHYSIOL, UNIV TEX HEALTH CTR, 83- *Mem:* Soc Gen Physiol. *Res:* Active ion transport in epithelium and biomembranes. *Mailing Add:* Dept Physiol Med Sch Univ Tex 7703 Floyd Curl Dr San Antonio TX 78284

SMITH, THOMAS CHARLES, b Elyria, Ohio, Dec 6, 25. PHARMACOLOGY. *Educ:* Oberlin Col, AB, 47; Harvard Univ, MA, 49, PhD, 52; Northwestern Univ, MD, 62. *Prof Exp:* Lab asst, Oberlin Col, 44-47; asst bot, Cambridge Jr Col, 48; teaching fel gen physiol, Harvard Univ, 49, teaching fel zool, 50 & endocrinol, 51; from instr to asst prof pharmacol, Sch Med, Boston Univ, 53-58; asst prof, Med Sch, Northwestern Univ, 59-62; intern, Univ Chicago Hosps, 62-63; asst prof gynec, obstet & pharmacol, Sch Med, Marquette Univ, 63-65; from asst dir to assoc dir div clin res, Ortho Res Found, 65-66, dir div pharmacol, 66-68; dir clin pharmacol, Parke, Davis & Co, 68-81; MED DIR & ATTEND STAFF, CLIN INVEST UNIT, BRONSON METHODIST HOSP, 81- *Concurrent Pos:* Assoc attend staff, Milwaukee County Hosp, Wis, 63-65, head gynec-endocrine lab & family planning clin, 64-65; vis lectr, Sch Med, Marquette Univ, 66-; spec attend staff, Somerset Hosp, Somerville, NJ, 67-68; lectr, Rutgers Univ, 67-68 & Univ Mich, 69-; attend staff, Chelsea Med Clin, Mich, 70- *Mem:* Am Soc Pharmacol & Exp Therapeut; Am Soc Clin Pharmacol & Therapeut. *Res:* Physiology of reproduction; mammary gland function; hormones affecting metabolism and tumor growth; drugs and adipose tissue; endocrine and clinical pharmacology. *Mailing Add:* Jasper Clin Invest Ctr Upjohn Co 526 Jasper St Kalamazoo MI 49007

SMITH, THOMAS DAVID, b Eng, Nov 25, 23; m 47. PHYSICAL CHEMISTRY. *Educ:* Univ London, BSc, 44, PhD(chem), 47. *Prof Exp:* Chemist, C A Parsons & Co, Ltd, Eng, 39-44; res assoc, Brit Coke Res Asn, 44-47; res assoc, Univ Southern Calif, 48-49; res chemist, Union Oil Co, Calif, 49-50; res fel, Cambridge Univ, 50-51; res supvr, E I du Pont de Nemours & Co, Inc, 51-57, res mgr, 57-64, lab dir, 64-65, asst plant mgr, 65-66, lab dir, 66-68, asst dir res & develop, 68-72, dir res, Imaging Systs Dept, 72-86; RETIRED. *Mem:* Am Chem Soc; Soc Photog Sci & Eng; Royal Photog Soc Gt Brit; Inst Elec & Electornic Engrs; Soc Photo-Optical Instrumentation Engrs; Am Acad Sci. *Res:* Physical and colloid chemistry, especially in photographic systems; electronic imaging systems. *Mailing Add:* RR 1 Box 1330 Greensboro VT 05841

SMITH, THOMAS ELIJAH, b North Augusta, SC, Apr 11, 33; m 53; c 2. BIOCHEMISTRY. *Educ:* Benedict Col, BS, 53; George Washington Univ, MS, 59, PhD(biochem), 62. *Prof Exp:* Chemist, Lab Exp Med & Clin Therapeut, Nat Heart Inst, 53-54, biochemist, Lab Clin Biochem, 56-62; NIH fel enzyme mech, Wash Univ, 62-63; sr biochemist, Biol & Med Div, Melpar, Inc, 63-65; sr biochemist, Lawrence Livermore Lab, Univ Calif, 65-74; assoc prof biochem, Univ Tex Health Sci Ctr Dallas, 74-80; PROF & CHMN, DEPT BIOCHEM, COL MED, HOWARD UNIV, 80- *Concurrent Pos:* Dir, Biomed Div Summer Teaching & Res Inst, Lawrence Livermore Lab, Univ Calif, Livermore, Calif, 72-74; consult, E I DuPont Co, Wilmington, Del, 72-74; asst dean, Grad Sch Biomed Sci, Univ Tex Health Sci Ctr, Dallas, Tex, 74-76; mem, test comt, Nat Bd Med Examiners, 87-91 & bd sci counselors, Nat Heart Lung & Blood Inst, NIH, 88-92. *Mem:* AAAS; Am Chem Soc; Am Soc Biol Chemists; Sigma Xi. *Res:* Enzyme mechanisms. *Mailing Add:* Dept Biochem Howard Univ Col Med 520 W St NW Washington DC 20059

SMITH, THOMAS GRAVES, JR, b Winnsboro, SC, Mar 22, 31; m 56. NEUROPHYSIOLOGY. *Educ:* Emory Univ, BA, 53; Oxford Univ, BA & MA, 56; Columbia Univ, MD, 60. *Prof Exp:* Intern, Bronx Munic Hosp, New York, 60-61; vis res assoc biol, Mass Inst Technol, 64-66; res med officer physiol, 64-68, CHIEF SECT SENSORY PHYSIOL, LAB NEUROPHYSIOL, NAT INST NEUROL COMMUN & NEUROL DIS & STROKE, 68- *Concurrent Pos:* Vis prof, John Cortin Sch Med, Australian Nat Univ, Canberra, 80. *Mem:* AAAS; Am Physiol Soc; Soc Neurophysiol. *Res:* Neurophysiology and biophysics of excitable membranes and of synaptic transmission between nerve cells; video microscopy and image processing. *Mailing Add:* Nat Inst Neurol Dis & Stroke Bldg 36 Rm 2C02 Bethesda MD 20892

SMITH, THOMAS HADWICK, mechanical & chemical engineering, for more information see previous edition

SMITH, THOMAS HARRY FRANCIS, b Paterson, NJ, Feb 15, 28; m 51; c 1. TECHNICAL MANAGEMENT. *Educ:* Fordham Univ, BS, 47; Philadelphia Col Pharm, MSc, 56, PhD(toxicol), 61. *Prof Exp:* Asst pharmacologist, Hoffmann-La Roche, Inc, 49-51 & Wallace Labs, 51-52; bacteriologist, Children's Hosp Philadelphia, 54-55; asst zool & bot & res pharmacologist, Philadelphia Col Pharm, 56-57, admin asst, 57-59; res pharmacologist & parasitologist, Vet Sch, Univ Pa, 57-58, lectr anat, physiol & microbiol, Sch Nursing, 58-61; exp pharmacologist & toxicologist, Wyeth Labs, Inc, 61-64; tech info coordr, Avon Prod Inc, NJ, 64-67; dir sci serv, Lehn & Fink Div, Sterling Drug Inc, 67-71; dir qual assurance, Lanvin Charles of the Ritz, 71-76; dir prod integrity, Norda, 76-80; corp toxicologist, IBM Corp, 81-89; PROG MGR, CHEM UTILIZATION TECHNOL, 90- *Concurrent Pos:* Instr anesthesiol, Misericordia Hosp, 57-59. *Mem:* Am Acad Clin Toxicol; Soc Toxicol. *Res:* Pharmacology; dermatotoxicology; psychopharmacology; semiconductor toxicology; environmental toxicology. *Mailing Add:* IBM Corp Hq 2000 Purchase St Purchase NY 10577

SMITH, THOMAS HENRY, b Lackawanna, NY, Aug 14, 47; m 77; c 2. ORGANIC CHEMISTRY. *Educ:* Niagara Univ, BS, 69; Ariz State Univ, PhD(org chem), 74. *Prof Exp:* Res asst org chem, Ariz State Univ, 69-74; fel, Stanford Res Inst, 74-75; org chemist, SRI INT, 75-87; mgr, chem process develop, 87-89, DIR, SYNTHESIS REAGENTS, AM BIONETICS, 89- *Mem:* AAAS; Am Chem Soc. *Res:* Synthetic organic chemistry; synthesis of biologically active compounds; drug design; nucleic and synthesis reagents. *Mailing Add:* 2041 Greenwood Dr San Carlos CA 94070

SMITH, THOMAS JAY, b Rochester, NY, May 9, 40; m 62; c 2. OCCUPATIONAL HEALTH & SAFETY, HUMAN FACTORS. *Educ:* Univ Wis-Madison, BA, 62, PhD(physiol), 77; Univ Calif, San Diego, MSc, 66. *Prof Exp:* Programmer systs, Planning Res Corp, San Diego, Calif, 66-67 & Comput Ctr, Univ Wis-Madison, 67-69; teaching asst physiol, Univ Wis-Madison, 69-75, teaching fel toxicol, 77-79; vis instr kinesiology, Simon Fraser Univ, 75-76, asst adj prof, 80-87; SUPVR HUMAN FACTORS RES, US BUR MINES, 87- *Concurrent Pos:* Vis prof physiol, dept biol, Beloit Col, 79; consult, 80- *Mem:* Am Physiol Soc; Human Factors Soc; AAAS. *Res:* Human health and performance effects of exposure to occupational-environmental hazards; behavioral cybernetic analysis of motor performance, growth and development, and safety and hazard management; ergonomic-human factors evaluation of occupational health and safety problems. *Mailing Add:* US Bur Mines Twin Cities Res Ctr 5629 Minnehaha Ave S Minneapolis MN 55147

SMITH, THOMAS JEFFERSON, b Atlanta, Ga, June 12, 30; m 57; c 2. GEOPHYSICS, MATHEMATICS. *Educ:* Emory Univ, BA, 51; Univ Wis, MS, 57, PhD(math), 61. *Prof Exp:* Asst prof math, Kalamazoo Col, 61-62; mem staff geophys, Carnegie Inst Wash, assoc prof, 70-74; PROF MATH, KALAMAZOO COL, 74- *Mem:* Am Math Soc. *Res:* Minkowskian and Finsler geometries; convex sets; numerical methods. *Mailing Add:* Dept Math Kalamazoo Col Kalamazoo MI 49007

SMITH, THOMAS LOWELL, MICROCIRCULATION, HEMODYNAMICS. *Educ:* Wake Forest Univ, PhD(cardiovasc physiol), 79. *Prof Exp:* ASST PROF PHYSIOL, BOWMAN GRAY SCH MED, WAKE FOREST UNIV, 82- *Mailing Add:* Dept Physiol Bowman Gray Sch Med Wake Forest Univ 300 S Hawthorne Rd Winston Salem NC 27103

SMITH, THOMAS PATRICK, zoology, for more information see previous edition

SMITH, THOMAS STEVENSON, b Hubbard, Ohio, Feb 8, 21; m 44; c 3. SOLID STATE PHYSICS. *Educ:* Kenyon Col, AB, 47; Ohio State Univ, PhD(physics), 52. *Hon Degrees:* LHD, Kenyon Col, 70, Cardinal Stritch Col, 80; DSc, Ripon Col, 71; LLD, Lawrence Univ, 80. *Prof Exp:* Instr physics, Kenyon Col, 46-47; asst, Ohio State Univ, 47-51, res fel, 51-52; from asst prof to prof, Ohio Univ, 52-69, asst to pres, 61-62, vpres acad affairs, 62-67, provost, 67-69; pres, Lawrence Univ, 69-79. *Concurrent Pos:* Chmn, Great Lakes Dist Selection Rhodes Scholar; educ consult-examr, Comn Cols & Univs, NCent Asn Cols & Sec Schs. *Mem:* Am Phys Soc; AAAS; Sigma Xi; Am Asn Physics Teachers. *Res:* Cryogenics; superconductivity; nuclear magnetic resonance; x-ray powder diffraction; vapor pressures at high temperature. *Mailing Add:* Rte 1 79A Pine River WI 54965

SMITH, THOMAS W, b Akron Ohio, Mar 29, 36; m 58; c 3. MEDICINE. *Educ:* Harvard Univ, AB, 58, MD, 65. *Prof Exp:* From asst prof to assoc prof, 71-79, PROF MED, SCH MED, HARVARD UNIV, 79-; CHIEF CARDIOVASC DIV, BRIGHAM & WOMEN'S HOSP, 74- *Mem:* Am Heart Asn; Am Col Cardiol; Am Soc Clin Invest; Am Fed Clin Res; Am Physiol Soc; Am Soc Pharmacol & Exp Therapeut; Asn Am Physicians. *Res:* Mechanism of action digitalis; mechanism of inotropic agents. *Mailing Add:* Cardiovasc Div Brigham & Women's Hosp 75 Francis St Boston MA 02115

SMITH, THOMAS WOODS, b Portsmouth, Ohio, Dec 16, 43; m 68; c 2. ORGANIC POLYMER CHEMISTRY. *Educ:* John Carroll Univ, BS, 69; Univ Mich, PhD(org chem), 73. *Prof Exp:* Chemist, Lubrizol Corp, 63-70; assoc scientist, 73-75, SCIENTIST, WEBSTER RES CTR, XEROX CORP, 75- *Concurrent Pos:* Consult, Environ Res Inst Mich, 72-73. *Mem:* Am Chem

Soc; Sigma Xi. *Res:* The synthesis of functional polymers; the mechanism of polymerization processes and the utilization of macromolecules as catalysts for chemical processes. *Mailing Add:* Xerox Corp 22 Hidden Meadow Penfield NY 14526

SMITH, THOR LOWE, b Zion, Ill, June 11, 20; m 49; c 2. POLYMER PHYSICS. *Educ:* Wheaton Col, BS, 42; Ill Inst Technol, MS, 44; Univ Wis, PhD(chem), 48. *Prof Exp:* Res chemist, Hercules Co, 48-54; sr res engr, Jet Propulsion Lab, Calif Inst Technol, 54-56, chief solid propellant chem sect, 56-59; chmn propulsion dept, Stanford Res Inst, 59-61, dir propulsion sci div, 61-64, sci fel, 64-68; prof chem, Tex A&M Univ, 68-69; RES STAFF MEM, IBM RES DIV, 69- *Concurrent Pos:* Mem, Eval Panel Nat Res Coun, Nat Bur Standards, 74-77; mem bd trustees, Gordon Res Conf, 78-84, chmn, 81-82; mem, Nat Mat Adv Bd Comts, 71-72 & 78-80. *Honors & Awards:* Bingham Medal, Soc Rheol, 78; Centennial scholars lectr, Case Western Reserve Univ, 80; Res Award, Soc Plastics Engrs, 83; Whitby Mem Lectr, Univ Akron, 74. *Mem:* Nat Acad Eng; Brit Soc Rheol; fel Am Phys Soc; Soc Rheol (pres, 67-69); Soc Plastics Engrs; Am Chem Soc. *Res:* Mechanical and other physical properties of polymer systems; deformation and fracture of polymeric materials; rheology of dispersions and polymers; rejuvenation and physical aging of polymeric glasses. *Mailing Add:* IBM Res Div Almaden Res Ctr 650 Harry Rd Mail Stop K93/801 San Jose CA 95120-6099

SMITH, TIM DENIS, b Eugene, Ore, Dec 30, 46; m 67; c 1. BIOLOGY, STATISTICS. *Educ:* Pac Lutheran Univ, BA, 69; Univ Wash, PhD(biomath), 73. *Prof Exp:* Res assoc oceanog, Univ Wash, 72-73; fisheries biologist, Nat Marine Fisheries Serv, 73-75; asst prof zool, Univ Hawaii, 75-78; fisheries biologist, Southwest Fisheries Ctr, 78-85; FISHERIES BIOLOGIST, NORTHEAST FISHERIES CTR, 85- *Concurrent Pos:* Mem sci comt, Int Whaling Comn, 74-; mem, Comt Sci Adv, US Marine Mammal Comn, 76-79, 89-; mem, Sci & Statist Comt, Western Pac Regional Fisheries Mgt Coun, 77-78. *Mem:* Soc Marine Mammalogy; Resource Modeling Asn. *Res:* Applied and theoretical population biology, especially of large mammals; management of living resources; dynamics of populations of fishes and large mammals, especially marine mammals; natural resource utilization; history of marine science. *Mailing Add:* Northeast Fisheries Ctr Woods Hole MA 02543

SMITH, TIMOTHY ANDRE, b LaCrosse, Wis, Jan 9, 37; m 64; c 5. DENTAL STRESS. *Educ:* Marquette Univ, BS, 58; Univ NC, MA, 61, PhD(psychol), 63. *Prof Exp:* Instr psychol, Fla State Univ, 62-63; assoc prof psychol, 69-73, dir, Learning Resources, Col Educ & Dent, 73-75, prof community dent, 76-87, PROF ORAL HEALTH SCI, UNIV KY COL DENT, 87- *Concurrent Pos:* Consult, Chattanooga State Tech Col, 75, Mass Inst Technol, 76, Univ Minn, 77, Sci Res Assoc, 77, Univ Wash, 79, Dept Educ, State Alaska, 80-83 & 85; fel, Harvard Sch Med, 76; vis prof, Univ Wash, 83-84, Royal Dent Col, Aarhus, Denmark, 90-91. *Mem:* Am Psychol Asn; Am Educ Res Asn; Am Asn Dental Sch; Int Asn Dental Res; Am Asn Pub Health Dent. *Res:* Improving the measurement of dental stress through physiological and self-report methods; reduction of dental stress through behavioral methods. *Mailing Add:* Dept Oral Health Sci Col Dent Univ Ky Lexington KY 40536-0084

SMITH, TODD IVERSEN, b Mobile, Ala, June 11, 40; m 70. PHYSICS. *Educ:* Cornell Univ, BA, 61; Rice Univ, MA, 63, PhD(physics), 65. *Prof Exp:* Res assoc physics, Stanford Univ, 65-68; asst prof physics & elec eng, Univ Southern Calif, 68-74; RES PHYSICIST, HANSEN LAB, STANFORD UNIV, 74- *Mem:* AAAS; Am Phys Soc; Inst Elec & Electronics Engrs. *Res:* Superconducting microwave cavities; Josephson tunneling between superconductors; fluctuation effects and flux flow effects in thin superconducting films; instrumentation using Josephson junctions as sensors. *Mailing Add:* High Energy Phys Lab Stanford Univ Stanford CA 94305

SMITH, TOWNSEND JACKSON, b West Union, WVa, Oct 15, 10; m 41; c 1. AGRONOMY. *Educ:* WVa Univ, BSA, 36; Ohio State Univ, PhD(agron), 40. *Prof Exp:* Agt, USDA & asst, Ohio Agr Exp Sta, 36-40; instr, Univ & asst agronomist, Exp Sta, Univ Ariz, 40-41, asst prof & asst agronomist, 41-46; assoc agronomist, Exp Sta, 46-48, prof, 48-78, EMER PROF AGRON, VA POLYTECH INST & STATE UNIV, 78- *Mem:* Am Soc Agron. *Res:* Physiology of cotton fibers; forage crops breeding; soybean breeding, culture and physiology; use of ionizing radiation in alfalfa improvement. *Mailing Add:* 1014 Allendale Ct SW Blacksburg VA 24060-5414

SMITH, TRUDY ENZER, b Eger, Czech, May 23, 24; nat US; m 53; c 6. PHYSICAL CHEMISTRY. *Educ:* Greensboro Col, AB, 44; Ohio State Univ, PhD(chem), 57. *Prof Exp:* Asst, Ohio State Univ, 45-50; res chemist, Aerojet Gen Corp, Gen Tire & Rubber Co, Calif, 51-53; phys chemist, US Naval Ord Test Sta, 53-60; instr chem, Univ Conn, 60-62; from asst prof to assoc prof, 62-77, PROF CHEM, CONN COL, 77- *Mem:* Am Chem Soc; Sigma Xi. *Res:* Chemical kinetics; mass spectrometry. *Mailing Add:* Dept Chem Conn Col New London CT 06320-4196

SMITH, VANN ELLIOTT, b Pensacola, Fla, June 28, 40; m 69; c 2. MARINE BIOLOGY. *Educ:* Fla State Univ, BS, 62, MS, 64; Scripps Inst Oceanog, PhD(marine biol), 68. *Prof Exp:* Coordr Lake & Marine Res, Cranbrook Inst, 71-85; PROJ MGR, OCEANOG SERVS, RAYTHEON SERV CO, 85- *Concurrent Pos:* Edison scholar, Cranbrook Inst Sci, 71-72. *Mem:* Sigma Xi; Int Asn Great Lakes Res. *Res:* Contaminants in Great Lakes waters and wildlife; comparative marine biochemistry; heavy metals and pesticides in lake ecosystems; remote sensing of lakes, lake watersheds and coral reef systems. *Mailing Add:* 2004 W Spinningwheel Bloomfield Hills MI 48013

SMITH, VELMA MERRILINE, b San Bernardino, Calif, Mar 11, 40; m 61. ALGEBRA. *Educ:* Calif State Col, San Bernardino, BA, 67; Univ Calif, Riverside, MA, 69, PhD(math), 72. *Prof Exp:* From asst prof to assoc prof, 72-81, PROF MATH, CALIF STATE POLYTECH UNIV, 81-, CHMN DEPT, 82- *Mem:* Am Math Soc; Math Asn Am. *Res:* Commutative algebra, especially ideal and ring theory; teacher education. *Mailing Add:* Dept Math Calif State Polytech Univ 3801 W Temple Ave Pomona CA 91768

SMITH, VICTOR HERBERT, b Lewistown, Mont, Aug 1, 25; m 50; c 5. HOSPITAL ADMINISTRATION, ENVIRONMENTAL HEALTH. *Educ:* Mont State Col, BS, 50; Ore State Col, PhD(chem), 55. *Prof Exp:* Biol scientist, Hanford Atomic Prod Oper, Gen Elec Co, 54-65; sr res scientist, Pac Northwest Labs, Battelle Mem Inst, 65-81; MGR, KENNEWICK PRIMARY CLINIC, 82- *Concurrent Pos:* AEC fel radiation chem & biophys, Univ Minn, 59-61; mem SC-37, Nat Comt Radiation Protection, 73- *Mem:* AAAS; Am Chem Soc; Radiation Res Soc; Health Physics Soc; Soc Exp Biol & Med. *Res:* Heterocyclics; radiation induced reactions and effects on organics and biological systems; radiation protection; removal of radioactive emitters; chelation therapy; effects and treatment of incorporated radionuclides, toxic metals, organometallics and combined insults. *Mailing Add:* 1007 W 27th Kennewick WA 99337-4308

SMITH, VICTORIA LYNN, b Dayton, Ohio, July 29, 59. PHYTOPATHOLOGY. *Educ:* Ohio Northern Univ, BS, 81; Ohio State Univ, MS, 83; NC State Univ, PhD(plant path & soil sci), 87. *Prof Exp:* Postdoctoral res assoc, Cornell Univ, 87-89; res plant pathologist, Beltsville Agr Res Ctr, USDA, 89-90; ASST SCIENTIST, CONN AGR EXP STA, 90- *Mem:* Am Phytopath Soc; Can Phytopath Soc; Sigma Xi. *Res:* Biological control of phytophthora cinnamomi on woody ornamental plants; biology and control of dogwood anthralnose. *Mailing Add:* Plant Path & Ecol 123 Huntington St New Haven CT 06504

SMITH, VINCENT C, b Albany, NY, Nov 4, 14; m 40; c 4. FLORICULTURE, MARKETING. *Educ:* Cornell Univ, BS, 37; NY Univ, MS, 47. *Prof Exp:* Teacher, Rockland County Voc Educ & Exten Bd, 40-47; asst prof floricult, 48-55, head dept, 56-75, prof ornamental hort, 56-80, EMER PROF, STATE UNIV NY AGR & TECH COL, 80- *Mem:* Am Soc Hort Sci. *Res:* Problems involved in marketing floral and ornamental horticulture crops. *Mailing Add:* 468 Cleveland Ave Cornell NY 14843

SMITH, VIVIAN SWEIBEL, health sciences, for more information see previous edition

SMITH, VIVIANNE C(AMERON), b Woodford, Eng, July 7, 38; US citizen; m 65; c 2. PSYCHOPHYSIOLOGY. *Educ:* Columbia Univ, BS, 62, MA, 64, PhD(psychol), 67. *Prof Exp:* From instr to assoc prof, 68-79, PROF OPHTHAL, UNIV CHICAGO, 79- *Concurrent Pos:* Mem, Int Res Group Colour Vision Deficiencies. *Honors & Awards:* Tyler Medal, 90. *Mem:* Fel Optical Soc Am; Asn Res Vision & Ophthal. *Res:* Mechanism of color vision in humans; theories of color vision; spatial and temporal factors in vision. *Mailing Add:* Visual Sci Ctr Univ Chicago 939 E 57th St Chicago IL 60637

SMITH, W JOHN, b Toronto, Ont, Dec 20, 34; m 64; c 1. BIOLOGY, ANIMAL BEHAVIOR. *Educ:* Carleton Univ, Can, BSc, 57; Univ Mich, MS, 58; Harvard Univ, PhD(biol), 61. *Prof Exp:* Asst prof zool, 63-68, assoc prof, 68-76, PROF BIOL & PSYCHOL, UNIV PA, 76- MEM, INST NEUROL SCI, 67- *Concurrent Pos:* Res assoc, Mus Comp Zool, Harvard Univ, 61-64; consult, Penrose Res Lab, Philadelphia Zool Soc, 65-; res assoc, Smithsonian Trop Res Inst, 66-; res assoc, Acad Natural Sci, Pa, 67- *Mem:* Am Ornithologists Union; Brit Ornithologists Union; Animal Behavior Soc. *Res:* Animal communication and social behavior; ecology; systematics; evolutionary theory. *Mailing Add:* Leidy Labs Univ of Pa Philadelphia PA 19104

SMITH, W(ILLIAM) P(AYNE), b Superior, Wis, Jan 5, 15; m 42; c 3. ELECTRICAL ENGINEERING. *Educ:* Univ Minn, BEE, 36, MS, 37; Univ Tex, PhD, 50. *Prof Exp:* Engr, Commonwealth Edison Co, Ill, 37-39; asst prof, Chicago Tech Col, 39-41; dean, Sampson Col, 46-50; prof elec eng, 50-80, chmn, dept elec eng, Sch Eng & Archit, 55-56, dean, 65-80, EMER DEAN, UNIV KANS, 80- *Concurrent Pos:* Lectr, Univ Tex, 48-50. *Mem:* Am Soc Eng Educ; Inst Elec & Electronics Engrs. *Mailing Add:* 6508 Overbrook Shawnee Mission KS 66208

SMITH, WADE KILGORE, b Paterson, NJ, Sept 7, 37; m 63; c 2. HEMATOLOGY, IMMUNOLOGY. *Educ:* Oberlin Col, AB, 59; Sch Med, Johns Hopkins Univ, MD, 63. *Prof Exp:* Intern, Mt Sinai Hosp, New York, 63-64, resident, 64-68, chief res, 68-69, fel hematol, 69; res asst immunol, Med Ctr, Duke Univ, 70-71, instr, 71-72, assoc med & immunol, 72-74; chief, hemat-oncol sect, Vet Admin Med Ctr, Richmond, 81-89; asst prof, 75-80, ASSOC PROF MED, MED COL VA, 75- *Concurrent Pos:* Instr med, Mt Sinai Sch Med, 68-69; Nat Cancer Inst Spec fel, Med Ctr, Duke Univ, 70-72; mem, Med Col Va/Va Commonwealth Univ Cancer Ctr, 76-; curric coord, Sch Med, Med Col Va, 82-87, chmn, Inst Animal Care & Use Comt, 86-89; dir, Hunter Holmes McGuire Vet Admin Med Ctr Comprehensive Cancer Ctr. *Mem:* Am Soc Histocompatibility & Immunogenetics; Am Soc Hematol; Am Soc Microbiol; Int Soc Hematol; NY Acad Sci. *Res:* Leukocyte antigens and immune destruction of leukocytes; humoral factors suppressing immune responses in tumor bearing or normal graft bearing hosts; clinical cancer chemotherapy trials. *Mailing Add:* Med Col Va VA Commonwealth Univ Richmond VA 23298-0162

SMITH, WALDO E(DWARD), b New Hampton, Iowa, Aug 20, 00; m 27; c 2. HYDRAULIC ENGINEERING, GEOPHYSICS. *Educ:* Univ Iowa, BS, 23, MS, 24. *Prof Exp:* Asst engr, Burns & McDonnell Eng Co, Mo, 24-26 & Black & Veatch, 26-27; instr theoret & appl mech, Univ Ill, 27-28; assoc prof civil eng & actg head dept, Robert Col, Istanbul, 28-31; asst prof, NDak State Univ, 31-35; hydraul engr, Muskingum Watershed Conserv Dist, Ohio, 35-39; mem staff flood control, US Eng Off, WVa, 39-40; soil conserv serv, USDA, 40-41, head sect hydrol land use, Hydrol Div, Off Res, 41-43; hydraul engr, US Pub Rd Admin, Washington, DC, 43-44; exec dir & ed Trans, Am Geophys Union, 44-70, ed, Geophys Monograph Series, 58-70, EMER EXEC DIR, AM GEOPHYS UNION, 70-; CONSULT, 70- *Concurrent Pos:* Dep state engr, NDak, 32-34; collabr, Soil Conserv Serv, USDA, 37-40; prof lectr, George Washington Univ, 46-61; specialist, Res & Develop Bd, US Dept Defense, DC, 47-53; US del, Int Union Geod & Geophys, Oslo, 48, Brussels, 51, Rome, 54, Toronto, 57, Helsinki, 60, Berkeley, 63, Switz, 68,

Tokyo, 71, Moscow, 72, Canberra, 79 & Hamburg, 83; asst to pres, Nat Grad Univ, 70-78. *Mem:* Fel AAAS; fel Am Soc Civil Engrs; fel Am Geophys Union. *Res:* Geophysical education; hydrology; environmental and natural resources problems. *Mailing Add:* 1330 Massachusetts Ave NW Apt 20005 Washington DC 20005

SMITH, WALKER O, JR, b Buffalo, NY, Nov 21, 50; c 1. BIOLOGICAL OCEANOGRAPHY, PHYCOLOGY. *Educ:* Univ Rochester, BS, 72; Duke Univ, PhD(bot), 76. *Prof Exp:* Res asst oceanog, Duke Univ, 72-76; from asst prof to assoc prof, 76-86, PROF BOT, UNIV TENN, 86- *Mem:* AAAS; Phycol Soc Am; Am Soc Limnol & Oceanog; Am Geophys Union. *Res:* Flux of carbon and nitrogen in polar systems. *Mailing Add:* Dept Bot Univ Tenn 588 Dabney Hall Knoxville TN 37916

SMITH, WALLACE BRITTON, b Parrish, Ala, Jan 21, 41; m 60; c 2. APPLIED PHYSICS. *Educ:* Jacksonville State Univ, BS, 67; Auburn Univ, MS, 69, PhD(physics), 72. *Prof Exp:* Asst prof physics, Appalachian State Univ, 72-73, res physicist, 73-74; head physics sect, Southern Res Inst, 74-77, head physics div, 77-84, assoc dir, 84-88; PRES, SOUTHERN RES TECHNOL, INC, 88- *Mem:* Inst Elec & Electronics Engrs; Sigma Xi; Am Asn Aerosol Res; Adv Pollution Control Asn; Ges Aerosolforschung. *Res:* Electrical breakdown in insulators and semiconductors; particle sizing techniques and instruments; physics of the electrostatic precipitation; fabric filtration processes. *Mailing Add:* PO Box 114 Trussville AL 35173

SMITH, WALTER LAWS, b London, Eng, Nov 12, 26; m 50; c 2. MATHEMATICAL STATISTICS. *Educ:* Cambridge Univ, BA, 47, MA, 50, PhD(math statist), 53. *Prof Exp:* Statistician, Med Sch, Cambridge Univ, 53-54, lectr math, 56-58; from asst prof to assoc prof, 53-62, PROF STATIST, UNIV NC, CHAPEL HILL, 62- *Mem:* Am Math Soc; fel Am Statist Asn; fel Inst Math Statist; fel Royal Statist Soc; Int Statist Inst. *Res:* Probability theory; operations research. *Mailing Add:* Dept Statist 318 Phillips Hall Univ of NC Chapel Hill NC 27514

SMITH, WALTER LEE, b Siler City, NC, Nov 12, 48; m 72. PHYSICS. *Educ:* NC State Univ, BS, 71; Harvard Univ, PhD(appl physics), 76. *Prof Exp:* physicist optical & mat physics, Lawrence Livermore Lab, 76- 83; DIR, THERMA-WAVE INC, FREMONT, CALIF, 83- *Mem:* Am Inst Physics; Am Phys Soc; Mat Res Soc; Electrochem Soc. *Res:* Nonlinear optics; laser physics; ultraviolet materials properties; absolute measurement techniques; laser-induced breakdown physics; semiconductor fabrication, thermal wave physics. *Mailing Add:* 2339 Chateau Way Livermore CA 94550

SMITH, WALTER THOMAS, JR, b Havana, Ill, Feb 28, 22; m 45; c 2. ORGANIC CHEMISTRY. *Educ:* Univ Ill, BS, 43; Ind Univ, PhD(org chem), 46. *Prof Exp:* Fels fund postdoctoral fel, Univ Chicago, 46-47; from instr to asst prof org chem, Univ Iowa, 47-53; assoc prof, 53-56, PROF ORG CHEM, UNIV KY, 56- *Concurrent Pos:* Chemist, Mallinckrodt Chem Works, St Louis, 43 & 44; Fulbright lectr & dept head, Univ Libya, Tripoli, 62-63; Fulbright lectr, Am Univ Beirut, 64-65; vis prof, Univ Maine, 80-86; consult-legal expert, var orgn. *Mem:* Fel AAAS; fel Am Inst Chemists; Am Chem Soc. *Res:* Organic analysis; medicinal chemistry of anticancer compounds and interferon inducers; graft polymers as synthetic nucleotides; detoxification of chemical weapons; enzyme activity in mixed solvents. *Mailing Add:* Dept Chem Univ Ky Lexington KY 40506

SMITH, WALTON RAMSAY, b Asheville, NC, Apr 26, 48. WOOD SCIENCE & TECHNOLOGY, WOOD PHYSICS. *Educ:* NC State Univ, BS, 71; Univ Calif, Berkeley, MS, 75, PhD(wood sci & technol), 81. *Prof Exp:* Consult, Walton R Smith Consult, 71-73; res asst, Univ Calif, Berkeley, 73-77; researcher, Ctr Technique du Bois, Paris, 77-78; dir grad prog, Int Trade Forest Prod, Univ Wash, coordr, Int Wood Construct Res Prog & assoc prof wood physics, 78-90; DIR & ASST PROF WOOD SCI, APPALACHIAN EXPORT CTR HARDWOODS, 90- *Concurrent Pos:* Consult, Dept Wood Physics, Ctr Technique du Bois, Paris, 81, consortium FAO, UN in Rome, Italy, Grenada, WI, Malawi, Africa, Third World Wood Res Labs. *Mem:* Forest Prod Res Soc; Soc Woods Sci & Technol; Int Union Forestry Res Orgn; Am Forestry Asn. *Res:* Development and coordination of both fundamental and applied programs in wood physics, technical and economic issues in international wood construction, international specifications for forest products; biomass fuel laboratory to fully characterize wood and other biomass fuel types; wood quality laboratory for analysis and characterization of woody materials with respect to potential end uses; product performance in international markets; thermal properties of wood; dimensional stability of wood and wood products. *Mailing Add:* Appalachian Export Ctr Hardwoods 4000 Hampton Ctr Suite B PO Box 6061 Morgantown WV 26506-6061

SMITH, WARREN DREW, b Tampa, Fla, Dec 22, 42; m 71. BIOMEDICAL SIGNAL PROCESSING, BIOMEDICAL MODELING & SIMULATION. *Educ:* Princeton Univ, BS, 64; Univ NMex, MS, 68; Univ Okla, PhD(elec eng), 71. *Prof Exp:* From asst prof to assoc prof, 73-82, PROF BIOMED ENG, ELEC & ELECTRONIC ENG, CALIF STATE UNIV, SACRAMENTO, 82- *Concurrent Pos:* Consult, Food & Drug Sect, Calif Health Dept, 75-76, Sutter Community Hosps, 76-77 & Lawrence Livermore Lab, 78-79; proj dir, Found Calif State Univ, 75-77, Sutter Hosps Med Res Found, 76-78 & Nat Inst Gen Med Sci, 86-88. *Mem:* Inst Elec & Electronics Engrs; Inst Elec & Electronics Engrs Eng Med & Biol Soc; Am Soc Eng Educ; Biomed Eng Soc; Asn Advan Med Instrumentation. *Res:* Developing anesthesia monitor that processes human brain waves to display the level of anesthesia of a patient during surgery; author of numerous technical publications. *Mailing Add:* Dept Elec & Electronic Eng Calif State Univ 6000 J St Sacramento CA 95819

SMITH, WARREN HARVEY, b Brooklyn, NY, Oct 6, 35; m 60; c 3. PHYSICAL CHEMISTRY. *Educ:* City Col NY, BS, 58; Syracuse Univ, PhD(phys chem, kinetics), 64. *Prof Exp:* Sr res chemist, Monsanto Res Corp, Miamisburg, 64-67, group leader plutonium chem, 67-69, plutonium fuels develop mgr, 69-71, isotope separation mgr, 71-73, applied physics mgr, Mound Lab, 73-85; MGR, NUCLEAR TECHNOL, MOUND LAB, EG&G INC, 85- *Mem:* AAAS; Am Chem Soc; Am Inst Chemists; Sigma Xi. *Res:* Gas phase kinetics; physical-inorganic chemistry of the actinide elements; use of plutonium-238 as a fuel for heat sources. *Mailing Add:* 5413 Coppermill Pl Dayton OH 45429

SMITH, WARREN JAMES, optical design, geometrical optics, for more information see previous edition

SMITH, WARREN LAVERNE, b Wayne, Nebr, July 6, 24; m 48; c 5. PHYSICS, ELECTRICAL ENGINEERING. *Educ:* Univ Wis, BSEE, 45. *Prof Exp:* Staff mem, 54-62, supvr eng, Bell Tel Labs, Ins, 62-87; RETIRED. *Concurrent Pos:* Tech adv to TC-49, IEC, 73- *Honors & Awards:* CB Sawyer Mem Award. *Mem:* Fel Inst Elec & Electronics Engrs; AAAS. *Res:* Development and design of precision frequency standards, quartz crystal units and monolithic crystal filters. *Mailing Add:* 3046 Meadowbrook Circle N Allentown PA 18103

SMITH, WAYNE EARL, b Franklin, Ind, Jan 7, 27; m 53; c 3. MATHEMATICS. *Educ:* Pomona Col, BA, 49; Univ Calif, Los Angeles, MA, 53, PhD(math), 58. *Prof Exp:* Asst math, Univ Calif, Los Angeles, 50-53, assoc math & jr res mathematician, 54-58; asst prof math, Occidental Col, 58-62; asst prof appl math, Univ Colo, 62-63; vis asst prof biostatist, 63-65, lectr biostatist & asst res statistician, 65-68, ADMINISTRATIVE ANALYST, UNIV CALIF, LOS ANGELES, 68- *Mem:* Math Asn Am; Asn Instnl Res; Sigma Xi. *Res:* Numerical analysis; probability and statistics; simulation of university processes. *Mailing Add:* Off Acad Planning & Budget Univ Calif Los Angeles CA 90024-1405

SMITH, WAYNE H, b Marianna, Fla, Aug 10, 38; m 62. FORESTRY. *Educ:* Univ Fla, BSA, 60; Miss State Univ, MS, 62, PhD(soils), 65. *Prof Exp:* Asst soils, Miss State Univ, 63-64; from asst prof to assoc prof forestry, Univ Fla, 64-78, asst dir res, 70-78, prof forest resources & conserv & dir, Environ & Natural Sci Progs, 78-84, PROF FORESTRY & DIR, BIOMASS ENERGY SYSTS, UNIV FLA, 80-, DIR, ENERGY EXTEN SERV, 90- *Concurrent Pos:* Fac develop leave, Coop State Res Serv, USDA, Washington, DC, 73-74; dir, Fed Agency Liaison, Wash, DC, 85. *Mem:* Soc Am Foresters; Am Soc Agron; AAAS; Sigma Xi; Soil Sci Soc Am. *Res:* Nutritional problems of forest trees, particularly nitrogen metabolism and forest soil-plant relationships; environmental effects of forest practices, biomass energy production; bioenergy conversions; energy conservation. *Mailing Add:* Ctr Biomass Energy Systs Univ Fla-IFAS Gainesville FL 32611

SMITH, WAYNE HOWARD, b Pittsburgh, Pa, July 18, 46; m 65; c 3. ELECTROCHEMISTRY. *Educ:* Univ Pittsburgh, BS, 71; Univ Tex, Austin, PhD(chem), 74. *Prof Exp:* Fel, Calif Inst Technol, 74-76; asst prof chem, Tex Tech Univ, 76-83; STAFF MEM, LOS ALAMOS NAT LAB, 83- *Concurrent Pos:* Consult, Monogram Indust, 75-76, Westvaco, 79- & Mikro Environ Lab, 80- *Mem:* Am Chem Soc; Electrochem Soc. *Res:* Electroorganic synthesis; homogeneous transition metal catalysis via electrochemically generated organometallic; kinetics and mechanisms of reactions initiated electrochemically. *Mailing Add:* Los Alamos Nat Lab Mail Stop E 501 Los Alamos NM 87545

SMITH, WAYNE LEE, b Oneonta, NY, Jan 29, 36; m 59; c 3. INORGANIC CHEMISTRY, PHYSICAL CHEMISTRY. *Educ:* Hartwick Col, BA, 57; Pa State Univ, PhD(chem), 63. *Hon Degrees:* MA, Colby Col, 83. *Prof Exp:* Res assoc chem, Univ Mich, 63-64; res chemist, Allied Chem Corp, 64-66; asst prof chem, Carnegie-Mellon Univ, 66-67; from asst prof to assoc prof, 67-83, chmn dept, 82-89, PROF CHEM, COLBY COL, 83- *Concurrent Pos:* Vis prof, Univ Mich, Ann Arbor, 74-75, Dartmouth Col, 81-82 & Univ NC, Chapel Hill, 89-90. *Mem:* Am Chem Soc; Royal Soc Chem. *Res:* Coordination compounds of the nontransition metal elements; organometallics; heteroborane chemistry; chemical education. *Mailing Add:* Dept Chem Colby Col Waterville ME 04901

SMITH, WENDELL VANDERVORT, b Caldwell, Idaho, Apr 16, 12; m 38; c 3. PHYSICAL CHEMISTRY. *Educ:* Col Idaho, BS, 33; Univ Calif, PhD(phys chem), 37. *Prof Exp:* Res chemist, Gen Labs, US Rubber Co, 37-59 & Res Ctr, 59-72; res chemist corp res & develop, Oxford Mgt & Res Ctr, Uniroyal Inc, 72-77; RETIRED. *Mem:* Am Chem Soc. *Res:* Ionic entropies; new rubber products; theory of emulsion polymerization; physical properties of rubbers; radiation chemistry of polymers. *Mailing Add:* Three Nettleton Ave Newton CT 06470

SMITH, WENDY ANNE, b Pittsburgh, Pa, July 12, 54. CELLULAR ENDOCRINOLOGY, DEVELOPMENTAL BIOLOGY. *Educ:* New Col, Sarasota, Fla, BA, 75; Duke Univ, PhD(zool), 81. *Prof Exp:* Teaching fel, dept zool, Duke Univ, 77-81; NRSA res fel, dept pharmacol, Duke Univ, 81-83 & dept biol, Univ NC, Chapel Hill, 83-85; ASST PROF PHYSIOL, DEPT BIOL, NORTHEASTERN UNIV, BOSTON, 85- *Concurrent Pos:* Prin investr, NIH, dept biol, Northeastern Univ, Boston, 86. *Mem:* Am Soc Zoologists; Sigma Xi; Soc Cell Biol; Soc Neurosci. *Res:* Regulation of endocrine cell function in insects and of insect molting and metamorphosis; cellular mechanisms of action of the cerebral molt-stimulating peptide; prothoracicotropic hormone. *Mailing Add:* Dept Biol Northeastern Univ 414 Mugar Hall Boston MA 02115

SMITH, WESLEY R, b Allentown, Pa, Nov 5, 28; m 55; c 6. FLUID PHYSICS, MOLECULAR PHYSICS. *Educ:* Lehigh Univ, BS, 50, MS, 51; Princeton Univ, PhD(physics), 57. *Prof Exp:* Instr, Princeton Univ, 56-58; from asst prof to assoc prof, 58-74, PROF PHYSICS, LEHIGH UNIV, 74- *Mem:* AAAS; Am Phys Soc; Sigma Xi. *Res:* Application of shock tubes to measurements of chemical and physical properties of gases, liquids and solids; studies of acoustic waves in solids. *Mailing Add:* Dept of Physics Lehigh Univ Bethlehem PA 18015

SMITH, WILBUR S, engineering; deceased, see previous edition for last biography

SMITH, WILLARD NEWELL, b Wellington, Kans, Jan 27, 26. CELL PHYSIOLOGY. *Educ:* Univ Md, BS, 50, MS, 53, PhD, 66. *Prof Exp:* Asst parasitologist, Ga Exp Sta, 54-57; parasitologist, Animal Dis & Parasite Res Div, USDA, 57-62 & Sch Dent, Univ Md, 62-66; assoc prof dent, Univ Tex Dent Sci Inst, Houston, 66-81; RETIRED. *Concurrent Pos:* Nat Inst Dent Res spec fel, Univ Tex Dent Br Houston, 67-70. *Mem:* Int Asn Dent Res; Am Soc Parasitol; Am Soc Microbiol. *Res:* Isolation and characterization of virus-like particles from oral microorganisms; parasites of South American primates. *Mailing Add:* 5206 Sanford Rd Houston TX 77035

SMITH, WILLIAM ADAMS, JR, b Parkersburg, WVa, July 13, 29; c 5. QUALITY ENGINEERING & ASSURANCE, INTEGRATED INFORMATION SYSTEM DEVELOPMENT. *Educ:* Naval Acad, BS, 51; Lehigh Univ, MS, 57; NY Univ, DEngSc, 66. *Prof Exp:* Instr indust eng, Lehigh Univ, 55-57, dir, Comput Lab, 57-67, prof indust eng, 67-73; prof & head, Dept Indust Eng, 73-82, dir, productivity res & extension prog, 75-84, PROF INDUST ENG & COORDR, ADVAN PROG DEVELOP INDUST EXTEN, NC STATE UNIV, 84- *Concurrent Pos:* Alcoa Professorship, Lehigh Univ, 68-69; Ford Found residency eng, Am Soc Eng Educ, Smith Kline Corp, 69-70; consult, IBM, Air Prod, du Pont, Western Elec Co, Corning, Gen Elec & Northern Telecom; chmn, Asn Coop Eng, 79-80, Pub Affairs Coun, Am Asn Eng Socs, 83-85 & Nat Productivity Network, 84-85; advan automation engr, Northern Telecom Integrated Network Systs, 84-86; pres, NC Qual Leadership Found, 89. *Mem:* Fel Am Inst Indust Engrs (pres, 75-76); Sigma Xi; Inst Mgt Sci; Am Soc Eng Educ; Soc Mfg Engrs; Am Soc Qual Control. *Res:* Management systems engineering; source data automation; electronics assembly and test performance; organizational productivity and quality measurement; total quality management; technology management. *Mailing Add:* Box 7906 NC State Univ Raleigh NC 27695-7906

SMITH, WILLIAM ALLEN, b Ashland, Ky, June 26, 40; m 66; c 2. MATHEMATICS, NUMERICAL ANALYSIS. *Educ:* Mass Inst Technol, BS, 62, PhD(math), 66. *Prof Exp:* Asst prof math, Univ SC, 66-70; ASSOC PROF MATH, GA STATE UNIV, 70- *Mem:* Math Asn Am. *Res:* difference equations; numerical analysis. *Mailing Add:* Dept Math & Comp Sci Ga State Univ Atlanta GA 30303-3083

SMITH, WILLIAM BOYCE, b Port Arthur, Tex, Sept 7, 38; m 63; c 3. MATHEMATICAL STATISTICS. *Educ:* Lamar Univ, BS, 59; Tex A&M Univ, MS, 60, PhD(statist), 67. *Prof Exp:* Asst prof math, Lamar State Col, 62-64; from asst prof to assoc prof, 66-73, asst dean col sci, 72-77, assoc dean col sci, 84-85, head dept, 77-86, PROF STATIST, TEX A&M UNIV, 73- *Concurrent Pos:* Vis prof, Southern Methodist Univ, 70, Nat Agr Exp Sta, Argentina, 77 & 87; vis scholar, Japanese Soc for Prom Sci, 80 & 86; invited prof, Ecole Nat Supiemredes Télécommunications, Paris France, 87. *Honors & Awards:* Hartley Award, 82. *Mem:* Biomet Soc; Am Statist Asn; Math Asn Am; Int Statist Inst. *Res:* Statistical estimation theory with incomplete observation vectors; legal statistics methods; multivariate analysis. *Mailing Add:* 1040 Rose Circle College Station TX 77840

SMITH, WILLIAM BRIDGES, b Washington, DC, Feb 13, 44; m 65; c 2. COMPUTER SCIENCE, ELECTRICAL ENGINEERING. *Educ:* Univ Md, BS, 62; Princeton Univ, MS, 63; Univ Pa, PhD(elec eng), 67. *Prof Exp:* mem tech staff, Prog Design, AT&T Bell Labs, 62-67, supvr, No 4 Electronic Switching Syst Design, 67-70, dept head, toll network studies, 70-74, dir, Opers Systs Dev, 74-78, exec dir, No 5 Electronic Switching Syst Div, 79-82, exec dir, Opers Technol Div, 86-90, EXEC DIR, COMMUN SERV NETWORK DIV, AT&T BELL LABS, 91-; GEN DIR, ITT EUROPE. *Concurrent Pos:* Instr, Ill Inst Technol Grad Sch, 68-70; bd of overseers, Armor Col Eng, Ill Inst Technol, 80-82; Gov Thompson's Technol Task Force, 81; vpres & gen tech dir, ITT, Europe, 82-86. *Mem:* Sr mem Inst Elec & Electronic Engrs. *Res:* Management of large software and hardware systems development; telecommunications networks. *Mailing Add:* Rm 1A380 Bell Labs Red Hill Rd Middletown NJ 07748

SMITH, WILLIAM BURTON, b Muncie, Ind, Dec 13, 27; m 53; c 2. SYNTHETIC ORGANIC & NATURAL PRODUCT CHEMISTRY. *Educ:* Kalamazoo Col, BA, 49; Brown Univ, PhD(chem), 54. *Prof Exp:* Res assoc chem, Fla State Univ, 53-54 & Univ Chicago, 54-55; from asst prof to assoc prof, Ohio Univ, 55-61; prof chem & chmn dept, Tex Christian Univ, 61-81; RETIRED. *Concurrent Pos:* Partic fel, Oak Ridge Assoc Univs, 55-; Welch vis scientist, Tex Christian Univ, 60-61; vis prof, Univ Sussex, UK, 81. *Honors & Awards:* W T Doherty Award, Am Chem Soc, 90. *Mem:* Am Chem Soc; Sigma Xi. *Res:* Physical organic chemistry of carbonium ions and free radicals; nuclear magnetic resonance; synthesis of biomimetic molecules. *Mailing Add:* Dept Chem Tex Christian Univ Ft Worth TX 76129

SMITH, WILLIAM CONRAD, b Cisco, Tex, May 20, 37; m 59; c 2. PHYSICS. *Educ:* NTex State Univ, BS, 60, MS, 62; Iowa State Univ, PhD(physics), 71. *Prof Exp:* Instr physics & math, Decatur Baptist Col, 61-62; asst prof physics, Howard Payne Col, 62-64 & Mankato State Col, 70-73; from asst prof to assoc prof physics, Tex Womans Univ, 73-83. *Mem:* Am Asn Physicists in Med; Am Phys Soc; Am Asn Physics Teachers. *Res:* Hyperfine fields in magnetic metallic compounds; nuclear magnetic resonance; effect of magnetic fields on axon signals; design of microprocessor-based instruments. *Mailing Add:* 705 Sundown Ct Las Cruces NM 88001

SMITH, WILLIAM EDGAR, experimental pathology, for more information see previous edition

SMITH, WILLIAM EDMOND, b Wilmington, NC, Nov 16, 39; m 67; c 4. PULP CHEMISTRY, PAPER CHEMISTRY. *Educ:* NC State Univ, MS, 65, PhD(wood & paper sci), 69; Univ SC, MBA, 73. *Prof Exp:* Res forest prod technologist, US Forest Prod Lab, 64-69; sr res chemist, Res Lab, 69-72, dir tech serv, Paper Div, Sonoco Prod Co, 72-85; VPRES & TECH DIR, EZE PROD INC, 85- *Mem:* Tech Asn Pulp & Paper Indust; Soc Wood Sci & Technol. *Res:* Product development; stress analysis of structures produced from paper and plastics; basic failure criteria of materials; process control; secondary fiber containment dispersion, paper machine press section optimization. *Mailing Add:* EZE Prod Inc PO Box 5744 Greenville SC 29606

SMITH, WILLIAM EDWARD, b Philadelphia, Pa, May 30, 38; m 63; c 2. INDUSTRIAL CHEMICAL ENGINEERING. *Educ:* La Salle Col, BS, 65; Purdue Univ, Lafayette, PhD(chem), 69. *Prof Exp:* NIH fel, Mass Inst Technol, 69-70; res chemist, Gen Elec Res & Develop Corp, 70-74, mgr, Catalytic Processes Unit, 74-79, mgr, Chem Eng Br, 79-83, mgr, Inorg Mat Lab, 83-85, mgr, Phys Chem Lab, 86-88; DIR CHEM RES, POLAROID CORP, 88- *Mem:* Am Chem Soc; Catalysis Soc. *Res:* Materials and process research and development; homogeneous and heterogeneous catalysis; monomer synthesis. *Mailing Add:* 117 Grove St Wellesley MA 02181-7803

SMITH, WILLIAM FORTUNE, b Vancouver, BC, Oct 11, 31; US citizen; m 58; c 3. MATERIALS SCIENCE, PHYSICAL METALLURGY. *Educ:* Univ BC, BA, 52; Purdue Univ, MS, 55; Mass Inst Technol, ScD(phys metall), 68. *Prof Exp:* Res engr, Metall Res Labs, Reynolds Metals Co, Va, 57-62; res engr, Metall Res Labs, Kaiser Aluminum Co, Washington, 65-67; assoc prof, 68-71, PROF ENG, FLA TECHNOL UNIV, 71- *Mem:* Am Soc Metals; Am Inst Mining, Metall & Petrol Engrs. *Res:* Precipitation reactions in the solid state; physical metallurgy of aluminum alloys; stress corrosion cracking. *Mailing Add:* 1150 Willa Vista Trail Maitland FL 32751

SMITH, WILLIAM GRADY, b Dover, Ark, Mar 29, 37; m 59; c 3. BIOCHEMISTRY. *Educ:* Univ Ark, BS, 59, MS, 60; Okla State Univ, PhD(biochem), 64. *Prof Exp:* Asst prof biochem & path, 64-66, from asst prof to assoc prof biochem, 66-77, PROF BIOCHEM, SCH MED, UNIV ARK, LITTLE ROCK, 77- *Concurrent Pos:* Fel biochem, Univ Minn, 63-64; res grants, NSF, 65-75 & NIH, 67-70; Lederle med fac award, 68-71. *Mem:* Am Soc Biol Chem. *Res:* Amino acid metabolism; metabolic control. *Mailing Add:* Dept Biochem Univ Ark Col Med Med Sci 4301 W Markham Little Rock AR 72201

SMITH, WILLIAM H, b Kingston, Okla, Jan 25, 29; m 54; c 2. ANIMAL SCIENCE. *Educ:* Okla State Univ, BS, 56; Purdue Univ, MS, 57, PhD(animal nutrit), 59. *Prof Exp:* Prof animal nutrit & mgt, Purdue Univ, West Lafayette, 59-76; RESIDENT DIR RES, TEX A&M UNIV, 76- *Concurrent Pos:* Sabbatical, Univ Calif, Davis, 66. *Mem:* Am Soc Animal Sci; Sigma Xi; Am Soc Agron. *Res:* Nutrient requirements of beef cows; value of cornstalks; methods of preventing grass tetany; value of liquid supplements; total digestible nutrients for heifers; value of forage quality. *Mailing Add:* 3305 Ridgemont St Irving TX 75062

SMITH, WILLIAM HAYDEN, b Paducah, Ky, Oct 19, 40; m 65. ASTRONOMY. *Educ:* Univ Ky, BS, 62; Princeton Univ, MA, 63, PhD(phys chem), 66. *Prof Exp:* Res assoc chem physics, Princeton Univ, 66; res assoc surface physics, Univ Ky, 66-67; res assoc astrophys, 68, res staff, 69-73, res astronr, Princeton Univ Observ, 73-76; PROF DEPT CHEM & DEPT EARTH & PLANETARY SCI, WASH UNIV, 76- *Concurrent Pos:* NATO sr fel, Inst Physics, Stockholm, 71; NSF res grant, Princeton Univ Observ, 71-; NATO sr fel, 75; vis prof, Inst Astron, Univ Hawaii, 75-76; fel, McDonnell Ctr Space Sci, 75- *Mem:* Am Chem Soc; Am Phys Soc; Am Astron Soc; Int Astron Union. *Res:* Planetary atmospheres; atomic and molecular physics. *Mailing Add:* Dept Chem & Earth Planetary Sci Wash Univ One Brookings Dr PO Box 1134 St Louis MO 63130

SMITH, WILLIAM HULSE, b Trenton, NJ, May 9, 39; m 63, 83; c 3. PLANT PATHOLOGY. *Educ:* Rutgers Univ, BS, 61, PhD(plant path), 65; Yale Univ, MF, 63. *Prof Exp:* Asst prof forestry, Rutgers Univ, 64-66; from asst prof to assoc prof, 66-78, from asst dean to actg dean, Sch Forestry & Environ Studies, 71-83, PROF FOREST BIOL, YALE UNIV, 79- *Mem:* AAAS; Am Phytopath Soc; Ecol Soc Am; Soc Am Foresters; Sigma Xi. *Res:* Chemistry and biology of the rhizosphere; influence of gaseous and particulate air contaminants on woody plant health. *Mailing Add:* Sch Forestry & Environ Studies Yale Univ New Haven CT 06511

SMITH, WILLIAM K, b Danville, Pa, Mar 8, 20; m 47; c 2. MATHEMATICS. *Educ:* Bucknell Univ, AB, 41, MA, 46; Univ Mich, PhD(math), 53. *Prof Exp:* Instr math, Bucknell Univ, 41-42 & 46-47, asst prof, 51-56; assoc prof, Antioch Col, 56-57; prof, Bucknell Univ, 57-64, New Col, 64-66 & Hobart & William Smith Cols, 66-67; prof math, New Col, Fla, 67-76; prof math & chmn dept, Ill Wesleyan Univ, 76-; RETIRED. *Mem:* Am Math Soc; Math Asn Am. *Res:* Functional analysis. *Mailing Add:* 736 Forestview Dr Sarasota FL 34232-2456

SMITH, WILLIAM KIRBY, b Greensboro, NC, Mar 6, 47; m 80; c 2. BIOPHYSICAL ECOLOGY, ENVIRONMENTAL PHYSIOLOGY. *Educ:* San Diego State Col, BS, 67, MS, 71; Univ Calif, Los Angeles, PhD, 77. *Prof Exp:* ASST PROF, DEPT BOT, UNIV WYO, 77- *Mem:* Am Soc Plant Physiologists; Ecol Soc Am; AAAS; Am Inst Biol Sci; Bot Soc Am. *Res:* Biophysical and physiological ecology; plant and animal adaptations in harsh or unusual environments; photosynthesis, water relations and growth physiology. *Mailing Add:* Dept Bot Univ Wyo PO Box 3165 Laramie WY 82071

SMITH, WILLIAM LEE, b Providence, RI, June 3, 22; m 57; c 3. EXPLORATION GEOLOGY, REMOTE SENSING. *Educ:* Columbia Univ, AB, 49; Rutgers Univ, MS, 51; Pac Western Univ, PhD, 82. *Prof Exp:* Petrologist, US Geol Surv, Washington, DC, 52-56; prin geologist, Battelle Mem Inst, 56-67; geologist, Bellcomm, Inc, Washington, DC, 67-72; geologist, Syst Planning Corp, 72-75 & ERIM, Washington, DC, 75-78; vpres, Spectral Data Corp, 78-88; MGR, REMOTE SENSING APPLICATIONS, DECISION SCI APPLICATIONS, INC, 88- *Mem:* Am Soc Photogram. *Res:* Mineralogy of ores of radioactive and rarer elements; economic geology; remote sensing for mineral deposits; systems analysis and planning for earth resources observation from satellites. *Mailing Add:* Decision Sci Applications Suite 400 1110 N Glebe Rd Arlington VA 22201

SMITH, WILLIAM LEE, b Tulsa, Okla, Oct 28, 45; m 68; c 3. BIOCHEMISTRY. *Educ:* Univ Colo, Boulder, BA, 67; Univ Mich, Ann Arbor, PhD(biol chem), 71. *Prof Exp:* NIH fel biochem, Univ Calif, Berkeley, 71-74; sr scientist biochem, Mead Johnson & Co, Bristol-Myers Co, 74-75; asst prof, 75-79, PROF BIOCHEM, MICH STATE UNIV, 79- *Concurrent Pos:* Estab investr, Am Heart Asn; adj prof physiol, Mich State Univ, 84- *Mem:* AAAS; Am Soc Biol Chemists. *Res:* Regulation of prostaglandin metabolism; mechanism of prostaglandin action; prostaglandins; kidney. *Mailing Add:* Dept Biochem Mich State Univ East Lansing MI 48824

SMITH, WILLIAM MAYO, b Fredericksburg, Va, Nov 30, 17; m 40; c 4. ENVIRONMENTAL SCIENCES. *Educ:* Va Mil Inst, BS, 38; Univ Ala, MS, 41; Univ Md, PhD(org chem), 46. *Prof Exp:* Cellulose chemist, Sylvania Indust Corp, Va, 38-43; instr chem, Univ Md, 43-46; sr res chemist, Firestone Tire & Rubber Co, 47-54, group leader, Defense Res Div, 54-56; asst dir res & develop, Escambia Chem Corp, Conn, 56-58, vpres & dir res, 58-67; vpres, Air Reduction Co, Inc, NY, 67-69; tech dir polymers & plastics, Air Prod & Chem, Inc, 69-71, group res & develop coordr, Chem Group, 71-79, dir sci affairs, 79-81; RETIRED. *Concurrent Pos:* Consult, Chem, Plastics, Polymers, Environ Health & Safety, Cancer & Environ, 81- *Mem:* Am Chem Soc; Chem Soc London; NY Acad Sci; Chem Indust Inst Toxicol; Indust Res Inst. *Res:* Plastics; polymerizations; petrochemicals; organic reactions; chemical carcinogens and toxic substances; industrial research management; health and safety. *Mailing Add:* 19 Painted Bunting Amelia Island FL 12034

SMITH, WILLIAM NOVIS, JR, b Chicago, Ill, May 21, 37; m 58. ORGANIC CHEMISTRY, INORGANIC CHEMISTRY. *Educ:* Mass Inst Technol, BS, 59; Univ Calif, Berkeley, PhD(org chem), 63. *Prof Exp:* Res chemist, Org Chem Dept, E I du Pont de Nemours & Co, 62-64; res assoc chem, Foote Mineral Co, 64-71, mgr chem res, 72-74; asst to dir, Eastern Res Ctr, Stauffer Chem Co, 74-76; asst dir corp res, Air Prod, 76-77, asst dir contract res, 77-79; mgr develop progs, Reentry Systs Div, Gen Elec Co, 80-82; PRES, R K CARBON FIBERS, 82-; PRES, AM HYPERFORM, 82- *Mem:* Am Chem Soc; Soc Plastics Indust; Inst Elec & Electronic Engrs; Am Inst Chem Engr; Am Ceramic Soc; Soc Advan Mat & Process Eng. *Res:* Catalysis, inorganic and organic lithium chemistry; organometallic chemistry; extractive metallurgy, catalysts, polymers, anionic polymerization; polyolefin catalysts; composites; armor; textiles; coatings; 42 US patents, 4 books, and 12 technical publications. *Mailing Add:* 412 S Perth Philadelphia PA 19147-1322

SMITH, WILLIAM OGG, b Shawnee, Okla, July 17, 25; m 48; c 3. MEDICINE. *Educ:* Harvard Med Sch, MD, 49. *Prof Exp:* Intern, Univ Chicago, 49-50; resident med, Vet Admin Hosps, Boston, 52-54 & Oklahoma City, 54-55; chief resident & clin asst med, 55-56, from instr to assoc prof, 56-66, vchmn dept med, 67-75, PROF MED, MED SCH, UNIV OKLA, 66- *Concurrent Pos:* Asst dir radioisotope serv, Vet Admin Hosp, Oklahoma City, 56-60, from assoc chief to chief med serv, 60-71. *Mem:* Fel Am Col Physicians; Soc Exp Biol & Med; Am Soc Nephrol. *Res:* Renal and electrolyte physiology; magnesium metabolism. *Mailing Add:* Dept of Med Smith/111A 921 NE 13th St Univ of Okla Oklahoma City OK 73104

SMITH, WILLIAM OWEN, b Louisville, Ky, Sept 2, 41; m 64; c 2. PHOTOBIOLOGY. *Educ:* Univ Ky, BS, 67, PhD(plant physiol), 75. *Prof Exp:* PLANT PHYSIOLOGIST, RADIATION BIOL LAB, SMITHSONIAN INST, 75- *Mem:* Am Soc Plant Physiologists; Am Soc Photobiol. *Res:* Molecular aspects of plant photomorphogenesis; biochemistry of phytochrome. *Mailing Add:* 2817 Rhoderick Rd Frederick MD 21701

SMITH, WILLIAM R, b Lyman, Okla, June 26, 25. STRUCTURAL DYNAMICS, MATHEMATICS. *Educ:* Bethany Nazarene Col, BA, 48; Wichita State Univ, MA, 50; Univ Calif, Los Angeles, PhD(biophys), 67. *Prof Exp:* Engr, Beech Aircraft Corp, 51-53; sr group engr, McDonnell Aircraft Corp, 53-60; asst prof math & physics, Pasadena Col, 60-62; sr engr, Lockheed Aircraft Corp, 62-63; sr engr-scientist, McDonnell Douglas Aircraft Corp, 66-71; teacher math, Glendale Col, Calif, 72; asst prof math & physics, Mount St Mary's Col, 72-73; tech staff, Rockwell Int Corp, 73-86; CDI Corp, 86-88; SR ENGR SCIENTIST, MCDONNELL DOUGLAS AIRCRAFT CORP, 88- *Honors & Awards:* Cert Recognition Award, NASA, 82. *Mem:* AAAS; Sigma Xi; NY Acad Sci; Am Inst Aeronaut & Astronaut. *Res:* Mathematical analysis of the electrical activity of the brain; mathematical modeling; time series analysis; engineering dynamics; digital signal processing; image processing. *Mailing Add:* 2405 Roscomare Rd Los Angeles CA 90077-1839

SMITH, WILLIAM ROBERT, b San Antonio, Tex, Jan 11, 35; m 63. NUCLEAR PHYSICS. *Educ:* Univ Tex, BS, 57, BA, 58, PhD(physics), 63. *Prof Exp:* Res assoc nuclear physics, Nuclear Physics Lab, Univ Tex, 63 & Neutron Physics Div, Oak Ridge Nat Lab, 63-65; sr res officer, Nuclear Physics Lab, Oxford Univ, 65-66; res assoc nuclear physics, Nuclear Physics Lab, Univ Southern Calif, 66; assoc prof physics, Trinity Univ, Tex, 67-83; LAND DEVELOP, 83- *Concurrent Pos:* Ed low energy nuclear physics, Comput Physics Commun, 68- *Mem:* Am Phys Soc. *Res:* Low energy nuclear reaction theory. *Mailing Add:* 63 E Craig St San Antonio TX 78212

SMITH, WILLIAM ROBERT, modeling, numerical analysis, for more information see previous edition

SMITH, WILLIAM RUSSELL, b Denton, Tex, Jan 13, 17; m 39; c 2. MICROBIOLOGY. *Educ:* NTex State Univ, BS, 37, MS, 38; Univ Tex, PhD(bact), 55. *Prof Exp:* From asst prof to prof, 46-73, REGENTS PROF BIOL, LAMAR UNIV, 73- *Concurrent Pos:* Res scientist, Univ Tex, 51-53. *Mem:* Am Soc Microbiol; Sigma Xi; NY Acad Sci. *Res:* Chemical and heat activation of bacterial spores; bacteriology of foods; medical bacteriology; general, food and medical microbiology. *Mailing Add:* 4785 Dellwood Lane Beaumont TX 77706

SMITH, WILLIAM S, b Greenwich, Conn, July 29, 18; m; c 4. ORTHOPEDIC SURGERY. *Educ:* Univ Mich, AB, 40, MD, 43; Am Bd Orthop Surg, dipl, 53. *Prof Exp:* Instr orthop surg, Univ Mich, 50; from instr to prof, Ohio State Univ, 52-63; PROF ORTHOP SURG, UNIV MICH, ANN ARBOR, 63- *Concurrent Pos:* Res grants, Easter Seal Found, 56-59, Orthop Res & Educ Found, 60 & NIH, 61-63. *Mem:* Fel Am Acad Orthop Surg; Orthop Res Soc; Clin Orthop Soc; Am Orthop Asn; Am Asn Surg of Trauma. *Res:* Congenital dislocation of the hip. *Mailing Add:* 1405 E Ann St Ann Arbor MI 48109

SMITH, WILLIAM WALKER, b Duncan, Okla, Sept 26, 40; div; c 3. MATHEMATICS. *Educ:* Southeastern State Col, BS, 61; La State Univ, MS, 63, PhD(math), 65. *Prof Exp:* From asst prof to assoc prof, 65-79, chmn dept, 76-81, PROF MATH, UNIV NC, CHAPEL HILL, 79- *Mem:* Math Asn Am; Am Math Soc; Nat Coun Teachers Math. *Res:* Algebra; commutative ring and ideal theory; mathematics education. *Mailing Add:* Dept Math Univ NC Chapel Hill NC 27514

SMITH, WILLIAM WARD, medical entomology, economic entomology; deceased, see previous edition for last biography

SMITH, WILLIS DEAN, b Ipava, Ill, Aug 5, 42; m 66. SOLID STATE PHYSICS. *Educ:* Bradley Univ, BS, 64; Washington Univ, MA, 66, PhD(physics), 70. *Prof Exp:* Tech staff mem physics, Sandia Labs, NMex, 70-75; sci staff mem, Comt Sci & Technol, US House Rep, 75-77; prof staff mem, Comt on Energy & Natural Resources, US Senate, 77-; MGR STRATEGIC PLANNING, BOEING AEROSPACE CO. *Concurrent Pos:* Energy consult, Comt Interior & Insular Affairs, US Sen, 74-75; Inst Elec & Electronics Engrs cong sci fel, 74-75. *Mem:* AAAS; sr mem Inst Elec & Electronics Engrs. *Res:* Ultrasonics; electron-phonon interactions; surface waves; ferroelectric ceramics; information processing and display; photoconductors; electrooptics and photovoltaic materials and devices; solar energy. *Mailing Add:* Strategic Planning Dept Boeing Aerospace Co MS 84-22, Box 3999 Seattle WA 98006

SMITH, WINFIELD SCOTT, b Detroit, Mich, Nov 1, 41; m 63. OPTICS. *Educ:* Oakland Univ, BA, 63; Univ Ariz, MS, 67, PhD(physics), 70. *Prof Exp:* Asst prof physics, Univ Nev, Reno, 70-71; res assoc, Univ Ariz, 70-71, prog mgr, Optical Sci Ctr, 72-79; CONTRAVES-GOERZ CORP, PITTSBURGH, 79- *Mem:* Optical Soc Am; Sigma Xi. *Res:* Atmospheric optics; optical testing and fabrication. *Mailing Add:* 727 Old Mill Rd Pittsburgh PA 15238

SMITH, WINTHROP WARE, b New York, NY, Aug 4, 36; m 65; c 1. EXPERIMENTAL ATOMIC PHYSICS. *Educ:* Amherst Col, BA, 58; Mass Inst Technol, PhD(physics), 63. *Prof Exp:* Nat Acad Sci-Nat Res Coun res assoc physics, Nat Bur Standards, 63-65; instr, Columbia Univ, 65-66, asst prof, 66-69; assoc prof, 69-75, PROF PHYSICS, UNIV CONN, 75- *Concurrent Pos:* Mem, Joint Inst Lab Astrophys, Boulder, Colo, 63-65, vis fel, 75-76; lectr physics, Univ Colo, 64-65; res partic, Oak Ridge Nat Lab, 69- *Mem:* AAAS; fel Am Phys Soc. *Res:* Beam-foil spectroscopy; low energy nuclear physics; atomic physics and collisions; atomic hyperfine structure and lifetimes of excited states; laser spectroscopy. *Mailing Add:* Dept Physics Univ Conn Storrs CT 06269

SMITH, WIRT WILSEY, b Colorado Springs, Colo, Nov 2, 20; m 50; c 2. MEDICINE. *Educ:* Rice Inst, BA, 42; Univ Tex, MD, 51. *Prof Exp:* Anal chemist, Dow Chem Co, 42-44; intern, Univ Wis Hosps, 51-52; resident path, Univ Tex M D Anderson Hosp & Tumor Inst, 52-56; assoc exp surg, 58-59, asst prof, 59-65, ASSOC PROF EXP SURG, SCH MED, DUKE UNIV, 65- *Concurrent Pos:* Hite fel, Univ Tex M D Anderson Hosp & Tumor Inst, 53-56; USPHS fel, Sch Med, Duke Univ, 56-57; Nat Inst Neurol Dis & Blindness clin trainee, 57-58; mem, hyperbaric prog proj, Duke Univ, Med Ctr, 63- *Mem:* AAAS; Am Fedn Clin Res. *Res:* Hyperbaric oxygenation and physiology. *Mailing Add:* Dept Surg Duke Univ Med Ctr Box 3823 Durham NC 27706

SMITHBERG, MORRIS, b Brooklyn, NY, Aug 28, 24; m 54; c 3. EMBRYOLOGY, NEUROANATOMY. *Educ:* Univ Rochester, PhD(zool), 53. *Prof Exp:* Asst biol, Univ Rochester, 48-52; fel, Jackson Mem Lab, 52-57; asst prof anat, Univ Fla, 57-60; from asst prof to assoc prof, 60-69, actg head dept, 75-77, PROF ANAT, UNIV MINN, MINNEAPOLIS, 69- *Mem:* Am Asn Anat. *Res:* Development in frogs; pregnancy in prepubertal mice; teratology in mice and fish. *Mailing Add:* Dept Anat Univ Minn Minneapolis MN 55455

SMITHCORS, JAMES FREDERICK, veterinary medicine, for more information see previous edition

SMITHER, ROBERT KARL, b Buffalo, NY, July 18, 29; m 55; c 3. MATERIAL SCIENCE, X-RAY OPTICS SYNCHROTRON RADIATION. *Educ:* Univ Buffalo, BA, 51; Yale Univ, MS, 52, PhD(physics), 56. *Prof Exp:* Instr physics, Yale Univ, 55-56; PHYSICIST, ARGONNE NAT LAB, 56- *Mem:* Am Phys Soc; AAAS; Sigma Xi; Am Archaeol Soc; Am Astrophys Soc. *Res:* Structure of new materials using Synchrotron radiation. *Mailing Add:* 537 N Washington St Hinsdale IL 60521

SMITHERMAN, RENFORD ONEAL, b Randolph, Ala, Aug 26, 37; m 59; c 1. FISH BIOLOGY. *Educ:* Auburn Univ, BS, 59, PhD(fisheries), 64; NC State Col, MS, 61. *Prof Exp:* Asst fish cult, Auburn Univ, 61-64; leader, La Coop Fishery Unit, US Bur Sportfish & Wildlife, 64-67; coordr fisheries res, 67-72, assoc prof fisheries & allied aquacult, 72-77, PROF AQUACULT, AUBURN UNIV, 77- *Concurrent Pos:* Chief party, AID-Auburn-Univ-Repub of Panama Aquacult Proj, 72-73. *Mem:* World Maricult Soc; Catfish Farmers of Am; Am Fisheries Soc. *Res:* Fish genetics, hybridization, pathology and ecology; biological weed control with fishes; crawfish ecology; polyculture of fishes; aquaculture. *Mailing Add:* Dept Fisheries & Allied Aquacult Auburn Univ Auburn AL 36830

SMITH-EVERNDEN, ROBERTA KATHERINE, b Los Angeles, Calif. MICROPALEONTOLOGY, SOIL CONSERVATION. *Educ:* Univ Alaska, BA, 57; Univ Calif, Berkeley, MA, 60; Univ BC, PhD(geol), 66. *Prof Exp:* Geologist, Smithsonian Inst, 65-73; asst prof lectr, George Washington Univ, 67-68; asst prof, Howard Univ, 68-70; LECTR & RES ASSOC, UNIV CALIF, SANTA CRUZ, 75-, CONSULT GEOL, 75- *Concurrent Pos:* Mem, tech adv comt, Calif Bd Forestry. *Mem:* Soc Econ Paleont & Mineralogists; Soc Woman Geogrs; Sigma Xi; Asn Eng Geologists; AAAS; NAm Micropaleont Soc; Asn Women Geoscientists. *Res:* Ecology and paleoecology of living and fossil benthonic and planktonic foraminifera; Tertiary biostratigraphy; environmental geology; mass wasting and soil loss. *Mailing Add:* Inst Marine Sci Univ of Calif Santa Cruz CA 95060

SMITH-GILL, SANDRA JOYCE, b Chicago, Ill, Jan 8, 44; m 67. DEVELOPMENTAL BIOLOGY, ENDOCRINOLOGY. *Educ:* Univ Mich, BS, 65, MS, 66, PhD(zool), 71. *Prof Exp:* Asst prof biol, Swarthmore Col, 71-74 & George Washington Univ, 74-76; vis asst prof, 76-77, assoc prof zool, Univ Md, 77-87; MICROBIOLOGIST, NAT CANCER INST, 87- *Concurrent Pos:* Microbiologist, Nat Cancer Inst, 80-87. *Mem:* Int Soc Develop Biol; Soc Develop Biol; Int Pigment Cell Soc; Am Soc Zoologists; Tissue Cult Asn. *Res:* Protein-protein interactions; genetic and structural basis of antibody recognition of protein antigens; antigenic structure of the mycoprotein. *Mailing Add:* Lab Genetics NCI Bldg 37 Rm 2810 Bethesda MD 20892

SMITHIES, OLIVER, b Halifax, Eng, July 23, 25; US citizen. PATHOLOGY. *Educ:* Oxford Univ, Eng, PhD(biochem), 51. *Prof Exp:* Postdoctoral fel phys chem, Univ Wis-Madison, 51-53; res asst & assoc, Connaught Med Res Lab, Toronto, Can, 53-60; from asst prof to prof genetics & med genetics, Univ Wiis-Madison, 60-63, Leon J Cole prof, 71-80, Hilldale prof, 80-88; EXCELLENCE PROF PATH, UNIV NC, CHAPEL HILL, 88- *Concurrent Pos:* Merkel scholar, 61; mem, Nat Adv Med Sci Coun, NIH, 85. *Honors & Awards:* William Allen Mem Award, Am Soc Human Genetics, 64; Karl Landsteiner Mem Award, Am Asn Blood Banks, 84; Gairdner Found Int Award, 90. *Mem:* Nat Acad Sci; Am Acad Arts & Sci; Genetics Soc Am (vpres, 74, pres, 75); fel AAAS. *Res:* Targetted modification of specific genes in living animals; author of various publications. *Mailing Add:* Dept Path Univ NC Chapel Hill Chapel Hill NC 27514

SMITH-SOMERVILLE, HARRIETT ELIZABETH, b Guntersville, Ala, Jan 5, 44; m 82. CELL BIOLOGY, ULTRASTRUCTURE. *Educ:* Univ Ala, BS, 66; Univ Tex, Austin, PhD(biol sci), 70. *Prof Exp:* Fel biophysics, Univ Chicago, 70-72, res assoc physiol, 73-76; from asst prof to assoc prof, 76-86, PROF BIOL SCI, DEPT BIOL SCI, UNIV ALA, 86- *Concurrent Pos:* Teaching assoc biol, Dept Biol, Univ SFla, 73; vis asst prof, Dept Biol Sci, Univ Ill, 75-76. *Mem:* Am Soc Cell Biol; Soc Protozoologists; Am Micros Soc; AAAS; Sigma Xi. *Res:* Oral apparatus structure, food vacuole formation and membrane recycling in Tetrahymena Vorax; cell motility; microtubules; microfilaments; microfilament-membrane interactions. *Mailing Add:* Dept Biol Sci Univ Ala Box 870344 Tuscaloosa AL 34587-0344

SMITHSON, GEORGE RAYMOND, JR, b New Vienna, Ohio, Mar 2, 26; m 50; c 3. ENVIRONMENTAL SCIENCE, RESEARCH ADMINISTRATION. *Educ:* Wilmington Col, BS, 49; Miami Univ, MS, 50. *Prof Exp:* Asst chem, Miami Univ, 49-50; prof phys sci, Rio Grande Col, 50-52; prin chemist extractive metall div, Batelle-Columbus Labs, Batelle Mem Inst, 52-60, proj leader, 60-61, sr scientist, 61-65, assoc chief minerals & metall waste technol div, 68-70, chief waste control & process technol div, 70-71, asst mgr environ systs & processes dept, 71-74, mgr environ technol prog off, Environ Energy Res Dept, 74-80, mgr environ progs off, Chem Dept, 80-85; RETIRED. *Concurrent Pos:* Vpres & prog dir, Metcalf & Eddy Ohio, Inc, 85-87; vpres, Lawhon Assocs, 86-88; environ consult, 86- *Mem:* Fel AAAS; emer fel Am Inst Chem; Sigma Xi. *Res:* Waste management and control; process technology; fluidized-bed technology; thermodynamics of extractive metallurgical systems; electrowinning; electrodialysis; sorption technology. *Mailing Add:* 3068 Kingston Ave Grove City OH 43123

SMITHSON, SCOTT BUSBY, b Oak Park, Ill, Oct 28, 30; m 53; c 2. GEOPHYSICS, PETROLOGY. *Educ:* Univ Okla, BS, 54; Univ Wyo, MA, 59; Univ Oslo, DSc(petrol geophys), 63. *Prof Exp:* Asst seismologist, Shell Oil Co, 54-57; analyst, Geotech Corp, 57-58; Royal Norweg Coun Sci & Indust Res fel, 63-64; from asst prof to prof geophys, 64-77, PROF GEOL, UNIV WYO, 77- *Mem:* Geol Soc Am; Soc Explor Geophys; Am Geophys Union; Mineral Soc Am; Norweg Geol Soc. *Res:* Solid earth geophysics and petrology; structure and composition of the continental crust of the earth. *Mailing Add:* Dept Geol Univ Wyo PO Box 3006 Laramie WY 82071

SMITH-SONNEBORN, JOAN, b Albany, NY, Nov 5, 35; div; c 2. MOLECULAR BIOLOGY. *Educ:* Bryn Mawr Col, BA, 57; Ind Univ, PhD(zool, biochem), 62. *Prof Exp:* Fel biochem, Brandeis Univ, 61-62; fel virol & microbiol, Univ Calif, Berkeley, 62-64; res assoc zool, Univ Wis-Madison, 64-71; ASSOC PROF ZOOL, UNIV WYO, 71- *Mem:* Am Soc Cell Biol; fel Gerontol Soc Am; NY Acad Sci; AAAS. *Res:* Extranuclear DNA of organelles; mutagenesis and repair; cellular aging; interaction of cell components and external environment on the determination of modulation of gene expression during development and aging as well as genotoxicology; the Paramecium system of cellular aging will be used with the new biotechnology available to try to alter the aging process; transformation of these cells with DNA coding for DNA repair genes will be explored and the effect of introduced genes on normal life span determined. *Mailing Add:* Dept Zool Univ Wyo Box 3166 Laramie WY 82071

SMITH-THOMAS, BARBARA, b Palo Alto, Calif, Oct 2, 42; c 1. OPERATING SYSTEM SECURITY, DISTRIBUTED COMPUTATION. *Educ:* Reed Col, BA, 64; Carnegie-Mellon Univ, MS, 70, PhD(math), 73; Ga Inst Technol, MS, 82. *Prof Exp:* Mellon fel, Univ Pittsburgh, 72-73; from asst prof to assoc prof math, Memphis State Univ, 73-79; vis assoc prof math, Univ Ala, Birmingham, 79-80; asst prof comput sci, 82-86, assoc prof comput sci, Univ NC, Greensboro, 86-87; AT&T BELL LABS, 87- *Concurrent Pos:* Consult, AT&T Bell Labs, 84-86. *Mem:* Asn Women Math; Asn Comput Mach. *Res:* Multi-level security enhancements to Unix system; secure windowing graphics terminals to Unix system; secure distributed computing. *Mailing Add:* AT&T Guilford Ctr W3f-58 PO Box 20046 Greensboro NC 27420

SMITHWICK, ELIZABETH MARY, b Casco, Wis, Jan 20, 28. PEDIATRICS, IMMUNOLOGY. *Educ:* Univ Wis, BS, 48, MD, 55; Am Bd Allergy & Immunol, dipl, 77. *Prof Exp:* Intern, Kings Co Hosp, NY, 55-56 & Bellevue Hosp, 56-57; resident pediat, Metrop Hosp & Babies Hosp, 57-58; sr house officer, Queen Charlotte's Hosp, London, Eng, 59-60; from instr to assoc prof, State Univ NY Downstate Med Ctr, 60-73; assoc mem & assoc prof pediat, Sloan-Kettering Inst Cancer Res, Cornell Univ, 73-82; PROF PEDIAT, UNIV CALIF, DAVIS, 82- *Concurrent Pos:* USPHS grant, Southwestern Med Sch, Univ Tex Dallas, 63-64. *Mem:* AAAS; Soc Pediat Res; Am Acad Pediat; Am Rheumatism Asn. *Res:* Immune deficiency diseases; neutrophil and monocyte function; rheumatology. *Mailing Add:* Dept Pediat, Univ Calif Davis 2516 Stockton Blvd Sacramento CA 95817

SMITS, FRIEDOLF M, b Stuttgart, Ger, Nov 10, 24; US citizen; m 55; c 3. PHYSICS. *Educ:* Univ Freiburg, PhD(physics), 50. *Prof Exp:* Res assoc physics, Univ Freiburg, 50-54; mem tech staff device develop, Bell Tel Labs, 54-62; mgr dept radiation physics, Sandia Corp, 62-65; dept head, Device Develop, Bell Tel Labs, 65-68, dir, Semi- Conductor Device Lab, 68-71, Mos Tech & Memory Lab, 71-75, Integrated Circuit Support Lab, 75-86; RETIRED. *Mem:* Fel Inst Elec & Electronics Eng. *Res:* Geological age determinations; physics of semiconductor devices; physics of radiation damage in semiconductors; ultrasonic and optical memories. *Mailing Add:* 2079 Greenwood Rd Allentown PA 18103

SMITS, TALIVALDIS I(VARS), b Riga, Latvia, Sept 18, 36; US citizen; m 67. DETECTION & ESTIMATION THEORY, UNDERWATER ACOUSTICS. *Educ:* Univ Minn, BS, 58, MSEE, 62, PhD(elec eng), 66. *Prof Exp:* Res asst elec eng, Univ Minn, 58-62, res assoc, 66-67, asst prof, 66-68; assoc prof, Cath Univ Am, 68-78; mem tech staff, 77-82, sect head, 82-85, STAFF ENG, TRW, 85- *Concurrent Pos:* Assoc ed, J Acoust Soc Am, 74-86; consult, Anal Adv Group, McLean, Va, 73-75, Undersea Res Corp, Falls Church, Va, 75-76 & Planning Systs Inc, 76; lectr elec eng, Cath Univ Am, 78- *Mem:* Acoust Soc Am; Inst Elec & Electronics Engrs; Am Sci Affil. *Res:* Random signal processing; passive and active sonar-radar detection and estimation; statistical communication theory; pattern recognition; statistical methods in system analysis; non-Gaussian random processes. *Mailing Add:* 2811 Crest Ave Cheverly MD 20785-2965

SMITTLE, BURRELL JOE, b Paola, Kans, July 13, 34; m 55; c 2. ENTOMOLOGY. *Educ:* Univ Ark, BS, 55, MS, 56; Rutgers Univ, PhD(entom), 64. *Prof Exp:* Med entomologist, US Army, 56-59; med entomologist, Insects Affecting Man & Animal Res Lab, Agr Res Serv, USDA, 59-89; MGR, FLA LINEAR ACCELERATOR IRRADIATION FACIL, 89- *Concurrent Pos:* Courtesy prof entom, Univ Fla, 65- *Mem:* Entom Soc Am; Am Soc Testing & Mat. *Res:* Irradiation of agricultural commodities. *Mailing Add:* 1621 NW 71st St Gainesville FL 32605

SMITTLE, DOYLE ALLEN, b Bradley, Ark, Feb 27, 39; m 57; c 2. OLERICULTURE, PLANT PHYSIOLOGY. *Educ:* Univ Ark, BS, 61, MS, 65; Univ Md, PhD(hort), 69. *Prof Exp:* Res asst hort, Univ Ark, 61-66; asst prof, Wash State Univ, 68-73; from asst prof to assoc prof, 73-81, PROF HORT, UNIV GA, 81- *Mem:* Sigma Xi; Am Soc Hort Sci. *Res:* Soil-plant-water relations and post-harvest handling of cucurbits, onions and edible legumes. *Mailing Add:* Coastal Plain Exp Sta Univ of Ga Tifton GA 31794

SMITTLE, RICHARD BAIRD, b New Martinsville, WVa, Mar 5, 43; m 67; c 2. MICROBIAL FOOD ECOLOGY. *Educ:* WVa Univ, AB, 67, MS, 68; NC State Univ, PhD(food sci), 73. *Prof Exp:* Microbiologist, US Food & Drug Admin, 68-70; chief microbiologist, CPC Int, Best Foods, 73-77; dir, Silliker Labs NJ, 77-88 & Silliker Labs Pa, 88-91; vpres opers, 90-91, EXEC VPRES & INFO SER ED, SILLIKER LABS GROUP INC, 91- *Concurrent Pos:* Adj prof, Kean Col NJ, 74-75 & NY Univ, 85-87; comt mem, Adv Bd Mil Personnel Supplies, Nat Res Coun, 81-83. *Mem:* Inst Food Technologists; Am Soc Microbiol; Int Asn Milk Food & Environ Sanitarians; Soc Appl Bact. *Res:* Storage and survival of lactic streptococci and lactobacilli to starvation and frozen conditions; growth and survival in high acid and high osmotic foods and environments of bacteria, yeasts and molds. *Mailing Add:* Silliker Labs Group Inc 1304 Hallsted St Chicago Heights IL 60411

SMOAKE, JAMES ALVIN, b Langdale, Ala, Oct 5, 42. PHYSIOLOGY, BIOCHEMISTRY. *Educ:* Jacksonville State Univ, BA, 65; Univ Tenn, Knoxville, MS, 66, PhD(zool), 69. *Prof Exp:* Nat Cancer Inst spec cancer res trainee biochem, St Jude Children's Res Hosp, 72-73; MEM FAC BIOL, N MEX INST MINING & TECHNOL, 73- *Concurrent Pos:* Sabbatical leave, Endocrine & Metab Sect, Vet Admin Med Ctr, Memphis, 79-80. *Mem:* AAAS; Sigma Xi. *Res:* Action of insulin on liver cells. *Mailing Add:* Dept Biol NMex Inst Mining & Technol Socorro NM 87801

SMOCK, DALE OWEN, b Cochranton, Pa, Feb 13, 15; m 42. ELECTRICAL ENGINEERING. *Educ:* Grove City Col, BS, 42; Carnegie-Mellon Univ, BS, 48; Purdue Univ, MS, 62. *Prof Exp:* Instr pre-radar, US Naval Training Sch, Grove City Col, 42-45; jr engr, Westinghouse Elec Corp, Md, 45; instr eng, 45-49, from asst prof to assoc prof, 49-56, actg head, Eng Dept, 72-77, prof, 56-80, chmn, Eng Dept, 77-80, EMER PROF ELEC ENG, GROVE CITY COL. *Mem:* Am Soc Eng Educ; Inst Elec & Electronics Engrs. *Res:* Electronics; electromagnetic theory; transmission circuits; electrical measurements; linear systems. *Mailing Add:* Dept of Eng Grove City Col Grove City PA 16127

SMOKE, MARY E, b Charlestown, WVa, Sept 30, 31; m 65. MATHEMATICAL STATISTICS. *Educ:* Am Univ, BS, 55; Stanford Univ, MS, 58, PhD(math statist), 64. *Prof Exp:* Asst biostatistician, Med Ctr, Univ Calif, San Francisco, 61-65; from asst prof to assoc prof, 65-74, PROF MATH, CALIF STATE UNIV, LONG BEACH, 74- *Mem:* Inst Math Statist; Am Statist Asn. *Res:* Non-parametric inference. *Mailing Add:* Dept of Math Calif State Univ 1250 Bellflower Blvd Long Beach CA 90840

SMOKE, WILLIAM HENRY, b Battle Creek, Mich, Nov 7, 28; m 65. MATHEMATICS. *Educ:* Univ Mich, BA, 58, MA, 60; Univ Calif, Berkeley, PhD(math), 65. *Prof Exp:* Asst prof, 65-74, ASSOC PROF MATH, UNIV CALIF, IRVINE, 74- *Mem:* Am Math Soc. *Res:* Algebra. *Mailing Add:* Dept Math Univ Calif Irvine CA 92717

SMOKER, WILLIAM ALEXANDER, b Ishpeming, Mich, July 28, 15; m 41; c 3. FISHERIES, FORESTRY. *Educ:* San Jose State-Univ Calif, BS, 38; Univ Wash, PhD, 55. *Prof Exp:* Asst fisheries lab, Univ Wash, 46-47; fisheries biologist, Wash Dept Fish, 47-51, asst supvr res, 51-56; sr biologist, Alaska Dept Fish & Game, 56-59, chief biol res div, 59-60; asst lab dir, Auke Bay Biol Lab, Bur Com Fisheries, US Fish & Wildlife Serv, 61-67, lab dir, Auke Bay Fisheries Lab, Nat Marine Fisheries Serv, 67-82; RETIRED. *Concurrent Pos:* Adj prof fisheries, Univ Alaska, Juneau. *Mem:* AAAS; Am Fisheries Soc; Am Inst Fishery Res Biol. *Res:* Ecological factors determining abundance of fresh water and marine fishes in Alaska. *Mailing Add:* Auke Bay Fisheries Lab Box 210155 Nat Marine Fisheries Serv Auke Bay AK 99821

SMOKER, WILLIAM WILLIAMS, b Washington, DC, Sept 6, 45; m 75; c 2. FISH & WILDLIFE SCI. *Educ:* Carleton Col, BA, 67; Ore State Univ, MS, 70, PhD(fisheries), 82. *Prof Exp:* ASSOC PROF SCH FISHERIES & OCEAN SCI, UNIV ALASKA, FAIRBANKS, 78- *Concurrent Pos:* Assoc dir, Alaska Sea Grant Col, 82-83; vis assoc prof, Fac Fisheries, Univ Hokkaido, Hakodate, 88- 89. *Mem:* Am Fisheries Soc; Genetics Soc Am; Am Soc Limnol & Oceanog; Am Inst Fisheries Res Biologists; Int Asn Genetics in Aquacult. *Res:* Genetics of Pacific salmon, particularly as applied to fish culture; technology of the culture of Pacific salmon. *Mailing Add:* Juneau Ctr Fisheries & Ocean Sci 11120 Glacier Hwy Juneau AK 99801

SMOKOVITIS, ATHANASSIOS A, b Thessaloniki, Greece, June 2, 35; m 65. ANIMAL PHYSIOLOGY. *Educ:* Aristotelian Univ, dipl vet med, 57, dipl biol, 66, PhD(physiol), 68. *Prof Exp:* Vis investr res, Inst Med Res, Mitchell Found, Washington, DC, 73-75; res assoc, Sch Med, Ind Univ, 75-76; vis investr & adv res, Gaudius Inst, Health Res Orgn, Holland, 76-77; vis investr & lectr res, Dept Physiol, Med Sch, Vienna, Austria, 77-81; lectr physiol teaching, 70-73, PROF & HEAD, DEPT PHYSIOL RES, FAC VET MED, ARISTOTELIAN UNIV, THESSALONIKI, GREECE, 81- *Concurrent Pos:* Hon prof physiol, Univ Vienna, Austria, 79. *Mem:* Am Physiol Soc; Am Heart Asn. *Res:* Physiology of fibrinolysis; author of various publications. *Mailing Add:* Kon Melenikou 27 Thessaloniki 54635 Greece

SMOL, JOHN PAUL, b Montreal, Que, Oct 10, 55. PALEOLIMNOLOGY. *Educ:* McGill Univ, BSc, 77; Brock Univ, MSc, 79; Queen's Univ, PhD(paleolimnol), 82. *Prof Exp:* Vis scientist paleolimnol, Nat Sci & Eng Res Coun, Geol Surv Can, 83-84; fel, Nat Sci & Eng Res Coun, 82-83, ASST PROF BIOL, QUEEN'S UNIV KINGSTON, 84- *Mem:* Am Soc Limnol & Oceanog; Int Asn Theoret & Appl Limnol; Freshwater Biol Asn; Int Phycol Asn; Am Phycol Asn; Soc Can Limnologists. *Res:* Limnology and paleoecology of lakes; lake acidification; entrophication; high Arctic lakes; alpine lakes. *Mailing Add:* Dept Biol Queen's Univ Kingston ON K7L 3N6 Can

SMOLANDER, MARTTI JUHANI, b Helsinki, Finland, Sept 8, 55; c 1. EXERCISE PHYSIOLOGY, WORK PHYSIOLOGY. *Educ:* Univ Jyv10skyl10, BA, 77, MSc, 82; Univ Knopie, PhD(physicol), 87. *Prof Exp:* Asst researcher, 81-83, researcher, 84-87, SPECIALIZED RESEARCHER, INST OCCUP HEALTH, 88- *Concurrent Pos:* Vis scholar, Johns Hopkins Univ, 86-87; vis scientist, Swed Inst Occup Health, 87; docent exercise physiol, Univ Knopio, 89- *Mem:* Am Physiol Soc; Am Col Sports Med; Int Comn Occup Health; Scand Physiol Soc; Finnish Physiol Soc. *Res:* Applied physiology; assessment of work capacity in different occupational groups; evaluation of physiological responses due to protective clothing and equipment; evaluation of mechanisms and individual factors producing heat or cold stress. *Mailing Add:* Inst Occup Health Laajaniityntie 1 Vantaa 01620 FinlandAd

SMOLENSKY, MICHAEL HALE, b Chicago, Ill, May 10, 42; m 80; c 3. MEDICAL CHRONOBIOLOGY, ENVIRONMENTAL PHYSIOLOGY. *Educ:* Univ Ill, Urbana-Champaign, BS, 64, MS, 66, PhD(physiol), 71. *Prof Exp:* From asst prof to assoc prof, 70-87, PROF ENVIRON PHYSIOL, DIV PULMONARY MED, GRAD SCH BIOMED SCI, DEPT PHARMACEUT, UNIV TEX HEALTH SCI CTR, HOUSTON, 87- *Concurrent Pos:* Res assoc, Tex Allergy Res Found, 71- & McGovern Allergy Clin, Houston, 71-; co-ed, Chronobiol Int, 84-, Ann Rev Chronopharmacol; secy-treas, organizing comt, Int Conf Chronopharmacol, 84- *Mem:* AAAS; Int Soc Study Chronobiol; Soc Menstrual Cycle Res; NY Acad Sci. *Res:* Shift work, occupational health; chronopharmacology, chronotoxicology; chronobiology; public health; allergic asthma; investigation of human biological rhythmic phenomena relative to the diagnosis and treatment (chronopharmacology) of humans, especially heart, allergic asthma and cancer disease, as well as their cause or exacerbation due to environmental factors. *Mailing Add:* Sch Pub Health Univ Tex Health Sci Ctr PO Box 20186 Houston TX 77225

SMOLIAR, STEPHEN WILLIAM, b Philadelphia, Pa, July 8, 46. INFORMATION SCIENCE. *Educ:* Mass Inst Technol, BS, 67, PhD(appl math), 71. *Prof Exp:* Instr comput sci, Israel Inst Technol, 71-73; asst prof comput & info sci, Univ Pa, 73-78; mem tech staff, Gen Res Corp, 78-; MEM PROF STAFF, SCHLUMBERGER-DOLL RES. *Mem:* Asn Comput Mach; Inst Elec & Electronics Engrs. *Res:* Distributed data processing. *Mailing Add:* USC/Info Sci Inst 4676 Admiralty Way Marina Del Rey CA 90292

SMOLIK, JAMES DARRELL, b Rapid City, SDak, Mar 28, 42. PLANT NEMATOLOGY. *Educ:* SDak State Univ, BS, 65, MS, 69, PhD(plant path), 73. *Prof Exp:* Foreman pest control, M L Warne Chem & Equip Co, 66; asst plant path, 67-69, from res asst to res assoc, 70-75, ASST PROF PLANT NEMATOL, SDAK STATE UNIV, 75- *Mem:* Sigma Xi; Soc Nematologists. *Res:* Effect of nematodes on productivity of row, field and legume crops; nematode ecology studies in native range. *Mailing Add:* 1206 Fifth St Brookings SD 57006

SMOLIN, LEE, b New York, NY, June 6, 55. QUANTIZATION OF THE GRAVITATIONAL FIELD, QUANTUM DESCRIPTION OF SPACE & TIME. *Educ:* Hampshire Col, BA, 75; Harvard Col, MA, 78, PhD(physic), 79. *Prof Exp:* Postdoctoral, Inst Theoret Physics, Univ Calif, Santa Barbara, 80-81; mem, Inst Advan Study, 81-83; postdoctoral, Envico Frumi Inst, Univ Chicago, 83-84; asst prof, physics, Yale Univ 84-88; ASSOC PROF, PHYSICS, SYRACUSE UNIV, 88- *Concurrent Pos:* Vis scientist, Inst Theoret Physics. *Res:* Reconciling quantum mechanics with general relativity; elementary particle theory and the problem of the self-organization of biological systems. *Mailing Add:* Dept Physics Syracuse Univ Syracuse NY 13244

SMOLINSKY, GERALD, b Philadelphia, Pa, Feb 25, 33; m 79; c 2. DIELECTRIC FILMS, PHOTO RESISTS. *Educ:* Drexel Inst, BS, 55; Univ Calif, Berkeley, PhD(org chem), 58. *Prof Exp:* Fel, Columbia Univ, 58-59; MEM TECH STAFF, BELL LABS, 59- *Mem:* AAAS; Am Chem Soc; Am Vacuum Soc. *Res:* Dielectric planarization of integrated circuit; dielectric materials from spin-on glasses. *Mailing Add:* AT&T Bell Labs Murray Hill NJ 07974-2070

SMOLLER, JOEL A, b New York, NY, Jan 2, 39; m 60; c 3. MATHEMATICS. *Educ:* Brooklyn Col, BS, 57; Ohio Univ, MS, 58; Purdue Univ, PhD(math), 63. *Prof Exp:* Instr math, Univ Mich, 63-64; vis mem, Courant Inst Math Sci, NY Univ, 64-65; from asst prof to assoc prof, Univ Mich, Ann Arbor, 65-69; vis mem, Courant Inst Math Sci, NY Univ, 69-70; PROF, UNIV MICH, ANN ARBOR, 70- *Concurrent Pos:* Vis prof, Math Res Ctr, 72-73; fel Guggenheim, 80; vis prof, Ecole Norwale Supericure, Paris, 85, Harvard Univ, 88-89. *Mem:* Am Math Soc. *Res:* Partial differential equations; geometry. *Mailing Add:* Dept of Math Univ of Mich Ann Arbor MI 48109

SMOLLER, SYLVIA WASSERTHEIL, b Poland, Feb 24, 32; US citizen; m 71; c 2. BIOSTATISTICS, EPIDEMIOLOGY. *Educ:* Syracuse Univ, BS, 53, MA, 55; NY Univ, PhD(statist), 69. *Prof Exp:* Engr human factors, IBM, 58-61; statistician ment health, Astor Home Children, 62-64; asst prof math, State Univ NY Col New Paltz, 64-69; PROF, EPIDEMIOL & SOCIAL MED, ALBERT EINSTEIN COL MED, 69-, HEAD DIV EPIDEMIOL & BIOSTATIST, 85- *Concurrent Pos:* Consult, Int Proj, Asn Vol Sterilization, 72-; fel coun epidemiol, Am Heart Asn, 76. *Mem:* Am Pub Health Asn; Soc Epidemiol Res; Am Statist Asn; fel NY Acad Sci; fel Am Col Epidemiol; Soc Clin Trials; Am Heart Asn. *Res:* Epidemiological studies of hypertension, cardiovascular disease, cancer and clinical traits; computer applications. *Mailing Add:* Dept Epidemiol & Social Med Albert Einstein Col of Med 1300 Morris Park Ave Bronx NY 10461

SMOLUCHOWSKI, ROMAN, b Zakopane, Austria, Aug 31, 10; nat US; m 51; c 2. SOLID STATE PHYSICS, ASTROPHYSICS. *Educ:* Univ Warsaw, MA, 33; Univ Groningen, Holland, PhD(physics), 35. *Prof Exp:* Mem, Inst Adv Study, 35-36; res assoc, Univ Warsaw & head,physics sect, Inst Metals, Warsaw Inst Technol, 36-39; instr & res assoc, Princeton Univ, 40-41; res physicist, Gen Elec Co, 41-46; assoc prof & mem staff,Metals Res Lab, Carnegie Inst Technol, 46-50, prof physics & metall, 50-56, prof physics, 56-60; prof solid state sci & dir,Solid State Lab, Princeton Univ, 60-78; prof astron & physics, 78-87, EMER PROF, UNIV TEX, AUSTIN, 87- *Concurrent Pos:* Mem,Solid State Panel, Res & Develop Bd, US Dept Defense, 49, tech adv bd aircraft nuclear propulsion, 50; secy,Solid State Sci Adv Panel, Nat Res Coun, 50-61, chmn comt solids, 61-67, chmn,Div Phys Sci, 69-75,; chmn adv comt magnetism, Off Naval Res, 52-56; Fulbright prof, Univ Paris, 55-56, exchange prof, 65-66; lectr, Univ Liege, Belg, 56; vis prof, Nat Res Coun Brazil, 58-59 & Tech Univ Munich, Ger, 74; mem adv comt metall, Oak Ridge Nat Lab, 60-62; mem physics surv comt, Nat Acad Sci, 64-66 & 70-72, space sci bd, 69-75, comt planetary and lunar expl, 80-83; John Simon Guggenheim Mem Fel, 74; fel, Churchill Col Cambridge Univ Eng, 74; assoc ed, Fundamentals Cosmic Physics. *Mem:* Fel Am Phys Soc; Am Astron Soc; Am Crystallog Asn; fel Am Acad Arts & Sci; AAAS; Brasilian Acad Sci; Finnish Acad Sci; Mex Acad Sci. *Res:* Condensed matter in astrophysics; lattice imperfections; radiation effects; phase transformation; magnetism. *Mailing Add:* Dept Astron & Physics Univ Tex Austin TX 78712

SMOOK, MALCOLM ANDREW, b Seattle, Wash, Aug 22, 24; m 45; c 2. ORGANIC CHEMISTRY. *Educ:* Univ Calif, BS, 45; Ohio State Univ, PhD(chem), 49. *Prof Exp:* Chemist, E I Du Pont de Nemours & Co, Inc, 49-52, res supvr, 52-53, div head, 53-58, from asst lab dir to lab dir, 58-63, asst dir res & develop div, Elastomers Dept, 63-68, asst dir res & develop div, Plastics Dept, 68-76, gen lab dir, Plastics Prod & Resins Dept, 76-80, mgr patents & regulatory affairs, Polymer Prod Dept, 80-85. *Concurrent Pos:* Mem res & technol adv comt, Materials & Structures, NASA, 75-78; consult, 85- *Mem:* Am Chem Soc; Sigma Xi; Soc Rheol. *Res:* Organic fluorine chemistry; polymer and rubber chemistry. *Mailing Add:* 59 Rockford Rd Wilmington DE 19806

SMOOKE, MITCHELL D, b Hartford, Conn, Aug 10, 51; m 77; c 1. NUMERICAL ANALYSIS, COMPUTATIONAL COMBUSTION. *Educ:* Rensselaer Polytech Inst, BS, 73; Harvard Univ, MS, 74, PhD (appl math), 78; Univ Calif, Berkeley, MBA, 83. *Prof Exp:* Staff scientist, Sandia Nat Labs, 78-84; asst prof, 84-86, ASSOC PROF MECH ENG, YALE UNIV, 86- *Concurrent Pos:* Vis prof, Catholic Univ, Holland, 85, Ecole Centrale, France, 88; consult, Gen Motors, 87-, United Technol, 87-, Gen Elec, 89-; comt propellant res, Army Res Off, 88. *Mem:* Soc Indust & Appl Math; Asn Comput Mach; Combustion Inst. *Res:* The development and application of adaptive numerical algorithms for problems in combustion. *Mailing Add:* Dept Mech Eng Yale Univ New Haven CT 06520

SMOOT, CHARLES RICHARD, b Marmet, WVa, Nov 15, 28; m 51; c 6. RESEARCH ADMINISTRATION. *Educ:* Charleston Univ, BS, 51; Purdue Univ, PhD(phys chem), 55. *Prof Exp:* Chemist, FMC Co, 47-51; chemist, Du Pont Co, 55-59, from res supvr to sr res supvr, 59-72, lab supt, 72-82, res mgr, 82-90, TECHNOL MGR, E I DU PONT DE NEMOURS & CO INC, 90- *Concurrent Pos:* Asst prof, WVa Univ Br, Parkersburg, 61-63. *Mem:* Am Chem Soc; Sigma Xi. *Res:* Applied research on polymers; research administration. *Mailing Add:* Du Pont Co-Polymers Exp Sta Lab 174-303 Wilmington DE 19898

SMOOT, EDITH L, b Ft Worth, Tex, Aug 21, 51; div; c 1. PALEOBOTANY, PLANT ANATOMY. *Educ:* Ohio State Univ, BSc, 76, MSc, 78, PhD(bot), 83. *Prof Exp:* Teaching asst & res asst bot, dept bot, Ohio State Univ, 77-82; fel, Am Asn Univ Women, 82-83; ASST PROF BIOL, HOPE COL, 83- *Concurrent Pos:* Teaching assoc bot, dept bot, Univ Tex, 79; researcher, Inst Polar Studies, Ohio State Univ, 85-86. *Mem:* Bot Soc Am; Int Orgn Paleobot; Int Asn Wood Anatomists; AAAS; fel Linnean Soc. *Res:* Phloem anatomy of Paleozoic and Mesozoic plants; investigation of Permian and Triassic silicified plants from the Central Transantarctic Mountains, Antarctica. *Mailing Add:* Dept Bot Ohio State Univ 1735 Neil Ave Columbus OH 43210

SMOOT, GEORGE FITZGERALD, b Wetumpka, Ala, Jan 16, 22; m 43; c 2. HYDROLOGY, INSTRUMENTATION. *Educ:* Auburn Univ, BS, 50. *Prof Exp:* Eng technician, Ala, 48-50, hydraul engr, 50-52, hydraul engr, Alaska, 52-56 & Ohio, 56-62, res hydrologist, Wash, DC, 62-66, coordr res on instrumentation, Wash, DC, US Geol Surv, 66-77. *Concurrent Pos:* Chmn working group tech comt, Int Orgn Standardization & mem hydrometry comt, Int Asn Sci Hydrol, 68-; mem, Interagency Adv Comt Nat Oceanog Instrumentation Ctr; chmn, Task Group Velocity Measurements, Am Soc Testing & Mats; consult, World Bank. *Mem:* Am Soc Civil Eng; Am Soc Testing & Mats; Am Geophys Union. *Res:* Research, design and development of instrumentation for use in hydrologic investigations. *Mailing Add:* 32600 River Rd Orange Beach AL 36561

SMOOT, GEORGE FITZGERALD, III, b Yukon, Fla, Feb 20, 45. ASTROPHYSICS, COSMIC RAY PHYSICS. *Educ:* Mass Inst Technol, BS(math) & BS(physics), 66, PhD(physics), 70. *Prof Exp:* Res physicist, Mass Inst Technol, 70; RES PHYSICIST, UNIV CALIF, BERKELEY, 71-, RES PHYSICIST, LAWRENCE BERKELEY LAB, 74- *Mem:* Am Phys Soc; Am Astron Soc; Sigma Xi; Int Astron Union. *Res:* Measurements of cosmic background radiation as a cosmological probe of the early universe; satellite and balloon-borne superconducting magnetic spectrometer experiments on the charged cosmic rays; remote sensing using microwave radiometers. *Mailing Add:* 10 Panoramic Way Berkeley CA 94704

SMOOT, LEON DOUGLAS, b Provo, Utah, July 26, 34; m 53; c 4. CHEMICAL ENGINEERING. *Educ:* Brigham Young Univ, BS & BEngS, 57; Univ Wash, MS, 58, PhD(chem eng), 60. *Prof Exp:* Lab asst chem, Brigham Young Univ, 54-55, instr math, 55-56, res asst chem eng, 56-57; consult engr heat transfer, 58-59; asst chem eng, Univ Wash, 57-60; asst prof, Brigham Young Univ, 60-63; sr tech specialist res & develop, Lockheed Propulsion Co, 63-67; assoc prof, 67-70, chmn dept, 70-77, PROF CHEM ENG, BRIGHAM YOUNG UNIV, 70-, DEAN ENG & TECHNOL, 77- *Concurrent Pos:* Indust partic, US-UK-Can Tech Coop Prog, 64-72; vis asst prof, Calif Inst Technol, 66-67; chmn ad hoc hybrid combustion comt, Int Agency Chem Rocket Propulsion Group, 66; consult, several companies and agencies in the US & Europe, 70- *Mem:* Am Inst Chem Engrs; Am Inst Aeronaut & Astronaut; Sigma Xi; Am Soc Eng Educ; Int Combustion Inst. *Res:* Combustion; energy; fossil fuels. *Mailing Add:* 1811 N 1500 East Provo UT 84604

SMOSNA, RICHARD ALLAN, b Chicago, Ill, Nov 3, 45; m 67; c 2. STRATIGRAPHY. *Educ:* Mich State Univ, BS, 67; Univ Ill, MS, 70, PhD(geol), 73. *Prof Exp:* Instr geol, Hanover Col, 71-72; petrol geologist, WVa Geol Surv, 72-78; PROF GEOL & GEOG, WVA UNIV, 78- *Concurrent Pos:* Adj asst prof, WVa Univ, 74- *Honors & Awards:* Levorsen Award, Am Asn Petrol Geologist, 74; Distinguished Tech Commun Award, Soc Tech Commun, 75. *Res:* Determination of paleoenvironments, paleoecology, stratigraphy and petroleum potential of Silurian-aged carbonate rocks of central Appalachians. *Mailing Add:* Dept Geol & Geog WVa Univ 425 White Hall Morgantown WV 26506

SMOTHERS, JAMES LLEWELLYN, b Jackson, Tenn, Aug 30, 30; m 64. ANIMAL PHYSIOLOGY, ENDOCRINOLOGY. *Educ:* Lambuth Col, BS, 52; Univ Tenn, MS, 53, PhD(zool), 61. *Prof Exp:* Nat Heart Found fel marine biol, Inst Marine Sci, Univ Miami, 61-62; from asst prof to assoc prof, 62-71, PROF BIOL, UNIV LOUISVILLE, 71- *Mem:* Am Soc Zoologists; Sigma Xi; AAAS; Int Oceanog Found. *Res:* Effects of dietary and hormonal factors on mitochondrial structure and function, mechanisms of actions of hormones; comparative physiology of respiratory enzyme activities and respiration of animals and tissues. *Mailing Add:* Dept of Biol Univ of Louisville Louisville KY 40292

SMOTHERS, WILLIAM JOSEPH, b Poplar Bluff, Mo, Mar 17, 19; m 43; c 2. ENGINEERING. *Educ:* Univ Mo, BS, 40, MS, 42, PhD(ceramic eng), 44. *Prof Exp:* Chem analyst, Mo Portland Cement Co, 39; lab asst physics, Mo Sch Mines, 40-42; engr, Mo Exp Sta, Rolla, 42-44; res engr, Bowes Elec Ceramic Corp, 44-50; assoc prof, Inst Sci & Technol, Univ Ark, 50-53; dir ceramic res, Ohio Brass Co, 54-63; sect mgr refractories, Homer Res Labs, Bethlehem Steel Corp, 63-82; RETIRED. *Honors & Awards:* Toledo Glass & Ceramic Award, Am Ceramic Soc, 75. *Mem:* Fel AAAS; fel Am Ceramic Soc (vpres, 64-65, pres, 71-72); Am Chem Soc. *Res:* Refractories research; differential thermal analysis; solid state; ceramic materials. *Mailing Add:* 2700 Woodside Rd Bethlehem PA 18017-3607

SMOUSE, PETER EDGAR, b Long Beach, Calif, Apr 17, 42. HUMAN GENETICS, STATISTICS. *Educ:* Univ Calif, Berkeley, BS, 65; NC State Univ, PhD(genetics), 70. *Prof Exp:* NSF grant, Univ Tex, Austin, 70-72; from asst prof to assoc prof human genetics, 72-83, PROF BIOL SCI, UNIV MICH, ANN ARBOR, 85- *Concurrent Pos:* Mem, Comt Quant Genetics & Common Dis, Nat Inst Gen Med Sci, NIH, 78, study sect mammalian genetics, 81; assoc ed, Theoret Pop Biol, 79-81; mem rev panel, Pop Biol & Physiol Ecol, NSF, 80-82. *Mem:* Int Soc Genetics; Soc Study Evolution; Genetics Soc Am; Am Soc Human Genetics; Am Soc Naturalists. *Res:* Biometry; genetics; ecology; demography; epidemiology; anthropology; taxonomy. *Mailing Add:* Theoret & Appl Genetics Cook Col Rutgers Univ New Brunswick NJ 08903

SMOUSE, THOMAS HADLEY, b Cumberland, Md, July 10, 36; m 59; c 3. FOOD SCIENCE & TECHNOLOGY, AGRICULTURAL & FOOD CHEMISTRY. *Educ:* Pa State Univ, BS, 58; Rutgers Univ, MS, 64, PhD(food sci), 65. *Prof Exp:* Analytical chemist, Nabisco Res Ctr, 58-61; res fel fats & oils, Rutgers Univ, 61-65; sr res chemist, Campbell's Inst Food Res, 65-67; res assoc, Anderson Clayton Foods, 67-77; mgr lipid sci, Ralston Purina Co, 77-88; MGR OIL PROCESS RES, ARCHER DANIEL MIDLAND CO, 88- *Mem:* Sigma Xi; Am Oil Chemists Soc; Am Chem Soc; Inst Food Technol. *Res:* Exploratory and applied research in fat and oil constituents and their interactions with foods to produce flavor effects. *Mailing Add:* 810 Stephens Creek Lane Decatur IL 62526-9712

SMOYER, CLAUDE B, b Pittsburgh, Pa, Jul 18, 34; m 59; c 4. RESEARCH ADMINISTRATION. *Educ:* Univ Rochester, BS, 59, MS, 64. *Prof Exp:* Res asst, Univ Rochester, 59-60; res staff mem, IBM, T J Watson Res Ctr, 60-65; scientist, 65-70, admin asst, 70-74, mgr admin, 74-79, MGR OPER, XEROX WEBSTER RES CTR, 80- *Mailing Add:* Xerox Webster Res Ctr 800 Phillips Rd, W128-28E Webster NY 14580

SMUCKER, ARTHUR ALLAN, b Dhamtari, India, Nov 27, 23; US citizen; m 48; c 6. BIOCHEMISTRY, COMPUTER-INSTRUMENT INTERFACING. *Educ:* Goshen Col, BA, 49; Univ Ill, MS, 51, PhD(chem), 54. *Prof Exp:* From asst to instr chem, Univ Ill, 49-53; from instr to assoc prof, 54-74, dir comput serv, 84-88, PROF CHEM, GODSHEN COL, 74-; RETIRED. *Concurrent Pos:* Consult, Miles Labs, Inc, 59-61; fels, Nat Inst Arthrities & Metab Dis, Univ Calif, Berkeley, 63-64 & Univ Iowa, 72-73. *Mem:* Am Chem Soc; Am Sci Affil. *Res:* Enzyme purification, kinetics and structure. *Mailing Add:* 701 College Ave Goshen IN 46526-4913

SMUCKER, SILAS JONATHAN, b Goshen, Ind, Dec 31, 04; m 35; c 2. SOIL CONSERVATION. *Educ:* Goshen Col, AB, 30; Purdue Univ, MS, 32. *Prof Exp:* Asst plant pathologist, Div Forest Path, USDA, 34-44, soil conservationist, Soil Conserv Serv, 45-62; agriculturist, AID, 62-69; CONSULT AGR SERV, 70- *Mem:* Am Phytopath Soc; Soil Conserv Soc Am. *Res:* Elm tree diseases; wood decay fungi and wood preservatives; soil and water conservation; wildlife biology; tropical agriculture. *Mailing Add:* 1801 Greencroft Blvd No 215 Goshen IN 46526

SMUCKLER, EDWARD AARON, experimental pathology, biochemistry; deceased, see previous edition for last biography

SMUDSKI, JAMES W, b Greensburg, Pa, Oct 31, 25; m 49; c 3. PHARMACOLOGY, DENTISTRY. *Educ:* Univ Pittsburgh, BS, 50, DDS, 52, Univ Calif, San Francisco, MS, 61, PhD(pharmacol), 65. *Prof Exp:* Pvt pract dent, 52-58; asst prof pharmacol, 63-64, assoc prof & head dept, 64-67, dir div grad & post grad educ, 70-73, prof pharmacol & physiol & head dept, Sch Dent, Univ Pittsburgh, 67-76; dean Sch Dent, Univ Detroit, 76-82; dean dent med, Univ Pittsburgh, 83-88; RETIRED. *Concurrent Pos:* Nat Inst Dent Res res-teacher trainee, 58-62 & career develop award, 62-64; consult, Oakland Vet Admin Hosp, Pittsburgh, Pa, 64-76; consult, Nat Bd Dent Exam, 65-75. *Mem:* Am Dent Asn; Am Inst Oral Biol; Int Asn Dent Res. *Res:* Pharmacology of agents affecting the central, autonomic and peripheral nervous systems. *Mailing Add:* Dent Med 257 Tech Rd Pittsburgh PA 15205

SMUK, JOHN MICHAEL, b Biwabik, Minn, Aug 16, 32; m 56; c 3. CHEMICAL ENGINEERING. *Educ:* Univ Wis, MS, 56, PhD(chem eng), 60. *Prof Exp:* Res engr, Forest Prod Lab, USDA, 60-64, proj leader, 64-66; consult engr waste treatment, Ruble-Miller Assocs, 66-69; SR RES ENGR, POTLATCH CORP, 69-, MGR PROCESS ENG, 81- *Concurrent Pos:* Consult,67. *Mem:* Am Chem Soc; Am Inst Chem Engrs; Tech Asn Pulp & Paper Indust; Instrument Soc Am. *Res:* Furfural plant and process design; acid decomposition of simple sugars; kinetic studies; chemistry of wood; process control; secondary fiber process design; oxygen bleaching; computer applications; pulp and paper plant design. *Mailing Add:* 321 E Faribault Duluth MN 55803

SMULDERS, ANTHONY PETER, b Oss, Netherlands, July 6, 42. PHYSIOLOGY. *Educ:* Loyola Univ, Los Angeles, BS, 66; Univ Calif, Los Angeles, PhD(physiol), 70. *Prof Exp:* from asst prof to assoc prof, 70-81, PROF BIOL, LOYOLA MARYMOUNT UNIV, 81-, ASSOC DEAN SCI & ENG, 72- *Concurrent Pos:* Res physiologist, Univ Calif, Los Angeles, 70-; comnr, Los Angeles County Narcotics and Dangerous Drugs Comn, 73-, Calif State Adv Bd Drug Progs, 82- *Mem:* AAAS; Sigma Xi; Biophys Soc; Nat Asn Adv Health Prof. *Res:* Transport phenomena, the movement of ions and non-electrolytes across biological and artificial membranes; improvemnt of university science teaching; drug abuse and prevention. *Mailing Add:* Col Sci & Eng Loyola Marymount Univ Los Angeles CA 90045

SMULLIN, LOUIS DIJOUR, b Detroit, Mich, Feb 5, 16; m 39; c 4. ELECTRICAL ENGINEERING. *Educ:* Univ Mich, BSE, 36; Mass Inst Technol, SM, 39. *Prof Exp:* Draftsman, Swift Elec Welder Co, Mich, 36; engr, Ohio Brass Co, 36-38, Farnsworth TV Corp, 39-40 & Scintilla Magneto Div, Bendix Aviation Corp, 40-51; sect head radiation lab, Mass Inst Technol, 41-46; head microwave tube lab, Fed Telecommun Labs Div, Int Tel & Tel Corp, NJ, 46-48; head tube lab, Res Lab Electronics, Mass Inst Technol, 48-

50, div head, Lincoln Lab, 50-55, from assoc prof to prof, 55-74, chmn dept, 66-74, Dugald Caleb Jackson prof elec eng, 74- 86, EMER PROF & SR LECTR ELEC ENG, MASS INST TECHNOL, 86- *Concurrent Pos:* Mem steering comt, Kanpur Indo-Am Prog, 61-65; vis prof, Indian Inst Technol, Kanpur, 65-66; NSF Working Group Sci & Eng Instr, India; mem comt telecommun, Nat Acad Eng; bd govs, Israel Inst Technol. *Mem:* Nat Acad Eng; Am Phys Soc; fel Inst Elec & Electronics Engrs; fel Am Acad Arts & Sci. *Res:* Plasma physics; technology assessment. *Mailing Add:* Dept Elec Eng Mass Inst Technol 38-294 Cambridge MA 02139

SMULOW, JEROME B, b New York, NY, July 29, 30; m 68; c 1. ORAL PATHOLOGY, PERIODONTOLOGY. *Educ:* NY Univ, AB, 51, DDS, 55; Tufts Univ, MS, 61, cert, 64. *Prof Exp:* From instr to assoc prof, 60-68, PROF PERIODONT, SCH DENT MED, TUFTS UNIV, 69- *Concurrent Pos:* Fulbright-Hays fel, Iran, 72-73. *Mem:* AAAS; Tissue Cult Asn; Am Dent Asn; Int Asn Dent Res. *Res:* Histopathology; tissue cultures. *Mailing Add:* 673 Boylston Brookline MA 02146

SMULSON, MARK ELLIOTT, b Baltimore, Md, Mar 25, 36; m 66; c 2. BIOCHEMISTRY. *Educ:* Washington & Lee Univ, AB, 58; Cornell Univ, MNS, 61, PhD(biochem), 71. *Prof Exp:* Fel biochem, Albert Einstein Med Ctr, 64-65; USPHS fel, Nat Cancer Inst, 65-67; from asst prof to assoc prof, 67-78, PROF BIOCHEM, SCHS MED & DENT, GEORGETOWN UNIV, 78- *Mem:* Am Asn Cancer Res; Am Soc Biol Chemists. *Res:* Molecular biology; poly adenosine diphosphoribose polymerase in control of DNA replication and in nucleosomal structure of chromatin; carcinogens interaction with nucleosomes. *Mailing Add:* Dept Biochem Schs Med & Dent Georgetown Univ Washington DC 20007

SMULYAN, HAROLD, b Philadelphia, Pa, Jan 2, 29; m 52; c 3. INTERNAL MEDICINE, CARDIOLOGY. *Educ:* Univ Pa, AB, 49; Univ Buffalo, MD, 53. *Prof Exp:* From instr to assoc prof, 59-72, PROF MED, STATE UNIV NY UPSTATE MED CTR, 71-; CHIEF CARDIOL, VET ADMIN MED CTR HOSP, SYRACUSE, 78- *Mem:* Am Fedn Clin Res; NY Acad Sci; Am Heart Asn. *Res:* Hypertension; circulatory control; exercise physiology. *Mailing Add:* Dept Med 750 E Adams St State Univ NY Upstate Med Ctr Syracuse NY 13210

SMURA, BRONISLAW BERNARD, b Solvay, NY, Aug 9, 30; m 52; c 3. CHEMICAL ENGINEERING. *Educ:* Syracuse Univ, BChE, 52, MChE, 54, PhD(chem eng), 68. *Prof Exp:* Res engr, Indust Chem Div, Allied Chem Corp, 57-79; mgr process eng, Linden Chem & Plastics Corp, 80-89; ADV, PRESSURE VESSEL SERV, 89- *Mem:* Am Inst Chem Engrs. *Res:* Industrial inorganic chemicals with recent major emphasis on electrolytic production of chlorine and caustic soda. *Mailing Add:* 4051 S St Marcellus NY 13108

SMUTNY, EDGAR JOSEF, b New York, NY, Apr 20, 28. ORGANIC CHEMISTRY. *Educ:* Univ Colo, BA, 48; Univ Minn, PhD(chem), 53. *Prof Exp:* Mem staff, Allied Chem & Dye Corp, 48-49; asst org chem, Univ Minn, 50-52; res fel, Calif Inst Technol, 53-55; res chemist, 55-68, res supvr, 68-72, SR STAFF CHEMIST, SHELL DEVELOP CO, 72- *Mem:* Am Chem Soc; The Chem Soc. *Res:* Small ring compounds; strain energy; photochemistry; free radical chemistry; heterocyclic chemistry; organometallic chemistry; sulfur compounds; homogeneous palladium catalysis; heterogeneous catalysis; fuels and lubricant research. *Mailing Add:* Shell Develop Co Chem Dept PO Box 1380 Houston TX 77001

SMUTS, MARY ELIZABETH, b Waterbury, Conn, Mar 15, 48; m 72. DEVELOPMENTAL BIOLOGY. *Educ:* Albertus Magnus Col, BA, 70; Temple Univ, PhD(develop biol), 75; Harvard Sch Pub Health, MS, 84. *Prof Exp:* Res fel, Lab Develop Biol & Anomalies, Nat Inst Dent Res, NIH, 74-76; asst prof develop biol, Cath Univ Am, 76-78; asst prof biol, Wheaton Col, 78-83; REGIONAL TOXICOLOGIST, REGION I, US ENVIRON PROTECTION AGENCY, 83- *Mem:* Soc Develop Biol; Am Soc Cell Biol; Am Soc Zoologist; Soc Risk Anal. *Res:* Cranio-facial development. *Mailing Add:* 45 S Washington Norton MA 02766

SMUTZ, MORTON, b Twin Falls, Idaho, Jan 10, 18; m 45, 70. CHEMICAL ENGINEERING. *Educ:* Kans State Col, BS, 40, MS, 41; Univ Wis, PhD(chem eng), 50. *Prof Exp:* Chem engr, Monsanto Chem Co, 41; asst prof chem eng, Bucknell Univ, 49-51; assoc prof, Iowa State Univ, 51-55, prof & head dept, 55-61; asst dir, Ames Lab, US Atomic Energy Comn, 55-64, dep dir, 64-69; chmn, Coastal & Oceanog Eng Dept, Col Eng, Univ Fla, 75-78, assoc dean eng res & prof chem eng, 69-79; sr prof engr, Nat Oceanic & Atmospheric Admin, 79-85; RETIRED. *Concurrent Pos:* Dir, Div Marine Sci, Instrument Soc Am, 82-84. *Mem:* Am Chem Soc; Am Soc Eng Educ; Am Inst Chem Engrs; Coastal Soc (secy, 77). *Res:* Laser raman spectroscopy, optical fibers. *Mailing Add:* 9901 Montrose Rd Rockville MD 20852

SMYERS, WILLIAM HAYS, b Pittsburgh, Pa, Jan, 9, 01; m 26; c 4. PARAPOL-A STYRENE-ISOBUTYLENE COPOLYMER. *Educ:* Univ Pittsburgh, BChem, 24; Southeastern Univ, LLB, 36. *Prof Exp:* Asst chemist blast furnace, Carnegie Steel Co, 18-19; five fels, Mellon Inst, Pittsburgh, 19-25; dir res, Richardson Corp, Rochester, NY, 24-25, Duquesne Slag Prod Co, Pittsburgh, 25-29; mem patent dept, Aluminum Co Am, 29-32; polymer group head patent dept, Exxon Res & Eng Co, 32-63; RETIRED. *Mem:* Am Chem Soc; fel Am Inst Chemists; Nat Soc Inventors; NY Acad Sci; Johnson O'Connor Res Found. *Res:* Patent soliciting; motor fuel octane improvement. *Mailing Add:* 229 Sylvania Pl Westfield NJ 07090-3123

SMYLIE, DOUGLAS EDWIN, b New Liskeard, Ont, June 22, 36; wid; c 4. GEOPHYSICS. *Educ:* Queen's Univ, Ont, BSc, 58; Univ Toronto, MA, 59, PhD(physics), 63. *Prof Exp:* Fel geophys, Univ Toronto, 64; asst prof, Univ Western Ont, 64-68; from asst prof to assoc prof, Univ BC, 68-72; PROF EARTH SCI, YORK UNIV, 72- *Concurrent Pos:* Nat Res Coun Can operating grant, 65- *Mem:* Am Geophys Union; fel Royal Astron Soc. *Res:* Rotation of the earth; Chandler wobble; main magnetic field; elasticity theory of dislocations; dynamics of the earth's core. *Mailing Add:* Dept Earth & Atmospheric Sci York Univ 4700 Keele St Downsview ON M3J 1P6 Can

SMYLIE, ROBERT EDWIN, b Lincoln Co, Miss, Dec 25, 29; c 3. MECHANICAL ENGINEERING. *Educ:* Miss State Univ, BSc, 52, MSc, 56; Mass Inst Technol, MSc, 67. *Prof Exp:* Indust engr, Ethyl Corp, Tex, 52-54; instr mech eng, Miss State Univ, 54-56; lead engr, Skybolt Missile Syst Thermo-Conditioning Systs, Douglas Aircraft Co, Calif, 56-62; chief, Apollo Support Off, Crew Systs Div, Manned Spacecraft Ctr, Goddard Space Flight Ctr, 62-66, asst chief div, 67-68, actg chief, 68-70, chief, Crew Systs Div, 70-76, dep dir, 76-87; RETIRED. *Concurrent Pos:* Mem US deleg engaged in discussions with USSR to establish common docking systems for spacecraft of the two countries. *Honors & Awards:* Except Serv Medal, NASA, 69; Victor Prather Award, 71. *Mem:* Am Inst Aeronaut & Astronaut. *Res:* Analyses, design and development in specific advanced system areas such as space suits, extravehicular activity support hardware and environmental and thermal control subsystems. *Mailing Add:* 15101 Interlachen Dr Silver Springs MD 20906

SMYRL, WILLIAM HIRAM, b Brownfield, Tex, Dec 12, 38; m 64; c 2. PHYSICAL CHEMISTRY, ELECTROCHEMISTRY. *Educ:* Tex Tech, BS, 61; Univ Calif, Berkeley, PhD(chem), 66. *Prof Exp:* Asst prof pharmaceut chem, Univ Calif, San Francisco, 66-68; mem tech staff, Boeing Sci Res Labs, 68-72; MEM TECH STAFF, SANDIA LABS, 72- *Mem:* Electrochem Soc; Sigma Xi. *Res:* Molten salts; corrosion science; modeling of corrosion and electrochemical processes; photoelectrochemistry; digital measurement of Faradaic impedance of electrochemical and corrosion reactions. *Mailing Add:* 2637 13th Terr NW New Brighton MN 55112

SMYRNIOTIS, PAULINE ZOE, biochemistry, for more information see previous edition

SMYTH, CHARLES PHELPS, physical chemistry, atomic & molecular physics; deceased, see previous edition for last biography

SMYTH, DONALD MORGAN, b Bangor, Maine, Mar 20, 30; m 51; c 2. SOLID STATE CHEMISTRY. *Educ:* Univ Maine, BS, 51; Mass Inst Technol, PhD(inorg chem), 54. *Prof Exp:* Sr engr, Sprague Elec Co, 54-61, head solid state res, 61-71; assoc prof metall, mat eng & chem, 71-73, prof, 73-88, DIR MAT RES CTR, LEHIGH UNIV, 71-, PAUL B REINHOLD PROF, MAT SCI, ENG & CHEM, 88- *Honors & Awards:* Battery Div Res Award, Electrochem Soc, 60; Edward C Henry Award, Am Ceramic Soc, 87, Kraner Award, 90. *Mem:* Am Chem Soc; Electrochem Soc; Am Inst Chem; Am Ceramic Soc; fel Mat Research Soc. *Res:* Defect chemistry of complex metal oxides; effect of composition, impurities and nonstoichiometry on properties of insulating, semiconducting and superconducting oxides. *Mailing Add:* Mat Res Ctr Bldg 5 Lehigh Univ Bethlehem PA 18015

SMYTH, JAY RUSSELL, b Trenton, NJ, Apr 24, 39; m; c 5. CERAMICS SCIENCE, MATERIALS SCIENCE. *Educ:* Rutgers Univ, BS, 61, MS, 63; Pa State Univ, PhD(ceramics sci), 74. *Prof Exp:* Develop engr, Western Elec, Inc, 63-66; supvr prod eng, Mitronic, Inc, 66-68; eng mgr, Nat Berylia Corp, 68-71; from asst prof to assoc prof, Iowa State Univ, 74-81; sr mat engr, Garrett Turbine Engine Co, 81-85; SR ENG SUPVR, GARRETT AUXILIARY POWER DIV, ALLIED SIGNAL, 85- *Mem:* Fel Am Ceramic Soc; Nat Inst Ceramic Engrs; Nat Asn Parliamentarian; Am Soc Metals; Am Inst Parliamentarians. *Res:* Mechanics properties of materials including fracture and deformation; advanced turbine engines; ceramics for turbine engine applications. *Mailing Add:* 223 E Garfield St Tempe AZ 85281

SMYTH, JOSEPH RICHARD, b Louisville, Ky, Oct 10, 44. GEOLOGY, MINERALOGY. *Educ:* Va Polytech Inst, BS, 66; Univ Chicago, SM, 68, PhD(mineral), 70. *Prof Exp:* Res fel geol, Harvard Univ, 70-72; vis fel, Lunar Sci Inst, 72-74, res scientist, 74-76; mem staff, Los Alamos Sci Lab, 76-84; PROF GEOL, UNIV COLO, BOULDER, 84- *Concurrent Pos:* Vis sr lectr, Univ Cape Town, 75. *Mem:* Mineral Soc Am; Mineral Soc Japan; Am Geophys Union; Meteoritical Soc; Geol Soc Am. *Res:* Crystal chemistry of rock-forming silicates; igneous petrology; radioactive waste isolation. *Mailing Add:* Dept Geol Univ Colo Boulder CO 80309

SMYTH, MICHAEL P(AUL), b Albany, NY, Oct 2, 34; m 77; c 1. ELECTRICAL ENGINEERING, SYSTEMS ANALYSIS. *Educ:* Syracuse Univ, BS, 57, MS, 59; Univ Pa, PhD(elec eng), 63. *Prof Exp:* Elec engr, Gen Elec Co, 57; res asst, Radar Display, Syracuse Univ, 57-59; from instr to asst prof elec eng, Univ Pa, 59-67; assoc prof, 67-71, dir eng, 71-74, PROF, WIDENER UNIV, 71- *Concurrent Pos:* Ed consult, Bell Tel Co, Pa, 61-; mem, Franklin Inst, 62-; consult, Gen Elec Co, 65-67 & Philadelphia Elec Co, 74. *Honors & Awards:* Ralph R Tetor Award, Soc Automative Engrs, 74. *Mem:* Inst Elec & Electronics Engrs; Am Soc Eng Educ; Soc Automative Engrs. *Res:* Methods of systems analysis; industrial educational methods. *Mailing Add:* Sch Eng Widener Univ Chester PA 19013

SMYTH, NICHOLAS PATRICK DILLON, b Dublin, Ireland, Apr 1, 24; nat US; m 55; c 5. SURGERY. *Educ:* Nat Univ Ireland, BSc, 46, MSc, 48, MB, BCh, 49; Univ Mich, MS, 54. *Prof Exp:* From instr to assoc prof, 58-68, assoc clin prof, 68-83, clin prof surg, Sch Med, George Washington Univ, 83-; dir surg res, Washington Hosp Ctr, 68-, consult, 70-; RETIRED. *Concurrent Pos:* Chief surg, St Elizabeth's Hosp, 60-63, consult, 63-; consult, DC Gen Hosp, 60- & NIH, 70-; chmn dept surg, Washington Hosp Ctr, 63-68; consult thoracic surg, NIH & Walter Reed Army Med Ctr & Vet Admin Hosp. *Mem:* Am Heart Asn; Am Col Surgeons; Am Asn Thoracic Surg; Am Col Chest Physicians; Am Fedn Clin Res. *Res:* Thoracic and cardiovascular surgery. *Mailing Add:* 4041 Gulfshore Blvd N No 809 Naples FL 33940

SMYTH, THOMAS, JR, b Binghamton, NY, May 12, 27. INSECT PHYSIOLOGY. *Educ:* Princeton Univ, AB, 48; Johns Hopkins Univ, PhD(biol), 52. *Prof Exp:* Res assoc & instr biol, Tufts Univ, 52-55; from asst prof to assoc prof, 55-73, PROF ENTOM, PA STATE UNIV, 73- *Mem:* AAAS; Biophys Soc; Am Soc Zool; Entom Soc Am. *Res:* Neuromuscular and sensory physiology of arthropods; spider venom neurotoxins. *Mailing Add:* Dept Entom Pa State Univ University Park PA 16802

SMYTHE, CHEVES MCCORD, b Charleston, SC, May 25, 24; m 49; c 6. MEDICINE. *Educ:* Harvard Univ, MD, 47. *Prof Exp:* From asst prof to assoc prof, Med Col SC, 57-66, dean sch med, 62-64; dir, Asn Am Med Cols, 66-70; dean, 70-75, PROF MED, UNIV TEX MED SCH, HOUSTON, 76- *Concurrent Pos:* Teaching fels med, Harvard Univ, 48-49 & 54-55; teaching fel, Columbia Univ, 50-52, Am Col Physicians & Life Ins Med Res Fund fels, 51-52; Markle fel med, 55-60; dir gen commissioning & opers & prof med, Aga Khan Hosp & Med Col, Karachi, Pakistan, 82, prof med & dean fac Health Sci, 82-85. *Mem:* AMA; Am Fedn Clin Res; Am Col Physicians. *Res:* medical education. *Mailing Add:* Univ Tex Sch Med MSMB 1150 PO Box 20708 Houston TX 77225

SMYTHE, RICHARD VINCENT, b Philadelphia, Pa, June 27, 39; m 62; c 2. ENTOMOLOGY. *Educ:* Col Wooster, BA, 61; Univ Wis-Madison, MS, 63, PhD(entom), 66. *Prof Exp:* Entomologist, Southern Forest, Exp Sta, 66-69, proj leader entom, 69-74, staff entomologist, Forest Serv, USDA, 74-76, staff asst dep chief res, 76-77; asst dir continuing res, 77-81, DEPT DIR, N CENT FOREST EXP STA, FOREST SERV, USDA, 81- *Concurrent Pos:* Consult, Nat Pest Control Asn, 71-76. *Mem:* AAAS; Entom Soc Am; Sigma Xi; Soc Am Foresters. *Res:* Feeding behavior, physiology and ecology of wood products insects, chiefly subterranean termites. *Mailing Add:* 7910 Oak Hollow Lane Fairfax Station VA 22039

SMYTHE, ROBERT C, b Orlando, Fla. CIRCUIT THEORY, CRYSTAL FILTERS & RESONATORS. *Educ:* Rice Univ, BA, 52, BS, 53; Univ Fla, MS, 57. *Prof Exp:* Asst vpres, Syst, Inc, 62-65; dir, Sawtek, Inc, 79-87; FROM VPRES TO SR VPRES, PIEZO TECHNOL, 65-, DIR, 70- *Mem:* Inst Elec & Electronic Engrs; AAAS; Acoust Soc Am. *Res:* Linear and nonlinear theory of piezoelectric resonators and filters especially monolithic filters; precision resonator measurement. *Mailing Add:* Piezo Technol Inc PO Box 547859 Orlando FL 32854

SMYTHE, WILLIAM RODMAN, b Calif, Jan 6, 30; m 54; c 4. PHYSICS. *Educ:* Calif Inst Technol, BS, 51, MS, 52, PhD(physics), 57. *Prof Exp:* Engr, Microwave Lab, Gen Elec Co, 56-57; res assoc physics, Univ Colo, Boulder, 57-58, from asst prof to assoc prof, 58-67, chmn nuclear physics lab, 67-69, 81-83, 90-92, PROF PHYSICS, UNIV COLO, BOULDER, 67- *Mem:* Am Phys Soc. *Res:* Nuclear physics and particle accelerators. *Mailing Add:* Dept Physics Univ Colo Boulder CO 80309-0390

SNADER, KENNETH MEANS, b Harrisburg, Pa, Apr 6, 38; m 60; c 3. CHROMATOGRAPHIC TECHNIQUES. *Educ:* Philadelphia Col Pharm & Sci, BSc, 60; Mass Inst Technol, PhD(org chem), 68. *Prof Exp:* Sr investr med chem, Smith, Kline & French Labs, 68-80, asst dir, 80-84; head, Dept Med Chem, Seapharm, 84-87. *Concurrent Pos:* Lectr, Philadelphia Col Pharm, & Sci, 75-81. *Mem:* Am Chem Soc; AAAS; Am Soc Microbiol. *Res:* Natural products isolation and structure determination; marine natural products; microbial metabolites, antibiotics and mycotoxins; synthetic medicinal chemistry; anti-inflammatory oxygen heterocycles; immunological RNA; nuclear magnetic resonance and mass spectroscopy interpretations; high performance liquid chromatography. *Mailing Add:* Natural Prod Br Frederick Cancer Res & Develop Ctr Nat Cancer Inst Bldg 1052 Frederick MD 21702-1201

SNAPE, WILLIAM J, b Camden, NJ, July 18, 12. GASTROINTESTINAL PHYSIOLOGY. *Educ:* Thomas Jefferson Univ, MD, 40. *Prof Exp:* Prof med, Univ Calif, Los Angeles, 43-86; RETIRED. *Mailing Add:* 261 Moore Lane Haddonfield NJ 08033

SNAPE, WILLIAM J, JR, b Camden, NJ, Aug 24, 43; m; c 2. GASTROENTEROLOGY. *Educ:* Jefferson Med Col, Philadelphia, Pa, MD, 69. *Prof Exp:* PROF MED, SCH MED, UNIV CALIF, LOS ANGELES, 82- *Mem:* Am Soc Clin Invest; Am Physiol Soc; Biophys Soc; Am Fedn Clin Res. *Res:* Smooth muscle physiology. *Mailing Add:* 2650 Elm Ave Suite 201 Long Beach CA 90806

SNAPER, ALVIN ALLYN, b Hudson Co, NJ, Sept 9, 27; m 49; c 3. RESEARCH ADMINISTRATION. *Educ:* McGill Univ, BS, 49. *Prof Exp:* Sr chemist, Bakelite Div, Union Carbide Corp, 50-52; chief chemist, McGraw Colorgraph Co, 52-55; vpres, Marcal Electro-Sonics Co, 55-61 & Houston Fearless Corp, 61-63; consult, Marquardt Corp, 63-64; dir res, Fed Res & Develop Corp, 64-66; vpres, Advan Patent Technol, Inc, 69-80; pres, Nicoa Corp, 81-82; CONSULT, AEROSPACE CORP, 66-; PRES, NEO-DYNE RES CORP, 82- *Concurrent Pos:* Consult, US Libr Cong, 67- & US Air Force Missile Command, 68-; corp staff consult, Telecommunications Industs Inc, 72-, Multi-Arc Vacuum Systs, Inc, 82- & Am Methyl Corp, 82-87. *Mem:* Sr mem AAAS; sr mem Am Ord Asn; sr mem Soc Photo-Optical Instrument Eng; sr mem Instrument Soc Am. *Res:* Basic research in ultrasonics for environmental waste treatment and biological effects on bacteria and virus. *Mailing Add:* 2800 Cameo Circle Las Vegas NV 89107

SNAPP, THOMAS CARTER, JR, b Suffolk, Va, Aug 23, 38; m 60; c 1. ORGANIC CHEMISTRY, ANALYTICAL CHEMISTRY. *Educ:* E Tenn State Col, BS, 59; Univ Miss, PhD(org & anal chem), 64. *Prof Exp:* Develop assoc, 74-78, dept head, 70-81, CHEMIST, TEX EASTMAN CO, 63-, ASST DIV HEAD, 81- *Mem:* Catalysis Soc; Am Chem Soc. *Res:* Cyclodehydrogenation reactions; organic syntheses by heterogenious catalytic vapor phase reactions; surface catalysis; epoxidation and organic peracid chemistry; chemistry of lactones. *Mailing Add:* 713 Kay Dr Box 7444 Longview TX 75601

SNAPPER, ERNST, b Groningen, Neth, Dec 2, 13; nat US; m 41; c 2. PHILOSOPHY OF MATHEMATICS. *Educ:* Princeton Univ, MA, 39, PhD(math), 41. *Prof Exp:* Instr math, Princeton Univ, 41-45; from asst prof to prof, Univ Southern Calif, 45-55; Andrew Jackson Buckingham prof, Miami Univ, 55-58; prof, Ind Univ, 58-63; prof, 63-71, B P CHENEY PROF MATH, DARTMOUTH COL, 71- *Concurrent Pos:* Vis assoc prof, Princeton Univ, 49-50, vis prof, 54-55; NSF fel, Harvard Univ, 53-54. *Honors & Awards:* Carl B Allendoerfer Award, Math Asn Am, 80. *Mem:* AAAS; Am Math Soc; Math Asn Am. *Res:* Algebra; geometry; combinatorial theory; philosophy of mathematics. *Mailing Add:* Dept of Math Dartmouth Col Hanover NH 03755

SNAPPER, JAMES ROBERT, b Los Angeles, Calif, May 23, 48; m 72; c 2. PULMONARY MEDICINE, CRITICAL CARE MEDICINE. *Educ:* Princeton Univ, AB, 70; Dartmouth Med Sch, BMS, 72, Harvard Med Sch, MD, 74. *Prof Exp:* Intern, Mass Gen Hosp, 74-75, resident, 75-76, clin fel internal med, 76-79; res fel, Peter Bent Brigham/Harvard Sch Pub Health, 76-79, resident, 78-79; asst prof, 79-83, ASSOC PROF MED, VANDERBILT UNIV, 83- *Concurrent Pos:* Sr investr,Pulmonary Res, Ctr Lung Res, Vanderbilt Univ, 86- *Mem:* Am Physiol Soc; Am Thoracic Soc; Am Col Chest Physicians; Am Heart Asn; AAAS; Am fed Clin Res. *Res:* Role of lipid mediators and the pathophysiologic mechanisms responsible for acute lung injury and altered airway responsiveness. *Mailing Add:* B1308 Med Ctr N Vanderbilt Univ Hosp Nashville TN 37232

SNARE, LEROY EARL, b Garden City, Mo, Nov 6, 31; m 60; c 3. AERONAUTICS, ASTRONAUTICS. *Educ:* Univ Mo-Kansas City, BA, 53, MS, 59; Mass Inst Technol, MS, 62. *Prof Exp:* Gen res physicist, Systs Analysis Div, Naval Avionics Ctr, 59-62, chief, Dynamic Analysis & Simulation Br, 62-72, res physicist, 72-76, dir, Systs Analysis Div, 76-80, dep dir, Appl Res Dept, 80-84, dep dir, Eng Dept, 84-86, res coordr for avionics, 86-91; PHYSICS INSTR, IND VOC TECH COL, 91- *Mem:* Inst Elec & Electronics Engrs; Nat Asn Unmanned Vehicles. *Res:* Alignment of inertial navigation systems; design, development and testing of alignment and filtering programs for airborne computers; auxiliary equipment for aligning inertial navigation systems aboard ships; analysis and conceptual design of air-to-surface missile systems and electronic intelligence systems. *Mailing Add:* Ind Voc Tech Col Instrnl Support Serv Div One W 26th St Indianapolis IN 46206-1763

SNARR, JOHN FREDERIC, b Cincinnati, Ohio, Jan 3, 39; m 60; c 2. PHYSIOLOGY. *Educ:* Univ Cincinnati, EE, 61; Drexel Inst, MS, 62; Northwestern Univ, PhD(physiol), 67. *Prof Exp:* Asst prof, 67-73, assoc dean student affairs, 75-90, ASSOC PROF PHYSIOL, MED SCH, NORTHWESTERN UNIV, 73-, ASSOC DEAN STUDENT PROGS, 90- *Res:* Quantification of nutrient supply system operation and regulation. *Mailing Add:* Assoc Dean Student Progs Med Sch Northwestern Univ Chicago IL 60611

SNAVELY, BENJAMIN BRENEMAN, b Lancaster, Pa, Jan 6, 36; m 61; c 2. QUANTUM ELECTRONICS, SOLID STATE PHYSICS. *Educ:* Swarthmore Col, BS, 57; Princeton Univ, MSE, 59; Cornell Univ, PhD(eng physics), 62. *Prof Exp:* Sr res physicist, Eastman Kodak Co, 62-65, res assoc solid state physics, Res Labs, 65-69, head solid state & molecular physics lab, 69-73; assoc div leader, Laser Div, Lawrence Livermore Lab, 73-75; asst dir, Physics Div, 75-81, tech asst to dir res, Res Lab, 81-83, asst dir, Image Rec Div, 83-85, ASST GEN MGR, ADVAN TECHNOL PROD, FED SYSTS DIV, EASTMAN KODAK CO, 85- *Concurrent Pos:* Vis prof, Phys-Chem Inst, Univ Marburg, 68-69; assoc prof, Inst Optics, Univ Rochester. *Mem:* Am Phys Soc; fel Optical Soc Am. *Res:* Photoconductivity in silver halides and II-VI compounds; electronic and optical properties of thin films; electroluminescence; organic dye lasers; tunable lasers; laser induced photochemistry; electro-optical imaging systems. *Mailing Add:* 27 Countryside Rd Fairport NY 14450

SNAVELY, DEANNE LYNN, b Columbus, Ohio, Nov 16, 51; m 82; c 1. PHYSICAL CHEMISTRY. *Educ:* Ohio State Univ, BS, 77; Yale Univ, PhD(phys chem), 83. *Prof Exp:* Res scholar, Stanford Univ, 83-85; assoc res scientist, Yale Univ, 85; ASST PROF PHYS CHEM, BOWLING GREEN STATE UNIV, 85- *Mem:* Am Phys Soc. *Res:* Infrared absorption spectoscopy of supersonic jets of polyatomic molecules, vibrational and rotational analysis; photoacoustic absorbtion spectroscopy of highly excited vibrational states in gas phase molecules; kinetic studies of reactions initiated by laser pumping of highly excited vibrational states in gas phase molecules; vibrational spectroscopy of molecules; photoinitiated reaction kinetics. *Mailing Add:* Chem Dept Bowling Green State Univ Bowling Green OH 43403

SNAVELY, EARL SAMUEL, JR, b Brackettville, Tex, Apr 10, 27; m 53; c 1. PHYSICAL CHEMISTRY. *Educ:* Agr & Mech Col, Tex, BS, 47; Univ Tex, MA, 50, PhD, 58. *Prof Exp:* Chemist, Oyster Mortality Proj, Res Found, Agr & Mech Col, Tex, 47-48; chemist, Southern Alkali Corp, 50-51; res scientist, Defense Res Lab, Univ Tex, 51-58; chemist, Oak Ridge Nat Lab, 58-60; dir chem res, Tracor, Inc, 60-66; sr res chemist, Mobile Oil Corp, Tex, 66-68, res assoc, Field Res Lab, 68-80, eng consult, Mobile Oil Res & Develop Corp, 80-87; RETIRED. *Concurrent Pos:* Ed, Corrosion Div, J Electrochem Soc. *Mem:* Sigma Xi. *Res:* Electrochemistry; corrosion; surface chemistry; environmental science. *Mailing Add:* 2610 Oak Cliff Lane Arlington TX 76012

SNAVELY, FRED ALLEN, inorganic chemistry; deceased, see previous edition for last biography

SNAVELY, PARKE DETWEILER, JR, b Yakima, Wash, Apr 7, 19; m 42; c 3. GEOLOGY. *Educ:* Univ Calif, Los Angeles, BA, 41, MA, 51. *Prof Exp:* From jr geologist to supvry geologist, 42-53, supvr, Pac Region, 53-59, res geologist, 59-60, chief, Pac Coast Br, 60-66, chief, Off Marine Geol & Hydrol, 66-69, asst chief geologist, 69-71, SR RES GEOLOGIST, OFF MARINE GEOL, US GEOL SURV, 71- *Concurrent Pos:* Chmn marine geol panel, US-Japan Coop Prog Natural Resources, 70-; res assoc, Univ Calif, Santa Barbara, 69-76. *Mem:* Fel Geol Soc Am; Am Asn Petrol Geologists. *Res:* Tertiary geology and mineral resource potential of western Oregon and Washington and adjacent continental shelf; relation of plate tectonics to structural, stratigraphic, and igneous history of Pacific coast states. *Mailing Add:* 1210 Larnel Pl Los Altos CA 94024

SNAZELLE, THEODORE EDWARD, b Richmond, Ind, Aug 30, 41; m 61; c 2. MICROBIOLOGY. *Educ:* Belmont Col, BS, 65; Purdue Univ, MS, 68, PhD(plant path), 70. *Prof Exp:* Instr biol, Rock Valley Col, Rockford, Ill, 70-72; asst prof biol, Univ Tenn, Nashville, 72-74, assoc prof,74-79, coordr, 75-79, prof, 79; prof biol, Tenn State Univ, 79-80; PROF BIOL, MISS COL, CLINTON, 80- *Concurrent Pos:* Vis researcher, Gulf Coast Res Lab, Ocean Springs, Miss, 75. *Mem:* Am Soc Microbiol; Sigma Xi. *Res:* Pigment production in Bacillus cereus; Narcissus diseases and pests. *Mailing Add:* Dept Biol Sci Box 4045 Miss Col Clinton MS 39058

SNEAD, CLARENCE LEWIS, JR, b Richmond, Va, Sept 25, 36; m 60; c 3. SOLID STATE PHYSICS. *Educ:* Univ Richmond, BS, 59; Univ NC, PhD(physics), 65. *Prof Exp:* Res assoc physics, Univ NC, 65; res assoc mat sci, Northwestern Univ, 65-67; from asst physicist to assoc physicist, Brookhaven Nat Labs, 67-71, ed, Sect A, Phys Rev & Res Collabr, 71-74, assoc physicist, 74-80, physicist, Mat Sci Dept, 80-84, PHYSICIST, DEPT NUCLEAR ENERGY, BROOKHAVEN NAT LAB, 84-, DIV HEAD, NEUTRAL BEAM DIV, 89- *Concurrent Pos:* Consult ed, Phys Rev, 74-90. *Mem:* Am Phys Soc; Am Inst Mech Engrs; Metall Soc. *Res:* Radiation effects, especially in type II superconductors and metals; positron annihilation studies in defects in metals; internal-friction studies of defects; high-energy proton irradiation effects. *Mailing Add:* Bldg 830 Brookhaven Nat Lab Upton NY 11973

SNEAD, O CARTER, III, b Princeton, WVa, Oct 24, 43. CHILD NEUROLOGY. *Educ:* Univ WVa, MD, 70. *Prof Exp:* PROF PEDIAT NEUROPHYSIOL, SCH MED, UNIV ALA, BIRMINGHAM, 84- *Mailing Add:* Neurol Div Childrens Hosp 450 Sunset Blvd Los Angeles CA 90027

SNEADE, BARBARA HERBERT, b Altoona, Pa, Nov 15, 47; m 69; c 2. ANALYTICAL CHEMISTRY. *Educ:* Bridgewater Col, BA, 65; Nova Univ, MBA, 85. *Prof Exp:* Res assoc, Am Tobacco Co, 69-71; lab dir, Lee County Sheriff's Dept, 73-77; teacher math & sci, Lee County Schools, 71-73 & 77-80; TECH DIR, HF SCI, INC, 80- *Mem:* Am Chem Soc; Am Asn Clin Chem; Am Soc Qual Control. *Res:* Turbidity; characterization of zero and development of permanent standards. *Mailing Add:* 3170 Metro Pkwy Ft Myers FL 33916-7597

SNECK, HENRY JAMES, (JR), b Schenectady, NY, Nov 9, 26; m 52; c 3. MECHANICAL ENGINEERING. *Educ:* Rensselaer Polytech Inst, BME, 51, PhD(mech eng), 63; Yale Univ, MEng, 52. *Prof Exp:* Jr engr, Eastman Kodak Co, 51; test engr, Gen Elec Co, 52-53; from instr to assoc prof, 53-76, PROF MECH ENG, RENSSELAER POLYTECH INST, 76- *Concurrent Pos:* Consult, Corp Res & Develop Ctr, Gen Elec Co, 53- *Mem:* AAAS; Am Soc Mech Engrs; Sigma Xi. *Res:* Bearings, seals, lubrication, atmospheric thermal pollution. *Mailing Add:* 21 Bolivar Ave Troy NY 12180

SNECKENBERGER, JOHN EDWARD, b Hagerstown, Md, Aug 17, 37; m 68; c 3. MANUFACTURING ENGINEERING. *Educ:* WVa Univ, BS, 64, MS, 66, PhD(eng), 69. *Prof Exp:* From asst prof to assoc prof, 70-81, PROF MECH ENG, WVA UNIV, 81- *Concurrent Pos:* Mem res adv bd, Nelson Industs, Inc, Wis, 78-; eng, IBM, 83; consult, DuPont, 84-89. *Mem:* Am Soc Mech Engrs; Am Soc Eng Educ; Soc Mfg Engrs. *Res:* Engineering systems design, automation and control; concurrent engineering. *Mailing Add:* Mech & Aero Eng WVa Univ PO Box 6101 Morgantown WV 26506-6101

SNEDAKER, SAMUEL CURRY, b Long Beach, Calif, May 22, 38; m 68; c 4. ECOLOGY. *Educ:* Univ Fla, BSA & BSF, 61, MS, 63, PhD(ecol), 70. *Prof Exp:* Res assoc ecol, 68-69, asst prof, 70-73, asst prof ecol & environ eng sci, Resource Mgt Systs Prog, 73-74, asst prof aquatic sci, Univ Fla, 71-74, asst prof ecol, Inst Food & Agr Sci, 74-75; prof biol & living resources, Univ Miami, 75-82, prof marine affairs, 82-86, prof biol & living resources, 86-89, PROF MARINE BIOL & FISHERIES, UNIV MIAMI, 89- *Concurrent Pos:* Res fel, East-West Ctr, Honolulu, Hawaii. *Mem:* AAAS; Am Inst Biol Sci; Asn Trop Biol; Ecol Soc Am. *Res:* Structure and function of tropical lowland and coastal ecosystems with respect to their relationship to man. *Mailing Add:* Sch Marine & Atmospheric Sci 4600 Rickenbacker Causeway Miami FL 33149-1098

SNEDDEN, WALTER, b Renton, Scotland, Feb 4, 36; Can citizen; m 60; c 5. EXPERIMENTAL MEDICINE, MASS SPECTROMETRY. *Educ:* Glasgow Univ, BSc, 56, PhD (chem), 59. *Prof Exp:* Scientist, Shell Res Ltd, Eng, 59-62, sr scientist, 62-67; sr lectr biochem, St Bartholomews Hosp, London, 67-76, reader biochem, 76-77; assoc prof, 77-78, PROF BIOCHEM, MEM UNIV MED SCH, NFLD, CAN, 78- *Concurrent Pos:* Sr lectr, Carlett Park Col Technol, 64-66; consult metab dis, Inst Child Health, London, 72-76; consult biochem metab, Janeway Childrens Hosp, Nfld, Can, 85- *Mem:* Can Soc Clin Invest. *Res:* Pharmacology of antihypertensive drugs, using them as probes to explore the physiological and biochemical mechanisms which control blood pressure in man; the effects of these drugs on intermediary metabolism. *Mailing Add:* Medicorp Labs Mem Med Sch Health Sci Ctr St Johns NF A1B 3V6 Can

SNEDDON, LEIGH, b New South Wales, Australia. VULNERABILITY OF SPREAD SPECTRUM COMMUNICATIONS. *Educ:* Univ Sydney, BSc, 74; Univ S Wales, MSc, 75; Oxford Univ, PhD(physics), 78. *Prof Exp:* Res assoc, Oxford Univ, 78-79, & Princeton Univ, 79-82; asst prof physics, Brandeis Univ, 82-89; TECH STAFF, TASC, 89- *Concurrent Pos:* Consult, Bell Labs, 81-82; vis scientist, Philip Morris Inc, 84-85; vis scientist, Inst Laue Langevin, France, 86-87. *Mem:* Am Phys Soc. *Res:* Vulnerability of spread spectrum communications; non-linear electronic materials; collective phenomena; scientific applications of computers; solid state physics. *Mailing Add:* TASC 55 Walkers Brook Dr Reading MA 01867

SNEDECOR, JAMES GEORGE, b Ames, Iowa, June 9, 17; m 44; c 2. PHYSIOLOGY. *Educ:* Iowa State Univ, BS, 39; Ind Univ, PhD(zool), 47. *Prof Exp:* Asst prof zool, La State Univ, 47-48; from asst prof to assoc prof, 48-57, PROF PHYSIOL, UNIV MASS, AMHERST, 57- *Concurrent Pos:* NIH sr res fel, 55-56. *Mem:* Endocrine Soc; Am Soc Zoologists; Am Physiol Soc. *Res:* Carbohydrate metabolism; avian thyroid physiology. *Mailing Add:* 49 Fairfield Amherst MA 01002

SNEDEGAR, WILLIAM H, b Ward, WVa, Aug 31, 26; m 48; c 2. NUCLEAR PHYSICS. *Educ:* WVa Univ, AB, 48, MS, 49; Univ Ky, PhD(physics), 58. *Prof Exp:* Physicist, Nat Bur Stand, 49-52; instr physics, Wis Col, Superior, 52-53; asst prof, Univ Ky Aid Prog to Univ Indonesia, 57-61; assoc prof physics, Eastern Ky State Col, 61-63; prof & chmn dept, Parsons Col, 63-67; prof physics & chmn dept, Clarion Univ Pa, 67-88; RETIRED. *Mem:* Am Asn Physics Teachers. *Mailing Add:* 1057 L Main St Clarion PA 16214

SNEDEKER, ROBERT A(UDLEY), b New York, NY, Aug 3, 28; m 52; c 3. CHEMICAL ENGINEERING. *Educ:* Mass Inst Technol, SB, 50, SM, 51; Princeton Univ, PhD, 56. *Prof Exp:* Engr, Photo Prod Dept, E I du Pont de Nemours & Co, 55-58, res supvr, 58-67; tech dir, Scott Paper Co, 67-70; vpres mfg, Merrimac Paper Co, 70-83; SR VPRES RES & DEVELOP, ARNOX CORP, 87- *Mem:* Am Chem Soc; Am Inst Chem Engrs; Sigma Xi; TAPPI; Paper Indust Mgt Asn. *Res:* Polymer fabrication; photopolymerization; coating and drying; papermaking techniques; pollution control; fire retardants. *Mailing Add:* Seven Mashie Way North Reading MA 01864-3423

SNEE, RONALD D, b Wash, Pa, Dec 11, 41; m 67; c 2. STATISTICS. *Educ:* Wash & Jefferson Col, BA, 63; Rutgers Univ, MS, 65, PhD(statist), 67. *Prof Exp:* Asst prof statist, Rutgers Univ, 66-68; from statistician to sr statistician, 68-75, consult, 75-76, consult supvr statistician, 76-87, QUAL SYST MGR, DEPT ENG, DUPONT CO, 87- *Concurrent Pos:* Statist fac, Univ Del, 70-75, 84, 88- *Honors & Awards:* Brumbaugh Award, 71, Shewell Prize, 72, Wilcoxon Prize, 72, 75 & 81, Youden Prize, 74 & 77, Ellis R Ott Award, 80, Am Soc Qual Control Shewhart Medal, 85, William G Hunter Award, 90. *Mem:* Fel Am Statist Asn; fel Am Soc Qual Control; Biometrics Soc; fel Am Asn Adv Sci. *Res:* Design and analysis of experiments; data analysis; graphical methods; mixture experiments; model building; statistical thinking; scientific problem solving, quality management, and technology. *Mailing Add:* 109 Red Pine Circle Newark DE 19711

SNEEN, RICHARD ALLEN, b Menomonie, Wis, July 19, 30. ORGANIC CHEMISTRY. *Educ:* St Olaf Col, BA, 52; Univ Ill, PhD(chem), 55. *Prof Exp:* Fel & res asst, Univ Calif, Loa Angeles, 55-56; from instr to assoc prof chem, 56-72, PROF CHEM, PURDUE UNIV, LAFAYETTE, 72- *Mem:* Am Chem Soc; Royal Soc Chem. *Res:* Physical organic chemistry; reaction mechanisms; kinetics; stereochemistry; solvolysis reactions; ion-pair intermediates. *Mailing Add:* 5808 North 75 East West Lafayette IN 47906

SNEIDER, ROBERT MORTON, b Asbury Park, NJ, Mar 2, 29; m 56; c 3. PETROLEUM. *Educ:* Rutgers Univ, BS, 51; Univ Wis, PhD(geol), 62. *Prof Exp:* Res geologist, Shell Develop Co, 57-65, res assoc, 65-66, from staff prod geologist to sr staff prod geologist, 66-71, res sect leader geol eng, Shell Develop Co, 71-74; pres, Sneider & Meckel Assocs, Inc, 74-81; PRES, ROBERT M SNEIDER EXPLOR INC, 81-; PARTNER, RICHARDSON, SANGREE & SNEIDER, 86- *Concurrent Pos:* Distinguished lectr, Soc Petrol Engrs, 77-78, am Asn Petrol Geol, 88-89. *Honors & Awards:* Distinguished Serv Award, Am Asn Petrol Geologists, 91. *Mem:* AAAS; Am Geol Inst; Am Asn Petrol Geol; Soc Econ Paleontologists & Mineralogists; Can Well Logging Soc; NY Acad Sci. *Res:* Exploration and reservoir geology and petrophysics of petroleum reservoirs in the Gulf Coast, California, Canadian Rocky Mountains, Alaska, Australia. *Mailing Add:* Robert M Sneider Explor, Inc 11767 Katy Fwy Suite 330 Houston TX 77079

SNEIDER, THOMAS W, b Fremont, Ohio, Apr 19, 38; m 65; c 3. BIOCHEMISTRY, MOLECULAR BIOLOGY. *Educ:* Univ Detroit, BSc, 61; Marquette Univ, MSc, 63, PhD(physiol), 65. *Prof Exp:* Instr oncol, Univ Wis-Madison, 67-69; from asst prof to assoc prof pharmacol, Baylor Col Med, 69-75; assoc prof, 75-81, PROF BIOCHEM, COLO STATE UNIV, 81- *Concurrent Pos:* NSF fel biochem & oncol, McArdle Lab, Univ Wis-Madison, 65-67; NIH res career develop award, 73-78; chmn, Grad Fac, Cellular & Molecular Biol, Colo State Univ, 78-80; NIH Fogarty Ctr sr int fel, Inst Molecularbiol II, Univ Zurich, 82-83; mem, Nat Res Coun Life Sci Review Panel, 82-85, Nat Sci Found Biochem, Biophys & Molec Biol Review Panel, 84-87. *Mem:* Am Soc Biol Chemists; Am Asn Cancer Res. *Res:* Chemical carcinogenesis; mechanisms and functions of eukaryotic DNA modifications; control of cellular differentiation. *Mailing Add:* Dept of Biochem Colo State Univ Ft Collins CO 80523

SNELGROVE, JAMES LEWIS, b Cookeville, Tenn, Jan 9, 42; m 65; c 2. REACTOR PHYSICS, REACTOR FUEL DEVELOPMENT & TESTING. *Educ:* Tenn Polytech Inst, BS, 64; Mich State Univ, MS, 66, PhD(physics), 68. *Prof Exp:* PHYSICIST, ARGONNE NAT LAB, 68- *Mem:* Am Phys Soc; Am Nuclear Soc; Sigma Xi; AAAS. *Res:* Development and use of fuels for research and test reactors, including design of reactor cores and testing of new high-density fuels. *Mailing Add:* Eng Physics Div Argonne Nat Lab 9700 S Cass Ave Argonne IL 60439

SNELL, A(BSALOM) W(EST), b Parler, SC, Apr 29, 24; m 51; c 4. AGRICULTURAL ENGINEERING. *Educ:* Clemson Univ, BS, 49; Iowa State Univ, MS, 52; NC State Univ, PhD(agr eng), 64. *Prof Exp:* Asst prof agr eng, Clemson Univ, 49-51; instr & res fel, Iowa State Univ, 51-52; from asst prof to prof & head dept, 52-75, chmn directorate, Water Resources Res Inst, 64-75, ASSOC DIR, SC AGR EXP STA, CLEMSON UNIV, 75- *Mem:* Am Soc Agr Engrs; Sigma Xi. *Res:* Water resources engineering; irrigation; drainage; water movement in soils; agricultural research; administration. *Mailing Add:* 116 Lewis Rd Clemson SC 29631

SNELL, ARTHUR HAWLEY, b Montreal, Que, Mar 10, 09; nat US; m 37, 41; c 3. NUCLEAR PHYSICS, ATOMIC PHYSICS. *Educ:* Univ Toronto, BA, 30; McGill Univ, MSc, 31, PhD(physics), 33. *Prof Exp:* Res assoc, McGill Univ, 33-34; Brit 1851 Exhib scholar, Univ Calif, 34-37, res assoc, Radiation Lab, 37-38; res instr physics, Univ Chicago, 38-42, sr physicist & chief cyclotron sect, Metall Lab, Manhattan Proj, Chicago, 42-44; sr physicist, sect chief & group leader, Oak Ridge Nat Lab, 44-48, dir physics div, 48-57, from asst dir to assoc dir lab, 57-73, dir fusion energy div, 58-67, consult, 73-87; RETIRED. *Concurrent Pos:* Chmn subcomt nuclear intruments & tech, Nat

Acad Sci-Nat Res Coun, 54-61; adv & mem US deleg Atoms for Peace Conf, Geneva, Switz, 55, 58 & 68; mem bd dirs, Oak Ridge Inst Nuclear Studies, 59-62. *Honors & Awards:* Gov Gen Can Res Medal, 33. *Mem:* Fel AAAS; fel Am Phys Soc; fel Royal Soc Arts; Sigma Xi. *Res:* Fast neutron effects in massive uranium metal; decay curve and intensity of delayed neutrons for control of chain reaction; identification of delayed neutron emitters; radioactive decay of neutron; neutrino recoil spectrometry; atomic electron loss following radioactive decay. *Mailing Add:* Rt 4 Box 222 James Ferry Rd Kingston TN 37763

SNELL, CHARLES MURRELL, b Johnson City, Tenn, Aug 19, 46. COMPUTATIONAL PHYSICS, GEOPHYSICS. *Educ:* Vanderbilt Univ, BA, 67; Univ Ariz, MS, 69. *Prof Exp:* Physicist explosives eng, Explosive Excavation Res Off, US Army Engr Waterways Exp Sta, 71-73; physicist comput physics, Lawrence Livermore Lab, 73-78; physicist comput physics & geophys, 78-80, PHYSICIST, PARTICLE-IN-CELL NUMERICAL MODELING, LOS ALAMOS NAT LAB, 80- *Res:* Physical properties of solid materials; constitutive relations of rocks and soils; material equations of state; underground and underwater explosions; computer modeling of material dynamics; particle-in-cell numerical modeling; simulation of intense charged-particle beams. *Mailing Add:* 2021-C22 Los Alamos NM 87544

SNELL, ESMOND EMERSON, b Salt Lake City, Utah, Sept 22, 14; m 41; c 4. BIOCHEMISTRY. *Educ:* Brigham Young Univ, BA, 35; Univ Wis, MA, 36, PhD(biochem), 38. *Hon Degrees:* DSc, Univ Wis, 82. *Prof Exp:* Res assoc chem, Univ Tex, 39-41, from asst prof to prof, 41-56; from assoc prof to prof biochem, Univ Wis, 45-51; prof biochem, Univ Calif, Berkeley, 56-76; prof & chmn, Dept Microbiol, 76-80, Ashbel Smith prof, 80- 90, EMER PROF MICROBIOL & CHEM, UNIV TEX, AUSTIN, 90- *Concurrent Pos:* Ed, Ann Rev Biochem, Ann Rev, Inc, 69-83, pres, 72-76; Guggenheim fels, 54, 62, 70. *Honors & Awards:* Lilly Award, Am Soc Bacteriologists, 45; Mead Johnson Award, Am Inst Nutrit, 46 & Osborne Mendel Award, 51; Kenneth A Spencer Award, Am Chem Soc, 74; William C Rose Award, Am Soc Biol Chemists, 85. *Mem:* Nat Acad Sci; AAAS; Am Soc Microbiol; Am Soc Biol Chem (pres, 61-62); Am Chem Soc; fel Am Inst Nutrit. *Res:* Metabolism and mechanism of action of vitamin B6; vitamin metabolism and transport; pyruvoyl enzymes; pyridoxal phosphate enzymes. *Mailing Add:* Dept Microbiol Univ Tex Austin TX 78712

SNELL, FRED MANGET, b Soochow, China, Nov 11, 21; m 46; c 3. BIOPHYSICS. *Educ:* Maryville Col, Tenn, AB, 42; Harvard Univ, MD, 45; Mass Inst Technol, PhD(biochem), 52. *Prof Exp:* Intern pediat, Children's Hosp, 45-46, from jr asst resident to asst resident, 48-49; res assoc biol, Mass Inst Technol, 52-54; assoc biochem, Harvard Med Sch, 54-57, asst prof, 57-59; prof biophys & chmn dept, 59-70, dean grad sch, 67-69, master, Col A, 68-71, PROF BIOPHYS SCI, SCH MED, STATE UNIV NY BUFFALO, 70- *Concurrent Pos:* Nat Found Infantile Paralysis fel, Mass Inst Technol, 52-54; Nat Found Infantile Paralysis fel, Children's Med Ctr, 52-54; Palmer sr fel, Harvard Med Sch, 54-57; US Navy rep & mem atomic bomb casualty comn, Comt Atomic Casualties, Nat Res Coun; mem biophys sci training comt, Nat Inst Gen Med Sci, 62-66, chmn, 65-66; mem biophys panel, President's NIH Study Comt, 64; mem interdisciplinary panel, Comn Undergrad Educ Biol Sci, 65, chmn biomath subpanel, 66-68; consult, Comn Undergrad Progs in Math, 65-68; mem adv comt on NIH training progs, Nat Acad Sci-Nat Res Coun; ed, Biophys J, 66-69; vis prof, atmospheric sci, Ore State Univ, Corvallis, 77-78. *Mem:* AAAS; Biophys Soc (pres, 71). *Res:* membrane processes; global thermodynamics; instructional computer simulations. *Mailing Add:* Dept Biophys Sci Sch Med State Univ NY Health Ctr Buffalo NY 14214

SNELL, GEORGE DAVIS, b Bradford, Mass, Dec 19, 03; m 37; c 3. GENETICS, TISSUE TRANSPLANTATION. *Educ:* Dartmouth Col, BS, 26; Harvard Univ, MS, 28, ScD(genetics), 30. *Hon Degrees:* MD, Charles Univ, Prague, 68; ScD, Dartmouth Col, 74 & Univ Maine & Gustavus Aldolphus Col, 81, Bates Col, 82, Ohio State Univ, 84; LLD, Colby Col, 82. *Prof Exp:* Instr zool, Dartmouth Col, 29-30 & Brown Univ, 30-31; Nat Res Coun fel, Univ Tex, 31-33; asst prof, Wash Univ, 33-34; res assoc, Jackson Lab, 35-56, sci adminstr, 49-50, sr staff scientist, 57-68, retired assoc, 69, emer sr staff scientist, 69-73; RETIRED. *Concurrent Pos:* Guggenheim fel, Univ Tex, 53-54; mem allergy & immunol study sect, NIH, 58-62. *Honors & Awards:* Corecipient, Nobel Prize in Physiol & Med, 80; Gairdner Found Award, 76; Wolf Prize Med Res, 78. *Mem:* Nat Acad Sci; AAAS; Am Soc Nat; foreign assoc Fr Acad Sci; Am Philos Soc. *Res:* Genetics of the house mouse; radiation genetics; genetics and immunology of tissue transplantation; immunogenetics. *Mailing Add:* 21 Atlantic Ave Bar Harbor ME 04609

SNELL, JAMES LAURIE, b Wheaton, Ill, Jan 15, 25; m 52; c 2. PROBABILITY. *Educ:* Univ Ill, BS, 47, MA, 48, PhD(math), 51. *Prof Exp:* Fine instr, Princeton Univ, 51-54; from asst prof to assoc prof, 54-62, PROF MATH, DARTMOUTH COL, 62- *Mem:* Am Math Soc; Math Asn Am. *Res:* Probability theory. *Mailing Add:* Dept of Math Dartmouth Col Hanover NH 03755

SNELL, JOHN B, b Waterbury, Conn, May 10, 36; m 66; c 3. PLASTICS CHEMISTRY. *Educ:* Univ Wis, BSChE, 59; Inst Paper Chem, PhD(phys chem), 64. *Prof Exp:* Sr chemist, Tape Div, 3M Co, 64-68, res supvr, 68-71, res specialist, Indust Spec Div, 71-84, SR TECH SERV SPECIALIST, SCOTCHLITE GLASS BUBBLE PRODS, 84- *Res:* Epoxy resins and curing agents for reinforced plastics and advanced composites; polymers for vibration dampening; surface treatments for glass bubbles. *Mailing Add:* Bldg 230-1F-02 3M Ctr St Paul MN 55101

SNELL, JUNIUS FIELDING, b Lovell, Wyo, Feb 6, 21; m 45; c 2. BIOCHEMISTRY. *Educ:* Univ Tex, BS, 43; Univ Wis, MS, 44, PhD(biochem), 49. *Prof Exp:* Res scientist, Chas Pfizer & Co, Inc, 45-46, dir radiobiochem lab, 52-61; actg chmn dept biochem, 66-67, PROF BIOCHEM, BIOPHYS & MICROBIOL, OHIO STATE UNIV, 61- *Concurrent Pos:* Dir, Pfizer Therapeut Inst, 55-61. *Mem:* Soc Nuclear Med; Am Soc Microbiol; NY Acad Sci. *Res:* Phosphorous metabolism; yeast growth; antibiotic fermentations; mode of action of antibiotics; non-specific immunity; reticuloendothelial system. *Mailing Add:* Dept of Biochem Vivian Hall Ohio State Univ 464 W 12th St Columbus OH 43210

SNELL, RICHARD SAXON, b Richmond, Eng, May 3, 25; m 49; c 5. ANATOMY. *Educ:* Univ London, MB, BS, 49, PhD(med), 55, MD, 61. *Prof Exp:* House surgeon, King's Col Hosp, London, 48-49, jr lectr anat, King's Col, London, 49-53, lectr anat & hist, 53-59; lectr anat, Univ Durham, 59-63; from asst prof to assoc prof, Yale Univ, 63-67; prof & chmn dept, NJ Col Med & Dent, 67-69; prof, Univ Ariz, 70-72; PROF ANAT & CHMN DEPT, MED SCH, GEORGE WASHINGTON UNIV, 72- *Concurrent Pos:* Vis prof, Yale Univ, 69 & Harvard Univ, 70-71. *Mem:* Am Asn Anatomists; Anat Soc Gt Brit & Ireland. *Res:* Pigmentation of mammalian skin and its control; light and electron microscopic appearances of the skin; histochemistry of cholinesterase in the peripheral and central parts of the nervous system. *Mailing Add:* Dept Anat George Washington Univ Sch Med 2300 Eye St NW Washington DC 20037

SNELL, ROBERT ISAAC, b Hancock, Mich, Mar 16, 37; m 65; c 1. MATHEMATICS. *Educ:* Northern Mich Univ, BS, 59; Univ Mich, MS, 60; Univ Colo, PhD(appl math), 68. *Prof Exp:* Instr math, Mich Tech Univ, 60-64; asst prof math, Univ Puget Sound, 68-73, assoc prof, 73-80; CONSULT. *Mem:* Math Asn Am; Soc Indust & Appl Math. *Res:* Continued fractions; real and complex analysis; differential equations. *Mailing Add:* 112 Bon Bluff Rd Fox Island WA 98333

SNELL, ROBERT L, b El Dorado Springs, Mo, Jan 28, 25; m 55; c 2. ORGANIC CHEMISTRY. *Educ:* Drury Col, BS, 48; Mo Sch Mines, MS, 52; Tex Tech Col, PhD(chem), 59. *Prof Exp:* Chemist, Dowell, Inc, 54-55; from asst prof to assoc prof, 59-66, PROF CHEM, E TENN STATE UNIV, 66- *Concurrent Pos:* Adj prof pharmacol, Quillen-Dishner Col Med. *Mem:* Am Chem Soc; Sigma Xi. *Mailing Add:* Dept Chem E Tenn State Univ Johnson City TN 37601

SNELL, ROBERT ROSS, b St John, Kans, Apr 17, 32; m 52; c 2. CIVIL ENGINEERING. *Educ:* Kans State Univ, BS, 54, MS, 60; Purdue Univ, Lafayette, PhD(civil eng struct), 63. *Prof Exp:* Civil engr, Kans State Hwy Comn, 54-55; from instr to assoc prof civil eng, Kans State Univ, 57-67; Ford Found resident, Rust Eng Co, Pa, 67-68; PROF CIVIL ENG, KANS STATE UNIV, 68-, HEAD DEPT, 72- *Mem:* Am Soc Eng Educ; Am Soc Civil Engrs; Nat Soc Prof Engrs. *Res:* Structural analysis and design; systems optimization; structural modeling. *Mailing Add:* Dept of Civil Eng Seaton Hall Kans State Univ Manhattan KS 66506

SNELL, RONALD LEE, b Salina, Kans, May 15, 51; m 78. ASTRONOMY. *Educ:* Univ Kans, BA, 73; Univ Tex, Austin, MA, 75, PhD(astron), 79. *Prof Exp:* Teaching asst astron, Univ Kans, 72-73; teaching asst & res asst, Univ Tex, 73-79, instr 76-79; res assoc, 79-84, asst prof, 84-87, ASSOC PROF ASTRON, UNIV MASS, 87- *Mem:* Am Astron Soc; Int Astron Union. *Res:* The study of the structure and dynamics of interstellar clouds and star formation through radio frequency observations of the radiation emitted by atoms and molecules in space. *Mailing Add:* GRC Tower B Five Col Radio Astron Observ Univ Mass Amherst MA 01003

SNELL, WILLIAM J, b LaSalle, Ill, Oct 11, 46; m 68; c 2. SIGNAL TRANSDUCTION, FERTILIZATION. *Educ:* Univ Ill, BS, 68; Yale Univ, PhD(cell & develop biol), 75. *Prof Exp:* PROF CELL BIOL & NEUROSCI, SOUTHWESTERN MED CTR, UNIV TEX, DALLAS, 77- *Concurrent Pos:* Prin investr, NIH & NSF grants. *Mem:* Am Soc Cell Biol; Sigma Xi; AAAS. *Res:* Cell-cell interactions during fertilization in chlamydomonas; signal transduction induced by cell contact; activation of adenylyl cyclose by a novel mechanism. *Mailing Add:* Dept Cell Biol & Neurosci Univ Tex Southwestern Med Ctr Dallas TX 75235-9039

SNELLING, CHRISTOPHER, b Hartford, Conn, Nov 8, 35; m 59; c 1. ELECTRICAL ENGINEERING, MATHEMATICS. *Educ:* Union Col, NY, BEE, 57; Univ Rochester, MS, 64. *Prof Exp:* Assoc physicist, Haloid Corp, 57-60, physicist, Haloid-Xerox, 60-62, sr physicist, Xerox Corp, 62-65, scientist, 65-68; proj engr, Hamco Mach & Electronics Corp, 68-72; scientist, 72-76, tech specialist/proj mgr I, 76-80, MEM RES STAFF, XEROX CORP, 81- *Mem:* Inst Elec & Electronics Engrs; Soc Photographic Scientists & Engrs; Sigma Xi. *Res:* Measurement and analysis of xerographic photoreceptors; xerographic process studies; electrostatics; photoconductivity; radiometry; direct current instrumentation; development of Czochralski silicon crystal growing furnaces for semiconductor production. *Mailing Add:* Five High Meadow Dr Penfield NY 14526

SNELLINGS, WILLIAM MORAN, b Norfolk, Va, May 7, 47; m 70; c 3. TOXICOLOGY, INHALATION TOXICOLOGY. *Educ:* Va Polytech Inst, BS, 69; Univ Mich, PhD(toxicol), 76. *Prof Exp:* Sr technician, Hazelton Labs, 69-71; toxicologist, Carnegie-Mellon Inst Res, 76-80; mgr inhalation toxicol, Bushy Run Res Ctr, 80-81, asst dir, 81-85, assoc dir, 85-89; DIR, IND CHEM DIV, UNION CARBIDE CHEM & PLASTICS, INC, 89- *Concurrent Pos:* Prin investr Ethylene Oxide Toxicity Prog worldwide consortium Ethylene Oxide Toxicity producers, 76-81. *Mem:* Soc Toxicol; Soc Toxicol Inhalation Specialty Sect. *Res:* Toxicity evaluation of various industrial chemicals; assessment of the oncogenic, developmental, reproductive and general toxic effects of test chemicals; design and development of inhalation chambers, vapor generators, aerosol generators, and atmospheric sampling systems. *Mailing Add:* Indust Chem Div Union Carbide Chem & Plastics Inc 39 Ridgebury P4 Danbury CT 06817-0001

SNELLMAN, LEONARD W, b Lansford, Pa, June 27, 20; m 48; c 4. METEOROLOGY. *Educ:* Kenyon Col, AB, 43. *Prof Exp:* Forecaster, US Weather Bur, 46-51; civilian consult meteorol, US Air Force-Hq Air Weather Serv, 53-65; chief, Sci Serv Div, WRN RGN Nat Weather Serv, Oceanic & Atmospheric Admin, 65-82; RETIRED. *Concurrent Pos:* Lectr, Univ Utah,

66-68, adj asst prof, 68-; NCR comt assignments, Acad Sci, 87- *Honors & Awards:* Silver Medal, Dept of Com, 69, Gold Medal, 78; Spec Award for Voyager Flight Support AMS, 87. *Mem:* Fel Am Meteorol Soc. *Res:* Synoptic and satellite meteorology especially weather forecasting. *Mailing Add:* 4278 S 2700 E Salt Lake City UT 84124

SNELSIRE, ROBERT W, b Pittsburgh, Pa, May 8, 33; m 57; c 3. ELECTRICAL ENGINEERING. *Educ:* Bethany Col, BA, 56; Carnegie Inst Technol, BS, 56, MS, 58, PhD(elec eng), 64. *Prof Exp:* Instr elec eng, Carnegie Inst Technol, 58-63; sr engr, Westinghouse Defense Ctr, 63-64; asst prof elec eng, State Univ NY Buffalo, 64-67; ASSOC PROF ELEC ENG, CLEMSON UNIV, 67- *Concurrent Pos:* Consult, Bell Aerosysts Co, 66-67 & Wachovia Bank & Trust Co, 66- *Mem:* Inst Elec & Electronics Engrs; Am Soc Eng Educ. *Res:* Computer science; simulation of human behavior. *Mailing Add:* Dept Elec & Comput Eng Clemson Univ Hwy 76 Clemson SC 29631

SNELSON, ALAN, b Manchester, Eng, Oct 17, 34; m 77; c 1. PHYSICAL CHEMISTRY, THERMODYNAMICS. *Educ:* Univ Manchester, Eng, BSc, 57, MSc, 58, PhD(chem), 60. *Prof Exp:* Fel chem, Univ Calif, Berkeley, 60-62; SR CHEMIST, IIT RES INST, 62- *Concurrent Pos:* Lectr, Ill Inst Technol, 67- *Mem:* Am Chem Soc. *Res:* Thermochemistry; spectroscopy; kinetics; cryogenics atmospheric chemistry. *Mailing Add:* 935 W Argyle Chicago IL 60640

SNELSON, FRANKLIN F, JR, b Richmond, Va, June 13, 43; c 3. ICHTHYOLOGY. *Educ:* NC State Univ, BS, 65; Cornell Univ, PhD(vert zool), 70. *Prof Exp:* From asst prof to assoc prof, chmn biol sci, 81-88, PROF BIOL, UNIV CENT FLA, 81- *Concurrent Pos:* NASA grant, 72-79, US Fish Wildlife grant, 80-81; assoc ed, American Midland Naturalist, 82- *Mem:* Am Soc Ichthyologists & Herpetologists; Am Inst Biol Sci; Ecol Soc Am; Soc Syst Zool; Am Soc Zool. *Res:* Systematics and ecology of fishes; biology of sharks and stingrays; reproductive ecology of livebearing fishes; ecology of coastal marine fishes in Florida; systematics of minnows. *Mailing Add:* Dept Biol Univ Cent Fla Orlando FL 32816-0990

SNELSON, SIGMUND, b Santa Paula, Calif, June 22, 32; m 60; c 2. GEOLOGY. *Educ:* Univ Redlands, BS, 53; Univ Wash, MS, 55, PhD(geol), 57. *Prof Exp:* Fulbright fel, Univ Graz, 57-58; geologist, Shell Oil Co, 59-66, sr geologist, 66-69, staff geologist, Shell Develop Co, 69-70 & Shell Oil Co, 70-74, sr staff geologist & geol adv, Geol Sect, Shell Develop Co, 75-87, GEOL CONSULT, SHELL OIL CO, 87- *Honors & Awards:* A I Levorsen Mem Award, Am Asn Petrol Geologists, 72. *Mem:* Geol Soc Am; Am Asn Petrol Geologists; fel AAAS. *Res:* Tectonics of North Alaska; structural evolution of California; oil accumulations in coastal California; Appalachian geology; South American and Caribbean geology; evolution of the Gulf of Mexico basin, continental margins and rift basins. *Mailing Add:* Shell Develop Co PO Box 481 Houston TX 77001

SNETSINGER, DAVID CLARENCE, b Barrington, Ill, Apr 22, 30; m 53; c 4. POULTRY NUTRITION. *Educ:* Univ Ill, BS, 52, MS, 57, PhD(poultry nutrit), 59. *Prof Exp:* Asst poultry, Univ Ill, 55-59; from asst prof to assoc prof poultry sci, Univ Minn, St Paul, 59-68; mgr, Com Layer Res Div, Ralston Purina Co, 68-70, mgr, com egg & breeder res div, 70-72, mgr, Gen Poultry Res Div, 72-76, dir, Poultry res & Mkt Dept, 76-85, dir, poultry bus group, 85-87, vpres res, 87-90, SR ADV, PURINA MILLS, 91- *Mem:* Fel, Poultry Sci Asn; Am Inst Nutrit; World Poultry Sci Asn; Animal Sci Asn; Dairy Sci Asn. *Res:* Amino acid and mineral nutrition and metabolism. *Mailing Add:* Res & Mkt Purina Mills PO Box 66812 St Louis MO 63166-6812

SNETSINGER, KENNETH GEORGE, b San Francisco, Calif, Feb 21, 39; m 78. GEOCHEMISTRY, MINERALOGY. *Educ:* Stanford Univ, BS, 61, MS, 62, PhD(mineral), 66. *Prof Exp:* Nat Acad Sci-Nat Res Coun resident res assoc, 66-69, RES SCIENTIST, NASA AMES RES CTR, 69- *Mem:* Mineral Soc Am; Mineral Soc Can; NY Acad Sci. *Res:* Mineralogy and geochemistry of meteorites, lunar samples, platinum metals; composition and mineralogy of atmospheric aerosols. *Mailing Add:* 668 Bancroft St Santa Clara CA 95051

SNETSINGER, ROBERT J, b Diamond Lake, Ill, Mar 6, 28; m 60; c 2. ECOMONIC ENTOMOLOGY, ARACHNOLOGY. *Educ:* Univ Ill, Urbana, BS, 52, MS, 53, PhD(entom), 60. *Prof Exp:* Asst econ entom, Ill Nat Hist Surv, 55-60; from asst prof to assoc prof entom, 60-71, PROF ENTOM, PA STATE UNIV, UNIVERSITY PARK, 71- *Concurrent Pos:* Ed, Int Mushroom Cong, 62, Entom Soc Pa, 64-77 & Pa Pest Control Quart, 67-75; vis prof, Univ PR, 83-84. *Mem:* Entom Soc Am; Entom Soc Can; Arachnids Soc Am. *Res:* Biology and control of animal pests of mushrooms; structural pest control and urban ecology; biology and control of arachnids; history of pest control; mushroom culture. *Mailing Add:* Dept Entom 106 Patterson Pa State Univ University Park PA 16802

SNIDER, ALBERT MONROE, JR, b Hoffman, NC; m 66; c 2. POLYMERS, ANALYTICAL CHEMISTRY. *Educ:* Univ NC, Chapel Hill, BA, 59, MEd, 62; Appalachian State Univ, MA, 69; Univ Pittsburgh, PhD(chem), 74. *Prof Exp:* Teaching asst chem, Appalachian State Univ, 67-69; chemist, Mellon Inst, 70; instr, Community Col Allegheny County, 75; res asst, Dept Chem, Sch Eng, Univ Pittsburgh, 69-74, res assoc polymer sci, 75-76; lab dir, K Tator Assocs, 76-77; group leader, Spectros Lab, Carnegie-Mellon Inst Res, 78-79; group leader polymer characteriation, Merck & Co, 79-81; MGR, CHEM & NONMETALS LAB, GEN ELEC CO, 81- *Mem:* Am Chem Soc; Soc Appl Spectros; Coblentz Soc; Sigma Xi. *Res:* Molecular spectroscopy and polymer science particularly utilizing infrared, raman, nuclear magnetic resonance, photoelectron, and mass spectroscopies, x-ray diffraction and electron microscopy. *Mailing Add:* 602 Orchard Hill Dr Pittsburgh PA 15238

SNIDER, ARTHUR DAVID, b Richmond, Va, Oct 7, 40. MATHEMATICS. *Educ:* Mass Inst Technol, BS, 62; Boston Univ, MA, 66; NY Univ, PhD(math), 71. *Prof Exp:* Analyst, Instrumentation Lab, Mass Inst Technol, 62-66; asst prof, 70-77, ASSOC PROF MATH, UNIV S FLA, 77- *Concurrent Pos:* Math consult, Honeywell Aerospace Corp, 74. *Mem:* Soc Indust & Appl Math; Am Math Soc; Math Asn Am. *Res:* Applied mathematics; numerical analysis; differential equations; plasmas. *Mailing Add:* 14701 Oak Lake Pl Lutz FL 33529

SNIDER, BARRY B, b Chicago, Ill, Jan 13, 50; m 75; c 1. NATURAL PRODUCTS SYNTHESIS. *Educ:* Univ Mich, BS, 70; Harvard Univ, PhD(chem), 73. *Prof Exp:* Fel chem, Columbia Univ, 73-75; asst prof, Princeton Univ, 75-81; assoc prof, 81-85, PROF CHEM, BRANDEIS UNIV, 85- *Concurrent Pos:* Alfred P Sloan Found fel, 79; Dreyfus Teacher Scholar, 82-87. *Mem:* Am Chem Soc; Royal Soc Chem. *Res:* Synthetic methods development; Ene reactions; Lewis acid catalysis; alkene carbofunctionalization; intramolecular cycloadditions and cyclizations; total synthesis of natural products. *Mailing Add:* Dept Chem Brandeis Univ Waltham MA 02254

SNIDER, BILL CARL F, b Cedar Rapids, Iowa, July 11, 20; m 54; c 1. STATISTICS. *Educ:* Univ Wichita, BA, 42, MA, 51; Univ Iowa, PhD, 55. *Prof Exp:* Res assoc prev ment health, Child Welfare Res Sta, Univ Iowa, 56-65, res assoc comput ctr, 62-65, asst prof statist, Col Educ & Comput Ctr, 65-69, assoc prof educ, 69-74, PROF EDUC, PSYCHOL, MEASUREMENT & STATIST, UNIV IOWA, 74- STATIST CONSULT, COMPUT CTR, 74- *Mem:* Am Educ Res Asn. *Res:* Mental health; statistics and preventive mental health. *Mailing Add:* Col Educ & Univ Comput Ctr Univ Iowa Iowa City IA 52242

SNIDER, DALE REYNOLDS, b Cincinnati, Ohio, Mar 21, 38; m 61; c 2. ELECTROMAGNETISM, HIGH ENERGY PHYSICS. *Educ:* Ohio State Univ, BS & MS, 61; Univ Calif, San Diego, PhD(physics), 68. *Prof Exp:* Instr nuclear eng, US Navy Nuclear Power Sch, Calif, 61-65; theoret physicist, Lawrence Radiation Lab, Univ Calif, Berkeley, 68-70; asst prof, 70-77, ASSOC PROF PHYSICS, UNIV WIS-MILWAUKEE, 77- *Concurrent Pos:* Vis prof physics, Univ Ill, Urbana, 74-75. *Mem:* Am Phys Soc. *Res:* Theoretical high energy physics; strong interaction theory; electromagnetic devices; nonlinear dynamic; surface studies; metallic clusters; superconductivity. *Mailing Add:* 4254 N Ardmore Ave Shorewood WI 53211

SNIDER, DIXIE EDWARD, JR, b Frankfort, Ky, Jan 16, 43; m 66; c 2. MEDICINE. *Educ:* Western Ky State Col, BS, 65; Univ Louisville, MD, 69. *Prof Exp:* Intern internal med, Barnes Hosp, 69-70, resident, 70-71; resident, Vanderbilt Univ, 71-72; fel allergy & clin immunol, Wash Univ, 72-73; med officer tuberc, USPHS, 73-74; chief, 74-85, DIR, RES & DEVELOP BR, TUBERCULOSIS CONTROL DIV, BUR STATE SERV, CTR DIS CONTROL, USPHS, 85- *Concurrent Pos:* Fel allergy & clin immunol, Wash Univ, 75-76. *Mem:* Am Thoracic Soc; Int Union Against Tuberc; Am Col Physicians; Am Acad Allergy; Am Pub Health Asn. *Res:* Tuberculin skin testing; mycobacterial drug resistance; preventive therapy of tuberculosis; treatment of tuberculosis; prostaglandins and asthma; lymphocyte cyclic nucleotide metabolism. *Mailing Add:* USPHS Ctr Dis Control 1600 Clifton Rd Mail Stop E-10 Atlanta GA 30333

SNIDER, DONALD EDWARD, b Lakewood, Ohio, Sept 12, 44; m 78. ATMOSPHERIC PHYSICS. *Educ:* Ohio State Univ, BS, 66, MS, 68, PhD(physics), 71. *Prof Exp:* RES PHYSICIST, BALLISTIC RES LABS, 71-, ATMOSPHERIC SCI LAB, 76- *Mem:* Optical Soc Am; Am Meteorol Soc. *Res:* Investigations of atmospheric effects on electro optical systems; attenuation by atmospheric gases; attenuation and scattering by natural and man made aerosols; refraction and scintillation due to turbulance; atmospheric optics; propagation of electro-magnetic energy. *Mailing Add:* Atmospheric Sci Lab SLCAS AE E WIN2AA 5297-9500 Zonel White Sands Missile Range NM 88002

SNIDER, GORDON LLOYD, b Toronto, Ont, Apr 11, 22; US citizen; m 45; c 3. PULMONARY DISEASES. *Educ:* Univ Toronto, MD, 44; Am Bd Internal Med, dipl, 53; Am Bd Pulmonary Dis, dipl, 58. *Prof Exp:* Intern, Toronto Gen Hosp, 44-45; resident med, Bronx Hosp, New York, 46-47, resident path, Mass Mem Hosp, Boston, 47-48; fel med, Lahey Clinic, Boston, 48-49; resident pulmonary med, Trudeau Sanitarium, NY, 49-50; asst dir chest dept, Michael Reese Hosp, 50-61; chief div thoracic med, Mt Sinai Hosp, 61-66, actg chmn dept med, 65-66; chief pulmonary dis sect, Wood Vet Admin Hosp, 66-68; prof med, Sch Med & head respiratory sect, Univ Hosp, Boston Univ, 68-76; chief pulmonary dis sect, Boston Vet Admin Hosp, 68-87, CHIEF MED SERV, BOSTON VA MED CTR, 86- *Concurrent Pos:* Attend physician, Winfield Hosp, Ill, 52-61; consult physician & dir pulmonary function lab, Munic Tuberc Sanitarium, Chicago, 52-68; med consult, Social Security Admin, 58-66; from asst prof to prof med, Chicago Med Sch, 58-66, actg chmn dept, 65-66; consult, West Side Vet Admin Hosp, Chicago, 64-66; prof, Sch Med, Marquette Univ, 66-68; attend physician, Milwaukee County Gen Hosp & med consult, Mt Sinai Hosp, Milwaukee, 66-68; mem respiratory dis comt, Tuberc Inst Chicago & Cook County; mem med adv bds, Suburban Cook County Tuberc Sanitarium Dis & Asthma & Allergy Res Found, Greater Chicago; mem med adv comt, Chicago Chap, Cystic Fibrosis Res Found; consult comt, Div Sanatoria & Tuberc Control, Mass Dept Pub Health; mem pulmonary dis adv comt, Nat Heart, Lung & Blood Inst, 80-84; pulmonary dis adv coun, Vet Admin, 79-81, pres Vet Pulmonary Physicians Asn, 80-81; mem adv coun, Spec Ctr Res Chronic Obstructive Lung Dis, Harvard Sch Pub Health, 80-84; chmn, Fed Lung Prog Comt, Am Thoracic Soc, 85-; Int adv, Aspen Lung Conf, 85-90; chmn, NIH Safety & Data Monitoring Bd, 89-; Maurice B Strauss prof med, Boston Univ & Tufts Univ Sch Med, 86- *Honors & Awards:* Simon Rodbard Mem Lectr, Am Col Chest Physicians, 85; Alton Ochsner Award, 90; Parker B Francis Lectr, 91. *Mem:* Fel Am Col Chest Physicians; fel Am Col Physicians; Am Fedn Clin Res; Am Thoracic Soc. *Res:* Clinical pulmonary disease and physiology; experimental pulmonary diseases. *Mailing Add:* 24 Holly Rd Waban MA 02168

SNIDER, JERRY ALLEN, b Danville, Ill, Feb 17, 37; m 67; c 2. BRYOLOGY, CYTOLOGY. *Educ:* Southern Ill Univ, BA, 67; Univ NC, Chapel Hill, MA, 70; Duke Univ, PhD(bot), 73. *Prof Exp:* Asst prof biol, Baylor Univ, 73-74; from asst prof to assoc prof, 74-89, HERBARIUM CUR, DEPT BIOL SCI, UNIV CINCINNATI, 74-, PROF BIOL SCI, 90- *Mem:* Am Bryol & Lichenological Soc; Sigma Xi; British Bryol Soc; Int Asn Plant Taxon; Int Asn Bryol; Norweg Bryol Soc. *Res:* Cytology and taxonomy of bryophytes; morphological development in bryophytes. *Mailing Add:* Dept Biol Sci Univ Cincinnati Cincinnati OH 45221-0006

SNIDER, JOHN WILLIAM, b Middleport, Ohio, Sept 13, 24; m 49; c 1. PHYSICS. *Educ:* Miami Univ, AB, 49, MA, 51; Ohio State Univ, PhD, 57. *Prof Exp:* Res assoc, Ohio State Univ, 57-59; from asst prof to assoc prof, 53-73, PROF PHYSICS, MIAMI UNIV, 73- *Concurrent Pos:* Sr res physicist, Mound Lab, Monsanto Chem Co, 58-69; consult, Ohio River Div, Army Corps Engrs, 63-65. *Mem:* Am Asn Physics Teachers. *Res:* Low temperature physics, especially superconductivity and thermal transpiration; calorimetry. *Mailing Add:* Dept of Physics Miami Univ Oxford OH 45056

SNIDER, JOSEPH LYONS, b Boston, Mass, June 10, 34. ATOMIC PHYSICS, ASTROPHYSICS. *Educ:* Amherst Col, BA, 56; Princeton Univ, PhD(physics), 61. *Prof Exp:* Instr & res fel physics, Harvard Univ, 61-64, asst prof, 64-69; assoc prof, 69-75, PROF PHYSICS, OBERLIN COL, 75- *Mem:* Am Phys Soc; Am Astron Soc; Am Asn Physics Teachers; Am Asn Univ Profs. *Res:* Solar physics; atomic physics; relativity and gravitation. *Mailing Add:* Dept Physics Oberlin Col Oberlin OH 44074

SNIDER, NEIL STANLEY, b Schenectady, NY, May 25, 38. THEORETICAL CHEMISTRY. *Educ:* Purdue Univ, BSc, 59; Princeton Univ, MA, 61, PhD(chem), 64. *Prof Exp:* NSF fel, 64-65; res assoc chem, Cornell Univ, 65 & Yale Univ, 65-66; asst prof, 66-72, ASSOC PROF CHEM, QUEEN'S UNIV, ONT, 72- *Mem:* Am Chem Soc; Can Asn Physicists; Chem Inst Can. *Res:* Theory of rates of homogeneous gas phase reactions; statistical mechanics of classical fluids. *Mailing Add:* Dept of Chem Queen's Univ Kingston ON K7L 3N6 Can

SNIDER, PHILIP JOSEPH, b Richmond, VA, Apr 5, 29; m 52; c 2. GENETICS. *Educ:* Richmond Univ, BS, 52; Harvard Univ, AM, 55, PhD, 57. *Prof Exp:* Res assoc biol div, Genetics Sect, Oak Ridge Nat Lab, 58-59; asst prof bot, Univ Calif, Berkeley, 59-63; dir univ honors prog, 65-70, ASSOC PROF BIOL, UNIV HOUSTON, 63- *Concurrent Pos:* Consult mem, Biol Sci Curriculum Studies, 64-70. *Mem:* AAAS. *Res:* Microbial and molecular genetics, especially regulatory genetics of reproductive processes. *Mailing Add:* Dept Biol Univ Houston 4800 Calhoun Rd Houston TX 77204

SNIDER, RAY MICHAEL, b Los Angeles, Calif, Dec 28, 48; m 82; c 2. PEPTIDE RECEPTOR PHARMACOLOGY, DRUG DISCOVERY. *Educ:* Calif State Univ, Northridge, BA, 74; Ohio State Univ, PhD(pharmacol), 82. *Prof Exp:* Grad res assoc pharmacol, Ohio State Univ, 76-81; postdoctoral fel pharmacol, Mayo Clinic & Found, 82-83; res investr neurochem, Univ Mich, 83-86; sr scientist immunol, T Cell Sci, 86-88; SR RES SCIENTIST PHARMACOL, PFIZER CENT RES, 88- *Honors & Awards:* Individual Nat Res Serv Award, NIMH, 82. *Mem:* AAAS; Am Soc Pharmacol & Exp Therapeut. *Res:* Discovery and pharmacological characterization of novel prototype molecules interactions with receptors; drug discovery; peptide and cytokine receptors; cholinergic and biogenic amine receptors; second messenger pharmacology. *Mailing Add:* Exploratory Med Chem Dept Pfizer Central Res Eastern Point Rd Groton CT 06340

SNIDER, ROBERT FOLINSBEE, b Calgary, Alta, Nov 22, 31; div; c 4. THEORETICAL CHEMISTRY. *Educ:* Univ Alta, BSc, 53; Univ Wis, PhD(theoret chem), 58; FRSC; FCIC. *Prof Exp:* Fel appl chem, Nat Res Coun Can, 58; from instr to assoc prof chem, 58-69, PROF CHEM, UNIV BC, 69- *Mem:* Can Asn Physicists; Am Inst Phys; Chem Inst Can. *Res:* Statistical mechanics; transport properties of gases; collisions of non-spherical molecules. *Mailing Add:* Dept Chem 2036 Main Mall Vancouver BC V6T 1Y6 Can

SNIDER, THEODORE EUGENE, b Pittsburg, Kans, Apr 22, 43; m 68; c 2. ORGANIC CHEMISTRY. *Educ:* Pittsburg State Univ, BS, 65, MS, 67; Okla State Univ, PhD(org chem), 72. *Prof Exp:* Instr chem, Eastern Okla State Col, 67-68; from instr to assoc prof, 68-78, PROF CHEM, CAMERON UNIV, 78- *Concurrent Pos:* Chmn & prof, Phys Sci Dept, Cameron Univ, 89- *Mem:* Am Chem Soc. *Res:* Synthesis of folic acid inhibitors and the calculation and significance of group electronegativity. *Mailing Add:* Phys Sci Dept Cameron Univ 2800 Gore Blvd Lawton OK 73505

SNIECKUS, VICTOR A, b Kaunas, Lithuania, Aug 1, 37; Can citizen; m 66; c 2. ORGANIC CHEMISTRY. *Educ:* Univ Alta, BSc, 59; Univ Calif, Berkeley, MS, 61; Univ Ore, PhD(chem), 65; FCIC, 78. *Prof Exp:* Nat Res Coun Can fel chem, 65-66; from asst prof to assoc prof, 66-79, fel, 66-67, PROF CHEM, UNIV WATERLOO, 79- *Concurrent Pos:* H C Orsted Found Fel, Univ Copenhagen, Denmark, 73; vis prof, Univ Geneva, Switzerland, 76-77; Can-Japan exchange fel, 81; France-Can exchange fel, 85. *Mem:* Am Chem Soc; Chem Inst Can; Int Soc Heterocyclic Chem. *Mailing Add:* Dept Chem Univ Waterloo Waterloo ON N2L 3G1 Can

SNIPES, CHARLES ANDREW, b Tampa, Fla, Nov 1, 36. PHYSIOLOGY. *Educ:* Western Carolina Univ, AB, 57; Duke Univ, PhD(physiol, pharmacol), 67. *Prof Exp:* From instr to asst prof pediat, Johns Hopkins Univ, 62-68, asst dir res training prog pediat endocrinol, 62-68; asst prof, 68-71, ASSOC PROF PHYSIOL, SCH MED, HAHNEMANN UNIV, 71- *Mem:* AAAS; Endocrine Soc; Am Physiol Soc; Lawson Wilkins Pediat Endocrine Soc; Am Soc Zoologists. *Res:* Neuroendocrinology; metabolism of steroid hormones; growth hormone and insulin on amino acid accumulation; control of development. *Mailing Add:* Dept Physiol & Biophys Hahnemann Univ 230 N Broad St Philadelphia PA 19102

SNIPES, DAVID STRANGE, b Hartsville, SC, Apr 16, 28; m 53; c 4. GEOLOGY. *Educ:* Wake Forest Col, BS, 50; Univ NC, PhD(geol), 65. *Prof Exp:* Geologist, Standard Oil Co Calif, 56-59; assoc prof geol, Furman Univ, 63-68; assoc prof, 68-79, PROF GEOL, CLEMSON UNIV, 79- *Concurrent Pos:* Consult var co & pvt individuals. *Res:* Ground water exploration in fractured igneous and metamorphic rocks; fault zones and dolerite dikes; X-ray analysis of minerals. *Mailing Add:* Dept of Geol Clemson Univ Clemson SC 29634-1905

SNIPES, MORRIS BURTON, b Clovis, NMex, Oct 29, 40; m 58; c 1. RADIATION BIOLOGY. *Educ:* Univ NMex, BS, 67, MS, 68; Cornell Univ, PhD(phys biol), 71. *Prof Exp:* ASSOC SCIENTIST RADIOBIOL, INHALATION TOXICOL RES INST, 71- *Mem:* Sigma Xi; Radiation Res Soc; Health Physics Soc. *Res:* Metabolism of radionuclides and radiation dosimetry. *Mailing Add:* 782 Highway 66 E Tijeras NM 87059

SNIPES, WALLACE CLAYTON, b Graham, NC, Oct 11, 37; m 60; c 3. PHARMACOLOGY. *Educ:* Wake Forest Col, BS, 60; Duke Univ, PhD(phsics), 64. *Prof Exp:* From asst prof to prof biophys, Pa State Univ, University Park, 72-88; VPRES RES & DEVELOP, ZETACHRON INC, 84- *Concurrent Pos:* Vis prof, Univ Calif, Berkeley, 69 & Univ Calif, Santa Cruz, 74-75. *Mem:* Am Phys Soc; Am Asn Pharmaceut Scientists; Biophys Soc. *Res:* Drug delivery systems. *Mailing Add:* Zetachron Inc 100 N Sci Park Rd State Col PA 16803

SNIPP, ROBERT LEO, b Omaha, Nebr, Aug 13, 36; m 63; c 2. PHYSICAL CHEMISTRY. *Educ:* Creighton Univ, BS, 58, MS, 60; Univ Iowa, PhD(phys chem), 65. *Prof Exp:* Asst prof, 64-71, chmn dept, 74-77, ASSOC PROF CHEM, CREIGHTON UNIV, 71- *Mem:* Am Chem Soc. *Res:* Physical chemistry of macromolecules, particularly polyelectrolytes in solution. *Mailing Add:* Dept Chem Creighton Univ Omaha NE 68178-0002

SNITGEN, DONALD ALBERT, b St John's, Mich, Feb 25, 36; m 59; c 3. BIOLOGY, SCIENCE EDUCATION. *Educ:* Cent Mich Univ, BS, 60; Mich State Univ, MS, 64, PhD(sci educ), 71. *Prof Exp:* Teacher high sch, Mich, 62-66; from instr to assoc prof, 66-79, PROF BIOL, NORTHERN MICH UNIV, 79- *Res:* Distribution of stream bottom fauna, especially insects; preservice elementary school teachers' attitudes toward biological science; development of audio-tutorial program for non-majors in biological science; feasibility study to develop a regional environmental education center in Upper Peninsula of Michigan. *Mailing Add:* Dept of Biol Northern Mich Univ Marquette MI 49855

SNITZER, ELIAS, b Lynn, Mass, Feb 27, 25; m 50; c 5. PHYSICS. *Educ:* Tufts Univ, BS, 45; Univ Chicago, MS, 50, PhD(physics), 53. *Prof Exp:* Res physicist, Minneapolis-Honeywell Regulator Co, 54-56; assoc prof electronics eng, Lowell Tech Inst, 56-58; res assoc, Mass Inst Technol, 59; res physicist, Am Optical Corp, 59-68, dir basic res, 68-75 & corp res, 75-77; mgr, tech planning, United Technol Corp Res Ctr, 77-79, mgr, Appl Physics Lab, 79-84; dir, Fiber & Integrated Optics Group, Polaroid Corp, 84-88; PROF, CERAMIC SCI & ENG, RUTGERS UNIV, 89- *Honors & Awards:* George W Morey Award, Am Ceramic Soc, 71; Quantum Electronics Award, Inst Elec & Electronics Engrs, 79. *Mem:* Nat Acad Eng; Am Phys Soc; Optical Soc Am; Am Ceramic Soc; Inst Elec & Electronics Engrs. *Res:* Physical optics; glass technology; solid state physics; materials and instrument research in optics. *Mailing Add:* Ceramic Sci & Eng Dept Rutgers Univ PO Box 909 Piscataway NJ 08855-0909

SNIVELY, LESLIE O, b Laramie, Wyo, Mar 11, 53; m 75. MAGNETIC INTERACTIONS. *Educ:* Colo State Univ, BS, 75; Mont State Univ, MS, 77, PhD(physics), 81. *Prof Exp:* Res assoc, Mont State Univ, 81-82; PRES, INTERLINK TECH, 82- *Mem:* Am Phys Soc. *Res:* Experimental and theoretical investigation of supererchange interaction in magnetically lower-dimentsional metal-halide compounds using magnetic susceptibility and magnetization measurement. *Mailing Add:* One Heather Lane Amherst NH 03031

SNOBLE, JOSEPH JERRY, b Center Point, Iowa, Feb 11, 31; m 55; c 2. PHYSICS, SCIENCE EDUCATION. *Educ:* Iowa State Teachers Col, BA, 57, MA, 61; State Univ Iowa, PhD(sci educ), 67. *Prof Exp:* Teacher sci, Kingsley Pub Sch, 57-59; instr, State Univ Iowa, 61-67; PROF PHYSICS, CENT MO STATE UNIV, 67- *Mem:* Nat Sci Teachers Asn; Sch Sci & Math Asn; Am Asn Physics Teachers; Sigma Xi. *Res:* Teacher education in science, especially elementary, secondary and collegiate teaching procedures and methods; meaningful demonstration and laboratory activities and experiments. *Mailing Add:* Route 5 Box 174 Warrensburg MO 64093

SNODDON, W(ILLIAM) J(OHN), chemical engineering, for more information see previous edition

SNODDY, EDWARD L, b Kelso, Tenn, Mar 6, 33; m 52; c 2. MEDICAL ENTOMOLOGY, ECOLOGY. *Educ:* Mid Tenn State Univ, BS, 62; Auburn Univ, PhD(med entom), 66. *Prof Exp:* NDEA fel med entom, Auburn Univ, 62-65; prof med entom, Coastal Plain Exp Sta, Univ Ga, 65-76; MED ENTOMOLOGIST, WATER QUALITY & ECOL BR, TENN VALLEY AUTHORITY, ALA, 76- *Concurrent Pos:* Consult, US Air Force Hosp, Robins AFB, Warner Robins, Ga, 69-76; mem sci adv panel, WHO, 74-; secy-treas, Ga Mosquito Control Asn, 75-; Ga dir, Mid-Atlantic Mosquito Control Asn, 75-; consult, Armed Forces Pest Mgt Bd, Dept Defense, 76-; mem Fed Interagency Comt, Forest Integrated Pest Mgt, 81- *Mem:* Entom Soc Am; Ecol Soc Am; Am Mosquito Control Asn. *Res:* Ecology and taxonomy of medically important arthropods; attractants and repellents for insects of medical importance, particularly Simuliidae, Tabanidae, Ceratopogonidae and Ixodidae. *Mailing Add:* 309 Meadow Hill Rd Muscle Shoals AL 35661

SNODGRASS, HERSCHEL ROY, physics; deceased, see previous edition for last biography

SNODGRASS, MICHAEL JENS, microscopic anatomy; deceased, see previous edition for last biography

SNODGRASS, REX JACKSON, b St Louis, Mo, Feb 24, 34. ALTERNATIVE ENERGY, INFORMATION SCIENCE. *Educ:* Harvard Univ, AB, 56; Univ Md, MS, 60, PhD(physics), 63. *Prof Exp:* Physicist, Harry Diamond Labs, 53-59; physicist, Nat Bur Stand, 60-66; fel, Inst Mat Sci & mem fac physics, Univ Conn, 66-72, mgr tech serv, New Eng Res Appl Ctr, 72-80; WITH ENVIRON SCI INFO CTR, NAT OCEANIC & ATMOSPHERIC ADMIN, 80- *Concurrent Pos:* Nat Bur Stand training fel, Univ Paris, 63-64. *Mem:* Am Phys Soc; AAAS; Am Soc Info Sci. *Res:* Information management; energy related problems, especially solar energy research; energy conservation efficiencies. *Mailing Add:* PO Box 236 Weaverville NC 28787

SNOEYENBOS, GLENN HOWARD, b Glenwood City, Wis, Sept 16, 22; m 49; c 3. VETERINARY MEDICINE. *Educ:* Mich State Col, DVM, 45. *Prof Exp:* Sta vet, Univ Minn, 45-46; PROF VET MED, UNIV MASS, AMHERST, 47- *Concurrent Pos:* Grants, USPHS, 65-68, Fats & Proteins Res Found, 66-69 & USDA, 66-71. *Mem:* Am Vet Med Asn; Am Asn Avian Path (secy treas, 61-70); Poultry Sci Asn; Conf Res Workers Animal Dis. *Res:* Infectious diseases of poultry; methods of preventing salmonellosis as a public health problem. *Mailing Add:* Dept Animal Sci Univ Mass Amherst MA 01003

SNOEYINK, VERNON LEROY, b Kent Co, Mich, Oct 10, 40; m 64; c 2. ENVIRONMENTAL ENGINEERING. *Educ:* Univ Mich, BS, 64, MS, 66, PhD(water resources eng), 68. *Prof Exp:* Engr, Metcalf & Eddy Engrs, 68-69; from asst prof to assoc prof sanit eng, Dept of Civil Eng, 69-77, PROF ENVIRON ENG, DEPT CIVIL ENG, UNIV ILL, 77- *Mem:* Am Soc Civil Engrs; Am Water Works Asn; Water Pollution Control Fedn; Asn Environ Eng Prof. *Res:* Water purification using adsorption processes; water chemistry; drinking water purification. *Mailing Add:* Dept Civil Eng Univ Ill Urbana IL 61801

SNOKE, ARTHUR WILMOT, b Baltimore, Md, Oct 5, 45; m 66; c 2. GEOLOGY, STRUCTURAL GEOLOGY. *Educ:* Franklin & Marshall Col, AB, 67; Stanford Univ, PhD(geol), 72. *Prof Exp:* Nat Res Coun assoc, US Geol Surv, 71-73; from asst prof to assoc prof geol, Univ SC, 74-84; PROF, UNIV WYO, 84- *Concurrent Pos:* Lectr, Humboldt State Univ, 74. *Mem:* Geol Soc Am; Am Geophys Union. *Res:* Structural geology and tectonics of orogenic belt; petrogenesis of mylonitic rocks; structural analysis of polyphase-deformed terranes, northeastern Nevada, Wyoming foreland, Tobago West Indies, and southern Alpine basement (northern Italy). *Mailing Add:* Dept Geol & Geophys Univ Wyo PO Box 3006 Laramie WY 82071-3006

SNOKE, J ARTHUR, b Rochester, NY, Mar 3, 40; m 64; c 2. SEISMOLOGY. *Educ:* Stanford Univ, BS, 63; Yale Univ, MS, 64, PhD(physics), 69. *Prof Exp:* Asst prof physics, Mid East Tech Univ, Turkey, 69-72; res fel seismol, Dept Terrestrial Magnetism, Carnegie Inst, Washington, DC, 72-77; from asst prof to assoc prof, 77-87, PROF GEOPHYSICS, VA POLYTECH INST & STATE UNIV, BLACKSBURG, 87- *Mem:* Am Geophys Union; Seismol Soc Am; AAAS. *Res:* Subducting plates: structure and dynamics; earthquake source: models and barameter estimates; seismicity patterns: temporal and spatial. *Mailing Add:* Dept Geol Sci Va Polytech Inst & State Univ Blacksburg VA 24061-0420

SNOKE, ROY EUGENE, b Shippensburg, Pa, Aug 6, 43; m 67; c 2. ENZYMOLOGY. *Educ:* Shippensburg State Col, BS, 65; Univ NDak, MS, 67, PhD(biochem), 70. *Prof Exp:* NIH fel, Inst Enzyme Res, Univ Wis-Madison, 70-72, asst prof biochem, 72; sr res chemist, 72-79, RES ASSOC, EASTMAN KODAK CO, 79- *Mem:* Am Soc Microbiol; Am Chem Soc. *Res:* Regulation of gluconeogenesis; enzyme mechanisms and adaptations involved in biological control, specifically phosphoenolpyruvate carboxykinase; microbial enzyme isolation and characterization; use of enzymes for biotransformation; design of enzyme analytical systems; clinical diagnostic analysis systems. *Mailing Add:* 1085 Marigold Dr Webster NY 14580-8727

SNOOK, JAMES RONALD, b Seattle, Wash, Oct 2, 30; m 52; c 4. GEOLOGY. *Educ:* Ore State Univ, BS, 52, MS, 57; Univ Wash, PhD(geol), 62. *Prof Exp:* From instr to asst prof geol, Ore State Univ, 59-62; res geologist, Humble Oil & Ref Co, 62-64, prod geologist, 64-65; explor geologist, Stauffer Chem Co, 65-67; from asst prof to assoc prof, 67-72, chmn dept, 68-71, PROF GEOL, EASTERN WASH UNIV, 72- *Concurrent Pos:* Consult, US Borax, 77-81. *Mem:* Am Inst Mining, Metall & Petrol Eng; Am Inst Prof Geol; fel Geol Soc Am; Am Asn Petrol Geol; Nat Asn Geol Teachers. *Res:* Igneous and metamorphic petrology; economic geology. *Mailing Add:* Dept of Geol Eastern Wash Univ Cheney WA 99004

SNOOK, THEODORE, b Titusville, NJ, Apr 14, 07; m 33; c 1. HISTOLOGY, EMBRYOLOGY. *Educ:* Rutgers Univ, BSc, 29, MSc, 30; Cornell Univ, PhD(histol), 33. *Prof Exp:* Asst zool, Rutgers Univ, 29-30; instr histol & embryol, Cornell Univ, 30-34; from instr histol & embryol to asst prof anat, Col Med, Syracuse Univ, 34-46; asst prof, Tulane Univ, 46-49; assoc prof, Sch Med, Univ Pittsburgh, 49-53; chmn dept, 67-72, from assoc prof to prof, 53-77, EMER PROF ANAT, SCH MED, UNIV NDAK, 77- *Mem:* AAAS; Am Asn Anatomists; Biol Photog Asn; Microcirc Soc; Sigma Xi. *Res:* Development of pharyngeal tonsil; spleen vascular connections and lymphatics; comparative mammalian spleen morphology. *Mailing Add:* 343 Sheridan Rd Racine WI 53403

SNOPE, ANDREW JOHN, b Paterson, NJ, Jan 19, 39; m 57; c 3. GENETICS, CYTOGENETICS. *Educ:* Del Valley Col, BS, 60; Rutgers Univ, MS, 62; Ind Univ, PhD, 66. *Prof Exp:* Asst prof biol, Univ Md, Baltimore, 66-70; from asst prof to assoc prof biol, Essex Community Col, 70-77, chmn, Div Sci & Math, 82-88, assoc dean iNstr, 88-90, PROF BIOL, ESSEX COMMUNITY COL, 77-, DEAN INSTR, 90- *Res:* Chromosome structure and behavior; human cytogenetics; instructional methods in college biology teaching. *Mailing Add:* Essex Community Col Baltimore MD 21237

SNOVER, JAMES EDWARD, b Troy, NY, Nov 23, 20; m 42; c 4. MATHEMATICS. *Educ:* State Univ NY, BA, 41; Syracuse Univ, MA, 50, PhD, 55. *Prof Exp:* Asst prof math, Assoc Cols Upper NY, 46-49; asst prof, 54-71, ASSOC PROF MATH, FLA STATE UNIV, 71- *Mem:* Am Math Soc; Math Asn Am. *Res:* Analysis. *Mailing Add:* 1704 Myrick Rd Fla State Univ Tallahassee FL 32303-4334

SNOVER, KURT ALBERT, b Albany, NY, Apr 26, 43; m 64; c 1. PHYSICS. *Educ:* Fla State Univ, BS, 64; Stanford Univ, MS, 68, PhD(physics), 69. *Prof Exp:* Res assoc nuclear physics, State Univ NY Stony Brook, 69-71, asst prof physics, 71-72; SR RES ASSOC NUCLEAR PHYSICS, UNIV WASH, 72-, RES ASST PROF PHYSICS, 77- *Mem:* Am Phys Soc. *Res:* Low energy nuclear physics. *Mailing Add:* Dept Physics Univ Wash Seattle WA 98195

SNOW, ADOLPH ISAAC, b Providence, RI, Oct 8, 21; m 44; c 2. CHEMISTRY. *Educ:* Brown Univ, ScB, 43; Iowa State Col, PhD(chem), 50. *Prof Exp:* From asst to res assoc inst atomic res, Iowa State Col, 43-50; instr inst study metals, Univ Chicago, 50-52; head phys chem sect, Sinclair Res, Inc, 52-56, dir radiation lab, 56-59, radiation div, 59-66 & radiation & instrumenation div, 66-69; mgr phys res, Atlantic Richfield Co, 69-72, mgr phys & environ res, 72-78, sr consult, 78-82, sr res adv, 82-85; RETIRED. *Concurrent Pos:* Consult air & water qual control, Alyeska Pipeline Serv Co, 72-; consult, A I Snow, 85- *Mem:* Am Chem Soc; Am Soc Metals; Am Crystallog Asn; Sigma Xi; AAAS. *Res:* X-ray crystallography; uranium and thorium alloy phase diagrams; metallurgy of thorium and alloys; neutron diffraction; bonding in solids; physical properties of catalysts; radiation and tracer chemistry; petroleum processing and instrumentation; environment; fate of oil in water; air and water pollution. *Mailing Add:* 731 Dunbar PO Box 487 Beecher IL 60401

SNOW, ANNE E, b Spokane, Wash, Apr 9, 43; m 80. PHYSIOLOGY. *Educ:* Univ Ore, BA, 66; Univ Wash, PhD(pharmacol), 77. *Prof Exp:* Res technician pharmacol, Univ Wash, 66-73, res asst, 73-74, NIH trainee, 74-77; res assoc pharmacol, Va Commonwealth Univ, 77-80, res assoc physiol, Med Col Va, 80-83, res assoc physiol, Pharm Sch, 84-87; COORDR CLIN SERV, PHARM, RICHMOND MEM HOSP, 87- *Concurrent Pos:* Consult, A H Robins Pharmaceut Co, 79-80. *Mem:* Soc Neurosci; AAAS; Intersci Res Found. *Res:* Mechanisms by which stress produces anti-nocioception; effect of stress in altering responses to pharmacologic agents; effects of opiates and endogenous peptides on serotonergic systems; role of calcium in excitation-contraction coupling; role of calcium in fertilization and cell division. *Mailing Add:* Pharm Richmond Mem Hosp 1300 Westwood Ave Richmond VA 23227

SNOW, BEATRICE LEE, b Boston, Mass, June 9, 41. MAMMALIAN GENETICS, MICROCOMPUTERS. *Educ:* Suffolk Univ, AB, 62; Univ NH, MS, 64, PhD(zool), 71. *Prof Exp:* From instr to assoc prof biol, 65-74, actg chmn dept, 72-73, chmn dept biol, 73-78, COORD MED TECHNOL PROG, SUFFOLK UNIV, 68-, PROF BIOL, 74- *Concurrent Pos:* Allied health adv, Brookline Pub Schs, 68-70; dir, Marine Sci Prog, NH Col & Univ Coun, 75-76; coordr, Biol-Comput Prog, Suffolk Univ, 81- *Mem:* Am Soc Human Genetics; Am Genetic Asn; Am Inst Biol Sci; Am Soc Med Technol; Am Soc Zool. *Res:* Alkaline phosphatase activity and siren mutation expressivity in the mouse. *Mailing Add:* Suffolk Univ Beacon Hill 41 Temple St Boston MA 02114

SNOW, CLYDE COLLINS, b Ft Worth, Tex, Jan 7, 28; m 55; c 5. PHYSICAL ANTHROPOLOGY, FORENSIC ANTHROPOLOGY. *Educ:* Eastern NMex Univ, BS, 50; Tex Tech Col, MS, 55; Univ Ariz, PhD, 67; Am Bd Forensic Anthrop, dipl. *Prof Exp:* Res asst anat, Med Col SC, 60-61; res anthropologist, 61-65; chief, Appl Biol Sect, Civil Aeromed Inst, Fed Aviation Agency, 65-69, chief phys anthrop res, 69-79; FORENSIC ANTHROP CONSULT, 79- *Concurrent Pos:* From adj instr to adj asst prof anthrop, Univ Okla, 62-80, adj prof, 80-, res assoc, Sch Med, 64-; trustee, Forensic Sci Found, 73-79; forensic anthrop consult, Okla State Med Examr, 78- & Med Examr, Cook County, Ill, 79-; consult, select comt assassinations, US House Rep, 78-79; pres, Forensic Sci Educ, Inc, 82-86. *Mem:* Am Acad Forensic Sci (vpres, 78-79); Am Anthrop Asn; Am Asn Phys Anthrop; Soc Study Human Biol; Am Soc Forensic Odontol; Sigma Xi. *Res:* Forensic anthropology; study of human skeletal remains to establish personal identification and cause of death. *Mailing Add:* Okla State Med Examiner 901 N Stonewall Oklahoma City OK 73117

SNOW, DAVID BAKER, b Albuquerque, NMex, Nov 15, 41; m 70; c 2. MATERIALS SCIENCE. *Educ:* Mass Inst Technol, BS, 63, ScD(metall), 71. *Prof Exp:* Res metallurgist, Refractory Metals Prod Dept, Gen Elec Co, 70-77; SUPV, TEM & LIGHT MICROS, UNITED TECHNOL RES CTR, 77- *Mem:* Am Soc Metals; Metall Soc; Electron Microscopy Soc Am; Sigma Xi; Mat Res Soc. *Res:* Physical metallurgy of rapidly-solidified metals and alloys; development of advanced titanium-based alloys; recovery and recrystallization; applications of electron microscopy to materials science. *Mailing Add:* United Technol Res Ctr Silver Lane (MS 129-26) E Hartford CT 06108-1096

SNOW, DONALD L(OESCH), b Cleveland, Ohio, Apr 10, 17; m 48; c 3. ENVIRONMENTAL HEALTH ENGINEERING. *Educ:* Case Western Reserve Univ, BS, 39; Univ Wis, MS, 41; Environ Engrs Intersoc, dipl. *Prof Exp:* Sanit engr, Pan-Am Sanit Bur, USPHS, 43-48, sr sanit engr, 48-51, chief res facilities planning br, 51-54, chief sanit eng br, 54-60, lab design documentation proj, NIH, 61-62, ed radiol health data, 62-64, chief radiation surveillance ctr, Bur Radiol Health, 64-67 & Standards & Intel Br, 68-69, dir, Off Criteria & Standards, 69-71; DIR, NAT CTR TOXICOL RES PROG OFF, UNIV ARK, 71- *Concurrent Pos:* Ed, J Inter-Am Asn Sanit Eng, 46-48; pres, Fed Conf Sanit Engrs, 64; sanit eng dir, USPHS, 43-52. *Honors & Awards:* Hemispheric Award, Inter-Am Asn Sanit Engrs, 54. *Mem:* Am Soc Civil Engrs; Am Acad Environ Engrs; fel Am Pub Health Asn; Inter-Am Asn Sanit Eng (secy, 46-48); AAAS; Sigma Xi. *Res:* Medical research facilities planning; environmental health engineering; radiological health data program management; radiation standards; environmental standards. *Mailing Add:* 13723 Rivercrest Dr Little Rock AR 72212

SNOW, DONALD RAY, b Los Angeles, Calif, Mar 19, 31; m 58; c 6. MATHEMATICS, COMPUTERS IN MATHEMATICS. *Educ:* Univ Utah, BSME & BA, 59; Stanford Univ, MSME, 60, MS, 62, PhD(math), 65. *Prof Exp:* Res asst comput ctr, Stanford Univ, 61-62; res engr res labs, Lockheed Missiles & Space Co, 62-64; res assoc math, Univ Minn, Minneapolis, 64-66; asst prof, Univ Colo, Boulder, 66-69; assoc prof, 69-74, PROF MATH, BRIGHAM YOUNG UNIV, 74- *Concurrent Pos:* Vis prof, Fulbright-Hayes sr lectureship to Peru, 74 & vis res prof, Dept Appl Math, Univ Waterloo, Ontario, 76-77; chmn bd dir, Rocky Mountain Math Consortium, 77-78; lectr, Math Asn Am, 78-; Atomic Energy Comn fel, Stanford Univ; mem bd dirs, Utah Coun Comput in Educ, 82-85; vis res prof, Imp Col, Univ London, 90. *Honors & Awards:* Hamilton Watch Award. *Mem:* AAAS; Am Math Soc; Math Asn Am; Soc Indust & Appl Math; Nat Coun Teachers Math; Am Math Asn; Sigma Xi. *Res:* Calculus of variations; functional equations; combinatorics; partial and ordinary differential and integral equations and inequalities; history of math; computers in math instruction and research; numerical analysis. *Mailing Add:* Dept Math Brigham Young Univ Provo UT 84602

SNOW, DOUGLAS OSCAR, b Port Maitland, NS, Nov 27, 17; m 51. MATHEMATICS. *Educ:* Acadia Univ, BSc, 43, MA, 46; Brown Univ, MSc, 52; Queen's Univ, Can, PhD(math), 56. *Prof Exp:* Asst prof math, Mt Allison Univ, 46-47; from asst prof to assoc prof, 47-56, PROF MATH, ACADIA UNIV, 56- *Mem:* Am Math Soc; Math Asn Am; Can Math Cong. *Res:* Analysis; integration in abstract spaces. *Mailing Add:* PO Box 583 Wolfville NS B0P 1X0 Can

SNOW, EDWARD HUNTER, b St George, Utah, June 26, 36; m 56; c 3. SOLID STATE PHYSICS. *Educ:* Univ Utah, BA, 58, PhD(physics), 63. *Prof Exp:* Mem tech staff, Physics Dept Res & Develop Lab, Semiconductor Div, Fairchild Camera & Instrument Corp, 63-68, mgr, Physics Dept, 68-71; VPRES & DIR OPERS, EG&G RETICON CORP, 71- *Concurrent Pos:* Lectr, Univ Santa Clara. *Honors & Awards:* Cert of Merit, Franklin Inst, 75. *Mem:* Fel Inst Elec & Electronics Engrs; Am Phys Soc. *Res:* Optoelectronics; electrical properties of insulators and semiconductors; semiconductor device physics; properties of interfaces between metals; insulators. *Mailing Add:* Eg&G Inc 345 Potrero Ave Sunnyvale CA 94086

SNOW, ELEANOUR ANNE, b Portland, Ore, Apr 4, 60; m 89. MINERAL KINETICS. *Educ:* Pomona Col, BA, 82; Brown Univ, ScM, 84, PhD (geol), 87. *Prof Exp:* ASST PROF MINERAL, UNIV ARIZ, 87- *Mem:* Am Geophys Union; Geol Soc Am; Mineral Soc Am; Sigma Xi. *Res:* The effect of deformation on the mechanisms and kinetics of mineral reactions. *Mailing Add:* Dept Geosci Univ Ariz Tucson AZ 85721

SNOW, GEORGE ABRAHAM, b New York, NY, Aug 24, 26; m 48; c 3. PHYSICS. *Educ:* City Col, BS, 45; Princeton Univ, MS, 47, PhD(physics), 49. *Prof Exp:* Jr physicist, Brookhaven Nat Lab, 48-51, assoc physicist, 51-55; physicist, US Naval Res Lab, 55-58; assoc prof physics, 58-61, actg chmn dept physics & astron, 70-71, PROF PHYSICS, UNIV MD, COLLEGE PARK, 61- *Concurrent Pos:* Mem, Inst Advan Study, 52-53; vis lectr, Univ Wis, 55; NSF sr fel, Europ Orgn Nuclear Res, Geneva, 61-62, sci assoc, 80; John S Guggenheim fel & Fulbright res scholar, Univ Rome, 65-66; vis prof, Univ Paris, 72-73 & Tohoku Univ, 79; consult, Argonne Nat Lab, Fermilab, Brookhaven Nat Lab & Prentice Hall Publ Co; mem bd trustees, Univ Res Asn, 73-78, vchmn, 74, chmn Sci Comt, 75-77; vchmn div particles & fields, Am Phys Soc, 75, chmn, 76; gen res fel, Univ Md, 86; vis sci, Univ Bologna, 86. *Mem:* Fel AAAS; Am Asn Physics Teachers; Fel Am Phys Soc; Fedn Am Sci; European Phys Soc. *Res:* Experimental and theoretical high energy physics; neutrino interactions; e-plus e-minus and muon interactions; experimental test of Pauli principle. *Mailing Add:* Dept Physics & Astron Univ Md College Park MD 20742

SNOW, GEORGE EDWARD, b Denver, Colo, Aug 6, 45; m 83. ZOOLOGY. *Educ:* Rockhurst Col, Kansas City, MO, BA, 67; Univ Colo, Boulder, MA, 74, PhD(biol), 77. *Prof Exp:* Instr biol sci, Community Col Denver, 72-77; asst prof, 77-84, ASSOC PROF BIOL SCI, MT ST MARY'S COL, 84- *Concurrent Pos:* Proj dir, Local Course Improv Grant, NSF, 79-82, Undergrad Curric & Course Develop Grant, 91-; Danforth assoc, Danforth Asn, 81; lectr, Calif Polytech State Univ, 85-89; instr, Cuesta Col, 86-89, Allan Hancock Col, 87-89. *Mem:* AAAS; Am Soc Zoologists; Sigma Xi; Nat Asn Biol Teachers. *Res:* Control mechanisms of thermoregulation in reptiles. *Mailing Add:* Mt St Mary's Col 12001 Chalon Rd Los Angeles CA 90049

SNOW, JAMES BYRON, JR, b Oklahoma City, Okla, Mar 12, 32; m 54; c 3. OTOLARYNGOLOGY. *Educ:* Univ Okla, BS, 53; Harvard Univ, MD, 56. *Prof Exp:* Asst prof otorhinolaryngol, Med Ctr, Univ Okla, 62-64, prof & head dept, 64-72; PROF OTORHINOLARYNGOL & CHMN DEPT, SCH MED, UNIV PA, 72- *Mem:* Am Acad Otolaryngol Head & Neck Surg; Am Col Surgeons; Soc Univ Otolaryngol; Am Otol Soc; Am Laryngol Asn; Sigma Xi; Triological Soc. *Res:* Clinical disorders of smell and taste. *Mailing Add:* Dept Otorhinolaryngol Univ Pa Philadelphia PA 19104

SNOW, JEAN ANTHONY, b Richmond, Ind, Apr 18, 32; m 62, 76; c 2. MYCOLOGY, AEROBIOLOGY. *Educ:* DePauw Univ, AB, 54; Pa State Univ, PhD(plant path, genetics), 64. *Prof Exp:* Asst prof plant path, Univ Mass, 63-67; chief plant pathologist, Standard Fruit Co, 67-70; res assoc, Ctr Air Environ Studies, Pa State Univ, University Park, 70-77; CONSULT, 77- *Concurrent Pos:* Consult, Municipal Tree Restoration Prog, Eastern States, Penn State Univ, 88- *Mem:* Int Soc Plant Path; Int Asn Aerobiol; Sigma Xi. *Res:* Epidemiology and biometeorology of fungal diseases of plants; aerobiology, especially occurrence and dispersal of fungus air spora and aeroallergens. *Mailing Add:* 720E W Beaver Ave State College PA 16801-3920

SNOW, JOEL A, b Brockton, Mass, Apr 1, 37; m 59; c 2. NATIONAL LABORATORY MANAGEMENT. *Educ:* Univ NC, BS, 58; Wash Univ, MS, 63, PhD(physics), 67. *Prof Exp:* Prog dir theoret physics, NSF, 68-71, head, Off Interdisciplinary Res, 70-71, dep asst dir sci & technol, Res Appln, 71-74, dir, Off Planning & Resources Mgt, 74-76, dir, Div Policy Res & Anal, 76; sr policy analyst, Off Sci & Technol Policy, Exec Off Pres, 76-77; assoc dir res policy, US Dept Energy, 77-81, dir sci & technol affairs, 81-88; ASSOC VPRES RES, ARGONNE NAT LAB & UNIV CHICAGO, 88- *Concurrent Pos:* Instr physics & electronics, US Navy Nuclear Power Sch, 58-61; res assoc, dept physics & fel, Ctr Advan Study, Univ Ill, 67-68. *Honors & Awards:* William S Jump Found Award, 73; Arthur S Flemming Award, 74. *Mem:* Fel Am Asn Advan Sci; Am Phys Soc; Sigma Xi; World Future Soc. *Res:* Statistical mechanics, super conductivity, transport and field theory; technology assessment; science policy; environment and energy problems. *Mailing Add:* 5503 S Kenwood Chicago IL 60637

SNOW, JOHN ELBRIDGE, b Marion, Ohio, June 4, 15; m 41; c 3. ORGANIC CHEMISTRY. *Educ:* Oberlin Col, AB, 38; Cornell Univ, MS, 40, PhD(org chem), 42. *Prof Exp:* Group leader org synthesis res dept, Heyden Chem Corp, 42-56; mgr chem & plastics res, Res Div, Curtiss-Wright Corp, 56-61; res dir, Rap-in-Wax Co, 61-65; mgr process & prod develop, Packages Co Div, Champion Papers Co Div, US Plywood Champion Papers Co, 65-70, dir appl res, Champion Packages Co, Div Champion Int, 70-80; CONSULT, FLEXIBLE PACKAGING, 80- *Concurrent Pos:* Lectr, Dept Food Sci, Univ Minn, 70-81, vis prof, 72. *Mem:* Am Chem Soc; Packaging Inst. *Res:* Organic chemicals; pentaerythritols; resins and plastics; flexible packaging meterials; films; extrusion coating and laminating; adhesives; paper technology; adhesion; coextrusion coating; surface chemistry. *Mailing Add:* 4750 Dona Lane Minneapolis MN 55422

SNOW, JOHN THOMAS, b St Petersburg, Fla, Dec 29, 43; m 66. ORGANIC CHEMISTRY. *Educ:* Earlham Col, AB, 65; Middlebury Col, MS, 67; Univ Calif, Davis, PhD(chem), 70. *Prof Exp:* Res assoc chem, Univ Calif, Davis, 70-71; chief chemist, US Sugar Corp, 71-74; Nat Res Coun assoc, USDA, Western Regional Res Lab, Berkeley, Calif, 74-75; res scientist, 75-76, MGR BIOCHEM, MKT DEPT, CALBIOCHEM, LA JOLLA, CALIF, 77- *Mem:* Am Chem Soc; Sigma Xi; Int Food Technologists. *Res:* Mechanistic and synthetic organic chemistry. *Mailing Add:* PO Box 31 Del Mar CA 92014

SNOW, JOHN THOMAS, b St Louis, Mo, Dec 14, 45; m 69. MESOMETEOROLOGY, GEOPHYSICAL FLUID DYNAMICS. *Educ:* Rose Polytech Inst, BS, 68, MS, 69; Purdue Univ, PhD(atmospheric sci), 77. *Prof Exp:* Assoc prof, Dept Geosci, 77-89, PROF ATMOSPHERIC SCI, DEPT EARTH & ATMOSPHERIC SCI, PURDUE UNIV, 89- *Mem:* Sigma Xi; Am Meteorol Soc; Royal Meteorol Soc. *Res:* Fluid dynamics of mesoscale meteorological phenomena; severe local storms; tornadoes and other geophysical vortices; gravity waves; lake and sea breeze circulations. *Mailing Add:* Dept Earth & Atmospheric Sci Purdue West Lafayette IN 47906

SNOW, JOHNNIE PARK, b Abilene, Tex, July 12, 42; m 68; c 2. PLANT PATHOLOGY. *Educ:* McMurry Col, BA, 65; Univ Ark, MS, 67; Tex A&M Univ, PhD(plant path), 70. *Prof Exp:* Fel plant path, NC State Univ, 70-72; from asst prof to assoc prof, 72-76, PROF PLANT PATH, LA STATE UNIV, BATON ROUGE, 81- *Mem:* Sigma Xi; Am Phytopath Soc. *Res:* Basic and applied research on soybean diseases. *Mailing Add:* Dept of Plant Path La State Univ Baton Rouge LA 70803

SNOW, JOSEPH WILLIAM, b Scarborough, Maine, Apr 3, 39. WIND & SOLAR ENERGY. *Educ:* Boston Col, BS, 61; Univ Utah, BS, 64; Univ Wis, Madison, MS, 75; Univ Va, PhD(environ sci), 81. *Prof Exp:* Weather officer, Air Weather Serv, US Air Force, 62-67; proj mgr, E G & G Inc, 68-69; supvr meteorologist, Panama Canal Co, Balboa, 70-71; res assoc, Dept Environ Sci, Univ Va, 80-81; asst prof meteorol, Lyndon State Col, 81-; RES SCI, HANSCON AFB, 81- *Mem:* Am Meteorol Soc; Am Geophys Union; Am Inst Aeronaut & Astronaut. *Res:* Analytical explanation of area climates and the modifications effected by man; coastal wind power and wind shear within the atmospheric boundary layer. *Mailing Add:* 146 Pine Point Rd Portland ME 04074

SNOW, LOUDELL FROMME, b Kansas City, Mo, July 17, 33; m 60; c 1. MEDICAL ANTHROPOLOGY. *Educ:* Univ Colo, BA, 59; Univ Ariz, MA, 70, PhD(anthrop), 71. *Prof Exp:* Mem fac, Col Human Med & asst prof anthrop, 71-73 & community med, 74-78, ASSOC PROF ANTHROP, MICH STATE UNIV, 78- *Mem:* AAAS; Am Anthrop Asn; Soc Med Anthrop; Sigma Xi; Am Folklore Soc. *Res:* Folk medical systems; folk practitioners as psychotherapists; witchcraft beliefs; behavioral science in the medical school curriculum; spirit possession and trance states; impact of cultural background on beliefs and attitudes concerning female reproductive cycle. *Mailing Add:* Dept Anthrop Mich State Univ 354 Baker Hall East Lansing MI 48824

SNOW, MICHAEL DENNIS, b Sacramento, Calif, Nov 9, 42; m 66; c 2. PHYTOPATHOLOGY, MICROBIAL ECOLOGY. *Educ:* Sacramento State Col, BA, 65; Wash State Univ, PhD(phytopath), 74. *Prof Exp:* From asst prof to assoc prof biol, 70-84, chmn dept phys & life sci, 77-80, PROF BIOL, UNIV PORTLAND, 85- *Concurrent Pos:* NSF fac develop fel, 81; res assoc, Corvallis Environ Res Lab, 81-82; vis scientist, US Environ Protection Agency, 91. *Honors & Awards:* Nat Tech Achievement Award Ecol, US Environ Protection Agency, 86. *Mem:* Am Soc Plant Physiologists; AAAS; Sigma Xi. *Res:* Stress physiology of crop plants. *Mailing Add:* Phys/Life Sci Univ Portland 5000 N Williamette Blvd Portland OR 97203-5798

SNOW, MIKEL HENRY, b Three Rivers, Mich, Sept 18, 44. MUSCLE REGENERATION, MUSCLE PLASTICITY. *Educ:* Olivet Col, BA, 66; Univ Mich, PhD(anat), 71. *Prof Exp:* Instr anat, Sch Med, Univ Miami, 71-72, asst prof, 72-75; asst prof, 75-79, ASSOC PROF ANAT, UNIV SOUTHERN CALIF, 79- *Concurrent Pos:* Vis prof, Sch Med, Univ Miami,

80; ed, Anat Record, 81- *Mem:* Am Asn Anatomists; Develop Biol. *Res:* The role of satellite cells in muscle regeneration and denervation; muscle adaptation to chronic and acute exercise in mammals. *Mailing Add:* 1883 Peterson Ave South Pasadena CA 91030

SNOW, MILTON LEONARD, b Providence, RI, Feb 16, 30; m 58; c 4. PHYSICAL CHEMISTRY. *Educ:* Brown Univ, ScB, 51; Princeton Univ, MA, 53, PhD(phys chem), 56. *Prof Exp:* Chemist, Davison Chem Co, WR Grace Co, 56-59; proj supvr, 76, SR CHEMIST, APPL PHYSICS LAB, JOHNS HOPKINS UNIV, 59- *Mem:* Sigma Xi. *Res:* Heterogeneous catalysis; gas phase reaction kinetics; supersonic ramjet performance analysis and prediction; hypersonic ramjet design; air pollution analysis; chemical anti-submarine warfare. *Mailing Add:* 1329 Winding Way Lane Wheaton MD 20902-1449

SNOW, PHILIP ANTHONY, b Baltimore, Md, July 19, 51. HYDROLOGY, MINERALOGY. *Educ:* Univ Md, BS, 73, MS, 77, PhD(soil, mineral), 81. *Prof Exp:* Res asst soil chem, 73-77, soil & mineral, 78-81, lectr soil physics, 81-82, res assoc soil mineral, Univ Md, 82-83; CHEMIST, GEO-SCI CONSULTS, 83- *Honors & Awards:* Bausch & Lomb Sci Award; T Blair Chem Award. *Mem:* Am Soc Agron; Soil Sci Soc Am; Soc Appl Spectros. *Res:* Rectification of domestic and industrial waste waters by land application; reclamation of acid mine and dredge spoil materials; development of x-ray spectroscopy techniques for the analysis of soil and geologic materials; research and development in chemical and biological warfare. *Mailing Add:* Box 53 Henderson MD 21640

SNOW, RICHARD HUNTLEY, b Worcester, Mass, Apr 26, 28; m 52; c 3. CHEMICAL ENGINEERING. *Educ:* Harvard Univ, AB, 50; Va Polytech Inst, MS, 52; Ill Inst Technol, PhD(chem eng), 56. *Prof Exp:* Res fel chem eng, 52-56, res engr, 56-63, sr engr, 63-73, mgr chem eng res, 77-83, dir, Nat Inst Petrol & Energy Res, 83-85, eng adv, Iit Res Inst, 73-; interim dir, Ctr Hazardous Waste Mgt, 87-88; RETIRED. *Mem:* AAAS; fel Am Inst Chem Engrs; Am Chem Soc; Soc Mining Engrs; Am Inst Mining, Metall & Petrol Engrs; Acad Hazardous Waste Mgt. *Res:* Chemical process development; chemical kinetics; thermodynamics; process simulation; hazardous waste control and minimization; particle processing. *Mailing Add:* IIT Res Inst 10 W 35th St Chicago IL 60616

SNOW, RICHARD L, b Salt Lake City, Utah, Jan 27, 30; m 53; c 5. PHYSICAL CHEMISTRY. *Educ:* Univ Utah, BS, 53, PhD(phys chem), 57. *Prof Exp:* From asst prof to assoc prof, 57-66, PROF CHEM, BRIGHAM YOUNG UNIV, 66- *Concurrent Pos:* NSF sci fac fel, Brown Univ, 63-64; Oak Ridge Assoc Univs fel, Savannah River Lab, E I du Pont de Nemours & Co, Inc, 71-72; fac fel, Battelle Northwest Norcus, 83. *Mem:* Sigma Xi; Am Chem Soc. *Res:* Quantum chemistry; theory of liquids. *Mailing Add:* Dept Chem Brigham Young Univ Provo UT 84601

SNOW, SIDNEY RICHARD, b Los Angeles, Calif, June 22, 29. GENETICS. *Educ:* Univ Calif, Los Angeles, BS, 54, PhD(bot), 58. *Prof Exp:* From instr to assoc prof, 57-70, PROF GENETICS, UNIV CALIF, DAVIS, 70- *Mem:* AAAS; Genetics Soc Am; Bot Soc Am; Am Soc Microbiol; Sigma Xi. *Res:* Genetics of fungi, especially yeast; genetic recombination; microbial genetics; cyto-genetics. *Mailing Add:* Dept Genetics Univ Calif Davis CA 95616

SNOW, THEODORE PECK, b Seattle, Wash, Jan 30, 47; m 69; c 3. ASTRONOMY. *Educ:* Yale Univ, BA, 69; Univ Wash, MS, 70, PhD(astron), 73. *Prof Exp:* Res assoc astrophys sci, Princeton Univ Observ, 73-76, mem res staff, 76-77; asst prof physics & astrophysics, 77-80, fel, Lab Atmospheric & Space Physics, 77-85, assoc prof astrophys, 80-87, PROF ASTROPHYS, 87-, DIR, CTR ASTROPHYS & SPACE ASTRON, 86- *Concurrent Pos:* Mem, many NASA & Am Astron Soc adv panels & comts. *Mem:* Int Astron Union; Am Astron Soc; Sigma Xi. *Res:* Visible wave length, infrared and ultraviolet space-borne spectroscopy of hot stars, stellar winds and interstellar gas and dust. *Mailing Add:* Ctr for Astrophysics & Space Astronomy Univ Colo-Campus Box 391 Boulder CO 80309

SNOW, THOMAS RUSSELL, b Danville, Va, Mar 29, 44. CARDIOVASCULAR PHYSIOLOGY. *Educ:* Carnegie-Mellon Univ, BS, 65; Duke Univ, PhD(physics), 71. *Prof Exp:* Instr biomed eng, Baylor Col Med, 71-73; res assoc physics, Duke Univ, 69-71, NIH fel physiol, 73-75, asst prof, 78-82, asst mem physiol, 83-90, assoc prof, 84- 90; DEPT SURG, UNIV SFLA, 90- *Mem:* Biophys Soc; Int Soc Heart Res; Am Physiol Soc. *Res:* Determination of the relevant factors controlling bio-energetics of the mammalian myocardium. *Mailing Add:* Dept Surgery Univ S Fla USF-MDC Box 16 Tampa FL 33612

SNOW, WILLIAM ROSEBROOK, b New York, NY, Jan 6, 30; m 51; c 2. ATOMIC PHYSICS, MOLECULAR PHYSICS. *Educ:* Stanford Univ, BS, 52; Univ Wash, MS, 65, PhD(physics), 66. *Prof Exp:* Rotational trainee reactor physics, Hanford Atomic Prod Oper, Gen Elec Corp, 52-54; physicist, Precision Technol, Inc, Calif, 54-58; staff assoc afterglows & atomic beams, Gen Atomic, 58-62; mem res staff satellite mass spectrometry, Aerospace Corp, 66-68; asst prof physics, Univ Mo-Rolla, 68-73, assoc prof, 73-78; mem staff, 78-81, DIR RES & DEVELOP, PAC WESTERN SYST, INC, 81- *Mem:* Am Phys Soc; Electrochem Soc; Am Vacuum Soc; Sigma Xi. *Res:* Ion-neutral reactions in plasmas and afterglows; negative ion charge transfer; ionosphere reactions; mass spectrometry; molecular scattering; plasma enhanced chemical vapor deposition. *Mailing Add:* 505 E Evelyn Ave Pac Western Syst Inc Mountain CA 94041

SNOW, WOLFE, b New York, NY, May 17, 38; m 60; c 3. MATHEMATICS. *Educ:* Brooklyn Col, BS, 59; NY Univ, MS, 61, PhD(math), 64. *Prof Exp:* From instr to asst prof, 64-83, ASSOC PROF MATH, BROOKLYN COL, 84- *Mem:* Soc Actuaries; Math Asn Am. *Res:* Stability for differential-difference equations. *Mailing Add:* Dept Math Brooklyn Col Brooklyn NY 11210

SNOWDEN, DONALD PHILIP, b Los Angeles, Calif, Sept 9, 31. SOLID STATE PHYSICS. *Educ:* Calif Inst Technol, BS, 53; Univ Calif, Berkeley, MA, 55, PhD(physics), 59. *Prof Exp:* Staff mem, Gen Atomic Div, Gen Dynamics Corp, 59-67; staff mem, Gulf Gen Atomic, 67-73; prin physicist, IRT corp, 73-82; SR SCIENTIST MISSION RES CORP, 82- *Mem:* Am Phys Soc; Sigma Xi. *Res:* Application of high current superconductors; semiconductor and device radiation effects; physical and electrical surface characterization; vacuum vapor deposition; photovoltaic solar cells; fiber optics communications; electromagnetic pulse (EMP) experiment. *Mailing Add:* 6656 Glidden St San Diego CA 92111

SNOWDEN, JESSE O, b McComb, Miss, Oct 19, 37; m 75; c 3. CLAY MINERALOGY, COASTAL GEOLOGY. *Educ:* Millsaps Col, BS, 59; Univ Mo, MA, 61, PhD(geol), 66. *Prof Exp:* Instr geol, Millsaps Col, 62-63; asst prof geol, Miss State Univ, 64-66; assoc prof geol, Millsaps Col, 66-69; from asst prof to prof geol, Univ New Orleans, 69-85, chmn, 85-90; DEAN, COL SCI & TECHNOL, SOUTHEAST MO STATE UNIV, 90- *Concurrent Pos:* Vis asst prof, Univ Mo, Columbia, 65-66; vis assoc prof, Univ Miss, 67, Gulf Coast Res Lab, 68-74, Univ Mich, 79. *Mem:* Am Asn Petrol Geologists; Clay Mineral Soc; Geol Soc Am; Soc Sedimentary Geol. *Res:* Research in evolution of the emergent Mississippi Delta and the barrier islands of Mississippi-Alabama. *Mailing Add:* Col Sci & Technol Southeast Mo State Univ One University Plaza Cape Girardeau MO 63701

SNOWDEN, WILLIAM EDWARD, materials science, for more information see previous edition

SNOWDON, CHARLES THOMAS, b Pittsburgh, Pa, Aug 8, 41. ANIMAL BEHAVIOR, PHYSIOLOGICAL PSYCHOLOGY. *Educ:* Oberlin Col, BA, 63; Univ Pa, MA, 64, PhD(psychol), 68. *Prof Exp:* Fel, Inst Neurol Sci, Univ Pa, 68-69; asst prof psychol, 69-74, assoc prof, 74-79, PROF PSYCHOL & ZOOL, UNIV WIS-MADISON, 79- *Concurrent Pos:* Affil scientist, Wis Regional Primate Res Ctr, 72; fac fel, NSF, 75-76; res scientist develop award, Nat Inst Mental Health, 77-; ed, Animal Behaviour, 85-88. *Mem:* Animal Behav Soc (pres-elect, pres, 88-92); Am Soc Zool; Psychonomics Soc; Am Soc Primatologists; Int Primatological Soc; Am Psychol Asn; Soc Conserv Biologists. *Res:* Communication and social behavior; evolution of language; social development and parental care; reproductive physiology and communication of reproductive status; captive breeding of endangered primates; field studies of endangered primates. *Mailing Add:* Dept Psychol Univ Wis-Madison 1202 W Johnson St Madison WI 53706-1296

SNOWDOWNE, KENNETH WILLIAM, b Homestead, Pa, June 2, 47; m 68. CALCIUM BIOLOGY. *Educ:* Clarion State Col, BS, 70; Univ Pittsburgh, PhD(pharmacol), 77. *Prof Exp:* NIH fel, Mayo Found, 77-80; res assoc, Univ Pittsburgh, 80-82, res asst prof, 82-85; ADJ ASST PROF, SCH DENT, UNIV PAC, 86- *Mem:* Biophys Soc; AAAS; Am Physiol Soc. *Res:* Cellular calcium homeostasis; role of calcium in control of cellular processes such as contraction, secretion, mitotic division and motility. *Mailing Add:* Dept Biochem Sch Dent Univ Pac 2155 Webster St San Francisco CA 94115

SNOWMAN, ALFRED, b London, Eng, Jul 11, 36; US Citizen; m 61; c 2. NON FERROUS METALS, PACKAGING MATERIAL. *Educ:* London Univ, BSc, 58; Farleigh Dickinson Univ, MBA, 73. *Prof Exp:* Res eng, Philco Corp, 58-60; mgr, Accurate Specialties Hackensack, 60-64; tech consult, Assoc Metals, 64-67; tech dir, Potters Indust, 79-83; tech mgr, Semi Alloys Inc, 83-89; MGR, MARGOLIN CONSULTS, 89- *Mem:* Am Inst Mining Metall Petrol Engrs; Am Soc Metals; Wire Asn; Am Standard Testing Mat; Wire Asn. *Res:* High purity precious metal; alloys used in soldering of semiconductor component. *Mailing Add:* 121 Huguenot Ave Englewood NJ 07631

SNUDDEN, BIRDELL HARRY, b Elkhorn, Wis, Nov 20, 35; m 64; c 2. BACTERIOLOGY, FOOD SCIENCE. *Educ:* Univ Wis, BS, 57, MS, 61, PhD(bact, food sci), 64. *Prof Exp:* Fel food sci, Mich State Univ, 64-66; from asst prof to assoc prof, 66-80, PROF BIOL, UNIV WIS-EAU CLAIRE, 80- *Mem:* Am Soc Microbiol; Int Asn Milk, Food & Environ Sanitarians. *Res:* Aquatic microbiology; food and water borne diseases. *Mailing Add:* Dept of Biol Univ of Wis Eau Claire WI 54702-4004

SNUSTAD, DONALD PETER, b Bemidji, Minn, Apr 6, 40; m 64; c 1. GENETICS. *Educ:* Univ Minn, BS, 62; Univ Calif, Davis, MS, 63, PhD(genetics), 65. *Prof Exp:* From asst prof to assoc prof genetics, 65-74, PROF GENETICS & CELL BIOL, UNIV MINN, ST PAUL, 74- *Mem:* Genetics Soc Am; Am Genetic Asn; Am Soc Microbiol; Am Inst Biol Sci. *Res:* Plant molecular genetics; virus-host cell interactions. *Mailing Add:* Dept Genetics Univ Minn 250 Biol Sci Ctr St Paul MN 55108

SNYDER, ALBERT W, b Halifax, Pa, Dec 28, 25. HIGH LEVEL RADIOACTIVE WASTE MANAGEMENT. *Educ:* Franklin & Marshall Col, BS, 51; Iowa State Univ, MS, 53. *Prof Exp:* DIR EXPLOR NUCLEAR POWER DEVELOP, SANDIA NAT LAB, ALBUQUERQUE, NMEX, 53- *Mem:* Fel Inst Elec & Electronics Engrs; Am Nuclear Soc. *Mailing Add:* Dir Explor Nuclear Power Develop Sandia Nat Labs PO Box 5800 Albuquerque NM 87185

SNYDER, ANDREW KAGEY, b Philadelphia, Pa, May 19, 37; m 59; c 2. MATHEMATICS. *Educ:* Swarthmore Col, BA, 59; Univ Colo, MA, 61; Lehigh Univ, PhD(math), 65. *Prof Exp:* Instr math, Lehigh Univ, 64-65 & Mass Inst Technol, 65-67; asst prof, 67-69, ASSOC PROF MATH, LEHIGH UNIV, 69- *Mem:* Am Math Soc; Math Asn Am; Sigma Xi. *Res:* Functional analysis; sequence spaces; summability. *Mailing Add:* Dept Math 14 Lehigh Univ Bethlehem PA 18015

SNYDER, ANN C, b Lansing, Mich, July 16, 51. SKELETAL MUSCLE, ENDOCRINE PHYSIOLOGY. *Educ:* Purdue Univ, PhD(exercise physiol), 82. *Prof Exp:* Asst prof med phys educ, Ball State Univ, 82-86; ASST PROF HUMAN KINTICS, UNIV WIS-MILWAUKEE, 86- *Res:* Exercise physiology. *Mailing Add:* Human Kinetics Dept Univ Wis-Milwaukee Milwaukee WI 53201

SNYDER, ANN KNABB, b West Reading, Pa, Aug 1, 44; m 65; c 1. MEDICAL RESEARCH, ENDOCRINOLOGY & METABOLISM. *Educ:* Pa State Univ, BS, 65; Univ Ill, MS, 68, PhD(physiol), 71. *Prof Exp:* Med technologist, Pediat Immunol Lab, Duke Univ Med Ctr, 70-71, Dept Immunol, Rush-Presby-St Lukes Med Ctr, 71-73; med technologist endocrinol, Med Res Prog, Vet Admin Med Ctr, North Chicago, 74-83; RES ASST PROF, DEPT MED, CHICAGO MED SCH, NORTH CHICAGO, ILL, 81-; RES PHYSIOL, 88- *Mem:* Sigma Xi; AAAS. *Res:* Influence of ethanol on carbohydrate metabolism in mammals, especially in respect to its interaction with the effects of thyroxine and insulin. *Mailing Add:* 505 E North Ave Lake Bluff IL 60044

SNYDER, ARNOLD LEE, JR, b Washington, DC, Oct 12, 37; m 63; c 3. SPACE PHYSICS, AURORAL & IONOSPHERIC PHYSICS. *Educ:* George Washington Univ, BCE, 60; Univ Colo, MS, 66; Univ Alaska, PhD(geophysics), 72. *Prof Exp:* Weather officer, Detachment 15, 1st Weather Wing, US Air Force, 61-62, weather officer, NY Air Defense Sector, 62-65, solar forecaster, Space Environ Support Ctr, 66-69, sect chief, Global Weather Cent, 72-76, br chief, Geophysics Lab, 76-80, test dir, 80-81, prog dir, Electronic Systs Div, US Air Force, 81-87; tech dir & adj prof, Univ Lowell, 87-89; SCIENTIST, MITRE CORP, 89- *Concurrent Pos:* Lectr, Western New Eng Col, 78-79. *Honors & Awards:* Res & Develop Award, US Air Force, 81; Prog Mgr Value Eng Award, Dept Defense, 84. *Mem:* Am Geophys Union; Am Meteorol Soc; Sigma Xi; Int Test & Evaluation Asn. *Res:* Space forecasting techniques. *Mailing Add:* RR 2 Box 135 Orrington ME 04474

SNYDER, BENJAMIN WILLARD, b Albion, Mich, July 5, 39; m 65; c 1. ENDOCRINOLOGY, PHARMACOLOGY. *Educ:* Albion Col, BA, 62; Univ Mich, MS, 67, PhD(zool), 70. *Prof Exp:* Fel reproductive biol, Sch Hygiene & Pub Health, Johns Hopkins Univ, 70-71 & Harvard Med Sch, 72-73; assoc prof biol, Swarthmore Col, 73-80; sr res biol, Sterling-Winthrop Res Inst, 80-89; assoc prof, Vassar Col, 89-91; PROJ LEADER, TSI-MASON RES INST, 91- *Mem:* Sigma Xi; Am Soc Zoologists; Endocrine Soc. *Res:* Reproductive biology; regulation of gonadotropin secretion in primates; maintenance of pregnancy in primates; endometriosis; contraception. *Mailing Add:* Dept Reproductive Physiol TSI-Mason Res Inst Worcester MA 01608

SNYDER, CARL EDWARD, chemistry; deceased, see previous edition for last biography

SNYDER, CARL HENRY, b Pittsburgh, Pa, Sept 18, 31; m 53; c 2. ORGANIC CHEMISTRY. *Educ:* Univ Pittsburgh, BS, 53; Ohio State Univ, PhD(org chem), 58. *Prof Exp:* Res chemist, Eastman Kodak Co, 58-59; res asst chem, Purdue Univ, 59-61; from asst prof to assoc prof, 61-74, PROF CHEM, UNIV MIAMI, 74- *Mem:* Am Chem Soc; Sigma Xi. *Res:* Reactions in dipolar, aprotic solvents; stereochemistry of carbonyl reductions; biomolecular eliminations; organoborane chemistry. *Mailing Add:* 6890 SW 78 Terr Miami FL 33143

SNYDER, CHARLES THEODORE, b Powell, Wyo, July 19, 12; m 73; c 2. HYDROLOGY. *Educ:* Univ Ariz, BSc, 48. *Prof Exp:* Geologist/Hydrologist, US Geol Survey, 46, 48-75; vis scientist, Carter County Museum, Ekalaka, Mont, 83-84; res assoc, Calif Acad Sci, 85-86; INDEPENDENT RES, 86- *Mem:* AAAS; Arctic Inst NAm; Soc Vert Paleontologists. *Res:* Hydrology and climatology of Ice Age and modern lakes in western United States; stream environmental studies; high altitude, cold weather or arid zone field operations safety and survival; sand dune morphology. *Mailing Add:* 552-17 Bean Creek Rd Scotts Valley CA 95066

SNYDER, CHARLES THOMAS, b Belle Plaine, Kans, July 2, 38; m 60; c 4. FLIGHT DYNAMICS & CONTROL, AERODYNAMICS. *Educ:* Univ Wichita, BS, 62; Stanford Univ, MS, 69, Engr, 76. *Prof Exp:* Aerospace eng flight dynamics, Flight & Systs Simulation Br, Ames Res Ctr, NASA, 62-65, proj engr, 65-70, group leader, 70-74, chief, Flight Systs Res Div, 74-80, dir aeronaut & flight systs, 80-85, DIR AEROSPACE SYSTS, AMES RES CTR, NASA, 85- *Concurrent Pos:* Dryden mem fel, Nat Space Club, 72; mem, NASA Adv Subcomt Aviation Safety & Oper Systs, 75-77 & Adv Subcomt Avionics & Controls, 77-80; assoc dir, Stanford/Ames Joint Inst Aeronaut & Acoust, 77-80; bd dirs, Am Helicopter Soc, 83-85. *Mem:* Assoc fel Am Inst Aeronaut & Astronaut; Am Helicopter Soc. *Res:* Flight mechanics; stability and control; handling qualities; guidance and navigation; avionics systems; aerodynamics; aircraft operating problems; simulation technology. *Mailing Add:* NASA-Ames Res Ctr Mail Stop 200-3 Moffett Field CA 94035

SNYDER, CLIFFORD CHARLES, b Ft Worth, Tex, Feb 16, 16; m 39; c 1. PLASTIC SURGERY. *Educ:* Univ Tenn, BS, 40, MD, 44; Am Bd Surg, dipl; Am Bd Plastic Surg, dipl. *Prof Exp:* Asst prof surg, Sch Med, Univ Tex, 52-54; from asst prof to assoc prof, Sch Med, Univ Miami, 54-67; EMER PROF SURG & ASSOC DEAN, SCH MED, UNIV UTAH, 67-, SURG ENDOWED PROF, 90- *Concurrent Pos:* Attend physician, Jackson Mem, Doctor's & Cedars of Lebanon Hosps; consult, Vet Admin Hosp, 54-67 & Nat Surg Consult, 69-; chief staff, Variety Children's Hosp, 58-67; mem, Am Bd Plastic Surg, 63-87; abstr ed, J Plastic & Reconstruct Surg, 67-73, co-ed, 73-; mem, Vet Admin Nat Adv Comt, 74- *Mem:* Am Soc Plastic & Reconstruct Surg (asst secy, 59-60, vpres, 66-67); Am Soc Surg of Hand; Am Asn Plastic Surg (pres, 74); Am Col Surgeons; hon mem Am Col Vet Surgeons. *Res:* Plastic and reconstructive surgery; transplantation of organs; snake bite; regeneration of nerves; wound healing. *Mailing Add:* Dean's Off Univ Utah Sch Med Salt Lake City UT 84132

SNYDER, CONWAY WILSON, b Kirksville, Mo, Jan 24, 18; m 43; c 3. PHYSICS. *Educ:* Univ Redlands, AB, 39; Univ Iowa, MS, 41; Calif Inst Technol, PhD(physics), 48. *Hon Degrees:* DSc, Univ Redlands, 68. *Prof Exp:* Jr physicist, US Naval Ord Lab, Washington, DC, 41-42; mem staff, Off Naval Res, Washington, DC, 48-49; with Fairchild Engine & Airplane Corp Nuclear Engine Propulsion Aircraft Proj, Ky, 49-51; Oak Ridge Nat Lab, 51-54; asst prof physics, Fla State Univ, 54-56; sr res engr, Jet Propulsion Lab, Calif Inst Technol, 56-59, scientist specialist, 59-63, staff scientist, 63-69, Viking orbiter scientist, 69-77, Viking proj scientist, 77-80, asst proj mgr sci, Infrared Astron Satellite Proj, 81-84; RETIRED. *Honors & Awards:* Medal for Except Sci Achievement, NASA, 68, 70, 73 & 77. *Mem:* Am Phys Soc; Sigma Xi. *Mailing Add:* 21206 Seep Willow Way Canyon Country CA 91351

SNYDER, DANA PAUL, b Winnipeg, Man, Apr 29, 22; nat US; m 52; c 4. ECOLOGY. *Educ:* Univ Ill, BS, 47, MS, 48; Univ Mich, PhD(zool), 51. *Prof Exp:* Instr biol, Pa Col Women, 52-53; res mammalogist, Carnegie Mus, 51-55; vis lectr, Smith Col, 56; from instr to assoc prof, 55-85, EMER PROF ZOOL, UNIV MASS, AMHERST, 85- *Mem:* Am Soc Mammal; Ecol Soc Am; Soc Syst Zool; Wildlife Soc; Am Inst Biol Sci; Am Ornith Union. *Res:* Systematics and ecology of mammals, especially Tamias striatus; geographic variation; population biology. *Mailing Add:* Dept Zool Univ Mass Amherst MA 01003

SNYDER, DANIEL RAPHAEL, b Detroit, Mich, July 5, 40; m 61; c 2. NEUROPSYCHOLOGY, PRIMATOLOGY. *Educ:* Wayne State Univ, BS, 62; Univ Mich, Ann Arbor, MS, 64, PhD(physiol psychol), 70. *Prof Exp:* Res assoc psychiat, 69-70, NIMH Biol Sci Training Prog fel biol psychiat, 70-71, asst prof lab animal sci, 71-73 & comp med & anthrop, 73-76, ASSOC PROF COMP MED & ANTHROP, SCH MED, YALE UNIV, 76-, HEAD NEUROBEHAV & PRIMATE RES FAC, 74- *Concurrent Pos:* Consult, Am Asn Accreditation Lab Animal Care, 74-; sci adv, Inst Biol Sci, Univ Islamabad, Pakistan, 75-; vet neurologist, Comp Med Referral Clin, Sch Med, Yale Univ, 76- *Mem:* Soc Neurosci; Psychonomic Soc; Int Primatol Soc; AAAS; Animal Behav Soc. *Res:* Neural mechanisms of social and emotional behavior; biomedical primatology & primate behavior; neuropsychology; behavioral & neurotoxicology. affective behavior; behavioral toxicology. *Mailing Add:* 47 Griffing Pond Rd Branford CT 06405-6410

SNYDER, DAVID HILTON, b Giles Co, Va, June 24, 38; m 54; c 5. VERTEBRATE BIOLOGY, ETHOLOGY. *Educ:* Univ Mo-Columbia, BA, 58, MA, 62; Univ Notre Dame, PhD(biol), 71. *Prof Exp:* Teacher, Berkeley Pub Schs, Mo, 58-59; from instr to assoc prof, 62-77, PROF BIOL, AUSTIN PEAY STATE UNIV, 77- *Mem:* Am Soc Ichthyologists & Herpetologists; Soc Study Amphibians & Reptiles; Am Soc Zoologists; Ecol Soc Am; Am Soc Mammal; Sigma Xi. *Res:* Reproductive behavior of amphibians; taxonomy of American amphibians and reptiles. *Mailing Add:* Dept Biol Austin Peay State Univ Clarksville TN 37044

SNYDER, DEXTER DEAN, b Toledo, Ohio, Feb 6, 42; m 68; c 3. PHYSICAL CHEMISTRY, ELECTROCHEMISTRY. *Educ:* Wabash Col, AB, 64; Mass Inst Technol, PhD(phys chem), 68. *Prof Exp:* Prof staff phys chem, Arthur D Little, Inc, 68-71; mem staff, Bendix Res Labs, 71-72; sr assoc res chem, 72-75, sr res chemist, 75-80, staff res scientist, 80-82, SR STAFF RES SCIENTIST, GEN MOTORS RES LABS, 82- *Mem:* Am Chem Soc; Electrochem Soc; Sigma Xi; Am Inst Chem Engrs; Mat Res Soc. *Res:* Thermodynamics of solids; cooperative and membrane phenomena; recycling and pollution control; electrochemical power sources , alloy electrodeposition, corrosion science, wear resistant surface treatments. *Mailing Add:* Gen Motors Res Labs Phys Chem Dept Gen Motors Corp Warren MI 48090-9055

SNYDER, DONALD BENJAMIN, b North Manchester, Ind, Oct 6, 35; m 65; c 2. ORNITHOLOGY. *Educ:* Manchester Col, BS, 57; Ohio State Univ, MS, 59, PhD(zool), 63. *Prof Exp:* Asst prof biol, Cent Wesleyan Col, 63-64; asst prof biol, Geneva Col, 64-69; PROF BIOL, EDINBORO UNIV PA, 69- *Mem:* Wildlife Soc; Asn Field Ornithologists. *Res:* Animal behavior; wildlife ecology. *Mailing Add:* Dept Biol & Health Serv Edinboro Univ Pa Edinboro PA 16444

SNYDER, DONALD DUWAYNE, b Mich, Apr 11, 28; m 51; c 2. PHYSICS. *Educ:* Andrews Univ, BA, 48; Mich State Univ, PhD(physics), 57. *Prof Exp:* Asst physics, Mich State Univ, 54-56, lectr, 57; sr res physicist labs, Gen Motors Corp, 57-59; from assoc prof to prof physics, Andrews Univ, 59-67, chmn dept, 59-67; assoc prof, 67-70, chmn div arts & sci, 68-78, PROF PHYSICS, IND UNIV, SOUTH BEND, 70-, CHMN, 81- *Mem:* Am Asn Physics Teachers. *Res:* Solid state and metal physics; compound semiconductors. *Mailing Add:* Dept Physics Ind Univ 1700 Mishawaka Ave PO Box 7111 South Bend IN 46634

SNYDER, DONALD LEE, b Bridgeport, Ohio, Sept 3, 43; m 67; c 3. ELECTROCHEMISTRY, CORROSION. *Educ:* Cleveland State Univ, BES, 66; Case Western Reserve Univ, MS, 68, PhD(phys chem), 70; John Carrell, MBA, 85. *Prof Exp:* RES ASSOC METAL FINISHING, HARSHAW CHEM CO, 70-, GROUP LEADER, 73-, RES & DEVELOP MGR, 82-, RES DIR, M&T HARSHAW, 83-, DIR TENHNOL, 87- *Mem:* Am Chem Soc; Am Electroplaters Soc; Electrochem Soc; Am Soc Testing & Mat; Inst Metal Finishing; Soc Automotive Eng; Nat Asn Corrosion Eng. *Res:* Electrochemical research primarily on the deposition of metals and the study of corrosion. *Mailing Add:* M & T Harshaw 23800 Mercantile Rd Cleveland OH 44122

SNYDER, EVAN SAMUEL, b Lehighton, Pa, Aug 24, 23; m 48; c 3. PHYSICS. *Educ:* Ursinus Col, BS, 44; Univ Pa, MS, 51, PhD(physics), 57. *Prof Exp:* From instr to assoc prof, 46-69, PROF PHYSICS & DEPT CHMN, URSINUS COL, 69- *Concurrent Pos:* Vis prof, NSF Inst, NMex State Univ, 59, 64-69; res partic, Oak Ridge Nat Lab, 60, 62; NSF sci fac fel, Princeton Univ, 68-69. *Mem:* Am Phys Soc; Am Asn Physics Teachers. *Mailing Add:* 80 Linfield Rd Trappe Collegeville PA 19426

SNYDER, FRANKLIN F, b Holgate, Ohio, Nov 11, 10; m; c 3. CIVIL ENGINEERING. *Educ:* Ohio State Univ, BCE, 32, CE, 42. *Prof Exp:* Surv foreman, Ohio Div Forestry, Rockbridge, Ohio, 33-34; jr hydraul engr, US Geol Surv, Washington, DC, 34-35 & Tenn Valley Authority, Knoxville, 35-37; hydraul engr, Pa Dept Forests & Waters, 37-40; assoc hydraul engr, US Weather Bur, Pittsburgh, Pa, 40 & Washington, DC, 40-42; supvry hydraul engr & asst chief, Hydrol & Hydraul Br, Corps Engrs, Washington, DC; partner, Nunn, Snyder & Assocs, 72-78; RETIRED. *Concurrent Pos:* Consult var co, 56-82. *Mem:* Nat Acad Eng; fel Am Soc Civil Engrs; Sigma Xi; Am Geophys Union; Am Meteorol Soc. *Res:* Author of various publications. *Mailing Add:* 1516 Laburnum St McLean VA 22101

SNYDER, FRED CALVIN, b Valley View, Pa, Apr 9, 16; m 42. AGRICULTURE. *Educ:* Pa State Univ, BS, 39, MS, 47, PhD(agr ed), 55. *Prof Exp:* Teacher pub schs, Pa, 39-41, 46-47; actg dir short courses, Pa State Univ, 56-58, dir short courses & chmn corresp agr & home econ, 58-79; RETIRED. *Mem:* AAAS. *Mailing Add:* 3048 Morningside Blvd Port St Lucie FL 34952

SNYDER, FRED LEONARD, b New Ulm, Minn, Nov 22, 31; m 55, 78; c 3. BIOCHEMISTRY. *Educ:* St Cloud State Col, BS, 53; Univ NDak, MS, 55, PhD(biochem), 58. *Hon Degrees:* DSc, Univ NDak, 83. *Prof Exp:* From res scientist to chief scientist, 58-79, from asst chmn to assoc chmn, 75-88, VCHMN MED & HEALTH SCI DIV, OAK RIDGE ASSOC UNIVS, 88- *Concurrent Pos:* Prof biochem, Med Units, Univ Tenn, Memphis, 64-86; prof med chem, Univ NC, Chapel Hill, 66-; assoc ed, Cancer Res, 71-78; ed, Handbook of Lipid Res, 87-; adj prof, Univ Tenn-Oak Ridge Grad Sch Biomed Sci, 72-; exec ed, Archives Biochem & Biophys, 87. *Mem:* Soc Exp Biol & Med; Am Soc Biol Chemists; Am Asn Cancer Res; Sigma Xi. *Res:* Metabolism and chemistry of lipids; cancer and pulmonary disorders; membranes; bioactive phospholipids; separation techniques. *Mailing Add:* Med Sci Div Oak Ridge Assoc Univs PO Box 117 Oak Ridge TN 37831

SNYDER, FREEMAN WOODROW, b Philadelphia, Pa, Dec 6, 17; m 38; c 2. PLANT PHYSIOLOGY. *Educ:* Univ Idaho, BS, 38; Cornell Univ, PhD(plant physiol), 50. *Prof Exp:* Eng aide, Soil Conserv Serv, USDA, 42-43, agr engr, 46-47; asst bot, Cornell Univ, 48-50; asst agronomist, Univ Ark, 50-53; plant physiologist, Sugar Beet Invests, Plant Sci Res Div, Agr Res Serv, USDA, 53-64, res plant physiologist, 64-68, plant physiologist, North Cent Region, 68-75, plant physiologist, Northeastern Region, 75-78, mem staff, Sci & Educ Admin, 78-83; RETIRED. *Concurrent Pos:* Adj assoc prof, Crop & Soil Sci, Mich State Univ, 70-75; actg lab chief, Light & Plant Growth Lab, Beltsville Agr Res Ctr, 81; coordr, Photosynthesis Prog, Org Econ Coop & Develop, 80-83. *Mem:* AAAS; Bot Soc Am; Am Soc Plant Physiol; Soc Sugar Beet Technol; Am Soc Agron. *Res:* Role of environmental and genetic factors in growth, development and yield of crop plants. *Mailing Add:* PO Box 169 Manor Dr Dublin PA 18917-0169

SNYDER, GARY DEAN, b New Castle, Pa, Aug 22, 47; m 72; c 2. BIOCHEMICAL ENDOCRINOLOGY, MEMBRANE RECEPTORS. *Educ:* Pa State Univ, BS, 69, MS, 72, PhD(biol), 75. *Prof Exp:* Fel endocrinol, Univ Ill Col Med, 75-77; fel endocrinol, 77-79, ASST PROF PHYSIOL, UNIV TEX HEALTH SCI CTR, DALLAS, 79- *Res:* Biochemical mechanism of action of hypothalamic releasing hormones on the anterior pituitary; regulation of hormone secretion from the anterior pituitary. *Mailing Add:* Dept Physiol Univ Tex Health Sci Ctr 5323 Harry Hines Blvd Dallas TX 75235

SNYDER, GARY JAMES, b Lebanon, Pa, Oct 11, 59. HIGH-SPIN MOLECULES, PHOTOACOUSTIC CALORIMETRY. *Educ:* Univ NMex, BS, 81; Calif Inst Technol, PhD(chem), 88. *Prof Exp:* Res assoc, Univ Colo, 88-90, lectr org chem, 89-90; ASST PROF, UNIV CHICAGO, 90- *Concurrent Pos:* Presidential young investr, NSF, 91. *Mem:* Am Chem Soc; AAAS; Sigma Xi. *Res:* Design, synthesis and study of high-spin organic molecules and compounds with novel electronic structures; application of time-resolved photoacoustic calorimetry to the study of high-energy organic transients. *Mailing Add:* Dept Chem Univ Chicago 5735 S Ellis Ave Chicago IL 60637

SNYDER, GARY WAYNE, b Blue Creek, WVa, May 26, 54; m 75; c 2. BOTANY-PHYTOPATHOLOGY. *Educ:* Glenville State Col, BS, 76; Miami Univ, Oxford, Ohio, MS, 79, PhD(bot), 84. *Prof Exp:* Instr bot, Miami Univ, Hamilton, 81-82; asst prof biol, Glenville State Col, 82-90; COORDR, NAT ENVIRON POLICY ACT PROG, URANIUM ENRICHMENT PLANT, MARTIN MARIETTA ENERGY SYS, 90- *Res:* Comparison of phenological events in forest stands and relating event to successional patterns; testing and utilization of pesticides and fungicides in orchards. *Mailing Add:* 767 Hopetown Rd Apt B-1 Chillicothe OH 45601

SNYDER, GEORGE HEFT, b Evanston, Ill, July 19, 39; m 69; c 2. AGRONOMY, SOIL CHEMISTRY. *Educ:* Ohio State Univ, BS, 62, MS, 64, PhD(agron), 67. *Prof Exp:* Asst instr agron, Ohio State Univ, 63-64; from asst prof to assoc prof, 67-79, PROF SOILS, EVERGLADES RES & EDUC CTR, UNIV FLA, 79- *Mem:* Am Soc Agron; Int Turfgrass Soc; Sigma Xi. *Res:* Soil chemistry; nutrient uptake by plants; wetland agriculture; nutrient leaching and transformations. *Mailing Add:* Agr Res & Educ Ctr Univ Fla PO Box 8003 Belle Glade FL 33430-8003

SNYDER, GEORGE RICHARD, b Grand Junction, Colo, July 26, 29; c 6. MARINE & FRESHWATER FISH ECOLOGY. *Educ:* Colo State Univ, BS, 58, MS, 60; Univ Idaho, PhD(fish sci), 80. *Prof Exp:* Planning biologist, US Bur Com Fisheries, 61-63; prog leader, Fish Passage Res, 63-69, dep dir, Environ-Conserv Div, 69-79, DIR, AUKE BAY FISHERIES LAB, NAT MARINE FISHERIES SERV, 81- *Concurrent Pos:* Adj prof, Univ Alaska, Fairbanks, 83- *Mem:* Am Fisheries Soc; Am Inst Fisheries Res Biologists. *Res:* Marine and freshwater fish ecology; life history salmonids; ecology of salmonids in Alaska; habitat effects on biota in Alaska; high seas driftnet interceptions; US and Canada northern border salmonid interceptions; salmonid enhancement; effects of Exxon Valdez oil spill on marine biota, water, sediments, fish. *Mailing Add:* 9454 Herbert Pl Juneau AK 99801

SNYDER, GLENN J(ACOB), b Akron, Ohio, Aug 1, 23; m 51; c 4. MECHANICAL ENGINEERING. *Educ:* Ohio Univ, BSME, 49. *Prof Exp:* Tool designer, Gun Mount Div, Firestone Tire & Rubber Co, 42-43; engr, Ohio Boxboard Co, 49-51; sr engr, Aerospace Div, Goodyear Tire & Rubber Co, 51-55; sr engr, Nuclear Power Div, Babcock & Wilcock Co, 55-60, supvr reactor vessel, internals control rod drives, 60-77, mgr & prin engr, Nuclear Power Generation Div, 77-84; VPRES ENG, SM/MS INC, 84- *Concurrent Pos:* Prof adj fac, Cent Va Community Col, 84-86. *Honors & Awards:* Centennial Medallion, Am Soc Mech Engrs, 80. *Mem:* Nat Soc Prof Engrs; Am Soc Mech Engrs; Soc Mfg Engrs; Am Nuclear Soc. *Res:* Author of several technical papers, holds patents. *Mailing Add:* SM/MS Inc PO Box 3305 Lynchburg VA 24503-3305

SNYDER, GREGORY KIRK, b Chicago, Ill, Sept 27, 39; m 75; c 2. COMPARATIVE PHYSIOLOGY, EVOLUTIONARY BIOLOGY. *Educ:* Humboldt State, BS, 63; Calif State Univ, San Diego, MS, 65; Univ Calif, Los Angeles, PhD(zool), 70. *Prof Exp:* Postdoctoral fel physiol, Univ Fla, Gainesville, 70-72; asst prof biol, Univ Calif, Riverside, 72-75; PROF BIOL, UNIV COLO, BOULDER, 75- *Mem:* Am Physiol Soc; Am Soc Zoologists; Fedn Am Socs Exp Biol. *Res:* Comparative physiology of cardiorespiratory function; adaptations to unique environment, especially high altitudes and diving; developmental regulation of the microvascular supply to tissues. *Mailing Add:* Dept EPO Biol Univ Colo Campus Box 334 Boulder CO 80309-0334

SNYDER, HAROLD LEE, b Denver, Colo, Dec 30, 52; m 77; c 2. POLYMER BLENDS, PHASE TRANSITIONS. *Educ:* Lewis & Clark Col, BS, 75; Univ Calif, Santa Barbara, PhD(phys chem), 80. *Prof Exp:* Staff scientist, E I Du Pont de Nemours & Co, Inc, 80-84, res supvr & mgr, 84-89, bus mgr specialty elastomers, 89-90, BUS MGR, ENGR POLYMERS, E I DU PONT, 91- *Concurrent Pos:* Lectr, Morgan State Univ, 81-82. *Res:* Dynamical aspects of chemilumenescent reactions in crossed molecular beams; dynamical aspects of phase transitions in polymer blends; structure and property relations on polymers. *Mailing Add:* Five Chemin Des Voirets Plan-Les-Ouates Geneva 1228 Switzerland

SNYDER, HARRY E, b Peoria, Ill, Jan 14, 30; m 52; c 3. FOOD SCIENCE, BIOCHEMISTRY. *Educ:* Univ Calif, Berkeley, AB, 51; Univ Calif, Davis, PhD(microbiol), 59. *Prof Exp:* From asst prof to assoc prof food technol, Iowa State Univ, 59-68, prof, 68-79; PROF FOOD SCI, UNIV ARK, 79- *Mem:* AAAS; Inst Food Technologists; Am Chem Soc; Am Asn Cereal Chemists; Am Soc Biol Chemists; Sigma Xi. *Res:* Extraction and refining of soybean oil; world food problems; possibility of making use of soy protein in human diets. *Mailing Add:* Dept Food Sci Univ Ark 272 Young Ave Fayetteville AR 72703

SNYDER, HARRY RAYMOND, JR, b Lawrence, Mass, Jan 19, 24; m 49; c 2. ORGANIC CHEMISTRY, MEDICINAL CHEMISTRY. *Educ:* Brown Univ, ScB, 49; Boston Univ, MA, 52, PhD(org chem), 58. *Prof Exp:* Res chemist, R J Reynolds Tobacco Co, 54-56; sr res chemist, Morton-Norwich Prod, Inc, 56-61, unit leader org chem, 61-73, sci assoc, Norwich-Eaton Pharmaceut Div, 73-83; mgr lab serv & safety, Chem Dept, Cornell Univ, Ithaca, NY, 84-90; RETIRED. *Concurrent Pos:* Counr, Norwich Sect, Am Chem Soc, 71-87; vis lectr org chem, Vassar Col, NY, 84- *Mem:* Am Chem Soc. *Res:* Synthesis of nitrofurans and other heterocycles for possible medicinal uses. *Mailing Add:* PO Box 371 Norwich NY 13815

SNYDER, HERBERT HOWARD, b Ravenswood, WVa, Feb 26, 27; m 66; c 1. APPLIED MATHEMATICS, MATHEMATICAL PHYSICS. *Educ:* Marietta Col, AB, 49; Lehigh Univ, MA, 51, PhD(math), 65; Univ SAfrica, PhD(appl math), 71. *Prof Exp:* Instr math, Lehigh Univ, 49-53; develop engr, ITT Fed Labs Div, Int Tel & Tel Corp, 53-63; instr math, Newark Col Eng, 63-64; instr math, Lehigh Univ, 64-65; asst prof, Drexel Inst, 65-66; assoc prof, 66-72, PROF MATH, SOUTHERN ILL UNIV, CARBONDALE, 72- *Concurrent Pos:* Vis assoc prof, Univ Ariz, 71-72; ed-in-chief, Handbuch der Electrotechnic; trustee, Ind Technol Univ. *Mem:* Am Math Soc; Ger Math Asn. *Res:* Function-theory on linear algebras; partial differential equations; electromagnetic theory; guided wave propagation and non-linear electron-wave interactions. *Mailing Add:* Dept Math Southern Ill Univ Carbondale IL 62901

SNYDER, HOWARD ARTHUR, b Lehighton, Pa, Mar 7, 30; m 75. FLUIDS, SOLID STATE PHYSICS. *Educ:* Rensselaer Polytech Inst, BS, 52; Univ Chicago, SM, 56, PhD(physics), 61. *Prof Exp:* Asst prof physics, Brown Univ, 61-68; assoc prof, 68-89, PROF AEROSPACE ENG SCI, UNIV COLO, BOULDER, 89- *Concurrent Pos:* NSF res grants. *Mem:* Am Phys Soc; Am Geophys Union. *Res:* Hydrodynamics and acoustics of liquid helium; laboratory modeling experiments of the atmosphere and the oceans; stability of fluid flow; low gravity fluid management. *Mailing Add:* Aerospace Box 429 Univ of Colo Boulder CO 80309

SNYDER, HUGH DONALD, b Norwalk, Conn, Sept 2, 23; m 52, 66, 69; c 3. PAPER INDUSTRY, MICROBIOLOGY. *Educ:* Rutgers Univ, BS, 48, PhD(plant path), 59. *Prof Exp:* Mycologist, Nuodex Prod Co, Inc, 49-54; chief microbiol lab, Troy Chem Co, 56-61; vpres & tech dir, Cosan Chem Corp, 62-66; tech dir indust biocides, Velsicol Chem Corp, 66-69; tech rep, Merck Chem Div, Merck & Co, Inc, 70-74; mgr tech serv, Vinings Chem Co, 74-89; PRES, SPECIALTY PROD DIV, ISCCO, 89-; VPRES INDUST SLIME CONTROL CONSULT, 89- *Mem:* Am Phytopath Soc; Soc Indust Microbiol; Bot Soc Am; Am Soc Microbiol; Tech Asn Pulp & Paper Indust; Sigma Xi. *Res:* Industrial microbiology and biocides; paper slime and deposit control methods. *Mailing Add:* 6140 Blackwood Circle Norcross GA 30093

SNYDER, J EDWARD, JR, b Grand Fork, NDak, Oct 23, 24. OCEANOGRAPHY. *Educ:* Naval Acad, BS, 44; Mass Inst Technol, MS, 55. *Prof Exp:* Oceanogr, US Navy; RETIRED. *Mem:* Nat Acad Eng. *Mailing Add:* 1224 Perry William Dr McLean VA 22101

SNYDER, JACK AUSTIN, b Lansing, Mich, Oct 21, 27; m 49; c 2. BIOCHEMISTRY. *Educ:* Mich State Univ, BS, 49; Univ Wis, MS, 51, PhD(biochem), 53. *Prof Exp:* From res scientist to res assoc, 58-66, res supvr, 66-71, res assoc, 71-73, tech serv mgr, Pharmaceut Div, Biochem Dept, 73-

81, CONSULT PHARMACEUT RES, E I DU PONT DE NEMOURS & CO, INC, 81 - *Mem:* Am Chem Soc. *Res:* Synthesis; structure in relation to activity; process development; drug candidate evaluation. *Mailing Add:* Rd 4 Coffee Run Apt B-5-C Hockessin DE 19707

SNYDER, JACK RUSSELL, zoology, for more information see previous edition

SNYDER, JAMES NEWTON, physics, computer science; deceased, see previous edition for last biography

SNYDER, JOHN CRAYTON, b Salt Lake City, Utah, May 24, 10 005 WH; m 42; c 3. MICROBIOLOGY, POPULATION POLICY. *Educ:* Stanford Univ, AB, 31; Harvard Univ, MD, 35; dipl, Am Bd Prev Med, 49. *Hon Degrees:* LLD, Harvard Univ, 64. *Prof Exp:* Fel surg, Mass Gen Hosp, 36-37; Soc Fels jr fel, Harvard Univ, 39-40; mem staff, Int Health Div, Rockefeller Found, 40-46; USA Typhus Comn, 42-46, Lt Col US Army Med Corps, 42-45; prof pub health bact, Harvard Univ, 46-50 & microbiol, 50-61, dean fac, 54-71, Henry Pickering Walcott prof, 61-71, prof pop & pub health & med dir, Ctr Pop Studies, Sch Pub Health, 71-76; chief, Div Pop Policy, Pathfinder Fund, 78; lectr, Mass Inst Technol & assoc dir, Int Pop Initiatives, 79-82; CONSULT, 82- *Concurrent Pos:* Chmn expert comt, Trachoma, World Health Org, 61; consult, Univ Assocs for Int Health, 74-83 & Med in Pub Interest, 76-80; mem, bd dirs, Theobald Smith Res Inst, 85-91. *Honors & Awards:* Order of the Nile, Egypt, 44. *Mem:* Am Pub Health Asn; Am Soc Trop Med & Hyg; Am Epidemiol Soc; Asn Am Physicians; Am Acad Arts & Sci. *Res:* Typhus fever and other rickettsial diseases; chemotherapy and immunology; trachoma and related diseases of the eye; human fertility control and population problems. *Mailing Add:* 112 Hugh Cargill Rd Concord MA 01742

SNYDER, JOHN L, b Lansing, Mich, June 23, 30; m 65. PHYSICAL GEOLOGY. *Educ:* Mich State Univ, BS, 51; Dartmouth Univ, AM, 53; Northwestern Univ, PhD(geol), 57. *Prof Exp:* Instr geol, Univ Tex, 57, asst prof, 57-62; dir educ, Am Geol Inst, 62-69; assoc prog dir undergrad educ div, NSF, 69-72, prog mgr, Div Higher Educ in Sci, 72-78, prof dir, local course improv prog, 78-82, dir volcanology & mantle geochem prog, 82-89, DIR PETROL & GEOCHEM PROG, NSF, 89- *Concurrent Pos:* Mem steering comt, Geo-Study & mem steering comt & adv bd, Earth Sci Curriculum Proj; vis prof, Dept Geol, Univ Mo-Columbia, 79. *Mem:* Geol Soc Am; AAAS; Nat Asn Geol Teachers; Am Geophys Union. *Res:* Igneous petrology; science education. *Mailing Add:* Nat Sci Found 1800 G St NW Washington DC 20550

SNYDER, JOHN WILLIAM, b Oakhill, WVa, May 12, 40. PHYSICAL OPTICS, SPECTROSCOPY. *Educ:* Ohio State Univ, BS, 63, MS, 64, PhD(physics), 68. *Prof Exp:* Mem fac, 68-78, chmn dept physics, 77-83, PROF PHYSICS, SOUTHERN CONN STATE COL, 81- *Mem:* Am Asn Physics Teachers; Am Optical Soc. *Res:* Fourier transform spectroscopy; microprocessors. *Mailing Add:* 342 Norton St New Haven CT 06511

SNYDER, JOSEPH QUINCY, b Joplin, Mo, Aug 7, 20; m 42; c 4. CHEMISTRY. *Educ:* Univ Okla, BS, 42, MS, 51, PhD(chem, chem eng), 54. *Prof Exp:* Org chemist & asst dir res, Samuel Roberts Noble Found, Okla, 48-51; instr chem, Univ Okla, 51-52 & 53-54, instr chem eng, 54-55; dir res, DanCu Chem Co, 55-56; sr res chemist, Monsanto Co, 56-64, sr res group leader, 64-70, sr res specialist, 71-85; RETIRED. *Mem:* Am Chem Soc. *Res:* Free radical copolymerizations; hydrocarbon reactions; catalytic conversions of olefins. *Mailing Add:* 1151 Tompkins St Charles MO 63301-2618

SNYDER, JUDITH ARMSTRONG, b Washington, DC, Nov 11, 46; m 72. CELL BIOLOGY. *Educ:* Univ Calif, Berkeley, AB, 68, PhD(bot), 73. *Prof Exp:* Res asst bot, Univ Calif, Berkeley, 72; res assoc cell biol, Univ Colo, Boulder, 73-78; from asst prof biol sci to assoc prof, 78-89, PROF BIOL SCI, UNIV DENVER, 89- *Concurrent Pos:* Res assoc, NIH fel, 75-78. *Honors & Awards:* Barton L Weller Professorship. *Mem:* AAAS; Am Soc Cell Biol. *Res:* Isolation and characterization of the intact mitotic apparatus from mammalian tissue culture cells and investigation of factors controlling mitotic spindle assembly and chromosome movement. *Mailing Add:* Dept Biol Sci Univ of Denver Denver CO 80208

SNYDER, LAWRENCE CLEMENT, b Ridley Park, Pa, Apr 16, 32; m 58, 89; c 3. CHEMICAL PHYSICS. *Educ:* Univ Calif, Berkeley, BS, 53; Carnegie Inst Technol, MS, 54, PhD(chem), 59. *Prof Exp:* Mem tech staff, Bell Labs, 59-80 & 82-84; PROF & CHMN DEPT CHEM, STATE UNIV NY, ALBANY, 80-82 & 84- *Concurrent Pos:* Lectr chem, Columbia Univ, 65-67; lectureship in chem, Robert A Welch Found, 71. *Mem:* Fel AAAS; Am Chem Soc; fel Am Phys Soc; fel Am Inst Chem. *Res:* Electronic structure of molecules; structure and thermochemistry of silicon hydrides; silicon crystal surface reconstruction; chemistry and physics of defects in semiconductors. *Mailing Add:* Chem Dept State Univ NY Albany NY 12222

SNYDER, LEON ALLEN, genetics; deceased, see previous edition for last biography

SNYDER, LEWIS EMIL, b Ft Wayne, Ind, Nov 26, 39; m 62; c 2. ASTROPHYSICS, MOLECULAR PHYSICS. *Educ:* Ind State Univ, BS, 61; Southern Ill Univ, MA, 64; Mich State Univ, PhD(physics), 67. *Prof Exp:* Res assoc astrophys, Nat Radio Astron Observ, 67-69; from asst prof to assoc prof astron, Univ Va, 69-75; PROF ASTRON, UNIV ILL, 75- *Concurrent Pos:* Mem, Ctr Advan Studies, Univ Va, 69-75; mem radio & radar astron comn, Int Sci Radio Union, 69-; mem radio astron subcomt, Comt Radio Frequencies, Nat Res Coun, 71-74; vis fel, Joint Inst Lab Astrophys, Boulder, Colo, 73-74; Alexander von Humboldt Found sr US scientist award, 83-84. *Mem:* Sigma Xi; Am Phys Soc; Am Astron Soc; Int Astron Union; AAAS; Astron Soc Pac. *Res:* Spectral line radio astronomy and chemical composition of the interstellar medium, comets and evolved stars. *Mailing Add:* 103 Astron Bldg Univ Ill 1002 W Green St Urbana IL 61801

SNYDER, LLOYD ROBERT, b Sacramento, Calif, July 30, 31; m 52; c 4. ANALYTICAL CHEMISTRY. *Educ:* Univ Calif, BS, 52, PhD, 54. *Prof Exp:* Res chemist, Shell Oil Co, 54-56 & Technicolor Corp, 56-57; from sr res chemist to sr res assoc, Union Oil Co Calif, 57-71; dir separations res, 71-72, VPRES CLINICAL CHEM, TECHNICON CORP, 72- *Honors & Awards:* Petrol Chem Award, Am Chem Soc, 70, Chromatography Award, 84; Dal Nogare Award, 76; Palmer Award, 85; Martin Award, 89. *Mem:* Am Chem Soc. *Res:* Preparative separations; high speed liquid chromatography; adsorption and adsorption chromatography; analytical separations; computer simulation. *Mailing Add:* 26 Silverwood Ct Orinda CA 94563

SNYDER, LOREN RUSSELL, b Milwaukee, Wis, June 19, 41; m 86; c 2. MOLECULAR BIOLOGY. *Educ:* Univ Minn, Duluth, BA, 63; Univ Chicago, PhD(biophys), 68. *Prof Exp:* Jane Coffin Childs Mem Fund Med Res fel, Int Lab Genetics & Biophys, Naples, 68-69, fac sci, Univ Paris, 69-70; from asst prof to assoc prof, 70-79, actg chair, 87-88, PROF MICROBIOL, MICH STATE UNIV, 79- *Concurrent Pos:* NIH res grants, 74-77, 80-83 & 83-86; NSF res grants, 78-80, 80-83 & 86-89; vis prof, Harvard Univ, 84-85. *Mem:* Am Soc Microbiol. *Res:* Molecular basis for control of gene expression in bacteria. *Mailing Add:* Dept Microbiol Mich State Univ East Lansing MI 48824

SNYDER, LOUIS MICHAEL, b Boston, Mass, May 10, 35; m 58; c 3. HEMATOLOGY. *Educ:* Brown Univ, AB, 57; Chicago Med Sch, MD, 62; Am Bd Internal Med, Hematol Bd, dipl. *Prof Exp:* NIH grant hemat, Mass Gen Hosp, 65-66; DIR DIV HEMAT, ST VINCENT HOSP, 68-; PROF INTERNAL MED & PEDIAT, MED SCH, UNIV MASS, 79- *Concurrent Pos:* Mem med adv bd, New Eng Hemophilia Asn, 72-; chmn med adv bd, Cent Mass, Leukemia Soc of Am, 73-78. *Mem:* Am Fedn Clin Res; Am Soc Hemat; fel Am Col Physicians; NY Acad Sci. *Res:* Red cell metabolism; red cell membrane structure and function. *Mailing Add:* Lab Med St Vincent Hosp Worcester MA 01604

SNYDER, MELVIN H(ENRY), JR, b Pittsburgh, Pa, Sept 22, 21; m 46, 59; c 6. THERMODYNAMICS. *Educ:* Carnegie Inst Technol, BS, 46; Wichita State Univ, MS, 50; Okla State Univ, PhD, 67. *Prof Exp:* From instr to assoc prof aeronaut eng, 46-58, head dept, 51-58, asst dean, 58-67, PROF AERONAUT ENG, WICHITA STATE UNIV, 58-, CHAIRPERSON DEPT, 77- *Concurrent Pos:* Consult, 51-; vis prof, Von Karman Inst Fluid Dynamics, Belg, 70-71. *Mem:* Am Inst Aeronaut & Astronaut. *Res:* Drag of bodies in sheared-flow fields; lift, drag and pitching moment of delta wings; use of power to aerothermodynamics. *Mailing Add:* Dept of Aeronaut Eng Wichita State Univ Wichita KS 67208

SNYDER, MERRILL J, b McKeesport, Pa, May 25, 19; m 42; c 3. CLINICAL MICROBIOLOGY. *Educ:* Univ Pittsburgh, BS, 40; Univ Md, MS, 50, PhD(bact), 53; Am Bd Med Microbiol, dipl. *Prof Exp:* Clin chemist, McKeesport Hosp, Pa, 38-41; med bacteriologist, Dept Virus & Rickettsial Dis, US Army Med Ctr, DC, 45-49; instr bact & med, Sch Med, Univ MD, 49-53, asst prof med in clin bact, 53-57, from asst prof to assoc prof microbiol, 55-65, assoc dir div infectious dis, 57-74, from assoc prof to prof med in clin microbiol, 59-83, res prof med, 83-86, EMER PROF MED, SCH MED, UNIV MD, BALTIMORE CITY, 86- *Concurrent Pos:* Head diag microbiol & serol, Univ Md Hosp, 59-71, hosp epidemiologist, 73-77. *Mem:* Am Soc Microbiol; Infectious Dis Soc Am; AAAS. *Res:* Infectious diseases. *Mailing Add:* Dept Med Univ Md Sch Med Baltimore MD 21201

SNYDER, MILTON JACK, chemistry, construction materials; deceased, see previous edition for last biography

SNYDER, MITCHELL, b Philadelphia, Pa, Nov 4, 38; m 63; c 5. STATISTICS. *Educ:* Yeshiva Univ, BA & BHL, 60; NY Univ, MS, 62; Univ Chicago, PhD(statist), 66. *Prof Exp:* Consult, Biol Sci Comput Ctr, Univ Chicago, 62-65; mem tech staff, Bell Tel Labs, 65-68; sci dir comput ctr, 69-83, LECTR MATH, BAR-ILAN UNIV, ISRAEL, 68- *Concurrent Pos:* Lectr, Roosevelt Univ, 64-65; corp statistician, Tadiràn, Ltd, 83- *Mem:* Inst Math Statist; Am Statist Asn; Asn Comput Mach; Israel Statist Asn; Info Processing Asn Israel; Am Soc Qual Control; Israel Soc Qual Assurance. *Res:* Multivariate analysis; industrial statistics; computer applications in data analysis. *Mailing Add:* Dept Math Bar-Ilan Univ Ramat Gan Israel

SNYDER, NATHAN W(ILLIAM), b Montreal, Que, Apr 21, 18; nat US; m 44; c 2. ENERGY CONVERSION, HEAT & MASS TRANSFER. *Educ:* Univ Calif, Berkeley, BS, 41, MS, 44, PhD(mech eng, math), 47. *Prof Exp:* Instr & res scientist, Univ Calif, Berkeley, 42-47, asst prof mech eng, 47-53, assoc prof process eng & chmn dept nuclear eng, 53-57, prof nuclear eng & chmn dept, 57-58; sr staff scientist in space technol, chmn, Space Power & Energy Conversion Panel & adv, Propulsion Panel, Inst Defense Anal, 58-61; vpres res & eng, Royal Res Corp, 61-62; chief scientist, Kaiser Aerospace & Electronics Corp, 62-64; Neely Prof nuclear eng, Ga Inst Technol, 64-66, Neely Prof aerospace eng, 66-68; asst sr vpres res & eng, N Am Rockwell Corp, 68-71; pres, N W Snyder Assocs, 71-72; adj prof, Energy & Kinetics, Univ Calif, Los Angeles, 73-74; chief scientist, 72-78, tech dir, 78-80, MGR TECHNOL DEPT, RALPH M PARSONS CO, 81- *Concurrent Pos:* Consult various govt agencies and pvt industs, 44-; mem adv comt nuclear systs in space, NASA, 59-61, biotechnol & human res adv comt, 69-71 & life sci comt, 71-74; energy conversion adv to Air Force, 62-70, mem Air Force Sci Adv Bd, 67-70; mem space technol panel, President's Sci Adv Comt, 64-67; mem adv comt isotopes & radiation develop, Atomic Energy Comn, 66-67; mem, Environ Impacts Panel, Am Inst Biol Sci, 74-78; chmn transp comt, Calif Intersoc Legis Adv Comn, 77-; chmn tech comt, Dept Energy Strategic Petrol Reserve Prog, 78-81. *Honors & Awards:* George Washington Award & Engr of Year, Inst Advan Eng, 77; Skylab Achievement Award, NASA, 74. *Mem:* Am Phys Soc; Am Inst Aeronaut & Astronaut; Am Nuclear Soc; Acoust Soc Am; Am Inst Chem Engrs. *Res:* Energy conversion; physics of fluids and heat; mass transfer; space technology; nuclear power; space power; acoustics; physics of boiling, originated thin film or microfilm theory of boiling; thin film thermal instruments; sea water desalination; nuclear and electric propulsion;

environmental control and life support; solid waste conversion and resource recovery; chemical conversion of wastes to energy; interbasin transfer of water from Alaska to the United States, Canada and Mexico. *Mailing Add:* 14901 La Cumbre Dr Pacific Palisades CA 90272

SNYDER, PATRICIA ANN, b Batavia, NY, Sept 24, 40; c 1. VACUUM ULTRAVIOLET SPECTROSCOPY. *Educ:* Syracuse Univ, BS, 62; Univ Calif, San Diego, PhD(chem), 70. *Prof Exp:* Teaching asst chem, Syracuse Univ, 61-62; chemist, Allied Chem, 61-62 & E I Du Pont de Nemours & Co, 62-64; res asst chem, Univ Calif, San Diego, 64-70; res assoc, Ore State Univ, 70-73, instr, 73-74; asst prof, Baylor Univ, 74-75; from asst prof to assoc prof, 75-83, PROF CHEM, FLA ATLANTIC UNIV, 83-; PROF CHEM, GRAD FAC, UNIV FLA, 83- *Concurrent Pos:* From asst prof to assoc prof chem, grad fac, Univ Fla, 77-83, vis prof, 79; vis scientist, Brookhaven Nat Lab, 80, 82 & 83, consult, 85. *Mem:* Am Chem Soc; Sigma Xi. *Res:* Vacuum ultraviolet natural circular dichroism; magnetic circular dichroism; absorption spectroscopy with synchrotron radiation; electronic and geometric structure of molecules; spectroscopy with a crossed field undulator; a storage ring insertion device to produce intense variably polarized light. *Mailing Add:* Dept Chem Fla Atlantic Univ Boca Raton FL 33431

SNYDER, R L, b Pittsburgh, Pa, Nov 1, 11. ENGINEERING. *Educ:* Lehigh Univ, BS, 34. *Prof Exp:* Asst prof elec eng, Univ Pa, Philadelphia, 47-50; chief comput res, Aberdeen Proving Ground, Md, 50-53; consult elec eng, Richard Snyder, Moorstown, NJ, 53-59; mgr info syst group, Hughes Aircraft Lab, Malibu, Calif, 59-63; RETIRED. *Mem:* Fel Inst Elec & Electronics Engrs; fel Inst Radio Engrs. *Mailing Add:* 4625 Van Kleek Dr New Smyrna Beach FL 32169

SNYDER, RICHARD GERALD, b Northampton, Mass, Feb 14, 28; m 49; c 6. INJURY BIOMECHANICS, PHYSICAL ANTHROPOLOGY. *Educ:* Univ Ariz, BA, 56, MA, 57, PhD(phys anthrop), 59; Am Bd Forensic Anthropology, dipl. *Prof Exp:* Asst anthrop, Univ Ariz, 57-59, from assoc res engr to res phys anthropologist, Appl Res Lab, 59-60; chief phys anthrop, Civil Aeromed Res Inst, Fed Aviation Agency, 60-66; mgr biomech dept, Automotive Safety Res Off, Ford Motor Co, Mich, 66; assoc prof anthrop, Mich State Univ, 66-68; from assoc prof to prof anthrop, Univ Mich, Ann Arbor, 68-85, head, Biomed Dept & res scientist, Hwy Safety Res Inst, Inst Sci & Technol, 69-84, dir, NASA Ctr Excellence Man-Syst Res, Transp Inst, 84-85, EMER PROF ANTHROP, UNIV MICH, ANN ARBOR, 85-, EMER RES SCIENTIST, HWY SAFETY RES INST, 89-; PRES, BIODYNAMICS INT, 86- *Concurrent Pos:* Mem staff, Ariz Transp & Traffic Inst, 59-60; assoc prof syst eng, Univ Ariz, 60; adj assoc prof, Univ Okla, 61-66; res assoc, Univ Chicago, 63-66; consult, USAF, US Navy, NASA, US Dept Transp, Southwest Res Inst, US Army, Dept Health, Educ & Welfare & Am Inst Biol Scientists; mem biodynamics comt aerospace med panel, Adv Group Aeronaut Res & Develop-NATO, 63-, planning comt, Int Meeting on Impact, Portugal, 71, adv comt, Stapp Car Crash Conf, 69- & adv panel grad prog systs safety eng, NC State Univ, 69-; mem comt on hearing, bioacoust & biomech, Nat Acad Sci-Nat Res Coun, 70-72, Trauma Res Comt, 84-85; mem fac, Bioeng Prog, Univ Mich, Ann Arbor, 70; mem med adv bd, Prof Race Pilots Asn, 74-, bd dirs, Snell Mem Found, 90- *Honors & Awards:* Nat Safety Coun Metrop Life Award, 70; Arch T Colwell Merit Award, Soc Automotive Engrs, 73; Harry G Moselsy Award, Aerospace Med Asn, 75. *Mem:* Fel AAAS; fel Am Anthrop Asn; Aerospace Indust Life Sci Asn (vpres, 74); fel Aerospace Med Asn; fel Am Acad Forensic Sci; assoc fel Am Inst Aeronaut & Astronaut. *Res:* Human biology; aviation and automotive medicine; biomedical sciences; human tolerances to impact trauma; occupant restraint systems; dental morphology; forensic medicine. *Mailing Add:* 3720 N Silver Dr Tucson AZ 85749

SNYDER, ROBERT, b Brooklyn, NY, Jan 17, 35; m 57; c 2. BIOCHEMICAL PHARMACOLOGY. *Educ:* Queens Col, NY, BS, 57; State Univ NY, PhD(biochem), 61. *Prof Exp:* Trainee pharmacol, Col Med, Univ Ill, 61-63; prof, Jefferson Med Col, Pa, 63-81; PROF TOXICOL & DIR JOINT GRAD PROG TOXICOL, COL PHARM, RUTGERS UNIV, 81-, PROF PHARMACOL & CHEM RES TOXICOL, & CHMN DEPT PAHRMACOL & TOXICOL, 82-; DIR, HEALTH EFFECTS ASSESSMENT DIV, NJ INST TECHNOL, 84- *Concurrent Pos:* Vis prof toxicol, Univ Tubingen, Germany, 71-72; adj prof, Thomas Jefferson Univ, Pa, 81- & Univ Med & Dent NJ, 82- *Mem:* Am Col Toxicol; Am Asn Univ Prof; Am Soc Pharmacol & Exp Therapeut; Soc Toxicol; Int Soc Biochem Pharmacol. *Res:* Metabolic conversion of xenobiotics to reactive metabolites which are ultimately responsible for toxicological or carcinogenic processes; the mechanism(s) by which benzene produces bone marrow damage leading to aplastic anemia or leukemia; enzymes which metabolize benzene; characterization of the metabolites of benzene which result in the formation of adducts to DNA; target cells for benzene in bone marrow; mechanism of neurotoxicity of monochloroacetic acid; strategies for investigation of the toxicology of complex mixtures such as those found in leachates from chemical waste dumps. *Mailing Add:* Dept Pharmacol & Toxicol Col Pharm Rutgers Univ PO Box 789 Piscataway NJ 08854

SNYDER, ROBERT DOUGLAS, b Lancaster, Pa, Apr 15, 34; m 55; c 3. ENGINEERING MECHANICS, APPLIED MATHEMATICS. *Educ:* Ind Inst Technol, BSME, 55; Clemson Univ, MSME, 59; WVa Univ, PhD(theoret & appl mech), 65. *Prof Exp:* Servo engr, Bell Aircraft Corp, 55; instr mech eng, Ind Inst Technol, 55-57; from instr to asst prof mech, Clemson Univ, 57-60; from instr to prof, WVa Univ, 62-75; chmn, Dept Eng Sci, Mech & Mat, 75-76, DEAN ENG, UNIV NC, CHARLOTTE, 76- *Mem:* Nat Soc Prof Engrs; Am Soc Mech Engrs; Sigma Xi. *Res:* Continuum mechanics. *Mailing Add:* Col Eng Univ NC Charlotte NC 28223

SNYDER, ROBERT GENE, b Boise, Idaho, July 4, 29; m 53; c 2. MOLECULAR SPECTROSCOPY. *Educ:* Ore State Univ, BA, 51, MA, 53, PhD(chem), 55. *Prof Exp:* Fel vibrational spectros, Univ Minn, 55-56; chemist, Shell Develop Co, 56-63; res fel, Polytech Inst Indust Chem, Milan, Italy, 63-64; chemist, Shell Develop Co, 64-72; res chemist, Western Regional Res Lab, 72-75; res scientist, Midland Macromolecular Inst, 75-77; RES FEL, DEPT CHEM, UNIV CALIF, BERKELEY, 77- *Honors & Awards:* Cobentz Soc Award, 65. *Mem:* Am Phys Soc; Am Chem Soc; Soc Appl Spectros; Biophys Soc. *Res:* Vibrational spectroscopy; spectra and structure and phase behavior of chain molecules, polymers, and biopolymers. *Mailing Add:* Dept Chem Univ Calif Berkeley CA 94720

SNYDER, ROBERT L(EON), b Albion, NY, Sept 3, 34; m 60; c 2. MATERIALS SCIENCE, ENGINEERING. *Educ:* Rochester Inst Technol, BS, 56; Iowa State Univ, PhD(metall), 60. *Prof Exp:* Res engr, Metall, Res & Eng Ctr, Ford Motor Co, 60-65; res scientist, Res Div, Am Standard, Inc, 65-67; assoc prof mat, 67-70, PROF MECH ENG, ROCHESTER INST TECHNOL, 70- *Mem:* Am Soc Metals; Am Soc Eng Educ; Am Soc Mech Engrs; Nat Soc Prof Engrs. *Mailing Add:* Dept Mech Eng Rochester Inst Technol One Lomb Mem Dr Rochester NY 14623

SNYDER, ROBERT LEROY, b Ellwood City, Pa, Apr 24, 26; m 49; c 4. COMPARATIVE PATHOLOGY, VERTEBRATE ECOLOGY. *Educ:* Pa State Univ, BS, 50, MS, 52; Johns Hopkins Univ, ScD(hyg), 60. *Prof Exp:* Res aide wildlife ecol, US Fish & Wildlife Serv, 50-52; biologist, Pa State Game Comn, 52-56; res asst vert ecol, Johns Hopkins Univ, 56-59; from res assoc to assoc dir, Penrose Res Lab, Zool Soc Philadelphia, 59-69; asst instr path, Univ Pa, 61-62, assoc comp path, 62-66, asst prof, 66-70, dir, Penrose Res Lab, Zool Soc Philadelphia, 69-89, ASSOC PROF PATH, DIV GRAD MED, UNIV PA, 70- *Concurrent Pos:* Adj prof, Beaver Col, 89- *Mem:* Am Asn Lab Animal Sci; Am Asn Zool Parks & Aquariums; Wildlife Dis Assoc. *Res:* Comparative pathology, viral hepatitis and population ecology; study of chronic viral diseases and their role in the development of cancer; woodchuck hepatitis virus and similar agents of importance in cancer research. *Mailing Add:* 245 Chapel Lane King of Prussia PA 19406

SNYDER, ROBERT LYMAN, b Plattsburg, NY, June 5, 41; m 63; c 2. HIGH TEMPERATURE SUPERCONDUCTIVITY, XRAY CRYSTALLOGRAPHY. *Educ:* Marist Col, BS, 63; Fordham Univ, PhD(phys chem), 68. *Prof Exp:* Res asst & guest assoc, Brookhaven Nat Lab, NY, 66-68; from asst prof to assoc prof, 70-77, PROF CERAMIC SCI, NY STATE COL CERAMICS, ALFRED UNIV, 83-, DIR, INST CERAMIC SUPERCONDUCTIVITY, 87- *Concurrent Pos:* NIH fel, Crystallog Lab, Univ Pittsburgh, Pa, 68; Nat Res Coun fel, Electronics Res Ctr, NASA, Cambridge, Mass, 69; vis scientist, Lawrence Livermore Lab, Livermore, Calif, 77 & 78 & US Nat Bur Standards, Gaithersburg, Md, 80 & 81; vis prof, Siemens Cent Res Lab, Munich, Ger, 83 & 91; chmn tech comt, Int Centre Diffraction Data JCPDS, 86-90; chmn, Appl Crystallog Div, Am Crystallog Asn, 89-91; mem, US Nat Comt for Crystallog, Nat Acad Sci, 91-94. *Mem:* Am Crystallog Asn; Am Ceramic Soc; Joint Comt Powder Diffraction Standards; Mat Res Soc; Nat Inst Ceramic Eng. *Res:* Establishing structure-property relationships in technologically important ceramic materials; broad range of analytical techniques with a particular emphasis on x-ray, neutron and electron diffraction and thermal analysis. *Mailing Add:* Inst Ceramic Superconductivity NY State Col Ceramics Alfred Univ Alfred NY 14802

SNYDER, RUTH EVELYN, b Canadian, Tex, May 21, 11; m 42; c 3. MEDICINE, RADIOLOGY. *Educ:* Park Col, BA, 32; Univ Tex, MD, 36; Am Bd Radiol, cert, 43. *Prof Exp:* Intern, NY Infirmary, Women & Child, 36-37; fel, Strang Clinic, 37-38; clin fel radiation ther, Mem Hosp, 39-42; asst radiologist, NY Hosp, 42-45; asst roentgenologist, Mem Hosp, 42-45; assoc roentgenologist & radiation ther, Hosp Spec Surg, 44-47; roentgenologist, Strang Clinic, Mem Hosp, 48-51; asst roentgenologist, 51-52; assoc roentgenologist, Mem Hosp, 52-77; CONSULT RADIOL, NY INFIRMARY, 54-; attend roentgenologist & sr staff, 77-81, CONSULT, MEM SLOAN-KETTERING CTR, 81- *Concurrent Pos:* Instr radiol, Cornell Univ Med Col, 52-61, clin instr, 61-63, clin asst prof, 64- *Mem:* AMA; Am Women's Med Asn; Radiol Soc NAm; Soc Surg Oncol; Asn Women Sci; Am Col Radiol. *Res:* Mammography. *Mailing Add:* 222 E 68th St New York NY 10021

SNYDER, SOLOMON H, b Washington, DC, Dec 26, 38; m 62; c 2. NEUROPHARMACOLOGY. *Educ:* Georgetown Univ, MD, 62. *Hon Degrees:* DSc, Northwestern Univ, 81, Georgetown Univ, 86 & Ben-Gurlon Univ, 90. *Prof Exp:* Intern med, Kaiser Found Hosp, 62-63; res assoc pharmacol, NIMH, 63-65; from asst prof pharmacol to prof psychiat & pharmacol, 66-77, DISTINGUISHED SERV PROF PSYCHIAT & PHARMACOL, MED SCH, JOHNS HOPKINS UNIV, 77-, DIR DEPT NEUROSCI, 80- *Concurrent Pos:* Asst resident psychiat, Johns Hopkins Hosp, 65-68. *Honors & Awards:* John Jacob Abel Award, Am Soc Pharmacol & Exp Therapeut, 70, Goodman & Gilman Award, 80; A E Bennett Award, Soc Biol Psychiat, 70; Hofheimer Prize, Am Psychiat Asn, 72 & Spec Presidential Commendation, 85, Distinguished Serv Award, 89; Gaddum Prize, Brit Pharmacol Soc, 74; Francis O Schmitt Award, 74; Daniel Efron Award, Am Col Neuropsychopharmacol, 74; Salmon Award, 78; Lasker Prize, 78; Harvey Lectr, 78; Wolf Prize Med, Israel, 83; George Cotzias Award, Am Acad Neurol, 85; Sci Achievement Award, AMA, 85; J Allyn Taylor Prize, 90. *Mem:* Nat Acad Sci; fel Inst Med; Am Col Neuropsychopharmacol; Am Soc Pharmacol & Exp Therapeut; assoc Neurosci Res Prog; Int Soc Neurochem; fel Am Acad Arts & Sci. *Res:* Neurotransmitters; mechanism of action of psychotropic drugs. *Mailing Add:* Dept Neurosci Med Sch Johns Hopkins Univ 725 N Wolfe St Baltimore MD 21205

SNYDER, STANLEY PAUL, b Rifle, Colo, Sept 11, 42; m 66; c 2. VETERINARY PATHOLOGY, ONCOLOGY. *Educ:* Colo State Univ, DVM, 66, MS, 67; Univ Calif, Davis, PhD(comp path), 71. *Prof Exp:* Am Cancer Soc fel, Univ Calif, Davis, 71-72; asst prof vet med, Ore State Univ, 72-74; asst prof path, 74-78, ASSOC PROF PATH, COLO STATE UNIV, 78- *Mem:* Am Vet Med Asn; Am Col Vet Path; Am Asn Cancer Res; Vet Cancer Soc. *Res:* Viral and comparative oncology; pathogenesis of viral diseases; leprology. *Mailing Add:* Dept of Path Col of Vet Med Colo State Univ Ft Collins CO 80523

SNYDER, STEPHEN LAURIE, b Herkimer, NY, Oct 2, 42; m 66; c 2. BIOLOGICAL CHEMISTRY. *Educ:* Hobart Col, BS, 64; State Univ NY Binghamton, MA, 67; Univ Vt, PhD(chem), 70. *Prof Exp:* Res assoc biochem, Univ Colo, Boulder, 70-72; Nat Res Coun fel, Agr Res Serv, New Orleans, La, 72-75; biochemist, Armed Forces Radiobiol Res Inst, Bethesda, Md, 75-78; mem fac, Dept Chem, US Naval Acad, Annapolis, Md, 78-81; mem staff, Naval Med Res Inst, 81-83; PROG MGR BIOTECHNOL, OFF NAVAL RES, 89- *Mem:* Am Chem Soc. *Res:* Reaction mechanisms; enzymology; charge-transfer complexes; radiation biology; pathophysiology of endotoxins; role of lysosomal hydrolases in inflammation. *Mailing Add:* PO Box 447 Clinton MD 20735

SNYDER, THOMA MEES, b Baltimore, Md, May 21, 16; m 58; c 2. PHYSICS. *Educ:* Johns Hopkins Univ, PhD(physics), 40. *Prof Exp:* Instr physics, Princeton Univ, 40-42, res assoc, Off Sci Res & Develop contract, 42-43; res assoc, Los Alamos Sci Lab, 43-45; res assoc, Res Lab, Gen Elec Co, 46-47, proj head preliminary pile assembly & mem intermediate breeder reactor staff, Knolls Atomic Power Lab, 47-49, asst mgr, physics sect, 49-52, mgr reactor sect, 52-54, mgr phys sect, 54-56, mgr res oper, 56-57, mgr phys sect, Vallecitos Atomic Lab, 57-64, consult, Res & Eng Prog, 64-69, CONSULT SCIENTIST, NUCLEAR ENERGY DIV, GEN ELEC CO, 70- *Concurrent Pos:* Mem cross sect adv group, US AEC, 48-56, secy, 48, mem adv comt reactor physics, 50-72, chmn, 54; tech adv, US AEC, Geneva Conf Peaceful Uses of Atomic Energy, 55 & 58; mem, Mission Atomic Energy, Eng & Belg, 56. *Mem:* Fel Am Nuclear Soc; fel Am Phys Soc. *Res:* Nuclear energy technology; processes and materials; nuclear and reactor physics; solid state and plasma physics; radiation effects. *Mailing Add:* 208 Kalkar Santa Cruz CA 95060

SNYDER, VIRGIL W(ARD), b Midland, Mich, Mar 4, 34; div; c 2. STRUCTURAL DYNAMICS, FINITE ELEMENTS. *Educ:* Mich Technol Univ, BSCE, 56, MSCE, 62; Univ Ariz, PhD(aerospace eng), 68. *Prof Exp:* Stress engr, Northrup Aircraft, Inc, 56-58; struct engr, NAm Aviation, Inc, 58-60; instr eng mech, Mich Technol Univ, 60-62; consult, Kitt Peak Nat Observ, Univ Ariz, 63-65; PROF ENG MECH, MICH TECHNOL UNIV, 65- *Concurrent Pos:* Vis prof, Monash Univ, Australia, 84-85. *Mem:* Am Soc Eng Educ; Soc Exp Mech; Am Acad Mech. *Res:* Vibrations, dynamics, finite elements and rock mechanics; computer software developed for mechanics problems. *Mailing Add:* Dept Mech Eng-Eng Mech Mich Technol Univ Houghton MI 49931

SNYDER, VIRGINIA, b Coldwater, Ohio, July 30, 57. ELECTRON MICROSCOPY. *Educ:* Defiance Col, BS, 79. *Prof Exp:* RES TECHNICIAN, MED COL OHIO, 85- *Mem:* Am Soc Cell Biol; Am Asn Anatomists; Am Asn Electron Micros. *Mailing Add:* Dept Biol Univ Wis Rm 254 Gardner Platteville WI 53818-3099

SNYDER, WALTER STANLEY, b Oakland, Calif, Jan 17, 49; m 76. GEOLOGY. *Educ:* Stanford Univ, BS, 72, MS, 73, PhD(geol), 77. *Prof Exp:* Fel, 77-78, res assoc geol, Lamont-Doherty Geol Observ, Columbia Univ, 78-81; RES GEOL, PHILLIPS PETROL, 81- *Mem:* Geol Soc Am; Am Inst Mining Engrs. *Res:* Stratigraphy, tectonics and mineral resources. *Mailing Add:* 1400 84th Ave N St Petersburg FL 33702

SNYDER, WARREN EDWARD, b Hutchinson, Kans, Feb 24, 22; m 43; c 3. MECHANICAL ENGINEERING. *Educ:* Univ Kans, BS, 43; Univ Minn, MS, 48, PhD(mech eng), 50. *Prof Exp:* Mech engr, US Naval Res Lab, 43-46; instr mech eng & asst head, Univ Minn, 46-50; assoc prof & head, Univ Kans, 50-52; sr res engr, Res Labs, Gen Motors Corp, 52-57; dir eng div, Midwest Res Inst, 57-62; vpres eng, Cummins Eng Co, Ind, 62-66; vpres eng & res div, Am Bosch Arma Corp, Mass, 66-70; VPRES ENG, WAUKESHA ENG DIV, DRESSER INDUST, INC, 70- *Mem:* Am Soc Mech Engrs; Am Soc Eng Educ; Soc Automotive Engrs. *Res:* Design; analysis; mathematics; management. *Mailing Add:* 13800 Watertoen Plank Rd Elm Grove WI 53122

SNYDER, WESLEY EDWIN, b Orlando, Fla, Nov 11, 46; m 68; c 3. ELECTRICAL ENGINEERING. *Educ:* NC State Univ, BS, 68; Univ Ill, Urbana-Champaign, MS, 70, PhD(elec eng), 75. *Prof Exp:* Vis asst prof elec eng, Univ Ill, Urbana-Champaign, 75; asst prof, 76-81, ASSOC PROF ELEC ENG, NC STATE UNIV, 81- *Concurrent Pos:* Fel, Langley Res Ctr, NASA, 76; consult, UN, 75 & 77, IBM, Westinghouse, Gen Elec, & Res Triangle Inst; vis scientist, WGer Space Agency, 79. *Mem:* Sr mem Inst Elec & Electronics Engrs; Soc Mech Engrs; Asn Comput Mach; Robotics Inst Am. *Res:* Computer image analysis; machine vision; robotics. *Mailing Add:* 3603 Octavia St Raleigh NC 27606

SNYDER, WILBERT FRANK, b Marion, Ohio, Apr 19, 04; m 33, 76; c 2. PHYSICS, HISTORY OF SCIENCE. *Educ:* NCent Col, BA, 26; Univ Ill, AM, 27. *Prof Exp:* From jr physicist to physicist, Nat Bur Stardards, 27-46, asst chief, Microwave Stand Sect, 46-54, asst to chief, Radio Stand Div, 54-56, asst chief, Electronic Calibration Ctr, 56-62, coordr calibration serv, Radio Stand Eng Div,62-69, annuitant, Electromagnetics Div, 69-72, guest worker, 72-87; RETIRED. *Honors & Awards:* Bronze Medal for Superior Serv, Dept of Com, 68. *Mem:* Fel Acoust Soc Am; Sigma Xi; sr mem Inst Elec & Electronics Eng; sr mem Instrument Soc Am. *Res:* Acoustics of buildings and sound; standards and testing of hearing aids, audiometers and sixteen millimeter sound motion picture projectors; radar countermeasures; microwave standardization; calibration of electronic standards; radio history. *Mailing Add:* 350 Ponca Pl Boulder CO 80303

SNYDER, WILLARD MONROE, b Lehighton, Pa, Sept 29, 18; m 48. HYDROLOGY. *Educ:* Ursinus Col, BS, 40; Mass Inst Technol, MS, 48. *Prof Exp:* Engr hydrol, Fed-State Flood Forecasting Serv, Pa, 47-50; engr hydrol, Hydraul Data Br, Tenn Valley Authority, 50-55, head statist analytical unit, Hydrol Sect, 55-57, head hydrol sect, Hydraul Data Br, 57-60, staff res hydrologist, Off Tributary Area Develop, 60-62; prof hydrol, Ga Inst Technol, 63-69; res hydrol engr, Sci & Educ Admin-Agr Res, USDA, 69-75, res invest leader watershed eng, 70-73, res leader watershed hydrol, Southeast Watershed Res Ctr, Athens area, 74-80; RETIRED. *Concurrent Pos:* Consult, Oak Ridge Nat Lab, 67; consult hydrol, 80- *Mem:* Am Soc Civil Eng; Am Geophys Union; Am Water Resources Asn; Sigma Xi. *Res:* Formulation and evaluation of hydrologic models based on statistical analysis and on explicit and implicit solution of watershed process equations. *Mailing Add:* 275 Gatewood Circle Athens GA 30607

SNYDER, WILLIAM, electrical engineering, for more information see previous edition

SNYDER, WILLIAM JAMES, b Altoona, Pa, Nov 4, 41; m 64; c 1. CHEMICAL ENGINEERING. *Educ:* Pa State Univ, BS, 63, MS, 65, PhD(chem eng), 67. *Prof Exp:* Fel, Lehigh Univ, 67-68; asst prof chem eng, 68-74, assoc prof, 74-80, PROF CHEM ENG, BUCKNELL UNIV, 80- *Mem:* Am Inst Chem Engrs; Am Chem Soc; Am Soc Eng Educ. *Res:* Thermodynamic properties of solutions; heterogeneous catalysis; differential thermal analysis; polymers in solution; application of computers to chemical engineering plant design; mathematical modeling and simulation. *Mailing Add:* Dept Chem Eng Bucknell Univ Lewisburg PA 17837

SNYDER, WILLIAM RICHARD, b Brooklyn, NY, Jan 24, 47; m 69; c 3. ORGANIC CHEMISTRY, POLYMER CHEMISTRY. *Educ:* Hamline Univ, BA, 69; Northwestern Univ, MS, 70, PhD(org chem), 74. *Prof Exp:* RES SPECIALIST, 3M CO, 76- *Concurrent Pos:* Res fel, Calif Inst Technol, 74-76. *Mem:* Am Chem Soc. *Res:* Organic synthesis of biologically active molecules. *Mailing Add:* Kettelhack Riker Pharm GMBH PO Box 1340 Borken D4280 Germany

SNYDER, WILLIAM ROBERT, b Youngstown, Ohio, Mar 11, 46; m 77. PHOSPHOLIPASE MECHANISMS, MEMBRANE ENZYMOLOGY. *Educ:* Ohio State Univ, BS, 68; Univ Chicago, PhD(biochem), 72. *Prof Exp:* Res fel chem, Harvard Univ, 72-74; sr res scientist biochem, Armour Pharmaceut, 74-77; res fel neuropath, Ohio State Univ, 77-78; vis asst prof chem, Univ Ill, Chicago, 78-81; ASST PROF CHEM, NORTHERN ILL UNIV, 81- *Mem:* Am Chem Soc; AAAS; Sigma Xi. *Res:* Properties of enzymes involved in lipid metabolism; relationship of lipid hydrolysis to the structure and function of biological membranes; involvement of membrane alteration in biological processes. *Mailing Add:* 1923 Tanager Lane Geneva IL 60134-3153

SNYDER, WILLIAM THOMAS, b Knoxville, Tenn, Oct 18, 31; m 56; c 3. ENGINEERING MECHANICS. *Educ:* Univ Tenn, BS, 54; Northwestern Univ, MS, 56, PhD(mech eng), 58. *Prof Exp:* Asst prof mech eng, NC State Univ, 58-61; assoc prof thermal sci, State Univ NY, Stony Brook, 61-64; assoc prof aerospace eng, Space Inst, Univ Tenn, Tullahoma, 64-70; prof eng sci & mech & head dept, 70-83, DEAN ENG, UNIV TENN, KNOXVILLE, 83- *Mem:* Am Soc Eng Educ; Am Soc Heating, Refrig & Air Conditioning Engrs; Am Acad Mech; Energy Conserv Soc; Asn Energy Engrs; Soc Eng Mgt. *Res:* Combustion; lubrication; magnetohydrodynamics; energy conservation. *Mailing Add:* Col Eng Univ Tenn Knoxville TN 37996-2000

SNYDERMAN, RALPH, b New York, NY, Mar 13, 40; m 67. INTERNAL MEDICINE, RHEUMATOLOGY. *Educ:* Washington Col, BS, 61; State Univ NY, MD, 65. *Prof Exp:* Intern med, Med Ctr, Duke Univ, 65-66, resident, 66-67; res assoc immunol, Lab Microbiol, Nat Inst Dent Res, 67-69, sr investr, 69-72; assoc prof med, 72-77, assoc prof immunol, 72-79, PROF MED, DUKE UNIV MED CTR, 77-, CHIEF RHEUMATIC & IMMUNOL DIS, 75-, PROF IMMUNOL, 79-; CHIEF RHEUMATOLOGY, DURHAM VET HOSP, 72- *Concurrent Pos:* Howard Hughes med investr; sr vpres, med res & develop, Genentech, Inc, 87-89; James B Duke prof med, 89. *Honors & Awards:* Alexander von Humboldt Prize, 85. *Mem:* Am Asn Clin Investr; Am Asn Immunol; Am Acad Allergy; Am Rheumatism Asn; Am Fedn Clin Res; Asn Am Physicians; Am Asn Med Cols; Asn Acad Health Ctrs; Soc Med Adminr. *Res:* Investigation of the biological effectors of inflammation. *Mailing Add:* Duke Univ Med Ctr Box 3701 Durham NC 27710

SNYDERMAN, SELMA ELEANORE, b Philadelphia, Pa, July 22, 16; m 39; c 2. PEDIATRICS, MEDICINE. *Educ:* Univ Pa, AB, 37, MD, 40. *Prof Exp:* Fel pediat, 44-46, from instr to assoc prof, 46-47, PROF PEDIAT, SCH MED, NY UNIV, 67- *Concurrent Pos:* Assoc attend physician, NY Univ Hosp, 52-60, attend pediatrician, 60-; attend physician, Bellevue Hosp, 58-; career scientist, Health Res Coun, City of New York, 61-; mem nutrit study sect, NIH; dir, Metab Dis Ctr, NY Univ. *Honors & Awards:* Borden Award, Am Acad Pediat, 75. *Mem:* Soc Pediat Res; Am Pediat Soc; Am Acad Pediat; Am Inst Nutrit; Am Soc Clin Nutrit; Soc for Inherited Metab Dis; Soc Study Inborn Errors Metab. *Res:* Pediatric nutrition, especially amino acid metabolism and requirements. *Mailing Add:* Dept of Pediat NY Univ Sch of Med New York NY 10016

SNYGG, JOHN MORROW, b Oswego, NY, Dec 2, 37; m 65; c 3. APPLIED MATHEMATICS, CLIFFORD ALGEBRA. *Educ:* Harvard Univ, BA, 59; NY Univ, MA, 62, PhD(math), 67. *Prof Exp:* Lectr math, Hunter Col, 64-67; asst prof, 67-76, assoc prof, 76-87, FULL PROF MATH, UPSALA COL, 87- *Res:* Clifford algebra; quantum mechanics; population growth. *Mailing Add:* Dept Math & Physics Upsala Col East Orange NJ 07019

SO, ANTERO GO, b Davao City, Philippines, Jan 3, 32; US citizen; m 65; c 2. INTERNAL MEDICINE, HEMATOLOGY. *Educ:* Univ Santo Tomas, MD, 56; Univ Wash, PhD(biochem), 65. *Prof Exp:* USPHS fel, Western Reserve Univ, 60-62; USPHS trainee, Univ Wash, 62-65; Helen Hay Whitney Found res fel, Univ Geneva, 66-67; res instr biochem, Univ Wash, 67-68; asst prof biochem, 68-73, assoc prof med, 68-74, ASSOC PROF BIOCHEM, UNIV MIAMI, 73-, PROF MED, 74- *Concurrent Pos:* Estab investr, Am Heart Asn, 69, mem coun basic sci, 70; investr, Howard Hughes Med Inst, 74. *Mem:* Am Soc Clin Invest; Am Soc Biol Chemists. *Res:* Regulation of DNA and RNA synthesis in mammalian tissues. *Mailing Add:* Univ Miami Sch Med PO Box 01690 Miami FL 33101

SO, RONALD MING CHO, b Hong Kong, Nov 26, 39; US citizen; m 68; c 2. MECHANICAL ENGINEERING, AERONAUTICAL SCIENCES. *Educ:* Univ Hong Kong, BSc, 62; McGill Univ, MEng, 66; Princeton Univ, MA, 68, PhD(mech sci), 71. *Prof Exp:* Exec trainee, Shell Co, Hong Kong, 62-63; instr mech eng, Univ Hong Kong, 63-64; res scientist paper sci, Union Camp Corp, Res & Develop, 70-72; asst prof mech eng, Rutgers Univ, 72-76; mech engr res & develop, Gen Elec Corp, 76-81; assoc prof, 81-83, PROF MECH ENG, ARIZ STATE UNIV, 83- *Concurrent Pos:* Commonwealth scholar, 62-64; fluid physics consult, Res Cottrell Corp, 74-76; adj asst prof, Fairleigh Dickinson Univ, 74-76; adj assoc prof, Union Col, 77-78 & 79-81. *Mem:* Am Soc Mech Engrs; Am Phys Soc; Am Inst Aeronaut & Astronaut. *Res:* Fluid dynamics; energy and power generation research; wind power systems; combustion; nuclear reactors and gas turbines; flow induced vibrations; turbulent flows; heat transfer; atmospheric surface layers. *Mailing Add:* Dept Mech & Aero Eng Ariz State Univ Tempe AZ 85281

SOARE, ROBERT I, b Orange, NJ, Dec 22, 40; m 66; c 1. MATHEMATICS. *Educ:* Princeton Univ, AB, 63; Cornell Univ, PhD(math), 67. *Prof Exp:* From asst prof to prof math, Univ Ill, Chicago Circle, 67-75; Chmn Dept Comput Sci, 83-87, PROF MATH, UNIV CHICAGO, 75- *Concurrent Pos:* NSF grant recursive anal, 68-70; prin investr, NSF grant recursive function theory, 70-; sr fel, Grad Col, Univ Ill, Chicago Circle, 71; assoc ed, Proc Am Math Soc, 71-74. *Mem:* Asn Comput Mach; Am Math Soc; Asn Symbolic Logic. *Res:* Mathematical logic, particularly recursive functions. *Mailing Add:* Dept Math Univ Chicago 5734 University Ave Chicago IL 60637

SOARES, EUGENE ROBBINS, b New Bedford, Mass, Nov 22, 45; m 71. MAMMALIAN GENETICS. *Educ:* Univ RI, BS, 67, PhD(biol sci), 72. *Prof Exp:* NIH trainee, Jackson Lab, 72-73; staff fel, Nat Inst Environ Health Sci, 75-77; MAMMALIAN GENETICIST, CHEM INDUST INST TOXICOL, 76- *Mem:* AAAS; Environ Mutagen Soc; Genetics Soc Am; Am Genetic Asn; Sigma Xi. *Res:* Mammalian genetics; the genetic effects of chemical mutagens and electromagnetic radiation in mice; studies of chromosomal aberrations biochemical mutations, dominant and recessive lethal mutations and polygenic mutations in mice. *Mailing Add:* Seven Durham Woods Bucks Hill Rd Durham NH 03824-3202

SOARES, JOSEPH HENRY, JR, b Fall River, Mass, July 30, 41; div; c 2. NUTRITION, BIOCHEMISTRY. *Educ:* Univ Md, BS, 64, MS, 66, PhD, 69. *Prof Exp:* Animal nutritionist, Bur Com Fisheries, US Dept Interior, 68-69; res nutritionist, Nat Marine Fisheries Serv, US Dept Com, 69-72; from asst prof to assoc prof, 72-79, PROF NUTRIT, UNIV MD, COLLEGE PARK, 79-, CHMN, GRAD PROG NUTRIT SCI, 84- *Concurrent Pos:* Am Feed Mfrs res award, 77; vis prof, Human Nutrit Inst, USDA, 82, Dept Pediat Med, Univ SC, 89-90. *Honors & Awards:* Res Award, Am Feed Mfrs Asn, 77. *Mem:* AAAS; Am Inst Nutrit; Poultry Sci Asn; Am Soc Bone & Mineral Res. *Res:* Molecular nutrition; micro nutrient nutrition; gene regulation of calcification and vitamin D hormones; bone growth, metabolism and calcification as influenced by nutrition, aging and endocrine status; vitamin D metabolism and estrogen influences in females. *Mailing Add:* Dept Poultry Sci Univ Md College Park MD 20740

SOAVE, ROSEMARY, b New York, NY, Jan 23, 49; m 91. INFECTIOUS DISEASE, PARASITOLOGY-CRYPTOSPORIDIUM. *Educ:* Fordham Univ, BS, 70; Cornell Univ, MD, 76. *Prof Exp:* Chief resident med, Mem Hosp-Sloan Kettering Cancer Ctr, 79-80; med intern med, NY Hosp-Cornell Med Ctr, 76-77, med resident med, 77-79, fel infectious dis, 80-82, asst prof med, 82-89, asst prof pub health, 85-89, ASSOC PROF MED & PUB HEALTH, NY HOSP-CORNELL MED CTR, 89- *Concurrent Pos:* Lectr, Merck Sharp Dohme, 83-; prin investr, NIH res grant, 85-88. *Honors & Awards:* Arthur Palmer Award, 76; Jean Roughgarden Frey Award, 76. *Mem:* Am Col Physicians; Am Fedn Clin Res; NY Acad Sci; Infectious Dis Soc Am; Sigma Xi; AAAS. *Res:* Cryptosporidium; intestinal host defense mechanism against Cryptosporidium; treatment for cryptosporidiosis. *Mailing Add:* 525 E 68th St Box 125 New York NY 10021

SOBCZAK, THOMAS VICTOR, b Brooklyn, NY, Aug 6, 37; m 60; c 6. UNANTICIPATED INTELLIGENCE, RADIO FREQUENCY MANIPULATION. *Educ:* St John's Univ, BA, 59; Hofstra Univ, MBA, 65; Sussex Col, Eng, PhD(mgt), 72. *Prof Exp:* Analyst, Sperry Gyroscope Co, 59-60; mgt specialist, Kollsman Instrument Co, 60-63; consult, Airborne Instrument Lab, 63-67; adminr, Citibank, 67-68; dept head, Computer Opers, Pic Design Corp, 68-70 & Waldes Kohinoor Inc, 70-79; vpres, Little Peoples Prod Ctr, 79-86; EXEC VPRES, APPLN CONFIGURED COMPUTERS, 87- *Concurrent Pos:* Dir plans, Nat Defense Exec Reserve, Dept Com, 63-; adj prof, Hofstra Univ, 68-72 & NY Inst Technol, 73-76; aerospace appointee, NATO, 87- *Mem:* Fel Soc Mfg Engrs; fel Inst Advan Engrs; fel Inst Prod Engrs; Data Processing Mgt Asn. *Res:* Manipulated and integrated equipment architectures and software code to create platform level security optimizing any operating system; software based radio frequency driven non-lethal weapons which end war making potential. *Mailing Add:* PO Box 0433 Baldwin NY 11510-0433

SOBEL, ALAN, b New York, NY, Feb 23, 28; m 52; c 2. ELECTRONICS ENGINEERING, PHYSICS. *Educ:* Columbia Univ, BS, 47, MS, 49; Polytech Inst Brooklyn, PhD(physics), 64. *Prof Exp:* Engr, Telectro Indust Corp, NY, 49-50; asst chief engr, Electronic Workshop, Inc, 50-51; electronic engr, Freed Radio Corp, 51-53; chief electronics dept, Freed Electronics & Controls Corp, 53-55, head functional eng dept, Fairchild Controls Corp, 55-56; proj engr, Skiatron Electronics & TV Corp, 56-57; physicist, Zenith Radio Corp, 64-77; vpres, Lucitron Inc, 78-86, pres, 86-87; CONSULT, 88- *Concurrent Pos:* NSF Coop Grad Fel, 59-61; assoc ed, Inst Elec & Electronic Engrs Transactions Electron Devices, 70-77; contrib ed, Info Display, 90- *Mem:* Sr mem Inst Elec & Electronics Engrs; Am Phys Soc; fel Soc Info Display; Int Soc Optical Eng. *Res:* Flat-panel displays; display systems & devices; gas discharges; electronic devices and circuits. *Mailing Add:* 633 Michigan Ave Evanston IL 60202

SOBEL, EDNA H, b New York, NY, Nov 2, 18. MEDICINE, PEDIATRICS. *Educ:* Univ Wis, BA, 39, MA, 40; Boston Univ, MD, 43; Am Bd Pediat, dipl, 54. *Prof Exp:* Intern, Montefiore Hosp, New York, 44; clin fel pediat, Harvard Med Sch & Mass Gen Hosp, 44-47, res fel, Harvard Med Sch, 47-49, res assoc, 55-56, clin & res fel, Mass Gen Hosp, 48-49; instr, Col Med, Univ Cincinnati, 50-53; from asst prof to assoc prof, 56-68, PROF PEDIAT, ALBERT EINSTEIN COL MED, 68- *Concurrent Pos:* Commonwealth Fund fel advan med, Mass Gen Hosp, 49-50; Commonwealth Fund fel, 63-64; asst, Sch Pub Health, Harvard Univ, 44-46; vis physician, Children's Hosp, Cincinnati, Ohio, 50-53, asst physician, 55-56; res assoc, Children's Cancer Res Found, 53-56; asst vis pediatrician, Bronx Munic Hosp, Cent Res, 53-60, assoc vis pediatrician, 60-68, vis pediatrician, 68-; asst prof, Antioch Col & res assoc, Fels Res Inst, 51-53; consult, Misericordia Hosp, New York, 58-59; attend pediatrician, Lincoln Hosp, 59-70. *Mem:* AAAS; Am Pediat Soc; Endocrine Soc; Europ Soc Paediatric Endocrinol; Lawson Wilkins Pediat Endocrine Soc; Sigma Xi. *Res:* Endocrine function of normal and abnormal children. *Mailing Add:* Brookhaven Lexington 1010 Waltham St C 557 Lexington MA 02173

SOBEL, HENRY WAYNE, b Philadelphia, Pa. PARTICLE PHYSICS, COSMIC RAY PHYSICS. *Educ:* Rensselaer Polytech Inst, BS, 62; Case Inst Technol, PhD(physics), 69. *Prof Exp:* Asst res physicist, 69-74, assoc res physicist, 74-80, RES PHYSICIST, UNIV CALIF, 80- *Res:* Neutrino physics; fission. *Mailing Add:* 3891 Cedron St Irvine CA 92714

SOBEL, JAEL SABINA, b Israel, Nov 29, 35. CANCER RESEARCH, CELL MOTILITY. *Educ:* Cornell Univ, BA, 57; Columbia Univ, MA, 62; Univ Wis-Madison, PhD(zool), 66. *Prof Exp:* Fel cancer res, Sloan-Kettering Inst, 68-70; lectr embryol, Med Sch, Tel-Aviv Univ, 72-76; res fel, Lab Radiobiol, Univ Calif, San Francisco, 77-79; ASST PROF EMBRYOL & HISTOL, STATE UNIV NY BUFFALO, 79- *Concurrent Pos:* Consult, Lab Human Reproduction & Fetal Develop, Tel-Aviv Univ, 72-76. *Honors & Awards:* Rothschild Prize, Israel, 73. *Mem:* AAAS; Am Soc Cell Biol; Am Asn Anatomists. *Res:* Cell motility and characterization of cytoskeletal proteins during normal embryonic development and in developmental mutants; development of the trophoblast with emphasis on the regulation of invasive behavior. *Mailing Add:* Dept Anat State Univ NY 321 Farber Hall Buffalo NY 14214

SOBEL, KENNETH MARK, b Brooklyn, NY, Oct 3, 54. MULTIVARIABLE CONTROL, ADAPTIVE CONTROL. *Educ:* City Col NY, BSEE, 76; Rensselaer Polytech Inst, MEng, 78, PhD(comput & syst eng), 80. *Prof Exp:* Res asst, Rensselaer Polytech Inst, 76-79, instr syst eng, 79-80; RES SCIENTIST, LOCKHEED CALIF CO, 80- *Concurrent Pos:* Adj asst prof, Calif State Univ, Northridge, 81. *Mem:* Sr mem Inst Elec & Electronics Engrs; Sigma Xi. *Res:* Optimal output feedback; robust multivariable control; adaptive control; linear and nonlinear filtering; parameter identification. *Mailing Add:* Elec Eng Dept City Col NY 138th & Convent Ave New York NY 10031

SOBEL, MARK E, b Brooklyn, NY, Apr 14, 49. MOLECULAR BIOLOGY. *Educ:* Brandeis Univ, Mass, BA, 70; City Univ NY, PhD(biomed sci), 75; Mt Sinai Sch Med, MD, 75. *Prof Exp:* SR INVESTR PATH, NAT CANCER INST, 83- *Concurrent Pos:* Dir, Concepts Molecular Biol Course, Am Asn Pathologists, 87-91. *Honors & Awards:* Commendation Medal, USPHS, 89. *Mem:* Am Asn Pathologists; Am Soc Biochem & Molecular Biol; Am Soc Cell Biol; Am Soc Microbiol; AAAS. *Res:* Tumor invasion and metastasis; connective tissue gene regulation. *Mailing Add:* 9401 Bulls Run Pkwy Bethesda MD 20817-2405

SOBEL, MICHAEL I, b Brooklyn, NY, Feb 5, 39; m 59; c 2. SCIENCE POLICY. *Educ:* Swarthmore Col, BA, 59; Harvard Univ, MA, 61, PhD(physics), 64. *Prof Exp:* Res assoc physics, Northeastern Univ, 64; from asst prof to assoc prof, 64-72, PROF PHYSICS, BROOKLYN COL, 72- *Concurrent Pos:* NSF res grant, 65-72; res assoc, Harwell, Eng, 67-68; NATO fel, 73 & sr fel, 75; Woodrow Wilson Found fac develop award, 79. *Mem:* AAAS; Am Phys Soc; Am Asn Physics Teachers; NY Acad Sci. *Res:* Nucleon-nucleon interactions; heavy ion reactions; Author Non-Scientist publication; nuclear arms control. *Mailing Add:* Dept Physics Brooklyn Col Brooklyn NY 11210

SOBEL, ROBERT EDWARD, b New York, NY, Aug 8, 41; m 63; c 1. CLINICAL CHEMISTRY, BIOCHEMISTRY. *Educ:* Columbia Col, BA, 62; George Washington Univ, MS, 66, PhD(biochem), 69. *Prof Exp:* NIH fel path, Col Med, Univ Fla, 69-71; asst prof cell & molecular biol, Med Col Ga, 71-78, dir clin chem, Clin Path Labs, 71-79, assoc prof cell & molecular biol, Dept Path, 78-79; TECH DIR, NAT HEALTH LABS, VIENNA, VA, 79- *Mem:* AAAS; Am Chem Soc; Am Asn Clin Chem; Sigma Xi. *Res:* Methods in clinical chemistry; correlation of laboratory results with clinical findings; lipid metabolism; trace metals. *Mailing Add:* Nat Health Labs 1007 Electric Ave Vienna VA 22180

SOBELL, HENRY MARTINIQUE, b Los Angeles, Calif, Nov 7, 35; m 58; c 5. MOLECULAR BIOLOGY, X-RAY CRYSTALLOGRAPHY. *Educ:* Columbia Col, AB, 56; Med Sch, Univ Va, MD, 60. *Prof Exp:* Instr, genetics & develop biol, Mass Inst Technol, 60-61, res assoc, 61-65; assoc prof, dept chem, Col Arts & Sci, 65-73 & dept radiation biol & biophysics, Sch Med & Dent, 68-73, PROF DEPT RADIATION BIOL & BIOPHYSICS, SCH MED & DENT, UNIV ROCHESTER & DEPT CHEM, COL ARTS & SCI, 73- *Concurrent Pos:* Helen Hay Whitney fel, 61-65; NIH award, 66-71; vis investr & lectr, Rockefeller Univ, 67-68; vis prof pharmacol, Stanford Med Sch, 72-73. *Mem:* AMA; Am Crystallog Asn; AAAS; Am Chem Soc; Am Biophys Soc. *Res:* Drug-nucleic acid crystallography; actinomycin D-DNA binding; protein-nucleic acid interactions; wave phenomena in the DNA double-helix. *Mailing Add:* Dept Chem/Radiation Biol Univ Rochester Rochester NY 14642

SOBELL, LINDA CARTER, b Reno, Nev, May 8, 48; m 69; c 2. TREATMENT RESEARCH, SUBSTANCE ABUSE PROBLEMS. *Educ:* Univ Calif, Riverside, BA, 70, Irvine, MA, 74, Riverside, PhD(psychol), 76. *Prof Exp:* SR SCI ADDICTION RES FOUND, 84-; PROF PSYCHOL, UNIV TORONTO, 88- *Mem:* Asn Advan Behav Ther; Am Psychol Asn; Can Psychol Asn. *Res:* Published seventy articles in various journals and four books in the area of addictive behaviors research and treatment. *Mailing Add:* Addiction Res Found 33 Russell St Toronto ON M5S 2S1 Can

SOBELL, MARK BARRY, b Philadelphia, Pa, May 14, 44; US & Can citizen; m 69; c 2. PSYCHOLOGY, ADDICTIONS. *Educ:* Univ Calif, Los Angeles, AB, 66, Riverside, MA, 67, PhD(psychol), 70. *Prof Exp:* Res analyst II, Orange County Dept Mental Health, 72-74; assoc prof psychol, Vanderbilt Univ, 74-80; assoc prof, 80-87, PROF PSYCHOL & BEHAV SCI, UNIV TORONTO, 87- *Concurrent Pos:* Dir grad training on alcohol dependence, Dept Psychol, Vanderbilt Univ, 74-80, dir clin training, 79-80; sr scientist, Addiction Res Found, 80-; head, Sociobehav Res, Addiction Res Found, 88-90, chair, Treatment Res & Develop Dept, 90- *Mem:* Fel Am Psychol Asn; Asn Advan Behav Ther; Soc Psychologists in Addictive Behav. *Res:* Behavioral treatment of alcohol problems; conceptualizations of alcohol problems; models of addiction. *Mailing Add:* Addiction Resident Foundation 33 Russell St Toronto ON M5S 2S1 Can

SOBER, DANIEL ISAAC, b New York, NY, Sept 5, 42; m 73; c 1. EXPERIMENTAL NUCLEAR PHYSICS. *Educ:* Swarthmore Col, AB, 63; Cornell Univ, PhD(physics), 69. *Prof Exp:* Res asst physics, Princeton-Pa Accelerator, Princeton Univ, 68-70; adj asst prof, Univ Calif, Los Angeles, 70-75; from asst prof to assoc prof, 75-83, PROF PHYSICS, CATH UNIV AM, 83- *Concurrent Pos:* Vis prof, Polytech Inst Darmstadt, 81-82. *Mem:* Am Phys Soc. *Res:* Electromagnetic and weak interactions of elementary particles and nuclei; particle detectors; accelerators. *Mailing Add:* Dept Physics Cath Univ Am Washington DC 20064

SOBERMAN, ROBERT K, b NY, New York, Apr 8, 30; m 54; c 2. ENVIRONMENTAL PHYSICS, ASTRONOMY. *Educ:* City Col New York, BS, 50; NY Univ, MS, 52, PhD(physics), 56; Temple Univ, MBA, 72. *Prof Exp:* Res physicist, Vallecitos Atomic Lab, Gen Elec Co, 55-57; sr scientist res & advan develop div, Avco Corp, 57-59; assoc prof elec eng, Northeastern Univ, 59-60; chief meteor physics br, Air Force Cambridge Res Labs, 60-66; mgr environ progs, Gen Elec Space Sci Lab, 66-76; vpres progs, Univ City Sci Ctr, 76-78; dir, Appl Sci Dept, Franklin Res Ctr, 78-88; LECTR, ASTRON DEPT, UNIV PA, 88- *Concurrent Pos:* Adj assoc prof, Northeastern Univ, 60-64; adj prof, Drexel Univ, 68-; mem, Post Apollo Sci Eval Comt, 65-; vchmn comn 22B, Int Astron Union; sect noctilucent cloud subcomn, Int Union Geod & Geophys & mem Cosmic Dust Panel, Comt Space Res, Int Coun Sci Unions. *Mem:* AAAS; Am Astron Soc; Int Astron Union; Am Geophys Union. *Res:* Micrometeoroid flux and composition; rocket sampling of noctilucent clouds and cosmic dust; artificial meteors; recoverable and nonrecoverable spacecraft studies of meteoroids. *Mailing Add:* 2056 Appletree St Philadelphia PA 19103

SOBERON, GUILLERMO, biochemistry, for more information see previous edition

SOBEY, ARTHUR EDWARD, JR, b Shawnee, Kans, May 28, 24; m 48; c 4. PHYSICS, MATHEMATICS. *Educ:* Univ Tex, BS, 49, MA, 51, PhD(physics), 58. *Prof Exp:* Res scientist, Defense Res Lab, Univ Tex, 50-58; eng specialist, Chance Vought Aircraft Co, Tex, 58-59; mem tech staff, 59-66, mgr marine sci progs, 66-68, mgr signal processing progs, 69-71, mgr antisubmarine warfare surveillance progs, Tex Instruments Inc, 72-73; prog mgr, Electronic Sci, LTV Aerospace & Defense Co, 73-76, mgr electronics & optics res, 77-86, eng proj mgr, Vought Missiles & Advan Prog Div, 87-88, DEP PROG MGR, LTV MISSILES & ELECTRONICS GROUP MISSILES DIV, LTV AEROSPACE & DEFENSE CO & MGR, IR&D PLANNING, 88- *Res:* Underwater acoustics, including propagation and ambient noise; acoustic signal processing; space-time signal processing; infrared sensors and systems; electrooptic devices and subsystems; noise-cancelling microphones; atmospheric research; infrared scene simulation. *Mailing Add:* 914 Northlake Dr Richardson TX 75080-4914

SOBIESKI, JAMES FULTON, b Berlin, Wis, Mar 18, 40; m 76; c 4. PHOTORECEPTORS, TONERS. *Educ:* Univ Wis-Madison, BS, 61; Lawrence Univ, MS, 63, PhD(chem), 67. *Prof Exp:* sr res chemist, Microfilm Prod Lab, 68-85, PROCESS DEVELOP SPECIALIST, DOCUMENT SYSTEMS, 3M CO, ST PAUL, 85- *Mem:* Soc Photog Sci & Eng. *Res:* Applications of photoconductors to imaging systems; imaging systems. *Mailing Add:* 225 Hickory St Mahtomedi MN 55115

SOBIESZCZANSKI-SOBIESKI, JAROSLAW, b Wilno, Poland, Mar 11, 34; nat US; m 58; c 2. APPLIED MECHANICS, OPTIMIZATION METHODS. *Educ:* Warsaw Tech Univ, dipl aeronaut, 55, MS, 57, Dr Tech Sci(theory of thin shells), 64. *Prof Exp:* Asst aeronaut struct, Warsaw Tech Univ, 55-57, sr asst, 57 & 60-64, adj prof, 64-66; designer cranes & steel struct, Design Off Heavy Mach, Poland, 58-59; res fel, Inst Aeronaut, Norweg Inst Technol, 66; from asst prof to assoc prof aerospace eng, Parks Col Aeronaut Tech, St Louis, 66-71; Nat Acad Sci sr res fel, 70-71, aerospace engr, 71-73, sr res scientist, Struct Div, 73-80, br head multidisciplinary anal & optimization & struct dir, 80-84, DEP HEAD, INTERDISCIPLINARY RES OFF, STRUCT DIRECTORATE, LANGLEY RES CTR, NASA, 84- *Concurrent Pos:* Cert expert stress & vibration, Polish Eng Asn, 62-64; Norweg Govt fel, Inst Aeronaut, Norweg Inst Technol, 64-65; NASA res grant nonlinear struct anal, 68-70; consult, Polish Aviation Indust, 61-64; assoc prof lectr, George Washinton Univ, 71-80, prof lectr, 80-87, res sci, Space Div, NASA. *Honors & Awards:* Awards, Polish Soc Theoret & Appl Mech, 63-65; Medal for Except Eng Achievement, NASA, 88. *Mem:* Assoc fel Am Inst Aeronaut & Astronaut; assoc fel Royal Aeronaut Soc. *Res:* Experimental and numerical stress analysis; development of finite element methods for analysis of nonlinear structures; development of automated methods for interdisciplinary systems analysis and design; optimization of structures. *Mailing Add:* c/o NASA 246 Hampton VA 23665-5225

SOBIN, LESLIE HOWARD, b New York, NY, Feb 10, 34; m 62; c 1. PATHOLOGY. *Educ:* Union Col, NY, BS, 55; State Univ NY, MD, 59; Am Bd Path, dipl anat path, 64. *Prof Exp:* Res fel, Inst Cell Res, Karolinska Inst, Sweden, 58; asst path, Med Col, Cornell Univ, 60-62, from instr to asst prof, 62-66; prof, WHO, 65-68; assoc prof, Med Col, Cornell Univ, 68-70; pathologist, WHO, 70-81; PATHOLOGIST, ARMED FORCES INST PATH, 81-; PROF PATH, UNIFORMED SERV UNIV HEALTH SCI, 84- *Concurrent Pos:* Vis prof, Fac Med, Kabul, Afghanistan, 65-68; mem, WHO Expert Adv Panel Cancer, 81-; head, WHO Ctr Int Histol Classification Tumors, 83-; co-ed, Atlas Tumor Path, Armed Forces Inst Path, 84-, assoc dir, Sci Publ Armed Forces Inst Path, 87- *Mem:* Am Asn Path; Int Acad Path; fel Royal Col Pathologists. *Res:* Histological classification of tumors; gastrointestinal pathology. *Mailing Add:* Dept Gastrointestinal Path Armed Forces Inst Path Washington DC 20306-6000

SOBIN, SIDNEY S, b Bayonne, NJ, Jan 1, 14; m 59; c 2. PHYSIOLOGY. *Educ:* Univ Mich, BS, 35, MA, 36, PhD(physiol), 38, MD, 41; Am Bd Internal Med, cert, 51. *Prof Exp:* Asst physiol, Univ Mich, 34-38; resident med, Barnes Hosp, St Louis, Mo, 42-44; Nat Res Coun fel physiol, Harvard Med Sch, 44-46; assoc, Univ Southern Calif, 47-56, dir cardiovasc lab, Childrens Hosp, 49-56; res prof med, Sch Med, Loma Linda Univ, 56-66; PROF PHYSIOL, SCH MED, UNIV SOUTHERN CALIF, 66-; ADJ PROF PHYSIOL, UNIV CALIF, SAN DIEGO, 77- *Concurrent Pos:* Res Career Awardee, NIH, 62-, mem, cardiovasc A study sect, 66-70, exp cardiovasc sci, 82-86. *Honors & Awards:* Landis Award, Microcirculatory Soc, 80. *Mem:* Microcirculatory Soc; Soc Exp Biol & Med; Am Physiol Soc; fel Am Col Physicians. *Res:* Micro and peripheral circulation; pulmonary circulation and hypertension. *Mailing Add:* Dept Physiol Biophys Sch Med Univ Southern Calif 2025 Zonal Ave Los Angeles CA 90033

SOBKOWICZ, HANNA MARIA, b Warsaw, Poland, Jan 1, 31; m 72. NEUROLOGY. *Educ:* Med Acad, Warsaw, MD, 54, cert bd neurol, 59, PhD(med sci), 62. *Prof Exp:* From jr asst to sr asst neurol, Med Acad, Warsaw, 59-63; Nat Multiple Sclerosis Soc res fel tissue cult, Mt Sinai Hosp, 63-65; vis fel, Columbia Univ, 65-66; from asst prof to assoc prof, 66-79, PROF NEUROL, MED SCH, UNIV WIS-MADISON, 79- *Concurrent Pos:* Prin investr, NIH, 68- *Mem:* Soc Neurosci; Int Brain Res Orgn; Asn Res Otolaryngol; Int Soc Develop Neurosci; NY Acad Sci; Electron Micros Soc Am. *Res:* Development and regeneration of nervous system in culture; organ of Corti; spinal cord; spinal ganglia; cerebellum. *Mailing Add:* Dept Neurol Univ Wis Med Sch Madison WI 53706

SOBOCINSKI, PHILIP ZYGMUND, b Salem, Mass, Oct 29, 34; m 58; c 2. MICROBIOLOGY, RADIATION BIOLOGY. *Educ:* Tufts Univ, BS, 56; City Univ NY, MA, 64; Univ Rochester, Sch Med, PhD(radiation biol), 70. *Prof Exp:* Chief, clin lab, US Army Hosp, 59-61; asst dir biochem, First US Army Area Med Lab, 61-62; chief biochem, SEATO Med Res Lab, Bangkok, 63-65 & Armed Forces Radiobiol Res Inst, 70-74; chief phys sci, US Army Med Res Inst Infectious Dis, 74-80; dir res progs, 80-84, US Army Med Res & Develop Command, 80-84, dep comdr & dep asst surgeon gen, 84-86; ASST DIR, UNIV-INDUST RES PROG, UNIV WIS-MADISON, 87- *Concurrent Pos:* Consult biochem to USA Surg Gen, 81-87; pres, Int Biotechnics Corp, 86-; dir, Concept Develop & Commercialization Corp, 87- *Mem:* Sigma Xi; Soc Exp Biol & Med; NY Acad Sci. *Res:* Biological effects of ionizing radiation; host immunologic and metabolic responses to infectious diseases; trace metal metabolism; leukocyte physiology and function. *Mailing Add:* 118 Pipestem Pl Rockville MD 20854

SOBOCZENSKI, EDWARD JOHN, b Exeter, NH, July 2, 29; m 53; c 2. INSECT TOXICOLOGY. *Educ:* Univ NH, BS, 52, MS, 54; Ohio State Univ, PhD(chem), 56. *Prof Exp:* Res chemist, Exp Sta, E I du Pont de Nemours & Co, Inc, 56-65, patent liaison chem & law, 65-70, res biologist, 70, res supvr insecticides, 70-81, supvr info resources, 81-90, personnel & admin, 90; RETIRED. *Concurrent Pos:* Consult, 90- *Mem:* Am Chem Soc. *Res:* Discovery and development of agricultural chemicals; inventor of Venzar R herbicide sold for control of weeds in sugarbeets in Europe and Demosan R soil fungicide. *Mailing Add:* RD 1 Chadds Ford PA 19317

SOBOL, BRUCE J, b June 10, 23; US citizen; m 51; c 2. MEDICINE, PHYSIOLOGY. *Educ:* Swarthmore Col, BS, 47; NY Univ, MD, 50; Am Bd Internal Med, dipl. *Prof Exp:* Intern med, Third Med Div, Bellevue Hosp, New York, 50-51, asst resident, 51-52; resident cardiol, Vet Admin Hosp, Boston, 52-53; dir, Cardiopulmonary Lab, Westchester County Med Ctr, 59-78; res prof med, NY Med Col, 77-90, dir med res, 81-83; RETIRED. *Concurrent Pos:* Prof med, New York Med Col, 70-77; dir clin res, Boehringer Ingelheim Ltd, 78-81. *Mem:* Am Physiol Soc; fel Am Col Physicians; fel Am Col Chest Physicians; Am Heart Asn; fel NY Acad Sci. *Res:* Cardiac and pulmonary physiology. *Mailing Add:* 275 Ridgeburg Rd Ridgefield CT 06877

SOBOL, HAROLD, b Brooklyn, NY, June 21, 30; m 57; c 4. PHYSICAL ELECTRONICS, COMMUNICATIONS. *Educ:* City Col NY, BS, 52; Univ Mich, MSE, 55, PhD(elec eng), 60. *Prof Exp:* Res assoc radar, Willow Run Labs, Univ Mich, 52-55, res assoc phys electronics, Electron Physics Lab, 56-60; mem tech staff, Watson Res Ctr, Int Bus Mach Corp, 60-62; mem tech staff & group head, RCA Labs, 62-68, mgr microwave electronics, RCA Solid State Div, 68-70, mem, RCA Corp Res & Eng Staff, 70-72, group head, Commun Technol Res, 72-73; dir prod develop, Collins Transmission Systs Div, 73-85, vpres eng & tech, 85-88, ASSOC DEAN ENG & PROF ELEC ENG, ROCKWELL INT, 88- *Concurrent Pos:* Nat lectr, Inst Elec & Electronic Engrs, 70. *Mem:* Am Physics Soc; fel Inst Elec & Electronic Engrs, 70; Sigma Xi. *Res:* Radar propagation studies; electron devices; microwaves; superconductivity; plasmas; communications. *Mailing Add:* Univ Tex PO Box 19019 Arlington TX 76019

SOBOL, MARION GROSS, b New York, NY, Dec 2, 30; m 57; c 4. ECONOMICS OF COMPUTERIZATION. *Educ:* Syracuse Univ, BA, 51; Univ Mich, MBA, 57, PhD(econ), 61. *Prof Exp:* Lectr econ, Univ Col, Rutgers, 64-71; assoc prof statist & dept chmn, Rider Col, 71-74; PROF &

CHMN MGT INFO SYSTS, COX SCH BUS, SOUTHERN METHODIST UNIV, 74- *Concurrent Pos:* Vpres & bd dirs, Decision Sci Inst, 87-89. *Mem:* Inst Mgt Sci; Decision Sci Inst. *Mailing Add:* Cox Sch Bus Southern Methodist Univ Dallas TX 75275

SOBOL, STANLEY PAUL, b Boston, Mass, Oct 8, 37; m 63; c 2. FORENSIC SCIENCE, RESEARCH ADMINISTRATION. *Educ:* Tufts Univ, BS, 59. *Prof Exp:* Biochemist, Pharmacol Dept, Arthur D Little Inc, 59-61; from chemist to res coordr, US Food & Drug Admin, Boston Dist, 61-69; forensic chemist, Bur Narcotics & Dangerous Drugs, Lab Div, 69-70, chief chemist, 70-73; LAB DIR CHEM, DRUG ENFORCEMENT ADMIN, SPEC TESTING & RES LAB, 73- *Concurrent Pos:* Consult, UN Div Narcotics, 73- & Pakistan Narcotic Control Bd, 81; mem forensic subcomt, Joint Comt on Powder Diffraction Stand & Org Subcomt, 75-81; fel, US Dept Com Sci & Technol Fel Prog, 83-84; mem, ed bd, J Sci, 81- *Honors & Awards:* Spec Achievement Award, Bur Narcotics & Dangerous Drugs, 73; Exceptional Serv Award, Drug Enforcement Admin, 75, Excellence of Performance Award, 77, Outstanding Performance Award, 87. *Mem:* Am Acad Forensic Sci; Int Asn Toxicologists; Am Mgt Asn; Asn Off Analytical Chemists; Am Soc Crime Lab Dirs. *Res:* Trace organic analysis; computer assisted correlations of drug exhibits and establishment of data base. *Mailing Add:* 7704 Old Springhouse Rd McLean VA 22102-3494

SOBOLEV, IGOR, b Zlin, Czech, July 31, 31; nat US; m 53; c 2. ORGANIC CHEMISTRY. *Educ:* State Univ NY Col Forestry, Syracuse, BS, 54, MS, 55, PhD(org chem), 58. *Prof Exp:* Res chemist, Olympic Res Div, Rayonier, Inc, Wash, 58-61; chemist, Shell Develop Co, 61-64, sr technologist, Indust Chem Div, Shell Chem Co, NY, 64-66, chemist, 66-67, res supvr, Shell Develop Co, Calif, 67-70; SECT HEAD, CHEM, KAISER ALUMINUM & CHEM CORP, 70- *Mem:* Am Chem Soc; Tech Asn Pulp & Paper Indust. *Res:* Chemical process and product research and development involving raw materials and intermediates in aluminum production; reduction cell technology, organic polymers, fluorocarbons, atmospheric science, heterogeneous catalysis and flame retardants; fluorine chemistry. *Mailing Add:* C & P Technology Inc Five Rita Way Orinda CA 94563-4131

SOBOTA, ANTHONY E, b Bradenville, Pa, May 29, 38; m 62; c 1. BACTERIOLOGY. *Educ:* Ind Univ Pa, BSEd, 60; Univ Pittsburgh, MS, 63, PhD(biol), 66. *Prof Exp:* NIH fel, Purdue Univ, 66-67; USDA grant, 67-68; assoc prof, 68-77, PROF BIOL SCI, YOUNGSTOWN STATE UNIV, 77- *Mem:* Am Inst Biol Sci; Am Soc Microbiologists; Sigma Xi. *Res:* Importance of adherence in urinary tract infections. *Mailing Add:* Dept Biol Sci Youngstown State Univ Youngstown OH 44503

SOBOTA, WALTER LOUIS, b Detroit, Mich, Oct 30, 46; c 2. NEUROPSYCHOLOGY, COGNITIVE ASSESSMENT. *Educ:* Univ Detroit, BA, 68, PhD (clin psychol), 73; Am Bd Prof Psychologists, dipl, 84; Am Bd Neuropsychol, dipl, 84. *Prof Exp:* PSYCHOLOGIST, SINAI HOSP, DETROIT, 73-; ADJ INSTR, WAYNE STATE UNIV MED SCH, 82- *Concurrent Pos:* Psychologist pvt pract, 77-; coun rep, Am Psychol Asn, 88-91. *Mem:* Am Psychol Asn; Int Neuropsychol Soc; AAAS; Nat Acad Neuropsychologists. *Res:* Neuropsychological sequelae of resuscitation from cardiac arrest; neuropsychological sequelae following anesthesia in the elderly; neuropsychological deficits associated with systemic lupus and chronic fatigue syndrome. *Mailing Add:* Dept of Psychiat Sinai Hospital Detroit 14800 W McNichols Rd Suite 230 Detroit MI 48235

SOBOTKA, THOMAS JOSEPH, b Baltimore, Md, Aug 16, 42; m 64; c 2. NEUROBEHAVIORAL TOXICOLOGY, BEHAVIORAL TERATOLOGY. *Educ:* Loyola Col, BS, 64; Loyola Stritch Sch Med, MS, 67, PhD(pharmacol), 69. *Prof Exp:* Res pharmacologist, 69-78, actg chief, Whole Animal Toxicol Br, 81, SUPVRY PHARMACOLOGIST, LEADER NEUROBEHAVIORAL TOXICOL TEAM, 78- *Concurrent Pos:* Mem organizing comt, Conf Nutrit & Behav, Franklin Res Found, 79-80; exec secy, Interagency Collab Group Hyperkinesis, Dept Health & Human Serv, 81-82; rep Interagency Comt Learning Disabilities, Food & Drug Admin, 86-87, Interagency Comt Neurotoxicol, 89, deleg, OECD ad hoc meeting Neurotoxicity test guidelines, 89-90. *Honors & Awards:* Commemorative Medal, Polish Soc Internal Med, 76. *Mem:* Am Soc Pharmacol & Exp Therapeut; Soc Neurosci; Behav Pharmacol Soc; Neurobehav Toxicol Soc; Asn Govt Toxicol; Neurobehav Teratology Soc; Int Brain Res Orgn; World Fedn Neurosci. *Res:* Neurotoxicity hazard of chemicals found in foods; effects of chemicals on the developing nervous system. *Mailing Add:* Neurobehav Toxicol Team/HFF-162 Ctr for Food Safety & Appl Nutrit/ Food & Drug Admin Washington DC 20204

SOBOTTKA, STANLEY EARL, b Plum City, Wis, Dec 20, 30. PHYSICS. *Educ:* Univ Wis, BS, 55; Stanford Univ, MS, 57, PhD(physics), 60. *Prof Exp:* Mem tech staff, Sci Res Labs, Boeing Airplane Co, 59-60 & Watkins-Johnson Co, 60-63; assoc prof, 64-71, PROF PHYSICS, UNIV VA, 71- *Mem:* Am Phys Soc. *Res:* High energy electron scattering; electron beam-plasma interactions; lasers; pion and muon interactions with nuclei; nuclear particle detectors; x-ray diffraction; x-ray detectors. *Mailing Add:* Dept physics Univ Va Charlottesville VA 22901

SOBSEY, MARK DAVID, b Lakewood, NJ, Sept 5, 43; m 65, 82; c 2. VIROLOGY, ENVIRONMENTAL MICROBIOLOGY. *Educ:* Univ Pittsburgh, BS, 65, MS, 67; Univ Calif, Berkeley, PhD(environ health sci), 71. *Prof Exp:* Fel, Baylor Col Med, 71-72, from instr to asst prof, 72-74; from asst prof to assoc prof, 74-84, PROF ENVIRON MICROBIOL, SCH PUB HEALTH, UNIV NC, CHAPEL HILL, 84- *Concurrent Pos:* vis scientist, Lab Infectious Dis, NIH, Bethesda, Md, 81. *Mem:* Am Water Works Asn; Am Soc Microbiol; Water Pollution Control Fedn; AAAS. *Res:* Environmental microbiology; public health aspects of water and shellfish pollution; environmental virology. *Mailing Add:* CB No 7400 Rosenau Hall Univ NC Sch Pub Health Chapel Hill NC 27599-7400

SOCHA, WLADYSLAW WOJCIECH, b Paris, France, July 3, 26; m 56. IMMUNOLOGY, PATHOLOGY. *Educ:* Jagiellonian Univ, MD, 52; Cracow Acad Med, Poland, DMedS(genetics), 59. *Prof Exp:* Assoc path, Inst Oncol, Warsaw, Gliwice & Cracow, 52-61; asst prof forensic med, Cracow Acad Med, Poland, 55-64, dir inst pediat, 65-68; assoc dir, 77-80, DIR, PRIMATE BLOOD GROUP REF LAB & WHO COLLAB CTR HAEMATOL PRIMATE ANIMALS, 80-, RES PROF FORENSIC MED, LAB EXP MED & SURG PRIMATES, SCH MED, NY UNIV, 69- *Concurrent Pos:* Fr Asn Study Cancer fel, Regional Anticancer Ctr, Univ Montpellier, 59-60; US AID fel, 66; Fr Nat Inst Med Res fel, Ctr Hemotypology, Nat Ctr Sci Res, Toulouse, France, 71; mem comt human genetics, Polish Acad Sci, 65-68; vis scientist, Nat Inst Health & Med Res, Toulouse, France, 72; assoc ed, J Med Primatol, 73-77, ed, 77-; mem rev bd, J Human Evolution, 78-; vis scientist, Col France, Paris, 78, 84, 87; consult ed, Am Jour Primatol, 81- *Honors & Awards:* Polish Med Asn Award, 60; Polish Surg Asn Award, 60; Polish Acad Sci Award, 66. *Mem:* Am Soc Human Genetics; Am Soc Primatologists; Am Asn Lab Animal Sci; Int Primatol Soc; Int Soc Heart Transplant; NY Acad Sci. *Res:* Blood and serum groups; comparative serology; seroprimatology; population genetics; pathology of tumors; pathology of nonhuman primates; forensic pathology and serology. *Mailing Add:* Lab Exp Med & Surg Primates NY Univ Med Ctr New York NY 10016

SOCHER, SUSAN HELEN, b Chicago, Ill, June 19, 44. CELL BIOLOGY. *Educ:* Mt Mary Col, BS, 66; Case Western Reserve Univ, PhD(cell biol), 70. *Prof Exp:* Univ res fel & NIH trainee, Sch Med, Vanderbilt Univ, 70-72, res assoc cell biol, 72; ASST PROF CELL BIOL, BAYLOR COL MED, 72- *Mem:* AAAS; Am Soc Cell Biol; Develop Biol Soc. *Res:* Hormonal regulation of gene expression during mammary gland development and in mammary cancer; chromatin biochemistry. *Mailing Add:* Dept of Cell Biol Baylor Col of Med Houston TX 77030

SOCIE, DARRELL FREDERICK, b Toledo, Ohio, Oct 29, 48; m 77; c 2. MECHANICAL BEHAVIOR OF MATERIALS. *Educ:* Univ Cincinnati, BS, 71, MS, 73; Univ Ill, PhD(mechs), 77. *Prof Exp:* Engr, Struct Dynamics, 71-74; from asst prof to assoc prof, 77-85, PROF MECH ENG, UNIV ILL, 85- *Concurrent Pos:* Pres, Somat Corp, 82; guest prof, Fed Tech Col, Zurich, 85. *Honors & Awards:* Ralph Teetor Award, Soc Automotive Engrs, 80. *Mem:* Soc Automotive Engrs; Am Soc Metals; Nat Soc Prof Engrs; Am Soc Testing & Mats. *Res:* Fatigue life prediction methods for structural design; cyclic deformation of cast materials; multiaxial fatigue and creep. *Mailing Add:* RR 1 Box 4 St Joseph IL 61873

SOCOLAR, SIDNEY JOSEPH, b Baltimore, Md, Feb 10, 24; m 51; c 2. MEMBRANE PHYSIOLOGY, BIOPHYSICS. *Educ:* Johns Hopkins Univ, AB, 43, AM, 44, PhD(chem), 45. *Prof Exp:* Jr instr chem, Johns Hopkins Univ, 43-44, res chemist, 44-46; instr chem, Univ Ill, 47-48; from instr to asst prof phys sci, Univ Chicago, 50-57; math physicist, Heat & Mass Flow Analyzer Lab, Columbia Univ, 57-59, res assoc physiol, Col Physicians & Surgeons, 59-69, asst prof, 69-71; from asst prof to prof, 71-84, EMER PROF PHYSIOL & BIOPHYS, SCH MED, UNIV MIAMI, 85-; SR ASSOC, HEALTH ACTION RESOURCE CTR, NY, 86- *Concurrent Pos:* Phillips fel, Pa State Col, 46-47. *Mem:* Soc Gen Physiologists; Biophys Soc; Am Pub Health Asn. *Res:* Cell-to-cell membrane channels: formation, permeability and permeability regulation; electrophysiology; membrane physiology. *Mailing Add:* 606 W 116th St New York NY 10027-7011

SOCOLOFSKY, MARION DAVID, b Marion, Kans, Sept 23, 31; m 53; c 2. MICROBIOLOGY. *Educ:* Kans State Univ, BS, 53; Univ Tex, MA, 55, PhD(bact), 61. *Prof Exp:* From asst prof to assoc prof microbiol, 61-68, PROF MICROBIOL, LA STATE UNIV, BATON ROUGE, 68-, CHMN DEPT, 66-86, 88- *Mem:* AAAS; Am Soc Microbiol; Electron Micros Soc Am; Brit Soc Gen Microbiol. *Res:* Electron microscopy; bacterial ultrastructure. *Mailing Add:* Dept of Microbiol La State Univ Baton Rouge LA 70803

SOCOLOW, ARTHUR A, b New York, NY, Mar 23, 21; m 49; c 3. ECONOMIC GEOLOGY. *Educ:* Rutgers Univ, BS, 42; Columbia Univ, MA, 47, PhD(econ geol), 55. *Prof Exp:* Asst field geologist, State Geol Surv, Va, 42; photogram engr, US Geol Surv, 42 & 46, geologist, 52 & Eagle Picher Mex, 47; asst econ geol, Columbia Univ, 47-48; instr geol & dir geol field camp in Colo, Southern Methodist Univ, 48-50; from instr to asst prof geol, Boston Univ, 50-55; from asst prof to prof geol, Univ Mass, 55-57; econ geologist, 57-61, state geologist & dir, PA Geol Surv, 61-86; CONSULT GEOLOGIST, 86- *Concurrent Pos:* Photogram, US Army Air Corps, 42-46; geologist, Defense Minerals Explor Authority, 52; geol adv, Boston Mus Sci, 55-57, lectr, 56; lectr, Pa State Univ, 59-73; mem, NSF Earth Sci Conf, 59; dir annual field conf Pa Geol, 61-86; gov rep & past chmn, Res Comt & Environ Protection Comt, Interstate Oil Compact Comn, 72-; mem & past chmn, Am Comn Stratig Nomenclature; past chmn, Pa Water Resources Coord Comt; mem, Outer Continental Shelf Policy Comt, US Dept Interior; counr & fel, Geol Soc Am; ed, PA Geology, 69-86. *Mem:* Fel AAAS; fel Mineral Soc Am; fel Am Geophys Union; Soc Econ Geologists; Nat Asn Geol Teachers; Meteoritical Soc; Asn Am State Geologists (past pres); Sigma Xi. *Res:* Genesis and structural control of ore deposits; regional structure interpretation; alteration effects related to igneous rocks and ore deposits; geologic interpretation of aeromagnetic data; geologic impact on man's environment; geologic hazards; waste disposal siting. *Mailing Add:* 26 Salt Island Rd Gloucester MA 01930-1945

SOCOLOW, ROBERT H(ARRY), b New York, NY, Dec 27, 37; m 62, 86; c 2. ENERGY POLICY, ENVIRONMENTAL SCIENCES. *Educ:* Harvard Univ, BA, 59, MA, 61, PhD(physics), 64. *Prof Exp:* Asst prof physics, Yale Univ, 66-71, jr fac fel, 70-71; assoc prof environ sci, 71-77, PROF ENVIRON SCI, PRINCETON UNIV, 77-, DIR, CTR ENERGY & ENVIRON STUDIES, 78- *Concurrent Pos:* NSF fel, 64-66; mem, Inst Advan Studies, 71; Guggenheim & Ger Marshall Fund fels, Energy Res Group, Cavendish Lab, Univ Cambridge, 77-78. *Mem:* Fel Am Phys Soc; Fedn Am Sci; fel AAAS. *Res:* Energy utilization; regional and global constraints on growth. *Mailing Add:* Eng Quad D-214 Mech/Aerosp Eng Princeton Univ Princeton NJ 08540

SODAL, INGVAR E, b Norway, Feb 12, 34. CARDIOPULMONARY RESEARCH, PATIENT MONITORING IN ANESTHESIOLOGY & INTENSIVE CARE. *Educ:* Trondheim Tech Col, dipl elec eng, 59; Univ Colo, BS, 64. *Prof Exp:* Asst prof anesthesiol & head dept, Bioeng & Clin Res Div, Col Med, Ohio State Univ, 79-82; pres, Mastron Inc, 83-89; chief scientist, Paradym Sci & Technol, 89-90; PRES, MED PHYSICS, COLO, 91- *Mem:* Biomed Eng Soc; Asn Advan Med Instrumentation; Instrument Soc Am. *Mailing Add:* 1550 Moss Rock Pl Boulder CO 80304

SODANO, CHARLES STANLEY, b Newark, NJ, Nov 13, 39; c 5. NATURAL PRODUCTS CHEMISTRY. *Educ:* Seton Hall Univ, BS, 61, MS, 63; Ariz State Univ, PhD(org chem), 67. *Prof Exp:* Res chemist cancer res, Pfizer Inc, 66-69; res chemist, Nabisco Brands Inc, 69-75, mgr anal systs res, 75-83, dir methods develop, 83-90, MGR, STRATEGIC SERV, NABISCO BRANDS INC, 90- *Mem:* Am Chem Soc; Am Asn Cereal Chem; Am Soc Testing & Mat. *Res:* Structure elucidation of anti-tumor agents; fabricated foods development; catalytic hydrogenation; analytical methods for food analysis; process control; shelf life; lab robotics; technology assessment. *Mailing Add:* Nabisco Brands Inc PO Box 1943 East Hanover NJ 07936-1943

SODD, VINCENT J, b Toledo, Ohio, Nov 20, 34; m 56; c 4. NUCLEAR CHEMISTRY, NUCLEAR MEDICINE. *Educ:* Xavier Univ, Ohio, BS, 56, MS, 58; Univ Pittsburgh, PhD(nuclear chem), 64. *Prof Exp:* Asst chem, Xavier Univ, Ohio, 56-58; res chemist, Robert A Taft Sanit Eng Ctr, 58-60, nuclear chemist, 64-66, dep chief nuclear med lab, 66-71, CHIEF NUCLEAR MED LAB & CHIEF RADIOPHARMACEUT DEVELOP SECT, USPHS, 71- *Concurrent Pos:* Asst clin prof, Col Med, Univ Cincinnati, 68-74, assoc prof, 74-77, prof, 77- *Mem:* Sigma Xi; Am Chem Soc. *Res:* Nuclear medicine investigations involving clinic practice, radiation exposure reduction, dosimetry and instrumentation development; radiopharmaceutical production; cyclotron and linear accelerator research; activation analysis; semiconductor theory and use; development of analytical procedures for radionuclides regarded as being hazardous to our environment. *Mailing Add:* 5987 Turpin Hills Dr Cincinnati OH 45244

SODEMAN, WILLIAM A, b New Orleans, La, Mar 26, 36; c 2. GASTROENTEROLOGY. *Educ:* Univ Mo, BA, 56; Univ Pa, MD, 60. *Prof Exp:* Assoc prof, dept internal med & chief, gastroenterol, Univ Ark, Little Rock, 70-73, assoc dir, Clin Res Ctr, 71-73; assoc prof, dept med & chief, div gastroenterol, Med Col Ohio, Toledo, 73-75; vchmn, dept med, 75-76, actg chmn, dept comprehensive med, 76-77, med dir, Med Clins, 76-79, asst dir, Med Ctr, 80-84, dir, Pub Health Prog, 82-83, dep dean acad affairs, Col Med, 84-88, PROF, DEPT MED, COL MED, UNIV S FLA, TAMPA, 75-, PROF, DEPT COMPREHENSIVE MED, 76-, CHMN, 77-, ASSOC DEAN ACAD AFFAIRS, LSUMC-S COL MED, 88- *Mem:* Sigma Xi; fel Am Col Physicians; Am Gastroenterol Asn; Am Soc Trop Med & Hyg; Am Soc Gastrointestinal Endoscopy. *Res:* Pathologic physiology; clinical parasitology; medical malacology; intermediate hosts in ecology. *Mailing Add:* LSUMC Sch of Med in S'port PO Box 33932 Shreveport LA 71130-3932

SODERBERG, LEE STEPHEN FREEMAN, b Chicago, Ill, Apr 17, 46; m 83; c 2. HEMATOPOIESIS, DRUG ABUSE. *Educ:* Rutgers Univ, PhD(microbiol), 73. *Prof Exp:* res fel immunol, Harvard Med Sch, 73-77; asst prof, 77-84, ASSOC PROF IMMUNOL, COL MED, UNIV ARK MED SCI, 84- *Concurrent Pos:* Prin investr, NIH grant, 80-83, Alcohol, Drug Abuse & Mental Health Admin grant, 91-94; co-investr, NIH grants, 86- *Mem:* Am Asn Immunologists; Am Soc Microbiol; Sigma Xi; Soc Exp Biol Med. *Res:* Immunotoxicity of abused nitrite inhalants; immunotoxicology of commercial pesticides; immunity and vaccines to chlamydial genital infections; the role of bone marrow natural suppressor cells in immunodeficiencies. *Mailing Add:* Dept Microbiol & Immunol Univ Ark Col Med 4301 W Markham Slot 511 Little Rock AR 72205

SODERBERG, ROGER HAMILTON, b Congress Park, Ill, June 19, 36; m 59; c 2. ENVIRONMENTAL CHEMISTRY. *Educ:* Grinnell Col, AB, 58; Mass Inst Technol, PhD(coord chem), 63. *Prof Exp:* From instr to assoc prof, 62-75, PROF CHEM, DARTMOUTH COL, 75- *Mem:* Am Chem Soc; AAAS. *Res:* Metal coordination chemistry. *Mailing Add:* Dept of Chem Dartmouth Col Hanover NH 03755

SODERBLOM, LAURENCE ALBERT, b Denver, Colo, July 17, 44; m 68; c 3. PLANETARY GEOLOGY. *Educ:* NMex Inst Mining & Technol, BS(geol) & BS(physics), 66; Calif Inst Technol, PhD(planetary sci & geophysics), 70. *Prof Exp:* Geophysicist, 70-78, SUPVR PHYS SCIENTIST, US GEOL SURV, 78- *Concurrent Pos:* Assoc ed, J Geophys Res, 71-73; dep team leader, Voyager Imaging Sci Team, NASA, 72-, Comt Lunar & Planetary Explor, Space Sci Bd, 73-77, Viking Orbiter Imaging Team, 76-78, Galileo Near Infrared Mapping Spectrometer Team, 77- & Space Sci Adv Comt, 80- *Mem:* Am Geophys Union. *Res:* Global geologic histories of planets and satellites of the solar system employing earth-based and spacecraft remote-sensing data; established timescales for planetary evolution; computerized image processing. *Mailing Add:* 3940 N Paradise Rd Flagstaff AZ 86001

SODERLING, THOMAS RICHARD, b Bonners Ferry, Idaho, May 25, 44; m 65; c 3. PHYSIOLOGY, BIOCHEMISTRY. *Educ:* Univ Idaho, BS, 66; Univ Wash, PhD(biochem), 70. *Prof Exp:* NIH fel, Vanderbilt Univ, 71-72, Am Diabetes Asn fel, 72-73, from asst prof to prof physiol, Med Sch, 73-91; ASSOC DIR, VOLLUM INST & PROF BIOCHEM & MOLECULAR BIOL, ORE HEALTH SCI UNIV, 91- *Concurrent Pos:* Investr, Howard Hughes Med Inst, 76-89. *Honors & Awards:* Andrew Mellon Found Scientist-Educr Award, 74. *Mem:* Am Soc Biol Chemists; Soc Neurosci. *Res:* Regulation of protein phosphorglation in brain; mechanism of action of insulin; neuroscience. *Mailing Add:* Vollum Inst Advan Biomed Res Ore Health Sci Univ 3181 Sam Jackson Park Rd Portland OR 97201-3098

SODERLUND, DAVID MATTHEW, b Oakland, Calif, Oct 1, 50; m 72. ENTOMOLOGY, BIOCHEMICAL TOXICOLOGY. *Educ:* Pac Lutheran Univ, BS, 71; Univ Calif, Berkeley, PhD(entomol), 76. *Prof Exp:* Vis res fel, insecticide biochem, Rothamsted Exp Sta, Harpenden, Eng, 76-77; asst prof, 78-84, ASSOC PROF, DEPT ENTOMOL, NY STATE AGR EXP STA, CORNELL UNIV, 84- *Concurrent Pos:* Rockefeller Found fel, 76-77; consult, Crop Chem Res & Develop, Mobil Chem Co, 78-81 & Agrochem Div, Rhone-Poulene Inc, 82-85. *Mem:* Am Chem Soc; Entomol Soc Am; AAAS. *Res:* Biochemical and physiological interactions of insecticide chemicals and insect growth regulators in insects and mammals. *Mailing Add:* Dept of Entomol NY State Agr Exp Sta Geneva NY 14456

SODERMAN, J WILLIAM, b Helsinki, Finland, Oct 31, 35; US citizen; m 56; c 3. GEOLOGY. *Educ:* Columbia Univ, BA, 57; Univ Ill, MS, 60, PhD(geol), 62. *Prof Exp:* Geologist, Texaco Inc, 62-72, supvr geol res, 72-74, asst div geologist, 74-78; chief geologist, Monsanto Oil Co, 78-79, dir domestic explor, 79-82, vpres, Explor, 82-88, VPRES EXPLOR & PROD, BG EXPLOR AM INC, 88- *Mem:* Geol Soc Am; Am Asn Petrol Geol; Soc Econ Paleont & Mineral. *Res:* Geology of sedimentary rocks. *Mailing Add:* BG Explor Am Inc British Gas PLC 1100 Louisianna Suite 2500 Houston TX 77002

SODERQUIST, DAVID RICHARD, b Idaho Falls, Idaho, June 26, 36; m 84. PSYCHOACOUSTICS. *Educ:* Utah State Univ, BS, 61, MS, 63; Vanderbilt Univ, PhD(psychol), 68. *Prof Exp:* Sr human factors specialist, Syst Develop Corp, 63-65; asst prof, 68-72, assoc prof, 72-86, PROF PSYCHOL, UNIV NC, GREENSBORO, 86- *Mem:* Acoust Soc Am; Sigma Xi. *Res:* Psychological acoustics; signal detection; pitch perception and temporal masking. *Mailing Add:* Dept Psychol Univ NC 1000 Spring Garden Greensboro NC 27412

SODERSTROM, EDWIN LOREN, b Riverside, Calif, Feb 8, 31; m 60; c 3. ENTOMOLOGY. *Educ:* Calif State Polytech Col, BS, 57; Kans State Univ, MS, 59, PhD(entom), 62. *Prof Exp:* Res asst entom, Kans State Univ, 61-62; RES ENTOMOLOGIST, AGR RES SERV, USDA, 62- *Mem:* Entom Soc Am. *Res:* Entomological research on geographical populations of rice weevils; response of stored product insects to light; effects of pesticides on populations of dried fruit and tree nut insects; controlled atmosphere fumigation; insect attractants and repellents. *Mailing Add:* 3285 E Rialto Ave Fresno CA 93726

SODERSTROM, KENNETH G(UNNAR), b Red Bank, NJ, Apr 21, 36; m 62; c 2. MECHANICAL ENGINEERING, SOLAR ENERGY. *Educ:* Univ Fla, BME, 58, MSE, 59, PhD(mech eng), 72. *Prof Exp:* From asst prof to assoc prof mech eng, 61-76, chmn dept, 63-68, PROF MECH ENG, DEPT MECH ENG, UNIV PR, MAYAGUEZ, 76-, SR SCIENTIST, CTR ENERGY & ENVIRON RES, 73-, ASSOC DIR, 79- *Concurrent Pos:* Res assoc, PR Nuclear Ctr, Univ PR, 62- *Mem:* Int Solar Energy Soc; Am Soc Mech Engrs. *Res:* Nuclear engineering in fields of irradiation effects on emissivity of materials and convection heat transfer; solar energy, particularly experimental and analytical system studies; solar data measurements and modeling. *Mailing Add:* Dept Mech Eng Univ PR Mayaguez PR 00709

SODERWALL, ARNOLD LARSON, b Portland, Ore, Nov 13, 14; m 38; c 2. ENDOCRINOLOGY. *Educ:* Linfield Col, AB, 36; Univ Ill, AM, 37; Brown Univ, PhD(endocrinol), 41. *Prof Exp:* Asst zool, Univ Ill, 36-38; from instr to assoc prof zool, Univ Ore, 41-61, prof, 61-77, prof biol, 77-80; RETIRED. *Concurrent Pos:* AEC fel; vis prof, Cornell Univ, 54-55 & 65-66 & Univ Hawaii Sch Med, 72-73. *Mem:* Soc Study Reproduction; Am Asn Anat. *Res:* Gerontological studies related to reproductive capacities in rodents; x-irradiation effects on simulated aging; electrophoretic studies of blood gonadotropins; nature of litter size loss in senescent hamsters; preimplantation death in older females; induction of decidual cell response by air injections in young and senescent female hamsters. *Mailing Add:* 2493 Harris Eugene OR 97405

SODETZ, JAMES M, b Chicago, Ill, Oct 9, 48. PROTEIN BIOCHEMISTRY. *Educ:* Univ Notre Dame, PhD(biochem), 75. *Prof Exp:* ASSOC PROF BIOCHEM, UNIV SC, 83- *Mailing Add:* Dept Biochem Univ SC Columbia SC 29208

SODICKSON, LESTER A, b New York, NY, Oct 23, 37; m 63; c 3. PHYSICS, ANALYTICAL INSTRUMENTS. *Educ:* Mass Inst Technol, BS, 58, PhD(physics), 63. *Prof Exp:* Sr scientist & dir new prod, Am Sci & Eng Inc, 63-70, pres, Biotech Diag, 70-71; prog mgr, Damon Corp, 71-72, vpres, Res & Eng Div, 72-74, res & develop, IEC Div, 75-81 & Inst Res, Damon Biotech, 81-83; dir appl res & sr res assoc, Corning Med, 84; PRES, CAMBRIDGE RES ASN, 85- *Mem:* AAAS; Am Phys Soc; Am Asn Clin Chem. *Res:* Electromagnetic sensing of molecular species for clinical chemistry; environmental pollution; infrared and optical physics; computer science; microencapsulation of living cells; cell culture; biotechnology. *Mailing Add:* 263 Waban Ave Waban MA 02168

SODICOFF, MARVIN, b Brooklyn, NY, June 12, 37; m 60; c 3. ANATOMY, RADIATION BIOLOGY. *Educ:* Brooklyn Col, BS, 59; Univ Cincinnati, PhD(anat), 66. *Prof Exp:* PROF ANAT, MED SCH, TEMPLE UNIV, 66- *Honors & Awards:* Lindback Award. *Mem:* Radiation Res Soc; Am Asn Anatomists. *Res:* Radiation biology. *Mailing Add:* Dept of Anat Temple Univ Med Sch 3400 N Broad St Philadelphia PA 19140

SOECHTING, JOHN F, b Sept 27, 43; US citizen. BIOENGINEERING & BIOMEDICAL ENGINEERING. *Educ:* Lehigh Univ, BS, 65; Cornell Univ, PhD(mech), 69. *Prof Exp:* Res assoc biomech, Brown Univ, 69-72; assoc, 72-74, lectr, 74-75, from asst prof to assoc prof, 75-85, PROF NEUROPHYSIOL, UNIV MINN, MINNEAPOLIS, 85- *Concurrent Pos:* NIH spec fel, 75-76. *Mem:* Soc Neurosci; Sigma Xi. *Res:* Motor control. *Mailing Add:* Dept of Physiol Univ of Minn Minneapolis MN 55455

SOEDEL, WERNER, b Prague, Czech, Apr 24, 36; US citizen; m 61; c 4. MECHANICS, ENGINEERING MECHANICS. *Educ:* Frankfurt State Inst Eng, Ing Grad, 57; Purdue Univ, MSME, 65, PhD(mech eng), 67. *Prof Exp:* Proj engr mech eng, Adam Opel AG, 57-63; from asst prof to assoc prof, 67-75, PROF MECH ENG, PURDUE UNIV, 75- *Concurrent Pos:* NAm ed J Sound & Vibration, 89- *Honors & Awards:* Roe Award, Am Soc Eng Educ, 86. *Mem:* Am Acad Mech; Am Soc Mech Engrs; Am Acoustical Soc. *Res:* Vibrations of shell structures; dynamic interactions of solids and fluids; gas dynamics; acoustics; hydrodynamics. *Mailing Add:* 901 Allen St West Lafayette IN 47906

SOEDER, ROBERT W, b Philadelphia, Pa, Oct 5, 35; m 59. ORGANIC CHEMISTRY. *Educ:* Ursinus Col, BS, 57; Univ Del, MS, 59, PhD(org chem), 62. *Prof Exp:* Fel org chem, Univ Minn, 61-62; from asst prof to assoc prof, Wilkes Col, 62-67; assoc prof, 67-72, PROF CHEM, APPALACHIAN STATE UNIV, 72- *Mem:* Am Chem Soc. *Res:* Synthesis and reactions of heterocyclic compounds; reactions of B-diketones; chemical constituents of ferns. *Mailing Add:* Dept of Chem Appalachian State Univ Boone NC 28608

SOEIRO, RUY, b Boston, Mass, May 28, 32; m 66. INFECTIOUS DISEASES, CELL BIOLOGY. *Educ:* Harvard Univ, AB, 54; Tufts Univ, MD, 58. *Prof Exp:* From instr biochem to asst prof med, 67-73, assoc prof med & cell biol, 73-76, assoc prof immunol, 73-78, PROF MED & IMMUNOL, ALBERT EINSTEIN COL MED, 78-, CO-DIR DIV INFECTIOUS DIS, 73- *Concurrent Pos:* USPHS trainee bact, Harvard Med Sch, 62-65; USPHS spec fel biochem, Albert Einstein Col Med, 66-67; City New York career res scientist award, 68-73. *Mailing Add:* Dept Med/Cell Biol Albert Einstein Col of Med Bronx NY 10461

SOELDNER, JOHN STUART, b Boston, Mass, Sept 22, 32; m 62; c 3. DIABETES RESEARCH, GLUCOSE SENSOR. *Educ:* Tufts Univ, BSc, 54, Dalhousie Univ, MD, 59. *Prof Exp:* PROF MED, UNIV CALIF, DAVIS, SCH MED, 87- *Honors & Awards:* Upjohn Award, Am Diabetic Asn, 86. *Mem:* Am Physiol Soc; Am Diabetes Asn; Endocrine Soc; Am Soc Clin Invest. *Res:* Research pathogenesis of diabetes, type 1 and type 2; insulin secretions, genetics of type 1 and type 2 diabetes; study miniature implantable glucose sensor and implantable artificial beta cell for diabetes; study glycosylation proteins, particularly hemoglobin. *Mailing Add:* Univ Calif Davis Med Ctr 4301 "X" St-FOLB-2-C Sacramento CA 95817

SOERENS, DAVE ALLEN, b Sheboygan, Wis, Aug 26, 52; m 72; c 2. PRESSURE-SENSITIVE ADHESIVES, NONWOVENS. *Educ:* Calvin Col, BS, 74; Univ Wis-Milwaukee, PhD(chem), 78. *Prof Exp:* SR RES CHEMIST, 3M CO, 78- *Mem:* Am Chem Soc; NAm Thermal Anal Soc. *Res:* Adhesives research and development; thermal analysis; nonwovens. *Mailing Add:* 736 Kensington Rd Neenah WI 54956-4908

SOERGEL, KONRAD H, b Coburg, Ger, July 27, 29; US citizen; m 55; c 4. INTERNAL MEDICINE, GASTROENTEROLOGY. *Educ:* Univ Erlangen, MD, 54, DrMedSci, 57. *Prof Exp:* Res fel gastroenterol, Sch Med, Boston Univ, 58-60, instr med, 60-61; from asst prof to assoc prof, 61-69, PROF MED, MED COL WIS, 69-, CHIEF DEPT GASTROENTEROL, 61- *Concurrent Pos:* Consult gastroenterologist, Wood Vet Admin Hosp, 67; consult, Vet Admin Res Serv Rev Bd, 69-71 & 72-74; mem gen med A study sect, NIH, 76-80. *Mem:* Am Fedn Clin Res; Am Gastroenterol Asn; Am Soc Clin Invest. *Res:* Absorption of water, electrolyte and sugar from the human small intestine. *Mailing Add:* Dept Med Med Col Wis 9200 W Wisconsin Ave Milwaukee WI 53226

SOFER, SAMIR SALIM, b Teheran, Iran, Oct 10, 45. CHEMICAL ENGINEERING, BIOENGINEERING. *Educ:* Univ Utah, BS, 69; Tex A&M Univ, ME, 71; Univ Tex, PhD(chem eng), 74. *Prof Exp:* Process design engr, Celanese Chem Co, 69-72; res assoc, Univ Tex, 73-74; from asst prof to assoc prof, Univ Okla, 74-80, dir, Sch Chem Eng & Mat Sci, 75-80, prof chem eng, 80-86; PROF & SPONSOR CHMN, DEPT BIOTECHNOL, NJ INST TECHNOL, 86- *Concurrent Pos:* Fel, Clayton Found Biochem Inst, 74-75. *Honors & Awards:* First place, SCORE (Student Contest on Relevant Eng), 75. *Mem:* Am Inst Chem Engrs; AAAS; Am Soc Eng Educ. *Res:* Insolubilized enzyme technology and biochemical reactor design; reaction kinetics; process design. *Mailing Add:* NJ Inst Technol 161 Warren St Newark NJ 07102

SOFER, WILLIAM HOWARD, b Brooklyn, NY, Jan 14, 41; m 64; c 2. MOLECULAR GENETICS. *Educ:* Brooklyn Col, BS, 61; Univ Miami, PhD(cell physiol), 67. *Prof Exp:* NIH fel, Johns Hopkins Univ, 67-69, NSF fel, 69-71, asst prof biol, 69-75, assoc prof, 75-80; MEM FAC, WAKSMAN INST MICROBIOL, RUGTERS UNIV, 80- *Concurrent Pos:* NIH grant, 71-; Nat Inst Environ Health Sci grant, 77-; Dept of Energy contract, 76- *Mem:* AAAS; Genetics Soc Am. *Res:* Regulation of the activity of alcohol dehydrogenase in Drosophila; mechanisms of mutagenesis; aging. *Mailing Add:* Dept Biol Sci Rutgers Univ New Brunswick NJ 08903

SOFFEN, GERALD A(LAN), b Cleveland, Ohio, Feb 7, 26; m 79. BIOLOGY. *Educ:* Univ Calif, Los Angeles, BA, 49; Univ Southern Calif, MS, 56; Princeton Univ, PhD(biol), 60. *Prof Exp:* USPHS fel biol, Sch Med, NY Univ, 60-61; sr space scientist, Jet Propulsion Lab, Calif Inst Technol, 61-69; proj scientist, Viking, Langley Res Ctr, 69-78, DIR LIFE SCI, HQ, NASA, 78- *Mem:* AAAS. *Res:* Physiology; biochemistry; growth, metabolism and physiology of the cell; effects of ultraviolet light; transport mechanisms; exobiology; muscle biochemistry; science administration. *Mailing Add:* 617 4th Place SW Washington DC 20024

SOFFER, ALFRED, b South Bend, Ind, May 5, 22; m 56; c 3. MEDICINE. *Educ:* Univ Wis, BA, 42, MD, 45; Am Bd Internal Med, dipl. *Prof Exp:* Electrocardiographer, Genesee Hosp, NY, 51-58; dir cardiopulmonary lab, Rochester Gen Hosp, 59-62; assoc med, Northwestern Univ, 63-64; clin asst prof, 64-65, assoc prof, 65-68, PROF MED, CHICAGO MED SCH/UNIV HEALTH SCI, 68- *Concurrent Pos:* Sr ed jour, AMA, 62-67; ed-in-chief, Dis of the Chest, 68-; exec dir, Am Col Chest Physicians, 69-; consult on med jours to secy, Dept Health, Educ & Welfare, DC, 71-; ed, Heart & Lung, J Total Care, 72. *Mem:* Fel Am Col Physicians; fel Am Col Chest Physicians; fel Am Col Cardiol; fel Am Med Writers' Asn; Am Fedn Clin Res. *Res:* Medical administration; cardiology. *Mailing Add:* 911 Busse Hwy Park Ridge IL 60068

SOFFER, BERNARD HAROLD, b Brooklyn, NY, Mar 2, 31; m 56; c 1. PHYSICS. *Educ:* Brooklyn Col, BS, 53; Mass Inst Technol, MS, 58. *Prof Exp:* Staff mem, Lab Insulation Res, Mass Inst Technol, 58-59; res physicist, Hughes Res Lab, Calif, 59-61; res physicist, Appl Physics Lab, Quantatron Inc, 61-62, sr scientist, Optical Physics Div, Korad Corp, Calif, 62-69; mem tech staff, 69-80, sr staff physicist, 80-87, SR SCIENTIST, HUGHES RES LABS, 87- *Concurrent Pos:* Consult. *Mem:* Am Phys Soc; Sigma Xi; sr mem Inst Elec & Electronics Eng; fel Optical Soc Am. *Res:* Optical, infrared and spin resonance spectroscopy of solids; laser physics and laser materials; optical physics; image and information processing; optical computing; neural networks. *Mailing Add:* 665 Bienveneda Ave Pacific Palisades CA 90272

SOFFER, MILTON DAVID, b New York, NY, Dec 11, 14; m 45; c 3. SYNTHETIC ORGANIC CHEMISTRY, NATURAL PRODUCTS CHEMISTRY. *Educ:* Univ Ark, BS, 37; Harvard Univ, AM, 39, PhD(org chem), 42. *Prof Exp:* Indust chemist, Wm R Rogers, Inc, NY, 37; from instr to prof, 42-70, SOPHIA SMITH PROF CHEM, SMITH COL, 70- *Concurrent Pos:* Res grant, Res Corp, 45-52; Guggenheim fel, Oxford Univ, 50-51; NSF grant, 52-; NSF sr fel, Harvard Univ, 58-59; vis prof, Univ Mass, 62 & Hollins Col, 63; consult, Tex Co, 43-47. *Mem:* AAAS; Am Chem Soc; NY Acad Sci; Sigma Xi. *Res:* Synthetic and structural investigations of natural products; terpenes and alkaloids; porphyrins and chlorins; synthesis of high molecular weight and cyclic compounds. *Mailing Add:* 407 Fairway Village Leeds MA 01053

SOFFER, RICHARD LUBER, b Baltimore, Md, Oct 1, 32; m 68; c 2. BIOCHEMISTRY. *Educ:* Amherst Col, BA, 54; Harvard Univ, MD, 58. *Prof Exp:* Asst resident med, Sch Med, NY Univ, 61-62, resident, 64-65; fel biochem, Pasteur Inst, Paris, 62-64; asst mem enzymol, Inst Muscle Dis, New York, 65-67; from asst prof to assoc prof molecular biol, Albert Einstein Col Med, 72-76; PROF MED & BIOCHEM, MED COL, CORNELL UNIV, 76- *Concurrent Pos:* Career develop award, Nat Inst Arthritis & Metab Dis, 68; fac res award, Am Cancer Soc, 73. *Mem:* Am Soc Biol Chemists. *Res:* Post-translational protein modification catalyzed by aminoacyl-t RNA-protein transferases; angiotensin-converting enzyme and regulation of vasoactive peptides; enzymes involved in the metabolism of thyroid hormones by target cells; angiotensin receptors. *Mailing Add:* Dept Biochem & Med Cornell Univ Med Col 1300 York Ave New York NY 10021

SOFIA, R DUANE, b Ellwood City, Pa, Oct 8, 42; m 65; c 4. PHARMACOLOGY. *Educ:* Geneva Col, BS, 64; Fairleigh-Dickinson Univ, MS, 69; Univ Pittsburgh, PhD(pharmacol), 71. *Prof Exp:* Res biologist, Lederle Labs, NY, 64-67; res assoc pharmacol, Union Carbide Corp, 67-69; sr pharmacologist, Pharmakon Labs, Pa, 69; sr res pharmacologist, 71-73, dir, Dept Pharmacol & Toxicol, 73-76, vpres biol res, 76-80, vpres res & develop, 80-82, VPRES PRE-CLIN RES, WALLACE LABS, CRANBURY, 82- *Concurrent Pos:* Consult, Pharmakon Labs, 69-71. *Mem:* Am Soc Pharmacol & Exp Therapeut; Soc Toxicol; Soc Neurosci; Int Soc Study Pain; Am Rheumatism Asn. *Res:* Pharmacology and toxicology of various constituents of marihuana; development of new drugs for cardiovascular, pulmonary and central nervous system diseases and pain relief. *Mailing Add:* 11 Endwell Lane Willingboro NJ 08046

SOFIA, SABATINO, b Episcopia, Italy, May 14, 39; m 63; c 2. ASTROPHYSICS. *Educ:* Yale Univ, BS, 63, MS, 65, PhD(astrophys), 66. *Prof Exp:* Nat Acad Sci-Nat Res Coun res assoc astrophys, Goddard Inst Space Studies, NASA, 66-67; from assoc prof to prof astron, Univ SFla, 67-73; vis fel, Joint Inst Lab Astrophys, 73-74; sr res assoc, Univ Rochester, 74-75; staff scientist, Hq, NASA, 75-77, sr res assoc solar phys, Nat Acad Sci-Nat Res Coun, 77-79, space scientist, Goddard Space Flight Ctr, 79-85; PROF ASTRON, YALE UNIV, 85- *Concurrent Pos:* Adj prof astron, Univ Fla, 75-; mem, space & earth sci adv comt, NASA, 85-88. *Mem:* Am Astron Soc; Int Astron Union; Am Geophys Union. *Res:* Solar physics, variability and evolution; stellar evolution; interstellar matter. *Mailing Add:* Yale Univ Observ PO Box 6666 New Haven CT 06511

SOGAH, DOTSEVI YAO, b Ghana, West Africa, April, 19, 45; m 73; c 3. HOST-GUEST CHEMISTRY, BIOMATERIALS. *Educ:* Univ Ghana, BSc, first class, 70, Hons, 71; Univ Calif, Los Angeles, MS, 74, PhD(chem), 75. *Prof Exp:* Fel chem, Univ Calif, Santa Barbara, 75-77; asst res chemist, Univ Calif, Los Angeles, 78-79, asst prof bio-org chem, 79-80; res chemist polymer chem, E I du Pont de Nemours & Co, Inc, 81-83, group leader, 83-84, res supv, 84-90, res mgr, 90-91; PROF CHEM, CORNELL UNIV, 91- *Concurrent Pos:* Fel, African Am Inst, 71-75; mem, Bd Sci & Technol, Nat Res Coun, 88-89; mem, Nat Res Coun Briefing Panel Thin Films & Interfaces. *Honors & Awards:* Distinguished Bayer/Mobay Lectr, Cornell Univ; Waddell Prize, Univ Ghana, 74. *Mem:* Am Chem Soc; Sigma Xi; Int Soc African Scientists (pres, 87-88); NY Acad Sci; AAAS. *Res:* Synthesis and complexation of Macrocyclic Hosts; liquid-solid interface chemistry; asymmetric inductions and catalysis of Michael addition reactions; synthesis of optically active polymers; organosilicon chemistry; polymers for biomedical applications; drug delivery systems; group transfer polymerization; monolayers and surface interactions; fluoropolymers; living polymerizations. *Mailing Add:* Dept Chem Baker Lab Cornell Univ Ithaca NY 14853

SOGANDARES-BERNAL, FRANKLIN, b Panama, CZ, May 12, 31; div; c 3. PARASITOLOGY. *Educ:* Tulane Univ, BS, 54; Univ Nebr, MS, 55, PhD(zool), 58. *Prof Exp:* Parasitologist, Marine Lab, State Bd Conserv, Fla, 58-59; from instr to prof zool, Tulane Univ, La, 59-71, mem grad fac, 61-71, exec officer biol, 62-65, univ coord sci planning, 65-67, dir lab parasitol,

67-71; prof zool & chmn dept, Univ Mont, 71-72, prof microbiol, 72-74; chmn dept, 74-77, PROF BIOL, SOUTHERN METHODIST UNIV, 74- *Concurrent Pos:* Guest investr, Lerner Marine Lab, Am Mus Natural Hist, 57 & 60; mem adv panel syst biol, Biomed Div, NSF, 63-66; mem bd sci adv, Saltwater Fish Div, Fla State Bd Conserv, 64-; consult in path, Dept Path, Baylor Univ Med Ctr, 75-, med staff affil, 77-, dir, Ctr Infectious Dis Res, Baylor Res Fedn, 84-; asst to dean, Div Continuing Educ, Univ Tex Health Sci Ctr, 77-80; consult engrs, SputterTex Corp. *Honors & Awards:* Henry Baldwin Ward Medal, Am Soc Parasitol, 69. *Mem:* Am Asn Pathologists; Am Soc Zool; Am Soc Parasitol; Coun Biol Ed; Wildlife Dis Asn; Soc Social Biol. *Res:* Evolutionary biology of parasitism immunopathology; photobiology. *Mailing Add:* Dept Biol Southern Methodist Univ Dallas TX 75275

SOGIN, H(AROLD) H, b Chicago, Ill, Dec 14, 20; m 46; c 4. MECHANICAL ENGINEERING. *Educ:* Ill Inst Technol, BS, 43, MS, 50, PhD(mech eng), 52. *Prof Exp:* Asst prof mech eng, Ill Inst Technol, 53-55; from asst prof to assoc prof eng, Brown Univ, 55-60; prof, 60-89, dept head, 78-89, EMER PROF MECH ENG, TULANE UNIV, 89- *Concurrent Pos:* Tulane Res Coun award, Inst Mech Statist Turbulence, Univ Marseille, 65-66; NSF res grants, 62-64, 69-71 & 72-75. *Mem:* Am Soc Mech Engrs; Sigma Xi. *Res:* Heat transfer; convection; thermal instability; measurements of thermal properties. *Mailing Add:* Dept Mech Eng Tulane Univ New Orleans LA 70118

SOGN, JOHN ALLEN, b Buffalo, NY, May 11, 46; m 69; c 2. IMMUNOLOGY, BIOCHEMISTRY. *Educ:* Brown Univ, AB, 68; Rockefeller Univ, PhD(biochem), 73. *Prof Exp:* Res assoc immunol & biochem, Rockefeller Univ, 73-76, asst prof, 76-77; sr staff fel, 77-78, res chemist, Nat Inst Allergy & Infectious Dis, 78-87, prog dir, Cancer Immunol, Nat Cancer Inst, 87-90, ACTG CHIEF, CANCER IMMUNOL, NAT CANCER INST, NIH, 90- *Mem:* Am Chem Soc; Harvey Soc; Am Asn Immunol; Sigma Xi. *Res:* Basic studies in molecular and cellular immunology relevant to the immune response to cancer. *Mailing Add:* 9208 Cedarcrest Dr Bethesda MD 20814

SOGNEFEST, PETER WILLIAM, b Melrose Park, Ill, Feb 4, 41; m 64; c 3. GENERAL MANAGEMENT. *Educ:* Univ Ill, BSEE, 64, MS, 67. *Prof Exp:* Engr, Magnavox Co, 64-67; sr fel, Mellon Inst Sci, 67-71; gen mgr res & mfg, Essex Int Inc, 69-77; bus unit mgr, 77-80, vpres & gen mgr, Motorola, Inc, Ill, 80-84; CHMN & CHIEF EXEC OFFICER, DIGITAL APPLIANCE CONTROLS, INC, 84- *Concurrent Pos:* Vis fel, Mellon Inst Sci, 71- *Mem:* Inst Elec & Electronics Engrs. *Res:* Metal-Oxide-semiconductor integrated circuits as applied to automotive electrical systems; digital controls as applied to major appliances; microprocessor based instruments as applied to agriculture and construction equipment. *Mailing Add:* Four Back Bay Rd S Barrington IL 60011

SOGO, POWER BUNMEI, b San Diego, Calif, Feb 26, 25; m; c 3. PHYSICS. *Educ:* San Diego State Col, AB, 50; Univ Calif, PhD(physics), 55. *Prof Exp:* Physicist, Radiation Lab, Univ Calif, 55-59; asst prof physics, San Diego State Col, 59-62; from asst prof to assoc prof physics, Pomona Col, 62-66; from assoc prof to prof, Calif State Col San Bernardino, 66-71; prof physics, Univ Hawaii, Hilo, 71-86, chmn, natural sci div, 78- 80; RETIRED. *Res:* Electron paramagnetic resonance. *Mailing Add:* 1400 Kapiolani St Hilo HI 06720

SOH, SUNG KUK, b Korea, Mar 5, 51; m 80; c 1. POLYMER SCIENCE, POLYMER ENGINEERING. *Educ:* Seoul Nat Univ, BS, 73, MS, 76; Univ NH, PhD(chem eng), 81. *Prof Exp:* Chem engr, Res & Develop, Yuyu Indust Co, 73-74; ASST PROF CHEM ENG, MANHATTAN COL, 81- *Mem:* Am Inst Chem Engrs; Am Chem Soc; Sigma Xi. *Res:* Polymerization kinetics; emulsion polymerization and latex technology; polymer reactor modeling and control. *Mailing Add:* Dept Chem Eng Univ Detroit 4001 W McNichols Rd Detroit MI 48221

SOHACKI, LEONARD PAUL, b Bay City, Mich, Aug 21, 33; m 60; c 5. LIMNOLOGY, ZOOLOGY. *Educ:* Mich State Univ, BS, 61, MS, 65, PhD(limnol), 68. *Prof Exp:* Asst prof, 68-71, ASSOC PROF BIOL, STATE UNIV NY COL ONEONTA, 71- *Mem:* Am Soc Limnol & Oceanog; Sigma Xi; Water Pollution Control Fedn. *Res:* Eutrophication and productivity. *Mailing Add:* Dept of Biol State Univ NY Col Oneonta Oneonta NY 13820

SOHAL, GURKIRPAL SINGH, b Punjab, India, Oct 1, 48; m 75; c 1. NEUROEMBRYOLOGY. *Educ:* Punjab Univ, BS, 69; La State Univ, PhD(anat), 73. *Prof Exp:* Asst prof anat, Fla Int Univ, 73-75; asst prof anat, Med Col Ga, 75-77, assoc prof, 77-80; mem fac, dept biol sci, Fla Int Univ, 80-; AT DEPT ANAT, MED COL GEORGIA. *Concurrent Pos:* Res grant, Med Col Ga, 75; NIH res grant, 77- *Mem:* Am Asn Anatomists; AAAS; Soc Neurosci; Sigma Xi. *Res:* Factors responsible for cell death and cell differentiation in the developing brain. *Mailing Add:* Dept Anat Med Col Ga Augusta GA 30912

SOHAL, MANOHAR SINGH, b Ludhiana, India, June 1, 43; US citizen; m 75; c 2. THERMOFLUIDS, HEAT TRANSFER. *Educ:* Birla Inst Technol & Sci, India, BE, 65, ME, 67; Univ Houston, PhD(mech eng), 72. *Prof Exp:* Lectr mech eng & heat transfer thermodynamics, Birla Inst Technol & Sci, India, 67-68; res fel, Eindhoven Univ Technol, Netherlands, 72-73, Univ Strathclyde, Scotland, 73-74; spec res asst, Univ Manchester Inst Sci & Technol, England; res assoc, Solar Energy Lab, Univ Houston, 75-76; develop engr, Res & Develop Lab, M W Kellogg, Houston, Tex, 76-80; PRIN PROG SPECIALIST, IDAHO NAT ENG LAB, EG&G, IDAHO, INC, 80- *Concurrent Pos:* Lectr mech eng, Univ Houston, 76; proj mgr, Thermal Sci Prog. *Mem:* Am Soc Mech Engrs; Sigma Xi. *Res:* Analysis of transient thermohydraulic phenomena in nuclear reactors; heat transfer, fluid flow and two-phase flow problems; project management of advanced ceramic heat exchangers for high temperature applications. *Mailing Add:* Idaho Nat Eng Lab EG&G Idaho Inc PO Box 1625 Idaho Falls ID 83415-3527

SOHAL, PARMJIT S, b Panjaur, India, Apr 6, 59; Can citizen; m 87; c 1. BIOLOGICAL SCIENCES. *Educ:* Punjab Univ, BSc, 79; Punjab Agr Univ, MSc, 82; Univ Sask, PhD(biochem), 88. *Prof Exp:* Res fel biochem, Univ Delhi, India, 81-82; lab demonstr biochem, Univ Sask, 82-88; postdoctoral fel biochem, Simon Fraser Univ, Can, 88-90; postdoctoral fel biochem, 90-91, RES SCIENTIST NUTRIT BIOCHEM, UNIV ALTA, 91- *Honors & Awards:* Travel Award, Am Soc Biochem & Molecular Biol, 91. *Mem:* Can Biochem Soc; Can Soc Nutrit Sci; Am Soc Biochem & Molecular Biol; Am Inst Nutrit; Biochem Soc. *Res:* Regulation of lipid metabolism; nutritional and hormonal regulation of lipogenic enzymes; nutrition and lipid-dependent signal transduction systems; role of diet and hormones in postnatal metabolic development. *Mailing Add:* 533 Newton Res Bldg Univ Alta Edmonton AB T6G 2C2 Can

SOHAL, RAJINDAR SINGH, b Amritsar, India, July 1, 36; m 71. CELL BIOLOGY. *Educ:* Panjab Univ, BS, 60, MS, 61; Tulane Univ, PhD(biol), 65. *Prof Exp:* Asst prof biol, Xavier Univ, 65-66; from instr to asst prof cardiovasc res, Sch Med, Tulane Univ, 66-69; from asst prof to assoc prof, 69-79, PROF BIOL, SOUTHERN METHODIST UNIV, 79- *Concurrent Pos:* Sr vis scholar, Zool Dept, Cambridge Univ, 75 & 79 & Univ Dusseldorf, Dept of Biochem, 84; vis prof, dept path, Linkoping Univ, Sweden, 87, 88. *Mem:* Am Soc Cell Biol; fel Geront Soc. *Res:* Differentiation and aging of cells; free radical biochemistry; relationship between life span and metabolic rate. *Mailing Add:* Dept Biol Southern Methodist Univ Dallas TX 75275

SOHL, CARY HUGH, b Pittsfield, Mass. ACOUSTICS. *Educ:* Renssalaer Polytech Inst, BS, 74; Northwestern Univ, MS, 76, PhD(physics), 79. *Prof Exp:* PHYSICIST, E I DU PONT DE NEMOUR & CO INC, 79- *Res:* Plant process monitoring. *Mailing Add:* 32 Renee Lane Newark DE 19711-3421

SOHL, NORMAN FREDERICK, b Oak Park, Ill, July 14, 24; m 47; c 1. PALEONTOLOGY. *Educ:* Univ Ill, BS, 49, MS, 51, PhD(geol), 54. *Prof Exp:* Asst, State Geol Surv, Ill, 49-50; instr geol, Bryn Mawr Col, 52-53 & Univ Ill, 53-54; geologist, 54-68, chief br paleont & stratig, 68-73, RES PALEONTOLOGIST, US GEOL SURV, 73- *Concurrent Pos:* Pres bd dirs, Inst Malacol, 62-; vis prof, Univ Kans, 66; res assoc, Smithsonian Inst, 67-; mem bd overseers, Harvard Univ. *Honors & Awards:* Meritorious Serv Award, Dept Interior, 74, Distinguished Serv Award, 81. *Mem:* Soc Econ Paleont & Mineral; Soc Syst Zool; Soc Study Evolution; Paleont Soc Am (pres, 85-86). *Res:* Upper Cretaceous gastropoda; Mesozoic stratigraphy. *Mailing Add:* 10629 Marbury Rd Oakton VA 22124

SOHLER, ARTHUR, b New York, NY, Sept 23, 27; m 58; c 3. BIOCHEMISTRY, MICROBIOLOGY. *Educ:* City Col New York, BS, 51; St John's Univ, NY, MS, 54; Rutgers Univ, PhD(microbiol chem), 57. *Prof Exp:* Asst, St John's Univ, NY, 53-54; res assoc, Inst Microbiol, Rutgers Univ, 57-58, asst res specialist biochem, 58-59; biochemist, Bur Res in Neurol & Psychiat, NJ Neuropsychiat Inst, 58-73; BIOCHEMIST, BRAIN BIOCENTER, 73- *Mem:* AAAS; Am Chem Soc; Am Soc Microbiol. *Res:* Biochemistry of mental illness; clinical and microbial biochemistry; chemistry of natural products. *Mailing Add:* Princeton Brain Bio Ctr 862 Rte 518 Skillman NJ 08558

SOHLER, KATHERINE BERRIDGE, epidemiology; deceased, see previous edition for last biography

SOHMER, BERNARD, b New York, NY, July 16, 29; m 52; c 2. MATHEMATICS. *Educ:* NY Univ, BA, 49, MS, 51, PhD(math), 58. *Prof Exp:* Mathematician, Army Signal Corps, 51-52; instr math, NY Univ, 52-53; lectr, City Col New York, 53-57; instr, NY Univ, 57-58; from asst prof to assoc prof, 58-69, from assoc dean to dean students, 68-75, PROF MATH, CITY COL NY, 69- *Concurrent Pos:* Chair, fac senate, City Col NY. *Mem:* AAAS; Am Math Soc; Math Asn Am. *Res:* Structure theory of groups; rings algebras. *Mailing Add:* City Col of NY 139th St & Convent Ave New York NY 10031

SOHMER, SEYMOUR H, b Bronx, NY, Feb 27, 41; m 67; c 2. SYSTEMATIC BOTANY. *Educ:* City Col NY, BS, 63; Univ Tenn, MS, 66; Univ Hawaii, PhD(bot), 71. *Prof Exp:* Dir herbarium & assoc prof bot, Univ Wis-LaCrosse, 67-80; CHMN DEPT BOT, BERNICE P BISHOP MUS, HONOLULU, 80-, ASST DIR, RES & SCHOLARLY STUDIES, 85- *Concurrent Pos:* Res fel, Smithsonian Inst, 75-76 & partic Flora of Ceylon proj, 73-74; NSF assignment to assess status basic res trop biol, 77-78; forest botanist, div bot, Off Forests, Dept Primary Industs, Lae, Papua, New Guinea, 79-; chmn, Standing Comt Bot, Pac Sci Asn, 83-; sr biodiversity adv, Agency Int Develop, 90-91. *Mem:* AAAS; Soc Study Evolution; Asn Trop Biol; Int Asn Plant Taxon; Am Soc Plant Taxon; Asn Pac Systematists (secy/treas, 83-). *Res:* Systematic revisionary work with selected angiosperms; ascertaining the identities and relationships among complex groups of flowering plants, particularly Psychotria (Rubiaceae). *Mailing Add:* Dept Bot Bernice P Bishop Mus PO Box 19000-A Honolulu HI 96819

SOHN, DAVID, b Far Rockaway, NY, Dec 5, 26; m 62, 85; c 6. PATHOLOGY, TOXICOLOGY. *Educ:* Yeshiva Col, BA, 46; Columbia Univ, AM, 48; Polytech Inst Brooklyn, MA, 53; State Univ NY Downstate Med Ctr, MD, 57. *Prof Exp:* Asst attend path, Montefiore Hosp & Med Ctr, New York, 62-63; asst pathologist, Maimonides Hosp & Med Ctr, New York, 63-65; attend path, Ctr Chronic Dis & assoc prof, New York Med Col, 65-88; DIR LABS, GRACIE SQUARE HOSPITAL, 65- *Concurrent Pos:* Mem comt alcohol & drug abuse, Nat Safety Coun, 72-; mem toxicol resource comt, Col Am Pathologists, 73-81, chmn, 74-80, mem surv comt, 74-81; assoc dean, New York Med Col, 73-88, asst clin prof dermat, 74-; consult toxicol subcomt, Diag Devices Comt, US Food & Drug Admin, 75-, chmn, 80-81 & 85-86. *Mem:* Fel Col Am Pathologists; fel Am Soc Clin Pathologists; Am Asn Clin Chemists; fel Am Acad Forensic Sci; Am Chem Soc. *Res:* Methodology in the identification of drugs of abuse in biologic fluids, their quantitation and confirmed identification; quantitation of therapeutic drugs in body fluids. *Mailing Add:* 8 Muriel Ave Lawrence NY 11559

SOHN, HONG YONG, b Kaesung, Korea, Aug 21, 41; US citizen; m 71; c 2. EXTRACTIVE METALLURGY, REACTION ENGINEERING. *Educ:* Seoul Nat Univ, BS, 62; Univ NB, MSc, 66; Univ Calif, Berkeley, PhD(chem eng), 70. *Prof Exp:* Res engr chem eng, Cheil Sugar Co, 61-64 & E I du Pont de Nemours & Co, 73-74; from asst prof to assoc prof, 74-80, PROF METALL ENG, UNIV UTAH, 80- *Concurrent Pos:* Res assoc, State Univ NY, Buffalo, 71-73; consult, Lawrence Livermore Lab, 75-, Kennecott Co, Cabot Corp, 84-, Utah Power & Light Co, 87- & DuPont Co, 87-; adj assoc prof fuels eng, Univ Utah, 78-80, adj prof, 80- & adj prof chem eng, 87-; Dreyfus Found teacher-scholar award, 77; extractive metall lectr, Metall Soc, 90. *Honors & Awards:* Fulbright Distinguished Lectr, 83; Extractive Metall Award, Metall Soc, 90. *Mem:* Am Inst Mining, Metall & Petrol Engrs(dir, 83-84); Am Inst Chem Engrs; Am Chem Soc; Sigma Xi; NAm Thermal Analysis Soc. *Res:* Extractive metallurgy; self-propagating high temperature synthesis; oil shale conversion; gas-solid reactions; combustion of solids. *Mailing Add:* Dept Metall Eng Univ Utah Salt Lake City UT 84112-1183

SOHN, ISRAEL GREGORY, b Ukraine, Nov 12, 11; nat US; m 41; c 2. PALEONTOLOGY. *Educ:* City Col NY, BS, 35; Columbia Univ, AM, 38; Hebrew Univ, Israel, PhD, 66. *Prof Exp:* Preparator, 41-42, geologist, 42-85, CONSULT GEOLOGIST, US GEOL SURV, 85- *Concurrent Pos:* From assoc prof lectr to prof lectr, George Wash Univ, 58-68, adj prof, 69-; guest lectr, Hebrew Univ, Israel, 62-63 & Acad Sinica, Nanjing, Peoples Repub China, 79; res assoc, Smithsonian Inst, 68- *Mem:* Fel Geol Soc Am; Soc Econ Paleont & Mineral; Paleont Soc; Am Asn Petrol Geol. *Res:* Micropaleontology, especially post Devonian Ostracoda. *Mailing Add:* Rm E-308 Nat Mus of Natural Hist Washington DC 20560

SOHN, KENNETH S (KYU SUK), b Seoul, Korea, Aug 8, 33; m 62; c 2. ELECTRICAL ENGINEERING. *Educ:* Upsala Col, BS, 57; Stevens Inst Technol, MS, 59, ScD(elec eng), 67. *Prof Exp:* Instr elec eng, Stevens Inst Technol, 59-66; asst prof, 66-69, ASSOC PROF ELEC ENG, NJ INST TECHNOL, 69- *Concurrent Pos:* Consult, NY Tel Co, 67-73. *Mem:* Inst Elec & Electronics Engrs; Sigma Xi. *Res:* Determination of the electrical conductivity of semi-conductors by optical method; non-magnetic-DC-DC converters; ultrasonic array scanner for non-invasively visualizing blood vessels; medical instrumentation; investigation of planar edge-contact Josephson Junction radiation detector and mixer. *Mailing Add:* Dept Elec Eng Newark Inst Technol 323 High St Newark NJ 07102

SOHN, YUNG JAI, b Tokyo, Japan; US citizen; m 61; c 3. CARDIOVASCULAR PHARMACOLOGY. *Educ:* Univ Rochester, BA, 58; State Univ NY, MD, 62. *Prof Exp:* From asst prof to assoc prof anesthesiol & pharmacol, Univ Miami, 70-75; assoc prof, Univ Calif, San Francisco, 75-87; VIS PROF, YONSEI UNIV, KOREA, 85-, CLIN PROF, UNIV CALIF, SAN FRANCISCO, 87- *Concurrent Pos:* Sr scientist, Univ Gronigen, Neth, 80-81. *Res:* Neuromuscular pharmacology. *Mailing Add:* Dept Anesthesia Univ Calif Box 0648 San Francisco CA 94143

SOHR, ROBERT TRUEMAN, b Green Bay, Wis, Apr 4, 37. EARTH & MARINE SCIENCES, ENVIRONMENTAL HEALTH. *Educ:* Valparaiso Univ, BS, 60; Purdue Univ, BS, 61; Univ Chicago, MBA, 69. *Prof Exp:* Process engr, Natural Gas Pipeline Co Am, 60-67; cryogenic tech engr, Liquid Carbonic Div, Gen Dynamics, 67-69; regional mgr, US Stoneware Div, Norton, 69-72, Pollution Control Syst Div, Hormel, 72-75; exec vpres, Am Envirodyne Div, Pettibond, 75-78; US opers mgr, Tywood Industs, 78-79; tech dir, R T Sohr, PCHE, 79-89; TECH DIR, AIR DISTRIB ASSOCS, 89- *Mem:* Am Chem Soc; Am Inst Chemists; Am Inst Chem Engrs; Air Pollution Control Asn; Am Fisheries Soc. *Res:* Development of a filtration system for solvent and heat recovery, and dust to allow recirculation; state of the art chemical absorbtion systems for foundry coreroom, corn wet milling and many other industrial applications; develop rubber deflashing with liquid N2 and blow molding with liquid CO2. *Mailing Add:* 140 Lemoyne Pkwy Oak Park IL 60302

SOIFER, DAVID, b New York, NY, Sept 16, 37; m 60; c 2. CELLULAR & MOLECULAR NEUROBIOLOGY. *Educ:* Swarthmore Col, Columbia Univ, BS, 61; Cornell Univ, PhD(anat), 69. *Prof Exp:* AMA fel regulatory biol, Inst Biomed Res, 68-70; sr res scientist, 70-80, assoc res scientist, 77-86, DEP DIR, CSI/IBR CTR DEVELOP NEUROSCIS & HEAD, LAB CELL BIOL, INST BASIC RES DEVELOP DISABILITIES, 87- *Concurrent Pos:* Vis asst prof anat, Col Med & vis asst prof cell biol, Grad Sch Med Sci, Cornell Univ, 70-77; assoc prof anat cell biol, State Univ NY Downstate Med Ctr, 77-86; prof biol, CUNY Col Staten Island & Grad Ctr, 87- *Mem:* AAAS; Am Soc Cell Biol; Soc Neurosci; Am Soc Neurochem; Int Soc Neurochem. *Res:* Neurobiology; biology of microtubules; dynamics of the neuronal cytoskeleton; the molecular biology of neurofibrillary degeneration; function of cytoskeletal proteins in cells of the nervous system. *Mailing Add:* Inst for Basic Res Develop Disabilities 1050 Forest Hill Rd Staten Island NY 10314

SOIFER, HERMAN, US citizen. ENGINEERING. *Educ:* Cooper Union Sch Eng, BCE, 44. *Prof Exp:* From design engr to assoc, Consult Eng Firms, 46-71; PARTNER, ALPERN & SOIFER CONSULT ENGRS, 71- *Concurrent Pos:* Adj instr, C W Post Sch Eng, 66-68. *Mem:* Am Soc Civil Engrs; Nat Soc Prof Engrs. *Res:* Sanitary and environmental engineering. *Mailing Add:* 2635 Pettit Ave Bellmore NY 11710

SOIKE, KENNETH FIEROE, b Minneapolis, Minn, July 8, 27; m 78; c 3. VIROLOGY, CELL CULTURE. *Educ:* Univ Minn, BA, 49; Ore State Univ, PhD(microbiol), 55. *Prof Exp:* Res assoc, Sterling Winthrop Res Inst, 55-61; assoc prof microbiol, Albany Med Col, 61-73; assoc dir res, Primate Res Inst, 73-75; SR RES SCIENTIST, DELTA REGIONAL PRIMATE CTR, 75- *Concurrent Pos:* Adj prof, Primate Res Inst, NMex State Univ, 84- *Mem:* Am Soc Microbiol; Am Soc Virol; Int Soc Interferon Res; Int Soc Chemother; Int Soc Antiviral Res. *Res:* Evaluation of antiviral drugs and recombinant human interferons in the treatment of viral disease in nonhuman primates; pathogenesis of viral diseases in monkeys. *Mailing Add:* Delta Regional Primate Res Ctr Tulane Univ Three Rivers Rd Covington LA 70433

SOJKA, GARY ALLAN, b Cedar Rapids, Iowa, July 15, 40; m 62; c 2. MICROBIAL PHYSIOLOGY, GENETICS. *Educ:* Coe Col, BA, 62; Purdue Univ, MS, 65, PhD(microbiol), 67. *Prof Exp:* From res assoc to asst prof biol, Ind Univ, Bloomington, 67-72, assoc prof biol & assoc chmn dept, 72-78; prof biol, 77, chmn, 78-81, dean orto & sci, 81-84, PRES, BUCKNELL UNIV, 84- *Mem:* AAAS; Am Soc Biochemists; Am Soc Microbiol. *Res:* Control of metabolism at a molecular level in photosynthetic bacteria. *Mailing Add:* Biol Dept Jordan Hall Ind Univ Bloomington IN 47405

SOJKA, ROBERT E, b Chicago, Ill, Dec 28, 47; m 77; c 2. SOIL & WATER CONSERVATION, CROP STRESS MANAGEMENT. *Educ:* Univ Calif, Riverside, BA, 69, PhD(soil sci), 74. *Prof Exp:* Lab asst, Soils Dept, Univ Riverside, 66-70, res asst, 70-74; post doctorate res assoc, Agron Dept, Univ Ark, 74-76; asst prof soils, NDak State Univ, 76-78; soil scientist, USDA Agr Res Serv, SC, 78-86, SOIL SCIENTIST, USDA AGR RES SERV, ID, 86- *Concurrent Pos:* adj assoc prof, Clemson Univ, 78-86; adj prof, Univ Idaho, 86- *Mem:* Am Soc Agron; Soil Sci Soc Am; Crop Sci Soc Am; Int Soc Soil Sci; Inst Soil & Tillage Res Orgn; Sigma Xi. *Res:* Physical edaphology, the study of the effects of soils physical properties on plant response, including the effects of soil compaction, aeration, temperature, flooding, and drought. *Mailing Add:* USDA-ARS 3793 N-3600 E Kimberly ID 83341

SOJKA, STANLEY ANTHONY, b Buffalo, NY, Nov 6, 46; m 70; c 2. PHYSICAL ORGANIC CHEMISTRY. *Educ:* Canisius Col, BS, 68; Ind Univ, Bloomington, PhD(org chem), 72. *Prof Exp:* Res assoc spectros, Nat Res Coun, 72-74; res chemist, Naval Res Lab, 74-76; sr res chemist, Hooker Chem & Plastics Corp, 76-80; mgr environ technol, 80-85, DIR, NEW BUS DEVELOP, OCCIDENTAL CHEM CORP, 85- *Mem:* Am Chem Soc; Com Develop Asn. *Res:* Using carbon-13 nuclear magnetic resonance spectroscopy to solve chemical problems; carbon-13 chemically induced dynamic nuclear polarization developed to gain knowledge about mechanism and kinetics of reactions. *Mailing Add:* Occidental Chem Corp 360 Rainbow Blvd S Niagara Falls NY 14302

SOKAL, ROBERT REUVEN, b Vienna, Austria, Jan 13, 26; nat US; m 48; c 2. POPULATION BIOLOGY, TAXONOMY. *Educ:* St John's Univ, China, BS, 47; Univ Chicago, PhD(zool), 52. *Hon Degrees:* DSc, Univ Crete, Iraklion, 90. *Prof Exp:* From instr to assoc prof entom, Univ Kans, 51-61, prof statist biol, 61-69; prof biol sci, 68-72, chmn & dir grad studies, 80-83, actg vprovost res grad studies, 81-82, LEADING PROF ECOL & EVOLUTION, STATE UNIV NY, STONY BROOK, 72- *Concurrent Pos:* Watkins scholar, Univ Ill, 56; NSF sr fel, Galton Lab, Univ Col, London, 59-60; Fulbright vis prof zool, Hebrew & Tel-Aviv Univs, Israel, 63-64; NIH career investr, 64-69; NATO sr fel, Cambridge Univ, 75; vis prof, Inst Advan Studies, Oeiras, Portugal, 71-80; vis distinguished scientist & Guggenheim Found fel, Univ Mich, 75-76; vis prof zool, Univ Vienna, 77 & 78, Fulbright vis prof human biol & Guggenheim Found fel, 84; vis prof, Col France, Paris, 89; corresp, Nat Mus Natural History, Paris, 90- *Mem:* Nat Acad Sci; hon fel Linnean Soc London; hon mem Soc Syst Zool; Am Soc Naturalists (ed, 69-74, pres, 84); Classification Soc (pres, 69-71); Soc Study Evolution (vpres, 67, pres, 77); fel AAAS; fel Am Acad Arts Sci; Int Fedn Classification Socs (vpres, 87, 90, pres, 88-89). *Res:* Geographic variation analysis; numerical taxonomy; theory of systematics; spatial models; human variation; European ethnohistory. *Mailing Add:* Dept Ecol & Evolution State Univ NY Stony Brook NY 11794

SOKATCH, JOHN ROBERT, b Joliet, Ill, Dec 20, 28; m 57; c 3. BACTERIOLOGY. *Educ:* Univ Mich, BS, 50; Univ Ill, MS, 52, PhD(bact), 56. *Prof Exp:* Res assoc chem, Wash State Univ, 56-58; asst dean grad col, Univ Okla, 70, from asst prof to assoc prof, Sch Med, 58-67, assoc dean grad col, 71-77, assoc dir res admin, 73-77, prof microbiol, Sch Med, 67-84, PROF & CHMN, DEPT BIOCHEM & MOLECULAR BIOL, SCH MED, UNIV OKLA, 84- *Concurrent Pos:* Fulbright sr res scholar, Sheffield, Eng, 63-64; USPHS res career develop award, 62-72; Fogarty sr int fel, Cambridge Univ, 79; vis prof, GBF, Graunschweig, Fed Rep Ger. *Mem:* Am Soc Microbiol; Am Soc Biol Chemists; Am Acad Microbiol; Sigma Xi. *Res:* Metabolism of branched chain amino acids by bacteria and regulation of catabolic pathways. *Mailing Add:* Dept Microbiol Univ Okla Health Sci Ctr PO Box 26901 Oklahoma City OK 73190

SOKOL, HILDA WEYL, b St Louis, Mo, Dec 19, 28; m 51; c 3. NEUROENDOCRINOLOGY. *Educ:* Hunter Col, AB, 50; Radcliffe Col, AM, 51, PhD, 57. *Prof Exp:* Instr sci, Boston Univ, 54-55; instr zool, Wellesley Col, 55-58; res assoc physiol, Harvard Med Sch, 60-61; res assoc, 61-63, from instr to asst prof physiol, 63-75, ASSOC PROF PHYSIOL, DARTMOUTH MED SCH, 75- *Mem:* Fel AAAS; Am Soc Zoologists; Endocrine Soc; Asn Women Sci. *Res:* Comparative endocrinology; cytology and physiology of the pituitary gland and hypothalamus; releasing factors, anterior and posterior pituitary hormones; diabetes insipidus; sexual dimorphism. *Mailing Add:* Dept Physiol Dartmouth Med Sch Hanover NH 03756

SOKOL, ROBERT JAMES, b Rochester, NY, Nov 18, 41; m 64; c 3. OBSTETRICS & GYNECOLOGY, COMPUTER SCIENCE. *Educ:* Univ Rochester, BA, 63, MD, 66; Am Bd Obstet & Gynec, dipl, 72, cert maternal-fetal med, 75. *Prof Exp:* Intern & resident obstet & gynec, Barnes Hosp, Wash Univ, 66-70; from obstetrician & gynecologist to chief obstetrician & gynecologist, US Air Force Hosp, Ellsworth AFB, 70-72; asst prof obstet & gynec, Sch Med & Dent, Univ Rochester, 72-73; from asst prof to prof obstet & gynec, Case Western Reserve Univ, 73-83; PROF & CHMN OBSTET & GYNEC, WAYNE STATE UNIV, 83-, CHIEF OBSTET & GYNEC, HUTZEL HOSP, 83-; DIR, CS MOTT CTR HUMAN GROWTH & DEVELOP, 83- *Concurrent Pos:* Buswell fel maternal-fetal med, Strong Mem Hosp, Sch Med & Dent, Univ Rochester, 72-73; asst prog dir, Perinatal Clin Res Ctr, Cleveland Metrop Gen Hosp, 73-78, co-prog dir, 73-81, prog dir, 81-83 & assoc dir dept obstet & gynec, 81-83; fel maternal-fetal med, Cleveland Metrop Gen Hosp & Case Western Reserve Univ, 74-75; consult, Nat Inst Child Health & Human Develop, 78-79 & 84-, Nat Inst Alcohol Abuse & alcoholism, 79-, Ctr Dis Control, 81, Nat Inst Health, 82-83, Health

Resources & Serv Admin, 84; mem, Alcohol Psychosocial Res Rev Comt & Nat Inst Alcohol Abuse & Alcoholism, 82-, grad fac, Dept Physiol, Wayne State Univ, 84-; assoc examr, Am Bd Obstet & Gynec, 84- *Mem:* Perinatal Res Soc; Res Soc Alcoholism; Soc Gynec Invest; Soc Perinatal Obstetricians; Cent Asn Obstet & Gynecologists; Behav Teratology Soc; Sigma Xi. *Res:* Perinatal risk assessment; database management and statistical analysis; alcohol-related birth defects; low birth weight risks and outcomes; algorithmic diagnosis and management; fetal risks of ultrasound exposure; fetal alcohol syndrome. *Mailing Add:* Deans Off Sch Med 540 E Canfield Detroit MI 48201

SOKOL, RONALD JAY, b Chicago, Ill, July 18, 50. PEDIATRIC GASTROENTEROLOGY & NUTRITION, VITAMIN E RESEARCH. *Educ:* Univ Ill, Urbana, BS, 72; Univ Chicago Pritzker Sch Med, MD, 76. *Prof Exp:* Pediat resident, Univ Colo Health Sci Ctr, 76-79, chief resident, 79-80; fel pediat gastroenterol & nutrit, Childrens Hosp Res Found, Cincinnati, 80-83; asst prof pediat, 83-88, ASSOC PROF PEDIAT, UNIV COLO SCH MED, 88- *Concurrent Pos:* Grants Rev Comt, Am Liver Found, 85-88. *Mem:* Am Acad Pediat; Am Asn Study Liver Dis; Am Gastroenterol Asn; N Am Soc Pediat Gastroenterol & Nutrit. *Res:* Causes, mechanisms and treatment of human vitamin E deficiency states; investigations of the effect of vitamin E on the structure and function of the hepatocyte. *Mailing Add:* Dept Pediat Div Pediat Gastroenterol & Nutrit Box C228 Univ Hosp 4200 E Ninth Ave Denver CO 80262

SOKOLOFF, ALEXANDER, b Tokyo, Japan, May 16, 20; nat US; m 56; c 3. ECOLOGICAL GENETICS. *Educ:* Univ Calif, Los Angeles, AB, 48; Univ Chicago, PhD(ecol), 54. *Prof Exp:* Res assoc cancer, Univ Chicago, 54, instr biol, 55; from instr to asst prof biol, Hofstra Col, 55-58; geneticist, William H Miner Agr Res Inst, NY, 58-60; assoc res botanist, Univ Calif, Los Angeles, 60-61, assoc res geneticist, Univ Calif, Berkeley, 61-66; assoc prof natural sci div, 65-66, prof biol, 66-90, EMER PROF BIOL, CALIF STATE COL, SAN BERNARDINO, 90- *Concurrent Pos:* NSF res grant, Cold Spring Harbor Lab Quant Biol, 58-60; ed, Tribolium Info Bull, 60-; USPHS res grant, 61; res geneticist, Univ Calif, Berkeley, 66-68; NSF res grants, 67-75; assoc ed, Evolution, 72-74; chmn subcomt on insect stocks, Comt for Maintenance of Genetic Stocks, Genetics Soc Am, 74-85; res grant, Army Res Off, 74-79; dir, Tribolium Stock Ctr, 61-; assoc ed, J Advan Zool, India, 80-; mem adv bd, J Stored Prod Res, 61- *Mem:* Am Soc Zoologists; Am Soc Nat; Am Genetic Soc; Genetics Soc Am; Entom Soc Am; Soc Study Evolution; fel Royal Entom Soc London; Genetics Soc Can; Japanese Soc Pop Ecol; Sigma Xi. *Res:* Population ecological genetics of Tribolium; genetic control of flour beetles. *Mailing Add:* Dept Biol Calif State Univ San Bernardino CA 92407

SOKOLOFF, JACK, b New York, NY, July 28, 22; m 48; c 3. THEORETICAL PHYSICS. *Educ:* Univ Mich, BS, 48, MS, 49; Northwestern Univ, PhD(physics), 56. *Prof Exp:* Teaching asst physics, Univ Mich, 48-49; aerodynamicist, Bell Aircraft Corp, 49-50; teaching asst physics, Northwestern Univ, 52-54; student res assoc physics div, Argonne Nat Lab, 54-56; res scientist, Lockheed Palo Alto Res Lab, 56-68; prof physics, York Univ, 68-88; RETIRED. *Concurrent Pos:* Vis scientist, Lab Theoret Physics & Elem Particles, Univ Paris-Sud, Orsay, France, 75-76. *Mem:* Am Phys Soc. *Res:* Variation principles as applied to problems of atomic structure and atomic collisions. *Mailing Add:* 38 Valentine Dr Don Mills ON M3A 3J8 Can

SOKOLOFF, JEFFREY BRUCE, b New York, NY, Oct 7, 41; m 68; c 3. SOLID STATE PHYSICS. *Educ:* Queen's Col, NY, BS, 63; Mass Inst Technol, PhD(physics), 67. *Prof Exp:* Res assoc, Brookhaven Nat Lab, 67-69; PROF PHYSICS, NORTHEASTERN UNIV, 69- *Concurrent Pos:* Vis mem staff, Weitzmann Inst, 79-80; vis prof, Ariz State Univ, 88. *Mem:* Am Phys Soc. *Res:* Magnetic and transport properties of metallic ferromagnets and ferrites; charge density wave conductivity in one-dimensional conductors; theory of ideal friction between sliding solid surfaces; excitations in crystals with two incommensurate periods; vibrations of DNA in solution. *Mailing Add:* Dept Physics Northeastern Univ Boston MA 02115

SOKOLOFF, LEON, b Brooklyn, NY, May 9, 19; c 2. PATHOLOGY. *Educ:* NY Univ, BA, 38, MD, 44. *Prof Exp:* Asst prof path, NY Univ, 50-52; chief sect rheumatic dis, Lab Exp Path, Nat Inst Arthritis, Metab & Digestive Dis, 53-73; PROF PATH, STATE UNIV NY STONY BROOK, 73- *Concurrent Pos:* Mem, path study sect, NIH, 56-60, gen med A, 78-84; vis prof, Royal Soc Med, 85. *Honors & Awards:* Philip Hench Award, 65; Van Breemen Award, Dutch Rheumatism Asn, 66. *Mem:* AAAS; Harvey Soc; Am Asn Pathologists; master Am Col Rheumatism; hon mem Europ Soc Osteoarthrology. *Res:* Pathology of rheumatic diseases. *Mailing Add:* 25 View Rd Setauket NY 11733

SOKOLOFF, LOUIS, b Philadelphia, Pa, Oct 14, 21; m 47; c 2. PHYSIOLOGY, BIOCHEMISTRY. *Educ:* Univ Pa, BA, 43, MD, 46. *Hon Degrees:* MD, Univ Lund, 80; DSc, Albert Einstein Col Med, 82. *Prof Exp:* Intern, Philadelphia Gen Hosp, 46-47; res fel physiol, Grad Sch Med, Univ Pa, 49-51, instr physiol, 51-54, assoc, 54-56; assoc chief sect cerebral metab, 53-56, chief, lab clin sci, 56-68, CHIEF LAB CEREBRAL METAB, NIMH, 68- *Concurrent Pos:* Vis prof, Col France, 68-69. *Honors & Awards:* F O Schmitt Award, 80; Albert Lasker Clin Med Res Award, 81; Karl Spencer Lashley Award, 87; Nat Acad Sci Award Neurosci, 88. *Mem:* US Nat Acad Sci; AAAS; Am Soc Biol Chemists; Am Neurol Asn; Am Physiol Soc; Am Soc Neurochem. *Res:* Cerebral circulation and metabolism; neurochemistry; biochemical basis of hormone actions; protein biosynthesis; thyroxine. *Mailing Add:* Lab of Cerebral Metab NIMH Bldg 36 Rm 1A-05 Bethesda MD 20892

SOKOLOFF, VLADIMIR P, b Tomsk, Siberia, Nov 8, 04; nat US; m 33; c 1. GEOCHEMISTRY. *Educ:* Univ Calif, PhD(soil sci), 37. *Prof Exp:* Jr chemist, Citrus Exp Sta & asst prof microbiol, Univ Calif, 37-43; soil scientist, US Geol Surv, 43-50; vis prof, Johns Hopkins Univ, 50-53; consult geochemist, Makhtsavei, Israel, 53-56; consult, US Geol Surv, 56 & Shell Oil Co, 57; eastern Europe specialist & phys scientist, US Bur Mines, DC, 58-66; trans rev ed, Am Geol Inst, 66, mem int geol rev staff, 66-74; RETIRED. *Concurrent Pos:* Consult geochemist, Zinc Corp, Pty, Ltd, Australia, 48, Western Mining Corp, 49 & Conzinc-Riotinto, 65. *Honors & Awards:* Commendable Serv Medal, US Dept Interior, 66. *Mem:* Fel AAAS; fel Am Geog Soc; fel Am Inst Chem; NY Acad Sci. *Res:* Applied geochemistry; physiology and parasitism of nitrate-reducing microorganisms; bacterial leaching of ores; soils physics in terrain intelligence. *Mailing Add:* PO Box 9724 Washington DC 20016

SOKOLOSKI, MARTIN MICHAEL, b Freeland, Pa, Sept 7, 37; m 62; c 3. CONDENSED MATTER PHYSICS, MATHEMATICS. *Educ:* Bucknell Univ, BS, 59, MS, 60; Catholic Univ Am, PhD(physics), 69. *Prof Exp:* Mathematician, Dept Defense, Nat Security Agency, 60-61; aerospace technician, NASA Goddard Space Flight Ctr, 61-63, aerospace engr, 63-66; res assoc, Catholic Univ Am, 70-71; mem staff, Harry Diamond Labs, 71-79; MGR ELECTRONICS, NASA, 79- *Mem:* Am Phys Soc; AAAS. *Res:* Determination of the nature of interface electronic states; theoretical studies of the static and dynamic properties of disordered systems; many body problems; basic research management. *Mailing Add:* 5526 Phelps Luck Dr Columbia MD 21045

SOKOLOSKI, THEODORE DANIEL, b Philadelphia, Pa, July 10, 33; m 61; c 3. PHARMACY, PHYSICAL CHEMISTRY. *Educ:* Temple Univ, BS, 55; Univ Wis, Madison, MS, 59, PhD(pharm), 61. *Prof Exp:* Asst prof pharm, Wash State Univ, 61-64; from asst prof to assoc prof pharm, 64-73, PROF PHARMACEUT & PHARMACEUT CHEM, OHIO STATE UNIV, 73- *Mem:* Am Chem Soc; Am Pharmaceut Asn; Acad Pharmaceut Sci. *Res:* Application of physical chemistry to pharmaceutical systems. *Mailing Add:* 217 Lloyd Parks Hall Ohio State Univ Main Campus Columbus OH 43210

SOKOLOVE, PHILLIP GARY, b Los Angeles, Calif, Aug 24, 42; div; c 1. NEUROBIOLOGY, BIOPHYSICS. *Educ:* Univ Calif, Berkeley, AB, 64; Harvard Univ, PhD(biophysics), 69. *Prof Exp:* Actg asst prof neurobiol, Stanford Univ, 71-72; from asst prof to assoc prof, 72-83, PROF BIOL, UNIV MD, BALTIMORE COUNTY, 83-, ASSOC DEAN ARTS & SCI, 87- *Concurrent Pos:* Res assoc, Stanford Univ, 72-74; consult, SRI Int, 72-79; Carnegie Sci fel, 82-83. *Mem:* Am Soc Zoologists; Soc Gen Physiologists; Soc Neurosci; Am Physiol Soc; AAAS; Sigma Xi. *Res:* Biological circadian rhythms; reproductive neuroendocrinology in molluscs; behavioral neurobiology. *Mailing Add:* Off Dean Arts & Sci Univ Md Baltimore County 5401 Wilkens Ave Baltimore MD 21228

SOKOLOW, MAURICE, b New York, NY, May 19, 11; wid; c 2. MEDICINE. *Educ:* Univ Calif, AB, 32, MD, 36; Am Bd Internal Med & Cardiovasc Dis, dipl. *Prof Exp:* Intern, San Francisco Hosp, Calif, 35-36; asst resident med, Univ Calif Hosp, 36-37; resident physician, New Eng Med Ctr, Boston, Mass, 37-38; researcher cardiovasc dis, Michael Reese Hosp, 38-39; clin instr med, 40-45, lectr, 45-46, from asst prof to prof, 46-78, chief, Electrocardiogram Dept, Univ Hosp, 46-78, chief cardiovasc serv, Univ Hosp, 58-74, EMER PROF MED, SCH MED, UNIV CALIF, SAN FRANCISCO, 78- *Concurrent Pos:* Res fel med, Sch Med, Univ Calif, San Francisco, 39-40, vis physician, 40-47; attend cardiologist, Langley Porter Clin, 46-; consult, Vet Admin Hosps, San Francisco & Oakland, Calif, 46-; researcher, Nat Heart Hosp, London, 53-54; mem Coun Arteriosclerosis, Am Heart Asn. *Mem:* Fel Am Col Physicians; Am Soc Clin Invest; Asn Univ Cardiologists; hon fel Am Col Cardiol; Am Fedn Clin Res (vpres, 49). *Res:* Rheumatic fever; electrocardiography; hypertension; cardiac arrhythmias; cardiac failure. *Mailing Add:* M312 Box 0214 Univ Calif San Francisco CA 94143

SOKOLOWSKI, DANNY HALE, b Alton, Ill, June 1, 38; m 65; c 2. SOLID STATE PHYSICS. *Educ:* Southern Ill Univ, AB, 61; Univ Mo, Rolla, MS, 63; St Louis Univ, PhD(physics), 74. *Prof Exp:* Instr physics, Southern Ill Univ, 63-66, Marquette Univ, 69-71; assoc prof physics, Lewis & Clark Community Col, 71-81; corp mgr telecommun, Gen Dynamics Co, 83-89; MGR TELECOMMUN, DOW CORNING CORP, 89- *Mem:* Am Phys Soc; Sigma Xi; Am Asn Physics Teachers. *Res:* Electrical and thermoelectrical properties of semiconductors. *Mailing Add:* Dow Corning Co Midland MI 48686

SOKOLOWSKI, HENRY ALFRED, b Hamtramck, Mich, Jan 21, 23; m 55; c 3. PHYSICS, MATHEMATICS. *Educ:* Univ Pa, AB, 57. *Prof Exp:* Physicist, 51-56, chief propellant physics, 56-61, chief ballistics lab, 61-71, chief test instrumentation div, 71-74, chief test & eval div, 74-76, DIR TECH SUPPORT DIRECTORATE, FRANKFORD ARSENAL, 77- *Mem:* Sigma Xi. *Res:* Ballistic, environmental materials test and evaluation utilizing the disciplines of physics, mathematics, metallurgy, chemistry, electrical engineering, mechanical engineering and associated specialized scientific fields. *Mailing Add:* 2731 Kirkbride St Philadelphia PA 19137

SOKOLSKI, WALTER THOMAS, b Newark, NJ, Oct 29, 16; m 47; c 2. MICROBIOLOGY. *Educ:* Ind Univ, AB, 48; Purdue Univ, MS, 53, PhD(bact), 55. *Prof Exp:* Serologist, Venereal Dis Res Lab, USPHS, 45-46; chemist, Parke, Davis & Co, 48-51; microbiologist, Upjohn Co, 54-59, head spec microbiol methods, Control Div, 59-67, head microbiol res, 67-70, mem staff infectious dis res, 70-78; RETIRED. *Concurrent Pos:* Consult antibiotic fermentation, Panlabs Taiwan Inc, Taiwan, 80-81; clin microbiol consult. *Honors & Awards:* Award, Am Soc Microbiol, 52. *Mem:* Am Soc Microbiol; Am Soc Med Technol; Am Soc Clin Path; Soc Protozool; Soc Cryobiol. *Res:* Screening methods for new antibiotics; paper and column chromatography; microbiological assay for antibiotics; in vitro methodology in clinical research; environmental control. *Mailing Add:* 3304 Cranbrook Kalamazoo MI 49007

SOLAND, RICHARD MARTIN, b New York, NY, July 27, 40; m 79; c 5. OPERATIONS RESEARCH. *Educ:* Rensselaer Polytech Inst, BEE, 61; Mass Inst Technol, PhD(math), 64. *Prof Exp:* Mem tech staff, Advan Res Dept, Res Anal Corp, 64-71; assoc prof statist-opers res, Univ Tex, Austin, 71-76; assoc prof, Dept Indust Eng, Ecole Polytech, Univ Montreal, 76-78; PROF OPERS RES, GEORGE WASHINGTON UNIV, 78-, CHAIR,

DEPT OPERS RES, 89- *Concurrent Pos:* Asst prof lectr, Dept Bus Admin, George Washington Univ, 65-68, assoc prof lectr, Dept Eng Admin, 68-69; Fulbright lectr, Helsinki Sch Econ, Finland, 69-70; vis prof, Res Ctr, Inst d Admin des Enterprises, Univ Aix-Marseille, Aix-en-Provence, France, 73-74; vis prof, Carabobo Univ, Valencia, Venezuela, 75, Univ Copenhagen, Denmark, 82 & 88; conf chmn, Tenth Triennial Conf Oper Res, 84; consult, Inst Defense Anal, 83-85 & Anal Serv Inc, 86- *Mem:* Opers Res Soc Am; Inst Mgt Sci; Inst Elec & Electronics Engrs; Math Programming Soc; Can Opers Res Soc; Inst Indust Engrs. *Res:* Multiple criteria decision making; branch-and-bound methods in mathematical programming; applications of mathematical programming; mathematical modeling; facility location; decision analysis; Bayesian statistics; missile defense models. *Mailing Add:* Dept of Opers Res George Washington Univ Washington DC 20052

SOLANDT, OMOND MCKILLOP, b Winnipeg, Man, Sept 2, 09; m 41, 72; c 3. PHYSIOLOGY. *Educ:* Univ Toronto, BA, 31, MA, 32, MD, 36; Cambridge Univ, MA, 39; FRCP, 64. *Hon Degrees:* Eleven from Can univs, 46-68. *Prof Exp:* Lectr physiol, Cambridge Univ, 39; dir, SW London Blood Supply Depot, 40; dir tank sect, Army Oper Res Group, 42, from dept supt to supt, 43-45; chmn, Defence Res Bd, Dept Nat Defence, 46-56; asst vpres res & develop, Can Nat Rwy, 56, vpres, 57-63; vpres res & planning, De Havilland Aircraft Can Ltd, 63-66; chmn, Sci Coun Can, 66-72; consult, Mitchell, Plummer & Co, Ltd, 72-75; SR CONSULT, INST ENVIRON STUDIES, UNIV TORONTO, 76- *Concurrent Pos:* Chancellor, Univ Toronto, 65-71; chmn, Sci Adv Bd for Northwest Territories, 76-81, pub gov, Toronto Stock Exchange; consult, Int Ctr Agr Res in Dry Areas, Syria, 76-81, Int Ctr Insect Physiol & Ecol, 77-82, Int Ctr Diarrheal Dis Res, Bangladesh, 79-82 & Int Wheat & Maize Improv Ctr, Mex, 76-86, WAfrican Rice Develop Asn, 87. *Honors & Awards:* Companion, Order Can; Order Brit Empire, 46; US Medal of Freedom, 47. *Mem:* Am Physiol Soc; Can Physiol Soc; Can Oper Res Soc (pres, 58-60); fel Royal Soc Can; foreign hon mem Am Acad Arts & Sci. *Res:* Operational research. *Mailing Add:* RR 1 Bolton ON L7E 5R7 Can

SOLAR, SAMUEL LOUIS, organic chemistry; deceased, see previous edition for last biography

SOLARI, MARIO JOSE ADOLFO, b Rosario, Santa Fe, Arg, Sept 29, 48; m 74; c 3. METALLURGY, WELDING. *Educ:* Nat Col No 1 Rosario, Arg, BA, 65; Nat Univ Rosario, Engr, 72; Nat Univ del Sur, DrIng, 85. *Prof Exp:* Researcher metall, Nat Atomic Energy Comn, Arg, 74-79, head, Div Welding Technol, 79-83; head, Dept Mat, Atucha II Nuclear Power Proj, Empresa Nuclear Arg Centrales Electricas, SAm, 83-87; INDEPENDENT INVESTR MAT, CONSEJO NACIONAL DE INVESTIGATIONES CIENTIFICAS Y TEC, ARG, 88-; MGR, CTI CONSULTORES INGENIERIS, SRL, 88- *Concurrent Pos:* Vis prof univs, Chile, Brazil, Uruguay, Venezuela, Mex, Colombia & Peru, 74-91, Nat Univ Plata, Arg, 89-91; consult, var co, 74-91; postdoctoral fel solidification, Multinat Prog Metall, Orgn Am States, Nat AEC, Arg, 76, postdoctoral fel welding, 80; dir, Welding Technol Proj, Nat Atomic Energy Comn-SECYT, 78-83; mem, Nat Standards Inst, 80-91; deleg comt IX, Int Inst Welding, 80-91; acad dir, Postgrad Welding Eng Course, CNEA-VBA-IAS, 81-83; head, Welding Div, Lemit-La Plata, 88-91; prin prof, Univ Belgrano, Buenos Aires, Arg, 88-91. *Mem:* Int Inst Welding; Am Welding Soc. *Mailing Add:* Fla 274 3 of 31 Buenos Aires 1005 Argentina

SOLARO, R JOHN, PROTEIN PHOSPHORYLATION & REGULATION. *Educ:* Univ Pittsburgh Sch Med, PhD(physiol), 71. *Prof Exp:* PROF PHYSIOL, UNIV CINCINNATI SCH MED, 81- *Res:* Calcium binding proteins. *Mailing Add:* Dept Physiol M/C 901 Box 6998 Univ Ill Col Med Chicago IL 60680

SOLARZ, RICHARD WILLIAM, b Minneapolis, Minn, Dec 12, 47; m 68; c 1. CHEMICAL PHYSICS. *Educ:* Mass Inst Technol, SB, 69; Univ Chicago, PhD(chem physics), 74. *Prof Exp:* PHYSICIST LASER PHYSICS, LAWRENCE LIVERMORE LAB, 74- *Res:* Laser physics and photochemistry; spectroscopy of excited atoms and molecules; laser isotope separation. *Mailing Add:* Lawrence Livermore Lab Box 808-MS L-495 Livermore CA 94550

SOLBERG, JAMES J, b Toledo, Ohio, May 27, 42; m 66. INDUSTRIAL ENGINEERING, OPERATIONS RESEARCH. *Educ:* Harvard Col, BA, 64; Univ Mich, MA & MS, 67, PhD(indust eng), 69. *Prof Exp:* Asst prof indust eng, Univ Toledo, 68-72; assoc prof, 72-81, PROF INDUST ENG, PURDUE UNIV, 81- *Mem:* AAAS; Opers Res Soc Am; Inst Mgt Sci; Am Inst Indust Eng; Soc Mgf Eng. *Res:* Graph theory; queueing theory; scheduling; probability; computer aided manufacturing. *Mailing Add:* Sch Indust Eng Purdue Univ West Lafayette IN 47907

SOLBERG, MYRON, b Boston, Mass, June 11, 31; m 56; c 3. FOOD SCIENCE, FOOD MICROBIOLOGY. *Educ:* Univ Mass, BS, 52; Mass Inst Technol, PhD(food technol), 60. *Prof Exp:* Res asst food technol, Mass Inst Technol, 54-60; qual control mgr, Colonial Provision Co, 60-64; from asst prof to assoc prof food sci, 64-70, PROF FOOD SCI, RUTGERS UNIV, NEW BRUNSWICK, 70-, DIR, CTR ADVAN FOOD TECHNOL, 85- *Concurrent Pos:* Lectr, Meat Sci Inst, 65-72; vis prof food eng & biotechnol, Israel Inst Technol, 73-74; co-ed, J Food Safety, 77-87. *Honors & Awards:* Distinguished Food Scientist, NY, Inst Food Technologists, 81, Nicholal Appert Medalist, 89. *Mem:* Fel AAAS; fel Inst Food Technologists; Am Soc Qual Control; Am Soc Microbiol; Am Meat Sci Asn; fel Am Chem Soc. *Res:* Mode of action of microbial inhibition; microbial evaluation of protein quality; assurance of microbiological safety in mass-feeding; regulation of toxinogenesis in clostridium perfringens. *Mailing Add:* Ctr for Advan Food Technol Rutgers Univ New Brunswick NJ 08903

SOLBERG, RICHARD ALLEN, b Decorah, Iowa, Sept 21, 32; m 54; c 4. PLANT PATHOLOGY. *Educ:* Univ Mont, BA, 54; Wash State Univ, MS, 56; Univ Calif, Los Angeles, PhD(bot), 61. *Prof Exp:* Asst air pollution, Inst Technol, Wash State Univ, 54-56; res botanist, Univ Calif, Los Angeles, 56-57, sr lab technician plant path, 57-61; assoc prof, 61-70, dir biol sta, 61-67, assoc dean, 67-69, dean col arts & sci, 69-82, PROF BOT, UNIV MONT, 70-, ASSOC ACAD VPRES, 82- *Mem:* Bot Soc Am; Soc Develop Biol; Am Soc Limnol & Oceanog. *Res:* Plant anatomy and cytology; pathological cytology of virus infection; developmental anatomy; fine structure of cells; tissue differentiation and culture; phytotoxicants and virus cytology and their interrelationships. *Mailing Add:* Box 187 White Fish MT 59937

SOLBERG, RUELL FLOYD, JR, b Norse, Tex, July 27, 39; m 59; c 2. ELECTROMAGNETICS, RESEARCH & DEVELOPMENT. *Educ:* Univ Tex, Austin, BS, 62, MS, 67; Trinity Univ, MBA, 77. *Prof Exp:* Res engr underwater acoustics div, Appl Res Labs, Univ Tex, 62-65, asst supvr mech eng sect, 65-67; res engr, 67-70, sr res engr, 70-87, PRIN ENGR, DEPT ADVAN SYSTS ENG, SOUTHWEST RES INST, 87- *Concurrent Pos:* Tech asst, Appl Mech Rev, 80-83. *Honors & Awards:* Centennial Medallion, Am Soc Mech Engrs, 80, Bd gov Cert, 82, 83, 84, 85, 87 & 89; Clifford H Shumaker Award, 90. *Mem:* Sigma Xi; Am Soc Mech Engrs; Nat Soc Prof Engrs; Soc Allied Weight Engrs; Human Factors Soc; Instrument Soc Am. *Res:* Structural optimization; mechanical design; response of structures to periodic and impulsive loading; behavioral science; flexible automation; material fatigue; environmental effects; zero-gravity in space flight; oceanography; management science; human factors; corrosion; acoustics; mass measurement in microgravity. *Mailing Add:* Southwest Res Inst PO Drawer 28510 San Antonio TX 78228-0510

SOLBRIG, OTTO THOMAS, b Buenos Aires, Arg, Dec 21, 30; US citizen; m 56; c 2. PLANT ECOLOGY. *Educ:* Univ Calif, Berkeley, PhD(bot), 59. *Hon Degrees:* MA, Harvard Univ, 69. *Prof Exp:* Botanist, Harvard Univ, 59-61, from asst cur to assoc cur, 61-66; from assoc prof to prof, Univ Mich, Ann Arbor & biosystematist, Bot Gardens, 66-69; prof biol, 69-83, dir, Gray Herbarium & supvr, Bussey Inst, 78-83, Paul C Mangelsdorf prof natural sci, 83-87, BUSSEY PROF BIOL, HARVARD UNIV, 87- *Concurrent Pos:* Hon travel fel, Univ Calif, Berkeley, 59-60; NSF & Am Acad Arts & Sci grants, 59-; lectr, Harvard Univ, 64-66; secy gen, Int Orgn Plant Biosysts, 64-69; mem, Int Orgn Biosysts & Orgn Trop Studies; dir, Struct Ecosystems Prog, US/IBP, 70-75; mem, IUBS Comt, Nat Acad Sci, 75-80; dir, Decade of the tropics prog. *Mem:* Int Union Biol Sci (pres, 85-); Genetics Soc Am; Soc Study Evolution (secy, 73-78, pres, 81-82); Sigma Xi (secy-treas, 76-82); fel Am Acad Arts & Sci; fel AAAS; Latin Am Study Soc; Ecol Soc Am; Brit Ecol Asn. *Res:* Cytotaxonomical and cytogenetical studies of plant species; chemical and physiological studies of natural plant population; evolution of plants; plant population biology; resources and humans in Latin America; savana ecosystem. *Mailing Add:* Gray Herbarium Harvard Univ 22 Divinity Ave Cambridge MA 02138

SOLC, KAREL, b Nachod, Czech, July 25, 33; m 57; c 2. PHYSICAL CHEMISTRY. *Educ:* Inst Chem Technol, Prague, Czech, MSc, 56; Czech Acad Sci, PhD(macromolecular chem), 61. *Prof Exp:* From scientist to sr scientist, Inst Macromolecular Chem, Czech Acad Sci, 61-68; res instr chem, Dartmouth Col, 71; res scientist, 71-74, SR RES SCIENTIST, MICH MOLECULAR INST, 74-, PROF POLYMER CHEM, 84- *Concurrent Pos:* NSF vis fel, Dartmouth Col, 68-70; Mich Found Advan Res vis fel, 70-71; fel, Japan Soc Promotion Sci Res, 85. *Mem:* AAAS; Am Chem Soc; Am Phys Soc; Sigma Xi. *Res:* Physical chemistry of polymers; statistical mechanics and thermodynamics; chain statistics; chemical kinetics. *Mailing Add:* 4310 James Dr Midland MI 48640-3707

SOLDANO, BENNY A, b Utica, NY, Nov 17, 21; m 46; c 2. PHYSICAL CHEMISTRY. *Educ:* Alfred Univ, BS, 43; Univ Wis, PhD(phys chem), 49. *Prof Exp:* From chemist to sr chemist, Oak Ridge Nat Lab, Tenn, 49-71; prof chem, 71-77, PROF PHYSICS, FURMAN UNIV, 71- *Mem:* Am Chem Soc. *Res:* Ion exchange; thermodynamics; kinetics; solution chemistry. *Mailing Add:* 114 W Pasadana Lane Oakridge TN 37830-6301

SOLDAT, JOSEPH KENNETH, b Chicago, Ill, May 4, 26; m 52; c 3. HEALTH PHYSICS, RADIOLOGICAL PHYSICS. *Educ:* Univ Colo, BS, 48; Am Bd Health Physics, cert, 61. *Prof Exp:* Indust hyg engr, Med Ctr, Univ Colo, 48; from technician to sr engr, Gen Elec Co, Wash, 48-65; sr res scientist, 65-73, res assoc, 73-76, STAFF ENVIRON SCIENTIST, PAC NORTHWEST LABS, BATTELLE MEM INST, 77- *Concurrent Pos:* Bd dirs, Environ Radiation Sect, Health Physics Soc, 90-93. *Mem:* AAAS; Am Chem Soc; fel Health Physics Soc. *Res:* Human doses from environmental radiation sources; movement of radionuclides through the biosphere to man; radioactive waste management; surveillance of waste effluents and the environs for radioactive and nonradioactive materials. *Mailing Add:* Pac Northwest Labs Battelle Mem Inst PO Box 999 Richland WA 99352

SOLDATI, GIANLUIGI, b Bologna, Italy, Feb 17, 37; m 86; c 2. ORGANIC CHEMISTRY. *Educ:* Univ Bologna, DSc(org chem), 61, Columbia Univ, cosmetic Technol, dipl, 76-78. *Prof Exp:* Petrol Res Fund fel, Univ Mass, 62; lectr & res assoc org & anal chem, Univ Bologna, 63-64, phys & org chem, 64-65; sr res chemist, Agr Chem, Uniroyal, Inc, Conn, 65-70; sr synthetic chemist, 70-75, sr res chemist, 75-78, GROUP LEADER, CARTER PROD RES DIV, CARTER-WALLACE, INC, 78- *Mem:* Soc Cosmetic Chemists; Am Chem Soc. *Res:* antihypertensive agents; inorganic antiperspirans; depilatory agents; polymers; synthesis of antiperspirant salts and complexes; synthesis and study of new anticalculus and anticaries materials; anticholinergics and antihypertensive agents; patent writing and liaison with Legal Department; surfactants; emulsion technology; product development cosmetics and toiletries; antiperspirant product development. *Mailing Add:* Carter Prod Res-POB 1 Div Carter-Wallace Inc Cranbury NJ 08512

SOLDO, ANTHONY THOMAS, b New York, NY, Sept 11, 27; m 51; c 3. BIOCHEMISTRY, NUTRITION. *Educ:* Brooklyn Col, BS, 50, MA, 53; Ind Univ, PhD(biochem), 60. *Prof Exp:* Biochemist, Schering Corp, NJ, 59-62; res assoc, Inst Muscle Dis Inc, NY, 62-64; asst prof, 65-72, assoc prof, 72-81, PROF BIOCHEM, SCH MED, UNIV MIAMI, 81-; RES CHEMIST, VET ADMIN HOSP, 65- *Mem:* AAAS; Soc Protozool; NY Acad Sci; Am Inst Nutrit; Sigma Xi. *Res:* Nutrition and nucleic acid metabolism of Protozoa; biochemistry of endosymbiotes. *Mailing Add:* Res Div 151 Vet Admin Med Ctr 1201 NW 16th St Miami FL 33125

SOLDO, BETH JEAN, b Binghamton, NY, Sept 30, 48; m 75. DEMOGRAPHY, GERONTOLOGY. *Educ:* Fordham Univ, BA, 70; Duke Univ, MA, 74, PhD(demog), 77. *Prof Exp:* Asst dir, Ctr Demog Studies, Duke Univ, 74-77; sr res scholar, 77-81, asst prof, 81-85, ASSOC PROF, 85-, CHAIR DEPT DEMOG & DIR, CTR POP RES, GEORGETOWN UNIV, 86- *Concurrent Pos:* Sr res scholar, Ctr Pop Res, Georgetown Univ, 77-, sr res fel, Kennedy Inst Ethics, 81-; consult, US Senate Comt Aging, Health Care & Financing Admin, Nat Inst Aging. *Mem:* Geront Soc Am; Pop Asn Am; Am Pub Health Asn; Am Sociol Soc. *Res:* Implications of a changing age structure and disease profile for the organization, structure, and financing of health care facilities, particularly long-term care services. *Mailing Add:* Dept Demog Georgetown Univ 233 Poulton Hall Washington DC 20057

SOLE, MICHAEL JOSEPH, b Timmins, Ont, Mar 5, 40; m 64; c 2. NEUROCHEMISTRY, MOLECULAR BIOLOGY. *Educ:* Univ Toronto, BSc, 62, MD, 66; FRCP(C), 74. *Prof Exp:* Intern, Toronto Gen Hosp, 66-67, jr resident, 67-68, sr resident, 68-69, staff cardiologist, 74-89; cardiol fel, Cardiovascular Res Inst, 69-71 & Peter Bent Brigham Hosp, 71-74; res assoc nutrit, Mass Inst Technol, 73-74; from asst prof to assoc prof med, Univ Toronto, 74-83, prof med & physiol & dir cardiol res, 83-89, HEART & STROKE FOUND ONT DISTINGUISHED RES PROF & DIR, CTR CARDIOVASC RES, UNIV TORONTO, 89-; DIR CARDIOL, TORONTO HOSP, 89- *Concurrent Pos:* Fels, Ont Heart Found, 73-80, res assoc, 80-, Med Review Comt, 78-; Sci Review Comt, Can Heart Found, 76-; fels, Coun Clin Cardiol & Am Col Cardiol, 76-; Hon secy-treas, Banting Res Found, 78-81; staff, Inst Med Sci, Univ Toronto, 77-; vchmn, Can Heart Found, 80-; mem. Rev Comt, Gairdner Found, 80-; exec comt, Health Res Develop Coun, Ont, 84-86, exec, Basic Sci Coun Am Heart Asn, 87-, Dir, Heart & Stroke Found Ont, 86- *Honors & Awards:* Res Award, Can Cardiovascular Soc, 75; William Goldie Prize, 80; Res Achievement Award, Can Cardiovasc Soc, 89. *Mem:* Am Heart Asn; Can Cardiovascular Soc; Am Fedn Clin Res; Int Soc Heart Res; Am Soc Clin Invest; Asn Am Physicians. *Res:* Central and peripheral neurotransmitter metabolism in cardiovascular disease; molecular biology of myocardial hypertrophy and failure. *Mailing Add:* Toronto Hosp EN 13-208 200 Elizabeth St Toronto ON M5G 2C4 Can

SOLECKI, ROMAN, b Lwow, Poland, Apr 6, 25; US citizen; m 48. SOLID MECHANICS. *Educ:* Warsaw Polytech Inst, BS, 50, PhD(appl mech), 56; Inst Fund Technol Res, Warsaw, DSc, 60. *Prof Exp:* Asst prof civil eng, Warsaw Polytech Inst, 50-56, adj prof, 56-60; assoc prof continuum mech, Inst Fund Technol Res, Warsaw, 60-68; PROF MECH ENG, UNIV CONN, 68- *Concurrent Pos:* Royal Norwegian Coun Sci Res fels, 62- & 64-; NSF sr foreign sci fel, Univ Conn, 68- *Mem:* Am Soc Mech Engrs; Sigma Xi; Am Acad Mech; Soc Eng Sci. *Res:* Wear; fracture mechanics; theory of elasticity. *Mailing Add:* Dept of Mech Eng PO Box U-139 Storrs CT 06268

SOLED, STUART, b New York, NY, May 11, 48; m; c 1. SOLID STATE CHEMISTRY. *Educ:* City Col New York, BS, 69; Brown Univ, PhD(chem), 73. *Prof Exp:* Res assoc chem res, Brown Univ, 73-77; res chemist, Allied Chem Co, 77-80; Staff Exxon Res & Engr Co, 80-87; DIR MAT TECH, SUNSTONE INC, 87- *Concurrent Pos:* Res assoc, Lab Inorg Chem, Univ Paris, 74-75. *Mem:* Am Chem Soc; Am Crystallog Asn. *Res:* Preparation, structure and properties of materials in solid state chemistry. *Mailing Add:* Cooks Cross Rd Pittstown NJ 08867

SOLEM, G ALAN, invertebrate zoology; deceased, see previous edition for last biography

SOLEM, JOHNDALE CHRISTIAN, b Chicago, Ill, Nov 8, 41; m 65. LASERS, NUCLEAR EXPLOSIVE PHYSICS. *Educ:* Yale Univ, BS, 63, MS, 65, PhD(physics), 68, MPhil, 67. *Prof Exp:* Group leader, Thermonuclear Weapons Physics Group, Los Alamos Nat Lab, 73-76, Neutron Physics Group, 77-79, High Power Density Group, 78, alt div leader, Physics Div, 78-80, ASSOC DIV LEADER, THEORET DIV, LOS ALAMOS NAT LAB, 80- *Concurrent Pos:* Mem, US Air Force Sci Adv Bd, 72-77 & Munitions & Armament Panel, 73-77. *Mem:* Am Phys Soc; AAAS; Am Nuclear Soc. *Res:* X-ray and gamma-ray lasers; laser-driven shockwaves; nuclear physics; transport theory; plasma physics; nuclear explosive physics. *Mailing Add:* Los Alamos Nat Lab Theoret Div MS-B210 Box 1663 Los Alamos NM 87545

SOLENBERGER, JOHN CARL, b San Diego, Calif, Apr 2, 41; m 71. INDUSTRIAL CHEMISTRY. *Educ:* Univ NMex, BS, 63; Wash Univ, PhD(chem), 69. *Prof Exp:* Sr res chemist, plastics dept, E I du Pont de Nemours & Co, Inc, 69-80. *Mem:* Am Chem Soc; Sigma Xi. *Res:* Development of membranes for use as separators on chlor-alkali cells, polymeric coatings and binders, and general industrial process research. *Mailing Add:* Five Wood Rd Wilmington DE 19806-2021

SOLER, ALAN I(SRAEL), b Philadelphia, Pa, Dec 9, 36; m 60; c 2. ENGINEERING MECHANICS. *Educ:* Univ Pa, BS, 58, PhD(mech eng), 62; Calif Inst Technol, MS, 59. *Prof Exp:* Res scientist, Dyna Struct, Inc, Pa, 61-64; mem tech staff, Ingersoll Rand Res Ctr, NJ, 64-65; from asst prof to assoc prof mech eng, 65-77, PROF MECH ENG, UNIV PA, 77- *Concurrent Pos:* Adj prof, Drexel Inst Technol, 64-65; consult, Ingersoll Rand Res Ctr, 65- *Mem:* Am Inst Aeronaut & Astronaut. *Res:* Viscoelasticity and thermoviscoelasticity; engineering theory of thick walled shells; buckling of deep beams; cable dynamics; transportation dynamics. *Mailing Add:* Dept of Mech Eng Univ of Pa Philadelphia PA 19104

SOLEZ, KIM, b Washington, DC, June 20, 46; m 68; c 2. PATHOLOGY, RENAL MEDICINE. *Educ:* Oberlin Col, BA, 68; Univ Rochester, MD, 72. *Prof Exp:* Nat Kidney Found fel, 76-77; asst prof, 77-83, ASSOC PROF PATH & MED, JOHNS HOPKINS UNIV, 83-, PATHOLOGIST, JOHNS HOPKINS HOSP, 77- *Concurrent Pos:* NIH res career develop award. *Honors & Awards:* Res Career Develop Award, NIH. *Mem:* Am Soc Nephrology; Int Soc Nephrology. *Res:* Acute renal failure; renal circulation; renal transplantation; glomerul-onephritis; atherosclerosis. *Mailing Add:* Dept of Path Johns Hopkins Hosp Baltimore MD 21205

SOLI, GIORGIO, b Rome, Italy, Feb 3, 20; nat US; m 70; c 3. MICROBIOLOGY. *Educ:* Univ Rome, DSc(microbiol), 47. *Prof Exp:* Res asst microbiol, Med Sch, Univ Calif, Los Angeles, 49-50; res asst petrol explor & develop, Gen Petrol Corp, Calif, 51-53; microbiologist & consult, Soli Microbiol Labs, 53-59; microbiologist, US Naval Ord Test Sta, China Lake, 59-62, res microbiologist, Res Dept, 62-68; staff scientist, Naval Undersea Ctr, Hawaii, 68-70; res microbiologist, Michelson Labs, China Lake, 70-73; SCI CONSULT, 73 - *Concurrent Pos:* Guest scientist, Oceanog Mus, Monaco, 66-67; resident scientist, Oceanic Inst, Hawaii, 68-70. *Mem:* AAAS; Am Soc Microbiol; Am Soc Limnol & Oceanog; Sigma Xi; NY Acad Sci. *Res:* Consulting in microbiology applied to diversified problems in marine pollution, petroleum technology and agriculture. *Mailing Add:* PO Box 1679 Solvang CA 93464-1679

SOLIE, LELAND PETER, b Barron, Wis, July 19, 41; m 67; c 3. ACOUSTICS. *Educ:* Stanford Univ, BS, 64, MS, 67, PhD(appl physics), 71. *Prof Exp:* Res asst microwave acoust, Hansen Lab, Stanford Univ, 65-70; vis prof & res assoc, Norwegian Tech Inst, 71-72; mem tech staff microwareacoust, Sperry Res Ctr, 73-87; SR MEM TECH STAFF, ELECTRONIC DECISION INC, 87- *Mem:* Sigma Xi; Inst Elec & Electronics Engrs. *Res:* Signal processing with surface acoustic wave devices; particular emphasis on band pass filters, convolvers, surface wave amplifiers and wave propagation in layered media. *Mailing Add:* Electronic Decision 1776 E Washington St Urbana IL 61801

SOLIE, THOMAS NORMAN, b Spring Grove, Minn, Sept 16, 31; m 59; c 2. PHYSICAL CHEMISTRY, BIOPHYSICS. *Educ:* Univ Minn, Minneapolis, BA, 59; Univ Ore, MA, 63, PhD(chem), 65. *Prof Exp:* Instr biophys, Med Ctr, Univ Colo, 65; USPHS fel, 65-66; res assoc molecular biol, Vanderbilt Univ, 66; asst prof chem, Luther Col, Iowa, 66-67; asst prof, 67-74, ASSOC PROF BIOPHYS & CHEM, COLO STATE UNIV, 74- *Mem:* AAAS; Am Chem Soc; Biophys Soc; Sigma Xi; Inst Elec & Electronic Engrs. *Res:* Biophysical chemistry; instrumentation design; biological signal processing and spectral analysis. *Mailing Add:* Dept Physiol & Biophys Colo State Univ Ft Collins CO 80523

SOLIMAN, AFIFI HASSAN, b Cairo, Egypt, Feb 2, 31; m 62; c 5. TRANSPORTATION, PHOTOGRAMMETRY. *Educ:* Ain-Shams Univ, Cairo, BSc, 58; Ohio State Univ, MSc, 62, PhD(geod sci), 68. *Prof Exp:* Design engr, Suez Canal Authority, Egypt, 58-60; res asst geod sci, Res Ctr Found, Ohio State Univ, 62-64; chief engr, Deleuw Cather & Brill Eng Co, Columbus, Ohio, 64-66; asst prof civil eng, McMaster Univ, 66-69; from asst prof to assoc prof civil eng, 69-75, exec secy, Ctr Transp Studies, 70-73, PROF CIVIL ENG & DIR CE COOP STUDIES PROG, UNIV MAN, 87- *Concurrent Pos:* Nat Res Coun Can & Can Transp Comn grants, Ctr Transp Studies, Univ Man; mem, Hwy Res Bd, Nat Acad Sci-Nat Res Coun; examr, Asn Man Land Surveyors, 73; assoc prof engrs, Prov of Man, 83-; vis prof civil eng, Univ Tex, 80-81; prof civil eng, Concordia Univ, Montreal, 83-85, adj prof,__85-87; consult, Can Aero Serv Ltd, Can Transp Comn & Transp Develop Ctr, Dept Transp, Air Serv Constr, Eng & Arch Br; chmn Tech Adv Comt, Energy Conserv, Man Dept Indust, 79-80; chmn Transp Div, Can Soc Civil Eng, 86-; coord & Tech chmn, annual CSCE Transp Prog, 87-, mem steering comt, N Am Conf on Microcomputers Transp, 87-, Can Conf Eng Educ, 87, chmn, Tech prog, 88. *Mem:* Am Soc Photogram; Eng Inst Can; Can Roads & Transp Asn; Am Cong Surv & Mapping; Inst Transp Engrs; Can Soc Civil Eng; Am Soc Eng Educ. *Res:* Transportation planning, especially public transportation in urban areas; transportation growth and demand; forecasting, methods and techniques for determining potential demand for highways; transportation energy conservation; systems approach to transportation problems; freight transport. *Mailing Add:* Dept Civil Eng Univ Man Winnipeg MB R3T 2N2 Can

SOLIMAN, KARAM FARAG ATTIA, b Cairo, Egypt, Oct 15, 44; US citizen; m 73; c 4. NEUROENDOCRINOLOGY, ENDOCRINOLOGY. *Educ:* Cairo Univ, BS, 64; Univ Ga, MS, 71, PhD(endocrinol), 72. *Prof Exp:* Res asst physiol, Univ Ga, 68-72; asst prof, Sch Vet Med, Tuskegee Inst, 72-75, assoc prof, 75-79; PROF PHYSIOL, SCH PHARM, FLA A&M UNIV, 79-, CHMN DIV BASIC PHARMACEUT SCI, 81- *Concurrent Pos:* Prin investr grants, NASA, 76- & NIH, 76-81 & 83-88. *Mem:* Am Physiol Soc; Endocrine Soc; Neurosci Soc; Chronobiol Soc; Am Soc Pharmacol Exp Therap; Soc Exp Biol Med. *Res:* Investigate the role of peripheral nervous system in the regulation of the endocrine gland function; elucidate the physiology and the pharmacology of the role of the autonomic nervous system in the regulation of adrenal cortex function. *Mailing Add:* Col Pharm Fla A&M Univ Tallahassee FL 32307

SOLIMAN, MAGDI R I, b Alexandria, Egypt, June 30, 42; m 67; c 2. MOLECULAR PHARMACOLOGY, NEURO- & CHRONOPHARMACOLOGY. *Educ:* Alexandria Univ, Egypt, BSc, 64, MS, 68; Univ Ga, PhD(pharmacol), 72. *Prof Exp:* Instr pharmacol, Fac Pharm, Alexandria Univ, Egypt, 64-69; teaching asst, Sch Pharm, Univ Ga, 69-72; res assoc biomed pharmacol, Penn State Col Med, 72-74; from asst prof to assoc prof, Fac Pharm, Alexandria Univ, 74-80; vis scientist, 80-82, from asst prof to assoc prof, 82-85, PROF PHARMACOL, COL PHARM, FLA A&M UNIV, 85-, DIR, CHRONOPHARMACOL RES LAB, 82- *Concurrent Pos:* Res grants, NIH & NASA. *Mem:* Sigma Xi; Egyptian Pharmaceut Soc; Egyptian Pharmacol Soc; Am Soc Pharmacol & Exp Therapeut. *Res:* Chronobiotic drugs to alleviate the deleterious symptomatology associated with internal rhythm desynchronization; elucidation of the neurochemical basis of motion and space sickness; toxicology. *Mailing Add:* Col Pharm Fla A&M Univ Tallahassee FL 32307

SOLIN, STUART ALLAN, b Baltimore, Md, Sept 9, 42; m 64; c 3. SOLID STATE PHYSICS. *Educ:* Mass Inst Technol, BS, 63; Purdue Univ, MS, 66, PhD(physics), 69. *Prof Exp:* Asst physics, Purdue Univ, 64-69; from asst prof to assoc prof physics, Univ Chicago, 69-80; MEM FAC, DEPT PHYSICS, MICH STATE UNIV, 80-, DIR, CTR FUNDAMENTAL MAT RES, 86- *Mem:* Am Phys Soc. *Res:* Solid state physics; laser Raman spectroscopy of solids; fundamental properties of lasers; superconductivity. *Mailing Add:* NEC Res Inst Inc Four Independence Way Princeton NJ 08540

SOLINGER, ALAN M, b Nov 27, 48; m 83; c 2. IMMUNOLOGY, RHEUMATOLOGY. *Educ:* Columbia Univ, BA, 70; Univ Cincinnati, MD, 74. *Prof Exp:* From asst prof to assoc prof med, Col Med, Univ Cincinnati, 81-91; ASST DIR, ANTI-INFLAMMATORIES, DRUG DEVELOP DEPT, CIBA-GEIGY CORP, 91- *Mem:* AAAS; NY Acad Sci; Am Col Rheumatology; Am Col Physicians; Am Fedn Clin Res; Asn Clin Pathologists. *Res:* Clinical immunology; rheumatology; cellular immunology and its part in chronic synovitas. *Mailing Add:* Phamaceut Div Dev 4094 Ciba-Geigy Corp 556 Morris Ave Summit NJ 07901

SOLIS-GAFFAR, MARIA CORAZON, b Tacloban City, Philippines, Dec 1, 39, US citizen; m 70; c 1. BIOCHEMISTRY, MEDICINAL CHEMISTRY. *Educ:* Univ Santo Tomas, Manila, BS, 61; Mass Col Pharm, MS, 64, PhD(pharmaceut/biol), 67; Fairleigh Dickinson Univ, MBA, 84. *Prof Exp:* Qual control chemist, Inhelder Labs, 61; instr chem, Emmanuel Col, Mass, 64-67; sr res chemist, Colgate-Palmolive Co, 67-79; sr scientist, Johnson & Johnson Prod, Inc, 79-84, prin scientist, 84-86, group leader, Res Div, 86-88; Mgr, Johnson & Johnson Health Care Co, 89; ASST DIR, JANSSEN PHARMACEUTICA, 89- *Mem:* Am Chem Soc; Am Pharmaceut Asn; Am Asn Pharmaceut Scientists. *Res:* Structure-activity relationships; enzyme inhibition studies; pathological calcification; mouth odor studies; human clinical studies for evaluation of oral products in mouth odor reduction; biochemistry of sulfur metabolism in oral cavity; oral products and pharmaceutical product development and formulations; clinical research; dermatology; infectious diseases. *Mailing Add:* Janssen Pharmaceutica 40 Kingsbridge Rd Piscataway NJ 08855-3998

SOLISH, GEORGE IRVING, b Providence, RI, Jan 7, 20; m 46; c 3. OBSTETRICS & GYNECOLOGY, HUMAN GENETICS. *Educ:* Providence Col, BS, 41; Tufts Univ, 48, MD, 50; Univ Mich, MS, 61, PhD(human genetics), 68. *Prof Exp:* From instr to asst prof, 57-68, assoc prof, 68-79, PROF OBSTET & GYNEC, STATE UNIV NY DOWNSTATE MED CTR, 79-; DIR MED GENETICS SERVS, MAIMONIDES MED CTR, 86- *Concurrent Pos:* Res asst, Med Sch, Univ Mich, 60-63; consult, Margaret Sanger Res Bur, 64- *Mem:* Am Col Obstet & Gynec; Am Fertil Soc; Am Soc Human Genetics; Sigma Xi. *Res:* Population genetics; prezygotic selection and control of fertility; infertility; reproduction; prenatal genetic diagnosis. *Mailing Add:* Dept Obstet & Gynec Maimonides Med Ctr 4802 Tenth Ave Brooklyn NY 11219

SOLL, ANDREW H, b Mar 20, 45. CELL BIOLOGY, PHYSIOLOGY. *Educ:* Harvard Univ, MD, 70. *Prof Exp:* PROF CLIN & PRECLIN MED, MED SCH, UNIV CALIF, LOS ANGELES, 75- *Res:* Cellular mechanisms underlying regulation of secretion and growth in gastric mucosa. *Mailing Add:* Dept Med Wadsworth Vet Admin Hosp Bldg 115 Rm 203 Los Angeles CA 90073

SOLL, DAVID RICHARD, b Philadelphia, Pa, Apr 29, 42; c 1. DEVELOPMENTAL BIOLOGY. *Educ:* Univ Wis, BA, 64, MA, 68, PhD(zool), 70. *Prof Exp:* Fel develop biol, Univ Wis, 69-70, Brandeis Univ, 71-72; asst prof, 72-77, ASSOC PROF ZOOL, UNIV IOWA, 77- *Concurrent Pos:* Res grants, NSF, 74 & 76 & NIH, 78, 79 & 81; mem, Cell Biol Study Sect, NIH, 78-83. *Mem:* Soc Develop Biol; AAAS. *Res:* An analysis of the molecular mechanisms controlling cell differentiation and muticellular morphogenesis. *Mailing Add:* Dept Biol Univ Iowa Iowa City IA 52242

SLL, DIETER GERHARD, b Stuttgart, Ger, Apr 19, 35; US citizen; m 64; c 3. MOLECULAR BIOLOGY. *Educ:* Stuttgart Tech Univ, MSc, 60, PhD(chem), 62. *Prof Exp:* Fel, Inst Enzyme Res, Univ Wis, Madison, 62-65, from asst prof to assoc prof, 65-76, PROF MOLECULAR BIOPHYS, YALE UNIV, 76- *Concurrent Pos:* Guggenheim Found fel, 72 & 89. *Honors & Awards:* Humboldt Prize, 88. *Mem:* Am Soc Microbiol; fel AAAS; Am Soc Biol Chem; Am Chem Soc. *Res:* Regulation of gene expression; plant molecular biology. *Mailing Add:* Dept Molecular Biophys & Biochem Yale Univ PO Box 6666 New Haven CT 06511

SOLLA, SARA A, b Buenos Aires, Arg, June 30, 50; m 74. STATISTICAL MECHANICS, COMPUTATIONAL NEUROSCIENCE. *Educ:* Univ Buenos Aires, Arg, BS, 74; Univ Wash, PhD(physics), 82. *Prof Exp:* Lectr thermodyn, Nat Univ Technol, Buenos Aires, Arg, 74-76; teaching asst, Univ Wash, 76-77, res asst, 77-82; postdoctoral assoc, Lab Atomic & Solid State Physics, Cornell Univ, 82-84, lectr, 84; postdoctoral assoc, IBM Watson Res Ctr, 84-86; MEM TECH STAFF, AT&T BELL LABS, 86- *Concurrent Pos:* Vis scientist, Boston Univ, 85; vis prof, ENS & ENSEA, Paris, France, 86, Ctr Telecommun Res, Columbia Univ, 87-88 & Nordic Inst theoret Atomic Physics, Copenhagen, 90. *Mem:* Am Phys Soc; Int Neural Network Soc; NY Acad Sci; AAAS. *Res:* Neural networks; statistical models to describe learning and adaptation in computational systems which are biologically inspired and try to mimic the parallel computation used by the brain for tasks such as associative memory and pattern recognition; author of various publications. *Mailing Add:* AT&T Bell Labs Rm 4G-336 Crawford Corners Rd Holmdel NJ 07733-3030

SOLLBERGER, ARNE RUDOLPH, b Dresden, Ger, Mar 17, 24; m 54; c 2. BIOMETRY. *Educ:* Caroline Inst, Stockholm, Sweden, MB, 49, MD, 57. *Prof Exp:* From asst anat to assoc prof, Caroline Inst, Stockholm, Sweden, 48-62; prof pharmacol, Univ PR, 62-64; assoc med, Case Western Reserve Univ, 64-65; chief biomet, Eastern Res Supply Ctr, Vet Admin Hosp, West Haven, Conn, 65-67; assoc prof psychiat, Med Sch, Yale Univ, 68-72; prof, 72-88, VIS PROF PHYSIOL & INFO PROCESSING, MED SCH, SOUTHERN ILL UNIV, 88- *Concurrent Pos:* Lectr anat & physiol, two nursing schs & Sch Indust Art, Stockholm, 57-62; asst ward physician, Hosp Swedish Diabetes Found, Stockholm, 54-62; lectr biometrics, Yale Univ, 66-72; ed-in-chief, J Interdiscplinary Cycle Res, 78-89; chmn biol rhythms study group, Int Soc Biometeorol, 70-88; bd mem, Found Study Cycles, 75, pres, 76- *Honors & Awards:* Award, Biometeorol Res Found, 75. *Mem:* Hon mem Int Soc Chronobiol (secy, 55-67). *Res:* Cardiology and diabetes; biological rhythms, especially statistical problems; normal values in medicine; biomedical computer processing. *Mailing Add:* Lindegren Hall Southern Ill Univ Med Sch Carbondale IL 62901

SOLLBERGER, DWIGHT ELLSWORTH, biology; deceased, see previous edition for last biography

SOLLER, ARTHUR, b New York, NY, Aug 15, 36; m 65; c 2. INDUSTRIAL MICROBIOLOGY. *Educ:* City Col New York, BS, 57; Brandeis Univ, PhD(biol), 63. *Prof Exp:* NIH fel microbiol, Univ Milan, Italy, 63-64 & Int Lab Genetics & Biophys, Naples, 65-67; for consult genetics, 68-70; asst prof microbiol, Eppley Cancer Inst, Omaha, Nebr, 71 & Univ Ill, Chicago, 72; RES ASSOC, STAUFFER CHEM CO, 73- *Mem:* Am Soc Microbiologists. *Res:* Strain improvement of glutamic acid producing microorganisms via mutation and selection; development of phage resistant strains; investigation of microorganisms producing high yields of amino acids. *Mailing Add:* 1715 Julian Ct El Cerrito CA 94530

SOLLER, ROGER WILLIAM, b Bronxville, NY, Nov 18, 46. PHARMACOLOGY, NEUROPHARMACOLOGY. *Educ:* Colby Col, BA, 68; Cornell Univ, PhD(neurobiol), 73. *Prof Exp:* Fel pharmacol, Sch Med, Univ Pa, 73-75, instr & res assoc, 75-77, asst prof, 77-79; sci assoc pharmacol, 79-81, VPRES DIR SCI AFFAIRS, GLENBROOK LABS, DIV STERLING DRUG INC, 81- *Concurrent Pos:* Pharmaceut Mfrs Asn fel, 74-76; Pa plan scholar, Univ Pa, 76-79. *Mem:* Soc Neurosci; Sigma Xi. *Res:* Mechanisms of action of analgesics, hormones and neurotransmitter substances; clinical pharmacology. *Mailing Add:* Proprietary Assoc 1150 Connecticut Ave NW Washington DC 20036

SOLLERS-RIEDEL, HELEN, b Baltimore, Md, Sept 29, 11; wid. ENTOMOLOGY. *Educ:* Wilson Teachers Col, BS, 34. *Prof Exp:* With insect pest surv, Bur Entomol & Plant Quarantine, 37- 43, jr entomologist, 43-47, asst entomologist, Insects Affecting Man and Animals, 46-53, entomologist, Plant Pest Control Div, Agr Res Serv, USDA, 53-71; med entomologist, NIH grant, 71-85; RETIRED. *Mem:* Am Mosquito Control Asn; Royal Soc Trop Med & Hyg; Am Soc Trop Med & Hyg; Entom Soc Am; Entom Soc Can; NY Acad Sci; hon fel Indian Soc Malaria & Other Commun Dis. *Res:* Mosquito research and analysis. *Mailing Add:* PO Box 19009 Washington DC 20036

SOLLFREY, WILLIAM, b New York, NY, Mar 8, 25; m 49. PHYSICS. *Educ:* NY Univ, BA, 44, MS, 46, PhD(physics), 50. *Prof Exp:* Asst proj engr, Sperry Gyroscope Co, 44-47; res assoc, Math Res Group, NY Univ, 47-51; sr engr, W L Maxson Corp, 51-55; sr res engr, Chicago Midway Labs, 55-57; mgr syst anal, Mech Div, Gen Mills, Inc, 57-61; PHYS SCIENTIST, RAND CORP, 61- *Concurrent Pos:* Instr, Polytech Inst Brooklyn, 53-55. *Mem:* AAAS; Am Phys Soc. *Res:* Advanced analysis in military systems; mathematics; electrical engineering. *Mailing Add:* Rand Corp 1700 Main St Santa Monica CA 90407-2138

SOLLID, JON ERIK, b Denver, Colo, Oct 1, 39; m 65, 80, 87; c 10. OPTICS. *Educ:* Univ Mich, Ann Arbor, BS, 61; NMex State Univ, MS, 65, PhD(physics), 67. *Prof Exp:* Physicist, White Sands Missile Range, 61-62; res assoc plasma physics, Los Alamos Sci Lab, 65-66; sr res scientist, Convair Aerospace Div, Gen Dynamics, 67-72 & Sci Res Lab, Ford Motor Co, 72-74; staff scientist, Los Alamos Nat Lab, 74-81, proj leader, 81-85; vpres & gen mgr, Los Alamos Div, Newport Corp, 85-86; PRES & FOUNDER, SOLLID OPTICS, INC, 86- *Concurrent Pos:* Adj prof, Tex Christian Univ, 70-72; prof, Northern NMex Community Col, 78-80; adj prof, Los Alamos Br, Univ NMex, 80- *Mem:* Am Phys Soc; Am Asn Physics Teachers; Optical Soc Am; Soc Photo-Optical Instrumentation Engrs; Sigma Xi. *Res:* Optical diagnostics; holographic interferometry and coherent optics; applications in experimental mechanics and plasma physics; laser fusion and free electron lasers. *Mailing Add:* 365 Valle Del Sol Los Alamos NM 87544-3563

SOLLITT, CHARLES KEVIN, b Minneapolis, Minn, Aug 8, 43; m 67; c 2. OCEAN WAVES, OCEAN & COASTAL STRUCTURES. *Educ:* Univ Wash, BS, 66, MS, 68; Mass Inst Technol, PhD(civil eng), 72. *Prof Exp:* Res asst, Univ Wash, 66-68 & Mass Inst Technol, 68-72; asst prof, 72-78, ASSOC PROF, ORE STATE UNIV, 78-, DIR, OH HINSDALE WAVE RES LAB, 81- *Mem:* Am Soc Civil Engrs. *Res:* Analytical and experimental work in ocean wave-structure-foundation interaction; breakwater behavior; wave and current measurement systems analysis and interpretation; fate of contaminated sediments deposited at offshore disposal sites. *Mailing Add:* Civil Eng Dept Apperson 206 Ore State Univ Corvallis OR 97331

SOLLMAN, PAUL BENJAMIN, b Ft Branch, Ind, May 2, 20; m 41; c 4. ORGANIC CHEMISTRY. *Educ:* Univ Ind, BS, 47; Univ Minn, PhD(org chem), 51. *Prof Exp:* res chemist, G D Searle Co, 51-80; res chemist, Regis Chem Co, 80-88; RETIRED. *Mem:* Am Chem Soc. *Res:* Organic synthesis; medicinal chemistry. *Mailing Add:* 549 Jupiter Dr Ft Myers FL 33908

SOLLNER-WEBB, BARBARA THEA, b Washington, DC, Dec 21, 48; m 73. MOLECULAR BIOLOGY. *Educ:* Mass Inst Technol, BS, 70; Stanford Univ, PhD(biol), 76. *Prof Exp:* staff fel, Molecular Biol, Nat Inst Arthritis, Metab & Digestive Dis, NIH, 76-77; FEL, DEPT EMBRYOL, CARNEGIE INST, WASHINGTON, 77- *Res:* Structure and function of chromatin; nuclease protease and polymerase action on nucleoprotein and nuclei; ribosomal RNA transcriptional control regions of xenopus laevis; DNA sequencing. *Mailing Add:* Dept Embryol Rm 413 WBSB Johns Hopkins Univ Sch Med 725 N Wolfe St Baltimore MD 21205

SOLLOTT, GILBERT PAUL, b Philadelphia, Pa, July 12, 27; m 54; c 2. ORGANIC CHEMISTRY. *Educ:* Univ Pa, BA, 49; Temple Univ, MA, 56, PhD(org chem), 62. *Prof Exp:* Chemist, R M Hollingshead Corp, 50-52 & Betz Labs, 52-53; org chemist, Pitman-Dunn Lab, Frankford Arsenal, 53-62, chief org chem group, 62-77; res chemist, US Army Armament Res & Develop Ctr, 77-87; SR SCIENTIST, GEO CTR CORP, 87- *Honors & Awards:* Outstanding Achievement Award, US Dept Army, 64 & 82. *Mem:* AAAS; Am Chem Soc; Sigma Xi; NY Acad Sci; Royal Soc Chem. *Res:* Organic, organometallic and organometalloid chemistry including phosphorus, arsenic, boron, silicon and germanium; chemiluminescence research; new synthetic methods; mechanisms; ferrocene chemistry; polymers; nitrocompounds. *Mailing Add:* 618 Gawain Rd Plymouth PA 19462

SOLMAN, VICTOR EDWARD FRICK, b Toronto, Ont, May 24, 16; m 42; c 2. ZOOLOGY, ECOLOGY. *Educ:* Univ Toronto, BA, 38, MA, 39, PhD(biol), 42. *Prof Exp:* Asst zool, Univ Toronto, 36-42; limnologist, Nat Parks Bur, Dept Mines & Resources & Dom Wildlife Serv, 45-49, chief biologist, Dom Wildlife Serv, 49-50 & Can Wildlife Serv, 50-53, asst chief, 53-64, staff specialist, 64-81; RETIRED. *Honors & Awards:* Gold Medal, Prof Inst Pub Serv Can, 77; Kuhring Award, Bird Strike Comn, Europe, 86. *Mem:* Fel AAAS. *Res:* Cladocera of Costello Creek, Algonquin Park, Ontario; ecological relations of waterfowl, especially predatory fish; ecology; wildlife research and management; reduction of bird hazards to aircraft. *Mailing Add:* 614 Denbury Ave Ottawa ON K2A 2P1 Can

SOLMON, DONALD CLYDE, b Fall River, Mass, Mar 28, 45; m 70; c 2. COMPUTED TOMOGRAPHY. *Educ:* Southeastern Mass Tech Inst, BS, 67; Ore State Univ, MS, 73, PhD(math), 74. *Prof Exp:* Vis asst prof math, Univ Ore, 74-75; George William Hill res instr, State Univ NY at Buffalo, 75-77; asst prof, 77-81, ASSOC PROF MATH, ORE STATE UNIV, 81- *Concurrent Pos:* Vis lectr math, Univ des Saarlandes, 81. *Mem:* Am Math Soc; Math Asn Am. *Res:* Applications of analysis and functional analysis to obtain a deeper understanding of problems in medical radiology especially computed tomography. *Mailing Add:* Ore State Univ Corvallis OR 97331

SOLMSSEN, ULRICH VOLCKMAR, organic chemistry, for more information see previous edition

SOLN, JOSIP ZVONIMIR, b Zagreb, Yugoslavia, Mar 31, 34; m 66; c 1. THEORETICAL PHYSICS. *Educ:* Univ Zagreb, BSc, 57, PhD(physics), 60. *Prof Exp:* Res assoc parity violation in mu decay, Rudjer Boskovic Inst, Zagreb, 57-61, researcher particle physics & field theory, 62- 64; fel high energy physics, European Ctr Nuclear Res, Geneva, Switz, 61-62; res assoc broken symmetries, Univ Calif, Los Angeles, 64-65, asst prof in residence, 65-66; asst prof particle physics, Univ Wis, Milwaukee, 66-70; vis asst prof, Univ Ill, Chicago Circle, 70-71; res assoc, Inst Theoret Sci, Univ Ore, 71-72; PHYSICIST, NUCLEAR RADIATION EFFECTS LAB, HARRY DIAMOND LABS, 72- *Honors & Awards:* Scientific Achievement Award, Sigma Xi, 75. *Mem:* Sigma Xi; Am Phys Soc. *Res:* Nonconservation of parity in weak decays; quantum field theory; soluble models; particle production in pion-proton collision; high energy behavior of the scattering amplitude; broken symmetries; solid state devices; Cerenkov and stimulated radiations; free electron lasers; radiation propagation; unified gauge field theories; vacuum structure in Gauge Field Theories; supersymmetry; covariant perturbation theory; radiation shielding. *Mailing Add:* Harry Diamond Labs Nuclear Radiation Effects Lab Adelphi MD 20783

SOLNIT, ALBERT J, b Los Angeles, Calif, Aug 26, 19. PSYCHIATRY, PEDIATRICS. *Educ:* Univ Calif, BA, 40, MS, 42, MD, 43. *Hon Degrees:* MA, Yale Univ, 64. *Prof Exp:* Sterling prof, 66-90, EMER STERLING PROF PEDIAT & PSYCHIAT & SR RES SCIENTIST, CHILD STUDY CTR, YALE UNIV, 90- *Concurrent Pos:* Attend physician pediat & psychiat, Yale-New Haven Hosp, 52-; mem fac & supv analyst, Western New Eng Inst Psychoanal, 62 & NY Psychoanal Inst, 66-; fel, Branford Col, Yale Univ, 67-, chmn, Ctr Study Educ, Inst Social & Policy Studies, 71-73; managing ed, Psychoanal Study Child, 71-; vis prof psychiat & human develop, Ben-Gurion Univ Negev, 73-74; Sigmund Freud mem prof, Univ Col, London, 83-84; Sigmund Freud vis prof, Hebrew Univ, 85-87. *Honors & Awards:* William C Menninger Award, Am Col Physicians, 79; Agnes Purcell McGavin Award, Am Psychiat Asn, 80; Andrew Rackow Mem Lectr, Abington Hosp Ment Health Ctr, 80; Lindemann Distinguished Lectr, NY Hosp, 83; Peter Blos Biennial Lectr, 83; C Anderson Aldrich Award, Am Acad Pediat, 89; Simon Wile Award, Am Acad Child & Adolescent Psychiat, 91. *Mem:* Fel Inst Med-Nat Acad Sci; Inst Asn Child & Adolescent Psychiat (hon pres, 90-). *Res:* Child and adolescent psychiatry. *Mailing Add:* Dept Pediat Child Study Ctr Yale Univ New Haven CT 06510

SOLO, ALAN JERE, b Philadelphia, Pa, Nov 7, 33; m 63; c 2. MEDICINAL CHEMISTRY, ORGANIC CHEMISTRY. *Educ:* Mass Inst Technol, SB, 55; Columbia Univ, AM, 56, PhD(chem), 59. *Prof Exp:* Res assoc org chem, Rockefeller Inst, 58-62; from asst prof to assoc prof, 62-70, PROF MED CHEM, STATE UNIV NY BUFFALO, 70-, CHMN DEPT, 69- *Concurrent Pos:* Consult, Westwood Pharmaceut Inc, 71- *Mem:* Am Chem Soc; NY Acad Sci. *Res:* Synthesis and structure-activity relationships of steroid hormones; investigations of mechanism of action of steroid hormones; synthesis and qsar of dihydropyridine-type calcium channel antagonists. *Mailing Add:* Dept Med Chem Sch Pharm State Univ NY Buffalo NY 14260

SOLODAR, ARTHUR JOHN, b East Orange, NJ, Apr 18, 40; m 64; c 3. ORGANIC CHEMISTRY, CATALYSIS. *Educ:* Swarthmore Col, BA, 62; Yale Univ, MS, 63, PhD(chem), 67. *Prof Exp:* Nat Cancer Inst fel, Mass Inst Technol, 67-68; sr res chemist, 68-74, res specialist, 74-80, sr res specialist, 80-84, SR PROCESS CONSULT, MONSANTO CO, 84- *Mem:* Sigma Xi; Am Chem Soc. *Res:* Homogeneous catalysis; asymmetric synthesis; phase-transfer catalysis; exploratory process research. *Mailing Add:* 8135 Cornell Ct University City MO 63130

SOLODAR, WARREN E, b New York, NY, Sept 29, 25; m 50; c 2. ORGANIC CHEMISTRY. *Educ:* NY Univ, AB, 48; Stevens Inst Technol, MS, 53. *Prof Exp:* Assoc chemist, Hoffmann-La Roche, Inc, 48-54; chemist, Polaroid Corp, 54-64; scientist, Xerox Corp, 64-87; MEM STAFF, DX IMAGING, 87- *Concurrent Pos:* Res fel, Koor Chem Ltd, Beer Sheva, Israel, 74-75; vis prof, Hebrew Univ Jerusalem, 82. *Mem:* Am Chem Soc; Soc Photog Scientists & Engr. *Res:* Pharmaceuticals; vitamins; hypertensive agents; analgesics; azo and anthraquinone dyes; photographic developers; organic photoconductors; pigments; photoelectrophoretic imaging materials; liquid xerographic inks. *Mailing Add:* 480 Montgomery Ave Merion PA 19066

SOLOFF, BERNARD LEROY, b New York, NY, June 21, 31; m 61; c 2. ANATOMY. *Educ:* Univ Cincinnati, BS, 53, MS, 56; Rice Univ, PhD(biol), 61. *Prof Exp:* Res asst trace metals, M D Anderson Hosp & Tumor Inst, 61; asst prof biol, Stephen F Austin State Col, 61-62; res assoc anat, Med Units, Univ Tenn, Memphis, 63-64; RES PHYSIOLOGIST, LITTLE ROCK HOSP DIV, VET ADMIN, 65- *Concurrent Pos:* Nat Heart Inst fel & training grant, Marine Lab, Inst Marine Sci, Univ Miami, 62-63; USPHS trainee, Med Units, Univ Tenn, Memphis, 64-65; instr, Med Ctr, Univ Ark, 65-70, asst prof, 70- *Mem:* NY Acad Sci; AAAS; Am Asn Anatomists; Tissue Culture Asn; Electron Micros Soc Am. *Res:* Ultrastructure of leukocytes; ultrastructure of hemoglobin interactions within erythrocytes; ultrastructure of lung; ultrastructure of bacteria and bacteriophage; ultrastructure of heart. *Mailing Add:* Vet Admin Hosp 4300 W Seventh St Little Rock AR 72205

SOLOFF, LOUIS ALEXANDER, b Paris, France, Oct 2, 04; nat US; m 34; c 1. MEDICINE, CARDIOLOGY. *Educ:* Univ Chicago, MD, 30. *Prof Exp:* Chief labs, St Joseph's & St Vincent's Hosps & Eagleville Sanatorium, 34-45; chief dept cardiol, Episcopal Hosp, 50-56; chief div cardiol, Health Sci Ctr, 56-71, PROF MED, TEMPLE UNIV, 56-, BLANCHE P LEVY DISTINGUISHED SERV UNIV PROF, 71- *Mem:* Asn Univ Cardiol; Am Heart Asn; Am Col Physicians; Am Col Cardiol; Sigma Xi. *Res:* Diseases of the heart. *Mailing Add:* 1901 Walnut St Philadelphia PA 19103

SOLOFF, MELVYN STANLEY, b Los Angeles, Calif, Oct 6, 38; m 68; c 2. ENDOCRINOLOGY. *Educ:* Univ Calif, Los Angeles, AB, 62, MA, 64, PhD(zool), 68. *Prof Exp:* Res fel, Univ Calif, Los Angeles, 68-69; assoc res biologist, Sterling-Winthrop Res Inst, 69-70; asst prof biochem, 70-74, assoc prof, 74-79, PROF BIOCHEM, MED COL OHIO, 79- *Mem:* Endocrine Soc; Am Soc Biol Chemists. *Res:* Hormone receptors; mechanisms of hormone action. *Mailing Add:* Dept Biochem CS 10008 Med Col Ohio Toledo OH 43699

SOLOMON, ALAN, b New York, NY, May 16, 33; m 77; c 2. HEMATOLOGY, ONCOLOGY. *Educ:* Bucknell Univ, BS, 53; Duke Univ, BSMed, 56, MD, 57; Am Bd Internal Med, dipl, 64. *Prof Exp:* Intern, Mt Sinai Hosp, New York, 57-58, asst resident med, 59-60; asst resident, Montefiore Hosp, 58-59; clin assoc, Nat Cancer Inst, 60-62; chief resident, Mt Sinai Hosp, 62-63, asst attend physician, 65-66; PROF MED, DEPT MED, UNIV TENN MED CTR, KNOXVILLE, 66- *Concurrent Pos:* Res fel hemat, Mt Sinai Hosp, 63-65; Nat Inst Arthritis & Metab Dis spec fel, Rockefeller Inst, 63-65; USPHS res career develop award, 67-72; from asst physician to assoc physician, Rockefeller Inst, 63-66, guest investr, 63-66; prin investr, USPHS grant, 65-, mem review comt, Nat Cancer Inst, Clin Cancer Prog Proj, 78-83. *Honors & Awards:* Laszlo Mem Lectr, Montifiore Hosp, 84. *Mem:* Am Soc Clin Invest; fel Am Col Physicians; Am Soc Clin Oncol; Am Asn Cancer Educ; Am Soc Hemat. *Res:* Amyloidosis; human immunoglobulins; cancer; pathophysiology of the human light chain-associated renal and systemic diseases: myeloma (cast) nephropathy, light chain deposition disease and amyloidosis AL. *Mailing Add:* Univ Tenn Med Ctr 1924 Alcoa Hwy Knoxville TN 37920

SOLOMON, ALLEN M, b Mt Clemens, Mich, Apr 29, 43; m 66; c 2. PLANT ECOLOGY, PALYNOLOGY. *Educ:* Univ Mich, Ann Arbor, BA, 65; Rutgers Univ, New Brunswick, PhD(bot), 70. *Prof Exp:* Res asst bot, Rutgers Univ, New Brunswick, 68-70; asst prof geosci, Univ Ariz, 70-76; res assoc, Oak Ridge Nat Lab, 76-81, res staff ecologist, 81-87; proj leader, Int Inst Appl Syst Anal, Laxenburg, Austria, 87-90; PROF FOREST ECOL, MICH TECH UNIV, 89- *Concurrent Pos:* Ed, Amqua Newslett, 79-85; co-prin investr, Terrestrial Ecosysts, Climate & Global Carbon Cycle, NSF, 78-81, 81-84 & 84-87. *Mem:* AAAS; Ecol Soc Am; Am Quaternary Asn; Am Inst Biol Sci; Int Asn Veg Sci; Int Asn Ecol. *Res:* Plant ecology; global ecology; vegetation response to global environmental change; quaternary palynology; plant geography; modelling paleoenvironmental reconstruction. *Mailing Add:* Rte 1 Box 206C Calumet MI 49913

SOLOMON, ALVIN ARNOLD, b Chicago, Ill, Aug 17, 37. MATERIALS SCIENCE. *Educ:* Univ Ill, BS, 59, MS, 61; Stanford Univ, PhD(mat sci), 68. *Prof Exp:* Develop engr, Advan Systs Develop Div, IBM Corp, 61-64; postdoctoral res, French Atomic Energy Comn, Saclay, 68-69; metallurgist, Argonne Nat Lab, 69-74; PROF NUCLEAR ENG, PURDUE UNIV, 74- *Concurrent Pos:* Instr, San Jose State Col, 63-64; mem staff, Denver Res Inst, 66; Centre Nat de la Rechercha Scientifique, 80-81. *Honors & Awards:* Ceramographic Award, Am Ceramic Soc. *Mem:* Fel Am Ceramic Soc; Am Nuclear Soc. *Res:* High temperature materials. *Mailing Add:* Purdue Univ Sch Nuclear Eng West Lafayette IN 47907

SOLOMON, ARTHUR KASKEL, b Pittsburgh, Pa, Nov 26, 12; m; c 2. BIOPHYSICS. *Educ:* Princeton Univ, AB, 34; Harvard Univ, MA, 35, PhD(phys chem), 37; Cambridge Univ, PhD(physics), 47. *Hon Degrees:* ScD, Cambridge Univ, 64. *Prof Exp:* Res assoc physics & chem, Harvard Univ, 39-41, Exp Off, Brit Ministry Supply, 41-43 & Brit Admiralty, 43-45; mem staff, Radiation Lab, Mass Inst Technol, 45; asst prof physiol chem, 46-56, from assoc to prof biophys prof, 57-83, EMER PROF BIOPHYS, HARVARD MED SCH, 83- *Concurrent Pos:* Assoc, Peter Bent Brigham Hosp, 50-72; mem ed bd, J Gen Physiol, 58-; chmn comt on higher degrees in biophys, Harvard Univ, 59-80; NIH radiation study sect, 60-63, biophys sci training comn, 63-68, chmn, 66-68; secy-gen, Int Union Pure & Appl Biophys, 61-72; mem bd, Int Orgn & Progs, Nat Res Coun, Nat Acad Sci, 67-80, chmn, 77-79; sci policy adv to Thai govt, UNESCO, 68-72, mem, US Nat Comt, 69-74. *Honors & Awards:* Order Andres Bello, Govt of Venezuela, 74. *Mem:* Fel AAAS; Am Chem Soc; Am Physiol Soc; Biophys Soc; fel Am Acad Arts & Sci; Soc Gen Physiol. *Res:* Permeability of cellular membranes and model systems. *Mailing Add:* Biophys Lab Harvard Med Sch 25 Shattuck St Boston MA 02115

SOLOMON, DAVID EUGENE, b Milton, Pa, June 22, 31; m 50; c 3. GENERAL MANAGEMENT, MATERIALS SCIENCE. *Educ:* Susquehanna Univ, AB, 58; Bucknell Univ, MS, 60; Eastern Mich Univ, MBA, 75. *Prof Exp:* Sr engr, Electron Tube Lab, Westinghouse Elec Corp, Md, 59-65; sr res engr, Electron Physics Lab, Univ Mich, Ann Arbor, 65-67; chief engr, Electro Optics Div, Bendix Corp, Mich, 67-71, prog mgr Mars probe, Bendix Aerospace Systs Div, 71-72; dir laser fusion mat sci, KMS Fusion, Inc, 72-85, vpres opers, 80-85; PRES, SOLO HILL ENG, INC, 85- *Mem:* Am Vacuum Soc; fel Inst Elec & Electronics Engrs. *Res:* Broad research in the material sciences aimed at laser thermonuclear fuel pellets, encompassing glass synthesis and fabrication, polymers, copolymers, cryogenics, thin film deposition and others. *Mailing Add:* 3415 Woodlea Dr Ann Arbor MI 48103

SOLOMON, DAVID HARRIS, b Mass, Mar 7, 23; m 46; c 2. GERIATRICS, ENDOCRINOLOGY. *Educ:* Brown Univ, AB, 44; Harvard Med Sch, MD, 46. *Prof Exp:* House officer med, Peter Bent Brigham Hosp, 46-47, sr asst resident physician, 50-51; sr asst surgeon, NIH, 48-50; from instr to assoc prof, Sch Med, Univ Calif, Los Angeles, 52-66, chmn, Dept Med, 71-81, assoc dir, Multicampus Div Geriat Med, 82-90, PROF MED, SCH MED, UNIV CALIF, LOS ANGELES, 66-, DIR, CTR ON AGING, 90- *Concurrent Pos:* Res fel, Peter Bent Brigham Hosp, 47-48; fel endocrinol, New Eng Ctr Hosp, 51-52; attend physician, Harbor Gen Hosp, Torrance, 52-66, chief dept med, 66-71; attend physician, Vet Admin Ctr, 52-; mem, bd dirs, Am Geriat Soc & Am Fedn Aging Res. *Mem:* Endocrine Soc; Am Col Physicians; Inst Med-Nat Acad Sci; Am Thyroid Asn; Asn Am Physicians. *Res:* Thyroid hormone metabolism in animals and man; effect of illness and aging on thyroid hormone metabolism; appropriateness of surgical operations and other major procedures. *Mailing Add:* Dept Med Univ Calif Sch Med Los Angeles CA 90024

SOLOMON, DONALD W, b Detroit, Mich, Feb 6, 41; m 89; c 1. ANALYTICAL MATHEMATICS. *Educ:* Wayne State Univ, BS, 61, MA, 63, PhD(math), 66, MD, 68. *Prof Exp:* From teaching asst to instr math, Wayne State Univ, 63-66; from asst prof to assoc prof math, 66-74, assoc chmn dept, 75-78, PROF MATH, UNIV WIS-MILWAUKEE, 74-, CHMN, DIV NATURAL SCI, 76- *Concurrent Pos:* NSF res grants, 67-68 & 70-71; Univ Wis Grad Sch res grants, 69, 71-72 & 73-74. *Mem:* AAAS; Am Math Soc; Math Asn Am; Soc Indust & Appl Math; NY Acad Sci. *Res:* Measure, integration and differentiation. *Mailing Add:* Dept Math Univ Wis Milwaukee WI 53201

SOLOMON, EDWARD I, b New York, NY, Oct 20, 46; m; c 1. PHYSICAL INORGANIC CHEMISTRY, BIOINORGANIC CHEMISTRY. *Educ:* Rensselaer Polytech Inst, BS, 68; Princeton Univ, MA, 70, PhD(chem), 72. *Prof Exp:* Danish Nat Sci Found fel chem, H C Orsted Inst, Univ Copenhagen, 73-74; NIH fel, Noyes Lab, Calif Inst Technol, 74-75; A P Sloan res fel, 76; from ast prof to prof, Mass Inst Technol, 75-81; PROF CHEM, STANFORD UNIV, 82- *Concurrent Pos:* Sloan fel, 76-78; assoc ed, Inorg Chem; hon prof, Xiamen Univ, People's Repub China; invited prof, Univ Paris, Orsay. *Honors & Awards:* O K Rice Lectr; Reilly Lectr; World Bank Lectr; First Glen Seaborg Lectr. *Mem:* Am Chem Soc; Am Phys Soc; Sigma Xi; Am Asn Univ Profs; fel AAAS. *Res:* Inorganic spectroscopy and ligand field theory; spectral and magnetic studies on bioinorganic systems; interactions between metals in polynuclear complexes; spectroscopic studies of active sites in metalloprotein and heterogeneous catalysts; synchrotron spectroscopic studies of inorganic materials. *Mailing Add:* Dept Chem Stanford Univ Stanford CA 94305

SOLOMON, FRANK I, b Denver, Colo, Oct 7, 24; m 48; c 3. ELECTROCHEMISTRY. *Educ:* City Col New York, BChE, 47. *Prof Exp:* Asst vpres & tech dir, Yardney Elec Corp, 49-71; tech dir, Molecular Energy Corp, 71-74; PRES, ELECTROMEDIA, INC, 74- *Mem:* Electrochem Soc. *Res:* Storage batteries, alkaline; gas diffusion electrodes. *Mailing Add:* 8 Hampton Ct Lake Success NY 11020

SOLOMON, GEORGE E, b Seattle, Wash, July 14, 1925. AERONAUTICAL & ASTRONAUTICAL ENGINEERING. *Educ:* Univ Wash, BS, 49; Calif Inst Technol, MS, 50, PhD(aeronaut & physics), 53. *Prof Exp:* Res fel, Caltech, 53-54; tech consult re-entry body aerodynamics, Ramo-Wooldridge Corp, 53-54, mem tech staff, Guided Missile Res Div, 54-58, dir, syst res & anal div, 58-62, vpres, Space Technol Labs, 62-65; vpres & dir, Syst Labs, TRW Syst Group, TRW Inc, 65-68, vpres & dir, mkt & requirements anal, 68-71, vpres & gen mgr, TRW Defense & Space Syst Group, 71-81, exec vpres & gen mgr, electronics & defense sector, 81-88; RETIRED. *Mem:* Nat Acad Eng; Sigma Xi; Aerospace Indust Asn. *Mailing Add:* 220 Miramar Ave Montecito CA 93108

SOLOMON, GEORGE FREEMAN, b Freeport, NY, Nov 25, 31; m 79; c 2. PSYCHONEUROIMMUNOLOGY, FORENSIC PSYCHIATRY. *Educ:* Stanford Univ, AB, 52, MD, 55. *Prof Exp:* From asst to assoc prof psychiat, Stanford Univ, 62-73; chief psychiat, Valley Med Ctr Fresno, 74-83; prof, Univ Calif, San Francisco, 80-84; PROF PSYCHIAT, UNIV CALIF-LOS ANGELES, 84- *Concurrent Pos:* Chief psychiat training & res, Palo Alto Vet Admin Hosp, 62-70; chief substance abuse treat unit, Sepulveda Vet Admin Med Ctr, 84- *Mem:* Fel Am Psychiat Asn; fel Acad Behav Med; fel Int Col Psychosomatic Med; Am Acad Psychiat & Law; fel Soc Behav Med; Am Psychosomatic Soc. *Res:* Psychoneuroimmunology, especially of AIDS and aging criminal behavior, violence and aggression; post-traumatic stress syndrome in Vietnam veterans; normal behavior; psychosomatic medicine; immunity of schizophrenia. *Mailing Add:* 19054 Pac Coast Hwy Malibu CA 90265

SOLOMON, GORDON CHARLES, b Salida, Colo, Dec 5, 24; m 49; c 6. ANATOMY, PATHOLOGY. *Educ:* Colo State Univ, BS, 49, MS, 51, DVM, 55, PhD(path), 63. *Prof Exp:* Vet epidemiologist, Commun Dis Ctr, USPHS, 55-58, lab dir, Southwest Rabies Invests, 58-60, vet pathologist, 64-66; assoc prof anat, Colo State Univ, 66-86; RETIRED. *Concurrent Pos:* USPHS fel, Colo State Univ, 63-64; consult, Southwest Radiol Health Lab, USPHS, 66-68. *Mem:* Am Asn Vet Anat; World Asn Vet Anat; Wildlife Dis Asn. *Res:* Bone and connective tissue; wildlife anatomy and diseases. *Mailing Add:* 1309 Luke St Ft Collins CO 80524

SOLOMON, HARVEY DONALD, b New York, NY, Dec 14, 41; m 66; c 2. METALLURGY, MATERIALS SCIENCE. *Educ:* NY Univ, BS, 63; Univ Pa, PhD(metall), 68. *Prof Exp:* METALLURGIST & MAT SCIENTIST, RES & DEVELOP CTR, GEN ELEC CO, 68- *Concurrent Pos:* Adj asst prof, Union Col, 89- *Honors & Awards:* Joseph Vilella Award, Am Soc Testing & Mat, 79. *Mem:* Am Inst Mining, Metall & Petrol Engrs. *Res:* Physical and mechanical metallurgy; fatigue; fatigue crack propagation; creep-fatigue interactions; electrical interconnections; fatigue of solders; welding metallurgy; materials for advanced energy systems; superalloys; metallurgy of stainless steels; stress corrosion cracking. *Mailing Add:* Gen Elec Res & Develop Ctr PO Box 8 Schenectady NY 12345

SOLOMON, HERBERT, b New York, NY, Mar 13, 19; m 47; c 3. MATHEMATICAL STATISTICS. *Educ:* City Col NY, BS, 40; Columbia Univ, MA, 41; Stanford Univ, PhD(math statist), 50. *Prof Exp:* Asst res mathematician, Columbia Univ, 43-44, assoc math statistician, 44-46; instr, City Col NY, 46; asst prof math statist, Stanford Univ, 47; mathematician, Off Naval Intel, US Dept Navy, 48, Off Naval Res, DC, 49-52; prof math, Teachers Col, Columbia Univ, 52-59; prof, 58-89, EMER PROF STATIST, STANFORD UNIV, 89- *Concurrent Pos:* Guggenheim fel, 58. *Honors & Awards:* Wilks Medalist, Am Statist Asn, 75; Townsend Harris Medalist, City College, NY, 77; Distinguished Public Serv Medal, USN, 78. *Mem:* Am Statist Asn; Inst Math Statist (pres, 65). *Res:* Psychometrics; engineering statistics; operations research. *Mailing Add:* Dept of Statist Stanford Univ Stanford CA 94305

SOLOMON, HOWARD FRED, radiobiology, cancer biology, for more information see previous edition

SOLOMON, JACK, b Brooklyn, NY, July 26, 41; m 70; c 3. PHYSICAL CHEMISTRY. *Educ:* Mass Inst Technol, BS, 63; Columbia Univ, PhD(phys chem), 67. *Prof Exp:* Res scientist, 67-72, proj scientist, Linde Res Lab, Union Carbide Co, 72-74; sr proj engr, 74-77, supvr, Gas Prod Develop Lab, 77-80, process mgr, Gas Prod Mkt Develop, 80-83, prod mgr, New Electronics Applns, 83-87, assoc dir develop, 87-90, MGR APPLN TECHNOL PLANNING, UNION CARBIDE CO, 90- *Concurrent Pos:* Chmn tech sessions, Semicon West, 86; Reviewer Indust & Eng Chem Res; chmn, Tekcon, 88. *Mem:* Am Chem Soc; Sigma Xi; Semiconductor Equip & Mat Inst; Int Soc Hybrid Microelectronics. *Res:* Chemistry of oxygen in cryogenic, aqueous and other systems; atmospheres for carburizing; hardening and sintering; applications of oxygen in combustion; semiconductor fabrication technology; use of inert gas in polymer processing and other industrial processes; 4 US patents. *Mailing Add:* Linde Industrial Gas Develop Union Carbide Co Tarrytown NY 10591

SOLOMON, JAMES DOYLE, b Bee Branch, Ark, May 18, 34; m 61; c 1. FOREST ENTOMOLOGY. *Educ:* Univ Ark, BS, 56, MS, 60; Miss State Univ, PhD(entom), 71. *Prof Exp:* Res entomologist, Asheville, NC, 60-61 & Stoneville, Miss, 61-75, PRIN RES ENTOMOLOGIST, SOUTHERN FOREST EXP STA, STONEVILLE, MISS, 75- *Concurrent Pos:* Adj prof, Miss State Univ, 72-; mem, Interagency Task Force, Cross-Fla Barge Canal Environ Impact Statement, 72-73; mem, USDA task force, Long Range Forest Res Planning, Southern Region, 73-74. *Honors & Awards:* US Dept Agr, Cert of Merit Award for Outstanding Res, 85. *Mem:* Entom Soc Am; Sigma Xi. *Res:* Hardwood insects with emphasis on the Cossid, Cerambycid and Sesiid Borers of living trees and shrubs. *Mailing Add:* US Forest Serv PO Box 227 Stoneville MS 38776

SOLOMON, JAY MURRIE, b Washington, DC, June 6, 36; m 59; c 3. COMPUTATION FLUID DYNAMICS, APPLIED MATHEMATICS. *Educ:* Univ Md, BS, 58, MS, 60, PhD(appl math), 68. *Prof Exp:* Res asst, Univ Md, 58-60; aerospace engr res viscous flows, 60-67, MATHEMATICIAN APPL MATH, WHITE OAK LAB, NAVAL SURFACE WEAPONS CTR, 67- *Mem:* Soc Indust & Appl Math; Am Inst Aeronaut & Astronaut. *Res:* Development, analysis and application of finite difference methods to steady subsonic inviscid flows, incompressible Navier-Stokes equations and unsteady inviscid flows. *Mailing Add:* 2232 Hidden Valley Lane Silver Spring MD 20914

SOLOMON, JEROME JAY, b Brooklyn, NY, Apr 23, 45. PHYSICAL CHEMISTRY, MASS SPECTROMETRY. *Educ:* Brooklyn Col, BS, 66; Cornell Univ, PhD(phys chem), 72. *Prof Exp:* Res assoc phys chem & ion-molecule reactions, Rockefeller Univ, 72-75; assoc res scientist, NY Univ Med Ctr, 75-77, asst prof, Environ Med & Mass Spectrometry, Inst Environ Med, 77-83, res assoc prof, 83-90, PROF & DIR, LAB DNA CHEM & CARCINOGENESIS, INST ENVIRON MED, NY UNIV MED CTR, 90- *Mem:* Am Soc Mass Spectrometry; Sigma Xi; Am Asn Cancer Res; AAAS. *Res:* Development of the analytical capability of mass spectrometry for use in biomedical research; emphasis on DNA chemistry. *Mailing Add:* Environ Med NY Univ Med Ctr 550 First Ave New York NY 10016

SOLOMON, JIMMY LLOYD, b Milan, Tenn, Oct 3, 41; m 64; c 2. MATHEMATICS. *Educ:* Univ Miss, BS, 64; Miss State Univ, MS, 66; Tex A&M Univ, PhD(math), 72. *Prof Exp:* Asst mathematician, dept aerophys, Miss State Univ, 64-65, instr, 66-67; instr math, Tex A&M Univ, 71-72, asst prof, 72-75; assoc prof, 75-83, HEAD, DEPT MATH & STATIST, MISS STATE UNIV, 81-, PROF MATH, 83- *Concurrent Pos:* Consult. *Mem:* Am Math Soc; Math Assoc Am. *Res:* Fixed point theory, numerical analysis, and statistical pattern recognition. *Mailing Add:* Dept Math Drawer MA Miss State Univ Mississipi State MS 39762

SOLOMON, JOEL MARTIN, b Malden, Mass, Dec 25, 32; m 84; c 3. IMMUNOGENETICS, IMMUNOHEMATOLOGY. *Educ:* Boston Col, BS, 53; Johns Hopkins Univ, ScM, 57; Univ Wis, PhD(med genetics), 63. *Prof Exp:* Immunohematologist, NIH, 57-60; sr res assoc, Am Nat Red Cross, 63-64, sr res assoc & training dir, 64-67; dir blood bank, Brooklyn-Cumberland Med Ctr, 67-70; blood prod dir, E R Squibb & Sons, Inc, 70-73; dep dir, Div Blood & Blood Prod, Bur Biologics, Food & Drug Admin, 74-77, dir, 78-81; policy coordr, Off Secy, Dept Health & Human Serv, 81-83; scientist adminr, dept Transfusion Med, 83-85, spec asst to dir extramural affairs, 85-86, exec sec, HEM-2 study sect, NIH, 86-88; dir, Div Blood & Blood Prod, Ctr, Biologics, FDA, 88-91; EXEC DIR, AM ASN BLOOD BANKS, 91- *Concurrent Pos:* Clin instr pediat, Calif Col Med, 63-64; mem fac genetics & fac microbiol & immunol, Grad Sch, NIH, 64-67 & 74-; lectr, Sch Med, George Washington Univ, 66-67; clin assoc prof path, Sch Med, State Univ NY Downstate Med Ctr, 67-73. *Mem:* Int Soc Blood Transfusion; Am Asn Blood Banks. *Res:* Quantitative hemagglutination; genetics of human blood groups; immunochemistry of blood group antigens; parasite physiology. *Mailing Add:* 15714 Cherry Blossom Way North Potomac Gaithersburg MD 20878

SOLOMON, JOLANE BAUMGARTEN, b New York, NY, Sept 23, 27; m 57; c 3. PHYSIOLOGY, ENDOCRINOLOGY. *Educ:* Hunter Col, BS, 52; Radcliffe Col, MS, 55, PhD(physiol), 58. *Prof Exp:* Teaching fel, biol, Harvard Univ, 53-55, sci news writer, 55-57; teaching fel, anat, Harvard Med Sch, 57-59; res assoc nutrit, Harvard Sch Pub Health, 60-63; vis lectr, 63-72, assoc prof, 74-80, PROF BIOL, BOSTON COL, 80- *Concurrent Pos:* Dir, Off Resources, Boston Col, 70-72; Carnegie res fel, 72-74; grants, Nat Inst Drug Abuse, 75-76 & 76-78. *Mem:* AAAS; Am Diabetes Asn; Am Women Sci; Entomol Soc Am; Endocrine Soc. *Res:* Effect of THC on reproduction in male and female rats; effect of lighting regimens on growth of the American cockroach, P Americana; lipid metabolism in hyperglycemic ob/ob mice. *Mailing Add:* Dept Biol Boston Col 140 Commonwealth Ave Chestnut Hill MA 02167

SOLOMON, JOSEPH ALVIN, b New Kensington, Pa, July 25, 25; m 53; c 1. PHYSICAL CHEMISTRY. *Educ:* Westminster Col, BS, 49; Carnegie-Mellon Univ, MS, 58, PhD, 59. *Prof Exp:* Analytical chemist, US Steel Corp, NJ, 51-54; analyst, Gulf Res & Develop Co, 54-55; prof chem, St Joseph Col, Md, 58-61 & Marietta Col, 61-62; coal res engr, WVa Univ, 62-63; prof chem, St Joseph Col, Md, 63-65; from asst prof to assoc prof, 65-70, PROF CHEM, PHILADELPHIA COL PHARM & SCI, 70- *Mem:* AAAS; Am Chem Soc. *Res:* Coal research, especially removal of sulfur both prior to and following combustion. *Mailing Add:* 363 S Manoa Rd Havertown PA 19083

SOLOMON, JULIUS, b Brooklyn, NY, Apr 14, 36; m 63; c 3. HIGH ENERGY PHYSICS. *Educ:* Columbia Univ, AB, 57; Univ Calif, Berkeley, PhD(physics), 63. *Prof Exp:* Res assoc, Lawrence Radiation Lab, Univ Calif, 63; instr, Princeton Univ, 63-66; from asst prof to assoc prof, 66-82, PROF PHYSICS, UNIV ILL AT CHICAGO, 83- *Mem:* Am Phys Soc. *Res:* Experimental high energy physics; K zero meson decays; pion-proton scattering; neutron-proton scattering; k-zero regeneration; k-zero-k-zero bar mass differences; high transverse momentum scattering with Hadron beams; Dimuon productions with hadron beams. *Mailing Add:* Dept Physics Univ Ill at Chicago Box 4348 Chicago IL 60680

SOLOMON, KENNETH, b Brooklyn, NY, Oct 8, 47; m; c 2. GERIATRICS, SEXUALITY. *Educ:* NY Univ, BA, 67; State Univ NY, Buffalo, MD, 71. *Prof Exp:* Asst instr psychiat, Albany Med Col, 72-75, from instr to asst prof, 75-77; asst prof, Med Col Va, 77-79, asst dir residency training, 77-79; fac assoc geront, Va Ctr Aging, Va Commonwealth Univ, 78-79; asst prof psychiat, Univ Md, 79-81; assoc dir educ & planning, Levindale Hebrew Geriat Ctr & Hosp, 81-83; adj asst prof psychiat, Univ Md, 82-87, assoc mem grad fac, 80-89; ASSOC PROF PSYCHIAT, ST LOUIS UNIV, 89-; CHIEF, GERIATRIC PSYCHIAT, ST LOUIS DEPT VET AFFAIRS MED, 89- *Concurrent Pos:* Adj prof, Union Experimenting Cols & Univs, 81-90; assoc clin prof, geriat & psychiat, Univ Md, 87-90; chief, geriat & psychiat serv, Sheppard & Enoch Pratt Hosp, 83-89. *Mem:* Fel Am Psychiat Asn; fel Geront Soc Am; Am Pub Health Asn; fel Am Geriat Soc; AAAS. *Res:* Psychogeriatrics; depression; Alzheimer's disease; sterotyping the elderly; sexuality; men's issues. *Mailing Add:* Dept Psychiat St Louis Univ Sch Med 1221 S Grand Blvd St Louis MO 63104

SOLOMON, LAWRENCE MARVIN, b Montreal, Que, June 1, 31; m 59; c 2. DERMATOLOGY, MEDICAL EDUCATION. *Educ:* McGill Univ, BA, 53; Univ Geneva, MD, 59; FRCP(C), 64; Am Bd Dermat, dipl, 65. *Prof Exp:* Intern, Jewish Gen Hosp, Montreal, 59-60; resident med, Queen Mary Vet Hosp, 60-61; resident dermat, Grad Hosp, Univ Pa, 61-64; from asst prof to prof, 66-76,actg dir, Univ Hosp, 69-70, PROF DERMAT & HEAD DEPT, UNIV ILL MED CTR, 74- *Concurrent Pos:* Res fel, Jewish Gen Hosp, Montreal, 64-66; consult, Vet Admin Hosp, Hines, Ill, 66-, 68- & West Side Vet Admin Hosp, 69-; chmn, comt dermat agents, Food & Drug Admin, 76-78; chief ed, J Invest Dermat, 77-83. *Honors & Awards:* Gold Award, Am Acad Dermat & Am Acad Allergy, 64. *Mem:* Am Acad Dermat; Soc Invest Dermat; Int Soc Pediat Dermat (pres, 79-83); Am Soc Pediat Dermat; Am Dermat Asn. *Res:* Biochemical and pharmacological studies in atopic dermatitis; pharmacogenetic changes in hereditary skin diseases; catecholamines; congenital malformation of the skin. *Mailing Add:* Dept Dermat Univ Ill Med Ctr PO Box 6998 Chicago IL 60680

SOLOMON, LOUIS, b New York, NY, June 20, 31; m 60; c 2. MATHEMATICS. *Educ:* Harvard Univ, AB, 51, AM, 52, PhD(math), 58. *Prof Exp:* Asst prof math, Bryn Mawr Col, 58-59 & Haverford Col, 59-62, 63-64; vis mem, Inst Advan Study, 62-63; asst prof, Rockefeller Univ, 64-65; from assoc prof to prof, NMex State Univ, 65-69; PROF MATH, UNIV WIS-MADISON, 69- *Concurrent Pos:* Vis prof, Univ London, 71-72. *Mem:* Am Math Soc; Math Asn Am. *Res:* Finite groups, especially groups generated by reflections and linear groups over finite fields; combinatorics. *Mailing Add:* Dept Math Univ Wis 480 Lincoln Dr Madison WI 53706

SOLOMON, M MICHAEL, b Philadelphia, Pa, Sept 20, 24; m 48; c 2. ORGANIC CHEMISTRY, POLYMER CHEMISTRY. *Educ:* Temple Univ, BA, 45, MA, 47; Purdue Univ, PhD, 51. *Prof Exp:* Asst, Temple Univ, 45-47 & Purdue Univ, 47-48; res assoc endocrine chem, Worcester Found Exp Biol, 51-52; sr develop chemist, Silicone Prod Dept, 52-60, mgr liaison & info, Space Sci Lab, Missile & Space Div, 60-66, mgr eng mat & tech lab, Lab Oper, Power Transmission Div, Pa, 66-71, mgr metall & fabrication lab oper, Power Delivery Group, Group Tech Resources Oper, Pittsfield, Mass, 71-75; MGR SWITCHGEAR RESOURCES SUPPORT OPERS, SWITCHGEAR DISTRIB TRANSFORMER DIV, TECH RESOURCES OPERATORS, GEN ELEC CO, PHILADELPHIA, 75- *Concurrent Pos:* Mgr environ technol, Power Delivery Div, King of Prussia, Pa, 79-; consult mat & environ, Lawrence J Dove Assocs, Philadelphia, 85- *Mem:* Am Chem Soc; Am Inst Aeronaut & Astronaut; Inst Elec & Electronics Eng; Sigma Xi. *Res:* Management; metals; metal fabrication and equipment development; ceramics; electrochemistry for use in power delivery equipment; composite materials; dielectrics; silicone chemistry; biological metabolism of adrenocorticotrophic hormone and cortisone; organic chemistry; hazardous materials treatment and dispositions, fire and arson investigations. *Mailing Add:* 1871 Ambler Rd Abington PA 19001

SOLOMON, MALCOLM DAVID, b Swansea, Wales, Oct 16, 42. SOFTWARE SYSTEMS LAB MANAGEMENT & INSTRUMENTS. *Educ:* Univ London, BSc, 64, PhD(org chem), 67; Univ Calif, Berkeley, AB, 83. *Prof Exp:* Res chemist, Med Ctr, Univ Calif, San Francisco, 67-69; fel genetics, Med Ctr, Stanford Univ, 70-71; res chemist, Ultrachem Corp, Walnut Creek, 71-73, tech dir, Sci Res Info Serv Inc, San Francisco, 73-74; toxiologist, Hine Inc, San Francisco, 74-79; software eng, Oxbridge, Mountain View, Ca, 83-84; SOFTWARE ENG, HEWLETT-PACKARD, PALO ALTO, CA, 84- *Mem:* Am Chem Soc; Royal Soc Chem; Asn Comput Mach. *Res:* Synthesis and chemistry of natural products; mass spectrometry; gas-liquid chromatography; narcotics and dangerous drugs; computer graphics; laboratory information management systems. *Mailing Add:* 535 Everett Ave No 303 Palo Alto CA 94301-1517

SOLOMON, MARVIN H, b Chicago, Ill, Mar 11, 49. OPERATING SYSTEMS, PROGRAMMING LANGUAGES. *Educ:* Univ Chicago, BS, 70; Cornell Univ, MS, 74, PhD(comput sci), 77. *Prof Exp:* Vis instr comput sci, Aarhus Univ, 75-76; instr, 76-77, asst prof, 77-82, ASSOC PROF COMPUT SCI, UNIV WIS-MADISON, 82- *Concurrent Pos:* Vis scientist, IBM Corp, 84-85. *Mem:* Asn Comput Mach; Inst Elec & Electronics Engrs. *Res:* Theory of programming languages; distributed operating systems; graph theory as applied to multiple-computer systems; computer networks; electronic mail. *Mailing Add:* Dept Comput Sci Univ Wis 1210 W Dayton St Madison WI 53706

SOLOMON, NEIL, b Pittsburgh, Pa, Feb 27, 32; m 55; c 3. PHYSIOLOGY. *Educ:* Western Reserve Univ, AB, 54, MD & MS, 61; Univ Md, PhD(physiol), 65. *Prof Exp:* Instr, Sch Med, Johns Hopkins Univ, 63-68, asst prof psychiat, 63-; Secy, Md State Dept Health & Ment Hyg, 69-; ENDOCRINOLOGIST, NEIL SOLOMON & RICHARD LAYTON. *Concurrent Pos:* Am Heart Asn res fel, 65-67; intern med, Johns Hopkins Hosp, 61-62, asst resident, 62-63, asst, 63-64, instr, 64-69; vis physician & asst chief med, Baltimore City Hosp, 63-68; consult, Vet Admin Hosp, Perry Point, Md, 63-68; assoc prof physiol, Sch Med, Univ Md, Baltimore, 63-70; asst sr surgeon, Nat Inst Child Health & Human Develop, 64-65; vis physician, Univ Md Hosp, 65-68; clin prof pharmacol, Sch Med, Univ Miami, 78. *Mem:* Am Fedn Clin Res; Am Heart Asn; Am Physiol Soc; Fedn Am Socs Exp Biol; NY Acad Sci. *Res:* Aging and heart and endocrine function. *Mailing Add:* 901 Dulaney Valley Rd Suite 602 Towson MD 21204

SOLOMON, PETER R, b New York, NY, Feb 19, 39; m 60, 75; c 3. SOLID STATE PHYSICS, COAL SCIENCE. *Educ:* City Col NY, BS, 60; Columbia Univ, MA, 63, PhD(physics), 65. *Prof Exp:* Res asst physics, Watson Lab, IBM Corp, 63-65; exp physicist, United Technol Res Ctr, 65-68, prin scientist, 68-71, asst to dir res progs & technol, 71-73, prin physicist, 73-80; PRES, ADVANCED FUEL RES INC, 80- *Concurrent Pos:* Committeman, Nat Acad Sci; chmn, Fuel Chem Div, Am Chem Soc, 84-85. *Honors & Awards:* Richard A Glean Award, Am Chem Soc. *Mem:* Am Phys Soc; Sigma Xi; Am Chem Soc; Combustion Inst. *Res:* Low temperature physics; electrical instabilities in semiconductors; coal science; superconductivity; instabilities in solids. *Mailing Add:* Advanced Fuel Res Inc PO Box 18343 East Hartford CT 06118-0343

SOLOMON, PHILIP M, b New York, NY, Mar 29, 39; m 58; c 1. ASTROPHYSICS, MOLECULAR PHYSICS. *Educ:* Univ Wis, BS, 59, MS, 61, PhD(astron), 64. *Prof Exp:* Res assoc astrophys, Princeton Univ, 64-66; lectr & sr res assoc astron, Columbia Univ, 66-70; assoc prof astrophysics, Univ Minn, Minneapolis, 71-74; PROF ASTRON, STATE UNIV NY STONY BROOK, 74- *Concurrent Pos:* Vis scientist, Inst Theoret Astron, Univ Cambridge, 67-72. *Mem:* Am Astron Soc. *Res:* Molecular opacities; interstellar matter; planetary atmospheres; interstellar chemistry; radioastronomy, masers; quasi-stellar objects. *Mailing Add:* Box 2902 East Setauket NY 11733

SOLOMON, ROBERT DOUGLAS, b Delavan, Wis, Aug 28, 17; m 43; c 4. PATHOLOGY. *Educ:* Johns Hopkins Univ, MD, 42. *Prof Exp:* Pathologist, Kankakee State Hosp, 49-50; assoc dir, Terre Haute Med Lab, 50-54; assoc pathologist, Sinai Hosp, Baltimore, Md, 55-58; asst prof path, Univ Md, 58-60; assoc pathologist, City of Hope Med Ctr, 60-63, dir path res, 63-67; dir labs, Doctors' Hosp San Leandro, Calif, 67-75; dir labs, Edgewater Hosp, Chicago, 75-76; assoc pathologist, Wilson Mem Hosp, Johnson City, 78-85; clin prof path, State Univ NY Upstate Med Ctr, 79-87; ADJ PROF BIOL, UNIV NC, WILMINGTON, 88- *Concurrent Pos:* Fel cancer res, Michael Reese Hosp, 47-49; trainee, Nat Cancer Inst, 58-60; consult, Regional Off US Vet Admin, Md, 58-60; assoc prof, Univ Southern Calif, 61-; fel coun arteriosclerosis, Am Heart Asn. *Mem:* Fel Royal Soc Med; Am Soc Clin Path; Am Chem Soc; Am Col Physicians; Col Am Path; Asn Clin Scientists; Sigma

Xi. *Res:* Urinary pigments; nutritional influences on carcinogenesis; leukoplakia and vitamin A; experimental arteriosclerosis; vascular surgery; mechanisms of aging. *Mailing Add:* 113 S Belvedere Dr Hampstead NC 28443

SOLOMON, SAMUEL, b Brest Litovsk, Poland, Dec 25, 25; Can citizen; m 53; c 3. BIOCHEMISTRY, ENDOCRINOLOGY. *Educ:* McGill Univ, BSc, 47, MSc, 51, PhD(biochem), 53. *Prof Exp:* Res asst, McGill Univ, 51-53; from res asst to res assoc, Columbia Univ, 53-57, assoc biochem, 58-59, asst prof, 59-60; assoc prof, 60-67, PROF BIOCHEM OBSTETS & GYNECOL & EXP MED, MCGILL UNIV, 67-; DIR, ENDOCRINE LAB, ROYAL VICTORIA HOSP, 65- *Concurrent Pos:* Chem Inst Can fel, 65; Can Soc Clin Invest Schering traveling fel, 65 & 69; consult, Ayerst Labs, 65-79 & Ortho Pharmaceut Co, 68-78; mem sen, McGill Univ, 69-71 & 74-77 & bd gov, 74-77; mem sci adv comt, Connaught Res Inst, 79-82; dir, Res Inst, Royal Victoria Hosp, 82-85; chmn steering comt, Int Study Group Steroid Hormones, Italy, 83-; mem var site vis teams, NSF, NIH & others. *Honors & Awards:* Price Orator, Am Soc Obstet & Gynecol; McLaughlin Gold Medal, Royal Soc Can. *Mem:* AAAS; Soc Gynec Invest; fel Royal Soc Can; Perinatal Res Soc (pres, 75-76); Am Chem Soc; Sigma Xi; hon fel Am Gynec & Obstet Soc; fel Chem Inst Can; Endocrine Soc. *Res:* Hormones in pregnancy; endocrinology. *Mailing Add:* Endocrine Lab 687 Pine Ave W Montreal PQ H3A 1A1 Can

SOLOMON, SEAN CARL, b Los Angeles, Calif, Oct 24, 45; m 67; c 2. GEOPHYSICS. *Educ:* Calif Inst Technol, BS, 66; Mass Inst Technol, PhD(geophys), 71. *Prof Exp:* Fel, NSF, 71-72; from asst prof to assoc prof, 72-83, PROF GEOPHYS, MASS INST TECHNOL, 83- *Concurrent Pos:* Lunar Sample Analysis Planning Team, NASA, 74-76, mem, Venus Orbital Imaging Radar Sci Working Group, 77-78, mem, Lunar & Planetary Geosci Rev Panel, 80-88, chmn, 86-88, mem, Megellan Proj Sci Group, 81-91, chmn, Planetary Geol & Geophys Working Group, 84-86, mem, Space & Earth Sci Adv Comt, 84-87; assoc ed, J Geophys Res, 76-78; Comt Planetary & Lunar Exploration, Nat Acad Sci-Nat Res Coun, 76-79, mem, Space Sci Bd, 78-82, chmn, Comt Earth Sci, 79-82, mem, Bd Earth Sci, 85-88; Alfred P Sloan res fel, 77-81; mem, Lunar & Planetary Sci Coun, Univ Space Res Asn, 78-80, 91-93; assoc ed, Eos Trans Am Geophys Union, 78-81; mem, Tech Rev Panel Nuclear Test Ban Treaty Verification, Defense Advan Projs Agency, 81-87; John Simon Guggenheim mem fel, 82-83; vis fac, Univ Calif, Los Angeles, 82-83; assoc, Space Sci Working Group, Am Univ, 84-91, chmn, 87-89; assoc ed, Geophys Res Lett, 86-88; Standing Comt, Global Seismic Network, Inc Res Insts Seismol, 87-90, chmn, 88-90; vis assoc, Calif Inst Technol, 90-91; mem, Div Planetary Sci, Am Astron Soc. *Mem:* fel Am Geophys Union (pres-elect & pres, Planetology Sect, 84-88); Seismol Soc Am; AAAS; Sigma Xi; Geol Soc Am; Am Astron Soc. *Res:* Earthquake seismology; marine geophysics; planetary geology and geophysics. *Mailing Add:* 54-522 Mass Inst Technol Cambridge MA 02139

SOLOMON, SEYMOUR, b Milwaukee, Wis, May 27, 24. HEADACHE. *Educ:* Marquette Univ, MD, 47. *Prof Exp:* Chief neurol, Philadelphia Gen Hosp, 52-53; attend neurologist, Bronx Municipal City Hosp, 71-81; head EEG Dept, 55-71, ATTEND NEUROLOGIST, MONTEFIORE HOSP & MED CTR, 55-, DIR, HEADACHE UNIT, MED CTR, 80- *Concurrent Pos:* asst clin prof neurol, dept neurol, Col Physicians & Surgeons, Columbia Univ, 58-64; assoc prof neurol, 80-83, prof neurol, Albert Einstein Col Med, Yeshiva Univ, 83-; mem, Migraine Res Group, World Fedn Neurol, 85-; treas & vpres, Am Asn Study Headaches, 88- *Mem:* Am Acad Neurol; Assoc Res Nervous & Mental Diseases; Am Assoc Study Headache; Int Assoc Study Pain; Am Pain Soc; Int Headache Soc. *Res:* Clinical research in the field of headache, particularly migraine, tension-type and cluster headache. *Mailing Add:* Montefiore Med Ctr 111 East 210 St Bronx NY 10467

SOLOMON, SIDNEY, b Worcester, Mass, Feb 22, 23; m 47; c 2. PHYSIOLOGY. *Educ:* Univ Mass, BS, 48; Univ Chicago, PhD(physiol), 52. *Prof Exp:* From instr to assoc prof physiol, Med Col Va, 52-63; chmn dept, 63-78, PROF PHYSIOL, SCH MED, UNIV NMEX, 65- *Concurrent Pos:* Guggenheim fel, Berlin, Ger, 62-63; consult, Adv Panel Regulatory Biol, NSF, 65-67, consult metab biol, 80-; prof dir metab biol, NSF, 67-68; consult, Nat Bd Med Exam, 68-72; spec asst, Div Phys Molecular & Cell Biol, NSF, 78-79. *Honors & Awards:* Guggenheim Fel, 62-63. *Mem:* Soc Exp Biol & Med; Biophys Soc; Am Physiol Soc. *Res:* Renal and comparative physiology; active transport. *Mailing Add:* Dept Physiol Univ NMex Sch Med 915 Stanford NE Albuquerque NM 87131

SOLOMON, SOLOMON SIDNEY, b New York, NY, Dec 2, 36; m 62; c 2. MEDICINE, METABOLISM. *Educ:* Harvard Univ, AB, 58; Univ Rochester, MD, 62. *Prof Exp:* Intern internal med, Med Ctr, Tufts Univ, 62-63, resident, Univ & Boston City Hosp, 63-65; res & educ assoc, Vet Admin Hosp, Memphis, 69-71; from asst prof to assoc prof med, 69-77, PROF MED, UNIV TENN, MEMPHIS, 77-, CHIEF ENDOCRINOL & METAB, VET ADMIN HOSP, MEMPHIS, 71-, ASSOC DEAN, RES COL MED, CTR HEALTH SCI, UNIV TENN, 83- *Concurrent Pos:* Vet Admin Hosp career develop award, Univ Tenn, Memphis, 69-71; attend physician, City of Memphis Hosp, 71-; reviewer, Journals & grants, NIH & Vet Admin Hosp; chair, Am Diabetes Asn Metab, 83; chair, Air Force Systs Command Regulation & Soc Sci Citation Index, 77 & 88; prof pharmacol, 86- *Mem:* Am Soc Clin Invest; Am Fedn Clin Res; Am Diabetes Asn (pres, 76); Endocrine Soc; Am Soc Pharmacol & Exp Therapeut; Fedn Am Soc Exp Biol. *Res:* Diabetes; intermediary metabolism; mechanism of action of insulin; role of second messenger's cyclic adenosine monophosphate in adipose tissue in normal and diabetic conditions; cyclic adenosine monophosphate phosphodiesterase, lipolysis; hormonal receptors diabetic animal models. *Mailing Add:* Vet Admin Hosp 1030 Jefferson Ave Memphis TN 38104

SOLOMON, SUSAN, b Chicago, Ill, Jan 19, 56; m 88. PHOTOCHEMISTRY. *Educ:* Ill Inst Technol, BS, 77; Univ Calif, Berkeley, MS, 79 & PhD(chem), 81. *Prof Exp:* RES CHEMIST, NAT OCEANIC & ATMOSPHERIC ADMIN, 81- *Concurrent Pos:* Assoc ed, J Atmospheric Sci, 83-86; mem, comt solar & space physics, NASA, 83-86, space & earth sci adv comt, 85-88; head proj scientist, Nat Ozone Exped, McMurdo Sta, Antarctica, 86-87; adj fac, Univ Colo, Boulder, 83-; assoc ed, J Geophys Res, 85- *Honors & Awards:* J B MacElwane Award, Am Geophys Union, 85. *Mem:* Am Geophys Union; Royal Meteorol Soc. *Res:* Photochemistry; transport processes in the earth's stratosphere and mesosphere; polar ozone. *Mailing Add:* ERL Aeron Lab NOAA 325 Broadway Boulder CO 80303

SOLOMON, THOMAS ALLAN, b New Kensington, Pa, Apr 3, 41; m 68. PHARMACOLOGY, CARDIOVASCULAR PHYSIOLOGY. *Educ:* Westminster Col, Pa, BS, 64; WVa Univ, MA, 67; Univ Pittsburgh, PhD(pharmacol), 72. *Prof Exp:* From instr to asst prof psychiat & behav biol, Sch Med, 72-76, instr environ med, Sch Pub Health, 73-75, asst prof psychiat & behav biol, Sch Med, Johns Hopkins Univ, 74-76, instr environ med, Sch Pub Health, 73-75; head cardiovasc pharmacol, Pharmaceut Div, Sandoz, Inc, 74-80; assoc dir, Dept Clin Pharmacol, Revlon Health Care, 80-85, dir, 85-87; assoc dir, Dept Clin Pharmacol, 87-90, ASSOC DIR, DEPT MED IMAGING, BERLEX LABS, 90- *Mem:* NY Acad Sci; Johns Hopkins Med Surg Soc; AAAS; Am Heart Asn; Sigma Xi. *Res:* Circulation; cardiovascular system and its regulation; control mechanisms involved in hypertension; cardiovascular pharmacology. *Mailing Add:* 265 Old Mill Rd Chester NJ 07930

SOLOMON, VASANTH BALAN, b Nagercoil, Madras, India, Aug 8, 35; US citizen; m 60; c 2. APPLIED STATISTICS. *Educ:* Univ Madras, BSc, 58, MSc, 61; Iowa State Univ, PhD(statist), 70. *Prof Exp:* Statistician, Rubber Res Inst Ceylon, 62-64; biometrician, Dept Fisheries & Forestry, Govt of Can, 67-69; assoc prof statist, Drake Univ, 70-88; PROF STATIST, RADFORD UNIV, 88- *Mem:* Am Statist Asn; Sigma Xi. *Res:* Statistical research in epidemiological problems. *Mailing Add:* Dept Math & Statist Radford Univ Norwood St & Rte 11 Radford VA 24142

SOLOMONOW, MOSHE, b Tel-Aviv, Israel, Oct 24, 44; US & Israeli citizen; m; c 2. REHABILITATION ENGINEERING, NEUROSCIENCES. *Educ:* Calif State Univ, BS, 70, MS, 72; Univ Calif, Los Angeles, PhD(eng), 76. *Prof Exp:* Chief engr med eng, Calmag Electronics, 68-71; res engr neuromuscular eng, Rancho Los Amigos Hosp, 71-72; proj engr med eng, Clamag Electronics, 72-73; clin intern prosthetics, Child Amputee Clin, Univ Calif, Los Angeles, 75, res engr rehabilitative eng, 73-80; assoc prof, dept biomed eng, Tulane Univ, 80-83; PROF & DIR BIOENGINEERING, LA STATE UNIV MED CTR, NEW ORLEANS, 83- *Concurrent Pos:* Consult, Olivetti Am Inc, 74-75 & Child Amputee Clin, Univ Calif, Los Angeles, 75-76; prin engr, Bennett Respiration Prod Inc, 77; consult, Lida Inc, 75-, Perceptronics Inc, 75-76, Vet Admin Hosp, Brentwood, 77-78, Vet Admin Hosp, Sepulveda, 78-80 & Dept Health, La, 81-85, NIH, 80, NSF, 87-, Vet Admin, Washington, 87- *Mem:* Biomed Eng Soc; AAAS; Sigma Xi; Inst Elec & Electronics Engrs; Int Soc Biomechanics; Int Soc Electrophysiological Kinesiology; Orthopedic Res Soc. *Res:* Prosthetics, orthutics, biomechanics and electrophysiology of movement. *Mailing Add:* Dept Orthopaedics La State Univ Med 433 Bolivar St New Orleans LA 70112

SOLOMONS, CLIVE (CHARLES), b Johannesburg, SAfrica, June 6, 31; m 56; c 3. BIOCHEMISTRY, PEDIATRICS. *Educ:* Univ Witwatersrand, BS, 52, PhD(biochem), 56. *Prof Exp:* Biochemist, SAfrican Inst Med Res, 52-55; biochemist, Dent Res Univ, Coun Sci & Indust Res & Univ Witwatersrand, 55-61; asst prof biochem, McGill Univ, 61-63; assoc prof pediat, 63-75, PROF ORTHOP, DIR ORTHOP RES, ASSOC PROF ANAESTHESIOL, 75-, UNIV COLO MED CTR, DENVER, RES PROF, 85- *Concurrent Pos:* Fel radiation biol, Univ Rochester, 58-59; Can Med Res Coun grant, 61-63; NIH grant, 64-; NSF grant; Cystic Fibrosis Res Found grant. *Res:* Application of analytical biochemistry to clinical and basic science investigation, especially on the metabolism of connective tissue disorders; metals in the environment, cystic fibrosis and renal disease, and anaesthetic risk; Reyes syndrome; orthopedics anaesthesiology. *Mailing Add:* 164 S Fairfax Denver CO 80222

SOLOMONS, GERALD, b London, Eng, Feb 22, 21; US citizen; m 55; c 2. PEDIATRICS, CHILD GROWTH. *Educ:* Royal Col Physicians & Surgeons, Edinburgh, LRCP, LRCS, 43; Royal Col Physicians & Surgeons Eng, dipl child health, 48; Am Bd Pediat, dipl, 52. *Prof Exp:* Asst supt, Charles V Chapin Hosp, Providence, RI, 52; pvt pract, 53-59; dep dir pediat, Inst Health Sci, Brown Univ, 59-62, asst mem, 60-62; from asst prof to assoc prof, 62-69, PROF PEDIAT, UNIV IOWA, 69-, DIR CHILD DEVELOP CLIN, 63-, ACTG HEAD, INST CHILD BEHAV & DEVELOP, 75- *Concurrent Pos:* Consult, NIH, 59-; prog dir, Regional Ctr Child Abuse & Neglect, 75- *Mem:* Fel Am Acad Pediat; fel Am Acad Cerebral Palsy (pres, 77-78); Am Asn Ment Deficiency. *Res:* Child abuse; minimal brain damage, its diagnosis, drug therapy and effect on learning. *Mailing Add:* 319 Mullin Ave Iowa City IA 52246

SOLOMONS, NOEL WILLIS, b Boston, Mass, Dec 31, 44. CLINICAL NUTRITION, GASTROENTEROLOGY. *Educ:* Harvard Univ, AB, 66; Harvard Med Sch, MD, 70. *Prof Exp:* Instr med, Univ Chicago, 75-76, res assoc gastroenterol, 77-79; from asst to assoc prof clin nutrit, dept nutrit & food sci, Mass Inst Technol, 80-84; SR SCIENTIST, CTR STUDIES SENSORY IMPAIRMENT, AGING & NUTRITION, GUATEMALA CITY, 85- *Concurrent Pos:* Nutrit Found grant, 74-77; Josiah Macy Jr Found fac fel, 75-76; res assoc clin nutrit, Div Human Nutrit & Biol, Inst Nutrit Cent Am & Panama, 76-78, affil sci, 78-; mem, Comt Int Nutrit, Nat Acad Sci, 79-82. *Mem:* Am Gastroenterol Asn; Am Soc Clin Nutrit; Am Fedn Clin Res; Am Soc Nutrit; Latin Am Nutrit Soc. *Res:* Trace mineral nutrition; protein-energy malnutrition; trace mineral absorption, non-invasive and stable isotope technology in absorptive physiology. *Mailing Add:* CeSSIAM, Hosp de Ojos y Oidos Diagonal 21 y 19 Calle, Zona 11 Guatemala City Guatemala

SOLOMONS, THOMAS WILLIAM GRAHAM, b Charleston, SC, Aug 30, 34. ORGANIC CHEMISTRY. *Educ:* The Citadel, BS, 55; Duke Univ, PhD(chem), 59. *Prof Exp:* Sloan Found fel, Univ Rochester, 59-60; instr, 60-61, from asst prof to assoc prof, 61-73, PROF CHEM, UNIV S FLA, 73- *Mem:* Am Chem Soc. *Res:* Synthesis and reactions of heterocyclic aromatic compounds. *Mailing Add:* Hanging Birch House Hanging Birch Lane Horam Heathfield E Sussex TN21 OPA England

SOLOMONS, WILLIAM EBENEZER, b Ridgeland, SC, Oct 2, 43; m 65; c 2. MEDICINAL CHEMISTRY, ORGANIC CHEMISTRY. *Educ:* Berry Col, BA, 65; Univ Miss, PhD(pharmaceut chem), 70. *Prof Exp:* Res asst prof, Ctr Health Sci, Univ Tenn, Memphis, 70-73, res assoc prof med chem, 73-76; assoc prof, 76-83, PROF CHEM, UNIV TENN, MARTIN, 83- *Mem:* Am Chem Soc; Sigma Xi. *Res:* Synthetic organic chemistry; organic synthesis and structure determination; the relationships between molecular structure and biological activity; synthesis of novel analgesic and antipsychotic agents. *Mailing Add:* 107 Alberta Martin TN 38237

SOLOMONSON, LARRY PAUL, b Scarville, Iowa, June 26, 41; m 68; c 2. BIOCHEMISTRY. *Educ:* Luther Col, BA, 63; Univ Chicago, PhD(biochem), 69. *Prof Exp:* Res chemist res & develop, Borden Chem Co, 63-64; amanuensis, Physiol Inst, Univ Aarhus, Denmark, 69-70; scientist, Max Planck Inst, Berlin, Ger, 70-74; vis asst prof, Col Med, Univ Iowa, 74-76; asst prof, 76-79, assoc prof biochem, 80-86, prof & actg chmn, 86-88, CHMN, DEPT BIOCHEM & MOLECULAR BIOL, UNIV SOUTH FLA, 88- *Mem:* AAAS; Am Chem Soc; Am Soc Biol Chemists; Am Soc Plant Physiologists; Sigma Xi. *Res:* Mechanism and regulation of nitrate assimilation; molecular properties and functions of the sodium pump. *Mailing Add:* Dept Biochem & Molecular Biol Col Med Univ South Fla Tampa FL 33612-4799

SOLON, LEONARD RAYMOND, b White Plains, NY, Sept 11, 25; m 46; c 3. RADIOLOGICAL PHYSICS, RADIATION BIOLOGY. *Educ:* Hamilton Col, AB, 47; Rutgers Univ, MSc, 49; NY Univ, PhD(radiol health), 60. *Prof Exp:* Teaching asst physics, Rutgers Univ, 47-49; physicist, Nuclear Develop Assocs, 50-52; physicist radiation br, Health & Safety Lab, US Atomic Energy Comn, 52-54, asst chief, 54-59, chief, 59-60; dir appl nuclear tech, Tech Res Group, Inc, NY, 60-64; mgr res & develop, Del Electronics Corp, mem vpres & tech dir, Hadron Inc, Westbury, 67-75; DIR, BUR RADIATION CONTROL, DEPT HEALTH, NEW YORK, 75- *Concurrent Pos:* Lectr, Med Center, NY Univ, 56-60, adj asst prof, 60-62, adj assoc prof, 62-; Mem tech consults panel, Div Mil Appln, US Atomic Energy Comn, 57-60, consult, Health & Safety Lab, 62-65; prof health physics, US Merchant Marine Acad, 64. *Mem:* AAAS; Am Phys Soc; Sigma Xi; Health Physics Soc; Am Nuclear Soc. *Res:* Radiation protection and health physics; reactor and accelerator shielding; environmental radiation measurements; laser physics and applications; biomedical instrumentation; radiation dosimetry; stratospheric sampling; application of lasers to thermonuclear fusion. *Mailing Add:* 28 Pilgrim Ave Yonkers NY 10710

SOLONCHE, DAVID JOSHUA, b New York, NY, Apr 10, 45; m 66; c 2. BIOMEDICAL ENGINEERING, BIOMEDICAL COMPUTING. *Educ:* Yeshiva Col, BA, 66; Worcester Polytech Inst, PhD(biomed eng), 71. *Prof Exp:* Res asst, 71-72, from instr to asst prof orthod, Sch Dent Med, 72-79, DIR BIOENG, HEALTH CTR, UNIV CONN, 79- *Concurrent Pos:* Consult eng, Trinity Col, 74; assoc prof biomed eng, Hartford Grad Ctr, 74- *Mem:* Inst Elec & Electronics Eng; Biomed Eng Soc; Instrument Soc Am. *Res:* Application of computers to biomedical research; biological signal processing; bioelectronics; automated orthodontic diagnosis. *Mailing Add:* Dir Bioeng Univ Conn Health Ctr Farmington CT 06030

SOLORZANO, ROBERT FRANCIS, b New York, NY, May 21, 29; m 85; c 6. VIROLOGY, MEDICAL MICROBIOLOGY. *Educ:* Georgetown Univ, BS, 51; Pa State Univ, MS, 56, PhD(bact), 62. *Prof Exp:* Bacteriologist, Montefiore Hosp Chronic Dis, New York, 53-54; res asst virol, Children's Hosp, Philadelphia, Pa, 56-58; asst, Pa State Univ, 58-62; asst virologist, Coastal Plain Exp Sta, Univ Ga, 62-68; assoc prof, 68-78, PROF VET MICROBIOL, COL VET MED & SR VIROLOGIST, VET MED DIAG LAB, UNIV MO-COLUMBIA, 78- *Concurrent Pos:* Mem, Am Asn & NCent Conf Vet Lab Diagnosticians; Fulbright res fel, Mexico. *Mem:* AAAS; Am Soc Microbiol; Am Asn Vet Lab Diagnosticians; Am Leptosirosis Res Conf. *Res:* Japanese B encephalitis ecology; effect of sonic vibrations on Newcastle virus; entero-cytopathogenic human orphan virus serology; fluorescent antibody test for hog cholera; serology for leptospirosis; ecology of hog cholera virus; enteric virus diseases of swine; diagnostic virology, and serology; pseudorabies; turkey parvovirus. *Mailing Add:* Vet Med Diag Lab Col Vet Med Univ Mo Columbia MO 65211

SOLOTOROVSKY, MORRIS, b New York, NY, Oct 10, 13; m 45; c 4. BACTERIOLOGY. *Educ:* Univ Va, BS, 34; NY Univ, MS, 38; Columbia Univ, PhD(bact), 47. *Prof Exp:* Asst sanit sci, Col Physicians & Surgeons, Columbia Univ, 36-, instr epidemiol, 41-42; bacteriologist, Guggenheim Bros, NY, 40-41; res assoc chemotherapy, Merck Inst, 46-58; PROF BACT, RUTGERS UNIV, 58- *Mem:* AAAS; Am Soc Microbiol; Am Asn Immunol; Fel Am Acad Microbiol. *Res:* Host-parasite interaction; chemotherapy of tuberculosis and fungus diseases; cell-mediated immunity; bacterial vaccines. *Mailing Add:* Nelson Biol Labs Rutgers Univ New Brunswick NJ 08903

SOLOVAY, ROBERT M, LOGIC, SET THEORY. *Prof Exp:* PROF MATH, UNIV CALIF-BERKELEY, 86- *Mem:* Nat Acad Sci. *Mailing Add:* Dept Math Univ Calif Berkeley CA 94720

SOLOW, DANIEL, b Washington, DC, Nov 19, 49; m 80. MATHEMATICAL PROGRAMMING. *Educ:* Carnegie-Mellon Univ, BS, 70; Univ Calif, Berkeley, MS, 72; Stanford Univ, PhD(opers res), 78. *Prof Exp:* ASST PROF OPERS RES, CASE WESTERN RESERVE UNIV, 78- *Mem:* Opers Res Soc; Math Prog Soc; Am Math Soc; Math Asn Am. *Res:* Development of computational algorithms for solving mathematical problems arising in combinatorial optimization, mathematical programming and operations research. *Mailing Add:* Dept Opers Res Case Western Reserve Univ University Circle Cleveland OH 44106

SOLOW, MAX, b Philadelphia, Pa, Nov 20, 16; m 41; c 2. PHYSICS, METALLURGY. *Educ:* George Washington Univ, BEE, 43, MS, 50; Catholic Univ, PhD(physics), 57. *Prof Exp:* Radio engr, Nat Bur Stand, 46-49, electronic scientist, 49-53; physicist, US Naval Ord Lab, 53-60; sr scientist, Martin Co, 60-64; res coordr physics, US Navy Marine Eng Lab, 64-68 & Naval Ship Res & Develop Ctr, Annapolis, 68-71, sr res scientist/tech consult, 71-80; MEM STAFF, UNIV MD, COLLEGE PARK, MD, 81- *Mem:* Am Phys Soc; AAAS; Am Soc Metals; Inst Elec & Electronics Engrs; Am Inst Mining, Metall & Petrol Engrs; Sigma Xi. *Res:* Solid state physics; metals; electrochemistry; corrosion, materials; vacancy and dislocation technique and theory; random noise; explosion hydrodynamics; lasers; holography. *Mailing Add:* 823 Painted Post Ct Baltimore MD 21208

SOLOWAY, ALBERT HERMAN, b Worcester, Mass, May 29, 25; m 53; c 3. MEDICINAL CHEMISTRY. *Educ:* Worcester Polytech Inst, BS, 48; Univ Rochester, PhD(org chem), 51. *Prof Exp:* USPHS fel, Sloan-Kettering Inst, 51-53; res chemist, Eastman Kodak Co, 53-56; res assoc surg, Harvard Med Sch, 56-63; from asst chemist to assoc chemist, Mass Gen Hosp, 56-73; assoc prof med chem, Northeastern Univ, 66-71, chmn dept med chem & pharmacol, Col Pharm & Allied Health Professions, 71-74, prof med chem & chem, 71-77, dir grad sch, Pharm & Allied Health Professions, 73-77, dean, Col Pharm & Allied Health Professions, 75-77; prof med chem & dean, 77-88, PROF MED CHEM, COL PHARM, OHIO STATE UNIV, 88- *Mem:* Fel AAAS; Am Chem Soc; Am Asn Cols Pharm; Am Pharmaceut Asn; Am Asn Cancer Res. *Res:* Cancer therapy; development of drugs for chemoimmuno and chemoradiotherapy; use of Boron compounds in cancer. *Mailing Add:* Ohio State Univ Col Pharm 500 W 12th Ave Columbus OH 43210

SOLOWAY, HAROLD, b New York, NY, June 15, 17; m 47; c 2. ORGANIC CHEMISTRY, MEDICINAL CHEMISTRY. *Educ:* Brooklyn Col, BA, 38; Polytech Inst Brooklyn, MS, 48. *Prof Exp:* Chemist, George Washington Coffee Div, Am Home Foods, Inc, 46-48; asst chemist, Sterling-Winthrop Res Inst, 49-52; chemist, US Vitamin & Pharmaceut Corp, 52-55, sr res chemist, 55-66; sr chemist, Endo Labs, Inc, 66-73, group leader, 73-76; group leader, Biomed Dept, E I duPont de Nemours & Co Inc, 76-84; RETIRED. *Mem:* NY Acad Sci; AAAS; Am Chem Soc. *Res:* Synthesis and reactions of organic heterocyclic compounds; synthesis of organic medicinals; alkaloid chemistry. *Mailing Add:* 529A Heritage Hills Somers NY 10589-1907

SOLOWAY, S BARNEY, b New York, NY, Jan 21, 15; m 38; c 2. ORGANIC CHEMISTRY. *Educ:* City Col New York, BS, 36; Univ Colo, PhD(chem), 55. *Prof Exp:* Chemist, Div Insecticide Invests, USDA, 41-47; chemist & asst dir, Julius Hyman & Co, 47-52; supvr org res, Agr Res Div, Shell Develop Co, 52-64, head, Org Chem Div, Woodstock Agr Res Ctr, Shell Res Ltd, 64-67, head org chem, 67-80, asst to dir, Agr Res Div, Shell Develop Co, 80- 86; RETIRED. *Mem:* Am Chem Soc. *Res:* Synthesis; stereochemistry; agricultural chemicals. *Mailing Add:* 3401 Mansfield Lane Modesto CA 95350

SOLOWAY, SAUL, b New York, NY, Apr 12, 16; m 44; c 3. ORGANIC CHEMISTRY. *Educ:* City Col New York, BS, 36; Columbia Univ, AM, 38, PhD(chem), 42. *Prof Exp:* Hernschiem fel, Mt Sinai Hosp, New York, 40-41; mem staff, Nat Defense Res Comn, US Bur Mines, 41-43, Panel Chem Corp, 43-44 & Grosvenor Labs, 44-46; instr chem, City Col New York, 46-50, from asst prof to assoc prof, 50-73; CONSULT CHEMIST, 73- *Concurrent Pos:* Consult, Faberge, Inc, 56-, dir res, 58-; consult, Revlon, 60-64. *Mem:* Am Chem Soc; NY Acad Sci. *Res:* Chelation; cosmetics; perfume; encapsulation of liquids; lipids; organic analysis; polymerization; thermochromism. *Mailing Add:* 180 Broadview Ave New Rochelle NY 10804

SOLOYANIS, SUSAN CONSTANCE, b New York, NY, Jan 21, 52. GLACIAL GEOLOGY, PETROLEUM GEOLOGY. *Educ:* Smith Col, AB, 72; Univ Mass, Amherst, MS, 75, PhD(geol), 78. *Prof Exp:* Geologist environ geol, Conn Valley Urban Area Proj, US Geol Surv, 71-75; teaching asst geol & geog, Univ Mass, Amherst, 74-78; geologist petrol geol, Amoco Prod Co, 78-89; HYDROGEOLOGIST, MITRE CORP, 89- *Concurrent Pos:* Teaching asst geol, Univ Ill, Urbana, 72-73. *Mem:* Geol Soc Am; Am Geophys Union; Sigma Xi; Soc Econ Paleontologists & Mineralogists. *Res:* Pleistocene paleomagnetic stratigraphy; magnetization of sediments; glacial sedimentation; petroleum exploration. *Mailing Add:* 9611 Azalea Circle Garden Ridge TX 78266-2501

SOLSKY, JOSEPH FAY, b Corning, NY, June 9, 49; m 70; c 2. ANALYTICAL CHEMISTRY. *Educ:* State Univ NY, Buffalo, BA, 71, PhD(chem), 78. *Prof Exp:* Asst prof chem, Creighton Univ, 76-85; US ARMY CORPS ENGRS, 85- *Mem:* Am Chem Soc. *Res:* Investigations of stationary phases used in chromatographic systems including liquid crystal and permanently bound types. *Mailing Add:* US Army Corps Engrs 420 S 18th St Omaha NE 68102-2501

SOLT, DENNIS BYRON, ORAL PATHOLOGY. *Educ:* Temple Univ, PhD(exp path), 78. *Prof Exp:* ASSOC PROF PATH, NORTHWESTERN UNIV MED & DENT SCHS, 82- *Mailing Add:* Dept Path Northwestern Univ Med & Dent Schs 303 E Chicago Ave Chicago IL 60611

SOLT, PAUL E, b Allentown, Pa, Feb 23, 29; m 50; c 2. PNEUMATIC CONVEYING OF BULK MATERIALS. *Educ:* Lehigh Univ, BS, 50. *Prof Exp:* Serv engr, Fuller Co, 50-52, res engr, 54-56; Lt, USAF, Korea, 52-54; proj engr, Mack Trucks, 56-62; res engr, Fuller Co, GATX, 62-68, mgr res, 68-84; CONSULT, PNEUMATIC CONVEYING CONSULTS, 84- *Concurrent Pos:* Course dir, Ctr Prof Advan, 72- & Am Inst Chem Engrs, 84-; consult, Teltech Resource Network, 87-; consult ed, Power & Bulk Eng, 89- *Mem:* Am Inst Chem Engrs. *Res:* Pneumatic conveying of bulk materials, including design, troubleshooting, engineering lectures and courses, expert witness and system modifications. *Mailing Add:* 529 S Berks St Allentown PA 18104

SOLTAN, HUBERT CONSTANTINE, b Wilno, Poland, Dec 16, 32; Can nat; m 62; c 3. MEDICAL GENETICS. *Educ:* Univ Toronto, BA, 55, PhD(human genetics), 59; Univ Western Ont, MD, 70. *Prof Exp:* Res fel genetics, Hosp Sick Children, Toronto, Ont, 55-58; asst prof biol, St Mary's Univ, NS, 58-61; from asst prof to assoc prof human genetics, Fac Med, 61-77, clin assoc prof pediat, 71-77, PROF HUMAN GENETICS, UNIV

WESTERN ONT, 77-, CLIN PROF PEDIAT, 77-; MED GENETICIST, CHILDREN'S HOSP WESTERN ONT, 77- *Mem:* Genetics Soc Can; Asn Genetic Counr Ont; Can Col Med Geneticists; Am Soc Human Genetics. *Mailing Add:* Children's Hosp Western Ont 800 Commissioners Rd E London ON N6C 2V5 Can

SOLTANPOUR, PARVIZ NEIL, b Tehran, Iran, Mar 21, 37; m 60; c 4. SOIL FERTILITY, AGRONOMY. *Educ:* Am Univ Beirut, BS, 61, MS, 63; Univ Nebr, PhD(soil fertility), 66. *Prof Exp:* FROM ASST PROF TO PROF SOIL FERTILITY, COLO STATE UNIV, 66- *Concurrent Pos:* Consult, Egypt Water Mgt Proj, 78-83, Morocco Dryland Proj, 85-88, Asn Int Develop, Comn Int Develop & Colo State Univ; chmn, Coun on Soil Testing & Plant Anal, 83-84. *Mem:* Am Soc Agron; Soil Sci Soc Am; Int Soc Soil Sci; Sigma Xi. *Res:* Methods of soil testing for fertilizer recommendations; soil fertility and plant nutrition. *Mailing Add:* Agron Dept Colo State Univ Ft Collins CO 80523

SOLTER, DAVOR, b Zagreb, Yugoslavia, Mar 22, 41. DEVELOPMENTAL BIOLOGY. *Educ:* Univ Zagreb, MD, 65, MSc, 68, PhD(biol), 71. *Prof Exp:* Instr anat, Med Sch, Univ Zagreb, 66-68, instr biol, 68-72, asst prof, 72-73; assoc scientist, 73-75, assoc mem, 75-80, PROF, WISTAR INST, 81-; DIR, MAX PLANCK INST. *Concurrent Pos:* Europ Molecular Biol Orgn scholar, 71; Damon Runyon Mem Cancer Fund fel, 73; assoc ed, Develop Biol, 80-87; mem study sect human embryol & develop, NIH, 81-85; Wistar prof biol, fac arts & sci, Univ Pa, 84 -; ed bd, Cell, 87, Genes & Develop, 87. *Mem:* Soc Develop Biol. *Res:* Development of early mouse embryo; role of membrane molecules in development of early mouse embryo; cross-reacting antigens on embryos and tumor cells; regulation and differentiation of embryo derived teratocarcinomas; genetic control of development, nuclear transfer and trangenic animals. *Mailing Add:* Max Planck Inst Stubeweg 51 D7800 Freiburg Germany

SOLTERO, RAYMOND ARTHUR, b Milwaukee, Wis, July 20, 43; m 60; c 3. LIMNOLOGY, WATER POLLUTION. *Educ:* Mont State Univ, BS, 66, MS, 68, PhD(bot), 71. *Prof Exp:* from asst prof to assoc prof, 71-79, PROF BIOL, EASTERN WASH UNIV, 80- *Mem:* Am Soc Limnol & Oceanog; Mem: Int Soc Limnol; Sigma Xi. *Res:* Eutrophication of lakes, streams and reservoirs; lake restoration. *Mailing Add:* Dept Biol Eastern Wash Univ Cheney WA 99004

SOLTES, EDWARD JOHN, b Montreal, Que, Mar 25, 41; m; c 4. WOOD CHEMISTRY. *Educ:* McGill Univ, BSc, 61, PhD(carbohydrate chem), 65. *Prof Exp:* Fel, Ohio State Univ, 65-66, lectr, 66; sr res chemist, Tech Ctr, St Regis Paper Co, 66-76, asst to dir res & develop, 70-71, responsibility Sylvachem Res & Develop, 73-75, responsibility wood chem, 75-76; assoc prof forest sci, Tex A&M Univ, 76-81, prof wood chem, 81-84, prof forest sci & plant physiol, Agr Exp Sta, 84-89, interim dept head, 89-90, ASSOC HEAD ACAD AFFAIRS, TEX A&M UNIV, 90- *Concurrent Pos:* Chmn, Div Cellulose, Paper & Textile Chem, Am Chem Soc, 79, secy-gen, Macromolecular Secretariat, 82, counr, Am Chem Soc, 84-91, chmn, Tex A&M Sect, 86. *Honors & Awards:* Res Award, Tex Forestry Asn, 83. *Mem:* Am Chem Soc. *Res:* Wood chemistry, utilization of agricultural and forestry residues, pyrolysis, naval stores; physiology of tissue culture processes, molecular bases for host/pathogen interactions, photo bioreactor development and bioprocessing; resource sustainability. *Mailing Add:* Forest Sci Dept Tex A&M Univ College Station TX 77843-2135

SOLTYSIK, EDWARD A, b Newark, NJ, Aug 23, 29; m 58; c 3. ATOMIC PHYSICS. *Educ:* Lafayette Col, BS, 50; Ind Univ, MS, 52, PhD(nuclear physics), 56. *Prof Exp:* Lectr physics, Univ Nev, 55-56; physicist, Lawrence Radiation Lab, 56-62; assoc prof, 62-71, PROF PHYSICS, UNIV MASS, AMHERST, 71- *Concurrent Pos:* Consult, Lawrence Radiation Lab, 62-65 & Air Force Off Sci Res grants, 63- *Res:* Nuclear decay, shake off process and inner Bremsstrahlung; atomic physics, polarization of collisional radiation, especially radiation resulting from the collisions of electrons and protons on atoms. *Mailing Add:* Dept of Physics Univ of Mass Amherst MA 01003

SOLTYSIK, SZCZESNY STEFAN, b Zakopane, Poland, Mar 27, 29; m 55; c 2. ANIMAL BEHAVIOR, NEUROPHYSIOLOGY. *Educ:* Jagiellonian Univ, Poland, MD, 53; Polish Acad Sci, PhD(behav sci), 60, Docent Sci, 65. *Prof Exp:* Asst prof human physiol, Sch Med, Jagiellonian Univ, 50-53; asst prof neurophysiol, Nencki Inst Exp Biol, 54-64, docent, 65-69; assoc prof, Inst Psychoneurol, Warsaw, 65-71; asst res anatomist, Brain Res Inst, 62-64, assoc res anatomist, Sch Med, 71-75, assoc prof, 75-76, PROF PSYCHIAT, DEPT PSYCHIAT & NEUROPSYCHIAT INST, UNIV CALIF, LOS ANGELES, 76- *Mem:* Int Brain Res Orgn; Psychonomic Soc. *Res:* Blocking, protection from extinction; emotional behavior, classical and operant conditioning in normal and brain operated kittens at different ages. *Mailing Add:* Neuropsychiat Inst Rm 58-258 Univ Calif Los Angeles Los Angeles CA 90024

SOLTZ, DAVID LEE, b La Cross, Wis, Nov 7, 46; m 78; c 2. POPULATION ECOLOGY, ICHTHYOLOGY. *Educ:* Univ Calif, BA, 68, PhD(biol), 74. *Prof Exp:* from asst prof to assoc prof, 74-82, PROF BIOL, CALIF STATE UNIV, 82-, CHMN DEPT, 81- *Concurrent Pos:* NSF grant, 77-79; vis scientist, Univ Mich, 84; consult, Bur Land Mgt, US Fish & Wildlife Serv, Los Angeles County, Santa Barbara County, 78- *Mem:* AAAS; Am Soc Ichthyologists & Herpetologists; Ecol Soc Am; Soc Study Evolution. *Res:* Population biology; evolutionary and reproductive ecology of fish populations; community ecology of isolated freshwater habitats. *Mailing Add:* Dept Biol Calif State Univ 1250 Bellflower Blvd Long Beach CA 90840-3702

SOLTZBERG, LEONARD JAY, b Wilmington, Del, July 10, 44. PHYSICAL CHEMISTRY, CRYSTALLOGRAPHY. *Educ:* Univ Del, BS, 65; Brandeis Univ, MA, 67, PhD(phys chem), 69. *Prof Exp:* Nat Res Coun res assoc, Air Force Cambridge Res Lab, 69; asst prof, 69-73, assoc prof, 73-79, PROF CHEM, SIMMONS COL, 79- *Mem:* Sigma Xi; Am Crystallog Asn; Am Chem Soc. *Res:* Chemical crystallography; optical and x-ray crystallography; phase transitions; microscopy; pedagogical computer application. *Mailing Add:* Dept of Chem Simmons Col 300 The Fenway Boston MA 02115

SOLURSH, MICHAEL, b Los Angeles, Calif, Dec 22, 42; m 64; c 1. DEVELOPMENTAL BIOLOGY, CELL BIOLOGY. *Educ:* Univ Calif, Los Angeles, BA, 64; Univ Wash, PhD(zool), 69. *Prof Exp:* Teaching asst zool, Univ Wash, 64-66; from asst prof to assoc prof, 69-79, PROF BIOL, UNIV IOWA, 79- *Mem:* Am Soc Cell Biol; Am Soc Zool; Soc Develop Biol; Tissue Culture Asn; Am Asn Anatomists. *Res:* Extracellular materials in morphogenesis and migration of primary mesenchyme cell in sea urchin embryos; cartilage cell differentiation and limb morphogenesis (heterotypic and homotypic cell interaction during chondrogenesis). *Mailing Add:* Dept of Biol Univ of Iowa Iowa City IA 52242-1368

SOLVIK, R S(VEN), b Fauske, Norway, Apr 24, 24; m 49; c 2. CHEMICAL ENGINEERING, CHEMISTRY. *Educ:* Tech Univ Norway, MS, 49. *Prof Exp:* Prod supvr, Mjondalen Rubber Factory, 50-52; engr, Technol Sect, E I du Pont de Nemours & Co, 52-57; group leader, Polymer Pilot Plant, US Indust Chem Co Div, Nat Distillers & Chem Corp, 57, res supvr polyolefin res, 57-59, asst mgr, 59-61, mgr, 61-66; DIR RES, CHEMPLEX CO, ROLLING MEADOWS, 66- *Mem:* Am Chem Soc; Soc Plastics Engrs. *Res:* Mechanical and industrial rubber goods and plastics; process development in high and low density polyehtylene, copolymers and polypropylene; catalyst and exploratory research in olefin polymerization; applications research, polymer development and technical service. *Mailing Add:* 176 Club Circle LBS Lake Barrington Shore Barrington IL 60010-1611

SOM, PRANTIKA, b Silchar, Assam, India, Aug 31, 42; US citizen. NUCLEAR MEDICINE, VETERINARY MEDICINE. *Educ:* Univ Calcutta, ISc, 60, DVM, 65; Johns Hopkins Univ, ScM, 69. *Prof Exp:* Demonstr path, Bengal Vet Col, 65-66; investr, Marine Biol Lab, Wood's Hole, Mass, 67-68; asst pathobiol, Johns Hopkins Med Inst, 67-69, sr res fel, 73-74; from asst scientist to assoc scientist, 75-80, SCIENTIST, NUCLEAR MED, BROOKHAVEN NAT LAB, 80- *Concurrent Pos:* Reserve vet asst surgeon, Govt W Bengal, 65-66; jr res fel, Johns Hopkins Univ, 70-72, asst radiol, 73-74; mem vet serv comt, Brookhaven Nat Lab, 75-; mem educ comt, Soc Nuclear Med, 76-; res asst prof, State Univ NY, 79-86; consult, Vet Admin Hosp, Nathpat, 81-; res assoc prof, State Univ NY, 86- *Honors & Awards:* Raymond Star Gold Medal. *Mem:* Soc Nuclear Med; Am Vet Med Asn; Radiol Soc NAm; Am Asn Lab Animal Sci. *Res:* Radiopharmaceutical development; evaluations and studies on their pharmacokinetics, metabolism and toxicology. *Mailing Add:* Two Taylor Commons Shirley NY 11067

SOMA, LAWRENCE R, b New York, NY, Feb 2, 33; m 55; c 3. ANESTHESIOLOGY. *Educ:* Univ Pa, VMD, 57. *Prof Exp:* Intern vet med, Animal Med Ctr, NY, 57-58; fel anesthesiol, Sch Med, 60-62, instr, Sch Vet Med, 62-64, from asst prof to assoc prof, 64-72, PROF ANESTHESIOL, SCH VET MED, UNIV PA, 72-, CHMN DEPT CLIN STUDIES, 75- *Concurrent Pos:* NIH career develop award, 67-72, staff mem, Dept Anesthesiol, Sch Med, 71-; spec fel, Heart Lung Inst, 74-75. *Mem:* Am Vet Med Asn; AAAS; Am Soc Vet Physiologists & Pharmacologists; Am Thoracic Soc; Am Soc Anesthesiol. *Res:* Veterinary anesthesiology; anesthesia and pharmacology; effects of respiratory stimulants in the dog; cardiovascular effects of local anesthetics; effects of anesthetics on the fetus, pathophysiology of shock lung; physiology of bronchial circulation. *Mailing Add:* Dept Anesthesiol Univ Pa Sch Vet Med New Bolten Ctr Kennett Square PA 19348

SOMANI, ARUN KUMAR, b Beawar, Raj, India, July 16, 51; m 87; c 3. PARALLEL COMPUTER SYSTEMS, FAULT-TOLERANT COMPUTING. *Educ:* Birla Inst Technol & Sci, Pilani, India, BE Hons, 73; IIT, Delhi, ME, 79; McGill Univ, MSEE, 83, PhD(elec eng), 85. *Prof Exp:* Tech officer, Electronics Corps India, 73-74; sci officer, Dept Electronics, Systems Group, Govt India, 74-79, scientist D, 79-82; asst prof, 85-90, ASSOC PROF ELEC ENG, DEPT ELEC ENG, UNIV WASH, SEATTLE, 90- *Concurrent Pos:* Assoc prof, Dept Computer Sci & Eng, Univ Wash, 90-; consult, Boeing Com, 91- *Mem:* Sr mem Inst Elec & Electronics Engrs; Inst Elec & Electronics Engrs Computer Soc; Asn Comput Mach. *Res:* Design of fault tolerant parallel computer system; fault diagnosis algorithms; parallel computer algorithms; computer communication networks; modeling and analysis of computer systems. *Mailing Add:* 16609 126th Ave NE Woodinville WA 98072

SOMANI, PITAMBAR, b Chirawah, India, Oct 31, 37; m 60; c 3. CLINICAL PHARMACOLOGY, MEDICINE. *Educ:* G R Med Col, Gwalior, India, MD, 60; Marquette Univ, PhD(pharmacol), 65. *Prof Exp:* Demonstr pharmacol, Indian Inst Med Sci, New Delhi, 60-62; from instr to asst prof, Sch Med, Marquette Univ, 65-69; assoc prof, Med Col Wis, 69-71, assoc clin prof, 71-74; prof pharmacol, Sch Med, Univ Miami, 74-80; dir clin pharmacol, Med Col Ohio, 80-90; ASST DIR HEALTH, STATE OF OHIO, 91- *Concurrent Pos:* Wis Heart Asn res grants, 65-71; NIH res grants, 66-72 & 74-78; Fla Heart Asn grant, 75 & 78; consult, Selvi & Co, Italy, 65-66, Abbott Labs, 74-76, Riker Labs, 77 & Dupont Labs, 78; mgr gen pharmacol dept, Abbott Labs, 71-74. *Mem:* AAAS; Am Soc Pharmacol & Exp Therapeut; Am Fedn Clin Res; fel Am Col Clin Pharmacol; Am Med Asn. *Res:* Cardiovascular and autonomic pharmacology; drug-design; clinical pharmacology; public health. *Mailing Add:* Ohio Dept Health 246 N High St Columbus OH 43266-0588

SOMANI, SATU M, b India, Mar 14, 37; m 66; c 1. PHARMACOLOGY, BIOCHEMICAL PHARMACOLOGY. *Educ:* Osmania Univ, BSc, 56; Univ Poona, MSc, 59; Duquesne Univ, MS, 64; Univ Liverpool, PhD(biochem pharmacol), 69. *Prof Exp:* Lectr chem, Vivek Vardhini Col, Osmania Univ, India, 59-61; scientist, Nuclear Sci & Eng Corp, Pa, 64-67; from instr to asst prof pharmacol, Univ Pittsburgh, 71-74; assoc prof, 74-82, PROF PHARMACOL & TOXICOL, SCH MED, SOUTHERN ILL UNIV, SPRINGFIELD, 82- *Concurrent Pos:* Ellis T Davies fel, Liverpool Univ, England, 67-69; NIH fel, Univ Pittsburgh, 69-70; Health Res found grant, Univ Pittsburgh, 71; grant, EPA, 77-80, Am Heart Asn, 83-84 & Dept Army, 84- *Mem:* AAAS; Soc Toxicol; Am Soc Clin Pharmacol & Therapeut; Fedn Am Socs Exp Biol; NY Acad Sci. *Res:* Distribution, metabolism and excretion of anticholinesterases, caffeine, theophylline and pollutants in animals and man; competition of drugs for the plasma protein binding; biliary excretion of drugs; toxicology; analysis of water pollutants and mutagenicity; effects of exercise on pharmokinetics of drugs. *Mailing Add:* Dept Pharmacol SIll Univ Sch Med PO Box 3926 Springfield IL 62708

SOMASUNDARAN, P(ONISSERIL), b Annallur, India, June 28, 39; m 66; c 1. SURFACE & COLLOID CHEMISTRY. *Educ:* Univ Kerala, BS, 58; Indian Inst Sci, Bangalore, BE, 61; Univ Calif, Berkeley, MS, 62, PhD(eng), 64. *Prof Exp:* Sr lab asst biochem, Nat Chem Lab, Poona, India, 58-59; res asst metall & mat sci, Univ Calif, Berkeley, 61-64; sr mineral res engr, Int Minerals & Chem Corp, 64-67; res chemist, Res Dept-Basic Sci, R J Reynolds Industs Inc, 67-70; from assoc prof to prof mineral eng, Henry Krumb Sch Mines, 70-83, LA VON DUDDLESON KRUMB PROF, SCH ENG & APPL SCI, COLUMBIA UNIV, 83- *Concurrent Pos:* NSF grants; Am Iron & Steel Inst grants; consult, NIH, 73, Ill Inst Technol Res Inst, 74-77, Amoco Prod Co, 74-77, Int Paper Co, 75, NSF, 77, B F Goodrich Co, 77-81, Exxon Corp, 77-, Occidental Res, 77, Am Cyanamid, 78, Proctor & Gamble, 78-79, Union Carbide, 79, Colgate Palmolive, 79-, IBM, 84-85 & UNESCO, 82, DuPont, 88-; dir, Langmuir Ctr Colloid & Interfaces, 87-; hon prof, Central S Univ Technol, China, 87-; chmn, Henry Krumb Sch of Mines, 88-; Brahm Prakash chair, Indian Inst Sci, Bangalore, 90. *Honors & Awards:* Antoine M Gaudin Award, Soc Mining Engrs, Am Inst Mining Metall & Petrol Engrs, 82; Robert H Richards Award, Am Inst Mining Engrs, 87; Arthur F Taggart Award, Soc Mining Engrs, 87; Henry Krumb lectr, Am Inst Mining Metall & Petrol Engrs, 88. *Mem:* Nat Acad Eng; Am Inst Mining, Metall & Petrol Engrs; Am Inst Chem Engrs; Int Asn Surface & Colloid Scientists. *Res:* Surface and colloid chemistry; electrokinetics; flotation; flocculation; adsorption; mineral processing; enhanced oil recovery; superconductor processing. *Mailing Add:* Sch Eng & Appl Sci Columbia Univ New York NY 10027

SOMEKH, GEORGE S, b Brussels, Belg, Apr 3, 35; US citizen; m 60; c 2. CHEMICAL ENGINEERING. *Educ:* Mass Inst Technol, BS, 56, MS, 57. *Prof Exp:* Proj engr, Plastics Div, 57-60, eng scientist, Chem Div, 60-77, RES ENGR, CHEM DIV, UNION CARBIDE CORP, 77- *Mem:* Am Inst Chem Engrs. *Res:* Separation and purification processes in petro-chemistry, petroleum refining and water pollution abatement; solvent extraction, azeotropic and extractive distillation; Rankine cycle fluids, lubricants and systems design. *Mailing Add:* 43 Winding Brook Rd New Rochelle NY 10804

SOMERO, GEORGE NICHOLLS, b Duluth, Minn, July 30, 40; m 68. BIOCHEMISTRY, PHYSIOLOGY. *Educ:* Carleton Col, BA, 62; Stanford Univ, PhD(biol), 67. *Prof Exp:* NSF fel, Univ BC, 67-69; I W Killam fel, 69-70; from asst prof to prof marine biol, Scripps Inst Oceanog, Univ Calif, San Diego, 80-91; PROF ZOOL, ORE STATE UNIV, 91- *Concurrent Pos:* John Dove Isaacs prof natural philos, Univ Calif, San Diego, 84. *Mem:* Nat Acad Sci; AAAS; Am Soc Zool. *Res:* Comparative biochemistry of environmental adaptation. *Mailing Add:* Dept Zool Ore State Univ Corvallis OR 97331-2914

SOMERS, ANNE R, b 1913. ENVIRONMENTAL MEDICINE. *Educ:* Vassar Col, BA, 35. *Hon Degrees:* DSc, Med Col Wis, 75. *Prof Exp:* Adj prof, Dept Environ Community Med, Robert Wood Johnson Med Sch, 71-84; RETIRED. *Concurrent Pos:* Freelance writer & speaker, 54- *Mem:* Inst Med-Nat Acad Sci; fel Am Col Hosp Adminrs; hon mem Soc Teachers Family Med. *Res:* Co-author of one book. *Mailing Add:* G205 Penswood Village Newtown PA 18940

SOMERS, EMMANUEL, b Leeds, Eng, July 3, 27; Can citizen; m 51; c 2. ENVIRONMENTAL TOXICOLOGY. *Educ:* Univ Leeds, BSc, 48, MSc, 50, DSc(chem), 69; Bristol Univ, PhD(chem), 56. *Prof Exp:* Prin sci off pesticide chem, Long Ashton Res Sta, Bristol Univ, 51-67; sect head food contaminants, Food & Drug Directorate, 67-68, chief food div, Health Protection Br, 68-72, dir, Food Res Labs, 72-74, dir-gen, Environ Health Directorate, 74-87, DIR-GEN, DRUGS DIRECTORATE, CAN DEPT NAT HEALTH & WELFARE, 88- *Concurrent Pos:* Nat Res Coun fel, Pesticide Res Inst, London, Ont, 57-58; NSF grant, Conn Agr Exp Sta, 63-64; Consult, WHO, 72-; mgr, Int Prog Chem Safety, WHO Geneva, 80; dir, Toxicol Forum, Inst Risk Res, Univ Waterloo, 88- *Mem:* Fel Chem Inst Can; fel Royal Soc Chem; Int Acad Environ Safety. *Res:* Mode of action of agricultural fungicides; analysis, metabolism and biochemistry of food contaminants and additives; research management; environmental health; risk assessment, science policy. *Mailing Add:* Health Protection Br Can Dept Nat Health & Welfare Ottawa ON K1A 0L2 Can

SOMERS, GEORGE FREDRICK, JR, b Garland, Utah, July 9, 14; m 39; c 3. PLANT PHYSIOLOGY. *Educ:* Utah State Univ, BS, 35; Oxford Univ, BA, 38, BSc, 39; Cornell Univ, PhD(plant physiol), 42. *Prof Exp:* Instr biochem, Cornell Univ, 41-44, from asst prof to assoc prof, 44-51; assoc dir, Del Agr Exp Sta, 51-59, chmn dept agr biochem & food tech, 52-59, assoc dean, Sch Agr, 54-59, chmn dept biol, 59-71, H Fletcher Brown prof, 62-81, EMER PROF BIOL, UNIV DEL, 81- *Concurrent Pos:* Plant physiologist, Plant, Soil & Nutrit Lab, USDA, 44-51, asst dir lab, 49-51; mem comt effects of atomic radiation on agr & food supplies, Nat Acad Sci-Nat Res Coun, 56-60; vis prof, Philippines, 58-59; ed, Gen Biochem Sect, Chem Abstr, 63-71; vis s#ientist, Brookhaven Nat Lab, 71; distinguished fac lectr, Univ Del, 80. *Mem:* Fel AAAS; Am Soc Plant Physiol; Bot Soc Am. *Res:* Enzymes; cell wall chemistry; physiological ecol gy; halophytes as potential food plants. *Mailing Add:* 22 Minquil Dr Newark DE 19713

SOMERS, KENNETH DONALD, b Fremont, Mich, Mar 2, 38; m 61; c 3. MICROBIOLOGY. *Educ:* Univ Mich, BA, 60, MS, 62; Univ Chicago, PhD(microbiol), 69. *Prof Exp:* Res assoc virol, Ciba Pharmaceut Co, 63-65; res assoc biochem virol, Baylor Col Med, 69-70, asst prof, 70-74; assoc prof, 74-78, PROF MICROBIOL, EASTERN VA MED SCH, 78- *Mem:* AAAS; Am Soc Microbiol; Am Asn Cancer Res; Soc Exp Biol & Med. *Res:* Oncogenic RNA viruses; cancer biology. *Mailing Add:* Microbiol Dept Eastern Va Med Sch PO Box 1980 Norfolk VA 23501

SOMERS, MICHAEL EUGENE, b Astoria, NY, Aug 11, 29; m 54; c 4. NEUROBIOLOGY, HISTOLOGY. *Educ:* Univ Bridgeport, BA, 51; Clark Univ, MA, 55, PhD(animal morphol), 67. *Prof Exp:* Instr, Univ Bridgeport, 55-59, from asst prof to assoc prof, 60-69, prof biol & chmn, 70-, EMER PROF BIOL, UNIV BRIDGEPORT. *Mem:* Am Soc Zool; Am Soc Ichthyol & Herpet; Am Micros Soc; Am Fisheries Soc. *Res:* Neuroanatomy of Crustacea; fine structure of invertebrate nervous systems; fish olfactory system. *Mailing Add:* 925 Longbrook Ave Stratford CT 06497

SOMERS, PERRIE DANIEL, b Winona, Minn, Oct 18, 18; m 42; c 4. BIOCHEMISTRY. *Educ:* Wabash Col, AB, 41; Purdue Univ, MS, 43, PhD(biochem), 46. *Prof Exp:* Res chemist, 46-51, GROUP LEADER, LAB TECH CTR, INT MULTIFOODS CORP, 51- *Mem:* Am Chem Soc; Am Asn Cereal Chemists. *Res:* Enzymic reactions; biological food chemistry; new food product development; food process design; food product patents; new cereal products. *Mailing Add:* 13776 74th Pl N Osseo MN 55369

SOMERSCALES, EUAN FRANCIS CUTHBERT, b London, Eng, Jan 23, 31; US citizen; m 64; c 2. HEAT TRANSFER, FLUID MECHANICS. *Educ:* Univ London, BSc, 53; Rensselaer Polytech Inst, MME, 61; Cornell Univ, PhD(heat transfer), 65. *Prof Exp:* Apprentice, NBrit Locomotive Co, 53-55; instr mech eng, 58-59, asst prof, 64-68, ASSOC PROF, RENSSELAER POLYTECH INST, 68- *Concurrent Pos:* Sr vis fel, Univ Manchester Inst Sci Technol, 75-76; sr vis scientist, Nat Phys Lab, London, 83. *Honors & Awards:* Bengough Medal & Prize, Inst Metals, 88. *Mem:* Am Soc Mech Eng; Nat Asn Corrosion Engrs; Sigma Xi. *Res:* Fluid mechanics and heat transfer with application to free convection and the fouling of heat transfer surfaces. *Mailing Add:* Mech Eng Dept Rensselaer Polytech Inst Troy NY 12180-3590

SOMERSET, JAMES H, b Philadelphia, Pa, Apr 19, 38; m 63; c 2. MECHANICAL & AEROSPACE ENGINEERING. *Educ:* Drexel Inst Technol, BS, 61; Syracuse Univ, MS, 63, PhD(mech & aerospace eng), 65. *Prof Exp:* Engr, Scott Paper Co, 58-61; asst prof mech & aerospace eng, 65-69, assoc prof, 69-80, PROF MECH & AEROSPACE ENG, COL ENG, SYRACUSE UNIV, 80- *Concurrent Pos:* NSF grant, 66-68; consult, Singer Publ Co, 66- *Mem:* Am Inst Aeronaut & Astronaut; Sigma Xi. *Res:* Stochastic response of structures; dynamic response of structures to periodic and impulse loads; dynamics; vibrations; stability of systems; plate and shell structures; biomechanics. *Mailing Add:* 307 Bradford Pkwy Syracuse NY 13224

SOMERSON, NORMAN L, b Philadelphia, Pa, Dec 17, 28; m 55; c 6. MEDICAL MICROBIOLOGY. *Educ:* Marietta Col, BS, 50; Univ Pa, MS, 52, PhD, 54. *Prof Exp:* Asst, Univ Pa, 53-54; asst prof, Bucknell Univ, 55; bacteriologist, Philadelphia Gen Hosp, 55; res microbiologist, Merck & Co, 56-62; sr scientist, Nat Inst Allergy & Infectious Dis, 62-66; assoc prof med microbiol, 66-69, assoc prof pediat, 67-70, PROF MED MICROBIOL, OHIO STATE UNIV, 69-, PROF PEDIAT, 70- *Concurrent Pos:* NIH grant pulmonary physiol in infection, 68-73; contract antigenicity of Mycoplasma pneumoniae, NIH, 66-72; mem, Int Subcomt Nomenclature of Mycoplasmas, 66-72; consult & lab supvr, Ohio State Dept Health, 75-81 & 82-; distinguished vis prof biol, US Air Force Acad, 81-82. *Mem:* AAAS; Am Soc Microbiol; Soc Exp Biol Med. *Res:* Penicillin and glutamic acid fermentation process; microbial steroid conversions; mycoplasmas, including nucleic acid homology, serology, pathogenicity, vaccine process and lung changes in infection; male hybrid sterility in Drosophila. *Mailing Add:* Dept Med Microbiol Ohio State Univ Col of Med Columbus OH 43210-1239

SOMERVILLE, CHRISTOPHER ROLAND, b Kingston, Ontario, Can, Oct 11, 47; m 76. MOLECULAR GENETICS, LIPID BIOCHEMISTRY. *Educ:* Univ Alta, BS, 74, MS, 76, PhD(genetics), 78. *Prof Exp:* Res assoc genetics, Univ Ill, 78-80; asst prof genetics, Univ Alta, 80-82; assoc prof, 82-86, PROF MOLECULAR BIOL, MICH STATE UNIV, 86- *Honors & Awards:* Schull Award, Am Soc Plant Physiologists, 87; Young Presidential Investr Award, NSF, 84. *Mem:* Am Soc Plant Physiologists; Am Oil Chemists Soc. *Res:* Physiological genetics and biochemistry of lipid metabolism and membrane biogenesis in higher plants and molecular genetics of Arabidopsis. *Mailing Add:* DOE Plant Res Lab Mich State Univ East Lansing MI 48824

SOMERVILLE, GEORGE R, US citizen. CHEMICAL ENGINEERING. *Educ:* Tex A&M Univ, BS, 42. *Prof Exp:* Plant engr, Chem Warfare Serv, US Army, 43-45; process engr, Neches Butane Prod Co, 46-53; assoc chem engr, 55-56, sr chem engr, 56-59, sr indust chemist, 59, asst mgr org & biol chem, 59-60, mgr encapsulation sect, 60-61, mgr spec projs, 61-64, actg dir, 64-65, asst dir, 65-74, dir, San Antonio Labs, Dept Chem & Chem Eng, 74-76, DIR DEPT APPL CHEM & CHEM ENG, SOUTHWEST RES INST, 76- *Mem:* Am Chem Soc; Sigma Xi. *Res:* Development of the process, materials and techniques for encapsulating various materails for commercial and military purposes. *Mailing Add:* 2519 Cedar Falls San Antonio TX 78232-4221

SOMERVILLE, PAUL NOBLE, b Vulcan, Alta, May 7, 25; nat US; m 54; c 2. STATISTICS. *Educ:* Univ Alta, BSc, 49; Univ NC, PhD(statist), 53. *Prof Exp:* Teacher, Lethbridge Sch Div, Can, 42-44; assoc prof statist, Va Polytech Inst & assoc statistician exten serv, Agr Exp Sta, 53-55; vis prof math, Am Univ, 55-57; asst proj dir, C-E-I-R, Inc, Ariz, 58-61, mgr, Utah Off, 61-62; mgr tech eval, RCA Corp, Patrick AFB, 62-72; assoc prof, 72-79, PROF STATIST, UNIV CENT FLA, 79- *Concurrent Pos:* Guest scientist, Nat Bur Standards, 55-57; lectr, Univ Ariz, 58-61 & Brigham Young Univ, 62; chmn math dept, Fla Inst Technol, 63-72; adj prof, Univ Fla, Genesys, 68-72. *Mem:* Fel Am Statist Asn; Am Meteorol Soc; Int Statist Inst. *Res:* Statistics; climatology; education; computer simulation; model building; design of experiments; consulting. *Mailing Add:* Dept Statist Univ Cent Fla Box 25000 Orlando FL 32816-0370

SOMERVILLE, RICHARD CHAPIN JAMES, b Washington, DC, May 30, 41; m 65; c 2. METEOROLOGY, FLUID DYNAMICS. *Educ:* Pa State Univ, BS, 61; NY Univ, PhD(meteorol), 66. *Prof Exp:* Res meteorologist, Geophys Fluid Dynamics Lab, Environ Sci Serv Admin, 67-69; res scientist, Courant Inst Math Sci, NY Univ, 69-72; meteorologist, Inst Space Studies,

Goddard Space Flight Ctr, NASA, 71-74; scientist, Nat Ctr Atmospheric Res, 74-79; PROF METEOROL & HEAD, CLIMATE RES GROUP, SCRIPPS INST OCEANOG, UNIV CALIF, SAN DIEGO, 79- *Concurrent Pos:* Fel, Nat Ctr Atmospheric Res, 66-67; fel geophys fluid dynamics prog, Woods Hole Oceanog Inst, 67, staff mem, 70, 76; adj assoc prof, NY Univ, 71-73 & Columbia Univ, 71-74. *Honors & Awards:* Fel, Am Meteorol Soc, 87. *Mem:* Am Meteorol Soc; Am Geophys Union. *Res:* Theoretical dynamic meteorology; numerical fluid dynamics; thermal convection; atmospheric general circulation; numerical weather prediction; parameterization of small-scale processes; climate modeling. *Mailing Add:* Scripps Inst Oceanog Mail Code 0224 Univ Calif San Diego La Jolla CA 92093-0224

SOMERVILLE, RONALD LAMONT, b Vancouver, BC, Feb 27, 35; nat US; m 55; c 5. BIOCHEMISTRY, BIOTECHNOLOGY. *Educ:* Univ BC, BA, 56, MSc, 57; Univ Mich, PhD, 61. *Prof Exp:* Res assoc biochem, Univ Mich, 60-61; asst prof, Univ Mich, Ann Arbor, 64-67; assoc prof, 67-77, PROF BIOCHEM, PURDUE UNIV, WEST LAFAYETTE, 77- *Concurrent Pos:* Fel biol sci, Stanford Univ, 61-64. *Mem:* Am Soc Biol Chemists; Genetics Soc Am; Am Soc Microbiol. *Res:* industrial production of proteins and small molecules by bacterial fermentation; genetic analysis; DNA-mediated redesign of proteins. *Mailing Add:* Dept of Biochem Purdue Univ West Lafayette IN 47907

SOMES, GRANT WILLIAM, b Bloomington, Ind, Jan 30, 47; m 67; c 3. STATISTICS AS APPLIED TO MEDICALLY RELATED DATA, CATEGORICAL & NONPARAMETRIC STATISTICS. *Educ:* Ind Univ, AB, 68; Univ Ky, PhD(statist), 75. *Prof Exp:* Asst prof res design, Dept Commun Med, Univ Ky, 75-79, asst prof statist, Dept Statist, 76-79; postdoctoral epidemiol, Univ Minn, 76; assoc prof statist, Res Prog, E Carolina Univ, 79-84; assoc prof, 84-87, PROF STATIST, DEPT BIOSTATIST & EPIDEMIOL, UNIV TENN, MEMPHIS, 87-, CHMN DEPT, 84- *Concurrent Pos:* Statist consult, J Nuclear Med, 76-85; prin investr, Biomed Res Support grant, 76-79 & 81-84; co-prin investr, Nat Heart, Lung, Blood Inst & NIH, 85-; adj prof statist, Dept Math, Memphis State Univ, 88-, Dept Psychol, 90- *Mem:* Sigma Xi; Am Statist Asn; Biomet Soc. *Res:* Cardiovascular risk factors, epilepsy, psychosocial factors and illness, smoking and behavior, dentistry and nutrition; statistical theory, mainly in categorical data analysis and nonparametric statistics; author of numerous publications. *Mailing Add:* Dept Biostatist & Epidemiol Univ Tenn Health Sci Ctr Memphis TN 38163

SOMES, RALPH GILMORE, JR, b Melrose, Mass, Aug 15, 29; m 85; c 8. AVIAN GENETICS, HUMAN NUTRITION. *Educ:* Univ Mass, BS, 60, PhD(poultry genetics), 63. *Prof Exp:* From asst prof to prof nutrit & genetics, Univ Conn, 63-91; CONSULT, 91- *Mem:* Poultry Sci Asn; Am Genetic Asn; World Poultry Sci Asn. *Res:* Genetic investigations of feather pigment systems and new mutant traits in the domestic fowl; genetic-nutritional interaction. *Mailing Add:* Dept of Nutrit Sci Box U-17 Univ Conn 3642 Horsebarn Rd Ext Storrs CT 06268

SOMJEN, GEORGE G, b Budapest, Hungary, May 2, 29; m 76; c 4. PHYSIOLOGY, PHARMACOLOGY. *Educ:* Univ Amsterdam, MD, 56; Univ NZ, MD, 61. *Prof Exp:* Asst pharmacol, Univ Amsterdam, 53-56; lectr physiol, Univ Otago, NZ, 56-60, sr lectr, 61-62; res fel, Harvard Med Sch, 62-63; from asst prof to assoc prof, 63-71, PROF PHYSIOL & NEUROBIOL, DUKE UNIV, 71- *Concurrent Pos:* Consult, Nat Inst Environ Health Sci, 71-75; invited speaker, XXVIIIth Int Cong Physiol, 81; vis prof, London, 75 & 85, Ibadan, Nigeria, 78. *Mem:* Am Asn Univ Profs; Am Soc Pharmacol Exp Therapeut; Soc Neurosci; Am Physiol Soc; hon mem Hungarian Physiol Soc; Sigma Xi. *Res:* Reflex function of spinal cord; mechanism of seizures; properties of neurons; effects of drugs and ions on central nervous system and on peripheral junctions; blood-brain barrier; hypoxia of central nervous system-stroke. *Mailing Add:* Dept Cell Biol Box 3709 Duke Univ Med Ctr Durham NC 27710

SOMKAITE, ROZALIJA, b Lithuania, Feb 10, 25; US citizen. PHARMACEUTICAL CHEMISTRY, ANALYTICAL CHEMISTRY. *Educ:* St John's Univ, NY, BS, 54; Univ Wis, MS, 56; Rutgers Univ, PhD(pharmaceut sci), 62. *Prof Exp:* Assoc scientist, Warner-Chillcot Pharmaceut Co, 56-58; teaching asst, Rutgers Univ, 58-59, NIH res fel anal, 61-62; sr scientist, Ethicon, Inc, 62-70; mgr anal res dept, 70-74, DIR ANAL SERV, REHEIS CHEM CO, 74- *Mem:* Am Pharmaceut Asn; Am Chem Soc; Am Microchem Soc; Soc Appl Spectros. *Res:* Analytical research applying multiple technique systems. *Mailing Add:* 386 Hillside Pl South Orange NJ 07079-2903

SOMKUTI, GEORGE A, b Budapest, Hungary, Jan 6, 36; US citizen; m 59; c 2. FERMENTATION BIOCHEMISTRY, APPLIED GENETICS. *Educ:* Tufts Univ, BS, 59; Purdue Univ, MS, 63, PhD(microbiochem), 66. *Prof Exp:* NIH fel, Purdue Univ, 66-68; asst prof microbiochem & immunol, Duquesne Univ, 68-69; res assoc cell biol, Purdue Univ, 69-73; sr res scientist, Res & Develop, Lederle Labs, Am Cyanamid Co, 73-76; RES LEADER MICROBIOL & BIOCHEM, EASTERN REGIONAL RES CTR, USDA, 76- *Concurrent Pos:* Mem, NSF Curric Develop Comt Univ Tex, San Antonio, 74-75 & NIH Special Studies Sect, 76; ed, J Food Protection, 82- & J Indust Microbiol, 85-; NSF Spec Studies Sect, 83-90; mem, bd dirs, Soc Indust Microbiol, 84-87. *Honors & Awards:* SIM Chas Porter Award, 88. *Mem:* Am Soc Microbiol; Soc Indust Microbiol (pres, 85-86); NY Acad Sci; Inst Food Technol; Am Dairy Sci Asn. *Res:* Microbial physiology and metabolism; plasmid function; applied enzymology. *Mailing Add:* Eastern Regional Res Ctr USDA 600 E Mermaid Lane Philadelphia PA 19118

SOMLYO, ANDREW PAUL, b Budapest, Hungary, Feb 25, 30; US citizen; m 61; c 1. PHYSIOLOGY, PATHOLOGY. *Educ:* Univ Ill, Chicago, BS, 54, MS & Md, 56; Drexel Inst Technol, MS, 63. *Hon Degrees:* MA, Univ Pa, 81. *Prof Exp:* Intern, Philadelphia Gen Hosp, 56-57, resident, 57-58; asst resident med, Mt Sinai Hosp, New York, 58-59; sr asst resident, Bellevue Hosp, 59-60; asst physician, Columbia-Presby Med Ctr, 60-61; res assoc, Presby Hosp, 61-66; from asst prof to assoc prof, 64-71, PROF PATH, UNIV PA, 71-, PROF PHYSIOL, 73- *Concurrent Pos:* Heart Asn Southeast Pa res fel, Philadelphia Gen Hosp, 57-58; NIH spec res fel, Presby Hosp, Philadelphia, 61-66; USPHS res career prog award, Presby-Univ Pa Med Ctr, 66-73; dir, Pa Muscle Inst; prof physiol & path, Univ of Pa Sch Med, 67-88; sr res pathologist, Presby-Univ Pa Med Ctr, 67-80; Charles Slaughter prof & chmn dept physiol & prof med cardiol, Univ Va, Sch Med, 88- *Mem:* Microbeam Anal Soc; AAAS; Soc Gen Physiol; Am Physiol Soc; Biophys Soc; Am Soc Cell Biol. *Res:* Development and application of quantitative electron optical techniques in biology including electron probe analysis and electron energy loss analysis; ultrastructure and cell physiology of vascular smooth muscle and skeletal muscle; pharmacology. *Mailing Add:* Dept Physiol Univ Va Sch Med Box 449 Charlottesville VA 22908

SOMLYO, AVRIL VIRGINIA, b Sask, Can, Apr 9, 39; m 61; c 1. CELL PHYSIOLOGY. *Educ:* Univ Sask, BA, 58, MSc, 61; Univ Pa, PhD, 76. *Prof Exp:* Co prin investr, 68-79, from res assoc prof to res prof physiol, 82-89, PROF PHYSIOL, UNIV PA, 89- *Concurrent Pos:* Mem, Biol Instrumentation Panel, NSF, 79-83; mem, Pharmacol Study Sect, NIH, 81-84, Physiol Study Sect, 90-93; mem, Nat Inst Heart, Lung & Blood Cardiol Adv Comt, 85-89; coun, Biophysics Soc, 84-87, exec coun, 85-87; coun, Cell & Gen Physiol Sect, Am Physiol Soc, 85-88, chmn, 87-88; mem, US Nat Comt, Int Union Physiol Soc, 88-91, Nat Acad Res Coun Deleg Gen Assembly, 89; mem, Sci Prog Comt, Int Physiol Cong, 88-93, Glasgow, 93; mem, Cell Transport & Metabolism Res Study Comt, Am Heart Asn, 89-92. *Mem:* Am Soc Pharmacol & Exp Therapeut; Biophys Soc; Sigma Xi; Soc Gen Physiologists; Am Physiol Soc. *Res:* Basic function and structure of striated and vascular smooth muscle, including excitation-contraction coupling, contractile proteins; the role of the in situ distribution of elements, especially calcium, within organelles, using high spatial resolution electron probe x-ray microanalysis and electron energy loss analysis. *Mailing Add:* Pa Muscle Inst Univ Pa Sch Med B42 Anat-Chem Bldg G3 Philadelphia PA 19104-6083

SOMMER, ALFRED, b New York, NY, Oct 2, 42; m 63; c 2. CHILD SURVIVAL, BLINDNESS PREVENTION. *Educ:* Union Col, BS, 63; Harvard Med Sch, MD, 67; Johns Hopkins Univ, MHS, 73. *Prof Exp:* Med epidemiologist, Ctrs Dis Control, 69-72; dir & prin investr clin epidemiol, Helen Keller Int Blindness Prev, 76-80; found dir, Int Ctr Epidemiol & Prev Ophthal, 80-90; FROM ASST PROF TO PROF OPHTHAL, EPIDEMIOL & INT HEALTH, JOHNS HOPKINS UNIV, 80, DEAN, SCH HYG & PUB HEALTH, 90- *Concurrent Pos:* Med adv, Helen Keller Int, 73-; steering comt, Int Vitamin A Consultative Group, 75-; bd mem, Int Agency Prev Blindness, 78-; comt chmn, Nat Insts Health, 81-; bd dirs, Nat Soc Prev Blindness, 87-89; chmn, Prog Adv Group Blindness Prev, World Health Orgn, 88-90; comt mem, Inst Med & Nat Acad Sci, 89. *Honors & Awards:* Helen Keller Blindness Prevention Award, Helen Keller Int, 80; Distinguished Serv Award for Contrib to Vision Care, Am Pub Health Asn, 88; Charles A Dana Award, Pioneering Achievements in Health, Charles A Dana Found, 88; E V McCollum Int Lectr Nutrit, Am Inst Nutrit, Fedn Am Soc Exp Biol, 88; Award for Distinguished Contrib World Ophthal, Int Fedn Ophthal Socs, 90. *Mem:* Am Acad Ophthal; Am Ophthal Soc; Soc Epidemiol Res; Asn Res in Vision & Ophthal; Am Pub Health Asn; Am Inst Nutrit. *Res:* Epidemiologic assessment of blinding diseases; prevention of childhood mortality in developing countries and assessment of medical technology. *Mailing Add:* 1041 Sch Hyg & Pub Health 615 N Wolfe St Baltimore MD 21205

SOMMER, ALFRED HERMANN, b Frankfurt, Ger, Nov 19, 09; US citizen; m 38; c 3. ELECTRON EMISSION. *Educ:* Berlin Univ, Dr Phil(chem), 34. *Prof Exp:* Res engr photo multipliers, Baird TV Co, London, 36-46; res engr TV camera tubes, EMI-Res Lab, Eng, 46-53; res engr electron emission, RCA-Res Labs, Princeton, NJ, 53-74; res engr, Thermo-Electron Co, Waltham, Mass, 74-78; CONSULT, 78- *Honors & Awards:* Gaede-Langmuir Award, Am Vacuum Soc, 82. *Mem:* Am Phys Soc; fel Inst Elec & Electronics Engrs. *Res:* New photoemissive materials; secondary emission; thermionic emission; photo multipliers; television camera tubes; image intensifier tubes; thermionic energy conversion. *Mailing Add:* 37 Dogwood Lane Northampton MA 01060

SOMMER, CHARLES JOHN, b New York, NY, Jan 12, 51. STATISTICS, BIOMETRICS. *Educ:* Manhattan Col, BS, 72; State Univ NY Buffalo, MA, 73, PhD(statist sci), 77. *Prof Exp:* Biostatistician, Sidney Farber Cancer Inst, 77-78; asst prof statist, Temple Univ, 78-; ASST PROF, DEPT MATH & COMPUTER SCIENCE, SUNY, BROCKPORT, 84- *Mem:* Am Statist Asn. *Mailing Add:* Dept of Math & Comput Sci State Univ NY Brockport NY 14420

SOMMER, HARRY EDWARD, b Chatham, NY, July 25, 41; m 64; c 2. FOREST PHYSIOLOGY. *Educ:* Univ Vt, BSAgr, 63; Univ Maine, MS, 66; Ohio State Univ, PhD(bot), 72. *Prof Exp:* Res assoc tissue cult, Sch Forest Resources, Univ Ga, 72-74; scientist, Weyerhaauser Forestry Res Ctr, 74-76; ASST PROF TISSUE CULT, SCH FOREST RESOURCES, UNIV GA, 76- *Mem:* Bot Soc Am; Am Soc Plant Physiologists; Sigma Xi. *Res:* Tissue culture of trees. *Mailing Add:* Sch of Forest Resources Univ of Ga Athens GA 30602

SOMMER, HELMUT, b Ger, Aug 23, 22; nat US; m 46; c 6. ELECTRICAL ENGINEERING, ELECTRONICS. *Educ:* Agr & Mech Col, Tex, BS, 44, MS, 47, PhD, 50. *Prof Exp:* Electronic scientist, Nat Bur Stand, 49-53 & Diamond Ord Fuze Labs, US Dept Army, 53-57; res prof electronics, Univ Fla, 57-58; chief, Microwave Br, Diamond Ord Fuze Labs, US Dept Army, 58-62, chief, Systs Res Lab, Harry Diamond Labs, 62-66, assoc tech dir, Harry Diamond Labs, 66-80; CONSULT, 80- *Concurrent Pos:* Consult, Catholic Univ, 53-60. *Mem:* Inst Elec & Electronics Engrs. *Res:* Radar; microwaves; military electronics; proximity fuzes. *Mailing Add:* 9502 Hollins Ct Bethesda MD 20817

SOMMER, HOLGER THOMAS, b Wittgendorf, Ger, June 18, 50; m 77. COMBUSTION, FLUID-THERMO SCIENCE. *Educ:* Tech Univ Aachen, dipl ing, 74, Imp Col Sci & Technol, London, MS, 77; Tech Univ Aachen, Dr Ing, 79. *Prof Exp:* Instr thermo-fluid, Tech Univ Aachen, 74-75; consult air conditioning, Behr Eng, Aachen, 76; res engr combustion, Imp Col London, 76-77; sr scientist combustion-mech fluid, Tech Univ Aachen, 77-80; prin investr solar energy, Ger Sci Found, Desert Res Inst, 80; ASST PROF COMBUSTION-FLUID MECH, CARNEGIE-MELLON UNIV, 81- *Concurrent Pos:* Consult engr, Forensic Consult & Engrs, 81-; fel, Lilly Endowment, 81. *Honors & Awards:* R R Teetor Award, Soc Automotive Engrs, 82. *Mem:* Am Soc Mech Engrs; Soc Automotive Engrs; Combustion Inst. *Res:* Fluid mechanics of combustion; ignition of combustible mixtures; development of optical diagnostic instrumentation; numerical modeling of combustion systems; spray formation; coal-energy conversion processes. *Mailing Add:* Dept Mech Eng Carnegie-Mellon Univ 5000 Forbes Ave Pittsburgh PA 15213

SOMMER, JOACHIM RAINER, b Dresden, Ger, Apr 11, 24; nat US; m 51; c 2. PATHOLOGY. *Educ:* Univ Munich, MD, 50; Am Bd Path, dipl, 58, cert anat & clin path, 69. *Prof Exp:* Asst, Path Inst, Munich, Ger, 51-52; asst, Med Clin Munich, 52-53; intern, Garfield Mem Hosp, Washington, DC, 53-54; resident path, Garfield Mem & De Paul Hosps, 54-58; assoc, 58-59, from asst prof to assoc prof, 59-70, PROF PATH, MED CTR, DUKE UNIV, 70-, PROF PHYSIOL, 81- *Mem:* Sigma Xi. *Res:* Histochemistry; cardiac ultrastructure and function; electron microscopy, cryotechniques. *Mailing Add:* Dept Path Duke Univ Med Ctr PO Box 3548 Durham NC 27706

SOMMER, JOHN G, b Portsmouth, Ohio, Jan 28, 26; m 60; c 4. POLYMER PHYSICS, RUBBER STRUCTURE PROPERTY RELATIONSHIPS. *Educ:* Univ Dayton, BChE, 51; Univ Akron, MS, 65. *Prof Exp:* Lab mgr, Dayco Corp, 51-60; asst chief chemist, Precision Rubber Prod Corp, 60; sect head, Rubber Compounding & Processing, GenCorp Res, 60- 88. *Concurrent Pos:* Lectr & consult, 88- *Honors & Awards:* Melvin Mooney Distinguished Technol Award, Rubber Div, Am Chem Soc, 88. *Mem:* Am Chem Soc, Rubber Div. *Res:* Molding of rubber; rubber part design; rubber use in aerospace applications; rubber dynamic properties; rubber materials science; flammability of rubber; physical testing & physical properties. *Mailing Add:* 5939 Bradford Way Hudson OH 44236-3905

SOMMER, KATHLEEN RUTH, b Port Washington, Wis, June 2, 47. TOXICOLOGY. *Educ:* Ripon Col, BA, 69; Univ Iowa, PhD(biochem), 73; Am Bd Toxicol, dipl, 81. *Prof Exp:* Res assoc chem, Univ Wis-Milwaukee, 69; instr pharmacol, Baylor Col Med, 73-74; USPHS res fel, Baylor Col Med, 74-76; toxicologist, Shell Oil Co, 76-80; DIR RES, TOXICON CORP, 80- *Concurrent Pos:* Consult, Scientists Coop Indust, 72-76, Nat Adv Res Resources Coun, NIH, 74-78 & Div Res Resources, NIH, 78-; guest lectr, Med Sch, Univ Tex, 78-; dir, Reid Rd Municipal Utility District #1, 82- (vpres bd, 88-); EMT Paramedic, Tex, 85-89. *Mem:* Am Col Toxicol; Am Indust Health Coun; Am Chem Soc; AAAS; Sigma Xi. *Res:* Drug metabolism and toxicity; pesticide toxicology; mass spectroscopy; effects of toxins on reproduction; toxicology, waste water effluent, plant compliance; employee right-to-know compliance. *Mailing Add:* 10315 Crescent Moon Houston TX 77064

SOMMER, LEO HARRY, b New York, NY, Sept 21, 17; m 44; c 3. ORGANIC CHEMISTRY. *Educ:* Pa State Univ, BS, MS, 42, PhD(org chem), 45. *Prof Exp:* From instr to prof chem, Pa State Univ, 43-65; PROF CHEM, UNIV CALIF, DAVIS, 65- *Concurrent Pos:* Res fel, Harvard Univ, 50-51; Guggenheim fel, 60-61; consult, Dow Corning Corp, 47- *Honors & Awards:* F S Kipping Award, Am Chem Soc, 63. *Mem:* Am Chem Soc; Royal Soc Chem. *Res:* Stereochemistry and reaction mechanisms of silicon centers in organosilicon compounds; chemistry of multiple-bonded unsaturated organosilicon compounds. *Mailing Add:* Dept of Chem Univ of Calif Davis CA 95616

SOMMER, LEONARD SAMUEL, b Springfield, Mass, July 3, 24; c 2. CARDIOLOGY. *Educ:* Yale Univ, BS, 44; Columbia Univ, MD, 47; Am Bd Internal Med, dipl, 57 & 77; Am Bd Cardiovasc Dis, dipl, 75. *Prof Exp:* Intern med, Peter Bent Brigham Hosp, 47-48; instr med, Med Sch, Georgetown Univ, 51-52; resident med, Peter Bent Brigham Hosp, 48-49 & 53-54; res fel, Harvard Med Sch, 54-55, asst, 55-56; from asst prof to assoc prof, 56-74, PROF MED, SCH MED, UNIV MIAMI, 74- *Concurrent Pos:* Teaching fel, Harvard Med Sch, 48-49; Am Heart Asn res fel cardiol, Columbia-Presby Med Ctr & New York Hosp, Cornell Univ, 49-50; Nat Heart Inst res fel cardiol, Hammersmith Hosp, London, Eng, 52-53; fel cardiol, Cardiovasc Lab, Children's Med Ctr, Boston, 54-56; consult cardiologist, Adolescent Unit, Children's Med Ctr, Boston, Mass, 54-56; asst physician, Peter Bent Brigham Hosp, 54-56; investr, Howard Hughes Med Inst, 56-59; mem coun clin cardiol, Am Heart Asn; dir, Cardiovasc Lab, Jackson Mem Hosp, 56-75 & dir, Exercise Labs, 78- *Mem:* Am Heart Asn; fel Am Col Cardiol; fel Am Col Physicians; Am Fedn Clin Res; Sigma Xi. *Res:* Cardiovascular physiology and diseases. *Mailing Add:* Univ Miami Sch Med-D62 Div Cardiol PO Box 016960 Miami FL 33101

SOMMER, NOEL FREDERICK, b Scio, Ore, Jan 21, 20; m 46; c 1. PLANT PHYSIOLOGY, PATHOLOGY. *Educ:* Ore State Col, BS, 41; Univ Calif, MS, 52, PhD(plant path), 55. *Prof Exp:* County agr exten agent, Ore State Col, 46-51; res asst, Univ Calif, 52-55; plant pathologist, USDA, 55-56; asst pomologist, 56-63, lectr & assoc pomologist, 63-67, lectr & pomologist, 67-75, chmn dept pomol, 75-81, LECTR POMOL & POSTHARVEST PATHOLOGIST, UNIV CALIF, DAVIS, 81- *Honors & Awards:* Bronze Medal, Agr Chamber, Vaucluse, France. *Mem:* Am Phytopath Soc; Am Soc Hort Sci; Am Soc Microbiol; Mycol Soc Am; NY Acad Sci; AAAS. *Res:* Physiology and pathology of fruits and vegetables after harvest; mycotoxins. *Mailing Add:* Dept of Pomology Univ of Calif Davis CA 95616

SOMMER, SHELDON E, b New York, NY, Nov 3, 37; m 60. GEOCHEMISTRY. *Educ:* City Col New York, BS, 59; City Univ New York, MA, 61; Tex A&M, MS, 64; Pa State Univ, PhD(geochem), 69. *Prof Exp:* Sec sch teacher, Bd Ed, NY, 59-61; res asst geol, Kans Geol Surv, 61-62; oceanogr, Tex A&M, 62-63, res scientist, 63-64; asst geochem & mineral, Pa State Univ, 64-69; assoc prof geochem, 69-76, ASSOC PROF GEOL, UNIV MD, COLLEGE PARK, 76- *Mem:* AAAS; Geochem Soc; Mineral Soc Am; Soc Appl Spectros. *Res:* Geochemistry of marine sediments and sea water; low temperature mineral synthesis; study of geological materials by electron spectroscopy and electron microprobe spectroscopy. *Mailing Add:* 16806 Shepstow Ct Dallas TX 75248

SOMMERFELD, JUDE T, b Elmwood Place, Ohio, Feb 4, 36; c 4. CHEMICAL ENGINEERING. *Educ:* Univ Detroit, BChE, 58; Univ Mich, MSE, 60, PhD(chem eng), 63. *Prof Exp:* Sr systs engr, Monsanto Co, 63-65, eng specialist, 65-66; sr systs engr, Wyandotte Chem Corp, 66-67, mgr systs eng, 67-68, dir process eng, 68-70; assoc prof, 70-75, assoc dir chem eng, 81-88, PROF CHEM ENG, GA INST TECHNOL, 75- *Mem:* Am Chem Soc; Am Inst Chem Engrs; Nat Soc Prof Engrs. *Res:* Energy conservation; computer applications; applied mathematics; systems engineering; management science; thermodynamics; kinetics; catalysis. *Mailing Add:* Sch Chem Eng Ga Inst Technol Atlanta GA 30332-0100

SOMMERFELD, MILTON R, b Thorndale, Tex, Nov 24, 40; m 63; c 2. PHYCOLOGY. *Educ:* Southwest Tex State Col, BS, 62; Wash Univ, PhD(bot), 68. *Prof Exp:* Teaching asst biol, Southwest Tex State Col, 61-62; teaching asst bot, Wash Univ, 64-65, instr, 65; from asst prof to assoc prof bot, 68-77, chair dept, 81-88, PROF BOT, ARIZ STATE UNIV, 78-, ASSOC DEAN, 89- *Mem:* AAAS; Bot Soc Am; Phycol Soc Am; Int Phycol Soc. *Res:* Morphogenesis and development of the algae; systematics; morphogenesis; life cycles; ecology of the algae; water quality; endolithic algae. *Mailing Add:* Dept Bot Ariz State Univ Tempe AZ 85287

SOMMERFELD, RICHARD ARTHUR, b Chicago, Ill, July 4, 33. GEOCHEMISTRY, GEOLOGY. *Educ:* Univ Chicago, PhD(geophys), 65. *Prof Exp:* Micrometeorologist, Univ Wash, 61-64; fel geochem, Univ Calif, Los Angeles, 65-67, inst geophys fel, 66-67; assoc geologist, 67-76, RES GEOLOGIST, ROCKY MOUNTAIN FOREST & RANGE EXP STA, US FOREST SERV, 77- *Concurrent Pos:* NSF fel, 65-66. *Mem:* Int Glaciol Soc; AAAS; Am Geophys Union. *Res:* Physical chemistry of mineral reactions, particularly the reactions of quartz and water; metamorphism and solid mechanics of snow; ice crystallization from vapor; acoustic properties of snow; ice surface chemistry; snow chemistry. *Mailing Add:* US Forest Serv 240 W Prospect St Ft Collins CO 80526

SOMMERFELDT, THERON G, b Cardston, Alta, Can, May 27, 23; m 48; c 5. SOIL SCIENCE, PHYSICAL CHEMISTRY. *Educ:* Univ Alta, BSc, 50; Utah State Univ, MS, 52, PhD(soil chem), 61. *Prof Exp:* Asst agronomist, Can Sugar Factories, 51-53; asst soil scientist, NDak State Univ, 53-60; self employed, 60-61; asst soil scientist, Univ Idaho, 61-65; SOIL SCIENTIST, CAN DEPT AGR, 65- *Mem:* Am Soc Agron; Agr Inst Can; Can Soc Soil Sci; Prof Inst Pub Serv Can; Can Soc Agr Engrs. *Res:* Reclamation and drainage of saline and alkali soils; soil and water pollution from fertilizers and animal wastes; investigations, management and reclamation of dryland salinity; animal waste disposal and utilization. *Mailing Add:* 1705 20 St S Lethbridge AB T1K 2G1 Can

SOMMERFIELD, CHARLES MICHAEL, b New York, NY, Oct 27, 33; m 69; c 2. THEORETICAL PHYSICS, QUANTUM FIELD THEORY. *Educ:* Brooklyn Col, BS, 53; Harvard Univ, AM, 54, PhD(physics), 57. *Hon Degrees:* MA, Yale Univ, 67. *Prof Exp:* NSF fel physics, Univ Calif, 57-58, instr & jr res physicist, 58-59; res fel, Harvard Univ, 59-61, Corning lectr, 60-61; from asst prof to assoc prof, 61-67, PROF PHYSICS, YALE UNIV, 67- *Concurrent Pos:* Vis asst res mathematician, Univ Calif, 65; vis fel, Mass Inst Technol, 69-70; mem, Inst Advan Studies, 89. *Mem:* AAAS; Am Phys Soc. *Res:* Theories of quantized fields and elementary particle interactions. *Mailing Add:* Dept Physics Yale Univ PO Box 6666 New Haven CT 06511

SOMMERMAN, GEORGE, b Baltimore, Md, July 2, 09. PHYSICS. *Educ:* Johns Hopkins Univ, BEE, 29, DEng, 33. *Prof Exp:* Fel engr, Westinghouse Elec, 54-74, consult, 74-82; consult, Oak Space Ridge Nat Lab, 74-82; RETIRED. *Honors & Awards:* Alfred Noble Prize. *Mem:* Fel Inst Elec & Electronics Engrs. *Mailing Add:* 13801 York Rd Cockeysville MD 21030

SOMMERMAN, KATHRYN MARTHA, b New Haven, Conn, Jan 11, 15. ENTOMOLOGY. *Educ:* Univ Conn, BS, 37; Univ Ill, MS, 41, PhD(entom), 45. *Prof Exp:* Artist entom, Univ Ill, 37-38, artist & asst entom, Ill Natural Hist Surv, 39-45; instr biol, Wells Col, 45; asst prof zool, Eastern Ill Col Educ, 46; entomologist, Army Med Dept Res & Grad Sch, Washington, DC, 46-51; entomologist bur entom & plant quarantine, USDA, 51-53, collabr, Sect Insect Identification, Entom Res Br, Agr Res Serv, Md, 53-58; res entomologist, Arctic Health Res Ctr, 55-73, chief entom unit, 60-73; RETIRED. *Concurrent Pos:* Entomologist, Alaskan Insect Proj, US Dept Army, 48; fel, Univ Ill, Urbana, 49; res consult, 73-77. *Mem:* fel Entom Soc Am; Sigma Xi; Wilderness Soc. *Res:* Systematics and bionomics of Psocoptera. *Mailing Add:* SR 76 Box 384 Greenville ME 04441

SOMMERS, ARMIGER HENRY, b Clarksdale, Miss, June 15, 20; m 49; c 5. CHEMISTRY. *Educ:* Notre Dame Univ, BS, 42, MS, 43, PhD(org chem), 48. *Prof Exp:* Res chemist, Notre Dame Univ, 44-45 & Columbia Univ, 46; res chemist, Abbott Labs, 47-63, Licensing, 63-85; RETIRED. *Mem:* Am Chem Soc; Sigma Xi. *Res:* Organic synthesis of nitrogen compounds for medicinal use; new drug information and licensing. *Mailing Add:* 120 Wimbledon CT Lake Bluff IL 60044

SOMMERS, ELLA BLANCHE, b Lahoma, Okla, Mar 12, 08. PHARMACY. *Educ:* Univ Okla, BS, 30, MS, 31; Ohio State Univ, PhD, 54. *Prof Exp:* From assoc prof to prof pharm, Univ Okla, 42-78, asst dean pharm, 71-78, consult, Col Pharm & Off Develop, 78- *Mem:* Am Chem Soc; Am Pharmaceut Asn; Sigma Xi. *Res:* Freeze drying. *Mailing Add:* 1805 S Virginia Norman OK 73071

SOMMERS, HENRY STERN, JR, b St Paul, Minn, Apr 21, 14; m 38; c 4. PHYSICS. *Educ:* Univ Minn, AB, 36; Harvard Univ, PhD(physics), 41. *Prof Exp:* Instr, Harvard Univ, 41-42; mem staff, Mass Inst Technol, 42-45; asst prof physics, Rutgers Univ, 46-49; mem staff, Los Alamos Sci Lab, NMex, 49-54; fel, RCA Labs, RCA Corp, 54-84; RETIRED. *Concurrent Pos:* Fulbright lectr & Guggenheim fel, Hebrew Univ, Israel, 60-61. *Mem:* AAAS; fel Am Phys Soc; Fedn Am Sci. *Res:* Nuclear and semiconductor physics; instrumentation; cryogenics; photoconductivity; quantum electronics; experimental research on basic physics and control of power spectrum of injection lasers. *Mailing Add:* Pennswood Village No C201 Newtown PA 18940

SOMMERS, HERBERT M, b Colorado Springs, Colo, Sept 4, 25; m 55; c 4. PATHOLOGY. *Educ:* Northwestern Univ, BS, 49, MD, 52. *Prof Exp:* Instr, 59-61, assoc, 61-62, from asst prof to assoc prof, 62-71, PROF PATH, MED SCH, NORTHWESTERN UNIV, CHICAGO, 71-; DIR CLIN MICROBIOL, NORTHWESTERN MEM HOSP, 72- *Concurrent Pos:* Res fel path, Med Sch, Northwestern Univ, Chicago, 54-58; attend pathologist, Chicago Wesley Mem Hosp, 58-68, Passavant Mem Hosp, 68-73 & Northwestern Mem Hosp, 73-; consult, Vet Admin Res Hosp; trustee, Am Bd Path, 82-88. *Mem:* Am Soc Clin Path; Am Asn Pathologists; Am Soc Microbiol; Am Thoracic Soc; Col Am Path; Sigma Xi. *Res:* Experimental pathology of ischemic myocardium and mechanisms of ventricular fibrillation; improvement of methods in clinical microbiology; laboratory methods for mycobacterial susceptiblity testing. *Mailing Add:* Dept Path Med Sch Northwestern Univ 303 E Chicago Ave Chicago IL 60611

SOMMERS, JAY RICHARD, b Brooklyn, NY, May 19, 39; m 61; c 3. ORGANIC CHEMISTRY, TECHNICAL MANAGEMENT. *Educ:* Brooklyn Col, BS, 61; Univ Pittsburgh, PhD(org chem), 66. *Prof Exp:* Res chemist, Org Chem Dept, E I du Pont de Nemours & Co, 65-69; sr res scientist & mgr, Surg Specialty Div, Johnson & Johnson Co, 69-74, mgr surg apparel & fabrics develop, Surgikos, 74-76, dir prod develop, Surgikos, 76-80, mgr fiber technol, Johnson & Johnson Prod, Inc, 80-81; dir prod develop, Int Playtex Inc, 81-83; dir female care res & develop, 83-87, DIR RES CORP SCI & TECHNOL, KIMBERLY-CLARK, 87- *Mem:* Am Chem Soc; Asn Advan Med Instrumentation; Am Asn Textile Chem & Colorists; Sigma Xi; Asn Res Dirs. *Res:* Textile chemicals; nonwoven fabrics and finishes; disposable apparel; fabric flammability; medical/surgical products; biomedical devices; internal and external sanitary protection; health and beauty aids; basic and long range research and development. *Mailing Add:* 1985 Willed Creek Pt Marietta GA 30068

SOMMERS, LAWRENCE M, b Clinton, Wis; m 48; c 1. DEVELOPMENT GEOGRAPHY, SCANDINAVIAN GEOGRAPHY. *Educ:* Univ Wis, BS, 41, PhM, 46; Northwestern Univ, PhD(geog), 50. *Prof Exp:* Instr, 49-51, from asst prof to assoc prof, 51-55, head dept, 55-62, chmn, 62-79, PROF GEOG, MICH STATE UNIV, 55- *Concurrent Pos:* Mem, comt geog, adv geog br, off Naval Res-Nat Res Coun, 58-61; consult & examr, N Cent Accrediting Asn, 63-79; vis scientist, NSF & Asn Am Geographers, 68-70; ed, Denoyer Geppert Co, 75-86; prof, Environ Qual Ctr, Mich State Univ, 79-81, asst provost, 87-90; US Deleg, study group, Comn High Latitude Develop, Int Geog Union, 84- *Honors & Awards:* res awards, Social Sci Res Coun, Am Scand Found, Off Naval Res. *Mem:* AAAS; Asn Am Geographers; Am Geog Soc; Sigma Xi; Explorers Club. *Res:* Development issues in the state of Michigan, arid Southwest, and Scandanavian countries; Norwegian North Sea oil and gas developments; development problems and water quality in Norway; spatial analysis of Lake Michigan and the geography of Michigan. *Mailing Add:* Dept Geog Mich State Univ E Lansing MI 48824

SOMMERS, LEE EDWIN, b Beloit, Wis, July 30, 44; m 66, 79; c 3. SOIL MICROBIOLOGY. *Educ:* Wis State Univ-Platteville, BS, 66; Univ Wis-Madison, MS, 68, PhD(soil sci), 70. *Prof Exp:* Assoc prof, 70-80, PROF SOIL MICROBIOL, PURDUE UNIV, 80- *Mem:* AAAS; Am Soc Agron; Soil Sci Soc Am. *Res:* Effect of soil chemical and physical properties on microbial growth; microbial transformations of heavy metals; role of soils and sediments in eutrophication; plant nutrient and metal transformations in soils amended with industrial and municipal wastes. *Mailing Add:* Dept Agron Colo State Univ Ft Collins CO 80523

SOMMERS, RAYMOND A, b Marshfield, Wis, Nov 22, 31; m 54; c 14. ANALYTICAL CHEMISTRY, MICROCOMPUTERS. *Educ:* Univ Wis-Stevens Point, BS, 53; Lawrence Univ, MS, 59, PhD(chem), 63. *Prof Exp:* From asst prof to assoc prof, 62-76, PROF ANAL CHEM, UNIV WIS-STEVENS POINT, 76- *Concurrent Pos:* NSF sci faculty fel, Dept Chem, Mich State Univ, 68-69. *Mem:* Am Chem Soc. *Res:* microcomputers in teaching chemistry. *Mailing Add:* Dept Chem Univ Wis Stevens Point WI 54481

SOMMERS, SHELDON CHARLES, b Indianapolis, Ind, July 7, 16; m 43. PATHOLOGY. *Educ:* Harvard Univ, SB, 37, MD, 41. *Prof Exp:* Assoc prof path, Sch Med, Boston Univ, 53-61; clin prof, Univ Southern Calif, 62-88; from assoc prof path to prof, Columbia Univ, 63-68, clin prof path, Col Physicians & Surgeons, 68-88; pathologist & dir, 68-81, CONSULT, PATH LAB, LENOX HILL HOSP,81- *Concurrent Pos:* Res assoc, Cancer Res Inst, New Eng Deaconess Hosp, 50-61; lectr, Harvard Med Sch, 53-61; pathologist, Mass Mem Hosps, 53-61, Scripps Hosp, 61-63 & Delafield Hosp, NY, 63-68; ed, Path Ann, 66-86 & Path Decenn, 66-75; sci dir, Coun Tobacco Res, 81-87, consult, 88; consult path, 81- *Mem:* Am Soc Clin Path; Am Asn Path; Col Am Path. *Res:* Experimental pathology; intestinal disease; cancer; kidney disease; endocrine pathology. *Mailing Add:* NY Univ Med Ctr 550 First Ave Lenox Hill Hosp 100 E 77th St New York NY 10016

SOMMERS, WILLIAM P(AUL), b Detroit, Mich, July 22, 33; m 56, 78; c 5. ENGINEERING, RESEARCH MANAGEMENT. *Educ:* Univ Mich, BSE, 55, MSE, 56, PhD(mech eng), 61. *Prof Exp:* Engr, Martin Co, 56-57, sr engr, 57-58; res assoc aeronaut eng, Inst Sci & Technol, Univ Mich, 59-61; chief chem propulsion, Martin Co, 61-63; proj scientist mech & aeronaut eng, 63-65, res dir eng & sci mgt, 65-67, vpres & dir NASA progs, 67-71, pres & mem bd dirs, 71-73, pres, Technol Mgt Group, 73-79, SR EXEC VPRES, BOOZ ALLEN & HAMILTON INC, 79- *Concurrent Pos:* Consult, Ethyl Corp, 60-61. *Mem:* Assoc fel Am Inst Aeronaut & Astronaut; sr mem Am Astron Soc; Sigma Xi. *Res:* Detonative combustion; fluid dynamics; heat transfer; propulsion and aerospace sciences. *Mailing Add:* Booz Allen & Hamilton Inc 555 Montgomery 17th Floor San Francisco CA 94111

SOMMER SMITH, SALLY K, b Menomine, Mich, Oct 2, 53. CELLULAR BIOLOGY. *Educ:* Tufts Univ, PhD(cell biol & anat), 80. *Prof Exp:* ASST PROF SCI, BOSTON UNIV, 84- *Mailing Add:* Dept Sci Col Basic Studies Boston Univ Boston MA 02215

SOMMESE, ANDREW JOHN, b New York, NY, May 3, 48; m 71; c 2. TRANSCENDENTAL ALGEBRAIC GEOMETRY. *Educ:* Fordham Univ, BA, 69; Princeton Univ, PhD(math), 73. *Prof Exp:* Gibbs instr, Yale Univ, 73-75; asst prof, Cornell Univ, 75-79; assoc prof, 79-83, PROF MATH, UNIV NOTRE DAME, 83-, CO-DIR, CTR APPL MATH, 87-, CHMN DEPT MATH, 88- *Concurrent Pos:* Mem, Inst Advan Study, NJ, 75-76; guest prof, Univ Gottingen, 77, Univ Bonn, 78-79 & Max Planck Inst Math, Bonn, W Germany 84-85, & 87; Sloan Fel, Alfred P Sloan Found, 79; ed, Manuscripta Mathematica, 86-; consult, Gen Motors Res Lab, 86- *Mem:* Am Math Soc; Sigma Xi. *Res:* Projective classification of algebraic varieties; topology of algebraic varieties; adjunction theory of projective manifolds. *Mailing Add:* Dept Math Univ Notre Dame Notre Dame IN 46556

SOMOANO, ROBERT BONNER, b Houston, Tex, Sept 2, 40; m 62; c 3. SOLID STATE PHYSICS. *Educ:* Tex A&M, BS, 62, MS, 64; Univ Tex, PhD(physics), 69. *Prof Exp:* MEM TECH STAFF, JET PROPULSION LABS, 69- *Mem:* Am Inst Physics; Am Phys Soc. *Res:* Liquid metals; polymer physics; superconductivity. *Mailing Add:* 835 Old Landmark Lane La Canada CA 91011

SOMOGYI, LASZLO P, b Budapest, Hungary, June 1, 31; US citizen; m 51; c 2. PLANT PHYSIOLOGY, HORTICULTURE. *Educ:* Univ Agr Sci Hungary, BS, 56; Rutgers Univ, MS, 60, PhD(hort), 62. *Prof Exp:* Lab technician plant physiol, Cornell Univ, 57-58; jr res pomologist, Univ Calif, Davis, 62-64; proj leader, Hunt-Wesson Foods, Inc, 64-70; dir res & develop, Vacu-Dry Co, 70-74; tech dir, Biophys Res & Develop Corp, 74-76; sr food scientist, SRI Int, 76-79; vpres, Finn-Cal Prod, Inc, 79-81; pres, Etel, Inc, 81-89; SR CONSULT, SRI INT, 89- *Honors & Awards:* Outstanding Mem Award, Inst Food Technologist, Northern Calif, 84. *Mem:* Fel Inst Food Technologists; Am Soc Enologists. *Res:* Food processing and product development; food dehydration; technoeconomic market studies of food ingredients and additives; environmental impact of food processing operations; harvesting and storage of fruits and vegetables. *Mailing Add:* 12 Highgate Ct Kensington CA 94707

SOMORJAI, GABOR ARPAD, b Budapest, Hungary, May 4, 35; m 57; c 2. PHYSICAL CHEMISTRY. *Educ:* Budapest Tech Univ, ChE, 56; Univ Calif, PhD(chem), 60. *Hon Degrees:* Dr, Univ Paris, Budapest Tech Univ. *Prof Exp:* Mem res staff, Res Ctr, Int Bus Mach Corp, 60-64; from asst prof to assoc prof, 64-67, Miller prof, 77-78, PROF CHEM, UNIV CALIF, BERKELEY, 72-; DIR SURFACE SCI & CATALYSIS PROG, LAWRENCE BERKELEY LAB, 64- *Concurrent Pos:* Guggenheim fel, 69-70; vis fel, Emmanuel Col, Cambridge, Eng, 69; Unilever vis prof, Bristol Univ, 71-72; chmn, Div Colloid & Surface Chem, Am Chem Soc, 75; Royal Soc lectr, 83. *Honors & Awards:* Emmett Award, Am Catalysis Soc, 77; Baker Lectr, Cornell Univ, 77; Colloid & Surface Chem Award, Am Chem Soc, 81, Peter Debye Award, 89; Palladium Medal, 86; G N Lewis Lectr, Univ Calif, Berkeley, 87. *Mem:* Nat Acad Sci; fel Am Phys Soc; Am Chem Soc; fel AAAS; Am Acad Arts & Sci. *Res:* Chemistry of surfaces and solids; catalysis; surface science of energy conversion; mechanism of catalysis of hydrocarbon reactions by metals and otides; structure of surfaces, tribology. *Mailing Add:* Dept Chem Univ Calif Berkeley CA 94720

SOMORJAI, RAJMUND LEWIS, b Budapest, Hungary, Jan 21, 37; Can citizen; m 70; c 2. THEORETICAL BIOLOGY, BIOPHYSICS. *Educ:* McGill Univ, BSc, 60; Princeton Univ, PhD(physics, phys chem), 63. *Prof Exp:* NATO sci fel, Cambridge Univ, 63-65; RES OFFICER, NAT RES COUN CAN, 65- *Concurrent Pos:* Adj prof, Dept Physiol & Biophys, Univ Ill Med Ctr, 75- *Mem:* Chem Inst Can; Am Phys Soc; Can Asn Physics. *Res:* Approximation methods; calculation of the dynamics of protein folding and enzyme action; structure-function relationships in biology; properties of complex, hierarchical systems; nonequilibrium phenomena; nonlinear problems. *Mailing Add:* Inst Biol Sci Rm 1113 M54 Nat Res Coun Can Ottawa ON K1A 0R6 Can

SOMSEN, ROGER ALAN, b River Falls, Wis, May 4, 31; m 56; c 1. PULP CHEMISTRY. *Educ:* Univ Wis, BS, 53; Lawrence Col, Inst Paper Chem, MS, 55, PhD(pulp paper), 58. *Prof Exp:* Sr res chemist, Olin Mathieson Chem Corp, 58-59, asst supvr pulp & paper res, 59-60, supvr, 60-62, process eng res, 62-68, tech serv mgr, Pulp & Paper Div, 68-72, tech serv dir, 72-82, res & develop mgr, 82-84, tech serv mgr, 84-86, CUST SERV MGR, MANVILLE FOREST PROD CORP, INC, 86- *Concurrent Pos:* Adj prof, paper sci & technol, Northeast La Univ. *Mem:* Tech Asn Pulp & Paper Indust; Am Chem Soc. *Res:* Improvement of old grades and development of new grades of paper, including pulping, bleaching and papermaking; product development in cartons, bags and corrugated containers. *Mailing Add:* PO Box 488 West Monroe LA 71294-0488

SON, CHUNG HYUN, b Changyun, Korea, Mar 16, 17; US citizen; m 39; c 3. FOOD PRODUCTS DEVELOPMENT. *Educ:* Rutgers State Univ NJ, BS, 56, MS, 57, PhD(food sci), 59. *Prof Exp:* Sr scientist, Del Monte Corp Res Ctr, 59-83, consult food processing, 84-89; RETIRED. *Mailing Add:* 782 Tiffany Pl Concord CA 94518

SONAWANE, BABASAHEB R, b Nandgaon, India, April 5, 40; m 67; c 2. ENVIRONMENTAL TOXICOLOGY. *Educ:* Univ Peona, India, BS, 62, MS, 65; Univ Mo-Columbia, PhD(enthomol & toxicol), 71. *Prof Exp:* Lectr zool & entomol, Col Agr, Univ Peona, 63-67; res specialist toxicol & path, Univ Mo, Columbia, 71-72; fel environ & toxicol, Nat Inst Environ Health Sci, 72-75; sr res assoc pediat & pharm, Children's Hosp, Philadelphia, 75-81. *Concurrent Pos:* Vis fel, Nat Inst Environ Health Sci, 72-75. *Mem:* Teratology Soc Am; AAAS; Am Chem Soc; NY Acad Sci. *Res:* Developmental pharmacology-toxicology; mechanisms of xenobiotic toxicity and teratogenicity; regulation of drug metabolism and action during perinatal development and its modification by environmental factors such as nutrition, disease and exposure to chemicals. *Mailing Add:* 13204 Moran Dr Gaithersburg MD 20878

SONDAK, NORMAN EDWARD, b Cornwall, NY, Sept 1, 31; m 54; c 3. COMPUTER SCIENCE, INFORMATION SYSTEMS. *Educ:* City Col NY, BE, 53; Northwestern Univ, MS, 54; Yale Univ, DEng(eng), 58. *Prof Exp:* Sr technologist, Res Labs, Socony Mobil, 56-61; mgr data processing, Electronic Data Processing Div, RCA Corp, 61-63; vpres data processing, J Walter Thompson Co, NY, 63-68; prof comput sci & head dept, Worcester Polytech Inst, 68-78, dir comput ctr, 68-71; prof & chmn, 78-85, PROF, INFO & DECISION SYSTS, SAN DIEGO STATE UNIV, 85- *Concurrent Pos:* Affil prof, Clark Univ, 69-78; mem coop staff, Worcester Found Exp Biol, 70-78; res prof, Med Sch, Univ Mass, 75-78. *Honors & Awards:* Kellog lectr, 79. *Mem:* Data Processing Mgt Asn; Soc Indust & Appl Math; Asn Comput Mach; Inst Elec & Electronics Engrs; Am Soc Info Sci. *Res:* Programming languages; data base management systems; operating systems; social implications of computing; computer architecture; computer science education; computer networks; structured systems design; word processing; microcomputer systems; artificial intelligence; expert systems. *Mailing Add:* Info Systs Dept San Diego State Univ San Diego CA 92182

SONDEL, PAUL MARK, b Milwaukee, Wis, Aug 14, 50; m 73; c 2. TUMOR IMMUNOLOGY, IMMUNOGENETICS. *Educ:* Univ Wis-Madison, BS, 71, PhD(genetics), 75, Harvard Med Sch, MD, 77. *Prof Exp:* Res & teaching asst, Dept Genetics, Univ Wis-Madison, 71-72; res aide, Dept Immunol, Harvard Med Sch, 73-74; res assoc, Immunobiol Res Ctr, Univ Wis-Madison, 74-75; res fel tumor immunol, Sidney Farber Cancer Inst, 75-77; intern pediat, Univ Minn Hosp, 77-78; resident, Univ Wis Hosp, 78-80; asst prof pediat & human oncol, 80-84, asst prof med genetics, 81-84, assoc prof, 84-87, PROF, PEDIAT, HUMAN ONCOL & MED GENETICS, UNIV WIS-MADISON, 87- *Concurrent Pos:* Scholar, Leukemia Soc Am, 81-86; fel, G A & J L Hartford Found, 81-84. *Mem:* Transplantation Soc; Am Asn Immunologists; Am Asn Clin Histocompatibility; Am Fedn Clin Res; Am Soc Clin Invest; Soc Pediat Res. *Res:* Tumor and transplantation immunogenetics: the in vitro responses of human lymphocytes to normal and abnormal cell populations to better define the role of human leucocyte antigen factors in immunoregulation. *Mailing Add:* Univ Wis Clin Sci Ctr 660 Highland Ave Rm K4-448 Madison WI 53792

SONDER, EDWARD, b Ger, May 1, 28; nat US; m 53; c 2. PHYSICS, MATERIALS SCIENCE. *Educ:* Queens Col, BS, 50; Univ Ill, MS, 51, PhD(physics), 55. *Prof Exp:* Res assoc solid state physics, Iowa State Col, 55-56; PHYSICIST, OAK RIDGE NAT LAB, 56- *Concurrent Pos:* Vis prof physics, Okla State Univ, 74-75; adj prof, Vanderbilt Univ, 87- *Mem:* Am Phys Soc; Am Ceramic Soc. *Res:* Imperfections in solids, radiation effects in insulators and semiconductors; color centers in alkali halides; electrical properties of insulators; varistors; high temperature superconductors. *Mailing Add:* Oak Ridge Nat Lab PO Box 2008 Oak Ridge TN 37830-6056

SONDEREGGER, THEO BROWN, b Brimingham, Ala, May 31, 25; m 47; c 3. MEDICAL PSYCHOLOGY. *Educ:* Fla State Univ, BS, 46; Univ Nebr, Lincoln, MA, 48 & 60, PhD(clin psychol), 65. *Prof Exp:* Teaching asst psychol, Univ Nebr, Lincoln, 59-62, instr med psychol, Med Ctr, 65-69; asst prof psychol, Nebr Wesleyan Univ, 65-68; from asst prof to assoc prof, 69-78, PROF PSYCHOL, UNIV NEBR, LINCOLN, 76-, PROF MED PSYCHOL, UNIV NEBR MED CTR, 78- *Concurrent Pos:* Vis scholar, Dept Neurosci, Northwestern Univ, Evanston, 73-74; vis assoc res anatomist, Med Sch, Univ Calif, Los Angeles, 74, vis res psychol, Dept Psychiat, 79-80; vis res assoc biol, Calif Inst Technol, 79-80; vis prof, Brain Res Inst, Univ Calif, Los Angeles, 80-81. *Honors & Awards:* Outstanding Scientist Award, Sigma Xi, 91. *Mem:* Am Psychol Asn; Soc Neurosci; fel AAAS; Int Soc Develop Biol; Psychonomic Soc. *Res:* Neonatal narcotic addiction; fetal alcohol syndrome (animal model); intracranial self stimulation; catecholamines and the developing nervous system; psychology of women. *Mailing Add:* 1710 S 58th St Lincoln NE 68506

SONDERGAARD, NEAL ALBERT, b Schenectady, NY, Mar 20, 49; m 77. PHYSICAL CHEMISTRY, CHEMICAL PHYSICS. *Educ:* Marist Col, BA, 70; Brown Univ, MSc, 73, PhD(chem), 77. *Prof Exp:* Fel, Wash Univ, 77; fel phys chem, Johns Hopkins Univ, 77-80; CHEMIST, NAVAL SHIP RES & DEVELOP CTR, ANNAPOLIS, 80- *Concurrent Pos:* Fel, Johns Hopkins Univ, 80- *Mem:* Sigma Xi. *Res:* Physical and chemical phenomena of high current density sliding electric contacts; techniques include molecular beams, ion cyclotron resonance and mass spectroscopy. *Mailing Add:* 591 Treslow Glen Dr Severna Park MD 21146

SONDERGELD, CARL HENDERSON, b Brooklyn, NY, Nov 4, 47; m 69; c 2. GEOPHYSICS, ROCK MECHANICS. *Educ:* Queen's Col, NY, BA, 69, MA, 73; Cornell Univ, PhD(geophysics), 77. *Prof Exp:* Res assoc geothermal energy, Cornell Univ, 77; vis fel, Nat Oceanic & Atmospheric Admin, Univ Colo, 77-78, res assoc rock mech, Coop Inst Res Environ Sic, 78-81; sr res scientist, 81-83, staff res scientist, 83-88, RES ASSOC, AMOCO PROD CO, 88- *Concurrent Pos:* Adj prof, Univ Colo, 80-; vis scientist, Los Alamos Nat Lab. *Mem:* Am Geophys Union; Acoust Emission Soc; Sigma Xi. *Res:* Acoustic emissions in rock; elasticity of rocks and polycrystals; geothermal energy-two-phase convection in porous media; acoustic logging and shear wave anisotropy; acoustic magnetic and electrical properties of rock. *Mailing Add:* Amoco Prod Co PO Box 3385 Tulsa OK 74102

SONDHAUS, CHARLES ANDERSON, b San Francisco, Calif, Oct 5, 24; m 55; c 3. BIOPHYSICS, RADIOBIOLOGY. *Educ:* Univ Calif, Berkeley, AB, 50, PhD(biophys), 58. *Prof Exp:* Radiol physicist, Naval Radiol Defense Lab, 51-55; res asst biophys, Lawrence Radiation Lab, Univ Calif, Berkeley, 56-58, biophysicist, Donner Lab, 59-64, asst prof, 65-66, ASSOC PROF RADIOL SCI, UNIV CALIF, IRVINE, 66- *Concurrent Pos:* Nat Acad Sci-Nat Res Coun Donner fel biophys, Karolinska Inst & Hosp, Stockholm, Sweden, 58-59; vis lectr, Mont State Univ, 60-62; lectr, Div Med Physics, Univ Calif, Berkeley, 60-65; res collabr, Med Dept, Brookhaven Nat Lab, 60-; consult radiol physicist, Off Civil Defense, US Air Force, NASA, 62-64, US Vet Admin & Fed Aviation Agency, 65-, Los Angeles County Gen Hosp, 65-68 & Orange County Med Ctr, 68-; mem subcomt rel biol effectiveness, Nat Coun Radiation Protection, 60-66, biol effects high altitude cosmic radiation, Int Comn Radiol Protection, 63-65, high energy & space radiation dosimetry, Int Comn Radiation Units, 64- & space radiation study panel, Nat Acad Sci-Nat Res Coun, 64-68. *Mem:* Am Asn Physicists in Med; Am Phys Soc; Biophys Soc; Radiation Res Soc. *Res:* Applied nuclear & radiation physics, dosimetry; biophysical and optical microanalysis; cellular and mammalian radiobiology. *Mailing Add:* Radiol Sci Dept Col Med Univ Calif Irvine CA 92717

SONENBERG, MARTIN, b New York, NY, Dec 1, 20; m 56; c 2. ENDOCRINOLOGY, BIOCHEMISTRY. *Educ:* Univ Pa, BA, 41; NY Univ, MD, 44, PhD(biochem), 52. *Prof Exp:* Intern, Beth Israel Hosp, 44-45; asst resident med, Goldwater Hosp, 45-46; from instr to assoc prof, 53-72, PROF MED, MED COL, CORNELL UNIV, 72-, PROF BIOCHEM, CORNELL GRAD SCH MED SCI, 66- *Concurrent Pos:* Am Cancer Soc fel, Mem Ctr Cancer & Allied Dis, 52-57; Guggenheim fel, Carlsberg Lab, Copenhagen Univ, 57-58; clin asst, Mem Ctr Cancer & Allied Dis, 51-; assoc, Sloan-Kettering Inst Cancer Res, 52-60, assoc mem, 60-66, mem, 66-; assoc attend physician, Mem & James Ewing Hosps, 59-; attend physician, Mem Hosp, 69-; NIH endocrinol study sect, 83-87, chmn, 85-87. *Honors & Awards:* Van Meter award, Am Thyroid Asn, 52; Sloan Award, 68. *Mem:* AAAS; Biophys Soc; Am Soc Biol Chem; Am Soc Clin Invest; Am Thyroid Asn; Endocrine Soc. *Res:* Chemistry and physiology of pituitary hormones; protein chemistry; mechanism of hormone action. *Mailing Add:* Mem Sloan-Kettering Cancer Ctr New York NY 10021

SONENSHEIN, ABRAHAM LINCOLN, b Paterson, NJ, Jan 13, 44; m 67; c 2. MICROBIOLOGY, MOLECULAR BIOLOGY. *Educ:* Princeton Univ, AB, 65; Mass Inst Technol, PhD(biol), 70. *Prof Exp:* Am Cancer Soc fel, Inst Microbiol, Univ Paris, 70-72; from asst prof to assoc prof, 72-82, PROF MOLECULAR BIOL & MICROBIOL, SCH MED, TUFTS UNIV, 82- *Concurrent Pos:* Nat Inst Gen Med Sci res support grant, 72-92; NSF res grant, 79-81. *Mem:* AAAS; Am Soc Microbiol; Fedn Am Sci; Am Soc Biochem & Molecular Biol. *Res:* Bacterial sporulation; control of transcription; RNA polymerase; genetics and physiology of Bacillus subtilis; phage infection of Bacillus subtilis. *Mailing Add:* Dept Molecular Biol & Microbiol Sch Med Tufts Univ Boston MA 02111

SONENSHINE, DANIEL E, b New York, NY, May 11, 33; m 57. ZOOLOGY. *Educ:* City Col NY, BA, 55; Univ MD, PhD(zool), 59. *Prof Exp:* Asst zool, Univ Md, 55-58, asst instr, 58-59; instr biol, Univ Akron, 59-61; mem staff, Old Dom Univ, 61-74; PROF MICROBIOL, EASTERN VA MED SCH, NORFOLK, 74- *Mem:* Am Soc Parasitol; Sigma Xi. *Res:* Acarology; parasitology; ecology and life history of ticks; physiology. *Mailing Add:* Dept of Biol Old Dominion Univ Norfolk VA 23508

SONETT, CHARLES PHILIP, b Pittsburgh, Pa, Jan 15, 24; m 48; c 2. PHYSICS. *Educ:* Univ Calif, Los Angeles, PhD(physics), 54. *Prof Exp:* From asst to assoc prof physics, Univ Calif, 51-53; mem tech staff & head range develop group, Ramo Wooldridge Corp, 54-57; mem sr staff & head space physics, Space Technol Labs, Inc, 57-60; chief sci, off lunar & planetary prog, NASA, 60-62, chief space sci div, Ames Res Ctr, 62-70, dep dir astronaut, 71-73; head dept planetary sci & dir, Lunar & Planetary Lab, 73-77, prof, 73-90, REGENTS PROF PLANETARY SCI, LUNAR & PLANETARY LAB, UNIV ARIZ, 90- *Concurrent Pos:* Lectr eng, Univ Calif, Los Angeles, 55-58; Guggenheim fel, Imp Col, Univ London, 68-69; ed, Cosmic Electrodynamics, 70-72; co-ed, Astrophys & Space Sci, 73-80; mem, Space Sci Steering Comt, NASA, chmn subcomts lunar sci & planetary & interplanetary sci, 60-62, mem subcomts planetology, 62-65 & particles & fields, 62-63, mem outer planets sci adv group, 71-72, mem, Outer Planets Sci Working Group, Post-Apollo Sci Planning Conf & ad hoc working group on planetary remote sensing, 72, mem NASA/Jet Propulsion Lab terrestrial bodies sci working group, 76, chmn NASA/Univs Space Res Asn, tethered satellite exp rev panel, 84, mem Mars Observer exp rev panel, 85; mem, Comn 17 & 49, Int Astron Union; consult, Jet Propulsion Lab & Rockwell Int; mem bd trustees, Univ Space Res Asn, 77-83; co-ed, The Moon & Planets, 78-; mem Los Alamos br adv comn, Univ Calif Inst Geophys & Planetary Physics, 84-85; Carnegie fel, Univ Edinburgh, 85; fel, Inst Advan Studies, Univ Indiana, 90. *Honors & Awards:* Space Sci Award, Am Inst Aeronaut & Astronaut, 69; Exceptional Sci Achievement Medal, NASA, 69 & 72. *Mem:* Am Geophys Union; Sigma Xi; AAAS. *Res:* Planetary and interplanetary physics. *Mailing Add:* Dept Planetary Sci Univ Ariz Tucson AZ 85721

SONG, BYOUNG-JOON, b Mar 4, 50; US citizen; m; c 2. PROTEIN BIOCHEMISTRY, MOLECULAR BIOLOGY. *Educ:* Seoul Nat Univ, BS, 72, MS, 76; Univ Minn, PhD(pharmacol, biochem), 83. *Prof Exp:* Res asst pharmacol, Univ Minn, 77-82; postdoctoral fel biochem, Lab Molecular Carcinogenesis, Nat Cancer Inst, NIH, 83-86; sr staff fel, 86-90, SECT CHIEF, LAB METAB & MOLECULAR BIOL, NAT INST ALCOHOL ABUSE & ALCOHOLISM, 90- *Concurrent Pos:* Adj prof, Dept Pharmacol, Univ Md, 90- *Mem:* Am Soc Biochem & Molecular Biol; AAAS. *Res:* Molecular biology on ethanol inducible cytochrome P450; molecular biology on the pyruvate dehydrogenase multienzyme complex. *Mailing Add:* Lab Metab & Molecular Biol NIAAA 12501 Washington Ave Rockville MD 20852

SONG, CHANG WON, b Chun Chon City, Korea, Apr 10, 32. RADIOBIOLOGY, IMMUNOLOGY. *Educ:* Seoul Nat Univ, BS, 57; Univ Korea, MS, 59; Univ Iowa, PhD(radiation biol), 64. *Prof Exp:* Res asst radiation biol, Univ Iowa, 60-64; asst mem, Res Labs, Albert Einstein Med Ctr, 64-69; asst prof, Med Col Va, 69-70; from asst prof to assoc prof, 70-78, PROF & DIR RADIATION BIOL, MED SCH, UNIV MINN, MINNEAPOLIS, 78- *Concurrent Pos:* Consult, Vet Admin & Nat Can Inst. *Mem:* Cell Kinetic Soc; AAAS; Radiation Res Soc; Am Asn Cancer Res; Europ Soc Hyperthermic Oncol; N Am Hyperthermia Group. *Res:* Relationship between vascular changes and curability of tumors by radiotherapy or hyperthermia; effect of radiation on immune system and feasibility of combination radio- and immuno-therapy for treatment of cancer; radiosensitization and radioprotection of tumors. *Mailing Add:* Dept Therapeut Radiol Univ Minn Health Sci Ctr Box 494 MAYO 420 Delaware St SE Minneapolis MN 55455

SONG, CHARLES CHIEH-SHYANG, b Taiwan, China, Jan 12, 31; m 55; c 3. CIVIL ENGINEERING, FLUID MECHANICS. *Educ:* Nat Taiwan Univ, BS, 53; Univ Iowa, MS, 56; Univ Minn, Minneapolis, PhD(civil eng), 60. *Prof Exp:* From asst prof to assoc prof, 61-79, PROF CIVIL ENG, UNIV MINN, MINNEAPOLIS, 79- *Honors & Awards:* J C Stevens Award, 80. *Mem:* AAAS; Int Asn Hydraul Res; Soc Naval Archit & Marine Engrs; Am Soc Civil Engrs; Am Water Resources Asn. *Res:* Flows at large Reynolds numbers; computional hydrodymincs; turbulent flow modeling; drainage and water distribution system modeling; hydraulic transient; sediment transport; effect of ice on flooding. *Mailing Add:* St Anthony Falls Hydrology Lab Univ Minn Minneapolis MN 55414

SONG, JIAKUN, b Shanghai, China, Aug 27, 44. COMPARATIVE NEUROANATOMY, ONTOGENY & PHYLOGENY OF VERTEBRATES. *Educ:* Shanghai Col Fisheries, BS, 67; Univ Mich, Ann Arbor, MS, 82, PhD(biol sci), 89. *Prof Exp:* Sci & technol dir aquaculture, Biol Stat Bur Fisheries, Ning-De Region Fujian Prov, China, 70-75; asst researcher ichthyol, Inst Zool, Academia Sinica, Beijing, 75-80; res assoc fish taxon, Mus Zool, Univ Mich, Ann Arbor, 81-83 & teaching asst & lab instr vertebrate biol, Dept Biol, 83-86; res assoc neurobiol, Neurobiol Unit, Scripps Inst Oceanog, Univ Calif San Diego, 87-89; POSTDOCTORAL RES FEL NEUROSCI, SCH LIFE & HEALTH SCI, UNIV DEL, 89- *Mem:* AAAS; Am Soc Ichthyologists & Herpetologists; Am Soc Zoologists; Chinese Soc Zoologists; Int Brain Res Orgn; Soc Neurosci. *Res:* Lateral line system of fishes; application of novel studies on organization of cranial nerves to questions of phylogenetic relationships of fishes. *Mailing Add:* Sch Life & Health Sci Univ Del Newark DE 19716

SONG, JOSEPH, b Seoul, Korea, May 11, 27; nat US; m 58. PATHOLOGY. *Educ:* Seoul Nat Univ, MD, 50; Univ Tenn, MS, 56; Univ Ark, MD, 65. *Prof Exp:* Instr path, Med Sch, Univ Tenn, 52-56; instr, Sch Med, Boston Univ, 56-61; assoc prof, Sch Med, Univ Ark, 61-65; DIR DEPT PATH, MERCY HOSP, 65- *Concurrent Pos:* Assoc dir, RI State Cancer Cytol Proj, 56-59; sr instr, Med Sch, Tufts Univ, 59-61; assoc mem, Inst Health Sci, Brown Univ, 59-61; assoc pathologist, Providence Lying-In Hosp, 59-61; assoc med examr, State RI, 59-61; consult, St Joseph's Hosp, Providence, RI, 59-61 & Vet Admin Hosps, Little Rock & North Little Rock, Ark; clin prof, Sch Med, Creighton Univ. *Honors & Awards:* Martin Luther King Med Achievement Award, 72; Statesman in Health Care Award, 87. *Mem:* Fel Am Soc Clin Path; Am Asn Path & Bact; fel Am Col Path. *Res:* Hepatic pathology in sickle cell disease; exfoliative cytology in cancer of the cervix; splenic function and tumor growth, experimental cancer research. *Mailing Add:* Mercy Hosp Dept of Path Sixth & University Des Moines IA 50314

SONG, KONG-SOP AUGUSTIN, b Korea, 1934. SOLID STATE PHYSICS. *Educ:* Chunpuk Nat Univ, Korea, 56, MS, 57; Univ Paris, Dr 3e Cycle, 64; Univ Strasbourg, Dr es Sci(physics), 67. *Prof Exp:* Jr researcher, Nat Ctr Sci Res, France, 63-69; from asst prof to assoc prof, 69-80, PROF PHYSICS, UNIV OTTAWA, 80- *Concurrent Pos:* Instr, Univ Strasbourg, 65-69. *Mem:* Am Phys Soc. *Res:* Electronic and optical properties of semiconductors and insulators. *Mailing Add:* Dept Physics Univ Ottawa Ottawa ON K1N 6N5 Can

SONG, MOON K, b Taejon, Korea, July 9, 31; m 66; c 2. FORMULATE RESEARCH PROPOSAL. *Educ:* Univ Hawaii, BA, 64, MS, 66, PhD(genetics), 72. *Prof Exp:* Res asst sea water, Univ Hawaii, 63-66, jr res chemist, 66-69; postdoctoral lysosomal enzymes, Ind Univ, 73-74; res chemist, 74-83, CHIEF, MTR LAB, US DEPT VET AFFAIRS, 83- *Concurrent Pos:* Asst res prof, Univ Calif, Los Angeles, 82-87, assoc res prof, 87-; vis prof, Korean Inst Technol, 91. *Mem:* Am Inst Nutrit; Am Soc Clin Nutrit; Am Col Nutrit. *Res:* Elucidation of the mechanisms by which zin is absorbed and utilized by organ specific cells; treatment of patients suffering from various diseases such as diabetes, cancer, aging, skin disorders and hypertension. *Mailing Add:* 10922 Yolanda Ave Northridge CA 91326

SONG, PILL-SOON, b Osaka, Japan, Aug 5, 36; m 84; c 3. MOLECULAR BIOPHYSICS. *Educ:* Univ Seoul, BS, 58, MS, 60; Univ Calif, PhD(biochem), 64. *Prof Exp:* Res assoc biochem & biophys, Iowa State Univ, 64-65; from asst prof to prof, 65-75, PAUL W HORN PROF, TEX TECH UNIV, 75- *Concurrent Pos:* Robert A Welch Found grant photochem res, 66-, res grant, 72-75; NSF grant, 70-; Nat Cancer Inst grant, 72- *Mem:* Am Soc Biol Chemists; Am Soc Photobiol. *Res:* Photobiology of Phytochrome in plants and photosensory transduction in stentor; molecular spectroscopy and photochemistry of photoreceptor pigments, energy transduction, and quantum biology. *Mailing Add:* Dept Chem Univ Nebr Lincoln NE 68588-0304

SONG, SEH-HOON, b Seoul, Korea, June 29, 36; Can citizen; m 62; c 4. BIOPHYSICS, CARDIOPULMONARY PHYSIOLOGY. *Educ:* Yonsei Univ, MD, 60; State Univ NY Buffalo, MA, 69; Univ Western Ont, PhD(biophys), 72. *Prof Exp:* Instr physiol, Sch Med, Yonsei Univ, 60-62 & 66-67; res assoc, State Univ NY Buffalo, 67-69; lectr, 72-73, ASST PROF BIOPHYS, UNIV WESTERN ONT, 73- *Mem:* Biophys Soc; Am Physiol Soc; Can Physiol Soc. *Res:* Compartmentalization in the microcirculation of various organs, spleen, skeletal muscles and heart; transport of materials through the endothelial membranes. *Mailing Add:* Dept Biophys Fac Med Univ Western Ont London ON N6A 5C1 Can

SONG, SUN KYU, b Yonchon, Korea, May 15, 27; US citizen; m 56; c 3. NEUROPATHOLOGY. *Educ:* Yonsei Univ, Korea, MD, 49. *Prof Exp:* Asst prof, 65-71, ASSOC PROF NEUROPATH, MT SINAI SCH MED, 72- *Concurrent Pos:* USPHS spec fel, Mt Sinai Hosp, 59-63; asst attend neuropathologist, Mt Sinai Hosp, 63-; assoc attend physician, City Hosp Ctr, Elmhurst, NY, 64- *Mem:* Am Asn Neuropath; Histochem Soc; Am Asn Path & Bact; Am Soc Exp Path; Am Acad Neurol. *Res:* Histochemistry and electron microscopy of neuromuscular junction and pathology of neuromuscular diseases. *Mailing Add:* Dept Neuropath Mt Sinai Sch Med Fifth Ave & 100 St New York NY 10029

SONG, WON-RYUL, b Korea; US citizen; m 59; c 2. POLYMER CHEMISTRY. *Educ:* Yonsei Univ, Korea, BS, 52; McMaster Univ, MS, 58; Polytech Inst Brooklyn, PhD(polymer chem), 65. *Prof Exp:* Chemist, Am Cyanamide Co, 61-63; SR STAFF CHEMIST, EXXON CHEM CO, 65- *Mem:* Am Chem Soc. *Res:* Fundamental studies on polymeric lube oil additives. *Mailing Add:* 36 Dorset Lane Short Hills NJ 07078-3399

SONG, YO TAIK, b Korea, Feb 23, 32; m 60; c 3. NUCLEAR ENGINEERING. *Educ:* Yonsei Univ, Korea, BE, 54; Univ Ill, Urbana, MS, 62, PhD(nuclear eng), 68. *Prof Exp:* Nuclear engr, Korean Atomic Energy Res Inst, 59-60; nuclear physicist, US Naval Civil Eng Lab, 63-67; asst prof nuclear eng, Univ Tenn, Knoxville, 68-69; nuclear engr, Tenn Valley Authority, 69-70; assoc dir, Prof Adv Serv Ctr, Univ Colo, 70-72; res nuclear engr, US Naval Surface Weapons Ctr, White Oak Lab, 72-79; PROG MGR NUCLEAR ENERGY, US DEPT ENERGY, GERMANTOWN, MD, 79- *Mem:* Am Nuclear Soc. *Res:* Radiation shielding; fast reactor physics; fuel management; nuclear weapons; radiation, neutral and charged; transport through various media; reactor and weapons safety. *Mailing Add:* 14208 Woodwell Terr Silver Spring MD 20906

SONGER, JOSEPH RICHARD, b South Charleston, WVa, Dec 20, 26; m 48; c 7. MICROBIOLOGY. *Educ:* Eastern Nazarene Col, AB, 51; Iowa State Univ, MS, 65. *Prof Exp:* Bacteriologist, USDA, 51-60, vet microbiologist, Nat Animal Dis Ctr, Sci & Educ Admin, Agr Res Serv, 60-86; CONSULT, BIOHAZARD CONTROL, 86- *Mem:* Am Soc Microbiol; Am Soc Safety Eng; Am Indust Hyg Asn; Sigma Xi; Am Biol Safety Asn (pres, 88-89). *Res:* Biological laboratory safety, disinfection, sterilization, air filtration and airborne infection; animal disease research, vesicular diseases; hog cholera; equine infectious anemia; biological hazard assessment; contamination control and euthanasia. *Mailing Add:* 419 Ninth St Ames IA 50010

SONGSTER, GERARD F(RANCIS), b Darby, Pa, Aug 29, 27; m 53; c 2. ELECTRICAL ENGINEERING. *Educ:* Drexel Inst Technol, BSEE, 51; Univ Pa, MSEE, 65, PhD(elec eng), 62. *Prof Exp:* Res engr, Philco Corp, 51-52; instr digital comput, Moore Sch Elec Eng, Univ Pa, 52-56; asst prof elec eng, Drexel Inst Technol, 56-62, assoc prof & actg dir biomed eng prog, 63; sr scientist, Res Div, Melpar Inc, 63-64; elec engr, US Naval Res Lab, 64-65; NIH spec fel, Mass Inst Technol, 65-67; physiol studies, NASA Electronics Res Ctr, 67-70; prof elec eng & chmn dept, Old Dominian Univ, 70-75; elec engr, Naval Ship Eng Ctr, 75-; ELEC ENGR, GEN ELEC CO, TRANS DEVELOP ENG. *Mem:* AAAS; Am Soc Eng Educ; Sr mem, Inst Elec & Electronics Engrs; Sigma Xi. *Res:* Electrophysiology of nerve tissue; switching theory; computer simulation of living systems; underwater acoustics; automated measurement. *Mailing Add:* 203 Alton St Syracuse NY 13215

SONI, ATMARAM HARILAL, b Shihor, India, Oct 5, 35; m 64; c 3. MECHANICAL ENGINEERING. *Educ:* Univ Bombay, BSc, 57; Univ Mich, BS, 59, MS, 61; Okla State Univ, PhD(mech eng), 67. *Prof Exp:* Res asst comput prog, Univ Mich, 61-64; res asst mech eng, 64-67, from asst prof to assoc prof, 67-77, PROF MECH ENG, OKLA STATE UNIV, 77- *Concurrent Pos:* Prin investr, NSF grant, 68-69 & 70-72, dir appl mech conf, 69-71. *Mem:* Am Soc Mech Engrs; Am Soc Eng Educ. *Res:* Machine design; synthesis and analysis of mechanisms; fatigue; reliability. *Mailing Add:* Sch Mech Eng Univ Cincinnati Main Campus Cincinnati OH 45221

SONI, KUSUM, b Hoshiarpur, India, Nov 14, 30; US citizen; m 58; c 2. MATHEMATICAL ANALYSIS. *Educ:* Univ Panjab, India, BA, 49, MA, 51; Ore State Univ, PhD(math), 64. *Prof Exp:* Lectr math, Panjab Educ Serv, 52-59; asst prof, Ore State Univ, 66-67; from asst prof to assoc prof, 67-83, PROF MATH, UNIV TENN, KNOXVILLE, 83- *Concurrent Pos:* Vis mem, dept math sci, Univ Dundee, Scotland, 82; vis prof, Centre Math & Comput Sci, Amsterdam, 87, Indian Inst Sci Bangalore, 87- *Mem:* Am Math Soc; Math Asn Am. *Res:* Classical analysis, asymptotic expansions and approximation. *Mailing Add:* Univ Tenn Dept Math Knoxville TN 37996-1300

SONI, PREM SARITA, b Kisumu, Kenya, Nov 17, 48; Brit citizen; m 74. OPTOMETRY. *Educ:* Univ Manchester, BSc, 72; Ind Univ, OD, 75, MS, 79. *Prof Exp:* Optometrist, Eng, 72-75; lectr optom, 76-78, ASST PROF OPTOM, IND UNIV, 78- *Concurrent Pos:* Grant-in-aid, Ind Univ, 78-; Am Acad Optom grant, 78-; Wesley-Jassen, Inc grant, 78- *Mem:* Fel Brit Optical Asn; Am Acad Optom; Am Optom Asn; Contact Lens Educr Asn. *Res:* Corneal physiology and pathology with special reference to contact lens use. *Mailing Add:* Sch Optom 800 E Atwater Bloomington IN 47405

SONIN, AIN A(NTS), b Tallinn, Estonia, Dec 24, 37; US citizen; m 71; c 2. THERMO-FLUID SCIENCES. *Educ:* Univ Toronto, BASc, 60, MASc, 61, PhD(aerospace sci), 65. *Prof Exp:* From asst prof to assoc prof fluid mech, 65-74, PROF MECH ENG, MASS INST TECHNOL, 74- *Concurrent Pos:* Consult; sr scientist, Thermo Electron Corp, 81-82. *Mem:* AAAS; Am Phys Soc; Am Soc Mech Engrs; Am Nuclear Soc. *Res:* Fluid mechanics; thermodynamics; heat, mass and charge transport; electrochemistry. *Mailing Add:* Rm 3-256 Mass Inst of Technol Cambridge MA 02139

SONIS, MEYER, b Philadelphia, Pa, Jan 29, 19; m 44; c 3. PSYCHIATRY. *Educ:* Univ Pa, AB, 39; Hahnemann Med Col, MD, 43; Am Bd Psychiat & Neurol, dipl, 50, cert in child psychiat, 60. *Prof Exp:* Lectr psychiat, Sch Social Work, Univ Pa, 52-57, assoc, Sch Med, 52-61, instr, 55-61; assoc prof child psychiat, 61-69, prof psychiat, 69-76, CHIEF CHILD PSYCHIAT, SCH MED & EXEC DIR, PITTSBURGH CHILD GUID CTR, UNIV PITTSBURGH, 61-, CO-DIR POSTGRAD PEDIAT 62-, PROF CHILD DEVELOP & CARE, SCH HEALTH RELATED PROFESSIONS, 70-, PROF CHILD PSYCHIAT, SCH MED, 76- *Concurrent Pos:* Assoc physician, Children's Hosp Philadelphia, 51-61, sr physician & coordr child psychiat, 56-61; staff psychiatrist, Philadelphia Child Guid Clin, 51-61, supvr psychiat & sr psychiatrist, 54-61, coordr contract, Clin Eval Ctr, 57-61, dir training, 58-61, co-dir post-grad pediat, 60-62; vchmn sect content, Nat Conf Training Child Psychiat, 61-63; mem ad hoc comt outpatient studies, NIMH, 62-64, mem, Conf Planning Child Psychiat Serv, 64-66, mem planning comt, Conf Med Sch Child Psychiat, 65-66; mem bd examr, Am Bd Psychiat & Neurol, 62-66; mem, Joint Comn Ment Health for Children, 65-66; consult, Cath & Pub Schs, Pittsburgh, 65-66. *Mem:* AAAS; AMA; fel Am Psychiat Asn; fel Am Acad Child Psychiat; Asn Psychiat Clins for Children (pres, 65-67). *Mailing Add:* 3811 O'Hara St Dept Psychiatry Univ Pittsburgh Pittsburgh PA 15213

SONIS, STEPHEN THOMAS, b Oct 6, 45; m; c 2. PERIODONTOLOGY, ORAL ONCOLOGY. *Educ:* Norwich, BS, 67; Tufts, DMD, 72; Harvard, DMSc, 76. *Prof Exp:* CHIEF DENT SERV, BRIGHAM & WOMEN'S HOSP, BOSTON, 89-; PROF ORAL MED, HARVARD UNIV. *Mem:* Am Asn Periodontology; Am Acad Oral Med; Am Acad Oral Path; Am Dental Asn. *Res:* Wound healing; stomatitis; oral medicine. *Mailing Add:* Dept Surg Brigham & Women's Hosp 75 Francis St Boston MA 02115

SONLEITNER, FRANK JOSEPH, b Chicago, Ill, Jan 23, 32; div; c 3. POPULATION ECOLOGY. *Educ:* Univ Chicago, AB, 51, SB, 56, PhD(zool), 59. *Prof Exp:* Fel, Dept Zool, Univ Sydney, 59-61; lectr, Univ Calif, Berkeley, 61-62; asst prof entom, Univ Kans, Lawrence, 62-65; asst prof, 65-69, ASSOC PROF, DEPT ZOOL, UNIV OKLA, NORMAN, 69- *Mem:* AAAS; Am Inst Biol Sci; Ecol Soc Am; Entom Soc Am; Sigma Xi. *Res:* Computer simulation models of population dynamics and ecogenetics (natural selection). *Mailing Add:* Dept Zool Univ Okla Norman OK 73019

SONN, GEORGE FRANK, b Pt Pleasant, NJ, Jan 19, 36; m 60; c 2. COLOR TECHNOLOGY, FIBER SPECIALIST. *Educ:* NC State Univ, BS, 58; Rutgers Univ, MS, 60. *Prof Exp:* Tech mgr, Basf Inmont, 60-68; MKT DIR, MAGRUDER COLOR, 68- *Concurrent Pos:* Consult, Color & Fibers, 70- *Mem:* Soc Plastics Engrs; Am Asn Textile Chemists & Colorists. *Res:* Color technology; aqueous printing inks and synthetic pigmented fibers. *Mailing Add:* 141 Mundy Ave Edison NJ 08820

SONNEBORN, DAVID R, b Baltimore, Md, Oct 20, 36; m 62; c 2. DEVELOPMENTAL BIOLOGY, MICROBIOLOGY. *Educ:* Swarthmore Col, BA, 57; Brandeis Univ, PhD(biol), 62. *Prof Exp:* NIH fel virol, Univ Calif, Berkeley, 62-64; from asst prof to assoc prof, 64-72, PROF ZOOL, UNIV WIS-MADISON, 72- *Concurrent Pos:* Panel mem, Develop Biol Sect, NSF, 71-74. *Mem:* AAAS; Soc Develop Biol; Am Soc Microbiol; Am Soc Cell Biol. *Res:* Cell differentiation. *Mailing Add:* Zool Res Bldg Univ Wis 1117 W Johnson St Madison WI 53706

SONNEBORN, LEE MEYERS, b Baltimore, Md, Dec 27, 31; m 55; c 2. MATHEMATICS. *Educ:* Oberlin Col, BA, 51; Calif Inst Technol, PhD(math), 56. *Prof Exp:* Asst math, Calif Inst Technol, 53-56; Fine instr, Princeton Univ, 56-58; from asst prof to assoc prof, Univ Kans, 58-67; PROF MATH, MICH STATE UNIV, 67-, DIR GRAD STUDIES, 70- *Concurrent Pos:* Math Asn Am vis lectr & NSF res grant, 65-67. *Mem:* Am Math Soc; Math Asn Am. *Res:* Group theory; differential equation; topology. *Mailing Add:* Mich State Univ East Lansing MI 48824

SONNEMANN, GEORGE, b Munich, Germany, Feb 2, 26; nat US; m 54. ENGINEERING MECHANICS, INFORMATION MANAGEMENT. *Educ:* NY Univ, BS, 47, MS, 49; Univ Mich, PhD(eng mech), 55. *Prof Exp:* Instr physics, Newark Col Eng, 47-48; instr eng mech, Univ Detroit, 48-49; asst prof aeronaut eng, Drexel Inst Tech, 49-52; res assoc, Univ Mich, 52-54; sr engr, Westinghouse Elec Corp, 54-55, Univ Pittsburgh-Westinghouse Elec Corp fel prog, 55-57; from assoc prof to prof mech eng, Univ Pittsburgh, 57-61, Westinghouse prof & dir grad studies mech eng, 57-61; dir staff eng & tech asst to gen mgr, Fecker Div, Am Optical Co, 61-63; chief adv design, United Aircraft Corp Systs Ctr, 63-66, mgr prod eng, 66-67, eng mgr, 67-69; mgr adv progs, Raytheon Co, Sudbury, 69-75; vpres-MIS, Com Union Assurance Co, 75-77; vpres, Conn Gen Life Ins Co, 77-78; vpres planning & mgt info, Nationwide Ins Co, 78-84; PRES, INFO, FINANCE & TECHNOL INC, 84- *Concurrent Pos:* Engr, Franklin Inst, 50; consult, Westinghouse Elec Corp, Am Optical Soc & Copes-Vulcan Div, Blaw-Knox Co; adj prof, Ohio State Univ, 84-85. *Mem:* Am Soc Mech Engrs; Nat Asn Corp Dirs; Proj Mgt Inst. *Res:* Thermal stress analysis and fluid flow problems in reactor engineering; continuum mechanics; structural analysis; heat conduction; optical instrumentation; guidance systems; computer peripherals; manufacturing systems; management information systems; data processing; planning. *Mailing Add:* 543 Montgomery School Lane Wynnewood PA 19096

SONNENBERG, HARDY, b Schoensee, Ger, Apr 12, 39; Can citizen; m 64; c 2. EXPERIMENTAL PHYSICS, ENGINEERING PHYSICS. *Educ:* Univ Alta, BSc, 62; Stanford Univ, MS, 64, PhD(elec eng), 67. *Prof Exp:* Eng specialist, GTE Sylvania, 66-73; mgr res & develop, Optical Diodes, Inc, 73-74; mem sci staff, Xerox Res Ctr Can, Ltd, 75-78, mgr physics & eng, 78-85, mgr res opers, 86-88, MGR, TECHNOL & ENG SYSTS, XEROX RES CTR CAN, LTD, 88- *Concurrent Pos:* Referee, Am Inst Physics, 70-; Arpa proposal consult, US Govt, 74-; grant appl consult, Can Govt; mem, Task Force Univ Indust, Can Mfrs Asn, 81 & Indust Adv Coun, McMaster Univ, 87-; adv comt, Ryerson Polytech Inst; pres, Sheridan Res Park Asn, 87-88. *Honors & Awards:* Charles G Ives Eng Award, Soc Photog Scientists & Engrs. *Mem:* Sigma Xi; Inst Elec & Electronics Engrs; Am Phys Soc; Soc Photog Scientists & Engrs. *Res:* Investigations of the physics and systems aspects of photoactive-pigment-electrography and the coupling of such systems to high-speed channels; laser scanning in xerography; research management; management of technology. *Mailing Add:* Xerox Res Ctr 2660 Speakman Dr Mississauga ON L5K 2L1 Can

SONNENBLICK, EDMUND H, b New Haven, Conn, Dec 7, 32. CARDIOLOGY. *Educ:* Harvard Univ, MD, 58. *Prof Exp:* OLFMAN PROF MED & DIR, CARDIOVASC CTR, ALBERT EINSTEIN COL MED, 84-, CHIEF, CARDIOL DIV, 75- *Mailing Add:* Div Cardiol Albert Einstein Col Med 1300 Morris Park Ave Bronx NY 10461

SONNENFELD, GERALD, b New York, NY, Oct 14, 49; m 78; c 1. INTERFERON, LYMPHOKINES. *Educ:* City Col New York, BS, 70; Univ Pittsburgh, PhD(microbiol & immunol), 75. *Prof Exp:* Fel infectious dis & immunol, Sch Med, Stanford Univ, 76-78; asst prof microbiol & immunol, 78-83, assoc prof microbiol & immunol, 83-88, ASSOC PROF ORAL BIOL, SCH DENT, 84-, PROF MICROBIOL & IMMUNOL, SCH MED, UNIV LOUISVILLE, 88- *Concurrent Pos:* Assoc guest worker, NASA Ames Res Ctr, 76-78; Sect ed immunol, J Interferon Res, 81- *Mem:* Am Asn Immunologists; Am Soc Microbiol; Am Soc Gravitational Biol; Am Soc Virol; Int Soc Interferon Res. *Res:* Biological role of interferon; relationship of interferon to immune responses; resistance to infectious diseases and carcinogens. *Mailing Add:* Dept Microbiol & Immunol Sch Med Univ Louisville Louisville KY 40292

SONNENFELD, PETER, b Berlin, Ger, Jan 20, 22; Can citizen; m 59; c 2. GEOLOGY. *Educ:* Absolutorium, Univ Bratislava, 48; Dr rer nat(geol, geog), Charles Univ, Prague, 49. *Prof Exp:* Geologist, Falconbridge Nickel Mines, Nfld, 51-52; consult, Bennett & Burns, Sask, 52-53, Imp Oil Ltd, Alta, 53-58 & Shell Can Ltd, 58-63; asst prof geol & geog, Tex Col Arts & Indust, 63-66; from assoc prof to prof geol, 66-89, head dept, 68-73, EMER PROF GEOL, UNIV WINDSOR, 89- *Res:* Sedimentology; genesis of sedimentary rocks; dolomitization; evaporite formation; petroleum geology. *Mailing Add:* Dept Geol Univ Windsor Windsor ON N9B 3P4 Can

SONNENFELD, RICHARD JOHN, b Britton, Okla, Apr 29, 19; m 42, 75; c 4. ORGANIC CHEMISTRY, SYNTHETIC INORGANIC CHEMISTRY. *Educ:* Univ Pittsburgh, BS, 41; Univ Okla, MS, 55, PhD(chem), 56. *Prof Exp:* Foreman, Weldon Spring Ord Works, 41-43; anal chemist, Phillips Petrol Co, 46-49, rubber chemist, 49-53, sr group leader, 56-63, sect mgr, 63-82; RETIRED. *Mem:* Am Chem Soc; AAAS. *Res:* Synthetic rubber by emulsion and stereospecific polymerization; chemicals from petroleum; free radicals; organo-metallic compounds. *Mailing Add:* 842 S E Concord Dr Bartlesville OK 74003

SONNENSCHEIN, CARLOS, CELLULAR BIOLOGY. *Educ:* Univ Buenos Aires, Arg, MD, 58. *Prof Exp:* PROF CELLULAR BIOL & CANCER, SCH MED, TUFTS UNIV, 80. *Res:* Control of cellular proliferation; human breast tumors; diagnostic and therapeutic approaches. *Mailing Add:* Dept Anat & Cellular Biol Tufts Univ 136 Harrison Ave Boston MA 02111

SONNENSCHEIN, RALPH ROBERT, b Chicago, Ill, Aug 14, 23; m 52; c 3. PHYSIOLOGY. *Educ:* Northwestern Univ, BS, 43, MS, 46, MD, 47; Univ Ill, PhD(physiol), 50. *Prof Exp:* Asst physiol, Northwestern Univ, 44-46; intern, Michael Reese Hosp, Chicago, Ill, 46-47; res asst psychiat, Univ Ill, 49-51, res assoc, 51; from asst prof to assoc prof physiol, 51-62, PROF PHYSIOL, UNIV CALIF, LOS ANGELES, 62- *Concurrent Pos:* USPHS res fel, 57-58; Swed Med Res Coun fel, 64-65; liaison scientist, Off Naval Res, London, 71-72. *Mem:* AAAS; Microcirc Soc; Am Physiol Soc; Soc Exp Biol & Med; Sigma Xi; hon mem Hungarian Physiol Soc. *Res:* Peripheral circulation. *Mailing Add:* Dept Physiol Univ Calif Sch Med Los Angeles CA 90024-1751

SONNENWIRTH, ALEXANDER COLEMAN, medical microbiology; deceased, see previous edition for last biography

SONNER, JOHANN, b Munich, Ger, May 3, 24; nat US; m 57; c 2. MATHEMATICS. *Educ:* Univ Munich, Dr rer nat, 54. *Prof Exp:* Asst prof, State Sch Eng, Ger, 56-57; tech consult, Wright Air Develop Ctr, Ohio, 57-58; prof math, Univ SC, 58-67; PROF MATH, UNIV NC, CHAPEL HILL, 67- *Mem:* Am Math Soc; Math Asn Am; Ger Math Asn; Math Soc France. *Res:* Foundations of mathematics; general topology. *Mailing Add:* Dept Math Univ NC Chapel Hill NC 27599-3902

SONNERUP, BENGT ULF OSTEN, b Malmo, Sweden, July 7, 31; m 55; c 3. SPACE PHYSICS, FLUID MECHANICS. *Educ:* Chalmers Inst Technol, Sweden, BME, 53; Cornell Univ, MAE, 60, PhD(fluid mech), 61. *Prof Exp:* Proj engr, Stal-Laval Steam Turbine Co, Sweden, 54-56; proj engr, Bofors Co, Sweden, 56-58; fel, Ctr Radiophys & Space Res, Cornell Univ, 61-62; fel, Inst Plasma Physics, Royal Inst Technol, Sweden, 62-64; assoc prof, 64-70, prof, 70-81, SYDNEY E JUNKINS PROF ENG SCI, DARTMOUTH COL, 81- *Concurrent Pos:* Lectr, Uppsala Univ, 63; Europ Space Res Orgn fel, Europ Space Res Inst, Italy, 70-71; vis scientist, Max Planck Inst Extraterrestrial Physics, Garching, Fed Repub Ger, 78-79, 86-87; ed, J Geophys Res, 82-85. *Mem:* AAAS; Am Geophys Union; Am Inst Aeronaut & Astronaut. *Res:* Plasma physics and magnetohydrodynamics applied to problems in space physics, particularly the structure of the magnetopause current layer and boundary layer; the magnetosphere and the nature of magnetic field merging. *Mailing Add:* Thayer Sch Eng Dartmouth Col Hanover NH 03755

SONNET, PHILIP E, b New York, NY, Feb 6, 35; m 58; c 3. ORGANIC CHEMISTRY. *Educ:* Columbia Univ, AB, 56; Rutgers Univ, PhD(org chem), 63. *Prof Exp:* NIH fel org chem, Mass Inst Technol, 63-64; RES CHEMIST, AGR RES CTR, USDA, 64- *Concurrent Pos:* Instr, Univ Md, 72-76. *Mem:* Am Chem Soc; Entom Soc Am. *Res:* Insect pheromones, identification and synthesis; aliphatic synthesis. *Mailing Add:* 600 E Mermaid Lane Wyndmoor Philadelphia PA 19118-2551

SONNICHSEN, GEORGE CARL, b Chicago, Ill, Nov 15, 41; m 70. CHEMISTRY. *Educ:* DePauw Univ, BS, 63; Mich State Univ, PhD(chem), 67. *Prof Exp:* NSF fel chem, Univ Calif, Berkeley, 67-68, lectr, 68-69; res chemist, 69-75, res supvr, 75-78, RES ASSOC, E I DU PONT DE NEMOURS & CO, 78- *Mem:* Am Chem Soc. *Res:* Heterogeneous and homogeneous catalysis. *Mailing Add:* 614 Lindsay Rd Wilmington DE 19809

SONNICHSEN, HAROLD MARVIN, b Hancock, Minn, Apr 4, 12; m 39; c 2. ADHESIVES, ADHESIVE TAPE. *Educ:* Tex Col Mines, AB, 34; Harvard Univ, Phd(org chem), 39. *Prof Exp:* Res chemist, Electrochem Dept, E I du Pont de Nemours & Co, 39-40, supvr, Sales Res Sect, 40-43, plant supvr, 43-44; tech serv mgr, Permacel Tape Corp, 44-48, asst dir, 48-52, tech dir, 52-55, vpres, 55-60; dir fiber & saturant res, Dewey & Almy Chem Div, W R Grace & Co, 60-64, vpres, Precision Tech Prod, 65-75; PRES, H M SONNICHSEN & ASSOCS, 75- *Concurrent Pos:* Consult, Adhesive Tape, 65- *Mem:* Am Chem Soc; fel Am Inst Chem; Am Soc Testing & Materials; Tech Asn Pulp & Paper Indust. *Res:* Properties and applications of synthetic high polymers; pressure sensitive adhesives; structural adhesives; latex; paper; artificial leather; paper and nonwoven disposable products. *Mailing Add:* 37 Robin Hood Rd Arlington MA 02174

SONNINO, CARLO BENVENUTO, b Torino, Italy, May 12, 04; US citizen; m 49; c 3. ELECTROCHEMISTRY, ELECTROMETALLURGY. *Educ:* Univ Milano, PhD(chem eng), 27. *Prof Exp:* Dir res & mgr, Dept Flotation, Italian Aluminum Co, 28-33; mgr, Tonolli Co, 33-34; pres, LCI Consult, 34-39; mem bd & mgr, LAESA, 39-43; tech adv to bd dir, Boxal S A, Switzerland, 44-52; mgr & tech adv, Kreisler Co, 52-53; tech mgr, Alumacraft, St Louis, 53-56; MAT ENG MGR, EMERSON ELEC CO, ST LOUIS, 56-; PROF METALL ENG, UNIV MO, ROLLA, 68- *Concurrent Pos:* Prof metall & mat sci, Washington Univ, 60-67; tech adv, Thompson Brand, Paris & Rouen, 70-75; consult, Monsanto, Wagner Elec, & Amax, 75- *Honors & Awards:* Klixon Award, Am Soc Heating, Refrigeration, & Air Conditioning Engrs, 60; Knight Comdr, President Italian Repub, 77. *Mem:* Sigma Xi; Am Soc Testing Mat; fel Am Soc Metals; Soc Metal Engrs. *Res:* Synthetic cryolite; anodizing of aluminum alloys. *Mailing Add:* 7206 Kingsbury Blvd St Louis MO 63130

SONNTAG, BERNARD H, b Goodsoil, Sask, June 27, 40; m 63; c 3. AGRICULTURAL ECONOMICS. *Educ:* Univ Sask, BSA, 62, MSc, 65; Purdue Univ, PhD(agr econ), 71. *Prof Exp:* Economist, O William Carr & Assoc, Ottawa, 66-68; economist, Econ Br, Agr Can, Sask, 62-66 & 79-80 & Lethbridge, 68-79, dir, Res Br, Brandon, 80-86 & Swift Current, 86-89, DIR, RES BR, AGR CAN, LETHBRIDGE, 89- *Concurrent Pos:* Dir, Can Agr Econ & Farm Mgt Soc, 83-85. *Mem:* Agr Inst Can; Can Agr Econ & Farm Mgt Soc. *Res:* Production economics research in cereals, forages, special crops, beef, dairy, sheep, and soil and water resources. *Mailing Add:* Agr Can Research Sta, Lethbridge Lethbridge AB T1J 4B1

SONNTAG, NORMAN OSCAR VICTOR, b Brooklyn, NY, Sept 10, 19; m 47; c 2. ORGANIC CHEMISTRY, ANALYTICAL CHEMISTRY. *Educ:* Polytech Inst Brooklyn, PhD(chem), 51. *Prof Exp:* Res chemist, Polytech Inst Brooklyn, 49-51 & Colgate Palmolive Co, 51-55; chief chemist, Chem Div, Celanese Corp Am, 55-56; res chemist, Emery Industs, Inc, 56-59; assoc mgr, Res & Develop Div, Nat Dairy Prod Corp, Ill, 59-66; mgr process res, Glyco Chem, Inc, 66-68, dir res, 68-77; tech dir chem div, Southland Corp, 78-80; asst prof, Bishop Col, Dallas, Tex, 81-84; CONSULT, 80- *Mem:* Am Chem Soc; Am Oil Chem Soc (vpres, 78-79, pres, 79-80). *Res:* Reduction of highly arylated conjugated cyclic ketones; reactions of aliphatic acid chlorides; chemical utilization of fats; fatty chemicals; synthetic fatty acids; nitrogen derivatives; dibasic acids; hydantoin chemicals; agriculture and food chemistry. *Mailing Add:* 306 Shadowood Trail Red Oak TX 75154-1424

SONNTAG, RICHARD E, b Chicago, Ill, Apr 17, 33; m 57; c 2. MECHANICAL ENGINEERING, THERMODYNAMICS. *Educ:* Univ Mich, BSE, 56, MSE, 57, PhD(mech eng), 61. *Prof Exp:* From asst prof to assoc prof, 60-67, PROF MECH ENG, UNIV MICH, ANN ARBOR, 67- *Mem:* Am Soc Mech Engrs; Am Soc Eng Educ; Sigma Xi. *Res:* Low temperature thermodynamics; phase equilibria; pressure-volume-temperature behavior. *Mailing Add:* 3925 Penberton Dr Ann Arbor MI 48105

SONNTAG, ROY WINDHAM, b Cleburne, Tex, Nov 17, 29; m 53; c 4. ORGANIC CHEMISTRY. *Educ:* NTex State Col, BS, 53; Univ Tex, PhD(org chem), 59. *Prof Exp:* Res chemist, Monsanto Chem Co, 58 & Esso Res & Eng Co, 59-60; from asst prof to assoc prof, 60-69, PROF CHEM, MCMURRY COL, 69- *Mem:* Am Chem Soc. *Res:* Molecular rearrangements of organic systems; conformational analysis. *Mailing Add:* Dept Chem McMurray Col McMurry Sta Box 158 Abilene TX 79697

SONNTAG, WILLIAM EDMUND, b Waterbury, Conn, Jan 1, 50; m 73; c 1. AGING, ALCOHOLISM. *Educ:* Tufts Univ, BS, 72; Univ Bridgeport, MS, 74; Tulane Univ, PhD(physiol psychol), 79. *Prof Exp:* Asst prof, 84-89, ASSOC PROF PHYSIOL, BOWMAN GRAY SCH MED, WAKE FOREST UNIV, 89-, ASSOC DIR BASIC SCI RES, J PAUL STICHT CTR AGING, 90- *Mem:* Endocrine Soc; Geront Soc; AAAS; Am Fedn Aging Res. *Res:* Molecular neuroendocrinology; effects of age and/or alcohol on the regulation of growth hormone and insulin-like growth factors. *Mailing Add:* Dept Physiol Bowman Gray Med Sch Wake Forest Univ Winston-Salem NC 27103

SONODA, RONALD MASAHIRO, b Hilo, Hawaii, June 4, 39; m 66; c 2. PLANT PATHOLOGY. *Educ:* Sacramento State Col, AB, 63; Univ Calif, Davis, MS, 65, PhD(plant path), 69. *Prof Exp:* From asst prof to assoc prof, 69-81, PROF, INST FOOD & AGR SCI, UNIV FLA, 81- *Concurrent Pos:* Fac develop grant, Univ Calif, Davis, 79-80. *Mem:* Am Phytopath Soc; Am Soc Microbiol. *Res:* Diseases of citrus; diseases of tropical forage legumes and grasses; diseases of tomatoes. *Mailing Add:* 1014 Carribbean Ave Ft Pierce FL 34982

SONS, LINDA RUTH, b Chicago Heights, Ill, Oct 31, 39. MATHEMATICS. *Educ:* Ind Univ, AB, 61; Cornell Univ, MS, 63, PhD(math), 66. *Prof Exp:* from asst prof to assoc prof, 65-78, PROF MATH, NORTHERN ILL UNIV, 78- *Concurrent Pos:* NSF grant, 70-72 & 74-75; mem bd gov, Math Asn Am, Comt Undergrad Prog, chair, Subcomt Qual Literacy Requirements, Comt Undergrad Prog. *Mem:* Math Asn Am; Am Math Soc; Asn Women Math; Nat Coun Teachers Math. *Res:* Mathematical analysis, especially complex function theory. *Mailing Add:* Dept Math Northern Ill Univ De Kalb IL 60115

SONSTEGARD, KAREN SUE, developmental biology, experimental pathology, for more information see previous edition

SONTAG, EDUARDO DANIEL, b Buenos Aires, Arg, Apr 16, 51;,US citizen; m 81; c 2. CONTROL THEORY, SYSTEMS THEORY. *Educ:* Univ Buenos Aires, Lic, 72; Univ Fla, PhD(math), 76. *Prof Exp:* From asst prof to assoc prof, 77-87, PROF MATH, RUTGERS UNIV, NJ, 87-, MEM GRAD FAC COMPUT SCI, 85-, MEM GRAD FAC ELEC ENG, 88- *Concurrent Pos:* Prin investr, Air Force Off Sci Res, 78-, NJ Dept Higher Educ, 85-86, NSF, 85-87 & 88-, Ctr Comput Aids Indust Prod, 86- *Mem:* Am Math Soc; Soc Indust & Appl Math; sr mem Inst Elec & Electronic Engrs. *Res:* Systems and control theory with applications to robotics and neural networks; related problems in mathematics, particularly in algebra. *Mailing Add:* Dept Math Rutgers Univ New Brunswick NJ 08903

SONTHEIMER, RICHARD DENNIS, b Beaumont, Tex, Nov 3, 45. DERMATOLOGY. *Educ:* Univ Tex, MD, 72. *Prof Exp:* PROF DERMAT & INTERNAL MED, HEALTH SCI CTR, UNIV TEX, 84- *Mem:* Am Soc Clin Invest; Am Dermat Asn. *Mailing Add:* Dept Dermat Univ Tex Health Sci Ctr Dallas TX 75235

SOO, SHAO-LEE, b Peiping, China, Mar 1, 22; US citizen; m 52; c 3. MECHANICAL ENGINEERING. *Educ:* Nat Chiaotung Univ, BS, 45; Ga Inst Technol, MS, 48; ScD, Harvard Univ, 51. *Prof Exp:* Teaching fel appl physics, Harvard Univ, 51; instr mech eng, Princeton Univ, 51-52, lectr, 52-54, from asst prof to assoc prof, 54-59; PROF MECH ENG, UNIV ILL, URBANA, 59- *Concurrent Pos:* Indust consult, 51-; mem consult team Skylab I, Univ Space Res Assoc, NASA, 71-72; Agard lectr, NATO, 73; distinguished lectr, Fulbright-Hays Prog, Buenos Aires, Arg, 74; mem sci adv bd, US Environ Protection Agency, 76-78; adv energy transp, World Bank, 79; dir, S L Soo Assocs Inc, Urbana, 80-; Kumar Consult Inc, 86; Int Powder Inst, 76- *Honors & Awards:* Appl Mech Rev Award, Am Soc Mech Engrs, 72; Distinguished Lectr Award, Int Freight Pipeline Soc, 81; Alcoa Found Award, 85. *Mem:* Fel Am Soc Mech Engrs; Am Soc Eng Educ; Combustion Inst; Sigma Xi; Chinese Acad Sci. *Res:* Basic formulation of nonequilibrium fluid dynamics; experimental research in multi-phase flow; nonequilibrium ionized gases; gas-surface interaction; atmospheric transport of air pollutants and control by electrostatic precipitation; pneumatic conveying. *Mailing Add:* 123 Mech Eng Bldg Univ Ill 1206 W Green St Urbana IL 61801

SOOD, MANMOHAN K, b Manpur Nagaria, India, Apr 17, 41; nat US; m 65; c 2. PETROLOGY, GEOCHEMISTRY. *Educ:* Panjab Univ, India, BSc, 60, MSc, 63; Univ Western Ont, MSc, 68, PhD(geol), 69. *Prof Exp:* Tech asst geol, Govt Punjab, 63-64; res assoc, Univ Western Ont, 69-70; from asst prof to assoc prof, 70-80, chmn dept, 74-82, PROF EARTH SCI, NORTHEASTERN ILL UNIV, 80-, DIR UNIV HONORS PROG, 85- *Concurrent Pos:* Geochemist, Argonne Nat Lab, Ill, 82-83; geol consult, 79- *Mem:* Mineral Soc Am; Int Asn Advan Earth & Environ Sci(founding secy, 72-75). *Res:* Phase equilibrium related to alkaline igneous rocks; geochemistry and petzology of proterozoic granites, Wisconsin, USA; geological isolation of hazardous wastes; issues in science and math; education in schools and colleges. *Mailing Add:* Dept Earth Sci Northeastern Ill Univ Chicago IL 60625

SOOD, SATYA P, b Abohar, Punjab, India, Feb 4, 23; m 57; c 1. PHYSICAL CHEMISTRY, POLYMER CHEMISTRY. *Educ:* Forman Christian Col, Punjab, BSc, 42; Acton Tech Col, London, API, 53; State Univ NY, MS, 56; Univ Hawaii, PhD(chem), 63. *Prof Exp:* Chemist various chem concerns, India, 42-49; tech asst, Indian High Comn, London, Eng, 50-53; teaching asst chem, State Univ NY, 53-55; res asst, Tex Tech Col, 55-57; asst, 57-62, res assoc physics, 62-63, from asst prof to assoc prof, 63-71, chmn div sci, 71-80, PROF CHEM, UNIV HAWAII, HILO, 71- *Concurrent Pos:* Lectr, Bahawal Col, Pakistan, 46-47; res fel, McMaster Univ, 68-69. *Res:* Chemical kinetics; synthetic organic chemistry; ultraviolet spectroscopy; dipole moments; radiation chemistry. *Mailing Add:* 523 W Lani Kaula St Hilo HI 96720-4091

SOOD, VIJAY KUMAR, b New Delhi, India, May 13, 51; m 80. AGRICULTURAL & FOOD CHEMISTRY. *Educ:* Nat Dairy Res Inst, India, BS, 70, MS, 72; Cornell Univ, PhD(food sci), 78. *Prof Exp:* Sr technologist, Punjab Dairy Develop Corp, India, 72-73; asst dairy technologist, Punjab Coop Dept, 73-74; SR SCIENTIST FOOD RES, UNIVERSAL FOODS CORP, 78- *Concurrent Pos:* Vis scientist, Nat Inst Agron Res, France, 78. *Mem:* Inst Food Technologists; Am Dairy Sci Asn. *Res:* Develop new food products and processes. *Mailing Add:* 1800 Montee Ste Julie Varennes PQ J3X 1S1 Can

SOODSMA, JAMES FRANKLIN, b Hull, Iowa, Feb 3, 38; m 64; c 2. COMBUSTION ENGINEERING. *Educ:* Univ SDak, BA, 63; Univ NDak, MS, 65, PhD(biochem), 68; Univ Tulsa, MS. *Prof Exp:* chief scientist, William K Warren Med Res Ctr, Inc, 71-77; TEST ENGR, JOHN ZINK CO, 77- *Concurrent Pos:* AEC fel, Med Div, Oak Ridge Assoc Univs, 68-71. *Mem:* Am Chem Soc. *Res:* Studies on rat kidney glucose-6-phosphatase and associated phosphotransferase activities; rat liver enzyme system which cleaves glyceryl ethers; mammalian newborn phospholipid and carbohydrate metabolism; pollution control. *Mailing Add:* 6301 E 56th Pl Tulsa OK 74135

SOOHOO, RONALD FRANKLIN, b Kwangtung, China, Sept 1, 28; US citizen; m 57; c 2. ELECTRICAL ENGINEERING, PHYSICS. *Educ:* Mass Inst Technol, SB, 48; Stanford Univ, MS, 52, PhD(elec eng, physics), 56. *Prof Exp:* Asst engr, Pac Gas & Elec Co, 48-51; dir res anal, Cascade Res Corp, 54-58; res physicist, Lincoln Lab, Mass Inst Technol, 58-61; assoc prof eng & appl sci, Calif Inst Technol, 61-64; chmn dept elec eng, 64-70, PROF ELEC ENG, UNIV CALIF, DAVIS, 64- *Concurrent Pos:* Consult, Space Technol Labs, 62-64, Ampex Corp, 62-64, Bunker-Ramo Corp, 62-64, E&M Labs, 67- & Lawrence Livermore Lab, 69-; NATO fel, Nat Ctr Sci Res, Bellevue, France, 70. *Mem:* Fel Inst Elec & Electronics Engrs; Am Inst Physics. *Res:* Magnetism and magnetic materials; solid state physics; microwave electronics; computer devices and systems; quantum electronics. *Mailing Add:* 568 Reed Davis CA 95616

SOOKNE, ARNOLD MAURICE, b New York, NY, Oct 9, 15; m 39; c 2. TEXTILE CHEMISTRY, TEXTILE FLAMMABILITY. *Educ:* Brooklyn Col, BS, 35; George Washington Univ, MA, 42. *Prof Exp:* Res assoc, Am Asn Textile Chemists & Colorists, Washington, DC, 36-37, Nat Res Coun, 37-38 & Textile Found, Nat Bur Standards, 38-44; assoc dir, Harris Res Labs Div, Gillette Res Inst, Inc, 44-65, vpres labs, 65-68, vpres, Inst, 68-69; dir chem develop, res ctr, Burlington Industs Inc, 69-77, asst dir corp res & develop, 78-83; TEXTILE CONSULT, 83- *Concurrent Pos:* Vis prof, Univ NC, Greensboro, 79-89. *Honors & Awards:* Olney medal, Am Asn Textile Chem & Colorists, 60; Harold DeWitt Smith Award, Am Soc Test & Mat, 71; Inst Medal, Brit Textile Inst, 84. *Mem:* Am Chem Soc; Am Asn Textile Chem & Colorists; Fiber Soc; Brit Textile Inst. *Res:* Physical chemistry of textiles; dyeing and finishing of textiles; flammability of textiles. *Mailing Add:* 117 Batchelor Dr Greensboro NC 27410

SOOKY, ATTILA A(RPAD), b Rakoscsaba, Hungary, Aug 22, 32; US citizen; m 64; c 3. FLUID MECHANICS. *Educ:* Budapest Tech Univ, BS, 55, MS, 56; Purdue Univ, PhD(fluid mech), 64. *Prof Exp:* Asst prof, 64-71, ASSOC PROF POLLUTION CONTROL, UNIV PITTSBURGH, 71- *Mem:* Am Soc Civil Engrs; Water Pollution Control Fedn; Sigma Xi. *Res:* Pollution control; fate of pollution in natural waters; health effects of pollutants. *Mailing Add:* Dept Civil Eng Grad Sch Pub Health Univ Pittsburgh Pittsburgh PA 15261

SOONG, TSU-TEH, b Honan, China, Feb 10, 34; US citizen; m 59; c 3. CIVIL ENGINEERING, STRUCTURAL CONTROL & IDENTIFICATION. *Educ:* Univ Dayton, BS, 55; Purdue Univ, MS, 58, PhD(eng sci), 62. *Prof Exp:* Instr mech, Purdue Univ, 58-62; sr res engr, Jet Propulsion Lab, Calif Inst Technol, 62-63; from asst prof to assoc prof eng sci, 63-68, chmn dept, 70-80, prof eng & civil eng, 68-89, SAMUEL P CAPEN PROF ENG, STATE UNIV NY, BUFFALO, 89- *Concurrent Pos:* Lectr, Univ Calif, Los Angeles, 62-63; res mathematician, Cornell Aeronaut Lab, 64-; NSF res grants, 64-, sci faculty fel, Delft Technol Univ, 66-67. *Honors & Awards:* Humboldt Found Sr US Scientist Award, 87. *Mem:* Nat Soc Prof Engrs; Am Soc Civil Engrs. *Res:* Stochastic processes with applications to analysis of engineering systems; identification and control of mechanical and structural systems. *Mailing Add:* Dept Civil Eng Ketter Hall State Univ NY Buffalo NY 14260

SOONG, YIN SHANG, b Shanghai, Repub of China, July 14, 47; m; c 2. PHYSICAL OCEANOGRAPHY. *Educ:* Nat Taiwan Univ, BS, 69; Fla State Univ, MS, 74, PhD(phys oceanog), 78. *Prof Exp:* From asst prof to assoc prof, 77-89, PROF OCEANOG, MILLERSVILLE UNIV, 89- *Mem:* Am Meteorol Soc; Am Geophys Union. *Res:* Remote sensing; inertial currents; mesoscale oceanic phenomenon. *Mailing Add:* Dept Earth Sci Millersville Univ Millersville PA 17551

SOONPAA, HENN H, b Estonia, Mar 18, 30; nat US; m 59; c 2. SOLID STATE PHYSICS. *Educ:* Concordia Col, BA, 51; Univ Ore, MA, 53; Wayne State Univ, PhD(phys chem), 55. *Prof Exp:* Asst, Univ Ore, 51-52; res fel, Wayne State Univ, 52-54; res fel, Iowa State Univ, 56; sr scientist, Gen Mills, Inc, 57-58; assoc prof physics, Gustavus Adolphus Col, 58-59; sr & prin scientist, Gen Mills, Inc, 59-62 & Honeywell Corp Res Ctr, 62-66; assoc prof, 66-72, PROF PHYSICS, UNIV NDAK, 72- *Mem:* Am Phys Soc; Am Asn Physics Teachers; Am Vacuum Soc. *Res:* Solid state physics; transport phenomena; quantum size effects; two dimensional systems. *Mailing Add:* Univ NDak Box 8008 Grand Forks ND 58202

SOORA, SIVA SHUNMUGAM, b Madras, India, Mar 1, 57; US citizen; m 82; c 2. REFRACTORIES-ALUMINO SILICATE, FUSED. *Educ:* Banaras Hindu Univ, BTech, 79; Ga Inst Technol, MS, 81; Ga Southwestern Col, MBA, 85. *Prof Exp:* Process engr, 81-84, sr process engr, 84-88, DIR RES & DEVELOP, CE MINERALS, 88- *Concurrent Pos:* All India Glass Mfrs Fed Award for Achievement in Glass, 78-79. *Mem:* Am Ceramic Soc. *Res:* Alumina/alumino silicate and fused silica refractory materials; developing new products for the various industries and also in developing new uses for existing products for both materials. *Mailing Add:* CE Minerals PO Box 37 Andersonville GA 31711

SOOS, ZOLTAN GEZA, b Budapest, Hungary, July 31, 41; US citizen; m 66. PHYSICAL CHEMISTRY. *Educ:* Harvard Col, AB, 62; Calif Inst Technol, PhD(chem, physics), 65. *Prof Exp:* NSF fel, Stanford Univ, 65-66; from asst to assoc prof, 66-74, PROF CHEM, PRINCETON UNIV, 74- *Concurrent Pos:* Vis scientist, Sandia Corp, 71. *Mem:* Am Chem Soc. *Res:* Theory of molecular excitons; many-body methods in para magnetic crystals. *Mailing Add:* Dept of Chem Princeton Univ Princeton NJ 08544

SOOST, ROBERT KENNETH, b Sacramento, Calif, Nov 13, 20; m 49; c 3. GENETICS. *Educ:* Univ Calif, PhD(genetics), 49. *Prof Exp:* From asst geneticist to assoc, 49-65, chmn, Dept Plant Sci, 69-75, geneticist, 65-86, EMER GENETICIST, UNIV CALIF, RIVERSIDE, 86- *Mem:* fel Am Soc Hort Sci; Sigma Xi. *Res:* Citrus genetics and breeding. *Mailing Add:* PO Box 589 Inverness CA 94937

SOOY, FRANCIS ADRIAN, otolaryngology; deceased, see previous edition for last biography

SOOY, WALTER RICHARD, b Boston, Mass, Dec 28, 32; m 82; c 2. PHYSICS. *Educ:* Mass Inst Technol, BS, 56; Univ Southern Calif, MS, 58; Univ Calif, Los Angeles, PhD(physics), 63. *Prof Exp:* Mem tech staff, Hughes Aircraft Co, 56-62, staff physicist, 62-64, sr staff physicist, 64-66, sr scientist, 66-70, dept mgr, 68-70; supt optical sci div, Naval Res Lab, 70-75; vpres & chief scientist, Sci Applns Inc, 75-81; SR STAFF SCIENTIST, LAWRENCE LIVERMORE NAT LAB, 81- *Concurrent Pos:* Consult, Off Secy Defense & Naval Mat Command, 75- *Mem:* AAAS; Optical Soc Am; Am Phys Soc. *Res:* Superconductivity; microwave oscillators; nuclear physics; optics; lasers; systems. *Mailing Add:* 4890 Cobbler Ct Pleasonton CA 94566

SOPER, DAVISON EUGENE, b Milwaukee, Wis, Mar 21, 43; m 71; c 2. THEORETICAL ELEMENTARY PARTICLE PHYSICS, QUANTUM CHROMODYNAMICS. *Educ:* Amherst Col, AB, 65; Stanford Univ, PhD(physics), 71. *Prof Exp:* From instr to asst prof physics, Princeton Univ, 71-77; from asst prof to assoc prof, 77-82, PROF PHYSICS, UNIV ORE, 82- *Mem:* Am Phys Soc. *Res:* Quantum field theory and particle physics; classical field theory. *Mailing Add:* Inst of Theoret Sci Univ of Ore Eugene OR 97403

SOPER, GORDON KNOWLES, b Gunnison, Colo, July 25, 38; m 58; c 2. PLASMA PHYSICS. *Educ:* Univ Tenn, BS, 59, MS, 62, PhD(physics), 64. *Prof Exp:* From asst prof to assoc prof physics, US Air Force Inst Technol, 64-72; proj officer, HQ, 72-75, chief, Electronics Vulnerablity Div, 75-77, chief, Atmospheric Effects Div, 77-78, ASST TO DEP DIR SCI & TECHNOL EXP RES, DEFENSE NUCLEAR AGENCY, WASHINGTON, DC, 78- *Mem:* Am Phys Soc; Am Asn Physics Teachers. *Res:* Theoretical plasma physics, particularly stability theory. *Mailing Add:* 14824 N Ashdale Ave Woodbridge VA 22193

SOPER, JAMES HERBERT, b Hamilton, Ont, Apr 9, 16; m 46; c 4. BOTANY. *Educ:* McMaster Univ, BA, 38, MA, 39; Harvard Univ, PhD(biol), 43. *Prof Exp:* Botanist, Can Dept Agr, 45-46; spec lectr bot, Univ Toronto, 46-47, from asst prof to prof bot, 47-67; chief botanist, 67-81, EMER CUR, BOT DIV, NATURAL MUS NATURAL SCI, 81- *Concurrent Pos:* Curator, Herbarium Vascular Plants, Univ Toronto, 46-67. *Mem:* Royal Can Inst, (pres, 62-63); Can Bot Asn (pres, 82-83). *Res:* Flora of Ontario; distribution of vascular plants of North America; data-processing and automated cartography. *Mailing Add:* 621 Echo Dr Ottawa ON K1S 1P1 Can

SOPER, JON ALLEN, b Wyandotte, Mich, Mar 7, 36; m 58; c 5. ELECTRICAL ENGINEERING. *Educ:* Mich Technol Univ, BS, 57, MS, 61; Univ Mich, Ann Arbor, PhD(elec eng), 69. *Prof Exp:* Instr elec eng, Mich Technol Univ, 57-60, asst prof, 60-63; design and develop engr, Raytheon Mfg Co, 63-64; asst prof elec eng, Mich Technol Univ, 64-65; asst res engr, Univ Mich, 67-68; asst prof, 68-70, assoc prof, 70-79, PROF ELEC ENG, MICH TECHNOL UNIV, 79-, ASST DEPT HEAD, 84- *Concurrent Pos:* Actg assoc dean, Mich Tech Univ, 78-79, actg dept head, 90-91. *Mem:* Sr mem Inst Elec & Electronic Engrs; Am Soc Eng Educ. *Res:* Antenna theory & behavior; microwave networks; radar systems and navigation systems; electromagnetic interactions with snow; microwave coupling; electromagnetic compatibility and interference. *Mailing Add:* Dept Elec Eng Mich Technol Univ Houghton MI 49931

SOPER, QUENTIN FRANCIS, b Buhl, Minn, Dec 3, 19; m 46; c 4. PHARMACEUTICAL CHEMISTRY. *Educ:* Univ Minn, BChem, 40; Univ Ill, PhD(org chem), 43. *Prof Exp:* Asst chem, Univ Ill, 40-43, spec asst, Nat Defense Res Comt Contract, 43-44; sr org chem, Eli Lilly & Co, 44-65, head agr chem res, 65-72, agr sr assoc, 72-77, res adv, 77-84; RETIRED. *Honors & Awards:* John Scott Award, Am Inst Chemists. *Mem:* Am Chem Soc; Weed Sci Soc Am; Ctr Appln Sci & Technol. *Res:* Synthesis of new war gases; synthesis of new chemicals useful in the biosynthesis of new penicillins; quinoxaline formation and the ortho effect; hindrance at beta carbon atom; synthetic pharmaceuticals, herbicides and pesticides. *Mailing Add:* 2120 W 38th St Indianapolis IN 46208-3202

SOPER, RICHARD GRAVES, b Trenton, Mich, Dec 10, 50; m 81; c 5. SURGICAL PATHOLOGY, CYTOPATHOLOGY. *Educ:* Univ Mich, BS, 72; Univ Tenn, MD, 77. *Prof Exp:* Instr path, Med Sch, Univ Tenn, 77-81; med dir, Roche Biomed Labs, Inc, 82-86 & CytoDiagnostics, Univ Okla, 87-90; fel cytopath, Sch Med, Johns Hopkins Univ, 87; PATHOLOGIST MED TECHNOL, COLUMBIA STATE, MIDDLE TENN MED CTR, 90- *Concurrent Pos:* Consult pathologist, CytoDiagnostics, Oklahoma City, Okla, 90-, Nat Ref Lab, Nashville, Tenn, 90- & Nat Health Labs, Birmingham, Ala, 90- *Mem:* Col Am Pathologists; Am Soc Clin Pathologists; Am Acad Path; Am Soc Cytol; Int Soc Anal Cytol; Fedn Am Soc Exp Biol. *Res:* Immunohistochemistry and static/flow cytometry of human neoplastic disease tissues. *Mailing Add:* Middle Tenn Med Ctr 423 N University Ave Murfreesboro TN 37130

SOPER, ROBERT JOSEPH, b Weston, Ont, Aug 25, 27; m 57; c 2. SOIL SCIENCE. *Educ:* Univ Sask, BA & BSA, 53, MSc, 55; McGill Univ, PhD(agr chem), 59. *Prof Exp:* From asst prof to assoc prof, 58-69, PROF SOILS, FAC AGR, UNIV MAN, 69- *Concurrent Pos:* App sr officer P-5 head, Soils, Irrigation & Crop Prod Sect, Atomic Energy Food & Agr, Vienna, Austria. *Honors & Awards:* Queen's Jubilee Medal. *Mem:* Fel Can Soc Soil Sci; fel Agr Inst Can; Int Soc Soil Sci; Am Soc Agron. *Res:* Soil fertility and chemistry. *Mailing Add:* Dept of Soil Sci Univ of Man Fac of Agr Winnipeg MB R3T 2N2 Can

SOPER, ROBERT TUNNICLIFF, b Iowa City, Iowa, Sept 16, 25; m 51; c 6. MEDICINE, SURGERY. *Educ:* Cornell Col, BS, 49; Univ Iowa, MD, 52; Am Bd Surg, dipl, 59. *Prof Exp:* Intern med, Cleveland City Hosp, Ohio, 52-53; resident, 54-57, instr, 57-58, assoc, 58-59, from asst prof to assoc prof, 59-67, PROF SURG, UNIV IOWA HOSPS, 67- *Concurrent Pos:* Surg registr, Alder Hey Children's Hosp, Liverpool, Eng, 59-60. *Mem:* AMA; Am Col Surg; Brit Asn Pediat Surg; Am Pediat Surg Asn; Am Acad Pediat. *Res:* Clinical pediatric surgery. *Mailing Add:* Dept of Surg Univ of Iowa Hosps Iowa City IA 52242

SOPHER, ROGER LOUIS, b Long Beach, Calif, Oct 7, 36; m 72. PATHOLOGY. *Educ:* St Mary's Col, BS, 58; Johns Hopkins Univ, MD, 62. *Prof Exp:* Intern path, Med Ctr, Univ Calif, Los Angeles, 62-63, resident, 63-64; resident, Sch Med, Univ NMex, 64-66, from asst prof to assoc prof path, 68-72, vchmn dept, 68-83; chief lab serv, Albuquerque Vet Admin Hosp, 69-83; prof path & dir clin labs, Hahnemann Univ, Philadelphia, Pa, 83-85; chmn path & lab med, Mercy Cath Med Ctr, Philadelphia, Pa, 85-88; PROF & CHMN DEPT PATH, UNIV NDAK, 88- *Mem:* AAAS; Col Am Path; Am Soc Clin Path; Am Asn Pathologists & Bacteriologists; AMA. *Res:* Pulmonary effects of altered atmospheres; computer applications to biomedicine. *Mailing Add:* Dept Pathol Univ N Dak Grand Forks ND 58201

SOPHIANOPOULOS, ALKIS JOHN, b Athens, Greece, Aug 29, 25; US citizen; m 55; c 2. BIOCHEMISTRY, BIOPHYSICS. *Educ:* Drew Univ, AB, 53; Purdue Univ, MS, 57, PhD(chem), 60. *Prof Exp:* Trainee biophys, Dept Chem, Univ Ill, Urbana, 60-61; asst prof biochem, Univ Tenn Med Units, 61-68; ASSOC PROF BIOCHEM, EMORY UNIV, 68- *Honors & Awards:* Eli Lilly Med Fac Award, 65. *Mem:* Am Soc Biol Chemists; Biophys Soc; Am Chem Soc. *Res:* Physical chemistry of macromolecules; biophysical studies of relation of macromolecular structure to biological activity. *Mailing Add:* Dept Biochem Emory Univ Atlanta GA 30322

SOPORI, MOHAN L, b Kashmir, India, Dec 3, 42. IMMUNOLOGY. *Educ:* All India Inst Med Sci, PhD(biochem), 70. *Prof Exp:* ASST PROF MED MICROBIOL & IMMUNOL, MED CTR, UNIV KY, 80- *Mailing Add:* Dept Immunol Lovelace Biomed Ctr 400 Gibson Blvd SE Albuquerque NM 87108

SOPP, SAMUEL WILLIAM, b Hammond, Ind, Aug 28, 34; m 63; c 2. INORGANIC CHEMISTRY, PHYSICAL CHEMISTRY. *Educ:* Ind State Univ, BS, 57; Ariz State Univ, MS, 62; Univ Ill, PhD(inorg chem), 65. *Prof Exp:* Prod develop assoc, Merck & Co, 65-68, supvr & fel prod develop, 69-72, sr group leader, Marine Prod Develop Labs, 73-84; PARTNER, CHEMICON ASSOCS, 84- *Mem:* Am Chem Soc; Fine Particle Soc. *Res:* Characterization of transition metal complexes using nuclear magnetic resonance; evaluation of new product concepts and characterizing physical properties of inorganic materials; development of novel coating materials for electrical steels. *Mailing Add:* Barcroft Co 40 Cape Henlopen Dr PO Box 481 Lewes DE 19958-1117

SOPPER, WILLIAM EDWARD, b Slatington, Pa, Aug 16, 28; m 51; c 3. FORESTRY. *Educ:* Pa State Univ, BS, 54, MF, 55; Yale Univ, PhD(forest hydrol), 60. *Prof Exp:* Asst forestry, 54-55, instr, 55-60, from asst prof to assoc prof, 60-68, PROF FOREST HYDROL, FOREST RES LAB, PA STATE UNIV, 68-, HYDROLOGIST, INST RES LAND & WATER RESOURCES, 69- *Concurrent Pos:* Mem forest influences & watershed mgt sect, Int Union Forestry Res Orgns, 64-66. *Mem:* Soc Am Foresters; Water Pollution Contol Fedn. *Res:* Forest hydrology; watershed management; forest influences; land application of wastewater, strip mine reclamation with sludge. *Mailing Add:* 416 Outer Dr State College PA 16801

SORAUF, JAMES E, b Milwaukee, Wis, May 19, 31; m 62; c 2. GEOLOGY. *Educ:* Univ Wis-Madison, BS, 54, MS, 55; Univ Kans, PhD(geol), 62. *Prof Exp:* From asst prof to assoc prof, 62-75, chmn dept, 77-80, PROF GEOL, STATE UNIV NY, BINGHAMTON, 75-, CHMN DEPT, 86- *Concurrent Pos:* Trustee, Paleont Res Inst, 83-, vpres, 86-88, pres, 88- *Mem:* Am Asn Petrol Geol; Geol Soc Am; Soc Econ Paleont & Mineral. *Res:* Permian stratigraphy; paleontology of Devonian corals. *Mailing Add:* Dept of Geol Sci State Univ of NY Binghamton NY 13901

SORBELLO, RICHARD SALVATORE, b New York, NY, Aug 10, 42. THEORETICAL SOLID STATE PHYSICS. *Educ:* Mass Inst Technol, BS, MS, 65; Stanford Univ, PhD(appl physics), 70. *Prof Exp:* Res assoc low temperature physics, Swiss Fed Inst, Zurich, 71-73; from asst prof to assoc prof, 73-83, PROF PHYSICS, UNIV WIS-MILWAUKEE, 83- *Concurrent Pos:* Vis prof, Vrije Univ, Amsterdam, 80. *Mem:* Am Phys Soc; AAAS. *Res:* Transport theory and atomic diffusion in solids; electromigration and thermomigration in metals; electronic structure of solids; dielectric response and transport properties of metallic microstructures. *Mailing Add:* Dept of Physics Univ of Wis Milwaukee WI 53201

SORBER, CHARLES ARTHUR, b Kingston, Pa, Sept 12, 39; m 72; c 2. ENVIRONMENTAL ENGINEERING, CIVIL ENGINEERING. *Educ:* Pa State Univ, BS, 61, MS, 66; Univ Tex, Austin, PhD (environ eng), 71. *Prof Exp:* Proj engr, Harris, Henry & Potter, Inc, Pa, 65-66; chief, Gen Eng Br, USArmy Environ Hyg Agency, 66-69; comdr, US Army Med Environ Eng Res Univ, Md, 71-73; dir, Environ Qual Div, US Army Med Bioeng Res & Develop Lab, 73-75; asst dean, Col Sci & Math, Univ Tex, San Antonio, 76-77, assoc prof environ eng, Div Environ Studies, 75-80, acting dir, Div Earth & Phys Sci & dir, Ctr Appl Res & Technol, 76-80; prof civil eng & assoc dean, Col Eng, Univ Tex, Austin, 80-86, L B Meaders prof eng, 85-86; PROF CIVIL ENG & DEAN SCH ENG, UNIV PITTSBURGH, 86- *Concurrent Pos:* Prin investr, Fischer & Porter, Co, Pa, 76-78, Environmental Protection Agency grant, 76-86 & US Army & Mobility, Equip Res & Develop Command, 77-78; co-prin investr, Environmental Protection Agency grant, 77-79, NSF grant, 77-80 & Southwest Res Inst & Environmental Protection Agency grant, 78-83; grant, Environ Protection Agency, 84-86, 87-90 & 90-93; prin investr, MACME grant, 85-86, NSF grant, 89-91. *Honors & Awards:* Serv Award, Water Pollution Control Fedn, 85 & 89. *Mem:* Am Pub Health Asn; Am Soc Eng Educ; Am Soc Civil Engrs; Am Water Works Asn; Int Asn Water Pollution Res & Control; Nat Soc Prof Engrs; Water Pollution Control (vpres, 90-91). *Res:* Health effects associated with water and wastewater treatment processes including land application of wastewater and sludges; wastewater reuse including membrane processes and health effects; water-wastewater disinfection; kinetics and efficiency. *Mailing Add:* 240 Benedum Hall Univ Pa Pittsburgh PA 15261

SORBY, DONALD LLOYD, b Fremont, Nebr, Aug 12, 33; m 59; c 2. PHARMACY, PHARMACEUTICAL CHEMISTRY. *Educ:* Univ Nebr, BS, 55; Univ Wash, Seattle, MS, 58, PhD(pharm), 60. *Prof Exp:* From asst prof to prof pharm & pharmaceut chem, Sch Pharm, Univ Calif, San Francisco, 60-72; prof pharm & chmn dept pharm practice, Col Pharm, Univ Wash, 72-74; dean,Sch Of Pharm, Univ Mo, Kansas City, 74-84; DEAN, SCH PHARM, UNIV PAC, 84- *Concurrent Pos:* USPHS grant, 63-65. *Mem:* Am Pharmaceut Asn; Acad Pharmaceut Sci. *Res:* Interactions between drugs and adsorbent materials and how they affect action of various drug molecules; relationships between physical and chemical properties of drugs and their in vivo action. *Mailing Add:* 4362 Yacht Harbor Dr Stockton CA 95204

SORDAHL, LOUIS A, b Chicago, Ill, Aug 24, 36; m 62; c 2. PHYSIOLOGY, BIOCHEMISTRY. *Educ:* Rutgers Univ, AB, 58, MS, 61, PhD(biochem), 64. *Prof Exp:* Res asst biol, Rutgers Univ, 58-62, asst instr physiol, 63, instr, 63-64; from instr to asst prof pharmacol, Baylor Col Med, 66-72; assoc prof, 72-77, PROF BIOCHEM, UNIV TEX MED BR GALVESTON, 77- *Concurrent Pos:* NIH staff fel geront, Baltimore City Hosps, Md, 64-66; vis prof cardiovasc med, Mayo Clin, 76-; co-adj prof exp med, Baylor Col Med, 78-; mem, Va Cardiovasc Merit Preview Panel, 87-91, chmn, 89-91, dir, Grad Prog Human Biol Chem & Genetics, 88- *Honors & Awards:* Hektoen Gold Medal Award, AMA, 70. *Mem:* Int Soc Heart Res; fel Am Col Cardiol; Am Physiol Soc; Am Heart Asn; Biophys Soc; Am Soc Biochem & Mol Biol. *Res:* Intermediary metabolism and metabolic diseases; cardiac bioenergetics; oxidative phosphorylation; enzymes and mechanisms of calcium transport in mitochondria; experimental surgery; coronary thrombosis. *Mailing Add:* Dept Human Biol Chem & Genetics Univ Tex Med Br Galveston TX 77550-2774

SOREF, RICHARD ALLAN, b Milwaukee, Wis, June 26, 36; m 69. ELECTROOPTICS. *Educ:* Univ Wis, BS, 58, MS, 59; Stanford Univ, PhD(elec eng), 64. *Prof Exp:* Staff mem, Solid State Physics Div, Lincoln Lab, Mass Inst Technol, 64-65; res staff mem, appl physics dept, Sperry Res Ctr, 65-83; RES SCIENTIST, USAF ROME LAB, BEDFORD, 83- *Concurrent Pos:* Charles E Ryan award basic res, Rapid Area Distrib Support, USAF, 88. *Mem:* Am Phys Soc; Optical Soc Am; sr mem Inst Elec & Electronic Eng; Soc Photo-Optical Instrumentation Engrs. *Res:* Electro-optical modulation in semiconductors; integrated optics; fiber optics; electrooptic effects in liquid crystals; liquid crystal displays; infrared detectors; nonlinear optical effects in solids; optical communication; optical switching; sensors; quantum well and superlattice devices. *Mailing Add:* USAF Rome Lab (RL/ESO) Hanscom AFB MA 01731-5000

SOREIDE, DAVID CHRISTIEN, b Arlington, Va, July 20, 45; m 72. ENGINEERING PHYSICS. *Educ:* Univ Colo, BS, 67; Univ Wash, MS, 68, PhD(physics), 78. *Prof Exp:* Sr engr, Aerodyn Lab, Boeing Com Airplanes, 77-80, scientist, Laso Instrumentation Lab, 80-85, prin investr, 85-90, CHIEF SCIENTIST, LIDAR-AIR DATA, HIGH TECHNOL CTR, BOEING DEFENSE & SPACE GROUP, 90- *Mem:* Optical Soc Am; Int Soc Optical Eng; Am Phys Soc; Am Inst Aeronaut & Astronaut. *Res:* Lidar systems to sense air velocities for airborne applications. *Mailing Add:* Boeing Co PO Box 3999 MS 7J-05 Seattle WA 98105

SORELL, HENRY P, b Coeymans, NY, Nov 15, 23; m 70. ORGANIC CHEMISTRY. *Educ:* Rensselaer Polytech Inst, BS, 48, MS, 50, PhD(chem), 54. *Prof Exp:* Teaching asst, Org Labs, Rensselaer Polytech Inst, 48-51; sr chemist coated abrasives, Behr-Manning Div, Norton Co, 51-55; fel dent mat res, Mellon Inst, 55-56, sr fel, 56-57; vpres dent res & mfg, Luxene, Inc, 57-58; sect leader adhesives develop, Hughson Chem Co, 58-61; group leader pressure sensitive tapes, Mystik Tape Div, Borden Chem Co, Ill, 63-65, sect leader, 65-70; tech dir, Pipeline Tapes, Plicoflex, Inc, 70-72; vpres res & develop, Anchor Continental, Inc, 70-85; RETIRED. *Concurrent Pos:* Admin fel dent plastics, Mellon Inst, 57-58. *Mem:* Am Chem Soc. *Res:* Organic adhesives; adhesion and effect of environmental factors on organic adhesives and on adhesion. *Mailing Add:* 125 Calhoun St Johnston SC 29832-1310

SOREM, MICHAEL SCOTT, b Berkeley, Calif, Apr 27, 45; div; c 2. LASER PHYSICS, PHOTOCHEMISTRY. *Educ:* Stanford Univ, BS, 67, MS, 68, PhD(physics), 72. *Prof Exp:* Physicist elastic-plastic flow codes, Lawrence Livermore Lab, 67; res asst high resolution spectros, Stanford Univ, 67-72; physicist, Nat Bur Standards, 72-74; STAFF PHYSICIST LASER RES, LOS ALAMOS SCI LAB, 74- *Concurrent Pos:* Nat Res Coun fel, Nat Bur Standards & Joint Inst Lab Astrophys, 72-74. *Res:* Tunable laser source development; sub-Doppler high resolution atomic and molecular spectroscopy; optically pumped ir lasers; excited-state spectroscopy; high average power solid state lasers. *Mailing Add:* Four Piedra Ct White Rock NM 87544

SOREM, RONALD KEITH, b Northfield, Minn, June 18, 24; m 53; c 4. GEOLOGY, DEEP-SEA MINERAL RESOURCES. *Educ:* Univ Minn, BA, 46, MS, 48; Univ Wis, PhD, 58. *Prof Exp:* Asst, State Geol Surv, Minn, 44-46; geol asst, Univ Minn, 47; field asst, US Geol Surv, Alaska, 47, geologist, 48-55; asst, Univ Wis, 56-57, fel econ geol, 58-59; from asst prof to prof, 59-82, assoc dean sci, 80-81, EMER PROF GEOL, WASH STATE UNIV, 83-, DIR, MARINE MINERAL RES MUSEUM, 88- *Concurrent Pos:* Strategic minerals adv, US For Opers Admin, 53-55; sr vis res fel, Univ Manchester, 70; co-investr, Nat Sci Found Int Decade Ocean Explor, Seabed Assessment Prog, 72-75; co-investr & co-chief scientist at sea, Deep Ocean Mining Environ Study Proj, Nat Oceanic & Atmospheric Admin, deep sea expeds, Pac Ocean, 75-79; resources consult, 83-, vpres, comn on manganese, Int Asn Genesis Ore Deposits, 76-86, pres, 86-90; grant, Nat Oceanic & Atmospheric Admin, US Geol Surv & US Bur Mines, 76-80; consult, mineralogy, volcanic ash study, possible health hazards, eruption Mount St Helens Wash, Wash State Univ, 80-; vis res fel, Japan Soc Prom Sci, 81; field examination sulfide, tungsten & manganese deposits caucusus Region, USSR, 82. *Mem:* Sigma Xi; fel Geol Soc Am; fel Mineral Soc Am; Soc Econ Geol; Int Asn Genesis Ore Deposits. *Res:* Mineralogy and origin of manganese deposits; application of micro x-ray analysis to ore and petrographic microscope studies; properties, texture and composition of ore minerals; exploration origin and evaluation of marine manganese nodule deposits; conservation of natural resources; manganese resources of Philippines; Ni-laterites of Cuba. *Mailing Add:* Marine Mineral Resources Mus Wash State Univ Res & Technol Park Pullman WA 99163

SOREN, ARNOLD, b Vienna, Austria, Oct 30, 10; US citizen; m 61. ORTHOPEDIC SURGERY. *Educ:* Univ Vienna, MD, 34, PhD(comp morphol), 51. *Prof Exp:* Resident orthop surg, Gen Hosp, Vienna, 34-47; asst, Univ Vienna, 47-51; asst, Univ Munich, 53-54; docent, Med Fac, Univ Vienna, 55-56; asst prof, 63-67, assoc prof, 67-81, PROF ORTHOP SURG, SCH MED, NY UNIV, 81- *Concurrent Pos:* USPHS grant, Sch Med, NY Univ, 62-; assoc attend physician, Univ Hosp, NY Univ, 62-; attend physician, Vet Admin Hosp, New York, 62-; univ docent, Med Fac, Univ Vienna, 63-; asst admitting physician, Bellevue Hosp, New York, 64- *Honors & Awards:* City of Vienna Prize, 52; Fed Pres of Austria Prize, 55. *Res:* General orthopaedic surgery; rheumatic diseases and histopathology of arthritis. *Mailing Add:* Dept Orthop Surg NY Univ Sch Med New York NY 10016

SORENSEN, ANDREW AARON, b Pittsburgh, Pa, July 20, 38; m 68; c 2. MEDICAL EDUCATION. *Educ:* Univ Ill, BA, 59; Yale Univ, BD, 62, MPhil, 70, PhD(med sociol), 71; Univ Mich, MPub Health, 66. *Prof Exp:* Instr psychiat, Med Sch, Boston Univ, 70-71; asst prof community serv educ, Cornell Univ, 71-73; asst prof prev med, Univ Rochester, 73-76, assoc prof & assoc chmn prev med, Sch Med, 76-83; prof & dir, Sch Pub Health, Univ Mass, 83-90; PROVOST & VPRES ACAD AFFAIRS, UNIV FLA, GAINESVILLE, 90- *Concurrent Pos:* Vis assoc health serv res, Harvard Med Sch, 75-76; NSF fac sci fel, 75-76; vis fel, Univ Cambridge, 79-80; vis prof community med, Welsh Nat Sch Med, 81. *Mem:* Asn Teachers Prev Med; AAAS; Am Sociol Asn; Int Epidemiol Asn. *Res:* Health services; sociology of addictions. *Mailing Add:* Tigert Hall Rm 235 Univ Fla Gainesville FL 32611

SORENSEN, ARTHUR (SHERMAN), JR, engineering mechanics, for more information see previous edition

SORENSEN, CHRISTOPHER MICHAEL, b Omaha, Nebr, Oct 1, 47; c 1. PHASE TRANSITIONS. *Educ:* Univ Nebr, BS, 69; Univ Colo, MS, 73, PhD(physics), 77. *Prof Exp:* PROF PHYSICS, KANS STATE UNIV, 77- *Mem:* Am Phys Soc; Am Chem Soc; Am Asn Aerosol Res; German Soc Aerosol Res. *Res:* Phase transitions and critical phenomena; optics and light scattering; combustion generated particulates; aerosol dynamics. *Mailing Add:* Dept Physics Kans State Univ Manhattan KS 66506

SORENSEN, CRAIG MICHAEL, b May 29, 54. SUPPRESSOR T CELLS, IMMUNE RESPONSE GENES. *Educ:* Wash Univ, St Louis, PhD(immunol), 80. *Prof Exp:* Asst prof immunol, Jewish Hosp Wash Univ Med Ctr, 81-90; SR RES SCIENTIST ONCOGENE SCI, MANHASSET, NY, 90- *Mem:* Am Asn Immunologists; NY Acad Sci. *Res:* T cell regulation; oncogene expression; cell cycle regulation. *Mailing Add:* 350 Community Dr Manhasset NY 11030

SORENSEN, DALE KENWOOD, b Centuria, Wis, July 21, 24; m 48; c 3. VETERINARY SCIENCE. *Educ:* Kans State Col, DVM, 46; Univ Wis, MS, 50, PhD(virol path), 53. *Prof Exp:* Consult vet, UNRRA, 46-47; instr vet sci & head, Sect Clin Med, Univ Wis, 47-53; asst prof med, 53-57, head dept, 65-72, acting dean, 72-73, chmn, Dept Vet Clin Sci, 73-76, chmn, Dept Large Animal Clin Sci, 76-79, assoc dean, 80-87, PROF MED, COL VET MED, UNIV MINN, ST PAUL, 58-, ACTG DEAN, 72- *Concurrent Pos:* Med scientist, Brookhaven Nat Lab, 57-58; vet consult, AID, Dept of State, Philippine Island Mission, 66 & Indonesia, 74; prin vet, USDA-Coop State Res Serv, 85-86. *Mem:* Am Vet Med Asn. *Res:* Pneumonia of calves; radiation syndrome in dogs; diseases of swine and viral respiratory diseases of cattle; leukemia of cattle; animal pathology. *Mailing Add:* 460 Vet Hosp Bldg Univ Minn St Paul MN 55108

SORENSEN, DAVID PERRY, b Spring City, Utah, Nov 1, 30; m 52. ORGANIC CHEMISTRY. *Educ:* Univ Utah, BS, 52, PhD, 55. *Prof Exp:* Res chemist, M W Kellogg Co, Pullman, Inc, 55-57; sr res chemist, 57-60, res supvr, 60-63, res specialist, 63-65, res mgr, 65-67, tech dir, Imaging Res Lab, 67-71, tech dir, Printing Prod Div, 71-81, dir, Corp Technol Assessment, 81-89, EXEC DIR, CORP TECH PLANNING & COORD, 3M CO, 89- *Honors & Awards:* 3M Carlton Soc, 88. *Mem:* Am Chem Soc; Sigma Xi; Tech Asn Graphic Arts. *Res:* Imaging sciences; printing technology. *Mailing Add:* 4140 Lakewood Ave White Bear Lake MN 55110

SORENSEN, DAVID T, inorganic chemistry; deceased, see previous edition for last biography

SORENSEN, EDGAR LAVELL, b Mendon, Utah, Nov 26, 18; m 48; c 3. PLANT BREEDING. *Educ:* Utah State Univ, BS, 41, MS, 52; Univ Wis, PhD(agron), 55. *Prof Exp:* Soils technologist, Bur Reclamation, Utah, 46-47; state seed supvr, Utah State Dept Agr, 47-49; prog specialist, Agr Conserv, Prod & Mkt Admin, 49-51, RES AGRONOMIST, AGR RES SERV, USDA, 55- *Mem:* Am Soc Agron; Crop Sci Soc Am. *Res:* Breeding improved varieties of alfalfa; insect and disease resistance. *Mailing Add:* Dept Agron Waters Hall Kans State Univ Manhattan KS 66506

SORENSEN, ELSIE MAE (BOECKER), physiology, toxicology, for more information see previous edition

SORENSEN, FREDERICK ALLEN, b Pittsburgh, Pa, July 18, 26. MATHEMATICAL STATISTICS. *Educ:* Carnegie Inst Technol, BS, 47, MS, 49, PhD(math), 59. *Prof Exp:* Instr, Carnegie-Mellon Univ, 51-54, 59-71; RETIRED. *Concurrent Pos:* Asst math, Carnegie Inst Technol, 47-51; statistician, Westinghouse Elec Corp, 51-54; statistician, Res Lab, US Steel Corp, 56-61, head, Oper Res Sect, 61-64, res mathematician, 64-77, assoc res consult, Math Div, 77-85. *Mem:* AAAS; Am Soc Qual Control; Am Statist Asn; Inst Math Statist. *Res:* Theory of control charts; design and analysis of industrial and engineering experiments. *Mailing Add:* 1074 Findley Dr Pittsburgh PA 15221

SORENSEN, HAROLD C(HARLES), b Bancroft, Nebr, Dec 21, 34; m 60; c 2. STRUCTURAL ENGINEERING & DYNAMICS, ENGINEERING MECHANICS. *Educ:* Univ Nebr, BS, 57, MS, 62, PhD(eng mech), 66. *Prof Exp:* Jr engr, Dept Rds, State of Nebr, 57-59; instr eng mech, Univ Nebr, 59-65; asst prof civil eng, 66-75, ASSOC PROF CIVIL ENG, WASH STATE UNIV, 75- *Concurrent Pos:* Consult, Palouse Prod, 77, Weyerhaeuser Corp, 78-80 & Wash Pub Power Supply Syst, 80. *Mem:* Am Soc Eng Educ; Am Concrete Inst; Earthquake Eng Res Inst. *Res:* Structural dynamics; earth sheltered homes; lateral buckling of parallel chord trusses; soil-structural interaction-AFWL. *Mailing Add:* Dept Civil Eng Wash State Univ Pullman WA 99164-2910

SORENSEN, KENNETH ALAN, b Providence, RI, Aug 11, 44; m 69; c 3. ENTOMOLOGY. *Educ:* Univ RI, BS, 66; Kans State Univ, MS, 68, PhD(entom), 70. *Prof Exp:* Nat Defense Educ Act fel entom, Kans State Univ, 66-69, res asst, 69-70; exten asst prof, 70-75, EXTEN ASSOC PROF ENTOM, NC STATE UNIV, 75- *Mem:* Entom Soc Am. *Res:* Study of insect pest population dynamics and crop damage under grower conditions; evaluate new and review effectiveness of existing insecticides on vegetables and develop insect pest management programs for growers' use. *Mailing Add:* 3312 Gardner Hall NC State Univ Raleigh NC 27650

SORENSEN, LAZERN OTTO, b Dannebrog, Nebr, Nov 1, 27; wid; c 1. MARINE PHYCOLOGY. *Educ:* Nebr State Univ, BS, 50; Univ Nebr, Lincoln, MS, 52, PhD(bot), 56. *Prof Exp:* Assoc prof, Nebr Wesleyan Univ, 53-56; assoc prof, 56-63, dean sci & math, 68-75, head dept biol, 65-68, PROF BIOL, PAN AM UNIV, 63-, DIR MARINE LAB, PAN AM UNIV, 75- *Mem:* Phycol Soc Am. *Res:* Physiology of macroscopic marine algae. *Mailing Add:* Biol Pan Am Univ Brownsville TX 78520

SORENSEN, LEIF BOGE, b Odense, Denmark, Mar 25, 28; US citizen; m 68; c 1. MEDICINE, BIOCHEMISTRY. *Educ:* Copenhagen Univ, MD, 53, PhD(biochem), 60. *Prof Exp:* Instr anat, Copenhagen Univ, 50-51; resident, St Luke's Hosp, 54; intern, Copenhagen County Hosp, Hellerup, Denmark, 54-55; res asst geront, Med Sch, Wash Univ, 55-56; res asst med, Argonne Cancer Res Hosp, Univ Chicago, 56-57; resident, Copenhagen Munic Hosp, 57-58; resident, Copenhagen Univ Hosp, 58; from instr to assoc prof, 58-70, PROF MED, UNIV CHICAGO, 70-, ASSOC CHMN, DEPT MED, 70- *Concurrent Pos:* Fulbright scholar, Med Sch, Wash Univ, 55-56; lectr, Ill Acad Gen Pract, 62-63; fac mem, Am Col Physicians Postgrad Course, 62 & 69; sr fel, Fogarty Int Ctr, NIH, 80-81. *Mem:* AAAS; Am Rheumatism Asn; Am Soc Clin Invest; NY Acad Sci; Am Soc Geriat. *Res:* Gout; purine metabolism; aging of the immune system. *Mailing Add:* Dept Med Pritzker Sch Med Univ Chicago 5841 Maryland Ave Chicago IL 60637

SRENSEN, PAUL DAVIDSEN, b Seattle, Wash, Dec 4, 34; m 59; c 3. BOTANY, TAXONOMY. *Educ:* Univ Iowa, BA, 62, MS, 66, PhD(bot, plant taxon), 67. *Prof Exp:* Asst hort taxonomist, Arnold Arboretum, Harvard Univ, 67-68, asst cur, 68-70; asst prof, 70-75, ASSOC PROF PLANT TAXON, NORTHERN ILL UNIV, 76-; VPRES, ENVIRON CONSULTS & PLANNERS, ENCAP INC, 75- *Concurrent Pos:* Assoc prof bot, Univ Iowa, 78; vis scholar, The Gray Herbarium, Harvard Univ, 86-87. *Mem:* Int Asn Plant Taxon; Am Soc Plant Taxon; Natural Areas Asn; Nature Conservancy. *Res:* Taxonomy of vascular plants; systematics and ecology of flowering plants; distributional relationships of plants and habitats of the Upper Midwest; taxonomic studies of Mexican and Central American floras. *Mailing Add:* Dept Biol Sci Northern Ill Univ De Kalb IL 60115-2861

SORENSEN, RALPH ALBRECHT, b Lynwood, Calif, Apr 19, 45; m 84; c 3. DEVELOPMENTAL BIOLOGY, IMMUNOLOGY. *Educ:* Univ Calif, Riverside, BA, 67; Yale Univ, PhD(biol), 72. *Prof Exp:* Res fel, Dept Physiol & Anat, Sch Med, Harvard Univ, 72-74; asst prof biol sci, DePaul Univ, 74-77; asst prof, 77-84, ASSOC PROF BIOL, GETTYSBURG COL, 84- *Mem:* Soc Develop Biol; Soc Study Reproduction. *Res:* Early development of the sea urchin embryo; meiotic maturation and fertilization of the mammalian oocyte; preimplantation development of mammalian embryos. *Mailing Add:* Dept Biol Gettysburg Col Box 392 Gettysburg PA 17325-1486

SORENSEN, RAYMOND ANDREW, b Pittsburgh, Pa, Feb 27, 31; m 53; c 1. NUCLEAR PHYSICS. *Educ:* Carnegie Inst Technol, BS, 53, MS, 55, PhD(physics), 58. *Prof Exp:* NSF fel, Copenhagen Univ, 58-59; from instr physics to res assoc, Columbia Univ, 59-61; from asst prof to assoc prof, 61-68, chmn dept, 80-89, PROF PHYSICS, CARNEGIE-MELLON UNIV, 68- *Concurrent Pos:* NSF sr fel, Niels Bohr Inst, Copenhagen, Denmark, 65-66; Nordita prof, Res Inst Physics, Stockholm, 70-71 & 76-77; assoc ed, Nuclear Physics A, 72- *Mem:* Am Phys Soc; fel AAAS. *Res:* Theoretical nuclear structure physics. *Mailing Add:* Dept of Physics Carnegie-Mellon Univ Pittsburgh PA 15213

SORENSEN, RICARDO U, b Valdivia, Chile, Mar 13, 39; m 70; c 3. IMMUNOLOGY. *Educ:* Univ Chile, MD, 64. *Prof Exp:* Sect head, Immunol Dept, Pub Health Inst, Santiago, Chile, 68-70, dept head, 70-76; from asst prof to assoc prof pediat & immunol, Case Western Reserve Univ, 77-89; PROF PEDIAT & IMMUNOL, LA STATE UNIV MED CTR, 89- *Mem:* Am Asn Immunol; Pediat Res Soc. *Res:* Congenital immunodeficiency disease; development of immunity in humans; cellular immunity to pseudomonas; immune response to Bacillus Calmette-Guerin immunization in human newborns. *Mailing Add:* 1542 Tulane Ave New Orleans LA 70112

SORENSEN, ROBERT CARL, b Omaha, Nebr, July 24, 33; m 58; c 2. SOIL CHEMISTRY. *Educ:* Univ Nebr, BS, 55, MS, 57; Iowa State Univ, PhD(soil chem), 64. *Prof Exp:* From asst prof to assoc prof, 64-75, PROF AGRON, UNIV NEBR, LINCOLN, 75- *Mem:* Am Soc Agron; Soil Sci Soc Am; fel Nat Assoc Col Teachers Agr; Sigma Xi. *Res:* Reactions and movement of phosphorus in soils. *Mailing Add:* Dept Agron Univ Nebr Lincoln NE 68583

SORENSEN, THEODORE STRANG, b Dixonville, Alta, June 6, 34; m 66. PHYSICAL ORGANIC CHEMISTRY. *Educ:* Univ Alta, BSc, 56; Univ Wis, PhD(org chem), 60. *Prof Exp:* Imp Chem Industs fel, Univ Leicester, 60-62; from asst prof to assoc prof, 62-74, PROF CHEM, UNIV CALGARY, 74- *Mem:* AAAS; Am Chem Soc; Chem Inst Can. *Res:* Organic reaction mechanisms; stable aliphatic carbonium ions and their reactions; unusual organometallic compounds. *Mailing Add:* Dept of Chem Univ of Calgary Calgary AB T2N 1N4 Can

SORENSON, FRED M, b Brigham City, Utah, Feb 19, 27; m 50; c 2. DENTISTRY. *Educ:* Univ Utah, BS, 51; Univ Ore, DMD, 58, MSD, 63. *Prof Exp:* Asst radiol, Univ Utah, 51-54, res consult dent, 59; res asst & instr biochem, 59-60, from asst prof to assoc prof dent, 60-65, PROF DENT, DENT SCH, UNIV ORE, 65-, DIR CLINS, 65-, CHMN, DEPT ORAL RADIOL, 76- *Concurrent Pos:* Res consult, Vet Hosp, Portland, 66- *Mem:* AAAS; assoc Am Acad Oral Roentgenol. *Res:* Cancer; dental pulp physiology and pathology; dental materials and instrumentation in dental research. *Mailing Add:* 1193 Troon Rd Lake Oswego OR 97034

SORENSON, HAROLD WAYNE, b Omaha, Nebr, Aug 28, 36; m 58; c 3. CONTROL SYSTEMS ENGINEERING. *Educ:* Iowa State Univ, BS, 57; Univ Calif, Los Angeles, MS, 63, PhD(control systs eng), 66. *Prof Exp:* Sr res engr, Gen Dynamics/Astronaut, 57-62; head space systs group, AC Electronics Div, Gen Motors Corp, Calif, 63-66; guest scientist, Ger Exp Aerospace Facil, Inst Control Syst Technol, WGer, 66-67; asst prof systs dynamics & control, Univ Calif, San Diego, 68-71, from assoc prof to prof eng sci, 71-89; chief scientist, US Air Force, 85-88; VPRES BEDFORD GROUP & DIR AIR FORCE FED FUNDED RES & DEVELOP CTR, MITRE CORP, 90- *Concurrent Pos:* Consult, Adv Group Aerospace Develop, NATO, Paris, 66-67, Aerojet-Gen Corp, Azusa, Calif, 68-70 & Aerospace Corp, 71-75; pres, Orincon Corp, 74-81; mem, US Air Force Sci Adv Bd, 81-85. *Mem:* Fel Inst Elec & Electronics Engrs; Oper Res Soc Am; AAAS. *Res:* Control of stochastic and deterministic dynamical systems, including optimal deterministic control theory, numerical methods for optimal control, linear and nonlinear filtering for stochastic systems, optimal and suboptimal control of stochastic systems. *Mailing Add:* Mitre Corp Burlington Rd Bedford MA 01730

SORENSON, JAMES ALFRED, b Madison, Wis, Aug 21, 38; m 61; c 4. MEDICAL PHYSICS, NUCLEAR MEDICINE. *Educ:* Univ Wis-Madison, BS, 63, MS, 64, PhD(radiol sci), 71. *Prof Exp:* Physicist, Sect Nuclear Med, Univ Wis-Madison, 66-71, asst prof med physics, Dept Radiol, 71-73; assoc prof, 73-80, PROF RADIOLOGY, UNIV UTAH, 80- *Mem:* AAAS; Soc Nuclear Med; Am Asn Physicists Med. *Res:* In vivo determination of body composition and elemental concentrations by radiation transmission measurements; whole-body counting of radioactivity; contrast improvement in radiography; positron tomography; image perception. *Mailing Add:* 525 Shady Wood Way Madison WI 53714-2731

SORENSON, JOHN R J, b Sturgeon Bay, Wis, June 13, 34; m 59; c 7. MEDICINAL CHEMISTRY & PHARMACOLOGY. *Educ:* Univ Wis, BS, 60; Univ Kans, PhD(med chem), 65. *Prof Exp:* Sr res chemist, G D Searle & Co, 65-70; asst prof environ health, Col Med, Univ Cincinnati, 70-76, adj asst prof med chem, Col Pharm, 76-77; assoc prof, 77-81, PROF MED CHEM, COL PHARM & PHARMACOL & COL MED, UNIV ARK, LITTLE ROCK, 81- *Concurrent Pos:* Corresp mem, UNESCO Int Ctr Trace Element Study. *Mem:* NY Acad Sci; Asn Bioinorg Sci; Am Chem Soc. *Res:* Medicinal chemistry; pharmacology. *Mailing Add:* Col of Pharm Slot 522 4301 W Markham St Little Rock AR 72205

SORENSON, MARION W, b Salt Lake City, Utah, Dec 29, 26; m 48; c 3. ETHOLOGY, ZOOLOGY. *Educ:* Univ Utah, BS, 59, MS, 60; Univ Mo, PhD(zool), 64. *Prof Exp:* Res assoc zool, 64-65, asst prof, 65-70, ASSOC PROF BIOL SCI, UNIV MO-COLUMBIA, 70- *Mem:* AAAS; Animal Behav Soc; Am Soc Zool; Ecol Soc Am; Am Soc Mammal. *Res:* Social and reproductive behavior of vertebrates, especially small mammals. *Mailing Add:* 5700 E Highway WW Columbia MO 65201

SORENSON, ROBERT LOWELL, b Albert Lea, Minn, Aug 3, 40. ANATOMY. *Educ:* Univ Minn, BA, 62, PhD(anat), 67. *Prof Exp:* from instr to assoc prof, 67-85, PROF ANAT, UNIV MINN, MINNEAPOLIS, 85- *Concurrent Pos:* USPHS fel, Minn Med Res Found, 68-71; res fel, Rigshospitalet, Copenhagen, Denmark, 68; USPHS fel, Univ Minn, 71-; investr, Minn Med Res Found, 67-70. *Mem:* AAAS; Am Anat Asn. *Res:* Diabetes; islet cytology; protein synthesis and secretion autoimmunity. *Mailing Add:* 4901 Woodlawn Blvd Minneapolis MN 55417

SORENSON, WAYNE RICHARD, b St Paul, Minn, Dec 19, 26; m 54; c 3. POLYMER CHEMISTRY, ORGANIC & INORGANIC CHEMISTRY. *Educ:* Col St Thomas, BS, 49; Univ Md, PhD(org chem), 54. *Prof Exp:* Res chemist, E I du Pont de Nemours, 53-61; group leader Continental Oil Co, 61-64, sect leader, 64-67, mgr res & develop, 67-72, dir plastics res & develop, 72-77, coordr new ventures, 77-78; dir res & develop, Tenneco Chem, Inc, 78-79, vpres, 80-82; vpres res & develop, Church & Dwight Co Inc, 82-90; RETIRED. *Mem:* Am Chem Soc; Soc Plastics Engrs. *Res:* New polymer-forming reactions; properties of polymers; polymer applications; inorganic chemicals and detergents. *Mailing Add:* 29 Catskill Ct Belle Mead NJ 08502

SORENSON, WILLIAM GEORGE, b Albert Lea, Minn, Oct 30, 35; m 57; c 4. MYCOLOGY. *Educ:* Univ Iowa, BA, 58, MS, 62; Univ Tex, PhD(bot), 64. *Prof Exp:* res microbiologist, Northern Utilization Res & Develop Div, USDA, 63-66; asst prof biol, Oklahoma City Univ, 66-68; NIH fel, Univ Okla, 68-70, asst prof bot & microbiol, 70-77; RES MYCOLOGIST, APPALACHIAN LAB OCCUP SAFETY & HEALTH, 77-; ADJ PROF MICROBIOL & IMMUNOL, WVA UNIV, 77- *Mem:* Mycol Soc Am; Bot Soc Am; Am Soc Microbiol; Int Biodeterioration Res Group. *Res:* Physiology and taxonomy of fungi; mycotoxins. *Mailing Add:* NIOSH/DRDS 944 Chestnut Ridge Rd Morgantown WV 26505

SORGER, GEORGE JOSEPH, b Vienna, Austria, Sept 20, 37; Can citizen; m 61; c 2. BIOCHEMICAL GENETICS, MICROBIOLOGY. *Educ:* McGill Univ, BS, 59; Yale Univ, PhD(microbiol), 64. *Prof Exp:* Res assoc bot, Ore State Univ, 64-66; from asst prof to assoc prof, 66-78, PROF BIOL, MCMASTER UNIV, 78-, NAT RES COUN CAN FEL, 66- *Concurrent Pos:* Exchange scientist, Chem Bact, Nat Ctr Sci Res, Marseille, 73-74, Cold Spring Harbor Lab, 81-82. *Mem:* Can Fedn Biol Soc; Can Soc Biochem; Can Soc Plant Molecular Biol. *Res:* Regulation and mechanism of action of nitrate reductase; nitrite reductase, studied using a molecular-genetical approach; isolation and study of genes concerned with nitrate assimilation and its regulation in neurospora; regulation of nitrate reductase in differnet organs of maize; developmental connections. *Mailing Add:* 525 Life Sci Bldg Dept of Biol McMaster Univ Hamilton ON L8S 4L8 Can

SORIA, RODOLFO M(AXIMILIANO), b Berlin, Germany, May 16, 17; nat US; m 47; c 3. ELECTRICAL ENGINEERING. *Educ:* Mass Inst Technol, SB, 39, SM, 40; Ill Inst Technol, PhD(elec eng), 47. *Prof Exp:* Asst elec & radio eng, Ill Inst Technol, 40-42; instr electronics & microwave, US Army Sig Corps Training Prog, 42, lectr, 43, instr elec eng, 43-47, asst prof, 47; proj engr, Amphenol Corp, 46-49, dir res, 49-54, dir eng, 54-56, vpres res & eng, 56-68, dir corp res & eng, Bunker-Ramo Corp, 68, group vpres, Res & Eng, Amphenol Components Group, 69-71; MGT CONSULT, CONSULT INT LTD, 72- *Concurrent Pos:* Pres, Nat Electronics Conf, 54; consult, Adv Group Electronic Parts, Off Dir Defense Res & Eng, 55-61; chmn, Electronic Components Conf, 57; mem comt, Radio Frequency Cables & Connectors & US deleg, Int Electrotech Comn. *Mem:* Fel Inst Elec & Electronics Engrs; Sigma Xi. *Res:* Radio communications; wave propagation; microwaves; antennae; electronic components and systems. *Mailing Add:* 5028 Fair Elms Ave Western Springs IL 60558

SORIANO, DAVID S, b Jersey City, NJ, June 3, 53; m 78; c 2. HETEROCYCLIC SYNTHESIS & PHASE-TRANSFER CATALYSIS, CONFORMATIONAL ANALYSIS. *Educ:* Fairleigh Dickinson Univ, Teaneck, NJ, BS, 75, MS, 77; Univ Nebr-Lincoln, PhD(chem), 80. *Prof Exp:* Res chemist, Buffalo Res Lab, Allied Corp, 80-82, sr res chemist, 82-84; ASSOC PROF CHEM, UNIV PITTSBURGH, BRADFORD, PA, 84- *Mem:* Sigma Xi; Am Chem Soc. *Res:* Immobilized enzymes as catalysts for organic synthesis; asymmetric organic synthesis; fermentation processes leading to chiral synthons for use in organic synthesis. *Mailing Add:* Dept Chem Univ Pittsburgh Bradford PA 16701

SORIERO, ALICE ANN, b Brooklyn, NY, Oct 12, 47. REPRODUCTIVE PHYSIOLOGY, AGING. *Educ:* Hunter Col, BA, 68; State Univ NY Downstate Med Ctr, PhD(anat), 72. *Prof Exp:* Instr obstet, State Univ NY Downstate Med Ctr, 72-75; instr, Dent Sch, Univ Md, 75-76; ASST PROF ANAT, UNIV TEX MED BR, GALVESTON, 76- *Mem:* Sigma Xi; Am Asn Anatomists. *Res:* Reproductive physiology and aging of the female; biochemical effects of estrogen on uterine metabolism; histological changes in aging uterus. *Mailing Add:* Dept of Anat Univ of Tex Med Br Galveston TX 77550

SORKIN, HOWARD, b New York, NY, Aug 29, 33; m 57; c 2. ORGANIC CHEMISTRY, DISPLAY DEVICE CHEMISTRY. *Educ:* City Col New York, BS, 55; Cornell Univ, MS, 57, PhD(org chem), 59. *Prof Exp:* Sr res chemist, Cent Res Labs, Airco, 59-68; mem tech staff, Solid State Div, RCA Corp, Somerville, 68-76; sr chemist, Timex Components, Inc, Somerset, NJ, 76-80; PROG MGR, PHILIPS LABS, 80- *Concurrent Pos:* Adj prof, Rutgers Univ, 79-80. *Mem:* Am Chem Soc; Soc Info Display. *Res:* Synthesis and properties of liquid crystals and their application to electro-optic display devices; new polymers and polymerization processes; electrophoretic display devices; LCD device technology. *Mailing Add:* Philips Labs 345 Scarborough Rd Briarcliff Manor NY 10510

SORKIN, MARSHALL, b Chicago, Ill, July 12, 28; m 50; c 4. COSMETIC CHEMISTRY. *Educ:* Roosevelt Univ, BS, 50; Northwestern Univ, MS, 59. *Prof Exp:* Chemist, Rock Island RR, 50-51; asst plant mgr, S Buchsbaum & Co, 51-52; res & develop group leader, Helene Curtis Ind Inc, 52-61, res dir, Toiletries Div, Alberto-Culver Co, 61-68; dir toiletries res, 68-80, VPRES TOILETRIES DIV, CARTER PROD DIV, CARTER-WALLACE, INC, 80- *Mem:* Am Chem Soc; Soc Cosmetic Chem. *Res:* Cosmetic and proprietary drug formulations; skin and hair physiology. *Mailing Add:* Carter Prods Res PO Box 1 Cranbury NJ 08512-0001

SOROF, SAM, b New York, NY, Jan 24, 22; m 67; c 2. BIOCHEMISTRY. *Educ:* City Col New York, BS, 44; Univ Wis, PhD(physiol chem), 49. *Prof Exp:* Asst physiol chem, Univ Wis, 47-49, res assoc, 49-50; res biochemist, Vet Admin Hosp, New York, 51-52; res assoc, Inst Cancer Res & Lankenau Hosp Res Inst, 52-55, assoc mem, Inst, 55-61, SR MEM, INST CANCER RES, FOX CHASE CANCER CTR, PHILADELPHIA, 61- *Concurrent Pos:* Res fel phys biochem, Nat Cancer Inst, 50-51; vis scientist, Biochem Inst, Univ Uppsala, 56 & Salk Inst Biol Studies, 71-72; assoc ed, Cancer Res, 72-80 & Cancer Biochem & Biophys, 75-90; adj prof, Dept Path & Lab Med, Med Sch, Univ Pa, 81-90. *Mem:* AAAS; Am Soc Cell Biol; Am Soc Biol Chem; Am Asn Cancer Res. *Res:* Molecular biology of cells; biochemistry of cancer, carcinogenesis, cell differentiation, liver proteins, mammary gland in culture. *Mailing Add:* Inst Cancer Res Fox Chase Cancer Ctr 7701 Burholme Ave Philadelphia PA 19111

SOROFF, HARRY S, b Sydney, NS, Feb 2, 26. SURGERY. *Educ:* Temple Univ, MD, 48; Am Bd Surg, dipl, 60; Bd Thoracic Surg, dipl, 61. *Prof Exp:* Intern, Philadelphia Jewish Hosp, 48-49; asst resident surg, Montefiore Hosp, New York, 50-51; chief resident, Beth David Hosp, 51-52; chief metab div, Surg Res Univ, Brooke Army Med Ctr, Ft Sam Houston, Tex, 53-56; resident surg, Lakeside Hosp, Cleveland, Ohio, 56-57; fel thoracic surg, Peter Bent Brigham Hosp, 57-61, chief thoracic lab, 60-61; from asst prof to prof surg, Sch Med, Tufts Univ, 68-74, dir Tufts Surg Serv, Boston City Hosp, 70-74; CHMN DEPT SURG, MED SCH/HEALTH SCI CTR, STATE UNIV NY STONY BROOK, 74-, PROF SURG, 77- *Concurrent Pos:* Fel surg metab,

Columbia-Presby Med Ctr, 52-53; res fel, Peter Bent Brigham Hosp, Boston, 57-60; fel thoracic surg, Mt Auburn & Malden Hosps, 57-60; asst surgeon, Boston City Hosp, 61-64; estab investr, Am Heart Asn, 61-66; assoc dir clin study unit, New Eng Ctr Hosp, Boston, 61-, asst surgeon, 61-64, surgeon, 64-; sr consult, Lemuel-Shattuck Hosp, Jamaica Plain, Mass, 67- *Mem:* Int Soc Burn Injuries; Am Soc Artificial Internal Organs; Int Cardiovasc Soc; Am Asn Thoracic Surg; Am Burn Asn. *Res:* Thoracic and cardiovascular surgery; surgical metabolism. *Mailing Add:* Vet Admin Hosp Northport NY 11768

SOROKIN, PETER, b Boston, Mass, July 10, 31. QUANTUM ELECTRONICS. *Educ:* Harvard Univ, AB, 52, BS, 53, PhD(appl physics), 58. *Prof Exp:* Staff physicist, 57-68, IBM FEL, T J WATSON RES CTR, 68- *Mem:* Nat Acad Sci; NY Acad Sci; fel Am Optical Soc; fel Am Phys Soc Res Lasers; Am Acad Sci. *Mailing Add:* T J Watson Res Ctr IBM Corp PO Box 218 Yorktown Heights NY 10598

SOROKIN, SERGEI PITIRIMOVITCH, b Boston, Mass, Apr 13, 33. HISTOLOGY, EMBRYOLOGY. *Educ:* Harvard Univ, AB, 54, Harvard Med Sch, MD, 58. *Prof Exp:* Instr anat, Harvard Med Sch, 60-65, assoc & tutor, 65-69, asst prof, 69-70; asst prof, 70-77, ASSOC PROF CELL BIOL, SCH PUB HEALTH, HARVARD UNIV, 77- *Concurrent Pos:* Res fel path, Harvard Med Sch, 58-59, fel anat, 59-60; vis asst prof, Cornell Univ, 62-63; mem pulmonary res eval comt, Vet Admin, 69-70; lung cancer adv group, Nat Cancer Inst, 71-74. *Mem:* Histochem Soc; Soc Cell Biol; Am Asn Anat. *Res:* Cell and biology; in vitro culturing techniques; cytological differentiation as studied with aid of electron microscopy, histochemistry and autoradiography; physiology; pulmonary morphology. *Mailing Add:* Dept Anat Boston Univ Med 80 E Concord St Boston MA 02118

SOROOSHIAN, SOROOSH, b Kerman, Iran, July 2, 48; US citizen; m; c 2. HYDROLOGIC MODELLING, SURFACE HYDROLOGY. *Educ:* Calif State Polytech Univ, San Luis Obispo, BS, 71; Univ Calif Los Angeles, MS, 73, PhD(syst eng), 78. *Prof Exp:* Asst prof, Dept Systs Eng & Civil Eng, Case Western Reserve Univ, Cleveland, Ohio, 78-82; assoc prof, 83-87, PROF, DEPTS HYDROL & WATER RESOURCES & SYSTS & INDUST ENG, UNIV ARIZ, 87-, DEPT HEAD, HYDROL & WATER RESOURCES, 89- *Concurrent Pos:* Assoc ed, Water Resource Res, Am Geophys Union, 83-88, ed, 88-92; prog chmn, fall meetings, Hydrol Sect, Am Geophys Union, 84-87, mem, Hydrol Exec Comt, 84-; mem, Water Sci & Technol Bd, Comt Restoration Aquatic Ecosysts, Nat Res Coun, 90- *Mem:* Am Soc Civil Eng; Am Meteorol Soc; AAAS; Am Water Resources Asn. *Res:* Surface hydrology, including rainfall-runoff modeling, flood forecasting, application of remote sensing in hydrology of climate studies. *Mailing Add:* Dept Hydrol & Water Resources Univ Ariz Tucson AZ 85721

SORRELL, FURMAN Y(ATES), JR, b Wadesboro, NC, July 14, 38; div; c 2. ENVIRONMENTAL FLUID DYNAMICS. *Educ:* NC State Univ, BS, 60; Calif Inst Technol, MS, 61, PhD(aeronaut), 66. *Prof Exp:* Res engr, Pratt & Whitney Aircraft, 61-62; res fel, Univ Colo, Boulder, 66-67, asst prof aerospace eng sci, 67-68; from asst prof to assoc prof eng mech, 68-75, dir grad prog, 70-74, PROF MECH & AEROSPACE ENG, NC STATE UNIV, 76- *Concurrent Pos:* Prof marine sci faculty, NC State Univ, 72-, prof air conserv faculty, 77-; assoc, Perry Assoc Consult Engrs, 74-75; co-chmn ocean panel, Comt Applications Rev High Resolution Passive Satellites, NASA, 76-78; Nat Oceanic & Atmospheric Admin grant, Nearshore Ocean Currents & Mixing, 76-81, chmn, Panel Marine Waste Disposal & mem Steering Comt, Conf Marine Pollution, 79-80; grant Impact Off-Shore Pipelines, 80-82; mem prog review comt, Prog Phys & Chem Energy Storage, US Dept Energy, 81; tech dir, NC Alternative Energy Corp, 81-82; res contracts, Int Bus Mach, 82-84, USAF, 85-88, Sematech, 88-, Semiconductor Res Corp, 88-, Army Res Off, 89- & Off Naval Res, 90- *Mem:* Am Phys Soc; Am Geophys Union; Am Soc Mech Engrs. *Res:* Fluid dynamics; physics of fluids; models of fluid and heat transfer systems; laboratory and field measurements. *Mailing Add:* Dept Mech & Aerospace Eng NC State Univ Raleigh NC 27695-7910

SORRELL, GARY LEE, b Middletown, Ohio, Dec 7, 43; m 65; c 1. DECISION SUPPORT SYSTEMS, OPERATIONS RESEARCH. *Educ:* Park Col, BA, 65; Univ Okla, Norman, MA, 67. *Prof Exp:* Mathematician, Nat Security Agency, 67-73; opers dir, Teledyne Brown Eng, 73-74; sr opers analyst, US Postal Serv, 74-76; proj mgr, Genasys Corp, 76-77; sr assoc, J Watson Noah Assocs, 77-78; EXEC VPRES, MGT CONSULT & RES, INC, 78- *Mem:* Opers Res Soc Am; Inst Cost Anal; Mil Opers Res Soc. *Res:* Computer systems design and development; decision support systems applications; resource management; cost analysis and estimation; economic analysis. *Mailing Add:* 5111 Leesburg Pike Suite 514 Falls Church VA 22041

SORRELL, MICHAEL FLOYD, b St Louis, Mo, July 4, 35; m 57; c 4. GASTROENTEROLOGY. *Educ:* Univ Nebr, Omaha, BS, 57, MD, 59. *Prof Exp:* Intern med, Nebr Methodist Hosp, 59-60; pvt pract, 60-66; resident internal med, Col Med, Univ Nebr, Omaha, 66-68, fel gastroenterol, 68-69; NIH trainee liver dis & nutrit, Col Med & Dent, NJ, 69-71; from asst prof to assoc prof, 71-76, PROF MED, UNIV NEBR MED CTR, OMAHA, 76-, CHMN DEPT INTERNAL MED, 81- *Concurrent Pos:* NIH acad career develop award, 71-76; dir, Liver Study Unit, 71- *Mem:* Am Fedn Clin Res; Am Gastroenterol Asn; Am Asn Study Liver Dis; Int Asn Study Liver Dis (secy-treas, 81-); fel Am Col Physicians; Asn Am Physicians. *Res:* Toxic effects of alcohol and its metabolites on protein fabrication and membrane repair; drug metabolism in liver disease. *Mailing Add:* Dept of Internal Med Univ Nebraska Med Ctr Omaha NE 68198-2000

SORRELLS, FRANK DOUGLAS, b Toccoa, Ga, May 14, 31; m 54; c 1. DEVELOPMENT & DESIGN, STRESS ANALYSIS. *Educ:* Univ Tenn, BSME, 57, MS, 68. *Prof Exp:* Exec vpres res & develop, Charles A Lee Assocs, 67-76; CONSULT, DEVELOP & DESIGN, PVT PRACT, 76- *Concurrent Pos:* Mgr technol transfer, Valmet Paper Mach, 88- *Mem:* Nat Soc Prof Engrs; Am Soc Mech Engrs. *Res:* Machinery and apparatus for various manufacturing industries; stress analysis; applied mathematics; computer codes; awarded 22 patents. *Mailing Add:* 5516 Timbercrest Trail Knoxville TN 37909

SORRELLS, GORDON GUTHREY, b Dallas, Tex, Mar 5, 34. SEISMOLOGY, GEOMECHANICS. *Educ:* Southern Methodist Univ, BS, 55, MS, 61, PhD(geophysics), 71. *Prof Exp:* Res geophysicist seismol, Teledyne Geotech, 67-70; sr res assoc seismic measurement, Southern Methodist Univ, 70-71; prog mgr & prin investr geothermal, Teledyne Geotech, 71-74; dir, Senturion Sci, 74-75; consult hydraul fracturing, 75-76, TECH DIR GEOTHERMAL & HYDROCARBON, TELEDYNE GEOTECH, 76- *Concurrent Pos:* Consult, Dowell Div, Dow Chem Co, 75-76 & Dept of Energy, 78. *Mem:* Am Geophys Union; Seismol Soc Am. *Res:* Development of seismic techniques to assess and control environmental risk of induced seismicity associated with geothermal and hydrocarbon production. *Mailing Add:* 2714 Country Club Pkwy Garland TX 75043

SORRELLS, MARK EARL, b Hillsboro, Ill, Mar 23, 50. PLANT BREEDING, GENETICS. *Educ:* Southern Ill Univ, BS, 73, MS, 75; Univ Wis, PhD (plant breeding), 77. *Prof Exp:* Fel, Dept Agron, Univ Wis, 77-78; ASST PROF PLANT BREEDING, CORNELL UNIV, 78- *Mem:* Am Soc Agron; Crop Sci Soc Am; Genetic Soc Can; Am Genetic Asn; Sigma Xi. *Res:* Plant genetics; plant physiology. *Mailing Add:* Dept Plant Breeding & Biomet Cornell Univ 252 Emerson Hall Ithaca NY 14850

SORRELS, JOHN DAVID, b Poteau, Okla, July 5, 27; m 51; c 2. PHYSICS, COMPUTER SCIENCE. *Educ:* Mass Inst Technol, BS, 50; Rice Univ, MA, 51; Calif Inst Technol, PhD(physics), 56; Univ Juarez, MSc, 77. *Prof Exp:* Engr, Ramo-Wooldridge Corp, 55-62; GROUP DIR SPACECRAFT SYSTS ENG, AEROSPACE CORP, LOS ANGELES, 62- *Mem:* AAAS; Am Phys Soc; Sigma Xi. *Res:* Cosmic rays; satellite orbit determination; control systems; data processing systems development; medical science. *Mailing Add:* 2738 Vista Mesa Dr Rancho Palos Verdes CA 90274

SORRENTINO, SANDY, JR, b Buffalo, NY, Dec 23, 43; m 65; c 3. NEUROENDOCRINOLOGY. *Educ:* Canisius Col, AB, 65; Univ Tenn, PhD(anat), 69; Univ Rochester, MD, 75. *Prof Exp:* asst prof anat, 71-76, ASST CLIN PROF, SCH MED, UNIV ROCHESTER, 76- *Mem:* Am Asn Anat. *Mailing Add:* 1570 Long Pd Rd Rochester NY 14626

SORROWS, HOWARD EARLE, b Hewitt, Tex, Aug 10, 18; m 43; c 5. PHYSICS. *Educ:* Baylor Univ, BA, 40; George Wash Univ, EE & MA, 47; Cath Univ Am, PhD(physics, math, civil eng), 58. *Prof Exp:* Pub sch teacher, Tex, 40-41; jr physicist, Nat Bur Standards, 41-43, radio physicist, 43-45, physicist, 45-50; electronics engr & proj officer, Bur Ord, US Dept Navy, 50-53, solid state & supvry physicist, Off Naval Res, 53-59; from dir tech int, long range planning & new prod dir to mgr space & environ sci servs, Tex Instruments, Inc, 59-65; mgr off opers & planning, Inst Mat Res, 65-67, dep dir, Inst Mat Res, 67-69, dir, Inst Appl Technol, 69-70, assoc dir progs, 70-78, dir off res & technol appln, Nat Bur Standards, 78-88, STAFF DIR, BD ASSESSMENT, NAT INST STANDARDS & TECHNOL, NAT RES COUN, 88- *Concurrent Pos:* Adv coun elec eng & sci dept, Univ Pa, 77-83; invited expert mgt res, Nigerian Workshop, 81; adj prof, American Univ, 83-87; consult, La Col, 79, bd Regents, 88- *Mem:* AAAS; Inst Elec & Electronic Engrs; Am Phys Soc. *Res:* Solid state and surface physics; photoconductivity; precise measurement of electromagnetic power; voltage, current, impedance, antenna gain and field intensity at frequencies up to and including microwaves; electronic countermeasures; research and development management; forecasting; national science and technical policy. *Mailing Add:* 8820 Maxwell Dr Potomac MD 20854

SORSCHER, ALAN J, b Flint, Mich, Sept 19, 34; m 57, 83; c 6. MEDICINE. *Educ:* Wayne State Univ, MD, 59. *Prof Exp:* Attend physician, Hurley Med Ctr, McLaren Hosp, 64-66 & 68-86; chief, Gastrointestinal Sect, USAF Hosp, Keesler AFB, Biloxi, 66-68; MED DIR, HOUSE STAFF & OCCUP HEALTH, HOLY CROSS HOSP, DETROIT, MICH, 86- *Concurrent Pos:* Res asst, Univ Chicago, 60-61; clin instr, Univ Ill, 62-64; from clin asst prof to clin assoc prof, Col Human Med, Mich State Univ, 72-86; consult, Mich Atty Gen, 84 & Mich Dept Social Serv, 85-86. *Res:* Hormonal risk factors in breast carcinogenesis. *Mailing Add:* 351 Donegal Dr Rochester MI 48309

SORTER, PETER F, b Vienna, Austria, Feb 8, 33; US citizen; m 65. ORGANIC CHEMISTRY, INFORMATION SCIENCE. *Educ:* Lafayette Col, BA, 54; DePauw Univ, MA, 56; Univ Iowa, PhD(chem), 62. *Prof Exp:* Info scientist, 62-65, mgr Sci Lit Dept, 66-79, dir res serv, 80-84, dir Info Ctr, 85-90, ASST VPRES, HOFFMAN-LA ROCHE, 91- *Concurrent Pos:* Mem bd dirs, Documentation Abstr, Inc, 70- *Mem:* Am Chem Soc; Chem Notation Asn (vpres, 72, pres, 73); Am Soc Info Sci; Drug Info Asn. *Res:* Storage retrieval of chemical and biological information, especially chemical structures. *Mailing Add:* Hoffmann-La Roche Inc 150 W End Ave New York NY 10023

SOSA, OMELIO, JR, b Camaguey, Cuba, Feb 2, 39; US citizen; m 61; c 2. ENTOMOLOGY, HOST PLANT RESISTANCE. *Educ:* Okla State Univ, BS, 64; Purdue Univ, MS, 71, PhD(entom), 77. *Prof Exp:* Agr res tech emtom, 65-73, entomologist, 73-76, RES ENTOMOLOGIST, AGR RES SERV, USDA, 76- *Concurrent Pos:* Assoc ed, Fla Entomologist, 84-; hispanic employment prog mgr, USDA, ARS, South Atlantic Area (Va, NC & SC, Ga, Fla, PR, VI), 88; adj asst prof, Univ Fla; first vpres, Fla Div, Am Soc Sugar Cane Technol, 89-90; mem, Nat Plant Genetics Resources Bd, 91-92. *Mem:* Entom Soc Am; Am Soc Sugarcane Technol; Am Registry Prof Entom; Am Soc Agron. *Res:* Reduction of crop losses in sugarcane by controlling or suppressing insect population. *Mailing Add:* US Sugarcane Field Sta Star Rte Box Eight Canal Point FL 33438

SOSEBEE, RONALD EUGENE, b Abilene, Tex, July 2, 42; m 64; c 2. PLANT PHYSIOLOGY, ECOLOGY. *Educ:* Abilene Christian Col, BS, 64; NMex State Univ, MS, 66; Utah State Univ, PhD(plant physiol), 70. *Prof Exp:* Instr range sci, Utah State Univ, 69; from asst prof to assoc prof, 69-79, PROF RANGE MGT, TEX TECH UNIV, 79-, ASSOC CHMN, DEPT RANGE & WILDLIFE MGT, 80- *Mem:* Am Soc Plant Physiol; Ecol Soc Am; fel Soc Range Mgt; Weed Sci Soc Am; Sigma Xi. *Res:* Photosynthesis of native plants; plant-soil water relationships; soil temperature and plant growth; carbohydrate relationships in plants; noxious plant control. *Mailing Add:* Dept Range & Wildlife Mgt Tex Tech Univ Lubbock TX 79409

SOSINSKY, BARRIE ALAN, b New York, NY, Aug 27, 52. INORGANIC CHEMISTRY, ORGANOMETALLIC CHEMISTRY. *Educ:* Univ Ill, Chicago Circle, BS, 71; Bristol Univ, PhD(chem), 75. *Prof Exp:* Fel, Cornell Univ, 74-76 & Univ Calif, Los Angeles, 76-78; ASST PROF CHEM, RICE UNIV, 78- *Mem:* Am Chem Soc. *Res:* Low valent transition metal catalysis; metal clusters; prebiotic chemistry; organometallic reaction mechanisms. *Mailing Add:* NO Two 45 Lexington St West Newton MA 02165-1029

SOSKA, GEARY VICTOR, b Sewickley, Pa, June 30, 48; m 69; c 2. ROBOTICS. *Educ:* Southwestern Univ, BS, 84; Univ Akron, BS, 90. *Prof Exp:* Nucler missile syst specialist, USAF, 68-72; field serv engr, Unimation Inc, 72-73; automation specialist, robotics, Ford Motor Co, 73-79; mfg res & develop engr, John Deere, 79-81; dir mkt & sales, Cybotech Corp, 81-86; ENG CONSULT, GOODYEAR TIRE & RUBBER CO, 86-; INSTR ROBOTICS, STARK TECH COL, 89- *Concurrent Pos:* Consult & pres, Midwest Technologies, 80-; lectr, Purdue Univ, 85-88; mem bd adv, Robotics Int, Soc Mfg Engrs, 89-90. *Res:* Robotics and manufacturing engineering education. *Mailing Add:* 7388 Ashburton Circle NW North Canton OH 44720

SOSLAU, GERALD, b New York, NY, Jan 22, 44; m 66; c 4. BIOCHEMISTRY, HEMATOLOGY. *Educ:* Queens Col NY, BA, 65; Univ Rochester, PhD(biochem), 70. *Prof Exp:* Fel biochem, Med Sch, Univ Pa, 70-71, res assoc, 71-75; from asst prof to assoc prof, 75-89, PROF BIOCHEM & NEOPLASTIC DIS, HAHNEMANN MED COL, 89- *Mem:* Sigma Xi; Am Soc Biol Chemists; Am Soc Cell Biol; Am Soc Hematol. *Res:* Platelets and their receptors in hemostasis and vascular interactions; molecular biology of herpes simplex virus glycoproteins and their functional roles. *Mailing Add:* Hahnemann Med Col Broad & Vine Philadelphia PA 19102-1192

SOSNOVSKY, GEORGE, b Petersburg, Russia, Dec 12, 20; US citizen; m 44. ORGANIC CHEMISTRY. *Educ:* Univ Munich, dipl, 44; Univ Innsbruck, PhD(chem), 48. *Prof Exp:* Res assoc chem, Univ Innsbruck, 48-49; tech officer, Commonwealth Sci & Indust Res Orgn, Australia, 49-51; in-chg org process develop, Cent Res Lab, Imp Chem Industs, Ltd, 51-56; fel & res assoc, Univ Chicago, 56-59; sr scientist, Res Inst, Ill Inst Technol, 59-63, assoc prof chem, 63-66; lectr, 66-67, PROF CHEM, UNIV WIS-MILWAUKEE, 67- *Concurrent Pos:* Res consult, Ill Inst Technol Res Inst, 63-66; USPHS spec sr res fel, Univ Col, London & Univ Tuebingen, 67-68; ed, Synthesis, 69-85; regional dir, Nat Found Cancer Res, 80-85. *Mem:* Am Chem Soc; Royal Soc Chem. *Res:* Free radical chemistry; organometallic and organometalloid peroxides; metal ion-catalyzed and photochemical reactions of peroxides; phosphorus intermediates of biological interest; synthesis and biological applications of new phosphorus compounds containing a spin label; novel synthetic methods; medicinal chemistry; structure-activity relationship of anticancer drugs; syntheses and biological evaluation of new anticancer drugs; contrast enhancing agents for NMR-Imaging; NMR-Imaging for diagnostic assessment of tumors. *Mailing Add:* Dept Chem Univ Wis-Milwaukee Milwaukee WI 53201

SOSNOWSKI, THOMAS PATRICK, b Scranton, Pa, Aug 11, 36; m 64; c 3. OPTICS, GENERAL PHYSICS. *Educ:* Pa State Univ, BS, 62; Case Western Reserve Univ, MS, 65, PhD(eng), 67. *Prof Exp:* Mem tech staff, Bell Tel Labs, 68-80; tech mgr, GTE Labs, 80-84; dir eng, Eikonix, 84-86; CONSULT, SOSNOWSKI ASSOC, 86- *Concurrent Pos:* Ford Found fel, 62. *Mem:* Sr mem Inst Elec & Electronic Engrs Eng Mgt Soc; Int Technol Inst. *Res:* Visual communication research; microprocessors; communication systems research; electronic imaging systems; lasers; electro-optics. *Mailing Add:* Sosnowski Associates 58 Sears Rd Wayland MA 01778

SOSSONG, NORMAN D, b Wash, Mar 27, 39. ELEMENTARY PARTICLE PHYSICS, INTERNAL MEDICINE. *Educ:* Walla Walla Col, BS, 61; Univ Wash, MS, 66, PhD(physics), 69; Univ Chicago, MD, 75. *Prof Exp:* asst prof, Walla Walla Col, 69-72, assoc prof physics, Pacific Union Col, 79-83; PRIVATE PRATICE INTERN MED, 79- *Mem:* Am Phys Soc; Am Med Asn. *Mailing Add:* PO Box 430 Deer Park CA 94576

SOSULSKI, FRANK WALTER, b Weyburn, Sask, Dec 2, 29; m; c 3. FOOD SCIENCE & TECHNOLOGY, AGRICULTURAL CHEMISTRY. *Educ:* Univ Sask, BSA, 54; Wash State Univ, MS, 56, PhD(agron), 59. *Prof Exp:* Asst prof field husb, 58-66, assoc prof crop sci, 66-71, PROF CROP SCI, UNIV SASK, 71-, ASSOC MEM FOOD SCI, 90- *Honors & Awards:* Bronze Medal, Polish Acad Sci. *Mem:* Am Asn Cereal Chemists; Agr Inst Can; Inst Food Technol; Am Oil Chemists Soc. *Res:* Cereal, oilseed, legume quality and utilization; food chemistry. *Mailing Add:* Dept Crop Sci Univ Saskatchewan Saskatoon SK S7N 0W0 Can

SOTERIADES, MICHAEL C(OSMAS), b Istanbul, Turkey, Mar 25, 23; US citizen; m 62. CIVIL ENGINEERING, SOIL MECHANICS. *Educ:* Nat Tech Univ Athens, Dipl Eng, 48, DrEng, 52; Mass Inst Technol, ScD(soil mech), 54. *Prof Exp:* Consult struct & found, A Woolf & Assoc, Mass, 52-53, assoc engr, 56-57; res asst soil dynamics, Mass Inst Technol, 53-54; consult found & struct, Greece, 54-55; head design & specifications, Greek Govt, 55-56; asst to pres eng, Doxiadis Assoc, 58-59, vpres & treas, Doxiadis Assoc, Inc, Washington, DC, 59-61; PROF STRUCT, CATH UNIV AM, 61- *Concurrent Pos:* Consult, Bldg Res Adv Bd, Nat Acad Sci, 62. *Res:* Aseismic analysis and design; systems analysis; computer methods in structural analysis. *Mailing Add:* Dept Civil Eng Cath Univ Am 620 Mich Ave NE Washington DC 20064

SOTIRCHOS, STRATIS V, b Mytilene, Greece, Feb 8, 56; m 83; c 2. REACTION ENGINEERING, APPLIED NUMERICAL ANALYSIS. *Educ:* Nat Technol Univ Athens, dipl, 79; Univ Houston, PhD(chem eng), 82. *Prof Exp:* asst prof, 82-87, ASSOC PROF CHEM ENG, UNIV ROCHESTER, 87- *Mem:* Am Chem Soc; Am Inst Chem Engrs. *Res:* Applied mathematics and numerical analysis in chemical engineering; reaction engineering; combustion and gasification processes; transport and reaction in multiphase systems. *Mailing Add:* Dept Chem Eng Univ Rochester Rochester NY 14627

SOTO, AIDA R, b Havana, Cuba, Dec 3, 31; US citizen. ORGANIC CHEMISTRY, BIOCHEMISTRY. *Educ:* Univ Havana, BS, 53 & 55; Univ Miami, MS, 62, PhD(chem), 66. *Prof Exp:* Res chemist, Villanueva Univ, 55-58, asst prof, 58-61; fel, Dept Pharmacol, Univ Miami, 65-68, instr med, Sch Med, 68-69; supvr, Chem Res & Develop Dept, 69-72, group leader, Biol Res & Develop Dept, 72-74, group leader immunochem res & develop, 74-84, SECT HEAD, DADE DIV, BAXTER INT, 84- *Mem:* AAAS; Am Chem Soc; NY Acad Sci; Am Asn Clin Chemists. *Res:* Base promoted reactions of sulfonate esters in dipolar aprotic solvents; purification and characterization of proteolytic enzymes; clinical enzymology; radioimmunoassays; the use of immunologic techniques in clinical chemistry. *Mailing Add:* 3150 NW 19th Terr Miami FL 33125

SOTO, GERARDO H, b Havana, Cuba, Nov 23, 22; US citizen; div; c 1. AGRONOMY, SOIL CLASSIFICATION. *Educ:* Univ Havana, Agr Eng, 48. *Prof Exp:* Chief, Soils Dept, Agr & Indust Develop Bank, Cuba, 56-60; chief, Soils Dept, Agr Exp Sta, Cuba, 60-62, chief, Soils Dept, Inst Hydraulic Resources, Cuba, 62-63; sr soil scientist, Org Am States, 64-70, dir, Div III Regional Develop, 70-75; AGRICULTURIST, INT BANK RECONSTRUCT & DEVELOP, 75- *Concurrent Pos:* Tech ed, United Eng Ctr, New York, 63-64. *Mem:* Am Soc Agron; Int Soil Sci Soc; Soil Conserv Soc Am. *Res:* Aspects of rural development based on rational utilization of physical and human resources. *Mailing Add:* 7005 Barkwater Ctr Bethesda MD 20817

SOTOMAYOR, RENE EDUARDO, b Santiago, Chile, Jan 3, 37; m 66; c 1. GENETICS, MUTAGENESIS. *Educ:* Univ Chile, BHu, 56, PhL, 66. *Prof Exp:* Asst prof biol, Sch Vet Med, Univ Chile, 66-74, assoc prof, Sch Med & Sch Nursing, 67-71; consult, 74-75, RES ASSOC MUTAGENESIS, OAKRIDGE NAT LAB, 75- *Concurrent Pos:* Fel, Pan Am Health Orgn/WHO, 72-74. *Mem:* Environ Mutagen Soc; Genetics Soc Am; Int Genetics Fedn; Latin Am Soc Genetics. *Res:* Mechanisms of chemical mutagenesis in mammalian systems; DNA repair in the germ cells; relationships between repair and genetic damage; cytogenetics of induced chromosome abnormalities in the germ cells of mammals. *Mailing Add:* 3700 Sutherland Ave No J4 Knoxville TN 37919

SOTOS, JUAN FERNANDEZ, b Tarazona, Spain, May 18, 27; US citizen; c 6. PEDIATRICS. *Educ:* Univ Valencia, MD, 51. *Prof Exp:* Intern, Univ Valencia Hosp, 52-53; resident pediat path, St Christopher's Hosp, Philadelphia, Pa, 53-54, resident pediat, 54-55; resident pediat, Children's Hosp, Columbus, Ohio, 55-56, instr & chief resident, 56-57; instr, Mass Gen Hosp, 60-62; from asst prof to assoc prof, 62-67, PROF PEDIAT, COL MED, OHIO STATE UNIV, 67-, DIR DIV ENDOCRINOL & METAB, DEPT PEDIAT, 63- *Concurrent Pos:* Fel, Mass Gen Hosp, 57-60; dir, Clin Res Ctr, Children's Hosp, Columbus, 62-72. *Mem:* Endocrine Soc; Am Pediat Soc; Am Diabetes Asn; Soc Pediat Soc; Pediat Endocrine Soc. *Res:* Metabolic and endocrine disorders of children. *Mailing Add:* Children's Hosp 700 Childrens Dr Columbus OH 43205

SOTTERY, THEODORE WALTER, b Lebanon, Pa, Feb 8, 27; m 49; c 4. GENERAL CHEMISTRY TEACHING. *Educ:* Dartmouth Col, BNS, 46; Clark Univ, cert chem, 49; Univ Maine, MS, 56, PhD(chem), 66. *Prof Exp:* Res vol endocrine res, Harvard Med Sch, 46-47; sci storekeeper, Mass-Ft Devens, 47; lab asst chem, Columbia Univ, 49-50; qual control group supvr pigments, E I du Pont de Nemours & Co, 50-51; instr sci, Finch Jr Col, 51-52; asst chem, Univ Maine, Orono, 54-55, instr, 56-61; from asst prof to assoc prof, Univ S Maine, Portland, 61-73, prof chem, 73-; RETIRED. *Concurrent Pos:* Consult, Howell Labs, Bridgton, Maine, Fairchild Camera & Instrument, South Portland, Maine, 66-; chem reviewer, several publ co; bk reviewer, J Chem Educ, 75- *Res:* Failure of the Darzans reaction to produce alpha-phenyl-substituted glycidic esters; reaction of sodium metal with dimethyl formamide. *Mailing Add:* 24 Chamberlain Ave Portland ME 04101

SOUBY, ARMAND MAX, b Murfreesboro, Tenn, Jan 12, 17; m 47; c 4. HYDROGENATION & HYDROGENOLYSIS, CATALYTIC REACTIONS. *Educ:* Vanderbilt Univ, BS, 38. *Prof Exp:* res assoc, Exxon Res & Eng Co, 39-71; proj mgr, Univ NDak, 72-78, adj prof chem eng, 75-81, prin investr, 79-81; CONSULT, 81- *Concurrent Pos:* Consult synthetic fuels, 82- *Mem:* Fel AAAS; fel Am Inst Chemists; Am Inst Chem Engrs; Am Chem Soc; Sigma Xi; Nat Soc Prof Engrs. *Res:* Conversion of coal to premium solid, liquid or gaseous fuels. *Mailing Add:* 103 Nichols San Marcos TX 78666

SOUCIE, WILLIAM GEORGE, b Missoula, Mont, Mar 20, 42; m 66; c 4. FOOD CHEMISTRY, BIOTECHNOLOGY. *Educ:* Carroll Col, BA, 64; Incarnate Word Col, MS, 68; NC State Univ, PhD(biochem), 73. *Prof Exp:* Res assoc, Chem Dept, Univ Colo, 73-76; res scientist, Protein Prods Lab, Kraft Res & Develop, 76-77, group leader, 77-83, group leader, Basic Food Sci Lab, 83-85, sr group leader, 85-87, mgr, biotechnol dept, 87-88; ASSOC DIR, BIOTECHNOLOGY, KRAFT GEN FOODS, 88- *Concurrent Pos:* Instr, chem dept, Col Lake County, 81-83. *Mem:* Am Chem Soc; Inst Food Technol; Am Oil Chemists Soc. *Res:* Investigations into the electrical, physical and chemical properties of proteins as a basis for the use of proteins in human foods; colloid chemistry of food constituents; emulsion science; biotechnology applications to fats & oils, cheese and crops. *Mailing Add:* Kraft Gen Foods 1801 Maple Ave Evanston IL 60201

SOUDACK, AVRUM CHAIM, b July 5, 34; Can citizen; m 78; c 4. ELECTRICAL ENGINEERING. *Educ:* Univ Man, BScEE, 57; Stanford Univ, MS, 59, PhD(elec eng), 61. *Prof Exp:* Asst prof elec eng, Univ BC, 61-65; vis asst prof, Univ Calif, Berkeley, 65-66; assoc prof, 66-71, PROF ELEC ENG, UNIV BC, 71- *Concurrent Pos:* Vis assoc prof, Israel Inst Technol, 69-70; vis prof, Weizmann Inst Sci, Israel, 74-75. *Honors & Awards:* Marv Emerson Award, Soc Comput Simulation, 72. *Mem:* Simulation Coun; Inst Elec & Electronic Engrs. *Res:* Approximate solution of nonlinear differential equations; analog and hybrid simulation of nonlinear systems; stability of harvested predator-prey systems; analytical solutions of ecological models; choas in nonlinear dynamic systems. *Mailing Add:* Dept Elec Eng Univ BC Vancouver BC V6T 1W5 Can

SOUDEK, DUSHAN EDWARD, b Prague, Czech, May 4, 20; m 47; c 3. MEDICAL CYTOGENETICS. *Educ:* Univ Brno, Czech, MD, 49, CScbiol, 56. *Prof Exp:* Asst prof biol, Univ Brno, Czech, 53-62, privatdocent, 64, head dept genetics, 63-68; vis scientist dept anat, Univ Western Ont, 68-69; prof, 69-87, EMER PROF, DEPT PSYCHIAT & PEDIAT, QUEEN'S UNIV, 87- *Concurrent Pos:* Res assoc, Ont Ment Health Found, 69-85. *Honors & Awards:* S Moravian Province Prize, 62; Mendel Medal, Mendel's Mus, Brno, 65. *Mem:* Am Soc Human Genetics. *Res:* Human cytogenetics; mental defects, normal variants, chromosomal evolution; structure of chromosomes and cell nucleus. *Mailing Add:* 371 Elmwood St Kingston ON K7M 2Z2 Can

SOUDER, PAUL A, b New Jersey, May 26,44. ELEMENTARY PARTICLE PHYSICS, NUCLEAR PHYSICS. *Educ:* Wheaton Col, BS, 66; Princeton Univ, PhD(physics), 71. *Prof Exp:* Res assoc physics, Princeton Univ, 71-72; res staff physicist, Yale Univ, 72-75, from instr to assoc prof physics, 75-82; PROF PHYSICS, SYRACUSE UNIV, 88- *Concurrent Pos:* Vis assoc prof physics, Harvard Univ, 82-83. *Mem:* Am Phys Soc. *Mailing Add:* Dept Physics Syracuse Univ 201 Physics Bldg Syracuse NY 13210

SOUDER, WALLACE WILLIAM, b Columbus, Kans, June 12, 37; m 65; c 3. NUCLEAR PHYSICS, GEOPHYSICS. *Educ:* Kans State Col, BS, 60; Iowa State Univ, PhD(physics), 69. *Prof Exp:* SR RES PHYSICIST, RES CTR, PHILLIPS PETROL CO, 69- *Mem:* Am Phys Soc; Soc Explor Geophys; Soc Prof Well Log Analysts. *Res:* New techniques for mineral exploration. *Mailing Add:* 601 Oakridge Dr Bartlesville OK 74006

SOUHRADA, FRANK, b Sluknov, Czech, Sept 22, 37; m 62; c 2. CHEMICAL ENGINEERING. *Educ:* Inst Chem Technol, Prague, Dipl Ing, 61; Czech Acad Sci, PhD(chem eng), 64. *Prof Exp:* Res scientist, Inst Chem Process Fundamentals, Czech Acad Sci, 64-68; Nat Res Coun Can fel dynamic simulation, Univ NB, Fredericton, 68-70; res assoc fluidization, McMaster Univ, 70-71; process engr, Int Nickel Co, 71-74; sr res engr process develop, Gulf Can Res & Develop, 74-86; PRES, PROCESS ENGINEERING INC, 86- *Mem:* Can Soc Chem Engrs; Chem Inst Can. *Res:* Mass transfer; liquid-liquid extraction; ion exchange; fluidization; dynamic simulation; optimization; alternate energy sources; heavy oil and tar sands; heat and mass transfer; optimization; feasibility studies; new technology; flare systems. *Mailing Add:* Seven Totteridge Rd Islington ON M0A 1Y9 Can

SOUKUP, RODNEY JOSEPH, b Faribault, Minn, Mar 9, 39; m 65; c 3. ELECTRICAL ENGINEERING, PHYSICS. *Educ:* Univ Minn, Minneapolis, BS, 61, MSEE, 64, PhD(elec eng), 69. *Prof Exp:* Prin develop engr, Univac, Sperry Rand Corp, 69-71; instr elec eng, Univ Minn, 72; asst prof elec eng, Univ Iowa, 72-76; assoc prof, 76-80, PROF ELEC ENG, UNIV NEBR, 80-, CHMN DEPT, 78- *Concurrent Pos:* Rockwell Int; chmn Nat Elec Eng Dept Heads Asn, 90-91. *Mem:* Am Vacuum Soc; Inst Elec & Electronic Engrs; Am Soc Eng Educ. *Res:* Solar cells and scanning electron microscopy; physical electronics; thin film devices with a study of materials used and methods of fabrication. *Mailing Add:* Dept Elec Eng 212 N WSEC Univ Nebr Lincoln NE 68588-0511

SOULE, DAVID ELLIOT, b Norwalk, Conn, Feb 24, 25; m 49; c 3. PHYSICS. *Educ:* DePauw Univ, AB, 49; Northwestern Univ, MS, 51, PhD(physics), 54. *Prof Exp:* Res physicist, Union Carbide Lab, 54-66 & Douglas Aircraft Advan Res Lab, Calif, 66-71; PROF PHYSICS, WESTERN ILL UNIV, 71- *Concurrent Pos:* NSF fel, Royal Soc Mond Lab, Univ Cambridge, 61-62; sr vis Dept Sci & Indust Res fel, Dept Physics, Univ Sussex, 65. *Mem:* Am Phys Soc. *Res:* Solid state physics; transport properties; photoconductivity; susceptibility; low-temperature electronic properties such as deHaas-vanAlphen effect. *Mailing Add:* Dept Physics Western Ill Univ Adams St Macomb IL 61455

SOULE, DOROTHY (FISHER), b Lakewood, Ohio, Oct 8, 23; m 43; c 2. MARINE BIOLOGY. *Educ:* Miami Univ, BA, 45; Occidental Col, MA, 63; Claremont Grad Sch, PhD, 69. *Prof Exp:* Res assoc biochem, Allan Hancock Found, Univ Southern Calif, 45-47; cur & instr comp anat, biol & microbiol, Occidental Col, 61-63; asst prof embryol, invert biol & zool, Calif State Col, Los Angeles, 63-65; res assoc marine biol, Univ Southern Calif, 67-71, marine biol & pollution, 71-76, sr res scientist, 76-86, dir, Harbors Res Lab, 76-84, sr res scientist, 76-86, RES PROF BIOL SCI, HANCOCK INST MARINE SCI, UNIV SOUTHERN CALIF, 86-, DIR, HARBORS ENVIRON PROJS, 71-, CUR BRYOZOA, 64- *Concurrent Pos:* Independent consult, 61-; coordr, Environ Qual Projs, Univ Southern Calif-Sea Grant, 72-78, assoc dir, 74-76; mem, Eng Panel, Nat Acad Sci, 73-75; adj prof environ eng, Univ Southern Calif, 74-; mem, marine fish adv comt, Dept Com, 76-79, sci adv bd, Environ Protection Agency, 78-81, consult, 81-; vpres, SOS Environ, Inc, 78-; res assoc, Los Angeles County Mus Natural Hist, 85-, Santa Barbara Mus of Natural Hist, 87-; lectr physiol, Calif Col Med, 65-66. *Mem:* AAAS; Int Bryozool Asn; Am Soc Zool; Western Soc Naturalists; Sigma Xi; Pac Sci Asn. *Res:* Ecology and pollution in urban harbors, beaches and estuaries; effluent and ocean dumping pollution; environmental impact assessment and coordination; coral reef ecology, systematics and ecology of tropical, temperate Bryozoa; bryozoan development. *Mailing Add:* Allan Hancock Found Univ Southern Calif Los Angeles CA 90089-0371

SOULE, JAMES, b Bradford, Pa, Jan 3, 20; m 42. HORTICULTURE. *Educ:* Cornell Univ, BS, 41; Univ Miami, MS, 51; Univ Fla, PhD(hort), 54. *Prof Exp:* Chemist, Emulsion Res Lab, Eastman Kodak Co, 43-46; res asst, Univ Miami, 51; agent, Agr Mkt Serv, USDA, 54-56; from assoc prof to prof, 56-85, EMER PROF HORT SCI, UNIV FLA, 85- *Concurrent Pos:* Assoc ed, Am Soc Hort Sci, 73-88. *Mem:* AAAS; Am Chem Soc; fel Am Soc Hort Sci; Int Soc Hort Sci. *Res:* Tropical horticulture; rootstock-scion relationships; taxonomy; postharvest handling and propagation of horticultural crops. *Mailing Add:* 2141 Fifield Hall Univ Fla Gainesville FL 32611

SOULE, JOHN DUTCHER, b Moline, Ill, Oct 11, 20; m 43; c 2. ZOOLOGY. *Educ:* Miami Univ, AB, 42; Univ Southern Calif, MS, 48, PhD(zool), 52. *Prof Exp:* Asst zool, 47, from instr to assoc prof, 50-63, chmn dept histol, 53-77, PROF HISTOL & PATH, SCH DENT, UNIV SOUTHERN CALIF, 63-, PROF BIOL, 70- *Concurrent Pos:* Res assoc, Am Mus Natural Hist, 61-, Los Angeles County Mus Natural Hist, 85-, Santa Barbara Mus Nat Hist, 87-; Hancock Found res scholar, Univ Southern Calif, 52-, asst dean, Sch Dent, 78. *Mem:* Fel AAAS; Am Micros Soc; Am Soc Zool; Int Asn Dent Res; Am Inst Biol Sci; Sigma Xi. *Res:* Taxonomy, histogenesis, postlarval development, histology, anatomy, ecology and reef communities of Bryozoa; histology and histochemistry of tooth development in fish, amphibia and reptiles. *Mailing Add:* 2361 Hill Dr Los Angeles CA 90041

SOULE, OSCAR HOMMEL, b St Louis, Mo, Oct 6, 40; m 71; c 2. ECOLOGY, BIOLOGY. *Educ:* Colo Col, BA, 62; Univ Ariz, MS, 64, PhD(ecol), 69. *Prof Exp:* Acad dean, 72-73, MEM FAC ECOL, EVERGREEN STATE COL, 71- *Concurrent Pos:* Ford Found fel, Mo Bot Garden, 70-71; sr ecologist, HDR Ecosci, 77-80; partner, Lidman & Soule, Consults, 78-; vis prof, Colo Col, 79. *Mem:* AAAS; Ecol Soc Am; Brit Ecol Soc; Sigma Xi. *Res:* Terrestrial aspects of applied environmental studies; special interests in desert biology, urban ecology and environmental education. *Mailing Add:* Evergreen State Col Olympia WA 98505

SOULE, ROGER GILBERT, b Northport, NY, Feb 21, 35; m 59; c 3. EXERCISE PHYSIOLOGY. *Educ:* State Univ NY Col Cortland, BS, 57; Univ Ill, MS, 58; Wash State Univ, PhD(exercise physiol), 67. *Prof Exp:* Instr phys educ & health, Dutchess Community Col, 60-64; instr phys educ, Wash State Univ, 64-67; from asst prof to assoc prof exercise physiol, Sargent Col, Boston Univ, 71-76; prof, Liberty Baptist Col, Va, 76-79; PROF, BIOLA UNIV, LA MIRADA, 79- *Mem:* AAAS; Am Physiol Soc; Am Col Sports Med. *Res:* Energy cost of exercise; physical fitness levels of various populations; metabolic substrate utilization during exercise; control of temperature under exercise and environmental stress. *Mailing Add:* 14203 Figueras Rd La Mirada CA 90638

SOULE, SAMUEL DAVID, obstetrics & gynecology; deceased, see previous edition for last biography

SOULEN, JOHN RICHARD, b Milwaukee, Wis, June 19, 27; m 55; c 3. PHYSICAL CHEMISTRY. *Educ:* Carroll Col, Wis, BA, 50; Univ Wis, PhD(phys chem), 55. *Prof Exp:* Asst chem, Univ Calif, 50-52 & Univ Wis, 52-54; res chemist, Penwalt Chems Corp, 55-59, proj leader, inorg res dept, 59-63, group leader, contract res dept, 63-68, dir contract res, 68-73, assoc mgr res & develop, 73-82; PRES, TECH & MGT SERV, INC, 83- *Concurrent Pos:* Lectr, Univ Pa, 60-61. *Mem:* AAAS; Am Chem Soc. *Res:* Inorganic, high temperature and ultrahigh pressure chemistry; thermodynamics; spectroscopy; kinetics. *Mailing Add:* 5333 Hickory Bend Bloomfield Hills MI 48304

SOULEN, RENATE LEROI, b Berlin, Ger, June 10, 33; US citizen; m 55; c 3. MEDICINE, RADIOLOGY. *Educ:* NY Univ, BA, 53; Med Col Pa, MD, 57; Am Bd Radiol, dipl, 63. *Prof Exp:* Intern, Albert Einstein Med Ctr, 57-58; from resident to instr, Hosp, Jefferson Med Col, 59-63; from instr to prof radiol, Health Sci Ctr, Med, Sch, Temple Univ, 63-85; PROF RADIOL, JOHNS HOPKINS UNIV, 85- *Concurrent Pos:* Nat Cancer Inst fel, Hosp, Jefferson Med Col, 61-62; mem coun cardiovasc radio, Am Heart Asn. *Mem:* Fel Am Heart Asn; fel Am Col Radiol; Asn Univ Radiol; Radiol Soc NAm; Am Inst Ultrasonics in Med; Soc Magnetic Resonance Med. *Res:* Cardiovascular system. *Mailing Add:* 2833 Montclair Dr Ellicott City MD 21043

SOULEN, ROBERT J, JR, b Phoenixville, Pa, July 16, 40; m 63; c 2. CRYOGENIC PHYSICS. *Educ:* Rutgers Univ, BA, 62, PhD(physics), 66. *Prof Exp:* proj leader cryogenic physics, Nat Bur Standards, 67-86; HEAD SUPERCONDUCTING MAT SECT, NAVAL RES LABS, 86- *Mem:* Fel Am Phys Soc. *Res:* Very low temperature techniques, and low temperature thermometry; superconductivity. *Mailing Add:* Naval Res Lab Code 6344 Washington DC 20375-5000

SOULEN, ROBERT LEWIS, b Chicago, Ill, Jan 19, 32; m 54; c 3. ORGANIC CHEMISTRY. *Educ:* Baker Univ, AB, 54; Kans State Univ, PhD(org chem), 60. *Prof Exp:* From res chemist to sr res chemist, Austin Res Labs, Jefferson Chem Co Inc, Tex, 60-64; LILLIAN NELSON PRATT PROF CHEM & CHMN DEPT, SOUTHWESTERN UNIV, TEX, 64- *Concurrent Pos:* NSF grant, 66-67; Robert A Welch grant, 66-88; consult, Texas Res Inst; chmn local sect Activ Comt, Am Chem Soc, 85-87, chmn Nominations & Elections Comt, 90-, bd trustees, Mem Ins Prog, 90- *Mem:* Sigma Xi; fel AAAS; Am Chem Soc. *Res:* Exploratory and applications research in rigid and flexible polyurethane foams; polyolefine polymerization; vinyl halogen displacement reactions; synthesis of organofluorine derivatives; fluoropolymers. *Mailing Add:* Dept Chem Southwestern Univ Georgetown TX 78626

SOULEN, THOMAS KAY, b Waukesha, Wis, Apr 7, 35; m 58; c 3. BIOCHEMISTRY. *Educ:* Univ Wis, BA, 57, MS, 61, PhD(biochem), 63. *Prof Exp:* Asst prof bot, Univ Wis, 63-64; asst prof, Univ Minn, Minneapolis, 64-69, ASSOC PROF BOT, UNIV MINN, ST PAUL, 69- *Mem:* AAAS; Am Chem Soc; Am Soc Plant Physiologists; Sigma Xi. *Res:* Nitrogen metabolism of higher plants, especially with reference to development; growth and flowering of Lemnaceae. *Mailing Add:* Dept of Bot Univ of Minn St Paul MN 55108

SOULES, JACK ARBUTHNOTT, b Ashtabula, Ohio, Jan 26, 28; m 49, 70; c 3. PHYSICS. *Educ:* Ohio State Univ, BS, 48, MSc, 50, PhD(physics), 54. *Prof Exp:* Res assoc & asst instr, Ohio State Univ, 54-55; from asst prof to prof physics, NMex State Univ, 55-68; dean col arts & sci, 68-81, PROF PHYSICS, CLEVELAND STATE UNIV, 81- *Concurrent Pos:* Am Coun Educ fel acad admin, 65-66. *Mem:* Am Phys Soc. *Res:* Solid state physics; x-rays; biophysics; laser physics. *Mailing Add:* Physics Dept Cleveland State Univ Cleveland OH 44115

SOULSBY, MICHAEL EDWARD, b Montgomery, WVa, Sept 4, 41; m 60; c 3. MEDICAL PHYSIOLOGY, BIOPHYSICS. *Educ:* WVa Univ, AB, 63, MS, 68, PhD(biophysics), 71. *Prof Exp:* USPHS fel, Appalachian Lab Occup Respiratory Dis, 71-72; from instr to asst prof physiol, Va Commonwealth Univ, 72-76; asst prof physiol & biophys, 77-83, ASSOC PROF PHYSIOL, BIOPHYS & TOXICOL, UNIV ARK MED SCI, 84- *Concurrent Pos:* Consult, Gen Med Corp, 76-; prin investr, Heart, Lung & Blood Inst, NIH, 76-78 & Ark Br, Am Heart Asn, 78- *Mem:* Am Heart Asn; Am Physiol Soc; Biophys Soc; Sigma Xi. *Res:* Myocardial and vascular smooth muscle physiological and biophysical properties during ischemia, cardiovascular shock and hypertension. *Mailing Add:* Dept of Physiol & Biophysics 4301 W Markham Little Rock AR 72205

SOUNG, WEN Y, b Tainan, Taiwan, Feb 14, 45; m 75; c 3. SOLID-LIQUID SEPARATION, FLUIDIZATION. *Educ:* Nat Cheng-Kung Univ Taiwan, BS, 67, MS, 69; WVa Univ, PhD(chem eng), 73. *Prof Exp:* Sr process engr, Catalytic Inc, Philadelphia, Pa, 73-75; chem engr, Hydrocarbon Res Inc, Lawrenceville, NJ, 75-79; sr staff engr, Exxon Res & Eng Co, Bayton, Tex, 79-86; FEL ENG, WESTINGHOUSE ELEC CORP, 86- *Mem:* Am Inst Chem Eng; Asn Am Chinese Prof. *Res:* Process development and improvement for catalytical coal gasification and liquefaction processes, involving catalyst recovery from spent char, solid-liquid separation, ash utilization and new catalyst development for coal liquefaction. *Mailing Add:* 426 Oaklawn Dr Pittsburgh PA 15241

SOURES, JOHN MICHAEL, b Galati, Roumania, Jan 2, 43; US citizen; m 69; c 5. LASER-MATTER INTERACTION, ULTRA-HIGH-POWER LASER DEVELOPMENT. *Educ:* Univ Rochester, BS, 65, MS, 67, PhD(mech & aerospace sci), 70. *Prof Exp:* Res assoc, Univ Rochester, 70-72, group leader, Glass Laser Develop, 75-78, group leader, Laser Fusion Exp, 78-81, SR SCIENTIST, LAB LASER ENERGETICS, UNIV ROCHESTER, 72-, DIV DIR, EXP DIV, 79- *Concurrent Pos:* Consult, Oak Ridge Nat Lab, 70-71, Link Found, 83-84, Eastman Kodak, 84-85 & Lawrence Livermore Nat Lab, 88-90; dep dir, Lab Laser Energetics, Univ Rochester, 83- *Mem:* Am Phys Soc; Optical Soc Am; AAAS; NY Acad Sci. *Res:* Laser-driven inertial fusion, including plasma physics of high-temperature, high-density fusion plasmas; development of plasma diagnostic techniques; high-power solid-state laser development; laser-beam smoothing systems; non-linear optics. *Mailing Add:* Lab Laser Energetics 250 E River Rd Rochester NY 14623

SOURKES, THEODORE LIONEL, b Montreal, Que, Feb 21, 19; m 43; c 2. BIOCHEMISTRY. *Educ:* McGill Univ, BSc, 39, MSc, 46; Cornell Univ, PhD(biochem), 48. *Prof Exp:* Chemist, Gen Eng Co, Ont, 42-44; biochemist, Frank W Horner, Ltd, Que, 44-45; asst biochem, Cornell Univ, 46-48; asst prof pharmacol, Med Sch, Georgetown Univ, 48-50; sr res assoc, Merck Inst Therapeut Res, NJ, 50-53; sr res biochemist, 53-65, DIR LAB CHEM NEUROBIOL, ALLAN MEM INST PSYCHIAT & PROF PSYCHIAT, FAC MED, MCGILL UNIV, 65-, PROF BIOCHEM, 70- & PROF PHARMACOL, 90- *Concurrent Pos:* From instr to assoc prof psychiat, Fac Med, McGill Univ, 54-65; assoc scientist, Royal Victoria Hosp, 70 & assoc dean med, 72-75; sr fel Award, Parkinson's Dis Found, NY, 63-66; Raven Press lectr, Int Soc Neurochem, Singapore & India, 86. *Honors & Awards:* Heinz-Lehmann Award, Can Col Neuropsychopharmacol, 82, Medal, 90; Jasper Pub Lecture, Can Assoc for Neurosci, 87, Order Andres Bello, Govt Venezuela, 87. *Mem:* Am Soc Neurochem; Am Soc Biol Chem; Am Soc Pharmacol & Exp Therapeut; Can Biochem Soc; Royal Soc Can; Int Soc Neurochem; Int Brain Res Orgn; Soc Neurosci. *Res:* Catecholamines and other biogenic amines; amino acid decarboxylases; amine oxidases; biochemistry of extrapyramidal syndromes; central pathways of response to stress; copper metabolism; biochemistry of mental diseases; history of biochemistry; imaging of brain serotonin. *Mailing Add:* Dept Psychiat McGill Univ 1033 Pine Ave W Montreal PQ H3A 1A1 Can

SOURS, RICHARD EUGENE, b Baltimore, Md, Sept 5, 41; m 64; c 1. MATHEMATICAL ANALYSIS. *Educ:* Towson State Teachers Col, BS, 63; Mich State Univ, MS, 65; Univ Va, PhD(math), 71. *Prof Exp:* Instr math, 65-68 & 71-77, ASSOC PROF MATH & COMPUT SCI, WILKES COL, 77- *Mem:* Am Math Soc; Math Asn Am; Sigma Xi. *Res:* Some aspects of integral operators on hilbert spaces. *Mailing Add:* Dept Math Wilkes Col Wilkes-Barre PA 18766

SOUSA, LYNN ROBERT, b Oakland, Calif, Apr 14, 43; m 64; c 1. ORGANIC CHEMISTRY, PHOTOCHEMISTRY. *Educ:* Univ Calif, Davis, BSc, 66; Univ Wis, PhD(org chem), 71. *Prof Exp:* Fel org chem, Univ Calif, Los Angeles, 71-73; asst prof, Mich State Univ, 73-78; from asst prof to assoc prof, 78-85, PROF CHEM, BALL STATE UNIV, 85- *Mem:* Am Chem Soc. *Res:* Photochemical reaction mechanisms; development of photochemical techniques for organic synthesis and applications of complexation by crown ethers in organic chemistry. *Mailing Add:* Dept of Chem Ball State Univ Muncie IN 47306

SOUTAS-LITTLE, ROBERT WILLIAM, b Oklahoma City, Okla, Feb 25, 33; m 54, 75; c 5. BIODYNAMICS, TISSUE BIOMECHANICS. *Educ:* Duke Univ, BS, 55; Univ Wis, MS, 59, PhD(mech), 62. *Prof Exp:* assoc dir applied math, 65-67, chmn, dept mech eng, 72-77, chmn, dept biomech, 77-90, from asst prof to assoc prof, 65-70, PROF, DEPT BIOMECH & METALL, MECH & MAT SCI, 70-, DIR, BIOMECH EVAL LAB, MICH STATE UNIV, 89- *Concurrent Pos:* Consult, Lawrence Livermore Lab, Univ Calif, 77-78 & Biomech Interface, 84-; legal expert witness, 76- *Mem:* Sigma Xi; Am Soc Biomechanics; Soc Eng Sci; Am Soc Mech Eng; Am Col Sports Med. *Res:* Tissue mechanics; body dynamics; orthopedics; sports medicine and rehabilitation. *Mailing Add:* Dept Biomech Mich State Univ A439 E Fee Hal East Lansing MI 48824-1316

SOUTH, FRANK E, b Norfolk, Nebr, Sept 20, 24; m 46; c 2. PHYSIOLOGY. *Educ:* Univ Calif, AB, 49, PhD(physiol), 52. *Prof Exp:* Jr res physiologist, Univ Calif, 52-53; asst prof physiol, Univ PR, 53-54, Col Med, Univ Ill, 54-61; from asst prof to prof, Colo State Univ, 61-65; prof physiol & investr, Dalton Res Ctr, Univ Mo-Columbia, 65-77; dir, 77-82, PROF, SCH LIFE & HEALTH SCI, UNIV DEL, 82- *Concurrent Pos:* NIH sr res fel, 61-65; co-dir, Hibernation Info Exchange, Off Naval Res; mem bd dirs, Int Hibernation Soc, 65- *Mem:* Fel AAAS; Am Physiol Soc; Am Soc Zool; Soc Cryobiol; Soc Gen Physiol; Int Hibernation Soc. *Res:* Environmental physiology; hibernation, hypothermia, acclimatization and adaptations to extreme environments; neurophysiology, thermoregulation and physiology of marine mammals; history of physiology and medicine. *Mailing Add:* Sch Life & Health Sci Univ Del Newark DE 19711

SOUTH, GRAHAM ROBIN, b Thorpe, Eng, Oct 27, 40; m 66; c 2. PHYCOLOGY. *Educ:* Liverpool Univ, BS, 63, PhD(marine algal ecol), 66, DSc, 90. *Prof Exp:* NATO fel phycol, Univ BC, 66-67; from asst prof to assoc prof biol, Mem Univ Nfld, 67-76, assoc cur herbarium, 67-71, head dept biol, 76-84, cur herbarium, 71-, prof biol, 76-; dir, Huntsman Marine Sci Ctr, St Andrews, NB, Can, 85-90; DIR & PROF MARINE STUDIES, INST OF MARINE RESOURCES, UNIV S PAC, 90- *Concurrent Pos:* Res fel, Edward Percival Marine Lab, Univ Canterbury, NZ, 73-74; pres, Coun Can Univ Biol Chmn, 82-84; pres, Biol Coun Can, 82-84; chmn, Int Orgn Comt, Int Phycological Cong, 85-91; mem, Can Nat Comt, Int Union Biol Sci; assoc ed, Can J Bot, 83-89; ed, Phycologia, 77-82, assoc ed, 83-90. *Mem:* Phycol Soc Am; Int Phycol Soc; Marine Biol Soc UK; Brit Phycol Soc; Royal Soc NZ; hon foreign mem, Societas pro Fauna et Flora Fennica; Sigma Xi. *Res:* Ecology, distribution, taxonomy and biology of benthic marine algae; laboratory culture of marine algae for experimental, ecological and life-history studies; flora of marine algae of eastern Canada and the North Atlantic Ocean; aquaculture. *Mailing Add:* Univ SPac Inst Marine Resources Box 1168 Suva Fiji

SOUTH, HUGH MILES, b Houston, Tex, Nov 10, 47; m 76. DIGITAL SIGNAL PROCESSING, SONAR SYSTEMS ANALYSIS. *Educ:* Rice Univ, BA, 71; Johns Hopkins Univ, PhD(elec eng), 81. *Prof Exp:* Instr elec eng, Johns Hopkins Univ, 73-75, sr engr, Appl Physics Lab, 75-82, supvr, Span Lab, 79-85, PRIN ENGR, JOHNS HOPKINS UNIV, 82-, SUPVR, SIGNAL PROCESSING GROUP & LECTR ELEC ENG, 85-, MGR, AUTOMATED SURVEILLANCE PROJ, 90- *Concurrent Pos:* Mem, Underwater Acoust Tech Comt, Inst Elec & Electronic Engrs Acoust, Speech & Signal Processing Soc. *Mem:* Inst Elec & Electronic Engrs; Acoust Soc Am; Sigma Xi; Europ Asn Signal Processing. *Res:* Design of hardware and software systems for digital signal processing; beam forming; spectral estimation. *Mailing Add:* 11166 Woodelves Way Columbia MD 21044-1090

SOUTH, MARY ANN, b Portales, NMex, May 23, 33; m 83; c 2. PEDIATRICS, IMMUNOLOGY. *Educ:* Eastern NMex Univ, BA, 55; Baylor Univ, MD, 59. *Prof Exp:* Intern, Presby-St Luke's Hosp, Chicago, 59-60; resident pediat, Baylor Univ, 60-62, fel, Infectious Dis, Col Med, 62-64; fel & instr immunol, Univ Minn, Minneapolis, 64-66; from asst prof to assoc prof, Col Med, Baylor Univ, 66-73; assoc prof pediat, Univ Pa, 73-77; chmn, dept pediat, Sch Med, Tex Tech Univ, 78-79, res prof pediat, 79-82; med officer, Neurol Inst, NIH, 82-85; prof pediat, 86-89, DISTINGUISHED PROF PEDIAT, W K KELLOGG FOUND MEHARRY MED COL, 89- *Concurrent Pos:* USPHS Res career develop award, 68-73. *Mem:* Am Asn Immunol; Am Pediat Soc; Infectious Dis Soc Am; Am Med Women's Asn; Int Soc Exp Hemat; Soc Gnotobiol. *Res:* Pediatric immunology; immune deficiency diseases; congenital infections. *Mailing Add:* Dept Pediat Meharry Med Col 1005 D B Todd Blvd Nashville TN 37208

SOUTHAM, CHESTER MILTON, b Salem, Mass, Oct 4, 19; m 39; c 3. ONCOLOGY, GERIATRICS. *Educ:* Univ Idaho, BS, 41, MS, 43; Columbia Univ, MD, 47. *Prof Exp:* Intern med, Presby Hosp, NY, 47-48; instr, Med Col, Cornell Univ, 51-52, from asst prof to assoc prof, Sloan-Kettering Div, 52-71; head div med oncol, 71-79, PROF MED, JEFFERSON MED COL, 71- *Concurrent Pos:* Am Cancer Soc res fel, Mem Ctr Cancer & Allied Dis, 48-49, Damon Runyon Fund clin res fel, 49-51, sr res fel, 51-52; asst, Sloan-Kettering Inst, 49-52, assoc & head clin virol sect, 52-63, mem, 63-71; asst attend physician, Mem Hosp, New York City, 52-58, assoc attend physician, 59-71; asst vis physician, James Ewing Hosp, 52-59, assoc vis physician, 59-71; mem, bd dirs, Am Asn Cancer Res, 66-70. *Mem:* Am Col Physicians; Am Asn Cancer Res (pres, 68-69); Am Fedn Clin Res; Am Asn Immunol; Am Soc Exp Path. *Res:* Clinical oncology, immunology; chemotherapy of cancer; oncolytic and oncogenic viruses; transplantation and tissue culture of human cancer; cancer immunology; carcinogenesis. *Mailing Add:* Thomas Jefferson Univ Hosp 111 11th St Suite 4001 Philadelphia PA 19107

SOUTHAM, DONALD LEE, b Cleveland, Ohio, Aug 28, 29; m 52; c 2. MECHANICAL ENGINEERING. *Educ:* Case Western Reserve Univ, BS, 51, MS, 54. *Prof Exp:* Designer eng, The Yoder Co, 51-53; chief engr, TRW, Inc, 53-66; vpres eng, Harris Corp, 66-74; V PRES ENG, CAST EQUIP DIV, COMBUSTION ENG, INC, 74- *Mem:* Am Soc Mech Engrs; Am Foundry Soc. *Mailing Add:* 10325 Whitewood Rd Cleveland OH 44141

SOUTHAM, FREDERICK WILLIAM, b NS, July 2, 24; m 47; c 3. PHYSICAL CHEMISTRY. *Educ:* Queen's Univ, Can, BSc, 46, MSc, 47; Mass Inst Technol, PhD(phys chem), 50. *Prof Exp:* Group leader, Electrometall Div, Aluminum Labs, Ltd, 50-71; sect head res lab, 71-72, Aluminum Co Can, sr tech consult, 72-85; RETIRED. *Concurrent Pos:* Mem, Grants Comt Chem & Metall Eng, Nat Res Coun Can, 70-73; consult, 85- *Mem:* Fel Chem Inst Can; Can Soc Chem Eng. *Res:* Processes associated with production of aluminum; heat and mass transfer; high temperature reaction kinetics; environmental control. *Mailing Add:* 36 Van Order Dr Kingston ON K7M 1B7 Can

SOUTHAM, JOHN RALPH, b Youngstown, Ohio, Oct 30, 42. MARINE GEOLOGY. *Educ:* Purdue Univ, BSEE, 65, MSEE, 67; Univ Ill, MS, 69, PhD(physics), 74. *Prof Exp:* Aerospace scientist, Lewis Res Ctr, NASA, 67; fel, 74, ASSOC PROF MARINE GEOL, ROSENSTIEL SCH MARINE & ATMOSPHERIC SCI, UNIV MIAMI, 75- *Mem:* AAAS; Am Geophys Union; Europ Geophys Soc; Int Asn Math Geol; Am Phys Soc. *Res:* Dynamic modelling of marine systems incorporating chemical, physical and biological processes; deep sea sedimentation and sedimentation processes in lakes and enclosed seas. *Mailing Add:* Univ Miami 4600 Rickenbacker Causeway Virginia Key FL 33149

SOUTHARD, ALVIN REID, b Centertown, Ky, June 30, 26; m 50; c 4. SOIL SCIENCE, GEOLOGY. *Educ:* Utah State Univ, BS, 57, MS, 58; Cornell Univ, PhD(soil classification), 63. *Prof Exp:* From asst prof to assoc prof soils, Mont State Univ, 63-67; from assoc prof to prof soil sci & biometeorol, Utah State Univ, 67-89, head dept, 83-89, emer prof, 89-; RETIRED. *Concurrent Pos:* Exp sta rep, Nat Coop Soil Surv, Mont, 63-67 & Utah, 67-; conservationist, US Agency Int Develop, Ecuador; Fulbright sr res scholar, Australia, 81; consult OICD, Peoples Repub China, 82. *Mem:* Am Soc Agron. *Res:* Soil genesis and classification in Utah, New York and Montana; soils of the alpine tundra in Alaska; soils of wet and dry tropics in Ecuador, Brazil, Mauritania, Hawaii & Australia. *Mailing Add:* 940 River Heights Blvd Logan UT 84321-5623

SOUTHARD, JOHN BRELSFORD, b Baltimore, Md, May 21, 38; m 60; c 2. GEOLOGY. *Educ:* Mass Inst Technol, SB, 60; Harvard Univ, MA, 63, PhD(geol), 66. *Prof Exp:* NSF fel, Calif Inst Technol, 66-67; from asst prof to assoc prof, 67-85, PROF GEOL, MASS INST TECHNOL, 85- *Mem:* AAAS; Geol Soc Am; Am Geophys Union; Int Asn Sedimentol; Soc Econ Paleontologists & Mineralogists. *Res:* Physical sedimentology; mechanics of sediment transport; marine geology; fluvial geomorphology. *Mailing Add:* Dept of Earth Atmospheric & Planetary Sci Mass Inst of Technol Cambridge MA 02139

SOUTHARD, WENDELL HOMER, b Des Moines, Iowa, July 21, 27. BIOCHEMISTRY. *Educ:* Drake Univ, BS, 50; Univ Ill, MS, 53, PhD(biol chem), 60. *Prof Exp:* Asst pharm, Univ Ill, 50-51, instr mfg pharm, 51-55, pharm, 55-56; from asst prof to prof, 67-89, EMER PROF PHARMACEUT CHEM, COL PHARM, DRAKE UNIV, 89- *Mem:* Sigma Xi. *Res:* Carbohydrate and microbial metabolism; manufacturing pharmacy. *Mailing Add:* 1542 Wilson Ave Des Moines IA 50316

SOUTHARDS, CARROLL J, b Bryson City, NC, June 18, 32; m 56; c 3. NEMATOLOGY, PLANT PATHOLOGY. *Educ:* NC State Univ, BS, 54, MS, 61, PhD(plant path), 65. *Prof Exp:* Asst county agr agent, NC Agr Exten Serv, 57-59; res asst hort, NC State Univ, 59-61, plant path, 61-65; from asst prof to assoc prof plant path, 65-74, PROF AGR BIOL & HEAD DEPT, INST AGR, UNIV TENN, KNOXVILLE, 74- *Mem:* Am Soc Phytopathologists; Soc Nematologists; Sigma Xi; Entom Soc Am. *Res:* Host-parasite relationships of tobacco, soybeans and vegetable crops and root-knot and cyst nematodes; host resistance; variability of root-knot nematodes. *Mailing Add:* Dept Entom & Plant Path Univ Tenn PO Box 1071 Knoxville TN 37901

SOUTHERLAND, WILLIAM M, SULFER METABOLISM. *Educ:* Duke Univ, PhD(biochem), 77. *Prof Exp:* ASSOC PROF BIOCHEM, COL MED, HOWARD UNIV, 83- *Res:* Enzymology of sulfer metabolism; toxicity of sulfer oxides. *Mailing Add:* Dept Biochem Col Med Howard Univ 2400 Sixth St NW Washington DC 20059

SOUTHERN, BYRON WAYNE, b Toronto, Ont, June 13, 46; m 71; c 3. CONDENSED MATTER PHYSICS. *Educ:* York Univ, BSc, 69; McMaster Univ, MSc, 71, PhD(physics), 73. *Prof Exp:* Nat Res Coun Can fel physics, Imp Col Sci & Technol, London, Eng, 73-75; res physicist, Inst Laue Langevin, Grenoble, France, 75-79; PROF PHYSICS, UNIV MAN, 79- *Mem:* Can Asn Physicists; Am Phys Soc. *Mailing Add:* Dept Physics Univ Man Winnipeg MB R3T 2N2 Can

SOUTHERN, THOMAS MARTIN, b Beaumont, Tex, June 19, 42; m 66; c 2. TECHNICAL MANAGEMENT, QUALITY ASSURANCE STANDARDS. *Educ:* Lamar Univ, BS, 64; Tex Tech Univ, MS, 66; Univ Houston, PhD(anal chem), 69; Southern Methodist Univ, MBA, 77. *Prof Exp:* Chemist & qual control mgr, 69-80, mgr plastics prod develop & polyethylene qual mgr develop, 80-89, MGR POLYETHYLENE QUAL CONTROL & SR CHEMIST ANAL QUAL ASSURANCE, TEX EASTMAN CO, 89- *Concurrent Pos:* Pres & chmn, Cardinal Premium Finance, Longview, Tex; vpres & bd mem, Parish Premium Finance, Coushatta, La. *Mem:* Am Soc Qual Control. *Res:* Liquid chromatography; thermoanalytical chemistry; application of computers dedicated to analytical instruments and data reduction; hot melt adhesives-formulation and application; computers; quality systems and standards. *Mailing Add:* Sr Chemist Qual Assurance PO Box 7444 Longview TX 75601

SOUTHERN, WILLIAM EDWARD, b Wayne Co, Mich, Dec 22, 33; m; c 3. ORNITHOLOGY, WETLAND SCIENCE. *Educ:* Cent Mich Univ, BS, 55; Univ Mich, MA, 59; Cornell Univ, PhD(comp vert ethol, animal ecol, wildlife mgt), 67. *Prof Exp:* Pub sch teacher, Mich, 55-56 & 57-58; asst prof biol sci, Northern Ill Univ, 59-68,; assoc prof biol sci, 68-72, prof ornith, Univ Mich Biol Sta, 75-78; prof avian behav ecol, Northern Ill Univ, 72-90, dir coop educ prog, 84-90; PRES, ENCAP INC, 74- *Concurrent Pos:* Grants, Frank M Chapman Mem Fund, 61-64 & 65, Sigma Xi, 61, 63-64 & 68-69, Northern Ill Univ, 62, 65, 67-68, 69, 72 & 76-77, Brown Fund, Cornell Univ, 63-64, Max McGraw Wildlife Found, 68-70, NSF, 59, 62 & 71-73, sci fac fel, 63-64, desert biol fel, 65, Nat Park Serv, 74-82, Off Naval Res, 71-72 & Ill Dept Conserv, 73-74 & 81-85; US Air Force off sci res, 77-79; mem, Ill Endangered Species Protection bd, exec secy, 76-; US Dept Educ, 84-90. *Honors & Awards:* Hann lectr, Biol Sta, Univ Mich, 73; Ernest P Edwards Prize, Wilson Ornith Soc, 75. *Mem:* Sigma Xi; Soc Wetland Scientists; Am Ornithologists Union; Nat Asn Environ Professionals; Wilson Ornith Soc; Cooper Ornithologists Soc. *Res:* Bird/riveraft collision issues; avian behavior, ecology and population dynamics; environmental assessments; endangered species issues; wetland delineation; mitigation and management. *Mailing Add:* Encap Inc 400 E Hillcrest Suite 240 De Kalb IL 60115

SOUTHGATE, PETER DAVID, b Woking, Eng, July 20, 28; m 52; c 4. PHYSICS. *Educ:* Univ London, BSc, 48, MSc, 52, PhD(physics), 59. *Prof Exp:* Res scientist, Mullard Res Labs, Eng, 48-58; res physicist, Res Inst, Ill Inst Technol, 59-65; MEM TECH STAFF, DAVID SARNOFF RES CTR, 66- *Mem:* Am Phys Soc. *Res:* Luminescence and recombination processes in semiconductors and organic materials; non-linear optical interactions in crystals; ferroelectric and pyroelectric phenomena and their application to radiation detection; acoustoelectric interactions in semiconductors; analysis of television tube and videodisc manufacturing processes. *Mailing Add:* David Sarnoff Res Ctr Princeton NJ 08543-5300

SOUTHIN, JOHN L, b Brockville, Ont, Can, June 10, 39. GENETICS. *Educ:* Queen's Univ, Ont, BSc, 61; Univ Calif, Los Angeles, MA, 62, PhD(zool), 63. *Prof Exp:* From lectr to asst prof, 63-67, ASSOC PROF GENETICS, MCGILL UNIV, 67-, DIR RESIDENCES, 72- *Concurrent Pos:* Vis scholar, Univ Calif, Los Angeles, 64 & 65; vis prof, Univ Havana, 67 & 69, prof, 70- *Mem:* Genetics Soc Am. *Res:* Gene structure in Drosophila; chemically induced mutation and the problem of mosaicism in Drosophila. *Mailing Add:* Dept Biol McGill Univ Box 6070 St A Montreal PQ H3A 2T6 Can

SOUTHREN, A LOUIS, b New York, NY, Oct 12, 26; m 50; c 3. MOLECULAR BIOLOGY. *Educ:* NY Univ, AB, 49; Chicago Med Sch, MD, 55. *Prof Exp:* From asst prof med & asst attend physician to assoc prof med & assoc attend physician, 61-69, PROF MED & ATTEND PHYSICIAN, NEW YORK MED COL, 69-, RES PROF OPHTHAL, 79-, CHIEF DIV ENDOCRINOL/METABOLISM, 63-, PROF OBSTET & GYNEC (ENDOCRINOL). *Concurrent Pos:* Fel endocrinol, Mt Sinai Hosp, New York, 58-59; fel endocrine res, Jewish Hosp Brooklyn, 59-61; fel, training prog steroid biochem, Worcester Found Exp Biol, 61-62; USPHS res grants, 63-; career scientist, Health Res Coun of New York, 63-72 & 74-75; dir, USPHS Training Prog Endocrinol & Metab, 65-72; mem glaucoma panel, Nat Adv Eye Coun, 80-81. *Mem:* AAAS; fel Am Col Physicians; Endocrine Soc; Soc Gynec Invest; Am Fertil Soc. *Res:* Glucocorticoid metabolism in glaucoma. *Mailing Add:* Dept Med New York Med Col Valhalla NY 10595

SOUTHWARD, GLEN MORRIS, b Boise, Idaho, Oct 8, 27; c 2. STATISTICS, SAMPLING. *Educ:* Univ Wash, BS, 49, MS, 56, PhD, 66. *Prof Exp:* Biologist, Int Pac Halibut Comn, 55-67; asst prof statist & asst statistician, Wash State Univ, 67-70 & Univ Wis, 70-71; biometrician, Int Pac Halibut Comn, 71-75; assoc prof, 75-80, PROF EXP STATIST, NMEX STATE UNIV, 80- *Concurrent Pos:* Vis prof, dept statist, Univ Edinburgh, 81-82. *Mem:* Am Inst Fisheries Res Biologists; Am Statist Asn; Biomet Soc. *Res:* Biometry; experimental statistics, biomathematics, statistical consulting; sampling. *Mailing Add:* Dept Exp Statist NMex State Univ Col Agr Las Cruces NM 88003

SOUTHWARD, HAROLD DEAN, b Headrick, Okla, June 22, 30; m 54; c 4. NUCLEAR PHYSICS, SOLID STATE PHYSICS. *Educ:* West Tex State Univ, BS, 51; Univ Tex, MA, 57, PhD(physics), 58. *Prof Exp:* Engr, Aircraft Armaments, Inc, 54-55; sr res technologist, Mobil Oil Corp, 58-63; assoc prof, 63-69, dir bur eng res, 71-76, PROF ELEC ENG, UNIV NMEX, 69-, CHMN ELEC ENG & COMPUT SCI, 81- *Mem:* Am Phys Soc; Inst Elec & Electronics Eng. *Res:* Radiation effects of solid state electronics; radiation measurement; photovoltaic energy systems. *Mailing Add:* Dept Elec Eng & Comput Sci Tapy Hall Univ NMex Albuquerque NM 87131

SOUTHWELL, P(ETER) H(ENRY), b Rochdale, Eng, Nov 29, 24; m 46; c 4. MECHANICAL & AGRICULTURAL ENGINEERING. *Educ:* Royal Naval Eng Col, Eng, Engr, 45; Univ Sask, MSc, 60. *Prof Exp:* Exp officer, Nat Inst Agr Eng, Eng, 48-54; asst prof eng sci, 55-60, ASSOC PROF, SCH ENG, UNIV GUELPH, 61- *Concurrent Pos:* Chmn, Assoc Comt Agr & Forestry Aviation, Nat Res Coun Can, 66-72; mem, Energy & Agr Policy Comt, Ont Govt, 80-81. *Honors & Awards:* Pilcher Mem Prize, Royal Aeronaut Soc, 52. *Mem:* Can Soc Agr Engrs; Am Soc Agr Engrs; assoc fel Royal Aeronaut Soc; Inst Mech Engrs. *Res:* Combustion engines and turbines; terrain-vehicle systems and terra-mechanics; agricultural aviation; pesticides application systems; energy analysis and energy ratios of food production; biomass fuels. *Mailing Add:* RR 1 Rockwood ON N0B 2K0 Can

SOUTHWICK, CHARLES HENRY, b Wooster, Ohio, Aug 28, 28; m 52; c 2. PRIMATE ECOLOGY. *Educ:* Col Wooster, BA, 49; Univ Wis, MS, 51, PhD(zool), 53. *Prof Exp:* Asst prof biol, Hamilton Col, 53-54; NSF fel, Bur Animal Population, Oxford Univ, 54-55; from asst prof to assoc prof zool, Ohio Univ, 55-61; assoc prof pathobiol, Sch Hyg & Pub Health, Johns Hopkins Univ, 61-68; prof, 68-80, PROF, DEPT EPO BIOL, UNIV COLO, 79- *Concurrent Pos:* Mem numerous primate expeds, Panama, India, Nepal, Indonesia, Malaysia, Burma, China, Kenya, 51-89; Fulbright fel, Aligarh Muslim Univ, India, 59-60; mem, Calif Primate Res Ctr, 74-76 & Gov Sci Adv Coun, Md, 75-77; mem, Primate Adv Cmt, Nat Acad Sci, Nat Res Coun & Adv Bd, Caribbean Primate Res Ctr, Cmt Res & Explor, Nat Geog Soc. *Mem:* Fel AAAS; Ecol Soc Am; Am Soc Mammal; Am Soc Zoologists; Fel Animal Behav Soc (pres, 68); Am Soc Primate; Int Primate Soc; Sigma Xi; Fel Acad Zool. *Res:* Vertebrate population dynamics; stress physiology; sociobiology and animal behavior, especially behavioral ecology. *Mailing Add:* 6507 Baseline Rd Boulder CO 80303

SOUTHWICK, DAVID LEROY, b Rochester, Minn, Aug 30, 36; m 59; c 3. PRECAMBRIAN GEOLOGY, TECTONICS. *Educ:* Carleton Col, BA, 58; Johns Hopkins Univ, PhD(geol), 62. *Prof Exp:* Geologist, US Geol Surv, 62-68; from asst prof to prof geol, Macalester Col, 68-77; sr geologist, 77-89, ASST DIR & RES ASSOC, MINN GEOL SURV, 89- *Concurrent Pos:* Adj assoc prof, Univ Minn, 83- *Mem:* Am Geophys Union; fel Geol Soc Am; fel Geol Asn Can. *Res:* Petrology of metamorphic and igneous rocks; structural geology; stratigraphy and structural geology of the Precambrian rocks of Minnesota. *Mailing Add:* 2642 Univ Ave St Paul MN 55114

SOUTHWICK, EDWARD EARLE, b Northampton, Mass; m. PHYSIOLOGICAL ECOLOGY. *Educ:* Univ Mich, BSME, 65, MS, 67; Wash State Univ, PhD(zool), 71. *Prof Exp:* Asst prof biol, Duquesne Univ, 71-73; lectr physiol ecol, Grad Sch, Georgetown Univ, 74-75; dir, Chippewa Nature Ctr, Inc, 75-77; assoc prof, 77-85, PROF BIOL, STATE UNIV NY, BROCKPORT, 85- *Concurrent Pos:* prin investr, NSF res grant, 80; Mellon Found fel microclimate & pollination, 81-83; Alexander von Humboldt fel, 83, 84, 88; prin investr, NATO res, 88; fac exchange scholar, 90. *Mem:* Ecol Soc Am; Sigma Xi; Int Bee Res Asn; Am Soc Bot; Entomol Soc Am. *Res:* Melittology; nectar biology and nutrition; plant insect relationships; pollination ecology; microclimates; cold temperature physiology; animal energetics including energy balance; thermoregulation and metabolism, especially in social insects; superorganism attributes of social insects. *Mailing Add:* Dept Biol Sci State Univ NY Brockport NY 14420

SOUTHWICK, EVERETT WEST, b Providence, RI, Sept 19, 41; m 83; c 2. ORGANIC CHEMISTRY. *Educ:* Univ RI, BA, 63; Univ NH, PhD(org chem), 73. *Prof Exp:* Fel org chem, State Univ NY Buffalo, 72-74, Duke Univ, 73-74; sr chemist, Liggett & Myres Inc, 74-80; CHEMIST, PHILIP MORRIS USA, 80- *Mem:* Am Chem Soc; Am Chem Soc. *Res:* Synthesis, isolation, purification and characterization of organic compounds of potential value as flavorants; structure-activity; relations in olfaction. *Mailing Add:* Philip Morris USA Res Ctr POB 26583 Richmond VA 23261

SOUTHWICK, FRANKLIN WALLBURG, b Boston, Mass, May 29, 17; m 40; c 4. POMOLOGY. *Educ:* Mass State Col, BS, 39; Ohio State Univ, MS, 40; Cornell Univ, PhD(pomol), 43. *Prof Exp:* Asst prof pomol, Univ Conn, 43-45 & Cornell Univ, 45-48; head dept plant & soil sci, Univ Mass, Amherst, 64-77, dir grad prof, 73-80, prof pomol, 48-83; RETIRED. *Honors & Awards:* Gold Medal, Mass Soc Promoting Agr, 65. *Mem:* AAAS; fel Am Soc Hort Sci. *Res:* Fruit storage; respiration; growth regulating substances; nutrition. *Mailing Add:* 993 E Pleasant Univ of Mass Amherst MA 01002

SOUTHWICK, HARRY W, b Grand Rapids, Mich, Nov 21, 18; m 42; c 4. SURGERY. *Educ:* Harvard Univ, BS, 40, MD, 43; Am Bd Surg, dipl, 51. *Prof Exp:* Clin prof surg, Univ Ill Col Med, 58-71; prof surg, Rush Med Col & chmn dept gen surg, Rush-Presby-St Luke's Med Ctr, 71-84; RETIRED. *Concurrent Pos:* Pvt pract; head sect gen surg, Presby-St Luke's Hosp, 67-71. *Mem:* Am Col Surg; Am Surg Asn; Soc Head & Neck Surgeons; Soc Surg Oncol. *Res:* Oncology. *Mailing Add:* 8155 N Lost Lake Dr Sayner WI 54560

SOUTHWICK, LAWRENCE, b Worcester, Mass, Apr 6, 12; m 37; c 3. AGRONOMY, HORTICULTURE. *Educ:* Univ Mass, BS, 33, MS, 38. *Prof Exp:* Instr, Univ Mass, 35-45; bioprod res & develop chemist & tech ed, Dow Chem Co, 45-80; RETIRED. *Mem:* Am Soc Agron; fel Weed Sci Soc Am; Am Soc Hort Sci; Sigma Xi. *Res:* Chemical weed control; plant nutrition; soils; dwarf fruit trees; growth substances; pest control; environmental science. *Mailing Add:* 4504 Bond Ct Midland MI 48640

SOUTHWICK, PHILIP LEE, b Lincoln, Nebr, Nov 15, 16; m 42; c 1. CHEMISTRY. *Educ:* Univ Nebr, AB, 39, AM, 40; Univ Ill, PhD(org chem), 43. *Prof Exp:* Res chemist, Merck & Co, NJ, 43-46; from asst prof to assoc prof, 46-55, prof chem, 55-82, EMER PROF CHEM, CARNEGIE-MELLON UNIV, 82- *Concurrent Pos:* Mem, Fulbright Act Awards Chem Comt, Div Chem & Chem Technol, Nat Acad Sci-Nat Res Coun, 60-63. *Mem:* Fel AAAS; Am Chem Soc; fel NY Acad Sci. *Res:* Stability of enols; direct aromatic carboxymethylations; synthetic antimetabolites; chemistry of penicillins, pteridines, pyrimidines, pyrrolidines and indoles; stereochemistry of conjugate addition reactions and formation and reactions of aziridines and aziridinium ions; fluorescent diazepines, indolenines and cyanines; synthesis of first N-amidino pyrazine carboxamide diuretic leading to the drug amiloride. *Mailing Add:* Dept Chem Carnegie-Mellon Univ Pittsburgh PA 15213

SOUTHWICK, RICHARD ARTHUR, b White River Junction, Vt, Sept 16, 24; m 45; c 2. PLANT BREEDING. *Educ:* Univ Vt, BS, 50, MS, 54. *Prof Exp:* Instr agron, Univ Mass, Amherst, 54-58, asst prof, 58-67; assoc prof, State Univ NY Agr & Tech Col Cobleskill, 67-69, prof plant sci, 69-87; RETIRED. *Concurrent Pos:* Botanist, George Landis Arboretum, Esperance, 70-82, assoc dir, 82-84; mem, Adv Comt on Rare & Endangered Plants, State NY, 71-79. *Mem:* Am Hort Soc; Am Forestry Asn; Sigma Xi; NY Acad Sci. *Res:* Plant propagation; improvement, evaluation, and collection of ornamental plant materials. *Mailing Add:* Rd Two Box 648 Cobleskill NY 12043

SOUTHWICK, RUSSELL DUTY, b Woonsocket, RI, Dec 27, 31; m 54; c 3. PHYSICS. *Educ:* Rensselaer Polytech Inst, BS, 53; Univ Conn, MS, 55. *Prof Exp:* Asst, Univ Conn, 53-55; res physicist, Preston Labs, Inc, 55-60, asst dir res, 60-66, vpres & asst dir res, 66-76, vpres & dir res, 76-86, PRES, AM GLASS RES, INC, 86- *Mem:* Am Phys Soc; Am Ceramic Soc; Soc Exp Stress Anal. *Res:* Glass technology; mechanical properties of glass, especially strength and surface friction. *Mailing Add:* 112 Forresta Dr Box 149 Butler PA 16001

SOUTHWICK, WAYNE ORIN, b Lincoln, Nebr, Feb 6, 23; m 44; c 3. ORTHOPEDIC SURGERY. *Educ:* Univ Nebr, AB, 45, MD, 47; Am Bd Orthop Surg, dipl, 58. *Prof Exp:* Asst anat & histol, Col Med, Univ Nebr, 46-47; intern med, Boston City Hosp, 47-48, asst resident surg, Fifth Surg Div, 48-50; asst resident orthop surg, Hosp, Johns Hopkins Univ, 50-54, instr, Sch Med, Univ & chief resident, Hosp, 54-55, asst prof, Univ, 55-58; assoc prof, 58-61, chief sect, 58-77, PROF ORTHOP SURG, SCH MED, YALE UNIV, 61-, CONSULT ORTHOP, UNIV HEALTH SERV, 69- *Concurrent Pos:* Fel, Branford Col, Yale Univ, 70- *Mem:* Am Acad Orthop Surg; Am Orthop Asn; fel Am Col Surg; AMA; Orthop Res Soc. *Res:* Degenerative cervical disk disease; slipped epiphysis; experimental osteomyelitis; histology of cartilage. *Mailing Add:* PO Box 390 Old Lyme CT 06371-0390

SOUTHWORTH, HAMILTON, b New York, NY, Apr 7, 07; m 33; c 4. INTERNAL MEDICINE. *Educ:* Yale Univ, BA, 29; Johns Hopkins Univ, MD, 33; Am Bd Internal Med, dipl, 40. *Prof Exp:* Intern, Presby Hosp New York, 33-35; asst resident med, Johns Hopkins Univ, 35-37, asst, Sch Med, 35-37; from asst to assoc, Col Physicians & Surgeons, 37-46, from asst clin prof to clin prof, 47-69, prof clin med, 69-72, EMER PROF CLIN MED, COLUMBIA UNIV, 72- *Concurrent Pos:* US rep med subcomt, Europ Regional Adv, UNRRA, 44; attache, US Embassy, London, 44-45; Europ rep comt med res, Off Sci Res & Develop, 44-45; from asst attend physician to attend, Presby Hosp, 45-72; consult, Sta Hosp, US Mil Acad, 46-73 & Presby Hosp, 72-; mem med exam bd, Am Bd Internal Med, 64-70. *Honors & Awards:* Order of the Cedar Award, Govt Lebanon, 81. *Mem:* AMA; fel Am Col Physicians; Am Clin & Climat Asn; NY Acad Med. *Res:* Toxicity of sulfonamides; hemolytic and aplastic anemias; cardiac resuscitation. *Mailing Add:* 200 E 66th St New York NY 10021

SOUTHWORTH, RAYMOND W(ILLIAM), b North Brookfield, Mass, Oct 23, 20. CHEMICAL ENGINEERING. *Educ:* Worcester Polytech Inst, BS, 43; Yale Univ, MEng, 44, DEng, 48. *Prof Exp:* From asst instr to assoc prof chem eng, Yale Univ, 43-66; dir Comput Ctr, 66-81, prof math & comput sci, 66-85, EMER PROF MATH, COL WILLIAM & MARY, 85- *Concurrent Pos:* Mem staff, Brookhaven Nat Lab, 48-49; mem, Inst Math Sci, NY Univ, 60-61; dir, Southeastern Va Regional Comput Ctr, 69-81. *Mem:* Am Chem Soc; Am Inst Chem Engrs; Asn Comput Mach; Am Math Soc; Soc Indust & Appl Math. *Res:* Digital computing; numerical methods. *Mailing Add:* Dept Math Col William & Mary Williamsburg VA 23185

SOUTHWORTH, WARREN HILBOURNE, b Lynn, Mass, Feb 10, 12; m 37; c 2. PUBLIC HEALTH, PREVENTIVE MEDICINE. *Educ:* Univ Mass, BS, 34; Boston Univ, MS, 35; Mass Inst Technol, DrPH, 44. *Prof Exp:* Teacher biol, Whitman High Sch & Belmont High Sch, Mass, 36-39; res dir, Mass Dept Pub Health, 41-42; prof health sci, Panzer Col, 42-44; coordr sch health, Wis Dept Pub Instr, 44-48; coordr med team, Wis State Bd Health, 52-53; assoc prof, 44-52, prof prev med, 77-81, PROF CURRIC & INSTR, UNIV WIS-MADISON, 53-, EMER PROF PREV MED, 81- *Concurrent Pos:* Lectr pub health, NY Univ, 43-44; field rep, Fed Venereal Dis Prog, Am Social Hyg Asn, 44. *Honors & Awards:* William A Howe Award, 68. *Mem:* AAAS; Am Pub Health Asn; Am Sch Health Asn (treas, 60-74); Soc Pub Health Educrs. *Res:* Community health education; school health education; occupational health education; patient health education; drug education; family life education. *Mailing Add:* 3207 Stevens Madison WI 53705

SOUTO BACHILLER, FERNANDO ALBERTO, b Andujar, Spain, Mar 27, 51; m 74; c 3. PHYSICAL ORGANIC CHEMISTRY, PHOTOCHEMISTRY. *Educ:* Univ Granada, Spain, Lic Sc, 74; Univ Alta, Can, PhD(chem), 78; Ministerio Educación, Madrid, Spain, 88. *Prof Exp:* Res assoc chem, Imp Col, London, 78-79; from asst prof to assoc prof chem, 79-88, DIR LAB, CRIL, UNIV PR, 84-, PROF CHEM, 88- *Concurrent Pos:* Vis prof plant biol, Univ Granada, Spain, 87-88, vis prof chem, EPFL, Lausanne, Switz, 88-89 & Univ Malaga, Spain, 89-90. *Mem:* Sigma Xi; fel Royal Soc Chem; fel Am Inst Chemists; Electrochem Soc; Am Soc Pharmacog; Am Chem Soc. *Res:* Organic molecular photophysics; organic photochemistry; organic electrochemistry; tropical plant products, their isolation, characterization, photochemistry and in vitro biosynthesis; association of organic dyes in solution, equilibrium polymerization, absorption and emission of electronic excitation energy; excitation energy transfer; photochemistry of N-oxides. *Mailing Add:* Chem Dept Univ PR Mayaguez PR 00708

SOVEN, PAUL, b New York, NY, Sept 30, 39; m 61; c 3. PHYSICS. *Educ:* City Col New York, BS, 60; Univ Chicago, MS, 61, PhD(physics), 65. *Prof Exp:* Mem staff, Bell Tel Labs, 65-67; from asst prof to assoc prof, 67-76, PROF PHYSICS, UNIV PA, 77- *Mem:* Am Phys Soc. *Res:* Theory of metals. *Mailing Add:* Dept Physics Univ Pa Philadelphia PA 19174

SOVERS, OJARS JURIS, b Riga, Latvia, July 11, 37; US citizen; m 59. ASTRONOMY. *Educ:* Brooklyn Col, BS, 58; Princeton Univ, PhD(physics, phys chem), 62. *Prof Exp:* NSF fel chem, Oxford Univ, 61-62; fel, Columbia Univ, 62-63, res assoc, Watson Lab, 63-64; res engr, Gen Tel & Electronics Lab, Inc, 64-72; res engr, Sony Corp, 72-79; AT JET PROPULSION LAB, CALIF INST TECHNOL, 79- *Mem:* Am Phys Soc; AAAS; Am Geophys Union; Sigma Xi. *Res:* Applications of very long baseline radio interferometry to astrometric; geodetic measurements for spacecraft navigation. *Mailing Add:* 1367 La Solana Dr Altadena CA 91001

SOVIE, MARGARET D, NURSING ADMINISTRATION. *Educ:* Syracuse Univ, BS, 64, MS, 68, PhD(educ), 72. *Prof Exp:* Supvr & instr nursing, Good Shepherd Hosp & State Univ Hosp, 63-66; assoc prof nursing, educ dir & coord nursing serv, Upstate Med Ctr Univ Hosp, State Univ New York, 66-71, assoc dean & dir continuying educ, 72-76; assoc dir, assoc dean & prof nursing admin, Univ Rochester Med Ctr, 76-88; ASSOC EXEC DIR & CHIEF NURSING OFFICER, HOSP UNIV PA, 88-, JANE DELANO PROF NURSING ADMIN & ASSOC DEAN NURSING PRACT, 88- *Concurrent Pos:* Mem nursing serv, Am Nursing Asn, 86-; consult, Nat Ctr Health Serv Res, 87-; comt strategy design, qual rev & assurance med, Inst Med, 88- *Mem:* Inst Med-Nat Acad Sci. *Res:* Identifying the costs of providing nursing care to patients; creating and testing models ofnursing practice that assures quality care while controlling costs and examination of clinical practices to assure quality care. *Mailing Add:* Univ Pa Hosp 3400 Spruce St Philadelphia PA 19104-4283

SOVISH, RICHARD CHARLES, b Cleveland, Ohio, July 22, 25; m 54; c 3. ORGANIC CHEMISTRY, RADIATION CHEMISTRY. *Educ:* Ohio Univ, BS, 49; Case Western Reserve Univ, MS, 52, PhD(chem), 54. *Prof Exp:* Res chemist, Dow Chem Co, 54-62; res scientist, Lockheed Missile & Space Co, 62-63; staff mem, Raychem Corp, 63-67, sect leader, 67-70, mgr mfg compounding dept, 70-73, mgr eng, Thermofit Div, 73-75, tech mgr, Utilities Div, Raychem Belg, 75-78, int tech dir, Telecom Div, 78-80, tech dir Europe, 80-83, exec tech dir, Corp Res & Develop, 83-89, VPRES, CORP

TECHNOL, RAYCHEM CORP, 89- *Mem:* Am Chem Soc; Sigma Xi; Soc Plastics Engrs. *Res:* Irradiation effects on polymers; polymer cross-linking; graft copolymers; mechanical properties of polymers. *Mailing Add:* Raychem Corp 300 Constitution Dr Menlo Park CA 94025

SOVOCOOL, G WAYNE, b Cortland, NY, Oct 30, 42; m 71; c 2. PHYSICAL-ORGANIC CHEMISTRY, ANALYTICAL CHEMISTRY. *Educ:* Rochester Inst Technol, BS, 65; Cornell Univ, MS, 67, PhD(chem), 71. *Prof Exp:* Fel chem, Univ NC, Chapel Hill, 71-72; res chemist, 72-81, supvry phys scientist, 81-87, CHEM, US ENVIRON PROTECTION AGENCY, 87- *Concurrent Pos:* Adj assoc prof, Biochem Lab, Bot Dept, Univ NC, Chapel Hill, 81-82. *Mem:* Am Chem Soc; Am Soc Mass Spectrometry. *Res:* Structure determination and quantitative measurement of organic chemical compounds occurring in complex mixtures, as human tissue and environmental samples, through the use of mass spectrometry; quality assurance of analytical data; predictive methods for chemical and physical properties of compounds. *Mailing Add:* 3155 Highview Dr Henderson NV 89014-2131

SOWA, JOHN ROBERT, b South Bend, Ind, Aug 21, 34; m 61; c 4. ORGANIC CHEMISTRY. *Educ:* Univ Notre Dame, BS, 56; Univ Pa, PhD(org chem), 64. *Prof Exp:* Res asst chem, Sowa Chem Co, 58-59; res assoc, Univ Ariz, 64-66; from asst prof to assoc prof, 67-90, PROF CHEM, UNION COL, NY, 90-; asst prof, 67-77, ASSOC PROF CHEM, UNION COL, NY, 77- *Concurrent Pos:* Henry Busche teaching fel, Univ Pa, 62; consult, Sowa Chem Co, 66-, decontamination res div, Edgewood Arsenal, 67 & Schenectady Chem Co, 68-; vis prof, Rensselaer Polytech Inst, 73-74 & Univ Albany, 81. *Mem:* Am Chem Soc; NY Acad Sci; Am Inst Chemists; Sigma Xi. *Res:* Nuclear magnetic resonance applied to mechanisms in organic chemistry; silicon-carbon d-pi/p-pi bonding; mustard reactions with purines; vinyl polymerizations; organophosphorous chemistry; acetylenes; liquid crystals; diazonium ions; undergraduate organic laboratory experiments. *Mailing Add:* Dept of Chem Union Col Schenectady NY 12308

SOWA, WALTER, b Flin Flon, Man, Nov 29, 33; m 64; c 2. ORGANIC CHEMISTRY. *Educ:* Queen's Univ, Ont, BSc, 56, MSc, 58, PhD(org chem), 62; York Univ, MBA, 75. *Prof Exp:* Nat Res Coun Can fel, 61-63; res scientist, 63-73, sr res scientist, 73-78, asst dir, Dept Appl Chem, 75-79, prin res scientist, 80-81, SR PROJ OFFICER PROJ DEVELOP, ONT RES FOUND, 81- *Mem:* AAAS; Chem Inst Can; Am Inst Mining, Metall & Petrol Engrs; Am Chem Soc; Can Inst Mining & Metall. *Res:* Organic synthesis; chemistry of carbohydrates; antiradiation compounds; selective extractants for metals; chemistry of industrial solvent extraction of metals; forest products; chromotography. *Mailing Add:* Ortech Int 2395 Speakman Dr Mississauga ON L5K 1B3 Can

SOWDER, LARRY K, b Bedford, Ind, Mar 17, 38. MAHTEMATICS EDUCATION. *Educ:* Univ Wis, PhD(math), 69. *Prof Exp:* PROF, MATH SCI, SAN DIEGO STATE UNIV, 86- *Mailing Add:* Math Sci Dept San Diego State Univ San Diego CA 92182

SOWELL, JOHN BASIL, b Phoenix, Ariz, Oct 22, 58; m 83; c 3. PLANT PHYSIOLOGICAL ECOLOGY. *Educ:* Univ Calif, Davis, BS, 79; Univ Idaho, PhD(bot), 85. *Prof Exp:* ASST PROF BIOL, SOUTHWEST STATE UNIV, MINN, 85- *Mem:* Am Soc Plant Physiologists; Ecol Soc Am. *Res:* Physiological ecology of alpine timberline trees; winter water relations; effects of low soil temperature on plant functioning; ecological modeling. *Mailing Add:* Dept Biol Southwest State Univ Marshall MN 56258

SOWELL, JOHN GREGORY, b Knoxville, Tenn, Jan 22, 41; m 67; c 1. PHARMACOLOGY, OPHTHALMOLOGY. *Educ:* Murray State Univ, BS, 63; Univ Tenn, MS, 67, PhD, 69. *Prof Exp:* Instr pharmacol, Univ Tenn, 70-71; ASST PROF PHARMACOL & OPHTHAL, MED CTR, UNIV ALA, BIRMINGHAM, 73-; AT DEPT PHARMACOL, SAMFORD UNIV. *Concurrent Pos:* Fel, Univ Southern Calif, 71-73. *Mem:* Asn Res Vis & Ophthal; Endocrine Soc. *Res:* Mechanism of steroid action; glaucoma. *Mailing Add:* Dept Pharmacol Samford Univ 800 Lakeshore Dr Birmingham AL 35229

SOWELL, KATYE MARIE OLIVER, b Winston-Salem, NC, Apr 6, 34; wid; c 1. MATHEMATICS. *Educ:* Flora Macdonald Col, BA, 56; Univ SC, MS, 58; Fla State Univ, PhD(math educ), 65. *Prof Exp:* Asst prof math, Elon Col, 58-60; instr, Univ Southern Miss, 60-63; instr & res assoc math educ, Fla State Univ, 65; from asst prof to assoc prof, 65-71, PROF MATH, ECAROLINA UNIV, 72-, ACTG DEAN, COL ARTS & SCI, 90- *Concurrent Pos:* Dir, student teaching prog math & supvr student teachers, ECarolina Univ, 66-79 & math educ comput lab, 81-84; NSF grants, 68-74; consult, var bk publs & city & county bds of educ, Eastern NC, 67-; consult, Ctr Individualized Instr Systs, Durham, NC, 71-73; mem, Adv Coun Math, NC State Dept Pub Instr, 70-72, Metric Educ, 74-76; vis scholar, Univ Mich, 80 & 81; co-ed, The Centroid, 83-85; dir, Eastern Carolina Educ Consults, 84- *Honors & Awards:* W W Rankin Mem Award, 85. *Mem:* Math Asn Am; Nat Coun Teachers Math; Asn Women Math; Sigma Xi. *Res:* Mathematics education; geometry; number theory. *Mailing Add:* 103 Col Court Dr Greenville NC 27858-3926

SOWER, STACIA ANN, b Ft Belvoir, Va, Nov 16, 50. NEUROENDOCRINOLOGY, REPRODUCTION. *Educ:* Univ Utah, BA, 73; Ore State Univ, MS, 78, PhD(fisheries physiol), 81. *Prof Exp:* Postdoctoral fel endocrinol, Univ Wash, 80-83; asst prof zool, 83-87, ASSOC PROF ZOOL & BIOCHEM, UNIV NH, 87- *Concurrent Pos:* Consult, Sea Run, Inc, Maine, 85-; secy, Div Comp Endocrin, Am Soc Zoologists, 88-89; mem, Physiol Processes Panel, NSF, 88-92; prin investr, NSF grant, 89-94; mem, Tech Prog Comn for 2nd Int Fish Endocrinol Symp, France, 90-92. *Honors & Awards:* Career Develop Award, NSF, 88, Fac Award, Women Scientists & Engrs, 91. *Mem:* Am Soc Zoologists; Soc Neurosci; Am Fisheries Soc; AAAS. *Res:* Comparative reproductive physiology and endocrinology in fishes; structure and function of brain hormones, particularly lampreys. *Mailing Add:* Dept Zool Univ NH Durham NH 03824

SOWERS, ARTHUR EDWARD, b Chicago, Ill, Dec 20, 43; m 82. MEMBRANE BIOLOGY. *Educ:* Univ Ill, Urbana, BS, 66; Tex A&M Univ, PhD(biol), 77. *Prof Exp:* Res fel membranes, Med Sch, Univ NC, 78-82; SCIENTIST II MEMBRANES, AM RED CROSS RES LAB, 82- *Mem:* AAAS; Biophys Soc; Am Soc Cell Biol; Electron Micros Soc Am; Bioelectromagnetics Soc. *Res:* Electric field effects on membrane structure and function; electrofusion, electroporation and lateral diffusion of mobile membrane components; membrane structure-function relationships. *Mailing Add:* Holland Lab Biomed Sci Am Red Cross, 15601 Crabbs Br Way Rockville MD 20855

SOWERS, EDWARD EUGENE, b Crawfordsville, Ind, Nov 26, 42; m 67; c 2. INDUSTRIAL ORGANIC CHEMISTRY. *Educ:* Wabash Col, AB, 64; Tufts Univ, PhD(org chem), 70; Ind Univ Law Sch, JD, 90. *Prof Exp:* Staff chemist res, Reilly Tar & Chem Corp, 69-75, mgr prod develop, 75-80, SR SECT HEAD, REILLY INDUSTS INC, 81- *Mem:* Am Chem Soc. *Res:* Synthesis and product development of nitrogen heterocycles; synthesis and characterization of linear and crosslinked polymers. *Mailing Add:* Reilly Industs Inc 1500 S Tibbs Ave Indianapolis IN 46241

SOWERS, GEORGE F(REDERICK), b Cleveland, Ohio, Sept 23, 21; m 44; c 4. CIVIL & GEOLOGICAL ENGINEERING. *Educ:* Case Western Reserve Univ, BS, 42; Harvard Univ, MS, 47. *Prof Exp:* From assoc prof to prof, 47-65, REGENTS PROF CIVIL ENG, GA INST TECHNOL, 65- *Concurrent Pos:* Vis lectr, India, 59, 65; consult engr, Law Eng Testing Co, 47 & Washington, DC, 57- *Honors & Awards:* Middlebrook Award, Am Soc Civil Engrs, 77, Terzaghi lectr, 79. *Mem:* Am Soc Civil Engrs; Geol Soc Am; Am Soc Testing & Mat; Nat Soc Prof Engrs; Int Soc Soil Mech & Found Engrs; Sigma Xi. *Res:* Soils; engineering geology; rock mechanics. *Mailing Add:* Dept Civil Eng Ga Inst Technol Atlanta GA 30332

SOWINSKI, RAYMOND, b Hammond, Ind, Feb 8, 24; m 54. BIOCHEMISTRY. *Educ:* Ind Univ, BS, 49, PhD(biochem), 52. *Prof Exp:* Res asst phys chem, Yale Univ, 52-53; supvy biochemist, Mercy Hosp, Chicago, Ill, 53-55; res assoc biochem, Med Sch, Northwestern Univ, 55-58 & biophys, Univ Pittsburgh, 58-60; sr res assoc hemat, Hektoen Inst Med Res, Cook County Hosp, 60-63; asst prof microbiol, Albany Med Sch, 63-66; assoc prof biochem, Rochester Inst Technol, 66-87; RETIRED. *Mem:* AAAS; Am Chem Soc; NY Acad Sci. *Res:* Protein chemistry; physical biochemistry; neurochemistry. *Mailing Add:* 171 Brandywine Terr Rochester NY 14623-5251

SOWLS, LYLE KENNETH, b Darlington, Wis, Feb 28, 16; m 74; c 6. WILDLIFE BIOLOGY. *Educ:* Univ Wis, PhD(wildlife mgt), 51. *Prof Exp:* Biologist, Delta Waterfowl Res Sta, Wildlife Mgt Inst, Can, 46-50; leader coop Wildlife Res Unit, Univ Ariz, 50-86; RETIRED. *Concurrent Pos:* Fulbright vis lectr zool, Univ Col Rhodesia & Nyasaland, 62-63; wildlife consult, Food & Agr Orgn, Philippines, 67; mem, pigs & peccaries comt, Species Surv Comm, Int Union Conserv Nature. *Honors & Awards:* Terrestial Publ Award, Wildlife Soc, 55 & 80; Thomas E McCullough Award, Ariz Wildlife Fedn, 82. *Mem:* Wildlife Soc; Am Soc Mammal; Int Union Conserv Nature. *Res:* Game birds and mammals. *Mailing Add:* Rm 106 Biol Sci East Univ Ariz Tucson AZ 85721

SOWMAN, HAROLD G, b Murphys Boro, Ill, July 21, 23; m 45; c 2. CERAMIC ENGINEERING. *Educ:* Univ Ill, BS, 48, MS, 49, PhD(ceramic eng), 51. *Prof Exp:* Res assoc, Knoll Atomic Lab, Gen Elec Co, 52-57; mgr & supvr nuclear mat res, 3M Co, 57-65, sr res specialist, 66-70, corp scientist, 70-87; RETIRED. *Concurrent Pos:* Assoc ceramist, Titaniom Alloy Div, Nat Lead Co, 51-52. *Honors & Awards:* John Jeppson Medal & Award, Am Ceramic Soc, 85. *Mem:* Nat Acad Eng; fel Am Ceramic Soc; Nat Inst Ceramic Eng. *Res:* High temperature materials and ceramics; utilization of chemical ceramic on SOL-GEL technology for fibers, fabric, coatings and abrasive minerals. *Mailing Add:* 855 Towne Circle Stillwater MN 55082

SOYKA, LESTER F, b Chicago, Ill, Mar 12, 31; div; c 4. CARDIOVASCULAR CLINICAL RESEARCH. *Educ:* Univ Ill, MD, 61. *Prof Exp:* CLIN DIR THERAPEUT AREA, BRISTOL-MYERS RESEARCH CENTER, 82- *Mem:* Soc Pediat Res; Endocrine Soc; Am Soc Clin Res; Am Soc Pharmacol & Exp Therapeut. *Mailing Add:* Bristol-Myers Res Ctr Five Research Pkwy PO Box 5100 Wallingford CT 06492

SOZEN, M(ETE) A(VNI), b Istanbul, Turkey, May 22, 30; m 56. CIVIL ENGINEERING. *Educ:* Robert Col Istanbul, BS, 51; Univ Ill, MS, 52, PhD(civil eng), 57. *Hon Degrees:* Doctorate, Bogazici Univ, 89. *Prof Exp:* Engr, Kaiser Engrs, 52 & Hardesty & Hanover, 53; res assoc, 55-57, from asst prof to assoc prof, 57-63, PROF CIVIL ENG, UNIV ILL, URBANA, 63- *Honors & Awards:* Res Prize, Am Soc Civil Engrs, 63, R C Reese Prize, 70 & Moisseiff Prize, 72, Howard Award, 87, Base Award, 88; J W Kelly Award, Am Concrete Inst, 74, Bloem Award, 85; Drucker Award, 86. *Mem:* Nat Acad Eng; Am Soc Civil Engrs; Am Concrete Inst. *Res:* Reinforced and prestressed concrete structures; earthquake-resistant design. *Mailing Add:* Univ Ill 1245 NCEL Urbana IL 61801-2397

SPACH, MADISON STOCKTON, b Winston Salem, NC, Nov 10, 26; m; c 4. PEDIATRIC CARDIOLOGY. *Educ:* Duke Univ, AB, 50, MD, 54. *Prof Exp:* From instr to assoc prof, 57-68, PROF PEDIAT, SCH MED, DUKE UNIV, 68-, CHIEF PEDIAT CARDIOL, 60- *Concurrent Pos:* Chmn, Nat Heart, Lung & Blood Inst Manpower Rev Comt, 82-85. *Mem:* Soc Pediat Res (pres, 74); Am Acad Pediat; fel Am Col Cardiol; Asn Europ Pediat Cardiol; NY Acad Sci; Int Soc Heart Res; Biomed Eng Soc. *Res:* Electrophysiology, determining the mechanisms by which antisotropic structural complexities of the cardiac muscle alters the kinetics of ionic channels during the propogation of action potentials. *Mailing Add:* Box 3090 Duke Univ Med Ctr Durham NC 27706

SPACIE, ANNE, b Boston, Mass, Aug 19, 45. AQUATIC TOXICOLOGY, POLLUTION BIOLOGY. *Educ:* Mt Holyoke Col, BA, 67; Univ Calif, San Diego, MS, 69; Purdue Univ, PhD(limnol), 75. *Prof Exp:* Researcher aquatic biol, Union Carbide Corp, 69-73; asst prof, 75-81, ASSOC PROF FISHERIES, PURDUE UNIV, 81- *Concurrent Pos:* Vis prof, Savannah Rider Ecol Lab, 79; consult, US Environ Protection Agency; Alexander von Humboldt fel, Fed Repub Germany, 84-85. *Mem:* Am Fisheries Soc; Am Soc Limnol & Oceanog; AAAS; Am Soc Testing & Mat; Soc Environ Toxicol & Chem. *Res:* The accumulation and toxicity of synthetic organic compounds in fish and other aquatic organisms; effects of stream modifications on water quality and the distribution of fishes. *Mailing Add:* Forestry & Natural Resources Purdue Univ West Lafayette IN 47907-1159

SPACKMAN, DARREL H, b Morgan, Utah, July 18, 24; m 47; c 5. BIOCHEMISTRY, CANCER. *Educ:* Univ Utah, BA, 50, MA, 52, PhD, 54. *Prof Exp:* Res assoc, Rockefeller Inst, 54-59; sr biochemist, Spinco Div, Beckman Instruments, Inc, 59-62; res asst prof biochem, obstet & gynec, Univ Wash, 62-68; sr res biochemist, Pac Northwest Res Found, 68-82, chmn, Dept Microbiol, 82-85; RETIRED. *Concurrent Pos:* Res scientist, Fred Hutchinson Cancer Res Ctr, 75-85; asst prof, Rehab Med Dept, Univ Wash, 77-80. *Mem:* Am Asn Cancer Res; Am Soc Biol Chemists. *Res:* Amino acids, peptides and proteins of physiological fluids and tissues in hosts with malignancies; deprivation therapy in cancer research; methodology for automatic amino acid analysis. *Mailing Add:* 500 166th Ave NE Bellevue WA 98008

SPACKMAN, WILLIAM, JR, b Chicago, Ill, Sept 20, 19; m 42; c 3. PALEOBOTANY. *Educ:* Univ Ill, BS, 42; Harvard Univ, MA, 47, PhD(bot), 49. *Prof Exp:* Assoc biologist, US Naval Shipyard, Pa, 44-45; from asst prof to assoc prof, 49-61, prof paleobot, 61-86, dir coal res sect, Earth & Mineral Sci Exp Sta, 86-87, EMER PROF PALEOBOT, PA STATE UNIV, 86- *Concurrent Pos:* Mem, Int Comn Coal Petrol. *Honors & Awards:* Joseph Becker Award, Am Inst Mining, Metall & Petrol Engrs; G H Cady Award, Geol Soc Am; Reinhardt Thiessen Medal, Int Comn Coal Petrol. *Mem:* Geol Soc Am. *Res:* Tertiary floras; fossil woods; plant phylogeny; coal petrology; modern phytogenic sediments; peat to coal transformation. *Mailing Add:* Dept Geosci 510 Deike Bldg Pa State Univ University Park PA 16802

SPADAFINO, LEONARD PETER, b Jersey City, NJ, Oct 25, 31; m 56; c 2. ORGANIC CHEMISTRY. *Educ:* Univ Ga, BS, 58, PhD(chem), 63. *Prof Exp:* ORG CHEMIST, TENN EASTMAN CO, 63- *Mem:* Am Asn Textile Chemists & Colorists; Am Chem Soc. *Res:* Kinetics of the decomposition of peroxides in systems where stable free radicals function as scavengers of reactive radicals; anthraquinone and azo dyes for synthetic fibers and films. *Mailing Add:* 4413 Beechcliff Dr Kingsport TN 37664

SPADONI, LEON R, b Kent, Wash, Aug 11, 30; m 57; c 3. OBSTETRICS & GYNECOLOGY, REPRODUCTIVE ENDOCRINOLOGY & INFERTILITY. *Educ:* Univ Wash, BS, 53, MD, 57. *Prof Exp:* Intern med, Minn Gen Hosp, 57-58; resident obstet & gynec, Univ Integrated Hosp, 60-63, from instr to assoc prof, Sch Med, 63-74, PROF OBSTET & GYNEC, SCH MED, UNIV WASH, 74-, VCHMN, 77 - *Concurrent Pos:* Attend physician, Univ & Harborview Med Ctr, 63-; consult, Univ Wash Hall Health Ctr, 63 -; consult, Madigan Gen Hosp, Tacoma, 69; pres, Pac Coast Fertil Soc, 83; examr, Am Bd Obstet & Gynec, 85; chief of staff, Univ Wash Med Ctr, 89-91. *Mem:* Am Fertil Soc; Am Col Obstet & Gynec; Soc Reproductive Surgeons; Am Bd Obstet & Gynec. *Res:* Infertility; gynecologic endocrinology. *Mailing Add:* Dept Obstet & Gynec RH-20 Univ Wash Med Ctr Seattle WA 98195

SPAEDER, CARL EDWARD, JR, b Meadville, Pa, Mar 29, 35; m 62; c 3. METALLURGY. *Educ:* Pa State Univ, University Park, BS, 57; Carnegie-Mellon Univ, MS, 63; Univ Pittsburgh, PhD(metall), 70. *Prof Exp:* Sr res metallurgist, US Steel Res Ctr, 57-90; SR RES METALLURGIST, ARISTECH CHEM, 90- *Mem:* Am Soc Metals; Am Soc Mech Engrs. *Res:* Elevated temperature; properties of metals; cryogenic properties; formability; material characteristics. *Mailing Add:* 3905 Princess Ct Murraysville PA 15668

SPAEPEN, FRANS, b Mechelen, Belgium, Oct 29, 48; m 73; c 3. PHYSICAL METALLURGY, MATERIALS SCIENCE. *Educ:* Univ Leuven, Belgium, ME, 71; Harvard Univ, PhD(appl physics), 75. *Prof Exp:* Res fel, 75-77, from asst prof to assoc prof, 77-83, GORDON MCKAY PROF APPL PHYSICS, HARVARD UNIV, 83-, DIR MAT RES LAB, 90- *Concurrent Pos:* Vis prof, Univ Leuven, Belgium, 84-85; counr, Mat Res Soc, 86-89 & 90-93; chmn, Phys Metall Gordon Conf, 88. *Mem:* Am Soc Metals; fel Am Phys Soc; Mat Res Soc; Am Soc Mining & Metall Engrs; Böhmische Phys Soc. *Res:* Atomic transport (viscosity, diffusivity, phase transformations) in amorphous materials: metals, semiconductors and oxides; artificial multilayers: preparation, stability, mechanical properties; properties of interfaces: crystal/melt, amorphous/crystalline semiconductor, grain boundaries, grain growth, interface tension, interface stress; quasicrystals. *Mailing Add:* Div Appl Sci Harvard Univ Pierce Hall 29 Oxford St Cambridge MA 02138

SPAET, THEODORE H, b New York, NY, June 24, 20; m 41, 71; c 3. MEDICINE. *Educ:* Univ Wis, BA, 42; New York Med Col, MD, 45; Am Bd Internal Med, dipl. *Prof Exp:* Intern, Montefiore Hosp, New York, 45-46, asst resident med, 48-49; asst resident, Morrisania Hosp, 49-50; from instr to asst prof, Sch Med, Stanford Univ, 51-55; assoc prof path, Col Physicians & Surgeons, Columbia Univ, 55-65; PROF MED, ALBERT EINSTEIN COL MED, 65- *Concurrent Pos:* Damon Runyon clin res fel hemat, New Eng Ctr Hosp, Boston, 50-51; dir dept hemat, Montefiore Hosp, 55-85; consult, Manhattan Vet Admin Hosp, 58, St Luke's Hosp & Bronx Vet Admin Hosp; former mem exec comt, Coun Thrombosis, Am Heart Asn; mem coun arteriosclerosis, Am Heart Asn. *Mem:* AAAS; Asn Am Physicians; Am Soc Clin Invest; Am Physiol Soc; Am Soc Hemat; Am Fedn Clin Res; Int Soc Hemat. *Res:* Hemorrhagic diseases; thrombosis; hemolytic anemias. *Mailing Add:* Montefiore Med Ctr Albert Einstein Col Med 110 E 210 St Bronx NY 10467

SPAETH, GEORGE L, b Philadelphia, Pa, Mar 3, 32; m 58; c 3. OPHTHALMOLOGY. *Educ:* Yale Univ, BA, 54; Harvard Med Sch, MD, 59; Am Bd Ophthal, dipl, 65. *Prof Exp:* Intern, Univ Hosp, Ann Arbor, Mich, 60; resident ophthal, Wills Eye Hosp, 63; clin assoc, Nat Inst Neurol Dis & Blindness, 63-65; instr ophthal, Univ Pa, 65-68; clin instr ophthal, Temple Univ, 68, from assoc prof to prof, 68-74; PROF OPHTHAL, THOMAS JEFFERSON UNIV, 74- *Concurrent Pos:* Asst ophthalmologist, Grad Hosp, Univ Pa, 65, assoc ophthalmologist, 66; clin asst, Wills Eye Hosp, 65, sr asst surgeon, 66, assoc surgeon, 68, dir glaucoma serv, 68, attend surgeon, 73, pres med staff, 84; grants, Nat Soc Prev Blindness, 67 & 68, Nat Coun Combat Blindness, 68 & Nat Eye Inst, 72-75; attend surgeon ophthal, Chestnut Hill Hosp, 75-84; pres, Am Glaucoma Soc. *Honors & Awards:* Sr Hawr Award, Am Acad Ophthal. *Mem:* Fel Am Ophthal Soc; fel Am Acad Ophthal & Otolaryngol; fel Danish Ophthal Soc; fel Royal Soc Med; fel Am Col Surgeons; fel Royal Soc Health; fel Ophthal Soc UK; AMA; Asn Res Vision & Ophthal; AAAS; Pan-Am Ophtalmic Asn; Int Soc Pediat Ophthal. *Res:* Diagnosis, treatment and pathophysiology of glaucoma; metabolic diseases, their ocular aspects and treatment, especially homocystinuria; sociology of chronic disease; ocular surgery, especially of glaucoma. *Mailing Add:* 15 Laughlin Lane Philadelphia PA 19118

SPAETH, RALPH, b Cleveland, Ohio, Mar 21, 05; m 32; c 2. MEDICINE. *Educ:* Western Reserve Univ, AB, 27, MD, 31; Am Bd Pediat, dipl, 38. *Prof Exp:* From instr to prof, 36-72, emer clin prof, Col Med, Univ Ill, Chicago, 72-; emer clin prof pediat, Rush Med Col, 72-; RETIRED. *Concurrent Pos:* Supvr physician, East Off, Chicago Div, State of Ill Children & Family Serv, 37-75; mem, Nat Comn Venereal Dis, 71-72; regional medical coordr, Proj Head Start to 78. *Mem:* Am Acad Pediat; AMA; Sigma Xi; Am Venereal Dis Asn; NY Acad Sci. *Res:* Immunology and clinical management of tetanus; immunology and immunization against mumps; prevention of rabies; therapy of poliomyelitis with convalescent serum; active immunization against measles. *Mailing Add:* 9030 S Bell Ave Chicago IL 60620

SPAGHT, MONROE EDWARD, b Arcata, Calif, Dec 9, 09; m; c 3. PHYSICAL CHEMISTRY, ORGANIC CHEMISTRY. *Educ:* Stanford Univ, AB, 29, AM, 30, PhD(chem), 33. *Hon Degrees:* DSc, Rensselaer Polytech Inst, 58, Drexel Inst, 62; LLD, Univ Manchester, 64, Calif State Cols, 65, Milliken Univ, 67, Wesleyan Univ, 68; Colo Sch Mines, DEng, 71. *Prof Exp:* Res chemist & technologist, Shell Oil Co, 33-45, vpres, Shell Develop Co, 46-48, pres, 49-52, exec vpres, Shell Oil Co, 53-60, pres, 61-65, managing dir, Shell Group, 65-70; dir, Royal Dutch Petrol Co, 65-80; RETIRED. *Concurrent Pos:* Chmn bd dirs, Shell Oil Co, 65-70. *Honors & Awards:* Midwest Res Inst Award, 62; Johnson Lectr & Medallist, Stockholm, Sweden, 66; Brit Soc Chem Indust Medal, 66. *Mem:* Nat Acad Eng; fel AAAS; Am Chem Soc; fel Am Inst Chem Eng; Brit Soc Chem Indust (int pres, 63-64). *Res:* Thermodynamics of organic compounds; chemical derivatives of petroleum. *Mailing Add:* Shell Centre London SE1 7NA England

SPAHN, GERARD JOSEPH, b Baltimore, Md, May 4, 38; m 61; c 4. MICROBIOLOGY. *Educ:* Mt St Mary's Col, Md, BS, 60; St John's Univ, MS, 62; Univ Md, PhD(microbiol), 65. *Prof Exp:* Lab scientist rabies diag, Livestock Sanit Serv Lab, 65-66; virologist, Microbiol Assoc, Inc, Md, 66-72; sr scientist, Litton Bionetics, Inc, 72-76; virologist, Microbiol Assocs, Inc, 76-77; sr scientist, Environ Control, Inc, Rockville, Md, 77-79; DIR SAFETY, SALK INST BIOL STUDIES, SAN DIEGO, 79- *Mem:* Am Asn Lab Animal Sci; Am Soc Microbiol; Tissue Cult Asn; Am Biol Safety Asn; Sigma Xi. *Res:* Oncogenic virus expression in cell culture; tumorigenicity in vivo and transformation in vitro; studies on spontaneous neoplasms of rats and mice; environmental, occupational health, and safety program. *Mailing Add:* Salk Inst PO Box 85800 San Diego CA 92138

SPAHN, ROBERT JOSEPH, b Chicago, Ill, July 2, 36; m 61; c 4. APPLIED MATHEMATICS. *Educ:* Mich Technol Univ, BS, 58; Mich State Univ, PhD(physics), 63. *Prof Exp:* Engr, NAm Aviation, Inc, Ohio, 63-64; asst prof, 64-70, ASSOC PROF MATH, MICH TECHNOL UNIV, 70- *Mem:* Soc Indust & Appl Math. *Res:* Solutions of boundary value problems in partial differential equations. *Mailing Add:* Dept of Math Mich Technol Univ Houghton MI 49931

SPAHR, SIDNEY LOUIS, b Bristol, Va, Sept 5, 35; m 60, 75; c 2. AGRICULTURE, DAIRY SCIENCE. *Educ:* Va Polytech Inst, BSc, 58; Pa State Univ, MSc, 60, PhD(dairy sci), 64. *Prof Exp:* Instr dairy sci, Pa State Univ, 62-64; asst prof dairy husb, 64-70, from assoc prof to prof dairy sci, 72-85, PROF ANIMAL SCI, UNIV ILL, URBANA, 85- *Concurrent Pos:* Staff officer, Nat Acad Sci, Washington, DC, 70-72. *Honors & Awards:* Am Dairy Sci Asn, 87. *Mem:* AAAS; Am Dairy Sci Asn; Am Soc Animal Sci; US Animal Health Asn. *Res:* Dairy cattle management and nutrition; dairy automation; electronic animal identification; automatic data acquisition; application of data base management; expert systems and artificial intelligence techniques in dairy herd management. *Mailing Add:* 215 Animal Sci Lab Univ of Ill 1207 W Gregory Dr Urbana IL 61801

SPAHT, CARLOS G, II, b New Orleans, La, June 22, 42; m 64; c 2. MATHEMATICS, OPERATIONS RESEARCH. *Educ:* La State Univ, BS, 64, MS, 66, PhD(math), 70. *Prof Exp:* Asst math, La State Univ, Baton Rouge, 64-70, spec lectr, 70; from asst prof to assoc prof math, 72-85, prof math & comput sci, 81-85, chmn, Math & Comput Sci Dept, 85-88, PROF MATH & COMPUT SCI, LA STATE UNIV, SHREVEPORT, 88- *Concurrent Pos:* Instr & consult, Educ Ctr, Barksdale Air Force Base, 73-75. *Mem:* Am Math Asn; Am Math Soc; Nat Coun Teachers Math. *Res:* Abstract algebra; operation research field. *Mailing Add:* Dept of Math La State Univ 8515 Youree Dr Shreveport LA 71115

SPAID, FRANK WILLIAM, b Pocatello, Idaho, Mar 7, 38; m 64; c 2. FLUID DYNAMICS. *Educ:* Ore State Univ, BS, 59; Calif Inst Technol, MS, 61, PhD(mech eng), 64. *Prof Exp:* Assoc res engr, Jet Propulsion Lab, Pasadena, 59-60, res engr, 61; supvr, Douglas Aircraft Co, Inc, 64-67; asst prof aeronaut,

Univ Calif, Los Angeles, 67-72; CHIEF SCIENTIST, McDONNELL DOUGLAS RES LABS, 72- *Mem:* Am Inst Aeronaut & Astronaut. *Res:* The interaction of a liquid or gaseous jet with a supersonic flow; boundary layer separation; transonic fluid dynamics; hypersonic fluid dynamics. *Mailing Add:* McDonnell Douglas Res Labs PO Box 516 St Louis MO 63166

SPAIN, IAN L, b Saltwood, Eng, June 19, 40; wid; c 3. CHEMICAL PHYSICS. *Educ:* Univ London, BSc, 61, PhD(chem physics), 64, Royal Col Sci, ARCS, 61, Imp Col, dipl, 64. *Prof Exp:* Res assoc, Inst Study Metals, Univ Chicago, 64-66; asst prof molecular physics, Univ Md, Col Park, 66-70, from assoc prof to prof mat sci, 70-79, dir, lab high pressure sci, 72-79, dir eng mat prog, 75-79; PROF, DEPT PHYSICS, COLO STATE UNIV, 79- *Concurrent Pos:* Prin res fel, Admiralty Underwater Weapons Estab, Gt Brit, 74-75; foreign collabr, Nuclear Study Ctr of Grenoble, 77 & 78, 86, 87, 88 Max Planck Inst Solid State Sci, Stuttgart, 84; vis scientist, RSRE, Gt Mahrenn, Eng, 87 & 88. *Mem:* Am Inst Physics; fel Brit Inst Physics; Sigma Xi. *Res:* Electronic, optical and structural studies of solids at high pressure; growth, structural characterization and physical properties of carbon, grahite and carbon fibers; synthetic metals. *Mailing Add:* Dept Physics Colo State Univ Ft Collins CO 80523

SPAIN, JAMES DORRIS, JR, b Washington, DC, Feb 3, 29; m 52; c 3. BIOCHEMISTRY. *Educ:* Mich Technol Univ, BS, 51; Med Col Va, MS, 53; Stanford Univ, PhD(chem), 56. *Prof Exp:* Res fel biochem, Univ Tex, M D Anderson Hosp & Tumor Inst, 55-56; from asst prof to assoc prof chem, Mich Technol Univ, 56-62, head dept biol sci, 62-68, prof biochem, 62-84 DIR, INSTRNL COMPUT CTR, EASTERN MICH UNIV, 84-85. *Concurrent Pos:* Vis prof chem, Clemson Univ, 85- *Honors & Awards:* Mich Tech Fac Res Award, 65. *Mem:* Am Chem Soc; Sigma Xi. *Res:* Computerized instruction design; computer modeling of biological systems; mixing of water masses in large lakes; simulation of biological and chemical systems; liver damage and azo dye carcinogenesis, histochemistry; precipitation chromatography; physical and chemical limnology; computer search and retrieval. *Mailing Add:* 504 Hunters Glen Central SC 29630

SPALATIN, JOSIP, b Ston, Yugoslavia, Jan 29, 13; m 41; c 3. VIROLOGY. *Educ:* Univ Zagreb, BS, 38, DVM, 41; Univ Giessen, PhD(vet med), 44. *Prof Exp:* Teaching asst animal infectious dis, Univ Zagreb, 39-46; res assoc vaccine & sera prod, Vetserum Kalinovica, Yugoslavia, 46-53; from res assoc virol to assoc prof zoonoses, Sch Med, Univ Zagreb, 53-61; vis prof virol, Univ Sask, 61-63; RES ASSOC VIROL, UNIV WIS-MADISON, 63- *Concurrent Pos:* Yugoslav fel, Univ Wis-Madison, 56-59. *Mem:* Asn Yugoslav Microbiologists (treas, 40-45); Wildlife Dis Asn; Sigma Xi. *Res:* Epidemiology, diagnosis and immunology in groups of mixoviruses; arborviruses and psittacosis lymphogranuloma venereum agents. *Mailing Add:* 4505 Onaway Pass Madison WI 53711

SPALDING, DAN WESLEY, physics, for more information see previous edition

SPALDING, DONALD HOOD, plant pathology, for more information see previous edition

SPALDING, GEORGE ROBERT, b Lancaster, Pa, Dec 1, 27; m 52; c 3. ARCHITECTURAL ACOUSTICS, INDUSTRIAL NOISE. *Educ:* Pa State Univ, BS, 53, MS, 55. *Prof Exp:* Res asst meteorol, Univ Mich, 55; res physicist, 55-72, sr res scientist, 72-89, PRIN SCIENTIST, ARMSTRONG WORLD INDUST, INC, 89- *Mem:* Acoust Soc Am; Am Soc Testing & Mat. *Res:* Architectural and landscape office acoustics; industrial noise control; design and evaluation of systems for use in open plan office spaces; acoustical materials testing. *Mailing Add:* Armstrong World Indust Inc Box 3511 2500 Columbia Ave Lancaster PA 17604

SPALL, HENRY ROGER, b Newcastle upon Tyne, Eng, Oct 10, 38; US citizen. GEOPHYSICS. *Educ:* Univ London, BSc, 62, PhD(geophys), 70; Southern Methodist Univ, MS, 68. *Prof Exp:* Res asst geophys, Cambridge Univ, 62-64 & Southwest Ctr Advan Studies, Univ Tex, Dallas, 64-67; lectr geol, Southern Methodist Univ, 67-68; geologist, Mobil Res Labs, 68-69; Coop Inst Res Environ Sci fel, Univ Colo, Boulder, 70-71; geophysicist, Environ Res Labs, Nat Oceanic & Atmospheric Admin, 71-73; GEOPHYSICIST, US GEOL SURV, 73- *Concurrent Pos:* Mem working group 10, Comn Geodynamics, Int Union Geod & Geophys, 72-; ed, Geol, 73- & Earthquakes & Volcanoes, 75- *Mem:* Am Geophys Union; Seismol Soc Am; fel Geol Soc Am; fel Royal Astron Soc; Europ Asn Sci Ed. *Res:* Paleomagnetism and plate tectonics. *Mailing Add:* Off of Sci Publ US Geol Surv Reston VA 22092

SPALL, WALTER DALE, b Greeley, Colo, Apr 23, 43; m 70. ANALYTICAL CHEMISTRY. *Educ:* Colo Col, BA, 66; Univ NMex, PhD(anal chem), 70. *Prof Exp:* Res chemist, Chem Div, Uniroyal, Inc, 70-75; MEM STAFF, LOS ALAMOS SCI LAB, UNIV CALIF, NMEX, 75- *Concurrent Pos:* Part-time asst prof, Univ New Haven, 71-75. *Mem:* Am Chem Soc; Sigma Xi. *Res:* Chromatography; analytical instrumentation; computer automation of instrumentation; mass spectroscopy. *Mailing Add:* Group LSB-1 Mail Stop M880 Los Alamos Sci Lab Los Alamos NM 87545

SPALLHOLZ, JULIAN ERNEST, b Boston, Mass, Oct 8, 43; m 64; c 2. BIOCHEMISTRY, NUTRITION. *Educ:* Col State Univ, BS, 65, MS, 68; Univ Hawaii, PhD(biochem), 71. *Prof Exp:* Fel, Dept Biochem, Colo State Univ, 71-72, res assoc, 72-73, instr, 73-74; res chemist nutrit, Lab Exp Metab Dis, Vet Admin Hosp, 74-78; assoc res chemist, State Univ NY, Albany, 78; assoc prof, 78-84, PROF, DEPT FOOD & NUTRIT, TEX TECH UNIV, 84- *Concurrent Pos:* Interim dir, Inst Nutrit Sci, 81-84, dir, 85-; ed-in-chief, J Nutrit Immunol. *Mem:* Sigma Xi; Am Inst Nutrit. *Res:* Nutritional importance of trace metals, especially selenium; immunology toxicology; application of physical probes such as nitroxides, fluorescent molecules and radionuclide probes to biological, biochemical and immunological research. *Mailing Add:* Dept Food & Nutrit Tex Tech Univ Lubbock TX 79409

SPAMFORD, BRYANT, b Pittsburgh, Pa, Sept 10, 46. EXCERCISE PHYSIOLOGY. *Educ:* Univ Pittsburgh, PhD(excercise physiol & phys educ), 73. *Prof Exp:* DIR, EXERCISE PHYSIOL LAB, DIV ALLIED HEALTH, SCH MED, UNIV LOUISVILLE, 73-, PROF, 78-, ASST DEAN, GRAD SCH, 84-, DIR, HEALTH PROMOTION CTR. *Mem:* Am Physiol Soc; Sigma Xi; Am Col Sports Med; Am Alliance Health, Phys Educ & Recreation. *Mailing Add:* Health Promotion Ctr 102C Carmichael Bldg Univ Louisville Louisville KY 40292

SPANDE, THOMAS FREDERICK, b Madison, Wis, June 22, 37; m 68. ORGANIC CHEMISTRY. *Educ:* St Olaf Col, BA, 59; Princeton Univ, PhD(org chem), 65. *Prof Exp:* Staff fel org chem, 64-66, RES CHEMIST, NIH, 66- *Mem:* Am Chem Soc. *Res:* Steroid chemistry; amino acid and protein chemistry, particularly indole and tryptophan chemistry. *Mailing Add:* Lab 1A-18 Lab Bioorganic Chem Bldg Eight NIDDK Bethesda MD 20892

SPANDORFER, LESTER M, b Norfolk, Va, Oct 16, 25; m 56; c 2. ELECTRICAL ENGINEERING, COMPUTER SCIENCE. *Educ:* Univ Mich, BSEE, 47, MSEE, 48; Univ Pa, PhD(elec eng), 56. *Prof Exp:* Mem tech staff, Bell Tel Labs, 48-50; res asst dir comput ctr & proj mgr, Univ Pa, 50-57; sr engr, Sperry Univac Div, Sperry Rand Corp, 57-58, dept mgr, 58-60, staff consult, 60-65, dept mgr, 65-67, dir tech develop, 67-72, dir data entry, 72-77, off automation, 77-85; PRES, DESIGN DATA, INT, 85- *Mem:* Fel Inst Elec & Electronics Engrs. *Res:* Computer design; application of semiconductor and magnetic devices. *Mailing Add:* 8012 Ellen Lane Cheltenham PA 19012

SPANEL, LESLIE EDWARD, b St Louis, Mo, Mar 13, 37; m 61; c 2. SOLID STATE PHYSICS. *Educ:* Univ Mo-Rolla, BS, 59; Iowa State Univ, PhD(physics), 64. *Prof Exp:* Res asst physics, Ames Lab, AEC, Iowa State Univ, 60-64; res specialist, Microelectronics Orgn, Boeing Space Div, Boeing Co, 64-68; asst prof, 68-74, ASSOC PROF PHYSICS, WESTERN WASH UNIV, 74- *Mem:* Am Phys Soc. *Res:* Fermi surface of magnetic and nonmagnetic metals; transport properties of semiconductors. *Mailing Add:* Physics/Astron Dept Western Wash Univ Bellingham WA 98225

SPANG, ARTHUR WILLIAM, b Detroit, Mich, Aug 16, 17; m 44, 67; c 5. MICROCHEMISTRY. *Educ:* Wayne State Univ, BS, 41. *Prof Exp:* Org microanalyst & head lab, Parke, Davis & Co, 41-48, med detailing, 48-49; mem analytical staff, Upjohn Co, 49-50; head microanal lab, Olin Mathieson Chem Corp, 50, supvr res analytical lab, 51-54; OWNER, SPANG MICROANAL LAB, 54- *Mem:* Am Chem Soc; Am Microchem Soc; Royal Soc Chem. *Res:* New methods for the microanalysis of new types of organic and organometallic compounds. *Mailing Add:* Spang Microanal Lab Star Rte One Box 142 Eagle Harbour MI 49951-9711

SPANG, H AUSTIN, III, b New Haven, Conn, July 16, 34; m 57; c 3. CONTROL ENGINEERING. *Educ:* Yale Univ, BE, 56, MEng, 58, DEng(elec eng), 60. *Prof Exp:* Lab asst, Yale Univ, 56-58, res engr commun, 58-60; CONTROL ENGR, GEN ELEC RES & DEVELOP CTR, 60- *Concurrent Pos:* Instr, New Haven Col, 58-60; lectr, Univ Calif, Los Angeles, 66-69; assoc ed, Automatica, Int Fedn Automatic Control, 67-80; assoc ed appln, Inst Elec & Electronics Engrs Trans on Automatic Control, 77-79; ed, Automatica, Int Fedn Automatic Control, 80- *Mem:* Fel Inst Elec & Electronics Engrs; Soc Indust & Appl Math; Sigma Xi. *Res:* Applications of control and their digital implementation; multivariable control; real-time computer control; computer aided control system design. *Mailing Add:* 2525 Hilltop Rd Schenectady NY 12309

SPANGENBERG, DOROTHY BRESLIN, b Galveston, Tex, Aug 31, 31; m 58; c 1. ZOOLOGY, BIOCHEMISTRY. *Educ:* Univ Tex, BA, 56, MA, 58, PhD(zool), 60. *Prof Exp:* Dir, Spangenberg Labs, 60-62; res assoc, Med Ctr, Univ Ark, 62-65; assoc prof biol res, Univ Little Rock, 65-66; res scholar zool, Ind Univ, 66-69; res assoc, Water Resources Lab, Univ Louisville, 69-70, Dept Oral Biol, Sch Dent, 70-72; vis assoc prof molecular, cellular & develop biol, Univ Colo, Boulder, 72-77; assoc prof, 77-80, RES PROF, EASTERN VA MED SCH, 80- *Concurrent Pos:* Grants, NSF, 64-66, Sigma Xi, 65-66, NIH, 66-, NIH & Nat Inst Dent Res, 78, NIH & Nat Inst Child Health & Human Develop & Dept Energy Contract, 77-82 & NASA, 84- *Mem:* Am Soc Zoologists; Am Soc Cell Biol; Sigma Xi; Electron Micros Soc Am; AAAS; NY Acad Sci; Am Soc Gravity & Space Biol. *Res:* Development of coelenterate model systems for study of mechanisms of cellular and organismal development, especially metamorphosis, utilizing biochemical and cytological technics. *Mailing Add:* Eastern Va Med Sch Norfolk VA 23501

SPANGLER, CHARLES WILLIAM, b Philadelphia, Pa, Feb 12, 38; m 61; c 2. PHYSICAL ORGANIC CHEMISTRY. *Educ:* Mass Inst Technol, BS, 59; Northeastern Univ, MS, 61; Univ Md, PhD(org chem), 64. *Prof Exp:* Great Lakes Cols Asn teacher intern org chem, Ohio Wesleyan Univ, 64-65; from asst prof to assoc prof, 65-81, PROF ORG CHEM, NORTHERN ILL UNIV, 81- *Concurrent Pos:* Res Corp & NSF res grants; presidential res prof, Northern Ill Univ, 91- *Mem:* Am Chem Soc; Royal Soc Chem; Mat Res Soc. *Res:* Chemistry of conjugated polyenes; electrophilic substitution; new organic materials for nonlinear optics; electrocyclic reactions; photochemistry of polyunsaturated systems; sigmatropic migrations; conjugated conducting polymers and monomers. *Mailing Add:* Dept Chem Northern Ill Univ De Kalb IL 60115

SPANGLER, DANIEL PATRICK, b Meadows of Dan, Va, Apr 22, 34; m 61; c 1. GEOLOGY, HYDROLOGY & WATER RESOURCES. *Educ:* Berea Col, BA, 56; Univ Va, MS, 64; Univ Ariz, PhD(geol), 69. *Prof Exp:* Geologist, Va Hwy Dept, 57-59 & 61-63; asst, Univ Va, 63-65 & Univ Ariz, 65-67; geologist, Agr Res Serv, USDA, 67-69; asst prof geol, SFla Univ, 69-74; asst prof, 74-77, ASSOC PROF GEOL, UNIV FLA, 77- *Concurrent Pos:* Consult, hydrogeol, 69-; partic, NSF-Am Geol Inst Tenth Int Field Inst, Spain, 71; res award, Univ SFla & Penrose bequest res grant, Geol Soc Am, Tampa, Fla, 72; state & fed res grants, 74- *Mem:* Geol Soc Am; Am Water

Resources Asn; Nat Asn Geol Teachers; Nat Water Well Asn; Am Inst Prof Geologists; Sigma Xi. *Res:* Hydrogeologic systems; application of geophysics to hydrogeologic problems; engineering geology. *Mailing Add:* Dept Geol Univ Fla Gainesville FL 32611

SPANGLER, FRED WALTER, b Park Ridge, Ill, Feb 27, 18; m 41; c 2. CHEMISTRY. *Educ:* Carthage Col, AB, 40; Univ Ill, PhD(org chem), 44. *Prof Exp:* Asst chem, Univ Ill, 40-42; res chemist, Eastman Kodak Co, 44-52, tech assoc, 52-59, asst supt, Film Emulsion Div, 59-81; RETIRED. *Mem:* Am Chem Soc; Soc Photog Sci & Eng. *Res:* Grignard reactions involving the naphthalene nucleus; organic chemicals used in photography; anthraquinone and related dyes; detergents and wetting agents; photographic emulsions. *Mailing Add:* 121 Nob Hill Rochester NY 14617

SPANGLER, GEORGE RUSSELL, b Susanville, Calif, Oct 22, 42; m 62; c 4. AQUATIC ECOLOGY, POPULATION DYNAMICS. *Educ:* Humboldt State Col, BS, 64; Univ Toronto, MS, 66, PhD(zool), 74. *Prof Exp:* Res scientist fisheries, Ont Dept Lands & Forests, 68-71; scientist-in-chg, Lake Huron Res Unit, Ont Ministry Natural Resources, 72-78; ASSOC PROF FISHERIES, UNIV MINN, 78- *Mem:* Am Fisheries Soc; Int Asn Great Lakes Res; Biomet Soc; Am Inst Fishery Res Biologists. *Res:* Population dynamics of fish stocks; predator-prey interactions; effects of exploitation on fish communities; efficiency and selectivity of fishing gear. *Mailing Add:* Dept Fisheries & Wildlife Univ Minn 200 Hodson Hall St Paul MN 55108

SPANGLER, GLENN EDWARD, b York, Pa, June 20, 42; m 86. FORENSIC SCIENCE, ENVIRONMENTAL DETECTION TECHNOLOGY. *Educ:* Gettysburg Col, BA, 65; Univ Va, PhD(physics), 70. *Prof Exp:* Physicist thermometry, Nat Bur Stand, 70-71; res physicist spectrometry, US Army Mobility Equip Res & Develop Command, 71-78; PRIN SCIENTIST DETECTION, ETG INC, 78- *Concurrent Pos:* Consult, explosive vapor detection, 70-79. *Mem:* Am Phys Soc; Am Chem Soc; Am Soc Mass Spectrometry. *Res:* Ion mobility spectrometry; mass spectrometry; atmospheric pressure ionization; chemical and electrochemical vapor detection; membrane permeability; gas purification; environmental science. *Mailing Add:* Environ Technol Group Inc 1400 Taylor Ave Baltimore MD 21284-9840

SPANGLER, GRANT EDWARD, b Lebanon, Pa, Oct 17, 26; m 54; c 3. METALLURGICAL ENGINEERING, MATERIAL SCIENCES. *Educ:* Lehigh Univ, BS, 50; Univ Pa, MS, 53. *Prof Exp:* Res scientist metall, Westinghouse Atomic Power Div, 50-52, Res Lab, Air Reduction Co, 53-56 & Franklin Inst Res Labs, 56-64; res scientist, 64-65, sect dir, 56-66, 65-66, dept dir, 66-78, GEN DIR METALL RES DIV, REYNOLDS METALS CO, 78- *Mem:* Fel Am Soc Metals; Am Inst Mining, Metall & Petrol Engrs. *Res:* Fuel element development; treatment of molten metals by powder injection; purification of reactive metals by floating zone refining; alloy development, physical metallurgy and process metallurgy of aluminum and aluminum alloys. *Mailing Add:* 307 Coal Port Rd Richmond VA 23229

SPANGLER, HAYWARD GOSSE, b Redbank, NJ, July 6, 38; m 66; c 2. ENTOMOLOGY. *Educ:* La Sierra Col, BS, 61; Univ Ariz, MS, 63; Kans State Univ, PhD(entom), 67. *Prof Exp:* RES ENTOMOLOGIST, CARL HAYDEN BEE RES CTR, AGR RES SERV, USDA, ARIZ, 67- *Mem:* Entom Soc Am; Sigma Xi; Int Union Study Social Insects. *Res:* Insect behavior; insect bioacoustics. *Mailing Add:* Carl Hayden Bee Res Ctr USDA-ARS 2000 E Allen Rd Tucson AZ 85719

SPANGLER, JOHN ALLEN, b Morgantown, WVa, Jan 1, 18; m 48; c 3. CHEMISTRY. *Educ:* WVa Univ, AB, 39, PhD(chem), 42. *Prof Exp:* Res chemist, Am Viscose Corp, 42-44; assoc prof, 46-52, chmn dept, 55-58, PROF CHEM, SAN DIEGO STATE UNIV, 54- *Mem:* Am Chem Soc. *Res:* Multiple-junction thermocouples for low freezing-point measurements; physical constants related to easier methods of analysis; ternary systems; physical methods of analysis; fiber chemistry. *Mailing Add:* 4959 Catactin Dr San Diego CA 92115-2608

SPANGLER, JOHN DAVID, b Lincoln, Nebr, Nov 18, 36; m 58; c 5. PHYSICS. *Educ:* Kans State Univ, BS, 58; Duke Univ, PhD(physics), 61. *Prof Exp:* Res assoc physics, Duke Univ, 61-62; asst prof, DePauw Univ, 64-65; from asst prof to assoc prof, 65-80, PROF PHYSICS, KANS STATE UNIV, 80- *Mem:* Am Phys Soc; Soc Indust & Appl Math; Am Asn Physics Teachers. *Res:* Theoretical applied physics. *Mailing Add:* 601 Brevoort Lane Green Bay WI 54301

SPANGLER, MARTIN ORD LEE, b Roanoke, Va, Sept 17, 28; m 56; c 4. ORGANIC CHEMISTRY, BIOCHEMISTRY. *Educ:* Bridgewater Col, BA, 50; Va Polytech Inst & State Univ, MS, 53, PhD(chem), 59. *Prof Exp:* Res assoc anal biol fluids, Univ Mich Hosp, 53-55; instr chem, Va Polytech Inst & State Univ, 55-56; assoc prof, Waynesburg Col, 58-59 & King Col, 59-66; assoc prof, 66-68, chmn dept, 73-80, PROF CHEM, ELIZABETHTOWN COL, 68- *Concurrent Pos:* Vis prof, Hershey Med Ctr, Pa State Univ, 72-73; lectr, Ohio State Univ, 80-81. *Mem:* AAAS; Am Chem Soc; Sigma Xi. *Res:* Organic synthesis and mechanisms of organic reactions; synthesis of antitumor agents and antibiotics from sugar derivaties. *Mailing Add:* Dept Chem 210 Musset Hall Elizabethtown Col Elizabethtown PA 17022-2298

SPANGLER, PAUL JUNIOR, b York, Pa, Nov 21, 24; m 50; c 1. ENTOMOLOGY. *Educ:* Lebanon Valley Col, AB, 49; Ohio Univ, MS, 51; Univ Mo, PhD(entom), 60. *Prof Exp:* Mus asst entom, Univ Kans, 51-53; instr entom, Univ Mo, 53-57; fishery res biologist, US Fish & Wildlife Serv, 57-58; syst entomologist, Entom Res Div, Agr Res Serv, USDA, 58-62; ASSOC CUR, DIV COLEOPTERA, NAT MUS NATURAL HIST, SMITHSONIAN INST, 62- *Concurrent Pos:* Lectr, Grad Fac, Univ Md. *Mem:* Entom Soc Am; Sigma Xi. *Res:* Systematics, biology and zoogeography of aquatic beetles. *Mailing Add:* Dept Entom NHB-169 Nat Mus Natural Hist Smithsonian Inst Washington DC 20560

SPANGLER, ROBERT ALAN, b Celina, Ohio, Apr 10, 33; m 59. BIOPHYSICS. *Educ:* Harvard Univ, AB, 55, MD, 59; State Univ NY Buffalo, PhD(biophys), 64. *Prof Exp:* Res fel, 59-65, asst prof, 65-70, actg chmn dept, 70-77, ASSOC PROF BIOPHYS, SCH MED, STATE UNIV NY BUFFALO, 70- *Mem:* Biophys Soc; NY Acad Sci; AAAS. *Res:* Chemical kinetics in biological systems; non-equilibrium thermodynamics of biological systems; transport; models; medical diagnostic imaging. *Mailing Add:* Dept of Biophys Sci State Univ of NY at Buffalo Buffalo NY 14214

SPANGLER, STEVEN RANDALL, b Stamford, Conn, Sept 25, 50; m 70; c 2. RADIO ASTRONOMY, THEORETICAL ASTROPHYSICS. *Educ:* Univ Iowa, BA & MS, 72, PhD(physics), 75. *Prof Exp:* Res assoc space physics, Univ Iowa, 75-76; res assoc, 76-78, ASST SCIENTIST RADIO ASTRON, NAT RADIO ASTRON OBSERV, 78- *Mem:* Am Astron Soc; Am Phys Soc. *Res:* Observations of extragalactic radio sources and their interpretation in terms of hydrodynamics, statistical physics and radiation theory. *Mailing Add:* Dept Physics & Astron Univ Iowa Iowa City IA 52242

SPANIER, ARTHUR M, b New York, NY, July 5, 48; m 72; c 4. PHYSIOLOGY, BIOCHEMISTRY. *Educ:* Herbert H Lehman Col City Univ New York, BA, 70, MA, 77; Rutgers State Univ NJ, PhD(physiol), 77. *Prof Exp:* MDAA postdoc fel, dept med, Univ Chicago, 77-79, NIH postdoc fel, 79-80,; res assoc & instr biophys, Med Col Va, 80-81; asst mem, Okla Med Res Found, 81-85; res physiologist, 85-86, LEAD SCIENTIST & RES PHYSIOLOGIST, USDA/ARS SOUTHERN REGIONAL RES CTR, 86- *Concurrent Pos:* Prin investr, Univ Chicago, 77-79 & Okla Res Found, NIH, 83-86; co-investr, 3 NIH grants to Okla Med Res Found, 81-85, Med Col Va, 80-81 & Univ Chicago, 79-80; consult, George Washington Univ, 85-; adj asst prof physiol & biophys, Univ Okla Health Sci Ctr, 82-85. *Mem:* Am Physiol Soc; Am Chem Soc; Am Soc Cell Biol; Inst Food Technologists; Am Meat Sci Asn; Am Comt on Proteolysis. *Res:* Muscle & meat physiology; enzymology and cell biology; meat flavor chemistry and nutrition; protein, peptide and amino acid flavor; proteinase biochemistry; natural food product chemistry; lysosomes, lysosomal hydrolases of muscle and other tissue. *Mailing Add:* USDA/ARS-SRRC 1100 Robert E Lee Bldg New Orleans LA 70124

SPANIER, EDWARD J, b Philadelphia, Pa, May 13, 37; m 68; c 2. INORGANIC CHEMISTRY. *Educ:* La Salle Univ, BA, 59; Univ Pa, PhD(inorg chem), 64. *Prof Exp:* Res chemist, E I du Pont de Nemours & Co, 64-65; asst prof inorg chem, Seton Hall Univ, 65-72; asst dean sci & eng, 72-73, assoc dir planning for health affairs, 73-74, asst dean admin, Sch Med, 74-75, assoc dean admin, Sch Med, 75-80, asst vpres health affairs, 80-81, asst vpres financial serv, 81-85, TREAS, WRIGHT STATE UNIV, 84-, VPRES BUS & FINANCE, 85- *Concurrent Pos:* Adj assoc prof chem, Wright State Univ, 72- *Mem:* AAAS; Am Chem Soc; Sigma Xi. *Res:* Chemistry of the hydrides of boron, silicon and germanium; nuclear magnetic resonance; reactions of metal carbides; chemistry of group V elements. *Mailing Add:* Wright State Univ Dayton OH 45435

SPANIER, EDWIN HENRY, b Washington, DC, Aug 8, 21; div; c 3. MATHEMATICS. *Educ:* Univ Minn, BA, 41; Univ Mich, MS, 45, PhD(math), 47. *Prof Exp:* Mathematician, Signal Corps, US War Dept, 41-44; Jewett fel, Inst Advan Study, 47-48; from asst prof to prof math, Univ Chicago, 48-59; PROF MATH, UNIV CALIF, BERKELEY, 59- *Concurrent Pos:* Guggenheim fel, Univ Paris, 52-53; Fulbright distinguished lectr, Chile, 73. *Mem:* Am Math Soc; Math Asn Am. *Res:* Topology; formal languages. *Mailing Add:* Dept Math Univ Calif Berkeley CA 94720

SPANIER, JEROME, b St Paul, Minn, June 3, 30; m 52; c 3. MATHEMATICS. *Educ:* Univ Minn, BA, 51; Univ Chicago, MS, 52, PhD(math), 55. *Prof Exp:* Asst, Univ Minn, 50-51; mathematician, Bettis Atomic Power Lab, Westinghouse Elec Corp, Pa, 55-67; mem tech staff, Math Group, NAm Rockwell Corp, Calif, 67-70, group leader math group, Sci Ctr, 70-71; PROF MATH, CLAREMONT GRAD SCH, 71-, DEAN, 82-, VPRES & DEAN, 86- *Concurrent Pos:* Consult, Atomics Int Div, NAm Rockwell Corp, 71- *Mem:* Am Math Soc; Soc Indust & Appl Math; Math Asn Am; AAAS. *Res:* Monte Carlo methods; numerical analysis; random walk processes; transport theory; applications of numerical techniques to nuclear reactor design. *Mailing Add:* VPres Acad Affairs & Dean Claremont Grad Sch Claremont CA 91711-6160

SPANIOL, CRAIG, b Charleston, WVa, Feb 22, 44; m 63; c 3. ALTERNATIVE ENERGY SYSTEMS. *Educ:* WVa State Col, BS, 66; Ohio Univ, MS, 69; Rensselaer Polytech Inst, PhD(eng sci), 74. *Prof Exp:* Elec engr, E M Johnson Consult Eng, 62-66; elec engr, Chesapeake & Potomac Telephone Co, 66-70; nuclear engr, Gen Elec Co, 70-73; prin engr nuclear eng, Babcock & Wilcox Co, 73-76; asst prof elec eng, WVa Inst Technol, 76-81; CONSULT ENGR, MIDWEST TECH, INC, 81-; ASSOC PROF INDUST TECH, WVA STATE COL, 85- *Concurrent Pos:* Pvt legal consult, 76- *Mem:* Inst Elec & Electronics Engrs; Am Nuclear Soc. *Res:* High temperature furnace/autoclave development; high pressure gas encapulation; crystal technology; alternative energy systems. *Mailing Add:* Dept Indust Tech WVa State Col Institute WV 25112

SPANIS, CURT WILLIAM, b Barrie, Ont, May 6, 32; US citizen; m 65. NEUROPHYSIOLOGY, PSYCHOLOGY. *Educ:* Queen's Univ, Ont, BA, 57; Univ Calif, Los Angeles, MA, 60, PhD(physiol), 62; Alvarez Soc Med, Mex, dipl, 75. *Prof Exp:* Asst prof biol, San Diego State Col, 62-63 & microbiol, Inst Marine Sci, Univ Miami, 64; fel biol clocks, Scripps Inst Oceanog, Univ Calif, San Diego, 64-65; from asst prof to assoc prof biol, 65-70, chmn dept, 66-72, PROF BIOL, UNIV SAN DIEGO, 72-; RESEARCHER, DEPT PSYCHIAT, VET MEM HOSP, LA JOLLA, CALIF, 75- *Concurrent Pos:* NIH grants, Univ Miami, 64, Scripps Inst Oceanog, Univ Calif, San Diego, 64-65, Univ San Diego, 69-72; NSF grants, 64 & 75; vis prof, Biol Inst, Helgoland, Germany, 64; mem NSF grants rev bd. *Mem:* Fel Am Inst Chem; hon mem Mex Soc Biol Psychiat; Soc Neurosci; Brit Brain Res Asn; Am Soc Microbiol; Int Brain Res Orgn-World Fedn

Neuroscientists. *Res:* Sleep and the biochemistry of rapid eye movement sleep; memory; amnesia-electroshock; exercise physiology, sports medicine and nutrition. *Mailing Add:* Dept of Biol Univ of San Diego San Diego CA 92110

SPANN, CHARLES HENRY, b Brandon, Miss, Sept 11, 39; m 63; c 3. EXPERIMENTAL PATHOLOGY, HUMAN ANATOMY. *Educ:* Tougaloo Col BS, 62; Univ Miss Med Ctr, MS, 73, PhD(human anat & path), 74. *Prof Exp:* Instr biol & sci, Holtzclaw High Sch, 66-69 & Crystal Springs High Sch, Miss, 69-70; ASSOC PROF BIOL & DIR HEALTH CAREERS, JACKSON STATE UNIV, 74- *Concurrent Pos:* Consult pre-nursing prog, Meridian Jr Col, Miss, 76-77; res trainer, Minority Biomed Res Support Prog, Jackson State Univ, 77-, dir, 81-, dir, Health Careers Training Prog. *Mem:* Am Soc Anatomists; AAAS. *Res:* Relationship and susceptibility of scorbutic Guinea Pigs to endotoxic shock, with emphasis on the histopathological effects of the general viscera and treatment modalities. *Mailing Add:* Dept Biol Jackson State Univ 1400 John R Lynch St Jackson MS 39217

SPANN, JAMES FLETCHER, (JR), b Dothan, Ala, Nov 21, 35; m 56; c 2. CARDIOLOGY. *Educ:* Emory Univ, MD, 61. *Prof Exp:* Intern med, Mass Gen Hosp, 61, asst resident, 62; sr investr, Cardiol Br, attend physician, Inst & consult cardiologist & med coordr, Surg Br, Nat Heart Inst, 66-68; assoc prof med & physiol, chief cardiovasc diag & asst chief cardiovasc med, Sch Med, Univ Calif, Davis, 68-70; PROF MED, CHIEF CARDIOVASC SECT, HEALTH SCI CTR, TEMPLE UNIV, 70- *Concurrent Pos:* Fel, Cardiol Br, Nat Heart Inst, 63-65, spec fel, 65-66. *Mem:* AAAS; Am Col Cardiol; Am Fedn Clin Res; NY Acad Sci; Am Soc Pharmacol & Exp Therapeut. *Res:* Clinical and investigative cardiology; cardiovascular physiology and pathophysiology; cardiac hypertrophy and congestive heart failure. *Mailing Add:* Dept Med, Med Univ S Carolina 171 Ashley Ave Charleston SC 29425

SPANNINGER, PHILIP ANDREW, b Quakertown, Pa, May 31, 43; m 63; c 3. ORGANIC CHEMISTRY. *Educ:* Philadelphia Col Textiles & Sci, BS, 65; Clemson Univ, MS, 67, PhD(org chem), 70. *Prof Exp:* NIH fel, Univ Tex, Austin, 70-71; sr res chemist, Polyester Res, Goodyear, 71-74, proj mgr, joint ventures & licensing technol, 74-77, mgr, Int Chem Div, 77-80, dir, technol & ventures mgt, 80-88; VPRES, INT, GENCORP, INC, 88- *Concurrent Pos:* Sch Indust Mgt, Carnegie-Mellon Univ. *Mem:* Am Chem Soc. *Res:* Organometallic chemistry; stereochemistry; reaction mechanisms; catalysis; boron-nitrogen heteroaromatic compounds; high temperature polymers. *Mailing Add:* 2555 Olentangy Dr Akron OH 44313

SPANO, FRANCIS A, b New York, NY, Jan 6, 31; m 59; c 3. ORGANIC CHEMISTRY, BIOCHEMISTRY. *Educ:* City Col New York, BS, 53; Fordham Univ, PhD(org chem), 63. *Prof Exp:* Res chemist, Allied Chem Corp, 62-66; res specialist, Gen Aniline & Film Corp, 66; asst prof org chem, 66-72, PROF CHEM, MIDDLESEX COUNTY COL, 72-, DEAN DIV SCI, 73- *Mem:* Am Chem Soc. *Res:* Illucidation of ozone oxidation of heterocyclic aeromatic compounds. *Mailing Add:* Dept Chem Middlesex County Col 155 Mill Rd PO Box 3050 Edison NJ 08818-3050

SPANSWICK, ROGER MORGAN, b Eng, June 24, 39; m 63; c 2. BIOPHYSICS, PLANT PHYSIOLOGY. *Educ:* Univ Birmingham, BSc, 60; Univ Edinburgh, dipl biophys, 61, PhD(biophys), 64. *Prof Exp:* Asst lectr physics, Univ Edinburgh, 62-64; postdoctoral res fel plant biophys, Cambridge Univ, 64-67; asst prof, 67-73, assoc prof, 73-79, PROF PLANT PHYSIOL, CORNELL UNIV, 79- *Concurrent Pos:* Sci Res Coun sr vis fel, Cambridge Univ, 73-74; John Simon Guggenheim Mem fel, Univ Calif, Davis, 81-82. *Mem:* Brit Soc Exp Biol; Biophys Soc; Am Soc Plant Physiologists; AAAS. *Res:* Transport of ions across plant cell membranes; intercellular and long distance transport of ions in plants; partitioning of photosynthetic assimilates in relation to seed development in crop plants. *Mailing Add:* Sect Plant Biol Cornell Univ 228 Plant Sci Bldg Ithaca NY 14853-5908

SPAR, IRVING LEO, b New York, NY, July 6, 26; m 48; c 3. IMMUNOLOGY. *Educ:* George Washington Univ, BS, 47; Univ Rochester, PhD(biol), 52. *Prof Exp:* Jr scientist radiation biol, Sch Med & Dent, Univ Rochester, 52-53, from instr to assoc prof, 54-61; assoc prof radiol, Med Sch, Univ Ky, 62-63; assoc prof, 63-70, assoc grad studies, 75-85, PROF, RADIATION BIOL & BIOPHYS, SCH MED & DENT, UNIV ROCHESTER, 70-, SR ASSOC DEAN GRAD STUDIES & FINANCIAL AID, 85- *Mem:* Soc Nuclear Med; Am Asn Cancer Res; Radiation Biol Soc; Fedn Am Soc Exp Biol. *Res:* Iodine labeled antigens, antibodies and components of complement involved in inflammation and tumor rejection. *Mailing Add:* Dept Radiation Biol Univ Rochester Med Ctr Rochester NY 14642

SPAR, JEROME, b New York, NY, Oct 7, 18; m 45; c 2. METEOROLOGY. *Educ:* City Col New York, BS, 40; NY Univ, MS, 43, PhD(meteorol), 50. *Prof Exp:* From instr to prof meteorol, NY Univ, 46-73; prof meteorol, City Col, City Univ NY, 73-84. *Concurrent Pos:* Dir meteorol res, US Weather Bur, 64-65. *Mem:* Fel Am Meteorol Soc; Am Geophys Union; fel NY Acad Sci; Royal Meteorol Soc. *Res:* Atmospheric radioactivity; numerical weather prediction; cyclogenesis; applied meteorology; climatic variations; synoptic and dynamic meteorology; general circulation and air-sea interactions. *Mailing Add:* 18 Fieldmere Ave Glen Rock NJ 07452

SPARACINO, CHARLES MORGAN, b Charleston, WVa, Oct 18, 41; m 64; c 2. ORGANIC CHEMISTRY. *Educ:* Emory Univ, BS, 65, PhD(org chem), 69. *Prof Exp:* NIH fel, Worcester Found Exp Biol, 69-70; Nat Inst Gen Med Sci fel, 70-71, CHEMIST, RES TRIANGLE INST, RESEARCH TRIANGLE PARK, 71- *Mem:* Am Chem Soc. *Res:* Organic synthesis; natural product biosynthesis; drug metabolism. *Mailing Add:* 3209 Jenifer Dr Durham NC 27705

SPARACINO, ROBERT R, b New York, NY, Nov 6, 27; m 49; c 3. ELECTRICAL ENGINEERING, INSTRUMENTATION. *Educ:* City Col New York, BEE, 50; Polytech Inst Brooklyn, MEE, 55; Mass Inst Technol, ScD, 61. *Prof Exp:* Proj engr, Atlantic Electronics Corp, 50-54; chief engr & asst secy to corp, Penn-East Eng Corp, 54-58; res asst instrumentation, Mass Inst Technol, 58-59 & 60-61; sect head res & develop systs eng, AC Electronics Div, Gen Motors Corp, Wakefield, Mass, 61-62, lab dir, 62-63, dir res & develop, Los Angeles, 63-64, dir res & develop, Milwaukee, 64-68, dir eng, 68-70; vpres & mgr qual assurance dept, Bus Prod Group, Xerox Corp, 70-71, vpres & mgr prod design & eng, 71-73, vpres technol & eng, 73, sr vpres, Copier Duplicator Develop Div, 73-74, pres, Info Technol Group, 75-78, pres, Reprographics Tech Group, 78-80, corp vpres, 74-80, sr vpres, Info Prod Group, 80-82; PRES, SPARACINO ASSOCS, INC, 81-, SPARACINO MGT CO, INC, 83- *Mem:* Inst Elec & Electronics Engrs; Sigma Xi. *Mailing Add:* 175 Blackberry Dr Stamford CT 06903

SPARANO, BENJAMIN MICHAEL, animal pathology, toxicology; deceased, see previous edition for last biography

SPARAPANY, JOHN JOSEPH, b Albany, NY, Nov 11, 28; m 52; c 3. PHYSICAL ORGANIC CHEMISTRY. *Educ:* Buena Vista Col, BS, 51; NDak State Col, MS, 53; Okla State Univ, PhD(chem), 59. *Prof Exp:* Res chemist, E I du Pont de Nemours & Co, Inc, NY, 59-70; assoc scientist, 70-73, tech dir, 73-77, V PRES RES & DEVELOP, HYSOL DIV, DEXTER CORP, 77- *Mem:* Am Chem Soc. *Mailing Add:* 111 Nolan Dr Allegany NY 14706-1114

SPARBER, SHELDON B, b Brooklyn, NY, Sept 29, 38. NEUROPSYCHOPHARMACOLOGY, NEUROBEHAVIORAL TOXICOLOGY. *Educ:* Univ Minn, PhD(pharmacol), 67. *Prof Exp:* PROF PHARMACOL & PSYCHIAT & ADJ PROF PSYCHOL, MED SCH, UNIV MINN, 78- *Mem:* Am Soc Pharmacol & Exp Therapeut; AAAS; Soc Neurosci; Soc Develop Psychobiol; Soc Behav Teratology; Sigma Xi. *Mailing Add:* Dept Pharmacol 3-249 Millard Hall Univ Minn Med Sch 435 Delaware St SE Minneapolis MN 55455

SPARBERG, ESTHER BRAUN, b New York, NY, June 17, 22; m 44; c 2. HISTORY OF SCIENCE, CHEMISTRY. *Educ:* Univ NC, BS, 43; Columbia Univ, MA, 45, EdD(sci educ), 58. *Prof Exp:* Technician, Rockefeller Inst, 43-44; teacher high sch, NY, 46-47; spec instr chem, 59-63, instr, 63-66, from asst prof to assoc prof, chem & hist sci, 66-77, prof hist sci, 77-80, PROF CHEM, HOFSTRA UNIV, 77- *Concurrent Pos:* Consult, NSF Coop Col Sch Sci Progs, dir prog, Hofstra-Uniondale Schs, 70-72, Hofstra-New Hyde Park-Herricks Schs, 72-74; dir, NSF proj, Hofstra-Farmingdale, Glen Cove, Wantagh, Queens, NY Schs, 75-76; dir, NSF, Pre-Col Teacher Develop in Sci Proj, Hofstra Univ, 77-79 & Honors Workshop for Teachers, 84-85. *Mem:* Am Chem Soc; Hist Sci Soc; Am Asn Physics Teachers. *Res:* Plasma and serum studies with Tiselius electrophoresis equipment. *Mailing Add:* 25 Emerson Dr Great Neck NY 11023

SPARGO, BENJAMIN H, b Six Mile Run, Pa, Aug 18, 19; m 42; c 2. PATHOLOGY. *Educ:* Univ Chicago, BS, 48, MS & MD, 52. *Prof Exp:* From instr to assoc prof, 53-64, PROF PATH, SCH MED, UNIV CHICAGO, 64- *Concurrent Pos:* Chmn, comt Diag Electron Micros, Vet Admin, 75-; res career award, Heart & Lung Inst, NIH, 64- *Mem:* Am Soc Nephrology; Nat Kidney Found; AAAS. *Res:* Pathology of renal diseases. *Mailing Add:* Dept Pathol BH P312 Box 327 Univ Chicago 5841 S Maryland Ave Chicago IL 60637

SPARKES, ROBERT STANLEY, b Niagara Falls, NY, June 20, 30; m 71; c 2. MEDICINE, HUMAN GENETICS. *Educ:* Antioch Col, BS, 52; Univ Rochester, MD, 56. *Prof Exp:* Asso med, Sch Med, Univ Wash, 61-63; assoc physician, City of Hope Med Ctr, 63-64; PROF MED & MED GENETICS, SCH MED, UNIV CALIF, LOS ANGELES, 64-, VCHMN DEPT MED, 81- *Mem:* AAAS; Am Soc Human Genetics; Am Fedn Clin Res; Asn Am Physicians. *Res:* Human-medical cytogenetics; human biochemical genetics; genetic linkage; tissue culture. *Mailing Add:* Dept Med Univ Calif Sch of Med Los Angeles CA 90024

SPARKMAN, DENNIS RAYMOND, b Ennis, Tex, Jan 12, 54. ALZHEIMERS DISEASE & DEMENTIA. *Educ:* Univ Tex, Arlington, BS, 76; Tex A&M Univ, PhD(cell & molecular biol), 82. *Prof Exp:* Res assoc, Univ Tex Southwestern Med Ctr, 76-78, asst instr, 82-85, res instr neurol, Inst Path, 85-88, ASST PROF PATH, UNIV TEX SOUTHWESTERN MED CTR, DALLAS, 88- *Mem:* AAAS; Am Soc Cell Biol; Sigma Xi; Soc Neurosci; Am Asn Neuropath. *Res:* Pathology of neurofibrillary degeneration and cytoskeletal alterations that occur in Alzheimer's disease, related dementias and Parkinson's disease through isolation of antibodies, cellular and ultrastructural studies and molecular cloning. *Mailing Add:* Dept Path Univ Tex Southwestern Med Ctr 5323 Harry Hines Blvd Dallas TX 75235-9072

SPARKMAN, DONAL ROSS, b Seattle, Wash, June 7, 07; m 48; c 4. MEDICINE. *Educ:* Univ Wash, BS, 30; Univ Pa, MD, 34; Am Bd Internal Med, dipl, 47. *Prof Exp:* From clin asst prof to clin prof, 54-56, assoc prof med, 66-76, EMER ASSOC PROF, MED SCH UNIV WASH, 76- *Concurrent Pos:* Assoc dir, Cancer Control Prog, Fred Hutchinson Cancer Res Ctr, 76-79; at Am Cancer Soc, Seattle, 79-84. *Mem:* Am Heart Asn; Am Col Physicians. *Res:* Cardiac rehabilitation; relationship of stress to heart disease. *Mailing Add:* 6545 Park Point Way NE Seattle WA 98115

SPARKMAN, MARJORIE FRANCES, b McShan, Ala, Jan 25, 23. PHYSIOLOGY. *Educ:* Fla State Col Women, BM, 45; Univ Ala, BS, 61; Ohio State Univ, MS, 62, PhD(physiol), 68. *Prof Exp:* Instr nursing, Southern Baptist Hosp, New Orleans, La, 51-56; head nurse, Nursing Serv, Wichita Falls Gen Hosp, Tex, 56-57; actg dir nursing, Southern Baptist Hosp Sch Nursing, New Orleans, 58-60; instr nursing, Col Med, Ohio State Univ, 62-64, asst prof physiol, 68-72; assoc prof, 72-77, PROF PHYSIOL & DIR CONTRACT GRANT ADMIN, COL NURSING, FLA STATE UNIV, 77- *Mem:* Sigma Xi. *Res:* Effects of 100 percent oxygen at atmospheric pressure in rats. *Mailing Add:* 306 Tallwood Dr Tallahassee FL 32312

SPARKMAN, ROBERT SATTERFIELD, b Brownwood, Tex, Feb 18, 12; m 42. SURGERY. *Educ:* Baylor Univ, BA & MD, 35, Am Bd Surg, dipl, 48. *Hon Degrees:* LLD, Baylor Univ, 74. *Prof Exp:* Intern & resident, Cincinnati Gen Hosp, Ohio, 35-40; clin prof surg, Univ Tex Southwest Med Sch Dallas, 62-68; chief, 68-81, EMER CHIEF, DEPT SURG, MED CTR, BAYLOR UNIV, 82- *Concurrent Pos:* Mem attend staff, Baylor Hosp, 45-; chief surg consult, US Fifth Army, 48-73; chmn, Asn Prog Dirs Surg, 81-83; pres, Tex Surg Soc, 65, Southern Surg Asn, 78 & fel, 87. *Mem:* Am Surg Asn; fel Am Col Surg; Asn Prog Dirs Surg; Int Soc Surg; Int Biliary Asn. *Res:* Surgical disease of gallbladder, bile duct; surgical history. *Mailing Add:* 1004 N Washington Dallas TX 75204

SPARKS, ALBERT KIRK, b Wichita Falls, Tex, July 31, 23; m 43; c 1. MARINE BIOLOGY. *Educ:* Tex A&M Univ, BS, 47, MS, 49, PhD(biol oceanog), 57. *Prof Exp:* Asst biol, Tex A&M Univ, 47-49, instr, 49; asst prof, Sam Houston State Univ, 49-51; asst biol oceanog, Tex A&M Univ, 51-52, asst prof, 52-53, asst phys oceanog, Tex A&M Res Found, 53-56, chief biol & asst dir, Marine Lab, 56-58; assoc prof fisheries, Univ Wash, 58-63, prof, 63-70; dir, Bur Commercial Fisheries Biol Lab, Tex, 70-71; ctr dir, Gulf Coastal Fisheries Ctr, Nat Marine Fisheries Serv, 71-73, dep assoc dir resource res, Washington, DC, 73-76, invert pathologist, Northwest & Alaska Fisheries Ctr, 76-91; RETIRED. *Concurrent Pos:* Consult to numerous indust & state agencies, 56-70; consult, Res & Develop Div, Humble Oil & Refining Co, 57-62 & Hawaii Dept Fish & Game, 63; adv, Ministry Nat Resources & Wildlife, Kenya, 65; prof, Tex A&M Univ, 70-73; affil prof, Col Fish, Univ Wash, 70- *Mem:* AAAS; Am Soc Zoologists; Nat Shellfisheries Asn (vpres, 68, pres, 69); Soc Invert Pathologists (vpres, 66-68, pres, 68-70; Wildlife Dis Asn. *Res:* Invertebrate pathology; marine fisheries. *Mailing Add:* 9629 42nd NE Seattle WA 98105

SPARKS, ALTON NEAL, b Robert Lee, Tex, Jan 25, 32; m 53; c 4. ENTOMOLOGY. *Educ:* Tex Tech Col, BS, 58; Iowa State Univ, MS, 59, PhD(entom), 65. *Prof Exp:* Entomologist, Entom Res Div, Agr Res Serv, USDA, Ariz, 59- 61, Iowa, 61-65, Okla 65-66, res leader & dir, Southern Grain Inspects Res Lab, 66-83, res leader & nat tech adv field crop entom, Agr Res Serv, 83-88; RETIRED. *Concurrent Pos:* Assoc prof, Okla State Univ, 65-66, Univ Ga, 67- & Univ Fla, 81. *Honors & Awards:* Agr Entomologist Award, Am Registry Prof Entomologist, 84. *Mem:* Entom Soc Am. *Res:* Biology and ecology of cotton insects and grain insects; screening insecticides; European corn borer and insects attacking small grains; pheromones and their effects on the behavior of nocturnal insects; insect migration. *Mailing Add:* 1627 N Park Tifton GA 31794

SPARKS, ARTHUR GODWIN, b Savannah, Ga, Feb 10, 38; m 58; c 3. MATHEMATICS, COMPUTER SCIENCES. *Educ:* Ga Southern Col, BS, 60; Univ Ga, MEd, 62; Univ Fla, MA, 64; Clemson Univ, PhD(math), 69. *Prof Exp:* Instr math & physics, high sch, Ga, 60-61; instr math, Ga Southern Univ, 64-65, asst prof, 65-66 & 69-72, assoc prof math & comput sci, 72-80, prof math & comput sci, 80-88, PROF & HEAD, MATH & COMPUT SCI, GA SOUTHERN UNIV, 88- *Mem:* Am Math Soc; Math Asn Am; Soc Indust Appl Math; Asn Comput Mach; Sigma Xi. *Res:* Analysis; convexity; computer science. *Mailing Add:* Dept Math & Comput Sci Ga Southern Univ Statesboro GA 30460

SPARKS, CECIL RAY, b Lockwood, WVa, Nov 16, 30; m 56; c 3. ACOUSTICS, FLUID DYNAMICS. *Educ:* Univ Tex, BS, 53; Univ Pittsburgh, MS, 56. *Prof Exp:* Develop engr, New Prod Dept, Westinghouse Elec Corp, 53-57; asst dir dept appl physics, 57-74, dir eng physics, 74-85, VPRES, SOUTHWEST RES INST, 85- *Mem:* Acoust Soc Am. *Res:* Noise control; machinery and structure vibrations; fluid mechanics; instrumentation. *Mailing Add:* 10906 Janet Lee San Antonio TX 78230

SPARKS, CHARLES EDWARD, b Peoria, Ill, July 29, 40; m 77; c 3. LIPOPROTEIN METABOLISM & APOLIPOPROTEINS. *Educ:* Mass Inst Technol, BS, 63; Jefferson Med Col, MD, 68; Am Bd Path, cert; Am Bd Clin Chem, cert. *Prof Exp:* Asst prof, Med Col Pa, 75-77, assoc prof biochem & physiol, 77-82; assoc prof path, 82-87, PROF PATH & LAB MED, UNIV ROCHESTER, 88- *Concurrent Pos:* Fel, Coun Arteriosclerosis, Am Heart Asn & Am Diabetes Asn, 89-; vis assoc prof biochem, Med Col Pa, 82- *Res:* Hormonal regulation of lipoprotein metabolism-relationships to diabetes and atherosclerosis. *Mailing Add:* Dept Path & Lab Med Univ Rochester Med Ctr 601 Elmwood Ave Box 608 Rochester NY 14642

SPARKS, CULLIE J(AMES), JR, b Belpre, Ohio, May 8, 29; m 51; c 5. METALLURGY. *Educ:* Univ Ky, BS, 52, EngrD(metall), 57. *Prof Exp:* Res assoc metall, Univ Ky, 53-56; METALLURGIST & MAT SCIENTIST, OAK RIDGE NAT LAB, 58-, GROUP LEADER, 80- *Concurrent Pos:* Officer, USAF, Wright-Patterson AFB, Ohio-Mat Sci, 56-58. *Mem:* Am Soc Metals; Sigma Xi; Am Crystallog Asn; Am Phys Soc. *Res:* Relationship between the geometrical structure of materials and their physical and chemical behavior; x-ray scattering measurements of the short-range order and atomic displacements; crystallographic distributions of lattice defects and relationship to solid solution strengthening; atomic structure of thin films, surfaces, interfaces, epitaxial mismatch, growth imperfections and reaction rate. *Mailing Add:* 804 W Outer Dr Oak Ridge TN 37830

SPARKS, DARRELL, b Tipton Hill, NC, Apr 14, 38. PLANT PHYSIOLOGY, HORTICULTURE. *Educ:* NC State Col, BS, 61; Mich State Univ, MS, 62, PhD(hort), 65. *Prof Exp:* From asst prof to assoc prof, 65-76, PROF HORT RES, UNIV GA, 76- *Mem:* AAAS; Am Soc Hort Sci; Am Soc Plant Physiol; Bot Soc Am. *Res:* Applied ecology; mineral nutrition and general physiology of tree fruit crops. *Mailing Add:* Dept Hort Univ Ga Athens GA 30602

SPARKS, DAVID LEE, b Guntersville, Ala, Dec 22, 37; m 63; c 3. NEUROPHYSIOLOGY. *Educ:* Univ Ala, BA, 59, MA, 62, PhD(psychol), 63. *Prof Exp:* Instr psychol, Univ Ala, 62-63; USPHS fel neurosurg, Med Ctr, Univ Miss, 63-65; instr psychiat, Med Ctr, 65-67, from asst prof to prof psychol, 67-81, chmn dept, 69-74, PROF PHYSIOL & BIOPHYSICS, UNIV ALA, BIRMINGHAM, 81 - *Mem:* AAAS; Soc Neurosci; Asn Res Vision & Ophthal. *Res:* Sensory-motor function; neural control of eye movements. *Mailing Add:* Dept Physiol & Biophysics Univ Ala University Sta Birmingham AL 35294

SPARKS, HARVEY VISE, b Flint, Mich, June 22, 38; m 69; c 4. MEDICAL PHYSIOLOGY. *Educ:* Univ Mich, MD, 63. *Prof Exp:* USPHS fel physiol, Harvard Med Sch, 63-65 & Univ Goteborg, 65-66; from instr to prof, Univ Mich, 66-79; PROF & CHMN, DEPT PHYSIOL, MICH STATE UNIV, 79- *Concurrent Pos:* Mem, Coun on Circulation, Am Heart Asn; mem, Nat Bd Med Examiners; Markle Scholar Acad Med, John & Mary Markle Found, 67. *Mem:* Am Heart Asn; Microcirculatory Soc; Soc Exp Biol & Med; Am Physiol Soc (pres, 87-88); Am Col Sports Med. *Res:* Metabolic control of coronary and skeletal muscle blood flow using mathematical model simulations and experimental approaches. *Mailing Add:* Dept Physiol A110 E Fee Hall Mich State Univ East Lansing MI 48824-1101

SPARKS, JOSEPH THEODORE, physics, for more information see previous edition

SPARKS, MORGAN, b Pagosa Springs, Colo, July 6, 16; m 49; c 4. CHEMISTRY. *Educ:* Rice Univ, BA, 38, MA, 40; Univ Ill, PhD(chem physics), 43. *Hon Degrees:* DSc, Univ NMex, 80. *Prof Exp:* Mem staff, Nat Defense Res Comt, Univ Ill, 41-43; res chemist, Bell Tel Labs, Inc, 43-48, mem semiconductor group, 48-53, dept head semiconductor device feasibility, 53-55, dir solid state electronics res, 55-58, dir, Transistor Dept, 58-59, exec dir, Components & Solid State Div, 59-68, exec dir, Semiconductor Components Div, 68-69, vpres tech info & personnel, 69-71, vpres electronics technol, 71-72; pres, Sandia Labs, 72-81; dean, R O Anderson Sch Mgt, Univ NMex, 81-84; RETIRED. *Concurrent Pos:* Vpres, Western Elec Co, Inc, 72-81. *Honors & Awards:* Jack A Morton Award, Inst Elec & Electronics Engrs, 77. *Mem:* Nat Acad Eng; Am Chem Soc; fel Am Phys Soc; fel Inst Elec & Electronic Engrs; fel Am Inst Chemists. *Res:* Solid state physics and chemistry; electron device development; semiconductors; transistors; thin film devices; passive components; memory elements. *Mailing Add:* 904 Lamp Post Circle SE Southeast Albuquerque NM 87123

SPARKS, PETER ROBERT, b Bristol, Eng, July 29, 47; m 76; c 2. WIND ENGINEERING, EARTHQUAKE ENGINEERING. *Educ:* Univ Bristol, BSc, 68; Univ London, PhD(structural eng), 74. *Prof Exp:* Sci officer, Bldg Res Sta, Eng, 68-73, higher sci officer, 73-75, sr sci officer, 75-77; vis prof eng mech, Va Polytech Inst & State Univ, 77-79, assoc prof, 79-82; assoc prof, 82-87, PROF CIVIL ENG & ENG MECH, CLEMSON UNIV, 87- *Concurrent Pos:* Hurricane & Tornado damage surv, Nat Res Coun, 84-; Dir Wind Eng Res Coun, 88- *Mem:* Am Soc Civil Engrs; Am Soc Eng Educ; Sigma Xi. *Res:* The behavior of structures under wind and earthquake loading; full-scale and model investigations of structural performance and loading; architectural aerodynamics; mitigation of damage due to natural hazards. *Mailing Add:* Dept Civil Eng Clemson Univ Clemson SC 29631

SPARKS, RICHARD EDWARD, b Kingston, Pa, Apr 19, 42; m 66; c 2. AQUATIC BIOLOGY, AQUATIC TOXICOLOGY. *Educ:* Amherst Col, BA, 64; Univ Kans, MS, 68; Va Polytech Inst & State Univ, PhD(biol), 71. *Prof Exp:* Teacher gen sci & biol, US Peace Corps, Univ Nigeria, Methodist Higher Elem Teacher Training Col, Nigeria, 64-66; res assoc, Ctr Environ Studies, Va Polytech Inst & State Univ, 71-72; asst aquatic biologist, 72-77, assoc aquatic biologist, 77-80, AQUATIC BIOLOGIST, ILL NATURAL HIST SURV, 80- *Concurrent Pos:* Consult, US Army Corps Engrs, 74-81, Ill Power Co, 75-78 & Upper Miss River Basin Comn, 79-80; lectr, Bradley Univ, 75-; adj prof, Western Ill Univ, 76- *Mem:* AAAS; Am Fisheries Soc; Ecol Soc Am; NAm Benthological Soc; Sigma Xi; Am Inst Biol Sci. *Res:* Biological monitoring for pollution control; restoration of degraded aquatic ecosystems; ecology of Illinois River and Mississippi River; effects of toxicants and contaminants on aquatic organisms; relationships between annual cycle of flood and low flow and populations and productivity. *Mailing Add:* Ill Natural Hist Surv River Res Lab PO Box 599 Havana IL 62644

SPARKS, ROBERT D, b Newton, Iowa, 32. HEALTH ADMINISTRATION. *Educ:* Iowa State Univ, MD, 57. *Hon Degrees:* DHH, Creighton Univ. *Prof Exp:* Intern, Charity Hosp New Orleans, 57-58, resident, 58-59, asst vis physican, 59-63, vis physician, 63-72; attending physician chief, gastro-intestinal med, VA Hosp New Orleans, 72-76; asst dean, Affil Hosp Progs, 64-69, prof med, Tulane Univ, 69-72, VDean, 68-69, Dean, 69-72; prog dir, W K Kellogg Found, 76-81, vpres, 80-81, sr vpres, 81-82, chief prog officer, 82-86, pres & trustee, 82-88; assoc med dir, Addiction Treat Serv & dir educ & res, Battle Creek Adventist Hosp, 90-91; VPRES, DIV PROD SAFETY & COMPLIANCE, SYNTEX CORP, 91- *Concurrent Pos:* Consult, DePaul Hosp, 60-72, USPHS, VA Hosp New Orleans & Alexandria; mem, bd dirs, Nat Coun Alcoholism & Drug Dependence, 81, treas, 86-88, chmn, 89-90; mem, President Reagan's Bd Advisors Pvt Sector Initiatives, 85-89; founding mem bd dirs, Am Pharmaceut Inst & Inst Health Pract & Policy, Bard Col, 88 & Consumer Health Info Res Inst, 89; fel, US-China Educ Inst, 88; emer pres & sr consult, W K Kellogg Found, 88-; chmn, Comt Eval Treat Alcohol Probs, Inst Med, Nat Acad Sci, 90. *Mem:* Inst Med-Nat Acad Sci; fel Am Col Physicians; Asn Am Med Cols. *Res:* Internal medicine; alcohol and drug addictions; health planning. *Mailing Add:* Battle Creek Adventist Hosp 165 N Washington Ave Battle Creek MI 49016

SPARKS, ROBERT EDWARD, b Marshall, Mo, Sept 25, 30; m 55; c 3. CHEMICAL ENGINEERING. *Educ:* Univ Mo, BS, 52; Johns Hopkins Univ, DEng, 60. *Prof Exp:* Res engr, Esso Res & Eng Co, 60-62, sr engr, 62-63; from asst prof to prof chem eng, Case Western Reserve Univ, 63-72; PROF CHEM ENG, WASH UNIV, 72- *Concurrent Pos:* Consult, Nat Inst Arthritis & Metab Dis, 65-74 & Goodyear Tire & Rubber Co, 66-74. *Mem:* AAAS; Am Inst Chem Engrs; Am Soc Artificial Internal Organs. *Res:* Medical engineering; design of the artificial kidney; membrane transport; emulsion breaking; velocity profile control; mass transfer and fluid mechanics in chemical reactors; microencapsulation; controlled drug release; inventive reasoning. *Mailing Add:* Dept Chem Eng Urbauer Hall Wash Univ Lindell-Skinker Blvd St Louis MO 63130-4899

SPARKS, WALTER CHAPPEL, b New Castle, Colo, Aug 22, 18; m 42; c 3. HORTICULTURE. *Educ:* Colo State Univ, BS, 41, MS, 43. *Hon Degrees:* DSc, Univ Idaho, 84. *Prof Exp:* Instr agr, Pueblo Col, 41; asst hort, Colo State Univ, 41-43, from instr to assoc prof, 43-47; assoc horticulturist, 47-57, horticulturist, 57-68, res prof hort, 68-81, coordr potato progs, 76-81, co-dir, Postharvest Inst Perishables, 80-81, EMER PROF HORT, UNIV IDAHO, 81- *Concurrent Pos:* Actg supt, Aberdeen Br Exp Sta, 52, 56 & 65; Jenne res fel, Univ Idaho & rep, Nat Inter-Regional Potato Introd & Preserv Proj, 57; consult, Corporacion De La Produccion Santiago, Chile, 66, Australian Govt & Commonwealth Sci & Indust Res Orgn, Venezuelan Corp of Agr Mkt, 75, Japan, 75, 76 & 77, Repub S Africa, 77; exchange res prof, Res Inst Com & Indust Plants, Kolding, Denmark, 72-73; guest lectr, ten Europ countries, 72-73, Greece, Israel, Australia & NZ, 73, Europ Asn Potato Res, Poland, 78 & Ger, 79; adv, Israeli Veg Bd, 80, PEI, Can, 80 & Philippines, 81; Int Potato Ctr, Lima Peru, 82, Moscow, Russia, 83, Jamaica, 88. *Honors & Awards:* Potato Hall of Fame-Brussels, Belgium, 77; Eldred L Jenne Res Fel Award, 57. *Mem:* AAAS; hon mem Am Potato Asn (pres, 64-65); Am Soc Hort Sci; Am Inst Biol Sci; Europ Asn Potato Res. *Res:* Mechanical injury and storage; cultural practices of potatoes. *Mailing Add:* Res & Exten Ctr Univ Idaho Aberdeen ID 83210

SPARLIN, DON MERLE, b Joplin, Mo, Mar 29, 37; m 59; c 4. SOLID STATE PHYSICS, SEMICONDUCTORS-INSULATORS. *Educ:* Univ Kans, BS, 59; Northwestern Univ, PhD(physics), 64. *Prof Exp:* Instr physics, Case Western Reserve Univ, 64-65, asst prof, 65-68; from asst prof to assoc prof, 68-90, PROF PHYSICS, UNIV MO-ROLLA, 90- *Mem:* Am Asn Physics Teachers; Am Inst Physics; AAAS. *Res:* Electronic and magnetic properties of materials. *Mailing Add:* Dept Physics Univ Mo Rolla MO 65401

SPARLING, ARTHUR BAMBRIDGE, b Rossburn, Man, Jan 3, 30; m 55; c 5. SANITARY ENGINEERING. *Educ:* Univ Man, BSc, 53; Univ Toronto, MASc, 54; Wash Univ, DSc(environ & sanit eng), 68. *Prof Exp:* Pub health engr, Prov of Man, 54-67, chief engr, Clean Environ Comn, 67-71; ASSOC PROF CIVIL ENG, UNIV MAN, 71- *Mem:* Water Pollution Control Fedn; Am Water Works Asn. *Res:* Waste treatment; water pollution and treatment. *Mailing Add:* Dept Civil Eng Univ Manitoba Winnepeg MB R3T 2N2 Can

SPARLING, DALE R, b St Clair, Mich, Dec 19, 29; m 58; c 3. GEOLOGY, DEVONIAN CONODONTS. *Educ:* Univ Wyo, BS, 54; Wayne State Univ, MS, 56; Ohio State Univ, PhD(geol), 65. *Prof Exp:* Asst geol, Wayne State Univ, 54-56; petrol geologist, Creole Petrol Corp, 56-61; asst geol, Ohio State Univ, 62-65, instr, 65; lectr, West Wash State Col, 66; instr, Dayton Univ, 66-67; asst prof, Earlham Col, 67-68; from asst prof to assoc prof, 68-74, chmn earth sci prog, 71-76, chmn, dept biol & earth/space sci, 87-89, PROF GEOL, SOUTHWEST STATE UNIV, MINN, 74- *Mem:* Geol Soc Am; Am Asn Petrol Geologist; Ohio Acad Sci; Pander Soc; Paleont Soc. *Res:* Stratigraphy; sedimentology; conodont taxonomy and biostratigraphy. *Mailing Add:* Earth Sci Prog Southwest State Univ Marshall MN 56258

SPARLING, DONALD WESLEY, JR, b Chicago, Ill, Sept 20, 49; m 71; c 2. BIOMETRICS & BIOSTATISTICS, RESEARCH ADMINISTRATION. *Educ:* Southern Ill Univ, Carbondale, BA, 71, MS, 74; Univ NDak, Grand Forks, PhD(biol), 79. *Prof Exp:* Instr natural resource, Univ Minn Tech Col, Crookston, 74-78; asst dir, Coop Wildlife Res Lab, Southern Ill Univ, 78-79; asst prof biol & ecol, Dept Biol, Ball State Univ, Muncie, Ind, 79-82; statistician, Northern Prairie Wildlife Res Ctr, Jamestown, Md, 82-86, RES BIOLOGIST, PATUXENT WILDLIFE RES CTR, US FISH & WILDLIFE SERV, LAUREL, MD, 86- *Concurrent Pos:* Adj prof, Prince Georges County Community Col, 90- *Mem:* Am Ornithologists Union; Wilson Ornith Soc; Sigma Xi. *Res:* Population and ecological effects of contaminants on wildlife with emphasis on pesticides, acid precipitation; author of numerous technical publications. *Mailing Add:* Patuxent Wildlife Res Ctr US Fish & Wildlife Serv Laurel MD 20708

SPARLING, MARY LEE, b Ft Wayne, Ind, May 20, 34; m 56, 75; c 2. EMBRYOLOGY, MEMBRANES. *Educ:* Univ Miami, BS, 55; Duke Univ, MA, 58; Univ Calif, Los Angeles, PhD(embryol, zool), 62. *Prof Exp:* Part-time lectr gen zool, Univ Calif, Los Angeles, 62-64; lectr embryol, 66-68, from asst prof to assoc prof, 66-76, secy treas fac, 78-80, PROF BIOL, CALIF STATE UNIV, NORTHRIDGE, 76- *Concurrent Pos:* Consult, Oak Ridge Nat Lab, 58-59; NSF grant, 71-73 & 81-83, NIH grant, 87-89; nat bd dir, Sigma Xi, 78- *Mem:* Sigma Xi; Am Soc Cell Biol; Soc Develop Biol; Am Soc Zoologists. *Res:* Protein and lipid changes in cell membranes during early development in normal embryos and in those treated with agents producing abnormalities in development. *Mailing Add:* Dept Biol Calif State Univ Northridge CA 91330

SPARLING, PHILIP FREDERICK, b Evanston, Ill, Sept 10, 36; m 63; c 4. MEDICINE, BACTERIOLOGY. *Educ:* Princeton Univ, AB, 58; Harvard Univ, MD, 62. *Prof Exp:* Resident physician, Mass Gen Hosp, 62-64; officer, Comn Corps Venereal Dis Res, Ctr Dis Control, 64-66; fel bacteriol, Harvard Med Sch, 66-68; fel infectious dis, Mass Gen Hosp, 68-69; from asst prof to assoc prof, 69-75, PROF MED & MICROBIOL, UNIV NC, CHAPEL HILL, 75-, CHMN, DEPT MICROBIOL IMMUNOL, 81- *Concurrent Pos:* NIH res career develop award, 71-76; mem adv comt, Ctr Dis Control, 72-; reader bacteriol, Univ Bristol, 74-75; chief, Div Infectious Dis, Univ NC, 75-81; dir, NC Prog on Sexually Transmitted Dis, Sch Med, 78-; mem, microbiol comt, Nat Bd Med Examrs, 78-83, chmn, 81-83; mem, NIH Study Sect Bact, Mycol I, 80-84, chmn, 82-84. *Mem:* Am Soc Microbiol; Asn Am Phys; Am Clin Asn; Am Soc Clin Invest; Infectious Dis Am. *Res:* Infectious diseases; genetics and biochemistry of microbial antibiotic resistance; biochemical genetics of microbial pathogenicity; bacterial physiology; immunobiology of Neisseria gonorrhoeae. *Mailing Add:* 547 Clin Sci Bldg Mem Hosp Chapel Hill NC 27514

SPARLING, REBECCA HALL, b Memphis, Tenn, June 7, 10; m 35, 48; c 1. TROUBLE-SHOOTING PROBLEMS WITH MATERIALS, TECHNICAL WRITING & SPEAKING. *Educ:* Vanderbilt Univ, BA, 30, MS, 31. *Prof Exp:* Metallurgist prod, Am Cast Iron Pipe Co, 31-32 & Lakeside Malleable Castings Co, 32-34; tech writer, William H Baldwin, NY, 34-35; consult, self-employed, 36-44; chief mat processing engr design, Turbodyne Corp, 44-51; design spec materials, prod mat, Gen Dynamics, Pomona, Calif, 51-68; consult mat energy, self-employed, Laguna Hills, Calif, 68- 84; RETIRED. *Concurrent Pos:* Metallurgist invest, Naval Gunn Factory, Wash, 42; prof engr, Calif, 50-; speaker, US Navy, US Air Force, Army Missile Command, 50-68, Univ Southern Calif, Los Angeles, Calif State Univ, TV, radio, etc, 65-85; mem tech comt, Aircraft Industs Asn, 51-60; mem var comts, Am Soc Metals, 52-68, Soc Women Engrs, 57-80 & Am Soc Nondestructive Testing, 54-87; expert witness, Environ Protection Agency, Calif Air Resources Bd, Pub Utilities Comm, etc, 69-84; co-chmn, San Bernardino County Sci Comt, 72; Engr mem, San Bernardino County Air Pollution Bd, 73; intervenor, rep AAUW, AF, EI, Calif State Energy Comm, 75-85, Comt Power Plant Siting, 84. *Honors & Awards:* Achievement Award, Soc Women Engrs; Outstanding Engr Award, Inst Advan Eng. *Mem:* Fel Am Soc Metals; fel Soc Women Engrs; fel Inst Advan Eng; Am Soc Nondestructive Testing; Am Nuclear Soc. *Res:* Developed visible penetrant inspection for nondestructive examination of metal components and new techniques for short-time tests of structural elements at elevated temperatures. *Mailing Add:* 650 W Harrison Ave Claremont CA 91711

SPARLING, SHIRLEY, b Detroit, Mich, Oct 28, 29. PHYCOLOGY. *Educ:* Iowa State Univ, BS, 50, MS, 51; Univ Calif, PhD(bot), 56. *Prof Exp:* Instr bot, Cent Col, Iowa, 51-53, Univ BC, 56-59 & Univ Calif, Santa Barbara, 59-63; instr bot, 63-80, PROF BIOL SCI, CALIF POLYTECH STATE UNIV, SAN LUIS OBISPO, 80- *Mem:* Bot Soc Am. *Res:* Morphology, anatomy, reproduction and life cycles of marine algae, especially red algae. *Mailing Add:* Dept of Biol Sci Calif Polytech State Univ San Luis Obispo CA 93407

SPARNINS, VELTA L, b Riga, Latvia, May 16, 28. MEDICINE. *Educ:* Univ Minn, Minneapolis, BS, 65, MS, 68, PhD(biochem), 70. *Prof Exp:* Teaching asst, Univ Minn, 65-67, res specialist, 70-75, res fel, 75-77, res assoc, dept biochem, 77, scientist, 77-86, sr scientist, dept lab med & path, 86-89, LECTR, UNIV MINN, 89- *Honors & Awards:* JP Fridley Scholarship. *Mem:* Am Chem Soc; AAAS. *Res:* Inhibition of chemical carcinogenesis by chemicals and components of food. *Mailing Add:* 3220 Rankin Rd NE Minneapolis MN 55418

SPARROW, D(AVID) A, b Boston, Mass, June 30, 47. NUCLEAR REACTIONS AND SCATTERING. *Educ:* Princeton Univ, BA, 69; Mass Inst Technol, MS, 71, PhD(physics), 74. *Prof Exp:* Instr math, physics & chem, Univ Mass, 71-73; res assoc physics, Univ Colo, 74-77 & Univ Md, 77-78; asst prof physics, Univ Pa, 78-85; RES STAFF MEM, INST DEFENSE ANALYSIS, 86- *Concurrent Pos:* Vis asst prof, Univ Md, 78; vis assoc prof, Temple Univ, 85-86. *Res:* Theoretical physics: nuclear reactions and scattering, analytic methods; data-to-data (purely empirical) relations; isospin violation; antinucleon interaction; algebraic and analytic techniques in nuclear and molecular scattering; ultra-short wavelength lasers; air vehicle detection and engagement; military application of modeling and simulation; impact of technol on military systems. *Mailing Add:* Inst Defense Analyses Sci & Tech Div 1801 N Beauregard St Alexandria VA 22311

SPARROW, E(PHRAIM) M(AURICE), b Hartford, Conn, May 27, 28; m 52; c 1. MECHANICAL ENGINEERING, HEAT TRANSFER. *Educ:* Mass Inst Technol, BS, 48, MS, 49; Harvard Univ, MA, 50, PhD(mech eng), 56. *Hon Degrees:* Dr, Univ Brazil, 67. *Prof Exp:* Res engr, Oak Ridge Nat Lab, 49; mech engr, Raytheon Mfg Co, 52-53; res scientist, Lewis Res Ctr, Nat Adv Comt Aeronaut, 53-59; chmn fluid mech prog, 68-80, PROF MECH ENG, UNIV MINN, MINNEAPOLIS, 59- *Concurrent Pos:* Lectr, Commonwealth Sci & Indust Res Orgn, Australia, 65; chief-of-party, US Agency Int Develop Prog Grad Educ in Brazil, 66-67; vis prof, Israel Inst Technol, 69; consult, Solar Energy Panel, US Off Sci & Technol, 72; ed, J Heat Transfer, 72-80, US Sci Comt, Fifth Int Heat Transfer Conf, 74, chmn, 78; mem adv panel, US Cong, Off of Technol Assessment, 75-77; vis prof, Xian Jiaotong Univ, 84; prog dir, 86, div dir, NSF, 86-88; distinguished lectr, Am Soc Mech Engr, 86-91. *Honors & Awards:* Heat Transfer Mem Award, Am Soc Mech Engrs, 62, Centennial Award, 80, Charles Russ Richards Mem Award, 85; Max Jakob Award for Eminence in Heat Transfer Res, Am Soc Mech Engrs/Am Inst Chem Engrs, 77; Ralph Coats Roe Award for Eminence in Eng Educ, Am Soc Eng Educ, 78, Sr Res Award, 89; Hawkins Mem Lectr, Purdue Univ, 85; Worcester Reed Warner Medal, Am Soc Mech Engr, 86. *Mem:* Nat Acad Eng; Fel Am Soc Mech Engrs; Sigma Xi. *Res:* Analytical and experimental research in heat transfer and fluid mechanics. *Mailing Add:* Dept Mech Eng Univ Minn Minneapolis MN 55455-0111

SPARROW, ELENA BAUTISTA, b Col, Laguna, Philippines; m 72; c 2. SOIL MICROBIOLOGY, ENVIRONMENTAL MICROBIOLOGY. *Educ:* Univ Philippines, BS, 62; Cornell Univ, MS, 66; Colo State Univ, PhD(agron, soil microbiol), 73. *Prof Exp:* Res asst soil chem & microbiol, Int Rice Res Inst, 62-64, asst soil microbiologist, 66-69; fel microbiol ecol, dept agron, Colo State Univ, 73; independent microbiologist, Arctic Environ Res Lab, 75-76; microbiologist, US Environ Protection Agency, Arctic Environ Res Sta, 76-77 & US Arny Cold Regions Res & Eng Lab, Alaska Projs Off, 77-80; fel soil microbiol, Agr Forestry Exp Sta, Univ Alaska, 87; SOIL SCIENTIST, AGR RES SERV, SUBARCTIC AGR UNIT, USDA, 88- *Concurrent Pos:* Affil asst prof environ microbiol, Inst Water Resources, 75 & 81-83; adj researcher, soil sci sept, Univ Minn, 79; affil asst prof, Sch Agr & Land Resources Mgt, Univ Alaska, Fairbanks, 84-85, affil assoc prof soil microbiol, 85-, lectr, 86-; consult microbiologist, 81- *Mem:* Am Soc Microbiol; Int Soc Soil Sci; Soil Sci Soc Am; Sigma Xi; Asn Women Sci. *Res:* Ecology of microorganisms in terrestrial and freshwater environments; effects and degradation of organic pollutants; microbial transformations of minerals and nutrient cycling. *Mailing Add:* 1127 Park Dr Fairbanks AK 99709

SPATOLA, ARNO F, b Albany, NY, May 9, 44; m 82; c 1. BIO-ORGANIC CHEMISTRY. *Educ:* Cornell Univ, AB, 66; Univ Mich, MS, 69, PhD(chem), 71. *Prof Exp:* Lectr chem, Univ Mich, 70-71; assoc, Univ Ariz, 71-73; from asst prof to assoc prof, 73-83, PROF CHEM, UNIV LOUISVILLE, 83-, PROF BIOCHEM, 90- *Concurrent Pos:* Pres, Peptides Int, Inc, Louisville, Ky; vis prof, Univ Padova, 82; chmn-elect, Gordon Conf on Peptides, 90; counr & chmn, Pub Comt, Am Peptide Soc. *Honors & Awards:* Devoe-Raynolds Award, 90. *Mem:* AAAS; Am Chem Soc; Sigma Xi; Am Peptide Soc. *Res:* Polypeptide synthesis, solution, solid phase methods and catalytic transfer hydrogenation; hormones and hormone analogues incorporating novel amino acids and amide bond replacements; peptide antagonists of LH-RH as potential ovulation inhibitors; structure-function studies on peptide hormones; collagenase inhibitors; growth hormone releasing peptides; cholecystokinin analogs as appetite suppressants; enzyme mimetics. *Mailing Add:* Dept Chem Belknap Campus Univ Louisville Louisville KY 40292

SPATZ, DAVID MARK, b Pottstown, Pa, Oct 10, 46. ORGANIC CHEMISTRY, AGRICULTURAL CHEMISTRY. *Educ:* Clarkson Col, BS, 68; Univ Mich, Ann Arbor, PhD(med chem), 72. *Prof Exp:* Fel nucleotide synthesis, Stanford Univ, 72-74; sr res chemist, Dow Chem Pharmaceut Res & Develop, 74-77; res chemist, Agr Res Div, Am Cyanamid Co, 77-80; AT CHEVRON CHEM CO, 80- *Mem:* Am Chem Soc; AAAS. *Res:* Synthetic organic chemistry; medicinal and pesticidal chemistry; drug design; heterocyclic and natural product synthesis including nucleotides and terpenes. *Mailing Add:* Chevron Chem Co 15029 San Pablo Ave San Ramon CA 94583-0947

SPATZ, MARIA, NEUROCYTOBIOLOGY, TISSUE CULTURE. *Prof Exp:* LAB CHIEF, NAT INST NEUROL & COMMUN DIS, NIH, 85- *Mailing Add:* Nat Inst Neurol & Commun Dis NIH 900 Rockville Pk Bethesda MD 20892

SPATZ, SIDNEY S, b Pittsburgh, Pa, Jan 13, 24; m 46; c 3. ORAL SURGERY. *Educ:* Univ Pittsburgh, BS, 43, DDS, 45; Am Bd Oral Surg, dipl, 59. *Prof Exp:* PROF ORAL SURG & CHMN DEPT, SCH DENT MED, UNIV PITTSBURGH, 71-, ASST DEAN CLIN AFFAIRS, 85- *Concurrent Pos:* Head oral & maxillofacial surg & dent, Montefiore Hosp. *Mem:* Am Asn Oral & Maxillofacial Surg; Int Asn Oral Surg; fel Am Col Dent; fel Int Col Dent. *Res:* Evaluation of drugs and techniques in relation to clinical oral surgery. *Mailing Add:* 1207 Beechwood Ct Pittsburgh PA 15206

SPATZ, SYDNEY MARTIN, b New York, NY, June 9, 12; m 36; c 1. ORGANIC CHEMISTRY. *Educ:* Univ Iowa, BA, 35, MS, 37; Iowa State Col, PhD(org chem), 41. *Prof Exp:* Res assoc, Nat Defense Res Comt, Iowa State Col, 42-43; res chemist, Nat Aniline Div, Allied Chem & Dye Corp, 43-47; chief chemist, Polak's Frutal Works, Inc, 47-53; pres & mgr, Spatz Chem, Inc, 53-54; res supvr, Nat Aniline & Specialty Chem Div, Allied Chem Corp, 54-71; sr prod eng specialist, Mead Papers Div, Mead Corp, 72-74, res fel, Mead Cent Res Div, 74-77; Consult, 77-84. *Mem:* Am Chem Soc; Sigma Xi. *Res:* Dicarboxylic anhydride chemistry; synthetic antimalarials; organometallics of lithium; ultraviolet absorbers; epoxy curing agents; polyester fire-retardant resins; cationic dyestuffs; color precursors for copy systems; carbonless copy; N-heterocyclic chemistry; infringement searches. *Mailing Add:* 6698 Tenth Ave N Apt 321 Lake Worth FL 33467-1448

SPAULDING, HARRY SAMUEL, JR, b Waterbury, Vt, Dec 12, 30; m 56; c 5. ALLERGY, IMMUNOLOGY. *Educ:* Albany Col Pharm, NY, BS, 53; Duquesne Univ, MS, 55; Univ Vt, Burlington, MD, 59. *Prof Exp:* Post surgeon, 24th Med Detachment, 61-63; resident pediat, Walter Reed Army Med Ctr, 63-65; chief pediat, US Army Hosp, Ft Carson, 66-68; pvt pract, Beverly Hosp, Mass, 68-71; chief, Dept Clin Admin, Reynolds Hosp, Ft Sill, 71-72; chief, Gen Pediat Serv, Fitzsimons Army Med Ctr, Denver, 74-77; asst chief, Allergy-Immunol Serv, Fitzsimons Army Med Ctr, Aurora, Colo, 77-80, dep comdr, Dep Med Activ & Med Educ, 80-81, dep comdr clin serv & dir med educ, 86-88 & 88-89, comdr, 88, staff allergist, 89, CHIEF, ALLERGY-IMMUNOL & CONSULT TO SURGEON GEN, FITZSIMONS ARMY MED CTR, AURORA, COLO, 89-; ASST CLIN PROF PEDIAT, MED CTR, UNIV COLO, 75- *Concurrent Pos:* Consult pediat, Colo State Hosp, Pueblo, 66-68; clin instr pediat, Med Ctr, Tufts Univ, 68-71; fel allergy-immunol, Fitzsimons Army Med Ctr & Nat Jewish Hosp, Denver, 72-74; staff affil, Nat Jewish Hosp & Res Ctr, 77- *Mem:* Am Acad Pediat; fel Am Acad Allergy; Asn Mil Allergists. *Res:* Association between gastroesophageal reflux and asthma; pharmacology of aminophylline with respect to coagulation problems and sensitivity through its ethylene, diamine fraction. *Mailing Add:* Fitzsimons Army Med Ctr Allergy-Immunol Serv Denver CO 80240

SPAULDING, LEN DAVIS, b Spring Valley, Ill, Oct 31, 42. INORGANIC CHEMISTRY, ORGANOMETALLIC CHEMISTRY. *Educ:* Antioch Col, BSc, 65; Univ Cincinnati, PhD(chem), 72. *Prof Exp:* Res assoc, Ga Inst Technol, 72-74; assoc scientist chem, Brookhaven Nat Lab, 74-80; with Exxon Chem Corp, 80-; AT PLATINA CATALYST LAB INC. *Concurrent Pos:* Consult, Mad River Chem Co, 67-69; NIH fel, Ga Inst Technol, 73-74. *Mem:* Am Chem Soc; AAAS. *Res:* Catalysis; Fischer-Tropsch and related reaction; porphyrin chemistry; transition metal organometallic chemistry. *Mailing Add:* 732 Dixie Lane Plainfield NJ 07062

SPAULDING, MALCOLM LINDHURST, b Providence, RI, Feb 15, 47. OCEAN ENGINEERING. *Educ:* Univ RI, BS, 69, PhD(mech eng), 72; Mass Inst Technol, MS, 70. *Prof Exp:* Asst prof eng mech, Old Dominion Univ, 72-73; from asst prof to assoc prof, 73-83, PROF OCEAN ENG, UNIV RI, 83- *Concurrent Pos:* Consult var pvt industs, 75-; Royal Norweg Res Coun Fel, 82-83; Fulbright-Hayes fel, Leningrad, USSR; panel mem, Nat Res Coun, 87-90. *Mem:* Am Soc Civil Engrs; Am Soc Mech Engrs; AAAS; Am Geophys Union; Marine Technol Soc. *Res:* Numerical modelling of coastal and shelf processes to include circulation, temperature, salinity and pollutant transport; computational fluid mechanics; oil spill fates and impact modeling. *Mailing Add:* Dept of Ocean Eng Univ of RI Kingston RI 02881-0814

SPAULDING, STEPHEN WAASA, b San Francisco, Calif, Aug 24, 40; m 69; c 2. ENDOCRINOLOGY, MEDICAL RESEARCH. *Educ:* Pomona Col, BA, 62; McGill Univ, MD & CM, 66. *Prof Exp:* Intern & asst res med, Osler Serv, Johns Hopkins Univ, 66-68; clin assoc endocrinol, NIH Geront Ctr, 68-70; from fel to asst prof endocrinol, Sch Med, Yale Univ, 70-76; assoc prof, 76-81, PROF, STATE UNIV NY, BUFFALO, 81- *Concurrent Pos:* Attend physician, Yale New Haven Hosp & West Haven Vet Admin Hosp, 72-76; Am Col Physicians res scholar, 72; NIH spec res fel, 73-74; clin investr, Vet Admin, 74; Am Col Physicians traveling scholar endocrinol, 76; chief, Endocrine Unit, Buffalo Gen Hosp, 76-82; assoc chief staff res, Buffalo VA Med Ctr, 82-; Sr Int Fogarty fel, 84-85. *Mem:* Am Thyroid Asn; Endocrine Soc; Am Fedn Clin Res; Am Col Physicians; Am Soc Clin Invest. *Res:* Actions of thyrotropin on chromatin structure and function; hypothalamic-pituitary interrelationships; thyroid hormone metabolism. *Mailing Add:* 3495 Bailey Ave Buffalo NY 14215

SPAULDING, THEODORE CHARLES, PHARMACY, PHARMACEUTICALS. *Educ:* Univ NC, PhD(pharmacol), 73- *Prof Exp:* DIR PHARM & PHARMACOL, HEALTH CARE GROUP, BRIT OXYGEN CO, DIV ANIQUEST, 85- *Mailing Add:* Dept Pharmacol & Pharm Anaquest Murray Hill NJ 07974

SPAULDING, WILLIAM BRAY, clinical medicine, for more information see previous edition

SPAUSCHUS, HANS O, b Liedemeiten, Ger, June, 15, 23; US citizen; m 59; c 2. MATERIALS SCIENCE. *Educ:* Ill Col, AB, 46; Tulane Univ, MS, 48, PhD(phys chem), 50. *Prof Exp:* Chemist, Gen Elec Co, 50-53, mgr lab, 53-56, res assoc, 56-68, mgr lab, 68-80; DIR RES, GA INST TECHNOL, 80- *Concurrent Pos:* Consult indust, Spauschus Assocs, 80- *Honors & Awards:* Steinmetz Medal, Gen Elec Co, 73. *Mem:* AAAS; Inst Elec & Electronics Engrs; fel Am Soc Heating, Refrig & Air Conditioning Engrs. *Res:* Biomass conversion to fuels and chemicals; high temperature solar thermal systems; materials sciences; heat pumps; technology appraisal and transfer; vapor compression cooling science and technology. *Mailing Add:* Spaschus Assoc Inc 1575 Northside Dr Suite 410 Atlanta GA 30318

SPAYD, RICHARD W, b Reading, Pa, Dec 10, 32; m 58; c 3. ORGANIC CHEMISTRY. *Educ:* Albright Col, BS, 58; Univ Del, MS, 60, PhD(org chem), 62. *Prof Exp:* From res chemist to sr res chemist, Eastman Kodak Co, 62-67, lab head, 67-89; RETIRED. *Mem:* Am Chem Soc; Am Asn Clin Chem. *Res:* Photographic systems; use of radioactive isotopes to study reaction mechanisms; dry multilayer films for clinical analysis. *Mailing Add:* 8395 Cypress Hollow Dr Sarasota FL 34238-5500

SPAZIANI, EUGENE, b Detroit, Mich, July 22, 30; m 53; c 2. ENDOCRINOLOGY, REPRODUCTIVE PHYSIOLOGY. *Educ:* Univ Calif, Los Angeles, BA, 52, MA, 54, PhD(zool), 58. *Prof Exp:* Asst zool, Univ Calif, Los Angeles, 52-55; hon res asst physiol, Univ Col, London, 58-59; from instr to assoc prof, 59-68, chmn dept, 77-80, PROF ZOOL, UNIV IOWA, 68- *Concurrent Pos:* Lalor Found fel, 60; vis investr, Inst Biomed Res, AMA, 66-67; mem, Bd Examrs, Grad Record Exam, advan test in biol, Educ Testing Serv, Princeton, NJ, 66-70; mem, Panel Undergrad Sci Partic Prog, NSF & consult, Panel Preprof Training, Comn Undergrad Educ Biol Sci; vis prof physiol, Univ Calif, Sch Med, San Francisco, 81. *Mem:* AAAS; Am Soc Zoologists; Am Physiol Soc; Endocrine Soc; Soc Study Reproduction. *Res:* Mechanisms of steroid hormone action in vertebrate reproductive organs; hormonal control of cellular transport; pigmentation; invertebrate endocrinology. *Mailing Add:* Dept Biol Univ Iowa Iowa City IA 52242

SPEAKMAN, EDWIN A(ARON), electronics, for more information see previous edition

SPEAR, BRIAN BLACKBURN, b Los Angeles, Calif, July 1, 47; m 72; c 1. GENE STRUCTURE. *Educ:* Amherst Col, AB, 69; Yale Univ, MPhil, 70, PhD(biol), 73. *Prof Exp:* Res assoc, Dept Molecular Cell & Develop Biol, Univ Colo, 73-75; asst prof biol sci, Northwestern Univ, 76-82; res scientist, dept molecular biol, Abbott Labs, 82-; AT DEPT BIOL SCI, NORTHWESTERN UNIV. *Mem:* Sigma Xi; Am Soc Cell Biol; AAAS. *Res:* Gene and chromosome structure; polytene chromosome organization in Drosophila and protozoa; molecular biology of ciliated protozoa; structure and evolution of genes for ribosomal RNA and actin. *Mailing Add:* Abbott Labs D-91L 1400 Sheridan Rd North Chicago IL 60064

SPEAR, CARL D(AVID), b Salt Lake City, Utah, Dec 6, 27; m 58; c 6. METALLURGY. *Educ:* Univ Utah, BS, 55, PhD(metall), 60. *Prof Exp:* Res metallurgist, Corning Glass Works, 60-63; asst prof metall, Univ Idaho, 63-66; head dept mech eng, 66-77, PROF MECH ENG, UTAH STATE UNIV, 66- *Mem:* Soc Mfg Engrs; Am Soc Metals. *Res:* Behavior of materials in manufacturing processes; non-traditional processes. *Mailing Add:* Dept Mech Eng Utah State Univ Logan UT 84322

SPEAR, GERALD SANFORD, b Providence, RI, Mar 3, 28; m 64; c 3. PATHOLOGY. *Educ:* Harvard Univ, AB, 48; Johns Hopkins Univ, MD, 52; Am Bd Path, dipl, 59. *Prof Exp:* Asst med, Sch Med, Wash Univ, 52-53; asst, Johns Hopkins Univ, 53-54, instr, 54-56 & 58-59, from asst prof to assoc prof path, Sch Med, 64-77; PROF PATH & MEM MED STAFF, CALIF COL MED, UNIV CALIF, IRVINE, 77- *Concurrent Pos:* Intern, Barnes Hosp, St Louis, Mo, 52-53; from asst pathologist to asst resident, Johns Hopkins Hosp, 53-56, resident, 58-59, pathologist, 59-77; vis pathologist, Baltimore City Hosps, 59-60. *Mem:* Int Soc Nephrology; Am Asn Path; Am Soc Pediat Nephrology; Am Soc Nephrology; Int Acad Path; Soc Pediat Path. *Res:* renal and pediatric pathology. *Mailing Add:* Dept of Path Univ Calif Irvine CA 92717

SPEAR, IRWIN, b New York, NY, Jan 4, 24; m 49; c 5. PLANT PHYSIOLOGY, BIOLOGY. *Educ:* Cornell Univ, BS, 47; Harvard Univ, AM, 49, PhD(biol), 53. *Prof Exp:* Asst prof bot & physiologist, Plant Res Inst, 53-59, assoc prof bot, 59-69, PROF BOT, UNIV TEX, AUSTIN, 69- *Concurrent Pos:* Mem biol advan placement comt, Col Entrance Exam Bd. *Mem:* AAAS; Am Soc Plant Physiol; Bot Soc Am; Soc Exp Biol & Med; Scand Soc Plant Physiol; Sigma Xi. *Res:* Physiology of growth and development, especially flowering; social consequences of biological discoveries. *Mailing Add:* Biol Labs Univ Tex PO Box 7640 Austin TX 78713-7640

SPEAR, JOSEPH FRANCIS, b Baltimore, Md, May 3, 43; c 2. PHYSIOLOGY. *Educ:* Loyola Col, Md, BS, 65; Univ Pa, PhD(physiol), 69. *Prof Exp:* Instr physiol, Sch Med, 70, Sch Vet Med, 70-72, from asst prof to assoc prof, 72-79, PROF PHYSIOL, SCH VET MED, UNIV PA, 79- *Concurrent Pos:* Pa Heart Asn res fel, Univ Pa, 70-71 & res grant, 71-72, res fel physiol, Dept Med, Univ Pa Hosp, 71-72; mem, Coun Basic Sci, Am Heart Asn, 71, estab investr, 72. *Mem:* AAAS; fel Am Col Cardiol; Am Heart Asn; Soc Gen Physiologists; Cardiac Muscle Soc. *Res:* Cardiovascular physiology; cardiac electrophysiology. *Mailing Add:* Sch of Vet Med Univ of Pa Philadelphia PA 19104

SPEAR, JO-WALTER, b Bridgton, NJ, Nov 3, 42; c 2. HYDROGEOLOGY, LANDFILL DESIGN. *Educ:* Rutgers Univ, AB, 63; Univ Pa, MSc, 70. *Prof Exp:* Teacher chem, Vineland Sr High Sch, 69-72; proj mgr, Pandullo Quirk Assocs, 72-78; sanit engr, Del Basin Comn, 78-79; dir chem process div, John Sexton Contractors, 79-82, dir corp develop, 82-86; asst prof civil eng, Midwest Col Eng, 80-86; pres, Morgen Environmental, 86-87; sr civil eng, Rogers Golden & Halpern, 87-90; SR ENVIRON ENGR, CHZM HILL, 90- *Concurrent Pos:* Tech dir, Alternative Technol, 80-86. *Mem:* Nat Soc Prof Engrs; Am Inst Chem Engrs; Am Soc Civil Engrs; fel Am Inst Chemists; Am Pub Works Asn; Water Pollution Control Asn; Govt Refuse Control & Disposal Asn. *Res:* Numerical risk assessment of landfill liner design. *Mailing Add:* 135 Stenton Ave Plymouth Meeting PA 19462-1219

SPEAR, PATRICIA GAIL, b Chattanooga, Tenn, Dec 14, 42; m 83. VIROLOGY, CELL BIOLOGY. *Educ:* Fla State Univ, BA, 64; Univ Chicago, PhD(virol), 69. *Prof Exp:* USPHS trainee & res assoc virol, Dept Microbiol, Univ Chicago, 69-71; Arthritis Found fel & res assoc biochem, Rockefeller Univ, 71-73; from asst prof to assoc prof, 73-82, PROF MOLECULAR GENETICS & CELL BIOL, UNIV CHICAGO, 82- *Concurrent Pos:* USPHS res career develop award, 75; consult comt virol & cell biol, Am Cancer Soc, 75-78; consult human cell biol prog, NSF, 75-77; mem med adv bd, Leukemia Res Found, Inc, 80-84; consult, Microbiol & Infectious Dis Res Comn, Nat Inst Allergy & Infectious Dis, NIH, 83- *Mem:* AAAS; Am Soc Virol; Am Soc Microbiologists. *Res:* Virus-induced modifications of cell membranes. *Mailing Add:* Dept Molecular Genet & Cell Biol MKL 008 Univ Chicago 303 E Chicago Ave Chicago IL 60611

SPEAR, PAUL WILLIAM, b Baltimore, Md, Nov 3, 08; m 44; c 3. MEDICINE. *Educ:* Johns Hopkins Univ, BA, 30, MD, 34; Am Bd Internal Med, dipl, 41, recert, 74. *Prof Exp:* From asst to instr med, Johns Hopkins Univ, 37-41; chief med, Manhattan Beach Vet Admin Hosp, 47-50; from asst chief med to chief med, Brooklyn Vet Admin Hosp, 50-63; dir med, Montefiore-Morrisania Affil, 63-76; emer prof med, Albert Einstein Col Med, 75-; med dir, Queens County Div Island Peer Rev Orgn, 84-86; RETIRED. *Concurrent Pos:* Vis physician, Sinai Hosp, Baltimore, 37-41; clin assoc prof med, Col Med, State Univ NY Downstate Med Ctr, 51-63; attend physician, Kings County Hosp, 57-63, Maimonides Hosp, 58-63 & Montefiore Hosp, 63-; assoc prof med, Albert Einstein Col Med, 68-75. *Mem:* Am Soc Hemat; Int Soc Hemat; Am Fedn Clin Res; fel Am Col Physicians; fel NY Acad Med. *Res:* Hematology. *Mailing Add:* 55 Manhasset Woods Rd Manhasset NY 11030

SPEAR, ROBERT CLINTON, b Los Banos, Calif, June 26, 39; m 62; c 2. ENGINEERING. *Educ:* Univ Calif, Berkeley, BS, 61, MS, 63; Cambridge Univ, PhD(eng), 68. *Prof Exp:* Mech engr, US Navy, Calif, 63-65, 68-69; US Pub Health Serv fel, 69-70, from asst prof to assoc prof, 70-80, dir, Northern Calif Occup Health Ctr, 79-89, assoc dean, 89-91, PROF ENVIRON HEALTH SCI, SCH PUB HEALTH, UNIV CALIF, BERKELEY, 81- *Concurrent Pos:* Sr int fel, Fogarty Int Ctr, NIH, 77-78. *Mem:* AAAS; Am Soc Mech Engrs; Am Indust Hyg Asn. *Res:* Engineering aspects of occupational and environmental health. *Mailing Add:* Sch of Pub Health Univ of Calif Berkeley CA 94720

SPEARE, EDWARD PHELPS, b Springfield, Mass, Jan 12, 21; m 48; c 5. ZOOLOGY. *Educ:* Northland Col, BA, 48; Univ Mich, MA, 50; Mich State Univ, PhD(zool), 58. *Prof Exp:* Assoc prof, 50-72, PROF BIOL, OLIVET COL, 72- *Concurrent Pos:* From instr to assoc prof, Biol Sta, Mich State Univ, 54-64. *Honors & Awards:* Nat Wildlife Fedn Award, 58-59. *Mem:* Am Fisheries Soc; Am Soc Ichthyologists & Herpetologists; Nat Audubon Soc. *Res:* Ecological ichthyology; stream ecology; life history of percid fish. *Mailing Add:* Dept of Biol Olivet Col Olivet MI 49076

SPEARING, ANN MARIE, b Olean, NY, Jan 29, 47. PLANT BIOPHYSICS. *Educ:* State Univ NY Col Buffalo, BA, 69; State Univ NY Col Forestry, Syracuse Univ, MS, 71; Univ Md, PhD(plant biophys), 75. *Prof Exp:* Asst prof bot, Wheaton Col, 75-78; actg asst dir, Environ Prog, 78-81, asst dean, 81-84, ASSOC DEAN, GRAD COL, UNIV VT, 84- *Concurrent Pos:* Assoc prof forestry, Univ Vt, 83- *Mem:* AAAS; Am Inst Biol Sci; Am Soc Plant Physiologists; Ecol Soc Am; Sigma Xi; Asn Women Sci. *Res:* All aspects of physiological ecology, particularly responses of plants to light; effects of physical aspects of environment such as temperature and light on physiology of plants. *Mailing Add:* RR 3-Box 7200 Stow VT 05672

SPEARING, CECILIA W, b New York, NY, Jan 29, 27. BIOCHEMISTRY. *Educ:* Hunter Col, BA, 47; Columbia Univ, MA, 49; George Washington Univ, MS, 61. *Prof Exp:* Lab technician, Med Col, Cornell Univ, 47-50; lab technician, Col Physicians & Surgeons, Columbia Univ, 50; instr biol, Barber-Scotia Col, 50-52; biochemist, Walter Reed Army Inst Res, 52-59, NIH, 59-63 & Food & Drug Admin, 63-65; biol sci adminr, 65-76, STAFF ASSOC, NSF, 76- *Mem:* AAAS; Am Inst Biol Sci; Am Chem Soc. *Res:* Intermediary metabolism. *Mailing Add:* 11500 Fairway Dr No 402 Reston VA 22090-4434

SPEARS, ALEXANDER WHITE, III, b Grindstone, Pa, Sept 29, 32; m 51, 77; c 1. ORGANIC CHEMISTRY, PHYSICAL CHEMISTRY. *Educ:* Allegheny Col, BS, 53; Univ Buffalo, PhD(chem), 60. *Prof Exp:* Res assoc chem, Univ Buffalo, 56-58; instr, Millard Fillmore Col, 58-59; res assoc, Res Div, P Lorillard Co, 59-61, sr res chemist, 61-65, dir basic res, 65-68, dir res & develop, 68-71, vpres res & develop, 71-75, sr vpres, 75-77, EXEC VPRES OPERS & RES, LORILLARD CORP, 77- *Concurrent Pos:* Asst prof, Greensboro Div, Guilford Col, 61-65; adv, Nat Cancer Inst, 68-79; mem, Tech Study Group, Cigarette Safety Act, 84, ISO Comt 126, 88. *Honors & Awards:* Distinguished Achievement Award in Tobacco Sci, Philip Morris, Inc, 70. *Mem:* AAAS; Am Chem Soc. *Res:* Cancer chemotherapy; pyrolytic reactions and products; spectroscopy; chromatography. *Mailing Add:* Lorillard Inc PO Box 21688 Greensboro NC 27420-1688

SPEARS, BRIAN MERLE, b La Grande, Ore, Oct 5, 50; m 78; c 4. ENTOMOLOGY, RANGE SCIENCE. *Educ:* Ore State Univ, BS, 72; Tex Tech Univ, MS, 75; Univ Idaho, PhD(entom), 78. *Prof Exp:* Asst res scientist entom, Univ Ariz, 78-79; range mgr, Idaho Dept Lands, 80-81; RANGE CONSERVATIONIST, BUR INDIAN AFFAIRS, 81- *Mem:* Soc Range Mgt; Entom Soc Am; Sigma Xi. *Res:* Bionomics of Arizona range and forest insect pests. *Mailing Add:* 680 Eighth Madras OR 97741-1564

SPEARS, DAVID LEWIS, b Belvidere, Ill, July 22, 40; div; c 2. OPTOELECTRONICS, ELECTROOPTICS. *Educ:* Monmouth Col, Ill, BA, 62; Dartmouth Col, MA, 64; Purdue Univ, PhD(physics), 69. *Prof Exp:* Mem, res staff, 69-84, ASST HEAD, ELECTROOPTICAL DEVICES GROUP, LINCOLN LAB, MASS INST TECHNOL, 84- *Mem:* Am Phys Soc; Inst Elec & Electronic Engrs. *Res:* Acoustoelectric effect in GaAs; x-ray lithography; surface wave devices; integrated optics; optical waveguide modulators; heterodyne detection; infrared heterodyne radiometry; infrared detectors; diode lasers. *Mailing Add:* Lincoln Lab Mass Inst Technol PO Box 73 Lexington MA 02173

SPEARS, JOSEPH FAULCONER, b Moreland, Ky, Aug 1, 15; m 45; c 2. BIOLOGY. *Educ:* Univ Ky, BS, 38. *Prof Exp:* Asst to dir, Bernheim Natural Hist Found, Ky, 38-40; supvr archaeol, US Dept Interior, 40-42; asst proj leader golden nematode control, Bur Entom & Plant Quarantine, USDA, 46-51, proj leader golden nematode control, Plant Pest Control Div, 51-56, staff officer control opers, Plant Pest Control Div, Agr Res Serv, 56-59, chief staff officer, 59-68, assoc dir plant protection div, 68-71, asst dep adminr plant protection progs, Animal & Plant Health Serv, 71-78, asst dir plant protection & quarantine progs, 78-80; RETIRED. *Mem:* Entom Soc Am; Orgn Trop Nematologists; Soc Nematologists; Weed Sci Soc Am. *Res:* Control of nematodes, plant pests, insects, weeds and plant diseases. *Mailing Add:* 8719 Stockton Pkwy Alexandria VA 22308

SPEARS, KENNETH GEORGE, b Erie, Pa, Oct 23, 43; m 66; c 1. LASER PHOTOPHYSICS, MEDICAL IMAGING WITH LASERS. *Educ:* Bowling Green State Univ, BS, 66; Univ Chicago, MS, 71, PhD(chem), 71. *Prof Exp:* Fel chem physics, Nat Res Coun, Nat Oceanic & Atmospheric Admin, Boulder, Colo, 70-72; asst prof, Northwestern Univ, 72-78, assoc prof chem, 78-89, JOINT PROF BIOMED ENG, NORTHWESTERN UNIV, 87-, PROF CHEM, 89- *Concurrent Pos:* Alfred P Sloan Found fel, 74-76; mem adv bd, Midwest Bio-Laser Inst, 85-; consult, Laser Surg Syst, 81-84. *Mem:* Am Phys Soc; Am Chem Soc; AAAS; Int Soc Optical Eng. *Res:* Molecular photophysics, electron and energy transfer in solution; picosecond and other laser techniques; medical imaging and corneal applications of lasers; photophysical probes of vesicles and peptides; picosecond infrared spectroscopy. *Mailing Add:* Dept Chem Northwestern Univ 2145 Sheridan Rd Evanston IL 60208-3113

SPEARS, RICHARD KENT, b Brush, Colo, Mar 28, 37; m 57; c 5. ENGINEERING MATERIALS, ENGINEERING MECHANICS. *Educ:* Colo Sch Mines, BS, 59; Univ Denver, MS, 64; Univ Fla, PhD(eng sci), 77. *Prof Exp:* Engr metall, Martin Co, Denver, 59-64; applns engr, Honeywell, Inc, Minneapolis, 64-66; design engr, Gen Elec Co, St Petersburg, Fla, 66-82; PRES, LARGO SCI INC, FLA, 82- *Mem:* Am Soc Metals (treas, 77-78). *Res:* Ferroelectricity; glass ceramics; glass coated wire; polymer research; adhesive research; impact property of metals. *Mailing Add:* 3104 Roberta St Largo FL 33517

SPEARS, SHOLTO MARION, b Scottsville, Ky, Aug 29, 00; m 22; c 3. ENGINEERING. *Educ:* Univ Ky, BS, 22, CE, 33; Univ Mich, PhD(civil eng), 42. *Prof Exp:* Struct engr, Ogle Construct Co, Ill, 22-31; from asst prof to assoc prof civil eng, Ill Inst Technol, 31-42; dean sch eng, Fenn Col, 46-50; city engr, East Cleveland, 50-52; dir, Cleveland Mem Med Found, 53-57; prof civil eng, Univ Ark, Fayetteville, 57-68; vis prof, 68-73, EMER VIS PROF CIVIL ENG, UNIV MO-COLUMBIA, 73- *Honors & Awards:* Octave Chanute Medal, Western Soc Engrs, 42. *Mem:* Am Soc Civil Engrs; Am Soc Eng Educ; Nat Soc Prof Engrs. *Res:* Methods of highway traffic control; psychology in highway design. *Mailing Add:* 1423 Beechwood Terr Manhattan KS 66502

SPECHT, DONALD FRANCIS, b Harvey, NDak, Oct 15, 33; m 60; c 2. NEURAL NETWORKS, ELECTRICAL ENGINEERING & MEDICAL ELECTRONICS. *Educ:* Univ Santa Clara, BEE, 55; Carnegie-Mellon Univ, MS, 56; Stanford Univ, PhD(elec eng), 66. *Prof Exp:* Electronics engr, Radio Corp Am, NJ, 55-57; res engr, Lockheed Missiles & Space Co, 57-63, res specialist biomed data anal, 63-66, scientist, Lockheed Palo Alto Res Lab,

66-70; mgr prog develop, Gould, Inc, Palo Alto, 70-74; mgr res, Smithkline Instruments, 75-81; dir res, Ekoline Inc, 81-82; vpres, Xonics Imaging, 82-84; mgr, image processing, KLA Instruments, 84-88; SR MEM, LOCKHEED PALO ALTO RES LAB, 88- *Concurrent Pos:* Assoc ed, Inst Elec & Electronic Engrs Trans on Neural Networks. *Mem:* Inst Elec & Electronic Engrs; Int Neural Networks Soc. *Res:* Neural networks, sonar, and ultrasonic imaging; medical instrumentation; adaptive pattern-recognition techniques; nonlinear regression techniques; nonparametric probability estimators; radar target discrimination; automatic analysis of electrocardiograms; digital radiography. *Mailing Add:* 869 Terrace Dr Los Altos CA 94024

SPECHT, EDWARD JOHN, b Loveland, Colo, July 29, 15; m 38; c 2. GEOMETRY. *Educ:* Walla Walla Col, BS, 39; Univ Colo, MS, 41; Univ Minn, PhD(math), 49. *Hon Degrees:* DSc, Andrews Univ, 84. *Prof Exp:* Prof math, Andrews Univ, 47-72; prof, 72-86, EMER PROF, MATH, IND UNIV, SOUTH BEND, 86- *Concurrent Pos:* Vis lectr math, Univ Minn, 57. *Mem:* Fel AAAS; Math Asn Am; Sigma Xi; Am Math Soc. *Res:* Development of pasch geometry which follows from the incidence, betweeness, plane separation and least upper bound axioms. *Mailing Add:* 1023 Oakland South Bend IN 46615

SPECHT, HAROLD BALFOUR, b Schenectady, NY, May 13, 27; Can citizen; m 49; c 5. ENTOMOLOGY, ECOLOGY. *Educ:* McGill Univ, BSc, 48; Univ Wis, MSc, 51; Rutgers Univ, PhD(entom), 59. *Prof Exp:* RES SCIENTIST & ENTOMOLOGIST, KENTVILLE RES STA, CAN DEPT AGR, 48- *Mem:* Entom Soc Am; Entom Soc Can. *Res:* Fruit insect ecology; integrated control studies on insects affecting apples; factors affecting mite populations on apple trees; pea aphid ecology; apple aphid ecology; tobacco cutworm investigation. *Mailing Add:* Kentville Res Sta Can Dept Agr Kentville NS B4N 1J5 Can

SPECHT, HEINZ, PHYSIOLOGY. *Educ:* Johns Hopkins Univ, PhD(physiol), 33. *Prof Exp:* Administrator, NIH, 68-71; RETIRED. *Mailing Add:* Fairhaven C135 7200 Third Ave Sykesville MD 21784

SPECHT, JAMES EUGENE, b Scottsbluff, Nebr, Sept 12, 45; m 69. PLANT PHYSIOLOGY, PLANT BREEDING. *Educ:* Univ Nebr, BS, 67, PhD(genetics), 74; Univ Ill, MS, 71. *Prof Exp:* MEM FAC AGRON, INST AGR & NATURAL RESOURCES, UNIV NEBR, 74- *Mem:* AAAS; Am Soc Agron; Sigma Xi. *Res:* Development of physiological screening techniques and tools to aid in the breeding and improvement of soybean varieties. *Mailing Add:* 309 Keim Hall E Campus Univ Nebr Lincoln NE 68503-0915

SPECHT, LAWRENCE W, b Roscoe, NY, Aug 5, 28; m 51; c 6. GENETICS. *Educ:* Cornell Univ, BS, 51; Mich State Univ, MS, 55, PhD, 57. *Prof Exp:* PROF DAIRY SCI, PA STATE UNIV, 57- *Mem:* Am Dairy Sci Asn. *Res:* Dairy cattle genetics and breeding, especially progeny testing and sire selection in artificial insemination; electronic data processing of milk production records. *Mailing Add:* Dept Dairy & Animal Sci 310 Agr Sci Bldg Pa State Univ University Park PA 16802

SPECHT, ROBERT DICKERSON, b Seattle, Wash, May 11, 13; m 36; c 5. MATHEMATICS. *Educ:* Univ Fla, AB, 36, MS, 38; Univ Wis, PhD(math), 42. *Prof Exp:* Instr math, Univ Fla, 36-38; asst, Univ Wis, 38-41; instr, Univ Fla, 41-42; asst, Brown Univ, 42; from asst physicist to assoc mathematician, David Taylor Model Basin, US Dept Navy, 42-45; asst prof math, Univ Wis, 45-49; mathematician, Rand Corp, Santa Monica, 49-79; RETIRED. *Concurrent Pos:* Consult, Rand Corp, 79- *Res:* Applied mathematics; mechanics. *Mailing Add:* 14930 McKendree Ave Pacific Palisades CA 90272

SPECIAN, ROBERT DAVID, b Niagara Falls, NY, Apr 25, 50; m 70; c 2. CYTOLOGY, ANIMAL PHYSIOLOGY. *Educ:* Southern Methodist Univ, BS, 72, MS, 74; Tulane Univ, PhD(biol), 80. *Prof Exp:* Fel cell biol & anat, Med Sch, Harvard Univ, 78-81; from asst prof to assoc prof anat, 81-90, PROF CELLULAR BIOL & ANAT, SCH MED, LA STATE UNIV, SHREVEPORT, 90- *Concurrent Pos:* Adj prof, dept biochem & molecular biol, Sch Med, La State Univ, Shreveport, 85- *Mem:* AAAS; Am Asn Anatomists; Am Soc Cell Biol; Am Soc Parasitol; NY Acad Sci. *Res:* Regulation and mechanisms of mucin synthesis and secretion in the mammalian intestine; physiochemical behavior of high molecular weight glycoproteins and their interaction with the intestinal flora; ishemia/reperfusion and intestinal function. *Mailing Add:* Dept Anat La State Univ 1501 Kingshwy PO Box 33932 Shreveport LA 71130-3932

SPECK, DAVID RALPH, b Lindsay, Calif, Oct 31, 27; m 52; c 1. LASER SCIENCE, INERTIAL CONFINEMENT FUSION. *Educ:* Fresno State Col, BS, 51; Univ Calif, Berkeley, MA, 53, PhD(physics), 56. *Prof Exp:* PHYSICIST, LAWRENCE LIVERMORE NAT LAB, UNIV CALIF, 56- *Mem:* Am Phys Soc. *Res:* Development and use of high power glass laser systems for inertial confinement fusion research; plasma physics; interaction of high power optical radiation with materials including non-linear optics and optical damage to materials. *Mailing Add:* 12 Corwin Dr Alamo CA 94507

SPECK, JOHN CLARENCE, JR, b Indianapolis, Ind, Jan 6, 17; m 40; c 5. BIOCHEMISTRY. *Educ:* Univ Ill, BS, 39; Univ NC, PhD(org chem), 43. *Prof Exp:* Asst chem, Univ NC, 39-40; chemist, US Naval Res Lab, 41-43; res assoc chem, Ind Univ, 43-45; res assoc, 45-46, from instr to prof, 46-87, EMER PROF BIOCHEM, MICH STATE UNIV, 87- *Concurrent Pos:* Adj Prof Biochem, Mich State Univ, 88- *Mem:* Am Chem Soc; Am Soc Biol Chemists. *Res:* Chemistry of enzymes and other natural products. *Mailing Add:* Dept Biochem Mich State Univ East Lansing MI 48824

SPECK, JOHN EDWARD, b Toronto, Ont, May 22, 25; m 51; c 2. PERIODONTOLOGY. *Educ:* Univ Toronto, DDS, 49, dipl periodont, 52. *Prof Exp:* From instr to assoc prof, 52-71, PROF PERIODONT & CHMN DEPT, FAC DENT, UNIV TORONTO, 71- *Concurrent Pos:* Consult, Disabled Vet Admin, Sunnybrook Mil Hosp, 65-, Hosp for Sick Children, 67 & med-dent staff, Univ Toronto, Sunnybrook Hosp, 68-; consult, North York Gen Hosp, 74-; active staff, St Michaels Hosp, 84- *Mem:* Can Acad Periodont (secy-treas, 55-66, pres, 70-71); Am Acad Periodont; fel Am Col Dent; fel Royal Col Dent Can (registr-secy-treas, 66-). *Mailing Add:* Dept Periodont Univ Toronto Fac Dent 124 Edward St Toronto ON M5G 1G6 Can

SPECK, MARVIN LUTHER, b Middletown, Md, Oct 6, 13; m 40; c 3. FOOD SCIENCE & TECHNOLOGY. *Educ:* Univ Md, BS, 35, MS, 37; Cornell Univ, PhD(bact), 40. *Prof Exp:* Bacteriologist, Western Md Dairy, 35-36; instr bact, Univ Md, 40-41; asst chief bacteriologist, Nat Dairy Res Labs, 41-47; from assoc prof to prof dairy bact, 47-57, William Neal Reynolds prof, 57-79, EMER PROF FOOD SCI & MICROBIOL, NC STATE UNIV, 79- *Concurrent Pos:* Jr bacteriologist, USDA, 36; bacteriologist, Dairymen's League, 40; instr, Univ Md, 45; consult, USPHS, 50-51 & 53; WHO fel, Europe, 66; Consult, 79- *Honors & Awards:* Borden Award, Am Dairy Sci Asn, 59 & Pfizer Award, 67; Nordica Int Res Award, Am Cultured Dairy Prod Inst, 81; Nat Award for Agr Excellence in Sci, Nat Agr-Mkt Asn, 84. *Mem:* AAAS; Inst Food Technol; Am Soc Microbiol; Am Dairy Sci Asn; fel Inst Food Technologists. *Res:* Nutrition and metabolism of lactic acid bacteria; injury and destruction of bacteria by physical and chemical agents; uses and functions of intestinal lactobacilli. *Mailing Add:* 3204 Churchill Rd Raleigh NC 27607

SPECK, REINHARD STANIFORD, b Rockport, Mass, Apr 30, 22; m 59. MEDICAL BACTERIOLOGY. *Educ:* Middlebury Col, AB, 44; Boston Univ, MD, 48. *Prof Exp:* Intern, Boston City Hosp, 48-49; from instr to assoc prof, 49-68, vchmn dept, 63-76, PROF MICROBIOL, SCH MED, UNIV CALIF, SAN FRANCISCO, 68- *Mem:* Am Soc Microbiol; Am Asn Hist Med. *Res:* Antibiotic antagonism; plague immunity; experimental airborne infection; infectious diseases; intestinal bacterial flora; history of fevers and infectious diseases. *Mailing Add:* Dept Microbio & Immunol Univ Calif 513 Parnassus Ave San Francisco CA 94143

SPECK, RHOADS MCCLELLAN, b Glenside, Pa, Apr 12, 20; m 46; c 4. ORGANIC CHEMISTRY. *Educ:* Philadelphia Col Pharm, BSc, 42; Pa State Univ, MSc, 49, PhD(org chem), 52. *Prof Exp:* Asst chemist, Eastern Regional Res Ctr, Agr Res Serv, USDA, 42-44 & 46-47; asst anal chem, Pa State Univ, 47-48; res chemist, Res Ctr, Hercules Inc, Wilmington, 52-75, sr res chemist, 75-81, res scientist, 81-83; RETIRED. *Concurrent Pos:* Consult, Hercules Inc, 86-87. *Mem:* Am Chem Soc; Sigma Xi. *Res:* Syntheses and physical properties of high molecular weight hydrocarbons; syntheses of agricultural chemicals; rosin and fatty acid chemistry; emulsion polymerization; free radical reactions. *Mailing Add:* RD 1 PO Box 675 F Berkeley Springs WV 25411

SPECKMANN, ELWOOD W, b Brooklyn, NY, Jan 10, 36; m 62; c 4. PHYSIOLOGY, NUTRITION. *Educ:* Rutgers Univ, BS, 57; Mich State Univ, MS, 59, PhD(nutrit physiol), 62. *Prof Exp:* Res asst nutrit physiol, Mich State Univ, 58-60; res scientist, Biospecialties Br, Physiol Div, Aerospace Med Res Labs, Wright-Patterson AFB, Ohio, 62-65; from asst dir to dir nutrit res, Nat Dairy Coun, 65-83, interim educ dir, 67-88, vpres, 83-85, pres, 85-89, vpres, Nutrit Res & Tech Serv, United Dairy Indust Assoc, 89-90; DIR RES PROG, SHRINERS HOSPS CRIPPLED CHILDREN, 91- *Concurrent Pos:* Air Force liaison rep food & nutrit bd & working group nutrit & feeding probs, Man in Space Comt, Space Sci Bd, Nat Acad Sci-Nat Res Coun, 63-65; Salisbury res fel, 60-62; fel coun Arteriosclerosis, Am Heart Asn. *Mem:* Inst Food Technologists; Am Inst Nutrit; NY Acad Sci; Coun Agr Sci & Technol; fel Am Heart Asn; Am Soc Clin Nutrit (treas, 82-86); Fedn Am Soc Exp Biol (treas, 86-88); Fel Am Col Nutrit; NY Acad Sci; Soc Res Adminr. *Res:* Physiology of circulation and heart; arteriosclerosis; physiology of burns; physiology of digestion and metabolism of foods; nutrient interactions; bone health. *Mailing Add:* Shriners Hosps for Crippled Children Hq PO Box 31356 Tampa FL 33631-3356

SPECKMANN, GUNTER WILHELM-OTTO, b Ger, Oct 3, 34; Can citizen; m 62; c 3. VETERINARY BACTERIOLOGY. *Educ:* Vet Col Hannover, Ger, DMV, 67; Univ Guelph, DVM, 68. *Prof Exp:* Veterinarian poultry dis, 68-76, VETERINARIAN BACTERIOL, ANIMAL PATH DIV, AGR CAN, 76- *Concurrent Pos:* Drug evaluation. *Mem:* Am Asn Zoo Veterinarians. *Res:* Epidemiological studies of Salmonella and Yersinia carriers in livestock. *Mailing Add:* RR Three Stittsville ON K2S 1B8 Can

SPECTER, STEVEN CARL, b Philadelphia, Pa, June 4, 47; m 69; c 2. VIROLOGY. *Educ:* Temple Univ, BA, 69, PhD(microbiol), 75. *Prof Exp:* Res fel microbiol, Albert Einstein Med Ctr, Philadelphia, 74-76, asst dir, 76-79; asst prof, 79-84, ASSOC PROF MICROBIOL & IMMUNOL, COL MED, UNIV SFLA, 84- *Concurrent Pos:* Virol consult, Tampa Gen Hosp, 79. *Mem:* Am Soc Microbiol; AAAS; fel Am Acad Microbiol. *Res:* Immune suppression by viruses, tumors and drugs, most notably friend leukemia virus and marijuana; clinical virology studies on antivirals and herpesviruses. *Mailing Add:* Dept Med Microbio Univ SFla 12901 N 30th Tampa FL 33612

SPECTOR, ABRAHAM, b Nyack, NY, Jan 14, 26; m 83; c 2. BIOCHEMISTRY. *Educ:* Bard Col, AB, 47; NY Univ, PhD(biochem), 57. *Hon Degrees:* MD, Univ Repub Uruguay, 81; DSc, Bard Col, 85. *Prof Exp:* Res chemist, Lederle Labs, Am Cyanamid Corp, 48-52; from instr to assoc biochem, Howe Lab, Mass Eye & Ear Infirmary, Harvard Med Sch, 58-65; lectr biol chem, Northeastern Univ, 59-62; from asst prof to assoc prof of ophthal, 65-73, PROF OPHTHALMIC BIOCHEM, COL PHYSICIANS & SURGEONS, COLUMBIA UNIV, 73-, DIR, LAB BIOCHEM & MOLECULAR BIOL, 76- *Concurrent Pos:* Mem, Nat Eye Inst Vision Res, 70-71; mem, Nat Eye Inst Bd Sci Advisors, 74; mem, Nat Eye Inst, 76-80, chmn, 78-80; vis prof opthal, Univ Puerto Rico, 82-; vis prof biochem, Med Univ Shangai, 86-; John Simon Guggenheim fel, 71-72, Fulbright, 81. *Honors & Awards:* Int Award, Japanese Coop Cataract Res Group, 87; Bausch & Lomb Sci Medal, 44. *Mem:* Am Soc Biol Chem; Am Chem Soc; Asn Res Vision & Ophthal (pres, 76); AAAS; Harvey Soc. *Res:* Protein chemistry; ophthalmic biochemistry; enzymology; biosynthesis of proteins and nucleic acids; oxidative stress. *Mailing Add:* Col Physicians & Surgeons Dept Opthal Columbia Univ New York NY 10032

SPECTOR, ARTHUR ABRAHAM, b Philadelphia, Pa, May 14, 36; m 60; c 3. BIOCHEMISTRY, INTERNAL MEDICINE. *Educ:* Univ Pa, BA, 56, MD, 60. *Prof Exp:* Intern, Abington Mem Hosp, 60-61; res med officer, Nat Heart Inst, 63-68; from asst prof to assoc prof, 68-75, PROF BIOCHEM & MED & DIR ARTERIOSCLEROSIS SCOR CTR, UNIV IOWA, 75- *Concurrent Pos:* NIH res career develop award, 69-74; Nat Heart Inst fel biochem, 63-65; mem coun arteriosclerosis, Am Heart Asn; mem metab study sect, NIH, 73-77 & rev comt, Ischemic Heart Dis Ctr, 78-79; chmn res comt & mem bd dirs, Iowa Heart Asn, 75-77; chmn & mem, Great Plains Regional Res Comn, Am Heart Asn, 77-80; mem & chmn biomed adv comt, Oak Ridge Assoc Univ, 78-82. *Mem:* AAAS; Tissue Cult Asn; Am Asn Cancer Res; Am Soc Biol Chemists; Am Soc Clin Invest. *Res:* Lipid metabolism; membranes; fatty acids; prostadlanging. *Mailing Add:* Dept Biochem 4-550 Bowen Sci Bldg Univ Iowa Iowa City IA 52242

SPECTOR, BERTRAM, b New York, NY, Nov 1, 21; m 45; c 2. ENVIRONMENTAL HEALTH, MEDICAL SYSTEMS. *Educ:* City Col New York, BEE, 45; Hunter Col, MS, 57; Cornell Univ, PhD(med sci), 61. *Prof Exp:* Vpres, Seversky Electronatom, NY, 60-63; dean acad affairs, NY Inst Technol, 63-64, chmn life sci dept, 63-64 & 67-68, prof life sci, 63-87, vpres res, 66-87; vpres, Ctr Educ Technol, 85-87; CONSULT MED ENGR, 87- *Concurrent Pos:* Prin investr grants, Off Educ, Dept Health, Educ & Welfare, Carnegie Corp, Ford Found Fund for Advan Educ & Proj ULTRA, 64-; prin investr systs anal, Brevard County Schs, Fla, 64-; prin co-investr, Dept Health, Educ & Welfare grant, 68-; adj prof, Shaw Univ, 68-70; consult, Pan Am Airlines, 68, Nova Univ Advan Technol & Hofstra Univ, 69 & Wash Univ, 69-; mem NSF panel, 69; mem bd trustees, Affiliated Cols & Univs, Inc; mem bd dirs, Afro-Am Coun Higher Educ; mem bd dirs & exec comt, Cancirco, Inc; chmn, Environ Control Comn, Town of Oyster Bay, 74- & Comt Energy & Natural Resources, 73-; vpres res, New York Chiropractic Col, Glen Head, 76-; pres, New Ctr Wholistic Health & Res, 81-; Sloan fel. *Mem:* Fel Am Inst Biol Sci. *Res:* Physics; educational technology; environmental sciences and technology; pattern electromyography; biofeedback; moire contourography; computer technology. *Mailing Add:* 303 A Sea Oats Dr Juno Beach FL 33408

SPECTOR, CLARENCE J(ACOB), b New York, NY, June 19, 27; m 50; c 3. ENGINEERING PHYSICS, MATERIALS SCIENCE. *Educ:* Va Polytech Inst & State Univ, BS, 53; Stevens Inst Technol, MS, 57. *Prof Exp:* Mem tech staff, Bell Tel Labs, Inc, 53-62; sr engr & mgr display components develop, 62-76, MGR THERMAL TECHNOL, IBM SYST PROD, IBM CORP, 76- *Mem:* AAAS; Inst Elec & Electronics Engrs. *Res:* Physics of dielectrics, magnetic and semiconductor materials; passive and magnetic thin film devices; metallurgy. *Mailing Add:* IBM Gen Prod Div Dept E 19-028 5600 Little Rd San Jose CA 95193

SPECTOR, HAROLD NORMAN, b Chicago, Ill, Feb 26, 35; m 61; c 1. SOLID STATE PHYSICS. *Educ:* Univ Chicago, SB, 57, SM, 58, PhD(physics), 61. *Prof Exp:* NSF fel physics, Hebrew Univ, Israel, 61-62; asst prof, Case Inst Technol, 62-63; res physicist, IIT Res Inst, 63-66, assoc prof, 66-76, PROF, ILL INST TECHNOL, 76- *Concurrent Pos:* Vis prof, Hebrew Univ, Israel, 73-74; fel GTE, 83 & consult, 83-86. *Mem:* Fel Am Phys Soc; Sigma Xi. *Res:* Solid state theory, optical and electronic properties of semiconducting quantum well systems electronphonon interactions in solids, transport theory, effect of strong electric and magnetic fields on electronic processes in solids. *Mailing Add:* Dept Physics Ill Inst Technol Chicago IL 60616

SPECTOR, LEO FRANCIS, b Kansas City, Mo, Nov 10, 23; m 68; c 5. PLANT ENGINEERING, MECHANICAL DESIGN ENGINEERING. *Educ:* Univ Kans, BSME, 49. *Prof Exp:* Sr assoc ed mach design, Penton Publ Co, 51-62, regional ed, 67-70; ed assembly eng, Hitchcock Publ Co, 62-67; ed plant eng, Tech Publ Co, 70-86; vpres & ed dir plant eng, Cahners Publ Co, Div Reed Publ, USA, 86-90; CONSULT PLANT ENG, FREE LANCE, 90- *Concurrent Pos:* Secy, Comt Sintered Metal Brake Mat (Standardization), Am Soc Testing & Mat, 56-66; mem, Several Am Nat Standards Comts on Fasteners, 64-70; chmn, Am Nat Standards Comt on Interference Fits, 65-67 & Nat Ann Plant Eng Conf & Related Regional Confs, 75-89. *Honors & Awards:* Leo F Spector Award for Outstanding Contrib to Field of Plant Eng, 90. *Mem:* Am Soc Mech Engrs. *Res:* Technical editor and writer, written on subjects relating to plant engineering, design engineering and technical article preparation. *Mailing Add:* 1800 Wakeman Ct Wheaton IL 60187

SPECTOR, LEONARD B, b Newark, NJ, Dec 6, 18; div. BIOCHEMISTRY. *Educ:* Harvard Univ, PhD(chem), 50. *Prof Exp:* Assoc biochemist, Mass Gen Hosp, 50-60; asst prof, 60-62, ASSOC PROF BIOCHEM, ROCKEFELLER UNIV, 62- *Mem:* Am Soc Biol Chemists; Sigma Xi. *Res:* Purines; pyrimidines; urea synthesis; phosphorus compounds; enzyme reactions; covalent by enzymes. *Mailing Add:* Dept Biochem Rockefeller Univ New York NY 10021

SPECTOR, NOVERA HERBERT, b Cincinnati, Ohio, Aug 23, 19; m 41, 81; c 4. NEUROBIOLOGY, BIOPHYSICS. *Educ:* City Col New York, BS, 41; Univ Pa, PhD(physiol), 67. *Prof Exp:* Consult engr, 41-62; from res assoc to asst prof psychiat, Med Col Va, Va Commonwealth Univ, 66-68, asst prof physiol, 68-69; prof physiol, Fac Med, Univ Claude Bernard, France, 69-71; chief dept neurophysiol & sr res physiologist, Walter Reed Inst Res, 71-76; dir, Neurobiol Prog, NSF, 76-77; prof neurosci, Birmingham Med Sch, Univ Ala, 80-83; HEALTH SCI ADMINR, FUNDAMENTAL NEUROSCI PROG, NAT INST NEUROL DIS & STROKE, NIH, 77- *Concurrent Pos:* Consult, NASA, 68-71; adj prof physiol & biophys & adj prof anat, Med Ctr, Georgetown Univ, Washington, DC, 78-; vis prof psychiatry, Birmingham Med Ctr, Univ Ala, 80-83; adj prof, microbiol, Univ Ala, Birmingham, 85- *Honors & Awards:* Metal-Nikov Gold Medal, Italy, 90; Physiol Soc Medal, Poland, 90. *Mem:* Sigma Xi; Soc Exp Biol & Med; Am Physiol Soc; Soc Neurosci; Tissue Cult Asn; Am Soc Microbiol; Int Soc Neuroimmunomodulation (pres, 87-90); Asn des Physiologists, Int Brain Res Orgn. *Res:* Physiology; neuroimmuno modulation; neural data processing; biophysics of neurons in vivo and in vitro; neural substrates of sensation and behavior; epistemology; central nervous system influences on host responses to antigens and diseases; neuropharmacology; alcoholism; psychophysics; hypothalamic control mechanisms; neuroanatomy; neuroimmunogenesis; neuroimmunomodulation; energy balance in mammals; central nervous system regulation of automatic functions. *Mailing Add:* NIH NINDS Fed Bldg Rm 916 Bethesda MD 20892

SPECTOR, REYNOLD, b Boston, Mass, Nov 3, 40; m 73; c 2. PHARMACOLOGY, INTERNAL MEDICINE. *Educ:* Harvard Col, AB, 62; Yale Univ, MD, 66. *Prof Exp:* Sr med adv, Army Med Corps, US Army, Repub Korea, 68-70; instr med, Sch Med, Harvard Univ & Peter Bent Brigham Hosp, 71-74, from asst prof to assoc prof, 74-78, chief, div clin pharmacol, 76-78; dir, div gen med, 80-85, prof internal med & pharmacol & dir, div clin pharmacol, Col Med, Univ Iowa, Iowa City, 78-87, dir, div gen med, 80-85; dir clin res ctr & med dir, Poison Control Ctr, Hosps & Clins, 85-87; EXEC DIR CLIN SCI, MERCK SHARP & DOHME RES LAB, RAHWAY, NJ, 87- *Concurrent Pos:* Vis prof biochem, Stanford Univ, 83-84. *Mem:* Asn Am Physicians; Am Soc Clin Invest; Am Soc Pharmacol & Exp Therapeut; fel Am Col Physicians; Int Soc Neurochem. *Res:* Passage of drugs, vitamins and hormones in and out of brain; poison victim treatment; effect of diet on the pharmacokinetics of drugs. *Mailing Add:* Merck Sharp & Dohme Res Labs PO Box 2000 (WBD-268) Rahway NJ 07065

SPECTOR, RICHARD M, b St Louis, Mo, Jan 13, 38; div; c 3. THEORETICAL PHYSICS. *Educ:* Harvard Univ, BA, 59; Oxford Univ, PhD(physics), 62; Wayne State Univ, JD, 76. *Prof Exp:* Vis scientist physics, Saclay Nuclear Res Ctr, France, 62; res assoc, Univ Rochester, 62-64; prof assoc, NSF, 64-65; from asst prof to assoc prof physics, Wayne State Univ, 65-78; assoc, Dykema, Gossett, Spencer, Goodwin & Trigg, Detroit, 78-81; ASSOC, HONIGMAN, MILLER, SCHWARTZ & COHN, DETROIT, 81- *Mem:* Am Bar Asn; Am Phys Soc. *Res:* Singular potential theory; spectra of rapidly rotating stars; geophysical aspects of pleochroic halos; group theory of elementary particles. *Mailing Add:* 745 Coronado Ave Coral Gables FL 33143

SPECTOR, SAMUEL, b Brooklyn, NY, Mar 11, 14; m 43; c 3. PEDIATRICS. *Educ:* Columbia Univ, BS, 34; Long Island Col Med, MD, 37. *Prof Exp:* Intern, Beth El Hosp, 37-38; intern, Kingston Ave Hosp, 38; resident pediat, Willard Parker Hosp, 39-41; resident, Univ Hosp, Univ Mich, 41-42, instr, Med Sch, 42-43; from asst prof to prof, Sch Med, Case Western Reserve Univ, 46-70; prof pediat & chmn dept, Univ Chicago, 70-79; prof, 79-90, EMER PROF PEDIAT, UNIV CALIF, SAN DIEGO, 90-; EMER PROF PEDIAT, UNIV CHICAGO, 82- *Concurrent Pos:* From assoc pediatrician to assoc dir pediat, Babies & Children's Hosp, 46-66; dir pediat, Children's Hosp of Akron, 67-70; dir pediat, Wyler Children's Hosp, Univ Chicago Hosps & Clins, 70-79 & LaRabida Children's Hosp & Res Ctr, 73-78; attending pediat, Univ Hosp, Med Ctr, Univ Calif, San Diego. *Mem:* Soc Pediat Res; Am Pediat Soc; Am Acad Pediat; Sigma Xi. *Res:* Metabolic and endocrine problems of childhood. *Mailing Add:* 225 Dickinson St Med Ctr Univ Calif H814L San Diego CA 92103

SPECTOR, SHELDON LAURENCE, b Detroit, Mich, Feb 13, 39; m 66; c 3. ALLERGY, IMMUNOLOGY. *Educ:* Wayne State Univ, MD, 64. *Prof Exp:* Fel allergy & clin immunol, 69-71, clin coordr, 71-72, HEAD SECT ALLERGY & CLIN IMMUNOL, NAT JEWISH HOSP & RES CTR, 72- *Concurrent Pos:* Asst med, Mt Sinai Hosp Sch Med, 65-66; asst prof, Med Sch, Univ Colo, 71-77, assoc prof, 77-; vis prof, Hebrew Univ, Jerusalem, 78; Lady Davis fel allergy, Hebrew Univ, William Beaumont Soc, 78. *Mem:* Am Thoracic Soc; Am Soc Internal Med; fel Am Acad Allergy; fel Am Col Physicians; fel Am Col Chest Physicians. *Res:* Bronchial inhalation challenge techniques; new modalities of treatment of asthma and rhinitis including unmarketed preparation; how certain substances in the environment affect bronchial and/or nasal reactivity; how commonly used medications affect asthmatic patients. *Mailing Add:* 11645 Wilshire Blvd Suite 600 Los Angeles CA 90025

SPECTOR, SYDNEY, b New York, NY, Oct 28, 23; m 48; c 2. PHARMACOLOGY. *Educ:* Univ Denver, BS, 48, MS, 50; Jefferson Med Col, PhD(pharmacol), 56. *Prof Exp:* Asst physiol, Univ Denver, 47-50; asst pharmacol, Sch Med, Wash Univ, 50-52; res assoc, Wyeth Inst Med Res, Pa, 52-55; pharmacologist, Nat Heart Inst, Md, 56-68; HEAD PHYSIOL CHEM & PHARMACOL, ROCHE INST MOLECULAR BIOL, 68- *Concurrent Pos:* Instr, Hahnemann Med Col, 54; adj prof pharmacol, Howard Med Sch, 60- & New York Med Col, 68-; adj prof pharmacol & anesthesiol, Col Physicians & Surgeons, Columbia Univ, 70- *Honors & Awards:* Amer Soc Exp Therapeut Award, 79; P K Smith Award, 87. *Mem:* AAAS; Am Soc Pharmacol & Exp Therapeut (pres, 79); Am Col Neuropsychopharmacol. *Res:* Biochemical pharmacoloy; correlation between pharmacological effects of drugs and chemical changes, particularly of the central nervous system; development of antibodies toward drugs. *Mailing Add:* Dept Pharmacol & Psych Vanderbilt Univ Med Ctr Rm AA2232 Nashville TN 37232

SPECTOR, THOMAS, b New Haven, Conn, July 20, 44; m 69; c 1. BIOCHEMISTRY, PHARMACOLOGY. *Educ:* Univ Vt, BA, 66; Yale Univ, PhD(pharmacol), 70. *Prof Exp:* Fel biochem, Univ Mich, 70-72; PRIN SCIENTIST, WELLCOME RES LABS, 72- *Concurrent Pos:* Adj assoc prof, dept pharmacol, Univ NC, 76- *Mem:* Am Soc Biol Chemists. *Res:* Enzymology, mechanisms of inhibition and substrate catalysis; inhibitor and substrate specificities; studies of the mechanisms of interactions of drugs and drug matabolites with chemotherapeutically important enzymes of the purine and pyrimidine metabolic pathways; studies of enzymes from human immunodefficiency and herpes viruses; chemotherapy. *Mailing Add:* Burroughs Wellcome Co Research Triangle Park NC 27709

SPEDDEN, H RUSH, b Colville, Wash, May 31, 16; m 51; c 4. MINING ENGINEERING. *Educ:* Univ Wash, BS, 39; Mont Sch Mines, MS, 40. *Hon Degrees:* MinDrE, 64. *Prof Exp:* Asst, Mass Inst Technol, 40-41, instr mineral dressing, 41-42, asst prof, 46-52 res engr & head minerals res dept, Metals Res Lab, Union Carbide Corp, 52-57, dir res, Union Carbide Ore Co, 57-64; res dir, Kennecott Copper Corp, Metal Mining Div, Res Ctr, 64-74, dir tech admin, 74-77; CONSULT MINERAL PROCESSING ENGR, 77- *Concurrent Pos:* Prod specialist, Foreign Econ Admin, 42-44; adj prof, Univ Utah. *Honors & Awards:* Robert H Richards Award, Am Inst Mining, Metall & Petrol Engrs, 71. *Mem:* Am Inst Mining, Metall & Petrol Engrs; Soc Mining Engrs (pres, 70); Mining & Metall Soc Am. *Res:* Minerals beneficiation; extractive metallurgy. *Mailing Add:* 4131 Cumorah Dr Salt Lake City UT 84124-4040

SPEDDING, ROBERT H, b Lockport, NY, Feb 8, 31; m 59; c 3. DENTISTRY. *Educ:* Ind Univ, AB, 53, DDS, 60, MSD, 63; Am Bd Pedodont, dipl, 67. *Prof Exp:* Teaching asst pedodont, Sch Dent, Ind Univ, 62-63; from instr to assoc prof, 63-72, PROF PEDIAT DENT, COL DENT, UNIV KY, 72- *Mem:* Am Dent Asn; Am Acad Pediat Dent; Am Asn Dent Schs; Am Asn Univ Prof. *Res:* Effects of various materials on primary tooth pulps and periodontal tissues. *Mailing Add:* Med Ctr 0841 Rm D 141 Col Dent Univ Ky 800 Rose St Lexington KY 40536

SPEECE, HERBERT E, b Meadowlands, Minn, Oct 29, 14; m 45; c 2. MATHEMATICS. *Educ:* York Col, AB, 38; Tex Christian Univ, MA, 43; NC State Col, MS, 51; Univ NC, PhD(math), 56. *Prof Exp:* Head, Dept Sci & Math Educ, 72-80, prof, 47-80, EMER PROF MATH, NC STATE UNIV, 80- *Concurrent Pos:* Dir, Nat Acad Sci-NSF Inserv Insts, NC, 59-, assoc dir, NSF Acad Year Inst, 65-67; dir, Eng Concepts Curriculum Proj Implementation Ctr Southeast, NSF, 71-72 & 72-73; chmn adv bd, NC Student Acad Sci; chmn selection comt, NC Jr Sci & Humanities Symp; dir, Comput Educ Ctr, Dept Math-Sci Educ, NC State Univ, 73-78. *Res:* Tensors and differential geometry. *Mailing Add:* 3408 Wade Ave Raleigh NC 27607

SPEECE, SUSAN PHILLIPS, b Chicago, Ill, Aug 13, 45; div; c 2. AIDS EDUCATION. *Educ:* Purdue Univ, BS, 67, MS, 71; Ball State Univ, EdD, 78. *Prof Exp:* Dept chair sci, Gosport Sch, 67-68; instr biol, Purdue Univ, 68-70; dept chair sci, Wes Del High Sch, 71-76; from asst prof to assoc prof, 77-84, PROF & CHAIR BIOL, ANDERSON UNIV, 84- *Concurrent Pos:* Adj fac biol, Ball State Univ, 80-84; dist dir, Nat Sci Teacher's Asn, 88-90; mem, Ind Pesticide Rev Bd, 88-, Col Comt, Nat Sci Teacher's Asn, 91-93; prin investr, Comn Higher Educ grant for AIDS Educ, 89-90; mem AIDS Adv Bd, Ind Dept Educ, 90- *Honors & Awards:* Distinguished Serv Award, Hosier Asn Sci Teachers, 89. *Mem:* Nat Asn Biol Teachers (pres, 89); Hoosier Asn Sci Teachers (pres, 86-89); AAAS; Int Soc AIDS Educ; Indian Acad Sci; Nat Asn Res Sci Teaching. *Res:* AIDS research and its interpretation for education; effective science teaching strategies. *Mailing Add:* Biol Dept Anderson Univ Anderson IN 46012

SPEED, EDWIN MAURICE, b Enterprise, Miss, Aug 17, 18; m 42; c 2. DENTISTRY, ANATOMY. *Educ:* Birmingham Southern Col, BA, 52; Univ Ala, DMD, 54, MS, 65. *Prof Exp:* Pvt pract, 54-61; resident periodont, Vet Admin Hosp, Birmingham, Ala, 63-65; mem, Fac Dent & asst to dean, Sch Dent, Univ Ala, Brimingham, 65-66, prof dent & asst dean, 66-79; RETIRED. *Concurrent Pos:* Consult, US Army, Ft Benning, Ga, 71-72; mem coun, Nat Bd Dent Examr, 75-77 & Dent Hyg Nat Bd, 75-76. *Mem:* Am Dent Asn; Am Acad Periodont; fel Am Col Dent; Am Asn Dent Schs (vpres, 74, pres, 75). *Res:* Wound healing. *Mailing Add:* 744 River Haven Circle Birmingham AL 35244-1242

SPEED, RAYMOND A(NDREW), b Muldoon, Tex, Sept 30, 22; m 53; c 2. CHEMICAL ENGINEERING. *Educ:* Univ Tex, BSChE, 49. *Prof Exp:* From jr chemist to res specialist, Humble Oil & Refining Co, 49-63; res specialist, Esso Res & Eng Co, 63-71; sr staff engr, Enjay Chem Co, 71-78, eng assoc, Exxon Chem Co USA, 78-82; RETIRED. *Mem:* Am Chem Soc. *Res:* Polyolefin polymers; separations. *Mailing Add:* 16110 Peach Bough Lane Houston TX 77095-4061

SPEED, ROBERT CLARKE, b Los Angeles, Calif, June 20, 33; m 54; c 2. GEOLOGY, GEOPHYSICS. *Educ:* Univ Colo, BS, 54; Stanford Univ, MS, 58, PhD(geol), 61. *Prof Exp:* Res supvr, Jet Propulsion Lab, Calif Inst Technol, 60-66; from asst prof to assoc prof, 66-74, PROF GEOL, NORTHWESTERN UNIV, EVANSTON, 74- *Res:* Tectonics and structural geology; structure and evolution of accretionary prisms and forearcs; tectonics of southeastern Caribbean and US cardillera; fold and thrust belts; seismicity. *Mailing Add:* Dept Geol Northwestern Univ 633 Clark St Evanston IL 60208

SPEEDIE, MARILYN KAY, b Salem, Ore, Nov 13, 47; m 68; c 2. BIOTECHNOLOGY, APPLIED MICROBIOLOGY. *Educ:* Purdue Univ, BSPh, 70, PhD(med chem & pharmacog), 73. *Prof Exp:* Asst prof, Sch Pharm, Ore State Univ, 73-75; from asst prof to assoc prof pharmacog, 75-88, CHMN, DEPT BIOMED CHEM, SCH PHARM, UNIV MD, 88- *Concurrent Pos:* Vis assoc prof, Stanford Univ, 84-85; Found Microbiol lectr, Am Soc Microbiol, 83-84. *Mem:* Am Soc Pharmacog; Am Soc Microbiol; Soc Indust Microbiol; Am Chem Soc; Sigma Xi; AAAS. *Res:* Regulation and enzymology of secondary metabolism in Streptomycetes; expression and secretion of proteins from heterologous genes cloned into streptomycetes. *Mailing Add:* Dept Biomed Chem 20 N Pine Baltimore MD 21201

SPEEN, GERALD BRUCE, b Philadelphia, Pa, Oct 2, 30; c 3. HEAT TRANSFER, ENERGY CONSERVATION. *Educ:* Univ Del, BS, 52; Univ Calif, Los Angeles, MS, 54. *Prof Exp:* Asst physics, Univ Del, 51-52; mem tech staff, Res & Develop Labs, Hughes Aircraft Co, 52-54; res physicist, Micronics Inc Div, Zenith Plastics Corp, 54-55; res engr, Summers Gyroscope Co, 55-56; exec engr, ITT Fed Labs, Int Tel & Tel Corp, 56-64; gen mgr, Western Develop Ctr, Conductron Corp, Subsid McDonnell-Douglas Corp, 64-68; pres, Data Instruments Co, 68-73; pres, Lanco-Supreme, Inc, Hyatt Corp, 73-75, pres, Supreme Aire & Elmet Corp, Santa Fe Springs, 73-77; group vpres, Elsters Inc, 77-78; PRES, G SPEEN & ASSOCS, CONSULTS IN ENERGY FIELD, 78- *Concurrent Pos:* Mem tech coord comt gas lubrication, Off Naval Res, 58-66. *Honors & Awards:* Inst Elec & Electronics Engrs Award, 63. *Mem:* Am Phys Soc; Am Inst Aeronaut & Astronaut; sr mem Inst Elec & Electronics Engrs. *Res:* Guidance and control systems; sensor design; gyroscopes; accelerometers; gas lubrication and bearing design; pneumatic systems; data acquisition; environmental control systems; energy conservation and recovery systems; heat and solar actuated systems; thermodynamics. *Mailing Add:* 17339 Halsted St Northridge CA 91325

SPEER, CLARENCE ARVON, b Lamar, Colo, Feb 14, 45; m 77; c 2. COCCIDIOSIS, SARCOCYSTOSIS. *Educ:* Colo State Univ, BS, 67; Utah State Univ, MS, 70, PhD(zool), 72. *Prof Exp:* Asst prof histol, Univ Tex, Houston, 72-73; res assoc malaria res, Univ NMex, 73-75; from asst prof to assoc prof microbiol, Univ Mont, 75-83; assoc prof, 83-86, PROF & HEAD VET SCI, MONT STATE UNIV, 86- *Concurrent Pos:* Consult, Nat Res Inst Amazon, Manaus, 77-83. *Mem:* Am Asn Immunologists; Am Soc Parasitologists; Can Soc Zoologists; Soc Protozoologists. *Res:* In vitro cultivation and biochemical, physiological and ultrastructural aspects of protozoan parasites that cause coccidiosis, malaria and Chagas' disease. *Mailing Add:* Dept Parasitol Marsh Lab Mont State Univ Bozeman MT 59717

SPEER, FRIDTJOF ALFRED, b Berlin, Germany, Aug 23, 23; US citizen; m 51; c 3. PHYSICS. *Educ:* Tech Univ Berlin, Dipl Ing, 50, Dr Ing(physics), 53. *Prof Exp:* Asst prof physics, Tech Univ Berlin, 50-55; div chief missile develop, US Army, 55-60; div chief, spacecraft develop, 60-65, mgr, mission opers, 65-71, mgr, sci projs, 71-83, assoc dir sci, Marshall Space Flight Ctr, NASA, 83-86; DIR, CTR ADV SPACE PROPULSION, TULLAHOMA, TENN, 87- *Concurrent Pos:* Ed, Sci Abstr Periodical, Berlin, 53-55. *Honors & Awards:* Except Serv Medals, NASA, 69; Holger Toftoy Award, Am Inst Aeronaut & Astronaut, 79. *Mem:* Assoc fel Am Inst Aeronaut & Astronaut; fel Am Astronaut Soc. *Res:* Science policy; high energy astronomy. *Mailing Add:* 1214 Chandler Rd Huntsville AL 35801

SPEER, VAUGHN C, b Milford, Iowa, Apr 5, 24; m 47; c 4. NUTRITION, BIOCHEMISTRY. *Educ:* Iowa State Univ, BS, 49, MS, 51, PhD, 57. *Prof Exp:* Asst nutrit, Iowa State Univ, 49-51; nutritionist, Ralston Purina Co, 51-53; assoc animal husb, 53-57, from asst prof to assoc prof, 58-66, prof, 66-90, EMER PROF NUTRIT, IOWA STATE UNIV, 91- *Mem:* Am Inst Nutrit; Am Soc Animal Sci. *Res:* Swine nutrition; nutritional effects on swine reproduction. *Mailing Add:* Dept Animal Sci Iowa State Univ 337 Kildee Hall Ames IA 50011

SPEERS, GEORGE M, b State Center, Iowa, Dec 10, 40; m 63; c 2. POULTRY NUTRITION. *Educ:* Iowa State Univ, BS, 63, MS, 65, PhD(poultry nutrit), 68. *Prof Exp:* Grad asst, Iowa State Univ, 63-65, res assoc, 65-68; from asst prof to assoc prof animal nutrition, Univ Minn, St Paul, 68-74; poultry prod mgr, 74-84, DIR, ANIMAL PROD, LAND O'LAKES, INC, 84- *Mem:* Poultry Sci Asn; World Poultry Sci Asn; Sigma Xi. *Res:* Amino acid and mineral metabolism in poultry. *Mailing Add:* Land-O-Lakes PO Box 1395 Minneapolis MN 55440

SPEERS, LOUISE (MRS HENRY CROIX), b Nanking, China, Dec 30, 19; US citizen; m 51. ORGANIC CHEMISTRY. *Educ:* Vassar Col, AB, 41; Columbia Univ, PhD(org chem), 49. *Prof Exp:* Bacteriologist, Typhus Vaccine Prod, Lederle Labs, Am Cyanamid Co, 41-42; chemist, Plastics Dept, E I du Pont de Nemours & Co, 42-44; proj leader med prod, Cent Res Labs, Airco Inc, 49-82; RETIRED. *Mem:* Am Chem Soc; Sigma Xi. *Res:* Organic synthesis in pharmaceuticals; anesthetics, analgesics, muscle relaxants, acetylenics and fluoro aliphatic compounds. *Mailing Add:* 50 High Summit NJ 07901

SPEERS, WENDELL CARL, SURGICAL PATHOLOGY, MEDICAL EDUCATION. *Educ:* Johns Hopkins Univ, MD, 71. *Prof Exp:* ASSOC PROF PATH, UNIV COLO HEALTH & SCI CTR, 79- *Mailing Add:* Dept Pathol Univ Col Med Ctr Porter Mem Hosp 2525 S Downing Denver CO 80220

SPEERT, ARNOLD, b Bronx, NY, June 19, 45; m 67; c 2. PHYSICAL ORGANIC CHEMISTRY. *Educ:* City Col New York, BS, 66; Princeton Univ, PhD(chem), 71. *Prof Exp:* From asst prof to assoc prof chem, William Paterson Col, 70-80, asst to vpres acad affairs, 71-78, assoc dean, 78-79, vpres, 79-85, PROF CHEM, WILLIAM PATERSON COL, NJ, 80-, PRES, 85- *Mem:* AAAS; Am Chem Soc. *Res:* Aromaticity; nuclear magnetic resonance spectroscopy; stereochemistry; iron carbonyl complexes. *Mailing Add:* William Paterson Col Wayne NJ 07470

SPEES, STEVEN TREMBLE, JR, b Earl Park, Ind, May 12, 33; m 53; c 3. INORGANIC CHEMISTRY. *Educ:* Purdue Univ, BS, 56; Univ Southern Calif, PhD(phys chem), 61. *Prof Exp:* Instr chem, Ohio State Univ, 61-62; asst prof inorg chem, Univ Minn, Minneapolis, 62-67; assoc prof inorg chem, 67-77, PROF CHEM, LYMAN BRIGGS COL, MICH STATE UNIV, 77- *Mem:* AAAS; NY Acad Sci; Am Chem Soc; The Chem Soc; Sigma Xi. *Res:* Chemistry of coordination compounds; synthesis; molecular and electronic structures; kinetics; optical activity; stereochemistry; photochemistry; nuclear magnetic resonances. *Mailing Add:* Lyman Briggs Col Holmes Hall Mich State Univ East Lansing MI 48824

SPEHRLEY, CHARLES W, JR, b Pottsville, Pa, July 16, 44. ELECTROMECHANICAL SYSTEM DESIGN, MACHINE DESIGN. *Educ:* Dartmouth Col, AB, 66, BEng, 67, ME(mech eng design), 70. *Prof Exp:* Proj engr, Rohm & Haas Co, 66-68; proj engr, Creare Inc, 68-72, dir, Mach Systs Eng Div, 72-76, vpres, Creare Innovations Inc, 76-86; VPRES, SPECTRA INC, 86- *Mem:* Am Soc Mech Engrs. *Res:* Xerographic and ink-jet copiers and printers; precision dynamic drives and mechanisms; electromechanical and digital servo systems; paper handling, feeding and collation; optomechanical and raster imaging systems. *Mailing Add:* Spectra Inc PO Box 68-C Hanover NH 03755

SPEICH, G(ILBERT) R(OBERT), physical metallurgy; deceased, see previous edition for last biography

SPEICHER, BENJAMIN ROBERT, b Swatow, China, Jan 23, 09; US citizen; m 32. ZOOLOGY. *Educ:* Denison Univ, AB, 29; Univ Pittsburgh, MS, 31, PhD(genetics), 33. *Hon Degrees:* ScD, Colby Col, 69. *Prof Exp:* Asst, Univ Pittsburgh, 29-33; visitor, Carnegie Inst Technol, 33-35; asst, Amherst Col, 35; Nat Res Coun fel, Columbia Univ, 35-36; from instr to prof, 37-74, actg head dept, 42-45, head dept, 45-63, EMER PROF ZOOL, UNIV MAINE, ORONO, 74- *Concurrent Pos:* Consult, Oak Ridge Nat Lab, 55-65. *Mem:* AAAS; Genetics Soc Am; Am Soc Zool; Sigma Xi. *Res:* Genetics of Hymenoptera; cytology of parthenogenesis. *Mailing Add:* 4357 Plass Drive Napa CA 94558

SPEICHER, CARL EUGENE, b Carbondale, Pa, Mar 21, 33; m 58; c 3. PATHOLOGY. *Educ:* Kings Col, BS, 54; Univ Pa, MD, 58. *Prof Exp:* Intern med surg, obstet & pediat, Hosp Univ Pa, 58-59; resident path, Hosp Univ PA, 59-63; pathologist, US Air Force, 63-70; fel path, Upstate Med Ctr, State Univ NY, 70-71; pathologist, US Air Force, 71-77; PROF & DIR, CLIN LAB, OHIO STATE UNIV HOSP, 77- *Concurrent Pos:* Clin assoc prof path, Univ Tex Health Sci Ctr, San Antonio, 71-77; chmn, Dept Path, Wilford Hall, Med Ctr, Lackland AFB, Tex, 75-77. *Mem:* Col Am Pathologists; Am Soc Clin Pathologists; Am Med Asn; Acad Clin Lab Physicians & Scientists. *Res:* Application of laboratory medicine to patient care, using problem solving approach and computer assistance. *Mailing Add:* Rm N-343 Ohio State Univ Hosp 410 W Tenth Ave Columbus OH 43210

SPEIDEL, DAVID H, b Pottsville, Pa, Aug 10, 38; m 62. GEOCHEMISTRY, RESOURCES. *Educ:* Franklin & Marshall Col, BS, 60; Pa State Univ, PhD(geochem), 64. *Prof Exp:* Res assoc geochem, Pa State Univ, 64-66; from asst prof to assoc prof geol, 66-70, from assoc dean to dean sci fac, 70-78, chmn dept geol, 80-88, PROF GEOL, QUEENS COL, NY, 70- *Concurrent Pos:* Vis scholar, Cong Res Serv, 77-78; section head, Major Proj, Earth Sci Div, NSF, Washington, DC, 88- 89. *Mem:* AAAS; Am Ceramic Soc; Geol Soc Am; Mineral Soc Am; Am Geophys Union; Soc Environ Geochem & Health; NY Acad Sci. *Res:* Resource analysis of inorganic materials; environmental geochemistry; determination of composition of coexisting minerals and the study of their change as a function of environment, especially oxygen fugacity. *Mailing Add:* Dept Geol Queens Col Flushing NY 11367-0904

SPEIDEL, EDNA W, b Indianapolis, Ind, June 14, 08; m 34; c 3. BIOCHEMISTRY. *Educ:* Butler Univ, BS, 29; Univ Mich, MA, 30; Univ Iowa, PhD(biochem), 34. *Prof Exp:* Res fel biochem, Univ Tenn, 34-36; res assoc anat, Univ Minn, Minneapolis, 58-77; RETIRED. *Res:* Biochemistry of diabetes mellitus; microanalytical techniques; mineral metabolism; radioactive turnover. *Mailing Add:* 5443 41st Pl NW Washington DC 20015

SPEIDEL, JOHN JOSEPH, b Iowa City, Iowa, Sept 17, 37; m 67; c 1. POPULATION BIOLOGY, PUBLIC HEALTH. *Educ:* Harvard Univ, AB, 59, MD, 63, MPH, 65. *Prof Exp:* Intern med, St Luke's Hosp, New York, 63-64; resident pub health, City of New York Dept Health, 65-67, dep dir maternal & infant care proj, 66-67; chief develop group, Off Surgeon Gen, US Army, 67-69; dep chief res div, Off Pop, AID, 69-70, chief res div, 70-77, assoc dir, 77-78, dep dir, 78-83; VPRES, POP CRISIS COMT, 83- *Mem:* Am Pub Health Asn; Population Asn Am; Brit Soc Study Fertil. *Res:* Population research including demograph, social science, operational and contraceptive development. *Mailing Add:* Pop Crisis Comt 1120 19th St NW Washington DC 20036

SPEIDEL, T(HOMAS) MICHAEL, b Memphis, Tenn, Apr 17, 36; div; c 2. ORTHODONTICS. *Educ:* State Univ Iowa, BA, 58; Loyola Univ, DDS, 63; Univ Minn, MSD, 67; Am Bd Orthod, dipl, 72. *Prof Exp:* Teaching asst, 64-65, res assoc, 65-66, instr, 66-68, asst prof, 68-71, assoc prof, 71-77, PROF ORTHOD & DENT, SCH DENT, UNIV MINN, 77- *Mem:* Am Asn Orthodontists; Am Dent Asn; Int Asn Dent Res. *Res:* Quantitation of occlusal function. *Mailing Add:* Dept Ortho 6-320 Moos Tower Univ Minn Minneapolis MN 55455

SPEIER, JOHN LEO, JR, b Chicago, Ill, Sept 29, 18; m 44; c 6. CHEMISTRY. *Educ:* St Benedict's Col, BS, 41; Univ Fla, MS, 43; Univ Pittsburgh, PhD(chem), 47. *Prof Exp:* Asst, Univ Fla, 41-43; sr fel organo-silicon chem, Mellon Inst, 47-56; res supvr, 56-65, mgr org res, 65-69, scientist, 70-75, SR SCIENTIST, DOW CORNING CORP, 75- *Honors & Awards:* Scientist of the Year Award, Indust Res & Develop, 78; Frederick Stanley Kipping Award, Am Chem Soc, 90. *Mem:* Sigma Xi; AAAS; Am Chem Soc. *Res:* Resin acids in pine tree oleoresins; polymerization of silicones; organo-silicon compounds, especially synthesis, derivatives and properties; synthesis and applications of carbon functional silicones. *Mailing Add:* Res Dept Dow Corning Corp Midland MI 48640

SPEIGHT, JAMES G, b Durham, Eng, June 24, 40; m 63; c 1. ORGANIC CHEMISTRY. *Educ:* Univ Manchester, BSc, 61, PhD(chem), 65. *Prof Exp:* Imp Chem Indust res fel chem, Univ Manchester, 65-67; res officer, Res Coun Alta, 67-80; res assoc, Exxon Res & Eng Co, 80-84; RES ASSOC, WESTERN RES INST, 84- *Mem:* Fel Chem Inst Can; fel Royal Soc Chem; assoc Royal Inst Chem; Am Chem Soc; Sigma Xi. *Res:* Naturally occurring high molecular weight organic residues, especially coal, asphalt, and petroleum. *Mailing Add:* Western Res Inst PO Box 3395 Univ Sta Laramie WY 82071

SPEIL, SIDNEY, b Revere, Mass, Feb 21, 17; m 40, 79; c 2. CERAMICS. *Educ:* Mass Inst Technol, BS, 36, DSc(ceramics), 39. *Prof Exp:* Asst ceramics, Mass Inst Technol, 37-39; res engr, Ideal Tooth, Inc, Mass, 39-40; engr nonmetals, US Bur Mines, Tenn, 40-46; sr res engr, Johns-Manville Corp, Denver, 46-52, chief aviation & spec thermal insulations res, 52-64, basic chem res, 64-66, dir corp res & develop, 67-76, dir appl technol & int div res, 76-80, sr scientist & vpres, 80-82; RETIRED. *Honors & Awards:* Electrochem Soc Award, 41. *Mem:* Am Ceramic Soc; Am Inst Aeronaut & Astronaut; Am Chem Soc. *Res:* Home, industrial and aerospace thermal insulations; cryogenic thermal insulation; fiberization of glass wool; high temperature ceramic compositions; fiber reinforcements and fiber reinforced composites; synthetic and natural silicates; asbestos. *Mailing Add:* 3425 S Race Englewood CO 80110

SPEISER, ROBERT DAVID, b New York, NY, Aug 28, 43; m 72; c 2. PURE MATHEMATICS. *Educ:* Columbia Col, AB, 65; Cornell Univ, PhD(math), 70. *Prof Exp:* Res assoc psychol, Ctr Res in Educ, Cornell Univ, 70-71; asst prof math, Univ Tex, Austin, 71-73; ASST PROF MATH, ILL STATE UNIV, 73- *Concurrent Pos:* Vis assoc prof math, Univ Minn, Minneapolis, 78-79. *Res:* Algebraic geometry; commutative algebra. *Mailing Add:* 799 E 3800 N Provo UT 84604

SPEISER, THEODORE WESLEY, b Del Norte, Colo, Nov 23, 34; m 56; c 3. ASTROPHYSICS, GEOPHYSICS. *Educ:* Colo State Univ, BS, 56; Calif Inst Technol, MS, 59; Pa State Univ, PhD(physics), 64. *Prof Exp:* Res physicist, Nat Bur Standards, 59, 60-61; Nat Acad Sci res assoc earth-sun rels, Goddard Space Flight Ctr, NASA, Md, 64-66; from lectr to assoc prof astro-geophys, 67-85, PROF ASTROPHYSICAL, PLANETARY & ATMOSPHERIC SCI, UNIV COLO, BOULDER, 85- *Concurrent Pos:* Fel, Imp Col, Univ London, 66-67; awardee, US Spec Prog, Alexander von Humboldt Found, 77-78. *Mem:* Am Geophys Union. *Res:* Theories of the aurora and magnetosphere configuration; particle motion and acceleration; magnetospheric and solar wind plasma dynamics. *Mailing Add:* Dept Astrophysical Planetary & Atmospheric Sci Univ Colo Boulder CO 80309

SPEISMAN, GERALD, b New York, NY, Feb 27, 30; m 57; c 1. THEORETICAL PHYSICS. *Educ:* City Col New York, BS, 51; Calif Inst Technol, PhD(physics), 55. *Prof Exp:* Mem sch math, Inst Adv Study, 55-56; asst prof, 56-69, ASSOC PROF PHYSICS, FLA STATE UNIV, 69- *Concurrent Pos:* Physicist, Avco-Everett Res Lab, Avco Corp, 56-57. *Mem:* Am Phys Soc. *Res:* Quantum field theory; many-body problem; statistical mechanics; mathematical physics. *Mailing Add:* Dept of Physics Fla State Univ Tallahassee FL 32306

SPEIZER, FRANK ERWIN, b San Francisco, Calif, June 8, 35; m 57; c 4. EPIDEMIOLOGY, ENVIRONMENTAL MEDICINE. *Educ:* Stanford Univ, BA, 57, MD, 60. *Hon Degrees:* AM, Harvard Univ, 89. *Prof Exp:* Actg instr med, Sch Med, Stanford Univ, 65-66; vis scientist epidemiol, Brit Med Res Coun, Statist Res Unit, 66-68; assoc prof med, Sch Med, Harvard Univ, 70-76, chief, Div Clin Epidemiol, Channing Lab & assoc prof med, 76-85, assoc physician, Throndike Lab, 68-77, PROF MED & ENVIRON SCI, SCH MED, HARVARD UNIV, 87-, CO-DIR, CHANNING LAB, 89- *Concurrent Pos:* Assoc vis physician, Boston City Hosp, 68-77 & Peter Brent Brigham Hosp, 77-; Edmund Livingston Traudeu fel, Am Thoracic Soc, 68-70; career develop award, Nat Inst Environ Health Sci, 70-76; mem prog comt, 2nd Task Force Res Plans Environ Res, NIH, mem, Task Force Epidemiol Lung Dis & consult, 79-81; chmn epidemiol sect, Workshops on Environmentally Related Non-Oncogenic Lund Dis, US Task Force on Environ Cancer & Heart & Lung Dis, 82-83; consult, Sci Adv Bd for Long Range Planning Environ Protection Agency, Wash, DC, 84-; working group mem, clin appl, Nat Heart, Lung & Blood Inst, Wash, DC, 84-86; mem, Policy Bd, Honolulu Heart Study, Nat Heart, Lung & Blood Inst, Wash, DC, 85-; mem, Task Force on Asthma Morbidity and Mortality, Nat Heart, Lung & Blood Inst, Wash, DC, 86-87. *Mem:* Am Epidemiol Soc; Am Soc Clin Invest; fel Am Col Epidemiol; fel Am Col Chest Physicians. *Res:* Epidemiological studies of chronic diseases associated with environmental exposure, particularly heart, lung and cancer. *Mailing Add:* Dept Med Harvard Med Sch 180 Longwood Ave Boston MA 02115

SPEJEWSKI, EUGENE HENRY, b East Chicago, Ind, Sept 15, 38; m 63; c 4. NUCLEAR PHYSICS. *Educ:* Univ Notre Dame, BS, 60; Ind Univ, Bloomington, PhD(exp physics), 66. *Prof Exp:* Res assoc physics, Ind Univ, Bloomington, 65-67; res assoc, Princeton Univ, 67-69, instr, 69-71; asst prof, Oberlin Col, 71-72; dir univ isotope separator proj, 72-85, proj mgr, Navy SDS proj, Oak Ridge Nat Lab, Assoc Univs, 85-87, CHMN SPEC PROJ DIV, 87- *Concurrent Pos:* Consult, Oak Ridge Nat Lab, 71-72; vis prof physics, Univ Tenn, 81-82. *Mem:* AAAS; Am Phys Soc; Sigma Xi. *Res:* Nuclear structure. *Mailing Add:* PO Box 117 Oak Ridge Assoc Univs Oak Ridge TN 37831-0117

SPELIOTIS, DENNIS ELIAS, b Kalamata, Greece, Nov 27, 33; US citizen; m 58; c 3. SOLID STATE PHYSICS, MAGNETISM. *Educ:* Univ RI, BS, 55; Mass Inst Technol, MS, 57, EE, 58; Univ Minn, PhD(magnetism), 61. *Prof Exp:* Staff physicist, Int Bus Mach Develop Labs, 61-63, mgr recording physics, 63-66, adv physicist, 66-67; assoc prof elec eng, Univ Minn, Minneapolis, 67-69; dir eng, Micro-Bit Corp, 69-76; PRES, ADVAN DEVELOP CORP, 77- *Concurrent Pos:* Consult, Kodak, BASF, Hitachi, Gen Elec, Mitsubishi Kasei, Toda, Orient Chem, Nashua, Digital Equip, Seagate, NKK, TDK, 3M, Cabot, Vermont Res, Polaroid. *Honors & Awards:* Fel, Inst Elec & Electronic Engrs. *Mem:* Inst Elec & Electronics Engrs. *Res:* Hard magnetic materials and their applications to bulk magnetic storage devices; magnetic recording; electron beam addressable memories; digital computer memory architecture. *Mailing Add:* Advan Develop Corp Eight Ray Ave Burlington MA 01803

SPELKE, ELIZABETH SHILIN, b New York, NY, May 28, 49; m 88; c 2. COGNITIVE SCIENCE. *Educ:* Harvard Univ, BA, 71; Cornell Univ, PhD(psychol), 78. *Prof Exp:* From asst prof to assoc prof psychol, Univ Pa, 77-86; PROF PSYCHOL, CORNELL UNIV, 86- *Concurrent Pos:* Vis scientist, Ctr Cognitive Sci, Mass Inst Technol, 82-83 & Nat Ctr Sci Res, Paris, 84-85; Fulbright fel, 84; Boyd McCendress young investr award, Am Psychol Asn, 85; John Simon Guggenheim fel, 88; mem, Adv Panel Human Perception & Cognition, NSF, 90-92. *Mem:* AAAS; Soc Res Child Develop; Soc Philos & Psychol; Sigma Xi. *Res:* Perceptual development, cognitive development and conceptual change in human infants and children; physical phenomena. *Mailing Add:* Dept Psychol Cornell Univ Uris Hall Ithaca NY 14853

SPELL, ALDENLEE, b Rayne, La, Feb 9, 20; m 52; c 2. PHYSICAL CHEMISTRY. *Educ:* Southwestern La Inst, BS, 41; Tulane Univ, MS, 43; Brown Univ, PhD(chem), 52. *Prof Exp:* Chemist, Shell Develop Co, 43-48; scientist, Signal Corps, US Dept Army, 51-52; res supvr, Rohm & Haas Co, 52-82; RETIRED. *Mem:* Am Chem Soc; Soc Appl Spectros. *Res:* Infrared absorption and reflection spectroscopy; fractionation and analysis of polymers; gas chromotography. *Mailing Add:* 514 Portsmouth Ct Doylestown PA 18901-2510

SPELLACY, WILLIAM NELSON, b St Paul, Minn, May 10, 34; m 81; c 3. OBSTETRICS & GYNECOLOGY. *Educ:* Univ Minn, BA, 55, BS, 56, MD, 59; Am Bd Obstet & Gynec, dipl, 66, maternal & fetal med cert, 75. *Prof Exp:* Intern, Minneapolis Gen Hosp, Minn, 59-60; from instr to asst prof obstet & gynec, Univ Minn, 63-67; from assoc prof to prof, Med Sch, Univ Miami, 67-74; prof obstet & gynec & chmn dept, Col Med, Univ Fla, 74-79; PROF & HEAD, DEPT OBSTET & GYNEC, COL MED, UNIV ILL, 79 - *Concurrent Pos:* Fel obstet & gynec, Univ Minn, 60-63; NIH, Pop Coun & Food & Drug Admin grants, 64-69; Josiah Macy Jr Found fel, 66-69; examr, Am Bd Obstet & Gynec. *Mem:* Nat Acad Sci; Endocrine Soc; Am Fertil Soc; Asn Prof Gynec & Obstet; Am Fedn Clin Res. *Res:* Metabolism of pregnant woman and fetus; effects of ovarian steroids on carbohydrate and lipid metabolism; studies of placental function and fetal maturity; endocrinology of reproduction. *Mailing Add:* Harbourside Med Tower Four Columbia Dr Suite 4 Tampa FL 33606

SPELLENBERG, RICHARD (WILLIAM), b San Mateo, Calif, June 27, 40; m 64; c 2. PLANT TAXONOMY. *Educ:* Humboldt State Col, BA, 62; Univ Wash, PhD(bot), 68. *Prof Exp:* From asst prof to assoc prof, 68-77, PROF BIOL, NMEX STATE UNIV, 77- *Concurrent Pos:* Consult, endangered & threatened plant species. *Mem:* Am Soc Plant Taxon; Int Asn Plant Taxon. *Res:* Systematics of Gramineae and Nyctaginaceae. *Mailing Add:* Dept of Biol Box 3 AF NMex State Univ Las Cruces NM 88001

SPELLER, STANLEY WAYNE, b Victoria, BC, June 6, 42; m 67; c 1. WILDLIFE BIOLOGY. *Educ:* Univ Victoria, BS, 65; Carleton Univ, MS, 68; Univ Sask, PhD(mammal), 72. *Prof Exp:* Biologist, Can Wildlife Serv, Environ Can, 72-75; biologist, Environ Assessment Sect, Dept Indian Affairs & Northern Develop, 75-77; chief, Wildlife Res & Interpretation, Can Wildlife Serv, Environ Can, 77-81; supvr, 81-89, MGR ENVIRON, SAFETY & INDUST HYG, EXPLOR DEPT, PETROCANADA, 89- *Mem:* Can Soc Environ Biol. *Res:* Management of research programs on effects of insecticides on wildlife; wildlife and limnology studies for parks Canada; wildlife research of rare and endangered species; wildlife interpretation programs in Atlantic Region. *Mailing Add:* Petrocanada Box 2844 Calgary AB T2P 3E3 Can

SPELLMAN, CRAIG WILLIAM, b Longview, Wash, Dec 15, 46; m 75; c 2. TUMOR BIOLOGY, CELLULAR IMMUNOLOGY. *Educ:* Univ Wash, BS, 69; Mont State Univ, Bozeman, MS, 76; Univ Utah, PhD(path), 78. *Prof Exp:* NIH teaching fel, 78-80, res asst prof path, 80-81, ASST PROF PATH & CELL BIOL, SCH MED, UNIV NMEX, 81- *Concurrent Pos:* Consult, Becton Dickinson Monoclonal Ctr, 81; prin investr, Nat Cancer Inst RO-1 Grants, NIH, 81-, ad hoc reviewer, 84- *Honors & Awards:* Wilson S Stone Mem Award, Syst Cancer Ctr, M D Anderson Hosp, Univ Tex, Houston, 79. *Mem:* Am Asn Pathologists; AAAS. *Res:* T-cell immune circuits operative in anti-tumor immunity; UV-induced syngeneic skin tumors and UV-irradiated hosts exhibiting defined states of tumor susceptibility. *Mailing Add:* Dept Immunol & Microbiol Tex Col Osteop Med 3500 Camp Bowie Blvd Ft Worth TX 76107-2690

SPELLMAN, JOHN W, b Ft Worth, Tex, Oct 3, 41; m 59; c 2. MATHEMATICS. *Educ:* Tex Lutheran Col, BA, 63; Emory Univ, MA, 65, PhD(math), 68. *Prof Exp:* Fel math, Univ Fla, 68-69; asst prof, Tex A&M Univ, 60-71; assoc prof math, Pan Am Univ, 71-80, head dept, 75-80; PROF MATH, SOUTHWEST TEX STATE UNIV, 80- *Mem:* Am Math Soc. *Res:* Functional analysis; semigroups of operators; real analysis. *Mailing Add:* Dept Math SW Tex State Univ 601 University Dr San Marcos TX 78666

SPELLMAN, MITCHELL WRIGHT, b Alexandria, La, Dec 1, 19; m 47; c 8. SURGERY. *Educ:* Dillard Univ, AB, 40; Howard Univ, MD, 44; Univ Minn, PhD(surg), 55; Am Bd Surg, dipl, 53. *Hon Degrees:* DSc, Georgetown Univ, 74; DSc, Univ Fla, 77; LLD, Dillard Univ, 83. *Prof Exp:* From intern to asst resident surg, Cleveland Metrop Gen Hosp, Ohio, 44-46; asst resident, Freedmen's Hosp, Howard Univ, 46-47, chief resident thoracic surg, 47-48 & surg, 49-50, asst physiol, Col Med, 48-49 & surg, 50-51; res asst, Exp Surg Lab, Univ Minn, 51-53, sr resident surg, Univ Hosp, Univ Minn, 53-54; from asst prof to prof, Col Med, Howard Univ, 54-68; prof & asst dean Sch Med, Univ Calif, Los Angeles & dean, Charles R Drew Postgrad Med Sch, 69-78; prof surg & dean med serv, Med Sch & exec vpres, Med Ctr, 78-90, dir, Int Med Progs, 83-90, EMER PROF SURG, EMER DEAN MED SERV & EMER DEAN INT PROJS, MED SCH HARVARD UNIV, 90- *Concurrent Pos:* Dir, Exp Surg Lab, Col Med, Howard Univ, 54-61, res asst prof, Grad Fac Physiol, 55-69, chief med officer, Howard Univ Div, DC Gen Hosp, 61-68; Mem, DC Bd Exam Med & Osteop, 55-68; exec vpres & mem bd dirs, Nat Med Asn Found, 68-70; mem nat rev comt, Regional Med Progs, 68-70; mem spec adv group, Vet Admin, 69-73, nat surg consult, Cent Off, 69-73; clin prof surg, Sch Med, Univ Southern Calif, 69-78; mem bd visitors, Med Ctr, Duke Univ, 70-75; mem, Comn Study of Accreditation of Selected Health Educ Progs, 70-72; bd dirs, Sun Valley Forum Nat Health, 70; bd trustees, Occidental Col, 71-78, Kaiser Found Health Plan, Inc & Kaiser Found Hosps, 71- & Lloyds Bank Calif, 74-; mem bd overseer's comt visit univ health serv, Harvard Col, 72-78; mem vis comt, Sch Med, Stanford Univ, 72-73 & Univ Mass Med Ctr, 74-75; bd regents, Georgetown Univ, 72-78; chmn adv comt, Med Devices Appln Br, Nat Heart & Lung Inst, 72-73; fel, Ctr Advan Study Behav Sci, Stanford, 75-76; vis prof surg, Sch Med, Stanford Univ, 75-76; mem, Epcot Life & Health Pavilion Adv Bd, 81-; mem bd dirs, Monogram Industs, Inc, 81-83; bd dirs, Georgetown Univ, 86-; mem, Transitional Coun, United Arab Emirates Univ, Fac Med & Health Sci, 87-; hon sr surgeon, Beth Israel Hosp, 90- *Honors & Awards:* Sinkler Award Surg, Nat Med Asn, 68; Warfield Award, Freedmen's Hosp, 69. *Mem:* Inst Med-Nat Acad Sci; Am Asn Univ Professors; Soc Univ Surg; AMA; Am Surg Asn; Nat Med Asn. *Res:* Radiation biology; cardiovascular physiology; evaluation of methods of closure of bronchial stump; blood volume. *Mailing Add:* Harvard Med Sch 25 Shattuck St Boston MA 02115

SPELMAN, MICHAEL JOHN, b Rochester, NY, Mar 28, 39; m 67; c 3. METEOROLOGY, ATMOSPHERIC SCIENCE. *Educ:* LeMoyne Col, BS, 62; Pa State Univ, MS, 69. *Prof Exp:* RES METEOROLOGIST, GEOPHYS FLUID DYNAMICS LAB, NAT OCEANIC & ATMOSPHERIC ADMIN, 69- *Mem:* Am Meteorol Soc. *Res:* Investigation of the structure and circulations of the atmosphere and oceans through numerical modeling on super computers. *Mailing Add:* Geophys Fluid Dynamics Lab PO Box 308 Princeton NJ 08542

SPELSBERG, THOMAS COONAN, b Clarksburg, WVa, July 6, 40; m 67; c 3. GENETICS, BIOCHEMISTRY. *Educ:* WVa Univ, AB, 63, PhD(genetics, biochem), 67. *Prof Exp:* Fel biochem, Univ Tex M D Anderson Hosp & Tumor Inst, 67-68, res asst, 68-69, asst biochemist, 69-70; asst prof obstet, gynec & biochem, Sch Med, Vanderbilt Univ, 70-74; assoc prof biochem, Mayo Med Sch & Mayo Grad Sch Med, 74-77; prof & head biochem sect, 79-83, MEM STAFF, DEPT CELL BIOL, MAYO CLIN, 74-, CHMN DEPT BIOCHEM & MOLECULAR BIOL, 88-; PROF BIOCHEM, MAYO MED SCH & MAYO GRAD SCH MED, 77- *Concurrent Pos:* Nat Genetics Found fel; distinguished lectr, Univ Conn, Univ NJ Med Sch & Oral Roberts Med Sch; distinguished investr, Mayo Grad Sch Med. *Mem:* AAAS; Am Soc Biol Chemists; Am Soc Cell Biol; Endocrine Soc; Am Soc Reprod Biol; Am Soc Bone & Mineral Res. *Res:* Role of nuclear proteins and steroid hormones in regulation of gene activity; DNA-protein interactions; steroid receptor interaction with chromatin; steroid action in human bone cells. *Mailing Add:* Dept Biochem & Molecular Biol Mayo Clin Rochester MN 55901

SPENADEL, LAWRENCE, b Brooklyn, NY, Apr 1, 32; m 55; c 2. PHYSICAL CHEMISTRY, POLYMER CHEMISTRY. *Educ:* Queens Col, NY, BS, 53; Univ Cincinnati, MS, 54, PhD(phys chem), 57. *Prof Exp:* Res chemist, Esso Res & Eng Co, 56-65, sr chemist, Enjay Polymer Labs, 66-70, RES ASSOC, EXXON CHEM CO, 70- *Mem:* Inst Elec & Electronic Engrs; Am Chem Soc; Soc Plastics Eng. *Res:* Compounding of ethylene propylene terpolymer and polyethylenes for wire and cable applications; development of thermoelastic rubbers; combustion; formulation and testing of high energy solid propellants; dispersion measurements on platinum catalysts. *Mailing Add:* Exxon Chem Co PO Box 5200 Baytown TX 77522

SPENCE, ALEXANDER PERKINS, b St Louis, Mo, Apr 5, 29; m 55; c 3. COMPARATIVE ANATOMY, EMBRYOLOGY. *Educ:* Univ Mo, BSEd, 60, MST, 61; Cornell Univ, PhD(biol), 69. *Prof Exp:* From asst prof to assoc prof, 61-78, PROF ANAT & EMBRYOL & CHMN DEPT, STATE UNIV NY, CORTLAND, 78- *Mem:* Sigma Xi; Am Soc Zoologists. *Res:* Ultrastructure of spermatogenesis in Rana pipiens; sperm-egg chemotaxis in amphibians. *Mailing Add:* Dept Biol Sci State Univ NY PO Box 2000 Cortland NY 13045

SPENCE, DALE WILLIAM, b Beaumont, Tex, Apr 8, 34; m 55; c 3. EXERCISE PHYSIOLOGY. *Educ:* Rice Inst, BS, 56; NTex State Univ, MS, 59; La State Univ, EdD(phys educ), 66. *Prof Exp:* Instr phys educ, NTex State Univ, 58-59 & Hardin-Simmons Univ, 59-62; asst, La State Univ, 62-63; from instr to assoc prof, 63-74, PROF HUMAN PERFORMANCE & HEALTH SCI, RICE UNIV, 74- *Concurrent Pos:* Fel, Baylor Col Med, 68-69, vis assoc prof, 71-80, prof, 80-86; dir exercise rehab & res, St Joseph Hosp, 74-76; vis scientist, Manned Spacecraft Ctr, 69-70; mem staff, Houston Cardovasc Rehabilitation Ctr, 80-86; consult scientist, Sch Aerospace Med, 80-81; adj prof med, Baylor Col Med, 87- *Mem:* fel Am Col Sports Med; Aerospace Med Asn. *Res:* Cardiovascular physiology and cardiovascular rehabilitation. *Mailing Add:* Human Performance & Health Sci Rice Univ Box 1892 Houston TX 77251-1892

SPENCE, DAVID, b Halifax, Eng, Sept 23, 41; m 62; c 4. ATOMIC PHYSICS. *Educ:* Univ Durham, BSc, 63; Univ Newcastle-upon-Tyne, PhD(physics), 67. *Prof Exp:* Res staff appl scientist physics, Yale Univ, 67-71; asst physicist, 71-74, PHYSICIST, ARGONNE NAT LAB, 74- *Res:* Atomic and molecular spectroscopy; physical and gaseous electronics. *Mailing Add:* Argonne Nat Lab Argonne IL 60439

SPENCE, GAVIN GARY, b St Paul, Minn, July 23, 42; m 65; c 3. ORGANIC CHEMISTRY. *Educ:* Williams Col, BA, 64; Princeton Univ, AM, 67, PhD(chem), 68. *Prof Exp:* Res chemist, Hercules Inc, 68-75, sr res chemist, 75-79, res scientist, 79-87; DIR POLYMER RES & DEVELOP, CALLAWAY CHEM CO, 87- *Mem:* Am Chem Soc. *Res:* Synthetic organic chemistry, in particular synthesis and evaluations of organic polymers for use in paper, adhesives, textiles. *Mailing Add:* 7113 Stillwater Dr Columbus GA 31904-1958

SPENCE, HARLAN ERNEST, b Winchester, Mass, June 14, 61; m 83; c 1. MAGNETOSPHERIC & IONOSPHERIC PHYSICS. *Educ:* Boston Univ, BA, 83; Univ Calif, Los Angeles, MS, 85, PhD(geophys & space physics), 89. *Prof Exp:* Off Naval Res grad fel, space physics res, Dept Earth & Space Sci, Univ Calif, Los Angeles, 83-87, grad res asst, Inst Geophys & Planetary Physics, 87-89; MEM TECH STAFF SPACE PHYSICS, SPACE PARTICLES & FIELDS DEPT, AEROSPACE CORP, 89- *Concurrent Pos:* Co-investr grants, NASA, 89-; assoc ed, J Geophys Res-Space Physics, 91-; mem, Geospace Environ Modelling Prog Adv Panel, NSF, 91- *Mem:* Am Geophys Union. *Res:* Magnetospheric magnetic field modelling; magnetospheric convection and dynamics of the magnetotail; substorm signatures in the inner magnetosphere; coherence of ionospheric auroral precipitation structures; ultra-low frequency waves in the magnetosphere. *Mailing Add:* Aerospace Corp MS M2-260 PO Box 92957 Los Angeles CA 90009

SPENCE, HILDA ADELE, b Chattanooga, Tenn, Oct 13, 29. MICROBIOLOGY. *Educ:* Univ Tenn, Chattanooga, BS, 51; La State Univ, MS, 66, PhD(microbiol), 71. *Prof Exp:* Med technologist, Charity Hosp LA, New Orleans, 51-59; instr med technol, 59-64, asst prof microbiol, 64-74, ASSOC PROF MICROBIOL, LA STATE UNIV MED CTR, 89- *Concurrent Pos:* Mem rev bd, Nat Accrediting Agency Clin Lab Sci, 72-77. *Honors & Awards:* Am Soc Med Technol Res Award, 63. *Mem:* Am soc Microbiol; Am Soc Med Technol (pres, 78-79); Am Soc Virol. *Res:* Neurotropic strains of influenza viruses; clinical microbiology; microcomputers in medical education and research. *Mailing Add:* Dept of Microbiol La State Univ Med Ctr 1901 Perdido St New Orleans LA 70112

SPENCE, JACK TAYLOR, b Salt Lake City, Utah, Nov 16, 29; m 51; c 3. ANALYTICAL CHEMISTRY, INORGANIC CHEMISTRY. *Educ:* Univ Utah, BS, 51, PhD(chem), 57. *Prof Exp:* Fel, Univ Ore, 57-58; from asst prof to prof chem, Utah State Univ, 58-89, dept head, 76-81; RETIRED. *Concurrent Pos:* USPHS res career develop award, Nat Inst Gen Med Sci, 68-73. *Mem:* AAAS; Am Chem Soc. *Res:* Organic chelating agents; coordination compounds; inorganic photochemistry; inorganic biochemistry; mechanisms of enzyme reactions. *Mailing Add:* PO Box 142 Teasdale UT 84773

SPENCE, JOHN EDWIN, b Fall River, Mass, Oct 26, 34; m 58; c 3. ELECTRICAL ENGINEERING. *Educ:* Bradford Durfee Col Technol, BS, 57; Univ Wis, MS, 60, PhD(elec eng), 62. *Prof Exp:* Mem staff, Digital & Analog Comput Labs, Allis-Chalmers Mfg Co, 57-59; asst, Univ Wis, 59-62; assoc prof elec eng, Univ RI, 62-67; tech dir antennas & propagation group, Electronics & Commun Div, Atlantic Res Corp, Va, 67-68; assoc prof elec eng, 68-76, PROF ELEC ENG, UNIV RI, 76- *Concurrent Pos:* Consult, Amecom Div, Litton Systs, Inc, Md, 65. *Mem:* Inst Elec & Electronics Engrs. *Res:* Electromagnetic theory; wave propagation. *Mailing Add:* Dept of Elec Eng Univ of RI Kingston RI 02881

SPENCE, KEMET DEAN, b Portland, Ore, Jan 10, 37; m 58; c 4. MICROBIOLOGY, BIOCHEMISTRY. *Educ:* Ore State Univ, BS, 60, MS, 62, PhD(microbiol, biochem), 65. *Prof Exp:* Microbiologist, Ore Fish Comn, 62-64; res fel, Argonne Nat Lab, 65-68; from asst prof to assoc prof, 68-80, PROF MICROBIOL, WASH STATE UNIV, 81- *Mem:* Am Soc Microbiol; Soc Invert Path. *Res:* Biochemical and applied studies of microbial pathogens and immunity of insects. *Mailing Add:* Dept Microbiol Washington State Univ Pullman WA 99164

SPENCE, LESLIE PERCIVAL, b St Vincent, WI, Aug 16, 22; m 53; c 2. MICROBIOLOGY. *Educ:* Bristol Univ, MB, ChB, 50; Univ London, dipl trop med & hyg, 51; FRCP, 72. *Prof Exp:* Med officer, Trinidad Govt Med Serv, 51-62; dir regional virus lab, Trinidad & prof virol, Univ West Indies, 62-68; prof microbiol, McGill Univ, 68-72; chmn dept, 82-88, PROF MICROBIOL, UNIV TORONTO, 72- *Concurrent Pos:* Rockefeller Found fel, Rockefeller Found Virus Labs, NY, 55-56; consult, Nat Inst Allergy & Infectious Dis, 69-73 & Pan-Am Health Orgn, 69-70. *Mem:* Am Soc Trop Med & Hyg; Can Soc Microbiol; Am Soc Microbiol. *Res:* Arboviruses; viral gastroenteritis. *Mailing Add:* Dept Microbiol Banting Inst Rm 215 101 College Toronto ON M2M 1C2 Can

SPENCE, MARY ANNE, b Tulsa, Okla, Sept 8, 44; m 72. HUMAN GENETICS. *Educ:* Grinnell Col, BA, 66; Univ Hawaii, PhD, 69. *Prof Exp:* Asst prof, 70-75, assoc prof, 75-80, PROF PSYCHIAT & BIOMATH, SCH MED, UNIV CALIF, LOS ANGELES, 80- *Concurrent Pos:* NIH fel genetics curriculum, Univ NC, Chapel Hill, 69-70; mem, Ment Retardation Res Ctr, Neuropsychiat Inst, Univ Calif, Los Angeles, 74-, assoc dean, Grad Div, 88- *Honors & Awards:* Woman of Sci Award, Univ Calif, 79. *Mem:* Am Soc Human Genetics; Genetics Soc Am; Behav Genetics Asn. *Res:* Mathematical and computer models for family data analysis; applications for genetic counseling. *Mailing Add:* 1652 Benedict Canyon Dr Beverly Hills CA 90210

SPENCE, ROBERT DEAN, b Bergen, NY, Sept 12, 17; m 42; c 4. ENGINEERING. *Educ:* Cornell Univ, BS, 39; Mich State Col, MS, 42; Yale Univ, PhD(physics), 48. *Prof Exp:* Asst physics, Mich State Col, 41-42; instr elec commun, Mass Inst Technol, 42-45; from asst prof to prof physics & astron, 47-76, actg head dept physics, 56-57, prof physics, 76-86, EMER PROF PHYSICS, MICH STATE UNIV, 86- *Concurrent Pos:* Vis prof & Guggenheim fel, Bristol Univ, 55-56; vis prof, Eindhoven Technol Univ, 64 & State Univ Leiden, 71. *Mem:* Nat Soc Prof Engrs; Am Soc Heating Refrig & Air Conditioning Engrs. *Res:* Mathematical and crystal physics; nuclear magnetic resonance. *Mailing Add:* 1849 Ann St East Lansing MI 48823

SPENCE, SYDNEY P(AYTON), b Yonkers, NY, Dec 30, 21; m 44; c 5. CHEMICAL ENGINEERING. *Educ:* Univ Rochester, BS, 44. *Prof Exp:* Jr tech rep, Halowax Prods Div, 43-44 & 46, mem tech staff, Bakelite Co Div, 46-51, chem engr, Res & Develop Dept, 51-56, proj engr, 56-57, group leader process res, Union Carbide Plastics Co Div, 57-67, process tech mgr phenolic & epoxy resins, Coatings Intermediates Div, 67-71, DEVELOP SCIENTIST, UNION CARBIDE CORP, BOUND BROOK, 71- *Mem:* Am Inst Chem Engrs. *Res:* Development of commercial processes for epoxy resins, bisphenol A, phenolic resin, acrylonitrile-butadiene-styrene resins, di-paraxylylene, polyester resins, chlorinated hydrocarbons. *Mailing Add:* 2159 Bayberry Lane Westfield NJ 07090

SPENCE, THOMAS WAYNE, b Washington, Pa, Sept 14, 38; m 68; c 3. OCEANOGRAPHY. *Educ:* Duquesne Univ, BA, 60; Univ Chicago, PhD(geophys sci), 73. *Prof Exp:* Meteorologist, US Air Force, 60-63, DeNardo & McFarland, Inc, 63-65; res asst hydrodynamics, Univ Chicago, 66-73; from asst prof to assoc prof oceanog, Tex A&M Univ, 73-83; sci prog officer, Off Naval Res, 81-83, prog mgr, 83-87; PROG DIR, NSF, 87- *Concurrent Pos:* Sci fel, Senate Appropriations Comm. *Mem:* Oceanog Soc; Am Geophys Union; Am Meteorol Soc; Sigma Xi. *Res:* Geophysical fluid dynamics including atmospheric and oceanic dynamics; laboratory models of geophysical fluid motions; instability; large and mesoscale ocean dynamics; science management. *Mailing Add:* 6245 N Kensington McLean VA 22101

SPENCE, WILLARD LEWIS, b Providence, RI, Mar 16, 35; m 58; c 3. BOTANY. *Educ:* Colby Col, BA, 57; Univ Iowa, MS, 59; Univ Calif, Berkeley, PhD(bot), 63. *Prof Exp:* Asst cur bot, NY Bot Gardens, 63-64; from asst prof to assoc prof biol, 64-69, PROF BIOL, FRAMINGHAM STATE COL, 69- *Mem:* Am Soc Plant Taxon; Bot Soc Am; Sigma Xi. *Res:* Vascular plant taxonomy; local flora; economically valuable plants. *Mailing Add:* Dept of Biol Framingham State Col Framingham MA 01701

SPENCE, WILLIAM J, b Peoria, Ill, July 11, 37; wid; c 1. TECTONOPHYSICS, EARTHQUAKE PREDICTION. *Educ:* State Univ NY, Albany, BS, 59, MS, 60; Pa State Univ, PhD(geophys), 73. *Prof Exp:* Instr phys chem, Spencer Cent High Sch, 61-62; res gophysicist, US Coast & Geol Surv, 62-70, Environ Res Lab, 71, Nat Oceanic & Atmospheric Admin, 72; RES GEOPHYSICIST, US GEOL SURV, 73-; adj prof, Colo Sch of Mines, Golden, 85-87. *Concurrent Pos:* Prin investr, Seismicity & Tectonics Proj, comp earthquake & tsunami hazard for zones, Circum-Pac Proj & deep hole desalinization of Dolores River, Colo Proj; pres, Front Range Chap, Am Geophys Union, 89. *Mem:* Seismol Soc Am; Am Geophys Union; AAAS; Geol Soc Am; Sigma Xi. *Res:* Causes and consequences of great earthquakes; plate tectonics; earthquakes induced by reservoirs or by fluid injection into substrata; aftershocks; seismic siting of critical facilities; tectonic development of western United States. *Mailing Add:* Nat Earthquake Info Ctr Us Geol Surv Box 25046 MS967 Denver CO 80225

SPENCER, ALBERT WILLIAM, b Omaha, Nebr, Jan 1, 29; m 56; c 4. ZOOLOGY. *Educ:* Colo State Univ, BS, 57, MS, 62, PhD(zool), 65. *Prof Exp:* Asst prof zool, Eastern NMex Univ, 64-65; asst prof zool, 65-74, ASSOC PROF BIOL, FT LEWIS COL, 74- *Mem:* Am Soc Mammal; Soc Study Evolution; Ecol Soc Am; Wildlife Soc; Genetics Soc Am. *Res:* Vertebrate population biology, particularly speciation. *Mailing Add:* Dept Biol Sci Ft Lewis Col Durango CO 81301

SPENCER, ALEXANDER BURKE, b San Antonio, Tex, Dec 28, 32; m 63; c 2. GEOLOGY. *Educ:* Tex Western Col, BSc, 55; Univ Okla, MS, 61; Univ Tex, Austin, PhD(geol), 66. *Prof Exp:* From instr to asst prof chem & geol, Carnegie Inst Technol, 65-67; res geologist, Mobile Res & Develop Corp, 67-71, sr res geologist, 71-74, mem staff, Mobil Oil Libya, Ltd, 75-77, mem staff, 77-81, explor supvr, Mobil Explor & Producing Serv, Inc, 81-89; CONSULT, 89- *Mem:* Soc Explor Geophysicists; Am Asn Petrol Geol; Soc Econ Paleont & Mineral. *Res:* Petrology of sandstone, international petroleum exploration. *Mailing Add:* 7708 Turnberry Lane Dallas TX 75248

SPENCER, ANDREW NIGEL, b Fulmer, Eng, Feb 13, 45; m 66; c 3. INVERTEBRATE PHYSIOLOGY. *Educ:* Univ London, BSc, 67; Univ Victoria, PhD(zool), 71. *Prof Exp:* Sci Res Coun fel zool, Univ Bristol, 71-72; vis asst prof zool, 72-73; lectr biol, Inst Biol, Univ Odense, 73-75; PROF ZOOL & ASSOC CHMN DEPT, UNIV ALTA, 75- *Concurrent Pos:* Consult, New Can Encyclopedia; vis prof, Univ Pet M Curie, Villefranche, 74-75; grant selection comt, animal biol, Nat Sci & Eng Res Coun Can. *Honors & Awards:* McCalla Professorship. *Mem:* Am Soc Zoologists; Can Soc Zoologists; Brit Soc Exp Biol; Soc Neurosci. *Res:* The behavioural neurophysiology of hydrozoans; central control of rhythmical behavior in invertebrates; function of neuropeptides; voltage-clamp analysts. *Mailing Add:* Dept Zool Univ Alta Edmonton AB T6G 2E9 Can

SPENCER, ANDREW R, US Citizen. METALLURGICAL ENGINEERING, FORENSIC SCIENCE. *Educ:* US Merchant Marine Acad, BS, 45; Wayne State Univ, BS, 48. *Prof Exp:* Serv metallurgist, Steel Sales Corp, 49-55; pres & tech dir, Precision Testing Labs, Inc, 56-57; sr scientist, Metall Res Dept, Chrysler Corp, 58-61; staff metall engr, Bendix Res Lab, 61-73; mgr porous metal prod, Filter Prod Div, Facet Enterprises Inc, 73-77; TECH DIR, METALLURGICAL ADVISORS CO, 77- *Mem:* Am Soc Metals; Am Inst Mining, Metall & Petrol Engrs; Nat Asn Corrosion Engrs; Soc of Automotive Engrs; Am Soc Testing & Mat; Am Acad Forensic Sci. *Res:* Failure analysis; applying engineering materials for use in hostile environments; powder metallurgy; diffusion bonding; oxidation resistance of metals and super alloys; lubrication; friction and wear; manufacturing engineering. *Mailing Add:* Metall Adv Co 4900 Leafdale Royal Oak MI 48073

SPENCER, ARMOND E, b Crandon, Wis, Oct 1, 33; m 58; c 4. MATHEMATICS. *Educ:* Mich State Univ, BS, 58, MS, 61, PhD(math), 67. *Prof Exp:* High sch instr, Mich, 58-60; instr math, Lansing Community Col, 62-65; asst prof, Western Mich Univ, 66-67 & Univ Ky, 67-71; assoc prof, 71-76, PROF MATH, STATE UNIV NY COL POTSDAM, 76- *Mem:* Am Math Soc; Math Asn Am. *Res:* Finite group theory. *Mailing Add:* Dept of Mathematics Potsdam Col State Univ NY Potsdam NY 13625

SPENCER, ARTHUR COE, II, b Pittsburgh, Pa, Dec 16, 39; m 60; c 2. APPLIED MATHEMATICS, ENGINEERING SCIENCE. *Educ:* Allegheny Col, BS, 61; Univ Pittsburgh, MS, 64. *Prof Exp:* Res mathematician, PPG Industs, Inc, 61-67; sr engr, 67-73, fel engr, 73-76, MGR THERMAL HYDRAUL METHODS, WESTINGHOUSE ELEC CORP, 76- *Mem:* Soc Indust & Appl Math. *Res:* Computational fluid dynamics; numerical methods for partial differential equations; two phase flow; heat transfer. *Mailing Add:* 2513 Col Park Rd Allison Park Pittsburgh PA 15101

SPENCER, ARTHUR MILTON, JR, b Salt Lake City, Utah, Jan 6, 20; m 48; c 7. FUEL ENGINEERING, SURFACE CHEMISTRY. *Educ:* Univ Utah, BS, 49, MS, 51, PhD(fuels eng), 62. *Prof Exp:* Tech asst, explosives res group, Univ Utah, 53-59; sr res chemist, Allegany Balistics Lab, Hercules Inc, Md, 62-63; sr res assoc oil well stimulation & cement, West Co, Tex, 64-71; chief chemist, Petrol Technol Corp, 71-78; sr res scientist, Rocket Res Co, 78-79; chief chemist, Petrol Technol Corp, 79-85; RETIRED. *Concurrent Pos:* Fel, Petrol Res Fund, 59-62; teacher, high sch math & sci, 89- *Mem:* Am Chem Soc; Soc Petrol Engrs. *Res:* Explosives for oil well stimulation; acid corrosion at high temperatures; fuel and water gels; high temperature retarders for oil well cementing; high temperature explosives for geothermal wells. *Mailing Add:* 13719 115th Ave NE Kirkland WA 98034-2165

SPENCER, BROCK, b Horton, Kans, Sept 25, 39; m 64; c 2. INORGANIC CHEMISTRY. *Educ:* Carleton Col, BA, 61; Univ Calif, Berkeley, PhD(chem), 65. *Prof Exp:* From instr to assoc prof, 65-76, PROF CHEM, BELOIT COL, 76-, CHMN DEPT, 80- *Concurrent Pos:* Vis prof, Case Western Reserve Univ, 67-68, Uppsala Univ, 71-72, Univ Calif, Berkeley, 79 & Univ Wis-Madison, 85. *Mem:* Am Chem Soc; AAAS. *Res:* Molecular spectroscopy and x-ray diffraction determination of molecular structure and bonding. *Mailing Add:* Dept of Chem Beloit Col Beloit WI 53511

SPENCER, CHARLES WINTHROP, b Cambridge, Mass, Dec 25, 30; m 54; c 3. GEOLOGY. *Educ:* Colby Col, AB, 53; Univ Ill, MS, 55. *Prof Exp:* Res asst, Clay Mineral, State Geol Surv, Ill, 53-55; geologist, US Geol Surv, 55-59; geologist, Texaco, Inc, Mont, 59-66, from asst dist geologist to dist geologist, 66-73, div lab mgr, Colo, 73-74; PROG CHIEF, US GEOL SURV, 74- *Honors & Awards:* A I Levorsen Award. *Mem:* Am Asn Petrol Geologists; Soc Econ Paleontologists & Mineralogists; Soc Prof Well Log Analysts; Soc Petrol Engrs. *Res:* Geology of mineral deposits; petroleum exploration; stratigraphy of Paleozoic and Cretaceous; hydrodynamics; environmental interpretation of sandstones and carbonates; petrol geology of southern Brazil; geology of low permeability (tight) gas reservoirs; origins of overpressured and underpressured gas reservoirs. *Mailing Add:* 13528 W Alaska Dr Lakewood CO 80228

SPENCER, CHERRILL MELANIE, b Derbyshire, Eng, Feb 17, 48; m 89. DESIGN MAGNETS FOR PARTICLE ACCELERATORS & MAGNETIC RESONANCE IMAGERY, DESIGN PARTICLE DETECTORS FOR ELEMENTARY PARTICLE PHYSICS. *Educ:* Univ London, BSc 69, Univ Oxford, DPhil, 72. *Prof Exp:* Royal Soc Europ fel elem particle physics, Italian Nat Lab, 72-74; res assoc elem particle physics, Univ Wis, Madison, 74-77; res assoc, Fla State Univ, 77-79; staff scientist, Sci Applns Inc, 79-84; physicist, Resonex Inc, 84-88; PHYSICIST, STANFORD LINEAR ACCELERATOR CTR, 88- *Mem:* Am Phys Soc; AAAS; Inst Physics UK; Asn Women Sci. *Res:* Design & build iron cored electromagnets for particle accelerators and magnetic resonance imaging; designed & built non-destructive coal analyzer using prompt neutron activation analysis; Magnetics resonance resonance imagery, design particle detectors for elementary particle physics. *Mailing Add:* Bin 12 Stanford Linear Accelerator Ctr PO Box 4349 Stanford CA 94309

SPENCER, CHESTER W(ALLACE), b Greeley, Kans, Nov 2, 24; m 48; c 4. METALLURGY. *Educ:* Univ Kans, BS, 49, MS, 50; Univ Wis, PhD(metall), 52. *Prof Exp:* Sr engr atomic energy div, Sylvania Elec Prod, Inc, 52-54; res metallurgist, Carnegie Inst Technol, 54-56; asst prof metall eng, Cornell Univ, 56-58, assoc prof, 58-62; mgr mat res & develop, Res & Adv Develop Div, Avco Corp, Mass, 62-64; vpres, Chase Brass & Copper Co, 64-77; EXEC DIR, NAT MAT ADV BD, NAT ACAD SCI, 77- *Concurrent Pos:* Consult, Res & Adv Develop Div, Avco Corp, 59- *Mem:* Am Soc Metals; Am Inst Mining, Metall & Petrol Engrs; Brit Inst Metals. *Res:* Eutectoid and peritectoid transformations in alloys; electrical and physical properties of semiconducting intermetallic compounds; reactions between liquids and solids. *Mailing Add:* RD 6 Box 41-A Christianburg VA 24079-9109

SPENCER, CLAUDE FRANKLIN, b Athens, Pa, Feb 14, 19; m 44; c 3. ORGANIC CHEMISTRY. *Educ:* Univ Mich, BS, 42; Mass Inst Technol, PhD(chem), 50. *Prof Exp:* Res chemist, Merck & Co, Inc, 42-46 & 50-59; sr res chemist, Norwich Pharmacol Co, 59-65, group leader, 65-80, res assoc, Norwich-Eaton Pharmaceut, 80-83; RETIRED. *Mem:* Am Chem Soc; fel Am Inst Chemists; Sigma Xi; Int Soc Heterocyclic Chemists; AAAS. *Res:* Synthesis and structure determination of natural products; synthesis and transannular rearrangements in eight-membered ring compounds; synthesis of medicinal and veterinary products; design and synthesis of biologically active compounds. *Mailing Add:* Box 244 A Hall Quarry Mt Desert ME 04660-9802

SPENCER, DAVID R, b New York, NY, Apr 24, 42; m 68; c 2. COMPUTER PUBLISHING, ELECTRONIC PRINTING. *Educ:* Mass Inst Technol, BSEE, 64, MSEE, 68. *Prof Exp:* Mgr, Graphics Eng, EG&G, Inc, 62-63; pres, Datalog Div, Litton Systs, Inc, 73-82; pres, Muirhead NAm, Inc, 82-83; chmn, Data Recording Systs, Inc, 83-89; MANAGING PARTNER, DAVID R SPENCER & ASSOCS, 89- *Concurrent Pos:* Div, Long Island Forum Technol, 78-82; dir, Long Island Ventura Group, 88- *Mem:* Inst Elec & Electronic Engrs. *Res:* Image processing, compression and hardcopy recording; high resolution xerography; color printing technology, including imaging engines and controller/rips. *Mailing Add:* Three Giffard Way Melville NY 11747

SPENCER, DEREK W, b South Shields, Eng, May 2, 34; m 57; c 3. OCEANOGRAPHY, MARINE GEOLOGY. *Educ:* Univ Manchester, BSc, 54, PhD(geochem), 57. *Prof Exp:* Geochemist, Imp Oil Ltd, Can, 57-65; sr scientist, 65-78, assoc dir res, 78-80, SR SCIENTIST, WOODS HOLE OCEANOG INST, 78- *Mem:* AAAS; Am Geophys Union; Am Asn Petrol Geologists; Geochem Soc. *Res:* Trace element geochemistry of sediments and ocean water; chemical oceanography; ocean circulation and time scale of ocean mixing. *Mailing Add:* Woods Hole Oceanog Inst Woods Hole MA 02543

SPENCER, DOMINA EBERLE (MRS PARRY MOON), b New Castle, Pa, Sept 26, 20; m 61; c 1. MATHEMATICS, PHYSICS. *Educ:* Mass Inst Technol, SB, 39, MS, 40, PhD(math), 42. *Prof Exp:* Asst illum eng, Mass Inst Technol, 42; asst prof physics, Am Univ, 42-43, Tufts Col, 43-47 & Brown Univ, 47-50; assoc prof math, 50-60, PROF MATH, UNIV CONN, 60- *Honors & Awards:* Illum Eng Soc Gold Medal, 74. *Mem:* Am Math Soc; fel Optical Soc Am; fel Illum Eng Soc; Math Asn Am. *Res:* Application of tensors to physics, field theory; nomenclature, color, calculation of illumination; design of lighting for vision; foundations of electrodynamics; mathematics of nutrition. *Mailing Add:* Dept of Math Univ of Conn Storrs CT 06268

SPENCER, DONALD CLAYTON, b Boulder, Colo, Apr 25, 12; m 36, 51; c 3. MATHEMATICS. *Educ:* Univ Colo, BA, 34; Mass Inst Technol, BSc, 36; Cambridge Univ, PhD(math), 39, ScD, 63. *Hon Degrees:* DSc, Purdue Univ, 71. *Prof Exp:* Instr math, Mass Inst Technol, 39-42; from assoc prof to prof, Stanford Univ, 42-50; from assoc prof to prof, Princeton Univ, 50-63; prof, Stanford Univ, 63-68; prof, 68-72, Henry Burchard Fine prof, 72-78, EMER HENRY BURCHARD FINE PROF MATH, PRINCETON UNIV, 78- *Honors & Awards:* Bocher Prize, Am Math Soc, 48; Nat Medal Sci, 89. *Mem:* Nat Acad Sci; Am Math Soc; Am Acad Arts & Sci. *Res:* Differential geometry; partial differential equations. *Mailing Add:* 943 County Rd 204 Durango CO 81301

SPENCER, DONALD JAY, b Salt Lake City, Utah, Apr 1, 28; m 50; c 7. LASER PHYSICS, AEROTHERMODYNAMICS. *Educ:* Univ Utah, BS, 54; Univ Calif, MS, 56; Univ Southern Calif, PhD(quantum electronics), 81. *Prof Exp:* Electronics engr, Hughes Aircraft Co, 54-57; res physicist, Ramo-Wooldridge Corp, 57-58, Space Technol Labs, 58-60; sr scientist, 60-90, CASUAL SR SCIENTIST, AEROSPACE CORP, 90- *Concurrent Pos:* Consult, Donald J Spencer Assocs, 90- *Mem:* Am Phys Soc; Am Inst Aeronaut & Astronaut. *Res:* Chemical lasers research and atmospheric transmissions of laser beams; electric arc driven wind tunnels; plasma and flow diagnostics; reentry simulation; aerothermodynamic materials testing; experimental spectroscopy. *Mailing Add:* Aerospace Corp Mail Sta M5/741 PO Box 92957 Los Angeles CA 90009-2957

SPENCER, DONALD LEE, solar energy, thermal sciences; deceased, see previous edition for last biography

SPENCER, DWIGHT LOUIS, b Harveyville, Kans, June 24, 24; m 48; c 6. ANIMAL ECOLOGY. *Educ:* Kans State Teachers Col, BS, 52, MS, 55; Okla State Univ, PhD(zool), 67. *Prof Exp:* Instr high sch, Kans, 53-60; from lectr to prof, 72-86, assoc chmn dept, 74-86, EMER PROF BIOL, EMPORIA KANS STATE UNIV, 86- *Mem:* Am Soc Mammal. *Res:* Mammalian ecology and speciation; ecological speciation study of Neotoma floridana and Neotoma micropus in Kansas and Oklahoma. *Mailing Add:* Dept Biol Sci Emporia State Univ 1200 Coml St Emporia KS 66801

SPENCER, E MARTIN, b Cleveland, Ohio, Dec 6, 29; c 3. MEDICINE. *Educ:* Dartmouth Col, AB, 52; Harvard Univ, MD, 56; Rockefeller Univ, PhD(biochem), 69. *Prof Exp:* Res assoc, Beth Israel Hosp, Sch Med, Harvard Univ, 59-60; attend physician, Harlem Hosp, New York, 69-70; guest investr, Rockefeller Univ, 69-70; ASSOC CLIN PROF MED, UNIV CALIF, SAN FRANCISCO, 70-; DIR, LAB GROWTH & DEVELOP, CHILDREN'S HOSP, SAN FRANCISCO, 80- *Concurrent Pos:* Rotating intern, San Francisco Gen Hosp, Univ Calif Serv, 56-57; sr resident med, Bellevue Hosp, Columbia Univ Serv, 60-61, Univ Calif Med Ctr, San Francisco, 62-63; vis asst, Cardiol Ctr, Cantonal Hosp & 1 Univ, Geneva, Switz, 61-62; policy bd mem, Sickle Cell Vaso-Occlusive Clin Trials, Nat Heart & Lung Inst, 72, mem, Ad Hoc Comt, Studies Sickle Cell Dis, 73-75; chmn, Workshop Extracorporeal Treatment Sickle Cell Dis, NIH, 74-75. *Mem:* Fel Am Col Physicians; Western Soc Clin Res; Am Soc Bone & Mineral Res. *Res:* Somatomedin, its role in the control of cellular proliferation and its physiologic role and regulation; hormonal regulation of vitamin D metabolism. *Mailing Add:* Dept Res Oper 622 Children's Hosp San Francisco CA 94118

SPENCER, EDGAR WINSTON, b Monticello, Ark, May 27, 31; m 58; c 2. GEOLOGY. *Educ:* Washington & Lee Univ, BS, 53; Columbia Univ, PhD(geol), 57. *Prof Exp:* Lectr geol, Hunter Col, 54-56, instr, 57; asst prof & actg chmn dept, 57-59, assoc prof, 59-63, PROF GEOL, WASHINGTON & LEE UNIV, 63-, CHMN DEPT, 59- *Concurrent Pos:* Prin investr, NSF grant, 59-62, sci fac fel tectonics in NZ & Australia, 65-66; res grant, Switz & Spain, 71-72; pres, Rockbridge Area Conserv Coun, 78; res grant, Am Chem Soc, 81-82, Greece, 82 & Western NAm, 90. *Mem:* Fel AAAS; fel Geol Soc Am; Am Asn Petrol Geologists; Nat Asn Geol Teachers; Am Inst Prof Geologists; Am Geophys Union. *Res:* Tectonics; regional structure; land use planning; author of six books. *Mailing Add:* Dept Geol Washington & Lee Univ Lexington VA 24450

SPENCER, EDWARD G, b Lynchburg, Va, July 21, 20; m 46; c 2. MATERIALS SCIENCE ENGINEERING. *Educ:* George Washington Univ, BSE, 45; Boston Univ, MA, 50. *Prof Exp:* Physicist, US Naval Res Lab, 43-46 & 49-53; physicist, Cambridge Air Force Res Lab, 46-49; physicist, Diamond Ordinance Fuze Lab & Dept Defense, 53-58 & AT&T Bell Labs, 58-87; RETIRED. *Concurrent Pos:* Co-ed, Conf Magnetism & Magnetic Mat, New York City, 64-68. *Mem:* Fel Am Phys Soc. *Res:* Microwave radar; upper atmospheric physics; semiconductor and ferrimagnetic materials for scanning phase array microwave radars; magnetic spin resonance; dielectric materials for electrooptic; elastooptic and microwave ultrasonic applications; superconducting metals and alloys. *Mailing Add:* 76 Roland Rd Murray Hill NJ 07974

SPENCER, ELAINE, b Portland, Ore, Aug 6, 19; m 42; c 5. BIOCHEMISTRY. *Educ:* Linfield Col, BA, 40; Mass Inst Technol, MS, 48; Univ Ore, PhD(biochem), 61. *Prof Exp:* Instr chem, Ore State Col, 46-47, instr math, 47; instr, 56-59, from instr to asst prof chem, 59-74, ASSOC PROF CHEM, PORTLAND STATE UNIV, 74- *Mem:* Sigma Xi; Am Chem Soc. *Res:* Oxidative enzymes; proteins; enzyme kinetics. *Mailing Add:* 4835 NE Broadway Portland OR 97213

SPENCER, ELVINS YUILL, b Edmonton, Alta, Oct 2, 14; m 42; c 2. ORGANIC CHEMISTRY, AGRICULTURAL CHEMISTRY. *Educ:* Univ Alta, BSc, 36, MSc, 38; Univ Toronto, PhD(chem), 41. *Prof Exp:* Chief chemist, Fine Chem Can, 41-42; res chemist, Gelatin Prod Corp, Ont & Mich, 42-43; res engr, Consol Mining & Smelting Co, BC, 43-46; res chemist, E B Eddy Paper Co, Que, 46; assoc prof chem, Univ Sask, 46-51; prin chemist, Res Inst, Can Dept Agr, Univ Western Ont, 51-60, dir Res Inst, 60-78, sr scientist dept agr, 78-79, chmn, Int Expert Comt Dioxins, Ont Ministry

Environ, Can Dept Agr, Univ Western Ont, 83-90; RETIRED. *Concurrent Pos:* Coordr res, Sask Res Coun, 49-51; hon lectr, Univ Western Ont, 51-61, hon prof, 61-91; consult, Dept Natural Resources, Sask on Potash & Sabbatical at Cambridge Univ, 56-57; mem expert comt pesticide residues, Food & Agr Orgn; vis scholar, Rockefeller Conf Ctr, Bellagio, Italy, 76; ed, Pesticide Biochem & Physiol, 78-90; mem, Fed Pest Mgt Adv Bd, 85-89; fel, Agrochem Div, Am Chem Soc & Chem Inst Can. *Mem:* Am Chem Soc; Chem Inst Can; Agr Inst Can; Can Biochem Soc. *Res:* Flotation agent; pharmaceuticals; cereal chemistry; synthetic polypeptides; oils and fats; organic chemistry and biochemistry of pesticides; pesticide biochemistry and physiology. *Mailing Add:* Seven Westview Dr London ON N6A 2Y2 Can

SPENCER, FRANK, b Rochester, Kent, Eng, 41. PHYSICAL ANTHROPOLOGY. *Educ:* FIMLS, London, 65; Univ Windsor, BA, 73; Univ Mich, Ann Arbor, MA, 74, PhD(anthrop), 79. *Prof Exp:* Chief med tech path, St Bartholomew's Hosp, Kent, Eng, 65-69; tech dir, Hotel Dieu Hosp, Windsor, Ont, 69-73; PROF & CHMN DEPT ANTHROP, QUEENS COL, 79- *Mem:* Am Asn Phys Anthropologists; Am Anthrop Asn; Am Soc Hist Med. *Res:* History of physical anthropology and medicine; paleoanthropology; Plio Pleistocene hominid evolution. *Mailing Add:* Dept Anthrop Queens Col Kissena Blvd Flushing NY 11367

SPENCER, FRANK COLE, b Haskell, Tex, Dec 21, 25; c 3. MEDICINE. *Educ:* NTex State Col, BS, 44; Vanderbilt Univ, MD, 47; Am Bd Surg & Bd Thoracic Surg, dipl. *Prof Exp:* Intern surg, Johns Hopkins Hosp, 47-48; asst res surgeon, Univ Calif Med Ctr, Los Angeles, 49-50; from asst resident surg to surgeon, Johns Hopkins Hosp, 53-65, from instr to assoc prof surg, Sch Med, Johns Hopkins Univ, 54-61; prof, Sch Med, Univ Ky, 61-65; PROF SURG & CHMN DEPT, SCH MED, NY UNIV, 65- *Concurrent Pos:* Fel, Sch Med, Johns Hopkins Univ, 48-49; USPHS fel cardiovasc surg, Sch Med, Univ Calif, Los Angeles, 51; Markle scholar; consult, Walter Reed Army Hosp, 57. *Mem:* Soc Univ Surg; Soc Clin Surg; Am Asn Thoracic Surg; Am Surg Asn. *Res:* Cardiovascular and thoracic surgery. *Mailing Add:* Dept Surg NY Univ Sch Med 550 First Ave New York NY 10016

SPENCER, FREDERICK J, b Newcastle-on-Tyne, Eng, June 30, 23; US citizen; m 54; c 2. PREVENTIVE MEDICINE, PUBLIC HEALTH. *Educ:* Univ Durham, MB, BS, 45; Harvard Univ, MPH, 58. *Prof Exp:* Health dir, Va State Dept Health, 56-62, dir bur epidemiol, 62; assoc prof, Med Col Va, Va Commonwealth Univ, 62-63, prof prev med, 63-81, chmn dept, 62-81, asst dean student activ, Sch Med, 78-81, EMER PROF, MED COL VA, VA COMMONWEALTH UNIV, 81- *Concurrent Pos:* Walter Reed lectr, Richmond Acad Med, 66. *Mem:* Am Pub Health Asn. *Res:* Epidemiology and its application in administration of medical care; history of medicine. *Mailing Add:* 560 Caroline Dr Ruther Glen VA 22546

SPENCER, GORDON REED, b Ithaca, NY, June 30, 25; m 63. ELECTRICAL ENGINEERING. *Educ:* Cornell Univ, BSEE, 46. *Prof Exp:* Instr elec eng, Cornell Univ, 47-48; from eng specialist to res specialist electron devices, Philco Corp, 48-60; PRIN ENGR, RAYTHEON CO, 60- *Mem:* AAAS; Am Phys Soc; Inst Elec & Electronic Engrs; Soc Info Display; Sigma Xi. *Res:* Electron optics; application of electron optics to computer-driven displays and electronic imaging; application of spectroscopy to display contrast enhancement and filter design. *Mailing Add:* Res & Develop Dept Raytheon Co 465 Ctr St Quincy MA 02169

SPENCER, GUILFORD LAWSON, II, b Natick, Mass, Feb 21, 23; m 51; c 1. MATHEMATICS. *Educ:* Williams Col, BA, 43; Mass Inst Technol, MS, 48; Univ Mich, PhD(math), 53. *Prof Exp:* Instr math, Univ Md, 51-53; from asst prof to prof, 53-70, FREDERICK LATIMER WELLS PROF MATH, WILLIAMS COL, 70- *Res:* Topology and hyperbolic systems of partial differential equations. *Mailing Add:* 37 Harwood St Williamstown MA 02167

SPENCER, HAROLD GARTH, b Avon Park, Fla, May 19, 30; m 56; c 2. PHYSICAL CHEMISTRY. *Educ:* Univ Fla, BS, 52, MS, 58, PhD(phys chem), 59. *Prof Exp:* From asst prof to assoc prof, 59-68, head dept chem & geol, 66-77, prof chem, 68-88, ALUMNI PROF, CLEMSON UNIV, 88- *Concurrent Pos:* Vis scientist, Imperial Col, 74; AID adv, Instituto Universitario da Beira Interior, Portugal, 82; consult, Vicellon Inc, 74-77, CARRE Inc, 77-88 & Du Pont, 89- *Mem:* Am Chem Soc; Sigma Xi; AAAS; NAm Membrane Soc. *Res:* Physical chemistry of polymers and polymer membranes, structure and properties; transport in polymers; thin film coatings. *Mailing Add:* Dept Chem Clemson Univ Clemson SC 29634-1905

SPENCER, HARRY EDWIN, b Friendship, NY, June 8, 27; m 53; c 4. PHYSICAL CHEMISTRY. *Educ:* Syracuse Univ, BA, 50; Univ Calif, PhD, 54. *Prof Exp:* Chemist, Navy Ord Div, 53-59, res assoc, Res Labs, 59-72, sr res assoc, Res Labs, Eastman Kodak Co, 72-85; VIS PROF & AFFIL SCHOLAR, OBERLIN COL, 85- *Concurrent Pos:* Adj fac chem, Rochester Inst Technol, 63-80; mem, NY State Rating Comt for PhD Prog in Chem, 73-74. *Honors & Awards:* Hon mem, Soc Photographic Sci & Technol Japan. *Mem:* Am Chem Soc; Am Phys Soc; Royal Photog Soc; fel Am Inst Chemists; fel Soc Photog Scientists. *Res:* Radiation chemistry; photoconductivity; infrared detectors; theory of photography; photochemistry. *Mailing Add:* 138 Hollywood St Oberlin OH 44074

SPENCER, HERBERT W, III, b Louisville, Ky, June 12, 45; m 67, 84; c 2. PHYSICS, AIR POLLUTION. *Educ:* Vanderbilt Univ, BA, 67; Auburn Univ, MS, 69, PhD(physics), 74. *Prof Exp:* Teaching asst physics, Auburn Univ, 67-73; res physicist electrostatic precipitators, Southern Res Inst, 74-77; mem staff, Joy Mfg Co, 77-80, mgr, Adv Tech Dept, Western Precipitation Div, 80-88; EXEC VPRES, EC&C TECHNOL, 89-; PRES, HWS ENG & RES CO. *Mem:* Am Phys Soc; Sigma Xi; Air & Waste Mgt Asn; Am Inst Chem Engrs. *Res:* Expert cleaning industrial gases for particulates SO2, NOx using scrubber, electrostatic precipitators and fabric filters. *Mailing Add:* 23629 Mill Valley Valencia CA 91355

SPENCER, HERTA, METABOLISM. *Educ:* Case Western Reserve Univ, MD, 46. *Prof Exp:* CHIEF METAB SECT, VET ADMIN HOSP, 61-; PROF INTERNAL MED, SCH MED, LOYOLA UNIV, 64- *Res:* Mineral and trace element metabolism. *Mailing Add:* Vet Admin Hosp Box 35 Hines IL 60302

SPENCER, HUGH MILLER, b Winfield, Mo, Nov 24, 97; m 36; c 3. PHYSICAL CHEMISTRY. *Educ:* Univ Mo, AB, 19, AM, 21; Univ Calif, PhD(chem), 24. *Prof Exp:* Asst chem, Univ Mo, 18-20, instr phys chem, 20-21; instr, Yale Univ, 24-27; asst prof chem, 27-45, ASSOC PROF CHEM, UNIV VA, 45- *Mem:* AAAS; Am Chem Soc; fel Am Inst Chemists. *Res:* Thermodynamics of solutions; phase equilibria; constant humidity systems; galvanic cells; heat capacity equations; isotopes and nuclear transformations; thermodynamic functions of gases; heat capacities of solids; history of chemistry. *Mailing Add:* New Chem Bldg Univ of Va Charlottesville VA 22901

SPENCER, JACK T, b Mantua, Ohio, Sept 23, 12; m 34; c 1. BOTANY, AGRONOMY. *Educ:* Kent State Univ, BS, 35; Univ Wis, MSc, 36; Ohio State Univ, PhD(bot, agron), 39. *Prof Exp:* Agt maize breeding, Bur Plant Indust, USDA, 36-40; res agronomist forage crops, Univ Ky, 40-42 & 46-49; consult foreign res, USDA, 49-61; prog dir, NSF, 61-68; exec dir, Orgn Trop Studies, 68-72; dir develop, Shippensburg Univ, 72-77; FED LIAISON CONSULT COLS & UNIVS, 78- *Mem:* AAAS; Am Inst Biol Sci; Bot Soc Am; Orgn Trop Studies. *Res:* Plant genetics; tropical science; science administration. *Mailing Add:* 1303 Azalea Lane De Kalb IL 60115-2329

SPENCER, JAMES ALPHUS, b Clayton, Okla, Nov 5, 30; c 3. PLANT PATHOLOGY. *Educ:* Univ Ark, BS, 53, MS, 62; NC State Univ, PhD(plant path), 66. *Prof Exp:* Res asst plant path, Univ Ark, 57-62; agr res technician, NC State Univ, USDA, 62-66; from asst prof to assoc prof, 66-76, PROF PLANT PATH, MISS STATE UNIV, 76- *Mem:* Am Phytopath Soc; Am Soc Hort Sci; Can Plant Path Soc; Am Rose Soc. *Res:* Host-parasite relationships; mycology; woody ornamental plants; diseases; rose disease. *Mailing Add:* Dept Plant Path Miss State Univ PO Box 5328 Mississippi State MS 39762

SPENCER, JAMES BROOKES, b Canton, China, July 16, 26; US citizen; m 48; c 4. HISTORY OF SCIENCE. *Educ:* Lawrence Col, BS, 48; Univ Wis, MS, 56, PhD(hist sci), 64. *Prof Exp:* Res physicist & proj leader, Bjorksten Res Labs, Wis, 54-57; asst prof physics, Augustana Col, Ill, 57-59; asst prof, 63-70, ASSOC PROF HIST SCI, ORE STATE UNIV, 70- *Concurrent Pos:* Vis asst prof, Johns Hopkins Univ, 65-66; NSF res grant, 65-69; vis asst prof, Univ Wis, 69; amanuensis, Niels Bohr Inst, Copenhagen, Denmark, 70-71. *Mem:* Hist Sci Soc. *Res:* History of 19th and early 20th century physical science, particularly magnetooptics and the structure of matter spectroscopy. *Mailing Add:* 5745 NW Oak Creek Dr Corvallis OR 97330

SPENCER, JAMES EUGENE, b Kansas City, Mo, Jan 2, 38; m 64. NUCLEAR & ACCELERATOR PHYSICS, NEURAL NETWORKS & CONTROL THEORY. *Educ:* Mass Inst Technol, BS, 64, PhD(physics), 69. *Prof Exp:* Fel physics, Stanford Univ, 69-71; staff mem, Los Alamos Sci Lab, 71-78; STAFF MEM, STANFORD LINEAR ACCELERATOR CTR, 78- *Concurrent Pos:* Consult, Lawrence Livermore Lab, 70-72; mem tech adv panel, Los Alamos Physics Facil, 73-75; reviewer, Phys Res & Phys Rev Letts, 74-; mem prog adv comt, Ind Cyclotron Lab, 76-78. *Mem:* Am Phys Soc; Sigma Xi. *Res:* Particle physics; magnetic optics; synchrotron radiation; quantum electronics; storage rings; adaptive control theory. *Mailing Add:* Stanford Linear Accelerator Ctr PO Box 4349 Stanford CA 94309

SPENCER, JAMES NELSON, b Rainelle, WVa, Nov 11, 41. PHYSICAL CHEMISTRY. *Educ:* Marshall Univ, BS, 63; Iowa State Univ, PhD(phys chem), 67. *Prof Exp:* Student chemist, Int Nickel Co, WVa, 61-63; assoc prof chem, Lebanon Valley Col, 67-80; MEM FAC, CHEM DEPT, FRANKLIN & MARSHALL COL, 80- *Mem:* Am Chem Soc. *Res:* Hydrogen bonding; thermodynamic properties of solutions. *Mailing Add:* 107 S Duke St Lancaster PA 17602

SPENCER, JAMES W(ENDELL), b Ithaca, NY, Aug 3, 27; m 46; c 3. EDUCATION ADMINISTRATION. *Educ:* Cornell Univ, BCE, 49, MCE, 51; Stanford Univ, PhD, 67. *Prof Exp:* Instr civil eng, 49-51, from asst prof to prof, 51-87, vdir coop exten, 70-73, assoc dean, NY State Col Agr & Life Sci, 73-78, spec asst to pres, 78-79, vprovost, 79-87, EMER PROF AGR ENG, CORNELL UNIV, 87- *Concurrent Pos:* Field engr, D J Belcher & Assocs, NY, 54; lectr & assoc res engr, Inst Transp & Traffic Eng, Calif, 57-58; consult, NY State Temp Comn Agr, 58 & Ford Found, Colombia, SAm, 63; NSF sci faculty fel, Stanford Univ, 64-65. *Res:* Highway engineering; engineering-economic planning. *Mailing Add:* 1071 Taughannock Blvd Ithaca NY 14850

SPENCER, JESSE G, b Farmville, NC, Apr 10, 35; m 73; c 3. PHYSICAL CHEMISTRY, INORGANIC CHEMISTRY. *Educ:* Univ NC, BS, 57; Univ Va, MS, 59, PhD(chem), 62. *Prof Exp:* Res assoc chem, Univ NC, 61-62; from asst prof to assoc prof, Univ Charleston, 62-68, head dept, 65-76, prog dir, Med Lab Technol, 75-79, chmn, Div Health Serv, 76-79, prof chem, 68-84, dir, comput serv & records, 81-84; PROF CHEM & HEAD DEPT, VALDOSTA STATE COL, 84- *Mem:* Am Chem Soc; Sigma Xi. *Res:* Solution thermochemistry and polarography of transition metals in aqueous and nonaqueous media; trace analysis. *Mailing Add:* Dept Chem Valdosta State Col Valdosta GA 31698

SPENCER, JOHN EDWARD, b Panama, CZ, Mar 22, 49; US citizen; m 75; c 1. PHYSICAL CHEMISTRY. *Educ:* Millsaps Col, BS, 71; Rice Univ, PhD(chem), 75. *Prof Exp:* Res assoc chem, Univ Calif, Irvine, 75-77; res chemist, Lighting Bus Group, Gen Elec Co, 77-80; MEM TECH STAFF, TEX INSTRUMENTS INC, 80- *Mem:* Am Chem Soc; AAAS; Electrochem Soc. *Res:* Plasma etching of semiconductor thin films; plasma spectroscopy; high temperature chemistry and combustion. *Mailing Add:* 1213 Balboa Circle Plano TX 75075

SPENCER, JOHN FRANCIS THEODORE, b Magrath, Alta, Jan 18, 22; m 45, 73; c 6. MICROBIOLOGY, MICROBIAL GENETICS OF YEASTS. *Educ:* Univ Alta, BSc, 49, MSc, 51; Univ Sask, PhD(chem), 55. *Prof Exp:* Asst res officer, 51-59, assoc res officer, 59-63, sr res officer, Eng & Process Develop Sect, Prairie Regional Lab, Nat Res Coun Can, 63-74; RES ASSOC, DEPT BIOL, GOLDSMITH'S COL, UNIV LONDON & THAMES POLYTECH, LONDON, 74- *Concurrent Pos:* Mem, Int Yeast Coun, 69- *Mem:* AAAS; Am Soc Microbiol; Can Soc Microbiol. *Res:* Genetic improvement of industrial yeasts using protoplast fusion as a principal tool; intergenetic fusion of baking yeasts with osmotolerent species; role of mitochondria as well as nuclear genomes in yeast performance; yeast genetics & molecular biology. *Mailing Add:* PRO1M1 Ave Belgrano Y Pasajecasteros San Miguel De Tucuman 4000 Argentina

SPENCER, JOHN HEDLEY, b Stapleford, Eng, Apr 10, 33; Can citizen; m 58; c 3. BIOCHEMISTRY. *Educ:* St Andrews Univ, BSc, 55, Hons, 56; McGill Univ, PhD(biochem), 60. *Prof Exp:* Res asst biochem, McGill Univ, 56-59; Damon Runyon Mem Fund Cancer Res vis fel, Columbia Univ, 59-61, res assoc, 61; lectr & teaching fel, McGill Univ, 61-63, from asst prof to prof biochem, 63-78; head dept, 78-90, PROF BIOCHEM, QUEEN'S UNIV, 78- *Concurrent Pos:* Sci Officer biochem comt, Med Res Coun Can, 73-79; mem grants panel, Nat Cancer Inst, 75-79; mem, adv panel Collab Res Grants Prog, NATO, 86-89; chmn, 88-89, mem, Comité Scientifique, Programme Des Actions Structurantes, FONDS, FCAR. *Honors & Awards:* Ayerst Award, Can Biochem Soc, 72. *Mem:* Can Biochem Soc (treas, 66-69, vpres, 78-79, pres, 79-80); Am Soc Biochem & Molecular Biol; AAAS; Brit Biochem Soc; Can Fed Biol Soc (vpres, 80-81, pres, 81-82); Sigma Xi; fel Royal Soc Can. *Res:* Chemistry and primary structure of DNA; gene expression transcription and control. *Mailing Add:* Dept Biochem Queen's Univ Kingston ON K7L 3N6 Can

SPENCER, JOHN LAWRENCE, b Sanford, Fla, Sept 10, 32; m 54, 79; c 2. TECHNICAL SERVICE, QUALITY CONTROL. *Educ:* DePauw Univ, AB, 54; Univ Mich, MS, 56, PhD(isoxazolines), 58. *Prof Exp:* Org chemist, Lederle Lab, Am Cyanamid Co, 58-60; sr org chemist, 60-67, res scientist, 67-71, mgr antibiotic prod technol, 71-78, mgr sterile operations, 78-80, mgr tech serv & qual control, 78-85, res assoc, 85-86, sr res scientist, Eli Lilly & Co, 86-88; prof chem, 88-90, JESSE BALL DUPONT PROF NAT SCI, FLA SOUTHERN COL, 90- *Mem:* Am Chem Soc; Parenteral Drug Asn. *Res:* Synthesis of heterocyclic systems, particularly isoxazolines, oxadiazoles, benzodiazepines and quinazolines; antibiotic modifications, particularly tetracyclines, penicillins and cephalosporins; antibiotic manufacturing including fermentation, purification, and bulk parenteral operations; analytical supervision and quality control supervisions; vancomycin process improvements production scale high pressure liquid chromatograph. *Mailing Add:* Dept Chem Fla Southern Col 111 Lake Hollingsworth Dr Lakeland FL 33801-5698

SPENCER, JOSEPH WALTER, b Salt Lake City, Utah, May 24, 21; m 48; c 5. GEOPHYSICS. *Educ:* Brigham Young Univ, BS, 47; Pa State Univ, PhD(physics), 52. *Prof Exp:* Asst math, Brigham Young Univ, 47-48; asst physics, Pa State Univ, 48-49, asst acoust, 49-50; res physicist, Calif Res Corp, Standard Oil Co Calif, 52-54, proj leader geophys, 54-56, group supvr, 56-60; sr geophysicist, Calif Oil Co, 60-65, staff geophysicist, Western Div, Chevron Oil Co, 65-69, chief geophysicist, Western Div, Chevron Oil Co, 69-77, CHIEF GEOPHYSICIST, CENT REGION, CHEVRON USA, 77- *Mem:* Soc Explor Geophysicists; Am Geophys Union. *Res:* Fluid dynamics; viscous behavior of high molecular weight hydrocarbons; wave propagation in earth materials; interpretation of geophysical data; electronic computer applications. *Mailing Add:* 7505 S Steele St Littleton CO 80122

SPENCER, LARRY T, b Palo Alto, Calif, Oct 15, 41; m 64; c 3. AQUATIC ECOLOGY, INVERTEBRATE ZOOLOGY. *Educ:* Brigham Young Univ, BS, 63; Ore State Univ, MA, 65; Colo State Univ, PhD(zool), 68; Univ Calif, MLS, 75. *Prof Exp:* Teaching asst zool, Brigham Young Univ, 62-63, Ore State Univ, 65-67 & Colo State Univ, 65-67; from instr to assoc prof biol, 67-78, PROF BIOL, PLYMOUTH STATE COL, 78- *Concurrent Pos:* Vis prof, Univ Hawaii, 82. *Mem:* AAAS; Am Soc Limnol & Oceanog; Ecol Soc Am; Am Soc Zoologists; Hist Sci Soc. *Res:* Population biology of marine and fresh water invertebrates; biology of cephalopod mollusks; history of biology and American science; exploration and settlement of the Trans-Mississippi West; evolution, its impact on biological and intellectual thought; data base management and information retrieval in the biological sciences. *Mailing Add:* Dept Natural Sci Plymouth State Col Plymouth NH 03264

SPENCER, LORRAINE BARNEY, b Ogden, NY, Jan 26, 24; m 42; c 4. PHYCOLOGY. *Educ:* Guilford Col, BS, 66; Wake Forest Univ, MA, 70, PhD(biol), 73. *Prof Exp:* Asst instr, Wake Forest Univ, 68-72, res fel, 72-73; ASSOC PROF BIOL, ST AUGUSTINE'S COL, 74- *Concurrent Pos:* Adj prof biol, Guilford Col Urban Ctr, 74. *Mem:* Am Inst Biol Sci; Sigma Xi. *Res:* Biosystematics of Zephyranthes. *Mailing Add:* 315 White Oak Dr Cary NC 27513

SPENCER, MARY STAPLETON, b Regina, Sask, Oct 4, 23; m 46; c 1. BIOCHEMISTRY. *Educ:* Univ Sask, BA, 45; Bryn Mawr Col, MA, 46; Univ Calif, PhD(agr chem), 51. *Prof Exp:* Chemist, Ayerst, McKenna & Harrison, Ltd, 46-47; chemist, Nat Canners Asn, Calif, 48-49; asst food chem, Univ Calif, 50-51, instr, 51-53; from asst prof to assoc prof biochem, 53-64, actg head dept biochem, 60-61, prof plant sci, 64- 83, PROF, UNIV ALTA, 84- *Concurrent Pos:* Mem, Nat Res Coun Can, 70-76; mem, Natural Sci & Eng Res Coun Can, 86-; mem bd gov, Univ Alta, 76-79; mem, Task Force on Post-Secondary Educ, Alberta Govt Comn on Educ Planning, 70-72. *Mem:* Am Soc Plant Physiol Biochem Soc. *Res:* Metabolism of aging tissue; biology of ethylene, its effects on plants, animals, microorganisms, its relationship to aging and to plant productivity, biogenesis; cyanide-resistant respiration; post-harvest physiology of fruits and vegetables. *Mailing Add:* Dept Plant Sci Univ Alta Edmonton AB T6G 2M7 Can

SPENCER, MAX M(ARLIN), b Rocky Ford, Colo, Jan 10, 35; m 55; c 3. MECHANICAL ENGINEERING. *Educ:* Okla State Univ, BS, 56, MS, 57, PhD(eng), 60. *Prof Exp:* Res asst & lectr eng, Okla State Univ, 56-60; assoc engr, Boeing Co, Kans, 57, faculty assoc adv design group, 58, res specialist, 60; res specialist & stress consult, Ballistics Res Labs, Aberdeen Proving Ground, Md, 60-61; res specialist, Boeing Co, 61-62, stress res group chief, 62-63, res specialist & fatigue group head, 63-69, sr eng supvr, 69-78, struct technol mgr & stress unit chief, 78-80, struct technol chief, 80-87, tech chief, 87-88, CHIEF ENGR STRUCT, BOEING CO, 88- *Concurrent Pos:* Guest lectr, Univ Wichita, 62; designated eng rep, Fed Aviation Admin, 74. *Mem:* Am Soc Mech Engrs. *Res:* Aircraft structural analysis; stress and fatigue. *Mailing Add:* 12039 SE 20th Bellevue WA 98005

SPENCER, MERRILL PARKER, b Pawnee, Okla, Feb 27, 22; m 44; c 4. CARDIOVASCULAR PHYSIOLOGY. *Educ:* Baylor Univ, MD, 45. *Prof Exp:* Intern, Herman Hosp, Tex, 45-46; med resident, Crile Vet Admin Hosp, 50-51; instr physiol & pharmacol, Bowman Gray Sch Med, 51-54, asst prof physiol, 54-59, assoc prof physiol & pharmacol, 59-63; dir, Va Mason Res Ctr, 63-71; pres & dir, Inst Appl Physiol & Med, 71-86; MED DIR, PAC VASCULAR INC, 87- *Concurrent Pos:* USPHS fel, Western Reserve Univ, 48-50; mem coun circulation & coun basic sci, Am Heart Asn; pres, Oceanographic Inst Wash. *Mem:* Am Physiol Soc; Am Heart Asn. *Res:* Medical electronics; cardiopulmonary medicine. *Mailing Add:* Inst Appl Physiol & Med 701 16th Ave Seattle WA 98122

SPENCER, PAUL ROGER, b Madison, Wis, Dec 24, 41; wid; c 2. HEAD-DISK INTERFACE RELIABILITY & FAILURE ANALYSIS, COMPUTER PRINTING TECHNOLOGY. *Educ:* Wash State Univ, BS, 63; Univ Ill, MS, 65, PhD(physics), 69. *Prof Exp:* Sci co-worker electron-nuclear double resonance, Second Phys Inst, Stuttgart, Ger, 69-70; mem res staff, electrophotog, Xerographic Technol Lab & Cent Res Lab, Xerox Corp, Webster, NY, 70-79; MEM TECH STAFF, ELECTROPHOTOG DISK DRIVE ENG, BOISE PRINTER DIV & DISK MECHANISM DIV, HEWLETT-PACKARD CORP, BOISE, IDAHO, 79- *Mem:* Am Phys Soc. *Res:* Electrophotographic printing, ink jet printing, specialized sensor development and head-disk interface reliability and failure analysis; four professional articles and eight patents in electrophotography and ink jet printing. *Mailing Add:* 6045 Becky Dr Meridian ID 83642-5333

SPENCER, PETER SIMNER, b London, Eng, Nov 30, 46; US citizen; m 69; c 2. NEUROBIOLOGY, NEUROTOXICOLOGY. *Educ:* Univ London, BSc, 68, PhD(path), 71. *Prof Exp:* Res asst, Nat Hosp Nervous Dis, Univ London, UK, 68-70, res fel, Royal Free Hosp Sch Med, 70-71; fel path, Albert Einstein Col Med, 71-73, asst prof, 73-81, assoc prof neurosci, 77-83, prof neurosci, 83-, assoc prof path & dir, Inst Neurotoxicol, 79-; DIR, ORE HEALTH SCI UNIV. *Concurrent Pos:* Joseph P Kennedy Jr Found fel, 74-76; consult, Nat Inst Occup Safety & Health, 76-77 & Environ Protection Agency, 77-; assoc ed, J Neurocytol, 77-; chmn adv bd, J Neurotoxicol, 78-; mem adv bd, Rutgers Univ Toxicol Prog, 84, Howe & Assocs, 85 & Peripheral Nerve Repair & Regeneration, 85; mem, Bd Toxicol & Environ Health Hazards, Nat Acad Sci, 84, Safe Drinking Water Comt, 85; secy, Third World Med Res Found, 85- *Honors & Awards:* Weil Award, Am Asn Neuropathologists, 76. *Mem:* Am Asn Neuropathologists; Am Soc Cell Biol; AAAS; Anat Soc Gt Brit & Ireland; British Neuropath Soc; World Fedn Neurol; Royal Col Pathologists; hon mem Pan-Am Neuroepidiology Found. *Res:* Cellular relationships in the nervous system and the effects of neurotoxic chemicals. *Mailing Add:* Oregon Health Sci Univ 3181 SW Sam Jackson Park Rd L606 Portland OR 97201

SPENCER, RALPH DONALD, b Kolambugan, Philippines, July 22, 20; US citizen; m 47; c 2. ORGANIC CHEMISTRY. *Educ:* Col Wooster, BA, 41; Stanford Univ, MA, 42; Cornell Univ, PhD(org chem), 47. *Prof Exp:* Du Pont fel chem, Cornell Univ, 47-48, Goodrich Tire & Rubber Co Proj res assoc, 50-51; chemist, Pineapple Res Inst, Hawaii, 48-50; chemist, E I du Pont de Nemours & Co, 51-57; assoc prof, 67-85, SR FEL, MELLON INST, CARNEGIE-MELLON UNIV, 57-, EMER PROF CHEM, 85- *Mem:* Am Chem Soc. *Res:* New approach to desalination of sea water; daylily research and hybridizing. *Mailing Add:* 1207 N Western Dr Monroeville PA 15146-4403

SPENCER, RANDALL SCOTT, b Sept 29, 37; US citizen; m 66; c 2. PALEONTOLOGY, STRATIGRAPHY. *Educ:* Univ Wis-Madison, BS, 60; Univ Kans, MS, 62, PhD(geol), 68. *Prof Exp:* From asst prof to assoc prof, Old Dominion Univ, 66-77, asst chmn dept, 74-76, assoc dean, 78-81, PROF GEOL SCI, OLD DOMINION UNIV, 78-, CHMN DEPT, 81- *Concurrent Pos:* NSF grant, 70-; fel Cushman Found. *Mem:* Fel Geol Soc Am; Soc Econ Paleontologists & Mineralogists. *Res:* Upper Paleozoic brachiopods; Cenozoic foraminifera of Atlantic Coast; statistical studies in brachiopod evolution; marine Pleistocene stratigraphy and fauna of the mid-Atlantic seaboard. *Mailing Add:* Dept Geol Sci Old Dominion Univ Norfolk VA 23508

SPENCER, RICHARD L, b Dunlap, Iowa, Feb 11, 34; m 59; c 3. BIOCHEMISTRY. *Educ:* Fresno State Col, AB, 56; San Jose State Col, MS, 63; Univ Calif, Davis, PhD(biochem), 66. *Prof Exp:* Asst chemist, Calif Chem Co, 56-59; res assoc biochem, Univ Ill, Urbana, 65-66; NIH fel, Univ Minn, 66-68; from asst prof to assoc prof, 68-74, PROF CHEM, SOUTHWEST STATE UNIV, 74- *Mem:* Am Chem Soc. *Res:* Enzyme chemistry. *Mailing Add:* Dept of Chem Southwest State Univ Marshall MN 56258

SPENCER, RICHARD PAUL, b New York, NY, June 7, 29; m 56; c 3. NUCLEAR MEDICINE, BIOCHEMISTRY. *Educ:* Dartmouth Col, AB, 51; Univ Southern Calif, MD, 54; Harvard Univ, MA, 58, PhD(biochem), 61. *Hon Degrees:* MA, Yale Univ, 68. *Prof Exp:* From asst prof to assoc prof biophys, Univ Buffalo, 61-63; from assoc prof to prof nuclear med, Sch Med, Yale Univ, 68-74; PROF NUCLEAR MED & CHMN DEPT, SCH MED, UNIV CONN HEALTH CTR, FARMINGTON, 74 - *Concurrent Pos:* NSF fel, Harvard Univ, 57-58, Helen Hay Whitney fel, 58-60. *Mem:* AAAS; Am

Physiol Soc; Soc Nuclear Med; Biophys Soc. *Res:* Organ structure and function as studied by radiopharmaceuticals; models of biological growth and differentiation; intestinal metabolism and transport. *Mailing Add:* Dept of Nuclear Med Univ of Conn Health Ctr Farmington CT 06030

SPENCER, SELDEN J, b Towanda, Pa, Apr 28, 23; m 51; c 2. BIOLOGICAL SCIENCES. *Educ:* Mansfield State Col, BS, 48; Pa State Univ, MEd, 52, DEd(biol sci), 62. *Prof Exp:* High sch teacher, Pa, 48-49; chem technician, Sylvania Elec Prod, Inc, 50-52; admin asst, Educ Off, Pa State Col, 52-53, admin asst educ film res prog, 53-54; high sch teacher, Pa, 54-57; mem, Inst Sci Teachers, Pa, 57-58; prof biol, Goddard Col, 58-63; from asst prof to assoc prof, State Univ NY Col New Paltz 63-88 chmn dept, 72-78; RETIRED. *Res:* Bird banding research; bank swallow nesting sites; Arctic wilderness exploration; West Indian cushion starfish. *Mailing Add:* 55 DuBois Rd New Paltz NY 12561

SPENCER, TERRY WARREN, b Los Angeles, Calif, Feb 10, 30; m 53; c 2. GEOPHYSICS. *Educ:* Univ Calif, Los Angeles, AB, 51; Calif Inst Technol, PhD(geophys), 56. *Prof Exp:* Sr res physicist, Chevron Res Co, Calif, 56-66; chmn dept, 66-77, PROF GEOPHYS, TEX A&M UNIV, 66- *Mem:* Am Geophys Union; Soc Explor Geophys. *Res:* Elastic wave propagation theory; petroleum scesmology. *Mailing Add:* Dept Geophys Tex A&M Univ College Station TX 77843

SPENCER, THOMAS, b Dec 24, 46. MATHEMATICS. *Educ:* Univ Calif, Berkeley, BA, 68; NY Univ, PhD, 72. *Prof Exp:* Vis mem, Inst Advan Study, Princeton Univ, 72-74; fel, Harvard Univ, 74-75; assoc prof, Rockefeller Univ, 75-77; prof, Rutgers Univ, 78-80; prof, Courant Inst Math, NY Univ, 80-86; PROF, INST ADVAN STUDY, PRINCETON UNIV, 86- *Concurrent Pos:* Sloan fel. *Mailing Add:* RD 1 Box 425A Bergen Ave Princeton NJ 08540

SPENCER, THOMAS A, b Orange, NJ, Mar 31, 34; m 56; c 4. ORGANIC CHEMISTRY. *Educ:* Amherst Col, AB, 56; Univ Wis, PhD(chem), 60. *Prof Exp:* Res assoc chem, Univ Wis, 60; from instr to prof & chmn dept, 60-72, NEW HAMPSHIRE PROF CHEM, DARTMOUTH COL, 72- *Concurrent Pos:* Alfred P Sloan Found res fel, 65-68; mem Grants Prog Adv Comt, Res Corp, 72-78. *Mem:* AAAS; Am Chem Soc. *Res:* Organic chemical synthesis; natural products, particularly terpenoids; steroid biosynthesis; biochemical reaction mechanisms. *Mailing Add:* Dept Chem Dartmouth Col Hanover NH 03755

SPENCER, THOMAS H, METALLURGY, HEAT TREATMENTS. *Educ:* Bradley Univ, BS. *Prof Exp:* Prod apprentice & HT supv, Caterpillar Tractor Co, East Peoria Plant, 36-48, heat treatment supt, 48-52, asst plant metallurgist, 52-55, heat treatment mgr, 55, plant metallurgist, 55-57, qual control mgr, 57-70, mfg & mat develop mgr, 70-79; CONSULT, CAM CONSULT INC, 80- *Concurrent Pos:* Mem Iron & Steel Tech Comt Panel D, Soc Automotive Engrs, 55-77, chmn, 64-65, Mfg Standards Comt, Soc Mfg Engrs, 72-74, ad hoc educ planning comt, 75, eng educ comt, 76, Comt Computeraided Mfg, Nat Acad Sci, 77-80; chmn, Iron & Steel Tech Comt, Div 13, 66-76, Mat Coun, 76; adv, Air Force Integrater Computeraided Mfg Prog, 77, Nat Prod Ctr, Washington, DC, 78-79, Soft-Tech, 78. *Mem:* Fel Am Soc Metals; fel Soc Automotive Engrs; Am Welding Soc; Am Soc Qual Control; Am Inst Metall Engrs; Soc Mfg Engrs. *Res:* Materials development; ferrous & non-ferrous, plastics, powders and coatings; management information systems. *Mailing Add:* Rancho Viejo 17 Cortez Brownsville TX 78520

SPENCER, WALTER WILLIAM, b Mansfield, Ohio, Nov 10, 33; m 64; c 2. CLINICAL CHEMISTRY. *Educ:* Heidelberg Col, BS, 55; Purdue Univ, West Lafayette, MS, 58, PhD(biochem), 60. *Prof Exp:* Purdue Res Found fel, Purdue Univ, West Lafayette, 60-61; CLIN CHEMIST, ST ELIZABETH MED CTR, 61-, ADMIN DIR LAB SERV, 85- *Concurrent Pos:* Clin asst prof, Univ Dayton, 70-74, clin assoc prof, 74-80; treas, Clin Chem Consult, Inc, 78; consult, Va Hosp, Dayton, 84- *Honors & Awards:* Katchman Award, Ohio Valley Sect, Am Asn Clin Chem, 78. *Mem:* Fel AAAS; Am Asn Clin Chemists; Sigma Xi. *Res:* Development of new procedures for use in the field of clinical chemistry. *Mailing Add:* Clin Lab St Elizabeth Med Ctr Dayton OH 45408

SPENCER, WILLIAM ALBERT, b Oklahoma City, Okla, Feb 16, 22; m 45; c 2. REHABILITATION, PEDIATRICS. *Educ:* Georgetown Univ, BS, 42; Johns Hopkins Univ, MD, 46; Am Bd Pediat, dipl, 55. *Prof Exp:* Intern, Hopkins Hosp, 46-47, resident, 47-48; pres, Inst Rehab & Res, Tex Med Ctr, 59-87; from instr to asst prof pediat, Baylor Col Med, 50-57, asst prof physiol, 54-57, prof rehab & chmn dept, 57-89, EMER PROF DEPT REHAB, BAYLOR COL MED, 89. *Concurrent Pos:* Med dir, Southwestern Poliomyelitis Respiratory Ctr, 50-59; Horowitz vis prof, Inst Phys Med & Rehab, 64; asst attend physician, Ben Taub Gen Hosp; mem active staff, Tex Children's Hosp; mem consult staff, M D Anderson Hosp & Tumor Inst Houston; mem courtesy staff, St Luke's Hosp; chmn spec med adv group to Vet Admin, 74-75; dep dir & actg dir, Nat Inst Handicapped Res, Washington, DC, 79-80, intermittent consult, 79-; mem, Panel Testing Handicapped, Nat Acad Sci, 80-81, comt health care racial-ethnic minorities & handicapped persons, 80-81; mem sci adv bd, Paralyzed Vet Am Technol & Res Found, 81-; mem, Sci Merit Rev Bd, Vet Admin Rehab, Res & Develop, 81. *Honors & Awards:* Gold Medal, Int Cong Phys Med, 72; Gold Key Award, Am Cong Rehab Med, 72, Coulter Award, 78. *Mem:* Inst Med-Nat Acad Sci; AMA; Am Acad Pediat; Am Physiol Soc; Am Cong Rehab Med (pres, 68-69); AAAS; Nat Rehab Asn; NY Acad Sci; Asn Comput Mach; Sigma Xi. *Res:* Development of principles of rehabilitation medicine; application of electronic technology to research in disabling chronic disease and injuries; planning health services for disabled at community and national level. *Mailing Add:* Dept Rehab Baylor Col Med Houston TX 77025

SPENCER, WILLIAM F, b Carlinville, Ill, Mar 4, 23; m 46; c 3. SOIL CHEMISTRY. *Educ:* Univ Ill, BS, 47, MS, 50, PhD(agron), 52. *Prof Exp:* Asst soil physics, Univ Ill, 48-49; asst chemist, Citrus Exp Sta, Univ Fla, 51-54; soil scientist, Agr Res Serv, USDA, Wyo, 54-55 & Calif, 55-57; assoc soil chemist, Citrus Exp Sta, Univ Fla, 57-62; SUPVRY SOIL SCIENTIST, AGR RES SERV, USDA, 62- *Concurrent Pos:* Consult, Cent Univ Venezuela, 59. *Honors & Awards:* Fel AAAS; Fel Am Soc Agron. *Mem:* AAAS; Am Chem Soc; fel Soil Sci Soc Am; Am Soc Agron; Soc Environ Toxicol Chem; Ctr Appln Sci & Technol. *Res:* Soil chemistry of pesticides, nutrient enrichment, waste disposal on land as related to water quality and vapor behavior of pesticides and other toxic organic chemicals in the environment. *Mailing Add:* USDA Agr Res Serv Univ Calif 1278 Geol Bldg Riverside CA 92521

SPENCER, WILLIAM J, b Kansas City, Mo, Sept 25, 30; m 53; c 2. SOLID STATE PHYSICS. *Educ:* William Jewell Col, AB, 52; Kans State Univ, MS, 56, PhD(physics), 60. *Hon Degrees:* Dr, William Jewel Col, 69. *Prof Exp:* Mem tech staff, Bell Tel Labs, Pa, 59-60, supvr, Piezoelec Devices Group, 60-68, head Piezoelec Devices Dept, 68-72, dir, univ rels & tech employ, 72-73,; dir microelectronics, Sandia Labs, 73-78, systs develop, Livermore, 78-81; mgr, Integrated Circuit Lab, Xerox Corp, Palo Alto Res Ctr, 81-82, vpres, Sci Ctr, 82-86, mgr, 83-86, group vpres & sr tech officer, Corp Res Group, 86-90; RES PROF MED, SCH MED, UNIV NMEX, 78-; PRES & CHIEF EXEC OFFICER, SEMATECH, 90- *Concurrent Pos:* Pres, Solid State Circuits Coun, Inst Elec & Electronics Engrs, 78-79; mem, Computer Sci & Technol bd, 88-90 & Executone Bd Dirs meeting, 89- *Honors & Awards:* Electronic 100 Award, 63; C B Sawyer Award, 72. *Mem:* Nat Acad Eng; Am Phys Soc; fel Inst Elec & Electronics Engrs; AAAS; Sigma Xi. *Res:* Integrated circuits design and processing; biomedical applications. *Mailing Add:* Sematech 2706 Montopolis Dr Austin TX 78741

SPENDLOVE, JOHN CLIFTON, b Provo, Utah, Dec 24, 25; m 44; c 5. BACTERIOLOGY. *Educ:* Brigham Young Univ, BS, 49, MS, 50; Ohio State Univ, PhD(bact), 53. *Prof Exp:* Bacteriologist & chief agents biol br, Biol Warfare Assessment Lab, Chem Corps, Dugway Proving Ground, 53-57, opers res analyst, Opers Res Group, Army Chem Ctr, 57-62, tech dir planning & eval directorate, Deseret Test Ctr, 62-72, chief biol defense div, Plans & Studies Directorate, 72-75, chief Environ & Life Sci Div, Materiel Test Directorate, US Dept Army, Dugway Proving Ground, 75-82,; INDEPENDENT CONSULT, 87- *Mem:* AAAS; Am Soc Microbiologists; Sigma Xi; NY Acad Sci; Soc Indust Microbiol. *Res:* Aerobiology; decontamination; pathogenic bacteriology; biological and chemical warfare operations research; hazardous microbial aerosols in the environment. *Mailing Add:* 4166 Fortuna Way Holladay UT 84124

SPENDLOVE, REX S, b Hoytsville, Utah, Apr 29, 26; m 49; c 5. VIROLOGY, IMMUNOLOGY. *Educ:* Brigham Young Univ, BS, 50, MS, 52; Ohio State Univ, PhD, 55; Am Bd Med Microbiol, dipl. *Hon Degrees:* Doctorate, Utah State Univ, 89. *Prof Exp:* Instr microbiol, Univ Conn, 55-58; res microbiologist, Viral & Rickettsial Dis Lab, Calif State Dept Pub Health, 58-66; head dept bact & pub health, Utah State Univ, 66-73, prof virol, 66-81; RETIRED. *Concurrent Pos:* Mem ed bd, Excerpta Medica; mem reovirus study group, Vert Virus Subcomt, Int Comn Nomenclature Viruses; pres, HyClone Labs, Inc, 75- *Mem:* AAAS; Am Soc Microbiol; NY Acad Sci; Soc Exp Biol & Med; Am Asn Immunol. *Res:* Reovirus replication and genetics; affinity of reovirus for host cell microtubules; enhancement of reovirus infectivity by capsid removal; effect of proteolytic enzymes on viral structure; viral pollution of water; Rotavirus Gastroenteritis. *Mailing Add:* 725 S State Hwy 89-91 Logan UT 84321

SPENGER, ROBERT E, b Oakland, Calif, Sept 20, 24; m 59; c 1. ORGANIC CHEMISTRY. *Educ:* Univ Calif, Berkeley, AB, 54; Univ Calif, Los Angeles, PhD(org chem), 62. *Prof Exp:* Chemist, Radiation Lab, Univ Calif, 54-57; asst, Univ Calif, Los Angeles, 57-62, asst res chemist, Univ Calif, Riverside, 62-64; asst prof, 64-74, PROF CHEM, CALIF STATE UNIV, FULLERTON, 74- *Mem:* AAAS; Am Chem Soc. *Res:* Synthesis of isotopically-labelled compounds; synthesis of organo-arsenic compounds. *Mailing Add:* Dept of Chem Calif State Univ Fullerton CA 92634

SPENGLER, KENNETH C, b Harrisburg, Pa. METEOROLOGY. *Educ:* Dickinson Col, BA, 36; Mass Inst Technol, MS. *Hon Degrees:* DSc, Univ Nev, 66. *Prof Exp:* Exec dir, 46-88, EMER EXEC DIR, AM METEOROL SOC, 88- *Mem:* AAAS; Coun Eng & Sci Soc; Am Inst Aeronaut & Astronaut; Am Geophys Union. *Mailing Add:* Am Meteorol Soc 45 Beacon St Boston MA 02108

SPENGOS, ARIS C(ONSTANTINE), engineering, for more information see previous edition

SPENNER, FRANK J(OHN), b Riverside, Iowa, July 4, 01; m 39; c 1. ELECTRICAL ENGINEERING. *Educ:* Univ Iowa, BS, 24, MS, 27. *Prof Exp:* Asst foreman, Potter Condenser Co, Ill, 27-28; asst engr, Western Elec Co, 28-32; surveyor, US Coast & Geod Surv, Iowa, 33-34 & Johnson Co, Iowa, 36-39; instr eng, drawing & math, Trinidad State Jr Col, 41-47; from asst prof to assoc prof eng, 47-63, assoc prof elec eng, 63-71, EMER ASSOC PROF ENG DRAWING, UNIV WYO, 71- *Mem:* Inst Elec & Electronics Engrs. *Res:* Design, inspection and development on equipment for measuring electrical characteristics of cables, coils and condensers; basic circuits. *Mailing Add:* 561 N Seventh St Laramie WY 82070

SPENNY, DAVID LORIN, b Covington, Ky, Nov 5, 43; m 82. PHYSICS EDUCATION, UNDERGRADUATE RESEARCH. *Educ:* Wittenberg Univ, BS, 65; Univ Colo, Boulder, PhD(physics), 70. *Prof Exp:* Lectr physics, Univ Colo, Denver, 70-71; asst prof phys sci, Univ Colo, Boulder, 71-74; asst prof, NMex Highlands Univ, 74-75; asst prof physics, Bemidji State Univ, 76-79; assoc prof, 80-89, PROF PHYSICS, UNIV SOUTHERN COLO, 89- *Mem:* Am Asn Physics Teachers; AAAS. *Res:* Supervise undergraduate student research in fields of physics, especially optics. *Mailing Add:* Dept of Physics Univ Southern Colo Pueblo CO 81001-2034

SPENSER, IAN DANIEL, b Vienna, Austria, June 17, 24; m 51; c 2. BIO-ORGANIC CHEMISTRY. *Educ:* Univ Birmingham, BSc, 48; Univ London, PhD(biochem), 52, DSc, 69. *Prof Exp:* Demonstr biochem, Kings Col, Univ London, 48-52, asst lectr biochem & chem, Med Col, St Bartholomew's Hosp, 52-54, lectr, 54-57; from asst prof to assoc prof biochem, 57-64, PROF CHEM, MCMASTER UNIV, 64- *Concurrent Pos:* Fel, Nat Res Coun Can, 53-54; vis prof, Lab Org Chem, Eidgenössische Tech Hochschule, Zürich, 71, 89; vis prof, Inst Org Chem Technol Univ Denmark, 77, Inst Org Chem Univ Karlsruhe, Fed Rep Ger, 81 & Inst Pharmaceut Biol Univ Bonn, Fed Rep Ger, 89. *Honors & Awards:* Sr Scientist Award, NATO, 80; Can (Nat Sci & Eng Res Coun Can)-Japan (Japan Soc Prom Sci) Exchange Award, 82 & 83; John Labatt Ltd Award, Chem Inst Can, 82 & 83. *Mem:* Am Soc Biol Chemists; fel Chem Inst Can; Royal Soc Chem; Brit Biochem Soc; fel Royal Soc Can; Phytochem Soc NAm; Am Soc Pharmacog. *Res:* Biosynthesis of alkaloids and of B vitamins; chemistry and metabolism of amino acids. *Mailing Add:* Dept of Chem McMaster Univ Hamilton ON L8S 4M1 Can

SPERA, FRANK JOHN, b Philadelphia, Pa, Dec 6, 50; m 77. MAGMA TRANSPORT, IRREVERSIBLE THERMODYNAMICS. *Educ:* Franklin & Marshall Col, BA, 72; Univ Calif, Berkeley, BA, 74, PhD(geol), 77. *Prof Exp:* Asst prof, 77-81, ASSOC PROF THERMODYNAMICS & PETROL, PRINCETON UNIV, 82- *Concurrent Pos:* Vis lectr, Univ Calif, Los Angeles, 81-82, vis res geophysicists, 81-82; prin investr, NSF, 77-82. *Mem:* Am Geophys Union; Geol Soc Am. *Res:* Application of thermodynamics and fluid dynamics to magnatic processes; eruption and ascent of magma; experimental rheology of magma; origin of compositional zonation in magma chambers. *Mailing Add:* Dept Goel Univ Calif Santa Barbara CA 93106

SPERANDIO, GLEN JOSEPH, b Glen Carbon, Ill, May 8, 18; m 46; c 1. CLINICAL PHARMACY. *Educ:* St Louis Col Pharm, BS, 40; Purdue Univ, MS, 47, PhD(pharm), 50. *Prof Exp:* Anal chemist, Grove Labs, 40-42, chief control chemist, 44-46 & United Drug Co, 42-43; asst dept mgr, William R Warner, Inc, 43-44; from instr to assoc prof pharm, 46-60, head dept clin pharm, 71-78, assoc dean, 78-83, PROF PHARM, PURDUE UNIV, WEST LAFAYETTE, 60-; EXEC DIR, IND SOC HOSP PHARM, 83- *Concurrent Pos:* Consult, Vet Admin Hosps, 69-79, Surgeon Gen, US Army, 74-80. *Honors & Awards:* Glen J Sperandio Award for Advan of Pharm, 84. *Mem:* Am Soc Hosp Pharmacists; Am Pharmaceut Asn; Soc Cosmetic Chem; Am Asn Col Pharm; Sigma Xi; assoc AMA. *Res:* Product formulation; tablets; dermatological medication; cosmetics; pharmaceuticals; hospital pharmacy. *Mailing Add:* Dept Clin Pharm Purdue Univ West Lafayette IN 47907

SPERANZA, GEORGE PHILLIP, b Johnston City, Ill, Aug 27, 24; m 44; c 4. ORGANIC CHEMISTRY, POLYMER CHEMISTRY. *Educ:* Southern Methodist Univ, BS, 48; Univ Ill-Urbana, MS, 49, PhD(org chem), 51. *Prof Exp:* Res chemist, Jefferson Chem Co, 51-56, supvr, 56-68; mgr, 68-82, res fel, 82-90, SR RES FEL, TEXACO CHEM CO, 90- *Mem:* Am Chem Soc. *Res:* Synthetic organic chemistry involving petroleum based chemicals; urethane chemistry; exploratory research and supervisory positions. *Mailing Add:* 2800 Silverleaf Circle Austin TX 78757

SPERATI, CARLETON ANGELO, b Fergus Falls, Minn, Sept 1, 18; m 41; c 3. FLUOROPOLYMER SYSTEMS, POLYMER CHEMISTRY. *Educ:* Luther Col, AB, 38; Univ Ill, MA, 39, PhD(org chem), 41. *Prof Exp:* Res chemist, Plastics Dept, E I du Pont de Nemours & Co, 41-52, res supvr, Polychem Dept, 52-55, sr res supvr, 55-60, sr res chemist, 60-62, res assoc, 62-69, res fel, Plastics Prod & Resins Dept, 69-79; C Paul Stocker prof eng, Ohio Univ, 79-80; CONSULT, 81- *Concurrent Pos:* C Paul Stocker adj prof chem eng, Ohio Univ; vis prof, interim prog, Luther Col, Decorah, Iowa, 88; chmn subcomt terminology, Am Soc Testing & Mat, 84-, chmn sect fluoropolymers, 86-, mem comt terminology, 86-; mem, US tech adv group, tech comt plastics, Int Orgn Standardization, 86-, Standards for fluoropolymers, 88-, convenor, gen vocabulary, 89-, chmn US tech adv group, terminology, 89- *Mem:* AAAS; Am Chem Soc; Soc Plastics Eng; Am Soc Testing & Mat; Sigma Xi. *Res:* Steric hindrance; stable vinyl alcohols; low reflection coatings; laminating resins; condensation polymers; photochromic systems; computers in polymer studies; synthesis conditions and molecular structure versus properties of fluorocarbon and other polymers; thermal analysis of polymers; plastics terminology; principles of terminology. *Mailing Add:* 23 Mustang Acres Parkersburg WV 26104

SPERBER, DANIEL, b Vienna, Austria, May 8, 30; m 63; c 1. PHYSICS. *Educ:* Hebrew Univ, Israel, MSc, 54; Princeton Univ, PhD(physics), 60. *Prof Exp:* Teaching asst physics, Hebrew Univ, Israel, 53-54 & Israel Inst Technol, 54-55; asst, Princeton Univ, 55-60; instr, Ill Inst Technol, 61-62, lectr, 62-64, assoc prof, 64-67; assoc prof, 67-72, PROF PHYSICS, RENSSELAER POLYTECH INST, 72- *Concurrent Pos:* From assoc physicist to sr physicist, IIT Res Inst, 60-66, sci adv, 66-67; Nordita prof, Niels Bohr Inst, Univ Copenhagen, 73-74; NATO res fel, 75-76, Neils Bohr Inst, 75-76; vis prof, GSI Darmstadt, WGermany, 83; Fulbright res scholar, Saha Inst Nuclear Physics, Calcutta, India, 87-88. *Mem:* Fel Am Phys Soc; Phys Soc Israel; NY Acad Sci. *Res:* Nuclear structure, reactions and decay modes; physics of fission and heavy ions, nuclear equation of state; atomic spectroscopy; application of group theory to quantum mechanics. *Mailing Add:* Dept of Physics Rensselaer Polytech Inst Troy NY 12180-3590

SPERBER, GEOFFREY HILLIARD, b Bloemfontein, SAfrica, Dec 26, 33; Can citizen; m 63; c 3. ANATOMY, DENTISTRY. *Educ:* Univ Witwatersrand, BSc, 54, Hons, 58, BDS, 56, PhD, 74; Univ Rochester, MSc, 62. *Hon Degrees:* For Assoc RSSAF, 88. *Prof Exp:* Jr lectr anat, Med Sch, Univ Witwatersrand, 57-58; from asst prof to assoc prof anat & oral surg, 61-72, PROF ORAL BIOL, FAC DENT, UNIV ALTA, 72- *Concurrent Pos:* Nat Res Coun Can res grants, 64-66; Nat Res Coun Can sr res fel, Univ Witwatersrand, 69-70; ed, Asn Can Fac Dent Newslett, 72-85; fel, Can Fund Dent Educ, 85; pres, Midwest Sect, Can Asn Dent Res, 84-85; McCalla prof, Univ Alta, 90-91. *Mem:* Can Asn Anatomists; Can Asn Phys Anthrop; Can Dent Asn; Int Asn Dent Res; Int Dent Fedn; Am Asn Anatomists; AAAS; Craniofacial Genetics Soc; Am Cleft Palate Craniofacial Asn. *Res:* Dental science; physical anthropology; embryology; skull growth; oral pathology and teratology; comparative odontology. *Mailing Add:* Fac Dent Univ Alta Edmonton AB T6G 2N8 Can

SPERBER, STEVEN IRWIN, b Brooklyn, NY, May 25, 45; m 73. GEOMETRY. *Educ:* Brooklyn Col, BA, 66; Univ Pa, MA & PhD(math), 75. *Prof Exp:* Instr math, York Col, City Univ New York, 71-73, adj fac, Lehman Col, 74-75; lectr math, Univ Ill, Urbana, 75-77, asst prof, 77-80; MEM FAC, DEPT MATH, UNIV MINN, 80- *Mem:* Am Math Soc; Sigma Xi. *Res:* A study of the p-adic cohomology of the generalized hypergeometric functions and the associated Frobenius structure of the deformation equation. *Mailing Add:* Dept Math Univ Minn 127 Vincent Hall Minneapolis MN 55455

SPERBER, WILLIAM H, b Sturgeon Bay, Wis, Feb 15, 41; m 63; c 2. MICROBIOLOGY, BIOCHEMISTRY. *Educ:* Univ Wis-Madison, BS, 64, MS, 67, PhD(bact), 69. *Prof Exp:* Chief microbiol sect, Best Foods Res Ctr Div, CPC Int, Inc, 69-72; scientist microbiol sect, Res & Develop Ctr, 72-74, sr scientist, 74-77, RES ASSOC, CORP MICROBIOL, PILLSBURY CO, 77- *Mem:* Am Soc Microbiologists; NY Acad Sci; Soc Appl Bacteriol; Inst Food Technol; AAAS. *Res:* Food microbiology, lactics, osmophilics, food poisoning organisms, evolution, philosophy of science. *Mailing Add:* 5814 Oakview Cir Minnetonka MN 55345

SPERELAKIS, NICK, b Joliet, Ill, Mar 3, 30; m 60; c 6. CARDIOVASCULAR PHYSIOLOGY. *Educ:* Univ Ill, BS, 54, MS, 55, PhD(physiol), 57. *Prof Exp:* Asst physiol, Univ Ill, 54-57; from instr to assoc prof, Western Reserve Univ, 57-66; prof physiol, sch med, Univ Va, 66-83; PROF & DIR, DEPT PHYSIOL, COL MED, UNIV CINCINNATI, 83- *Concurrent Pos:* Estab investr, Am Heart Asn, 61-66; hon res assoc biophys, Univ Col, Univ London; vis prof, Ctr Advan Studies, Mex, 72 & Univ St Andrews, Scotland, 72-73; assoc ed, Circulation Res J, 70-75; chmn, Steering Comt Cell & Gen Physiol Sect, Am Physiol Soc, 81-82. *Mem:* Am Physiol Soc; Soc Gen Physiologists; Int Soc Heart Res; Cardiac Muscle Soc; Biophys Soc. *Res:* Electrophysiology of nerve, muscle and muscle ultrastructure; transmission of excitation in cardiac and smooth muscles; excitation-contraction coupling; hormone-membrane interaction; active ion transport; membrane properties; mechanism of action of calcium-antagonistic drugs; electrophysiology of cultured heart cells; developmental changes in electrical properties of the heart; electrical properties of myocardial slow channels. *Mailing Add:* Dept Physiol Univ Cincinnati Col Med Cincinnati OH 45267

SPERGEL, DAVID NATHANIEL, b Rochester, NY, Mar 25, 61. ASTROPHSYICS, ELEMENTARY PARTICLE PHYSICS. *Educ:* Princeton Univ, AB, 82; Harvard Univ, MA, 84, PhD(astron), 85. *Prof Exp:* Res assoc astron, Harvard Univ, 86; ASST PROF ASTRON, PRINCETON UNIV, 87- *Concurrent Pos:* mem staff, Inst Advan Study, 86-88; Sloan fel, Alfred P Sloan Found, 88. *Res:* Stellar dynamics; early universe; cosmic strings; dark matter detection. *Mailing Add:* Dept Astron Princeton Univ Princeton NJ 08540

SPERGEL, MARTIN SAMUEL, b New York, NY, Sept 13, 37; m 59; c 3. ASTROPHYSICS, HIGH ENERGY PHYSICS. *Educ:* Rensselaer Polytech Inst, BS, 59; Univ Rochester, MA, 61, PhD(physics), 64. *Prof Exp:* Physicist, Indust Nucleonics Corp, 56 & Xerox Corp, 60; recitation instr basic physics, Univ Rochester, 59-61, res asst elem particles, 61-63; res scientist, Grumman Aircraft Corp, 63-67; assoc prof, 67-80, PROF PHYSICS, YORK COL, GRAD SCH & UNIV CTR, CITY UNIV NEW YORK, 80-, CHMN DEPT, 85- *Concurrent Pos:* Adj asst prof, C W Post Col, 65-; vis lectr astron, State Univ NY Stony Brook, 66-67; vis res scientist, Brookhaven Nat Labs, 75-; prin investr, NASA, 78-81. *Mem:* AAAS; Am Phys Soc; Am Geophys Union; Inst Elec & Electronics Eng. *Res:* Planetary physics; interactions of cosmic rays with interplanetary and interstellar matter; molecules in space; radiation environment of solar system; solid state. *Mailing Add:* Dept Physics York Col Jamaica NY 11451

SPERGEL, PHILIP, b New York, NY, Mar 5, 26; m 48; c 2. INSTRUMENTATION, ELECTRICAL ENGINEERING. *Educ:* City Col New York, BEE, 48; NY Univ, MEE, 51. *Prof Exp:* Proj engr, Sperry Gyroscope Co, 48-54; chief engr, Indust Nucleonics Corp, 54-57 & Epsco, Inc, 57-61; mem staff, Mitre Corp, 61-62; dir eng, Baird Atomic, Inc, 62-67; vpres res & develop, 67-74, VPRES CORP QUAL ASSURANCE, INSTRUMENTATION LAB, INC, 74- *Mem:* Am Soc Qual Control; Inst Elec & Electronics Engrs; Asn Advan Med Instrumentation. *Res:* Development of optical, electronic, nuclear, mechanical and chemical instruments to meet specific and general purpose applications. *Mailing Add:* Seven Wainwright Rd No 17 Ledges Winchester MA 01890

SPERLEY, RICHARD JON, b Staples, Minn, May 28, 39. ORGANIC CHEMISTRY. *Educ:* Concordia Col, BA, 61; Univ Minn, Minneapolis, PhD(org chem), 66. *Prof Exp:* Res chemist, Res Ctr, Uniroyal Inc, NJ, 66-72, SR RES SCIENTIST, TIRE DIV, UNIROYAL INC, 72- *Mem:* Am Chem Soc; Sigma Xi. *Res:* Polymer and elastomer degradation. *Mailing Add:* 5168 Hale Ct Troy MI 48098-3404

SPERLING, FREDERICK, Jan 16, 13. TOXICOLOGY, TERATOLOGY. *Educ:* Univ Chicago, PhD(zool), 52. *Prof Exp:* Emer prof pharmacol & toxicol, Col Med, Howard Univ; RETIRED. *Concurrent Pos:* Vis prof pharmacol, Hedassah Med Sch; emer scientist, Soc Exp Biol Med, Wash, DC; toxicologist, Sperling Lab, Dept Pharm, Howard Univ Col Med. *Honors & Awards:* Educ Award, Soc Toxicol. *Mem:* Fel AAAS; Soc Toxicol; Soc Exp Biol Med; Am Soc Pharmacol & Exp Therapeut; Am Soc Zool; Am Chem Soc. *Res:* Toxicology; methodology; teratology. *Mailing Add:* 5902 Mt Eagle Dr No 407 Alexandria VA 22303

SPERLING, GEORGE, US citizen. VISION, HUMAN INFORMATION PROCESSING. *Educ:* Univ Mich, BS(math & biophys), 55; Columbia Univ, MA, 56; Harvard Univ, PhD(psychol), 59. *Prof Exp:* Res asst psychol, Harvard Univ, 57-59; vis assoc prof, dept psychol, Wash Sq Col, NY Univ, 62-63; adj assoc prof, dept psychol, Columbia Univ, 64-65; actg assoc prof, dept psychol, Univ Calif, Los Angeles, 67-68; fel, John Simon Guggenheim Mem Found, 69-70; adj prof, 70-80, PROF PSYCHOL, GRAD SCH ARTS & SCI, NY UNIV, 80-, DIR, HUMAN INFO PROCESSING LAB, 80-

Concurrent Pos: Mem, tech res staff, Acoust & Behav Res Ctr, AT&T Bell Labs, 59-70; hon res assoc, dept psychol, Univ Col, Univ London, 69-70; vis prof, dept psychol, Univ Western Australia, 72 & Univ Wash, Seattle, 77; mem, steering comt, Soc Computers Psychol, 74-78, 85-86, bd dirs, Eastern Psychol Asn, 82-85, sci adv bd, USAF, 88-; founder & organizer, Ann Interdisciplinary Conf, 75-; mem, exec bd, Soc Math Psychol, 79-85, chmn, 83-84; William James fel, Am Psychol Soc; vis scholar, dept psychol, Stanford Univ, 84. *Honors & Awards:* Distinguished Sci Contrib Award, Am Psychol Asn, 88. *Mem:* Nat Acad Sci; fel AAAS; Asn Res Vision & Ophthal; fel Am Psychol Asn; Soc Comput in Psychol; fel Optical Soc Am; Am Psychol Soc; Psychonomic Soc; Soc Exp Psychologists; Soc Math Psychol. *Res:* Vision and visual perception; mathematical and theoretical psychology; computational vision and computer image processing; human information processing. *Mailing Add:* Dept Psychol & Neural Sci NY Univ Six Washington Pl Rm 980 New York NY 10003

SPERLING, HARRY GEORGE, b New York, NY, Aug 26, 24; m 50; c 2. VISION, PSYCHOPHYSICS. *Educ:* Univ Pa, AB, 44; New Sch Social Res, MSc, 46; Columbia Univ, PhD(psychol), 53. *Prof Exp:* Jr instr psychol, Johns Hopkins Univ, 47-48; res psychologist, US Naval Med Res Lab, 48-52, chief psychophys res sect, 52-58; chief colorimetry sect, 58-59; from sr scientist to mgr manned systs sci systs & res div, Honeywell Inc, 59-67; PROF NEUROSCI & OPHTHAL, MED SCH & DIR, SENSORY SCI CTR, UNIV TEX HEALTH SCI CTR, HOUSTON, 67-, PROF NEURAL SCI, GRAD SCH, 80- *Concurrent Pos:* Consult, Int Comn Illum, 59-; clin assoc prof, Univ Minn, 61-67; mem, Armed Forces-Nat Res Coun Comt Vision & chmn working group laser-eye effects, 66-70; adj prof, Baylor Col Med, 67- & Rice Univ, 72-; mem, Nat Adv Eye Coun, NIH, 75-79. *Mem:* Assoc Res Vision & Ophthal; fel Optical Soc Am; Psychonom Soc; Soc Neurosci; Int Res Group Color Vision Deficiencies. *Res:* Psychophysical, electrophysiological and anatomical studies of color and brightness vision; intense light effects on retina. *Mailing Add:* Sensory Sci Ctr Univ Tex Grad Sch Biomed Sci Houston TX 77025

SPERLING, JACOB L, b Linz, Austria, Jan 3, 49; US citizen; m 78; c 2. PLASMA PHYSICS, MICROWAVE & RADIO FREQUENCY COMMUNICATIONS & RADAR. *Educ:* Columbia Univ, BS, 71; Princeton Univ, MA, 73, PhD(plasma physics), 75. *Prof Exp:* Sr scientist plasma physics, Gen Atomic Co, 75-78; staff scientist, 78-80, sr scientist, 80-83, prin scientist energy, 83-88, DIV MGR SYSTS SURVIVABILITY GROUP, JAYCOR, 88- *Mem:* Am Phys Soc. *Res:* Theoretical and applied plasma physics; health physics aspects of radioactivity; coal physics; microwave and radio frequency communication; radar propagation. *Mailing Add:* Jaycor PO Box 85154 San Diego CA 92138

SPERLING, LESLIE HOWARD, b Yonkers, NY, Feb 19, 32; m 57; c 2. POLYMER SCIENCE. *Educ:* Univ Fla, BS, 54; Duke Univ, MA, 57, PhD, 58. *Prof Exp:* Res chemist, Buckeye Cellulose Corp, Procter & Gamble Co, 58-65; res assoc phys chem, Princeton Univ, 65-67; from asst prof to assoc prof, 67-77, PROF, DEPT CHEM ENG & MAT RES CTR, LEHIGH UNIV, 77-, PROF, MAT SCI ENG DEPT. *Concurrent Pos:* Mem, Ctr Polymer Sci Eng, Lehigh Univ. *Mem:* Am Chem Soc; Am Inst Chem Engrs; Soc Plastics Eng. *Res:* Physical chemistry of cellulose; physical and mechanical properties of polymers; polymer blends, particularly interpenetrating polymer networks; triglyceride oil-based interpenetrating polymer networks; noise damping polymer systems; interpenetrating polymer network nomenclature and block copolymers; small-angle neutron scattering from polymer latexes; chain conformation within latex particles; author. *Mailing Add:* Whitaker No 5 Lehigh Univ Bethlehem PA 18015

SPERLING, MARK ALEXANDER, b Lodz, Poland, Sept 6, 38; Australian citizen; m 66; c 2. ENDOCRINOLOGY, DIABETES. *Educ:* Univ Melbourne, MB & BS, 62; Am Bd Pediat, dipl, 70. *Prof Exp:* Prof pediat & Dir, Div Endocrin & Diabetes, Univ Cincinnati Col Med, Children's Hosp Med Ctr, 78-89; CHMN, DEPT PEDIAT, UNIV PITTSBURGH SCH MED, 89-; PEDIAT-IN-CHIEF, CHILDREN'S HOSP PITTSBURGH, 89- *Concurrent Pos:* Intern, resident & sr resident pediat, Royal Children's Hosp, Australia, 63-68; fel, Pediat Endocrin & Metabolism, Children's Hosp, Pittsburgh, 68-70; asst prof Pediat, Univ Calif Los Angeles, Harbor Gen Hosp Campus, 70-75, assoc prof pediat & chief, Div Pediat Endocrin, 75-78; comt mem Endocrine Soc Comt on Sci Prog, 84-87, Maternal & Child Health Res Comt, Nat Inst Child Health, 84-89; vchmn clin Res Children's Med Ctr, Cincinnati, 87-89; mem, Nat Diabetes Adv Bd, NIH, 90-94. *Mem:* Am Soc Clin Invest; Endocrine Soc; Am Diabetes Asn; Soc Pediat Res; Am Pediat Soc; Am Fedn Clin Res. *Res:* Hormonal control of carbohydrate metabolism; perinatal glucose homeostasis; insulin and glucagon receptors; endocrinology of hypertension; endocrinology of growth and development; numerous publications including 110 original papers, approximately 40 book chapters, editor and co-editor 2 books and approximately 150 published abstracts. *Mailing Add:* Dept Pediat Children's Hosp Pittsburgh 3705 Fifth Ave at Desoto St Pittsburgh PA 15213-2583

SPERO, CAESAR A(NTHONY), JR, b Newport, RI, Oct 3, 21. MECHANICAL & SYSTEMS ENGINEERING. *Educ:* Mass Inst Technol, BS, 44; Univ Northern Colo, MS, 77. *Prof Exp:* Proj engr equip design, Owens-Corning Fiberglas Corp, 46-50, eng mgr, 50-52; staff sci asst, Naval Underwater Ord Sta, 52-55, head, Eng Dept, 55-60, head, Testing & Eval Dept, 60-61, head, Develop Dept, 61-64, head, Shipborne Equip Dept, 64-65, assoc dir, Systs Develop, 65-71, chief, Res & Develop, Newport Lab, 71, assoc dir, Weapons, 71-72, dir, Systs Develop, 72-76, assoc tech dir, Prod Lines, 76-78, dep tech dir, Naval Underwater Syts Ctr, 78-80; vpres, OSD Gould Inc, 80-86; RETIRED. *Concurrent Pos:* Consult. *Mem:* Math Asn Am. *Res:* Complex system development from conceptual stage through actual manufacture, installation and operational testing. *Mailing Add:* 325 Mail Coach Rd Portsmouth RI 02871

SPERO, LEONARD, b New York, NY, May 30, 21; m 43; c 4. BIOCHEMISTRY. *Educ:* City Col New York, BS, 41; Univ Wis, MS, 43, PhD(biochem), 48. *Prof Exp:* Asst, Univ Wis, 42-44 & 46-48; biochemist, US Army Med Res Inst Infectious Dis, 48-63, chief, Chem Br, 63-71, biochemist, Path Div, 71-75, asst chief, Path Div, 75-81; CONSULT, 81- *Concurrent Pos:* Secy Army res & study fel, 60; lectr, Georgetown Univ, 65. *Mem:* Am Soc Biol Chem. *Res:* Protein chemistry; isolation and purification; reactive groups; immunochemistry; bacterial toxins. *Mailing Add:* 635 Schley Ave Frederick MD 21701

SPERONELLO, BARRY KEVEN, b Passaic, NJ, July, 29, 50; m 75. HETEROGENEOUS CATALYSIS, MATERIALS SCIENCE. *Educ:* Rutgers Univ, BS, 72, MS, 75, PhD(ceramic eng), 76. *Prof Exp:* Res & proj leader, 76-81, RES GROUP LEADER, ENGELHARD CORP, 81- *Mem:* Am Ceramic Soc; NAm Catalysis Soc. *Res:* Synthesis and properties of oxide materials; physical and catalytic properties of oxide catalysts. *Mailing Add:* Res Dept Engelhard Corp 101 Wood Ave Iselin NJ 08830-0770

SPERRY, CLAUDE J, JR, b Greenwood, SC, Aug 8, 25; m 48; c 1. ELECTRICAL ENGINEERING. *Educ:* Clemson Col, BEE, 48; Univ Ill, MS, 54. *Prof Exp:* From instr to assoc prof, 48-65, PROF ELEC ENG, TULANE UNIV, 65-, RES ASSOC PHYSIOL, 51- *Mem:* Inst Elec & Electronics Engrs; Sigma Xi. *Res:* Electricity in medical research, especially remote recording of subcortical potentials. *Mailing Add:* 33 Stilt St New Orleans LA 70124

SPERRY, PHILIP ROGER, b Mountainview, NJ, July 6, 20; m 42; c 4. METALLURGY. *Educ:* Univ Notre Dame, BS, 52. *Prof Exp:* Metallographer, Univ Notre Dame, 47-51; res metallurgist, Kaiser Aluminum & Chem Corp, 52-57; res metallurgist, mgr, Physics Sect & sr res scientist, Metals Res Labs, Olin Corp, 58-76; mgr phys metall & chief metallurgist, Consol Aluminum Corp, 76-85; RETIRED. *Concurrent Pos:* Mem, Metals Handbk Comt, Am Soc Metals, 70-72, Comt Revise Aluminum, 81-83; mem, Welding & Joining Comt, Aluminum Asn, 77-82, rep, Tech Comt, 82-84; mem, Comt Asc, Am Welding Soc, 83. *Honors & Awards:* Matthewson Gold Medal, Am Inst Mech Engrs, 53. *Mem:* Am Soc Metals Int; Metall Soc. *Res:* Non-ferrous metals, mainly aluminum; alloy development; corrosion; metal physics; metallographic interpretation of microstructure related to metal processing and service experience; author or co-author of 26 publications; granted 29 patents. *Mailing Add:* 185 Elmwood Dr Meriden CT 06450

SPERRY, ROGER WOLCOTT, b Hartford, Conn, Aug 20, 13; m 49; c 2. NEUROBIOLOGY. *Educ:* Oberlin Col, AB, 35, MA, 37; Univ Chicago, PhD(zool), 41. *Hon Degrees:* DSc, Univ Cambridge, 72, Univ Chicago, 76, Kenyon Col, 79, Rockefeller Univ, 80, Oberlin Col, 82. *Prof Exp:* Res fel, Harvard Univ, 41-46; from asst prof anat to assoc prof psych, Univ Chicago, 46-53; Hixon prof, 54-84, EMER PROF PSYCHOBIOL, CALIF INST TECHNOL, 84- *Concurrent Pos:* Sect chief, neurol diseases & blindness, NIH, 52-53; mem, Fel Comt, NSF, 63-64, Exp Psychol Study Sect, 66-70, chmn, 69-70; mem corp vis comt psychol, Mass Inst Technol, 69-76. *Honors & Awards:* Nobel Prize in Med, 81; Howard Crosby Warren Medal, Soc Exp Psychologists, 69; William Thompson Wakeman Res Award, Nat Paraplegia Found, 72; Passano Award in Med Sci, 73; Karl Lashley Award, Am Philos Soc, 76; Wolf Prize in Med, 79; Ralph Gerard Award, Soc Neurosci, 79; Albert Lasker Med Res Award, 79; Mentor Soc Award, 87; Nat Medal of Sci, 89. *Mem:* Nat Acad Sci; AAAS; Am Psychol Asn; Am Physiol Soc; Am Philos Soc; hon mem Am Neurol Asn; foreign mem Royal Soc; foreign mem USSR Acad Sci. *Res:* Brain organization and neural mechanisms; neural plasticity tested by nerve and muscle transplantation; selective patterning in growth of nerve connections; cytospecificity and chemoaffinity theory; neural mechanism in perception and memory; split-brain approach to cerebral organization; hemispheric specialization; mind-brain concepts and the human value implications. *Mailing Add:* Calif Inst Technol Div Biol 156-29 Pasadena CA 91125

SPERRY, THEODORE MELROSE, b Toronto, Ont, Feb 20, 07; US citizen; m 35, 77. BOTANY, ECOLOGY. *Educ:* Butler Univ, BS, 29; Univ Ill, MS, 31, PhD(bot), 33. *Prof Exp:* Asst bot, Univ Ill, 29-32; timber cruiser & estimator, US Forest Serv, Ill, 33-34, tech foreman, 35-36; sr foreman ecol, Nat Park Serv, Wis, 36-41; consult, Curtis Prairie Univ Wis, Madison, 36-90; from asst prof to prof bot & ecol, 46-74, EMER PROF BOT & ECOL & CUR HERBARIUM, PITTSBURG STATE UNIV, KANS, 74- *Concurrent Pos:* Weather forecaster, US Army Air Corps, 41-45; consult, Nat Inst Study Agr Belg Congo, 51-52; secy-treas, Grasslands Res Found, 54-58. *Mem:* Nat Parks Asn; Am Soc Plant Taxon; Wilderness Soc; Nature Conserv; hon mem Soc Ecol Restoration, 90. *Res:* Grassland ecology; plant taxonomy; environmental conservation; prairie restoration (world's oldest). *Mailing Add:* 1413 S College Pittsburg KS 66762

SPERRY, WILLARD CHARLES, b Dunsmuir, Calif, Nov 29, 31; m 66; c 2. NUCLEAR PHYSICS. *Educ:* Stanford Univ, BS, 54; Univ Calif, Davis, MA, 67, PhD(physics), 68. *Prof Exp:* Physicist, Aerojet-Gen Corp, 56-60; asst prof, 66-68, ASSOC PROF PHYSICS, CENT WASH UNIV, 68- *Res:* Nuclear structure by means of mesic atoms; improvement of undergraduate physics laboratories. *Mailing Add:* Dept of Physics Cent Wash Univ Ellensburg WA 98926

SPERTI, GEORGE SPERI, biophysics, for more information see previous edition

SPESSARD, DWIGHT RINEHART, b Westerville, Ohio, July 6, 19; m 43; c 2. ORGANIC CHEMISTRY. *Educ:* Otterbein Col, BS, 41; Western Reserve Univ, PhD(inorg & org chem), 44. *Prof Exp:* Group leader, Lubrication Sect, Chem Div, Naval Res Lab, 44-47; res assoc chem, Gen Elec Res Lab, 47-49; from asst prof to assoc prof, Muskingum Col, 49-53; from asst prof to assoc prof, 53-60, chmn dept, 58-61, PROF CHEM, DENISON UNIV, 60-, WICKENDEN CHAIR, 66- *Mem:* Am Chem Soc. *Res:* Organophosphorus, organosilicon and organofluorine chemistry. *Mailing Add:* 228 Granview Rd Granville OH 43023-1248

SPESSARD, GARY OLIVER, b Orange, Calif, Sept 27, 44; m 68; c 1. SYNTHETIC ORGANIC CHEMISTRY. *Educ:* Harvey Mudd Col, BS, 66; Univ Wis-Madison, MS, 68; Wesleyan Univ, PhD(org chem), 71. *Prof Exp:* Fel org chem, Univ Alta, 70-72; vis res assoc, Ohio State Univ, 72-73; from asst prof to assoc prof, 73-86, PROF CHEM, ST OLAF COL, 86- *Concurrent Pos:* vis assoc prof chem, Univ Utah, 79-80; vis prof chem, Ore State Univ, 86-87. *Mem:* Am Chem Soc. *Res:* Synthetic organic chemistry of small ring compounds; synthetic organic and natural products chemistry; chemistry of phytoalexins. *Mailing Add:* Dept of Chem St Olaf Col 1520 St Olaf Ave Northfield MN 55057-1098

SPETNAGEL, THEODORE JOHN, b Chillicothe, Ohio, May 26, 48; m 70; c 2. ENVIRONMENTAL SCIENCES, COMPUTER SCIENCES. *Educ:* Clemson Univ, BS, 70; Ga Inst Technol, MS, 72. *Prof Exp:* Struct engr, Appalachian Consult Engrs, 67-70; res asst, Ga Inst Technol, 70-71; struct engr, Atlantic Bldg Systs, Inc, 71-78; Civil engr, hq Ft McPherson, 78-79, hq US Army Forces Command, 79-84, DEP ENGR, HQ SECOND US ARMY, 84- *Mem:* Am Soc Civil Engrs; Nat Soc Prof Engrs; Soc Am Mil Engrs. *Res:* Structural engineering. *Mailing Add:* 855 Kipling Dr NW Atlanta GA 30318-1634

SPEYER, JASON L, b Boston, Mass, Apr 30, 38. GUIDANCE CONTROL AEROSPACE. *Educ:* Mass Inst Technol, BS, 60, MS, 65; Harvard Univ, PhD(appl math), 68. *Prof Exp:* Res engr, Child Stock Draper Lab, Cambridge, Mass, 70-76; PROF, SYST THEORY & GUID CONTROL AEROSPACE, UNIV TEX-AUSTIN, 76- *Mem:* Fel Am Inst Aeronaut & Astronaut; fel Inst Elec & Electronic Engrs. *Res:* Deterministic and stochastic optimum control theory. *Mailing Add:* 3830 Avenida Del Sol Studio City CA 91604-4024

SPHON, JAMES AMBROSE, b Luxor, Pa, Nov 4, 39; m 67; c 3. MASS SPECTROMETRY. *Educ:* St Vincent Col, BS, 66; Wayne State Univ, PhD(org chem), 77. *Prof Exp:* Chemist, 65-67, res chemist, 67-75, SUPVY CHEMIST, MASS SPECTROMETRY LAB, FOOD & DRUG ADMIN, 75- *Mem:* Am Chem Soc; Am Soc Mass Spectrometry. *Res:* Application of mass spectrometry to structure elucidation and method development for components of foods: pesticides, mycotoxins, direct & indirect food additives and veterinary drugs. *Mailing Add:* 9900 Worrell Ave Glenn Dale MD 20769-9260

SPIALTER, LEONARD, b Newark, NJ, Jan 18, 23; m 46; c 2. ORGANIC CHEMISTRY, COMPUTER SYSTEMS. *Educ:* Rutgers Univ, BS, 44, PhD(chem), 49; Polytech Inst Brooklyn, MS, 48. *Prof Exp:* Res chemist, Montclair Res Corp, NJ, 44-47; instr org chem, Univ Col, Rutgers Univ, 48-49; fel free radicals, Harvard Univ, 49-51; from res chemist to sr scientist & head org sect, Chem Res Lab, Aerospace Res Labs, Wright-Patterson AFB, 51-75; DIR, INSTRUMENTORS I-V, 75- *Mem:* AAAS; Am Chem Soc; Am Inst Chemists; The Chem Soc; Electrochem Soc; Sigma Xi. *Res:* Organosilanes; animes; molecular rearrangements; free radicals; laboratory automation; information storage-retrieval; computer-based nomenclature; liquid fuel/byproducts from agricultural residues; computer-based information systems; computer-aided manufacturing. *Mailing Add:* 2536 England Ave Dayton OH 45406

SPICER, CLIFFORD W, b LaGrange, Ohio, Mar 9, 19. METALLURGY. *Educ:* Univ Cincinnati, BS, 42. *Prof Exp:* Metallurgist, Eastern Steel Div, US Steel Corp, 42-83; RETIRED. *Mem:* Fel Am Soc Metals Int; Am Inst Mining Metall & Petrol Engrs. *Mailing Add:* 40 Rookery Way Hilton Head Island SC 29928

SPICER, DONALD Z, b St Paul, Minn, Mar 15, 37; m 68; c 3. MATHEMATICS, COMPUTER SCIENCE. *Educ:* Univ Minn, BA, 59, PhD(math), 65; Columbia Univ, MA, 60; dipl, Cambridge Univ, 84. *Prof Exp:* From actg asst prof to asst prof math, Univ Calif, Los Angeles, 65-67; asst prof, Univ Ky, 67-70; from asst prof to assoc prof math, Vassar Col, 70-86, assoc dean col, 80-83; dir acad comput & adj prof, Dartmouth Col, 85-88, ASST PROVOST UNIV COMPUT & PROF MATH, NOTRE DAME UNIV, 88- *Concurrent Pos:* Proj dir, Fund Improvement Postsecondary Educ, HEW; vis lectr, Univ Kent, Canterbury, Eng, 84-85. *Mem:* Am Math Soc; Math Asn Am; Asn Comput Mach. *Res:* Computer algebra. *Mailing Add:* Off Univ Comput Univ Notre Dame Notre Dame IN 46556

SPICER, LEONARD DALE, b Detroit, Mich, July 7, 42; m 68; c 2. PHYSICAL BIOCHEMISTRY, BIOMACROMOLECULAR STRUCTURE. *Educ:* Univ Mich, BSCh, 64; Yale Univ, PhD(phys chem), 68. *Prof Exp:* Assoc chem kinetics, Univ Wash, 68-69; from asst prof to prof phys chem, Univ Utah, 69-83, assoc dean, Grad Sch, 80-83; PROF BIOCHEM & RADIOL, DUKE UNIV, 83-, DIR DUKE NUCLEAR MAGNETIC RESONANCE CTR, 86- *Concurrent Pos:* Dreyfus Found fel, 71-77. *Mem:* AAAS; Am Chem Soc; Am Phys Soc; Fedn Am Socs Exp Biol; Soc Nuclear Med. *Res:* Biophysical nuclear medical resonance; protein structure and function; hot atom chemistry; high energy and thermal kinetics; photoassisted catalysis; intermolecular vibrational energy transfer; atmospheric chemistry; nuclear medicine with positron. *Mailing Add:* Dept Biochem Duke Univ Med Ctr Box 3711 Durham NC 27710

SPICER, SAMUEL SHERMAN, JR, b Denver, Colo, Aug 12, 14; m 41; c 3. EXPERIMENTAL PATHOLOGY, HISTOCHEMISTRY & CELL BIOLOGY. *Educ:* Univ Colo, BS, 36, MD, 39. *Hon Degrees:* MD, Univ Linkoping, Sweden. *Prof Exp:* Intern, Univ Hosp, Univ Wis, 39-40; from asst surgeon to med dir, Nutrit Lab, Lab Phys Biol & Lab Exp Path, NIH, 40-55, chief sect biophys histol, 61-66; PROF PATH, MED UNIV SC, 66- *Honors & Awards:* Hon Mem, Histochem Soc; Pioneer in Histochem Award, 8th Int Congress Histochem & Cytochem. *Mem:* Histochem Soc; Am Soc Cell Biol; Am Soc Exp Path. *Res:* Nutrition; folic acid deficiency; malariology; industrial toxicology; biochemistry of erythrocytes; muscle protein chemistry; histochemistry and ultrastructural cytochemistry of carbohydrates; basic proteins and enzymes; immunocytochemistry-hormones and cell enzymes; experimental pathology of genetic diseases including cystic fibrosis; cochlear abnormality in presbyacusis; salivary gland structure and function; histology and histochemistry of developing and adult lung. *Mailing Add:* Dept Path 171 Ashley Ave Charleston SC 29425-2645

SPICER, WILLIAM EDWARD, b Baton Rouge, La, Sept 7, 29; m 51; c 3. SOLID STATE PHYSICS. *Educ:* Col William & Mary, BS, 49; Mass Inst Technol, SB, 51; Univ Mo, MS, 53, PhD(physics), 55. *Hon Degrees:* Dr Technol, Univ Linkoping, Sweden, 75. *Prof Exp:* Res physicist, RCA Labs, 55-62; from assoc prof to prof elec eng, 62-78, dep dir, Stanford Synchrotron Radiation Proj, 73-75, PROF ELEC ENG & MAT SCI, STANFORD UNIV, 72-, CONSULT DIR, STANFORD SYNCHROTRON RADIATION LAB, 75-, STANFORD W ASCHERMAN PROF ENG, 78- *Concurrent Pos:* Consult, Varian Assocs; Japan Soc Prom Sci fel, 72; mem adv group electron devices, Dept Defense, 73; Guggenheim fel, 78-79; overseas fel, Churchill Col, Cambridge Univ, 79 & 84; vis prof, Fudan Univ, Shanghai, 83; ed, J Crystal Growth. *Honors & Awards:* Oliver E Buckley Solid State Physics Prize, Am Phys Soc, 80; Medard W Welch Award, Am Vacuum Soc, 84. *Mem:* AAAS; fel Am Phys Soc; fel Inst Elec & Electronics Engrs; Sigma Xi; Am Vacuum Soc. *Res:* Electronic structure and optical properties of solids and surfaces; photoelectric emission; semiconductors; alloys; amorphous solids; surface and interface states; surface science and catalysis. *Mailing Add:* Stanford Electronics Lab Stanford Univ Stanford CA 94305

SPICHER, JOHN L, b Belleville, Pa, Sept 12, 35; m 59; c 3. MEDICAL TECHNOLOGY. *Educ:* Eastern Mennonite Col, BS, 58; Geisinger Med Ctr, MT, 64. *Prof Exp:* Res asst endocrinol, Med Col Va, 60; res asst chromatog, Geisinger Med Ctr, 62-64, asst instr med technol, 64-66; res asst path, Sch Med, Univ Pittsburgh, 66-68; SR ENGR HUMAN SCI, WESTINGHOUSE RES & DEVELOP, WESTINGHOUSE ELEC CORP, 68- *Mem:* Am Asn Clin Chemists; Am Soc Clin Pathologists. *Res:* Development of planning methods for health care systems which relate to consumer need, demands and available resources. *Mailing Add:* Rd Ten Irwin PA 15642

SPICHER, ROBERT G, b Pittsburgh, Pa, Apr 24, 35; m 59, 81; c 2. SANITARY & ENVIRONMENTAL ENGINEERING. *Educ:* Cornell Univ, BCE, 58; Univ Calif, Berkeley, MS, 59; Wash Univ, St Louis, ScD(environ & sanit eng), 63. *Prof Exp:* Prod engr, Shell Oil Co, Calif, 59-60; asst prof sanit & civil eng, Univ Miami, 63-65; assoc dean grad studies, 79-80, PROF SANIT & CIVIL ENG, SAN JOSE STATE UNIV, 65- *Honors & Awards:* Lincoln Arc Welding Struct Nat Grand Award, 58. *Mem:* Am Soc Civil Engrs; Am Water Works Asn; Water Pollution Control Fedn. *Res:* Environmental engineering; cannery waste treatment; sanitary engineering aspects of shelters; treatment of photographic wastes; sanitary landfill stabilization; radioactive contamination of water; fuel gas production from biomass. *Mailing Add:* Dept of Civil Eng San Jose State Univ San Jose CA 95192

SPICKERMAN, WILLIAM REED, b Council Bluffs, Iowa, Dec 28, 25; m 57. APPLIED MATHEMATICS. *Educ:* Univ Omaha, BA, 49, MS, 53; Xavier Univ, MS, 58; Univ Ky, PhD(curriculum), 65. *Prof Exp:* Teacher high sch, Iowa, 49-50; engr aid, Omaha Dist, Mo River Div, Corps of Engr, 51-52; teacher high sch, Iowa, 52-56; tech engr, Jet Engine Dept, Gen Elec Co, Ohio, 56-57; teacher high sch, Ohio, 57-58; engr, Avco Corp, 58-61; specialist-engr, Goodyear Aircraft Corp, 61-62; sr scientist, Spindletop Res Inc, Ky, 65-67; assoc prof, 67-72, PROF MATH, EAST CAROLINA UNIV, 72- *Mem:* Am Math Soc; Math Asn Am; Soc Indust & Appl Math; assoc Opers Res Soc Am; Sigma Xi; Nat Coun Teachers Math. *Res:* Mathematics education; recursive sequences. *Mailing Add:* Dept Math East Carolina Univ Greenville NC 27834

SPIEGEL, ALLEN DAVID, b New York, NY, June 11, 27; m 55; c 3. PUBLIC HEALTH, COMMUNICATIONS. *Educ:* Brooklyn Col, AB, 47; Columbia Univ, MPH, 54; Brandeis Univ, PhD(social welfare), 69. *Prof Exp:* Health educr to chief, Radio & TV Unit, New York City Health Dept, 51-61; health educ assoc, Med Found, Inc, 61-66; PROF PREV MED & COMMUNITY HEALTH, STATE UNIV NY DOWNSTATE MED CTR, 69- *Concurrent Pos:* WHO fel, Israel Med Schs, 74; consult to comnr, Social & Rehab Serv, Dept Health, Educ & Welfare, 70; commun consult, Health Info Systs, Inc, 70-71; curriculum consult, Grad Prog Health Care Admin, Baruch Col, 70-; health manpower consult, NJ Regional Med Prog, 72; adj prof, St Francis Col, 74-76; fac, Staff Col, NIMH, 79; consult, Cancer Proj, Urban Health Inst, 79 & Home Care Proj, Temple Univ, 81, VA Med Ctr, Booklyn Home Care, 83, Long Term Care Greater Dept Med Record Asn, 84, Clin Prev, Robt Wood Johnson Med Sch, Piscataway, NJ, 87, Home Health Care Tele Med, Inc, 88, High Tech Home Care, Case Western Reserve Med Sch, 88 & Patient Educ, Hutchinson, Black, Mill & Cook, 90. *Mem:* Am Pub Health Asn; Health Educ Media Asn; Am Social Asn; Am Teachers Prev Med; Soc Pub Health Educators. *Res:* Medical sociology; medical communications; public health education; patient education; curriculum development; mass media health program; health care administration. *Mailing Add:* Dept Prev Med & Commun Health State Univ NY Downstate Med Ctr 450 Clarkson Ave Box 43 Brooklyn NY 11203

SPIEGEL, ALLEN J, b New York, NY, Sept 17, 32. PHARMACY, PHARMACEUTICAL CHEMISTRY. *Educ:* Columbia Univ, BS, 53, MS, 55; Univ Fla, PhD(pharm), 57. *Prof Exp:* Res assoc, Pharmaceut Res & Develop Dept, Chas Pfizer & Co, Inc, 57-66, mgr res coordr, New Prod Dept, Pfizer Int, 66-69, patent agt, Legal Div, 69-76, mgr foreign patents, 76-81, asst dir, 81-86, DIR FOREIGN PATENTS, PFIZER INC, 86- *Mem:* Am Pharmaceut Asn; Am Chem Soc; NY Acad Sci. *Res:* Pharmaceutical product development; research administration; patent law. *Mailing Add:* Legal Div Pfizer Inc 235 E 42nd St New York NY 10017

SPIEGEL, ALLEN M, PATHOPHYSIOLOGY. *Prof Exp:* CHIEF, MOLECULAR PATHOPHYSIOL BR, NAT INST DIABETES DIGESTIVE & KIDNEY DIS, NIH, 86- *Mailing Add:* NIH Nat Inst Diabetes Digestive & Kidney Dis Molecular Pathophysiol Br Bldg 10 Rm 8D17 Bethesda MD 20892

SPIEGEL, EDWARD A, b New York, NY, Mar 7, 31. ASTROPHYSICS, APPLIED MATHS. *Educ:* Univ Calif, Los Angeles, BA, 52; Univ Mich, MS, 54, MA, 56, PhD(astron), 58. *Prof Exp:* From instr to asst prof astron, Univ Calif, 58-60; res scientist, Inst Math Sci, NY Univ, 60-65; from assoc prof to prof physics, 65-69; prof astron, 69-80, RUTHERFORD PROF ASTRON, COLUMBIA UNIV, 80- *Concurrent Pos:* Peyton advan fel, Princeton Univ, 59-60; assoc, Woods Hole Oceanog Inst, 59-; consult, Goddard Inst Space Studies, NASA, 60-69 & Int Coun Asn Sci Educ; NSF sr fel, 66-67. *Mem:* Int Astron Union; NY Acad Sci; Royal Astron Soc. *Res:* Astrophysical fluid dynamics; chaos. *Mailing Add:* Dept Astron Columbia Univ New York NY 10027

SPIEGEL, EUGENE, b Brooklyn, NY, Sept 16, 41; m 68; c 3. MATHEMATICS. *Educ:* Brooklyn Col, BS, 61; Mass Inst Technol, PhD(math), 65. *Prof Exp:* Instr math, Mass Inst Technol, 64-65; Bateman res fel, Calif Inst Technol, 65-66, instr, 66-67; from asst prof to assoc prof, 67-78, dept head, 81-84, PROF MATH, UNIV CONN, 78- *Concurrent Pos:* Vis prof, Ecole Polytechnique Federal, Lausanne, 73, Weizmann Inst, Rehovot, 81. *Mem:* Am Math Soc. *Res:* Algebra; combinatorics. *Mailing Add:* Dept Math Univ Conn Storrs CT 06268

SPIEGEL, EVELYN SCLUFER, b Philadelphia, Pa, Mar 20, 24; m 55; c 2. DEVELOPMENTAL BIOLOGY. *Educ:* Temple Univ, BA, 47; Bryn Mawr Col, MA, 51; Univ Pa, PhD(zool), 54. *Prof Exp:* Asst prog dir regulatory biol, NSF, 54-55; instr, Colby Col, 55-59; res assoc prof, 62-78, RES PROF BIOL, DARTMOUTH COL, 78- *Concurrent Pos:* Lalor fel, Univ Pa, 52-53; Am Cancer Soc fel, 61-62; vis res assoc, Calif Inst Technol, 64-65; vis assoc res biologist, Univ Calif, San Diego, 70-71; vis res scientist, Nat Inst Med Res, Eng, 71; NIH guest investr, 75-76; mem, Corp Marine Biol Lab, 75-, bd trustees, 81-85, 88-92; vis prof, Bioctr, Univ Basel, 79-82, 85. *Mem:* Soc Develop Biol. *Res:* Ultrastructural and immunocytochemical studies of cell adhesion; extracellular matrix and microvilli. *Mailing Add:* Dept Biol Sci Dartmouth Col Hanover NH 03755

SPIEGEL, HERBERT ELI, b New York, NY, July 7, 33; m 58; c 6. BIOCHEMISTRY, CLINICAL CHEMISTRY. *Educ:* Brooklyn Col, BS, 56; George Wash Univ, MS, 61; Rutgers Univ, New Brunswick, PhD(biochem, physiol), 66, Fairleigh Dickinson Univ, MBA, 81. *Prof Exp:* Sr asst health officer biochem pharmacol, NIH, 57-62; sr biochemist, Schering Corp, 62-64; clin chemist, Mountainside Hosp, 64-67; chief clin chemist, Hoffmann-La Roche Inc, 67-72, dir, Dept Clin Biochem, 72-80, dir, Dept Clin Lab Res, 80-85; chief sci off, Quadretek Assocs, 85-87; dir, BurChem, Hearst Corp, 87-89; CHIEF CLIN CHEM, ST VINCENT'S HOSP & MED CTR, 89- *Concurrent Pos:* Mem comn toxicol, Int Union Pure & Appl Chem, 74-; spec adv majority state senate NJ, 83-, cong comt sci & technol, 87; chmn, NJ State Comn Cancer Res, 83-86, vchmn, 86-, mem comn sci & technol, 85-; mem panel kits & devices, Food & Drug Admin, 89- *Honors & Awards:* Am Asn Clin Chemists Awards, 72 & 77. *Mem:* Am Chem Soc; Fel Am Asn Clin Chemists; Fel Am Inst Chemists; Fel Nat Acad Clin Biochemists (pres, 81-); Am Soc Clin Path. *Res:* Biochemical pharmacology, especially catecholamines; clinical chemistry methods, especially fluorimetry and radioactivity; drug effects on clinical chemistry. *Mailing Add:* 39 Greendale Rd Cedar Grove NJ 07009

SPIEGEL, JUDITH E, muscle biochemistry, for more information see previous edition

SPIEGEL, LEONARD EMILE, b New York, NY, Sept 12, 24; m 50; c 5. ECOLOGY. *Educ:* Drew Univ, AB, 48; Northwestern Univ, MS, 50; Cornell Univ, PhD(wildlife mgt), 54. *Prof Exp:* Asst wildlife mgt, State Dept Conserv, NY & Cornell Univ, 51-53; game mgt supvr, Div Wildlife, Ohio Dept Natural Resources, 53-55; instr biol, Alpena Community Col, 55-57; from asst prof to assoc prof, Cent Mich Univ, 57-63; asst prof, Cornell Univ, 63; Chmn Dept, 74-81, PROF BIOL, MONMOUTH COL, NJ, 63- *Concurrent Pos:* Environ consult, 67-; mem, NJ State Mosquito Control Comn, 76-; admin asst, NJ Sea Grant, 83-, assoc dir, 88- *Mem:* Wildlife Soc; Am Inst Biol Sci. *Res:* Ecology of game animals; plant ecology; plant soil wildlife interrelationships. *Mailing Add:* 56 Golf St West Long Branch NJ 07764

SPIEGEL, MELVIN, b New York, NY, Dec 10, 25; m 55; c 2. BIOLOGY. *Educ:* Univ Ill, BS, 48; Univ Rochester, PhD(zool), 52. *Hon Degrees:* MA, Dartmouth Col, 66. *Prof Exp:* Res fel zool, Univ Rochester, 52-53; USPHS res fel biol, Calif Inst Technol, 53-55; asst prof, Colby Col, 55-59; from asst prof to assoc prof zool, 59-66, chmn dept biol sci, 72-74, PROF BIOL, DARTMOUTH COL, 66- *Concurrent Pos:* Mem, NIH Cell Biol Study Sect, 66-70; vis sr res biologist, Univ Calif, San Diego, 70-71; vis prof, Nat Inst Med Res, Eng, 71; bd corp mem & mem bd trustees, Marine Biol Lab, 75-79; prog dir develop biol, NSF, 75-76; mem exec comt, Bd Trustees, Marine Biol Lab, 78-81; vis prof, Bioctr, Univ Basel, 79-82, 85. *Mem:* Am Soc Cell Biologists; Sigma Xi; Am Soc Zoologists; fel AAAS; Int Soc Develop Biologists (secy-treas, 77-81); Soc Develop Biol. *Res:* Developmental biology; protein synthesis; fertilization; cell reaggregation; specificity of cell adhesion. *Mailing Add:* Dept Biol Sci Dartmouth Col Hanover NH 03755

SPIEGEL, ROBERT, b Brooklyn, NY, Dec 28, 28; m 50; c 3. CONTROL SYSTEMS, ENERGY CONSERVATION SYSTEMS. *Educ:* Polytech Inst Brooklyn, BEE, 50, MEE, 54. *Prof Exp:* Chief systs engr, Polarad Electronics, 56-62; prin engr, Gen Precision, 62-67; eng mgr, Micro Power, 67-70; dir eng, Trygon Electronics, 71-74; PRES, ECONOWATT CORP, 75- *Mem:* Inst Elec & Electronics Engrs; Instrument Soc Am. *Res:* Mathematical modeling of heat transfer in commercial buildings; published 11 papers in electronics and energy management; three patents in electronics; energy management. *Mailing Add:* 85 Storer Ave Pelham NY 10803

SPIEGEL, STANLEY LAWRENCE, b New York, NY, Oct 27, 35; m 72; c 3. SPACECRAFT CHARGING. *Educ:* NY Univ, BS, 57; Harvard Univ, AM, 59, PhD(physics), 66. *Prof Exp:* Fel, meteorol dept, Mass Inst Technol, 66-68; res assoc, math dept, Northweastern Univ, 69-73; PROF MATH, UNIV LOWELL, 73- *Concurrent Pos:* Sr scientist, EG&G Environ Consult, 78-79, consult, 79-82; fac res fel, Air Force Off Sci Res, 81 & 82, prin investr, 82-83 & 85-88. *Mem:* Am Geophys Union; Am Meteorol Soc; AAAS; NY Acad Sci; Sigma Xi. *Res:* Derivation, analysis and testing of computer algorithms for the automatic real time determination of space vehicle potentials in various plasma environments, such as at geosynchronous and low earth orbits; numerical modeling of geophysical fluids; electrostatic analyzer measurements. *Mailing Add:* Dept Math Univ Lowell Lowell MA 01854

SPIEGEL, ZANE, b Middletown, NY, Nov 6, 26; m 59; c 2. HYDROLOGY, GEOLOGY. *Educ:* Univ Chicago, BS, 49, MS, 52; NMex Inst Mining & Technol, PhD(earth Sci), 62. *Prof Exp:* Geologist hydrol, US Geol Surv, 49-53; water resources engr, NMex State Engr Off, 54-58; vis prof eng, Imp Col, Univ London, 63-64; proj mgr hydrol, UN Spec Fund, Argentina, 64-66; water resources engr, NMex State Eng Off, 66-71; CONSULT HYDROL, 71- *Concurrent Pos:* Fulbright lectr grant, Univ de San Agustin, Arequipa, Peru, 58-59; water resources res fel, Water Resources Dept, Harvard Univ, 62-63; vis lectr, Univ Minn, Minneapolis, 67-68; vis assoc prof, NMex Inst Mining & Technol, 71; course coordr, Continuing Educ Dept, Col Santa Fe, 73-77, 85; US Environ Protection Agency Extramural Grant Reviewer, 73-75; hydrologist, Ohio State Univ, 80-82; instr physics, Col Santa Fe, 88. *Mem:* Fel Geol Soc Am; fel Am Soc Civil Engrs; Nat Water Well Asn; Am Geophys Union. *Res:* Fundamental concepts of hydrology; impacts of wells on streamflow and estuary salinity; movement and removal of contaminating brines from fresh ground waters; Cenozoic geohydrology; environmental impact analysis; aquifer performance test analysis. *Mailing Add:* PO Box 222 New Canaan CT 06840-0222

SPIEGELBERG, HANS L, b Basel, Switz, Jan 8, 33; c 3. ALLERGY. *Educ:* Univ Basel, MD, 58. *Prof Exp:* Mem staff, Scripps Clin & Res Found, 63-90; PROF & HEAD DEPT PEDIAT, PEDIAT IMMUNOL & ALLERGY DIV, UNIV CALIF SAN DIEGO SCH MED, 90- *Mem:* Am Soc Clin Invest; Am Soc Exp Path; Am Asn Immunologists; Int Col Allergy; Am Acad Allergy; Swiss Soc Allergy & Immunol. *Res:* Immunology. *Mailing Add:* Dept Pediat 0609D Univ Calif San Diego Sch Med La Jolla CA 92093-0609

SPIEGELBERG, HARRY LESTER, b New London, Wis, Apr 24, 36; m 60; c 4. PAPER CHEMISTRY. *Educ:* Univ Wis, BSChE, 59; Inst Paper Chem, MS, 63, PhD(mech, physics), 66; Univ Chicago, MBA, 80. *Prof Exp:* Instr mech, Univ Wis, 57-59; design engr, Kimberly Clark Corp, 59-61, res chemist, 65-68, mgr corp res & eng, New Concepts Lab, 68-73, dir contract res, 72-73, DIR RES & DEVELOP, FEMININE CARE PROD, CONSUMER BUS DIV, KIMBERLY-CLARK CORP, 73-, DIR RES & DEVELOP, CONSUMER & SERV TISSUE PROD, 79-, VPRES RES, CONSUMER TISSUE, 85- *Concurrent Pos:* Chmn, Gordon Res Conf Chem & Physics of Paper, 71 & Indust Liason Coun, Univ Wis. *Res:* Mechanical properties of pulp fibers; mathematical analysis of screening systems; innovative process; long range invention of new products; absorbent materials. *Mailing Add:* 3624 S Barker Lane Appleton WI 34915

SPIEGELHALTER, ROLAND ROBERT, b Dubuque, Iowa, May 31, 23; m 46; c 2. ANALYTICAL CHEMISTRY. *Educ:* Carroll Col, PhB, 48; Univ Kans, MA, 50. *Prof Exp:* Anal chemist, Commercial Solvents Corp, 50-56; sr develop chemist, Chemstrand Corp, 56-64; asst plant chemist, Escambia Chem Corp, 64-69, chief chemist, Eschabia Plant, Air Prod & Chem, Inc, 69-84; RETIRED. *Mem:* Am Chem Soc. *Res:* Titrimetry in nonaqueous solvents. *Mailing Add:* 8411 Country Walk D Pensacola FL 32514-4632

SPIEGELMAN, BRUCE M, b Bayshore, NY, Nov 14, 52. CELL DIFFERENTIATION, CELLULAR DEVELOPMENT. *Educ:* Princeton Univ, PhD(biochem), 78. *Prof Exp:* Fel biol, Mass Inst Technol, 78-82; ASST PROF PHARMACOL, DANA FARBER CANCER INST, MED SCH, HARVARD UNIV, 82- *Mem:* Am Soc Cell Biol; AAAS. *Mailing Add:* Dana Farber Cancer Inst Harvard Univ Med Sch 44 Binney St Boston MA 02115

SPIEGELMAN, GERALD HENRY, b New York, NY, Oct 22, 38; m 60; c 3. ORGANIC CHEMISTRY, INDUSTRIAL CHEMISTRY. *Educ:* City Col New York, BS, 59; Columbia Univ, MA, 60; Stevens Inst Technol, PhD(chem), 69. *Prof Exp:* Chemist, Ultra Div, 60-64, sr res chemist, 67-72, group leader, 72-76, mgr res & develop, 76-79, ASST DIR CORP RES & DEVELOP, WITCO CHEM CORP, 79- *Mem:* Am Chem Soc; Am Oil Chemists Soc; Am Soc Testing & Mat; Am Textile Chemists & Colorists Soc. *Res:* Surfactants; detergents; organometallics; unit processes; process development; analytical chemistry. *Mailing Add:* 211 Indian Rd Wayne NJ 07470-4915

SPIEGELMAN, MARTHA, b New York, NY, May 22, 36; m 64. EMBRYOLOGY, CYTOLOGY. *Educ:* Albertus Magnus Col, BA, 58; Columbia Univ, PhD(biol), 71. *Prof Exp:* Res fel develop genetics, Dept Anat, Med Col, Cornell Univ, 70-71, instr micros anat, 72-74, asst prof anat, 74-76; assoc develop genetics, Mem Sloan-Kettering Cancer Ctr, 76-82; LECTR, SMITH COL, 82- *Concurrent Pos:* Adj assoc prof anat, Med Col, Cornell Univ, 78-81; vis investr, Mem Sloan-Kettering Cancer Ctr, 82- *Mem:* Soc Develop Biol; Am Soc Zoologists; Sigma Xi. *Res:* Fine structural analysis of genetic abnormalities in mouse embryos, especially cellular motility and cell-cell interactions during development. *Mailing Add:* Biol Sci Smith Col Northampton MA 01063

SPIEGLER, KURT SAMUEL, b Vienna, Austria, May 31, 20; nat US; m 46; c 3. CHEMISTRY. *Educ:* Hebrew Univ, Israel, MSc & PhD(chem), 44. *Prof Exp:* Develop chemist, Anglo-Iranian Oil Co, 44-46; res physicist, Palestine Potash Co, 46-47; actg head water purification proj, Weizmann Inst Sci, 48-50; Weizmann Inst & AEC fel, Mass Inst Technol, 50-52; res chemist, Geol Div, Gulf Res & Develop Co, 53-55, sect head phys geochem, 55-59; prof chem, Israel Inst Technol, 59-62; sr scientist, Pratt & Whitney Aircraft Div, United Aircraft Corp, 62-64; prof mech eng in residence, Univ Calif, Berkeley, 64-78; prof chem eng, Mich Technol Univ, 78-81; EMER PROF

MECH ENG, UNIV CALIF, BERKELEY, 78-; PROF, FROMM INST LIFELONG LEARNING, UNIV SAN FRANCISCO, 89- *Concurrent Pos:* Consult, Bur Reclamation, 75-80. *Honors & Awards:* Sr Res Award, Japan Soc Prom Sci, 80. *Mem:* Asn Energy Engrs; Am Chem Soc. *Res:* Ion exchange; electrochemistry; thermodynamics; water purification; fuel cells; membrane physics; geochemistry. *Mailing Add:* Dept Mech Eng Univ Calif Berkeley CA 94720

SPIEKER, ANDREW MAUTE, b Columbus, Ohio, Aug 15, 32; m 61; c 2. HYDROLOGY. *Educ:* Yale Univ, BS, 54; Stanford Univ, MS, 56, PhD(geol), 65. *Prof Exp:* Geologist, US Geol Surv, Ohio, 57-65, hydrologist, Ill, 65- 67, NY, 67-68, staff hydrologist, Water Resources Div, Washington, DC, 68-70, dep proj dir, San Francisco Bay Region Environ & Resources Planning Study, 70-75, western region rep, Lan Info & Anal Off, 76-79, asst dist chief, 79-81, STAFF HYDROLOGIST, WATER RESOURCES DIV, WESTERN REGION, US GEOL SURV, 81- *Honors & Awards:* W R Boggess Award, Am Water Resources Asn, 74. *Mem:* AAAS; Geol Soc Am; Am Geophys Union. *Res:* Hydrology of the urban environment; environmental geology; geology and hydrology of ground water; application of earth sciences to urban and regional planning; stratigraphy. *Mailing Add:* 341 Linfield Dr Menlo Park CA 94025

SPIELBERG, NATHAN, b Philadelphia, Pa, Feb 2, 26; m 47; c 3. PHYSICS. *Educ:* Emory Univ, AB, 47; Ohio State Univ, MS, 48, PhD(physics), 52. *Prof Exp:* Asst physics, Ohio State Univ, 47-49, res assoc, 51-53, asst prof welding eng, 53-54; assoc physicist, Philips Labs, NAm Philips Co, Inc, 54-58, sr physicist, 58-60, staff physicist, 60-65, res physicist, 65-69; PROF PHYSICS, KENT STATE UNIV, 69- *Concurrent Pos:* Adj prof geosciences, Tex Tech Univ, 78; mem, Liquid Crystal Inst, Kent State Univ, 79-; vis prof, Weizmann Inst, 83-84. *Mem:* Am Phys Soc; Am Asn Physics Teachers; Am Crystallog Asn; Electron Probe Analysis Soc Am; Sigma Xi. *Res:* X-ray spectrochemical analysis; high resolution x-ray spectroscopy; x-ray physics; crystal perfection; x-ray interferometry; structure of liquid crystals. *Mailing Add:* Dept Physics Kent State Univ Kent OH 44242

SPIELBERG, STEPHEN E, b Philadelphia, Pa, June 7, 34. MATHEMATICS. *Educ:* Univ Pa, BA, 56; Univ Minn, MA, 58, PhD(math), 63. *Prof Exp:* Asst prof, 63-77, ASSOC PROF MATH, UNIV TOLEDO, 77- *Mem:* Math Asn Am; Am Math Soc. *Res:* Probability; statistics; Wiener integrals. *Mailing Add:* Dept Math Univ Toledo 2801 W Bancroft St Toledo OH 43606

SPIELBERGER, CHARLES DONALD, b Atlanta, Ga, Mar 28, 27; m 71; c 3. CLINICAL PSYCHOLOGY, BEHAVIORAL MEDICINE. *Educ:* Ga Inst Technol, BS, 49; Univ Iowa, BA, 51, MA, 53, PhD(psychol), 54. *Prof Exp:* Asst prof psychiat, Med Sch, Duke Univ, 55-58, from asst prof to assoc prof psychol, 55-63; prof psychol, Vanderbilt Univ, 63-67; training specialist, NIMH, 65-67; prof psychol, Flat State Univ, 67-72; prof psychol, 72-85, DISTINGUISHED UNIV RES PROF PSYCHOL, UNIV SFLA, 85- *Concurrent Pos:* Dir, Ctr Res Behav Med & Health Psychol, Univ SFla, 77-; ed, Am J Community Psychol, 71-78. *Honors & Awards:* Distinguished Contribution Community Psychol, Am Psychol Asn, 82, Distinguished Contribution Clin Psychol, 89; Distinguished Contribution Personality Assessment, Soc Personality Assessment, 90. *Mem:* Fel Am Psychol Asn (treas, 87-90, pres, 91-92); Soc Personality Assessments (pres, 86-); Int Coun Pyschologists (pres, 86-87); Int Soc Test Anxiety Res (pres, 82-84). *Res:* Nature and assessment of anxiety, anger, curiosity and job stress; anger expression and control; cross-cultural research on emotion; stress management, behavioral medicine and health psychology; mental health consultation; community psychology. *Mailing Add:* Dept Psychol Univ SFla Tampa FL 33620-8200

SPIELER, HELMUTH, b Irvington, NJ, Aug 25, 45; m 71; c 2. PHYSICS OF SEMICONDUCTOR DEVICES. *Educ:* Tech Univ Munich, Ger, dipl physics, 71, Dr Rer Nat, 74. *Prof Exp:* Staff scientist, Tech Univ Munich, 73-75 & GSI Darmstadt, 75-82; staff scientist, 82-87, SR STAFF SCIENTIST, LAWRENCE BERKELEY LAB, 87- *Mem:* Am Phys Soc; Inst Elec & Electronic Engrs. *Res:* Radiation detectors and electronics; semiconductor detector systems for high energy physics; physics of semiconductor devices; design and fabrication of integrated circuits. *Mailing Add:* Lawrence Berkeley Lab 50B-6208 One Cyclotron Rd Berkeley CA 94720

SPIELER, RICHARD ARNO, b Syracuse, NY, Apr 8, 32. GENETICS, ZOOLOGY. *Educ:* Univ Chicago, BA, 52, PhD(zool), 62. *Prof Exp:* Asst prof zool, Ill Inst Technol, 62-68; from asst prof to assoc prof, 68-77, PROF BIOL, CALIF STATE UNIV, FRESNO, 77- *Mem:* AAAS. *Res:* Genetics of meiosis; parental care by reptiles and amphibians. *Mailing Add:* Dept of Biol Calif State Univ Fresno CA 93740

SPIELER, RICHARD EARL, b Washington, DC, Mar 11, 42; m 87; c 2. CHRONOBIOLOGY, ICHTHYOLOGY. *Educ:* Univ Md, BA, 63; Ark State Univ, BS, 70, MS, 71; La State Univ, PhD(marine sci), 75. *Prof Exp:* CUR FISHES, MILWAUKEE PUB MUS, 75- *Concurrent Pos:* Instr anat, fisheries biol & ichthyol, Ark State Univ, 70-71; adj prof, Univ Wis, Milwaukee, 75-; grants, numerous orgn, 76-; instr zool, Milwaukee Inst Art & Design, 85-87; actg head vertebrate zool, Milwaukee Pub Mus, 87-89. *Mem:* AAAS; Am Fisheries Soc; Am Soc Zoologists; Int Soc Chronobiol; World Aquacult Soc. *Res:* Temporal integration of fishes with emphasis on both basic and applied aspects; author of numerous publications. *Mailing Add:* Milwaukee Pub Mus Milwaukee WI 53233

SPIELHOLTZ, GERALD I, b New York, NY, Mar 12, 37; m 78; c 1. ANALYTICAL CHEMISTRY. *Educ:* City Col New York, BS, 58; Univ Mich, MS, 60; Iowa State Univ, PhD(anal chem), 63. *Prof Exp:* Assoc anal chem, Iowa State Univ, 63; from instr to asst prof chem, Hunter Col, 63-68; from asst prof to assoc prof, 68-75, PROF CHEM, LEHMAN COL, 75- *Mem:* Am Chem Soc; Sigma Xi. *Res:* Atomic absorption spectroscopy; wet oxidation of materials prior to analysis; analytical chemistry applied to anthropology. *Mailing Add:* Dept of Chem Herbert H Lehman Col Bronx NY 10468

SPIELMAN, ANDREW, b New York, NY, Feb 24, 30; m 55; c 3. MEDICAL ENTOMOLOGY. *Educ:* Colo Col, BS, 52; Johns Hopkins Univ, ScD(med entom), 56. *Prof Exp:* Biologist, Tenn Valley Authority, 53; from instr to assoc prof, 59-80, PROF TROP PUB HEALTH, SCH PUB HEALTH, HARVARD UNIV, 80- *Honors & Awards:* Medal of Honor, Am Mosquito Control Asn, 88. *Mem:* Am Soc Trop Med & Hyg; Entom Soc Am; Am Mosquito Control Asn; Am Soc Zool; AAAS. *Res:* Epidemiology of lyme disease, babesiosis and arthropod-borne diseases; physiology of salivation and reproduction in mosquitoes. *Mailing Add:* Dept Trop Pub Health 665 Huntington Ave Harvard Sch Pub Health Boston MA 02115

SPIELMAN, BARRY, b Chicago, Ill, Oct 29, 42; m 66; c 2. EDUCATIONAL ADMINISTRATION, RESEARCH ADMINISTRATION. *Educ:* Ill Inst Technol, BS, 64; Pa State Univ, MS, 67; Syracuse Univ, PhD(elec eng), 71. *Prof Exp:* Res electronics engr, Microwave Integrated Circuits, Naval Res Lab, 71-73, head, Millimeter Wave Tech Sect, 78-84, Solid State Circuits Sect, 78-84, Microwave Technol Br, 84-87; PROF ELECTROMAGNETICS & CHMN, DEPT ELEC ENG, WASH UNIV, ST LOUIS, 87-, DIR, MICROELECTRONIC SYSTS LAB, 90- *Concurrent Pos:* Consult, McDonnell-Douglas Cent Res Lab, 88-89; rep, Nat comt Superconductivity, Inst Elec & Electronics Engrs, Microwave Theory & Tech Soc, 89-90, mem & chmn, Adcom Tech Comt, MTT-6. *Honors & Awards:* Distinguished Serv Award, Inst Elec & Electronics Engrs Microwave Theory & Tech Soc, 89. *Mem:* Sigma Xi; Inst Elec & Electronics Engrs Theory & Tech Soc (secy-treas 73, vpres 87, pres 88). *Res:* Microwave and millimeter-wave planar superconducting components including directional couplers, filters, phase shifters, modulators and switches; applied numerical solutions to electromagnetic problems pertinent to the specific components; experimental measurements verify computations. *Mailing Add:* Wash Univ One Brookings Dr Box 1127 St Louis MO 63130

SPIELMAN, HAROLD S, b Philadelphia, Pa, Dec 11, 14; m 41; c 1. ELECTRONICS, SCIENCE EDUCATION. *Educ:* City Col New York, BS, 34, MS, 35; Columbia Univ, EdD(sci ed), 50. *Prof Exp:* Teacher high sch, Pa, 36-42; instr radio theory, Cent Signal Corps Sch, 42-44; instr high sch, NY, 46-50; PROF SCI EDUC, CITY COL NEW YORK, 50-, CHMN DEPT SEC EDUC, 68- *Concurrent Pos:* Nat Sci Teachers Asn liason rep, Nat Asn Indust-Ed Coop, 64-; consult, Sci Teachers Workshop, Prentice-Hall Inc, 65- *Mem:* Fel AAAS; Asn Res Sci Teaching; Nat Sci Teachers Asn. *Res:* Electronics. *Mailing Add:* 531 Main St Roosevelt Island NY 10044

SPIELMAN, RICHARD SAUL, b New York, NY, Feb 25, 46; div; c 1. HUMAN GENETICS, QUANTITATIVE VARIATION. *Educ:* Harvard Col, AB, 67; Univ Mich, Ann Arbor, PhD(human genetics), 71. *Prof Exp:* Res assoc human genetics, Med Sch, Univ Mich, Ann Arbor, 71-74; from asst prof to assoc prof, 74-89, PROF HUMAN GENETICS, MED SCH, UNIV PA, 89- *Concurrent Pos:* Vis scholar, Imp Cancer Res Fund, London, 82-83. *Mem:* Sigma Xi; AAAS; Genetics Soc Am; Am Soc Human Genetics; Am Diabetes Asn. *Res:* Human variation; biometric genetics; genetics of disease susceptibility. *Mailing Add:* Dept Human Genetics Sch Med Univ Pa 422 Curie Blvd Philadelphia PA 19104-6145

SPIELMAN, WILLIAM SLOAN, b Tulsa, Okla, Aug 7, 47; m 70; c 2. CARDIOVASCULAR & RENAL PHYSIOLOGY, CELL & MOLECULAR BIOLOGY. *Educ:* Westminster Col, BA, 69; Univ Mo, Columbia, PhD(physiol), 74. *Prof Exp:* Fel physiol, Univ NC, 75-77; fel & instr, Mayo Med Sch, 77-78, asst prof, 78-80; from asst to assoc prof, 80-87, PROF PHYSIOL, MICH STATE UNIV, 87-, DIR GRAD STUDIES, 89- *Concurrent Pos:* NIH career develop award, 81. *Mem:* Am Physiol Soc; Sigma Xi; Am Fedn Clin Res; Am Soc Nephrology. *Res:* Normal and abnormal function of the kidney; receptors and signalling mechanisms. *Mailing Add:* Dept Physiol 502 Biochem Bldg Mich State Univ East Lansing MI 48824-1319

SPIELVOGEL, BERNARD FRANKLIN, b Ellwood City, Pa, Apr 23, 37; m 63, 81; c 3. INORGANIC CHEMISTRY. *Educ:* Geneva Col, BS, 59; Univ Mich, PhD(chem), 63. *Prof Exp:* From instr to asst prof chem, Univ NC, 63-67; chief, Inorg Br, Chem Div, US Army Res Off, 67-88; PRES, BORON BIOLOGICALS, INC, 86- *Concurrent Pos:* Vis sr res assoc, Duke Univ, 67-72, adj assoc prof chem, 72-81, adj prof chem, 81- *Honors & Awards:* NC Distinguished Chemist Award, Am Inst Chem, 84. *Mem:* Am Chem Soc (chmn NC sect, 84). *Res:* Boron hydride chemistry; synthesis of Boron analogs of amino acids; peptides, DNA and other biologically important molecules. *Mailing Add:* Boron Biologicals Inc 2118 O'Berry St Raleigh NC 27607

SPIELVOGEL, LAWRENCE GEORGE, b Newark, NJ, June 2, 38. FORENSIC ENGINEERING. *Educ:* Drexel Univ, BS, 62. *Prof Exp:* Assoc, Robert G Werden & Assoc, Inc, 59-70; PRES, LAWRENCE G SPIELVOGEL, INC, 70- *Concurrent Pos:* Asst post eng, Walter Reed Med Ctr, 63-65; eng, Utility Surv Corp, 65-66; adj instr, Evening Col, Drexel Univ, 68-84; instr, plumbing design course, Am Soc Sanit Engrs, 69-78; adj asst prof, Col Eng Technol, Temple Univ, 71-74; lectr, Grad Sch Fine Arts, Univ Pa, 71-78 & Sch Continuing Educ, NY Univ, 76-79; vis lectr, Grad Sch Archit, Yale Univ, 75-82. *Honors & Awards:* Crosby Field Award, Am Soc Heating, Refrig & Air Conditioning Engrs, 81. *Mem:* Am Soc Heating, Refrig & Air Conditioning Engrs; Illuminating Eng Soc; Am Consult Engrs Coun; Am Soc Mech Engrs; Chartered Inst Bldg Serv Eng. *Res:* Energy in buildings. *Mailing Add:* Wyncote House Wyncote PA 19095-1499

SPIELVOGEL, LESTER Q, b Brooklyn, NY, June 27, 37; m 69; c 2. ABSTRACT WAVE THEORY, OCEANOGRAPHY. *Educ:* Cooper Union, BME, 59; Columbia Univ, MS, 62; NY Univ, PhD(math), 69. *Prof Exp:* Mech engr, Sperry Rand Corp, 59-69; asst prof civil eng, Univ Hawaii, 69-70; asst physics, Nat Oceanic & Atmospheric Admin, US Dept Com, 70-71, physicist oceanog, 71-81; CHIEF SCIENTIST, SEACO, 81- *Concurrent Pos:* Adj prof oceanog, Univ Hawaii, 74-83. *Mem:* Soc Indust Appl Math; AAAS. *Res:* Ocean waves and abstract wave theory; engineering analysis; engineering physics; mathematics. *Mailing Add:* 619 Kumukahi Pl Honolulu HI 96825

SPIER, EDWARD ELLIS, b Los Angeles, Calif, Aug 21, 21; m 45; c 3. SEMI-EMPIRICAL ANALYSIS OF COMPOSITES, TESTING OF COMPOSITE MATERIALS. *Educ:* San Diego State Univ, BS, 66. *Prof Exp:* Eng specialist struct anal, Gen Dynamics Convair Div, 58-84; ENG STAFF SPECIALIST, RES & DEVELOP STRUCT, ROHR INDUST INC, 84- *Concurrent Pos:* Mem, Mat Adv Bd, Nat Acad Sci, 64-66. *Res:* Authored or co-authored approximately 25 papers, most of which involved analysis and testing of carbon/epoxy and metal matrix composite structures; wrote reports on semi-empirical analysis of carbon/epoxy short-column stiffened panels. *Mailing Add:* 5245 Joan Ct San Diego CA 92115

SPIERER, PIERRE, b Geneva, Switz, Aug 26, 48; m 74; c 3. CHROMOSOME ORGANIZATION. *Educ:* Univ Geneva, BS, 71, MS, 72, PhD(biol), 75. *Prof Exp:* Fel biochem, Univ Mass, 75-76; fel molecular genetics, Univ Stanford, Calif, 77-79; RES SCIENTIST MOLECULAR GENETICS, UNIV GENEVA, 80- *Res:* Molecular genetics of development of the fruit fly Drosophila; analysis of chromosome organization and of genes controlling development. *Mailing Add:* Dept Animal Biol Genetics Lab Univ Geneva 154 Rte de Malagnou 1224 Chene Bougeries Switzerland

SPIERS, DONALD ELLIS, b Richmond, Va, 48; m 84; c 2. TEMPERATURE REGULATION, ENVIRONMENTAL PHYSIOLOGY. *Educ:* Va Polytech Inst & State Univ, BS, 70, MS, 72; Mich State Univ, PhD(physiol), 80. *Prof Exp:* Asst fel, 82-90, ASSOC FEL, JOHN B PIERCE FOUND, 90-; ASST PROF EPIDEMIOL, YALE UNIV, 84- *Concurrent Pos:* Biologist, Peace Corps, 73-75. *Mem:* Am Physiol Soc; Am Soc Zool; Am Inst Biol Sci; Bioelectromagnetics Soc; Sigma Xi; NY Acad Sci. *Res:* Environmental physiology; development of temperature regulation in birds and mammals; neonatal responses to stress. *Mailing Add:* John B Pierce Found 290 Congress Ave New Haven CT 06519

SPIERS, JAMES MONROE, b Wiggins, Miss, July 31, 40; m 65; c 3. PLANT PHYSIOLOGY, AGRONOMY. *Educ:* Miss State Univ, BS, 63, MS, 66; Tex A&M Univ, PhD(agron, crop physiol), 69. *Prof Exp:* Res plant physiologist, 69-71, res plant physiologist-in-charge fruit & forage res, 71-73, RES HORTICULTURIST, LOCATION LEADER & RES LEADER, SMALL FRUIT RES STA, SCI & EDUC ADMIN-FED RES, USDA, 73- *Mem:* Int Soc Hort Sci; Am Soc Agron; Am Soc Hort Sci; Crop Sci Soc Am. *Res:* Nutrition and cultural requirements of blueberries, strawberries and blackberries; hormonal regulation of rooting, flowering and growth of blueberries. *Mailing Add:* USDA Small Fruit Res Sta PO Box 287 Poplarville MS 39470

SPIES, HAROLD GLEN, b Mountain View, Okla, Mar 30, 34; m 90; c 1. NEUROENDOCRINOLOGY, REPRODUCTIVE BIOLOGY. *Educ:* Okla State Univ, BS, 56; Univ Wis, MS, 57, PhD(animal sci & genetics), 59. *Prof Exp:* Asst, Univ Wis, 56-59; from asst prof to assoc prof animal physiol, Kans State Univ, 59-66; res assoc anatomist, Univ Calif, Los Angeles, 67-68; res scientist, Delta Regional Primate Res Ctr, 68-72, chmn reproductive physiol, Ore Regional Primate Res Ctr, 72-83, ASSOC DIR RES, ORE REGIONAL PRIMATE RES CTR, 83- *Concurrent Pos:* NIH spec fel neuroendocrinol res, Univ Calif, Los Angeles, 66-67; assoc prof, Tulane Univ, 68-72; prof, Med Sch, Univ Ore, 73- *Mem:* Endocrine Soc; Am Asn Anat; Soc Study Reproduction (vpres, 79-80, pres, 80-81); Am Physiol Soc; Am Soc Animal Sci; Neurosci Soc; Soc Study Exp Biol Med. *Res:* Reproductive physiology and endocrinology of laboratory animals and primates; hypothalamo-hypophysial-gonadal interrelationships; neural regulation of endocrine changes in the menstrual cycle. *Mailing Add:* Ore Regional Primate Res Ctr 505 NW 185th Ave Beaverton OR 97006

SPIES, JOSEPH REUBEN, ANALYTICAL CHEMISTRY, IMMUNOCHEMISTRY. *Educ:* Univ Md, College Park, PhD(org chem), 34. *Hon Degrees:* LHD, Univ SDak, 87. *Prof Exp:* Head Allergen Invests, USDA, 36-73; RETIRED. *Mailing Add:* 507 N Monroe St Arlington VA 22201

SPIES, ROBERT BERNARD, b Palo Alto, Calif, May 21, 43; m 63; c 3. MARINE BIOLOGY, MARINE ECOLOGY. *Educ:* St Mary's Col, BS, 65; Univ Pac, MS, 69; Univ Southern Calif, PhD(marine biol), 71. *Prof Exp:* Sr res officer marine ecol, Fisheries & Wildlife Dept, Victoria, Australia, 70-73; SR ENVIRON SCIENTIST, ENVIRON DIV, LAWRENCE LIVERMORE LAB, UNIV CALIF, 73- *Concurrent Pos:* Ed, Marine Environ Res, 87-; pres, Appl Marine Sci Inc, Livermore, Calif; chief scientist, Exxon Valdez Oil Spill Assessment, US Govt, Alaska. *Mem:* AAAS; Am Soc Limnol & Oceanog; Western Soc Naturalists; Am Chem Soc; Soc Environ Toxicol Chem. *Res:* Effects of xenobiotic compounds on reproduction in estuarine fish; effects of petroleum hydrocarbons in the marine benthos; dynamics of petroleum hydrocarbons, trace elements and radionuclides in benthic organisms; dynamic processes in marine benthos. *Mailing Add:* PO Box 5507 L453 Livermore CA 94550

SPIESS, ELIOT BRUCE, b Boston, Mass, Oct 13, 21; m 51; c 2. GENETICS. *Educ:* Harvard Univ, AB, 43, AM, 47, PhD(genetics), 49. *Prof Exp:* Instr biol, Harvard Univ, 49-52; from asst prof to prof biol, Univ Pittsburgh, 52-66; prof biol sci, 66-89, EMER PROF BIOL SCI, UNIV ILL, CHICAGO, 89- *Concurrent Pos:* Am Acad Arts & Sci grant in aid, 53; AEC res grant, 55-72; dir NIH grad training grant genetics, Univ Pittsburgh, 63-66; NSF res grant, 72-83; ed, Evolution, Soc Study Evol Jour, 75-78; pres, Am Soc Naturalists, 81. *Mem:* Behav Genetics Asn; fel AAAS; Genetics Soc Am; Soc Study Evolution; Am Soc Naturalists (pres, 81). *Res:* Genetics of adaptive mechanisms in populations of Drosophila; behavior genetics. *Mailing Add:* Dept Biol Sci Box 4348 Univ Ill Chicago Chicago IL 60680

SPIESS, FRED NOEL, b Oakland, Calif, Dec 25, 19; m 42; c 5. OCEANOGRAPHY. *Educ:* Univ Calif, AB, 41, PhD(physics), 51; Harvard Univ, MS, 46. *Prof Exp:* Nuclear engr, Knolls Atomic Power Lab, Gen Elec Co, 51-52; res physicist, 52-61, dir, Marine Phys Lab, 58-80, prof oceanog, 61-90, actg dir, Scripps Inst Oceanog, 61-63, chmn dept oceanog, 63-64 & 76-77, dir, 64-65, assoc dir, 65-80, dir, Inst Marine Resources, Univ Calif, San Diego, 80-88, EMER PROF OCEANOG, INST MARINE RESOURCES, UNIV CALIF, SAN DIEGO. *Honors & Awards:* Ewing Medal, Am Geophys Union-US Navy, 83; Pioneers of Underwater Acoust Medal, Acoust Soc Am, 85; Marine Technol Soc-Lockheed Award, 85; Sigma Xi, 49; Franklin Inst, John Price Wetherill Medal, 65; AAAS Newcomb Cleveland Prize, 80; Conrad Medal, USN, 74. *Mem:* Nat Acad Eng; fel Acoust Soc Am; fel Am Geophys Union; fel Marine Technol Soc. *Res:* Underwater acoustics; marine geophysics; ocean technology. *Mailing Add:* Scripps Inst Oceanog Univ Calif San Diego La Jolla CA 92093-0205

SPIESS, JOACHIM, b Ludenscheid, Ger, April 20, 40; m 67; c 3. PROTEIN CHEMISTRY. *Educ:* Univ Munchen, MD, 73, PhD, 76. *Prof Exp:* Wiss asst, Max Planck Inst Biochem, 73-76; fel, Salk Inst, 76-77, from res assoc to assoc res prof, 78-83; assoc res prof, Salle Inst, 83-87; PRES, SCI MEM & DIR, MAX PLANCK INST EXP MED, GOETTINGEN, 87- *Concurrent Pos:* Adj prof, Salle Inst, 90. *Mem:* Ger Chem Soc; Endocrine Soc; Am Chem Soc; Am Soc Biol Chemists. *Res:* Conducting biochemistry of hormonal peptides. *Mailing Add:* Dept Molecular Neuroendocrinol Max Planck Inst Exper Med Hermann Rein Str 3 Goettingen 3400 Germany

SPIESS, LURETTA DAVIS, developmental biology, for more information see previous edition

SPIETH, HERMAN THEODORE, b Charlestown, Ind, Aug 21, 05; m 31; c 1. ZOOLOGY. *Educ:* Ind Cent Col, AB, 26; Ind Univ, PhD(zool), 31. *Hon Degrees:* LLD, Ind Cent Col, 58. *Prof Exp:* Asst zool, Ind Univ, 26-30, instr, 31-32; from instr to assoc prof biol, City Col New York, 32-53; prof zool & chmn div life sci, Univ Calif, Riverside, 53-56, provost, 56-58, chancellor, 58-64; prof zool, 64-73, chmn dept, 64-71, EMER PROF ZOOL, UNIV CALIF, DAVIS, 73- *Concurrent Pos:* Lectr, Columbia Univ, 38-53; res assoc, Am Mus Natural Hist, NY, 43-56. *Mem:* Am Soc Naturalists; Soc Study Evolution; Am Soc Zool; fel Entom Soc Am; fel Animal Behav Soc. *Res:* Biology and taxonomy of Ephemeroptera; sexual behavior in Drosophila. *Mailing Add:* Dept of Zool Univ of Calif Davis CA 95616

SPIETH, JOHN, MOLECULAR & DEVELOPMENTAL GENETICS, GENE EXPRESSION. *Educ:* Univ Wash, PhD(zool), 78. *Prof Exp:* RES ASSOC, DEPT BIOL, IND UNIV, 78- *Mailing Add:* Dept Biol Jordan Hall Ind Univ Bloomington IN 47405

SPIETH, PHILIP THEODORE, b New York, NY, June 10, 41; m 63; c 4. POPULATION GENETICS, EVOLUTION. *Educ:* Univ Calif, Berkeley, AB, 62; Univ Ore, PhD(biol), 70. *Prof Exp:* Ford Found fel pop biol, Univ Chicago, 70-71; asst prof & asst geneticist, 71-76, ASSOC PROF & ASSOC RES GENETICIST, UNIV CALIF, BERKELEY, 76-, ASSOC DEAN, STUDENT AFFAIRS, COL NAT RES, 80- *Mem:* Genetics Soc Am; Soc Study Evolution; Mycol Soc Am. *Res:* Genetic variation of natural populations; empirical population genetics of fungi; microevolutionary processes of speciation. *Mailing Add:* Dept Plant 147 Hilgard Hall Univ Calif 2120 Oxford St Berkeley CA 94720

SPIGARELLI, STEVEN ALAN, b Highland Park, Ill, Mar 26, 42. FISH ECOLOGY, POLLUTION BIOLOGY. *Educ:* Northwestern Univ, BA, 64; Univ Ill, Urbana, MS, 66; Mich State Univ, PhD(aquatic ecol), 71. *Prof Exp:* Asst ecologist, Argonne Nat Lab, 71-74, ecologist aquatic ecol, 74-82; DIR CTR ENVIRON STUDIES, BEMIDJI STATE UNIV, 82- *Concurrent Pos:* Mem res comt, Int Atomic Energy Agency, 75-79; assoc ed, J Great Lakes Res, 77- *Mem:* Am Fisheries Soc; Fisheries Res Bd Can; Int Asn Great Lakes Res; Am Chem Soc. *Res:* Thermal ecology; radioecology; fish behavior; stress ecology. *Mailing Add:* Dept Environ Studies Bemidji State Unvi Bemidji MN 56601

SPIGHT, CARL, physics, for more information see previous edition

SPIKE, CLARK GHAEL, b Ypsilanti, Mich, Sept 15, 21; m 47; c 2. INORGANIC CHEMISTRY. *Educ:* Mich State Norm Col, BS, 44; Univ Mich, PhD(chem), 52. *Prof Exp:* Instr chem, Mich State Norm Col, 46-48; res chemist, Ethyl Corp, 52-58; assoc prof chem, 58-61, head dept, 61-77, interim dean, Col Arts & Sci, 77-79, actg assoc vpres acad affairs, 79-80, PROF CHEM, EASTERN MICH UNIV, 61- *Mem:* AAAS; Am Chem Soc; Sigma Xi. *Res:* Coordination complexes; metallo-organic compounds. *Mailing Add:* 18580 Grass Lake Rd Manchester MI 48158

SPIKER, STEVEN L, b Omaha, Nebr, Nov 1, 41; m 71; c 1. PLANT CHROMATIN STRUCTURE. *Educ:* Univ Iowa, BS, 64, MS, 67, PhD(plant physiol), 70. *Prof Exp:* Asst prof biol, Am Univ Beirut, 72-74; asst prof molecular biol, Dept Bot & Plant Path, Ore State Univ, 78-81; ASSOC PROF MOLECULAR GENETICS, NC STATE UNIV, 81- *Mem:* Am Soc Plant Physiologists; Genetics Soc Am; Am Soc Biochem & Molecular Biol. *Res:* Isolation, characterization and evolution of plant chromosomal proteins; physical and chemical studies of protein-protein interactions in the nucleosome; role of chromosomal proteins in forming transcribable or inert chromatin structure. *Mailing Add:* Genetics Dept NC State Univ Raleigh NC 27695-7614

SPIKES, JOHN DANIEL, b Los Angeles, Calif, Dec 14, 18; m 42; c 3. PHOTOBIOLOGY. *Educ:* Calif Inst Technol, BS, 41, MS, 46, PhD(chem, embryol), 48. *Prof Exp:* From asst prof to assoc prof biol, Univ Utah, 48-55, head dept exp biol, 54-62, dean, Col Lett & Sci, 64-67, chmn dept biol, 84-85, prof biol, 55-88, EMER PROF BIOL, UNIV UTAH, 89- *Concurrent Pos:* Cell physiologist, US AEC, Washington, DC, 58-60, consult, 60-65; counr, Smithsonian Inst, 66-72; mem comt photobiol, Nat Acad Sci-Nat Res Coun, 72-75; vis prof, Univ Padua, Italy. *Honors & Awards:* distinguished Res Prof Award, Univ Utah, 72; Medal, Europe Soc Photobiol, 89. *Mem:* Am Chem Soc; Inter-Am Photochem Soc; Am Soc Photobiol (pres, 74-75); Europ Photochem Asn; Europ Soc Photobiol; Am Cornea Soc. *Res:* Photobiology; photosensitized reactions; mechanisms of the sensitized photooxidation of biomolecules; sensitized photo effects on cells; photosensitized reactions as a tool in biology and medicine. *Mailing Add:* Dept Biol Univ Utah Salt Lake City UT 84112

SPIKES, JOHN JEFFERSON, toxicology, clinical chemistry, for more information see previous edition

SPIKES, PAUL WENTON, b Ft Worth, Tex, Mar 22, 31; m 53; c 2. MATHEMATICAL ANALYSIS. *Educ:* Miss Southern Col, BS, 53, MA, 57; Auburn Univ, PhD(math), 70. *Prof Exp:* Instr math, Copiah-Lincoln Jr Col, 57-58; assoc prof, William Carey Col, 58-65; chmn, Div Sci & Math, Alexander City State Jr Col, 66-68, chmn, Eve Div, 69-70; asst prof, 70-73, assoc prof, 73-78, PROF MATH, MISS STATE UNIV, 78- *Mem:* Am Math Soc; Math Asn Am. *Res:* Qualitative theory of ordinary differential equations. *Mailing Add:* Dept Math Box MA Miss State Univ Mississippi State MS 39762

SPILBURG, CURTIS ALLEN, b Cleveland, Ohio, May 27, 45; m 79; c 3. ENZYMOLOGY, LIPIDS. *Educ:* Carnegie Inst Technol, BS, 67; Northwestern Univ, MS, 69, PhD(phys chem), 72. *Prof Exp:* Fel biol chem, Harvard Med Sch, 72-76; res specialist, 76-78, group leader protein chem, Monsanto Co, 78-85; ASST PROF, JEWISH HOSP, WASHINGTON UNIV SCH MED, 85- *Concurrent Pos:* NIH fel, Harvard Med Sch, 72-74. *Res:* Enzyme isolation and characterization; structure function studies; chemical modification of proteins; lipids. *Mailing Add:* Dept Cardiol Jewish Hosp 216 S Kings Hwy St Louis MO 63110

SPILHAUS, ATHELSTAN FREDERICK, b Cape Town, SAfrica, Nov 25, 11; nat US; m 79; c 5. METEOROLOGY, OCEANOGRAPHY. *Educ:* Univ Cape Town, BSc, 31, DSc, 48; Mass Inst Technol, SM, 33. *Hon Degrees:* DSc, Coe Col, 61, Univ RI, 68, Hahnemann Med Col, 68, Philadelphia Col Pharm & Sci, 68, Hamilton Col, 70, Southeastern Mass Univ, 70, Univ Durham, 70, Univ SC, 71, Southwestern at Memphis, 72; LLD, Nova Univ Advan Technol, 70 & Univ Md, 79. *Prof Exp:* Vol engr, Junkers Airplane Works, Ger, 31-32; res engr, Sperry Gyroscope Co, NY, 33; asst meteorol, Mass Inst Technol, 34-35; asst dir tech serv, Dept Defence, Union SAfrica, 35-36; asst, Woods Hole Oceanog Inst, 36-37, phys oceanogr, 38-60; prof geophys, Univ Minn, Minneapolis, 66-67, dean, Inst Sci, 49-66; pres, Franklin Inst, Pa, 67-69 & Aqua Int, Inc, 69-71; fel, Woodrow Wilson Int Ctr Scholars, Smithsonian Inst, 71-74; spec asst to adminr, Nat Oceanic & Atmospheric Admin, US Dept Com, 74-80; PRES, PAN/GEO, INC, 84- *Concurrent Pos:* From asst prof to prof, NY Univ, 37-48, chmn dept meteorol, 38-47, dir res eng & phys sci, 46-48; mem subcomt meteorol, Nat Adv Comt, Aeronaut, 41-56; consult, Div Ten, Nat Defense Res Comt, 42-43, SAfrican Govt, 47, Brookhaven Nat Lab, 47-49, US Weather Bur, 47-56, Air Materiel Command & Sci Adv Bd, US Dept Air Force, 48-58, US Dept Defense & Res & Develop Adv Coun, Signal Corps, US Dept Army & Nat Oceanic & Atmospheric Admin, 80-; sci dir weapons effects, Atomic Tests, Nev, 51; cem, Baker Mission, Korea, 52; US rep exec bd, UNESCO, 54-58; comnr, US Sci Exhib, Seattle World's Fair, 61-63; mem comt pollution, Nat Acad Sci; mem, Nat Sci Bd, 66-; mem bd trustees, Aerospace Corp, El Segundo, Sci Serv, Inc, Int Oceanog Found & Pac Sci Ctr Found; chmn sci adv comt, Am Newspaper Publ Asn; mem bd trustees, Sea Educ Asn, 74-; vis prof, Tex A&M Univ, 74-75; Phi Beta Kappa lectr, 76-77; distinguished vis prof, Univ Tex, 77-78; Annenberg scholar, Univ Southern Calif, 81, vis scholar, Inst Marine & Coastal Studies, 81-83. *Honors & Awards:* Berzelius Medal, Sweden, 62; Proctor Prize, Sci Res Soc Am, 68; Compass Award, Marine Tech Soc, 81. *Mem:* Fel AAAS(pres, 70, chmn, 71); fel Am Geog Soc; Am Inst Aeronaut & Astronaut; fel Geog Soc; fel Royal Meteorol Soc; fel Am Geophys Union; Am Philos Soc. *Res:* Spilhaus space clock; bathythermograph; aircraft, meteorological and oceanographic instruments; physical oceanography. *Mailing Add:* PO Box 1063 Middleburg VA 22117

SPILHAUS, ATHELSTAN FREDERICK, JR, b Boston, Mass, May 21, 38; m 60; c 3. ASSOCIATION MANAGEMENT, GEOPHYSICS. *Educ:* Mass Inst Technol, SB, 59, SM, 60, PhD(oceanog), 65. *Prof Exp:* Oceanogr, US Govt, 65-67; asst exec dir, 67-70, EXEC DIR, AM GEOPHYS UNION, WASHINGTON, DC, 70- *Concurrent Pos:* Dir & vpres, Oceanic Educ Found, 70-74; treas, Renewable Natural Resources Found, 72-74, dir, 72-; secy, US Nat Comt, Union Geodesy & Geophys, 72-; chmn, Conv Liaison Coun, 81-82; dir, Asn Women Geoscientists Found, 84-88. *Mem:* Fel AAAS; Am Geophys Union; Am Soc Limnol & Oceanog; Coun Eng & Sci Soc Exec (pres, 80-81); Asn Earth Sci Ed (pres, 77); Soc Scholarly Publ (secy, 78-80). *Res:* Use of optical measurements in oceanography. *Mailing Add:* 10900 Picasso Lane Potomac MD 20854

SPILKER, BERT, b Washington, DC, July 3, 41; m 67; c 2. DRUG DEVELOPMENT. *Educ:* Univ Pa, AB, 62; Downstate Med Ctr, State Univ NY, PhD(pharmacol), 67; Univ Miami, MD, 77. *Prof Exp:* Asst pharmacol, Med Sch, Univ Calif, 67-68; sr pharmacologist, Pfizer Ltd, Kent, Eng, 68-70, Philips-Duphar, Weesp, Holland, 70-72 & Sterling-Winthrop Res Inst, 72-75; resident, Med Sch, Brown Univ, 77-78; sr med consult, JRB Assocs, Inc, McLean, Va, 78-79; sr clin scientist, 79-83, HEAD DEPT PROJ COORD, BURROUGHS WELLCOME CO, 83-; CLIN PROF, DEPT MED, MED SCH, UNIV NC, 79- *Concurrent Pos:* Adj prof, dept pharmacol, Med Sch, Univ NC, 79-, clin prof, Sch Pharm. *Mem:* Am Soc Clin Pharmacol & Therapeut; Am Soc Pharmacol & Exp Therapeut; Am Epilepsy Soc. *Res:* Drug development. *Mailing Add:* 2556 Booker Creek Rd Chapel Hill NC 27514

SPILKER, CLARENCE WILLIAM, organic chemistry; deceased, see previous edition for last biography

SPILLER, EBERHARD ADOLF, b Halbendorf, Ger, Apr 16, 33; m 64; c 2. X-RAY OPTICS. *Educ:* Univ Frankfurt, MSc, 60, PhD(physics), 64. *Prof Exp:* Asst prof physics, Univ Frankfurt, 65-68; STAFF MEM, T J WATSON RES CTR, IBM CORP, 68- *Mem:* Fel Optical Soc Am; Ger Phys Soc; AAAS; Int Soc Optical Eng. *Res:* Solid state physics; coherence of light; lasers; holography; nonlinerar optics; thin films; x-ray optics. *Mailing Add:* Watson Res Ctr IBM Corp PO Box 218 Yorktown Heights NY 10598

SPILLER, GENE ALAN, b Milan, Italy, Feb 19, 27; US citizen; m 81. CHOLESTEROL & DIETARY FIBER RESEARCH & WRITING. *Educ:* Univ Milan, Dr(chem), 49; Univ Calif, Berkeley, MS, 68, PhD(nutrit), 72. *Prof Exp:* Res Chemist, Univ Calif, Berkeley, 66-67, assoc specialist physiol, 68-72; prin scientist, Syntex Res, Palo Alto, Calif, 73-80; PRES, HEALTH RES & STUDIES CTR, LOS ALTOS, CALIF, 88-; PRES, SPHERA FOUND, LOS ALTOS, CALIF. *Concurrent Pos:* Lectr nutrit, Mills Col, Oakland, Calif, 71-81; consult, Clin Nutrit Res, 81-; auth med bks, 74-; lectr, Foothill Col, Los Altos, Calif, 74- *Mem:* Am Inst Nutrit; Am Soc Clin Nutrit; Am Asn Cereal Chemists; Am Diabetes Asn; Mediterranean Group; Am Col Nutrit. *Res:* Effects of dietary fiber on health and disease; effects of carbohydrates and fats in human physiology and health; effects of dietary patterns on diseases of aging; role of lesser known food components on human health; pharmacological effects of nutrients and methylxanthines; design and execution of clinical studies. *Mailing Add:* Health Res & Studies Ctr PO Box 338 Los Altos CA 94023

SPILLERS, WILLIAM R, b Fresno, Calif, Aug 4, 34; c 3. CIVIL ENGINEERING, ENGINEERING MECHANICS. *Educ:* Univ Calif, Berkeley, BS, 55, MS, 56; Columbia Univ, PhD(continuum mech), 61. *Prof Exp:* Struct designer, John Blume Assoc, Calif, 56-57; from instr to assoc prof civil eng, Columbia Univ, 59-68, prof civil eng & eng mech, 68-76; prof civil eng, Rensselaer Polytech Inst, 76-90; PROF & CHMN, DEPT CIVIL & ENVIRON ENG, NJ INST TECHNOL, 90- *Concurrent Pos:* NSF grant, 64; Guggenheim fel, NY Univ, 68-69; NSF fel, Univ Calif, Berkeley, 75-76. *Mem:* Am Soc Civil Engrs; Int Asn Bridge & Struct Engrs. *Res:* Problems of structural mechanics; optimization; fatigue of buried power transmission cables; environmental design of housing; design theory. *Mailing Add:* Dept Civil & Environ Eng NJ Inst Technol Newark NJ 17102

SPILLETT, JAMES JUAN, b Idaho Falls, Idaho, Oct 21, 32; m 61; c 2. ANIMAL ECOLOGY, WILDLIFE RESOURCES. *Educ:* Utah State Univ, BS, 61, MS, 65; Johns Hopkins Univ, ScD(animal ecol), 68. *Prof Exp:* Res asst pronghorn antelope, Utah State Univ, 62-64; field biologist, Indian Wildlife Surv, World Wildlife Fund, Morges, Switz, 66; res asst ecol res, Johns Hopkins Univ, 64-67; asst leader, Utah Coop Wildlife Res, Utah State Univ, 67-76; WILDLIFE BIOLOGIST, US FOREST SERV, 78- *Mem:* Wilderness Soc; Wildlife Soc; Brit Fauna Preserv Soc; Wildlife Preserv Soc India. *Res:* Ecuadorian mammals, taxonomy and distribution effects of livestock fences on pronghorn antelope movements; status of Indian wildlife, particularly the Indian rhino; ecology of the lesser bandicoot rat. *Mailing Add:* Box 298 Rockland ID 83271

SPILLMAN, CHARLES KENNARD, b Lawrence County, Ill, Feb 26, 34; m 59; c 1. AGRICULTURAL ENGINEERING. *Educ:* Univ Ill, Urbana, BS, 60, MS, 63; Purdue Univ, PhD(agr eng), 69. *Prof Exp:* Asst waste mgt, Univ Ill, Urbana, 60-62; exten agr engr, Mich State Univ, 62-66; from asst prof to assoc prof, Kans State Univ, 69-79, prof struct & environ, 79-82, head dept agr eng, 82-87, PROF GRAIN PROCESSING, KANS STATE UNIV, 87- *Honors & Awards:* Metal Bldgs Mfg Award, Am Soc Agr Engrs. *Mem:* Fel Am Soc Agr Engrs; Am Asn Cereal Chemists; Am Soc Eng Educ; Sigma Xi. *Res:* Physical properties of cereal grains. *Mailing Add:* Dept of Agr Eng Seaton Hall Kans State Univ Manhattan KS 66506

SPILLMAN, GEORGE RAYMOND, b Holdenville, Okla, Oct 21, 34; m 66; c 2. PHYSICS. *Educ:* Univ Okla, BS, 56, Univ Calif, Berkeley, MA, 61, PhD(physics), 64. *Prof Exp:* Assoc engr, Gen Dynamics/Convair, 56; proj officer, Res Directorate, Spec Weapons Ctr, Kirtland AFB, US Air Force, 56-59, proj officer, Weapons Lab, 63-67; staff mem, Los Alamos Sci Lab, 67-71, group leader, 71-74, assoc div leader, 74-76, alt div leader, 76-79, asst div leader, 79-83, prog mgr, 83-87, STAFF MEM, LOS ALAMOS SCI LAB, 87- *Concurrent Pos:* Consult, 81- *Res:* Nuclear explosion phenomenology; nuclear weapons effects; plasma physics; atomic processes in plasma; radiative transfer; spectral absorption coefficients; hydrodynamics. *Mailing Add:* Los Alamos Sci Lab PO Box 1663 Los Alamos NM 87545

SPILLMAN, RICHARD JAY, b Tacoma, Wash, Sept 13, 49. FAULT TOLERANT COMPUTING. *Educ:* Western Wash Univ, BS, 71; Univ Utah, MA, 73; Utah State Univ, PhD(elec eng), 78. *Prof Exp:* Asst prof elec eng, Univ Calif, Davis, 78-80; specialist eng, Boeing Co, 80-81; ASST PROF COMPUT SCI, PAC LUTHERN UNIV, 81- *Mem:* Inst Elec & Electronics Engrs; Am Soc Comput Mach. *Res:* Development of highly fault tolerant comput systems; analysis of system testability. *Mailing Add:* Dept Comput Sci Pac Luthern Univ Tacoma WA 98447

SPILLMAN, WILLIAM BERT, JR, b Charleston, SC, Jan 21, 46; m 74; c 2. FIBER OPTIC SENSORS, DEVICE MODELLING. *Educ:* Brown Univ, AB, 68; Northeastern Univ, MS, 72, PhD(physics), 77. *Prof Exp:* Res asst, Northeastern Univ, 72-77; tech staff mem, Sperry Res Ctr, 77-83; sr scientist, Geo-Centers, Inc, 83-84; mgr advan develop, 84-87, DIR RES, SIMMONDS PRECISION AIRCRAFT SYSTS, 87-; PRES, CATAMOUNT SCI INC, 91- *Concurrent Pos:* Adj assoc prof elec eng, Univ Vt, 87-90; mem adv bd, Smart Struct Inst, Univ Strathclyde, UK, 91- *Mem:* Optical Soc Am; Am Phys Soc; Inst Elec & Electronic Engrs; sr mem Instrument Soc Am; Int Soc Optical Eng. *Res:* Fiber optic sensing; fiber optic sensor multiplexing; ultrasonic sensing; magneto-optic materials; smart structures and skins for aerospace and civil applications; computer modelling of sensor and sensor systems. *Mailing Add:* Simmonds Precision Aircraft Systs Vergennes VT 05491

SPILMAN, CHARLES HADLEY, b Westerly, RI, Mar 30, 42; m 62, 85; c 2. ATHEROSCLEROSIS, LIPOPROTEINS. *Educ:* Clark Univ, AB, 65; Univ Mass, Amherst, PhD(physiol), 69. *Prof Exp:* Res asst, Univ Mass, 65-69; res assoc, Cornell Univ, 69-71 & Worcester Found Exp Biol, 71-72; res scientist II, 72-76, sr res scientist III, 76-85, SR SCIENTIST IV, UPJOHN CO, 85- *Concurrent Pos:* Vis scientist, Gladstone Found Labs, 85-86. *Mem:* NY Acad Sci; AAAS; Soc Exp Biol Med. *Res:* Discovery and development of drugs to treat atherosclerosis; lipoprotein binding to receptors; apolipoprotein binding domains; lipoprotein metabolism. *Mailing Add:* Metabolic Dis Res Upjohn Co Kalamazoo MI 49001

SPINAR, LEO HAROLD, b Colome, SDak, Feb 20, 29; m 56; c 4. PHYSICAL CHEMISTRY. *Educ:* Univ SDak, BA, 51; Univ Wis, MS, 53, PhD(chem), 58. *Prof Exp:* From instr to asst prof chem, Colo State Univ, 57-62; assoc prof, Univ Mo, 62-66; assoc prof, 66-69, PROF CHEM, SDAK STATE UNIV, 69- *Concurrent Pos:* Dir planning, program & budget, SDak State Univ, 73-82. *Mem:* Am Chem Soc. *Res:* Physical inorganic chemistry; high temperature properties of materials; thermodynamics; vapor pressure studies. *Mailing Add:* 211 Shepard Hall SDak State Univ Brookings SD 57007-0896

SPINDEL, ROBERT CHARLES, underwater acoustics, electrical engineering, for more information see previous edition

SPINDEL, WILLIAM, b New York, NY, Sept 9, 22; m 42, 67; c 2. CHEMISTRY, SCIENCE POLICY. *Educ:* Brooklyn Col, BA, 44; Columbia Univ, MA, 47, PhD(chem), 50. *Prof Exp:* Jr scientist, Manhattan Proj, Los Alamos Sci Lab, Univ Calif, 44-45; asst chem, Columbia Univ, 46-49; instr, Polytech Inst Brooklyn, 49-50; asst prof, State Univ NY Teachers Col, Albany, 50-54; from assoc prof to prof, Rutgers Univ, 57-64; prof, Belfer Grad Sch Sci, Yeshiva Univ, 64-74; exec secy, Off Chem & Chem Technol, 74-82, staff dir, Bd Chem Sci & Technol, 83-88, PRIN STAFF OFF SPEC PROJ, NAT ACAD SCI-NAT RES COUN, 88- *Concurrent Pos:* Res assoc, Columbia Univ, 54-56, vis assoc prof, 56-57, vis prof, 62-70, sr lectr, 70-74; NSF vis scientist, Yugoslavia, 71-72; Guggenheim fel, 61-62. *Honors & Awards:* Prof Staff Award, Nat Res Coun, 85. *Mem:* Fel AAAS; Am Chem Soc; Am Phys Soc. *Res:* Separation of stable isotopes; physical and chemical properties of isotopes; mass spectrometry. *Mailing Add:* 6503 Dearborn Dr Falls Church VA 22044

SPINDLER, DONALD CHARLES, analytical chemistry, for more information see previous edition

SPINDLER, MAX, b Antwerp, Belg, Dec 19, 38; US citizen; m 67; c 1. CIVIL & AERONAUTICAL ENGINEERING. *Educ:* Cooper Union, BCE, 61; Northwestern Univ, Ill, MS, 63, PhD(civil eng), 68. *Prof Exp:* Eng specialist, LTV Aerospace Corp, Tex, 67-70; asst prof civil eng, 70-77, ASSOC PROF CIVIL ENG, UNIV TEX, ARLINGTON, 77- *Mem:* Am Soc Civil Engrs; Am Inst Aeronaut & Astronaut; NY Acad Sci; Sigma Xi. *Res:* Noise spectra due to flow through stenosed heart valves; fluid mechanics; hydraulics; biomedical engineering. *Mailing Add:* 1708 Park Ridge Terr Arlington TX 76012

SPINDT, RODERICK SIDNEY, b Waupaca, Wis, Mar 5, 19; m 44; c 2. ORGANIC CHEMISTRY. *Educ:* Ripon Col, BA, 41; Univ Wis, MS, 44; Univ Pittsburgh, PhD(chem), 49. *Prof Exp:* Jr chemist org synthesis, Gulf Res & Develop Co, 44-45; fel org anal, Mellon Inst, 45-56; staff asst, Gulf Res & Develop Co, 56-58, sect head, 58-60, sr chemist, 60-65, from res assoc to sr res assoc, 65-80; RETIRED. *Concurrent Pos:* Lectr, Univ Pittsburgh, 52- *Mem:* Am Chem Soc; Soc Automotive Engrs; Air Pollution Control Asn. *Res:* Characterization of sulfur compounds; mechanism of engine deposit formation; mechanisms of combustion in engines; air pollution research; vehicle emissions; composition of gasoline. *Mailing Add:* 3957 Parkview Lane Allison Park PA 15101-3522

SPINELLI, JOHN, b Seattle, Wash, July 23, 25; m 49; c 1. ANALYTICAL CHEMISTRY, FOOD CHEMISTRY. *Educ:* Univ Wash, BS, 49. *Prof Exp:* Consult chemist, Food Chem & Res Labs, Wash, 49-62; res chemist, Technol Lab, Nat Marine Fisheries Serv, 62-80; DIR, UTILIZATION RES DIV, NORTHWEST & ALASKA FISHERIES CTR, SEATTLE, 80- *Mem:* NY Acad Sci; AAAS; Am Chem Soc; Inst Food Technol; Pac Fisheries Technologists. *Res:* Food process quality control, product analysis and development; biochemical changes in fish postmortem; protein isolates of marine origin; protein and nutritional requirements of salmonids; food uses for underutilized species; improvement of quality and safety of fishery products. *Mailing Add:* 10002 63rd Ave S Seattle WA 98178

SPINGOLA, FRANK, b Brooklyn, NY, Aug 3, 37; m 67. CHEMISTRY. *Educ:* Adelphi Univ, AB, 59; Polytech Inst Brooklyn, MS, 63, PhD(chem), 68. *Prof Exp:* Asst prof, 68-76, ASSOC PROF CHEM, DOWLING COL, 76- *Res:* Physical chemistry of aqueous solutions. *Mailing Add:* 14 Haight St Deer Park NY 11729

SPINING, ARTHUR MILTON, III, animal nutrition, biochemistry; deceased, see previous edition for last biography

SPINK, CHARLES HARLAN, b Platteville, Wis, Apr 9, 36; wid; c 1. ANALYTICAL CHEMISTRY, PHYSICAL CHEMISTRY. *Educ:* Univ Wis, BS, 58; Pa State Univ, PhD(phys chem), 62. *Prof Exp:* Fel, Univ Wash, 62-63; asst prof analytic chem, Juniata Col, 63-67; from asst prof to assoc prof, 67-72, PROF ANALYTIC CHEM, STATE UNIV NY COL, CORTLAND, 72- *Concurrent Pos:* Am Chem Soc-Petrol Res Fund grant, 63-64; USPHS res grant, 65-68; NY State Res Found fel & grant-in-aid, 69-72; res assoc, Lund Univ, Sweden, 73-74, Yale Univ, 80-81; USPHS res grant, 77-79, 80-87,; Nat Sci Found Res Award, Cornell Univ, 87-88. *Mem:* Biophys Soc; Am Chem Soc; Sigma Xi. *Res:* Thermochemical studies on solutes in bile salt solutions; thermochemical analysis of mixed organic-aqueous mixtures; heat capacities of model biochemical compounds; scanning calorimetry of micelles and lipid-detergent mixtures. *Mailing Add:* Dept Chem State Univ NY PO Box 2000 Cortland NY 13045

SPINK, D(ONALD) R(ICHARD), b Buffalo, NY, Mar 11, 23; m 46; c 6. EXTRACTIVE METALLURGY, CHEMICAL ENGINEERING. *Educ:* Univ Mich, BS, 45; Univ Rochester, MS, 49; Iowa State Univ, PhD(chem eng), 52. *Prof Exp:* Chem engr, Gen Elec Co, 46-48; asst, Iowa State Univ, 49-52; res engr, E I du Pont de Nemours & Co, 52; sr res engr, Carborundum Metals Climax, Inc, NY, 52-59, asst to mgr, Tech Br & mgr, Res & Develop Dept, 59-61, mgr, Tech Br, 61-65, vpres technol, 65-68; prof chem eng, Univ Waterloo, 68-88; FOUNDER & PRES, AIR POLLUTION CONTROL EXPERTS, TURBOTAK INC, 76- *Mem:* Am Inst Chem Engrs; Can Soc Chem Engrs; Am Inst Mining, Metall & Petrol Engrs; Soc Chem Indust; Can Inst Mining & Metall. *Res:* Extractive metallurgy of zirconium, hafnium and titanium; solvent extraction; extractive metallurgy; air pollution research; low energy scrubber development; coal treatment (sulfur removal); fly ash treatment (recovery of vanadium and nickel); zinc concentrate roasting without formation of ferrites; sulphur dioxide removal process. *Mailing Add:* Turbotak Inc 550 Parkside Dr Suite A-14 Waterloo ON N2L 5V4 Can

SPINK, GORDON CLAYTON, b Lansing, Mich, Jan 6, 35; m 60; c 2. MEDICAL EDUCATION, FAMILY MEDICINE. *Educ:* Mich State Univ, BS, 57, PhD(bot, cytol), 66, DO, 75. *Prof Exp:* Instr, Univ, 63-66, asst prof, Biol Res Ctr, 66-68, dir, Electron Micros Lab, 67-72, asst prof entom, Univ, 68-71, prof staff scientist electron micros, Pesticide Res Ctr, 71-72, instr & dir, Electron Micros Lab, 72-75, clin asst prof, Col Osteopath Med, 75-76, asst prof, Dept Family Med, 76-78, unit III coordr, 77-79, co-dir, Preceptor Prog, Dept Family Med, 78-80, actg asst dean grad & continuing educ, 80, ASSOC PROF, DEPT FAMILY MED, MICH STATE UNIV, 78- *Concurrent Pos:* Res collabr, Biol Dept, Brookhaven Nat Lab, 69; intern, Flint Osteopath Hosp, Flint, Mich, 75-76, dir med educ, 80-81; dir med educ, Lansing Gen Hosp, 82-84; dir med educ, Ingham Med Ctr, Lansing, Mich, 87-; med prac fel, Chicago Col Osteopathic Med, Chicago, Ill, 87-88. *Mem:* AAAS; Electron Micros Soc Am; Asn Hosp Med Educ; Asn Osteop Dirs Med Educ; Am Osteop Asn; Am Heart Asn; Am Med Soccer Asn. *Res:* Medical education. *Mailing Add:* 3910 Sandlewood Okemos MI 48864

SPINK, WALTER JOHN, b Hackensack, NJ, May 4, 33; m 57; c 2. STRATIGRAPHY, STRUCTURAL GEOLOGY. *Educ:* Lehigh Univ, BS, 57; Rutgers Univ, MS, 63, PhD(geol), 67. *Prof Exp:* Geologist, NJ Geol Surv, 60-66; from instr to asst prof, Rider Col, 66-69, chmn dept, 69-76, assoc prof geol, 69-80, chmn dept, 80-83; INKEEPER, THE INN, CRYSTAL LAKE, 86-; GEOL CONSULT, 86- *Mem:* AAAS; Geol Soc Am; Asn Prof Geol Scientists. *Res:* Areal geologic mapping and gravity survey of northwestern New Jersey; structural geology; stratigraphy and sedimentation. *Mailing Add:* Inn at Crystal Lake PO Box 12 Rte 153 Eaton Center NH 03832

SPINK, WESLEY WILLIAM, internal medicine; deceased, see previous edition for last biography

SPINKA, HAROLD M, b Chicago, Ill, Apr 2, 45; m 73; c 2. SPIN PHYSICS, NUCLEON-NUCLEON INTERACTIONS. *Educ:* Northwestern Univ, BA, 66; Calif Inst Technol, PhD(physics), 70. *Prof Exp:* Fel, 70-73, physicist, 76-87, SR PHYSICIST, ARGONNE NAT LAB, 87- *Concurrent Pos:* Adj asst prof physics, Univ Calif, Los Angeles, 73-76. *Mem:* Am Phys Soc; Sigma Xi. *Res:* Strong interactions using polarized beams and targets and nuclear beams and targets; nucleon-nucleon interactions. *Mailing Add:* High Energy Physics Div Argonne Nat Lab Bldg 362 Argonne IL 60439

SPINKS, DANIEL OWEN, b Dallas, Ga, Sept 5, 18; m 40; c 3. SOIL CHEMISTRY. *Educ:* Univ Ga, BS, 39, MSA, 47; NC State Col, PhD, 53. *Prof Exp:* Instr soils & physics, Abraham Baldwin Agr Col, 39-44; from instr to assoc prof, 47-61, prof soils & soil chemist, Agr Exp Sta, Univ Fla, 61-84, assoc dean resident instr, Inst Food & Agr Sci, 69-84; RETIRED. *Mem:* Nat Asn Cols & Teachers Agr; Soil Sci Soc Am. *Res:* Effect of organic matter on the availability of fixed soil phosphorus. *Mailing Add:* Rt One Box 136 C Lewisburg NC 27549

SPINKS, JOHN LEE, b Central City, Ky, June 19, 24; m 51; c 2. CONSULTING. *Educ:* Univ Ky, BSME, 51. *Hon Degrees:* PhD(eng), World Univ, 84. *Prof Exp:* Dep dir, Eng Div & Lt Col, Space Div, USAF, 61-73; supervising engr, South Coast Air Qual Mgr Dist, 56-83; PRES, ENVIRON EMISSIONS ENG CO, 83- *Concurrent Pos:* Hon mem, Nat Adv Bd, Am Biog Inst; instr, rock & ice mountaineering; lectr, marathon running; consult, Govt Air Qual Agencies. *Honors & Awards:* US Presidential Sports Award. *Mem:* Am Acad Environ Engrs; Inst Advan Eng; Am Soc Mech Engrs; Air Pollution Control Asn; Nat Soc Prof Engrs; Inst Environ Sci; Soc Environ Engrs; Am Soc Eng Educ; Soc Eng Sic; AAAS. *Res:* Pioneered development of engineering principles and technology for air pollution control techniques and adapted by other similar agencies throughout the world; developed air & water pollution control programs for US Air Force. *Mailing Add:* 26856 Eastvale Rd Rolling Hills CA 90274-4007

SPINKS, JOHN WILLIAM TRANTER, b Methwold, Eng, Jan 1, 08; m 39. CHEMISTRY. *Educ:* Univ London, BSc, 28, PhD(photochem), 30. *Hon Degrees:* DSc, Univ London, 57; LLD, Carleton Univ, 58 & Assumption Col, 62. *Prof Exp:* Asst prof chem, 30-39, prof phys chem, 39-74, head dept, 48-59, dean grad col, 49-59, pres, 59-74, EMER PROF PHYS CHEM & EMER PRES, UNIV SASK, 74- *Honors & Awards:* Order Brit Empire; Companion Can. *Mem:* Am Chem Soc; fel Royal Soc Can; fel Royal Inst Chem. *Res:* Photochemistry; molecular structure; radioactive tracers; radiation chemistry. *Mailing Add:* 932 Univ Dr 1011 Saskatoon SK S7N 0K1 Can

SPINNER, IRVING HERBERT, b Toronto, Ont, Dec 29, 22; m 44; c 3. CHEMICAL ENGINEERING. *Educ:* Univ Toronto, BSc, 51, MSc, 53, PhD(chem eng), 54. *Prof Exp:* Res consult, Stanley Mfg Co, 46-51, res chemist, 54-56; assoc prof chem eng, 56-70, PROF CHEM ENG, UNIV TORONTO, 70- *Concurrent Pos:* Indust chemist, 39-42; consult indust chem, 47-54 & 64. *Mem:* Fel Chem Inst Can. *Res:* Ion exchange, redox polymers and novel mass transfer techniques. *Mailing Add:* Dept Chem Eng Univ Toronto 200 College St Toronto ON M5S 1A4 Can

SPINNLER, JOSEPH F, b Greenwood, SC, July 8, 31; m 62; c 2. PHYSICAL CHEMISTRY. *Educ:* Lafayette Col, BS, 53; Yale Univ, MS, 58, PhD(chem), 60. *Prof Exp:* Scientist ballistics sect, Rohm and Haas Co, 59-65, sr scientist chem sect, 65-70; mem explor develop group, Micromedic Systs Inc, 70-71; SR SCIENTIST, ROHM AND HAAS CO, 71- *Mem:* Sigma Xi. *Res:* Development of hydrogen fluorine chemical laser; automation of spectrophotometric and electrophoretic clinical laboratory procedures;

metabolite and residue analysis using liquid chromatography; gas chromatography/mass spectrometry; thin layer chromatography and electrophoresis. *Mailing Add:* Rohm and Haas Co Norristown Rd Spring House PA 19477

SPINOSA, CLAUDE, b Italy, July 17, 37; US citizen; m 63; c 3. PALEONTOLOGY, GEOLOGY. *Educ:* City Col, New York, BS, 61; Univ Iowa, MS, 65, PhD(geol), 68. *Prof Exp:* Asst prof geol, Ind Univ, Southeast, 68-70; assoc prof, 70-76, chmn dept, 82-88, PROF GEOL, BOISE STATE UNIV, 76- *Mem:* Geol Soc Am; Paleont Soc. *Res:* Permian ammonoids; Permian stratigraphy; nautilus; biology. *Mailing Add:* Dept Earth Sci Boise State Univ Boise ID 83725

SPINRAD, BERNARD ISRAEL, b New York, NY, Apr 16, 24; m 51, 83; c 4. PHYSICS, NUCLEAR ENGINEERING. *Educ:* Yale Univ, BS, 42, MS, 44, PhD(phys chem), 45. *Prof Exp:* Sterling fel, Yale Univ, 45-46; phsyicist, Clinton Labs, Tenn, 46-49; from assoc physicist to sr physicist, Argonne Nat Lab, 49-72, dir, Reactor Eng Div, 57-63; dir, Div Nuclear Power & Reactors, Int Atomic Energy Agency, Vienna, 67-70; ad hoc prof, Univ Wis-Parkside, 71; chm dept nuclear eng, Iowa State Univ, 83-90; prof, 72-83, EMER PROF NUCLEAR ENG, ORE STATE UNIV, 91- *Concurrent Pos:* Adv US deleg, Conf Peaceful Uses of Atomic Energy, Geneva, 55-58; consult, Int Atomic Energy Agency, 61, 63; mem, Europ-Am Reactor Physics Comt, 61-66, chmn, 61-62; vis prof Univ Ill, 64; mem, Comt Nuclear & Alternative Energy Systs, Nat Acad Sci, 75-80; res scholar, Int Inst Appl Systs Anal, 78-79; mem comt Univ res reactors, Nat Res Coun, 86-88. *Mem:* AAAS; fel Am Phys Soc; fel Am Nuclear Soc; Sigma Xi; Am Chem Soc. *Res:* Physics of nuclear reactors; nuclear systems; energy systems and economics; nuclear reactor shutdown power; nuclear fuel cycle and nuclear safeguards. *Mailing Add:* 18803 37th Ave NE Seattle WA 98155

SPINRAD, HYRON, b New York, NY, Feb 17, 34; m 58; c 3. ASTRONOMY. *Educ:* Univ Calif, Berkeley, PhD(astron), 61. *Prof Exp:* Sr scientist, Jet Propulsion Lab, 61-64; from asst prof to assoc prof, 64-68, PROF ASTRON, UNIV CALIF, BERKELEY, 68- *Honors & Awards:* Dannie Heineman Prize, Atrophys. *Mem:* Nat Acad Sci; Am Atron Soc. *Res:* Study of planetary atmospheres; spectroscopic investigations of old stars and nuclei of galaxies; astrophysics; spectroscopy of faint, distant radio and cluster galaxies. *Mailing Add:* Dept Astron Univ Calif Berkeley CA 94720

SPINRAD, RICHARD WILLIAM, b New York, NY, Apr 6, 54; m 80; c 1. OPTICAL OCEANOGRAPHY, FLOW CYTOMETRY. *Educ:* Johns Hopkins Univ, BA, 75; Ore State Univ, MS, 78, PhD(geol oceanog), 82. *Prof Exp:* Pres mfg, Sea Tech, Inc, 84-86; prog mgr ocean optics, 87-88, DIV DIR, OFF NAVAL RES, 88-; RES SCIENTIST OCEANOG, BIGELOW LAB OCEAN SCI, 82- *Concurrent Pos:* Consult, Calgary Dept Waterworks, 89- *Mem:* Am Geophys Union; Oceanog Soc; Am Soc Limnol & Oceanog; Optical Soc Am; AAAS. *Res:* Optical properties of marine particles; light scattering characteristics of phytoplankton and bacteria. *Mailing Add:* Off Naval Res 800 N Quincy St Arlington VA 22217-5000

SPINRAD, ROBERT J(OSEPH), b New York, NY, Mar 20, 32; m 54; c 2. COMPUTER SCIENCE. *Educ:* Columbia Univ, BS, 53, MS, 54; Mass Inst Technol, PhD, 63. *Prof Exp:* Assoc engr, Bulova Res & Develop Lab, 54-55; from asst elec engr to assoc elec engr, Brookhaven Nat Lab, 55-63, elec engr, 63-66, head comput systs group, 65-68, sr elec engr, 66-67, sr scientist, 67-68; vpres, Sci Data Systs, 68-70, vpres Xerox Data Systs, 70-71, dir info sci, 71-76, vpres, Systs Develop Div, 76-78, vpres res, Parc, 78-83, dir, Systs Technol, 83-86, DIR, CORP TECHNOL, XEROX CORP, 87- *Concurrent Pos:* Consult, Bell Telephone Labs, 62-67, Rand Corp, 77-79 & Int Inst Appl Syst Anal, 78-80; educ comt, Comput Sci & Eng Bd, Nat Acad Sci, 68-70, mem, software intellectual property panel, 89; mem comput in elec eng comt, Nat Acad Eng, 69-72; mem math dept vis comt, Mass Inst Technol, 70-74; gen chmn, 1972 Fall Joint Comput Conf, 71-72; mem comut sci adv comt, Stanford Univ, chmn, 72 & 83; mem, Eng Adv Coun, Univ Calif, 77-85; Overseers Comt Info Technol, Harvard Univ, 79-85; Nat Res Coun Panel, Nat Bur Standards, 80-83; bd trustees, Educom, 82-89; mem adv group, CSNET, 82-83; mem Info Technol Workshop, NSF, 83; dir, Digital Pathways Inc, 83-86; AAAS panel, NSF Bilateral Progs, 84-85; consult ed-comput, McGraw-Hill Encyclopedia Sci & Technol, 87-; mem technol assessment adv comt, Comn Preserv & Access, 89-; mem Info Sci & Technol Study Group, Defense Advan Res Projs Agency, 90- *Mem:* Inst Elec & Electronic Engrs; Asn Comput Mach. *Res:* Electronics; computers; computer systems. *Mailing Add:* Xerox Corp 3333 Coyote Hill Rd Palo Alto CA 94304

SPIRA, ARTHUR WILLIAM, b New Britain, Conn, Oct 18, 41; m 65; c 3. DEVELOPMENTAL NEUROSCIENCE. *Educ:* City Col New York, BS, 62; Univ Mich, MS, 64, USPHS fel & PhD(anat), 67. *Prof Exp:* USPHS fel anat, McGill Univ, 67-68; asst prof, Univ BC, 68-73; from asst prof to assoc prof, 75-85, PROF ANAT, UNIV CALGARY, 85- *Concurrent Pos:* Mem, Grants Review Comt, Med Res Coun Can, 78-81; actg dir, Lions Sight Ctr, 85-86. *Mem:* Soc Neurosci; Am Asn Anat; Can Asn Anat; Asn Res Vision Ophthal. *Res:* Structure and function of the retina; ocular development; retinal histogenesis; retinal neurotransmitters; cytochemistry; neurochemistry; transmission and scanning electron microscopy. *Mailing Add:* Dept Anat Univ Calgary 3330 Hospital Dr NW Calgary AB T2N 4N1 Can

SPIRA, JOEL SOLON, b New York, NY, Mar 1, 27; m 54; c 3. ENGINEERING PHYSICS. *Educ:* Purdue Univ, BS, 48. *Prof Exp:* Jr engr, Glenn L Martin Co, 48-52; engr, Reeves Instrument Corp, 52-54, sr engr, 54-56, sr proj engr, 56-59; prin systs analyst, ITT Commun Systs Inc, 59-61; CHMN & DIR RES, LUTRON ELECTRONICS CO, INC, 61- *Mem:* Fel AAAS; Am Phys Soc; sr mem Inst Elec & Electronic Engrs. *Res:* Supersonic aerodynamics; microwaves; computers; electronic instruments; missile technology; weapons systems analysis; nuclear and military strategy; electronic and general technology; light dimming and electron power control; energy conservation. *Mailing Add:* 7200 Suter Rd Coopersburg PA 18036

SPIRA, MELVIN, b Chicago, Ill, July 3, 25; m 52; c 3. MEDICINE, PLASTIC SURGERY. *Educ:* Northwestern Univ, DDS, 47, MSD, 51; Med Col Ga, MD, 56. *Prof Exp:* From instr to assoc prof, 61-70, PROF PLASTIC SURG, BAYLOR COL MED, 70-, HEAD, DIV PLASTIC SURG, 76- *Mem:* Am Soc Plastic & Reconstruct Surg; Am Soc Maxillofacial Surg (pres, 74-); fel Am Col Surg; Am Asn Plastic Surg; Plastic Surg Res Coun. *Res:* Maxillofacial and microvascular surgery. *Mailing Add:* 5114 Glenmeadow Houston TX 77096

SPIRITO, CARL PETER, b Hartford, Conn, Apr 7, 41; m 64; c 3. NEUROBIOLOGY, ETHOLOGY. *Educ:* Cent Conn State Col, BA, 65; Univ Conn, PhD(biol eng), 69. *Prof Exp:* Nat Inst Child Health & Human Develop trainee, Univ Miami, 69-70; asst prof biol, Univ Va, 70-77; assoc prof neurophysiol, Ohio Univ, 77-80; chmn, 83-88, MEM FAC DEPT PHYSIOL, UNIV NEW ENG, 80- *Mem:* Soc Neurosci; Soc Exp Biol & Med. *Res:* Invertebrate neurobiology and behavior; neural control of locomotion. *Mailing Add:* Dept Physiology Univ New Eng Osteop Med 605 Pool Rd Biddeford ME 04005

SPIRO, CLAUDIA ALISON, b Castro Valley, Calif, March 4, 56. ARITHMETIC FUNCTIONS, PRIME NUMBER THEORY. *Educ:* Calif Inst Technol, BS & MS, 77; Univ Ill, Urbana, PhD(math), 81. *Prof Exp:* Grad teaching asst math, Univ Ill, Urbana, 77-81; George William Hill & Emmy Noether Instr, 81-83, ASST PROF MATH, STATE UNIV NY, BUFFALO, 83- *Mem:* Am Math Soc; Math Asn Am; Asn Women Math; Sigma Xi. *Mailing Add:* Dept Math 314 Diefendorf Hall Suny Health Sci Ctr 3435 Main St Buffalo NY 14214

SPIRO, HERZL ROBERT, b Burlington, Vt, Apr 22, 35; m 55; c 3. PSYCHIATRY, SOCIAL PSYCHOLOGY. *Educ:* Univ Vt, BA, 55, MD, 60. *Prof Exp:* Intern internal med, Cornell Univ, 60-61; resident psychiat, Johns Hopkins Univ, 61-64, from instr to assoc prof psychiat, 64-71; prof psychiat, Med Sch, prof social psychol, Grad Fac & dir ment health ctr, Rutgers Univ, 71-76; PROF PSYCHIAT & CHMN DEPT, MED COL WIS, 76-; PROF PSYCHIAT, SINAI SAMARITAN MED CTR, 88- *Concurrent Pos:* Fel, Johns Hopkins Univ, 61-64; assoc physician-in-charge psychiat liaison serv, Johns Hopkins Hosp, 64-66, psychiatrist-in-charge Henry Phipps outpatient serv, 66-70, dir outpatient & community ment health progs & div group process, 69-71; consult, Bur Disability Ins, Social Security Admin, 64-69; consult, NIMH, 69-, mem ment health serv res rev comt, 71-, mem task force health maintenance orgn, 71-72; mem, Nat Task Force Psychiat Res, 72-73. *Mem:* AAAS; fel Asn Psychiat Asn; fel Am Col Psychiat; fel Am Pub Health Asn; Am Psychosom Soc. *Res:* Social psychiatry including epidemiology of and attitudes towards mental illness; small group theory and practice; health and mental health service delivery systems; psychosomatic medicine and liaison psychiatry. *Mailing Add:* Sinai Samaritan Med Ctr 2000 W Kilbourn Milwaukee WI 53233

SPIRO, HOWARD MARGET, b Cambridge, Mass, Mar 23, 24; m 51; c 4. MEDICINE. *Educ:* Harvard Univ, BA, 43, MD, 47. *Hon Degrees:* Yale Univ, MA, 67. *Prof Exp:* From asst prof to assoc prof, 56-67, PROF MED, SCH MED, YALE UNIV, 67- *Mem:* Am Gastroenterol Asn; Am Soc Clin Invest. *Res:* Gastroenterology. *Mailing Add:* Yale Univ Sch Med 333 Cedar St Box 3333 New Haven CT 06510

SPIRO, IRVING J, b Chicago, Ill, Sept 20, 13; m 40; c 2. PHYSICS, OPTICAL ENGINEERING. *Educ:* Ill Inst Technol, BS, 36; Univ Calif, Los Angeles, MS, 61. *Prof Exp:* Design analyst, Int Harvester-Tractor Works, 36-38; chief draftsman, Graf Optical Co, 38-39; design engr, Lockheed Aircraft Corp, 38-45; eng supvr, Mission Appliance Corp, 45-47; gen mgr, Roxmar Optical Co, 47-51; chief engr, Borman Eng, Inc, 51-55, treas, 55-56; proj mgr, Aerophys Develop Corp, 56-58; sect head, Space Tech Labs, 58-60; sect head, 60-63, staff engr, 63-70, mgr, 70-85, SR ENGR, AEROSPACE CORP, 85- *Concurrent Pos:* Instr, Univ Calif, Los Angeles, 42-45. *Honors & Awards:* Governor's Award, Soc Photo-Optical Instrumentation Engrs, 81, President's Award, 87. *Mem:* Am Soc Mech Engrs; fel Optical Soc Am; Am Inst Aeronaut & Astronaut; Sigma Xi; fel Soc Photo-Optical Instrumentation Engrs (secy, 65-66, vpres, 66-67, pres, 68-70). *Res:* Military infrared systems, especially atmospheric transmission; thermal radiation properties of materials; constituents of smog and smog measuring instruments; servo systems utilizing man-optics combinations. *Mailing Add:* 4924 Mammoth Ave Sherman Oaks CA 91423-1320

SPIRO, JULIUS, b New York, NY, Nov 20, 21; m 46; c 2. ELECTRONICS ENGINEERING. *Educ:* City Col New York, BS, 53. *Prof Exp:* Electronic engr, Nevis Cyclotron, Columbia Univ, 47-53, Hudson Labs, 53-54; sr elec engr, Accelerator Dept, 54-86, CONSULT, BROOKHAVEN NAT LAB, 88- *Mem:* Sr mem Inst Elec & Electronics Engrs. *Res:* Electronic engineering applied to high energy particle accelerator design; logic and control systems for particle accelerators. *Mailing Add:* Brookhaven Nat Lab Upton NY 11973

SPIRO, MARY JANE, b Syracuse, NY, Nov 15, 30; m 52; c 2. BIOCHEMISTRY. *Educ:* Syracuse Univ, AB, 52, PhD, 55. *Prof Exp:* Res assoc biochem, Col Med, State Univ NY Upstate Med Ctr, 55-56; from res assoc to prin assoc, 60-84, ASSOC PROF MED, HARVARD MED SCH, 84- *Concurrent Pos:* Res fel, Harvard Med Sch, 56-60; sr investr, Joslin Res Lab, 74- *Mem:* Am Soc Biol Chemists; Soc Complex Carbohydrates; Am Diabetes Asn. *Res:* Glycoprotein biosynthesis, structure and change in disease states; structure and metabolism of thyroglobulin; extracellular matrix changes in diabetes. *Mailing Add:* Joslin Res Lab One Joslin Pl Boston MA 02215

SPIRO, ROBERT GUNTER, b Berlin, Germany, Jan 5, 29; nat US; m 52; c 2. BIOCHEMISTRY. *Educ:* Columbia Col, AB, 51; State Univ NY, MD, 55. *Hon Degrees:* AM, Harvard Univ, 75. *Prof Exp:* Intern, Syracuse Med Ctr, NY, 55-56; res assoc med, 60-63, from assoc to assoc prof biol chem, 64-74, PROF BIOL CHEM, HARVARD MED SCH, 74- *Concurrent Pos:* Am Cancer Soc res fel biochem, Harvard Med Sch, 56-58 & res fel med, Mass Gen Hosp, 58-60; USPHS res fel, 58-59; Am Heart Asn advan res fel, 59-61; estab investr, Am Heart Asn, 61-66; sr investr & chief sect, Complex Carbohydrates & Biomembranes, Joslin Res Lab, 61- *Honors & Awards:* Lilly Award, Am Diabetes Asn, 68; Claude Bernard Award, Europ Asn Study Diabetes, 75. *Mem:* Am Diabetes Asn; Am Soc Biol Chem; Am Chem Soc; Soc Complex Carbohydrates (pres, 78). *Res:* Chemical structure and biosynthesis of glycoproteins; biochemistry and biology of cell surfaces and basement membranes; biochemistry of diabetes mellitus; regulatory action of insulin; chemistry of connective tissues and basement membranes. *Mailing Add:* Joslin Res Lab One Joslin Place Boston MA 02215

SPIRO, THOMAS, b St Louis, Mo, May 29, 47. NEUROANATOMY, ANATOMY. *Educ:* Univ Mo, St Louis, BA, 71; St Louis Univ, MS(R), 76, PhD(anat), 78. *Prof Exp:* Res assoc anat, St Louis Univ, 78-; AT DEPT ANAT, SAINT MARY HEALTH CTR, ST LOUIS. *Mem:* Soc Neurosci; Sigma Xi. *Res:* Neuroanatomy with emphasis on modern hodological methodology. *Mailing Add:* 2809 Tall Oak St Louis MO 63129

SPIRO, THOMAS GEORGE, b Aruba, Netherlands Antilles, Nov 7, 35; m 59; c 2. CHEMISTRY. *Educ:* Univ Calif, Los Angeles, BS, 56; Mass Inst Technol, PhD(chem), 60. *Prof Exp:* Fulbright student, Copenhagen, 60-61; res chemist, Calif Res Corp, 61-62; NIH fel, Royal Inst Technol Sweden, 62-63; from instr to assoc prof, 63-64, chmn dept, 79-88 PROF CHEM, PRINCETON UNIV, 74-,. *Concurrent Pos:* Guggenheim fel, 89. *Honors & Awards:* Bomem Michelson Award, 86; Merit Award, NIH, 89. *Mem:* AAAS; Am Chem Soc; Am Soc Biol Chemists; Biophys Soc. *Res:* Resonance Raman spectroscopy; applications to biological structure; role of metals in biology; bonding in inorganic molecules; chemically modified electrodes. *Mailing Add:* Dept Chem Princeton Univ Washington Rd Princeton NJ 08540

SPIROFF, BORIS E N, b Waukegan, Ill, Dec 5, 25. BIOLOGY, EMBRYOLOGY. *Educ:* Loyola Univ, Ill, BS, 46; Univ Chicago, MS, 49; Northwestern Univ, PhD(biol), 53; Aquinas Inst Philos, MA, 69. *Prof Exp:* From instr to asst prof biol, Loyola Univ, Ill, 52-62; asst prof, Bard Col, 62-64; assoc prof, Canisius Col, 64-66; ASST PROF BIOL, LOYOLA UNIV, CHICAGO, 66- *Mem:* Am Soc Zool; Am Inst Biol Sci; Sigma Xi. *Res:* Developmental biology; history and philosophy of science; history of biology. *Mailing Add:* Dept of Biol Sci 6525 N Sheridan Rd Chicago IL 60626

SPITALNY, GEORGE LEONARD, b Philadelphia, Pa, Mar 7, 47; c 3. IMMUNOBIOLOGY. *Educ:* Pa State Univ, BS, 69; NY Univ, PhD(immunol, parasitol), 73. *Prof Exp:* Asst res sci, Sch Med, NY Univ, 73, from instr to asst prof immunobiol, 73-75; res assoc, 75-77, asst mem immunobiol staff, Trudeau Inst, 77-82; DIR IMMUNOL, BRISTOL-MYERS CO, 83- *Concurrent Pos:* Nat Inst Allergy & Infectious Dis fel, Sch Med, NY Univ, 74-75; fel, Trudeau Inst, 75-76, fel, Cancer Res Inst, 77; prin investr, Trudeau Inst, Nat Cancer Inst grant, 78-81; prin investr, Grant Nat Inst Alergy & Infectious Dis, 81-84; coprin investr, Grant Nat Cancer Inst, 81-85; assoc mem, Trudeau Inst, 82-83. *Mem:* Am Assoc Immunol; AAAS; Am Assoc Cancer Res; Am Assoc Microbiol; Reticuloendothelial Soc. *Res:* Mechanisms of immunity to infectious and neoplastic diseases; tumor induced suppressor cells; regulation of immunity; delivery of cytotoxic drug via linkage to monoclonal antibodies. *Mailing Add:* Dir Immunol Bristol-Myers Co Five Res Pkwy Wallingford CT 06492

SPITLER, LYNN E, b Grand Rapids, Mich, Sept 28, 38; m 67; c 2. IMMUNOLOGY, INTERNAL MEDICINE. *Educ:* Univ Mich, MD, 63. *Prof Exp:* Intern, Highland-Alameda County Hosp, 63-64; resident, Med Ctr, Univ Calif, San Francisco, 64-66; fel immunol, NY Univ, 66-67; fel immunol, 67-69, instr, 69-71, ASST PROF MED, MED CTR, UNIV CALIF, SAN FRANCISCO, 71- *Concurrent Pos:* Res assoc, Cancer Res Inst, 73-; assoc ed, J Immunol, 74-78; dir res, Children's Hosp, San Francisco, 75-; mem allergy & immunol res comt, Nat Inst Allergy & Infectious Dis, NIH, 76-; mem immunol rev comt, Vet Admin, Washington, DC, 77- *Mem:* Am Asn Immunologists; AAAS; Am Fedn Clin Res. *Res:* Immunopotentiator therapy; transfer factor; levamisole; multiple sclerosis immunology; immunotherapy of malignant melanoma; monoclonal antibodies. *Mailing Add:* Northern California Melanoma Ctrs 1895 Mountain View Dr Tiburon CA 94920

SPITLER, MARK THOMAS, b Rockford, Ill, Oct 19, 50. PHYSICAL CHEMISTRY. *Educ:* Stanford Univ, BS, 72; Univ Calif, Berkeley, PhD(phys chem), 77. *Prof Exp:* Guest scientist, Fritz Haber Inst, Max Planck Soc, 77-78; asst prof chem, Mount Holyoke Col, 79-84; AT POLAROID CORP, 84- *Concurrent Pos:* Vis asst prof chem, Amherst Col, 78-79. *Mem:* Sigma Xi; Am Chem Soc; AAAS; Electrochem Soc. *Res:* Photoelectrochemistry; semiconductor electrochemistry; photochemical energy conversion photochemistry at electrified interfaces; amorphous semiconductors. *Mailing Add:* 110 Tarbell Spring Rd Concord MA 01742-4023

SPITSBERGEN, JAMES CLIFFORD, b Washington, DC, Sept 1, 26; m 80; c 2. THERMOSET POLYMER TECHNOLOGY. *Educ:* George Washington Univ, BS, 49; Univ Del, MS, 59, PhD(phys polymer chem), 62. *Prof Exp:* Chemist, Eng Res & Develop Labs, Army Eng Corp, 51; chemist, Elec Hose & Rubber Co, 51-57; sr chemist, 57-61; sr chemist, Elastomers Lab, E I du Pont de Nemours & Co, 62-68; proj leader, polymer res & develop, Corp Res & Develop Lab, Witco Chem Corp, 68-83; polymer consult, Unitrode Corp, 83-85; POLYMER CONSULT, EPOXY CONSULT, 85- *Mem:* Am Chem Soc; Soc Plastics Eng; Int Electronics Packaging Soc; Soc Plastics Indust; Int Soc Hydrid Microelectronics. *Res:* Elastomer technology, particularly compositions for hose; molecular weight distribution-rheology relationships of elastomers, particularly neoprene; synthesis and characterization of thermosetting polymers, particularly epoxy resins; structure-property relationships of polymers; potting; transfer molding; powder coating; adhesion and laminating of electrical and electronic components. *Mailing Add:* 696 Knollwood Rd Franklin Lakes NJ 07417-1710

SPITTELL, JOHN A, JR, b Baltimore, Md, Apr 7, 25; m 49; c 5. INTERNAL MEDICINE, CARDIOVASCULAR DISEASES. *Educ:* Franklin & Marshall Col, BS, 44; Univ Md, MD, 49; Univ Minn, MS, 55. *Prof Exp:* From asst prof to prof, Mayo Med Sch, Univ Minn, 62-80, vchmn educ, Dept Med, Mayo Clin, 72-76, assoc dir continuing educ, Mayo Found, 78-84, Mary Lowell Leary prof med, 80-89; RETIRED. *Concurrent Pos:* NIH grant, 64-68; consult, Mayo Clin, 56-; mem spec ad hoc comt, Food & Drug Admin, 63; Nat Cardiovasc Conf Peripheral Vascular Dis, 64; bd regents, Am Col Physicians, 80-89. *Mem:* Fel Am Col Physicians; fel Am Col Cardiol. *Res:* Peripheral vascular disease; relationship of changes of blood coagulation and intravascular thrombosis; mechanism of action of Coumarin anticoagulants; aneurysmal disease; aortic dissection. *Mailing Add:* Mayo Clin Rochester MN 55901

SPITTLER, ERNEST GEORGE, b Cleveland, Ohio, May 4, 28. CHEMISTRY. *Educ:* Loyola Univ, Ill, AB, 51, PhL(philos), 53, ThL(theol), 63; Cath Univ Am, PhD(chem), 59. *Prof Exp:* Instr physics, Loyola Acad, Ill, 53-54; asst prof, 65-76, ASSOC PROF CHEM, JOHN CARROLL UNIV, 76- *Concurrent Pos:* Res assoc, Bushy Run Radiation Lab, Mellon Inst, 64-66; secy bd trustees, John Carroll Univ, 69-71, mem bd trustees, 80- *Mem:* Am Chem Soc; Hist Sci Soc. *Res:* Mercury-photosensitized reactions of hydrocarbon systems; use of carbon-14 tagged molecules as tracers in studying gas phase reactions; chemical effects of lasers and ultra-sound; photochemistry of inorganic complexes; history of periodic table; development of theories of chemistry. *Mailing Add:* Dept Chem John Carroll Univ, Univ Heights Cleveland OH 44118-3895

SPITTLER, TERRY DALE, b Buffalo, NY, Apr 29, 43; m 74; c 3. ORGANIC CHEMISTRY, PESTICIDE CHEMISTRY. *Educ:* Bowling Green State Univ, BA, 65; State Univ NY, Buffalo, MS, 68; State Univ NY, Albany, PhD(org chem), 74. *Prof Exp:* Res assoc paper & pulp, State Univ NY Col Environ Sci & Forestry, 74-75; res assoc coal & asphalt, Mont State Univ, 75-77; LAB COORDR, PESTICIDE RESIDUES, NY STATE AGR EXP STA, CORNELL UNIV, 77- *Mem:* Am Chem Soc. *Res:* Analytical organic methods development in pesticide residues; carbon and proton nuclear magnetic resonance; groundwater pollution and quality determination; pulp bleaching with alkaline hydrogen peroxide; pesticide worker exposure to second disposal. *Mailing Add:* Anal Div Cornell Univ Geneva NY 14456

SPITZ, IRVING MANFRED, b July 9, 39, Johannesburg, SAfrica; Israeli citizen; m 64; c 2. ENDOCRINOLOGY, METABOLISM. *Educ:* Witwatersrand Med Sch, Johannesburg, MB, BCh, 62, MD, 71; Royal Col Physicians, London, MRCP, 65. *Prof Exp:* Temp chief physician, dept chem endocrinol, Hadassah Univ Hosp, Jerusalem, 70-73, permanent chief physician, 73-74, actg chief dept, 75-76; assoc prof endocrinol, Hebrew Univ, 78-82; DIR CLIN RES, CTR BIOMED RES, POP COUN, NY, 82- *Concurrent Pos:* Head dept endocrinol & metab, Shaare Zedek Med Ctr, Jerusalem, 77-82. *Honors & Awards:* Albelheim Prize, 63. *Mem:* Endocrine Soc; Am Soc Andrology; NY Acad Sci; AAAS; Am Soc Clin Invest. *Res:* Endocrine control of human reproduction, hormonal aspects of contraception; role of prolactin in reproduction; hormonal regulation of hypogonadal states; inappropriate TSH secretion. *Mailing Add:* Ctr Biomed Res Pop Coun 1230 York Ave New York NY 10021

SPITZ, WERNER URI, b Stargard, Ger, Aug 22, 26; US citizen; c 3. PATHOLOGY, FORENSIC MEDICINE. *Educ:* Hebrew Univ, Jerusalem, MD, 53; Am Bd Path, dipl & cert path anat, 61, cert forensic path, 65. *Prof Exp:* Resident path, Tel-Hashomer Govt Hosp, Israel, 53-56; resident forensic med, Hebrew Univ, Jerusalem, 56-59; asst forensic path, Free Univ Berlin, 61-63; assoc med examr, Md Med-Legal Found, 63-65; asst med examr, Off Chief Med Examr, Md, 65-69, dep chief med examr, 69-72; CHIEF MED EXAMR, WAYNE COUNTY, MICH, 72-; PROF PATH, SCH MED, WAYNE STATE UNIV, 72- *Concurrent Pos:* Res fel forensic path, Univ Md, Baltimore City, 59-61; Nat Inst Gen Med Sci training & res grant forensic path, Md State Med Examr, 62-, NIH grant, 64-66; consult path, Israel Ministry of Health, 57-59; lectr, Johns Hopkins Univ, 66, assoc prof, Sch Hyg & Pub Health, 67-72, consult, appl physics lab, 72-; asst prof, Sch Med, Univ Md, 66, clin assoc prof, 69-72, mem grad fac, Col Park, 70-; dir res & training, Md Med-Legal Found, 67-; mem ed bd, J Forensic Sci, J Legal Med, Excerpta Medica-Forensic Sci; adj prof chem, Univ Windsor, Ont, Can, 78- *Mem:* Fel Col Am Path; fel Am Soc Clin Path; AMA; Soc Exp Biol & Med; Nat Asn Med Examrs; hon mem Latin Am Asn Legal Med, 82- *Res:* Mechanism of death by drowning; pathology of vehicular trauma; wound patterns by firearms and other agents. *Mailing Add:* Off Chief Med Examr 400 E Lafayette St Detroit MI 48226-2995

SPITZBART, ABRAHAM, b New York, NY, Oct 13, 15. MATHEMATICS. *Educ:* City Col New York, BS, 35; Harvard Univ, AM, 36, PhD(math), 40. *Prof Exp:* Instr math, Harvard Univ, 37-40, City Col New York, 40-41 & Univ Minn, 42; prof, Col of St Thomas, 42-43; from instr to assoc prof, 45-61, PROF MATH, UNIV WIS-MILWAUKEE, 61- *Mem:* Am Math Soc; Math Asn Am. *Res:* Approximation theory in complex variables; numerical analysis. *Mailing Add:* Dept Math Univ Wis Milwaukee WI 53201

SPITZE, LEROY ALVIN, b Ford Co, Kans, Sept 7, 17; m 42; c 5. CHEMISTRY. *Educ:* Southwestern Col, Kans, AB, 39; Rensselaer Polytech Inst, MS, 41, PhD(phys chem), 42. *Prof Exp:* Asst, Rensselaer Polytech Inst, 39-42; sr phys chemist, Owens-Corning Fiberglas Corp, Ohio, 42-47; prof chem, Southwestern Col, Kans, 47-55, chmn div natural sci, 52-55; assoc prof, 55-60, PROF CHEM, SAN JOSE STATE UNIV, 60- *Concurrent Pos:* Fulbright lectr, Vidyodaya Univ Ceylon, 65-66; NASA res grant, 70-; consult, FMC Corp, IBM Corp & Lockheed Missiles & Space Co. *Mem:* Am Chem Soc. *Res:* Adsorption; surface properties; reverse osmosis and hyperfiltration; diffusion; polymer application; transport parameters in waste water purification; superoxides; numeral separation. *Mailing Add:* 17840 Holiday Dr Morgan Hill CA 95037-4696

SPITZER, ADRIAN, b Dec 21, 27; m; c 1. NEPHROLOGY. *Educ:* Med Sch Bucharest, Rumania, MD, 52. *Prof Exp:* PROF PEDIAT & DIR DEPT NEPHROLOGY, ALBERT EINSTEIN COL MED, 73- *Concurrent Pos:* Vis prof, Wilhelmena Children's Hosp, Ufrecht, Holland & Dept Biochem, Oxford, Eng. *Mem:* Soc Pediat Res; Am Pediat Soc; Am Soc Nephrol; Am Soc Pediat Nephrol; Am Physiol Soc; Am Fedn Clin Res. *Res:* Developmental renal physiology. *Mailing Add:* Dept Pediat Albert Einstein Col Med 1410 Pelham Pkwy S Bronx NY 10461

SPITZER, CARY REDFORD, b New Hope, Va, Jul 31, 37; m 60; c 1. ELECTRONICS ENGINEERING, EARTH SCIENCES. *Educ:* Va Polytechnic Inst & State Univ, BS, 59; George Washington Univ, MS, 70. *Prof Exp:* Aerospace technologist, instrumentation, 62-69, exp mgr, planetary missions, 69-78, MGR PROG PLANS FLIGHT RES, NASA LANGLEY RES CTR, 78- *Concurrent Pos:* Lectr, Navigation Technol Seminars Inc, 88- *Honors & Awards:* Centennial Medal, Inst Elec & Electronics Engrs, 84. *Mem:* Am Asn Advan Sci; Am Inst Aeron & Astron; Inst Elec & Electronics Engrs. *Res:* Flight research and development and validation of advanced aircraft flight control laws; electronic displays; operating procedures. *Mailing Add:* 3409 Foxridge Rd Williamsburg VA 23188-2499

SPITZER, FRANK L, b Vienna, Austria, July 24, 26; nat US; m 51; c 2. MATHEMATICS. *Educ:* Univ Mich, PhD(math), 53. *Prof Exp:* From instr to asst prof math, Calif Inst Technol, 53-56; assoc prof, Univ Minn, 58-61; PROF MATH, CORNELL UNIV, 61- *Concurrent Pos:* NSF sr fel, 60-61; Guggenheim fel, 65-66. *Mem:* Nat Acad Sci; Am Math Soc. *Res:* Probability theory, mathematical analysis. *Mailing Add:* Dept Math Cornell Univ Ithaca NY 14850

SPITZER, IRWIN ASHER, b Los Angeles, Calif, July 4, 22; m 53; c 2. SYSTEMS ENGINEERING, MECHANICAL ENGINEERING. *Educ:* Univ Calif, BS, 44. *Prof Exp:* Sr engr, Kaiser Steel Corp, 47-56; prin engr, Grand Cent Rocket Co, 56-62; sr prog mgr, Lockheed Propulsion Co, 62-73; asst dir procurement, Amecom Div, Litton Indust, 73-75; SR STAFF ENGR, BALLISTIC MISSILE DIV, TRW SYSTS, 75- *Mem:* Am Soc Qual Control; Am Inst Aeronaut & Astronaut; Air Force Asn. *Res:* Solid rocket propulsion systems; advanced inter-continental ballistic missile basing concepts. *Mailing Add:* 306 Marcia St Redlands CA 92373

SPITZER, JEFFREY CHANDLER, b Malden, Mass, Dec 1, 40; m 67; c 1. ORGANIC CHEMISTRY, SPECTROSCOPY. *Educ:* Mass Inst Technol, BS, 61; Univ Ariz, PhD(chem), 66. *Prof Exp:* Res assoc org chem, Univ Calif, 66-67; asst ed, 67-69, sr assoc indexer, 69-79, SR ED, CHEM ABSTRACTS SERV, 79- *Res:* Terpene and polyacetylene structure determination; synthesis of pyrrole derivatives. *Mailing Add:* 2279 Canterbury Rd Columbus OH 43221

SPITZER, JOHN J, b Baja, Hungary, Mar 9, 27; m 51; c 2. PHYSIOLOGY. *Educ:* Univ Munich, MD, 50. *Prof Exp:* Demonstr physiol, Sch Med, Univ Budapest, 47-49; lectr, Sch Med, Dalhousie Univ, 51-52; asst prof, Fla State Univ, 52-54; res scientist, Div Labs & Res, NY State Dept Health, 54-57; from asst prof to prof physiol, Hahnemann Med Col, 57-73; PROF PHYSIOL & HEAD DEPT, LA STATE UNIV MED CTR, NEW ORLEANS, 73- *Concurrent Pos:* Vis scientist, Lab Physiol, Oxford Univ; Burroughs Wellcome Professorships, 83. *Honors & Awards:* Christian R & Mary F Lindback Award, 61. *Mem:* AAAS; Am Physiol Soc; Soc Exp Biol & Med; Am Heart Asn; NY Acad Sci. *Res:* Substrate metabolism in vivo; carbohydrate metabolism; hepatic non-parenchymal cells; shock and metabolism; sepsis-induced metabolic changes; oxygen free radical production. *Mailing Add:* Dept Physiol La State Univ Med Ctr New Orleans LA 70112

SPITZER, JUDY A, b Budapest, Hungary, Feb 25, 31; US citizen; m 51; c 2. PHYSIOLOGY, BIOCHEMISTRY. *Educ:* Fla State Univ, BA, 53; Albany Med Col, MS, 55; Hahnemann Med Col, PhD(microbiol, immunol), 63. *Prof Exp:* Res asst physiol, Fac Med, Dalhousie Univ, 51-52; asst biochem, Fla State Univ, 53-54; biochemist, Div Labs & Res, NY State Dept Health, 54-57; res assoc physiol, Hahnemann Med Col, 57-61 & 62-70, res asst prof physiol & biophys, 70-72; res assoc prof, 72-73, assoc prof med & physiol, 73-79, PROF PHYSIOL MED, LA STATE UNIV MED CTR, NEW ORLEANS, 79- *Concurrent Pos:* Prin investr, Off Naval Res Contract, 73-82 & NIH grants, 82-; chmn, Am Physiol Soc Educ Ctr, 82-87, Fedn Am Soc Exp Biol Educ Ctr, 84-86; mem, Surg, Anesthesiol & Trauma Study Sect, NIH, 84-89. *Mem:* NY Acad Sci; Endotoxin Soc; Am Physiol Soc; Shock Soc (secy, 85-); Soc Exp Biol Med; Am Soc Biochem & Molecular Biol. *Res:* Signal transduction mechanisms; metabolic and endocrine changes in shock. *Mailing Add:* Dept Physiol La State Univ Med Ctr New Orleans LA 70112-1393

SPITZER, LYMAN, JR, b Toledo, Ohio, June 26, 14; m 40; c 4. INTERSTELLAR MATTER, STELLAR DYNAMICS. *Educ:* Yale Univ, BA, 35; Princeton Univ, MA, 37, PhD(astrophys), 38. *Hon Degrees:* DSc, Yale Univ, 58, Case Inst Technol, 61 & Harvard Univ, 75; LLD, Univ Toledo, 63; DSc, Princeton Univ, 84. *Prof Exp:* Nat Res Coun fel, Harvard Univ, 38-39; instr physics, Yale Univ, 39-41, instr astron & physics, 41-42; scientist spec studies group, Div War Res, Columbia Univ, 42-44, dir sonar anal group, 44-46; assoc prof astrophys, Yale Univ, 46-47; prof, Princeton Univ, 47-52, Young prof astron, 52-82, chmn dept & dir observ, 47-79; RETIRED. *Concurrent Pos:* Dir, Proj Matterhorn, Princeton Univ, 53-61, chmn exec comt, Plasma Physics Lab, 61-66 & Univ Res Bd, 67-72, chmn, Asn Univ Res Astron Space Telescope Inst Coun, 81-90. *Honors & Awards:* Rittenhouse Medal, Franklin Inst, 57; Bruce Gold Medal, Astron Soc Pac, 73; Henry Draper Medal, Nat Acad Sci, 74; James Clerk Maxwell Prize, Am Phys Soc, 75; Distinguished Pub Serv Medal, NASA, 76; Gold Medal, Royal Astron Soc, 78; Jules Janssen Medal, Soc Astron de France, 80; Franklin Medal, Franklin Inst, 80; Nat Medal of Sci, 80; Crafoord Prize, 85. *Mem:* Nat Acad Sci; Am Astron Soc (pres, 60-62); foreign corresp Royal Soc Sci, Liege; foreign mem Royal Soc London; foreign assoc Royal Astron Soc England; Sigma Xi; Am Acad Arts & Sci; Am Philos Soc. *Res:* Interstellar matter; dynamics of stellar systems; physics of fully ionized gases; controlled thermonuclear research; space astronomy. *Mailing Add:* Peyton Hall Princeton Univ Observ Princeton NJ 08544

SPITZER, NICHOLAS CANADAY, b New York, NY, Nov 8, 42; m 67; c 2. DEVELOPMENTAL NEUROBIOLOGY. *Educ:* Harvard Univ, BA, 64, PhD(neurobiol), 69. *Prof Exp:* From asst prof to assoc prof biol, 73-82, chmn dept, 88-90, PROF BIOL, UNIV CALIF, SAN DIEGO, 82- *Mem:* Soc Neurosci; AAAS; Biophys Soc; Soc Develop Biol. *Res:* Embryonic development of neuronal membrane properties in vivo and in culture characterizing the order of appearance of phenotypes and defining the roles of RNA and protein synthesis in their expression; neurophysiology. *Mailing Add:* 0322 Biol Dept Univ Calif San Diego 9600 Gilman Dr La Jolla CA 92093-0322

SPITZER, RALPH, b New York, NY, Feb 9, 18; m 41; c 1. CHEMISTRY, PATHOLOGY. *Educ:* Cornell Univ, AB, 38; Calif Inst Technol, PhD(chem), 41; Univ Man, MD, 57; FRCPath. *Prof Exp:* Assoc phys chemist, Nat Adv Comt Aeronaut, 42-43; res assoc, Woods Hole Oceanog Inst, 43-45; assoc prof chem, Ore State Col, 46-50 & Univ Kans City, 50-53; res assoc med, Univ Man, 53-54; clin instr, 58-65, assoc prof, 65-82, clin prof, 82-85, EMER CLIN PROF CHEM PATH, UNIV BC, 85- *Concurrent Pos:* Dir biochem labs, Royal Columbian Hosp, 58-81, chem pathologist, 81-88; vis sr lectr, Univ Otago, NZ, 62-63. *Mem:* Acad Lab Physicians & Scientists; Can Soc Clin Chem; Am Soc Clin Res; sr mem Can Med Asn. *Res:* Chemical biology and pathology; endocrine effects of tumors; clinical enzymology; psychopharmacology. *Mailing Add:* 1911 Knox Rd Vancouver BC V6T 1S5 Can

SPITZER, ROBERT HARRY, b Chicago, Ill, July 25, 29. BIOCHEMISTRY. *Educ:* Valparaiso Univ, BA, 51; Loyola Univ, Ill, MS, 53, PhD(biochem), 55. *Prof Exp:* Instr exp surg, Sch Med, Wash Univ, 55-56; res chemist, Gillette Co, 56-61; from res assoc med to res assoc enzymol & exp hypersensitivity, 61-65, from asst prof enzymol & exp hypersensitivity to asst prof biochem, 65-70, assoc prof, 70-75, PROF BIOCHEM, CHICAGO MED SCH, 75- *Mem:* Am Asn Immunologists. *Res:* Cell biology; metabolism in lower vertebrates; immunobiology. *Mailing Add:* 3835 Brittany Rd Northbrook IL 60062

SPITZER, ROGER EARL, b Washington, DC, June 20, 35; m 62; c 3. PEDIATRICS, IMMUNOLOGY AND NEPHROLOGY. *Educ:* George Washington Univ, BS, 58; Howard Univ, MD, 62; Am Bd Pediat, dipl, 68, dipl pediat nephrology, 73. *Prof Exp:* Intern, Gen Hosp, Cincinnati, 62-63; resident pediat, Children's Hosp, 63-65; asst prof pediat, Col Med, Univ Cincinnati, 69-73; assoc prof, 73-77, PROF PEDIAT, STATE UNIV NY UPSTATE MED CTR, 77- *Concurrent Pos:* NIH spec fel nephrology-immunol, Children's Hosp Res Found, 67-69; res assoc immunol, Children's Hosp Res Found & attend pediatrician, Children's Hosp Med Ctr, 69-73, attend nephrologist, 72-73; clinician pediat, Gen Hosp, 59-73; consult, pediatrician, Good Samaritan Hosp, 71-73. *Mem:* Fel Am Acad Pediat; Am Soc Pediat Nephrology; Am Soc Nephrology; Soc Pediat Res; Am Asn Immunol. *Res:* Biology of complement; pediatric renal disease; biology of antibody formation. *Mailing Add:* Dept Pediat State Univ NY Health Sci Ctr Syracuse NY 13210

SPITZER, WALTER O, b Asuncion, Paraguay, Feb 19, 37; Can citizen. EPIDEMIOLOGY. *Educ:* Univ Toronto, MD, 62; Univ Mich, MHA, 66; Yale Univ, MHP, 70; FRCP(C). *Prof Exp:* Gen dir, Int Christian Med Soc, 66-69; assoc mem, Dept Family Med, McMaster Univ, 69-75, asst prof clin epidemiol & biostatist, 69-75; prof family med, 75-83, PROF EPIDEMIOL & HEALTH, MCGILL UNIV, 75-, PROF MED, 83-, CHMN, DEPT EPIDEMIOL & BIOSTATIST, 84-, STRATHCONA PROF PREV MED, 84- *Concurrent Pos:* Prin investr grants, Ont Min Health, 70-74, 71-72, Can Arthritis & Rheumat Soc, 72-74, Nat Cancer Inst Can, 72-74, 80-81 & 82, Health & Welfare Can, 76-78, Commonwealth of Australia, 79-80, Can Res Soc, 82, Toronto Dept Health, 83 & Govt Alta, 85-86; attend physician, McMaster Univ Med Ctr, 71-75; co-investr grants, Health & Welfare Can, 71-74, 72-74, NIH, 86-; mem, Task Force Demonstration Models, Comt Health Res, Ont Coun Health, 73-76; mem, Res Eval Panel Clin Res, Nat Cancer Inst Can, 74-78, Rev Panel Epidemiol Res, 80-81; sr physician, Dept Med, Montreal Gen Hosp, 75-, Royal Victoria Hosp, 85-; vis prof epidemiol, Univ Buenos Aires, Arg, 75, Sydney Univ, Australia, 79-80, Newcastle Univ, Australia, 79-80; vis fac, Int Ctr Res Cancer, WHO, Lyons, France, 75; vis prof clin epidemiol, Nat Univ Chile, 75; chmn, Task Force Eval of Periodic Health Exam, Health & Welfare Can, 76-85, mem, 85-, nat vis health scientist to Australia, 79-80; vis prof, Western Australia Inst Technol, 77, Univ Cologne, 87 & Shanghai Univ, Peoples Repub China, 87; dir, W K Kellogg Ctr Advan Studies in Primary Care, McGill Univ & Montreal Gen Hosp, 77-83, Div Clin Epidemiol, Dept Med, Montreal Gen Hosp, 79-86; John F McCleary vis prof, Univ BC, 78; prof, McGill Cancer Ctr, 78-, assoc dir, 79-; vis prof med & epidemiol, Royal North Shore Hosp, Sydney, Australia, 79-80; mem, Comt to Study Pain, Disability & Chronic Illness Behav, Inst Med-Nat Acad Sci, 85-; ed, J Clin Epidemiol. *Mem:* Inst Med-Nat Acad Sci; fel Am Col Epidemiol; Can Oncol Soc (pres, 88-89); Soc Epidemiol Res; Int Epidemiol Asn; Col Family Physicians Can; Am Fedn Clin Res; Am Epidemiol Soc. *Res:* Health manpower studies; causality in biomedical phenomena; epidemiology of clinical phenomena in primary care; clinical epidemiology of cancer; clinical epidemiology of rheumatic and arthritic complaints; randomized controlled trials; systematic error and bias in clinical and epidemiologic research; epidemiology of environmental health hazards; quality of life measurement and validation; pharmacoepidemiology; author or co-author of numerous publications. *Mailing Add:* McGill Univ Purvis Hall 1020 Pine Ave W Montreal PQ H3A 1A2 Can

SPITZER, WILLIAM CARL, b Chicago, Ill, Sept 15, 14; m 42; c 1. ORGANIC POLYMER CHEMISTRY. *Educ:* Univ Chicago, BS, 36, PhD(org chem), 40. *Prof Exp:* Org chemist, Sherwin-Williams Co, 40-47; org chemist, Paint Res Assoc Inc, 48-66, dir res, 66-75, dir res, PRA Labs, Inc, 75-81; RETIRED. *Concurrent Pos:* Instr, Ill Inst Technol, 50-51; lectr, DePaul Univ, 64; consult, 82-87. *Mem:* Am Chem Soc; Am Oil Chemists Soc; Am Soc Testing & Mat. *Res:* Organic coatings; synthetic resins; drying oils. *Mailing Add:* 221 White Fawn Trail Downers Grove IL 60516

SPITZER, WILLIAM GEORGE, b Los Angeles, Calif, Apr 24, 27; m 49; c 2. PHYSICS. *Educ:* Univ Calif, Los Angeles, BA, 49; Univ Southern Calif, MS, 52; Purdue Univ, PhD(physics), 57. *Prof Exp:* Mem tech staff, Hughes Aircraft Co, 52-53, Bell Labs, Inc, 57-62 & Bell & Howell Res Ctr, 62-63; chmn dept mat sci, 67-69 & 78-81, dept physics, 69-72, dean natural sci, 72-73, PROF PHYSICS, ELEC ENG & MAT SCI, UNIV SOUTHERN CALIF, 63- *Mem:* Fel Am Phys Soc; Inst Elec & Electronics Engrs; Sigma Xi. *Res:* Solid state and semiconductor physics; infrared properties of semiconductors and dielectrics. *Mailing Add:* Dept Mat & Sci Univ S Calif Vivian Hall Eng MC 0241 Los Angeles CA 90089

SPITZIG, WILLIAM ANDREW, b Cleveland, Ohio, Sept 12, 31; m 59; c 4. MECHANICAL METALLURGY, STRUCTURE-PROPERTY CORRELATIONS. *Educ:* Cleveland State Univ, BS, 60; Case Western Reserve Univ, MS, 62, PhD(mat sci), 65. *Prof Exp:* Metallurgist, Thompson-Ramo-Wooldridge Inc, 60-62; mat engr, Lewis Res Ctr, NASA, 62-66; sr scientist, US Steel Res Labs, 66-84; SR METALLURGIST, AMES LAB, IOWA STATE UNIV, 84- *Concurrent Pos:* Lectr, Univ Pittsburgh, 68-83; adj prof MS & E, Iowa State Univ, 84- *Honors & Awards:* Energy Res Award, Dept Energy, 87. *Mem:* Am Soc Metals; Am Inst Mining, Metall & Petrol Engrs; Mats Res Soc. *Res:* Structure-property relationships in body centered cubic single crystals, steels, titanium alloys, aluminum alloys, metal and ceramic composites and polymers; determination of quantitative correlations between the geometric properties of second phase populations and mechanical properties; computer controlled automatic image analysis techniques. *Mailing Add:* Ames Lab 211A Metals Develop Iowa State Univ Ames IA 50011-3020

SPITZNAGEL, EDWARD LAWRENCE, JR, b Cincinnati, Ohio, Sept 4, 41. MATHEMATICAL STATISTICS, BIOMETRICS. *Educ:* Xavier Univ, BS, 62; Univ Chicago, MS, 63, PhD(math), 65. *Prof Exp:* From instr to asst prof math, Northwestern Univ, Ill, 65-69; assoc prof, 69-80, PROF MATH, WASH UNIV, 80- *Mem:* Sigma Xi; Math Asn Am. *Res:* Statistics; finite group theory. *Mailing Add:* Dept Math Wash Univ St Louis MO 63130

SPITZNAGEL, JOHN A, b Pittsburgh, Pa, June 27, 41; m 72; c 4. PULSE POWER ENGINEERING, DEFECTS IN CRYSTAL GROWTH. *Educ:* Carnegie Inst Technol, BS, 63, MS, 64; Carnegie-Mellon Univ, PhD(metall & mat sci), 69. *Prof Exp:* Mem staff soil mech, US Army Waterways Exp Sta, 68-70; sr engr, 70-74, fel engr, 74-80, ADV ENGR, WESTINGHOUSE RES & DEVELOP CTR, 80-; PRES, ADJUNCT PROF, UNIV OF PITTSBURGH , 84- *Concurrent Pos:* Prin investr ion beam effects in solids, NSF, 74-; prin investr irradiation response mat fusion, Dept Energy, 77-81; mem, Damage Analysis & Fundamental Studies Task Group, Dept Energy, 77-; ed, Advan Techniques Characterizing Microstruct, Am Inst Mining, Metall & Petrol Engrs, 80-82; adj prof, dept mat sci & eng, Univ Pittsburg, 84-88. *Mem:* Am Inst Mining, Metall & Petrol Engrs; Am Soc Metals; Mats Res Soc. *Res:* Fundamental processes of ion beam, plasma and neutron interactions with solids; microstructural and microchemical effects of irradiation in metals and semiconductors; modification of surfaces by ion implantation; control of thermal stress induced defects during crystal growth. *Mailing Add:* Westinghouse Sci & Technol Ctr 1310 Beulah Rd Pittsburgh PA 15235

SPITZNAGEL, JOHN KEITH, b Peoria, Ill, Apr 11, 23; m 47; c 5. MICROBIOLOGY, MEDICINE. *Educ:* Columbia Univ, BA, 43, MD, 46; Am Bd Internal Med, dipl, 53. *Prof Exp:* Asst instr basic sci, US Army Med Sch, 47-49; asst med, Wash Univ, 49-52; vis investr, Rockefeller Inst, 52-53; from chief infectious dis serv to chief med serv, US Army Hosp, Ft Bragg, 53-57; from lectr to prof bact, immunol & med, Univ NC, Chapel Hill, 57-79; PROF MICROBIOL & IMMUNOL & CHMN DEPT, EMORY UNIV, 79- *Concurrent Pos:* USPHS sr res fel, 58-68; vis investr, Nat Inst Med Res, Eng, 67-68; consult mem, Bact & Mycol Study Sect, USPHS-Dept Health, Educ & Welfare; chmn, Bact Mycol Study Sect, NIH, 77-79, consult mem, 85- *Mem:* Infectious Dis Soc; Am Soc Microbiol; Sigma Xi; fel Am Col Physicians; Am Asn Immunol. *Res:* Role of cationic proteins in oxygen-independent antimicrobial capacity of neutrophil granulocytic granules. *Mailing Add:* Dept Microbiol & Immunol Sch Med Emory Univ 502 Woodruff Mem Bldg Atlanta GA 30322

SPITZNAGLE, LARRY ALLEN, b Lafayette, Ind, Oct, 17, 43; m 61; c 2. RADIOPHARMACEUTICAL CHEMISTRY. *Educ:* Purdue Univ, BS, 65, MS, 66, PhD(bionucleonics), 69. *Prof Exp:* Asst prof bionucleonics, Sch Pharm, Univ Wash, 69-75; asst prof nuclear med, Sch Med, Univ Conn Health Ctr, 75-79; assoc prof med chem, Sch Pharm, Univ Md, Baltimore, 79-81; ASSOC PROF NUCLEAR MED, SCH MED, UNIV CONN HEALTH CTR, 81- *Concurrent Pos:* NIH res grants, 71-74 & 75-85. *Mem:* AAAS; Am Chem Soc; Am Pharmaceut Asn; Soc Nuclear Med. *Res:* Development of new and improved radio-pharmaceuticals; Environmental and radiological protection in the research establishment. *Mailing Add:* Dept of Nuclear Med Univ of Conn Health Ctr Farmington CT 06032-9984

SPITZNOGLE, FRANK RAYMOND, physics, for more information see previous edition

SPIVACK, HARVEY MARVIN, b Brooklyn, NY, Feb 3, 48; m 72; c 1. NAVAL SYSTEMS ANALYSIS, OCEAN ACOUSTICS. *Educ:* Brooklyn Col, BS, 68; Purdue Univ, MS, 70, PhD(physics), 76. *Prof Exp:* Res assoc physics, Purdue Univ, 76; asst physicist, Brookhaven Nat Lab, 77; MEM PROF STAFF SYSTS ANAL, CTR NAVAL ANAL, 77- *Mem:* Am Phys Soc. *Res:* Cost and effectiveness of military systems; neutrino physics; theory of weak interactions. *Mailing Add:* 3219 Parkwood Terr Falls Church VA 22042

SPIVAK, JERRY LEPOW, b New York, NY, Jan 5, 38; m 67; c 2. HEMATOLOGY. *Educ:* Princeton Univ, AB, 60; Cornell Univ, MD, 64; Am Bd Int Med dipl, 71, cert hemat, 75. *Prof Exp:* Sr resident med, New York Hosp, 68-69; intern, Johns Hopkins Hosp, 64-65, asst resident, 65-66, chief resident, 71-72; from asst prof to assoc prof med, 72-88, DIR, DIV HEMAT, 80-, PROF MED, 88-, PROF ONCOL, SCH MED, JOHNS HOPKIN UNIV, 90- *Concurrent Pos:* Clin assoc, Nat Cancer Inst, 66-68; fel hemat, Johns Hopkins Hosp, 69-71; investr med, Howard Hughes Med Inst, 72-77. *Honors & Awards:* Borden Award; Res Career Develop Award, Merit Award, USPHS. *Mem:* Am Fedn Clin Res; Am Soc Hemat; Soc Exp Biol & Med; fel Am Coi Physicians; Int Soc Exp Hemat; Am Clin Climat Asn. *Res:* Erythropoietin and the regulation of erythropoiesis. *Mailing Add:* Sch of Med Johns Hopkins Univ 600 N Wolfe St Baltimore MD 21205

SPIVAK, STEVEN MARK, b New York, NY, Oct 11, 42; div. TEXTILES, TEXTILE ENGINEERING. *Educ:* Philadelphia Col Textiles & Sci, BS, 63; Ga Inst Technol, MS, 65; Univ Manchester, PhD(polymer & fiber sci), 67. *Prof Exp:* Asst prof textiles & apparel res assoc, Philadelphia Col Textiles & Sci, 67-70; from asst prof to assoc prof, 70-83, PROF TEXTILES & CONSUMER ECON, UNIV MD, 83- *Concurrent Pos:* Tech adv, ASCR Int, Annapolis Junction, MD, 75-; dir, Atex Consults, Washington, DC, 76- *Mem:* Am Asn Textile Chemists & Colorists; fel Standards Engrs Soc; Sigma Xi; Am Chem Soc; Fiber Soc; Am Soc Test Mat. *Res:* End-use performance aspects of textiles and related materials, including interior textiles (carpet, upholstery, draperies), flammability (seams, apparel, extinguishability), weathering, gaseous pollutants, clothing comfort and standards engineering. *Mailing Add:* Dept Textiles & Consumer Econ Univ Md College Park MD 20742-7531

SPIVEY, BRUCE ELDON, b Cedar Rapids, Iowa, Aug 29, 34; m 56; c 2. OPHTHALMOLOGY, MEDICAL EDUCATION. *Educ:* Coe Col, BA, 55; Univ Iowa, MD, 59, MS, 64; Univ Ill, MEd, 69; Am Bd Ophthal, dipl, 65. *Prof Exp:* Intern, Highland-Alameda County Hosp, Oakland, Calif, 59-60; resident, Univ Hosps, Iowa City, 60-63; res assoc ophthal, Col Med, Univ Iowa, 63-64, from asst prof to assoc prof, 66-71; dean, Sch Med, 71-76, PROF OPHTHAL & HEAD DEPT, PAC MED CTR, UNIV PAC, 71- *Concurrent Pos:* Co-dir, NIH grants, 67 & 72. *Mem:* AMA; Am Col Surgeons; Am Acad Ophthal (vpres, 77-); Pan-Am Asn Ophthal; Asn Res Vision & Ophthal. *Res:* Strabismus; ophthalmologic genetics. *Mailing Add:* Pacific Presbyn Med Ctr Clay & Webster Sts Box 7999 San Francisco CA 94120

SPIVEY, GARY H, b Midland, Tex, Dec 3, 43; m 65; c 2. ENVIRONMENTAL EPIDEMIOLOGY & OCCUPATIONAL EPIDEMIOLOGY. *Educ:* Univ Calif, Davis, BA, 65; Med Sch, Univ Calif, San Francisco, MD, 69; Johns Hopkins Univ, Baltimore, MPH, 75. *Prof Exp:* Gen Med Officer, Indian Health Serv, USPHS, 70-71, serv unit dir, Whiteriver Serv Unit, 71-72, onsite proj dir, Proj Apache, Maternal & Child Health, 72-73; asst prof, 75-80, assoc prof epidemiol, Sch Pub Health, 80-83, ADJ ASSOC PROF, 83-, MGR OF EPIDEMIOL & ENVIRON MED, UNOCAL, LOS ANGELES, 83- *Concurrent Pos:* Bd counr, Int Soc Environ Epidemiol, 89-91; gov coun, Soc Occup & Environ Health, 90-92. *Mem:* Am Pub Health Asn; Soc Epidemiol Res; Soc Occup & Environ Health; Am Col Epidemiol. *Res:* Health surveillance; health promotion and community impact of environmental pollutants. *Mailing Add:* Med Dept Unocal PO Box 7600 Los Angeles CA 90051

SPIVEY, HOWARD OLIN, b Gainesville, Fla, Dec 10, 31; m 59; c 3. PHYSICAL BIOCHEMISTRY. *Educ:* Univ Ky, BS, 54; Harvard Univ, PhD(biochem), 63. *Prof Exp:* Res assoc phys chem, Rockefeller Univ, 62-64; NIH fel chem, Mass Inst Technol, 64-65; asst prof, Univ Md, College Park, 65-67; from asst prof to assoc prof, 67-75, PROF BIOCHEM, OKLA STATE UNIV, 75- *Mem:* Am Chem Soc; Sigma Xi; Am Soc Biochem & Molecular Biol; AAAS. *Res:* Physical biochemistry; heterologous associations among the enzymes - studies of extents of formation, substrate channeling and other catalytic properties. *Mailing Add:* Dept of Biochem 454 PS-II Okla State Univ Stillwater OK 74078-0454

SPIVEY, ROBERT CHARLES, b Jacksboro, Tex, Apr 22, 09; m 29; c 2. GEOLOGY. *Educ:* Tex Tech Univ, AB, 31; Univ Iowa, MS, 36, PhD(paleont), 38. *Prof Exp:* Asst geol, Univ Iowa, 34-38 & instr, 38-40; geologist & paleontologist, Shell Oil Co, 40-50, sr geologist, 50-61, staff geologist, 61-67; CONSULT GEOLOGIST, 67- *Mem:* Geol Soc Am; Am Asn Petrol Geol. *Res:* Stratigraphy. *Mailing Add:* Rte 2 Box 363 Springdale AR 72764

SPIVEY, WALTER ALLEN, b Wilmington, NC, July 24, 26; m 52; c 2. STATISTICS. *Educ:* Univ NC, AB, 50, MA, 52, PhD(statist), 56. *Prof Exp:* From asst prof to assoc prof, 57-62, PROF STATIST, UNIV MICH, ANN ARBOR, 62- *Concurrent Pos:* NSF fel, Stanford Univ, 57; vis assoc prof, Harvard Univ, 59-60; vis prof, London Sch Econ, 65-66, Doshisha Univ, 67. *Mem:* Fel Royal Statist Soc; fel Am Statist Asn; Soc Indust & Appl Math; Math Asn Am; Inst Mgt Sci. *Res:* Statistics and data analysis; optimization theory and applications; statistical forecasting. *Mailing Add:* Dept Bus Admin & Stats Univ Mich Ann Arbor MI 48109

SPIZIZEN, JOHN, b Winnipeg, Man, Feb 7, 17; US citizen; m 43, 68; c 1. MICROBIOLOGY, BIOCHEMISTRY. *Educ:* Univ Toronto, BA, 39; Calif Inst Technol, PhD(bact), 42. *Prof Exp:* Asst biol, Univ Toronto, 38-39; Nat Res Coun fel med sci, Vanderbilt Univ, 42-43; instr bact, Med Sch, Loyola Univ, Ill, 43; assoc virus res, Sharp & Dohme, Inc, 46-54; from asst prof to assoc prof microbiol, Sch Med, Western Reserve Univ, 54-61; prof microbiol & head dept, Univ Minn, Minneapolis, 61-65; chmn dept microbiol, Scripps Clin & Res Found, 65-76, mem dept cellular biol, Res Inst Scripps Clin, 76-79; prof & head dept, 79-89, PROF EMER MICROBIOL & IMMUNOL, UNIV ARIZ, 89- *Concurrent Pos:* Mem, Life Sci Comt, NASA; mem recombinant DNA adv comt, NIH; fel, Nat Res Coun, 42-43; res career develop award, NIH, 56-61; mem, Am Cancer Soc Coun, 74-, Nat Acad Sci & Res Coun, Am Cancer Soc, 74-; adj prof, Univ Calif, San Diego; consult, various co. *Mem:* Am Soc Microbiol; Am Soc Biol Chemists; Am Asn Advan Sci. *Res:* Cloning of genes in bacillus species; identification and cloning of genes for insecticidal toxins. *Mailing Add:* Dept Microbiol & Immunol Univ Ariz Sch Med Tucson AZ 84724

SPJUT, HARLAN JACOBSON, b Salt Lake City, Utah, May 3, 22; m; c 5. PATHOLOGY. *Educ:* Univ Utah, BS, 43, MD, 46; Am Bd Path, dipl. *Prof Exp:* Intern, Jackson Mem Hosp, Miami, Fla, 46-47; asst resident path, Salt Lake Vet Hosp, Utah, 49-50; from asst resident to instr, Univ Utah, 50-53; from instr to assoc prof surg path & path, Sch Med, Wash Univ, 54-62; prof path, 62-83, actg chmn dept, 69-72 & 87-88, IRENE & CLARENCE PROF PATH, BAYLOR COL MED, 83- *Concurrent Pos:* Am Cancer Soc fel surg path, Sch Med, Wash Univ, 53-54; attend pathologist, St Louis Vet Hosp, 55-59; assoc pathologist, Barnes Hosp, 54-62; vis asst, Karolinska Hosp, Stockholm, Sweden, 59-60; attend pathologist, St Luke's Hosp, 71-80; consult, Houston Vet Admin Hosp; sr attend pathologist, Methodist Hosp, 80- *Mem:* Col Am Path; AMA; Int Acad Path. *Res:* Carcinoma; bone and large bowel tumors; cytology in diagnosis of carcinoma of various sites. *Mailing Add:* Dept of Path Baylor Col Med One Baylor Plaza Houston TX 77030

SPLETTSTOESSER, JOHN FREDERICK, b Waconia, Minn, Oct 17, 33; m 56; c 2. GEOLOGY. *Educ:* Univ Minn, 62. *Prof Exp:* Ed, Am Geol Inst, 62-63; head, Sci & Tech Div, Libr Cong, 64-67; asst dir, Inst Polar Studies, Ohio State Univ, 67-69, assoc dir, 69-74; admin dir, Ross Ice Shelf Proj Mgt Off, Univ Nebr, 74-77; PROG MGR MINN GEOL SURV, UNIV MINN, ST PAUL, 77- *Concurrent Pos:* Sci coordr US remote field camps, Antarctica, 78-; staff lectr geol, tourist ships to Antarctica, 83- *Honors & Awards:* Soviet Polar Medal, 74. *Mem:* AAAS; Am Inst Mining, Metall & Petrol Eng; Soc Mining Eng; Arctic Inst NAm; Am Geophys Union; Geol Soc Am. *Res:* Geology of Antarctica; mining geology; scientific editing; mineral resource potential of Antarctica. *Mailing Add:* One Jameson Point Rd Rockland ME 04841

SPLIES, ROBERT GLENN, b Bird Island, Minn, Oct 2, 25; m 52; c 1. ORGANIC CHEMISTRY. *Educ:* Univ Wis, BS, 47, MS, 48, PhD, 51. *Prof Exp:* Instr, Wis State Col, Milwaukee, 51-52; chemist, Solvay Process Div, Allied Chem Corp, NY, 52-53; coordr, Bjorksten Res Lab, Wis, 53-55; chemist, Oscar Mayer & Co, 55-57; asst prof org chem, Univ NDak, 57-59; from asst prof to assoc prof, Univ Wis-Milwaukee, 59-67; from assoc prof to prof org chem, Univ Wis-Waukesha, 67-88; RETIRED. *Mem:* Am Chem Soc. *Res:* Nitro and amino derivatives of aromatic hydrocarbons; color reactions of alkaloids. *Mailing Add:* Dept Chem Univ Wis 1500 Univ Dr Waukesha WI 53186

SPLIETHOFF, WILLIAM LUDWIG, b Matamoras, Pa, Apr 8, 26; m 49, 71; c 3. ORGANIC CHEMISTRY. *Educ:* Pa State Univ, BS, 46, MS, 48; Mich State Univ, PhD(org chem), 53. *Prof Exp:* Asst fuel tech, Pa State Univ, 46-48; asst gen org & phys chem, Mich State Univ, 50-52; res chemist textile fibers dept, E I du Pont de Nemours & Co, 52-60; dir mkt res chem div, Gen Mills, Inc, Ill, 60-62, mgr com develop, 62-67; asst managing dir, Polymer Corp Ltd, Sydney, Australia, 67-69; dir opers indust chem, Chem Div, Gen Mills Chem, Inc, 69-70, vpres, 70-77; exec vpres & dir, Henkel Corp, 77-86; srvpres & dir, Henkel Am, 81-86; MGT CONSULT, 86- *Concurrent Pos:* Dir & officer subsid, Mex, Japan, Ireland, SAfrica & Venezuela, 70-86. *Honors & Awards:* Com Develop Asn Honor Award, 82. *Mem:* Am Chem Soc; Com Develop Asn; Chem Mkt Res Asn. *Res:* Mechanism and kinetics of racemization of optically active halides by phenols; condensation polymers; new uses for synthetic fibers; market research and commercial development; general industrial chemical management. *Mailing Add:* 113 Sandy Hook Rd Chanhassen MN 55317

SPLINTER, WILLIAM ELDON, b North Platte, Nebr, Nov 24, 25; m 53; c 4. ENGINEERING. *Educ:* Univ Nebr, BSc, 50; Mich State Univ, MSc, 51, PhD(agr eng), 55. *Prof Exp:* Instr agr eng, Mich State Univ, 53-54; res assoc prof, NC State Univ, 54-61, prof, 61-68; prof & head dept, Univ Nebr, Lincoln, 68-84, George Holmes distinguished prof & head dept agr eng, 84-88, assoc vchancellor for res, 88-90, INTERIM VICE CHANCELLOR FOR RES & DEAN GRAD STUDIES, Univ Nebraska, 90. *Concurrent Pos:* Consult, Southern Rhodesia, Univ So Africa, 63, Ford Found, IIT Kharagpur, India, 66, Columbia, 68, Chile, 78, Peru, 78, Russia & Mexico, 80, China, 81 & 86, Germany, 83, Univ Melbourne, Australia, 85, Morocco, 86, Ireland, 89; mem, Nat Coun Res Adminrs, 88- *Honors & Awards:* Massey Ferguson Medal, Am Soc Agr Engrs, 78. *Mem:* Fel AAAS; fel Am Soc Agr Engrs (vpres, 76-77, pres, 78-79); Soc Automotive Engrs; Nat Acad Eng; Am Asn Eng Socs; Am Soc Agr Eng Found (pres, 87-89). *Res:* Bioengineering of plant systems; systems engineering of crop production; electrostatic application of agricultural pesticides; mathematical modeling of plants; machine design and development; human factors engineering. *Mailing Add:* 302 Admin Bldg Univ of Nebr Lincoln NE 68588-0433

SPLITTER, EARL JOHN, b Lorraine, Kans, June 29, 20; m 43; c 3. VETERINARY PARASITOLOGY. *Educ:* Kans State Col, DVM, 43, MS, 50. *Prof Exp:* Jr state vet, NC State Dept Agr, 43-46; from asst prof to assoc prof path, Kans State Col, 46-57; group leader & asst dep adminr, Coop State Res Serv, USDA, 77-81, prin vet, 57-86; RETIRED. *Honors & Awards:* Res Award, Am Feed Indust Asn, 86. *Mem:* Am Vet Med Asn; Nat Asn Fed Vets. *Res:* Blood parasitic diseases of domestic animals; research administration; veterinary pathology and medicine. *Mailing Add:* 7053 SE Bunker Hill Dr Hobe Sound FL 33455-7321

SPLITTER, GARY ALLEN, b Lumberton, NC, July 19, 45; m 67. VETERINARY MEDICINE, IMMUNOPATHOLOGY. *Educ:* Kans State Univ, BS, 67, DVM, 69, MS, 70; Wash State Univ, PhD(path), 76. *Prof Exp:* Instr path, Kans State Univ, 69-70; captain, Sch Aerospace Med, US Air Force, 70-72; NIH fel, Dept Vet Path, Wash State Univ, 72-76; ASST PROF IMMUNOPATH, DEPT VET SCI, UNIV WIS-MADISON, 76- *Concurrent Pos:* Ed reviewer, Am Vet Med Asn, 78- *Mem:* Am Vet Med Asn; Am Col Vet Toxicologists. *Res:* Pathology; immunology; mechanisms of host defense in chronic and viral diseases. *Mailing Add:* Vet Sci Univ Wis 1655 Linden Dr Madison WI 53706

SPLITTGERBER, GEORGE H, b Van Tassel, Wyo, Jan 25, 18; m 42; c 3. INORGANIC CHEMISTRY. *Educ:* Univ Nebr, BSc, 39, MSc, 40; Kans State Univ, PhD(chem), 60. *Prof Exp:* Chemist, Victor Chem Works, 40-42 & Sinclair Res & Develop Co, 42-48; from instr to prof chem, Colo State Univ, 48-88; RETIRED. *Mem:* Am Chem Soc; AAAS; Sigma Xi. *Res:* Antioxidants and corrosion inhibitors; nonaqueous polarography. *Mailing Add:* 709 Birky Rd Ft Collins CO 80526

SPLITTSTOESSER, CLARA QUINNELL, b Miles City, Mont, Jan 19, 29; m 59. BACTERIOLOGY, INSECT PATHOLOGY. *Educ:* Mont State Univ, BS, 50; Univ Wis, MS, 51, PhD(bact), 56. *Prof Exp:* Fel, Univ Wis, 56-57; assoc exp surg, Med Ctr, Univ Calif, 57-58; experimentalist, Cornell Univ, 59-62, res assoc insect path, NY State Exp Sta, 62-77. *Mem:* Soc Invert Path. *Res:* Physiology of microorganisms; viral and bacterial pathogens of insects. *Mailing Add:* Dept Entom NY State Agr Exp Geneva NY 14456

SPLITTSTOESSER, DON FREDERICK, b Norwalk, Wis, 27; m 59. BACTERIOLOGY. *Educ:* Univ Wis, BS, 52, MS, 53, PhD, 56. *Prof Exp:* Proj assoc bact, Univ Wis, 55-56; from asst prof to assoc prof bact, 58-69, chmn dept, 82-88, PROF MICROBIOL, CORNELL UNIV, 69- *Concurrent Pos:* Mem, Ad Hoc Subcomt Food Microbiol, Food Protection Comt, Nat Acad Sci-Nat Res Coun, 63 & Comt Microbiol Food, Adv, Bd Mil Personnel Supplies, Nat Res Coun, 72-75, chmn, Food Protection Comt, 82-86. *Mem:* Am Soc Microbiol; Inst Food Technol; Int Asn Milk, Food & Environ Sanitarians; Am Soc Enologists. *Res:* Sanitation in food processing; microbiology of frozen foods; physiology of spore germination; wine fermentation. *Mailing Add:* Dept of Food Sci & Technol Cornell Univ Geneva NY 14456

SPLITTSTOESSER, WALTER E, b Claremont, Minn, Aug 27, 37; m 60; c 3. PLANT PHYSIOLOGY, BIOCHEMISTRY. *Educ:* Univ Minn, BS, 58; SDak State Univ, MS, 60; Purdue Univ, PhD(plant biochem), 63. *Prof Exp:* Plant physiologist, Shell Develop Co, 63-64; biochemist, Univ Calif, Davis, 64-65; plant biochemist, 65-74, head, Div Veg Crops, 73-76, PROF VEG CROPS, UNIV ILL, URBANA, 75-, asst head, Dept Hort, 82-88. *Concurrent Pos:* NIH fel, 64-65; prof bot & microbiol, Univ Col, London, 72-73; prof soil sci, Rothamsted Exp Sta, Harpenden, Eng, 80; distinguished vis prof, Nagoya Univ, Japan, 82-; biotechnologist, Univ Col, Dublin, 87. *Honors & Awards:* J H Gourley Award, Am Soc Hort Sci, 74; Outstanding Grad Educator, Am Soc Hort Sci, 90. *Mem:* Am Soc Plant Physiol; fel Am Soc Hort Sci; Weed Sci Soc Am; Japan Soc Plant Physiol; Plant Growth Reg Soc Am. *Res:* Plant metabolism; amino acids in germinating seedlings; weed control; metabolism of herbicides in plants; environmental factors affecting herbicide action. *Mailing Add:* 201 Veg Crops Bldg Univ Ill 1103 W Dorner Dr Urbana IL 61801-4777

SPOCK, ALEXANDER, b Shamokin, Pa, May 5, 29. PEDIATRICS, ALLERGY. *Educ:* Loyola Col, Md, BS, 51; Univ Md, MD, 55. *Prof Exp:* From intern to resident pediat, Geisinger Mem Hosp, Danville, Pa, 55-58; from instr to assoc prof, 59-77, PROF PEDIAT, MED CTR, DUKE UNIV, 77- *Concurrent Pos:* Fel pediat allergy, Med Ctr, Duke Univ, 60-62. *Mem:* Am Acad Pediat; Am Acad Allergy; Am Col Allergists; Am Thoracic Soc. *Res:* Immunology; pulmonary physiology; lung disease and allergic problems in pediatric patients; cystic fibrosis. *Mailing Add:* Box 2994 Dept Pediat Duke Univ Med Ctr Durham NC 27706

SPODICK, DAVID HOWARD, b Hartford, Conn, Sept 9, 27; m 51, 69; c 2. CARDIOLOGY. *Educ:* Bard Col, AB, 47; New York Med Col, MD, 50. *Prof Exp:* From instr to prof med, Tufts Univ, 57-76; PROF MED, UNIV MASS, 76- *Concurrent Pos:* Nat Heart Inst spec fel, WRoxbury Vet Admin Hosp, NY, 56-57; Am Col Physicians Brower Traveling Scholar, 64; sr physician, Lemuel Shattuck Hosp, Boston, 57-76, chief cardiol div, 62-76; chief cardiac diag & res ctr, Boston Eve Clin, 60-; assoc med, Boston City Hosp, 65-, lectr, Sch Med, Boston Univ, 66- & Sch Med, Tufts Univ, 76-; attend cardiologist, Univ Mass Hosp, 76-; dir cardiol div, St Vincent Hosp, 76-84, dir clin cardiol, 85-; ed, Am J Noninvasive Cardiol, 85- *Mem:* Am Col Chest Physicians; Am Fedn Clin Res; Am Col Cardiol; Am Col Physicians; Am Heart Asn. *Res:* Noninvasive polycardiography; clinical pharmacology; exercise physiology; physical diagnosis; diseases of pericardium; electrocariography echo-doppler. *Mailing Add:* Cardiol Div St Vincent Hosp Worcester MA 01604

SPOEHR, ALBERT FREDERICK, b Milwaukee, Wis, Feb 24, 18; m 47. CHEMISTRY. *Educ:* Univ Wis, BS, 42. *Prof Exp:* Lab supvr, Hercules Powder Co, Del, 42-45; res chemist, Am Anode, Inc, 45-53; mgr tech servs, Latex Compounding Div, Polson Rubber Co, Garretsville, 53-70, tech dir, 70-79; prod mgr, Bearfoot Inc, Wadsworth, 79-82,; RETIRED. *Mem:* Am Chem Soc. *Res:* Rubber and latex compounding; explosives. *Mailing Add:* 1912 Phelps Ave Cuyahoga Falls OH 44223

SPOELHOF, CHARLES PETER, b Hackensack, NJ, Aug 6, 30; m 53; c 4. PHYSICS, MATHEMATICS. *Educ:* Univ Mich, BS(eng physics) & BS(eng math), 53, MS, 54. *Prof Exp:* Engr, EKCo, Camera Works, Navy Ord Div & Kodak Appartus Div, 54-62; tech asst to dir res & develop, Eastman Kodak Co, 62-64, proj mgr, 64-65, prog mgr, 66-68, asst to dir res & develop, 68-72, mgr govt prod, 72-73, dir res & eng, 73-75, mgr bus & prof prod, 75-82, vpres & asst gen mgr, 82-85, vpres & dir, tech assessment com & info sysyts, 85-86; RETIRED. *Concurrent Pos:* Mem sci adv comt, Defense Intel Agency. *Mem:* Nat Acad Eng; Optical Soc Am. *Res:* Optical and photographic systems; image evaluation; optical design; imaging sensors; optical measurements. *Mailing Add:* Five Mullet Dr Pittsford NY 14534

SPOEREL, WOLFGANG EBERHART G, b Stuttgart, Ger, July 11, 23; Can citizen; m 51; c 2. ANESTHESIOLOGY, PHARMACOLOGY. *Educ:* Univ Frankfurt, MD, 49; FRCP(C), 56. *Prof Exp:* Resident med, Hosp, Univ Frankfurt, 49-51; demonstr pharmacol & physiol, 53-55, from instr to sr assoc anesthesia, 57-58, from asst prof to clin prof, 60-65, assoc prof pharmacol, 66-72, head dept anesthesia & hon lectr pharmacol, 72-83, PROF ANESTHESIA, UNIV WESTERN ONT, 66- *Concurrent Pos:* Fel anesthesia, Mayo Found, 55-56; chief anesthesia, Victoria Hosp, London, 58-66 & Univ Hosp, Univ Western Ont, 72-83. *Honors & Awards:* Can Anesthesists Soc Medal, 86. *Mem:* Can Anesthetists Soc; Can Med Asn; Can Soc Clin Invest; Royal Col of Physician of Can. *Res:* Epidural anesthesia; anesthetic breathing circuits; cerebral anoxia; respiratory insufficiency. *Mailing Add:* Dept Anesthesiol Univ West Fac Med London ON N6A 5C1 Can

SPOERLEIN, MARIE TERESA, b Dormont, Pa, Nov 3, 25. PHARMACOLOGY, PHYSIOLOGY. *Educ:* Seton Hill Col, BA, 47; Rutgers Univ, MS, 54, PhD, 59. *Prof Exp:* Asst pharmacologist, Schering Corp, 47-54 & Maltbie Labs, Wallace & Tiernan, Inc, 54-56; asst physiol, Col Pharm, Rutgers Univ, Newark, 56-58, from asst prof to assoc prof pharmacol, 59-68; PROF PHARMACOL, COL PHARM, RUTGERS UNIV, NEW BRUNSWICK, 68- *Mem:* AAAS; Am Soc Pharmacol & Exp Therapeut; Am Pharmaceut Asn; NY Acad Sci; Sigma Xi. *Res:* Biochemical pharmacology, especially mechanism of drug action and nervous system pharmacology. *Mailing Add:* 252 Carol Jean Way Rutgers Univ PO Box 789 Somerville NJ 08876

SPOFFORD, JANICE BROGUE, b Chicago, Ill, Nov 14, 25; m 51; c 2. GENETICS. *Educ:* Univ Chicago, PhB, 44, SB, 46, PhD(zool), 55. *Prof Exp:* Instr natural sci, Univ Col, 48-51, asst prof biol, 55-61, res assoc zool, 56-70, ASSOC PROF BIOL, 61- *Mem:* AAAS; Genetics Soc Am; Soc Study Evolution; Am Soc Human Genetics; Am Soc Naturalists (secy, 71-74, pres, 79). *Res:* Population genetics; mechanism of position-effect variegation; multi-gene families in Drosophila development and evolution. *Mailing Add:* Dept of Ecol & Evol Univ of Chicago Chicago IL 60637

SPOFFORD, SALLY HOYT, b Williamsport, Pa, Apr 11, 14; m 42, 64. ORNITHOLOGY. *Educ:* Wilson Col, AB, 35; Univ Pa, MS, 36; Cornell Univ, PhD(ornith), 48. *Prof Exp:* Asst biol, Wilson Col, 37-39; med technician, Stark Gen Hosp, Charleston, SC, 42-44 & Kennedy Gen Hosp, Memphis, Tenn, 44-45; admin asst, Lab Ornith, Cornell Univ, 55-69, res collabr, 69-80; RETIRED. *Mem:* Wilson Ornith Soc; Nat Audubon Soc; Western Field Ornith; Am Geog Soc; Am Ornith Union. *Res:* Life history and ecology of pileated woodpecker; population and distribution studies of Southeastern Arizona birds; public education in ornithology; conservation; food habits and behavior of roadrunners. *Mailing Add:* Aguila-Rancho Box J Portal AZ 85632

SPOFFORD, WALTER O, JR, b Swampscott, Mass, May 9, 36; m 61; c 3. ENVIRONMENTAL SCIENCES & ENGINEERING. *Educ:* Northeastern Univ, BS, 59; Harvard Univ, MS, 60, PhD(water resources eng), 65. *Prof Exp:* Res asst, Harvard Water Resources Group, Harvard Univ, 61-65, res fel water resources mgt, Sch Pub Health, 65-66; res assoc, Resources for the Future, 68-74, dir qual environ div, 74-80, dir hazardous waste mgt prog, 85-88, SR FEL, RESOURCES FOR THE FUTURE, 78-, DIR, ENVIRON & DEVELOP PROG, 89- *Concurrent Pos:* Ford Found consult, Aswan Reg Develop Proj, Cairo, 65-66; WHO consult, Czech Res & Develop Ctr for Environ Pollution Control, 72-74 & Environmental Pollution Abatement Ctr, Poland, 76-78; mem panel on marine ecosyst anal, Nat Acad Sci-Nat Acad Eng, Sci & Eng Comt Adv to Nat Oceanic & Atmospheric Admin, 72-73; mem fac systs anal for environ pollution control, NATO Advan Study Inst, Baiersbronn, Ger, 72; consult, Los Alamos Sci Lab, 73-78, NSF, 73, World Bank, 74- 75, 79 & 89-, Ministry Conserv, Victoria, Australia, 74, Int Inst Appl Systs Anal, 78-79 & Asian Develop Bank, Manila, 80, 83, 86; res scholar water resources, Int Inst Appl Systs Anal, Austria, 74; mem subcomt on water resources adv comt, Int Inst Appl Systs Anal, Nat Acad Sci, 74-76, mem & chmn, Liaison Subcomt on Resources & Environ, 78-82 & mem, Int Coop in Systs Anal Res Exec Subcomt, 78-82; mem, Metro Study Task Force, Washington Ctr Metro Studies, Washington, DC, 77-78; mem bd dirs, Roy F Weston, Inc, Pa, 78-89; mem, US Nat Comt Scientific Hydrology, 78-; consult, Asian Develop Bank, Repub Korea 80, 83, Malaysia, 86, 87; consult, World Bank, Peoples Repub China, 89- *Mem:* Fel Am Inst Chemists; Am Soc Civil Engrs; Am Geophys Union; Pub Works Hist Soc; Sigma Xi; Asn Environ & Resource Economists. *Res:* Environmental economics and management; civil and sanitary engineering; water resources engineering and management; public health. *Mailing Add:* 3348 Beech Tree Lane Falls Church VA 22042

SPOFFORD, WALTER RICHARDSON, II, b Hackensack, NJ, Nov 25, 08; m 36, 64; c 5. ZOOLOGY, ANATOMY. *Educ:* Tufts Col, BS, 31; Yale Univ, PhD(zool), 38. *Prof Exp:* From asst to instr anat, Med Col, Cornell Univ, 35-40; from asst prof to assoc prof, Sch Med, Vanderbilt Univ, 40-49; assoc prof neuroanat, Col Med, State Univ NY Upstate Med Ctr, 49-70; RES AFFIL ORNITH, LAB ORNITH, CORNELL UNIV, 70- *Concurrent Pos:* Res affil zool, Brigham Young Univ, 71- *Mem:* Soc Study Evolution; Am Asn Anat; Wilson Ornith Soc; Am Ornith Union; Brit Ornith Union; Sigma Xi. *Res:* Neuroanatomy; axon terminals; experimental embryology; posterior neural plate mesoderm; avian systematics; egg-white proteins as evolutionary characters; falconiformes; arctic ecology. *Mailing Add:* Rancho Aguila Box J Portal AZ 85632

SPOHN, HERBERT EMIL, b Berlin, Ger, June 10, 23; US citizen; m 73; c 2. EXPERIMENTAL PSYCHOPATHOLOGY. *Educ:* City Col New York, BSS, 49; Columbia Univ, PhD, 55; Am Bd Prof Psychologists, dipl clin psychol, 62. *Prof Exp:* Lectr, City Col New York, 49-52; res assoc, Sarah Lawrence Col, Bronxville, 50-54; res psychologist, Franklin D Roosevelt Vet Admin Hosp, Montrose, 55-61, chief, Res Sect, 61-64; sr res psychologist, 65-80, DIR, RES DEPT, MENNINGER FOUND, 81- *Concurrent Pos:* Prin investr & res scientist, USPHS res grants, 66-76; mem, Ment Health Small Grants Rev Comt, NIMH, 72-76; mem rev comt, Treat, Develop & Assessment Res, NIMH, 83-86, chair, 86-87. *Mem:* Sigma Xi; Am Psychol Asn; AAAS; Soc Res Psychopath. *Res:* Experimental psychopathology; clinical and experimental psychopharmacology; schizophrenia; treatment evaluation research; mechanisms in schizophrenia. *Mailing Add:* Menninger Found PO Box 829 Topeka KS 66601

SPOHN, RALPH JOSEPH, b New York, NY; m 89; c 2. HOMOGENEOUS CATALYSIS, ORGANOMETALLIC CHEMISTRY. *Educ:* Providence Col, BS, 65; Mass Inst Technol, PhD(inorg chem), 70; Rutgers Univ, MBA, 75. *Prof Exp:* Chemist, Enjay Chem Lab, 70-72, res chemist, Corp Res Lab, 72-75, staff chemist & group leader, Chem Intermediate Technol Div, 75-79, sr staff chemist, 79-82, head synthesis gas process res, New Ventures Technol Div, 82-84, HEAD, CHEM ANAL LAB, EXXON CHEM CO, LINDEN, 84- *Mem:* Am Chem Soc; NY Acad Sci. *Res:* Homogeneous and heterogeneous catalysis, especially that of the reactions of carbon monoxide and/or hydrogen with organic substrates and their industrial applications. *Mailing Add:* Ten Kings Ct Woodcliff Lake NJ 07675-8022

SPOHN, WILLIAM GIDEON, JR, b Lancaster, Pa, Mar 8, 23; m 87; c 4. MATHEMATICS. *Educ:* St John's Col, Md, BA, 47; Univ Calif, Berkeley, MA, 50; Univ Pa, PhD(math), 62. *Prof Exp:* Instr math, Temple Univ, 52-54, Univ Del, 54-56 & Bowling Green State Univ, 56-59; William S Parsons fel, Johns Hopkins Univ, 66-67, sr staff mathematician, Appl Physics Lab, 59-84; RETIRED. *Mem:* Math Asn Am. *Res:* Analytical mathematics; applied mathematics; mathematical analysis; number theory; numerical analysis; system analysis. *Mailing Add:* 5423 Storm Drift Columbia MD 21045

SPOHR, DANIEL ARTHUR, b Meadville, Pa, Sept 13, 27; m 66; c 2. PHYSICS. *Educ:* Allegheny Col, BS, 49; Oxford Univ, DPhil(physics), 58. *Prof Exp:* Res physicist, Cryogenics Br, US Naval Res Lab, 49-54 & 58-68, consult physicist, 68-89; RETIRED. *Mem:* AAAS; Am Phys Soc; Sigma Xi; Soc Photo-Optical Instrumentation Engrs. *Res:* Infrared, fiber optic and optical systems; data and signal processing; instrumentation systems analysis and development; computer application to structural, thermal and circuit analysis; applied superconductivity; low temperature physics; nuclear cooling and orientation; application of advanced materials. *Mailing Add:* 4477 Que St NW Washington DC 20007

SPOKAS, JOHN J, b Lisle, Ill, Oct 15, 28; m 52; c 8. RADIATION PHYSICS, DOSIMETRY. *Educ:* St Procopius Col, BS, 52; Univ Ill, MS, 54, PhD(physics), 58. *Prof Exp:* Mem tech staff, RCA Labs, 57-61; assoc prof, 61-70, chmn dept, 67-72, prof physics & dir, Phys Sci Lab, 70-86, PROF PHYSICS, ILL BENEDICTINE COL, 86- *Concurrent Pos:* Tech dir, Exradin, Inc, 87- *Mem:* Am Phys Soc; Am Asn Physics Teachers; Radiation Res Soc; Am Asn Physicists Med. *Res:* Radiation dosimetry; nuclear instrumentation; charge transport in insulators; conducting plastics; nuclear magnetic resonance relaxation. *Mailing Add:* Dept Physics Ill Benedictine Col Lisle IL 60532

SPOKES, ERNEST M(ELVERN), b Philadelphia, Pa, Feb 10, 16; m 41; c 2. MINE SAFETY, MINING ENVIRONMENT. *Educ:* Lafayette Col, BS, 36, EM, 46; Univ Ky, MS, 49; Pa State Univ, PhD(mineral prep), 56. *Prof Exp:* Mining engr, Bethlehem Steel Co, 36-40; foreman, sintering plant, Nat Lead Co, 45-47, gen engr, 47-48; asst instr mining & metall eng, Univ Ky, 48-49, from asst prof to assoc prof, 49-57, prof mining eng, 57-63; chmn, Dept Mining & Petrol Eng, Univ Mo-Rolla, 63-69, prof mining eng, 68-83, chief USAID, Univ Proj Eng Educ, Saigon, 69-71, head sect, 71-80, actg dean, Sch Mines & Metall, 80-81, emer prof mining eng, 83-89; RETIRED. *Concurrent Pos:* Consult eng, self employed, 50-89; consult, Ky Water Pollution Control Comn, 57-63; mem panel bituminous coal mining, US Bur Mines Report on Energy Resources of Sci Adv to President, 63; consult, Caterpillar Tractor Co, 66, US Bur Mines, 67, Monterey Coal Co, 72-73 & Consolidation Coal Co, 75; dir, McNally Pittsburg Mfg Corp, 73-78; chmn, Comt Underground Coal Mine Safety, Nat Res Coun, 80-82. *Mem:* Am Inst Mining, Metall & Petrol Engrs (vpres, 67-68, 69-70); Soc Mining Engrs (vpres, 67-70); Engrs Coun Prof Develop. *Res:* Mineral and coal preparation; mining methods; mine safety; mining management; mineral industry economics; mine systems analysis; mining environments; mine roof control, mine ventilation control of pollution from mine refuse, ventilation of coal land product liability in mine accidents; control of acid pollution by mine run-off, coal strip mine planning including all legal requirements for safety and environmental protection. *Mailing Add:* Box 331 Modesto CA 95353

SPOKES, G(ILBERT) NEIL, b Isleworth, Eng, July 18, 35; m 69; c 4. RESEARCH DIRECTION, ANALYTICAL CHEMISTRY. *Educ:* Univ London, BSc, 56, PhD(flame spectra), 59. *Prof Exp:* Res assoc chem, Univ Mich, 59-60 & Yale Univ, 60-61; chem physicist, Stanford Res Inst, 61-72; dir eng res & develop dept, Hycel Inc, 72; prin engr & dir res, Technicon Instruments Corp, 73-76; vpres, Chem Serv Div, US Testing Co, Inc, Hoboken, NJ, 76-81, group vpres, Chem Serv Group, 81-87; TECH DIR, CHEMETRICS INC, CALVERTON, VA, 88- *Mem:* Am Soc Testing & Mat; Am Chem Soc; Am Soc Quality Control. *Res:* Emission and absorption spectra of flames and of negative ions; gaseous electronics; ion sampling; gas kinetics and thermochemistry; pyrolysis and oxidation of gases and solids using mass spectrometry; biomedical diagnostic instrumentation; Inorganic analytical instrumentation; environmental chemistry; water analysis. *Mailing Add:* 7267 Laurel Brook Lane Marshall VA 22115

SPOLJARIC, NENAD, b Zagreb, Yugoslavia, July 3, 34; US citizen; m 64; c 1. SEDIMENTARY PETROLOGY, ENVIRONMENTAL GEOLOGY. *Educ:* Univ Ljubljana, GE, 60; Harvard Univ, MA, 65; Bryn Mawr Col, PhD(sedimentary petrol), 70. *Prof Exp:* Explor geologist, Proizvodnja Nafte Co, Yugoslavia, 60-61; petrol geologist, Petrol Inst, Yugoslavia, 62-63; SR SCIENTIST, DEL GEOL SURV, UNIV DEL, 65- *Concurrent Pos:* Mem Int Petrol Tech Deleg To People's Repub of China, 83 & Australia & Indonesia, 84. *Honors & Awards:* Autometric Award, Am Soc Photogram, 76. *Mem:* Am Asn Petrol Geologists. *Res:* Study of glauconitic sediments; geographic distribution, stratigraphy, correlation and origin of these sediments; study of clay-mineralogy of the Mid-Atlantic Coastal Plain, USA; study of New Zealand glauconitic sediments. *Mailing Add:* Del Geol Surv Univ of Del Newark DE 19716

SPOLSKY, CHRISTINA MARIA, b Reute, Austria, Mar 3, 45; Can citizen; m 75; c 2. MITOCHONDRIAL DNA, EVOLUTION. *Educ:* Univ Toronto, BS, 67; Yale Univ, PhD(microbiol), 73. *Prof Exp:* Fel, Univ Pa, 73-75, Am Cancer Soc, 74; res assoc cell biol, Wistar Inst, 75-76; RES SCIENTIST, ACAD NATURAL SCI, PHILADELPHIA, 78-; AT DEPT ECOL & EVOLUTION, UNIV ILL. *Concurrent Pos:* Prin invstr, Whitehall Fedn grant, 81-84. *Mem:* Sigma Xi; Tissue Cult Asn; Am Soc Cell Biol; Am Soc Microbiol. *Res:* The relationship of chemical carcinogenesis to mutagenesis in cell cultures; the information content and biogenesis of mitochondrial DNA; use of mitochondrial DNA to determine toxonomic relationships of species; rate of evolution of mitochondrial DNA in vertebrates; origin of mitochondria as determined by rDNA sequences. *Mailing Add:* Dept Ecol & Evolution Univ Ill 505 Goodwin Ave Urbana IL 61801

SPOLYAR, LOUIS WILLIAM, b Detroit, Mich, May 6, 08; m 35; c 3. TOXICOLOGY. *Educ:* DePauw Univ, AB, 31; Ind Univ, MD, 36; Am Bd Prev Med, dipl, 50. *Prof Exp:* Lectr indust med, Sch Med, Ind Univ, Indianapolis, 40-46, asst prof pub health, 46-78; ASST HEALTH COMNR, IND STATE BD HEALTH, 69- *Concurrent Pos:* Dir div indust hyg, Ind State Bd Health, 37-56 & bur prev med, 56-68, asst comnr med opers, 68-69; dir prev med br, State Civil Defense, Ind, 50-, chmn opers br, Med Health Serv, 55; consult, Surgeon Gen, USPHS, 57. *Mem:* Am Med Asn; Indust Med Asn; Am Conf Govt Indust Hygienists (vpres, 48, pres 49). *Res:* Toxicology of cadmium and lead; toxicology and generation of arsine; industrial toxicology; chemotherapy of tuberculosis; laboratory determination of sickle cell anemia. *Mailing Add:* 6737 E Ninth St Indianapolis IN 46219

SPOMER, GEORGE GUY, b Denver, Colo, Mar 2, 37; m 60; c 4. PLANT PHYSIOLOGY, ECOLOGY. *Educ:* Colo State Univ, BS, 59, MS, 61, PhD(bot sci), 62. *Prof Exp:* From instr to asst prof bot, Univ Chicago, 62-68; from asst prof to assoc prof, Wash State Univ, 68-72; ASSOC PROF BOT, UNIV IDAHO, 72- *Concurrent Pos:* NSF res grant, 65-67, 75-76 & 79-82; Nat Geog Soc res grant, 75. *Mem:* AAAS; Ecol Soc Am; Am Soc Plant Physiol. *Res:* Plant eco-physiology, environmental analysis, alpine plant ecology, water relations, tree physiology; sagebrush eco-physiology. *Mailing Add:* Dept of Biol Univ of Idaho Moscow ID 83843

SPOMER, LOUIS ARTHUR, b Apr 17, 40; US citizen; m 62. PLANT PHYSIOLOGY, SOIL SCIENCE. *Educ:* Colo State Univ, BS, 63; Cornell Univ, MS, 67, PhD(plant sci), 69. *Prof Exp:* Meteorologist, Deseret Test Ctr, US Army, 69-71, phys scientist, US Dept Defense, 71-72; asst prof plant physiol & hort, 72-75, asst prof, 75-77, ASSOC PROF HORT, UNIV ILL, URBANA, 77- *Mem:* Am Soc Plant Physiol; Am Soc Agron; Am Soc Hort Sci; Soil Sci Soc Am. *Res:* Soil-plant-water relationships; water stress and crop growth; plant and crop water requirement. *Mailing Add:* Dept Hort Univ Ill 1301 W Gregory Dr Urbana IL 61801

SPONGBERG, STEPHEN ALAN, b Rockford, Ill, Oct 15, 42; m 72. SYSTEMATIC BOTANY. *Educ:* Rockford Col, BA, 66; Univ NC, Chapel Hill, PhD(bot), 71. *Prof Exp:* Asst cur bot, 70-76, CURATORIAL TAXONOMIST, ARNOLD ARBORETUM, HARVARD UNIV, 76- *Concurrent Pos:* Ed bd, J Arnold Arboretum, 71-, ed, 79- *Mem:* Bot Soc Am; Am Soc Plant Taxonomists; Int Asn Plant Taxon; Linnean Soc London; Int Dendrol Soc. *Res:* Taxonomic revisions of woody angiosperm genera of eastern Asiatic-eastern North American distribution, particularly genera of ornamental importance; taxonomy; nomenclature of woody plants cultivated in the North Temperate Zone. *Mailing Add:* Harvard Univ-Herbaria 22 Divinity Ave Cambridge MA 02138

SPONSELLER, D(AVID) L(ESTER), b Canton, Ohio, Oct 2, 31; m 55; c 7. METALLURGICAL ENGINEERING. *Educ:* Univ Notre Dame, BS, 53; Univ Mich, MSE, 58, PhD(metall eng), 62. *Prof Exp:* Instr marine eng, US Naval Acad, 55-57; res asst, Res Inst, Univ Mich, 57-60, instr metall eng, 60-62; asst prof, Univ Notre Dame, 62-65; staff metallurgist, Molybdenum Co, Mich, 65-87, res group leader, 67-70, res supvr, 70-79,; RES ENGR, ERIM TRANSP & ENERGY MAT CTR, 87- *Concurrent Pos:* Vis res metallurgist, Edgar C Bain Lab, US Steel Corp, 63. *Mem:* Am Soc Metals; Am Inst Mining, Metall & Petrol Engrs; Nat Asn Corrosion Engrs; Am Soc Mech Engrs; Soc Automotive Engrs. *Res:* Physical metallurgy and alloy development of steels and alloys for oil production and for elevated temperature service; corrosion in oil field environments; failure analysis of metals; differential thermal analysis; manufacturing metallurgy; mechanical testing; oxidation and sulfidation of metals; steel desulfurization; casting of metals; embrittlement of metals; issued five US patents. *Mailing Add:* 2648 Antietam Dr Ann Arbor MI 48105

SPONSLER, GEORGE C, b Dec 2, 27; m 55; c 3. MATHEMATICAL MODELS. *Educ:* Princeton Univ, BSE, 49, MA, 51, PhD, 52; George Washington Univ, JD, 81. *Prof Exp:* Chief scientist & dir, Tech Analysis & Oper Res, US Navy, 60-63; dir, Ctr for Exp Studies, IBM/FSD, 63-68; exec secy, Nat Acad Sci, Div Eng, 68-70; PRES, LAW, MATH & TECHNOL INC, 70- *Concurrent Pos:* Congressional fel, 87-88. *Mem:* Fel Am Phys Soc; fel AAAS; Sigma Xi; sr mem Inst Elec & Electronic Engrs. *Res:* Subjective probability; cold fuoron. *Mailing Add:* 7804 Old Chester Rd Bethesda MD 20817

SPOONER, ARTHUR ELMON, agronomy; deceased, see previous edition for last biography

SPOONER, BRIAN SANDFORD, b St Louis, Mo, Dec 27, 37; m 63; c 2. DEVELOPMENTAL BIOLOGY, CELL BIOLOGY. *Educ:* Quincy Col, BS, 63; Temple Univ, PhD(biol), 69. *Prof Exp:* Teaching asst biol, Temple Univ, 63-65; USPHS trainee, Univ Wash, 69; NIH fel, Stanford Univ, 69-71; asst prof biol, 71-75, assoc dir res, Div Biol, 75-77, assoc prof, 75-79, PROF BIOL, KANS STATE UNIV, 79- *Concurrent Pos:* Nat Inst Gen Med Sci grants, 72-75 & 75-80, Nat Heart, Lung & Blood Inst, 80-82 & Am Heart Asn, 80-82. *Mem:* Soc Develop Biol; Am Soc Cell Biol. *Res:* Control of differentiation during embryonic development; mechanism of cell movement; regulation of cytodifferentiation and morphogenesis; interactions in organogenesis; stability of the differentiated state. *Mailing Add:* Div of Biol Kans State Univ Manhattan KS 66506

SPOONER, CHARLES EDWARD, JR, b Boston, Mass, July 25, 32; m 62; c 2. NEUROPHARMACOLOGY, NEUROPHYSIOLOGY. *Educ:* Univ Calif, Los Angeles, BA, 56, MS, 61, PhD(neuropharmacol), 64. *Prof Exp:* Res pharmacologist, Riker Labs, Calif, 51-59; asst res pharmacologist, Med Sch, Univ Calif, Los Angeles, 65-68; from asst prof to assoc prof neurosci, 68-74, asst dean spec curricula, 69-71, asst dean admis & student affairs, 71-74, PROF NEUROSCI, SCH MED, UNIV CALIF, SAN DIEGO, 74-, ASSOC DEAN ADMIS, 74- *Concurrent Pos:* NIMH fel, 63-65; consult, Psychobiol Labs, Sepulveda Vet Hosp, Calif & Neurochem Sect, Space Biol Labs, Univ Calif, Los Angeles, 68-80; vis prof, Harvard Med Sch, 76; vis prof, Oxford Univ Sch Med, 87; nat chair, group student affairs, Asn Am Med Col, 89-90. *Mem:* AAAS; Am Soc Pharmacol & Exp Therapeut; Asn Am Med Col; Soc Exp Biol & Med; Am Educ Res Asn. *Res:* medical education. *Mailing Add:* Med Admis Off M-021 Univ Calif San Diego La Jolla CA 92093

SPOONER, ED THORNTON CASSWELL, b Blandford, Dorset, Eng, June 16, 50; m 72. GEOLOGY. *Educ:* Univ Cambridge, BA, 71, MA, 75; Oxford Univ, MA, 75; Univ Manchester, PhD(geol), 76. *Prof Exp:* Demonstr mineral, Oxford Univ, 73-77; ASST PROF GEOL, UNIV TORONTO, 77- *Concurrent Pos:* Lectr geol, Oriel & Pembroke Cols, Oxford Univ, 74-77; Can rep to Comn on Ore Forming Fluids in Inclusions, 78-; Natural Sci & Eng Res Coun Can grants, 78- *Mem:* Can Inst Mining & Metall; Brit Geol Asn. *Res:* Mineral deposits in geology, especially hydrothermal; geochemical methods of exploration for economic mineral deposits. *Mailing Add:* Dept Geol Univ Toronto Toronto ON M5S 1A1 Can

SPOONER, GEORGE HANSFORD, b Henderson, NC, Feb 24, 27; m 53; c 3. CLINICAL CHEMISTRY. *Educ:* Univ Miami, BS, 50; Univ NC, PhD(biochem), 58. *Prof Exp:* Res asst sanit eng, Sch Pub Health, Univ NC, 54-56, res asst, Sch Dent, 56-57, res assoc biochem, Sch Med, 57-58, instr, 58-61, USPHS trainee microbiol, 61-62, asst prof biochem, 62-65; asst prof path, Sch Med, Duke Univ, 65-73; ASSOC PROF PATH, MED UNIV SC, 73-; CLIN CHEMIST, VET ADMIN HOSP, CHARLESTON, 73- *Concurrent Pos:* Res scientist, State Sanitorium Syst NC, 58-61; biochemist, Clin Res Unit, NC Mem Hosp, 62-65; clin chemist, Vet Admin Hosp, Durham, 65-73. *Res:* Serum enzyme levels in the diagnosis of disease; continuous flow kinetics. *Mailing Add:* Lab Serv Vet Admin Hosp Charleston SC 29403

SPOONER, JOHN D, b Hillsborough Co, Fla, Dec 18, 35; m 58; c 5. ZOOLOGY, ENTOMOLOGY. *Educ:* Ga State Univ, BS, 60; Univ Fla, MS, 62, PhD(entom), 64. *Prof Exp:* Asst prof biol, Ga Southern Col, 64-66 & Augusta Col, 66-70; chmn div natural sci, 70-76, actg acad dean, 76, from asst prof to assoc prof, 70-78, PROF BIOL, UNIV SC, AIKEN, 78- *Concurrent Pos:* NSF grant, 66-68. *Mem:* Pan Am Acridiological Soc; Sigma Xi; Am Soc Zoologist; Entom Soc Am; Entom Soc Am. *Res:* Acoustical pair forming systems of Orthoptera, particularly phaneropterine katydids; geographic variation in orthopteran acoustical behavior; life history studies of phaneropterine katydids. *Mailing Add:* Dept Biol Univ SC 171 University Pkwy Aiken SC 29801

SPOONER, M(ORTON) G(AILEND), b Eau Claire, Wis, Jan 16, 24; m 50; c 4. ELECTRICAL ENGINEERING. *Educ:* Univ Wis, BS, 48, MS, 54, PhD(elec eng), 56. *Prof Exp:* Elec engr, Standard Oil Co, Ind, 48-52; from instr to asst prof, Univ Wis, 52-56; res electronics engr, 56-76, sr vpres tech opers, Cornell Aeronaut Lab Inc, 56-76; DIR STRATEGIC PLANNING, GARLAND DIV, E SYSTS, 76- *Mem:* Inst Elec & Electronic Engrs. *Res:* Technical management in high speed special purpose digital processing systems and large scale software systems; strategic planning. *Mailing Add:* E Systs Box 660023 Dallas TX 75266-0023

SPOONER, PETER MICHAEL, b Newport, RI, Nov 11, 42; m 90. ENDOCRINOLOGY, LIPID BIOCHEMISTRY. *Educ:* Bates Col, BS, 64; Univ Ill, Urbana-Champaign, MS, 66, PhD(physiol & biophysics), 70. *Prof Exp:* Sr staff fel, Lipid Res, Nat Inst Arthritis, Metab & Digestive Dis, Nat Heart, Lung & Blood Inst, NIH, 74-79, physiologist & biochemist, 79-80, health sci adminr, 80-85, chief prog rev, 85-90, CHIEF CARDIAC FUNCTIONS BR, NAT HEART, LUNG & BLOOD INST, NIH, 90- *Concurrent Pos:* Vis investr, Imperial Cancer Res Found Labs, London, 72-74; adj asst prof, Uniformed Serv, Univ Health Sci, 78-85; physiologist & biochemist, Nat Inst Arthritis, Metabolic & Digestive Dis, NIH, 77-79; lectr, Found Advan Educ Sci, 80-84; mem, Budget Comt, Basic Sci Coun, Am Heart Asn. *Mem:* Am Soc Biol Chemists; Sigma Xi; Am Heart Asn. *Res:* Endocrine control of uptake, metabolism and deposition of lipids into cells in vivo and vitro using model cell culture systems, ion channels and transport systems. *Mailing Add:* Rm 304 Fed Bldg Nat Heart, Lung & Blood Inst NIH 7550 Wisconsin Ave Bethesda MD 20892

SPOONER, ROBERT BRUCE, b Cleveland, Ohio, Aug 7, 20; wid; c 4. BIOMEDICAL ENGINEERING. *Educ:* Hiram Col, BA, 41; Northwestern Univ, PhD(physics), 49. *Prof Exp:* Asst, Northwestern Univ, 46-48; head, Thermodyn Analysis Sect, Lewis Flight Propulsion Lab, Nat Adv Comt Aeronaut, 49-53; mgr, Adv Nuclear Design Dept, Martin Co, 53-55; mgr assoc res, Koppers Co, Inc, 55-62; coordr sci & res adv group, Regional Indust Develop Corp, 63-65; pres, Impac, 65-74; dir, Med Instrumentation Ctr, MPC Corp, 74-75; SR PROJ ENGR, EMERGENCY CARE RES INST, 75- *Concurrent Pos:* Lectr, Case Western Reserve Univ, 50-52; chmn, Annual Res Conf Instrumentation Sci, 69-; vis lectr, Biomed Instrumentation, Bosphorus Univ, Istanbul, Turkey, 83. *Mem:* Am Phys Soc; sr mem Instrument Soc Am; sr mem Inst Elec & Electronics Engrs; Asn Advan Med Instrumentation; Am Soc Test & Mat. *Res:* Radioisotope and radiation applications; instrumentation and control; medical instrumentation; anesthesia and breathing systems; evaluations, investigations of accidents and incidents; study and correction of hazards, review and writing standards for equipment. *Mailing Add:* Emergency Care Res Inst 5200 Butler Pike Plymouth Meeting PA 19462

SPOONER, STEPHEN, b Worcester, Mass, Apr 2, 37; m 59; c 2. MATERIALS SCIENCE, METALLURGY. *Educ:* Mass Inst Technol, BS, 59, ScD(metall), 65. *Prof Exp:* Res asst, Mass Inst Technol, 59-65; from asst prof to assoc prof, 65-75, prof metall, 75-81, RES SCIENTIST, ENG EXP STA, GA INST TECHNOL, 65-; RES SCIENTIST, OAK RIDGE NAT LAB, 82- *Concurrent Pos:* Consult, Oak Ridge Nat Lab, 73-77; res scientist, Ga Inst Technol, 82. *Mem:* Am Phys Soc; Am Inst Mining, Metall & Petrol Engrs; Am Crystallog Asn. *Res:* Materials science; phase transformations; magnetic materials. *Mailing Add:* Solid State Div Oak Ridge Nat Lab Oak Ridge TN 37831

SPOOR, RYK PETER, b Albany, NY, June 30, 35; m 57; c 2. PHARMACOLOGY, PHYSIOLOGY. *Educ:* State Univ NY Albany, BS, 57; Union Univ, NY, PhD(pharmacol), 62. *Prof Exp:* Instr physiol & pharmacol, Sch Med, Creighton Univ, 62-64; asst prof, Univ SDak, 64-67; fel pharmacol, Emory Univ, 67-69; from asst prof to assoc prof pharmacol, Albany Med Col, 69-82, ASSOC PROF BIOL, ALBANY COL PHARM, 82- *Mem:* AAAS; Sigma Xi. *Res:* Muscle; cardiovascular pharmacology and physiology; phytotoxins. *Mailing Add:* 1052 Brierwood Blvd Schenectady NY 12308

SPOOR, WILLIAM ARTHUR, b New York, NY, Dec 14, 08; m 34; c 2. ZOOLOGY, PHYSIOLOGY. *Educ:* Univ Wash, BS, 31; Univ Wis, PhD(zool), 36. *Prof Exp:* From instr to prof zool, Dept Biol Sci, Univ Cincinnati, 36-68, head dept biol sci, 58-64; res aquatic biologist, Environ Res Lab, 68-82; RETIRED. *Concurrent Pos:* Mem aquatic life adv comt, Ohio River Valley Water Sanit Comn, 52-68; consult physiol of aquatic animals, Nat Water Qual Lab, Dept Interior, Minn, 66-68; mem nat tech comt on water qual requirements for aquatic life, Fed Water Pollution Control Admin, 67-68; mem summer staff, F T Stone Lab, Ohio State Univ, 48-62. *Mem:* Sigma Xi. *Res:* Environmental requirements and oxygen requirements of fish; activity detectors for aquatic animals; physiology of aquatic animals. *Mailing Add:* 1053 Jackson Rd Park Hills KY 41011

SPOREK, KAREL FRANTISEK, b Bohumin, Czech, Oct 12, 19; US citizen; m 51; c 1. ANALYTICAL CHEMISTRY, ORGANIC CHEMISTRY. *Educ:* St Andrews Univ, MA, 47. *Prof Exp:* Res chemist, Nobel Div, Imp Chem Indust Ltd, Scotland, 47-54, group leader anal res, Plant Protection Div, Eng, 54-57; res chemist, Eldorado Mining & Refining, Can, 57-58; head anal dept, Bioferm Corp, Calif, 58-60; res chemist, Tech Ctr, 60-61, chief org anal chem, 61-73, RADIATION OFFICER, TECH CTR, OWENS-ILL INC, 73- *Mem:* Am Chem Soc; Am Nuclear Soc; Am Soc Testing & Mat; Sigma Xi; Am Soc Safety Engrs. *Res:* High explosives; detonators; fuses; cellulose derivatives; insecticides; fungicides; fertilizers; polymers; polyethelene; silicones; pharmaceuticals; vitamin B-12; uranium; radiation chemistry. *Mailing Add:* 7142 Erie St Sylvania OH 43560-1134

SPORER, ALFRED HERBERT, b New York, NY, May 28, 29; m 55; c 2. INK CHEMISTRY, PHOTO CHEMISTRY. *Educ:* City Col, BA, 51; Univ Calif, Los Angeles, MS, 53, PhD(phys org chem), 56. *Prof Exp:* Res chemist, Esso Res & Eng Corp, 56; fel photochem, Univ Southern Calif, 56-57; staff chemist, 57-73, mgr appl sci, Res Div, 73-79, mem, Corp Tech Comt, 79-81, res staff mem, IBM Res, San Jose, 81-85, MGR, INK TECHNOL, IBM RES, IBM CORP, SAN JOSE, 85- *Mem:* Am Chem Soc; Soc Imaging Sci & Technol; Sigma Xi; AAAS. *Res:* Electrophotography; organic photoconductors; physical organic chemistry; photochemistry of complex ions, chelates and organic compounds in condensed phases; mechanisms of organic reactions; chromatography; technical management of programs in electrophotography; photoconductors; non-impact printing; magnetic recording media; ink jet inks. *Mailing Add:* IBM Almaden Res Ctr K41-803 650 Harry Rd San Jose CA 95120-6099

SPORN, EUGENE MILTON, toxicology, research management; deceased, see previous edition for last biography

SPORN, MICHAEL BENJAMIN, b New York, NY, Feb 15, 33; m 56; c 2. CANCER, BIOCHEMISTRY. *Educ:* Univ Rochester, MD, 59. *Prof Exp:* Intern med, Sch Med, Univ Rochester, 59-60; staff mem, Lab Neurochem, Nat Inst Neurol Dis & Blindness, 60-64; staff mem, 64-70, head lung cancer unit, 70-73, chief, Lung Cancer Br, 73-78, CHIEF LAB CHEMOPREVENTION, NAT CANCER INST, 78- *Mem:* Am Asn Cancer Res; Am Soc Biol Chem; Am Soc Pharmacol & Exp Therapeut. *Res:* Nucleic acids and cancer, vitamin A and related compounds; carcinogenesis studies; retinoids and cancer prevention; peptide growth factors, transforming growth factor-beta. *Mailing Add:* Nat Cancer Inst NIH Bldg 41 Rm C629 Bethesda MD 20892

SPORNICK, LYNNA, b Oct 6, 47; m. PHYSICS. *Educ:* Carnegie-Mellon Univ, BS, 69; Rutgers Univ, PhD(physics), 75; Johns Hopkins Univ, MS(comput sci), 81. *Prof Exp:* Fel, Dept Physics, Colo State Univ, 75-77; SR STAFF PHYSICIST, APPL PHYSICS LAB, JOHNS HOPKINS UNIV, 77- *Concurrent Pos:* Lectr, appl physics dept, Whiting Sch Eng, Johns Hopkins Univ, 85- *Mem:* Am Phys Soc; Am Asn Physics Teachers; Soc Comput Simulations. *Res:* Computer modeling of physical systems. *Mailing Add:* Appl Physics Lab Johns Hopkins Univ Johns Hopkins Rd Laurel MD 20723

SPOSITO, GARRISON, b Los Angeles, Calif, July 29, 39; m 76; c 6. PHYSICAL CHEMISTRY, SOIL CHEMISTRY. *Educ:* Univ Ariz, BS, 61, MS, 63; Univ Calif, PhD(soil sci), 65. *Prof Exp:* From asst prof to prof physics, Sonoma State Univ, 65-74; from asst prof to prof soil sci, Univ Calif, Riverside, 74-88, chmn div environ sci, 75-78; PROF SOIL PHYS CHEM, UNIV CALIF, BERKELEY, 88- *Concurrent Pos:* Fulbright fel, 73; vis fel, Nat Inst Agron Res, France, 81 & St Cross Col, Oxford Univ, Eng, 84; Guggenheim fel, 84. *Honors & Awards:* Soil Sci Award, 82; Horton Award, 90. *Mem:* Am Chem Soc; fel Am Geophys Union; fel Soil Sci Soc Am; Am Phys Soc; Hist Sci Soc; fel Am Soc Agron. *Res:* Environmental physical chemistry, statistical mechanics, thermodynamics of soils and clays; surface chemistry of soils, transport in porous media. *Mailing Add:* Dept Soil Biol Univ Calif Berkeley CA 94720

SPOSITO, VINCENT ANTHONY, b Pittsburg, Calif, Nov 15, 36; m 63; c 2. OPERATIONS RESEARCH, STATISTICS. *Educ:* Calif State Univ, Sacramento, BA, 65; Iowa State Univ, MS, 67, PhD(statist), 70. *Prof Exp:* Lab technician metall, Aerojet-General, Sacramento, 58-65; instr statist, Calif State Univ, Sacramento, 65; res assoc, 66-70, from asst prof to assoc prof, 70-78, PROF STATIST, IOWA STATE UNIV, 78- *Mem:* Am Statist Asn; Math Prog Soc. *Res:* Mathematical programming; linear and nonlinear programming. *Mailing Add:* Dept Statist Iowa State Univ 102C Snedecor Ames IA 50011

SPOTNITZ, HENRY MICHAEL, b New York, NY, July 7, 40; m 77; c 2. MEDICAL SCIENCES, PHYSIOLOGY. *Educ:* Harvard Univ, BA, 62; Columbia Univ, MD, 66. *Prof Exp:* Intern surg, Bellevue Hosp, NY, 66-67; staff assoc cardiol, Nat Heart Inst, 67-69; resident surg, Presby Hosp, NY, 69-75; asst prof med sci, 75-80, asst attend surgeon, Presby Hosp, NY, 75-80, ASSOC PROF SURGERY, COLUMBIA UNIV, 80-, LAB DIR CARDIOVASC SURG, 76-; ASSOC ATTEND SURGEON, PRESBY HOSP, NY, 80- *Concurrent Pos:* Estab investr, Am Heart Asn, 76-81; prin investr, Nat Heart Lung & Blood Inst, NIH res grant, 78-81; asst attend surgeon, Presby Hosp, NY, 75-80. *Mem:* AAAS; Am Heart Asn; fel Am Col Cardiol; fel Am Col Surgeons; Soc Univ Surgeons; Am Asn Thoracic Surg; Soc Thoracic Surg; Am Surg Asn. *Res:* Human left ventricular compliance and systolic mechanics; mechanical and pharmacologic circulatory support; open heart surgery; intraoperative echocardiography. *Mailing Add:* Dept Surg Columbia Univ 630 W 168th St New York NY 10032

SPOTTISWOOD, DAVID JAMES, b Melbourne, Australia, Aug, 28, 44; m 73; c 2. MINERAL PROCESSING, HYDROMETALLURGY. *Educ:* Univ Melbourne, Australia, 65; Colo Sch Mines, PhD(metall eng), 70. *Prof Exp:* Chem engr, Commonwealth Serum Labs, 65-66; asst prof mining eng, Queen's Univ, Can, 70-75; assoc prof metall eng, Mich Technol Univ, 75-79; ASSOC PROF METALL ENG, COLO SCH MINES, 79- *Concurrent Pos:* Consult, numerous co US, Can, Australia & Sam, 70-; prin investr, various projs, 70-; vis prof, Univ Conception, Chile, 74; vis lectr, McGill Univ, Can, K U Leuven, Belgium, Auckland Univ, NZ; vpres, Eng Systs Res Inc, 78- *Mem:* Am Inst Mining Metall & Petrol Engrs; Australasian Inst Mining & Metall; Can Inst Mining & Metall; Instrument Soc Am; Sigma Xi. *Res:* Mineral processing and hydrometallurgy, with emphasis on mathematical analysis for improved design, operation and automatic control; application of surface chemistry to mineral separations. *Mailing Add:* Seven Bennetts Pl Kalgoorlie WA 6430 Australia

SPOTTS, CHARLES RUSSELL, b Phoenix, Ariz, Oct 14, 33; m 54; c 4. MICROBIOLOGY. *Educ:* Univ Calif, Berkeley, BA, 55, PhD(microbiol), 61. *Prof Exp:* NIH fel, 61-63; instr microbiol, Univ Wash, 63, asst prof, Sch Med, 63-69; assoc prof, 69-74, PROF BIOL, CALIF STATE UNIV, NORTHRIDGE, 74- *Mem:* Am Soc Microbiol. *Res:* Bacterial metabolism; physiology of bacterial sporulation. *Mailing Add:* Dept of Biol Calif State Univ Northridge CA 91324

SPOTTS, JOHN HUGH, b Lauratown, Ark, Nov 2, 27; m 54; c 3. GEOLOGY. *Educ:* Univ Mo, BA, 50, MA, 51; Univ Western Australia, MSc, 56; Stanford Univ, PhD, 59. *Prof Exp:* Res geologist, Shell Develop Co, 52-53; geologist & geophysicist, Standard Oil Co, Calif, 53-56, geologist, 58-59, res geologist, La Habra Lab, Chevron Res Co, 59-68, mgr geol res, 68-70, div geologist, Standard Oil Co, Calif, 70-77, chief geologist, Chevron Resources Co, San Francisco, 77-81, VPRES EXPLORATION RES, CHEVRON OIL FIELD RES CO, LA HABRA, CALIF, 81- *Mem:* Mineral Soc Am; Am Asn Petrol Geologists; Soc Econ Paleontologists & Mineralogists; Int Asn Sedimentologists. *Res:* Mineralogy; sedimentary petrology; heavy minerals; geochemistry; carbonate petrography; petrofabrics. *Mailing Add:* 290 N Palisade Dr Orem UT 84057

SPOTTS, M(ERHYLE) F(RANKLIN), b Battle Creek, Iowa, Dec 5, 95; m 47; c 2. MACHINE DESIGN. *Educ:* Ohio Northern Univ, BS, 23; Ohio State Univ, MA, 33; Univ Mich, PhD(appl mech), 38. *Hon Degrees:* DEng, Ohio Northern Univ, 80. *Prof Exp:* Engr, Brown Steel Co, Ohio, 27-32; designer, Jeffrey Mfg Co, 33-35; assoc mech eng, Johns Hopkins Univ, 38-41; from asst prof to prof, 43-77, EMER PROF MECH ENG, TECHNOL INST, NORTHWESTERN UNIV, EVANSTON, 77- *Honors & Awards:* Worcester Reed Warner Medal, Am Soc Mech Engrs, 68; Century II Medallion, Am Soc Mech Engrs, 81; Machine Design Award, Am Soc Mech Engrs, 81. *Mem:* Fel Am Soc Mech Engrs. *Res:* Applied mechanics; mechanical vibrations; stress analysis. *Mailing Add:* Technol Inst McCormick Sch Eng & Appl Sci Northwestern Univ Evanston IL 60201

SPOTTS, ROBERT ALLEN, b Philadelphia, Pa, June 10, 45; m 69; c 2. PLANT PATHOLOGY. *Educ:* Colo State Univ, BS, 67, MS, 69; Pa State Univ, PhD(plant path), 74. *Prof Exp:* Chemist, Colo Dept Health, 69-71; asst prof plant path, Ohio Agr Res & Develop Ctr, 74-78; PROF PLANT PATH, ORE STATE UNIV, 78- *Mem:* Am Phytopath Soc. *Res:* Epidemiology, physiology and control of diseases of fruit crops. *Mailing Add:* Mid-Columbia Exp Sta Hood River OR 97031

SPRADLEY, JOSEPH LEONARD, b Baker, Ore, Oct 30, 32; m 55; c 4. ENGINEERING PHYSICS, HISTORY OF SCIENCE. *Educ:* Univ Calif, Los Angeles, BS, 54, MS, 55, PhD(eng physics), 58. *Prof Exp:* Mem tech staff, Hughes Aircraft Co, 54-58; assoc prof, Wheaton Col, 59-72, chmn dept, 68-70, chmn sci div, 90-91, PROF PHYSICS, WHEATON COL, ILL, 72- *Concurrent Pos:* Consult, Sunbeam Corp, 60-61; prof, Haigazian Col, Lebanon, 65-68; US Agency Int Develop sci specialist, Ahmadu Bello Univ, Nigeria, 70-72. *Mem:* Am Asn Physics Teachers. *Res:* Microwave antenna arrays; laser communications; history of science; prewar Japanese particle physics. *Mailing Add:* Dept Physics Wheaton Col Wheaton IL 60187

SPRADLIN, JOSEPH E, b Bloom, Kans, July 12, 29; m 48; c 3. BEHAVIOR ANALYSIS. *Educ:* Univ Kans, BA, 51; Ft Hays State Col, MS, 54; George Peabody Col, PhD(psychol), 59. *Prof Exp:* Teaching asst psychol, Ft Hays Col, 53-54; clin psychologist, Winfield State Sch, 54-56; res fel psychol, George Peabody Col, 56-58; res assoc, Bur Child Res, Univ Kans, 58-59, dir, Parsons Res Ctr, 59-69 & Kans Univ Affil Prog, 78-88, PROF HUMAN DEVELOP, UNIV KANS, 69-, DIR, PARSONS RES CTR, 87- *Concurrent Pos:* Consult ed, Am J Ment Retardation, 73-75 & J Speech & Hearing Dis, 75-78; mem, Ment Retardation Comt, Nat Inst Child Health & Human Develop, 74-78. *Mem:* Fel Am Psychol Asn; fel Am Psychol Soc. *Res:* Behavior analysis, with a special emphasis on stimulus control with persons with retardation. *Mailing Add:* 2601 Gabriel St Parson KS 67357

SPRADLIN, WILFORD W, b Bedford Co, Va, Oct 4, 32; m 58; c 2. PSYCHIATRY. *Educ:* Univ Va, BA, 53, MD, 57. *Prof Exp:* Intern, Royal Victoria Hosp, McGill Univ, 58; resident psychiat, Eastern State Hosp, 58-59 & Med Ctr, Duke Univ, 60-62; staff psychiatrist, Vet Admin Hosp, Durham, 62; assoc psychiat, Med Ctr, Duke Univ, 62-63; asst chief, Vet Admin Hosp, Durham, 63-64; asst prof, Med Ctr, Duke Univ, 64-67, chief psychiat day unit, 65-67, asst head psychiat inpatient serv, 65-67; prof psychiat & chmn dept behav med & psychiat, Med Ctr, WVa Univ, 67-78; PROF PSYCHIAT & CHMN DEPT, MED CTR, UNIV VA, 78- DEPT BEHAV MED & PSYCHIAT, MED CTR, W VA UNIV, 67- *Mem:* AAAS; fel Am Psychiat Asn; AMA; Sigma Xi. *Mailing Add:* Box 170 Ivy VA 22945

SPRADLING, ALLAN C, GENETICS. *Prof Exp:* STAFF MEM, HOWARD HUGHES MED INST, CARNEGIE INST WASHINGTON. *Mem:* Nat Acad Sci. *Mailing Add:* Howard Hughes Med Inst Carnegie Inst Washington 115 W University Pkwy Baltimore MD 21210

SPRAFKA, ROBERT J, b Chicago, Ill, Nov 24, 38; div. TECHNICAL MANAGEMENT. *Educ:* Purdue Univ, BS, 59, PhD(physics), 65. *Prof Exp:* Res assoc high energy physics, Purdue Univ, 64-66; physicist, Lawrence Radiation Lab, Calif, 66-67; from asst prof to assoc prof physics, Mich State Univ, 67-74, assoc prof physics & off health serv educ & res, 74-76, assoc prof community health sci, 76-82; sr assoc, E F Technol Inc, 81-84; dir, Systs Design & Implementation, LAM Consult Inc, 84-88; STAFF MEM, EAST OHIO GAS CO, 88- *Mem:* Am Phys Soc; Sigma Xi; Am Soc Heating, Refrigerating & Airconditioning Engrs. *Res:* Microcomputers as energy use monitors; computer applications in public health. *Mailing Add:* Tech Mkt Support East Ohio Gas Co 1717 E Ninth St Cleveland OH 44114

SPRAGG, JOCELYN, b New York, NY, Sept 16, 40. IMMUNOPHARMACOLOGY. *Educ:* Smith Col, AB, 62; Radcliffe Col, MA, 65; Harvard Univ, PhD(bact & immunol), 69. *Prof Exp:* IMMUNOLOGIST, DEPT RHEUMATOLOGY & IMMUNOL, BRIGHAM & WOMEN'S HOSP, 82-; ASSOC PROF MED & IMMUNOL, HARVARD MED SCH, 84- *Concurrent Pos:* Mem Coun Kidney & Cardiovasc Dis & Coun Thrombosis, Am Heart Asn; course developer & prin lectr, Radcliffe Summer Prog Sci, Radcliffe Col, 89- *Mem:* Am Asn Immunol; NY Acad Sci; AAAS; Am Heart Asn. *Res:* Human kallikrein-kinin systems. *Mailing Add:* Seeley Mudd Bldg Rm 625 Harvard Med Sch 250 Longwood Ave Boston MA 02115

SPRAGGINS, ROBERT LEE, b Sedalia, Mo, Feb 18, 39; m 63, 87; c 3. MASS SPECTROMETRY. *Educ:* La Tech Univ, BS, 63, MS, 66; Univ Okla, PhD(org chem), 70. *Prof Exp:* Chemist, Cities Serv Oil Co, 63-64; res fel, Alza Corp, Calif, 70-71; sea grant, Stevens Inst Technol, 71-72, res scientist, 72-74; res scientist, Ctr Trace Characterization, Tex A&M Univ, 75-77; sr scientist, 77-78, sr scientist & group leader mass spectros, Radian Corp, 79-81; SR SCIENTIST & MGR ANAL CHEM, SVMX CORP, AUSTIN, TEX, 81-; SR RES CHEMIST & MASS SPECTROMETRIST, MANVILLE RES & DEVELOP CO. *Honors & Awards:* Sigma Xi Res Award, 70. *Mem:* Am Chem Soc; Sigma Xi; Am Soc Mass Spectrometry. *Res:* Mass spectroscopy; biomedical and natural products; analytical chemistry; environmental chemistry. *Mailing Add:* 7885 W Walker Dr Littleton CO 80123

SPRAGINS, MELCHIJAH, b Mitchellville, Md, Jan 4, 19; m 49; c 4. PEDIATRICS. *Educ:* Johns Hopkins Univ, AB, 41, MD, 44; Am Bd Pediat, dipl, 49. *Prof Exp:* Instr pediat, Univ Buffalo, 47-48; from instr to asst prof, 48-84, EMER ASST PROF, PEDIAT, SCH MED, JOHNS HOPKINS UNIV, 84- *Concurrent Pos:* Instr pediat, Hosp, Johns Hopkins Univ, 48-50, pediatrician, Nursery, 57-64, instr pediat, Sch Med, 66-67, asst prof, 67-84; consult, Kernan's Hosp, Baltimore, 49-54; active staff mem, Union Mem Hosp, 50-; active staff mem, Women's Hosp of Md, 57-64, asst chief pediat, Women's Hosp of Md, 64-66; asst chief, 64-66, chief pediat, Greater Baltimore Med Ctr, 66-86. *Mem:* AMA; Am Acad Pediat. *Mailing Add:* 1506 Long Quarter Ct Lutherville MD 21093

SPRAGUE, BASIL SHELDON, b Hartford, Conn, Aug 3, 20; m 44; c 2. MATERIALS SCIENCE, POLYMER PHYSICS. *Educ:* Swarthmore Col, AB, 42; Polytech Inst Brooklyn, MChE, 44. *Hon Degrees:* ScD, Lowell Univ, 69. *Prof Exp:* Res engr plastics, Celanese Corp, 44-48, res engr textiles, 48-50, group leader textiles phys res, 50-52, head textile eval res, 52-56, head fiber physics & eval res, 56-64, mgr physics res dept, 64-65, mgr mat sci res dept, 65-68, dir mat sci res, 68-76, sr res fel, 76-82; RETIRED. *Honors & Awards:* H DeWitt Smith Medal, 76. *Mem:* AAAS; Am Chem Soc; Fiber Soc (vpres, 65, pres, 66). *Res:* Relationship of chemical constitution and morphology to physical properties of polymers and fibers; dyeing of synthetic fibers; materials research. *Mailing Add:* 356 Timber Dr Berkeley Heights NJ 07922

SPRAGUE, CHARLES CAMERON, b Dallas, Tex, Nov 16, 16; m 41; c 1. INTERNAL MEDICINE, HEMATOLOGY. *Educ:* Southern Methodist Univ, BBA & BS, 40; Univ Tex, MD, 43; Am Bd Internal Med, dipl. *Hon Degrees:* DSc, Southern Methodist Univ, 66, Univ Dallas, 83, Tulane Univ, 91. *Prof Exp:* Intern, US Naval Med Ctr, Md, 43-44; from asst to prof med, Sch Med, Tulane Univ, 47-67, prof hemat, 54-63, dean div, 63-67; prof med & dean, 67-72, pres, Univ Tex Health Sci Ctr Dallas, 72-86; pres, 86-88, CHMN SOUTHWESTERN MED FOUND, 88-; EMER PROF MED, UNIV TEX HEALTH SCI CTR, 86- *Concurrent Pos:* Fel, Sch Med, Tulane Univ, 48-49; Commonwealth res fel hemat, Sch Med, Wash Univ, 50-52 & Sch Med, Oxford Univ, 52; asst resident, Charity Hosp of La, 47-48, sr vis physician, 52-67; chmn, Gov Task Force Health Manpower, 81, Gov Med Educ Mgt Effectiveness Comt & Allied Health Educ Adv Comt, Coord Bd, Tex Col & Univ Syst; mem, coord bd, Tex Col & Univ, 88-, vchmn, 89- *Mem:* Inst Med-Nat Acad Sci; Am Soc Hemat (pres, 67); Am Fedn Clin Res; Int Soc Hemat; fel Am Col Physicians. *Res:* Hemoglobinopathies; leukemia; cancer chemotherapy. *Mailing Add:* Po Box 45708 Dallas TX 75245-0708

SPRAGUE, CLYDE HOWARD, mechanical engineering, for more information see previous edition

SPRAGUE, ESTEL DEAN, b Leavenworth, Kans, Oct 17, 44; m 67; c 2. PHYSICAL CHEMISTRY. *Educ:* Asbury Col, BA, 66; Univ Tenn, Knoxville, PhD(phys chem), 71. *Prof Exp:* Res kinetics & radiation chem, Max Planck Inst Coal Res, 71-73, Dept Chem Univ Wis-Madison, 73-74; asst prof, 74-80, ASSOC PROF PHYS CHEM, UNIV CINCINNATI, 80- *Mem:* Am Chem Soc; Sigma Xi; Biophys Soc. *Res:* Micelle chemistry; polymer models for micelles; electron spin resonance; radiation chemistry; kinetics of free-radical reactions; tunneling in hydrogen atom transfer reactions. *Mailing Add:* Dept of Chem Univ of Cincinnati Cincinnati OH 45221

SPRAGUE, G(EORGE) SIDNEY, b Lexington, Ky, Sept 9, 18; m 42, 74; c 3. POLYMER CHEMISTRY. *Educ:* Lehigh Univ, BS, 40; Univ Wis, MS, 43; NY Univ, PhD(chem), 50. *Prof Exp:* Res chemist, Sharples Chem Inc, 43-45 & Deering Milliken Res Trust, 47-49; from res chemist to sr res chemist, Am Cyanamid Co, 49-72; sr chemist, 73-90, CONSULT, LOCTITE CORP, 91- *Mem:* Am Chem Soc; Sigma Xi. *Res:* Adhesives and sealants; thermoplastics; solid rocket propellant and explosive binders; water soluble polymers; monomer synthesis. *Mailing Add:* Eight B Lyle Ct Farmington CT 06032-3531

SPRAGUE, GEORGE FREDERICK, b Crete, Nebr, Sept 3, 02; m; c 4. AGRONOMY. *Educ:* Univ Nebr, BSc, 24, MS, 26; Cornell Univ, PhD(genetics), 30. *Prof Exp:* Jr agronomist, Bur Plant Indust, USDA, DC, 24-28, from asst agronomist to agronomist, 28-42, from sr agronomist to prin agronomist, 42-58, head corn & sorghum sect, Crops Res Div, Agr Res Serv, 58-72; prof, 73-87, EMER PROF AGR, UNIV ILL, URBANA, 87- *Concurrent Pos:* Prof, Iowa State Univ, 48-58. *Mem:* Nat Acad Sci; fel Am Soc Agron (vpres, 59-60); Am Genetics Asn; Genetics Soc Am; Biomet Soc. *Res:* Corn breeding and genetics; statistics. *Mailing Add:* Dept Agr Univ Ill Urbana IL 61801

SPRAGUE, HOWARD BENNETT, b Cortland, Nebr, Dec 11, 98; m 44; c 1. AGRONOMY. *Educ:* Univ Nebr, BS, 21, MS, 23; Rutgers Univ, PhD(plant physiol), 26. *Prof Exp:* Asst agron, Univ Nebr, 21-23; instr, Rutgers Univ, 23-26, assoc prof, 27-31, prof, 31-42, head dept, Univ & agronomist, Exp Sta, 27-42; asst prof, Univ Minn, 26-27; head, Agr Res Div, Tex Res Found, 46-51; prof agron & head dept, Pa State Univ, 53-64; exec secy, Agr Bd, Nat Acad Sci, 64-69; agr consult, Aid & Var Int Agencies, 69-83; RETIRED. *Honors & Awards:* Medallion Award, Am Forage & Grassland Coun, 64. *Mem:* Am Forage & Grassland Coun (pres, 53-56); Soil Conserv Soc Am; fel AAAS (vpres & chmn sect O, 56, secy, 57-65); fel Am Soc Agron (pres, 64); Crop Sci Soc Am (pres, 60). *Res:* Land classification and use; soil fertility and management; plant nutrition; ecology of crop plants; plant breeding; grassland management; turf culture; agricultural and natural resource development. *Mailing Add:* 560 Ridge Ave State College PA 16803

SPRAGUE, ISABELLE BAIRD, b Manila, PI, May 30, 16. BIOLOGY. *Educ:* Mt Holyoke Col, AB, 37, MA, 39; Univ Kans, PhD(entom), 53. *Prof Exp:* From instr to prof zool, 45-64, chmn dept, 63-66, prof biol sci, 76-80, DAVID B TRUMAN PROF ZOOL & EMER PROF BIOL SCI, MT HOLYOKE COL, 80- *Concurrent Pos:* NSF fac fel, 58-59. *Mem:* Ecol Soc Am; Entom Soc Am. *Res:* Aquatic biology; endocrinology of insects; semi-aquatic hemiptera. *Mailing Add:* Dept Biol Sci Mt Holyoke Col South Hadley MA 01075

SPRAGUE, JAMES ALAN, b Cleveland, Ohio, Aug 24, 43; m 68. SURFACE MODIFICATION, THIN FILMS. *Educ:* Rice Univ, BA, 65, BS, 66, PhD(mat sci), 70. *Prof Exp:* RES METALLURGIST, NAVAL RES LAB, 71-; ADJ PROF, GEORGE WASHINGTON UNIV, 83- *Concurrent Pos:* NSF fel, Max Planck Inst Metall Res, 69-70. *Mem:* Am Inst Mining, Metall & Petrol Engrs; Am Ceramics Soc; Electron Micros Soc Am; Sigma Xi. *Res:* Electron microscopy of defects in solids; ion beam surface modification of materials; thin film processing, microstructure and properties. *Mailing Add:* Naval Res Lab Code 4670 Washington DC 20375

SPRAGUE, JAMES CLYDE, b Gibbons, Alta, Aug 4, 28; m 52; c 1. INDUSTRIAL ENGINEERING. *Educ:* Univ Okla, BSc, 60; Iowa State Univ, MSc, 67, PhD(indust eng), 69. *Prof Exp:* Chief economist, Hu Harries & Assocs, 60-61; chief engr, BJ Serv of Can, 61-65; dir eng eval, Gamma Eng, 65-66; asst prof indust eng, Iowa State Univ, 68-69; assoc prof mech eng, 69-77, PROF MECH ENG, UNIV ALTA, 77- *Mem:* Am Soc Eng Educ; Am Inst Indust Engrs. *Res:* Engineering economy and capital budgeting, design of industrial systems; mass production of homes. *Mailing Add:* Dept Mech Eng Univ Alberta Edmonton AB T6G 2E2 Can

SPRAGUE, JAMES MATHER, b Kansas City, Mo, Aug 31, 16; c 1. ANATOMY, METALLURGY & PHYSICAL METALLURGICAL ENGINEERING. *Educ:* Univ Kans, AB, 38, AM, 40; Harvard Univ, PhD(biol), 42. *Hon Degrees:* MA, Univ Pa, 71. *Prof Exp:* From asst mus mammals to asst instr zool, Univ Kans, 36-40; from asst to asst prof anat, Sch Med, Johns Hopkins Univ, 42-50; from asst prof to prof anat, 50-73, mem, Inst Neurol Sci, 54-73, assoc dir, 57-60, chmn dept, 67-76, dir Inst Neurol Sci, 73-80, JOSEPH LEIDY PROF ANAT, SCH MED, UNIV PA, 73- *Concurrent Pos:* Guggenheim fel, Cambridge Univ & Oxford Univ, 48-49;

Macy Fac Scholar award, 74-75; vis investr, Med Sch, Northwestern Univ, 48, Rockefeller Inst, 55, Cambridge Univ, 56 & Univ Pisa, 66 & 74, Univ Leuven, Belgium; consult, NIH, 57-58. *Honors & Awards:* Lindbach Found Award, 66. *Mem:* Nat Acad Sci; AAAS; Int Brain Res Orgn; Soc Neurosci; Am Asn Anat (vpres, 76-78). *Res:* Taxonomy and comparative anatomy of mammals; neuroanatomy of spinal cord; neurophysiology of brain stem and spinal cord; anatomy and physiology of brain stem and cerebellum; neural mechanisms of vision and visual behavior. *Mailing Add:* Dept Anat Sch Med Univ Pa Philadelphia PA 19104-6058

SPRAGUE, JOHN BOOTY, b Woodstock, Ont, Feb 16, 31; m 53; c 5. BIOLOGY. *Educ:* Univ Western Ont, BSc, 53; Univ Toronto, MA, 54, PhD(zool), 59. *Prof Exp:* Scientist-in-chg pollution studies, Biol Sta, Fisheries Res Bd, Can, 58-70; from assoc prof to prof zool, Univ Guelph, 70-88; CONSULT, 88- *Res:* Aquatic biology; effects of pollution on fish and other aquatic organisms; bioassays and water quality criteria. *Mailing Add:* Sprague Assocs Ltd 166 Maple St Guelph ON N1G 2G7 Can

SPRAGUE, LUCIAN MATTHEW, b Salt Lake City, Utah, Apr 14, 26. FISHERIES POLICY STUDIES. *Educ:* Univ Calif, AB, 50, PhD, 57. *Prof Exp:* Res asst, Univ Calif, 52-56; geneticist biol lab, Bur Commercial Fisheries, US Fish & Wildlife Serv, 56-60, chief subpop invest, 60-62, dep dir, Hawaii Area, 62-67; assoc dir med & natural sci, Rockefeller Found, 67-69; prof oceanog & dir, Int Ctr Marine Resource Develop, Univ RI, 69-72; sr fisheries specialist, Agr & Rural Develop Dept, Int Bank Reconstruction & Develop 72-88, Fisheries Consult, World Bank, 88-90; ADJ PROF FISHERIES, DEPT FISHERIES & AQUACULT, SEA GRANT COL, U FAL, GAINESVILLE, 90- *Concurrent Pos:* Res fel, Univ Uppsala, 66-67. *Mem:* AAAS; Genetic Soc Am; Sigma Xi. *Res:* Blood groups and genetics of natural populations of vertebrates, particularly teleosts; international fisheries resources and policy studies. *Mailing Add:* 4486 Occoquan View Ct Woodbridge VA 22192

SPRAGUE, MILTON ALAN, b Washburn, Wis, June 16, 14; m 44; c 4. AGRONOMY, CROP PHYSIOLOGY. *Educ:* Northland Col, BA, 36; Univ Wis, MS, 38, PhD(agron & plant phys), 41. *Prof Exp:* Asst agron, Univ Wis, 36-41; asst prof agron, Univ Ark, 41-46; asst prof agron, Rutgers Univ, 46-50, assoc prof & assoc res specialist, 50-56, chmn dept farm crops, 55-61, from prof to distinguished prof, 56-84, EMER PROF, RUTGERS UNIV, NEW BRUNSWICK, 84-, RES SPECIALIST, 56- *Concurrent Pos:* Dir, Am Soc Agron, 56-59, 69-72; lectr, Columbia Univ, 56-71; dir, Am Forage & Grassland Coun, 58-61, 69-72; agr consult, Latin Am, Yucatan, 66-72, 81, Am Heritage Dict, 71, NJ, 84; mem, Coun Agr Sci & Technol. *Honors & Awards:* Res Award, Am Soc Agron, 77. *Mem:* Fel Am Soc Agron; Am Soc Plant Physiologists; Sigma Xi; Crop Sci Soc Am; Soil Sci Soc Am. *Res:* Physiology forages; winter killing; seedling establishment; pasture and forage management; accumulation of respiratory by-products in alfalfa and the injurious effects of ice contact; factors affecting silage quality; microclimate affected by slope and its effect on the biosphere; no-tillage agriculture. *Mailing Add:* Dept Crop Sci Rutgers Univ Lipman Hall Box 231 New Brunswick NJ 08903

SPRAGUE, NEWTON G, b Indianapolis, Ind, Feb 8, 14; m 37; c 2. PHYSICS, ASTRONOMY. *Educ:* Butler Univ, BS, 35; Ind Univ, MS, 51, EdD(educ psychol), 55. *Prof Exp:* Asst physics, Butler Univ, 47-48; phys chemist, Indust Oils Lab, 48-49; teacher pub sch, Ind, 49-51, asst visual prod, 51-56, consult, 56-60; from asst prof to prof physics & astron, 60-78, dir, Univ Observ & Planetarium, 65-78, EMER PROF PHYSICS & ASTRON, BALL STATE UNIV, 78- *Mem:* AAAS; emer mem Am Astron Soc. *Res:* Spectroscopic and photometric stellar measurements. *Mailing Add:* 1212 N Ridge Rd Muncie IN 47304

SPRAGUE, PETER WHITNEY, b Rochester, NY, Oct 4, 41; m 63; c 2. ORGANIC CHEMISTRY. *Educ:* Western Reserve Univ, BA, 63, PhD(org chem), 66. *Prof Exp:* Fel, Ind Univ, 66-67; from asst prof to assoc prof chem, Calif State Col, San Bernardino, 67-72; res investr, 72-74, sr res investr, 74-76, group leader, 76-80, SECT HEAD ORG CHEM, E R SQUIBB & SONS, INC, 80- *Mem:* Am Chem Soc; AAAS. *Res:* Medicinal chemistry in areas of cardiovascular and antiinflammatory drugs. *Mailing Add:* E R Squibb & Sons Inc PO Box 4000 Princeton NJ 08543-4000

SPRAGUE, RANDALL GEORGE, b Chicago, Ill, Sept 22, 06; m 39; c 4. INTERNAL MEDICINE, ENDOCRINOLOGY. *Educ:* Northwestern Univ, Ill, BS, 30, MB & MS, 34, MD, 35, Mayo Grad Sch Med, Univ Minn, PhD(med), 42; Am Bd Internal Med, dipl, 42. *Hon Degrees:* LLD, Univ Toronto, 64. *Prof Exp:* Instr physiol, Med Sch, Northwestern Univ, Ill, 33-34; first asst med, Mayo Clin, 39-40, from instr to prof, Mayo Grad Sch Med, 42-71, emer prof med, Mayo Grad Sch Med, Univ Minn, 71-85; RETIRED. *Concurrent Pos:* Consult, Mayo Clin, 40-63, sr consult, 63-71, emer consult, 71-; mem metab & endocrinol study sect, NIH, 47-51; mem, Nat Adv Dent Res Coun, 63-67; consult, Rochester State Hosp, 72-81. *Mem:* Am Soc Clin Invest; Asn Am Physicians; Am Diabetes Asn (pres, 53-54); master Am Col Physicians; hon mem Royal Soc Med; corresp mem Royal Acad Med Belg. *Res:* Clinical investigation of metabolic and endocrine diseases; relation of the adrenal cortex to carbohydrate metabolism; endocrinology. *Mailing Add:* 410 SW Sixth Ave M-2 Rochester MN 55902

SPRAGUE, RICHARD HOWARD, b Cincinnati, Ohio, Nov 9, 24. MATHEMATICS. *Educ:* Maryville Col, BS, 49; Univ KY, MA, 52, PhD(math), 61. *Prof Exp:* Asst math, Ohio State Univ, 49-50; asst, Univ Ky, 50-52, instr, 53-56; asst prof, NMex State Univ, 58-60; asst prof, 61-67, ASSOC PROF MATH, IOWA STATE UNIV, 67- *Mem:* Am Math Soc. *Res:* Univalent functions; geometry; complex variables. *Mailing Add:* Dept Math Iowa State Univ 400 Carver Ames IA 50011

SPRAGUE, ROBERT ARTHUR, US citizen. OPTICS. *Educ:* Univ Rochester, BS, 67, PhD(optics), 71. *Prof Exp:* Sr scientist, Itek Corp Cent Res Labs, 71-74, staff scientist, 74-76; mem res staff, 76-80, RES AREA MGR, PALO ALTO RES CTR, XEROX CORP, 80- *Concurrent Pos:* Comt mem, US Nat Comt, Int Comn Optics, 73-75. *Mem:* Optical Soc Am; Soc Photo-Optical Instr Engrs; Am Inst Physics. *Res:* Electro-optics, input/output systems, optical signal processing, coherent optical processing and acousto-optics. *Mailing Add:* Xerox Palo Alto Res Ctr 3333 Coyote Hill Palo Alto CA 94304

SPRAGUE, ROBERT C, b Aug 3, 00. ELECTRONICS ENGINEERING. *Educ:* US Naval Post-Grad Sch, BS, 22; Mass Inst Technol, SM, 24. *Prof Exp:* DIR, FOUNDER & HON CHMN, SPRAGUE ELEC CO, 26- *Honors & Awards:* Medal of Honor, Radio-Electronics TV Manufacturers Asn, 54; Gold Knight Award, Nat Mgt Asn, 58; Medal of Honor, Electronic Indust Asn, 59. *Mem:* Nat Acad Eng; fel Inst Elec & Electronics Engrs; fel Am Acad Arts & Sci. *Mailing Add:* Sprague Elec Co PO Box 662 Williamstown MA 02167

SPRAGUE, ROBERT HICKS, b Rochester, NY, Mar 9, 14; m 36; c 3. ORGANIC CHEMISTRY. *Prof Exp:* Asst res lab, Eastman Kodak Co, NY, 31-34, res chemist, 35-52; group leader sensitizing dye res, Remington Rand, Inc, Conn, 52-57; asst head chem dept, Horizons, Inc, 57-58, head, 58-64; mgr dye chem dept, Itek Corp, 65-75; consult, Exxon Res & Eng Co, 75-76; consult, Res Triangle Inst, Olivetti Co Am, 76-81; consult, Ricoh Syst Inc, 81-85; RETIRED. *Honors & Awards:* Kosar Award, Soc Photog Sci & Eng, 68. *Mem:* AAAS; fel Soc Photog Sci & Eng; Am Chem Soc; NY Acad Sci. *Res:* Sensitizing dyes for photographic emulsions; dyes for color photography; pharmaceuticals; antibiotics; diuretics; tranquilizers; non-silver photographic systems; organic photoconductors. *Mailing Add:* 1041 Hillside Dr Chapel Hill NC 27514-2607

SPRAGUE, ROBERT W, b Omaha, Nebr, Aug 1, 23; m 78; c 2. ENVIRONMENTAL SCIENCE. *Educ:* Univ Calif, Los Angeles, BS, 44; Ohio State Univ, PhD(inorg chem), 57. *Prof Exp:* Sales engr, R E Cunningham & Son, 46-48; chemist, US Naval Ord Test Sta, Calif, 48-57; teaching asst & res fel chem, Ohio State Univ, 54-57; sr chemist, Minn Mining & Mfg Co, 57-58; res specialist propulsion, Rocketdyne Div, NAm Aviation, Inc, 58-60; res scientist, Aeronutronic Div, Philco Corp, 60-65; sr res chemist, 65-83, SR SCIENTIST, US BORAX RES CORP, ANAHEIM, 83- *Mem:* Fel AAAS; Am Chem Soc; Am Soc Test & Mat; Nat Fire Protection Asn. *Res:* Inorganic chemistry of nonmetals; chemistry of boron oxides, sulfides, halides; chemistry of oxide systems; environmental chemistry; fire retardance. *Mailing Add:* 5753 Wildriar Dr Rancho Palos Verdes CA 90274-1752

SPRAGUE, VANCE GLOVER, JR, b Bellefonte, Pa, Oct 28, 41; m 89. PHYSICAL OCEANOGRAPHY. *Educ:* Penn State Univ, BS, 63; Salve Regina Col, MS, 83; Naval War Col, MS, 83. *Prof Exp:* Oceanogr, Naval Oceanog Off, 65-72, sr scientist, 72-77, head, Analysis Sect, Phys Oceanog Br, 77-85, head, Phys Oceanog Br, 80-85, DIR, PHYS OCEANOG DIV, NAVAL OCEANOG OFF, 85- *Mem:* Am Geophys Union. *Res:* Airborne and shipboard field programs; design and construction of oceanographic data bases and mathematical models; transition of numerical ocean models for operational use and implementation of satellite remote sensing techniques. *Mailing Add:* Nat Oceanog Off Stennis Space Center MS 39522

SPRAIN, WILBUR, science education; deceased, see previous edition for last biography

SPRAKER, HAROLD STEPHEN, b Cedar Bluff, Va, May 13, 29; m 54; c 2. MATHEMATICS. *Educ:* Roanoke Col, BS, 50; Univ Va, MEd, 55, DEd(math educ), 60. *Prof Exp:* Teacher high sch, Va, 53-55, asst prin, 55-57; res assoc, Univ Va, 56-60, instr math, 59-60; from asst prof to assoc prof 60-65, PROF MATH, MID TENN STATE UNIV, 65-, CHMN DEPT, 67- *Concurrent Pos:* Apprentice coordr, Va State Dept Labor, 53-57; dir, NSF In-Serv Inst & vis scientist lectr. *Mem:* Math Asn Am; Nat Coun Teachers Math. *Res:* Mathematical education; geometry; algebra; statistics. *Mailing Add:* Dept Math & Statist Mid Tenn State Univ Box 34 Murfreesboro TN 37132

SPRATLEY, RICHARD DENIS, b Vancouver, BC, Apr 18, 38; m 64; c 3. PHYSICAL CHEMISTRY. *Educ:* Univ BC, BSc, 61; Univ Calif, Berkeley, PhD(chem), 65. *Prof Exp:* Res assoc chem, Brookhaven Nat Lab, 65-67; asst prof chem, 67-72, res adminr, 72-83, DIR RES SERVS, UNIV BC, 83- *Mem:* Can Asn Univ Res Adminr; Soc Univ Patent Adminr. *Res:* Infrared spectroscopy; x-ray and neutron diffraction; molecular structure and bonding. *Mailing Add:* Off Res Servs Univ BC Vancouver BC V6T 1Z3 Can

SPRATT, JAMES LEO, b Chicago, Ill, Jan 27, 32; c 2. PHARMACOLOGY. *Educ:* Univ Chicago, AB, 53, PhD(pharmacol), 57, MD, 61. *Prof Exp:* Res assoc pharmacol, Argonne Cancer Res Hosp, Univ Chicago, 57-61; from asst prof to assoc prof, 61-71, PROF PHARMACOL, UNIV IOWA, 71- *Concurrent Pos:* USPHS res career develop award, 63-68; Markle scholar, 63-68. *Mem:* AAAS; Am Soc Pharmacol & Exp Therapeut. *Res:* Therapeutics; radioisotopic tracer methods in cardiac glycoside research; cardiac glycosides and neurotoxicity; biochemical neuropharmacology. *Mailing Add:* Dept of Pharmacol Univ of Iowa Iowa City IA 52240

SPRATT, JOHN STRICKLIN, b San Angelo, Tex, Jan 3, 29; m 51; c 3. SURGERY. *Educ:* Univ Tex, Dallas, MD, 52; Univ Mo-Columbia, MSPH, 70; Southern Methodist Univ, BS, 76; Am Bd Surg, dipl, 60. *Prof Exp:* Asst physiol, Univ Tex Southwestern Med Sch Dallas, 52; intern surg, Barnes Hosp, 52-53, from asst resident to resident, 55-59; from instr to assoc prof, Sch Med, Washington Univ, 59-66; prof surg, Sch Med, Univ Mo-Columbia, 66-76, prof community health & med pract, 71-76; PROF SURG ONCOL, UNIV LOUISVILLE, 76- *Concurrent Pos:* USPHS cancer res fel radiother & surg, Mallinckrodt Inst Radiol, St Louis, 57-58; Am Cancer Soc fel, Barnes Hosp, 58-59; Am Cancer Soc advan clin fel, 60-63; from asst prof to assoc prof, Sch Med, Univ Mo, 61-66, dir clin res & mem sci adv comt, Cancer Res Ctr, 64, dir ctr, 65-76; med dir dept surg, Ellis Fischel State Cancer Hosp, Columbia, 61-76; med adv bur hearings & appeals, Soc Security Admin, 64-; mem rev comt sr clin traineeships surg, Cancer Control Prog, USPHS, 64-68; coordr cancer control, State of Mo; mem study sect, Supportive Serv Rev, 75-77; prof clin oncol, Am Cancer Soc, 76- *Mem:* Am Surg Asn; Am Asn Cancer Res; Soc Head & Neck Surg; Soc Surg Oncol. *Res:* Statistical analyses of the natural history of human cancer and the influence of therapy upon the natural history; use of roentgen and surgical therapy for cancer; cytokinetics of human cancer, application of operations research methods to clinical decisions; the role of patient and family education in rehabilitation; cancer control. *Mailing Add:* Dept Surg & Cmty Health James Graham Brown Cancer Ctr 529 S Jackson St Louisville KY 40202

SPRATTO, GEORGE R, b Waterbury, Conn, July 28, 40; m 68; c 2. PHARMACOLOGY. *Educ:* Fordham Univ, BS, 61; Univ Minn, PhD(pharmacol), 66. *Prof Exp:* Pharmacologist, Food & Drug Admin, 66-68; from asst prof to assoc prof pharmacol, 68-79, assoc head, Dept Pharmacol & Toxicol, 78-83, PROF PHARMACOL, PURDUE UNIV, 79-, ASSOC DEAN PROF PROGS, SCH PHARM, 84- *Concurrent Pos:* mem, Instnl Rev Bd, Pharmadynamics, Inc, 79-87; adj prof, Sch Med, Ind Univ, 81-; secy, coun deans, Am Pharmaceut Asn, 87-90, chmn, coun faculties, 79-80. *Honors & Awards:* Merck Sharp & Dohme Award Outstanding Achievement Prof Pharm, 89; Distinguished Serv Award, Am Sch Health Asn, 74. *Mem:* Am Soc Pharmacol & Exp Therapeut; Am Asn Col Pharm; Am Soc Hosp Pharmacists; Am Asn Pharmaceut Scientists; Am Pharmaceut Asn. *Res:* Assessment of the interation of acute or subchronic administration of acetylcholinesterase inhibitors and stress on endocrine parameters and glucose; interaction of central nervous system drugs in animals treated acutely or chronically with narcotics. *Mailing Add:* Off Dean Sch Pharm & Pharmacol Sci Purdue Univ West Lafayette IN 47907

SPRAWLS, PERRY, JR, b Williston, SC, Mar 2, 34; m 61; c 1. MEDICAL PHYSICS, BIOMEDICAL ENGINEERING. *Educ:* Clemson Univ, BS, 56, MS, 60, PhD, 68. *Prof Exp:* Engr, Bell Tel Labs, 56-58; physicist, Savannah River Labs, AEC, 59-60; from instr to assoc prof,59-77, PROF RADIOL, EMORY UNIV, 77- *Mailing Add:* Dept Radiol Emory Univ Woodruff Mc Admin Bldg Atlanta GA 30322

SPRAY, CLIVE ROBERT, b Oxford, Eng, Sept 8, 53; m 83. METABOLIC STUDIES WITH RADIOLABELED SUBSTRATES. *Educ:* Univ Hull, BSc, 77; Univ Bath, PhD(org chem), 81. *Prof Exp:* Res assoc chem, Univ Bristol, Eng, 80-81; RES ASSOC BIOL, UNIV CALIF, LOS ANGELES, 81- *Mem:* Am Chem Soc; Am Soc Plant Physiologists. *Res:* Biosynthesis and metabolism of the plant growth hormones; gibberellins, including studies on the molecular biology of dwarfing genes in maize; analysis of the gibberellin biosynthetic pathway in maize isolation and purification of gibberellin biosynthetic enzymes. *Mailing Add:* Dept Biol Univ Calif Los Angeles CA 90024

SPRAY, DAVID CONOVER, b Pittsburgh, Pa, June 7, 46; div; c 1. NEUROPHYSIOLOGY. *Educ:* Transylvania Col, BS, 68; Univ Fla, PhD(physiol), 73. *Prof Exp:* Res fel, 73-77, ASST PROF NEUROSCI, ALBERT EINSTEIN COL OF MED, 77- *Concurrent Pos:* Trainee, Ctr Neurosci, 69-73; mem corp, Marine Biol Lab, Woods Hole, 74- *Mem:* Am Physiol Soc; Biophys Soc; Soc Neurosci; Soc Gen Physiologists; Sigma Xi. *Res:* General neurophysiology, especially the physiology of chemical and electrical synapses, electro- and cutaneous receptors, cellular excitability and intracellular communication and excitability during development. *Mailing Add:* Dept Neurosci Albert Einstein Col Med 1300 Morris Park Ave New York NY 10461

SPRECHER, DAVID A, b Saarbrucken, Ger, Jan 12, 30; US citizen; m 79; c 2. MATHEMATICS. *Educ:* Univ Bridgeport, AB, 58; Univ Md, PhD(math), 63. *Prof Exp:* Instr math, Univ Md, 61-63; asst prof, Syracuse Univ, 63-66; assoc prof, 66-71, chmn, Dept Math, 72-75, actg dean, Col Lett & Sci, 78-79, dean, 79-80, PROF MATH, UNIV CALIF, SANTA BARBARA, 71-, PROVOST & DEAN, COL LETT & SCI, 81- *Concurrent Pos:* NSF grant, 65-67. *Mem:* Am Math Soc; Math Asn Am. *Res:* Structure of functions of several variables; superposition of functions and approximation theory. *Mailing Add:* Col Letter Sci Univ Calif Santa Barbara CA 93106

SPRECHER, HOWARD W, b Sauk City, Wis, Oct 13, 36; m 64. BIOCHEMISTRY. *Educ:* NCent Col, BA, 58; Univ Wis, PhD(biochem), 64. *Prof Exp:* Fel biochem, Hormel Inst, Univ Minn, 63-64; from asst prof to assoc prof physiol chem, 64-72, PROF PHYSIOL CHEM, OHIO STATE UNIV, 72- *Mem:* AAAS; Am Chem Soc; Am Oil Chem Soc; Am Soc Biol Chemists. *Res:* Organic synthesis, metabolism and characterization of lipids. *Mailing Add:* Dept Physiol Chem Ohio State Univ 337 Hamilton Hall 1645 Neil Ave Columbus OH 43210

SPREITER, JOHN R(OBERT), b Oak Park, Minn, Oct 23, 21; m 53; c 4. FLUID MECHANICS, SPACE PHYSICS. *Educ:* Univ Minn, BAeroE, 43; Stanford Univ, MS, 47, PhD(eng mech), 54. *Prof Exp:* Aeronaut engr, Flight Res Br, Ames Aeronaut Lab, Nat Adv Comt Aeronaut, 43-46, res scientist, Theoret Aerodyn Br, 47-58, res scientist, Theoret Br, Ames Res Ctr, NASA, 58-62, chief theoret studies br, Space Sci Div, 62-69; lectr, 50-68, PROF APPL MECH & AERONAUT & ASTRONAUT, STANFORD UNIV, 68- *Concurrent Pos:* Mem ionospheres & radio physics subcomt, Space Sci Steering Comt, NASA, 60-64; mem various comts, Inst Asn Geomag & Aeronomy, 64-, Am Inst Aeronaut & Astronaut, 68-74; consult, Neilson Eng & Res, Inc, 68-85, Gen Motors, 75-76, RMA Aerospace, 85- *Honors & Awards:* NASA Group Achievement Award, 83; Am Geophys Union Excellence in Reviewing, 88. *Mem:* AAAS; Am Geophys Union; Am Phys Soc; fel Am Inst Aeronaut & Astronaut, 71; Royal Astron Soc; Sigma Xi; Planetary Soc. *Res:* Geomagnetism; solar wind; cosmic fluid dynamics; space physics; magnetohydrodynamics; transonic flow theory; aerodynamics and fluid mechanics; subsonic, transonic and supersonic flow about wings and bodies; space plasma physics (solar wind and its interaction with the Earth, Moon, planets and comets). *Mailing Add:* 1250 Sandalwood Lane Los Altos CA 94024

SPREITZER, WILLIAM MATTHEW, b Highland Park, Mich, Aug 14, 29; m 52; c 2. AERONAUTICAL ENGINEERING. *Educ:* Univ Detroit, BAeE, 51. *Hon Degrees:* AeE, Univ Detroit, 57. *Prof Exp:* Eng draftsman, Dept Aeronaut, State Mich, 49-51; from jr res engr to sr res engr, Eng Develop Dept, Gen Motors, 51-61, sr liaison engr, Exec Dept, 61-66, head transp res dept, 66-72 & 79-85, transp & urban anal dept, 72-78, operating systs res, 85-87, mgr planning, 87-89, MGR, VEHICLE/HWY SYSTS COORD, GEN MOTORS RES LABS, 89- *Concurrent Pos:* Mem, Comt on Transp, Nat Res Coun Assembly Eng, 70-81 & Bay Area Rapid Transit Impact Adv Comt, 72-79; mem, Transp Develop Adv Comt, Hwy Users Fedn Safety & Mobility, 70-72; deleg transp panel, White House Conf on Aging, 71; mem panel on urbanization, transp & commun, Nat Acad Sci-Nat Res Coun Study for 79 UN Conf on Sci & Technol for Develop, 77-78; chmn, div A group 5, Transp Res Bd, Nat Acad Sci, 86- & subcomt Automotive Navig Aids, Soc Automotive Engrs, 85-; mem steering comt, Intel Vehicle Hwy Soc Am. *Honors & Awards:* Roy W Crum Award, Trans Res Bd Nat Res Coun, 84. *Mem:* Assoc fel Am Inst Aeronaut & Astronaut; Soc Automotive Engrs; Opers Res Soc Am. *Res:* Transportation and traffic science; automotive gas turbine engine research, development and applications; research administration. *Mailing Add:* Mgr Vehicle/Hwy Systs Coord Warren MI 48090-9055

SPREMULLI, GERTRUDE H, b Bucyrus, Ohio, Dec 21, 12; m 37; c 3. BIOCHEMISTRY. *Educ:* Heidelberg Col, BS, 33; Western Reserve Univ, MS, 38; Pa State Univ, PhD(agr biochem), 42. *Prof Exp:* Res chemist, Ranger Aircraft Engines, 42-44; res assoc, Columbia Univ, 44-45; asst prof, 56-72, prof, 72-77, chmn, Div Natural Sci & dean admin, 72-76, Dana prof, 76-77, EMER DANA PROF CHEM, ELMIRA COL, 78- *Concurrent Pos:* Lectr, Univ NC, 82-84. *Mem:* Am Chem Soc; Sigma Xi. *Res:* Biophysical chemistry; biochemistry, especially enzymes; physical properties. *Mailing Add:* 605 Kenmore Rd Chapel Hill NC 27514

SPREMULLI, LINDA LUCY, b Corning, NY, Sept 6, 47. CHEMISTRY. *Educ:* Univ Rochester, BA, 69; Mass Inst Technol, PhD(biochem), 73. *Prof Exp:* Assoc chem, Univ Tex, Austin, 73-74, fel, 74-76; asst prof, 76-81, ASSOC PROF CHEM, UNIV NC, 81- *Concurrent Pos:* Mem, Biomed Sci Study Sect, NIH, 82-86, Biochem study sect, 87- *Mem:* Am Soc Biol Chemists; Am Chem Soc; AAAS; Asn Women Sci; Am Soc Microbiol; Int Soc Plant Molecular Biol. *Res:* Characterization of mammalian mitochondrial protein synthesis; characterization of the ribosomes and auxiliary factors required for chloroplast protein synthesis and induction of this system by light. *Mailing Add:* 605 Kenora Rd Chapel Hill NC 27514

SPRENG, ALFRED CARL, b Alliance, Ohio, Feb 2, 23; m 49; c 3. STRATIGRAPHY. *Educ:* Col Wooster, AB, 46; Univ Kans, AM, 48; Univ Wis, PhD(geol), 50. *Prof Exp:* Asst, Univ Wis, 48-50; from asst prof to prof geol, 50-85, chma dept geol & geophys, 71-75, EMER PROF GEOL, UNIV MO-ROLLA, 85- *Concurrent Pos:* Consult, limestone & shale raw mats, 55- *Mem:* Paleont Soc; Am Asn Petrol Geologists; Geol Soc Am; Soc Econ Paleontologists & Mineralogists; Am Inst Prof Geologists. *Res:* Stratigraphic paleontology; carbonate petrology. *Mailing Add:* Dept Geol & Geophysics Univ Mo Rolla MO 65401-0249

SPRENKEL, RICHARD KEISER, b York, Pa, July 10, 43; m 65. ENTOMOLOGY. *Educ:* Pa State Univ, BS, 65, MS, 67; Univ Ill, PhD(entom), 73. *Prof Exp:* Res assoc entom, NC State Univ, 73-79; ASST PROF ENTOM, UNIV FLA, 79- *Mem:* Entom Soc Am; Sigma Xi. *Res:* Development of integrated pest management programs on row crops in Florida. *Mailing Add:* Rte 3 Box 4370 Quincy FL 32351

SPRESSER, DIANE MAR, b Welch, WVa, Dec 12, 43. GRAPH THEORY & ALGORITHMS. *Educ:* Radford Col, BS, 65; Univ Tenn, Knoxville, MA, 67; Univ Va, PhD(math sci & educ), 77. *Prof Exp:* From instr to assoc prof math, 67-80, actg head, dept math, 78-79, assoc prof math & comput sci, 80-82, PROF MATH & COMPUT SCI, JAMES MADISON UNIV, 82-, HEAD DEPT, 79- *Concurrent Pos:* Lectr, Vis Scientists Prog, Va Acad Sci, 74- & Asn Women in Math, 86; partic, Nat Identification Prog Advan of Women Higher Educ Admin, Am Coun Educ/Am Asn State Cols & Univs, 81-83. *Mem:* Am Math Soc; Asn Comput Mach; Math Asn Am; Asn Women in Math; Nat Coun Teachers Math. *Res:* Properties of graphs and analysis of related computer algorithms, with emphasis on time complexity; mathematical and computing education at the collegiate level. *Mailing Add:* Dept Math & Comput Sci James Madison Univ Harrisonburg VA 22807

SPRIGGS, ALFRED SAMUEL, b Houston, Tex, Aug 1, 22; m 49; c 4. ORGANIC CHEMISTRY. *Educ:* Dillard Univ, AB, 42; Howard Univ, MS, 44; Washington Univ, PhD(chem), 54. *Prof Exp:* Asst prof chem, Tenn Agr & Indust State Col, 47-51; prof, Lincoln Univ, Pa, 54-55; PROF CHEM & CHMN DEPT, CLARK COL, 55- *Mem:* Fel AAAS; Am Chem Soc; Sigma Xi. *Res:* Isotope tracers with carbon 14; carbohydrates; organic synthesis; radiochemistry; chromatography. *Mailing Add:* 4629 Boulder Park Dr SW Atlanta GA 30331

SPRIGGS, RICHARD MOORE, b Washington, Pa, May 8, 31; m 53; c 3. MATERIALS RESEARCH, CERAMIC ENGINEERING. *Educ:* Pa State Univ, BS, 52; Univ Ill, MS, 56, PhD(ceramic eng), 58. *Prof Exp:* Asst ceramic eng, Univ Ill, 54-56; sr res engr ceramics, Ferro Corp, Ohio, 58-59; sr scientist, Res & Advan Develop Div, Avco Corp, Mass, 59-60, staff scientist, 60-62, sr staff scientist & ceramics res group leader, 62-64; assoc dir, Mat Res Ctr & dir, Phys Ceramics Lab, Lehigh Univ, 64-70, from assoc prof to prof metall & mat sci, 64-80, admin asst to pres, 70-71, asst vpres admin, 71-72, vpres admin, 72-78; vis sr staff assoc, 79-80, sr staff officer/staff scientist, Nat Mat Adv Bd, 80-87, staff dir bd assessment NBS Progs, 84-87; JOHN FRANCIS MCMAHON PROF CERAMIC ENG & DIR, CTR ADV CERAMIC TECHNOL, NY STATE COL CERAMICS, ALFRED UNIV, 87-, DIR, SPONSORED RES ACTIV, 88- *Concurrent Pos:* Am Coun Educ fel, Lehigh Univ, 70-71; consult to var corps & govt labs; foreign mem, Serbian Acad Sci & Arts, 86-; consult prof, Univ Belgrade, Yugoslavia, 85- *Honors & Awards:*

Ross Coffin Purdy Award, Am Ceramic Soc, 67; Hobart M Kramer Award, Am Ceramic Soc, 80; Orton Mem Lectr & McMahon Mem Lectr, Am Ceramic Soc, 88. *Mem:* Fel Brit Inst Ceramics; fel Am Ceramic Soc (treas, 80-82, vpres, 82-83, pres-elect, 83-84, pres, 84-85); Nat Inst Ceramic Engrs; Brit Ceramic Soc; Int Inst Sci Sintering; Am Soc Eng Educ; Am Soc Testing & Mat; AAAS; NY Acad Sci. *Res:* Physical ceramics; materials science; correlations among processing, internal structure and physical and mechanical properties of dense polycrystalline refractory ceramic oxide systems; author or coauthor of over 100 technical articles. *Mailing Add:* Alfred Univ Ctr Adv Ceramic Technol Alfred NY 14802

SPRINCE, HERBERT, b Lewiston, Maine, Dec 18, 12; wid. BIOCHEMISTRY, PHARMACOLOGY. *Educ:* Bates Col, BS, 34; Harvard Univ, MA, 35, PhD(cellular physiol), 39. *Prof Exp:* Lab asst physiol, Harvard Univ, 35-38; Parker fel, Med Col, Cornell Univ, 39-40; asst, Huntington Mem Hosp, 41; asst, Harvard Univ, 41-42; res assoc, Mass Inst Technol, 42-43; Nutrit Found fel, Rockefeller Inst, 43-45; dir div nutrit & microbiol, Ortho Res Found, Johnson & Johnson, 45-52; dir res, Elizabeth Biochem Lab, NJ, 52-54; chief res biochem, Vet Admin Med Ctr, Coatesville, Pa, 54-84; assoc prof pharmacol & psyciat, Jefferson Med Col, 68-85; RETIRED. *Concurrent Pos:* Vis lectr, New York Med Col, 53-60; res assoc psychiat, Univ Pa, 55-68; asst prof biochem, Grad Sch Med, Univ Pa, 61-67; hon assoc prof pharmacol & psychiat, Jefferson Med Col, 85- *Mem:* AAAS; Am Chem Soc; Am Asn Clin Chem; Soc Biol Psychiat; Am Inst Nutrit; NY Acad Sci. *Res:* Nutritional biochemistry; behavioral pharmacology; indoles, amino acids in schizophrenia and vitamin c in alcoholism; strucutre-activity relationships in depressant and excitatory behavior; xanthurenic acid in toxemias of pregnancy; animal, bacterial and protozoal growth factors. *Mailing Add:* 450 Apple Dr Exton PA 19341-2169

SPRING, BONNIE JOAN, b Hackensack, NJ, Oct 9, 49. PSYCHIATRY, PHARMACOLOGY. *Educ:* Bucknell Univ, BA, 71; Harvard Univ, MA, 75, PhD(psychol), 77. *Prof Exp:* From asst prof to assoc prof psychol, Harvard Univ, 77-84; prof & dir psychol, Tex Tech Univ, Lubbock, 84-88; CLIN TRAINING PROF PSYCHOL, CHICAGO MED SCH, 88- *Concurrent Pos:* Vis lectr, dept nutrit & food sci, Mass Inst Technol, 79-84, lectr dept psychiat, Columbia Col, NY, 79-85; res assoc prof, Univ Md Sch Med, 84-; comt mem, Nat Needs Biomed & Behav Res Personnel, Inst Med, Nat Acad Sci, 84-86, Nat Plan Res Schizophrenia Treat Panel, Nat Inst Mental Health, 87; mem, Coun Univ Dir Clin Psychol; field ed N & S Am, Psychopharmacol, Human Exp Studies, 87. *Mem:* Fel Am Psychol Asn; Soc Biol Psychiat; Am Col Neuropharmacol; Am Psychopathol Asn; Soc Exp Psychopath; Soc Behav Med; AAAS; Sigma Xi. *Res:* State and trait aspects of brain-behavior relationships as manifested in the areas of psychopathology and health psychology; attempting to distinguish between disturbances of information processing that mark an enduring trait of vulnerability to schizophrenia, versus those that mark the episodic psychotic state; study the capacity of food constituents to produce transient, "drug-like" effects on the brain and behavior. *Mailing Add:* Dept Psychol Univ Health Sci Chicago Med Sch 3333 Green Bay Rd North Chicago IL 60064-3095

SPRING, JEFFREY H, b Galt, Ont, Mar 14, 50; m 91; c 1. INSECT ENDOCRINOLOGY, INSECT EXCRETORY PHYSIOLOGY. *Educ:* Univ Waterloo, Can, BSc, 73, MSc, 76; Univ BC, Can, PhD(zool), 79. *Prof Exp:* NATO postdoctoral zool, Univ Cambridge, Eng, 79-81; univ res fel zool, La Trobe Univ, Melbourne, Australia, 82; asst prof, 83-87, ASSOC PROF BIOL, UNIV SOUTHWESTERN LA, 87- *Mem:* AAAS; Am Soc Zoologists. *Res:* Salt and water balance in insects; structure of malpighian tubules and rectum; physiology of primary urine formation and reabsorptive processes; source, structure, release and function of neurohormones controlling diuresis; isolation and identification of neuropeptides. *Mailing Add:* Dept Biol Univ Southwestern La Box 42451 Lafayette LA 70504-2451

SPRING, RAY FREDERICK, b Cincinnati, Ohio, Mar 28, 25; m 49; c 2. MATHEMATICS. *Educ:* Univ Cincinnati, BS, 48; Univ Ill, MS, 52, PhD(math), 55. *Prof Exp:* Chem engr, US Playing Card Co, 48-50; asst math, Univ Ill, 52-54; from asst prof to assoc prof, 55-66, PROF MATH, OHIO UNIV, 66- *Mem:* Am Math Soc. *Res:* Modern abstract algebra; group and lattice theories; digital computer programming; characterization and classification of metabelian p-groups and other groups by means of their subgroup lattices. *Mailing Add:* PO Box 269 Athens OH 45701

SPRING, SUSAN B, b New York, NY, May 1, 43. VIROLOGY, IMMUNOCHEMISTRY. *Educ:* Univ Chicago, PhD(microbiol), 68. *Prof Exp:* Exec secy, Microbiol & Infectious Dis Res Comt, Nat Inst Allergy & Infectious Dis, 81-84, PROG DIR DNA VIRUS STUDIES I, BIOL CARCINOGENESIS BR, DIV CANCER ETIOLOGY, NAT CANCER INST, NIH, 84- *Mem:* Am Soc Immunol; Am Soc Virol; Am Soc Microbiol; Sigma Xi; NY Acad Sci; Tissue Cult Asn. *Mailing Add:* NIH NIAID DMID V8 Westwood Bldg Rm 736 Bethesda MD 20892

SPRINGBORN, ROBERT CARL, b Geneva, Ill, Oct 19, 29; m 51; c 2. POLYMER SCIENCE, BIORESEARCH. *Educ:* Univ Ill, BS, 51; Cornell Univ, PhD(org chem), 54. *Prof Exp:* Res chemist, Monsanto Chem Co, 54-58; tech dir, Marbon Chem Div, Borg-Warner Corp, 58-63; vpres & tech dir, Ohio Rubber Co, 63-65; gen mgr, Ionics, Inc, 65-67; vpres, W R Grace & Co, 67-69; chmn & pres, Gen Econ Corp, 69-71; CHMN & CHIEF EXEC OFFICER, SPRINGBORN GROUP, INC, 85- *Concurrent Pos:* Mem, White House Conf Small Bus Com, 78- *Mem:* Am Chem Soc; AAAS; Soc Plastics Engrs; Plastics Inst Am; Nat Asn Life Sci Industs. *Res:* Polymer science including polymeric synthesis and processing; medical and health sciences, particularly related to bioresearch. *Mailing Add:* 4320 Gulfshore Blvd N Suite 216 Naples FL 33940-2662

SPRINGER, ALAN DAVID, b Linz, Austria, Jan 6, 48; US citizen; m 69; c 2. NEUROSCIENCE. *Educ:* Brooklyn Col, BS, 69; City Univ New York, PhD(psychol), 73. *Prof Exp:* Scholar neurosci, Univ Mich, 73-77; asst prof physiol, Univ Ill Med Ctr, 77-79; assoc prof, 79-84, PROF ANAT, NY MED COL, 84- *Concurrent Pos:* Prin investr, Nat Inst Aging, NIH grant & NSF grant, 78-81 & Nat Eye Inst, NIH grant, 81-90. *Mem:* Soc Neurosci; Asn Res Vision & Ophthal; Am Asn Anatomists; NY Acad Sci; AAAS. *Res:* Vision and optic nerve regeneration in vertebrates, including conditions leading to abnormal and normal patterns of regeneration and the role of various brain structures in mediating vision; retinal development and retinal regeneration. *Mailing Add:* Dept Anat New York Med Col Valhalla NY 10595

SPRINGER, ALLAN MATTHEW, b Baraboo, Wis, Oct 2, 44; m 67. CHEMICAL ENGINEERING, PULP & PAPER TECHNOLOGY. *Educ:* Univ Wis-Madison, BS, 66; Lawrence Univ, MS, 69, PhD(chem eng), 72. *Prof Exp:* Process engr, Olin Mathieson Chem Corp, 67-68; res engr, Nat Coun Paper Indust Air & Stream Improvement, 72-76; from asst prof to assoc prof pulp & paper technol, 76-86, PROF PAPER SCI & ENG, MIAMI UNIV, 86- *Concurrent Pos:* Sr Fulbright lectr, Univ Pertanian Malaysia, 79-80; sr Fulbright prof, Univ Sao Paulo, Brazil, 85. *Honors & Awards:* Environ Div Award, Tech Asn Pulp & Paper Indust, 89. *Mem:* Tech Asn Pulp & Paper Indust; Am Inst Chem Engrs; Am Asn Environ Eng Prof; Sigma Xi; Int Asn Water Pollution Res & Control. *Res:* Water pollution abatement through process modification; wastewater treatment optimization; resource recovery and recycling in the pulp and paper industry. *Mailing Add:* Dept Paper Sci & Eng Miami Univ Oxford OH 45056

SPRINGER, BERNARD G, b New York, NY, Feb 26, 35; div; c 2. SCIENCE POLICY. *Educ:* Univ Chicago, BA, 54, MS, 57, PhD(physics), 64. *Prof Exp:* Res assoc solid state physics, Univ Chicago, 64; Nat Acad Sci vis res fel, Univ Tokyo, 64-65; asst prof, Univ Southern Calif, 66-69; mem staff, Boeing Co, 59-60 & Hughes Aircraft Co, 69-72; SR PHYS SCIENTIST, RAND CORP, 72- *Mem:* Am Phys Soc. *Mailing Add:* PO Box 18182 Encino CA 91416

SPRINGER, CHARLES EUGENE, b Storm Lake, Iowa, Oct 25, 03; m 30; c 1. MATHEMATICS. *Educ:* Univ Okla, AB, 25, AM, 26; Oxford Univ, BSc, 40; Univ Chicago, PhD(math), 38. *Prof Exp:* Instr math, Univ Okla, 26-27 & Iowa State Col, 30; from instr to prof, Univ Okla, 30-61, chmn, Dept Math, 46-55, form David Boyd prof to David Ross Boyd emer prof math, 61-77; RETIRED. *Concurrent Pos:* Chmn dept math, Oklahoma City Univ, 70-72. *Res:* Differential geometry; dual geodesics on a surface; metric geometry of surfaces by use of tensor analysis and in four-dimensional space; union curves and curvature. *Mailing Add:* 1617 Jenkins St Norman OK 73072

SPRINGER, CHARLES SINCLAIR, JR, b Houston, Tex, Nov 2, 40; m 63; c 2. BIOPHYSICAL CHEMISTRY. *Educ:* St Louis Univ, BS, 62; Ohio State Univ, MSc, 64, PhD(chem), 67. *Prof Exp:* Res chemist, Aerospace Res Labs, 65-68; from asst prof to assoc prof, 68-85, PROF CHEM, STATE UNIV NY STONY BROOK, 85- *Concurrent Pos:* Vis assoc, Calif Inst Technol, 76-77; vis assoc prof, Med Sch, Harvard Univ, 83-84. *Honors & Awards:* US Air Force Res & Develop Award, 67. *Mem:* AAAS; Am Chem Soc; NY Acad Sci; Biophys Soc; Int Soc Magnetic Resonance; Soc Magnetic Resonance in Med; Sigma Xi. *Res:* Nuclear magnetic resonance and electron paramagnetic resonance studies of living systems and biological membranes; physical properties, and ionophore- and protein- catalyzed metal ion membrane transport, metal ion binding to membrane surfaces, physical chemistry of micelle and inverse micelle solutions. *Mailing Add:* 296 Sheep Pasture Rd East Setauket NY 11733

SPRINGER, DONALD LEE, b Hampton, Iowa, Mar 15, 33; m 55; c 3. SEISMOLOGY, PHYSICS. *Educ:* Univ Calif, Santa Barbara, BA, 56. *Prof Exp:* PHYSICIST SEISMOL, LAWRENCE LIVERMORE NAT LAB, 56- *Concurrent Pos:* Mem ground shock tech working group, Canal Studies, AEC, 66-70, mem ground shock subcomt, 69-70; mem seismic rev panel, US Air Force Tech Appl Ctr, 74-79 & Off Sci & Technol Policy, White House, 77-79; mem, US deleg, Ad Hoc Group Sci Experts Comt Disarmament, UN, 80- *Mem:* Seismol Soc Am. *Res:* Observational seismology; explosion seismology; geophysics; earth structure; earthquake prediction; seismic energy. *Mailing Add:* 5271 Irene Way Livermore CA 94550

SPRINGER, DWIGHT SYLVAN, b Harrisburg, Pa, Oct 8, 43; m 64; c 3. CHEMICAL ENGINEERING, CHEMISTRY EDUCATION. *Educ:* Univ Del, BChE, 65; Univ Minn, PhD(chem eng), 71; Long Island Univ, MS, 90. *Prof Exp:* Chem engr power supplies, Harry Diamond Labs, US Army, Washington, DC, 72-74; instr, US Mil Acad, 76-77, asst prof chem, 77-79; chem officer, US Army, Berlin, 80-81; ASSOC PROF CHEM, US MIL ACAD, 81- *Concurrent Pos:* Vis scholar chem, Stanford Univ, 90-91. *Honors & Awards:* Herbert W Alden Award, Am Defense Preparedness Asn, 75. *Mem:* Am Inst Chem Engrs; Am Chem Soc; Am Soc Eng Educ. *Res:* Chemical and conventional ammunition; chemical defense material; phosphorus and sulfur chemistry; lasers in chemical education. *Mailing Add:* Dept Chem US Mil Acad West Point NY 10996-1785

SPRINGER, EDWARD L(ESTER), b Baraboo, Wis, July 12, 31; m 61; c 2. WOOD HYDROLYSIS, WOOD STORAGE. *Educ:* Univ Wis, BS, 53, MS, 58, PhD(chem eng), 61. *Prof Exp:* Chem engr, Kimberly-Clark Corp, Wis, 55-56; CHEM ENGR, FOREST PROD LAB, USDA, 58- *Concurrent Pos:* Fulbright fel, Finland, 61-62. *Mem:* Tech Asn Pulp & Paper Indust; Am Chem Soc. *Res:* Wood preservation; preservation of pulp chips; kinetics of wood hydrolysis and of the delignification of wood; nonconventional pulping and pulp bleaching. *Mailing Add:* Forest Prod Lab USDA Forest Serv 1 Gifford Pinchot Dr Madison WI 53705-2398

SPRINGER, GEORG F, b Berlin, Ger, Mar 1, 24; nat US; m 51; c 3. IMMUNOCHEMISTRY. *Educ:* Univ Heidelberg, MA, 47; Univ Basel, MD, 51. *Prof Exp:* Res fel pediat, Sch Med, Univ Pa, 53, Woodward fel physiol chem, 52-53, asst instr path, Sch Med, 52-55, assoc clin path, 55-58, asst prof immunol, 56-61, assoc prof, 61-62, mem, Pepper Lab, 55-62; PROF MICROBIOL & IMMUNOL, MED SCH, NORTHWESTERN UNIV, EVANSTON, 63-, DIR IMMUNOCHEM RES, EVANSTON HOSP, 63- *Concurrent Pos:* Mem germ free res unit, Walter Reed Army Med Ctr, 54-55; in-chg blood bank & serol, Philadelphia Gen & Univ Hosps; ed various sci jour; Am Heart Asn estab investr, 58-63; John G Gibson, II lect, 66; mem, Northwestern Univ Cancer Ctr, Evanston Hosp Res & Educ Comt & Protection of Human Subjects Comt; mem med adv bd, Leukemia Res Found, Inc. *Honors & Awards:* Oehlecker Prize, Ger Soc Blood Transfusion, 66. *Mem:* AAAS; Am Soc Microbiol; Am Chem Soc; Am Heart Asn; NY Acad Sci. *Res:* Immunochemistry of blood-group active substances; carbohydrate chemistry; virus action on blood groups; immunology of human breast cancer; infectious mononucleosis; shock; physical chemistry of antigen-antibody interactions; tumor virus receptors. *Mailing Add:* Sch Med Univ Health Sci Med 333 Green Bay Rd North Chicago IL 60064

SPRINGER, GEORGE, b Cleveland, Ohio, Sept 3, 24; m 50; c 3. MATHEMATICS. *Educ:* Case Inst, BS, 45; Brown Univ, MS, 46; Harvard Univ, PhD(math), 49. *Prof Exp:* Moore instr math, Mass Inst Technol, 49-51; asst prof, Northwestern Univ, 51-54; vis prof & Fulbright lectr, Univ Münster, 54-55; from assoc prof to prof, Univ Kans, 55-64; assoc dean res & develop, 73-80, actg dean res & grad develop, 80-82, PROF MATH, IND UNIV, BLOOMINGTON, 64-, PROF COMPUT SCI, 87- *Concurrent Pos:* Vis prof, Univ Sao Paulo, 61; vis prof & Fulbright lectr, Univ Würzburg, 61-62; ed, J Math & Mech, Ind Univ, 65-; vis prof, Imp Col, Univ London, 71-72; consult ed, McGraw Hill Book Co, 71-; prog dir math sci sect, NSF, Washington, DC, 78-79. *Mem:* Am Math Soc; Math Asn Am; Asn Comput Mach. *Res:* Theory of functions of one and several complex variables; harmonic functions; conformal and quasiconformal mapping; programming languages. *Mailing Add:* Computer Sci Dept Lindley Hall Ind Univ Bloomington IN 47405

SPRINGER, GEORGE HENRY, b Bristol, RI, Jan 16, 18; m 41; c 1. GEOLOGY. *Educ:* Brown Univ, AB, 38, ScM, 40. *Prof Exp:* Geologist, Tenn Valley Authority, 41; from instr to assoc prof, 46-56, DISTINGUISHED SERV PROF GEOL, UNIV DAYTON, 84- *Mem:* AAAS; Nat Asn Geol Teachers; Int Glaciol Soc; Sigma Xi. *Res:* Structural geology; petrography. *Mailing Add:* 2373 Shelterwood Dr Dayton OH 45409

SPRINGER, GEORGE S, b Budapest, Hungary, Dec 12, 33; US citizen; m 63; c 2. AERONAUTICAL & ASTRONAUTICAL ENGINEERING. *Educ:* Univ Sydney, BE, 59; Yale Univ, MEng, 60, MS, 61, PhD(mech eng), 62. *Prof Exp:* Ford Found fel & instr mech eng, Mass Inst Technol, 62-63, asst prof, 63-67; from assoc prof to prof mech eng, UnivMich, Ann Arbor, 67-83; PROF DEPT AERONAUT & ASTRONAUT, STANFORD UNIV, 83-, CHMN DEPT, 90- *Honors & Awards:* Ralph E Teetor Award, Soc Automotive Engrs, 78; Pub Serv Achievement Award, NASA, 88. *Mem:* fel Am Soc Mech Engrs; Am Phys Soc; fel Am Inst Aeronaut & Astronaut; Soc Automotive Engrs; Soc Adv Mech & Process Eng. *Res:* Composite materials. *Mailing Add:* Dept Aeronaut & Astronaut Stanford Univ Stanford CA 94305

SPRINGER, JOHN KENNETH, b Trenton, NJ, Mar 26, 29; m 60; c 2. PLANT PATHOLOGY, NEMATOLOGY. *Educ:* Rutgers Univ, BS, 61, MS, 63, PhD(plant path), 66. *Prof Exp:* Supvr plant pest survs, NJ Dept Agr, 59-63; exten assoc, 63-66, asst exten specialist, 66-70, assoc exten specialist, 70-75, EXTEN SPECIALIST PLANT PATH, RUTGERS UNIV, 75- *Mem:* Am Phytopath Soc; Soc Nematol. *Res:* Survey of plant parasitic nematodes; mechanisms involved in Verticillium wilt syndrome; effect of soilborne diseases on production of crops. *Mailing Add:* Rutgers Res & Develop Ctr RD No 5 Box 232 Bridgeton NJ 08302-9499

SPRINGER, JOHN MERVIN, b Peoria, Ill, Apr 19, 41; m 66; c 3. CHEMICAL PHYSICS. *Educ:* Knox Col, BA, 63; Vanderbilt Univ, MS, 65, PhD(physics), 72. *Prof Exp:* Res assoc chem physics, 71-73, asst prof, 73-77, res assoc physics, 77-80, ASSOC PROF PHYSICS, FISK UNIV, 81- *Mem:* Sigma Xi; Am Asn Physics Teachers. *Res:* Crystal structure determinations via infrared and Raman spectroscopy; optical analysis of crystal defects. *Mailing Add:* 815 Kendall Dr Nashville TN 37209

SPRINGER, KARL JOSEPH, b San Antonio, Tex, Apr 14, 35; m 57; c 3. MECHANICAL ENGINEERING. *Educ:* Tex A&M Univ, BS, 57; Trinity Univ, MS, 66; Am Acad Engrs, dipl. *Prof Exp:* Res engr auto engines, Southwest Res Inst, 57-58; proj engr jet engines, Wright Air Develop Ctr, US Air Force, 58-60; field engr, E I du Pont de Nemours & Co, Inc, 60-62; proj engr, Automotive Res Assocs, 62-63; sr engr, US Army Fuels & Lubricants Res Lab, 63-67, mgr emissions res lab, Dept Automotive Res, 67-72, asst dir automotive res, 72-74, DIR DEPT EMISSIONS RES, ENGINES, EMISSIONS & VEHICLES RES DIV, SOUTHWEST RES INST, 74- *Concurrent Pos:* Mem odor & particulate subpanels, Diesel Emission Comt, Coord Res Coun, 68- *Mem:* Fel Am Soc Mech Engrs; Soc Automotive Engrs; Sigma Xi; Combustion Inst. *Res:* Emissions from diesel and gasoline vehicles; control of emissions from diesels and measurement of combustion odor. *Mailing Add:* 111 Shalimar Dr San Antonio TX 78213

SPRINGER, MARTHA EDITH, b Mountain View, Calif, Jan 24, 16. BIOLOGY. *Educ:* Stanford Univ, AB, 35, AM, 36; Univ Mich, PhD(bot), 44. *Prof Exp:* Teacher high sch, Calif, 36-40 & 41-42; instr bot & cur herbarium, Ind Univ, 44-45 & 46-47; instr bot, Conn Col, 45-46; assoc prof biol, Williamette Univ, 47-53, actg chmn dept, 48-50, prof, 53-81, cur, Peck Herbarium, 67-81, EMER PROF BIOL, WILLAMETTE UNIV, 81- *Mem:* AAAS; Bot Soc Am; Mycol Soc Am; Sigma Xi. *Res:* Taxonomy of aquatic phycomycetes, flowering plants and bryophytes; a morphologic and taxonomic study of the genus Monoblepharella. *Mailing Add:* PO Box 5000 Apt 521 Salem OR 97304

SPRINGER, MAXWELL ELSWORTH, b Bourbon, Mo, Oct 21, 13; m 53; c 3. SOIL MORPHOLOGY. *Educ:* Univ Mo, BS, 35, AM, 46; Univ Calif, Berkeley, PhD(soils), 53. *Prof Exp:* Asst agr econ, Univ Mo, 36-37, asst soils, 37-40, instr, 40-42 & 46-49, asst prof, 53-57; assoc prof agron, Univ Tenn, Knoxville, 57-67, prof plant & soil sci, 67-79; CONSULT, 79- *Concurrent Pos:* Soil surv specialist, Natural Resources Sect, Gen Hq, Supreme Comdr Allied Powers, Tokyo, 46-47; Fulbright Award, Univ Ghent, 66-67. *Mem:* Fel AAAS; Am Soc Agron. *Res:* Soil formation and classification; physical, chemical and mineralogical studies of soils. *Mailing Add:* 1600 Autry Way Knoxville TN 37909

SPRINGER, MELVIN DALE, b Saybrook, Ill, Sept 12, 18; m 48; c 1. MATHEMATICAL STATISTICS. *Educ:* Univ Ill, BS, 40, MS, 41, PhD(math statist), 47. *Prof Exp:* Asst math, Univ Ill, 41-44 & 46-47, instr, 47-48; asst prof, Mich State Col, 48-50; math statistician, Res Dept, US Naval Ord Plant, Ind, 50-56; sr opers analyst, Tech Opers, Inc, Va, 56-59; sr res statistician, Defense Res Labs, Gen Motors Corp, 59-67, dir reliability res & educ, A C Electronics Div, 67-68; prof, 68-84, EMER PROF INDUST ENG, UNIV ARK, FAYETTEVILLE, 84- *Mem:* Am Statist Asn; Math Asn Am. *Res:* Reliability theory and analysis; Bayesian statistics; experimental design; sampling theory; integral transforms in stochastic models; algebra of random variables. *Mailing Add:* Dept Indust Eng Univ Ark Fayetteville AR 72701

SPRINGER, PAUL FREDERICK, b Chicago, Ill, Apr 25, 22; m 49; c 4. WILDLIFE RESEARCH. *Educ:* Univ Ill, AB, 43; Univ Wis, MS, 48; Cornell Univ, PhD(wildlife conserv), 61. *Prof Exp:* Waterfowl res biologist, State Natural Hist Surv, Ill, 47-48; wildlife res biologist, US Fish & Wildlife Serv, 48-58, chief sect wetland ecol, Patuxent Wildlife Res Ctr, 58-63, leader, SDak Coop Wildlife Res Unit, 63-67, asst dir, Northern Prairie Wildlife Res Ctr, US Bur Sport Fisheries & Wildlife, 67-72, BIOLOGIST-IN-CHG, WILDLIFE RES FIELD STA, US FISH & WILDLIFE SERV, 73- *Concurrent Pos:* Mem comt agr pests, Agr Bd, Nat Res Coun, 56-58; secy, Nat Mosquito Control-Fish & Wildlife Mgt Coord Comt, 60-63, mem, 72-73; vpres, Raptor Res Found, 67-68; mem vector control comt, Water Resources Coun, 72-73; adj prof wildlife mgt, Humboldt State Univ, 73- *Mem:* Wildlife Soc; Am Ornithologists Union; Cooper Ornith Soc; Ecol Soc Am; Wilson Ornith Soc. *Res:* Waterfowl and wetland ecology and management; effects of mosquito control and chemical pesticides on wildlife; wildlife-estuarine relationships. *Mailing Add:* 1610 Panorama Dr Arcata CA 95521

SPRINGER, ROBERT HAROLD, b Downsville, Wis, Nov 7, 32; m 54; c 7. GASEOUS ELECTRONICS. *Educ:* Univ Minn, BS, 58, MS, 60, PhD(elec eng), 65. *Prof Exp:* RES PHYSICIST, LIGHTING RES & TECH SERV OPER, GEN ELEC CO, 65- *Mem:* Am Phys Soc. *Res:* All aspects of electrical discharges in gases related to light production; specializing in electrodes. *Mailing Add:* 6524 Duneden Ave Cleveland OH 44139

SPRINGER, TIMOTHY ALAN, LEUKOCYTE ADHESION. *Educ:* Harvard Univ, PhD(biochem & molecular biol), 76. *Prof Exp:* ASSOC PROF PATH & CHIEF, MEMBRANE IMMUNOCHEM LAB, SCH MED, HARVARD UNIV, 83- *Mailing Add:* Dept Pathol Harvard Med Sch Ctr Blood Res 800 Huntington Ave Boston MA 02115

SPRINGER, VICTOR GRUSCHKA, b Jacksonville, Fla, June 2, 28; m 65; c 2. BIOLOGY, MARINE SCIENCES. *Educ:* Emory Univ, AB, 48; Univ Miami, MS, 54; Univ Tex, PhD(vert zool), 57. *Prof Exp:* Ichthyologist, Marine Lab, State Bd Conserv, Fla, 57-61; res assoc, 61-62, assoc cur, 63-66, supvr, 70-71 & 84-86, CUR, DIV FISHES, NAT MUS NATURAL HIST, SMITHSONIAN INST, 67- *Concurrent Pos:* Ed, Proc Biol Soc Wash, 65-67; res assoc, Moore Lab, Occidental Col, 71-72 & Bishop Mus, 84-; Nat Geog Soc grant, 73-74; Max & Victoria Dreyfus Found Grant, 80, 82 & 86; bd dirs, Nat Aquarium Baltimore, 79-85. *Honors & Awards:* Stoye Award, 57. *Mem:* Fel AAAS; Am Soc Ichthyologists & Herpetologists (treas, 65-67); Soc Syst Zool (treas, 78-80); Sigma Xi. *Res:* Systematics; zoogeography; ecology; life histories of tropical marine fishes. *Mailing Add:* Div Fishes US Nat Mus Natural Hist Washington DC 20560

SPRINGER, WAYNE RICHARD, b Milwaukee, Wis, Nov 16, 46; m 72; c 2. ADHESION MECHANISMS. *Educ:* Northwestern Univ, BA, 68; Univ Calif, Berkeley, PhD(biochem), 77. *Prof Exp:* Fel, 77-79, RES BIOCHEMIST, UNIV CALIF, SAN DIEGO, 79-; RES BIOCHEMIST, 79-, RES SAFETY MGR, VET ADMIN MED CTR, SAN DIEGO, 88- *Mem:* Am Soc Biochem & Molecular Biol; Am Soc Cell Biol. *Res:* Mechanisms of cell to cell and cell to substrate adhesion; using the cellular slime mold as a model system. *Mailing Add:* Vet Admin Med Ctr 151 3350 La Jolla Village Dr San Diego CA 92161

SPRINGETT, BRIAN E, b Chatham, Eng, Apr 24, 36; m 63; c 2. PHYSICS, MATERIALS SCIENCE ENGINEERING. *Educ:* Cambridge Univ, BA, 60, MA, 64; Univ Chicago, MS, 63, PhD(physics), 66. *Prof Exp:* Res assoc physics, Univ Chicago, 66-67; asst prof, Univ Mich, Ann Arbor, 67-72; vis prof, Univ Quebec, 72-73 & Oakland Univ, 73-74; scientist, 74-77, TECH MGR, XEROX CORP, 77- *Mem:* AAAS; Am Phys Soc; Soc Photographic Scientists & Engrs. *Res:* Low temperature physics, gas discharges, ion and electron transport in dielectric media, amorphous photoconductors; xerographic sciences,; imaging and printing technologies; research administration. *Mailing Add:* Xerox Corp Bldg 103 800 Phillips Rd Webster NY 14580

SPRINGETT, DAVID ROY, b London, Ont, Apr 24, 35; m 58; c 4. MECHANICAL ENGINEERING. *Educ:* Univ Toronto, BASc, 58; Queen's Univ, Ont, MS, 62, PhD(mech eng), 64; Harvard Univ, dipl bus, 70. *Prof Exp:* Instr process control, Dept Mech Eng, Queen's Univ, Ont, 61-63; res engr, Burrough's Corp, 63-64; sr develop engr, Xerographic Systs, Explor Develop Dept, Xerox Corp, 64-65, staff asst to vpres eng, Off Prod Develop Dept, 65-66, mgr planning & admin, Bus Prod Div, 66-68, mgr div planning, Bus Prod & Systs Div, 68-69, prog mgr, Advan Develop Dept, 69-71, mgr advan copier develop, Info Technol Group, 71-73, mgr, Prod Technol Prog Off, 73-75; dir, Advan Bus Concepts Div, 75-77, dir, Major Progs Div, Rank Xerox Corp, 77-90; PRES, STRATEGIC MKT ASSOC, 90- *Concurrent Pos:* Fel, Ont Res Found, Queen's Univ, Ont, 61-62, Nat Res Coun Can, 61-63. *Res:* Engineering management; nonlinear control systems; systems engineering and simulation. *Mailing Add:* Strategic Mkt Assoc 2785 Pacific Coast Hwy Suite 251 Torrence CA 90505

SPRINGFIELD, HARRY WAYNE, b Dayton, Ohio, Sept 24, 20; m 50. RANGE ECOLOGY. *Educ:* Univ NMex, BS, 42; Univ Ariz, MS, 49; Agr & Mech Col, Tex, PhD(range mgt), 59. *Prof Exp:* Range conservationist, US Forest Serv, 47-52; agrostologist, Foreign Agr Serv, Iraq, 52-54; range scientist, US Forest Serv, 54-75; RETIRED. *Mem:* Soc Range Mgt; Am Soc Agron. *Res:* Ecological studies; game forage revegetation; germination characteristics of shrub seeds; mulching to establish shrub seedlings; mine spoil reclamation. *Mailing Add:* 13822 108th Dr Sun City AZ 85351

SPRINGGATE, CLARK FRANKLIN, biochemistry, for more information see previous edition

SPRINKLE, JAMES (THOMAS), b Arlington, Mass, Sept 2, 43; m 68; c 2. INVERTEBRATE PALEONTOLOGY. *Educ:* Mass Inst Technol, SB, 65; Harvard Univ, MA, 66, PhD(geol), 71. *Prof Exp:* Nat Res Coun-US Geol Surv assoc, Paleont & Stratig Br, US Geol Surv, Denver, 70-71; from asst prof to assoc prof, 71-83, PROF GEOL, UNIV TEX, AUSTIN, 83- *Concurrent Pos:* Prin investr, NSF grant, 77-80, 89-91. *Honors & Awards:* Schuchert Award, Paleont Soc, 82. *Mem:* AAAS; Paleont Soc; Geol Soc Am; Palaeont Asn England; Soc Syst Zoologists; Soc Econ Paleont & Min. *Res:* Primitive echinoderms; blastoids; Paleozoic stratigraphy and invertebrate paleontology; echinoderm biology and evolution. *Mailing Add:* Dept Geol Sci Univ Tex Austin TX 78713-7909

SPRINKLE, JAMES KENT, JR, b Cambridge, MA, Nov 1, 52; m 82; c 2. APPLIED NUCLEAR PHYSICS. *Educ:* State Univ NY, BS, 74; Univ Rochester, 76, MS, 77. *Prof Exp:* Res assoc, Argonne Nat Lab, 77-78; STAFF MEM, LOS ALAMOS NAT LAB, 78- *Mem:* Am Phys Soc. *Res:* Develop instrument based on nuclear radiation which determine the quantity (mass) of radioactive isoptope in various containers and matrics; instrument uranium and plutonium. *Mailing Add:* Los Alamos Nat Lab MS E540 Group Q-1 Los Alamos NM 87545

SPRINKLE, PHILIP MARTIN, b Greensboro, NC, Aug 5, 26; m 55; c 2. OTOLARYNGOLOGY. *Educ:* Univ Va, MD, 53. *Prof Exp:* Intern, Virginia Mason Hosp, Seattle, Wash, 53-54; pvt pract, Va, 54-60; resident gen surg, Watts Hosp, Durham, NC, 60-61; resident otolaryngol, Hosp, Univ Va, 61-64, asst prof, 64-65; assoc prof, 65-68, PROF OTOLARYNGOL & CHMN DEPT, MED CTR, W VA UNIV, 68- *Concurrent Pos:* Physician consult, Vet Admin Hosp & WVa Rehabil Ctr, 69- *Honors & Awards:* Prof Dr Ignacio Barroquer Mem Award. *Mem:* AMA; Am Acad Gen Pract; Am Acad Ophthal & Otolaryngol; Am Col Surgeons; Royal Soc Med; Sigma Xi. *Mailing Add:* RR 6 Box 10 Martinsville VA 24112-8806

SPRINKLE, ROBERT SHIELDS, III, b Martinsville, Va, Apr, 8, 35; m 61; c 3. ORGANOLEPTIC EVALUATION. *Educ:* Emory and Henry Col, BS, 57. *Prof Exp:* Chemist, 57-63, mgr prod develop, 63-65, supvr new prod div, 65-68, coordr res & develop, 68-76, dep dir res & develop, 76-78, dir, 78-80, vpres res & develop, 80-86, SR VPRES, RES & DEVELOP, AM TOBACCO CO, 86- *Mem:* Am Chem Soc. *Res:* Chemistry and composition of tobacco and tobacco smoke; pyrolytic products of combustion; applications of radioactive assay techniques for identification of particulates and gas phase constituents of tobacco smoke; spectroscopy; gas and liquid phase chromatography. *Mailing Add:* 9217 Groomfield Rd Richmond VA 23236

SPRINSON, DAVID BENJAMIN, b Russia, Apr 5, 10; nat US; m 43; c 3. BIOCHEMISTRY. *Educ:* City Col New York, BS, 31; NY Univ, MS, 36; Columbia Univ, PhD(biochem), 46. *Hon Degrees:* ScD, Columbia Univ, 91. *Prof Exp:* Asst thyroid biochem, Chem Lab, Montefiore Hosp, 31-42; from res assoc to prof, 48-78, EMER PROF BIOCHEM & MOLECULAR BIOPHYSICS, COL PHYSICIANS & SURGEONS, COLUMBIA UNIV, 78-; BIOCHEMIST, DEPT MED, ROOSEVELT HOSP, NY, 79- *Concurrent Pos:* Fulbright fel, Univ Paris, 52; Guggenheim fels, Stanford Univ, 57 & Univ Oxford, 60-61; career investr, Am Heart Asn, 58-75; vis scientist, NIH, 65; Brown-Hazen lect, NY State Dept Health, Albany, 69. *Mem:* Nat Acad Sci; Am Soc Biochem & Molecular Biol; Brit Biochem Soc; Am Chem Soc. *Res:* Intermediary metabolism of amino acids; biosynthesis of methyl groups and purines, aromatic compounds and sterols; regulation of metabolic pathways; mechanism of enzymic reactions. *Mailing Add:* St Luke's Roosevelt Hosp Ctr 428 W 59 St New York NY 10019

SPRINZ, HELMUTH, b Berlin, Ger, May 29, 11; nat US; m 59. PATHOLOGY. *Educ:* Univ Berlin, Dr med, 36. *Prof Exp:* Chief lab serv, 98th Gen Hosp, Med Corps, US Army, 49-53, chief path sect, Walter Reed Army Hosp, 53-59, dir div exp path, Walter Reed Army Inst Res, 59-71; prof, Med Sch, Univ Mo-Kansas City, 71-75; dir prof affairs, Kansas City Gen Hosp & Med Ctr, 71-73; PROF, UNIV KANS MED CTR, 76- *Concurrent Pos:* Consult, Midwest Res Inst, Kansas City, 77- *Honors & Awards:* Walter Reed Medal; Surgeon Gen's Medal; Stitt Award, Asn Mil Surg US. *Mem:* Am Asn Path; Int Acad Path; Am Col Physicians; Am Gastroenterol Asn; Sigma Xi. *Res:* Pathology and pathogenesis of infections; gastrointestinal diseases; general experimental and neuropathology of intoxications. *Mailing Add:* Dept Path WHW227 Univ Kans Col Med 39th St & Rainbow Blvd Kansas City KS 66103

SPRITZ, NORTON, b Baltimore, Md, June 19, 28; c 1. BIOCHEMISTRY. *Educ:* Johns Hopkins Univ, AB, 48; Univ Md, MD, 52. *Prof Exp:* Asst med, Med Col, Cornell Univ, 52-54, from instr to assoc prof, 56-66; assoc prof, Rockefeller Univ, 66-69; PROF MED, NY UNIV, 69- *Concurrent Pos:* Intern, 2nd Cornell Med Div, Bellevue Hosp, 52-53, asst res, 53-54, fel cardiol, 56-57, chief res, 57-58, asst vis physician, 58-63, attend cardiorenal lab, 58-60, dir lipid metab lab, 63-66, vis physician, 64-; estab investr, Health Res Coun New York, 59-; clin asst, Mem Hosp, 60-; asst vis physician, James Ewing Hosp, 60-; asst attend, NY Hosp, 60-65, assoc attend, 65-; guest investr & asst physician, Rockefeller Univ, 61-63, assoc physician, 66-; chief med, NY Vet Admin Hosp, 69-; chief med serv, Manhattan Vet Hosp, 69- *Mem:* Am Soc Clin Invest; Am Fedn Clin Res; Am Diabetes Asn. *Res:* Lipid metabolism as related to human disorders and particularly atherosclerosis. *Mailing Add:* Manhattan Vet Admin Hosp First Ave & 24th St New York NY 10010

SPRITZ, RICHARD ANDREW, b Philadelphia, Pa, Dec 19, 50; div. MEDICAL GENETICS, PEDIATRICS. *Educ:* Univ Wis, BS, 72; Pa State Univ, MD, 76. *Prof Exp:* Intern pediat, Children's Hosp Philadelphia, Univ Pa, 76-77, resident, 77-78; fel, Dept Human Genetics, Sch Med, Yale Univ, 78-80; asst prof, 81-86, ASSOC PROF MED GENETICS & PEDIAT, SCH MED, UNIV WIS, 86- *Mem:* Am Soc Human Genetics; AAAS. *Res:* Molecular aspects of the structure, organization and control of human genes; molecular mechanisms of RNA processing; the molecular basis of human genetic disorders. *Mailing Add:* Dept Pediat Univ Wis Clin Sci Ctr 600 Highland Ave Madison WI 53792

SPRITZER, ALBERT A, b Brooklyn, NY, Apr 2, 27; m 53; c 3. MEDICINE. *Educ:* Col Wooster, BS, 48; Albany Med Col, MD, 52; Univ Pittsburgh, MPH, 56. *Prof Exp:* ASST PROF OCCUP HEALTH GRAD SCH PUB HEALTH, UNIV PITTSBURGH, 57-, PROF RADIATION HEALTH, 73- *Concurrent Pos:* Dept Health, Educ & Welfare res grant, 65-; consult, Babcox & Wilcox, Duquesne Light; med dir, Nuclear Energy Systs, Westinghouse Elec Corp, 70- *Mem:* Am Indust Hyg Asn; Health Physics Soc; AMA; Am Occup Health Asn. *Res:* Radiation biology; industrial radiation health practice; occupational health and radiation health research in pulmonary clearance; physiology and radiation hazard evaluation. *Mailing Add:* 9 Churchill Rd Pittsburgh PA 15235

SPRITZER, MICHAEL STEPHEN, b New York, NY, July 15, 39; m 64; c 2. ANALYTICAL CHEMISTRY. *Educ:* Polytech Inst Brooklyn, BS, 60; Univ Mich, MS, 62, PhD(chem), 65. *Prof Exp:* Instr, Univ Mich, 65-66; asst prof, 66-77, PROF CHEM, VILLANOVA UNIV, 77- *Mem:* AAAS; Am Chem Soc; Sigma Xi. *Res:* Electrochemical analysis; electrochemistry in nonaqueous media; organic polarography and voltammetry; electrochemical and photoelectrochemical energy storage. *Mailing Add:* Dept Chem Villanova Univ Villanova PA 19085

SPROKEL, GERARD J, b Valkenburg, Netherlands, Aug 14, 21; US citizen; m 49. PHYSICAL CHEMISTRY. *Educ:* State Univ Utrecht, PhD(phys chem), 52. *Prof Exp:* Res chemist, Am Viscose Corp, Pa, 54-58; adv chemist, Int Bus Mach Corp, 58-65, adv solid state, 65-70, adv, Components Div, 70-74, MEM RES STAFF, RES DIV, IBM CORP, 74- *Mem:* AAAS; Am Chem Soc; Electrochem Soc; Inst Elec & Electronics Engrs. *Res:* Diffusion and surface properties in semiconductors; semiconducting and scintillation counters; injection lasers; liquid crystals; materials research. *Mailing Add:* 2831 Castle Dr San Jose CA 95125

SPROTT, DAVID ARTHUR, b Toronto, Ont, May 31, 30. MATHEMATICAL STATISTICS. *Educ:* Univ Toronto, BA, 52, MA, 53, PhD, 55. *Prof Exp:* Asst, Comput Ctr, Univ Toronto, 52-53, Defence Res Bd, 54 & Galton Lab, Eng, 55-56; assoc prof, 58-70, chmn dept & dean fac, 67-72, PROF MATH, UNIV WATERLOO, 70- *Mem:* Am Math Soc; Inst Math Statist; Math Asn Am. *Res:* Mathematical genetics; experimental design; statistical inference. *Mailing Add:* Dept Statist Univ of Waterloo Waterloo ON N2L 3G1 Can

SPROTT, GORDON DENNIS, b Badjeros, Ont, Feb 27, 45; m 68; c 2. BIOCHEMISTRY. *Educ:* Univ Guelph, BS, 68, MS, 70; McGill Univ, PhD(microbiol), 73. *Prof Exp:* Fel, 73-75, asst res officer, 75-81, SR RES OFFICER, NAT RES COUN CAN, 88- *Concurrent Pos:* Sect ed, Can J Microbiol, 86-89; adj prof, Ottawa Univ, Univ Guelph. *Honors & Awards:* Can Soc Microbiol Award, 89. *Mem:* Am Soc Microbiol; Can Soc Microbiol. *Res:* Physiology of methanogenic bacteria, including measurement of electrical and chemical potentials, energetics of ion transport, cell permeability, structure of ether-linked membrane lipids; metabolic pathways. *Mailing Add:* Inst Biol Sci 100 Sussex Dr Nat Res Coun Ottawa ON K1A 0R6 Can

SPROTT, JULIEN CLINTON, b Memphis, Tenn, Sept 16, 42; m 65. PLASMA PHYSICS. *Educ:* Mass Inst Technol, BS, 64; Univ Wis, MS, 66, PhD(physics), 69. *Prof Exp:* Lectr elec eng & proj assoc physics, Univ Wis-Madison, 69-70; physicist, Thermonuclear Div, Oak Ridge Nat Lab, 70-72; from asst prof to assoc prof, 72-79, PROF PHYSICS, UNIV WIS-MADISON, 79- *Concurrent Pos:* Consult, Oak Ridge Nat Lab, 72, McDonnell Douglas Corp, 77-80, Elec Power Res Inst, 78, TRW, 78-79, Argonne Nat Lab, 79-80 & Honeywell, 81; prin investr, Plasma Physics Contract, Univ Wis, US Dept Energy, 80-86. *Mem:* Fel Am Phys Soc. *Res:* Plasma confinement and heating in toroidal and magnetic mirror fields; toroidal multipoles; tokamaks; bumpy torii, reversed field pinches; computer simulation of plasmas; ionospheric and extra-terrestrial plasmas and cosmic rays, chaos. *Mailing Add:* Dept Physics Univ Wis Madison WI 53706

SPROTT, RICHARD LAWRENCE, b Tampa, Fla, Aug 9, 40; m 65; c 2. AGING, BEHAVIOR GENETICS. *Educ:* Univ NC, AB, 62, MA, 64, PhD(psychol), 65. *Prof Exp:* Fel behav genetics, Jackson Lab, 65-67, assoc staff scientist, 69-71, staff scientist, 71-80; asst prof psychol, Oakland Univ, 67-69; health scientist admin & aging, 80-81, br chief aging, 81-84, ASSOC DIR AGING, NAT INST AGING, NIH, 84- *Concurrent Pos:* Mem, Comt Animal Modes for Aging, Nat Res Coun, 78-79; head, Off Res Resources, Nat Inst Aging, 81-85. *Mem:* Am Psychol Asn; Behav Genetics Asn; Geront Asn Am. *Res:* Study of genetic determinants of behavior from maturity to senescence; stimulation and development of biological research on aging and the development of animal models for such research. *Mailing Add:* 11514 Regency Dr Potomac MD 20854

SPROUL, GORDON DUANE, b Edgewood, Md, Mar 6, 44; m 75; c 3. COORDINATION POLYMERS. *Educ:* Harvey Mudd Col, BS, 66; Univ Ill, MS, 69, PhD(chem), 71. *Prof Exp:* Fel chem, Tulane Univ, 71-72 & Univ SC, 72-75; from asst prof to assoc prof, 75-86, PROF CHEM, UNIV SC, BEAUFORT, 86- *Concurrent Pos:* Consult, 82-; pres, SC Acad Sci, 90-91. *Mem:* Am Chem Soc; Sigma Xi. *Res:* Design and synthesis of trans-coordinating bidentate monomer liqands and of comparable tetradentate liqands for coordination polymer formation with appropriate metal ions. *Mailing Add:* 980 Edith Lane Beaufort SC 29902

SPROUL, OTIS J, b Dover Foxcroft, Maine, July 9, 30; m 52; c 1. SANITARY ENGINEERING. *Educ:* Univ Maine, BS, 52, MS, 57; Wash Univ, St Louis, ScD(sanit eng), 61. *Prof Exp:* Instr civil eng, Univ Maine, 55-57, asst prof, 57-59; trainee sanit eng, Wash Univ, St Louis, 59-61; from assoc prof to prof civil eng, Univ Maine, 61-77; PROF & CHMN DEPT CIVIL ENG, OHIO STATE UNIV, 77- *Honors & Awards:* Rudolph Hering Award, Am Soc Civil Engrs, 71. *Mem:* Am Water Works Asn; Water Pollution Control Fedn; Am Soc Civil Engrs; Am Soc Eng Educ; Nat Soc Prof Engrs. *Res:* Virus inactivation by water and wastewater treatment processes; industrial air and water pollution. *Mailing Add:* Kingsbury Hall Col of Eng/ Phys Sci Univ NH Durham NH 03824

SPROUL, WILLIAM DALLAS, b Fitchburg, Mass, Mar 14, 43; m 68; c 2. REACTIVE SPUTTERING, UNBALANCED MAGNETRON SPUTTERING. *Educ:* Brown Univ, ScB, 66, ScM, 68, PhD(mat eng), 75. *Prof Exp:* Scientist, Am Can Corp, 75-77; sr engr, Borg-Warner Corp, 77-87; GROUP LEADER, BIRL, NORTHWESTERN UNIV, 87- *Concurrent Pos:* Co-ed, Physics & Chem Protective Coatings, 85; gen chmn, Int Conf Metall Coatings, 87-88 & 90; bd dir, Am Vacuum Soc, 90-91, chmn, Vacuum Metall Div, 91. *Mem:* Am Soc Metals Int. *Res:* Sputtering of hard, wear and corrosion resistant coatings; invented the high-rate reactive sputtering process. *Mailing Add:* BIRL Northwestern Univ 1801 Maple Ave Evanston IL 60201

SPROULE, BRIAN J, b Calgary, Alta, Oct 31, 25; m 55; c 4. MEDICINE, THORACIC DISEASES. *Educ:* Univ Alta, BSc, 49, MD, 51, MSc, 55; FRCPS(C). *Prof Exp:* Instr med, Univ Tex Southwestern Med Sch, Dallas, 55-59; from instr to assoc prof, 59-70, PROF MED & HEAD DIV RESPIRATORY DIS, UNIV ALTA, 70- *Concurrent Pos:* Consult, Can Dept Vet Affairs, 60-; gov, Am Col Physicians, 79-83; gov, Am Col Chest Phys, 76-78, chmn dept med, 75-76. *Mem:* Fel Am Col Chest Physicians; fel Am Col Physicians; Am Fedn Clin Res; Can Soc Clin Invest; Royal Col Physicians & Surgeons Can (vpres). *Res:* Pulmonary mechanics; blood gas derangements in chronic lung disease. *Mailing Add:* Dept Med Univ Alta Sch Med 2E436 WC McKenzie Ctr Edmonton AB T6G 2B7 Can

SPROULL, ROBERT FLETCHER, b Ithaca, NY, June 6, 47; m 71. COMPUTER SCIENCE. *Educ:* Harvard Col, AB, 68; Stanford Univ, MS, 70, PhD(comput sci), 77. *Prof Exp:* Staff programmer artificial intel, Stanford Univ, 69-70; comput specialist, Div Comput Res & Technol, NIH, 70-72; mem res staff comput sci, Xerox Palo Alto Res Ctr, 73-77; from asst prof to assoc prof, Carnegie-Mellon Univ, 77-84; VPRES, SUTHERLAND, SPROUL & ASSOC 79- *Concurrent Pos:* Mem tech adv coun, R R Donnelly & Sons, 81-; adj prof comput sci, Carnegie-Mellon Univ, 84. *Mem:* Asn Comput Mach; Inst Elec & Electronic Engrs. *Res:* Computer graphics; large-scale integrated circuits. *Mailing Add:* Sutherland Sproull & Assoc 4516 Henry St Pittsburgh PA 15213

SPROULL, ROBERT LAMB, b Lacon, Ill, Aug 16, 18; m 42; c 2. GENERAL PHYSICS. *Educ:* Cornell Univ, BA, 40, PhD(exp physics), 43. *Hon Degrees:* LLD, Nazareth Col, 83. *Prof Exp:* Physicist, RCA Labs, NJ, 43-46; from asst prof to prof physics, Cornell Univ, 46-68, dir lab atomic & solid state physics, 59-60, dir mat sci ctr, 60-63, vpres acad affairs, 65-68; vpres & provost, 68-70, pres, 70-84, chief exec officer, 74-84, EMER PRES & PROF PHYSICS, UNIV ROCHESTER, 84- *Concurrent Pos:* Part-time instr, Princeton Univ & Univ Pa, 43-45; physicist, Oak Ridge Nat Lab, 52 & Europ Res Assocs, Belg, 58-59; ed, J Appl Physics, 54-57; trustee, Assoc Univs, Inc, 62-63; dir, Advan Res Projs Agency, 63-65; mem bd dirs, John Wiley & Sons, Inc, NY, 65-89; trustee, Deep Springs Col, 67-74 & 82-86; mem bd dirs, United Technol Corp, 72-89, Xerox Corp, 76-89, Sybron Corp, 72-85 & Bausch & Lomb Corp, 82-89; trustee, Cornell Univ, 72-77; pres, Telluride Asn, 45-47; mem solid state sci adv panel, Off Naval Res & later Nat Acad Sci, 50-68; mem lab mgt coun, Oak Ridge Nat Lab, 65-75, chmn coun, 71-73; mem, Defense Sci Bd, 66-70, chmn bd, 68-70; mem statutory vis comt, Nat Bur Standards, 66-71, chmn comt, 68-71; mem sci adv comt, Gen Motors Corp, 71-80, chmn, 73-80; mem bd dir, Commonwealth Fund, 79-89 & Inst Defense Analysis, 85-91. *Mem:* Fel AAAS; Am Phys Soc; Am Acad Arts & Sci. *Res:* Thermionic electron emission; microwave radar; experimental solid state physics; imperfections in nonmetallic crystals, especially in barium oxide; low temperature physics; phonon scattering. *Mailing Add:* Univ Rochester Bausch & Lomb Bldg Rochester NY 14627

SPROULL, WAYNE TREBER, b Racine, Wis, Aug 3, 06; m 34; c 2. AIR POLLUTION. *Educ:* Univ Akron, BS, 27; Lehigh Univ, MS, 29; Univ Wis, PhD(physics), 33. *Prof Exp:* Asst physics, Lehigh Univ, 27-29 & Univ Wis, 29-32; res physicist, Res Labs Div, Gen Motors Corp, Mich, 33-46; res physicist, Res Lab, Lockheed Aircraft Corp, Calif, 46-47; chief liquid rockets sect, Jet Propulsion Lab, Calif Inst Technol, 47-48; head elec res dept, Western Precipitation Corp, 48-60, chief physicist, Western Precipitation Div, Joy Mfg Co, Calif, 60-65; sr staff physicist, Nat Eng Sci Co, 65-68; assoc, Petroff & Assocs, 68-69; consult, Western Precipitation Div, Joy Mfg Co, Los Angeles, 69-72; consult, Elec Power Res Inst, Palo Alto, Calif & Carolina Power & Light, Raleigh, NC, 72-90; RETIRED. *Concurrent Pos:* Staff physicist, Environ Resources, Inc, Calif, 68-69; mem ed staff air pollution criteria, US Govt Publ. *Honors & Awards:* US Off Sci Res & Develop. *Mem:* AAAS; Am Phys Soc; Air Pollution Control Asn. *Res:* Air pollution control; effects of dust clouds on gaseous discharges and gas flow; x-rays; improvements in the technology of industrial gas cleaning and electrical precipitation. *Mailing Add:* 3015 San Gabriel Ave Glendale CA 91208

SPROUSE, GENE DENSON, b Litchfield, Ill, May 7, 41; m 63; c 2. ATOMIC & MOLECULAR PHYSICS. *Educ:* Mass Inst Technol, BS, 63; Stanford Univ, MS, 65, PhD(physics), 68. *Prof Exp:* Res assoc physics, Stanford Univ, 67-69, asst prof, 69-70; from asst prof to assoc prof, 73-77, PROF PHYSICS, STATE UNIV NY STONY BROOK, 77- *Concurrent Pos:* Fel, Alfred P Sloan Found, 72-74; dir, Nuclear Struct Lab, 84-86, chmn, 90- *Honors & Awards:* Humboldt Prize, Am Phys Soc. *Mem:* Fel Am Phys Soc. *Res:* Hyperfine interactions; perturbed angular correlations; recoil implantation; laser spectroscopy. *Mailing Add:* Dept Physics State Univ NY Stony Brook NY 11794

SPROWLES, JOLYON CHARLES, b Columbia, SC, July 6, 44; m 68; c 3. ELECTROPLATING, PHYSICAL INORGANIC CHEMISTRY. *Educ:* Princeton Univ, AB, 66; Cornell Univ, PhD(inorg chem), 73. *Prof Exp:* Asst prof chem, Williams Col, 70-72; res assoc, Purdue Univ, 72-74; vis asst prof, Univ Mo-Columbia, 74-75; asst prof chem, Bates Col, 75-82; PROCESS ENGR, MAINE ELECTRONICS, INC, 82- *Mem:* Am Chem Soc. *Res:* Electroplating; uses of plasma in manufacturing; vibrational spectroscopic studies of equilibria and structure in solutions of organometallic cations and weak Lewis base anions. *Mailing Add:* 200 Heroax Blvd No 1907 Cumberland RI 02864

SPROWLS, DONALD O(TTE), b Arnold, Pa, Sept 9, 19; m 44; c 4. CHEMICAL ENGINEERING, METALLURGY. *Educ:* Drexel Inst Technol, BS, 43. *Prof Exp:* Technician, Chem Metall Div, Alcoa Res Labs, Pittsburgh, 36-43, res engr, 43-64, head stress corrosion sect, 64-77, assoc engr, Alcoa Tech Ctr, 77-82; RETIRED. *Concurrent Pos:* Consult aluminum corrosion, 83- *Honors & Awards:* Sam Tour Award, Am Soc Testing & Mat. *Mem:* Am Soc Testing & Mat. *Res:* Corrosion and stress corrosion of aluminum alloys. *Mailing Add:* 4419 7th St Alcoa Center PA 15069

SPROWLS, RILEY CLAY, b Medina, NY, July 22, 21; m 50; c 2. STATISTICS. *Educ:* Univ Chicago, PhD(statist), 51. *Prof Exp:* Instr statist, Univ Chicago, 49-51; from asst prof statist to prof bus statist, 51-71, PROF COMPUT & INFO SYSTS, UNIV CALIF, LOS ANGELES, 71- *Mem:* Am Statist Asn. *Res:* Business statistics; electronic computers. *Mailing Add:* Dept Mgmt 3250 GSM Univ of Calif 405 Hilgard Ave Los Angeles CA 90024

SPRUCH, GRACE MARMOR, b Brooklyn, NY, Nov 19, 26; m 50. SCIENCE EDUCATION, SCIENCE WRITING. *Educ:* Brooklyn Col, BA, 47; Univ Pa, MS, 49; NY Univ, PhD(physics), 55. *Prof Exp:* Res asst physics, Univ Pa, 47-48, asst instr, 48-49; res asst, NY Univ, 52-55, assoc res scientist, 55-56; instr physics, Cooper Union, 57-58; assoc res scientist, NY Univ, 58-63; Am Asn Univ Women fel, Oxford Univ, 63-64; vis assoc prof, Rutgers Univ, 64-65; assoc res scientist, NY Univ, 65-67, res scientist, 67-68; assoc prof physics, 69-75, PROF PHYSICS, RUTGERS UNIV, 75- *Concurrent Pos:* Secy, Int Conf Luminescence, NY Univ, 61; writer, ed & translator; hon res assoc appl sci, Harvard Univ, 77-78; hon assoc, Neiman Found for Jour, 77-78; fel, Ctr Energy & Environ Studies, Princeton Univ, 81; mem, interview team for china, US Physics exam & appln prog, Peoples Repub China, 85-86. *Mem:* Am Phys Soc. *Res:* Luminescence, photoconductivity and applications to biophysics; light scattering; science writing. *Mailing Add:* Dept Physics Rutgers Univ 101 Warren St Newark NJ 07102

SPRUCH, LARRY, b Brooklyn, NY, Jan 1, 23; m 50. THEORETICAL PHYSICS. *Educ:* Brooklyn Col, BA, 43; Univ Pa, PhD(physics), 48. *Prof Exp:* From asst instr to instr physics, Univ Pa, 43-46, Tyndale fel, 46-48; Atomic Energy Comn fel, Mass Inst Technol, 48-50; from asst prof to assoc prof, 50-61, PROF PHYSICS, FAC ARTS & SCI, NY UNIV, 61- *Concurrent Pos:* Consult, Lawrence Radiation Lab, 59-66; NSF sr fel, Univ London & Oxford Univ, 63-64; vis prof, Inst Theoret Phys, Univ Colo, 61 & 68; correspondent, Comments on Atomic & Molecular Physics, 72-; mem, Inst Advan Study, 81-82; deleg, China US Physics Exam & Appln, 85 & 86; mem adv bd, Inst Theoret Atomic & Molecular Physics, Harvard-Smithsonian Ctr Astrophys, 89-91. *Honors & Awards:* von Humboldt Sr Award, 85 & 88. *Mem:* Fel Am Phys Soc. *Res:* Beta decay; nuclear moments; isomeric transitions; internal conversion; atomic and nuclear scattering; variational principles; astrophysics; charge transfer; Thomas-Fermi theory; radiative corrections; atoms in magnetic fields; Levinson's theorem; casimir interactions; semi-classical radiation theory. *Mailing Add:* Dept of Physics Meyer Bldg NY Univ 4 Washington Pl New York NY 10003

SPRUGEL, DOUGLAS GEORGE, b Ames, Iowa, Feb 18, 48; m 84; c 1. PLANT ECOLOGY. *Educ:* Duke Univ, BS, 69; Yale Univ, MPhil, 71, PhD(plant ecol), 74. *Prof Exp:* Lectr ecol, Univ Pa, 73-74; res assoc, Argonne Nat Lab, 74-76, asst ecologist, 76-79; asst prof, dept forestry, Mich State Univ, 79-82; sr res assoc, 83-87, res assoc prof, 87-90, PROF, COL FOREST RES, UNIV WASH, 90- *Honors & Awards:* Mercer Award, Ecol Soc Am, 77. *Mem:* Ecol Soc Am (vpres, veg sect, 88, pres, 89); Am Inst Biol Sci; AAAS; Int Asn Veg Sci. *Res:* Effects of natural and human disturbance on natural ecosystems; air pollution effects on plants; tree ecophysiology; woody-tissue respiration. *Mailing Add:* Col Forest Res AR-10 Univ Wash Seattle WA 98195

SPRUGEL, GEORGE, JR, b Boston, Mass, Sept 26, 19; m 45; c 1. ZOOLOGY, ECOLOGY. *Educ:* Iowa State Col, BS, 46, MS, 47, PhD(econ zool), 50. *Prof Exp:* From instr to asst prof zool & entom, Iowa State Col, 46-54; spec asst to asst dir biol & med sci, NSF, 53-54, prog dir environ biol, 54-64; chief scientist, Nat Park Serv, 64-66; chief, 66-80, EMER CHIEF, ILL NATURAL HIST SURV, 80- *Concurrent Pos:* Asst & actg head, Biol Br, Off Naval Res, 51-53; mem adv comt environ biol, NSF, 65 & Nat Res Coun, 68-71; prog dir conserv ecosysts, US Int Biol Prog, 69-72; mem life sci comt, NASA, 73-78. *Honors & Awards:* Distinguished Serv Citation, Ecol Soc Am, 76. *Mem:* AAAS (vpres, biol sci, 71); Am Soc Zoologists (secy, 70-72); Ecol Soc Am (vpres, 68); Am Inst Biol Sci (vpres, 73, pres, 74). *Res:* Aquatic ecology; fish growth; animal population dynamics. *Mailing Add:* 2710 S 1st St Champaign IL 61820

SPRUIELL, JOSEPH E(ARL), b Knoxville, Tenn, Oct 13, 35; m 58; c 2. MATERIALS SCIENCE, ENGINEERING. *Educ:* Univ Tenn, BS, 58, MS, 60, PhD(metall eng), 63. *Prof Exp:* From asst prof to assoc prof metall eng, 63-71, PROF METALL ENG & POLYMER ENG, UNIV TENN, KNOXVILLE, 71-, HEAD DEPT MAT SCI & ENG, 85- *Concurrent Pos:* Consult, Metals & Ceramics Div, Oak Ridge Nat Lab, 60-77. *Mem:* Fel Am Soc Metals; Sigma Xi; Soc Plastics Engrs; Int Polymer Processing Soc; Fiber Soc; Soc Rheology. *Res:* X-ray diffraction; physical metallurgy; polymer science; polymer processing. *Mailing Add:* Dept Mat Sci & Eng Univ of Tenn Knoxville TN 37996

SPRUILL, NANCY LYON, b Takoma Park, Md, Mar 24, 49; m 69. PLANNING, PROGRAMMING & BUDGETING. *Educ:* Univ Md, BS, 71; George Washington Univ, MA, 75, PhD(math statist), 80. *Prof Exp:* Proj dir, Ctr Naval Analyses, 71-83; sr analyst planning, prog & budgeting, Manpower, Installations & Logistics, 83-85, ASSOC DIR INTERGOVERNMENTAL AFFAIRS, FORCE MGT & PERSONNEL, OFF ASST SECY DEFENSE, 85- *Concurrent Pos:* Assoc prof/lectr, Statist Dept, George Washington Univ, 78-79. *Honors & Awards:* Jerome Cornfield Award, 80. *Mem:* Am Statist Asn. *Res:* Resource implications of changes in Department of Defense policies and procedures, including changes recommended by Congress and organizations outside of the Department; measures of defense capability; confidentiality of different kinds of data releases. *Mailing Add:* Rm 2E 314 Pentagon Washington DC 20301

SPRULES, WILLIAM GARY, b Hamilton, Ont, Nov 5, 44; m 67; c 3. AQUATIC ECOLOGY. *Educ:* Queen's Univ, Ont, BSc, 66; Princeton Univ, MA, 68, PhD(ecol), 70. *Prof Exp:* From asst prof to assoc prof, 70-84, vprin res & grad studies, 86-89, PROF ZOOL, ERINDALE COL, UNIV TORONTO, 84- *Concurrent Pos:* Operating grant, Nat Sci & Eng Res Coun Can, Donner Can Found, Wildlife Toxicol Fund. *Mem:* Ecol Soc Am; Am Soc Limnol & Oceanog; Int Asn Theoret & Appl Limnol; Freshwater Biol Asn, Eng; Can Soc Zoologists; Soc Can Limnologists. *Res:* Size structure of aquatic plankton communities; aquatic food webs; zooplankton behavior. *Mailing Add:* Erindale Col Univ of Toronto Mississauga ON L5L 1C6 Can

SPRUNG, DONALD WHITFIELD LOYAL, b Kitchener, Ont, June 6, 34; m 58; c 2. NUCLEAR PHYSICS. *Educ:* Univ Toronto, BA, 57; Univ Birmingham, PhD(physics), 61, DSc, 77. *Prof Exp:* Instr physics, Cornell Univ, 61-62; from asst prof to assoc prof physics, 62-71, dean fac sci, 75-84 & 89, PROF PHYSICS, MCMASTER UNIV, 71- *Concurrent Pos:* Mem res staff, Lab Nuclear Sci, Mass Inst Technol, 64-65; C D Howe fel, Orsay, France, 69-70; guest prof, Univ Tuebingen, Ger, 80-81; exchange prof, Univ Mainz, Ger, 90-91; vis prof, Univ Barcelona, Spain, 91. *Honors & Awards:* Herzberg Medal, 72. *Mem:* Am Phys Soc; Can Asn Physicists; Brit Inst Physics; fel Royal Soc Can. *Res:* Nucleon-nucleon interaction; effective force in finite nuclei; nuclear structure and forces theory; theory of mesoscopic systems. *Mailing Add:* Dept Physics ABC 348 McMaster Univ 1280 Main St W Hamilton ON L8S 4M1 Can

SPRUNG, JOSEPH ASHER, b Wahpeton, NDak, Dec 25, 15; m 44; c 1. ORGANIC CHEMISTRY. *Educ:* Univ Minn, BChem, 38, MS, 39, PhD(org chem), 43. *Prof Exp:* Res chemist, Cent Res Lab, GAF Corp, Pa, 43-47, group leader photog sect, 47-51, sr res specialist, Photog Div, 51-61, mgr, Photog Emulsion Tech Dept, 61-62, assoc dir res & develop, 62-64, sr scientist, Photog Div, 64-80, tech consult photog sci & technol, 80; RETIRED. *Mem:* Fel AAAS; Am Chem Soc; Soc Photog Sci & Eng. *Res:* Synthesis of organic compounds required in studying ortho effect; synthesis of vitamin E and vitamin A intermediates; color photography processes; photographic emulsions. *Mailing Add:* 16 Devon Blvd Binghamton NY 13903

SPRUNT, EVE SILVER, b Brooklyn, NY, July 9, 51; m 73; c 2. GEOPHYSICS, GEOLOGY. *Educ:* Mass Inst Technol, SB, 72, SM, 73; Stanford Univ, PhD(geophys), 77. *Prof Exp:* Res assoc, Stanford Univ, 76-78; RES ASSOC, MOBIL RES & DEVELOP CORP, 78- *Mem:* Am Geophys Union; Geol Soc Am; Soc Explor Geophysicists; Soc Petrol Engrs; Soc Rock Mech; Soc Prof Well Log Analysts; Soc Core Analysts. *Res:* Rock physics, specifically solution transfer; quartz cathoduluminescence; porosity; permeability; velocity; scanning electron microscopy; hydraulic fracturing; core analysis. *Mailing Add:* Mobil Res & Develop Corp PO Box 819047 Dallas TX 75381

SPRY, ROBERT JAMES, b Dayton, Ohio, Feb 12, 38; m 69; c 4. SOLID STATE PHYSICS, OPTICS. *Educ:* Univ Ill, Urbana, BS, 60, MS, 62, PhD(solid state physics), 67. *Prof Exp:* Res asst solid state physics, Univ Ill, Urbana, 62-67; RES PHYSICIST, AIR FORCE MAT LAB, WRIGHT-PATTERSON AFB, 67- *Concurrent Pos:* ADJ PROF, WRIGHT STATE UNIV, 86- *Mem:* Am Phys Soc; Sigma Xi. *Res:* Optical properties of semiconductors; radiation damage; infrared spectroscopy; optical sensors; laser technology; low temperature physics; electro-optics, patent law. *Mailing Add:* Worley Rd Tripp City OH 45371

SPUDICH, JAMES ANTHONY, b Collinsville, Ill, Jan 7, 42; m 64; c 2. BIOCHEMISTRY. *Educ:* Univ Ill, Urbana, BS, 63; Stanford Univ, PhD(biochem), 67. *Prof Exp:* USPHS trainee, Stanford Univ, 68; US Air Force Off Sci Res fel, Cambridge Univ, 69 & NSF fel, 70; asst prof biochem, Univ Calif, San Francisco, 70-74, assoc prof, 74-; PROF, DEPT CELL BIOL, SCH MED, STANFORD UNIV. *Concurrent Pos:* Am Cancer Soc res grant, Univ Calif, San Francisco, 70. *Mem:* Nat Acad Sci; AAAS. *Res:* Molecular basis of mitosis, amoeboid movement and other forms of cell mobility. *Mailing Add:* Cell Biol D141 Fairchild Bldg Stanford Univ Stanford CA 94305

SPULLER, ROBERT L, b Shelbyville, Ind, Aug 21, 37; m 59; c 2. PROTOZOOLOGY, PARASITOLOGY. *Educ:* Purdue Univ, BS, 59, MS, 60; Univ Mich, MS, 63, PhD(zool), 68. *Prof Exp:* Teacher high sch, Ind, 60-62; asst prof, 68-70, assoc prof, 70-74, chmn div natural sci & math, 75-79, actg dean acad affairs, 77-78, PROF BIOL, LENOIR-RHYNE COL, 75-, CHMN DEPT, 68- *Mem:* AAAS; Soc Protozool; Sigma Xi; Am Soc Zool. *Res:* Electron microscopy of ciliated protozoa; ecological relationships in the protozoa. *Mailing Add:* Dean of Acad Affairs Lenoir-Rhyne Col Hickory NC 28603

SPURGEON, WILLIAM MARION, b Quincy, Ill, Dec 5, 17; m 41; c 3. PHYSICAL CHEMISTRY. *Educ:* Univ Ill, BS, 38; Univ Mich, MS, 39, PhD(phys chem), 41. *Prof Exp:* Res chemist, Tex Co, NY, 41-42 & 46; asst prof appl sci, Univ Cincinnati, 46-48; res dir & vpres, Am Fluresit Co, Ohio, 47-54; mgr phys chem unit, Flight Propulsion Lab Dept, Gen Elec Co, 54-59; mgr mat & processes dept, Res Labs, Bendix Corp, 59-73, dir mfg qual control, Home Systs Res, Res Labs, 73-78, sr res planner, 78-80; dir prod res prog, NSF, 80-85; DIR MFG ENG PROG, UNIV MICH, DEARBORN, 85- *Concurrent Pos:* Mem, Nat Mat Adv Bd, 78-80. *Honors & Awards:* Colwell Award, Soc Automotive Engrs, 67; Siegel Award, Soc Mfg Engrs, 81. *Mem:* Am Chem Soc; Am Soc Metals; fel Soc Mfg Engrs; Sigma Xi; Soc Auto Engrs. *Res:* Catalysis; solubility; fluid flow; building materials; ophthalmology; corrosion; coatings; thermal properties; semiconductors; friction materials; composites; separation processes; manufacturing processes; unit operations of manufacturing. *Mailing Add:* 24799 Edgemont Rd Southfield MI 48034

SPURLIN, HAROLD MORTON, b Atlanta, Ga, Apr 4, 05; m 33, 44; c 5. PHYSICAL CHEMISTRY. *Educ:* Ga Inst Technol, BS, 25; Yale Univ, PhD(chem), 28. *Prof Exp:* Res chemist, Res Ctr, Hercules Inc, 28-39, group leader, 39-45, tech asst to dir, 45-70; CONSULT, 70- *Honors & Awards:* Anselme Payen Award, 64. *Mem:* AAAS; Am Chem Soc; Am Inst Chem Engrs. *Res:* Cellulose; polymers; naval stores; propellants; process design. *Mailing Add:* 2704 Duncan Rd Wilmington DE 19808

SPURLOCK, BENJAMIN HILL, JR, mechanical engineering; deceased, see previous edition for last biography

SPURLOCK, CAROLA HENRICH, b Detroit, Mich, July 24, 26; m 54; c 2. CHROMATOGRAPHY, SPECTROPHOTOMETRY. *Educ:* Univ Detroit, BS, 48, MS, 50. *Prof Exp:* Asst chemist phys chem, Parke Davis & Co, 50-69, assoc chemist, 69-78; scientist anal serv, Warner-Lambert/Parke Davis, 78-86; RETIRED. *Mem:* AAAS; Am Chem Soc. *Res:* Ultraviolet-visible spectrophotometry, chromatography, potentiometric titration and instrumental analytical methods. *Mailing Add:* PO Box 122 Hawks MI 49743

SPURLOCK, JACK MARION, b Tampa, Fla, Aug 16, 30; m 52; c 4. CHEMICAL ENGINEERING, BIOMEDICAL ENGINEERING. *Educ:* Univ Fla, BChE, 52; Ga Inst Technol, MSChE, 58, PhD(chem eng), 61. *Prof Exp:* Qual control engr, Auto-Lite Battery Co, Ga, 54-55; res engr, Eng Exp Sta, Ga Inst Technol, 55-58, asst prof chem eng, Sch Chem Eng, 58-62; mgr, Aerospace Sci Lab, Orlando Div, Martin Co, 62-64; chief, Eng Res Group, Atlantic Res Corp, 64-69; pres, Health & Safety Res Inst, 69-74; prin res engr, Eng Exp Sta & assoc dir, Appl Sci Lab, Ga Inst Technol, 74-79, dir, Off Interdisciplinary Prog, 70-83, assoc vpres res, 83-85; pres, Cetrest Corp, 85-88; PRES, S & A AUTOMATED SYSTS, INC, 90- *Concurrent Pos:* Biomed eng consult, T A Jones Assocs, 71-74. *Honors & Awards:* M A Ferst Res Award, 61. *Mem:* Am Inst Chem Eng; fel Royal Soc Health; Am Chem Soc; Aerospace Med Asn; fel Am Inst Chem. *Res:* Acoustical effects on transport phenomena; conduction and convection heat transfer; vacuum environmental effects on liquid propellants; biomedical transport phenomena and instrumentation; chemical process economics; energy conservation; research management; alcohol fuels technology; systems design and analysis; rehabilitation technology for the disabled; computer-based training; spacecraft life support systems. *Mailing Add:* PO Box 176 Boca Raton FL 33429-0176

SPURLOCK, LANGLEY AUGUSTINE, b Charleston, WVa, Nov 9, 39. ORGANIC CHEMISTRY. *Educ:* WVa State Col, BS, 59; Wayne State Univ, PhD(org chem), 63. *Prof Exp:* Res chemist, Gen Chem Div, Allied Chem Corp, NJ, 63-65, sr res chemist, Nitrogen Div, 66; asst prof org chem, Temple Univ, 66-69; assoc prof org chem, Brown Univ, 69-76; fel, US Dept Health Educ & Welfare, 76-77; spec asst to dir, Off Audit & Oversight, NSF, 77-80, staff assoc spec proj, 80-81, sr staff assoc oper math & phys sci dir, 81-82; dir biomed & environ, Spec Progs Div, 82-89, DIR CHEMSTAR DIV, CHEM MFRS ASN, 89- *Concurrent Pos:* Asst to president, Am Coun Educ, 73-76. *Mem:* Am Chem Soc; AAAS; Am Soc Asn Execs. *Res:* Mechanistic organic chemistry, bridged polycyclic compounds, molecular rearrangements, free radical and ionic additions; synthetic organic chemistry; conformational analysis; organic ultrasonic chemistry; sonochemistry. *Mailing Add:* 3718 Van Ness St NW Washington DC 20016

SPURR, ARTHUR RICHARD, b Glendale, Calif, July 21, 15; m 42; c 4. PLANT MORPHOLOGY, PLANT PHYSIOLOGY. *Educ:* Univ Calif, Los Angeles, BS, 38, MA, 40; Harvard Univ, AM, 42, PhD(biol), 47. *Prof Exp:* Instr biol, Harvard Univ, 47-48; instr truck crops, 48-53, from jr olericulturist to assoc olericulturist, 48-73, from asst prof to prof veg crops, 54-84, OLERICULTURIST, EXP STA, UNIV CALIF, DAVIS, 73-, EMER PROF VEG CROPS, 84- *Concurrent Pos:* NIH spec res fel, 60-61; NSF res grant, 71-72; French govt res grant, 77-78. *Mem:* AAAS; Electron Micros Soc Am; Microbeam Anal Soc; Am Inst Biol Scientists; Bot Soc Am. *Res:* Cytology and ultrastructure of vascular plants; electron microscopy and analytical techniques; pathology of physiological disorders; mineral nutrition; electron probe x-ray analysis; ion probe microanalysis; responses to salinity and boron toxicity. *Mailing Add:* 617 Elmwood Dr Davis CA 95616

SPURR, CHARLES LEWIS, b Sunbury, Pa, Nov 20, 13; m 40; c 2. MEDICINE. *Educ:* Bucknell Univ, BS, 35; Univ Rochester, MS, 38, MD, 40. *Prof Exp:* From intern to asst med, Clin, Univ Chicago, 40-43, from instr to asst prof, 43-48; chief dept med, Univ Tex M D Anderson Hosp & Tumor Clin, 48-49; assoc prof, Col Med, Baylor Univ, 49-57; prof med & dir oncol res ctr, Bowman Gray Sch Med, Wake Forest Univ, 57-82; chmn, Clin Trials Group, Piedmont Oncol Asn, 76-85; CHMN, SOUTHEAST CANCER CONTROL CONSORTIUM INC, 86- *Concurrent Pos:* Chief med res, Vet Hosp, Houston, Tex, 49-57; chmn of bd, NC Div, Am Cancer Soc; Charles L Spurr professorship. *Mem:* AAAS; AMA; fel Am Col Physicians; Am Soc Clin Oncol; Am Soc Hematol; Sigma Xi. *Res:* Hematology; cancer chemotherapy. *Mailing Add:* Bowman Gray Sch of Med Wake Forest Univ Winston-Salem NC 27103

SPURR, DAVID TUPPER, b Notikewin, Alta, May 21, 38; m 62; c 3. STATISTICS, BIOLOGY. *Educ:* Univ Alta, BSc, 61, MSc, 65; Ore State Univ, PhD(genetics), 68. *Prof Exp:* Pub lands appraiser, Lands Br, Alta Dept Lands & Forests, 61-63; asst animal sci, Univ Alta, 63-65; asst genetics, Ore State Univ, 66-68; res assoc animal genetics, Comput Ctr, Univ Ga, Athens,

69-70; res scientist statist, Statist Res Serv, 70-75, STATISTICIAN, RES STA, AGR CAN, 75- *Mem:* Am Statist Asn. *Res:* Biometrical genetics; experimental design theory; bioassay; sampling. *Mailing Add:* 3005 East View Saskatoon SK S7J 3J1 Can

SPURR, GERALD BAXTER, b Cambridge, Mass, June 1, 28; m 52; c 4. PHYSIOLOGY, NUTRITION. *Educ:* Boston Col, BS, 50; Univ Iowa, PhD(physiol), 54. *Prof Exp:* From res asst to instr physiol, Univ Iowa, 51-54; from instr clin physiol to prof physiol & biophys, Col Med, Univ Tenn, Memphis, 56-68; PROF PHYSIOL, MED COL WIS, 68- *Concurrent Pos:* Am Heart Asn fel, Univ Iowa, 54-56; Markle scholar med sci, 58-63; vis prof, Univ Valle, Colombia, 61-62 & 79-; consult, Res Serv, Vet Admin Ctr, Wood, Wis, 68- *Mem:* AAAS; Am Physiol Soc; Soc Exp Biol & Med; fel Am Col Sports Med; Sigma Xi; Am Inst Nutrit. *Res:* Environmental and cardiovascular physiology; cold heart; hypothermia; hyperthermia; electrolyte and fluid metabolism; peripheral circulation; respiration; exercise; nutrition; nutritional status; malnutrition. *Mailing Add:* Dept Physiol Med Col Wis Ctr 8701 Watertown Plank Rd Milwaukee WI 53226

SPURR, HARVEY WESLEY, JR, b Oak Park, Ill, June 8, 34; m 56; c 3. PLANT PATHOLOGY. *Educ:* Mich State Univ, BS, 56, MS, 58; Univ Wis, PhD(plant path), 61. *Prof Exp:* NIH fel plant path, Univ Wis, 61-63; plant pathologist, Agr Res Sta, Union Carbide Corp, NC, 63-69; assoc prof plant path, 69-74, PROF PLANT PATH, NC STATE UNIV, 74-; RES PLANT PATHOLOGISTS, OXFORD TOBACCO LAB, USDA, 69-, RES LEADER & LAB DIR, 88- *Mem:* Am Phytopath Soc; Soc Nematol. *Res:* Plant disease control research; biological and chemical control; biochemistry of plant disease. *Mailing Add:* RR 3 Box 320 Oxford NC 27565

SPURR, ORSON KIRK, JR, b Cambridge, NY, Sept 4, 30; m 53; c 2. PHYSICAL CHEMISTRY. *Educ:* Dartmouth Col, BA, 52; Cornell Univ, MS, 56, PhD(phys chem), 58. *Prof Exp:* Res polymer chemist, Chem & Plastics Div, Union Carbide Corp, 58-79, res polymer chemist, Specialty Chem & Plastic Div, 79-85; consult, Amoco Performance Prod Inc, 86-88; CONSULT, 88- *Mem:* Am Chem Soc; Sigma Xi. *Res:* Epoxy resins. *Mailing Add:* 1270 Cornell Rd Bridgewater NJ 08807-2301

SPURR, STEPHEN HOPKINS, forestry, ecology; deceased, see previous edition for last biography

SPURRELL, FRANCIS ARTHUR, b Independence, Iowa, Apr 13, 19; m 42; c 1. VETERINARY MEDICINE, RADIATION BIOLOGY. *Educ:* Univ Wis, BS, 41; Iowa State Col, DVM, 46; Univ Minn, PhD(vet med), 55; Am Bd Vet Radiol, dipl. *Prof Exp:* Instr vet anat, Univ Minn, 47-49, vet obstet, 49-55, assoc prof vet radiol, 55-62, dir summer inst radiation biol, 60-64, prof vet radiol, 62-68, PROF THERIOGENOL & GENETICS, UNIV MINN, ST PAUL, 68- *Concurrent Pos:* County livestock agt, 46-47. *Mem:* Am Soc Animal Sci; Am Vet Radiol Soc; Am Vet Med Asn; Educators Vet Radiol Sci; Sigma Xi. *Res:* Genetics; heritability of fertility in dairy cattle; gaiting inheritance in racing thoroughbreds, with applied computer data base management procudures; veterinary radiology; heredity and diseases of animals; theriogenology computer systems; "bleeders" in racing horses. *Mailing Add:* PO Box 8137 St Paul MN 55108

SPURRIER, ELMER R, b Ava, Mo, Aug 1, 20; m 41; c 3. MICROBIOLOGY, PUBLIC HEALTH. *Educ:* Univ Mo, BS, 49; Univ Minn, MPH, 54; Univ NC, MSPH, 62, DrPH, 64. *Prof Exp:* Technologist, Landon-Meyer Labs, Ohio, 49-50; supvr serologist, Pub Health Lab, 50-53, supvr serol & virol, 54-60, asst lab dir, 64-66, dir labs, Mo Dept Health, 66-87; RETIRED. *Concurrent Pos:* Lectr microbiol, Sch Med, Univ Mo, 64- *Mem:* Am Soc Microbiol; Am Pub Health Asn; Conf Pub Health Lab Dirs; Asn State & Territorial Pub Health Lab Dirs. *Res:* Antigenic relationships among parainfluenza viruses. *Mailing Add:* Lookout Point Rte 3 Eldon MO 65026

SPURRIER, WILMA A, HIBERNATION RESEARCH. *Prof Exp:* DIR, DIV NEUROL SURG, MED CTR, LOYOLA UNIV, 76- *Mailing Add:* Dept Neurol Surg Bldg 54 Rm 248 Med Ctr Loyola Univ 2160 First Ave Maywood IL 60153

SPYHALSKI, EDWARD JAMES, b Chase, Wis, Apr 13, 25; m 55; c 3. ENTOMOLOGY. *Educ:* Univ Wis, BS, 50, MA, 51; NC State Col, PhD(entom), 59. *Prof Exp:* Res entomologist, Am Cyanamid Co, 51-55; mgr tech serv, Niagara Chem Div, FMC Corp, NY, 59-69; res biologist, Air Prod & Chem, Inc, 69-71; sr biochem field specialist, PPG Industs, Inc, 71-88, Valent USA Corp, 89-90; RETIRED. *Mem:* Weed Sci Soc Am; Entom Soc Am; Soc Nematol. *Res:* Relationship of chemicals to pesticidal activity; pesticide formulation research; crop culture research; pest control procedures. *Mailing Add:* 348 Penn Ave Floyd VA 24091

SQUIBB, ROBERT E, b Sacramento, Calif, Sept 30, 42; m 71; c 2. TOXICOLOGY. *Educ:* Rutgers Univ, BA, 73, MS, 75, PhD(toxicol), 77. *Prof Exp:* Sr staff fel, Nat Inst Environ Health Sci, 77-81; SR PRIN SCIENTIST, SCHERING-PLOUGH CO, 82- *Mem:* Soc Toxicol; Am Soc Pharmacol & Exp Therapeut; Am Col Toxicol; NY Acad Sci. *Res:* Design, conduct and evaluate preclinical toxicity studies in support of drug discovery and development programs. *Mailing Add:* 551 E Shore Trail Sparta NJ 07871

SQUIBB, SAMUEL DEXTER, b Limestone, Tenn, June 20, 31; m 51; c 2. ORGANIC CHEMISTRY. *Educ:* ETenn State Univ, BS, 52; Univ Fla, PhD(chem), 56. *Prof Exp:* Assoc prof chem, Western Carolina Univ, 56-60; from asst prof to assoc prof chem, Eckerd Col, 60-64, dir chem prog, 60-64; PROF CHEM & CHMN DEPT, UNIV NC, ASHEVILLE, 64- *Concurrent Pos:* Vis prof chem, Univ NC, Chapel Hill, 76-87. *Honors & Awards:* Charles H Stone Award for Chem Educ Achievements, Carolina-Piedmont Sect, Am Chem Soc, 79; Distinguished NC Chemist Award, NC Inst of Chemist, 86. *Mem:* Am Chem Soc; fel Am Inst Chem; Sigma Xi; Nat Col Sci, Teachers Asn. *Res:* Allyl-type optically active quaternary ammonium salts; aqueous solution chemistry. *Mailing Add:* Dept of Chem Univ of NC Asheville NC 28804

SQUIER, DONALD PLATTE, b Des Moines, Iowa, Aug 16, 29. MATHEMATICS. *Educ:* Stanford Univ, BS, 51, PhD(math), 55. *Prof Exp:* Mathematician, Remington Rand Univac Div, Sperry Rand Corp, 55-57; asst prof math, San Diego State Col, 57-59; reservoir engr, Calif Res Corp, 59-64; assoc prof math, Colo State Univ, 64-69; prof math & statist, Univ West Fla, 69-83; RETIRED. *Mem:* Am Math Soc; Math Asn Am. *Res:* Elliptic partial differential equations; numerical analysis. *Mailing Add:* 7230 Lanier Dr No B Pensacola FL 32504

SQUIERS, EDWIN RICHARD, b Bath, NB, May 15, 48; US citizen. PLANT ECOLOGY, FIELD BOTANY. *Educ:* State Univ NY, Binghamton, BA, 70; Rutgers Univ, MS, 73; Ohio Univ, PhD(bot), 76. *Prof Exp:* Res assoc terrestrial veg, Jack McCormick & Assocs, Ecol Consults, 70-71, consult, 71-76; PROF BIOL & ENVIRON SCI, TAYLOR UNIV, 76-, DIR ENVIRON SCI PROG, 80- *Mem:* Ecol Soc Am; Am Inst Biol Sci; Sigma Xi; Bot Soc Am. *Res:* Organization and dynamics of secondary successsion succession systems; systems modeling; floristics; phenology of successional species; biogeography; wetlands survey and mapping. *Mailing Add:* Ctr Environ Studies Taylor Univ Upland IN 46989

SQUILLACOTE, MICHAEL EDWARD, b Washington, DC, Dec 27, 50. PHOTOCHEMISTRY, NUCLEAR MAGNETIC RESONANCE. *Educ:* Univ Chicago, BS, 72; Univ Calif, Los Angeles, PhD(chem), 78. *Prof Exp:* Res fel chem, Calif Inst Technol, 78-80; ASST PROF CHEM, BROWN UNIV, 80- *Mem:* Am Chem Soc; Sigma Xi. *Res:* Low temperature photochemical and nuclear magnetic resonance studies of reactive intermediates; potential surfaces of excited states; routes of intramolecular vibrational decay. *Mailing Add:* Dept Chem Auburn Univ Auburn AL 36849

SQUINTO, STEPHEN P, b Cook Co, Ill, Sept 19, 56; m 91; c 1. MOLECULAR BIOLOGY. *Educ:* Loyola Univ BA, 78, PhD(biochem & biophysics), 84. *Prof Exp:* Fel biochem, Loyola Univ, 78-84; fel Molecular biol, Northwestern Univ Med Sch, 84-86; asst prof biochem & molecular biol, La State Univ Med Ctr, 86-89; STAFF SCIENTIST CELL & MOLECULAR BIOL, REGENERON PHARMACEUT, 89- *Concurrent Pos:* Lectr natural sci, Loyola Univ, 84-86; ind nat res, NIH, 84-86; cancer res award, United Way & Cancer Asn, 87-89, Leukemia Soc, 88-89; consult, Nat Inst Drug Abuse, 89-90, NIH, 91-92. *Mem:* AAAS; Am Soc Microbiol; Am Soc Biochem & Molecular Biol; Am Soc Neurosci. *Res:* Discovery of novel neurotropic factor related to nerve growth factor; discovery that BNDF protects nerves from the neurotoxins that cause Parkinson's disease. *Mailing Add:* Dept Molecular & Cell Biol Regeneron Pharmaceut 777 Old Saw Mill Rd Tarrytown NY 10591

SQUIRE, ALEXANDER, b Dumfrieshire, Scotland, 17; m 45; c 9. ENGINEERING ADMINISTRATION, NUCLEAR POWER. *Educ:* Mass Inst Tech, SB, 39; Columbia Univ, PhD(exec mgt prog), 58. *Prof Exp:* Proj mgr, Bettis Automic Power Lab, 50-62; mgr, Plant Apparatus Div, Westinghouse Corp, 62-69, dir purchasing & traffic, Westinghouse Elec Corp, 69-71, pres, Westinghouse Hanford Co, 71-80; dep mgr dir, Wash Pub Power Supplies Syst, 80-85; CONSULT, NUCLEAR POWER & CONTRACT SETTLEMENT, 85- *Concurrent Pos:* Dir, Tri City Nuclear Coun, Wash, dir, Old Nat Bank, United Way; mem bd, Grad Ctr Richmond, Wash. *Mem:* Nat Acad Eng; Am Nuclear Soc; Am Soc Metals; Am Defense Preparedness Asn. *Res:* Development and construction of naval nuclear power plants; development and construction of breeder reactor facilities including fast flux test facility; construction of commercial nuclear power plants. *Mailing Add:* 2415 Winburn Ave Durham NC 27704

SQUIRE, DAVID R, b Bartlesville, Okla, Mar 11, 35; m 56; c 4. POLYMER CHEMISTRY, PHYSICAL CHEMISTRY. *Educ:* Southern Methodist Univ, BS, 57; Rice Univ, PhD(theoret chem), 61. *Prof Exp:* Assoc chem, Duke Univ, 61-62; assoc prof chem & head dept, Nicholls State Col, 61-62; chief phys chem br, US Army Res Off, 62-77, assoc dir chem div, 77-88; mem staff, Defense Advan Res Proj Agency, 88-90; ADJ & DIR RES & DEVELOP, LOKER HYDROCARBON RES INST, UNIV SOUTHERN CALIF, LOS ANGELES, 90- *Concurrent Pos:* Vis asst prof, Duke Univ, 62-72; adj prof, NC State Univ, 70. *Mem:* Am Chem Soc; Am Phys Soc. *Res:* Theoretical chemistry; fundamental investigations in polymer chemistry; radiation chemistry of polymers; statistical mechanics of solids and liquids. *Mailing Add:* 5316 Key Blvd No 1202 Arlington VA 22209-1542

SQUIRE, EDWARD NOONAN, chemistry, for more information see previous edition

SQUIRE, LARRY RYAN, b Cherokee, Iowa, May 4, 41; c 1. NEUROSCIENCE, NEUROPSYCHOLOGY. *Educ:* Oberlin Col, BA, 63; Mass Inst Technol, PhD(psychol), 68. *Prof Exp:* From asst prof to assoc prof, 73-81 PROF PSYCHIAT, UNIV CALIF, SAN DIEGO, 81-; RES CAREER SCIENTIST, VET ADMIN MED CTR, 80- *Concurrent Pos:* NIMH interdisciplinary fel, Albert Einstein Col Med, 68-70; clin investr, Vet Admin Med Ctr, 73-76. *Mem:* Fel Am Psychol Asn; Int Neuropsychol Soc; Soc Neurosci; Psychonomic Soc; fel AAAS; Int Brain Res Orgn. *Res:* Organization and neurological foundations of memory in man and non-human primate; neural plasticity; memory disorders in man; electroconvulsive therapy and memory. *Mailing Add:* Dept Psychiat M-003 Univ Calif San Diego Med Sch La Jolla CA 92093

SQUIRE, RICHARD DOUGLAS, b New York, NY, Oct 9, 40; m 80. GENETICS, RADIATION BIOLOGY. *Educ:* Hofstra Univ, BA, 63, MA, 69; NC State Univ, PhD(genetics), 69. *Prof Exp:* Asst prof biol, Long Island Univ, Brooklyn Ctr, 69-73, adj assoc prof, 74-75, 76-78; res fel, Med Ctr, Cornell Univ, 75-76; from asst prof to assoc prof, 78-87, PROF BIOL, UNIV PR, MAYAGUEZ, 87- *Concurrent Pos:* Adj assoc prof, Staten Island Community Col, 74-75, 77 & St John's Univ, 77; consult, Gamont-Doherty Geol Observ, 73. *Mem:* AAAS; Genetics Soc Am; Soc Study Evolution; Am Genetic Asn. *Res:* Radiation genetics, ecological genetics, developmental genetics, cytogenetics and ecology of Artemia; chemically-induced polyploidy in animals; animal and human cytogenetics; genetics of color patterns in birds and fish. *Mailing Add:* Dept Biol Univ PR Mayaguez Campus Mayaguez PR 00708

SQUIRE, ROBERT ALFRED, b Dobbs Ferry, NY, July 1, 30; m 50; c 3. COMPARATIVE PATHOLOGY. *Educ:* Univ Vt, BS, 52; Cornell Univ, DVM, 56, PhD(vet path), 64; Am Col Vet Path, dipl. *Prof Exp:* Private practice, Vt, 56-60; asst path, Cornell Univ, 60-61, instr, 62-64; asst prof, 64-68, ASSOC PROF PATH, SCH MED, JOHNS HOPKINS UNIV, 68-, ASSOC PROF COMP MED, 77- *Concurrent Pos:* Mem adv comt, Registry Comp Path, 68-; chmn comt lab animal dis, Nat Acad Sci-Nat Res Coun, 69-; mem adv coun, Morris Animal Found, 71-; chmn adv coun, NY State Vet Col, Cornell Univ, 71-; lectr, Armed Forces Inst Path; dir comp path, Johns Hopkins Univ, 66-76. *Mem:* Int Acad Path; Am Asn Cancer Res; Am Col Vet Path; Am Vet Med Asn. *Res:* Pathology of hematopoietic tissues; animal models of human disease, particularly lymphomas and immunologic diseases. *Mailing Add:* Dept Comp Med Johns Hopkins Univ 720 Rutland Ave Baltimore MD 21205

SQUIRE, WILLIAM, b New York, NY, Sept 22, 20; m 48; c 2. APPLIED MATHEMATICS, FLUID MECHANICS. *Educ:* City Col New York, BS, 41; Univ Buffalo, MA, 59. *Prof Exp:* Inspector, Philadelphia Signal Corps Inspection Zone, US Dept Army, 42-43; jr physicist, Nat Bur Standards, 43-45; asst physicist, 45-48; assoc physicist, Cornell Aeronaut Lab, Inc, 48-57; aerodynamicist, Bell Aircraft Corp, 57-59; sr res engr, Southwest Res Inst, 59-61; prof, 61-86, EMER PROF, AEROSPACE ENG, WVA UNIV, 86- *Concurrent Pos:* Vis lectr, UTA, 87-88. *Mem:* assoc fel Inst Aeronaut & Astronaut. *Res:* Turbulence; high temperature gas dynamics; boundary layer theory; numerical integration. *Mailing Add:* Dept Mec & Aerospace Eng WVa Univ Morgantown WV 26506-6101

SQUIRES, ARTHUR MORTON, b Neodesha, Kans, Mar 21, 16. CHEMICAL ENGINEERING. *Educ:* Univ Mo, AB, 38; Cornell Univ, PhD(phys chem), 47. *Prof Exp:* Asst chem, Univ Mo, 38 & Cornell Univ, 38-41; lab technician, E I du Pont de Nemours & Co, NY, 41; phys chemist, M W Kellogg Co, 42-43 & Kellex Corp, 43-46; asst head, Process Develop Dept, Hydrocarbon Res, Inc, 46-51, head, 51-59; prof, 67-74, City Col New York, chmn dept, 71-74, distinguished prof chem eng, 74-76; Frank C Vilbrandt prof chem eng, 76-82, distinguished prof chem eng, 78-86, EMER DISTINGUISHED PROF, VA POLYTECH INST & STATE UNIV, 86- *Concurrent Pos:* process consult, 59-67. *Honors & Awards:* Storch Award, Am Chem Soc, 73. *Mem:* Nat Acad Eng; Am Soc Mech Eng; Am Chem Soc; Am Inst Chem Eng; Am Acad Arts & Sci. *Res:* Physical chemistry of solutions; multistage fractionation; flow-properties of fluid-solids systems; hydrocarbon synthesis and cracking; iron ore reduction; coal and oil gasification; hydrogen production; power generation; dust collection; low-temperature processes; small-scale coal combustion; heat transfer in shallow vibrated and gas-fluidized beds of particles; government management of technological change; vibrated-bed microreactors. *Mailing Add:* Dept Chem Eng Va Polytech Inst & State Univ Blacksburg VA 24061

SQUIRES, CATHERINE L, b Sacramento, Calif, Apr 9, 41; m 66; c 2. MOLECULAR GENETICS. *Educ:* Univ Calif, Santa Barbara, PhD(molecular biol & biochem), 72. *Prof Exp:* Postdoctoral fel, Stanford Univ, 71-74; asst prof, Darmouth Col, 75-77; from asst prof to assoc prof, 77-87, PROF, DEPT BIOL SCI, COLUMBIA UNIV, 87- *Mem:* Sigma Xi; AAAS; Am Soc Microbiol; Harvey Soc; Am Soc Biol Chemists. *Res:* Expression of the transciption and translation machinery in prokaryotes. *Mailing Add:* Dept Biol Sci Columbia Univ 710 Fairchild New York NY 10027

SQUIRES, DALE EDWARD, b San Diego, Calif, Aug 28, 50; m 86; c 1. RESOURCE ECONOMICS, ECONOMICS OF REGULATION. *Educ:* Univ Calif, Berkeley, BSc, 73, MSc, 78; Cornell Univ, PhD(resource econ), 84. *Prof Exp:* Asst prof econ theory, Fac Econ & Mgt, Univ Agr Malaysia, 75-77; sr economist, Dept Fisheries, Sabah Malaysia, 77-78; intern, Agency Int Develop, 79; INDUST ECONOMIST, NAT MARINE FISHERIES SERV, 82-; ASSOC ADJ PROF, ECON OCEAN RESOURCES, DEPT ECON, UNIV CALIF, SAN DIEGO, 89- *Concurrent Pos:* Scholastic fel, Foreign Lang Area Studies, Indonesia, 78-79 & Sea Grant Scholar, 80-81; mem, Working Group Ltd Access Alternatives, Pac Fishery Mgt Coun, 84-86, Ltd Entry Tech Adv Group, 87-89, Ltd Entry Oversight Comt, 89-90 & Groundfish Mgt Team, 89-; vis prof, Dept Econ, Univ Queensland, Brisbane, Qld, Australia, 90; assoc adj prof, Pub Policy Anal, Grad Sch Int Rels & Pac Studies, Univ Calif, San Diego, 91- *Mem:* Am Econ Asn; Am Agr Econ Asn; Asian Fisheries Soc. *Res:* Analysis of markets for tradable quotas and property rights; productivity measurement; economics of quotas; measurement of capacity utilization. *Mailing Add:* Southwest Fisheries Sci Ctr PO Box 271 La Jolla CA 92038-0271

SQUIRES, DONALD FLEMING, b Glen Cove, NY, Dec 19, 27; m 51; c 2. GENERAL MARINE & ENVIRONMENTAL SCIENCES. *Educ:* Cornell Univ, AB, 50, PhD, 55; Univ Kans, MA, 52. *Prof Exp:* Asst cur paleont, Am Mus Natural Hist, 55-61, assoc cur, 61-62; assoc cur marine invert, Mus Natural Hist, Smithsonian Inst, 62-63, cur-in-chg, 63-64, chmn dept invert zool, 65, dep dir, 66-68; actg assoc provost grad studies & res, 72-73, prof biol sci, earth & space sci & dir, Marine Sci Res Ctr, State Univ NY Stony Brook, 68-85; PROF MARINE SCI & DIR, MARINE SCI INST, UNIV CONN, 85- *Concurrent Pos:* Fulbright res fel, NZ, 59; dir, NY Sea Grant Inst, 71-85. *Honors & Awards:* Secy's Gold Medal, Smithsonian Inst. *Mem:* Fel AAAS; Sigma Xi; Marine Technol Soc. *Res:* Marine and coastal policy. *Mailing Add:* Univ Conn Marine Sci Inst Box U-6 Rm 215 438 Whitney Rd Ext Storrs CT 06269

SQUIRES, LOMBARD, ENGINEERING. *Prof Exp:* Asst gen mgr, E I du Pont de Nemours & Co; RETIRED. *Mem:* Nat Acad Eng. *Mailing Add:* 100 Moorings Park Dr No F101 Naples FL 33940

SQUIRES, PAUL HERMAN, b Sewickley, Pa, July 14, 31; m 53; c 6. CHEMICAL ENGINEERING, POLYMER PROCESSING. *Educ:* Rensselaer Polytech Inst, BChE, 53, PhD(chem eng), 56; Univ Wis, MS, 54. *Prof Exp:* Engr, Plastics Dept, 56-59, supvr, 59-63, sr res engr, 63-67, res assoc, Eng Dept, 67-76, prin consult, Eng Dept, E I Du Pont De Nemours & Co, Inc, 76-; RETIRED. *Honors & Awards:* Presidents' Cup, Soc Plastics Engrs, 89, Distinguished Serv Award, 89. *Mem:* Fel Soc Plastics Engrs; Am Inst Chem Engrs. *Res:* Energy and momentum transfer in viscous materials, especially translation of theory into design and development of plastics processing equipment; laminated safety glass interlayer; extrusion. *Mailing Add:* 1235 E Lake Shore Dr Landrum SC 29356

SQUIRES, RICHARD FELT, b Sparta, Mich, Jan 15, 33; m 70; c 1. NEUROPHARMACOLOGY, NEUROCHEMISTRY. *Educ:* Mich State Univ, BS, 58. *Prof Exp:* Res biochemist, Pasadena Found Med Res, 61-62; head biochemist, res dept, Ferrosan A/S, Soeborg, Denmark, 63-78; group leader CNS biol, nerv res neurochem, Lederle Labs, Am Cyanamid, 78-79; PRIN RES SCIENTIST, NATHAN KLINE INST PSYCHIAT RES, 79- *Mem:* Soc Neurosci; Europ Neurosci Asn; Am Soc Neurochem; Int Soc Neurochem; Am Soc Biol Chemists; Am Soc Pharmacol & Exp Therapeut. *Res:* Gamma aminobutyric acid; benzodiazepine; picrotoxin receptors in brain; biochemistry; characterization of receptors in central nervous system; development of novel psychotropic drugs; neurological and psychiatric disorders. *Mailing Add:* Nathan Kline Inst Orangeburg NY 10962

SQUIRES, ROBERT GEORGE, b Sewickley, Pa, Oct 1, 35; m 57; c 3. CHEMICAL ENGINEERING. *Educ:* Rensselaer Polytech Inst, BChE, 57; Univ Mich, Ann Arbor, MSE, 58, MS, 60, PhD(chem eng), 63. *Prof Exp:* From asst prof to assoc prof, 62-72, PROF CHEM ENG, PURDUE UNIV, WEST LAFAYETTE, 72- *Mem:* Am Inst Chem Engrs; Am Chem Soc; Am Soc Eng Educ; Catalysis Soc. *Res:* Heterogeneous catalysis; adsorption; reaction kinetics. *Mailing Add:* 807 N Salisbury St West Lafayette IN 47906-2715

SQUIRES, ROBERT WRIGHT, b Barberton, Ohio, Aug 25, 21; m 48; c 2. MICROBIOLOGY. *Educ:* Kent State Univ, BS, 48; Purdue Univ, MS, 50, PhD(microbiol), 54. *Prof Exp:* Instr gen bact, Purdue Univ, 49, instr food bact, 50-53; sr microbiologist, Eli Lilly & Co, 54-65, res scientist, 65-66, asst mgr pilot plant opers, 66-67, mgr antibiotic opers, 67-69, eng coordr antibiotic fermentations, 69-80; tech dir biomfg, Searle Chem, Inc, 81-83, dir biochem process technol, Nutrasweet Group, G D Searle, 84-89; CONSULT, 90- *Honors & Awards:* Charles Porter Award, Soc Indust Microbiologists, 66. *Mem:* Fel Soc Indust Microbiologists (treas, 61-66, pres, 68-69); Am Inst Biol Scientists; Sigma Xi; Am Soc Microbiol; Instrument Soc Am. *Res:* Antibiotic fermentation pilot plant operations; fermentation development and equipment design; continuous fermentation; process scaleup and fermentation process control dynamics; fermentor reactor design large airlift as well as agitated tanks. *Mailing Add:* 10105 Dorsey Hill Rd Louisville KY 40223

SRAMEK, RICHARD ANTHONY, b Baltimore, Md, June 5, 43. RADIO ASTRONOMY. *Educ:* Mass Inst Technol, BS, 65; Calif Inst Technol, PhD(astron), 70. *Prof Exp:* Res assoc, Nat Radio Astron Observ, 70-72, asst scientist, 72-74, assoc scientist, 74-75; res assoc, Arecibo Observ, 75-78; VLA SCIENTIST, NAT RADIO ASTRON OBSERV, 78- *Mem:* Am Astron Soc; Int Union Radio Sci. *Res:* Extra-galactic astronomy and experimental tests of general relativity. *Mailing Add:* 1005 Calle Del Sol Socorro NM 87801

SRB, ADRIAN MORRIS, b Howells, Nebr, Mar 4, 17; m 40; c 3. GENETICS. *Educ:* Univ Nebr, AB, 37, MS, 41; Stanford Univ, PhD(genetics), 46. *Hon Degrees:* DSc, Univ Nebr, 69. *Prof Exp:* Asst prof biol, Stanford Univ, 46-47; from assoc prof to prof plant breeding, 48-65, prof, 65-76, Jacob Gould Schurman prof genetics, 76-84, JACOB GOULD SCHURMAN EMER PROF GENETICS, CORNELL UNIV, 84- *Concurrent Pos:* Nat Res Coun fel, Calif Inst Technol, 46-47; Guggenheim fel & Fulbright res fel, Univ Paris, 53-54, NSF sr res fel, 60-61, Univ Edinburgh, 67-68; dir, NIH pre- & postdoctoral training prog genetics, Cornell Univ, 62-67; Darling lectr, Allegheny Col, 64; mem genetics study sect, NIH, 64-67 & genetics training comt, 69-; vis scholar, Va Polytech Inst, 65; co-chmn prog comt, XI Int Bot Cong; trustee, Cornell Univ, 75-80; mem educ adv bd, John Simon Guggenheim Mem Found, 76- *Mem:* Nat Acad Sci; fel AAAS; Genetics Soc Am; Am Soc Naturalists; fel Am Acad Arts & Sci. *Res:* Physiological genetics of fungi; mutagenesis; extranuclear heredity; developmental genetics. *Mailing Add:* Genetics & Develop Cornell Univ, 219 Bradfield Hall Ithaca NY 14853

SREBNIK, HERBERT HARRY, b Berlin, Ger, Mar 25, 23; nat US; m 51; c 2. ANATOMY. *Educ:* Univ Calif, BA, 50, MA, 55, PhD(anat), 57. *Prof Exp:* Asst anat, 53-57; from instr to prof anat, 57-89, chmn, Dept Biol, 87-89, EMER PROF ANAT, UNIV CALIF, BERKELEY, 90- *Mem:* AAAS; Soc Exp Biol & Med; Am Inst Nutrit; Am Asn Anat; Endocrine Soc. *Res:* Endocrine-nutrition interrelationships; physiology of reproduction; anterior pituitary function; investigations into hormonal and nutritional factors influencing the course and outcome of gestation in laboratory animals. *Mailing Add:* Dept Molecular & Cell Biol Univ Calif Berkeley CA 94720

SREBRO, RICHARD, b New York, NY, Jan 9, 36; m 67; c 3. PHYSIOLOGY. *Educ:* Wash Univ, MD, 59. *Prof Exp:* Intern, State Univ NY Upstate Med Ctr, 59-60; resident ophthal, Sch Med, Wash Univ, 60-62; sr asst surgeon, Lab Phys Biol, NIH, 62-64; res scientist biophys, Walter Reed Army Inst Res, 65-68; from res asst prof to assoc prof physiol, State Univ NY Buffalo, 68-77; assoc prof ophthal & physiol, 77-87, PROF OPHTHAL, SOUTHWESTERN MED SCH, DALLAS, 87- *Concurrent Pos:* Nat Inst Neurol Dis & Blindness fel, 64-65; sr sci investr award, res to prevent blindness, 88. *Mem:* Am Asn Artificial Intel; Asn Res Vision & Ophthal; Soc Neurosci. *Res:* Physiology of vision. *Mailing Add:* Dept Ophthal Southwestern Med Sch 5323 Harry Hines Blvd Dallas TX 75235

SREEBNY, LEO MORRIS, b New York, NY, Jan 8, 22; m 45; c 2. PATHOLOGY, BIOCHEMISTRY. *Educ:* Univ Ill, AB, 42, DDS, 45, MS, 50, PhD(path), 54. *Prof Exp:* Asst therapeut, Col Dent, Univ Ill, 49-50, instr, 51-53, from asst prof to assoc prof oral path, 53-57; from assoc prof to prof oral biol, Sch Dent, Univ Wash, 57-75, chmn dept, 57-75, prof path, Sch Med, 65-75; dean, Sch Dent Med, 75-79, PROF ORAL BIOL, SCH DENT MED, STATE UNIV NY, STONY BROOK, 75- *Concurrent Pos:* Fulbright lectr & advan res award, Hebrew Univ Jerusalem, 63-64; dir, Ctr Res Oral Biol, Sch Med, Univ Wash, 68-75; mem bd dirs, Am Asn Dent Res, 81; chmn, Sci Prog, Int Dent Fed, 70-88. *Honors & Awards:* Anat Sci Award, Int Asn Dent Res, 69; Silver Medal Dent Soc, Paris, France, 79; List of Honor, Fed Dent Int, 89. *Mem:* AAAS; Int Asn Dent Res; Am Dent Asn; Int Dent Fedn. *Res:* Secretory mechanism of salivary secretions; pathophysiology of diseases of the oral cavity; nutrition and oral diseases. *Mailing Add:* Sch Dent Med State Univ NY Stony Brook NY 11794

SREE HARSHA, KARNAMADAKALA S, b India, May 25, 36; m 67; c 2. MATERIALS SCIENCE. *Educ:* Univ Mysore, BSc, 55; Indian Inst Sci, Bangalore, dipl metall, 57; Univ Notre Dame, MS, 60; Pa State Univ, PhD(metall), 64. *Prof Exp:* Res fel, Iowa State Univ, 65-67; asst prof mat sci, 67-77, PROF MAT SCI, SAN JOSE STATE UNIV, 77-, CHMN, DEPT MAT ENG, 78- *Mem:* Am Soc Metals; Am Inst Mining, Metall & Petrol Engrs; Mat Res Soc. *Res:* Structure and transformations of solids; theory of dislocation; surfaces; thermodynamics; semiconductors. *Mailing Add:* Dept Mat Sci San Jose State Univ Washington Sq San Jose CA 95192

SREENIVASAN, KATEPALLI RAJU, b Kolar, India, Sept 30, 47; m 80; c 1. SCIENCE & SOCIETY, TEACHING OF SCIENCE & TECHNOLOGY. *Educ:* Bangalore Univ, BE, 68; Indian Inst Sci, ME, 70, PhD(aeronaut eng), 75. *Prof Exp:* JRD Tata fel, Indian Inst Sci, 72-74, proj asst, 74-75; fel, Univ & Sydney, 75, Univ Newcastle, 76-77; res assoc, Johns Hopkins Univ,77-79; from asst prof to assoc prof, 82-85, PROF MECH ENG, YALE UNIV, 85-, CHMN DEPT, 87-, & HAROLD W CHEEL PROF, 88- *Concurrent Pos:* Vis scientist, Indian Inst Sci, 79, vis prof, 82; vis sci, DFVLR, Gottingen, W Ger, 83; vis prof, Cal Inst Tech, Pasadena, 86; actg chmn, Coun Eng, 87. *Honors & Awards:* Narayan Gold Medal, Indian Inst Sci, 75; fel, Humboldt Found, 83. *Mem:* ASME; Am Inst Astronaut & Aeronaut; fel Am Inst Physics; Am Math Soc; Sigma Xi. *Res:* Origin and dynamics of turbulence; control of turbulent flows; chaotic dynamics; fractals. *Mailing Add:* Mech Eng M6 ML Yale Univ New Haven CT 06520

SREENIVASAN, SREENIVASA RANGA, b Mysore, India, Oct 20, 33; m 63; c 5. PHYSICS, ASTROPHYSICS. *Educ:* Univ Mysore, BS, 50, BS, 52; Gujarat Univ, India, PhD(physics), 58. *Prof Exp:* Res fel, Harvard Univ, 59-61; Nat Acad Sci res assoc, Goddard Inst Space Studies, New York, 61-64; vis scientist, Max Planck Inst Physics & Astrophys, 64-66; from asst prof to assoc prof physics, 67-75, PROF PHYSICS, UNIV CALGARY, 75- *Concurrent Pos:* Vis prof, Royal Inst Technol, Univ Stockholm & vis scientist, Swed Natural Sci Res Coun, 74-75. *Mem:* Am Astron Soc; Am Geophys Union; Am Phys Soc; fel Royal Astron Soc; Int Astron Union; Am Meteorol Soc; Astron Soc Japan. *Res:* Theoretical astrophysics; theoretical plasma physics; general relativity; nonlinear phenomena in physics. *Mailing Add:* Dept Physics & Astron Univ Calgary Calgary AB T2N 1N4 Can

SREEVALSAN, THAZEPADATH, b Kanjiramattom, India, Jan 25, 35. MICROBIOLOGY, VIROLOGY. *Educ:* Univ Kerala, BSc, 53, MSc, 56; Univ Tex, PhD(microbiol), 64. *Prof Exp:* Teacher, St Ignatius High Sch, India, 53-54; res asst virol, Pasteur Inst, Coonoor, 56-61 & Univ Tex, 61-63, assoc res scientist, 64-66; res scientist virol, Cent Res Sta, E I du Pont de Nemours & Co, Inc, 66-69; asst prof, 69-73, assoc prof, 73-80, PROF MICROBIOL, MED & DENT SCH, GEORGETOWN UNIV, 80- *Mem:* AAAS; Am Soc Microbiol. *Res:* Viruses, replication of animal viruses, molecular biology of animal virus development, mode of RNA replication; interfrons their biological activity and mode of action in inhibiting cell growth and viral multiplication; control of proliferation in animal cells. *Mailing Add:* Dept Microbiol Georgetown Univ Sch Med 3900 Reservoir NW Washington DC 20007

SRERE, PAUL ARNOLD, b Davenport, Iowa, Sept 1, 25; m 53; c 4. ENZYMOLOGY, METABOLISM. *Educ:* Univ Calif, Los Angeles, BS, 47; Univ Calif, Berkley PhD(comp biochem), 51. *Hon Degrees:* Dr, Pecs Univ Hungary, 90. *Prof Exp:* Asst physiol, Univ Calif, 47-51; asst biochemist, Mass Gen Hosp, 51-53; from asst prof to assoc prof biochem, Univ Mich, 56-63; biochemist, Biomed Div, Lawrence Radiation Lab, Univ Calif, 63-66; PROF BIOCHEM, SOUTHWESTERN MED SCH, UNIV TEX HEALTH SCI CTR DALLAS, 66- *Concurrent Pos:* Childs Found Med Res fel, Yale Univ, 53-54; USPHS fels, Pub Health Res Inst, New York, 54-55 & Max Planck Inst Cell Biol, Ger, 55-56; chief basic biochem unit, Gen Med Res, Vet Admin Hosp, 69-72. *Honors & Awards:* William S Middleton Award, 74. *Mem:* Am Chem Soc; Fedn Am Soc Exp Biol. *Res:* Intermediary metabolism; enzymology; metabolic regulation. *Mailing Add:* Vet Admin Med Ctr 4500 S Lancaster Rd Dallas TX 75216

SRETER, FRANK A, b Szanda, Hungary, Oct 2, 21; US citizen; m 44; c 2. BIOCHEMISTRY. *Educ:* Budapest Tech Univ, MS, 43, PhD(animal nutrit), 44; Vet Sch Budapest, DVM, 49; Med Univ Budapest, MD, 50. *Prof Exp:* From instr to assoc prof, Budapest Tech Univ, 44-51; assoc prof, E-tv-s Lorand Univ, Budapest, 51-56; from instr to asst prof, Univ BC, 57-63; res assoc, Dept Muscle Res, Retina Found, 63-72; SR RES SCIENTIST, DEPT MUSCLE RES, BOSTON BIOMED RES INST, 72-; ASSOC, DEPT NEUROPATH, HARVARD MED SCH, 72- *Concurrent Pos:* Can Muscular Dystrophy, Inc fel, 63-66; estab investr, Am Heart Asn, 66-71; assoc biochemist, Mass Gen Hosp. *Mem:* Am Physiol Soc; Biophys Soc. *Res:* Muscle physiology and biochemistry. *Mailing Add:* Dept Muscle Res Boston Biomed Res Inst 20 Staniford St Boston MA 02114

SRIBNEY, MICHAEL, b Alta, Can, Feb 5, 27; m 54; c 2. BIOCHEMISTRY. *Educ:* Univ Alta, BSc, 52; McMaster Univ, MSc, 53; Univ Chicago, PhD, 57. *Prof Exp:* Fel, Enzyme Inst, Univ Wis, 58-60; asst prof biochem & psychiat, Yale Univ, 60-69; ASSOC PROF BIOCHEM, QUEEN'S UNIV, ONT, 69- *Mem:* AAAS; Am Chem Soc. *Res:* Lipids. *Mailing Add:* Dept of Biochem Queen's Univ Kingston ON K7L 3N6 Can

SRIDARAN, RAJAGOPALA, b Papireddipatti, India, Feb 22, 50; m 78; c 2. REPRODUCTIVE ENDOCRINOLOGY. *Educ:* Univ Madras, India, BS, 70, MS, 72; Univ Health Sci, Chicago Med Sch, PhD(physiol), 77. *Prof Exp:* Instr zool, Madras Christian Col, 72-73; res assoc endocrinol, Univ Nebr Med Ctr, Omaha, 77-78, Univ Ill Med Ctr, 78-81; asst prof, 81-87, ASSOC PROF PHYSIOL, MOREHOUSE SCH MED, ATLANTA, 87-; ADJ PROF BIOL, GA STATE UNIV, ATLANTA, 89- *Mem:* Am Physiolol Soc; Endocrin Soc; Soc Neurosci; Soc Study Reproduct; AAAS; Soc Exp Biol Med. *Res:* Corpus luteum function and maintenance of pregnancy; mechanism of antifertility action of gonadotropin-releasing hormone agonisk (GnRM-sq) during pregnancy in the rat; circadian rhythms in reproductive endocrinology. *Mailing Add:* Dept Physiol Morehouse Sch Med 720 Westview Dr SW Atlanta GA 30310-1495

SRIDHAR, CHAMPA GUHA, Indian citizen. SOLID STATE PHYSICS. *Educ:* Calcutta Univ, BSc, 63; Jadavpur Univ, Calcutta, MSc, 65; Northeastern Univ, MS, 68; Univ Conn, PhD(solid state physics), 73. *Prof Exp:* Teaching asst physics, Northeastern Univ, 66-68; asst, Univ Conn, 68-73, fel liquid crystal, 73; fel solid state physics, Stanford Univ, 74; Nat Res Coun fel theoret solid state physics, Ames Res Ctr, NASA, Moffett Field, 74-77; sr process engr, SX-70 film, Polaroid Corp, 77; DEVELOP ENGR PRECISION FREQUENCY STANDARD, HEWLETT PACKARD CO, 78- *Mem:* Am Phys Soc; Sigma Xi. *Res:* Ban structure and optical properties calculation of solids; optical and electron spin resonance studies of radiation induced colour centers in solids; magnetic properties of liquid crystals. *Mailing Add:* 834 Hierra Ct Los Altos CA 94022

SRIDHAR, RAJAGOPALAN, b Trichinopoly, Madras, India, July 29, 41; Can citizen. BIOCHEMISTRY, CHEMISTRY. *Educ:* Univ Delhi, BSc, 61; Kurukshetra Univ, India, MSc, 63; Univ London, PhD(org chem), & DIC, 68. *Prof Exp:* Fel org chem, Res Inst for Med & Chem, Cambridge, 68-71; res assoc, Johns Hopkins Univ, 71-72; assoc biochem, Univ Western Ont, 73-75; res fel radiobiol, Ont Cancer Treatment & Res Found, 75-78; ASST MEM BIOMEMBRANE RES, OKLA MED RES FOUND, 78-; PROF, UNIV SC. *Concurrent Pos:* Consult biochem, Ont Cancer Treatment & Res Found, London Clin, Victoria Hosp, 78-; adj asst prof, Dept Radiol Sci, Okla Univ Health Sci Ctr, 81- *Mem:* Assoc mem Radiation Res Soc; Am Soc Photobiol; Biophys Soc; Am Asn Cancer Res; Sigma Xi. *Res:* Electron spin resonance studies in biological systems; DNA and membrane damage due to carcinogens and pharmaceuticals; radiosensitizers and other cytotoxic agents specific for hypoxic cells; multicell spheroids as a solid tumor model; nuclear magnetic resonance studies in biological systems. *Mailing Add:* Dept Radiother Howard Univ Hosp 2041 Georgia Ave NW Washington DC 20060

SRIDHARA, S, HORMONE ACTION, REGULATION TRANSCRIPTION. *Educ:* Indian Inst Sci, Bangalora, India, PhD(biochem), 65. *Prof Exp:* ASSOC PROF BIOCHEM, HEALTH & SCI CTR, TEX TECH UNIV, 82- *Res:* Gene activity during insect metamorphosis. *Mailing Add:* Dept Biochem Tex Tech Univ Health Sci Ctr 3601 Fourth St Lubbock TX 79430

SRIDHARAN, NATESA S, b Madras, India, Oct 2, 46; US citizen; m; c 1. COGNITIVE MODEL BUILDING. *Educ:* Indian Inst Technol, BTech, 67; State Univ NY, Stonybrook, MS, 69, PhD(comput sci), 71. *Prof Exp:* Res assoc, Stanford Univ, 71-74; vis scientist, Technische Univ, Munich, 74; from asst prof to assoc prof comput sci, Rutgers Univ, 74-84; div scientist, BBN Labs, 84-86; MGR, ARTIFICIAL INTEL, FMC CORP, 86- *Concurrent Pos:* Prin investr, NSF grants, 79-; sci adv, Nat Ctr Telecommun Study, France, 82-86; prin investr, KRNL/Defense Adv Res Projs Agency, 86. *Honors & Awards:* Siemen's Medal, Indian Inst Technol, 67. *Mem:* Am Asn Artificial Intel; Inst Elec & Electronic Engrs; Asn Comput Mach. *Res:* Building logical, computer-based models of human reasoning, including those of scientist, lay people and lawyers; development of tools to assist in such modelling; large-scale software design; implementation of A1 solutions to real problems. *Mailing Add:* 271 Old Adobe Rd Los Gatos CA 95030

SRIHARI, SARGUR N, b Bangalore, India, May 7, 50; m 77. ARTIFICIAL INTELLIGENCE, RECOGNITION. *Educ:* Bangalore Univ, BSc, 68; Indian Inst Sci, BE, 70; Ohio State Univ, PhD(comput & info sci), 76. *Prof Exp:* Res asst, Ohio State Univ, 70-75; asst prof, Wayne State Univ, 76-78; from asst prof to assoc prof, 78-87, PROF COMPUT SCI, STATE UNIV NY, BUFFALO, 87-, DIR, CTR DOCUMENT ANAL & RECOGNITION, 91- *Concurrent Pos:* Assoc ed, Pattern Recognition J, 83-; prin investr, USPS Proj, 84-; consult, Xerox Corp, 85-; co-prin investr, RADC-AI Consortium, 85- *Mem:* Asn Comput Mach; sr mem Inst Elec & Electronic Engrs; Pattern Recognition Soc; Am Asn Artificial Intelligence. *Res:* Pattern recognition and artificial intelligence; document image analysis; expert systems. *Mailing Add:* Dept Comput Sci State Univ NY at Buffalo Amherst NY 14260

SRINATH, MANDYAM DHATI, b Bangalore, India, Oct 12, 35; m 65; c 2. ELECTRICAL ENGINEERING. *Educ:* Univ Mysore, BSc, 54; Indian Inst Sci, DIISc, 57; Univ Ill, Urbana, MS, 59, PhD(elec eng), 62. *Prof Exp:* Asst prof elec eng, Univ Kans, 62-64; asst prof, Indian Inst Sci, 64-67; assoc prof info & control sci, 67-76, PROF ELEC ENG, SOUTHERN METHODIST UNIV, 76- *Mem:* Inst Elec & Electronics Engrs; Soc Indust & Appl Math. *Res:* Control and estimation theory; digital signal processing; identification. *Mailing Add:* Dept of Elec Eng Southern Methodist Univ Dallas TX 75275

SRINIVASAN, ASOKA, b Bangalore, India, May 13, 39, US citizen; m 67; c 2. PHEROMONES. *Educ:* Univ Mysore, India, BSc, 63; Univ Calif, Berkeley, PhD(entom), 70; MBA, Millsaps Col, 84. *Prof Exp:* Asst prof biol, 69-74, assoc prof, 74-80, chmn, Biol Dept, 71-81, chmn, nat sci div, 81-83, DIR BIOMED RES, TOUGALOO COL, 73-, PROF BIOL, 80- *Concurrent Pos:* Asst to pres, Miss Bio Asn, 87-, secy/treas, 88- *Mem:* Tissue Cult Asn; Entom Soc Am; Am Inst Biol Sci; AAAS; Sigma Xi. *Res:* Development and activity of the sex pheromone gland in Lepidoptera; tissue culture of sex pheromone gland of insects. *Mailing Add:* 1948 Cherokee Dr Jackson MS 39211

SRINIVASAN, BHAMA, b Madras, India, Apr 22, 35; nat US. MATHEMATICS. *Educ:* Univ Madras, BA, 54, MSc, 55; Univ Manchester, PhD(math), 59. *Prof Exp:* Lectr math, Univ Keele, 60-64; Nat Res Coun Can fel, Univ BC, 65-66; reader math, Univ Madras, 67-70; assoc prof math, Clark Univ, 70-80; PROF MATH, UNIV ILL CHICAGO CIRCLE, 80- *Mem:* Am Math Soc; London Math Soc; Indian Math Soc; Asn Women Math (pres, 81-83). *Res:* Representations of finite Chevalley groups. *Mailing Add:* Dept Math M-C 249 Univ Ill Chicago Circle Box 4348 Chicago IL 60680

SRINIVASAN, G(URUMAKONDA) R, b Mysore, India; m 68; c 4. THEORETICAL MODELING, RADIATION EFFECTS. *Educ:* Univ Mysore, BS, 54, Hons, 56; Indian Inst Sci, dipl metall, 58; Colo Sch Mines, MS, 61; Univ Ill, Urbana, PhD(metall), 66. *Prof Exp:* Res scholar metall, Indian Inst Sci, 58-59; res asst, Colo Sch Mines, 59-61; res asst, Univ Ill, Urbana, 61-66; res assoc mat sci, Cornell Univ, 66-68; from asst prof to assoc prof mat sci, Cath Univ Am, 68174; adv engr, 74-82, SR ENG MGR, GEN TECHNOL DIV, IBM CORP, 82- *Concurrent Pos:* Ed jour, Electronics Div, Electrochem Soc, 84-90; tech planning comt, Electrochem Soc, 83-, Europ affairs comt, 82- *Mem:* Metall Soc; Electrochem Soc. *Res:* Theoretical modeling; epitaxy; semiconductor devices; phase transformations; electron microscopy; x-ray diffraction and scattering; device physics; head of theoretical modeling group covering transistor processes and devices; cosmic ray and radiation effects; stress effects; ion implantation channeling theory and energy loss; epitaxy and diffusion; computer-aided device design; dislocation modeling; device physics. *Mailing Add:* 15 Mark Vincent Dr Poughkeepsie NY 12603

SRINIVASAN, MAKUTESWARAN, b Tiruchirapalli, India, Jan 26, 45; m 71; c 2. MATERIALS SCIENCE, PHYSICAL METALLURGY. *Educ:* Univ Madras, India, BSc, 64; Indian Inst Sci, BE, 67; Univ Wash, MSinMetD, 69, PhD(metall), 72. *Prof Exp:* Res assoc mat sci, Univ Wash, 72-74; staff scientist graphite, Union Carbide Corp, 74-78; develop assoc silicon carbide, Carborundum Co, 78-79, sr res assoc struct ceramic, 79-83; mgr mat characterization & properties res, 84-90, RES MGR, STRUCT & REFRACTORY PROD, SOHIO ENGINEERED MAT CO, 90- *Mem:* Am Soc Metals; Am Ceramic Soc; Am Soc Nondestructive Testing. *Res:* Mechanical properties of ceramics, mainly high temperature, high performance ceramics such as silicon carbide and silicon nitride; tribology; corrosion; advanced nondestructive evaluation; reliability and ceramic design; process yield improvement; materials selection. *Mailing Add:* Mat Solutions PO Box 663 Grand Island NY 14072-0663

SRINIVASAN, P R, b Villupuram, India, Nov 24, 27. BIOCHEMISTRY. *Educ:* Univ Madras, BSc, 46, PhD(biochem), 53; Banaras Hindu Univ, MSc, 48. *Prof Exp:* Asst res officer biochem, Indian Coun Med Res, 52-53; Fulbright-Smith Mundt fel, 53-54, res fel, 53-57, res assoc, 57-58, from instr to assoc prof, 58-70, PROF BIOCHEM, COL PHYSICIANS & SURGEONS, COLUMBIA UNIV, 70- *Mem:* Am Chem Soc; Harvey Soc; Am Soc Microbiol; NY Acad Sci; Am Soc Biol Chemists. *Res:* Biological function of methylated bases in nucleic acids; regulatory mechanisms and transformation of normal cells by animal viruses. *Mailing Add:* Columbia Univ 630 W 168th St New York NY 10032

SRINIVASAN, RAMACHANDRA SRINI, b Madurai, India, Mar 16, 39; US citizen; m 71; c 1. MATHEMATICAL MODELING & COMPUTER SIMULATION OF PHYSIOLOGICAL SYSTEMS, SPACEFLIGHT PHYSIOLOGY. *Educ:* Madras Univ, India, BE, 60; Indian Inst Sci, ME, 62; Purdue Univ, MSEE, 65; Calif Inst Technol, PhD(elec eng), 69. *Prof Exp:* Res assoc, Elec Eng Dept, Rice Univ, 70-72 & Baylor Col Med, Tex Med Ctr, 72-75; biomed engr, Abbott Labs, Chicago, 75-77; asst prof systs analysis, Dept Med & Biomet, Med Univ SC, 77-80; proj scientist & biomed engr, Gen Elec Govt Serv, 80-87; SR RES SCIENTIST, KRUG LIFE SCI, INC, HOUSTON, 87- *Concurrent Pos:* Instr, Elec Eng Dept, Rice Univ, 74-75 & Univ Houston, 87-; prin investr, NASA, 87-, co-investr, 91- *Mem:* Sr mem Biomed Eng Soc; Inst Elec & Electronics Engrs; Sigma Xi. *Res:* Mathematical modeling and computer simulation of physiological systems and cardiovascular systems; spaceflight biomedical data. *Mailing Add:* Krug Life Sci Inc 1290 Hercules Dr Suite 120 Houston TX 77058

SRINIVASAN, RANGASWAMY, b Madras, India, 1929; US citizen. OPTICS, CHEMICAL DYNAMICS. *Educ:* Univ Madras, BSc Hons, 49, MSc, 51; Univ Southern Calif, PhD(phys chem), 56. *Prof Exp:* Mgr, T J Watson Res Ctr, IBM, 61-90; PRES, UV TECH, 90- *Concurrent Pos:* Guggenheim fel, 65; vis prof, Ohio State Univ, 66-67, Columbia-Presby Med Ctr, 84-, Wellman Lab, Harvard Med Sch, 86-88. *Mem:* Fel AAAS; fel Am Phys Soc; fel Am Soc Laser Surg & Med; Am Chem Soc. *Res:* Photochemistry; laser interaction with matter; laser surger. *Mailing Add:* UV Tech 2508 Dunning Dr Yorktown Heights NY 10598

SRINIVASAN, SATHANUR RAMACHANDRAN, b Madras, India, July 16, 38; m 67; c 1. CHEMISTRY, BIOCHEMISTRY. *Educ:* Univ Madras, BSc, 58, BSc(tech), 60, MSc, 62, PhD(leather technol), 65. *Prof Exp:* Res assoc, 67-72, from asst prof to assoc prof, 72-81, PROF DEPT MED & BIOCHEM, SCH MED, LA STATE UNIV MED CTR, NEW ORLEANS, 81- *Concurrent Pos:* Fed Repub Ger acad exchange fel, Darmstadt Tech Univ, 66; mem, Coun Arteriosclerosis, Am Heart Asn. *Mem:* Am Heart Asn; Am Soc Biochem & Molecular Biol. *Res:* Cardiovascular connective tissue and its relation to the pathogenesis of atherosclerosis; role of lipoproteins and lipoprotein-proteoglycans complexes in atherosclerosis; lipoprotein metabolism. *Mailing Add:* Dept of Med La State Univ Med Ctr New Orleans LA 70112

SRINIVASAN, VADAKE RAM, b Ponnani, India, Nov 18, 25; US citizen; m 57; c 2. BIOCHEMISTRY, MICROBIOLOGY. *Educ:* Univ Madras, MA, 48, PhD(biol chem), 51; Univ Mainz, Dr rer Nat(org chem), 55. *Prof Exp:* Res assoc microbiol, Univ Ill, Urbana, 56-59; asst res prof biochem, Univ Pittsburgh, 59-60; res asst prof microbiol, Univ Ill, Urbana, 60-65; assoc prof, 65-70, PROF MICROBIOL, LA STATE UNIV, BATON ROUGE, 70- *Concurrent Pos:* Res grants, NIH, 66-, Am Cancer Soc, 66- & Am Sugarcane League, 68; partic, NSF Int Prog, 72; guest prof, Max Planck Inst Biochem, 72-73. *Mem:* Am Soc Microbiol; Am Chem Soc; Brit Biochem Soc; fel Am Inst Chem; NY Acad Sci. *Res:* Microbial biochemistry and molecular biology; intracellular differentiation in bacteria, control of macromolecular synthesis; single cell protein from cellulose wastes. *Mailing Add:* Dept Microbio La State Univ 602 Life Sci Bldg Baton Rouge LA 70803-0100

SRINIVASAN, VAKULA S, b Madras, India, Mar 25, 36; nat US; m 67; c 1. ELECTROCHEMISTRY, ANALYTICAL CHEMISTRY. *Educ:* Univ Madras, BSc, 56, MA, 58; La State Univ, Baton Rouge, PhD(chem), 65. *Prof Exp:* Archaeol chemist, Govt India, 57; scientist, Indian Atomic Energy Estab, 58-61; vis asst prof chem, La State Univ, Baton Rouge, 65; res fel, Case Inst Technol, 65-67; mem sci staff, TRW Systs, Calif, 67-71; res fel, Calif Inst Technol, 71; assoc prof, 77-78, PROF CHEM, BOWLING GREEN STATE UNIV, 78- *Concurrent Pos:* Res fel, Purdue Univ, 73; consult, Gen Atomics Corp, 74, Energy Conversion Devices, Mich, Vulcan Mat, Ohio, Dinner Bell Inc, Henry Filters, Capitol Plastics, Ohio; vis prof, Tohoku Univ, Japan, 81; fac fel, NASA, 84-85, 85-86; res fel, Argonne Nat Lab, Ill, 84. *Mem:* Sigma Xi; Am Chem Soc; Royal Soc Chem. *Res:* Energy conversion; electrochemistry; space power systems; mixed valence complexes; enzyme electrodes. *Mailing Add:* Dept Chem Bowling Green State Univ Bowling Green OH 43403

SRINIVASAN, VIJAY, b Tamil Nadu, India, Oct 30, 54; m 84; c 1. COMPUTER AIDED DESIGN & MANUFACTURING, MODELING & SIMULATION. *Educ:* Indian Inst Technol, BTech, 76, PhD(mech eng), 80. *Prof Exp:* Postdoctoral fel mech eng, Okla State Univ, 80-83; RES STAFF MEM, INT BUS MACH CORP RES DIV, T J WATSON RES CTR, 83-, MGR, 85- *Concurrent Pos:* Adj assoc prof, Dept Mech Eng, Columbia Univ, 86-90, adj prof, 91- *Mem:* Soc Indust & Appl Math. *Res:* Computer modeling and simulation of engineering systems in design and manufacturing; theories and algorithms in computer aided design and computer aided manufacturing. *Mailing Add:* IMB Res Div Rm 2-150 T J Watson Res Ctr Yorktown Heights NY 10598-0218

SRINIVASARAGHAVAN, RENGACHARI, b Madras, India, Aug 25, 48; m 81. ENVIRONMENTAL ENGINEERING, SANITARY ENGINEERING. *Educ:* Univ Madras, India, BTech, 70; Rose-Hulman Inst Technol, MS, 72; Okla State Univ, PhD(environ eng), 74. *Prof Exp:* process engr, 75-80, ASSOC, GREELEY & HANSEN, ENGRS, 80- *Mem:* Water Pollution Control Fedn; Am Soc Civil Engrs; Tech Asn Paper & Pulp Indust. *Res:* Industrial and municipal wastewater pollution control; resource recovery; conservation and production of energy from waste material. *Mailing Add:* Greeley & Hansen Engrs 222 S Riverside Plaza Chicago IL 60626

SRIPADA, PAVANARAM KAMESWARA, b India, Jan 17, 33; US citizen; m 64; c 3. SYNTHESIS OF MEMBRANE CONSTITUENTS. *Educ:* Anbhra Univ, BS,52, MSc,53,DSc(org chem), 58. *Hon Degrees:* Fel Royal Inst Chem, London, Eng, 69. *Prof Exp:* Lectr, Univ Toronto, 68-72; res assoc, Univ Rhode Island, 72-76 & Univ Conn Health Ctr, 76-81; res biochemist, Biophys Inst, Boston Univ Med Sch, 81-89; CONSULT, 89- *Concurrent Pos:* Vis scientist, Dept Org Chem, Nagarjuna Univ, India, 85-86. *Mem:* Am Chem Soc; NY Acad Sci; royal Inst Chem. *Res:* Synthesizing various labeled and unlabeled components of biomembranes to facilitate physical studies of their reconstituted molecular assemblies to probe into the mechanism of the progression of atherosclerosis. *Mailing Add:* 18 O'Rourke Path Newton MA 02159

SRIVASTAV, RAM PRASAD, b Khairabad, India, Oct 13, 34; m 59; c 3. APPLIED MATHEMATICS. *Educ:* Univ Lucknow, BSc, 53, MSc, 55, PhD(math), 58; Univ Glasgow, PhD(appl math), 63, DSc, 72. *Prof Exp:* Lectr math, Indian Inst Technol, Kanpur, 60-64, asst prof, 64-66; asst prof, Duke Univ, 66-67; assoc prof math, State Univ NY, Stony Brook, 67-73, actg chmn, 76-77, dir grad studies, 86-90, PROF APPL MATH, STATE UNIV NY, STONY BROOK, 73- *Concurrent Pos:* Assoc ed, J Appl Math, Soc Indust & Appl Math, 70-76; NSF grants, State Univ NY Stony Brook, 72, 74 & 89; vis mem math res ctr, Univ Wis-Madison, 73-74; US Army Res Off grant, 76-; vis fel, Princeton Univ, 81; IPA mathematician, US Army Res Off, 84-86; vis mem, Courant Inst Math Sci, NY Univ, 88; assoc ed, Computers & Math with Applications, 84- , Appl Numerical Math, 85- & Appl Math Lett, 87-; consult, UN Develop Prog, IIT, Madras, India, 89; US Army Res Off grant, 90- *Mem:* Am Math Soc; Soc Indust & Appl Math; fel Indian Nat Acad Sci; Math Asn Am; Am Acad Mech. *Res:* Integral equations; mixed boundary value problems in elasticity; numerical analysis and scientific computing; fracture mechanics. *Mailing Add:* 53 Twisting Dr Lake Grove NY 11755-1827

SRIVASTAVA, ASHOK KUMAR, b Basti, India, July 5, 51; m 78; c 2. GLYCOGEN METABOLISM, PROTEIN KINASES. *Educ:* Lucknow Univ, India, BSc, 68, MSc, 70; Kanpur Univ, India, PhD(biochem), 74. *Prof Exp:* Fel, Univ Southern Calif, Los Angeles, 74-77; res assoc, Vanderbilt Univ, 77-80; SR INVESTR CLIN RES INST, MONTREAL, 81-; ASST PROF, UNIV MONTREAL, 82- *Mem:* Soc Exp Biol & Med; Can Biochem Soc; Am Soc Pharmacol & Exp Therapeut; Am Soc Biochem & Molecular Biol. *Res:* Regulation of cell function involving reversible protein phosphorylation dephosphorylation mechanisms; signal transduction in pathophysiological states. *Mailing Add:* 110 Pine Ave W Montreal PQ H2W 1R7 Can

SRIVASTAVA, BEJAI INDER SAHAI, b Shahjahanpur, India, June 1, 32; m 62; c 3. BIOCHEMISTRY, MOLECULAR BIOLOGY. *Educ:* Agra Univ, BSc, 52; Univ Lucknow, MSc, 54; Univ Sask, PhD(plant physiol), 60. *Prof Exp:* Res asst plant physiol, Main Sugar Cane Res Sta, India, 54-55; lectr bot, Ramjas Col, Delhi, 55-57; res biochemist, Grain Res Lab, Winnepeg, Can, 61-63; assoc prof plant physiol, Carver Res Found, Tuskegee Inst, 63-65; sr cancer res scientist, Roswell Park Mem Inst, 65-68, assoc cancer res scientist, 68-77; asst res prof biol, 66-74, ASSOC RES PROF BIOCHEM, STATE UNIV NY BUFFALO, 75-; CANCER RES SCIENTIST V, ROSWELL

PARK MEM INST, 77- *Concurrent Pos:* Res fel, Univ Sask, 60-61; NSF grants, 63-69 & 69-72; AEC grant, 66-69; NIH grant, 73-81. *Mem:* Am Soc Plant Physiol; Am Soc Biol Chemists; Am Asn Cancer Res; NY Acad Sci. *Res:* Biochemical markers for the differential diagnosis of leukemias; gene analysis for leukemia diagnosis; cytokines and cytokine receptors in human leukemias; human T-cell leukemia viruses. *Mailing Add:* Dept Lab Med Roswell Park Mem Inst Buffalo NY 14263

SRIVASTAVA, HARI MOHAN, b Ballia, India, July 5, 40; m 78; c 2. MATHEMATICS, MATHEMATICAL PHYSICS. *Educ:* Univ Allahabad, BSc, 57, MSc, 59; Univ Jodhpur, PhD(math), 65. *Hon Degrees:* Fel Royal Astron Soc, London, 68; FNASc, India, 69; fel Inst Math & Applns, UK, 75. *Prof Exp:* Lectr math, D M Govt Col, Gauhati Univ, India, 59-60 & Univ Roorkee, 60-63; lectr math, Univ Jodhpur, 63-68, reader, 68-69; asst prof, WVa Univ, 67-69; assoc prof, 69-74, PROF MATH, UNIV VICTORIA, BC, 74- *Concurrent Pos:* UGC grant, India, 65; vis prof numerous univs, sci acad & math insts around the world, 67-; grants, Natural Sci & Eng Res Coun Can, 69-; ed, Jnanabha, 72-; regional ed, Pure & Appl Math Sci, 76-; reviewer, Math Rev, Zentralblatt für Math & Appl Mech Rev. *Mem:* Am Math Soc; fel Inst Math & Its Appl, UK; fel Royal Astron Soc London; fel Indian Nat Acad Sci; Can Math Soc. *Res:* Special functions; operational calculus and related areas of differential and integral equations; Fourier analysis; combinatorial analysis; applied methematics; queuing theory; fractional calculus, complex analysis, mathematical physics and astrophysical applications; author, co-author or co-ed of 7 research monographs/books and over 300 journal articles. *Mailing Add:* Dept Math & Statist Univ Victoria Victoria BC V8W 3P4 Can

SRIVASTAVA, JAGDISH NARAIN, b Lucknow, India, June 20, 33; m 51; c 3. STATISTICS, MATHEMATICS. *Educ:* Univ Lucknow, BS, 51, MS, 54; Indian Statist Inst, Calcutta, dipl, 58; Univ NC, PhD(math statist), 61. *Prof Exp:* Statistician, Indian Inst Sugarcane Res, Lucknow, 55-57 & Indian Coun Agr Res, New Delhi, 58-59; res assoc, Univ NC, 61-63; assoc prof math, Univ Nebr, 63-66; PROF STATIST & MATH, COLO STATE UNIV, 66- *Concurrent Pos:* Consult, Lincoln State Hosp, Nebr, 63-64; res grants, Aerospace Res Labs, Air Force Base, Dayton, Ohio, 65-71, 74-, Nat Bur Standards, 69-72, 74-, Air Force Off Sci Res, 71- & NSF, 71-80; dir, Vis Lectr Prog Statist, US & Can, 73-75; founder & ed-in-chief, J Statist Planning & Inference, 75-84, chair gov bd, 84-; sessional pres, Indian Soc Agr Statist, 77. *Honors & Awards:* Award, J Indian Soc Agr Statist, 61. *Mem:* Fel Am Statist Asn; fel Inst Math Statist; Int Statist Inst; Indian Statist Inst; Forum Interdisciplinary Math (vpres, 75-77). *Res:* Multivariate analysis; combinatorial mathematics; weather modification statistics; foundations of statistics; sampling; design of experiments; co-authored one book, edited four books and published more than 100 papers. *Mailing Add:* Dept Statist Colo State Univ Denver CO 80523

SRIVASTAVA, KRISHAN, b Kampur, India, July 9, 31; Can citizen; m; c 5. HIGH VOLTAGE ENGINEERING. *Educ:* Agra Univ, India, BSc, 49; Roorkee Univ, India, BE, 52; Glasgow Univ, PhD(elec), 57. *Prof Exp:* Res assoc, Strathclyde Univ, Glasgow, 55-56; res engr, A Reyrolle & Co, 57-58; lectr, Univ Roorkee, India, 58-59; head, Elec Eng Dept, Eng Col, Univ Jodhpur, India, 59-60; sr res engr, Brush Elec Co, 61-62; PROF, DEPT ELEC ENG, UNIV WATERLOO, ONT, CAN, 66-, VPRES, STUDENT & ACAD SERVS, 83- *Concurrent Pos:* assoc chmn, 69-71, chmn, Dept Elec Eng, Univ Waterloo, 72-78. *Mem:* Fel Inst Elec & Electronics Engrs. *Res:* High Voltage Engineering; electrical insulation engineering; gaseous discharges. *Mailing Add:* Dept Elec Eng Univ BC Main Mall Mcleod Bldg 2075 Westbrook Pl Vancouver BC V6T 1W5 Can

SRIVASTAVA, LALIT MOHAN, b Gonda, India, Sept 7, 32; m 64. BIOLOGY. *Educ:* Univ Allahabad, BSc, 50, MSc, 52; Univ Calif, Davis, PhD(bot), 62. *Prof Exp:* Mercer res fel, Harvard Univ, 61-64, Maria Moors Cabot res fel, 64-65; from asst prof to assoc prof, Simon Fraser Univ, 65-71, acad vpres, 69-70, chmn, 85-90, PROF BIOL SCI, SIMON FRASER UNIV, 71- *Concurrent Pos:* Mem, Mgt Adv Coun, Ministry Educ, BC, 78-81; pres, Enmar Res Corp, 81- *Mem:* Can Soc Plant Physiol; Can Fedn Biol Socs; Am Soc Plant Physiol; Plant Growth Regulator Soc Am. *Res:* Cambium, xylem and phloem; cell growth; gibberellins, receptors, mode of action; physiology of growth, nutritional requirements and chemical constituents of seaweeds. *Mailing Add:* Dept Biol Sci Simon Fraser Univ Burnaby BC V5A 1S6 Can

SRIVASTAVA, LAXMI SHANKER, b Deoria, Uttar Pradesh, India, Mar 2, 38; m 60; c 3. ENDOCRINOLOGY. *Educ:* Bihar Univ, BVSc & AH, 59; Univ Mo-Columbia, MS, 61, PhD(animal breeding), 64. *Prof Exp:* Vet surgeon, Bihar Govt, India, 59-60; instr endocrinol, Univ Mo-Columbia, 64-65, res assoc, Space Res Ctr, 67-68; from asst prof to assoc prof, 69-78, PROF EXP MED, DEPT INTERNAL MED, MED CTR, UNIV CINCINNATI, 78-, DIR ENDOCRINOL & METAB LABS, 69- *Concurrent Pos:* Univ Res Fund grant, St Louis Univ, 65-66; NIH grant, Wash Univ, 66-67. *Mem:* Endocrine Soc; Am Fedn Clin Res; Am Asn Clin Chemists. *Res:* Mammary cancer; neuroendocrine control of pituitary function; mammalian reproductive physiology. *Mailing Add:* 1868 Loisview Lane Cincinnati OH 45255

SRIVASTAVA, MUNI SHANKER, b Gonda, India, Jan 20, 36; m 64; c 4. MATHEMATICAL STATISTICS. *Educ:* Univ Lucknow, BSc, 56, MSc, 58; Stanford Univ, PhD(statist), 64. *Prof Exp:* From asst prof to assoc prof math, 63-72, PROF MATH, UNIV TORONTO, 72- *Concurrent Pos:* Vis res staff, Princeton Univ, 65-66; assoc prof, Univ Conn, 70-71; vis prof, Univ Wis & Indian Statist Inst, 77-78. *Mem:* Fel Inst Math Statist; fel Am Statist Asn; fel Royal Statist Soc; fel Int Statist Inst. *Res:* Multivariate statistics. *Mailing Add:* Dept Statist Univ Toronto Toronto ON M5S 1A7 Can

SRIVASTAVA, PRAKASH NARAIN, b Allahabad, India, Dec 7, 29; m 55; c 4. REPRODUCTIVE BIOCHEMISTRY. *Educ:* Lucknow Univ, India, BSc, 49, MSc, 51; Cambridge Univ, PhD(biochem), 65. *Prof Exp:* Asst res officer hormones, Indian Vet Res Inst, 58-68; asst prof, 69-75, assoc prof, 75-81, PROF BIOCHEM, UNIV GA, 82- *Mem:* Am Soc Biol Chemists; Am Physiol Soc; Soc Study Reproduction; Soc Study Fertil UK. *Res:* Sperm enzymes and their inhibitors in fertilization. *Mailing Add:* Dept Biochem Univ Ga Boyd Grad Studies Bldg Rm 626 Athens GA 30602

SRIVASTAVA, REKHA, b Chikati, Orissa, India, Feb 15, 45; m 78; c 2. MATHEMATICS. *Educ:* Utkal Univ, India, BSc, 62; Banaras Hindu Univ, MSc, 65, PhD(math), 67. *Prof Exp:* Lectr math, Khallikote Col, Berhampur Univ, 68-69, Rourkela Sci Col, 69-70, Women's Col, Berhampur Univ, 70-71 & Ravenshaw Col, Utkal Univ, India, 71-72; res fel, Univ Victoria, 72-73, res assoc & vis scientist math, 73-77, sessional lectr math, 78-89, ADJ ASSOC PROF MATH, UNIV VICTORIA, 89- *Concurrent Pos:* Reviewer, Mathematical Reviews; Zentralblatt für Mathematik. *Mem:* Am Math Soc; Vijnana Parishad India; Asn Women Math; Indian Math Soc; Indian Sci Cong Asn. *Res:* Special functions; operational calculus including integral transforms and related areas of integral equations; Fourier analysis; G, H and the generalized Lauricella functions of several variables; fractional calculus; statistical applications; author or co-author of over 40 research papers in professional journals. *Mailing Add:* Dept Math & Statist Univ Victoria Victoria BC V8W 3P4 Can

SRIVASTAVA, SATISH KUMAR, b Rae Bareli, India, July 21, 37; m 62; c 3. BIOCHEMISTRY, GENETICS. *Educ:* Univ Lucknow, India, BS, 56, MS, 58, PhD(biochem), 62. *Prof Exp:* Tutor biochem, Postgrad Med Sch, Chandigarh, India, 64-66; res scientist, City of Hope Nat Med Ctr, 66-74; asst prof pharmacol, Med Sch, Univ Southern Calif, 70-74; PROF HUMAN BIOL CHEM & GENETICS, UNIV TEX MED BR GALVESTON, 74- *Concurrent Pos:* Coun Sci & Indust Res India fel, Univ Lucknow, 62-64; NIH res grant, 71. *Mem:* Am Soc Hemat; Asn Res Vision & Ophthal. *Res:* Genetics of glycolipid storage diseases; glutathione metabolism in red cells and lens; enzyme kinetics and red cell metabolism; biochemical alterations in senile cataract formation. *Mailing Add:* C-55 Child Health Center Univ Tex Med Br Galveston TX 77550

SRIVASTAVA, SURAT PRASAD, b Allahabad, India, July 1, 37; m. GEOPHYSICS. *Educ:* Indian Inst Technol, Kharagpur, BSc, 58, MTech, 60; Univ BC, PhD(physics), 63. *Prof Exp:* Nat Res Coun Can res fel, Dom Observ, Can, 63-64; asst prof geophys, Univ Alta, 64-65; RES SCIENTIST, ATLANTIC GEOSCI CENTRE, BEDFORD INST OCEANOG, 65- *Concurrent Pos:* Mem, Can Subcomt Geomagnetism, Assoc Comt Geod & Geophys, 66. *Mem:* Am Geophys Union. *Res:* Tectonic implications of the subsurface structures across the continental slope and margin obtained using gravity, magnetic and seismic measurements; application of magnetotelluric method on land and sea. *Mailing Add:* Marine Geophys Div Atlantic Geosci Centre Bedford Inst Dartmouth NS B2Y 4A2 Can

SRIVASTAVA, SURESH CHANDRA, b Aligarh, India, Jan 1, 39; US citizen; m 68; c 2. NUCLEAR MEDICINE, RADIOPHARMACEUTICALS. *Educ:* Agra Univ, BS, 55, MS, 57; Univ Allahabad, PhD(chem), 60. *Prof Exp:* AEC fel, La State Univ, New Orleans, 62-65; res assoc, Brookhaven Nat Lab, 65-67; vis scientist, Sch Chem, Univ Paris, 67-69; res assoc, Ga Inst Technol, 69-71; chemist, Res Triangle Inst, NC, 71-74; clin asst prof radiol, Downstate Med Ctr, Brooklyn, NY, 74-75; assoc scientist, 75-78, scientist, 78-79, HEAD, RADIONUCLIDE & RADIOPHARMACEUT RES DIV, MED DEPT, BROOKHAVEN NAT LAB, 83-, SR SCIENTIST, 90- *Concurrent Pos:* consult, Northport Vet Admin Hosp, NY, 82-, Mem Sloan Kettering Cancer Res Ctr, NY, 83-86, Cremascoli Inc, Milan, 86-, Mallinckrodt Med Inc, 90-; int expert, Int Atomic Energy Agency, 83-, bd dir, Radiopharmaceut Sci Coun, 85-87; prin investr radionuclide & radiopharmaceut res prog, 83-; ed antibody spec issue, Int J Nucl Med Biol, 86; ed bd, Int J Biol Markers, 86; subcomt nucl med, Dept of Energy, Off Health Environ Res Adv Comt, 87-88; ed, NATO Advan Study Inst, Proc on Monoclonal Antibodies, 88. *Honors & Awards:* NATO Advan Study Inst Award, 85; Spec Recognition Award, Chilean Soc Biol Nuclear Med, 89; Fed Lab Consortium Award, 88. *Mem:* Am Chem Soc; Soc Nuclear Med; Radiopharmaceut Sci Coun (bd dirs, 85-87); AAAS; Indo-Am Soc Nuclear Med (vpres, 87-88, pres, 89-90). *Res:* Radiopharmaceuticals; novel diagnostic reagents and therapeutic agents for in vitro and in vivo applications; radiolabeled monoclonal antibodies for imaging and therapy of cancer; chemistry and production of short-lived gamma and position- emitting radionuclides and of beta emitters of interest to nuclear medicine; technetium chemistry and radio pharmaceuticals; blood cell labeling. *Mailing Add:* Dept Med Brookhaven Nat Lab Upton NY 11973

SRIVASTAVA, TRILOKV N, b Lucknow, India, June 1, 36; Can citizen; m 77; c 4. PURE MATHEMATICS, MATHEMATICAL STATISTICS. *Educ:* Lucknow Univ, BS, 57, MS, 59; Gorakhpur Univ, PhD(math), 69; Sheffield Univ, PhD(statist). *Prof Exp:* Sr lectr math, Loyola Col Montreal, 63-64, from asst prof to assoc prof, 64-71; vis prof, Univ Isfahahan, Iran, 71-72; ASSOC PROF MATH, CONCORDIA UNIV, 72- *Res:* Study of differential geometry of special Kawaguchi manifold, generalized statistical distributions and distributions of general functions of random variables; published 30 research articles in refereed journals; reviewer for mathematical reviews. *Mailing Add:* Dept Math Concordia Univ 1455 Demaisonneuve Blvd W Montreal PQ H3G 1M8 Can

SRIVASTAVA, UMA SHANKER, b Lucknow, Uttar Pradesh, India, Mar 13, 34; Can citizen; m 70; c 1. MOLECULAR BIOLOGY, BIOCHEMISTRY. *Educ:* Lucknow Univ, India, BSc, 55, MSc, 57; Laval Univ, Can, DSc, 65. *Prof Exp:* From asst prof to assoc prof, 68-79, PROF NUTRIT & TOXICOL, DEPT NUTRIT, UNIV MONTREAL, 79-, PROF, PROG TOXICOL, 80- *Concurrent Pos:* Exchange fel, France & Can, 73; prof toxicol, Div Toxicol, Johns Hopkins Univ, 80; sr res scientist biochem & appl nutrit, Cent Food & Tehcnol Res Inst, India, 80-81; prof pharmacol & physiol, San Luis, Patosi, Mexico, 89-90. *Mem:* NY Acad Sci; Am Inst Nutrit; fel Am Col Nutrit; Soc Environ Geochem & Health; Can Biochem Soc; Fr-Can Asn Advan Sci; Can Soc Nutrit. *Res:* Biochemical and molecular aspects of malnutrition in the development of the brain; modulation, translation and decoding of coding sequences in the messenger RNA of the brain of well fed and dietary restricted rats; toxic effects of heavy metals in the nutrition of animals and man; etiology of progressive muscular dystrophy; nutrients and minerals in North American diet; role of chemically defined diet on the growth of pea aphids. *Mailing Add:* Dept Nutrit Univ Montreal Montreal PQ H3C 3J7 Can

SRIVATSAN, TIRUMALAI SRINIVAS, b Madras, India, July 14, 57; US citizen; m 89. MATERIALS TESTING, SOLID & FRACTURE MECHANICS. *Educ:* Univ Bangalore, BEng, 80; Ga Inst Technol, MS, 81, PhD(mech eng), 84. *Prof Exp:* Res fel, Ctr Computational Studies, 84-85; res fel & instr mat processing, Ga Tech Res Inst, 85-86; proj engr & mgr, Mat Modification Inc, 86-87; ASST PROF MAT SCI & MECH, UNIV AKRON, 87- *Concurrent Pos:* Mem, Struct Mat Comt, Metals, Minerals & Mat Soc, 87-; distinguished lectr, Am Soc Mat Int, 88; ed-in-chief, Mat & Mfg Processes Int J, 89- *Mem:* Am Soc Mech Engrs; Am Soc Mat Int; Metals Minerals & Mat Soc. *Res:* Mechanical behavior of engineering materials; relationship between micristructure and mechanical properties; mechanical life-time failure prediction; thermal-mechanical fatigue; materials processing and characterization; electron microscopy; fracture mechanics; stress analysis; composite materials; author of over 100 publications. *Mailing Add:* Dept Mech Eng Univ Akron Akron OH 44325-3903

SRNKA, LEONARD JAMES, b Cleveland, Ohio, Nov 17, 46; div; c 1. INVERSE THEORY, GEOPHYSICAL APPLICATIONS. *Educ:* Purdue Univ, BS, 68; Univ Newcastle, Eng, PhD(physics), 74. *Prof Exp:* Sci officer plasma physics, Culham Lab, UK Atomic Energy Authority, Eng, 70-73; fel, Lunar Sci Inst, Univ Space Res Asn, 74-75, staff scientist, Lunar & Planetary Inst, 75-79; MEM STAFF, LONG RANGE RES DIV, EXXON PROD RES CO, 79- *Concurrent Pos:* Vis scientist, Lunar & Planetary Inst, 79-; group leader, wave equation methods, Long Range Res Div, Exxon Prod Res Co, 84- *Mem:* Am Geophys Union; Am Phys Soc; Soc Explor Geophys. *Res:* Electromagnetic fields in planetary interiors; exploration geophysics, seismic and electromagnetic inversion. *Mailing Add:* Exxon Prod Res Co PO Box 2189 Houston TX 77001

SROLOVITZ, DAVID JOSEPH, b Milwaukee, Wis, Mar 13, 57; m 78; c 3. MICROSTRUCTURAL EVOLUTION, STRUCTURE OF DEFECTS. *Educ:* Rutgers Univ, BS, 78; Univ Pa, MSE, 80, PhD (mats sci & eng), 81. *Prof Exp:* Postdoctoral, Exxon Res & Eng Co, 82-84; staff mem, Los Alamos Nat Lab, 84-87; ASSOC PROF MATS SCI & APPL PHYSICS, UNIV MICH, 87- *Concurrent Pos:* Mem, supercomputing mats sci comt, Nat Mats Adv Bd, Nat Res Coun, 86-88, Defense Advan Res Projs Agency, Mat Res Coun. *Mem:* Am Inst Mech Eng- MetallSoc; Mats Res Soc; Am Phys Soc. *Res:* Microstructural evolution; dislocation dynamics; structure and thermodynamics of interfaces; mechanics of defected materials; film growth. *Mailing Add:* Dept Mat Sci & Eng Univ Mich Dow Bldg Ann Arbor MI 48109-2136

SROOG, CYRUS EFREM, b New York, NY, Mar 25, 22; m 43; c 2. ORGANIC CHEMISTRY, POLYMER CHEMISTRY. *Educ:* Brooklyn Col, BA, 42; Univ Buffalo, PhD(chem), 50. *Prof Exp:* Instr chem, Univ Buffalo, 46-50; res chemist, E I Du Pont De Nemours & Co, Inc, 50-54, res supvr, 54-61, develop mgr, 62-64, res mgr, 64-76; OWNER, POLYMER CONSULT, INC, 86- *Concurrent Pos:* Res fel, NY Acad Sci, 76-86. *Mem:* Am Chem Soc; NY Acad Sci; The Chem Soc. *Res:* High temperature polymers; thermally stable polymers; heterocyclic, organic nitrogen, organic sulfur and metalloorganic compounds; aromatic polyimides; heterocyclic polymers. *Mailing Add:* 3227 Coachman Rd Surrey Park Wilmington DE 19803

SROUR, JOSEPH RALPH, b Tampa, Fla, Jan 7, 41; m 69. SOLID STATE ELECTRONICS, RADIATION EFFECTS. *Educ:* Cath Univ Am, BEE, 63, MEE, 66, PhD(elec eng), 68. *Prof Exp:* Mem res tech staff radiation effects, 68-76, mgr, 76-78, MGR SOLID STATE ELECTRONICS, NORTHROP RES & TECHNOL CTR, 78- *Concurrent Pos:* Lectr elec eng, Loyola Univ Los Angeles, 70-72; vchmn publ radiation effects comt, Inst Elec & Electronic Engrs, 74-76, vchmn radiation effects comt, 79-82. *Mem:* Inst Elec & Electronic Engrs; Am Phys Soc; Sigma Xi. *Res:* Radiation effects on electronic materials, devices and circuits; semiconductor device physics; integrated circuits; radiation hardening. *Mailing Add:* Northrop Res & Technol Ctr PO Box 3070 Manhatten Beach CA 90266

SRYGLEY, FLETCHER DOUGLAS, b Nashville, Tenn, Mar 27, 38; m 68; c 3. MATERIALS SCIENCE ENGINEERING. *Educ:* David Lipscomb Col, BA, 60; Duke Univ, PhD(physics), 66. *Prof Exp:* Asst prof physics, Stetson Univ, 66-73; assoc prof, 73-79, PROF PHYSICS, DAVID LIPSCOMB COL, 79- *Mem:* Sigma Xi; Am Phys Soc; Am Asn Physics Teachers. *Res:* Electron spin resonance studies of radiation damage in single crystals; electron spin resonance and ultraviolet studies of color centers in magnesium oxide; fourier analysis of x-ray line shapes. *Mailing Add:* Dept of Physics Box 4114 David Lipscomb Univ Nashville TN 37204-3951

STAAL, GERARDUS BENARDUS, b Assen, Neth, Aug 19, 25; m 57; c 2. ENTOMOLOGY, BIOLOGY. *Educ:* State Agr Univ, Wageningen, Ing, 57, PhD(insect physiol), 61. *Prof Exp:* Sr res officer, Neth Orgn Appl Sci Res, 61-68; dir biol res, 68-80, DIR INSECT RES, SANDOZ CROP PROTECTION RES DIV, 80- *Concurrent Pos:* Neth Orgn Pure Res fel biol, Harvard Univ, 62-63. *Mem:* AAAS; Entom Soc Am; Am Chem Soc; Am Soc Zoologists; Am Orchid Soc. *Res:* Comparative insect endocrinology; insect bioassay of toxicants; juvenile hormone analogs; antagonists and other principles affecting growth and development of insects; neurohormones; neurotransmitters. *Mailing Add:* Sandoz Corp Protection Res Div 975 California Ave Palo Alto CA 94304

STAAT, ROBERT HENRY, b Denver, Colo, Apr 2, 42; m 79; c 3. PATHOGENIC MECHANISMS, MICROBIAL ADHERENCE. *Educ:* Univ NMex, BS, 65, MS, 68; Univ Minn, PhD(microbiol), 75. *Prof Exp:* Res assoc, Univ NMex, 67-69; scientist, Sch Dent, Univ Minn, 71-73, res fel, 73-75; asst prof microbiol, Sch Dent, Med Univ SC, 75-76; from asst to assoc prof, 76-83, PROF ORAL HEALTH, SCH DENT, UNIV LOUISVILLE, 83-, ASSOC PROF MICROBIOL, SCH MED, 76- *Concurrent Pos:* Prin investr, Sugar Asn grant, 79-82 & Nat Inst Dent Res, NIH, 76-83; dir, Sterilizer Monitoring Prog. *Mem:* AAAS; Am Soc Microbiol; Am Asn Dent Res; Sigma Xi; Am Asn Dent Sch. *Res:* Determination of the pathogenic mechanisms of oral Streptococci, specifically, definition of the adherence reaction of streptococcus mutans to the tooth surface and purification of the adherence factors for use in a dental caries vaccine; infection control principles as applied to dentistry. *Mailing Add:* Dept Oral Health Sch Dent Univ Louisville Louisville KY 40292

STAATS, GUSTAV W(ILLIAM), b Forest Park, Ill, Nov 30, 19; m 49; c 2. ELECTRICAL ENGINEERING. *Educ:* Ill Inst Technol, BS, 42, MS, 48, PhD(elec eng), 56. *Prof Exp:* Engr, Motor & Generator Dept, Allis-Chalmers Mfg Co, 42-56, staff engr, Thermal Power Dept, 56-63, res engr, Res Div, 63-65; from asst prof to assoc prof elec eng, 65-77, PROF ELEC ENG, UNIV WIS-MILWAUKEE, 77- *Mem:* Fel Inst Elec & Electronics Engrs. *Res:* Rotating electrical machinery; high strength magnetic fields; electric power systems. *Mailing Add:* 6124 N Lake Dr Whitefish Bay WI 53217

STAATS, PERCY ANDERSON, b Belleville, WVa, Feb 20, 21; m 44; c 4. PHYSICAL CHEMISTRY, PHYSICS. *Educ:* Marietta Col, AB, 43; Univ Minn, MS, 49. *Hon Degrees:* DS, Fisk Univ, 74. *Prof Exp:* Instr physics, Marietta Col, 43; tech supvr, Tenn Eastman Corp, 43-46; chemist, Rohm and Haas Co, Pa, 49-52; chemist, Oak Ridge Nat Lab, 52-85; RETIRED. *Concurrent Pos:* Guest lectr & lab dir, Infrared Spectros Inst, Fisk Univ, 57-75; traveling lectr, Oak Ridge Inst Nuclear Studies, 59-60, 62-63 & 65-66. *Mem:* Am Chem Soc. *Res:* Molecular structure by infrared spectroscopy; infrared spectra of gases as solids at low temperatures; inorganic ions in solid solution; isotopes, especially tritium; gas lasers; plasma diagnostics using far infrared submillimeter lasers. *Mailing Add:* 119 Manchester Rd Oak Ridge TN 37830

STAATS, WILLIAM R, b Chicago, Ill, Sept 10, 35; m 80; c 2. CHEMICAL ENGINEERING. *Educ:* Ill Inst Technol, BS, 57, MS, 60, PhD, 70. *Prof Exp:* Chem engr, Inst Gas Technol, 57-58; proj develop officer, Rome Air Develop Ctr, 59-62; chem engr, Inst Gas Technol, 62-69; vpres eng res, Polytech, Inc, 70-75; assoc dir, Inst Gas Technol, 75-79; DIR PHYS SCI, GAS RES INST, 79- *Mem:* AAAS; Am Inst Chem Engrs; Combustion Inst; Am Chem Soc. *Res:* Gas technology. *Mailing Add:* 705 Timber Trail Dr Naperville IL 60565-2705

STAATZ, MORTIMER HAY, b Kalispell, Mont, Oct 20, 18; m 52; c 3. ECONOMIC GEOLOGY. *Educ:* Calif Inst Technol, BS, 40; Northwestern Univ, MS, 42; Columbia Univ, PhD(geol), 52. *Prof Exp:* Asst geol, Northwestern Univ, 41-42; GEOLOGIST, US GEOL SURV, 42-44 & 46- *Mem:* Geol Soc Am; Soc Econ Geologists; Mineral Soc Am. *Res:* Pegmatites of Colorado and South Dakota; geology of eastern Great Basin and Washington; beryllium, fluorspar and phosphate deposits; vein-type uranium and thorium deposits; thorium and rare earth resources in United States. *Mailing Add:* 13435 Braun Rd Golden CO 80401

STAATZ, WILLIAM D, b Glendale, Calif; c 3. INTEGRINS. *Educ:* Univ Puget Sound, BS, 67; Wash State Univ, MS, 69; Univ Edinburgh, PhD(zool), 76. *Prof Exp:* Res asst, Max Planck Inst, 69-71; asst res scientist, City Hope Med Ctr, 76-78; res fel, Eye Inst, Med Col Wis, 78-80; res assoc develop biol, Univ Wis-Madison, 80-84; asst prof genetics & develop biol, State Univ NY, Fredonia, 84-89; RES INSTR, SCH MED, WASH UNIV, 89- *Concurrent Pos:* Fel, Robert E Cook Res Fund, 77; res fel, Wash Univ Sch Med, 87-89. *Mem:* Am Soc Cell Biol; Soc Develop Biol; AAAS. *Res:* Role of cell adhesion receptors and extracellular matrix molecules in normal and pathologic development and function of animal systems; protein chemistry. *Mailing Add:* Div Lab Med Wash Univ Sch Med 660 S Euclid Box 8118 St Louis MO 63110

STABA, EMIL JOHN, b New York, NY, May 16, 28; m 54; c 5. PHARMACOGNOSY. *Educ:* St John's Univ, NY, BS, 52; Duquesne Univ, MS, 54; Univ Conn, PhD(pharmacog), 57. *Prof Exp:* Prof pharmacog & chmn dept, Univ Nebr, Lincoln, 57-68; asst dean, Univ Minn, Minneapolis, 74-78, prof pharmacog & chmn dept, Col Pharm, 68-85; RETIRED. *Concurrent Pos:* Consult var indust & govt agencies; NSF sr foreign scientist, Poland, Hungary & Czech, 69; Fulbright-Hays res fel, Ger, 70; Coun Sci & Indust Res vis scientist, India, 73; partic, US-Repub China Coop Sci Prog, Plant Cell & Tissue Culture, 74; mem, US Pharmacopeia Comt Rev-Natural Prod, 80- & Life Sci Adv Comt, NASA, 84-87. *Honors & Awards:* Lunsford-Richardson Award, 58. *Mem:* AAAS; Am Soc Pharmacog (pres, 71-72); Tissue Culture Asn; Am Pharmaceut Asn; Soc Econ Bot. *Res:* Cultivation, extraction and tissue culture of medicinal plants; herbal teas. *Mailing Add:* Dept Med Chem Col Pharm Univ Minn Minneapolis MN 55455

STABENFELDT, GEORGE H, b Shelton, Wash, June 26, 30; m 53; c 4. PHYSIOLOGY, ENDOCRINOLOGY. *Educ:* Wash State Univ, BA, 55, DVM, 56, MS, 62; Okla State Univ, PhD, 68. *Prof Exp:* Pvt pract, Ore, 56-57, Idaho, 57-58 & Wash, 58-60; instr vet path, Wash State Univ, 60-62; asst prof physiol, Okla State Univ, 62-68; assoc prof clin sci & assoc res physiologist, Nat Ctr Primate Biol, 68-75, PROF REPRODUCTION, DEPT REPRODUCTION, SCH VET MED, UNIV CALIF, DAVIS, 75- *Mem:* Am Vet Med Asn; Am Soc Vet Physiol & Pharmacol; Am Physiol Soc; Soc Study Fertility; Soc Study Reproduction. *Res:* Endocrinology of female reproductive cycle, including the estrous and menstrual cycles, pregnancy and parturition. *Mailing Add:* Dept Reproduction Univ Calif Sch Vet Med Davis CA 95616

STABLEFORD, LOUIS TRANTER, b Meriden, Conn, July 30, 14; m 41; c 3. BIOLOGY. *Educ:* Univ Va, BS, 37; Yale Univ, PhD(zool), 41. *Prof Exp:* Lab asst, Univ Va, 34-37 & Yale Univ, 37-40; from instr to prof, 41-72, chmn dept, 58-78, Dana prof, 72-79, EMER PROF BIOL, LAFAYETTE COL, 79- *Honors & Awards:* Christian & Mary Lindbach Found Award, 65. *Mem:* Am Soc Zoologists; Soc Develop Biol. *Res:* Experimental embryology; early amphibian development; aging of connective tissue. *Mailing Add:* Dept Biol Lafayette Col Easton PA 18042

STABLER, TIMOTHY ALLEN, b Port Jervis, NY, Sept 27, 40. DEVELOPMENTAL BIOLOGY, ENDOCRINOLOGY. *Educ:* Drew Univ, BA, 62; DePauw Univ, MA, 64; Univ Vt, PhD(zool), 69. *Prof Exp:* Asst prof biol, Hope Col, 69-71; asst prof biol & health professions, 73-76, actg chmn biol, 81 & 87, chmn biol, 83-85, ASSOC PROF BIOL & HEALTH PROFESSIONS ADV, IND UNIV NORTHWEST, 76- *Concurrent Pos:* NIH fel, Univ Minn, 68-69; mem, NSF workshop develop biol, Univ Calif, San Diego, 71; NIH trainee reproductive endocrinol, Sch Med, Boston Univ, 71-73; adj asst prof physiol, Northwest Ctr Med Educ, Ind Univ Sch Med, 74-76, adj assoc prof physiol, 76- *Mem:* AAAS; Am Soc Zool; Am Inst Biol Sci; NY Acad Sci; Tissue Culture Asn; Nat Asn of Advisors for Health Prof; Cent Asn Adv Health Prof (treas, 80-); Int Electrophoresis Soc. *Res:* Steroid receptor biochemistry; tissue culture of steroid-producing tissues; tow-dimensional electrophoresis of steroid tissues. *Mailing Add:* Dept of Biol Ind Univ Northwest Gary IN 46408

STACE-SMITH, RICHARD, b Creston, BC, May 2, 24; m 51. PLANT PATHOLOGY. *Educ:* Univ BC, BSA, 50; Ore State Col, PhD(plant path), 54. *Prof Exp:* Asst plant pathologist, 50-54, assoc plant pathologist, 54-58, plant pathologist, 58-81, HEAD PLANT PATH SECT, CAN DEPT AGR, 81- *Mem:* Can Phytopath Soc; Agr Inst Can. *Res:* Rubus virus diseases; virus purification and properties. *Mailing Add:* 6660 NW Marine Dr Vancouver BC V6T 1X2 Can

STACEY, JOHN SYDNEY, b June 15, 27; US citizen; m 54; c 3. GEOPHYSICS. *Educ:* Univ Durham, BSc, 51; Univ BC, MASc, 58, PhD(physics), 62. *Prof Exp:* Engr, Marconi Wireless Tel Co, 51-55; lectr elec eng, Univ BC, 57-58; PHYSICIST, ISOTOPE GEOL BR, US GEOL SURV, 62- *Mem:* Geol Soc Am; Am Soc Mass Spectrometry. *Res:* Mass spectrometry in geologic studies and related data processing techniques; lead isotope and U-Pb zitcon geochronology for ore genesis and crustal evolution. *Mailing Add:* PO Box 34 El Granada CA 94018

STACEY, LARRY MILTON, b Greensboro, NC, July 30, 40. PHYSICS. *Educ:* Univ NC, Chapel Hill, BS, 62, PhD(physics), 68. *Prof Exp:* Fel physics, Rutgers Univ, New Brunswick, 67-70; chem engr, Calif Inst Technol, 70-71; ASST PROF PHYSICS, ST LOUIS UNIV, 71- *Mem:* Am Phys Soc; Am Asn Physics Teachers. *Res:* Applications of magnetic resonance to the study of solids and fluids; critical point phenomena. *Mailing Add:* Dept of Physics St Louis Univ 221 N Grand Blvd St Louis MO 63103

STACEY, WESTON MONROE, JR, b US; div; c 3. REACTOR PHYSICS, FUSION PLASMA THEORY. *Educ:* Ga Inst Technol, BS, 59, MS, 63; Mass Inst Technol, PhD(nuclear eng), 66. *Prof Exp:* Nuclear engr, Knolls Atomic Power Lab, 62-64, mgr reactor kinetics, 66-69; sect head reactor theory, Argonne Nat Lab, 69-72, assoc dir, Appl Physics Div, 72-77, dir fusion prog, 73-77; CALLAWAY PROF NUCLEAR ENG, GA INST TECHNOL, 77- *Honors & Awards:* Distinguished Assoc Award, Dept of Energy, 90. *Mem:* Fel Am Nuclear Soc; fel Am Phys Soc. *Res:* Nuclear reactor theory, fusion reactor technology, plasma physics; fusion reactor design. *Mailing Add:* Sch Nuclear Eng Ga Inst Technol Atlanta GA 30332

STACH, JOSEPH, b Wallington, NJ, Aug 21, 38; m 63; c 2. SOLID STATE ELECTRONICS, ELECTRICAL ENGINEERING. *Educ:* Newark Col Eng, BS, 60; Pa State Univ, MS, 62, PhD(elec eng), 66. *Prof Exp:* Instr elec eng, Pa State Univ, 62-65; mem tech staff, Bell Tel Labs, 66-67; from asst prof to prof elec eng, Pa State Univ, University Park, 67-91; EXEC VPRES, PLASMA-THERM, 91- *Concurrent Pos:* Consult, Air Prod & Chem Carborunchem. *Res:* Investigation of avalanche breakdown in metal barrier diodes and surface properties of insulators on semiconductors; boron nitride processing, HCI oxidations and plasma etching. *Mailing Add:* Plasma-Therm I P Inc 9509 International Ct St Petersburg FL 33761

STACH, ROBERT WILLIAM, b Chicago, Ill, Feb 12, 45; m 66; c 1. NEUROBIOCHEMISTRY, BIOCHEMISTRY. *Educ:* Ill Wesleyan Univ, BA, 67; Univ Wis-Madison, PhD(org chem), 72. *Prof Exp:* Trainee neurobiochem, Depts Genetics & Biochem, Sch Med, Stanford Univ, 72-74; from asst prof to assoc prof biochem, molecular biol, anat & cell biol, State Univ NY Upstate Med Ctr, 74-87, assoc mem fac, Ctr Neurobehav Sci, 79-87; DIR RES & PROF CHEM, UNIV MICH, FLINT, 87- *Concurrent Pos:* Biomed res support grants, 74-75, 79, 81, 84-85 & 88-89; NIH res grants, 75-85 & 87-90. *Mem:* Am Soc Neurochem; AAAS; NY Acad Sci; Soc Neurosci; Am Soc Biochem & Molecular Biol; Int Soc Neurochem. *Res:* Factors involved in the growth, development and regeneration of the nervous system with special emphasis on the mechanism of the nerve growth. *Mailing Add:* Dir Res Rm 210 CROB Univ Mich 303 E Kearsley Flint MI 48502-2186

STACHEL, JOHANNA, b Muenchen, Ger, Dec 3, 54. RELATIVISTIC HEAVY ION PHYSICS, LOW & INTERMEDIATE ENERGY HEAVY ION PHYSICS. *Educ:* Johannes-Gutenberg Univ Mainz, Ger, dipl chem, 77, PhD(nuclear physics), 82. *Prof Exp:* Res assoc nuclear physics, Univ Mainz, Ger, 79-83; res assoc, State Univ NY, Stony Brook, 83-84, vis assoc, 84- 85, asst prof, 85-89, ASSOC PROF NUCLEAR PHYSICS, STATE UNIV NY, STONY BROOK, 89- *Concurrent Pos:* Feoda-Lynen fel, Alexander von Humboldt Found, 83-85; A P Sloan fel, A P Sloan Found, 86-90; prin investr, NSF grants, 87-, presidential young investr, 88-93. *Mem:* Am Phys Soc. *Res:* Experiments in relativistic heavy ion physics to study nuclear matter at extreme condition of high temperature and density; search for phase transition to quark-gluon-plasma. *Mailing Add:* Dept Physics State Univ NY Stony Brook Stony Brook NY 11794-3800

STACHEL, JOHN JAY, b New York, NY, Mar 29, 28; m 53; c 3. PHYSICS. *Educ:* City Univ New York, BS, 56; Stevens Inst Technol, MS, 59, PhD(physics), 62. *Prof Exp:* Instr physics, Lehigh Univ, 59-61; instr, Univ Pittsburgh, 61-62, res assoc, 62-64; from asst prof to assoc prof, 64-72, PROF PHYSICS & DIR CTR EINSTEIN STUDIES, BOSTON UNIV, 72- *Concurrent Pos:* Vis res assoc, Inst Theoret Physics, Warsaw, 62; vis prof, King's Col, Univ London, 70-71; vis sr res fel, Dept Physics, Princeton Univ, 77-84; ed, Collection Papers Albert Einstein, Princeton Univ Press, 77-88. *Res:* General relativity; foundations of quantum theory; history and philosophy of physics. *Mailing Add:* Ctr Eisntein Studies Boston Univ 590 Commonwealth Ave Boston MA 02215

STACK (STACHIEWICZ), B(OGDAN) R(OMAN), b Lwow, Poland, Sept 16, 24; nat US; m 55; c 4. ELECTRONICS ENGINEERING. *Educ:* Bristol Univ, BSEE, 47; McGill Univ, MEE, 53. *Prof Exp:* Engr, Radio Eng Prod, Ltd, Can, 47-52; proj engr, Lenkurt Elec Co, Calif, 52-55; asst sect head, Stromberg-Carlson Div, Gen Dynamics Corp, NY, 55-57; assoc lab dir, Int Tel & Tel Fed Labs, Calif, 57-62; dept mgr, Philco Corp, Ford Motor Co, 62-64; prog mgr, Stanford Res Inst, 64-68; mgr, Systs Design Dept, WDL Div, Philco-Ford Corp, Palo Alto, 68-78; mgr, Aydin Satellite Commun, 77-80, vpres eng, 80-82, vpres & gen mgr, Aydin Systs Div, 82-86; PRES, BRS ASSOCS, 86- *Concurrent Pos:* Expert witness, Dept Justice, ITT-ABC merger proc; vis lectr commun, Univ Alexandria, Egypt, lectr, Telecommun Cert Prog, San Francisco State Univ. *Mem:* Sr mem Inst Elec & Electronics Engrs. *Res:* Space and ground communications systems; signal analysis and detection; 6 US patents in the field of telecommunications. *Mailing Add:* 358 Toyon Ave Los Altos CA 94022

STACK, JOHN D, b Los Angeles, Calif, July 24, 38; m 63; c 1. THEORETICAL PHYSICS. *Educ:* Calif Inst Technol, BSc, 59; Univ Calif, Berkeley, PhD(physics), 65. *Prof Exp:* Actg asst prof physics, Univ Calif, Berkeley, 65-66; from asst prof to assoc prof, 66-81, PROF PHYSICS, UNIV ILL, URBANA, 82- *Concurrent Pos:* Vis assoc, Calif Inst Technol, 69-70, Standford Linear Accelerator Ctr, 73. *Mem:* Am Phys Soc. *Res:* Field theory; lattice gauge theory; Feynman path integrals. *Mailing Add:* Dept Physics-237D Univ Ill 1110 W Green St Urbana IL 61801

STACK, STEPHEN M, b Monahans, Tex, Feb 12, 43; m 65; c 2. CYTOLOGY. *Educ:* Univ Tex, Austin, BAS, 65, PhD(cytol, bot), 69. *Prof Exp:* Asst prof, 69-74, ASSOC PROF BOT & PLANT PATH, COLO STATE UNIV, 74- *Mem:* Am Soc Cell Biol; AAAS; Bot Soc Am. *Res:* Structure and function of chromosomes. *Mailing Add:* Dept Bot & Plant Path Colo State Univ Plant Sci Bldg Ft Collins CO 80523

STACKELBERG, OLAF PATRICK, b Munich, Ger, Aug 2, 32; US citizen; m 54; c 3. MATHEMATICS. *Educ:* Mass Inst Technol, BS, 55; Univ Minn, MS, 60, PhD(math), 63. *Prof Exp:* Teaching asst math, Univ Minn, 58-63; from asst prof to assoc prof, Duke Univ, 63-76; PROF MATH & CHMN DEPT, KENT STATE UNIV, 76- *Concurrent Pos:* Alexander V Humboldt fel, Stuttgart Tech Univ, 65-66; vis assoc prof, Univ Ill, Urbana, 69-70 & Univ London, 74; ed, Duke Math J, 71-74. *Mem:* Am Math Soc; Math Asn Am; Inst Math Statist. *Res:* Probability; metric number theory. *Mailing Add:* Dept Math Sci Kent State Univ Kent OH 44242

STACKMAN, ROBERT W, b Dayton, Ohio, June 29, 35; m 62; c 3. RESEARCH ADMINISTRATION, POLYMER CHEMISTRY. *Educ:* Univ Dayton, BS, 57; Univ Fla, PhD(org chem), 61. *Prof Exp:* Res asst cyclopolymers of silanes, Univ Fla, 58-61; res chemist, Summit Labs, Celanese Corp, 61-65, sr res chemist, Celanese Res Co, 65-72, res assoc, 72-74, res supvr polymer flammability res, 74-76, res supvr polymer & specialty chem res, 76-80, res assoc, Celanese Res Co, 80-84; sr group leader, Adv Mat Corp Res & Develop, 84-89, SR SECT MGR, PIONEERING POLYMER RES, S C JOHNSON & SON INC, 89- *Mem:* AAAS; Am Chem Soc; fel Am Inst Chem. *Res:* Condensation polymerization; emulsion polymerization; high temperature polymers; cyclopolymerization; organic synthesis; organosilicon compounds; addition polymerization; polymer modification and stabilization; flammability of polymers; water soluble polymers; fermentation processes; liquid crystal polymers. *Mailing Add:* SC Johnson & Son Inc 1525 Howe St Mail Sta 117 Racine WI 53403-5011

STACKPOLE, JOHN DUKE, b Boston, Mass, Dec 28, 35; m 60; c 3. METEOROLOGY. *Educ:* Amherst Col, BA, 57; Mass Inst Technol, MS, 59, PhD(meteorol), 64. *Prof Exp:* Res meteorologist, 64-73, SUPVRY RES METEOROLOGIST, NAT WEATHER SERV, NAT OCEANIC & ATMOSPHERIC ADMIN, 73- *Res:* Numerical weather prediction. *Mailing Add:* 7305 Kipling Pkwy District Heights MD 20747-1863

STACY, CARL J, b Joplin, Mo, Jan 20, 29; m 51; c 3. POLYMER SCIENCE. *Educ:* Kans State Col Pittsburg, BA, 51; Purdue Univ, PhD(phys chem), 56. *Prof Exp:* Res fel starch, Purdue Univ, 55-56; res physicist, Phillips Petrol Co, 56-61, sr res chemist, 62-86; INDEPENDENT CONSULT, POLYMERS, PLASTICS, RUBBER, 86- *Mem:* AAAS; Am Chem Soc; Sigma Xi. *Res:* Polymer molecular weight distribution and structure by light scattering, gel permeation chromatography, ultracentrifuge and other physical techniques. *Mailing Add:* 2929 Sheridan Rd Bartlesville OK 74006

STACY, DAVID LOWELL, b Kansas City, Mo, Oct 7, 50. ANGIOGENESIS, INFLAMMATION. *Educ:* Oral Roberts Univ, BA, 78; Univ Nebr Med Ctr, PhD(physiol), 86. *Prof Exp:* Grad res assoc physiol, Univ Nebr Med Ctr, 80-85; postdoctoral fel physiol, Eastern Va Med Sch, 85-88, res asst prof, 88; sr scientist, 88-89, RES INVESTR, GLAXO INC, 89- *Concurrent Pos:* Travel award, Japanese Soc Microcirculation, 87. *Mem:* Microcirculation Soc; Am Phys Soc; Am Diabetes Asn. *Res:* Physiology, pharmacology, and pathology of small blood vessels; therapeutics of hypertension, inflammation, diabetes and angiogenesis; compounds which inhibit tumor angiogenesis. *Mailing Add:* Dept Pharmacol Glaxo Res Inst Five Moore Dr Research Triangle Park NC 27709

STACY, GARDNER W, b Rochester, NY, Oct 29, 21; m 47, 67; c 5. ORGANIC CHEMISTRY, CHEMICAL EDUCATION. *Educ:* Univ Rochester, BS, 43; Univ Ill, PhD(org chem), 46. *Prof Exp:* Asst, Off Sci Res & Develop Proj, Ill, 43-46; fel biochem, Med Col, Cornell Univ, 46-48; from asst prof to assoc prof, 48-60, prof, 60-88, EMER PROF, WASH STATE UNIV, 88- *Concurrent Pos:* Dir region VI, Am Chem Soc, 70-77, mem coun, 56-, nat pres off & dir, 78-80, pres, 79, mem, Comt Chem & Pub Affairs, 81-

86, subcomt energy, 82-86, consult to comt, 88- *Honors & Awards:* PRF Int Award, Australia & New Zealand, 63-64. *Mem:* Am Chem Soc; fel AAAS; fel Am Inst Chemist. *Res:* Sulfur-containing heterocyclic tautomeric systems and ring-chain tautomerism; prospective antimalarials; medicinal chemistry. *Mailing Add:* 5024 S Stone St Spokane WA 99223

STACY, RALPH WINSTON, environmental health, research administration; deceased, see previous edition for last biography

STACY, T(HOMAS) D(ONNIE), b Houston, Tex, Jan 13, 34; m 54; c 4. PETROLEUM ENGINEERING. *Educ:* La Polytech Inst, BS, 57, MS, 62; Miss State Univ, PhD(eng), 66. *Prof Exp:* Petrol engr, Pan-Am Petrol Corp, 57-58 & 62-63; asst petrol eng, La Polytech Inst, 61-62; from instr to asst prof, Miss State Univ, 63-68; area engr, Pan Am Petrol Corp, 68-76; mgr res, Amoco Prod Co, 76-80, mgr prod, Amoco Int, 80-81, mgr prod serv, vpres prod, Amoco, PRES & CHMN AMOCO CAN. *Mem:* AAAS; Am Inst Mining, Metall & Petrol Engrs; Soc Petrol Engrs (treas); Soc Petrol Engrs (pres-elect, 81). *Res:* Petroleum engineering, surface chemistry and gas adsorption. *Mailing Add:* Amoco Canada Petro Co Ltd 444 7th Ave SW Calgary AB T2P 0Y2 Can

STADELMAIER, H(ANS) H(EINRICH), b Stuttgart, Ger, Nov 14, 22; nat US; m 46; c 3. PHYSICAL METALLURGY. *Educ:* Univ Stuttgart, Dipl, 51, Dr rer nat, 56. *Prof Exp:* Interpreter, US Mil Govt, Ger, 45-47; from res assoc to res prof, 52-80, PROF METALL, NC STATE UNIV, 80- *Concurrent Pos:* Vis scientist, Max Planck Inst Metall, Stuttgart, Ger, 85-90. *Mem:* Fel Am Soc Metals; Ger Mat Soc; Metall Soc; Assoc Inst Mining Engrs. *Res:* Alloy phases; x-ray crystallography; electronic materials; permanent magnet materials. *Mailing Add:* Mat Sci & Eng Dept NC State Univ Box 7907 Raleigh NC 27695-7907

STADELMAN, WILLIAM JACOB, b Vancouver, Wash, Aug 8, 17; m 42; c 2. FOOD SCIENCE. *Educ:* Wash State Univ, BS, 40; Pa State Univ, MS, 42, PhD(biochem), 48. *Prof Exp:* Asst poultry husb, Pa State Univ, 40-42; asst prof, Wash State Univ, 48-52, assoc prof, 52-55; from assoc prof to prof poultry sci, 55-62, prof, 62-83, EMER PROF FOOD SCI, PURDUE UNIV, WEST LAFAYETTE, 83- *Concurrent Pos:* Consult, Food & Poultry Int Indust; mem, Tech Adv Comt, Poultry & Egg Inst Am, 57-83; mem bd dirs, Res & Develop Assocs, Food & Container Inst, 66-69, 72-75; mem, Sci Adv Comt, Refrig Res Found, 67-; mem adv bd mil personnel supplies, Food Irradiation Comt, Nat Acad Sci, 67-69; mem, Tech Adv Comt, Nat Turkey Fedn, 71-83, Tech Adv Comt, Am Egg Bd, 74-83. *Honors & Awards:* Christie Award, Poultry & Egg Nat Bd, 55; Res Award, Am Egg Bd, 75; Sci Award, Inst Food Technol, 77. *Mem:* Inst Food Technol; Poultry Sci Asn (pres, 77-78); Am Soc Heat, Refrig & Air-Conditioning Eng; World Poultry Sci Asn; Am Meat Sci Asn; Int Inst Refrig. *Res:* Effects of refrigeration and freezing on quality preservation of protein rich foods; poultry products quality evaluation and preservation; new product development. *Mailing Add:* Dept Food Sci Smith Hall Purdue Univ West Lafayette IN 47907

STADELMANN, EDUARD JOSEPH, b Graz, Austria, Sept 24, 20; m. PLANT PHYSIOLOGY, CELL PHYSIOLOGY. *Educ:* Innsbruck Univ, PhD(bot, philos), 53; Univ Freiburg, Venia Legendi, 57. *Hon Degrees:* Dr, Agr Univ Vienna, Austria, 89. *Prof Exp:* Asst bot, Freiburg Univ, 54-61, privat docent, 57, sr asst, 62-64; from asst prof to assoc prof, 64-72, PROF PLANT PHYSIOL, UNIV MINN, ST PAUL, 72- *Concurrent Pos:* Muellhaupt scholar biol, Ohio State Univ, 58-59; res assoc, Univ Minn, 63-64; Humboldt award, Bonn, Ger, 73-75; Agr, Univ Vienna, Austria, 87; vis prof, dept bot, Seoul Nat Univ, Korea, 78-79, vis scientist, Shijiazhuang, Hebei, China, 85, 86, 87; Fulbright award, 79 & 87. *Mem:* Ger Bot Soc; Swiss Bot Soc; Swiss Soc Natural Sci; Austrian Zool-Bot Soc; Am Inst Biol Sci; Sigma Xi; Am Soc Plant Physiologists. *Res:* Permeability; cytomorphology; salt resistance; protoplasmatology; radiation effects; desiccation resistance. *Mailing Add:* Dept Hort Sci 228 Univ Minn St Paul MN 55108-1011

STADLER, DAVID ROSS, b Columbia, Mo, May 24, 25; m 52; c 4. GENETICS. *Educ:* Univ Mo, AB, 48; Princeton Univ, MA, 50, PhD, 52. *Prof Exp:* Instr biol, Univ Rochester, 52; Gosney res fel genetics, Calif Inst Technol, 52-53, USPHS fel, 53-55; instr bot, 56-57, asst prof, 57-59, from asst prof to assoc prof genetics, 59-67, PROF GENETICS, UNIV WASH, 67- *Concurrent Pos:* Ed, Genetics, Genetics Soc Am, 73-76. *Mem:* Genetics Soc Am (treas, 69-71). *Res:* Genetics of microorganisms; mutation, recombination and development. *Mailing Add:* Dept of Genetics Univ of Wash Seattle WA 98195

STADLER, LOUIS BENJAMIN, b Monroe, Mich, Feb 26, 26; m 51; c 3. PHARMACEUTICAL CHEMISTRY, ANALYTICAL CHEMISTRY. *Educ:* Univ Mich, BS, 48, MS, 50, PhD(pharmaceut chem), 54. *Prof Exp:* Sr anal chemist, Parke, Davis & Co, 53-63, mgr anal standards, 63-64; asst head qual control, William S Merrell Co Div, Richardson-Merrell Inc, 64-65, head qual control, 66-71, Merrell Nat Labs, 71-73, Master Documents Admin, 73-75, mgr, Qual Opers Records, systs & planning, Merrell-Nat Labs, 75-81, qual opers tech proj mgr, 81-83, qual assurance compliance coordr & qual opers compliance mgr, Merrell Dow Pharmaceut, 83-90; RETIRED. *Concurrent Pos:* Mem adv panel steroids, Nat Formulary, 60-65, comt specifications, 66-75, panel trypsin & chymotrypsin, 70-75; mem rev comt, US Pharmacopeia, 70-80, asst, 80-90. *Mem:* Am Pharmaceut Asn; Am Chem Soc; Am Soc Qual Control. *Res:* Analytical methodology for testing drug substances, pharmaceutical dosage forms and associated standards; improved control techniques for pharmaceuticals; technical management. *Mailing Add:* 511 Laramie Trail Cincinnati OH 45215-2503

STADNICKI, STANLEY WALTER, JR, b Norwich, Conn, Sept 30, 43; m 65; c 4. RESEARCH ADMINISTRATION. *Educ:* Assumption Col, Mass, BA, 65; Clark Univ, Mass, MA, 70; Worcester Polytech Inst, PhD(biomed eng), 76. *Prof Exp:* Res scientist, E G & G Mason Res Inst, 67-76; ASST DIR TOXICOL, MED RES LABS, PFIZER, INC, 76- *Mem:* Soc Toxicol; Am Soc Pharmacol & Exp Therapeut; Am Col Toxicol; Soc Toxicol Can; Inst Elec & Electronics Engrs. *Res:* Directing research for the purpose of testing the toxicological effects of new drug candidates; interpretation, analysis and documentation of research results. *Mailing Add:* Pfizer Inc Central Res Eastern Pt Rd Groton CT 06340

STADTER, JAMES THOMAS, b Baltimore, Md. COMPUTER SIMULATION. *Educ:* Loyola Col, BS, 59; Univ Md, MA, 64; Am Univ, PhD(math), 75. *Prof Exp:* Assoc engr, 60-65, sr engr, 65-81, PRIN PROF STAFF, APPL PHYSICS LAB, JOHNS HOPKINS UNIV, 81-, INSTR COMPUT SCI & APPLIED MATH, 80- *Res:* Aeroelasticity and structural analysis; applied mathematics; eigenvalue estimation procedure. *Mailing Add:* Appl Physics Lab Johns Hopkins Univ Johns Hopkins Rd Laurel MD 20723

STADTHERR, LEON, b New Ulm, Minn, Nov 27, 42. BIOINORGANIC CHEMISTRY, ENGINEERING. *Educ:* St John's Univ, Minn, BS, 65; Univ NDak, PhD(chem), 70. *Prof Exp:* Res assoc chem, Univ Va, 70-72; res assoc chem, Iowa State Univ, 73-76; process develop spec, Gen Resource Corp, 77-84. *Mem:* Am Chem Soc; Sigma Xi. *Res:* Coordination compounds; stereochemistry of transition metal ion and lanthanide ion complexes; ion exchange. *Mailing Add:* PO Box 61 Gibbon MN 55335-0061

STADTHERR, MARK ALLEN, b Austin, Minn, May 15, 50. CHEMICAL ENGINEERING. *Educ:* Univ Minn, BChE, 72; Univ Wis-Madison, PhD(chem eng), 76. *Prof Exp:* ASST PROF CHEM ENG, UNIV ILL, URBANA-CHAMPAIGN, 76- *Mem:* Am Inst Chem Engrs; Am Chem Soc; Am Soc Eng Educ; Soc Indust & Appl Math. *Res:* Chemical process simulation, optimization and design; sparse matrix computations; resource management. *Mailing Add:* Dept Chem Eng Univ Ill 1209 W California St Urbana IL 61801-3731

STADTHERR, RICHARD JAMES, b Gibbon, Minn, Nov 24, 19. PLANT PHYSIOLOGY, PLANT BREEDING. *Educ:* Univ Minn, BS, 49, MS, 51, PhD(hort), 63. *Prof Exp:* Teaching asst, Univ Minn, 48-51, asst prof hort, Exten, 53-54, instr in charge of res turf & nursery crops, 54-61; res asst, Cornell Univ, 51-52; asst prof nursery crops res, Univ Mass, 52-53; assoc prof hort, Exten, NC State Univ, 63-67; from asst prof to prof hort, La State Univ, Baton Rouge, 67-82; RETIRED. *Honors & Awards:* Burpee Award, 49. *Mem:* Am Soc Hort Sci; Int Plant Propagators Soc; Am Magnolia Soc; Am Hort Soc; Am Rhododendron Soc. *Res:* Testing-breeding of azaleas. *Mailing Add:* 1373 First Ave Box 214 Gibbon MN 55335

STADTMAN, EARL REECE, b Carrizozo, NMex, Nov 15, 19; m 43. BIOCHEMISTRY. *Educ:* Univ Calif, Berkeley, BS, 42, PhD(comp biochem), 49. *Hon Degrees:* DSc, Univ Mich, 87. *Prof Exp:* Res asst, Dept Food Technol, Univ Calif, 43-46, Leopold Wrasse res assistantship, Div Plant Nutrit, 46-47, Leopold Wrasse grad res fel & sr lab technician, 47-78, res asst 48-49; AEC res fel, Mass Gen Hosp, 49-50; chemist, Lab Cellular Physiol & Metab, 50-58, chief, Enzyme Sect, 58-62, CHIEF, LAB BIOCHEM, NAT HEART, LUNG & BLOOD INST, NIH, 62- *Concurrent Pos:* Lectr, USDA Grad Sch, 54-, Georgetown Univ, 56-58 & Univ Md, 59-; lectr var socs & univs, 56-; vis scientist, Max Planck Inst, Ger, 59-60 & Pasteur Inst, France, 60; ed, J Biol Chem, 60-; exec ed, Archives Biochem, Biophys, 60-; mem adv comt, Oak Ridge Nat Lab, 63-66; chmn, Div Biol Chem, Am Chem Soc, 63-64, 81-82; chmn biochem div, Found Advan Educ in Sci, 64-; chmn, Comt Policy & Procedures, Am Soc biol Chemists, 65-66, coun, 72-85; NIH lectr, 66; vis prof var univs, 67-; adv bd, Biochemistry, 69-76 & Trends in Biochem Res, 76-79; coun, Int Union Biochem, 76-82, chmn, Interest Group Comt, 79-84; Nat Acad Sci deleg, Int Union Biochem, 76. *Honors & Awards:* Lewis Award, Am Chem Soc, 53; Burroughs-Wellcome Lectr, Mass Gen Hosp, 68; Microbiol Award, Nat Acad Sci, 70; Distinguished Serv Award, HEW, 70; Plenary Lectr, Am Soc Biol Chemists, 76; Nat Medal Sci, 79; Distinguished Camille & Henry Dreyfus Lectr, Northwestern Univ, 80; Presidential Rank Award Sr Exec Serv, 81; Merck Award, Am Soc Biol Chemists, 83; Kamen Lectr, Univ Calif, 85; John Muntz Mem Lectr, Union Univ, 85; Seventh Robert E Olson Lectr, St Louis Univ, 90; Robert A Welch Award Chem, 91. *Mem:* Nat Acad Sci; Am Chem Soc; Am Soc Biol Chemists (pres, 82-83); Am Soc Microbiol; Am Acad Arts & Sci; Int Union Biochem; Biophys Soc; Protein Soc; fel Oxygen Soc; Geront Soc. *Res:* Microbial and intermediary metabolism; enzyme chemistry; biochemical function of vitamin B12 and ferredoxin; metabolic regulation of biosynthetic of biosynthetic pathways; membrane transport; oxygen radical mediated modification of enzymes; role of protein modification in aging; oxygen toxicity; author of various publications. *Mailing Add:* Lab Biochem Nat Heart Lung & Blood Inst NIH Bethesda MD 20014

STADTMAN, THRESSA CAMPBELL, b Sterling, NY, Feb 12, 20; m 43. BIOCHEMISTRY, MICROBIOLOGY. *Educ:* Cornell Univ, BS, 40, MS, 42; Univ Calif, PhD(microbiol), 49. *Prof Exp:* Asst nutrit, Agr Exp Sta, Cornell Univ, 42-43; res assoc food microbiol, Univ Calif, 43-46; asst, Harvard Med Sch, 49-50; BIOCHEMIST, NAT HEART INST, 50- *Concurrent Pos:* Whitney fel, Oxford Univ, 54-55; Rockefeller grant, Inst Cell Chem, Univ Munich, 59-60; French Govt fel, Inst Biol & Phys Chem, France, 60. *Honors & Awards:* Hillebrand Award, 79; Rose Award, 87. *Mem:* Nat Acad Sci; Am Soc Biol Chemists (secy, 78-81); Brit Biochem Soc; Am Chem Soc; Am Soc Microbiol. *Res:* Amino acid intermediary metabolism; one-carbon metabolism; methane formation; microbial biochemistry; selenium biochemistry. *Mailing Add:* Nat Heart Lung & Blood Inst Bethesda MD 20892

STAEBLER, DAVID LLOYD, b Ann Arbor, Mich, Apr 25, 40; m 61; c 2. KINESCOPE DESIGN, PHOTOVOLTAIC CELLS. *Educ:* Pa State Univ, BS, 62, MS, 63; Princeton Univ, MA, 67, PhD(elec eng), 70. *Prof Exp:* Mem tech staff, 63-81, HEAD, KINESCOPE SYSTS GROUP, RCA LABS, 81- *Concurrent Pos:* Vis prof, Inst Phys Chem Sao Carlos, Univ San Paulo, Brazil, 74-75; vis mem tech staff labs, RCA Ltd, Zurich, Switz, 79-80. *Mem:* Inst Elec & Electronics Eng; AAAS. *Res:* Electron gun and kinescope design; photovoltaic properties of amorphous silicon; hologram storage in electro-optic materials; photochromic and electrochromic phenomena; hologram storage in electro-optic materials; optical and electronic properties of amorphous silicon. *Mailing Add:* RR1 362-C2 Pennington NJ 08534

STAEHELIN, LUCAS ANDREW, b Sydney, Australia, Feb 10, 39; m 65; c 3. CELL BIOLOGY, PHOTOSYNTHESIS. *Educ:* Swiss Fed Inst Technol, DiplNatw, 63, PhD(biol), 66. *Prof Exp:* Res scientist, Dept Sci & Indust Res, NZ, 66-69; res fel cell biol, Harvard Univ, 69-70; asst prof, 70-73, assoc chmn dept molecular cell & develop biol, 72-73, assoc prof, 73-78, PROF CELL BIOL, UNIV COLO, BOULDER, 79- *Concurrent Pos:* Nat Inst Gen Med Sci grant, 71-; study sect cell biol, NIH, 80-84; vis prof, Inst Biol & Microbiol, Univ Freiburg, Germany, 78, cell biol, Swiss Fed Inst Technol, 84. *Honors & Awards:* Humboldt Award, 78. *Mem:* AAAS; Am Soc Cell Biol; Am Soc Plant Physiol. *Res:* Structure and function of biological membranes; freeze-etch electron microscopy; photosynthesis; plant cell walls; plant cell secretion. *Mailing Add:* Dept Molecular Cell Develop Biol Univ Colo Box 347 Boulder CO 80309

STAEHLE, ROGER WASHBURNE, b Detroit, Mich, Feb 4, 34; div; c 4. METALLURGICAL ENGINEERING, CORROSION. *Educ:* Ohio State Univ, BMetE & MS, 57, PhD(metall eng), 65. *Prof Exp:* Res asst corrosion, Ohio State Univ, 61-65, from asst prof to assoc prof metall eng, 65-70, prof, 70-79,; pres & chmn, Automated Transp Systs, Inc, Minneapolis, 84-86; dean, Inst Technol, 79-83, prof chem eng & mat sci, 83-84, ADJ PROF, UNIV MINN, 88- *Concurrent Pos:* Consult, 3M Co, Oak Ridge Nat Lab, Monsanto Co, Int Nickel Co, Inc, NUS Corp & Parameter Inc; mem adv panel, Mat Div, Nat Bur Standards; Int Nickel prof corrosion sci & eng, 71-76; ed, Corrosion J, 73-79; dir, Fontana Corrosion Ctr, 75-79; indust consult, North Oaks, Minn, 86- *Honors & Awards:* Willis Rodney Whitney Award, Nat Asn Corrosion Engrs, 80. *Mem:* Nat Acad Eng; fel Am Soc Metals; Electrochem Soc; Sigma Xi; Am Concrete Inst; Am Soc Testing & Mat; Nat Asn Corrosion Eng. *Res:* Passivity of metals; stress corrosion cracking; process of fracture; fatigue; optical properties of surfaces; surface chemistry; author of various publications; granted one patent. *Mailing Add:* 22 Red Fox Rd North Oaks MN 55127

STAELIN, DAVID HUDSON, b Toledo, Ohio, May 25, 38; m 62; c 3. RADIO ASTRONOMY, METEOROLOGY. *Educ:* Mass Inst Technol, SB, 60, SM, 61, ScD(elec eng), 65. *Prof Exp:* From instr to assoc prof, 65-76, PROF ELEC ENG, MASS INST TECHNOL, 76- *Concurrent Pos:* Ford fel eng, 65-67; vis asst scientist, Nat Radio Astron Observ, 68-69; dir, Environ Res & Technol, Inc, 69-79; pres, Pictel Corp, 84, chmn 84-87; chmn, Nat Acad Sci Comn Radio Frequency Req Res, 83-86; mem, Space Appln Adv Comt, NASA, 83-86, prin invest spaceflight exp, Nimbus-E Microwave Spectrometer, 72, Scanning Microwave Spectrometer, 75, Tech Adv Comt, Comsat, 84-87; asst dir, Lincoln Lab, Mass Inst Technol, 90- *Honors & Awards:* NASA Award, Voyager Sci Invest - Planetary Radio Astron, Uranus, 86, Neptune, 90; Alan Berman Res Publ Award, Naval Res Lab, 88. *Mem:* AAAS; Am Astron Soc; Am Geophys Union; Inst Elec & Electronic Eng; Am Meteorol Soc. *Res:* Planetary atmospheres; pulsars; space-based and ground-based meteorological observations using passive microwave techniques; microwave and optical instrumentation; atmospheric sensing; communications satellites; video image processing. *Mailing Add:* Dept Elec Eng & Comput Sci Mass Inst Technol Cambridge MA 02139

STAETZ, CHARLES ALAN, b North Platte, Nebr, July 12, 45; m 68; c 2. ECONOMIC ENTOMOLOGY. *Educ:* Chadron State Col, BS, 67; Univ Nebr, MS, 72, PhD(entom), 75. *Prof Exp:* Entomologist, Velsicol Chem Corp, 75-78; MEM STAFF, AGR CHEM GROUP, FMC CORP, 78- *Mem:* Entom Soc Am. *Res:* New insecticides; insecticide resistance-detection and countermeasures. *Mailing Add:* 85 Hickory Lane Newtown PA 18940

STAFFORD, BRUCE H(OLLEN), b North Platte, Nebr, Aug 25, 22; m 47; c 2. ELECTRICAL ENGINEERING, OPERATIONS RESEARCH. *Educ:* Univ Nebr, BSc, 43; Univ Md, MSc, 49. *Prof Exp:* Head dir systs sect, Oper Res Br, US Naval Res Lab, 43-54; opers analyst, Strategic Air Command, US Dept Air Force, 54-57, chief opers anal, 8th Air Force, 57-59, chief oper capability div, Opers Anal, 59-61, dep chief, Opers Anal, Strategic Air Command, 61-63, chief opers anal, HQ, 63-71, chief sci & res, 71-74; RETIRED. *Res:* Circuit analysis; servo-mechanisms; weapon control systems, including radar and computers; nuclear physics; electronic countermeasures; guided missiles; aircraft; operations research. *Mailing Add:* 3803 Chiswell Ct Greensboro NC 27410

STAFFORD, DARREL WAYNE, b Parsons, Kans, Mar 11, 35; m 57; c 3. ZOOLOGY, MOLECULAR BIOLOGY. *Educ:* Southwest Mo State Col, BA, 59; Univ Miami, Fla, PhD(cellular physiol), 64. *Prof Exp:* NIH fel, Albert Einstein Col Med, 64-65; asst prof, 65-70, assoc prof, 70-77, PROF ZOOL, NUTRIT & BIOCHEM, UNIV NC, CHAPEL HILL, 78- *Mem:* AAAS. *Res:* Cell division and protein synthesis. *Mailing Add:* Dept Biol Univ NC 203 Wilson Hall 046A Chapel Hill NC 27514

STAFFORD, FRED E, b New York, NY, Mar 30, 35; m 63. PHYSICAL INORGANIC CHEMISTRY, SOLID STATE CHEMISTRY. *Educ:* Cornell Univ, AB, 56; Univ Calif, Berkeley, PhD(chem), 59. *Prof Exp:* NSF fel, Free Univ Brussels, 59-61; from asst prof to assoc prof chem, Northwestern Univ, 61-74; prog officer sci develop progs, 74-75, prof dir solid state chemistry, NSF, 75-87, DIR OF SPEC PROJS, UNIV OF CHICAGO, 87- *Concurrent Pos:* Mem, Comt on High Temp Sci & Technol, Res Coun Can-Nat Acad Sci, 75-78. *Mem:* Am Chem Soc; Am Phys Soc; Am Soc Pub Admin; Fed Exec Inst Alumni Asn. *Res:* Mass spectrometry and spectroscopy of simple inorganic systems; high temperature chemistry; solid state chemistry: administers research activities in a broad interdiciplinary area. *Mailing Add:* Univ Chicago Res & Admin 970 E 58th St Chicago IL 60637-1432

STAFFORD, HELEN ADELE, b Philadelphia, Pa, Oct 9, 22. PLANT PHYSIOLOGY. *Educ:* Wellesley Col, BA, 44; Conn Col, MA, 48; Univ Pa, PhD, 51. *Prof Exp:* Instr bot & res assoc biochem, Univ Chicago, 51-54; from asst prof to assoc prof, 54-65, PROF BOT, REED COL, 65- *Concurrent Pos:* Guggenheim fel, Harvard Univ, 58-59; NSF sr fel, Univ Calif, Los Angeles, 63-64. *Mem:* Bot Soc Am; Am Soc Plant Physiol; Am Soc Biol Chemists; Phytochem Soc NAm (pres, 77-78). *Res:* Plant biochemistry; metabolism of phenolic compounds; regulation of and metabolism of phenolic compounds in higher plants. *Mailing Add:* Dept Biol Reed Col 3203 SE Woodstock Blvd Portland OR 97202

STAFFORD, JOHN WILLIAM, b New York, NY, Mar 11, 32; m 58; c 2. ELECTRONIC PHYSICAL DESIGN & SEMICONDUCTOR PACKAGING, ENGINEERING MECHANICS-STRESS ANALYSIS. *Educ:* Mass Inst Technol, BS, 54; Brooklyn Polytech Inst, MS, 59 & 70; Fairleigh Dickinson Univ, MBA, 79. *Prof Exp:* Struct engr, Grumman Aircraft, Bethpage, NJ, 57-60 & Knolls Atomic Power Lab, Schenectady, 60-61; tech supvr & MTS, Bell Tel Labs, Murray Hill, NJ, 61-82 & AT&T Info Systs, Holmdel, NJ, 82-84; tech dir & founder, Lytel Inc, Branchburg, NJ, 84-87; LAB HEAD & SR MTS, MOTOROLA INC, SCHAUMBURG, ILL, 88- *Concurrent Pos:* Chmn, Prog Subcomt Packaging, Electronic Component & Technol Conf, 80-89; dir, Int Electronic Packaging Soc, 83-86; mem, Component Hybrids & Mfg Technol Admin Comn, Int Elec & Electronics Eng, 86-89; prog chmn, 41st Electronic Component & Technol Conf, 90-91. *Mem:* Sr mem Inst Elec & Electronics Engrs. *Res:* Electronics systems physical design and semiconductor packaging; development of precision equipment for electronic manufacturing; high volume low manufacturing cost technology; author/coauthor 24 scientific publications; two patents awarded and six pending. *Mailing Add:* 77 Highgate Course St Charles IL 60174

STAFFORD, MAGDALEN MARROW, b Sharon, Pa, May 20, 42; m 75. GERONTOLOGY, REHABILITATION. *Educ:* Duquesne Univ, BSN, 64; Case Western Reserve Univ, MSN, 68, PhD(nursing), 81. *Prof Exp:* Staff nurse geriat, Pittsburgh Vet Admin Hosp, 64-65; staff nurse metabolics, Cleveland Vet Admin Hosp, 65-68; clin nurse specialist geront & rehab, Cuyahoga County Hosps, 68-81; assoc prof nursing, Youngstown State Univ, Ohio, 81-86, chmn dept nursing, 82-84; asst prof & prof-in-charge nursing, Pa State Univ Extended Nursing Prog, Pa State Univ, 86-87; assoc exec dir nursing serv & educ, 87-91, VPRES NURSING, ST ELIZABETH HOSP MED CTR, YOUNGSTOWN, OHIO, 91- *Concurrent Pos:* Consult geriat nursing, Butler Vet Admin Med Ctr, 84- *Mem:* Am Heart Asn; Am Nurses Asn. *Res:* Perceived health needs of elderly clients; gerontological nursing; rehabilitation nursing. *Mailing Add:* St Elizabeth Hosp Med Ctr 1044 Belmont Ave Youngstown OH 44501

STAFFORD, ROBERT OPPEN, b Milwaukee, Wis, Jan 28, 20; m 61; c 3. ENDOCRINOLOGY, NAUTICAL ARCHEOLOGY. *Educ:* Univ Wis, BA, 41, MA, 48, PhD(zool), 49. *Prof Exp:* Res scientist, Upjohn Co, 49-60, asst dir biol res, 60-62, biochem res, 62-68, asst to exec vpres pharmaceut div, 68-71, chmn & chief exec officer, Upjohn Healthcare Serv, 76-82, vpres corp planning, Upjohn Co, 70-82; RETIRED. *Mem:* Inst Nautical Archeol; Soc Ocean Studies; Int Oceanograph Found. *Res:* Pharmacology, virology, pathology; metabolic diseases; management of research. *Mailing Add:* PO Box 125 Key Colony Beach FL 33051

STAFFORD, THOMAS P(ATTEN), aeronautics, astronautics, for more information see previous edition

STAFSUDD, OSCAR M, JR, b Allison Park, Pa, Nov 10, 36; m 67; c 1. SOLID STATE SPECTROSCOPY. *Educ:* Univ Calif, Los Angeles, BA, 59, MS, 62, PhD(physics), 67. *Prof Exp:* Physicist, Atomics Int Div, NAm Aviation, Inc, 60-64 & Hughes Res Labs, 64-67; asst prof eng, 67-72, assoc prof eng, 74-80, assoc prof appl sci, 74-80, PROF ELEC ENG, UNIV CALIF, LOS ANGELES, 80- *Concurrent Pos:* Consult, Hughes Res Labs, 67- *Mem:* Am Phys Soc; Optical Soc Am; Am Soc Eng Educ. *Res:* Laser technology; crystal growth; solid state electronics. *Mailing Add:* Dept Elec Eng 7732 Boelter Univ Calif 405 Hilgard Ave Los Angeles CA 90024

STAGEMAN, PAUL JEROME, b Persia, Iowa, June 21, 16; m 37; c 1. BIOCHEMISTRY. *Educ:* Univ Omaha, AB, 39; Univ Iowa, MS, 50; Univ Nebr, PhD, 63. *Prof Exp:* Res chemist, Cudahy Packing Co, 39-41; from asst prof to prof chem, 41-80, EMER PROF CHEM, UNIV NEBR, OMAHA, 80- *Mem:* AAAS; Am Chem Soc; Am Inst Chemists; Sigma Xi. *Res:* Ultracentrifugation; lipoproteins; atherosclerosis; plant pigments. *Mailing Add:* 308 W Oak St Council Bluffs IA 51503

STAGER, CARL VINTON, b Kitchener, Ont, June 10, 35; m 62; c 4. SOLID STATE PHYSICS. *Educ:* McMaster Univ, BSc, 58; Mass Inst Technol, PhD(physics), 61. *Prof Exp:* Mem res staff, Francis Bitter Nat Magnet Lab, Mass Inst Technol, 60-63; from asst prof to assoc prof, 63-72, PROF PHYSICS, MCMASTER UNIV, 72- *Mem:* Can Asn Physicists (treas, 64-68); Am Phys Soc. *Res:* Magnetism of insulating crystals; crystal fields spectra; electron paramagnetic resonance. *Mailing Add:* Dept Physics McMaster Univ 1280 Main St W Hamilton ON L8S 4L8 Can

STAGER, HAROLD KEITH, b Gardena, Calif, Dec 5, 21; m 49, 87; c 4. GEOLOGY. *Educ:* Univ Calif, Los Angeles, BA, 48. *Prof Exp:* Asst assayer, Golden Queen Mining Co, 41; geologist, Mineral Deposits Br, US Geol Surv, 48-63, Geol Br, 63-64, Base Metals Br, 64-65, field officer, Off Minerals Explor, 65-82; RETIRED. *Concurrent Pos:* Consult mining geologist, 82- *Mem:* Geol Soc Am. *Res:* Mining geology; mineral deposits; strategic and rare metals. *Mailing Add:* PO Box 1197 Bodega Bay CA 94923

STAGG, RONALD M, b Brooklyn, NY; m 52; c 5. PHYSIOLOGY, ENDOCRINOLOGY. *Educ:* Tusculum Col, BA, 50; Brooklyn Col, MA, 55; Rutgers Univ, PhD(endocrinol), 62. *Prof Exp:* Asst adminr, Willard F Greenwald, Med & Chem Consult, NY, 51-53; asst to med dir admin, Warner-Chilcott Lab Div, Warner-Lambert Pharmaceut Co, NJ, 53-56; instr zool, Drew Univ, 59-60; instr physiol, Med Col Va, 60-65; assoc prof, 65-74, PROF BIOL, HARTWICK COL, 74- *Concurrent Pos:* NIH grant, 63-65. *Mem:* Am Soc Zool. *Res:* Physiology, specifically mammalian; endocrinology, specifically the relationship between hormones and nutrition. *Mailing Add:* Dept of Biol Hartwick Col Oneonta NY 13820

STAGG, WILLIAM RAY, b Lexington, Ky, Sept 15, 37; m 62; c 2. PHYSICAL INORGANIC CHEMISTRY. *Educ:* Univ Ky, BS, 59; Iowa State Univ, PhD(chem), 63. *Prof Exp:* Res chemist, FMC Corp, NJ, 63-64; res assoc chem, Univ Ill, 66-67; asst prof, Colgate Univ, 67-72; assoc prof chem, Randolph-Macon Women's Col, 72-77; SR RES CHEMIST,

BABCOCK & WILCOX CO, 77- *Mem:* Am Nuclear Soc; Am Chem Soc. *Res:* Complex equilibria of lanthanide elements; heteroatom ring systems of sulfur, nitrogen and phosphorous; environmental chemistry; nuclear reactor coolant chemistry. *Mailing Add:* Babcock & Wilcox Naval Nuclear Div PO Box 785 Mt Athos Rd Lynchburg VA 24505

STAGNO, SERGIO BRUNO, b Santiago, Chile, Oct 31, 41; m 68; c 1. PEDIATRICS, INFECTIOUS DISEASES. *Educ:* Univ Chile, Bachelor, 60, MD, 67. *Prof Exp:* Instr pediat & parasitol, Sch Med, Univ Chile, 70-71; res assoc pediat, 72-73, asst prof, 73-77, asst prof microbiol, 75-81, assoc prof pediats, 77-80, assoc prof microbiol, 81-85, PROF PEDIAT, UNIV ALA, 80-, PROF MICROBIOL, 85- *Concurrent Pos:* Fel, Univ Ala, 71-72. *Mem:* Am Soc Microbiol; Soc Pediat Res; Chilean Pediat Soc; Chilean Parasitol Soc; Am Acad Pediats. *Mailing Add:* Dept Pediat Univ Ala Sch Med Univ Sta Birmingham AL 35294

STAHEL, EDWARD P(AUL), b New York, NY, June 3, 34; m 57, 65; c 5. CHEMICAL ENGINEERING. *Educ:* Princeton Univ, BSE, 55; Univ Notre Dame, MS, 57; Ohio State Univ, PhD(chem eng), 61. *Prof Exp:* Res engr, E I Du Pont de Nemours & Co, 61-62; from asst prof to assoc prof, 62-74, grad adminr, 66-72, PROF CHEM ENG, NC STATE UNIV, 74- *Mem:* AAAS; Am Chem Soc; Am Inst Chem Engrs; Am Soc Eng Educ. *Res:* Chemical engineering kinetics and reactor design; transport phenomena; polymer chemical engineering; biotechnological engineering. *Mailing Add:* Dept Chem Eng NC State Univ Box 7905 Raleigh NC 27695

STAHL, BARBARA JAFFE, b Brooklyn, NY, Apr 17, 30; m 51; c 4. COMPARATIVE ANATOMY, EVOLUTION. *Educ:* Wellesley Col, BA, 52; Radcliffe Col, AM, 53; Harvard Univ, PhD(biol), 65. *Prof Exp:* PROF BIOL, ST ANSELM COL, 54- *Mem:* AAAS; Soc Vert Paleont. *Res:* Evolution of holocephali and early vertebrates. *Mailing Add:* Dept Biol St Anselm Col Manchester NH 03102-1310

STAHL, C(HARLES) D(REW), b Altoona, Pa, Aug 28, 23; m 48; c 1. PETROLEUM ENGINEERING. *Educ:* Pa State Univ, BS, 47, MS, 50, PhD(petrol eng), 55. *Prof Exp:* Asst, 47-48, res assoc, 49-53, from asst prof to assoc prof, 53-61, PROF PETROL ENG, PA STATE UNIV, UNIVERSITY PARK, 61-, HEAD DEPT, 62- *Concurrent Pos:* Consult, Minerals Div, Pa State Dept Forests & Waters & Socony Mobil Oil Co, Venezuela, 58. *Honors & Awards:* Am Asn Oilwell Drilling Contractors Award, 56. *Mem:* Am Inst Mining, Metall & Petrol Engrs. *Res:* Displacement of immiscible fluids in porous media. *Mailing Add:* Col Mineral Industs Pa State Univ University Park PA 16802

STAHL, FRANKLIN WILLIAM, b Boston, Mass, Oct 8, 29; m 55; c 2. GENETICS. *Educ:* Harvard Univ, AB, 51; Univ Rochester, PhD(biol), 56. *Hon Degrees:* DSc, Oakland Univ, 66 & Univ Rochester, 82. *Prof Exp:* Grad teaching asst & res asst, Univ Rochester, 51-54, NSF fel, 54-55; fel, NSF-Nat Res Coun, Div Med Sci, Calif Inst Technol, 55-57, res fel, 57-58; assoc prof zool, Univ Mo, 58-59; assoc prof biol & res assoc, Inst Molecular Biol, 59-63, actg dir, 73-74, PROF BIOL & MEM, INST MOLECULAR BIOL, UNIV ORE, EUGENE, 63-; RES PROF MOLECULAR GENETICS, AM CANCER SOC, 85- *Concurrent Pos:* Vol scientist, Div Molecular Genetics, Med Res Coun, Cambridge, Eng, 64-65; mem, Virol Study Sect, NIH, 68-71, spec fel, 69; vis scientist, Molecular Gentics Unit, Med Res Coun, Univ Edinburgh, Scotland & Int Lab Genetics & Biophys, Naples, 69-70; Guggenheim fel, 75 & 85; Lady Davis vis prof, Dept Genetics, Hebrew Univ, Jerusalem, 75-76; MacArthur fel, 85- *Mem:* Nat Acad Sci; Am Acad Arts & Sci. *Res:* Genetics of bacteriophage. *Mailing Add:* Inst Molecular Biol Univ Ore Eugene OR 97403

STAHL, FRIEDA A, b Brooklyn, NY, May 27, 22; m 42; c 2. THERMAL PHYSICS. *Educ:* Hunter Col, BA, 42; Hofstra Col, MA, 57; Claremont Grad Sch, PhD(educ), 69. *Prof Exp:* Jr physicist, US Army Signal Corps, NJ & Ala, 42-44 & Petty Labs, Petty Geophys Eng Co, Tex, 44-46; physicist, Hillyer Instrument Corp, NY, 46-48; sr physicist, Sylvania Res Labs, NY, 48-52; lectr physics, 58-59, from asst prof to assoc prof, 59-73, assoc dean acad planning, 70-75, PROF PHYSICS, CALIF STATE UNIV, LOS ANGELES, 73- *Concurrent Pos:* NSF sci fac fel, Harvey Mudd Col & Claremont Grad Sch, 66-68; res assoc physics, Harvey Mudd Col, 69-70, 75-76 & 90-91; consult ed, College Teaching, 89- *Honors & Awards:* Coler-Maxwell Prize, Int Soc for the Arts, Sci & Technol, 87. *Mem:* Am Phys Soc; Am Asn Physics Teachers; Sigma Xi; Asn Women Sci. *Res:* Electro-optical behavior in semiconductors; ultrasound propagation in solid methane and deuteromethane as a function of temperature, with particular interest in the lambda-type phase transitions of these substances; electro-optical phenomena in thin-film metal-insulator-metal structures. *Mailing Add:* Dept Physics & Astron Calif State Univ Los Angeles Los Angeles CA 90032

STAHL, GLENN ALLAN, b Snyder, Tex, Mar 14, 45; m 70; c 2. POLYMER CHEMISTRY. *Educ:* Univ Houston, PhD(polymer chem), 75. *Prof Exp:* Robert A Welch fel polymers, Univ Houston, 75-76; fel, Univ Ala, Tuscaloosa, 76-77; res chemist, Res & Develop Ctr, B F Goodrich Co, 77-80; sr chemist, Phillips Petrol Co, 80-87; SR STAFF CHEMIST, EXXON CHEM CO, 87- *Concurrent Pos:* Grant adv, Paint Res Inst, 77-79. *Mem:* Am Chem Soc; Soc Plastic Engrs. *Res:* Water soluble polymers; polymers for oil revovery; science education through industry-academic cooperation; anchored organic reagents; application of polypropylene in textiles. *Mailing Add:* 8111 Hurst Forest Dr Humble TX 77346

STAHL, JOEL S, b Youngstown, Ohio, June 10, 18; wid; c 1. PLASTICS IN BUILDING. *Educ:* Ohio State Univ, BChE, 39. *Prof Exp:* Mgr spec prod, Ashland Oil, Inc, 39-50; pres, Cool Ray Co, 50-51; PRES, STAHL INDUSTS, INC, 51- *Concurrent Pos:* Plastics-chem expert, US Trade Develop Mission, 63 & US Indust Develop Mission, 66. *Mem:* Soc Plastics Engrs; Soc Plastics Indust; Am Soc Testing & Mat; Asn Consult Chemists & Chem Engrs; NY Acad Sci. *Res:* Fire-resistant polymers for use in construction applications; development of systems and processes for low-cost housing using plastic-foam cores and thin-skins sandwich construction. *Mailing Add:* 530 E Central #1504 Orlando FL 32801

STAHL, JOHN BENTON, b Columbus, Ohio, Mar 28, 30; m 74; c 1. LIMNOLOGY. *Educ:* Iowa State Univ, BS, 51; Ind Univ, AM, 53, PhD(zool), 58. *Prof Exp:* Sessional lectr biol, Queen's Univ, Ont, 58-59; asst prof, Thiel Col, 59-63; asst prof, Wash State Univ, 63-66; asst prof, 66-72, ASSOC PROF BIOL, SOUTHERN ILL UNIV, 72- *Mem:* N Am Benthological Soc; Am Soc Limnol & Oceanog; Ecol Soc Am; Int Asn Theoret & Appl Limnol. *Res:* Chironomidae and Chaoborus. *Mailing Add:* Dept Zool Southern Ill Univ Carbondale IL 62901-6501

STAHL, JOHN WENDELL, b Wilkinsburg, Pa, Aug 24, 56; m 77; c 3. ANALYTICAL CHEMISTRY. *Educ:* Geneva Col, BS, 79; Pa State Univ, PhD(chem), 83. *Prof Exp:* Asst prof chem, Bloomsburg Univ, Pa, 83-85; ASSOC PROF CHEM, GENEVA COL, 85- *Concurrent Pos:* Assoc mem, Comn VI, Int Union Pure & Appl Chem, 85-; vis prof, Univ Pittsburgh, 88. *Mem:* Am Chem Soc; Am Sci Affil. *Res:* Environmental analysis; inorganic sulfur chemistry; enthalpimetry; electrochemistry; computerization of instrumentation. *Mailing Add:* Dept Chem Geneva Col Beaver Falls PA 15010

STAHL, LADDIE L, b Terre Haute, Ind, Dec 23, 21; m 42; c 3. FLUID MECHANICS. *Educ:* Purdue Univ, BS, 42; Johns Hopkins Univ, MS, 50. *Prof Exp:* Mgr prod planning & market res, Guided Missile Dept, Gen Elec Co, 54-55, Missile & Space Vehicle Dept, 55-59, mgr tech planning, 59-60, tech rels, 60-61, mgr spec prod, Gen Eng Lab, 61-62, adv technol appln, 62-64, mgr, Info Eng Lab, Adv Technol Labs, 64-65, res & develop appln, 65-70, adminr progs & systs, Res & Develop Ctr, 68-70, mgr res & develop appln, 70-74, planning & resources, electronics sci & eng, 74-76, Electronics Systs Progs Oper, electronics sci & eng, 76-84, mgr spec progs, proj develop oper res & develop appln oper, Corp Res & Develop, 84-90; RETIRED. *Concurrent Pos:* Alt mem, Gen Staff Comt Army Reserve Policy, 63; mem, electronics adv group, US Army Electronics Command, 70-74, chmn, 71-74; mem Secy Labor Nat Adv Comt Jobs Ret, 71- & Nat Adv Coun Employer Support of Guard & Reserve, 72-; consult, US Army Aviation Systs Command sci adv group aviation systs, 73-75; mem, Army Reserve Forces Policy Comt, 73-76, chmn, Army Reserve Subcomt, 76; mem, US Army Sci Bd, 78-87; dir, Technol Transfer Prog, Data Storage Systs Ctr, Carnegie Mellon Univ, 90- *Mem:* Sr mem Inst Elec & Electronics Engrs; Am Inst Aeronaut & Astronaut; Am Defense Preparedness Asn. *Res:* Velocity fields induced by supersonic lifting surfaces; weapon system design and selection. *Mailing Add:* 29 Fairway Lane Rexford NY 12148

STAHL, NEIL, b Sheridan, Ind, June 11, 42; m 67; c 1. MATHEMATICAL ANALYSIS. *Educ:* Ind Univ, Bloomington, AB, 64; Brown Univ, PhD(appl math), 70. *Prof Exp:* Asst prof ecosysts anal, Univ Wis-Green Bay, Marinette Campus, 69-72, asst prof math, Univ Wis Ctr, Marinette Campus, 72-76; ASSOC PROF MATH, UNIV WIS CTR, FOX VALLEY CAMPUS, 76- *Mem:* Soc Indust & Appl Math; Am Math Asn; Asn Comput Mach. *Res:* Differential equations and their applications; educational software. *Mailing Add:* 921 Whittier Dr Appleton WI 54914

STAHL, PHILIP DAMIEN, b Wheeling, WVa, Oct 4, 41; m 68; c 3. PHYSIOLOGY, CELL BIOLOGY. *Educ:* WLiberty State Col, BS, 64; WVa Univ, PhD(pharmacol), 67. *Prof Exp:* From asst prof to assoc prof, 71-81, PROF PHYSIOL, MED SCH, WASH UNIV, 82-, HEAD, 84- *Concurrent Pos:* Fel, Space Sci Res Ctr, Univ Mo, 67; Arthritis Found fel molecular biol, Vanderbilt Univ, 68-70. *Mem:* Brit Biochem Soc; Am Chem Soc; Am Physiol Soc; Am Soc Biol Chemists. *Res:* Lysosomes. *Mailing Add:* Dept Cell Biol Wash Univ Med Sch St Louis MO 63110

STAHL, RALPH HENRY, b Berlin, Ger, Dec 29, 26; US citizen; m 55; c 3. EXPERIMENTAL PHYSICS. *Educ:* Harvard Univ, AB, 49, MA, 50, PhD(nuclear physics), 54. *Prof Exp:* Mem tech staff, Radiation Lab, Univ Calif, 54-56; mem tech staff, Gen Atomic Div, Gen Dynamics Corp, 56-68; secy-treas, Systs, Sci & Software, 68-72; mem tech staff, IRT Corp, 72-77; VPRES, JAYCOR, 77- *Concurrent Pos:* Vis prof, Univ Ill, Urbana, 61. *Mem:* Am Phys Soc; Am Nuclear Soc. *Res:* Nuclear weapons effects on military systems; electronics and electronics components; nuclear reactor physics. *Mailing Add:* 3060 Cranbrook Ct La Jolla CA 92037

STAHL, ROLAND EDGAR, b Northumberland, Pa, Sept 2, 25; m 55; c 2. ORGANIC CHEMISTRY. *Educ:* Bucknell Univ, BS, 50; Cornell Univ, PhD(chem), 54. *Prof Exp:* Asst org chem, Cornell Univ, 50-53; res chemist, Am Cyanamid Co, 54-56; res chemist, E I DuPont De Nemours & Co Inc, NY, 56-60, Tenn, 60-65, res chemist, 65-75, mkt res rep, 75-77, toxicol & regulatory affairs coordr, 77-78, sr regulatory affairs specialist, 73-83, regulatory affairs consult, Del, 83-90; RETIRED. *Mem:* Am Chem Soc. *Res:* Peroxide and radical chemistry; polymer applications; adhesives; textile finishing and applications. *Mailing Add:* 2619 Skylark Rd Wilmington DE 19808

STAHL, S SIGMUND, b Berlin, Ger, June 16, 25; nat; m 47; c 1. PERIODONTOLOGY, ORAL PATHOLOGY. *Educ:* Univ Minn, DDS, 47; Univ Ill, MS, 49; Am Bd Periodont, dipl. *Prof Exp:* Res assoc periodont, 49-50, from instr to prof, 50-71, PROF PERIODONT & ORAL MED & CHMN DEPT, COL DENT, NY UNIV, 71-, ASSOC DEAN ACAD AFFAIRS, 78- *Concurrent Pos:* Attend, Vet Admin Hosp, Brooklyn, 50-53; consult, 58-; exec secy, Guggenheim Found Inst Dent Res, NY Univ, 64- *Honors & Awards:* Int Asn Dent Res Award, 71; Am Acad Periodont Award, 76. *Mem:* AAAS; Sigma Xi; Am Dent Asn; fel Am Med Writers Asn; fel Am Col Dent. *Res:* Periodontal pathology; clinical treatment evaluation; wound healing. *Mailing Add:* 1110 Second Ave New York NY 10022

STAHL, SAUL, b Antwerp, Belg, Jan 23, 42; US citizen; m 72. MATHEMATICS. *Educ:* Brooklyn Col, BA, 63; Univ Calif, Berkeley, MA, 66; Western Mich Univ, PhD(math), 75. *Prof Exp:* Systs programmer, Int Bus Mach, 69-73; asst prof math, Wright State Univ, 75-77; asst prof, 77-80, ASSOC PROF MATH, UNIV KANS, 80- *Mem:* Math Asn Am; Am Math Soc. *Res:* Graph theory; combinatorial topology. *Mailing Add:* Dept Math 25 C St Univ Kans Lawrence KS 66045

STAHL, WILLIAM J, b New York, NY, Jan 3, 39; m 64; c 3. BIOCHEMISTRY. *Educ:* Merrimack Col, AB, 60; Fordham Univ, MS, 61; St John's Univ, NY, PhD(biochem), 69. *Prof Exp:* Res asst biochem, Albert Einstein Med Ctr, 62-64; res chemist, Tenneco Chem Inc, 64-65; asst prof, 65-74, ASSOC PROF BIOCHEM, JOHN JAY COL CRIMINAL JUSTICE, 74- *Concurrent Pos:* Chief Coroner, Putnam County, NY, 79- *Mem:* AAAS; Am Soc Microbiol; Am Chem Soc. *Res:* Enzymes associated with invasive microbes and biochemical intermediates; development of vaccine for heroin and related alkaloids; ria-digoxin, amerod immunology. *Mailing Add:* Dept Biochem John Jay Col Criminal Justice 445 W 59th St New York NY 10019

STAHL, WILLIAM LOUIS, b Glen Dale, WVa, Aug 2, 36; m 59; c 2. BIOCHEMISTRY, NEUROCHEMISTRY. *Educ:* Univ Notre Dame, BS, 58; Univ Pittsburgh, PhD(biochem), 63. *Prof Exp:* Res assoc biochem, NIH, 65-67; res asst prof physiol & med, 67-71, assoc prof physiol & med, 71-77, PROF PHYSIOL & BIOPHYS & MED, SCH MED, UNIV WASH, 77-; CHIEF NEUROCHEM, VET ADMIN HOSP, SEATTLE, 67- *Concurrent Pos:* United Cerebral Palsy Res & Educ Found fel, biochem dept, Inst Psychiat, Maudsley Hosp, Univ London, 63-65. *Mem:* Am Chem Soc; Am Soc Biochem & Molecular Biol; Soc Neurosci; Am Soc Neurochem; Int Soc Neurochem; Histochem Soc. *Res:* Cation transport systems; structure and function of biological membranes. *Mailing Add:* Neurochem Lab GMR-151 VA Med Ctr 1660 S Columbian Way Seattle WA 98108

STAHLEY, WILLIAM, b Stuttgart, Ark, Sept 30, 28; m 53; c 2. ELECTRICAL ENGINEERING. *Educ:* Okla State Univ, BS, 57, MS, 59, PhD(eng), 64. *Prof Exp:* Staff mem, Sandia Corp, 57-58; instr elec eng, Okla State Univ, 58-60; sr engr, Autonetics Div, NAm Aviation, Inc, 61-63, tech specialist, NAm Rockwell Corp, 63-65, chief avionic data processing, 65-68; SR SCI ADV, NAVCOM DEFENSE ELECTRONICS INC, 73- *Mem:* Inst Elec & Electronics Engrs. *Res:* Digital computational requirements for various military and industrial control systems; derivation of mechanization equations, specification of hardware and definition of programming requirements. *Mailing Add:* 9335 Tanager Ave Fountain Valley CA 92708

STAHLMAN, CLARENCE L, dairy industry, for more information see previous edition

STAHLMAN, MILDRED, b Nashville, Tenn, July 31, 22. PEDIATRICS, PHYSIOLOGY. *Educ:* Vanderbilt Univ, BA, 43, MD, 46; Am Bd Pediat, dipl, 54. *Hon Degrees:* MD, Univ Goteborg, Sweden, 73; Univ Nancy, France, 82. *Prof Exp:* From instr to asst prof pediat, 51-59, from instr to asst prof physiol, 54-60, assoc prof pediat, 64-70, PROF PEDIAT, SCH MED, VANDERBILT UNIV, 70-, PROF PATHOL, 82- *Concurrent Pos:* Lederle med fac award, 61-62; USPHS career develop award, 64-68; mem human embryol & develop study sect, USPHS, 64-68; perinatal biol, Infant Mortality Study Sect, 69-73; Adv Child Health & Human Develop Coun, 76-80. *Honors & Awards:* Thomas Jefferson Award, 80; Apgar Award, 87. *Mem:* Inst Med-Nat Acad Sci; Am Pediat Soc; Am Physiol Soc; Am Fedn Clin Res; AAAS; Soc Pediat Res. *Res:* Newborn cardiorespiratory and fetal physiology; cardiology; rheumatic fever. *Mailing Add:* Dept Pediat Vanderbilt Univ Sch Med Nashville TN 37232

STAHLMAN, PHILLIP WAYNE, b Shattuck, Okla, Jan 4, 48; m 69; c 2. WEED SCIENCE, AGRONOMY. *Educ:* Panhandle State Univ, BS, 70; NDak State Univ, MS, 73; Univ Wyo, PhD, 89. *Prof Exp:* Asst agronomist, NCent Br, NDak Agr Ecp Sta, 72-75; supt agron res, Harvey County Exp Field Agron, 75-76, RES WEED SCIENTIST, FT HAYS BR, KANS AGR EXP STA, KANS STATE UNIV, 76- *Mem:* Weed Sci Soc Am; Am Soc Agron; Coun Agr Sci & Technol. *Res:* Weed control in dryland wheat and grain sorghum; chemical fallow and reduced tillage systems; control of field bindweed. *Mailing Add:* Ft Hays Exp Sta 1232 240th Ave Hays KS 67601

STAHLY, DONALD PAUL, b Columbus, Ohio, May 29, 37; m 59; c 2. MICROBIOLOGY. *Educ:* Ohio State Univ, BS, 59, MS, 61; Univ Ill, PhD(microbiol), 65. *Prof Exp:* NIH fel, Univ Minn, 65-66; from asst prof to assoc prof, 66-79, PROF MICROBIOL, UNIV IOWA, 79- *Concurrent Pos:* Sabbatical, Scripp's Clin & Res Found, 77-78. *Mem:* AAAS; Am Soc Microbiol. *Res:* Bacterial sporulation; plasmids and bacteriophages in Bacillus species and gene cloning; Bacillus species pathogenic for insects. *Mailing Add:* Dept Microbiol Univ Iowa Iowa City IA 52242

STAHLY, ELDON EVERETT, organic chemistry; deceased, see previous edition for last biography

STAHMANN, MARK ARNOLD, b Spanish Fork, Utah, May 30, 14; m 41; c 2. BIOCHEMISTRY, BOTANY-PHYTOPATHOLOGY. *Educ:* Brigham Young Univ, BA, 36; Univ Wis, PhD(biochem), 41. *Prof Exp:* Asst chem, Rockefeller Inst, 42-44; res assoc org chem, Mass Inst Technol, 44-45; res assoc biochem, 46-47, from asst prof to prof, 47-82, EMER PROF BIOCHEM, UNIV WIS-MADISON, 82- *Concurrent Pos:* Guggenheim fel, Pasteur Inst, Paris, 55; Fulbright scholar, Nagoya, 67; FAO consult, Biol Inst, Sao Paulo, 74. *Mem:* AAAS; Am Chem Soc; Am Soc Biol Chemists; Soc Exp Biol & Med; Am Phytopath Soc. *Res:* Anticoagulant 4-hydroxycoumarins; warfarin; biochemistry of plant diseases; synthetic polypeptides; polypeptidyl proteins; virus diseases; plant proteins; molecular pathology of atherosclerosis. *Mailing Add:* 742 Hickman Dr Melbourne FL 32901

STAHNKE, HERBERT LUDWIG, b Chicago, Ill, June 10, 02; m 29; c 2. SCORPIOLOGY, SYSTEMATICS. *Educ:* Univ Chicago, SB, 28; Univ Ariz, MA, 34; Iowa State Univ, PhD(zool), 39. *Prof Exp:* Teacher pub schs, Ariz, 28-40; assoc prof sci, 41-47, prof zool & dir poisonous animals res lab, 47-72, head div life sci, 50-62, EMER PROF ZOOL & EMER DIR POISONOUS ANIMALS RES LAB, ARIZ STATE UNIV, 72- *Mem:* AAAS; Soc Syst Zool; Fel Explorers Club; Sigma Xi; Am Inst Biol Sci. *Res:* Scorpion taxonomy and biogeography; scorpion anti-venom. *Mailing Add:* 2625 E Southern Ave Cot 65 Tempe AZ 85282

STAHR, HENRY MICHAEL, b White, SDak, Dec 10, 31; m 52; c 5. VETERINARY TOXICOLOGY, MASS SPECTROSCOPY. *Educ:* SDak State Univ, Brookings, BS, 56; Union Col, Schnectady, NY, MS, 61; Iowa State Univ, Ames PhD(food chem), 76. *Prof Exp:* Anal develop chemist anal chem, Gen Elec Co, 56-65; sr scientist anal chem, Philip Morris Res, 65-69; PROF ANAL TOXICOL, IOWA STATE UNIV, 69- *Mem:* Am Chem Soc; Soc Appl Spectros; Am Microchem Soc; Am Col Vet & Comp Toxicol; Am Asn Vet Lab Diagnostician; Asn Off Anal Chemists; Sigma Xi; Soc Toxicol. *Res:* Analytical toxicology; develop techniques to analyze biological and environmental samples for toxic substances; natural intoxicants; man made chemicals; food contamination; safety considerations of contamination in foods; development of simple tests for field application and applying quantitative analytical procedures to provide quantitative data for toxicological assessments; veterinary medicine; mutagenic testing of substances-natural and synthetic. *Mailing Add:* Vet Diag Lab Col Vet Med Iowa State Univ Ames IA 50011

STAIB, JON ALBERT, b Toledo, Ohio, Mar 23, 40; m 67; c 2. COSMIC RAY PHYSICS, PHYSICS OF MUSIC. *Educ:* Univ Toledo, BS, 63; Case Western Reserve Univ, MS, 67, PhD(physics), 69. *Prof Exp:* Asst prof, 69-70, ASSOC PROF PHYSICS, JAMES MADISON UNIV, VA, 70- *Concurrent Pos:* Consult, Case Western Reserve Univ, 69-73; lectr, Univ Ore, 81-82. *Mem:* Am Asn Physics Teachers; Int Planetarium Soc. *Res:* Gamma ray astronomy; atmospheric gamma radiation. *Mailing Add:* Dept Physics James Madison Univ Harrisonburg VA 22807

STAIFF, DONALD C, b Everett, Wash, Feb 26, 36; m 59; c 2. PHARMACEUTICAL CHEMISTRY, BIONUCLEONICS. *Educ:* Univ Wash, BS, 59, PhD(pharmaceut chem), 63. *Prof Exp:* Asst prof pharmaceut chem, Ohio Northern Univ, 63-64; asst prof pharmaceut chem & bionucleonics, NDak State Univ, 64-67; anal res chemist, Western Pesticide Res Lab, Nat Commun Dis Ctr, USPHS, Wash, 67-72; CHIEF CHEMIST, WENATCHEE RES STA, ENVIRON PROTECTION AGENCY, 72- *Concurrent Pos:* Mead-Johnson Labs grant, 64-65; NSF inst grant, 65-66; Soc Sigma Xi grant-in-aid res, 65-66; guest lectr, Training Prog, Perrine Primate Res Lab, Environ Protection Agency, Fla. *Mem:* Am Pharmaceut Asn; Am Chem Soc; Health Physics Soc. *Res:* Conformational and configurational studies of some substituted phenyl-cyclohexane compounds by modern instrumental methods; metabolism studies including use of radiotracer techniques; effect of pesticides on health and persistence in the environment. *Mailing Add:* 2017 N Western Ave Wenatchee WA 98801

STAIGER, ROGER POWELL, b Trenton, NJ, Nov 23, 21; m 44; c 1. ORGANIC CHEMISTRY. *Educ:* Ursinus Col, BS, 43; Univ Pa, MS, 48, PhD, 53. *Prof Exp:* From instr to assoc prof, Ursinus Col, 43-63, prof & chmn dept chem, 63-87; RETIRED. *Concurrent Pos:* Consult, Maumee Chem Co, 55-64; vis prof, Temple Univ, 63-82 & Alexandria Hosp, Nevis, 68-73. *Mem:* Am Chem Soc; Sigma Xi. *Res:* Synthesis of organic heterocyclic compounds. *Mailing Add:* 707 Chestnut St Collegeville PA 19426

STAIKOS, DIMITRI NICKOLAS, b Piraeus, Greece, Dec 18, 19; m 47; c 2. ELECTROCHEMISTRY. *Educ:* Nat Univ Athens, dipl, 42; Western Reserve Univ, MS, 50, PhD(chem), 51. *Prof Exp:* Off Naval Res asst, Western Reserve Univ, 49-50; res chemist, Pa, 50-56, res engr, Eng Res Lab, 56-60, sr res phys chemist, 60-64, sr res phys chemist, Cent Res Dept, Exp Sta, E I Du Pont De Nemours & Co, Inc, 64-82; RETIRED. *Concurrent Pos:* Instr, St Joseph's Col, Pa, 54-55. *Mem:* Am Chem Soc; Electrochem Soc. *Res:* Fundamentals of electrochemical processes; corrosion; electronic instrumentation; ultrasonics; surface tension; fused salts; electrochemistry in nonaqueous solutions; electroless deposition; software systems programming. *Mailing Add:* 1306 Quincy Dr Green Acres Wilmington DE 19803-5146

STAINER, DENNIS WILLIAM, b Liverpool, Eng, Aug 25, 32; m 57, 84; c 3. BIOCHEMISTRY, MICROBIOLOGY. *Educ:* Univ Liverpool, BSc, 54, Hons, 55, PhD(biochem), 58. *Prof Exp:* Asst dir, 60-83, DIR, CONNAUGHT LABS, LTD, 83- *Concurrent Pos:* Nat Res Coun Can fel biochem, Food & Drug Directorate, 57-59. *Mem:* Am Soc Microbiol; Int Asn Biol Standardization. *Res:* Production of diphtheria and tetanus toxoids and pertussis vaccine and studies on their immunogenicity; development of new bacteriological media and their application in bacterial fermentations; immunology. *Mailing Add:* 95 Elgin Mills Rd W Richmond Hill ON L4C 4M1 Can

STAINS, HOWARD JAMES, b Frenchtown, NJ, Apr 16, 24; m 54; c 3. ZOOLOGY. *Educ:* NC State Col, BS, 49, MS, 52; Univ Kans, PhD(zool), 55. *Prof Exp:* Lab instr econ zool, NC State Col, 48, res biologist, 49-51; res biologist, Univ Kans, 51-54, instr biol, 54-55; from asst prof to assoc prof, 55-71, PROF ZOOL, UNIV SOUTHERN ILL, 71- *Mem:* Wildlife Soc; Ecol Soc Am; Am Soc Mammal; Soc Study Evolution. *Res:* Furbearing mammals; osteology and ecology of mammals; wildlife techniques. *Mailing Add:* Dept Zool & Life Sci 0317 Southern Ill Univ Carbondale IL 62901

STAINSBY, WENDELL NICHOLLS, b New York, NY, Nov 14, 28; m 52; c 4. PHYSIOLOGY. *Educ:* Bucknell Univ, AB, 51; John Hopkins Univ, ScD, 55. *Prof Exp:* From instr to assoc prof, 57-69, PROF PHYSIOL, COL MED, UNIV FLA, 69- *Concurrent Pos:* NIH res grant, 58- *Honors & Awards:* Citation Award, Am Col Sports Med, 86. *Mem:* Am Physiol Soc; Am Col Sports Med. *Res:* Circulatory and muscle physiology; muscle metabolism and circulation; tissue gas transport. *Mailing Add:* Dept of Physiol Univ of Fla Col of Med Gainesville FL 32610

STAIR, PETER CURRAN, b Pasadena, Calif, Jan 25, 50; m 83; c 1. PHYSICAL CHEMISTRY. *Educ:* Stanford Univ, BS, 72; Univ Calif, Berkeley, PhD(chem), 77. *Prof Exp:* From asst prof to assoc prof, 77-87, PROF CHEM, NORTHWESTERN UNIV, 87- *Mem:* Am Vacuum Soc; Am Phys Soc; Am Chem Soc. *Res:* Structure and chemistry of metal and metal oxide surfaces, catalysis, and corrosion. *Mailing Add:* Dept of Chem Northwestern Univ Evanston IL 60208

STAIR, WILLIAM K(ENNETH), b Clinton, Tenn, Oct 1, 20; m 45; c 2. MECHANICAL ENGINEERING. *Educ:* Univ Tenn, BS, 48, MS, 49. *Prof Exp:* Asst engr, Tenn Rwy Co, 39-41; eng aide, Tenn Valley Authority, 41-43; instr mech eng, Univ Tenn, 48-49; res participant nuclear eng, Oak Ridge Inst Nuclear Studies, 49-50; from asst prof to assoc prof mech eng, 50-62, assoc dir eng exp sta, 70-76, asst dean res, 70-72, PROF MECH ENG, UNIV TENN, KNOXVILLE, 62-, ASSOC DEAN RES, COL ENG, 72- *Mem:* Fel Am Soc Lubrication Engrs; Am Soc Mech Engrs. *Res:* Lubrication and fluid dynamics; heat transfer; combustion phenomena. *Mailing Add:* 2317 Lakemoor Dr Knoxville TN 37920

STAIRS, GORDON R, b Millville, NB, May 18, 32; m 54; c 4. INSECT PATHOLOGY. *Educ:* Univ NB, BSc, 54; McGill Univ, MSc, 58, PhD(entom), 63. *Prof Exp:* Res officer forest entom, Can Dept Agr, 54-58; res scientist, Insect Path Res Inst, 58-65; from asst prof to assoc prof, 65-73, PROF ENTOM, OHIO STATE UNIV, 73- *Res:* Natural population biology and the effects of microorganisms on these populations; possible utilization of microorganisms in the control of pest populations; forest entomology and insect ecology. *Mailing Add:* Dept Entom 103 Bot & Zool Bldg Ohio State Univ 1735 Neil Ave Columbus OH 43210

STAIRS, ROBERT ARDAGH, b Montreal, Que, June 10, 25; m 48; c 2. SOLUTION PROPERTIES, ACID PRECIPITATION. *Educ:* McGill Univ, BSc, 48; Univ Western Ont, MSc, 51; Cornell Univ, PhD(inorg chem), 55. *Prof Exp:* Instr chem, Cornell Univ, 53-55; lectr, Queen's Univ, Ont, 55-58, asst prof, 58-64; actg chmn dept, Trent Univ, 66-67, from assoc prof to prof chem, 64-90, chmn dept, 86-89, EMER PROF, TRENT UNIV, 90- *Concurrent Pos:* Pres, R & R Labs, Ltd, 85-86. *Mem:* Fel Chem Inst Can; AAAS; Am Chem Soc; Sigma Xi. *Res:* Physical properties of electrolyte solutions; viscosity of solutions; analysis acid precipitation; analytical methods. *Mailing Add:* Dept Chem Trent Univ Peterborough ON K9J 7B8 Can

STAKE, PAUL ERIK, b Grandy, Minn, Jan 15, 44; m 66; c 3. ANIMAL NUTRITION. *Educ:* Univ Minn, BS, 68; SDak State Univ, MS, 71; Univ Ga, PhD(nutrit biochem), 74. *Prof Exp:* Lab supvr dairy sci, SDak State Univ, 68-71; res asst animal nutrit, Univ Ga, 71-74; asst prof, 74-79, ASSOC PROF NUTRIT SCI, UNIV CONN, 79-, ACTG DEPT HEAD, 86 & 87- *Mem:* Am Dairy Sci Asn; Am Chem Soc; Am Soc Animal Sci; Nutrit Today Soc; Am Inst Nutrit. *Res:* Comparative metabolism of dietary essential and non-essential trace elements in animals and humans. *Mailing Add:* Dept Nutrit Sci Univ Conn Main Campus U-17 3624 Horsebarn Storrs CT 06268

STAKER, DONALD DAVID, b Wheelersburg, Ohio, Jan 16, 26; m 47; c 4. ORGANIC CHEMISTRY. *Educ:* Ohio Univ, BS, 47, MS, 48; Ohio State Univ, PhD(chem), 52. *Prof Exp:* Asst, Ohio Univ, 44-45, 47-48; asst, Ohio State Univ, 48-49, res fel, 49-52; res chemist, Monsanto Chem Co, 52-56, res proj leader, 56-57, res group leader, 57-61; res sect leader, 61-67, res sect mgr, fatty acid div, 67-78, RES DEPT, EMERY CHEM DIV, NAT DIST & CHEM CORP, 78- *Mem:* Am Chem Soc; Am Oil Chem Soc. *Res:* Fatty acid chemistry, production processes, utilization; reaction mechanisms; rubber chemicals; polymer properties; lubricant additives; process development; metalworking lubricants. *Mailing Add:* 591 Abilene Trail Cincinnati OH 45215-2554

STAKER, MICHAEL RAY, b Dayton, Ohio, Nov 25, 47; m 76. METALLURGY, MECHANICAL ENGINEERING. *Educ:* Univ Dayton, BME, 70; Mass Inst Technol, MS, 71, PhD(metall), 75. *Prof Exp:* Mat engr, Gen Elec Aircraft Engine Group, 75-78; METALLURGIST, ARMY MAT & TECHNOL LAB, 78- *Concurrent Pos:* Lectr, Dept Mech Eng, Northeastern Univ, 80 & Dept Civil Eng, Tufts Univ, 84. *Mem:* Am Soc Metals; Am Soc Testing & Mat. *Res:* Metal defects and structure by electron microscopy; structure and mechanical property relations in metals; fractography, fatigue, fracture; dislocations; high temperature deformation of metals; super alloys; failure analysis; high strain rate deformation and fracture. *Mailing Add:* 40 Beaver Rd Reading MA 01867

STAKER, ROBERT D, b Newport, RI, July 3, 45; m 85; c 2. PHYCOLOGY, OPERATIONS RESEARCH. *Educ:* Univ Dayton, BS, 67, MBA, 81, MA, 84; Univ Ariz, MS, 71, PhD(bot), 73. *Prof Exp:* Res asst phycol, Dept Biol, Univ PEI, 73-75; asst res scientist phytoplanktology, NY Ocean Sci Lab, 75-79; admin asst, Univ Dayton, 79-81; MGT ANALYST, WRIGHT-PATTERSON AFB, 81- *Mem:* Am Soc Limnol & Oceanog; Phycol Soc Am; Int Soc Limnol; Sigma Xi; Nat Estimating Soc. *Res:* Marine and freshwater algal taxonomy and ecology; resource management; modeling using regression analysis and develop cost factors for planning and budget purposes; cost estimating for Air Force program office MIS modernization. *Mailing Add:* 3542 Woodgreen Dr Xenia OH 45385

STAKER, WILLIAM PAUL, b Aberdeen, SDak, Apr 9, 19; m 49; c 2. REACTOR PHYSICS. *Educ:* Ill State Univ, BS, 40; Univ Iowa, MS, 42; NY Univ, PhD(physics), 50. *Prof Exp:* Asst, Univ Iowa, 41-42; physicist & chief ballistic engr, Burnside Lab, E I du Pont de Nemours & Co, 42-46; res assoc, NY Univ, 46-50, proj dir, Res Div, Col Eng, 47-50; assoc physicist, Argonne Nat Lab, 50-52, group leader, 52; sr proj physicist, Eng Res Dept, Standard Oil Co, Ind, 52-56; sr physicist, Nuclear Div, Combustion Eng, Inc, 56-57, mgr exp physics, 57-59, mgr physics, Naval Reactors Div, 59-61, mgr eng & physics, 61-67, mgr fast breeder develop, Nuclear Power Dept, 67-73, mgr prod eng & develop, analysis, Nuclear Power Systs, 73-75, mgr, C-E/KWU coord, 75-79, proj mgr liquid metal fast breeder reactor develop, 79-82; RETIRED. *Concurrent Pos:* Consult, 82-84. *Mem:* Am Phys Soc; Am Nuclear Soc; Sigma Xi. *Res:* Neutron physics; reactor engineering; tracer studies and radioisotope applications. *Mailing Add:* 53 Walnut Farms Dr Farmington CT 06032-2115

STAKES, DEBRA SUE, b Winnsboro, Tex, Jan 7, 51; m 89. MARINE GEOCHEMISTRY, GEOLOGY. *Educ:* Rice Univ, BA(geol) & BA(chem), 73; Ore State Univ, PhD(oceanog), 78. *Prof Exp:* Res assoc marine geochem, Mass Inst Technol, 78-80; res fel, Calif Inst Technol, 80-84; prog mgr, NSF, 83-84; asst prof, 84-91, ASSOC PROF, UNIV SC, 91- *Mem:* Am Geophys Union; Geol Soc Am; Union Concerned Scientists. *Res:* Chemical and mineralogical evidence of seawater interacting with oceanic rocks; origin and structure of oceanic crust. *Mailing Add:* Dept Geol Univ SC Columbia SC 29208

STAKGOLD, IVAR, b Oslo, Norway, Dec 13, 25; nat US; m 64; c 1. APPLIED MATHEMATICS. *Educ:* Cornell Univ, BME, 45, MME, 46; Harvard Univ, MA, 48, PhD(appl math), 49. *Prof Exp:* From instr to asst prof appl math, Harvard Univ, 49-56; head math & logistics brs, US Off Naval Res, 56-59; assoc prof eng sci, Northwestern Univ, Evanston, 60-64, prof eng sci & math, 64-75, chmn dept eng sci, 69-75; prof & chmn dept math sci, Univ Del, 75-91; RETIRED. *Concurrent Pos:* Vis asst prof, Stanford Univ, 53-54; liaison scientist, US Off Naval Res, London, 67-69; vis prof, Oxford Univ, 73-74, Univ Col, London, 78, Victoria Univ, Wellington, NZ, 81 & Polytech Inst, Lausanne, Switz, 81 & 84, Massey Univ, NZ, 87, Univ Bari, Italy, 87; consult, various indust & govt agencies; assoc ed, J Integral Equations, Int J Eng Sci, J Math Appl Anal; chair ethics comn, Counc Sci Soc Pres, 90, Conf Bd Math Sci, 90-92; mem Joint Policy Bd Math. *Mem:* Am Math Soc; Soc Indust & Appl Math (pres, 89-90); Math Asn Am; London Math Soc; Soc Natural Philos; Am Acad Mech. *Res:* Nonlinear boundary value problems. *Mailing Add:* Dept of Math Sci Univ of Del Newark DE 19716

STAKLIS, ANDRIS A, b Valmiera, Latvia, Feb 4, 39; US citizen; m 61; c 1. ORGANIC CHEMISTRY. *Educ:* Univ Nebr, BS, 61, PhD(org chem), 65. *Prof Exp:* Asst head met develop sect, Manned Spacecraft Ctr, NASA, 65-67; sr scientist, Bell Aerospace Co Div, Textron Inc, mgr mat develop, New Orleans Oper, 71-77; mgr eng mat, Brake & Steering Div, 77-83, MGR FRICTION MAT DIV, BENDIX CORP, 84- *Mem:* Am Chem Soc; Am Soc Test & Mat; Soc Automotive Engrs; Am Soc Metals; Sigma Xi. *Res:* Direct materials development for air cushion ships; development and evaluation of nonmetallic materials; corrosion and stress corrosion protection of metallic materials; cavitation-erosion damage; brake lining development and processing. *Mailing Add:* Bendix Friction MTRLS Div PO Box 238 Troy NY 12181

STAKNIS, VICTOR RICHARD, b Bridgewater, Mass, June 14, 20; m 42. MATHEMATICS. *Educ:* Bridgewater Teachers Col, BS, 42; Mass Inst Technol, BS, 46; Boston Univ, MA, 50, PhD(math), 53. *Prof Exp:* Instr math, Ft Devens Br, Univ Mass, 46-49; lectr, Boston Univ, 51-52, instr, 52-53; from asst prof to assoc prof math, Northeastern Univ, 53-85; RETIRED. *Mem:* Am Math Soc. *Res:* Topology. *Mailing Add:* 90 Stoneleigh Rd Watertown MA 02172

STAKUTIS, VINCENT JOHN, b Boston, Mass, June 20, 20; m 47; c 4. ATMOSPHERIC PHYSICS, OPTICS. *Educ:* Boston Col, BS, 43; Brown Univ, MS, 50, PhD(physics), 54. *Prof Exp:* Lab instr optics, Boston Col, 42-43; instr & asst to headmaster, Marianapolis Prep Sch, Conn, 46-48; asst optics, photog, atomic physics & acoust, Brown Univ, 48-53; res physicist, Geophys Res Dir, Air Force Cambridge Res Labs, 53-57, supvry physicist, 57-58, br chief atmospheric optics, 58-60; mem tech staff, 60-63, mem sr tech staff, Mitre Corp, Bedford, 63-87; RETIRED. *Concurrent Pos:* Private consult. *Mem:* AAAS; Sigma Xi; Am Geophys Union; Optical Soc Am; Acoust Soc Am. *Res:* Ultrasonic attenuation in aqueous suspensions; electromagnetic propogation; scattering processes in the atmosphere; high altitude sky luminance and albedo measurements; vision in the atmosphere; satellite reconnaissance systems; nuclear weapon environmental effects. *Mailing Add:* 160 Grant St Lexington MA 02173

STALCUP, MARVEL C, b Mandan, NDak, Dec 30, 31; c 6. PHYSICAL OCEANOGRAPHY, GEOLOGY. *Educ:* Univ Idaho, BS, 61. *Prof Exp:* RES SPECIALIST OCEANOG, WOODS HOLE OCEANOG INST, 61- *Concurrent Pos:* Consult oceanog, ECO-Zist, Teheran, Iran, 77-78. *Mem:* AAAS. *Res:* Identification and description of water masses and currents of the Atlantic and Indian Oceans and Caribbean Sea. *Mailing Add:* 456 Locustfield Rd East Falmouth MA 02536

STALEY, DAVID H, b Columbus, Ohio, Jan 30, 30; m 58; c 3. MATHEMATICS. *Educ:* Oberlin Col, AB, 52; Ohio Univ, MS, 54; Ohio State Univ, PhD(math), 63. *Prof Exp:* Instr math, Henry Ford Community Col, 56-57 & Oberlin Col, 60-61; from asst prof to assoc prof, 61-70, PROF MATH, OHIO WESLEYAN UNIV, 70- *Concurrent Pos:* Consult legal math questions, USDA; mem, N Cent Eval Team, USDA. *Mem:* Math Asn Am; Nat Coun Teachers Math. *Res:* Commutability of operators in a topological space; mathematical models of insect populations; mathematical models for temperature and heat accumulation. *Mailing Add:* Dept Math Ohio Wesleyan Univ E Campus Delaware OH 43015

STALEY, DEAN ODEN, b Kennewick, Wash, Oct 18, 26; m 63; c 5. METEOROLOGY. *Educ:* Univ Wash, BS, 50, PhD(meteorol), 56; Univ Calif, Los Angeles, MA, 51. *Prof Exp:* From instr to asst prof, Univ Wis, 55-59; assoc prof & assoc meteorologist, 59-65, PROF ATMOSPHERIC SCI & RES PROF, INST ATMOSPHERIC PHYSICS, UNIV ARIZ, 65- *Mem:* Am Meteorol Soc; Am Geophys Union. *Res:* Dynamic and synoptic meteorology. *Mailing Add:* Inst Atmospheric Physics Univ Ariz Tucson AZ 85721

STALEY, JAMES T, b Pittsburgh, Pa, Oct 15, 34; m; c 5. ALUMINUM ALLOYS. *Educ:* Univ Pittsburgh, BS, 62, MS, 67; Drexel Univ, PhD(mat eng), 89. *Prof Exp:* CHIEF SCIENTIST AEROSPACE & INDUST PROD GROUP & CORP CONSULT HIGH STRENGTH ALUM PROD, ALCOA LABS. *Mem:* Fel Am Soc Metals Int; Metall Soc Am Inst Mining Metall & Petrol Engrs; Sigma Xi. *Res:* Physical metallurgy; development of aluminum alloys and processes; applications of aluminum alloy mill products in the aerospace industry; over 40 publications; six patents. *Mailing Add:* Alcoa Tech Ctr Aluminum Co Am Bldg C-8-221-2044 Kensington PA 15069

STALEY, JAMES TROTTER, b Brookings, SDak, Mar 14, 38; m 63; c 2. BACTERIAL TAXONOMY, MICROBIAL ECOLOGY. *Educ:* Univ Minn, Minneapolis, BA, 60; Ohio State Univ, MSc, 63; Univ Calif, Davis, PhD(bact), 67. *Prof Exp:* Instr microbiol, Mich State Univ, 67-69; asst prof environ sci & eng, Univ NC, Chapel Hill, 69-71; from asst prof to assoc prof microbiol, 71-82, PROF MICROBIOL, UNIV WASH, 82- *Concurrent Pos:* Mem, Bergey's Manual Trust, 76-; ed, Bergey's Manual Syst Bacteriol, vol 3, 83-87; mem, Int Symp Environ Biochem, 85-; assoc ed, Intern J System Bacteriol, 91. *Mem:* Am Soc Microbiol; AAAS. *Res:* Biology and taxonomy of Ancalomicrobium, Prosthecomicrobium and other prosthecate, budding and gas vacuolate bacteria; aquatic bacteriology; fresh water microbiology; microbial ecology; bacteriology of drinking water; microbiology of desert varnish formation; chitin degradation in polar environments; sea ice bacteriology; bacterial taxonomy. *Mailing Add:* Dept Microbiol Univ Wash Seattle WA 98195

STALEY, JOHN M, b Three Rivers, Mich, Sept 12, 29; m 80; c 4. PLANT PATHOLOGY, FORESTRY. *Educ:* Univ Mont, BS, 51; WVa Univ, MS, 53; Cornell Univ, PhD(plant path), 62. *Prof Exp:* Asst forest path, WVa Univ, 51-53; plant pathologist, Pac Northwest Forest & Range Exp Sta, US Forest Serv, 53, Northeastern Forest Exp Sta, 56-62, res plant pathologist, Rocky Mountain Forest & Range Exp Sta, 62-82; AGR RES TECHNICIAN, DEPT PLANT PATH, WASH STATE UNIV, 83- *Concurrent Pos:* Res asst, Cornell Univ, 56-62; affil prof, Grad Fac, Colo State Univ, 67- *Mem:* Am Phytopath Soc; Mycol Soc Am. *Res:* Complex diseases of forest trees; vascular wilt diseases of forest trees; foliage diseases of coniferous trees; rust diseases of cereals. *Mailing Add:* Dept Plant Path Wash State Univ Puyallup WA 98373

STALEY, L(EONARD) M(AURICE), b Dodsland, Sask, Feb 11, 26; m 53; c 2. AGRICULTURE, ENGINEERING. *Educ:* Univ BC, BASc, 51; Univ Calif, MS, 56. *Prof Exp:* Asst prof agr eng, Ont Agr Col, 56-57; from asst prof to assoc prof agr eng, 51-71, prof, 71-75, prof bio-resource eng, 75-81, HEAD, DEPT BIO-RESOURCE ENG, UNIV BC, 81- *Concurrent Pos:* Lectr, Inner Mongolian Col Eng, China, 85. *Honors & Awards:* Maple Leaf Award, Can Soc Agr Engrs. *Mem:* Am Soc Agr Engrs; Can Soc Agr Engrs (pres, 83-84). *Res:* Solar and thermal environmental control in greenhouses and confinement housing systems; methods of energy conservation in agriculture. *Mailing Add:* Bio-Resource Eng Dept Univ BC Vancouver BC V6T 1W5 Can

STALEY, RALPH HORTON, b Boston, Mass, Mar 15, 45; m 83; c 2. PHYSICAL CHEMISTRY. *Educ:* Dartmouth Col, AB, 67; Calif Inst Technol, PhD(chem), 76. *Prof Exp:* Res physicist, Feldman Res Labs, Picatinny Arsenal, 68-71; asst prof phys chem, Mass Inst Technol, 75-81; RESEARCH SUPVR, E I DU PONT DE NEMOURS & CO, INC, 81- *Mem:* Am Chem Soc; AAAS. *Res:* Advanced inorganic materials; ceramics. *Mailing Add:* Cent Res & Develop Dept E356 Exp Sta E I Du Pont de Nemours & Co Inc Wilmington DE 19880-0356

STALEY, ROBERT NEWTON, b Canova, SDak, Oct 15, 35; m 70; c 2. ORTHODONTICS, PHYSICAL ANTHROPOLOGY. *Educ:* Univ Minn, Minneapolis, BS, 57, DDS, 59; Univ Chicago, MA, 67; State Univ NY Buffalo, cert orthod, 69, MS, 70; Am Bd Orthod, dipl, 83. *Prof Exp:* Intern dent, Zoller Mem Dent Clin, Univ Chicago Hosp & Clins, 59-60, mem staff, 62-65; from asst prof to assoc prof, 70-85, PROF ORTHOD, COL DENT, UNIV IOWA, 85- *Concurrent Pos:* Captain, Dent Corp, US Army, 60-62. *Mem:* Int Asn Dent Res; Am Dent Asn; Am Asn Orthodont; Am Cleft Palate Asn; Am Asn Phys Anthrop. *Res:* Investigates the relations of craniofacial morphology to clinical orthodontics, growth and genetics; orthodontic biomechanics. *Mailing Add:* Dept of Orthod Univ of Iowa Col of Dent Iowa City IA 52242

STALEY, STUART WARNER, b Pittsburgh, Pa, July 11, 38; m 63; c 2. PHYSICAL & ORGANIC CHEMISTRY, STRUCTURE CHEMISTRY. *Educ:* Williams Col, BA, 59; Yale Univ, MS, 61, PhD (phys org chem), 64. *Prof Exp:* Res assoc phys org chem, Univ Wis, 63-64; from asst prof to prof chem, Univ Md, College Park, 64-78; prof chem, Univ Nebr-Lincoln, 78-85; PROF CHEM, CARNEGI MELLON UNIV, 86- *Concurrent Pos:* Vis prof, Swiss Fed Inst Technol, Zurich, 71-72, & Univ Calif, Riverside, 84. *Mem:* Am Chem Soc. *Res:* Photoelectron, electron transmission, and nuclear magnetic resonance spectroscopy of organic compounds; reaction mechanisms; carbanion chemistry; molecular orbital calculations; synthesis of theoretically interesting molecules; molecular electronics; determination of molecular structure. *Mailing Add:* Dept Chem Carnegie Mellon Univ 4400 Fifth Ave Pittsburgh PA 15213

STALEY, THEODORE EARNEST LEON, veterinary anatomy, veterinary physiology, for more information see previous edition

STALFORD, HAROLD LENN, b Avery, Okla, July 22, 42; m 63; c 5. APPLIED MATHEMATICS. *Educ:* Okla State Univ, BS, 65; Univ Calif, Berkeley, MS, 66, PhD(appl mech), 70. *Prof Exp:* Asst res engr, Univ Calif, 70; opers res analyst, Radar Div, US Naval Res Lab, 70-71, math analyst, 71-76; sr analyst, Dynamics Res Corp, 76-80; PRES, PRACTICAL SCI, INC, 80- *Concurrent Pos:* Sr analyst, Dynamics Res Corp, 76- *Mem:* Sigma Xi. *Res:* Estimation, control, system identification and differential games; design and evaluation of optimum survivability maneuvers for airborne applications; non-linear track filter sand predictors for shipboard applications; statistics; system identification and classification. *Mailing Add:* 495 Berry Patch Lane Marietta GA 30067-5072

STALHEIM, OLE H VIKING, b Garretson, SDak, Sept 23, 17; m 42; c 4. VETERINARY HISTORY. *Educ:* Tex A&M Univ, DVM, 41; Univ SDak, MA, 61; Univ Wis, PhD(bact), 63. *Prof Exp:* Vet practitioner, Vermillion, SDak, 41-58; NIH fel, Univ Wis, 60-63; res vet, Nat Animal Dis Crt, Sci & Educ Admin-Agr Res, USDA, 63-; ASSOC PROF & COLLABR, DEPT HIST, IOWA STATE UNIV, AMES, 85- *Concurrent Pos:* Fulbright-Hays award, 80-81 & 87-89. *Honors & Awards:* E A Pope Award, 79; 12th Int Vet Congress Prize, Am Vet Med Asn, 84. *Mem:* Am Vet Med Asn; Am Soc Microbiol; Conf Res Workers Animal Dis; US Animal Health Asn. *Res:* Microbial nutrition, metabolism and virulence; immunity to bacterial infection; veterinary history; chemotherapy. *Mailing Add:* 603 Ross Hall Iowa State Univ Ames IA 50011

STALICK, WAYNE MYRON, b Oregon City, Ore, Aug 24, 42; m 67; c 1. ORGANIC CHEMISTRY. *Educ:* Univ Ore, BA, 64; Northwestern Univ, PhD(org chem), 69. *Prof Exp:* Asst prof org chem, Calif State Univ, San Jose, 69-70; fel, Ohio State Univ, 70-72, lectr, 72; from asst prof to assoc prof, 72-87, PROF ORG CHEM, GEORGE MASON UNIV, 87- *Concurrent Pos:* Sabbatical, Naval Res Lab, 85-86. *Mem:* Sigma Xi; Am Chem Soc. *Res:* Chemistry of petroleum products, in particular, catalytic reactions of hydrocarbons, photochemistry and pryrolosis of long chain alkylaromatic compounds; synthetic technique in organic chemistry. *Mailing Add:* Dept Chem George Mason Univ Fairfax VA 22030

STALKER, ARCHIBALD MACSWEEN, b Montreal, Que, June 29, 24; m 51; c 4. GLACIAL GEOLOGY, PALEOENVIRONMENTS. *Educ:* McGill Univ, BA, 45, MSc, 48, PhD(geol), 50. *Hon Degrees:* DSc, Univ Lethbridge, 84. *Prof Exp:* Geologist, Geol Surv Can, 50-87; RETIRED. *Concurrent Pos:* Distinguished vis scholar, Univ Lethbridge, 87-90, adj prof, 90- *Mem:* Can Quaternary Asn; fel Geol Soc Am; Geol Asn Can; Am Quaternary Asn; Quaternary Res Asn (UK). *Res:* Glacial geology; geomorphology; relations of Cordilleran and Laurentide glaciations; preglacial drainage; early man in New World; climate, stratigraphy and mammals of Quaternary; Quaternary vertebrate paleontology. *Mailing Add:* 2126 Strathmore Blvd Ottawa ON K2A 1M7 Can

STALKER, HAROLD THOMAS, b Pittsburgh, Pa, Oct 27, 50; m 72; c 2. PLANT CYTOGENETICS, PEANUT BREEDING. *Educ:* Univ Ariz, BS, 72, MS, 73; Univ Ill, PhD(genetics), 77. *Prof Exp:* Res assoc cytogenetics, 77-79, asst prof crop sci, 79-83, ASSOC PROF, NC STATE UNIV, 83-, ASSOC PROF CROP SCI & BIOTECHNOL, 87- *Mem:* Crop Sci Soc Am; Am Genetic Asn; Am Peanut Res & Educ Soc; Soc Econ Bot; Botanical Soc Am. *Res:* Cytogenetics; genetics; speciation and biosystematics of wild and cultivated species of the genus Arachis. *Mailing Add:* 1206 Ivy Lane Cary NC 27511

STALKER, HARRISON DAILEY, b Detroit, Mich, July 3, 15; m 41; c 1. GENETICS, EVOLUTIONARY BIOLOGY. *Educ:* Col Wooster, BA, 37; Univ Rochester, PhD(zool), 41. *Prof Exp:* Asst zool, Univ Rochester, 37-42; from instr to assoc porf, 42-56, PROF ZOOL, WASH UNIV, 56- *Concurrent Pos:* NSF sr fel, 61; mem genetics adv panel, NSF, 59-62. *Mem:* AAAS; Genetics Soc Am; Soc Study Evolution; Am Soc Nat; Am Soc Human Genetics. *Res:* Cytogenetics and physiology of Drosphila species and populations. *Mailing Add:* 12505 Village Circle Dr St Louis MO 63127

STALKUP, FRED I(RVING), JR, b Temple, Tex, Feb 3, 36; m 65; c 1. CHEMICAL ENGINEERING. *Educ:* Rice Univ, BA, 57, PhD(chem eng), 62. *Prof Exp:* Sr res engr, Atlantic-Richfield Co, 61-65, prin res engr, 65-67, dir, Process Develop Res, 67-69, dir, Reservoir Eng Res, 69-71, dir, Reservoir Math Res, 71-77, dir recovery res, 77-79; consult, 79-83; SR RES ADV, 83- *Honors & Awards:* Sigma Xi Res Award, 62; Lester C Uren Award, Soc Petrol Engrs, 85, Distinguished Lectr, 85. *Mem:* Nat Acad Eng; Soc Petrol Engrs. *Res:* Hydrocarbon phase equilibria; miscible, immiscible and thermal methods of oil recovery; reservoir engineering; secondary and tertiary recovery; mathematical modeling; author, SPE monograph: misable displacement 21 publications. *Mailing Add:* Arco Oil & Gas Co 2300 W Plano Pkwy Plano TX 75075

STALL, ROBERT EUGENE, b Leipsic, Ohio, Dec 11, 31; m 52; c 2. PLANT PATHOLOGY. *Educ:* Ohio State Univ, BSc, 53, MSc, 54, PhD(bot, plant path), 57. *Prof Exp:* Res asst plant path, Ohio Agr Exp Sta, 57-63, assoc prof plant path & assoc plant pathologist, 63-69, PROF PLANT PATH, UNIV FLA, 69- *Concurrent Pos:* Vis prof, Nat Inst Agr Technol, 78-79. *Mem:* AAAS; fel Am Phytopath Soc. *Res:* Bacterial phytopathology. *Mailing Add:* Dept Plant Path Univ Fla Gainesville FL 32611

STALL, WILLIAM MARTIN, b Bluffton, Ohio, Apr 14, 44; m 69. HORTICULTURE, WEED SCIENCE. *Educ:* Ohio State Univ, BSA, 67; Univ Fla, MSA, 69, PhD(vegetable crops), 73. *Prof Exp:* Teacher biol, Southwestern City Sch, Grove City, Ohio, 69-71; exten agent veg, Fla Coop Exten Serv, 74-80; PROF, EXTEN VEG SPECIALIST, VEG CROPS DEPT, UNIV FLA, 80- *Mem:* Am Soc Hort Sci; Weed Sci Soc Am; Nat Agr Plastics Asn. *Res:* Integrated pest management; herbicide efficacy and phytotoxicity; weed crop interference; herbicide and growth regulator residues for minor crop clearance. *Mailing Add:* Veg Crops Dept Univ Fla Gainesville FL 32611

STALLARD, RICHARD E, b Eau Claire, Wis, May 30, 34; m; c 3. ANATOMY, PERIODONTOLOGY. *Educ:* Univ Minn, BS, 56, DDS, PhD(anat), 62. *Prof Exp:* Co-dir periodont res prog, EDent Dispensary, 62-65; from assoc prof to prof periodont & chmn dept, Univ Minn, Minneapolis, 65-68; asst dir, Eastman Dent Ctr, NY, 68-70; prof periodont & anat, Med Ctr & dir, Clin Res Ctr, Sch Grad Dent, Boston Univ, 70-74, asst dean, Sch Grad Dent, 72-74; dent dir & head dept periodont, Group Health Plan, Inc, 75-80; prof pub health, Univ Minn, Minneapolis, 76-; CONSULT & CHIEF DENT SERV, HAMAD GEN HOSP, DOHA, QATAR. *Concurrent Pos:* Consult, Cambridge State Sch & Hosp, 61-63; mem grad fac dent & grad fac anat, Univ Minn, Minneapolis, 65-68; consult, Vet Admin Hosp, 67-70, Univ Minn Hosp, 67-69 & consult, Wright-Patterson Air Force Base, 68-73; mem training grant comt, Nat Inst Dent Res, 69-73; consult, US Naval Hosp, Chelsea, Mass, 71-74, Boston Univ Hosp, 71-74 & Fairview Hosp, Minneapolis, 75-84; mem, Grants & Allocations Comt, Am Fund Dent Health, 74-80; consult, Hamad Gen Hosp, 84- *Mem:* Fel AAAS; Am Acad Periodont (pres, 74); Am Dent Asn; Am Acad Oral Path; Int Col Oral Implantologists (pres, 80-81). *Res:* Microcirculation; occlusion; implantology and etiology of periodontal disease. *Mailing Add:* Hamad Gen Hosp PO Box 3050 Doha Qatar

STALLCUP, MICHAEL R, b Dallas, Tex, Nov 6, 47. MOLECULAR BIOLOGY, GENETICS. *Educ:* Yale Univ, BA, 69; Univ Calif, Berkeley, PhD(biochem), 74. *Prof Exp:* Fel biochem & biophysics, Univ Calif, San Francisco, 74-79; asst prof biol, Univ SC, 80-85; ASSOC PROF, DEPT PATH, UNIV SOUTHERN CALIF, SCH MED, 85- *Concurrent Pos:* NSF grad trainee award, 69-72; fel, Am Cancer Soc, 74-75, Nat Res Serv Award, NIH, 77-79; prin investr, NIH res grant, 80-95, Am Cancer Soc res grant, 88-92; Res Career Develop Award, NIH, 83-88; mem Molecular Biol study sect, NIH, 84-88. *Mem:* Am Soc Cell Biol; Endocrine Soc. *Res:* Biochemical and genetic studies on regulation of gene expression in mammalian cells by steroid hormones. *Mailing Add:* Dept Path HMR 301 Univ Southern Calif Sch Med 2011 Zonal Ave Los Angeles CA 90033

STALLCUP, ODIE TALMADGE, b Paragould, Ark, Dec 2, 18; m 47; c 3. NITROGEN METABOLISM, PHYSIOLOGY OF REPRODUCTION. *Educ:* Univ Ark, BSA, 43; Univ Mo, AM, 47, PhD(nutrit/physiol), 50. *Prof Exp:* Instr dairy sci, Dept Dairy Husb, Univ Mo, 47-50; from assoc prof to prof, 50-85, univ prof animal sci, 85-88, EMER UNIV PROF ANIMAL SCI, UNIV ARK, 89- *Concurrent Pos:* Instr dairy sci, Dept Animal Indust, Univ Ark, 45-46. *Mem:* Am Soc Animal Sci; Am Dairy Sci Asn; Soc Study Reproduction; Histochem Soc; Am Inst Nutrit; fel AAAS; Int Fedn Soc Histochem & Cytochem. *Res:* Nutrition and physiology; nutritive value of forages for cattle; nitrogen in ruminants; histochemical studies on bovine embryo growth and reproduction in cattle. *Mailing Add:* Dept Animal Sci Univ Ark Fayetteville AR 72701

STALLCUP, WILLIAM BLACKBURN, JR, b Dallas, Tex, Oct 18, 20; m 42; c 5. VERTEBRATE ZOOLOGY. *Educ:* Southern Methodist Univ, BS, 41; Univ Kans, PhD(zool), 54. *Prof Exp:* Instr biol, Southern Methodist Univ, 45-50; asst instr, Univ Kans, 50-53; from instr to assoc prof, 53-62, chmn dept, 63-67, assoc dean sch humanities & sci, 71-74, assoc provost, 74-80 & 81-83, provost Ad Interim, 80-81, actg provost, 86, pres ad interim, 86-87, PROF BIOL, SOUTHERN METHODIST UNIV, 62- *Concurrent Pos:* Spec asst to pres, Southern Methodist Univ, 82-83. *Mem:* Am Soc Mammal; Am Ornith Union. *Res:* Comparative myology and serology of birds; vertebrate natural history. *Mailing Add:* PO Box 1257 Ranchos De Taos NM 87557

STALLEY, ROBERT DELMER, b Minneapolis, Minn, Oct 25, 24; m 50; c 4. NUMBER DENSITY, ADDITIVE MODULAR NUMBER THEORY. *Educ:* Ore State Col, BS, 46, MA, 48; Univ Ore, PhD(math), 53. *Prof Exp:* Instr math, Univ Ariz, 49-51; instr, Iowa State Col, 53-54, asst prof, 54-55; instr, Fresno State Col, 55-56; from asst prof to prof, 56-89, EMER PROF MATH, ORE STATE UNIV, 89- *Concurrent Pos:* Prin investr, Nat Sci Found res contract, Addition Theorems Density Spaces, 67-71. *Mem:* Am Math Soc. *Res:* Has obtained extensions of the fundamental metric theorems of additive number theory, results for density spaces and results concerning the cardinality of zero sum subsets (submultisets) of certain sets (multisets) of residue classes. *Mailing Add:* Dept Math Ore State Univ Corvallis OR 97331-4605

STALLING, DAVID LAURENCE, b Kansas City, Mo, Oct 24, 41; m 62; c 3. ANALYTICAL CHEMISTRY, ORGANIC CHEMISTRY. *Educ:* Mo Valley Col, BS, 62; Univ Mo-Columbia, MS, 64, PhD(biochem), 67. *Prof Exp:* Instr agr chem, Univ Mo-Columbia, 66-68; res scientist environ chem & trace chem anal, Columbia Nat Fisheries Res Lab, 67-88; SR VPRES RES & DEVELOP, ABC LABS, INC, 88- *Concurrent Pos:* Chem consult, Regis Chem Co, 66-69; co-founder, Anal Bicohem Labs, Inc, 67; assoc referee for contaminants in aquatic biota, Asn Official Anal Chemists, 75-84; co-founder, ABC Labs, 67. *Mem:* Am Chem Soc; Asn Off Anal Chemists; Chemometrics Soc; Soc Environ Toxicol & Chem. *Res:* Development of rapid methods of analysis of biologically important compounds, especially amino acids, purine and pyrimidine bases, organic pollutants, trace environmental contaminants, chlorinated dibenzofurans, biphenyls, dibenze-p-dioxins and pesticides by gas-liquid chromatography and combined gas chromatography-mass spectrometry computer techniques; biochemical effects of pesticides and organic contaminants on fish; gas-chromatography mass-spectrometry computer studies on environmental contaminants; organic residues in national pesticide monitoring programs; development and application of multivariate statistical methods and pattern recognition techniques for characterizing complex data from environmental contaminants in fish and the acquatic ecosystem; development of automated methodology and systems for environmental pollutants. *Mailing Add:* 2505 Shephard Blvd Columbia MO 65201

STALLINGS, CHARLES HENRY, b Durham, NC, Dec 28, 41; m 65; c 2. PLASMA PHYSICS. *Educ:* NC State Univ, BS, 63, MS, 64; Univ Wis-Madison, PhD(physics), 70. *Prof Exp:* Sr physicist, 70-77, dept mgr, 77-78, dir prog off, 79-81, dir prog develop, 81-82, VPRES, PHYSICS INT CO, 82- *Mem:* Am Phys Soc; Inst Elec & Electronics Eng. *Res:* Generation and propagation of intense relativistic electron beams and their interaction with background on target plasma; high density imploding plasmas and x-ray diagnostics. *Mailing Add:* Physics Int Co 2700 Merced St San Leandro CA 94577-0599

STALLINGS, JAMES CAMERON, b Denton, Tex, Jan 16, 19; m 49; c 2. ORGANIC CHEMISTRY. *Educ:* Univ Tex, PhD(org chem), 49. *Prof Exp:* Prof chem, Sam Houston State Col, 49-57; sr res chemist, Celanese Corp Am, 57-59; prof chem & dir dept, 59-78, dean, Col Sci, 78-81, EMER DEAN, SAM HOUSTON STATE UNIV, 81- *Mem:* Am Chem Soc. *Res:* Steric hindrance; synthetic hypnotics; free radicals in solution; organic synthesis; chemical education. *Mailing Add:* 1912 18th St Huntsville TX 77340-4211

STALLINGS, JOHN ROBERT, JR, b Morrilton, Ark, July 22, 35. TOPOLOGY. *Educ:* Univ Ark, BS, 56; Princeton Univ, PhD(math), 59. *Prof Exp:* NSF fel math, Oxford Univ, 59-60; from instr to assoc prof, Princeton Univ, 60-67; PROF MATH, UNIV CALIF, BERKELEY, 67- *Concurrent Pos:* Sloan Found fel, 62-65. *Honors & Awards:* Frank Nelson Cole Prize, Am Math Soc, 70. *Mem:* Am Math Soc. *Res:* Three-manifolds; geometric topology; group theory from topological and homological viewpoints. *Mailing Add:* 1107 Keith Ave Berkeley CA 94708

STALLKNECHT, GILBERT FRANKLIN, b Spooner, Minn, Sept 21, 35; m 58; c 4. PLANT PHYSIOLOGY, PLANT BIOCHEMISTRY. *Educ:* Univ Minn, BS, 62, MS, 66, PhD(plant physiol), 68. *Prof Exp:* Agr res technician, USDA Sugar Beet Invests, Univ Minn, St Paul, 63-67; asst prof plant physiol, Univ Idaho, 68-72, assoc prof, 72-; AT SOUTHERN AGR RES CTR, MONT STATE UNIV. *Mem:* Am Soc Plant Physiol; Scand Soc Plant Physiol. *Res:* Fungus physiology; metabolic studies in host-parasite physiology; physiology of tuberization in potatoes; productivity and physiological age of potato tubers; physiology of seed crops. *Mailing Add:* Agr Exp Sta Mont State Univ Rte 1 Box 131 Huntley MT 59037

STALLMANN, FRIEDEMANN WILHELM, b Koenigsberg, Germany, July 29, 21; m 53; c 2. MATHEMATICS. *Educ:* Stuttgart Tech Univ, Dipl math, 49; Univ Giessen, Dr rer nat, 53. *Prof Exp:* Asst math, Univ Giessen, 53-55, lectr, 55-59; asst, Brunswick Tech Univ, 59-60; chief math sect, Spec Res Unit Med Electronic Data Processing, Vet Admin, DC, 60-64; assoc prof, 64-69, PROF MATH, UNIV TENN, KNOXVILLE, 69- *Concurrent Pos:* Consult, Oak Ridge Nat Lab, 64- *Mem:* Am Math Soc. *Res:* Numerical analysis; conformal mapping and differential equations, lead field theory of electrocardiogram; computer analysis of electrocardiograms; complex variables. *Mailing Add:* Univ Tenn 207-A Ayres Hall Knoxville TN 37916

STALLMEYER, J(AMES) E(DWARD), b Covington, Ky, Aug 11, 26; m 53; c 6. STRUCTURAL ENGINEERING. *Educ:* Univ Ill, BS, 47, MS, 49, PhD(civil eng), 53. *Prof Exp:* Res asst prof, 53-57, assoc prof, 57-60, PROF CIVIL ENG, UNIV ILL, URBANA, 60- *Honors & Awards:* Adams Mem Award, Am Welding Soc. *Mem:* Am Soc Civil Engrs; Am Soc Mech Engrs; Am Soc Testing & Mat; Soc Exp Stress Anal; Am Concrete Inst; Sigma Xi; Am Ry Eng Asn. *Res:* Fatigue of metals and structures; welded structures; brittle fracture; structural analysis, design and dynamics; research on fatigue strength of welded connections and plate girder bridges. *Mailing Add:* Dept Civil Eng Univ Ill 205 N Mathews St Urbana IL 61801

STALLWOOD, ROBERT ANTONY, b Oxbow, Sask, June 15, 25. NUCLEAR PHYSICS. *Educ:* Univ Toronto, BASc, 49, MA, 50; Carnegie Inst Technol, PhD(physics), 56. *Prof Exp:* Physicist, Nuclear Sci Sect, Gulf Res & Develop Co, 56-64; physicist, Gen Elec Space Sci Ctr, 64-67; assoc prof physics, Thiel Col, 67-80. *Concurrent Pos:* Vis prof physics, Cleveland State Univ, 80-82 & Case Western Reserve Univ, 83- *Mem:* Am Phys Soc. *Res:* High energy nuclear physics; gamma ray and x-ray spectroscopy; neutron physics. *Mailing Add:* 12 Eagle St Greenville PA 16125

STALNAKER, CLAIR B, b Parkersburg, WVa, July 21, 38; m 63; c 2. FISH BIOLOGY, GENETICS. *Educ:* WVa Univ, BSF, 60; NC State Univ, PhD(zool), 66. *Prof Exp:* Res asst fisheries biol, NC State Univ, 60-66; asst prof, Utah State Univ, 66-72, assoc prof fisheries biol, 72-76, asst unit leader, Utah Coop Fishery Unit, 66-75; LEADER, COOP INSTREAM FLOW SERV GROUP, US FISH & WILDLIFE SERV, FT COLLINS, 76- *Mem:* AAAS; Am Fisheries Soc; Wildlife Soc; Am Soc Ichthyol & Herpet; Soc Am Nat; Sigma Xi. *Res:* Physiological-genetic studies of fishes; administration of multi-agency, interdisciplinary programs; physical aspects of stream ecology; aquatic environmental interactions. *Mailing Add:* 742 Cottonwood Dr Ft Collins CO 80524-1517

STALTER, RICHARD, b Jan 16, 42; US citizen; m 68. BOTANY, PLANT ECOLOGY. *Educ:* Rutgers Univ, BS, 63; Univ RI, MS, 66; Univ SC, PhD(biol), 68. *Prof Exp:* Asst prof biol, High Point Col, 68-69 & Pfeiffer Col, 69-70; from asst prof to assoc prof, 71-83, dir environ studies prog, 75-85, PROF BIOL, ST JOHN'S UNIV, NY, 83- *Mem:* Torrey Bot Club; Sigma Xi. *Res:* Barrier island ecology; flora of barrier islands; water relations of dune vegetation. *Mailing Add:* Dept of Biol St John's Univ Jamaica NY 11439

STAM, JOS, b Rotterdam, Netherlands, Apr 3, 24; nat US; m 59; c 4. ENVIRONMENTAL LAW, TOXICOLOGY. *Educ:* Tech Col, Dordrecht, BSc, 46; Delft Univ Technol, MS(chem eng), 52. *Prof Exp:* Chief engr, Unie Chem, Inc, 52-56; res engr, Carothers Res Lab & Textile Res Lab, E I du Pont de Nemours & Co, 56-65, tech assoc, 65-70, staff engr, Netherlands, 70-72, prod environ mgr, 72-85; INDEPENDENT CONSULT, ENVIRON AFFAIRS, 85- *Res:* Toxicology, ecology and health legislation of chemicals. *Mailing Add:* Ten Chemin Sous-Le-Cret Troinex 1256 Switzerland

STAMATOYANNOPOULOS, GEORGE, b Athens, Greece, Mar 11, 34; m 64; c 1. MEDICAL GENETICS, HEMATOLOGY. *Educ:* Nat Univ Athens, MD, 58, DSc, 60. *Prof Exp:* Asst med, Nat Univ Athens, 58-59, asst med & hemat, 61-64; res assoc med, 64-65, instr, 65-66, res asst prof, 65-69, assoc prof, 69-72, PROF MED, DIV MED GENETICS, UNIV WASH, 72- *Concurrent Pos:* Royal Hellenic Res Found fel, 61-64. *Mem:* Am Soc Human Genetics; Am Soc Clin Invest; Europ Soc Human Genetics; Genetics Soc Am; Asn Am Physicians. *Res:* Developmental genetics-genetic hematology. *Mailing Add:* 3336 Cascadia Ave Seattle WA 98144

STAMBAUGH, EDGEL PRYCE, b Blain, Ky, Aug 31, 22; m 50; c 1. HYDROTHERMAL HYDROMETALLURGY TECHNOLOGY. *Educ:* Ohio State Univ, BS, 50, MS, 51. *Prof Exp:* Res chemist, N L Indust, 51-56; prin chemist, Battelle Columbus Labs, 56-57; proj leader, Nat Distillers & Chem Corp, 57-59; proj leader, 59-65, sr chemist, 65-79, assoc mgr, 79-82, RES LEADER, BATTELLE COLUMBUS LABS, 82- *Concurrent Pos:* Prin investr, N L Indust, 53-56; liaison, Zirconium Plant, Nat Distillers & Chem Corp, 57-59; proj mgr, US Environ Protection Agency, 75-80; proj mgr, Energy Prog, Battelle Columbus Labs, 73-76, ultra-fine ceramic oxides, synthetic rutile & strategic metals, 78-82. *Mem:* Sigma Xi; Am Inst Mining, Metall & Petrol Engrs. *Res:* Development of new improved commercial processes, based on hydrothermal technology, for combating feedstock shortages, rising energy costs, pollution abatement regulations, and inflation pressures in the inorganic chemicals, mineral processing, strategic metals, ceramic and electronic industries. *Mailing Add:* 921 Evening St Worthington OH 43085

STAMBAUGH, JOHN EDGAR, JR, b Everrett, Pa, Apr 30, 40; m 61; c 4. ONCOLOGY, CLINICAL PHARMACOLOGY. *Educ:* Dickinson Col, BS, 62; Jefferson Med Col, MD, 66, Thomas Jefferson Univ, PhD(pharmacol), 68. *Prof Exp:* From instr to asst prof, 68-74, ASSOC PROF PHARMACOL, JEFFERSON MED COL, THOMAS JEFFERSON UNIV, 74- *Concurrent Pos:* Resident med & AMA spec scholar, Thomas Jefferson Univ Hosp, 68-70, fel oncol, 70-72; staff physician, Cooper Hosp, Camden, NJ, 72- & Underwood Hosp, Woodbury, 73- *Mem:* Am Asn Cancer Res; Am Soc Clin Oncol; Am Soc Pharmacol & Exp Therapeut; Am Soc Clin Pharmacol. *Res:* Clinical drug metabolism and drug interactions; clinical oncology. *Mailing Add:* 341 Station Ave Haddonfield NJ 08033

STAMBAUGH, RICHARD L, b Mechanicsburg, Pa, Aug 16, 36; m 56; c 2. BIOCHEMISTRY. *Educ:* Albright Col, BS, 53; Univ Pa, PhD(biochem), 59. *Prof Exp:* Res assoc biochem, Philadelphia Gen Hosp, 58-59; dir biochem, Elwyn Res & Eval Ctr; instr biochem in pediat, Univ Pa, 59-63; sr res investr, Fels Res Inst & instr biochem, Sch Med, Temple Univ, 63-66; ASSOC PROF, DIV REPRODUCTIVE BIOL, SCH MED, UNIV PA, 66- *Concurrent Pos:* Consult, Penrose Res Lab, Zool Soc Philadelphia & Elwyn Res & Eval Ctr, 63-67. *Mem:* AAAS; Soc Study Reproduction; Fedn Am Socs Exp Biol. *Res:* Enzymology; biochemistry of reproduction. *Mailing Add:* Div Reproductive Biol Univ Pa Sch Med G3 Rm 307 Philadelphia PA 19104

STAMBAUGH, WILLIAM JAMES, b Allenwood, Pa, Dec 1, 27; m 52; c 3. FOREST PATHOLOGY. *Educ:* Pa State Univ, BS, 51, MS, 52; Yale Univ, PhD(forest path), 57. *Prof Exp:* Instr bot, Pa State Univ, 53-57, asst prof forest path, 57-61; from asst prof to assoc prof, 61-72, PROF FOREST PATH, DUKE UNIV, 72-, ASSOC DEAN ACAD PROG, 84- *Honors & Awards:* Southern Forest Path Achievement Award, 82. *Mem:* AAAS; Am Phytopath Soc; Sigma Xi; Am Inst Biol Sci. *Res:* Diseases of forest trees, with emphasis on forest pest management and biocontrol; microbiology of forest soils. *Mailing Add:* Sch of Forestry & Environ Studies Duke Univ Durham NC 27706

STAMBROOK, PETER J, b London, Eng, July 24, 41; m 81; c 1. BIOLOGY. *Educ:* Rensselaer Polytech Inst, BSc, 63; Syracuse Univ, MSc, 65; State Univ NY Buffalo, PhD(biol), 69. *Prof Exp:* Fel cell biol, Med Ctr, Univ Ky, 69-71; investr cell, develop & molecular biol, Dept Embryol, Carnegie Inst Washington, 71-74; asst prof cell, develop & molecular biol, Case Western Reserve Univ, 74-80; PROF ANAT & BIOCHEM, CELL & MOLECULAR BIOL, COL MED, UNIV CINCINNATI, 80-, PROF MOLECULAR GENETICS, BIOCHEM & MICROBIOL, 86- *Concurrent Pos:* Assoc dir, Barrett Cancer Ctr; Sr Int Fogarty fel, 81-82. *Mem:* AAAS; Am Soc Cell Biol. *Res:* Regulation of cell cycle with particular focus on DNA replication; gene expression and mechanisms of mutation in eukaryotic cells and animals. *Mailing Add:* Dept Anat & Cell Biol Col Med Univ Cincinnati Cincinnati OH 45267

STAMER, JOHN RICHARD, b Plankinton, SDak, May 19, 25; m 58; c 3. MICROBIOLOGY. *Educ:* Dakota Wesleyan Univ, BA, 50; SDak State Col, MS, 53; Cornell Univ, PhD(bact), 62. *Prof Exp:* Res assoc bact, Univ Ill, 53-56; jr scientist, Smith Kline & French Labs, 56-58; asst bact, Cornell Univ, 58-62, NIH fel, 62-64; from asst prof to assoc prof bact, 64-77, PROF MICROBIOL, NY STATE AGR EXP STA, CORNELL UNIV, 77- *Mem:* AAAS; Am Soc Microbiol. *Res:* Microbial physiology and nutrition. *Mailing Add:* Dept of Food Sci NY State Agr Exp Sta Geneva NY 14456

STAMER, PETER ERIC, b New York, NY, June 4, 39; m 68. PHYSICS. *Educ:* Stevens Inst Technol, BS, 61, MS, 63, PhD(physics), 66. *Prof Exp:* Instr physics, Upsala Col, 65-66; from asst prof to assoc prof, 66-78, PROF PHYSICS, SETON HALL UNIV, 78- *Concurrent Pos:* Jr res assoc physics, Stevens Inst Technol, 66-68, res assoc, 68- *Mem:* AAAS; Am Asn Physics Teachers; Am Phys Soc. *Res:* Experimental elementary particle physics; pi-P, P-P, K-P interactions at 147 GeV/C and neutrino interactions at 2.0-5.0 GeV/C. *Mailing Add:* Box 347 Sparta NJ 07871

STAMEY, THOMAS ALEXANDER, b Rutherfordton, NC, Apr 26, 28; m 56; c 5. UROLOGY. *Educ:* Vanderbilt Univ, AB, 48; Johns Hopkins Univ, MD, 52; Am Bd Urol, dipl, 61. *Prof Exp:* Intern, Johns Hopkins Hosp, 52-53; mem, Brady Urol House Staff Residency Prog, Johns Hopkins Univ, 53-56; urol consult, US Armed Forces, UK, 56-58; from asst prof to assoc prof urol, Johns Hopkins Univ, 58-61; assoc prof surg, 61-64, PROF SURG, SCH MED, STANFORD UNIV, 64-, CHMN DEPT UROL, 61- *Concurrent Pos:* Mem comt renal dis & urol training grants, NIH, 67-72, chmn, 71-72; mem sci adv bd, Nat Kidney Found, sci adv coun, Coop Study Pyelonephritis, USPHS; mem sci adv comt, Hosp Sick Children, Toronto, Can; mem, Study of Res in Nephrology & Urol, NIH; assoc ed, Campbell's Urol. *Honors & Awards:* Hugh Hampton Young Award, Am Urol Asn, 72; Sheen Award, Am Col Surgeons, 90. *Mem:* Inst Med Nat Acad Sci; Am Urol Asn; Am Surg Asn; Soc Univ Urol. *Res:* Renal physiology and disease and urinary tract infections; microbiology and hypertension; cancer of the prostate. *Mailing Add:* Dept Urol Stanford Univ Sch Med Stanford CA 94305

STAMEY, WILLIAM LEE, b Chicago, Ill, Oct 19, 22; m 45; c 3. MATHEMATICS. *Educ:* Univ Northern Colo, AB, 47; Univ Mo, MA, 49, PhD(math), 52. *Prof Exp:* From asst instr to instr math, Univ Mo, 47-52; asst prof, Ga State Univ, 52-53; from asst prof to assoc prof, Kans State Univ, 53-62, assoc dean, 63-69, dean, Col Arts & Sci, 69-87, PROF MATH, KANS STATE UNIV, 62- *Concurrent Pos:* Secy-treas, Coun Cols Arts & Sci, 78-87. *Mem:* Am Math Soc; AAAS; Math Asn Am. *Mailing Add:* 416 Edgerton Ave Manhattan KS 66502-3712

STAMLER, JEREMIAH, b New York, NY, Oct 27, 19; m 42; c 1. PREVENTIVE MEDICINE, PUBLIC HEALTH. *Educ:* Columbia Univ, AB, 40; State Univ NY, MD, 43. *Prof Exp:* Intern, Long Island Col Med Div, Kings County Hosp, 44; res assoc, Cardiovasc Dept, Med Res Inst, Michael Reese Hosp, Chicago, 49-55, asst dir dept, 55-58; dir heart dis control prog, Chicago Bd Health, 58-74, dir div adult health & aging, 63-74; from asst to assoc prof, 59-71, chmn Dept, Community Health & Prev Med, 72-86, Mem Hosp, 73-85, DINGMAN PROF CARDIOL, MED SCH, NORTHWESTERN UNIV, CHICAGO, 71-73, PROF MED 71- *Concurrent Pos:* Fel path, Long Island Col Med, 47; res fel, Cardiovasc Dept, Med Res Inst, Michael Reese Hosp, Chicago, 48; Am Heart Asn estab investr, 52-58, fel coun arteriosclerosis, 63-64 & coun epidemiol, 64-66; dir chronic dis div, Chicago Bd Health, 61-63; chmn coun arteriosclerosis, Am Heart Asn, 63-64, mem exec comt, Coun Epidemiol, 64-66 & coun high blood pressure res; western hemisphere ed, Atherosclerosis, 63-75; exec dir, Chicago Health Res Found, 63-72; consult, St Joseph Hosp, 64-, Rush-Presby-St Luke's Hosp, 64- & Atherosclerosis Cardiol Drug Lipid Coop Study & Cardiovasc Res Prog Eval Comt, Vet Admin, 65; prof lectr med, Div Biol Sci, Pritzker Sch Med, Univ Chicago, 70-; vis prof, Dept Internal Med, Rush Presby-St Luke's Med Ctr, 72-; attend physician, Northwest Mem Hosp, 72-; sponsor, Nat Health Educ Comt; mem, Worcester Found Exp Biol; specialist clin nutrit, Am Bd Nutrit; chmn coun epidemiol & prev, Int Soc Cardiol, 74-78. *Honors & Awards:* Med J Award, Lasker Found, 65; Blakeslee Award, Am Heart Asn, 64; Donald Reid Medal, London Sch Hgy & Royal Col Physicians, London, England, 88. *Mem:* Fel AAAS; fel Am Col Cardiol; Am Diabetes Asn; Am Fedn Clin Res; Asn Teachers Prev Med. *Res:* Cardiovascular physiology, medicine, epidemiology and preventive medicine, particularly atherosclerosis and hypertension; chronic disease, preventive medicine and public health. *Mailing Add:* Dept Community Health & Prev Med Northwestern Univ Med Sch 680 N Lake Shore Dr Suite 1102 Chicago IL 60611

STAMM, ROBERT FRANZ, b Mt Vernon, Ohio, Mar 28, 15; m 64. SPECTROSCOPY, OPTICAL PHYSICS. *Educ:* Kenyon Col, AB, 37; Iowa State Univ, PhD(phys chem), 42. *Prof Exp:* Asst chem, Iowa State Univ, 37-42; res physicist, Am Cyanamid Co, 42-54, group leader, Basic Res Dept, 54-59 & Phys Res Dept, 59-61, res assoc, Chem Dept, 61-66, res fel, 66-72; sr res investr, Clairol, Inc, Div Bristol Myers Co, 73-82; CONSULT, CUBE-CORNER RETROREFLECTIVE SHEET HWY SIGNS, 87- *Mem:* Am Chem Soc; Am Phys Soc; Optical Soc Am; NY Acad Sci. *Res:* Raman spectroscopy; light scattering; fluorescence; radiation chemistry and sterilization; neutron activation analysis; spectroscopy of triplet molecules and excited transients; photochromism; flash photolysis; kinetic spectroscopy; retroreflectors; optical properties of human hair fibers; FTIR spectroscopy. *Mailing Add:* 158 Rufous Ln Sedona AZ 86336-7116

STAMM, STANLEY JEROME, b Seattle, Wash, July 14, 24; m; c 3. MEDICINE. *Educ:* Seattle Univ, BS, 48; St Louis Univ, MD, 52; Am Bd Pediat, dipl, 58. *Prof Exp:* Intern, King County Hosp, Seattle, Wash, 52-53; resident pediat, Univ Wash, 53-55; instr, 57-58; dir cardiac diag lab, 58-59, co-dir dept cardiol & attend, 59-63, co-dir cystic fibrosis clin, 63, dir cardiopulmonary res lab, 62-67, dir cardiol dept, 67-70, DIR CARDIOPULMONARY DEPT, CHILDREN'S ORTHOP HOSP, 70- *Concurrent Pos:* NIH cardiac trainee, Children's Orthop Hosp, 55-56, hosp cardiac fel, 56-57; clin assoc prof, Univ Wash, 66-69. *Mem:* Fel Am Acad Pediat; Am Heart Asn; fel Am Col Chest Physicians; fel Am Col Angiol. *Mailing Add:* Children's Orthop Hosp & Med Ctr 4800 Sand Point Way NE Seattle WA 98105

STAMMER, CHARLES HUGH, b Indianapolis, Ind, Apr 1, 25; m 47; c 2. ORGANIC CHEMISTRY, AMINO ACID CHEMISTRY. *Educ:* Univ Ind, BS, 48; Univ Wis, PhD(org chem), 52. *Prof Exp:* Res chemist, Merck & Co, Inc, NJ, 52-62; assoc prof, 62-80, PROF CHEM, UNIV GA, 80-, ASSOC DIR, SCH CHEM SCI, 85- *Mem:* Am Chem Soc; AAAS. *Res:* Synthesis of cyclopropane amino acids and peptides. *Mailing Add:* 718 Riverhill Dr Athens GA 30606

STAMPER, EUGENE, b New York, NY, Mar 24, 28; m 53; c 2. MECHANICAL ENGINEERING. *Educ:* City Col New York, BME, 48; NY Univ, MME, 52. *Prof Exp:* Aeronaut res scientist, Nat Adv Comt Aeronaut, 48-49; design engr, S Schweid & Co, 49-50, Karp Metal Prod Co, 50-51 & Seelye, Stevenson, Value & Knecht, 51-52; assoc prof mech eng, NJ Inst Technol, 52-69, prof, 69-88, asst dean acad affairs, 72-85; consult engr, J R Loring & Assoc, 88-90; CONSULT, 90- *Concurrent Pos:* Consult, 56- *Honors & Awards:* Distinguished Serv Award, Am Soc Heating, Refrig & Air-Conditioning Engrs. *Mem:* Am Soc Mech Engrs; Am Soc Eng Educ; fel Am Soc Heating, Refrig & Air-Conditioning Engrs; NY Acad Sci; Sigma Xi; fel Am Soc Heating Refrigerating Air-Conditioning Engr. *Res:* Heat transfer; fluid mechanics; thermodynamics; refrigeration; air conditioning; building energy studies. *Mailing Add:* 73 Cranford Pl Teaneck NJ 07666

STAMPER, HUGH BLAIR, b Warren, Ohio, Dec 13, 43; m 72; c 2. RESEARCH ADMINISTRATION. *Educ:* Ohio State Univ, BSc, 67, MSc, 68, PhD(microbiol), 72. *Prof Exp:* Bacteriologist, Clin Lab, Licking County Mem Hosp, Newark, Ohio, 68-69; asst prof microbiol & immunol, Biol Dept, Old Dominion Univ, Norfolk, Va, 72-75; asst instr & res assoc, dept microbiol & immunol, Downstte Med Ctr, State Univ NY Brooklyn, 75-77; health scientist adminr, Div Lung Dis, Nat Heart, Lung & Blood Inst, 77-83, exec secy, Spec Rev Sect, Div Res Grants, 83-85, exec secy, Immunol Sci Study Sect, Div Res Grants, NIH, USPHS, 85-87; CHIEF, BIOL SCI REV SECT & ASST CHIEF, REFERRALS, & REV BR, DIV RES GRANTS, NIH, USPHS, 87- *Mem:* AAAS; Am Asn Immunologists; Sigma Xi. *Res:* Research administration; lymphocyte biology. *Mailing Add:* 6604 Hollingsworth Terr Derwood MD 20855

STAMPER, JAMES HARRIS, b Richmond, Ind, Sept 10, 38; div. MATHEMATICAL MODELING, HUMAN EXPOSURE TO PESTICIDES. *Educ:* Miami Univ, BA, 60; Yale Univ, MS, 62, PhD(physics), 65. *Prof Exp:* Asst prof physics, Elmira Col, 62-63, dir math & physics, 65-66; asst prof, Univ Fla, 67-70; prof & chmn, Dept Physics & Chem, Fla Southern Col, 70-79; RES ASSOC, UNIV FLA, 77- *Concurrent Pos:* Consult, Battelle Mem Inst, 68-70; reader, advan placement exam physics, Educ Testing Serv, Princeton, NJ, 80-85; Consult, Tex Tech Univ Health Sci Ctr, San Benito, Tex, 84; Consult, Duphar BV, Crop Protection Div, Amsterdam, Holland, 86-

Mem: Am Asn Physics Teachers. *Res:* Atomic and molecular physics; environmental sciences; theory of atomic collisions and quantum effects of interatomic exchange forces; mathematical modeling; environmental fate of pesticides, farm and greenhouse worker exposure to pesticides; pesticide drift. *Mailing Add:* 98 Imperial Southgate Lakeland FL 33803

STAMPER, JOHN ANDREW, b Middletown, Ohio, Mar 28, 30; m 59; c 3. PLASMA PHYSICS. *Educ:* Ohio State Univ, BS, 53; Univ Ky, MS, 58; Univ Md, PhD(physics), 68. *Prof Exp:* Mem tech staff semiconductor physics, Tex Instruments, Inc, 58-63; RES PHYSICIST, NAVAL RES LAB, DC, 68- *Honors & Awards:* E O Hulburt Award, US Naval Res Lab, 74. *Mem:* Fel Am Phys Soc. *Res:* Experimental plasma physics, including physics of laser-matter interactions; semiconductor physics, including thermal, thermoelectric and thermomagnetic effects; laser-produced plasmas; interaction of laser radiation with plasmas. *Mailing Add:* Rt 2 Box 71-E Indian Head MD 20640

STAMPER, MARTHA C, b Dawson Springs, Ky, May 7, 25. ORGANIC CHEMISTRY. *Educ:* DePauw Univ, AB, 47; Univ Wis, PhD(org chem), 52. *Prof Exp:* ORG CHEMIST, PROCESS RES DIV, ELI LILLY & CO, 52- *Mem:* Am Chem Soc. *Res:* antibiotics and pharmaceuticals. *Mailing Add:* 6238 Harbridge Rd Indianapolis IN 46220

STAMPF, EDWARD JOHN, JR, b Evergreen Park, Ill, May 5, 49; m 72. INORGANIC CHEMISTRY. *Educ:* Northern Ill Univ, BS, 72; Univ SC, PhD(chem), 76. *Prof Exp:* Fel, Univ SC, 76-77; ASST PROF CHEM, LANDER COL, 77- *Mem:* Am Chem Soc; Sigma Xi. *Res:* Boron hydrides and organoboranes; compounds are synthesized using high vacuum technology and studied by nuclear magnetic resonance spectroscopy. *Mailing Add:* 202 Lynn St Greenwood SC 29646

STAMPFER, JOSEPH FREDERICK, b Dubuque, Iowa, Mar 15, 30; m 53; c 2. PERSONAL CHEMICAL PROTECTIVE EQUIPMENT. *Educ:* Dartmouth Col, AB, 52; Univ NMex, PhD(chem), 58. *Prof Exp:* Assoc prof chem, Univ Mo-Rolla, 67-77; staff mem, 58-67 & 77-90, ASSOC STAFF MEM, LOS ALAMOS NAT LAB, 91- *Mem:* AAAS; Am Chem Soc; Am Meteorol Soc; Am Indust Hyg Asn. *Res:* Personal protective clothing chemical permeation; respiratory protection; atmospheric chemistry; sorbent efficiency; self-contained breathing apparatus; generation, sampling and characterization of vapors and aerosols; surfactants; surface to atmosphere exchange; isotope separation; aircraft sampling. *Mailing Add:* Group HSE-5 MSK 499 Los Alamos Nat Lab Los Alamos NM 87545

STAMPFL, RUDOLF A, b Vienna, Austria, Jan 21, 26; US citizen; m 56; c 2. ELECTRICAL ENGINEERING, ELECTRONIC COMMUNICATIONS. *Educ:* Vienna Tech Univ, BS, 48, MSEE, 51, PhD(electronic commun), 53. *Prof Exp:* Sr engr, US Army Res & Develop Lab, NJ, 53-57, dep br head electronic instruments, 57-59; head instrumentation br, Goddard Space Flight Ctr, NASA, 59-64; vis lectr instrumentations systs, Univ Calif, Los Angeles, 64-65; chief systs div, Goddard Space Flight Ctr, NASA, Greenbelt, 65-67, dept asst dir advan projs, 67-73; dir, Systs Directorate, Naval Air Develop Ctr, 73-77, dir, Commun Navig Technol Diretorate, 77-79, dir, 79-; AT RCA ASTRO ELEC DIV. *Concurrent Pos:* Mem, CCIR Group IV, 63- *Honors & Awards:* Harry Diamond Award, Inst Elec & Electronics Engrs. *Mem:* Fel Inst Elec & Electronics Engrs. *Res:* Aerospace systems; electronics. *Mailing Add:* Inst Elec & Electronics Engrs 445 Hoes Lane Box 1331 Piscataway NJ 08855-1331

STAMPFLI, JOSEPH, b Rochester, NY, Aug 9, 32; m 64; c 3. SPECTRA SETS, HYPNONORMAL OPERATORS. *Educ:* Univ Rochester, BA, 54; Univ Mich, MA, 55, PhD(math), 59. *Prof Exp:* Instr math, Yale Univ, 59-61; from asst prof to assoc prof, NY Univ, 61-67; assoc prof, 67-69, chmn dept, 80-83, PROF MATH, IND UNIV, 69- *Concurrent Pos:* res fel, Off Naval Res, 64-65; prin investr, Nat Sci Found, 65-; Sherman Fairchild distinguished scholar, Calif Inst Technol, 74-75. *Mem:* Am Math Soc; AAAS; Am Asn Artificial Intelligence. *Res:* Operators on Hilbert Space; hyponormal operators, spectral sets, and local spectral theory. *Mailing Add:* Ind Univ Bloomington IN 47405

STAMPS, JUDY ANN, b San Francisco, Calif, Mar 13, 47. ANIMAL BEHAVIOR, ECOLOGY. *Educ:* Univ Calif, Berkeley, BA, 69, MA, 71, PhD(zool), 74. *Prof Exp:* Actg asst prof, 73-74, ASST PROF ZOOL, UNIV CALIF, DAVIS, 74- *Concurrent Pos:* NSF fel, 76-78. *Mem:* Animal Behav Soc; Am Soc Ichthyologists & Herpetologists; Ecol Soc Am; AAAS; Am Soc Zoologists. *Res:* Evolution of social systems, ecological determinates of variability of social behavior; parent-offspring conflict; ontogeny of social behavior; lizard social behavior. *Mailing Add:* Dept of Zool Univ of Calif Davis CA 95616

STANA, REGIS RICHARD, b Greensburg, Pa, Sept 7, 41; m 68; c 4. DATA ANALYSIS, STATISTICS. *Educ:* Univ Pittsburgh, BS, 63, MS, 65, PhD(chem eng), 67, PE, 88. *Prof Exp:* Fel engr water & waste treat, Westinghouse Res, 67-78; adv engr uranium recovery, Wyo Mineral Corp, 78-81; SR CONSULT ENG URANIUM RECOVERY PHOSPHATE FERTILIZERS, INT MINERALS & CHEM, 81- *Mem:* Am Inst Chem Engrs. *Res:* Reverse osmosis membranes; fabrication and utilization; process development for recovery of uranium from secondary sources, phosphate fertilizers process improvement; on-line expert systems; artificial intelligence. *Mailing Add:* 935 Heathercrest Lakeland FL 33813

STANABACK, ROBERT JOHN, b Weehawken, NJ, Dec 24, 30; m 57; c 1. ORGANIC CHEMISTRY. *Educ:* Rutgers Univ, BA, 53; Seton Hall Univ, MS, 64, PhD(chem), 66. *Prof Exp:* Assoc scientist, Warner-Lambert Res Inst, 56-67; sr chemist, Tenneco Chem, Inc, Piscataway, 67-78; sr chemist org pigments res & develop, AM Cyanamid Co, 78-83; sr chem, Inmont Corp, 83-85; SR CHEMIST, MAGRUDER COLOR CO, 85- *Mem:* Am Chem Soc. *Res:* Synthetic organic medicinals; thyroxine analogs; central nervous depressants; biocides; plasticizers; synthetic polymers; vinyl chloride technology and additives. *Mailing Add:* Nine Union Grove Rd Gladstone NJ 07934

STANACEV, NIKOLA ZIVA, b Milosevo, Yugoslavia, July 17, 28. BIOLOGICAL CHEMISTRY. *Educ:* Univ Zagreb, Chem E, 53, PhD(chem), 58. *Prof Exp:* Asst prof med, Univ Zagreb, 55-58; fel div biosci, Nat Res Coun Can, Ottawa, 58-59; res assoc chem & chem eng, Univ Ill, Urbana, 59-61; res assoc cell chem lab, Dept Biochem, Columbia Univ, 61-62; res assoc, Banting & Best Dept Med Res, Univ Toronto, 62-64, lectr, 64-65; res assoc biol chem, Harvard Med Sch, 65-67; assoc prof, 67-74, PROF CLIN BIOCHEM, UNIV TORONTO, 74- *Res:* Organic biochemistry; chemistry and biochemistry of membrane lipids; isolation, determination of constitution and biosynthesis of complex lipids of membranes of animal and bacterial origin. *Mailing Add:* Dept Clin Biochem Univ Toronto 100 College St Toronto ON M5G 1L5 Can

STANAT, DONALD FORD, b Jackson, Miss, Jan 10, 37; m 58; c 3. COMPUTER SCIENCE. *Educ:* Antioch Col, BS, 59; Univ Mich, Ann Arbor, MS, 62, PhD(commun sci), 66. *Prof Exp:* Assoc res mathematician, Univ Mich, 66-67; asst prof, 67-72, assoc prof, 72-82, PROF COMPUT SCI, UNIV NC, CHAPEL HILL, 82- *Concurrent Pos:* Consult, IBM Corp, 67-71, Naval Res Lab, 81-84; vis scientist, IBM Corp, 79-80. *Mem:* AAAS; Asn Comput Mach; Sigma Xi. *Res:* Algorithm analysis; data structures and models of computation; parallel computation and cellular computers; research centers on the execution of functional language programs on highly parallel computers; algorithm design; programming language sematics; architecture of parallel computers. *Mailing Add:* Dept Comput Sci Univ NC Chapel Hill NC 27599-3175

STANBERRY, LAWRENCE RAYMOND, b Detroit, Mich, Aug 27, 48; m 69; c 2. VIROLOGY, PEDIATRIC INFECTIOUS DISEASES. *Educ:* Southwestern Univ Georgetown, BS, 70; Univ Ill Med Ctr, Chicago, MD, 77, PhD (pharmacol), 79. *Prof Exp:* Intern pediat, Childrens Med Ctr, Dallas, 77-78; fel oncol & exp therapeut, Univ Ill Med Ctr, 78-79; resident pediat, Univ Utah Med Ctr, Salt Lake, 79-80, fel ped infections dis, 80-82; asst prof, 82-87, ASSOC PROF PEDIAT, CHILDRENS HOSP RES FOUND, UNIV CINCINNATI COL MED, 87- *Concurrent Pos:* Fel John Hartford Found, 84-87; chmn spec review comt vitro antiviral screen, Nat Inst Allergy & Infectious Dis, 88-; mem Epidemiol & Dis Control #2 study sect, Nat Inst Health, 89-; prin investr, Nat Inst Allergy & Infectious Dis, 85-; Procter & Gamble Univ Explor grant, 85-88. *Mem:* Infectious Dis Soc Am; Soc Pediat Res; Int Soc Antiviral Res; Am Soc Virol; Pediat Infectious Dis Soc; Am Soc Microbiol. *Res:* Human herpes viruses, their vaccine development and anti-viral evaluation; molecular analysis of latency and pathogenesis. *Mailing Add:* 3475 Whitfield Ave Cincinnati OH 45220

STANBRIDGE, ERIC JOHN, b London, Eng, May 28, 42; m 71; c 2. CELL BIOLOGY, MICROBIOLOGY. *Educ:* Brunel Univ, HNC, 62; Stanford Univ, PhD(med microbiol), 71. *Prof Exp:* Tech officer virol, Nat Inst Med Res, UK, 60-65; res asst cell biol, Wistar Inst Anat & Biol, 65-67; mem sci staff cell biol, Nat Inst Med Res, UK, 68-69; instr med microbiol, Sch Med, Stanford Univ, 73-75; from asst prof to assoc prof microbiol, 75-82, PROF MICROBIOL, COL MED, UNIV CALIF, IRVINE, 82- *Concurrent Pos:* Spec fel, Leukemia Soc Am, 76-78; res career develop award, Nat Cancer Inst, 78-83; Eleanor Roosevelt Int Fel, 83,84. *Mem:* Am Soc Microbiol; Tissue Cult Asn; Int Orgn Mycoplasmologists; NY Acad Sci. *Res:* Cancer biology; somatic cell genetics; mycoplasmology. *Mailing Add:* Dept of Microbiol Univ of Calif - Irvine Irvine CA 92717

STANBRO, WILLIAM DAVID, b St Louis, Mo, Nov 29, 46; m 69; c 4. CHEMICAL SYSTEMS ANALYSIS, CHEMICAL SENSORS. *Educ:* George Washington Univ, BS, 68, PhD(chem), 72; Johns Hopkins Univ, MS, 85. *Prof Exp:* Res asst, Geophys Lab, Carnegie Inst Wash, 69-70; NSF presidential intern, 72-73, chemist & sr prof staff, Appl Physics Lab, Johns Hopkins Univ, 73-86; vpres res, Biotronic Systs Corp, 86-89; STAFF MEM, LOS ALAMOS NAT LAB, 89- *Concurrent Pos:* Patent comt, Am Inst Chemists, 89-90. *Mem:* Am Chem Soc; AAAS; Int Union Pure & Appl Chem; fel Am Inst Chemists. *Res:* Chemical dynamics and photochemistry of oxidants in condensed media; development of chemical and biological sensors; development of arms control verification procedures. *Mailing Add:* Los Alamos Nat Lab MS E541 Los Alamos NM 87545

STANBROUGH, JESSE HEDRICK, JR, b Ruston, La, May 1, 18; c 1. PHYSICS. *Educ:* Univ Tex, BS, 49, MA, 50. *Prof Exp:* Asst geophys, Univ Tex, 48-50, res physicist, Defense Res Lab, 50-60; tech staff asst, Naval Underwater Syst Ctr, Newport, RI, 79-88; corp dir, Benthos Corp, 68-88; exec secy, Undersea Warfare Res & Develop Planning Coun, Woods Hole Oceanog Inst, 60-61, tech asst to dir, 61-67, exec secy Joint Oceanog Insts Deep Earth Sampling Prog, 66-67, exec asst, Ocean Eng Dept, 67-76, RES PHYSICIST, WOODS HOLE OCEANOG INST, 60. *Concurrent Pos:* Co-founder, Tracor, Inc, 55-60; tech asst, Comt Undersea Warfare, Nat Acad Sci, 56; consult variable depth sonar, US Navy, 58-60, Washington Anal Servs Ctr Inc & EG&G Inc, 79-81. *Res:* Oceanography; underwater sound. *Mailing Add:* 36 Riddle Hill Rd Falmouth MA 02540

STANBURY, DAVID MCNEIL, b Boston, Mass, May 9, 52. INORGANIC CHEMISTRY. *Educ:* Duke Univ, BA, 74; Univ Southern Calif, PhD(chem), 78. *Prof Exp:* Fel, Stanford Univ, 78-80; ASST PROF CHEM, RICE UNIV, 80- *Mem:* Am Chem Soc; Sigma Xi; AAAS. *Res:* Kinetics and mechanisms of inorganic redox reactions. *Mailing Add:* Dept Chem Rice Univ PO Box 1892 Houston TX 77251

STANBURY, JOHN BRUTON, b Clinton, NC, May 15, 15; m 45; c 5. EXPERIMENTAL MEDICINE. *Educ:* Duke Univ, BA, 35; Harvard Med Sch, MD, 39; Am Bd Internal Med, dipl, 49. *Hon Degrees:* MD, Univ Leiden, 75. *Prof Exp:* House officer, Mass Gen Hosp, Boston, 40-41; asst in med, Harvard Med Sch, from asst to assoc clin prof, 56-66, lectr med, 66-; prof exp med, Mass Inst Technol, 66-81; RETIRED. *Concurrent Pos:* Res fel pharmacol, Harvard Med Sch, 47-48; chief med resident, Mass Gen Hosp, 48-49, asst in med, 49-50, chief thyroid clin & lab, 49-66, from asst physician to physician, 50-66, consult physician, 66-, sr physician, 81-86. *Mem:* Endocrine Soc; Am Soc Clin Invest; Asn Am Physicians; Am Thyroid Asn. *Res:* Endocrinology; metabolism; genetics; metabolic disease. *Mailing Add:* 43 Circuit Rd Chestnut Hill MA 02167

STANCAMPIANO, CHARLES VINCENT, b Brooklyn, NY, Oct 27, 48; m 69; c 1. ELECTRICAL ENGINEERING, SUPERCONDUCTIVITY. *Educ:* Rensselaer Polytech Inst, BS, 69; Univ Rochester, MS, 71, PhD(elec eng), 76. *Prof Exp:* Res assoc & asst prof, 75-76, ASST PROF ELEC ENG, UNIV ROCHESTER, 77-, SCIENTIST, LAB LASER ENERGETICS, 80- *Concurrent Pos:* Consult, Dept Radiation Biol & Biophys, Univ Rochester, 76-80; co-investr, Ctr Naval Anal grant, 78- *Mem:* Inst Elec & Electronics Engrs; Electron Devices Soc; Am Phys Soc. *Res:* Microwave applications of the Josephson effect; nonequilibrium superconductivity. *Mailing Add:* Eastman Kodak Co Kodak Res Lab Bldg 81 Device Dev Lab Rochester NY 14650

STANCEL, GEORGE MICHAEL, b Chicago, Ill, Dec 29, 44; m 72. BIOCHEMISTRY, ENDOCRINOLOGY. *Educ:* St Thomas Col, BS, 66; Mich State Univ, PhD(biochem), 70. *Prof Exp:* ASSOC PROF PHARMACOL, UNIV TEX MED SCH HOUSTON, 72- *Concurrent Pos:* NIH fel endocrinol, Univ Ill, 71-72. *Mem:* Endocrine Soc; NY Acad Sci; Am Chem Soc; Tissue Cult Asn. *Res:* Biochemical endocrinology; steroid hormone action; hormone receptors; estrogen regulation of uterus and pituitary. *Mailing Add:* Dept Pharmacol Univ Tex Health Sci Ctr PO Box 20036 Houston TX 77025

STANCER, HARVEY C, b Toronto, Ont, Mar 6, 26; m 58; c 2. PSYCHIATRY, NEUROCHEMISTRY. *Educ:* Univ Toronto, BA, 50, PhD(path chem), 53, MD, 55; Royal Col Physicians Can, cert psychiat, 62, fel, 72. *Prof Exp:* Head neurochem, Toronto Psychiat Hosp, 62-66; assoc prof, 69-72, chief clin invest univ & head neurochem, 66-76, prof psychiat res, 74-81, chief Affective Disorders Unit, Clarke Inst Psychiat, 78-84, vchmn, dept psychiat, 80-85, PROF PSYCHIAT, FAC MED, UNIV TORONTO, 72- *Concurrent Pos:* McLean fel, Maudsley Inst, Univ London, 58-59; NY State Dept Hyg fel, Columbia Univ-Presby Med Ctr, 59-61; McLellan fel, Univ Toronto, 61-62; assoc, Med Res Coun Can, 62-64; vis prof, Univ Calif, Los Angeles, 83- *Honors & Awards:* Clarke Inst Prize, Univ Toronto, 70 & 74; McNeil Award, 72. *Mem:* Int Soc Neurochem; Neurochem Soc; Soc Biol Psychiat; Psychiat Res Soc; Can Endocrinol Soc & Metab. *Res:* Clinical psychiatric investigation and animal behavioral investigation of brain biogenic amines; genetics of affective disorders; psychopharmacology. *Mailing Add:* Clarke Inst of Psychiat Univ Toronto Toronto ON M5T 1R8 Can

STANCHFIELD, JAMES ERNEST, b Salem, Mass, June 22, 52; m 73; c 1. MOLECULAR BIOLOGY, BIOCHEMISTRY. *Educ:* Univ Mass, Amherst, BS, 74; Dartmouth Col, PhD(biol), 78. *Prof Exp:* staff fel molecular biol, Nat Cancer Inst, 78-80; SALES DIR, BETHESDA RES LABS INC, 80- *Mem:* AAAS; Soc Develop Biol; Am Soc Cell Biol. *Res:* Transcriptional and translational control mechanisms in eucaryotes. *Mailing Add:* Life Technol Inc 8717 Grovemont Circle PO Box 6009 Gaithersburg MD 20877

STANCL, DONALD LEE, b Oak Park, Ill, Feb 21, 40; m 66. MATHEMATICS, OPERATIONS RESEARCH. *Educ:* Knox Col, AB, 62; Univ Ill, PhD(math), 66; Nichols Col, MBA, 78. *Prof Exp:* Instr math, Princeton Univ, 67-69; asst prof, Univ Kans, 69-72; from assoc prof to prof math & statist, Nichols Col, 72-83; sr software engr, Sanders Assocs, 84-85; PROF MATH, ST ANSELM COL, 85- *Mem:* Math Asn Am; Am Math Soc. *Res:* Operations research and general applications of mathematics to problems of management. *Mailing Add:* 40 Briar Hill Rd New Boston NH 03070-9375

STANCL, MILDRED LUZADER, b Parkersburg, WVa; m 66. TOPOLOGY. *Educ:* Marietta Col, AB, 49; Univ Ill, AM, 62, PhD(math), 69. *Prof Exp:* Systs analyst, Sperry-Rand Corp, 54-57 & Radio Corp Am, 57-60; teaching asst math, Univ Ill, 61-66; asst prof, Trenton State Col, 68-69 & Kans Univ, 69-72; assoc prof, 72-76, PROF MATH, NICHOLS COL, 76- *Mem:* Am Math Soc; Math Asn Am; Sigma Xi. *Res:* Isomorphisms of smooth manifolds surrounding polyhedra. *Mailing Add:* 40 Briar Hill Rd New Boston NH 03070-9375

STANCYK, STEPHEN EDWARD, b Denver, Colo, Apr 8, 46; m 78. MARINE ECOLOGY, INVERTEBRATE ZOOLOGY. *Educ:* Univ Colo, Boulder, BA, 68; Univ Fla, MS, 70, PhD(zool), 74. *Prof Exp:* Instr biol, Dept Zool, Univ Fla, 74-75; ASST PROF MARINE SCI & BIOL, UNIV SC, 75- *Mem:* Am Soc Zoologists; AAAS; Ecol Soc Am; Southeastern Estuarine Res Soc; Sigma Xi. *Res:* Reproductive ecology of marine invertebrates; estuarine zooplankton dynamics; marine turtle conservation; systematics of Phoronids. *Mailing Add:* Dept of Biol Univ of SC Columbia SC 29208

STANCZYK, FRANK ZYGMUNT, b Montreal, Que, July, 4, 36; m 70; c 1. PERINATAL PHYSIOLOGY, STEROID BIOCHEMISTRY. *Educ:* Western Ill Univ, BS, 61; McGill Univ, MS, 67, PhD(exp med), 72. *Prof Exp:* Fel reproductive biol, Obstet & Gynec Dept, Univ Southern Calif, 72-74, from instr to asst prof, 74-76, asst prof, Obstet & Gynec Dept & Physiol Dept, 76-80; asst scientist perinatal physiol, Ore Regional Primate Res Ctr, 80-86; asst prof, Dept Obstet & Gynec, Ore Health Sci Univ, 80-86; ASSOC PROF, DEPT OBSTET & GYNEC, UNIV SOUTHERN CALIF, LOS ANGELES, 86- *Mem:* Soc Gynec Invest; Endocrine Soc; Am Chem Soc; Can Biol Soc; Sigma Xi. *Res:* In vivo and in vitro studies of steroid hormone metabolism in pregnancy and endocrinopathies; pharmacokinetics and endocrine effects of contraceptive steroids; prediction and detection of ovulation; regulation of parturition and placental production of hormones. *Mailing Add:* Dept Obstet & Gynec Univ Southern Calif 1240 N Mission Rd Los Angeles CA 90033

STANCZYK, MARTIN HENRY, b Jersey City, NJ, Jan 26, 30; m; c 7. EXTRACTIVE METALLURGY. *Educ:* Univ Ariz, BS, 57, MS, 58. *Prof Exp:* Bur Mines fel, US Bur Mines, Ariz, 57-58, extractive metallurgist, 57-60 & Ala, 60-61, res extractive metallurgist, 61-68, supvry metallurgist, College Park Metall Res Ctr, 68-72, res dir, Tuscaloosa Metall Res Ctr, 73-88. *Concurrent Pos:* Adj prof, Univ Ala, 84-88; vpres, Tech Dev-wTe Corp, 88-90; vpres res, SERI Corp, 90- *Honors & Awards:* Silver & Gold Medals from the Dept of the Interior for Meritorious & Distinguished Serv, 80, 87. *Mem:* Am Inst Mining, Metall & Petrol Engrs; Am Inst Mining Engrs; Sigma Xi. *Res:* Extractive metallurgy of nonmetallic minerals, including beneficiation studies, dewatering mineral wastes and developing new or improved ceramic and refractory materials; recycling solid wastes from municipal and cerebride collection programs. *Mailing Add:* 4941 Red Oak Lane Tuscaloosa AL 35405

STANDAERT, FRANK GEORGE, b Paterson, NJ, Nov 12, 29; m 59; c 3. PHARMACOLOGY. *Educ:* Harvard Univ, AB, 51; Cornell Univ, MD, 55. *Prof Exp:* Intern med, Johns Hopkins Hosp, Baltimore, Md, 55-56; from instr to assoc prof pharmacol, Med Col, Cornell Univ, 59-67; prof pharmacol & chmn dept, Sch Med & Dent, Georgetown Univ, 67-86; adad vpres & dean med sch, 86-89, ASSOC VPRES RES, MED COL OHIO, 90- *Concurrent Pos:* Res fel pharmacol, Med Col, Cornell Univ, 56-57; USPHS career develop award, 61-67; guest scientist, Naval Med Ctr, Bethesda, Md; mem, Nat Res Coun & Am Asn Dent Schs; chmn publ comt, Fedn Am Socs Exp Biol; mem comt toxicol, Nat Acad Sci-Nat Res Coun; mem neurol dis prog proj rev comt & pharmacol & toxicol comt, NIH; ed, Neuropharmacol, J Pharmacol & Exp Therapeut; mem merit rev bd neurobiol, Vet Admin; mem basic pharmacol adv comt, Pharmaceut Mfrs Asn Found; mem bd dirs, Washington Heart Asn; secy & pres, Asn Med Sch Pharmacol. *Mem:* Am Soc Clin Pharmacol & Therapeut; Am Soc Pharmacol & Exp Therapeut (pres, 90-91); Soc Exp Biol & Med; Soc Toxicol; Peripatetic Soc. *Res:* Pharmacology of neuromuscular transmission; toxicology, neuropharmacology. *Mailing Add:* Assoc Vpres Res Med Col Ohio C S 10008 Toledo OH 43699

STANDEFER, JIMMY CLAYTON, b Stanton, Tex, Mar 2, 41; m 68; c 2. BIOCHEMISTRY, CLINICAL CHEMISTRY. *Educ:* Univ Kans, BA, 63, PhD(biochem), 67; Nat Registry Clin Chem, cert, 70. *Prof Exp:* Biochemist, Walter Reed Army Inst Res, 67-70; instr biochem, Sch Nursing, Univ Md, 69-70; ASST PROF CLIN CHEM, SCH MED, UNIV N MEX, 70- *Mem:* Am Asn Clin Chem; Am Acad Forensic Sci. *Res:* Membrane chemistry; isoenzymes; lipids. *Mailing Add:* Dept Path Univ NMex Sch Med 915 Stanford NE Albuquerque NM 87131

STANDER, JOSEPH W, b Covington, Ky, Dec 2, 28. ALGEBRA. *Educ:* Univ Dayton, BS, 49; Cath Univ, MS, 57, PhD(math), 59. *Prof Exp:* Teacher, Hamilton Cath High Sch, 49-50 & Col Ponce, 50-55; from asst prof to assoc prof math, Univ Dayton, 60-74, dean, Grad Studies & Res, 67-74, vpres, Acad Affairs & Provost, 74-89, PROF MATH, UNIV DAYTON, 74- *Mem:* Math Asn Am; Sigma Xi. *Res:* Matrix theory. *Mailing Add:* Univ Dayton Dayton OH 45469

STANDFORD, GEOFFERY BRIAN, b London, Eng, Mar 29, 16; m 59; c 3. ENVIRONMENTAL MANAGEMENT. *Educ:* Royal Col Physicians & Surgeons, MRCS & LRCP, 39; dip med radiol, 47. *Prof Exp:* Vis prof, Sch Archit & Environ Planning Calif State Polytech Col, San Luis Obispo. 70; adj prof, Environ Studies Ctr, Antioch Col, 71; vis prof & dir, Environics Ctr, St Edwards Univ, 71-72; biomed res scientist, Urban Health Module Sch Pub Health, Univ Tex, Houston, 72-74; VIS PROF & RESOURCE RECOVERY PLANNING SPECIALIST, INST URBAN STUDIES, UNIV TEX, ARLINGTON, 74-; PRES, ARGO-CITY, INC, 74- *Concurrent Pos:* Adj prof, Sch Archit, Rice-Univ, 72-74; mem environ & eco-systs planning comt, Prep Planning Group, UN Environ Prog for 1976 Habitat Cont, 74; UN Environ Prog deleg, Sump on Develop Patterns of Humans Settlements in Developing Countries for Yr 2000, 75; proj dir res into effects of landmix, Environ Protection Agency; dir, Greenhills Environ Studies Ctr; trustee, Environic Found Int. *Mem:* Archit Asn Gt Brit; Am Soc Testing & Mat; fel Royal Photog Soc Gt Brit; Soil Asn; AAAS. *Res:* Use of municipal waste resources for restoring soil fertility and water characteristies; measurement of the effects of crop yield, water quality and climate application to regional environmental management. *Mailing Add:* 7575 Wheatland Rd Cedar Hill TX 75104

STANDIFER, LEONIDES CALMET, JR, b Gulfport, Miss, Apr 24, 25; m 57; c 2. PLANT PHYSIOLOGY. *Educ:* Miss State Univ, BS, 50, MS, 54; Univ Wis, PhD(bot), 59. *Prof Exp:* Plant physiologist, Firestone Plantations Co, 54-61; from asst prof to assoc prof, 61-74, prof bot, 74-77, PROF HORT, LA STATE UNIV, BATON ROUGE, 78- *Mem:* Bot Soc Am; Am Soc Plant Physiol; Weed Sci Soc Am. *Res:* Patterns of plant recovery from flame injury; histological responses of certain plants to herbicides, and physiology of herbicidal action. *Mailing Add:* 1244 Pasture View Dr Baton Rouge LA 70810

STANDIFER, LONNIE NATHANIEL, b Itasca, Tex, Oct 28, 26; div. ENTOMOLOGY, PARASITOLOGY. *Educ:* Prairie View Agr & Mech Col, BS, 49; Kans State Col, MS, 51; Cornell Univ, PhD(med & vet entom & parasitol), 54. *Prof Exp:* Instr biol sci & supvr campus pest control, Tuskegee Inst, 51-52; asst livestock insect control, Cornell Univ, 53-54; asst prof biol sci, Southern Univ, 54-56; res scientist & apiculturist, USDA, 56-70, dir, Bee Res Lab, 70-85, res leader, Honey Bee Pollination Lab, 72-85, tech adv apicult, Western Region, 73-85; RETIRED. *Honors & Awards:* Award 120, Excellence Biol & Life Scis, Nat Consortium Black Prof Develop. *Mem:* AAAS; Entom Soc Am; Am Soc Parasitol; Am Beekeeping Fedn. *Res:* Medical and veterinary entomology and parasitology; control of insects of public health importance; insect physiology and nutrition; botany and plant pathology; honey bee physiology and nutrition, protein and lipids; pollen chemistry, fatty acids, sterols and hydrocarbons. *Mailing Add:* 5401 Overtone Ridge Blvd Apt 1001 Ft Worth TX 76132

STANDIL, SIDNEY, b Winnipeg, Man, Oct 19, 26; m 50; c 4. PHYSICS. *Educ:* Queen's Univ, Ont, BSc, 48, MSc, 49; Univ Man, PhD, 51. *Prof Exp:* From asst prof to assoc prof, 51-63, dean fac grad studies, 73-79, prof physics, 63-86, SR SCHOLAR, UNIV MAN, 86- *Mem:* Am Phys Soc. *Res:* Cosmic ray and space physics. *Mailing Add:* 772 Campbell St Winnipeg MB R3N 1C6 Can

STANDING, CHARLES NICHOLAS, b Minneapolis, Minn, Dec 24, 43; m 68. CHEMICAL ENGINEERING, FOOD SCIENCE. *Educ:* Univ Minn, BS, 65, PhD(chem eng), 70. *Prof Exp:* Scientist/engr, Pillsbury Co, Minneapolis, 70-74, sr scientist/engr, 74-75, group leader, 75-78, sect mgr res & develop, 78-79; dept head, Processes Res & Develop, Gen Mills, 79-83, dir Res & Develop, 83-86; GEN MGR, PROCESSED FOODS, JW ALLEN & CO, WHEELING, ILL, 86- *Mem:* Am Inst Chem Engrs; Inst Food Technologists. *Res:* Process development; food product development and exploratory food research; technomic feasibility analyses; process engineering, process scaleup and plant engineering; biochemical engineering and dynamics of microbiological populations. *Mailing Add:* J W Allen & Co 555 Allendale Dr Wheeling IL 60090

STANDING, KEITH M, b Ogden, Utah, Aug 2, 28; m 56; c 7. VERTEBRATE ZOOLOGY. *Educ:* Brigham Young Univ, BS, 53, MS, 55; Wash State Univ, PhD(zool, bot), 60. *Prof Exp:* Assoc prof, 58-69, Chmn Dept, 69-72, PROF BIOL, CALIF STATE UNIV, FRESNO, 69-, CHMN DEPT, 82- *Concurrent Pos:* NSF res grants, 64-67; mem, NSF Conf Histochem-Its Appl in Res & Teaching, Vanderbilt Univ, 65; mem gov bd, Moss Landing Marine Labs; bd dirs, Ctr Urban & Regional Studies; vis scholar, Univ Calif, Berkeley & Univ Calif, Davis. *Mem:* AAAS; Am Soc Zool. *Res:* Histological analysis of reproductive organs of blue grouse; comparative histology of nephron units of kangaroo rats; isolation of nephron units of Dipodomys by various techniques; cytotaxonomy of Dipodomys; embryonic kidney development. *Mailing Add:* Dept Biol Calif State Univ Fresno CA 93740

STANDING, KENNETH GRAHAM, b Winnipeg, Man, Apr 3, 25; div; c 4. MASS SPECTROMETRY. *Educ:* Univ Man, BSc, 48; Princeton Univ, AM, 50, PhD(physics), 55. *Prof Exp:* From asst prof to assoc prof, 53-64, PROF PHYSICS, UNIV MAN, 64- *Concurrent Pos:* Nuffield Found Dom traveling fel, Wills Physics Lab, Bristol Univ, 58-59; Nat Res Coun Can sr res fel, Univ Grenoble, 67-68; vis prof, Inst Nuclear Physics, Orsay, 85-86. *Mem:* Am Phys Soc; Can Asn Physicists; Am Soc Mass Spectrometry. *Res:* Mass spectrometry of biomolecules and molecular clusters; nuclear physics and applications. *Mailing Add:* Dept Physics Univ Man Winnipeg MB R3T 2N2 Can

STANDING, MARSHALL B, ENGINEERING. *Prof Exp:* RETIRED. *Mem:* Nat Acad Eng. *Mailing Add:* 5434 Via Carrizo Laguna Hills CA 92653

STANDISH, CHARLES JUNIOR, b Triangle, NY, Nov 10, 26. MATHEMATICS. *Educ:* Hamilton Col, NY, BA, 49; Johns Hopkins Univ, MA, 51; Cornell Univ, PhD(math), 54. *Prof Exp:* Instr math, Hamilton Col, NY, 51-52; asst, Cornell Univ, 52-54; asst prof, Union Univ, NY, 54-57; mathematician, IBM Corp, 57-84; RETIRED. *Concurrent Pos:* Vis assoc prof, NC State Univ, 60-61; vis lectr, Sch Adv Technol, State Univ NY, Binghamton, 74-78, adj lectr, Dept Elec Eng, 84- *Mem:* Am Math Soc; Soc Indust & Appl Math. *Res:* Kalman filtering; control theory. *Mailing Add:* RD Two Box 672 Greene NY 13778-9405

STANDISH, E MYLES, JR, b Hartford, Conn, Mar 5, 39; m 68; c 3. ASTRONOMY. *Educ:* Wesleyan Univ, BA, 60, MA, 62; Yale Univ, PhD(astron), 68. *Prof Exp:* Asst prof astron, Yale Univ, 68-72; MEM TECH STAFF, JET PROPULSION LAB, 72- *Mem:* Am Astron Soc-Div Dynamic Astron; Int Astron Union. *Res:* Celestial mechanics; numerical analysis; continuous improvement of the planetary, lunar and natural satellite ephemerides. *Mailing Add:* Jet Propulsion Lab 301-150 4800 Oakgrove Dr Pasadena CA 91109

STANDISH, NORMAN WESTON, b Marion, Iowa, Apr 4, 30; m 56; c 3. TECHNICAL MANAGEMENT, ORGANIC CHEMISTRY. *Educ:* Beloit Col, BS, 52; Purdue Univ, MS, 57, PhD(org chem), 59. *Prof Exp:* Chemist, Selectron Div, Pittsburgh Plate Glass Co, 52-53; res assoc biochem, Standard Oil Co, 60-65, tech dir plastics, Prophylactic Brush Div, 65-67, res supvr, Cleveland, 67-70, supvr develop, Tech Serv & Polymers, 70-75, mgr, Tech Serv, 75-80, lab dir, Explor Prod, 80-84, mgr, Strat Planning, 84-87; mgr prog develop, 88-89, SR RES ASSOC, BP AM, INC, CLEVELAND, 90- *Concurrent Pos:* Mem adv bd, Col Sci, Tex A&M Univ; chmn & nat coun, Cleveland Sect, Am Chem Soc. *Mem:* Am Chem Soc; Soc Plastics Engrs; Soc Nematol; fel Am Inst Chemists; Soc Plastics Indust; Soc Petrol Engrs. *Res:* Computer modelling of fluid flow in reservoirs, enhanced oil recovery systems; polymer processing and fabrication plastic market development; technology strategic planning; technical program development and commercialization. *Mailing Add:* 32250 Burlwood Dr Solon OH 44139

STANDISH, SAMUEL MILES, b Campbellsburg, Ind, July 6, 23; m 49; c 2. ORAL PATHOLOGY, FORENSIC ODONTOLOGY. *Educ:* Ind Univ, DDS, 45, MS, 56; Am Bd Oral Path, dipl, 59; Am Bd Forensic Odontol, dipl, 77. *Prof Exp:* Instr dent, 52-57, from asst prof to assoc prof oral path, 57-67, asst dean grad & postgrad educ, 69-74, prof oral path, 67-88, assoc dean grad & postgrad educ, 74-88, EMER PROF, DENT DIAG SCI & ORAL PATH, SCH DENT, IND UNIV, INDIANAPOLIS, 88- *Concurrent Pos:* Mem, Clin Cancer Educ Comt, Nat Cancer Inst, 69-79; consult, Coun Dent Educ, Am Dent Asn, 71-77; mem, Nat Bd Test Constructors Comt, Am Dent Asn, 66-75; pres, Am Bd Oral Path, 78-79 & Orgn Teachers Oral Diag, 87-88. *Honors & Awards:* Odontol Award, Am Acad Forensic Sci. *Mem:* Am Dent Asn; fel Am Acad Oral Path (pres, 72-73); Int Asn Dent Res; fel Am Acad Forensic Sci. *Res:* Salivary gland pathophysiology and experimental carcinogenesis; inflammatory mechanisms; striated muscle regeneration; muscle diseases; clinical oral pathology. *Mailing Add:* Ind Univ Sch of Dent - DS S 110 1121 W Michigan St Indianapolis IN 46202

STANDLEE, WILLIAM JASPER, b Zybach, Tex, May 2, 29; m 58; c 3. POULTRY NUTRITION. *Educ:* Tex Tech Univ, BS, 54, MS, 55; Tex A&M Univ, PhD(poultry sci), 63. *Prof Exp:* Animal nutritionist, Standard Milling Co, Tex, 55-57; salesman, Van Waters & Rogers, 57-58; res asst, Tex Agr Exp Sta, 58-63; dir nutrit & res, Darragh Co, Ark, 63-65; dir nutrit, Burrus Feed Mills, Tex, 65-68; dir nutrit & res, Food Div, Valmac Industs, Inc, Ark, 68-71; dir res & nutrit, B & D Mills, 71-79; CONSULT POULTRY & ANIMAL NUTRIT, 79- *Res:* Nutrition and feeding management of turkey breeders; broiler chicken breeders, market turkeys and broilers and egg production chickens. *Mailing Add:* 815 N Lucas Dr Grapevine TX 76051-5063

STANDLEY, ROBERT DEAN, b Findlay, Ill, Aug 25, 35; m 59; c 2. ELECTRICAL ENGINEERING. *Educ:* Univ Ill, BS, 57; Rutgers Univ, MS, 60; Ill Inst Technol, PhD(elec eng), 66. *Prof Exp:* Assoc engr, IIT Res Inst, 60-62, res engr, 62-64, asst mgr microwaves & antennas, 64-65, mgr electromagnetic compatibility, 65-66; MEM TECH STAFF COHERENT OPTICS RES, BELL TEL LABS, INC, 66- *Mem:* Inst Elec & Electronics Engrs. *Res:* Microwave filters; antennas; electromagnetic compatibility; avalanche transit time diode oscillators; optical modulators; optical integrated circuits; fiber optics; satellite communications. *Mailing Add:* Four Sunnybank Dr Shrewsbury NJ 07701

STANEK, ELDON KEITH, b Novinger, Mo, Dec 12, 41; m 69; c 2. ELECTRICAL ENGINEERING. *Educ:* Ill Inst Technol, BSEE, 64, MS, 65, PhD(elec eng), 69. *Prof Exp:* Asst prof elec eng, Ill Inst Technol, 68-70; from asst prof to assoc prof elec eng, 70-77, prof, 77-80; PROF & HEAD ELEC ENG, MICH TECH UNIV, 80- *Concurrent Pos:* NSF grant, WVa Univ, 71-73, Bur Mines grant, 72-79; consult, Dept Energy, 77- & Union Carbide Corp, 77- *Mem:* Inst Elec & Electronics Engrs; Am Soc Elec Engrs; Sigma Xi. *Res:* Electrical power systems; digital simulation and mathematical modeling; simulation of switching transients; inductive interference; induced voltage in cables. *Mailing Add:* Dept Elec Eng Univ Mo Rolla MO 65401-0249

STANEK, KAREN ANN, b Orofino, Idaho, Oct 19, 50. HYPERTENSION. *Educ:* Univ Idaho, BS, 73; Univ Wis-Madison, MS, 75, PhD(physiol), 78; Univ Miss-Jackson, MD, 88. *Prof Exp:* Fel physiol, 78-79, instr, 79-80, asst prof, 80-84, sr res assoc physiol, Med Ctr, Univ Miss, 84-88; intern med, Univ SDak, 88-89; RESIDENT PHYS MED & REHAB, UNIV COLO HEALTH SCI CTR, DENVER, 89- *Honors & Awards:* Sci Res Award, AOA, 85-86. *Mem:* Am Physiol Soc; AMA; Asn Acad Physiatrists. *Res:* Quantitating changes in blood flow; resistance patterns in the development and treatment of hypertensive rats; radioactive microsphere techiques; hi-rider wheel chair research. *Mailing Add:* 11113 East Alameda Ave No 102 Aurora CO 80012

STANEK, PETER, b Chicago, Ill, Dec 3, 37; m 60; c 2. MATHEMATICS, SYSTEMS ANALYSIS. *Educ:* Univ Chicago, MS, 58, PhD(math), 61. *Prof Exp:* Mem, Inst Defense Anal, 61-62; analyst, Opers Eval Group, 62-63; asst prof math, Univ Southern Calif, 63-65; sr scientist, Jet Propulsion Lab, 65-68 & Lear Siegler Inc, 68-72; mem staff, Systs Applns Inc, 72-78; MEM STAFF, KETRON INC, 78- *Mem:* Am Math Soc; Soc Indust & Appl Math; Human Factors Soc; Am Inst Aeronaut & Astronaut. *Res:* Algebra; operations research; communications engineering; human factors engineering; computer applications. *Mailing Add:* 1281 Idylberry Rd San Rafael CA 94903

STANFEL, LARRY EUGENE, operations research, computer science, for more information see previous edition

STANFIELD, JAMES ARMOND, b Covington, Ky, Aug 28, 17; m 42; c 3. ORGANIC CHEMISTRY. *Educ:* Eastern Ky State Col, BS, 40; Univ Tenn, MS, 42, PhD(phys org chem), 47. *Prof Exp:* Instr chem, Univ Ky, 41-42; instr, Univ Tenn, 42-46; from asst prof to prof chem, 47-56-, asst dir sch chem, 65-85, res assoc, res inst, 51-85, EMER PROF, GA INST TECHNOL, 85- *Mem:* Am Chem Soc. *Res:* Organic synthesis; catalytic hydrogenation kinetics; spirobarbituric acids; chemistry of uramil. *Mailing Add:* 1065 Ferncliff Rd NE Atlanta GA 30324-2522

STANFIELD, KENNETH CHARLES, b Los Angeles, Calif, Sept 21, 42; c 1. EXPERIMENTAL HIGH ENERGY PHYSICS. *Educ:* Univ Tex, BS, 64; Harvard Univ, AM, 67, PhD(physics), 69. *Prof Exp:* Res assoc physics, Univ Mich, 69-71; asst prof physics, Purdue Univ, 71-77; assoc head, Fermi Nat Accelerator Lab, 77-79, head, Proton Dept, 79-81, Exp Areas Dept, 82, Bus Sect, 84, HEAD, RES DIV, FERMI NAT ACCELERATOR LAB, 85- *Concurrent Pos:* Mem prog adv comt, Zero Gradient Synchrotron, Argonne Nat Lab, 75- *Mem:* Sigma Xi. *Res:* Experimental research, using electronic techniques, into the nature of elementary particle properties. *Mailing Add:* 38 W Elliott Ct St Charles IL 60175

STANFIELD, MANIE K, b St Petersburg, Fla, Feb 15, 31. ORGANIC CHEMISTRY, BIOCHEMISTRY. *Educ:* Univ Chicago, BA, 54, MS, 57; Univ Calif, Los Angeles, PhD(org chem), 62. *Prof Exp:* Asst org chem, Mass Inst Technol, 62-63 & Rockefeller Univ, 63-65; ASST PROF BIOCHEM, SCH MED, TULANE UNIV, 65- *Mem:* AAAS; Am Chem Soc. *Res:* Organic syntheses of small biologically interesting molecules; stability and polymerization of beta-lactam antibiotics. *Mailing Add:* Dept Biochem Tulane Univ Med Sch New Orleans LA 70112

STANFORD, AUGUSTUS LAMAR, JR, b Macon, Ga, Jan 20, 31; m 52; c 2. SOLID STATE PHYSICS. *Educ:* Ga Inst Technol, BS, 52, MS, 57, PhD(physics), 58. *Prof Exp:* Sr staff consult, Sperry Rand Corp, 58-64; assoc prof, 64-74, PROF PHYSICS, GA INST TECHNOL, 74- *Concurrent Pos:* NASA res grant, 64- *Mem:* Am Phys Soc. *Res:* Nuclear spectroscopy; ferroelectrics; pyroelectrics; phonon in solids. *Mailing Add:* Dept Physics Ga Inst Technol 225 North Ave NW Atlanta GA 30332

STANFORD, GEOFFREY, b March 29,16. RESEARCH ADMINISTRATION. *Educ:* Royal Col Surgeons,MRCS,39,Royal Col Physicians,LRCP,39,Univ London,DMR(radiotherapy),46. *Hon Degrees:* Royal Photographic Soc,FRPS,44. *Prof Exp:* Vis prof,Sch Archit, Rice Univ, 73-75; assoc prof, Sch Pub Health, 73-75,adj prof, Inst Urban Studies, Univ Tx, 75-76; dir ,Greenhills Environ Ctr, 76-87; DIR EMER ENVIRON, DALLAS NATURE CTR, 87-; PRES,AGRO-CITY INC, 74- *Concurrent Pos:* Vis prof,Sch Archit & Engr,Calif Polytech, 70; vis prof,Antioch Col, 71; dir,Environ Inst,St Edwards Univ, 72-73. *Mem:* Dallas Nature Ctr(dir emer

& trustee0; Archit Asn Gt Brit. *Res:* Energy and materials recovery from refuse and sewage by innovative (A/I) methods; biogas generation by advanced procedures; biomass cultivation by coppicing; regional economic resourses co-ordination. *Mailing Add:* Dallas Nature Ctr 7171 Mountain Creek Pkwy Dallas TX 75249

STANFORD, GEORGE STAILING, b Halifax, NS, July 23, 28; US citizen; m 56; c 3. REACTOR PHYSICS. *Educ:* Acadia Univ, BSc, 49; Wesleyan Univ, MA, 51; Yale Univ, PhD(nuclear energy levels), 56. *Prof Exp:* Proj engr infrared instrumentation, Perkin-Elmer Corp, 55-59; PHYSICIST, ARGONNE NAT LAB, 59- *Mem:* Am Asn Advan Sci; fel Am Scientists. *Res:* Experimental reactor physics; verification of arms-control treaties. *Mailing Add:* Argonne Nat Lab D208 9700 S Cass Ave Argonne IL 60439

STANFORD, JACK ARTHUR, b Delta, Colo, Feb 18, 47; m; c 2. LIMNOLOGY. *Educ:* Colo State Univ, BS, 69, MS, 71; Univ Utah, PhD(limnol), 75. *Prof Exp:* Asst fish & wildlife, Colo State Univ, 65-69, asst zool, 69-72; res limnol, Univ Utah & Univ Mont, 72-74; asst prof limnol & biol, NTex State Univ, 74-79; RES PROF ZOOL, UNIV MONT, 79-, DIR, BIOL STA, 80-, BIERMAN PROF, 87- *Concurrent Pos:* Dir, Flathead Res Group, Biol Sta, Univ Mont, 77-; consult, Nature Conserv Nordic Coun Ecol; vis lectr, Norway, 80; vis lectr, SAfrica, 90. *Mem:* Sigma Xi; Int Soc Theoret & Appl Limnol; Am Soc Limnol & Oceanog; Ecol Soc Am; NAm Benthological Soc. *Res:* All aspects of limnological study in lakes and streams with special interest in nutrient cycling by algae and heterotrophic bacteria; benthic ecology and life histories of the Plecoptera; hyporheic ecology. *Mailing Add:* Biol Sta Univ Mont Polson MT 59860

STANFORD, JACK WAYNE, b Eldorado, Tex, Dec 21, 35; m 58; c 2. PLANT TAXONOMY. *Educ:* Baylor Univ, BA, 58; Tex Tech Univ, MS, 66; Okla State Univ, PhD(bot), 71. *Prof Exp:* Teacher jr high sch, Tex, 60-62, high sch, 62-66; from asst prof to assoc prof, 66-74, PROF BIOL, HOWARD PAYNE UNIV, 74- *Res:* Pollen morphology of the Mimosoideae; floristic studies of central Texas. *Mailing Add:* Dept Biol Howard Payne Univ Brownwood TX 76801

STANFORD, MARLENE A, FOOD CHEMISTRY. *Educ:* Ind Univ, BS, 74; Northwestern Univ, MS, 78, PhD(phys chem), 80. *Prof Exp:* Lab technician & res asst, Gibbs Labs, Wilmette, Ill, 75-76; res fel & teaching asst, dept chem, Northwestern Univ, Evanston, 76-80; mem staff, new prod develop, Pet Foods Res & Develop, 80-82 & Cent Res & Develop, 82-84, SCIENTIST, TECHNOL & NEW BUS RES & DEVELOP, QUAKER OATS CO, BARRINGTON, ILL, 84- *Mem:* Am Chem Soc; Inst Food Technologists. *Res:* Product concepts, formulations and processes for foods; new technologies to support new cereals, snacks and beverages; product concepts, formulations and processes for all pet food categories. *Mailing Add:* Kraft Inc Res & Develop 801 Waukegan Rd Glenview IL 60025-4391

STANG, ELDEN JAMES, b Victoria, Kans, Jan 23, 40; m 63; c 3. APPLIED PHYSIOLOGY, TEACHING. *Educ:* Kans State Univ, BS, 67; Iowa State Univ, MS, 69, PhD(hort), 73. *Prof Exp:* Res asst hort, Iowa State Univ, 67-69, res assoc, 69-73, instr, 73; exten horticulturist pomol, Ohio State Univ, 73-78; PROF POMOL, UNIV WIS-MADISON, 78- *Concurrent Pos:* Peace Corps, Chile, 61-63; assoc ed, Am Soc Hort Sci, 80-84; Fulbright fel, Finland, 87. *Mem:* Am Soc Hort Sci; Int Soc Hort Sci. *Res:* Production of tree fruit and small fruit; weed control; plant nutrition; plant growth regulators. *Mailing Add:* Univ of Wis Dept of Hort 1575 Linden Dr Madison WI 53706

STANG, LOUIS GEORGE, b Portland, Ore, Oct 25, 19; m 43; c 3. RADIOCHEMISTRY, NUCLEAR WASTE MANAGEMENT. *Educ:* Reed Col, BA, 41. *Prof Exp:* Res chemist, Nat Defense Res Comt, Northwestern Univ, 42-43, Calif Inst Technol, 43, Clinton Labs, Tenn, 43-44, Metall Lab, Univ Chicago, 44-45, Monsanto Chem Co, Ohio, 45 & Universal Oil Prod Co, Ill, 45-47; div head, 47-80, chemist, Brookhaven Nat Lab, 47-82; PROPRIETOR, DORILU ASSOC, 87- *Concurrent Pos:* USAEC consult, Yugoslavia & Israel, 60 & rep, regional meetings utilization res reactors, Int Atomic Energy Agency, Manila, 63, Bombay, 64; Indian Atomic Energy Estab & Int Atomic Energy Agency lect prod radioisotopes, Bombay, 64. *Honors & Awards:* Distinguished Serv Award, Am Nuclear Soc, 69; Cert Appreciation, Am Nuclear Soc, 78. *Mem:* Emer mem Am Chem Soc; assoc mem Sigma Xi. *Res:* Production of radioisotopes; spallation reactions; radionuclide generators; design of radioactive laboratories and equipment; use of teeth as indicators of concentrations of trace elements in the human body; health effects of photovoltaic materials. *Mailing Add:* 13769 Exotica Lane Wellington FL 33414

STANG, PETER JOHN, b Nurnberg, Ger, Nov 17, 41; US citizen; m 69; c 2. ORGANIC CHEMISTRY. *Educ:* DePaul Univ, BS, 63; Univ Calif, Berkeley, PhD(chem), 66. *Prof Exp:* NIH fel chem, Princeton Univ, 66-68, instr, 68-69; from asst prof to assoc prof, 69-79, PROF CHEM, UNIV UTAH, 79-, CHMN DEPT, 89- *Concurrent Pos:* Assoc ed, J Am Chem Soc, 82; Lady Davis Fel, Haifa, Israel; Fulbright sr scholar, Yugoslavia, 87-88. *Honors & Awards:* Alexander von Humboldt US sr scientist award, 77; Mendeleev Lectr, USSR, 89. *Mem:* Am Chem Soc; Chem Soc; fel AAAS; fel Japan Soc Prom Sci. *Res:* Generation, nature and chemistry of unsaturated reactive intermediates (vinyl cations, carbenes, ylides and strained ring compounds; mechanism of metal mediated vinylic couplings, novel transition metal complexes; preparation and uses of alkynyl esters, and alkynyliodonium species, enzyme inhibition (new suicide substrates and their mode of action), novel antitumor agents. *Mailing Add:* Dept Chem Univ Utah Salt Lake City UT 84112

STANG, ROBERT GEORGE, b Los Angeles, Calif, June 20, 38; m 64. MATERIALS SCIENCE ENGINEERING. *Educ:* Long Beach State Col, BS, 61; Univ Calif, Los Angeles, MS, 65; Stanford Univ, PhD(mat sci & eng), 72. *Prof Exp:* Instr mech eng, Long Beach State Col, 65-66; res asst mat sci, Dept Mat Sci & Eng, Stanford Univ, 66-71; asst prof, Inst Mil Engenharia, Rio de Janeiro, 71-72; res assoc, Dept Mat Sci & Eng, Stanford Univ, 72-73; asst prof, 73-79, ASSOC PROF, DEPT MAT SCI & ENG, UNIV WASH, 79- *Concurrent Pos:* Inco fel, Dept Mat Sci & Eng, Stanford Univ, 66-68; sr Fulbright-Hayes researcher & lectr, Montanuniversitat, Austria, 80-81; consult, US Coast Guard, 78, Hewlett Packard, Boise Div, 79 & Battelle Pac Northwest Labs, 78-83; assoc prog dir, metall prog, div mat res, Nat Sci Found, Washington DC, 84-85, prog dir, 85-86. *Mem:* Am Soc Metals; Metall Soc Am; Inst Mining Metall & Petrol Engrs; Sigma Xi; Mat Res Soc; Am Soc Eng Educ. *Res:* Structure-property relationships in materials; deformation at ambient and high temperatures in metals, alloys and ceramics; fatigue and fracture; effect of micro-structure on deformation and fracture. *Mailing Add:* Dept Mat Sci & Eng FB-10 Univ Wash Seattle WA 98195

STANGE, HUGO, b Elizabeth, NJ, June 24, 21; m 42; c 5. ORGANIC CHEMISTRY, INORGANIC CHEMISTRY. *Educ:* Northwestern Univ, BS, 42, PhD(chem), 50. *Prof Exp:* Chemist, Pa Ord Works, US Rubber Co, 42; chemist res dept, Olin Mathieson Chem Corp, 50-52, sect leader, 52-55; mgr org res, FMC Corp, 55-60, mgr org & polymer res, 60-62, res mgr, 62-65, asst dir, Cent Res Dept, 65-72, dir, Princeton Ctr Tech Dept, 72-82; RETIRED. *Concurrent Pos:* Consult, 83-; pres, Princeton Chap Sigma Xi, 88-89. *Mem:* Asn Res Dirs (pres, 82-82); Am Chem Soc; Royal Soc Chem; Am Inst Chemists; Am Inst Chem Eng; Sigma Xi. *Res:* Thianaphthene and boron chemistry; industrial process and product development in organic and inorganic chemistry; agricultural pesticides; polymers; chemical research management; general technical management; government research contract management. *Mailing Add:* 19 Hamilton Ave Princeton NJ 08542

STANGE, LIONEL ALVIN, b Los Angeles, Calif, June 27, 35; m 67; c 2. SYSTEMATIC ENTOMOLOGY. *Educ:* Univ Calif, Berkeley, BS, 58; Univ Calif, Davis, MS, 60, PhD(entom), 65. *Prof Exp:* Prof entom, Nat Univ Tucuman, Arg, 65-78; TAXON ENTOMOLOGIST, BUR ENTOM, DIV PLANT INDUST & CONSUMER SERV, DEPT AGR, FLA, 78 - *Concurrent Pos:* Investr entom, Miguel Lillo Found, Tucuman, Arg, 65-78; grants, Sigma Xi, 68, Nat Coun Res Technol, Buenos Aires, 70-75 & Nat Geog Soc, 75-77; vis curator, Mus Comp Zool, Harvard Univ, 70; vis prof, North East Univ Corrientes, Arg, 74; Smithsonian fel, 89. *Honors & Awards:* Ellsworth Award, Am Mus Natural Hist, NY, 60; Nat Geog Soc Grant, 74-76. *Mem:* Nat Geog Soc; Sigma Xi. *Res:* Biosystematics of the Neuroptera especially Myrmeleontidae (world) and of Hymenoptera (Eumenidae and Megachilidae) of the western hemisphere. *Mailing Add:* Bur Entom Fla Dept Agr Gainesville FL 32602

STANGEBY, PETER CHRISTIAN, b Can, Sept 6, 43; m 64; c 2. FUSION ENERGY. *Educ:* Univ Toronto, BSc, 66, MSc, 67; Oxford Univ, dipl sci, 68, DPhil(plasma physics), 71. *Prof Exp:* PROF PLASMA PHYSICS & FUSION, INST AEROSPACE STUDIES, UNIV TORONTO, 72- *Concurrent Pos:* Sci consult, Joint Europ Tours Fusion Energy Proj Europ Community, 84- & Princeton Plasma Physics Lab, 83-84 & 90-91. *Mem:* Am Vacuum Soc; Can Asn Physicists. *Res:* Fusion energy particularly edge studies of tokamaks; modeling impurity behavior. *Mailing Add:* Inst Aerospace Studies Univ Toronto 4925 Dufferin St Downsview ON M3H 5T6 Can

STANGER, ANDREW L, b Boulder, Colo, Apr 12, 48; m. COMPUTER SCIENCE, SOLAR PHYSICS. *Educ:* Univ Colo, BA, 71. *Prof Exp:* Comput programmer signal processing, Naval Undersea Ctr, 71-72; res asst nuclear eng, Gen Atomic Co, 72-74; engr & scientist sci comput, TRW Systs Group, Inc, 74-75; sci programmer solar physics, 75-85, ASSOC SCIENTIST SOLAR PHYSICS, HIGH ALTITUDE OBSERV, NAT CTR ATMOSPHERIC RES, 85- *Honors & Awards:* Res support Award, Solar Maximum Mission Satellite, Nat Ctr Atmospheric Res, 80. *Res:* Image processing; scientific analysis of solar corona images; spacecraft control software. *Mailing Add:* High Altitude Observ Nat Ctr Atmospheric Res 1850 Table Mesa Dr PO Box 3000 Boulder CO 80307

STANGER, PHILIP CHARLES, b Newark, NJ, Nov 11, 20; m 43; c 2. ASTRONOMY. *Educ:* Montclair State Teachers Col, AB, 42; Okla Agr & Mech Col, MS, 49; Ohio State Univ, MA, 54. *Prof Exp:* Instr math, Okla Agr & Mech Col, 46-48; instr, Ohio Univ, 48-50; from instr to asst prof, 52-59, from asst prof astron to assoc prof, 59-74, PROF ASTRON, OHIO WESLEYAN UNIV, 74- CHMN DEPT, 59- *Mem:* AAAS; Am Astron Soc. *Res:* Spectroscopic binaries; stellar atmospheres. *Mailing Add:* 22 Gniswald St Delaware OH 43015

STANGHELLINI, MICHAEL EUGENE, b San Francisco, Calif, Mar 21, 40; m 66; c 2. PLANT PATHOLOGY. *Educ:* Univ Calif, Davis, BA, 63; Univ Hawaii, MS, 65; Univ Calif, Berkeley, PhD(plant path), 69. *Prof Exp:* Asst prof, 69-72, assoc prof, 72-81, PROF PLANT PATH, UNIV ARIZ, 81- & RES SCIENTIST PLANT PATH, AGR EXP STA, 77- *Mem:* Am Phytopath Soc; Sigma Xi. *Res:* Soil borne fungal pathogens. *Mailing Add:* 750 Ko Vaya Dr Tucson AZ 85704

STANIFORTH, DAVID WILLIAM, agronomy; deceased, see previous edition for last biography

STANIFORTH, RICHARD JOHN, b Sidmouth, Eng, Oct 2, 46; Brit & Can citizen; m 71; c 4. WEED SCIENCE, PLANT REPRODUCTION. *Educ:* Univ Col NWales, BSc, 68; Univ Western Ont, PhD(plant sci), 75. *Prof Exp:* Lectr biol, Univ Western Ont, 73-75; asst prof, 75-82, ASSOC PROF ECOL, UNIV WINNIPEG, 82- *Concurrent Pos:* Sci dir field sta, Churchill Northern Studies Ctr, Man, 80-81; adj prof, dept bot, Univ Man, 80-; vis prof, Nfld Forest Res Ctr, Environ Can, St Johns, 81-82. *Mem:* Can Bot Asn. *Res:* Seed ecology of Poygonum (smart weed) species; flouride air pollutants and seed production in boreal forest plants; reproduction and survival in Opuntia fragilis; ecology of subarctic estuaries. *Mailing Add:* Dept Biol Univ Winnipeg 515 Portage Ave Winnipeg MB R3B 2E9 Can

STANIFORTH, ROBERT ARTHUR, b Cleveland, Ohio, Oct 5, 17; m 44; c 3. INORGANIC CHEMISTRY. *Educ:* Case Western Reserve Univ, BA, 39; Ohio State Univ, MS, 42, PhD(inorg chem), 43. *Prof Exp:* Asst chem, Ohio State Univ, 39-43; res chemist, Monsanto Chem Co, Ohio, 44-46, group leader, 47, sect chief, AEC, Mound Lab, 48, res dir, 48-54, mgr chem develop, Inorg Chem Div, 54-59, asst dir develop, 59-62; mgr prod planning, Monsanto Indust Chem Co, 62-69, mgr commun & info, 69-75; RETIRED. *Concurrent Pos:* Res chemist, Gen Aniline & Film Corp, Pa, 43. *Mem:* Am Chem Soc; Electrochem Soc. *Res:* Ultramicrobalances; chelate compounds of the rare earth metals; radiochemistry; metals; semiconductors. *Mailing Add:* 1215 Walnut Hill Farm Dr St Louis MO 63005

STANIONIS, VICTOR ADAM, b New York, NY, Dec 24, 38; m 60; c 2. APPLIED MATHEMATICS. *Educ:* Iona Col, BS, 60; NY Univ, MS, 64; Queen's Col, MA, 70; Columbia Univ, PhD(math), 75. *Prof Exp:* Instr math & Physics, Iona Col, 61-66, from asst prof to assoc prof physics, 66-89, chmn dept, 75-82, PROF PHYSICS, IONA COL, 90- *Mem:* Am Asn Physics Teachers; Math Asn Am; Nat Coun Teachers Math. *Res:* Computers in physics teaching; nature of problem solving in physics and mathematics. *Mailing Add:* Dept Physics Iona Col 715 North Ave New Rochelle NY 10801

STANISLAO, BETTIE CHLOE CARTER, b Alexandria, La, June 12, 34; m 60. NUTRITION, FOOD SERVICE SYSTEMS. *Educ:* Northwestern State Univ, La, BS, 56; Pa State Univ, MSc, 60; Case Western Reserve Univ, DPhil, 76. *Prof Exp:* Therapeut dietitian, Baptist Hosp, Alexandria, 56-57; asst hotel & inst admin, Pa State Univ, 57-59; dietetic intern, Barnes Hosp, St Louis, Mo, 59-60; chief therapeut dietitian, Pawtucket Mem Hosp, 61-63; therapeut dietitian, Good Samaritan Hosp, Phoenix, 63-64; nutrit asst, Coop Exten Serv, RI, 64-65; asst prof food & nutrit, Univ RI, 65-71; asst nutrit, Case Western Reserve Univ, 72-73; assoc prof food & nutrit & chairperson dept, 76-80, SYSTS ANALYST & COMPUT SUPVR, FOOD SERVS, NDAK STATE UNIV, 84- *Concurrent Pos:* Nutrit adv, Child Develop Ctr, Univ RI, 65-71; support serv contract, Food & Nutrit Serv, Nutrit Educ & Training Prog, USDA, 78-80; dietition, Fargo Diabetes Educ Ctr, 80- *Mem:* Am Dietetic Asn; Am Home Econ Asn; Soc Nutrit Educ; Am Asn Univ Women; Am Diabetes Asn; Am Asn Diabetes Educrs. *Res:* Food habits of college men; effectiveness of nutrition counseling in changing food habits of pedodontic patients; nutrition in preventive dentistry; nutrition in diabetic care; computer food service systems. *Mailing Add:* 3520 Longfellow Rd Fargo ND 58102

STANISLAO, JOSEPH, b Manchester, Conn, Nov 21, 28; m 60. INDUSTRIAL ENGINEERING. *Educ:* Tex Tech Col, BS, 57; Pa State Univ, MS, 59; Columbia Univ, DEngSc, 70. *Prof Exp:* Asst prof indust eng, NC State Col, 59-61; dir res & develop, Darlington Fabrics Corp, 61-62, actg plant mgr, 62; from asst prof to assoc prof indust eng, Univ RI, 63-71; prof & chmn dept, Cleveland State Univ, 71-75; DEAN & PROF, COL ENG & ARCHIT, NDAK STATE UNIV, 75- *Concurrent Pos:* Res grants, Am Soc Mfg & Tool Eng, US Steel Co & Gen Elec Co, 65-66, Naval Air Syst Command, 68-71; lectr, Indust Eng Dept, Columbia Univ, 66-67; consult, Asian Productivity Orgn, 72. *Mem:* AAAS; sr mem Am Inst Indust Engrs; Am Soc Metals; Am Soc Mech Engrs; Am Soc Eng Educ; Sigma Xi. *Res:* Manufacturing engineering, technical aspects, economic considerations and organizational theory; machinability, instrumentation and nondestructive testing techniques. *Mailing Add:* Col of Eng & Archit NDak State Univ Sta Fargo ND 58102

STANISZ, ANDRZEJ MACIEJ, b Cracow, Poland, Nov 20, 61; Can citizen; m 78; c 2. NEUROIMMUNOLOGY, IMMUNOPHARMACOLOGY. *Educ:* Univ Cracow, MS, 74, PhD(immunol), 77; Med Acad Cracow, MD. *Prof Exp:* Asst prof cell biol, Univ Cracow, Poland, 77-79; res assoc immunol, Dept Micro & Immunol, Wash Univ, 79-83; asst prof med, 83-89, ASSOC PROF IMMUNOL & PATH, MCMASTER UNIV, 89- *Mem:* Am Asn Immunologists; NY Acad Sci; Int Soc Immunopharmacol; Int Soc Neuroimmunomodulation; Soc Mucosal Immunol; Can Asn Gastroenterol. *Res:* Interactions between nervous and immune system in health and diseases; gastrointestinal and rheumatic diseases. *Mailing Add:* Dept Path Intestine Dis Res Unit McMaster Univ 1200 Main St W Hamilton ON L8N 3Z5 Can

STANITSKI, CONRAD LEON, b Shamokin, Pa, May 3, 39; m 63; c 2. CHEMICAL EDUCATION. *Educ:* Bloomsburg State Col, BSEd, 60; State Col Iowa, MA, 64; Univ Conn, PhD(inorg chem), 71. *Prof Exp:* Teacher high sch, Pa, 60-63 & Goshen Cent Sch, 64-65; instr chem, Edinboro State Col, 65-67; teaching fel, Univ Conn, 70-71; asst prof, Ga State Univ, 71-75; assoc prof chem, Kennesaw Jr Col, 75-76; assoc prof chem, 76-80, CHMN DEPT, RANDOLPH-MACON COL, 76-, PROF, 80-; DEAN, MT UNION COL, 88. *Concurrent Pos:* W Nelson Gray Distinguished Prof, 83. *Honors & Awards:* Gustav Ohauv Award, Creative Col Sci Teaching, 73. *Mem:* Am Chem Soc; Sigma Xi; AAAS. *Res:* Solid state hydride synthesis and reaction studies; chemical education. *Mailing Add:* Mt Union Col Alliance OH 44601

STANKIEWICZ, RAYMOND, b Brooklyn, NY, Sept 3, 32; m 55; c 3. INDUSTRIAL ENGINEERING DESIGN. *Educ:* Allied Inst Technol, BS, 67; Am Western Univ, MS, 82. *Prof Exp:* Founder, owner, engr & model & pattern maker, Am Eng Model Co, 59-84; prog mgr, tool design mgr & machine shop mgr, Russell Plastics Technol, 84-87; sr develop & tooling engr & qual control mgr, Symbol Technol Inc, 87-88; CONSULT, DESIGN & MANUFACTURE NEW PRODS, LONG ISLAND, MANHATTAN, CONN & NJ, 88- *Honors & Awards:* Gen Elec Innovation Award, 87. *Mem:* Inst Indust Engrs; Soc Plastic Engrs; Soc Mech Engrs; Soc Am Mil Engrs; Soc Mfg Engrs. *Res:* Mold design for injection, compression, rotational, blow, bag, high-temp, etc; pattern making and model making. *Mailing Add:* Am Eng Model Co 866 Bohemia Pkwy Bohemia NY 11716

STANKO, JOSEPH ANTHONY, b Wilkes-Barre, Pa, July 2, 41; m 62; c 3. INORGANIC CHEMISTRY. *Educ:* King's Col, BS, 62; Univ Ill, PhD(inorg chem), 66. *Prof Exp:* Asst prof chem, Pa State Univ, 66-73; ASSOC PROF CHEM, UNIV S FLA, TAMPA, 73- *Mem:* Am Chem Soc; The Chem Soc; Am Crystallog Asn. *Res:* X-ray crystallography; chemistry of platinum anti-cancer drugs; synthesis of 1-dimensional conductors. *Mailing Add:* Dept of Chem Univ of SFla Tampa FL 33620

STANKOVICH, MARIAN THERESA, b Houston, Tex, Nov 14, 47. ELECTROCHEMISTRY, BIOCHEMISTRY. *Educ:* Univ St Thomas, Tex, BA, 70; Univ Tex, Austin, PhD(anal chem), 75. *Prof Exp:* Scholar biochem, Univ Mich, 75-77; asst prof anal chem, Univ Mass, Amherst, 77-80; MEM FAC, CHEM DEPT, UNIV MINN, 80- *Concurrent Pos:* Fac res grant, Univ Mass, 77-78; Cottrell Corp res grant, 78-79. *Mem:* Am Chem Soc; Electrochem Soc. *Res:* Spectral and electrochemical study of electron transfer in flavoproteins, riboflavin, and flavin analogs; parameters studied are redox potentials; number of electrons transferred in a reaction; kinetics of electron transfer. *Mailing Add:* 207 Pleasant St SE Chem Dept Univ Minn Minneapolis MN 55455-0431

STANLEY, DANIEL JEAN, b Metz, France, Apr 14, 34; US citizen; m 60; c 3. MARINE GEOLOGY. *Educ:* Cornell Univ, BSc, 56; Brown Univ, MSc, 58; Univ Grenoble, DSc, 61. *Prof Exp:* Res geologist, French Petrol Inst, 58-61; geologist, Pan-Am Petrol Corp, 61-62; asst to dir geol, US Army Engrs Waterways Exp Sta, 62-63; asst prof sedimentol, Univ Ottawa, 63-64; asst prof marine geol, Dalhousie Univ, 64-66; assoc cur sedimentol, 66-68, supvr div, 68-71, cur sedimentol, 68-71, geol oceanogr, 71-79, SR SCIENTIST, SMITHSONIAN INST, 79- *Concurrent Pos:* post-doctoral, Woods Hole Oceangraphic Inst, 63; Nat Res Coun Can travel award, USSR, 66; founder & ed, Maritime Sediments J, 64-66; Nat Acad Sci exchange award, Poland, 67 & Romania, 76; adj prof, Univ Maine, Orono, 74-, Nat Sch; Petrol, France, 78-; adv, Int Court Justice, Hague, 81. *Honors & Awards:* Francis Shepard Medal Excellence Marine Geol, 90. *Mem:* Fel AAAS; fel Geol Soc Am; Soc Econ Paleontologists & Mineralogists; Am Mineralogists; Am Asn Petrol Geologists; corresp mem Geol Soc Belg; Sigma Xi. *Res:* Sedimentology of Nile Delta flysch in Alps, Carpathians and Caribbean; modern slope, canyon and deep sea fan deposits; marine geology studies of Nova Scotian shelf, Bermuda, northwestern Atlantic and the Mediterranean; directs multi-national studies of Nile delta of Egypt; author or co-editior of 8 books and over 200 articles in scientific journals. *Mailing Add:* Div Sedimentol E-109 NMNH Smithsonian Inst Washington DC 20560

STANLEY, DAVID WARWICK, b Muncie, Ind, Oct 12, 39. FOOD SCIENCE. *Educ:* Univ Fla, BS, 62, MS, 63; Univ Mass, PhD(food sci), 67. *Prof Exp:* Res fel, Smith Col, 67-68; asst prof food sci, Univ Toronto, 68-70; from asst prof to assoc prof, 70-79, PROF FOOD SCI, UNIV GUELPH, 79- *Concurrent Pos:* Ed, Can Inst Food Sci & Tech Jour, 76-82. *Mem:* Inst Food Technol; fel Can Inst Food Sci & Technol. *Res:* Animal protein systems including meat texture and muscle protein biochemistry; cell membranes including isolation, composition, structure and function; plant protein systems including food uses of plant proteins; food analysis; postharvest physiology of fruits and vegetables. *Mailing Add:* Dept Food Sci Univ Guelph Guelph ON N1G 2W1 Can

STANLEY, EDWARD LIVINGSTON, b Orange, NJ, Sept 6, 19; m 43; c 2. CHEMISTRY. *Educ:* Princeton Univ, AB, 40, MA, 43, PhD(chem), 47. *Prof Exp:* Asst, Princeton Univ, 40-41, res assoc anal chem, Off Sci Res & Develop & Manhattan Dist Proj, 41-43; supvr, Anal Lab, Rohm & Haas Co, 43-50, lab head, Res Div, 50-57, foreign area supvr, 57-76; CONSULT, 76- *Mem:* AAAS; Am Chem Soc; Sigma Xi. *Res:* Analytical chemistry; facility is mass-burning, generating electricity. *Mailing Add:* 124 Plymouth Rd Gwynedd Valley PA 19437

STANLEY, EVAN RICHARD, b Sydney, Australia, 1944; m; c 2. CELL BIOLOGY, MEDICAL RESEARCH. *Educ:* Univ Western Australia, BSc, 67; Univ Melbourne, PhD(med biol), 70. *Prof Exp:* Fel med biol, Walter & Eliza Hall Inst Med Res, Melbourne, Australia, 70-72; lectr cell biol, Dept Med Biophys, Univ Toronto, 72-73, asst prof, 73-76; from asst prof to prof, depts microbiol, immunol & cell biol, 76-87, PROF & CHMN DEPT DEVELOP BIOL & CANCER, ALBERT EINSTEIN COL MED, 87- *Concurrent Pos:* Mem sr sci staff, Ont Cancer Inst, 72-76. *Res:* Biochemical and genetic studies on hemopoietic growth factor action. *Mailing Add:* Dept Biol & Cancer Albert Einstein Col Med 1300 Morris Park Ave Bronx NY 10461

STANLEY, GEORGE GEOFFREY, b Palmerton, Pa, May 2, 53; m; c 1. HOMOGENEOUS CATALYSIS. *Educ:* Univ Rochester, BS, 75; Tex A&M Univ, PhD(chem), 79. *Prof Exp:* Fel, Univ Louis Pasteur, France, 79-81; ASST PROF INORG, WASHINGTON UNIV, ST LOUIS, 81-; ASST PROF, LOUISIANA STATE UNIV. *Concurrent Pos:* Fel, NATO, 79 & Nat Ctr Sci Res, France, 81. *Mem:* Am Chem Soc. *Res:* Synthesis, structure and reactivity of bi-and poly-metallic transition metal compounds with particular emphasis on homogeneous catalytic reactions involving hydrogen, carbon monoxide, nitrogen and organic substrates. *Mailing Add:* Dept Chem La State Univ Baton Rouge LA 70803-1804

STANLEY, GEORGE M, b Detroit, Mich, Mar 15, 05. HISTORY OF GREAT LAKES, DEATH VALLEY NAT PARK. *Educ:* Univ Mich, BS, 28, MS, 32 & PhD(geol), 36. *Prof Exp:* Prof geol, Univ Mich, 32-48; prof geol, Univ Calif, Fresno, 48-67; RETIRED. *Mem:* Geol Soc Am; Sigma Xi. *Mailing Add:* 545 E Buckingham Way Fresno CA 93704

STANLEY, GERALD R, b Niles, Mich, Nov 14, 43; m 66; c 2. COMPUTER SIMULATION POWER CIRCUITS & APPLICATIONS, DESIGNER OF TIME DELAY SPECTROMETRY INSTRUMENTATION. *Educ:* Mich State Univ, S, 65; Univ Mich, MSE, 66. *Prof Exp:* MGR RES & RESOURCES, CROWN INT, 66- *Mem:* Inst Elec & Electronics Engrs; Acoust Soc Am; Asn Comput Mach; Audio Eng Soc. *Res:* Circuit designer of power amplifiers and instrumentation for audio and industrial applications; invented/developed technology used for gradient amplifiers used in whole-body magnetic resonance imaging. *Mailing Add:* 1718 W Mishawaka Rd Elkhart IN 46517

STANLEY, H(ARRY) EUGENE, b Norman, Okla, Mar 28, 41; m 67; c 3. STATISTICAL MECHANICS, BIOLOGICAL PHYSICS. *Educ:* Wesleyan Univ, BA, 62; Harvard Univ, PhD(physics), 67. *Prof Exp:* Staff mem solid state theory group, Lincoln Lab, Mass Inst Technol, 67-68; fel physics, Miller Inst Basic Res Sci, Univ Calif, Berkeley, 68-69; from asst prof physics to assoc prof, Mass Inst Technol, 71-76, Hermann von Helmholtz assoc prof health sci & technol, 73-76; PROF PHYSICS, BOSTON UNIV, 76-, PROF PHYSIOL, SCH MED & DIR, CTR POLYMER STUDIES, 78-, UNIV PROF, 79- *Concurrent Pos:* NSF fel theoret physics, 62-66; consult, Lincoln Lab, Mass Inst Technol, 69-71; vis prof physics, Osaka Univ, 75, Univ Toronto, 77, Sch Physics & Chem, 79 & Peking Normal Univ & Nanking Univ, 81; John Simon Guggenheim Mem fel, 79-81. *Honors & Awards:* Joliot-Curie Medal, 79. *Mem:* AAAS; fel Am Phys Soc; Biophys Soc; NY Acad Sci. *Res:* Phase transitions and critical phenomena; biomedical physics; polymer physics; physics of random media; percolation; liquid state physics; cooperative functioning of polymers and other systems with no underlying lattice; fractals in biology and medicine. *Mailing Add:* Dept Physics Boston Univ Boston MA 02215

STANLEY, HAROLD RUSSELL, b Salem, Mass, June 26, 23; m 46; c 3. ORAL PATHOLOGY. *Educ:* Univ Md, DDS, 48; Am Univ, BS, 52; Georgetown Univ, MS, 53; Am Bd Oral Path, dipl, 57. *Prof Exp:* Intern, Marine Hosp, USPHS, Baltimore, Md, 48-49; resident oral path, Armed Forces Inst Path, 51-53; mem staff, Nat Inst Dent Res, 53-66, clin dir, 66-68; PROF PATH, UNIV FLA, 81-, CHMN DEPT ORAL MED, 70- *Concurrent Pos:* Hon prof, San Carlos, Univ Guatemala, 60- *Honors & Awards:* Sci Award, Int Asn Dent Res, 78. *Mem:* Hon fel Am Asn Endodont; Int Asn Dent Res; Am Dent Asn; Am Acad Oral Path (pres, 67). *Res:* Diseases of the human dental pulp; periodontium; oral mucous membranes. *Mailing Add:* 2C Sea Oats Ormond Beach FL 32176

STANLEY, HUGH P, b Modesto, Calif, July 14, 26; m 59; c 2. ELECTRON MICROSCOPY, CELL BIOLOGY. *Educ:* Univ Calif, Berkeley, BA, 51; Ore State Univ, MA, 58, PhD(zool), 61. *Prof Exp:* NIH fel zool, Zool Sta, Naples, Italy, 61-63 & Cornell Univ, 63 & sr fel biol struct, Univ Wash, 63-65; asst prof anat, Univ Minn, 65-66; from asst prof to assoc prof zool, 66-76, prof biol, 76-87, EMER PROF BIOL, UTAH STATE UNIV, 87- *Mem:* AAAS; Am Soc Zool. *Res:* Ultrastructure of developing cell systems, especially vertebrate spermatid differentiation. *Mailing Add:* 3310 D Bailer Hill Rd Friday Harbor WA 98250

STANLEY, JOHN PEARSON, b Washington, DC, Dec 17, 15; m 41; c 7. BIOCHEMISTRY. *Educ:* Cath Univ Am, BS, 37, MS, 39, PhD(biochem), 42. *Prof Exp:* Chemist & spec agt, Fed Bur Invest, Washington, DC, 41-44; res chemist, Gelatin Prod Co, Mich, 44, head bact labs, 44-46, asst chief control dept & res & develop labs, 46; dir res & develop, R P Scherer Corp, 46-68, tech dir, 68-79; RETIRED. *Mem:* Fel AAAS; Am Chem Soc; Am Pharmaceut Asn; fel Am Inst Chemists; NY Acad Sci; Sigma Xi. *Res:* Soft gelatin capsules; gelatin; vitamins; nutrition; research and development administration; product development; technical service. *Mailing Add:* 1032 Yorkshire Rd Grosse Pointe MI 48230

STANLEY, JON G, b Edinburg, Tex, Oct 28, 37; m 65; c 3. AQUACULTURE. *Educ:* Univ Mo, AB, 60, BS, 63, PhD(zool), 66. *Prof Exp:* Asst prof biol, DePaul Univ, 66-69; from asst prof to assoc prof, Univ Wis-Milwaukee, 69-72; fisheries biologist & leader, Maine Coop Fishery Res Unit, Univ Maine, Orono, 72-; CTR DIR, US FISH & WILDLIFE SERV, NAT FISHERIES RES CTR, GREAT LAKES. *Concurrent Pos:* Nat Acad Sci exchange scholar, Czech. *Mem:* AAAS; Am Soc Zoologists; Am Fisheries Soc. *Res:* Polyploidy and genetics in aquaculture; biology of Chinese fishes such as Grass Carp; gynogenesis and breeding of freshwater fish; environmental effects of pesticides on fishes; acid rain effects on fishes. *Mailing Add:* Zool Dept Univ Maine Orono ME 04473

STANLEY, KENNETH EARL, b Auburn, NY, Nov 7, 47; m 71; c 2. BIOSTATISTICS. *Educ:* Alfred Univ, BA, 69; Bucknell Univ, MA, 70; Univ Fla, PhD(statist), 74. *Prof Exp:* Res asst prof statist, State Univ Ny, Buffalo, 75-77; BIOSTATISTICIAN, SIDNEY FARBER CANCER INST, 77-; ASST PROF BIOSTATIST, HARVARD UNIV, 77- *Concurrent Pos:* Statistician, Ludwig Lung Cancer Study Group, 77-; coord statistician, Eastern Coop Oncol Group, 78-80; mem expert adv panel cancer, WHO, consult, 81-; co-dir, Collaborating Ctr Cancer Biostatistics Eval, Harvard Sch Pub Health, WHO. *Mem:* Am Statist Asn; Biomet Soc; Soc Clin Trials; Int Asn Study Lung Cancer. *Res:* Clinical trials in cancer. *Mailing Add:* 104 Waterhill St Lynn MA 01905

STANLEY, LUTICIOUS BRYAN, JR, b Atlanta, Ga. TECHNICAL MANAGEMENT, REGIONAL PROJECT ADMINISTRATION. *Educ:* Southern Tech Inst, BCET, 74, BMET, 82; Ga State Univ, MS, 86. *Prof Exp:* Field engr civil eng, Jordan Jones & Goulding, Inc, 74-79; proj engr, Mayes, Sudderth & Etheredge, Inc, 79-82; asst regional proj mgr, Westinghouse Elec Corp, 82-88; PRIN, L B S INTERPRICES, 89- *Mem:* Am Soc Mech Engrs. *Res:* Improvement of the operation of turbines at nuclear power plants. *Mailing Add:* PO Box 5386 WSB Gainesville GA 30501

STANLEY, MALCOLM MCCLAIN, b Henderson, Ky, Mar 2, 16; m 43; c 2. MEDICINE. *Educ:* Centre Col, AB, 37; Univ Louisville, MD, 41; Am Bd Internal Med & Am Bd Gastroenterol, dipl. *Prof Exp:* From intern to asst resident med, Gallinger Munic Hosp, DC, 41-43; from asst resident to chief resident, Evans Mem Hosp, Boston, 43-46; from instr to assoc prof, Med Sch, Tufts Univ, 46-57; prof exp med, Sch Med, Univ Louisville, 57-59, prof med, 59-62; prof med, Univ Ill Col Med, 62-82; prog dir, Clin Res Ctr, 62-70, SECT CHIEF GASTROENTEROL, HINES VET ADMIN HOSP, 70-; PROF MED, STRITCH SCH MED, LOYOLA UNIV, 82- *Concurrent Pos:* Am Cancer Soc fel, Joseph H Pratt Diag Hosp, Boston, 46-48; res assoc, Pratt Diag Hosp, Boston, 48, mem staff, 49-57. *Mem:* Am Soc Clin Invest; Am Asn Study Liver Dis; Am Physiol Soc; Am Gastroenterol Asn; Cent Soc Clin Res. *Res:* Gastroenterology, especially intestinal absorption and secretion; bile salt metabolism; liver cirrhosis; ascites. *Mailing Add:* Hines Vet Admin Hosp PO Box 23 Hines IL 60141

STANLEY, MELISSA SUE MILLAM, b South Bend, Wash, June 23, 31; div. MEDICAL TECHNOLOGY. *Educ:* Univ Ore, BS, 53, MA, 59; Univ Utah, PhD(zool, entom), 65. *Prof Exp:* Med technologist, Hosps & Labs, 53-57; teaching asst biol, Univ Ore, 57-58; med technologist, Hosps & Labs, 58-59; instr, Westminster Col, Utah, 59-61, asst prof, 61-63, actg chmn dept, 59-60, chmn dept, 60-63; res assoc, Pioneering Lab Insect Path, USDA, 65-67; from asst prof to assoc prof biol, 67-74, prog coordr, 68-69, PROF BIOL, GEORGE MASON UNIV, 74- *Mem:* Am Soc Zoologists; AAAS. *Res:* Arthropod tissue culture; textbook author on general biology. *Mailing Add:* Dept Biol George Mason Univ Fairfax VA 22030

STANLEY, NORMAN FRANCIS, b Rockland, Maine, May 6, 16; m 63; c 2. CHEMISTRY. *Prof Exp:* Res chemist, Algin Corp Am, 40-53, res dir, 53-59; asst tech dir, Marine Colloids, Inc, 59-64, res chemist, 64-75; sr scientist, Marine Colloids Div, FMC Corp, 74-85; RETIRED. *Mem:* AAAS; Am Chem Soc; Soc Rheol; Am Inst Aeronaut & Astronaut. *Res:* Polysaccharide chemistry; chemistry and technology of marine algae, algal products and watersoluble gums; design and analysis of experiments. *Mailing Add:* PO Box 723 Rockland ME 04841

STANLEY, PAMELA MARY, b Melbourne, Australia, Mar 25, 47; m 70; c 2. CARBOHYDRATE STRUCTURES, SOMATIC CELL GENETICS. *Educ:* Univ Melbourne, BSc Hons, 68, PhD(virol), 72. *Prof Exp:* Fel somatic cell, Univ Toronto, Int, 72-75, res assoc, 75-77; from asst prof to assoc prof, 77-86, PROF CELL BIOLOGY, ALBERT EINSTEIN COL MED, NY, 86- *Concurrent Pos:* Mem grant rev panel, Am Cancer Soc & Palhobiochem study sect, NIH. *Mem:* Am Soc Biol Chemists; Int Asn Women Biochemists. *Res:* Generation of animal cell mutants which express altered carbohydrates at the cell surface to isolate genes coding for glycosylation enzymes, to delineate glycosylation pathways and to study structure; function relationships of cell surface carbohydrates; glycosyltransferase genes; molecular genetics. *Mailing Add:* Dept Cell Biol Albert Einstein Col Med 1300 Morris Park Ave Bronx NY 10461

STANLEY, PATRICIA MARY, b Oneonta, NY, Mar 28, 48; m 77. MICROBIOLOGY, MICROBIAL ECOLOGY. *Educ:* Cornell Univ, BS, 70; Univ Wash, MS, 72, PhD(microbiol), 75. *Prof Exp:* Res specialist, Dept Microbiol, Univ Minn, 76-79; prin microbiologist, 79-86, SCIENTIST, ECOLAB, INC, 86- *Mem:* Am Soc Microbiol; Soc Indust Microbiol; Am Soc Testing & Mat. *Res:* In situ metabolism of nitrifying and heterotrophic bacteria; biology of nitrifying bacteria; use of fluorescent antibody staining in microbial ecology; bacterial adhesion to surfaces; antimicrobial activity of biocides and disinfectants; industrial enzymology. *Mailing Add:* Ecolab Inc Osborn Bldg St Paul MN 55102

STANLEY, RICHARD PETER, b New York, NY, June 23, 44; m 71; c 2. ALGEBRAIC COMBINATORICS, ENUMERATIVE COMBINATORICS. *Educ:* Calif Inst Technol, BS, 66; Harvard Univ, PhD(math), 71. *Prof Exp:* Miller fel math, Miller Inst Basic Res Sci, 71-73; Moore instr, 70-71, from asst prof to assoc prof, 73-79, PROF APPL MATH, MASS INST TECHNOL, 79- *Concurrent Pos:* Res scientist & consult, Jet Propulsion Lab, 65-72; consult, Bell Tel Labs, 73-; Guggenheim, fel, 83-84. *Honors & Awards:* Polya Prize, Soc Indust & Appl Math, 75. *Mem:* Am Math Soc; Math Asn Am. *Res:* Development of a unified foundation to combinatorial theory; interactions between algebra and combinatorics. *Mailing Add:* Dept Math Mass Inst Technol Cambridge MA 02139

STANLEY, RICHARD W, b Milesburg, Pa, Dec 16, 28; m 52; c 3. NUTRITION, BIOCHEMISTRY. *Educ:* Pa State Univ, BS, 56, MS, 58, PhD(dairy sci), 61. *Prof Exp:* Instr nutrit, Pa State Univ, 57-61; from asst prof to assoc prof, Univ Hawaii, 61-70, chmn, Dept Animal Sci, 68-85, prof nutrit, Col Trop Agr, 70-85, prof nutrit & biochem, 85-88, EMER PROF NUTRIT & BIOCHEM, UNIV HAWAII, 85- *Concurrent Pos:* Fel, Univ Mo, 67-68. *Mem:* Am Dairy Sci Asn; Am Soc Animal Sci. *Res:* Ruminant nutrition, including utilization of metabolites formed in the rumen as they influence the productive performance of domestic cattle. *Mailing Add:* 1052 Lunaanela St Kailua HI 96734

STANLEY, ROBERT LAUREN, b Seattle, Wash, Dec 30, 21; m 47; c 2. MATHEMATICS. *Educ:* Univ Wash, BS, 43, MA, 47; Harvard Univ, PhD, 51. *Prof Exp:* Guest lectr philos, Univ BC, 51-52, guest lectr math, 52-54; asst prof, Univ SDak, 54-57; asst prof, Wash State Univ, 57-61; assoc prof, 61-66, PROF MATH, PORTLAND STATE UNIV, 66- & HEAD DEPT MATH, 78- *Mem:* Am Math Soc; Asn Symbolic Logic; Math Asn Am. *Res:* Mathematical logic and foundations; logical analysis in philosophy of science. *Mailing Add:* Dept Math Portland State Univ Box 751 Portland OR 97201

STANLEY, ROBERT LEE, JR, b Dodge Co, Ga, Mar 7, 40; m 68; c 2. AGRONOMY. *Educ:* Univ Ga, BSA, 63, PhD(agron), 69; Clemson Univ MSA, 64. *Prof Exp:* Asst prof, 68-74, ASSOC PROF AGRON, UNIV FLA, 74- *Mem:* Am Soc Agron; Crop Sci Soc Am; Soc Range Mgt. *Res:* Forage crops management and utilization. *Mailing Add:* North Fla Res Educ Ctr Rt 3 Box 4370 Quincy FL 32351

STANLEY, ROBERT WEIR, physics; deceased, see previous edition for last biography

STANLEY, ROLFE S, b Brooklyn, NY, Nov 4, 31; m 52; c 4. GEOLOGY. *Educ:* Williams Col, BA, 54; Yale Univ, MS, 55, PhD(geol), 62. *Prof Exp:* Geologist, Shell Oil Co, 57-59; NSF fel & lectr geol, Yale Univ, 62-64; from asst prof to assoc prof, 64-72, instnl res grant, 64-66, chmn dept, 64-78, PROF GEOL, UNIV VT, 72- *Concurrent Pos:* Res assoc, Ctr Technophysics, Tex A&M Univ, 71 & 72; prin investr, Northern Vt Serpentinite belt; res grant, Nat Res Coun, Repub China. *Mem:* Fel Geol Soc Am. *Res:* Structural geology; structural petrology; regional geology of western New England; compilation of the geological map of Massachusetts; metamorphic core of the central mountains of Taiwan. *Mailing Add:* Dept Geol Univ Vt 85 S Prospect St Burlington VT 05405

STANLEY, RONALD ALWIN, b Edinburg, Tex, June 18, 39; div; c 4. PLANT PHYSIOLOGY. *Educ:* Univ Ark, BS, 61, MS, 63; Duke Univ, PhD(plant physiol), 70; Univ Southern Calif, MPA, 81. *Prof Exp:* Botanist, Tenn Valley Authority, 64-75; asst prof biol, Univ SDak, Springfield, 75; PLANT PHYSIOLOGIST, ENVIRON PROTECTION AGENCY, 76- *Concurrent Pos:* Consult, WHO, 86-87. *Mem:* Ecol Soc of Am; Soc Wetland Scientists; Asn Aquatic Vascular Plant Biologists. *Res:* Synergistic interactions of chemical, physical and biological factors in the environment with aquatic macrophytes; adaptation to the aquatic environment; assessment of risk from toxic substances. *Mailing Add:* 2 Meadow Grass Court Gaithersburg MD 20878

STANLEY, STEVEN MITCHELL, b Detroit, Mich, Nov 2, 41; m 69. PALEONTOLOGY, EVOLUTION. *Educ:* Princeton Univ, AB, 63; Yale Univ, PhD(geol), 68. *Prof Exp:* Asst prof paleont, Univ Rochester, 67-69; from asst prof to assoc prof, 69-74, PROF PALEOBIOL, JOHNS HOPKINS UNIV, 74- *Honors & Awards:* Charles Schuchert Award, Paleont Soc, 77. *Mem:* Paleont Soc; Geol Soc Am; Soc Study Evolution; Am Acad Arts & Sci. *Res:* Functional morphology and evolution of bivalve mollusks and other taxa with fossil records; rates and patterns of evolution and extinction; benthic marine ecology and paleoecology; biomechanics. *Mailing Add:* Dept Earth & Planetary Sci Johns Hopkins Univ 34th & Charles St Baltimore MD 21218

STANLEY, THEODORE H, b New York, NY, Feb 4, 40. ANESTHESIOLOGY. *Educ:* Columbia Univ, MD, 65. *Prof Exp:* PROF ANESTHESIOL, COL MED, UNIV UTAH, 79- *Mailing Add:* Dept Anesthesiol & Surg Univ Utah 50 N Medical Dr Salt Lake City UT 84112

STANLEY, WENDELL MEREDITH, JR, b New York, NY, Nov 9, 32; m 58; c 3. MOLECULAR BIOLOGY, BIOCHEMISTRY. *Educ:* Univ Calif, Berkeley, AB, 57; Univ Wis-Madison, MS, 59, PhD(biochem), 63. *Prof Exp:* From instr to asst prof biochem, Sch Med, NY Univ, 65-67; asst prof, 67-70, ASSOC PROF BIOCHEM, UNIV CALIF, IRVINE-CALIF COL MED, 70- *Concurrent Pos:* USPHS grant, Sch Med, NY Univ, 63-65; assoc dean, Undergrad Affairs, Sch Biol Sci, Univ Calif, Irvine, 80. *Mem:* AAAS; Am Soc Biol Chem. *Res:* Control of protein biosynthesis in eukaryotes. *Mailing Add:* Dept Molecular Biol & Biochem Univ Calif Irvine CA 92717

STANLEY, WILLIAM DANIEL, b Bladenboro, NC, June 13, 37; m 62; c 1. ELECTRICAL ENGINEERING TECHNOLOGY. *Educ:* Univ SC, BS, 60; NC State Univ, MS, 62, PhD(elec eng), 63. *Prof Exp:* Develop engr, Electro-Mech Res, Inc, 63; asst prof elec eng, Clemson Univ, 64-66; assoc prof, Old Dominion Univ, 66-72, chmn, Dept Eng Technol, 70-74, dir, Div Eng Technol, 74-76, grad prog dir, Dept Elec Eng, 76-79, chmn, Dept Elec Eng Technol, 79-90, PROF, OLD DOMINION UNIV, 72-, EMINENT PROF, 85-, CHMN, DEPT ENG TECHNOL, 91- *Concurrent Pos:* Consult, NASA, 67-69. *Mem:* Inst Elec & Electronic Engrs. *Res:* Communications systems analysis and design; network synthesis; radiometer measurements. *Mailing Add:* Dept Eng Tech Old Dominion Univ Col Eng & Tech Norfolk VA 23529

STANLEY, WILLIAM LYONS, b Teh Chou, China, May 30, 16; US citizen; m 41; c 3. ORGANIC CHEMISTRY. *Educ:* Marietta Col, AB, 39; Univ Calif, PhD(chem), 48. *Prof Exp:* Chemist & asst group leader, Carbide & Carbon Chem Corp, WVa, 39-45; res chemist, Western Regional Res Lab, USDA, 48-51 & Res & Develop Ctr, Union Oil Co, 51-54; prin chemist, Fruit & Veg Chem Lab, USDA, 54-60, chief fruit lab, Western Utilization Res & Develop Div, Agr Res Serv, Calif, 60-68; UN develop prog citrus res technologist, Food Inst, Centre Indust Res, Haifa, Israel, 68-69; chief res chemist, Fruit & Veg Processing Lab, Western Regional Lab, 70-76, EMER CONSULT COLLABR, SCI & EDUC ADMIN-AGR RES, 76- *Concurrent Pos:* Mem comt fruit & veg prod, Adv Bd Mil Personnel Supplies, Nat Res Coun-Nat Acad Sci; consult, Almond Bd Calif & Dried Fruit Asn, Calif. *Mem:* Fel Am Chem Soc; Phytochem Soc NAm. *Res:* Synthetic organic chemistry; petrochemicals; chemistry of natural products; flavor components of citrus fruits; immobilized enzymes in food processing. *Mailing Add:* 8545 Carmel Valley Rd Carmel CA 93923

STANNARD, CARL R, JR, b Syracuse, NY, July 24, 35; m 67; c 2. IN-SERVICE EDUCATION, WRITING MATERIALS AT VARIOUS LEVELS. *Educ:* Syracuse Univ, BS, 56, PhD(physics), 64; Cornell Univ, MS, 60. *Prof Exp:* Res assoc physics, Syracuse Univ, 63-64; asst prof, 64-70, ASSOC PROF PHYSICS, STATE UNIV NY, BINGHAMTON, 70- *Concurrent Pos:* IBM prin investr, State Univ NY, Binghamton, 67-69, dir undergrad physics prog, 70-78, proj dir, Physics Technol Ctr, 71- 75, chair physics dept, 78-81; dir, Southern Tier Educators (K-12) Prog Understanding Physics (STEP UP), NY State Dept Educ, 88- *Mem:* Am Phys Soc; Sigma Xi; Am Asn Physics Teachers; Nat Sci Teachers Asn; Am Asn Univ Professors. *Res:* Pedagogy of physics: improvement of methods and structures to provide education in physics to a broader population; at K-12 levels: increased awareness and utility of principles of physics among teachers and students; encouragement of interest in physics engineering technology among women, minorities, disabled, etc; at college level: courses in practical physics literacy for non-science students; inservice training for K-12 teachers to help them become more knowledgeable in science and help them make science more exciting for their students, especially those underrepresented in science. *Mailing Add:* Dept Physics State Univ NY PO Box 6000 Binghamton NY 13902-6000

STANNARD, J NEWELL, b Owego, NY, Jan 2, 10; wid; c 1. TOXICOLOGY. *Educ:* Oberlin Col, BA, 31; Harvard Univ, MA, 34, PhD(gen physiol & biophys), 35. *Prof Exp:* Asst prof pharmacol, Emory Univ Med Sch, 39-41; pharmacologist, USPHS, NIH, Bethesda, Md, 41-44, sr pharmacologist, 46-47; naval officer, Res Div, Bur Med & Surg, Navy Dept, Wash, DC, 44-46; asst & instr physiol, Sch Med & Dent, Univ Rochester, 35-39, asst dir educ, Atomic Energy Proj, 48-52, chief, Radioactive Inhalation Sect, 52-59, assoc dir educ, 59-69, assoc dean grad studies, Med Sch, 59-75, EMER PROF RADIATION BIOL, BIOPHYSICS, PHARMACOL & TOXICOL, SCH MED & DENT, UNIV ROCHESTER; ADJ PROJ COMMUNITY MED & RADIOL, UNIV CALIF, SAN DIEGO, 78- *Concurrent Pos:* Consult, health physics, radiation protection, radiation biol & toxicol, several orgn; mem, Radiation Exposure Adv Comt, US Environ Protection Agency, Nat Res Coun Ocean Affairs Bd-Workshop Transuranics in Environ, Nat Acad Sci & Adv Comt Reactor Safeguards, US Nuclear Regulatory Comn, Wash, DC; chmn, Sci Comt-57, Nat Coun Radiation Protection & Measurements; lectr radiobiol, Traveling Lect Prog, Am Inst Biol Sci. *Honors & Awards:* Distinguished Achievement Award, Health Physics Soc, 77; Parker Lectr, Battelle Pac Northwest Labs, 88; Taylor Lectr, Nat Coun Radiation Protection & Measurements, 90. *Mem:* Am Physiol Soc; Soc Pharmacol & Exp Therapeut; Radiation Res Soc; fel AAAS; Soc Gen Physiologists; Am Indust Hyg Asn; fel Health Physics Soc (pres-elect, 68-69, pres, 69-70); Biophys Soc; Soc Toxicol; Sigma Xi. *Res:* Cellular radiobiology and metabolism and biological effects of radioisotopes in the body; setting of radiation protection standards for radioisotopes in workers, the general population, and the environment, particularly alpha emitters such as plutonium and uranium; writing a history of radioactivity and health; author of over 100 publications. *Mailing Add:* Dept Community Med Univ Calif 17441 Plaza Animado No 132 San Diego CA 92128

STANNARD, WILLIAM A, b Whitefish, Mont, Sept 5, 31; m 51; c 4. MATHEMATICS. *Educ:* Univ Mont, BA, 53; Stanford Univ, MA, 58; Mont State Univ, EdD, 66. *Prof Exp:* Instr math, Northern Mont Col, 60-62; instr, Mont State Univ, 62-66; from assoc prof to prof math, Eastern Mont Col, 66-84, chmn dept, 68-77; prof math, Calif State Univ, Bakersfield, 84-88, Fresno, 88-89, prof educ, Hayward, 89-90; INSTR MATH, NEWARK MEM HIGH SCH, 90- *Concurrent Pos:* Partic NSF Math Inst, Rutgers Univ, 67. *Res:* Teaching undergraduate mathematics. *Mailing Add:* 40384 Imperio Pl Fremont CA 94539

STANNERS, CLIFFORD PAUL, b Sutton Surrey, Eng, Oct 19, 37; Can citizen; m 59; c 3. MOLECULAR BIOLOGY, CELL BIOLOGY. *Educ:* McMaster Univ, BSc, 58; Univ Toronto, MSc, 60, PhD(med biophys), 63. *Prof Exp:* Fel molecular biol, Mass Inst Technol, 62-64; from asst prof to prof med biophys, Univ Toronto, 64-82; sr scientist biol res, Ont Cancer Inst, 64-82; PROF BIOCHEM, MCGIL UNIV 82-, DIR,CANCERCENTRE, 88- *Concurrent Pos:* Grants, Med Res Coun & Nat Cancer Inst Can, 65-, US Nat Cancer Inst, 73-79 & Multiple Sclerosis Soc Can, 79-; assoc ed, J Cell Physiol, 73-, Cell, 75-84; mem grants panel, Nat Cancer inst Can, 76-81; sci adv, Amyotropic Lateral Sclerosis Soc Can, 77-79. *Mem:* Can Biochem Soc; Can Soc Cell Biol. *Res:* Growth control of animal cells; protein synthesis somatic cell genetics; molecular genetics; cell virus interactions; persistent infection with vesicular stomatitis virus; molecular genetics; human cancer; human carcinoembryonic antigen. *Mailing Add:* Cancer Centre McGill Univ McIntyre Sci Bldg Sherbrooke St Montreal PQ H3A 2M5 Can

STANNETT, VIVIAN THOMAS, b Langley, Eng, Sept 1, 17; nat US; m 46; c 1. POLYMER CHEMISTRY. *Educ:* London Polytech Inst, BSc, 39; Polytech Inst Brooklyn, PhD(chem), 50. *Prof Exp:* Plant chemist, Brit Celanese Co, 39-41; chief chemist, Utilex, Ltd, 44-47, dir, 50-51; asst group leader polymers, Koppers Co, 51-52; asst prof polymer chem, State Univ NY Col Forestry, Syracuse Univ, 52-56, prof, 56-61; assoc dir, Camille Dreyfus Lab, Res Triangle Inst, 61-67; prof, 67-69, vprovost & dean grad sch, 75-82, CAMILLE DREYFUS PROF CHEM ENG, NC STATE UNIV, 69- *Concurrent Pos:* Res assoc, Mellon Inst, 51. *Honors & Awards:* Borden Award & Payen Award, Am Chem Soc, 74; Int Award & Gold Medal, Soc Plastics Eng. *Mem:* Am Chem Soc; Tech Asn Pulp & Paper Indust; fel NY Acad Sci; Soc Chem Indust; fel Royal Soc Chem. *Res:* Physical chemistry and engineering properties of plastics; cellulosic plastics; plastics-paper combinations; radiation chemistry of polymers. *Mailing Add:* Dept Chem Eng NC State Univ Box 7905 Raleigh NC 27695-7905

STANOJEVIC, CASLAV V, b Belgrade, Yugoslavia, June 23, 28; US citizen; m 70; c 1. MATHEMATICS. *Educ:* Univ Belgrade, BS, 52, MS, 54, PhD(math), 55. *Prof Exp:* From asst prof to assoc prof math, Univ Belgrade, 58-61; from assoc prof to prof math, Univ Detroit, 62-68; PROF MATH, UNIV MO-ROLLA, 68- *Concurrent Pos:* Vis prof, Ohio State Univ, 67-68 & La State Univ, New Orleans, 71-52. *Mem:* Am Math Soc; Math Asn Am; Inst Math Statist. *Res:* Fourier analysis; geometry of quantum states and normed linear spaces; applied probability; analysis; theory of probability. *Mailing Add:* Dept Math Univ Mo PO Box 249 Rolla MO 65401

STANONIS, DAVID JOSEPH, b Louisville, Ky, Mar 19, 26. ORGANIC CHEMISTRY. *Educ:* Univ Ky, BS, 45; Northwestern Univ, PhD(chem), 50. *Prof Exp:* Instr chem, Northwestern Univ, 48-49; asst prof, Clark Univ, 49-50 & Loyola Univ, Ill, 50-54; res chemist, Southern Regional Res Lab, USDA, 56-79; VPRES, BENDAL, INC, METAIRIE, LA, 80- *Mem:* Am Chem Soc; Sigma Xi; Am Inst Chem; Fiber Soc. *Res:* Stereochemistry; mechanisms of organic reactions; cellulose chemistry. *Mailing Add:* Bendal Inc - 306-C 7809 Airline Hwy Metairie LA 70003

STANONIS, FRANCIS LEO, b Louisville, Ky, July 9, 31; c 2. MINERALOGY, PETROLOGY. *Educ:* Univ Ky, BS, 51, MS, 56; Pa State Univ, PhD(mineral & petrol), 58. *Prof Exp:* Geologist, Carter Oil Co, 58-60 & George A Hoffman Co, 60-62; pres, Mitchell & Stanonis Inc, 62-67 & Int Pollution Control, Inc, 67-69; assoc prof geol & geog, 69-74, chmn, Div Sci & Math, 73-79, PROF GEOL & GEOG, UNIV SOUTHERN IND, EVANSVILLE, 74-, DEAN SCH SCI & ENG TECHNOL, 88- *Concurrent Pos:* Owner, Red Banks Oil & Gas Co, 60-69; pres, Enviro-Sci Corp, 70-76; mem environ comt, Interstate Oil Compact Comn, 76-; owner, Stanonis Mineral Explor Co, 69-; comnr environ protection, Commonwealth, Ky, 75-77; explor mgr, Wiseroil Co, Sisterville, WVa, 84-85. *Mem:* Geol Soc Am; Am Inst Prof Geol. *Res:* Archaeoastronomy. *Mailing Add:* Box No 150 Henderson KY 42420

STANOVSKY, JOSEPH JERRY, b Galveston, Tex, Mar 4, 28; wid; c 4. ENGINEERING MECHANICS, CIVIL ENGINEERING. *Educ:* Southern Methodist Univ, BSCE, 48; Univ Tex, Austin, MSCE, 51; Pa State Univ, PhD(eng mech), 66. *Prof Exp:* Steel detailer, Austin Bros Steel Co, Tex, 48-49; instr civil eng, Univ Tex, Austin, 50-51; sr struct test engr, Convair, Tex, 51-53; design supvr, Austin Co, Tex, 53-54; design engr supvr, Fluor Corp, Tex, 54-58; asst prof civil eng, Tex Technol Col, 58-59; design engr, Boeing Airplane Co, Wash, 59-60; instr eng mech, Pa State Univ, 61-66; ASSOC PROF ENG MECH, UNIV TEX, ARLINGTON, 66- *Concurrent Pos:* Vis assoc prof civil eng, Univ Petrol & Minerals, Dhahran, Saudi Arabia, 74-76. *Honors & Awards:* Ralph R Teetor Award, Soc Automotive Engrs, 74. *Mem:* Soc Automotive Engrs; Soc Petrol Engrs; Am Inst Aeronaut & Astronaut; AAAS. *Res:* Shock response on nonlinear structures; structural dynamics; experimental mechanics; structures; plasticity. *Mailing Add:* Dept Aerospace Eng Univ Tex Box 19088 Uta Sta Arlington TX 76019

STANSBERY, DAVID HONOR, b Upper Sandusky, Ohio, May 5, 26; m 48; c 4. ZOOLOGY, HYDROBIOLOGY. *Educ:* Ohio State Univ, BS, 50, MS, 53, PhD, 60. *Prof Exp:* Fel, Stone Lab Hydrobiol, Ohio State Univ, 53-55, asst instr gen zool, 56-57, instr animal ecol, 58-60, from asst prof to assoc prof, 61-71, PROF ZOOL, OHIO STATE UNIV, 71- *Concurrent Pos:* Cur natural hist, Ohio State Mus, 61-72; dir mus zool, Ohio State Univ, 71-77. *Mem:* AAAS; Ecol Soc Am; Soc Syst Zool; Soc Study Evolution; Am Malacol Union (pres, 70-71); Sigma Xi. *Res:* Zoogeography, ecology, evolution and taxonomy of freshwater forms, especially bivalve molluscs and decapod crustaceans. *Mailing Add:* Mus Zool Ohio State Univ 1813 N High St Columbus OH 43210

STANSBREY, JOHN JOSEPH, b St Louis, Mo, Dec 30, 18; m 63; c 2. PHYSICAL CHEMISTRY. *Educ:* Washington Univ, AB, 41, MS, 43, PhD(chem), 47. *Prof Exp:* Asst chem, Washington Univ, 41-45, lectr, US Army Training Prog, 43-44; res chemist, Am Can Co, Ill, 45-47; res physicist, Anheuser-Busch, Inc, 47-53; mem tech staff, Bell Tel Labs, Inc, 53-56; mem group staff, Res Develop & Mfg Div, NCR Corp, 56-73; scientist, Glidden Coatings & Resins Div, D P Joyce Res Ctr, SCM Corp, 78-86; CONSULT STATISTICIAN, RICERCA, INC, 88- *Concurrent Pos:* Instr, Webster Col, 44-45, Cuyahoga Community Col W, 86, Ashland Col, 87 & Kent State Univ, 87-88; instr exten div, Univ Cincinnati, 56-57. *Mem:* Am Chem Soc; Am Statist Asn; Sigma Xi; Inst Mgt Sci. *Res:* Colloid chemistry; thixotropy; electrochemistry; conductance; electrophoresis; ultracentrifuge; light scattering and viscosity of protein solutions; emission and absorption spectra of biological materials; solid state physics and transistors; mathematical programming; statistical analysis and experimental design; computer automation of laboratory and production processes. *Mailing Add:* 137 Parmelee Dr Hudson OH 44236

STANSBURY, E(LE) E(UGENE), b Indianapolis, Ind, Dec 14, 18; m; c 4. PHYSICAL METALLURGY. *Educ:* NC State Col, BChE, 40; Univ Cincinnati, MS, 42, PhD(metall), 46. *Prof Exp:* Asst, Univ Cincinnati, 40-42, from instr to asst prof metall, 42-47; assoc prof, 47-52, PROF METALL, UNIV TENN, KNOXVILLE, 52-, ALUMNI DISTINGUISHED SERV PROF, 76- *Concurrent Pos:* Consult, Oak Ridge Nat Lab, 47- *Mem:* Am Soc Metals; Metall Soc; Am Soc Eng Educ; Soc Hist Technol; Sigma Xi. *Res:* Thermodynamics of metal systems; kinetics of phase transformations; corrosion. *Mailing Add:* 435 Dougherty Hall Univ Tenn Knoxville TN 37916

STANSBURY, EDWARD JAMES, physics, for more information see previous edition

STANSBURY, HARRY ADAMS, JR, b Morgantown, WVa, Sept 14, 17; m 44; c 4. CHEMISTRY. *Educ:* WVa Univ, AB, 40, MS, 41; Yale Univ, PhD(org chem), 44. *Prof Exp:* Group leader, Union Carbide Corp, WVa, 44-71; state dir comprehensive health planning, Gov Off, 71-76; EXEC SECY, WVA ASN SCH ADMIN, 76- *Mem:* Am Chem Soc. *Res:* Development of new pesticides; registration of pesticides; synthetic organic chemistry; residue analyses. *Mailing Add:* 806 Montrose Dr Charleston WV 25303-2609

STANSBY, MAURICE EARL, b Cedar Rapids, Iowa, Apr 25, 08; m 38; c 1. NUTRITION. *Educ:* Univ Minn, BChem, 30, MSc, 33. *Prof Exp:* Jr chemist, US Bur Commercial Fisheries, Mass, 31-35, Md, 35-37 & Wash, 38-40, technologist chg fishery prod lab, Alaska, 40-42, dir tech lab, 42-66, dir food sci, Pioneer Res Lab, Wash, 66-71, dir environ conserv div, 72-75, SCI CONSULT, NORTHWEST FISHERIES CTR, NAT MARINE FISHERIES SERV, 75- *Concurrent Pos:* Lectr, Sch Fisheries, Univ Wash, 38-89. *Mem:* Am Chem Soc; Inst Food Technologists; Am Oil Chemists Soc. *Res:* Analysis, preservation and processing of fish; chemistry and nutritional properties of fish oils; effects of contaminants in the environment upon fish. *Mailing Add:* 2725 Montlake Blvd E Seattle WA 98112

STANSEL, JOHN CHARLES, b Spring Canyon, Utah, Nov 18, 35; m 60; c 7. ENGINEERING, FUEL TECHNOLOGY. *Educ:* Univ Utah, BS, 60; Calif Inst Technol, MS, 62. *Prof Exp:* Res engr rockets, Jet Propulsion Lab, 61; mem tech staff fluid flow, 62-66, sect head nuclear, 66-72, dept mgr lasers, 72-78, asst lab mgr lasers, 78-80, chief engr combustion & gasification, TRW Inc, 80-84; mgr, Eng & Res & Develop, CBU, 84-89; DIR PROG MGT, TRW VEHICLE SAFETY SYSTS INC, 89- *Concurrent Pos:* Chmn nuclear space safety, Atomic Indust Forum, 67-70. *Res:* Application of physics and engineering principles to development of advanced chemical combustion and laser devices. *Mailing Add:* TRW Vehicle Safety Systs Inc 4505 W 26 Mile Rd Washington MI 48094

STANSFIELD, BARRY LIONEL, b Toronto, Ont, June 10, 42; m 67; c 3. PLASMA PHYSICS. *Educ:* Univ Toronto, BASc, 65; Univ BC, MASc, 67, PhD(plasma physics), 71. *Prof Exp:* Fel, 71-72, PROF PLASMA PHYSICS, INST NAT SCI RES, 72- *Mem:* Can Asn Physicists; Am Phys Soc. *Res:* Confinement of plasmas using electric as well as magnetic fields; development of plasma diagnostics; laser-produced plasmas and laser diagnostics. *Mailing Add:* Quebec Univ Inst Nat Sci Res Energy CP 1020 Varennes PQ J0L 2P0 Can

STANSFIELD, ROGER ELLIS, b Sanford, Maine, July 16, 26; m 51; c 2. ORGANIC CHEMISTRY. *Educ:* Northwestern Univ, BS, 50; Carnegie Inst Technol, PhD(chem), 55. *Prof Exp:* from asst prof to assoc prof, 56-65, chmn dept, 71-74, PROF CHEM, BALDWIN-WALLACE COL, 65- *Concurrent Pos:* Vis prof, Forman Christian Col, W Pakistan, 64-66; Fel, Duke Univ, 54-56. *Mem:* AAAS; Am Chem Soc; Am Asn Univ Prof. *Res:* Synthesis of peptides; alkaloids of nicotiana; esterification; chemistry of pyroles; computer interfacing. *Mailing Add:* Dept of Chem Baldwin-Wallace Col Berea OH 44017

STANSFIELD, WILLIAM D, b Los Angeles, Calif, Feb 7, 30; m 53; c 3. GENETICS. *Educ:* Calif State Polytech Col, BS, 52, MA, 58; Univ Calif, Davis, MS, 61, PhD(animal breeding), 63. *Prof Exp:* INSTR BIOL, CALIF POLYTECH STATE UNIV, SAN LUIS OBISPO, 63- *Mem:* Genetics Soc Am; Am Genetic Asn; Soc Study Evolution; Sigma Xi. *Res:* Immunogenetics. *Mailing Add:* Dept of Biol Sci Calif Polytech State Univ San Luis Obispo CA 93407

STANSLOSKI, DONALD WAYNE, b Big Rapids, Mich, June 22, 39; m 59; c 4. CLINICAL PHARMACY, MICROCOMPUTERS. *Educ:* Ferris State Col, BS, 61; Univ Nebr-Lincoln, MS, 69, PhD(pharmaceut sci), 70. *Prof Exp:* Asst prof pharm, Col Pharm, Univ Nebr-Lincoln, 70-72; CLIN COORDR, RAABE COL PHARM, OHIO NORTHERN UNIV, 72-, DEPT CHMN, 75- *Concurrent Pos:* Consult, Vet Admin Hosp, Lincoln, Nebr, 71, Ohio Dept Pub Welfare, 74-78 & various comput systs, 75- *Honors & Awards:* Merck Award. *Mem:* Am Pharmaceut Asn; Am Asn Cols Pharm. *Res:* Chemistry of mesoionic compounds and role of pharmacist in the provision of drug therapy; computer applications to health care; expanding the use of computers in the provision of health care. *Mailing Add:* 3551 County Rd 44 Ada OH 45810

STANSLY, PHILIP GERALD, ANTIBIOTIC RESEARCH, VIRAL ONONOGY. *Educ:* Univ Minn, PhD(biochem), 44, Univ Wis, PD,(enzyme chem), 54. *Prof Exp:* Prog dir biochem, Nat Cancer Inst, NIH, 68-80; RETIRED. *Mailing Add:* 2301 Tarleton Dr Charlottesville VA 22901

STANTON, BRUCE ALAN, b Providence, RI, Mar 31, 52. RENAL PHYSIOLOGY, ION TRANSPORT. *Educ:* Univ Maine, BS, 74; Yale Univ, MS, 76, PhD(physiol), 80. *Prof Exp:* Teaching fel physiol, Sch Med, Yale Univ, 80-83, assoc res scientist, 83-84; asst prof, 84-88, ASSOC PROF PHYSIOL, DARTMOUTH MED SCH, 88- *Concurrent Pos:* Prin investr, NIH grant, 85-88; mem, Kidney Coun, Am Heart Asn; established investr, Am Heart Asn; ed bd, Am J Physiol. *Mem:* Am Soc Nephrology; Int Soc Nephrology; Soc Gen Physiologists; Am Physiol Soc; Biophys Soc; Am Heart Asn. *Res:* Renal physiology; cellular mechanisms of ion transport studied by electrophysiological and ultrastructural techniques; regulation of sodium, potassium and hydrogen; ion transport by adrenal corticosteroids and acid-base disorders. *Mailing Add:* Dept Physiol Dartmouth Med Sch 615 Remsen Bldg Hanover NH 03756

STANTON, CHARLES MADISON, b San Diego, Calif, July 2, 42; m 71. MATHEMATICS, COMPUTER SCIENCE. *Educ:* Wesleyan Univ, BS, 64; Stanford Univ, PhD(math), 69. *Prof Exp:* Lectr, Wesleyan Univ, 68-69, asst prof math, 69-76; asst prof math, Fordham Univ, 76-80; consult programmer, Comput Ctr, Univ Notre Dame, 84-85; VIS ASSOC PROF COMPUT SCI, IND UNIV, SOUTH BEND, 85- *Concurrent Pos:* Vis assoc prof math, Wesleyan Univ, 80-81; adj assoc prof math, Univ Notre Dame, 81-82, IHES, 82 & Max Planck Inst Math, 83. *Mem:* Am Math Soc; Math Asn Am; Asn Comput Mach; Inst Elec & Electronics Engrs. *Res:* Complex analysis; algebras of analytic funtions; Riemann surfaces; symbolic manipulation. *Mailing Add:* Dept Computer Sci Ind Univ 1700 Mishawaka Ave South Bend IN 46634

STANTON, GARTH MICHAEL, b Cleveland, Ohio, June 15, 33; m 57, 87; c 5. ORGANIC CHEMISTRY. *Educ:* Univ Detroit, BS, 56, MS, 58; Purdue Univ, PhD(chem), 62. *Prof Exp:* Res chemist, 62-69, sr res chemist, 69-80, SR RES ASSOC, CHEVRON RES CO, STANDARD OIL CO, CALIF, 80- *Mem:* AAAS; Am Soc Testing & Mat. *Mailing Add:* Chevron Res & Technol Co 100 Chevron Way Richmond CA 94802

STANTON, GEORGE EDWIN, b Danville, Pa, Mar 28, 44; m 65; c 2. AQUATIC ECOLOGY, INVERTEBRATE ECOLOGY. *Educ:* Bucknell Univ, BS, 66; Univ Maine, Orono, PhD(zool, entom), 69. *Prof Exp:* USDA res asst, Univ Maine, Orono, 66-69; from asst prof to assoc prof, 69-76, PROF BIOL, COLUMBUS COL, 76-, BIOL DEPT CHAIRPERSON, 82- *Concurrent Pos:* Adj assoc prof, Ga State Univ, 72-73; vis assoc prof biol, Mt Lake Biol Sta, 75; NSF sci fac fel, Auburn Univ, 76-78; Danforth fel, 79; mem, Muscogee County Bd Educ, 76-80, 82-87, first vpres, 85, pres, 86-87. *Mem:* Am Inst Biol Sci; Ecol Soc Am; Entom Soc Am; AAAS; Am Benthological Soc. *Res:* Ecology of small watershed systems and their impoundments; crayfish ecology; dynamics and succession of macroinvertebrate communities associated with carrion decomposition. *Mailing Add:* Dept Biol Columbus Col Columbus GA 31993

STANTON, HUBERT COLEMAN, b Orofino, Idaho, May 3, 30; m 50; c 2. PHARMACOLOGY. *Educ:* Idaho State Col, BS, 51; Ore State Col, MS, 53; Univ Iowa, PhD(pharmacol), 58. *Prof Exp:* Asst pharmacol, Univ Iowa, 55-58; instr, Sch Med, Univ Colo, 58-60; sr pharmacologist & group leader, Mead Johnson & Co, 60-65; asst prof pharmacol, Col Med, Baylor Univ, 65-68; head, Dept Animal Physiol, Biol Res Ctr, Shell Develop Co, 68-79; dir biol res, Mead Johnson & Co, 79-82; dir cardiovascular preclin res, Pharmaceut Res & Develop, Bristol Myers Corp, 82-84; CONSULT & MED WRITER, 87- *Mem:* AAAS; fel Am Col Vet Pharmacologists; Soc Exp Biol & Med; Am Soc Pharmacol & Exp Therapeut; Am Asn Med Writers. *Res:* Autonomic pharmacology; neonatal physiology; cardiovascular pharmacology. *Mailing Add:* 11028 Sycamore Ct Auburn CA 95603

STANTON, K NEIL, ENGINEERING. *Educ:* Univ New South Wales, BE, 59, ME, 61, PhD(elec eng), 64. *Prof Exp:* Trainee engr, Australian Iron & Steel Co, 52-58; teaching asst, Univ New South Wales, Australia, 59-64; asst prof sch elec eng, Purdue Univ, 65-70; sr res eng, Syst Control, Inc, Palo Alto, Ca, 70-76; supvr, Consorcio Hidroservice-Sci, 76-78; pres, ENSYSCO, 76-78; pres & chief exec officer, 79-85, vpres, 85-90, CHMN BD, ESCA CORP, 90- *Mem:* Fel Inst Elec Electronics Engrs; Inst Elec & Electronics Engrs Power Eng Soc. *Res:* Author of several books and articles; developed medium term load forecasting programs for several electric utilities companies; developed the dynamic energy balance technique for simulation of long term power system dynamics. *Mailing Add:* ESCA Corp 11120 NE 33rd Pl Bellevue WA 98004

STANTON, MEARL FREDRICK, b Staunton, Ill, Aug 14, 22; m 51; c 4. MEDICINE. *Educ:* St Louis Univ, MD, 48. *Prof Exp:* Sr instr path, Sch Med, St Louis Univ, 50-54; PATHOLOGIST, NAT CANCER INST, 56-, ED-IN-CHIEF JOUR, 68- *Mem:* AMA; Am Soc Exp Path. *Res:* Host-parasite relationships of obligate intracellular parasites; experimental and clinical cancer research; pulmonary diseases. *Mailing Add:* Lab Path Nat Cancer Inst Bethesda MD 20892

STANTON, NANCY KAHN, b San Francisco, Calif, Mar 23, 48; m 71; c 2. MATHEMATICS. *Educ:* Stanford Univ, BS, 69; Mass Inst Technol, PhD(math), 73. *Prof Exp:* Instr math, Mass Inst Technol, 73-74; lectr, Univ Calif, Berkeley, 74-76; Ritt asst prof math, Columbia Univ, 76-81; assoc prof, 81-85, PROF MATH, UNIV NOTRE DAME, 85- *Concurrent Pos:* Mem, Inst Advan Study, 79-80; Sloan fel, 81-85; vis, Institut des Hautes 'Etudes Scientifiques, 82-83; Gast prof, Max Planck Inst, 83; vis prof, Univ Mich, 86-87. *Mem:* Am Math Soc; Math Asn Am; Asn Women Math. *Res:* Spectrum of the Laplacian on complex manifolds, geometry of complex manifolds with boundary. *Mailing Add:* Dept Math Univ Notre Dame Notre Dame IN 46556

STANTON, NANCY LEA, b Casper, Wyo, Jan 13, 44; m 70; c 2. COMMUNITY ECOLOGY, SOIL BIOLOGY. *Educ:* Creighton Univ, BS, 66; Univ Chicago, PhD(biol), 72. *Prof Exp:* Asst prof, 72-80, actg dept head, 81-82, dept head, 82-85, ASSOC PROF ECOL, DEPT ZOOL, UNIV WYO, 81- *Concurrent Pos:* Consult, Environ Protection Agency & NSF, 80-83; prog dir ecol, NSF. *Mem:* Ecol Soc Am; Sigma Xi; Soc Nematologists; AAAS; Am Inst Biol Sci; Am Soc Naturalists. *Res:* Community and evolutionary ecology including soil microarthropods and nematodes, plant and animal interactions, specifically on grasslands; pollination biology; parasite ecology. *Mailing Add:* Dept Zool & Physiol Box 3166 Univ Wyo Laramie WY 82071-3166

STANTON, NOEL RUSSELL, b Dover, NJ, Dec 29, 37; m 62; c 1. HIGH ENERGY PHYSICS. *Educ:* Rutgers Univ, BA, 60; Cornell Univ, PhD(exp physics), 65. *Prof Exp:* Res assoc exp high energy physics, Univ Mich, 65-68; asst prof, 68-73, assoc prof, 73-77, PROF PHYSICS, OHIO STATE UNIV, 78- *Mem:* Am Phys Soc. *Res:* Strong interactions of elementary particles at high energy. *Mailing Add:* Dept Physics 2094 Smith Lab Ohio State Univ 174 W 18th Ave Columbus OH 43210

STANTON, RICHARD EDMUND, b Brooklyn, NY, Aug 31, 31; m 57; c 5. THEORETICAL CHEMISTRY. *Educ:* Niagara Univ, BS, 52; Univ Notre Dame, PhD(phys chem), 57. *Prof Exp:* Fel, Cath Univ Am, 56-57; from asst prof to assoc prof, 57-69, PROF PHYS CHEM, CANISIUS COL, 69-, CHMN, CHEM DEPT, 81- *Concurrent Pos:* Consult, Union Carbide Res Inst, 61-63, Occidental Chem, 90-; Sloan fel, 69-71; vis prof, Univ Manchester, 70-71; vis scientist, Brookhaven Nat Lab, 85-86; Sloan fel, 69-71. *Mem:* Am Chem Soc; Am Phys Soc. *Res:* Quantum chemistry, especially self-consistent field convergence theory, methodology and electron correlation theory; group theory of transition states in chemical kinetics; relativistic techniques; carbon clusters. *Mailing Add:* Dept of Chem Canisius Col Buffalo NY 14208

STANTON, ROBERT E, b Philadelphia, Pa, Dec 5, 47; m 77. RADIATION ONCOLOGY, MACHINE CALIBRATION & QUALITY ASSURANCE. *Educ:* Univ Pa, BA, 69, MS, 69; Drexel Univ, PhD(biomed eng), 82. *Prof Exp:* Radiologist physicist, Univ Pa, 71-73; from jr physicist to sr physicist, Mem Sloan Kettering Hosp, 73-75; SR CLIN PHYSICIST, COOPER HOSP/UNIV MED CTR, 75- *Concurrent Pos:* Lectr, Gwynedd-Mercy Col Sch Radiation Tech, 78-85; partic consult, Am Col Radiol, 78-; mem ad hoc comt, Proposal Eval, Nat Cancer Inst, 83-85; adj assoc prof, Drexel Univ, 85-89; consult, Emer Care Res Inst, 85-; adj asst prof, Univ Med & Dent NJ-Robert Woods Johnson Med Sch, 85- *Mem:* Am Asn Physicists Med; Am Col Radiol; Health Physics Soc; Soc Magnetic Resonance Imaging. *Res:* Clinical medical physics; radiation oncology; machine parameter measurement; quality assurance. *Mailing Add:* Cooper Hosp Univ Med Ctr Camden NJ 08103

STANTON, ROBERT JAMES, JR, b Los Angeles, Calif, June 17, 31; m 53; c 2. GEOLOGY, PALEONTOLOGY. *Educ:* Calif Inst Technol, BS, 53, PhD(geol), 60; Harvard Univ, MA, 56. *Prof Exp:* Res geologist, Shell Develop Co, 59-67; from assoc prof to prof, 67-85, head dept, 78-82, RAY C FISH PROF GEOL, TEX A&M UNIV, 85- *Concurrent Pos:* Vis prof, Univ Erlangen-Nurnberg, WGer, 84. *Mem:* Fel Geol Soc Am; Soc Econ Paleontologists & Mineralogists; Paleont Soc. *Res:* Paleoecology-development of criteria and techniques for determining ancient environments from the fossil record; research focus on cenozoic of California; modern of Texas gulf coast, triassic reefs of Austrian Alps. *Mailing Add:* Dept of Geol Tex A&M Univ Col Sta TX 77843

STANTON, ROBERT JOSEPH, b Pottsville, Pa, Feb 5, 47; m 70; c 2. MATHEMATICS. *Educ:* Drexel Inst Technol, BS, 69; Cornell Univ, MA, 71, PhD(math), 74. *Prof Exp:* Asst prof math, Rice Univ, 69-80; MEM FAC, DEPT MATH, OHIO STATE UNIV, 80- *Concurrent Pos:* NSF grant, 75-; mem, Sch Math, Inst Advan Study, 77-78. *Mem:* Am Math Soc. *Res:* Analysis on lie groups. *Mailing Add:* Dept Math Ohio State Univ 231 W 18th Ave Columbus OH 43210

STANTON, THADDEUS BRIAN, b Cincinnati, Ohio, Nov 11, 51; m 72; c 2. ANAEROBIC BACTERIOLOGY, MICROECOLOGY. *Educ:* Thomas More Col, BA, 71; Univ Mass, PhD(microbiol), 80. *Prof Exp:* NIH postdoctoral fel, Dept Microbiol, Univ Ill, 80-82; res microbiologist, Pfizer Cent Res, 82-83; res microbiologist, 83-91, LEAD SCIENTIST SWINE DYSENTERY, PHYSIOPATH RES UNIT, NAT ANIMAL DIS CTR, AGR RES SERV, USDA, 91- *Concurrent Pos:* Consult, Pfizer Cent Res, Upjohn Co; asst prof, collabr, Microbiol Dept, Iowa State Univ, 87- *Mem:* Am Soc Microbiol; Conf Res Workers Animal Dis; AAAS; Soc Microbiol Ecol & Dis. *Res:* Phylogenetic analysis of spirochetes based on 16S rRNA comparisons; DNA probes for swine pathogen Serpulina hyodysenteriae; identification of virulence-associated enzymes and proteins of S hyodysenteriae; one US patent. *Mailing Add:* Nat Animal Dis Ctr Agr Res Serv USDA PO Box 70 Ames IA 50010

STANTON, TONI LYNN, b Johnstown, Pa, July 21, 44. NEUROPEPTIDES, PINEAL MELATONIN. *Educ:* Univ Md, BS, 68, MS, 71; Thomas Jefferson Univ, PhD(pharmacol), 81. *Prof Exp:* Res asst, Dept Physiol, Sch Med, Univ Pa, 72-78; res asst, 78-81, ASST RES SCIENTIST, ALFRED I DU PONT INST, 81- *Concurrent Pos:* Lectr, Col Nursing, Univ Del; consult pychoneuroendocrinol, Michael R Babitts Fund, 81- *Mem:* Soc Neurosci; Int Hibernation Soc. *Res:* Behavioral and physiological aspects of neuropeptide action in the mammalian central nervous system; role of pineal melatonin in mechanisms that control the state of hibernation; control of central activity state. *Mailing Add:* 19481 Pompano Lane 109 Huntington Beach CA 92648

STANTON, WILLIAM ALEXANDER, b Washington, DC, Sept 9, 15; m 42; c 3. ORGANIC CHEMISTRY. *Educ:* Univ Md, BS, 36, PhD(org chem), 41. *Prof Exp:* Res chemist, Tech Div, Photo Prod Dept, NJ, E I du Pont de Nemours & Co Inc, 41-45, group leader, 46-49, chief supvr, Plant Process Dept, Prod Div, 49-50, plant process supt, 50-52, prod supt, 53-56, asst plant mgr, NY, 57-58, dir, Parlin Res Lab, NJ, 58-63, mgr prod mkt, Del, 64-65, dir printing & indust sales, 66, dir int opers, 67-80; RETIRED. *Honors & Awards:* J Award, Soc Motion Picture & TV Engrs, 50. *Mem:* AAAS; Am Chem Soc; Sigma Xi. *Res:* Natural products; photographic emulsions and processing solutions; synthetic color-forming polymers for photographic emulsions. *Mailing Add:* 726 Loveville Rd No 28 Hockessin DE 19707-1504

STANWICK, GLENN, b Milwaukee, Wis, Oct 17, 28. SOLID STATE PHYSICS, NUCLEAR PHYSICS. *Educ:* Northwestern Univ, BSc, 50, MSc, 51. *Prof Exp:* Res scientist, Johnsons Control, 51-53; ENGR, ACCURATE AUTOMATIC INC, 53- *Mem:* Am Phys Soc; Soc Manufacturing Engrs. *Mailing Add:* 1325 Valley Ridge Rd Brookfield WI 53005

STAPELBROEK, MARYN G, b Veghel, Netherlands, Aug 3, 47; US citizen; m 79; c 2. INFRARED DETECTOR DEVELOPMENT, DETECTOR PHYSICS. *Educ:* Univ Wis, Oshkosh, BS, 72; Univ Conn, MS 74, PhD (solid state physics), 76. *Prof Exp:* Resident res assoc, Naval Res Lab, Washington DC, 76-78; MEM TECH STAFF, RES & DEVELOP, SCI CTR, ROCKWELL INT, 78- *Mem:* Am Phys Soc; AAAS; Sigma Xi. *Res:* Invention and development of highly sensitive infrared detectors, based on impurity band conduction, specifically the blocked impurity band (BIB) detector and the solid state photomultiplier (SSPM). *Mailing Add:* M/S-BC17 Rockwell Int Sci Ctr Anaheim CA 92803

STAPH, HORACE E(UGENE), b Petrolia, Tex, Jan 8, 21; m 50; c 3. MECHANICAL ENGINEERING. *Educ:* Rice Univ, BSME, 43; Univ Tex, MSME, 51; Univ Minn, PhD(mech eng), 59. *Prof Exp:* Design engr hydraul, Douglas Aircraft Co, Inc, 43-44; asst prof mech eng, Univ Tex, 46-60; sr res engr, Southwest Res Inst, 60-85; RETIRED. *Mem:* Am Soc Mech Engrs; Am Soc Testing & Mats. *Res:* Lubrication; wear; friction phenomena, especially wet-brake studies. *Mailing Add:* Dept Eng St Mary's Univ One Camino Santa Maria San Antonio TX 78284

STAPLE, PETER HUGH, b Tonbridge, Eng, Oct 15, 17; m 52; c 2. PHYSIOLOGY. *Educ:* Univ London, BDS, 40, BSc, 49, PhD(sci, histochem), 52. *Prof Exp:* Mem sci staff, Med Res Coun, Eng, 51-57; lectr physiol, Univ Birmingham, 57; res assoc, Med Res Labs, Charing Cross Hosp, Univ London, 57-59; instr pharmacol, Univ Ala, 59-60, assoc prof, 60-63, assoc prof dent, Med Col & Sch Dent, 59-63; from assoc prof to prof, 63-87, EMER PROF ORAL BIOL, SCH DENT MED, STATE UNIV NY, BUFFALO, 87- *Concurrent Pos:* Hon vis assoc prof, Univ BC, 71; vis prof, Univ Ill, 81-82; Nuffield Found Dental Res Fel, 49-51. *Mem:* Histochem Soc; Int Asn Dent Res. *Res:* Histochemistry in relation to dental disease; phenytoin sodium and adjuvant-induced arthritis; glycosaminoglycans in gingival sulcus fluid. *Mailing Add:* Dept Oral Biol State Univ NY Buffalo Sch Dent Med Buffalo NY 14214

STAPLE, TOM WEINBERG, b Hamburg, Ger, May 6, 31; US citizen; m 64; c 1. MEDICINE, RADIOLOGY. *Educ:* Univ Ill, Chicago, BS, 53, MD, 55. *Prof Exp:* From instr to prof radiol, Mallinckrodt Inst Radiol, Sch Med, Wash Univ, 73-75; adj prof radiol, Univ Calif, Irvine, 75; STAFF MEM, DEPT RADIOL, MEM HOSP, LONG BEACH, 75- *Concurrent Pos:* Consult, VA Hosp, Long Beach, Calif. *Mem:* Fel Am Col Radiol; Asn Univ Radiol; assoc mem Am Acad Orthop Surg. *Res:* Bone growth and arthrography. *Mailing Add:* 2801 Atlantic Ave Box 1428 Long Beach CA 90801

STAPLES, BASIL GEORGE, b Eliot, Maine, Aug 13, 14; m 35; c 2. CORROSION TESTING, ELECTRICAL INSPECTION. *Educ:* Univ Maine, BS, 35, MS, 36. *Prof Exp:* Foreman, Gen Chem Co, 36-42; supt, US Rubber Co, 42-45; chemist, The Pfaudler Co, 45-64, engr, 64-79; RETIRED. *Mem:* Am Chem Soc; fel Am Ceramic Soc (vpres, 74-75); fel Am Soc Testing & Mat; Nat Asn Corrosion Engrs; Am Inst Chemists; Nat Inst Ceramic Engrs. *Res:* Glass coated vessels and other equipment, with special emphasis on practical design details, such as closures and gaskets. *Mailing Add:* 275 Colwick Rd Rochester NY 14624

STAPLES, JON T, b Waterville, Maine, Sept 14, 38; m 62; c 3. ORGANIC CHEMISTRY, POLYMER CHEMISTRY. *Educ:* Bowdoin Col, AB, 61; Univ NC, Chapel Hill, PhD(org chem), 66. *Prof Exp:* Res assoc org chem, Mass Inst Technol, 66-67; sr res chemist, 67-71, RES ASSOC, EASTMAN KODAK LABS, ROCHESTER, 71-, HEAD LAB, 75- *Mem:* Am Chem Soc. *Res:* Protecting group chemistry; peptide synthesis; monomer and polymer synthesis; photographic science. *Mailing Add:* 14 Ithaca Dr Pittsford NY 14534

STAPLES, LLOYD WILLIAM, b Jersey City, NJ, July 8, 08; m 41; c 3. MINING ENGINEERING. *Educ:* Columbia Univ, AB, 29; Univ Mich, MS, 30; Stanford Univ, PhD(mineral), 35. *Prof Exp:* With Mich State Geol Surv, 30-31; instr geol, Mich Col Mining & Technol, 31-33; res assoc mineral, Stanford Univ, 35-36; instr geol, Ore State Col, 36-37; from instr to prof, 39-74, head dept, 58-68, EMER PROF GEOL, UNIV ORE, 74- *Concurrent Pos:* Chief geologist, Horse Heaven Mines, Sun Oil Co, 37-41 & Cordero Mining Co, Nev, 41-45; Guggenheim fel, Mex, 60-61; consult, UNESCO, Paris, 68-71; mem, State Ore Bd Geologist Examrs, 77-80. *Honors & Awards:* Legion of Honor, Soc Mining Engr, 90. *Mem:* Fel Geol Soc Am; fel Mineral Soc Am; distinguished mem, Am Inst Mining, Metall & Petrol Engrs. *Res:* Mineralogy and crystallography; economic geology of quicksilver; microchemistry of minerals; mineral determination by microchemical methods; field and x-ray study of zeolites. *Mailing Add:* Dept Geol Univ Ore Eugene OR 97403

STAPLES, RICHARD CROMWELL, b Hinsdale, Ill, Jan 29, 26; m 54; c 3. PHYTOPATHOLOGY. *Educ:* Colo State Univ, BS, 50; Columbia Univ, AM, 54, PhD(plant physiol), 57. *Prof Exp:* Fel biochem, 52-57, from asst biochemist to assoc biochemist, 57-64, PLANT BIOCHEMIST, BOYCE THOMPSON INST PLANT RES, INC, 64-, PROG DIR PHYSIOL OF PARASITISM, 66- *Honors & Awards:* Humboldt Sr Scientist Award, 81. *Mem:* Am Soc Plant Physiologists; fel Am Phytopath Soc; Am Inst Biol Sci. *Res:* Responses of rust fungi to the environment; mechanisms by which these fungi sense chemical and physical stimuli which induce formation of the infection structures. *Mailing Add:* Boyce Thompson Inst Tower Road Ithaca NY 14853

STAPLES, ROBERT, b Philadelphia, Pa, Dec 9, 16; m 43; c 3. ENTOMOLOGY. *Educ:* Univ Mass, BS, 40; Cornell Univ, PhD, 48. *Prof Exp:* Asst entomologist, Conn Agr Exp Sta, 49-50; assoc prof entom, 50-74, PROF ENTOM, UNIV NEBR, LINCOLN, 74-, ASSOC ENTOMOLOGIST, 50- *Mem:* Entom Soc Am. *Res:* Arthropod transmission of plant viruses; economic entomology. *Mailing Add:* 1040 N 65th St Lincoln NE 68505

STAPLES, ROBERT EDWARD, b Cobourg, Ont, Dec 5, 31; m 57; c 2. TERATOLOGY, DEVELOPMENTAL & REPRODUCTIVE TOXICOLOGY. *Educ:* Univ Sask, BSA, 54, MSc, 56; Cornell Univ, PhD(reproductive physiol), 61. *Prof Exp:* Asst animal husb, Univ Sask, 54-56 & Cornell Univ, 56-61; sect head, Endocrinol Dept, William S Merrell Co, 61-63; staff scientist, Worcester Found Exp Biol, 63-67; head, Unit Teratology & Reproduction, Merck Inst Therapeut Res, 67-71; head sect exp teratology, Environ Toxicol Br, Nat Inst Environ Health Sci, 71-75, asst br chief teratology & dir, Environ Teratology Info Ctr, 75-78; STAFF TERATOLOGIST, HASKELL LAB TOXICOL & INDUST MED, E I DU PONT DE NEMOURS & CO, 78- *Concurrent Pos:* Prin investr, USPHS grant, 64-67; consult, William S Merrell Co, 63-67; adj assoc prof, Pharmacol Dept, Med Sch, Univ NC, Chapel Hill, 73-78 & Dept Anat, Jefferson Med Col, Philadelphia, 80-; US rep, US-USSR Collab Environ Health, 72-79; secy-treas, Specialty Sect Reprod & Develop Toxicol, Soc Toxicol, 83-85, pres, 91-92; consult, Am Petrol Inst, 80-90, Reproductives Effects Assessment Group, Off Res & Develop, USEPA, 80; panel mem, Off Technol Assessment, Cong US, 84-85; invited pres, Japan Terat Soc, Nagoya, Japan, 86; nominating comt, Am Col Toxicol, 87-88; consult, Chem Mfrs, Asn, Pharmaceut Mfr Asn, Inst de la vie, EPA, FDA, WHO; cong task force, Effects Pesticides Human Health, 88 & Contraceptive Develop Br, NIH, 90; chmn, Int Fedn Teratology Soc, 83-85; mem, Mid-Atlantic Reproduction & Tentative Asn Steering Comt, Behav Teratology Soc, 81-84, chmn, Const Comt, 83-84. *Mem:* Teratology Soc (secy, 73-77, pres, 83-84); Soc Study Reproduction; Soc Toxicol; Europ Teratology Soc; Int Fedn Teratology Socs; Behav Teratology Soc. *Res:* Indentification of chemicals and environmental agents that represent reproductive or developmental hazards; detection through development of test methodology, the conduct of tests and evaluation and extraportion of the results. *Mailing Add:* Haskell Lab Toxicol & Indust Med E I du Pont de Nemours & Co PO Box 50 Elkton Rd Newark DE 19714

STAPLETON, HARVEY JAMES, b Kalamazoo, Mich, Dec 22, 34; m 57; c 3. MAGNETIC RESONANCE. *Educ:* Univ Mich, BS, 57; Univ Calif, Berkeley, PhD(physics), 61. *Prof Exp:* From asst prof to assoc prof, 61-69, PROF PHYSICS, UNIV ILL, URBANA, 69-, ASSOC DEAN, GRAD COL, 80-, ASSOC VCHANCELLOR RES, 87- *Mem:* Sigma Xi; fel Am Phys Soc; Biophys Soc. *Res:* Paramagnetic resonance; electron spin-lattice relaxation; nuclear orientation; physics of solids, surfaces and biomolecules. *Mailing Add:* Univ Ill 402 Swanlund Admin Bldg 601 E John St Champaign IL 61820

STAPLETON, JAMES H, b Royal Oak, Mich, Feb 8, 31; m 63; c 3. MATHEMATICAL STATISTICS. *Educ:* Eastern Mich Univ, AB, 52; Purdue Univ, MS, 54, PhD(math statist), 57. *Prof Exp:* Statistician, Gen Elec Co, 57-58; from asst prof to assoc prof, 58-73, chmn, dept statist & probability, 69-75, PROF STATIST, MICH STATE UNIV, 73- *Concurrent Pos:* NSF fac sci fel, Univ Calif, Berkeley, 66-67; vis prof, Sch Econ, Univ Philippines, 78-79. *Mem:* Am Math Asn; Inst Math Statist; Am Statist Asn. *Res:* Linear models. *Mailing Add:* Dept Statist & Probability Mich State Univ East Lansing MI 48823

STAPLETON, JOHN F, b Brooklyn, NY, Jan 25, 21; m 50; c 5. MEDICINE. *Educ:* Fordham Univ, AB, 42; Georgetown Univ, MD, 45; Am Bd Internal Med, dipl, 53; Am Bd Cardiovasc Dis, dipl, 57. *Hon Degrees:* DSc, Georgetown Univ, 83. *Prof Exp:* Intern, Providence Hosp, Washington, DC, 45-46; resident med, Georgetown Univ, 49-51, clin instr med, Hosp, 52-54, from instr to asst prof, 54-65; assoc prof med & chief cardiol, Woman's Med Col Pa, 65-67; PROF MED & ASSOC DEAN, SCH MED, GEORGETOWN UNIV, 67-, MED DIR, UNIV HOSP, 67- *Concurrent Pos:* Nat Heart Inst res fel, Georgetown Univ, 51-52; dir med educ, St Vincent Hosp, Worcester, Mass, 54-65; attend physician, Philadelphia Vet Admin Hosp, 65-67; consult, Vet Admin Hosp, Wilmington, Del, 65-67. *Mem:* Am Heart Asn; Am Col Physicians; AMA. *Res:* Clinical cardiology; medical education; hospitals. *Mailing Add:* Dean Med NW 104 Med-Dent Georgetown Univ 3800 Reservoir Rd NW Washington DC 20007

STAPLEY, EDWARD OLLEY, b Brooklyn, NY, Sept 25, 27; m 49; c 3. MICROBIOLOGY. *Educ:* Rutgers Univ, BS, 50, MS, 54, PhD(microbiol), 59. *Prof Exp:* From jr microbiologist to sr microbiologist, Merck Sharp & Dohme Res Labs, 50-66, res fel microbiol, 66-69; asst dir basic microbiol res, 69-74, dir microbial chemotherapeut, 74-76, dir, 76-78, sr dir Basic Microbiol, 78-83, EXEC DIR, MERCK INST THERAPEUT RES, 84- *Honors & Awards:* Waksman Award, 90. *Mem:* AAAS; Am Acad Microbiol; Am Soc Microbiol; Soc Indust Microbiol (vpres, 74-75, pres, 76-77); NY Acad Sci; Infectious Dis Soc Am. *Res:* Isolation of microorganism; mutation; fermentation; ergosterol production by yeasts; microbial transformations of steroids; isolation and utility of antibiotic-resistant microorganisms; detection, characterization and evaluation of new antibiotics; microbial transformations of sulfur; detection and production of microbiol chemotherapeutics. *Mailing Add:* 110 Highland Ave Metuchen NJ 08840

STAPP, HENRY P, b Cleveland, Ohio, Mar 23, 28. PHYSICS. *Educ:* Univ Mich, BS, 50; Univ Calif, MA, 52, PhD, 55. *Prof Exp:* Theoret physicist, Lawrence Berkeley Lab, Univ Calif, 55-58 & Inst Theoret Physics, Swiss Fed Inst Technol, 58; THEORET PHYSICIST, LAWRENCE BERKELEY LAB, UNIV CALIF, 59- *Res:* Elementary particle physics. *Mailing Add:* Lawrence Berkeley Lab 1 Cyclotron Blvd Univ of Calif Berkeley CA 94720

STAPP, JOHN PAUL, b Bahia, Brazil, July 11, 10; US citizen; m 57. BIOPHYSICS. *Educ:* Baylor Univ, BA, 31, MA, 32; Univ Tex, PhD(biophys), 40; Univ Minn, BM & MD, 44; Am Bd Prev Med, dipl, 56. *Hon Degrees:* DSc, Baylor Univ, 56, NMex State Univ, 79. *Prof Exp:* Instr zool, Decatur Col, 32-34; proj officer, Aero Med Lab, Wright Field, US Air Force, 46-53, chief, Aero Med Field Lab, Holloman Air Force Base, 53-58, chief, Aero Med Lab, Wright Air Force Base, 58-60, asst to comdr aerospace med, Aerospace Med Ctr, Brooks Air Force Base, 60-65, resident in biophys, Armed Forces Inst Path, 65-67; prin med scientist, Nat Hwy Safety Bur, 67-72; adj prof & consult, Safety & Systs Mgt Ctr, Univ Southern Calif, 72-76; CHMN, SPACE CTR COMN, ALAMAGORDO, NMEX. *Concurrent Pos:* Vpres, Int Astron Fedn, 60; consult, Nat Acad Sci, Nat Traffic Safety Agency, Gen Serv Admin & Nat Bur Stand; permanent chmn, Annual Stapp Car Crash Conf, Soc Automotive Engrs, 55- *Honors & Awards:* Nat Medal of Technol, 91; Cheney Award Valor, 54; Gorgas Medal, Mil Surg Asn, 57; Cresson Medal, Franklin Inst, 73; Excalibur Award, Safety Adv Coun, US Dept Transp, 75; Honda Award Automotive Safety, Am Soc Mech Engrs, 85; FISITA Annual Medal, 90. *Mem:* Fel Am Inst Aeronaut & Astronaut (pres, 59); fel, Soc Automotive Engrs; Aerospace Med Asn (vpres, 57); AMA; Civil Aviation Med Asn (pres, 68); Sigma Xi. *Res:* Aerospace and industrial medicine; biodynamics of crashing and ditching; impact injury; medical biophysics. *Mailing Add:* PO Box 553 Alamogordo NM 88310-0553

STAPP, WILLIAM B, b Cleveland, Ohio, June 17, 29; m 55; c 3. ENVIRONMENTAL SCIENCES. *Educ:* Univ Mich, Ann Arbor, BA, 51, MA, 58, PhD(conserv), 63. *Prof Exp:* Instr sci, Cranbrook Sch Boys, Mich, 51-52, instr biol, 54-58; conservationist, Aullwood Audubon Ctr, Ohio, 58-59; instr conserv, 59-61, lectr, 63-64, from asst prof to assoc prof, 64-72, PROF NATURAL RESOURCES & CHAIRPERSON BEHAV & ENVIRON PROG, SCH NAT RESOURCES, UNIV MICH, ANN ARBOR, 72- *Concurrent Pos:* Res assoc, Cranbrook Inst Sci, Mich, 55-57; consult conserv, Ann Arbor Pub Schs, 61-68, youth progs, Nat Audubon Soc, New York, 65-66; consult, Int Film Bur, Chicago, Kalamazoo Nature Ctr, Mich & Creative Visuals, Tex, 66-68; consult environ educ prog, University City Pub Schs, Mo, 66-67, DeKalb Pub Schs, Ill, 67-68, Grand Haven Pub Schs, Mich, 67-, Raleigh County Sch Syst, WVa, 68-69, Toledo Bd Educ, Ohio, 70- & State of Alaska, 72-; consult, Nat Youth Movement Natural Beauty & Conserv, Washington, DC, 67-69, NJ Environ Educ Prog, 67-69, High Rock Interpretive Ctr, NY, 68-70, Seven Ponds Nature Ctr, 69- & Tapes Unlimited, Div Educ Unlimited Corp, 72-; consult environ interpretive ctr, Dept of Interior, 67-69 & environ educ, Dept HEW, 68 & div col support, Off Educ, 70-; consult audio-cassette series ecol, Am Soc Ecol Educ, 72- & proj man & environ, Nat TV Learning Systs, Miami, Fla, 72-; mem conserv comt, Mich Dept Pub Instr, 65-68; mem bd dirs, Drayton Plains Interpretive Ctr, Pontiac, Mich, 67-; prog dir, Ford Found grant, 68-70; mem working comt, Ann Arbor Environ Interpretive Ctr, 69-; mem bd dirs, Mich Pesticide Coun, 69-71; mem comn educ, Int Union Conserv Nature & Natural Resources, 69-; mem bd adv, Gill Inst Environ Studies, NJ, 70-; mem, Ecol Ctr Commun Coun, Washington, DC, 71- & Educ Resources Info Ctr Sci, Math & Environ Educ, Columbus, Ohio, 71-; fac adv, Ecol Ctr Ann Arbor, 71-, vpres & mem bd dirs, 72-; mem, Pub Sanit Systs, Los Angeles, Calif, 72- & Mich Pop Coun, 72-; chmn Gov Task Force Develop State Environ Educ Plan, 72-; mem task force estab guidelines environ educ elem & sec schs, Mich Dept Educ, 72-; dir environ educ, UNESCO, Paris, France, 74-76; mem, US Deleg World Conf Environ Educ, Tbilisi, USSR, 77; consult, UN Environ Prog, Nairobi, Kenya, 77, Asian Conf Environ Educ, Sri Lanka, 78, Nat Leadership Conf Environ Educ, Washington, DC, 78, Nat Audubon Soc, 78, Consumer Dynamics, Inc, 79, Int Union Conserv Nature & Natural Resources, Morges, Switz, 79-81, Unesco, 79-80, Morton Arboretum, 80, Environ Educ Plan Sri Lanka, Nat Park Ser, Int Br, 86-87, Annapurna Proj, King Mahendra Trust Nature Conserv, Katmandu, Nepal, 87, Mahaweli Environ Proj, Nat Park Serv,

Colombo, Sri Lanka, 87, adv, govt Venezuela, Nat Environ Educ Plan, Caracas, 78; team dir, Nat Univ Benin, W Africa, 87; adv comt, Int Conf Environ Educ Teachers & Students, Hague, Neth, 87; proj dir, Kellogg Found Grant, 87-89, Ford Motor Co Grant, 87-89, Ohio Pub Interest Found Grant, 85-87. *Honors & Awards:* Samuel Trask Dana Award Conserv, 62; Key Man Award, Conserv Educ Asn, 71; Lorado Taft Spec Recognition Award, 77; Walter Jeskie Award, 88. *Mem:* Am Nature Study Soc (vpres, 66-67, pres, 69-70); Conserv Educ Asn; Nat Audubon Soc; Asn Interpretive Naturalists (vpres, 69-71). *Res:* Environmental education and ecology; programs directed at helping man to develop a fuller understanding of environmental resource problems and his role in helping to resolve them. *Mailing Add:* Samuel T Dana Bldg Univ Mich Ann Arbor MI 48109

STAPPER, CHARLES HENRI, b Amsterdam, Neth, Mar 27, 34; US citizen; m 58. ELECTRICAL ENGINEERING, SOLID STATE PHYSICS. *Educ:* Mass Inst Technol, BS, 59, MS, 60; Univ Minn, Minneapolis, PhD(elec eng, physics), 67. *Prof Exp:* Coop student elec eng, Gen Radio Co, 57-59; elec engr, IBM Corp, 60-65; teaching assoc elec eng, Univ Minn, Minneapolis, 66-67; engr-mgr elec eng & physics, 67-69, SR ENGR, IBM CORP, 70- *Mem:* Sr mem Inst Elec & Electronics Engrs; Sigma Xi. *Res:* Application of mathematical theory to practical engineering and physics problems; statistical models for semiconductor devices, yields and manufacturing processes. *Mailing Add:* IBM Corp IBM GenTech Div Dept A20 Bldg 861-1 Essex Jct VT 05452

STAPRANS, ARMAND, b Riga, Latvia, Feb 28, 31; nat US; m 55; c 3. ELECTRICAL ENGINEERING. *Educ:* Univ Calif, BS, 54, MS, 55, PhD(elec eng), 59. *Prof Exp:* Asst elec eng, Univ Calif, 54-55, res asst, Microwave Tube Lab, 55-58; engr, 57-59, sr eng mgr, Super Power Opers, 69-71, mgr eng, High Power Microwave Opers, 71-75, mgr, Coupled Cavity Traveling Wave Tube Opers, 75-78, CHIEF ENGR, PALO ALTO MICROWAVE TUBE DIV, VARIAN ASSOCS, 78- *Mem:* Inst Elec & Electronics Engrs; Sigma Xi. *Res:* Microwave electronics; space-charge waves in periodic beams; electron optics; linear beam; super power tubes; coupled cavity traveling wave tubes; high power microwave windows; insulation of high voltages in vacuum; gyrotrons. *Mailing Add:* 445 Knoll Dr Los Altos CA 94022

STAR, AURA E, b New York, NY, Mar 15, 30; m 50; c 2. BOTANY, NATURAL PRODUCTS CHEMISTRY. *Educ:* Hunter Col, BA, 49; Mt Holyoke Col, MA, 51; Rutgers Univ, PhD(cytogenetics), 67. *Prof Exp:* Chemist, Baltimore Light & Power Co, Md, 51-52; instr biol, Morgan State Col, 52-53; from asst prof to assoc prof, 67-75, PROF BIOL, TRENTON STATE COL, 75- *Concurrent Pos:* Sigma Xi res grant, 63. *Mem:* AAAS; Bot Soc Am; Am Inst Biol Sci; Phytochem Soc; Torrey Bot Club. *Res:* Biochemical systematics of ferns and grasses; flavonoid chemistry; chemical biogeography of Pityrogramma; physiology of flavonoids in ferns; carotenogenesis in algae. *Mailing Add:* 26 White Pine Lane Princeton NJ 08540

STAR, JEFFREY L, b New York, NY, 1953; m 85; c 2. OCEANOGRAPHY. *Educ:* Mass Inst Technol, SB, 75; Univ Calif, San Diego, PhD(oceanog), 81. *Prof Exp:* DEVELOP ENGR, GEOG DEPT, UNIV CALIF, SANTA BARBARA, 83-; INSTR ENG, GEORGE WASHINGTON UNIV, 89- *Concurrent Pos:* res fel geog, Nat Ctr Geog Info Anal, 89-; mem, Comn Radiation Epidemiol Res Progs, Nat Res Coun, 91- *Mem:* AAAS; Am Soc Photogram & Remote Sensing; Asn Comput Mach. *Mailing Add:* Geog Dept Univ Calif Santa Barbara CA 93106

STAR, JOSEPH, b Far Rockaway, NY, Sept 2, 16; m 46; c 2. ELECTRONIC SYSTEMS ENGINEERING. *Educ:* Univ NC, BS, 37. *Prof Exp:* Engr, Radio Develop & Res Corp, 37-40; radio engr, Ft Monmouth Signal Lab, War Dept, 40-43; proj engr, Lab for Electronics, Inc, 46-48; staff engr, Hillyer Instrument Co, Inc, 48-51; chief electronics develop, Astrionics Div, Fairchild Engine & Aircraft Corp, 52-59; vpres eng, Instrument Systs Corp, 59-63; vpres & corp dir res & develop, Lundy Electronics & Systs, Inc, Glen Head, 63-73; INDEPENDENT CONSULT, 73- *Mem:* Sr mem Inst Elec & Electronics Engrs; Water Pollution Control Fedn; Int Asn Pollution Control. *Res:* Complex electronics for military and commercial purposes; derivation and investigation of new electronic and electromechanical devices; pollution control devices and systems; environmental systems engineering. *Mailing Add:* 186 Parkway Dr Roslyn Heights NY 11577

STAR, MARTIN LEON, b Brooklyn, NY, May 3, 28; m 55; c 1. COMPUTER SCIENCE, APPLIED STATISTICS. *Educ:* City Col New York, BBA, 48. *Prof Exp:* Qual control supvr, Sonotone Corp, 52-55; comput programmer, Remington Rand Univac, 55-56; systs analyst, Underwood Corp, 56-57; opers res analyst, Stevens Inst Technol, 57-59; asst programming mgr, Teleregister Corp, 59-61; programming supvr on-line systs, Nat Cash Register Co, 61-67; ASSOC DIR MGT SERV, S D LEIDESDORF & CO, NEW YORK, 67- *Concurrent Pos:* Partner, Eisner & Lubin, 78- *Mem:* Data Processing Mgt Asn (treas, 70-). *Res:* Statistical techniques in auditing; on-line systems and programming techniques. *Mailing Add:* 25 Willow Lane Great Neck NY 11023

STARACE, ANTHONY FRANCIS, b New York, NY, July 24, 45; m 68; c 2. ATOMIC PHYSICS. *Educ:* Columbia Univ, AB, 66; Univ Chicago, MS, 67, PhD(physics), 71. *Prof Exp:* Res assoc physics, Imp Col, Univ London, 71-72; from asst prof to assoc prof, 73-81, PROF PHYSICS, UNIV NEBR-LINCOLN, 81-, CHMN DEPT PHYSICS & ASTRON, 84- *Concurrent Pos:* Alfred P Sloan Found fel, 75-79; prin investr, Dept Energy res contract, 76-85, res grant, 85-; Alexander von Humboldt res fel, Freiburg Univ, Fed Repub Ger, 79-80; prin investr, NSF res grant, 81-; mem, Nat Res Coun Comt Atomic, Molecular & Optical Sci, 86- 89, 90-91; chmn, Div Atomic Molecular & Optical Physics, Am Phys Soc, 90-91. *Mem:* Fel Am Phys Soc; Brit Inst Physics. *Res:* Theory of atomic photoabsorption and photoionization, atomic collisions, atoms in high magnetic fields, and multiphoton ionization process. *Mailing Add:* Dept Physics & Astron Univ Nebr Lincoln NE 68588-0111

STARBIRD, ALFRED D, engineering; deceased, see previous edition for last biography

STARBIRD, MICHAEL PETER, b Los Angeles, Calif, July 10, 48; m 78; c 2. TOPOLOGY. *Educ:* Pomona Col, BA, 70; Univ Wis-Madison, MA, 73, PhD(math), 74. *Prof Exp:* From asst prof to assoc prof, 74-88, PROF MATH, UNIV TEX, AUSTIN, 88-, ASSOC DEAN, COL NATURAL SCI, 89- *Concurrent Pos:* Vis mem, Inst Advan Study, 78-79, Jet Propulsion Lab, 85-86. *Mem:* Am Math Asn; Math Asn Am. *Res:* Geometric topology. *Mailing Add:* Dept Math Univ Tex Austin TX 78712

STARCHER, BARRY CHAPIN, b Los Angeles, Calif, Dec 1, 38; m 60; c 3. NUTRITION, BIOCHEMISTRY. *Educ:* Univ Calif, Davis, BS & MS, 62; NC State Univ, PhD(biochem), 65. *Prof Exp:* Asst mem biochem, Inst Biomed Res, 66-70; asst prof path, Univ Colo Med Ctr, 70-72; asst prof biochem, Med Ctr, Univ Ala, Birmingham, 72-74; res asst prof, Pulmonary Div, Sch Med, Washington Univ, 74-80; mem fac, Dept Home Econ, Austin, 80-84, PROF BIOCHEM, UNIV TEX HEALTH CTR, TYLER, 84- *Mem:* Am Nutrit Soc. *Res:* Studies on the biochemistry of copper and zinc metabolism; enzyme induction in relation to stress, and the chemistry of the crosslinking amino acids in elastin; connective tissue components of lung; the role of elastin in calcification and arteriosclerosis. *Mailing Add:* Dept Biochem Univ Tex Health Ctr PO Box 2003 Tyler TX 77510

STARCHMAN, DALE EDWARD, b Wallace, Idaho, Apr 16, 41; m 69; c 4. MEDICAL PHYSICS. *Educ:* Pittsburg State Univ, BS, 63; Univ Kans, MS, 65, PhD(radiation biophys), 68; Am Bd Radiol, cert; Am Bd Health Physics, cert. *Prof Exp:* Chief health physicist, Ill Inst Technol Res Inst & radiol physicist, Inst Radiation Therapy Mercy Hosp & Med Ctr, Chicago, 68-71; PRES, MED PHYSICS SERV, INC, 71- *Concurrent Pos:* Consult, Aultman Hosp & Timken Mercy Hosp, Canton, Ohio & Northeast Ohio Conjoint Radiation Oncol Ctr, 71-; mem bd, Mideast Region Radiol Physics Ctr Bd Adv & prof, Univ Akron, 73-; prof & chmn radiation biophys, Northeastern Ohio Univ Col Med, 74-; mem, Adv Staff, Akron Gen Med Ctr, 75-; bd dirs, Am Asn Physicists Med, 84-86. *Mem:* Am Asn Physicists Med; Soc Nuclear Med; Health Physics Soc; Am Asn Therapeut Radiologists; Am Col Radiol; Sigma Xi. *Res:* Electron beam perturbation by cavities; radiation dosimetry; post irradiation atrophic changes of bone; information optimization with dose minimization in diagnostic radiology; radiation oncology treatment development. *Mailing Add:* Med Physics Serv Inc 5942 Easy Pace Circle NW Canton OH 44718

STARE, FREDRICK J, b Columbus, Wis, Apr 11, 10; c 3. NUTRITION. *Educ:* Univ Wis, PhD, 34; Univ Chicago, MD, 41. *Hon Degrees:* DSc, Suffolk Univ, 63, Trinity Col, Dublin, Ireland, 74, Muskingum Col, 77. *Prof Exp:* prof nutrit & chmn dept, 42-76, EMER PROF NUTRIT, SCH PUB HEALTH, HARVARD UNIV, 80- *Concurrent Pos:* Ed, Nutrit Rev, 42-68; co-founder & dir, Am Coun Sci & Health. *Honors & Awards:* Goldberger Award, AMA, 61. *Mem:* Am Chem Soc; Am Soc Biol Chemists; Am Inst Nutrit; Biochem Soc; Am Soc Clin Invest; hon mem Am Dietetic Asn; Soc Nutrit Educ. *Res:* Diet in relation to heart disease; exposing food faddism; heart disease and obesity. *Mailing Add:* 267 Cartwright Rd Wellesley MA 02181

STARFIELD, BARBARA HELEN, b Brooklyn, NY, Dec 18, 32; m 55; c 4. PEDIATRICS. *Educ:* Swarthmore Col, BA, 54; State Univ NY Downstate Med Ctr, MD, 59; Johns Hopkins Univ, MPH, 63. *Prof Exp:* Teaching asst anat, State Univ NY Downstate Med Ctr, 54-57; from intern to resident pediat, Harriet Lane Home, Johns Hopkins Hosp, 59-62; from instr to asst prof pediat, Sch Med, John's Hopkins Univ, 63-73, instr pub health admin, Sch Hyg & Pub Health, 65-66, from asst prof to assoc prof med care & hosps, 66-75, PROF & HEAD DIV HEALTH POLICY, SCH HYG & PUB HEALTH, JOHNS HOPKINS UNIV, 75-, JOINT APPOINTMENT PEDIAT, 73- *Concurrent Pos:* Nat Ctr Health Serv Res & Develop res scientist develop award; med dir, Community Nursing Proj, Dept Pediat & dir, Pediat Med Care Clin, Johns Hopkins Hosp, 63-66, asst dir community health, Comprehensive Child Care Proj & mem, Comt Planning & Develop, 65-67, pediatrician, dir, Pediat Clin Scholars Prog, 71-76; mem spec rev comt, Exp Med Care Rev Orgns, Dept Health, Educ & Welfare, mem, Health Serv Res Study Sect, 74-78, Nat Ctr Health Serv Res & Develop, 82-85; mem, Nat Prof Standards Rev Coun, 80-81; chmn health serv develop grants study sect, 86-91; mem, Nat Adv Coun, Agency, Health Care Policy & Res, 90-; First Annual Res Award, Am Pediat Asn, 90. *Honors & Awards:* Award, Enuresis Found, 67; George Armstrong Award, Am Pediat Asn, 83. *Mem:* Inst Med Nat Acad Sci; Am Pub Health Asn; Sigma Xi; Am Pediat Soc; Int Epidemiol Asn; Am Pediat Asn (pres, 80). *Res:* Cost effectiveness of health care; care of vulnerable population subgroups; epidemiology of child health and measurement of health status; primary care; health services research. *Mailing Add:* Sch Hyg & Pub Health Johns Hopkins Univ 624 N Broadway Baltimore MD 21205

STARICH, GALE HANSON, ENDOCRINOLOGY, BIOCHEMISTRY. *Educ:* Univ Nev, PhD(biochem), 81. *Prof Exp:* ASST PROF ENDOCRINOL, ANDERSON MED SCH, UNIV NEV, 84- *Mailing Add:* Internal Med Anderson Bldg Rm 108 Nev Med Sch Reno NV 89557

STARK, ANTONY ALBERT, b Seattle, Wash, Oct 29, 53; m 76. ATMOSPHERIC CHEMISTRY & PHYSICS. *Educ:* Calif Inst Technol, BS(physics) & BS(astron), 75; Princeton Univ, MA, 77, PhD(astrophys), 79. *Prof Exp:* Physicist, Lawrence Livermore Lab, 75-76; MEM TECH STAFF, AT&T BELL LABS, 79- *Concurrent Pos:* Vis lectr, Princeton Univ, 80- *Mem:* Int Astron Union; Am Astron Soc. *Res:* Galactic dynamics; interstellar medium; sub-millimeter-wave observations; constructing a submillimeter telescope and remote observatory at the South Pole. *Mailing Add:* AT&T Bell Labs Rm L231 Crawford Hill Holmdel-Keyport Rd Holmdel NJ 07733-0400

STARK, BENJAMIN CHAPMAN, b Saginaw, Mich, Nov 22, 49; m 79; c 2. RNA PROCESSING, PHYSIOLOGY OF RECOMBINANT BACTERIA. *Educ:* Univ Mich, Ann Arbor, BS, 71; Yale Univ, MPh, 74, PhD(biol), 77. *Prof Exp:* Fel biochem, Dept Bot, Wash State Univ, 77-79; res assoc, 79-82, vis asst prof, Dept Biol, Ind Univ, 82-83; asst prof, 83-88, ASSOC PROF, DEPT BIOL, ILL INST TECH, 88- *Concurrent Pos:* NSF energy related fel,

Wash State Univ, 77-78. *Mem:* AAAS; Sigma Xi. *Res:* RNA precursor processing and processing enzymes; small RNAs in higher plants; physiology and biochemistry of recombinant bacteria; detection of carcinogens in food. *Mailing Add:* Dept Biol Ill Inst Tech ITT Ctr Chicago IL 60616

STARK, DENNIS MICHAEL, b Baltimore, Md, May 16, 42. LABORATORY ANIMAL SCIENCE, INVITRO TOXICOLOGY. *Educ:* Univ Ga, DVM, 66; Cornell Univ, PhD(immunol), 69. *Prof Exp:* Res asst immunol, Cornell Univ, 66-69; asst prof, C W Post Col, Long Island Univ, 69-73; assoc prof path & dir animal facil, Med Ctr, NY Univ, 73-76; ASSOC PROF & DIR LAB ANIMAL RES CTR, ROCKEFELLER UNIV, 76- *Mem:* AAAS; Am Soc Microbiol; Am Asn Lab Animal Sci (pres, 87); Am Vet Med Asn; fel NY Acad Sci; Soc Toxicol. *Res:* Immunology of diseases of laboratory animals; in vitro measurement of cytotoxicity. *Mailing Add:* 500 E 63rd St No 23A New York NY 10021-6399

STARK, EGON, b Vienna, Austria, Sept 28, 20; nat US; m 48; c 3. MICROBIOLOGY. *Educ:* Univ Man, BS, 47, MS, 48; Purdue Univ, PhD(microbiol), 51. *Prof Exp:* Asst org chem, Univ Man, 44-47, asst microbiol, 45-48; asst bact, Purdue Univ, 48-51, Purdue Res Found Indust fel & res assoc microbiol, 51-53; consult microbiol, 53-54; sr res scientist, Joseph E Seagram & Sons, Ky, 54-66; prof biol, 66-87, EMER PROF BIOL, ROCHESTER INST TECHNOL, 87- *Mem:* AAAS; Am Soc Microbiol; Am Chem Soc. *Res:* Microbiology of bacteria, yeasts, fungi; taxonomy, physiology, enzymology, ecology, fermentations, water pollution, waste disposal; process of producing a heat-stable bacterial amylase. *Mailing Add:* Dept Biol Rochester Inst Technol Rochester NY 14623

STARK, FORREST OTTO, b Bay City, Mich, Mar 31, 30; m 58; c 6. ORGANOSILICON CHEMISTRY, SILICONE MATERIALS. *Educ:* Univ Pittsburgh, BS, 55; Pa State Univ, University Park, PhD(phys & organic chem), 61. *Prof Exp:* Chemist, Anal Labs, 51-61, Phys Chem Labs, 61-71, mgr, Med Tech Serv & Develop, 71-76, Resins & Chem Res, 76-79 & Elastomers Res, 79-81, dir, Silicone Res, 81-86, DIR HEALTH & ENVIRON SCI, DOW CORNING CORP, 86- *Concurrent Pos:* Mem, adv bd, Mich Molecular Inst, 84-85. *Mem:* Am Chem Soc; Soc Chem Indust. *Res:* Synthesis and characterization of silicone, silicon and ceramic materials for a wide range of applications in high technology industries such as aerospace, automotive, construction, medical and electronics. *Mailing Add:* 5311 Sunset Midland MI 48640-2535

STARK, FRANCIS C, JR, b Drumright, Okla, Mar 19, 19; m 41; c 2. HORTICULTURE. *Educ:* Okla State Univ, BS, 40; Univ Md, MS, 41, PhD(hort), 48. *Prof Exp:* From asst prof to prof veg crops, Univ Md, 45-64, prof hort & head dept, 64-74, chmn food sci fac, 66-73, provost, Div Agr & Life Sci, 74-80, actg vchancellor acad affairs, 81 & 82, EMER PROF HORT, UNIV MD, COLLEGE PARK, 80-, SPEC ASST TO VCHANCELLOR ACAD AFFAIRS, 82- *Concurrent Pos:* Chmn, Gov Comn on Migratory Labor, Md, 63-77; trustee, Lynchburg Col, 70-79; dir, Coun Agr Sci & Technol, 76-79. *Mem:* Fel AAAS; fel Am Soc Hort Sci. *Res:* Nutrition, physiology, breeding and culture of vegetable crops. *Mailing Add:* 7318 Radcliffe Dr College Park MD 20740-3024

STARK, GEORGE ROBERT, b New York, NY, July 4, 33; m 56; c 2. BIOCHEMISTRY. *Educ:* Columbia Univ, BA, 55, MA, 56, PhD(chem), 59. *Prof Exp:* Res assoc biochem, Rockefeller Inst, 59-61, asst prof, 61-63; from asst prof to prof biochem, Sch Med, Stanford Univ, 63- 83; ASSOC DIR, IMPERIAL CANCER RES FUND, LONDON, 83- *Concurrent Pos:* Guggenheim fel, 70-71. *Mem:* Nat Acad Sci; Am Chem Soc; Am Soc Biol Chem. *Res:* Chemistry and reactions of proteins; control of mammalian gene expression; structure-function relationships of enzymes; proteins of DNA tumor viruses; gene amplification; response to interferon. *Mailing Add:* Imp Cancer Res Fund PO Box 123 Lincolns Inn Fields London WC2A 3PX England

STARK, HAROLD EMIL, b San Diego, Calif, July 26, 20; m 44, 71; c 4. ENTOMOLOGY. *Educ:* San Diego State Col, BA, 43; Univ Utah, MS, 48; Univ Calif, PhD, 65. *Prof Exp:* Asst, Univ Utah, 46-48; jr entomologist, USPHS, 48-49, med entomologist, 50-63, trainin officer health mobilization, 63-64, ecol & chief vert-vector unit, Commun Dis Ctr, Ga, 64-68, entomologist, Walter Reed Army Inst Res, US Army Med Component/ SEATO, Bangkok, Thailand, 68-70, res entomologist, Ecol Invest, Ctr Dis Control, USPHS, 70-73; res zoologist, Environ & Ecol Br, Dugway Proving Ground, Utah, 73-79; RETIRED. *Mem:* Entom Soc Am. *Res:* Systematics of Siphonaptera; ecology of small wild rodents and fleas in relation to natural occurrence of plague and tularemia; preparation of training literature and audiovisuals for vector-borne diseases; preparation of environmental impact assessments and statements. *Mailing Add:* 1205N-400W Trenton UT 84338

STARK, HAROLD MEAD, b Los Angeles, Calif, Aug 6, 39; m 64. NUMBER THEORY. *Educ:* Calif Inst Technol, BS, 61; Univ Calif, Berkeley, MA, 63, PhD(math), 64. *Prof Exp:* From instr to asst prof math, Univ Mich, Ann Arbor, 64-66; asst prof, Univ Mich, Dearborn Ctr, 66-67; from asst prof to assoc prof, Univ Mich, Ann Arbor, 67-68; assoc prof, 69-72, PROF MATH, MASS INST TECHNOL, 72- *Concurrent Pos:* Off Naval Res fel, 67-68; Sloan fel, 68-70. *Mem:* Am Math Soc; Math Asn Am. *Res:* Analytic and elementary number theory with emphasis on zeta functions and applications to quadratic fields. *Mailing Add:* Dept Math Mass Inst Technol 77 Mass Ave Cambridge MA 02139

STARK, HENRY, b Antwerp, Belg, May 25, 38; US citizen; m 60; c 2. COHERENT OPTICS, PATTERN RECOGNITION. *Educ:* City Col New York, BS, 61; Columbia Univ, MS, 64, DrEngSc(elec eng), 68. *Prof Exp:* Asst proj engr, Bendix Corp, 61-62; res engr, Columbia Univ, 62-69; sr lectr elec eng, Israel Inst Technol, 69-70; from asst prof to assoc prof, Yale Univ, 70-77; assoc prof, 78-82, PROF, RENSSELAER POLYTECH INST, 82- *Concurrent Pos:* Lectr, City Col New York, 67 & 69; jr fel, Weizmann Inst Sci, 69; Frederick Gardner Cottrell grant, Res Corp, 71-72, NSF, 73- & Air Force Res & Develop Command, 78-; consult, Rome Air Develop Ctr & Gen Elec Corp Res, 78- *Mem:* Fel Optical Soc Am; Inst Elec & Electronics Engrs; NY Acad Sci; Sigma Xi. *Res:* Information science, coherent optics, image and data processing; systems; electrical communications; mathematical statistics. *Mailing Add:* Dept Elec & Systs Eng Jonsson Ctr Rensselaer Polytech Inst Troy NY 12180

STARK, J(OHN) P(AUL), JR, b Des Moines, Iowa, Nov 9, 38; m 59. PHYSICS, METALLURGY. *Educ:* Univ Okla, BS, 60, PhD(metall), 63. *Prof Exp:* From asst prof to assoc prof, 63-72, PROF MECH ENG, UNIV TEX, AUSTIN, 72- *Concurrent Pos:* Consult, Humble Oil & Refining Co, Tex, 63-67 & Tracor, Inc, 64-65; NSF res grant, 66-68 & 74-; Air Force Off Sci Res grant, 72-76. *Mem:* Am Inst Mining, Metall & Petrol Engrs; Am Soc Metals; Am Phys Soc. *Res:* Diffusion in solids; thermodynamics; phase transformations. *Mailing Add:* Dept Mech Eng Univ Tex Austin TX 78712

STARK, JAMES CORNELIUS, b Port Jefferson, NY, Sept 1, 41; m 63; c 3. ORGANIC CHEMISTRY, BIOCHEMISTRY. *Educ:* Eastern Nazarene Col, BS, 63; Purdue Univ, Lafayette, PhD(org chem), 69. *Prof Exp:* ASSOC PROF CHEM, EASTERN NAZARENE COL, 68- *Mem:* Am Chem Soc. *Res:* Preparation and reactions of polyhalo-organic compounds. *Mailing Add:* Dept Chem Eastern Nazarene Col Quincy MA 02170-2999

STARK, JEREMIAH MILTON, b Norfolk, Va, Apr 1, 22; m 49, 62; c 3. MATHEMATICS. *Educ:* US Coast Guard Acad, BS, 44; NTex State Col, BS, 46; Mass Inst Technol, SM, 49, PhD(math), 54. *Prof Exp:* Instr math, Mass Inst Technol, 49-52, mathematician instrumentation lab, 54-56; head dept, 56-77, PROF MATH, LAMAR UNIV, 56- *Concurrent Pos:* NSF sci fac fel, Stanford Univ, 63-64. *Mem:* AAAS; Am Math Soc; Math Asn Am; Soc Indust & Appl Math. *Res:* Analysis; complex variables. *Mailing Add:* Dept Math Lamar Univ PO Box 10047 Beaumont TX 77004

STARK, JOEL, b New York, NY, Nov 18, 30; m 50; c 2. SPEECH PATHOLOGY. *Educ:* Long Island Univ, BA, 50; Columbia Univ, MA, 51; NY Univ, PhD(speech), 56. *Prof Exp:* Instr speech, Long Island Univ, 51-54; asst prof, City Col New York, 54-65; assoc prof speech path, Sch Med, Stanford Univ, 65-68; assoc prof, 68-72, PROF COMMUN ARTS & SCI & DIR SPEECH & HEARING CTR, QUEENS COL, NY, 72- *Concurrent Pos:* Nat Inst Neurol Dis & Blindness fel, 62-64; mem coun except children. *Mem:* Am Speech & Hearing Asn; Am Asn Ment Deficiency. *Res:* Communications disorders; language development and disorders in children. *Mailing Add:* Speech & Hearing Ctr Queens Col Flushing NY 11367

STARK, JOHN, JR, b Headland, Ala, Aug 26, 21; m 58; c 2. CHEMICAL ENGINEERING. *Educ:* Univ Ala, BS, 48, MS, 61, PhD(chem eng), 64. *Prof Exp:* Chemist, Astilleros Dominicanos, 55-60; asst prof chem eng, Univ Ala, 64-65; assoc prof, Univ SAla, 65-77, prof chem eng & chmn dept, 77-86; MEM FAC, UNIT TEX, AUSTIN, 86- *Concurrent Pos:* Consult, NASA, Ala, 64-68. *Mem:* Am Chem Soc; Am Inst Chem Engrs. *Res:* Turbo grid-plate efficiencies; distribution of noncondensable gases in liquids; effect of surface waves on evaporation rates. *Mailing Add:* Univ Tex Austin TX 78712

STARK, JOHN HOWARD, b Port Jefferson, NY, Sept 1, 41; m 63; c 3. PAPERMAKING CHEMISTRY. *Educ:* Eastern Nazarene Col, BS, 63; Purdue Univ, MS, 65, PhD(biochem), 69. *Prof Exp:* SR RES ASSOC MAT RES, INT PAPER, 87- *Honors & Awards:* Texaco Res Award, Am Chem Soc. *Mem:* Am Chem Soc; Sigma Xi. *Res:* Papermaking chemistry; application of new materials to improve product performance. *Mailing Add:* Int Paper Long Meadow Rd Tuxedo NY 10987

STARK, LARRY GENE, b Abilene, Kans, Dec 31, 38; m 76; c 2. PHARMACOLOGY. *Educ:* Univ Kans, BS, 61, MS, 63; Stanford Univ, PhD(pharmacol), 68. *Prof Exp:* From asst prof to assoc prof, 69-82, asst dean curricular affairs, 80-83, PROF PHARMACOL, SCH MED, UNIV CALIF, DAVIS, 82-, CHMN DEPT, 83- *Concurrent Pos:* NIH fel, Univ Chicago, 68-69; guest prof, Pharmacol Inst, Univ Bern, Switz, 83-84. *Mem:* Soc Neurosci; Am Soc Pharmacol & Exp Therapeut; Am Epilepsy Soc. *Res:* Anticonvulsant drugs and models of epilepsy. *Mailing Add:* 850 Burr St Davis CA 95616

STARK, LAWRENCE, b New York, NY, Feb 21, 26; m 49; c 3. BIOENGINEERING. *Educ:* Columbia Univ, AB, 45; Albany Med Col, MD, 48; Am Bd Psychiat & Neurol, dipl, 57. *Hon Degrees:* ScD(hon), SUNY, 88. *Prof Exp:* Intern, US Naval Hosp, St Albans, 48-49; res asst biochem, Oxford Univ & mem, Trinity Col, 49-50; res asst physiol, Univ Col, London, 50-51; asst prof physiol & pharmacol & res assoc neurophysiol & neuromuscular physiol, NY Med Col, 51; from instr to asst prof neurol & assoc physician, Yale Univ, 55-60; head neurol sect, Ctr Commun Sci, Res Lab Electronics & Electronic Systs Lab, Mass Inst Technol, 60-65; prof bioeng, neurol & physiol & chmn biomed eng dept, Univ Ill, Chicago Circle, 65-68; PROF PHYSIOL OPTICS, UNIV CALIF, BERKELEY, 68-, PROF ENGR SCI, 69- *Concurrent Pos:* Fel neurol & EEG, Neurol Inst, Columbia Univ & Presby Hosp, 51-52; fel neurol, Sch Med, Yale Univ, 54-55; fel, Mass Inst Technol, 60-65; fel neurol, Mass Gen Hosp, 60-65; Guggenheim fel, 68-70; vis app, Univ Col, London, 50-51, Nobel Inst Neurophysiol, Stockholm, 57, Harvard Univ, 63-65, Univ Calif, Los Angeles, 65 & Stanford Univ, 68 & 75; dir, Biosysts, Inc, Cambridge, 62-67 & Biocontacts, Inc, Berkeley, 69-73; chmn neurosci work session, Math Concepts of Cent Nerv Syst, 64, Gordon Conf Biomath, 65 & bioeng training comt, Nat Inst Gen Med Sci, 67-68; consult, NIH, NSF & var indust companies; assoc ed or ed bd mem, Math Biosci, Inst Elec & Electronics Eng-SMC, Brain Res, J Appl Physiol, J Neurosci & Comput in Biol & Med; prof neurol, Univ Calif, San Francisco, 74- *Honors & Awards:* Morlock Award in Biomed Eng, 77; Taylor Award, 89. *Mem:* Am Physiol Soc; Biophys Soc; Am Acad Neurol; fel Inst Elec & Electronic Engrs; Asn Comput Mach. *Res:* Application of communication and information theory to neurophysiology; normal and abnormal neurological control systems; cybernetics; pattern recognition and artificial intelligence; information flow in biological evolution and economic theory; neurological control theory, especially applied to ocular motor systems; robotic manipulation, locomotion and vision compared with biological mechanisms. *Mailing Add:* 481 Minor Hall Univ Calif Berkeley CA 94720

STARK, MARVIN MICHAEL, b Mich, Mar 14, 21; c 3. DENTISTRY. *Educ:* Univ Calif, Los Angeles, AB, 48, DDS, 52. *Prof Exp:* PROF OPER DENT & ORAL BIOL, SCH DENT, UNIV CALIF, SAN FRANCISCO, 53-; CHIEF DENT OFFICER, STATE OF CALIF DEPT HEALTH, 75- *Concurrent Pos:* Res fel dent med, Sch Dent Med, Harvard Univ, 52-53. *Mem:* Am Asn Endodont; fel Int Col Dent; fel Am Col Dent; Am Dent Asn; Int Asn Dent Res. *Mailing Add:* Dept Dent Univ Calif San Francisco CA 94143

STARK, NATHAN JULIUS, b Minneapolis, Minn, Nov 9, 20; m 43; c 4. HEALTH LAW. *Educ:* US Merchant Marine Acad, BS, 43; Chicago Kent Col Law, JD, 48. *Hon Degrees:* LLD, Park Col, 69, Univ Mo; LHD, Hahnemann Univ. *Prof Exp:* Plant mgr, Englander Co, Inc, Chicago, 49-51; partner law firm, Downey, Abrams, Stark & Sullivan, Kansas City, 52-53; vpres, Rival Mfg Co, Kansas City, 54-59; sr vpres opers, Hallmark Cards, Inc, Kansas City, 59-74; vchancellor, Schs Health Professions & pres, Univ Health Ctr, Univ Pittsburgh, 74-79, prof health sci & sr vchancellor, Grad Sch Pub Health, 80-86; undersecy, US Dept Health & Human Serv, 79-80; RETIRED. *Concurrent Pos:* Secy, Eddie Jacobson Mem Found, 60-; pres & chmn, Kansas City Gen Hosp & Med Ctr, 62-74; vchmn health ins benefits adv comt, HEW, 65-70, secy task force Medicaid, 69-70, chmn adv comn incentive reimbursement exp, 68-70, chmn capital investment conf, HEW-Health Resources Admin, 76; dir, Woolf Bros, Inc, ERC Corp & Nat Fidelity Ins Co Hallmark Continental Ltd, Ireland, 71-73; pres & chmn bd, Crown Ctr Redevelop Corp, Kansas City, 72-74; chmn community hosp-med staff group pract prog, Robert Wood Johnson Found, 74-79; mem med malpract adv comt, Inst of Med of Nat Acad Sci, 75-79; mem bd, Am Nurses Found, 75-79; mem tech bd, Milbank Mem Fund, 76-79; fel, Hastings Ctr, mem, bd trustees, 81; mem, exec bd, Nat Bd Med Examiners. *Honors & Awards:* Trustee Award, Am Med Asn, 74. *Mem:* Inst of Med of Nat Acad Sci; hon mem Am Hosp Asn; hon mem Am Col Hosp Adminrs; AMA. *Mailing Add:* 1401 New York Ave NW Washington DC 20005

STARK, NELLIE MAY, b Norwich, Conn, Nov 20, 33; m 62. FOREST ECOLOGY, SOIL ECOLOGY. *Educ:* Conn Col, BA, 56; Duke Univ, MA, 58, PhD(plant ecol, bot), 62. *Prof Exp:* Botanist, Pac Southwest Forest & Range Exp Sta, US Forest Serv, 58-64; res assoc, Lab Atmospheric Physics, Desert Res Inst, Univ Nev, Reno, 64-72; assoc prof, 72-79, PROF FORESTRY, UNIV MONT, 79- *Concurrent Pos:* Mem, Alph Helix Res Exped for Desert Res Inst, Brazil & Peru, 67; Int Biol Prog grants & NSF grants, 72-73. *Mem:* Ecol Soc Am; Bot Soc Am; Am Inst Biol Sci; Soc Am Foresters; Int Soc Trop Foresters. *Res:* Nutrient cycling and soil ecology in tropical and temperate forests; fire and logging ecology; applied concept of the biological life of a soil to land use management; forest ecology; soil chemistry; xylem sap chemistry; chemical perturbation of forest ecosystems; aging in trees. *Mailing Add:* Sch Forestry Univ Mont Missoula MT 59812

STARK, PAUL, b Philadelphia, Pa, Feb 1, 29; m 52; c 3. PHARMACOLOGY, PHYSIOLOGY. *Educ:* McGill Univ, BSc, 49; Univ Rochester, PhD(pharmacol), 63; Sch Law, Ind Univ, JD, 77. *Prof Exp:* Chief prod biochem, Gerber Prod Co, 52-55; res technician, Stromberg-Carlson Co, 55-57; res chemist, Allerton Chem, 57-60; sr pharmacologist, Eli Lilly & Co, 63-66, res scientist, 67-72, res assoc, 72-84; PRES, INT CLIN RES CORP, 84- *Concurrent Pos:* Assoc prof, Sch Med, Ind Univ, Indianapolis. *Mem:* Int Col Neuropsychopharmacol; Am Soc Pharmacol & Exp Therapeut; Am Physiol Soc; Am Bar Asn; Soc Neurosci. *Res:* Neuropharmacological and psychopharmacological techniques in the study of neuro-transmitters within the central nervous system; clinical evaluation of psychotropic drugs. *Mailing Add:* 160 Olde Mill Circle S Dr Indianapolis IN 46260

STARK, PHILIP HERALD, b Iowa City, Iowa, Mar 2, 36; m 81; c 1. GEOLOGY. *Educ:* Univ Okla, BS, 58; Univ Wis, MS, 61, PhD(geol), 63. *Prof Exp:* Explor geologist, Mobil Oil Corp, 63-65, sr explor geologist, 65-66, regional comput coordr, 66-69; mgr geol applns, 69-74, dir tech applns, 74-77, vpres spec projs, 77-84, exec vpres Int Opers, 84-88, VPRES TECH MKT, PETROL INFO CORP, 88- *Concurrent Pos:* Lectr, Continuing Educ Prog, Am Asn Petrol Geologists. *Mem:* Soc Econ Paleontologists & Mineralogists; Am Asn Petrol Geologists. *Res:* Stratigraphy and micropaleontology of Paleozoic flysch facies; computer applications in geology for petroleum exploration. *Mailing Add:* Petrol Info Corp PO Box 2612 Denver CO 80202

STARK, RICHARD B, b Conrad, Iowa, Mar 31, 15; m 67. PLASTIC SURGERY. *Educ:* Stanford Univ, AB, 36; Cornell Univ, MD, 41; Am Bd Plastic Surg, dipl, 52 & 78. *Prof Exp:* Intern, Peter Bent Brigham & Children's Hosps, Boston, Mass, 41-42; resident, Children's Hosp, 42; plastic surgeon, Northington Gen Hosp, Tuscaloosa, Ala, 45-46 & Percy Jones Gen Hosp, Battle Creek, Mich, 46; surgeon, Kingsbridge Vet Hosp & NY Hosp, 47-50; from inst to assoc prof clin surg, Cornell Med Col, 50-55; from instr to assoc prof, 55-73, PROF CLIN SURG, COL PHYSICIANS & SURGEONS, COLUMBIA UNIV, 73- *Concurrent Pos:* Fel, Med Sch, Stanford Univ, 46-47; plastic surgeon, NY Hosp, 48; attend surgeon chg plastic surg, St Luke's Hosp, 58-77; vis prof, Univ Tex, 65, Univ Mich, 66, Walter Reed Med Ctr, 70, Univ Man, 71 & Univ Pa, 82, SAM, 73, France & Spain, 80; from vpres to pres, Am Bd Plastic Surg, 66-68; ed, Annals Plastic Surg, 77-82. *Honors & Awards:* Res Prize, Am Soc Plastic & Reconstruct Surg Found, 51; Medal of Honor, Vietnam, 67 & 69; Order of San Carlos, Colombia, 69. *Mem:* AAAS; Soc Univ Surg; AMA; Am Asn Plastic Surg; Am Col Surgeons; Am Surg Asn. *Res:* Circulation in skin grafts; homologous transplants of skin; pathogenesis of harelip and cleft palate; aesthetic surgery. *Mailing Add:* 35 E 75th St New York NY 10021

STARK, RICHARD HARLAN, b Ozawkie, Kans, Dec 5, 16; m 42; c 4. COMPUTER SCIENCE. *Educ:* Univ Kans, AB, 38; Northwestern Univ, MS, 42, PhD(math), 46. *Prof Exp:* Jr physicist, US Naval Ord Lab, 42-43 & Los Alamos Sci Lab, 44-45; instr math, Northwestern Univ, 46-48; mathematician, Los Alamos Sci Lab, 48-51; mgr math anal, Knolls Atomic Power Lab, Gen Elec Co, 52-56, mgr math & comput oper, Atomic Power Equip Dept, 56-61, consult analyst, Comput Dept, 62-64; from assoc prof to prof math & info sci, Wash State Univ, 64-69; PROF COMPUT SCI, NMEX STATE UNIV, 69- *Mem:* Soc Indust & Appl Math; Asn Comput Mach. *Res:* Proofs of program validity. *Mailing Add:* Dept Comput Sci NMex State Univ Las Cruces NM 88001

STARK, ROBERT M, b New York, NY, Feb 6, 30; m 55; c 4. OPERATIONS RESEARCH, CIVIL ENGINEERING. *Educ:* Johns Hopkins Univ, AB, 51; Univ Mich, MA, 52; Univ Del, PhD(appl sci), 65. *Prof Exp:* Instr physics, Rochester Inst Technol, 55-57; asst prof math & asst dean col eng, Cleveland State Univ, 57-62; from ast prof to assoc prof civil eng, statist & comput sci, Univ Del, 62-72; vis assoc prof civil eng, Mass Inst Technol, 72-73; PROF CIVIL ENG & MATH SCI, UNIV DEL, 76- *Concurrent Pos:* Res physicist, Bausch & Lomb Optical Co, 55-56; consult various industs & govt agencies. *Mem:* Fel AAAS; Opers Res Soc Am; Am Soc Civil Engrs; Am Soc Eng Educ; Nat Coun Teachers Math. *Res:* Civil engineering systems; applied probability; engineering management. *Mailing Add:* Dept Math Sci Univ Del Newark DE 19716

STARK, RONALD WILLIAM, b Can, Dec 4, 22; nat US; m 44; c 2. FOREST ENTOMOLOGY. *Educ:* Univ Toronto, BScF, 48, MA, 51; Univ BC, PhD(forest entom), 58. *Prof Exp:* Agr res officer, Div Forest Biol, Sci Serv, Can Dept Agr, 48-59; asst prof entom & asst entomologist, Agr Exp Sta, Univ Calif, Berkeley, 59-61, from assoc prof to prof, 61-70, vchmn dept entom & parasitol, 68-70, entomologist, 61-70; grad dean, coordr res, 70-77, PROF FORESTRY & ENTOM, UNIV IDAHO, 78-, EMER PROF. *Concurrent Pos:* NSF sr fel, 67-68; collabr, Pac Southwest Forest & Range Exp Sta, US Forest Serv; Am rep & chmn working group forest entom, Int Union Forest Res Orgn; proj leader, Pest Mgt Prog Pine Back Beetle Ecosyst, Int Biol Prog; mem, USDA comt scientists, 76-78, dep prog mgr, USDA Expanded Douglas-fir Tussock Moth Prog, Portland, Ore, 77-78; prog mgr, Int Spruce Budworms Prog, Western Component, US Forest Serv. *Honors & Awards:* Gold Medalist, Entom Soc Can. *Mem:* AAAS; Soc Am Foresters; fel Entom Soc Am; Ecol Soc Am; Entom Soc Can. *Res:* Population dynamics; integrated pest management; research management. *Mailing Add:* 520 S First Sandpoint ID 83864-1206

STARK, ROYAL WILLIAM, b Wellington, Ohio, Apr 30, 37; m 62; c 2. SOLID STATE PHYSICS, LOW TEMPERATURE PHYSICS. *Educ:* Case Inst Technol, BS, 59, MS, 61, PhD(physics), 62. *Prof Exp:* Res assoc solid state physics, Case Inst Technol, 62; from instr to prof, Univ Chicago & Inst Study Metals, 63-72; PROF PHYSICS, UNIV ARIZ, 72- *Concurrent Pos:* Alfred P Sloan res fel, 64-70. *Mem:* Am Phys Soc. *Res:* Electronic properties of metals; magnetic breakdown; Fermi surface and band structure; ferromagnetism; plasma effects. *Mailing Add:* 1285 N Speedway Pl Tucson AZ 85715

STARK, RUTH E, b Philadelphia, Sept 22, 50. NUCLEAR MAGNETIC RESONANCE. *Educ:* Cornell Univ, AB, 72; Univ Calif, San Diego, PhD(phys chem), 77. *Prof Exp:* Res assoc fel, Nat Magnet Lab, Mass Inst Technol, 77-79; asst prof chem, Amherst Col, 79-85; ASSOC PROF CHEM, COL STATEN ISLAND, CITY UNIV NEW YORK, 85- *Mem:* Am Chem Soc; Sigma Xi; Am Women Sci; Biophys Soc. *Mailing Add:* Dept Chem Amherst Col Amherst MA 01022

STARK, WALTER ALFRED, JR, b San Antonio, Tex, Aug 30, 40; m 68; c 1. PHYSICAL CHEMISTRY. *Educ:* Princeton Univ, AB, 62; Univ Calif, Berkeley, PhD(chem), 67; Univ NMex, MBA, 82. *Prof Exp:* Staff mem mat sci, Sandia Labs, 67-73; STAFF MEM MATS SCI, LOS ALAMOS NAT LAB, 73- *Mem:* Am Phys Soc. *Res:* Transport properties of materials, especially diffusion and permeation; high temperature materials compatibility. *Mailing Add:* 275 Kimberly Los Alamos NM 87544

STARK, WILLIAM POLSON, b French Camp, Calif, Dec 30, 43; m 65; c 2. AQUATIC ENTOMOLOGY, SYSTEMATICS. *Educ:* Southeastern Okla State Univ, BS, 65; NTex State Univ, MS, 72; Univ Utah, PhD(biol), 74. *Prof Exp:* Asst prof, 76-80, assoc prof, 80-86, PROF BIOL, MISS COL, 86- *Concurrent Pos:* Adj prof biol, NTex State Univ, 77-87; res assoc, Fla Collection Arthropods, 77-87. *Mem:* Entom Soc Am; NAm Benthological Soc; Soc Syst Zool. *Res:* Systematics and biology of Nearctic Plecoptera. *Mailing Add:* 1603 Laurelwood Dr Clinton MS 39056

STARK, WILLIAM RICHARD, b Lexington, Ky, Apr 28, 45; m 68; c 1. DISTRIBUTED COMPUTATION & ALGEBRAIC ASPECTS OF COMPUTATION. *Educ:* Univ Ky, BS, 68; Univ Wis-Madison, PhD(math), 75. *Prof Exp:* Fel, Univ Wis-Madison, 69-74; instr math, Univ Tex, Austin, 74-78; asst prof, Calif State Univ, San Jose, 78-79; assoc prof math, Univ SFla, 78-85; mem, Tech Staff, AT&T Bell Labs, 85-87; ASSOC DIR, INST CONSTRUCT MATH, UNIV SFLA, 87- *Concurrent Pos:* Consult, US Army, 84-85. *Mem:* Am Math Soc; Asn Symbolic Logic; Asn Comput Mach; Math Asn Am; Inst Elec & Electronic Engrs Computer Soc; Union Concerned Scientists. *Res:* Algebraic aspects of asynchronized, asynchronous distributed computation; artificial life; computational semantics; logics of knowledge. *Mailing Add:* Math Univ SFla Tampa FL 33620

STARKE, ALBERT CARL, JR, b Cleveland, Ohio, Jan 14, 16; m 41; c 3. ORGANIC CHEMISTRY, BIOCHEMISTRY. *Educ:* Fla Southern Col, BS, 36; Northwestern Univ, PhD(chem), 40. *Prof Exp:* Lab instr, Dent Sch, Northwestern Univ, 37-40, Nat Defense Res Comt fel, 40-42; res chemist, GAF Corp, 42-46, patent searcher, 46-47, patent liaison, 48-50, patent agt, 50-55, supvr tech info serv, 55-66, mgr tech info, 66-72; res specialist, Univ Conn, 72-81; info mgr, New Eng Res Appln Ctr, 72-81; RETIRED. *Concurrent Pos:* Mem, Franklin Inst, Chem Abstr communicator, 66- *Mem:* Fel AAAS; Am Chem Soc; Am Soc Info Sci; fel Am Inst Chemists. *Res:* Physiological and synthetic organic chemistry; color photography; patent soliciting and prosecution; warfare agents; polymers; storage and retrieval of technical information; computer systems design; documentation; computer searching. *Mailing Add:* 10 Carriage Cove Way Sanford FL 32773-6012

STARKE, EDGAR ARLIN, JR, b Richmond, Va, May 10, 36; m 61; c 2. PHYSICAL METALLURGY. *Educ:* Va Polytech Inst, BS, 60; Univ Ill, MS, 61; Univ Fla, PhD(metall), 64. *Prof Exp:* Res metallurgist, Savannah River Lab, E I du Pont de Nemours & Co, Inc, 61-62; from asst prof to prof metall, Ga Inst Technol, 64-82, dir, Fracture & Fatigue Res Lab, 78-82; EARNEST OGLESBY PROF MAT SCI, UNIV VA, 83-, DEAN, SCH ENG & APPL SCI, 84- *Concurrent Pos:* Consult, Lockheed Co, Northrop Corp, Reynolds Metals Co, Arco Metals Co, Kaiser Aluminum; vis scientist, Oak Ridge Nat Lab, 67 & Max Planck Inst Metall Res, 71; Sigma Xi res award, 70, Nonferrous Div Wire Asn, 72; mem res coun, Defense Advan Res Projs Agency, acad adv comt, Aluminum Asn, bd rev, Metall Trans Asn. *Mem:* Am Inst Mining, Metall & Petrol Engrs. *Res:* Strengthening mechanisms; alloy theory; fracture and fatigue; aluminum alloys. *Mailing Add:* Sch Eng & Appl Sci Univ Va Charlottesville VA 22901

STARKEY, EUGENE EDWARD, b Yakima, Wash, July 14, 26; m 54; c 3. DAIRY SCIENCE. *Educ:* Calif State Polytech Col, BS, 52; Univ Wis, MS, 54, PhD(dairy husb, genetics), 58. *Prof Exp:* Asst dairying, Univ Wis, 52-55; dairy husbandman, Dairy Husb Res Br, Agr Res Serv, USDA, 55-57; asst prof dairy prod, Utah State Univ, 57-60; prof dairy prod, Univ Wis-Madison, 60-78; PROF & HEAD DEPT DAIRY SCI, CALIF POLYTECH STATE UNIV, 78- *Mem:* Am Dairy Sci Asn. *Res:* Dairy cattle breeding sire selection and evaluation of environmental influences on production. *Mailing Add:* 1730 Portola San Luis Obispo CA 53711

STARKEY, FRANK DAVID, b Indianapolis, Ind, Aug 6, 44; m 67; c 2. ORGANIC CHEMISTRY. *Educ:* Wabash Col, AB, 66; Brown Univ, PhD(org chem), 73. *Prof Exp:* Asst prof chem, Ill Wesleyan Univ, 71-77, assoc prof, 77-80, head dept, 79-80. *Concurrent Pos:* Vis res assoc, Univ Minn, 77-78. *Mem:* Am Chem Soc; AAAS. *Res:* Carbonium ion chemistry; synthesis and reaction of various substrates that give carbonium ions. *Mailing Add:* 1274 Regent St Schenectady NY 12309-5351

STARKEY, JOHN, b Manchester, Eng, Aug 11, 36; m 58; c 2. STRUCTURALGEOLOGY, PETROFABRIC ANALYSIS. *Educ:* Univ Liverpool, BSc, 57, PhD(geol), 60. *Prof Exp:* NATO & Dept Sci & Indust Res Gt Brit res fels geol, Inst Crystallog & Petrol, Swiss Fed Inst Technol, 60-62; Miller res fel, Univ Calif, Berkeley, 62-65; from asst prof to assoc prof, 65,79, PROF GEOL, UNIV WESTEN ONT, 79- *Concurrent Pos:* Leverhulme Europ fel, 60-61; Royal Soc bursary, Imp Col, Univ London, 71-72; vis prof geochem, Fed Univ Bahia, Salvador, Brazil, 75-77; vis prof geol, Monash Univ, Australia, 78; vis prof geol, Eidgenoessiche Technische Hochschule, Zurich, Switzerland, 78-79. *Mem:* Brit Mineral Soc; Am Mineral Soc; Mineral Soc Can; Geol Asn Can; Sigma Xi. *Res:* Petrofabric analysis of rocks, primarily by x-ray techniques image analysis, with particular reference to the microstructures of deformed rocks; crystallography of plagioclase feldspars and their twinning; crystal chemistry of rock forming minerals. *Mailing Add:* Dept Geol Univ Western Ont London ON N6A 3B7 Can

STARKEY, PAUL EDWARD, b Fultonham, Ohio, Dec 9, 20; m 42; c 4. PEDODONTICS, DENTISTRY. *Educ:* Ind Univ, DDS, 43; Am Bd Pedodont, dipl, 58. *Prof Exp:* Instr pedodontics, Ohio State Univ, 55-56; assoc prof, Ind Univ, Indianapolis, 59-63, chmn clin div, 61 & dept pedodontics, 68, prof pedodontics, Sch Dent, 63-88, EMER PROF PEDIAT DENT, 88- *Concurrent Pos:* Pvt pract, 46-59; exam mem, Am Bd Pedodont, 69. *Honors & Awards:* Frederick Bachman Lieber Distinguished Teaching Award, Ind Univ, 68. *Mem:* Am Soc Dent for Children (pres, 67); Am Acad Pedodont; Am Dent Asn; Asn Pedodontics Diplomates; Am Asn Dent Schs. *Res:* Clinical children's dentistry; educational research. *Mailing Add:* 1760 E 110th Indianapolis IN 46280

STARKEY, WALTER L(EROY), b Minneapolis, Minn, Oct 5, 20; m 49; c 2. MECHANICAL ENGINEERING. *Educ:* Univ Louisville, BME, 43; Ohio State Univ, MSc, 47, PhD(mech eng), 50. *Prof Exp:* Instr mech eng, Univ Louisville, 43-46; from instr to prof, 47-78, EMER PROF MECH ENG, OHIO STATE UNIV, 78- *Concurrent Pos:* Consult, 50- *Honors & Awards:* Mach Design Award, Am Soc Mech Engrs, 71. *Mem:* Fel Am Soc Mech Engrs. *Res:* Fatigue of metals; mechanics of materials; machine dynamics; mechanical design of machinery. *Mailing Add:* 7000 Coffman Rd Dublin OH 43017

STARKOVSKY, NICOLAS ALEXIS, b Alexandria, Egypt, Jan 15, 22; US citizen; m 59; c 2. ORGANIC CHEMISTRY. *Educ:* Univ Cairo, BS, 46, MS, 54, PhD(chem), 56. *Prof Exp:* Res chemist, Memphis Chem Co, Egypt, 51-60; fel, Columbia Univ, 60-61; res chemist, Dow Chem Co, 61-64; dir res & develop, Collab Res, Inc, Mass, 64-70; dir proj develop, Ortho Res Found, NJ, 71-72; sr scientist, Wampole Div, Carter Wallace, Inc, Cranbury, 73-80; dir pro develop, Meloy Labs, Va, 80-82; RETIRED. *Concurrent Pos:* Vol ESL teacher, Fairfax Co Pub Schs Adult Ed, N Va Community Col & N Va Literary Coun, 82- *Mem:* Am Chem Soc. *Res:* Immunology; clinical medicine. *Mailing Add:* 6352 Silas Burke Burke VA 22015-3447

STARKS, AUBRIE NEAL, JR, b Dermott, Ark, Aug 20, 46. ANALYTICAL CHEMISTRY. *Educ:* Southern Ill Univ, Carbondale, BA, 67; Univ Ark, Fayetteville, PhD(chem), 75. *Prof Exp:* Instr chem, Univ Ark, Fayetteville, 73-74; intern, Hendrix Col, 74-75; asst prof chem, Thiel Col, 75-80; MEM FAC, CHEM DEPT, NORTHEAST LOUISIANA UNIV, 80- *Mem:* Am Chem Soc. *Res:* Photovoltammetric investigation of transition metal complexes for photocurrents generated in optically-shielded electrode systems. *Mailing Add:* 2012 Elton Ft Worth TX 76117-6507

STARKS, KENNETH JAMES, b Ft Worth, Tex, July 27, 24; m 51; c 2. ENTOMOLOGY. *Educ:* Univ Okla, BS, 50, MS, 51; Iowa State Univ, PhD(entom), 54. *Prof Exp:* Asst prof, Univ Ky, 53-61; mem staff, USDA, Uganda, 61-69; prof entom, Okla State Univ, 69; res & location leader, Agr Res Serv, USDA, 69-85; RETIRED. *Mem:* Entom Soc Am. *Res:* Grain insects investigations. *Mailing Add:* 724 W Ute Ave Stillwater OK 74075

STARKS, THOMAS HAROLD, b Owatonna, Minn, Aug 19, 30; m 59; c 2. STATISTICS. *Educ:* Mankato State Col, BA, 52; Purdue Univ, MS, 54; Va Polytech Inst, PhD(statist), 59. *Prof Exp:* Spec serv engr, E I du Pont de Nemours & Co, 59-61; ASSOC PROF MATH, SOUTHERN ILL UNIV, CARBONDALE, 61- *Mem:* AAAS; Am Statist Asn; Int Asn Math Geol; Inst Math Statist. *Res:* Design of experiments; statistical inference; geostatistics. *Mailing Add:* Dept Math Neckers C 0279 Southern Ill Univ Carbondale IL 62901

STARKS, THOMAS LEROY, b Muskegon, Mich, June 20, 47. PLANT PHYSIOLOGY, TRACEMETAL BIOAVAILABILITY. *Educ:* Ferris State Col, BS, 69; Cent Mich Univ, MS, 76; Univ NDak, PhD, 79. *Prof Exp:* Res scientist, Domestic Mining & Mineral Inst, Univ NDak, 79-81, fel, 81-82; TEACHING FEL, USDA/ARS HUMAN NUTRIT RES CTR, GRAND FORKS, NDAK, 82- *Mem:* Sigma Xi; AAAS. *Res:* Intrinsically labeling of plants with radio and stable isotopes and the resulting effect on plant proteins. *Mailing Add:* USDA/ARS Human Nutrit Res Ctr Box 7166 Univ Sta Grand Forks ND 58202

STARKWEATHER, GARY KEITH, b Lansing, Mich, Jan 9, 38; m 61; c 2. PHYSICAL OPTICS, ELECTROOPTICS. *Educ:* Mich State Univ, BS, 60; Univ Rochester, MS, 66. *Prof Exp:* Engr, Bausch & Lomb, Inc, 62-64; area mgr optical systs, Xerox Palo Alto Res Ctr, 64-80, sr res fel, 80-88; APPLE FEL, ADVAN TECHNOL GROUP, APPLE COMPUTER INC, 88- *Concurrent Pos:* Instr optics, Monroe Community Col, 68-69. *Mem:* Optical Soc Am; Soc Photog Inst Engr. *Res:* Optics and electronics and their specific system interaction, involving display and hard copy image systems. *Mailing Add:* 13325 Paramount Dr Saratoga CA 95070

STARKWEATHER, HOWARD WARNER, JR, b Cambridge, Mass, July 20, 26; m 48; c 3. PHYSICAL CHEMISTRY. *Educ:* Haverford Col, AB, 48; Harvard Univ, AM, 50; Polytech Inst Brooklyn, 50-52, PhD(chem), 53. *Prof Exp:* Chemist, Rayon Dept, 47, res chemist, Ammonia Dept, 48-49 & Plastics Dept, 52-57, sr res chemist, 57-66, res assoc, plastics dept, 66-76, CENT RES & DEVELOP DEPT, E I DU PONT DE NEMOURS & CO, INC, 76- *Mem:* Am Chem Soc; Am Phys Soc; fel NAm Thermal Anal Soc. *Res:* Polymer chemistry; polymerization kinetics; polymer properties and molecular structure; polymer crystallography. *Mailing Add:* Cent Res & Develop Dept E I du Pont de Nemours & Co Inc Wilmington DE 19880-0356

STARKWEATHER, PETER LATHROP, b Glen Ridge, NJ, Nov 7, 48; m 72; c 1. AQUATIC ECOLOGY, INVERTEBRATE PHYSIOLOGY. *Educ:* Union Col, NY, BS, 70; Dartmouth Col, PhD(biol sci), 76. *Prof Exp:* Teaching asst zool & ecol, State Univ NY Albany, 70-72; teaching fel biol sci, Dartmouth Col, 72-76, res assoc ecol, 76-78; from asst prof to assoc prof, 78-88, PROF AQUATIC ECOL, UNIV NEV, 88-, CHMN DEPT, 84- *Concurrent Pos:* Co-prin investr biol sci, NSF grant, Dartmouth Col, 78-81, vis asst prof, 79 & 80; prin investr, NSF grant, Univ Nev, 81-84; Am Coun Educ fel, 87-88. *Mem:* Am Soc Limnol & Oceanog; Am Soc Zoologists; Ecol Soc Am; AAAS; Sigma Xi; Am Asn Higher Educ. *Res:* Feeding biology and behavior of microcrustacean zooplankton and rotifers; biological rhythms and invertebrate ecology; examination and analysis of feeding behavior and ecology of zooplankton, both freshwater and marine. *Mailing Add:* Dept Biol Sci Univ Nev Las Vegas Maryland Pkwy Las Vegas NV 89154

STARLING, ALBERT GREGORY, b Joiner, Ark, Feb 24, 39. MATHEMATICS, COMPUTER SCIENCE. *Educ:* Univ Ark, BSEE, 61, MS, 64, PhD(math), 69. *Prof Exp:* Jr engr electronics, Int Bus Mach Corp, 61-62, assoc engr & mathematician appl math, 64-65; asst prof math, Univ Mass, Amherst, 69-71; ASSOC PROF MATH, WESTERN CAROLINA UNIV, 71-; CHMN, COMP SCI DEPT, UNIV ARK, 88- *Concurrent Pos:* Fel, Univ Mass, 69-71. *Mem:* Am Math Soc. *Res:* Directed graphs of finite groups. *Mailing Add:* Comp Sci Dept Sci Eng Bldg Univ Ark Fayetteville AR 72701-8331

STARLING, JAMES LYNE, b Henry Co, Va, Aug 16, 30; m 68; c 1. PLANT BREEDING, STATISTICS. *Educ:* Va Polytech Inst, BS, 51; Pa State Univ, MS, 55, PhD(agron), 58. *Prof Exp:* Asst forage crop breeding, 54-57, from instr to assoc prof agron, 57-69, prof agron & head dept, 69-85, ASSOC DEAN ADMIN, COL AGR, PA STATE UNIV, 85- *Mem:* AAAS; Am Soc Agron; Sigma Xi. *Res:* Forage crop breeding; genetics and cytogenetics of forage crop species; experimental design. *Mailing Add:* Col Agr 201 Agr Admin Bldg Pa State Univ University Park PA 16802

STARLING, JANE ANN, b Waco, Tex, Jan 4, 46. METABOLISM, PARASITE PHYSIOLOGY. *Educ:* Rice Univ, BA, 67, PhD(physiol), 72. *Prof Exp:* Fel zool, Univ Mass, 72-75; vis asst prof biol, Univ Pittsburgh, 75-76; asst prof, 76-83, ASSOC PROF BIOL, UNIV MO-ST LOUIS, 83- *Concurrent Pos:* NIH fel, 73-75. *Mem:* AAAS; Am Micros Soc; Am Soc Parasitol; Sigma Xi. *Res:* Trehalose metabolism; energy metabolism in intestinal helminths; host parasite integration; carbohydrate and amino acid transport in helminths; anaerobic energy metabolism. *Mailing Add:* Univ Mo Dept Biol 8001 Natural Bridge Rd St Louis MO 63121

STARLING, KENNETH EARL, b Corpus Christi, Tex, Mar 9, 35; m 60; c 3. CHEMICAL ENGINEERING. *Educ:* Tex A&I Univ, BS, 57 & 58; Ill Inst Technol, MS, 60, PhD(gas technol), 62. *Prof Exp:* Res engr, Inst Gas Technol, Ill Inst Technol, 62-63; Robert A Welch Found fel chem, Rice Univ, 63-64; sr res engr, Standard Oil Co, NJ, 64-66; from asst prof to assoc prof, 62-77, PROF CHEM ENG, UNIV OKLA, 77- *Concurrent Pos:* Consult, Standard Oil Co, NJ, 68-, Inst Gas Technol, Ill Inst Technol, 70- & J F Pritchard & Co, 71-; fel, Inst Low Temperature Physics, Cath Univ Louvain, 72-73. *Mem:* AAAS; Am Inst Chem Engrs; Am Chem Soc; Soc Petrol Engrs. *Res:* Energy conversion; fossil energy processes; thermodynamics; correlation of fluid properties. *Mailing Add:* Sch Chem Eng Univ Okla Norman OK 73019

STARLING, THOMAS MADISON, b Loneoak, Va, Aug 12, 23; m 61; c 2. PLANT BREEDING. *Educ:* Va Polytech Inst, BS, 44; Iowa State Univ, MS, 47, PhD, 55. *Prof Exp:* From asst agronomist to assoc agronomist, Va Polytech Inst & State Univ, 44-60, prof agron, 60-70 & 71-85, assoc dean grad sch, 70-71, W G Wysor prof agr, 85-88, EMER PROF VA POLYTECH INST & STATE UNIV, 88- *Mem:* Fel Am Soc Agron; fel Crop Sci Soc Am. *Res:* Plant breeding and genetics of winter barley and wheat. *Mailing Add:* 618 Woodland Dr NW Blacksburg VA 24060

STARMER, C FRANK, b Greensboro, NC, Sept 4, 41; m 63; c 4. PHARMACOLOGY, CARDIOLOGY. *Educ:* Duke Univ, BSEE, 63, MSEE, 65; Rice Univ, 65-66; Univ NC, PhD(biomath, bioeng), 68. *Prof Exp:* Res assoc med, 63-65, assoc biomath, 66-68, asst prof Med & community health sci, 68-71, assoc prof comput sci, 71-77, assoc prof med, Med Ctr, 77-90, PROF COMPUT SCI, DUKE UNIV, 77-, PROF EXP MED, 90- *Concurrent Pos:* NIH career develop award, Duke Univ, 72-77; mem coun, Div Res Resources, NIH, 84-88; US-USSR exchange (sudden death), 87-90. *Mem:* AAAS; Asn Comput Mach; Am Heart Asn; Am Col Med Informatics. *Res:* Computer science; applied biostatistics; communication in cellular systems; pharmacology. *Mailing Add:* Duke Univ Med Ctr Box 3181 Durham NC 27710

STARNES, WILLIAM HERBERT, JR, b Knoxville, Tenn, Dec 2, 34; m 86. POLYMER DEGRADATION & STABILIZATION, ORGANIC REACTION MECHANISMS. *Educ:* Va Polytech Inst, BS, 55; Ga Inst Technol, PhD(chem), 60. *Prof Exp:* Res chemist, Humble Oil & Refining Co, Esso Res & Eng Co, 60-62, sr res chemist, 62-65, res specialist, 65-67, res assoc, 67-71; res assoc & instr, Dept Chem, Univ Tex, Austin, 71-73; mem tech staff, AT&T Bell Labs, 73-85; prof chem, Polytech Univ, Brooklyn, 85-89, head, Dept Chem & Life Sci, 85-88, assoc dir, Polymer Durability Ctr, 87-89; GOTTWALD PROF CHEM, COL WILLIAM & MARY, 89- *Concurrent Pos:* Sect head, Humble Oil & Refining Co, Esso Res & Eng Co, 64; vis scientist, Tex Acad Sci, 64-67; consult, numerous indust co, 85-; adv bd & bd rev, J Vinyl Technol, 81-83; chmn, chem subpanel, AAAS proj 2061. *Honors & Awards:* MA Ferst Award, Sigma Xi, 60. *Mem:* Fel AAAS; fel Am Inst Chemists; Am Chem Soc; NY Acad Sci. *Res:* Degradation, stabilization, microstructure, flammability and polymerization chemistry of synthetic polymers; physical organic studies of polymer reactions; free-radical chemistry; liquid-phase autoxidation; organic synthesis; applied carbon-13 nuclear magnetic resonance spectroscopy. *Mailing Add:* Dept Chem Col William & Mary Williamsburg VA 23185

STAROS, JAMES VAUGHAN, b May 20, 47; US citizen; m 76; c 3. PROTEIN CHEMISTRY. *Educ:* Dartmouth Col, AB, 69; Yale Univ, PhD(molecular biophysics & biochem), 74. *Prof Exp:* Helen Hay Whitney fel, Dept Chem, Harvard Univ, 74-77; from asst prof to prof biochem, 78-91, interim chmn, Dept Biochem, 88-91, PROF MOLECULAR BIOL & BIOCHEM & CHAIR, DEPT MOLECULAR BIOL, VANDERBILT UNIV, 91- *Concurrent Pos:* Mem, Grad Fel Eval Panel in Biochem & Biophys, Nat Res Coun, 87, Cellular & Molecular Basis of Dis Rev Comn, NIH, 88-92, chair, 90-92; chair, Nat Inst Diabetes & Digestive & Kidney Dis Workshop on Membrane Protein Struct, 90. *Mem:* Am Chem Soc; Biophys Soc; Fedn Am Scientists; AAAS; Am Soc Biochem & Molecular Biol; Protein Soc. *Res:* Protein chemistry; structure and function of biomembrane proteins; design and synthesis of new chemical probes of biomembrane protein structure and function. *Mailing Add:* Dept Molecular Biol Vanderbilt Univ Nashville TN 37235

STARR, ALBERT, b New York, NY, June 1, 26; m 55; c 2. THORACIC SURGERY. *Educ:* Columbia Col, BA, 46; Columbia Univ, MD, 49. *Prof Exp:* Asst surg, Columbia Univ, 56-57; from instr to assoc prof surg, 57-64, PROF CARDIOPULMONARY SURG, MED SCH, UNIV ORE, 64-, HEAD DIV, 63- *Honors & Awards:* Award, Am Heart Asn, 63; Rene Le Riche Award Cardiovasc Surg, 65. *Mem:* Am Surg Asn; Am Asn Thoracic Surg; Am Col Cardiol; Int Cardiovasc Soc. *Res:* Prosthetic values for cardiac surgery. *Mailing Add:* Dept Surg Ore Health Sci Univ 3181 SW Sam Jackson Park Rd Portland OR 97201

STARR, ARNOLD, b New York, NY, Aug 5, 32; c 3. COGNITIVE DISORDERS. *Educ:* Kenyon Col, Gambier, Ohio, AB, 53; NY Univ, MD, 57. *Prof Exp:* Asst prof med, Sch Med, Stanford Univ, 64-71; from assoc prof to prof med, Univ Calif, Irvine, 71-77, chief, Div Neurol, 73-77, founding chmn, Dept Neurol, 77-86, PROF NEUROL, PSYCHOBIOL & COGNITIVE SCI, UNIV CALIF, IRVINE, 77-, PROF PSYCHIAT, 83- *Concurrent Pos:* Dir, Residency Prog, Univ Calif, Irvine, 73-90; vis prof, Dept Psychol, Univ London Nat Hosp, 85-86, 87 & 88, Dept Neurol, Univ Hanzhou, China, 86 & Dept Neurol, Vienna, Austria, 89; mem, Bd Sci Counr, NIH, 88-92, Comt Mapping & Evoked Potentials, Int Fed Soc EEG, 89 & Therapeut & Technol Assessment Subcomt, Am Acad Neurol, 90. *Mem:* Am EEG Soc; Am Neurol Asn; fel Am Acad Neurol; Soc Neurosci; fel Acoust Soc Am; Am Physiol Soc. *Res:* Auditory brainstem evoked potentials; magnetic stimulation; multiple sclerosis; memory and cognitive disorders; author of over 130 publications. *Mailing Add:* Dept Neurol Med Surg Rm 154 Univ Calif Irvine CA 92717

STARR, C DEAN, b Tulare, SDak, Apr 24, 21; m 46; c 4. METALLURGY. *Educ:* SDak Sch Mines, BS, 43; Univ Utah, MS, 48, PhD(metall), 49. *Prof Exp:* Jr metallurgist, AC Spark Plug Co Div, Gen Motors Corp, 43-44, spectrographer, 46-47; asst res prof metall, Univ Calif, Berkeley, 49-53; chief res metallurgist, Wilbur B Driver Co, Gen Tel & Electronics Corp, 54-60, tech dir eng, 60-66, vpres eng & tech dir eng, 66-71, vpres eng & res, 71-79; vpres eng & res, Amax Specialty Metals Corp, 79-84; PRES, C DEAN STARR INC, 84- *Concurrent Pos:* Vis sr lectr, Grad Sch Metall, Stevens Inst Technol, 61-84. *Honors & Awards:* Sam Tour Award, Am Soc Testing & Mat, 66. *Mem:* Am Soc Metals; Am Inst Mining, Metall & Petrol Engrs; Electrochem Soc. *Res:* Alloy development for resistance, particularly heat resisting, thermocouple, corrosion and high strength alloys; thermodynamic and kinetic studies of simple and complex alloy systems. *Mailing Add:* 1621 Farr Rd Wyomissing PA 19610

STARR, CHAUNCEY, b Newark, NJ, Apr 14, 12; m 38; c 2. ENGINEERING PHYSICS. *Educ:* Rensselaer Polytech Inst, EE, 32, PhD(physics), 35. *Hon Degrees:* DrEng, Rensselaer Polytech Inst, 64; DrEng, Swiss Inst Technol, Switz, 80; DSc, Tulane Univ, 86. *Prof Exp:* Coffin fel & res fel physics, Harvard Univ, 35-37; res physicist, P R Mallory Co, Ind, 37-38; res assoc phys chem, Mass Inst Technol, 38-41; physicist, D W Taylor Model Basin, Bur Ships, US Dept Navy, 41-42, Radiation Lab, Univ Calif, 42-43 & Manhattan Dist, Oak Ridge, 43-46; dir atomic energy res dept, N Am Aviation, Inc, 46-55, vpres, 55-66, gen mgr, Atomics Int Div, 55-60, pres, 60-66; dean sch eng & appl sci, Univ Calif, Los Angeles, 67-72; pres, 72-78, vchmn, 78-87, EMER PRES, ELEC POWER RES INST, 87- *Concurrent Pos:* Consult, US Off Sci & Technol, NASA, Atomic Energy Comn & US Air Force; dir, Atomic Indust Forum; mem, Rockefeller Univ Coun, Rockefeller Univ, NY. *Honors & Awards:* Legion of Honor, French Govt, 78; Walter H Zinn Award, Am Nuclear Soc, 79; Rockwell Medal, 88; Pres Nat Med Tech, President US, 90. *Mem:* Nat Acad Eng (vpres); Am Inst Aeronaut & Astronaut; Sigma Xi; fel Am Phys Soc; Am Nuclear Soc (pres, 58-59); Royal Acad Eng Sci, Sweden. *Res:* Semiconductors; thermal conductivity of metals; high pressures; cryogenics; magnetic susceptibilities at low temperatures; gas discharge phenomena, solid state; Atomic energy and nuclear reactors; risk analysis. *Mailing Add:* Elec Power Res Inst 3412 Hillview Ave Palo Alto CA 94303

STARR, DAVID WRIGHT, b Anna, Tex, Dec 8, 12; wid; c 2. MATHEMATICS. *Educ:* Southern Methodist Univ, AB, 33; Univ Ill, AM, 37, PhD(math), 40. *Prof Exp:* High sch instr, Tex, 33-37, prin, 34; asst math, Univ Ill, 38-40; instr ground sch aviation, Southern Methodist Univ, 40-43, from instr to prof math, 40-78, coordr war training serv, Civil Aeronaut Admin, 41-44, chmn dept, 63-77, EMER PROF MATH, SOUTHERN METHODIST UNIV, 78- *Mem:* Am Math Soc; Math Asn Am. *Res:* Analysis; Schrodinger wave equation from the point of view of singular integral equations. *Mailing Add:* Dept Math Southern Methodist Univ Dallas TX 75275

STARR, DUANE FRANK, b Pasadena, Calif, Oct 20, 42; m 65; c 2. PHYSICAL CHEMISTRY. *Educ:* Wesleyan Univ, BA, 64; Ore State Univ, PhD(phys chem), 73. *Prof Exp:* Resident res assoc lasers, Naval Res Lab, Nat Res Coun, 73-75; staff chemist propellant chem, Allegany Ballistics Lab, Hercules Inc, 75-77; ENGR URANIUM ENRICHMENT, NUCLEAR DIV, OAK RIDGE GASEOUS DIFFUSION PLANT, UNION CARBIDE CORP, 77- *Mem:* Am Chem Soc; Am Phys Soc. *Res:* Economic assessment of advanced isotope separation methods. *Mailing Add:* 109 Woodridge Ln Oak Ridge TN 37830-8242

STARR, E(UGENE) C(ARL), electrical power engineering; deceased, see previous edition for last biography

STARR, JAMES LEROY, b Almont, Mich, Aug 14, 39; m 60; c 2. SOIL PHYSICS. *Educ:* Mich State Univ, BS, 61, MS, 70; Eastern Baptist Theol Sem, MA, 66; Univ Calif, Davis, PhD(soil sci), 73. *Prof Exp:* Voc Agr teacher, Carson City Community Schs, Mich, 61-62 & Mayville Community Schs, 62-64; teaching asst, Mich State Univ, 67-68, instr soils & dir audiotutorial lab, 68-70; res asst, Univ Calif, 70-72, staff res assoc, 72-74; from asst scientist to assoc scientist soil physics, Conn Agr Exp Sta, 74-78; RES SOIL SCIENTIST, USDA, 79- *Mem:* Sigma Xi; Am Soc Agron; Soil Sci Soc Am; Int Soc Soil Sci; Am Geophys Union. *Res:* Soil water nitrogen relations; infiltration; movement of nutrients and agrochemicals to ground water. *Mailing Add:* USDA Agr Res Serv BARC-W B007, RM 253 Beltsville MD 20705

STARR, JASON LEONARD, b Chelsea, Mass, Aug 13, 28; m 51; c 3. ONCOLOGY, BIOCHEMISTRY. *Educ:* Harvard Univ, AB & AM, 49, MD, 53. *Prof Exp:* Intern med, Beth Israel Hosp, Boston, Mass, 53-54; asst prof, Med Sch, Northwestern Univ, 61-65; from assoc prof to prof, Col Med, Univ Tenn, Memphis, 65-72; prof med, Sch Med, Univ Calif, Los Angeles, 73-74; CLIN PROF MED, UNIV TENN, MEMPHIS, 74-; DIR ONCOL, BAPTIST MEM HOSP, MEMPHIS, 74- *Concurrent Pos:* Fel, Mayo Found, Univ Minn, 56-58; fel biochem, Sch Med, Western Reserve Univ, 58-61; Am Cancer Soc fel, 61-63 & scholar, 71-72; sr investr, Arthritis Found, 63-68. *Mem:* Am Soc Clin Invest; Am Asn Immunologists; Am Chem Soc. *Res:* Genetic control of antibody synthesis; biochemistry of neoplastic cells; cancer chemotherapy; immuno-oncology. *Mailing Add:* 9715 Checkerboard St Houston TX 77016

STARR, JOHN EDWARD, b St Louis, Mo, July 12, 39; m 61; c 2. PHOTOGRAPHY. *Educ:* Colo Col, BA, 61; Stanford Univ, PhD(org chem), 65. *Prof Exp:* Sr chemist, 65-69, lab head, Res Labs, 69-78, RES ASSOC, EASTMAN KODAK CO, 78- *Res:* Spectral sensitization of photographic emulsions by sensitizing dyes; color reproduction. *Mailing Add:* Res Labs Eastman Kodak Co Kodak Park Rochester NY 14650

STARR, MATTHEW C, CARDIOVASCULAR, MICROCIRCULATION. *Educ:* Univ Southern Calif, PhD(psychol), 74. *Prof Exp:* HEALTH SCI ADMINR, NAT HEART LUNG & BLOOD INST, NIH, 81- *Mailing Add:* 3709 35th St NW Washington DC 20016

STARR, MORTIMER PAUL, b New York, NY, Apr 13, 17; m 44; c 3. APPLIED MICROBIOLOGY. *Educ:* Brooklyn Col, BA, 38; Cornell Univ, MS, 39, PhD(bact, biochem, plant path), 43. *Prof Exp:* Tutor biol, Brooklyn Col, 39-44, from instr to asst prof, 44-47; from asst prof bact & asst bacteriologist to assoc prof bact & assoc bacteriologist, 47-58, spec asst to chancellor res grants & contracts, 63-67, PROF BACT EXP STA, UNIV CALIF, DAVIS, 58- *Concurrent Pos:* Nat Res Coun fel, Hopkins Marine Sta, Stanford Univ, 44-46; cur, Int Collection of Phytopathogenic Bacteria, 47-; vis specialist fac agron, Nat Univ Colombia, 49; NIH spec fels, Cambridge & Ghent Univs, 53-54, Plant Dis Div, Univ Auckland, NZ, 62; Guggenheim Mem Found fel, Max Planck Inst Med Res, 58, guest, 59; vis prof, Chile, 66; Guggenheim Mem Found fel, Swiss Fed Inst Technol, Zurich, 68-69; vis prof,

Univ Hamburg, 72-73, Univ Bordeaux, 75-76, Univ Gottingen, 77; partic, Int Cong Microbiol, Rome, 53, Stockholm, 58, Montreal, 62, Moscow, 66, Mexico, 70; partic, Int Bot Cong, Paris, 54, Montreal, 59 & Edinburgh, 64; mem, Int Conf Sci Probs Plant Protection, Budapest, 60, Int Conf Cult Collections, Toronto, 62, Tokyo, 68, Brno, 81; mem, Conf Global Impacts Appl Microbiol, Stockholm, 63, Addis Ababa, 67, Bombay, 69; mem, Int Conf Phytopathogenic Bacteria, Harpenden, 64, Lisbon, 67, Angers, 78; Alexander von Humboldt Award, 77. *Honors & Awards:* Bernardo O'Higgins Medal, First Class, Repub of Chile, 67; Silver Medal, Purkyne Univ, Brno, 68. *Mem:* AAAS; Soc Gen Microbiol; Am Phytopath Soc; Am Soc Microbiol. *Res:* Biochemistry, metabolism, genetics, ecology and taxonomy of phytopathogenic bacteria; industrial microbiology; microbial pigments; pectin metabolism; philosophical grounds of taxonomy and ecology; bacterial morphogenesis; bacterial diversity and ecology; aquatic bacteria; international microbiology; university administration. *Mailing Add:* 751 Elmwood Dr Davis CA 95616

STARR, NORMAN, b Scranton, Pa, Apr 14, 33; m 65. MATHEMATICAL STATISTICS. *Educ:* Univ Mich, BA, 55, MA, 60; Columbia Univ, PhD(math statist), 65. *Prof Exp:* Assoc math, Evans Res & Develop Corp, 61-65; asst prof statist, Univ Minn, 65-66; from asst prof to assoc prof, Carnegie-Mellon Univ, 66-68; assoc prof, Univ Mich, Ann Arbor, 68-73, prof math, Dept Statist, 68-86; CONSULT, 86- *Mem:* Inst Math Statist; Am Math Soc; Am Statist Asn. *Res:* Sequential analysis; optimal stopping; statistical allocation and theory; applied probability. *Mailing Add:* 4775 Waters Rd Ann Arbor MI 48103

STARR, NORTON, b Kansas City, Mo, June 18, 36; m 59; c 2. MATHEMATICS. *Educ:* Harvard Col, AB, 58; Mass Inst Technol, PhD(math), 64. *Prof Exp:* Instr math, Mass Inst Technol, 64-66; asst prof, 66-71, assoc prof, 71-78, PROF MATH, AMHERST COL, 78- *Concurrent Pos:* Vis asst prof, Univ Waterloo, 72-73. *Mem:* Am Math Soc; Math Asn Am. *Res:* Operator limit theory. *Mailing Add:* Dept Math Amherst Col PO Box 2239 Amherst MA 01002

STARR, PATRICIA RAE, b Hood River, Ore, Feb 28, 35. MICROBIOLOGY, HISTORY & PHILOSOPHY OF SCIENCE. *Educ:* Ore State Univ, BS, 57, MS, 62; Univ Ore, Med Sch, PhD(microbiol), 68. *Prof Exp:* Instr microbiol, Dent Sch, Univ Ore, 68-69; assoc, Ore State Univ, 69-71; asst prof, Univ Ill, Urbana, 71-75; res assoc, Providence Hosp, 75-77; chair, 80-83, INSTR SCI DIV, MT HOOD COMMUNITY COL, 77- *Mem:* AAAS; Am Soc Microbiol; Hist Sci Soc. *Res:* Relationship of sterol synthesis to respiratory adaptation in yeast; amino acid uptake systems in yeast; white blood cell function. *Mailing Add:* 5049 NE Simpson St Portland OR 97218

STARR, PATRICK JOSEPH, b St Paul, Minn, Oct 24, 39. MECHANICAL ENGINEERING, INTELLIGENT SYSTEMS. *Educ:* Univ Minn, Minneapolis, BME, 62, MSME, 66, PhD(mech eng), 70. *Prof Exp:* Develop eng, Honeywell Inc, 62-64, 68; teaching assoc, Univ Minn, Minneapolis, 64-70; control syst engr, Northern Ord Div, FMC Corp, 70-71; asst prof, 71-77, ASSOC PROF, INDUST ENG & OPERS RES DIV, DEPT MECH ENG, UNIV MINN, 77- *Res:* modeling of large scale socio-industrial systems; system dynamics; technology assessment; expert simulation; manufacturing system analysis. *Mailing Add:* Dept of Mech Eng Univ of Minn Minneapolis MN 55455

STARR, PHILLIP HENRY, b Poland, Nov 16, 20; nat US; div; c 3. PSYCHIATRY. *Educ:* Univ Toronto, MD, 44. *Prof Exp:* Asst prof neuropsychiat & pediat, Sch Med, Wash Univ, 52-55, dir community child guid clin, 52-56; ASSOC PROF NEUROL & PSYCHIAT, SCH MED, UNIV NEBR, OMAHA, 57- *Concurrent Pos:* Chief psychiat consult, St Louis Children's Hosp, 51-56; chief children's outpatient serv, Nebr Psychiat Inst, 56-64, consult, 64-; consult, Offut Air Base Hosp & Immanuel Ment Health Ctr. *Mem:* Fel Am Psychiat Asn; fel AMA; Int Asn Child Psychiat; fel Am Acad Child Psychiat. *Res:* Child and adult psychiatry. *Mailing Add:* PO Box 5973 Scottsdale AZ 85261

STARR, RICHARD CAWTHON, b Greensboro, Ga, Aug 24, 24. PHYCOLOGY. *Educ:* Ga Southern Col, BS, 44; George Peabody Col, MA, 47; Vanderbilt Univ, PhD(biol), 52. *Prof Exp:* From instr to prof bot, Ind Univ, Bloomington, 52-76; PROF BOT, UNIV TEX, AUSTIN, 76-, DIR CULT COLLECTION ALGAE, 76- *Concurrent Pos:* Guggenheim fel, 59; US sr award, Alexander von Humboldt Found, 72. *Honors & Awards:* Gilbert Morgan Smith Medal, Nat Acad Sci, 85. *Mem:* Nat Acad Sci; Bot Soc Am; AAAS; Phycol Soc Am. *Res:* Morphology and cultivation of green algae; genetics and development of Volvox. *Mailing Add:* Dept Bot Univ Tex Austin TX 78712

STARR, ROBERT I, b Laramie, Wyo, Dec 11, 32; m 56; c 2. CHEMISTRY, ENVIRONMENTAL SCIENCES. *Educ:* Univ Wyo, BS, 56, MS, 59, PhD(plant physiol), 72. *Prof Exp:* Res biochemist, US Fish & Wildlife Serv, Colo, 60-63; plant physiologist, Colo State Univ, 63-64; anal chemist, US Food & Drug Admin, 64-65; chemist, Colo State Univ, 65-69; res chemist, Bur Sport Fisheries & Wildlife, US Dept Interior, 69-74; environ scientist, US Geol Surv, 74-77, chief, Environ-Tech Unit, 77-78; chief, Biol-Ecol Sci Br, US Dept Interior, 78-81; pvt consult environ chem, 81-84; consult environ chem, US Dept Interior, 84-89; SR RES CHEMIST, PESTICIDES, USDA, 89- *Concurrent Pos:* Res biochemist, Wildlife Res Ctr, US Fish & Wildlife Serv, Colo, 68-69. *Mem:* AAAS; Am Chem Soc; fel Am Inst Chemists. *Res:* Pesticide chemistry as related to plant soil and water systems, including method development studies; plant, soil, and chemistry matters relating to mining operations; trace metals in soil and water systems. *Mailing Add:* 404 Tulane Dr Ft Collins CO 80525

STARR, THEODORE JACK, b Plainfield, NJ, Aug 22, 24; m 54; c 3. MICROBIOLOGY. *Educ:* City Col New York, BS, 49; Univ Mass, MS, 51; Univ Wash, PhD(microbiol), 53. *Prof Exp:* Res assoc microbiol, Haskins Labs, 48-49; instr biol, Univ Ga, 54-55; fishery res biologist, US Fish & Wildlife Serv, 55-57; McLaughlin fel virol, Med Br, Univ Tex, 57-60; assoc res scientist, Lab Comp Biol, Kaiser Found Res Inst, 60-62; assoc prof microbiol, Univ Notre Dame, 62-68; prof biol sci & assoc head dept, Col Arts & Sci, Univ Ill, Chicago Circle, 68-70; asst vpres acad affairs, Grad Studies & Res, 75-77, head dept, 70-80, PROF BIOL SCI, STATE UNIV NY COL BROCKPORT, 70- *Concurrent Pos:* Fac exchange scholar, State Univ NY, 75. *Mem:* Am Soc Microbiol; Soc Exp Biol & Med. *Res:* Cytochemistry; virology; marine and space biology; gnotobiology. *Mailing Add:* Dept of Biol Sci State Univ of NY Brockport NY 14420

STARR, THOMAS LOUIS, b Cincinnati, Ohio, Mar 22, 49; m 70; c 3. ANALYTICAL CHEMISTRY, PHYSICAL CHEMISTRY. *Educ:* Univ Detroit, BS, 70; Univ Louisville, PhD(phys chem), 76. *Prof Exp:* Anal chemist, Major appliance labs, Gen Elec Co, 77-80; SR RES SCIENTIST, GA INST TECHNOL, 80- *Concurrent Pos:* Pres scholar, Univ Detroit, 67-70; J B Speed fel, Univ Louisville, 76. *Honors & Awards:* J M Houchens Prize, Univ Louisville, 77. *Mem:* Am Chem Soc; Am Ceramic Soc. *Res:* Chemistry and physics of materials processing; ceramics and ceramic composites; high performance coatings; molecular and microstructure modeling; acid rain effects on materials. *Mailing Add:* Georgia Tech Res Inst Ga Inst Technol Atlanta GA 30332

STARR, WALTER LEROY, b Portland, Ore, Feb 9, 24; m 44; c 2. PHYSICS. *Educ:* Univ Southern Calif, BS, 50; Calif Inst Technol, MS, 51. *Prof Exp:* Physicist, US Naval Missile Test Ctr, 50; res physicist, US Naval Civil Eng Lab, 51-55; res scientist, Phys Sci Lab, Lockheed Missiles & Space Co, 55-67; physicist, Ames Res Ctr, NASA, 67-88; RETIRED. *Mem:* AAAS; Am Phys Soc; Am Geophys Union. *Res:* Atomic and molecular physics; atmospheric processes; ionization and excitation; absorption cross sections. *Mailing Add:* 13124 Byrd Lane Los Altos CA 94022

STARRATT, ALVIN NEIL, b Paradise, NS, Sept 18, 36; m 67; c 2. NATURAL PRODUCTS CHEMISTRY. *Educ:* Acadia Univ, BSc, 59; Univ Western Ont, PhD(org chem), 63. *Prof Exp:* Brit Petrol Co res fel, Imp Col, Univ London, 63-64; res fel, Res Inst Med & Chem, Mass, 64-65; RES SCIENTIST, RES INST, CAN AGR, 65- *Mem:* Am Chem Soc; fel Chem Inst Can; Entom Soc Can. *Res:* Natural products influencing the behavior of insects and insect neuropeptides. *Mailing Add:* Res Ctr Agr Can 1400 Western Rd London ON N6G 2V4 Can

STARRETT, ANDREW, b Greenwich, Conn, Mar 18, 30; m 51; c 3. MAMMALOGY. *Educ:* Univ Conn, BS, 51; Univ Mich, MS, 55, PhD(zool), 58. *Prof Exp:* Instr zool, Univ Mich, 56-57; from instr to asst prof biol, Univ Southern Calif, 57-64; asst prof, Northeastern Univ, 64-65; assoc prof, 65-69, PROF BIOL, CALIF STATE UNIV, NORTHRIDGE, 69- *Concurrent Pos:* Res assoc, Los Angeles County Mus Natural Hist. *Mem:* AAAS; Soc Syst Zool; Soc Study Evolution; Am Soc Mammalogists. *Res:* Vertebrate and mammalian evolution and distribution; mammalian morphology and systematics, particularly Chiroptera. *Mailing Add:* Dept Biol Calif State Univ Northridge CA 91330

STARRETT, RICHMOND MULLINS, b Gardner, Mass, Oct 1, 43; m 66; c 3. BIOTECHNOLOGY, INDUSTRIAL MICROBIOLOGY. *Educ:* Univ NC, BA, 66; Iowa State Univ, PhD(org chem), 70. *Prof Exp:* Res assoc, R A Welch Found, Tex Tech Univ, 71-72; res chemist, 72-75, sr res chemist, 75-79, RES ASSOC, DEPT RES & DEVELOP, ETHYL CORP, 79- *Mem:* Am Chem Soc; Org Reactions Catalysis Soc; Nat Asn Advan Sci; Soc Indust Microbiol. *Res:* Development of biological processes for industrial chemicals; investigation of fundamental microbiological methodology, including recombinant DNA techniques; study of novel synthetic and catalytic processes for chemical manufacture. *Mailing Add:* 939 Edmund Hawes Rd Baton Rouge LA 70810

STARRFIELD, SUMNER GROSBY, b Los Angeles, Calif, Dec 29, 40; m 66; c 3. THEORETICAL ASTROPHYSICS. *Educ:* Univ Calif, Berkeley, BA, 62; Univ Calif, Los Angeles, MA, 65, PhD(astron), 69. *Prof Exp:* Lectr astron, Yale Univ, 67-69, asst prof, 69-71; scientist, Thomas J Watson Res Ctr, IBM Corp, 71-72; from asst prof to assoc prof, 72-79, PROF ASTROPHYS, ARIZ STATE UNIV, 80- *Concurrent Pos:* Res grants, NSF, 74 & 75-89, NASA, 80-89; collabr, Los Alamos Sci Lab, 74-; fels, Assoc Western Univ, 85 & Joint Inst Lab Astrophys, 86. *Honors & Awards:* Philips lectr, Haverford Col, 78. *Mem:* Royal Astron Soc; Int Astron Union; Am Astron Soc; Am Phys Soc. *Res:* Stellar structure and evolution; hydrodynamical studies of novae; stellar pulsation; ultraviolet studies of novae. *Mailing Add:* Dept Physics Ariz State Univ Tempe AZ 85287

STARUSZKIEWICZ, WALTER FRANK, JR, b Ellwood City, Pa, Jan 31, 39; m 63; c 3. ANALYTICAL CHEMISTRY, FOOD CHEMISTRY. *Educ:* Geneva Col, BS, 60; Univ Hawaii, MS, 65. *Prof Exp:* Res asst biochem, Pineapple Res Inst Hawaii, 64-66; chemist, Del Monte Corp, 66-67; RES CHEMIST ANAL CHEM FOODS, US FOOD & DRUG ADMIN, 67- *Mem:* Am Chem Soc; fel Asn Off Anal Chemists. *Res:* Development of analytical methods for the detection of decomposition in foods; applications of gas and liquid chromatography for the determination of histamine and other biogenic amines in seafoods. *Mailing Add:* US Food & Drug Admin HFF-423 200 C St SW Washington DC 20204

STARY, FRANK EDWARD, b St Paul, Minn, Jan 3, 41; m 64. EPOXY POLYMER CHEMISTRY-CHEMILUMINESCENCE. *Educ:* Univ Minn, BChem, 63; Univ Cincinnati, PhD(inorg chem), 69. *Prof Exp:* Res asst nuclear magnetic resonance, Univ Cincinnati, 64-68 & Univ Calif, Irvine, 68-72; res assoc, Univ Mo, St Louis, 72-74; from asst prof to assoc prof, 74-82, PROF CHEM, MARYVILLE COL, ST LOUIS, 82-, CHMN DEPT, 74- *Mem:* Am Chem Soc; Sigma Xi. *Res:* Mercaptan dissociation constants; organometallic electrochemistry; ozonation and singlet oxygen; pulsed nuclear magnetic resonance of solids to investigate molecular motions, mainly in plastic crystals. *Mailing Add:* 13550 Conway Rd Creve Coeur MO 63141

STARZAK, MICHAEL EDWARD, b Woonsocket, RI, Apr 21, 42; m 67; c 2. BIOPHYSICAL CHEMISTRY. *Educ:* Brown Univ, BS, 63; Northwestern Univ, PhD(chem), 68. *Prof Exp:* From actg instr to actg asst prof chem & grant, Univ Calif, Santa Cruz, 68-70; from asst prof to assoc prof, 70-88, PROF CHEM, STATE UNIV NY, BINGHAMTON, 88- *Concurrent Pos:* Corp mem, Marine Biol Lab, 74- *Mem:* Sigma Xi; Am Chem Soc; Am Phys Soc; Biophys Soc; NY Acad Sci. *Res:* Excitable membrane phenomena; photochemistry; stochastic processes; membrane channels. *Mailing Add:* Dept Chem State Univ NY Binghamton NY 13902-6000

STARZL, THOMAS E, b Le Mars, Iowa, Mar 11, 26; m 54; c 3. SURGERY. *Educ:* Westminister Col, Mo, BA, 47; Northwestern Univ, MA, 50, PhD(anat) & MD, 52. *Hon Degrees:* Dr, Westminister Col, Mo, New York Med Col, Univ Wyo & Westmar Col. *Prof Exp:* Intern surg, Johns Hopkins Hosp, 52-53, asst resident, 55-56; resident, Sch Med, Univ Miami, 56-58; resident & instr, Northwestern Univ, 58-59, assoc, 59-61, asst prof, 61; assoc prof surg, Sch Med, Univ Colo, Denver, 62-64, prof, 64-81, chmn dept, 72-81; PROF SURG, UNIV PITTSBURGH, 81- *Honors & Awards:* Int Soc Surg Medal, 65; Eppinger Prize (Freiburg), 70; Brookdale Award in Med, 74; Middleton Award, 68; Mod Med Distinguished Achievement Award, 69; David M Hume Mem Award, Nat Kidney Found, 78. *Mem:* Am Col Surg; fel Am Acad Arts & Sci; Soc Univ Surg; Am Surg Asn; Soc Vascular Surg. *Res:* General and thoracic surgery; neurophysiology, cardiac physiology; transplantation of tissues and organs. *Mailing Add:* Dept Surg 1084 Scaife Hall Univ Pittsburgh 4200 5th Ave Pittsburgh PA 15260

STARZYK, MARVIN JOHN, b Chicago, Ill, Feb 3, 35; m 58; c 4. MICROBIOLOGY. *Educ:* Loyola Univ, Chicago, BS, 57; Univ Wis-Madison, PhD(microbiol), 62. *Prof Exp:* Asst prof natural sci, Northern Ill Univ, 61-64; group leader microbiol, Res Dept, Brown & Williamson Tobacco Corp Ky, 64-65, asst sect leader biol sci, 65-66; asst prof, 66-71, assoc prof, 71-85, chmn dept biol sci, 84, PROF MICROBIOL, NORTHERN ILL UNIV, 85- *Concurrent Pos:* Consult, Brown & Williamson Tobacco Corp, 66-67. *Honors & Awards:* Fulbright Sr Lectr, USSR, 83. *Mem:* Am Soc Microbiol; AAAS; Sigma Xi. *Res:* Aquatic Microbiology, the pathological ecology of microorganisms associated with contaminated waters. *Mailing Add:* Dept Biol Sci 311 Montgomery Hall Northern Ill Univ De Kalb IL 60115

STASHEFF, JAMES DILLON, b New York, NY, Jan 15, 36; m 59; c 2. MATHEMATICS, PHYSICAL MATHEMATICS. *Educ:* Univ Mich, BA, 56; Princeton Univ, MA, 58, PhD(math), 61; Oxford Univ, DPhil(math), 61. *Prof Exp:* Moore instr math, Mass Inst Technol, 60-62; from asst prof to prof, Univ Notre Dame, 62-70; prof math, Temple Univ, 70-78; PROF MATH, UNIV NC, 76- *Concurrent Pos:* NSF grants, 64-; mem Inst Advan Study, 64-65, 87, Sloan fel, 69-70; vis prof, Princeton Univ, 68-69, Univ Pa, 83, Rutgers Univ, 87. *Mem:* Am Math Soc; Math Asn Am. *Res:* Algebraic topology, especially homotopy theory; homological algebra, fibre spaces, H-spaces and characteristic classes with applications to theoretical physics. *Mailing Add:* Math Univ NC Chapel Hill NC 27599-3250

STASIW, ROMAN OREST, b Ukraine, May 3, 41; US citizen; m 68; c 1. CLINICAL CHEMISTRY, BIOCHEMISTRY. *Educ:* Univ Rochester, BS, 63; State Univ NY Buffalo, PhD(inorg chem), 68. *Prof Exp:* Analyst, E I du Pont de Nemours & Co, summers 62 & 63; asst scientist, Cancer Res Ctr, Columbia, Mo, 68-73; SCIENTIST, TECHNICON, 73- *Mem:* Am Asn Clin Chem. *Res:* Inorganic and synthetic organic chemistry; enzymology; clinical automation. *Mailing Add:* 98 N Grant Ave Congers NY 10920

STASKIEWICZ, BERNARD ALEXANDER, b Monessen, Pa, Aug 20, 24; m 49; c 5. PHYSICAL CHEMISTRY. *Educ:* Washington & Jefferson Col, AB, 46; Carnegie Inst Technol, MS, 50, PhD(chem), 53. *Prof Exp:* Instr chem, Washington & Jefferson Col, 46-51; res chemist, Esso Res & Eng Co, Standard Oil Co, NJ, 53-56 & Rayonier, Inc, 56-58; assoc prof, 58-62, PROF CHEM, WASHINGTON & JEFFERSON COL, 62-, chmn dept, 67-, CHMN, DIV SCI & MATH, 88- *Mem:* Am Chem Soc. *Res:* Thermodynamics; cellulose chemistry; automotive lubricants. *Mailing Add:* Dept of Chem Washington & Jefferson Col Washington PA 15301

STASKO, AIVARS B, b Riga, Latvia, May 22, 37; Can citizen; m 63; c 3. AQUATIC ECOLOGY & FISHERIES, RESEARCH MANAGEMENT. *Educ:* Univ Toronto, BASc, 60, PhD(zool), 69. *Prof Exp:* Res assoc limnol, Univ Wis-Madison, 67-70; res scientist, St Andrews, 70-79, head, Crustaceans Sect, 77-79, ASSOC DIR, FISHERIES RES PROG ANALYSIS, DEPT FISHERIES & OCEANS, OTTAWA, ONT, 80- *Concurrent Pos:* Ed, Underwater Telemetry Newslett, 71-76. *Res:* Crab and lobster biology and fisheries; underwater biotelemetry; responses of fish to environmental factors. *Mailing Add:* Dept Fisheries & Oceans Sta 1256 200 Kent St Ottawa ON K1A 0E6 Can

STASZAK, DAVID JOHN, b Milwaukee, Wis, Mar 29, 44; m 65; c 2. ANIMAL PHYSIOLOGY, BIOCHEMISTRY. *Educ:* Iowa State Univ, BS, 66, MS, 68, PhD(physiol), 71. *Prof Exp:* Res asst insect physiol, Iowa State Univ, 66-68, teaching asst human physiol, 68-69, res asst insect physiol, 69-71; asst prof biochem, Ill Col, 71-72; assoc prof physiol, Ga Col, 72-76, assoc prof biol & dir res servs, 76-80; PROF BIOL & DEAN GRAD STUDIES, UNIV WIS-STEVENS POINT, 80- *Concurrent Pos:* USDA grant, Iowa State Univ, 66-71; asst prof, Dept Biol, MacMurray Col, 71-72; consult, Biochem Sect, Res Dept, Regional Ment Health Ctr, Cent State Hosp, 75-79. *Mem:* Am Inst Biol Sci; Sigma Xi; AAAS; Nat Coun Univ Res Adminrs. *Res:* Influence of low temperature on animals; chill-coma; thermal acclimation. *Mailing Add:* Off Dean Grad Studies & Coordr Res Univ Wis Stevens Point WI 54481

STASZEKY, FRANCIS M, b Wilmington, Del, Apr 16, 18. MECHANICAL ENGINEERING. *Educ:* Mass Inst Technol, BS & MS, 43. *Prof Exp:* Mech engr, Union Oil Co, Calif, 43-45, E I du Pont de Nemours, Wilmington, 46-48 & Boston Edison Co, 48-57; supt, eng & consult, 57-64; asst vpres, Boston Edison Co, 64-67, exec vpres, 67-69, pres & chief operating officer, 79-83, dir, 68-83; CONSULT, 83- *Mem:* Nat Acad Eng; Inst Elec & Electronics Engrs; Am Soc Mech Engrs. *Mailing Add:* 144 Chestnut Circle Lincoln MA 01773

STATE, DAVID, b London, Ont, Nov 13, 14; nat US; m 45; c 5. SURGERY. *Educ:* Univ Western Ont, BA, 36, MD, 39; Univ Minn, MS, 45, PhD(surg), 47; Am Bd Surg, dipl, 46; Bd Thoracic Surg, dipl, 52. *Prof Exp:* Intern, Victoria Hosp, Ont, 39-40; from intern to sr resident surg, Univ Minn Hosp, 41-45, res asst, Univ, 45-46, from instr to assoc prof, 46-52, dir cancer detection ctr, Univ Hosp, 48-52; clin assoc prof surg, Sch Med, Univ Southern Calif, 52-58; prof, Albert Einstein Col Med, 58-71, chmn dept, 59-71; PROF SURG & VCHMN DEPT, UNIV CALIF, LOS ANGELES & CHMN DEPT SURG, HARBOR GEN HOSP, 71- *Concurrent Pos:* Fel path, St Luke's Hosp, Chicago, 40-41; dir surg, Cedars of Lebanon Hosp, Los Angeles, Calif, 53-58, Bronx Munic Hosp Ctr, New York, 59-71 & Hosp of Albert Einstein Col Med, 66-71. *Mem:* AAAS; Soc Exp Biol & Med; Am Thoracic Soc; Soc Univ Surg; AMA. *Res:* General, thoracic and open heart surgery; gastrointestinal physiology. *Mailing Add:* Harbor/Calif LA Med Ctr 1000 W Carson St Torrance CA 90502

STATE, HAROLD M, b Washington, Mo, Apr 15, 10; m 39; c 1. ANALYTICAL CHEMISTRY, INORGANIC CHEMISTRY. *Educ:* Cent Col, Mo, AB, 32; Princeton Univ, AM, 35, PhD(chem), 36. *Prof Exp:* Asst prof chem, Culver-Stockton Col, 36-37; from instr to prof, 37-75, EMER PROF CHEM, ALLEGHENY COL, 75- *Concurrent Pos:* Vis asst prof, Univ Ill, 41-42. *Mem:* Am Chem Soc. *Res:* Application of Werner complexes to analysis; higher valent complexes of nickel; chemistry of coordination compounds. *Mailing Add:* RD 4 Box 320 Saegertown PA 16433

STATEN, RAYMOND DALE, b Stillwater, Okla, May 17, 22; m 46; c 4. AGRONOMY, BOTANY. *Educ:* Okla State Univ, BS, 47; Univ Nebr, MS, 49, PhD(agron), 51. *Prof Exp:* Asst prof agron, Univ Ark, 51-56; asst prof, Tex A&M Univ, 56-60, assoc prof agron, 60-87, EMER PROF AGRON, 87- *Mem:* Am Soc Agron. *Res:* Forage crop breeding and improvement; pasture management; grain and fiber crops production; morphology. *Mailing Add:* Dept Agron Tex A&M Univ College Station TX 77843

STATES, JACK STERLING, b Laramie, Wyo, Nov 6, 41; m 65; c 2. MICROBIAL ECOLOGY, MYCOLOGY. *Educ:* Univ Wyo, BAEd, 64, MSc, 66; Univ Alta, PhD(bot), 69. *Prof Exp:* Res assoc bot, Univ Wyo, 69-70; from asst prof to assoc prof, 70-84, PROF BIOL, NORTHERN ARIZ UNIV, 84- *Concurrent Pos:* High sch instr, Wyo, 69-70; chmn comt teaching, Mycol Soc Am, 80-81; southwest regional dir, nat bd dirs, Sigma Xi, 86-88, mem nat comt membership, 86-88. *Mem:* Mycol Soc Am; Sigma Xi; Regist Forensics Experts. *Res:* Soil microfungi; ecological studies and effects of industrial pollutants; mycorrhizal fungi; mycophagy by mammals and insects. *Mailing Add:* Dept Biol Northern Ariz Univ Flagstaff AZ 86011

STATHOPOULOS, THEODORE, b Athens, Greece, Sept 30, 47; Can citizen; m 79; c 2. WIND ENGINEERING. *Educ:* Nat Tech Univ, Athens, dipl, 70; Univ Western Ont, MS, 76, PhD(wind eng), 79. *Prof Exp:* Engr struct design, Stefanou & Assoc, Athens, 70-73; res assoc, 79, from asst prof to assoc prof, 79-87, PROF, CONCORDIA UNIV, 87-, ASSOC DIR, CTR BLDG STUDIES, 83- *Mem:* Tech Chamber Greece; Am Soc Civil Engrs; Can Soc Wind Eng; Wind Engr Res Coun. *Res:* Wind loads on buildings; wind tunnel testing techniques; economical measurements of area averaged wind loads on structures (pneumatic averaging technique); wind environmental problems. *Mailing Add:* Ctr Bldg Studies Concordia Univ 1455 De Maisonneuve Blvd W Montreal PQ H3G 1M8 Can

STATLER, IRVING C(ARL), b Buffalo, NY, Nov 23, 23; m 53; c 2. NEUROSCIENCES, BIOMATHEMATICS. *Educ:* Univ Mich, BS(aeronaut eng) & BS(eng math), 45; Calif Inst Technol, PhD(aeronaut, math), 56. *Prof Exp:* Res engr, Cornell Aeronaut Lab, Inc, 46-53; sr res engr, Jet Propulsion Lab, Calif Inst Technol, 53-55; prin engr, Cornell Aeronaut Lab, Inc, 56-57, asst head, Appl Mech Dept, 57-63, head, 63-70; res scientist, US Army Mobility Res & Develop Lab, Ames Res Ctr, NASA, 70-72, dir, Dept Defense, Aeromechanics Lab, 72-85; dir, Adv Group Aerospace & Res Develop, NATO, 85-88; DIR ENG, HUMAN FACTORS RES DIV, AMES RES CTR, NASA, 88- *Concurrent Pos:* Lectr, Univ Buffalo, 56-57; mem flight mech panel, Adv Group Aerospace Res & Develop, NATO. *Mem:* AAAS; fel Am Inst Aeronaut & Astronaut; Am Helicopter Soc; fel Royal Aeronaut Soc. *Res:* Aerodynamics; dynamic stability and control; aeroelasticity; rotary wing aerodynamics; applied mathematics; human factors; aeronautical & astronautical engineering. *Mailing Add:* Ames Res Ctr NASA Mail Stop 262-1 Moffett Field CA 94035-5000

STATON, ROCKER THEODORE, JR, b McComb, Miss, Dec 6, 20; m 41; c 3. MECHANICAL & INDUSTRIAL ENGINEERING. *Educ:* Miss State Col, BS, 41; Ga Inst Technol, MS, 49; Johns Hopkins Univ, PhD(indust eng), 55. *Prof Exp:* Instr mech eng, Miss State Col, 46-48; asst prof indust eng, Ga Inst Technol, 48-51 & Johns Hopkins Univ, 51-54; assoc prof, Ga Inst Technol, 54-58, asst dean col eng, 56-61, assoc dean, 61-66, dean undergrad div, 66-77, prof indust eng, 58-81, dir inst res, 77-81; RETIRED. *Mem:* Am Soc Mech Engrs; Am Soc Eng Educ. *Res:* Academic administration. *Mailing Add:* 10660 Parsons Rd Duluth GA 30136

STATT, TERRY G, b Rochester, NY, Apr 19, 53. REFRIGERATION & AIR CONDITIONING TECHNOLOGIES, DEVELOPMENT OF NEW TECHNOLOGIES. *Educ:* Stevens Inst Technol, BE & ME, 75. *Prof Exp:* Assoc engr, Hittman Assocs, Inc, 75-77; sr engr, Automation Industs, Inc, 77-78; sr assoc, PRC Energy Analysis Co, 78-80; mech engr, Energy Applications, Inc, 80-83; mech engr, 83-85, PROG MGR REFRIG SYSTS, US DEPT ENERGY, 85- *Concurrent Pos:* Mem, Refrig Comt, Am Soc Heating, Refrig & Air Conditioning Engrs, 88-90, Task Group Halocarbon Emissions & Tech Comt Unitary Air Conditioners & Heat Pumps; mem bd dirs, US Nat Comt, Int Inst Refrig, 91- *Honors & Awards:* Citation of Excellence, UN Envrion Prog, 89. *Mem:* Am Soc Heating Refrig & Air Conditioning Engrs; Int Inst Refrig. *Res:* Managing and coordinating research activities among industry, government and universities; international environmental studies. *Mailing Add:* 10207 Raleigh Tavern Lane Ellicott City MD 21042

STATTON, GARY LEWIS, b New Brighton, Pa, Nov 4, 37; m 58; c 6. POLYMER CHEMISTRY. *Educ:* Geneva Col, BS, 59; Univ Fla, PhD(org chem), 64. *Prof Exp:* Res chemist, 64-66, SR RES CHEMIST, ATLANTIC RICHFIELD CO, 66- *Mem:* Am Chem Soc; Royal Soc Chem. *Res:* Organometallics; polymers; polyurethanes. *Mailing Add:* 1392 Bittersweet Lane West Chesterfield PA 19380

STATZ, HERMANN, b Herrenberg, Ger, Jan 9, 28; nat US; m 53; c 2. PHYSICS, SOLID STATE PHYSICS. *Educ:* Stuttgart Tech Univ, MS, 49, Dr rer nat(physics), 51. *Prof Exp:* Res assoc, Max Planck Inst Metal Res, Ger, 49-50; Ger Res Asn fel physics, Stuttgart Tech Univ, 51-52; mem solid state & molecular theory group, Mass Inst Technol, 52-53; group leader, 53-58, asst gen mgr, 58-69, ASST GEN MGR & TECH DIR, 69-, DIV GEN MGR, RAYTHEON RES DIV, RES DIV, RAYTHEON CO, 88- *Mem:* Nat Acad Eng; fel Inst Elec & Electronic Engrs; fel Am Phys Soc. *Res:* Semiconductor physics, surfaces and devices; noise in electronic devices, device modeling ferromagnetism; paramagnetic resonance; exchange interactions in solids; masers and lasers. *Mailing Add:* 10 Barney Hill Rd Wayland MA 01778-3602

STATZ, JOYCE ANN, b Minn, July 21, 47. COMPUTER SCIENCES. *Educ:* Col St Benedict, Minn, BA, 69; Syracuse Univ, MA, 71, PhD(comput sci), 73. *Prof Exp:* Res assoc comput sci, Syracuse Univ, 72-73; asst prof comput sci, Bowling Green State Univ, 73-78; SYSTS PROGRAMMER, TEX INSTRUMENTS, 78- *Concurrent Pos:* Ed, SIGCUE Bull, Asn Comput Mach, 74-78. *Mem:* Asn Comput Mach; Inst Elec & Electronics Engrs Comput Soc. *Res:* Computer education; operating systems development; logo. *Mailing Add:* 5305 Valburn Circle Austin TX 78731

STAUB, FRED W, b Apr 5, 28; US citizen; m 65; c 3. HEAT TRANSFER, FLUID DYNAMICS. *Educ:* Rensselaer Polytech Inst, BME, 52, MME, 53. *Prof Exp:* Heat transfer engr, Gen Eng Lab, 53-61, proj engr, Res & Develop Ctr, 61-67, mgr two phase processes, Gen Eng Lab, 67-68, mgr heat transfer unit, 68-82, CONSULT ENGR, CORP RES DEVELOP CTR, GEN ELEC CO, 82- *Concurrent Pos:* US Atomic Energy Comn-Europ Atomic Energy Comn personnel exchange rep, France, 66; vis fel, Cambridge Univ, 79; Coolidge Fel, Gen Elec Co, 76. *Honors & Awards:* Melville Medal, Am Soc Mech Engrs, 80. *Mem:* Fel Am Soc Mech Engrs; Am Inst Chem Engrs. *Res:* Applied research and development in convective heat transfer and fluid flow radiation exchange processes; adiabatic and diabatic two phase flow processes. *Mailing Add:* 1186 Godfrey Lane Schenectady NY 12309

STAUB, HERBERT WARREN, b Brooklyn, NY, Aug 31, 27; m 55; c 1. NUTRITIONAL BIOCHEMISTRY. *Educ:* Syracuse Univ, AB, 49; Rutgers Univ, MS, 57, PhD(biochem, physiol), 60. *Prof Exp:* Asst, Rutgers Univ, 57-60; sr res specialist, Gen Foods Corp, 60-80, prin scientist, Nutrit Tech Ctr, 80-90; CONSULT, 90- *Concurrent Pos:* Mem coun arteriosclerosis, Am Heart Asn; adj prof nutrit, Pace Univ Westchester, 75-81. *Mem:* Am Inst Nutrit; fel Am Col Nutrit; Am Chem Soc; Soc Nutrit Educ; Inst Food Technologists; fel Am Inst Chemists. *Res:* Nutritional biochemistry; atherosclerosis; proteins; carbohydrates; enzymology; relationship of dietary carbohydrates to metabolic activity; protein quality evaluation and protein nutrition; dietary fiber. *Mailing Add:* 41 Clover Lane Hightstown NJ 08520

STAUB, NORMAN CROFT, b Syracuse, NY, June 21, 29; m 53; c 5. PHYSIOLOGY. *Educ:* Syracuse Univ, AB, 50; State Univ NY, MD, 53. *Prof Exp:* Intern, Walter Reed Army Med Ctr, Washington, DC, 54; instr physiol, Grad Sch Med, Univ Pa, 57-58; vis asst prof, 58-59, asst res physiologist, 59-60, from asst prof to assoc prof physiol, 60-70, PROF PHYSIOL, CARDIOVASC RES INST, MED CTR, UNIV CALIF, SAN FRANCISCO, 70-, MEM SR STAFF, 58- *Concurrent Pos:* Res fel physiol, Grad Sch Med, Univ Pa, 56-58. *Mem:* AAAS; Am Physiol Soc; Microcirc Soc (pres, 78-79); Int Soc Lymphology; Am Thoracic Soc. *Res:* Pulmonary physiology; pulmonary structure-function relations; kinetics of reaction of oxygen and hemoglobin; diffusion of oxygen and carbon monoxide; pulmonary capillary bed; pulmonary edema and blood flow; pulmonary lymph and lymphatics. *Mailing Add:* Dept Physiol CVRI Box 0130 Univ Calif San Francisco CA 94143

STAUB, ROBERT J, b Chicago, Ill, Jan 29, 22. ECOLOGY, BOTANY. *Educ:* St Mary's Col, Minn, BS, 43; Univ Minn, Minneapolis, MS, 49, PhD(ecol), 66. *Prof Exp:* Teacher high schs, Mo, Minn, Tenn & Ill, 43-50; instr biol, 50-53, assoc prof, 59-61, chmn dept, 61-85, PROF BIOL, CHRISTIAN BROS UNIV, 70- *Concurrent Pos:* Res grants, Dept of Interior, 67-69 & Environ Protection Agency, 70-72. *Mem:* Ecol Soc Am; Bot Soc Am; Am Water Resources Asn; Am Inst Biol Sci; Am Bryol & Lichenological Soc. *Res:* Plant variation; water pollution and its effects on phytoplankton; aquatic ecology. *Mailing Add:* Dept Biol Christian Brothers Univ 650 E Parkway Memphis TN 38104-5519

STAUBER, WILLIAM TALIAFERRO, b East Orange, NJ, June 15, 43; m 70; c 1. PHYSIOLOGY. *Educ:* Ithaca Col, BS, 67; Rutgers Univ, MS, 69, PhD(physiol), 72. *Prof Exp:* NSF fel, Univ Iowa, 72-73, Muscular Dystrophy Asn fel physiol, 74-75, assoc physiol, 76-79; NSF fel, 72-73, muscular dystrophy asn fel physiol, 74-75, assoc physiol, Univ Iowa, 76-79; asst prof, 79-81, assoc prof, 79-85, PROF, WVA UNIV, 85- *Mem:* Sigma Xi; Am Physiol Soc. *Res:* Physiology-pathology of skeletal muscle protein breakdown; muscle injury and repair; cumulative trauma disorders; eccentric muscle action. *Mailing Add:* Dept Physiol WVa Univ Health Sci Ctr N Morgantown WV 26506

STAUBITZ, WILLIAM JOSEPH, b Buffalo, NY, Mar 19, 15; m 44; c 4. MEDICINE, UROLOGY. *Educ:* Gettysburg Col, AB, 38; Univ Buffalo, MD, 42. *Prof Exp:* Chmn urol, Roswell Park Mem Inst, 49-60; PROF UROL & CHMN DEPT, SCH MED, STATE UNIV NY COL BUFFALO, 60- *Concurrent Pos:* Chmn dept urol, Buffalo Gen Hosp, Buffalo Children's Hosp & Edward J Meyer Mem Hosp, 60-; consult, Roswell Park Mem Inst, 60-; consult & mem dean's comt, Vet Admin Hosp, 65-; mem, Residency Rev Comt Urol, 68- *Mem:* Can Urol Asn; Am Urol Asn; Am Col Surg; Am Acad Pediat; Am Asn Genito-Urinary Surg. *Res:* Carcinoma of the prostate; carcinoma of the testes; urinary tract infections. *Mailing Add:* Vet Admin Hosp 3495 Bailey Ave Buffalo NY 14215

STAUBUS, ALFRED ELSWORTH, b San Jose, Calif, Nov 20, 47; m 72; c 1. PHARMACOKINETICS, FORENSIC TOXICOLOGY. *Educ:* Univ Calif, San Francisco, PharmD, 71, PhD(pharmaceut chem), 74. *Prof Exp:* ASSOC PROF PHARMACEUT & PHARMACEUT CHEM, COL PHARM, OHIO STATE UNIV, 74- *Concurrent Pos:* Dir clin pharmacokinetic lab, Interdisciplinary Oncol Unit, Ohio State Univ Comprehensive Cancer Ctr, 77-; vis prof, Abbott Labs, North Chicago, 76. *Mem:* Am Pharmaceut Asn; Acad Pharmaceut Sci; Am Soc Hosp Pharmacists; Am Asn Cancer Res; Am Asn Pharmaceut Scientists; Am Acad Forensic Sci. *Res:* Forensic pharmacokinetics and toxicology; clinical pharmacology of phase I-II. *Mailing Add:* Col Pharm Ohio State Univ 500 W 12th Ave Columbus OH 43210

STAUBUS, JOHN REGINALD, b Cissna Park, Ill, Mar 21, 26; m 51; c 1. DAIRY SCIENCE. *Educ:* Univ Ill, BS, 50, MS, 56, PhD(dairy sci), 59. *Prof Exp:* Asst dairy sci, Univ Ill, 54-59, res assoc, 59-60; from asst prof to assoc prof dairy scci, Ohio State Univ, 60-69, prof dairy sci, 69-87, exten specialist, 60-87, emer prof dairy sci, 87; RETIRED. *Mem:* AAAS; Am Dairy Sci Asn; Sigma Xi; Am Soc Animal Sci. *Res:* Nutrition in dairy science; ruminant nutrition and physiology; bacteriology of silage; forage plant physiology and composition. *Mailing Add:* 915 Brentford Dr Columbus OH 43220

STAUDENMAYER, RALPH, b July 28, 42; US citizen. METALLURGICAL ENGINEERING. *Educ:* Univ Calif, Los Angeles, BS, 66; Univ Ariz, MS, 68; Univ Ark, PhD(chem), 73. *Prof Exp:* Chief chemist metall, TRW Inc, Wendt Sonis, 73-80; DIR ENG, HUGHES TOOL CO, 80- *Concurrent Pos:* Cert Calif Jr Col Instr. *Mem:* Am Powder Metall Inst; Am Soc Testing & Mat. *Res:* Powder metallurgy and gas deposition on cemented carbides. *Mailing Add:* 15711 Four Leaf Dr Houston TX 77084

STAUDENMAYER, WILLIAM J(OSEPH), b Rochester, NY, Jan 4, 36; m 63; c 2. CHEMICAL ENGINEERING. *Educ:* Clarkson Col Technol, BS, 57; Cornell Univ, PhD(chem eng), 63. *Prof Exp:* Develop engr, Mfg Exp Div, 57-59, res engr, Res Labs, 62-69, res assoc, 69-86, SR RES ASSOC, EASTMAN KODAK, 86-, LAB HEAD, IMAGE FIXING, 89- *Mem:* Soc Photog Sci & Eng. *Res:* Electrophotography; photo receptors; high temperature elastomers. *Mailing Add:* 47 Greylock Ridge Pittsford NY 14534

STAUDER, WILLIAM, b New Rochelle, NY, Apr 23, 22. GEOPHYSICS, SEISMOLOGY. *Educ:* St Louis Univ, AB, 43, MS, 48; Univ Calif, PhD(geophys), 59. *Prof Exp:* Instr, Marquette Univ High Sch, 48-49; res asst geophys, Univ Calif, 57-59; from instr to assoc prof geophys, 60-66, chmn dept earth & atmospheric sci, 72-75, dean Grad Sch/Univ Res Adminr, 75-88, PROF GEOPHYS, ST LOUIS UNIV, 66-, ASSOC ACAD VPRES, 89- *Concurrent Pos:* Mem geophys adv panel, Air Force Off Sci Res, 61-71; mem panel seismol, Comt Alaska Earthquake, Nat Acad Sci-Nat Res Coun, 64-72; mem adv panel, Nat Ctr Earthquake Res, 66-76; mem ad hoc comt triggering of earthquakes, AEC, 69-72. *Mem:* Fel Am Geophys Union; Seismol Soc Am (vpres, 64, pres, 65). *Res:* Focal mechanism of earthquakes; crustal structure in central United States; seismicity of southeastern Missouri. *Mailing Add:* 3601 Lindell Blvd St Louis MO 63108

STAUDHAMMER, JOHN, b Budapest, Hungary, Mar 15, 32; US citizen; m 60; c 2. ELECTRICAL ENGINEERING, COMPUTER GRAPHICS. *Educ:* Univ Calif, Los Angeles, BS, 54, MS, 56, PhD(eng), 63. *Prof Exp:* From asst to assoc eng, Univ Calif, Los Angeles, 54-59; sr syst engr, Syst Develop Corp, 59-64; prof eng, Ariz State Univ, 64-67; prof elec eng, NC State Univ, 67-80; PROF ELEC ENG, UNIV FLA, 80- *Concurrent Pos:* Consult, various industs, 57-; designer, Douglas Aircraft Co, 59; tech adv, US Army Comput Systs Command, 76-77; nat lectr, Asn Comput Mach, 76-79; comput engr, US Army Res Off, 78-79; adv prof, Zhejiang Univ, Hangzhou, PRC, 85-; expert witness, computer graphics & systs, 87- *Mem:* Am Soc Eng Educ; sr mem Inst Elec & Electronics Engrs; Asn Comput Mach. *Res:* Use of computers in circuit design; design and analysis of computer systems; system engineering of graphics displays. *Mailing Add:* Dept Elec Eng Univ Fla Gainesville FL 32611

STAUDHAMMER, PETER, b Budapest, Hungary, Mar 4, 34; US citizen; m 58; c 3. ENGINEERING, PHYSICAL CHEMISTRY. *Educ:* Univ Calif, Los Angeles, BS, 55, MS, 56, PhD(eng, phys chem), 57. *Prof Exp:* Res engr, Univ Calif, Los Angeles, 55-57; sr res engr, Jet Propulsion Lab, Calif Inst Technol, 57-59; head, Chem Sect, TRW, Inc, 59-60, mgr, Propulsion Res Dept, 60-66, mgr, Design & Develop Lab, 66-73, mgr, Res Lab, 75-81, mgr energy systs opers, 81-87, vpres, Defense Proj Div, 87-90, VPRES, CTR AUTOMOTIVE TECHNOL, TRW INC, 90- *Concurrent Pos:* Magnetic Fusion Adv Comt, 86- *Honors & Awards:* Engr Achievement Award for Viking Biol Inst, Inst Advan Eng, 76; Group Achievement Award for Pioneer Venus Sci Team, NASA, 80, Award for Voyager Jupiter-Saturn Ultra-violet Spectrometer, 81; NASA Group Achievement Award for Pioneer Venus Sci Team, 80. *Mem:* Assoc fel Am Inst Aeronaut & Astronaut; Combustion Inst; Soc Automotive Engrs. *Res:* After burning of automobile exhaust; regenerable fuel cells; combustion and chemical kinetics of rocket propellants; developer of Apollo lunar module descent engine; space science instruments; fusion research; spacecraft and ground systems engineering; automotive technology. *Mailing Add:* 5060 Rolling Meadows Rd Rolling Hills Estates CA 90274

STAUFFER, ALLAN DANIEL, b Kitchener, Ont, Mar 11, 39; m 62; c 2. ATOMIC PHYSICS. *Educ:* Univ Toronto, BSc, 62; Univ London, PhD(appl math), 66. *Prof Exp:* Asst lectr math, Royal Holloway Col, 64-66; fel, 66-67, asst prof, 67-71, assoc prof, 71-80, PROF PHYSICS, YORK UNIV, 81- *Concurrent Pos:* Vis prof, Royal Holloway Col, London, 74-75; vis fel, Joint Inst Lab Astrophysics, Boulder, Co, 89-90. *Mem:* Am Phys Soc; Can Operational Res Soc; Opers Res Soc Am; Brit Inst Physics; Can Asn Physicists. *Res:* Theoretical atomic collisions; atomic structure problems. *Mailing Add:* Dept Physics York Univ 4700 Keele St Downsview ON M3J 1P3 Can

STAUFFER, CHARLES HENRY, b Harrisburg, Pa, Apr 17, 13; m 39; c 3. PHYSICAL CHEMISTRY. *Educ:* Swarthmore Col, AB, 34; Harvard Univ, AM, 36, PhD(chem), 37. *Prof Exp:* Lab asst org chem, Harvard Univ, 34-36, from instr to assoc prof chem, Worcester Polytech Inst, 37-58; prof & head dept, St Lawrence Univ, 58-65; prof & chmn Div Natural Sci, 65-77, EMER PROF, BATES COL, 77- *Concurrent Pos:* Dir chem kinetics data proj, Nat Acad Scis, 54-64; mem adv com, Off Critical Constants, 61-64. *Mem:* Am Chem Soc; fel AAAS; Sigma Xi. *Res:* Enolization of unsymmetrical ketones; gaseous formation and decomposition of tertiary alkyl halides; reaction kinetics; experimental and theoretical calculations of rates of reaction in gas and liquid phases. *Mailing Add:* 10 Champlain Ave Lewiston ME 04240

STAUFFER, CLYDE E, b Duluth, Minn, Nov 8, 35; m 58; c 2. BIOCHEMISTRY. *Educ:* NDak State Univ, BS, 56, MS, 58; Univ Minn, PhD(biochem), 63. *Prof Exp:* Res chemist, Procter & Gamble Co, 63-76; dir, Kroger Baked Foods Res & Develop, 76-81; dir res & develop, Colso Prods Inc, 81-82; OWNER, TECH FOOD CONSULTS, 82- *Mem:* AAAS; Am Asn Cereal Chemists; Am Soc Biol Chemists; Sigma Xi; Inst Food Technologists. *Res:* Protein biophysical chemistry; enzymology; surface and interfacial adsorption from solution; immunochemistry. *Mailing Add:* 631 Christopal Dr Cincinnati OH 45231

STAUFFER, EDWARD KEITH, b Logan, Utah, July 6, 41; m 65, 79; c 2. MEDICAL PHYSIOLOGY. *Educ:* Utah State Univ, BS, 64, MS, 69; Univ Ariz, PhD(physiol), 74. *Prof Exp:* Assoc, Col Med, Univ Ariz, 74-75; ASSOC PROF PHYSIOL, SCH MED, UNIV MINN, DULUTH, 75- *Mem:* Am Physiol Soc; Soc Neurosci; Sigma Xi; AAAS. *Res:* Neurophysiological studies of motor control with emphasis on afferent, central and efferent mechanisms found in the spinal cord; electrophysiology of neurons in tissue culture. *Mailing Add:* Sch of Med Univ of Minn 10 Univ Dr Duluth MN 55812-2487

STAUFFER, GARY DEAN, b Wenatchee, Wash, Feb 26, 44; m 68; c 2. FISHERIES. *Educ:* Univ Wash, BS, 66, MS, 69, PhD(fisheries & statist), 73. *Prof Exp:* Fishery biologist salmon res, Quinault Resource Develop Proj, Quinault Tribal Coun, 71-72; FISHERY BIOLOGIST NAT MARINE FISHERY SERV, 73- *Res:* Stock assessment and fishery evaluation of pacific coast fisheries for developing management information including groundfish species of Alaska and small pelagic species off the coast of California. *Mailing Add:* 7600 Sand Point Way NE BIN C15700 Seattle WA 98115

STAUFFER, GEORGE FRANKLIN, b Hanover, Pa, Oct 23, 07; m 31. ASTRONOMY. *Educ:* Millersville State Col, BS, 32; Univ Pa, MS, 38, EducD, 63. *Prof Exp:* Teacher pub sch, Pa, 26-27 & high schs, 29-57; prof astrom, Millersville State Col, 57-80; RETIRED. *Mem:* AAAS; Am Astron Soc; Nat Sci Teachers Asn. *Mailing Add:* B118 211 Willow Valley Sq Lancaster PA 17602

STAUFFER, HOWARD BOYER, b Philadelphia, Pa, Aug 10, 41; c 1. APPLIED MATHEMATICS, BIOMETRICS. *Educ:* Williams Col, BA, 64; Univ Calif, Berkeley, PhD(math), 69. *Prof Exp:* Fel, Univ BC, 69-70; asst prof math, Calif State Univ, Hayward, 70-80; biomathematician, BC Ministry Forests, 80-83; PROF COMPUT SCI, HUMBOLDT STATE UNIV, 83- *Concurrent Pos:* Fulbright prof, Nat Univ Malaysia, 74-75; res fel, Pac Forest Res Ctr, Victoria, BC, 75-76; vis instr, Univ BC, 78-80; mgr, image processing, NASA-Ames Res Ctr. *Mem:* Am Math Soc; Asn Comput Mach. *Res:* Mathematics, statistics, operations research and computer science applications; forestry applications. *Mailing Add:* Math Dept Humboldt State Univ Arcata CA 95521

STAUFFER, JACK B, b Newton, Iowa, May 19, 28; m 53; c 3. TRANSPORTATION SYSTEM DEVELOPMENT. *Educ:* Univ Ill, BSME, 54. *Prof Exp:* Design engr, ACF Industs, Inc, 55-59, proj engr, 59-69; proj mgr, US AEC, 62-69 & Westinghouse Elec Corp, 69-72; dir, Transp Test Ctr, Us Dept Transp, 72-76; asst dir indust eng, Consol Rail Corp, 77-79, dir, Technol Serv Lab, 79-82 & appl res, 82-87; DIR TEST ENG, TRANSP TEST CTR, ASN AM RAILROADS, 87- *Concurrent Pos:* Pvt consult, 76-77. *Mem:* Nat Soc Prof Engrs. *Res:* Development of nuclear, space, underwater and transportation projects. *Mailing Add:* Dir Tech Oper Transport Test Box 11130 Pueblo CO 81001

STAUFFER, JAY RICHARD, JR, b Lancaster, Pa, Apr 8, 51. AQUATIC ECOLOGY, ICHTHYOLOGY. *Educ:* Cornell Univ, BS, 72; Va Polytech Inst & State Univ, PhD, 75. *Prof Exp:* Asst prof, 75-80, assoc prof aquatic ecol, Appalachian Environ Lab, Univ Md, 80-84; assoc prof, 84-88, PROF FISHERY SCI, PENN STATE UNIV, 88- *Concurrent Pos:* Mem, Pa Rare & Endangered Fishes Coun, 77- *Mem:* Am Inst Fishery Res Biologists; Am Fisheries Soc; Am Soc Ichthyologists & Herpetologists. *Res:* Zoogeography of freshwater fishes; status of rare and endangered fishes; temperature behavior of fishes; assessment of environmental stresses; systematics of African cichlids. *Mailing Add:* Sch Forest Resources 88 Ferguson Bldg Penn State Univ University Park PA 16802

STAUFFER, JOHN RICHARD, b Findlay, Ohio, Oct 27, 52; m 80. STELLAR EVOLUTION, SPECTROSCOPY. *Educ:* Case Western Reserve Univ, BS, 74; Univ Calif, Berkeley, MS, 77, PhD(astron), 82. *Prof Exp:* FEL, HARVARD-SMITHSONIAN CTR ASTROPHYSICS, 81-; AT DOMINION ASTROPHYS OBSERV. *Mem:* Am Astron Soc. *Res:* Emperical pre-main sequence evolutionary tracks for low mass stars; observational constraints on the physical process at work in active galaxy nuclei. *Mailing Add:* 11332 Rocoso Rd Lakeside CA 92040

STAUFFER, MEL R, b Edmonton, Alta, July 16, 37; m 58, 72; c 6. STRUCTURAL GEOLOGY. *Educ:* Univ Alta, BSc, 60, MSc, 61; Australian Nat Univ, PhD(geol), 64. *Prof Exp:* From asst prof to assoc prof, 65-75, PROF STRUCT GEOL, UNIV SASK, 75- *Concurrent Pos:* Vis lectr, Univ Alta, 64-65; Nat Res Coun fel, Univ BC, 65-66. *Mem:* Geol Asn Can. *Res:* Structures in rocks, both primary and secondary. *Mailing Add:* Dept of Geol Sci Univ of Sask Saskatoon SK S7N 0W0 Can

STAUFFER, ROBERT ELIOT, b Chicago, Ill, June 9, 13; m 34; c 4. PHYSICAL CHEMISTRY. *Educ:* Mt Union Col, BA, 32; Harvard Univ, MA, 34, PhD(phys chem), 36. *Hon Degrees:* DSc, Mt Union Col, 58. *Prof Exp:* Asst electrochem, Harvard Univ, 34-36; res assoc, Eastman Kodak Co, 36-54, head, Emulsion Res Div, 54-71, asst dir, Kodak Res Labs, 71-78; RETIRED. *Concurrent Pos:* Instr, Rochester Inst Technol, 47-49. *Mem:* AAAS; Am Chem Soc; NY Acad Sci. *Res:* Theory of electrolytes; protein electrochemistry and composition; photographic emulsions; viscosities of strong electrolytes; photographic chemistry. *Mailing Add:* 353 Oakridge Dr Rochester NY 14617

STAUFFER, THOMAS MIEL, b Edmore, Mich, June 24, 26; m 54; c 2. FISH BIOLOGY. *Educ:* Mich State Univ, Lansing, BS, 49, MS, 66. *Prof Exp:* From fisheries technol fish res to supvr sea lamprey res, Mich Dept Conserv, 50-64; BIOLOGIST IN CHARGE FISH RES, MARQUETTE FISHERIES RES STA, 64- *Concurrent Pos:* Head, Great Lakes Res, Mich Dept Natural Resources, 64-72, anadromous fisheries res, 72-; assoc ed, Transactions of Am Fisheries Soc, 77-; mem bd tech experts, Great Lakes Fishery Comn, 80-81. *Mem:* Am Fisheries Soc; Am Inst Fisheries Res Biologists. *Res:* Determination of the cause of reproductive failure of planted lake trout and assessment of reproduction by coho and chinook salmon in the Great Lakes. *Mailing Add:* 193 Lakewood Lane Harvey MI 49855

STAUFFER, TRUMAN PARKER, SR, b Illmo, Mo, May 29, 19; m 45; c 1. PHYSICAL GEOGRAPHY. *Educ:* Univ Kansas City, BA, 61; Univ Mo, Kansas City, MA, 64; Univ Nebr, PhD(geog), 72. *Prof Exp:* From teacher geog to admin supt aide, Ft Osage Sch Dist, 61-68; from asst prof to assoc prof, 75-77, PROF GEOSCI, PROF GEOG, UNIV MO, KANSAS CITY, 77- *Concurrent Pos:* Coun mem, Underground Construct Res Coun, Am Soc Civil Engrs, 74-; consult, Union Carbide of AEC, 75. *Mem:* Asn Am Geogr; fel Geog Soc Am; Sigma Xi; Nat Coun Geog Educ; Int Conf Bldg Off. *Res:* Utilization and economic development of underground space for the conservation of space and energy by planned excavation and conversion of mined areas preserving the qualities of the surface. *Mailing Add:* Dept Geosci Univ Mo 5100 Rockhill Rd Kansas City MO 64110

STAUGAARD, BURTON CHRISTIAN, b Paterson, NJ, Aug 6, 29; m 53; c 4. EMBRYOLOGY. *Educ:* Brown Univ, AB, 50; Univ RI, MS, 54; Univ Conn, PhD(embryol), 64. *Prof Exp:* Dir med photog dept, RI Hosp, 50-52; sales rep, x-ray dept, Gen Elec Co, 54-58; instr zool, Univ NH, 61-64, asst prof, 64-67; res fel anat, Sch Med, Vanderbilt Univ, 67-68, asst prof, 68-70; PROF SCI & BIOL, UNIV NEW HAVEN, 70- *Mem:* AAAS; Am Soc Zoologists; Biol Photog Asn. *Res:* Morphological and functional changes of the developing mammalian embryo in preparation for independent existence; developmental biochemistry and anatomy of kidney, liver and placenta. *Mailing Add:* Dept Biol Univ New Haven 300 Orange Ave New Haven CT 06516

STAUM, MUNI M, b New York, NY, Oct 30, 21; m 46; c 2. RADIOCHEMISTRY, PHARMACEUTICAL CHEMISTRY. *Educ:* City Col New York, BS, 42; Columbia Univ, BS, 51; Univ Fla, PhD(pharmaceut chem), 61. *Prof Exp:* Develop chemist, Am Cyanamid Co, 53-57; sr res scientist, Olin Mathieson Chem Corp, 61-67; asst prof radiol, Sch Med, Univ Pa, 67-87; RETIRED. *Concurrent Pos:* Am Found Pharmaceut Educ fel. *Mem:* Am Chem Soc; Am Pharmaceut Asn; Soc Nuclear Med. *Res:* Organic reaction mechanisms; pharmaceutical drug development; development of radioactive pharmaceuticals for diagnostic nuclear medicine. *Mailing Add:* Two Saddle Lane Cherry Hill NJ 08002

STAUNTON, JOHN JOSEPH JAMESON, b Binghamton, NY, July 4, 11; m 39; c 6. INSTRUMENTATION, PATENTS. *Educ:* Univ Notre Dame, BSEE, 32, MS, 34, EE, 41. *Hon Degrees:* DEng, Midwest Col Eng, 69. *Prof Exp:* Jr engr mfg, Bantam Ball Bearings Co, Ind, 35-36; head, physics dept, DePaul Univ, 36-38; engr instruments, Coleman Elec Co, Maywood, Ill, 38-44, dir res, Coleman Instruments, Inc, 44-56, sr staff scientist, 56-64; sr staff scientist instruments, Perkin-Elmer Corp, Ill, 64-78; TECH CONSULT 78- *Mem:* Fel Inst Elec & Electronic Engrs; Optical Soc Am; Sigma Xi. *Res:* Optical, electronic, thermal control and electrochemical instrumentation for clinical and chemical analysis. *Mailing Add:* 310 Wesley Ave Oak Park IL 60302

STAUSS, GEORGE HENRY, b East Orange, NJ, Mar 25, 32; m 59; c 2. PHYSICS. *Educ:* Princeton Univ, AB, 53; Stanford Univ, MS, 58, PhD(physics), 61. *Prof Exp:* PHYSICIST, US NAVAL RES LAB, 61- *Mem:* Am Phys Soc. *Res:* Nuclear magnetic resonance and electron paramagnetic resonance, principally in magnetically ordered compounds and semiconductors. *Mailing Add:* 7701 Tauxemont Rd Alexandria VA 22308

STAUT, RONALD, b New York, NY, Mar 30, 41; m 73; c 4. CERAMICS, PHYSICAL CHEMISTRY. *Educ:* Rutgers Univ, BS, 63, MS, 66, PhD(ceramics), 67. *Prof Exp:* Res assoc inorg chem & ceramics, Mat Res Group, Gen Refractories Co, 67-73, mgr, 73-81, dir corp res & develop, 81-82, vpres res & develop, US Refractories Div, 82-83; CONSULT, 83- *Mem:* Am Ceramic Soc; Can Ceramic Soc. *Res:* Inorganic chemistry; glass-ceramics; refractories; glass; technical and electronic ceramics. *Mailing Add:* 17 Sunny Hill Dr Pittsburgh PA 15228

STAVCHANSKY, SALOMON AYZENMAN, b Mexico City, Mex, May 7, 47; m 70; c 2. PHARMACY, PHARMACEUTICS. *Educ:* Nat Univ Mex, BS, 69; Univ Ky, PhD(pharmaceut sci), 74. *Prof Exp:* Anal chemist, Nat Med Ctr, Mex, 68-69; develop pharmacist, Syntex Labs, Mex, 69-70; vis scientist, Sloan Kettering Inst Cancer Res, 74; asst prof, 74-80, ASSOC PROF PHARM, UNIV TEX, AUSTIN, 80-, BIOPHARMACEUT COORDR, DRUG DYNAMICS INST, 75- *Concurrent Pos:* Consult, Alcon Labs, 75- & Dept Health, Educ & Welfare, 76- *Mem:* Am Pharmaceut Asn; Am Chem Soc; Mex Pharmaceut Asn. *Res:* Analytical chemistry of pharmaceutical systems; protein binding; application of short lived isotopes for the identification of neoplastic tumors. *Mailing Add:* Dept Pharm Univ of Tex Austin TX 78712

STAVELY, JOSEPH RENNIE, b Wilmington, Del, May 28, 39; m 65; c 1. PLANT PATHOLOGY. *Educ:* Univ Del, BS, 61; Univ Wis-Madison, MS, 63, PhD(plant path, bot), 65. *Prof Exp:* Fel plant path, Univ Wis-Madison, 65-66; res plant pathologist, Tobacco Lab, Plant Genetics & Germplasm Inst, 66-80, RES PLANT PATHOLOGIST, MICROBIOL & PLANT PATH LAB, PLANT SCI INST, AGR RES SERV, USDA, 80- *Concurrent Pos:* Pres, Potomac Div Am Phytopath Soc, 79-80; assoc ed, Phytopathology, 87-90. *Mem:* Am Genetic Asn; Crop Sci Soc Am; Am Phytopath Soc; Am Soc Hort Sci. *Res:* Disease resistance in Phaseolus beans; bean diseases, especially rust, pathogenic specialization, genetics of resistance, development of comprehensive and stable resistance for United States green, wax and dry Phaseolus beans. *Mailing Add:* Beltsville Agr Res Ctr West Rm 252 Bldg 011A Agr Res Serv USDA Beltsville MD 20705

STAVER, ALLEN ERNEST, b Scribner, Nebr, Dec 5, 23; m 65; c 4. SYNOPTIC METEOROLOGY. *Educ:* Univ Nebr, Omaha, BGen Ed, 56; NY Univ, MS, 59; Univ Wis-Madison, PhD(meteorol), 69. *Prof Exp:* Weather officer, Air Weather Serv, US Air Force, 43-67; asst prof, 69-72, ASSOC PROF METEOROL, NORTHERN ILL UNIV, 72- *Mem:* Am Meteorol Soc; Nat Weather Asn; Sigma Xi. *Res:* Dynamic and synoptic meteorology utilizing satellite data; computerized meteorologicalmodels. *Mailing Add:* Dept of Geog Northern Ill Univ De Kalb IL 60115

STAVINOHA, WILLIAM BERNARD, b Temple, Tex, June 11, 28; m 56, 67; c 4. PHARMACOLOGY, TOXICOLOGY. *Educ:* Univ Tex, BS, 51, MS, 54, PhD(pharmacol), 59. *Prof Exp:* From instr to asst prof pharmacol & toxicol, Med Br, Univ Tex, 58-60; chief toxicol res, Civil Aeromed Inst, Fed Aviation Agency, Okla, 60-68; assoc prof pharmacol, 68-72, PROF PHARMACOL, UNIV TEX MED SCH SAN ANTONIO, 72- *Concurrent Pos:* Asst res prof, Med Ctr, Univ Okla, 60, adj prof, 62. *Honors & Awards:* Sigma Xi res award, Univ Tex Med Br, Galveston. *Mem:* Soc Neurochem; Am Soc Pharmacol & Exp Therapeut; Int Soc Neurochem. *Res:* Neurochemistry; insecticides; adaptive mechanisms; study of rapidly metabolized compounds in the CNS; anti-inflammatory drug development. *Mailing Add:* Dept Pharmacol Univ Tex Health Sci Ctr San Antonio TX 78284-7764

STAVIS, GUS, b New York, NY, June 5, 21; m 44; c 3. ELECTRICAL ENGINEERING. *Educ:* City Col New York, BEE, 41; Fairleigh Dickinson Univ, MBA, 79. *Prof Exp:* Mem staff, Fed Telecommun Labs, Int Tel & Tel Corp, 41-44, assoc head, Air Navig Dept, 46-52; res assoc, Radio Res Lab, Harvard Univ, 45-46; mgr radar advan develop, Kearfott Div, Singer Co, 69-85; RETIRED. *Mem:* Fel Inst Elec & Electronics Engrs; Inst Navig. *Res:* Navigation and radar electronics techniques and systems, including microwave, sonar and optical radiation devices, propagation, transmitters, receivers and signal processing. *Mailing Add:* 1710 SW 22nd Boynton Beach FL 33426

STAVITSKY, ABRAM BENJAMIN, b Newark, NJ, May 14, 19; m 42; c 2. IMMUNOLOGY. *Educ:* Univ Mich, AB, 39, MS, 40; Univ Minn, PhD(bact-immunol), 43; Univ Pa , VMD, 46. *Prof Exp:* Asst bact, Med Sch, Univ Minn, 42; bacteriologist, Dept Pediat, Univ Pa, 44-46; asst prof immunol, Sch Med, Case Western Reserv Univ, 47-49, from asst prof to prof microbiol, 49-83, prof molecular biol & microbiol, 83-89, EMER PROF, SCH MED, CASE WESTERN RESERV UNIV, 89- *Concurrent Pos:* Res fel immunochem, Calif Inst Technol, 46-47; NSF fel, Nat Inst Med Res, Eng, 58-59; bacteriologist, State Dept Health, Minn, 42 & Children's Hosp, Philadelphia, Pa, 44-46; estab investr, Am Heart Asn, 54-59; mem microbiol fel panel, USPHS, 60-63; expert comts immunochem & teaching immunol, WHO, 63-; ed, J Cellular Physiol, Wistar Inst, 66-74 & J Immunol Methods, Immunopharmacology; mem microbiol test comt, Nat Bd Med Examr, 70-74. *Mem:* Fel AAAS; Am Soc Microbiol; Am Asn Immunol; Am Soc Trop Med Hyg. *Res:* Induction and regulation of cellular and humoral immunity in schistosomiasis japonica. *Mailing Add:* Dept Molecular Biol & Microbiol Case Western Reserve Univ Cleveland OH 44106

STAVN, ROBERT HANS, b Palo Alto, Calif, July 30, 40. OCEANOGRAPHY, GEOPHYSICS. *Educ:* San Jose State Col, BA, 63; Yale Univ, MS, 65, PhD(ecol), 69. *Prof Exp:* Lectr biol, City Univ New York, 67-70, instr, 70-71; asst prof, 71-77, ASSOC PROF BIOL, UNIV NC, GREENSBORO, 77- *Concurrent Pos:* Grant-in-aid, Univ NC, Greensboro, Res Coun, 71-85, univ res assignment leave, 79; res grant, NC Bd Sci & Technol, 74-75; consult hydrospheric optics, Water Qual & Watershed Res Lab, USDA, Durant, Okla, 83; vis prof, Univ Southern Calif, 86, Naval Ocean Res & Develop Activ, 87-89, Naval Oceanic & Atmospheric Res Lab, 90; researcher, Univ NC, Greensboro & Off Naval Res, 86-87; Off Naval Res grants, 88-92. *Mem:* Am Soc Limnol & Oceanog; Am Geophys Union; Ecol Soc Am; Biomet Soc; Optical Soc Am; Oceanog Soc. *Res:* Aquatic ecology; physiological ecology; oceanographic optics; theory of the ecological niche; transmission and absorption of light by the ocean; energy-balance of air water interface. *Mailing Add:* Dept Biol Univ NC Greensboro NC 27412

STAVRIC, BOZIDAR, b Skopje, Yugoslavia, Oct 31, 26; Can citizen; m 58; c 2. FOOD TOXICOLOGY, NATURAL PRODUCTS. *Educ:* Univ Zagreb, BSc, 50, PhD(org chem), 58. *Prof Exp:* Lectr org chem, Univ Zagreb, 50-63, asst prof, 63; res scientist biochem, Health Protection Br, 65-72, res scientist toxicol, 72-80, RES SCI FOOD RES, FOOD DIRECTORATE, HEALTH PROTECTION BR, HEALTH & WELFARE CAN, 80- *Concurrent Pos:* Nat Res Coun Can fel biosci, 63-65; vis prof, Univ Ottawa, 86- *Mem:* AAAS; Am Chem Soc; Soc Exp Biol & Med; Soc Toxicol; Am Col Toxicol; NY Acad Sci. *Res:* Isolation and identification of mutagens and other naturally occuring toxic components in foods; experimentally induced hyperuricemia in animals for studies in the fields of hyperuricemia and hyperuricosuria; isolation of biologically active impurities in food additives; caffeine metabolism in monkey; humans exposure to polyaromatic hydrocarbons; bioavailability of benzo(a)pyrene from foods; toxicity of methylxanthines (caffeine, theoffiline) to humans. *Mailing Add:* Food Res Div Hlth Protect Br Health & Welfare Can Ottawa ON K1A 0L2 Can

STAVRIC, STANISLAVA, b Celje, Yugoslavia, Nov 13, 33; Can citizen; m 58; c 2. BIOCHEMISTRY. *Educ:* Univ Zagreb, BSci, 59, PhD(biochem), 62. *Prof Exp:* Technician org synthesis, Fac Biochem & Pharm, Univ Zagreb, 56-57, technician radiobiol, Inst Rudjer Boskovic, 57-59, res assoc, 59-63; fel org synthesis, Dept Chem, Univ Ottawa, 64-65; RES SCIENTIST, BUR MICROBIOL HAZARDS, HEALTH PROTECTION BR, HEALTH & WELFARE CAN, 65- *Mem:* Can Soc Microbiol; Soc Appl Microbiol UK; Inst Food Techologists; Int Soc Toxinology. *Res:* Effects of irradiation on the metabolism of nucleic acids in bacteria; detection methods for bacterial enterotoxins; isolation and characterization of bacterial enterotoxins; pathogenicity of campylobacter species; competitive exclusion of salmonella from young chicks; isolation and characterization of chickens gut flora; development of defined mixture of bacterial isolates for protection of poultry against salmonella. *Mailing Add:* Bur Microbiol Hazards Health Protection Br Ottawa ON K1A 0L2 Can

STAVROLAKIS, J(AMES) A(LEXANDER), b Storrs, Utah, Oct 1, 21; c 5. MATERIALS SCIENCE, CHEMISTRY. *Educ:* Rutgers Univ, BS, 43; Mass Inst Technol, ScD, 49. *Prof Exp:* Engr, Gen Elec Co, 49-52; supvr, Armour Res Found, Ill Inst Technol, 52-55; asst mgr, Mallinckrodt Chem Works, 55-56; develop mgr, Crucible Steel Co, 56-61; mgr mat res & eng, Am-Standard Corp, 61-64, vpres develop & eng, 64-67, gen mgr, 67-70; pres, Glasrock Prod, Inc, 70-72 & StanBest, Inc, 73-77; DIR, GERBER PLUMBING FIXTURES CORP, 78- *Mem:* Am Soc Metals; Am Chem Soc; Am Ceramic Soc; Sigma Xi; Am Inst Chemists. *Res:* Refractory materials; physical testing; nucleonics; brass metallurgy; grey iron; vitreous china technology and compositions. *Mailing Add:* 4656 W Touhy Ave Chicago IL 60646

STAVROUDIS, ORESTES NICHOLAS, b New York, NY, Feb 22, 23; m 49; c 2. MATHEMATICS, OPTICS. *Educ:* Columbia Univ, AB, 48, MA, 49; Imp Col, dipl & Univ London, PhD, 59. *Prof Exp:* Asst math, Rutgers Univ, 50-51; mathematician, US Dept Navy, 51; mathematician, Nat Bur Stand, 51-54, in chg lens anal & design, 57-67; prof, Optical Sci Ctr, Univ Ariz, 67-88; SR STAFF SCIENTIST, FAIRCHILD SPACE CO, 88- *Concurrent Pos:* Fac fel, Stanford Univ, 76; vis prof, Nat Chiao Tung Univ, Hsinchu, Taiwan, 82-83. *Mem:* Fel AAAS; Soc Hist Technol; Am Math Soc; Soc Indust & Appl Math; Math Asn Am; Hist Sci Soc; Optical Soc Am. *Res:* Geometric and physical optics; differential equations; differential geometry; diffraction; micro computers. *Mailing Add:* 1060 Reed Ave No 36 Sunnyvale CA 94086

STAY, BARBARA, b Cleveland, Ohio, Aug 31, 26. INSECT MORPHOLOGY, PHYSIOLOGY. *Educ:* Vassar Col, AB, 47; Radcliffe Col, MA, 49, PhD(biol), 53. *Prof Exp:* Asst biol, Harvard Univ, 52; Fulbright Scholar, Commonwealth Sci & Indust Res Orgn, Australia, 53-54; entomologist, Qm Res & Eng Ctr, US Dept Army, 54-59; Lalor fel, Harvard Univ, 59; vis asst prof zool, Pomona Col, 60; asst prof biol, Univ Pa, 61-67; assoc prof, 67-77, PROF BIOL, UNIV IOWA, 77- *Mem:* Entom Soc Am; Am Soc Zoologists; Am Soc Cell Biol. *Res:* Histochemistry of blowfly during metamorphosis and larval blowfly midgut; histology of scent glands, physiology and fine structure of accessory reproductive glands in cockroaches; control of reproduction in cockroaches; regulation of corpora allata. *Mailing Add:* Dept Biol Univ Iowa Iowa City IA 52242

STEAD, EUGENE ANSON, JR, b Atlanta, Ga, Oct 6, 08; m 40; c 3. MEDICINE. *Educ:* Emory Univ, BS, 28, MD, 32. *Prof Exp:* Intern med, Peter Bent Brigham Hosp, Boston, 32-33, intern surg, 34-35; instr, Univ Cincinnati, 35-37; asst, Harvard Med Sch, 37-39, instr, 39-41, assoc, 41-42; prof, Sch Med, Emory Univ, 42-46, dean, 45-46; Florence McAlister prof, 47-78, EMER PROF MED, SCH MED, DUKE UNIV, 78- *Concurrent Pos:* Fel, Harvard Univ, 33-34; from asst resident to resident, Cincinnati Gen Hosp, 35-37; resident physician, Thorndike Mem Lab & asst, Boston City Hosp, 37-39; assoc med, Peter Bent Brigham Hosp, 39-42, actg physician-in-chief, 42; physician-in-chief, Univ Div, Grady Hosp, 42-46 & Duke Hosp, 47-67; distinguished physician, Vet Admin, 78-85; ed, NC Med J, 82. *Honors & Awards:* Abraham Flexmer Award, Am Med Cols, 70; Gold Heart Award, Am Heart Asn, 76; Kaber Medal, Asn Am Physicians, 80. *Mem:* Inst Med-Nat Acad Sci; Am Soc Clin Invest (secy, 46-48); Asn Am Physicians (secy, 62-67, pres, 71-72); Am Fedn Clin Res. *Res:* Cardiovascular studies. *Mailing Add:* Duke Hosp Box 3910 Durham NC 27710

STEAD, FREDERICK L, b Toledo, Ohio, Dec 20, 23; m 47; c 4. NATURAL GAS EXPLORATION & DEVELOPMENT, DISPOSAL OF HAZARDOUS WASTE. *Educ:* Col Wooster, BA, 47; Univ Tex-Austin, MA, 50. *Prof Exp:* Instr geol, Univ Tex, 48-50; staff geologist, Continental Oil Co, Midland, Tex, 50-53, dist geologist, 53-54; div mgr, Ada Oil Co, Midland, Tex, 54-56; chief geologist, McAlester Fuel Co, Magnolia, Ark, 56-60; pres, Great Lakes Gas Corp, Dallas & Houston, Tex, 63-67; pres, Coastline Petrol Corp, Los Angeles, 70-72; pres, Malibu Mining Corp, Los Angeles, 72-74; pres, Helmet Petrol Corp, Denver, Colo, 75-76; consult, 76-79; PRES, F L STEAD & ASSOC INC, DALLAS, TEX, 79-; PRES, STEAD MGT SERV, INC, 79- *Concurrent Pos:* Pres, Tri-Coast Petrol Corp, Los Angeles, 70-72; pres coun, Am Inst Mgt, 77. *Mem:* Am Asn Petrol Geologists; Am Inst Prof Geologists (vpres, 76-77); fel Geol Soc Am; Soc Mining Eng; Am Inst Mining Eng-Soc Mining Engrs. *Res:* Exploration in frontier areas, preferably basin margins, looking for major deposits of natural gas for clients. *Mailing Add:* 14803 Le Grande Dr Dallas TX 75244

STEAD, WILLIAM WALLACE, b Durham, NC, Aug 23, 48; m 77; c 1. MEDICAL INFORMATICS, HOSPITAL INFORMATION SYSTEMS. *Educ:* Duke Univ, BA, 70, MD, 73; Am Bd Internal Med, dipl, 77 & 80. *Prof Exp:* Intern med, 73-74, asst resident, 74-75, fel nephrology, 75-77, assoc med & nephrology, 77-80, asst prof, 80-83, ASSOC PROF MED & NEPHROLOGY, DUKE UNIV MED CTR, 84-, ASSOC PROF COMMUNITY & FAMILY MED, 85- *Concurrent Pos:* Dir med ctr info systs, Duke Univ Med Ctr, 85-; dir ctr dialysis, Durham Vet Admin Med Ctr, 77-83, chief nephrology, 83-85; mem biomed libr rev comt, NIH, 87-; prin investr, Nat Libr Med Grant, 87- *Mem:* Am Col Med Informatics; Am Soc

Nephrology; Int Soc Nephrology; Am Fedn Clin Res. *Res:* Medical informatics database design, computer-based medical records, practice and hospital information systems and consultation systems; co-developer of the TMR medical information system; testing a model of an integrated academic information management system based upon integration of distributed resources. *Mailing Add:* Dept Med Ctr Info Systs Box 3900 Duke Univ Med Ctr Durham NC 27710

STEAD, WILLIAM WHITE, b Decatur, Ga, Jan 4, 19; m 75; c 1. INTERNAL MEDICINE, PULMONARY DISEASES. *Educ:* Emory Univ, AB, 40, MD, 43. *Prof Exp:* Resident med, Emory Univ, 44-45, Univ Cincinnati, 46-47 & Univ Minn, 48-49; chief of serv pulmonary dis, Vet Admin Hosp, Minneapolis, Minn, 54-57; assoc prof, Col Med, Univ Fla, 57-60; prof, Med Col Wis, 60-72; chief pulmonary dis, Vet Admin Hosp, Little Rock, Ark, 72-73; DIR TUBERC PROG, ARK DEPT HEALTH, 73-; PROF PULMONARY DIS, UNIV ARK, LITTLE ROCK, 72- *Concurrent Pos:* Fel cardiol, Univ Cincinnati, 47-48; med dir, Muirdale Sanatorium, Milwaukee, 60-72; consult, Dept Med, Vet Admin Hosp, Little Rock, 73- & Arthur D Little Co, Mass, 74. *Honors & Awards:* James D Bruce Award, Am Col Physicians, 88; Edward Livingston Trudeau Medal, Am Thoracic Soc, 88. *Mem:* AAAS; Am Soc Clin Invest; Am Fedn Clin Res (secy, 55-58, vpres, 58-59, pres, 59-60); Am Thoracic Soc; Am Col Chest Physicians. *Res:* Pulmonary physiology, development of spirometers; clinical and public health aspects of tuberculosis. *Mailing Add:* Ark Dept Health 4815 W Markham St Little Rock AR 72201

STEADMAN, JAMES ROBERT, b Cleveland, Ohio, Feb 7, 42; m 64, 89; c 4. PLANT PATHOLOGY. *Educ:* Hiram Col, BA, 64; Univ Wis-Madison, MS, 68, PhD(plant path), 70. *Prof Exp:* From asst prof to assoc prof, 69-75, FULL PROF PLANT PATH, UNIV NEBR-LINCOLN, 75- *Concurrent Pos:* Bean Indust, USDA & Chem Co grants, 80-; consult, Latin Am, Africa & Australia, 80-91; US AID title XII grants-bean improv, Dominican Repub, 81-, Honduras, 87-, & Jamaica, 89- *Mem:* Am Phytopath Soc; Sigma Xi; Int Soc Plant Path. *Res:* Epidemiology; vegetable diseases; white mold disease; bean rust; plant disease and microclimate interaction; disease resistance; disease management. *Mailing Add:* Dept of Plant Path Univ of Nebr Lincoln NE 68583-0722

STEADMAN, JOHN WILLIAM, b Cody, Wyo, Oct 13, 43; m 83; c 2. ELECTRICAL ENGINEERING, BIOENGINEERING. *Educ:* Univ Wyo, BS, 64, MS, 66; Colo State Univ, PhD(elec eng), 71. *Prof Exp:* Res engr life sci res, Convair Div, Gen Dynamics Corp, Calif, 66-68; from asst to assoc prof, 71-81, PROF BIOENG & ELEC ENG, UNIV WYO, 81-, ASSOC DEAN ENGR, 83- *Mem:* Inst Elec & Electronics Engrs; Am Soc Eng Educ. *Res:* Machine analysis of electroencephalograms; information processing in the nervous system; digital system design; microprocessor and microcomputer systems; electrical safety; microprocessor design. *Mailing Add:* Eng Col Univ Sta Box 3295 Laramie WY 82071

STEADMAN, ROBERT GEORGE, b Sydney, NSW, Nov 8, 39; m 70; c 3. TEXTILES. *Educ:* Univ New South Wales, BSc, 61, PhD(textile physics), 65. *Prof Exp:* Officer in charge, Cotton Fiber Lab, NSW Dept Agr, 64-66; textile mgr, Australian Wool Testing Authority, 66-68; assoc prof clothing & textiles, Univ Man, 68-71; exec vol, Can Exec Serv Overseas, Nigeria, 71-72; asst prof clothing & textiles, Tex Tech Univ, 72-78; assoc prof, Dept Textiles & Clothing, Colo State Univ, 78-81; prof, textile res engr, Textile Res Ctr, Tex Tech Univ, 81-88; PROF RES, AUSTRALIAN WOOL TESTING AUTHORITY, 88- *Mem:* Am Asn Textile Technol; Metric Asn; Am Meteorol Soc. *Res:* Textiles as thermal insulators and application to human biometeorology; economics and physiology of clothing and textiles; consumer problems in textiles. *Mailing Add:* Australian Wool Testing Authority Australian Wool Corp 369 Royal Parade Parkville 3052 Australia

STEADMAN, THOMAS REE, b Erie, Pa, Mar 15, 17; m 41; c 2. ORGANIC CHEMISTRY, HIGH PERFORMANCE FIBERS. *Educ:* Rensselaer Polytech Inst, BS, 37; Harvard Univ, AM, 38, PhD(org chem), 41. *Prof Exp:* Sr chemist org chem, B F Goodrich Co, 41-51; res assoc process res, Nat Res Corp, 51-57; mgr org chem res admin, W R Grace & Co, 57-68 & Allied Chem Corp, 68-72; sr chem economist consult, Battelle Mem Inst, 72-73; RETIRED. *Concurrent Pos:* Hormel Found fel, Univ Minn, 40-41; dir org chem res, Signal Oil & Gas Co, 68; adj prof chem eng, Ohio State Univ, 74. *Mem:* Am Chem Soc; Sigma Xi. *Res:* Beta-propiolactone; chemistry of formaldehyde; synthesis of amino acids; methionine and tryptophane; polyvinyl chloride additives; process chemistry; catalysts; water and air pollution; chemical economics; composite materials; high performance fibers; plastic composits. *Mailing Add:* 6400 Flotilla Dr No 74 Holmes Beach FL 34217

STEAR, EDWIN BYRON, b Peoria, Ill, Dec 8, 32; div; c 2. SYSTEMS SCIENCE, BIOMEDICAL ENGINEERING. *Educ:* Bradley Univ, BSME, 54; Univ Southern Calif, MS, 56; Univ Calif, Los Angeles, PhD(control & info systs), 61. *Prof Exp:* Mem tech staff missile syst design, Hughes Aircraft Co, 54-59; assoc res engr, Univ Calif, Los Angeles, 59-61; mgr, Control & Commun Lab, Lear Siegler, Inc, 63-64; assoc prof info systs, Univ Calif, Los Angeles, 64-69; prof elec eng, Univ Calif, Santa Barbara, 69-, chmn dept elec eng & comput sci, 75-; EXEC DIR, WASH TECHNOL CTR, UNIV WASH. *Concurrent Pos:* Mem, Am Automatic Control Coun, 63-; consult, several indust orgn, 64-; sr consult, US Air Force Space & Missile Test Ctr, 69-; mem, Sci Adv Bd, 71- *Mem:* Am Inst Aeronaut & Astronaut; fel Inst Elec & Electronics Engrs; NY Acad Sci. *Res:* Control systems theory; optimum filtering and data processing; biological control systems; computer analysis of electroencephalogram signals; technology for continuing education. *Mailing Add:* 14010 SE 44th Pl Bellevue WA 98006

STEARMAN, ROEBERT L(YLE), b Burley, Idaho, May 12, 23; m 47; c 2. SYSTEMS ANALYSIS. *Educ:* Ore State Univ, BS, 47, MS, 49; Johns Hopkins Univ, ScD(biostatist), 55. *Prof Exp:* US Pub Health Serv fel biostatist & virol, Johns Hopkins Univ, 55-56; biometrician, Nat Inst Arthritis & Metab Diseases, 56-57; sr biostatistician & supvr comput serv, Booz-Allen Appl Res, Inc, 57-61; prin statistician, Statist Sci Dept, Inst Adv Studies, Sci & Prof Serv Group, C-E-I-R, Inc, 61-68; chief, Systs Analysis Div, Ft Detrick, Md, 68-71; chief, Analytical Sci Off, US Army Biol Defense Res Labs, 71-72; chief, Systs Analysis Div, US Army Edgewood Arsenal, 72-74; math statistician, US Army, Exp Systs Div, 74-77, Test Design & Analysis Div, 77-82, opers res analysis, 82-84, Tech Analysis & Info Off, Dugway Proving Ground, 84-89; RETIRED. *Mem:* Am Mgt Asn; Nat Contract Mgt Asn; fel Royal Statist Soc. *Res:* Experimental design and analysis; biometry; statistical-mathematical modeling; computer applications and systems analysis; operations research; medical and biological laboratory and assay procedures; administration; contract management. *Mailing Add:* 449 Marvista Lane Tooele UT 84074

STEARMAN, RONALD ORAN, b Wichita, Kans, June 8, 32; m 57; c 2. AEROSPACE ENGINEERING. *Educ:* Okla State Univ, BS, 55; Calif Inst Technol, MS, 56, PhD(aeronaut), 61. *Prof Exp:* Res fel aeronaut, Calif Inst Technol, 61-62; sr analyst math & physics, Midwest Res Inst, 62-66; assoc prof mech & aerospace eng, Univ Kans, 64-66; assoc prof aerospace eng, 66-77, PROF AEROSPACE ENG & ENG MECH, UNIV TEX, AUSTIN, 77- *Concurrent Pos:* Air Force Off Sci Res res grant, 66- *Mem:* Am Inst Aeronaut & Astronaut; Soc Exp Stress Analysis; Am Helicopter Soc; Am Soc Engr Educ. *Res:* Aeroelastic and structural dynamics; reliability engineering. *Mailing Add:* Aerospace Dept Univ Tex Austin TX 78712

STEARN, COLIN WILLIAM, b Bishops Stortford, Eng, July 16, 28; m 53; c 3. PALEONTOLOGY, STRATIGRAPHY. *Educ:* McMaster Univ, BSc, 49; Yale Univ, PhD(geol), 52. *Prof Exp:* From asst prof to prof geol, McGill Univ, 52-68, asst dean fac grad studies & res, 60-63, chmn, Dept Geol Sci, 69-74 & 80-84, LOGAN PROF GEOL, MCGILL UNIV, 68- *Honors & Awards:* Billings Medal, Geol Asn Can. *Mem:* Geol Soc Am; Paleont Soc; Geol Asn Can; Royal Soc Can; Can Soc Petrol Geologists; Soc Sedimentary Geol. *Res:* Lower Paleozoic stratigraphy and paleontology of Canada; historical geology of North America; fossil stromatoporoids; organisms of Caribbean and Paleozoic reefs. *Mailing Add:* Dept Geol Sci McGill Univ 3450 University St Montreal PQ H3A 2A7 Can

STEARNER, SIGRID PHYLLIS, b Chicago, Ill, Jan 10, 19. CARDIOVASCULAR PHYSIOLOGY, RADIOBIOLOGY. *Educ:* Univ Chicago, BS, 41, MS, 42, PhD(zool), 46. *Prof Exp:* Biologist, Div Biol & Med Res, Argonne Nat Lab, 46-; RETIRED. *Mem:* AAAS; NY Acad Sci. *Res:* Late effects of ionizing radiations on the heart and vascular system, physiological and ultrastructural studies; other physiological effects of radiations on biological systems; pigmentation changes. *Mailing Add:* 1141 Iroquois Ave Apt 114 Naperville IL 60563

STEARNS, BRENTON FISK, b Chicago, Ill, July 28, 28; m 57, 74, 78; c 2. ENERGY CONVERSION. *Educ:* Pomona Col, BA, 49; Wash Univ, PhD(physics), 56. *Prof Exp:* Asst prof physics, Univ Ark, 54-57; from asst prof to assoc prof, Tufts Univ, 57-68; chmn dept physics, 68-74, assoc provost, 74-75, PROF PHYSICS, HOBART & WILLIAM SMITH COLS, 68- *Mem:* AAAS; Am Phys Soc; Am Asn Physics Teachers. *Res:* Applications of energy storage and efficient use. *Mailing Add:* 478 Wash St Geneva NY 14456

STEARNS, CHARLES EDWARD, b Billerica, Mass, Jan 20, 20; wid; c 6. GEOLOGY. *Educ:* Tufts Univ, AB, 39; Harvard Univ, MA, 42, PhD(geol), 50. *Hon Degrees:* LLD, Southeastern Mass Technol Inst, 62. *Prof Exp:* Asst geol, Tufts Univ, 41, instr, 41-42, 45, 46-48, asst prof, 48-51; asst prof, Harvard Univ, 51-54; assoc prof, 54-57, dean col lib arts, 54-67 & 68-69, actg provost, 66-67, prof, 57-87, EMER PROF GEOL, TUFTS UNIV, 87- *Mem:* AAAS; Geol Soc Am; Sigma Xi. *Res:* Pleistocene stratigraphy; shoreline geomorphology. *Mailing Add:* 381 Boston Rd Billerica MA 01821

STEARNS, CHARLES R, b McKeesport, Pa, May 2, 25. METEOROLOGY. *Educ:* Univ Wis, BS, 50, MS, 52, PhD(meteorol), 67. *Prof Exp:* Asst meteorol, Univ Wis, 55-56; chief physicist, Winzen Res, Inc, 56-57; res assoc meteorol, 57-65, asst prof, 65-69, chmn, Inst Environ Studies, 72-74, PROF METEOROL, UNIV WIS-MADISON, 69- *Concurrent Pos:* Consult, Aberdeen Proving Ground, Md, 69-; consult, Red Stone Arsenal, Ala; consult, Argonne Nat Lab, Ill. *Mem:* Am Meteorol Soc; Am Geophys Union; AAAS. *Res:* Micrometeorology, particularly boundary layer problems; evaporation from lakes; diffusion from power plants; antarctic meteorology; automatic weather stations. *Mailing Add:* Dept Meteorol Univ Wis 1325A Met-Space Sci Madison WI 53706

STEARNS, DAVID WINROD, b Muskegon, Mich, Mar 23, 29; m 48; c 5. STRUCTURAL GEOLOGY. *Educ:* Univ Notre Dame, BS, 53; SDak Sch Mines & Technol, MS, 55; Tex A&M Univ, PhD, 69. *Prof Exp:* Geologist, Shell Oil Co, 55-56 & Shell Develop Co, 56-66; from assoc prof to prof geol, Tex A&M Univ, 67-80, head dept, 71-80; DISTINGUISHED PROF, UNIV OKLA, 80- *Concurrent Pos:* Consult, Major Oil Co. *Mem:* Am Geophys Union; Geol Soc Am; Am Asn Petrol Geologists. *Res:* Structural geology and mechanical behavior of layered sedimentary rocks. *Mailing Add:* Col Geosci Univ Okla Norman OK 73019

STEARNS, DONALD EDISON, b Columbus, Ohio, Nov 1, 48; div; c 1. ZOOPLANKTON BEHAVIORAL ECOLOGY, ESTUARINE ECOLOGY. *Educ:* Dartmouth Col, AB, 70; Univ NH, MS, 74; Duke Univ, PhD(zool), 83. *Prof Exp:* Catedrático biol, Escuela Super Cieucias Marinas, Univ Autónoma Baja Calif, 74-76; postdoctoral res assoc, Skidaway Inst Oceanog, 84-86; res assoc, Dauphin Island Sea Lab, 86-88; res scientist assoc, Marine Sci Inst, Univ Tex, 88-89; ASST PROF BIOL, DEPT BIOL, RUTGERS UNIV, 89- *Concurrent Pos:* Adj res assoc & adj asst prof, Dept Biol, Univ Ala, Birmingham, 86-88; adj asst prof, Dept Zool & Wildlife Sci, Auburn Univ, 86-88; grad fac assoc, Univ S Ala, 86-88; prin investr, Nat Oceanic & Atmospheric Admin, 88-89. *Mem:* Am Soc Limnol & Oceanog; Nat Marine Educators Asn; Estuarine Res Fedn. *Res:* Marine and estuarine plankton ecology; zooplankton and ichthyoplankton behaviors, photobehavior, feeding, migration, egg production, endogenous rhythms; predator-prey interactions and sublethal pollution detection. *Mailing Add:* Dept Biol Rutgers Univ Camden NJ 08102

STEARNS, EDWIN IRA, b Matawan, NJ, Sept 3, 11; m 34; c 3. PHYSICAL CHEMISTRY. *Educ:* Lafayette Col, BS, 32; Rensselaer Polytech Inst, MS, 33; Rutgers Univ, PhD(phys chem), 45. *Prof Exp:* Physicist, Am Cyanamid Co, NJ, 33-43, chief physicist, 44-45, asst dir physics res, 45-51, mgr prod improv, Dyestuff Dept, 52-54, asst mgr, Midwest Territory, 54-59, tech mgr, Dyes Dept, 59-63, mgr sales develop, Dyes & Textile Chem Dept, 64-69, res assoc, 69-72; head dept textile sci, Clemson Univ, 72-77, PRES & CONSULT, E I STEARNS INC, 78- *Concurrent Pos:* Instr, Cooper Union, 38-39 & Adult Sch, Bound Brook, 40. *Honors & Awards:* Olney Medal, Am Asn Textile Chemists & Colorists, 67; Godlove Medal, Inter-Soc Color Coun, 67. *Mem:* Am Chem Soc; Am Asn Textile Chemists & Colorists (pres, 71-72); Inter-Soc Color Coun. *Res:* Photochemistry; phase rule; visual and infrared spectrophotometry; instrumentation; optical properties of pigments; spectrophotometer improvements; instrumentation in chemical processes. *Mailing Add:* 321 Woodland Way Clemson SC 29631-1547

STEARNS, EUGENE MARION, JR, b Evanston, Ill, May 3, 32; wid; c 3. BIOCHEMISTRY, MANAGEMENT. *Educ:* Denison Univ, BA, 54; Purdue Univ, West Lafayette, MS, 61, PhD(biochem), 65. *Prof Exp:* Res fel, Hormel Inst, Univ Minn, 65-67, res assoc lipid biochem, 67-70, asst prof lipid biochem, 70-76; sect leader, 76-81, group mgr biochem, 81-83, PROD DEVELOP MGR, LIFE SCI, CONKLIN CO INC, 83- *Mem:* AAAS; Am Chem Soc; Plant Growth Regulator Soc Am; Am Inst Biol Sci. *Res:* Animal health and nutrition; microbial- or enzyme-containing materials for home, agricultural and industrial uses; plant tissue culture; microbial biochemistry; plant growth regulators; plant protection. *Mailing Add:* Conklin Co Inc 889 Valley Park Dr PO Box 155 Shakopee MN 55379

STEARNS, FOREST, b Milwaukee, Wis, Sept 10, 18; m 43, 56; c 4. ECOLOGY, BOTANY. *Educ:* Harvard Univ, AB, 39; Univ Wis, PhM, 40, PhD(bot), 47. *Prof Exp:* Asst bot, Univ Wis, 40-42 & 46-47; instr bot exp sta, Purdue Univ, 47-49, asst prof, 49-57; botanist, Vicksburg Res Ctr, US Forest Serv, 57-60, proj leader forest wildlife habitat res, NCent Forest Exp Sta, 61-68; prof, 68-87, EMER PROF BOT, UNIV WIS, MILWAUKEE, 87- *Mem:* AAAS; Ecol Soc Am (pres, 75-76); Bot Soc Am; Wildlife Soc; Am Inst Biol Sci (pres, 81-82). *Res:* Autecology of trees and shrubs; seed germination; early succession and productivity; wetland, urban and landscape ecology and phenology. *Mailing Add:* Dept Biol Sci Univ Wis PO Box 413 Milwaukee WI 53201

STEARNS, H(ORACE) MYRL, b Kiesling, Wash, Apr 24, 16; m 39; c 4. ENGINEERING. *Educ:* Univ Idaho, BS, 37; Stanford Univ, EE, 39. *Hon Degrees:* DSc, Univ Idaho, 60. *Prof Exp:* Asst to chief engr, Gilfillan Bros, 39-41; mem eng staff, Sperry Gyroscope Co, 41-43, head Doppler radar develop prog, 43-45; res engr in charge klystron res & develop mfg, 45-48; exec vpres & gen mgr, 48-57, pres, 57-64, DIR & CONSULT TO BD DIRS, VARIAN ASSOCS, 48- *Concurrent Pos:* Dir Idaho Res Found, Univ Idaho, 90- *Honors & Awards:* Fel AAAS:. *Mem:* Am Mgt Asn; fel Inst Elec & Electronics Engrs. *Res:* Automatic frequency control and ranging; radar; microwave tubes; engineering management. *Mailing Add:* 246 La Cuesta Dr Menlo Park CA 94028

STEARNS, JOHN WARREN, b Santa Barbara, Calif, June 3, 33; m 67; c 1. NEGATIVE ION PRODUCTION, FUSION PLASMA DIAGNOSTICS. *Educ:* Univ Calif, Berkeley, AB, 59, MS, 67. *Prof Exp:* PHYSICIST ATOMIC & PLASMA PHYSICS, LAWRENCE BERKELEY LAB, UNIV CALIF, 59- *Res:* Low to high energy atomic and molecular cross-sections; low to medium energy surface interactions; beam-plasma-photon interactions; neutral beam development as related to magnetically confined fusion-plasma problems; beam-foil interactions; spin polarized neutral beams; laser/ spectroscopic VUV diagnostics; high-energy, high charge-state atomic physics; "Polarization of Fast Particle Beams by Collision Pumping". *Mailing Add:* Lawrence Berkeley Lab MS 4-230 Univ of Calif Berkeley CA 94720

STEARNS, MARTIN, b Philadelphia, Pa, Aug 16, 16; m 48; c 2. PHYSICS. *Educ:* Univ Calif, Los Angeles, BA, 43; Cornell Univ, PhD(physics), 52. *Prof Exp:* Res assoc, Carnegie Inst Technol, 52-57; staff scientist, Gen Atomic Div, Gen Dynamics Corp, 57-60; prof physics & chmn dept, 60-62, DEAN COL LIBERAL ARTS & PROF PHYSICS, WAYNE STATE UNIV, 62- *Concurrent Pos:* Consult, Ramo-Wooldridge Corp, 55. *Mem:* Fel Am Phys Soc. *Res:* High energy nuclear physics; plasma physics; bremsstrahlung and pair production; photoproduction of mesons; mesonic x-rays; nuclear reactors. *Mailing Add:* Dept Physics-135 Physics Wayne State Univ Detroit MI 48202

STEARNS, MARY BETH GORMAN, b Minneapolis, Minn; m 48; c 2. MAGNETISM, MULTILAYER FILMS. *Educ:* Univ Minn, BS, 46; Cornell Univ, PhD(physics), 52. *Prof Exp:* Asst physics, Cornell Univ, 47-51; res physicist, Carnegie Inst Technol, 52-56, Univ Pittsburgh, 57 & Gen Atomic Div, Gen Dynamics Corp, 58-60; sr scientist, Sci Lab, Ford Motor Co, 60-77, prin scientist, 77-81; PROF PHYSICS, ARIZ STATE UNIV, 81-; ASSOC DEAN RES, COL LIB ARTS & SCI, 90- *Concurrent Pos:* Mem, Exec Bd Condensed Matter Div, Am Phys Soc, 72-74, counr at large, 78-82, exec bd, 79-80, chmn elect nominating comt, 81-82; solid state panel, Nat Acad Sci, 74-78; mem adv panel, Condensed Matter Sci, NSF, 77-78; mem, Solid State Sci Rev Comt, Argonne Nat Lab, 79-80, chmn, 81; mem, Selection Comt, Fulbright Hays Scholar Physics, 78-79, chmn, 80-81; mem, Prog Adv Comt, Los Alamos Nat Lab, 85-87; mem, Magnetism & Magnetic Mat Adv Comt, 87-90; mem, Comn Magnetism, Int Union Pure Appl Sci, 75-78, sec, 78-81, chair, 81-84. *Honors & Awards:* Alexander Von Humboldt Sr Award, 84; Regents' Prof, 88. *Mem:* Fel Am Phys Soc; AAAS; Inst Elec & Electronics Engrs. *Res:* Photonuclear reactions; meson spectroscopy; thermoelectricity; solids; low energy nuclear physics; magnetism; M-ssbauer effect and pulsed nuclear magnetic resonance studies; electron scattering; extended x-ray absorption fine structure studies; electronic structure of transition metals; modulated multilayer structures. *Mailing Add:* Ariz State Univ Tempe AZ 85287

STEARNS, RICHARD EDWIN, b Caldwell, NJ, July 5, 36; m 63; c 2. COMPUTATIONAL COMPLEXICITY, ALGORITHRIS. *Educ:* Carleton Col, BA, 58; Princeton Univ, PhD(math), 61. *Prof Exp:* Mathematician, Res Lab, Gen Elec Corp, 61-65, mathematician, Res & Develop Ctr, 65-71, mathematician, Corp Res & Develop, 71-78; PROF COMPUT SCI, STATE UNIV NY, ALBANY, 78- *Concurrent Pos:* Vis prof, Hebrew Univ, 75, Math Sci Res Inst, 85; ed, Siam J Comput, 72- *Mem:* Math Asn Am; Asn Comput Mach. *Res:* Computational Complexity; algorithms; game theory. *Mailing Add:* Dept Comput Sci Univ Albany 1400 Washington Ave Albany NY 12222

STEARNS, RICHARD GORDON, b Buffalo, NY, Apr 28, 27; m 50; c 2. GEOLOGY, GEOPHYSICS. *Educ:* Vanderbilt Univ, AB, 48, MS, 49; Northwestern Univ, PhD(geol), 53. *Prof Exp:* Asst state geologist, State Div Geol, Tenn, 53-61; from asst prof to assoc prof, 61-68, chmn dept, 67-76, PROF GEOL, VANDERBILT UNIV, 68- *Mem:* AAAS; Geol Soc Am; Am Asn Petrol Geol; Am Geophys Union. *Res:* Stratigraphy; structure; geophysics; hydrogeology. *Mailing Add:* Box 1615 Sta B Vanderbilt Univ Nashville TN 37235

STEARNS, ROBERT INMAN, b Atlanta, Ga, Feb 26, 32; m 66; c 2. INORGANIC CHEMISTRY. *Educ:* Loyola Univ, La, BS, 53; Tulane Univ, MS, 55, PhD(inorg chem), 58. *Prof Exp:* Res specialist, Cent Res Dept, Monsanto Co, Mo, 59-68; from asst prof to assoc prof chem, Eve Div, Univ Mo-St Louis, 66-78; DIR RES, LORVIC CORP, 68-; PROF CHEM, EVE DIV, UNIV MO-ST LOUIS, 78- *Concurrent Pos:* Asst prof chem, Eve Div, Univ Mo-St Louis, 66-72, assoc prof, 72-78, prof 78- *Res:* Physical chemistry of fluorides in preventive dentistry; dental materials, cements and polymers; semiconductor materials research, particularly vapor phase depositon of single crystal thin films. *Mailing Add:* 2396 Wesglen Est Dr Maryland Heights MO 63043

STEARNS, ROBERT L, b New Haven, Conn, July 28, 26; m 58; c 2. PHYSICS. *Educ:* Wesleyan Univ, BA, 50; Case Inst Technol, MS, 52, PhD(physics), 55. *Prof Exp:* Instr physics, Case Inst Technol, 52-55 & Queens Col, 55-58; dean freshman, 74-77, from asst prof to assoc prof, 58-68, PROF PHYSICS, VASSAR COL, 68-, CHMN DEPT, 62-64, 66-69, 78-81 & 85- *Concurrent Pos:* Res collabr, Brookhaven Nat Lab, 57-, vis assoc physicist, 64-65; vis scientist, Europ Orgn Nuclear Res, Geneva, 70-71; vis staff mem, Los Alamos Nat Lab, 72-80. *Mem:* Am Phys Soc; Am Asn Physics Teachers. *Res:* Neutron physics; scattering of cold neutrons from cyrstals and liquids; nuclear structure physics using high energy proton scattering; nuclear structure-mesic atoms and hypernuclear physics. *Mailing Add:* Dept of Physics Vassar Col Poughkeepsie NY 12601

STEARNS, S(TEPHEN) RUSSELL, b Manchester, NH, Feb 28, 15; m 39; c 3. CIVIL ENGINEERING, EDUCATION ADMINISTRATION. *Educ:* Dartmouth Col, AB, 37, CE, 38; Purdue Univ, MS, 49. *Hon Degrees:* AM, Dartmouth Col. *Prof Exp:* Jr engr, Gannett, Eastman & Fleming, Pa, 38-40; jr prof engr, Navy Yard, Philadelphia, 40-41, engr, Dry Docks Assoc, 41-43; instr, 43-45, asst prof civil eng, 45-53, assoc dean, 73-75 & 80-82, prof, 53-80, EMER PROF CIVIL ENG, THAYER SCH ENG, DARTMOUTH COL, 80- *Concurrent Pos:* Field engr, Boston Univ, Alaska, 53; chief appl res br, Snow, Ice, Permafrost Res Estab, US Dept Army, 54-55, consult, 55-60; sr res engr, Oper Res Inc, 62-64; mem, NH Transp Comn, 66, 68; chmn Lebanon, NH Airport Authority, 66-69; mem, NH Tomorrow Exec Comn, 70-72; mem hwy res bd, Nat Acad Sci; dir, NH Bd Regist Prof Engrs, 73-82 & Am Soc Civil Engrs, 78-81; consult engr, UN Develop Prog, Poland, 74, 78, 80- & TAMS, 74-82. *Mem:* Nat Soc Prof Engrs; Am Soc Eng Educ; fel Am Soc Civil Engrs (pres, 83-84). *Res:* Transportation engineering, environmental planning and design; soil mechanics and foundations; permafrost engineering. *Mailing Add:* Thayer Sch Eng Dartmouth Col Hanover NH 03755

STEARNS, STEPHEN CURTIS, b Kapaau, Hawaii, Dec 12, 46; m 71; c 2. LIFE-HISTORY EVOLUTION, EVOLUTIONARY ECOLOGY. *Educ:* Yale Univ, BS, 67; Univ Wis, MS, 71; Univ BC, PhD(zool), 75. *Prof Exp:* Miller fel, Univ Calif, Berkeley, 75-76, vis asst prof zool, 76-77, Miller fel, 77-78; asst prof biol, Reed Col, Ore, 78-83; PROF ZOOL, UNIV BASEL, SWITZ, 83- *Concurrent Pos:* Founding mem, European Soc Evolutionary Biol; managing ed, J Evolutionary Biol. *Mem:* Am Soc Naturalists; Soc Study Evolution; Ecol Soc Am; Brit Ecol Soc. *Res:* Evolution and ontogeny of life-history traits-age at maturity, fecundity, reproductive effort, size of young and longivity including theoretical work, field studies and laboratory experiments. *Mailing Add:* Zool Inst Univ Basel Rheinsprung 9 Basel CH 4051 Switzerland

STEARNS, THOMAS W, b New York, NY, June 17, 09; wid; c 2. BIOCHEMISTRY. *Educ:* Univ Fla, BS, 34, MS, 37; Univ Minn, PhD(biochem), 40. *Prof Exp:* Asst, Univ Minn, 38-40; asst prof vet res, Iowa State Col, 40-46; asst prof chem, 46-49, assoc prof agr chem, 49-55, prof chem & asst chmn dept, 55-74, EMER PROF CHEM & ASST CHMN DEPT, UNIV FLA, 74- *Mem:* Am Chem Soc. *Res:* Physical chemistry bacteria; biochemistry foods; biosynthesis riboflavin. *Mailing Add:* Chem Bldg 300 Lei Bldg Univ Fla Gainesville FL 32603

STEBBING, NOWELL, b Copenhagen, Denmark, Sept 5, 41; Brit citizen; m 74; c 4. CANCER RESEARCH, ARTHRITIS RESEARCH. *Educ:* Univ Edinburgh, Scotland, BSc, 64, PhD(cell biol), 68. *Prof Exp:* Dir biol, G D Searle Res Lab, 69-79 & Genentech Inc, 79-82; vpres, Amgen Inc, 82-86; GEN MGR, ICI PHARMACEUT, 86- *Mem:* Am Soc Pharmacol & Exp Therapeut; Endocrine Soc; Am Soc Immunol. *Res:* Pharmaceutical research to find innovative treatments in cancer, arthritis and infection. *Mailing Add:* ICI Pharmaceut Alderley Park Macclesfield Cheshire SK10 4TG England

STEBBINGS, JAMES HENRY, b Grand Rapids, Mich, 1937; m 89; c 1. ENVIRONMENTAL EPIDEMIOLOGY, OCCUPATIONAL EPIDEMIOLOGY. *Educ:* St Louis Univ, BS, 60; Johns Hopkins Univ, ScD(epidemiol), 69. *Prof Exp:* Asst prof epidemiol, Univ Minn, 70-73; staff scientist, US Environ Protection Agency, Res Triangle Park, NC, 74-77; group leader epidemiol, Los Alamos Nat Lab, 77-79; assoc prof, Univ Minn, 80-81; group leader epidemiol, Argonne Nat Lab, 81-90; CHIEF SCIENTIST, MIDWEST EPIDEMIOL ASSOCS, 91- *Concurrent Pos:* Adj asst prof environ med, NY Univ, 68-70; asst epidemiol, Sch Pub Health & Admin Med, Columbia Univ, 68-69, instr, 69-70; adj assoc prof, Dept Epidemiol, Sch Pub Health, Univ NC, 75-77; clin assoc, Dept Family, Community & Emergency Med, Sch Med, Univ NMex, 77-78; mem, Inhalation Toxicol Comt, Nat Ctr Toxicol Res, Jefferson, Ark, 77-78, Sci Adv Group, Power Plant Siting Dept, Environ Qual Bd, St Paul, Minn, Sci Adv Panel, Overhead Power Lines Proj, Dept Health, Albany, NY, 81-86 & Adv Bd Great Lakes Ctr Occup Safety & Health, Col Med, Univ Ill, Chicago, 86-; consult, Dept Health & Environ Sci, Helena, Mont, 79-80, Comt Fed Res on Biol & Health Effects of Ionizing Radiation, Nat Res Coun, NIH, 80 & Nat Comn Air Qual, Wash, DC, 81; adj assoc prof, Epidemiol & Biomet Prog, Sch Pub Health, Univ Ill, Chicago, 82-, adj full mem, Grad Col, 82- & practicum supvr, Occup Med Residency Prog, Great Lakes Ctr Occup Safety & Health, Col Med, Univ Ill, Chicago, 82- *Mem:* Soc Epidemiol Res; Int Epidemiol Asn; Soc Environ Geochem & Health; Health Physics Soc. *Res:* Environmental and occupational epidemiology; health effects of radium and other radionuclides. *Mailing Add:* Midwest Epidemiol Assocs PO Box 743 Joliet IL 60434

STEBBINGS, RONALD FREDERICK, b London, Eng, Mar 20, 29; m 52; c 2. PHYSICS. *Educ:* Univ Col, Univ London, BSc, 52, PhD(atomic physics), 56. *Prof Exp:* Scientist, Atomic Physics Lab, San Diego, Calif, 58-65; reader physics, Univ Col, Univ London, 65-68; PROF PHYSICS & ASTRON, RICE UNIV, 68-, VPRES, STUDENT AFFAIRS, 83- *Concurrent Pos:* Chmn, dept Space Sci, 69-74. *Mem:* Fel Am Phys Soc; Am Geophys Union. *Res:* Experimental atomic physics, particularly as it relates to problems of astrophysical or aeronomic interest. *Mailing Add:* Rice Univ Box 1892 Houston TX 77251

STEBBINGS, WILLIAM LEE, b Orange Co, Calif, Mar 1, 45; m 68; c 1. ANALYTICAL CHEMISTRY. *Educ:* Iowa State Univ, BS, 66; Univ Wis, PhD(org chem), 72. *Prof Exp:* Sr chemist, 72-77, res specialist, 77-82, ANALYSIS MGR, 3M CO, 82- *Mem:* Am Chem Soc; Am Soc Mass Spectrometry; Am Asn Artificial Intel. *Res:* Applications of mass spectrometry in analytical chemistry; diagnostic applications of expert systems; prediction of economic consequences of research and development. *Mailing Add:* 717 Nightingale Stillwater MN 55082-5239

STEBBINS, DEAN WALDO, b Billings, Mont, Jan 14, 13; m 37; c 1. PHYSICS, ACADEMIC ADMINISTRATION. *Educ:* Mont State Col, BS, 35; Iowa State Univ, PhD(appl physics), 38. *Prof Exp:* Instr physics, State Col Wash, 38-39 & Agr & Mech Col Tex, 39-41; asst prof, Lehigh Univ, 46-47; from assoc prof to prof, Iowa State Univ, 47-60; physicist, Rand Corp, Calif, 60-63; prof physics & head dept, Mich Technol Univ, 63-65, dean fac, 65-66, vpres acad affairs, 66-76; RETIRED. *Concurrent Pos:* Consult, Opers Anal Off, Hq, US Dept Air Force, 50-60, Radiation Lab, Univ Calif, 56-58, Westinghouse Elec Corp, Pa, 57 & Ramo-Wooldridge Corp, 59. *Mem:* AAAS; Am Phys Soc; Am Asn Physics Teachers. *Res:* Classical physics; geophysics; presence and distribution of matter; interplanetary and interstellar space. *Mailing Add:* RR 4 Box 169 Ames IA 50010-9319

STEBBINS, GEORGE LEDYARD, b Lawrence, NY, Jan 6, 06; m 31, 58; c 3. BOTANY. *Educ:* Harvard Univ, AB & AM, 28, PhD(biol), 31. *Hon Degrees:* Dr, Univ Paris, 62, Ohio State Univ, 82 & Carleton Col, 83. *Prof Exp:* Asst bot, Harvard Univ, 29-31; instr biol, Colgate Univ, 31-35; jr geneticist, 35-39, from asst prof to prof, 39-73, EMER PROF GENETICS, UNIV CALIF, DAVIS, 73- *Concurrent Pos:* Jesup lectr, Columbia Univ, 46; Guggenheim fels, 54 & 60; secy gen, Int Union Biol Sci, 59-64; fac res lectr, Univ Calif, Davis, 62; hon fel, Smithsonian Inst, 82. *Honors & Awards:* Lewis Prize, Am Philos Soc, 60; Nat Medal Sci, 80. *Mem:* Nat Acad Sci; Am Soc Naturalists (pres, 69); Bot Soc Am (pres, 62); Soc Study Evolution (vpres, 47, pres, 48); Am Philos Soc. *Res:* Cytogenetics of parthenogenesis in the higher plants; production of hybrid and polyploid types of forage grasses; natural selection, developmental genetics and morphogenesis of higher plants; mechanisms of evolution. *Mailing Add:* Dept Genetics Univ Calif Davis CA 95616

STEBBINS, RICHARD GILBERT, b Providence, RI, May 20, 43; m 77; c 3. PHYSICAL CHEMISTRY. *Educ:* Wesleyan Univ, BS, 65; Tex A&M Univ, PhD(phys chem), 70. *Prof Exp:* Asst prof, 70-76, ASSOC PROF CHEM, BETHANY COL W VA, 76- *Concurrent Pos:* Vis prof, Mont State Univ, 75-76. *Res:* Analysis of trace organics by API mass spectrometry and the electron capture detector. *Mailing Add:* Dept Chem Univ Southern Maine 96 Falmouth St Portland ME 04103

STEBBINS, ROBERT CYRIL, b Chico, Calif, Mar 31, 15; m 41; c 3. ZOOLOGY. *Educ:* Univ Calif, MA, 41, PhD(zool), 43. *Prof Exp:* From instr to prof, 58-78, EMER PROF ZOOL, UNIV CALIF, BERKELEY, 78-, CUR HERPET, MUS VET ZOOL, 48- *Concurrent Pos:* Guggenheim fel, 49; ed, Am Soc Ichthyol & Herpet J, 55; NSF sr fel, 58-59. *Mem:* Soc Syst Zool; Am Soc Ichthyol & Herpet; fel Am Acad Zool. *Res:* Natural history and factors in the evolution of amphibians and reptiles; population studies of amphibians and reptiles; function of pineal apparatus; research development of biological science topics for schools; scientific illustrations. *Mailing Add:* 601 Plateau Dr Berkeley CA 94708

STEBBINS, ROBERT H, b Boston, Mass, Nov 14, 24; m 50; c 3. GENERAL EARTH SCIENCES. *Educ:* Mass Inst Technol, BS, 50; Columbia Univ, MA, 57. *Prof Exp:* Explor geologist, US Steel, 54-59; gen mgr, Hunting Geophys Surv Inc, 59-63; pres, Stebbins Mineral Surv Inc, 63-66; vpres, Gulf Resources & Chem Corp, 67-72; mgr, Minerals Div, Exxon Corp, 73-76; consult, Robert H Stebbin, 76-80; sr vpres, Int Energy Corp, 80-84; vpres, O'Connor Res Inc, 84-86; FOUNDER, FORUM RESOURCES, 79-; PRES, SCIS BANK INC, 86- *Mem:* Am Asn Advan Sci; Am Asn Petrol Geologists; Geol Soc Am; Soc Econ Geologists. *Res:* Development and application of models ore deposit formation to exploration programs world wide; application of rubber sheet algorithms to transference of data from one map or engineering drawing drawing to another. *Mailing Add:* 100 Sleepy Hollow Rd Richmond VA 23229-7831

STEBBINS, ROBIN TUCKER, b Philadelphia, Pa, July 9, 48; m 85; c 1. SOLAR PHYSICS, PHYSICS. *Educ:* Wesleyan Univ, BA, 70; Univ Colo, Boulder, MS, 73, PhD(physics), 75. *Prof Exp:* Fel, Advan Study Prog, Nat Ctr Atmospheric Res, 75-76; res assoc, Sacramento Peak Nat Observ, 76-77, asst astronr, 77-82, assoc astronr solar physics, 82-86; SR RES ASSOC, JOINT INST LAB ASTROPHYS, UNIV COLO, 86- *Mem:* Am Phys Soc; Optical Soc Am; Am Astron Soc; AAAS; Int Astron Union. *Res:* Global properties of the sun; relativity; fundamental tests. *Mailing Add:* JILA-Campus Box 440 Univ Colo Boulder CO 80309-0440

STEBBINS, WILLIAM COOPER, b Watertown, NY, June 6, 29; m 53; c 3. BIOACOUSTICS, COMPARATIVE PERCEPTION. *Educ:* Yale Univ, BA, 51; Columbia Univ, MA, 54, PhD(psychol), 57. *Prof Exp:* Res assoc otol, NY Univ Med Ctr, 57; asst prof psychol, Hamilton Col, 57-61; fel neurophsyiol, Med Sch, Univ Wash, 61-63; from asst prof to assoc prof, 63-70, PROF PSYCHOL & OTORHINOLARYNGOL, LITERARY COL & MED SCH, UNIV MICH, ANN ARBOR, 70- *Concurrent Pos:* Prin investr, Sigma Xi res grants, 60-61; prin investr, NIH res grants, 60-61 & 64-; fel, Univ Wash, 61-63; prin investr, NSF res grants, 74-; mem commun disorders rev comt, Nat Inst Neurol & Commun Disorders & Stroke, 76-80; mem, Comt Hearing, Bioacoust & Biomech, Nat Res Coun, 80- *Mem:* Fel AAAS; fel Acoust Soc Am; Int Primatol Soc; Asn Res Otolaryngol (pres elect, pres, past pres, 83-86); fel Am Psychol Soc. *Res:* Comparative bioacoustics and the evolution of hearing, animal psychophysics, hearing and auditory perception in nonhuman primates. *Mailing Add:* 340 Orchard Hills Dr Ann Arbor MI 48104

STEBELSKY, IHOR, b Krakow, Poland, Sept 6, 39; Can citizen; m 63; c 3. ENVIRONMENTAL SCIENCES, AGRICULTURAL GEOGRAPHY OF THE SOVIET UNION. *Educ:* Univ Toronto, BA, 62, MA, 64; Univ Wash, PhD(geog), 67. *Prof Exp:* Res asst geog, Univ Wash, 65-67, res assoc, 68; from asst prof to assoc prof, 68-82, chmn, Dept Geog, 82-89, PROF, DEPT GEOG, UNIV WINDSOR, 82- *Concurrent Pos:* Russian & Far Eastern Inst res assoc, Univ Wash & Moscow & Lenningrad, USSR, 68; Ont Dept Univ Affairs grant, Univ Windsor, 70-71; Can Coun res grant, 74; Can Coun-Acad Sci USSR travel grant, 76; External Affairs travel grant, United Kingdom, 77; Woodrow Wilson Int Ctr for Scholars fel, 83; Geog Ed, Encyclopedia of Ukraine, 88- *Mem:* Asn Am Geog; Can Asn Geog; Can Asn Slavists. *Res:* Geography of agricultural resources, food production and consumption in the Soviet Union; historical geography of the Soviet Union, with emphasis on population migration to siberia; land use and occupance in Ukraine; environmental impact in Ukraine; Ukrainian post World War II refugees and their immigration to Canada. *Mailing Add:* Dept Geog Univ Windsor Windsor ON N9B 3P4 Can

STEBEN, JOHN D, b Hinsdale, Ill, Feb 27, 36; m 59; c 2. PHYSICS, COMPUTER SCIENCE. *Educ:* Univ Ill, BS, 58, MS, 59, PhD(physics), 65. *Prof Exp:* Physicist, Midwest Univs Res Asn, 65-67; physicist, Phys Sci Lab, Univ Wis-Madison, 67-74, lectr nuclear eng, 70-74; asst prof radiation ther, Thomas Jefferson Univ, 74-77, lectr radiation technol, 78, sr systs analyst mgr serv, 76-84; SR MEM ENG STAFF, RCA-ESD SYSTS ENG, MOORESTOWN, NJ, 84- *Concurrent Pos:* Adj asst prof, Dept Neurol, Thomas Jefferson Univ, 79- *Mem:* Inst Elec & Electronic Engrs; Sigma Xi; Am Phys Soc. *Res:* Nuclear physics, particle accelerator physics; plasma and medical physics, neurologic studies; computer methods in these areas. *Mailing Add:* Gen Elec Corp 108-102 Marne Hwy Moorestown NJ 08057

STEBER, GEORGE RUDOLPH, b West Milwaukee, Wis, Sept 25, 38; m 64; c 2. ELECTRICAL ENGINEERING. *Educ:* Univ Wis-Milwaukee, BS, 63, MS, 66; Marquette Univ, PhD(elec eng), 69. *Prof Exp:* From instr to asst prof, 63-71, ASSOC PROF ELEC ENG, UNIV WIS-MILWAUKEE, 71-, ASST DEAN SCH ENG & APPL SCI, 77- *Concurrent Pos:* Grants, NSF & Wis Dept Natural Resources. *Mem:* Inst Elec & Electronics Engrs; Simulation Coun; Am Soc Eng Educ. *Res:* Control theory; hybrid computers and systems; electronic circuits; minicomputer interfacing; air pollution control. *Mailing Add:* Dept Elec Eng & Comput Sci Univ of Wis PO Box 413 Milwaukee WI 53201

STECHER, EMMA DIETZ, b Brooklyn, NY, 05; m 44. ORGANIC CHEMISTRY. *Educ:* Columbia Univ, BA, 25, MA, 26; Bryn Mawr Col, PhD(chem), 29. *Prof Exp:* Res chemist, Harvard Univ, 29-34; Am Asn Univ Women Berliner fel, Univ Munich, 34-35; res chemist, Exp Sta, Hercules Powder Co, Del, 35-37; lectr, Moravian Col Women, 38-41; res chemist, Merck & Co, Inc, NJ, 41; asst prof, Conn Col, 41-43; res chemist, Gen Aniline & Film Corp, Pa, 43-45; from instr to prof org chem, 45-71, EMER PROF ORG CHEM, BARNARD COL, COLUMBIA UNIV, 71- *Concurrent Pos:* Adj prof, Pace Univ, 71-83; alt coun, Am Chem Soc, 68-92. *Mem:* AAAS; Sigma Xi; Am Chem Soc; NY Acad Sci. *Res:* Microanalysis; diazotype paper; chlorophyll; unsaturated ketoacids and lactones; synthesis and oxidation potentials of benzanthraquinones. *Mailing Add:* 423 W 120th St Apt 74 New York NY 10027

STECHER, MICHAEL, b Milwaukee, Wis, Feb 8, 42; m 64; c 3. MATHEMATICS. *Educ:* Univ Wis-Milwaukee, BS, 64, MS, 65; Ind Univ, PhD(math), 73. *Prof Exp:* ASST PROF MATH, TEX A&M UNIV, 73- *Mem:* Am Math Soc; Soc Indust & Appl Math. *Res:* Partial differential equations; integral equations. *Mailing Add:* Dept Math Tex A&M Univ College Station TX 77843

STECHER, THEODORE P, b Kansas City, Mo, Dec 15, 30; m 56; c 4. ASTRONOMY. *Educ:* Univ Iowa, BA, 53, MS, 56. *Prof Exp:* Head, Observ Astron Br, 72-77, astronomer, 59-76, discipline scientist for astron, Space Shuttle Spacelab Proj, 76-82, SR SCIENTIST, NASA GODDARD SPACE FLIGHT CTR, 77- *Concurrent Pos:* Mem space sci sub-comt astron, NASA, 68-70; independent res fel, Goddard Space Flight Ctr, 71-72; vis fel, Joint Inst Lab Astrophys, Univ Colo/Nat Bur Standards, 71-72; US proj scientist, Astron Netherlands Satellite; prin investr, Ultraviolet Imaging Telescope for Astro Missions, Spacelab, 79-82 & mission scientist, OSS-3 payload & flight, 81-82. *Honors & Awards:* John C Lindsay Mem Award, NASA, 66,. *Mem:* Am Astron Soc; Int Astron Union; fel Royal Astron Soc. *Res:* Ultraviolet stellar spectrophotometry from rockets; stellar physics; interstellar grains and molecules; space instrumentation; gum nebula; gaseous nebulae; galaxies; globular clusters. *Mailing Add:* 10812 Margate Rd Silverspring MD 20901

STECHSCHULTE, AGNES LOUISE, b Owosso, Mich, Jan 9, 24. BIOLOGY, MICROBIOLOGY. *Educ:* Siena Heights Col, BS, 47; Detroit Univ, MS, 53; Cath Univ, PhD(biol), 61. *Prof Exp:* Chmn dept, 61-72, FROM INSTR TO PROF BIOL, BARRY COL, 60- *Concurrent Pos:* NIH res grant, 62-65. *Mem:* AAAS; Am Soc Microbiol; Nat Asn Biol Teachers; NY Acad Sci. *Res:* Lysozyme resistant mutants. *Mailing Add:* 11300 NE Second Ave Barry Univ Miami FL 33161

STECK, DANIEL JOHN, b Calumet, Mich, Mar 2, 46; m 70. RADON & NATURAL RADIOACTIVITY. *Educ:* Univ Mich, BS, 68; Univ Wis-Madison, MS, 70, PhD(physics), 76. *Prof Exp:* Staff physics, Los Alamos Sci Lab, 68; asst, Univ Wis-Madison, 68-69, res asst nuclear physics, 69-76; asst prof, 77-83, ASSOC PROF PHYSICS, ST JOHNS UNIV, 83- *Mem:* Am Phys Soc; Health Physics Soc; Am Asn Physics Teachers. *Res:* Environmental radioactivity; radon; solar and energy efficient housing. *Mailing Add:* 31148 County Rd 50 Avon MN 56310

STECK, EDGAR ALFRED, b Philadelphia, Pa, Dec 24, 18; m 47; c 2. ORGANIC CHEMISTRY, PARASITOLOGY. *Educ:* Temple Univ, AB, 39; Univ Pa, MS, 41, PhD(org chem), 42. *Prof Exp:* Asst bact, Temple Univ, 36-39; sr res org chemist, Winthrop Chem Co, 42-46; assoc mem, Sterling-Winthrop Res Inst, 46-56, mem, 56-58; med res group leader, Res Ctr, Johnson & Johnson, 58-60; dir res, Wilson Labs, 60-61; sr scientist, Nalco Chem Co, 61-65; dir res, McKesson Labs, 65-67; proj dir parasitic dis, Walter Reed Army Inst Res, 67-81; RETIRED. *Concurrent Pos:* Consult, chemotherapy of parasitic dis, WHO, 77- *Mem:* Am Soc Trop Med & Hyg; Am Chem Soc; The Chem Soc; Royal Soc Trop Med & Hyg; Swiss Chem Soc. *Res:* Chemotherapy of parasitic diseases; liposomes; nitrogen heterocyclic compounds. *Mailing Add:* 1913 Edgewater Pkwy Silver Spring MD 20903-1207

STECK, THEODORE LYLE, b Chicago, Ill, May 3, 39; m 61, 82; c 2. BIOCHEMISTRY. *Educ:* Lawrence Col, BS, 60; Harvard Univ, MD, 64. *Prof Exp:* Intern med, Beth Israel Hosp, Boston, 64-65; res fel, Sch Med, Harvard Univ, 65-66, 68-70 & Mass Gen Hosp, 68-70; res assoc, Nat Cancer Inst, 66-68; asst prof med, 70-73, asst prof biochem, 73-74, from assoc prof to prof biochem & med, 74-84, chmn biochem, 79-84, PROF BIOCHEM & MOLECULAR BIOL, UNIV CHICAGO, 84- *Concurrent Pos:* Schweppe Found fel, 71-74; fac res award, Am Cancer Soc, 75-80, mem adv comt biochem & chem carcinogenesis, 75-78; res council, NY Heart Asn, 84-86; bd sci counr, Nat Heart, Lung & Blood Inst, NIH, 84-88; coun invest awards, Am Cancer Soc, 86-89; publ comt, Am Soc Biochem & Molecular Biol, 83-86, awards comt, 88-90, nominating comt, 89-91, coun, 90-92. *Honors & Awards:* Robert A Welch Found Lectr, 89. *Mem:* AAAS; Am Soc Biol Chemists. *Res:* Membrane biochemistry; cell biology of the erythrocyte and of dictyostelium. *Mailing Add:* 920 E 58th St Chicago IL 60637

STECK, WARREN FRANKLIN, b Regina, Sask, May 10, 39; m 63; c 2. ORGANIC CHEMISTRY, ENTOMOLOGY. *Educ:* McGill Univ, BEng, 60; Univ Sask, PhD(org chem), 64. *Prof Exp:* Res assoc, Okla Univ Res Inst, 63-64; asst res officer, Nat Res Coun Can, 64-70, assoc res officer, 70-76, sr res officer, 76-80, asst dir, 80-81, assoc dir, 82-83, dir plant biotech inst, 83-90, DIR GEN PLANT BIOTECH INST, NAT RES COUN CAN, 91- *Mem:* Phytochem Soc NAm; Can Asn Conifer Biotech; Int Asn Plant Tissue Cult; Int Soc Chem Ecol; Can Res Mgt Asn. *Res:* Insect sex attractants and pheromones; chemical ecology. *Mailing Add:* 1326 Conn Saskatoon SK S7H 3L1 Can

STECKEL, RICHARD J, b Scranton, Pa, Apr 17, 36; m 61; c 2. MEDICAL & HEALTH SCIENCES. *Educ:* Harvard Univ, BA, 57, MD, 61. *Prof Exp:* Clin res assoc, Nat Cancer Inst, 65-67; PROF RADIOL SCI, UNIV CALIF, LOS ANGELES, 67- *Concurrent Pos:* Pres & chmn bd, Asn Am Cancer Insts. *Mem:* Fel Am Col Radiol; Radiol Soc NAm; Am Roentgen Ray Soc; Asn Univ Radiologists; Asn Am Cancer Insts; Soc Cancer Imaging. *Res:* Diagnostic imaging in cancer management. *Mailing Add:* 248 24th St Santa Monica CA 90402

STECKER, FLOYD WILLIAM, b New York, NY, Aug 12, 42; m 65; c 2. PHYSICS, ASTRONOMY. *Educ:* Mass Inst Technol, SB, 63; Harvard Univ, AM, 65, PhD(astrophys), 68. *Prof Exp:* Res assoc astrophys, NASA-Nat Res Coun, 67-68; astrophysicist, Lab Theoret Studies, 68-71, & Lab Space Physics, 71-77, SR ASTROPHYSICIST, LAB HIGH ENERGY ASTROPHYS, GODDARD SPACE FLIGHT CTR, NASA, 77- *Concurrent Pos:* Lectr, Univ MD, 85-; mem, Comt Cosmology & Galactic Structure, Int Astron Union. *Mem:* Am Astron Soc; fel Am Phys Soc; Int Astron Union. *Res:* High-energy astrophysics; cosmic-ray physics; gamma-ray astronomy and cosmology; infrared astrophysics; neutrino astrophysics; galaxy structure. *Mailing Add:* High Energy Astrophys Lab Goddard Space Flight Ctr NASA Greenbelt MD 20771

STECKL, ANDREW JULES, US citizen. SEMICONDUCTOR DEVICES, INTEGRATED CIRCUITS. *Educ:* Princeton Univ, BSE, 68; Univ Rochester, MS, 70, PhD(eng), 73. *Prof Exp:* Sr res engr, Honeywell Radiation Ctr, 72-73; mem tech staff, Rockwell Electronics Res Ctr, 73-76; ASSOC PROF IR DETECTORS & SOLID STATE DEVICES, RENSSELAER POLYTECH INST, 76- *Concurrent Pos:* Fac fel, T J Watson Res Ctr, IBM, 77. *Mem:* Am Phys Soc; Inst Elec & Electronics Engrs; Electron Devices Soc. *Res:* Semiconductors and solid state devices; integrated circuits and infrared detectors. *Mailing Add:* Dept Elec Eng Univ Cinn Main Campus 899 Rhodes Hall MC No 30 Cincinnati OH 45221

STECKLER, BERNARD MICHAEL, b Hebron, NDak, Jan 23, 32; m 54; c 4. ORGANIC CHEMISTRY, EDUCATIONAL ADMINISTRATION. *Educ:* St Martins Col, BS, 53; Univ Wash, PhD(org chem), 57. *Prof Exp:* Chemist, Nat Bur Standards, Washington, DC, 53, Northwest Labs, 54 & Shell Develop Co, 57-61; assoc prof, 61-75, PROF CHEM, SEATTLE UNIV, 75- *Mem:* AAAS; Am Chem Soc. *Res:* History and philosophy of science; interdisciplinary approaches to teaching physical science; integration of humanities and science disciplines; non-traditional studies curriculum development; phosphorus in delocalized pi-electron systems. *Mailing Add:* Dean Matteo Ricci II Seattle Univ Seattle WA 98122-2181

STEDINGER, JERY RUSSELL, b Oakland, Calif, June 22, 51; m 73; c 2. WATER RESOURCES PLANNING. *Educ:* Univ Calif, Berkeley, AB, 72; Harvard Univ, AM, 74, PhD(eng), 77. *Prof Exp:* PROF ENVIRON ENG, CORNELL UNIV, 77- *Honors & Awards:* Huber Res Prize, Am Soc Civil Engrs, 89. *Mem:* Am Geophys Union; Am Soc Civil Engrs; Inst Mgt Sci; Sigma Xi. *Res:* Application of statistics and scientific management techniques to problems in environmental engineering and water resources planning; reservoir operation and management; analysis of groundwater resources. *Mailing Add:* Sch Civil & Environ Eng Hollister Hall Cornell Univ Ithaca NY 14853-3501

STEDMAN, DONALD HUGH, b Dundee, Scotland, Feb 8, 43; m 64; c 3. ATMOSPHERIC CHEMISTRY. *Educ:* Cambridge Univ, BA, 64; Univ EAnglia, MSc, 65, PhD(chem), 67. *Prof Exp:* US Dept Health Educ & Welfare grant, Kans State Univ, 67-69; sr res scientist air pollution chem, Sci Res Labs, Ford Motor Co, 69-72; vis lectr atmospheric chem, Inst Environ Qual, 72-73, asst prof, 73-80, ASSOC PROF CHEM & ATMOSPHERIC & OCEANIC SCI, UNIV MICH, ANN ARBOR, 80- *Mem:* AAAS; The Chem Soc; Am Chem Soc; Am Phys Soc. *Res:* Gas phase chemical kinetics and spectroscopy of small molecules, particularly as related to aeronomy, atmospheric chemistry and air pollution; trace analysis of atmospheric pollutants. *Mailing Add:* 4910 E Princeton Ave Englewood CO 80110-5018

STEDMAN, JAMES MURPHEY, b Lockhart, Tex, July 6, 38; m 61; c 4. CLINICAL PSYCHOLOGY. *Educ:* Rockhurst Col, BA, 61; St Louis Univ, MA, 62, PhD (psych), 66; Am Bd Prof Psychol, dipl. *Prof Exp:* from asst prof psych to assoc prof psychol, 69-85, PROF & COORD, PSYCHOL TRAINING PROG, HEALTH SCI CTR, SAN ANTONIO, 85- *Concurrent Pos:* Secy & treas, Asn Psychol Internship Ctrs, 87-88; regional chair, Am Asn Psychiat Invest Children, 77-79. *Mem:* Fel Am Psychol Assoc; Behavior Therapy & Res Soc. *Res:* Children and family clinical issues. *Mailing Add:* Dept Psychiat Univ Tex Health Sci Ctr 7703 Floyd Curl Dr San Antonio TX 78284-7792

STEDMAN, ROBERT JOHN, b Marlow, Eng, Jan 28, 29; wid; c 2. ORGANIC CHEMISTRY. *Educ:* Cambridge Univ, BA, 49, MA & PhD(chem), 52. *Prof Exp:* Fel chem, Nat Res Coun Can, 52-54; res assoc, Med Col, Cornell Univ, 54-56; res assoc, Banting Inst, Univ Toronto, 57-58; res chemist, Chas Pfizer & Co, Conn, 58-60 & Smith Kline & French Labs, Pa, 60-69; assoc prof phys org chem, 69-76, PROF MED CHEM, SCH PHARM, TEMPLE UNIV, 76- *Concurrent Pos:* Consult, Smith Kline & French Labs, Pa, 79- *Mem:* AAAS; Am Chem Soc; The Chem Soc. *Res:* Natural products and medicinals; nuclear magnetic resonance spectroscopy. *Mailing Add:* Health Sci Ctr Temple Univ Sch Pharm 3307 N Broad St Philadelphia PA 19140-5101

STEEGE, DEBORAH ANDERSON, b Boston, Mass, Oct 2, 46; m 82. BIOCHEMISTRY, MOLECULAR BIOLOGY. *Educ:* Stanford Univ, BA, 68; Yale Univ, PhD(molecular biophys, biochem), 74. *Prof Exp:* Fel molecular biophys & biochem, Yale Univ, 74-76, fel biol, 76-77; asst prof, 77-83, ASSOC PROF BIOCHEM, DUKE UNIV, 83- *Concurrent Pos:* Am Cancer Soc fel, 74-76. *Mem:* Sigma Xi; Am Soc Microbiol; AAAS; Am Soc Biochem & Molecular Biol. *Res:* Post-transcriptional controls of gene expression; ribonucleic acid, messenger processing/stability; translational coupling; control strategies. *Mailing Add:* Dept of Biochem Duke Univ Med Ctr Durham NC 27710

STEEGMANN, ALBERT THEODORE, JR, b Cleveland, Ohio, Aug 15, 36; m 63; c 2. BIOLOGICAL ANTHROPOLOGY, PHYSICAL ANTHROPOLOGY. *Educ:* Univ Kans, BA, 58; Univ Mich, MA, 61, PhD(anthrop), 65. *Prof Exp:* From instr to asst prof anthrop, Univ Mo-Columbia, 64-66; from asst prof to assoc prof, 66-74, chmn, Anthrop Dept, 79-86, PROF ANTHROP, STATE UNIV NY BUFFALO, 74-; CONSULT, 86- *Concurrent Pos:* Vis colleague, Univ Hawaii, 67-68; NSF res grants, 67-70, 73-75 & 78-80; State Univ NY Buffalo fac res grants, 69-72; res assoc, Royal Ont Museum, 70-; vis prof, Inst Occup Med, Beijing, 89. *Mem:* AAAS; Am Anthrop Asn; Am Asn Phys Anthrop (secy & treas, 85-89); Soc Study Human Biol; Human Biol Coun (secy & treas, 74-77). *Res:* Human cold response, physiological and behavioral; cranio-facial evolution; American sub-arctic; nutritional anthropology; behavior; health and work capacity. *Mailing Add:* Dept Anthrop State Univ NY Buffalo Amherst NY 14261

STEEL, COLIN, b Aberdeen, Scotland, Feb 7, 33; m 58; c 3. PHYSICAL CHEMISTRY. *Educ:* Univ Edinburgh, BSc, 55, PhD(chem), 58. *Prof Exp:* Res assoc chem, State Univ NY Col Forestry, Syracuse Univ, 58-59; res assoc, Brandeis Univ, 59-60; asst prof, Univ Toronto, 60-61; res scientist, Itek Corp,

61-63; from asst prof to assoc prof, 63-77, PROF CHEM, BRANDEIS UNIV, 77- *Mem:* Am Chem Soc; Royal Soc Chem. *Res:* Reaction kinetics and photochemistry. *Mailing Add:* Dept Chem Brandeis Univ South St Waltham MA 02154-2700

STEEL, COLIN GEOFFREY HENDRY, b London, Eng, June 15, 46; m 70. INVERTEBRATE PHYSIOLOGY, COMPARATIVE ENDOCRINOLOGY. *Educ:* Univ Cambridge, BA, 67, MA, 71; Queen's Univ, PhD(zool), 71; Univ London, DIC, 75. *Prof Exp:* Fel insect physiol, Imp Col, Univ London, 71-72, res fel, 72-75; res assoc, 75-78, asst prof, 78-82, ASSOC PROF, DEPT BIOL, YORK UNIV, 82- *Concurrent Pos:* Sci consult, Ont Educ Commun Authority. *Mem:* Fel Royal Entom Soc London; Soc Exp Biol; Europ Soc Comp Endocrinol; Am Soc Zoologists; Can Soc Zool. *Res:* Neurosecretion in insects and crustacea; nervous and hormonal mechanisms controlling development; integration of behavior and development; circadian rhythms, photoperiodism and endocrine aspects. *Mailing Add:* Dept Biol York Univ Downsview ON M3J 1P3 Can

STEEL, HOWARD HALDEMAN, b Philadelphia, Pa, Apr 17, 21; m 64; c 4. ORTHOPEDIC SURGERY. *Educ:* Colgate Univ, BA, 42; Temple Univ, MD, 45, MS, 51; Am Bd Orthop Surg, dipl, 52; Univ Wash, PhD(anat), 65. *Prof Exp:* Resident, Temple Univ Hosp & Shriners Hosp Crippled Children, 48-52; chief orthop surgeon, Shriners Hosp Crippled Children, 65-86, PROF ORTHOP SURG, MED CTR, TEMPLE UNIV, 65-; CHIEF EMER, SHRINERS HOSP CRIPPLED CHILDREN, 86- *Concurrent Pos:* Staff surgeon, Med Ctr, Temple Univ, 51-; assoc prof, Div Grad Med, Univ Pa, 55-; clin prof, Med Sch, Univ Wash, 64-65; consult, Vet Admin Hosp, Philadelphia, 67- & Walson Army Hosp, Ft Dix, NJ, 67-; attend surgeon, St Christopher's Hosp for Children; clin prof ortho surg, Med Col PA, 85- *Mem:* Orthop Res Soc; AMA; Am Acad Orthop Surg; Am Orthop Asn; Am Fedn Clin Res; Eastern Orthop Asn; Scoliosis Res Soc. *Res:* Clinical investigation of hip problems in the child; clinical and bacteriological investigations of nosocomial infections; studies of the C1-C2 articulations in humans; studies on Protrusio acetabuli in the child with closure of the triraplate epiphysis; effect of gluteus medius and minimus advancement in cerebral palsied patient; rib resection in scoliosis; correlation of appearance of the face with skeletal diseases; studies on etiology of palsy in lower extremity after proximal tibial osteotomy. *Mailing Add:* Shriners Hosp Crippled Children 8400 Roosevelt Blvd Philadelphia PA 19152

STEEL, R KNIGHT, b New York, NY, Dec 1, 39; m 65; c 1. INTERNAL MEDICINE. *Educ:* Yale Univ, BA, 61; Columbia Univ, MD, 65. *Prof Exp:* From intern to chief resident med, Univ NC, Chapel Hill, 65-71, asst prof, 71-72; asst prof med, Univ Rochester, 72-77; assoc dir, Monroe Community Hosp, 72-77; ASSOC PROF MED, BOSTON UNIV, 77-, CHIEF GERIAT & DIR GERONT CTR, 77- *Mem:* Sigma Xi. *Res:* Medical education; geriatrics; health care delivery. *Mailing Add:* c/o Boston Univ Med Ctr 75 E Newton St Boston MA 02118

STEEL, ROBERT, b Winnipeg, Man, Mar 17, 23; m 52; c 2. MICROBIOLOGY. *Educ:* Univ Man, BS, 49, MS, 51; Univ Manchester, PhD(microbiol, biochem), 56. *Prof Exp:* Jr res officer, Div Appl Biol, Nat Res Coun Can, 51-54; Imp Chem Industs res fel, Univ Manchester, 55-58; res assoc, 58-71, HEAD MICROBIOL & CHEM SERV, UPJOHN CO, 71-, PROD MGR FERMENTATION OPERS, 76-, ASSOC DIR BIOCHEM ENG, 83- *Mem:* Am Soc Microbiol; Can Soc Microbiol; Am Chem Soc. *Res:* Steroid bioconversions; utilization of agricultural wastes by fermentation; production of 2, 3-butanediol, citric acid; biochemical engineering; agitation aeration studies in fermentation; mixing and scale-up of antibiotic fermentations. *Mailing Add:* 3505 Pinegrove Ln Kalamazoo MI 49008

STEEL, ROBERT GEORGE DOUGLAS, b St John, NB, Sept 2, 17; m 41; c 2. STATISTICAL ANALYSIS. *Educ:* Mt Allison Univ, BA, 39, BSc, 40; Acadia Univ, MA, 41; Iowa State Univ, PhD(statist), 49. *Prof Exp:* Asst prof math, Univ Wis & statistician, Agr Exp Sta, 49-52; assoc prof biol statist, Cornell Univ, 52-60; prof statist & grad adminr, 60-82, EMER PROF STATIST, NC STATE UNIV, 83- *Concurrent Pos:* Mem math res ctr, US Dept Army, Univ Wis, 58-59. *Mem:* Fel Am Statist Asn. *Res:* Nonparametric statistics; experimental design; data analysis. *Mailing Add:* 2106 Coley Forest Pl Raleigh NC 27607

STEEL, WARREN G, b New York, NY, Feb 16, 20; m 43; c 2. GEOLOGY. *Educ:* Univ NC, BS, 46, MS, 49. *Prof Exp:* Asst prof geol, NC State Col, 48-55; prof geol, 55-72, E B Andrews Prof Natural Sci, 72-82, chmn dept geol, 55-82, EMER E B ANDREWS PROF NATURAL SCI, MARIETTA COL, 82- *Concurrent Pos:* Geologist, US Geol Surv, 47-48 & 51 & NC Dept Conserv & Develop, 49-50 & 52; geologist-petrogr, Rare Minerals Br, US Bur Mines, 53-55; consult geologist, 55- *Mem:* Geol Soc Am; Nat Asn Geol Teachers. *Res:* Petrology; structural geology; geomorphology. *Mailing Add:* Adams Township Three Lowell OH 45744

STEELE, ARNOLD EDWARD, b Estherville, Iowa, June 21, 25; m 54; c 3. ZOOLOGY, NEMATOLOGY. *Educ:* Iowa State Univ, BA, 53, MS, 57. *Prof Exp:* Parasitologist, Animal Parasite & Dis Div, USDA, 55-56, zoologist plant nematol, Tifton, Ga, 55-59, ZOOLOGIST PLANT NEMATOL, SCI & EDUC ADMIN-FED RES, USDA, CALIF, 59- *Concurrent Pos:* Assoc ed, Soc Nematologists, 75-77. *Mem:* Soc Nematologists; Am Phytopath Soc; Sigma Xi. *Res:* Biology; host-parasite relationships and control of nematodes affecting production of sugarbeet and vegetable crops. *Mailing Add:* 1118 Briarwood Pl Salinas CA 93901

STEELE, CHARLES RICHARD, b Royal, Iowa, Aug 15, 33; m 69; c 4. APPLIED MECHANICS, BIO-MEDICAL ENGINEERING. *Educ:* Tex A&M Univ, BS, 56; Stanford Univ, PhD(appl mech), 60. *Prof Exp:* Eng specialist aircraft struct, Chance-Vought Aircraft, Dallas, 59-60; res scientist shell theory, Lockheed Res Lab, Palo Alto, 60-66; assoc prof, 66-71, PROF APPL MECH, STANFORD UNIV, 71- *Concurrent Pos:* Lectr, Univ Calif, Berkeley, 64-65; vis prof, Swiss Fed Inst Technol, Zurich, 71-72, Univ of Lulen, Sweden, 82 & Chung Kung Univ, Taiwan, 85; tech dir, Shelltech Assoc; chmn exec comt, appl mech div, Am Soc Mech Engrs, 83-84; ed-in-chief, Int J Solids Structures, 85- *Mem:* Am Inst Aeronaut & Astronaut; fel Am Soc Mech Engrs; Acoust Soc Am; fel Am Acad Mech (pres, 89-90). *Res:* Asymptotic analysis in mechanics; thin shell theory; mechanics of the inner ear; noninvasive determination of bone stiffness. *Mailing Add:* Div of Appl Mech Stanford Univ Stanford CA 94305

STEELE, CRAIG WILLIAM, b Port Arthur, Tex, Mar 22, 54. ICHTHYOLOGY, AQUATIC BEHAVIORAL TOXICOLOGY. *Educ:* Pa State Univ, BS, 76; Tex A&M Univ, MS, 78, PhD(zool), 86. *Prof Exp:* Fel aquatic toxicol, Dept Zool, Miami Univ, Ohio, 86-89; ASST PROF BIOL, DEPT BIOL & HEALTH SERV, EDINBORO UNIV, 90- *Concurrent Pos:* Adj asst prof, Greenwich Univ, Hawaii, 85-; vis scientist, Santa Fe Inst, 89. *Mem:* AAAS; Am Soc Zoologists; Animal Behav Soc; Brit Ecol Soc; Int Asn Ecol; Soc Environ Toxicol & Chem. *Res:* Ethology and behavioral ecology of aquatic animals, primarily fishes; chemoreception; hierarchical organization of food search and feeding behavior; behavioral toxicology; development of methodologies for the quantitative study of behavior; mathematical modelling of behavioral processes. *Mailing Add:* Dept Biol & Health Serv Edinboro Univ Edinboro PA 16444

STEELE, DAVID GENTRY, b Beeville, Tex, Feb 8, 41; m 80; c 1. ZOOARCHAEOLOGY, PHYSICAL ANTHROPOLOGY. *Educ:* Univ Tex, Austin, BA, 67; Univ Kans, PhD(anthrop), 70. *Prof Exp:* Fel anthrop, Smithsonian Inst, 70-71; from asst prof to assoc prof, Univ Alta, Edmonton, 71-79; ASSOC PROF ANTHROP, TEX A&M UNIV, 79- *Mem:* Am Soc Phys Anthrop; Soc Am Archaeologists. *Res:* Predator/prey relationships of man; human adaptations to the Texas coast; animal remains from a Roman farm site in southern Italy; human osteology; mammalian paleontology of the Texas Quaternary. *Mailing Add:* Dept Anthropol Tex A&M Univ College Station TX 77843

STEELE, DONALD HAROLD, b London, Ont, Nov 5, 32; m 59; c 1. ZOOLOGY. *Educ:* Univ Western Ont, BSc, 54; McGill Univ, MSc, 56, PhD(zool), 61. *Prof Exp:* Technician, Biol Sta, St Andrews, NB, 55-56; lectr biol, Sir George Williams Univ, 60-62, asst prof, 62; from asst prof to assoc prof, 62-75, PROF BIOL, MEM UNIV NFLD, 75- *Mem:* AAAS; Ecol Soc Am; Brit Ecol Soc; Int Asn Ecol; Can Soc Zool. *Res:* Marine ecology; zoogeography; systematics of marine amphipoda. *Mailing Add:* Dept Biol Mem Univ Nfld St John's NF A1B 3X9 Can

STEELE, EARL L(ARSEN), b Denver, Colo, Sept 24, 23; m 53; c 6. SOLID STATE PHYSICS, ELECTRICAL ENGINEERING. *Educ:* Univ Utah, BS, 45; Cornell Univ, PhD(physics), 52. *Prof Exp:* Lab asst physics, Univ Utah, 44-45; asst, Cornell Univ, 45-51; res physicist, Gen Elec Co, 51-56; chief res dept, Semiconductor Div, Motorola, Inc, Ariz, 56-58; asst lab mgr, Semiconductor Div, Hughes Aircraft Co, 58-59, lab mgr, 59-63; staff scientist, Res & Eng Div, Autonetics Div, N Am Aviation, Inc, Calif, 63-69; chmn dept, 71-80, PROF ELEC ENG, COL ENG, UNIV KY, 69- *Concurrent Pos:* Ed, Trans Electron Devices, Inst Elec & Electronics Engrs, 54-61; assoc prof, Ariz State Univ, 57-58; lectr, Univ Calif, Los Angeles, 59; phys sci coordr, Southern Calif Col, 62-63; affil prof & lectr, Univ Calif, Irvine, 67-69; officer, Southeastern Ctr Elec Eng Educ, 74- *Mem:* Am Phys Soc; Am Soc Eng Educ; Sigma Xi; Am Asn Physics Teachers; fel Inst Elec & Electronics Engrs; Int Soc Hybrid Microelectronics. *Res:* Semiconductor p-n junction theory and device design; solid state theoretical studies of band structure of barium oxide; transistors and parametric devices; microelectronics; lasers and electrooptics; computer aided electronic circuit design; quantum electronics. *Mailing Add:* Dept Elec Eng Univ Ky Lexington KY 40506-0046

STEELE, IAN MCKAY, b Syracuse, Ill, June 19, 44; div; c 4. LEAD CHEMICALS, COMPUTER CONTROL OF INSTRUMENTS. *Educ:* Rensselaer Poly Inst, BS, 66; Univ Ill, PhD(geol), 71. *Prof Exp:* RES ASSOC GEOPHYSICS, UNIV CHICAGO, 71- *Concurrent Pos:* consult, Hammond Lead Corp, 86- *Mem:* fel Mineral Soc Am; Geochem Soc; fel Meteoritical Soc; Am Geophys Union; Mineral Soc Great Brit; Microbeam Anal Soc. *Res:* Crystallograph, mineralogy and chemical processes of phases in lead acid batteries; analytical techniques of chemical and structural analysis; computer application to analysis techniques. *Mailing Add:* Univ Chicago 5734 S Ellis Ave Chicago IL 60637

STEELE, JACK, b Indianapolis, Ind, Jan 22, 42; m 68; c 3. INORGANIC CHEMISTRY, PHYSICAL CHEMISTRY. *Educ:* DePauw Univ, BA, 64; Univ Ky, PhD(inorg chem), 68. *Prof Exp:* Am Chem Soc Petrol Res Fund grant & teaching intern, Wash State Univ, 68-70; from asst prof to assoc prof, 70-80, chmn, dept chem & physics, 81-85, PROF CHEM, ALBANY STATE COL, 80- *Concurrent Pos:* NSF col sci improv prog mem, Albany State Col, 72-73, minority sch biomed support prog mem, 72-76; Minority Access to Res Career, 88- *Mem:* Am Chem Soc. *Res:* Stereochemistry of metal chelates of biologically important compounds; science education; analysis of environmental samples. *Mailing Add:* Dept Natural Sci Albany State Col Albany GA 31705

STEELE, JACK ELLWOOD, b Lacon, Ill, Jan 27, 24; m 55; c 2. BIONICS, INTELLIGENT SYSTEMS. *Educ:* Northwestern Univ, BM, 49, MD, 50; Wright State Univ, MS, 77. *Prof Exp:* Intern, Cincinnati Gen Hosp, 49-50; fel neuroanat, Med Sch, Northwestern Univ, 50-51; ward officer, US Air Force, 2750 US Air Force Hosp, Wright-Patterson AFB, 51-53, proj officer, Aerospace Med Lab, 53-71; pvt pract, 71-73; physician, Dayton Mental Health Ctr, 73-75, med dir, Drug Treatment Unit, 75-78; PRES, GEN BIONICS CORP, 79- *Concurrent Pos:* Physician, Buda Narcotics Clinic, 78-81; med dir, Nova House, 81-90. *Mem:* Inst Elec & Electronic Engrs; NY Acad Sci; AMA; Asn Comput Mach; Am Soc Clin Hypnosis. *Res:* Analysis and design of systems with lifelike behavior, intelligence in particular; logic of human mind; protologic. *Mailing Add:* 2313 Bonnieview Ave Dayton OH 45431-1987

STEELE, JAMES HARLAN, b Chicago, Ill, Apr 3, 13; m 41, 69; c 3. VETERINARY MEDICINE, PUBLIC HEALTH. *Educ:* Mich State Univ, DVM, 41; Harvard Univ, MPH, 42; Am Bd Vet Pub Health, dipl, Am Col Vet Med, dipl. *Prof Exp:* State Health Dept, Ohio, 42-43; vet, USPHS, 43-45, chief vet pub health, Nat Commun Dis Ctr, 45-68, asst surgeon gen vet affairs, 68-71; prof, Univ Tex Sch Pub Health, Houston, 71-83, emer prof environ sci, 83; RETIRED. *Concurrent Pos:* Consult, Pan-Am Sanit Bur, 44, WHO, 50-, Food & Agr, 60-, & White House Comt Consumer Protection; founder, Am Bd Vet Pub Health, 50-52; chmn, WHO-Food & Agr Orgn&Expert Comt Zoonoses, 3rd Report, 67; ed-in-chief, CRC Handbk Zoonoses, 78-84; hon dipl, Tenth World Vet Cong, 75; Conf Emer Mem, Pub Health Asn; consult Ger Health Serv, 86-88. *Honors & Awards:* Centennial Award, Am Pub Health Asn, 72; Int Vet Award, Am Vet Med Asn, 84; Int Award, contrib to world health, Ger Health Serv, 88. *Mem:* Am Soc Trop Med & Hyg; Asn Mil Surg US; Am Vet Med Asn; Am Vet Epidemiol Soc (pres, 88-90); hon mem World Vet Cong; hon mem World Vet Asn. *Res:* Veterinary public health; epidemiology of zoonoses and chronic diseases common to animals and man; cost benefits of international veterinary public health programs; tuberculosis in animals; food irradiation and hygiene. *Mailing Add:* Inst Environ Health Univ Tex Sch Pub Health PO Box 20186 Houston TX 77225

STEELE, JAMES PATRICK, b Louisville, Ky, Dec 6, 19; m 44; c 2. RADIOLOGY. *Educ:* Univ Louisville, MD, 43; Am Bd Radiol, dipl, 50. *Prof Exp:* PROF RADIOL, SCH MED, UNIV S DAK, VERMILLION, 49- *Concurrent Pos:* Radiologist, Sacred Heart Hosp, Yankton, 49-; prof, Sch Med, Univ Nebr; pres, SDak Health Res Inst. *Mem:* Am Roentgen Ray Soc; Radiol Soc NAm; AMA; Am Col Radiol. *Res:* Night vision as related to fluoroscopy; psychological effects of total body radiation. *Mailing Add:* 8122 Montero Dr Prospect KY 40059

STEELE, JOHN EARLE, b St John's, Nfld, Jan 29, 32; m 57; c 3. ENDOCRINOLOGY, CELL BIOLOGY. *Educ:* Dalhousie Univ, BSc, 54; Univ Western Ont, MSc, 56; Univ Sask, PhD(biol), 59. *Prof Exp:* Res officer, Can Dept Agr, 59-64; from asst prof to assoc prof, 64-75, PROF ZOOL, UNIV WESTERN ONT, 75- *Mem:* Can Soc Zool; Can Soc Cell Biol; Entom Soc Can. *Res:* Hormonal control of metabolism, water transport and growth and development in insects. *Mailing Add:* Dept Zool Univ Western Ont London ON N6A 5B7 Can

STEELE, JOHN H, b Edinburgh, UK, Nov 15, 26; m 56; c 1. BIOLOGY. *Educ:* Univ Col, London, BSc, 46, DSc, 64. *Prof Exp:* Scientist, Marine Lab, Scotland, 51-77, dep dir, 73-77; dir, 77-89, PRES, WOODS HOLE OCEANOG INST, 83-; DIR, EXXON CORP, 89- *Concurrent Pos:* Trustee, Bermuda Biol Sta Res Inc, 77-, Univ Corp Atmospheric Res, 87- & Rob Wood Johnson Found, 90-; mem, bd govs, Joint Oceanog Inst Inc, 77-89, Univ Corp Atmospheric Res, 78- & Arctic Res Comt, 87-; chmn ocean sci bd, Nat Res Coun, Nat Acad Sci, 81-82, mem, 78-88. *Honors & Awards:* Alexander Agassiz Medal, Nat Acad Sci, 73. *Mem:* Fel Royal Soc; fel Royal Soc Edinburgh; Am Acad Arts & Sci; fel AAAS. *Res:* Dynamics of marine ecosystems. *Mailing Add:* Woods Hole Oceanog Inst Woods Hole MA 02543

STEELE, JOHN WISEMAN, b Motherwell, Scotland, May 27, 34; m 58; c 4. PHARMACEUTICAL CHEMISTRY. *Educ:* Glasgow Univ, BSc, 55, PhD(pharmaceut chem), 59. *Prof Exp:* Lectr, 58-59, from asst prof to prof, 59-81, DEAN FAC PHARM, UNIV MAN, 81- *Concurrent Pos:* Fel, Chelsea Col Sci & Technol, 65-66; mem, Med Res Coun Can, 70-72; vis scientist, Med-Chem Inst, Univ Bern, 72-73; mem, Man Drug Standards & Therapeut Comt, 78-86. *Mem:* Can Pharmaceut Asn; assoc Royal Inst Chem; The Chem Soc; Asn Faculties Pharm Can (pres, 75-76). *Res:* Drug metabolism, especially of anabolic steroids and other drugs likely to be abused by athletes; methods of drug analysis, including gas-liquid chromatography. *Mailing Add:* Fac of Pharm Univ of Man Winnipeg MB R3T 2N2 Can

STEELE, KENNETH F, b Statesville, NC, Jan 16, 44; m 66; c 2. WATER QUALITY, WATER CHEMISTRY. *Educ:* Univ NC, Chapel Hill, BS, 62, PhD(geol), 71. *Prof Exp:* From instr to assoc prof, 70-83, PROF GEOL, UNIV ARK, FAYETTEVILLE, 83-, DIR ARK WATER RESOURCES RES CTR, 88- *Concurrent Pos:* Mem S cent mgt bd, Geol Soc Am, 80-82, 84-86, prog chmn, 85; bd dirs, Ark Ground Water Asn, 89-90, vpres, 91; mem mgt bd, Geol Soc Am, 80-82 & 84-86. *Honors & Awards:* Oak Ridge Assoc Univs fel. *Mem:* Geol Soc Am; Soc Environ Geochem & Health; Am Water Resources Asn; Asn Explor Geochemists; Asn Groundwater Sci & Engrs. *Res:* Major and trace element geochemical investigations applied to exploration and also environmental geochemistry. *Mailing Add:* Ark Water Resources Res Ctr 113 Ozark Hall Univ Ark Fayetteville AR 72701

STEELE, LAWRENCE RUSSELL, b Manhattan, Kans, Nov 7, 35; m 59; c 3. CHEMICAL ENGINEERING. *Educ:* Ohio State Univ, BChE & MSc, 58, PhD(chem eng), 62; Fairleigh-Dickinson Univ, MBA, 83. *Prof Exp:* Sr res engr, NAm Aviation, Inc, 62-63; res scientist, Columbia Univ, 63-66; SR RES ENGR, E R SQUIBB & SONS, INC, 66-, SECT HEAD, 75-, ASST DEPT HEAD, 81-, DIR, 83- *Mem:* Am Inst Chem Engrs; Am Chem Soc. *Res:* Chemical process development of pharmaceuticals. *Mailing Add:* 55 Cherrybrook Dr Princeton NJ 08540

STEELE, LENDELL EUGENE, b Kannapolis, NC, May 5, 28; m 49; c 4. NUCLEAR ENGINEERING, MATERIALS SCIENCE. *Educ:* George Washington Univ, BS, 50; Am Univ, MA, 59. *Prof Exp:* Chemist phys sci, Res Mgt, Agr Res Ctr, 49-50; res & develop officer radiol safety, US Air Force, 51-53; metall eng, US Atomic Energy Comn, 66-67; chemist, Naval Res Lab, 50-51, anal chemist, 53-57, sect head & br head, 57-66, br head & assoc supt mat sci & technol, 67-80, BR HEAD, RES MGR, NAVAL RES LAB, 80- *Concurrent Pos:* Consult, Metal Properties Coun, 67-; US deleg, Int Atomic Energy Agency, Vienna, 67-; task group leader, Metals Properties Coun, ed, 67-; consult nuclear engr, 86- *Honors & Awards:* Wash Acad Sci Eng Award, 62; Appl Sci Award, Naval Res Labs-Sigma Xi, 66; Spec Annual Prize Award, Am Nuclear Soc, 72; Dudley Medal, Am Soc Testing & Mat, 73, Award of Merit, 78. *Mem:* Fel Am Soc Metals; Am Nuclear Soc; fel Am Soc Testing & Mat; Res Soc Am; Fed Mat Soc (vpres, 83 pres, 84). *Res:* Fundamental and applied research on materials for advanced energy conversion systems, especially gas turbine materials but including response of nuclear structural material to nuclear effects for light water, breeder and fusion reactors as well. *Mailing Add:* 7624 Highland St Springfield VA 22150

STEELE, LEON, b Ill, Apr 8, 15; m 41; c 3. PLANT BREEDING. *Educ:* Ill Wesleyan Univ, BS, 40, DSc, 67. *Prof Exp:* Res assoc, Michael-Leonard Seed Co, 36-40; mgr dept res, Funk Seeds Int, 40-52, assoc res dir, 52-57, res dir, 57-78, vpres, 63-78, res consult, 78-88; consult, USAID & World Bank, 78-88,; RETIRED. *Concurrent Pos:* Pres, Funk Seeds Int Ltd, Can, 67-72. *Mem:* AAAS; Am Soc Agron; Genetics Soc Am; Am Genetic Asn. *Res:* Commercial and hybrid corn breeding; physiology of corn plant; corn diseases and their control through breeding for resistance. *Mailing Add:* 1214 Towanda Pl Bloomington IL 61701

STEELE, MARTIN CARL, b New York, NY, Dec 25, 19; m 41; c 4. SOLID STATE ELECTRONICS. *Educ:* Cooper Union, BChE, 40; Univ Md, MS, 49, PhD, 52. *Prof Exp:* Physicist & chief cryomagnetics sect, US Naval Res Lab, 47-55; res physicist, Res Lab, Radio Corp Am, 55-58, head semiconductor res group, 58-60, dir res labs, Japan, 60-63, head solid state electron physics group, 63-72; head semiconductor mat & device res, 72-81, staff res scientist, Electronics Dept, Res Labs, Gen Motors Corp, 81-85; DIR, INST AMORPHOUS STUDIES, 85- *Concurrent Pos:* Vis lectr, Princeton Univ, 65-66; adj prof elec eng, Wayne State Univ, 76-85. *Mem:* Fel Am Phys Soc; Electrochem Soc; fel Inst Elec & Electronics Engrs. *Res:* Solid state physics; superconductivity; galvanomagnetic effects in metals and semiconductors; high electric field effects in semiconductors; solid state plasma effects; microwave devices; infrared detection; integrated circuits; MOS devices; semiconductor surfaces. *Mailing Add:* 1098 Welsh Rd Huntington Valley PA 19006-6024

STEELE, RICHARD, b Charlotte, NC, Sept 6, 21; m 49; c 2. CHEMISTRY, POLYMER CHEMISTRY. *Educ:* Univ NC, SB, 42; Princeton Univ, MA, 48, PhD(chem), 49. *Prof Exp:* Res chemist, Rohm and Haas Co, 42-46; res chemist & head phys org chem sect, Textile Res Inst, 50-53; lab head, Rohm and Haas Co, 53-65; dir appln & prod develop, Celanese Int Co, NY, 65-66, vpres & tech dir, 66-71, sr vpres technol & admin, Celanese Fibers Mkt Co, 71- 73 & Celanese Fibers Co, 73-76, sr vpres mfg & technol, Celanese Fibers Int Co, 76-79, exec vpres, 80 -82; CONSULT, 82- *Honors & Awards:* Olney Medal, Am Asn Textile Chemists & Colorists, 64; Harold DeWitt Smith Award, Am Soc Testing & Mat, 78. *Mem:* AAAS; Am Chem Soc; Am Asn Textile Chemists & Colorists; Brit Textile Inst; Fiber Soc. *Res:* Structure of natural and synthetic fibers; chemistry of textile wet-finishing processes; cellulose chemistry. *Mailing Add:* Strafford Rd Tunbridge VT 05077

STEELE, RICHARD HAROLD, b Buffalo, NY, Aug 1, 19; m 52; c 3. BIOCHEMISTRY. *Educ:* Univ Ala, BS, 48; Tulane Univ, PhD(biochem), 53. *Prof Exp:* Vis investr, Inst Muscle Res, Marine Biol Lab, 54-57; PROF BIOCHEM, TULANE UNIV, 57- *Concurrent Pos:* Lederle Med Fac Award, 57-60; NIH sr res fels, 60 & 65. *Mem:* Am Chem Soc; Am Soc Biol Chemists. *Res:* Energy generation and transfer; spectroscopy; chemiluminescence and bioluminescence; copper metabolism; alcholism. *Mailing Add:* 3905 Cleveland Pl Metairie LA 70003

STEELE, ROBERT, b Scotland, Jan 16, 29; m 55; c 2. PREVENTIVE MEDICINE, EPIDEMIOLOGY. *Educ:* Univ Edinburgh, DPH, 56; Univ Sask, MD, 60; FRCP(C); FFCM; FRCP(ED). *Prof Exp:* Asst prof prev med, Col Med, Univ Sask, 58-62; med officer, Scottish Health Dept, 62-64; assoc prof prev med, Fac Med & dir res, 64-68, PROF COMMUNITY HEALTH & EPIDEMIOL & HEAD DEPT, FAC MED, QUEEN'S UNIV, ONT, 68- *Concurrent Pos:* Consult, Kingston Gen Hosp, Ont. *Mem:* Asn Teachers Prev Med; fel Int Epidemiol Asn; fel Am Pub Health Asn; Royal Med Soc; Can Asn Teachers Social & Prev Med. *Res:* Cancer; medical care; community health, Aids, international health. *Mailing Add:* Dept Epidemiol-Pub Health Queen's Univ Fac Med Kingston ON K7L 3N6 Can

STEELE, ROBERT DARRYL, b New Eagle, Pa, Dec 5, 46; m 79. INTERMEDIARY METABOLISM, PROTEIN METABOLISM. *Educ:* Univ Ariz, BS, 70, MS, 73; Univ Wis, Madison, PhD(nutrit & biochem), 78. *Prof Exp:* PROF DEPT NUTRIT SCI, UNIV WIS. *Honors & Awards:* Bio-Serve Award, Am Inst Nutrit. *Mem:* Am Inst Nutrit; Am Physiol Soc; Sigma Xi. *Res:* Investigating the central role of the liver in amino acid and protein metabolism in mammals; blood-brain barrier transport; folic acid metabolism. *Mailing Add:* Dept Nutrit Sci Univ Wis 1415 Linden Dr Madison WI 53706

STEELE, ROBERT WILBUR, b Denver, Colo, Aug 13, 20; m 42, 61; c 6. FOREST MANAGEMENT. *Educ:* Colo State Univ, BSF, 42; Univ Mich, MSF, 49; Colo State Univ, PhD(forest fire sci), 75. *Prof Exp:* Forest guard, US Forest Serv, Ore, 42-43, forester, Pac Northwestern Exp Sta, 46-55; forest mgr, SDS Lumber Co, 55-56; asst prof forestry, Univ Mont, 56-67, assoc prof, 67-70, prof, 70-81; CONSULT, FORESTRY ASSOC INT INC, 81- *Mem:* Soc Am Foresters; Am Meteorol Soc. *Res:* Forest fire control; development of techniques and machinery for fire detection and control; use and effects of prescribed fire in the forest; forest fire science; land surveying; sagebrush burning. *Mailing Add:* 1165 Hamilton Heights Corvallis MT 59828

STEELE, RONALD EDWARD, b Pittsburgh, Pa, May 19, 43; m 69. PHYSIOLOGY, ENDOCRINOLOGY. *Educ:* Pa State Univ, BS, 65; Univ Ky, PhD(physiol), 70. *Prof Exp:* Fel physiol, Worcester Found Exp Biol, 69-72; asst prof pediat endocrinol, Johns Hopkins Hosp, 72-74; sr scientist endocrinol, 74-86, admin mgr, 86-89, DIR ARTHROSCOLORIS, CIBA-GEIGY CORP, 89- *Mem:* Endocrine Soc; Am Soc Andrology. *Res:* Male and female reproductive endocrinology and corticosteroid physiology. *Mailing Add:* Pharmaceut Div Ciba-Geigy Corp 556 Morris Ave Summit NJ 07901

STEELE, SIDNEY RUSSELL, b Toledo, Ohio, June 30, 17; m 44; c 2. CHEMISTRY. *Educ:* Univ Toledo, BS, 39; Ohio State Univ, PhD(chem), 43. *Prof Exp:* Res chemist, Girdler Corp, Ky, 43-47; from assoc prof to prof chem, Eastern Ill Univ, 47-84, head dept, 67-77 & 78-79; RETIRED. *Mem:* Am Chem Soc. *Res:* Polarography; abnormal diffusion currents; water gas-shift catalysis; methanation of carbon monoxide. *Mailing Add:* Dept Chem Eastern Ill Univ Charleston IL 61920

STEELE, THEODORE KARL, b Brooklyn, NY, Oct 27, 22; m 52; c 2. ENERGY MANAGEMENT. *Educ:* City Col New York, BME, 43; NY Univ, MME, 49, EngScD, 51. *Prof Exp:* Design engr instruments, Bendix Aviation Corp, 43-44; test engr power plants, US Navy, 44-46; test engr accessories, Stratos Corp, 46-47; test engr rocket motros, M W Kellogg Co, 47-48; instr mech eng, NY Univ, 48-51; vpres, Bulova Res & Develop Labs, 51-64; SR VPRES, NY INST TECHNOL, 64- *Concurrent Pos:* Pvt consult, 48-; consult, res projs, NSF, NASA, Dept of Environ, US Navy, 64-; Consult & Designers Inc, 64-66. *Mem:* Am Inst Aeronaut & Astronaut; Am Soc Mech Engrs; Sigma Xi; Am Soc Eng Educ; Soc Mfg Eng. *Res:* Electromechanical instruments; fuzing and safing systems; computer aided instruction; energy systems; management science; heat power. *Mailing Add:* 22 Embassy Ct Great Neck NY 11021

STEELE, TIMOTHY DOAK, b Muncie, Ind, Apr 12, 41; div; c 2. HYDROLOGY, RESOURCE MANAGEMENT. *Educ:* Wabash Col, AB, 63; Stanford Univ, MS, 65, PhD(hydrol), 68; USDA Grad Sch, advan cert acct, 73. *Prof Exp:* Res hydrologist, Water Resources Div, Menlo Park, US Geol Surv, Colo, 66-68, res hydrologist, Systs Lab Group, Washington, DC, 68-72, hydrologist, Qual Water Br, 72-74, proj chief & hydrologist, Yampa River Basin Assessment Study, 75-80; sr proj hydrologist & chief, Water Qual Group, Woodward-Clyde Consult, Denver, 80-83; water resources mgr, In-Situ Inc, Denver, 83-89; MGR, PHYS SCI GROUP, ADVAN SCI INC, LAKEWOOD, 89- *Concurrent Pos:* Water qual specialist, US AID, Pakistan, 72; Alex von Humboldt res fel, Univ Bayreuth, WGer, 79; affiliated fac mem & guest lectr, Colo State Univ, Ft Collins, 79- *Mem:* Am Geophys Union; Int Asn Hydrol Sci; Int Water Res Asn; Am Chem Soc; Am Inst Hydrol. *Res:* Design of hydrologic data-collection networks; statistical analysis of data; hydrologic simulation and modeling; water resources planning and systems analysis; hydrogeochemistry; water quality; assessments of environmental impacts of energy-resource development; regional water-resources assessments; ground water contamination and resource conservation and recovery act regulations. *Mailing Add:* 326512 Meadow Mountain Rd Evergreen CO 80439

STEELE, VERNON EUGENE, b Blairsville, Pa, July 23, 46; m 68; c 2. RADIOBIOLOGY. *Educ:* Bucknell Univ, BS, 68; Univ Rochester, MS, 74, PhD(radiation biol), 75. *Prof Exp:* Investr carcinogenesis, Biol Div, Oak Ridge Nat Lab, 75-77; mem staff, Nat Inst Environ Health Sci, NIH, 77-82; res supvr, Northrop Serv, Inc, 82-; PROG DIR, DIV CANCER PREV & CONTROL, NAT CANCER INST, NIH, 89- *Mem:* Tissue Cult Asn; Sigma Xi; Am Asn Cancer Res. *Res:* Effects of radiation, chemical carcinogens and promoters on cell and tissue kinetics, morphology and physiology; teratology; chemoprevention. *Mailing Add:* Div Cancer Prev & Control Nat Cancer Inst NIH 9000 Rockville Pike Bethesda MD 20892

STEELE, VLADISLAVA JULIE, b Prague, Czech, July 8, 34; m 59; c 1. INVERTEBRATE PHYSIOLOGY, HISTOLOGY. *Educ:* McGill Univ, BSc, 57, MSc, 59, PhD(zool), 65. *Prof Exp:* Lectr histol, McGill Univ, 60-61; vis lectr, Mem Univ Nfld, 62-63, lectr histol & embryol, 63-65, asst prof, 65-72, dept head biol, 80-82, ASSOC PROF HISTOL DEVELOP BIOL CELL BIOLOGY, MEM UNIV NFLD, 72- *Mem:* Am Inst Biol Sci; Can Soc Zool; Nutrit Today Soc; Crustacean Soc Can; Soc Cell Biologists; Am Soc Zool. *Res:* Photoperiod, neurosecretion and steroid production in marine amphipods; influence of environmental factors on the reproduction of boreo-arctic intertidal amphipods; sensory receptors crustaceans. *Mailing Add:* Dept of Biol Mem Univ Nfld St John's NF A1B 3X9 Can

STEELE, WARREN CAVANAUGH, b Pocatello, Idaho, Oct 25, 29; m 55. PHYSICAL CHEMISTRY. *Educ:* Ore State Col, BA, 51, PhD(phys chem), 56. *Prof Exp:* Res chemist, Dow Chem Co, 56-58; res assoc chem, Tufts Univ, 58-60 & 62-64; res fel, Harvard Univ, 60-62; sr staff scientist, Space Systs Div, Avco Corp, Wilmington, 64-75; prin scientist, Energy Resources Co Inc, Cambridge, 75-78; proj leader, Foremost Res Ctr, Foremost-McKesson, Inc, Dublin, 78-86; lab dir, 85-87, SR CONSULT, CLAYTON ENVIRON CONSULTS, INC, 87- *Mem:* AAAS; Am Chem Soc. *Res:* Mass spectrometry; gas-surface reaction kinetics; high temperature thermochemistry; environmental chemistry. *Mailing Add:* 1854 San Ramon Ave Berkeley CA 94707

STEELE, WILLIAM A, b St Louis, Mo, June 4, 30; m 55; c 2. PHYSICAL CHEMISTRY. *Educ:* Wesleyan Univ, BA, 51; Univ Wash, PhD(phys chem), 54. *Prof Exp:* Fel, Cryogenic Lab, 54-55, from asst prof to assoc prof, 55-66, PROF PHYS CHEM, PA STATE UNIV, UNIVERSITY PARK, 66- *Concurrent Pos:* NSF fel, 57-58, sr fel, 63-64; mem comt colloid & surface chem, Nat Acad Sci-Nat Res Coun, 66-72; mem adv comt, Chem Div, NSF, 72-76; Unilever vis prof, Univ Bristol, 77; Guggenheim fel, 77; Fulbright fel, Univ Vienna, 79; assoc ed, J Phys Chem, 80- *Mem:* Am Chem Soc; Am Phys Soc. *Res:* Thermodynamics and statistical mechanics of liquids and physical adsorption of gases on solids. *Mailing Add:* Dept Chem 152 Davey Lab Pa State Univ University Park PA 16802-6302

STEELE, WILLIAM F, b Quincy, Mass, Mar 14, 20; m 54; c 2. MATHEMATICS. *Educ:* Boston Univ, AB, 51, MA, 52; Univ Pittsburgh, PhD(math), 61. *Prof Exp:* From instr to assoc prof math, Heidelberg Col, 52-63, prof, 63-86 & 88; RETIRED. *Concurrent Pos:* Lectr, Univ Pittsburgh, 58-68; NSF vis scholar, Mass Inst Technol, 71-72. *Mem:* Am Math Soc; Math Asn Am. *Res:* Summability of sequences by matrix methods. *Mailing Add:* 181 Clinton Ave Tiffin OH 44883

STEELE, WILLIAM JOHN, b Philadelphia, Pa, Mar 31, 29; m 54; c 3. BIOCHEMICAL PHARMACOLOGY. *Educ:* Univ Pa, AB, 51, PhD(biochem), 58. *Prof Exp:* Res assoc cancer, Univ Pa, 57-60, Chester Beatty Res Inst, Royal Cancer Hosp, 60-62; asst prof, Col Med, Baylor Univ, 62-67; from asst prof to assoc prof, 67-74, PROF PHARMACOL, COL MED, UNIV IOWA, 74- *Mem:* Am Asn Cancer Res; Am Chem Soc; Am Soc Pharmacol & Exp Therapeut; Am Soc Cell Biol; Brit Biochem Soc. *Res:* Biochemical basis of opiate and alcohol addiction; mode of action of protein synthesis inhibitors; regulation of translation on free and membrane-bound polysomes. *Mailing Add:* Dept Pharmacol Univ Iowa Col Med Iowa City IA 52242

STEELE, WILLIAM KENNETH, b Ft Wayne, Ind, Nov 2, 42; m; c 1. GEOLOGY, GEOPHYSICS. *Educ:* Case Western Reserve Univ, BS, 65, PhD(geol), 70. *Prof Exp:* From asst prof to assoc prof, 70-82, PROF GEOL, EASTERN WASH UNIV, 82- *Mem:* Am Geophys Union; AAAS; Sigma Xi. *Res:* General geophysics, especially gravity and magnetic modeling, paleomagnetism. *Mailing Add:* Dept Geol Eastern Wash Univ Chenev WA 99004

STEELINK, CORNELIUS, b Los Angeles, Calif, Oct 1, 22; m 49; c 2. ORGANIC CHEMISTRY. *Educ:* Calif Inst Technol, BS, 44; Univ Southern Calif, MS, 50; Univ Calif, Los Angeles, PhD(chem), 56. *Prof Exp:* Lectr chem, Univ Southern Calif, 49-50 & Orange Coast Col, 50-53; asst, Univ Calif, Los Angeles, 53-56; res fel, Univ Liverpool, 56-57; from asst prof to assoc prof chem, 57-70, PROF CHEM, UNIV ARIZ, 70- *Mem:* Am Chem Soc. *Res:* Structure of lignin; electron spin resonance studies on naturally-occurring compounds; isolation and structural elucidation of plant terpenoids; structures and reactions of aquatic humic acids. *Mailing Add:* Dept Chem Univ Ariz Tucson AZ 85721-0002

STEELMAN, CARROL DAYTON, b Vernon, Tex, Dec 9, 38; m 60; c 5. VETERINARY ENTOMOLOGY. *Educ:* Okla State Univ, BS, 61, MS, 63, PhD(entom), 65. *Prof Exp:* From asst prof to assoc prof, 65-73, PROF MED & VET ENTOM, LA STATE UNIV, BATON ROUGE, 73-,PROF ENTOM, 79- *Concurrent Pos:* USDA res grant, 67-69; asst dir, La Agr Exp Sta, 83, Ark Agr Exp Sta, 87; prof, Dept Entom, Univ Ark. *Mem:* Entom Soc Am; Am Mosquito Control Asn; Sigma Xi. *Res:* External parasites of domestic animals and poultry; disease-vector-host biological, ecological and control relationships; effects of insect parasites on animal hosts and resistance of animals to arthropod ectoparasites. *Mailing Add:* Dept Entom Rm 320 Agr Bldg Univ Ark Fayetteville AR 72701

STEELMAN, SANFORD LEWIS, b Hickory, NC, Oct 11, 22; m 45; c 2. CLINICAL PHARMACOLOGY. *Educ:* Lenoir-Rhyne Col, BS, 43; Univ NC, PhD(biol chem), 49. *Prof Exp:* Biochemist, Armour & Co Labs, 49-50, head endocrinol sect, 51-53, head dept biochem res, 53-56; assoc prof biochem, Baylor Col Med, 56-58; dir endocrinol, Merck Inst Therapeut Res, 68-70, sr clin assoc, Merck Sharp & Dohme Res Labs, Rahway, 70-76, dir clin pharmacol, 76-78, sr investr, 78-86; RETIRED. *Concurrent Pos:* Assoc prof, Postgrad Sch Med, Univ Tex, 56-58. *Mem:* AAAS; Am Chem Soc; Soc Exp Biol & Med; Endocrine Soc; Am Soc Exp Therapeut. *Res:* Isolation and biological and physicochemical properties of protein and peptide hormones; physiology, pharmacology and bioassay of steroidal hormones; analgesics; hypothalamic hormones; clinical pharmacology. *Mailing Add:* PO Box 5358 Viewmont Sta Hickory NC 28603-4002

STEEN, EDWIN BENZEL, b Wheeling, Ind, July 23, 01; m 27; c 2. PARASITOLOGY. *Educ:* Wabash Col, AB, 23; Columbia Univ, AM, 26; Purdue Univ, PhD(zool), 38. *Prof Exp:* Instr zool, Wabash Col, 23-25, actg head dept, 26-27; asst biol, NY Univ, 25-26; instr zool, Univ Cincinnati, 27-31; instr zool, Purdue Univ, 31-34, instr fish & game, Univ & Exp Sta, 34-36; tutor biol, City Col New York, 38-40; from asst prof to prof, 41-72, head dept, 64-65, EMER PROF BIOL, WESTERN MICH UNIV, 72- *Concurrent Pos:* Mem, Mich Bd Examnrs in Basic Sci, 61-63. *Mem:* AAAS; Sigma Xi; Am Inst Biol Sci. *Res:* Mammalian anatomy and physiology; medical and biological lexicography; zoology. *Mailing Add:* 1344 Orlando Dr Haslett MI 48840

STEEN, JAMES SOUTHWORTH, b Vicksburg, Miss, Oct 26, 40; m 60; c 2. BIOLOGY, IMMUNOLOGY. *Educ:* Delta State Col, BS, 62; Univ Miss, MS, 64, PhD(biol), 68. *Prof Exp:* From asst prof to assoc prof, 68-77, PROF BIOL, DELTA STATE COL, 77- *Mem:* Am Soc Microbiol. *Res:* Carbohydrate metabolism in bacteria; tissue transplantation and immunosuppression as related to the enhancement phenomenon in inbred strains of mice. *Mailing Add:* Dept of Biol Delta State Col Cleveland MS 38733

STEEN, LYNN ARTHUR, b Chicago, Ill, Jan 1, 41; m 63; c 2. SCIENCE WRITING. *Educ:* Luther Col, BA, 61; Mass Inst Technol, PhD(math), 65. *Hon Degrees:* DSc, Luther Col, 86. *Prof Exp:* From asst prof to assoc prof, 65-75, PROF MATH, ST OLAF COL, 75- *Concurrent Pos:* NSF sci faculty fel, Mittag-Leffler Inst, Sweden, 71-72; assoc ed, Am Math Monthly, 70-; ed, Math Mag, 76-80; contrib ed, Sci News, 77-82; proj dir, NSF comput grants, 78-83; assoc dir acad comput, St Olaf Col, 82-84; mem, Math Sci Educ Bd, Nat Res Coun, 85-; chmn, Conf Bd Math Sci, 88-90, Coun Sci Soc Pres, 89. *Honors & Awards:* Lester R Ford Award, Math Asn Am, 73 & 75. *Mem:* Am Math Soc; Math Asn Am (vpres, 80-81, pres, 85-86); AAAS; Conf Bd Math Sci; Coun Sci Soc Pres. *Res:* Analysis; function algebras; mathematical logic; general topology; mathematics education. *Mailing Add:* Dept Math St Olaf Coll Northfield MN 55057

STEEN, ROBERT FREDERICK, b Atlanta, Ga, Aug 19, 42; m 65; c 3. ELECTRICAL ENGINEERING, COMPUTER SCIENCE. *Educ:* Univ Rochester, BS, 64; NC State Univ, PhD(elec eng), 73. *Prof Exp:* Engr, IBM Corp, 64-70, staff engr, 70-71, adv engr, 73-77, TECH ADVR TO VPRES & CHIEF SCIENTIST, IBM CORP, 77- *Concurrent Pos:* Lectr & consult, Off Technol Assessment, US Cong, 78. *Mem:* Sr mem Inst Elec & Electronics

Engrs; AAAS; Asn Comput Mach. *Res:* Computer data communication; information sciences; data communication systems and protocols; coding theory. *Mailing Add:* Ctr D'etudes Et Recherches IBM France Dept 3778 Bp 30 La Gaude 06610 France

STEEN, STEPHEN N, b London, Eng, Sept 6, 23; US citizen; wid. MEDICINE. *Educ:* Mass Inst Technol, SB, 43; Univ Geneva, ScD(med biochem), 51, MD, 52; Am Bd Anesthesiol, dipl, 60. *Prof Exp:* Intern Abbot Hosp, Minneapolis, Minn, 53-54; res anesthesiol, Columbia-Presby Ctr, 54-56; instr, Albert Einstein Med Sch, 56-60; from asst prof to assoc prof, State Univ NY Downstate Med Ctr, 61-69; physician-in-chief anesthesia res, Cath Med Ctr Brooklyn & Queens, Inc, NY, 69-71; dir training & res, Harbor Gen Hosp, 71-75, PROF ANESTHESIOL & DIR RES, SCH MED, UNIV CALIF, LOS ANGELES, 71- *Concurrent Pos:* Actg dir, Delafield Hosp, NY, 56, attend anesthesiologist, 56-61; instr, Bronx Munic Hosp Ctr, 56-60 & Columbia Univ, 57-61; physician, Beth Israel Hosp, 57-61; attend anesthesiologist, Cent Islip State Hosp & St Francis Hosp, 58-, Misericordia Hosp, 58-62, Vet Admin 62- & St Barnabas Hosp, Bronx, NY, 56-61, dir anesthesiol, 60-61; vis attend anesthesiologist, Brooklyn Vet Admin Hosp, 61-66; from assoc vis anesthesiologist to vis anesthesiologist, Kings County Hosp Ctr, 61-; attend anesthesiologist, Harbor Gen Hosp & Torrance Mem Hosp, Univ Calif, Los Angeles; dir, Pulmonary Function Labs, Meditrina Med Ctr, 79-; physicians specialist anesthesia, Los Angeles City Col-Univ Southern Calif, 81- *Mem:* Fel Am Col Anesthesiol; sr mem Am Chem Soc; AMA; Am Soc Anesthesiol; Asn Am Med Cols. *Res:* Anesthesiology. *Mailing Add:* 1900 Ocean Blvd 1802 Long Beach CA 90802

STEENBERGEN, JAMES FRANKLIN, b Glasgow, Ky, May 11, 39; m 82. MICROBIOLOGY. *Educ:* Western Ky State Col, BS, 62; Ind Univ, MA, 65, PhD(microbiol), 68. *Prof Exp:* Res assoc marine microbiol, Ore State Univ, 68-69, vis asst prof microbiol, 69-70; asst prof, 70-75, assoc prof, 75-79, PROF MICROBIOL, SAN DIEGO STATE UNIV, 80- *Mem:* Am Soc Microbiol; World Mariculture Soc. *Res:* Microbial ecology and physiology; diseases of crustaceans and fish; invertebrate immunology. *Mailing Add:* Dept of Biol San Diego State Univ San Diego CA 92182

STEENBURG, RICHARD WESLEY, b Aurora, Nebr, Feb 3, 25; m 50; c 2. SURGERY. *Educ:* Harvard Univ, MD, 48; Am Bd Surg, dipl. *Prof Exp:* Assoc prof surg, Johns Hopkins Univ, 65-69; prof surg, Col Med, Univ Nebr, Omaha, 69; RETIRED. *Concurrent Pos:* Surgeon in chief, Baltimore City Hosps, 67-69. *Mem:* Soc Univ Surg. *Res:* General surgery; surgical endocrinology; vascular disease; renal physiology. *Mailing Add:* 6983 SE Harbor Circle Stuart FL 34996-1934

STEEN-MCINTYRE, VIRGINIA CAROL, b Chicago, Ill, Dec 3, 36; m 67. TEPHROCHRONOLOGY. *Educ:* Augustana Col, Ill, 59; Wash State Univ, MS, 65; Univ Idaho, PhD(geol), 77. *Prof Exp:* Asst geologist, George H Otto, Consult Geologist, Chicago, 59-61; jr geologist, Lab Anthrop, Wash State Univ, 64-66; phys sci technician, US Geol Surv, Denver, 70-75; res affil dept anthrop, Colo State Univ, 77-83; CONSULT TEPHROCHRONOLOGIST, 77- *Concurrent Pos:* Corresp mem, Int Asn Quaternary Res Comn Tephrochronology, 73-77; tephrochronologist, Valsequillo Early Man Proj, Mex, 66- & El Salvador Protoclassic Proj, 75-; guest lectr, NATO Advanced Studies Inst, Iceland, 80. *Mem:* Sigma Xi; Am Asn Quaternary Res. *Res:* Volcanic ash chronology; archaeologic site stratigraphy; petrography of friable Pleistocene deposits; tephra hydration dating; weathering of volcanic ejecta; human orgins; christian metaphysics, occult. *Mailing Add:* Box 1167 Idaho Springs CO 80452

STEENSEN, DONALD H J, b Clinton, Iowa, Apr 26, 29; m 54. FOREST ECONOMICS. *Educ:* Iowa State Univ, BS, 58; Duke Univ, MF, 60, PhD(forest econ), 65. *Prof Exp:* Asst prof forest econ & sampling, Auburn Univ, 60-65; asst prof, 65-73, ASSOC PROF FOREST ECON & MENSURATION, SCH FOREST RESOURCES, NC STATE UNIV, 73- *Mem:* Soc Am Foresters; Sigma Xi. *Res:* Forest mensuration. *Mailing Add:* Dept Forest Mgt-Wood Sci & Tech Box 8002 NC State Univ Sch Forest Resour Raleigh NC 27695

STEENSON, BERNARD O(WEN), b Crosby, NDak, Aug 2, 22; m 51; c 3. ELECTRICAL ENGINEERING. *Educ:* Ill Inst Technol, BS, 44; Calif Inst Technol, MS, 48, PhD(elec eng), 51. *Prof Exp:* Lab technician, Calif Inst Technol, 50-51; res physicist, Hughes Aircraft Co, 51-58, sr staff engr, 58-62, sr scientist, Space Systs Div, 62-71, Space & Commun Group, 71-88, chief scientist, 88-89; RETIRED. *Res:* Radar systems and communication satellites design and analysis. *Mailing Add:* 2878 W 230th St Torrance CA 90505

STEEPLES, DONALD WALLACE, b Hays, Kans, May 15, 45; m 67; c 2. GEOPHYSICS. *Educ:* Kans State Univ, BS, 69, MS, 70; Stanford Univ, MS, 74, PhD(geophys), 75. *Prof Exp:* Geophysicist seismol, US Geol Surv, 72-75; RES ASSOC GEOPHYS, STATE GEOL SURV KANS, 75- *Concurrent Pos:* Chmn, Geophysics Prog, Univ Kans; pres, Great Plains Geophysical, Kans; ed, Geophysics, J Soc of Explor Geophysicists. *Mem:* Am Geophys Union; Soc Explor Geophysicists; Seismol Soc Am. *Res:* Crust and upper mantle structure of central North America; kimberlites of Kansas; Pleistocene drainage of Kansas; passive seismic methods in exploration for geothermal energy; use of seismic methods for shallow exploration. *Mailing Add:* 2913 Westdale Rd Lawrence KS 66044

STEER, MARTIN WILLIAM, b Chelmsford, Essex, UK, Aug 19, 42; m 84; c 3. TIP GROWTH, SECRETION-ENDOCYTOSIS. *Educ:* Univ Bristol, UK, BSc Hons, 63, DSc(plant cell biol), 86; Queen's Univ Belfast, UK, PhD(cell biol), 66. *Prof Exp:* Res assoc bot, Univ Wis-Madison, 66-68; res asst bot, Queen's Univ Belfast, 63-66, lectr, 68-79, reader, 79-84; PROF & DEPT HEAD BOT, UNIV COL DUBLIN, 84- *Concurrent Pos:* Mem, Finance Comt & Cell Biol Comt, Soc Exp Biol, 77-80; consult, Gallagher's Tobacco Co, Belfast, 80-84; assoc ed, J Exp Bot, 83-89; dir, Electron Micros Lab, Univ Col Dublin, 86-90. *Mem:* Fel Royal Micros Soc; Am Soc Cell Biol; Soc Exp Biol. *Res:* Structure and function of plant cells, using mainly light and electron microscope techniques to record effects of physiological and biochemical perturbations; quantitative and stereological image analysis; secretory and endocytotic mechanisms and pollen tube tip growth. *Mailing Add:* Dept Bot Univ Col Dublin Nat Univ Ireland Dublin 4 Ireland

STEER, MAX DAVID, b New York, NY, June 14, 10; m 42. SPEECH & HEARING SCIENCES. *Educ:* Long Island Univ, 32, LLD, 57; Univ Iowa, MA, 33, PhD(psychol), 38. *Prof Exp:* Asst speech path, Univ Iowa, 33-35; from instr to prof speech sci, 35-70, dir, Speech & Hearing Clin, 46-70, head dept 63-70, Hanley distinguished prof, 70-77, DISTINGUISHED PROF EMER AUDIOL & SPEECH SCI, PURDUE UNIV, WEST LAFAYETTE, 77- *Concurrent Pos:* Consult, State of Ind Hearing Comn, 66-, US Off Educ, NIH, NSF & Ind State Training Sch Ment Retarded; consult, Neurol & Sensory Dis Control Prof, USPHS, Pan-Am Health Orgn, 72-, Latin Am Fedn Logopedics, Phoniatrics, Audiol, 72- & Univ Bogota, 74; mem nat res adv comt, Bur Educ Handicapped, US Off Educ, 72-; consult & vis lectr, Nat Rehab Inst, Panama, 74. *Mem:* AAAS; Acoust Soc Am; Am Psychol Asn; fel Am Speech & Hearing Asn (vpres, 49, pres, 51); Int Asn Logopedics & Phoniatrics (vpres, 63-65, 71-77 & 80-83); Sigma Xi. *Res:* Speech disorders and acoustics; audiology; clinical psychology; neurology; physiology and psychology of communication. *Mailing Add:* 342 W View Circle West Lafayette IN 47906

STEER, MICHAEL BERNARD, b Brisbane, Australia, Apr 26, 55; m 80; c 2. MICROWAVE ENGINEERING, ANALOG COMPUTER AIDED DESIGN. *Educ:* Univ Queensland, BE, 78, PhD(elec eng), 83. *Prof Exp:* PROF ELEC ENG, NC STATE UNIV, 83- *Concurrent Pos:* NSF presidential young investr award, 87. *Mem:* Sr mem Inst Elec & Electronic Engrs. *Res:* Simulation, measurement and computer aided design of nonlinear microwave circuits, high speed multi-chip modules, parameter extraction, and millimeter-wave quasi-optical techniques; author of more than 25 publications. *Mailing Add:* Elec & Computer Eng Dept NC State Univ Raleigh NC 27695-7911

STEER, RONALD PAUL, b Regina, Sask, Mar 7, 43; m 64; c 2. LASER CHEMISTRY, PHOTOCHEMISTRY. *Educ:* Univ Sask, BA, 64, PhD(chem), 68. *Prof Exp:* USPHS fel, Univ Calif, Riverside, 68-69; from asst prof to assoc prof, 69-78, PROF CHEM, UNIV SASK, 78- *Concurrent Pos:* Vis fel, Univ Southampton, 75-76; vis scientist, Nat Res Coun Can, 84-85. *Mem:* Fel Chem Inst Can; Int Photochem Soc. *Res:* Laser chemistry, photochemistry, photophysics and spectroscopy of small molecules; fluorescence probe techniques in biological systems; SCF MO calculations of excited state potential surfaces. *Mailing Add:* Dept Chem Univ Sask Saskatoon SK S7N 0W0 Can

STEERE, WILLIAM CAMPBELL, botany; deceased, see previous edition for last biography

STEERS, EDWARD, JR, b Bethlehem, Pa, May 21, 37; m 57; c 3. BIOLOGICAL CHEMISTRY. *Educ:* Univ Pa, AB, 59, PhD(biol), 63. *Prof Exp:* Staff fel biochem, 63-65, RES BIOLOGIST, LAB CHEM BIOL, NAT INST ARTHRITIS, METABOLISM & DIGESTIVE DIS, 65-, DEP DIR DIV INTRAMURAL RES, 85- *Concurrent Pos:* Lectr, George Washington Univ, 66- *Mem:* Am Soc Biol Chem; Am Soc Cell Biol. *Res:* Protein chemistry; relationship of structure to function in proteins; cell organelles. *Mailing Add:* Div Intramural Res Bldg 10 Rm 9N222 Nat Inst Diabetes & Digestive & Kidney Dis Washington DC 20052

STEEVES, HARRISON ROSS, III, b Birmingham, Ala, July 2, 37; m 57; c 4. HISTOCHEMISTRY, TAXONOMY. *Educ:* Univ of the South, BS, 58; Univ Va, MS, 60, PhD(biol), 62. *Prof Exp:* Instr zool, Univ Va, 62; fel histol, Med Ctr, Univ Ala, 62-65, instr, 65-66; asst prof histol & histochem, 66-68, ASSOC PROF ZOOL, VA POLYTECH INST & STATE UNIV, 68- *Honors & Awards:* Andrew Fleming Award Biol Res, 61. *Mem:* AAAS. *Res:* Invertebrate histochemistry; taxonomy, ecology and physiology of cave crustaceans and histochemistry of digestion in these forms. *Mailing Add:* Dept Zool Va Polytech Inst & State Univ Blacksburg VA 24061

STEEVES, JOHN DOUGLAS, b Calgary, Alta, Mar 25, 52. NEUROBIOLOGY, ZOOLOGY. *Educ:* Univ Manitoba, BSc, 73, PhD(physiol), 79. *Prof Exp:* Fel neurophysiol, dept physiol, Univ Alta, 78-79; asst prof, 79-85, ASSOC PROF NEUROBIOL, DEPT ZOOL, UNIV BC,85-, ASSOC PROF, GRAD PROG NEUROSCI, 85- *Concurrent Pos:* Consult, Can Broadcasting Corp, 81-83; vis prof, Jinan Univ, Guagzhou (Canton), People's Repub China, 85; assoc mem, dept anat, Univ BC, 87- *Mem:* Soc Neurosci; Can Asn Neurosci. *Res:* Central nervous system mechanisms which initiate and modify motor behavior in vertebrates, specifically birds and mammals; spinal cord injury and repair processes. *Mailing Add:* Dept Zool Univ BC 6720 University Blvd Vancouver BC V6T 1Z4 Can

STEEVES, RICHARD ALLISON, b Fredericksburg, Va, Feb 2, 38; m 65; c 3. ONCOLOGY, VIROLOGY. *Educ:* Univ Western Ont, MD, 61; Univ Toronto, PhD(med biophys), 66. *Prof Exp:* From sr cancer res scientist to assoc cancer res scientist, Roswell Park Mem Inst, 67-72; assoc prof develop biol & cancer, 72-77, vis assoc prof genetics, Albert Einstein Col Med, 77-80; asst prof, 80-83, ASSOC PROF HUMAN ONCOL, UNIV WIS-MADISON, 83- *Concurrent Pos:* Nat Cancer Inst Can fel, Dept Biol, McMaster Univ, 66-67; resident therapeut radiol, Albert Einstein Col Med, 77-80. *Mem:* Am Asn Cancer Res; Radiol Soc NAm; Am Endocurietheray Soc; Am Soc Therapeut Radiol. *Res:* Interaction of radiation and hyperthermin in cancer therapy; genetic control of target cells for murine leukemia viruses. *Mailing Add:* Dept Radiation Oncol Univ Wis-Madison 600 Highland Ave Madison WI 53792

STEEVES, TAYLOR ARMSTRONG, b Quincy, Mass, Nov 29, 26; m 56; c 3. BOTANY. *Educ:* Univ Mass, BS, 47; Harvard Univ, AM, 49, PhD(biol), 51. *Hon Degrees:* DCnL, Emmanuel Col, Saskatoon. *Prof Exp:* Jr fel, Soc Fels, Harvard Univ, 51-54, asst prof bot, Biol Labs, 54-59; from assoc prof to prof, 59-85, head dept, 76-81, RAWSON PROF BIOL, UNIV SASK, 85- *Concurrent Pos:* Ed, Bot Gazette, 69-75, Can J Bot, 79- *Honors & Awards:* Lawson Medal, Can Bot Asn, 78. *Mem:* Bot Soc Am; fel Royal Soc Can; Int Soc Plant Morphologists; Soc Develop Biol; Can Bot Asn (pres, 72-73). *Res:* Morphogenesis of vascular plants; plant tissue culture and growth hormones; plant architecture. *Mailing Add:* Dept Biol Univ Sask Saskatoon SK S7N 0W0 Can

STEFAN, HEINZ G, b Landskron, CSR, June 12, 36; m 62; c 3. HYDROMECHANICS, WATER RESOURCES ENGINEERING. *Educ:* Munich Tech Univ, Dipl Ing, 59; Univ Toulouse, Ing Hydraulicien, 60, DrIng(hydromech), 63. *Prof Exp:* Res fel hydraul, Univ Minn, 63-64; chief engr, Inst Water Resources & Hydraul Eng, Tech Univ, Berlin, 65-67; from asst prof to assoc prof, 67-77, PROF CIVIL ENG, UNIV MINN, MINNEAPOLIS, 77- *Concurrent Pos:* Consult, Power Co; UN expert, India, 81; lectr, foreign countries; assoc dir, St Anthony Falls Hydrol Lab, 74- *Mem:* Am Soc Civil Engrs; Am Geophys Union; Int Asn Hydrol Res; Am Water Resources Asn; Int Water Resources Asn. *Res:* River and lake hydromechanics; thermal pollution; aquatic systems; water resources; hydraulic structures. *Mailing Add:* Dept Civil Eng Univ Minn Minneapolis MN 55455

STEFANAKOS, ELIAS KYRIAKOS, b Athens, Greece, Sept 30, 40; m 68; c 1. ENGINEERING SCIENCE, SOLAR ENERGY. *Educ:* Wash State Univ, BS, 64, MS, 65, PhD(eng sci), 69. *Prof Exp:* Res asst elec eng, Wash State Univ, 66-68; from asst prof to assoc prof elec eng, Univ Idaho, 68-77; assoc prof, 77-80, PROF ELEC ENG, NC A&T STATE UNIV, 80- *Concurrent Pos:* Fel, 71-72, int travel grant, NSF, 78; vis prof, Greek Atomic Energy Comn, 75. *Mem:* AAAS; Inst Elec & Electronics Engrs; Am Soc Eng Educ; Int Solar Energy Soc. *Res:* Semiconductor materials and devices; photovoltaics; solar energy utilization. *Mailing Add:* 19202 Blount Rd Lutz FL 33549

STEFANCSIK, ERNEST ANTON, b Brooklyn, NY, Sept 10, 23; m. ORGANIC CHEMISTRY. *Educ:* St John's Univ, NY, BS, 43; NY Univ, MS, 47, PhD(chem), 53. *Prof Exp:* Chemist, Am Cyanamid Co, 43-46; teaching fel, NY Univ, 47-49, asst, 50-52; from res chemist to res assoc, Pigments Dept, E I du Pont de Nemours & Co, Inc, Newark, 52-82; RETIRED. *Mem:* AAAS; Am Chem Soc; Sigma Xi. *Res:* Physics and chemistry of organic pigments. *Mailing Add:* 162 Voorhees Corner Rd Flemington NJ 08822

STEFANESCU, DORU MICHAEL, b Sibiu, Romania, Nov 15, 42; m 77; c 2. METAL CASTING, SOLIDIFICATION OF ALLOYS. *Educ:* Polytech Inst Bucharest, BEng, 65, DEng, 73. *Prof Exp:* Jr researcher metal casting, Technol Inst Hot Processes, Bucharest, 68-70, sr res, 70-72, group head, 72-80; vis prof , Univ Wis-Madison, 80; assoc prof metall, 80-84, PROF METALL, UNIV ALA, 84-, UNIV RES PROF, 88- *Concurrent Pos:* Asst prof, Polytech Inst Bucharest, 73-80. *Mem:* Metall Soc; Am Soc Metals; Am Foundrymens Soc. *Res:* Solidification processing; physical chemistry of surface and interface reactions; metal matrix composites, superconductivity; numerical modeling of solidification; thermal analysis of alloys; cast metals technology. *Mailing Add:* Dept Metall Eng Col Eng Univ Ala PO Box G Tuscaloosa AL 35487

STEFANI, ANDREW PETER, b Cyprus, July 10, 26; US citizen; m 55; c 1. ORGANIC CHEMISTRY, PHYSICAL CHEMISTRY. *Educ:* Mich State Univ, BA, 56; Univ Colo, PhD(chem), 60. *Prof Exp:* Res assoc chem, State Univ NY Col Forestry, Syracuse Univ, 60-62; asst prof, Purdue Univ, 62-63; from asst prof to assoc prof, 63-68, PROF CHEM, UNIV MISS, 68-, CHMN DEPT, 77- *Concurrent Pos:* Res grants, Res Corp, 64, Petrol Res Fund & NSF, 64-68. *Mem:* AAAS; Am Chem Soc. *Res:* Chemical kinetics; free radical reactivity; solvent effects in chemical reactions; high pressure chemistry. *Mailing Add:* Box 174 University MS 38677

STEFANI, STEFANO, b Trieste, Italy, Apr 19, 29; US citizen; m 58; c 2. RADIOTHERAPY, RADIOBIOLOGY. *Educ:* Univ Trieste, BS, 48; Univ Perugia, MD, 54. *Prof Exp:* Intern, Univ Trieste, 54-55; resident radiol, Med Sch, Univ Padua, 55-56, asst prof, 56-57; resident radiother, G Roussy Inst, Paris, 57-59; asst prof, Med Sch, Univ Paris, 59-60; resident, Univ Md Hosp, 61-62; res radiobiologist, Med Sch, Northwestern Univ, 62-63; from instr to asst prof radiol, Stritch Sch Med, Loyola Univ Chicago, 63-69; CHIEF THERAPEUT RADIOL SERV, VET ADMIN HOSP, HINES, 69-; PROF RADIOTHER, RUSH MED SCH, CHICAGO, 77-; DIR RADIOTHER DEPT, MT SINAI HOSP MED CTR, CHICAGO, 73- *Concurrent Pos:* Fels radiother, Curie Found, Paris, 57 & radiobiol, Nat Inst Nuclear Sci & Technol, Saclay, France, 58-59; consult, WHO, Geneva, 61, Argonne Nat Lab, AEC, 64-65; clin investr, Vet Admin Hosp, Hines, 64-69, staff physician, 67-68; abstractor, J Surg, Gynec & Obstet, 66-; prof radiother, Chicago Med Sch, 69-77; consult radiother, 72; dir, Radiother Dept, Northwest Community Hosp, Arlington Heights, 75-78. *Mem:* AAAS; AMA; Radiation Res Soc; Transplantation Soc; Radiol Soc NAm. *Res:* Radiation potentiators; total body irradiation and radioprotection; physiology and pathology of lymphocytes; bone marrow radiosensitivity and transfusion. *Mailing Add:* 5617 S Dorchester 7N Chicago IL 60637

STEFANOU, HARRY, b New York, NY, June 16, 47; m 69; c 1. POLYMER PHYSICS. *Educ:* City Col New York, BS, 69; City Univ New York, PhD(phys chem), 73. *Prof Exp:* Res assoc polymer physics, Princeton Univ, 72-73; sr res chemist, Pennwalt Corp, 73-79, res scientist, 79-82, proj leader, 82, tech mgr, 82-85; mgr mat res, 85-86, MGR POLYMER APPLN, ARCO CHEM, 86- *Mem:* AAAS; Am Chem Soc; Am Phys Soc; Sigma Xi. *Res:* The areas of polymer crystallization kinetics; polymer viscoelasticity and solution thermodynamics; piezo- and pyro-electric polymers. *Mailing Add:* 201 Country Gate Rd Strafford PA 19087-5321

STEFANSKI, RAYMOND JOSEPH, b Buffalo, NY, July 22, 41; m 69; c 1. PHYSICS. *Educ:* State Univ NY Buffalo, BA, 63; Yale Univ, PhD(physics), 69. *Prof Exp:* Res assoc physics, Yale Univ, 68-69; physicist, Fermi Nat Accelerator Lab, 69-90; STAFF PHYSICIST, SUPERCONDUCTING SUPERCOLLIDER LAB, 90- *Mem:* Am Phys Soc; Opers Res Soc Am; Europ Phys Soc. *Res:* High energy physics and cosmic rays; neutrino interactions. *Mailing Add:* Superconducting Supercollider Lab 2550 Beckleymead Ave Mail Sta 2001 Dallas TX 75237

STEFANSSON, BALDUR ROSMUND, b Vestfold, Man. PLANT BREEDING. *Educ:* Univ Man, BSA, 50, MS, 52, PhD(plant sci), 66. *Prof Exp:* Res assoc, 52-66, assoc prof, 66-74, PROF PLANT SCI, UNIV MAN, 74- *Honors & Awards:* Grindley Medal, Agr Inst Can, 78. *Mem:* Fel Agr Inst Can. *Res:* Plant breeding and related research with oilseed crops, formerly soybeans, currently rapeseed; compositional changes in rapeseed, including low erucic acid, low glucosinolate and low fibre content, induced by breeding. *Mailing Add:* Dept Plant Sci Univ Man Winnipeg MB R3T 2N2 Can

STEFFAN, WALLACE ALLAN, b St Paul, Minn, Aug 10, 34; m 66; c 1. ENTOMOLOGY. *Educ:* Univ Calif, Berkeley, BS, 61, PhD(entom), 65. *Prof Exp:* Entomologist, Bernice P Bishop Mus, 64-85, head diptera sect, 68-85; DIR, IDAHO MUS NATURAL HIST, 85 - *Concurrent Pos:* NIH grant, 67-72; partic island ecosyst int res prog, US Int Biol Prog, 70; mem affiliated grad fac, Univ Hawaii; coop scientist, USDA, 84 -; res assoc, Bernice P Bishop MVS, 85 -; chmn, systs resources comt, Entom Soc Am, 85-87; mem, Axillary Fac, Idaho State. *Honors & Awards:* Grace M Griswold Lectr, Cornell Univ, 70. *Mem:* AAAS; Am Entom Soc; Am Mosquito Control Asn; Asn Sci Mus Dirs. *Res:* Systematics of Culicidae and Sciaridae; ecology of Culicidae; cytogenetics of Sciaridae; information management in museums. *Mailing Add:* 907 Yukon Dr Fairbanks AK 99709

STEFFEK, ANTHONY J, b Milwaukee, Wis, Aug 6, 35; m 60; c 3. DENTISTRY, PHARMACOLOGY. *Educ:* Marquette Univ, DDS, 62, MS, 63, PhD(physiol), 65. *Prof Exp:* Res assoc pharmacol, Nat Inst Dent Res, 65-70; ASST PROF & DIR DIV DEVELOP BIOL, WALTER G ZOLLER MEM DENT CLIN, UNIV CHICAGO, 75-, RES ASSOC, DEPT ANAT, 75- *Concurrent Pos:* Am Dent Asn grant pharmacol, Nat Inst Dent Res, 65-; vis prof pharmacol, Charles Univ, Prague, Czechoslovakia. *Mem:* Am Dent Asn; Am Physiol Soc. *Res:* Metabolism and pharmacological disposition of environmental agents producing experimentally-induced oral-facial malformations, specifically cleft lip and palate; teratology. *Mailing Add:* 4726 North Dover Chicago IL 60640

STEFFEN, ALBERT HARRY, b Menomonee Falls, Wis, May 24, 14; m 39; c 2. FOOD CHEMISTRY. *Educ:* Univ Wis, BS, 40. *Prof Exp:* Anal food chemist fats & oils, Armour & Co, 40-46, res food chemist, 46-52, fat & oil researcher, 52-56, asst prod mgr, Refining Div, 56-57, tech dir & qual control mgr, Lookout Oil & Refining Co Div, 57-67, prod mgr by-prod & mem staff, Cent Qual Control, Nebr, 67-68, mem cent qual assurance staff, Ill, 68-71, tech specialist, qual assurance staff, ariz, 71-79; CONSULT, 79- *Mem:* Am Oil Chem Soc; Am Soc Qual Control; Inst Food Technol. *Res:* Meat and meat by-products; fats and oils. *Mailing Add:* 4105 W Mission Ln Phoenix AZ 85051

STEFFEN, DANIEL G, b St Louis, Mo, Feb 9, 48; div; c 4. RESEARCH ADMINISTRATION. *Educ:* Univ Notre Dame, BS, 70; Univ Mo, PhD(physiol), 74. *Prof Exp:* Postdoctorate swine lipid metab, Shell Develop Co, 74-76; res scientist nutrit, Gen Foods Corp, 76-90; MGR SCI REL, KRAFT GEN FOODS, 90- *Mem:* Inst Food Technologists; Am Inst Nutrit; Sigma Xi. *Res:* Age and diet effects on physiological systems; application of technical strategies to development of regulatory policies. *Mailing Add:* Kraft Gen Foods 250 North St White Plains NY 10625

STEFFEN, JUERG, b Zurich, Switz, Dec 11, 42; m 70; c 4. LASERS, LASER-MATERIALS PROCESSING. *Educ:* Swiss Fed Inst Technol, Zurich, dipl, 66; Univ Berne, Switz, PhD(physics), 70. *Prof Exp:* Group leader lasers, Inst Appl Physics, Univ Berne, 70-72; head laser appln, Res Inst, Pierres Holding SA, 72-74; res & develop mgr, Lasag AG, Thun, 74-81; group leader lasers & appln, Asulab SA, Neuchâtel, 81-89; MANAGING DIR LASERS & APPLN, OERLIKON-PRC LASER SA, GLAND, 89- *Concurrent Pos:* Vis prof, Inst Appl Physics, Univ Darmstadt, 78-84, Physics Dept, Univ Kaiserslautern, 81-82. *Mem:* Swiss Phys Soc; Europ Phys Soc; fel Optical Soc Am; Int Soc Optical Eng. *Res:* Development of industrial Nd-YAG- and C02-lasers; development of applications procedures of lasers in materials processing. *Mailing Add:* Le Bon Pirouz Vich CH-1267 Switzerland

STEFFEN, ROLF MARCEL, b Basel, Switz, June 17, 22; m 49; c 2. NUCLEAR PHYSICS. *Educ:* Cantonal Col, Zurich, BS, 41; Swiss Fed Inst Technol, MS, 46, PhD, 48. *Prof Exp:* From asst prof to prof, 49-82, EMER PROF PHYSICS, PURDUE UNIV, WEST LAFAYETTE, 82- *Concurrent Pos:* Vis staff mem, Los Alamos Sci Lab, 64-; Sigma Xi Res Award, 64. *Mem:* Fel Am Phys Soc; Swiss Phys Soc. *Res:* Nuclear spectroscopy; angular correlations of nuclear radiation; influence of extranuclear fields on angular correlation; beta decay; nuclear structures studies by muonic x-rays; muonic atoms; hyperfine fields; heavy ion nuclear reactions. *Mailing Add:* PO Box 220 Tesuque NM 87574

STEFFENS, GEORGE LOUIS, b Bryantown, Md, June 13, 30; m 59; c 2. PLANT PHYSIOLOGY, PLANT GROWTH REGULATION. *Educ:* Univ Md, BS, 51, MS, 53, PhD(agron), 56. *Prof Exp:* Agron biochemist, Gen Cigar Co, Inc, Pa, 58-61; plant physiologist, Coastal Plain Exp Sta, USDA, Ga, 61-63, plant physiologist, 63-74, lab chief, 74-82, PLANT PHYSIOLOGIST, FRUIT LAB, PLANT SCI INST, BELTSVILLE AGR RES CTR, AGR RES SERV, USDA, 82- *Concurrent Pos:* Vis prof, agron dept, Cornell Univ, 81-82; chmn, Plant Growth Reg Soc Am, 86-87. *Honors & Awards:* Philip Morris Award, 72; Cert Merit, USDA, 82. *Mem:* Fel AAAS; Am Soc Plant Physiologists; Am Chem Soc; Scand Soc Plant Physiologists; Plant Growth Regulator Soc Am; Am Soc Hort Sci. *Res:* Natural and synthetic plant growth regulating chemicals, their development and physiological effects, especially tree crops (apple). *Mailing Add:* Fruit Lab Plant Sci Inst Beltsville Agr Res Ctr USDA Beltsville MD 20705

STEFFENSEN, DALE MARRIOTT, b Salt Lake City, Utah, Apr 17, 22; m 50, 70; c 3. GENETICS. *Educ:* Univ Calif, Los Angeles, AB, 48; Univ Calif, Berkeley, PhD(genetics), 52. *Prof Exp:* Asst res geneticist, Univ Calif, 52; from assoc geneticist to geneticist, Brookhaven Nat Lab, 52-61; prof genetics & develop, PROF AGRON, UNIV ILL, URBANA, 84-, PROF CELL BIOL, 87- *Concurrent Pos:* USPHS spec fel, Naples, Italy, 67-68. *Mem:* Genetics Soc Am; Am Soc Cell Biol; Soc Develop Biol. *Res:* Nuclear structure, developmental genetics and cell biology; biochemistry of nuclear structures; gene mapping on chromosomes by RNA-DNA hybridization; molecular development of maize. *Mailing Add:* Dept Cell Biol 505 S Goodwin Ave, 17 Morris Hall Urbana IL 61801

STEFFEY, EUGENE P, b Reading, Pa, Oct 27, 42; m 67; c 4. ANESTHESIOLOGY. *Educ:* Univ Pa, VMD, 67; Univ Calif, Davis, PhD(comp path), 73. *Prof Exp:* VET ANESTHESIOL ASST TO PROF & CHMN DEPT SURG, SCH VET MED, UNIV CALIF, DAVIS, 74- *Concurrent Pos:* Intern, Univ Calif, Davis, 68; pvt pract, Reading, Pa, 69. *Mem:* Am Physiol Soc; Am Soc Pharmacol & Exp Therapeut; Int Anesthesia Res Soc; Am Soc Anesthesiol; Am Vet Med Asn; Am Col Vet Anesthesiol. *Res:* Comparative anesthesiology; pharmacology of inhalation anesthetics and opioids; comparative cardiopulmonary pathophysiology. *Mailing Add:* Dept Surg Sch Vet Med Univ Calif Davis CA 95616

STEFFEY, ORAN DEAN, b Billings, Okla, May 19, 21; m 43; c 2. INDUSTRIAL HYGIENE, HEALTH PHYSICS. *Educ:* Phillips Univ, AB, 48; Okla State Univ, MS, 50, PhD(bot), 53. *Prof Exp:* Asst zool, Phillips Univ, 42, asst bot & zool, 47-48; res assoc, Res Found, 50-53, from instr to asst prof bot & plant path, 53-56; radiol health physics officer, Radiation Lab, Continental Oil Co, 56-72, indust hygienist, med dept, Conoco Inc, 72-85; RETIRED. *Concurrent Pos:* Mem, Okla State Radiation Adv Comt, 72-85. *Res:* Variations of the fruits of Quercus Macrocarpa; development of an autoradiographic technique for use in botanical investigations; cytomorphogenetic studies in sorghum. *Mailing Add:* 2017 N 6th Ponca City OK 74601

STEFFGEN, FREDERICK WILLIAMS, b San Diego, Calif, Nov 21, 26; m 49; c 2. FUEL SCIENCE, SURFACE CHEMISTRY. *Educ:* Stanford Univ, BS, 47; Northwestern Univ, Evanston, MS, 49; Univ Del, PhD(org chem), 53. *Prof Exp:* Res chemist, Cutter Labs, Calif, 49-50; res chemist, Esso Res & Eng Co, Standard Oil Co (NJ), La, 52-56; sr res chemist, Union Oil Res, Union Oil Co Calif, 56-60; res assoc, Richfield Res Ctr, Atlantic Richfield Co, 60-69; mgr process develop, Antox, Inc, WVa, 70-71; supvry res chemist, Pittsburgh Energy Res Ctr, US Bur Mines, 71-75, res supvr chem, Pittsburgh Energy Res Ctr, US Energy Res & Develop Admin, 75-77 & US Dept of Energy, 77-79, DIR, ADVAN RES CONTRACTS PROJ, MGT DIV, PITTSBURGH ENERGY TECH CTR, DEPT ENERGY, 79- *Mem:* Am Chem Soc; Am Inst Chem Engr; Catalysis Soc. *Res:* Heterogeneous catalysis, processes and catalyst preparation; conversions to produce low sulfur, clean fuels from coal, petroleum and organic wastes; support of grants and contracts in advanced coal research. *Mailing Add:* 2363 Caminito Eximio San Diego CA 92107-1524

STEFKO, PAUL LOWELL, b Middletown, Conn, June 29, 15; m 40; c 1. EXPERIMENTAL SURGERY, PHARMACOLOGY. *Educ:* Univ Va, BS, 44; Cornell Univ Med Col, MD, 47. *Prof Exp:* Tech surg asst, NY Hosp, New York, 38-43; exp surgeon & pharmacologist, Hoffman-La Roche, Inc, 47-61, sr pharmacologist & exp surgeon, 61-77; RETIRED. *Concurrent Pos:* Exp surg consult, USAEC & Off Naval Res & Develop, 40-43, cert, 45. *Mem:* AAAS; Sigma Xi; Am Indust Hyg Asn; Am Asn Lab Animal Sci; NY Acad Sci. *Res:* Vascular surgery; blood vessel transposition; gastroenterology; spasmolytics; antitussives. *Mailing Add:* Hoffmann-La Roche Inc 38 Blue Hill Rd Clifton NJ 07013

STEG, L(EO), b Vienna, Austria, Mar 30, 22; nat US; m 47; c 3. ENGINEERING MECHANICS, PHYSICS. *Educ:* City Col New York, BS, 47; Univ Mo, MS, 48; Cornell Univ, PhD(mech, math, physics), 51. *Prof Exp:* Chief engr, Fed Design Co, 46-47; instr mech eng, Univ Mo, 47-48; from instr to asst prof mech & mat, Cornell Univ, 48-55; syst engr, Missile & Space Div, Gen Elec Co, 55-56, mgr, GE Space Sci Lab, 56-80; sr vpres, Res Inst Div, Univ City Sci Ctr, 81-82, sci & pub fel, Brooking Inst, Wash DC, 83-84; PRES, STEG, RAY & ASSOC, 82- *Concurrent Pos:* Consult, Fed Design Co, 48-53, Lincoln Lab, Mass Inst Technol, 52-55, Ramo-Wooldridge Corp, 54-55 & US Dept Defense, 67-71; res engr, Boeing Airplane Co, 54; ed-in-chief, Am Inst Aeronaut & Astronaut Jour, 62-67; adj prof, Drexel Univ, 70-; mem, Bd Managers, Franklin Inst, 72-82; chmn, Comt Space Res Working Group-Mat Sci Space, 75-80; chmn, Gordon Conf Fundamentals of Cybernetics, 84; chmn, Gordon Conf Gravitational Effects Mat Separation & Living Systs, 85; lectr, United States Info Agency, 83-84. *Mem:* Fel AAAS; Am Inst Aeronaut & Astronaut; Sigma Xi. *Res:* Nonlinear mechanics; stability in dynamical systems; applied mechanics; space and reentry technology and applications; research administration; science policy. *Mailing Add:* 1616 Hepburn Dr Villanova PA 19085

STEGELMANN, ERICH J, b Litjenburg, Ger, Mar 28, 14; m 51; c 3. ELECTROOPTICS. *Educ:* Tech Univ, Berlin, Dipl Ing, 38, Dr Ing, 39. *Prof Exp:* Develop engr, Rheinmetall-Borsig, Ger, 44-45; scientist, Tech Off USSR, Berlin, 46-48; sr res engr, Can Aviation Electronics, Montreal, 52-54; design & res specialist, Lockheed Calif Corp, 54-60, sr res scientist, 60-63, mem tech staff, Northrop Space Lab, 63-64; sr tech specialist, Space Div, NAm Aviation, Inc, Calif, 64-68, mem tech staff, Autonetics Div, NAm Rockwell Corp, 68-70; mem tech staff, Electro-Optical Labs, Hughes Aircraft Co, Culver City, 72-80. *Mem:* Optical Soc Am. *Res:* Space physics; propagation of light in atmosphere. *Mailing Add:* 18559 Chatsworth St Northridge CA 91326-3141

STEGEMAN, GEORGE I, b Edinburgh, Scotland, Aug 4, 42; Can citizen; m 67; c 3. NONLINEAR OPTICS. *Educ:* Univ Toronto, BSc, 65, MSc, 66, PhD(physics), 69. *Prof Exp:* Prof physics, Univ Toronto, 69-80; prof optics, Univ Ariz, 80-90; PROF PHYSICS, UNIV CENT FLA, 90- *Concurrent Pos:* Vis prof, Univ Calif, Irvine, 79-80 & Stanford Univ, 80. *Honors & Awards:* Hertzberg Medal, Can Asn Physicists, 80. *Mem:* Fel Optical Soc Am; Can Asn Physicists; Inst Elec & Electronics Engrs; Am Phys Soc. *Res:* Propagation charcteristics of various waves guided by surfaces or films, their nonlinear interactions with matter and one another and their potential applications. *Mailing Add:* Creol Univ Cent Fla Orlando FL 32826

STEGEN, GILBERT ROLLAND, b Long Beach, Calif, Aug 19, 39; m 66. FLUID MECHANICS. *Educ:* Mass Inst Technol, BS, 61; Stanford Univ, MS, 62, PhD(aeronaut, astronaut), 67. *Prof Exp:* Develop engr, United Tech Ctr, United Aircraft Corp, 62-63; vis sr res fel, Col Aeronaut Eng, 67; asst res engr, Univ Calif, San Diego, 67-69; asst prof civil eng, Colo State Univ, 69-70; asst prof geol & geophys sci, Princeton Univ, 70-74; div mgr & sr res scientist, Flow Res Co, 74-76; mgr, 76-80, ASST VPRES, SCI APPLICATIONS INC, 80- *Mem:* Am Meteorol Soc; Sigma Xi; Am Phys Soc; Am Geophys Union. *Res:* Experimental fluid mechanics; experimental studies in atmospheric and oceanic turbulence. *Mailing Add:* Sci Applns Inc 13400 B Northup Way Apt 36 Bellevue WA 98005

STEGER, RICHARD WARREN, b Richmond, Calif, Aug 4, 48; m 79; c 3. NEUROENDOCRINOLOGY, NEUROCHEMISTRY. *Educ:* Univ Wyo, BA, 70, PhD(physiol), 74. *Prof Exp:* Asst obstet & gynec, Wayne State Univ, 74-77; asst physiol, Mich State Univ, 77-79; asst prof gynec, Health Sci Ctr, Univ Tex, 80-85; assoc prof physiol, 85-90, PROF PHYSIOL, MED SCH, SOUTHERN ILL UNIV, 90- *Mem:* Endocrine Soc; Soc Study Reproduction; Am Physiol Soc; Sigma Xi; Int Soc Neuroendocrinology; Soc Neuroscience. *Res:* Transduction of environmental signals into hormonal signals; age related changes in neuroendocrine function; prolactin & gonadotropin regulation; diabetes and neuroendocrine function. *Mailing Add:* Dept Physiol Southern Ill Univ Med Sch Carbondale IL 62901

STEGINK, LEWIS D, b Holland, Mich, Feb 8, 37; m 62; c 2. BIOLOGICAL CHEMISTRY. *Educ:* Hope Col, BA, 58; Univ Mich, MS & PhD(biol chem), 63. *Prof Exp:* Fel biochem, 63-65, asst prof pediat, 65-71, asst prof pediat & biochem, 68-71, assoc prof, 71-76, PROF PEDIAT & BIOCHEM, UNIV IOWA, 76- *Honors & Awards:* Mead Johnson Award, Am Inst Nutrit, 76. *Mem:* Soc Pediat Res; Am Chem Soc; Am Inst Nutrit; Am Soc Biol Chemists; Am Pediat Soc; Sigma Xi. *Res:* Biochemistry of normal and abnormal growth and development; amino acids; parental nutrition; acetylated proteins. *Mailing Add:* Dept of Biochem/Pediat Univ Iowa Iowa City IA 52242

STEHBENS, WILLIAM ELLIS, b Australia, Aug 6, 26; m 61; c 5. PATHOLOGY. *Educ:* Univ Sydney, MB & BS, 50, MD, 62; Oxford Univ, DPhil(path), 60; FRCP(A), 62; FRCPath, 61. *Prof Exp:* From lectr to sr lectr path, Univ Sydney, 53-62; from assoc prof to prof, Wash Univ, 66-68; prof, Albany Med Col, 68-74; PROF PATH, WELLINGTON SCH MED, UNIV OTAGO, NZ, 74- *Concurrent Pos:* Teaching fel path, Univ Sydney, 52; Nuffield Dominion Travelling fel, Univ Oxford, 58; sr res fel, Australian Nat Univ, 62-66, sr fel, 66; pathologist-in-chief & dir dept path & lab med, Jewish Hosp St Louis, 66-68; dir electron micros unit, Vet Admin Hosp, Albany, 68-74; dir, Malaghan Inst Med Res, 74- *Honors & Awards:* R T Hall Res Prize, Cardiac Soc Australia & NZ, 65. *Mem:* Royal Col Pathologists UK; Royal Col Australasia; Int Col Angiol. *Res:* Relationship of intimal thickening, thrombosis and hemodynamics to the pathogenesis of atherosclerosis and cerebral aneurysms; epidemiology of coronary heart disease. *Mailing Add:* Dept Path Wellington Sch Med PO Box 7343 Wellington South New Zealand

STEHLE, PHILIP MCLELLAN, b Philadelphia, Pa, Mar 3, 19; m 42; c 3. QUANTUM ELECTRONICS. *Educ:* Univ Mich, AB, 40, AM, 41; Princeton Univ, PhD(physics), 44. *Prof Exp:* Asst, Univ Mich, 40-41; instr physics, Princeton Univ, 41-44; instr, Harvard Univ, 46-47; asst prof, Univ Pittsburgh, 47, dept chmn, 70-75 & 82-85, prof, 53-89, EMER PROF PHYSICS, UNIV PITTSBURGH, 89- *Concurrent Pos:* Fulbright prof, Univ Innsbruch, 59-60, 65 & 85; vis prof, Univ Munich, 69. *Mem:* Am Phys Soc. *Res:* Quantum theory; quantum optics. *Mailing Add:* Dept Physics & Astron Univ of Pittsburgh Pittsburgh PA 15260

STEHLI, FRANCIS GREENOUGH, b Montclair, NJ, Oct 16, 24; m 48; c 4. GEOLOGY. *Educ:* St Lawrence Univ, BS, 49, MS, 50; Columbia Univ, PhD(geol), 53. *Prof Exp:* Asst prof invert paleont, Calif Inst Technol, 53-56; res engr, Pan Am Petrol Corp, 57, tech group supvr, 58-60; prof geol, Case Western Reserve Univ, 60-74, chmn dept, 61-73, Samuel St John prof earth sci, 74-80, actg dean sci, 75- 76, actg dean sci & eng, 76-77, dean sci & eng, 77-79; pres, Dicar Corp of Case Western Reserve, 78-80; dean grad studies & res, Univ Fla, 80-82; dean, Col Geosci, 82-86, dir, Weather Ctr, Univ Okla, 84-86; pres, Appl Systs Inst, Norman Okla, 83-86; CHMN, SCI ADV COMT, DOSECC INC, GAINESVILLE, FLA, 86- *Concurrent Pos:* Mem, NSF Earth Sci Adv Panel, 67-69, chmn, 69; chmn comt ocean drilling, Nat Res Coun, 81-82, chmn coun on Continental Sci Drilling, 83-86; mem, Energy Res Adv Bd, Dept of Energy, 84- *Mem:* Paleont Soc (pres, 77-80); fel Geol Soc Am; Geochem Soc; fel AAAS; Am Soc Eng Educ. *Res:* Paleozoic brachiopods; Mesozoic stratigraphy of Gulf Coast and western Mexico; paleoecology; carbonate rock formation and diagenesis; continental drift, polar wandering and paleoclimatology. *Mailing Add:* 7711 SW 103rd Ave Gainesville FL 32608-6214

STEHLING, FERDINAND CHRISTIAN, b Fredericksburg, Tex, Feb 18, 30; c 4. POLYMER PHYSICS, PHYSICAL CHEMISTRY. *Educ:* St Mary's Univ, Md, BS & BA, 50; Univ Tex, MA, 57, PhD(phys chem), 58. *Prof Exp:* Res assoc, 58-80, SR RES ASSOC POLYMERS, EXXON CHEM CO, 80- *Honors & Awards:* Prof Progress Award, Soc Prof Engrs & Chemists, 73. *Mem:* Am Chem Soc; Sigma Xi. *Res:* Polymer morphology and properties; mass spectroscopy. *Mailing Add:* 214 Post Oak St Baytown TX 77520

STEHLY, DAVID NORVIN, b Bethlehem, Pa, Oct 3, 33; m 61; c 2. INORGANIC CHEMISTRY. *Educ:* Moravian Col, BS, 59; Lehigh Univ, MS, 62, PhD(inorg chem), 67. *Prof Exp:* From instr to assoc prof, 60-75, PROF CHEM, MUHLENBERG COL, 75- *Mem:* Am Chem Soc. *Res:* Metal chelates of substitued amides; trace metal ions in surface waters. *Mailing Add:* 1254 Lehigh St Allentown PA 18103-3861

STEHNEY, ANDREW FRANK, b Chicago, Ill, May 4, 20; m 43; c 3. RADIOCHEMISTRY, RADIOBIOLOGY. *Educ:* Univ Chicago, BS, 42, PhD(chem), 50. *Prof Exp:* Instr chem, Univ Chicago, 49-50; assoc chemist, 50-71, assoc div dir, 76-84, SR CHEMIST, ARGONNE NAT LAB, 71-, SCIENTIST, 84- *Concurrent Pos:* Lectr, Univ Chicago, 56-61; vis scientist, Europ Orgn Nuclear Res, 61-63; mem subcomt radiochem, Nat Res Coun, 69-77; mem, Safe Drinking Water Comt, Nat Acad Sci, 76; mem subcomt, radiation adv comt, Environ Protection Agency, 86- *Mem:* AAAS; Am Chem Soc; Am Phys Soc; Radiation Res Soc; Health Physics Soc. *Res:* Toxicity of radium and other internal emitters; applications of radiochemistry to environmental studies; nuclear reactions. *Mailing Add:* Biol & Med Res Div Argonne Nat Lab Argonne IL 60439

STEHNEY, ANN KATHRYN, b Oak Ridge, Tenn, June 30, 46; c 2. CRYPTOLOGY. *Educ:* Bryn Mawr Col, AB, 67; State Univ NY, Stony Brook, MA, 69, PhD(math), 71. *Prof Exp:* From asst prof to prof math, Wellesley Col, 71-85, chmn dept, 77-78 & 80-81; RES STAFF MEM, CTR COMMUN RES, INST DEFENSE ANALYSIS, 83- *Concurrent Pos:* Vis scholar, Enrico Fermi Inst, Univ Chicago, 74-75; vis assoc prof, State Univ NY, Stony Brook, 81-82. *Mem:* Am Math Soc; Math Asn Am; Soc Indust & Appl Math. *Res:* Cryptology, analysis and statistical studies in communication. *Mailing Add:* Ctr Commun Res Thanet Rd Princeton NJ 08540

STEHOUWER, DAVID MARK, b Grand Rapids, Mich, July 14, 43; m 68; c 3. ORGANIC CHEMISTRY. *Educ:* Hope Col, BA, 65; Univ Mich, Ann Arbor, MS, 67, PhD(org chem), 70. *Prof Exp:* Sr chemist res labs, Texaco Inc, 70-76; sr res scientist, Gen Motors Res Labs, 76-81; tech adv fuels & lubricants, 81-85, MGR FUELS, LUBRICANTS & ORGANIC MATS, CUMMINS ENGINE CO, 85- *Mem:* Am Soc Testing & Mat; Soc Automotive Engrs; Am Chem Soc. *Res:* Lubricant additive chemistry; mechanisms of lubricant performance; corrosion mechanisms; engine oil wear protection; engine oil consumption; oil conservation and recycling. *Mailing Add:* Fleetguard Inc Box 3005 MC 41502 Columbus IN 47202-3005

STEHR, FREDERICK WILLIAM, b Athens, Ohio, Dec 23, 32; m 59; c 2. SYSTEMATICS, BIOCONTROL. *Educ:* Univ Ohio, BS, 54; Univ Minn, MS, 58, PhD(entom), 64. *Prof Exp:* Res fel entom, Univ Minn, 62-65; from asst prof to assoc prof, 65-76, PROF ENTOM, MICH STATE UNIV, 76-, ASST CHMN DEPT, 79- *Honors & Awards:* Karl Jordan Medal, Lepidop Soc, 74. *Mem:* Entom Soc Am; Soc Syst Zool; Lepidop Soc; Entom Soc Can. *Res:* Systematics of Lepidoptera and immature insects; biological control; ecology. *Mailing Add:* Dept Entom Mich State Univ East Lansing MI 48824-1115

STEHSEL, MELVIN LOUIS, b Long Beach, Calif, Oct 3, 24; m 57; c 2. ORGANIC CHEMISTRY, PLANT PHYSIOLOGY. *Educ:* Univ Calif, Berkeley, BS, 45, MS, 47, PhD(plant physiol), 50. *Prof Exp:* Off Naval Res fel biochem genetics, 50-51; French Govt fel, 51; res chemist, Socony Mobil Oil Co, 51-56; chief phys lab, Aerojet-Gen Corp, 56-65; assoc prof biol, 65-72, PROF BIOL, PASADENA CITY COL, 72- *Mem:* AAAS; Bot Soc Am; Am Chem Soc. *Res:* High temperature mechanical properties of graphite for nuclear rockets; action of natural and synthetic plant growth hormones; stimulation and cause of cell differentiation in plant tissue culture; plant biochemistry. *Mailing Add:* 1570 E Colorado Blvd Pasadena City Col Pasadena CA 91106

STEIB, RENE J, plant pathology; deceased, see previous edition for last biography

STEICHEN, RICHARD JOHN, b Wichita, Kans, July 1, 44; m 68; c 2. ENVIRONMENTAL ENGINEERING, COMPUTER APPLICATIONS. *Educ:* Rockhurst Col, BA, 66; Univ Kans, PhD(anal chem), 71. *Prof Exp:* Asst prof chem, Rockhurst Col, 71-73; Dir res, Goodyear Tire & Rubber Co, 73-87; DIR POLYESTER RES & DEVELOP, 87- *Concurrent Pos:* Mgr environ eng & anal serv. *Mem:* Am Chem Soc. *Mailing Add:* 130 Johns Ave Akron OH 44305

STEIDEL, ROBERT F(RANCIS), JR, b Goshen, NY, July 6, 26; m 46; c 4. MECHANICAL ENGINEERING. *Educ:* Columbia Univ, BS, 48, MS, 49; Univ Calif, DEng, 55. *Prof Exp:* From instr to asst prof mech eng, Ore State Col, 49-54; assoc, 54-55, from asst prof to assoc prof, 55-62, chmn div mech design, 61-64, chmn dept, 69-74, PROF MECH ENG, UNIV CALIF, BERKELEY, 62-; ENG ADV, JOHN WILEY & SONS, 73- *Concurrent Pos:* Consult, Bonneville Power Admin, 50-53, Missile & Space Div, Lockheed Aircraft Corp, 59-, Jet Propulsion Lab, Calif, 66-68 & Sandia Corp, 67; consult, Lawrence Radiation Lab, Univ Calif, 55-, proj engr, 58; fac athletic rep, 72-, assoc dean of eng, Univ Calif, 81-86. *Honors & Awards:* Chester F Carlson Award, 74. *Mem:* Am Soc Mech Engrs; Am Soc Eng Educ; Brit Inst Mech Engrs. *Res:* Engineering mechanics; mechanical vibration, systems analysis and design. *Mailing Add:* 1877 San Pedro Ave Berkeley CA 94707

STEIDTMANN, JAMES R, b Toledo, Ohio, Oct 14, 38. GEOLOGY. *Educ:* Bowling Green State Univ, BS, 60; Dartmouth Univ, MA, 62; Mich State Univ, PhD(geol), 68. *Prof Exp:* PROF GEOL & GEOPHYS, UNIV WYO, 68- *Honors & Awards:* A I Leaderson Mem Award, Am Asn Petrol Geologists. *Mem:* Soc Econ Paleontologists & Mineralogists; Geol Soc Am; Int Asn Sedimentologist. *Mailing Add:* Dept Geol & Geophysics Univ Wyo Box 3006 Laramie WY 82071

STEIER, WILLIAM H(ENRY), b Kendallville, Ind, May 25, 33; m 55; c 4. ELECTRICAL ENGINEERING, OPTICS. *Educ:* Evansville Col, BS, 55; Univ Ill, MS, 57, PhD(elec eng), 60. *Prof Exp:* Asst elec eng, Univ Ill, Urbana, 55-60, asst prof, 60-62; mem tech staff, Bell Tel Labs, NJ, 62-68; assoc prof, 68-76, co-chmn dept, 70-84, PROF ELEC ENG, UNIV SOUTHERN CALIF, 76- *Concurrent Pos:* Consult, Space Technol Labs, 61-62 & Northrop Corp Labs, 68- *Mem:* Fel Inst Elec & Electronics Engrs; AAAS. *Res:* Propagation of electromagnetic energy; millimeterwave detectors and transmission lines; optical transmission systems, modulators and components; lasers; optical signal processing. *Mailing Add:* Dept Elec Eng Univ Southern Calif MC 0483 Los Angeles CA 90008-0483

STEIGELMANN, WILLIAM HENRY, b Vineland, NJ, Jan 19, 35; m 58; c 2. ENGINEERING. *Educ:* Drexel Univ, BS, 56; Oak Ridge Sch Reactor Technol, dipl, 58; Union Col, NY, MS, 59. *Prof Exp:* Engr, RCA Corp, 56-57, NY Shipbldg Corp, 57-58 & Res Labs, Franklin Inst, Pa, 58-63; dept head nuclear eng, Kuljian Corp, 63-67; proj engr, NUS Corp, 67-69; lab mgr nuclear eng & energy studies, Res Labs, Franklin Inst, Pa, 69-76; vpres, Solar Energy Systs, Inc, 77-78; vpres energy technol, Synergic Resources Corp, 80-83; vpres, Assoc Utility Serv Inc, 83-86; pres, En-Save, 78-86; PRES, TECHPLAN ASSOC INC, 86- *Concurrent Pos:* Adj prof, Drexel Univ, 76- *Mem:* Am Nuclear Soc; Am Soc Mech Engrs; Int Solar Energy Soc. *Res:* Energy conversion; solar energy systems; energy conservation; co-generation; energy price and demand forecasting. *Mailing Add:* Tech Plan Assoc Inc 117 Partree Rd Cherry Hill NJ 08003

STEIGER, FRED HAROLD, b Cleveland, Ohio, May 11, 29; m 52, 91; c 2. APPLIED CHEMISTRY, ABSORBENT STRUCTURES. *Educ:* Univ Pa, BA, 51; Temple Univ, MA, 56. *Prof Exp:* Res chemist, Rohm and Haas Co, Pa, 51-60; group leader sanit protection, Johnson & Johnson, 60-62, sr res chemist, 62-68, sr res scientist, 68-74, sr res assoc, 74-76, prod develop mgr, Personal Prod Co Div, 76-88; CONSULT. *Concurrent Pos:* Abstractor, Chem Abstr, 53-68, ed textile sect, 68-85; vpres prof affairs, NJ Inst Chemists, 75-77. *Mem:* Am Chem Soc; Am Asn Textile Chem & Colorists; fel Am Inst Chem; Am Soc Test & Mat; Fiber Soc. *Res:* Textile chemistry; fibers and polymers; sanitary protection; absorption of liquids; cosmetics and toiletries; household products. *Mailing Add:* 10 Tompkins Rd East Brunswick NJ 08816

STEIGER, ROGER ARTHUR, b Potosi, Wis, Dec 29, 39; m 67; c 3. HIGH TEMPERATURE CHEMISTRY. *Educ:* Wis State Univ-Platteville, BS, 61; Univ Iowa, PhD(phys chem), 67. *Prof Exp:* sr res chemist indust chem div, 67-79, res assoc, 79-81, SR RES ASSOC, PPG INDUSTS INC, 81- *Mem:* AAAS; Am Chem Soc; Am Ceramic Soc. *Res:* Thermodynamics of refractory compounds; plasma chemical reactions; powder metallurgy of fine-grained cemented carbides; preparation, fabrication and properties of sinterable submicron ceramic powders. *Mailing Add:* 3215 Ridgeway Rd Greensburg PA 15601-3818

STEIGER, WALTER RICHARD, b Colo, Sept 4, 23; m 46; c 2. PHYSICS. *Educ:* Mass Inst Technol, BS, 48; Univ Hawaii, MS, 50; Univ Cincinnati, PhD(physics), 53. *Prof Exp:* Asst physics, Univ Hawaii, 48-50; instr, Univ Cincinnati, 51-52; from asst prof to assoc prof, 53-65, PROF PHYSICS, UNIV HAWAII, 65-, CHMN DEPT PHYSICS & ASTRON, 72- *Concurrent Pos:* Vis researcher high altitude observ, Univ Colo, 59-60; Fulbright res scholar, Tokyo Astron Observ, 66-67. *Mem:* AAAS; Am Phys Soc; Am Asn Physics Teachers. *Res:* Upper atmosphere physics, ionosphere; airglow. *Mailing Add:* Caltech Submillimeter Observ PO Box 4339 Hilo HI 96720

STEIGER, WILLIAM LEE, b New York, NY, Nov 17, 39; m 71; c 2. LINEAR OPTIMIZATION. *Educ:* Mass Inst Technol, SB, 61, SM, 63; Australian Nat Univ, PhD(statist), 69. *Prof Exp:* assoc prof, 74-84, PROF COMPUT SCI, RUTGERS UNIV, 84- *Concurrent Pos:* Assoc prof, Princeton Univ, 79-81, vis prof, 83-84; res fel, Australian Nat Univ, 79-81. *Mem:* Am Math Soc; Inst Math Statist; Australian Math Soc. *Res:* Computer sciences, theory; statistical computing; linear optimization. *Mailing Add:* Dept Comp Sci Rutgers Univ New Brunswick NJ 08903

STEIGLITZ, KENNETH, b Weehawken, NJ, Jan 30, 39; m 65; c 1. COMPUTER SCIENCE, ELECTRICAL ENGINEERING. *Educ:* NY Univ, BEE, 59, MEE, 60, EngScD(elec eng), 63. *Prof Exp:* From asst prof to prof elec eng, 63-85, PROF COMPUT SCI, PRINCETON UNIV, 85- *Concurrent Pos:* Assoc ed, J Asn Comput Mach, 78-81 & Networks, 80- *Honors & Awards:* Tech Achievement Award, Inst Elec & Electronics Engrs, 81, Centennial Medal, 84; Soc Award, Acoust, Speech & Signal Processing Soc, 86. *Mem:* Fel Inst Elec & Electronics Engrs; Asn Comput Mach. *Res:* Algorithms; digital signal processing. *Mailing Add:* Dept Comput Sci Princeton Univ Princeton NJ 08544

STEIGMAN, GARY, b New York, NY, Feb 23, 41; div. ASTROPHYSICS, COSMOLOGY. *Educ:* City Col New York, BS, 61; NY Univ, MS, 63, PhD(physics), 68. *Prof Exp:* Instr physics, New York Univ, 67-68; vis fel, Inst Theoret Astron, Cambridge, Eng, 68-70; res fel, Calif Inst Technol, 70-72; asst prof astron, Yale Univ, 72-78; assoc prof, Univ Del, 78-80, prof, Bartol Res Found, 80-86; PROF PHYSICS & ASTRON, OHIO STATE UNIV, 86- *Concurrent Pos:* Vis scientist, Nat Radio Astron Observ, 75; vis scholar, Stanford Univ, 79; res scientist, Inst Theoret Physics, Santa Barbara, 81; George Ellery Hale Distinguished Vis Prof, Enrico Fermi Inst, Univ Chicago, 83; vis theorist, Fermi Nat Accelerator Lab, 84, 85. *Mem:* Am Astron Soc; Int Astron Union. *Res:* Cosmology: primarily the early evolution of the universe; connections between elementary particle physics, cosmology, and astrophysics; big bang nucleosynthesis; astrophysical constraints on the properties of elementary particle candidates for the dark matter in the universe. *Mailing Add:* Dept Physics Ohio State Univ 174 W 18th Ave Columbus OH 43210

STEIGMANN, FREDERICK, b Austria, Apr 25, 05; nat US; m 37; c 3. MEDICINE. *Educ:* Univ Ill, BS, 28, MD, 30, MS, 38. *Prof Exp:* Asst, 33-34, from instr to assoc prof, 34-72, clin prof med, Univ Ill Col Med, 72-; RETIRED. *Concurrent Pos:* Pvt pract, 33-; dir depts therapeut & gastroenterol & attend physician, Cook County Hosp, 40- *Mem:* Fel Soc Exp Biol & Med; fel Am Soc Pharmacol & Exp Therapeut; fel Am Fedn Clin Res; fel Am Col Physicians; Am Gastroenterol Asn; Sigma Xi. *Res:* Gastroenterology; liver; vitamin A; protein metabolism. *Mailing Add:* 1205 W Kirby Ave Champaign IL 61821

STEILA, DONALD, b Cleveland, Ohio, Sept 26, 39; div; c 2. CLIMATOLOGY. *Educ:* Kent State Univ, BS, 65, MA, 66; Univ Ga, PhD(geog), 71. *Prof Exp:* Instr geog, Univ Ga, 70-71; asst prof, Univ Ariz, 71-72; from asst prof to assoc prof geog, E Carolina Univ, 72-81; PROF EARTH SCI, UNIV NC, CHARLOTTE, 82- *Concurrent Pos:* Dir, Math & Sci Educ Ctr, 83-87. *Mem:* Asn Am Geogr; Am Meteorol Soc; Soil Sci Soc Am; Am Soc Agron. *Res:* Quantitative identification of drought intensity; the impact of drought upon vegetation and the temporal and spatial patterns of drought occurrence; the spatial characteristics of soil bodies. *Mailing Add:* Dept Geol Earth Sci Univ NC Charlotte NC 28223

STEIMAN, HENRY ROBERT, b Winnipeg, Man, Aug 2, 38; m 73; c 2. DENTISTRY, PHYSIOLOGY. *Educ:* NDak State Univ, BS, 64; Wayne State Univ, MS, 67, PhD(physiol), 69; Univ Detroit, DDS, 73; Indiana Univ, MSD, 79. *Prof Exp:* Asst prof, 69-73, chmn dept physiol, 70-77, dir div biol sci, 73-77, ASSOC PROF PHYSIOL, DENT SCH, UNIV DETROIT, 73-, CHMN DEPT ENDODONTICS, 80- *Concurrent Pos:* Mich Asn Regional Med Progs grant, Dent Sch, Univ Detroit, 74-76; consult, Hypertension Coordinating & Planning Comt, Southeastern Mich, 74-77; consult, Detroit Receiving Hosp, 80-, Vet Admin Hosp, Allen Park, 81- *Mem:* Am Dent Asn; Am Asn Endodontics. *Mailing Add:* Dept Endodontics Univ Detroit-Dent 2985 E Jefferson Ave Detroit MI 48207

STEIMLE, TIMOTHY C, b Benton Harbor, Mich, May 6, 51; m 74. ATOMIC PHYSICS, MOLECULAR PHYSICS. *Educ:* Mich State Univ, BS, 73; Univ Calif, Santa Barbara, PhD(chem), 78. *Prof Exp:* Fel, Rice Univ, 78-80; fel, Univ Southampton, Eng, 80-81; res asst, Univ Ore, 81-85; ASST PROF CHEM, ARIZ STATE UNIV, 85- *Mem:* Am Phys Soc. *Res:* High resolution spectroscopy of gas phase free radials and ionic species. *Mailing Add:* Dept Chem Ariz State Univ Tempe AZ 85287

STEIN, ABRAHAM MORTON, b Chicago, Ill, Aug 9, 23; m 43; c 1. BIOCHEMISTRY. *Educ:* Univ Calif, Los Angeles, AB, 49, MA, 51; Univ Southern Calif, PhD(biochem), 57. *Prof Exp:* NIH res fel biochem, Brandeis Univ, 57-59; res assoc, Univ Pa, 59-65, sr res investr, 65-67; assoc prof col med, Univ Fla, 67-71; chmn dept, 71-74, PROF BIOL SCI, FLA INT UNIV, 71- *Concurrent Pos:* Adj prof col med, Univ Miami, 71- *Mem:* AAAS; Am Soc Biol Chem; Am Chem Soc; NY Acad Sci. *Res:* Chemical carcinogenesis; monoclonal antibodies to DHA adducts. *Mailing Add:* Dept Biol Sci Fla Int Univ NE 151st St & Biscayne Blvd Miami FL 33181

STEIN, ALAN H, b New York, NY, Apr 2, 47; m 69; c 1. ANALYTIC NUMBER THEORY, MICROCOMPUTERS. *Educ:* Queen's Col, BA, 68; NY Univ, MS, 70, PhD(math), 73. *Prof Exp:* From instr to asst prof, 72-80, ASSOC PROF MATH, UNIV CONN, 80- *Mem:* Am Math Soc; Math Asn Am. *Res:* Analytic number theory; additive number theory; binary numbers. *Mailing Add:* Dept Math Univ Conn 32 Hillside Ave Waterbury CT 06710

STEIN, ALLAN RUDOLPH, b Edmonton, Alta, Nov 14, 38. PHYSICAL ORGANIC CHEMISTRY. *Educ:* Univ Alta, BSc, 60; Univ Ill, Urbana, PhD(org chem), 64. *Prof Exp:* Asst org chem, Univ Ill, Urbana, 60-61; res scientist, Domtar Cent Res Labs, Senneville, Que, 64-65; from asst prof to assoc prof chem, 65-75, PROF CHEM, MEM UNIV, NFLD, 75- *Concurrent Pos:* Nat Res Coun Can res grants, 65-69, 75-78 & 78-81; vis prof, King's Col, Univ London, Univ Umea, Sweden, 72-73, Univ Alta, 86; res grant, Swedish Res Coun, 73; hon prof chem, Univ Auckland, New Zealand, 80-81; scientist, Atomic Energy Can, 81. *Mem:* Am Chem Soc; Brit Chem Soc; Chem Inst Can. *Res:* Organic reaction mechanism studies, especially reactions of ambident ions, the isonitriles and of phenol alkylation; ion-pair mechanism of nucleophilic displacement; set process. *Mailing Add:* Dept Chem Mem Univ St John's NF A1C 5S7 Can

STEIN, ARTHUR, b New York, NY, Aug 5, 18; m 41; c 2. MATHEMATICAL STATISTICS. *Educ:* City Col New York, BS, 38; Columbia Univ, MA, 44. *Prof Exp:* Statistician, Ballistic Res Labs, Aberdeen Proving Ground, US Dept Army, 41-44, mathematician, Ord Ballistic Team, 44-46, supvry ballistician, 46-51, dir qual control eng sect, Ord Ammunition Command, 52-55; prin engr & asst head, Opers Res Dept, Cornell Aeronaut Lab, Inc, 55-65, head, Systs Res Dept, 65-73, assoc dir, appl technol group, 70-74; vpres & dir Buffalo Facil, Falcon Res & Develop Co, 74-83; CONSULT, INST DEFENSE ANALYSIS, 84- *Concurrent Pos:* Lectr, Univ Chicago, 53-55; lectr, Univ Buffalo, 56; mem, comt on energetic mats, Nat Res Coun, 85, 86. *Mem:* Nat Acad Sci; Inst Math Statist; Am Statist Asn; Am Soc Qual Control; Am Defense Preparedness Asn; Sigma Xi; Opers Res Soc; fel Mil Opers Res Soc. *Res:* Operations research; weapons effectiveness; aircraft vulnerability; terminal ballistics; sample inspection and quality control; environmental systems analysis; reliability; probability; arms control; applied mathematics; mathematical modeling. *Mailing Add:* 30 Chapel Woods Ct Williamsville NY 14221

STEIN, ARTHUR A, b Toronto, Ont, Mar 12, 22; nat US; m 48; c 3. MEDICINE, PATHOLOGY. *Educ:* Univ Toronto, MD, 45; Am Bd Path, dipl, 52. *Prof Exp:* Intern, Victoria Hosp, London, Ont, 45-46; asst resident med, Jewish Hosp, St Louis, Mo, 47; asst resident path, City Hosp, 47-48; resident, Sch Trop Med, Univ PR, 48-49; instr path & bact, 49-52, from asst prof to assoc prof path, 52-59, dir res inst exp path & toxicol, 65-69, PROF PATH, ALBANY MED COL, 59- *Concurrent Pos:* Surg pathologist, Albany Hosp, 55-; sci adv, Ky Tobacco & Health Bd, Commonwealth of Ky, 71-; dir Micros Biol Res, Inc, 67- *Mem:* Am Soc Clin Path; AMA; Col Am Path; Am Acad Clin Toxicol; Soc Toxicol. *Res:* Toxicology. *Mailing Add:* 38 Colonial Ave Albany NY 12203

STEIN, BARRY EDWARD, b New York, NY, Dec 3, 44; m 68. NEUROPHYSIOLOGY, DEVELOPMENTAL PHYSIOLOGY. *Educ:* Queens Col, BA, 66, MA, 69; City Univ New York, PhD(neuropsychol), 70. *Prof Exp:* Fel neurophysiol & neuroanat, Univ Calif, Los Angeles, 70-72, asst res anatomist, 72-75; asst prof, 75-76, assoc prof, 76-82, PROF PHYSIOL, MED COL VA, VA COMMONWEALTH UNIV, 82- *Mem:* Sigma Xi; Int Brain Res Orgn; AAAS; Am Psychol Asn; Soc Neurosci; Am Physiol Soc. *Res:* Multisensory integration; the ontogenesis of sensory systems; neurophysiological, neuroanatomical and behavioral changes during early life. *Mailing Add:* Dept Physiol Med Col Va Commonwealth Univ Box 551 MCV Sta Richmond VA 23298-0551

STEIN, BARRY FRED, b Philadelphia, Pa, Nov 2, 37; m 62; c 3. SOLID STATE PHYSICS. *Educ:* Univ Pa, BA, 59, MS, 61, PhD(physics), 65. *Prof Exp:* Staff physicist, Univac Div, Sperry Rand Corp, 65-79, mgr, magnetic device res, Sperry Univac Div, 79-82; dir, device res, Univ City Sci Ctr, 83, spec asst to pres, 84, dir res & develop, 84, vpres, res & develop, 88-89, SR VPRES, RES & DEVELOP, BEN FRANKLIN TECH CTR, UNIV CITY SCI CTR, 89- *Concurrent Pos:* Adj prof, Grad Exten, Pa State Univ, 65-76. *Mem:* Am Phys Soc; Inst Elec & Electronic Engrs. *Res:* Electrical properties of gallium arsenide; chemical vapor deposition and magnetic and optical properties of gadolinium iron garnet; liquid phase epitaxial growth of garnets; magnetic bubble device fabrication; high resolution x-ray lithography; Josephson junction devices. *Mailing Add:* Ben Franklin Technol Ctr Univ City Sci Ctr 3624 Market St Philadelphia PA 19104

STEIN, BENNETT M, b New York, NY, Feb 2, 31; m 55, 87; c 3. NEUROSURGERY. *Educ:* Dartmouth Col, BA, 52; McGill Univ, MD, 55; Am Bd Neurol Surg, cert, 66. *Prof Exp:* Intern, US Naval Hosp, St Albans, NY, 58-59; surg resident, Columbia Presby Hosp, New York, 59-60, from asst resident to chief resident neurosurg, 60-64; asst prof, Neurol Inst, Columbia Univ, 68-71; prof neurosurg & chmn dept, New Eng Med Ctr, Tufts Univ, 71-80; PROF, CHEM DEPT NEUROSURG, COLUMBIA PRESBYTERIAN MED CTR, COL PHYSICIANS & SURGEONS, COLUMBIA UNIV, 80- *Concurrent Pos:* Fulbright scholar neurol, Nat Inst, Queens Sq, London, Eng, 58-59; NIH spec fel neuroanat, Columbia Univ, 64-66; consult, US Naval Hosp, Chelsea, Lemuel Shattuck Hosp, Boston & Vet Admin Hosp, Boston, 71- *Mem:* Fel Am Col Surg; Am Asn Anat; Soc Neurol Surg; Cong Neurol Surg; Am Acad Neurol Surg; Sigma Xi. *Res:* Cerebrovascular reactions, specifically cerebrovasospasm in response to subarachnoid hemorrhage; neuroanatomical problems. *Mailing Add:* 710 W 168th St New York NY 10032

STEIN, BLAND ALLEN, b New York, NY, Feb 27, 34; m 56; c 3. MATERIALS ENGINEERING, METALLURGICAL ENGINEERING. *Educ:* City Col New York, BME, 56; Va Polytech Inst & State Univ, MMetE, 64. *Prof Exp:* Mat res engr, 56-68, head metals sect, 68-74, asst head, Mat Res Br, 74-81, head advan mat br, 81-82, head polymeric mat br, 82-84, HEAD APPL MAT BR, LANGLEY RES CTR, NASA, 84- *Concurrent Pos:* Failure anal consult var govt agencies, 63-; educ pub mgt fel, Univ Va, 77-78; asst prof & lectr, George Washington Univ, 77-; adj prof, Christopher Newport Col, 80- *Mem:* Am Soc Testing & Mat; Sigma Xi; Am Soc Metals; Am Ceramic Soc; assoc fel Am Inst Aeronaut & Astronaut. *Res:* Metalls, ceramics, polymers and fiber reinforced composite materials for aerospace structural applications with special emphasis on the effect of service environments on mechanical and physical properties and degradation mechanisms. *Mailing Add:* 732 Jouett Dr Newport News VA 23602

STEIN, CAROL B, b Columbus, Ohio, Jan 1, 37. ZOOLOGY, FRESHWATER MALACOLOGY. *Educ:* Lake Erie Col, AB, 58; Ohio State Univ, MSc, 63, PhD(zool), 73. *Prof Exp:* Asst, Dept Bot, Ohio State Univ, 60, asst, Dept Zool & Entom, 60-61; mus technician natural hist, Ohio Hist Soc, Ohio State Mus, 64-66; teaching assoc, Biol Core Prog, Ohio State Univ, 68-69; mus technician natural hist, Ohio Hist Soc, Ohio State Mus, 69-70; asst curator, 70-72, CURATOR, MUS ZOOL, OHIO STATE UNIV, 72- *Mem:* Am Malacological Union; Am Inst Biol Sci; Sigma Xi; Nature Conservancy. *Res:* Systematics, zoogeography, and life history of the freshwater mollusks, especially the endangered species of eastern North America. *Mailing Add:* 13592 Johnstown-Utica Rd Johnstown OH 43031

STEIN, CHARLES M, MATHEMATICS. *Prof Exp:* Emer prof math, Stanford Univ; RETIRED. *Mem:* Nat Acad Sci. *Mailing Add:* 821 Santa Fe Stanford CA 94305

STEIN, CHARLES W C, b Philadelphia, Pa, Apr 28, 14; m 38. ORGANIC CHEMISTRY. *Educ:* Univ Pa, BS, 36, MS, 39, PhD(org chem), 42. *Prof Exp:* Asst instr chem, Lehigh Univ, 37; asst instr, Drexel Inst, 37-42; res chemist, Gen Aniline & Film Corp, 42-46; process develop chemist, Calco Div, Am Cyanamid Co, 46-51; tech assoc, GAF Corp, 51-79; RETIRED. *Mem:* AAAS; emer mem Am Chem Soc. *Res:* Dyestuff chemistry; pigments; organic synthesis; plastics. *Mailing Add:* 910 Summit Ave Westfield NJ 07090

STEIN, DALE FRANKLIN, b Kingston, Minn, Dec 24, 35; m 58; c 2. METALLURGY, SURFACE CHEMISTRY. *Educ:* Univ Minn, BS, 58; Rensselaer Polytech Inst, PhD(metall), 63; Cent Mich Univ, PhD, 85. *Prof Exp:* Asst plant metallurgist, Metall Inc, 57-58; prog metallurgist, Gen Elec Co, 58-59, metallurgist, Res & Develop Ctr, 59-67; assoc prof Sch Mineral & Metall Eng, Univ Minn, Minneapolis, 67-70, assoc prof mech eng, chem eng & mat sci, 70-71; prof metall eng & dept head, 71-77, vpres acad affairs, 77-79, PRES MICH TECHNOL UNIV, HOUGHTON, 79- *Concurrent Pos:* Consult, NSF Adv Comt, Mat Res Labs, 76-, & Dept Energy; Energy Res Adv Bd, 87; mem, Nat Mat Adv Bd. *Honors & Awards:* Hardy Gold Medal, Am Inst Mining, Metall & Petrol Engrs, 65; Giesler Award, Am Soc Metals, 67. *Mem:* Nat Acad Eng; fel Am Inst Mining, Metall & Petrol Engrs (pres elect, 79); Am Soc Metals; Sigma Xi; fel Metall Soc; fel AAAS. *Res:* Application of Auger spectroscopy to metallurgical problems, including brittle fracture, corrosion and structure stability, cleavage fracture, dislocation dynamics and high purity metals; author of over 60 publications. *Mailing Add:* Mich Technol Univ 1400 Townsend Dr Houghton MI 49931-1295

STEIN, DARYL LEE, b Canton, Ohio, Aug 27, 49; m 71. ORGANIC CHEMISTRY. *Educ:* Bowling Green State Univ, BS, 71, MS, 73; Mich State Univ, PhD(chem), 78. *Prof Exp:* Res scientist, Continental Oil Co, 77-80; res scientist, Lucidol Div, Pennwalt Corp, 80-89; sr res chemist, Akzo Chem, 89-91; RES SPECIALIST, QUANTUM CHEM, 91- *Mem:* Am Chem Soc. *Res:* Catalysts for polymers; additives for polymers. *Mailing Add:* 9667 Roundhouse Dr Westchester OH 45069

STEIN, DAVID MORRIS, b Johannesburg, SAfrica; m 75. OPERATIONS RESEARCH, COMPUTER SCIENCE. *Educ:* Univ Witwatersrand, BSc, 72, MSc, 74; Harvard Univ, PhD(eng), 77. *Prof Exp:* Res scientist comput sci, 77-81, MEM TECH PLANNING STAFF, IBM CORP, 81- *Res:* Application of mathematical techniques to industrial problems; optimization, decision and control methods applied to transportation problems; performance and storage organizations for future computer systems. *Mailing Add:* Hirst Rd Briarcliff Manor NY 10510

STEIN, DIANA B, b New York, NY, July 5, 37; m 58; c 4. BOTANY, MOLECULAR BIOLOGY. *Educ:* Barnard Col, AB, 58; Univ Mont, MA, 61; Univ Mass, Amherst, PhD(molecular biol), 76. *Prof Exp:* Instr bot, Univ mass, 64-69, instr plant physiol, 78, res assoc molecular biol, 76-79; asst prof, 80-86, ASSOC PROF BIOL SCI, MOUNT HOLYOKE COL, S HADLEY, MASS, 86- *Concurrent Pos:* Mem, DNA Study Comn Amherst, 77-78, Biohazards Comn, Amherst Col, 78- & Amherst Bd Health, 84- *Mem:* Bot Soc Am; Am Soc Plant Physiol; Sigma Xi; AAAS; Soc Plant Molecular Biol; Soc Develop Biol. *Res:* DNA sequence comparisons; DNA protein interactions; evolution of plant groups. *Mailing Add:* Dept Biol Sci Mt Holyoke Col South Hadley MA 01075

STEIN, DONALD GERALD, b New York, NY, Jan 27, 39; m 60; c 2. PSYCHOPHYSIOLOGY, BIOPSYCHOLOGY. *Educ:* Mich State Univ, BA, 60, MS, 62; Univ Ore, PhD(psychol), 65. *Prof Exp:* NIMH res fel, Mass Inst Technol, 65-66; prof neurol, univ mass med ctr, worcester, 78-87; asst prof psychol 7 co-dir animal lab, 66-69, assoc prof psychol, 69-73, PROF PSYCHOL, CLARK UNIV, 73-, DIR, BRAIN RES FACIL , 73-; DEAN GRAD SCH & ASSOC PROVOST RES, RUTGERS UNIV, NEWARK, 88- *Concurrent Pos:* USPHS res contract, 67; NSF res contract, 67-71; USPHS biomed sci grant, 69- Fulbright awards, 71-75; NIMH res career develop award, 72-78; prof, Univ Nice, France, 77; vis scientist, Nat Inst Health & Med Res, Lyon, 75-76, Paris, 79; France Nat Inst Aging res contract, 76-79; Vis scientist, Nat Asn Health & Med Res, Paris, 78, Univ L Pasteur, Strasbourg, France, 81-; Congressional fel sci & eng, AAAS, 80-81. *Mem:* Sigma Xi; Soc Neurosci; Psychonomic Soc; Europ Brain & Behav Soc; Int Soc Neuropsychol; fel AAAS. *Res:* Recovery from brain damage; aging and brain function; nerve growth factor and behavior; neuroplasticity. *Mailing Add:* Rutgers Univ Newark NJ 07102

STEIN, ELIAS M, b Antwerp, Belg, Jan 13, 31; nat US; m 59; c 2. MATHEMATICS. *Educ:* Univ Chicago, AB, 51, MS, 53, PhD, 55. *Prof Exp:* Instr, Mass Inst Technol, 56-58; mem fac, Univ Chicago, 58-62, assoc prof math, 61-62; mem, Inst Advan Study, 62-63; chmn dept, 68-71, PROF MATH, PRINCETON UNIV, 63- *Concurrent Pos:* Sloan Found res fel, 61-63; NSF sr fel, 62-63 & 71-72; sr vis fel, Sci Res Coun Gt Brit, 68. *Mem:* Nat Acad Sci; Am Math Soc. *Res:* Topics in harmonic analysis related to the Littlewood-Paley theory; singular integrals and differentiality properties of functions. *Mailing Add:* Dept Math Princeton Univ Princeton NJ 08544

STEIN, FRANK S, b Lancaster, Pa, Jan 11, 21; m 47; c 3. PHYSICS. *Educ:* Franklin & Marshall Col, BS, 42; Columbia Univ, MA, 47; Univ Buffalo, PhD(physics), 51. *Prof Exp:* Res physicist, Manhattan Proj, Columbia Univ, 44-46; instr physics, Univ Buffalo, 48-51; res physicist res labs, Westinghouse Elec Corp, 51-55, mgr semiconductor dept electronic tube div, 55, mgr power devices develop sect semiconductor dept, 55-60; mgr eng semiconductor div, Gen Instrument Corp, NJ, 60-61, dir res appl res lab, 61-63; sr scientist semiconductor dept, Delco Radio Div, 63-66, chief engr solid state prod, Div, 66-75, chief engr advan eng, 75-83, mgr res & devel, Delco Electronics Div, Gen Motors Corp, 83-86; CONSULT, 87- *Concurrent Pos:* Chmn joint electron device eng coun, Solid State Prod Eng Coun, 69-70 & 73-75; mem, Nat Acad Sci/Nat Acad Eng/Nat Res Coun Eval Panel Electronic Technol Div, Nat Bur Standards, 75-79. *Honors & Awards:* Excellence Award, Electronic Indust Asn, 85. *Mem:* Inst Elec & Electronics Eng. *Res:* Mass spectrometry; properties of semiconductors and semiconductor devices; solid state electronics; microelectronics. *Mailing Add:* 3204 Tallyho Dr Kokomo IN 46902

STEIN, FRED P(AUL), b Dallastown, Pa, Nov 22, 34; m 56; c 3. THERMODYNAMIC PROPERTIES. *Educ:* Lehigh Univ, BS, 56; Univ Mich, MSE, 57, PhD(chem eng), 61. *Prof Exp:* Sr chem engr, Air Prod & Chem Inc, 60-61; from asst prof to assoc prof, 63-71, assoc chmn, 83-89, PROF CHEM ENG, LEHIGH UNIV, 71- *Concurrent Pos:* Consult, Picatinny Arsenal, 63-65, Gardner Cryogenics, 65-69, Hershey Foods Corp, 71-72 & Air Prod & Chem, 76-; vis prof, Monash Univ, Melbourne, Australia, 73 & Univ Queensland, Brisbane, 74 & Univ Canterbury, Christchurch, NZ, 89-90. *Mem:* Am Inst Chem Engrs; Am Chem Soc. *Res:* Phase equilibria at cryogenic temperatures and at high temperatures; thermodynamic properties of mixtures; air preheaters in power plants; equations of state for electrolyte mixtures. *Mailing Add:* Dept of Chem Eng 111 Research Dr Lehigh Univ Bethlehem PA 18015-4791

STEIN, FREDERICK MAX, b Wyaconda, Mo, Feb 17, 19; m 43; c 2. MATHEMATICS. *Educ:* Iowa Wesleyan Col, AB, 40; Univ Iowa, MS, 47, PhD(math), 55. *Prof Exp:* Instr high schs, Iowa, 40-43; instr math, Univ Iowa, 43-44, asst, 46-47; assoc prof, Iowa Wesleyan Univ, 47-53; instr math, Univ Iowa, 53-54; assoc prof, 55-63, PROF MATH, COLO STATE UNIV, 63- *Mem:* Am Math Soc; Math Asn Am. *Res:* Approximation; orthogonal functions; differential and integro-differential equations; Sturm-Liouville systems. *Mailing Add:* 1212 W Olive St Ft Collins CO 80521

STEIN, GARY S, b Brooklyn, NY, July 30, 43; m 74. GENE EXPRESSION. *Educ:* Hofstra Univ, BA, 65, MA, 66; Univ Vt, PhD(cell biol), 69. *Prof Exp:* Res assoc biochem, Temple Univ, 71-72; from asst prof to assoc prof biochem, 72-78, assoc chmn dept, 81-87, PROF BIOCHEM & MOLECULAR BIOL, SCH MED, UNIV FLA, 78-; PROF & CHMN, DEPT CELL BIOL, UNIV MASS MED CTR, 87- *Concurrent Pos:* NIH fel, Sch Med, Temple Univ, 69-71; Damon Runyon Mem Fund cancer res grant, Sch Med, Univ Fla, 71-73; Am Cancer Soc, NSF & NIH grants, 74- *Mem:* AAAS; Am Soc Biol Chemists; Am Asn Cancer Res; Am Soc Cell Biol; Orthop Res Soc; Am Soc Bone & Mineral Res. *Res:* Molecular, biochemical, and cellular approaches to control of cell proliferation and differentiation in normal and tumor cells. *Mailing Add:* Dept Cell Biol Univ Mass Med Ctr 55 Lake Ave N Worcester MA 01655

STEIN, GEORGE NATHAN, b Philadelphia, Pa, Aug 11, 17; m 48; c 3. RADIOLOGY. *Educ:* Univ Pa, BA, 38; Jefferson Med Col, MD, 42; Am Bd Radiol, dipl, 49. *Prof Exp:* Intern, Jewish Hosp, Philadelphia, 43; resident radiol, Grad Hosp, 47-49, assoc radiologist, 49-51, from instr to assoc prof, Div Grad Med, 51-61, assoc dir dept, Ctr, 67-71, dir dept radiol, Presby-Univ Pa Med Ctr, 71-84, prof, 61-88, EMER PROF RADIOL, SCH MED, UNIV OF PA, 88- *Concurrent Pos:* Hon prof, Pontif Univ Javeriana, Colombia, 60. *Mem:* Radiol Soc NAm; Am Roentgen Ray Soc; fel Am Col Radiol. *Res:* Gastrointestinal radiology. *Mailing Add:* Dept Radiol Presby-Univ Pa Med Ctr 51 N 39th St Philadelphia PA 19104

STEIN, GRETCHEN HERPEL, b Asbury Park, NJ, Mar 27, 45; m 66; c 1. CELL BIOLOGY. *Educ:* Brown Univ, AB, 65; Stanford Univ, PhD(molecular biol), 71. *Prof Exp:* NIH fel cell biol, Sch Med, Stanford Univ, 71-73, res fel cell biol & molecular biol, 73-74; res assoc cell biol & molecular biol, 74-81, ASSOC PROF, ATTEND RANK, DEPT MOLECULAR, CELLULAR & DEVELOP BIOL, UNIV COLO, 82- *Concurrent Pos:* Vis scientist, Cancer Biol Prog, Frederick Cancer Res Ctr, 80-81; consult health scientist adminr & mem, aging planning panel, Nat Inst Aging, 81, mem aging rev comt, 81-83, cell adv comt, 82- *Mem:* Am Soc Cell Biol; Tissue Cult Asn; Int Cell Cycle Soc. *Res:* Control of cellular proliferation in normal and neoplastic human cells; mechanism for cessation of proliferation in senescent cells. *Mailing Add:* Dept Biol Campus Box 347 Univ Colo Boulder CO 80309-0347

STEIN, HARVEY PHILIP, b Brooklyn, NY, May 4, 40; m 65; c 3. ORGANIC CHEMISTRY, PUBLIC HEALTH. *Educ:* Queens Col, NY, BS, 61; Mass Inst Technol, PhD(org chem), 67. *Prof Exp:* Chemist, Stamford Res Labs, Am Cyanamid Co, summer 61; asst, Mass Inst Technol, 61-65; instr chem, Pa State Univ, 65-67, asst prof, 67-68; asst prof, Trenton State Col, 68-71; prof, Western Col, 71-75; sr scientist, 75-83, SCIENTIST DIR, USPHS, 83- *Res:* Reaction mechanisms; use of isotopes; biomedical effects of ethanol; public health; identification of previously unrecognized occupational hazards. *Mailing Add:* 11705 Silent Valley Lane Gaithersburg MD 20878-2433

STEIN, HERBERT JOSEPH, b Evanston, Ill, Mar 20, 28; m 65; c 3. ELECTRICAL ENGINEERING. *Educ:* Univ Ill, Urbana, BS, 58, MS, 59, PhD(elec eng), 64. *Prof Exp:* From instr to res assoc elec eng, Univ Ill, Urbana, 60-64; asst prof info eng, 64-66, actg head dept, 64-69, assoc prof elec eng & assoc dean col eng, 69-81, HEAD DEPT SYSTS ENG, UNIV ILL, CHICAGO CIRCLE, 81- *Concurrent Pos:* Consult, Honeywell Corp, Ill, 67-68. *Mem:* Sr mem Inst Elec & Electronics Engrs; Am Soc Eng Educ; found mem Am Soc Eng Mgt. *Res:* Engineering education and administration; implementation of interdisciplinary programs. *Mailing Add:* Dept Elec Eng & Comput Sci Univ Ill Box 4348 Chicago IL 60680

STEIN, HERMAN H, b Chicago, Ill, May 27, 30; m 51; c 2. BIOCHEMISTRY, PHARMACOLOGY. *Educ:* Univ Ill, BS, 51; Univ Minn, MS, 53; Northwestern Univ, PhD(chem), 56. *Prof Exp:* Lab asst, Northwestern Univ, 53-54; res chemist, Toni Co Div, Gillette Co, 56-61; sr res chemist, Abbott Labs, 61-66, group leader, 66, sect head, 67-72; assoc fel, 72-76, RES FEL, PHARMACEUT PROD DIV, ABBOTT LABS, 76- *Mem:* Am Chem Soc; Am Soc Pharmacol & Exp Therapeut. *Res:* Enzymology; automated metabolic and enzymic analyses; antianginal agents; cyclic adenosine monophosphate metabolism; pharmacology of nucleosides; beta-adrenergic blocking agents; lipid metabolism; inhibitors of renin. *Mailing Add:* Abbott Labs D-47B AP-10 Abbott Park IL 60064

STEIN, HOWARD JAY, b Baltimore, Md, May 28, 33; m 57; c 5. RESEARCH & DEVELOPMENT, CELL PHYSIOLOGY. *Educ:* Temple Univ, BA, 54; Univ Mich, MA, 58, PhD(bot), 61. *Prof Exp:* From asst prof to assoc prof biol, Kans State Col, Pittsburg, 60-65; assoc prof, 65-71, chmn dept, 66-68 & 71-75, PROF BIOL, GRAND VALLEY STATE COL, 71-, DIR RES & DEVELOP, 79 - *Concurrent Pos:* NSF grant, 62-64; staff biologist, Off Biol Educ, Am Inst Biol Sci, 69-71, vis biologist, 71-72, curric consult bur, 71-74; NSF Grants, 68-70, 68-69, 79-80, 80-81; EESA Title II Grants, 86, 87, 88. *Mem:* AAAS; Nat Sci Teachers Asn; Sigma Xi; Nat Asn Biol Teachers; Fedn Am Scientists. *Res:* Amino acid metabolism in plant roots; metabolism in plant mitochondria; development of geoglossum. *Mailing Add:* Dept Biol Grand Valley State Univ Allendale MI 49401

STEIN, IRVING F, JR, b Chicago, Ill, July 6, 18; m 50; c 2. SURGERY. *Educ:* Dartmouth Col, AB, 39; Northwestern Univ, MS, 41, MD, 43, PhD, 51; Am Bd Surg, dipl, 50. *Prof Exp:* ASST PROF SURG, MED SCH, NORTHWESTERN UNIV, 54- *Concurrent Pos:* Attend surgeon, Cook County Hosp; chief surg, Highland Park Hosp. *Mem:* AAAS; fel AMA; fel Am Col Surg; Am Fedn Clin Res. *Res:* Gastrointestinal and surgical research. *Mailing Add:* 625 Roger Williams Ave Highland Park IL 60035

STEIN, IVIE, JR, b Orange, Calif, Dec 31, 40; m 79. MATHEMATICS. *Educ:* Long Beach State Univ, BS, 62, MA, 63; Univ Calif, Los Angeles, CPhil, 70, PhD(math), 71. *Prof Exp:* Asst prof, 71-75, ASSOC PROF MATH, UNIV TOLEDO, 75- *Concurrent Pos:* Pac Missle Ctr, 66, 67, 68 & 74;

Argonne Nat Lab, 72. *Mem:* Am Math Soc; Soc Appl & Indust Math; Sigma Xi. *Res:* Calculus of variations; optimal control theory; numerical analysis; differential equations; optimization. *Mailing Add:* Dept Math Univ Toledo Toledo OH 43606

STEIN, JACK J(OSEPH), b New York, NY, Feb 14, 38; m 59; c 3. ELECTRICAL ENGINEERING. *Educ:* City Col New York, BEE, 59; Columbia Univ, MSEE, 60; NY Univ, DrEngSci, 65. *Prof Exp:* Res staff mem, IBM Res Div, Int Bus Mach Corp, 60-62; lectr, City Col New York, 62-63; instr, NY Univ, 63-65; systs engr, IBM Data Processing Div, Int Bus Mach Corp, 65-66; engr, Hughes Aircraft Co, 66-67; ASSOC PROF ELEC ENG, GT VALLEY GRAD CTR, PA STATE UNIV, 67- *Concurrent Pos:* Consult, Gen Elec, 68-69, Burroughs, 73-74. *Mem:* Inst Elec & Electronics Engrs. *Res:* Network theory; digital computers; solid state devices; biomedical simulation. *Mailing Add:* Pa State Great Valley 30 E Swedesford Rd Malvern PA 19355

STEIN, JAMES D, JR, b New York, NY, Aug 29, 41. MATHEMATICS. *Educ:* Yale Univ, BA, 62; Univ Calif, Berkeley, MA & PhD(math), 67. *Prof Exp:* Asst prof math, Univ Calif, Los Angeles, 67-74, NSF grant, 70-74; ASSOC PROF MATH, CALIF STATE UNIV, LONG BEACH, 74- *Mem:* Am Math Soc. *Res:* Banach algebras; continuity and boundedness problems in Banach spaces; measure theory. *Mailing Add:* 13930 NW Passage Marina Del Rey CA 90291

STEIN, JANET LEE SWINEHART, b Danville, Pa, Apr 3, 46; m 74. GENE EXPRESSION, TRANSCRIPTION. *Educ:* Elizabethtown Col, BS, 68; Princeton Univ, MA, 71, PhD(chem), 75. *Prof Exp:* Res assoc biochem, 74-76, from asst prof to prof immunol & med microbiol, Univ Fla, 76-87; PROF CELL BIOL, UNIV MASS MED CTR, 87- *Concurrent Pos:* Res grants, NSF, 75-90, Am Cancer Soc, 77-78, March of Dimes, 78- & NIH, 88-; mem, Physiol Chem Sect, NIH, 86-90, chair, 88-90; ed, Critical Rev Eukamyotic Gene Expression, 90- *Mem:* Am Soc Cell Biol; Am Chem Soc; AAAS; Am Soc Biochem Molecular Biol. *Res:* Regulation of gene expression, especially at the transcriptional level, in eukaryotic cells; structure and regulation of human histone genes. *Mailing Add:* Dept Cell Biol Univ Mass Med Ctr 55 Lake Ave N Worcester MA 01655

STEIN, JERRY MICHAEL, b Brooklyn, NY, Jan 9, 52; m 74. CONTACT LENSES, CARDIOVASCULAR PHYSIOLOGY. *Educ:* Brooklyn Col, BA, 73; Syracuse Univ, MS & PhD(psychol), 77. *Prof Exp:* Res assoc physiol & biophys, Reg Primate Res Ctr, Univ, Wash, 78-79; res assoc, Cardiovasc Ctr & Dept Psychol, Univ Iowa, Iowa City, 79-81; sr clin res assoc, 81-85, ASST DIR CLIN SCI, ALCON LABS, FT WORTH, TEX, 86- *Concurrent Pos:* Teaching asst, Syracuse Univ, 77; lectr psychol, Univ Wash, 78-, fel, Regional Primate Res Ctr, 78-79. *Mem:* Soc Neurosci; Assocs Clin Pharmacol. *Res:* Clinical testing of medical devices and ethical drugs; central nervous system control of cardiovascular physiology and renal functions. *Mailing Add:* Alcon Labs PO Box 6600 Ft Worth TX 76115

STEIN, JOHN MICHAEL, b Vienna, Austria, May 29, 35; US citizen; m 69; c 4. SURGERY. *Educ:* Harvard Univ, AB, 57, MD, 61; Am Bd Surg, dipl, 68. *Prof Exp:* Assoc surg, New York Hosp-Cornell Med Ctr, 62-63; asst instr, Albert Einstein Col Med, 66-67; chief burn study br, US Army Inst Surg Res, Tex, 67-69; assoc prof surg & burns, Albert Einstein Col Med, 69-79; dir, Burn Unit, Aricopa County Hosp, 79-89; DIR, BURN & TRAUMA CTR ARIZ, 89- *Concurrent Pos:* NIH fel surg, Albert Einstein Col Med, 64-65; pres med bd, Bronx Munic Hosp Ctr, 74-76; co-chmn burn comt & mem bd dirs Regional Emergency Med Servs, Coun of New York, 77- *Mem:* Am Burn Asn; NY Surg Soc; fel Am Col Surg; Am Trauma Soc; Asn Acad Surg. *Res:* Surgical training and research, expecially metabolic care of surgical and burned patients. *Mailing Add:* Burn & Trauma Ctr Ariz 1130 E McDowell Rd Suite B1 Phoenix AZ 85006

STEIN, KATHRYN E, IMMUNOCHEMISTRY, BIOCHEMISTRY. *Educ:* Albert Einstein Col Med, PhD(microbiol & immunol), 76. *Prof Exp:* RES CHEMIST, OFF BIOL RES REV, US FOOD & DRUG ADMIN, 80- *Mailing Add:* Div Bact Prods Bur Biol Bldg 29 Rm 124 US Food & Drug Admin 8800 Rockville Pike Bethesda MD 20892

STEIN, LARRY, b New York, NY, Nov 10, 31; m 60. NEUROSCIENCE, BEHAVIOR. *Educ:* NY Univ, BA, 52; Univ Iowa, MA, 53, PhD, 55. *Prof Exp:* Res psychologist, Walter Reed Army Inst Res, 55-57; res psychologist, Vet Admin Res Labs Neuropsychiat, 57-59; sr res scientist, Wyeth Labs, 59-64, mgr, Dept Psychopharmacol, 64-79; PROF & CHMN, DEPT PHRAMACOL, COL MED, UNIV CALIF, IRVINE, 79- *Concurrent Pos:* Res assoc, Bryn Mawr Col, 61-72, adj prof, 72-; adj prof med sch, Univ Pa. *Honors & Awards:* Bennett Award, Soc Biol Psychiat, 61. *Mem:* Am Psychol Asn; Am Physiol Soc; Am Soc Pharmacol & Exp Therapeut; Soc Neurosci. *Res:* Cellular mechanisms of reward and punishment, psychopharmacology, biological basis of schizophrenia and depression. *Mailing Add:* Dept Pharm Col Med MSII 360 Univ Calif Irvine CA 92717

STEIN, LAWRENCE, b Hampton, Va, July 21, 22; m 52; c 2. INORGANIC CHEMISTRY. *Educ:* George Washington Univ, BS, 48; Univ Wis, PhD(chem), 52. *Prof Exp:* Anal chemist, Nat Bur Standards, 48; asst, Univ Wis, 50-51; CHEMIST, ARGONNE NAT LAB, 51- *Concurrent Pos:* Adv panelist comt biol effects atmospheric pollutants, Nat Res Coun, 71-72; consult, Nat Inst Occupational Safety & Health, 73; consult, Gould, Inc, 83. *Mem:* Fel AAAS; Am Chem Soc. *Res:* Fluorine chemistry; interhalogen compounds; chemistry of noble gases, particularly radon; environmental radiation; biologic effects of atmospheric pollutants; infrared spectroscopy; chemistry of actinide elements; laser raman spectroscopy. *Mailing Add:* 1223 Gilbert Ave Downers Grove IL 60515-4516

STEIN, MARJORIE LEITER, b New York, NY. PROBLEM SOLVING, NETWORK ANALYSIS. *Educ:* Barnard Col, AB, 68; Princeton Univ, MA, 71, PhD(math), 72. *Prof Exp:* Res assoc, Math Res Ctr, Univ Wis, Madison, 72-73, lectr computer sci, Univ Wis, 73; res assoc, Nat Bur Standards, 73-75; sr math statistician, US Postal Serv, 75-76, sr opers res analyst, 76-77, mgt analyst/prog mgr, 77-79, prin economist, 79-86, PROG DIR, OPERS RES, US POSTAL SERV, 86- *Concurrent Pos:* Vis lectr, Math Asn Am, 75-78; docent, Nat Mus Natural Hist, Smithsonian Inst, 76-; gov-at-large, Math Asn Am, 77-80 & 88-91; mem coun, Asn Women Math, 78- *Mem:* Math Asn Am; Asn Women Math. *Res:* Combinatorial theory; networks; linear programming; applications to economics; manages development of decision support systems for postal management. *Mailing Add:* US Postal Serv 475 L'Enfant Plaza West SW Washington DC 20260-8114

STEIN, MARVIN, b St Louis, Mo, Dec 8, 23; m 50; c 3. PSYCHIATRY. *Educ:* Wash Univ, BS, 45, MD, 49. *Prof Exp:* Intern, St Louis City Hosp, 49-50; asst resident psychiat, Sch Med, Wash Univ, 50-51; asst instr, Sch Med, Univ Pa, 53-54, res assoc, 54-56, from asst prof to assoc prof, 56-63; prof, Med Col, Cornell Univ, 63-66; prof & chmn dept, State Univ NY Downstate Med Ctr, 66-71; prof psychiat & chmn dept, Mt Sinai Sch Med & Psychiatrist-in-Chief, 71-87; KLINGENSTEIN PROF, MT SINAI SCH MED, 87- *Concurrent Pos:* USPHS fel clin sch, Sch Med, Univ Pittsburgh, 51-53; fel psychiat, Sch Med, Univ Pa, 53-54; ment health career investr, NIMH, 56-61, mem ment health fels rev panel, 61-64, ment health res career award comt, 63-65, chmn, 65-67; mem behav med study sect, 81-83 & Geriat Rev comt, 86-88, Nat Insts Health; chmn spec rev comt, Ment Health Aspects of AIDS, 87-88, & chmn, Ment Health AIDS Res Rev Comt, Nat Inst Ment Health. *Mem:* Am Psychiat Asn (chmn res coun, 81-84); Asn Res Nerv & Ment Dis; Soc Biol Psychiat. *Res:* Brain and behavior; immunity. *Mailing Add:* Dept Psychiat Mt Sinai Sch Med New York NY 10029

STEIN, MARVIN L, b Cleveland, Ohio, July 15, 24; m 44; c 3. MATHEMATICS, COMPUTER SCIENCE. *Educ:* Univ Calif, Los Angeles, BA, 47, MA, 49, PhD, 51. *Prof Exp:* Asst math, Univ Calif, Los Angeles, 47-48, mathematician inst numerical anal, 48-52; sr res engr, Consol-Vultee Corp, 52-55; from asst prof to prof math, Univ Minn, Minneapolis, 55-70, dir, Univ Comput Ctr, 58-70, actg head, Comput Info & Control Sci Dept, 70-71, dir grad studies, 87-90, PROF COMPUT SCI, UNIV MINN, MINNEAPOLIS, 70- *Concurrent Pos:* Lectr, Univ Calif, Los Angeles, 54-55; Guggenheim fel, 63-64; vis prof, Tel Aviv Univ & Hebrew Univ, Jerusalem, 71-72. *Mem:* Am Math Soc; Soc Indust & Appl Math; Asn Comput Mach. *Res:* Numerical analysis; applications of super computers; parallel computer systems. *Mailing Add:* Comput Sci Dept Univ Minn Minneapolis MN 55455

STEIN, MICHAEL ROGER, b Milwaukee, Wis, Mar 21, 43; m 67; c 2. ALGEBRAIC K-THEORY. *Educ:* Harvard Univ, BA, 64; Columbia Univ, PhD(math), 70. *Prof Exp:* Asst prof math, 70-74, assoc prof, 74-80, PROF MATH, NORTHWESTERN UNIV, EVANSTON, 80- *Concurrent Pos:* Fel, Hebrew Univ, Jerusalem, 72-73; Sci Res Coun sr vis fel, King's Col, Univ London, 77-78; vis scientist, Weizmann Inst Sci, 78; vis scholar, Univ Chicago, 84. *Mem:* Am Math Soc. *Res:* Algebraic K-theory; algebra. *Mailing Add:* Dept Math Northwestern Univ 2003 Sheridan Rd Evanston IL 60208

STEIN, MYRON, b East Boston, Mass, May 27, 25; m 53; c 4. PHYSIOLOGY, MEDICINE. *Educ:* Dartmouth Col, BA, 48; Tufts Univ, MD, 52. *Prof Exp:* Instr med, Harvard Med Sch, 57-64, assoc, 64-65; from assoc prof to prof med sci, Brown Univ, 69-73; PROF MED, UNIV CALIF, LOS ANGELES, 73-; DIR PULMONARY DIV, BROTMAN MEM HOSP, 73- *Concurrent Pos:* Consult, Mass Rehab Comt, 60-65, Vet Admin Hosps, West Roxbury, 63- & Davis Park, RI, 65- *Mem:* Am Fedn Clin Res; Am Physiol Soc. *Res:* Pulmonary physiologic effects of pulmonary embolism; relationship of acid-base states and thyroid hormone transport; physiologic studies in clinical lung diseases. *Mailing Add:* Brotman Mem Ctr Pulmonary Med Div 3828 Delmas Terr Culver City CA 90230

STEIN, OTTO LUDWIG, b Augsburg, Ger, Jan 14, 25; nat US; m 58; c 4. PLANT MORPHOGENESIS. *Educ:* Univ Minn, BS, 49, MS, 52, PhD(bot), 54. *Prof Exp:* Asst bot, Univ Minn, 48-53; instr, Univ Mo, 55; USPHS res fel, Brookhaven Nat Lab, 55-58; from asst prof to assoc prof bot, Mont State Univ, 58-64; head dept, 70-74, from assoc prof to prof, 64-90, EMER PROF BOT, UNIV MASS, AMHERST, 90- *Concurrent Pos:* Res collab, Brookhaven Nat Lab, 58-70; vis asst prof, Univ Calif, 61-62; sr NATO res fel, Imp Col, Univ London, 71-72; dir, Univ Mass-Univ Freiburg, WGer, Exchange Prog, 79. *Mem:* Bot Soc Am; Soc Study Develop Biol; Soc Exp Biol & Med; Linnaean Soc, London. *Res:* Genetics; cytology; developmental anatomy of apical meristems and their derivatives. *Mailing Add:* Dept of Bot Univ of Mass Amherst MA 01003

STEIN, PAUL DAVID, b Cincinnati, Ohio, Apr 13, 34; m; c 3. CARDIOLOGY. *Educ:* Univ Cincinnati, BS, 55, MD, 59. *Prof Exp:* Fel cardiol, Col Med, Univ Cincinnati, 62-63 & Mt Sinai Hosp, New York, 63-64; res fel med, Peter Bent Brigham Hosp, Harvard Med Sch, Boston, 64-66; asst dir, catheterization lab, Baylor Univ Med Ctr, Dallas, 66-67; asst prof med, Creighton Univ, Omaha, 67-69; assoc prof, Col Med, Univ Okla, 69-73, res prof, 73-76; DIR CARDIOVASC RES, HENRY FORD HOSP, DETROIT, 76- *Concurrent Pos:* Adj prof physics, Oakland Univ, Rochester, Mich, 85- *Mem:* Fel Am COl Cardiol; fel Am Col Chest Physicians; Am Heart Asn; Am Physiol Soc; Laennec Soc; Cent Soc Clin Res; fel Am Col Physicians; Am Soc Mech Engrs. *Res:* Cardiovascular research and bioengineering; left ventricular function; mechanisms of heart sounds; disturbances of fluid flow; bioprosthetic valves; pulmonary embolism. *Mailing Add:* Cardiovasc Res Henry Ford Hosp 2799 W Grand Blvd Detroit MI 48202

STEIN, PAUL JOHN, b Pittsburgh, Pa, Sept 28, 50; m 74; c 3. BIOCHEMISTRY, INORGANIC CHEMISTRY. *Educ:* Bethany Col, WVa, BS, 72; Duke Univ, PhD(bioinorg chem), 76. *Prof Exp:* Res biochem, Inst Cancer Res, Fox Chase Cancer Inst, Philadelphia, 76-77; ASSOC PROF CHEM, COL ST SCHOLASTICA, 78- *Mem:* Am Chem Soc. *Res:* Fluorescence and nuclear magnetic resonance studies of enzymes; structure-function relationships of lectins. *Mailing Add:* Col St Scholastica 1200 Kenwood Ave Duluth MN 55811

STEIN, PAUL S G, b New York, NY, Apr 3, 43. NEUROBIOLOGY, MOTOR CONTROL. *Educ:* Harvard Univ, BA, 64; Univ Calif, Berkeley, 65; Stanford Univ, PhD(neurosci), 70. *Prof Exp:* Fel neurosci, Univ Calif, San Diego, 69-71; from asst prof to assoc prof, 71-86, PROF BIOL, WASHINGTON UNIV, 86- *Mem:* Soc Neurosci. *Res:* Spinal cord control of limb movement; scratch reflex in turtles; neuronal pattern generation. *Mailing Add:* Dept Biol Washington Univ St Louis MO 63130-4899

STEIN, PHILIP, b New York, NY, Apr 28, 32. PHYSIOLOGY. *Educ:* Brooklyn Col, BA, 53; George Washington Univ, MS, 54; Columbia Univ, MA, 59; Univ Geneva, PhD(biochem), 61. *Prof Exp:* Instr chem, Brooklyn Col, 60-62; instr, New York Community Col, 62-64; asst prof biol, Fairleigh Dickinson Univ, 64-68; ASSOC PROF BIOL, STATE UNIV NY COL, NEW PALTZ, 68- *Concurrent Pos:* USPHS grant, 62-64; NSF grant, 64-; consult biochemist, Nat Sugar Industs; consult, Sugar Refinery, Pepsi Cola, 72- *Mem:* Fel Am Inst Chem; Am Chem Soc; Am Soc Biol Chemists. *Res:* Chemical composition of the thyrotropic hormone secreted by the anterior pituitary gland; role of hypothalamus in regulation of thyroid function. *Mailing Add:* Dept of Biol State Univ NY Col New Paltz NY 12561

STEIN, REINHARDT P, b New York, NY, Dec 19, 35; m 62; c 2. ORGANIC CHEMISTRY. *Educ:* Rensselaer Polytech Inst, BS, 58; Ohio State Univ, PhD(org chem), 63. *Prof Exp:* Res chemist, Dow Chem Co, 63-64; res chemist, Wyeth Labs, Inc, 64-87, RES SCIENTIST, WYETH AYERST RES INC, 87- *Mem:* Am Chem Soc; Sigma Xi; Molecular Graphics Soc. *Res:* Total synthesis of natural products; new totally synthetic steroids; structural elucidation of natural products; synthesis of new drugs; design of new drugs; computer graphics of drugs; molecular modeling of drugs; computer systems management; computational chemistry. *Mailing Add:* Wyeth-Ayerst Res Inc CN 8000 Princeton NJ 08543

STEIN, RICHARD ADOLPH, b Edmonton, Alta, May 17, 37. ELECTRICAL ENGINEERING. *Educ:* Univ Alta, BS, 58; Univ Ill, Urbana, MS, 61; Univ BC, PhD(elec eng), 68. *Prof Exp:* Asst prof elec eng, Univ Alta, 61-65; assoc prof, 68-74, assoc dean, Fac Eng, 75-79, PROF ELEC ENG, UNIV CALGARY, 74- *Mem:* Inst Elec & Electronics Engrs; Am Soc Eng Educ. *Res:* Electrical circuit theory; electrical filter design; digital image processing. *Mailing Add:* Dept Elec Eng Univ Calgary 2500 Univ Dr NW Calgary AB T2N 1N4 Can

STEIN, RICHARD BERNARD, b New Rochelle, NY, June 14, 40; m 62; c 2. NEUROPHYSIOLOGY, BIOPHYSICS. *Educ:* Mass Inst Technol, BS, 62; Oxford Univ, MA & DPhil(physiol), 66. *Hon Degrees:* DSc, Univ Waterloo. *Prof Exp:* Res fel med res, Exeter Col, Oxford Univ, 65-68; assoc prof physiol, 68-72, PROF PHYSIOL, UNIV ALTA, 72- *Concurrent Pos:* USPHS fel, 66-68. *Mem:* Brit Physiol Soc; Can Physiol Soc; Neurosci Soc. *Res:* Motor control; information processing by nerve cells; sensory feedback; neural models. *Mailing Add:* Dept of Physiol Univ of Alta Edmonton AB T6G 2G7 Can

STEIN, RICHARD JAMES, b Palmerton, Pa, Aug 10, 30; m 53. POLYMER CHEMISTRY. *Educ:* Pa State Univ, BS, 58; Univ Akron, MS, 60, PhD(polymer chem), 67. *Prof Exp:* Res chemist, Goodyear Tire & Rubber Co, 60-63, sr res chemist, 66-71; polymer specialist, Insulating Mat Dept, Gen Elec Co, 71-85, staff chemist, 85-88; staff chemist, 88-89, CHIEF CHEMIST, INSULATING, MAT INC, 89- *Mem:* Am Chem Soc. *Res:* Polymer synthesis and properties. *Mailing Add:* 239 Pinewood Dr Schenectady NY 12303

STEIN, RICHARD JAY, b New York, NY, Jan 22, 46. GLOW-DISCHARGE PROCESSES, EQUPMENT DESIGN. *Educ:* Mass Inst Technol, SB, 67; Polytechnic Inst Brooklyn, MS, 70, PhD(physics), 72. *Prof Exp:* Scientist, Nat Bur Standards, 72-76; sr staff, Tex Instruments Res, 76-80 & GTE Labs, 80-81; mgr res & develop, Balzers Inc, 82-83; sr scientist, Perkin-Elmer Corp, 84-86; OWNER, R J STEIN ASSOCS, 86- *Res:* Microelectronic and coating equipment and process. *Mailing Add:* PO Box 252 West Redding CT 06896

STEIN, RICHARD STEPHEN, b Far Rockaway, NY, Aug 21, 25; m 51; c 4. POLYMER CHEMISTRY. *Educ:* Polytech Inst Brooklyn, BS, 45; Princeton Univ, MA, 48, PhD(phys chem), 49. *Prof Exp:* Asst, Polytech Inst Brooklyn, 45; asst, Princeton Univ, 49-50; asst prof, 50-57, from assoc prof to prof, 57-61, commonwealth prof, 61-80, GOESSMANN PROF CHEM & DIR, POLYMER RES INST, UNIV MASS, AMHERST, 80- *Concurrent Pos:* Fulbright vis prof, Kyoto Univ, 68; Ecole Super physics & chem, Paris, 86 & Universität Ulm, 88; vis prof, Syracuse Univ, 88, Laral Univ, 52. *Honors & Awards:* Int Award, Soc Plastics Eng, 69; Borden Award, Am Chem Soc, 72, Award in Polymer Chem, 83; Bingham Medal, Soc Rheol, 72; High Polymer Physics Award, Am Phys Soc, 76. *Mem:* Nat Acad Sci; Nat Acad Eng; Soc Rheol; Soc Polymer Sci Jap; Am Chem Soc; AAAS; Mat Res Soc; Am Phys Soc. *Res:* Molecular structure; light scattering; mechanical and optical properties of high polymers. *Mailing Add:* Lederlie Grad Res Tower 701 Univ of Mass Amherst MA 01002

STEIN, ROBERT ALFRED, b Chicago, Ill, Feb 20, 33; m 56; c 2. WEAPON SYSTEM ANALYSIS, ENGINEERING MECHANICS. *Educ:* Univ Ill, BS, 55; Ohio State Univ, MS, 58, PhD, 67. *Prof Exp:* Prin physicist, Dept Physics & Metall, Columbus Labs, Battelle Mem Inst, 55-60, sr physicist, 60-63, prog mgr, 63-66, assoc div chief, 66-70, asst sect mgr, 72-73, sect mgr, Defense Anal, 73-76, div chief, 70-76, assoc dir, Advan Syst Lab, 76-86, PROG MGR, ADV TECHNOL OFF, BATTELLE MEM INST, 86- *Concurrent Pos:* Guest ed, J Defense Res, 77. *Mem:* AAAS; Am Defense Preparedness Asn. *Res:* Hypervelocity impact phenomena and development of hypervelocity accelerators; weapon systems analysis; non-nuclear kill mechanisms; naval ordnance systems; army combat vehicle systems; terminal ballistics. *Mailing Add:* 11521 Danville Dr Rockville MD 20852

STEIN, ROBERT FOSTER, b New York, NY, Mar 4, 35; m 58; c 2. ASTROPHYSICS. *Educ:* Univ Chicago, BS, 57; Columbia Univ, PhD(physics), 66. *Prof Exp:* res fel, Carnegie Inst Wash Mt Wilson & Palomar Observs, 66-67; res fel, Harvard Observ, 67-69; asst prof astrophys, Brandeis Univ, 69-76; assoc prof astrophys, 76-81, PROF ASTROPHYS, MICH STATE UNIV, 81- *Concurrent Pos:* Consult, Smithsonian Astrophys Observ, 69-78; vis fel joint inst lab astrophys, Nat Bur Standards & Univ Colo, 73-74; vis scientist, Observatoire de Nice, 81, Univ St Andrews, 84, Nordita, 86; consult, Jet Propulsion Lab, 83-85. *Mem:* Am Astron Soc; Int Astron Union. *Res:* Astrophysical fluid dynamics; solar chromosphere and corona; radiative hydrodynamics; convection. *Mailing Add:* Dept Physics & Astron Mich State Univ East Lansing MI 48824

STEIN, ROBERT JACOB, PATHOLOGY, LEGAL MEDICINE. *Educ:* Univ Innsbruck, Austria, MD, 50; Northwestern Univ, MS, 60. *Prof Exp:* CHIEF MED EXAMR, COOK COUNTY, ILL, 76- *Concurrent Pos:* Consult & lectr forensic path. *Mem:* Am Acad Forensic Sci; Am Col Legal Med; Am Soc Pathologists. *Res:* Experimental pathology; forensic pathology; forensic toxicology. *Mailing Add:* 1828 W Polk St Chicago IL 60612

STEIN, ROY ALLEN, b Warren, Ohio, Aug 28, 47; m 73. ZOOLOGY, AQUATIC ECOLOGY. *Educ:* Univ Mich, BS, 69; Ore State Univ, MS, 71; Univ Wis, PhD(zool), 75. *Prof Exp:* Partic scientist fisheries, Smithsonian Inst, 72; res assoc zool, Univ Wis, 76; ASST PROF ZOOL, OHIO STATE UNIV, 76- *Mem:* Am Fisheries Soc; Am Inst Biol Sci; AAAS; Ecol Soc Am; Sigma Xi. *Res:* Behavioral ecology with a major emphasis on intra- and inter-specific interactions among aquatic organisms, specifically fish; examining how two important processes (predation and competition) structure fresh water communities. *Mailing Add:* Dept of Zool Ohio State Univ 1735 Neil Ave Columbus OH 43210

STEIN, RUTH E K, b New York, NY, Nov 2, 41; m 63; c 3. MEDICINE. *Educ:* Columbia Univ, New York, BA, 62; Albert Einstein Col, New York, MD, 66. *Prof Exp:* From instr to assoc prof, 70-83, PROF, DEPT PEDIAT, ALBERT EINSTEIN COL MED, 83- *Concurrent Pos:* Dir, Pediat Ambulatory Care Treatment Study, Albert Einstein Col Med, 77-83, Prev Intervention Res Ctr Child Health, 83-; vis scholar Robert Woods Johnson Clin Scholars Prog, Sch Med, Yale Univ, New Haven, Conn, 86-87, vis prof pub health, Dept Epidemiol & Pub Health, 86-87; mem, Pediat Emer Serv Comt, Inst Med, 91-92. *Honors & Awards:* Jacob I Berman & Dora B Friedman Fel Pediat, Albert Einstein Col Med, 76, Schick Medalist, Dept Pediat, 90. *Mem:* Ambulatory Pediat Asn (pres, 87-88); Am Acad Pediat; Am Pub Health Asn; Am Pediat Soc; Soc Pediat Res; Soc Behav Pediat. *Res:* Medical and health sciences. *Mailing Add:* 91 Larchmont Ave Larchmont NY 10538

STEIN, SAMUEL H, b New York, NY, Jan 6, 37; m 57; c 3. PHOTOGRAPHIC SCIENCE, PHYSICAL ORGANIC CHEMISTRY. *Educ:* City Col New York, BS, 57; Boston Univ, PhD(org chem), 67. *Prof Exp:* Res chemist, Nat Cash Register Co, Ohio, 57-59; res chemist, Itek Corp, summer 60, sr res chemist & proj leader org chem, Lexington Res Labs, 65-68, mgr paper develop group, Lexington Develop Labs, 68-69, mgr sci staff negative lithographic plate group, 69-71, mgr positive lithographic plate group, 71-72, mgr lithographic systs res dept, Lithographic Technol Lab, 72-74; mgr emulsion res, Chemco PhotoProd, 74-79, mem staff, New Prod Comt & Strategic Planning Comt, 79-84, dir emultion res & develop, 80-84; consult, Sesame Assocs, 84-87; tech dir, Citiplate, Inc, 87-89; TECH DIR, THERIMAGE DIV, AVERY DENNISON, 89- *Mem:* AAAS; Am Chem Soc; Soc Photog Sci & Eng; Sigma Xi. *Res:* Photochemistry, including unconventional photographic process and silver halide photo processes; silver halide emulsions for positive and negative systems; photoconductors, photopolymers and heterogeneous catalysis; graphic arts. *Mailing Add:* Nine Mathieu Dr Westborough MA 01581-3560

STEIN, SAMUEL RICHARD, b New York, NY, Jan 7, 46; m 66. PHYSICS. *Educ:* Brown Univ, ScB & ScM, 66; Stanford Univ, PhD(physics), 74. *Prof Exp:* Physicist time & frequency, 74-80, CHIEF PROG OFF, NAT BUR STANDARDS, 80- *Concurrent Pos:* Nat Res Coun fel, Nat Bur Standards, 74-76. *Mem:* Am Phys Soc. *Res:* Frequency standards; frequency metrology; superconductivity; lasers; quartz crystal oscillators. *Mailing Add:* 555 Jack Pine Ct Boulder CO 80302

STEIN, SEYMOUR NORMAN, b Chicago, Ill, Nov 23, 13; m 36; c 1. PHYSIOLOGY. *Educ:* Univ Ill, BS, 41, MD, 43. *Prof Exp:* From res asst to res assoc neurophysiol, Univ Ill, 46-49, asst prof & res physiologist, 49-51; head neurophysiol br & submarine & diving med br, Naval Med Res Inst, 52-60, head physiol div, 53-57; dep bio-sci officer, Pac Missile Range, US Dept Navy, 61-63; chief life sci officer, 64-65, chief med officer, 66-71, GUEST SCIENTIST, NASA-AMES RES CTR, 71-; PROF PHYSIOL, SAN JOSE STATE UNIV, 77- *Concurrent Pos:* Consult, Surgeon Gen, US Dept Army, 47-, NIMH, 53-; mem panel underwater swimmers, Nat Res Coun, 54-55; ed, J Biol Photog Asn, 56; pres, The Perham Found, 74-82, bd dirs, 75- *Honors & Awards:* Aerospace Contribution to Soc Medal, Am Inst Aeronaut & Astronaut, 80. *Mem:* Fel AAAS; Am Physiol Soc; Soc Exp Biol & Med; Biol Photog Asn; assoc fel Am Inst Aeronaut & Astronaut; Sigma Xi. *Res:* Basic and clinical studies on convulsions; effects of acute and chronic exposure to hypernormal amounts of carbon dioxide and oxygen; space medicine; bioinstrumentation. *Mailing Add:* 13080 Lorene Ct Mountain View CA 94040

STEIN, SHERMAN KOPALD, b Minneapolis, Minn, Aug 11, 26; m 50; c 3. MATHEMATICS. *Educ:* Calif Inst Technol, BSc, 46; Columbia Univ, MA, 47, PhD(math), 52. *Hon Degrees:* DH, Marietta Col, 75. *Prof Exp:* PROF MATH, UNIV CALIF, DAVIS, 53- *Honors & Awards:* L R Ford Award, Math Asn Am, 75. *Mem:* Am Math Soc; Math Asn Am. *Res:* Algebraic applications to geometry. *Mailing Add:* Dept Math Univ Calif Davis CA 95616

STEIN, STEPHEN ELLERY, b New York, NY, Dec 13, 48; m 74. PHYSICAL CHEMISTRY, CHEMICAL KINETICS. *Educ:* Univ Rochester, BS, 69; Univ Wash, PhD(phys chem), 74. *Prof Exp:* Res assoc, SRI Int, 74-75, phys chemist, 75-76; asst prof, Dept Chem, WVa Univ, 76-81, assoc prof, 81-82; RES CHEMIST, NAT BUR STANDARDS, 82-; ACTG CHIEF, CHEM KINETICS DIV, NAT BUR STANDARDS. *Mem:* NAm Photochem Soc; Am Chem Soc. *Res:* Thermochemistry and kinetics of elementary chemical reactions; rate and thermochemical estimation methods; unimolecular reactions; high temperature free radical reactions; gas-surface reactions; liquid-phase pyrolysis; analysis of complex reacting systems; coal conversion chemistry; combustion; ignition processes. *Mailing Add:* Chem Kinetics Div A149/222 Nat Inst Studies & Technol Gaithersburg MD 20899

STEIN, T PETER, b London, Eng, Apr 27, 41; m 67. BIOCHEMISTRY. *Educ:* Univ London, BSc, 62, MSc, 63; Cornell Univ, PhD(chem), 67. *Prof Exp:* Asst chem, Cornell Univ, 63-67; instr res surg & biochem, Sch Med, Univ Pa, 69-72, assoc prof surg, 72-86; PROF SURG & NUTRIT, SCH MED, UNIV MED & DENT OF NJ, 86- *Concurrent Pos:* NIH fel biochem, Univ Calif, Los Angeles, 67-69; adj prof surg, Univ Pa; adj prof nutrit, Rutgers Univ. *Mem:* AAAS; Am Chem Soc; Royal Soc Chem; Am Inst Nutrit; Am Soc Clin Nutrit; Am Physiol Soc; Am Col Nutrit; Am Soc Parenteral & Enteral Nutrit. *Res:* Protein metabolism during spacelift; lipid metabolism, clinical nutrition, nutritional assessment; nitrogen-15 metabolism and rates of protein synthesis in man; lung biochemistry. *Mailing Add:* 401 Haddon Ave Camden NJ 08103

STEIN, TALBERT SHELDON, b Detroit, Mich, Jan 6, 41; m 63; c 2. PHYSICS. *Educ:* Wayne State Univ, BS, 62; Brandeis Univ, MA, 64, PhD(physics), 68. *Prof Exp:* Res assoc physics, Univ Wash, 67-70; from asst prof to assoc prof, 70-82, PROF PHYSICS, WAYNE STATE UNIV, 82- *Concurrent Pos:* Alfred P Sloan res fel, 76-80. *Mem:* Fel Am Phys Soc; Am Asn Physics Teachers. *Res:* Experimental atomic physics including low energy positron-atom interactions; precision measurement of the g factor of the free electron; studies of effects of electric fields on neutral atoms. *Mailing Add:* 26707 Humber Huntington Woods MI 48070

STEIN, THEODORE ANTHONY, b St Louis, Mo, Aug 30, 38. PHOSPHOLIPIDS, CELL SIGNAL TRANSMISSION. *Educ:* St Louis Univ, BS, 60; Southern Ill Univ-Carbondale, MS, 70; City Univ NY, PhD(biochem), 88. *Prof Exp:* From res asst to res instr biochem, Sch Med, Wash Univ, 72-75; res supvr, biochem, 75-77, res coordr, biochem/physiol, 77-88, RES SCIENTIST, BIOCHEM/PHYSIOL & SURG, LONG ISLAND JEWISH MED CTR, 88- *Concurrent Pos:* Res asst prof, State Univ NY Stony Brook, 78-89; asst prof, Albert Einstein Col Med, 89-91. *Mem:* Sigma Xi; Am Pub Health Asn; Am Asn Clin Chem; Am Gastroenterol Asn; Am Fedn Clin Res. *Res:* Role of prostaglandins and leukotrienes in inflammatory bowel disease, gastric ulcer and motility; regulation of liver regeneration after partial hepatic resection; changes in pancreatic physiology associated with pharmacological agents; development of chromatographic methodoloetes to measure chromotheraputic agents in tissue and tumor. *Mailing Add:* Ten Glamford Ave Port Washington NY 11050

STEIN, WAYNE ALFRED, b Minneapolis, Minn, Dec 6, 37. ASTROPHYSICS. *Educ:* Univ Minn, BPhys, 59, PhD(physics), 64. *Prof Exp:* Res assoc astrophys, Princeton Univ, 64-66; asst res physicist, Univ Calif, San Diego, 66-69; asst prof astrophys, 69-73, assoc prof, 73-74, PROF PHYSICS, UNIV MINN, 74- *Concurrent Pos:* Alfred P Sloan Found fel, Univ Minn, Minneapolis & Univ Calif, San Diego, 69- *Mem:* Am Astron Soc; Am Phys Soc. *Res:* Infrared astronomy. *Mailing Add:* Univ Minn 116 Church St SE Univ Minn Minneapolis MN 55455

STEIN, WILLIAM EARL, b Rochester, NY, May 30, 24; m 47; c 4. ELECTRON PHYSICS, NUCLEAR PHYSICS. *Educ:* Univ Va, BEE, 46; Stanford Univ, MS, 50; Univ NMex, PhD(physics), 62. *Prof Exp:* Physicist, Los Alamos Sci Lab, 49-82; PYHSICIST, KIRK MAYER, 82- *Mem:* Am Phys Soc. *Res:* Electron-photon interactions and production of nearly monochromatic soft x-rays by the inverse Compton effect. *Mailing Add:* 124 Monte Rey Dr Los Alamos CA 87544

STEIN, WILLIAM EDWARD, b Cleveland, Ohio, June 18, 46. OPERATIONS RESEARCH. *Educ:* Case Western Reserve Univ, BS, 68; Purdue Univ, MS, 70; Univ NC, PhD(opers res), 75. *Prof Exp:* Asst prof math, Univ Ill, Chicago, 74-77; assoc prof math, Tex Christion Univ, 77-82; ASSOC PROF BUS, TEX A&M UNIV, 82- *Mem:* Opers Res Soc; Inst Mgt Sci; Asn Comput Mach. *Res:* Applied probability; fuzzy sets; stochastic dominance. *Mailing Add:* Bus Anal Dept Tex A&M Univ College Station TX 77843-4217

STEIN, WILLIAM IVO, b Wurzburg, Ger, July 22, 22; nat US; m 48; c 12. REFORESTATION. *Educ:* Pac Col, BS, 43; Ore State Univ, BF, 48; Yale Univ, MF, 52, PhD, 63. *Prof Exp:* Forester timber mgt res, Pac Northwest Res Sta, US Forest Serv, 48-52, asst res forester, 52-63, prin plant ecologist, 63- 80, res forester, 80-90, VOL, PAC NORTHWEST RES STA, US FOREST SERV, 90- *Mem:* Soc Am Foresters; Nature Conservancy; Am Forestry Asn. *Res:* study of ecology and physiology of Pacific Northwest species for the purpose of improving reforestation practices; reforestation research. *Mailing Add:* Forestry Sci Lab 3200 SW Jefferson Way Corvallis OR 97331

STEINBACH, LEONARD, b New York, NY, Feb 26, 27; m 51; c 3. ORGANIC CHEMISTRY. *Educ:* City Col New York, BS, 47; Polytech Inst Brooklyn, MS, 54. *Prof Exp:* Group leader org res, 56-62, prod mgr, 63-64, dir res & develop, 65, dir corp develop, 66-67, gen mgr, 67-69, vpres, 69-74, VPRES CORP DIR PURCHASING, INT FLAVORS & FRAGRANCES, INC, HAZLET, 74- *Concurrent Pos:* Mem chm fac, Monmouth Col, 56-61. *Mem:* AAAS; Am Chem Soc; NY Acad Sci. *Res:* Organic synthesis and development in aromatic chemicals, perfumes and flavor materials. *Mailing Add:* 10 Ramsgate Rd Cranford NJ 07016

STEINBECK, KLAUS, b Munich, Ger, Dec 11, 37; US citizen; m 60; c 4. FORESTRY, SILVICULTURE. *Educ:* Univ Ga, BSF, 61, MS, 63; Mich State Univ, PhD(forestry), 65. *Prof Exp:* Res plant physiologist, US Forest Serv, 65-68; asst prof, 69-72, assoc prof, 72-81, PROF FOREST RESOURCES, UNIV GA, 81- *Mem:* Soc Am Foresters. *Res:* Biomass production of short-rotation forests. *Mailing Add:* Sch Forest Resources Univ Ga Athens GA 30601

STEINBERG, ALFRED DAVID, b New York, NY, Nov 4, 40; m; c 4. IMMUNOLOGY, RHEUMATOLOGY. *Educ:* Princeton Univ, AB, 62; Harvard Univ, MD, 66. *Prof Exp:* Intern & resident internal med, Bronx Munic Hosp Ctr, 66-68; clin assoc, 68-70, sr staff fel, 70-71, sr investr, 71-81, chief cellular immunol sect, Nat Inst Arthritis, Diabetes & Digestive & Kidney Dis, 81-86, CHIEF, IMMUNOL SECT, ARTHRITIS & RHEUMATISM BR, NAT INST ARTHRITIS, METAB & DIGESTIVE DIS, NIH, 86-; MED DIR, US PUB HEALTH SERV, 78- *Concurrent Pos:* Clin Res Comt, NIH, 71-72, Peer Rev Comt, 72-74; USA-USSR Coop in Rheumatol & Immunol, 74-87; Lupus Study Group, Arthritis Found, 82-89, pres, 86-87; vis prof, Yale Univ, 81, Univ Calif, 82, 84 & 88, Univ Vienna, Austria, 83. *Honors & Awards:* Philip Hench Award, 74; Hollister-Stiers lectr, Wash State Univ, 85; Nelson lectr, Univ Calif, 87. *Mem:* NY Acad Sci; Am Fedn Clin Res; Am Rheumatism Asn; Am Asn Immunologists; Am Soc Exp Path; Am Soc Clin Invest; fel Am Col Physicians; Soc Exp Biol & Med; Transplantation Soc. *Res:* Autoimmunity; immune regulation. *Mailing Add:* Bldg 10 Rm 9N-218 NIH Bethesda MD 20892

STEINBERG, ARTHUR GERALD, b Port Chester, NY, Feb 27, 12; m 39; c 2. MEDICAL GENETICS, HUMAN GENETICS. *Educ:* City Col New York, BSc, 33; Columbia Univ, MA, 34, PhD(zool), 41. *Prof Exp:* Lectr genetics, McGill Univ, 40-44; mem opers res group, Off Sci Res & Develop, US Dept Navy, 44-46; assoc prof genetics, Antioch Col & chmn dept genetics, Fels Res Inst, 46-48; consult div biomet & med statist, Mayo Clinic, 48-52; geneticist, Children's Cancer Res Found & res assoc, Children's Hosp, Boston, Mass, 52-56; prof biol, Case Western Reserve Univ, 56-72, from asst prof to assoc prof, Dept Prev Med, 56-70, prof human genetics, Dept Reproductive Biol, 70-75, Francis Hobart Herrick prof biol, 72-82, prof human genetics, Dept Med, 75-82, FRANCIS HOBART HERRICK EMER PROF BIOL, 82- *Concurrent Pos:* Mem permanent comt int human genetics cong, NIH, 66-71; chmn med adv bd, Nat Genetics Found, 68-81; consult ed, Transfusion; adj staff mem, Cleveland Clin, 83-; consult, WHO. *Mem:* Fel AAAS; Am Soc Human Genetics (pres, 64); Genetics Soc Am; Am Asn Immunol; hon mem Japanese Soc Human Genetics. *Res:* Immunogenetics; study of genetic control of human immunoglobulins; population genetics; genetics of diabetes. *Mailing Add:* 20300 N Park Blvd No 4B Shaker Heights OH 44118

STEINBERG, BERNARD ALBERT, b New York, NY, Oct 2, 24; m 46; c 3. MICROBIOLOGY, VIROLOGY. *Educ:* NY Univ, AB, 47; Univ Ill, MS, 48, PhD(bact), 50. *Prof Exp:* Head sect bact & mycol, Squibb Inst Med Res, NJ, 50-56; head virus & cancer res, Wm S Merrell Co, 56-63; GROUP LEADER CHEMOTHER SECT, STERLING-WINTHROP RES INST, 63- *Mem:* AAAS; Am Soc Microbiol; Sigma Xi. *Res:* Antiviral chemotherapy; upper respiratory viruses; veterinary viruses; virus vaccines and immunology of viruses; viral etiology of cancer. *Mailing Add:* Sterling Winthrop Res Inst Rensselaer NY 12144

STEINBERG, BERNHARD, b New York, NY, June 18, 97; m 31; c 2. PATHOLOGY. *Educ:* Boston Univ, MD, 22. *Prof Exp:* House officer, Hosp Div, Med Col Va, 22-23; asst physician, Boston Psychopath Hosp, 23-24; dir labs & res, Toledo Hosp, Ohio, 27-64, dir inst med res, 43-64; ASSOC PROF RES IN PATH, SCH MED, LOMA LINDA UNIV, 64- *Concurrent Pos:* Nat Res Coun fel, Sch Med, Western Reserve Univ, 24-26, Crile res fel, 26-27; pvt pract oncol & hemat, Pomona, 65- *Honors & Awards:* Am Soc Clin Path Silver Medal, 37; Cincinnati Proctol Soc Cancer Res Award, 50. *Mem:* Am Soc Clin Path; AMA; Am Asn Path & Bact; Am Soc Exp Path; Am Soc Hemat. *Res:* Peritoneal infections; lung diseases; leukemias; hematopoiesis; cancer; originator, immunology of leukocyte; systemic disease of fat; development, functions and diseases of bone marrow; immuno-radiation therapy of cancer. *Mailing Add:* PO Box 1016 Pacific Palisades CA 90272

STEINBERG, BETTIE MURRAY, b Price, Utah, June 13, 37; m 60; c 3. CELL-VIRUS INTERACTIONS. *Educ:* Univ Calif, Riverside, BA, 59; Adelphi Univ, MS, 67; State Univ NY Stony Brook, PhD(microbiol), 76. *Prof Exp:* Fel, Dept Microbiol, State Univ NY Stony Brook, 76-78; res assoc, Dept Biol Sci, Columbia Univ, 78-80; res scientist, 80-83, HEAD SECT OTOLARYNGOL RES, LONG ISLAND JEWISH MED CTR, 83- *Concurrent Pos:* Lectr, Dept Biol Sci, Columbia Univ, 79-80; adj asst prof, Dept Surg, State Univ NY Stony Brook, 80-90; assoc prof otolaryngol, Albert Einstein Col Med, 90- *Mem:* Am Soc Microbiol; AAAS; NY Acad Sci; Am Acad Otolaryngol-Head & Neck Surg; Asn Res Otolaryngol. *Res:* Interaction between human papillomavirus type 11 (HPV 11) and laryngeal epithelial cells; major focus is on cellular and viral regulation of viral expression, using cells cultured in vitro and various cloned viral DNA mutants. *Mailing Add:* Dept Otolaryngol Long Island Jewish Med Ctr New Hyde Park NY 11042

STEINBERG, DANIEL, b Windsor, Ont, July 21, 22; US citizen; wid; c 3. BIOCHEMISTRY. *Educ:* Wayne State Univ, BS, 42, MD, 44; Harvard Univ, PhD(biochem), 51. *Prof Exp:* Intern internal med, Boston City Hosp, 44-45; resident, Detroit Receiving Hosp, 45-46; instr physiol, Med Sch, Boston Univ, 47-48; res scientist, Sect Cellular Physiol, Nat Heart Inst, 51-54, from actg chief to chief sect metab, 54-68; PROF MED, HEAD, DIV ENDOCRINOL & METAB & DIR, SPECIALIZED CTR RES ARTERIOSCLEROSIS, SCH MED, UNIV CALIF, SAN DIEGO, 68- *Concurrent Pos:* Vis scientist, Carlsberg Labs, Copenhagen, 52-53; pres, Found Advan Educ in Sci, 59; ed-in-chief, J Lipid Res, 61-64; chmn coun arteriosclerosis, Am Heart Asn, 68-69. *Honors & Awards:* Mayo Soley Award, 84; Distinguished Achievement Award, Am Heart Asn, 88. *Mem:* Nat Acad Sci; AAAS; Am Soc Biol Chem; Am Chem Soc; Am Oil Chem Soc; Soc Exp Biol & Med. *Res:* Mechanisms of hormone action; biochemistry of lipid and lipoprotein metabolism and its relation to atherosclerosis. *Mailing Add:* Div Endocrinol & Metab Dept Med M013D Univ Calif Sch of Med La Jolla CA 92093-0613

STEINBERG, DANIEL J, b Washington, DC, Feb 13, 35; m 59; c 2. THERMODYNAMICS, HYDRODYNAMICS. *Educ:* Johns Hopkins Univ, AB, 55; Harvard Univ, MA, 57, PhD(physics), 61. *Prof Exp:* SR PHYSICIST, LAWRENCE LIVERMORE LAB, UNIV CALIF, 61- *Mem:* Am Physics Soc. *Res:* Thermodynamics and equations of state; computer modeling of shock wave physics. *Mailing Add:* Lawrence Livermore Lab L-35 PO Box 808 Livermore CA 94551

STEINBERG, DAVID H, b Bronx, NY, Nov 24, 29; m 52; c 4. ORGANIC CHEMISTRY. *Educ:* Yeshiva Univ, BA, 51; NY Univ, MS, 56, PhD(org chem), 60. *Prof Exp:* Res asst biochem, Montefiore Hosp, NY, 52-53; res assoc org chem res div, NY Univ, 53-59; res chemist res div, Geigy Chem Corp, 59-72, res assoc, 72-83, SR STAFF SCIENTIST ADDITIVES RES, CIBA-GEIGY CHEM CORP 80- RES MGR, 83- *Mem:* Am Chem Soc; Royal Soc Chem. *Res:* Organic chemistry encompassing synthesis, reaction mechanism, stereochemistry and structure-activity relationships. *Mailing Add:* 2216 Wilson Ave Bronx NY 10469-5811

STEINBERG, DAVID ISRAEL, b St Louis, Mo, July 18, 42; m 65; c 2. OPERATIONS RESEARCH, APPLIED MATHEMATICS. *Educ:* Washington Univ, BS, 64, MS, 66, DSc, 68. *Prof Exp:* Asst prof appl math, Washington Univ, 67-72; assoc prof, 72-78, PROF MATH, SOUTHERN ILL UNIV, EDWARDSVILLE, 78-, DIR ASSESSMENT, 90- *Concurrent Pos:* Affil assoc prof, Washington Univ, 74-77, vis assoc prof comput sci, 77-78, affil prof, 78- *Mem:* Opers Res Soc Am; Am Asn Univ Professors; Inst Mgt Sci; Am Asn Higher Educ. *Res:* Mathematical programming; numerical linear algebra. *Mailing Add:* Dept Math & Comput Sci Southern Ill Univ Box 1653 Edwardsville IL 62026-1653

STEINBERG, ELIOT, b New York, NY, June 5, 23; m 47; c 3. ORGANIC CHEMISTRY, RESEARCH ADMINISTRATION. *Educ:* Polytech Inst Brooklyn, BS, 43, MS, 47. *Prof Exp:* Res chemist, Johnson & Johnson, NJ, 43-44; res chemist, Chilcott Labs Div, Maltine Co, 47-52; res adminr, Warner-Chilcott Labs, 52-58, dir res admin, 58-77, dir admin opers, Warner-Lambert Co, 77-81; mgr mem serv, Indust Res Inst, 81-88; consult, Rutgers Univ, 88-91; RETIRED. *Mem:* AAAS; Am Chem Soc; Am Soc Asn Execs; Coun Eng & Sci Soc Execs. *Res:* Information retrieval. *Mailing Add:* 20 Sherwood Dr Morristown NJ 07960

STEINBERG, ELLIS PHILIP, b Chicago, Ill, Mar 26, 20; m 44; c 3. CHEMISTRY. *Educ:* Univ Chicago, SB, 41, PhD(chem), 47. *Prof Exp:* Jr chemist, Elwood Ord Plant, US War Dept, Ill, 41-43; jr chemist Manhattan dist metall lab, Univ Chicago, 43-46; consult, AEC, Argonne Nat Lab, 46-47, assoc chemist & asst group leader, 47-58, sr chemist & group leader, 58-74, sect head nuclear & inorg chem, 74-82, dir, Chem Div, 82-88, acting assoc phys res, Lab Div, 86-87; CONSULT, SCI EDUC, 88- *Concurrent Pos:* Guggenheim fel, 57-58; mem sci adv comt space radiation effects lab accelerator, Col William & Mary; mem subcomt nuclear instruments & tech, Nat Acad Sci-Nat Res Coun, 48-59 & subcomt radiochem, 59-70; Am Chem Soc rep, Am Nat Standards Inst Subcomt Nuclear Med, 70-74; mem, Prog Adv Comt to Clinton P Anderson Meson Physics Lab, 77-80. *Honors & Awards:* Nuclear Chem Award, Am Chem Soc, 87. *Mem:* AAAS; Am Chem Soc; Am Phys Soc; Sigma Xi. *Res:* Nuclear and radiochemistry; nuclear fission; high energy nuclear reactions, meson-induced reactions, heavy-ion reactions. *Mailing Add:* Chem Div Argonne Nat Lab 9700 S Cass Ave Argonne IL 60439

STEINBERG, GEORGE MILTON, PHARMACOLOGY. *Educ:* Purdue Univ, PhD(org chem), 45. *Prof Exp:* Head Pharmacol Prog, NIH, 80-85; RETIRED. *Mailing Add:* 1730 Sunrise Dr Potomac MD 20854

STEINBERG, GUNTHER, b Cologne, Ger, Apr 14, 24; nat US; m 49; c 2. SURFACE CHEMISTRY, PHYSICAL CHEMISTRY. *Educ:* Univ Calif, Los Angeles, BS, 48, MS, 50, PhD(physiol chem), 56. *Prof Exp:* Res assoc, Scripps Metab Clin, Calif, 50; biochemist atomic energy proj, Univ Calif, Los Angeles, 50-56; res chemist, Martinez Res Lab, Shell Oil Co, 56-61, res chemist, Shell Develop Co, 61-64; res chemist, Stanford Res Inst, 64-67; sr staff chemist, Memorex Corp, 67-77; mgr, media res & mgr advan develop, 72-76, PRES & DIR, STEINBERG ASSOCS, 77- *Mem:* Am Chem Soc; Int Asn Colloid & Interface Scientists; NAm Thermal Analysis Soc. *Res:* Surface and physical chemistry of polymer composites; physical and chemical measurements; magnetic media processing and surface calorimetry development; electrical contact phenomena; heterogeneous catalysis; radiotracer applications. *Mailing Add:* 95 Lerida Ct Menlo Park CA 94025

STEINBERG, HERBERT AARON, b Bronx, NY, Sept 19, 29; m 55; c 2. MATHEMATICS. *Educ:* Cornell Univ, BA, 50; Yale Univ, MA, 51, PhD(math), 55. *Prof Exp:* Aerodynamicist, Repub Aviation Corp, 55-56; digital systs engr, Sperry Gyroscope Co, 56; sr mathematician, TRG Div, Control Data Corp, 56-68; dir sci serv, Math Applns Group, Inc, 68-76, res & develop dir, 76-85; DIR RES & DEVELOP, INFO SYST DIV, CADAM, 85- *Concurrent Pos:* Adj prof math, Polytech Inst, NY, 81-82. *Mem:* Soc Indust & Appl Math; Asn Comput Mach; Inst Elec & Electronics Eng; Math Asn Am; Am Math Soc; Sigma Xi. *Res:* Systems engineering; signal processing; Monte Carlo methods; computer simulation; radiation transport; random noise; stochastic processes; numerical analysis, computer generated imagery; computer aided design. *Mailing Add:* 25 N Lake Rd Armonk NY 10504

STEINBERG, HOWARD, b Chicago, Ill, Aug 23, 26; m 46; c 3. ORGANIC CHEMISTRY. *Educ:* Univ Ill, BS, 48; Univ Calif, Los Angeles, PhD(chem), 51. *Prof Exp:* AEC fel, Mass Inst Technol, 51-52; res chemist, Aerojet-Gen Corp, 52; res assoc org synthesis, Univ Calif, Los Angeles, 52-53; collabr natural prod, USDA, 53-54; res chemist, Pac Coast Borax Co, 54, mgr org res, US Borax Res Corp, 55-58, asst dir chem res, 58, from assoc dir to dir, 59-63, vpres, 63-69, PRES, US BORAX RES CORP, 69-, VPRES, US BORAX & CHEM CORP, 69-, MEM BD DIRS, 73- *Mem:* Am Chem Soc; Soc Chem Indust; Indust Res Inst; Am Inst Mining, Metall & Petrol Engrs. *Res:* Boron and synthetic organic chemistry; reaction mechanisms; kinetics. *Mailing Add:* 1401 Miramar Dr Fullerton CA 92631-2041

STEINBERG, JOSEPH, b New York, NY, Mar 22, 20; m 49; c 2. MATHEMATICAL STATISTICS, PROBABILITY SAMPLING. *Educ:* City Col New York, BS, 39. *Prof Exp:* Statistician pop div, US Bur Census, 40-42; math statistician, Social Security Bd, 42-44; statistician statist res div, US Bur Census, 44-45, chief statist methods br pop & housing div, 45-59, chief statist methods off, 59-60, chief statist methods div, 60-63; chief math statistician, Social Security Admin, 63-72; dir off surv methods res & asst comnr surv design, Bur Labor Statist, 72-75; PRES, SURV DESIGN, INC, 75- *Concurrent Pos:* Lectr, USDA Grad Sch, 42-74; consult, Orgn Am States, Chile, 65 & 67; vis prof surv res ctr, Inst Social Res, Univ Mich, 68, 69, 70 & 72; mem assembly behav & soc sci, Nat Acad Sci-Nat Res Coun, 72-77, mem comt Fed Agency Eval Res, 71-75, mem comt energy consumption measurement, 75-77; mem comt eval res, Social Sci Res Coun, 77-79; assoc ed, Am Statistician, 77-84. *Honors & Awards:* Distinguished Serv Award, US Dept Health Educ & Welfare, 68. *Mem:* Hon fel AAAS; hon fel Am Statist Asn; Inst Math Statist. *Res:* Sample survey design and statistical analysis; evaluation of non-sampling errors; response variance and bias; data linkage; computer analysis; quality control; operations research; cost functions and optimization. *Mailing Add:* 1011 Roswell Dr Silver Spring MD 20901-2131

STEINBERG, M(ORRIS) A(LBERT), b Hartford, Conn, Sept 24, 20; m 51; c 3. METALLURGY. *Educ:* Mass Inst Technol, BS, 42, MS, 46, DSc(metall), 48. *Prof Exp:* Instr metall, Mass Inst Technol, 45-48; head, Metall Dept, Horizons, Inc, 48-58; with Micrometric Instrument Co, 49-51; chief metallurgist, Horizons Titanium Corp, 51-55; secy & dir, Diwolfram Corp, 51-58; consult scientist, Lockheed Missiles & Space Co, 58-59, mgr mat, propulsion & ord res, Lockheed Aircraft Corp, Burbank, 59-62, lab dir, Mat Sci Lab, 62-64, mgr mat sci lab, Res & Develop Div, 64-65, dep chief scientist, 65-72, dir technol applns, 72-83, vpres sci, 83-86; RETIRED. *Concurrent Pos:* S K Wellman fel, Mass Inst Technol, 47-48; mem adv comt, US Israeli Bi-Nat Res Found, 75-; mem, Nat Mat Adv Bd, Nat Res Ctr, Nat Acad Sci, 75-; adj prof mat sci dept, Sch Eng, Univ Calif Los Angeles, 77-; pres, PVD Corp; mem bd dirs, HCC Industs, Encino, Calif, 80-87, Cadam Inc, Dialog Info Serv, Inc & Dataplan, 80-85; mem, Aeronaut Space Eng Bd, 83-86; chmn, Panel Continuing Educ & Utilization Engr NRC, Nat Acad Eng, 83-84, Air Force Studies Bd NRC Comt Nat Shape Technol, 84-85 & Aeronant Space & Eng Bd Comt NASA/Univ Retionships Aero/Space, 84-; mem, Nat Acad Eng Panel Mfg Res, 84-; mem, Comt Mat & Structures Technol Conf, 84; mem, Nat Acad Eng Ad Hoc Comt Evaluate Prog NSF's Directorate Eng; mem, Aeronaut & Space Eng Bd Comt, Nat Res Coun, Nat Acad Eng, 87 & Air Force Specification Bulletin Comt, 87-89. *Mem:* Nat Acad Eng; fel Am Soc Metall; fel Am Inst Chem; fel Am Inst Astronaut & Aeronaut; fel Inst Advan Eng; fel AAAS; Sigma Xi. *Res:* Missile and spacecraft materials; extractive metallurgy of reactive metals; powder metallurgy; high strength steels. *Mailing Add:* 348 Homewood Rd Los Angeles CA 90049

STEINBERG, MALCOLM SAUL, b New Brunswick, NJ, June 1, 30; m 83; c 4. DEVELOPMENTAL BIOLOGY, CELL BIOLOGY. *Educ:* Amherst Col, BA, 52; Univ Minn, MA, 54, PhD(zool), 56. *Prof Exp:* Instr zool, Univ Minn, 55; fel embryol, Carnegie Inst, Washington, 56-58; from asst prof to assoc prof biol, Johns Hopkins Univ, 58-66; prof, 66-75, HENRY FAIRFIELD OSBORN PROF BIOL, PRINCETON UNIV, 75- *Concurrent Pos:* Instr-in-charge embryol course, Woods Hole Marine Biol Lab, 67-72, trustee, 69-77; chmn, Div Develop Cell Biol, Am Soc Zool, 83-84; chmn, Gordon Res Conf Cell Contact & Adhesion, 85; mem bd biol, Nat Res Coun, Nat Acad Sci, 86-92. *Mem:* Fel AAAS; Am Soc Cell Biol; Am Soc Zool; Soc Develop Biol (secy, 70-73); Int Soc Develop Biol. *Res:* Mechanisms of animal morphogenesis; identification and analysis of cell adhesion systems; organization, assembly and adhesive interactions of desmosomes. *Mailing Add:* Dept Molecular Biol Princeton Univ Princeton NJ 08544

STEINBERG, MARCIA IRENE, b Brooklyn, NY, Mar 7, 44. ION TRANSPORT, ENZYMOLOGY. *Educ:* Brooklyn Col, BS, 64, MA, 66; Univ Mich, PhD(biochem), 73. *Prof Exp:* From asst prof to assoc prof pharmacol, State Univ NY Health Sci Ctr, Syracuse, 85-90; PROG DIR, BIOCHEM PROG, NSF, 90- *Concurrent Pos:* Mem, Basic Res Coun. *Mem:* Am Soc Biochem & Molecular Biol; Asn Women in Sci; AAAS. *Res:* Mechanism of action of ion transport ATPases. *Mailing Add:* NSF Rm 325 1800 G St NW Washington DC 20550

STEINBERG, MARSHALL, b Pittsburgh, Pa, Sept 18, 32; m 62; c 3. PHARMACOLOGY, TOXICOLOGY. *Educ:* Georgetown Univ, BS, 54; Univ Pittsburgh, MS, 56; Univ Tex Med Br Galveston, PhD(pharmacol, toxicol), 66; Nat Registry Clin Chem, cert, 70; dipl, Acad Toxicol Sci, 81. *Prof Exp:* Asst dir trop testing, Univ Pittsburgh, 55-56; chief clin path lab, 97th Gen Hosp, Med Serv Corps, US Army, Frankfurt, Ger, 56-60, chief biochem & toxicol div, 4th Army Med Lab, San Antonio, Tex, 61-63, chief toxicol div, Environ Hyg Agency, 66-71, dir lab serv US Army Environ Hyg Agency, 72-75; consult to US Army surg gen, Lab Sci, 75-76; vpres & dir bioassay prog, Tracor Jitco, 77-78; vpres & dir sci oper, 78-83, VPRES & SCI DIR, HAZLETON LABS CORP, 83- *Concurrent Pos:* Liaison mem, Armed Forces Pest Control Bd, 67-75; mem pesticide monitoring panel, Fed Working Group Pesticide Mgt, Coun Environ Quality, 71-72 & safety panel, 72-; adj prof, Am Univ, 81; mem, Prof Accreditation Bd, Acad Toxicol Sci, 82-85; secy, Toxicol Lab Accreditation Bd, 84-85; pres, Nat Capital Area Chap, Soc Toxicol, 84-85. *Mem:* Soc Toxicol, (secy, 83-85); Am Soc Pharmacol & Exp Therapeut; Am Conf Govt Indust Hygienists; Am Col Toxicol (pres elect, 84-85); Am Indust Hyg Asn. *Res:* Applied research in industrial and environmental toxicology, insect repellants, pesticides and fire extinguishants, particularly in regard to hazards owing to skin penetration or irritation as well as toxic effects due to inhalation. *Mailing Add:* Hercules Inc One Hercules Plaza Wilmington DE 19894

STEINBERG, MARTIN, b Chicago, Ill, Apr 18, 20; m 42; c 3. PHYSICAL CHEMISTRY. *Educ:* Univ Ill, BS, 41; Univ Chicago, PhD(chem), 49. *Prof Exp:* Res chemist, Continental Carbon Co, 41-45; res assoc inst nuclear studies, Univ Chicago, 49-51; res chemist, Gen Elec Co, 51-56; sr scientist, Armour Res Found, 56-61; head chem physics, Delco Electronics Div, Gen

Motors Corp, 61-74; res chemist, Quantum Inst, Univ Calif, Santa Barbara, 74-90; RETIRED. *Mem:* AAAS; Am Chem Soc; Am Phys Soc; Combustion Inst. *Res:* Carbon black formation and properties; electrochemistry in fused salts; stable isotope geochemistry; reentry physics; high temperature kinetics; air pollution; chemical lasers; gaseous radiation and spectroscopy; flame chemistry. *Mailing Add:* 345 N Ontare Rd Santa Barbara CA 93105

STEINBERG, MARTIN H, b New York, NY, July 2, 36; m 73; c 1. MOLECULAR BIOLOGY, GENETICS. *Educ:* Cornell Univ, AB, 58; Tufts Univ, MD, 62. *Prof Exp:* Med intern, Cornell Med Serv, 62-63; med resident, New Eng Med Ctr, 66-68, fel hemat, 68-70; asst prof, 70-74, assoc prof, 74-77, PROF MED, MED SCH, UNIV MISS, 77- *Concurrent Pos:* Mem, Res Rev Comt, Am Heart Asn, 81-84; prin investr, NIH, Vet Admin, 73-; assoc chief staff res, Jackson Vet Admin Med Ctr, 73-; asst dean, Sch Med, Univ Miss, 73-; vis prof med, Tufts Univ, 87. *Mem:* Am Fedn Clin Res; Am Soc Clin Invest; Am Soc Hemat; Asn Am Physicians. *Res:* Intrinsic factors which may determine the clinical course of sickle cell anemia; interactions of structural variants of hemogoblin with the thalassemia syndromes. *Mailing Add:* Vet Admin Med Ctr 151 Jackson MS 39216

STEINBERG, MARVIN PHILLIP, b Philadelphia, Pa, Oct 4, 22; m 46; c 3. AGRICULTURAL & FOOD CHEMISTRY. *Educ:* Univ Minn, BS, 43, MS, 49; Univ Ill, PhD(food technol), 53. *Prof Exp:* Asst chem eng, Univ Minn, 45-49; asst food tech, 49-52, from instr to assoc prof, 52-63, PROF FOOD ENG, UNIV ILL, URBANA, 63-, PROF AGR ENG, 72- *Mem:* Am Inst Chem Eng; Am Soc Agr Eng; Am Chem Soc; Inst Food Technol; Soc Indust Microbiol. *Res:* Application of chemical engineering to food product and process development; states of water bound by food constituents; NMR quantification, characterization and application to immediate moisture foods; agricultural biomass and food waste conversion into feed and energy. *Mailing Add:* Dept Food Sci Univ Ill 1304 W Pennsylvania Ave Urbana IL 61803

STEINBERG, MELVIN SANFORD, b Canton, Ohio, Mar 28, 28; m 54; c 2. THEORETICAL PHYSICS. *Educ:* Univ NC, BS, 49, MS, 51; Yale Univ, PhD(physics), 55. *Prof Exp:* Asst physics, Yale Univ, 54-55; asst prof, Stevens Inst Technol, 55-59; assoc prof, Univ Mass, 59-62; assoc prof, 62-86, PROF PHYSICS, SMITH COL, 86- *Concurrent Pos:* Res assoc, Woods Hole Oceanog Inst, 56-58 & 62; res assoc, Air Force Cambridge Res Lab, 60-61; NSF sci fac fel, 66-67. *Mem:* Am Phys Soc; Sigma Xi; Am Asn Phys Teachers. *Res:* Theory of solids; acoustics; electrodynamics; physics education; student's problems of comprehension in electricity and mechanics; new apparatus and instructional strategies; comparison of student's and historical learning. *Mailing Add:* Dept of Physics Smith Col Northampton MA 01063

STEINBERG, MEYER, b Philadelphia, Pa, July 10, 24; m 50; c 2. CHEMICAL ENGINEERING, CHEMISTRY. *Educ:* Cooper Union, BChE, 44; Polytech Inst Brooklyn, MChE, 49. *Prof Exp:* Jr chem engr, Manhattan Dist, Kellex Corp, 44-46; asst chem engr process develop, Deutsch & Loonam, NY, 47-50; assoc chem engr, Guggenheim Bros, Mineola, NY, 50-57; CHEM ENGR & HEAD, PROCESS SCI DIV, DEPT ENERGY & ENVIRON, BROOKHAVEN NAT LAB, 57- *Mem:* Fel Am Inst Chem Engrs; Am Chem Soc; Sigma Xi; AAAS; fel Am Nuclear Soc. *Res:* New processes for energy conversion, nuclear, fossil, geothermal and solar; development of materials for conservation and energy storage; process research and development in energy conversion. *Mailing Add:* Process Sci Div Brookhaven Nat Lab Upton NY 11973

STEINBERG, MITCHELL I, b Philadelphia, Pa, Jan 22, 44; m 66; c 3. PHARMACOLOGY. *Educ:* Philadelphia Col Pharm & Sci, BSc, 66; Univ Mich, PhD(pharmacol), 70. *Prof Exp:* Res fel, Univ Conn Health Ctr, 70-71, instr pharmacol, 71-72; sr pharmacologist, Eli Lilly Res Labs, 72-77, res scientist, 77-80, res assoc, 80-89, RES ADV, ELI LILLY RES LABS, 90- *Mem:* AAAS; Am Soc Pharmacol & Exp Therapeut; NY Acad Sci. *Res:* Interactions of pharmacological agents with excitable membranes of mammalian cardiac & neuronal tissues; antiarrhythmic drug research; cardiac electrophysiology. *Mailing Add:* Dept Cardiovasc & Pharmacol MC 607 Lilly Res Labs 307 E McCarty St Indianapolis IN 46285

STEINBERG, PHILLIP HENRY, experimental high energy physics; deceased, see previous edition for last biography

STEINBERG, RICHARD, b Brooklyn, NY, Feb 22, 30; m 53; c 3. ELECTRICAL ENGINEERING. *Educ:* City Col New York, BEE, 52; Univ Pa, MSEE, 58. *Prof Exp:* Jr engr, Govt & Indust Div, Philco Corp, 52-54; engr, Decker Corp, 54-56; sr engr, Electronics Div, Parsons Corp, 56-59; mem tech staff, Hughes Aircraft Co, 59-61; asst sect head elec eng, TRW Systs Group, Calif, 61-71; dept staff engr, Western Develop Labs Div, Philco-Ford Corp, 71; staff engr, 72-80, SR STAFF ENGR, LOCKHEED MISSILES & SPACE CO, SUNNYVALE, 80- *Concurrent Pos:* Instr, Univ Calif, Los Angeles, 61-64 & Calif State Col, Los Angeles, 61-68. *Res:* Scattering from random surfaces; acoustical and electromagnetic waves; nonlinear differential equations. *Mailing Add:* 2714 Preston Dr Mountain View CA 94040

STEINBERG, ROBERT, b Stykon, Rumania, May 25, 22; m 52. MATHEMATICS. *Educ:* Univ Toronto, PhD(math), 48. *Prof Exp:* Lectr math, Univ Toronto, 47-48; from instr to assoc prof, 48-62, PROF MATH, UNIV CALIF, LOS ANGELES, 62- *Mem:* Nat Acad Sci; Am Math Soc; Math Asn Am. *Res:* Group representations; algebraic groups. *Mailing Add:* Dept Math 6364 Math Sci Bldg Univ Calif Los Angeles CA 90024

STEINBERG, RONALD T, b New York, NY, Apr 18, 29; m 50; c 3. CHEMICAL ENGINEERING. *Educ:* Polytech Inst Brooklyn, BChE, 49; Clarkson Col Technol, MChE, 51. *Prof Exp:* Process engr, Schwarz Labs, Inc, 51-52, plant engr, 52-61; plant engr, Alcolac Inc, 61-66, mgr process deveop, 66-84; RETIRED. *Concurrent Pos:* Environ consult. *Mem:* Am Chem Soc. *Res:* Process development; specialty surfactants and monomers; ethoxylations, sulfations, and esterifications; diffusional operations and solids handling. *Mailing Add:* 2107 Cedar Circle Dr Catonsville MD 21228

STEINBERG, ROY HERBERT, b New York, NY, Dec 9, 35; m 59; c 1. NEUROPHYSIOLOGY. *Educ:* Univ Mich, BA, 56, MA, 57; NY Med Col, MD, 61; McGill Univ, PhD(neurophysiol), 65. *Prof Exp:* Intern med, Mass Mem Hosp, 61-62; USPHS fel, NIMH, 62-65; head Neurophysiol Br, Neurol Sci Div, Naval Aerospace Med Inst, Fla, 68-69; asst res physiologist, 69-72, assoc prof physiol, 72-77, PROF PHYSIOL & OPHTHAL, SCH MED, UNIV CALIF, SAN FRANCISCO, 78- USPHS career develop award, 71-; William C Bryant, Bernard C Spiegel & Harpuder Awards, NY Med Col. *Mem:* Soc Neurosci; Asn Res Vision & Ophthal; Am Phys Soc. *Res:* Physiology of the nervous system; vision, especially physiology and anatomy of the retina. *Mailing Add:* Dept Physiol Univ Calif S-762 San Francisco CA 94143

STEINBERG, SETH MICHAEL, b Washington, DC, June 5, 58; m 89; c 1. CLINICAL TRIALS, DATA ANALYSIS. *Educ:* Johns Hopkins Univ, BA, 79; Univ NC, Chapel Hill, MS, 81, PhD(biostatist), 83. *Prof Exp:* Statistician, EMMES Corp, 83-86; actg head, 86-90, HEAD, BIOSTATIST & DATA MGT SECT, CLIN ONCOL PROG, NAT CANCER INST, 90- *Mem:* Am Statist Asn; Biometric Soc. *Res:* Design, monitoring, and evaluation of clinical trials for treatment of cancer and AIDS; identification of prognostic factors for patients treated on studies; development and use of methods for evaluation of data from clinical trials. *Mailing Add:* 10826 Antigua Terr Rockville MD 20852

STEINBERG, STANLY, b Traverse City, Mich, Mar 10, 40; m 60. APPLIED MATHEMATICS. *Educ:* Mich State Univ, BS, 62; Stanford Univ, PhD(math), 68. *Prof Exp:* Asst prof math, Purdue Univ, Lafayette, 67-74; ASSOC PROF MATH, UNIV NMEX, 74- *Res:* Partial differential equations. *Mailing Add:* Dept Math-Statist Univ NMex Albuquerque NM 87131

STEINBERG, STUART ALVIN, b Chicago, Ill, Feb 3, 41; m 66; c 3. MATHEMATICS. *Educ:* Univ Ill, Urbana, BS, 63, PhD(math), 70; Univ Chicago, MS, 65. *Prof Exp:* Asst prof math, Univ Mo-St Louis, 70-71; from asst prof to assoc prof, 71-80, PROF MATH, UNIV TOLEDO, 80- *Mem:* Am Math Soc. *Res:* Algebra, especially ring theory and ordered algebraic structures. *Mailing Add:* Dept Math Univ Toledo Toledo OH 43615

STEINBERGER, ANNA, b Radom, Poland, Jan 1, 28; US citizen; m 50; c 2. CELL BIOLOGY, IMMUNOLOGY. *Educ:* State Univ Iowa, MS, 52; Wayne State Univ, PhD(microbiol), 61. *Prof Exp:* Bacteriologist, State Univ Iowa, 53-55; res virologist, Parke, Davis & Co, Mich, 55-56 & 58-59; asst mem, Albert Einstein Med Ctr, 61-71; PROF REPRODUCTIVE BIOL, UNIV TEX MED SCH HOUSTON, 71- *Concurrent Pos:* USPHS res grant, Albert Einstein Med Ctr, 61-71. *Mem:* Endocrine Soc; Soc Study Reproduction; Tissue Cult Asn; NY Acad Sci; Am Soc Andrology (vpres, 84-85, pres, 85-86). *Res:* Endoeroue and paracrine regulation of spermatogenesis in male mammalian gonads; hormonal control of spermatogenesis; secretion of gonadotropins; cell interactions in the testis; tissue culture. *Mailing Add:* Univ Tex Med Sch 6431 Fannin Suite 3212 Houston TX 77030

STEINBERGER, JACK, b Bad Kissingen, Ger, May 25, 21; nat US; m 61; c 2. PHYSICS. *Educ:* Univ Chicago, BS, 42, PhD, 48. *Prof Exp:* Mem, Inst Advan Study, Princeton, 48-49; res asst, Univ Calif, Berkeley, 49-50; Higgins prof physics, Columbia Univ, 50-71; physicist, Europ Ctr Nuclear Res, 68-86, PROF, Schola Normale Sup, Pisa, Italy, 86- *Concurrent Pos:* Mem, Inst Advan Study, Princeton, 59-60; hon prof, Univ Heidelberg. *Honors & Awards:* Nobel Prize in Physics, 88. *Mem:* Nat Acad Sci; Am Acad Arts & Sci; Heidelberg Acad Sci. *Res:* Mesons, spin, parity and other properties of pions; particle, spins, other properties of strange particles; two neutrinos; CP violating properties of kaons; interactions of neutrinos at high energies; quark structure of nucleons. *Mailing Add:* European Ctr Nuclear Res (CERN) Geneva 23 Switzerland

STEINBRECHER, LESTER, b Philadelphia, Pa, Sept 17, 27; m 51; c 3. INORGANIC CHEMISTRY. *Educ:* Temple Univ, BA, 50; Drexel Inst, MS, 57. *Prof Exp:* Chemist, Socony Mobil Oil Corp, 52-58; res chemist, 58-70, dir res, 70-88, SR TECHNOL CONSULT, AMCHEM PROD INC, 88- *Mem:* Nat Asn Corrosion Eng; Am Soc Electroplaters. *Res:* Solid state reactions; inorganic metallic coatings; analytical chemistry; coatings for metals; twenty-one US patents. *Mailing Add:* Amchem Prod Inc Ambler PA 19002

STEINBRENNER, ARTHUR H, b New York, NY, July 23, 17; div; c 2. MATHEMATICS EDUCATION. *Educ:* Columbia Univ, AB, 40, AM, 41, PhD, 55. *Prof Exp:* Instr math & physics, Graham-Eckes Sch, Fla, 41-46; asst math, Teachers Col, Columbia Univ, 46-48; asst prof, US Naval Acad, 48-53; from instr to asst prof, 53-56, from asst prof to prof math & educ, 56-70, prof, 70-83, EMER PROF MATH, UNIV ARIZ, 84- *Concurrent Pos:* Coordr, Sch Math Study Group, 58-61, Ariz Ctr Minn Math & Sci Teaching Proj, 63-65; Fulbright lectr, Australia, 63. *Mem:* Math Asn Am; Sch Sci & Math Asn; Nat Coun of Teachers Mathematics. *Res:* Mathematics curricula experiments; math learning by the blind. *Mailing Add:* Dept Math Univ Ariz Tucson AZ 85721

STEINBRENNER, EUGENE CLARENCE, b St Paul, Minn, Sept 3, 21; m 44; c 4. FORESTRY, SOILS. *Educ:* Univ Minn, BS, 49; Univ Wis, MS, 51; Univ Wash, Seattle, PhD(forest soils), 54. *Prof Exp:* Asst, Univ Wis, 49-51; asst, State Conserv Dept, Wis, 49-50; forest soils specialist, Forestry Res Ctr, Weyerhaeuser Co, 52-81; RETIRED. *Concurrent Pos:* Weyerhaeuser fel, 51-52, affil prof forest soils, Univ Wash, 74- Bullard fel, Harvard Univ, 68. *Mem:* Fel Soil Sci Soc Am; Soc Am Foresters; fel Am Sci Affil. *Res:* Soil classification and mapping; nutrition; productivity; soil management for site protection, rehabilitation and improvement; nursery soils. *Mailing Add:* PO Box 368 Centralia WA 98531

STEINBRUEGGE, KENNETH BRIAN, b St Louis, Mo, Dec 9, 39; m 66; c 3. ADVANCED SENSORS & SMART SENSORS. *Educ:* Univ Mo, Rolla, BS, 62, MS, 63. *Prof Exp:* Assoc engr laser applications res & develop, Res & Develop Ctr, Westinghouse Elec Corp, 63-65, scientist laser mat & systs, 65-69, sr scientist laser-optical res & develop, 69-85, fel scientist optical-analysis instrumentation, 85-86, mgr sensor appl, optical-nuclear sensors, 82-86, mgr fiber optics, Mach Technol Div, 86-89, SR ADV ENGR HEALTH PHYSICS INSTRUMENTATION, MECH TECHNOL DIV, WESTINGHOUSE ELEC CORP, 89- *Concurrent Pos:* Technol advocate, Westinghouse Elec Corp, 81-87. *Mem:* Optical Soc Am; Health Physics Soc; Am Nuclear Soc; Am Soc Testing & Mat. *Res:* Development of advanced microprocessor based continuous air monitors for radiological monitoring; research and devlopment on networking radiological instrumentation, analysis of radiological data for improved sensitivity and trending; use of nuclear techniques for sensing and analysis, such as detection of plastic explosives; optical analytical instrumentation; fiber optic sensors and communications; 12 US patents. *Mailing Add:* 3496 Ivy Lane Murrysville PA 15668

STEINBRUGGE, KARL V, b Tucson, Ariz, Feb 8, 19; m 42; c 2. ENGINEERING SEISMOLOGY. *Educ:* Ore State Univ, BS, 41. *Prof Exp:* Struct designer, Austin Co, 42-47; sr struct engr, Calif Div Archit, 48-50; prof, 50-78, EMER PROF, STRUCT DESIGN, UNIV CALIF, BERKELEY, 78-; CONSULT ENGR, 80- *Concurrent Pos:* Chief engr, Earthquake Dept, Pac Fire Rating Bur, 50-71, mgr, Earthquake Dept, Insurance Serv Off, 71-80; vpres, Earthquake Eng Res Inst, 62, pres, 68-70; mem US comt, Int Asn Earthquake Eng, 69, US chmn, 66-73; chmn task force earthquake hazard reduction, Off Sci & Technol, Washington, DC, 70; chmn adv group, Calif Legis Joint Comt Seismic Safety, 70-73; chmn eng criteria rev bd, San Francisco Conserv & Develop Comn, 71-72; chmn earthquake hazard reduction, Exec Off Pres, 77-78; chmn, Calif Seismic Safety Comn, 74-77, mem, 77-80; invited keynote speaker, foreign international conferences, Turkey, 68, 73, Peru, 73, NZ, 75, 89, Malaysia, 83; chmn, US Working Group Earthquake Related Casualties, 89-90 & Working Group on Risk, Int Asn Seismol & Physics Earth's Interium, 88- *Honors & Awards:* Alfred E Alquist Medal, 87. *Mem:* Seismol Soc Am (vpres, 66-67, pres, 67-68); hon mem Earthquake Eng Res Inst (pres, 67-68). *Res:* Earthquake engineering; author or co-authored 98 published scientific papers on earthquake engineering and earthquake damage. *Mailing Add:* 6851 Cutting Blvd El Cerrito CA 94530

STEINDLER, MARTIN JOSEPH, b Vienna, Austria, Jan 3, 28; nat US; m 52; c 2. NUCLEAR FUEL CYCLE, RESEARCH ADMINISTRATION. *Educ:* Univ Chicago, PhB, 47, BS, 48, MS, 49, PhD(chem), 52. *Prof Exp:* Res asst, US Navy Inorg Proj, Univ Chicago, 48-52, consult, 53; Argonne Nat Lab, 53; assoc chemist, 53-74, sr chemist, 74-77, assoc dir, Chem Eng Div, 77-84, DIR CHEM TECH DIV, ARGONNE NAT LAB, 84- *Concurrent Pos:* Mem, Atomic Safety & Licensing Bd Panel, 72-90; consult, Adv Comt Reactor Safeguards, 67-87, Lawrence Livermore Lab, 78-, Oak Ridge Gas Diffusion Plant, 80-; mem, Adv Comt Reactor Safeguards, 87-88, Adv Comt Nuclear Waste, 88- *Honors & Awards:* Robert E Wilson Award, Am Inst Chem Eng, 90. *Mem:* Am Chem Soc; Am Nuclear Soc; Royal Soc Chemists; Sigma Xi; AAAS; Am Inst Chem Eng. *Res:* Nuclear fuel cycle; radiological safety; nuclear waste disposal; fluorine chemistry of the actinide elements and fission product elements; reactor fuel reprocessing; non-aqueous inorganic chemistry. *Mailing Add:* Argonne Nat Lab Bldg 205 RmA128 9700 S Cass Ave Argonne IL 60439-4837

STEINECK, PAUL LEWIS, b Yonkers, NY, Jan 20, 42; m 88. MICROPALEONTOLOGY, BIOLOGICAL SYSTEMATICS. *Educ:* NY Univ, BA, 63, MS, 66; La State Univ, PhD(geol), 73. *Prof Exp:* From asst prof to assoc prof, 71-87, PROF NATURAL SCI, STATE UNIV NY COL, PURCHASE, 87- *Concurrent Pos:* chmn, Div Nat Sci, Suny, Purchase, 81-85. *Mem:* Geol Soc Am; Paleont Soc; Soc Econ Paleontologists & Mineralogists; Int Paleont Union; Micropaleont Soc. *Res:* Deep sea ostracoda from the Pacific and Caribbean and Southern ocean; emphasis on taxonomy and pale oceanography and evolution; deep sea ostracoda from eutrophic habitat islands (taxonomy, origin and ecology); ostracoda of tidal wetlands in the Hudson River. *Mailing Add:* Div Natural Sci State Univ NY Purchase NY 10577

STEINER, ANDRE LOUIS, b Haguenau, France, June 21, 28. ZOOLOGY. *Educ:* Univ Strasbourg, BSc, 52; Univ Paris, DSc(animal behav), 60. *Prof Exp:* Res assoc animal biol, Nat Res Coun, Paris, France, 52-56; teaching asst zool, Univ Montpellier, 56-59, teaching asst psychophysiol, 59-61, asst prof psychophysiol & animal behav, 61-65; vis prof zool & animal behav, Univ Montreal, 65-66; assoc prof, 66-72, PROF ZOOL, UNIV ALTA, 72- *Concurrent Pos:* Mem, Int Union Conserv Nature & Natural Resources, Switz, 67. *Mem:* AAAS; Animal Behav Soc; Can Soc Zool; Ecol Soc Am. *Res:* Behavior, ecology and distribution of solitary wasps and of some vertebrates; ecology, especially behavioral aspects and field studies; animal distribution and dispersion; cine-photo analysis of behavior. *Mailing Add:* Dept Zool Univ of Alta Edmonton AB T6G 2E2 Can

STEINER, ANNE KERCHEVAL, b Warrensburg, Mo, Aug 5, 36. MATHEMATICS. *Educ:* Univ Mo, AB, 58, MA, 63; Univ NMex, PhD(math), 65. *Prof Exp:* Eng asst, Am Tel & Tel Corp, 58-59; asst prof math, Tex Tech Col, 65-66; asst prof, Univ NMex, 66-68; from asst prof to assoc prof, 68-72, PROF MATH, IOWA STATE UNIV, 72- *Concurrent Pos:* Vis assoc prof, Univ Alta, 70-71; chair, Math Dept, Iowa State Univ, 82-86. *Mem:* Am Math Soc; Math Asn Am; Sigma Xi. *Res:* Point set topology. *Mailing Add:* Dept Math Iowa State Univ Ames IA 50011

STEINER, BRUCE, b Oberlin, Ohio, May 14, 31; m 60; c 2. OPTICS. *Educ:* Oberlin Col, AB, 53; Princeton Univ, PhD, 57. *Prof Exp:* Res assoc physics, Univ Chicago, 58-61; SCIENTIST, NAT INST STANDARDS & TECHNOL, 61- *Mem:* AAAS; Am Phys Soc; Inst Elec & Electronic Engrs; Optical Soc Am. *Res:* Structure of materials, optical properties of materials; breakdown of molecules under impact; electron detachment phenomena in ions and molecules; optical radiation measurement. *Mailing Add:* Nat Inst Standards & Technol Bldg Mat Rm B309 Gaithersburg MD 20899

STEINER, DONALD FREDERICK, b Lima, Ohio, July 15, 30. BIOCHEMISTRY, ENDOCRINOLOGY. *Educ:* Univ Cincinnati, BS, 52; Univ Chicago, MS & MD, 56. *Hon Degrees:* DSc, Royal Univ Umea, Sweden, 73 & Univ Ill Chicago, 84. *Prof Exp:* Intern, King County Hosp, Seattle, Wash, 56-57; asst med, Univ Wash, 57-60, med resident, 59-60; from asst prof to prof biochem, Univ Chicago, Ill, 60-70, A N Pritzker prof biochem & med, 70-74, actg chmn, Dept Biochem, 72-73, chmn, 73-79, dir, Diabetes-Endocrinol Ctr, 74-78, assoc dir, Diabetes & Res Training Ctr, 77-81, A N PRITZKER DISTINGUISHED SERV PROF BIOCHEM & MOLECULAR BIOL & MED, UNIV CHICAGO, 84-, SR INVESTR, HOWARD HUGHES MED INST, 85- *Concurrent Pos:* Res fel med, Univ Wash, 57-59; res career develop award, USPHS, 62-72; mem, Metab Study Sect, USPHS, 65-70, Coun Am Soc Biochem & Molecular Biol, 86-; E F F Copp mem lect, La Jolla, Calif, 71; consult, Cetus Corp, Emeryville, Calif, 80-, Eli Lilly & Co, Indianapolis, Ind, 87-, Baxter Corp, Chicago, Ill, 87-, Abbott Labs, 89- & Novo Labs, 89- *Honors & Awards:* Bordon Award, Asn Am Med Cols, 56 & 80; Lilly Award, Am Diabetes Asn, 69, Banting Medal, 76; Ernst Oppenheimer Award, Endocrine Soc, 70, Fred C Koch Award, 90; Hans Christian Hagedorn Medal, Steensen Mem Hosp, Copenhagen, 70; Gairdner Award, Gairdner Found, Can, 71; Diaz-Cristobal Award, Span Soc Study Diabetes, 73; Elliott P Joslin Medal, New Eng Diabetes Asn, 76; Boris Pregal Award, NY Acad Sci, 76; Passano Found Award, 79; EFF Copp Mem Lectr, La Jolla, Calif, 71; Mellon Lectr, Pittsburgh, Pa, 79; Pachkis Lectr, Philadelphia Endocrinol Soc, 79; Banting Mem Lectr, Brit Diabetic Soc, 81; David Rumbough Award, Juv Diabetes Found, 82; Rolf Luft Award & Lectr, Stockholm, Sweden, 84; Wolf Found Prize Med, 85. *Mem:* Nat Acad Sci; AAAS; Am Soc Biol Chemists; Biochem Soc; Am Diabetes Asn; Am Acad Arts & Sci; Sigma Xi; Int Diabetes Fedn; Endocrine Soc; Protein Soc. *Res:* Discovery, isolation, structural analysis and biosynthesis of proinsulin; mechanism of conversion of proinsulin to insulin; insulin binding to tissues and mechanism of action; evolutionary development of insulin and related hormones; study of normal and mutant insulin and insulin receptor genes; author of various publications. *Mailing Add:* Dept Biochem & Howard Hughes Med Inst Univ Chicago 5841 S Maryland Ave Chicago IL 60637

STEINER, ERICH E, b Thun, Switz, Apr 9, 19; nat US; m 44; c 3. GENETICS. *Educ:* Univ Mich, BS, 40; Ind Univ, PhD(bot), 50. *Prof Exp:* From instr to asst prof, 50-58, assoc prof, 58-61, chmn dept, 68-71 & 79-81, dir, Matthaei Bot Gardens, 71-77, 89-91. PROF BOT, UNIV MICH, ANN ARBOR, 61- *Concurrent Pos:* NSF sr fel, 60-61. *Mem:* Bot Soc Am; Genetics Soc Am; Soc Study Evolution; Am Soc Naturalists; Soc Econ Bot. *Res:* Genetics and evolutionary biology of Oenothera; genetics of incompatability. *Mailing Add:* Dept Biol Univ Mich Ann Arbor MI 48109

STEINER, EUGENE FRANCIS, b St Louis, Mo, July 15, 34; m 63; c 1. MATHEMATICS. *Educ:* Univ Mo, BS, 56, MA, 60, PhD(math), 63. *Prof Exp:* Asst prof physics, Southwestern La Inst, 56-57; eng physicist, McDonnell Aircraft Corp, 58-59; asst prof math, Univ NMex, 63-65; assoc prof, Tex Tech Col, 65-66; assoc prof, Univ NMex, 66-68; assoc prof, 68-72, PROF MATH, IOWA STATE UNIV, 72- *Concurrent Pos:* Vis prof, Univ Alta, 70-71. *Mem:* Am Math Soc; Math Asn Am. *Res:* General topology. *Mailing Add:* Dept Math Iowa State Univ 1429 Carver Ames IA 50011

STEINER, GEORGE, b Czech, Mar 11, 36; Can citizen; m 66; c 2. ENDOCRINOLOGY & METABOLISM. *Educ:* Univ BC, BA, 56, MD, 60; FRCP(C), 65. *Prof Exp:* Resident med, Royal Victoria Hosp, McGill Univ, 60-62; fel, Harvard Med Sch, Peter Bent Brigham Hosp & Joslin Clin Res Lab, 62-64; resident med & endocrinol, Royal Victoria Hosp, McGill Univ, 64-66; from lectr to assoc prof med & physiol, Univ Toronto, 66-80; dir, Lipid Res Clin, 66-80, dir, Diabetes Clin, 66-83, DIR, DIV ENDOCRINOL & METAB, TORONTO GEN HOSP, 80- *Concurrent Pos:* Med Res Coun Can scholar, 67-72. *Honors & Awards:* MDS Award, Can Soc Clin Chem; Pfizer lectureship, Clin Res Inst, Montreal. *Mem:* Am Diabetes Asn; Am Fedn Clin Res; Am Physiol Soc; Endocrine Soc; Can Soc Clin Invest; Am Heart Asn. *Res:* Interaction of carbohydrate and lipid metabolism; hyperlipemia and atherosclerosis. *Mailing Add:* Dept Med Rm 7302 Med Sci Bldg E-11-N-225 Univ Toronto 200 Elizabeth St Toronto ON M5G 2C4 Can

STEINER, GEORGE, b Budapest, Hungary, Sept 11, 47; Can citizen; m 74; c 2. THEORY OF SCHEDULING, ALGORITHMIC ORDER THEORY. *Educ:* Eötvös Univ, Budapest, Hungary, MSc, 71; Univ Waterloo, Can, PhD(math), 82. *Prof Exp:* Systs analyst, Steel Co Can, 74-80; asst prof, 81-85, ASSOC PROF MGT SCI, MCMASTER UNIV, 85- *Mem:* Opers Res Soc Am; Soc Indust & Appl Math; Inst Mgt Sci; Asn Comput Mach; Can Oper Res Soc. *Res:* Scheduling; combinatorial optimization; algorithmic order theory; operations research; author of various publications. *Mailing Add:* Fac Bus McMaster Univ Hamilton ON L8S 4M4 Can

STEINER, GILBERT, b Moscow, USSR, Jan 19, 37; US citizen; m 71; c 2. MATHEMATICS, ALGEBRA. *Educ:* Univ Mich, BS, 58, MS, 59; Univ Calif, Berkeley, PhD(math), 62. *Prof Exp:* Instr math, Reed Col, 62-64; asst prof, Dalhousie Univ, 64-68; from asst prof to assoc prof, 68-79, PROF MATH, FAIRLEIGH DICKINSON UNIV, TEANECK CAMPUS, 79- *Mem:* Am Math Soc; Math Asn Am. *Res:* Functional analysis. *Mailing Add:* Dept Math Fairleigh Dickinson Univ 1000 River Rd Teaneck NJ 07666

STEINER, HENRY M, b San Francisco, Calif, Aug 20, 23; c 4. ENGINEERING ECONOMICS, HIGHWAY ECONOMICS. *Educ:* Stanford Univ, BA, 44, MS, 50, PhD(civil eng), 65. *Prof Exp:* Engr War Dept, Korea, 46-47, Utah Construct Co, 50-51 & Aramco, Saudi Arabia, 53-55; asst prof sci, Univ Americas, 59-63; prof econ, Esuela Administracion Graduados, 65-67; assoc prof mgt, Univ Tex, Austin, 68-76; PROF ENG ECON, GEORGE WASHINGTON UNIV, 76- *Concurrent Pos:* Vis prof, Stanford Univ, 67-68 & Univ Calif, Berkeley, 87; consult, World Bank, Ecuador, Panama & Mex, 69, 77 & 91, Army CEngr, 88-89 & many pvt co. *Mem:* Am Soc Civil Engrs; Transp Res Forum. *Res:* Engineering economics; surface transportation, particularly roads. *Mailing Add:* 5315 Wehawken Rd Bethesda MD 20816

STEINER, HERBERT M, b Goppingen, Ger, Dec 8, 27; nat US. PARTICLE PHYSICS. *Educ:* Univ Calif, Berkeley, BS, 51, PhD(physics), 56. *Prof Exp:* Lectr, 57-60, from asst prof to assoc prof, 61-67, PROF PHYSICS, UNIV CALIF, BERKELEY, 67-, PHYSICIST, LAWRENCE BERKELEY LABS, 53- *Concurrent Pos:* Guggenheim fel, 60-61; vis scientist, Europ Ctr Nuclear Res, 60-61, 64 & 68-69; Alexander von Humboldt sr scientist, 76-77; guest scientist, Max Planck Inst Physics & Astrophysics, Munich, 76-78; vis prof, Japan Soc Promotion Sci, 78. *Mem:* Am Phys Soc. *Res:* High energy physics; elementary particle interactions. *Mailing Add:* Dept of Physics Univ of Calif Berkeley CA 94720

STEINER, JAMES W(ESLEY), b Lexington, Ky, Oct 3, 26; m 51; c 3. ELECTRICAL ENGINEERING. *Educ:* Univ Ky, BSEE, 48; Purdue Univ, MSEE, 50; Columbia Univ, EE, 60; NY Univ, MBA, 68. *Prof Exp:* Asst proj engr flight simulators, Curtiss-Wright Corp, 50-52; sr engr Lacrosse Missile, Fed Labs Div, Int Tel & Tel Corp, 52-55, develop engr, 55-56, proj engr, 56-58, sr proj engr Dew Line, 58-60, exec engr courier satellite, 60-66, proj mgr commun, Telemetry & Command Subsyst Develop, Intelsat III Commun Satellite, 66-68; dept mgr systs design, Western Develop Labs Div, Philco-Ford Corp, 68-69, mgr systs design, 69-70, mgr systs eng activ, 70-78, mgr, equip prog activ, 78-80, MGR, SYSTS INTEGRATION ACTIV, ESD DIV, FORD AEROSPACE & COMMUN CORP, 80- *Mem:* Sr mem Inst Elec & Electronics Engrs. *Res:* Range measurement by continuous wave phase shift measurements; digital command for satellite communications; telemetry and master timing systems; satellite ground support systems; telemetry, tracking and commanding; digital communications. *Mailing Add:* 14195 Wild Plum Lane Los Altos Hills CA 94022

STEINER, JEAN LOUISE, b Red Cloud, Nebr, Dec 9, 51; m 88. AGROCLIMATOLOGY. *Educ:* Cornell Col, Mt Vernon, Iowa, BA, 74; Kans State Univ, MS, 79, PhD(agron), 82. *Prof Exp:* Res scientist, Commonwealth Sci & Indust Res Orgn Ctr Irrig Res, Griffith, NSW Australia, 82-83; RES SOIL SCIENTIST, USDA-AGR RES SERV, CONSERV & PROD RES LAB, BUSHLAND, TEX, 83- *Concurrent Pos:* USDA-Cooperator, Tex A&M Univ, 83-; vis grad fac, Tex Tech Univ, 89-; assoc ed, Agron J, Am Soc Agron, 89-; adj prof agr, WTex State Univ, 90- *Mem:* Am Soc Agron; Soil Sci Soc Am; Soil & Water Conserv Soc; AAAS; Coun Adv Sci & Technol. *Res:* Developing improved dryland and irrigated cropping systems and practices to optimize evapotranspiration and water use; improve understanding of impact of crop residues on surface water and energy balances and processes; developing improved soil water balance models for cropping systems simulation; developing crop residue decomposition model. *Mailing Add:* Agr Res Serv USDA PO Drawer 10 Bushland TX 79012

STEINER, JOHN EDWARD, b Seattle, Wash, Nov 7, 17; m 42; c 3. AERONAUTICAL ENGINEERING MANAGEMENT. *Educ:* Univ Wash, BS, 40; Mass Inst Technol, MS, 41. *Hon Degrees:* Summa Laude Dignatus, Univ Wash, 78. *Prof Exp:* Aerodynamist com & mil airplane, Boeing Airplane Co, 41-44, chief aerodynamist, 44-48, sr group engr, 48-55, proj engr, 55-58, prog mgr 727 prog, 58-60, chief proj engr, 60-65, chief engr all com opers, 65-66, vpres eng & prod develop, 66-68, vpres mkt, Boeing Com Airplane Co, 68-70, vpres & div gen mgr, 70-73, vpres all opers, 73-74, vpres prog develop, 74-76, vpres, corp prod develop, Boeing Co, 76-83; chmn, Aeronaut Policy Rev Comt, Exec Off President of US, 83-89; RETIRED. *Concurrent Pos:* Trustee & mem exec comt, Pac Sci Ctr, 75- *Honors & Awards:* Elmer A Sperry Award, 80; Thulin Medal, Sweden; Sir C Kingsford Smith Award, Australia; Wright Brothers Mem Lectr, 82. *Mem:* Nat Acad Eng; fel Am Inst Aeronaut & Astronaut; fel Royal Aeronaut Soc. *Res:* High technology research involving aerodynamic efficiency of swept wings and total commercial and military configurations; structural efficiency and durability; propulsion integration; computer aided productivity improvement research in engineering and manufacturing. *Mailing Add:* 3425 Evergreen Pt Rd Belleview WA 98004

STEINER, JOHN F, b Milwaukee, Wis, July 21, 08; m 49. ELECTROCHEMISTRY. *Educ:* Univ Wis, BS, 29, MS, 32, PhD(chem), 33. *Prof Exp:* Instr chem, Univ Wis, 29-33, Alumni Asn fel, 33-34; develop chemist, Milwaukee Gas Specialty Co, 35-36; res chemist, Globe-Union, Inc, Wis, 36-37 & Miner Labs, Ill, 38-53; head food lab, Guardite Corp, 53-56; dir, Chem Res Labs, 56-85; RETIRED. *Concurrent Pos:* Dir develop, Sound Recording Serv, 47-; mem staff, Univ Ill & Wilson Col, 59-; consult, Ill Dept Revenue, 64- & US Dept Internal Revenue, 65- *Mem:* AAAS; Am Chem Soc; Electrochem Soc. *Res:* Limnology; ceramics; dentifrices; cereals; food technology; sound recording; information storage-retrieval. *Mailing Add:* 2748 S Superior Milwaukee WI 53207

STEINER, KIM CARLYLE, b Alton, Ill, Nov 21, 48; m 70; c 3. FOREST GENETICS. *Educ:* Colo State Univ, BS, 70; Mich State Univ, MS, 71, PhD(forest genetics), 75. *Prof Exp:* PROF FOREST GENETICS, PA STATE UNIV, 87- *Mem:* Sigma Xi; Soc Am Foresters. *Res:* Genetics and ecology of forest trees; genetic adaptation of trees to environmental stresses; taxonomy, distribution and geographic variation of forest trees. *Mailing Add:* Forest Resources Lab, Sch Forest Resources Pa State Univ University Park PA 16802

STEINER, KURT EDRIC, b Roanoke, Va, July 24, 46; m 80; c 3. DIABETES, ATHEROSCLEROSIS. *Educ:* Col Wooster, BA, 68; Univ Calif, Davis, PhD(biochem), 76. *Prof Exp:* From instr to asst prof physiol, Vanderbilt Med Sch, 81-84; res scientist metabolic dis, Wyeth Labs, 84-88, sect head, Wyeth-Ayerst Res, 88, assoc dir, 88-91, DIR CARDIOVASC & METAB DIS, WYETH-AYERST RES, 91- *Mem:* Am Diabetes Asn; Am Heart Asn. *Res:* Hormonal regulation of metabolism; mechanisms of insulin resistance in diabetes mellitus and the discovery of novel pharmacological interventions. *Mailing Add:* Wyeth-Ayerst Res CN 8000 Princeton NJ 08543-8000

STEINER, LISA AMELIA, b Vienna, Austria, May 12, 33; US citizen. IMMUNOLOGY. *Educ:* Swarthmore Col, BA, 54; Radcliffe Col, MA, 56; Yale Univ, MD, 59. *Prof Exp:* From asst prof to assoc prof, 57-80, PROF BIOL, MASS INST TECHNOL, 80- *Concurrent Pos:* Helen Hay Whitney res fel microbiol med sch, Wash Univ, St Louis, 62-65; Am Heart Asn res fel immunol, Wright-Fleming Inst, London, 65-67; mem allergy & immunol study sect, NIH, 74-78; mem bd overseers, Rosenstiel Basic Med Sci Res Ctr, 76-80, chmn, 80-; mem personnel comt, Am Cancer Soc, 79-84; mem bd sci counsrs, Nat Cancer Inst, 80-83; bd trustees, Helen Hay Whitney Found, 84-; mem-at-large, Biol Sci Comt, AAAS, 85- *Mem:* Am Asn Immunol; Am Soc Biol Chem; fel AAAS. *Res:* Structure and function of immunoglobulins; phylogeny and ontogeny of immune response; protein chemistry. *Mailing Add:* Dept Immunol Mass Inst Technol 77 Massachusetts Ave Cambridge MA 02139

STEINER, MANFRED, PLATELETS, BIOCHEMISTRY. *Educ:* Univ Vienna, Austria, MD, 55; Mass Inst Technol, PhD(biochem), 67. *Prof Exp:* PROF HEMAT, BROWN UNIV, 68-; DIR DIV HEMAT & ONCOL, BROWN UNIV MEM HOSP, 82- *Concurrent Pos:* Vis scientist, Roche Fund. *Mem:* Am Physiol Soc; Am Soc Biochem & Molecular Biol; Am Asn Path; Am Soc Hemat. *Mailing Add:* Dept Med Brown Univ Brown Sta Providence RI 02912

STEINER, MARION ROTHBERG, b New York, NY, May 23, 41; m 63; c 2. BIOCHEMISTRY, ONCOLOGY. *Educ:* Smith Col, BA, 62; Univ Ky, PhD(biochem), 68. *Prof Exp:* Instr biochem, Univ Ky, 68-70; fel, Baylor Col Med, 71-73, asst prof virol, 73-78; asst prof exp path, 78-83, ASSOC PROF MICRO IMMUNOL, UNIV KY, 83- *Mem:* Am Chem Soc; Am Soc Microbiologists; Am Asn Cancer Res. *Res:* Examination of the structure and function of the surface membrane of oncogenic cells, utilizing mouse mammary carcinomas and dysplasias as a model system. *Mailing Add:* Dept Micro Immunol MS409 Col Med Univ Ky Lexington KY 40506-0084

STEINER, MORRIS, b Shenandoah, Pa, Mar 1, 04; m 29; c 2. PEDIATRICS. *Educ:* NY Univ, MD, 28. *Prof Exp:* From asst clin prof to assoc clin prof, 53-66, from assoc prof to prof, 66-74, EMER PROF PEDIAT, STATE UNIV NY DOWNSTATE MED CTR, 74- *Concurrent Pos:* NY Tuberc & Health Asn grant, State Univ NY Downstate Med Ctr, 66-70; consult, Jewish Hosp, Brooklyn, 69- & King's County Hosp Med Ctr, 69- *Mem:* Am Acad Pediat; Am Thoracic Soc. *Res:* Tuberculosis; asthma, chronic broncho-pulmonary disease in children. *Mailing Add:* 456 E 19 St Brooklyn NY 11226

STEINER, PINCKNEY ALSTON, III, b Athens, Ga, Apr 5, 38; m 60, 88; c 2. SOLID STATE PHYSICS, MAGNETIC RESONANCE. *Educ:* Univ Ga, BS, 59; Duke Univ, PhD(physics), 65. *Prof Exp:* Res fel chem physics, H C Orsted Inst, Copenhagen Univ, 64-66; asst prof, 66-77, ASSOC PROF PHYSICS, CLEMSON UNIV, 77-, ASSOC DEPT HEAD. *Mem:* Am Phys Soc; Sigma Xi. *Res:* Electron paramagnetic resonance applied to various problems in solid state physics and in biophysics. *Mailing Add:* Dept of Physics & Astron Clemson Univ Clemson SC 29634-1911

STEINER, RAY PHILLIP, b Bronx, NY, Apr 28, 41; m 72. NUMBER THEORY. *Educ:* Univ Ariz, BSEE, 63, MS, 65; Ariz State Univ, PhD(math), 68. *Prof Exp:* Asst prof, 68-72, assoc prof, 72-78, PROF MATH, BOWLING GREEN STATE UNIV, 78- *Mem:* Fibonacci Asn; Am Math Soc; Math Asn Am; Nat Coun Teachers Math. *Res:* Finding the units and class numbers of algebraic number fields by linear programming techniques, Diophantine equations and Fibonacci numbers; Diophantine approximation. *Mailing Add:* Dept of Math Bowling Green State Univ Bowling Green OH 43403

STEINER, ROBERT ALAN, b Chicago, Ill, Mar 9, 47. NEUROENDOCRINOLOGY, REPRODUCTIVE PHYSIOLOGY. *Educ:* Univ Pac, BA, 69; Univ Ore, PhD(physiol), 75. *Prof Exp:* Sr fel endocrinol, Univ Wash, 75-77, asst prof physiol, 77-80, assoc prof, 80-85, PROF OBSTET & GYNEC, PHYSIOL & BIOPHYS, UNIV WASH, 85- *Concurrent Pos:* Res affil, Regional Primate Res Ctr, Univ Wash, 77-, Diabetes Res Ctr, 78- *Mem:* Am Physiol Soc; Soc Study Reproduction; Am Soc Zoologists; Am Endocrine Soc; Intl Neuroendocrine Soc; Soc Neurosci. *Res:* Control of neuropeptide gene expression; neuroendocrine control of the onset of puberty and growth. *Mailing Add:* Dept Physiol & Biophys SJ-40 Univ Wash Sch Med Seattle WA 98195

STEINER, ROBERT FRANK, b Manila, Philippines, Sept 29, 26; US citizen; m 56; c 2. PHYSICAL BIOCHEMISTRY. *Educ:* Princeton Univ, AB, 47; Harvard Univ, PhD(phys chem), 50. *Prof Exp:* Fel, US Naval Med Res Inst, 50-51, phys chemist, 51-70; PROF CHEM, UNIV MD, BALTIMORE COUNTY, 70-, CHMN DEPT, 74- *Concurrent Pos:* Jewett fel, 50-51; lectr, Georgetown Univ, 57-58 & Howard Univ, 58-59 & 60-61; mem, US Civil Serv Bd Exam, 58-; mem molecular biol panel, NSF, 67-70; ed, Res Commun Chem Path & Pharmacol, 70- & Biophys Chem, 73-; mem biophys & biophys chem study sect, NIH, 76-80. *Mem:* Am Chem Soc; Biophys Soc; Am Soc Biol Chem; NY Acad Sci. *Res:* Light scattering; fluorescence; protein interactions; nucleic acids; synthetic polynucleotides; statistical thermodynamics. *Mailing Add:* Chem Dept Univ Md 5401 Wilkens Ave Baltimore MD 21228-5329

STEINER, RUSSELL IRWIN, b Lebanon, Pa, July 21, 27; m 56; c 4. INDUSTRIAL ORGANIC CHEMISTRY. *Educ:* Lebanon Valley Col, BS, 49; Univ Conn, MS, 52, PhD(chem), 55. *Prof Exp:* Res chemist, Nat Aniline Div, Allied Chem Corp, 55-63, group leader, 63-65, res supvr indust chem div, 65-67; res assoc, Crompton & Knowles, 67-74, group leader disperse dyes, 74-80, dir process develop, 80-86, res & develop, 86-89, VPRES RES & DEVELOP, DYES & CHEM DIV, CROMPTON & KNOWLES, 89- *Mem:* Am Chem Soc; Am Asn Textile Chem & Colorists; Sigma Xi. *Res:* Textile dyes; food colors. *Mailing Add:* Res & Develop Dept Dyes Chem Div Crompton & Knowles Corp Reading PA 19603

STEINER, SHELDON, b Bronx, NY, Apr 23, 40; m 63; c 2. BIOCHEMISTRY. *Educ:* Drew Univ, BA, 61; Univ Ky, MS, 64, PhD(microbiol), 67. *Prof Exp:* Instr, Baylor Col Med, 71-73, from asst prof to assoc prof virol, 73-78; assoc prof biol sci, 78-83, PROF BIOL SCI, UNIV KY, LEXINGTON, 83- *Concurrent Pos:* NIH grant, Univ Ky, 69-71; NIH spec fel, Baylor Col Med, 72-75; fac res award, Am Cancer Soc, 75-80. *Mem:* AAAS; Am Soc Biol Chemists; Am Soc Microbiol; Am Soc Cell Biol; Soc Complex Carbohydrates. *Res:* Structure and function of procaryotic and eucaryotic membranes; biochemical characterization of membranes of malignant cells; role of eicosanoids in myoblast differentiation. *Mailing Add:* Dept of Biol Sci Univ Ky Lexington KY 40506

STEINER, WERNER DOUGLAS, b Milwaukee, Wis, Oct 8, 32; m 61; c 3. ORGANIC CHEMISTRY. *Educ:* Univ Karlsruhe, BA, 56, MA, 58; Univ Pa, PhD(org chem), 64. *Prof Exp:* Res & develop chemist, Org Chem Dept, Jackson Lab, E I Du Pont De Nemours & Co, Inc, Deepwater, NJ, 63-65, process develop chemist, 67-68 & dyes & intermediates, 68-75, sr chemist, 62-80. *Mem:* AAAS; Am Chem Soc. *Res:* Research and development of dyes for natural and man-made fibers; research and development of fluorine compounds for use as aerosol propellants, hydraulic fluids, instrument fluids, convective coolants and working fluids; process development of dyes and fluorine compounds; dye process development, manufacture, intermediates. *Mailing Add:* 6827 Thames Dr Gates Four Fayetteville NC 28306-2523

STEINERT, LEON ALBERT, b Shattuck, Okla, May 2, 30; m 88. QUANTUM PHYSICAL SYSTEMS, ELECTROMAGNETIC RADIATION. *Educ:* La Sierra Col, BA, 52; Univ Colo, MS, 56, PhD(physics), 62. *Prof Exp:* Physicist, Nat Bur Standards, 53-65; theoret physicist, Lawrence Radiation Lab, 66-67; sr scientist, McDonnell Douglas Corp, 67-70; sr res engr-scientist, Lockheed Corp, 72-79; res engr, Systs Control, Inc, 79; consult, IRT Corp & Miss State Univ, 80-81; RES DIR, PHYS SYNERGETICS INST, 81- *Concurrent Pos:* Lectr, Loma Linda Univ, 68. *Mem:* Am Phys Soc; Am Math Soc; Soc Indust & Appl Math. *Res:* Theoretical physics of quantum physical systems; applications of quantum statistical condensed-matter physics theory in microelectronics; electromagnetic fields scattering theory. *Mailing Add:* PO Box 61072 Sunnyvale CA 94088-1072

STEINERT, PETER MALCOLM, EPIDERMIS, KERATIN. *Educ:* Univ Adelaide, Australia, PhD(biochem), 72. *Prof Exp:* SR INVESTR, NAT CANCER INST, 73- *Mem:* Am Soc Cell Biol; Am Soc Biochem & Molecular Biol; Soc Invest Dermat. *Res:* Studies the structure, function and expression of keratin intermediate filaments of the epidemus and their associated proteins, including filagrin and the cornified cell envelop. *Mailing Add:* Dermat Br Nat Cancer Inst Bldg 10 Rm 12N238 Bethesda MD 20892

STEINETZ, BERNARD GEORGE, JR, b Germantown, Pa, May 30, 27; m 49; c 3. ENDOCRINOLOGY. *Educ:* Princeton Univ, AB, 49; Rutgers Univ, PhD(zool), 54. *Prof Exp:* Asst org res chemist, Irving Varnish & Insulator Co, 47; asst, Rutgers Univ, 52-54; sr scientist physiol res, Warner-Lambert Res Inst, 54-62, sr res assoc, 62-67; head reproductive physiol & fel, Ciba Res, Ciba Pharmaceut Co, NJ, 67-71, mgr endocrinol & metab, Pharmaceut Div, Ciba-Geigy Corp, Ardsley, NY 71-84; ASSOC PROF FORENSIC MED, LAB EXP MED & SURG PRIMATES, MED CTR, NY UNIV, TUXEDO, NY, 84- *Concurrent Pos:* Res assoc prof, NY Univ Sch Med, 74- *Mem:* Endocrine Soc; Am Physiol Soc; Orthop Res Soc; fel NY Acad Sci; Soc Exp Biol & Med; Soc Study Reproduction. *Res:* Reproductive physiology; hormone metabolism; hormones and connective tissue; hormones and aging; hormone interactions; role of mediators in osteoarthritis. *Mailing Add:* 336 Longbow Dr Franklin Lakes NJ 07417

STEINFELD, JEFFREY IRWIN, b Brooklyn, NY, July 2, 40. PHYSICAL CHEMISTRY. *Educ:* Mass Inst Technol, BS, 62; Harvard Univ, PhD(chem), 65. *Prof Exp:* NSF fel, 65-66; asst prof, 66-70, assoc prof, 70-79, PROF CHEM, MASS INST TECHNOL, 80- *Concurrent Pos:* Alfred P Sloan res fel, 69-71; John Simon Guggenheim fel, Univ Calif, Berkeley & Kammerlingh-Onnes Lab, Leiden, 72-73; consult, Aerospace Corp, 74-80, Los Alamos Nat Lab, 75-85 & KOR, Inc, 80-85; mem, Sci Adv Bd, Laser Technics, Inc, 82-; co-ed, Spectroclinic Acta, 83-; vis prof, Joint Inst of Lab Astrophys, 83 & 86; prof invité, Univerité de Bourgogne, Dijon, France, 91. *Mem:* AAAS; Am Phys Soc; Fedn Am Sci; Sigma Xi. *Res:* Molecular spectroscopy; energy transfer in molecular collisions; applications of lasers to chemical kinetics, laser-induced surface reactions. *Mailing Add:* Dept Chem Rm 2-221 Mass Inst Technol Cambridge MA 02139

STEINFELD, JESSE LEONARD, b West Aliquippa, Pa, Jan 6, 27; m 53; c 3. CANCER, MEDICINE. *Educ:* Univ Pittsburgh, BS, 45; Western Reserve Univ, MD, 49; Am Bd Internal Med, dipl, 58. *Hon Degrees:* LLD, Gannon Col, 72. *Prof Exp:* Instr med, Univ Calif, 52-54 & George Washington Univ, 54-58; asst dir, Blood Hosp, City of Hope, Duarte, Calif, 58-59; from asst prof to prof med, Sch Med, Univ Southern Calif, 59-68, cancer coordr, 66-68; dep dir, Nat Cancer Inst, 68-69; dep asst secy health & sci affairs, Dept Health, Educ & Welfare, 69-72; surgeon gen, USPHS, 69-73; prof med & dir dept oncol, Mayo Clin & Mayo Med Sch, Rochester, Minn, 73-74; prof med, Univ Calif, Irvine, 74-76; chief med serv, Long Beach Vet Admin Hosp, 74-76; PROF MED & DEAN, SCH MED, MED COL VA, 76- *Concurrent Pos:* AEC fel med, Univ Calif, 52-53; clin investr, Nat Cancer Inst, 54-58, mem, Krebiozen Rev Comt, 63- & Clin Studies Panel, 64-66, consult, 66-, mem chemother adv comt, 67 & cancer spec prog adv comt, 67; consult, Vet Admin Hosp, Long Beach, Calif, 59-, City of Hope Med Ctr, 60- & Kern County Gen Hosp, Bakersfield, 64-; mem, Calif State Cancer Adv Coun, 61-68. *Mem:* Soc Nuclear Med; Am Asn Cancer Res; AMA; fel Am Col Physicians; Am Fedn Clin Res. *Res:* Cancer chemotherapy; hematology; health administration. *Mailing Add:* 18676 Avenida Cordilleva San Diego CA 92128

STEINFELD, LEONARD, b New York, NY, Nov 16, 25; m 65; c 4. MEDICINE. *Educ:* Hofstra Col, BA, 49; State Univ NY Downstate Med Ctr, MD, 53. *Prof Exp:* Intern, Los Angeles County Gen Hosp, 53-54; resident pediat, Mt Sinai Hosp, New York, NY, 54-56, instr pediat, Col Physicians & Surgeons, Columbia Univ, 58-68; assoc prof, 66-69, PROF PEDIAT, MT SINAI SCH MED, 69- *Concurrent Pos:* NIH fel pediat cardiol, Mt Sinai Hosp, New York, 56-57, NY Heart Asn fel, 57-58; from asst attend pediatrician to assoc attend pediatrician, Mt Sinai Hosp, 59-69, attend pediatrician, 69-; consult, USPHS Hosp, Staten Island, 60- & Perth Amboy Gen Hosp, NY, 70- *Mem:* Am Acad Pediat; Am Col Cardiol; Am Heart Asn; NY Acad Sci. *Res:* Heart disease in infants and children. *Mailing Add:* Dept Pediat Mt Sinai Sch Med 5th Ave 100th St New York NY 10029

STEINFINK, HUGO, b Vienna, Austria, May 22, 24; nat US; m 48; c 2. MATERIALS SCIENCE & ENGINEERING. *Educ:* City Col New York, BS, 47; Columbia Univ, MA, 48; Polytech Inst Brooklyn, PhD(chem), 54. *Prof Exp:* Res chemist, Shell Develop Co, 48-51 & 54-60; assoc prof, 60-63, PROF CHEM ENG, UNIV TEX, AUSTIN, 63, 63-, JEWEL MCALISTER SMITH PROF ENG, 81- *Mem:* Am Chem Soc; Mineral Soc Am; Am Crystallog Asn; Am Inst Chem Engrs; Mat Res Soc. *Res:* Crystal structures of silicate minerals and silicate-organic complexes; crystal chemistry and physical properties of semiconductor and superconductor materials; materials science research. *Mailing Add:* Dept Chem Eng Univ Tex Austin TX 78712

STEINGISER, SAMUEL, b Springfield, Mass, June 6, 18; m 46; c 3. CHEMISTRY. *Educ:* City Col New York, BS, 38; Polytech Inst Brooklyn, MS, 41; Univ Conn, PhD, 49. *Prof Exp:* Chemist, Rockefeller Inst, 41-42; asst atom bomb proj S A M Labs, Columbia Univ, 42-43; asst, Carbide & Carbon Chems Corp, NY, 44-46; res assoc, Metall Lab, Univ Chicago, 43-44; group leader phys chem, Publicker Industs, Pa, 46; res assoc, US Off Naval Res, Conn, 46-50; group leader & res scientist, Cent Res Dept, Monsanto Chem Co, 50-54; group leader, Mobay Chem Co, 54-59, asst res dir, 59-65; scientist, Monsanto Res Corp, Ohio, 66-70, sci fel, Lopac Proj, Monsanto Co, Bloomfield, Conn, 70-77, res mgr & sr fel, 77-83; COMPUTER CONSULT, 85- *Concurrent Pos:* Mem mat adv bd, Nat Res Coun, 61-63; mem, Int Standardization Orgn; ed, J Cellular Plastics, J Elastomer & Plastics. *Mem:* Fel AAAS; Am Chem Soc; Soc Plastics Indust; Am Soc Test & Mat; Soc Rheol. *Res:* Mass spectroscopy; corrosion; electrolysis; nuclear chemistry; high-vacuum phenomena; magneto-optics; magnetic susceptibility; high polymer physics; mechanical properties; foams and elastomers; advanced composites; instrumentation design; plastics processing and development; computer technology. *Mailing Add:* Five Fox Chase Rd Bloomfield CT 06002-2107

STEINGOLD, HAROLD, b Providence, RI, May 7, 29; m 57; c 3. ENGINEERING. *Educ:* Brown Univ, AB, 49; Univ Calif, Los Angeles, MS, 60, PhD(eng), 64. *Prof Exp:* Engr, Hughes Aircraft, 59-63; engr, Rand Corp, Calif, 63-70; engr, Visualtek, 70-72; eng specialist, Actron Corp, 72-77; eng survr, Hughes Helicopters, 77-82, SR SCIENTIST, HUGHES AIRCRAFT, 82- *Concurrent Pos:* Lectr, exten, Univ Calif, Santa Barbara, 64-, Los Angeles, 65 & San Fernando Valley State Col, 70- *Mem:* Sr mem Inst Elec & Electronics Engrs; Am Inst Mining Engrs. *Res:* Applications of computer technology and coherent electromagnetic sources to communications imaging and data processing; infrared and microwave imaging systems; protheses for the physically handicapped. *Mailing Add:* 407 16th St Santa Monica CA 90402

STEINHAGEN, WILLIAM HERRICK, b Dayton, Ohio, May 25, 48; m 68; c 2. INHALATION TOXICOLOGY TECHNOLOGY. *Educ:* NC State Univ, Raleigh, BS, 83. *Prof Exp:* Inhalation tech, Wright-Patterson Air Force Base, 68-74; inhalation toxicologist, Becton-Dickinson Res Ctr, 74-78; sr res asst inhalation toxicol, Chem Indust Inst Toxicol, 78-79, res assoc, 79-81, sr res assoc, 81-84, assoc scientist inhalation toxicol, 84-89; TECH SALES MGR, ALL TECH ASSOC, INC, 89- *Concurrent Pos:* Consult, Northrop Serv, Inc, 85. *Mem:* Soc Toxicol; Wildlife Soc; Waterfowl USA, Inc. *Res:* Inhalation technology associated with the generation and analysis of gas, vapor, liquid aerosol and particulate test atmospheres for inhalation toxicology studies; author or co- author of over 40 articles and abstracts in the field of inhalation toxicology. *Mailing Add:* 412 Oak Ridge Rd Cary NC 27511

STEINHARDT, CHARLES KENDALL, b Milwaukee, Wis, Mar 1, 35; m 70; c 2. ORGANIC CHEMISTRY, MATHEMATICAL STATISTICS. *Educ:* Univ Wis, BS, 57; Univ Ill, PhD(org chem), 63. *Prof Exp:* Instr chem, Univ Wash, Seattle, 63-65; chemist, Lubrizol Corp, Ohio, 65-71 & Napko Corp, Tex, 73-77; SR CHEMIST, MOBAY CHEM CORP, 77- *Mem:* Am Chem Soc; Am Soc Qual Control. *Mailing Add:* 11427 Dunlap Dr Houston TX 77035-2332

STEINHARDT, EMIL J, b Pittsburgh, Pa, Aug 19, 37; m 60. MECHANICAL ENGINEERING. *Educ:* Univ Pittsburgh, BS, 59, MS, 61, PhD(mech eng), 65. *Prof Exp:* Instr mech eng, Univ Pittsburgh, 63-65; asst prof, 65-69, assoc prof, 69-81, PROF MECH ENG, WVA UNIV, 81- *Mem:* Am Soc Eng Educ; Am Soc Mech Engrs; Am Inst Aeronaut & Astronaut. *Res:* Satellite systems design; engineering systems design; rural systems and housing systems design. *Mailing Add:* Dept Mech Eng WVa Univ Box 6101 Morgantown WV 26506

STEINHARDT, GARY CARL, b Lansing, Mich, Sept 13, 44; m 73; c 2. SOIL SCIENCE, AGRONOMY. *Educ:* Mich State Univ, BS, 66, MS, 68; Purdue Univ, PhD(agron), 76. *Prof Exp:* Asst soil sci, Mich State Univ, 66-68; asst, 71-74, from instr, to asst prof, 74-82, ASSOC PROF AGRON, PURDUE UNIV, 82- *Mem:* Am Soc Agron; Soil Conserv Soc Am; Coun Agr Sci & Technol. *Res:* Effects of soil management and tillage practices on the physical properties of soil; agricultural aspects of land use planning. *Mailing Add:* Dept of Agron Purdue Univ West Lafayette IN 47907

STEINHARDT, PAUL JOSEPH, b Washington, DC, Dec 25, 52; m 79; c 3. INFLATIONARY UNIVERSE, QUASICRYSTALS. *Educ:* Caltech Univ, BS, 74; Harvard Univ, MA, 75, PhD(physics), 78. *Prof Exp:* Jr fel Harvard Univ, 78-81; from asst prof to assoc prof, 81-85, prof physics, Univ Pa, 85-89, MARY ANN WOOD PROF PHYSICS, UNIV PA, 89- *Concurrent Pos:* Consult, IBM Res, Yorktown Heights, NY, 78-, Mass Inst Technol Res Estab Corp, 87-; Sloan fel, 82-86; Monell fel, Inst Adv Study, 89-90. *Mem:* Fel Am Phys Soc; Sigma Xi; Am Astron Soc; Mat Res Soc. *Res:* Particle cosmology, especially phase transitions in the early universe; quasicrystals (theory of physical properties); amorphous solids and glasses. *Mailing Add:* Dept Physics Univ Pa Philadelphia PA 19104

STEINHARDT, RICHARD ANTONY, b Washington, DC, Sept 23, 39; m 77; c 2. CELL BIOLOGY. *Educ:* Columbia Univ, AB, 61, PhD(biol sci), 66. *Prof Exp:* NSF fel, Agr Res Coun Inst Animal Physiol, Babraham & Plymouth Marine Sta, Eng, 66-67; from asst prof to assoc prof, 67-78, PROF ZOOL, UNIV CALIF, BERKELEY, 79-, PROF MOLECULAR & CELL BIOL, 89- *Concurrent Pos:* Overseas fel, Churchill Col, Cambridge, Eng, 81. *Honors & Awards:* Miller Res Prof, Univ Calif, Berkeley, 79. *Mem:* AAAS; Soc Develop Biol; Am Soc Cell Biol. *Res:* Cell biology related to intracellular signals and gene expression; developmental biology; ion transport; membrane permeability. *Mailing Add:* Dept Molecular & Cell Biol 391 LSA Univ of Calif Berkeley CA 94720

STEINHART, CAROL ELDER, b Cleveland, Ohio, Mar 27, 35; m 58; c 3. AGRICULTURAL & ENVIRONMENTAL SCIENCE. *Educ:* Albion Col, AB, 56; Univ Wis, PhD(bot), 60. *Prof Exp:* Biologist lab gen & comp biochem, NIMH, 61-66, sci analyst div res grants, NIH, Md, 66-68, biologist, 68-70; SCI WRITER & ED, 70- *Concurrent Pos:* Specialist, Biodata Sect, Dept Human Oncol, Sch Med, Univ Wis, 77-80, proj assoc, Water Resources Ctr, 80-81; res analyst, Wis Dept Agr, Trade & Consumer Protection, 81-83; tech writer & ed, Middleton Mem Vet Hosp, 83- *Res:* Growth, differentiation and nutrition of plant tissue cultures; hormonal control of enzyme synthesis in plants; ecology and environmental problems; environmental indices; soil erosion control; medical science. *Mailing Add:* 104 Lathrop St Madison WI 53705

STEINHART, JOHN SHANNON, b Chicago, Ill, June 3, 29; m 58; c 3. GEOPHYSICS, SCIENCE POLICY. *Educ:* Harvard Univ, AB, 51; Univ Wis, PhD, 60. *Prof Exp:* Proj leader, Woods Hole Oceanog Inst, 56; NSF fel, Carnegie Inst Wash Dept Terrestrial Magnetism, 60-61, mem staff, 61-68; tech asst, Resources & Environ, Off Sci & Technol, Exec Off of the Pres, 68-70; assoc dir, Marine Studies Ctr, 73-77; PROF GEOPHYS & ENVIRON STUDIES, UNIV WIS-MADISON, 70- & CHAIR, ENERGY ANAL & POLICY PROG, 82- *Mem:* AAAS; Am Geophys Union. *Res:* Oil and gas availability; energy policy; environmental policy research; science and public policy research and teaching. *Mailing Add:* 220 Weeks Hall Univ Wis Madison WI 53706

STEINHART, WILLIAM LEE, b Philadelphia, Pa, May 31, 42; m 67; c 1. VIROLOGY. *Educ:* Univ Pa, AB, 64; Johns Hopkins Univ, PhD(biochem), 68. *Prof Exp:* Res assoc, dept biol chem, Col Med, Pa State Univ, 71-75; asst prof, 75-81, ASSOC PROF GENETICS, BOWDOIN COL, 81- *Concurrent Pos:* Mem bd dirs, Found Blood Res, 80-82; grant reviewer, Am Heart Asn, 83-85; mem med adv comt, Bd Pesticides Control, State of Maine, 85- *Mem:* AAAS; Am Soc Plant Physiologists; Am Soc Microbiol. *Res:* molecular genetics of virus replication; genetics of disease resistance in plants. *Mailing Add:* Dept Biol Bowdoin Col Brunswick ME 04011

STEINHAUER, ALLEN LAURENCE, b Winnipeg, Man, Oct 17, 31; US citizen; m 58; c 2. ENTOMOLOGY, ECOLOGY. *Educ:* Univ Man, BSA, 53; Ore State Univ, MS, 55, PhD(entom), 58. *Prof Exp:* From asst prof entom to assoc prof entom, Univ Md, 58-66; assoc prof, Ohio State Univ, 66-69; assoc prof, 69-71, PROF ENTOM, UNIV MD, COLLEGE PARK, 71-, CHMN DEPT, 75- *Concurrent Pos:* Entom specialist, Ohio State Univ-US Agency Int Develop, Brazil, 66-69; ed, Environ Entom, 71-75. *Mem:* AAAS; Entom Soc Am; Int Orgn Biol Control; Brazilian Entom Soc. *Res:* Applied ecology; forage crop insects; biological control; pest management; insect behavior; graduate training. *Mailing Add:* Dept of Entom Univ of Md College Park MD 20742

STEINHAUER, PAUL DAVID, b Toronto, Ont, Nov 29, 33; m 56; c 4. CHILD PSYCHIATRY. *Educ:* Univ Toronto, MD, 57; FRCP(C), 62. *Prof Exp:* Dir, training child psychiat, 74-91, PROF PSYCHIAT, HOSP SICK CHILDREN, TORONTO, 65- *Concurrent Pos:* Psychiat consult, Halton County Children's Aid Soc, 63-75, Toronto Cath Children's Aid Soc, 64-75 & Children's Aid Soc Metrop Toronto, 65-; sr staff psychiatrist, Hosp Sick Children, Toronto, 65-; chmn, Sparrow Lake Alliance. *Mem:* Can Psychiat Asn; Can Med Asn; Can Acad Child Psychiat (pres, 81-82). *Res:* Development of process model of family functioning and the family assessment measure based on that model; relative effectiveness of two models of foster care on protecting children's adjustment and development and foster parent satisfaction; development of an instrument for assessing/predicting parenting capacity. *Mailing Add:* Med Suite 126 400 Walmer Rd Toronto ON M5P 2X7 Can

STEINHAUS, DAVID WALTER, b Neillsville, Wis, July 29, 19; m 49; c 4. ATOMIC PHYSICS, SPECTROCHEMISTRY. *Educ:* Lake Forest Col, AB, 41; Johns Hopkins Univ, PhD(physics), 52. *Prof Exp:* Physicist, Cent Sci Co, 41-45, Cenco Indust fel, 41-42; res asst inst coop res, Johns Hopkins Univ, 46-52; mem staff & physicist, Los Alamos Nat Lab, 52-73, sect leader, 69-79; assoc prof physics, Memphis State Univ, 83-84; ADJ ASSOC PROF PHYSICS & ASTRON, UNIV NMEX, 84- *Concurrent Pos:* Mem comt line spectra of the elements, Nat Acad Sci-Nat Res Coun, 66-70; consult, Los Alamos Nat Lab, 79-83; distinguished US prof physics, Southwestern Memphis Col (now Rhodes Col), 79-83. *Mem:* AAAS; Optical Soc Am; Am Asn Physics Teachers; Soc Appl Spectros. *Res:* Instrument development; visible and ultraviolet spectroscopy; time resolution of spectra from spark discharges; high resolution spectroscopy; optical spectra of the heavy elements; spectrochemical analysis. *Mailing Add:* 2925 Candelita Court NE Albuquerque NM 87112-2108

STEINHAUS, JOHN EDWARD, b Omaha, Nebr, Feb 23, 17; m 43; c 5. ANESTHESIOLOGY. *Educ:* Univ Nebr, BA, 40, MA, 41; Univ Wis, MD, 45, PhD(pharmacol), 50; Am Bd Anesthesiol, dipl, 59. *Prof Exp:* Asst prof pharmacol, Marquette Univ, 50-51; assoc prof, Univ Wis, 51-54, asst prof anesthesiol, 54-58; assoc prof, 58-59, prof & chmn dept, 59-85, EMER PROF ANESTHESIOL, SCH MED, EMORY UNIV, 87- *Honors & Awards:* Distinguished Serv Award, Am Soc Anesthesiol, 82. *Mem:* Am Soc Anesthesiol (pres, 70); Asn Univ Anesthetists (pres, 71); Am Soc Pharmacol & Exp Therapeut; AMA. *Res:* Drug reactions and intoxications; antiarrhythmic agents; cough mechanism and suppression; depression of respiratory reflexes. *Mailing Add:* Dept Anesthesiol Emory Univ Sch Med Atlanta GA 30322

STEINHAUS, RALPH K, b Sheboygan, Wis, June 21, 39; m 65; c 2. ANALYTICAL CHEMISTRY. *Educ:* Wheaton Col, BS, 61; Purdue Univ, PhD(anal chem), 66. *Prof Exp:* Asst prof chem, Wis State Univ-Oshkosh, 65-68; from asst prof to assoc prof, 68-82, PROF CHEM, WESTERN MICH UNIV, 82- *Concurrent Pos:* Res assoc, Ohio State Univ, 75-76. *Mem:* Am Chem Soc; Sigma Xi. *Res:* Kinetics and mechanisms of transition metal chelates; factors affecting stability of metal chelates. *Mailing Add:* Dept of Chem Western Mich Univ Kalamazoo MI 49008

STEINITZ, MICHAEL OTTO, b New York, NY, June 12, 44; m 65; c 2. SOLID STATE PHYSICS, MATERIALS SCIENCE. *Educ:* Cornell Univ, BE, 65; Northwestern Univ, Evanston, PhD(mat sci), 70. *Prof Exp:* Coop student comput logic, Philco Corp, 62-63, coop student radio propagation, 63; Nat Res Coun Can fel, dept physics, Univ Toronto, 70-72, proj scientist solid state physics, dept metall, 72-73, instr, Scarborough Col, 70-72; from asst prof to assoc prof, 73-83, PROF PHYSICS, ST FRANCIS XAVIER UNIV, 83- *Concurrent Pos:* Lady Davis prof physics, Israel Inst Technol, 77-78 & 80-81. *Mem:* Am Phys Soc; Can Asn Physicists. *Res:* Magnetic properties of metals; thermal expansion; magnetostriction; neutron diffraction; ultrasonic attenuation; phase transitions; chromium; defect structures in oxides; charge and spin density waves; layered structures; model biomembranes; rare-earth metals. *Mailing Add:* PO Box 154 Dept Physics St Francis Xavier Univ Antigonish NS B2G 1C0 Can

STEINKAMP, MYRNA PRATT, b Warren, Pa, Dec 23, 38. BOTANY. *Educ:* Hood Col, BA, 62; Northwestern Univ, Evanston, MS, 64; Univ Calif, Santa Cruz, PhD(bot), 73. *Prof Exp:* Lab supvr, Stoller Res Co, 69-74; microbiologist, Agr Res Serv, USDA, Calif, 74-75; res assoc, Beet Sugar Develop Found, Colo, 76-89. *Mem:* Bot Soc Am; Am Bryological & Lichenological Soc. *Res:* Cytology of plant host-pathogenic fungi and pathotoxins on host tissue; fine structure and development of bryophyte spores. *Mailing Add:* 4700 Venturi Lane Ft Collins CO 80525

STEINKE, FREDERICH H, b Wilmington, Del, Nov 26, 35; m 60; c 2. POULTRY NUTRITION. *Educ:* Univ Del, BS, 57, MS, 59; Univ Wis, PhD(biochem), 62. *Prof Exp:* Asst mgr poultry nutrit, Ralston-Purina Co, 62-63, mgr turkey res, 63-72, mgr nutrit res, Cent Res Lab, 72-91; RETIRED. *Mem:* Poultry Sci Asn; Am Inst Nutrit; Inst Food Technol. *Res:* Poultry nutrition including broilers, laying hens and turkeys; nutrition evaluation; human nutrition. *Mailing Add:* Ralston Purina Co 4RN Checkerboard Sq St Louis MO 63164

STEINKE, PAUL KARL WILLI, b Friedeberg, Ger, July 13, 21; nat US; m 44; c 3. AGRICULTURAL MICROBIOLOGY. *Educ:* Univ Wis, BS, 47, MS, 48, PhD(agr bact), 51. *Prof Exp:* Asst agr bact, Univ Wis, 47-51; food bacteriologist, Chain Belt Co, 51-56; bacteriologist, Paul-Lewis Labs, Inc, 56-61; mgr tech serv, Chas Pfizer & Co, Inc, NY, 61-63, mgr com develop, 64-69, prod mkt mgr, 70-71, tech dir, Pfizer, Inc, Milwaukee, 72-86; RETIRED. *Mem:* Am Soc Microbiol; Am Soc Brewing Chem; Master Brewers Asn Am. *Res:* Microbiology of meat products; food poisoning bacteria; thermal death time of food spoilage organisms; dehydration of foods; brewing bacteriology and technology; chemistry of hops; cheese cultures and media. *Mailing Add:* 6060 N Kent Ave Whitefish Bay WI 53217

STEINKER, DON COOPER, b Seymour, Ind, Oct 6, 36; m 59. PALEOBIOLOGY. *Educ:* Ind Univ, BS, 59; Univ Kans, MS, 61; Univ Calif, Berkeley, PhD(paleont), 69. *Prof Exp:* Res Paleontologist, Univ Calif, Berkeley, 62-63, teaching asst paleont, 62-65, instr, 66-67; asst prof geol, San Jose State Col, 65-66; lectr, Univ Calif, Davis, summer 66; from asst prof to assoc prof, 67-78, PROF GEOL, BOWING GREEN STATE UNIV, 78- *Concurrent Pos:* Managing ed, J Paleont, 88-; nat ed, Compass, 90- *Mem:* Paleont Soc; Nat Asn Geol Teachers; Sigma Xi. *Res:* Foraminiferal biology and ecology; paleobiology. *Mailing Add:* Dept Geol Bowling Green State Univ Bowling Green OH 43403

STEINKRAUS, DONALD CURTISS, b Ames, Iowa, Sept 25, 50; m 87; c 2. MYCOLOGY, INSECT PATHOLOGY. *Educ:* Cornell Univ, BA, 75, PhD(entom), 87; Univ Conn, MS, 79. *Prof Exp:* Res assoc, Dept Entom, Cornell Univ, 87-89; ASST PROF MORPHOL & BIOCONTROL, DEPT ENTOM, UNIV ARK, FAYETTEVILLE, 89- *Mem:* Entom Soc Am; Sigma Xi. *Res:* Insect pathogenic microorganisms, particularly entomopathogenic fungi; biological control of filth flies and litter beetles on poultry and dairy farms and of lepidopterous pests of cotton, soybean and grain sorghum. *Mailing Add:* Dept Entom Univ Ark Fayetteville AR 72701

STEINKRAUS, KEITH HARTLEY, b Bertha, Minn, Mar 15, 18; m 41; c 5. MICROBIOLOGY, BIOCHEMISTRY. *Educ:* Univ Minn, BA, 39; Iowa State Col, PhD(bact), 51. *Prof Exp:* Chemist, Am Crystal Sugar Co, Minn, 39; microbiologist, Jos Seagram & Sons, Inc, 42; instr electronics, US Army Air Force Tech Training Sch, SDak, 42-43; res microbiologist, Gen Mills, Inc, Minn, 43-47; res microbiologist, Pillsbury Mills, Inc, 51; from asst prof to prof bact, 51-88, EMER PROF MICROBIOL, NY STATE COL AGR, CORNELL UNIV, 88- *Concurrent Pos:* Food & agr specialist, US Nutrit Surv, Ecuador & Vietnam, 59, Burma, 61 & 64, Malaya, Thailand & Korea, 64; spec consult interdept comt nutrit for nat defense, NIH, 59-60, spec consult off int res, 67-68; vis prof microbiol col agr, Univ Philippines, 67-69 & Polytech South Bank, London, 72-73; vis prof, UNESCO-Int Cell Res Orgn-UN Environ Prog Training Course Appl Microbiol, Inst Technol, Bandung, Indonesia, 74 & Inst Microbiol, Univ Gottingen, Germany, 80; int organizer Symp Indigenous Fermented Foods, Bangkok, 77; vis prof UNESCO/ICRO Training Course Appl Microbiol, Kasetsart Univ, Bangkok, 76; res scientist, Nestle Prod Tech Assistance Co, Ltd, La-Tour-de-Peilz, Switzerland, 79-80; consult, UN Ind Develop Orgn, 83-84; mem panel microbiol, UN Environ Prog-UNESCO-Int Cell Res Orgn, 84-; team dir, Ind Study Mission Japan, Tech Transfer Inst, 85; vis res scientist, Eastreco, Nestle, Singapore, 86-87, consult, Nestec-Westreco (Nestle), New Milford, Conn, 88-91. *Honors & Awards:* Int Award, Inst Food Technologists, 85; Fel, Inst Food Technologists, 87. *Mem:* Fel AAAS; fel Am Acad Microbiol; Am Soc Microbiol; Inst Food Technol; Sigma Xi. *Res:* Biochemical, microbial and nutritional changes in fermented protein-rich foods; biological and chemical transformations by yeasts, molds and bacteria; biological control of insects with parasitic spore-forming bacteria; extraction of plant proteins; protein hydrolysis; nutrition. *Mailing Add:* Dept Entom Cornell Univ Ithaca NY 14853

STEINLAGE, RALPH CLETUS, b St Henry, Ohio, July 2, 40; m 62; c 3. MATHEMATICS, FUZZY SETS & SYSTEMS. *Educ:* Univ Dayton, BS, 62; Ohio State Univ, MS, 63, PhD(math), 66. *Prof Exp:* From asst prof to assoc prof, 66-79, PROF MATH, UNIV DAYTON, 79- *Concurrent Pos:* Woodrow Wilson Fac Develop grant, 82, fel, Ohio State Univ, 62-63. *Mem:* Math Asn Am; Am Math Soc. *Res:* Measure theory and topology; Haar measure on locally compact Hausdorff spaces; function spaces; conditions related to equicontinuity; non-standard analysis; fuzzy topological spaces; experiential training in mathematics programs; applications of fuzzy sets; author of various books. *Mailing Add:* Dept Math Univ Dayton Dayton OH 45469-2316

STEINLE, EDMUND CHARLES, JR, b Scranton, Pa, Feb 7, 24; m 47; c 3. ORGANIC CHEMISTRY. *Educ:* DePauw Univ, AB, 47; Univ Iowa, MS, 49, PhD(org chem), 52. *Prof Exp:* Chemist, Ethyl Corp Res Lab, 52-55; CHEMIST, PLASTICS & CHEM RES & DEVELOP LAB, UNION CARBIDE CORP, 55- *Mem:* AAAS; Am Chem Soc; Am Oil Chemists Soc; fel Royal Soc Chem; Sigma Xi. *Res:* Fatty alchohols; biodegradable surfactants; surfactant intermediates. *Mailing Add:* 1220 Ridge Dr South Charleston WV 25309

STEINMAN, CHARLES ROBERT, b New York, NY, Aug 3, 38. RHEUMATOLOGY, BIOCHEMISTRY. *Educ:* Princeton Univ, AB, 59; Columbia Univ, MD, 63. *Prof Exp:* From intern to resident med, Presby Hosp, Columbia Univ, 63-65 & 68-69; assoc biochem, Nat Inst Arthritis, Metab & Digestive Dis, NIH, 65-67; fel rheumatology, Presby Hosp, Columbia Univ, 67-68 & 69-70; vis fel rheumatology, The London Hosp, 70; asst prof med, Mt Sinai Sch Med, 70-77, assoc prof, 77-; DIR, DIV RHEUMATOL & PROF, DEPT MED, STATE UNIV NY, DOWNSTATE MED SCH. *Concurrent Pos:* Asst attend physician, Mt Sinai Hosp, 70-77, assoc attend physician, 77-; asst attend physician, Beth Israel Hosp, 71; consult rheumatol, Bronx Vet Admin Hosp, 77- *Mem:* Am Rheumatism Asn; Harvey Soc; Lupus Found Am; Am Fed Clin Res; AAAS. *Res:* Study of rheumatoid arthritis; systemic lupus erythematosus and related disorders to determine their pathogenesis by biochemical, immunological and microbiological approaches. *Mailing Add:* Med Sch State Univ NY 450 Clarkson Ave Box 42 Brooklyn NY 11203

STEINMAN, HARRY GORDON, b Trenton, NJ, Jan 5, 13; m 36; c 1. ORGANIC CHEMISTRY. *Educ:* Mass Inst Technol, BS, 33; Rutgers Univ, MS, 36; Columbia Univ, PhD(chem), 42. *Prof Exp:* From biochemist to head sect biochem, Lab Clin Invest, Nat Inst Allergy & Infectious Dis, 38-66, chief viral reagents, Nat Cancer Inst, 66-69, Off Res Safety, 69-77, CONSULT, NAT CANCER INST, 77- *Mem:* Am Chem Soc; Soc Exp Biol & Med; Am Soc Biol Chem; Am Soc Microbiol; Tissue Cult Asn. *Res:* Cell-mediated immunity; chemotherapy of infectious diseases; penicillins and penicillinases; drug-protein interactions; viral oncology. *Mailing Add:* 1040 Deer Ridge Dr Apt 413 Baltimore MD 21210

STEINMAN, HOWARD MARK, b Detroit, Mich, Feb 18, 44; m 81; c 2. OXYGEN TOXICITY, PROTEIN STRUCTURE. *Educ:* Amherst Col, BA, 65; Yale Univ, PhD(molecular biophys), 70. *Prof Exp:* Assoc biochem, Duke Univ, 72-75, asst med res prof, 75-76; asst prof, 76-81, ASSOC PROF BIOCHEM, ALBERT EINSTEIN COL MED, 81- *Concurrent Pos:* Fel, Dept Biochem, Duke Univ, 70-72. *Mem:* Sigma Xi; Am Soc Biochem & Molecular Biol. *Res:* Biochemical mechanisms of oxygen toxicity; structure, function and relationships among enzymes. *Mailing Add:* Dept Biochem Albert Einstein Col Med 1300 Morris Park Ave Bronx NY 10461

STEINMAN, IRVIN DAVID, b New York, NY, Nov 7, 24; m 54; c 2. MICROBIOLOGY. *Educ:* Brooklyn Col, AB, 48; Univ Chicago, MS, 49; Rutgers Univ, PhD(microbiol), 58. *Prof Exp:* Instr microbiol col dent, NY Univ, 51-54; asst instr biol, Rutgers Univ, 54-58; mem fac, Monmouth Jr Col, NJ, 58-59; dir biol serv, US Testing Co, 59-61; dir prof serv, White Labs, 61-67; head med commun, Bristol Labs Int Corp, 67-68; clin res assoc, Ciba Pharm Co, 68-69; mem staff, Fed Trade Comn, 69-74, actg asst dir, Div Sci Opinion, 74-76, spec asst sci affairs, 76-80, sci adv & res analyst, Bur Consumer Protection, 81-87, NW BR COORDR, INTERSTATE COMN POTOMAC RIVER BASIN, FED TRADE COMN, 88- *Concurrent Pos:* Pvt consult fed regulatory matters relating to sci. *Mem:* AAAS; Am Soc Microbiol; Am Med Asn; Asn Mil Surg US; fel Am Acad Microbiol. *Res:* Microbial genetics; metabolism; pigmentation and production of antibiotics; scientific evaluation of false and misleading advertising claims; health sciences administration. *Mailing Add:* 14601 Notley Rd Silver Spring MD 20905

STEINMAN, MARTIN, b Passaic, NJ, Feb 16, 37; m 61; c 2. MEDICINAL CHEMISTRY, ORGANIC CHEMISTRY. *Educ:* Rutgers Univ, BS, 58, MS, 62; Univ Kans, PhD(med chem), 65. *Prof Exp:* From chemist to sr chemist, Schering Corp, 65-70, prin scientist, 70-73, sect leader, 73-76, mgr, 76-79, assoc dir, 79-89, DIR, SCHERING-PLOUGH RES, SCHERING CORP, 89- *Concurrent Pos:* Vis assoc prof pharmaceut chem, Rutgers Univ. *Mem:* Am Chem Soc; NY Acad Sci; fel Am Inst Chemists; Sigma Xi. *Res:* Stereochemistry; new heterocyclic ring systems; synthesis of potentially useful medicinal agents; chemical and antibiotic process research; structure-activity relationships. *Mailing Add:* Schering Corp 86 Orange St Bloomfield NJ 07003

STEINMAN, RALPH R, b Asheville, NC, Nov 23, 10; m 38; c 3. DENTISTRY. *Educ:* Emory Univ, DDS, 38; Univ Mich, MS, 53. *Prof Exp:* From instr to assoc prof caries control, 53-65, prof oral med, 65-77, mem fac, 77-81, EMER PROF, DEPTS OF DENT ASSIST AND HYG & ORAL DIAG, RADIOL & PATH, SCH DENT, LOMA LINDA UNIV, 81- *Mem:* AAAS; Int Asn Dent Res. *Res:* Pathology of dental caries, especially as related to nutrition; hypothalamic-parotid endocrine axis and its relation to dental caries. *Mailing Add:* Loma Linda Univ Dept Oral Diag-Radiol-Pathol Loma Linda CA 92354

STEINMAN, ROBERT, b New York, NY, Mar 30, 18; m 39; c 4. INORGANIC CHEMISTRY, ORGANIC CHEMISTRY. *Educ:* Carnegie Inst Technol, BS, 39; Univ Ill, MS, 40, PhD(chem), 42. *Prof Exp:* Res chemist, Owens-Corning Corp, 42-47; dir res, Waterway Projs, 47-48; pres, Garan Chem Corp, Calif, 48-67; VPRES RES & DEVELOP, WHITTAKER CORP, 67- *Mem:* AAAS; Am Chem Soc; Soc Plastics Eng; Soc Plastics Indust. *Res:* Phosphorus nitrogen chemistry; surface chemistry of glass; reinforced plastics. *Mailing Add:* 11455 Thurston Circle Los Angeles CA 90049-2425

STEINMEIER, ROBERT C, b Glendale, Calif, Apr 16, 43; m 68, 87; c 2. HEMEPROTEINS, KINETICS. *Educ:* Univ Nebr, Lincoln, BS, 65, PhD(biochem), 75. *Prof Exp:* Immunochemist, Hyland Div, Travenol Labs, 67-69; res assoc bioenergetics, State Univ NY Buffalo, 74-76, Nat Cancer Inst fel, 75-76; asst prof, 78-83, ASSOC PROF CHEM, UNIV ARK, 83-; ASST PROF CHEM, UNIV ARK MED SCH, 78- *Mem:* Am Chem Soc. *Res:* Structure-function studies on hemeproteins and enzymes by various chemical and kinetic methods. *Mailing Add:* Dept Chem Univ Ark Little Rock AR 72204-1084

STEINMETZ, CHARLES, JR, zoology, for more information see previous edition

STEINMETZ, CHARLES HENRY, b Logansport, Ind, Oct 5, 29; m 71, 88; c 4. OCCUPATIONAL HEALTH. *Educ:* Ind Univ, AB, 50, PhD(comp physiol), 53; Univ Cincinnati, MD, 60; Johns Hopkins Univ, MPH, 72; Am Bd Prev Med, cert gen prev med, 73. *Prof Exp:* Asst physiol, Ind Univ, 49-50, anat, 50-51 & zool, 51-53; asst chief space biol, Aero Med Field Lab, Holloman AFB, NMex, 53-56; epidemiologist, Off of Dir, Robert A Taft Sanit Eng Ctr, USPHS, Ohio, 58-60; intern, Staten Island Marine Hosp, NY, 60-61; asst gen mgr, Life Sci Opers, NAm Aviation, Inc, 62-68; dir, Systemed Corp, Md, 68-72; vpres, Nat Health Serv, 72-74; clin instr public health, Med Col, Cornell Univ, 75-79; med dir, Marathon Oil Co, 79-87; MED DIR, INDIANA BELL TEL CO, 87- *Concurrent Pos:* Chmn, Occup Med Sect Methodist Hosp, 88-89. *Mem:* Fel Aerospace Med Asn; fel Am Acad Occup Med; fel Am Occup Med Asn; fel Am Col Prev Med. *Res:* General preventive medicine; occupational health; systems analysis. *Mailing Add:* Med Dir Ind Bell Tel Co Rm 195 240 N Meridan St Indianapolis IN 46204

STEINMETZ, MICHAEL ANTHONY, b Ft Belvoir, Va, Nov 9, 51; m 73. NEUROPHYSIOLOGY, CEREBRAL CORTEX. *Educ:* Univ Mich, BS, 73; Mich State Univ, MS, 74, PhD(physiol), 82. *Prof Exp:* Res fel physiol, Johns Hopkins Univ, 82-83, res fel neurosci, 83-85, asst prof, 85-89, ASSOC PROF, DEPT NEUROSCI, SCH MED, JOHNS HOPKINS UNIV, 89- *Mem:* AAAS; Am Physiol Soc; Soc Neurosci. *Res:* Neurophysiology of the cerebral cortex in primate behavior. *Mailing Add:* Dept Neurosci Sch Med Johns Hopkins Univ 725 N Wolfe St Baltimore MD 21205

STEINMETZ, PHILIP R, b De Bilr, Neth, 27; m; c 2. KIDNEY PHYSIOLOGY, EPITHELIAL TRANSPORT. *Educ:* Univ Leiden, Neth, MD, 75. *Prof Exp:* PROF MED, UNIV CONN, 81- *Concurrent Pos:* Fac, Harvard Med Sch, 65-73; prof med, Univ Iowa, 73-81. *Honors & Awards:* Homer Smith Award, Renal Physiol, 85. *Mem:* Am Soc Clin Invest; Am Soc Nephrology; Am Physiol Soc; Soc Gen Physiol; Asn Am Physicians. *Mailing Add:* Dept Med Sch Med Univ Conn Farmington CT 06032

STEINMETZ, WALTER EDMUND, b Washington, DC, Jan 12, 21; m 47; c 2. ORGANIC CHEMISTRY. *Educ:* St Ambrose Col, BS, 43; Univ Iowa, MS, 47, PhD(org chem), 49. *Prof Exp:* Res chemist, Nalco Chem Co, 48-49, group leader org chem, 49-53, dir, 53-57, sr tech adv, 57-60; sect leader, El Paso Natural Gas Prod Co, 60-65; sect leader, Pennzoil Co, 65-69, mgr chem res div, 69-75, sr tech adv, 75-83; RETIRED. *Mem:* Am Chem Soc. *Res:* Herbicides; oil treatment chemicals; petrochemicals. *Mailing Add:* 2070 Holly Oak Dr Shreveport LA 71118

STEINMETZ, WAYNE EDWARD, b Huron, Ohio, Feb 16, 45. PHYSICAL CHEMISTRY, MOLECULAR SPECTROSCOPY. *Educ:* Oberlin Col, AB, 67; Harvard Univ, AM, 68, PhD(chem), 73. *Prof Exp:* Instr phys sci, St Peter's Boys' High Sch, 69-70; lab instr, Oberlin Col, 70-71; from asst prof to assoc prof, 73-88, PROF CHEM, POMONA COL, 88- *Concurrent Pos:* Guest prof, Eidgenössische Tech Hochschule, Zurich, 79-80 & 87-88; Woodrow Wilson fel. *Mem:* Am Chem Soc; AAAS; Sigma Xi. *Res:* Molecular structure and spectroscopy; application of circular dichroism, nuclear magnetic resonance, and low resolution microwave spectroscopy to conformational analysis; NMR spectroscopy of peptides; laboratory automation. *Mailing Add:* Seaver Chem Lab Pomona Col Claremont CA 91711-6338

STEINMETZ, WILLIAM JOHN, b Wheeling WVa, Nov 14, 39; m 71. APPLIED MATHEMATICS. *Educ:* St Louis Univ, BS, 60; Ga Inst Technol, MS, 62; Rensselaer Polytech Inst, PhD(math), 70. *Prof Exp:* Res scientist aeronaut eng, NASA Ames Res Ctr, 61-66; from asst prof to assoc prof, 70-81, PROF MATH, ADELPHI UNIV, 81- *Mem:* Soc Indust & Appl Math; Sigma Xi. *Res:* Applied mathematics, in particular singular perturbations of ordinary and partial differential equations; stochastic differential equations. *Mailing Add:* Dept of Math Adelphi Univ Garden City NY 11530

STEINMULLER, DAVID, b New York, NY, June 17, 34. TRANSPLANTATION IMMUNOLOGY. *Educ:* Swarthmore Col, BA, 56; Univ Pa, PhD(zool), 61. *Prof Exp:* Mus cur, Wistar Inst Anat & Biol, 61-62; lectr embryol, Fac Med, Univ Valencia, 62-63; res instr med genetics, Sch Med, Univ Wash, 63-65, res instr path, 65-66; from instr to asst prof, Sch Med, Univ Pa, 66-68; asst prof, Col Med, Univ Utah, 68-71, from assoc prof to prof path & surg, 71-76, assoc res prof surg, 76-77; prog immunol, Mayo Med Sch & consult immunol, Mayo Clin, 77-; AT TRANSPLANT SOC MICH. *Concurrent Pos:* Res fel, Wistar Inst Anat & Biol, 61-62; Fulbright fel, Univ Valencia, 62-63; res assoc, Div Path, Inst Cancer Res, 66-68; mem dent res insts & prog adv comt, Nat Inst Dent Res, 71-75; mem grad fac microbiol, Univ Minn, 78-; mem, Surg, Anesthesiol & Trauma Study Sect, NIH, 79- *Mem:* Transplantation Soc; Am Asn Cancer Res; Am Asn Immunol. *Res:* Immunology and biology of tissue and organ transplants; cellular immunology; carcinogenesis and tumor immunity. *Mailing Add:* Dept Surg 352C Med Labs Univ Iowa Col Med Iowa City IA 52242

STEINRAUF, LARRY KING, b St Louis, Mo, June 8, 31; m 68; c 1. BIOCHEMISTRY, PHYSICAL CHEMISTRY. *Educ:* Univ Mo, BS & MA, 54; Univ Wash, PhD(biochem), 57. *Prof Exp:* Asst prof phys chem, Univ Ill, Urbana, 59-64; assoc prof biochem, 64-68, PROF BIOCHEM & BIOPHYS, SCH MED, IND UNIV, INDIANAPOLIS, 68- *Concurrent Pos:* Res fel chem, Calif Inst Technol, 57-58; USPHS fel crystallog, Cavendish Lab, Cambridge, 58-59. *Mem:* Am Crystallog Asn. *Res:* Relation of molecular structure to biological activity. *Mailing Add:* Dept Biochem Ind Univ Purdue Univ Med 1100 W Michigan St Indianapolis IN 46202-5122

STEINSCHNEIDER, ALFRED, b Brooklyn, NY, June 11, 29; m 50; c 2. PEDIATRICS. *Educ:* NY Univ, BA, 50; Univ Mo, MA, 52; Cornell Univ, PhD(psychol), 55; State Univ NY, MD, 61. *Prof Exp:* Asst psychol, Univ Mo, 50-52; asst, Cornell Univ, 52-54; engr, Advan Electronics Ctr, Gen Elec Co, NY, 54-57; res assoc, State Univ NY Upstate Med Ctr, 58-64, from asst prof to assoc prof pediat, 64-77; prof pediat & dir, Sudden Infant Death Syndrome Inst, Sch Med, Univ Md, 77-83; PRES, AM SUDDEN INFANT DEATH SYNDROME, ATLANTA, GA, PORTLAND, OR, 83- *Mem:* Am Psychosom Soc; Soc Res Child Develop; Soc Psychophysiol Res; Am Acad Pediat. *Res:* Sudden infant death syndrome; child development; psychophysiology. *Mailing Add:* Am Sudden Infant Death Syndrome Inst 275 Carpenter Dr Atlanta GA 30328

STEIN-TAYLOR, JANET RUTH, b Denver, Colo; m. BOTANY, HORTICULTURE. *Educ:* Univ Colo, BA, 51; Wellesley Col, MA, 53; Univ Calif, PhD(bot), 57. *Prof Exp:* Lab technician & cur, Univ Calif, 57-59; from instr to assoc prof bot, biol & phycol, Univ BC, 59-71, prof bot, 71-85, asst dean sci, 79-85; RETIRED. *Concurrent Pos:* Vis assoc prof, Univ Calif, 65; dir, Western Bot Serv, Ltd, 72-75; ed, J Phycol, Phycological Soc Am, 75-80; Kilham sr fel, Univ BC, 75. *Honors & Awards:* Darbaker Award, Bot Soc Am, 60; Mary Elliot Award, Can Bot Asn, 81. *Mem:* Bot Soc Am; Phycol Soc Am (ed, News Bull, 60-64, Newsletter, 65-66, pres, 65); Can Bot Asn (vpres, 69-70, pres, 70-71); Brit Phycol Soc; Int Phycol Soc (treas, 82-87). *Res:* Morphology, physiology, ecology and distribution of freshwater algae. *Mailing Add:* PO Box 371 Glencoe IL 60022-0371

STEINWACHS, DONALD MICHAEL, b Boise, Idaho, Sept 9, 46; m 72. PUBLIC HEALTH ADMINISTRATION. *Educ:* Univ Ariz, BS, 68, MS, 70; Johns Hopkins Univ, PhD(opers res), 73. *Prof Exp:* Res mgr, Health Serv Res & Develop Ctr, 72-79, asst prof, Dept Health Serv, 73-79, asst dir, 79-81, dept dir, 81-82, assoc prof, Dept Health Serv Admin, Johns Hopkins Univ, 79-86, DIR, HEALTH SERV RES & DEVELOP CTR, 82- *Concurrent Pos:* Prof Dept Health Policy & Mgt, 86- *Mem:* AAAS; Opers Res Soc Am; Inst Mgt Sci; Am Pub Health Asn; Asn Health Serv Res. *Res:* Primary medical care; effects of availability, access, continuity, organization and financing on cost and quality; information systems; impact of hospital cost containment strategies; models for health resource allocation; health manpower planning and evaluation. *Mailing Add:* One Hickory Ct Lutherville Towson MD 21093

STEITZ, JOAN ARGETSINGER, b Minneapolis, Minn, Jan 26, 41; m 66; c 1. BIOCHEMISTRY, MOLECULAR BIOLOGY. *Educ:* Antioch Col, BS, 63; Harvard Univ, MA, 67, PhD(biochem, molecular biol), 68. *Hon Degrees:* DSc, Lawrence Univ, 81 & Sch Med, Univ Rochester, 84. *Prof Exp:* Asst prof, 70-74, assoc prof, 74-78, PROF MOLECULAR BIOPHYS & BIOCHEM, YALE UNIV, 78-; INVESTR, HOWARD HUGHES MED INST, 86- *Concurrent Pos:* NSF fel, Med Res Coun Lab Molecular Biol, Cambridge, Eng, 68-69, Jane Coffin Childs Fund med res fel, 69-70; assoc mem, Europ Molecular Biol Orgn, 87. *Honors & Awards:* Passano Award, 75; Eli Lilly Award, 76; US Steel Award, 82; Lee Halley Jr Award, Arthritis Res, 84. *Mem:* Nat Acad Sci; Am Soc Biol Chemists; fel AAAS. *Res:* Control of transcription and translation; RNA and DNA sequence analysis; structure and function of small ribonucleoproteins from eukaryotes; RNA processing. *Mailing Add:* Dept Molecular Biophys & Biochem Yale Univ 333 Cedar St New Haven CT 06510

STEITZ, THOMAS ARTHUR, b Milwaukee, Wis, Aug 23, 40; m 66; c 1. MOLECULAR BIOLOGY. *Educ:* Lawrence Univ, BA, 62; Harvard Univ, PhD(molecular biol, biochem), 66. *Hon Degrees:* DSc, Lawrence Univ, 81. *Prof Exp:* NIH grant, Harvard Univ, 66-67; Jane Coffin Childs Mem Fund Med Res fel, Med Res Coun Lab Molecular Biol, Cambridge Univ, 67-70; from asst prof to assoc prof, 70-79, PROF MOLECULAR BIOPHYS & BIOCHEM, YALE UNIV, 79-; INVESTR, HOWARD HUGHES MED INST, 86- *Concurrent Pos:* Macy fel, Max Planck Inst Biophys Chem, Goettingen, Ger & MRC Lab Molecular Biol, Cambridge, Eng, 76-77; actg dir, Div Biol Sci, Yale Univ, 81; Fairchild scholar, Calif Inst Technol, 84-85. *Honors & Awards:* Pfizer Award, Am Chem Soc, 80. *Mem:* Nat Acad Sci; Am Soc Biol Chemists; Biophys Soc Am; Am Crystallog Asn; Protein Soc; Am Acad Arts & Sci. *Res:* X-ray crystallographic structure determination of biological macromolecules; relation of enzyme structure and mechanism; structure studies of protein-nucleic acid interaction; theoretical studies protein secretion and membrane protein folding. *Mailing Add:* Dept Molecular Biophys & Biochem Yale Univ PO Box 6666 260 Whitney Ave New Haven CT 06511

STEJSKAL, EDWARD OTTO, b Chicago, Ill, Jan 19, 32; m 57. PHYSICAL CHEMISTRY, NUCLEAR MAGNETIC RESONANCE. *Educ:* Univ Ill, BS, 53, PhD(chem), 57. *Prof Exp:* Asst phys chem, Univ Ill, 53-56; NSF fel, Harvard Univ, 57-58; from instr to asst prof phys chem Univ Wis, 58-64, Wis Alumni Res Found fel, 58-59; res specialist, 64-67, sr res specialist, 67-80, fel sci, Monsanto Co, 80-86; PROF CHEM, NC STATE UNIV, 86- *Mem:* AAAS; Am Phys Soc; Am Chem Soc. *Res:* Molecular structure; spectroscopy; nuclear magnetic resonance; molecular motion in solids, liquids and gases; nuclear spin relaxation and diffusion phenomena; high-resolution nuclear magnetic resonance in solids; fluid rheology; lubrication. *Mailing Add:* NC State Univ Box 8204 Raleigh NC 27695-8204

STEJSKAL, RUDOLF, b Budejovice, Czech, Apr 16, 31; m 68; c 2. PATHOLOGY. *Educ:* Charles Univ, Czech, DDS, 56; Univ Chicago, MS, 69; Chicago Med Sch-Univ Health Sci, PhD(path), 72. *Prof Exp:* Trainee path, Univ Chicago, 67-69; res assoc exp path, Mt Sinai Hosp Med Ctr, Chicago, 70-72; res assoc & asst prof path, Univ Chicago, 72-73; sr res investr, Searle Res & Develop, France, 73-75, sr res scientist path, Searle Labs, 75-78, dir path, 78-83; ASSOC DIR PATH, BIO-RES LABS, CAN, 83- *Concurrent Pos:* Clin assoc path, Chicago Med Sch, 70-72. *Mem:* Europ Soc Toxicol; Soc Toxicol Pathologists; Int Acad Path; Am Asn Path. *Res:* Fate of blood group substances in human cancer; drug-induced carcinogenesis; toxicologic pathology. *Mailing Add:* Bio-Res Labs Ltd 87 Senneville Rd Senneville PQ H9X 3R3 Can

STEKEL, FRANK D, b Hillsboro, Wis, Aug 26, 41; m 67; c 2. SCIENCE EDUCATION, PHYSICS. *Educ:* Univ Wis-La Crosse, BS, 63; Univ Wis-Madison, MS, 65; Ind Univ, Bloomington, EdD(sci educ), 70. *Prof Exp:* From instr to assoc prof, 65-77, PROF PHYSICS, UNIV WIS-WHITEWATER, 77- *Mem:* Fel AAAS; Am Asn Physics Teachers; Nat Asn Res Sci Teaching; Nat Sci Teachers Asn; Sch Sci & Math Asn. *Res:* Development, implementation and evaluation of physical science instruction at all levels, from the elementary school up to the college and university level; archaeostronomy. *Mailing Add:* Dept Physics Univ Wis Whitewater WI 53190

STEKIEL, WILLIAM JOHN, b Milwaukee, Wis, Jan 1, 28; m 55; c 2. BIOPHYSICS. *Educ:* Marquette Univ, BS, 51; Johns Hopkins Univ, PhD(biophys), 57. *Prof Exp:* USPHS fel & instr physiol, 57-60, asst prof biophys, 60-66, assoc prof, 66-77, PROF PHYSIOL, MED COL WIS, 77- *Concurrent Pos:* Res exchange prof, Med Univ, Budapest, 63; fel, coun circulation, Am Heart Asn, coun high blood pressure, Cardiovasc Sect, APS. *Mem:* Biophys Soc; Am Physiol Soc; Microcirc Soc. *Res:* Electrophysiology; circulatory physiology; hypertension. *Mailing Add:* Dept Physiol Med Col Wis 8701 Watertown Plank Rd Milwaukee WI 53226

STEKLY, Z J JOHN, b Czechoslovakia, Oct 11, 33; US citizen. MECHANICAL ENGINEERING. *Educ:* Mass Inst Technol, BA, MA(mech eng) & MA(elec eng), 55, ScD, 59. *Prof Exp:* Chmn bd & tech dir, Magnetic Corp Am, 69-86; VPRES ADVAN PROGS, INTERMAGNETICS GEN CORP, 87- *Concurrent Pos:* Mem adv comt, Off Coal Res, 73-; mem ad hoc comt, Superconducting Magnet Technol, ERDA, 75-; mem, Comt Magnetic Fusion, Nat Res Coun, 81-82, Comt Elec Energy Systs, 84-86, Comt Crit Mat, US Army, 84-86, Comt Space Based Power, 87-; mem peer rev comt & mem comt, Nat Acad Eng, 82-84. *Mem:* Nat Acad Eng; Inst Elec & Electronics Engrs; Am Asn Physicists Med; Sigma Xi. *Mailing Add:* c/o Field Effects Intermagnetics Gen Corp Six Eastern Rd Acton MA 01720

STEKOLL, MICHAEL STEVEN, b Tulsa, Okla, May 7, 47; m 76; c 4. BIOCHEMISTRY. *Educ:* Stanford Univ, BS, 71; Univ Calif, Los Angeles, PhD(biochem), 76. *Prof Exp:* Res fel marine pollution, Univ Alaska, Fairbanks, 76-78, ASSOC PROF CHEM & BIOCHEM, UNIV ALASKA, SE & FAIRBANKS, 78- *Concurrent Pos:* Asst, Univ Calif, Los Angeles, 72-76, res trainee, 72-76; res biochemist, Nat Marine Fisheries Serv, 79; assoc res biologist, Univ Calif, Santa Barbara, 86-87. *Mem:* AAAS; Psychol Soc Am; Am Fisheries Soc. *Res:* Purification and properties of phytoalexin elicitors; effects of marine pollution on invertebrates; algal physiology & ecology; salmon biochemistry and physiology. *Mailing Add:* Juneau Ctr Fisheries & Ocean Sci 11120 Glacier Hwy Juneau AK 99801

STELCK, CHARLES RICHARD, b Edmonton, Alta, May 20, 17; m 45; c 4. CRETACEOUS, MICROPALEONTOLOGY. *Educ:* Univ Alta, BSc, 37, MSc, 41; Stanford Univ, PhD(geol), 50. *Prof Exp:* Lab asst, Univ Alta, 37-41; well site geologist, BC Dept Mines, 40-42; field geologist, Canol Proj, 42-44 & Imp Oil Co, 44-48; lectr, 48-53, from assoc prof to prof, 54-82, EMER PROF, UNIV ALTA, 82- *Honors & Awards:* Logan Medal. Geol Asn Can, 81. *Mem:* Geol Soc Am; Paleont Soc; fel Royal Soc Can; Geol Asn Can. *Res:* Stratigraphic paleontology of western Canada. *Mailing Add:* Dept Geol Univ Alta Edmonton AB T6G 2E3 Can

STELL, GEORGE ROGER, b Glen Cove, NY, Jan 2, 33; m 52; c 1. STATISTICAL MECHANICS. *Educ:* Antioch Col, BS, 55; NY Univ, PhD(math), 61. *Prof Exp:* Instr physics, Univ Ill, Chicago, 55-56; assoc res scientist, Inst Math Sci, NY Univ, 61-64; assoc res scientist, Belfer Grad Sch Sci, Yeshiva Univ, 64-65; from asst prof physics to assoc prof physics,

Polytech Inst Brooklyn, 65-68; from assoc prof to prof mech, 68-77, prof mech eng, 77-85, prof chem, 79-85, LEADING PROF CHEM & MECH ENG, STATE UNIV NY, STONY BROOK, 85- *Concurrent Pos:* Consult, Lawrence Radiation Lab, Univ Calif, 63-66; vis prof, Lab Theoret & High Energy Physics, Nat Ctr Sci Res, France, 67-68; prin investr, NSF grants, 70-, Guggenheim Fel, 84-85; prin investr, Dept Energy contract, 79-87, Dept Energy grant, 87- *Mem:* Am Math Soc; Am Phys Soc. *Res:* Statistical mechanics; especially molecular theory of fluids and lattice systems; mathematics associated with statistical mechanics, especially graph theory; generating functionals and non-linear integral equations; theory of critical phenomena and thermodynamics; dielectric, structural and transport properties of fluids; percolation and composite-media theory. *Mailing Add:* Nine Bobs Lane Setauket NY 11733

STELL, WILLIAM KENYON, b Syracuse, NY, Apr 21, 39; m; c 2. VISUAL PHYSIOLOGY, NEUROBIOLOGY. *Educ:* Swarthmore Col, BA, 61; Univ Chicago, PhD(anat), 66, MD, 67. *Prof Exp:* Staff assoc, Lab Neurophysiol, Nat Inst Neurol Dis & Stroke, 67-68, staff assoc neurocytol, Lab Neuropath & Neuroanat Sci, 68-69 & Lab of the Dir, 69, sr staff fel, Off Dir Intramural Res, 69-71 & Lab Neurophysiol, Sect Cell Biol, 71-72; assoc prof ophthal, Jules Stein Eye Inst, Univ Calif Los Angeles, 72-76, prof, 76-80, assoc dir, 78-80; chmn, 80-85, PROF ANAT, DIR, LION'S SIGHT CTR, UNIV CALGARY, 80- *Mem:* AAAS; Asn Res Vision & Ophthal; Soc Neurosci. *Res:* Neurocytology, ultrastructure and functional interconnections in vertebrate retina; neuropeptide chemistry and function; developmental neurobiology. *Mailing Add:* Dept Anat Univ Calgary 3330 Hosp Dr NW Calgary AB T2N 4N1 Can

STELLA, PAUL M, b Hartford, Conn, Dec 13, 44; m 79; c 1. MATERIALS SCIENCE ENGINEERING, MECHANICAL ENGINEERING. *Educ:* Princeton Univ, BA, 66; Calif State Univ-Northridge, MS, 71. *Prof Exp:* Sr engr, Spectrolab, Inc, Hughes Aircraft Co, 67-78; TECH GROUP LEADER, JET PROPULSION LAB, CALIF INST TECHNOL, 78- *Concurrent Pos:* Aerospace Power Comt, Am Inst Aeronaut & Astronaut, 84-87. *Mem:* Inst Elec & Electronics Engrs; Am Inst Aeronaut & Astronaut. *Res:* Design and development of advanced photovoltaic solar array systems for space applications; development of appropriate photovoltaic cell technology to satisfy requirements for operation near sun and far sun. *Mailing Add:* Jet Propulsion Lab MS303-310 4800 Oak Grove Dr Pasadena CA 91109

STELLA, VALENTINO JOHN, b Melbourne, Australia, Oct 27, 46; m 69; c 3. PHARMACY. *Educ:* Victorian Col Pharm, Melbourne, BPharm, 67; Univ Kans, PhD(anal pharmaceut chem, pharmaceut), 71. *Prof Exp:* Pharmacist, Bendigo Base Hosp, Australia, 67-68; asst prof pharm, Univ Ill, Med Ctr, 71-73; from asst prof to assoc prof, 73-81, PROF PHARMACEUT CHEM, SCH PHARM, UNIV KANS, 81- *Concurrent Pos:* Consult, E R Squibb & Sons, 78-, G D Searle, 80-, Sterling-Winthrop, 85-, Lederle Labs, 85- *Honors & Awards:* Lederle Award, Lederle Labs, 72 & 75. *Mem:* Am Chem Soc; Am Pharmaceut Asn; Victorian Pharmaceut Soc; fel Acad Pharmaceut Sci; fel AAAS; fel AM Asn Pharmaceut Scientists. *Res:* Physical pharmacy; pro-drugs and drug latentiation; drug stability; ionization kinetics; biopharmaceutics and pharmacokinetics; preformulation of anticancer drugs. *Mailing Add:* Dept Pharmaceut Chem Univ Kans Lawrence KS 66045-1500

STELLAR, ELIOT, b Boston, Mass, Nov 1, 19; m 45; c 2. PHYSIOLOGICAL PSYCHOLOGY. *Educ:* Harvard Univ, AB, 41; Brown Univ, MSc, 42, PhD(psychol), 47. *Hon Degrees:* DSc, Ursinus Col, 78, Emory Univ, 90; LHD, Johns Hopkins Univ, 83. *Prof Exp:* From instr to asst prof psychol, Johns Hopkins Univ, 47-54; from assoc prof to prof physiol psychol, 54-65, dir, Inst Neurol Sci, 65-73, provost, 73-78, PROF PHYSIOL PSYCHOL, SCH MED, UNIV PA, 78-, CHMN, DEPT ANAT, 90- *Concurrent Pos:* Mem, Nat Comn Protection Human Subj Biomed & Behav Res, 74-78; chmn, Nat Acad Sci Comt on Human Rights, 84-; mem, Bd Health Prom & Dis Prev, Inst Med, 84; mem bd dirs, Zool Soc, Philadelphia & chmn res comt, 86-89. *Honors & Awards:* Warren Medal, Soc Exp Psychol, 67. *Mem:* Inst Med-Nat Acad Sci; Am Philos Soc; AAAS; Am Psychol Asn; Am Acad Arts & Sci; Soc Neurosci; Am Asn Anatomists; Soc Exp Psychologists; Int Brain Res Orgn; Soc Study Ingestive Behav. *Res:* Motivation; learning; physiology of motivation. *Mailing Add:* Sch Med Univ Pa Philadelphia PA 19104-6058

STELLER, KENNETH EUGENE, b Lancaster, Pa, Mar 14, 41; m 64; c 2. ORGANIC CHEMISTRY. *Educ:* Franklin & Marshall Col, BS, 63; Northwestern Univ, PhD(org chem), 67. *Prof Exp:* Res chemist, 67-74, SR RES CHEMIST, HERCULES INC, 74- *Mem:* Am Chem Soc. *Res:* Polyether polymerization; elastomers; polymer modification; thermoplastic elastomers; peroxide research and development. *Mailing Add:* 13 Lamatan Rd Newark DE 19711-2315

STELLINGWERF, ROBERT FRANCIS, b Hawthorne, NJ, Apr 22, 47. STELLAR STRUCTURE & STABILITY. *Educ:* Rice Univ, BA, 69; Univ Colo, MS, 71, PhD(astrophys), 74. *Prof Exp:* Res assoc, Columbia Univ, 74-77; asst prof astron, Rutgers Univ, 77-80; res scientist, 80-84, DIV LEADER, MISSION RES CORP, 84- *Concurrent Pos:* Vis staff mem, Los Alamos Sci Lab, 76-; prin investr, NSF, 78- *Mem:* Am Astron Soc. *Res:* Stellar structure and stability; pulsation theory; astrophysical gas flow. *Mailing Add:* Mission Res Corp 1720 Randolph Rd Albuquerque NM 87106

STELLWAGEN, EARLE C, b Joliet, Ill, June 14, 33; m 58; c 4. BIOCHEMISTRY. *Educ:* Elmhurst Col, BS, 55; Northwestern Univ, MS, 58; Univ Calif, Berkeley, PhD(biochem), 63. *Prof Exp:* NIH res fel biochem, Univ Vienna, 63-64; from asst prof to assoc prof, 64-72, PROF BIOCHEM, UNIV IOWA, 72- *Concurrent Pos:* NIH career develop award, 67-72; mem biophysics & biophysical study sect B, NIH, 70-, chmn, 72-; vis scientist, Bell Labs, 71-72. *Mem:* Am Soc Biol Chem. *Res:* Relationship of structure of proteins to their biological function. *Mailing Add:* Dept Biochem 4-612BSB Univ Iowa Iowa City IA 52242

STELLWAGEN, ROBERT HARWOOD, b Joliet, Ill, Jan 6, 41; m 63; c 2. BIOCHEMISTRY. *Educ:* Harvard Univ, AB, 63; Univ Calif, Berkeley, PhD(biochem), 68. *Prof Exp:* Res biochemist, Univ Calif, Berkeley, 68; staff fel molecular biol, Nat Inst Arthritis & Metab Dis, 68-69; from asst prof to assoc prof, 70-80, interim chmn, 80-86, PROF BIOCHEM, SCH MED, UNIV SOUTHERN CALIF, 80- *Concurrent Pos:* USPHS fel biochem, Univ Calif, San Francisco, 69-70; vis scientist, Nat Ins Med Res, London, 79. *Mem:* AAAS; Am Soc Biochem & Molecular Biologists. *Res:* Biochemical control mechanisms in animal cells; enzyme induction by hormones; control of protein degradation. *Mailing Add:* 30906 Rue de la Pierre Rancho Palos Verdes CA 90274

STELLY, MATTHIAS, b Arnaudville, La, Aug 7, 16; m 40; c 4. SOIL FERTILITY. *Educ:* Univ Southwestern La, BS, 37; La State Univ, MS, 39; Iowa State Univ, PhD(soil fertil), 42. *Prof Exp:* Asst agronomist, Exp Sta, La State Univ, 42-43; from assoc prof to prof soils, Univ Ga, 46-57; prof in chg soil testing serv, La State Univ, 57-59, prof soil chem, 59-61; exec secy & treas, ASA Publs, Am Soc Agron, 61-70, exec vpres & ed-in-chief, 70-82; RETIRED. *Concurrent Pos:* Soil specialist, USDA, EAfrica, 52-53; Am secy, Comn Soil Fertil & Plant Nutrit, 7th Cong, Int Soc Soil Sci, 58-60; prog dir transl & printing, Soviet Soil Sci; ed, Agron J, 61-82; dir Am Soc Agron vis scientist prog, 63-72 & Agron Sci Found, 67-82; consult, Int Inst Tropical Agr, Nigeria, 84-85; Tropical Agr Res Fund, 85. *Mem:* Fel AAAS; fel Am Soc Agron; Crop Sci Soc Am; Soil Sci Soc Am; Int Soc Soil Sci. *Res:* Soil chemistry; identification of inorganic soil phosphorus compounds; radioactive materials as plant stimulants; crop rotation; radioactive phosphorus; fertilizer requirements and chemical composition of Bermuda grasses; methodology of soil testing; chemical and mineralogical investigations of Louisiana soils. *Mailing Add:* 2113 Chamberlain Ave Madison WI 53705

STELOS, PETER, b Lowell, Mass, May 17, 23. IMMUNOLOGY, IMMUNOCHEMISTRY. *Educ:* Berea Col, AB, 48; Univ Chicago, PhD(microbiol), 56. *Prof Exp:* Res assoc microbiol, Univ Chicago, 56-57; cancer res scientist, Roswell Park Mem Inst, 59-60, from sr cancer res to assoc cancer res scientist, 60-65; assoc prof microbiol, Hahnemann Med Col, 65-73; assoc surg & immunol, Peter Bent Brigham Hosp & Harvard Med Sch, 73-79, assoc med & biochem, Harvard Med Sch, 80-88; CONSULT, 88- *Concurrent Pos:* Fel chem, Yale Univ, 57-59. *Mem:* AAAS; Am Asn Immunol; Am Chem Soc; Sigma Xi. *Res:* Separation and properties of immunoglobulins and urinary proteins. *Mailing Add:* 83 Garden St West Roxbury MA 02132-4929

STELSON, PAUL HUGH, b Ames, Iowa, Apr 9, 27; m 50; c 4. PHYSICS. *Educ:* Purdue Univ, BS, 47, MA, 48; Mass Inst Technol, PhD(physics), 50. *Prof Exp:* Asst physics, Purdue Univ, 47; mem staff div indust coop, Nuclear Sci & Eng Lab, Mass Inst Technol, 50-52; sr physicist, 52-73, DIR, PHYS DIV, OAK RIDGE NAT LAB, 73- *Concurrent Pos:* Ford Found prof physics, Univ Tenn. *Mem:* Fel Am Phys Soc. *Res:* Nuclear physics; accelerators. *Mailing Add:* Oak Ridge Nat Lab PO Box 2008 Oak Ridge TN 37830

STELSON, T(HOMAS) E(UGENE), b Iowa City, Iowa, Aug 24, 28; m 51; c 4. CIVIL ENGINEERING. *Educ:* Carnegie Inst Technol, BS, 49, MS, 50, DSc(civil eng), 52. *Prof Exp:* From asst prof to prof civil eng, Carnegie-Mellon Univ, 52-61, Alcoa prof, 61-71, actg head dept, 57-59, head dept, 59-71; dean, Col Eng & asst vpres acad affairs, 71-77, PROF CIVIL ENG, GA INST TECHNOL, 71-, VPRES FOR RES, 77-; VPRES & CHMN BD DIRS, SYSTS PLANNING CORP, 68- *Concurrent Pos:* NSF fel, Calif Inst Technol, 62-63; co-dir, Transp Res Inst, Pa, 67-71; dir projs, Governor's Comt Transp, Commonwealth Pa, 68-71; mem exec bd, Skidaway Inst Oceanog, 71- & US Task Force on Future Mankind & Role of Christian Churches in World of Sci-Based Technol, 71- *Mem:* AAAS; Am Soc Civil Engrs; Am Soc Eng Educ; Nat Soc Prof Eng; Am Concrete Inst. *Res:* Fluid and soil mechanics; hydraulic and foundation engineering; transportation and systems engineering; environmental conditions. *Mailing Add:* 225 North Ave NW Georgia Tech VP Res Atlanta GA 30332

STELTENKAMP, ROBERT JOHN, b Dayton, Ky, Sept 13, 36; m 57; c 2. ORGANIC CHEMISTRY. *Educ:* Xavier Univ, Ohio, BS, 58; Purdue Univ, PhD(org chem), 62. *Prof Exp:* Sr res chemist perfumery res, 62-67, sec head var res & develop sections, 67-80, res assoc res & develop, 80-84, sr assoc, 84-87, ASSOC RES FEL, COLGATE PALMOLIVE CO, NJ, 87- *Mem:* Am Chem Soc; Sigma Xi; NY Acad Sci. *Res:* Identification of the composition of essential oils; structural identification of natural components and the toxicology of fragrance raw materials; organic synthesis. *Mailing Add:* 92 Emerson Rd Somerset NJ 08873

STELTING, KATHLEEN MARIE, b Ottawa, Kans, Sept 9, 42; c 1. ANALYTICAL CHEMISTRY. *Educ:* Eastern NMex Univ, BS, 64; Univ Mo-Columbia, PhD(anal chem), 73. *Prof Exp:* Clin chemist, Hertzler Clin, Halstead, Kans, 64-65; from asst prof to assoc prof anal chem, Calif State Univ, Fresno, 73-77; mgr separations & automation develop, Dept Res, Rockwell Hanford Opers, Rockwell Int Corp, 77-80; DIR BIOORGANIC CHEM, MIDWEST RES INST,80- *Mem:* Am Chem Soc; Am Chem Soc (chair elect, Indust & Eng Chem Div, 88); Robotics Int SME. *Res:* Chemistry support for toxicology research; ion selective electrodes; spectroscopic techniques for complexants analysis; environmental measurements; pesticide registration, microencapsulation & laboratory robotics. *Mailing Add:* Hewlett-Packard Rte 41 & Starr Rd PO Box 900 Avondale PA 19311

STELTS, MARION LEE, b Oregon City, Ore, Mar 28, 40; m 61; c 2. NUCLEAR PHYSICS. *Educ:* Univ Ore, BA, 61; Univ Calif, Davis, MS, 70, PhD(appl sci), 75. *Prof Exp:* Physicist, Lawrence Livermore Lab, 61-75; physicist nuclear physics, Brookhaven Nat Lab, 75-80; PHYSICIST, LOS ALAMOS NAT LAB, 80- *Mem:* Am Phys Soc. *Mailing Add:* Los Alamos Nat Lab P-15 MS406 Los Alamos NM 87545

STELZER, LORIN ROY, b Bloomer, Wis, Mar 6, 31; m 58; c 2. RESEARCH ADMINISTRATION, ENTOMOLOGY. *Educ:* Wis State Col, Whitewater, BEd, 53; Univ Wis, MS, 55, PhD(entom), 57. *Prof Exp:* Field res specialist, 57-62, supvr, Southern Field Res Sta, Fla, 62-69, East East/South Field Res Sta, NJ, 69-71, mgr registrat & regulatory affairs, res & develop, 71-85, mgr develop res, Ortho Div, 85-88, MGR CHEM SERV, CHEVRON CHEM CO, CALIF, 88- *Mem:* Entom Soc Am. *Res:* Field research, evaluation and development of chemicals for agricultural and home use. *Mailing Add:* Agr Chem Div Chevron Chem Co 15049 San Pablo Ave Richmond CA 94804

STEMBRIDGE, VERNIE A(LBERT), b El Paso, Tex, June 7, 24; m 44; c 3. PATHOLOGY. *Educ:* Tex Col Mines, BA, 43; Univ Tex, MD, 48; Am Bd Path, cert anat & clin path, 53. *Prof Exp:* Intern, Marine Hosp, Norfolk, Va, 48-49; resident path, Med Br, Univ Tex, 49-52, assoc dir clin labs & from asst prof to assoc prof, 52-56; chief aviation path sect, Armed Forces Inst Path, 56-59; assoc prof, Univ Tex Southwestern Med Ctr, Dallas, 59-61, chmn dept, 66-88, actg dean, 88-91, PROF PATH, UNIV TEX SOUTHWESTERN MED CTR, DALLAS, 61- *Concurrent Pos:* Consult, Vet Admin Hosp, Dallas, 59-, Civil Air Surgeon, Fed Aviation Agency, DC, 59-70 & Surgeon Gen, US Air Force, 62-70; trustee, Am Bd Path, 69-80, secy, 77-79, pres, 80; mem sci adv bd, Armed Forces Inst Path, DC, 71-75, chmn, 74-75, Residency Rev Comt Path, 72-78, chmn, 74-75, 77-78; pres, Dallas County Med Soc, 85; secy, Am Registry Path, 86, pres, 89-90. *Honors & Awards:* Ward Burdick Outstanding Award, Am Soc Am Path, 81. *Mem:* AMA; Am Asn Path; Col Am Path; Am Soc Clin Path (pres, 77-78); Int Acad Path; Asn Clin Lab Physicians & Scientists; Am Registry Path. *Res:* Neoplasms, environment and immunopathology. *Mailing Add:* Dept of Path Univ of Tex Southwestern Med Ctr Dallas TX 75235-9072

STEMKE, GERALD W, b Watseka, Ill, Oct 7, 35; m; c 2. MICROBIOLOGY. *Educ:* Ill State Univ, BS, 57; Univ Ill, PhD(chem), 63. *Prof Exp:* Teaching asst, Univ Ill, 59-60; trainee microbiol, NY Univ Med Ctr, 60-62, res assoc, 62-63; NIH fel, Pasteur Inst, Paris, 63-64, asst prof biol, Univ Pittsburgh, 65-66, asst prof microbial & molecular biol, 66-67, assoc prof biophys & microbiol, 67-70; PROF MICROBIOL, UNIV ALTA, 70- *Concurrent Pos:* Nat Res Coun Can res grant, 70-74; Nat Sci Eng, Res Coun Can res grant. *Mem:* AAAS; Int Org Mycoplasmol; Am Soc Microbiol; Can Soc Immunol. *Res:* Mycoplasma; ureaplasma surface antigens. *Mailing Add:* Dept Microbiol Univ Alta Edmonton AB T6G 2E2 Can

STEMLER, ALAN JAMES, b Chicago, Ill, July 29, 43; m 85; c 2. PLANT PHYSIOLOGY. *Educ:* Mich State Univ, BS, 65; Univ Ill, Urbana, PhD(plant physiol), 74. *Prof Exp:* Instr plant physiol, Univ Ill, Urbana, 74-75; fel, Dept Plant Biol, Carnegie Inst Washington, 75-78; PROF, DEPT BOT, UNIV CALIF, DAVIS, 78- *Mem:* Am Soc Photobiol; Am Soc Plant Physiologists. *Res:* Photochemical events of photosynthesis. *Mailing Add:* Bot Dept Univ Calif-Davis Davis CA 95616

STEMMER, EDWARD ALAN, b Cincinnati, Ohio, Jan 20, 30; m 54; c 5. THORACIC SURGERY. *Educ:* Univ Chicago, BA, 49, MD, 53; Am Bd Surg, dipl, 62; Am Bd Thoracic Surg, dipl, 63. *Prof Exp:* Intern med, Univ Chicago Clins, 53-54, res asst surg, Sch Med, 54-55, from asst resident to sr resident, 54-60, instr, 59-60; chief resident, Stanford Univ, 60-61, clin teaching asst, 61; chief resident, Palo Alto Vet Admin Hosp, 61-62, asst chief surg serv, 62-64; asst prof surg, Univ Utah, 64-65; asst prof in residence, 66-70, assoc prof, 70-76, PROF SURG, UNIV CALIF, IRVINE, 76-; CHIEF SURG SERV, LONG BEACH VET ADMIN HOSP, 65- *Concurrent Pos:* Responsible investr, Vet Admin Hosp, 62-; attend surgeon, Salt Lake County Hosp, Utah, 64-65; prin investr, NIH, 64-70. *Mem:* Fel Am Col Surg; Am Asn Thoracic Surg; Soc Thoracic Surg; Am Surg Asn. *Res:* Vascular surgery; metabolism of plasma proteins; myocardial functions; control of regional blood flow. *Mailing Add:* Long Beach Vet Admin Hosp 5901 E Seventh St Long Beach CA 90822

STEMMERMANN, GRANT N, b Bronx, NY, Oct 28, 18; m 44, 77; c 4. PATHOLOGY. *Educ:* Trinity Col, Conn, 35-37; McGill Univ, MD, 43. *Prof Exp:* Lab dir path, Hilo Hosp, Hawaii, 51-58; LAB DIR PATH, KUAKINI HOSP, 58-; CLIN PROF PATH, SCH MED, UNIV HAWAII, MANOA, 66- *Concurrent Pos:* Prin investr, Honolulu Heart Study. *Mem:* AAAS; Col Am Path; Am Asn Path & Bact; Am Soc Clin Path; Int Acad Path; Sigma Xi. *Res:* Geographic pathology, particularly neoplastic and cardiovascular diseases in migrants; biology of elastic tissue and disease patterns in the feral mongoose. *Mailing Add:* 46-458 Haiku Plantation Dr Kaneohe HI 96744

STEMMLER, EDWARD J, b Philadelphia, Pa, Feb 15, 29; m 58; c 5. MEDICINE. *Educ:* La Salle Col, BA, 50; Univ Pa Sch Med, MD, 60; Am Bd Intern cert, 67; subspecialty Bd Pulmonary Disease, cert, 72. *Hon Degrees:* DSc, Ursinus Col, 77, La Salle Univ, 83, Philadelphia Col Pharm & Sci, 89; LHD, Rush Univ, 86. *Prof Exp:* Intern, 60-61, med resident, 61-63, fel cardiol, Hosp Pa, 63-64; instr med, 64-66, assoc med, 66-67, assoc physiol, Grad Div Med, 67-72, assoc prof med, 70-74, prof med, 74-81, ROBERT G DUNLOP PROF MED, UNIV PA, 81-; EXEC VPRES, ASN AM MED COLS, 90- *Concurrent Pos:* Chief med resident, Hosp Univ Pa, 66-67, chief med, Vet Admin, 67-73, assoc dean, Univ Hosp, 73, actg dean, Sch Med, 74-75, dean, 75-86, exec vpres & dean, Med Ctr, 86-88, exec vpres, 88-89, emer dean, Sch Med, 89-; mem, Lower Merion Towship Bd, Sch Dir, 73-74; dean comt, Philadelphia Vet Admin Hosp, 74-; mem gov comt, health educ, 74-77; comt dean, 75, chmn, Pa Sch Med; chmn invest comt, 75-81; nat fund med educ, educ policy comt, 75-78; dir Rorer Group Inc, 78-; mem sci affairs comt, 78; select comt med prospective prog, 78-80; mem task force support med educ, 77-81, chmn, 79-81; mem coun deans admin bd, 80, chmn, 82-84; chmn, mgt educ prog, 83-85, comt practice plan, 85-86; lect, higher educ & health, western higher educators, 84; mem Inst Med, Nat Acad Sci, 86-; vchmn, Nat Bd Med Examr, 87-; invitational lectr, Asn Med Dean Europe, 86, Japan Med Educ Found, Japan, 87; AMA, 87, Am Med Asn & Annenberg Ctr Health Sci, 88. *Honors & Awards:* Frederick A Packard Award, 60; Albert Einstein Med Ctr Staff Award, 60; Roche Award, 60; Laureate Award, Am Col Physicians; Roland Holroyd Lect, La Salle Univ, 80; Aaron Brown Lect, Univ Pittsburgh Sch Med, 83; Gus Carroll Mem Lectr, Asn Am Med Cols, 88; Kiskadden Lectr, Am Asn Plastic Surgeons, 91. *Mem:* Inst Med-Nat Acad Sci; Am Col Physicians (treas, 75-81); Am Fed Clin Res; Am Heart Asn; AMA; Am Thoracic Soc; Asn Am Med Col; Am Clin & Climat Asn; fel AAAS. *Res:* Author of numerous scientific publications. *Mailing Add:* Asn Am Med Cols One DuPont Circle NW Suite 200 Washington DC 20036

STEMNISKI, JOHN ROMAN, b Nanticoke, Pa, Apr 29, 33; m 59; c 3. ORGANIC CHEMISTRY. *Educ:* Fordham Univ, BS, 55; Carnegie Inst Technol, MS, 59, PhD(org chem), 60. *Prof Exp:* Res chemist, Monsanto Res Corp, 59-65; staff chemist, 65-69, prin chemist polymer prod, 69-73, chief mat & process control lab, 73-84, SR TECHNOLOGIST, C S DRAPER LAB, MASS INST TECHNOL, 84- *Concurrent Pos:* Mem United Nation Environ Prog. *Mem:* Am Chem Soc; Royal Soc Chem; Am Soc Testing and Mat; Sigma Xi. *Res:* Synthesis of medicinal compounds; synthesis of oil additives; lubricant systems; oxidation mechanism; high density fluids; gel permeation chromatography; fluorine containing fluids, materials science studies; fiber optics instrumentation; chloro fluoro carbon alternatives. *Mailing Add:* 19 Walnut Rd Swampscott MA 01907

STEMPAK, JEROME G, b Chicago, Ill, Dec 24, 31; m 61; c 2. ANATOMY. *Educ:* Roosevelt Univ, BS, 58; Univ Ill, MS, 60, PhD(teratology), 62. *Prof Exp:* Instr, 63-66, asst prof, 66-77, ASSOC PROF ANAT, STATE UNIV NY DOWNSTATE MED CTR, 77- *Concurrent Pos:* USPHS fel, 62-63. *Mem:* Am Asn Anat; Am Soc Cell Biol. *Res:* Electron microscopy of differentiating cells and tissues; electron microscopic investigations on generation of cell organelles. *Mailing Add:* Dept of Anat State Univ NY Downstate Med Ctr Brooklyn NY 11203

STEMPEL, ARTHUR, b Brooklyn, NY, June 8, 17; m 50; c 2. CHEMISTRY. *Educ:* City Col New York, BS, 37; Columbia Univ, MA, 39. PhD(chem), 42. *Prof Exp:* Asst, Col Physicians & Surgeons, Columbia Univ, 38-39, org chem, Col Pharm, 40-41, res assoc, Chem Labs, 42-43; tutor & fel chem, Queens Col, 41-42; sr chemist, Hoffman-LaRoche, Inc, 43-68, from res group chief to sr res group chief, chem res dept, 79-85; CONSULT, SCR ASSOC, 85- *Mem:* Am Chem Soc. *Res:* Isolation of natural products of animal and plant origin; antibiotics; organic synthesis of pharmaceuticals; investigations on loco weeds; synthesis of benzodiazepines; cholinesterase inhibitors; anticurare compounds. *Mailing Add:* 1341 River Rd Teaneck NJ 07666

STEMPEL, EDWARD, b Brooklyn, NY, Mar 7, 26; m 59; c 1. PHARMACY. *Educ:* Brooklyn Col Pharm, BS, 49; Columbia Univ, MS, 52, MA, 55, EdD, 56. *Prof Exp:* From instr to assoc prof, 49-64, chmn dept, 64-79, assoc dean, 79-83, PROF PHARM, ARNOLD & MARIE SCHWARTZ COL PHARM & HEALTH SCI, 64- *Mem:* Am Pharmaceut Asn; Am Asn Col Pharm. *Res:* Dispensing pharmacy; long-acting dosage forms. *Mailing Add:* Arnold & Marie Schwartz Col Pharm & Health Sci 75 DeKalb Ave Brooklyn NY 11201-5372

STEMPEL, ROBERT C, b 1933. AUTOMOTIVE PRODUCTION ADMINISTRATION. *Prof Exp:* Var positions, 58-90, CHMN & CHIEF EXEC OFFICER, GEN MOTORS CORP, 90- *Mem:* Nat Acad Eng. *Mailing Add:* Gen Motors Corp 14-132 Gen Motors Bldg 3044 W Grand Blvd Detroit MI 48202

STEMPEN, HENRY, b Phila, Pa, May 10, 24; m 54; c 4. MICROBIOLOGY. *Educ:* Phila Col Pharm, BS, 45; Univ Pa, PhD(microbiol), 51. *Prof Exp:* Res asst cytol, Lankenau Hosp Res Inst, Phila, Pa, 45-46; from instr to asst prof, bact, Jefferson Med Col, 50-57, assoc prof microbiol, 57-62; asst prof biol, 62-63, ASSOC PROF MICROBIOL, RUTGERS UNIV, CAMDEN, 63- *Concurrent Pos:* NIH grants, 59-60 & 64-66. *Mem:* AAAS; Am Soc Microbiol; Mycol Soc Am. *Res:* Bacterial cytology and genetics; morphogenesis of myxomycetes. *Mailing Add:* 2364 Geneva Rd Glenside PA 19038

STEMPIEN, MARTIN F, JR, b New Britain, Conn, Sept 2, 30. BIOCHEMISTRY. *Educ:* Yale Univ, BS, 52, MS, 53, PhD(org chem), 57; Cambridge Univ, PhD(org chem), 60. *Prof Exp:* Res assoc & sr investr bio-org chem, 60-66, ASST TO DIR & BIO-ORG CHEMIST, OSBORN LABS, MARINE SCI, NY AQUARIUM, NY ZOOL SOC, 66- *Concurrent Pos:* Consult, Off Naval Res, 68; fel, NY Zool Soc, 68- *Mem:* AAAS; Am Chem Soc; NY Acad Sci; Am Soc Zool; The Chem Soc; Sigma Xi. *Res:* Physiologically active materials from extracts of marine invertebrates, particularly Porifera and Echinodermata; bio-chemical taxonomy of Phylum Porifera. *Mailing Add:* 185 W Houston St New York NY 10014

STEMPLE, JOEL G, b Brooklyn, NY, Feb 3, 42; m 68. MATHEMATICS. *Educ:* Brooklyn Col, BS, 62; Yale Univ, MA, 64, PhD, 66. *Prof Exp:* Asst prof math, 66-70, ASSOC PROF MATH, QUEENS COL, NY, 70- *Res:* Theory of graphs. *Mailing Add:* 46 Old Brook Rd Dix Hills NY 11746

STEMSHORN, BARRY WILLIAM, b Montreal, Que, Dec 15, 47; m 80. INFECTIOUS DISEASES, SCIENCE ADMINISTRATION. *Educ:* McGill Univ, BSc, 69; Univ Montreal, DMV, 74; Univ Guelph, PhD(microbiol & immunol), 79. *Prof Exp:* Staff scientist, Animal Dis Res Inst, Ont, 75-81; dir, Animal Path Lab, Saskatoon, 81-83, DIR ANIMAL DISEASES RES INST, NEPEAN, AGR CAN, 83- *Concurrent Pos:* Assoc grad fac mem, Dr Vet Sci prog, Dept Vet Microbiol & Immunol, Univ Guelph, 81-; mem Brucellosis Sci Adv Comt, US Animal Health Asn & Subcomt on taxon the Genus Brucella; Consult, UN Univ; coordr, Caribbean Animal & Plant Health Prog, Inter-Am Inst Coop Agr, Trinidad & Tobago, 88-90. *Mem:* Am Vet Med Asn; Can Vet Med Asn; US Animal Health Asn; Am Asn Vet Lab Diagnosticians. *Res:* Management of diagnostic service and research programs related to livestock diseases; microbiological and serological methods for diagnosis of brucellosis and other bacterial diseases. *Mailing Add:* 2748 Howe Nepean ON K2B 6W9 Can

STENBACK, WAYNE ALBERT, b Brush, Colo, June 12, 29; m 54; c 1. MICROBIOLOGY, ELECTRON MICROSCOPY. *Educ:* Univ Colo, BS, 55; Univ Denver, MS, 57; Univ Mo, PhD(microbiol), 62. *Prof Exp:* Instr exp biol, 62-66, asst prof exp biol, Dept Surg, Baylor Col Med, 66-75; electron microscopist, Tex Children's Hosp, 75- AT DEPT PATH, BAYLOR COL MED,. *Concurrent Pos:* Mem, Int Asn Comp Res on Leukemia & Related Dis. *Mem:* AAAS; Am Soc Microbiol; Am Asn Cancer Res; Electron Micros Soc Am. *Res:* Density gradient studies of Newcastle disease virus; control of endemic microorganisms in mouse colonies; electron microscopy, biological and biophysical studies of viruses associated with neoplasms. *Mailing Add:* Dept Path Tex Childrens Hosp PO Box 20269 Houston TX 77030

STENBAEK-NIELSEN, HANS C, b Slagelse, Denmark. AURORAL PHYSICS. *Educ:* Tech Univ Denmark, MSc, 65. *Prof Exp:* PROF GEOPHYS, UNIV ALASKA, 67- *Mem:* Am Geophys Union. *Res:* Space and auroral physics; optical observations of auroras and in upper atmosphere chemical releases. *Mailing Add:* Geophys Inst Univ Alaska Fairbanks AK 99775-0800

STENBERG, CHARLES GUSTAVE, b Chicago, Ill, Apr 11, 35; m 66. PHYSICS, COMPUTER SCIENCE. *Educ:* Univ Ill, Urbana, BS, 59, MS, 60, PhD(physics), 68. *Prof Exp:* PHYSICIST REACTOR PHYSICS, ARGONNE NAT LAB, 69- *Mem:* Am Nuclear Soc; Sigma Xi. *Res:* Method and code development of models and computer codes for reactor physics calculations. *Mailing Add:* Reactor Physics-Nat Lab 9700 S Cass Ave Argonne IL 60439

STENBERG, PAULA E, b Montreal, Que, Can, June 25, 53. HEMATOLOGY. *Educ:* Univ Calif, San Francisco, PhD(pathol), 84. *Prof Exp:* FEL PATH, VET ADMIN MED CTR, 84 - *Mem:* Am Soc Women Sci; Am Soc Cell Biol. *Mailing Add:* Clin Pathol Serv (113A) Vet Admin Med Hosp 4150 Clement St San Francisco CA 94121

STENBERG, VIRGIL IRVIN, b Grygla, Minn, May 18, 35; m 56; c 4. ORGANIC CHEMISTRY. *Educ:* Concordia Col, Moorhead, Minn, BA, 56; Iowa State Univ, PhD(org chem), 60. *Prof Exp:* From asst prof to assoc prof, 60-67, PROF CHEM, UNIV NDAK, 67- *Concurrent Pos:* Res grants, Res Corp, 60-62, Petrol Res Fund, 61-63, NIH, 61-64, 68-71 & 74-76, career develop award, 70-75; NSF res grant, 66-68; vis prof, Imp Col, Univ London, 71-72; Energy Res Develop Admin contract, 75-78 & Dept Energy contract, 78-87. *Honors & Awards:* Sigma Xi Res Award, 74; Chester Fritz Prof Award, 77. *Mem:* Am Chem Soc; Sigma Xi. *Res:* Natural products; medicinal chemistry of inflammation. *Mailing Add:* Univ NDak PO Box 7185 Grand Forks ND 58202

STENCEL, ROBERT EDWARD, b Wausau, Wis, Apr 16, 50; m 77; c 1. ASTRONOMY, SPECTROSCOPY. *Educ:* Univ Wis-Madison, BS, 72; Univ Mich, Ann Arbor, MS, 74, PhD(astron), 77. *Prof Exp:* Res asst solar physics, Sacramento Peak Observ, 75; res assoc astron, NASA Johnson Space Ctr, 77-78; res assoc astron, Nat Acad Sci & NASA, Goddard Space Flight Ctr, 78-80; res assoc astron, Joint Inst Lab, Astrophys, Univ Colo, 80-82; staff scientist, NASA Hq, 82-85; MEM FAC & EXEC DIR, CTR ASTROPHYS SPACE ASTRON, UNIV COLO, BOULDER, 85- *Concurrent Pos:* Res assoc, Nat Res Coun/Nat Acad Sci, 77-; mem adv coun, Archeoastron Ctr, Univ Md, 78-82; NSF & NASA grants; NASA, 82-85. *Mem:* Am Phys Soc; Am Astron Soc; Int Astron Union. *Res:* High resolution spectroscopy of the outer atmospheres of low temperature supergiant stars; radiative transfer studies. *Mailing Add:* Ctr Astrophys Space Astron Univ Colo Boulder CO 80309-0391

STENCHEVER, MORTON ALBERT, b Paterson, NJ, Jan 25, 31; m 55; c 3. REPRODUCTIVE BIOLOGY, HUMAN GENETICS. *Educ:* NY Univ, AB, 51; Univ Buffalo, MD, 56; Am Bd Obstet & Gynec, cert 65 & 85. *Prof Exp:* Intern med, Mt Sinai Hosp, NY, 56-57; resident, Columbia-Presby Med Ctr, 57-60; from instr to assoc prof obstet & gynec, Case Western Reserve Univ, 64-70, dir tissue cult lab, 64-70, assoc, Dept Med Educ, 68-70; prof obstet & gynec & chmn dept, Univ Utah, 70-77; PROF & CHMN, DEPT OBSTET & GYNEC, UNIV WASH, 77- *Concurrent Pos:* NIH res training fel genetics, 62-64; Oglebey res fel, Case Western Reserve Univ, 64; chief obstet & gynec serv, Malmstrom AFB Hosp, Mont. *Mem:* AAAS; AMA; fel Am Col Obstet & Gynec; fel Am Soc Obstet & Gynec; Soc Cryobiol. *Res:* Male infertility; human reproduction. *Mailing Add:* Dept Obstet & Gynec Univ Wash Seattle WA 98195

STENDELL, REY CARL, b San Francisco, Calif, Aug 12, 41; m 66; c 2. POLLUTION BIOLOGY. *Educ:* Univ Calif, Santa Barbara, BA, 63, MA, 67; Univ Calif, Berkeley, PhD(zool), 72. *Prof Exp:* Res biologist, US Fish & Wildlife Serv, 72-73, res coordr, Patuxent Wildlife Res Ctr, 73-77, prog coordr pollution biol, Washington DC, 77,79; asst dir, Northern Prairie Wildlife Res Ctr, 79-81, dir, 81-89, DIR, NAT ECOL RES CTR, US FISH & WILDLIFE SERV, 89- *Mem:* Am Ornithologists Union; Cooper Ornithol Soc. *Res:* Evaluation of the effects of environmental pollutants on wildlife, particularly birds. *Mailing Add:* US Fish & Wildlife Serv 4512 McMurray Ave Ft Collins CO 80525-3400

STENESH, JOCHANAN, b Magdeburg, Ger, Dec 19, 27; US citizen; m 57; c 2. MOLECULAR BIOLOGY, MICROBIOLOGY. *Educ:* Univ Ore, BS, 53; Univ Calif, Berkeley, PhD(biochem), 58. *Prof Exp:* Res assoc biochem, Weizmann Inst, 58-60 & Purdue Univ, 60-63; from asst prof to assoc prof, Western Mich Univ, 63-71, prof chem, 71-90; RETIRED. *Concurrent Pos:* Res grants, Nat Inst Allergy & Infectious Dis, 64-70, Am Cancer Soc, 65-67. *Mem:* AAAS; Am Chem Soc; Am Soc Biochem & Molecular Biol; Am Soc Microbiol. *Res:* Physical biochemistry; enzymology; protein synthesis; DNA replication and transcription. *Mailing Add:* Dept Chem Western Mich Univ Kalamazoo MI 49008

STENGEL, ROBERT FRANK, b Orange, NJ, Sept 1, 38; m 61; c 2. AEROSPACE ENGINEERING, MECHANICAL ENGINEERING. *Educ:* Mass Inst Technol, BS, 60; Princeton Univ, MSE, 65, MA, 66, PhD(aerospace & mech sci), 68. *Prof Exp:* Aerospace technologist rockets, NASA, 60-63; mem tech staff & group leader guid & control, Draper Lab, Mass Inst Technol, 68-73; mem tech staff & sect leader, Analytical Sci Corp, 73-77; assoc prof, 77-82, PROF MECH & AEROSPACE ENG, PRINCETON UNIV, 82- *Concurrent Pos:* Mem, Aerospace Guid & Control Systs Comt, Soc Automotive Engrs; vchmn, Cong Aeronaut Adv Comt; assoc ed-at-large, Inst Elec & Electronic Engrs Transactions on Automatic Control, 91- *Mem:* Assoc fel Am Inst Aeronaut & Astronaut; sr mem Inst Elec & Electronics Engrs. *Res:* Atmospheric flight mechanics; optimal control and estimation; space flight engineering; human factors; artificial intelligence. *Mailing Add:* D-202 Eng Quadrangle Princeton Univ Princeton NJ 08544

STENGER, FRANK, b Veszprem, Hungary, July 6, 38; m 61; c 2. APPLIED MATHEMATICS. *Educ:* Univ Alta, BSc, 61, MSc, 63, PhD(math), 65. *Prof Exp:* Guest worker, Nat Bur Stand, Washington, DC, 63-64; asst prof comp sci, Univ Alta, 65-66; asst prof math, Univ Mich, 66-69; from assoc prof to prof math, 69-89, PROF COMPUT SCI, UNIV UTAH, 89- *Concurrent Pos:* NSF grant, Univ Utah, 70-71 & 76-78; vis math res ctr, Univ Montreal, 71-72; US Army res grant, Univ Utah, 74-76 & 77-80; distinguished vis prof, Univ Tsukuba, 87. *Mem:* Am Math Soc; Can Math Cong; Soc Indust & Appl Math; Asn Comput Mach. *Res:* Quadrature; numerical solution of differential and integral equations; asymptotic approximation of functions and integrals; asymptotic solution of differential equations; numerical solution of inverse probles in ultrasonic tomography and in geophysics; sine numerical methods, parallel computation. *Mailing Add:* Dept Comput Sci Univ Utah Salt Lake City UT 84112

STENGER, RICHARD J, b Cincinnati, Ohio, Dec 13, 27; m 51; c 2. PATHOLOGY, ELECTRON MICROSCOPY. *Educ:* Col of the Holy Cross, AB, 49; Univ Cincinnati, MD, 53. *Prof Exp:* Intern, Cincinnati Gen Hosp, 53-54; resident path, Mass Gen Hosp, 54-55, 57-58, fel 58-59, asst, 59-60; asst prof, Col Med, Univ Cincinnati, 60-64; from asst prof to assoc prof path, Case Western Reserve Univ, 64-68; prof, New York Med Col, 68-72; PROF PATH, MT SINAI SCH MED, 72-; DIR PATH & LABS, BETH ISRAEL MED CTR, 72- *Concurrent Pos:* Chief, Lab Serv, W Point Army Hosp, 55-57; Nat Inst Arthritis & Metab Dis res grant & Nat Inst Gen Med Sci career develop award, 64-68; instr, Harvard Med Sch, 59-60; attend pathologist, Cincinnati Gen Hosp, 60-64; assoc pathologist, Cleveland Metrop Gen Hosp, 64-68; attend pathologist, Flower & Fifth Ave Hosps, NY & vis pathologist, Metrop Hosp Ctr, NY, 68-72. *Mem:* AAAS; Am Asn Path; Am Asn Study Liver Dis. Int Acad Path; Sigma Xi. *Res:* Light and electron microscopic studies of the liver, especially with regards to toxic effects. *Mailing Add:* Path & Labs Dept Nassau County Med Ctr 2201 Hempstead Tpke East Meadow NY 11554

STENGER, VERNON ARTHUR, b Minneapolis, Minn, June 11, 08; m 33; c 5. ANALYTICAL CHEMISTRY, GEOCHEMISTRY. *Educ:* Univ Denver, BS, 29, MS, 30; Univ Minn, PhD(anal chem), 33. *Hon Degrees:* DSc, Univ Denver, 71. *Prof Exp:* Asst, Eastman Kodak Co, 29-30; anal chemist, Univ Minn, 33-35; anal res chemist, 35-53, dir spec serv lab, 53-61, anal scientist, 61-73, CONSULT, DOW CHEM CO, 73- *Honors & Awards:* Anachem Award, 70; Midland Sect Award, Am Chem Soc, 79. *Mem:* Am Chem Soc; Sigma Xi; Geochem Soc; NY Acad Sci; fel Am Inst Chem. *Res:* Purity of reagents; technology of bromine and its compounds; instrumentation for water analysis; methods for environmental analysis. *Mailing Add:* 1108 E Park Dr Midland MI 48640-4249

STENGER, VICTOR JOHN, b Bayonne, NJ, Jan 29, 35; m 62; c 2. PHYSICS. *Educ:* Newark Col Eng, BS, 56; Univ Calif, Los Angeles, MS, 58, PhD(physics), 63. *Prof Exp:* Mem tech staff, Hughes Aircraft Co, 56-59; asst physics, Univ Calif, Los Angeles, 59-63; assoc prof, 63-74, PROF PHYSICS & ASTRON, UNIV HAWAII, 74- *Mem:* Am Phys Soc. *Res:* Elementary physics. *Mailing Add:* Dept Physics-Astron Univ Hawaii Manoa 2500 Campus Rd Honolulu HI 96822

STENGER, WILLIAM, b Bayonne, NJ, Jan 25, 42; m 68; c 2. MATHEMATICS. *Educ:* Stevens Inst Technol, BS, 63; Univ Md, PhD(math), 67. *Prof Exp:* Asst comput, Davidson Lab, Stevens Inst Technol, 62-63; res asst, Univ Md, 66-67; asst prof math, American Univ, 67-68 & Georgetown Univ, 68-69; asst prof, 69-72, PROF MATH, AMBASSADOR COL, 72- *Concurrent Pos:* Res grants, Air Force Off Sci Res, 67-68 & NSF, 68-69. *Mem:* Math Asn Am; Am Math Soc; Soc Indust Appl Math. *Res:* Variational theory of eigenvalues and related inequalities. *Mailing Add:* Dept Math Ambassador Col Big Sandy TX 75755

STENGER, WILLIAM J(AMES), SR, b Wheeling, WVa, May 9, 26; m 51; c 8. OCCUPATIONAL HEALTH, ENVIRONMENTAL AFFAIRS. *Educ:* WVa Univ, BSChE, 50, MSChE, 51, PhD(chem eng), 53. *Prof Exp:* Res chem engr, Burnside Lab, E I Du Pont De Nemours, 53-57, Eastern Lab, 57, Repauno Process Lab, 58, Elchem Res Lab, 59, Carney's Point Develop Lab, 59-61, Eastern Lab, 61, Repauno Develop Lab, 62-67 & Eastern Lab, 67-72, asst tech supt, Cape Fear Plant, 72-73, environ control supvr, 73-78, safety health & environ affairs mgr, textile fibers, 78-85; RETIRED. *Concurrent Pos:* Consult, Stenger Assocs, 85-, DuPont, Philips & DuPont Optical, Concepts Unlimited, Munsingwear & Takeda. *Mem:* Am Chem Soc; Am Inst Chem Engrs; Air Pollution Control Asn; Water Pollution Control Fedn; Sigma Xi. *Res:* Heat transfer; fluidization; coal gasification; nitrocellulose; amino acid synthesis; resolution of optical isomers; crystallization; cellulose and starch derivatives; propellant chemicals; hydrocarbon oxidation; polymer intermediates; environmental; occupational health. *Mailing Add:* 2210 Marlwood Dr Wilmington NC 28403

STENGLE, THOMAS RICHARD, b Lancaster, Pa, Nov 25, 29; m 78; c 1. PHYSICAL CHEMISTRY. *Educ:* Franklin & Marshall Col, BS, 51; Univ Mich, MS, 53, PhD(chem), 61. *Prof Exp:* From instr to assoc prof, 59-73, PROF CHEM, UNIV MASS, AMHERST, 73- *Mem:* Am Chem Soc; Am Phys Soc. *Res:* Application of nuclear magnetic resonance techniques to fast reactions, inorganic reaction mechanisms, molecular biology, solute-solvent interactions, amorphous polymers, and food systems. *Mailing Add:* Dept Chem Univ Mass Amherst MA 01003

STENGLE, WILLIAM BERNARD, b Lancaster, Pa, Feb 21, 23; m 48; c 3. WOOD CHEMISTRY. *Educ:* Franklin & Marshall Col, BS, 43; State Univ NY Col Forestry, Syracuse, MS, 49. *Prof Exp:* Chemist, Animal Trap Co Am, 47; asst, State Univ NY Col Forestry, Syracuse, 49; res chemist, Crossett Co, Ark, 49-58, asst tech serv dir, Paper Mill Div, 58-62; TECH DIR, TENN RIVER PULP & PAPER CO, COUNCE, 62- *Concurrent Pos:* Mem, Tenn Air Pollution Control Bd, 76-80. *Mem:* Am Chem Soc; Tech Asn Pulp & Paper Indust; Air Pollution Control Asn. *Res:* Hydroxy acids; alkaline pulping; bleaching; paper manufacture; tall oil; corrugated paperboard; water and air pollution abatement. *Mailing Add:* 116 College Ave Lancaster PA 17603

STENKAMP, RONALD EUGENE, b Bend, Ore, May 14, 48; m 70. BIOINORGANIC CHEMISTRY, PROTEIN CRYSTALLOGRAPHY. *Educ:* Univ Ore, BA, 70; Univ Wash, MSc, 71, PhD(chem), 75. *Prof Exp:* Fel crystallog, Dept Molecular Biophys & Biochem, Yale Univ, 75-76; Am Cancer Soc fel, Dept Molecular Biophys & Biochem, Yale Univ & Dept Biol Struct, Univ Wash, 77; res assoc crystallog, 78-81, ASSOC PROF, DEPT BIOL STRUCT, UNIV WASH, 81- *Mem:* Am Crystallog Asn; Am Chem Soc; AAAS. *Res:* Crystallographic studies of molecules of biological interest, including metallo proteins, cellular recognition proteins, enzymes and nucleic acids. *Mailing Add:* Dept Biol Struct Univ Wash SM-20 Seattle WA 98195

STENN, KURT S, b Chicago, Ill, Apr 6, 40; m 64; c 2. BIOLOGY OF SKIN. *Educ:* Univ Chicago, BS, 61; Univ Rochester, NY, MD, 65. *Hon Degrees:* MS, Yale Univ, 83. *Prof Exp:* PROF DERMAT & PATH, YALE UNIV, 71- *Mem:* Am Acad Dermat; Soc Invest Res Med; Am Soc Cell Biol; Am Soc Dermat Path; Sigma Xi; AAAS. *Res:* Epithelial mesenebynal interactions and the control of hair growth. *Mailing Add:* Yale Sch Med 333 Cedar St New Haven CT 06510

STENSAAS, LARRY J, b Nov 13, 32; US citizen; m 62; c 1. NEUROANATOMY, PLANT PHYSIOLOGY. *Educ:* Univ Calif, Berkeley, BA, 55, MA, 57; Univ Calif, Los Angeles, PhD(neuroanat), 65. *Prof Exp:* Stratigrapher, Richmond of Columbia, 57-58; head paleont sect, Cuba Calif Oil Co, 58-59; micropaleontologist, Standard of Calif, 59-60; from asst prof to assoc prof, 68-80, PROF PHYSIOL, UNIV UTAH, 80- *Concurrent Pos:* Cerebral Palsy Educ & Res Found fel, 65-67; NIH fel, 67-68; distinguished scientist, Armed Forces Inst Path, 84-85. *Mem:* AAAS; Am Asn Anat; Soc Neurosci. *Res:* Light and electron microscopy of normal and regenerating central nervous tissue; fertilization of plant communities by municipal wastes and rock phosphate solubilized by thermophyllic microorganisms. *Mailing Add:* Dept Physiol Univ Utah Salt Lake City UT 84112

STENSAAS, SUZANNE SPERLING, b Oakland, Calif, Mar 15, 39; m 62; c 1. NEUROANATOMY. *Educ:* Pomona Col, BA, 59; Univ Calif, Los Angeles, MA, 62; Univ Utah, PhD(anat), 75. *Prof Exp:* Instr, 71-75, ASST PROF ANAT, COL MED, UNIV UTAH, 76- *Mem:* Soc Neurosci; AAAS. *Res:* Biological compatability of materials with the brain; development of neuroprostheses; effects of electrical stimulation on the brain; problems of development and regeneration of the central nervous system. *Mailing Add:* Dept Path Univ Utah Sch Med 50 N Medical Dr Salt Lake City UT 84132

STENSETH, RAYMOND EUGENE, b Ludlow, SDak, Aug 5, 31; m 65; c 1. ORGANIC CHEMISTRY, PHARMACY. *Educ:* Univ Mich, BS, 53, MS, 57, PhD(pharmaceut chem), 61. *Prof Exp:* Sr res chemist, Monsanto Co, 60-68, res specialist, 68-85; RETIRED. *Mem:* Am Chem Soc; Sigma Xi. *Res:* Organic syntheses; nitrogen heterocycles; organophosphorus chemistry; food chemicals; pharmaceuticals; bacteriostats; fungistats; chemical processes. *Mailing Add:* 805 Westwood Dr St Louis MO 63105

STENSON, WILLIAM F, b Rome, NY, Dec 2, 45; c 3. GASTROENTEROLOGY. *Educ:* Wash Univ, MD, 71. *Prof Exp:* Asst prof, 79-85, ASSOC PROF MED, SCH MED, WASH UNIV, 85- *Mem:* Am Asn Immunologists; Am Soc Clin Invest. *Res:* Intestinal inflammation; immunology and physiology. *Mailing Add:* Sch Med Wash Univ Box 8124 St Louis MO 63110

STENSRUD, HOWARD LEWIS, b Minneapolis, Minn, Nov 16, 36; div; c 2. GEOLOGY, GEOCHEMISTRY. *Educ:* Univ Minn, Minneapolis, BA, 58; Univ Wyo, MA, 63; Univ Wash, PhD(geol), 70. *Prof Exp:* Explor geologist, Humble oil & Ref Co, 60; asst prof geol, Univ Minn, Morris, 61-66; asst prof, 70-77, PROF GEOL, CALIF STATE UNIV, CHICO, 77- *Mem:* Geol Soc Am; Soc Mining Engrs; Nat Assoc Geol Teachers. *Res:* Metamorphic petrology and Precambrian geology. *Mailing Add:* Dept of Geol & Phys Sci Calif State Univ Chico CA 95929

STENSTROM, MICHAEL KNUDSON, b Anderson, SC, Nov 28, 48. ENVIRONMENTAL ENGINEERING. *Educ:* Clemson Univ, SC, BS, 71, MS, 72, PhD(environ systs eng), 76. *Prof Exp:* Instr environ eng, Clemson Univ, 75; res engr, Amoco Oil Co, Standard Oil, Ind, 75-77; from asst prof to assoc prof, 77-84, PROF ENVIRON ENG, UNIV CALIF, LOS ANGELES, 84- *Concurrent Pos:* Consult, Amoco Oil Co, 77-79, Exxon Res & Develop, 80, Sohio, 82-84, Chevron, 84-, City of Los Angeles, 87- *Honors & Awards:* Best Dissortation Prize, Asn Environ Eng Prof, 75, 76; Huber Res Prize, Am Soc Civil Engrs, 89. *Mem:* Am Soc Civil Engrs; Water Pollution Control Fedn; Asn Environ Eng Prof; Int Asn Water Pollution Res; Am Chem Soc; AAAS. *Res:* Wastewater and water treatment processes, including real time control of the activated sludge process, oxygen transfer, anaerobic treatment systems, hazardous waste control; urban runoff control. *Mailing Add:* 4173 Eng I Univ Calif-Los Angeles Los Angeles CA 90024-1600

STENSTROM, RICHARD CHARLES, b Elkhorn, Wis, June 19, 36; m 60; c 1. GEOLOGY. *Educ:* Beloit Col, BS, 58; Univ Chicago, MS, 62, PhD(geophys sci), 64. *Prof Exp:* Am Chem Soc res fel geophys sci, Univ Chicago, 64-65; from instr to assoc prof, 65-77, PROF GEOL, BELOIT COL, 77-, CHMN, 87- *Concurrent Pos:* Coun-at-large, Nat Asn Geol Teachers, 85-87. *Mem:* Nat Asn Geol Teachers; fel Geol Soc Am; Geochem Soc; Am Geophys Union. *Res:* Diffusion rates through sediments; hydrologic effects of urbanization; environmental geology; basin behavior and water quality; effects of urbanization versus agricultural use; sediment load, surface and subsurface character, general water quality; chemical analysis of geochemical systems, using field, AA, and EDS-SEM techniques. *Mailing Add:* Dept Geol Beloit Col Beloit WI 53511

STENT, GUNTHER SIEGMUND, b Berlin, Ger, Mar 28, 24; nat US; m 51; c 1. MOLECULAR BIOLOGY, NEUROBIOLOGY. *Educ:* Univ Ill, BS, 45, PhD(phys chem), 48. *Hon Degrees:* DSc, York Univ, Toronto, 84. *Prof Exp:* Asst chem, Univ Ill, 44-48; Merck fel, Calif Inst Technol, Nat Res Coun, 48-50, Am Cancer Soc fel, Univ Copenhagen, 50-51, Pasteur Inst, Paris, 51-52; asst res biochemist & lectr bact, Univ Calif, Berkeley, 52-56, assoc prof bact, 56-59, prof arts & sci 67-68, chmn molecular biol & dir, Virus Lab, 80-86, PROF MOLECULAR BIOL, UNIV CALIF, BERKELEY, 59-, CHMN MOLECULAR & CELL BIOL, 87- *Concurrent Pos:* Document analyst, Field Intel Agency Tech, 46-47; mem, Genetics Study Sect, NIH, 59-64; sr fel, NSF, Univ Kyoto & Cambridge, 60-61; mem, Genetic Biol Panel, NSF, 65-69; external mem, Max Planck Inst Molecular Genetics, 66-; Guggenheim fel, Med Sch, Harvard Univ, 69-70; hon prof, Fac Sci, Univ Chile, 80; mem, Adv Bd, Basel Inst Immunol, 80-85; fel, Inst Advan Study, Berlin, 85-90; Inst Adv Study fel, 85-; chmn, Neurobiol Sect, Nat Acad Sci, 86-89; Fogarty scholar residence, NIH, Bethesda, 90-91. *Honors & Awards:* Runnstrom Medal, 86; Urania Medal, 90. *Mem:* Nat Acad Sci; Soc Neurosci; Am Acad Arts & Sci; Am Philos Soc; fel AAAS. *Res:* Nervous control of behavior; developmental biology; philosophy of science; author of various publications. *Mailing Add:* Dept Molecular Biol Univ Calif Berkeley CA 94720

STENUF, THEODORE JOSEPH, b Vienna, Austria, Feb 27, 24; US citizen; m 53; c 2. CHEMICAL ENGINEERING. *Educ:* Syracuse Univ, BChE, 49, MChE, 51, PhD(chem eng), 53. *Prof Exp:* Res asst & assoc chem eng, Syracuse Univ, 59-53; res engr, E I Du Pont de Nemours & Co, Inc, 53-59; corrosion engr, Esso Res & Eng, Standard Oil Co, NJ, 59-60; from asst to prof, 60-76, DISTINGUISHED TEACHING PROF PAPER SCI & ENG, COL ENVIRON SCI & FORESTRY, STATE UNIV NY, 77- *Mem:* Am Inst Chem Engrs; Am Chem Soc; Tech Asn Pulp & Paper Indust. *Res:* Fluid mechanics; heat transfer; mass transfer; paper sheet formation. *Mailing Add:* Four Heather Woods Ct Skaneateles NY 13152-1410

STENZEL, KURT HODGSON, b Stamford, Conn, Nov 3, 32; m 57; c 3. NEPHROLOGY, MEDICINE. *Educ:* NY Univ, BA, 54; Cornell Univ, MD, 58. *Prof Exp:* Intern med, Second Cornell Med Div, Bellevue Hosp, NY, 58-59, asst resident, 59-60, cardio-renal resident, 62-63; res fel, 63-64, from instr to assoc prof, 64-75, PROF MED & BIOCHEM, CORNELL UNIV MED COL, CHIEF, DIV NEPHROLOGY, 75- *Concurrent Pos:* NY Heart Asn res fel, 63-66 & investr, 66-70; attend physician, NY Hosp, 75-; med dir, Rogosin Inst, 84- *Honors & Awards:* Hoening Award Excellence Renal Med, Nat Kidney Found. *Mem:* Am Soc Biol Chemists; fel Am Col Physicians; Am Fedn Clin Res; Am Soc Artificial Internal Organs. *Res:* Dialysis; biomaterials; transplantation immunology; lymphocyte activation. *Mailing Add:* Rogosin Inst & Dept Med Cornell Univ Med Col 1300 York Ave New York NY 10021

STENZEL, REINER LUDWIG, b Breslau, Germany, Feb 18, 40; m 67; c 3. LABORATORY PLASMAS, PLASMA DIAGNOSTICS. *Educ:* Calif Inst Technol, MS, 66, PhD(elec eng), 70. *Prof Exp:* Res fel plasma physics, Calif Inst Technol, 69-70; adj asst prof, Univ Calif, Los Angeles, 70-73; mem physics staff, TRW Systs, Calif, 72-77; assoc prof, 77-81, PROF PHYSICS, UNIV CALIF, LOS ANGELES, 81- *Concurrent Pos:* Vis prof, Univ Tokyo, 80. *Mem:* Am Geophys Union; Am Phys Soc. *Res:* Experimental plasma physics on basic properties of plasmas such as waves, instabilities, beam-plasma interactions, double layers and magnetic reconnection; development of new laboratory plasma devices and diagnostic techniques. *Mailing Add:* 519 Almar Pacific Palisades CA 90272

STENZEL, WOLFRAM G, b Berlin, Ger, May 24, 19; nat US; m 43; c 1. PHYSICS. *Educ:* City Col New York, BS, 39, MS, 41. *Prof Exp:* Asst biochem, Warner Inst Therapeut Res, 42; res scientist phys chem, Nuodex Prod Co, Inc, 46-48; instr physics & math, Bloomfield Col, 48-51; physicist, B G Corp, 51-55; prin engr physics, Ford Instrument Co Div, 55-69, SR ENGR, SYSTS MGT DIV, SPERRY RAND CORP, 69- *Concurrent Pos:* Lectr, Adelphi Col, 57-60. *Mem:* Am Phys Soc. *Res:* Nuclear reactors and effects; mathematical physics; thermionics; traffic control; rocket trajectories. *Mailing Add:* 77-21 250th St Bellerose NY 11426

STEPAN, ALFRED HENRY, b St Paul, Minn, Jan 2, 20; m 44; c 5. ORGANIC CHEMISTRY. *Educ:* Col St Thomas, BS, 42; Univ Nebr, MS, 45, PhD(org chem), 51. *Prof Exp:* Chemist, Tenn Eastman Corp, 44-46; sr chemist, Continental Oil Co, 48-56; sr res chemist, Minn Mining & Mfg Co, 56-60, supvr, 60-63, mgr, 63-81, chem specialist, 81-82; RETIRED. *Mem:* Am Chem Soc. *Res:* Carbonless papers; dielectric paper; imaging systems involving dry silver. *Mailing Add:* 1952 Oak Knoll Dr White Bear Lake MN 55110

STEPANISHEN, PETER RICHARD, b Boston, Mass, Jan 20, 42; m 70; c 2. ACOUSTICS. *Educ:* Mich State Univ, BS, 63; Univ Conn, MS, 66; Pa State Univ, PhD(eng acoust), 69. *Prof Exp:* Sonar systs engr, Elec Boat Div, Gen Dynamics Corp, 63-66, sr systs engr, 66-70, res specialist acoust, 70-74; from asst prof to assoc prof, 74-82, PROF OCEAN ENG, UNIV RI, 83- *Concurrent Pos:* Acoust consult, Naval Underwater Systs Ctr, 75- & Raytheon Co, 76-; res grant prin investr, NIH, 76-82, ONR, 83- *Honors & Awards:* A B Wood Medal & Prize, Inst Acoust, Eng, 77. *Mem:* Fel Acoust Soc Am; Sigma Xi; Inst Elec & Electronic Engrs. *Res:* Underwater acoustics; mechanical vibrations and wave phenomena; medical ultrasonics; system theory and signal analysis. *Mailing Add:* Dept Ocean Eng Univ RI Kingston RI 02881

STEPENUCK, STEPHEN JOSEPH, JR, b Salem, Mass, Oct 12, 37; m 68; c 3. ENVIRONMENTAL CHEMISTRY. *Educ:* Merrimack Col, BS, 59; Col of the Holy Cross, MS, 61; Univ NH, PhD(phys chem), 71. *Prof Exp:* Instr chem, Merrimack Col, 61-65; teaching fel, Univ NH, 66-68, instr, 69-70; asst prof, 70-73, assoc prof, 73-79, PROF CHEM, KEENE STATE COL, 79-*Concurrent Pos:* Consult, indust, environ chem & chem safety. *Mem:* Am Chem Soc; Sigma Xi. *Res:* Radiation chemistry; clinical and diagnostic analyses; environmental analyses; marine chemistry, occupational health. *Mailing Add:* Dept Chem Keene State Col 229 Main St Keene NH 03431

STEPHAN, DAVID GEORGE, b Columbus, Ohio, Feb 8, 30; m 51; c 3. ENVIRONMENTAL & CHEMICAL ENGINEERING. *Educ:* Ohio State Univ, BChE & MSc, 52, PhD(chem eng), 55. *Prof Exp:* Res assoc heat transfer, Battelle Mem Inst, 52-55; technologist, Nat Lead Co Ohio, 55; chief air pollution control equip res, USPHS, 55-60, chief extramural res unit, Adv Waste Treatment Res Prog, 60-64, dep chief prog, 64-65, dep chief basic & appl sci br, 65-66; dir div res, Fed Water Pollution Control Admin, 66-68, asst comnr res & develop, Fed Water Qual Admin, 68-70; dir res prog mgt, Environ Protection Agency, 71-75, dir, Indust Environ Res Lab, 75-84, dir, Hazardous Waste Eng Res Lab, 84-85, SR ENG ADV, HAZARDOUS WASTE ENG RES LAB, ENVIRON PROTECTION AGENCY, 86-*Concurrent Pos:* Mem water reuse comt, Water Pollution Control Fedn, 64-71; mem water panel & spec consult comt pollution, Nat Acad Sci, 65; mem comt water resources res, Off Sci & Technol, 68-71; mem policy bd, Nat Ctr Toxicol Res, 72-74; mem governing bd, Int Asn Water Pollution Res, 74-84; vchmn, USA Nat Comt, Int Asn Water Pollution Res, 74-80; adj prof chem eng, Univ Cincinnati, 76-89; mem bd dirs, Environ Div, Am Inst Chem Engrs, 83-85. *Honors & Awards:* Ann Environ Div Award, Am Inst Chem Engrs, 83. *Mem:* Am Inst Chem Engrs; Am Acad Environ Engrs. *Res:* Heat transfer from vibrating plates; fabric air filtration; air pollution control equipment; advanced waste treatment; waste water renovation and reuse; water pollution control and pollution prevention. *Mailing Add:* 6435 Stirrup Rd Cincinnati OH 45244-3924

STEPHANAKIS, STAVROS JOHN, b Salonica, Greece, Apr 15, 40; US citizen; m 66; c 2. PHYSICS, ELECTRICAL ENGINEERING. *Educ:* NC State Univ, BS, 63, MEE, 65, PhD(elec eng), 69. *Prof Exp:* electronics engr, 68-72, RES PHYSICIST, US NAVAL RES LAB, 72- *Mem:* Am Phys Soc; Inst Elec & Electronics Engrs; Sigma Xi. *Res:* Study of hot, dense plasma discharges, x-ray and neutron emissions from such discharges; diagnostics. *Mailing Add:* Naval Res Lab Code 4773 Washington DC 20375-5000

STEPHANEDES, YORGOS JORDAN, b Athena, Greece, Sept 15, 51; c 2. MODELING, SIMULATION & CONTROL DESIGN. *Educ:* Dartmouth Col, BA, 73, PhD(eng sci), 80; Carnegie-Mellon Univ, MS, 75. *Prof Exp:* Res asst, Carnegie-Mellon Univ, 73-74; res & teaching asst eng sci, Dartmouth Col, 75-78; asst profg, 78-84, ASSOC PROF CIVIL ENG, UNIV MINN, 84-*Concurrent Pos:* Res assoc, Oak Ridge Nat Labs, 75; mem, Transp Res Bd, Nat Acad Sci, 76-; prin investr, Univ Minn, 78-; consult, Hennepin County, 80, Metrop Transp Comn, 84-, Wilbur Smith Assoc, 87 & Regional Econometrics, 87; dir res & develop, United Eng, 82-86; ed, NCent, Inst Transp Engrs, 83-85; referee sci manuscripts, NSF, 84-, Am Soc Civil Engrs, 84-, Appl Math Model, 86-; mem, Minn Ctr Survey Res jAdv Comt, 86-, Minn State & Regional Res Ctr, 86-; vchmn, Comt Adv Technol Appln in Urban Transp, Am Soc Civil Engrs, 87- *Mem:* Inst Elec & Electronics Engrs; Am Soc Civil Engrs; Inst Transp Engrs; Int Asn Math & Comput Simulation; Fr Asn Cybernet, Econ & Technol. *Res:* Modeling, simulation, estimation and optimal control design in engineering systems; demand-supply dynamics in transportation operations and economic development; technology and resource policy impact analysis and assessment; interface microcomputer expert systems. *Mailing Add:* Dept Civil Eng 12 Civil Mining Bldg Univ Minn 500 Pillsbury Dr S Minneapolis MN 55455

STEPHANOU, STEPHEN EMMANUEL, b Crete, Greece, Dec 29, 19; US citizen; c 3. RESEARCH MANAGEMENT, SYSTEMS ENGINEERING. *Educ:* Mass Inst Technol, SB, 42; Univ Kans, PhD(phys chem), 49. *Prof Exp:* Res tech physics, Mass Inst Technol, 41-42; analytical chemist, Hercules Inc, 42-45; res scientist chem eng, Univ Kans Res Found, 46; staff mem phys chem, Los Alamos Sci Lab, 49-52; res scientist, E I du Pont de Nemours & Co, Inc, 52-56; pres, Newport Res Assoc, 56-58; res mgr, Ford Motor Co, 58-63; deputy br chief, N Am Rockwell, 63-66; br chief, McDonnell Douglas Corp, 66-69; from assoc prof to prof, Univ Southern Calif, 69-85; RETIRED. *Concurrent Pos:* Lectr, Univ Southern Calif, 66-69. *Mem:* Acad Mgt; Inst Mgt Sci; Am Asn Univ Profs; Proj Mgt Inst. *Res:* Performing research on research and development management. *Mailing Add:* 4060 Beethoven St Los Angeles CA 90066

STEPHANS, RICHARD A, b Newark, NJ, May 29, 35; m 60; c 3. SYSTEM SAFETY ENGINEERING ANALYSIS, QUALITY ASSURANCE ENGINEERING. *Educ:* Purdue Univ, BS, 57; NMex State Univ, MS, 62. *Prof Exp:* Mgr safety & qual assurance eng, BDM Int, 81-91; PRIN CONSULT, ERC ENVIRON & ENERGY SERV CO, 91- *Concurrent Pos:* Pvt consult, United Space Boosters, Inc, 81; mem, Risk Anal Task Force, Am Soc Mech Engrs, 87- *Mem:* Syst Safety Soc; Am Soc Qual Control; Am Soc Mech Engrs. *Res:* Safety analysis and assessments of chemical and nuclear facilities; hazards analysis of engineering development programs; quality audit of organizations; planning for systems engineering projects; technical proposal preparation; environmental assessments. *Mailing Add:* 13724 Pruitt NE Albuquerque NM 87112

STEPHANY, EDWARD O, b Rochester, NY, July 6, 16; m 41; c 1. MATHEMATICS, STATISTICS. *Educ:* Univ Rochester, AB, 37, AM, 38; Syracuse Univ, PhD(statist), 56. *Prof Exp:* Chmn, Dept Math, State Univ NY Col, Brockport, 47-77, prof math, 77-80; RETIRED. *Mem:* Nat Coun Teachers Math. *Res:* Technique for comparing factors obtained through factor analysis; mathematics education. *Mailing Add:* 229 West Brockport NY 14420

STEPHAS, PAUL, b New York, NY, Aug 31, 29; m 59; c 1. PHYSICS. *Educ:* Univ Wash, Seattle, BS, 56; Rensselaer Polytech Inst, MS, 59; Univ Ore, PhD(physics), 66. *Prof Exp:* Asst res lab, Gen Elec Co, 56-58; metallurgist & ceramist, Vallecitos Atomic Lab, 58-62; asst prof physics, Univ BC, 66-69; PROF PHYSICS, EASTERN ORE STATE COL, 69- *Mem:* Am Phys Soc; Am Asn Physics Teachers; AAAS; Sigma Xi. *Res:* Atomic effects associated with beta decay; low energy nuclear physics; special relativistic mechanics. *Mailing Add:* Dept of Physics Eastern Ore State Col La Grande OR 97850

STEPHEN, CHARLES RONALD, b Montreal, Que, Mar 16, 16; nat US; m 41; c 3. ANESTHESIOLOGY. *Educ:* McGill Univ, BSc, 38, MD & CM, 40. *Prof Exp:* Chief dept anesthesia, Montreal Neurol Inst, 46-47; asst prof anesthesia, McGill Univ, 48-50; prof, Sch Med & chief div, Univ Hosp, Duke Univ, 50-66; prof, Univ Tex Southwest Med Sch Dallas, 66-71; head, dept anesthesia, Barnes Hosp, St Louis, 71-80; chief anesthesiology, St Luke's Hosp, St Louis, 80-85; RETIRED. *Concurrent Pos:* Chief, dept anesthesia, Childrens Mem Hosp, Montreal, 47-50; ed, Surv Anesthesiol, 56-; anesthesiologist, Parkland Mem Hosp, Dallas, 66-71; chief anesthesia, Children's Med Ctr, Dallas, 66-71; fel fac anesthetists, Royal Col Surgeons, 64. *Mem:* Am Soc Anesthesiol; Acad Anesthesiol; Inst Anesthesia Res Soc. *Res:* Pharmacology and physiology of clinical anesthesiology. *Mailing Add:* 15801 Harris Ridge Ct Chesterfield MO 63017-8725

STEPHEN, FREDERICK MALCOLM, JR, b Oakland, Calif, May 11, 43; m 84; c 3. FOREST ENTOMOLOGY, INSECT ECOLOGY. *Educ:* San Jose State Univ, BA, 67; Univ Calif, Berkeley, PhD(entom), 74. *Prof Exp:* Res asst, Univ Calif, Berkeley, 67-74; from asst prof to assoc prof, 74-82, PROF ENTOM, UNIV ARK, 82- *Concurrent Pos:* Vis scientist, Oxford Univ, 87. *Honors & Awards:* A D Hopkins Outstanding Contrib to Southern Forest Entomol, 89. *Mem:* Entom Soc Am; Int Union Forestry Res Orgn; AAAS; Entom Soc Can. *Res:* Forest insect ecology. *Mailing Add:* Dept Entom A-320 Univ Ark Fayetteville AR 72701

STEPHEN, KEITH H, b Ft Wayne, Ind, Jan 3, 34; m 59; c 2. INORGANIC CHEMISTRY. *Educ:* Wabash Col, BA, 59; Northwestern Univ, PhD(inorg chem), 65. *Prof Exp:* SR RES CHEMIST, EASTMAN KODAK CO, 64-*Mem:* Am Chem Soc; Soc Photog Scientists & Engrs. *Res:* Coordination compounds; electron transfer reactions; photographic chemistry. *Mailing Add:* 186 Heritage Circle Rochester NY 14615

STEPHEN, MICHAEL JOHN, b Johannesburg, SAfrica, Apr 7, 33; m 66. PHYSICS. *Educ:* Univ Witwatersrand, BS, 52, MS, 54; Oxford Univ, PhD(phys chem), 56. *Prof Exp:* Ramsey fel, 56-58; res assoc chem, Columbia Univ, 58-60; Imp Chem Industs res fel math, Oxford Univ, 60-62; from asst prof to assoc prof, Yale Univ, 62-68; PROF PHYSICS, RUTGERS UNIV, 68- *Concurrent Pos:* Consult, Bell Tel Labs, 64-66; vis prof, Mass Inst Technol, 67-68. *Mem:* Am Phys Soc. *Res:* Low temperature, solid state and molecular physics. *Mailing Add:* Dept Physics Busch Campus Rutgers Univ New Brunswick NJ 08903

STEPHEN, RALPH A, b Toronto, Ont, Feb, 18, 51; m 74; c 2. SEISMOLOGY. *Educ:* Univ Toronto, BASc, 74; Univ Cambridge, PhD(geophys), 78. *Prof Exp:* Asst scientist, 78-82, assoc scientist, 82-90, SR SCIENTIST GEOPHYS, WOODS HOLE OCEANOG INST, 90- *Mem:* Royal Astron Soc; Am Geophys Union; Soc Explor Geophysicists; Seismol Soc Am. *Res:* Borehole seismic experiments in oceanic crust; seismic structure of oceanic crust and synthetic seismogram development. *Mailing Add:* PO Box 567 West Falmouth MA 02574

STEPHEN, WILLIAM PROCURONOFF, b St Boniface, Man, June 6, 27; nat US; m 52; c 4. ENTOMOLOGY. *Educ:* Univ Man, BSA, 48; Univ Kans, PhD, 52. *Prof Exp:* From asst entomologist to assoc entomologist, Sci Serv, Can Dept Agr, 48-53; from asst prof & asst entomologist to assoc prof & assoc entomologist, 53-63, PROF ENTOM, ORE STATE UNIV, 63- *Concurrent Pos:* Consult, Orgn Am States, Chile, 70-71 & Food & Agr Orgn, UN, Arg, 72-75; mem, Bee Res Inst. *Mem:* AAAS; Entom Soc Am; Animal Behav Soc; Soc Study Evolution; Soc Syst Zool; Sigma Xi. *Res:* Insect behavior; systematic zoology; development physiology; population genetics; pollination. *Mailing Add:* Dept Entom Ore State Univ Corvallis OR 97331

STEPHENS, ARTHUR BROOKE, b Dinuba, Calif, July 6, 42; m 68. NUMERICAL ANALYSIS. *Educ:* Univ Colo, BA, 64; Univ Md, PhD(math), 69. *Prof Exp:* Asst prof math, Univ Hawaii, 69-71; asst prof, 73-77, ASSOC PROF MATH, MT ST MARY'S COL, 77- *Mem:* Am Math Soc; Math Asn Am. *Res:* Optimization problems in functional analysis; estimates for complex eigenvalues of positive matrices. *Mailing Add:* Dept Math & Comput Sci Univ Md Baltimore County 5401 Wilkens Ave Catonsville MD 21228

STEPHENS, CHARLES ARTHUR LLOYD, JR, b Brooklyn, NY, Apr 4, 17; m 39, 64; c 2. INTERNAL MEDICINE. *Educ:* Cornell Univ, AB, 38, MD, 42; Am Bd Internal Med, dipl, 52. *Prof Exp:* DIR RES, SOUTHWESTERN CLIN & RES INST, INC, UNIV ARIZ, 46-, PRES, 71-, ASSOC PROF INTERNAL MED, COL MED, 70-, ADJ PROF MICROBIOL, 80- *Concurrent Pos:* Sr consult, Tucson Med Ctr, St Joseph's & Palo Verde Hosps, Ariz; pvt pract; chmn med adv comt, Southern Chap, Am Red Cross; mem bd dirs & pres, Southwest Chap, Arthritis Found; adj prof microbiol, Univ Ariz, 76- *Mem:* AMA; fel Am Col Physicians; Am Rheumatism Asn; NY Acad Sci; Am Heart Asn; Sigma Xi. *Res:* Tissue culture research in rheumatic diseases. *Mailing Add:* 5265 E Knight Dr Tucson AZ 85712-2147

STEPHENS, CHARLES W, defense & aerospace systems; deceased, see previous edition for last biography

STEPHENS, CHRISTINE TAYLOR, b Rochester, NY, July 30, 51; m 74; c 1. PLANT PATHOLOGY. *Educ:* Smith Col, AB, 73; Ohio State Univ, MA, 76, PhD(plant path), 78. *Prof Exp:* Res asst veg, Cornell Univ, 74; res asst ornamental corps, Ohio State Univ, 74-78; ASST PROF ORNAMENTALS & VEG CROPS, MICH STATE UNIV, EAST LANSING, 78- *Mem:* Am Phytopath Soc. *Res:* Damping off and root rots diseases caused by soil borne fungi; ecological and epidemiological studies; characterization and control of rhizoctonia solan. *Mailing Add:* Dept of Bot and Plant Path Mich State Univ East Lansing MI 48824

STEPHENS, CLARENCE FRANCIS, b Gaffney, SC, July 24, 17; m 42; c 2. MATHEMATICS. *Educ:* J C Smith Univ, BS, 38; Univ Mich, MS, 39, PhD(math), 43. *Hon Degrees:* DSc, J C Smith Univ, 54. *Prof Exp:* Instr math, Prairie View Col, 40-42, prof, 46-47; prof & head dept, Morgan State Col, 47-62; prof, State Univ NY Col Geneseo, 62-69; prof & chmn dept, 69-87, EMER PROF MATH, STATE UNIV NY COL POTSDAM, 87- *Concurrent Pos:* Ford fel & mem, Inst Advan Study, 53-54. *Mem:* Assoc mem Am Math Soc; assoc mem Math Asn Am; Sigma Xi. *Res:* Non-linear difference equations analytic in a parameter. *Mailing Add:* 7244 Rte 255 Conesus NY 14435-9536

STEPHENS, DALE NELSON, b Los Angeles, Calif, Dec 20, 41; m 64; c 2. ORGANIC CHEMISTRY. *Educ:* Westmont Col, BA, 63; Univ Ariz, PhD(org chem), 67. *Prof Exp:* Res chemist, Corn Prod Co, 67-68; asst prof, 68-71, assoc prof, 71-77, PROF CHEM, BETHEL COL, MINN, 77- *Mem:* AAAS; Am Chem Soc. *Res:* Urethanes; epoxies; natural product synthesis; x-ray crystallography; organic synthesis. *Mailing Add:* Dept Chem Bethel Col 3900 Bethel Dr Arden Hills St Paul MN 55112-6999

STEPHENS, DOUGLAS ROBERT, b Portland, Ore, May 10, 35; m 61; c 2. CHEMICAL ENGINEERING, MATERIALS SCIENCE. *Educ:* Univ Wash, BS, 57; Univ Ill, MS, 59, PhD(chem eng), 61. *Prof Exp:* PROJ LEADER, LAWRENCE LIVERMORE LAB, 61- *Honors & Awards:* Spec Award Excellence Technol Transfer, Fed Lab Consortium, 86; Fel, Am Inst Chem Engrs. *Mem:* Am Inst Chem Engrs; Am Chem Soc; Am Nuclear Soc; AAAS. *Res:* Experimental high pressure equation of state and phase transformation measurements; electrical properties of solids at high pressures; in-situ chemical processing. *Mailing Add:* Lawrence Livermore Lab PO Box 808 L-85 Livermore CA 94550

STEPHENS, EDGAR RAY, physical chemistry; deceased, see previous edition for last biography

STEPHENS, FRANK SAMUEL, b Ind, June 30, 31; m 59. NUCLEAR CHEMISTRY. *Educ:* Oberlin Col, AB, 52; Univ Calif, PhD(chem), 55. *Prof Exp:* RES CHEMIST, LAWRENCE RADIATION LAB, UNIV CALIF, BERKELEY, 55- *Concurrent Pos:* Ford Found grant, Inst Theoret Physics, Copenhagen, Denmark, 59-60; guest prof, Physics Sect, Univ Munich, 70-71. *Mem:* Am Phys Soc. *Res:* Coulomb excitation; nuclear structure; heavy ion physics. *Mailing Add:* Lawrence Berkeley Lab Univ Calif Berkeley CA 94720

STEPHENS, GEORGE ROBERT, b Springfield, Mass, Nov, 10, 29; m 51; c 7. FOREST ECOLOGY, FOREST-INSECT RELATIONS. *Educ:* Univ Mass, BS, 52; Yale Sch Forestry, Yale Univ, MF, 58, PhD(forestry), 61. *Prof Exp:* Asst forester, 61-65, assoc forester, 65-75, forester, 75-79, CHIEF FORESTER, CONN AGR EXP STA, 80- *Concurrent Pos:* Secy, Conn Tree Protection Exam Bd, 61-78, mem, 79- *Mem:* Soc Am Foresters. *Res:* Natural succession in forest; effects of insects on tree mortality; forest management. *Mailing Add:* Dept Forestry & Hort Conn Agr Exp Sta PO Box 1106 New Haven CT 06504

STEPHENS, GREGORY A, b Salina, Kans, Dec 20, 47. CARDIOVASCULAR PHYSIOLOGY, RENAL PHYSIOLOGY. *Educ:* Univ Kans, BA, 69, PhD(physiol & cell biol), 76. *Prof Exp:* Res fel, Dept Physiol, Sch Med, Univ Mo, 75-78; asst prof, 78-84, prog dir physiol & anat, 88-90, ASSOC PROF PHYSIOL, SCH LIFE & HEALTH SCI, UNIV DEL, 84-, ASSOC DIR, 90- *Concurrent Pos:* NIH grant. *Mem:* Am Physiol Soc; Am Soc Zoologists; Am Heart Asn. *Res:* Control and functions of the renin angiotensin system in nonmammalian vertebrates. *Mailing Add:* Sch Life & Health Sci Univ Del Newark DE 19716

STEPHENS, GROVER CLEVELAND, b Oak Park, Ill, Jan 12, 25; m 49; c 3. ZOOLOGY. *Educ:* Northwestern Univ, BS, 48, MA, 49, PHD(biol), 52. *Prof Exp:* Asst philos, Northwestern Univ, 48-49, asst zool, 49-51; instr biol, Brooklyn Col, 52-53; asst zool, Univ Minn, Minneapolis, 53-55, from asst prof to prof, 55-64; prof organismic biol, 64-69, chmn dept organismic biol, 64-69, PROF DEVELOP & CELL BIOL, UNIV CALIF, IRVINE, 69- *Concurrent Pos:* Mem corp, Marine Biol Lab, Woods Hole, 53-, instr, 53-60, in-charge invert zool, 58-60; NSF sr fel, 59-60; NATO sr fel, 74. *Mem:* AAAS; Am Physiol Soc; Mar Biol Asn, UK; Am Soc Zool; Am Soc Limnol & Oceanog. *Res:* Invertebrate physiology; biological rhythms; feeding mechanisms in invertebrates; algal physiology; amino acid transport. *Mailing Add:* Dean Biol Scis Univ of Calif Irvine CA 92717

STEPHENS, GWEN JONES, biology, medical sciences, for more information see previous edition

STEPHENS, HAROLD W, b Trenton, NJ, Mar 16, 19; m 46; c 1. MATHEMATICS. *Educ:* Trenton State Col, BS, 41; Columbia Univ, MA, 44, EdD, 64. *Prof Exp:* Head dept math, Farragut Naval Acad, 44-46; instr, Univ Fla, 46-48, Univ Md, 48-49 & McCoy Col, Johns Hopkins Univ, 49-50; asst prof, Ball State Univ, 52-55 & Univ Tenn, 55-60; from assoc prof to prof math, Memphis State Univ, 60-89; RETIRED. *Concurrent Pos:* Lectr, NSF insts & workshops, Univ Tenn, Memphis State, Murray State Univ, Univ Amc, Austin Peay State Univ & Southwestern at Memphis, 57- *Mem:* Math Asn Am; Am Math Soc. *Res:* Mathematics courses for teacher training; algebra. *Mailing Add:* 64 N Yates Rd Memphis TN 38119

STEPHENS, HEATHER R, b Montreal, Que, Feb 28, 49. MUSCULAR DYSTROPHY, IMMUNOCYTOCHEMISTRY. *Educ:* McGill Univ, Montreal, BSc, 70; Univ Montreal, MSc, 72, BA, 75, PhD(anat), 76. *Prof Exp:* Res fel, Jerry Lewis Muscle Res Ctr, Hammersmith Hosp, London, 76-79; adj researcher, 79-85, ASSOC RESEARCHER, DEPT ANAT, UNIV MONTREAL, 85- *Concurrent Pos:* Vis res colleague, dept neuropath, Inst Psychiat, de Crespigny Park, London, Eng; vis res career scientist, neurobiol res lab, Vet Admin Med Ctr, 84-85. *Mem:* Can Asn Anatomists; Am Soc Cell Biol. *Res:* Nerve-muscle interactions in normal and dystrophic conditions; acetylcholinesterase and basal lamina immunocytochemistry at the motor endplate; computer three dimensional reconstruction of muscle spindles in normal and dystrophic muscle. *Mailing Add:* Dept Anat Fac Med Univ Montreal CP6128 Succursale A Montreal PQ H3C 3J7 Can

STEPHENS, HOWARD L, b Akron, Ohio, Oct 9, 19; m 45; c 1. POLYMER CHEMISTRY. *Educ:* Univ Akron, BS, 49, MS, 50, PhD(polymer chem), 60. *Prof Exp:* Res chemist, Rubber Res Labs, 49-52, Inst Rubber Chem, 53-56 & Inst Rubber Res, 56-57, admin asst, 57-65, from instr to assoc prof chem, 57-73, mgr appl res, Inst Polymer Sci, 66-81, PROF CHEM & POLYMER SCI, UNIV AKRON, 73-, HEAD DEPT POLYMER SCI, 78- *Mem:* Am Chem Soc. *Res:* Polymer oxidation; preparation and structure of graft polymers; emulsion polymerization; vulcanization. *Mailing Add:* Dept Polymer Sci Univ Akron Akron OH 44325-3909

STEPHENS, JACK E(DWARD), b Eaton, Ohio, Aug 17, 23; m 48; c 4. HIGHWAY ENGINEERING, CONSTRUCTION MATERIALS. *Educ:* Univ Conn, BS, 47; Purdue Univ, MS, 55, PhD(civil eng), 59. *Prof Exp:* Mem, Mat Sci Inst, 68-, from asst prof to assoc prof civil eng, 50-62, head dept, 65-72, prof civil eng, 62-88, EMER PROF CIVIL ENG, UNIV CONN, 89- *Concurrent Pos:* Jr engr, State Hwy Dept, Conn, 48-50, consult, 64-66; owner, Jack E Stephens Soil Lab, 58-; mem comts, Transp Res Bd, Nat Acad Sci-Nat Res Coun; asphalt panel SHRP, 87-89. *Honors & Awards:* H Jackson Tibbet Award, Am Soc Civil Engrs, 72; George Westinghouse Award, Am Soc Eng Educ, 74. *Mem:* Am Soc Civil Engrs; Am Soc Eng Educ; Am Rd Builders Asn (pres, 77); Asn Asphalt Paving Technologists; Am Soc Photogram; Sigma Xi. *Res:* Highway pavement design; bituminous and concrete mixes; traffic, planning and design; soil mechanics; effect of aggregate factors on pavement friction. *Mailing Add:* Civil Eng Dept Univ Conn Storrs CT 06269-3037

STEPHENS, JAMES BRISCOE, b San Francisco, Calif, Mar 5, 36; m 67; c 2. ATMOSPHERIC PHYSICS, SPACE PHYSICS. *Educ:* Univ Okla, BS, 64, MS, 66, PhD(eng physics), 71. *Prof Exp:* Scientist remote sensing, 66-69, scientist statist physics, 70-72, TASK TEAM LEADER TERRESTRIAL DIFFUSION, MARSHALL SPACE FLIGHT CTR, NASA, 73- *Concurrent Pos:* Mem, Atmos Effects Panel, NASA Hq, 74- *Mem:* Am Phys Soc; Sigma Xi. *Res:* The environmental effects from aerospace effluents. *Mailing Add:* 6800 Jones Valley Huntsville AL 35802

STEPHENS, JAMES FRED, b Lexington, Tenn, Sept 29, 32; m 72; c 3. POULTRY SCIENCE, MICROBIOLOGY. *Educ:* Univ Tenn, BS, 54, MS, 59, PhD(bact), 64. *Prof Exp:* Asst poultry, Univ Tenn, 56-62; from asst prof to assoc prof poultry sci, Clemson Univ, 62-68; assoc prof, 68-74, PROF POULTRY SCI, OHIO STATE UNIV, 74- *Mem:* Poultry Sci Asn; Am Soc Microbiol; World Poultry Sci Asn. *Res:* Poultry disease; nutrition relationships; pathogenecity of Salmonellae; drug resistance in enteric bacteria. *Mailing Add:* Dept Poultry Sci Ohio State Univ Columbus OH 43210

STEPHENS, JAMES REGIS, b Pittsburgh, Pa, Mar 16, 25; m 55; c 4. ORGANIC CHEMISTRY, POLYMER CHEMISTRY. *Educ:* St Vincent Col, BS, 47; Univ Pittsburgh, MS, 49; Northwestern Univ, PhD(chem), 53. *Prof Exp:* Res chemist, Sinclair Ref Co, 50 & Am Cyanamid Co, 53-57; sr res scientist, Amoco Chem Corp, 57-67, group leader res dept, 67-70, sect leader res & develop, 70-87, sr res assoc, 80-87; RETIRED. *Honors & Awards:* Glycerine Award, Glycerine Producer's Asn, 54; Outstanding Achievement Award, Soc Plastics Engrs, 87. *Mem:* Am Chem Soc; Soc Plastics Engrs. *Res:* Organic nitrogen compounds; stereochemistry of dioxanes; alkyd and magnet wire enamels; heterocyclic and aromatic polymers for high temperature service. *Mailing Add:* 7-S-361 Arbor Dr Naperville IL 60540

STEPHENS, JEFFREY ALAN, b Louisville, Ky, Oct 14, 58. MOLECULAR PHOTOIONIZATION PROCESSES. *Educ:* Huntington Col, BA(chem) & BA(math), 78; Boston Univ, MA & PhD(chem), 84. *Prof Exp:* RES ASSOC, UNIV CHICAGO, 84- *Concurrent Pos:* Vis scientist, Argonne Nat Lab, 84- *Mem:* Am Phys Soc; AAAS. *Res:* Effects of electron correlations in molecular photoionization-photoexcitation processes; calculation of molecular photoionization cross sections and angular distributions; interactional degradation of slow electrons in condensed matter; measurement-interpretation of liquid structure factors and radial distributions; x-ray diffraction of liquid systems. *Mailing Add:* AA Noyes Lab/Calif Inst Technology MC 127-72 Pasadena CA 91125

STEPHENS, JESSE JERALD, b Oklahoma City, Okla, June 3, 33; m 55; c 3. PHYSICAL METEOROLOGY. *Educ:* Univ Tex, BS, 58, MA, 61; Tex A&M Univ, Nat Defense Educ Act fel & PhD(meteorol), 66. *Prof Exp:* Meteorologist, US Weather Bur, 57-59; lectr meteorol, Univ Tex, 59-61, from instr to asst prof, 64-66; assoc prof, Univ Okla, 66-67; assoc prof, 67-71, chmn dept, 75-77 & 81-85, dir, Univ Comput Ctr, 83, assoc dir, Supercomput Comput Res Inst, 85, PROF METEOROL, FLA STATE UNIV, 71- *Mem:* Fel Am Meteorol Soc. *Res:* Scattering processes in the atmosphere; geophysical data processing. *Mailing Add:* Dept Meteorol Fla State Univ Tallahassee FL 32306

STEPHENS, JOHN ARNOLD, organic chemistry, for more information see previous edition

STEPHENS, JOHN C(ARNES), b Attalla, Ala, Sept 22, 10; m 36; c 3. AGRICULTURE ENGINEERING, GEOLOGY. *Educ:* Univ Ala, BS, 31, Stanford Univ, 32. *Prof Exp:* Agr engr, soil conserv serv, USDA, Ala, 33-39, asst proj engr, Everglades Proj, 39-46; chief water control engr, Dade County, 46-49; supvr, Everglades Proj, Soil & Water Conserv Res Div, 49-61, leader regional invests, watershed eng, south br, Ga, 61-65, dir Southeast Watershed Res Ctr, Ga, 65-69, chief northwest br, Soil & Water Conserv-Agr Res Serv, USDA, Boise, ID, 69-72, area dir, Lower Miss Valley Area, 72-76; RETIRED. *Concurrent Pos:* Collabr, Sci & Educ Admin, Agr Res Serv, USDA, 76-; consult geohydrologist, US AID, Asia, Nat Res Coun, Washington, DC, pvt eng firms, Miami, FL, 76-88. *Honors & Awards:* John Deere Gold Medal Award, Am Soc Agr Eng. *Mem:* Am Soc Agr Eng; Am Geophys Union; Am Soc Civil Eng. *Res:* Peat and muck investigations; drainage; irrigation; weed control; hydrology of agricultural watersheds. *Mailing Add:* 1111 NE Second St Ft Lauderdale FL 33301

STEPHENS, JOHN STEWART, JR, b Los Angeles, Calif, May 12, 32; m 53; c 1. MARINE BIOLOGY, FISH BIOLOGY. *Educ:* Stanford Univ, BS, 54; Univ Calif, Los Angeles, MA, 57, PhD, 60. *Prof Exp:* Asst zool, Univ Calif, Los Angeles, 54-58, assoc biol, Santa Barbara, 58-69; from instr to prof biol, 59-74, JAMES IRVINE PROF ENVIRON BIOL, OCCIDENTAL COL, 74- *Concurrent Pos:* Dir, Vantuna Oceanog Prog, Occidental Col, 69- *Mem:* Am Soc Ichthyologists & Herpetologists; Soc Syst Zool; Am Fisheries Soc; Am Inst Fishery Res Biologists; Sigma Xi. *Res:* Systematics and distribution of blenniod fishes; especially Chaenopsidae; osteology of tropical blennies; ecology of nearctic fishes of California, including effects of pollution and habitat destruction. *Mailing Add:* Dept Environ Biol Occidental Col Los Angeles CA 90041

STEPHENS, KENNETH S, b Kutztown, Pa, Dec 11, 32; m 53; c 3. QUALITY CONTROL. *Educ:* LeTourneau Inst, BS, 55; Rutgers Univ, MS, 60, PhD(appl & math statist), 66. *Prof Exp:* Qual control engr, Western Elec Co, Inc, Pa, 55-58 & 59-60; asst & instr math, Rutgers Univ, 58-59; chief qual control eng dept, 60-62, chief eng personnel & staff & Kans City coord dept, 62-63, res leader appl math & statist group systs res & develop, NJ, 63-67; prof & coordr math & indust eng, Letourneau Col, 67-72; lectr, Sch Indust & Systs Eng, Ga Inst Technol & Res Consult, Econ Develop Lab, Eng Exp Sta, 74-79; ADV, UN INDUST DEVELOP ORGN, 80- *Concurrent Pos:* Mem, UN team statist qual control, India, 62-63 & UN Indust Develop Orgn adv standardization & qual control, Thai Indust Stand Inst, Bangkok, Thailand, 72-74, Nigerian Standards Orgn, Lagos, Nigeria, 77-78, State Planning Orgn, Turkey, 80-81, Mauritius Standards Bur, Reduit, 80-82. *Mem:* Fel Am Soc Qual Control; Am Statist Asn; Europ Orgn Qual Control; Indian Asn Qual & Reliability; Sigma Xi. *Res:* Industrial statistics; quality control systems; standardization and certification; industrial engineering. *Mailing Add:* IO/IIS/INFR UNIDO PO Box 400 Vienna A-1400 Austria

STEPHENS, LAWRENCE JAMES, b Chicago, Ill, Aug 11, 40; m 64; c 3. ORGANIC CHEMISTRY, SCIENCE EDUCATION. *Educ:* Loyola Univ Chicago, BS, 63; Univ Nebr, Lincoln, PhD(org chem), 69. *Prof Exp:* Res assoc, Stanford Univ, 68-69; asst prof chem, Findlay Col, 69-73; PROF CHEM, ELMIRA COL, 73- *Concurrent Pos:* Res chemist, Corp Res & Develop Ctr, Gen Elec Co, Schnectady, NY, 84 & 85. *Mem:* Am Chem Soc; Nat Sci Teachers Asn; Am Asn Univ Prof. *Res:* Curriculum development in the natural sciences; mangrove succession patterns. *Mailing Add:* Dept of Chem Elmira Col Elmira NY 14901

STEPHENS, LEE BISHOP, JR, b Atlanta, Ga, Oct 22, 25; m 58; c 3. EMBRYOLOGY. *Educ:* Morehouse Col, BS, 47; Atlanta Univ, MS, 50; Univ Iowa, PhD, 57. *Prof Exp:* Instr biol, Dillard Univ, 50-53; instr, NC Col Durham, 53-54; assoc prof, Southern Univ, 57-62; from asst prof to assoc prof, 62-70, PROF BIOL, CALIF STATE UNIV, LONG BEACH, 70-; EMER PROF BIOL, SCH NATURAL SCIS, 83- *Concurrent Pos:* Assoc dean, Sch Natural Scis, 75-83. *Mem:* Am Soc Zool; Am Micros Soc; Sigma Xi. *Res:* Neuroembryology; regeneration; endocrinology and development of the nervous system. *Mailing Add:* Dept Biol Calif State Univ 2500 Bell Flower Blvd Long Beach CA 90840

STEPHENS, MARVIN WAYNE, b Grand Rapids, Mich, Mar 24, 43; m 66; c 3. ENVIRONMENTAL CHEMISTRY. *Educ:* Cedarville Col, BS, 65; Univ Nebr, Lincoln, PhD(chem), 72. *Prof Exp:* From asst prof to assoc prof chem, Malone Col, 69-80; VPRES & TECH DIR, WADSWORTH ALERT LABS, 80- *Mem:* Asn Off Anal Chemists; Am Soc Testing & Mat. *Res:* Testing procedures for environmental analyses; method development. *Mailing Add:* Wadsworth/Alert Labs Inc 4101 Shuffel Dr NW North Canton OH 44720

STEPHENS, MAYNARD MOODY, petroleum engineering, economic geology; deceased, see previous edition for last biography

STEPHENS, MICHAEL A, b Bristol, Eng, Apr 26, 27; m 62; c 1. MATHEMATICAL STATISTICS, APPLIED STATISTICS. *Educ:* Bristol Univ, BSc, 48; Harvard Univ, AM, 49; Univ Toronto, PhD(math), 62. *Prof Exp:* Instr math, Tufts Col, 49-50; lectr, Woolwich Polytech, Eng, 52-53 & Battersea Col Technol, 53-56; instr, Case Western Reserve Univ, 56-59; lectr, Univ Toronto, 59-62, asst prof, 62-63; from asst prof to prof, McGill Univ, 63-70; prof, Univ Nottingham & Univ Grenoble, 70-72; prof, McMaster Univ, 72-76; PROF MATH & STATIST, SIMON FRASER UNIV, 76- *Concurrent Pos:* Consult, Can Packers Ltd, 62-63 & various Montreal Drs, 63-67; fel UK Sci Res Coun, 80-81; mem adv comt, Statist Can, 86- *Honors & Awards:* Gold Medal, Statist Soc Can, 89. *Mem:* Fel Am Statist Asn; Statist Soc Can (pres, 83); Int Statist Inst; fel Inst Math Statist; Royal Statist Soc. *Res:* Mathematical statistics, distributions on a circle or a hyper-sphere; analysis of continuous proportions; goodness of fit statistics, robustness, density approximations. *Mailing Add:* Dept Math & Statist Simon Fraser Univ Burnaby BC V5A 1S6 Can

STEPHENS, N(OLAN) THOMAS, b Mountainair, NMex, Oct 20, 32; m 53; c 3. ENVIRONMENTAL SCIENCES, GENERAL. *Educ:* Univ NMex, BS, 55; NMex State Univ, BS, 61; Univ Fla, MSE, 67, PhD(environ eng), 69. *Prof Exp:* Res engr, Rocketdyne Div, NAm Rockwell Inc, 61-64; mgr saline water conversion opers, Struthers Sci & Int Corp, 64-66; sr environ engr, Southern Res Inst, 69-70; prof civil eng & air pollution specialist, Va Polytech Inst & State Univ, 70-; AT DEPT CHEM & ENVIRON ENG, FLA INST TECHNOL. *Mem:* Air Pollution Control Asn. *Res:* Fundamental and applied research on causes, effects, and control of air and water pollutants; research and development on processes for saline water conversion. *Mailing Add:* Dept Chem & Environ Eng Fla Inst Techol 150 Univ Blvd Melbourne FL 32901-6988

STEPHENS, NEWMAN LLOYD, b Kanth, India, Feb 28, 26; Can citizen; m 67; c 2. PHYSIOLOGY, BIOSTATISTICS. *Educ:* Univ Lucknow, India, MB & BS, 50, DM, 53; FRCP, London, 84. *Prof Exp:* Resident med officer, King George's Med Col, Univ Lucknow, 50-53; head sect med, Clara Swain Hosp, Bareilly, India, 55-58; med registr, Univ Col Hosp, London, Eng, 59-61; from asst prof to assoc prof, 67-73, PROF PHYSIOL, FAC MED, UNIV MAN, 73- *Concurrent Pos:* Res fel cardiol, Res & Educ Hosp, Univ Ill, 62-64; res fel physiol, Sch Hyg, Johns Hopkins Univ, 64-65; res fel med, Winnipeg Gen Hosp, Man, 65-66; Can Heart Found scholar & Med Res Coun Can grant, Fac Med, Univ Man, 67-; mem, Soc Scholars, Johns Hopkins Univ, 84. *Mem:* Fel Royal Soc Med; Am Physiol Soc; Biophys Soc; Can Physiol Soc; Can Soc Clin Invest. *Res:* Smooth muscle, biophysics, biochemistry and ultrastructure of normal muscle, effects of acidosis and hypoxia on these parameters; airway smooth muscle in asthma. *Mailing Add:* Dept Physiol Fac Med 425 Basic Med Sci Bldg Univ Man 770 Bannatyne Ave Winnipeg MB R3E 0W3 Can

STEPHENS, NOEL, JR, b Richmond, Ky, Dec 27, 28; m 58; c 4. ANIMAL SCIENCE, ANIMAL NUTRITION. *Educ:* Univ Ky, BS, 55, MS, 56, PhD(animal sci), 64. *Prof Exp:* From instr to assoc prof, 56-69, PROF ANIMAL SCI, BEREA COL, 69- *Mem:* Am Soc Animal Sci. *Res:* Amino acids and trace mineral research in swine nutrition. *Mailing Add:* Dept Animal Sci Berea Col Berea KY 40404

STEPHENS, PETER WESLEY, b Evanston, Ill, Jan 30, 51; m 78; c 2. SOLID STATE PHYSICS. *Educ:* Univ Calif, Berkeley, BA, 73; Mass Inst Technol, PhD(physics), 78. *Prof Exp:* Asst, Mass Inst Technol, 78-80; asst prof, 80-86, ASSOC PROF PHYSICS, STATE UNIV NY, STONY BROOK, 86- *Res:* Use of x-ray and neutron scattering to study unusual states of condensed matter; phase transitions; surface structure. *Mailing Add:* Dept Physics State Univ NY Stony Brook NY 11794

STEPHENS, PHILIP J, b West Bromwich, Eng, Oct 9, 40; m 62; c 1. SPECTROSCOPY, TRANSITION-METAL CHEMISTRY. *Educ:* Oxford Univ, BA, 62, DPhil(chem), 64. *Prof Exp:* Res fel chem, Univ Copenhagen, 64-65; res fel, Univ Chicago, 65-67; from asst prof to assoc prof, 67-76, PROF CHEM, UNIV SOUTHERN CALIF, 76- *Concurrent Pos:* Sci Res Coun fel, 64-66; Alfred P Sloan res fel, 68-70; John Simon Guggenheim Fel, 84-85. *Mem:* Am Chem Soc; Royal Soc Chem; Am Phys Soc. *Res:* Magneto-optical and spectroscopic properties of matter. *Mailing Add:* Dept Chem Univ Southern Calif Los Angeles CA 90089-0482

STEPHENS, RALPH IVAN, b Chicago, Ill, June 3, 34; m 58; c 3. MECHANICS. *Educ:* Univ Ill, BS, 57, MS, 60; Univ Wis, PhD(eng mech), 65. *Prof Exp:* Asst gen eng, Univ Ill, 57-59, instr theoret & appl mech, 59-60; instr eng mech, Univ Wis, 60-65; from asst prof to assoc prof, 65-72, PROF MECH ENG, UNIV IOWA, 72- *Concurrent Pos:* Indust consult, prod viability, 63- *Mem:* Am Soc Testing & Mat; Soc Automotive Engrs. *Res:* Fracture mechanics; fatigue of engineering materials; mechanical behavior; mechanics of solids; failure analysis. *Mailing Add:* Mech Eng Dept Univ of Iowa Iowa City IA 52242

STEPHENS, RAYMOND EDWARD, b Pittsburgh, Pa, Mar 5, 40. CELL BIOLOGY, PROTEIN CHEMISTRY. *Educ:* Geneva Col, BS, 62; Univ Pittsburgh, MS, 63; Dartmouth Med Sch, PhD(molecular biol), 65. *Prof Exp:* Fel, Univ Hawaii, 66; NIH fel, Harvard Univ, 66-67; from asst prof to assoc prof biol, Brandeis univ, 67-77; INVESTR, MARINE BIOL LAB, 70- *Concurrent Pos:* Mem cell biol study sect, NIH, 71-75 & 83-87; vis prof biol, Univ Pa, 77; adj prof physiol, Sch Med, Boston Univ, 78- *Mem:* AAAS; Am Chem Soc; NY Acad Sci; Soc Gen Physiol; Am Soc Cell Biol; Biophys Soc. *Res:* Protein subunit association; bio-chemistry of cell division and cell movement; microtubules; comparative physiology of motile systems; ionic control of ciliary movement. *Mailing Add:* Marine Biol Lab Woods Hole MA 02543

STEPHENS, RAYMOND WEATHERS, JR, b Marietta, Ga, Apr 20, 28; m 51; c 2. PETROLEUM GEOLOGY. *Educ:* Univ Ga, BS, 51; La State Univ, MS, 56, PhD(geol), 60. *Prof Exp:* Geologist, Shell Oil Co, 59-66; dist geologist, Pubco Petrol Co, 66-72; asst prof, 72-74, ASSOC PROF EARTH SCI, UNIV NEW ORLEANS, 74- *Mem:* Am Asn Petrol Geologists. *Res:* Stratigraphic and paleontologic geology. *Mailing Add:* Dept Geol Univ New Orleans New Orleans LA 70148

STEPHENS, RICHARD HARRY, b Bradford, Pa, Sept 16, 45; m 67; c 2. CHEMICAL ENGINEERING. *Educ:* Univ Tex, Austin, BSChE, 68; Mass Inst Technol, PhD(chem eng), 71. *Prof Exp:* Res engr, Exp Sta, E I du Pont de Nemours & Co, Inc, 68; res asst chem eng, Mass Inst Technol, 68-71; staff assoc, Arthur D Little, Inc, 71-75; mgr energy systs, Energy Resources Co, Mass, 76-78; sr engr, Polaroid Corp, 78-80, prin engr, 80-85; MGR TECH DIV, MARKEM CORP, 85- *Mem:* Am Chem Soc; Am Asn Textile Chemists & Colorists. *Res:* Coating process development; management of research and development. *Mailing Add:* 56 Washington Dr Acton MA 01720

STEPHENS, ROBERT JAMES, CELL BIOLOGY, BIOCHEMISTRY. *Educ:* Univ Southern Calif, PhD(cell biol), 65. *Prof Exp:* DIR CELL BIOL, SRI INT, 66- *Mailing Add:* Dept Biol SRI Int 333 Ravenswood Ave Menlo Park CA 94025

STEPHENS, ROBERT LAWRENCE, b Cincinnati, Ohio, July 12, 21; m 41; c 4. BIOCHEMISTRY, MICROBIOLOGY. *Educ:* Univ Fla, BS, 49, MS, 52, PhD(biochem), 56. *Prof Exp:* Chemist organometallic, Ethyl Corp, 56-58; res chemist natural prods, NC State Univ, 58-60; xres 58-60 & Pulp Chem Asn, 60-62; prof chem, LeTourneau Col, 62-65, acad vpres admin, 65-72; prof biol, Biola Col, 73-77; PROF CHEM, LeTOURNEAU COL, 77- *Mem:* Am Chem Soc; AAAS; Sigma Xi. *Res:* Natural products; microbiology of soil; ultraviolet irradiation effects on microorganisms. *Mailing Add:* Dept Chem Letourneau Col Longview TX 75602

STEPHENS, ROGER, chemistry, for more information see previous edition

STEPHENS, STANLEY LAVERNE, b Niagara Falls, NY, Apr 23, 43; m 64; c 1. MATHEMATICS. *Educ:* Anderson Col, BA, 65; Lehigh Univ, MS, 67, PhD(math), 72. *Prof Exp:* Instr, Moravian Col, 68-71; asst prof, 71-80, ASSOC PROF MATH, ANDERSON COL, 80- *Mem:* Am Math Soc. *Res:* Prime power groups, particularly the automorphism group of p-groups. *Mailing Add:* Dept Math Anderson Col Anderson IN 46012

STEPHENS, TIMOTHY LEE, b Bellingham, Wash, June 27, 44; m 69; c 1. NUCLEAR WEAPONS EFFECTS, RESEARCH MANAGEMENT. *Educ:* Calif Inst Technol, BS, 66; Harvard Univ, AM, 67, PhD(physics), 71. *Prof Exp:* Res fel physics, Smithsonian Astrophys Observ, 69-70; physicist, Gen Elec Co-Tempo, 70-81; prog mgr, Kaman Sci Corp, 81-84, VPRES & DIR RES, PHYS RES, INC, 84- *Res:* Optical emission processes of atoms and molecules; energy partitioning during high energy electron deposition in air; chemical and hydrodynamic properties of the atmosphere; environmental effects of nuclear weapons. *Mailing Add:* Phys Res Inc PO Box 4068 Huntsville AL 35807

STEPHENS, TRENT DEE, b Wendell, Idaho, Aug 14, 48; m 71; c 5. DEVELOPMENTAL BIOLOGY, MORPHOGENESIS. *Educ:* Brigham Young Univ, BS, 73, MS, 74; Univ Pa, PhD(anat), 77. *Prof Exp:* Sr fel, 77-79, res assoc pediat & anat, Univ Wash, 79-81; asst prof, 81-86, ASSOC PROF ANAT & EMBRYOL, IDAHO STATE UNIV, 86- *Concurrent Pos:* Adj asst prof, Creighton Univ, 82- *Mem:* Teratology Soc; Am Asn Anatomists; Sigma Xi; AAAS; Soc Develop Biol. *Res:* Developmental biology; limb field induction; size and location of the limb field as related to overall body plan; characteristics of limbness; comparative and evolutionary morphogenesis; limb defects. *Mailing Add:* Dept Biol Sci Idaho State Univ Pocatello ID 83209

STEPHENS, WILLIAM D, b Paris, Tenn, Nov 17, 32; div; c 3. ORGANIC CHEMISTRY. *Educ:* Western Ky State Col, BS, 54; Vanderbilt Univ, PhD(chem), 60. *Prof Exp:* Group leader high energy oxidizers, Thiokol Chem Corp, 59-61, sect chief org chem, 61-63; group leader basic mat, Goodyear Tire & Rubber Co, 63-66; prin chemist, Thiokol Chem Corp, 66-78; chief chemist, Atlantic Res Corp, 78-83; DIR ROCKET PROPULSION, US ARMY MISSILE COMMAND, 83- *Concurrent Pos:* Adj assoc prof, Univ Ala, Huntsville, 70-78. *Mem:* Am Chem Soc. *Res:* Solid propellant research; explosives, burning-rate catalysts; organometallic, organic nitrogen, sulfur and cyclic compounds; polymer chemistry; adhesives; bonding agents; urethane catalysts; antioxidants. *Mailing Add:* PO Box 12652 Huntsville AL 35807

STEPHENS, WILLIAM LEONARD, b Covington, Ky, Apr 19, 29; m 57. MICROBIOLOGY. *Educ:* Sacramento State Col, BS, 57; Univ Calif, Davis, PhD(microbiol), 63. *Prof Exp:* Res asst bact, Univ Calif, Davis, 57-63; from asst prof to assoc prof, 63-70, chmn dept, 68-74, PROF BACT, CHICO STATE COL, 70-, DEAN, SCH NATURAL SCI, 77- *Mem:* Am Soc Microbiol. *Res:* Carotenoid pigments of bacteria. *Mailing Add:* Dept Biol Chico State Col Chico CA 95927

STEPHENS, WILLIAM POWELL, b Rio de Janeiro, Brazil, Feb 18, 48; US citizen; m 76. NEW SYNTHETIC METHODS. *Educ:* Nasson Col, BS; Univ Vt, PhD(chem), 79. *Prof Exp:* Chmn, dept nat sci, 81-83, asst prof, 79-87, dean sci & technol, 83-87, ASSOC PROF CHEM, INTERAM UNIV, PR, 87- *Mem:* Am Chem Soc. *Res:* New synthetic methods; novel closure to the oxazoline ring system, the possibility of chiral induction upon closure and the subsequent reactions of these compounds. *Mailing Add:* Dept Chem Interamerican Univ Box 5100 San German PR 00753

STEPHENS-NEWSHAM, LLOYD G, b Saskatoon, Sask, Apr 30, 21; m 50; c 2. BIOPHYSICS. *Educ:* Univ Sask, BA, 43; McGill Univ, PhD(nuclear physics), 48. *Prof Exp:* Asst prof physics, Dalhousie Univ, 48-51; from asst prof to assoc prof, Fac Med, McGill Univ, 52-66; from assoc prof to prof physiol, 66-74, prof, 74-86, EMER PROF PHARMACY & PHARMACEUT SCI, FAC PHARM, UNIV ALTA, 86- *Concurrent Pos:* Consult, Victoria Gen Hosp, 48-51; radiation physicist, Royal Victoria Hosp, Montreal, 52-66. *Mem:* Can Physiol Soc; Can Asn Physicists; Biophys Soc; Sigma Xi; fel Can Col Physicists Med. *Res:* Effects of ionizing radiation; neutron activation analysis. *Mailing Add:* 1791 Brymea Lane Victoria BC V8N 6B7 Can

STEPHENSON, ALFRED BENJAMIN, b Unity, Va, May 24, 12; m 41; c 3. POULTRY HUSBANDRY. *Educ:* Va Polytech Inst, BS, 33; Rutgers Univ, MS, 34; Iowa State Col, PhD, 49. *Prof Exp:* Asst poultry breeding, Iowa State Col, 46-49; from asst prof to assoc prof, Utah State Agr Col, 49-53; from assoc prof to prof poultry breeding, 53-82, EMER PROF, UNIV MO-COLUMBIA, 82- *Mem:* Poultry Sci Asn. *Res:* Quantitative inheritance in poultry breeding. *Mailing Add:* 21 Bingham Rd Columbia MO 65203

STEPHENSON, ANDREW GEORGE, b Marion, Ohio, Dec 4, 50; m 77; c 2. PLANT REPRODUCTIVE ECOLOGY. *Educ:* Miami Univ, BA, 73; Univ Mich, MS, 75, PhD(bot), 78. *Prof Exp:* From asst prof to assoc prof, 78-86, PROF BIOL, PA STATE UNIV, 86- *Concurrent Pos:* vis researcher, Found Agr Plant Breeding, Wageningen, Neth. *Mem:* Soc Study Evolution; Ecol Soc Am; Am Soc Naturalists; Am Soc Plant Taxonomists. *Res:* Microgametophyte competition; ecology and evolution of plant reproduction; plant-animal coevolution in pollination, herbivory and dispersal. *Mailing Add:* 208 Mueller Lab Pa State Univ University Park PA 16802

STEPHENSON, CHARLES BRUCE, b Little Rock, Ark, Feb 9, 29; m 52. ASTRONOMY. *Educ:* Univ Chicago, BS, 49, MS, 51; Univ Calif, PhD(astron), 58. *Prof Exp:* Asst astron, Dearborn Observ, Northwestern Univ, 51-53 & Univ Calif, 56-57; from instr to prof astron, 58-88, WORCESTER R & CORNELIA B WARNER PROF ASTRON, CASE WESTERN RESERVE UNIV, 88- *Mem:* AAAS; Am Astron Soc; Int Astron Union. *Res:* Stellar spectra; galactic structure; positional astronomy. *Mailing Add:* Dept Astron Case Western Reserve Univ Cleveland OH 44106

STEPHENSON, CHARLES V, b Centerville, Tenn, Oct 1, 24; m 48; c 3. SOLID STATE PHYSICS. *Educ:* Vanderbilt Univ, BA, 48, MA, 49, PhD(physics), 52. *Prof Exp:* Res physicist, Sandia Corp, 52-56; asst prof physics, Ala Polytech Inst, 56-58; head physics sect, Southern Res Inst, 58-62; chmn dept, 67-74, PROF ELEC ENG, VANDERBILT UNIV, 62- *Concurrent Pos:* Consult, Sandia Corp, 56-58. *Mem:* AAAS; fel Am Phys Soc; Acoust Soc Am; Am Asn Physics Teachers; sr mem Inst Elec & Electronics Eng. *Res:* Molecular spectroscopy; solid state physics. *Mailing Add:* Dept Elec Eng Vanderbilt Univ Nashville TN 37235

STEPHENSON, DANNY LON, b Ft Worth, Tex, Nov 7, 37; m 63; c 1. ORGANIC CHEMISTRY, SPECTROSCOPY. *Educ:* Tex Christian Univ, BA, 59, MA, 60; Rice Univ, PhD(org chem), 64. *Prof Exp:* Res chemist, Phillips Petrol Co, 64-65; prof chem, Howard Payne Univ, 65-80, head dept & chmn, Div Sci & Math, 74-80; ADMINR, BEREAN BAPTIST SCH, 80- *Mem:* Am Chem Soc. *Res:* Mechanistic study of various condensation reactions with zinc chloride as the catalyst; natural products. *Mailing Add:* 16218 Autumn Wind Dr Houston TX 77090

STEPHENSON, DAVID ALLEN, b Denver, Colo, Nov 23, 42; m 63; c 2. MOLECULAR PHYSICS. *Educ:* NMex State Univ, BS, 64; Univ Mich, Ann Arbor, MS, 65, PhD(physics), 68. *Prof Exp:* Nat Res Coun-Environ Sci Serv Admin res fel, Environ Sci Serv Admin Res Labs, Colo, 68-70; assoc sr res physicist, Gen Motors Res Labs, 70-78; MEM TECH STAFF, SANDIA NAT LABS, 78- *Mem:* Optical Soc Am. *Res:* Raman spectroscopy of gases; gas phase reactions. *Mailing Add:* 967 Lynn St Livermore CA 94550

STEPHENSON, DAVID TOWN, b Colfax, Wash, Jan 28, 37; m 57; c 3. ELECTRICAL ENGINEERING. *Educ:* Wash State Univ, BS, 58; Univ Ill, MS, 62, PhD(elec eng), 65. *Prof Exp:* Asst elec eng, Univ Ill, 60-65, res assoc, 65-66; asst prof, 66-70, ASSOC PROF ELEC ENG, IOWA STATE UNIV, 70- *Mem:* Inst Elec & Electronics Engrs; Am Soc Eng Educ. *Res:* Antennas; application to radio astronomy, spacecraft, and communications systems; microwave measurements; teaching in field theory and measurements. *Mailing Add:* 322 Hickory Dr Ames IA 50010

STEPHENSON, EDWARD JAMES, b Birmingham, Ala, Aug 13, 47; m 71; c 1. MEDIUM ENERGY REACTIONS, POLARIZATION. *Educ:* Rice Univ, BA, 69; Univ Wis-Madison, MS, 71, PhD(physics), 75. *Prof Exp:* Fel nuclear physics, Lawrence Berkeley Lab, 75-78; fel nuclear physics, Argonne Nat Lab, Ill, 78-79; ASSOC PROF, IND UNIV CYCLOTRON FAC, 79- *Mem:* Am Phys Soc; AAAS. *Res:* Polarization; accelerator mass spectrometry; medium energy nuclear physics. *Mailing Add:* Cyclotron Fac Ind Univ Milo B Sampson Lane Bloomington IN 47401

STEPHENSON, EDWARD LUTHER, b Calhoun, Tenn, May 5, 23; m 47; c 2. ANIMAL NUTRITION. *Educ:* Univ Tenn, BS, 46, MS, 47; State Col Wash, PhD(poultry nutrit), 52. *Prof Exp:* From asst prof to prof animal nutrit, Col Agr, Univ Ark, Fayetteville, 49-64, head dept, 64-87; RETIRED. *Mem:* Am Soc Animal Sci; Soc Exp Biol & Med; fel Poultry Sci Asn; Am Inst Nutrit. *Res:* Poultry nutrition. *Mailing Add:* Old Wire Rd N Fayetteville AR 72703

STEPHENSON, EDWARD T, b Atlantic City, NJ, Nov 7, 29; m 53; c 4. PHYSICAL METALLURGY. *Educ:* Lehigh Univ, BS, 51, PhD(metall), 65; Mass Inst Technol, MS, 56. *Prof Exp:* Res engr, 56-73, sr res engr, 73-80, sr scientist, 80-88, RES FEL, HOMER RES LABS, BETHLEHEM STEEL CORP, 88- *Honors & Awards:* Grossman Award, Am Soc Metals, 64, Stoughton Award, 88. *Mem:* Fel Am Soc Metals; Am Inst Mining, Metall & Petrol Engrs. *Res:* Relation of strength and toughness to composition, processing and microstructure; electron microstructure, electrical resistivity, mechanical and magnetic properties, and internal friction of steel. *Mailing Add:* Homer Res Labs Bethlehem Steel Corp Bethlehem PA 18016

STEPHENSON, ELIZABETH WEISS, b Newark, NJ, Apr 1, 27; m 46; c 3. MUSCLE, CALCIUM REGULATION. *Educ:* Univ Chicago, BS, 47; George Washington Univ, PhD(physiol), 64. *Prof Exp:* Res asst, Ill Neuropsychiat Inst, 47-49; from instr to asst prof physiol, Sch Med, George Washington Univ, 64-71; sr staff fel, Lab Phys Biol, Nat Inst Arthritis, Metab & Digestive Dis, 71-77, res biologist, 77-81; ASSOC PROF PHYSIOL, UNIV MED & DENT, NJ MED SCH, 81- *Concurrent Pos:* Mem coun & exec bd, Biophys Soc, 83- *Mem:* AAAS; Am Physiol Soc; Biophys Soc (treas, 79-92); Soc Gen Physiol. *Res:* Ion transport across cellular and intracellular membranes; excitation-contraction coupling in skeletal muscle. *Mailing Add:* Dept Physiol Univ Med & Dent NJ Med Sch Newark NJ 07103-2757

STEPHENSON, FRANCIS CREIGHTON, b Brantford, Ont, Mar 24, 24; m 48; c 4. PHYSICS. *Educ:* Univ Toronto, BASc, 49, MA, 51, PhD(physics), 54. *Prof Exp:* Res physicist, Lamp Develop Dept, Gen Elec Co, 53-65; asst prof, 65-67, ASSOC PROF PHYSICS, CLEVELAND STATE UNIV, 67-, CHMN, PHYSICS DEPT, 89- *Mem:* Am Asn Physics Teachers. *Res:* Molecular spectroscopy; incandescent radiation; vibration; gas discharge. *Mailing Add:* Dept Physics Cleveland State Univ Euclid Ave E 24 St Cleveland OH 44115

STEPHENSON, FREDERICK WILLIAM, b Tynemouth, Eng, Sept 25, 39; m 68; c 1. ELECTRONICS ENGINEERING. *Educ:* Kings Col, Univ Durham, BSc, 61; Univ Newcastle Upon Tyne, PhD(elec eng), 65. *Prof Exp:* Tech mgr, Microelectronics Div, Electrosil Ltd, 65-67; sr res assoc, Univ Newcastle Upon Tyne, 67-68; lectr, Univ Hull, UK, 68-75, sr lectr, 75-78;

assoc prof, 78-81, assoc dean, res grad studies, 86-90, PROF ELEC ENG, VA POLYTECH INST & STATE UNIV, 83-, DEPT HEAD, 90- *Concurrent Pos:* R T French vis prof, Univ Rochester, 76-77; consult, Frequency Devices Inc, Haverhill, Mass, 79-; vis prof, Univ Cape Town, summer, 82 & 84. *Mem:* Fel Inst Elec & Electronic Engrs; Int Soc Hybrid Microelectronics; fel Inst Elec Engrs. *Res:* Synthesis and sensitivity evaluation of active resistance capacitance and switched-capacitor filters; applications of hybrid microelectronics. *Mailing Add:* The Bradley Dept Elec Eng 340 Whittemore Hall Blacksburg VA 24061-0111

STEPHENSON, GERARD J, JR, b Yonkers, NY, Mar 4, 37; m 60; c 1. THEORETICAL NUCLEAR PHYSICS. *Educ:* Mass Inst Technol, BS, 59, PhD(physics), 64. *Prof Exp:* Res fel physics, Calif Inst Technol, 64-66; from asst prof to assoc prof physics, Univ Md, College Park, 69-74; group leader, 78-82 and 84-85, STAFF MEM, LOS ALAMOS SCI LAB, 74- DEP DIV LEADER, 85- *Concurrent Pos:* Guggenheim Mem fel, Los Alamos Sci Lab, 72-73. *Mem:* Am Phys Soc. *Res:* Neutrino physics, theoretical studies of nuclear structure and of low and intermediate energy nuclear reactions, quark models applied to low energy nuclear physics. *Mailing Add:* MS D434 LANL PO Box 1663 Los Alamos NM 87545

STEPHENSON, HAROLD PATTY, b Angier, NC, Dec 22, 25; m 56; c 2. MOLECULAR SPECTROSCOPY. *Educ:* Duke Univ, BSME, 47, MA, 49, PhD(physics), 52. *Prof Exp:* Instr physics, Duke Univ, 48-49 & 51-52; asst, Appl Physics Lab, Johns Hopkins Univ, 51; assoc prof physics, Ill Wesleyan Univ, 52-53, prof & chmn dept, 53-57; assoc prof mech eng, Duke Univ, 57-60; from assoc prof to prof physics, 60-90, head dept, 60-83, EMER PROF PHYSICS, PFEIFFER COL, 90- *Concurrent Pos:* Vis instr physics, Univ NC, Charlotte, 83 & Livingstone Col, 83-84. *Mem:* Am Asn Physics Teachers. *Res:* Near ultraviolet absorption spectra of poly-atomic molecules. *Mailing Add:* Pfeiffer Col PO Box 590 Misenheimer NC 28109-0590

STEPHENSON, HUGH EDWARD, JR, b Columbia, Mo, June 1, 22; m 64; c 2. THORACIC SURGERY, CARDIOVASCULAR SURGERY. *Educ:* Univ Mo, AB & BS, 43; Wash Univ, MD, 45; Am Bd Surg, dipl, 53; Bd Thoracic Surg, dipl, 63. *Prof Exp:* Instr surg, Sch Med, NY Univ, 51-53; from asst prof to assoc prof, 53-55, chmn dept, 56-60, PROF SURG, SCH MED, UNIV MO-COLUMBIA, 56-, CHIEF, GEN SURG DIV, 76-, CHIEF OF STAFF, UNIV HOSP & CLIN, 82- *Mem:* Soc Vascular Surg; AMA; Am Asn Surg Trauma; Am Col Surg; Am Col Chest Physicians. *Res:* Cardiovascular research; oncology. *Mailing Add:* N323 Health Sci Ctr Univ Mo Columbia MO 65212

STEPHENSON, J(OHN) GREGG, b Kansas City, Mo, Sept 21, 17; m 48; c 2. ELECTRONICS ENGINEERING. *Educ:* Yale Univ, BE, 39; Stanford Univ, Engr, 41. *Prof Exp:* Asst engr, Ohio Brass Co, 41-42; res assoc, Radio Res Lab, Harvard Univ, 42-45; engr, Airborne Instruments Lab, 45-55, sect head appl electronics, 55-57, sect head, Appl Res Div, 57-58, dep dir, Proj Star, 58-61, tech asst to dir, Res & Eng Div, 61-63, prog dir, 63-68, dep dir, Reconnaissance & Surveillance Div, Cutler-Hammer, Inc, 68-69, dir, Tech Support, AIL Div, Eaton Corp, 70-81; RETIRED. *Mem:* Sr mem Inst Elec & Electronic Engrs; Am Inst Aeronaut & Astronaut; Sigma Xi. *Res:* Ultrahigh frequency and microwave receiving and transmitting equipment; space technology; ionospheric propagation; systems management. *Mailing Add:* PO Box 328 Londonberry VT 05148

STEPHENSON, JOHN, b Chichester, Eng, 1939; m 65; c 3. THEORETICAL PHYSICS. *Educ:* Univ London, BSc, 61, PhD(theoret physics), 64. *Prof Exp:* Lectr math, Univ Adelaide, 65-68; res assoc physics, 68-70, vis asst prof, 70-71, asst prof, 71-74, assoc prof, 74-81, PROF PHYSICS, UNIV ALTA, 81- *Mem:* Can Asn Physicists; Am Phys Soc. *Res:* Statistical mechanics and critical phenomena in fluids and magnetic systems. *Mailing Add:* Dept Physics Univ Alta Edmonton AB T6G 2J1 Can

STEPHENSON, JOHN CARTER, US citizen. PHYSICAL CHEMISTRY. *Educ:* Mass Inst Technol, BS, 66; Univ Calif, PhD(phys chem), 71. *Prof Exp:* Scientist, Avco Everett Res Lab, 70-71; SCIENTIST, NAT BUR STANDARDS, 71- *Mem:* Am Chem Soc; InterAm Photochem Soc. *Res:* Lasers; chemical kinetics; spectroscopy; energy transfer; air and water pollution. *Mailing Add:* Molecular Spectros Div Nat Bur Standards Gaithersburg MD 20899

STEPHENSON, JOHN LESLIE, b Farmington, Maine, Dec 4, 21; m 46; c 3. BIOMATHEMATICS. *Educ:* Harvard Univ, BS, 43; Univ Ill, MD, 49. *Prof Exp:* Asst theoret physics, Metall Lab, Univ Chicago, 43-45; physicist, US Naval Ord Lab, 45; intern, Staten Island Marine Hosp, NY, 49-50; from res assoc to asst prof, Univ Chicago, 52-54; scientist, Nat Heart & Lung Inst, 73-83, 54-73, chief sect theoret biophys; PROF BIOMATH PHYSIOL, CORNELL UNIV MED COL, 84- *Concurrent Pos:* USPHS fel anat, Univ Chicago, 50-52; vis prof, Inst Fluid Dynamics & Appl Math, Univ Md, College Park, 73-74. *Mem:* Am Phys Soc; Am Physiol Soc; Int Soc Nephrology; Soc Math Biol (pres, 83-85); Biophys Soc; Am Soc Nephrology; Soc Gen Physiologists. *Res:* Mathematical theory of transport in biological systems; theory of renal function. *Mailing Add:* Dept Physiol & Biophysics Cornell Univ Med Col 1300 York Ave New York NY 10021

STEPHENSON, KENNETH EDWARD, b Fayetteville, Ark, June 22, 51; m 73; c 3. ACCELERATOR PHYSICS, NUCLEAR DETECTOR TECHNOLOGY. *Educ:* Rice Univ, BA, 73; Univ Wis-Madison, MS, 74, PhD(physics), 79. *Prof Exp:* Res assoc, Argonne Nat Lab, 79-82; res physicist, EMR Photoelec, 82-83, group leader, 83-84, sect mgr, 84-87; res scientist, 87-91, PROG LEADER, SCHLUMBERGER-DOLL RES, 91- *Mem:* Am Phys Soc; Inst Elec & Electronic Engrs. *Res:* Novel approaches to the use of neutrons, neutron generators and detectors in the oil service industry; awarded two US patents. *Mailing Add:* Schlumberger-Doll Res Old Quarry Rd Ridgefield CT 06877

STEPHENSON, LANI SUE, b Honolulu, Hawaii, July 31, 48; m 74. NUTRITION, PARASITOLOGY. *Educ:* Cornell Univ, BS, 71, MNS, 73, PhD(gen nutrit), 78. *Prof Exp:* Exten aide, Dept Human Nutrit Food, Col Human Ecol, Cornell Univ, 71; from res asst to res assoc, Cornell Univ, 73-80, vis prof, 80-83, asst prof, 83-88, ASSOC PROF INT NUTRIT, DIV NUTRIT SCI, CORNELL UNIV, 88- *Concurrent Pos:* Co-investr, World Bank, Brit Overseas Develop Ministry & Cornell Health & Nutrit Proj, Kenya, 78-81; prin investr, Clark Found Urinary Schistosomiosis Growth & Anemia Proj, Kenya, 81-83, 84- *Honors & Awards:* Student Res Award, Am Inst Nutrit, 78. *Mem:* Am Pub Health Asn; Am Soc Parasitologists; Am Dietetic Asn; Am Inst Nutrit; Am Soc Trop Med Hyg. *Res:* Maternal and child health; protein-calorie malnutrition in young children; relationships between intestinal parasites, schistosomiasis and nutritional status; dietary methodologies; international nutrition problems. *Mailing Add:* Savage Hall Cornell Univ Ithaca NY 14853

STEPHENSON, LEE PALMER, b Fresno, Calif, Oct 21, 23; m 48; c 3. GEOPHYSICS. *Educ:* Fresno State Col, AB, 47; Univ Ill, MS, 49, PhD(physics), 53. *Prof Exp:* Asst physics, Univ Ill, 47-53; res physicist, Calif Res Corp, Stand Oil Co, Calif, 53-57, group supvr, 57-59, res assoc geophys, 59-63, SR RES ASSOC, CHEVRON OIL FIELD RES CO, 63- *Mem:* AAAS; Am Phys Soc; Am Chem Soc; Soc Explor Geophys; Am Asn Petrol Geologists; Sigma Xi. *Res:* Exploration seismology; seismic signal detection, data processing and interpretation; physical properties of earth materials; compaction and cementation of clastic sediments; optics; astronomy; astronomical instrumentation. *Mailing Add:* 1248 N Stanford Ave Fullerton CA 92631

STEPHENSON, LOU ANN, b Logansport, Ind, Apr 7, 54. TEMPERATURE REGULATION. *Educ:* Ind Univ, PhD(human performance), 81. *Prof Exp:* RES PHYSIOLOGIST, US ARMY RES INST ENVIRON MED, 83- *Mem:* Am Physiol Soc. *Mailing Add:* US Army Res Inst Environ Med Kansas St Natick MA 01760-5007

STEPHENSON, MARY LOUISE, b Brookline, Mass, Feb 23, 21. MOLECULAR BIOLOGY. *Educ:* Conn Col, AB, 43; Radcliffe Col, PhD(biochem), 56. *Prof Exp:* Res fel, 56-59, asst biochemist, 59-66, assoc biol chem, 66-74, ASSOC BIOCHEMIST, MASS GEN HOSP, HARVARD MED SCH, 74-, PRIN RES ASSOC, HARVARD MED SCH, 69- *Mem:* Am Asn Biol Chemists; Am Asn Cancer Res; Am Soc Cell Biol. *Res:* Biosynthesis of proteins. *Mailing Add:* 308 Ocean Ave Marblehead MA 01945

STEPHENSON, NORMAN ROBERT, b Toronto, Ont, Mar 15, 17; m 43; c 3. BIOCHEMISTRY, SCIENCE ADMINISTRATION. *Educ:* Univ Toronto, BA, 38, MA, 40, PhD, 42; Carleton Univ, DPA, 71. *Prof Exp:* Asst chemist, Insulin Comt Lab, Univ Toronto, 38-39, asst med res, Banting Inst, 39-42; res chemist, Stand Brands, Ltd, 45-48 & Consumers Res Labs, 48-50; res chemist, Physiol & Hormones Sect, Food & Drug Labs, Dept Nat Health & Walfare, 50-66, head anti-cancer sect & sci adv, Div Med & Pharmacol, Drug Adv Bur, 66-73, chief, health care prod div, Planning & Eval Directorate, 73-78, sr sci adv, bur drugs, drugs directorate, Health Protection Br, 78-80; RETIRED. *Mem:* Fel Chem Inst Can. *Res:* Endocrinology; biological and chemical assays of hormones and adrenal corticosteroids; estrogens, androgens; thyroid hormone; anterior pituitary hormones; insulin. *Mailing Add:* 44 Kilbarry Crescent Ottawa ON K1K 0H1 Can

STEPHENSON, PAUL BERNARD, b Jena, La, Dec 16, 37; m 59; c 3. PHYSICS. *Educ:* La Polytech Inst, BS, 60, MS, 61; Duke Univ, PhD(physics), 66. *Prof Exp:* Engr, Tex Instruments Inc, 61-62; res asst physics, Duke Univ, 62-66; from asst prof to assoc prof, 66-75, PROF PHYSICS, LA TECH UNIV, 75- *Mem:* Am Asn Physics Teachers; Am Phys Soc. *Res:* Solid state physics, particularly luminescence of organic crystals. *Mailing Add:* Rte 2 Box 2585 Ruston LA 71270

STEPHENSON, RICHARD ALLEN, b Cleveland, Ohio, June 8, 31; m 52; c 3. COASTAL GEOMORPHOLOGY, ENVIRONMENTAL PLANNING. *Educ:* Kent State Univ, BA, 59; Univ Tenn, MS, 61; Univ Iowa, PhD(geog, geol), 67. *Prof Exp:* Asst prof phys geog, ECarolina Univ, 62-67; asst prof phys geog & earth sci, Univ Ga, 67-71; assoc prof phys geog, 71-74, dir, Inst Coastal & Marine Resources, 74-77, PROF DEPT GEOG & PLANNING, E CAROLINA UNIV, 77- *Mem:* Asn Am Geog; Geol Soc Am; Sigma Xi. *Res:* Geomorphology; water resources; environmental resources; environmental planning. *Mailing Add:* 1733 Beaumont Dr Greenville NC 27858-4353

STEPHENSON, ROBERT BRUCE, b Colfax, Wash, Feb 15, 46; m 68; c 3. HUMAN PHYSIOLOGY, CARDIOVASCULAR PHYSIOLOGY. *Educ:* Wash State Univ, BS, 68; Univ Wash, PhD(physiol & biophysics), 76. *Prof Exp:* Assoc res engr, Boeing Co, 68-70; fel, Univ Wash, 70-76 & Mayo Clin & Found, 77-79; ASST PROF PHYSIOL, MICH STATE UNIV, 79- *Res:* Cardiovascular physiology; neural control of the circulation, particularly reflex regulation of blood pressure in normotension and hypertension. *Mailing Add:* Dept Biol Sci Wayne State Univ Detroit MI 48202

STEPHENSON, ROBERT CHARLES, b Oxford, Ohio, Dec 27, 16; m 42; c 3. GEOLOGY. *Educ:* Miami Univ, AB, 38; Johns Hopkins Univ, PhD(geol), 43. *Prof Exp:* Geologist & mining engr, Titanium Div, MacIntyre Develop, Nat Lead Co, 42-43; field geologist, Union Mines Develop Corp, Colo, 43-44 & Rocky Mt Div, Union Oil Co, Calif, 44-45; sr geologist, State Topog & Geol Surv, Pa, 45-46, asst state geologist, 46-52; geologist, Woodward & Dickerson, Inc, 52-55; exec dir, Am Geol Inst, DC, 55-63; prof geol & exec dir res found, Ohio State Univ, 63-72; dir, Ctr Marine Resources, 72-74, prof mgt & spec progs adminr, Off Univ Res, Tex A&M Univ, 74-77; assoc res prof & assoc dir, Res Lab Eng Sci, Univ Va, 80-82; RETIRED. *Concurrent Pos:* Assoc Univ Progs, Div Univ Progs, Energy Res & Develop Admin, Washington, DC, 75-77; consult, Off Energy Res, US Dept Energy, 78-83. *Mem:* AAAS; fel Geol Soc Am; Soc Econ Geologists; Am Asn Petrol Geol; Am Inst Mining, Metall & Petrol Eng. *Res:* Economic geology of metalliferous deposits; petroleum geology; geology of nonmetallic mineral resources; management of marine and coastal resources. *Mailing Add:* 69 Spruce Circle Green Ridge Village Newville PA 17241

STEPHENSON, ROBERT E(LDON), b Nephi, Utah, Aug 7, 19; m 42; c 4. ELECTRICAL ENGINEERING. *Educ:* Univ Utah, BS, 41; Calif Inst Technol, MS, 46; Purdue Univ, PhD(elec eng), 52. *Prof Exp:* From instr to prof elec eng & comput sci, Univ Utah, 46-71, assoc dean, Col Eng, 71-; RETIRED. *Concurrent Pos:* Instr, Purdue Univ, 50-52; engr, Hughes Aircraft Co, 55; engr, Sperry Utah Eng Labs, 59; consult, Sandia Corp, 59-64, Utah Power & Light, 81- *Mem:* Inst Elec & Electronics Engrs. *Res:* Computers; data systems; computer simulation; electric power. *Mailing Add:* 730 Hilltop Rd Salt Lake City UT 84103

STEPHENSON, ROBERT L, b Pittsburgh, Pa, Feb 11, 13; m 38; c 3. RAW MATERIALS BENEFICATION, COKE MANUFACTURE. *Educ:* Princeton Univ, AB, 35. *Prof Exp:* Observer, Duquesne Works, US Steel Corp, 33-37, metallurgist, 37-40, lab foreman, 40-45, chief metallurist, 45-51; from res engr to chief res engr, US Steel Res Lab, 51-78; consult, 78-86; RETIRED. *Concurrent Pos:* Distinguished mem, Iron & Steel Soc, Am Inst Mining, Metall & Petrol Engrs, 83. *Honors & Awards:* T L Joseph Award, Iron & Steel Soc, Am Inst Mining, Metall & Petrol Engrs, 79. *Mem:* Fel Am Soc Metals; Am Inst Mining Metall & Petrol Engrs. *Res:* Developed slide rule for accurately predicting the hardenability of steel from its chemical composition; developed methods for injecting hydrocarbon fuels through blast furnace tuyeres to decrease coke consumption and increase hot-metal production rate; determined methods for eliminating the harmful effects of alkalies on blast furnace performance. *Mailing Add:* 1309 Shady Ave Pittsburgh PA 15217-1339

STEPHENSON, ROBERT MOFFATT, JR, b Atlanta, Ga, Dec 25, 40; m 65; c 2. MATHEMATICS. *Educ:* Vanderbilt Univ, BA, 62; Tulane Univ, MS, 65, PhD(math), 67. *Prof Exp:* Asst prof math, Univ NC, Chapel Hill, 67-73; assoc prof, 73-78, chmn dept, 76-79, PROF MATH,UNIV SC, 78- *Mem:* Am Math Soc. *Res:* General topology. *Mailing Add:* Dept Math Univ SC Columbia SC 29208

STEPHENSON, ROBERT STORER, b Corpus Christi, Tex, Apr 30, 43; m 69; c 2. NEUROPHYSIOLOGY, VISION RESEARCH. *Educ:* Princeton Univ, AB, 65; Mass Inst Technol, SM, 67, PhD(neurophysiol), 73. *Prof Exp:* Lectr physiol, Fac Sci, Rabat, Morocco, 73-76; res assoc physiol, Dept Biol Sci, Purdue Univ, 76-; ASSOC PROF, DEPT BIOL SCI, WAYNE STATE UNIV, 81- *Concurrent Pos:* Fel, Purdue Univ, 76. *Mem:* AAAS; Sigma Xi; Biophys Soc; Asn Res Vision Ophthalmol. *Res:* Invertebrate photoreceptors and phototransduction; nerve regeneration; morphogenesis. *Mailing Add:* Dept Biol Sci Wayne State Univ Detroit MI 48202

STEPHENSON, SAMUEL EDWARD, JR, b Bristol, Tenn, May 16, 26; m 50, 70; c 4. MEDICINE, ORGANIC CHEMISTRY. *Educ:* Univ SC, BS, 46; Vanderbilt Univ, MD, 50; Am Bd Surg & Bd Thoracic Surg, dipl, 57. *Prof Exp:* Asst surg, Sch Med, Vanderbilt Univ, 53-55; from instr to asst prof, 55-61, assoc prof, Sch Med & assoc dir, Clin Res Ctr, 61-67, dir, S R Light Lab Surg Res, 59-62; chmn dept surg, Univ Hosp Jacksonville, 67-78; PROF SURG, UNIV FLA, 67-; CHIEF GEN SURG, BAPTIST MED CTR, 79- *Concurrent Pos:* Consult, Regional Respiratory & Rehab Ctr, 58 & Thayer Vet Admin Hosp, 59. *Res:* Medical electronics, especially physiological control of respiration and cardiac rate; experimental atherogenesis; malignant disease; cardiovascular surgery and neoplasms. *Mailing Add:* Baptist Pavillion 836 Prudential Dr Apt 808 Jacksonville FL 33207

STEPHENSON, STANLEY E(LBERT), b Ogden, Utah, Mar 12, 26; m 49; c 6. ELECTRICAL & NUCLEAR ENGINEERING. *Educ:* Univ Colo, BS, 48; Tex A&M Univ, MS, 61, PhD(elec eng), 64. *Prof Exp:* Engr, Lago Oil & Transport, 48-52; sr engr, Standard Oil Co, Ohio, 52-54; engr, Am Petrofina, 54-59; asst supvr, Nuclear Sci Ctr, Tex A&M Univ, 59-61, asst res engr, Activation Anal Lab, 61-62; assoc prof, 64-69, PROF ELEC ENG, UNIV ARK, FAYETTEVILLE, 69- *Mem:* Am Soc Eng Educ; Inst Elec & Electronics Engrs. *Res:* Digital control systems; optimal control systems. *Mailing Add:* Dept of Elec Eng Univ of Ark Fayetteville AR 72701

STEPHENSON, STEPHEN NEIL, b Hayden Lake, Idaho, Feb 3, 33; m 53; c 4. BOTANY, ECOLOGY. *Educ:* Idaho State Univ, BS, 55; Rutgers Univ, MS, 63, PhD(bot), 65. *Prof Exp:* Park ranger, Nat Park Serv, 57-61; instr bot, Douglass Col, Rutgers Univ, 62-63; asst prof, 65-72, ASSOC PROF BOT, MICH STATE UNIV, 72- *Concurrent Pos:* Mem eastern deciduous forest biome coord comt, Int Biol Prog, 68- *Mem:* AAAS; Am Soc Mammal; Ecol Soc Am; Am Inst Biol Sci. *Res:* Community structure and organization; biosystematics of Gramineae; biogeography of North America, especially arid and semiarid regions. *Mailing Add:* Dept of Bot 162 Plant Bio Lab Mich State Univ East Lansing MI 48824

STEPHENSON, STEVEN LEE, b Washington, DC, Mar 28, 43; m 72; c 1. MYCOLOGY, FOREST ECOLOGY. *Educ:* Lynchburg Col, BS, 68; Va Polytech Inst & State Univ, MS, 70, PhD(bot), 77. *Prof Exp:* Sec teacher biol, Bedford County Pub Schs, 70-74; PROF BOT, FAIRMONT STATE COL, 76- *Concurrent Pos:* Adj prof, WVa Univ, 82-85; res fel, Univ Va, 82; Fulbright scholar, Himachal Pradesh Univ, India, 87; vis prof, Univ Alaska, 89. *Mem:* Mycol Soc Am; Brit Mycol Soc; Torrey Bot Club; NAm Mycol Asn. *Res:* Distribution and ecology of Myxomycetes in temperate forest ecosystems; upland forests of the mid-Appalachian region of eastern North America. *Mailing Add:* 1115 Morningstar Lane Fairmont WV 26554

STEPHENSON, THOMAS E(DGAR), b Dahlgren, Ill, Oct 19, 22; m 46; c 4. NUCLEAR ENGINEERING, PHYSICS. *Educ:* Southern Ill Univ, BS, 45; Univ Tenn, MS, 50. *Prof Exp:* Physicist, Manhattan Dist, Atomic Energy Comn, 46-47; physicist, Oak Ridge Nat Lab, 47-54; nuclear engr, Convair Div, Gen Dynamics Corp, 54-55; design engr, Nuclear Div, Martin Co, 55-56; physicist, Sci Res Staff, Repub Aviation Corp, 56-65; physicist, Brookhaven Nat Lab, 65-70; physicist, S M Stoller Corp, 70-73; nuclear applications engr, Va Elec & Power Co, 73-76; sr nuclear engr, Burns & Roe, Inc, 76-80; sr engr, Stone & Webster, 80-89; consult, physics & eng, 89-90; PROJ ENGR, BROOKHAVEN NAT LAB, 90- *Concurrent Pos:* Adj assoc prof, Long Island Univ, 68-70. *Mem:* Am Phys Soc; Am Nuclear Soc. *Res:* Neutron and reactor physics; radiation effects and shielding; power reactor safety analysis and environmental effects; neutron cross section measurement and evaluation; nuclear fuel evaluation; nuclear licensing; neutron and nuclear polarization; low temperature physics. *Mailing Add:* 16 Briarfield Lane Huntington NY 11743

STEPHENSON, WILLIAM KAY, b Chicago, Ill, Apr 6, 27; m 51; c 3. PHYSIOLOGY. *Educ:* Knox Col, AB, 50; Univ Minn, PhD, 55. *Prof Exp:* Phys chemist, Nat Bur Stand, 49-50; asst zool, Univ Minn, 50-53, instr, 54; from asst prof to assoc prof biol, 54-64, chmn dept, 64-77, PROF BIOL, EARLHAM COL, 64- *Mem:* AAAS; Am Soc Zool; Am Soc Cell Biol. *Res:* Ion distribution; active transport; bioelectric phenomena; cnidarian behavior. *Mailing Add:* Dept Biol Earlham Col Richmond IN 47374

STEPKA, WILLIAM, b Veseli, Minn, Apr 13, 17; m 48; c 1. PLANT PHYSIOLOGY, PLANT BIOCHEMISTRY. *Educ:* Univ Rochester, AB, 46; Univ Calif, PhD, 51. *Prof Exp:* Asst, Univ Rochester, 46-47 & Univ Calif, 48-49; res assoc bot, Univ Pa, 51-54, asst prof, 54-55; plant physiologist in chg, Radiol Nutriculture Lab, Med Col Va, Va Commonwealth Univ, 55-68, prof, 68-82, EMER PROF PHARMACOG, HEALTH SCI DIV, VA COMMONWEALTH UNIV, 82- *Concurrent Pos:* Plant physiologist & biochemist, Am Tobacco Co, 55-68. *Mem:* Am Soc Plant Physiol; Am Soc Pharmacognosy; Am Asn Cols Pharm. *Res:* Discovery and isolation of cardioactive, hypotensive and contraceptive compounds from natural sources. *Mailing Add:* 715 Glendale Dr Richmond VA 23229

STEPLEMAN, ROBERT SAUL, b New York, NY, Nov 2, 42; m 67; c 2. MATHEMATICS. *Educ:* State Univ NY Stony Brook, BS, 64; Univ Md, College Park, PhD(math), 69. *Prof Exp:* Res assoc numerical anal, Inst Fluid Dynamics & Appl Math, Univ Md, College Park, 69; asst prof appl math & comput sci, Sch Eng & Appl Sci, Univ Va, 69-73; mem tech staff, David Sarnoff Res Ctr, RCA, 73-80; group head sci comput, Exxon Res & Eng Co, 80-85, group head comput & info support, 85-86, sect head, res comput, 86-90, RES & TECH COMPUT ADV, EXXON CHEM, 90- *Concurrent Pos:* Adj assoc prof comput sci, Rutgers Univ, 78-81; ed, Appl Numerical Math, 84-91. *Mem:* Math Asn Am; Asn Comput Mach; Soc Indust & Appl Math. *Res:* Numerical analysis; convergence of numerical methods; solution of elliptic partial differential equations; solution of singular Fredholm integral equations of the first kind; stopping criteria for numerical processes. *Mailing Add:* Exxon Chem PO Box 45 Linden NJ 07036

STEPLEWSKI, ZENON, b Komorki, Poland, Sept 27, 29; m; c 2. CELL BIOLOGY, IMMUNOTHERAPY. *Educ:* Silesian Med Sch, Zabrze, Poland, MD, 60, DM, 63; Inst Immunol, Wroclaw, Poland, DSc, 70. *Prof Exp:* Res asst & sr asst, dept histol & embryol, Silesian Med Sch, Zabrze, Poland, 53-63; assoc prof, dept tumor biol, Inst Oncol, Gliwice, Poland, 63-66, assoc scientist, Head Tumor Virus Lab, 69-73; vis scientist, 66-68, assoc scientist, 73-77, assoc prof, 77-83, PROF IMMUNOL, WISTAR INST, PHILADELPHIA, PA,84- *Concurrent Pos:* Ed, Hybridoma, Monoclonal Antibody News; mem, Biol Response Modifiers Adv Comt, Nat Cancer Inst; recipient, NIH grant. *Mem:* Am Cancer Soc; Europ Asn Cancer Res; Am Asn Cancer Res; Nat Cancer Inst. *Res:* Human cancer immunology; applications of mab's for diagnosis and therapy of solid tumors; studies of the mechanisms of tumor cell lysis by mab's in complement and effector cell dependent fashion; analysis of tumor associated antigens defined by monclonal antibodies mab's. *Mailing Add:* Wistar Inst 36th St & Spruce St Philadelphia PA 19104

STEPONKUS, PETER LEO, b Chicago, Ill, Sept 18, 41; m 62; c 4. PLANT PHYSIOLOGY, HORTICULTURE. *Educ:* Colo State Univ, BSc, 63; Univ Ariz, MSc, 64; Purdue Univ, PhD(plant physiol), 66. *Prof Exp:* Asst prof hort & asst horticulturist, Univ Ariz, 66-68; asst prof, 68-72, assoc prof hort, 72-77, assoc prof crop physiol, 77-79, PROF CROP PHYSIOL, CORNELL UNIV, 79- *Honors & Awards:* Kenneth Post Award, Soc Cryobiol, 71. *Mem:* AAAS; Am Soc Hort Sci; Am Soc Plant Physiol; Scand Soc Plant Physiol; Soc Cryobiol. *Res:* Stress physiology; biochemical mechanisms of cold acclimation; freezing injury; hormonal controls in high temperature injury and senescence; drought resistance of cereals. *Mailing Add:* 236 Emerson Hall Agron Cornell Univ Ithaca NY 14853

STEPTO, ROBERT CHARLES, b Chicago, Ill, Oct 6, 20; m 42; c 2. OBSTETRICS & GYNECOLOGY, PATHOLOGY. *Educ:* Northwestern Univ, BS, 41; Howard Univ, MD, 44; Univ Chicago, PhD(path), 48. *Prof Exp:* Asst, Col Med, Univ Chicago, 42; clin instr obstet & gynec, Stritch Sch Med, Loyola Univ Chicago, 50-60; from clin asst prof to clin assoc prof, Univ Ill, 60-70; chmn dept, Chicago Med Sch, 70-74, prof obstet & gynec, 70-75; prof obstet & gynec, Rush Med Col, 75-79; vpres, Chicago Bd Health, 64-88, pres, 88-90; prof, 79-90, EMER PROF OBSTET & GYNEC, UNIV CHICAGO, 90- *Concurrent Pos:* USPHS fel, Inst Res, Michael Reese Hosp, 48-50; chmn, Dept Obstet & Gynec, Provident Hosp, 53-63 & Mt Sinai Hosp & Med Ctr, 70-; dir obstet & gynec, Cook County Hosp, 72-75; mem, Food & Drug Adv Comt Obstet & Gynec, 72-76; mem, Family Planning Coord Coun, 76; mem maternal & preschool nutrit comt, Nat Acad Sci, 75. *Mem:* Fel AMA; Am Col Obstet & Gynec; Am Col Surg; hon fel Int Col Surg; Am Fertil Soc. *Res:* Endocrine pathology; oncology; sex hormones influence on tissue synthesis; laser surgery; gynecological/urology. *Mailing Add:* 5201 S Cornell Ave Chicago IL 60615

STERANKA, LARRY RICHARD, NEUROPHARMACOLOGY, AMPHETAMINES. *Educ:* Vanderbilt Univ, PhD(pharmacol), 76. *Prof Exp:* SECT HEAD, CNS PHARMACEUT, INC, BALTIMORE, 84- *Res:* Opiates. *Mailing Add:* 1008 Cold Bottom Rd Sparks Glencoe MD 21152

STERBENZ, FRANCIS JOSEPH, b Queens, NY, May 11, 24; m 56; c 3. BIOCHEMICAL PHARMACOLOGY. *Educ:* St John's Univ, NY, BS, 50, MS, 52; NY Univ, PhD, 57. *Prof Exp:* Asst bacteriologist, New York City Dept Hosps, 51-52; res assoc protozool, St John's Univ, NY, 52-56; instr physiol, NJ Col Med & Dent, 56-58, instr microbiol, 58-59; sr res microbiologist, Squibb Inst Med Res, 59-65; asst dept head, Bristol-Myers Co, 65-70, dept head biochem, 70-75, sr res investr clin res, 75-85, SR RES INVESTR MED SERV, RES & DEVELOP LAB, BRISTOL-MYERS PROD, HILLSIDE, 85- *Mem:* AAAS; Soc Protozool; NY Acad Sci; Am Soc Microbiol; Sigma Xi. *Res:* Bio-availability and biochemistry of analgesic, sedative and related drugs; immunochemistry; mechanisms involved in microbial pathogenicity; chemotherapy; nutritional physiology of microorganisms. *Mailing Add:* 60 Drake Rd Somerset NJ 08873

STERE, ATHLEEN JACOBS, b Boston, Mass, Feb 1, 21; m 43; c 3. HISTOCHEMISTRY. *Educ:* Bryn Mawr Col, AB, 41; Radcliffe Col, MA, 42; Pa State Univ, University Park, PhD(biol), 71. *Prof Exp:* Res asst immunol, Sch Med, Boston Univ, 44-46; res asst microbiol, Res Div, Albert Einstein Med Ctr, Philadelphia, 59-63; from res asst to asst prof, 63-77, ASSOC PROF BIOL, PA STATE UNIV, UNIVERSITY PARK, 77- *Honors & Awards:* Christian R & Mary F Lindback Found Award, 80. *Mem:* AAAS; Sigma Xi. *Res:* Histochemistry; effect of oxygen deprivation on cellular metabolism. *Mailing Add:* Dept Biol Pa State Univ Altoona PA 16601-3760

STERGIS, CHRISTOS GEORGE, b Greece, Dec 22, 19; US citizen; m 48; c 2. PHYSICS. *Educ:* Temple Univ, AB, 42, AM, 43; Mass Inst Technol, PhD(physics), 48. *Prof Exp:* Physicist, Radiation Lab, Mass Inst Technol, 44-45, res asst physics, 45-48; asst prof, Temple Univ, 48-51; physicist, 51-58, chief, Space Physics Lab, 59-63, chief aeronomy div, 63-83, SR SCIENTIST, AIR FORCE GEOPHYS LAB, 83- *Concurrent Pos:* Hon res asst, Univ Col, Univ London, 64-65; mem, comt high altitude rocket & balloon res, Nat Acad Sci, 63-66, comt Int Quiet Sun Yr, 63-67, comt solar-terrestrial res, 71-78, Aeronomy Panel, Interdept Comt Atmospheric Sci, 71-77. *Mem:* Am Phys Soc; Am Geophys Union; Sigma Xi. *Res:* Structure of the earth's upper atmosphere by means of rockets and satellites; scattering of solar radiations by the atoms and molecules of the upper atmosphere; solar and atmospheric ultraviolet radiations. *Mailing Add:* Ionospheric Physics Div Geophys Directorate Hanscom AFB Bedford MA 01730

STERIADE, MIRCEA, b Bucharest, Romania, Aug 20, 24; c 2. NEUROPHYSIOLOGY. *Educ:* Col Culture, Bucharest, BA, 44; Fac Med, Bucharest, MD, 52; Inst Neurol, Acad Sci, Bucharest, DSc(neurophysiol), 55. *Prof Exp:* Sr scientist, Inst Neurol, Acad Sci, Bucharest, 55-62, head lab, 62-68; assoc prof, 68-69, PROF PHYSIOL, FAC MED, UNIV LAVAL, 69- *Honors & Awards:* Claude Bernard Medal, Univ Paris, 65; Distinguished Scientist Award, Sleep Res Soc, 89. *Mem:* AAAS; Int Brain Res Orgn; Can Physiol Soc; Fr Neurol Soc; Fr Asn Physiol; Soc Neurosci. *Res:* Neuronal circuitry of thalamic nuclei and cortical areas; responsiveness of thalamic and cortical relay cells; thalamic and cortical inhibitory mechanisms during sleep and waking; ascending reticular systems; neuronal organization and properties related to shifts in vigilance states; cellular bases of thalamic oscillations. *Mailing Add:* Dept Physiol Univ Laval Fac Med Quebec PQ G1K 7P4 Can

STERK, ANDREW A, b Budapest, Hungary, Jan 29, 19; US citizen; m 50; c 2. INSTRUMENTATION, EXPERIMENTAL PHYSICS. *Educ:* Milan Polytech Inst, MS, 43, PhD(elec eng), 46. *Prof Exp:* Res engr, Automatic Elec Co, 46-47; sr develop engr, Servo Corp Am, 47-49; asst chief engr, Fed Mfg & Eng Corp, 49-50; tech dir automatic control, Magnetic Amplifiers, Inc, 50-53; chief develop engr, Philips Electronic Instruments, 53-62; dept mgr space instruments, Am Mach & Foundry Co, 62-67; consult engr & mgr space instrumentation progs, 67-80, SR SYSTS ENGR, MIL & DATE SYST OPERS, GEN ELEC SPACE DIV, PHILA, 81- *Mem:* Sr mem Inst Elec & Electronics Engrs; Am Inst Aeronaut & Astronaut; Soc Photo-Optical Instrumentation Engrs. *Res:* Space instrumentation; x-ray spectrometers; extreme ultraviolet spectrometers; x-ray and extreme ultraviolet spectro-heliographs; x-ray polarimeter; x-ray scattering processes; laser; lidar interferometers; manufacturing and processing in space; space communications; optical communications; remote sensing. *Mailing Add:* Gen Elec Co Space Div Rm/Bldg 9 Rm 2163 PO Box 8555 Philadelphia PA 19101

STERKEN, GORDON JAY, organic chemistry; deceased, see previous edition for last biography

STERLING, ARTHUR MACLEAN, b Ronan, Mont, June 21, 38; m 61; c 4. COMBUSTION. *Educ:* Gonzaga Univ, Spokane, Wash, BS, 61; Univ Wash, Seattle, PhD(chem eng), 69. *Prof Exp:* NIH spec fel, Univ Wash Sch Med, 69-72; wetenschappelijk medewerker, Acad Hosp Leiden, Neth, 72-75; assoc prof chem eng, 75-80, actg chmn, 87-88, PROF DEPT CHEM ENG, LA STATE UNIV, 80-, ASSOC DEAN, COL ENG, 89- *Concurrent Pos:* Assoc res engr, Boing Co, 66; instr, Dept Urol, Univ Wash Sch Med, 69-71, res assoc, Dept Chem Eng, 71-72; head, Urodynamic Lab, Acad Hosp Leiden, Neth, 73-75. *Mem:* Am Inst Chem Engrs; Combustion Inst. *Res:* Combustion phenomena, especially multi-phase processes; hazardous waste incineration; fluid mechanics and heat transfer; biomechanics; computational fluid mechanics. *Mailing Add:* 3304 CEBA La State Univ Baton Rouge LA 70803

STERLING, CLARENCE, b Millville, NJ, Mar 25, 19; m 42; c 5. BOTANY. *Educ:* Univ Calif, AB, 40, PhD(bot), 44. *Prof Exp:* Asst forestry, Univ Calif, 40-41 & bot, 44; res asst, Univ Ill, 46-47; instr, Univ Wis, 47-50; from asst prof to prof, 50-81, EMER PROF FOOD SCI & TECHNOL, UNIV CALIF, DAVIS, 81- *Concurrent Pos:* Fulbright grant, 56-57; Guggenheim fels, 56-57 & 63-64. *Mem:* AAAS; Bot Soc Am; Linnean Soc London. *Res:* Plant anatomy and morphology; cytology; submicroscopic structure of gels; crystal structure. *Mailing Add:* Dept of Food Sci & Technol Univ of Calif Davis CA 95616

STERLING, LEON SAMUEL, b Melbourne, Australia, May 17, 55; m 82; c 2. LOGIC PROGRAMMING, EXPERT SYSTEMS. *Educ:* Univ Melbourne, BSc, 76; Australian Nat Univ, PhD(math), 81. *Prof Exp:* Res fel, Dept Artificial Intel, Univ Edinburgh, UK, 80-83; Don Biegun postdoctoral fel, Dept Appl Math & Computer Sci, Weizmann Inst Sci, Israel, 83-85; asst prof, 85-89, ASSOC PROF COMPUTER SCI, DEPT COMPUTER ENG & SCI, CASE WESTERN RESERVE UNIV, 89- *Concurrent Pos:* Vis lectr, Computer Sci, Univ Melbourne, 87; dir, Ctr Automation & Intel Systs Res, 89- *Mem:* Am Asn Artificial Intel; Asn Computer Mach; Inst Elec & Electronic Engrs; Asn Automated Reasoning. *Res:* Development of a methodology for prolerg programming; build expert system tools and expert system applications; explore alternative foundations for artificial intelligence. *Mailing Add:* Ctr for Automation & Intel Systs Res Case Western Reserve Univ 2040 Adelbert Rd Cleveland OH 44106

STERLING, NICHOLAS J, b Cooperstown, NY, Nov 7, 34; m 62; c 2. MATHEMATICS. *Educ:* Williams Col, BA, 56; Syracuse Univ, MS, 61, PhD(math), 66. *Prof Exp:* Asst prof, 66-70, ASSOC PROF MATH, STATE UNIV NY BINGHAMTON, 70- *Mem:* Am Math Soc; Sigma Xi. *Res:* Non-associative ring theory. *Mailing Add:* Dept of Math State Univ NY Binghamton NY 13901

STERLING, PETER, b New York, NY, June 28, 40; m 61; c 2. NEUROANATOMY, NEUROPHYSIOLOGY. *Educ:* Western Reserve Univ, PhD(bbiol), 66. *Prof Exp:* Asst prof, 69-74, ASSOC PROF NEUROANAT, SCH MED, UNIV PA, 74- *Concurrent Pos:* NSF fel, Med Sch, Harvard Univ, 66-68, NIH fel, 68-69; NSF grant, Sch Med, Univ Pa, 69-71, NIH grant, 71- *Honors & Awards:* C Judson Herrick Award, Am Asn Anat, 71. *Mem:* Am Asn Anat; Soc Neurosci. *Res:* Relation between form of nerve cells and their physiological functioning, particularly in the visuo-motor system. *Mailing Add:* Dept Anat & Chem G3 Univ Pa Sch Med 36th & Hamilton Philadelphia PA 19104

STERLING, RAYMOND LESLIE, b London, Eng, Apr 19, 49; m 70, 83; c 4. UNDERGROUND CONSTRUCTION, STRUCTURAL ENGINEERING. *Educ:* Univ Sheffield, BEng, 70; Univ Minn, Minneapolis, MS, 75, PhD(civil eng), 77. *Prof Exp:* Construct engr, Egil Wefald & Assoc, Minneapolis, 70-71, Husband & Co, Eng, 71-73 & Setter, Leach & Lindstrom, Inc, Minneapolis, 76-77; from asst prof to assoc prof civil eng, 77-83, DIR, UNDERGROUND SPACE CTR, UNIV MINN, 77-, SHIMIZU PROF CIVIL ENG, 88- *Concurrent Pos:* Prin investr res proj, Univ Minn, 77-; consult archit & eng firms, 77-; lectr, 77-; vchmn, US Nat Comt on Tunneling Technol. *Mem:* Am Soc Civil Eng; Inst Struct Engrs; Nat Soc Prof Engrs; Am Underground Space Asn; Inst Civil Engrs; US Nat Comt on Tunneling Technol. *Res:* Underground construction and underground space use; earth-sheltered building design. *Mailing Add:* Underground Space Ctr Univ Minn 790 Civ Min E 500 Pillsbury Dr SE Minneapolis MN 55455

STERLING, REX ELLIOTT, b Eldorado, Kans, Sept 5, 24; m 48; c 2. BIOCHEMISTRY. *Educ:* Cent Mo State Col, BS, 48; Univ Ark, MS, 49; Univ Colo, PhD(biochem), 53. *Prof Exp:* Clin biochemist, Los Angeles County Gen Hosp, 53-71; from instr to assoc prof, biochem, Univ southern Calif, 53-88; head clin biochemist, Los Angeles County Gen Hosp, 71-88; RETIRED. *Mem:* Am Asn Clin Chem. *Res:* Diabetes and carbohydrate in cataract formation; carbohydrate metabolism and adrenal cortical function; prophyrins and porphyria; clinical biochemical methodology and automation. *Mailing Add:* 454 S Woodward Blvd Pasadena CA 91107

STERLING, ROBERT FILLMORE, b Toledo, Ohio, Aug 19, 19; m 47; c 2. ORGANIC POLYMER CHEMISTRY. *Educ:* Carnegie Inst Technol, BS, 42; Univ Pittsburgh, cert bus mgt, 62. *Prof Exp:* Res metallurgist, Magnesium Div, Dow Chem Co, 42-44; res engr, Res Labs, 46-49, mat & process engr, 50-52, res engr, 52-55, adv scientist, Atomic Power Dept, 55-57, mgr chem sect, 57-62, FEL ENGR, MISSILE LAUNCHING & HANDLING DEPT, MARINE DIV, WESTINGHOUSE ELEC CORP, 62- *Mem:* Am Chem Soc. *Res:* Thermal insulation materials; synthetic resins of improved electrical properties; chemistry and materials of nuclear reactors; reliability, research and development of missile launching and handling equipment; biological oceanography and agricultural biochemistry. *Mailing Add:* 1457 Hollenbeck Ave Sunnyvale CA 94087-4237

STERLING, THEODOR DAVID, b Vienna, Austria, July 3, 23; nat US & Can; m 48; c 2. COMPUTER SCIENCE, BIOMETRY. *Educ:* Univ Chicago, AB, 49, MA, 53; Tulane Univ, PhD, 55. *Prof Exp:* Instr math, Univ Ala, 54-55, asst prof statist, 55-57; asst prof statist, Mich State Univ, 57-58; asst prof prev med, Col Med, Univ Cincinnati, 58-66, assoc prof biostatist, 61-63, prof & dir med comput ctr, 63-66; prof comput sci, Wash Univ, 66-72; prof comput sci & fac interdisciplinary studies, 72-81, UNIV RES PROF, SIMON FRASER UNIV, 81- *Concurrent Pos:* Consult, NSF, 67, Environ Protection Agency, 71 & Fed Trade Comn, 72; vis prof statist, Princeton Univ, 78. *Mem:* Asn Comput Mach; fel AAAS; fel Am Statist Asn; fel Am Col Epidemiol; Comput Sci Asn Can (pres, 75-80). *Res:* Humanizing effects of automation; errors and foibles in investigations, especially medical; artificial intelligence. *Mailing Add:* Dept Comput Sci Simon Fraser Univ Burnaby BC V5A 1S6 Can

STERLING, WARREN MARTIN, b Chicago, Ill, Jan 4, 47; m 77; c 1. OPTICS, COMPUTER SCIENCES. *Educ:* Univ Ill, BS, 68; Carnegie-Mellon Univ, MS, 70, PhD(elec eng), 74. *Prof Exp:* Engr numerical control, Westinghouse Elec Corp, 68-74; engr elec eng, Xerox Corp, El Segundo, Calif, 74-82; DIR ENG, TERADATA CORP, 82 - *Concurrent Pos:* Res instr elec eng, Carnegie-Mellon Univ, 72; instr eng, Univ Calif, Los Angeles, 77-82. *Mem:* Inst Elec & Electronics Engrs. *Res:* Design of high performance multi-microprocessor systems for very large relational databases; automated inspection and manufacturing; digital image processing; optical computing. *Mailing Add:* Teradata Corp 100 N Sepulveda El Segundo CA 90245

STERLING, WINFIELD LINCOLN, b Edinburg, Tex, Sept 18, 36; m 61; c 3. ENTOMOLOGY. *Educ:* Pan Am Col, BA, 62; Tex A&M Univ, MS, 66, PhD(entom), 69. *Prof Exp:* Res assoc entom, 64-66, asst prof, 69-74, assoc prof, 74-81, PROF ENTOM, TEX A&M UNIV, 81- *Concurrent Pos:* AID consult, Univ Calif, Berkeley, 74-75; post-doctoral fel, Univ Queensland, 75-76. *Mem:* Entom Soc Am; Entom Soc Can; Am Inst Biol Sci; Sigma Xi. *Res:* Insect ecology, pest management and population dynamics. *Mailing Add:* Dept of Entom Tex A&M Univ College Station TX 77843

STERMAN, MELVIN DAVID, b Brooklyn, NY, Sept 19, 30; m 56; c 4. COLLOID CHEMISTRY, POLYMER CHEMISTRY. *Educ:* City Col New York, BS, 51; Purdue Univ, PhD(phys chem), 55. *Prof Exp:* Teaching asst, Iowa State Univ, 51-52; from res chemist to sr res chemist, 55-62, res assoc, Res Labs, 63-78, RES ASSOC, MFG TECHNOL DIV, EASTMAN KODAK CO, 79- *Mem:* Am Chem Soc. *Res:* Characterization of polymers by physical chemical techniques; electrical properties of polymers; chemistry of cross-linking of polymers; properties of cross-linked polymer works; polymer adsorption on surfaces; preparation and stability of lyophobic colloids in non-aqueous solvents; electrophoretic mobility of colloidal particles. *Mailing Add:* 8 Widewaters Lane Pittsford NY 14534-1024

STERMAN, SAMUEL, b Buffalo, NY, June 6, 18; m 53; c 4. PHYSICAL CHEMISTRY. *Educ:* Univ Buffalo, BS, 39. *Prof Exp:* Develop chemist, Nat Carbon Co Div, Union Carbide Corp, 40-45, res chemist, Linde Co Div, 45-50, supvr spec prod develop, Silicone Div, 50-66, asst dir res & develop, Chem & Plastics, 66-73, assoc dir res & develop, Union Carbide Tech Ctr, 73-88; RETIRED. *Honors & Awards:* Award, Soc Plastics Indust, 61. *Mem:* Am Chem Soc; Soc Plastics Indust; AAAS. *Res:* Textile chemicals; protective coatings; water repellants; surface active agents; composites; interface bonding; organofunctional silanes; urethane foam; high temperature polymers; surface chemistry; elastomers; patent management. *Mailing Add:* 56 Commodore Rd Chappaqua NY 10514

STERMER, RAYMOND A, b Barclay, Tex, July 22, 24; m 48; c 2. AGRICULTURAL ENGINEERING. *Educ:* Tex A&M Univ, BS, 50, MS, 58, PhD, 71. *Prof Exp:* Conserv aid, Soil Conserv Serv, Agr Res Serv, USDA, 43-47, agr engr, 50-55, res engr, Agr Mkt Serv, 55-63 & Agr Res Serv, 63-65, invests leader qual eval, Mkt Qual Res Div, 65-73, res agr engr, Grain Qual & Instrumentation Res Group, 73-76, res agr engr, 76-89, RES AGR ENGR, TEX AGR EXP STA, TEX A&M UNIV, USDA, 89- *Mem:* Am Soc Agr Engrs; Sigma Xi. *Res:* Instruments or techniques for rapid, objective measurement of quality of agricultural products; biomedical radio telemetry for monitoring physiological parameters in cattle; control of bacteria in meats. *Mailing Add:* Tex Agr Exp Sta Rm 231 Agr Eng Bldg College Station TX 77843

STERMER, ROBERT L, JR, b Wilmington, Del, Jan 27, 35; m 57; c 4. ELECTRONICS ENGINEERING. *Educ:* Univ Va, BEE, 60, MEE, 65; Duke Univ, PhD(elec eng), 71. *Prof Exp:* PHYSICIST, ELECTRONICS MAT, LANGLEY RES CTR, NASA, 60- *Concurrent Pos:* Lectr, George Washington Univ, 73- *Mem:* Inst Elec & Electronics Engrs; Am Phys Soc; Sigma Xi. *Res:* Development of electronic device technology for spacecraft; early work in hybrid circuits using film technology and semiconductor devices; development of spacecraft memory systems using bubble technology. *Mailing Add:* 701 Wickwood Dr Chesapeake VA 23320-5872

STERMITZ, FRANK, b Thermopolis, Wyo, Dec 3, 28; m 54; c 5. ORGANIC CHEMISTRY. *Educ:* Univ Notre Dame, BS, 50; Univ Colo, MS, 51, PhD(chem), 58. *Prof Exp:* Res chemist, Merck & Co, Inc, NJ, 51-53 & Lawrence Radiation Lab, Univ Calif, Berkeley, 58-61; from asst prof to assoc prof chem, Utah State Univ, 61-67; assoc prof, 67-69, CENTENNIAL PROF CHEM, COLO STATE UNIV, 69- *Concurrent Pos:* USPHS res career develop award, 63-67; vpres, Elars Biores Labs, 74-76; Fulbright sr fel, Argentina, 73; Fogarty sr fel, Peru, 82. *Mem:* Am Chem Soc; Am Soc Pharmacog; Phytochem Soc NAm; Int Soc Chem Ecol. *Res:* Alkaloid and other natural product isolation, structure proof and biosynthesis; medicinal chemistry; ecology; chemotaxonomy. *Mailing Add:* Dept of Chem Colo State Univ Ft Collins CO 80523

STERN, A(RTHUR) C(ECIL), b Petersburg, Va, Mar 14, 09; m 38, 76; c 3. AIR POLLUTION CONTROL. *Educ:* Stevens Inst Technol, ME, 30, MS, 33. *Hon Degrees:* DrEng, Stevens Inst Technol, 75. *Prof Exp:* Res asst smoke abatement, Stevens Inst Technol, 30-33; engr, J G White Eng Co, 33; supt air pollution surv, New York Dept Health, 35-37; engr exam, Munic Civil Serv Comn, 39-42; chief eng unit, Div Indust Hyg, State Dept Labor, NY, 42-54; chief lab eng & phys sci, Div Air Pollution, USPHS, 55-62; asst dir, Nat Ctr Air Pollution Control, 62-68; prof, 68-78, EMER PROF AIR HYG, DEPT ENVIRON SCI & ENG, SCH PUB HEALTH, UNIV NC, CHAPEL HILL, 78- *Concurrent Pos:* Ed, Current Titles From Eng Jours, 38; lectr, NY Univ, 44-54 & Columbia Univ, 46-52; res assoc, Univ Cincinnati, 55-62, asst clin prof, 61-62; mem, Adv Coun Elec Power Res Inst, 74-79; mem, Bldg Environ Adv Bd, Nat Acad Sci, 77-83. *Honors & Awards:* Richard Beatty Mellon Award, Air Pollution Control Asn, 70; Gordon M Fair Award, Am Acad Environ Eng, 83; Christopher Bartel Award, Int Union Of Air Pollution Prev Asn, 83. *Mem:* Nat Acad Eng; Air Pollution Control Asn (pres, 75-76); Am Soc Mech Engrs; Am Indust Hyg Asn. *Res:* Air pollution; history of air pollution. *Mailing Add:* Dept Environ Sci & Eng Sch Pub Health Univ NC Chapel Hill NC 27514

STERN, AARON MILTON, b Detroit, Mich, Aug 12, 20; m 55; c 2. PEDIATRICS, CARDIOLOGY. *Educ:* Univ Mich, AB, 42, MD, 45; Am Bd Pediat, dipl, 52. *Prof Exp:* Intern, Saginaw Gen Hosp, Mich, 45-46; from asst resident to resident, Univ Hosp, Univ Mich, Ann Arbor, 46-48; resident, Saginaw Gen Hosp, Mich, 48-49; from instr to assoc prof, 49-70, PROF PEDIAT & COMMUN DIS, MED SCH, UNIV MICH, ANN ARBOR, 70- *Concurrent Pos:* Consult, Alpena County Rheumatic Fever Clin, Mich, 51-56. *Mem:* Am Heart Asn; Am Acad Pediat. *Res:* Contrast studies of cardiovascular lesions and effects of cardiac surgery on physiologic and mental status of patients. *Mailing Add:* 1313 E Ann St Ann Arbor MI 48109

STERN, ALBERT VICTOR, b New York, NY, Apr 26, 23; div; c 2. ASTRONOMY, SYSTEMS ENGINEERING. *Educ:* Univ Calif, Berkeley, AB, 47, PhD(astron), 50. *Prof Exp:* Sect head digital subsysts, Hughes Aircraft Co, Fullerton, 54-57, sr scientist, 57-59, lab mgr adv systs, 59-66, asst div mgr, 66-69, chief scientist, Syst Div, 69-73, prog mgr, Missile Systs Group, Canoga Park, 73-76, PROG MGR, RADAR SYSTS GROUP, HUGHES AIRCRAFT CO, CULVER CITY, 76- *Res:* Celestial mechanics; weapons systems analysis; systems engineering. *Mailing Add:* 279 Ravenna Dr Long Beach CA 90803

STERN, ARTHUR IRVING, b New York, NY, Dec 8, 30; m 62; c 3. PLANT PHYSIOLOGY, PHOTOBIOLOGY. *Educ:* City Col New York, BS, 53; Brandeis Univ, PhD(biol), 62. *Prof Exp:* NIH fel develop biol, Brandeis Univ, 62-63; Kettering fel photosynthesis, Weizmann Inst, 63-64, USPHS fel, 64-65; from asst prof to assoc prof, 65-88, PROF BOT, UNIV MASS, AMHERST, 89- *Concurrent Pos:* Secy-treas, NE sect, Am Soc Plant Physiol, 78-91. *Mem:* Am Soc Plant Physiol; Am Soc Photobiol. *Res:* Chloroplast structure and function; chloroplast development; photophosphorylation; proton excretion in plant protoplasts and intact cells; plasma membrane redox. *Mailing Add:* 119 Huntington Rd Hadley MA 01035

STERN, ARTHUR PAUL, b Budapest, Hungary, July 20, 25; nat US; m 52; c 3. ELECTRONICS ENGINEERING. *Educ:* Univ Lausanne, BS, 46; Swiss Fed Inst Technol, dipl, 48; Syracuse Univ, MEE, 56. *Prof Exp:* Res engr, Jaeger, Inc, Switz, 48-50; instr, Swiss Fed Inst Technol, 50-51; res engr electronics lab, Gen Elec Co, 51-52, proj leader semiconductor applns, 52-54, mgr adv circuits, 54-57, mgr electronic devices & applications lab, 57-61; dir eng, Electronics Div, Martin Marietta Corp, Md, 61-64; dir opers, Defense Systs Div, Bunker-Ramo Corp, 64-66; vpres & gen mgr, Magnavox Res Labs, Magnavox Advan Prod & Systs Co, 66-70, vpres & gen mgr, Advan Prod Div, 70-79, pres, 80-90; PRES, EASTERN BEVERLY HILLS CORP, 91- *Concurrent Pos:* Non-res staff mem, Mass Inst Technol, 56-59; mem, Int Solid State Circuits Conf, 58-, chmn, 59-60. *Honors & Awards:* Centennial Medal, Inst Elect & Electronics Engrs, 84. *Mem:* Am Astronaut Soc; fel Inst Elec & Electronics Engrs (secy, 72, treas, 73, vpres 74, pres, 75); Sigma Xi; fel AAAS. *Res:* Applications of modern solid state physics; design, application and electronic circuit behavior of solid state electronic components; transistor circuit engineering. *Mailing Add:* 606 N Oakhurst Dr Beverly Hills CA 90210

STERN, DANIEL HENRY, b Richmond, Va, June 18, 34; m 63, 83; c 4. LIMNOLOGY, ECOLOGY. *Educ:* Univ Richmond, BS, 55, MS, 59; Univ Ill, PhD(zool), 64. *Prof Exp:* Asst prof biol, Tenn Technol Univ, 64-66 & La State Univ, New Orleans, 66-69; from asst prof to assoc prof, 69-75, PROF BIOL & NURSING, UNIV MO-KANSAS CITY, 75-, CHMN DEPT, 86- *Mem:* Ecol Soc Am; Am Soc Limnol & Oceanog; Phycologicol Soc Am; Micros Soc Am; NAm Benthological Soc. *Res:* Ecology of aquatic organisms; invertebrate ecology; applied ecology and environmental impacts; asbestos analysis. *Mailing Add:* Dept Biol Univ Mo Kansas City MO 64110-2499

STERN, DAVID P, b Decin, Czech, Dec 17, 31; US citizen; m 61; c 3. SPACE PHYSICS. *Educ:* Hebrew Univ, Israel, MSc, 55; Israel Inst Technol, DSc(physics), 59. *Prof Exp:* Res assoc physics, Univ Md, 59-61; Nat Acad Sci-Nat Res Coun resident res assoc, 61-63, PHYSICIST, GODDARD SPACE FLIGHT CTR, NASA, 63- *Concurrent Pos:* Chmn, AGU Comt, Hist Geophysics, 82-88. *Mem:* Am Phys Soc; Am Geophys Union. *Res:* Geomagnetic field, its structure, configuration and dynamics; geomagnetic plasmas and particle motion; theory of magnetospheric electric fields; field-aligned currents and associated processes. *Mailing Add:* Code 695 Goddard Space Flight Ctr NASA Greenbelt MD 20771

STERN, EDWARD ABRAHAM, b Detroit, Mich, Sept 19, 30; m 55; c 3. CONDENSED MATTER SCIENCE. *Educ:* Calif Inst Technol, BS, 51, PhD(physics), 55. *Prof Exp:* Res fel solid state physics, Calif Inst Technol, 55-57; from asst prof to prof, Univ Md, 57-66; PROF PHYSICS, UNIV WASH, 66- *Concurrent Pos:* Guggenheim fel, 63-64; NSF sr res fel, 70-71; Fulbright fel, 86-87. *Honors & Awards:* Warren Diffraction Physics Award, Am Crystal Asn, 79. *Mem:* Fel Am Phys Soc; AAAS. *Res:* Electronic structure of metals and alloys; collective effects; magnetism; atomic structure of amorphous and biological matter; ferromagnetism. *Mailing Add:* Dept Physics FM-15 Univ Wash Seattle WA 98195

STERN, ELIZABETH KAY, b Lansing, Mich, Oct 28, 45; m 74. GENERAL ACADEMIC PEDIATRICS. *Educ:* Mich State Univ, BA, 67; Univ Mich, PhD(human genetics), 74; Columbia Univ, MD, 78. *Prof Exp:* Intern & resident pediat, 78-81, fel ambulatory & behav pediat, 81-82, instr, 82-83, ASST PROF PEDIAT, ALBERT EINSTEIN COL MED, 83- *Mem:* AMA; Ambulatory Pediat Asn; fel Am Acad Pediat. *Res:* General academic pediatrics. *Mailing Add:* 113 Surrey Lane Tenafly NJ 07670

STERN, ERIC WOLFGANG, b Vienna, Austria, Nov 4, 30; nat US; m 60; c 1. CHEMISTRY. *Educ:* Syracuse Univ, BS, 51; Northwestern Univ, PhD(chem), 54. *Prof Exp:* Res chemist, Texaco, Inc, 54-57; res chemist, M W Kellogg Co Div, Pullman, Inc, NJ, 58-63, res assoc, 63-70; sect head res & develop, 70-87, SR RES ASSOC, ENGELHARD CORP, EDISON, 87- *Concurrent Pos:* Co-adj prof, Rutgers Univ, 65-74. *Mem:* AAAS; Am Chem Soc; Catalysis Soc; NY Acad Sci; Royal Chem Soc. *Res:* Heterogeneous catalysis; homogeneous catalysis; coordination chemistry; reaction mechanisms; molecular structure; catalyst characterization; medical application of precious metals; organometallics. *Mailing Add:* 234 Oak Tree Rd Mountainside NJ 07092

STERN, ERNEST, b Wetter, Ger, June 5, 28; US citizen; m 53; c 4. SOLID STATE PHYSICS, ACOUSTICS. *Educ:* Columbia Univ, BS, 53. *Prof Exp:* Sr engr, Sperry Gyroscope Co, 55-57; mem staff, Electronics Lab, Gen Elec Co, 58-62; vpres, Microwave Chem Lab, 62-64; staff mem, 64-68, group leader acoustics, 68-82, ASSOC HEAD, SOLID STATE DIV, LINCOLN LAB, MASS INST TECHNOL, 82- *Concurrent Pos:* Consult, US Dept Defense. *Honors & Awards:* Fel, Inst Elec & Electronics Engrs. *Mem:* Sigma Xi; Am Phys Soc. *Res:* Gyromagnetic phenomena; microwave frequencies; nonlinear magnetic phenomena; surface acoustics and acousto-electric phenomena; components and devices; x-ray lithography; superconducting devices; electronic engineering. *Mailing Add:* Mass Inst Technol Lincoln Lab Lexington MA 02173-0073

STERN, FRANK, b Koblenz, Ger, Sept 15, 28; nat US; m 55; c 2. THEORETICAL SOLID STATE PHYSICS. *Educ:* Union Col, NY, BS, 49; Princeton Univ, PhD(physics), 55. *Prof Exp:* Physicist, US Naval Ord Lab, Md, 53-62; Zurich Res Lab, TJ Watson Res Ctr, 65-66, mgr semiconductor electronic properties, 73-81, mgr device theory & modeling, 85- 90, RES STAFF MEM, IBM RES DIV, TJ WATSON RES CTR, 62- *Concurrent Pos:* Lectr, Univ Md, 55-58, part-time prof, 59-62; chmn, Organizing Comt, Electron Properties Two-Dimensional Systs Int Conf, New London, NH, 81; vis scientist, Cavendish Lab, Cambridge, 71, Max Planck Inst Für Festk06rperforschung, 80 & 88; mem at large, Exec Comt, Am Phys Soc Div Condensed Matter Physics, 87-90. *Honors & Awards:* John Price Wetherill Medal, Franklin Inst, 81; Jack A Morton Award, Inst Elec & Electronics Engrs, 88. *Mem:* AAAS; fel Am Phys Soc; Sigma Xi. *Res:* Cohesive energy of iron; semiconductors; injection lasers; optical properties of solids; quantum effects and transport in two dimensional electron systems. *Mailing Add:* IBM Res Div T J Watson Res Ctr Yorktown Heights NY 10598-0218o

STERN, HERBERT, b Can, Dec 22, 18; m 53; c 3. CELL BIOLOGY. *Educ:* McGill Univ, BSc, 40, MSc, 42, PhD, 45. *Hon Degrees:* DSc, McGill Univ. *Prof Exp:* Royal Soc Can fel, Univ Calif, 46-48; lectr cell physiol, Med Sch, Univ Witwatersrand, 48-49; assoc, Rockefeller Inst Med Res, 49-55; head biochem cytol, Plant Res Inst, Can Dept Agr, 55-60; prof bot, Univ Ill, Urbana, 60-65; chmn dept, 67-76, PROF BIOL, UNIV CALIF, SAN DIEGO, 65- *Concurrent Pos:* Mem develop biol panel, NSF; mem cell biol panel, NIH; mem spec subcomt cellular & subcellular struct & function, Nat Acad Sci. *Mem:* Am Soc Plant Physiol; Soc Develop Biol (pres, 64-65); Am Soc Cell Biol; fel Am Soc Biol Chem; Genetics Soc Am; Sigma Xi. *Res:* Cell biology and biochemistry. *Mailing Add:* Dept Biol B-022 Box 109 Univ Calif at San Diego La Jolla CA 92093

STERN, IRVING B, b New York, NY, Sept 12, 20; c 3. DENTISTRY, CELL BIOLOGY. *Educ:* City Col New York, BS, 41; NY Univ, DDS, 46; Columbia Univ, cert, 56. *Prof Exp:* Lectr periodont, Sch Dent, Univ Wash, 59-60, from asst prof to prof, 60-75; prof periodont & chmn dept, Sch Dent Med, Tufts Univ, 75-77; RETIRED. *Concurrent Pos:* Spec res fel anatomy, Sch Dent, Univ Wash, 61-62; USPHS grant. *Mem:* AAAS; Am Dent Asn; Am Acad Periodont; Am Soc Cell Biol; Int Asn Dent Res. *Res:* Ultrastructure and biology of oral epithelium and epithelial derivatives; ultrastructure of dento-gingival junction and cementum. *Mailing Add:* 247 84th Ave N E Bellevue WA 98004

STERN, IVAN J, b Chicago, Ill, Jan 5, 30; m 55; c 2. BIOCHEMISTRY, BACTERIOLOGY. *Educ:* Univ Ill, BSc, 51, MSc, 53; Ore State Univ, PhD(bact), 58. *Prof Exp:* Sr res biochemist, Norwich Pharmacal Co, 58-62; sr res biochemist, Baxter Labs, 62-72, head drug metab, 72-75, mgr biochem sect, Travenol Labs, Inc, 75-; TECH DIR, SMITH LAB, INC. *Concurrent Pos:* Lectr med, Univ Ill, 66-; adj asst prof, Stritch Sch Med, Loyola Univ Chicago. *Mem:* AAAS; Am Soc Pharmacol & Exp Therapeut. *Res:* Biochemical mechanisms of drug toxicity; enzymic degradation of mucopolysaccharides and glycoproteins; immunoassay; drug metabolism; pharmacokinetics. *Mailing Add:* Ten Bradford Terr No 5 Brookline MA 02146

STERN, JACK TUTEUR, JR, b Chicago, Ill, Jan 18, 42; m 67; c 2. BIOMECHANICS, HUMAN EVOLUTION. *Educ:* Univ Chicago, PhD(anat), 69. *Prof Exp:* Instr anat, Univ Chicago, 69-70, asst prof, 70-74; assoc prof, 74-81, PROF ANAT, STATE UNIV NY STONY BROOK, 81- *Concurrent Pos:* Assoc ed, Anat Rec, 72-80, Am J Phys Anthrop, 81-87; USPHS res career develop award, 73. *Mem:* Am Asn Phys Anthrop; Am Asn Anat; AAAS; Sigma Xi. *Res:* Functional anatomy of primates; evolution of erect posture; biomechanics and evolution of muscles. *Mailing Add:* 15 Southgate Rd Setauket NY 11733

STERN, JOHN HANUS, b Brno, Czech, May 21, 28; nat US; m 49. PHYSICAL CHEMISTRY. *Educ:* Univ Calif, BS, 53; Univ Wash, MS, 54, PhD(chem), 58. *Prof Exp:* From asst prof to assoc prof, 58-67, PROF CHEM, CALIF STATE UNIV, LONG BEACH, 67- *Concurrent Pos:* Am Chem Soc-Petrol Res Found int fac fel, Univ Florence, 64-65; vis prof, Hebrew Univ, Jerusalem, 71-72, Univ London, 79-80. *Mem:* Am Chem Soc. *Res:* Thermodynamics of electrolytes and non-electrolytes in aqueous solutions. *Mailing Add:* 7151 Carlton Ave Westminster CA 92683

STERN, JOSEPH AARON, b New York, NY, Apr 24, 27; m 50; c 3. FOOD SCIENCE & TECHNOLOGY. *Educ:* Mass Inst Technol, SB, 49, SM, 50, PhD(food technol), 53. *Prof Exp:* Food technologist, Davis Bros Fisheries, Mass, 48-49; asst food technol, Mass Inst Technol, 50-53; from asst prof to assoc prof fisheries technol, Univ Wash, 53-58; chief biochem unit, Space Med Sect, Boeing Co, 58-59, res prog dir, Bioastronaut Sect, 59-61, mgr adv space prog, 61-65 & Voyager prog planetary quarantine, 65-66, adv interplanetary explor prog, 66; sterilization group supvr, Environ Requirements Sect, Jet Propulsion Lab, Calif, 66-67, asst sect mgr sterilization, 67-69; PRES, BIONETICS CORP, 69- *Concurrent Pos:* Mem comt animal food prod, NSF-Nat Res Coun, 58-60. *Mem:* AAAS; Am Inst Aeronaut & Astronaut; NY Acad Sci; Am Inst Biol Sci; Sigma Xi. *Res:* Spoilage and preservation of food products; biochemical systems in space flight and extraterrestrial missions; planetary quarantine; spacecraft sterilization. *Mailing Add:* 119 Meredith Ave 20 Research Dr Hampton VA 23669

STERN, JUDITH S, b Brooklyn, NY, Apr 25, 43; m 64; c 1. NUTRITION. *Educ:* Cornell Univ, BS, 64; Harvard Univ, MS, 66, ScD, 70. *Prof Exp:* From res assoc to asst prof, Rockefeller Univ, 69-74; from asst prof to assoc prof, 75-82, PROF NUTRIT, UNIV CALIF, DAVIS, 82-, DIR FOOD INTAKE LAB, 80-, PROF, DIV CLIN NUTRIT & METAB, DEPT INT MED, 86- *Concurrent Pos:* Mem metabolism study sect, NIH, 85-87, nutrit study sect, 87-89; mem IOM Comt, Nutrit Labeling, 89-90. *Mem:* Inst Food Tech; Am Inst Nutrit; Am Dietetic Asn; AAAS; Sigma Xi; Am Physiol Soc; Am Soc Clin Nutrit; NAm Asn Study Obesity. *Res:* Studies of some critical factors involved in the development of obesity which include adipose cellularity, food intake, diet composition, exercise, hyperinsulinemia and tissue resistance in muscle and adipose. *Mailing Add:* Dept Nutrit Univ Calif Davis CA 95616

STERN, KINGSLEY ROWLAND, b Port Elizabeth, SAfrica, Oct 30, 27; nat US; m 56; c 2. TAXONOMIC BOTANY. *Educ:* Wheaton Col, Ill, BS, 49; Univ Mich, MA, 50; Univ Minn, PhD(bot), 59. *Prof Exp:* Asst, Univ Mich, 49-51 & Univ Ill, 54-55; asst, Univ Minn, 55-56 & 57-58, instr bot, 58-59; instr biol, Hamline Univ, 57-58; from asst prof to assoc prof, 59-68, PROF BOT, CALIF STATE UNIV, CHICO, 68- *Concurrent Pos:* NSF res grants, 59, 60 & 63-71; consult, Bot Field Surveys; vis prof, Univ Hawaii, 87. *Mem:* Am Soc Plant Taxon; Bot Soc Am. *Res:* Taxonomy of vascular plants, especially pollen grains, anatomy, cytology and morphogenesis. *Mailing Add:* Dept of Biol Sci Calif State Univ Chico CA 95929-0515

STERN, KURT, b Vienna, Austria, Apr 3, 09; nat US; m 39; c 3. PATHOLOGY, CANCER. *Educ:* Univ Vienna, MD, 33. *Prof Exp:* Instr biochem, Inst Med Chem, Univ Vienna, 30-33, res assoc, 33-38; jr physician, State Inst Study & Treatment Malignant Dis, Buffalo, 43-45; asst pathologist, Mt Sinai Hosp, 45-48, from asst to assoc dir, Mt Sinai Med Res Found, 48-60, dir blood ctr, 50-60; prof & pathologist, Res & Educ Hosp, 60-70, EMER PROF PATH, UNIV ILL COL MED, 70-; RES PROF, LAUTENBERG CTR IMMUNOL, HEBREW UNIV-HADASSAH MED SCH, JERUSALEM, 81- *Concurrent Pos:* Res fel, New York Cancer Hosp & Div Cancer, Bellevue Hosp, 39-40; from assoc to assoc prof, Chicago Med Sch, 49-60; sci ed, Bull Am Asn Blood Banks, 60; prof life sci, Bar-Ilan Univ, Israel, 69-80. *Honors & Awards:* John Elliott Mem Award, Am Asn Blood Banks, 72. *Mem:* Fel Am Soc Clin Path; Soc Exp Biol & Med; Am Soc Exp Path; Am Asn Immunol; Am Asn Cancer Res. *Res:* Experimental cancer research; immunology; experimental pathology; blood groups and immunohematology; physiopathology of reticulo-endothelial system. *Mailing Add:* Lautenberg Ctr Immunol, Hebrew Univ Hadassah Med Sch POB 1172 Jerusalem 91000 Israel

STERN, KURT HEINZ, b Vienna, Austria, Dec 26, 26; nat US; m 60; c 2. PHYSICAL INORGANIC CHEMISTRY. *Educ:* Drew Univ, AB, 48; Univ Mich, MS, 50; Clark Univ, PhD(chem), 53. *Prof Exp:* Asst, Univ Mich, 50; instr chem, Clark Univ, 50-52; from instr to assoc prof, Univ Ark, 52-60; res chemist, Electrochem Sect, Nat Bur Stand, 60-68; sect head high temperature electrochem, 68-74, RES CHEMIST, INORG & ELECTROCHEM BR & CONSULT, NAVAL RES LAB, 74- *Concurrent Pos:* Res assoc, Nat Acad Sci-Nat Res Coun, 59-60; mem fac, Grad Sch, NIH, 63-88. *Honors & Awards:* Turner Prize, Electrochem Soc, 51, Blum Award, 71. *Mem:* Am Chem Soc; Electrochem Soc; Royal Soc Chem; AAAS. *Res:* High temperature electrochemistry; molten salts; vaporization and thermal decomposition of inorganic salts. *Mailing Add:* Surface Chem Br Naval Res Lab Washington DC 20375-5000

STERN, LEO, b Montreal, Que, Jan 20, 31; m 55; c 4. PERINATAL BIOLOGY, CLINICAL PHARMACOLOGY. *Educ:* McGill Univ, BSc, 51; Univ Man, MD, 56; FRCPS(C), 64; Brown Univ, MA, 74. *Hon Degrees:* Dr, Univ Nancy, 77. *Prof Exp:* Demonstr, McGill Univ, 62-66, lectr, 66-67, from asst prof to assoc prof pediat, 67-73; PROF PEDIAT & CHMN DEPT, BROWN UNIV, 73- *Concurrent Pos:* Mead Johnson res fel, Karolinska Inst, Sweden, 58-59, Nat Res Coun Can med res fel, 59-60; Queen Elizabeth II scientist for res in dis of children, McGill Univ, 66-72; mem, Comn Study Perinatal Mortality, Prov of Que, 67-73; dir, Dept Newborn Med, Montreal Childrens Hosp, 69-73; pediatrician-in-chief, RI Hosp, 73- *Honors & Awards:* Queen Elizabeth II Res Scientist Award, 66. *Mem:* Perinatal Res Soc; Soc Pediat Res; Am Pediat Soc; Am Soc Clin Nutrit; Am Soc Clin Pharmacol & Therapeut. *Res:* Development pharmacology; perinatal biology, adaptation to extrauterine life, thermoregulation and bilirubin metabolism in the new born, respiratory adaptation in the normal and abnormal newborn infant. *Mailing Add:* 202 President Ave RI Hosp Providence RI 02906-5631

STERN, MARSHALL DANA, b New York, NY, Mar 18, 49; m 74; c 1. RUMEN MICROBIOLOGY, PROTEIN NUTRITION. *Educ:* State Univ NY, Farmingdale, AAS, 70; Cornell Univ, BS, 72; Univ RI, MS, 75; Univ Maine, PhD(animal nutrit), 77. *Prof Exp:* Res assoc fel, Univ Wis-Madison, 77-81; ASSOC PROF RUMINANT NUTRIT, UNIV MINN, ST PAUL, 81- *Concurrent Pos:* Ed, J Animal Sci, 82-84. *Mem:* Am Soc Animal Sci; Am Dairy Sci Asn; Nutrit Soc. *Res:* Protein (amino acid) requirements and nitrogen utilization in high producing dairy cows; metabolism of nutrients in gastro intestinal tract of ruminants; factors affecting fermentation and microbiol populations in the rumen. *Mailing Add:* Dept Animal Sci 130 Haecker Hall Univ Minn St Paul MN 55108

STERN, MARTIN, b New York, NY, Jan 9, 33; m 69; c 1. ORAL SURGERY. *Educ:* Harvard Univ, DMD, 56; Am Bd Oral Surg, dipl, 63. *Prof Exp:* ATTEND SURGEON IN CHG ORAL SURG, LONG ISLAND JEWISH MED CTR/QUEENS HOSP CTR AFFILIATION, 67-, ASSOC DIR DENT, 72-; PROF ORAL SURG, SCH DENT MED, STATE UNIV NY STONY BROOK, 71- *Concurrent Pos:* Asst clin prof, Sch Dent, Columbia Univ, 68-70. *Mem:* Am Dent Asn; Am Soc Oral Surg. *Mailing Add:* Dept Oral Surg Queens Hosp Ctr 82-68 164th St Jamaica NY 11432

STERN, MARVIN, b New York, NY, Jan 6, 16; m 42; c 3. PSYCHIATRY. *Educ:* City Col NY, BS, 35; NY Univ, MD, 39. *Prof Exp:* From fel to prof psychiat, 40-79, Menas S Gregory prof psychiat, 79-86, exec chmn dept, 76-86, ATTEND PSYCHIATRIST, UNIV HOSP, 52-, PROF, 86- *Concurrent Pos:* Consult, US Vet Admin Regional Off, Brooklyn, 51-66, Manhattan Vet Admin Hosp, 66-, & Brookdale Hosp, Brooklyn, 76-; assoc vis neuropsychiatrist, Bellevue Hosp, 52-62, vis neuropsychiatrist, 62- *Mem:* Psychosom Soc; Am Psychopath Asn; Am Psychiat Asn. *Res:* Psychosomatic medicine; altered brain function in organic disease. *Mailing Add:* 184 Rugby Rd Brooklyn NY 11226

STERN, MAX HERMAN, b Sioux City, Iowa, Mar 23, 20; m 46; c 1. ORGANIC CHEMISTRY. *Educ:* Morningside Col, BA, 41; Univ Wis, MS, 43, PhD(org chem), 45. *Prof Exp:* Asst, Univ Wis, 42-45; res chemist, Distillation Prod Industs, Eastman Kodak Co, 45-65, res assoc, Res Labs, 65-83; RETIRED. *Mem:* Am Chem Soc. *Res:* Vitamins A and E; carotenoids; soysterols; terpenes; photochemistry; photographic addenda. *Mailing Add:* 715 Winton Rd S Rochester NY 14618

STERN, MELVIN ERNEST, b New York, NY, Jan 22, 29; m 56; c 2. HYDRODYNAMICS. *Educ:* Cooper Union, BEE, 50; Ill Inst Technol, MS, 51; Mass Inst Technol, PhD(meteorol), 56. *Prof Exp:* From res assoc meteorol to physicist, Woods Hole Oceanog Inst, 51-64; PROF OCEANOG, GRAD SCH, UNIV RI, 64- *Concurrent Pos:* Guggenheim fel, 70-71. *Mem:* Fel Am Acad Arts & Sci, 75. *Res:* Oceanic circulation and turbulence; non-linear stability theory. *Mailing Add:* Dept Oceanog Fla State Univ Tallahassee FL 32306

STERN, MICHELE SUCHARD, b Chicago, Ill, Mar 17, 43; div; c 1. HYDROBIOLOGY, PLANT PHYSIOLOGY. *Educ:* Univ Ill, Urbana, BS, 64; Tenn Technol Univ, MS, 66; Tulane Univ, PhD(biol), 69. *Prof Exp:* Asst prof, 69-75, ASSOC PROF BIOL, UNIV MO-KANSAS CITY, 75- *Concurrent Pos:* Captain & nuclear med sci officer, US Army, Fort Sam, Houston, Tex, 82-85. *Mem:* Health Physics Soc; Am Chem Soc; Ecol Soc Am; Am Soc Limnol & Oceanog; Am Inst Biol Sci; Soc Environ Toxicol & Chem. *Res:* Isolation and characterization of plant proteases and their relationship to senescence; aquatic entomology and limnology; water pollution; aquatic toxicology; ecology; environmental sci. *Mailing Add:* Dept Biol Univ Mo Kansas City MO 64110

STERN, MIKLOS, b Budapest, Hungary, May 31, 57; US citizen. FIBER-OPTIC COMMUNICATIONS, DIGITAL DATA COMMUNICATION. *Educ:* Polytech Inst NY, BS, 81; Columbia Univ, MS, 82, PhD(elec eng), 90. *Prof Exp:* Mem tech staff, Bell Tel Labs, 82-84, MEM TECH STAFF, BELL COMMUN RES, 84- *Mem:* Inst Elec & Electronic Engrs. *Res:* High-speed digital fiber-optic communication systems; gigabit computer system based on Sonet/ATM protocols. *Mailing Add:* 5225 14th Ave Apt C5 Brooklyn NY 11219

STERN, MILTON, b Boston, Mass, Apr 20, 27; m 49; c 3. CHEMICAL ENGINEERING, PHYSICAL METALLURGY. *Educ:* Northeastern Univ, BS, 49; Mass Inst Technol, MS, 50, PhD(phys metall, corrosion), 52. *Prof Exp:* Res scientist, Metals Div, Union Carbide Corp, NY, 54-60, mgr res, Linde Div, Ind, 60-65, mgr mat res, NY, 65-67, mgr corp res 67, dir technol, Mat Systs Div, 67-68, vpres, Electronics Div, 68-69, exec vpres, Mining & Metals Div, 69-73; vpres, Kennecott Corp, 73-76, sr vpres, 76-78, exec vpres, 78-82; vchmn, Stauffer Chem Co, 83-86; CONSULT, 86- *Honors & Awards:* Willis R Whitney Award, 63. *Mem:* Nat Asn Corrosion Eng; Electrochem Soc; Am Inst Mining, Metall & Petrol Eng. *Res:* Electrochemistry; corrosion; kinetics; plasma technology; crystal growth; welding. *Mailing Add:* 53 Balfour Rd Palm Beach Gardens FL 33418

STERN, MORRIS, b St Louis, Mo, Nov 26, 30; m 52; c 2. COMPUTATIONAL MECHANICS, SOLID MECHANICS. *Educ:* Wash Univ, BS, 52; Univ Ill, MS, 57, PhD(eng mech), 62. *Prof Exp:* Teaching & res assoc theoret appl mech, Univ Ill, 56, asst prof, 62-66; assoc prof, 66-80, PROF ENG MECH, UNIV TEX, AUSTIN, 80- *Concurrent Pos:* Vis asst prof, Univ Colo, 65-66. *Mem:* Am Soc Eng Sci; Am Acad Mech. *Res:* Solid mechanics; continuum mechanics; computational fracture mechanics; boundary element methods. *Mailing Add:* Dept Aerospace Eng & Eng Mech Univ of Tex Austin TX 78712

STERN, PAULA HELENE, b New Brunswick, NJ, Jan 20, 38; m 59. PHARMACOLOGY. *Educ:* Univ Rochester, BA, 59; Univ Cincinnati, MS, 61; Univ Mich, PhD(pharmacol), 63. *Prof Exp:* Instr pharmacol, Univ Mich, 65-66; from asst prof to assoc prof, 66-77, PROF PHARMACOL, MED SCH, NORTHWESTERN UNIV, 77- *Concurrent Pos:* Fel pharmacol, Rochester Univ, 63-64 & Marine Biol Lab, Woods Hole, Mass, 64; res career develop award, NIH; consult, Food & Drug Admin; Am Inst Biol Sci/NASA mem, Gen Med B Study Sect, NIH. *Mem:* Am Soc Pharmacol & Exp Therapeut; Endocrine Soc; Asn Women Sci; Am Soc Bone & Mineral Res; Soc Exp Biol & Med. *Res:* Calcium metabolism; mechanisms of action of drugs and hormones on bone. *Mailing Add:* Dept Pharmacol Northwestern Univ Med Sch 303 E Chicago Ave Chicago IL 60611

STERN, RAUL A(RISTIDE), b Bucarest, Romania, Dec 26, 28; US citizen; m 53; c 2. PLASMA PHYSICS, GAS DYNAMICS. *Educ:* Univ Wis, BS, 52, MS, 53; Univ Calif, Berkeley, PhD(aeronaut sci), 59. *Prof Exp:* Res assoc, Univ Calif, Berkeley, 59-60; mem tech staff, Bell Labs, 60-81; PROF ASTROPHYS SCI & PHYS, UNIV COLO, BOULDER, 78- *Concurrent Pos:* Vis prof, New York Univ, 69-70, Univ Calif, Los Angeles, 77-78, Ctr Res Plasma Physics, Fed Polytech Sch, Lausanne, Switz, 82-83 & Univ Calif, Irvine, 85; vis res physicist, Univ Calif, Irvine, 75-, Ctr Res Plasma Physics, 83-, Ecole Polytechnique, Palaiseau, France, 86-; assoc ed, Physics of Fluids, 84-87; Sherman Fairchild Found distinguished scholar, Calif Inst Technol, 86. *Mem:* Fel Am Phys Soc. *Res:* Plasma waves, instabilities, transport, turbulence; shock and detonation waves; gas discharge physics; atmospheric physics; microwave and laser interactions with plasmas; diagnostics; nonlinear plasma properties. *Mailing Add:* Dept Astrophys Planetary & Atmospheric Sci Univ Colo Boulder CO 80309

STERN, RICHARD, b Paterson, NJ, Nov 27, 29; m 58, 80. ACOUSTICS. *Educ:* Univ Calif, Los Angeles, BA, 52, MS, 56, PhD(physics), 64. *Prof Exp:* Asst res physicist, Univ Calif, Los Angeles, 64-65, from asst prof to assoc prof eng, 66-76, prof, 76-, asst dean undergrad studies, 75-; AT APPL RES LAB, PA STATE UNIV. *Concurrent Pos:* Exchange fel, Imp Col, Univ London, 64-65. *Mem:* Fel Acoust Soc Am; Inst Elec & Electronics Engrs. *Res:* Experimentation in physical, engineering and medical acoustics. *Mailing Add:* 1150 Lindenhall Rd Boalsburg PA 16827

STERN, RICHARD CECIL, b New York, NY, Jan 4, 42; m 64; c 1. CHEMICAL PHYSICS, ISOTOPE SEPARATION. *Educ:* Cornell Univ, AB, 63; Harvard Univ, AM, 65, PhD(chem), 68. *Prof Exp:* From asst prof to assoc prof chem, Columbia Univ, 68-74; CHEMIST, LAWRENCE LIVERMORE LAB, UNIV CALIF, 74- *Mem:* Am Chem Soc; Am Phys Soc. *Res:* Laser isotope separation; photochemical kinetics; scattering and chemical reactions of low energy electrons; molecular beam and time-of-flight technology. *Mailing Add:* Lawrence Livermore Lab Univ of Calif Livermore CA 94550

STERN, RICHARD MARTIN, JR, b New York, NY, July 5, 48; m 88; c 1. AUDITORY PERCEPTION, AUTOMATIC SPEECH RECOGNITION. *Educ:* Mass Inst Technol, SB, 70, PhD(elec eng), 77; Univ Calif, Berkeley, MS, 72. *Prof Exp:* asst prof elec & biomed eng, 77-82, ASSOC PROF ELEC ENG, CARNEGIE MELLON UNIV, 82-, ADJ PROF, COMPUTER SCI, 88- *Concurrent Pos:* Vis prof, speech & commun sci, Nippon Telegraph & Telephone, Tokyo. *Mem:* Acoust Soc Am; Inst Elec & Electronic Engrs; Audio Eng Soc. *Res:* Auditory perception of binaural and monaural sounds and relation of psychoacoustical results to peripheral auditory physiology; computer recognition of speech sounds; computer generation and performance of music. *Mailing Add:* Dept Elec Eng Carnegie Mellon Pittsburgh PA 15213

STERN, ROBERT, b Bad Kreuznach, Ger, Feb 11, 36; US citizen; m 63; c 3. PATHOLOGY, BIOCHEMISTRY. *Educ:* Harvard Univ, BA, 57; Univ Wash, MD, 62. *Prof Exp:* USPHS officer, Nat Inst Dent Res, 63-65; sr scientist, Nat Inst Dent Res, 67-77; resident anat path, Nat Cancer Inst, 74-76; ASSOC PROF, DEPT PATH, UNIV CALIF, SAN FRANCISCO, 77- *Concurrent Pos:* Nat Cancer Inst spec fel, Weizmann Inst Sci, 65-67. *Mem:* AAAS; Am Soc Biol Chem; Am Soc Microbiol. *Res:* Transcriptional, translational controls in animal cells; translation of collagen messenger RNA; anatomic pathology; pathologic fibrosis. *Mailing Add:* Dept Path Univ Calif Box 0506 HSW 501 San Francisco CA 94143

STERN, ROBERT LOUIS, b Newark, NJ, Apr 10, 35; m 58; c 3. ORGANIC CHEMISTRY. *Educ:* Oberlin Col, AB, 57; Johns Hopkins Univ, MA, 59, PhD(org chem), 64. *Prof Exp:* Asst prof chem, Northeastern Univ, 62-65, assoc prof, 65-68; ASSOC PROF CHEM, OAKLAND UNIV, 68-, CO-CHMN DEPT, 74- *Mem:* AAAS; Am Chem Soc; Sigma Xi. *Res:* Organoanalytical chemistry; organic reaction mechanisms; biosynthesis; separation mechanisms of structurally related organic molecules; organic photochemistry. *Mailing Add:* Dept Chem Oakland Univ Rochester MI 48309

STERN, RONALD JOHN, b Chicago, Ill, Jan 20, 47; m 85; c 2. TOPOLOGY. *Educ:* Knox Col, BA, 68; Univ Calif, Los Angeles, MA, 70, PhD(math), 73. *Prof Exp:* Mem, Inst Advan Study, 73-74; instr math, Univ Utah, 74-76, asst prof, 76; mem, Inst High Sci Studies, 77-78; from assoc prof to prof math, Univ Utah, 79-89; PROF MATH, UNIV CALIF, IRVINE, 89- *Concurrent Pos:* Assoc prof math, Univ Hawaii, 79. *Mem:* Am Math Soc; Math Asn Am; AAAS. *Res:* Geometrical topology emphasizing the structure of topological manifolds especially of low dimension. *Mailing Add:* Dept Math Univ Calif Irvine CA 92717

STERN, SAMUEL T, b Buffalo, NY, May 27, 28; m 57; c 3. MATHEMATICS. *Educ:* Univ Buffalo, BA, 57, MA, 60; State Univ NY Buffalo, PhD(math), 62. *Prof Exp:* Instr math, Univ Buffalo, 58-62; asst prof, 62-65, PROF MATH, STATE UNIV NY COL BUFFALO, 65- *Concurrent Pos:* State Univ NY fac res fel, 67 & 70; assoc mathematician, Cornell Aeronaut Lab, summer 62, 63. *Mem:* Math Asn Am; Sigma Xi. *Res:* Noncommutative number theory; modern algebra; theory of skew groups and skew rings. *Mailing Add:* Dept Math State Univ NY 1300 Elmwood Ave Buffalo NY 14222

STERN, SILVIU ALEXANDER, b Bucharest, Romania, June 18, 21; US citizen; m 73; c 2. CHEMICAL ENGINEERING, PHYSICAL CHEMISTRY. *Educ:* Israel Inst Technol, BS, 45; Ohio State Univ, MS, 48, PhD(phys chem), 52. *Prof Exp:* Res assoc chem eng, Ohio State Univ, 52-55; res engr, Linde Div, Union Carbide Corp, NY, 55-58, group leader, 58-61, res supvr, 61-67; prof chem eng, 67-90, DONALD GAGE STEVENS DISTINGUISHED PROF MEMBRANE SCI & ENG, SYRACUSE UNIV, 90- *Concurrent Pos:* Acad vis chem dept, Imp Col Sci & Technol, London, 75; adj prof chem, 79-92, Col Environ Sci & Forestry, assoc mem, Inst Polymer Res, Col Environ Sci & Forestry, State Univ NY, 79-; bd dirs, NAm Membrane Soc, 85-89; assoc dir, Res Ctr Membrane Eng & Sci, Syracuse Univ, 87-; co-chmn, Orgn Comt, Int Cong Membranes & Membrane Processes, 89-90, sci & orgn comt, 85-87; sci adv, Fourth & Fifth Int Conf, Pervaporation Processes Chem Indust, 88-89 & 90-91. *Mem:* Am Chem Soc; fel Am Inst Chem Engrs; AAAS; Sigma Xi; Am Asn Univ Prof; Int Union Pure & Appl Chem; NAm Membrane Soc. *Res:* Transport phenomena in polymers; separation processes, particularly membrane separation processes; surface phenomena; biomedical engineering. *Mailing Add:* Dept Chem Eng & Mat Sci 320 Hinds Hall Syracuse Univ Syracuse NY 13244-1190

STERN, THEODORE, b Frankfurt am Main, Ger, Aug 27, 29; nat US; m 51; c 2. NUCLEAR ENGINEERING. *Educ:* Pratt Inst, BME, 51; NY Univ, MS, 56. *Prof Exp:* Engr, Foster Wheeler Corp, NY, 52-55, proj mgr, 55-56, head res reactor sect, 56-58; asst to tech dir, Atomic Power Dept, Westinghouse Elec Corp, 58, mgr adv develop, 58-59, mgr plant develop, 59-62, mgr, Projs Dept, 62-66, gen mgr, Pressurized Water Reactor Plant Div, 66-71, vpres & gen mgr, Nuclear Fuel Div, 71-72 & Water Reactor Div, 72-74, exec vpres, 74-90, SR EXEC VPRES, WESTINGHOUSE ELEC CORP, 90- *Mem:* Nat Acad Eng; Am Soc Mech Engrs; Am Nuclear Soc. *Res:* Application of nuclear energy to commercial generation of electric power. *Mailing Add:* Westinghouse Elec Corp 11 Stanwix St Pittsburgh PA 15222

STERN, THOMAS WHITAL, b Chicago, Ill, Dec 12, 22; m 55; c 1. GEOCHRONOLOGY. *Educ:* Univ Chicago, SB, 47; Univ Tex, MA, 48. *Prof Exp:* Geologist, US Geol Surv, 48-68, 71-88, chief, Isotope Geol Br, 68-71; RETIRED. *Mem:* AAAS; Geol Soc Am; Mineral Soc Am; Am Geophys Union; Geochem Soc. *Res:* Geochemistry; mineralogy; lead-uranium age determinations; isotope geology; autoradiography. *Mailing Add:* 2400 Foxhall Rd NW Washington DC 20007

STERN, VERNON MARK, b Sykeston, NDak, Mar 28, 23; m 47; c 2. ENTOMOLOGY. *Educ:* Univ Calif, Berkeley, BS, 49, PhD, 52. *Prof Exp:* Res asst entom, Univ Calif, Berkeley, 49-52; entomologist, Producers Cotton Oil Co, Ariz, 52-56; asst entomologist, 56-62, assoc prof, 62-68, PROF

ENTOM, UNIV CALIF, RIVERSIDE, 68-, ASSOC RES ENTOMOLOGIST, LAB NUCLEAR MED & RADIATION BIOL, LOS ANGELES, 66- *Concurrent Pos:* Collabr, USDA, 53-56; vpres, Ariz State Bd Pest Control, 53-56; coordr, Producers Agr Found, 54-56; NSF res grants, 61-69; Cotton Producers Inst res grant, 63-69; consult, US AEC, 65-66 & UN Food & Agr Orgn, 66-; Cotton Inst res grant, 69-; USDA res grant, 71-; int biol prog, NSF res grant, 72- *Mem:* Entom Soc Am; Ecol Soc Am; Sigma Xi. *Res:* Insect ecology; integrated control of arthropod pests; environmental radiation; radioecology; arthropods; population dynamics; insect migration and biology of insects. *Mailing Add:* Dept of Entom Univ of Calif Riverside CA 92521

STERN, W EUGENE, b Portland, Ore, Jan 1, 20; m 46; c 4. SURGERY. *Educ:* Univ Calif, AB, 41, MD, 43. *Prof Exp:* Instr neurol surg, Univ Calif, 51-52, chief neurosurg div, 52-84, from asst prof to prof, 52-87, chmn dept surg, 81-87, EMER PROF SURG, SCH MED, UNIV CALIF, LOS ANGELES, 87- *Concurrent Pos:* Consult, Los Angeles Vet Admin Hosp, 52-; vchmn, Am Bd Neurol Surgeons. *Mem:* AMA; Am Surg Asn; Am Asn Neurol Surg (pres, 78); Soc Neurol Surg (past pres, 76); Am Col Surgeons (secy). *Res:* Cerebral swelling; intracranial circulatory dynamics and intracranial mass dynamics. *Mailing Add:* Neurosurg Div Univ Calif Sch Med 405 Hilgard Ave Los Angeles CA 90024

STERN, WARREN C, b Bronx, NY, June 1, 44; m 65; c 3. NEUROPHARMACOLOGY, NEUROPHYSIOLOGY. *Educ:* Brooklyn Col, BS, 65; Ind Univ, PhD(psychopharmacol), 69. *Prof Exp:* Staff scientist, Worcester Found Exp Biol, 70-74; sr scientist, Squibb Inst, 74-75; sect head, Burroughs Wellcome, 75-81, dir clin neurosci, 81-84, dir new prod, 84-85; pres & chief exec officer, Pharmatec, 85-90; VPRES, CATO RES, 90- *Concurrent Pos:* Consult, var orgn, 75-90; dir, Innovet, 88-91 & Direct Therapeut, 91-; pres, Res Triangle Pharmaceut, 91- *Mem:* AAAS; NY Acad Sci; Am Soc Pharmacol & Exp Therapeut; Drug Info Asn. *Res:* Neuropharmacology; neurophysiology; neurobiology. *Mailing Add:* Cato Res 4364 S Alston Ave Durham NC 27713

STERN, WILLIAM, b Berlin, Germany, Jan 27, 46; m 74; c 1. BIOCHEMISTRY. *Educ:* NY Univ, BA, 67; Univ Mich, MS, 69, PhD(biochem), 72. *Prof Exp:* Asst, Pub Health Res Inst, 72-76, assoc, 76-83; scientist, Warner-Lambert, 83-84, sr scientist, 84-85; sr scientist, Organon Teknika, 85-86; CONSULT, 86- *Mem:* Am Acad Sci; Am Chem Soc. *Res:* Purify Endoproteases and develop inhibitors of them. *Mailing Add:* 113 Surrey Lane Tenafly NJ 07670

STERN, WILLIAM LOUIS, b Paterson, NJ, Sept 10, 26; m 49; c 2. PLANT ANATOMY. *Educ:* Rutgers Univ, BS, 50; Univ Ill, MS, 51, PhD(bot), 54. *Prof Exp:* From instr to asst prof wood anat, Sch Forestry, Yale Univ, 53-60; cur, Samuel James Record Mem Collection, 53-60; cur, Div Plant Anat, Smithsonian Inst, 60-64, chmn dept bot, 64-67; prof bot, Univ Md, College Park, 67-79, cur herbarium, 73-76; chmn dept, 79-85, PROF DEPT BOT, UNIV FLA, GAINESVILLE, 79- *Concurrent Pos:* Ed, Trop Woods, 53-60, Plant Sci Bull, 61-64 & Biotropica, 68-73; expert, UN Food & Agr Orgn, Philippines, 63-64; mem sci adv comt, Nat Trop Bot Garden, 69-80 & H P du Pont Winterthur Mus, 73-86; ed, Memoirs, Torrey Bot Club, 72-75; mem comt, Visit Arnold Arboretum, Harvard Univ, 72, vchmn comt, 73; prog dir syst biol, NSF, 78-79; bd dir, Am Inst Biol Sci, 88. *Honors & Awards:* Merit Award, Bot Soc Am, 87. *Mem:* Bot Soc Am (pres, 85, 86); Am Soc Plant Taxon (pres, 81); Am Inst Biol Sci; AAAS; fel Linnean Soc; Soc Econ Bot (treas, 88-91). *Res:* Plant anatomy and its relationship to systematic botany; plant morphology and phylogeny; orchid and wood anatomy and systematics tropical dendrology; natural history of tropical plants; history of botany and horticulture. *Mailing Add:* Dept Bot Univ Fla Gainesville FL 32611-2009

STERNBACH, DANIEL DAVID, b Montclair, NJ, May 28, 49; m 78; c 3. ORGANIC CHEMISTRY. *Educ:* Univ Rochester, BS, 71; Brandeis Univ, PhD(org chem), 76. *Prof Exp:* Swiss Nat Sci Found res asst, Swiss Fed Inst Technol, 76-77; res fel, Harvard Univ, 77-79; asst prof, Duke Univ, 79-86; SR SCIENTIST, GLAXO INC, 86-; ADJ PROF, UNIV NC, CHAPEL HILL, 88-; PRIN RES INVESTR, GLAXO INC, 88- *Mem:* Am Chem Soc. *Res:* Synthesis of interesting and biologically significant organic compounds and investigation of new synthetic methods. *Mailing Add:* Glaxo Inc Res Triangle Park Durham NC 27709

STERNBERG, ELI, mechanics; deceased, see previous edition for last biography

STERNBERG, HILGARD O'REILLY, b Rio de Janeiro, Brazil, July 5, 17; m 42; c 5. ALLUVIAL GEOMORPHOLOGY, ENVIRONMENTAL IMPACT OF ECONOMIC DEVELOPMENT. *Educ:* Univ Brazil, Rio de Janeiro, BA, 40, Licenciado, 41, Dr, 58; La State Univ, PhD(geog), 56. *Hon Degrees:* Dr, Univ Toulouse, France, 64. *Prof Exp:* Teaching asst geog, Col Pedro II, Rio de Janeiro, 38-42; prof, Catholic Univ, Rio de Janeiro, 41-44; asst prof, Univ Brazil, 42-44, prof, 44-64; prof, Inst Rio Branco, 47-56; prof, 64-88, EMER PROF GEOG, UNIV CALIF, BERKELEY, 88- *Concurrent Pos:* Consult, Nat Geog Coun, Brazil, 50-66; dir, Ctr Geog Studies, Brazil, 51-64; mem, UNESCO adv comt Arid Zone Res, 55-56; vis prof, Univ Heidelberg, 61, Columbia Univ, 63-64, Univ Peking, 84; consult, Max-Planck Soc, 75-78; mem, Comt Res Priorities Trop Biol, Nat Res Coun, Nat Acad Sci, 78-80. *Honors & Awards:* Nat Order Merit, Fed Repub Brazil, 56, Order Rio Branco, 67. *Mem:* Brazilian Acad Sci; Royal Geog Soc; Geog Soc Finland; Int Geog Union (vpres, 52-60); Leopoldina German Acad Natural Researchers; Soc Earth Sci Berlin; Serbian Soc Geog; fel AAAS. *Res:* Melding physico-biotic, cultural and historical geography; interface of human communities and their environments, and dysfunctions that occur on that interface, particularly as a consequence of development policies in the Neotropics, Brazil and Amazonia. *Mailing Add:* Dept Geog Univ Calif Berkeley CA 94720

STERNBERG, JOSEPH, b Brooklyn, NY, Nov 24, 21; m 46; c 4. FLUID MECHANICS, AERODYNAMICS. *Educ:* Calif Inst Technol, BS, 42, MS, 43; Johns Hopkins Univ, PhD(aeronaut), 55. *Prof Exp:* Res supvr supersonic flow, Calif Inst Technol, 43-46; aerodynamicist, US Army Ballistic Res Labs, 46-49, chief, Supersonic Wind Tunnels Br, 50-58 & Exterior Ballistics Lab, 58-62; consult aerodyn, Baltimore Div, Martin Marietta Corp, 62-63, mgr res & develop, 63-65, asst dir eng, 65-66, dir adv systs, Aerospace Hq, 66-70; sci adv to supreme allied comdr Europe, Supreme Hq Allied Powers Europe, 71-76; dir, Aerospace Group, Martin Marietta Corp, 76-; AT DEPT NAT SECURITY AFFAIRS, NAVAL POSTGRAD SCH. *Concurrent Pos:* Mem subcomt fluid mech, Nat Adv Comt Aeronaut, 50-58; mem res adv comt fluid mech, NASA, 58-62, mem res adv subcomt fluid mech, 67-69; mem consult panel, Chief Naval Opers, Opers Eval Group, 60-62; mem fluid dynamics panel, Adv Group for Aeronaut Res & Develop, NATO, 60-64; mem ground warfare panel, President's Sci Adv Comt, 69-70; mem, Army Sci Adv Panel, 70. *Honors & Awards:* Arthur S Flemming Award, 59. *Mem:* Assoc fel Am Inst Aeronaut & Astronaut; Am Phys Soc. *Res:* Boundary layer phenomena; shock wave reflections and structure; turbulent shear flows; missile and reentry systems. *Mailing Add:* Dept Nat Security Affairs Naval Postgrad Sch Monterey CA 93943

STERNBERG, MOSHE, b Marculesti, Rumania, Sept 3, 29; m 55; c 2. FOOD SCIENCE, BIOCHEMISTRY. *Educ:* Parhon Univ, Bucharest, Rumania, MS 52, PhD(org chem), 61. *Prof Exp:* Lab chief, Chem Pharmaceut Res Inst, Bucharest, Rumania, 55-57; res fel, Israel Inst Technol, 61-62; dir protein & carbohydrate res, Miles Labs, Inc, 62-80; vpres res & develop, Cutter Labs Inc, 80-86; VPRES RES & DEVELOP, CUTTER BIOLOGICAL/MILES INC, 86- *Mem:* Am Chem Soc; AAAS; Am Soc Microbiol. *Res:* Separation of industrial enzymes; proteins separation and characterization; human plasma proteins; biotechnology. *Mailing Add:* Cutter Labs Inc Res & Develop 4th & Parker St Berkeley CA 94710

STERNBERG, RICHARD WALTER, b Mt Pleasant, Iowa, Nov 21, 34; m 57; c 3. GEOLOGICAL OCEANOGRAPHY, MARINE SEDIMENTATION. *Educ:* Univ Calif, Los Angeles, BA, 58; Univ Wash, MSc, 61, PhD(oceanog), 65. *Prof Exp:* Assoc oceanog, Univ Wash, 63-65, res asst prof, 65-66; fel, Geomorphol Lab, Uppsala Univ, 66; res geophysicist, Univ Calif, San Diego, 67-68; asst prof, 68-73, assoc prof, 73-75, actg chmn dept, 78-79, PROF OCEANOG, UNIV WASH, 75- *Concurrent Pos:* adj assoc prof environ studies, Univ Wash, 73-90. *Mem:* Am Geophys Union. *Res:* Geological oceanography, especially processes of sediment transport and boundary-layer flow near the seafloor. *Mailing Add:* Sch Oceanog Univ of Wash Seattle WA 98195

STERNBERG, ROBERT JEFFREY, b Newark, NJ, Dec 8, 49; div; c 2. EXPERIMENTAL PSYCHOLOGY. *Educ:* Yale Univ, BA, 72; Stanford Univ, PhD(psychol), 75. *Prof Exp:* From asst prof to prof, 75-86, IBM PROF PSYCHOL & EDUC, YALE UNIV, 86- *Concurrent Pos:* prin investr Contracts Naval Res, 77-82, 85-88, Off Naval Res & Army Res Inst, 82-85, Army Res Inst, 85-90, Spencer Found, 82-84, 88-91, McDonnell Found, 87-90, Dept Educ, Off Educ Res & Improv, 90-; sr fac fel, Yale Univ, 82-83, Guggenheim fel, 85-86; consult, Air Force Off Sci Res, 86-88, Psychol Corp, 86-89, Harcourt Brace Jovanovich, 89-, Nat Comn Coop Educ, 90-; ed, Psychol Bull, 90- *Honors & Awards:* Cattell Award, Soc Multivariate Exp Psychol, 82. *Mem:* fel Am Psychol Asn; AAAS; Psychonomic Soc; Soc Multivariate Exp Psychol; Soc Res Child Develop. *Res:* Cognitive and contextual bases of human intelligence; proposal a triarchic theory, well supported by empirical data that accounts for many of the phenomena underlying intelligence. *Mailing Add:* Dept Psychol Yale Univ Box 11A Yale Sta New Haven CT 06520-7447

STERNBERG, ROBERT LANGLEY, b Newark, NJ, Apr 9, 22; m 50; c 3. MATHEMATICS, ENGINEERING. *Educ:* Northwestern Univ, BS, 46, MA, 48, PhD(math), 51. *Prof Exp:* Asst instrument engr, Clinton Labs, Manhattan Proj, 44-45, jr physicist, 45-46; asst math, Northwestern Univ, 46-50, lectr, 50-51; mathematician, Lab for Electronics, Inc, 51-63; mem, Inst Naval Studies, 63-66; chief adv study group & staff scientist, Res Dept, Elec Boat Div, Gen Dynamics Corp, 66-71; prof math, Univ RI, 71-73, lectr eng, 73-76, MATHEMATICIAN & SCI ADMINR, OFF NAVAL RES, BOSTON BR, 76- *Concurrent Pos:* Consult, Army Res Off, 63, McGill Univ, 65 & 67 & Sanders Assocs, 66; mathematician, Naval Underwater Systs Ctr, 72-73 & 74-76, consult, 76-; lectr math, Univ New Haven, 74-76; lectr statist, Univ Conn, 75. *Mem:* Am Math Soc; Sigma Xi; fel Brit Interplanetary Soc; Int Math Asn Comput & Simulation; Soc Indust & Appl Math. *Res:* Systems of differential equations and functional differential equations; applied mathematics; bennet functions and bang-bang control theory; microwave lens antennas; ocean resources; operations research; acoustic antennas; astronautics; thermo-nuclear deterrence; applied mathematics and ocean engineering. *Mailing Add:* 113 Seneca Dr Noank CT 06340

STERNBERG, STEPHEN STANLEY, b New York, NY, July 30, 20; m 58; c 2. PATHOLOGY, ONCOLOGY. *Educ:* Colby Col, BA, 41; NY Univ, MD, 44. *Prof Exp:* Resident path, Sch Med, Tulane Univ, 47-49; ATTEND PATH, MEM HOSP, 49-; PROF PATH, MED COL, CORNELL UNIV, 79- *Concurrent Pos:* Mem, Sloan-Kettering Inst, 49-; mem, Sci Adv Comt, Sch Med, Stanford Univ, 78-81; consult pathologist, Food & Drug Admin, 71-73 & NSF, Div Problem-Focused Res, 75-79; ed-in-chief, Am J Surg Path, 76-; ed, Human Path, 77-81; pres med bd, Mem Hosp, NY, 78-81, pres gen staff, 81-83; sci adv panel, Environ Protection Agency, 82-84; bd dir, Am Coun Sci & Health, 83-85; chmn bd dir, Am Coun Sci, 85-89; mem, NY Sci Policy Asn, 86-; mem of bd sci adv world health orgn collaborating ctr for the prev of colorectal cancer, 88- *Mem:* Am Asn Cancer Res; Int Acad Path; Soc Toxicol; NY Acad Med. *Res:* Carcinogenesis; surgical pathology; toxicology of cancer chemotherapeutic agents. *Mailing Add:* Mem Sloan-Kettering Cancer Ctr 1275 York Ave New York NY 10021

STERNBERG, VITA SHLOMO, b New York, NY, Jan 20, 36; m; c 5. MATHEMATICS. *Educ:* Johns Hopkins Univ, BA, 53, MA, 55, PhD, 56. *Hon Degrees:* Dr, Univ Mannheim, WGer, 90. *Prof Exp:* Vis fel, Inst Math Sci, NY Univ, 56-57; instr, Univ Chicago, 57-59; asst prof, 59-63, chmn dept, 75-78, PROF MATH, HARVARD UNIV, 63-, GEORGE PUTNAM PROF PURE & APPL MATH, 81- *Concurrent Pos:* Fel, Mortimer & Raymond Sackler Inst Advan Studies, Tel Aviv Univ, 80- *Mem:* Nat Acad Sci; Am Acad Arts & Sci. *Res:* Author of numerous technical publications. *Mailing Add:* Dept Math Harvard Univ Cambridge MA 02138

STERNBERG, YARON MOSHE, b Tel Aviv, Israel, May 26, 36; m 61; c 2. GROUND WATER HYDROLOGY. *Educ:* Univ Ill, BS, 61; Univ Calif, MS, 63, PhD(eng), 65. *Prof Exp:* From asst prof to assoc prof geol, Ind Univ, Bloomington, 65-70; assoc prof, 70-74, PROF CIVIL ENG, UNIV MD, COLLEGE PARK, 74- *Mem:* Am Soc Civil Engrs; Am Geophys Soc; Am Inst Mining, Metall & Petrol Engrs. *Mailing Add:* Dept of Civil Eng Univ of Md College Park MD 20742

STERNBERGER, LUDWIG AMADEUS, b Munich, Ger, May 26, 21; nat US; m 62. MEDICINE. *Educ:* Am Univ, Beirut, MD, 45. *Prof Exp:* Sr med bacteriologist, Div Labs & Res, State Dept Health, NY, 50-52, sr med biochemist, 52-53; asst prof med, Med Sch & assoc dir, Allergy Res Lab, Northwestern Univ, 53-55; chief path br, US Army Chem Res & Develop Labs, 55-67, chief basic sci dept, Med Res Labs, Army Chem Ctr, 67-77; asst prof microbiol, Sch Med, Johns Hopkins Univ, 66-77; prof brain res, Sch Med & Dent, Univ Rochester, 78-86; PROF NEUROL, PATH & ANAT, UNIV MD SCH MED, 86- *Concurrent Pos:* Fel exp path, Mem Cancer Ctr, New York, 48-50; assoc surg res, Sinai Hosp Baltimore, 66-; consult, Univ Iowa; neurosci investr award, 84. *Honors & Awards:* Paul A Siple Award, 72; Laureate of Alexander Von Humboldt Prize, 81. *Mem:* Soc Exp Biol & Med; Am Asn Immunologists; Am Acad Allergy; Histochem Soc; Am Soc Neurochem; Am Soc Neuropathologists. *Res:* Immunocytochemistry; neuroscience. *Mailing Add:* Univ Maryland Sch Med 22 S Greene St Baltimore MD 21201

STERNBURG, JAMES GORDON, b Chicago, Ill, Feb 22, 19; m 54; c 3. ENTOMOLOGY. *Educ:* Univ Ill, AB, 49, MS, 50, PhD(entom), 52. *Prof Exp:* Res assoc entom, 52-54, from asst prof to prof, 54-88, EMER PROF ENTOM, UNIV ILL, URBANA, 88- *Mem:* Entom Soc Am; Lepidop Soc. *Res:* Insect physiology and toxicology of insecticides; enzymatic detoxication of dichloro-diphenyl-trichloro-ethane by resistance house flies; effects of insecticides on neuroactivity in insects; behavior of nearctic and neotropical Lepidoptera; mimicry by insects. *Mailing Add:* Dept Entom-216 C Morrill Hall Univ Ill Urbana IL 61801

STERNER, CARL D, b Wellman, Iowa, Oct 15, 35; c 3. INORGANIC CHEMISTRY, PHYSICAL CHEMISTRY. *Educ:* Kearney State Col, BS, 60; Univ Tex, Austin, MA, 67; Univ Nebr, Lincoln, PhD(chem), 73. *Prof Exp:* Chem dept, 80-86, PROF CHEM, KEARNEY STATE COL, 87- *Concurrent Pos:* NSF fel, 71-72. *Mem:* Am Chem Soc. *Res:* Inorganic syntheses; solid state chemistry; electron spectroscopy chemical applications; materials science; GC/MS. *Mailing Add:* Dept Chem Univ Nebr Kearney NE 68849-0532

STERNER, JAMES HERVI, b Bloomsburg, Pa, Nov 14, 04; m 32, 71; c 3. MEDICINE. *Educ:* Pa State Univ, BS, 28, Harvard Univ, MD, 32; Am Bd Prev Med, dipl, 55; Am Bd Indust Hyg, dipl, 60. *Prof Exp:* Intern, Lankenau Hosp, Philadelphia, 32-34, chief resident physician, 34-35; dir lab indust med, Eastman Kodak Co, NY, 36-49, from assoc med dir to med dir, 49-68; assoc dean, 68-71, prof, 68-77, EMER PROF ENVIRON & OCCUP HEALTH, UNIV TEX SCH PUB HEALTH, HOUSTON, 76-; CLIN PROF OCCUP MED, UNIV CALIF, IRVINE, 76- *Concurrent Pos:* Instr indust med, Univ Rochester, 40-50, assoc prof med, Sch Med & Dent, 51-58, clin assoc prof, 58-68, clin assoc prof prev med, 59-61, clin prof prev med & community health, 61-68; med consult, Holston Ord Works, Tenn, 41-45; med dir, Clinton Eng Works, Tenn Eastman Co, 43-45; mem interim med adv bd, Manhattan Proj, AEC, 45-47, mem radiol safety sect & medico-legal bd, Oper Crossroads, 46, consult, Off Oper Safety, AEC, 48-, mem adv comt biol & med, 60-66 & gen adv comt, 71-74; mem comt toxicol, Nat Acad Sci-Nat Res Coun, 47-55 & 71-74; mem comt environ physiol, 65-68; mem expert adv comt social & occup health, WHO, 51-75, chmn expert comt med supvn in radiation work, 59; vis lectr, Harvard Med Sch, 52-56; mem comt occup health & safety, Int Labor Off, 52-75; chief indust med staff, Rochester Gen Hosp, NY, 55-62, consult, 63-68; trustee, Am Bd Prev Med, 55-67, vchmn occup med, 59-60, chmn 61-69; mem main comt, Nat Coun Radiation Protection & Measurements, 55-68; mem, Cancer Control Comt, Nat Cancer Inst, 57-61; mem adv comt, Nat Health Surv, 57-61; spec consult & chmn comt radiation studies, USPHS, 57-61, consult, Nat Ctr Health Statist, 66-75; mem, Gen Adv Comt Atomic Energy, NY, 59-65; sr assoc physician, Strong Mem Hosp, NY, 58-68; mem, Am Adv Bd, Am Hosp, Paris, 58-; chmn forum occup health, Nat Health Coun, 59, pres, 61; mem, Permanent Comn & Int Asn Occup Health, 60-; mem environ health panel, Exec Off Sci & Technol, 61-65; mem, Nat Environ Health Comt, 64-67; chmn, Nat Air Conserv Comn, 67; mem, Nat Adv Dis Prev & Environ Control Coun, 67-68; actg city health dir, Houston, Tex, 70; mem sci adv bd, Environ Protection Agency, 75-81. *Honors & Awards:* Cummings Award, Am Indust Hyg Asn, 55; Knudsen Award, Indust Med Asn, 57; Award, Am Acad Occup Med, 59. *Mem:* Fel Am Pub Health Asn; Am Col Prev Med (pres, 59-60); Am Acad Occup Med (pres, 52-53); fel Royal Soc Health; Am Chem Soc. *Res:* Clinical and experimental toxicology in industrial hygiene and environmental health. *Mailing Add:* 3354-0 Monte Hermoso Laguna Hills CA 92653

STERNER, ROBERT WARNER, b Elmhurst, Ill, Jan 15, 58; m 83; c 1. LIMNOLOGY, PHYCOLOGY. *Educ:* Univ Ill-Urbana, BS, 80, Univ Minn, PhD(ecology), 86. *Prof Exp:* ASST PROF BIOL, UNIV TEX, ARLINGTON, 88- *Concurrent Pos:* fel, Nat Sci Found, 80, Max Planck Inst, WGer, 87, Nat Sci Found/NATO, 88. *Mem:* Am Soc Limnol & Oceanog; Ecol Soc Am. *Res:* Ecological and physiological aspects of zooplankton/phytoplankton dynamics, combining herbivory and resource competition into one conceptual framework. *Mailing Add:* Dept Biol Univ Tex Box 19498 Arlington TX 76019

STERNFELD, LEON, b Brooklyn, NY, June 15, 13; m 34; c 2. MEDICAL ADMINISTRATION, RESEARCH ADMINISTRATION. *Educ:* Univ Chicago, SB, 32, MD, 36, PhD(biochem), 37; Columbia Univ, MPH, 43. *Prof Exp:* Intern pediat, Johns Hopkins Univ, 38-39 & Sydenham Hosp, 39-40; asst res, Jewish Hosp, Brooklyn, 40-41; epidemiologist-in-training, State Dept Health, NY, 41-42, jr epidemiologist, 42, asst dist state health officer, 43-44, dir med rehab, 44-50, dist health officer, 50-51; asst dir, Tuberc Div, State Dept Pub Health, Mass, 51-52; assoc dir, Field Training Unit, Harvard Univ, 52-53, lectr, Sch Pub Health, 53-57, from asst clin prof to assoc clin prof maternal & child health, 58-70, vis lectr maternal & child health, 70-74; MED DIR, UNITED CEREBRAL PALSY ASNS, INC, 71- *Concurrent Pos:* Chief Pub Health Admin, Korean Civil Asst Command, 53-55; chief prev med, Ft Devons, US Army, 55; City health comnr, Cambridge, Mass, 55-61; lectr, Simmons Col, 56-69; assoc physician, Children's Med Ctr, Boston, 57-69; dep health comnr, Mass Dept Pub Health, 61-69. *Mem:* Am Pub Health Asn; Am Asn Ment Deficiency; Am Acad Cerebral Palsy & Develop Med. *Res:* Chemical properties of essential bacterial growth factor; essential fructosuria; pathophysiology; medical and public health aspects of cerebral palsy and mental retardation; public health methodology and community health. *Mailing Add:* 1385 York Ave New York NY 10021

STERNFELD, MARVIN, b Cleveland, Ohio, Feb 24, 27; m 50; c 3. ORGANIC CHEMISTRY. *Educ:* Western Reserve Univ, BS, 49, MS, 51, PhD(chem), 53. *Prof Exp:* Pres, Cleveland Chem Labs, 53-66; PRES, RES ORGANICS INC, 66- *Concurrent Pos:* Head, Chem Dept, Ohio Col Podiat Med, 53-69. *Mem:* Am Chem Soc; Sigma Xi. *Res:* Synthesis of biochemicals, including biological buffers, fluorescent labels, brain research biochemicals, enzyme substrates and test reagents; amino acid derivatives, peptides, special biological dyes. *Mailing Add:* 4353 E 49th St Cuyahoga Heights OH 44125

STERNGLANZ, ROLF, b Sewell, Chile, May 18, 39; US citizen; m 64. MOLECULAR BIOLOGY, BIOCHEMISTRY. *Educ:* Oberlin Col, AB, 60; Harvard Univ, PhD(phys chem), 67. *Prof Exp:* NIH res fel biochem, Sch Med, Stanford Univ, 66-68; asst prof, 69-76, ASSOC PROF BIOCHEM, STATE UNIV NY STONY BROOK, 76- *Concurrent Pos:* Am Cancer Soc res grants, State Univ NY Stony Brook, 69- *Mem:* AAAS. *Res:* Mechanism of DNA replication; physical chemistry of DNA. *Mailing Add:* Dept Biochem State Univ NY Health Sci Ctr Stony Brook NY 11794

STERNGLASS, ERNEST JOACHIM, b Berlin, Ger, Sept 24, 23; nat US; m 57; c 2. PHYSICS. *Educ:* Cornell Univ, BEE, 44, MS, 51, PhD(eng physics), 53. *Prof Exp:* Asst physics, Cornell Univ, 44, res assoc, 51-52; physicist, US Naval Ord Lab, 46-52; res physicist, Res Labs, Westinghouse Elec Co, Pa, 52-60, adv physicist, 60-67; prof radiation physics, Univ Pittsburgh, 67-80, EMER PROF RADIOLOGICAL PHYSICS, 80- *Concurrent Pos:* Assoc, George Washington Univ, 46-47; Westinghouse Res Lab fel, Inst Henri Poincare, Paris, 57-58; vis prof, Inst Theoret Physics, Stanford Univ, 66-67. *Mem:* AAAS; fel Am Phys Soc; Am Astron Soc; Fedn Am Sci; Am Asn Physicists in Med. *Res:* Secondary electron emission; physics of electron tubes; electron and elementary particle physics; electronic imaging devices for astronomy and medicine; radiation physics; biological effects of radiation. *Mailing Add:* 170 West End Ave Apt 27 H New York NY 10023

STERNHEIM, MORTON MAYNARD, b Scranton, Pa, July 19, 33; m 55; c 3. PHYSICS. *Educ:* City Col, NY, BS, 54; NY Univ, MS, 56; Columbia Univ, PhD(physics), 61. *Prof Exp:* Fel physics, Brookhaven Nat Lab, 61-63; res assoc lectr, Yale Univ, 63-65; from asst prof to assoc prof, 65-71, PROF PHYSICS, UNIV MASS, 71- *Concurrent Pos:* Prin investr, NSF res grant, 65; consult, Las Alamos Nat Lab, 67-; visitor, Brookhaven Nat Lab, 66 & 68. *Mem:* Am Phys Soc; Am Asn Physics Teachers. *Res:* Theoretical nuclear physics; pion scattering and production; exotic atoms; incoherent processes involving nucleons and mesons at intermediate energies; nucleon-nucleon interactions; skyrmion models. *Mailing Add:* Dept Physics & Astron Univ Mass Amherst MA 01003

STERNHEIMER, RUDOLPH MAX, b Saarbruecken, Ger, Apr 26, 26; nat US; m 52. ATOMIC PHYSICS, NUMBER THEORY. *Educ:* Univ Chicago, BS, 43, MS, 46, PhD, 49. *Prof Exp:* Jr scientist, Div War Res, Metall Lab, Columbia Univ, 45-46; instr physics, Univ Chicago, 46-48; asst & instr, Yale Univ, 48-49; mem staff, Los Alamos Sci Lab, 49-51; from assoc physicist to physicist, 52-65, SR PHYSICIST, BROOKHAVEN NAT LAB, 65- *Mem:* Fel Am Phys Soc; Math Asn Am; Fibonacci Asn. *Res:* Atomic and nuclear physics; theory of solids; theory of nuclear quadrupole coupling; theory of ionization loss and Cerenkov radiation; focusing magnets; polarization of nucleons; theory of meson production; electronic polarizabilities of ions; k-ordering of atomic and ionic energy levels; problems in number theory. *Mailing Add:* Dept Physics 510A Brookhaven Nat Lab Upton NY 11973

STERNICK, EDWARD SELBY, b Cambridge, Mass, Feb 10, 39; m 60; c 3. MEDICAL PHYSICS. *Educ:* Tufts Univ, BS, 60; Boston Univ, MA, 63; Univ Calif, Los Angeles, PhD(med physics), 68; Northeastern Univ, MBA, 85. *Prof Exp:* Res scientist biophys, Nat Aeronaut & Space Admin, 63-64; instr radiol, Dartmouth-Hitchcock Med Ctr, 68-72, asst prof clin med, 72-78; assoc clin prof therapeut radiol, 78-87, CLIN PROF RADIATION ONCOL & DIR MED PHYSICS DIV, TUFTS-NEW ENGLAND MED CTR, 87- *Concurrent Pos:* Prof, NH Voc Tech Inst, 72-78; adj asst prof bioeng, Thayer Sch Eng, Dartmouth Univ, 73-78; consult, Vet Admin Hosp, 74- *Mem:* Fel Am Asn Physicists in Med; Soc Nuclear Med; Sigma Xi; Health Physics Soc; fel Am Col Med Physics; fel Am Col Radiol. *Res:* Application of computer technology to radiation medicine. *Mailing Add:* Dept of Radiation Oncol 750 Washington St Boston MA 02111

STERNLICHT, B(ENO), b Poland, Mar 12, 28; nat US; m 55, 75; c 3. ENERGY, PROPULSION. *Educ:* Union Col, BSEE, 49; Columbia Univ, PhD(appl mech), 54. *Hon Degrees:* DSc, Union Col, 70. *Prof Exp:* Engr, Gen Elec Co, 47-50; off mgr, Ameast Distribr Corp, 50-51; develop engr, Gen Elec Co, 51-53; res engr, Atomic Energy Comn, 53-54; eng specialist, Gen Eng Lab, Gen Elec Co, 54-58, consult engr hydrodyn, 58-61; tech dir & chmn bd,

Mech Technol Inc, 61-85; PRES, AMEAST TRADING CORP, 76-; PRES, BENJOSH MGT CORP, 80- *Concurrent Pos:* Lectr, Union Col, 56-57 & Mass Inst Technol, 57, 59; mem mat panel, Nat Acad Sci; founder sci based indust, Mamash, Israel; pres & chmn bd, Vols in Tech Assistance, 66-72; chmn comt power & propulsion, NASA; adv to energy, Pres Carter & Pres Reagan; bd mem, Vol in Tech Assistance; mem, Energy Policy Task Force, 81. *Honors & Awards:* Mach Design Award, Am Soc Mech Engrs, 66. *Mem:* Nat Acad Eng; Am Soc Lubrication Engrs; Inst Elec & Electronics Engrs; Am Inst Aeronaut & Astronaut; Am Soc Mech Engrs. *Res:* Turbomachinery; energy conversion; conservation diagnostic systems; automotive propulsion; separation systems; energy conversion, propulsion, separation and enrichment systems. *Mailing Add:* 123 Partridge Run Schenectady NY 12309

STERNLICHT, HIMAN, b New York, NY, May 31, 36; m 58; c 3. PHYSICAL CHEMISTRY. *Educ:* Columbia Univ, BA, 57, BS, 58; Calif Inst Technol, PhD(chem), 63. *Prof Exp:* Mem tech staff, Bell Tel Labs, NJ, 63-65; asst prof chem, Univ Calif, Berkeley, 65-70; mem tech staff, Bell Labs, Inc, 70-76; ASSOC PROF PHARMACOL, CASE WESTERN RESERVE UNIV, 76- *Concurrent Pos:* NIH res grant, 66-69. *Mem:* Am Phys Soc; Am Chem Soc. *Res:* Magnetic resonance studies, including small and macromolecular systems. *Mailing Add:* Dept Pharmacol Case Western Reserve Univ 2119 Abington Rd Cleveland OH 44106

STERNLIEB, IRMIN, b Czernowitz, Rumania, Jan 11, 23; US citizen; m 53. GASTROENTEROLOGY, ELECTRON MICROSCOPY. *Educ:* Univ Geneva, MSc, 49, MD, 52. *Prof Exp:* Intern, Morrisania City Hosp, Bronx, NY, 52-53; resident internal med, Bronx Munic Hosp Ctr, NY, 55-57; asst instr, 56-57, instr & assoc, 57-61, from asst prof to assoc prof, 61-72, PROF MED, ALBERT EINSTEIN COL MED, 72- *Concurrent Pos:* Fel internal med & gastroenterol, Mt Sinai Hosp, New York, 53-55; USPHS fel, 57-60 & spec fel, Lab Atomic Synthesis & Proton Optics, Ivry, France, 64-65. *Mem:* Am Soc Clin Invest; Int Asn Study Liver; Am Gastroenterol Asn; Am Asn Study Liver Dis; Am Soc Cell Biol. *Res:* Clinical, genetic, biochemical, diagnostic and morphologic aspects of human and canine inherited copper toxicosis; electron microscopy of human liver. *Mailing Add:* Albert Einstein Col of Med 1300 Morris Park Ave Bronx NY 10461

STERNLING, CHARLES V, b Pocatello, Idaho, Nov 15, 24. ENGINEERING RESEARCH. *Prof Exp:* Sr res assoc, Shell Develop Co, 49-89; RETIRED. *Mem:* Nat Acad Eng; Am Inst Chem Engrs; Am Chem Soc. *Mailing Add:* 1400 Stony Lane North Kingstown RI 02852

STERNSTEIN, MARTIN, b Chicago, Ill, Apr 25, 45; m 80; c 2. MATHEMATICS. *Educ:* Univ Chicago, BS, 66; Cornell Univ, PhD(math), 71. *Prof Exp:* From asst prof to assoc prof, 70-82, chmn dept, 72-76, 81-83, PROF MATH, ITHACA COL, 82- *Concurrent Pos:* Vis lectr, Col VI, 78-79; Fulbright prof, Univ Liberia, 79-80, 83-84. *Mem:* Am Math Soc; Math Asn Am. *Res:* Algebraic topology. *Mailing Add:* Dept of Math Ithaca Col Ithaca NY 14850

STERNSTEIN, SANFORD SAMUEL, b New York, NY, June 19, 36; m 58; c 2. POLYMER PHYSICS, POLYMER ENGINEERING. *Educ:* Univ Md, BS, 58; Rensselaer Polytech Inst, PhD(chem eng), 61. *Prof Exp:* From asst prof to prof polymers, 61-73, WILLIAM WEIGHTMAN WALKER PROF POLYMER ENG, RENSSELAER POLYTECH INST, 73- *Concurrent Pos:* NSF & Inst Paper Chem Pioneering Res grants, 63-65; Nat Inst Dent Res grant, 65-70; NSF res grants, 73-78 & NASA grants, 79- *Mem:* Am Inst Chem Eng; Soc Rheol; fel Am Phys Soc; Am Chem Soc. *Res:* Rheology; fracture; dynamic mechanical properties of polymers; polymer-solvent interactions and crazing; polymer network mechanics and rubber elasticity; composites. *Mailing Add:* Dept Polymer Eng Rensselaer Polytech Inst Troy NY 12181

STERPETTI, ANTONIO VITTORIO, b Rome, Italy, Jan 11, 56; Italian. CARDIOVASCULAR SURGERY, VASCULAR SURGERY. *Educ:* Liceo A Torlonia, Avezzano, Italy, BS, 74; Univ Rome, MD, 80, Gen Surg, 85. *Prof Exp:* Asst res surg, clinical surgery, Univ Rome, 80-85; cardiovasc fel surg, Creighton Univ, Omaha, Nebr, 84-88, res gen surg, 88; MEM, IA CLINICA CHIRURGICA, 88- *Mem:* Italian Soc Surg Res (pres, 81-); Assoc Acad Surg; Italian Soc Surg; Am Med Assoc; Am Col Chest Physicians. *Res:* Cardiovascular surgery; development of a new small arteries vascular graft. *Mailing Add:* Ia Clinica Chirurgica Policlinico Umberto I Viale del Policlinico Rome 00167 Italy

STERRETT, ANDREW, b Pittsburgh, Pa, Apr 3, 24; m 48; c 2. MATHEMATICS. *Educ:* Carnegie Inst Technol, BS, 48; Univ Pittsburgh, MS, 50, PhD(math), 56. *Prof Exp:* Lectr math, Univ Pittsburgh, 48-50; instr, Ohio Univ, 50-53; from asst prof to assoc prof, 53-65, chmn dept math, 60-63 & 65-68, dir comt undergrad prog in math, 70-72, dean col, 73-78, PROF MATH, DENISON UNIV, 65- *Concurrent Pos:* NSF fac fel statist, Stanford Univ, 59-60; vis scholar statist, Univ Calif, Berkeley, 66-67 & Univ NC, 78-79; dir comt on undergrad prog in math, Math Asn Am, 70-72. *Mem:* Am Math Soc; Math Asn Am; Am Statist Asn; Sigma Xi; Nat Coun Teachers Math. *Mailing Add:* Dept Math Denison Univ Granville OH 43023

STERRETT, FRANCES SUSAN, b Vienna, Austria, Sept 25, 13; nat US; wid; c 2. ENVIRONMENTAL CHEMISTRY. *Educ:* Univ Vienna, PhD(chem), 38. *Prof Exp:* Res chemist, Lab, France, 38-39; asst biochem, Med Ctr, Columbia Univ, 39-40; res chemist, van Ameringen & Haebler, Inc, NJ, 40-41, Woburn Degreasing Co, 43 & Fritzsche Bros, Inc, NY, 43-49; lectr, 53-57, from instr to prof, 57-85, EMER PROF CHEM, HOFSTRA UNIV, 85- *Concurrent Pos:* Lectr biochem sec high sch teachers, NSF, 65-67; co-ed, Ann NY Acad Sci, 82, ed, 83; chmn, Coord Comt Pub Affairs, NY sect, Am Chem Soc, mem, Environ Chem Comt; chmn, Environ Sci sect & Sci & Pub Policy sect, NY Acad Sci, mem Sci & Soc Comt; consult ground water qual probs, qual control of water analysis reagents, textbk eval; adj prof chem, Hofstra Univ, 86- *Mem:* AAAS; Am Chem Soc; fel Am Inst Chemists; Sigma Xi; fel NY Acad Sci. *Res:* Chemistry and chemical reactions in the environment in reference to the atmosphere, hydrosphere, lithosphere and biosphere; microanalysis; aromatic chemicals; essential oils; inorganic and qualitative chemistry; organic chemistry; quantitative analysis; chemical contaminants in drinking water. *Mailing Add:* 64 Hathaway Dr Garden City NY 11530

STERRETT, JOHN PAUL, b Springfield, Ohio, Dec 14, 24; m 49; c 2. PLANT PHYSIOLOGY. *Educ:* Univ WVa, BS, 50; Va Polytech Inst, MS, 61, PhD(plant physiol), 66. *Prof Exp:* County forester, WVa Conserv Comn, 50-53; forester, Bartlett Tree Expert Co, 53-59; res asst plant physiol, Va Polytech Inst, 59-61, asst prof, 61-69; plant physiologist, Veg Control Div, Ft Detrick, US Army, 69-74; plant physiologist, Sci Res Lab, Agr Res Serv, USDA, 74-86, plant physiologist, Foreign Dis-Weed Res Lab, 86-90; CONSULT PLANT PHYSIOLOGIST, 90- *Mem:* Plant Growth Regulators Soc Am. *Res:* Plant growth regulators for the control of weeds; determine physiological responses of woody plants to growth inhibitors. *Mailing Add:* 1935 Beaver Dam Rd Union Bridge MD 21791

STERRETT, KAY FIFE, b McKeesport, Pa, May 20, 31; m 60; c 3. GEOPHYSICS, SIGNAL PROCESSING. *Educ:* Univ Pittsburgh, BS, 53, PhD(phys chem), 57; Dartmouth Col, MS(eng sci), 85. *Prof Exp:* Asst, Univ Pittsburgh, 53-57; phys chemist, Nat Bur Stand, 57-61; mem res staff, Northrop Space Labs, 62-64, head, Space Physics & Chem Lab, 64-66; head phys chem lab, Northrop Space Labs, 66-67; CHIEF, RES DIV, US ARMY COLD REGIONS RES & ENG LAB, 67- *Concurrent Pos:* Neth Govt fel, Kamerlingh Onnes Lab, Univ Leiden, 57-58; mem exten teaching staff, Univ Calif, Los Angeles, 65-66; mem Army res coun, Dept Army, 67-68; spec asst to cmndg gen, US Army Elec Res & Develop Command, 77; chief, Eng Div, US Army Cold Regions Res & Eng Lab, 81-, tech dir, 86. *Mem:* Fel AAAS; Am Chem Soc; Am Phys Soc; Royal Soc Chem; Sigma Xi; Inst Elec & Electronics Engrs. *Res:* Technical management; digital signal processing; automatic control; low temperature physics; cold regions environment; physics of snow, ice and frozen soil; heat transfer. *Mailing Add:* HQ DA SARD-TT Pentagon Rm 3E426 Washington DC 20310

STERZER, FRED, b Vienna, Austria, Nov 18, 29; nat US; m 64. ELECTRONICS ENGINEERING. *Educ:* City Col New York, BS, 51; NY Univ, MS, 52, PhD(physics), 55. *Prof Exp:* mem staff, RCA Corp, 54-87, dir, Microwave Technol Ctr, RCA Labs, 72-87; dir, med syst, David Sorwoff Res Ctr, 87-88; PRES, MMTC, INC, 88- *Mem:* Nat Acad Eng; fel Inst Elec & Electronics Engrrs; Am Phys Soc. *Res:* Microwave spectroscopy, tubes and solid state devices; microwave lujjrer thermion treatment of cancer. *Mailing Add:* MMTC Inc 12 Roszel Rd Suite A-203 Princeton NJ 08540

STESKY, ROBERT MICHAEL, b Toronto, Ont, July 27, 45; m 71; c 2. GEOLOGY, GEOPHYSICS. *Educ:* Univ Toronto, BSc, 68, MSc, 70; Mass Inst Technol, PhD(geophys), 75. *Prof Exp:* ASST PROF GEOL, ERINDALE COL, UNIV TORONTO, 74- *Mem:* Am Geophys Union; Geol Asn Can. *Res:* Physical properties of rocks and minerals under high pressure; fractures and faults and igneous and metamorphic rocks. *Mailing Add:* Earth & Planetary Sci Rm 3032 Erindale Col Univ Toronto 3359 Mississauga Rd Mississauga ON L5L 1C6 Can

STETLER, DAVID ALBERT, b Pasadena, Calif, June 17, 35; m 64; c 2. PLANT CYTOLOGY. *Educ:* Univ Southern Calif, BSc, 59; Univ Calif, Berkeley, PhD(bot), 67. *Prof Exp:* Asst prof bot, Univ Minn, 67-69; asst prof biol, Dartmouth Col, 69-73; asst prof, 73-77, ASSOC PROF BOT, VA POLYTECH INST & STATE UNIV, 77- *Mem:* Am Soc Plant Physiologists; Bot Soc Am; Tissue Cult Asn; Int Asn Plant Tissue Cult. *Res:* Organelle development in plant cells; ultrastructure of plant tissues; ultrastructure of stressed animal tissues in the environment. *Mailing Add:* Dept Botany Va Polytech Inst Blacksburg VA 24061

STETLER, DEAN ALLEN, b Beloit, Kans, Nov 25, 54; m 72; c 3. MOLECULAR BIOLOGY. *Educ:* Univ Kans, BA, 76, PhD(microbiol), 80. *Prof Exp:* Res assoc molecular pharmacol, Pa State Col Med, 80-82, asst prof, 82-85; asst prof biochem & molecular biol, 85-89, ASSOC PROF BIOCHEM, UNIV KANS, 89-, CHMN, GENETICS PROG, 86- *Concurrent Pos:* Consult, Immunodiagnostics & Molecular Genetics. *Res:* Control of gene transcription; role of poly(A) polymerase in ultimate gene expression; role of protein phosphorylation in formation of autoimmunogenic nuclear proteins in rheumatic disease. *Mailing Add:* Dept Bio Sci 3042 HAW Univ Kans Lawrence KS 66045

STETSON, ALVIN RAE, b San Diego, Calif, July 23, 26; m 47; c 2. HIGH-TEMPERATURE COATINGS. *Educ:* San Diego State Col, AB, 48. *Prof Exp:* Anal chemist, 48-50, phys chemist, 50-53, from staff engr to sr res staff engr, 53-66, chief process res, 66-72, chief mat engr, 72-80, chief mat technol, 80-85, CONSULT, HIGH TEMPERATURE, SOLAR TURBINE INC, CATERPILLAR TRACTOR CO, 85-, CHIEF SCIENTIST, ADV COATINGS & APPLN, 89- *Mem:* Am Chem Soc; Nat Asn Corrosion Eng; Am Soc Metals. *Res:* Fused salt plating; high temperature metallic and ceramic protective coatings; reaction of materials at high temperatures; reentry and gas turbine environment simulation; braze joining of dissimilar metals; plasma arc testing and spraying; abrasive and abradable turbine tip seals; materials research supervision; thirteen US patents in the proctive field. *Mailing Add:* 4834 Lucille Dr San Diego CA 92115

STETSON, HAROLD W(ILBUR), b Bristol, Pa, July 2, 26; m 52; c 3. CERAMICS, INORGANIC CHEMISTRY. *Educ:* Pa State Univ, BS, 50, MS 52, PhD(ceramics), 56. *Prof Exp:* Sr engr, Corning Glass Works, 56-59 & Radio Corp Am, 59-62; sr engr, Western Elec Co, Princeton, 62-66, res leader ceramics, 66-69; dir res, Ceramic Metal Systs, Inc, 69-70; sr engr, Eastern Res Labs, TRW, 70-86; PRES, CERAMIC SCI ASSOC, 86- *Concurrent Pos:* Vis assoc prof, Rutgers Univ, 71. *Honors & Awards:* S J Geijsbeek Award, Am Ceramic Soc, 88; Fel, Am Inst Chemists. *Mem:* Am Ceramic Soc; Nat Inst Ceramic Engrs. *Res:* Ceramic materials for electronic uses; sintering theory of oxides; application of modern ceramic technology to archeological problems. *Mailing Add:* 222 N Chancellor St Newtown PA 18940

STETSON, KARL ANDREW, b Gardener, Mass, Oct 16, 37; m 59. HOLOGRAM INTERFEROMETRY, VIBRATION THEORY. *Educ:* Lowell Technol Inst, BSEE, 59; Univ Mich, MSE, 60; Royal Inst Technol, Stockholm, Sweden, PhD(phys optics), 69. *Prof Exp:* Elec engr, Bell Aerosysts, 61-62; res assoc, Inst Sci & Technol, Univ Mich, 62-65; engr, GCA, 66-67; prin res fel, Nat Phys Lab, UK, 69-71; res scientist, Ford Motor Co Res Labs, 71-73; SR SCIENTIST, UNITED TECHNOL RES CTR, 73- *Concurrent Pos:* Privat docent, Royal Inst Technol, 69-; adj prof, Worcester Polytech Inst, 80- *Honors & Awards:* Hefenji Award, Soc Exp Mech, 79, B J Lazan Award, 83. *Mem:* Optical Soc Am; Soc Exp Mech. *Res:* Coherent optical metrology with emphasis on holography and laser methods. *Mailing Add:* Photo & Appl Physics MS-92 United Technol Res Ctr East Hartford CT 06108

STETSON, KENNETH F(RANCIS), b Winthrop, Maine, May 2, 24; m 48; c 2. AERODYNAMICS, AEROPHYSICS. *Educ:* Univ Maine, BS, 49; US Air Force Inst Technol, BS, 52; Ohio State Univ, MS, 56. *Prof Exp:* Proj engr, Aircraft Lab, Wright-Patterson AFB, Ohio, 50-51, aeronaut engr, Aeronaut Res Labs, 52-56; sr staff scientist, Everett Res Lab, Avco Corp, Mass, 56-61, sr proj engr, Res & Develop Div, 61-66, sr staff scientist, Space Systs Div, 66-68; Ohio State Univ Res Found vis res assoc, 68-69, aerospace engr, Aerospace Res Labs, 69-75, AEROSPACE ENGR, FLIGHT DYNAMICS LAB, WRIGHT-PATTERSON AFB, 75- *Mem:* Assoc fel Am Inst Aeronaut & Astronaut. *Res:* Hypersonic, aerodynamic and aerophysics research. *Mailing Add:* 7104 Hartcrest Lane Centerville OH 45459

STETSON, MILTON H, b Springfield, Mass, Nov 25, 43; m 66; c 2. REPRODUCTIVE ENDOCRINOLOGY, BIOLOGICAL RHYTHMS. *Educ:* Cent Conn State Col, BA, 65; Univ Wash, MS, 68, PhD(zool), 70. *Prof Exp:* NIH fel, 66-70; res fel reproduction, Univ Tex, Austin, 71-73; from asst prof to assoc prof biol & health sci, 73-80, prof life & health sci & assoc dir, grad progs & res, 81-86, DIR, SCH LIFE & HEALTH SCI, UNIV DEL, 87- *Mem:* Am Physiol Soc; Am Soc Zool; fel AAAS; Soc Study Reproduction; Am Asn Univ Profs; Endocrine Soc; Soc Res Biol Rhythms. *Res:* Role of the circadian system in the timing of reproductive events; neural and neuroendocrine generation of female reproductive cyclicity; comparative endocrinology of the thyroid gland; ontogeny of puberty; pineal physiology. *Mailing Add:* Sch of Life-Health Sci Univ of Del Newark DE 19716

STETSON, ROBERT F, b New York, NY, Oct 20, 28. METALLURGICAL ENGINEERING. *Prof Exp:* Tech specialist, Gen Atomics, 58-90; RETIRED. *Mem:* Fel Am Soc Metals Int. *Res:* Design and construction of specialty testing equipment for materials; awarded one patent in plasma orifice nossel. *Mailing Add:* 6918 Tanglewood Rd San Diego CA 92111

STETSON, ROBERT FRANKLIN, b Lewiston, Maine, Apr 17, 32; m 67. BIOMATHEMATICS, PLASMA PHYSICS. *Educ:* Bates Col, BS, 54; Wesleyan Univ, MA, 56; Univ Va, PhD(physics), 59. *Prof Exp:* Asst physics, Wesleyan Univ, 54-56; instr, Univ Va, 56-58; asst prof, Univ Fla, 59-64; assoc prof, 64-69, dir 4-year degree prog, 83-87, PROF PHYSICS, FLA ATLANTIC UNIV, 69-, DIR FAC SCHOLARS PROG, 71-,. *Concurrent Pos:* Proj scientist, Air Force Off Sci Res, 62-64; consult, Col Entrance Exam Bd; dir, South Regional Ctr for Excellence, 86- *Mem:* Am Phys Soc; Am Asn Physics Teachers; fel AAAS. *Res:* Angular correlation of gamma rays; neutron scattering; non-traditional higher education; computer simulation of plasma and thermodynamic problems; computer simulation in biomathematics. *Mailing Add:* Dept Physics Fla Atlantic Univ Boca Raton FL 33431

STETTEN, DEWITT, JR, medicine; deceased, see previous edition for last biography

STETTENHEIM, PETER, b New York, NY, Dec 27, 28; m 65; c 2. ORNITHOLOGY. *Educ:* Haverford Col, BS, 50; Univ Mich, MA, 51, PhD(zool), 59. *Prof Exp:* Res zoologist, USDA, Avian Anat Proj, 58-69; book rev ed, Wilson Bull, 70-74; Ed, Condor, 74-85, ED, BIOGRAPHIES OF NORTH AMERICAN BIRDS, 85- *Concurrent Pos:* Mem, Int Comt Avian Anat Nomenclature; secy, bd dirs, Montshire Mus Sci; bd dirs, Soc Protection NH Forests. *Honors & Awards:* Co-recipient, Tom Newman Mem Int Award, Brit Poultry Breeders & Hatcheries Asn, 73; Hon Mem, Cooper Ornith Soc, 85. *Mem:* Wilson Ornith Soc; Cooper Ornith Soc; fel Am Ornith Union; Brit Ornith Union; Ger Ornith Soc. *Res:* Organize and edit comprehensive summary accounts on the biology of all species of North American birds; growth and structure of feathers. *Mailing Add:* Meriden Rd #64-255 Lebanon NH 03766

STETTER, JOSEPH ROBERT, b Buffalo, NY, Dec 15, 46; m 72; c 3. PHYSICAL CHEMISTRY, ELECTROCHEMISTRY. *Educ:* State Univ NY Buffalo, BA, 69, PhD(phys chem), 75. *Prof Exp:* Res asst, Linde Div, Union Carbide Corp, 66-68, chemist, 69; sr res chemist, Becton Dickinson & Co, 74-77, dir chem res, 77-80; scientist & sect head & group leader, Argonne Nat Lab, 80-85; PRES, TRANSDUCER RES, INC, NAPERVILLE, IL, 83- *Concurrent Pos:* Trustee, Lakeland Cent Sch Dist, 77-80; mem, gov bd, Fedn Anal Chem & Spectros Soc, 80-; adj assoc prof chem, Ill Inst Technol, 85-; sci adv bd, Environ Protection Agency, 85-; comt mem, Nat Acad Sci, 87-91. *Honors & Awards:* IR-100 Awards for Instrument Develop, 77, 84; New Tech Award, NASA, 79. *Mem:* Am Chem Soc; AAAS; Int Soc Exposure Anal; Instrument Soc Am; Sigma Xi; Electrochem Soc; Am Conf Govt Indust Hygienists. *Res:* Chemical sensors, physical adsorption, chemisorption, heterogeneous catalytic systems, surface chemistry, environmental chemistry, electrochemistry, analytical chemistry and instrumentation. *Mailing Add:* 1228 Olympus Dr Naperville IL 60540

STETTLER, JOHN DIETRICH, b Cleveland, Ohio, Mar 15, 34; m 57; c 5. LASER PHYSICS, PROPAGATION. *Educ:* Univ Notre Dame, BS, 56; Mass Inst Technol, PhD(physics), 62. *Prof Exp:* Res asst physics, Mass Inst Technol, 57-60; asst prof, Univ Mo-Rolla, 60-64; res physicist, US Army Missile Command, 64-76, mgr laser signature measurements, 76-79, chief, Optics Group, 79-80; sr scientist, Appl Res Inc, 81-82; RETIRED. *Concurrent Pos:* Adj prof physics, Univ Ala, Huntsville, 70- *Mem:* AAAS; Am Phys Soc; Am Asn Physics Teachers. *Res:* Eximer lasers and Raman shifting; propagation of submillimeter to visible radiation through a turbulent atmosphere; application of lasers to ballistic missile defense; particle beam optics. *Mailing Add:* 2004 Roundleaf Green SE Huntsville AL 35803-1832

STETTLER, REINHARD FRIEDERICH, b Steckborn, Switz, Dec 27, 29; m 55; c 1. FOREST GENETICS. *Educ:* Swiss Fed Inst Technol, dipl, 55; Univ Calif, Berkeley, PhD(genetics), 63. *Prof Exp:* Res officer silvicult, Res Div, BC Forest Serv, Can, 56-58; res assoc forest mgt, Fed Inst Forest Res, Switz, 58-59; from asst prof to assoc prof, 63-74, PROF FOREST GENETICS, UNIV WASH, 74- *Concurrent Pos:* Alexander von Humboldt fel, Inst Forest Genetics, Schmalenbeck, Ger, 69-70; vis scholar, Abegg Found, Berne, 76; guest lectr, Dept Bot, Univ Nijmegen, Neth, 77. *Mem:* AAAS; Sigma Xi. *Res:* Genetic control of morphogenesis in higher plants; reproductive biology of forest trees; genetic & physiological studies in short-rotation culture for fiber and energy; induction of haploid parthenogenesis. *Mailing Add:* Col of Forest Resources Univ of Wash Seattle WA 98195

STEUCEK, GUY LINSLEY, b New Haven, Conn, Jan 22, 42; c 2. PLANT PHYSIOLOGY. *Educ:* Univ Conn, BS, 63, PhD(plant physiol), 68; Yale Univ, MF, 65. *Prof Exp:* Nat Res Coun Can fel, Forest Prod Lab, BC, 68-69; from asst prof to assoc prof, 69-77, PROF BIOL, MILLERSVILLE UNIV, 77- *Mem:* AAAS; Am Soc Plant Physiol; Scan Soc Plant Physiol; Ecol Soc Am; Am Inst Biol Sci. *Res:* Influence of mechanical stress on plant growth and development; phloem transport and mineral nutrition. *Mailing Add:* Dept Biol Millersville Univ Millersville PA 17551

STEUDEL, HAROLD JUDE, b Milwaukee, Wis, Jan 3, 45. TOTAL QUALITY MANAGEMENT, COMPUTER SIMULATION. *Educ:* Univ Wis-Madison, BS, 68, PhD(mech eng), 74, Milwaukee, MS, 71. *Prof Exp:* From asst prof to assoc prof mgt, Marquette Univ, 74-82; assoc prof indust eng, 82-85, assoc dir, Univ Indust Res, 82-91, PROF INDUST ENG, UNIV WIS-MADISON, 85- *Concurrent Pos:* Pres, H J Steudel & Assoc Inc, 74-; mem, Col-Indust Coun for Mat Handling Educ, 80-84. *Mem:* Sr mem Inst Elec & Electronics Engrs; sr mem Soc Mfg Engrs; Am Soc Qual Control. *Res:* Computer simulation and statistical modeling for the design and analysis of computer integrated manufacturing systems; design and evaluation of cellular manufacturing layout configurations for just-in-time operations; job sequencing and control strategies for flexible machining workcells. *Mailing Add:* Indust Eng Dept Rm 457 Univ Wis-Madison Madison WI 53706

STEUER, MALCOLM F, b Marion, SC, Dec 16, 28; m 58; c 3. ATOMIC & MOLECULAR PHYSICS, NUCLEAR PHYSICS. *Educ:* US Merchant Marine Acad, BS, 50; Clemson Col, MS, 54; Univ Va, PhD(physics), 57. *Prof Exp:* Res fel, Univ Va, 57-58; asst prof, 58-63, assoc prof, 63-77, PROF PHYSICS, UNIV GA, 77- *Concurrent Pos:* NSF sci fac fel, Univ Wis, 64-65; sci fac fel, Argonne Nat Lab, 80-81. *Mem:* Am Phys Soc. *Res:* Interactions of MeV projectiles with matter; interactions of neutrons with nuclei. *Mailing Add:* Dept of Physics Univ of Ga Athens GA 30602

STEUNENBERG, ROBERT KEPPEL, b Caldwell, Idaho, Sept 18, 24; m 47. INORGANIC CHEMISTRY. *Educ:* Col Idaho, BA, 47; Univ Wash, PhD(chem), 51. *Prof Exp:* Mem staff, 51-67, SR CHEMIST, ARGONNE NAT LAB, 67- *Mem:* Am Chem Soc; Sigma Xi; Am Nuclear Soc; Electrochem Soc. *Res:* Fluorocarbons; interhalogen compounds; pyrometallurgical methods for processing nuclear reactor fuels; nuclear technology; high-temperature batteries; energy conversion; molten salt chemistry. *Mailing Add:* 60 Golden Larch Dr Naperville IL 60540

STEVEN, ALASDAIR C, b Alyth, Scotland, June 27, 47; US citizen; m 71; c 3. ELECTRON MICROSCOPY, IMAGE PROCESSING. *Educ:* Edinburgh Univ, Scotland, MA, 69, Cambridge Univ, PhD(theoret physics), 73; Basel Univ, Switz, SKMB, 75. *Prof Exp:* Res asst, Basel Univ, 73-78; vis scientist, 78-85, sect chief, 85-90, LAB CHIEF, NIH, 90- *Mem:* Electron Micros Soc Am; Biophys Soc. *Res:* Structural basis of molecular biology; high resolution electron microscopy; computer image analysis and model building; assembly properties of proteins, nucleoproteins, viruses, crystals and polymers; virus structure; macromolecular assembly. *Mailing Add:* Bldg 6 Rm 114 NIH Bethesda MD 20892

STEVEN, JAMES R, civil engineering, for more information see previous edition

STEVENS, ALAN DOUGLAS, b Nashua, NH, Aug 17, 26; m 49; c 5. OCCUPATIONAL HEALTH, VETERINARY MEDICINE. *Educ:* Cornell Univ, DVM, 47. *Prof Exp:* Private practice, Ga, 47-50; prog officer res grants, Div Environ Eng & Food Protection, US Pub Health Serv, 63-66, chief res grants, 66-67, chief res & training grants rev, Nat Ctr Urban & Indust Health, 67-68, environ control admin, 68-69, chief res grants, Bur Safety & Occup Health, Environ Control Admin, 69-70, asst dir to dir extramural progs, 70-75, dir training & man power develop, Nat Inst Occup Safety & Health, Ctr Dis Control, Dept Health, Educ & Welfare, 75-86; RETIRED. *Mem:* AAAS; Am Inst Chemists; Am Conf Govt Indust Hygienists. *Res:* Food chemistry and microbiology; irradiation of foods; virology; laboratory animal medicine; research administration; information retrieval. *Mailing Add:* 100 Silver Ave Ft Mitchell KY 41017

STEVENS, ALDRED LYMAN, energy conversion, solid mechanics, for more information see previous edition

STEVENS, ANN REBECCA, b Huntington, WVa, July 22, 39; m 65, 77; c 2. BIOCHEMISTRY, CELL BIOLOGY. *Educ:* Univ Ala, Tuscaloosa, BS, 61; Univ Colo, Denver, PhD(biochem), 66. *Prof Exp:* Res assoc cell biol, Inst Cellular, Molecular & Develop Biol, Univ Colo, Boulder, 66-68; res investr & dir electron micros labs, Vet Admin Hosp, 68-81; from asst prof to assoc prof, 68-82, from asst dean to assoc dean, res div, sponsored res, 81-85, PROF

BIOCHEM, UNIV FLA, 82-; PROF BIOCHEM, EMORY UNIV, 85-, ASSOC VPRES ACAD AFFAIRS & RES. *Concurrent Pos:* Nat Inst Allergy & Infectious Dis res grants, 70-76; Fulbright awardee, Pasteur Inst, Lille, France; assoc vpres acad affairs & dir sponsored progs, Emory Univ, 85- *Mem:* AAAS; Am Soc Cell Biol; Soc Exp Biol & Med; NY Acad Sci; Am Soc Biol Chemists; Nat Soc Res Admin (secy-treas, 83-84); Soc Res Admin. *Res:* Aspects of nucleic acid metabolism during growth and differentiation in pathogenic and nonpathogenic strains of Acanthamoeba and Naegleria; biochemical mechanism of the pathogenicity of free-living amoebae. *Mailing Add:* 303-B Sch Dent Emory Univ Atlanta GA 30322

STEVENS, AUDREY L, b Leigh, Nebr, July 21, 32; m 64; c 2. BIOCHEMISTRY, MICROBIOLOGY. *Educ:* Iowa State Univ, BS, 53; Western Reserve Univ, PhD(biochem), 58. *Prof Exp:* NSF fel, 58-60; instr pharmacol, Sch Med, Univ St Louis, 60-62, asst prof, 62-63; from asst prof to assoc prof biochem, Sch Med, Univ Md, Balitmore City, 63-66; MEM RES STAFF, BIOL DIV, OAK RIDGE NAT LAB, 66- *Concurrent Pos:* NSF res grants, 60-66. *Mem:* Am Soc Biol Chem. *Res:* Nuclear acid biosynthesis. *Mailing Add:* Biol Div Oak Ridge Nat Lab Oak Ridge TN 37830

STEVENS, BRIAN, b South Elmsall, Eng, July 22, 24; m 53; c 2. PHOTOCHEMISTRY. *Educ:* Oxford Univ, BA & MA, 50, DPhil(phys chem), 53. *Hon Degrees:* DSc, Oxford Univ, 77. *Prof Exp:* Fel, Nat Res Coun Can, 53-55; res asst chem, Princeton Univ, 55-56, res assoc, 56-57; res assoc tech off chem eng, Esso Res & Eng Co, NJ, 57-58; lectr chem, Univ Sheffield, 58-65, reader photochem, 65-67; prof chem, 67-89, GRAD RES PROF, UNIV SFLA, 89- *Concurrent Pos:* Askounes-Ashford distinguished scholar, 86. *Mem:* Am Chem Soc; Am Soc Photobiol. *Res:* Molecular luminescence and electronic energy transfer in complex molecules; photosensitized peroxidation of unsaturated molecules and reduction of dyes; solute re-encounter effects; electron donor acceptor orbital correlations. *Mailing Add:* Dept Chem Univ South Fla Tampa FL 33620

STEVENS, BRUCE RUSSELL, b Ogden, Utah, Apr 1, 52; m 75. MEMBRANE TRANSPORT, GASTROINTESTINAL ABSORPTION. *Educ:* Valparaiso Univ, BS, 74; Ill State Univ, MS, 77, PhD(physiol), 80. *Prof Exp:* Res asst, Ill State Univ, 75-80, instr, 78; fel, Med Sch, Univ Calif, Los Angeles, 80-83, res physiologist, 83-84; asst prof physiol, dept physiol, 84, ASSOC PROF PHYSIOL & SURG, COL MED, UNIV FLA, 84- *Concurrent Pos:* Vis scientist, Pavlov Inst Physiol, Leningrad; mem ad hoc rev comt, NSF grant rev. *Mem:* Am Physiol Soc. *Res:* Biomembrane transport mechanisms for amino acids and sugars; enzymatic hydrolysis of peptides and carbohydrates. *Mailing Add:* Dept Physiol Univ Fla Col Med Box J-274 Gainesville FL 32610

STEVENS, CALVIN H, b Sheridan, Wyo, Apr 3, 34; m 61; c 2. PALEOECOLOGY, STRATIGRAPHY. *Educ:* Univ Colo, AB, 56, MA, 58; Univ Southern Calif, PhD(geol), 63. *Prof Exp:* Geologist, Res Lab, Humble Oil Co, 58-60; asst prof geol, San Jose State Col, 63-65 & Univ Colo, 65-66; from asst prof to assoc prof, 66-72, PROF GEOL, SAN JOSE STATE UNIV, 72- *Mem:* Am Asn Petrol Geologists; fel Geol Soc Am; Soc Econ Paleont & Mineral. *Res:* Late Paleozoic paleoecology and paleontology; Great Basin geology and stratigraphy. *Mailing Add:* 1263 Clark Way San Jose CA 95125

STEVENS, CALVIN LEE, b Edwardsville, Ill, Nov 3, 23; m 47; c 1. CHEMISTRY. *Educ:* Univ Ill, BS, 44; Univ Wis, PhD(org chem), 47. *Hon Degrees:* Dr, Univ Nancy, France, 82. *Prof Exp:* Asst org chem, Univ Wis, 44-47; Du Pont fel, Mass Inst Technol, 47-48; from asst prof to assoc prof, 48-54, PROF ORG CHEM, WAYNE STATE UNIV, 54- *Concurrent Pos:* Guggenheim fel, Univ Paris, 55-56; sci liaison officer, Off Naval Res, London, 59-60; Fulbright fels, Sorbonne, 64-65, Univ Paris, 71-72. *Mem:* Am Chem Soc; The Chem Soc; Swiss Chem Soc; Chem Soc France. *Res:* Organic chemistry; epoxyethers and nitrogen analogs of ketenes; natural products; amino-sugars; amino-ketone rearrangements. *Mailing Add:* Dept of Chem 221 Chem Bldg Wayne State Univ Detroit MI 48202

STEVENS, CHARLES DAVID, b Pittsburgh, Pa, Feb 1, 12; m 37; c 5. BIOCHEMISTRY. *Educ:* Univ Cincinnati, AB, 33, MSc, 34, PhD(biochem), 37. *Prof Exp:* Res assoc biochem, Cardiac Lab, Sch Med, Univ Cincinnati, 38-42 & Lab Aviation Med, 42-45; biochemist, Dow Chem Co, 45-46; res assoc biochem, Gastric Lab, Univ Cincinnati, 46-50, asst prof, Dept Prev Med & Indust Health, 50-65; from assoc prof to prof, 65-80, EMER PROF BIOMET, EMORY UNIV, 80- *Concurrent Pos:* Fel, Med Col Va, 61-62. *Mem:* Fel AAAS; Soc Exp Biol & Med; Am Asn Cancer Res. *Res:* Selective localization of chemicals in acidic cancer tissue; respiration; biomathematics; synthesis and metabolism of organolead compounds. *Mailing Add:* 1519 Thornhill Ct Dunwoody GA 30338-4227

STEVENS, CHARLES EDWARD, b Minneapolis, Minn, June 5, 27; c 4. VETERINARY PHYSIOLOGY. *Educ:* Univ Minn, BS, 51, DVM & MS, 55, PhD(vet physiol & pharmacol), 58. *Hon Degrees:* Hon Prof, San Marcos Univ, Peru, 72. *Prof Exp:* Instr vet anat, Univ Minn, 51-52, asst vet physiol, 52-55, res assoc, 58-60; vet physiologist, Agr Res Serv, US Dept Agr, 60-61; assoc prof vet physiol, NY State Vet Col, Cornell Univ, 61-66, prof, 66-79, chmn, dept physiol, biochem & pharmacol, 73-79; ASSOC DEAN RES & GRAD STUDIES, COL VET MED, NC STATE UNIV, 79- *Concurrent Pos:* NIH spec res fel, 62-63, dir training prog comp gastroenterol, 71-76; field rep grad physiol, Cornell Univ, 68-70; mem, gen med study sect, NIH, 69-73; Fulbright lectr, 72. *Mem:* Am Soc Vet Physiol & Pharmacol (pres, 67-68); Am Physiol Soc; Am Vet Med Asn; Conf Res Workers Animal Dis; Comp Gastroenterol Soc; Am Gastroenterol Asn. *Res:* Comparative physiology of the digestive system; mechanisms of secretion, absorption and digesta transit. *Mailing Add:* Col Vet Med NC State Univ Raleigh NC 27650

STEVENS, CHARLES F, b Chicago, Ill, Sept 1, 34; m 56; c 3. NEUROBIOLOGY. *Educ:* Harvard Univ, BA, 56; Yale Univ, MD, 60; Rockefeller Univ, PhD, 64. *Prof Exp:* From asst prof to prof physiol & biophys, Sch Med, Univ Wash, 63-75; prof physiol, 75-83, PROF & CHMN MOLECULAR NEUROBIOL, SCH MED, YALE UNIV, 83- *Honors & Awards:* Spencer Award. *Mem:* Nat Acad Sci; Soc Neurosci; Am Acad Arts & Sci. *Res:* Synaptic transmission; properties of excitable membranes. *Mailing Add:* Molecular Neurobiol Salk Inst 10010 N Torrey Pines Rd La Jolla CA 92037

STEVENS, CHARLES LE ROY, b Chicago, Ill, Aug 8, 31; c 4. BIOPHYSICAL CHEMISTRY. *Educ:* Valparaiso Univ, BA, 53; Univ Pittsburgh, MS, 60, PhD(biophys), 62. *Prof Exp:* Physicist, US Army Biol Labs, 55-56; NIH res fel, 62-64; asst prof, 64-67, ASSOC PROF BIOPHYS, UNIV PITTSBURGH, 67- *Concurrent Pos:* NATO sr fel, Univ Uppsala, 69. *Mem:* AAAS; NY Acad Sci; Biophys Soc. *Res:* Physical chemistry of proteins and nucleic acids; the role of water in the structure of biological macromolecules; self-association of proteins; cell motility. *Mailing Add:* Dept Biol Sci Univ Pittsburgh 234 Langley Hall Pittsburgh PA 15260

STEVENS, CLARK, b Richland, Tex, Mar 24, 21; m 44; c 2. MICROBIOLOGY, CELL PHYSIOLOGY. *Educ:* Harding Col, BS, 49; Univ Ark, MA, 51; Vanderbilt Univ, PhD(biol), 56. *Prof Exp:* Instr sci, Beebe Jr Col, 47-49; asst prof biol, Harding Col, 50-52, prof, 55-66; instr, Vanderbilt Univ, 54-55; PROF & HEAD DEPT BIOL, ABILENE CHRISTIAN COL, 66- *Concurrent Pos:* NIH res fel, 62-63. *Res:* Bacterial physiology and biochemistry; animal virology; microbiology of water. *Mailing Add:* 3585 Hunters Glen Rd Abilene TX 79605

STEVENS, DALE JOHN, b Ogden, Utah, June 27, 36; m 62; c 6. PHYSICAL GEOGRAPHY. *Educ:* Brigham Young Univ, BA, 61; Ind Univ, MA, 63; Univ Calif, Los Angeles, PhD(geog), 69. *Prof Exp:* Instr geog, Univ Wyo, 63-64; assoc prof, 66-80, PROF GEOG, BRIGHAM YOUNG UNIV, 80- *Concurrent Pos:* Univ develop grant, Brigham Young Univ, 71-72, 84 & Austria, 77, 85 & 91. *Mem:* Asn Am Geog; Natural Arch & Bridge Soc. *Res:* Morphometric analysis of land forms, natural arches and bridges; climatology; Utah geography. *Mailing Add:* Dept Geog Brigham Young Univ 690 SWKT Provo UT 84602

STEVENS, DAVID ROBERT, b Logan, Utah, Apr 4, 49; m 81; c 3. TECHNICAL MANAGEMENT, IMMUNOLOGY. *Educ:* Wash State Univ, BS, 72, DVM, 74; Univ Calif, Davis, PhD(comput path), 77; Am Col Vet Pathologists, dipl, 79. *Prof Exp:* Prin scientist immunol, Advan Genetics Res Inst, Agrion Corp, 82-85, dir res & develop, 85-86, vpres res & develop, Diamond Sci Co, Agrion Corp, 86-87; at Hills Pets Prod, 88-89; CONSULT, 89- *Concurrent Pos:* Mem, Nat Agr Res & Exten Users Adv Bd, 85-87; Nebr Res & Develop Authority, 86- *Mem:* AAAS; Am Vet Med Asn; Am Asn Vet Lab Diagnosticians. *Res:* Industrial research and development management; animal and human health products. *Mailing Add:* 4404 Turn Berry Rd Lawrence KS 66146

STEVENS, DEAN FINLEY, b Derby, Conn, Oct 19, 23; m 51; c 3. ZOOLOGY, CELL BIOLOGY. *Educ:* Boston Univ, AB, 49, AM, 50; Clark Univ, PhD(cell biol), 64. *Prof Exp:* Res asst biol, Boston Univ, 50-51; teaching master, Mt Hermon Sch Boys, 51-54; staff scientist, Worcester Found Exp Biol, 54-67; ASSOC PROF ZOOL, UNIV VT, 67- *Concurrent Pos:* Fels, USPHS, Am Cancer Soc & NIH; Ortho Res Found spec grant. *Mem:* Am Soc Cell Biol; Am Asn Cancer Res. *Res:* Vascular physiology; cancer; mechanisms of cell division. *Mailing Add:* Middle Rd Colchester VT 05446

STEVENS, DONALD KEITH, b Troy, NY, July 30, 22; m 45, 65, 74; c 2. SOLID STATE PHYSICS. *Educ:* Union Col, BS, 43; Univ NC, PhD(chem), 53. *Prof Exp:* Physicist, US Naval Res Lab, 43-49; consult radiation effects in solids, Oak Ridge Nat Lab, 49-51, physicist, 53-57; chief,Metall & Mat Br, Div Res, US AEC, 57-60, asst dir,Res Metall & Mat Progs, 60-74, asst dir res mat sci, ERDA-Dept Energy, 74-81; dep assoc dir, 81-85, ASSOC DIR FOR BASIC ENERGY SCI, OFF ENERGY RES, DEPT ENERGY, 85- *Concurrent Pos:* Mem mat adv bd, Nat Acad Sci-Nat Res Coun, 59-62. *Mem:* Fel, AAAS; Sci Res Soc Am; Am Phys Soc. *Res:* Radiation effects in solids. *Mailing Add:* Off Energy Res US Dept of Energy Washington DC 20585

STEVENS, DONALD MEADE, b Lynchburg, Va, May 9, 47; m 70. NUCLEAR PHYSICS, COMPUTER SCIENCE. *Educ:* Va Polytech Inst & State Univ, BS, 69, MS, 70, PhD(physics), 74. *Prof Exp:* Res asst physics, Va Polytech Inst, 70-74; sr res engr nuclear physics, 74-79, group supvr, Diagnostic Develop Group, 80-81, group supvr data processing & diagnostics, 81-85, MGR, NONDESTRUCTIVE METHODS & DIAGNOSTICS SECT, LYNCHBURG RES CTR, BABCOCK & WILCOX CO, 85- *Concurrent Pos:* Guest res asst, Brookhaven Nat Lab, 70-73; guest scientist, Fermi Nat Accelerator Lab, 73-74. *Mem:* Am Phys Soc; Am Soc Mech Engrs; Inst Elec & Electronics Engrs; AAAS. *Res:* Applications in monitoring nuclear power plants and designing computer systems; data processing and diagnostic systems; nondestructive methods and diagnostics; acoustics. *Mailing Add:* Lynchburg Res Ctr PO Box 11165 Lynchburg VA 24506-1165

STEVENS, ERNEST DONALD, b Calgary, Alta, July 5, 41; m 64; c 2. PHYSIOLOGY, ZOOLOGY. *Educ:* Victoria Univ, BSc, 63; Univ BC, MSc, 65, PhD(zool), 68. *Prof Exp:* Assoc prof zool, Univ Hawaii, 68-75; PROF ZOOL, UNIV GUELPH, 88- *Concurrent Pos:* Vis prof, St Andrews, 75, Tohoku, Japan, 82 & Tex A&M, 89. *Mem:* Can Soc Zool; Soc Exp Biol. *Res:* Physiology, primarily of fish; mechanisms of respiration, especially as affected by muscular exercise; comparative physiology of muscle contraction. *Mailing Add:* Dept Zool Univ Guelph Guelph ON N1G 2W1 Can

STEVENS, FRANK JOSEPH, b Peru, Ill, Mar 9, 19; c 2. CHEMISTRY. *Educ:* Univ Ill, BS, 41; Iowa State Col, PhD(bio-org chem), 47. *Prof Exp:* Instr chem, Iowa State Col, 42-47; from asst prof to prof chem, Auburn Univ, 47-85; RETIRED. *Concurrent Pos:* Chmn, Premed-Predent Adv Comt, Auburn Univ, 69- *Mem:* AAAS; Am Chem Soc; NY Acad Sci. *Res:* Organic and pharmaceutical chemistry; plant growth regulators; indole and pyridazine derivatives. *Mailing Add:* 650 Meadowbrook Dr Auburn AL 36830

STEVENS, FRED JAY, b St Paul, Minn, June 10, 49. MACROMOLECULAR INTERACTIONS, COMPUTER MODELING. *Educ:* Hamline Univ, BA, 71; Northwestern Univ, MS, 74, PhD(biophys), 76. *Prof Exp:* Res assoc, Mich State Univ, 76-77; res biochemist, Abbott Labs, 81-83; post doctoral, 77-81, asst scientist, 81, asst biophysicist, 84, BIOPHYSICIST, ARGONNE NAT LAB, 87- *Mem:* AAAS; Biophys Soc. *Res:* Interactions of macromolecules such as idiotype; anti-idotypic antibodies and protein; nucleic acid; computer simulation. *Mailing Add:* Div Biol & Med Res A-141 BIM 202 Argonne Nat Lab Argonne IL 60439

STEVENS, FRITS CHRISTIAAN, b Ghent, Belg, Sept 18, 38; m 65; c 2. BIOCHEMISTRY, PROTEIN CHEMISTRY. *Educ:* Univ Ghent, Lic chem, 59; Univ Calif, Davis, PhD(biochem), 63. *Prof Exp:* Asst biochem, Pharmaceut Inst, Univ Ghent, 59-60 & Univ Calif, Davis, 60-63; sr researcher, Univ Brussels, 63-64; fel, Univ Calif, Los Angeles, 65-67; from asst prof to assoc prof, 67-78 assoc dean, 84-87, PROF BIOCHEM, UNIV MAN, 78-, DEPT HEAD, 87- *Concurrent Pos:* Exec dir, Manitoba Health Res Coun, 82- *Mem:* Can Biochem Soc; Am Soc Biochem & Molecular Biol. *Res:* Structure-function relationships in proteins. *Mailing Add:* Dept Biochem Univ Man Fac Med Winnipeg MB R3E 0W3 Can

STEVENS, GEORGE RICHARD, b Norfolk, Va, May 28, 31; c 4. STRUCTURAL GEOLOGY, TECTONICS. *Educ:* Johns Hopkins Univ, AB, 54, MA, 55, PhD(geol, tectonics), 59. *Prof Exp:* Asst prof geol, Lafayette Col, 57-66, asst acad dean, 64-65; head dept, 66-81, PROF GEOL, ACADIA UNIV, 66- *Concurrent Pos:* Consult, 55-67; vis, Univ Bergen, Norway, 73-74, Korea Inst Energy Res, 81-82. *Mem:* AAAS; fel Geol Soc Am; Geol Asn Can; Sigma Xi. *Res:* Structural vulcanology, metamorphic recrystallization; fabric of deformed rock and of igneous rock; remote sensing structural features; meteorite impact effects. *Mailing Add:* Dept Geol Acadia Univ Wolfville NS B0P 1X0 Can

STEVENS, GLADSTONE TAYLOR, JR, b Brockton, Mass, Dec 16, 30; m 57; c 2. ECONOMIC ANALYSIS, QUALITY CONTROL. *Educ:* Univ Okla, BS, 56; Case Inst Technol, MS, 62, PhD(indust eng & mgt), 66. *Prof Exp:* Proj engr, E I du Pont de Nemours & Co, 56-59; res engr, Thompson-Ramo-Wooldridge Co, 60-62; asst prof mech eng, Lamar State Col, 62-64; from asst prof to assoc prof indust eng & mgt, Okla State Univ, 66-75; PROF AND CHMN DEPT INDUST ENG, UNIV TEX, ARLINGTON, 75- *Concurrent Pos:* NASA res grant, 67-68; consult, Pub Serv Co of Okla, 73-75 & Standard Mfg Co of Dallas, 76- *Honors & Awards:* Eugene L Grant Award, Am Soc Eng Educ, 74. *Mem:* Am Inst Indust Engrs. *Res:* Allocation of capital funds; probabilistic models; in-plant service courses in areas of production control, operations research and quality control. *Mailing Add:* Dept Indust Eng Univ Tex Arlington TX 76019

STEVENS, HAROLD, b Salem, NJ, Oct 18, 11; m 38; c 2. NEUROLOGY. *Educ:* Pa State Univ, BS, 33; Univ Pa, AM, 34, PhD, 37, MD, 41. *Prof Exp:* PROF NEUROL, SCH MED, GEORGE WASHINGTON UNIV, 54- *Concurrent Pos:* Consult pediat neurol, DC Health Dept, 46-; sr attend neurologist, Children's Hosp, 51-; consult, Vet Hosp, 54- & Walter Reed Hosp & NIH, 57-; nat consult to Surg Gen, US Air Force; consult, FDA; distinguished prof neurol, Uniformed Serv Univ Health Sci. *Mem:* Am Asn Neurol Surg; Am Neurol Asn; Am Electroencephalog Soc; fel Am Col Physicians; fel Am Acad Neurol. *Res:* Clinical and pediatric neurology. *Mailing Add:* 4835 Del Ray Ave Bethesda MD 20814

STEVENS, HENRY CONRAD, b Vienna, Austria, Apr 17, 18; nat US; m 41; c 2. ORGANIC CHEMISTRY. *Educ:* Columbia Univ, BS, 41; Western Reserve Univ, MS, 49, PhD(chem), 51. *Prof Exp:* Res chemist, H Kohnstamm & Co, 41-42; res supvr, Chem Div, Pittsburg Plate Glass Co, 42-72, sr res supvr, Chem Div, 72-77, mgr, Univ & Govt Res Develop, 77-84, sr res assoc, PPG Indust, Inc, Barberton, 84-86; ADJ PROF, UNIV AKRON, 86- *Mem:* Am Chem Soc. *Res:* Chemistry of phosgene derivatives; free radical polymerization; polycarbonate resins; cycloadditions; tropolone syntheses; peroxides; epoxides; phase transfer catalysis; technology transfer; academic-industrial interface. *Mailing Add:* 1990 Brookshire Rd Akron OH 44313-5350

STEVENS, HERBERT H(OWE), JR, b Gardiner, Maine, May 12, 13; m 46; c 3. THEORETICAL PHYSICS. *Educ:* Ga Inst Technol, BSME, 36; New Sch Social Res, MALS, 69; Southeastern Mass Univ, MS, 86. *Prof Exp:* Engr & dir res aircraft seats, W McArthur Corp, 39-41; consult res & develop engr, 41-47; chief engr rolling steel doors, W Balfour Co, 47-53; supv engr elec shaver res, Schick, Inc, 53-55; chief engr paint sprayer develop, Champion Implement Co, 55-56; consult, 56-58; sr statist engr, M & C Nuclear, Inc, Div Tex Instruments, Inc, 58-65; consult engr, Walter Balfour & Co, Long Island City, 65-80; CONSULT RES & DEVELOP, 80- *Mem:* Am Soc Mech Engrs; AAAS; Philos Sci Asn. *Res:* Air-supported roofs; cooperative housing; philosophy; finitism; computer programming; theoretical physics; professional writing. *Mailing Add:* 218 Hix Bridge Road Westport MA 02790

STEVENS, HOWARD ODELL, (JR), b Canonsburg, Pa, May 29, 40; m 62; c 2. PHYSICS, ELECTRICAL ENGINEERING. *Educ:* Carnegie Inst Technol, BS, 62; Univ Md, MS, 67. *Prof Exp:* Physicist sensors & systs, 62-65, sr proj engr magnetic countermeasures, 65-69, sr proj engr superconducting mach, 69-79, head, Elec Propulsion & Mach Syst Br, 79-84, HEAD, ELEC SYSTS DIV, DAVID W TAYLOR NAVAL SHIP RES & DEVELOP CTR, 84- *Honors & Awards:* Solberg Award, Am Soc Naval Engrs, 84. *Mem:* Am Soc Naval Engrs. *Res:* Superconducting and advanced electrical machinery; high current switchgear; advanced current collection systems; superconducting magnets and cryogenic systems; electrical power and distribution systems. *Mailing Add:* 228 Wilt Shire Lane Severna Park MD 21146

STEVENS, J(AMES) I(RWIN), b Valley Station, Ky, July 15, 20; m 47; c 4. CHEMICAL ENGINEERING. *Educ:* Univ Louisville, BChE, 42, MChE, 43. *Prof Exp:* Instr, Univ Louisville, 43-44, res assoc, Inst Indust Res, 45-46; instr, Univ Del, 46-48; asst prof chem eng, Vanderbilt Univ, 48-52; engr, Phillips Petrol Co, Okla, 52-56, group leader, Idaho, 56-59, sect head, 59-62; chem engr Infilco/Fuller, 62-66, tech dir, 66-67; staff engr, 67, sr engr, 67-76, mgt staff assoc, 77-80, mgt staff, Arthur D Little, Inc, 80-87; PVT CONSULT, 87- *Mem:* Am Chem Soc; fel Am Inst Chem Engrs; Air & Waste Mgt Asn; Water Pollution Control Fedn. *Res:* Technical and social aspects of environmental management. *Mailing Add:* 9 Glen Terrace Bedford MA 01730-2037

STEVENS, JACK GERALD, b Port Angeles, Wash, Nov 3, 33; m 84; c 2. ANIMAL VIROLOGY, EXPERIMENTAL PATHOLOGY. *Educ:* Wash State Univ, DVM, 57; Colo State Univ, MS, 59; Univ Wash, PhD(virol), 62. *Prof Exp:* Asst prof microbiol, Wash State Univ, 62-63; from asst prof to assoc prof med microbiol & immunol, 63-73, prof microbiol, immunol & neurol, 73-80, PROF MICROBIOL, IMMUNOL, NEUROBIOL & NEUROL, SCH MED, COL LETTERS & SCI, UNIV CALIF, LOS ANGELES, 80-, CHMN, DEPT MICROBIOL & IMMUNOL, 81- *Concurrent Pos:* Mem, Infectious Dis Merit Rev Bd, Vet Admin Med Res Serv, 76-79; mem, Virol Study Sect, Div Res Grants, NIH, 78-82; mem, Fel Review Bd, Nat Multiple Sclerosis Soc, 81-86; mem nat bd, Med Exam, Microbiol & Immunol, 85-89. *Mem:* AAAS; Am Soc Microbiol; Am Asn Immunologists; Am Soc Exp Pathologists; Am Soc Virol. *Res:* Viral pathogenesis, particularly latent infections; diseases of the nervous system; neoplasms. *Mailing Add:* Dept Med Microbiol & Immunol Univ Calif Sch Med 405 Hilgard Ave Los Angeles CA 90024

STEVENS, JAMES EVERELL, b Ann Arbor, Mich, May 4, 50; m 72; c 2. PLASMA PROCESSING OF MATERIALS, RADIO FREQUENCY HEATING OF PLASMAS. *Educ:* Univ Mich, BS, 72, PhD(elec eng), 80. *Prof Exp:* Engr, Shared Applications, Inc, Ann Arbor, Mich, 73-79; RES PHYSICIST, PLASMA PHYSICS LAB, PRINCETON UNIV, 80- *Honors & Awards:* Award for Excellence in Plasma Physics Res, Div Plasma Physics, Am Phys Soc, 84. *Mem:* Am Phys Soc; Inst Elec & Electronics Engrs; Sigma Xi. *Res:* Plasma processing of materials; plasma etching; investigation of the interaction of radio frequency power with plasmas. *Mailing Add:* Plasma Physics Lab Princeton Univ-Forrestal Campus Princeton NJ 08544

STEVENS, JAMES LEVON, b Startex, SC, Dec 17, 47; m 76. ALUMINUM FOIL TECHNOLOGY, PASSIVE COMPONENTS. *Educ:* Wofford Col, BS, 70; Univ SC, PhD(physics), 75. *Prof Exp:* Res assoc physics, Case Western Reserve Univ, 76-77; asst prof physics, N Ga Col, 77-78; develop physicist, Mepco-Electra Inc, 78-85, sr physicist, Mepco-Centralab Inc, 85-89; SR PHYSICIST, PHILIPS COMPONENTS, 89- *Mem:* Am Phys Soc; Electrochem Soc. *Res:* Materials and processes for aluminum electrolytic capacitors; the technologies of etching, anodizing, fabrication and theoretical modeling. *Mailing Add:* 2820 Kinnerly Rd Irmo SC 29063

STEVENS, JAMES T, b Wellsboro, Pa, June 23, 46; m; c 3. INHALATION TOXICOLOGY, BIOCHEMICAL PHARMACOLOGY. *Educ:* WVa Univ, PhD(pharmacol), 72. *Prof Exp:* DIR TOXICOL, CIBA GEIGY CORP, 89- *Mem:* Soc Toxicol; Am Soc Pharmacol & Exp Therapeut. *Res:* General toxicology; risk assessment and biokinetics; 40 publications. *Mailing Add:* Ciba-Geigy Corp PO Box 18300 Greensboro NC 27419

STEVENS, JANICE R, b Portland, Ore; m 46; c 2. NEUROLOGY, PSYCHIATRY. *Educ:* Reed Col, BA, 44; Boston Univ, MD, 49. *Prof Exp:* Intern med, Mass Mem Hosp, 49-50; resident neurol, Boston City Hosp, 50-51; researcher neurol & assoc physician, Sch Med, Yale Univ, 51-54; resident, 54-55, from instr to assoc prof, 55-71, assoc prof psychiat, 74-77, PROF NEUROL, MED SCH, UNIV ORE, PORTLAND, 71-, PROF PSYCHIAT, 77- *Concurrent Pos:* Vis prof psychiat, Harvard Med Sch & Mass Gen Hosp, 71-73; NIMH guest worker, St Elizabeth's Hosp, Washington, DC, 75-76, mem staff, 80- *Mem:* AAAS; Am Acad Neurol; Am Electroencephalog Soc (pres, 73-74); Soc Neurosci; Am Epilepsy Soc; Am Pub Health Asn; Soc Biol Psychiat. *Res:* Neurology and electroencephalography of behavior; epilepsy; schizophrenia. *Mailing Add:* Dept Neurol & Psychol Ore Health Sci Univ 3181 SW Sam Jackson Park Rd Portland OR 97201

STEVENS, JOHN A(LEXANDER), b Baltimore, Md, Mar 25, 21; m 53; c 3. CIVIL ENGINEERING. *Educ:* Princeton Univ, BS, 43; Univ Miami, BSCE, 50; Putney Grad Sch Teacher Ed, MA, 51; Pa State Univ, MSCE, 52. *Prof Exp:* Proj engr, Am Dist Tel Co, NY, 46-49; from asst prof to assoc prof civil eng, Univ Miami, 52-, dir, Soils Eng Lab, 56-; RETIRED. *Mem:* Am Soc Civil Engrs; Am Soc Photogram; Am Forestry Asn. *Res:* Soils engineering; engineering geology; foundation engineering structure and properties of Florida marls and lime muds; air photo interpretation. *Mailing Add:* 9430 SW 93rd Ave Miami FL 33176-2924

STEVENS, JOHN BAGSHAW, b Toronto, Ont, June 26, 41; m 72; c 3. VETERINARY MICROBIOLOGY. *Educ:* Univ Toronto, BSc, 64; Univ Guelph, DVM, 69, MSc, 71; Iowa State Univ, PhD(vet microbiol), 75. *Prof Exp:* Res asst vet bacteriol, Dept Vet Microbiol & Immunol, Univ Guelph, 69; res assoc vet microbiol, Vet Med Res Inst, Iowa State Univ, 71-74; res scientist swine dis, 74-83, HEAD MICROBIOL SERV, HEALTH ANIMALS DIRECTORATE, ANIMAL PATH DIV, ANIMAL DIS RES INST, AGR CAN, 83- *Mailing Add:* Animal Dis Res Inst 3851 Fallowfield Rd Ottawa ON K2H 8P9 Can

STEVENS, JOHN CHARLES, magnetohydrodynamics, x-ray astronomy, for more information see previous edition

STEVENS, JOHN G, b Kansas City, Mo, Aug 7, 43; m 66; c 2. APPLIED MATHEMATICS, COMPUTER SCIENCE. *Educ:* Ind Univ, BS, 65; NY Univ, PhD(math), 72. *Prof Exp:* From asst prof to assoc prof, 69-80, PROF MATH, MONTCLAIR STATE COL, 80- *Concurrent Pos:* Consult, Exxon Res & Eng Co, 75- *Mem:* Am Math Soc; Soc Indust & Appl Math; Soc Comput Simulation. *Res:* Mathematical modeling of physical systems: neuro computing applications. *Mailing Add:* Dept Math & Comput Sci Montclair State Col Valley Rd Upper Montclair NJ 07043

STEVENS, JOHN GEHRET, b Mount Holly, NJ, Dec 16, 41; m 63; c 3. PHYSICAL CHEMISTRY, INFORMATION SCIENCE. *Educ:* NC State Univ, BS, 64, PhD(chem), 69. *Prof Exp:* From asst prof to assoc prof, 63-79, PROF CHEM, UNIV NC, ASHEVILLE, 79- *Concurrent Pos:* Mem ad hoc comt Mössbauer spectros data & conv, Nat Acad Sci, 70-73; ed Mössbauer Effect Data Index, Univ NC & Nat Bur Standards, 70-78; res assoc, Max Planck Inst Solid State Physics, 73; dir, Mössbauer Effect Data Ctr, 74-; dir, Mössbauer Effect Data Control, 74-; res prof, Inst Molecular Spectroscopy, Univ Nijmegen, Neth, 76-77, 78, 79, 80 & 81; co-ed, Mössbauer Effect Reference & Data J, 78-; recipient over 25 grants, NSF, Nat Standard Ref Data Syst, NATO Off Sci Res, Am Chem Soc, Res Corp; dir undergrad res prog, Int Comn Appln Mössbauer Effect; chem counr, counn undergrad res, 88- *Mem:* AAAS; Am Chem Soc; Am Phys Soc; Sigma Xi; Fedn Am Scientists. *Res:* Mössbauer spectroscopy; antimony chemistry; information sciences; evaluation of data. *Mailing Add:* Dept of Chem Univ of NC Asheville NC 28804

STEVENS, JOHN JOSEPH, b London, Eng, July 16, 41. CANCER. *Educ:* Univ Buenos Aires, MD, 64. *Prof Exp:* Res physician, Inst Biol & Exp Med, Buenos Aires, 64-67, Nat Acad Med, Arg, 65-67; res fel, Res Inst, Hosp Joint Dis, New York, 67-70, res assoc, 70-; RES ADMINR, DEPT RES, AM CANCER SOC. *Concurrent Pos:* Instr, Dept Biochem, Mt Sinai Sch Med, City Univ New York, 70-73, res asst prof, 73-; spec fel, Leukemia Soc Am, 74-76, scholarship, 76-81. *Mem:* Endocrine Soc; Am Asn Cancer Res. *Res:* Mechanism of steroid hormone action; studies on glucocorticoid-induced lymphocytolysis of malignant lymphocytes; chemotherapy of cancer. *Mailing Add:* Am Cancer Soc Inc 1599 Clifton Rd NE Altanta GA 30329

STEVENS, JOSEPH ALFRED, b Cleveland, Ohio, Jan 3, 27. MEDICAL MYCOLOGY, MEDICAL MICROBIOLOGY. *Educ:* Univ Dayton, BS, 49; Mich State Univ, MS, 53, PhD(microbiol), 57. *Prof Exp:* Res instr & fel, Mich State Univ, 57-59, res assoc, 59-61; from instr to assoc prof, 61-71, actg chmn dept, 70-74, PROF MICROBIOL, CHICAGO COL OSTEOP MED, 71- *Concurrent Pos:* Consult microbiol & pub health, Nat Bd Exam, Osteop Physicians & Surgeons Inc, 79- *Mem:* Am Soc Microbiol; Mycol Soc Am; NY Acad Sci; Int Soc Human & Animal Mycol; Am Asn Univ Professors; Sigma Xi. *Res:* Fungal serology; immune responses to major fungal pathogens; development of fungal antigens and antisera; fungal diagnostic-serologic tests. *Mailing Add:* Dept Microbiol Chicago Col Osteop Med 1122 E 53rd St Chicago IL 60615

STEVENS, JOSEPH CHARLES, b Grand Rapids, Mich, Feb 28, 29. PSYCHOPHYSICS, SENSORY PSYCHOLOGY. *Educ:* Calvin Col, AB, 51; Mich State Univ, MA, 53; Harvard Univ, PhD(psychol), 57. *Prof Exp:* From instr to asst prof psychol, Harvard Univ, 57-66; res assoc & lectr, 66-77, SR RES SCIENTIST & LECTR PSYCHOL, YALE UNIV, 77- *Concurrent Pos:* Fel, John B Pierce Found, 66- *Mem:* Soc Neurosci; Acoust Soc Am; Optical Soc Am; fel AAAS; fel NY Acad Sci. *Res:* Psychophysics of sensory and perceptual processes, especially somatosensory and kinesthetic sensory modalities. *Mailing Add:* John B Pierce Found Lab 290 Congress Ave New Haven CT 06519

STEVENS, KARL KENT, b Topeka, Kans, Jan 24, 39; m 60; c 3. ENGINEERING MECHANICS. *Educ:* Kans State Univ, BS, 61; Univ Ill, MS, 63, PhD(theoret & appl mech), 65. *Prof Exp:* Staff mem, Sandia Corp, NMex, 61-62; from asst prof to prof eng mech, Ohio State Univ, 65-78; prof ocean eng, 78-83, chmn dept, 81-83, PROF MECH ENG, FLA ATLANTIC UNIV, 83- *Concurrent Pos:* Vis scientist, US Army Ballistic Res Labs, 72-73. *Mem:* Am Soc Mech Engrs; Am Soc Eng Educ; Soc Naval Archit & Marine Engrs; Soc Exp Mechs. *Res:* Modal analysis, vibrations, finite element methods, and viscoelasticity. *Mailing Add:* Dept Mech Engr Fla Atlantic Univ Boca Raton FL 33431

STEVENS, KENNETH N(OBLE), b Can, Mar 23, 24; m 57; c 4. ACOUSTICS. *Educ:* Univ Toronto, BASc, 45, MASc, 48; Mass Inst Technol, ScD(elec eng), 52. *Prof Exp:* Instr appl physics, Univ Toronto, 46-48; asst elec eng, Mass Inst Technol, 48-51, instr, 51-52, mem res staff commun acoust, 52-54, from asst prof to assoc prof, 54-63, PROF ELEC ENG, MASS INST TECHNOL, 63-, CLARENCE JOSEPH LEBEL PROF, 76- *Concurrent Pos:* Consult & engr, Bolt Beranek & Newman, 52-; Guggenheim fel, 62-63; NIH spec fel & vis prof, Univ Col, Univ London, 69-70; mem, Nat Adv Coun Neurol & Commun Dis & Stroke, NIH, 81-85; vpres, Sensimetric Inc, 87- *Honors & Awards:* Silver Medal, Acoust Soc Am. *Mem:* Nat Acad Eng; Acoust Soc Am (pres, 76-77); Inst Elec & Electronic Engrs; Am Acad Arts & Sci. *Res:* Speech communication; psycho-acoustics. *Mailing Add:* Dept Elec Eng Mass Inst Technol Cambridge MA 02139

STEVENS, LEROY CARLTON, JR, b Kenmore, NY, June 5, 20; m 42; c 3. DEVELOPMENTAL BIOLOGY. *Educ:* Cornell Univ, BS, 42; Univ Rochester, PhD, 52. *Prof Exp:* Asst, Univ Rochester, 48-52, instr, Univ Sch, 51-52; res fel, 52-55, res assoc, 55-57, staff scientist, 57-67, SR STAFF SCIENTIST, JACKSON LAB, 67- *Concurrent Pos:* Guggenheim fel, Exp Embryol Lab, Col of France, 61-62. *Res:* Experimental embryology; cancer; mammalian embryology and teratocarcinogenesis. *Mailing Add:* Pretty Marsh Rd Desert ME 04660

STEVENS, LEWIS AXTELL, b Butte, Mont, Nov 17, 13; m 35; c 3. BIOPHYSICS. *Educ:* San Jose State Col, AB, 50. *Prof Exp:* Meteorol aid, Sci Serv Div, US Weather Bur, 51-52; physicist, Aviation Ord Dept, US Naval Ord Test Sta, 52-56, electronic scientist, 56-57, electronic scientist, Fuze Eval Div, Test Dept, 57-60; gen engr & head, Measurements Br, Propulsion Develop Dept, 60-61, gen proj engr, 62, res physicist, Explosives & Pyro-Tech Div, 62-70; CONSULT PHYSICIST, 70- *Concurrent Pos:* Mem fuze field tests subcomt, Joint Army-Navy-Air Force, 61-64. *Mem:* AAAS; Am Phys Soc; Am Inst Aeronaut & Astronaut. *Res:* Development of new medical tools and techniques; nuclear physics; meteorological aspects of health physics; technology of high speed aerial tow targets; technology of soft lunar landings; effects of microwave radiation on enzyme systems in living organisms; in vivo pathology of varying magnetic fields. *Mailing Add:* LASTEV Lab 725 Randall St Ridgecrest CA 93555

STEVENS, LLOYD WEAKLEY, b Philadelphia, Pa, Jan 14, 14; m 71; c 3. SURGERY. *Educ:* Univ Pa, AB, 33, MD, 37; Am Bd Surg, dipl, 44. *Prof Exp:* Assoc surg, Grad Sch Med, 46-49, assoc prof, Sch Med, 53-60, prof, 60-79, EMER PROF CLIN SURG, SCH MED, UNIV PA, 79- *Concurrent Pos:* Assoc prof, Women's Med Col Pa, 46-49; dir surg, Presby-Univ Pa Med Ctr & assoc surgeon, Univ Hosp; chief surg, Philadelphia Gen Hosp. *Mem:* Am Col Surgeons; Soc Surg Alimentary Tract. *Res:* Acute cholecystitis; peptic ulcer; ulcerative colitis. *Mailing Add:* The Hermitage 1204 Round Hill Rd Bryn Mawr PA 19010

STEVENS, MALCOLM PETER, b Birmingham, Eng, Apr 3, 34; US citizen; m 60; c 2. ORGANIC POLYMER CHEMISTRY. *Educ:* San Jose State Col, BS, 57; Cornell Univ, PhD(org chem), 61. *Prof Exp:* Res chemist, Chevron Res Co, Standard Oil Co Calif, 61-64; asst prof chem, Robert Col, Istanbul, 64-67; asst prof, Univ Hartford, 67-68; from asst prof to assoc prof, Am Univ Beirut, 68-71; assoc prof, 71-78, chmn dept, 78-81, PROF CHEM, UNIV HARTFORD, 78- *Concurrent Pos:* Vis prof, Univ Sussex, Eng, 77, Colo State Univ, 85; coun advan & support educ. *Mem:* Sigma Xi; Am Chem Soc. *Res:* Photopolymerization; thermal polymerization; polymer modification; synthesis of novel polymer systems, historical writing. *Mailing Add:* Dept Chem Univ Hartford West Hartford CT 06117-0395

STEVENS, MARION BENNION, CLINICAL NUTRITION & FOOD SCIENCE. *Educ:* Univ Wis, PhD(food sci), 56. *Prof Exp:* INSTR HEALTH OCCUPATIONS, EL PASO COMMUNITY COL, 82- *Mailing Add:* 2320 Gene Littler Dr El Paso Community Col El Paso TX 79936

STEVENS, MERWIN ALLEN, plant genetics, for more information see previous edition

STEVENS, MICHAEL FRED, b Urbana, Ill, May 17, 41; m 65; c 2. INDUSTRIAL CHEMISTRY, TOXICOLOGY. *Educ:* Eastern Ill Univ, BS, 64; Univ Ill, Urbana, MS, 66; Univ Nebr, Lincoln, PhD(org chem), 70. *Prof Exp:* From res assoc to sr res assoc, Appleton Papers Inc, 70-74, staff res assoc, 74-86, proj mgr prod develop, 86-89, PROD SAFETY SPECIALIST, APPLETON PAPERS INC, 89- *Mem:* Am Chem Soc; TAPPI. *Res:* Corporate liaison with government relative to toxic substances in control act matters; management of corporate toxicity testing programs; corporate consultant on chemical health and safety matters; air sampling work in plant mills for contaminants; advisor in product safety issues. *Mailing Add:* Appleton Papers Inc PO Box 359 Appleton WI 54912

STEVENS, PETER FRANCIS, b Teignmouth, Eng, Nov 13, 44. SYSTEMATIC BOTANY. *Educ:* Oxford Univ, BA, 66, MA, 72; Univ Edinburgh, PhD(bot), 70. *Prof Exp:* Forest botanist, Dept Forestry, Lae, Papua, New Guinea, 70-73; asst cur Arnold Arboretum, 73-80, assoc cur, 80-83, from asst prof to assoc prof, 77-83, PROF BIOL & CUR ARNOLD ARBORETUM & GRAY HERBARIUM, HARVARD UNIV, 83- *Mem:* Fel Linnean Soc London; Soc Study Evolution; Bot Soc Am; Int Asn Plant Taxonomists. *Res:* Morphology; systematics and evolution of Indo-Malesian plants and the Ericaceae and Clusiaceae of the world; biogeography; tropical ecology; theory and history of systematics. *Mailing Add:* Dept Biol Harvard Univ Herbaria 26 Oxford St Cambridge MA 02138

STEVENS, RICHARD EDWARD, b Washington, DC, Oct 30, 32; m 57; c 3. MICROSCOPY. *Educ:* Washington Col, BS, 54; Pa State Univ, MS, 56; Rensselaer Polytech Inst, PhD(phys chem), 75. *Prof Exp:* Asst chemist, Pa State Univ, 54-56 & Univ Colo, 56-59; scientist-chemist, Rocky Flats Div, Dow Chem Co, Colo, 59-62, develop chemist, 62-64; res chemist, Am Cyanamid Co, 64-67; microscopist, Ernest F Fullam, Inc, NY, 67-74; sr res microscopist, Walter C McCrone Assocs, 74-76; sr chemist, Nalco Chem Co, 76; mgr image anal appln, Bausch & Lomb, Inc, 76-78; res assoc ophthal, Park Ridge Hosp, 78-80; SR CHEMIST, AEROSPACE DIV, HERCULES, INC, 80-; sr scanning electron micros div, Carbon Composites Propellants, 88-91; RETIRED. *Mem:* Microbeam Anal Soc; Am Chem Soc; Electron Micros Soc Am. *Res:* Chemical and electron microscopy; crystallography; ultramicroanalysis; image analysis. *Mailing Add:* 4759 Bon Air St No 5 Salt Lake City UT 84117

STEVENS, RICHARD F, b Columbus, June 5, 02. POWER ENGINEERING DESIGN. *Educ:* Univ Wash, BSEE, 24, MSEE, 52. *Prof Exp:* Chief design, Bonneville Power Admin, Portland, 38-69; consult engr. Commonwealth Asn & Monasa Eng Co, Brazil, 71- 80; RETIRED. *Mem:* Fel Inst Elec & Electronics Engrs. *Res:* Design of AC/DC high voltage. *Mailing Add:* 4319 Chambers Lake Dr Olympia WA 98503

STEVENS, RICHARD JOSEPH, b Rochester, NY, Oct 31, 41; m 65; c 3. NEUROSCIENCES. *Educ:* Univ Rochester, BS, 63; Univ Ill, Urbana, MS, 65, PhD(biophysics), 69. *Prof Exp:* Aerospace technologist, NASA-Lewis Res Ctr, 63; res asst, Dept Physics, Univ Ill, 64-65, teaching asst human & cellular physiol, 66-67; res fel neuroanat, Dept Anat, Brain Res Inst, Univ Calif, Los Angeles, 69-70; asst prof, 70-75, ASSOC PROF HUMAN ADAPTABILITY, COL HUMAN BIOL, UNIV WIS-GREEN BAY, 75-

Concurrent Pos: Consult, Med Col Wis Pain Clin; President's teaching improvement grant, Univ Wis, 72; consult, Green Bay Childbirth Educ Asn, 75- *Mem:* AAAS; Sigma Xi. *Res:* Neurophysiology of vision, neuro-behavioral aspects of environmental contaminants; neuro-behavioral aspects of pain perception; innovative teaching of biology. *Mailing Add:* Col Human Biol Univ Wis Green Bay WI 54301

STEVENS, RICHARD S, b Cranston, RI, Mar 22, 25; m 52; c 3. MARINE GEOLOGY, PHYSICAL OCEANOGRAPHY. *Educ:* Brown Univ, AB, 50. *Prof Exp:* Oceanogr, US Naval Oceanog Off, 52-62, Off Naval Res, 62-80; assoc sea grant dir, NJ Marine Sci Consortium, 80-85. *Mem:* Am Geophys Union; Sigma Xi. *Res:* Geological and physical oceanography; ocean circulation, ocean bottom processes and their effects on sediment distribution and structure. *Mailing Add:* 75 Van Houten Fields West Nyack NY 10994

STEVENS, ROBERT E, forest entomology, for more information see previous edition

STEVENS, ROBERT EDWARD, b Lexington, Ky, Sept 19, 24; m 51; c 3. FISHERY BIOLOGY. *Educ:* Col William & Mary, BA, 50; NC State Univ, BS, 53, PhD(zool), 70. *Prof Exp:* Biologist fisheries, NC Wildlife Resources Dept, 53-56, SC Wildlife Resources Dept, 56-66; asst unit leader, NC Coop Fish Unit, 66-70; sr fisheries scientist, Marine Protein Corp, 70-75; chief coastal fisheries, Tex Parks & Wildlife Dept, 75-76; endangered species adminr, Nat Marine Fisheries Serv, 76-77; aquacult coordr, 77-78, CHIEF FISHERIES RES, US FISH & WILDLIFE SERV, 78- *Concurrent Pos:* Consult, Consol Edison, 75; chmn, Joint Subcomt Aquacult, Fed Coord Coun Sci & Eng Technol, 83-84. *Mem:* Am Fisheries Soc. *Res:* Hormone induced ovulation in fish; reservoir and pond culture; intensive silo culture of fresh, anadronous and salt water fish. *Mailing Add:* 125 Cameron Mews Alexandria VA 22314

STEVENS, ROGER TEMPLETON, b Syracuse, NY, Jan 11, 27; m 48, 79; c 2. ELECTRICAL ENGINEERING, SYSTEMS ENGINEERING. *Educ:* Union Col, BA, 49; Boston Univ, MA, 59; Blackstone Sch Law, LLB, 56; Va Polytech Inst & State Univ, MEng, 76; Calif Western Univ, PhD(elec eng), 78. *Prof Exp:* Tech writer, Raytheon Mfg Co, 50-51; engr, Lab Electronics Inc, 51-55; sr engr electronic design, Spencer Kennedy Labs, 55-56, AVCO Mfg Co, 56-57 & Electronics Systs Inc, 57-60; sect supvr, Sanders Assoc Inc, 60-65; group leader systs engr, Mitre Corp, 65-67; leading scientist systs engr, Dikewood Indust Inc, 67-70; group leader systs engr, Mitre Corp, 70-74; leading scientist systs engr, Dikewood Indust Inc, 74- 81; eng specialist II, EG & G, Inc, 81-83; MEM TECH STAFF, MITRE CORP, 83- *Concurrent Pos:* Author of numerous books. *Mem:* Soc Old Crows. *Res:* Operational test and evaluation; computer design and software development; display design; radar system design. *Mailing Add:* Mitre Corp PO Box 5520 Kirtland Air Force Base NM 87185

STEVENS, RONALD HENRY, b Philadelphia, Pa, Dec 3, 46. CELLULAR IMMUNOLOGY, HUMAN IMMUNOBIOLOGY. *Educ:* Ohio Wesleyan Univ, BA, 68; Harvard Univ, PhD(microbiol), 71. *Prof Exp:* Fel immunol, Nat Inst Med Res, London, 71-74; ASSOC PROF MICROBIOL & IMMUNOL, UNIV CALIF, LOS ANGELES, 74- *Mem:* Am Asn Immunologists; Am Fedn Clin Res; AAAS; NY Acad Sci. *Res:* Cellular and molecular interactions responsible for the successful initiation, maintenance, and termination of normal and abnormal humoral immune responses in humans. *Mailing Add:* Dept Microbiol 43-319 CHS Sch Med Univ Calif Los Angeles CA 90024

STEVENS, ROSEMARY ANNE, b Bourne, Eng, Mar 18, 35; US citizen; c 2. HISTORY OF MEDICINE, PUBLIC HEALTH. *Educ:* Oxford Univ, BA, 57, MA, 61; Univ Manchester, dipl social admin, 59; Yale Univ, MPH, 63, PhD(epidemiol), 68. *Hon Degrees:* LHD, Hahnemann Univ, 88. *Prof Exp:* Res asst pub health, Sch Med, Yale Univ, 62-65, res assoc, 66-68, from asst prof to prof, 68-76; prof health systs mgt, Sch Pub Health & Trop Med, Tulane Univ, 76-79; prof hist & sociol sci, 79-89, chmn dept, 80-83, SR FEL, LEONARD DAVIS INST HEALTH ECON, UNIV PA, 80-, CHMN, DEPT HIST & SOCIOL SCI, 86-, UPS FOUND PROF SOCIAL SCI, 90- *Concurrent Pos:* Hon res officer, London Sch Econ & Polit Sci, 62-63, vis lectr, 63-64 & 73-74; lectr, Sch Pub Health, Johns Hopkins Univ, 67-68; guest scholar, Brookings Inst, 67-68; consult, Bur Budget, Off President, 69, Ctr Res & Develop, AMA, 72, WHO, 74 & Brit Nat Health Serv, 81; mem, Comt Design Health Care Delivery Systs, Am Sociol Asn, 73, Subcomt Health, Comt Ways & Means, Adv Panel Nat Health Ins, US House Rep, 75, Comt Vital & Health Statist, 78-80 & US Nat Comt Int Union Hist & Philos Sci, Nat Acad Sci, 81-87; Rockefeller humanities fel, 82-83; Guggenheim fel, 84-85; Bellagio Study & Conf Ctr scholar, 84; Frohlich prof, Royal Soc Med, London, 86. *Honors & Awards:* Fulton Lectr, Yale Univ, 87; Arthur Viseltear Award, Am Pub Health Asn, 90; Welch Medal, Am Asn Hist Med, 90. *Mem:* Inst Med-Nat Acad Sci; fel Am Pub Health Asn; Am Sociol Asn; AMA; Sigma Xi. *Res:* History of medicine; comparative studies in health care policy; history of hospitals; medical education and manpower policies. *Mailing Add:* Dept Hist & Sociol Sci Univ Pa D6 Philadelphia PA 19104

STEVENS, ROY HARRIS, b New York, NY, Jan 8, 48; m 71; c 2. ORAL MICROBIOLOGY, DENTAL PULP BIOLOGY. *Educ:* Adelphi Univ, BA, 69; Rutgers Univ, MS, 72; Columbia Univ, DDS, 76. *Prof Exp:* Fel, Univ Pa, 77-79, from res assoc to res asst prof, Dept Microbiol, Sch Dent Med, 80-89; ASSOC PROF, SCH DENT & ORAL SURG, COLUMBIA UNIV, 89- *Concurrent Pos:* Co-prin investr, Nat Inst Dent Res Grants, 79-; ed consult, J Infectious Dis, 85, Oral Microbiol & Immunol, Endodontics & Dent Traumatology; vis teacher, Albert Einstein Med Ctr, 87-89. *Mem:* Am Soc Microbiol; Int Asn Dent Res; Am Asn Endodontists. *Res:* Oral microbiology; epidemiology, taxonomy, virulence and ecological interactions of oral bacteria. *Mailing Add:* Five Coleman Dr East Williston NY 11596

STEVENS, ROY WHITE, b Troy, NY, Sept 4, 34; m 56; c 2. MEDICAL MICROBIOLOGY, LABORATORY METHODS. *Educ:* State Univ NY Albany, BS, 56, MS, 58; Albany Med Col, PhD(microbiol), 65; Am Bd Med Microbiol, dipl, 71. *Prof Exp:* Bacteriologist, NY State Dept Health, 58-61, sr bacteriologist, 62-65, assoc bacteriologist, 65-67, sr assoc prin res scientist immunol, 67-73, prin res scientist immunol, 73-79, dir labs diag immunol, 79-89, DIR LABS RETROVIROL/IMMUNOL, NY STATE DEPT HEALTH, 89- *Concurrent Pos:* Adj assoc prof, dept microbiol & immunol, Albany Med Col, NY, 82-; ed, Diag Devices Manual & Directory; assoc prof, Biomed Sci Div, Sch Public Health, State Univ NY, Albany, 89- *Mem:* AAAS; Am Soc Microbiol; fel Am Acad Microbiol; Asn Med Lab Immunologists (pres, 89). *Res:* Diagnostic immunology, serology; medical microbiology; immunodiagnosis of infectious diseases and automation of clinical laboratory methods. *Mailing Add:* Wadsworth Ctr Labs & Res NY State Dept Health Empire State Plaza Albany NY 12201

STEVENS, STANLEY EDWARD, JR, b Ringgold, Tex, June 25, 44; m 69; c 3. GENETICS, ENZYMOLOGY OF NITROGEN ASSIMILATION. *Educ:* Univ Tex, Austin, BA, 66, MA, 68, PhD(bot), 71. *Prof Exp:* Teaching asst microbiol, Univ Tex, Austin, 67-68, res asst, 68, environ health trainee, 69-70, NIH postdoctoral fel protein chem, 71-73 & photosynthesis, 74-75; from asst prof to prof microbiol & molecular biol, Pa State Univ, 75-88, dir, Coop Prog Recombinant DNA Technol, 82-86, assoc dir, Biotechnol Inst, 84-87, dir, Coop Prog Biotechnol & chmn, Marine Sci Prog, 86-88; PROF BIOL & W HARRY FEINSTONE CHAIR EXCELLENCE MOLECULAR BIOL, MEMPHIS STATE UNIV, 88- *Concurrent Pos:* Vis scientist, Los Alamos Sci Lab, 75-76; consult, US Environ Protection Agency, 79, Gulf & Western Corp, 80-83, Standard Oil Co & Coulter Immunol, 84-86, Aluminum Co Am, 86 & Celgene Corp, 87; mem, NSF Grad Fel Panel & CRGO Biol Nit-Fix Panel, USDA, 83-85, Dept Energy Res Instr Panel, 85, NIH Acad Res Enhancement Panel, 86, NIH Biomed Res Shared Instr Panel, 90 & NIH Microbiol Physiol & Genetics Study Sect, 91; prog mgr, CRGO Biol Nit-Fix, USDA, 87. *Mem:* Fel Am Acad Microbiol; AAAS; Am Soc Biochem & Molecular Biol; Am Soc Microbiol; Am Soc Plant Physiologists; Phycol Soc Am; Sigma Xi. *Res:* Molecular biology of nitrogen assimilation and of the light harvesting antenna in cyanobacteria; fouling of marine structures by algae; biotechnology of cyanobacteria and algae; biological desulfurization of coal. *Mailing Add:* Dept Biol Memphis State Univ 509 Life Sci Bldg Memphis TN 38152

STEVENS, SUE CASSELL, b Roanoke, Va. BIOCHEMISTRY. *Educ:* Goucher Col, BA, 30; Columbia Univ, MA, 31, PhD(chem), 40. *Prof Exp:* Res biochemist, NY Skin & Cancer Hosp, New York, 32-35; biochemist, Fifth Ave Hosp, 35; res chemist, Col Physicians & Surgeons, Columbia Univ, 35-39, NY Orthop Hosp, 40-41 & Calif Milk Prod Co, 41-43; res dairy chemist, Golden State Co, Ltd, 43-46 & Swift & Co, 46-47; dir res & qual control, Steven Candy Kitchens, 47-48; assoc prof chem & biol, MacMurray Col, 48-49; chief biochemist, US Vet Admin Ctr, Dayton, Ohio, 49-52, res biochemist, 52-56, supvr res lab Hosp, Lincoln, Nebr, 56-65; dir, Div Endocrine Chem, Jewish Hosp St Louis, 65-79; RETIRED. *Concurrent Pos:* Asst prof path, Sch Med, Wash Univ, 67-79. *Mem:* Fel AAAS; fel Am Inst Chem; Am Soc Qual Control; NY Acad Sci; Am Chem Soc; Sigma Xi. *Res:* Clinical chemistry methods; electrolytes in biological fluids; steroids; hormones; automation. *Mailing Add:* PO Box 30206 Lincoln NE 68503

STEVENS, THOMAS MCCONNELL, b Plainfield, NJ, May 25, 27; m 54; c 4. VIROLOGY, ENTOMOLOGY. *Educ:* Haverford Col, BA, 50; Rutgers Univ, MS, 55, PhD(entom), 57. *Prof Exp:* Fel microbiol, St Louis Univ, 57-58, from instr to asst prof, Sch Med, 58-63; asst prof, Med Sch, Rutgers Univ, New Brunswick, 63-66; assoc prof exp med & from assoc dir to dir teaching labs, 66-72, PROF MICROBIOL, RUTGERS MED SCH, COL MED & DENT NJ, 72-, ASST DEAN, 68- *Mem:* Am Soc Microbiol. *Res:* Physical and chemical nature of the togaviruses using dengue virus as a model. *Mailing Add:* PO Box 103 Heathesville VA 22473-0103

STEVENS, TRAVIS EDWARD, b Leigh, Nebr, Dec 22, 27; m 61; c 4. ORGANIC CHEMISTRY. *Educ:* Wayne State Col, AB, 51; Iowa State Col, PhD(chem), 55. *Prof Exp:* Asst, Iowa State Col, 51-53; sr res chemist, 55-75, RES SECT MGR, ROHM & HAAS CO, 75- *Concurrent Pos:* Vis prof, Ind Univ, 65. *Mem:* Am Chem Soc. *Res:* Synthesis and properties of high-energy compounds; molecular rearrangements; polymer synthesis; paper chemicals; coatings and textile chemistry. *Mailing Add:* Rohm & Haas Co Spring House PA 19477

STEVENS, VERNON LEWIS, b Tacoma, Wash, Oct 10, 30; m 54; c 4. BIOCHEMISTRY, ANALYTICAL CHEMISTRY. *Educ:* Cent Wash State Col, BS, 57; Ore State Univ, MS, 60. *Prof Exp:* Res asst & biochemist, William S Merrell Co Div, Richardson-Merrell, Inc, 59-67; head anal chem, Enzomedic Lab, Inc, Wash, 67-69; CHEMIST, PUGET SOUND PLANT, TEXACO INC, 69- *Res:* Development of analytical procedures for gas-liquid and thin layer chromatography, autoanalyzer, radioisotopes and spectronic equipment; lipid synthesis in animals; nucleotides; clinical, environmental and petroleum chemistry. *Mailing Add:* 18252 NE Highland Lane Suquamish WA 98392

STEVENS, VINCENT LEROY, b Boston, Mass, July 14, 30; m 58. BIOCHEMISTRY. *Educ:* Univ Calif, Berkeley, AB, 53, PhD(biochem), 57. *Prof Exp:* Jr res biochemist, Med Ctr, Univ Calif, San Francisco, 57-59; asst prof chem & biochem, 59-62, assoc prof chem, 62-67, PROF CHEM, EASTERN WASH STATE COL, 67-, CHMN DEPT, 70-, DEAN, DIV HEALTH SCI, 74- *Concurrent Pos:* Consult, Deaconess Hosp, Spokane, 62- *Mem:* Am Chem Soc. *Res:* Organic and physical chemistry of nucleic acids and their derivatives. *Mailing Add:* 3124 S Lamonte St Spokane WA 99203

STEVENS, VIOLETE L, b China, June 2, 42; US citizen; m 70. PAINT & COATINGS, PHOTOCHEMISTRY. *Educ:* Viterbo Col, BS, 65; San Jose State Univ, MS, 67. *Prof Exp:* Res chemist, Stephan Chem Co, 65; res asst & teaching asst chem, San Jose State Col, 66-67; INDUST DEVELOP MGR,

DOW CHEM CO, 67- *Concurrent Pos:* Mem, bd dirs & vpres, Prof Develop Coun, AFP, Soc Mfg Engrs, 86-88; mem, Nat Paint & Coatings Sci Comt, 87- *Mem:* Nat Paint & Coatings Asn; Soc Mfg Engrs; Fedn Soc Coatings Technol; Chem Coaters Asn. *Res:* Photochemical decomposition of chlorinated compounds; synthesis of biocides; organo silanes; polyethers; radiation curable resins and diluents; latexes for paint and coatings; author of over 28 technical publications; holder of 15 US patents. *Mailing Add:* 4311 Andre St Midland MI 48642

STEVENS, WALTER, b Salt Lake City, Utah, Dec 6, 33; div; c 4. ANATOMY, RADIOBIOLOGY. *Educ:* Univ Utah, BS, 56, PhD(anat, radiobiol), 62. *Prof Exp:* From instr to assoc prof, 62-74, head chem group, Radiobiol Lab, 70-82, PROF ANAT, UNIV UTAH, 74-, ASSOC DEAN RES, 81-, INTERIM DEAN, 88- *Concurrent Pos:* Dir, Nat Inst Gen Med Sci Training Grant, 74-77; vis prof, Stanford Univ, 77. *Mem:* Endocrine Soc; Am Asn Anatomists; Soc Neurosci; Am Physiol Soc; Radiation Res Soc. *Res:* Mechanism of action of glucocorticoids in lymphoid tissues, central nervous system and lung; interaction of transuranic elements with biological systems; effect of radioactive fallout in humans; relationship to thyroid disease and leukemia. *Mailing Add:* Deans Office Sch Med Univ Utah Salt Lake City UT 84132

STEVENS, WALTER JOSEPH, b Atlantic City, NJ, Apr 29, 44; m 66; c 2. THEORETICAL & PHYSICAL CHEMISTRY, CHEMICAL PHYSICS. *Educ:* Drexel Univ, BS, 67; Ind Univ, Bloomington, PhD(chem physics), 71. *Prof Exp:* NSF fel, Argonne Nat Lab, 71-72, lab fel, 72-73; physicist, Lawrence Livermore Lab, 73-75; mem staff, Time & Energy Div, Nat Bur Standards, 75-77, mem staff, Molecular Spectros Div, 77-78; ASSOC DIR, CTR ADV RES BIOTECHNOL, 88- *Honors & Awards:* Silver Medal, Dept Com, 84, Gold Medal, 90. *Mem:* Am Chem Soc; Am Phys Soc. *Res:* Quantum chemistry; ab initio calculation of molecular wavefunctions and properties; theoretical biochemistry; biophysics. *Mailing Add:* Ctr Adv Res Biotechnol 9600 Gudelsky Dr Rockville MD 20850

STEVENS, WARREN DOUGLAS, b Long Beach, Calif, Sept 15, 44; m 63; c 1. BOTANY. *Educ:* Humboldt State Col, AB, 68; Mich State Univ, MS, 71, PhD(bot), 76. *Prof Exp:* Asst bot & plant path, Mich State Univ, 71-74; consult, Cyrus William Rice Div, NUS Corp, 75-77; res assoc bot & plant path, Mich State Univ, 76-78; B A KRUKOFF CUR CENT AM BOT, MO BOT GARDEN, 77- *Concurrent Pos:* Collabr, Smithsonian Hassan Flora Proj, 69; consult, Ingham County Circuit Court, 75-77. *Mem:* Am Soc Plant Taxonomists; Asn Trop Biol; Bot Soc Am; Int Asn Plant Taxon; Sigma Xi. *Res:* Flora of Nicaragua; systematics of Asclepiadaceae. *Mailing Add:* Mo Bot Garden PO Box 299 St Louis MO 63166

STEVENS, WILLIAM D, ENGINEERING. *Prof Exp:* emer dir, 83-88, CONSULT, FOSTER WHEELER CORP, 88- *Mem:* Nat Acad Eng. *Mailing Add:* Four Stony Brook Dr North Caldwell NJ 07006

STEVENS, WILLIAM F(OSTER), b Detroit, Mich, Oct 7, 22; m 62; c 5. CHEMICAL ENGINEERING. *Educ:* Northwestern Univ, BS, 44; Univ Wis, MS, 47, PhD(chem eng), 49. *Prof Exp:* Chem engr, Res Ctr, B F Goodrich Co, 49-51; from res assoc to assoc prof, 51-64, assoc dean, Grad Sch, 65-72, chmn dept, 76-79, prof chem eng, 64-, EMER PROF CHEM ENG, NORTHWESTERN UNIV, EVANSTON. *Concurrent Pos:* Consult, Vern E Alden Co, Ill, 52-59, Pure Oil Co, 58-63, Argonne Nat Lab, 59-61, Chicago Bridge & Iron Co, 64-70 & TecSearch, Inc, 65-71. *Mem:* Am Soc Eng Educ; Am Chem Soc; Am Inst Chem Engrs. *Res:* Applied mathematics and computers; process control and dynamics; process optimization. *Mailing Add:* Dept Chem Eng Northwestern Univ Evanston IL 60201

STEVENS, WILLIAM GEORGE, b Champaign, Ill, Sept 20, 38; m 61; c 3. ELECTROCHEMISTRY, ANALYTICAL CHEMISTRY. *Educ:* Mass Inst Technol, BS, 61; Univ Wis-Madison, PhD(chem), 66, PE(corrosion), 78. *Prof Exp:* Sr chemist, Corning Glass Works, 66-69; res specialist nonaqueous batteries, Res & Develop Div, Whittaker Corp, 69-72, res specialist anal chem, 72-76; SR RES ENGR, SOLAR GROUP, INT HARVESTER, 76-, CHIEF CHEMIST, 78- *Concurrent Pos:* QA mgr, S-Cubed, 81; consult, 88- *Mem:* Am Chem Soc; Electrochem Soc; Inst Elec & Electronics Engrs. *Res:* Polymer characterization and physical properties of materials; hot corrosion, QA (EPA protocols). *Mailing Add:* PO Box 2157 Julian CA 92036

STEVENS, WILLIAM Y(EATON), b South Portland, Maine, Nov 5, 31; m 66; c 2. COMPUTER SCIENCE, ENGINEERING PHYSICS. *Educ:* Bates Col, BS, 53; Cornell Univ, MS, 55, PhD(eng physics), 58. *Prof Exp:* Physicist, Gen Elec Co, 53; asst elec eng, Cornell Univ, 54-58; assoc engr, Int Bus Mach Corp, 58-60, staff systs planner, 60-63, adv engr, 63-69, SR ENGR, IBM CORP, 69- *Mem:* AAAS; Asn Comput Mach; Inst Elec & Electronics Engrs. *Res:* System design of digital computing and data processing systems; data communications, system reliability and maintainability. *Mailing Add:* IBM Corp PO Box 390 Poughkeepsie NY 12602

STEVENSON, BRUCE R, b Buffalo, NY, Sept 28, 52. EPITHELIAL CELL BIOLOGY. *Educ:* Harvard Univ, PhD(med sci), 83. *Prof Exp:* ASSOC BIOL, YALE UNIV, 83- *Mem:* Am Soc Cell Biol. *Res:* Biochemical characterization of tight junction. *Mailing Add:* Dept Anat & Cell Biol Univ Alta Edmonton AB T6G 2H7 Can

STEVENSON, CHARLES EDWARD, organic chemistry; deceased, see previous edition for last biography

STEVENSON, DAVID AUSTIN, b Albany, NY, Sept 6, 28; m 58; c 3. MATERIALS SCIENCE. *Educ:* Amherst Col, BA, 50; Mass Inst Technol, PhD(phys chem), 54. *Prof Exp:* Res assoc metall, Mass Inst Technol, 53-54, asst prof, 55-58; Fulbright scholar, Univ Munich, 54-55; PROF MAT SCI, STANFORD UNIV, 58- *Concurrent Pos:* Fulbright sr res fel, Max Planck Inst Phys Chem, 68, 69. *Mem:* Am Soc Metals; Am Inst Mining, Metall & Petrol Eng; Electrochem Soc; Sigma Xi. *Res:* Synthesis and properties of semiconducting materials and device applications; solid state electrochemistry; diffusion in compound semiconductors. *Mailing Add:* Dept of Mat Sci Stanford Univ Stanford CA 94305

STEVENSON, DAVID JOHN, b Wellington, New Zealand, Sept 2, 48. PLANETARY PHYSICS. *Educ:* Victoria Univ, BSc, 71, MS, 72; Cornell Univ, PhD(physics), 76. *Prof Exp:* Res fel, Earth Sci, Australian Nat Univc, 76-78; asst prof, Univ Calif, 78-80; assoc prof, 80-84, PROF PLANETARY SCI, CALIF INST TECHNOL, 84-, CHMN, DIV GEOL & PLANETARY SCI, 89- *Honors & Awards:* Fullbright Scholar, 71-76; Urey Prize, 84. *Mem:* Fel, Am Geophys Union, 86. *Res:* Origin, evolution and structure of planets, including Earth. *Mailing Add:* Caltech 17025 Pasadena CA 91125

STEVENSON, DAVID MICHAEL, b Shafton, Eng, Nov 1, 38; m 60; c 4. ELECTRICAL ENGINEERING, PHYSICS. *Educ:* Univ Leeds, BSc, 60; Cornell Univ, MS, 64, PhD(elec eng), 68. *Prof Exp:* Mem sci staff, Hirst Res Centre, Gen Elec Co Ltd, Eng, 60-63; consult, Cornell Aeronaut Lab, NY, 67-68; mem tech staff, David Sarnoff Res Ctr, RCA Labs, 38-73; eng mgr, 73-81, TECH DIR, SOLID STATE COMPONENTS OPER, VARIAN ASSOCS, 81- *Mem:* Inst Elec & Electronics Engrs; Am Phys Soc; Int Soc Hybrid Microelectronics. *Res:* Microwave electronics. *Mailing Add:* 41 Timberlane Topsfield MA 01983

STEVENSON, DAVID P(AUL), b Alameda, Calif, Jan 14, 14; m 40; c 2. CHEMISTRY. *Prof Exp:* Nat Res Coun Fel, Calif Inst Tech, 38-39; Asphalt Inst Fel, 39-40; fel, Westinghouse Elec & Mfg Co, 40-42; chemist, Shell Develop Co, 42-52; head dept chem physics, 52-57, dir fundamental res, 57-61, BASIC & GEN SCI RES, EMERYVILLE RES CTR, 61- *Mem:* NY Acad Sci. *Res:* Electron diffraction of gases; quantum mechanics of molecular structure; theory and analytical application of mass spectrometry; catalysis. *Mailing Add:* 390 Port Royal Ave San Mateo CA 94404

STEVENSON, DAVID STUART, b Virden, Man, Jan 23, 24; m 46; c 2. SOIL PHYSICS. *Educ:* Univ BC, BSA, 51; Ore State Univ, MSc, 56, PhD(soils), 63. *Prof Exp:* Res officer, Dom Exp Farm, Can Dept Agr, Sask, 56-57, Agr Res Sta, Alta, 62-66, res scientist, 66-78, sect head, Soil Sci & Agr Eng, Summerland Res Sta, 78-89; RETIRED. *Concurrent Pos:* Assoc ed, Can Jour Soil Sc, 79-84; mem Can Expert Comt on Soil & Water, 81-87. *Mem:* Sigma Xi; Can Soc Soil Sci. *Res:* Irrigation; soil-water-plant growth relationship. *Mailing Add:* 198 Dafoe Pl Penticton BC V2A 7E6 Can

STEVENSON, DENNIS A, b Mt Holly, NJ, Jan 25, 44; m 66; c 3. HEALTH PHYSICS, ENVIRONMENTAL MONITORING. *Educ:* Gettysburg Col, BA, 66; Univ Del, MS, 68, PhD(physics), 72; Am Bd Health Physics, Cert, 80. *Prof Exp:* Teaching res asst physics, Univ Del, 66-72; res assoc biophys, Univ Pittsburgh, 72-73; asst prof physics, Northeast La Univ, 73-77; asst health physics officer, Walter Reed Army Med Ctr, 77-80, health physics officer, 80-81; health physics officer/radiation protection officer, Dwight D Eisenhower Army Med Ctr, 81-83; process physicist, Westinghouse Savannah River Co, 83, area suprv, 83-84, tech supvr, 84-86, chief supvr, 86-88, mgr, environ monitoring, 88-89, MGR, HEALTH PROTECTION OPERS, WESTINGHOUSE SAVANNAH RIVER CO, 89- *Concurrent Pos:* Mem health physics, ANSI Standard Comt Multiple Dosimetry. *Mem:* Health Physics Soc; Am Acad Health Physics; Sigma Xi; Sci Res Soc NAm; Am Phys Soc. *Res:* Physical studies of biologically important macromolecules, protein-nucleic acid interactions, virology, effects of various ionizing radiations on macromolecules and living systems; applied health physics; radiation accident emergency preparedness; environmental monitoring. *Mailing Add:* 1316 Martinique Dr Augusta GA 30909

STEVENSON, DON R, b Syracuse, NY, Dec 19, 44; m 66; c 3. FLAME RETARDANTS. *Educ:* Syracuse Univ, BA, 66; Univ Ariz, PhD(organic polymers), 71. *Prof Exp:* fel, Univ mass, 71-72; researcher plastics, Glidden, 72-76; tech dir, Union Camp, 76-79; TECH DIR DEVELOP, DOVER CHEM, 79- *Mem:* Am Chem Soc. *Res:* Polymers and industrial polymer applications for reverse osmosis membrances, low profile plastics and coatings and flame retarded additives. *Mailing Add:* ICC Indust Inc Dover Chem Corp 15th & Davis Sts PO Box 40 Dover OH 44622-9712

STEVENSON, DONALD THOMAS, b Washington, DC, Sept 8, 23; m 46; c 3. SOLID STATE PHYSICS. *Educ:* Cornell Univ, AB, 44; Mass Inst Technol, PhD(physics), 50. *Prof Exp:* Asst physics, 49-50, res assoc, 50-51, mem staff, Lincoln Lab, 51-53, asst group leader solid state physics, 53-57, group leader, 57-61, asst dir, 60-88, VIS SCIENTIST, FRANCIS BITTER NAT MAGNET LAB, MASS INST TECHNOL, 88- *Mem:* AAAS; Am Phys Soc. *Res:* Semiconductors; high magnetic fields. *Mailing Add:* Francis Bitter Nat Magnet Lab Mass Inst Technol Rm NW14-3214 Cambridge MA 02139

STEVENSON, ELMER CLARK, b Pine City, Wash, Aug 20, 15; m 39; c 6. HORTICULTURE. *Educ:* Univ Md, BS, 37; Univ Wis, PhD(agron, plant path), 42. *Prof Exp:* Asst plant path, Univ Wis, 38-42; from asst plant pathologist to assoc plant pathologist, Drug Plant Invests, US Dept Agr, 42-48; from assoc prof to prof hort, Purdue Univ, 48-67, head dept, 58-67; prof, assoc dean agr & dir resident instruct, 67-80, EMER PROF HORT, EMER ASSOC DEAN & EMER DIR RESIDENT INSTR AGR, ORE STATE UNIV, 80- *Concurrent Pos:* Consult, US Dept Agr & Univ Ky, 58, US Agency Int Develop, Brazil, 62 & US Dept Agr & Miss State Univ, 64. *Mem:* Fel Am Soc Hort Sci. *Res:* Corn diseases and breeding; diseases of medicinal and special crops; mint breeding and production; vegetable breeding and genetics. *Mailing Add:* 8240 NW Chaparral Dr Corvallis OR 97330

STEVENSON, ENOLA L, b Feb 20, 39; US citizen. PLANT PHYSIOLOGY. *Educ:* Southern Univ, BS, 60; Univ NH, MS, 62, PhD(plant physiol), 68. *Prof Exp:* Res asst plant physiol, Univ NH, 60-62, 67-68; instr bot, Southern Univ, 62-64; asst prof, 68-72, ASSOC PROF BIOL, ATLANTA UNIV, 72- *Res:* Effects of light quality and intensity on plant growth and metabolism. *Mailing Add:* Rte 1 Box 60 Zachary LA 70791

STEVENSON, EUGENE HAMILTON, organic chemistry, nutrition; deceased, see previous edition for last biography

STEVENSON, EVERETT E, b Buffalo, NY, Jan 14, 23; m 45; c 3. MATHEMATICS. *Educ:* State Univ NY Col Buffalo, BS, 44; Univ Houston, MEd, 52; Ohio State Univ, PhD(math, math ed), 61. *Prof Exp:* From instr to assoc prof math, US Air Force Acad, 56-67, chief enrichment br, 64-66, exec officer, 66-67; fac mem, Indust Col Armed Forces, 67-68; assoc chmn dept, 69-79 & 80-83, chmn, 79-80, PROF MATH SCI, MEMPHIS STATE UNIV, 69-, PROF, 83- *Mem:* Nat Coun Teachers Math; Math Asn Am. *Res:* Mathematics education. *Mailing Add:* Dept Math Sci Memphis State Univ Memphis TN 38152

STEVENSON, F DEE, b Ogden, Utah, June 7, 33; m 51; c 5. CHEMICAL ENGINEERING. *Educ:* Univ Utah, BS, 55; Ore State Univ, PhD, 62. *Prof Exp:* Asst eng, Calif Res Corp, Standard Oil Co Calif, 55-57; from asst prof to prof chem eng, Iowa State Univ, 62-74; prof term with Chem Off, Div Phys Res, AEC, 72-74; prog mgr, Mat Sci & Molecular Sci Off, Div Basic Energy Sci, ERDA, 74-77; BR CHIEF, CHEM SCI DIV, OFF ENERGY RES, DEPT OF ENERGY, 77- *Mem:* Am Chem Soc; Am Inst Chem Engrs. *Res:* Kinetics of reactions; statistical application to data analysis and sequential experimental design; anhydrous separation and purification metals; thermodynamics of solutions, including liquid metals; high temperature and vacuum processing. *Mailing Add:* Div Chem Sci ER142 Off Energy Res Dept Engergy MS G-226 Washington DC 20545

STEVENSON, FORREST FREDERICK, b Kismet, Kans, Nov 12, 16; m 47; c 1. PLANT MORPHOLOGY. *Educ:* Cent Mo State Col, BS, 46; Univ Mo, MA, 48; Univ Mich, PhD(bot), 56. *Prof Exp:* Instr biol, Univ Kans City, 48-50 & McCook Jr Col, 50-51; from asst prof to prof biol, Ball State Univ, 55-82. *Mem:* Bot Soc Am; Am Bryol & Lichenological Soc. *Res:* Experimental plant morphology. *Mailing Add:* 1806 N Forest Ave Muncie IN 47304

STEVENSON, FRANK JAY, b Logan, Utah, Aug 2, 22; m 56; c 3. SOILS. *Educ:* Brigham Young Univ, BS, 49; Ohio State Univ, PhD(agron), 52. *Prof Exp:* From asst prof to assoc prof, 53-62, PROF SOIL CHEM, UNIV ILL, 62- *Honors & Awards:* Agron Res Award, Am Soc Agron, 80; Soil Sci Res Award, Soil Sci Soc Am, 83. *Mem:* Soil Sci Soc Am; Am Soc Agron; Int Soil Sci Soc. *Res:* Biochemical properties of soils; chemistry of soil organic matter. *Mailing Add:* Dept Agronomy Univ Ill Urbana IL 61801

STEVENSON, FRANK ROBERT, b Brooklyn, NY, Aug 29, 31; m 56; c 6. SOLID STATE PHYSICS. *Educ:* Polytech Inst Brooklyn, BS, 53, MS, 59. *Prof Exp:* Physicist, Sperry Gyroscope Co, NY, 53-56 & Curtis Wright Corp, Pa, 56-58; res assoc, RIAS Div, Martin Co, Md, 58-63; physicist, Lewis Res Ctr, NASA, Ohio, 63-71; environ scientist, Ford Motor Co, 71-90; ENVIRON CONSULT, 90- *Concurrent Pos:* Lectr, Goucher Col, 61-62 & John Carroll Univ, 64-65. *Mem:* Am Phys Soc. *Res:* Radiation damage; microwave electronics; low temperature physics; accelerators; air pollution; industrial noise control; energy conservation; water pollution management. *Mailing Add:* 18807 Ironwood Ave Cleveland OH 44110

STEVENSON, G W, b LaPorte, Ind, Sept 7, 51. PEDIATRIC ANESTHESIA. *Educ:* DePauw Univ, BA, 73; Ind Univ, Indianapolis, MD, 77; Am Bd Med Examr, dipl, 78; Am Bd Anesthesiol, dipl, 81. *Prof Exp:* Assoc, 81-86, ASST PROF CLIN ANESTHESIA, NORTHWESTERN UNIV MED SCH, 86- *Concurrent Pos:* Fel pediatric anesthesia, Children's Mem Hosp Chicago, 80-81 & Philadelphia, 81; provisional attend staff anesthesiologist, Children's Mem Hosp, Chicago, 81-82, Med co-dir, Outpatient Surg, 83-, dir anesthesia res, 87-, mem, Disaster Planning Comt, 85, Code Comt, 85 & Operating Room Mgt Comt, 86-; mem, Resident Eval Comt, Northwestern Univ Med Sch, 85-89; fel Am Acad Pediat, 89. *Mem:* Int Anesthesia Res Soc; Am Soc Anesthesiologists; Soc Ambulatory Anesthesia; Soc Pediat Anesthesia; Soc Leukocyte Biol; Am Fedn Clin Res; Am Asn Immunologists. *Res:* Effects of anesthetic agents on immunologic function; co-author of numerous publications. *Mailing Add:* Dept Anesthesia Children's Mem Hosp 2300 Children's Plaza Box 19 Chicago IL 60614

STEVENSON, GEORGE FRANKLIN, b St Thomas, Ont, Sept 13, 22; US citizen; m 45, 77; c 2. PATHOLOGY. *Educ:* Univ Western Ont, BA, 44, MD, 45. *Prof Exp:* Prof clin path, dean, Sch Allied Health Sci & dir, Med Technol Prog, Med Col SC, 66-71; dep comnr med technol, 65-67, comnr continuing educ, 67-72, exec vpres, 71-74, SR VPRES, AM SOC CLIN PATHOLOGISTS, 74-; PROF PATH, MED SCH, NORTHWESTERN UNIV, CHICAGO, 71- *Honors & Awards:* Burdick Award, Am Soc Clin Pathologists, 84. *Mem:* Int Acad Path; Am Asn Pathologists; Col Am Pathologists; Am Soc Clin Pathologists. *Res:* Pathology and administrative medicine. *Mailing Add:* Am Soc Clin Pathologists 2100 W Harrison St Chicago IL 60612

STEVENSON, HARLAN QUINN, b Waynesboro, Pa, Apr 1, 27; m 60; c 2. CYTOGENETICS, RADIOBIOLOGY. *Educ:* Pa State Univ, BS, 50; Univ Fla, PhD(radiation biol), 63. *Prof Exp:* Asst bot, Pa State Univ, 50-51 & Cornell Univ, 51-56; res assoc biol, Brookhaven Nat Lab, 56-60; asst prof, Univ Fla, 63-64; from asst prof to assoc prof, 64-72, chmn dept, 75-84, PROF BIOL, SOUTHERN CONN STATE UNIV, 72- *Mem:* AAAS; Soc Study Evolution; Am Inst Biol Sci; Am Soc Human Genetics; Genetics Soc Am; NY Acad Sci. *Res:* Chemical and radiation induced chromosomal aberrations; genetic and radiation effects in plant tumors; evolutionary and practical significance of multiple allopolyploidy; genetic counseling; bioethics. *Mailing Add:* Dept of Biol Southern Conn State Univ New Haven CT 06515

STEVENSON, HENRY C, b Long Island, NY, Sept 18, 48. CANCER IMMUNOLOGY. *Educ:* Stanford Univ, MD, 75. *Prof Exp:* SR INVESTR, FCRF, NAT CANCER INST, 80- *Mem:* Am Col Physicians; Am Col Allergists; Soc Biol Ther; Am Fedn Clin Res. *Mailing Add:* 10114 Thornwood Dr Kensington MD 20895

STEVENSON, HENRY MILLER, b Birmingham, Ala, Feb 25, 14; m 39; c 4. ORNITHOLOGY. *Educ:* Birmingham-Southern Col, AB, 35; Univ Ala, MS, 39; Cornell Univ, PhD(ornith), 43. *Prof Exp:* Lab asst geol, Birmingham-Southern Col, 35-36; lab asst biol, Univ Ala, 38-39; lab asst bot, Vanderbilt Univ, 40-41; lab asst ornith, Cornell Univ, 42-43; actg asst prof biol, Univ Miss, 43-44; assoc prof, Emory & Henry Col, 44-46; from asst prof to prof, 46-75, EMER PROF ZOOL, FLA STATE UNIV, 75-; RES FEL, TALL TIMBERS RES STA, 75- *Concurrent Pos:* Consult, Conserv Consults, Inc, 73-76; ed, Fla Field Naturalist, 73-76; Dept Game & Fresh Water Fish grant, 85-90. *Mem:* Am Ornith Union. *Res:* Avian taxonomy; quantitative field studies of birds; geographical distribution and migration of birds. *Mailing Add:* Tall Timbers Res Sta Tallahassee FL 32312

STEVENSON, IAN, b Montreal, Que, Oct 31, 18; nat US; m 47. MEDICINE, PSYCHIATRY. *Educ:* McGill Univ, BSc, 40, MD, CM, 43. *Prof Exp:* Intern & asst resident med, Royal Victoria Hosp, Montreal, 44-45; from intern to resident, St Joseph's Hosp, Phoenix, Ariz, 45-46; fel internal med, Ochsner Med Found, New Orleans, La, 46-47; Commonwealth fel med, Med Col, Cornell Univ, 47-49; asst prof med & psychiat, Sch Med, La State Univ, 49-52, assoc prof psychiat, 52-57; prof & chmn dept neurol & psychiat, 57-67, CARLSON PROF PSYCHIAT, SCH MED, UNIV VA, 67- *Concurrent Pos:* Consult, New Orleans Parish Sch Bd, 49-52, State Dept Pub Welfare, 50-52 & Southeast La State Hosp, Mandeville, 52-57; vis physician, Charity Hosp, New Orleans, 52-57; hon mem staff, DePaul Hosp, 52-57; psychiatrist-in-chief, Univ Va Hosp, 57-67. *Mem:* AAAS; Am Psychosom Soc; Am Soc Psychical Res; Am Psychiat Asn; AMA; Soc Sci Explor. *Res:* Experimental psychoses; psychotherapy; paranormal phenomena. *Mailing Add:* Dept Psychiat Univ Va Sch Med Charlottesville VA 22903

STEVENSON, IAN LAWRIE, b Hamilton, Ont, Dec 28, 26; m 53; c 2. AGRICULTURAL MICROBIOLOGY, CYTOLOGY. *Educ:* Ont Agr Col, BSA, 49; Univ Toronto, MSA, 51; Univ London, PhD(microbiol), 55. *Prof Exp:* From bacteriologist to sr bacteriologist, Microbiol Res Inst, Can Dept Agr, 51-67, head, Physiol & Nutrit Unit, 59-67, Cytol & Physiol Unit, Chem & Biol Res Inst, Ont, 67-72, assoc dir, Res Sta, Lethbridge, Alta, 72-74, prin res scientist, Chem & Biol Res Inst, Can Dept Agr, 74-86; RETIRED. *Concurrent Pos:* Lectr, Univ Ottawa, 55-59; vis scientist, Nat Inst Med Res, Mill Hill, 63-64. *Mem:* Am Soc Microbiol; Can Soc Microbiol; Brit Soc Gen Microbiol. *Res:* Physiology and growth of micro-organisms; microbiology of the soil; electron microscopy; cytology. *Mailing Add:* 14 Marielle Ct Ottawa ON K2B 8L6 Can

STEVENSON, IRONE EDMUND, JR, b Linthicum, Md, Apr 21, 30; m 60; c 2. INTERMEDIARY METABOLISM. *Educ:* Univ Md, BS, 53; Univ Pa, PhD(biochem), 61. *Prof Exp:* Asst instr biochem, Univ Pa, 54-58; asst zool, Yale Univ, 60-63; chemist, 63-81, sr res chemist, 81-87, RES ASSOC, E I DU PONT DE NEMOURS, 87- *Mem:* Am Soc Biol Chem. *Res:* Degradation of cholesterol by mammalian enzymes; intermediary metabolism of insects; metabolism and environmental fate of pesticides. *Mailing Add:* Agr Prod Dept E I du Pont de Nemours Wilmington DE 19880

STEVENSON, J(OSEPH) ROSS, b Canton, China, Sept 4, 31; US citizen; m 54; c 3. EFFECTS OF STRESS. *Educ:* Oberlin Col, BA, 53; Northwestern Univ, MS, 55, PhD, 60. *Prof Exp:* Asst zool & chem, Oberlin Col, 52-53; asst biol, Northwestern Univ, 53-55; instr, Chatham Col, 56-59; res assoc zool, Univ Wash, 59-60; from instr to assoc prof, 60-71, assoc dean grad col, 73-74, PROF BIOL SCI, KENT STATE UNIV, 71- *Concurrent Pos:* Jacques Loeb assoc, Rockefeller Univ, 63-64. *Mem:* AAAS; Am Soc Zool; Am Soc Cell Biol. *Res:* Effects of stress and hormones on the immune system; lymphocyte functions. *Mailing Add:* Dept Biol Sci Kent State Univ Kent OH 44242

STEVENSON, JAMES FRANCIS, b Greenville, Pa, July 15, 43; m 71; c 2. POLYMER PROCESSING, COMPOSITE MATERIALS. *Educ:* Rensselaer Polytech Inst, BChE, 65; Univ Wis, Madison, MS, 67, PhD(chem eng), 70. *Prof Exp:* NIH fel, Columbia Univ, 70-71; from asst prof to assoc prof chem eng, Cornell Univ, 71-77; res scientist, 77-79, group leader, 79-81, sect head, 81-86, MGR GENCORP, INC, 86- *Mem:* Am Inst Chem Engrs; Am Chem Soc; Soc Plastics Engrs; Soc Rheol; Polymer Processing Soc (treas, 85-88); Soc Advan Mat Process Eng. *Res:* Innovation, optimization, and control of processes for thermoplastic and thermoset composite materials; processes include extrusion, coating, injection transfer and compression, and molding; invention of net shape processes for efficient manufacturing; polymer rheology. *Mailing Add:* 2990 Gilchrist Rd Akron OH 44305

STEVENSON, JAMES RUFUS, b Trenton, NJ, May 19, 25; m 55; c 3. SURFACE PHYSICS. *Educ:* Mass Inst Technol, SB, 50; Univ Mo, PhD(physics), 58. *Prof Exp:* Res participant, Oak Ridge Nat Lab, 55; asst prof, 55-62, actg dir, 68-69, ASSOC PROF PHYSICS, GA INST TECHNOL, 62-, DIR SCH PHYSICS, 69- *Concurrent Pos:* Physicist, US Naval Res Lab, DC, 58, consult, 60-67; Fulbright-Hays vis prof, Univ Sci & Technol, Ghana, 65-66; mem comt applns physics, Am Phys Soc, 75-78; chmn comt educ, Am Phys Soc, 78. *Mem:* Am Phys Soc; Am Asn Physics Teachers; Optical Soc Am; Am Soc Eng Educ; Sigma Xi. *Res:* Synchrotron radiation, Auger spectroscopy and optical surface studies of metals, metal oxides, and semiconductors with applications to corrosion. *Mailing Add:* Sch of Physics Ga Inst of Technol Atlanta GA 30332

STEVENSON, JEAN MOORHEAD, b Circleville, Ohio, Oct 2, 04; m 40; c 3. SURGERY. *Educ:* Miami Univ, AB, 26; Univ Cincinnati, MB, 30, MD, 31. *Prof Exp:* Resident surg, Cincinnati Gen Hosp, 31-33 & 34-37 & Univ Calif, 33-34; from instr to assoc prof, 37-61, PROF SURG, COL MED, UNIV CINCINNATI, 61- *Mem:* Soc Univ Surgeons; Soc Clin Surgeons; AMA; Am Col Surgeons; Int Soc Surg; Sigma Xi. *Res:* Wound healing; development of technics for the management of wounds of violence; care of tissues in all surgical wounds. *Mailing Add:* 6345 Grand Vista Ave Pleasant Ridge OH 45242

STEVENSON, JEFFREY SMITH, b Salt Lake City, Utah, June 15, 51; m 74. REPRODUCTIVE PHYSIOLOGY, REPRODUCTIVE ENDOCRINOLOGY. *Educ:* Utah State Univ Logan, BS, 75; Mich State Univ, Lansing, MS, 77; NC State Univ, Raleigh, PhD(animal physiol), 80. *Prof Exp:* Grad res asst dairy sci, Mich State Univ, 75-77, animal sci, NC State Univ, 77-80; asst prof, 80-86, ASSOC PROF ANIMAL SCI, KANSAS STATE UNIV, 86- *Concurrent Pos:* Prin investr, Kans Agr Exp Sta, 80-; physiol comt, Am Dairy Sci Asn, 84-86. *Honors & Awards:* Young Scientist Award, Am Dairy Sci Asn, 90. *Mem:* Am Soc Animal Sci; Am Dairy Sci Asn; Soc Study Reproduction; Sigma Xi. *Res:* Physiologic and endocrinological factors associated with postpartum fertility of dairy cattle including estrous behavior, estrous detection, function of corpus luteum and other hormone interactions and treatments that may improve conception rates. *Mailing Add:* Dept Animal Sci Kansas State Univ Manhattan KS 66506-0201

STEVENSON, JOHN CRABTREE, b Everett, Wash, Feb 24, 37; m 60; c 3. MATHEMATICS. *Educ:* NY Univ, BA, 63, MS, 63; Adelphi Univ, PhD(math), 70. *Prof Exp:* From instr to assoc prof, 68-74, chmn dept, 72-78, PROF MATH, C W POST COL, LONG ISLAND UNIV, 74- *Concurrent Pos:* C W Post Col grant, dept physics, Imp Col, Univ London, 70-71. *Mem:* AAAS; Math Asn Am; Am Math Soc; Soc Indust & Appl Math. *Res:* Numerical solution of hyperbolic partial differential equations; plasma physics in the solar atmosphere and magnetosphere; multiple pool analysis of metabolic pathways. *Mailing Add:* 266 Southdown Rd Huntington NY 11743

STEVENSON, JOHN DAVID, b St Louis, Mo, Aug 12, 50; m 88. HEAVY-ION REACTION MECHANISMS, EXOTIC NUCLEI. *Educ:* Univ Ill, Urbana, BS, 72; Univ Calif, Berkeley, PhD(physics) 77. *Prof Exp:* ASST PROF PHYSICS, PHYSICS & ASTRON DEPT & NAT SUPERCONDUCTING CYCLOTRON LAB, MICH STATE UNIV, 84- *Mem:* Am Phys Soc. *Res:* Heavy ion reaction mechanisms at intermediate energies; study of the properties of nuclei far from stability. *Mailing Add:* Dept Physics & Astron Mich State Univ Cyclotron Lab East Lansing MI 48824

STEVENSON, JOHN RAY, b Ordway, Colo, May 10, 43; m. IMMUNOLOGY, MEDICAL MICROBIOLOGY. *Educ:* Kans State Teachers Col, BA, 65, MS, 67; Case Western Reserve Univ, PhD(microbiol), 72. *Prof Exp:* Asst prof biol & med, Biol Dept, Univ Mo, Kansas City, 74-80; vis asst prof, 80-83, asst prof, 83-87, ASSOC PROF MICROBIOL, MIAMI UNIV, OHIO, 87- *Mem:* AAAS; Am Soc Microbiol; Sigma Xi; Am Asn Immunologists. *Res:* Mechanisms of immunodeficiency induced by protein and/or zinc malnutrition; immune system development; cytokines in immune responses; cell-mediated immunity and host-parasite relationships; heparin activity. *Mailing Add:* Dept Microbiol Miami Univ Oxford OH 45056

STEVENSON, KENNETH EUGENE, b Modesto, Calif, June 3, 42; m 69; c 1. FOOD MICROBIOLOGY. *Educ:* Univ Calif, Davis, BS, 64, PhD(microbiol), 70. *Prof Exp:* Instr biol, Napa Col, Calif, 71; res microbiologist, Univ Calif, Davis, 71; asst prof, 71-76, ASSOC PROF FOOD MICROBIOL, MICH STATE UNIV, 76- *Concurrent Pos:* Sci adv, US Food & Drug Admin, Detroit, 72- *Mem:* Int Asn Milk, Food & Environ Sanit; Am Soc Microbiol; Inst Food Technol. *Res:* Microbiological analyses of foods; food poisoning microorganisms; microbiological aspects of plant sanitation and waste disposal; use of microorganisms in the production of food. *Mailing Add:* 2721 Fleetwood Dr San Bruno CA 94066

STEVENSON, KENNETH JAMES, b Calgary, Alta, Apr 16, 41; m 64; c 2. PROTEIN CHEMISTRY. *Educ:* Univ Alta, BSc, 62, PhD(biochem), 66. *Prof Exp:* Med Res Coun fel, Lab Molecular Biol, Cambridge Univ, 66-67; Killam fel, Univ BC, 67-69; from asst prof to assoc prof, 69-82, PROF BIOCHEM, UNIV CALGARY, 82- *Mem:* Can Biochem Soc; Brit Biochem Soc; Am Soc Biol Chemists; Sigma Xi; Protein Soc; NY Acad Sci. *Res:* Structure and function of dithiol reductases from Protozoa and archaebacteria; trivalent arsenicals. *Mailing Add:* Dept Biol Sci Univ Calgary Calgary AB T2N 1N4 Can

STEVENSON, KENNETH LEE, b Ft Wayne, Ind, Aug 1, 39; m 59; c 2. PHOTOCHEMISTRY. *Educ:* Purdue Univ, BS, 61, MS, 65; Univ Mich, PhD(phys chem), 68. *Prof Exp:* Teacher high schs, Ind & Mich, 61-65; from asst prof to assoc prof chem, 68-78, actg dean, Sch Sci & Humanities, 86-87, PROF CHEM, PURDUE UNIV, FT WAYNE, 78-, CHMN CHEM DEPT, 79-86 & 87- *Concurrent Pos:* Fel, Chem Dept, NMex State Univ, 75-76; sabbatical vis, Solar Energy Res Inst, Colo, 80. *Mem:* Sigma Xi; Am Chem Soc; Inter-Am Photochem Soc. *Res:* Photochemistry and spectra of coordination compounds of copper (I); kinetics and thermodynamics of ligand exchange reactions; induction of optical activity using light; photochemical conversion of solar energy. *Mailing Add:* Dept Chem Ind Univ-Purdue Univ Ft Wayne IN 46805-1499

STEVENSON, L HAROLD, b Bogalusa, La, Mar 18, 40; m 61; c 1. MICROBIOLOGY. *Educ:* Southeastern La Col, BS, 62; La State Univ, MS, 64, PhD(microbiol), 67. *Prof Exp:* Asst prof, 67-71, ASSOC PROF BIOL, UNIV SC, 71-, ASSOC PROF MARINE SCI & MICROBIOL, 73-; AT DEPT DIOL & ENVIRON SCI, MCNEESE STATE UNIV. *Mem:* AAAS; Am Soc Microbiol. *Res:* Bacterial ecology; distribution, activity and taxonomy of estuarine bacteria; marsh ecology. *Mailing Add:* Dept of Biol & Environ Sci Mc Neese State Univ Lake Charles LA 70609-2000

STEVENSON, LOUISE STEVENS, b Seattle, Wash, July 28, 12; Can citizen; wid; c 2. MINERALOGY. *Educ:* Univ Wash, BS, 32; Radcliffe Col, AM, 33. *Prof Exp:* Res asst climat, US Weather Bur, Seattle, 34; lectr geol & geog, Victoria Col, BC, 48-49; mus assoc geol, 51-57, cur, 57-80, HON CUR GEOL, REDPATH MUS, McGILL UNIV, 80- *Concurrent Pos:* Convener sect 17, Int Geol Cong, 70-72. *Mem:* Fel Geol Asn Can; Mineral Soc Am; Mineral Asn Can; Sigma Xi (hon secy, 73-84). *Res:* Petrogenesis of rare minerals; mineralogy applied to medicine and dentistry; petrology of siliceous lavas; adult education in mineralogy; geological education through university museums; origin of the Sudbury, Ontario orebody. *Mailing Add:* Redpath Mus McGill Univ 859 Sherbroket W Montreal PQ H3A 2K6 Can

STEVENSON, MARY M, b Philadelphia, Pa, Sept 10, 51; m 90. GENETIC CONTROL OF HOST RESISTANCE. *Educ:* Hood Col, BA, 73; Cath Univ, MS, 77, PhD(microbiol), 79. *Prof Exp:* Res asst, Nat Cancer Inst, NIH, 74-79; res fel, Montreal Gen Hosp Res Inst, 79-81, res assoc, 81-82; asst prof, 82-88, ASSOC PROF, DEPT MED, MCGILL UNIV, 88-, ASSOC MEM, DEPT PHYSIOL, 84- *Concurrent Pos:* Scholar, Med Res Coun, Can, 82-87. *Mem:* Sigma Xi; Reticuloendothelial Soc; Am Soc Microbiol; Am Asn Immunologists; Am Asn Trop Med & Hyg. *Res:* Genetic control of resistance to infection with intracellular pathogens, using as a model inbred, recombinant and congenic strains of mice; determination of the underlying mechanisms leading to resistance to malaria, Plasmodium chabaudi; immunology of host response to malaria. *Mailing Add:* Montreal Gen Hosp Res Inst 1650 Cedar Ave Montreal PQ H3G 1A4 Can

STEVENSON, MERLON LYNN, b Salt Lake City, Utah, Oct 31, 23; m 48; c 5. PARTICLE PHYSICS. *Educ:* Univ Calif, AB, 48, PhD(physics), 53. *Prof Exp:* From asst to lectr, 48-58, from asst prof to assoc prof, 58-64, PROF PHYSICS, UNIV CALIF, BERKELEY, 64-, FAC SR SCIENTIST, LAWRENCE BERKELEY LAB, 51- *Concurrent Pos:* NSF sr fel & vis prof physics, Inst High Energy Physics, Univ Heidelberg, 66-67. *Mem:* Am Phys Soc; AAAS; NY Acad Sci. *Res:* Neutrino physics and new particle search; electron-positron physics. *Mailing Add:* Dept Physics Univ Calif Berkeley CA 94720

STEVENSON, MICHAEL GAIL, b Little Rock, Ark, Jan 10, 43; m 64. NUCLEAR ENGINEERING. *Educ:* Univ Tex, Austin, BEngSc, 64, PhD(nuclear eng), 68. *Prof Exp:* Reactor safety analysis, Babcock & Wilcox Co, 68-71; mem staff & sect mgr, Argonne Nat Lab, 71-74; mem staff & group leader fast reactor safety, Los Alamos Nat Lab, 74-76, asst energy div leader, 76-79, assoc leader reactor safety, 79-81, dep energy div leader, 81-83, energy div leader, 83-86, DEP ASSOC DIR, ENERGY & TECHNOL, LOS ALAMOS NAT LAB, 87- *Mem:* Am Nuclear Soc. *Res:* Development of computational methods for analysis of reactor accidents; design and analysis of experiments related to reactor safety analysis; research program planning and management. *Mailing Add:* 1 Mariposa Ct White Rock NM 87544

STEVENSON, NANCY ROBERTA, b Vinton, Iowa, Feb 14, 38; m 73; c 4. PHYSIOLOGY, NUTRITION. *Educ:* Univ Northern Iowa, BS, 60; Rutgers Univ, MS, 63, PhD(nutrit), 69. *Prof Exp:* Nat Inst Arthritis, Metab & Digestive Dis fel, 69-71, instr physiol, 71-72, asst prof, 72-78, ASSOC PROF PHYSIOL, UNIV MED & DENT NJ, RW JOHNSON MED SCH, 78- *Mem:* AAAS; Am Gastroenterol Asn; Am Dietetic Asn; Am Physiol Soc; Int Soc Chronobiol. *Res:* Gastrointestinal digestion and absorption, intestinal blood flow; circadian rhythms. *Mailing Add:* Dept Physiol & Biophysics Univ Med & Dent NJ RW Johnson Med Sch Piscataway NJ 08854

STEVENSON, PAUL MICHAEL, b Denham, Eng, Oct 10, 54. RENORMALIZATION, NONPERTURBATIVE QUANTUM FIELD THEORY. *Educ:* Cambridge Univ, BA, 76; Imperial Col, London, DIC, 79; London Univ, PhD(theoret physics), 79. *Prof Exp:* Res assoc, dept physics, Univ Wis, Madison, 79-81 & 83-84; CERN fel, Europ Orgn Nuclear Res, 81-83; asst prof, 84-89, ASSOC PROF PHYSICS, PHYSICS DEPT, RICE UNIV, 89- *Res:* Theoretical elementary particle physics; optimization of perturbative quantum; chromodynamics calculations using renormalization-group invariance; nonperturbative studies of quantum field theory models using the Gaussian effective potential approach. *Mailing Add:* Dept Physics Rice Univ PO Box 1892 Houston TX 77251-1892

STEVENSON, RALPH GIRARD, JR, b Jersey Shore, Pa, Feb 14, 25; m 52; c 4. MINERALOGY, PETROLOGY. *Educ:* Univ NMex, BS, 49, MS, 50; Ind Univ, PhD(geol), 65. *Prof Exp:* Geol engr, Water Resources Div, US Geol Surv, 50-51; geologist, Skelly Oil Co, 51-55 & Shell Develop Co, 55-61; res geologist, Gulf Res & Develop Co, 64-66, staff geologist, Tech Serv Ctr, Gulf Oil Corp, 66-68; from asst prof to assoc prof, Univ S Fla, 68-78; sr proj geologist, Gulf Res & Develop Co, 78-80, dir, Reservoir Petrol Sect, 80-; RETIRED. *Mem:* Am Asn Petrol Geologists; Asn Prof Geol Scientists; Am Inst Mining, Metall & Petrol Eng. *Res:* Mineralogy and petrology of polymetamorphic and sedimentary rocks; crystal chemistry and geochemistry of phosphate minerals and clay minerals at atmospheric conditions. *Mailing Add:* 171 Horseshoe Lane Houston TX 77236

STEVENSON, RICHARD MARSHALL, b Detroit, Mich, July 2, 23; m 43; c 2. ELECTROPLATING, ORGANIC CHEMISTRY. *Educ:* Detroit Inst Technol, BS, 48. *Prof Exp:* Mfg chemist, Parke Davis & Co, 48-52; analytical chemist, Difco Labs, 52-53, Cadillac Motor Car Div, Gen Motors Corp, 53-58; sr org res chemist, div oxy metal industs, Udylike Corp, 58-75, McGean Chem Co, 75-77; RES DIR, DETROIT PLASTIC MOLDING CO, 78- *Concurrent Pos:* Pres, Richard M Stevenson & Co Consult Chemists, 77- *Mem:* Fel Am Inst Chemists; Am Chem Soc; Am Electroplaters Finishers Soc. *Res:* Electroplating plastics; electroless copper and nickel processes; electroplating acid cooper, nickel and zinc. *Mailing Add:* 2179 Allard Ave Grosse Point MI 48236-1911

STEVENSON, ROBERT EDWIN, b Columbus, Ohio, Dec 2, 26. MICROBIOLOGY, SCIENCE ADMINISTRATION. *Educ:* Ohio State Univ, BSc, 47, MSc, 50, PhD(bact), 54; Am Bd Microbiol, dipl. *Prof Exp:* Res assoc, US Pub Health Serv, 52-54, virologist, 54-58; head tissue cult div, Tissue Bank Dept, US Naval Med Sch, 58-60; head, Cell Cult & Tissue Mat Sect, Virol Res Resources Br, Nat Cancer Inst, 60-62, actg chief, 62-63, chief, 63-66, chief, Viral Carcinogenesis Br, 66-67; mgr biol sci, Develop Dept, Union Carbide Res Inst, NY, 67-71; vpres, Litton Bionetics & gen mgr, Nat Cancer Inst-Frederick Cancer Res Ctr, 72-80; DIR, AM TYPE CULT COLLECTION, 80- *Concurrent Pos:* Mem, Nat Inst Allergy & Infectious Dis, bd virus reference reagent, 63-65; cell cult comt, Int Asn Microbiol Soc, 63-67; comt transplantation, Nat Acad Sci-Nat Res Coun, 66-70; chmn cell cult comt, Am Type Cult Collection, 71-75, mem bd trustees, 72-; founding mem Am Asn Tissue Banks, chmn Cell & Tumor Coun, 77-81; fel Hastings Inst; vpres, World Fedn Cult Collections, 81-; chmn, Biotechnol Tech Adv

Comt, US Dept Commerce, 87- *Honors & Awards:* IR 100 Award, 70. *Mem:* AAAS; Am Soc Microbiol; Soc Cryobiol; Tissue Cult Asn; Am Asn Tissue Banks; Fed Culture Collections (pres, 88-90); Tissue Culture Asn (pres, 88-90). *Res:* Viral oncology; biomedical instrumentation; biological standardization. *Mailing Add:* Am Type Cult Collection 12301 Parklawn Dr Rockville MD 20852

STEVENSON, ROBERT EVANS, b Des Moines, Iowa, May 5, 16; m 48; c 2. OCEANOGRAPHY. *Educ:* Univ Hawaii, BS, 39; State Col Wash, MS, 42; Lehigh Univ, PhD(geol), 50. *Prof Exp:* Asst, Univ Hawaii, 39; lab asst, State Col Wash, 39-42; geologist, Wash State Div Geol, 42-44; field geologist, Venezuelan Atlantic Ref Co, 44-46; instr geol, Lehigh Univ, 46-50; geologist, State Geol Surv, SDak, 50-51; from asst prof to prof geol, Univ SDak, 51-71, chmn dept geol, 57-67, prof earth sci, 71-80, cur geol, Mus, 73-80. *Concurrent Pos:* Geologist, NY State Sci Serv, 47-48. *Mem:* AAAS; fel Geol Soc Am; Am Asn Petrol Geol; Paleont Soc. *Res:* Stratigraphy, sedimentation and paleontology of South Dakota; paleoecology. *Mailing Add:* 1225 Valley View Dr Vermillion SD 57069

STEVENSON, ROBERT EVERETT, b Fullerton, Calif, Jan 15, 21; m; c 2. OCEANOGRAPHY. *Educ:* Univ Calif, AB, 46, AM, 48; Univ Southern Calif, PhD(marine geol), 54. *Prof Exp:* Instr geol, Compton Col, 47-49; lectr, Univ Southern Calif, 49-51, dir inshore res oceanog, Hancock Found, 53-59, 60-61; res scientist, Off Naval Res, London, 59; dir marine lab, Tex A&M Univ, 61-63; assoc prof meteorol & geol, Fla State Univ, 63-65; res oceanogr & asst dir biol lab, US Bur Commercial Fisheries, 65-70; sci liaison officer, Off Naval Res, 70-85, dep dir, space oceanog & fleet liaison, 85-88, CONSULT OCEANOG, SCRIPPS INST OCEANOG, 88- *Concurrent Pos:* Distinguished lectr, Am Asn Petrol Geologists, 69-72; secy gen, Int Asn Phys Sci Oceans, 87- *Honors & Awards:* Skylab Sci Except Achievement Award, 74. *Mem:* Fel Geol Soc Am; Am Geophys Union. *Res:* Space oceanography; surface layer oceanography as related to meteorology and climatology. *Mailing Add:* PO Box 689 Del Mar CA 92014-1161

STEVENSON, ROBERT JAN, b Cleveland, Ohio, Jan 3, 52; m 74. PHYCOLOGY, STREAM BIOLOGY. *Educ:* Bowling Green State Univ, BS, 74, MS, 76; Univ Mich, PhD(natural resources), 81. *Prof Exp:* Scientist, Nalco Environ Sci, 76-77; res asst, Great Lakes Res Div, Univ Mich, 77-81; asst prof, 81-88, ASSOC PROF BIOL, UNIV LOUISVILLE, 88- *Mem:* Phycol Soc Am; Ecol Soc Am; NAm Benthological Soc; Am Soc Limnol & Oceanog; Am Inst Biol Sci. *Res:* Algae systematics and ecology; benthic diatom systematics; utilization of algae for environmental monitoring; studies of benthic algal ecology. *Mailing Add:* Dept Biol Univ Louisville Louisville KY 40292

STEVENSON, ROBERT LOUIS, b Princeton, NJ, Jan 15, 32; m 57; c 3. MATHEMATICS, COMPUTER SCIENCE. *Educ:* Hobart Col, BS, 54; Rutgers Univ, MEd, 60; NY Univ, PhD(math educ). *Prof Exp:* Teacher pub sch, NJ, 56-66; assoc prof, 66-80, PROF MATH, WILLIAM PATERSON COL, 80- *Res:* Number theory and coding theory. *Mailing Add:* RD 2 Box 475 Lafayette NJ 07848

STEVENSON, ROBERT LOVELL, b Long Beach, Calif. ANALYTICAL CHEMISTRY. *Educ:* Reed Col, BA, 63; Univ Ariz, PhD(chem), 66. *Prof Exp:* Sr chemist, Shell Develop Co, 66-69; sr chemist, Varian Aerograph, 69-75, mgr liquid chromatography-res & develop, Varian Assocs, 75-77; V PRES RES, ALTEX SCIENTIFIC, 77- *Mem:* Am Chem Soc. *Res:* Managing a research and development group developing high speed liquid chromatographs and accessories. *Mailing Add:* 3338 Carlyle Lab 201 Lafayette CA 94549-5202

STEVENSON, ROBERT THOMAS, b Washington, DC, July 23, 16; m 43; c 4. BIOLOGY. *Educ:* Am Univ, BA, 38; Univ Wis, MPh, 40, PhD(zool), 43. *Prof Exp:* Asst, Univ Wis, 38-43; asst prof biol & physiol, Univ Utah, 46-48; assoc prof, 48-68, head dept sci, 62-68, head dept life sci, 68-77, PROF BIOL, SOUTHWEST MO STATE UNIV, 68-, PROF LIFE SCI, 77- *Concurrent Pos:* Parasitologist, Neiman Stephenson Co, 41-42. *Res:* Parasitology; comparative histology and ultrastructure. *Mailing Add:* 727 E Kingsbury Springfield MO 65807

STEVENSON, ROBERT WILLIAM, b Philadelphia, Pa, Oct 22, 30; m 55; c 3. ORGANIC CHEMISTRY. *Educ:* Univ Pa, BS, 54; Ga Inst Technol, PhD(org chem), 58. *Prof Exp:* Res chemist, Celanese Corp Am, 58-61; sr res chemist, Mobil Chem Co, 62-69; chmn, Dept Phys Sci, Cheyney Univ, 74-75, 85-88, asst chmn, Dept Sci & Allied Health, 88-89, PROF SCI, CHEYNEY UNIV, 69- *Mem:* Am Chem Soc; Sigma Xi. *Res:* Synthesis of linear polyamides and polyesters, polyacetals and polyolefins; modification of fats. *Mailing Add:* Dept Sci & Allied Health Cheyney Univ Cheyney PA 19319

STEVENSON, ROBIN, b Concepcion, Chile, Mar 4, 23; US citizen; m 48; c 1. AERONAUTICAL ENGINEERING, SPACE SYSTEMS INTEGRATION. *Educ:* Mass Inst Technol, BS, 47, MS, 48. *Prof Exp:* Proj officer, Air Force Air Mat Command, 48-52, chief engr, B-36 Prog Off, Wright-Patterson AFB, 52-53, chief, Prod Div, Air Force Plant Rep Off, Gen Dynamics/Convair, Tex, 53-55, chief ground systs, Atlas Prog Off, Western Develop Div, Air Force, 55-59; vpres eng, Nat Aeronaut & Space Eng Inc, 59-61; asst div mgr, Missile & Space Systs Eng Div, Ling-Temco-Vought, Inc, 61-62, mem tech staff, Space Tech Labs Div, Thompson-Ramo-Wooldridge, Inc, 62-63; assoc dir standard launch vehicles, Aerospace Corp, 63-72, dir-systs eng, 72-87; CONSULT, 87 - *Mem:* AAAS; Am Inst Aeronaut & Astronaut. *Res:* Systems engineering and operations research studies in support of classified military missile systems and space vehicle development. *Mailing Add:* 5003 Kingspine Rd Rolling Hills Estates CA 90274

STEVENSON, ROBIN, b Falkirk, Scotland, Dec 25, 46; m 75. METALLURGY. *Educ:* Glasgow Univ, Scotland, BSc, 67; Mass Inst Technol, PhD(metall), 72. *Prof Exp:* Staff res engr, Gen Motors Res Labs, 73-83, supvr, Sheet Forming GP, 83-85, prog mgr, Gen Motors Advan Eng Staff, 85-88, SR STAFF RES ENG, GEN MOTORS RES LABS, 88- *Concurrent Pos:* Mem, Nat Mat Adv Bd Comt, Unified Life Cycle Eng. *Mem:* Am Soc Metals; Am Inst Mining Metall & Petrol Engrs. *Res:* Deformation mechanisms in solids; transmission electron microscopy; physical metallurgy; sheet metal deformation and manufacturing; physics and mechanics of machining processes. *Mailing Add:* Eng Mech Dept No 15 GM Res Labs Warren MI 48090-9055

STEVENSON, STUART SHELTON, b Bridgeport, Conn, Nov 11, 14. PEDIATRICS. *Educ:* Yale Univ, BA, 35, MD, 39; Harvard Univ, MPH, 44. *Prof Exp:* Instr pediat, Sch Med, Yale Univ, 41-43; Rockefeller fel, Harvard Univ, 43, asst maternal & child health, Sch Pub Health, 44, assoc child health, 46-47, asst prof, 47-49; res prof pediat, Sch Med, Univ Pittsburgh, 49-59; prof & chmn dept, Seton Hall Col Med, 59-64; clin prof, Col Physicians & Surgeons, Columbia Univ, 64-72, prof pediat, 72-74; dir pediat & attend pediatrician, St Luke's Hosp Ctr, 64-74. *Concurrent Pos:* Staff health comnr, Rockefeller Found, 44-46. *Mem:* Soc Pediat Res; Am Pediat Soc; fel Am Acad Pediat; NY Acad Med. *Res:* The newborn, especially carbonic anhydrase, hyaline membrane, thyroid, congenital malformations, nutrition, growth failure and prematurity. *Mailing Add:* 2 Fifth Ave New York NY 10011

STEVENSON, THOMAS DICKSON, b Columbus, Ohio, Sept 23, 24; m 52; c 3. MEDICINE. *Educ:* Ohio State Univ, BA, 45, MD, 48; Am Bd Internal Med, dipl, 56; Am Bd Path, dipl, 64. *Prof Exp:* Intern med, Johns Hopkins Hosp, 48-49; asst resident, Univ Minn Hosps, 50-51; from asst resident to resident, Ohio State Univ Hosps, 51-53; investr clin gen med & exp therapeut, Nat Heart Inst, 53-55; asst prof med & assoc dir div hemat, Sch Med, Univ Louisville, 55-61; assoc prof, 61-71, PROF PATH, COL MED, OHIO STATE UNIV, 71- *Mem:* Am Fedn Clin Res; Am Soc Clin Path; Am Soc Cytol. *Res:* Biochemical aspects of erythropoiesis; vitamin B-12 metabolism. *Mailing Add:* Dept of Path Col of Med Ohio State Univ 410 W 10th Ave Columbus OH 43210

STEVENSON, WALTER ROE, b Cortland, NY, Sept 16, 46; m 69; c 2. PLANT PATHOLOGY. *Educ:* Cornell Univ, BS, 68; Univ Wis-Madison, PhD(plant path), 72. *Prof Exp:* Res asst plant path, Univ Wis, 68-72; asst prof, Purdue Univ, 72-77, assoc prof, 77-79; assoc prof, 79-84, PROF PLANT PATH, UNIV WIS-MADISON, 84- *Concurrent Pos:* Ind liaison rep, Interregional Proj No 4, 74-79; mem assessment team, Pentachloronitrobenzene, Nat Agr Pesticide Impact Assessment Prog, 77-80; mem, Tech-Adv Comt, North Cent Comput Inst, 82-86; mem, ECOP-IPM Task Force, 84-87, Ecop-Database Task Force, 87-88. *Honors & Awards:* Exten Award, Am Phytopath Soc, 89. *Mem:* Am Phytopath Soc; Potato Soc Am. *Res:* Diseases of vegetable crops, including potato and mint crops; development of disease resistant cultivars; chemical control; epidemiology. *Mailing Add:* Dept Plant Path Univ Wis Madison WI 53706

STEVENSON, WARREN H, b Rock Island, Ill, Nov 18, 38; m 59; c 3. APPLIED OPTICS. *Educ:* Purdue Univ, BSME, 60, MSME, 63, PhD(mech eng), 65. *Prof Exp:* Engr, Martin Co, Colo, 60-61; from asst prof to assoc prof, 65-74, PROF APPL OPTICS, SCH MECH ENG, PURDUE UNIV, LAFAYETTE, 74- *Concurrent Pos:* Pres-elect, Laser Inst Am, 88, pres, 89. *Honors & Awards:* US Sr Scientist Award, Alexander von Humboldt Found WGer, 73. *Mem:* Am Soc Mech Engrs; Am Soc Eng Educ; Optical Soc Am; Laser Inst Am. *Res:* Application of advanced optical measurement techniques such as laser velocimetry and holography in the fields of fluid mechanics, heat transfer, combustion and automated manufacturing. *Mailing Add:* Sch Mech Eng Purdue Univ Lafayette IN 47907

STEVENSON, WILLIAM CAMPBELL, b Brooklyn, NY, Jan 22, 31; m 55; c 3. BIOCHEMISTRY. *Educ:* St John's Col, BS, 52. *Prof Exp:* Chemist, Quaker Maid Co, 54-57 & Nat Biscuit Co, 57-58; chemist, Mead Johnson Co, 58-67, sr scientist, 67-87; RETIRED. *Res:* Drug metabolism, isolation and identification of metabolites; assay of drugs in tissue. *Mailing Add:* 3520 Laurel Lane Evansville IN 47720

STEVENSON, WILLIAM D(AMON), JR, electrical engineering; deceased, see previous edition for last biography

STEVER, H GUYFORD, b Corning, NY, Oct 24, 16; m 46; c 4. AERONAUTICAL ENGINEERING. *Educ:* Colgate Univ, AB, 38, ScD, 58; Calif Inst Technol, PhD, 41. *Hon Degrees:* Numerous from US univs, 58-81. *Prof Exp:* Mem staff, radiation lab, Mass Inst Technol, 41-42, from asst prof to prof aero eng, 46-65; pres, Carnegie Mellon Univ, 65-72; dir, NSF, 72-76; sci adv to Pres US, 73-77; dir, White House Off Sci & Technol Policy, 76-77; CORP DIR & SCI CONSULT, 77- *Concurrent Pos:* Assoc dean eng, Mass Inst Technol, 56-59, head dept mech eng, naval archit & marine eng, 61-65; foreign secy, Nat Acad Eng, 84- *Honors & Awards:* Nat Medal of Sci, 91. *Mem:* Nat Acad Eng; Nat Acad Sci; AAAS; Am Phys Soc; Am Acad Arts & Sci. *Mailing Add:* Carnegie Comn Sci Tech & Govt 1616 P St Suite 400 Washington DC 20036

STEVERMER, EMMETT J, b Wells, Minn, Aug 13, 32; m 70; c 3. ANIMAL SCIENCE, BIOCHEMISTRY. *Educ:* Univ Wis, BS, 58, MS, 60, PhD(animal sci, biochem), 62. *Prof Exp:* From asst prof to assoc prof, 62-74, PROF ANIMAL SCI, IOWA STATE UNIV, 74- *Honors & Awards:* Extension Award, Am Soc Animal Sci. *Mem:* Am Soc Animal Sci. *Res:* Swine nutrition and reproductive physiology. *Mailing Add:* Dept of Animal Sci Iowa State Univ Ames IA 50011

STEVINSON, HARRY THOMPSON, b Passburg, Alta, Mar 5, 15; m 44; c 3. ELECTRICAL ENGINEERING. *Educ:* Univ Alta, BSc, 44. *Prof Exp:* Officer, Signal Div, Can Navy, 44-45; res officer flight, Nat Res Coun Can, 45-79; RETIRED. *Honors & Awards:* Can Aeronaut & Space Inst Baldwin Award, 56, McCurdy Award, 66. *Mem:* Can Aeronaut & Space Inst. *Res:* Communications; flight instrumentation; specialized aerial delivery problems; aircraft crash recovery systems. *Mailing Add:* 3558 Revelstoke Dr Ottawa ON K1V 7C1 Can

STEWARD, FREDERICK CAMPION, b London, Eng, June 6, 04; m 29; c 1. BOTANY, CELL BIOLOGY. *Educ:* Univ Leeds, BSc, 24, PhD(bot), 26; Univ London, DSc, 36. *Hon Degrees:* DSc, Univ Delhi, 74, Col William & Mary & Univ Guelph. *Prof Exp:* Demonstr bot, Univ Leeds, 26-27, asst lectr, 29-33; Rockefeller fel, Cornell Univ & Univ Calif, 27-29; Rockefeller fel, Univ Calif & Carnegie Inst Washington, 33-34; reader, Univ London, 34-47; dir aircraft equip, Ministry of Aircraft Prod, 40-45; res assoc, Univ Chicago, 45-46; vis prof bot & chmn dept, Univ Rochester, 46-50; prof bot, 50-65, Charles A Alexander prof, 65-73, EMER PROF BIOL SCI, CORNELL UNIV, 73- *Concurrent Pos:* Guggenheim fel, 64; Pauli lectr, Zurich, 68; Sir CV Raman vis prof, Madras, 74; Cecil & Ida Green vis prof, Univ BC, 75. *Honors & Awards:* Merit Award, Bot Soc Am, 61; Stephen Hales Award, Am Soc Plant Physiol, 64. *Mem:* Bot Soc Am; Am Soc Plant Physiol; fel Royal Soc; fel Am Acad Arts & Sci; fel Indian Sci Acad. *Res:* Plant physiology and biochemistry; respiration; salt intake; metabolism; protein synthesis; chromatography of amino acids; cell and tissue culture; morphogenesis. *Mailing Add:* 4947 Woodland Forrest Dr Tuscaloosa AL 35405-5761

STEWARD, JOHN P, b Huntington Park, Calif, Oct 9, 27. MEDICAL MICROBIOLOGY, IMMUNOLOGY. *Educ:* Stanford Univ, AB, 48, MD, 55. *Prof Exp:* Nat Inst Allergy & Infectious Dis fel med microbiol, Sch Med, Stanford Univ, 58-60, from instr basic med sci to asst prof exp med, 60-70, asst dean, Sch Med, 64-65, asst dir, Fleischmann Labs Med Sci, 64-70, sr lectr med microbiol, Sch Med & actg dir, Fleischmann Labs Med Sci, 70-74, assoc dean, Sch Med, 71-90, ADJ PROF MED MICROBIOL, SCH MED, STANFORD UNIV, 74-, EMER PROF MICROBIOL & IMMUNOL, 90- *Concurrent Pos:* Lectr, Sch Pub Health, Univ Calif, Berkeley, 63. *Mem:* Am Asn Immunologists; Am Soc Microbiol. *Res:* Host-parasite relationship between enterobacteriaceae and experimental animals. *Mailing Add:* 2070 Webster St Palo Alto CA 94301

STEWARD, KERRY KALEN, b Skowhegan, Maine, June 2, 30; m 56; c 1. WEED SCIENCE. *Educ:* Univ Conn, BS, 58, MS, 62, PhD(bot), 66. *Prof Exp:* res plant physiologist, Agr Res Serv, USDA, 66-88; CONSULT, 88- *Concurrent Pos:* Res leader & dir Aquatic Weed Control Lab. *Mem:* Am Inst Biol Sci; Bot Soc Am; Weed Sci Soc Am; AAAS; Aquatic Plane Mgt Soc. *Res:* Physiology of aquatic plants; mineral nutrition of aquatic plants; effects of environment on growth of aquatic plants. *Mailing Add:* Agr Res Serv USDA 3205 Col Ave Ft Lauderdale FL 33314

STEWARD, OMAR WADDINGTON, b Woodbury, NJ, May 28, 32; m 58; c 3. ORGANOMETALLIC CHEMISTRY, INORGANIC CHEMISTRY. *Educ:* Univ Del, BS, 53; Pa State Univ, PhD(chem), 57. *Prof Exp:* Proj leader fluorine & organosilicon chem, Dow Corning Corp, 57-62; NSF fel, Univ Leicester, 62-63; instr inorg chem, Univ Ill, 63; asst prof, Southern Ill Univ, 63-64; from asst prof to assoc prof, 64-72, PROF INORG CHEM, DUQUESNE UNIV, 72- *Mem:* Am Chem Soc; Sigma Xi. *Res:* Structure, bonding and reaction mechanisms of group 14 organometallic compounds; bonding and reaction mechanisms of coordination compounds. *Mailing Add:* Dept of Chem Duquesne Univ Pittsburgh PA 15282

STEWARD, ROBERT F, b Springboro, Pa, June 2, 23; m 46; c 2. MATHEMATICS. *Educ:* Wheaton Col, Ill, BS, 47; Rutgers Univ, MS, 49; Auburn Univ, PhD(math), 61. *Prof Exp:* Instr math, Va Mil Inst, 49-53; asst prof, Drexel Inst Technol, 53-57; instr, Auburn Univ, 57-58, 59-60; assoc prof, Western Carolina Col, 60-61, prof & chmn dept, 61-63; chmn dept, 67-73, PROF MATH, R I COL, 63- *Mem:* Math Asn Am. *Res:* Numerical analysis. *Mailing Add:* 24 Observatory Ave North Providence RI 02911

STEWARD, W(ILLIS) G(ENE), b Hastings, Nebr, June 11, 30; m 58; c 4. MECHANICAL ENGINEERING. *Educ:* Univ Colo, BS, 52, MS, 58; Colo State Univ, PhD, 69. *Prof Exp:* Engr, Gas Turbine Div, Gen Elec Co, 52-54; engr, Nat Bur Standards, 58-85; FLUID, THERM CO. *Concurrent Pos:* Consult energy cryog. *Mem:* Sigma Xi; Am Soc Mech Engrs. *Res:* Thermodynamics; heat transfer; fluid mechanics; cryogenics; solar energy. *Mailing Add:* Sugarloaf Rd 169 S Teak Lane Boulder CO 80302

STEWART, ALBERT CLIFTON, b Detroit, Mich, Nov 25, 19; m 49. RADIATION CHEMISTRY. *Educ:* Univ Chicago, SB, 42, SM, 48; St Louis Univ, PhD(chem), 51. *Prof Exp:* Chemist, Sherwin-Williams Paint Co, Ill, 43-44; asst inorg chem, Univ Chicago, 47-49; instr & res assoc, St Louis Univ, 49-51; sr chemist, Oak Ridge Nat Lab, 51-56; group leader, Res Lab, Nat Carbon Co Div, Union Carbide Corp, 56-59, asst dir res, Consumer Prod Div, 60-63, asst develop dir, 63-65, planning mgr new mkt develop, 65-66, mkt develop mgr, Chem & Plastics Develop Div, 66-69, mkt mgr rubber chem, Mkt Area, 69-71, mkt mgr chem coatings solvents, 71-73, int bus mgr, Chem & Plastics Div, 73-77, dir sales, Chem & Plastics Div, 77-79, nat sales mgr, Solvents & Intermediates Div, 79-82; dir, univ relations, 82-84, assoc dean & assoc prof, 84-87, ACTG DEAN, AN CELL SCH BUS, WESTERN CONN UNIV, 87- *Concurrent Pos:* Prof, Knoxville Col, 53-56; lectr, John Carroll Univ, 56-63; consult, Pub Affairs Div, Ford Found, 63; adminstr officer, NASA, 63 & Agency Int Develop, 64-69; treas, NY State Dormitory Authority, 71-76; consult, Union Carbide Corp, 84- *Mem:* AAAS; Am Chem Soc; Am Nuclear Soc; Radiation Res Soc. *Res:* Physical inorganic and radiation chemistry; research, development and general administration; marketing management; sales management; professor marketing. *Mailing Add:* 28 Hearthstone Dr Brookfield CT 06805

STEWART, ALEC THOMPSON, b Can, June 18, 25; m 60; c 3. SOLID STATE PHYSICS. *Educ:* Dalhousie Univ, BSc, 46, MSc, 49; Cambridge Univ, PhD(physics), 52. *Hon Degrees:* LLD, Dalhousie, 86. *Prof Exp:* From asst res officer to assoc res officer, Atomic Engr Can Ltd, 52-57; assoc prof physics, Dalhousie Univ, 57-60; from assoc prof to prof, Univ NC, Chapel Hill, 60-68; head dept, 68-74, PROF PHYSICS, QUEEN'S UNIV, ONT, 68- *Concurrent Pos:* J S Guggenheim fel & Kenan travelling prof, 65-66; consult, res granting agencies, US & Can. *Mem:* Fel Am Phys Soc; Can Asn Physicists (pres, 72-73); fel Royal Soc Can (pres, 84-87); fel NATO. *Res:* Solid state by positron annihilation in matter and neutron inelastic scattering. *Mailing Add:* Dept Physics Queen's Univ Kingston ON K7L 3N6 Can

STEWART, ARTHUR VAN, b Buffalo, NY, July 25, 38; m 65; c 3. DENTISTRY. *Educ:* Univ Pittsburgh, BS, 60, MEd, 64, DMD, 68, PhD(educ admin), 73. *Prof Exp:* USPHS postdoctoral fel, Sch Dent, Univ Pittsburgh, 68-70; chair Dept Community Dent & Dir Learning Res, Continuing Educ & Auxiliary training, Fairleigh Dickinson Univ, 70-75, asst dean for students & exec asst dean, 70-75; asst dean acad affairs & dir grad affairs, 75-88, spec asst to Univ Provost, 84-89, PROF DEPT GROWTH & SPECIAL CARE DENT, SCH DENT, UNIV LOUISVILLE, 88- *Concurrent Pos:* Consult, Headstart Prog, 70-75, Luthern Nursing Home, 70-75, Ringwood Dent Prog, 72-75, Patterson Child Dent Care Prog, 73-75, Am Dent Asn Comn Dent Accreditation, 90-; mem task force Curric, Am Asn Dent Sch, 90-, task force Outcomes Assess, 88-; dir Univ Accreditation, 84-87; chair Spec Int Group Geriatrics, Am Asn Dent Sch, 88-, Univ Senate Planning Comn, 88- *Mem:* NY Acad Sci; Am Dent Asn; Am Asn Dent Schs; Am Soc Geriat Dent; Geront Soc Am; Sigma Xi; fel Am Col Dent. *Res:* Dental education; health manpower; preventive dentistry; community health programs; health manpower development; gerontology/geriatrics. *Mailing Add:* Dept Growth & Special Care Univ Louisville Sch Dent Louisville KY 40292

STEWART, BARBARA YOST, b Johnstown, Pa, Oct 12, 32; div; c 2. INTRODUCTORY BIOLOGY LABS & COURSES. *Educ:* Swarthmore Col, BA, 54; Bryn Mawr Col, MA, 72, PhD(biochem), 75. *Prof Exp:* From lectr to asst prof, 75-86, ASSOC PROF BIOL, SWARTHMORE COL, 86-, ASSOC CHMN DEPT, 85- *Concurrent Pos:* Evaluator, Ursinus Col Biol Dept, 85 & 90; vis prof, Lafayette Col, 88 & 89; NSF reviewer, 90. *Mem:* Sigma Xi. *Res:* Effect of temperature on fatty acids; isozymes of lactate dehydrogenase in fish; investigative laboratories; writing in the sciences. *Mailing Add:* Dept Biol Swarthmore Col Swarthmore PA 19081

STEWART, BOBBY ALTON, b Erick, Okla, Sept 26, 32; m 56; c 3. WATER MANAGEMENT, SOIL CONSERVATION. *Educ:* Okla State Univ, BS, 53, MS, 57; Colo State Univ, PhD(soil sci), 61. *Prof Exp:* Soil scientist, Agr Exp Sta, Stillwater, Okla, 53-57, soil scientist, Agr Res Serv, Fort Collins, Colo, 57-68, DIR & SOIL SCIENTIST, AGR RES SERV, CONSERV & PROD RES LAB, US DEPT AGR, 68- *Concurrent Pos:* Instr, Colo State Univ, 62 & Tex A&M Univ, 72; ed, Advan in Soil Sci, 84- *Mem:* Fel Soil Conserv Soc Am; fel Am Soc Agron; fel Soil Sci Soc Am (pres, 81); Int Soil Sci Soc; Coun Agr Sci & Technol. *Res:* Conservation and production problems associated with agriculture in the Southern Great Plains of the United States; water-use efficiency; control of soil erosion. *Mailing Add:* Conserv & Prod Res Lab US Dept Agr PO Drawer 10 Bushland TX 79012

STEWART, BONNIE MADISON, b Loveland, Colo, July 10, 14; m 40; c 2. MATHEMATICS. *Educ:* Univ Colo, BA, 36; Univ Wis, PhM, 37, PhD, 40. *Prof Exp:* Asst math, Univ Wis, 38-40; instr, Mich State Univ, 40-42; asst prof, Denison Univ, 42-43; from asst prof to assoc prof, 43-53, prof, 53-80, EMER PROF MATH, MICH STATE UNIV, 80- *Mem:* Am Math Soc; Math Asn Am. *Res:* Matrix theory; number theory; graph theory; Euclidian geometry. *Mailing Add:* 4494 Wausau Rd Okemos MI 48864

STEWART, BRADLEY CLAYTON, b Ann Arbor, Mich, Feb 2, 54. SPEECH ROCOGNITION & SYNTHESIS, SIGNAL PROCESSING. *Educ:* Univ Calif, Santa Barbara, BS, 77. *Prof Exp:* Proj engr, Sonatech Inc, 77-78, E2 Tech, 78-80; contract eng, Santa Barbara Res Ctr, 80-81; VPRES ENG, COVOX INC, 81- *Res:* Implementation of computer voice recognition & synthesis. *Mailing Add:* Covox Inc 675 D Conger St Eugene OR 97402

STEWART, BRENT SCOTT, b Fairbanks, Alaska, Nov 19, 54; m 89. DEMOGRAPHY & FORAGING ECOLOGY OF MARINE MAMMALS & SEABIRDS. *Educ:* Univ Calif, Los Angeles, BA, 77, PhD(biol), 89; San Diego State Univ, MS, 81. *Prof Exp:* SR STAFF SCIENTIST, HUBBS SEA WORLD RES INST, 77- *Concurrent Pos:* Lectr, Soc Expeds, 88-; res assoc, Univ San Diego, 88-; adj fac mem, San Diego State Univ, 89- *Mem:* Fel Explorers Club; Ecol Soc Am; AAAS; Am Soc Zoologists; Am Soc Mammalogists; Soc Marine Mammal. *Res:* Population biology and ecology of marine mammals and seabirds; migrations, foraging ecology, demography, diving behavior and physiology of marine mammals and seabirds. *Mailing Add:* Hubbs Sea World Res Inst 1700 S Shores Rd San Diego CA 92109

STEWART, CARLETON C, b Schenectady, NY, July 13, 40; m 63; c 2. IMMUNOLOGY, BIOPHYSICS. *Educ:* Hartwick Col, BA, 62; Univ Rochester, MS, 64, PhD(radiation), 67. *Prof Exp:* Atomic Energy Proj res asst & instr radiation physics, Univ Rochester, 62-67; instr immunol, Univ Pa, 67-69; sr scientist, Smith Kline & French Labs, 69-70; asst, Sch Med, Wash Univ, 70-78, assoc prof cancer biol in radiol, 78-81; group leader, Exp Path, Los Alamos Nat Labs, 81-88; DIR FLOW CYTOMETRY, ROSWELL PARK CANCER INST, 88- *Concurrent Pos:* USPHS fel & grant, 67-69; Am Cancer Soc grant, 68-69; Cancer Ctr grant, Nat Cancer Inst, 70-78; NIH grants, 78 - *Mem:* AAAS; Soc Analytical; Am Asn Exp Pathologists; NY Acad Sci; Soc Leukocyte Biol; Sigma Xi. *Res:* Cellular immunology; tumor immunology; flow cytometry. *Mailing Add:* Roswell Park Cancer Inst Elm & Carlton St Buffalo NY 14263-0001

STEWART, CECIL R, b Monmouth, Ill, Mar 11, 37; m 58; c 2. PLANT PHYSIOLOGY. *Educ:* Univ Ill, BS, 58; Cornell Univ, MS, 63, PhD(plant physiol), 67. *Prof Exp:* NIH fel plant physiol, Purdue Univ, 66-68; asst prof 68-71, assoc prof, 71-76, PROF BOT, IOWA STATE UNIV, 76- *Mem:* AAAS; Am Soc Plant Physiol. *Res:* Plant metabolism. *Mailing Add:* Dept Bot 353 Bessey Iowa State Univ Ames IA 50011

STEWART, CHARLES JACK, b Rawlins, Wyo, June 17, 29; m 56; c 3. BIOCHEMISTRY. *Educ:* San Diego State Col, BA, 50; Ore State Univ, MS, 52, PhD(biochem), 55. *Prof Exp:* Fulbright grant biochem, Inst Org Chem, Univ Frankfurt, 54-55; from instr to assoc prof, 55-65, chmn, chem dept, 67-70, actg chmn, 80-81, chmn, Chem Dept, 86-89, chair-fac senate, 90-91, PROF CHEM, SAN DIEGO STATE UNIV, 65- *Concurrent Pos:* NIH res grant, 62-, spec fel, 63-64; guest prof chem, Max Planck Inst Med Res, Heidelberg, Ger, 75-76, res fel, 79-80. *Mem:* AAAS; Am Soc Biol Chemists; Am Chem Soc. *Res:* Mechanism of enzymes and antimetabolites; synthesis and enzymatic properties of coenzyme A analogs. *Mailing Add:* Dept of Chem San Diego State Univ San Diego CA 92182

STEWART, CHARLES NEIL, b Albany, NY, May 6, 45; m 68; c 3. SURFACE PHYSICS, GASEOUS ELECTRONICS. *Educ:* Union Col, NY, BS, 67; Univ Ill, Urbana-Champaign, MS, 69, PhD(physics), 74. *Prof Exp:* Asst physics, Dudley Observ, NY, 65-67; lab instr, Union Col, NY, 67; asst, Univ Ill, Urbana-Champaign, 67-73; res physicist, Lamp Phenomena Res Lab, 73-83, sr physicist, Lighting Systs Prod Eng Dept, 83-88, TECH LEADER, LIGHTING TECHNOL DIV, GEN ELEC CO, 88- *Mem:* Am Phys Soc; Sigma Xi. *Res:* Gas-solid interactions; field emission; low pressure discharges; oxide cathodes. *Mailing Add:* Gen Elec Nela Park 3432 Cleveland OH 44112

STEWART, CHARLES NEWBY, b New Westminster, BC, Can, May 16, 31; m 55; c 5. SCIENCE EDUCATION, PSYCHOPHARMACOLOGY. *Educ:* Seattle Pac Univ, BA, 53; Univ Ore, MS, 56, PhD(psychol), 63. *Prof Exp:* Assoc prof, 65-72, PROF PSYCHOL, FRANKLIN & MARSHALL COL, 72-, CHARLES A DANA CHAIR PSYCHOL, 89- *Concurrent Pos:* Affiliated scientist, Monell Chem Senses Ctr, 81-; assoc dean, Franklin & Marshall Col, 91- *Mem:* Sigma Xi; Psychonomics Soc; Asn Chemoreception Sci; AAAS; Soc Study Ingestive Behav; Am Inst Nutrit. *Res:* Chemical senses; influence of molecular structure on sweet taste perception; influence of nutritional state upon salt appetite; hormonal factors in sensory processes. *Mailing Add:* Deans Off Franklin & Marshall Col Lancaster PA 17604-3003

STEWART, CHARLES RANOUS, b La Crosse, Wis, Aug 6, 40; m 88. MICROBIAL GENETICS. *Educ:* Univ Wis, BS, 62; Stanford Univ, PhD(genetics), 67. *Prof Exp:* Am Cancer Soc fel biochem, Albert Einstein Col Med, 67-69; from asst prof to prof biol, 69-89, PROF BIOCHEM & CELL BIOL, RICE UNIV, 89- *Res:* Genetics and biochemistry of Bacillus subtilis and its virulent bacteriophages. *Mailing Add:* Dept Biochem & Cell Biol Rice Univ Houston TX 77251

STEWART, CHARLES WINFIELD, SR, b Wilmington, Del, Jan 27, 40; m 62; c 3. THEORETICAL CHEMISTRY. *Educ:* Univ Del, BS, 62, PhD(chem), 66. *Prof Exp:* Res chemist, Elastomer Chem Dept, 66-80, RES ASSOC POLYMER PROD DEPT, E I DU PONT DE NEMOURS & CO, INC, WILMINGTON, 80- *Res:* Polymer physics. *Mailing Add:* Four Jobs Lane Post Crossing Newark DE 19711

STEWART, CHESTER BRYANT, b Norboro, PEI, Dec 17, 10; m 42; c 2. MEDICINE, EPIDEMIOLOGY. *Educ:* Dalhousie Univ, BSc, 36, MD, CM, 38; Johns Hopkins Univ, MPH, 46, DrPH, 53; FRCP(C), 62. *Hon Degrees:* LLD, Univ Prince Edward Island, 73 & Dalhousie Univ, 79, Mt Allison, 83; DSc, St Francis Xavier Univ, 77. *Prof Exp:* Asst secy assoc comt med res, Nat Res Coun Can, 38-40; dean med, Dalhousie Univ, 54-71, prof epidemiol, 46-76, vpres health sci, 71-76; RETIRED. *Concurrent Pos:* Aviation med res, Med Br, Royal Can Air Force, 40-45. *Honors & Awards:* Centennial Medal, 67; Officer, Order of Can, 72; Queen Elizabeth Jubilee Medal, 77. *Mem:* Fel Am Pub Health Asn; Asn Can Med Cols (pres, 61-63); Can Pub Health Asn (pres, 68); Can Med Asn. *Res:* Aviation medicine; decompression sickness and anoxia; Bacillus Calmette-Guerin vaccination; tuberculosis; immunity; delivery of health care; hospital and medical insurance; physician manpower. *Mailing Add:* 6008 Oakland Rd Halifax NS B3H 1N8 Can

STEWART, DANIEL ROBERT, b New Kensington, Pa, July 25, 38; m 60; c 2. GLASS TECHNOLOGY, CERAMICS. *Educ:* Pa State Univ, BS, 60, MS, 62, PhD(ceramic technol), 64. *Prof Exp:* Sr scientist glass res, Owens-Ill, Inc, 64-67, sect chief glass sci, 67-70, dir glass & ceramics res, 70-72, dir corp res labs, 72-73, vpres corp staff & dir glass & ceramic technol, 73-; PRES, DURO TEMPCORP. *Mem:* Fel Am Ceramic Soc; Sigma Xi; Brit Soc Glass Technol. *Res:* Glass and ceramic materials and processing; research and development. *Mailing Add:* 1750 Eber Rd Holland OH 43666

STEWART, DAVID BENJAMIN, b Springfield, Vt, July 18, 28; m 52, 80; c 3. GEOLOGY, MINERALOGY. *Educ:* Harvard Univ, AB, 51; AM, 52, PhD(petrol), 56. *Prof Exp:* Geologist, US Geol Surv, 51-81, chief br exp geochem & mineral, 76-80, prog coordr radioactive waste mgt, 78-80, sr policy analyst, high level radioactive waste, State Planning Coun Radioactive Waste Mgt, 80-81, RES GEOLOGIST, US GEOL SURV, 81- *Concurrent Pos:* Guest prof, Univ Toronto, 68 & Swiss Fed Inst Technol, 71; mem, Lunar Sample Rev Bd, 70-72; prin investr, lunar feldspar Apollo 11-15, 69-72, lunar metamorphism, 73-76. *Mem:* Fel Geol Soc Am; fel Mineral Soc Am (pres, 88); Am Geophys Union. *Res:* Crustal structure of Maine by reflection seismology and digitized geoscience information systems; radioactive waste management; metamorphic recrystallization of lunar and terrestrial minerals; geochemistry of Maine Devonian coastal volcanic belt; crystal chemistry and phase relations of feldspar, silica and rock-forming silicates. *Mailing Add:* Nat Ctr 959 US Geol Surv Reston VA 22092

STEWART, DAVID PERRY, b Summersville, WVa, Mar 14, 16; m 43; c 2. GEOMORPHOLOGY. *Educ:* WVa Univ, AB, 38; Mich State Univ, MS, 48; Syracuse Univ, PhD(geol), 54. *Prof Exp:* Instr phys sci, Mich State Univ, 46-49; assoc prof geol, Marshall Col, 49-56; from asst prof to prof, 56-82, EMER PROF GEOL, MIAMI UNIV, 82- *Mem:* Geol Soc Am; Sigma Xi. *Res:* Glacial geology of Ohio and Vermont; glacial history of New England and midwestern United States. *Mailing Add:* Dept of Geol Miami Univ Oxford OH 45056

STEWART, DONALD BORDEN, b Sask, Can, Mar 15, 17; nat US; m 52; c 2. ENVIRONMENTAL MANAGEMENT. *Educ:* Univ Wash, BS, 39. *Prof Exp:* Chemist, B F Goodrich Co, 39-41, mgr, Gen Chem Lab, 41-42, oper mgr, Res Div, 42-48, opers mgr, Res Ctr, 48-56; bus mgr, Cent Labs, Gen Foods Corp, 56-57, dir admin serv, Res Ctr, 57-61; vpres, Sterling Forest Corp, 61-66; admin officer, Palisades Interstate Park Comn, 66-67, supt, 67-74, asst gen mgr, 74-78, dep gen mgr, 78-90; RETIRED. *Mem:* AAAS; Am Chem Soc; fel Am Inst Chem. *Res:* Analytical methods; industrial safety and hygiene. *Mailing Add:* 14 Ondaora Pkwy Highland Falls NY 10928

STEWART, DONALD CHARLES, b Salt Lake City, Utah, Dec 15, 12; m 48; c 2. RADIOCHEMISTRY. *Educ:* Univ Calif, Los Angeles, AB, 35; Univ Southern Calif, MS, 40; Va Polytech Inst & State Univ, BS, 44; Univ Calif, Berkeley, PhD(biochem), 50. *Prof Exp:* Chemist, Knudsen Creamery, 35-42; group leader, Metall Lab, Univ Chicago, 44-45, asst sec chief, 45-46; chemist, Radiation Lab, Univ Calif, 46-52; asst dir, 52-54, assoc chemist, 54-59, assoc dir chem div, Argonne Nat Lab, 59-77; RETIRED. *Concurrent Pos:* Co-ed, Prog in Nuclear Energy, Series IX, Pergamon Press, 63-74; consult anal, 78-83; secy, Am Chem Soc, Nuclear Sci Technol Div, 68-71. *Mem:* Fel AAAS; Am Chem Soc; Sigma Xi. *Res:* Analytical and inorganic chemistry of rare earth and actinide elements; author of 2 books on radioactivity and radioactive waste management, 81 and 85. *Mailing Add:* 17220 Tamara Lane Watsonville CA 95076

STEWART, DONALD GEORGE, b Pocatello, Idaho, Jan 9, 33. MATHEMATICS. *Educ:* Univ Utah, BA, 59, MS, 61; Univ Tenn, PhD, 63. *Prof Exp:* Asst prof math, Univ Tenn, 63-64; asst prof, 64-72, ASSOC PROF MATH, ARIZ STATE UNIV, 72- *Mem:* Am Math Soc. *Res:* Point-set topology. *Mailing Add:* Dept of Math Ariz State Univ Tempe AZ 85287

STEWART, DORATHY ANNE, b Beech Grove, Ind, June 2, 37. METEOROLOGY, PHYSICS. *Educ:* Univ Tampa, BS, 58; Fla State Univ, MS, 61, PhD(meteorol), 66. *Prof Exp:* Teacher sci, Suwannee High Sch, 58-59; asst meteorol, Fla State Univ, 59-66; res physicist atmospheric physics, 66-89, METEOROLOGIST, MISSILE COMMAND, 89- *Mem:* Am Meteorol Soc; Am Geophys Union; AAAS; Sigma Xi. *Res:* Evaluate atmospheric conditions which affect storage and performance of military equipment. *Mailing Add:* PO Box 12067 Huntsville AL 35815-1067

STEWART, DORIS MAE, b Sandsprings, Mont, Dec 12, 27; m 56; c 2. ZOOLOGY, PHYSIOLOGY. *Educ:* Univ Puget Sound, BS, 48, MS, 49; Univ Wash, PhD(zool), 53. *Prof Exp:* NIH fel, Univ Wash, 54; from instr to asst prof zool, Univ Mont, 54-57; asst prof biol, Univ Puget Sound, 57-58; head dept sci, Am Col Girls, Istanbul, 58-62; res asst prof zool, Univ Wash, 63-67, res assoc prof, 67-69; assoc prof biol, Cent Mich Univ, 70-72; res assoc prof zool, Univ Wash, 72-73; assoc prof, 73-81, PROF, SCI DEPT, UNIV BALTIMORE, 81- *Mem:* Sigma Xi; Am Physiol Soc; NY Acad Sci. *Res:* Muscle atrophy and hypertrophy and circulation and molting physiology in the spider. *Mailing Add:* Dept Sci Univ Baltimore Baltimore MD 21201

STEWART, EDWARD WILLIAM, b Cardiff, Mo, Sept 17, 31; m 54; c 3. METALLURGICAL ENGINEERING, CHEMISTRY. *Educ:* Wash Col, BS, 52; Lehigh Univ, MS, 54. *Prof Exp:* Tech supvr silicon, E I Du Pont de Nemours & Co, Inc, 58-62, tech supt, 62-65, prof mgr titanium dioxide, 65-68, dist sales mgr, 68-73, lab dir chem, 73-83, bus mgr, 83-85. *Mem:* Sigma Xi. *Res:* Titanium dioxide and color pigment product development. *Mailing Add:* 229 Plymouth Rd Wilmington DE 19803

STEWART, ELWIN LYNN, b Ellensburg, Wash, July 22, 40; m 64; c 1. MYCOLOGY. *Educ:* Eastern Wash State Col, BA, 69; Ore State Univ, PhD(mycol), 74. *Prof Exp:* Fel mycol, Dept Bot & Plant Path, Ore State Univ, 74-75; asst prof, 75-80, ASSOC PROF MYCOL, DEPT PLANT PATH, UNIV MINN, 80- *Mem:* Mycol Soc Am; Am Phytopath Soc; Brit Mycol Soc. *Res:* Fungal systematics; selection and utilization of mycorrhizal fungi in harsh site revegetation. *Mailing Add:* Dept Plant Path Univ Minn 405 Borlaugh Hall 1991 Upper Buford Circle St Paul MN 55108

STEWART, FRANK EDWIN, b Dallas, Tex, July 9, 41; m 72. PHYSICS, CHEMISTRY. *Educ:* Univ Tex, Arlington, BS, 61; Tex A&M Univ, MS, 64, PhD(physics), 66. *Prof Exp:* Instr physics, Tex A&M Univ, 64-66; Nat Acad Sci resident res assoc chem physics, Jet Propulsion Lab, Univ Calif, 66-67; asst prof physics, Northeast La Univ, 67-71; prof math, physics & astron & dir planetarium, Cooke County Jr Col, 71-77, DIR DATA PROCESSING, COOKE COUNTY COL COMPUTER CTR, 77- *Concurrent Pos:* Nat Defense Title IV fel, Tex A&M, 61-66, RA Welch Found Postdoctoral fel, Chem dept, 66. *Mem:* Am Asn Physics Teachers; Am Phys Soc; Sigma Xi; Soc Physics Students. *Res:* Electron paramagnetic resonance; charge-transfer complexes. *Mailing Add:* 1525 W California St Cooke County Col Gainesville TX 76240-4699

STEWART, FRANK MOORE, b Beirut, Lebanon, Dec 27, 17; US citizen; m 46; c 1. BIOMATHEMATICS. *Educ:* Princeton Univ, AB, 39; Harvard Univ, MA, 41, PhD(math), 47. *Prof Exp:* From instr to assoc prof, 47-61, PROF MATH, BROWN UNIV, 61- *Mem:* AAAS. *Res:* Mathematical models in biology; population genetics. *Mailing Add:* Dept of Math Box 1917 Brown Univ Providence RI 02912

STEWART, GARY FRANKLIN, b Okmulgee, Okla, Apr 3, 35; m 56; c 3. GEOLOGY. *Educ:* Okla State Univ, BS, 57; Univ Okla, MS, 63; Univ Kans, PhD(geol), 73. *Prof Exp:* Geologist, Humble Oil & Refining Co, 58-60; geologist, Kans State Geol Surv, Univ Kans, 62-71; asst prof, 71-73, ASSOC PROF GEOL, OKLA STATE UNIV, 73- *Concurrent Pos:* Consult, Oak Ridge Nat Lab, 70-72. *Mem:* Am Asn Petrol Geologists; Sigma Xi. *Res:* Geomorphology; stratigraphy; geologic mapping for environmental purposes; depositional environments of sedimentary rocks. *Mailing Add:* Dept Geol Okla State Univ Stillwater OK 74074

STEWART, GEORGE HAMILL, biomedical engineering; deceased, see previous edition for last biography

STEWART, GEORGE HUDSON, b Brooklyn, NY, May 13, 25; m 58; c 5. PHYSICAL CHEMISTRY. *Educ:* Univ Calif, Berkeley, BS, 49; Univ Utah, PhD(phys chem), 58. *Prof Exp:* Res asst chem, Univ Utah, 58-59; from instr to asst prof, Gonzaga Univ, 59-64, assoc prof chem & chmn dept chem & chem eng, 64-70, dean grad sch, 67-70; chmn dept, 70-76, PROF CHEM, TEX WOMAN'S UNIV, 70- *Mem:* AAAS; Sigma Xi; Am Chem Soc. *Res:* Physical chemistry of chromatography; dynamics of gas-liquid interface; flow in porous media. *Mailing Add:* 2003 W Oak St Denton TX 76201

STEWART, GEORGE LOUIS, b Washington, DC, Oct 30, 44; m 69. PARASITOLOGY. *Educ:* Tulane Univ, BS, 69; Rice Univ, PhD(parasitol), 73. *Prof Exp:* Fel parasitol, Rice Univ, 73-74, res assoc, 74, lectr, 75-77; ASST PROF PARASITOL, UNIV TEX, ARLINGTON, 77- *Concurrent Pos:* Consult, Phillips-Roxane, Inc, 73-78, Bellaire Blvd Animal Clin, 73-78 & Fielder Animal Clin, 77-; res grant, Phillips-Roxane, Inc, 73-78; fac res grant, Univ Tex, Arlington, 77-; NIH res grant, 79-82. *Mem:* Am Soc Parasitologists; Am Heartworm Soc. *Res:* Pathophysiology; host-parasite interactions; veterinary parasitology. *Mailing Add:* Dept Biol Univ Tex Arlington TX 76019

STEWART, GERALD WALTER, b Hamilton, Ohio, Oct 8, 44; m 82; c 2. PHYSICAL CHEMISTRY. *Educ:* Wilmington Col, Ohio, BS, 65; SDak Sch Mines & Technol, Rapid City, MS, 67; Univ Idaho, Moscow, PhD(phys chem), 71. *Prof Exp:* Res assoc, Washington Univ, St Louis, 71-73, Mass Inst Technol, 73-74; asst prof chem, WVa Univ, 74-77; chief, Supporting Res, US Dept Energy, 77-79; dir, Ctr Chem & Environ Physics, Aerodyne Res, Inc, 79-84, PRES, AERODYNE PROD CORP, 84- *Concurrent Pos:* Adj assoc prof chem, WVa Univ, 77-81, Boston Col, 82-; mem, res comt corrosion & deposits, Am Soc Mech Engrs, 84- *Mem:* Am Chem Soc; Sigma Xi; Combustion Inst. *Res:* Theoretical and experimental investigations on the chemistry of coal combustion; evaluation and measurement of kinetic and thermodynamic parameters controlling pollutant formation; chemiluminescence studies of reactions involving inorganic hydrides with strong oxidizing agents; ion-molecule reactions by ion cyclotron resonance spectroscopy. *Mailing Add:* Aerodyne Products Corp 76 Treble Cove Rd North Billerica MA 01862

STEWART, GLENN ALEXANDER, b Ellensburg, Wash, Jan 14, 41; m 62; c 2. SOLID STATE PHYSICS, SURFACE PHSYICS. *Educ:* Amherst Col, BA, 62; Univ Wash, MSE, 65, PhD(physics), 69. *Prof Exp:* Fel physics, Univ Wash, 69-70; res fel physics, Calif Inst Technol, 70-72; asst prof physics, 72-76, ASSOC PROF PHYSICS & DIR HONORS PROG, 76- *Mem:* Am Phys Soc. *Res:* Phase transitions in surface films, particularly in physically absorbed noble gas monolayers. *Mailing Add:* 1521 Williamsburg Pl Pittsburgh PA 15235

STEWART, GLENN RAYMOND, b Riverside, Calif, Feb 7, 36; div; c 2. VERTEBRATE ZOOLOGY, NATURAL HISTORY. *Educ:* Calif State Polytech Col, BS, 58; Ore State Univ, MA, 60, PhD(zool), 64. *Prof Exp:* From asst prof to assoc prof, 63-73, PROF ZOOL, CALIF STATE POLYTECH UNIV, 73- *Honors & Awards:* Ralph W Ames Res Award, 85. *Mem:* AAAS; Am Soc Ichthyol & Herpet; Am Soc Mammal; Am Inst Biol Sci; Soc Study Amphibians & Reptiles. *Res:* Ecology, taxonomy and behavior of reptiles, amphibians and mammals; status of endangered and rare species. *Mailing Add:* Dept Biol Sci Calif State Polytech Univ Pomona CA 91768

STEWART, GORDON ARNOLD, b Denver, Colo, July 2, 34; m 62; c 2. DAIRY SCIENCE. *Educ:* Univ Mo, BS, 56, MS, 58, PhD(dairy), 60. *Prof Exp:* Assoc prof agr, Southeast Mo State Col, 60-65; from asst prof to assoc prof dairy sci, La Tech Univ, 65-77, prof animal sci & dairy herd mgr, 77-86; DAIRY CONSULT, BEREND BROTHERS, INC, 86- *Concurrent Pos:* Assoc prof animal sci, Col of Agr, Haile Sellassie Univ, Alemaya, Ethiopia, 73-75. *Mem:* Nat Asn Cols & Teachers Agr (treas, 65-73); Am Dairy Sci Asn. *Res:* Dairy cattle nutrition, breeding and management. *Mailing Add:* Berend Brothers Inc PO Box 5164 Wichita Falls TX 76307

STEWART, GORDON ERVIN, b San Bernardino, Calif, June 25, 34; m 59; c 3. MICROWAVE PHYSICS, PLASMA PHYSICS. *Educ:* Univ Calif, Los Angeles, BS, 55, MS, 57; Univ Southern Calif, PhD(elec eng), 63. *Prof Exp:* Mem tech staff, Hughes Aircraft Co, 55-62; mem tech staff, Plasma Res Lab, 62-70, SECT HEAD, AEROSPACE CORP, 70- *Concurrent Pos:* Asst prof, Univ Southern Calif, 62-66; mem comn 6, Int Union Radio Sci. *Mem:* Inst Elec & Electronic Engrs; Am Phys Soc. *Res:* Radar scattering; antenna theory; wave propagation in plasmas; acoustic holography. *Mailing Add:* Aerospace Corp PO Box 95085 Bldg 120 Rm 1221 Los Angeles CA 90045

STEWART, GREGORY RANDALL, b Glendale, Calif, Apr 27, 49; m; c 2. HEAVY FERMION SYSTEMS. *Educ:* Calif Inst Tech, BS, 71; Stanford, MS, 73, PhD, 75. *Prof Exp:* Res asst, Univ Konstanz, WGer, 76-77; staff mem, Los Alamos Nat Lab, 77-85; PROF, PHYSICS, UNIV FLA, 85- *Concurrent Pos:* Vis prof, Inst fur Tech Physik, Kernforschungszentrum Karlsruhe, 83, 85, Tech Hochschule, Darmstadt/Transuranium Inst, Karlsruhe, WGer, 87; consult, Westinghouse, Res & Develop, 78, 80, MIT, 84, 85, Los Alamos Nat Lab, 85- *Mem:* Am Phys Soc. *Res:* Superconductivity; heavy fermions; specific heat; actinides. *Mailing Add:* Dept Physics 215 Williamson Hall Univ Fla Gainesville FL 32611

STEWART, GWENDOLYN JANE, Nov 12, 26; m 48; c 4. CELL ADHESION. *Educ:* WVa Univ Med Sch, PhD(microbiol & biochem), 62. *Prof Exp:* PROF, TEMPLE UNIV MED SCH, 71- *Mailing Add:* Thrombosis Res Ctr Temple Univ Med Sch Philadelphia PA 19140

STEWART, H(OMER) J(OSEPH), b Elba, Mich, Aug 15, 15; m 40; c 3. AERONAUTICAL ENGINEERING. *Educ:* Univ Minn, BAeroE, 36; Calif Inst Technol, PhD(aeronaut), 40. *Prof Exp:* Asst, 36-38, instr meteorol, 38-40, from instr to asst prof aeronaut & meteorol, 40-46, chief res anal sect, Jet Propulsion Lab, 44-56, assoc prof aeronaut, 46-49, chief liquid propulsion systs Div, Jet Propulsion Lab, 56-58, spec asst to dir lab, 60-62, mgr advan studies off, 62-68, prof aeronaut, 49-80, advan tech studies adv, Jet Propulsion Lab, 68-76, EMER PROF AERONAUT, CALIF INST TECHNOL, 80- *Concurrent Pos:* Mem tech eval group, Guided Missiles Comt, Res & Develop Bd, 48-52; mem sci adv bd, US Air Force, 49-55, 58-64; mem adv group artificial cloud nucleation, US Dept Defense, 51-55, chmn adv group on spec capabilities, 55-58; consult, Aerojet-Gen Corp, 51-58, 60-70, 75-84, Preparedness Invest Subcomt, US Senate, 57-58 & Rand Corp, 60-68; mem sci adv comt, Ballistics Res Lab, 58-77; mem bd dirs, Meteorol Res Inc, 62-66 & Sargent Industs, Inc, 66-79; dir off prog planning & eval, NASA, 58-60. *Mem:* Am Meteorol Soc; Am Inst Aeronaut & Astronaut. *Res:* Dynamic meteorology; theoretical aerodynamics; fluid and supersonic flows; guided missiles; space and planetary exploration systems. *Mailing Add:* Dept of Aeronaut Eng Calif Inst of Technol Pasadena CA 91125

STEWART, HAROLD BROWN, b Chatham, Ont, Can, Mar 9, 21; m 50; c 1. BIOCHEMISTRY. *Educ:* Univ Toronto, MD, 44, PhD, 50; Cambridge Univ, PhD, 55. *Prof Exp:* From assoc prof to prof, 55-86, chmn dept, 65-72, dean fac grad studies, 72-86, EMER PROF BIOCHEM, UNIV WESTERN ONT, 86- *Concurrent Pos:* Med Res Coun vis scientist, Cambridge Univ, 71-72. *Mem:* Am Soc Biol Chem; Can Physiol Soc; Can Biochem Soc; Brit Biochem Soc. *Res:* Intermediary metabolism in animals and microorganisms. *Mailing Add:* 118 Base Line Rd E London ON N6C 2N8 Can

STEWART, HAROLD L, b Houtzdale, Penn, Aug 6, 99; m 29; c 2. EXPERIMENTAL CARCINOGENICS. *Educ:* Jefferson Med Col, MD, 26, DSc, 64; Univ Perugia, MD, 65; Univ Turku, Finland, MD, 70; Am Bd Path, dipl; Pan Am Med Asn, dipl. *Prof Exp:* Intern, Fitzsimons Gen Hosp, Denver, 26-27, med staff, 27-29; asst pathologist, Philadelphia Gen Hosp, 29-37; from instr to asst prof path, Jefferson Med Col, 30-37, res fel, 29-30; pathologist, Off Cancer Invest, Harvard Univ & USPHS, 37-39; chief lab path, Nat Cancer Inst, USPHS, 39-69; chief dept path & anat, Clin Ctr, NIH, 54-69; CLIN PROF PATH, GEORGETOWN UNIV, 65-; ORGANIZER REGISTRY EXP CANCERS, NIH, 70-, SCIENTIST EMER, 76- *Concurrent Pos:* Chmn, US Nat Comn, Int Union Against Cancer, 53-59; mem study group, World Health Orgn, 57; consult, Food & Drug Admin, 69-71, Nat Cancer Inst, 70-76. *Honors & Awards:* Ward Burdick Award, Am Soc Clin Pathologists, 57; Distinguished Serv Award, Dept Health, Educ & Welfare, 66; Lucy Wortham James Award, James Ewing Soc, 67; F K Mostofi Award, Int Acad Path, 76; Gold-headed Cane Award, Am Asn Pathologists, 78; Fund for Exp Path & lectureship named in Honor, Uniformed Serv Univ Health Sci, Bethesda, Md, 86. *Mem:* Am Soc Clin Pathologists; Am Asn Cancer Res (pres, 58-59); Am Soc Exp Path (hon pres, 55); Am Asn Pathologists (pres, 50-51); Col Am Pathologists; Int Acad Path (pres, 53-55); Int Union Against Cancer (vpres, 62); Int Coun Socs Path (pres, 62); Int Soc Geog Path. *Mailing Add:* Nat Cancer Inst Bldg 41 Rm D311 Bethesda MD 20892

STEWART, HARRIS BATES, JR, b Auburn, NY, Sept 19, 22; m 59; c 2. OCEANOGRAPHY. *Educ:* Princeton Univ, AB, 48; Univ Calif, MS, 52, PhD(oceanog), 56. *Prof Exp:* Hydrographer, US Naval Hydrographic Off, 48-50; instr, Hotchkiss Sch, Conn, 50-51; res asst oceanog, Scripps Inst Oceanog, Univ Calif, 51-56; chief oceanogr, US Coast & Geod Surv, 57-65; dir, Inst Oceanog, Environ Sci Serv Admin, 65-70, dir, Atlantic Oceanog & Meteorol Labs, Nat Oceanic & Atmospheric Admin, 70-78; dir, Ctr Marine Studies, Old Dominion Univ, 80-85; RETIRED. *Concurrent Pos:* chmn, Ocean Surv Panel, Interagency Comt Oceanog, 62-66 & Int Progs Panel, 65-67; US nat coordr, Coop Invest of Caribbean & Adjacent Regions, 68-75; chmn adv coun, Dept Geol & Geophys Sci, Princeton Univ, 73-76; pres, Dade Marine Inst, 77-78; vchmn, Intergovt Oceanog Comn, Regional Asn for Caribbean, 78-82, US nat assoc, 76- *Mem:* Fel AAAS; fel Geol Soc Am; Am Geophys Union; fel Marine Technol Soc (vpres, 74-76); fel Int Oceanog Found (vpres, 82-86). *Res:* Coastal lagoons; marine geology; physical oceanography. *Mailing Add:* 644 Alhambra Circle Coral Gables FL 33134

STEWART, HERBERT, b Stanton, Ky, July 18, 28; m 53; c 3. SCIENCE EDUCATION, PLANT PHSIOLOGY. *Educ:* Univ Conn, BA, 54, MS, 56; Columbia Univ, EdD, 58. *Prof Exp:* Teacher pub sch, Ky, 51-53; instr bot, Univ Conn, 54-56; instr biol, Teachers Col, Columbia Univ, 57-58; prof sci educ & biol, Md State Teachers Col, Towson, 58-59; asst prof biol, Sch Com, NY Univ, 59-60; asst prof sci educ, Rutgers Univ, 60-61; assoc prof, Univ SFla, 61-67; PROF SCI EDUC, FLA ATLANTIC UNIV, 67- *Concurrent Pos:* Sci Manpower fel, Columbia Univ, 59-60; dir, NSF In-Serv Inst, 66-67; biol consult, Inst Univ Kerala, Trivandrum, S India, 66. *Mem:* Fel AAAS; Asn Comput Math & Sci Teaching; Asn Educ Teachers Sci; Asn Comput Math & Sci Teaching; Nat Asn Biol Teachers. *Res:* Synthesis and function of polymeric plant growth regulators. *Mailing Add:* Dept of Biol Sci Fla Atlantic Univ Boca Raton FL 33432

STEWART, IVAN, b Stanton, Ky, July 24, 22; m 47; c 3. PLANT CHEMISTRY. *Educ:* Univ Ky, BS, 48, MS, 49; Rutgers Univ, PhD(soils), 51. *Prof Exp:* From asst biochemist to assoc biochemist, Univ Fla, 51-61, biochemist, Citrus Exp Sta, 61-89. *Res:* Mineral nutrition of plants. *Mailing Add:* 1851 Peninsular Dr Haines City FL 33844

STEWART, J(AMES) R(USH), JR, b Orange, Tex, Dec 28, 26; m 51; c 4. CHEMICAL ENGINEERING. *Educ:* La Polytech Inst, BS, 50; Ill Inst Technol, MGT, 52. *Prof Exp:* Engr, Pennzoil Co, 52-56, sect supvr chem eng, Res Dept, 56-60, admin asst res, 60-68, sr res assoc, 68-84, mgr process res, 84-90; RETIRED. *Concurrent Pos:* Instr, Centenary Col, 57-62. *Mem:* Am Chem Soc; Am Inst Chem Engrs; Am Ornith Union; Cooper Ornith Soc. *Res:* Hydrocarbon and chemical processing; gas engineering; petrochemicals; fuel cells; solar energy; fertilizers; lube oil refining. *Mailing Add:* 519 Pine Edge Dr The Woodlands TX 77380

STEWART, J W, b Neosho, Missouri, Jan 18, 46; m 73; c 2. PSYCHOPHARMALOCOGY. *Educ:* Swarthmore Col, BA, 67; Yale Univ, MD, 71. *Prof Exp:* Staff psychiatrist, Brookdale Hosp, 76-78; asst prof, Columbia Univ Col Physicians, 80-88; RES PSYCHIATRIST, NY STATE PSYCHIATRIC INST, 78-; ASSOC PROF, COLUMBIA UNIV, 88- *Mem:* Am Psychiat Asn. *Res:* Nosology, biology and psychopharmacology of affective disorders. *Mailing Add:* 127 Berkeley Pl Brooklyn NY 11217

STEWART, JACK LAUREN, b Covington, Okla, Apr 3, 24; m 48; c 4. DENTISTRY. *Educ:* Univ Kansas City, DDS, 52. *Prof Exp:* Pvt pract, 52-62; from asst prof to prof dent, Univ Mo-Kans City, 63-70, coordr res, 67-70, asst dean, 70-82, assoc dean, Sch Dent, 82-89; RETIRED. *Concurrent Pos:* Investr, US Army res contract, 63-67, co-responsible investr, 67-70; consult, Leavenworth Vet Admin Ctr, 64- & Kansas City Vet Admin Hosp, 67-;

abstractor, Oral Res Abstr, Am Dent Asn; chmn sect comput appln, Am Asn Dent Schs, 68-69, from vchmn to chmn sect learning resources, 69-72. *Mem:* Am Asn Dent Schs; Am Dent Asn; Int Asn Dent Res. *Res:* Research administration; maxillofacial injuries; oral lesions. *Mailing Add:* 9724 Russell Overland Park MO 66212

STEWART, JAMES A, b Burnaby, BC, May 30, 20; m 50; c 2. ELECTRONICS ENGINEERING. *Educ:* Univ BC, BASc, 50; Stanford Univ, MSc, 59. *Prof Exp:* Prod engr, Lenkurt Elec Co Can, 51-54; sr electronics engr, Avro Aircraft, Ltd, 54-55; engr, Westinghouse Elec Corp, 55-56; res engr, Lenkurt Elec Co, Inc, 56-60; mem tech staff, West Coast Lab, Gen Tel & Electronics Labs, 60-63; sr staff engr, GTE Lenkurt Inc, 64-80; mem staff, Data Terminals Div, Hewlett-Packard, 80-; PRIN ENGR, GTE LENKURT, INC. *Mem:* Sr mem Inst Elec & Electronics Engrs. *Res:* Electronic circuitry; communication system design. *Mailing Add:* 839 Chesterton St Redwood City CA 94061

STEWART, JAMES ALLEN, b Pembroke, Ont, Can, Jan 7, 27; m 81; c 3. PHYSICAL CHEMISTRY. *Educ:* Queen's Univ, Ont, BA, 51, MA, 53; Univ Ottawa, PhD(phys chem), 59. *Prof Exp:* Anal res chemist, Dept Nat Health & Welfare, Can, 54-59; from asst prof to prof phys chem, Univ NDak, 59-87, emer prof chem, 87-; RETIRED. *Mem:* Am Chem Soc; Sigma Xi. *Res:* Chemical kinetics of hydrolytic enzyme systems, ester hydrolyses and excited alkali metal reactions; solvent isotope effects on reaction rates. *Mailing Add:* 596 Elizabeth St Pembroke ON K8A 1X2 Can

STEWART, JAMES ANTHONY, b Manchester, NH, Aug 2, 38; m 63; c 3. BIOCHEMISTRY. *Educ:* St Anselm Col, BA, 63; Univ Conn, PhD(biochem), 67. *Prof Exp:* Investr, Biol Div, Oak Ridge Nat Lab, 67-68; from asst prof to assoc prof, Univ NH, 68-78, prof & chmn dept, 78-85, coord, 85-86, int vpres res, 90-91, ASSOC DEAN, RES COL LIFE SCI & AGR, UNIV NH, 86- *Concurrent Pos:* Vis prof, Univ Tex Med Br, Galveston, 81-82; assoc dir, NH Agr Exp Sta, 86-; coordr Biol Sci prog, 85. *Mem:* AAAS; Soc Develop Biol; Am Inst Biol Sci; Am Biochem & Molecular Biol. *Res:* Regulation of protein and nucleic acid synthesis during development and differentiation of the mouse central nervous system. *Mailing Add:* Taylor Hall Univ of NH Durham NH 03824

STEWART, JAMES DREWRY, b Toronto, Ont, Mar 29, 41. MATHEMATICAL ANALYSIS. *Educ:* Univ Toronto, BSc, 63, PhD(math), 67; Stanford Univ, MS, 64. *Prof Exp:* Nat Res Coun Can fel, Univ London, 67-69; asst prof, 69-74, ASSOC PROF MATH, MCMASTER UNIV, 74- *Mem:* Am Math Soc; Math Asn Am; Can Math Soc. *Res:* Abstract harmonic analysis, functional analysis, history of mathematics. *Mailing Add:* Dept of Math McMaster Univ Hamilton ON L8S 4L8 Can

STEWART, JAMES EDWARD, b Anyox, BC, Aug 3, 28; m 67; c 1. BACTERIAL PHYSIOLOGY, MICROBIOLOGY. *Educ:* Univ BC, BSA, 52, MSA, 54; Univ Iowa, PhD, 58. *Prof Exp:* Scientist, Fisheries Res Bd Can, Maritimes Region, Dept Fisheries & Oceans, 58-74, Res & Develop Directorate, 74-76 & Resource Br, 76-80, dir, Fisheries Res Br, 80-87 & Biol Sci Br, 87-88, RES SCIENTIST, BIOL SCI BR, SCOTIA-FUNDY REGION, DEPT FISHERIES & OCEANS, NS, 88- *Mem:* Soc Invert Path; Sigma Xi; Can Soc Microbiol. *Res:* Microbial oxidation of hydrocarbons; enzymes; bacterial metabolism; defense mechanisms and diseases of marine animals; research management, microbial ecology and marine biotoxins. *Mailing Add:* Habitat Ecol Div-Biol Sci Br Scotia-Fundy Region Dartmouth NS B2Y 4A2 Can

STEWART, JAMES JOSEPH PATRICK, b Glasgow, Scotland, July 26, 46; US citizen; m 84. DEVELOPMENT OF MOPAC. *Educ:* Univ Strathclyde, Glasgow, Scotland, BSc, 69, PhD(chem), 72. *Prof Exp:* Asst prof, Glasgow Univ, Scotland, 72-75; assoc prof, Univ Strathclyde, Scotland, 75-85; res chemist, USAF, 87-91; CONSULT, STEWART COMPUTATIONAL CHEM, 91- *Concurrent Pos:* Res fel, Univ Tex Austin, 80-84; assoc, Nat Res Coun, 84-87. *Mem:* Am Chem Soc. *Res:* Semiempirical quartum chemistry software for predicting chemical properties. *Mailing Add:* 15210 Paddington Circle Colorodo Springs CO 80921

STEWART, JAMES LLOYD, b Chengtu, China, Jan 5, 18; nat US; m 45; c 1. PHYSICS. *Educ:* Univ Sask, BA, 38, MA, 40; Johns Hopkins Univ, PhD(physics), 43. *Prof Exp:* Lab asst physics, Univ Sask, 35-38, technician, Radon Plant, Sask Cancer Comn, 38-40; jr instr physics, Johns Hopkins Univ, 40-43; sci officer, Ballistics Lab, Can Armament Res & Develop Estab, 43-45; lectr physics, Queen's Univ, Ont, 45-46; asst prof, Rutgers Univ, 46-51; physicist, US Navy Electronics Lab, 51-67; physicist, Naval Undersea Ctr, 67-76, MEM STAFF, NAVAL OCEAN SYSTS CTR, 76- *Mem:* Sr mem Inst Elec & Electronics Engrs; fel Acoust Soc Am; Sigma Xi. *Res:* Slow neutrons; ballistics; ultrasonics; underwater acoustics; signal processing theory. *Mailing Add:* 9538 Summerfield St PO Box 887 Spring Valley CA 92077

STEWART, JAMES MCDONALD, b Taft, Tenn, Sept 22, 41; m 64; c 3. PLANT PHYSIOLOGY. *Educ:* Okla State Univ, BS, 63, PhD(plant physiol), 68. *Prof Exp:* PLANT PHYSIOLOGIST COTTON PHYSIOL, AGR RES SERV, USDA, 68-; ASSOC PROF PLANT & SOIL SCI, UNIV TENN, KNOXVILLE, 68-; AT DEPT AGRON, UNIV ARK. *Mem:* Sigma Xi; Am Soc Plant Physiologists; Crop Sci Soc Am; Agron Soc Am; AAAS. *Res:* Basic and applied research on cotton fiber and seed development and maturation with correlated studies concerning the effects of environment and heritable traits thereon. *Mailing Add:* Dept Plant & Soil Sci Univ Tenn PO Box 1071 Knoxville TN 37901

STEWART, JAMES MONROE, b Chicago, Ill, Feb 26, 46; m 69; c 2. SPEECH PERCEPTION, STATISTICS. *Educ:* Howard Univ, BA, 70, MA, 71; Ohio Univ, PhD(hearing & speech sci), 76. *Prof Exp:* Instr speech, Howard Univ, 70-71; teaching & res asst speech sci, Ohio Univ, 71-74; res assoc speech sci, Univ Tex Health Sci Ctr, Houston, 75-76; mem staff speech path & audiol, Tenn State Univ, 76-79, asst prof hearing & speech sci, 79-83; ASSOC PROF SPEECH, TENN TECH UNIV, 85- *Concurrent Pos:* Teacher eng & math, Washington DC Pub Sch, 70-73; teacher, Nashville Pub Sch, 84-85. *Mem:* Int Soc Phonetic Sci; Am Asn Phonetic Sci; Phonetic Soc Japan; Acoust Soc Am; Inst Acoust. *Res:* Psychological reality of speech perception; saliency of perceptual judgments; prevalence of communicative disorders. *Mailing Add:* PO Box 278 Cookeville TN 38503

STEWART, JAMES RAY, b Beeville, Tex, Aug 5, 37; m 68; c 2. CELL BIOLOGY, MICROBIOLOGY. *Educ:* NTex State Univ, BS, 59; Univ Ala, Tuscaloosa, MS, 65; Univ Tex, Austin, PhD(biol sci), 70. *Prof Exp:* NIH fel, dept biochem, Univ Tex Health Sci Ctr, San Antonio, 70-74; from asst prof to assoc prof, 74-86, PROF BIOL & CHEM, UNIV TEX, TYLER, 86- *Mem:* Am Soc Microbiol; Phycol Soc Am. *Res:* Bacterial and algae physiology; bacterial anatomy and taxonomy; microbial differentiation. *Mailing Add:* Dept Biol Univ Tex-Tyler 3900 University Blvd Tyler TX 75701

STEWART, JAMES T, b Birmingham, Ala, Dec 1, 38; m 63; c 3. PHARMACEUTICAL CHEMISTRY. *Educ:* Auburn Univ, BS, 60, MS, 63; Univ Mich, PhD(pharmaceut chem), 67. *Prof Exp:* From asst prof to assoc prof, 67-78, PROF & HEAD MED CHEM, UNIV GA, 78- *Concurrent Pos:* Mead-Johnson res grant, 67-68; NIH biomed sci grant, 68-69; mem, US Pharmacopeial Revision Comt, 80-85; Food & Drug Admin contracts, 77-80 & 81-84; grants, Knoll Pharmaceut, 78 & 79, Boots Pharmaceut, 80 & Hoffman-La Roche, Inc, 81; mem, US Pharmacopeial Revision Comt, 85-90; res grants, Glaxo Inc, 86, Upjohn, 86 & Burroughs Wellcome, 87. *Mem:* Am Chem Soc; Am Pharmaceut Asn; Acad Pharmaceut Sci; Am Asn Pharmaceut Scientists. *Res:* Fluorometric analysis of pharmaceuticals; liquid chromatography. *Mailing Add:* 575 Forest Heights Dr Athens GA 30606

STEWART, JENNIFER KEYS, b Rome, Ga, May 15, 47. ENDOCRINOLOGY. *Educ:* Emory Univ, BS, 68, MS, 69, PhD(physiol), 75. *Prof Exp:* Instr biol, Mercer Univ, 69-71; res fel endocrinol, Harborview Med Ctr, 75-78; res assoc, Univ Wash & Howard Hughes Med Inst, 79-80, res asst prof, 80-81; asst prof, 81-85, ASSOC PROF BIOL, VA COMMONWEALTH UNIV, 85- *Mem:* Am Physiol Soc; Am Endocrine Soc; AAAS; Asn Women Sci; Am Diabetes Asn. *Res:* Metabolic and endocrine physiology; control of hormone secretion; modulation of brain neurotransmitters. *Mailing Add:* Dept Biol Va Commonwealth Univ Richmond VA 23284-2012

STEWART, JOAN GODSIL, MARINE BOTANY. *Educ:* Pomona Col, BA, 53; Calif State Univ, San Diego, MA, 67; Univ Calif, Irvine, PhD(biol), 73. *Prof Exp:* Instr marine bot, Calif State Univ, San Diego, 67-68; res fel marine algae, Scripps Inst Oceanog, Univ Calif, 73- 75; scientist intertidal ecol, Lockheed Marine Biol Lab, 76; consult algal develop, 77-78; ASSOC RES MARINE BIOLOGIST MARINE ALGAE, SCRIPPS INST OCEANOG, UNIV CALIF, 78- *Mem:* Int Phycol Soc; Phycol Soc Am. *Res:* Developmental morphology and nearshore ecology of Rhodophyta. *Mailing Add:* A-002 Scripps Inst Oceanog Univ Calif San Diego CA 92093

STEWART, JOHN, b Redding, Calif, Dec 10, 29; m 56; c 4. ELECTRICAL ENGINEERING. *Educ:* Univ Calif, BSEE, 52. *Prof Exp:* Proj engr, Univac Div, Sperry-Rand Corp, Minn, 54-56, sr engr, 56-60; sr scientist, Systs Res Labs, Inc, 60-61, chief engr, 61-68, vpres res & develop, 68-72, group vpres, 72-; PRES, QUAL QUAN, INC. *Mem:* Inst Elec & Electronics Engrs. *Res:* Digital computers and data processing systems. *Mailing Add:* Qual Quan Inc 18 S West St Bellbrook OH 45305

STEWART, JOHN ALLAN, b Saskatoon, Sask, Feb 18, 24; m 48; c 5. SOIL SCIENCE. *Educ:* Univ BC, BSA, 50, MSA, 53; Univ Wis, PhD, 64. *Prof Exp:* Lectr soils, Univ BC, 50-51 & 52-54; plant physiologist, Can Dept Agr, 54-65; agronomist, Can Int Minerals & Chem Corp, 65-67, res agronomist, Ill, 67-68, mgr fertilizers & cropping systs res, Res & Develop Div, 68-70, mgr agr res, 70-72, dir res & develop, 72-78, dir agron serv, 79-87; AGRON CONSULT, 87- *Concurrent Pos:* Mem, Coun Agr Sci & Technol. *Mem:* Fel Am Soc Agron; Can Soc Soil Sci; Agr Inst Can; AAAS; fel Soil Sci Soc Am. *Res:* Mineral nutrition of agricultural crops; environmental aspects of agricultural technology; industrial minerals applications. *Mailing Add:* 15158 W Redwood Libertyville IL 60048

STEWART, JOHN CONYNGHAM, b New York, NY, Feb 10, 30; m 56; c 2. GEOLOGY. *Educ:* Trinity Col, Conn, BA, 52; Princeton Univ, MA, 56, PhD(geol), 57. *Prof Exp:* Geologist, Harvard NSF res proj, Dordogne dist, France, 57; geophys interpreter, Mobil Oil Co, Venezuela, 58-61; from instr to assoc prof, 61-74, chmn dept, 68-80, PROF GEOL, BROOKLYN COL, 74- *Concurrent Pos:* Danforth assoc. *Mem:* AAAS; Geol Soc Am; Soc Econ Paleont & Mineral; Am Asn Petrol Geol; Sigma Xi. *Res:* Stratigraphy. *Mailing Add:* Dept Geol Brooklyn Col Brooklyn NY 11210

STEWART, JOHN HARRIS, b Berkeley, Calif, Aug 7, 28; m 62; c 2. GEOLOGY. *Educ:* Univ NMex, BS, 50; Stanford Univ, PhD, 61. *Prof Exp:* GEOLOGIST, US GEOL SURV, 51- *Mem:* Geol Soc Am. *Res:* Stratigraphy; sedimentology; regional stratigraphy of Triassic rocks in Utah, Colorado, Nevada, Arizona and New Mexico and of late Precambrian and Cambrian in Nevada and California; regional and local mapping in Nevada; compilation of geologic map of Nevada. *Mailing Add:* 17324 Sierra Vista Mountainview CA 94043

STEWART, JOHN JOSEPH, b Paterson, NJ, July 27, 46. PHARMACOLOGY. *Educ:* Duquesne Univ, BS, 69; Univ Wis-Madison, MS, 72, PhD(pharm), 75. *Prof Exp:* USPHS fel physiol, Med Sch, Univ Tex, Houston, 75-77; assoc prof, 77-86, PROF PHARMACOL, MED CTR, LA STATE UNIV, 86- *Concurrent Pos:* Nat Inst Gen Med Sci grant, Med Sch, La State Univ, Shreveport, 78-81. *Mem:* Sigma Xi; Am Soc Pharmacol & Exp Therapeut; Soc Neurosci. *Res:* Central nervous system control of gastrointestinal function. *Mailing Add:* 7612 Brookhaven Way Shreveport LA 71105

STEWART, JOHN JOSEPH, JR, plasma physics, computational physics, for more information see previous edition

STEWART, JOHN L(AWRENCE), b Pasadena, Calif, Apr 19, 25; m 51; c 2. ELECTRICAL ENGINEERING. *Educ:* Stanford Univ, BS, 48, MS, 49, PhD(elec eng), 52. *Prof Exp:* Res engr, Jet Propulsion Lab, Calif Inst Technol, 49-51; res assoc, Stanford Univ, 52-53; asst prof elec eng, Univ Mich, 53-56; assoc prof, Calif Inst Technol, 56-57 & Univ Southern Calif, 57-60; prof, Univ Ariz, 60-62; pres, Santa Rita Technol Inc, Calif, 62-71; pres, Av-Alarm Corp, 71-89; pres, Couox, Inc, 82-89; PRES, SET, INC, 89- *Concurrent Pos:* Vert pest control res, 65-88. *Mem:* AAAS; sr mem Inst Elec & Electronic Engrs; Acoust Soc Am. *Res:* Electronic network simulations of animal sensory systems; bionics; speech and hearing; speech processing and recognition; acoustic control and cuing; artificial intelligence; signal analysis. *Mailing Add:* SET Inc 105 E Reserve St Vancouver WA 98661

STEWART, JOHN MATHEWS, b Vermillion, SDak, Apr 5, 20; m 43; c 1. ORGANIC CHEMISTRY. *Educ:* Univ Mont, BA, 41; Univ Ill, PhD(org chem), 44. *Prof Exp:* Res chemist, War Prod Bd, Univ Ill, 43-45 & Calif Res Corp, 45-46; from asst prof to prof chem, Univ Mont, 46-77, chmn dept, 59-67, dean grad sch, 68-77, actg acad vpres, 75-77; RETIRED. *Concurrent Pos:* Asst acad vpres, Univ Mont, 78- *Mem:* Am Chem Soc; Sigma Xi. *Res:* Reactions of olefin sulfides; additions to unsaturated nitriles; participation of cyclopropane rings in conjugation, ring-opening reactions of cyclopropanes; use of diazomethane in synthesis of heterocyclic compounds. *Mailing Add:* 111 Crestline Missoula MT 59803

STEWART, JOHN MORROW, b Guilford Co, NC, Oct 31, 24; m 49; c 3. PEPTIDE CHEMISTRY & BIOLOGY, ENDOCRINOLOGY. *Educ:* Davidson Col, BS, 48; Univ Ill, MS, 50, PhD(org chem), 52. *Prof Exp:* Instr chem, Davidson Col, 48-49; asst, Rockefeller Univ, 52-57, from asst prof to assoc prof biochem, 57-68; PROF BIOCHEM, MED SCH, UNIV COLO, DENVER, 68- *Mem:* Am Chem Soc; Am Soc Pharmacol & Exp Therapeut; NY Acad Sci; Am Soc Biol Chemists; Endocrine Soc; Soc Neuroscience. *Res:* Chemistry and pharmacology of peptide hormones, methods of peptide synthesis, antimetabolites and amino acids; synthetic organic chemistry. *Mailing Add:* Dept Biochem Univ Colo Med Sch Denver CO 80262

STEWART, JOHN WESTCOTT, b New York, NY, Nov 15, 26; m 54; c 1. PHYSICS. *Educ:* Princeton Univ, AB, 49; Harvard Univ, MA, 50, PhD(physics), 54. *Prof Exp:* Res fel, 54-56, asst prof, 56-60, ASSOC PROF PHYSICS, UNIV VA, 60-, ASST DEAN COL, 70- *Mem:* Fel Am Phys Soc; Am Asn Physics Teachers. *Res:* Properties of matter under combined field of high pressure and low temperature; meteorology. *Mailing Add:* Dept Physics J W Beams Lab Univ Va Charlottesville VA 22903

STEWART, JOHN WOODS, b Henderson, Tenn, Apr 5, 42; m 65; c 2. NUCLEAR ENGINEERING. *Educ:* Univ Tenn, BS, 65, MS, 67, PhD(nuclear eng), 69. *Prof Exp:* Nuclear engr, E I Dupont De Nemours & Co, Inc, 68-71, res supvr, 71-75, res mgr, 75-80, dept supt, 78-81, gen supt employee rels, Savannah River Plant, 81-82, sect dir, 83, tech mgr, Petrochem Dept, Atomic Energy Div, 84-87, design mgr, Eng Dept, 87-88, mgr, advan technol, Imaging Systs Dept, 88-90, CONSULT MGR, DUPONT ENG, E I DUPONT DE NEMOURS & CO, INC, 90- *Concurrent Pos:* Res assoc nuclear eng, Mass Inst Technol, 74-75. *Mem:* Am Nuclear Soc. *Res:* Nuclear reactor physics and engineering; equipment development; computer applications. *Mailing Add:* PO Box 6090 Newark DE 19714-6090

STEWART, JOHN WRAY BLACK, b Coleraine, NIreland, Jan 16, 36; Can citizen; m 65; c 2. SOIL SCIENCE, CHEMISTRY. *Educ:* Queen's Univ, Belfast, BSc, 58, BAgr, 59, PhD(soil sci), 63, DSc, 88. *Prof Exp:* From sci officer to sr sci officer soil sci, Chem Res Div, Ministry Agr, NIreland, 59-64; fel, Univ Sask, 64-65, from asst prof to assoc prof, 65-76, prof soil sci, 76-81, dir, Sask Inst Pedology & head, dept soil sci, 81-89, DEAN, COL AGR, UNIV SASK, 89- *Concurrent Pos:* Tech expert, Int Atomic Energy Agency, Vienna, 71-72; proj coordr, Can Int Develop Agency, Brazil, 76-; tech expert UN Brazil proj, 74-75; secy-gen, Scope/ICSU, 88-; fel, Berlin Inst Advan Studies, 89. *Mem:* Agr Inst Can; Brit Soc Soil Sci; Int Soil Sci Soc; fel Am Soc Agron; fel Can Soc Soil Sci; fel Soil Sci Soc Am. *Res:* Soil chemistry and fertility; cycling of macro and micro nutrients and heavy metals in the soil plant system. *Mailing Add:* Col Agr Univ Sask Saskatoon SK S7N 0W0 Can

STEWART, JOSEPH LETIE, b Salida, Colo, Aug 2, 27; m 50; c 3. ANTHROPOLOGY. *Educ:* Univ Denver, BA, 49, MA, 50; Univ Iowa, PhD(speech path, audiol, anthrop), 59. *Prof Exp:* Res assoc speech path, Univ Iowa, 58-59; asst prof audiol & dir hearing ctr, Univ Denver, 59-65; consult audiol & speech path, Neurol & Sensory Dis Control Prog, Nat Ctr Chronic Dis Control, USPHS, 65-70, CHIEF SENSORY DISABILITIES PROG, INDIAN HEALTH SERV, USPHS, 70- *Mem:* Fel Am Acad Audiol. *Res:* Epidemiology of otitis media world wide; effects of auditory deprivation; prevention of deaf-mutism. *Mailing Add:* Indian Health Serv 2401 12th St NW Albuquerque NM 87104

STEWART, KENNETH WILSON, b Walters, Okla, Mar 5, 35; m 56; c 4. ENTOMOLOGY, AQUATIC ECOLOGY. *Educ:* Okla State Univ, BS, 58, MS, 59, PhD(entom, zool), 63. *Prof Exp:* Entomologist, Rocky Mt Forest & Range Exp Sta, US Forest Serv, 58-59; head, dept biol, Coffeyville Col, 60-61; from instr to prof biol, Univ NTex, 61-79, prof & chmn biol sci, 79-83, chmn Div Environ Sci, 88-90, FAC RES GRANTS, UNIV NTEX, 63- *Concurrent Pos:* NIH Res grant, 66-88; consult investr, US Corps Engrs; NSF res grants, 77- *Mem:* Entom Soc Am; Am Entom Soc; NAm Benthological Soc (pres, 78-79). *Res:* Stream benthos community structure and dynamics; North American Plecoptera nymphs; passive dispersal of Algae and Protozoa by aquatic insects; food habits and life histories of aquatic insects and spiders; drumming behavior of Plecoptera. *Mailing Add:* Dept Biol Sci Univ NTex Denton TX 76203

STEWART, KENT KALLAM, b Omaha, Nebr, Sept 5, 34; m 56; c 4. ANALYTICAL CHEMISTRY, BIOCHEMISTRY. *Educ:* Univ Calif, Berkeley, AB, 56; Fla State Univ, PhD(chem), 65. *Prof Exp:* USPHS guest investr biochem, Rockefeller Univ, 65-67, res assoc, 67-68, asst prof, 68-69; res chemist, 70-75, lab chief, Nutrient Compos Lab, Nutrit Inst, Agr Res Serv, USDA, 75-82; head, Food Sci Dept, 82-85, PROF, DEPT BIOCHEM & NUTRIT, VA POLYTECH INST, 85- *Concurrent Pos:* Ed, J Food Compos & Anal, 87- *Mem:* fel AAAS; Am Chem Soc; fel Inst Food Technologists; Asn Off Anal Chemists. *Res:* Automated analyses, especially flow injection analyses; nutrient composition of foods; naturally occuring toxic materials in foods; protein chemistry. *Mailing Add:* Dept Biochem & Nutrit Va Polytech Inst & State Univ Blacksburgh VA 24061-0308

STEWART, KENTON M, b Withee, Wis, Aug 28, 31; m 54; c 3. ZOOLOGY, LIMNOLOGY. *Educ:* Wis State Univ, Stevens Point, BS, 55; Univ Wis-Madison, MS, 59, PhD(zool), 65. *Prof Exp:* Asst limnol, ecol & invert zool, Univ Wis-Madison, 58-61, asst limnol, 61-65, fel, 65-66; asst prof, 66-71, ASSOC PROF BIOL, STATE UNIV NY BUFFALO, 71- *Concurrent Pos:* Coop Inst Res Environ Sci Fel, Univ Colo. *Honors & Awards:* Chandler Meisner Award, Inst Asn Great Lakes Res. *Mem:* Am Soc Limnol & Oceanog; Ecol Soc Am; Int Asn Theoret & Appl Limnol; Int Asn Gt Lakes Res. *Res:* Physical limnology and eutrophication; comparative limnology of Finger Lakes of New York. *Mailing Add:* Dept Biol Sci State Univ of NY Buffalo NY 14260

STEWART, LAWRENCE COLM, b Mineola, NY, July 12, 55. DATA COMPRESSION, SIGNAL PROCESSING. *Educ:* Mass Inst Technol, SB, 76; Stanford Univ, MS, 77, PhD(elec eng), 81. *Prof Exp:* Res staff mem, Xerox Palo Alto Res Ctr, 77-84; res staff mem, Digital Equip Systs Res Ctr, 84-89, RES STAFF MEM, DIGITAL EQUIP CAMBRIDGE RES LAB, 89- *Concurrent Pos:* Vis lectr, Stanford Univ, 82; vis scientist, Mass Inst Technol, 89- *Mem:* Inst Elec & Electronic Engrs; Asn Comput Mach. *Res:* Multiprocessor computer systems; voice input and voice output systems for computers and multimedia. *Mailing Add:* Digital Equip Corp One Kendall Sq Bldg 700 Cambridge MA 02139

STEWART, LELAND TAYLOR, b San Francisco, Calif, Nov 24, 28; m 56; c 2. STATISTICS. *Educ:* Stanford Univ, BS, 51, MS, 57, PhD(statist), 65. *Prof Exp:* Res engr, Autonetics Div, NAm Aviation, Inc, 51-56 & Electronic Defense Labs, Sylvania Elec Prod, Inc, 58-61; statistician, C-E-I-R, Inc, 61-65; STAFF SCIENTIST, LOCKHEED MISSILES & SPACE CO, 65- *Mem:* Am Statist Asn. *Res:* Bayesian statistics and decision theory. *Mailing Add:* 152 Ferne Ct Palo Alto CA 94306

STEWART, MARGARET MCBRIDE, b Greensboro, NC, Feb 6, 27; m 69. VERTEBRATE ECOLOGY, HERPETOLOGY. *Educ:* Univ NC, AB, 48, MA, 51; Cornell Univ, PhD(vert zool), 56. *Prof Exp:* Lab instr anat & physiol, Woman's Col, Univ NC, 50-51; instr biol, Catawba Col, 51-53; asst bot & taxon, Cornell Univ, 53-56; from asst prof to prof vert biol, 56-77, DISTINGUISHED TEACHING PROF, STATE UNIV NY ALBANY, 77- *Concurrent Pos:* Res Found grant-in-aid, 58-61, 65-71, 73-74 & 84-85; grants, Am Philos Soc, 75 & 81 & NSF, 78-80; fac partic, Oak Ridge Assoc Univs, 83-91; ed gen herpet, Am Soc Ichthyol & Herpet, 83-85. *Mem:* Ecol Soc Am; Asn Trop Biol; Am Soc Ichthyol & Herpet; Soc Study Amphibians & Reptiles (pres, 79); fel Herpetologists' League; Soc Study Evolution. *Res:* Competition in tropical and temperate frogs; pattern polymorphism; population dynamics of Adirondack frogs; ecology and behavior of Eleutherodactylus, Puerto Rico and Jamaica; population ecology and behavior of frogs; Eleutherodactylus in Puerto Rico, ranid frogs in upstate New York; fire ecology in the Albany Pine Bush, an endangered community. *Mailing Add:* Dept Biol Sci State Univ NY Albany NY 12222

STEWART, MARK ARMSTRONG, b Yeovil, Eng, July 23, 29; US citizen; m 55; c 3. BIOCHEMISTRY, PSYCHIATRY. *Educ:* Cambridge Univ, BA, 52; Univ London, LRCP & MRCS, 56. *Prof Exp:* Asst psychiat, Sch Med, Wash Univ, 57-61, instr, 61-63, asst prof psychiat & pediat, 63-67, from assoc prof to prof psychiat, 67-72, assoc prof pediat, 68-72; IDA P HALLER PROF CHILD PSYCHIATRY, COL MED, UNIV IOWA, 72- *Concurrent Pos:* NIMH res career develop award, 61-71; dir psychiat, St Louis Children's Hosp. *Mem:* Int Soc Res Aggression; Am Soc Biol Chem; fel Am Acad Child Psychiat; fel Am Psychiat Asn; Soc Res Child Develop. *Res:* Genetic influences on children's aggressive and antisocial behavior. *Mailing Add:* Dept Psychiat Univ Iowa Col Med Iowa City IA 52242

STEWART, MARK THURSTON, b Montclair, NJ, Apr 27, 48. HYDROGEOLOGY. *Educ:* Cornell Univ, AB, 70; Univ Wis-Madison, MS(geol) & MS(water resources mgt), 74, PhD(geol), 76. *Prof Exp:* PROF GEOL, UNIV SFLA, TAMPA, 76-, CHAIR GEOL DEPT, 89- *Mem:* Asn Ground Water Scientists & Engrs; Soc Exploration Geophys; Geol Soc Am. *Res:* Applications of geophysical techniques to ground water resource investigations; interactions of hydrologic and geologic systems; karst hydrogeology. *Mailing Add:* Dept Geol SCA 203 Univ SFla Tampa FL 33620

STEWART, MARY E, b Wilmington, Del, Oct 24, 39. LIPIDS. *Educ:* Univ Del, BS, 61; Purdue Univ, MS, 64, PhD(biochem), 68. *Prof Exp:* Post doctorate trainee, Boston Biomed Res Found, 67-69; fel biochem, Boston Univ Sch Med, 69-71, res assoc, dept dermat, 71-78; asst res scientist, 78-81, ASSOC RES SCIENTIST, DEPT DERMAT, UNIV IOWA COL MED, 81- *Mem:* Soc Invest Dermat; Am Oil Chemists Soc; AAAS. *Res:* Investigating variations in human skin lipid composition, especially differences in levels of linoleate, to determine how lipid composition is influenced by age, heredity, hormonal drugs, etc. *Mailing Add:* 270 ML Univ Iowa Iowa City IA 52242

STEWART, MELBOURNE GEORGE, b Detroit, Mich, Sept 30, 27; m 54; c 3. PHYSICS. *Educ:* Univ Mich, AB, 49, MS, 50, PhD(physics), 55. *Prof Exp:* Instr physics, Univ Mich, 54-55; res assoc, Iowa State Univ, 55-56, from asst prof to assoc prof physics, 56-63; chmn dept, 63-73, assoc provost, 73-86, PROF PHYSICS, WAYNE STATE UNIV, 63- *Concurrent Pos:* Res assoc, Dept Physics & Astron, Univ Col London, UK, 86-87. *Mem:* Am Phys Soc. *Res:* Solid state physics. *Mailing Add:* Dept Physics Wayne State Univ Detroit MI 48202

STEWART, PAUL ALVA, b Leetonia, Ohio, June 24, 09; m 47; c 2. ECOLOGY, ORNITHOLOGY. *Educ:* Ohio State Univ, BS, 52, MS, 53, PhD(zool), 57. *Prof Exp:* Admin asst, State Dept Conserv, Ind, 58-59; wildlife res biologist, US Bur Sport Fisheries & Wildlife, 59-65; res entomologist, Agr Res Serv, USDA, 65-73; WILDLIFE CONSULT, 73- *Mem:* Am Soc Mammal; Wildlife Soc; Cooper Ornith Soc; Wilson Ornith Soc; Am Ornith Union; Brit Ornithol Union; Am Ornithol Union. *Res:* Ecology and management of the wood duck; ecology of blackbird congregations; biological control of insect pests; biology and management of black vulture. *Mailing Add:* 203 Mooreland Dr Oxford NC 27565-2852

STEWART, PETER ARTHUR, b Sask, May 12, 21; m 52; c 2. BIOPHYSICS, PHYSIOLOGY. *Educ:* Univ Man, BSc, 43; Univ Minn, MS, 49, PhD(biophys), 51. *Prof Exp:* Instr physiol, Univ Ill, 51-53, asst prof neurophysiol, Neuropsychiat Inst, Col Med, 53-54; from asst prof to assoc prof physiol, Emory Univ, 54-65, assoc prof physics, 61-65; prof med sci, 65-, EMER PROF MED SCI, BROWN UNIV. *Concurrent Pos:* Markle scholar, 56-61. *Mem:* AAAS; Am Physiol Soc; Biophys Soc; Biomed Eng Soc; Am Asn Physics Teachers. *Res:* Electrical parameters nerve membrane; protoplasmic movement and structure in slime molds; biological control systems analysis; computers in biomedicine; theoretical biology. *Mailing Add:* Brown Univ Box 6 Providence RI 02912

STEWART, REGINALD BRUCE, b Moose Jaw, Sask, May 30, 28; m 50; c 1. ANALYTICAL CHEMISTRY. *Educ:* Univ Man, BSc, 50. *Prof Exp:* Res chemist, Hudson Bay Mining & Smelting Co, 50-55; asst chief chemist, Noranda Mines Ltd, 55-62; res off analytical chem, Atomic Energy Can, 62-69, head analytical sci br, 69-85; ANALYTICAL CHEM CONSULT, 85- *Concurrent Pos:* Analytical chem consult, 85-, lab mgt consult, World Bank, Peoples Repub China, 87. *Mem:* Fel Chem Inst Can. *Res:* Analytical sciences, particularly nuclear power research and development and environmental monitoring. *Mailing Add:* One McWilliams Pl Pinawa MB R0E 1L0 Can

STEWART, RICHARD BYRON, b Waterloo, Iowa, Aug 22, 24; m 44; c 2. MECHANICAL ENGINEERING. *Educ:* Univ Iowa, BSME, 46, MS, 48, PhD(mech eng), 66; Univ Colo, ME, 59. *Prof Exp:* Instr mech eng, Univ Iowa, 46-48; from asst prof to assoc prof, Univ Colo, 48-60; supvry mech engr, Cryogenics Div, Nat Bur Standards, 60-66; prof mech eng, Worcester Polytech Inst, 66-69; chmn dept, 69-74, prof, 69-87, EMER PROF MECH ENG, UNIV IDAHO, 87- *Concurrent Pos:* Fulbright lectr, Col Eng, Univ Baghdad, 56-57. *Mem:* Am Soc Heat, Refrig & Air-Conditioning Engrs. *Res:* Thermodynamic properties and processes; cryogenics; thermodynamic properties of cryogenic fluids. *Mailing Add:* 1415 Chinook St Moscow ID 83843

STEWART, RICHARD CUMMINS, virology, vaccine development, for more information see previous edition

STEWART, RICHARD DONALD, b Lakeland, Fla, Dec 26, 26; m 52; c 3. INTERNAL MEDICINE, TOXICOLOGY. *Educ:* Univ Mich, AB, 51, MD, 55, MPH, 62, MA, 79; Am Bd Internal Med, dipl, 74; Am Bd Med Toxicol, dipl, 76. *Prof Exp:* Resident internal med, Med Ctr, Univ Mich, 59-62; staff physician, Med Dept, Dow Chem Co, 56-59, dir med res sect, Biochem Res Lab, 62-66; asst prof internal med & assoc prof prev med & toxicol,Sch Med, Med Col Wis, 66-69, prof environ med in internal med & toxicol & environ med, 69-78, chmn dept environ med, 66-78; CORP MED DIR, S C JOHNSON & SON, 78- *Concurrent Pos:* Corp med adv, S C Johnson & Son, 71-78; clin prof pharmacol & toxicol, Med Col Wis, 78-; adj prof, Univ Wis-Parkside, 78-; vis prof, Med Sch, Univ Hawaii, 80-; sect ed, Clin Med. *Honors & Awards:* Weisfeldt Mem Award, 75. *Mem:* Fel Am Col Physicians; Am Soc Artificial Internal Organs; Soc Toxicol; Am Acad Clin Toxicol; fel Am Occup Med Asn. *Res:* Human toxicology; development of the hollow fiber artificial kidney; air pollution epidemiological studies; experimental human exposures to artificial environments; tropical diseases; author of 125 publications. *Mailing Add:* 5337 Windpoint Rd Racine WI 53403

STEWART, RICHARD JOHN, b Duluth, Minn, May 30, 42; m 67. GEOLOGY. *Educ:* Univ Minn, BA, 65; Stanford Univ, PhD(geol), 70. *Prof Exp:* Asst prof, 69-77, ASSOC PROF GEOL, UNIV WASH, 77- *Concurrent Pos:* Geologist, Olympia Mts, US Geol Surv, 70; sedimentologist deep sea drilling proj, NSF, 71, res grant, Univ Wash, 72-73. *Mem:* Mineral Soc Am; Geol Soc Am; Am Asn Petrol Geologists; Soc Econ Paleontologists & Mineralogists; Am Geophys Union; Sigma Xi. *Res:* Sedimentary petrology; structural geology; geological and tectonic history of the northeast Pacific Ocean and its continental margin. *Mailing Add:* Dept Geol Sci Univ Wash Seattle WA 98195

STEWART, RICHARD WILLIAM, b Ames, Iowa, Oct, 22, 46; m 71; c 2. SYSTEM SURVIVABILITY. *Educ:* Mich State Univ, BS, 70; Iowa State Univ, MS, 73, PhD(elec eng), 77. *Prof Exp:* Staff engr, Johns Hopkins Appl Physics Lab, 77-79; group leader, IRT Corp, 79-89; div vpres, 89-90, VPRES & GROUP MGR, DIV MAXWELL LABS, SQ CORP, 90- *Mem:* Inst Elec & Electronics Engrs; Soc Photo-Optical Instrumentation Engrs. *Res:* Evaluation and enhancement of the survivability of military and civil systems operating in hostile environments; electromagnetic coupling, nonlinear propagation and stochastic estimation. *Mailing Add:* Div Maxwell Labs SQ Corp 3020 Callan Rd San Diego CA 92121

STEWART, RICHARD WILLIS, b Atlanta, Ga, Dec 27, 36; m 64; c 2. ATMOSPHERIC PHYSICS. *Educ:* Univ Fla, BS, 60; Columbia Univ, MA, 63, PhD(physics), 67. *Prof Exp:* Res assoc atmospheric physics, 67-69, staff scientist, Goddard Inst Space Studies, 69-78, RES SCIENTIST, NASA GODDARD SPACE FLIGHT CTR, 78- *Concurrent Pos:* Asst prof, Rutgers Univ, 68 & City Col New York, 69- *Mem:* Am Meteorol Soc; Am Geophys Union. *Res:* Aeronomy; planetary atmospheres. *Mailing Add:* 5359 Red Lake Columbia MD 21045-2433

STEWART, ROBERT ARCHIE, II, b Houston, Miss, Jan 23, 42; m 74; c 2. PLANT ECOLOGY. *Educ:* Miss State Univ, BS, 65, MS, 67; Ariz State Univ, PhD(bot), 71. *Prof Exp:* Partic, NSF advan seminar trop bot, Univ Miami, 68; from asst prof to assoc prof, 70-80, PROF BIOL, DELTA STATE UNIV, 80- *Mem:* Am Inst Biol Sci. *Res:* Plant ecology; flora of Mississippi. *Mailing Add:* Dept Biol Sci Delta State Univ Cleveland MS 38732

STEWART, ROBERT BLAYLOCK, b Stilwell, Okla, Feb 10, 26; m 48; c 2. PLANT PATHOLOGY. *Educ:* Okla State Univ, BS, 50; Tex A&M Univ, MS, 53, PhD(plant path), 57. *Prof Exp:* Instr agron, Okla State Univ, 54-56; asst prof plant path, Tex A&M Univ, 56-58; from assoc prof to prof bot, Sam Houston State Univ, 58-74, prof biol, 74-90. *Concurrent Pos:* Assoc prof from Okla State Univ, Imp Ethiopian Col Agr & Mech Arts, 59. *Mem:* AAAS; Mycol Soc Am; Soc Econ Bot. *Res:* Paleoethnobotany. *Mailing Add:* 3348 Winton Way Huntsville TX 77340

STEWART, ROBERT BRUCE, b Toronto, Ont, May 2, 26; m 45; c 2. ANIMAL VIROLOGY. *Educ:* Mt Allison Univ, BSc, 49; Queen's Univ, Ont, MA, 51, PhD(bact), 55. *Prof Exp:* Res officer, Defence Res Bd, Can, 51-55; from instr to asst prof bact, Sch Med & Dent, Univ Rochester, 55-63; assoc prof, 63-66, PROF BACT, QUEEN'S UNIV, ONT, 66- *Concurrent Pos:* head dept, Queen's Univ, Ont, 71-86. *Mem:* Am Soc Microbiol; Can Soc Microbiol; Can Soc Microbiologists (pres, 85-86). *Res:* Virology; regulation of virus growth; infectious disease. *Mailing Add:* Dept Microbiol & Immunol Queen's Univ Kingston ON K7L 3N6 Can

STEWART, ROBERT CLARENCE, b Sharon, Pa, Sept 23, 21; m 59. MATHEMATICS. *Educ:* Washington & Jefferson Col, BA, 42, MA, 44; Yale Univ, MA, 48. *Prof Exp:* Instr math, Washington & Jefferson Col, 42-44, 45-46; asst, Yale Univ, 46-50; from instr to prof, 50-76, CHARLES A DANA PROF MATH, TRINITY COL, CONN, 76- *Mem:* Am Math Soc; Math Asn Am. *Res:* Modern algebra; matrix theory; differential equations. *Mailing Add:* Dept Math Trinity Col Hartford CT 06106

STEWART, ROBERT DANIEL, b Salt Lake City, Utah, June 15, 23; m 47; c 4. PHYSICAL CHEMISTRY, METALLURGICAL CHEMISTRY. *Educ:* Univ Utah, BS, 50; Univ Wash, PhD(phys chem), 54. *Prof Exp:* Sr res chemist, Am Potash & Chem Corp, 54-56, group leader, 56-58, sect head, 58-68; group leader, Garrett Res & Develop Co, 68-75; group leader, Occidental Res Corp, 75-82; SUPVR, CYPRUS INDUST MINERALS, 83- *Mem:* Am Chem Soc; Electrochem Soc; Am Inst Mining, Metall & Petrol Engrs. *Res:* Gas phase kinetics; thermodynamics; inorganic polymers; extractive hydrometallurgy; minerals beneficiation. *Mailing Add:* 17052 El Cajon Ave Yorba Linda CA 92686

STEWART, ROBERT EARL, b Campbellton, NB, Feb 17, 35; m 60; c 2. FLUID MECHANICS, MICROMETEOROLOGY. *Educ:* NS Tech Col, BEng, 57; Univ BC, MEng, 63; Univ Waterloo, PhD(mech), 66. *Prof Exp:* Design engr, BC Hydro, 57-58, res engr, 58-61; demonstr mech eng, Univ BC, 61-63; res asst, Univ Waterloo, 63-66; asst prof environ eng, Univ Fla, 66-70; ASSOC PROF MECH & ENVIRON ENG, UNIV LOUISVILLE, 70- *Mem:* Am Meteorol Soc; Can Meteorol Soc; fel Royal Meteorol Soc. *Res:* Diffusion of gases and particulates released into the atmosphere. *Mailing Add:* Dept of Mech Eng Univ of Louisville Box 35260 Louisville KY 40292

STEWART, ROBERT F, b Seattle, Wash, Dec 31, 36; m 59; c 2. PHYSICAL CHEMISTRY. *Educ:* Carleton Col, AB, 58; Calif Inst Technol, PhD(chem), 63. *Prof Exp:* NIH fel, Univ Wash, 62-64; fel, 64-69, ASSOC PROF CHEM, CARNEGIE-MELLON UNIV, 69- *Concurrent Pos:* Alfred P Sloan fel, 70-72. *Res:* Ultraviolet absorption of single crystals; valence structure from x-ray scattering; x-ray diffraction. *Mailing Add:* Dept of Chem Carnegie-Mellon Univ 5000 Forbes Ave Pittsburgh PA 15213

STEWART, ROBERT FRANCIS, b Birmingham, Ala, Oct 31, 26; m 58; c 4. NUCLEAR CHEMISTRY, FUEL TECHNOLOGY. *Educ:* Univ Ala, BS, 49, MS, 50. *Prof Exp:* Jr chemist, Nat Southern Prod Corp, 50-52; head radioisotope lab, 63-69, RES CHEMIST, US BUR MINES, 54-, RES SUPVR, MORGANTOWN ENERGY RES CTR, 69- *Mem:* Am Chem Soc; Am Soc Testing & Mat; Instrument Soc Am. *Res:* Nuclear methods of continuous analysis of bulk materials for process control based on neutron interactions in matter. *Mailing Add:* Rte Eight Box 228E Morgantown WV 26505-9024

STEWART, ROBERT HENRY, b York, Pa, Dec 26, 41; m 86; c 1. PHYSICAL OCEANOGRAPHY, RADIO SCIENCE. *Educ:* Univ Tex, Arlington, BS, 63; Univ Calif, San Diego, PhD(oceanog), 69. *Prof Exp:* Asst res oceanogr, Scripps Inst Oceanog, 69-77, assoc res oceanogr & assoc adj prof, 77-83, res oceanogr & adj prof, 83-89; sr res scientist, Jet Propulsion Lab, Calif Inst Technol, 83-89; PROF, TEX A&M UNIV, 89- *Concurrent Pos:* Consult, NASA, 75-77; Topex proj scientist, 79-88; comt on earth sciences, Nat Acad Sci, Nat Res Coun, 87-89; res scientist, Jet Propulsion Lab, Calif Inst Technol, 79-83. *Mem:* Am Geophys Union; Int Union Radio Sci; Am Soc Photogrammetry & Remote Sensing. *Res:* Satellite oceanography; currents; ocean waves; radio scatter from the sea; numerous publications in general field of satellite oceanography plus experience in designing the Topex/Poseidon satellite mission. *Mailing Add:* Dept Oceanog Tex A&M Univ College Station TX 77843-3146

STEWART, ROBERT MURRAY, JR, b Washington, DC, May 6, 24; m 45; c 2. COMPUTER SCIENCE. *Educ:* Iowa State Col, BS, 45, PhD(physics), 54. *Prof Exp:* Instr elec eng, 46-48, res assoc physics, 48-54, from asst prof to assoc prof physics, 54-58, engr in charge, Cyclone Comput Lab, 56-67, assoc prof elec eng, 58-60, chmn dept comput sci, 69-83, PROF PHYSICS & ELEC ENG, IOWA STATE UNIV, 60-, ASSOC DIR COMPUT CTR, 63-, PROF COMPUT SCI, 69- *Concurrent Pos:* Sr physicist, Ames Lab, AEC; mem bd ed consults, Electronic Assocs, Inc; consult, Midwest Res Inst; mem educ comt, Am Fedn Info Processing Socs, 68-; mem, Comput Sci Bd, 72-; vchmn, Comput Sci Conf Bd, 75-76, chmn, 76-77. *Mem:* Am Phys Soc; Asn Comput Mach; Inst Elec & Electronics Eng; AAAS. *Res:* Design of logical control systems and digital computer systems; pattern recognition and adaptive logic. *Mailing Add:* 3416 Oakland St Ames IA 50010

STEWART, ROBERT WILLIAM, b Smoky Lake, Alta, Aug 21, 23; div; c 3. PHYSICS. *Educ:* Queen's Univ, Ont, BSc, 45, MSc, 47; Cambridge Univ, PhD(physics), 52. *Hon Degrees:* DSc, McGill Univ, 72. *Prof Exp:* Lectr physics, Queen's Univ, Ont, 46; defence sci serv officer, Pac Naval Lab Can, 50-55; from assoc prof to prof physics, Univ BC, 55-70; dir-gen, Pac Region, Ocean & Aquatic Sci, Can dept Fisheries & Oceans, 70-79; HON PROF PHYSICS, UNIV BC, 70- *Concurrent Pos:* Vis prof, Dalhousie Univ, 60-61 & Harvard Univ, 64; distinguished vis prof, Pa State Univ, 64; Commonwealth vis prof, Cambridge Univ, 67-68; chmn joint organizing comt, Global Atmospheric Res Prog, 72-78; mem, Comt Climatic Changes & the Ocean, 78-, chmn, 83-87; dep min, BC Ministry Univ, Sci & Comm, 79-84; pres, Alberta Res Coun, 84-87; interim dir, Ctr Earth & Ocean Physics, Univ Victoria, BC, 87- *Honors & Awards:* Patterson Medal, Can Meteorol & Oceanog Soc, 73; Sverdrup Gold Medal, Am Meteorol Soc, 76. *Mem:* Fel Royal Soc Can; fel Royal Soc; Can Asn Physicists; Can Meteorol Soc; Int Asn Phys Sci Ocean (pres, 77). *Res:* Turbulence; physical oceanography; air-sea interaction. *Mailing Add:* Dept Physics & Astron Univ Victoria PO Box 3055 Victoria BC V8W 3P6 Can

STEWART, ROBERTA A, b Rochester, NH, Aug 24, 23. ORGANIC CHEMISTRY. *Educ:* Univ NH, BS, 44; Smith Col, MA, 46, PhD, 49. *Prof Exp:* Res assoc chem, Smith Col, 48-49; instr, Wellesley Col, 49-53; from asst prof to assoc prof, 53-70, chmn dept chem, 63-66 & 69-73, chmn div natural sci & math, 67-73 & 85-88, asst to pres, 69-75, dean col, 75-83, PROF CHEM, HOLLINS COL, 70 - *Concurrent Pos:* NSF fel, Radcliffe Col, 60-61; Am Coun on Educ fel acad admin, Univ Del, 66-67. *Mem:* Sigma Xi; Am Chem Soc. *Res:* Synthetic experiments in direction of morphine; some reactions of beta tetralone; derivatives of cyclohexanone. *Mailing Add:* Box 9685 Hollins College VA 24020-1685

STEWART, ROBIN KENNY, b Ayr, Scotland, May 22, 37; m 64; c 3. ANIMAL ECOLOGY. *Educ:* Glasgow Univ, BS, 61, PhD(entom), 66. *Prof Exp:* Asst lectr agr zool, Glasgow Univ, 63-66; from asst prof to assoc prof, 66-77, coordr biol sci div, 72-75, PROF ENTOM & ZOOL, MACDONALD COL, MCGILL UNIV, 77-, CHMN DEPT ENTOM, 75- *Concurrent Pos:* Consult, Can Pac Investment, 74-75 & UN Develop Proj, 75- *Mem:* Can Entom Soc; British Ecol Soc. *Res:* Agricultural zoology; ecology; entomology; integrated control. *Mailing Add:* Dept Entom MacDonald Col 21111 Lakeshore Rd Ste Anne de Bellevue PQ H9X 1C0 Can

STEWART, ROSS, b Vancouver, BC, Mar 16, 24; m 46; c 2. PHYSICAL ORGANIC CHEMISTRY. *Educ:* Univ BC, BA, 46, MA, 48; Univ Wash, PhD, 54. *Prof Exp:* Lectr chem, Univ BC, 47-49; lectr, Can Serv Col, Royal Roads, 49-51, from asst prof to assoc prof, 51-55; from asst prof to assoc prof, 55-62, PROF CHEM, UNIV BC, 62- *Mem:* Fel Royal Soc Can; fel Chem Inst Can. *Res:* Organic oxidation mechanisms; protonation of weak organic bases; general acid catalysis; ionization of weak acids in strongly basic solution. *Mailing Add:* Dept Chem Univ BC 2075 Westbrook Pl Vancouver BC V6T 1W5 Can

STEWART, RUTH CAROL, b Englewood, NJ, Dec 18, 28. MATHEMATICS. *Educ:* Rutgers Univ, AB, 50, MA, 63, EdD(math educ), 69. *Prof Exp:* Dir music, High Sch, NJ, 50-53; instr, Wiesbaden Am High Sch, Ger, 54-57; chmn dept math, Frankfurt Am High Sch, Ger, 58-62 & 63-64; PROF MATH, MONTCLAIR STATE COL, 64- *Mem:* Math Asn Am. *Res:* Mathematics in areas of algebra and analysis; mathematics education in areas of curriculum and instruction. *Mailing Add:* Dept Math Montclair State Col Upper Montclair NJ 07043

STEWART, SHEILA FRANCES, neuroendocrinology; deceased, see previous edition for last biography

STEWART, SHELTON E, b Sanford, NC, Oct 1, 34; m 65; c 1. BOTANY, ZOOLOGY. *Educ:* ECarolina Col, BS, 56; Univ NC, MA, 59; Univ Ga, PhD(bot), 66. *Prof Exp:* Teacher, Pine Forest High Sch, 56-57; prof sci, Ferrum Jr Col, 59; prof biol, Lander Col, 59-63; instr bot, Univ Ga, 64-65; PROF BIOL, LANDER COL, 66- *Mem:* AAAS. *Res:* Plant taxonomy; plant biosystematics. *Mailing Add:* Dept of Biol Lander Col Greenwood SC 29646

STEWART, T BONNER, b Sao Paulo, Brazil, Nov 24, 24; US citizen; m 56; c 4. VETERINARY PARASITOLOGY. *Educ:* Univ Md, BS, 49; Auburn Univ, MS, 53; Univ Ill, Urbana, PhD(vet med sci), 63. *Prof Exp:* Parasitologist, USDA, Ala, 50-53 & Ga Coastal Plain Exp Sta, 53-60; fel, Univ Ill, Urbana, 60-62; res parasitologist, Animal Parasite Res Lab, Ga Coastal Plain Exp Sta, USDA, 63-64, supvry zoologist, 64-79; PROF PARASITOL, SCH VET MED, LA STATE UNIV, 79- *Concurrent Pos:* Vis prof, Rural Fed Univ, Rio de Janeiro, 86, Fed Univ, Rio Grande du Sol, Brazil, 86. *Mem:* AAAS; World Asn Advan Vet Parasitol; Am Soc Parasitol; Soc Protozool; Wildlife Dis Asn. *Res:* Life history of Cooperia punctata; gastrointestinal parasites of cattle; eradication of the kidneyworm of swine; beetles as intermediate hosts of nematodes; strongyloides ransomi of swine; ecology of swine parasites; perinatal infection of host by nematode parasites; anthelmintics for swine; host nutrition and parasite interaction. *Mailing Add:* Dept Vet Microbiol & Parasitol Sch Vet Med La State Univ Baton Rouge LA 70803

STEWART, TERRY SANFORD, b West Palm Beach, Fla, Feb 28, 51; m 74; c 3. ANIMAL BREEDING, SYSTEMS ANALYSIS. *Educ:* Univ Fla, BSA, 72, MSA, 74; Tex A&M Univ, PhD(animal genetics), 77. *Prof Exp:* Grad asst animal genetics, dept animal sci, Univ Fla, 73-74; res asst, Animal Sci Dept, Tex A&M Univ, 74-77; PROF ANIMAL SCI, PURDUE UNIV, 77- *Mem:* Am Soc Animal Sci; Am Genetics Asn; Biomet Soc; Am Regist Cert Animal Scientists. *Res:* Animal systems analysis; animal genetics; biometrics; species, beef cattle and swine. *Mailing Add:* Dept of Animal Sci Purdue Univ West Lafayette IN 47907

STEWART, THOMAS, b Leith, Scotland, Nov 25, 40. ORGANIC CHEMISTRY. *Educ:* Heriot-Watt Univ, BS, 63; Glasgow Univ, PhD(org chem), 66. *Prof Exp:* Fel org synthesis, Calif Inst Technol, 66-67 & Glasgow Univ, 67-68; SR SCIENTIST ORG SYNTHESIS, RES DIV, ROHM & HAAS CO, 68- *Concurrent Pos:* Res sect mgr, Polymer Technol Res. *Mem:* Am Chem Soc. *Res:* Synthesis of acrylic monomers and polymerization inhibition. *Mailing Add:* Gayman Rd RD Seven Doylestown PA 18901-9807

STEWART, THOMAS HENRY MCKENZIE, b Hertfordshire, Eng, Aug 17, 30; Can citizen; m 60; c 4. INTERNAL MEDICINE, IMMUNOLOGY. *Educ:* Univ Edinburgh, MB, ChB, 55; FRCP(C), 62. *Prof Exp:* House officer surg, All Saints Hosp, Chatham, Eng, 55-56; house officer med, Eastern Gen Hosp, Edinburgh, Scotland, 56 & Westminster Hosp, London, Ont, 58-60; resident, Ottawa Gen Hosp, 61-62; lectr nuclear med, Univ Mich, Ann Arbor, 63-64; lectr, 64-66, asst prof, 66-72, assoc prof, 72-77, PROF MED, UNIV OTTAWA, 77- *Concurrent Pos:* Teaching fel, Path Inst, McGill Univ, 60-61; res fel hemat, Univ Ottawa, 62-63. *Mem:* Can Soc Immunol; Can Soc Clin Invest; Soc Nuclear Med; Am Asn Cancer Res. *Res:* Immunology of cancer; host-tumor relationships; immunochemotherapy of human cancer; immunology of inflammatory bowel disease. *Mailing Add:* One Mt Pleasant Ottawa ON K1S 0L6 Can

STEWART, THOMAS WILLIAM WALLACE, acoustics, electronics; deceased, see previous edition for last biography

STEWART, W(ARREN) E(ARL), b Whitewater, Wis, July 3, 24; m 47; c 6. TRANSPORT PHENOMENA, NUMERICAL METHODS. *Educ:* Univ Wis, BS, 45, MS, 47; Mass Inst Technol, ScD(chem eng), 51. *Prof Exp:* Instr chem eng, Mass Inst Technol, 48; proj chem engr, Sinclair Res Labs, Inc, 50-56; from asst prof to assoc prof, 56-61, chmn dept, 73-78, PROF CHEM ENG, UNIV WIS-MADISON, 61- *Concurrent Pos:* Consult & vis prof, Univ Nac de La Plata, 62; assoc ed, J Comput & Chem Eng, 77-; mem, Math Res Ctr, Univ Wis, 78-86; McFarland-Bascom Prof, 83- *Honors & Awards:* Alpha Chi Sigma Res Award, Am Inst Chem Engrs, 81, Comput Chem Eng Award, 85; Chem Eng Div Lectr Award, Am Soc Eng Educ, 83; E V Murphree Award in Indust & Eng Chem, Am Chem Soc, 89. *Mem:* Am Chem Soc; fel Am Inst Chem Engrs; Sigma Xi; Am Soc Eng Educ. *Res:* Transport phenomena; chemical reactor modelling; multicomponent mass transfer; modelling of distillation systems; weighted residual methods; parameter estimation algorithms & software; boundary layer theory; pulmonary ventilation/perfusion distribution estimation; fusion reactor engineering. *Mailing Add:* 2004 Eng Bldg 1415 Johnson Dr Univ Wis Madison WI 53706

STEWART, WELLINGTON BUEL, b Chicago, Ill, June 18, 20; m 45; c 3. PATHOLOGY. *Educ:* Univ Notre Dame, BS, 42; Univ Rochester, MD, 45. *Prof Exp:* Intern path, Strong Mem Hosp, Rochester, NY, 45-46; Rockefeller Found fel, Univ Rochester, 48-49, Veteran fel, 48-50; assoc, Columbia Univ, 50-51, asst prof, 51-54; assoc prof, Col Physicians & Surgeons, 54-60; prof & chmn dept, Col Med, Univ Ky, 60-70; dir med comput ctr, 70-75, PROF PATH, UNIV MO-COLUMBIA, 70-, DIR LABS, 75- *Concurrent Pos:* Asst pathologist, Presby Hosp, New York, 50-54, assoc attend pathologist, 54-60; chmn, Bd Registry Med Technol, 64-67. *Mem:* Am Soc Clin Pathologists; Soc Exp Biol & Med; Harvey Soc; AMA; Col Am Pathologists. *Res:* Iron metabolism; fatty liver; red cell physiology; computers in medicine. *Mailing Add:* Dept Path-M646 Med Sci Univ Mo Sch Med-M228 Med Sci Bldg Columbia MO 65212

STEWART, WILLIAM ANDREW, b Liberty Center, Ohio, Apr 6, 33; m 60; c 5. MEDICINE. *Educ:* Miami Univ, AB, 54; Ohio State Univ, MD, 58. *Prof Exp:* Intern, 58-65, ASST PROF NEUROSURG, STATE UNIV NY UPSTATE MED CTR, 67- *Concurrent Pos:* Reader neurosurg, Fac Health Sci, Univ Ife, Nigeria, 74-75. *Mem:* Cong Neurol Surgeons; fel Am Col Surgeons; Am Asn Neurol Surgeons. *Res:* Trigeminal nerve; head injury. *Mailing Add:* 725 Irving Ave Syracuse NY 13210

STEWART, WILLIAM CHARLES, statistical analysis, operations research, for more information see previous edition

STEWART, WILLIAM HENRY, b Okalahoma City, Okla, Feb 9, 49; m 72; c 3. APPLIED STATISTICS. *Educ:* Univ Okla, BA, 71, MA, 74; Ore State Univ, PhD(statist), 79. *Prof Exp:* Asst prof statist, Okla State Univ, 78-84. *Mem:* Am Statist Asn; Biomet Soc. *Res:* Development of statistical methods for survival experiments with grouped data; general methods for categorical data with ordered classifications. *Mailing Add:* 8500 W 64 Terr Merriam KS 66202

STEWART, WILLIAM HOGUE, JR, b Mullins, SC, Dec 25, 36; m 59; c 3. SOLID STATE PHYSICS. *Educ:* The Citadel, BSEE, 59; Univ Cincinnati, MS, 61; Clemson Univ, PhD(physics), 64. *Prof Exp:* Res physicist, 66-71, sr res physicist, 71-74, res assoc, 74-79, SR SCIENTIST, MILLIKEN RES CORP, 79- *Mem:* Bioelectromagnetics Soc. *Res:* Fiber physics; static electricity and ion physics; paramagnetic resonance spectroscopy; solid state diffusion; x-ray diffraction and spectroscopy; nonlinear servomechanisms; digitally controlled machine design; Millitron ink jet printers. *Mailing Add:* PO Box 1927 Spartanburg SC 29304-1927

STEWART, WILLIAM HUFFMAN, b Minneapolis, Minn, May 19, 21; m 46; c 2. MEDICINE. *Educ:* Univ Minn, 39-41; La State Univ, MD, 45; Am Bd Pediat, dipl, 52. *Hon Degrees:* Numerous from US univs & insts, 66-69. *Prof Exp:* Resident pediatrician, Charity Hosp, New Orleans, 48-50; pvt pract, 50-51; epidemiologist, Commun Dis Ctr, USPHS, 51-53, actg chief heart dis control prog, 54-55, chief, 55-56, asst dir, Nat Heart Inst, 56-57, asst to surgeon gen, 57-58, chief div pub health methods, Off Surgeon Gen, 57-61 & div community health serv, 61-63, asst to spec asst to secy health & med affairs, 63-65, dir, Nat Heart Inst, 65, surgeon gen, 65-69; chancellor, 69-74, prof pediat & head dept, 73-77, HEAD DEPT PREV MED & PUB HEALTH, MED CTR, LA STATE UNIV, NEW ORLEANS, 77- *Concurrent Pos:* Mem tech adv bd, Milbank Fund & adv med bd, Leonard

Wood Mem; secy, Dept of Health & Human Resources, State of La, 74-77. *Mem:* AAAS; Am Pub Health Asn; Am Acad Pediat; Am Col Prev Med; Asn Teachers Prev Med. *Res:* Epidemiology; medical administration. *Mailing Add:* 219 W Gatehouse Dr Apt B Metairie LA 70001

STEWART, WILSON NICHOLS, botany, for more information see previous edition

STEYERMARK, JULIAN ALFRED, taxonomic botany, for more information see previous edition

STEYERT, WILLIAM ALBERT, physics, cryogenic refrigerators; deceased, see previous edition for last biography

STIBBS, GERALD DENIKE, b Schreiber, Ont, Apr 25, 10; m 55; c 3. DENTISTRY. *Educ:* Univ Ore, BS & DMD, 31. *Prof Exp:* Chmn dept oper dent, dir dent operatory & clin coordr, Sch Dent, 48-70, exec officer, Dept Fixed Partial Dentures, 50-57, chmn dept oper dent grad prog, 50-70, prof oper dent, Sch Dent, 48-70, prof fixed partial dentures, 54-70, spec asst to dean, 70-73, prof restorative dent, 73-76, EMER PROF RESTORATIVE DENT, SCH DENT, UNIV WASH, 76-, MEM STAFF, GRAD SCH, 50- *Concurrent Pos:* Mem assoc comt dent res, Nat Res Coun Can, 45-48; consult, Madigan Army Hosp, 48-52, coun dent health, Am Dent Asn, 54-55 & Pac Northwest Labs, Battelle Mem Inst, 70-74. *Mem:* Am Dent Asn; Am Acad Restorative Dent; Am Acad Gold Foil Opers; Int Asn Dent Res; Acad Operative Dent; Sigma Xi. *Res:* Restorative dentistry; dental materials. *Mailing Add:* 6227 51st Ave NE Seattle WA 98115

STIBITZ, GEORGE ROBERT, b York, Pa, Apr 30, 04; m 30; c 2. MEDICAL RESEARCH. *Educ:* Denison Univ, PhB, 26; Union Col, NY, MS, 27; Cornell Univ, PhD(physics), 30. *Hon Degrees:* DSc, Denison Univ, 66, Keene Col, 78, Dartmouth Col, 86. *Prof Exp:* Res mathematician, Bell Tel Labs, 30-41; tech aide, Nat Defense Res Comt, 41-45; math consult, 45-66; prof, 66-73, EMER PROF PHYSIOL, DARTMOUTH MED SCH, 73- *Honors & Awards:* Harry Goode Award, Am Fedn Info Processing Socs; Emanuel Piore Award, Inst Elec & Electronic Engrs, 78; Babbage Computer Pioneer Award, Computer Soc. *Mem:* Nat Acad Eng. *Res:* Computing devices; automatic control and stability; dynamic testing; logical design of computers; electronic music; mathematical and computer models of biomedical systems; computer programs for radiation therapy dosage; mathematical and computer models of physiological systems; passive electrical properties of cardiac cells; computer model of molecular motion in slits of capillaries. *Mailing Add:* Dept Physiol Dartmouth Med Sch Hanover NH 03756

STICH, HANS F, b Prague, Czech, Dec 24, 27; Ger citizen; m 55; c 1. CELL BIOLOGY, GENETICS. *Educ:* Univ Wurzburg, PhD(zool), 49. *Prof Exp:* Res assoc cell biol, Max Planck Inst Marine Biol, 50-57; assoc prof cancer res unit, Nat Cancer Inst Can, Univ Sask, 57-60; from assoc prof to prof genetics, Queen's Univ, Kingston, Ont, 60-66; prof biol & chmn dept, McMaster Univ, Hamilton, Ont, 66-68; PROF ZOOL, UNIV BC, 68-, HEAD ENVIRON CARCINOGENIC UNIT, BC CANCER CTR, VANCOUVER, 78- *Concurrent Pos:* Fulbright grants, Case Western Reserve Univ, 55-56, Univ Wis, 55-57; vis prof, MD Anderson Hosp & Tumor Inst, Houston, Tex, 63 & Roswell Park Mem Inst, Buffalo, NY, 64. *Honors & Awards:* Terry Fox Cancer Res Scientist, Nat Cancer Inst Can. *Mem:* Genetic Soc Can; Am Asn Cancer Res; NY Acad Sci; Can Nat Comt Sci Problems Environ. *Res:* Identification of population groups at high risk for cancer, identification of main causes of cancer among these groups; reduction of human exposure to cancer-causing agents; design and implementation of cancer-protective measures through intervention trials. *Mailing Add:* BC Cancer Res Ctr 601 W Tenth Ave Vancouver BC V5Z 1L3 Can

STICHA, ERNEST AUGUST, b Baltimore, Md, Mar 27, 11; m 39; c 2. METALLURGICAL ENGINEERING. *Educ:* Ill Inst Technol, BS, 33; Univ Mich, MSE, 36. *Prof Exp:* Supv engr, Crane Co, 36-59; chief metallurgist, Edward Valves, Inc Div, Rockwell Mfg Co, 59-65; sr res engr, Am Oil Co, 65-76; RETIRED. *Concurrent Pos:* Consult, 76- *Mem:* Am Soc Testing & Mat; Am Soc Metals. *Res:* Creep of metals; metallurgy of high temperature materials. *Mailing Add:* 715 N Kensington Ave LaGrange Park IL 60525

STICHT, FRANK DAVIS, b Plattsburg, Miss, June 14, 19; m 41; c 2. PHARMACOLOGY. *Educ:* Univ Miss, BS Pharm, 48; Baylor Univ, DDS, 56; Univ Tenn, Memphis, MS, 65. *Prof Exp:* From instr to prof, 61-84, EMER PROF PHARMACOL, UNIV TENN, MEMPHIS, 84- *Mem:* Am Soc Pharmacol & Exp Therapeut. *Res:* Autonomic and cardiovascular pharmacology; influence of drugs on the blood pressure within the tooth pulp; relationship of prostaglandins to periodontal disease; actions of calcium antagonists on veins. *Mailing Add:* 6092 Ridgewyck Dr Apt One Memphis TN 38115

STICKEL, DELFORD LEFEW, b Falling Waters, WVa, Dec 12, 27; m 52; c 1. SURGERY. *Educ:* Duke Univ, AB, 49, MD, 53; Am Bd Surg, dipl, 63; Bd Thoracic Surg, dipl, 63. *Prof Exp:* Asst surg, 57-59, from instr to assoc prof, 59-72, PROF SURG, MED CTR, DUKE UNIV, 72-, ASSOC DIR HOSP, 72- *Concurrent Pos:* Markle scholar & NIH career develop award, 62; attend physician, Durham Vet Admin Hosp, 65-66, chief surg serv, 66-68, chief staff, 70-72; consult, Watts Hosp, Durham, 66-70 & NC Eye & Human Tissue Bank, 69. *Mem:* AMA; Soc Univ Surgeons; Am Col Surgeons; Southern Surg Asn; Transplantation Soc; Am Soc Transplant Surg. *Res:* Clinical renal transplantation. *Mailing Add:* Dept Surg Duke Univ Med Ctr Box 3917 Durham NC 27710

STICKELS, CHARLES A, b Detroit, Mich, Apr 6, 33. METALLURGICAL ENGINEERING. *Educ:* Univ Mich, BS, 56, MS, 60, MA, 62, PhD(metall eng), 63. *Prof Exp:* Mem staff, Ford Motor Co, 63-85, prin res engr, 85-91; CONSULT, ENVIRON RES INST MICH, 91- *Mem:* Fel Am Soc Metals Int; Am Inst Mining Metall & Petrol Engrs; Sigma Xi. *Mailing Add:* 2410 Newport Rd Ann Arbor MI 48103

STICKER, ROBERT EARL, b New York, NY, Mar 4, 30; m 57; c 3. ORGANIC CHEMISTRY. *Educ:* Cornell Univ, AB, 53; Columbia Univ, AM, 57; Univ Kans, PhD(pinane chem), 65. *Prof Exp:* Res chemist, Eastman Kodak Co, 57-60; res chemist, 65-76, SR RES CHEMIST, NIAGARA CHEM DIV, FMC CORP, 76- *Mem:* Am Chem Soc; Int Soc Heterocyclic Chem. *Res:* Terpenes; heterocycles; surfactants; herbicides; plant regulators; fungicides. *Mailing Add:* 4252 Freeman Rd Middleport NY 14105-9640

STICKLAND, DAVID PETER, b Norwich, Eng, Aug 22, 54; m 79; c 1. ELEMENTARY PARTICLE PHYSICS. *Educ:* Univ Bradford, BTech, 77; Univ Bristol, PhD(physics), 81. *Prof Exp:* Res asst, 80-81, instr, 81-82, ASST PROF PHYSICS, PRINCETON UNIV, 82- *Mem:* Am Phys Soc. *Mailing Add:* Dept Physics Princeton Univ Princeton NJ 08544

STICKLE, GENE P, b New Castle, Pa, Apr 11, 29; m 54; c 2. PROCESS ENGINEERING, PROJECT ENGINEERING. *Educ:* Univ Tenn, BS, 53, MS, 54. *Prof Exp:* Res engr, Squibb Inst Med Res, Olin Mathieson Chem Corp, 54-61, sect head, Fermentation Mfg Dept, 61-62, sect head microbiol develop pilot plant, Squibb Inst Med Res, 62-66, sect head chem develop pilot plant, 67-69, asst dept dir chem develop, 69-74, antibiotics mfg develop mgr, 74-77, dir antibiotics process eng, 77-81, TECH ENG DIR, E R SQUIBB & SONS, INC, 81- *Concurrent Pos:* Lectr, Ctr Prof Advan, 81-85. *Mem:* Am Chem Soc; Am Inst Chem Eng; Sigma Xi. *Res:* Fermentation technology, scaleup, process design and development; plant start-up and manufacture of antibiotics, steroids, vitamins and enzymes; technical liaison; organic synthetics manufacture; pharmaceutical process engineering. *Mailing Add:* 11 Eaton Ave Spotswood NJ 08884

STICKLER, DAVID BRUCE, b Taunton, Mass, Nov 17, 41; m 64; c 3. COMBUSTION, FLUID MECHANICS. *Educ:* Mass Inst Technol, SB & SM, 64, PhD(hybrid combustion), 68. *Prof Exp:* Asst prof aeronaut & astronaut, Mass Inst Technol, 68-73; res scientist, 73-78, chmn aerophys, 78-82, CHIEF SCIENTIST, ENERGY TECHNOL, 82-, DIR, INNOVATIVE RES, AVCO RES LAB, INC, 87- *Concurrent Pos:* Consult lasers & energy, 68-73. *Mem:* Combustion Inst; assoc fel Am Inst Aeronaut & Astronaut. *Res:* Heterogeneous combustion; turbulent combustion and flow; coal combustion and gasification; hybrid combustion; slag flow in power systems; polymer pyrolysis; pollution control; glass manufacture; metal refining. *Mailing Add:* 2385 Revere Beach Pkwy Everett MA 02149

STICKLER, DAVID COLLIER, b Piqua, Ohio, Apr 12, 33; c 4. ACOUSTICS, APPLIED MATHEMATICS. *Educ:* Ohio State Univ, BSc, 56, MSc, 59, PhD(elec eng), 64. *Prof Exp:* Res assoc elec eng, Ohio State Univ, 55-65; mem tech staff, Bell Tel Labs, Inc, 65-72; sr engr, Systs Control, Inc, 72-73; sr res assoc, Pa State Univ, 73-77; sr res scientist, 77-80, RES PROF, NY UNIV, 80- *Res:* Electromagnetic theory; heat transfer; thermoelastic effects mechanics. *Mailing Add:* Math Dept NJ Inst Technol 323 Martin Luther King Blvd Newark NJ 07102

STICKLER, FRED CHARLES, b Villisca, Iowa, Dec 11, 31; m 55; c 3. AGRONOMY, CROP ECOLOGY. *Educ:* Iowa State Univ, BS, 53, PhD, 58; Kans State Univ, MS, 55. *Prof Exp:* From asst prof to assoc prof agron, Kans State Univ, 58-64; res agronomist, 64-71, prod planner, 71-73, mgr agr equip planning, 73-76, dir, Tech Ctr, 76-80, DIR, PROD & MKT PLANNING, DEERE & CO, 80- *Mem:* Sigma Xi. *Res:* Ecological aspects of crop production and management; crop management for improved mechanization. *Mailing Add:* 7108 36th Ave B Ct Moline IL 61265

STICKLER, GUNNAR B, b Peterskirchen, Ger, June 13, 25; US citizen; m 56; c 2. MEDICINE, PEDIATRICS. *Educ:* Univ Vienna, MD, 49; Univ Minn, PhD(pediat), 58; Am Bd Pediat, cert, 57, cert pediat & nephrol, 74. *Prof Exp:* Resident clin path, Krankenhaus III Orden, Munich, 50; resident path, Univ Munich, 50-51; intern, Mountainside Hosp, Montclair, 51-52; resident fel pediat, Mayo Found Grad Sch, 53-56; chief pediat, US Army Hosp, Munich, 56; sr res scientist, Roswell Park Mem, Buffalo, 56-57; instr pediat, Univ Buffalo, 57; asst to staff, Mayo Clin, 57-58, from instr to prof pediat, Mayo Found Grad Sch, 58-73, sect head pediat, 69-74, prof pediat & chmn dept, 74-80, consult pediat, Mayo Clin, 58-89; RETIRED. *Mem:* Am Acad Pediat; Soc Pediat Res; Am Soc Nephrol; Int Soc Nephrol; Am Soc Pediat. *Res:* Hypophosphatemic and various other forms of rickets; nephrotic syndrome; immunology of renal disease; growth failure in various disease processes; treatment of acute otitis media; hereditary bone diseases; urinary tract infection; natural history studies in chronic diseases; inflammatory bowel disease; study of anxieties in parents and children. *Mailing Add:* Emer Rm Mayo Found Mayo Grad Sch Med Rochester MN 55905

STICKLER, MITCHELL GENE, b Fairmont, WVa, Sept 19, 34; m 60; c 5. COMPUTER SCIENCE, SOFTWARE SYSTEMS FOR HOSPITALS. *Educ:* Univ W Va, BS, 60; Univ Pittsburgh, MS, 62. *Prof Exp:* Engr, Westinghouse Elec Corp, 60-62; mem tech staff, Bell Tel Labs, 62-64, supvr electronics, 64-66, info systs, 66-68; vpres & mem bd dirs, Virtual Comput Serv, Inc, NJ, 68-70; systs consult, Pentamation Enterprises Inc, Bethlehem, 70-77, dir eng, 77-80, vpres, 85-90; VPRES, FERRANTI HEALTHCARE SYSTS CORP, HUNT VALLEY, MD, 90- *Mem:* Inst Elec & Electronics Engrs. *Res:* Semiconductor device theory; integrated circuit design; information systems; process and inventory control; time sharing operating systems; telecommunications; multiprocessor design and microcoding; real-time hospital data base systems; manufacturing systems. *Mailing Add:* 527 E Lake Vista Circle Cockeysville MD 21030

STICKLER, WILLIAM CARL, b Stuttgart, Ger, Jan 25, 18; nat US; m 42, 58, 68; c 4. ORGANIC CHEMISTRY. *Educ:* Columbia Univ, AB, 41, AM, 44, PhD(chem), 47. *Prof Exp:* Asst chem, Columbia Univ, 41-44, 46-47; from asst prof to assoc prof chem, Univ Denver, 47-63, actg chmn dept, 71-72 & 83-84, prof, 63-83, mem teaching fac, 83-86, EMER PROF CHEM, UNIV DENVER, 83- *Concurrent Pos:* Instr, Sarah Lawrence Col, 43-44 & Hofstra Col, 46-47; vis prof & lectr, Univ Munich & Munich Tech Univ, 56-58; NSF fac fel, 57-58; vis prof, Univ Hamburg, 71 & 80; adj fac, Front Range

Community Col, Westminster, Colo, 88- *Mem:* Am Chem Soc; Royal Soc Chem; Soc German Chem; Sigma Xi. *Res:* Organic nitrogen chemistry; stereochemistry; reaction mechanisms; natural products and physiologically important compounds; sterane synthesis. *Mailing Add:* Dept Chem Univ Denver Denver CO 80208

STICKLEY, C(ARLISLE) MARTIN, b Washington, DC, Oct 30, 33; m 58; c 3. ELECTRO-OPTICS, MATERIALS. *Educ:* Univ Cincinnati, BSEE, 57; Mass Inst Technol, MSEE, 58; Northeastern Univ, PhD, 64. *Prof Exp:* Mem, Transistor Circuits Br, Commun Sci Lab, Air Force Cambridge Res Lab, Laurence G Hanscom Field, Mass, 58-62 & Laser Physics Br, 62-65, chief, Laser Physics Br, Optical Physics Lab, 65-71; dir mat sci off, Defense Advan Res Proj Agency, Dept Defense, 71-76; dir inertial confinement fusion, US Dept Energy, 76-79; VPRES & GEN MGR ADVAN TECHNOL, BDM CORP, 79- *Concurrent Pos:* Vis prof, Univ Rio Grande do Sul, Brazil, 66; assoc mem spec group optical masers, comt laser coord, Dept Defense, 64-71; prog chmn, Third Classified Conf Laser Tech, 67; co-chmn, Conf Laser Eng & Appln, 71; chmn, Inst Elec & Electronics Engrs, OSA Conf Inertial Fusion, 78; subcomt prog chmn, Optical Mats at CLEO, 89. *Mem:* Fel Inst Elec & Electronics Engrs; Sigma Xi; Soc Photo-Optical Instrumentation Engrs. *Res:* Optical fibers, sources and couplers; inertial confinement fusion; electro optical and semiconductor materials; laser devices; optical processing; communications technol; optical fiber sensors; advanced nuclear reactors. *Mailing Add:* 8108 Horse Shoe Lane Potomac MD 20854

STICKNEY, ALAN CRAIG, b Columbus, Ohio, Apr 28, 47; m 70; c 2. EDUCATIONAL COMPUTER SOFTWARE. *Educ:* Mich State Univ, BS, 69; Univ Mich, MS, 70, PhD(math), 75. *Prof Exp:* Instr math, Mich State Univ, 75-77; honors lectr math, Univ Del, 77-79; asst prof, 79-84, ASSOC PROF MATH, WITTENBERG UNIV, 84-, CHMN, DEPT MATH & COMPUT SCI, 82- *Mem:* Am Math Soc; Math Asn Am; Sigma Xi. *Res:* Educational software for college-level mathematics; undergraduate mathematics and computer science education; mathematics placement; computer-assisted instruction. *Mailing Add:* Dept Math & Comput Sci Wittenberg Univ Springfield OH 45501

STICKNEY, ALDEN PARKHURST, b Providence, RI, Sept 7, 22; m 51; c 2. MARINE ECOLOGY. *Educ:* Univ RI, BSc, 48; Harvard Univ, MA, 51. *Prof Exp:* Fishery aide, US Fish & Wildlife Serv, 51-53; res asst, Stirling Sch Med, Yale Univ, 53-54; chief, Atlantic Salmon Invest, US Fish & Wildlife Serv, 54-60, fishery res biologist, Fishery Biol Lab, Bur Com Fisheries, Maine, 60-72 & Biol Lab, Nat Marine Fisheries Serv, 72-73; marine resources scientist, Maine Dept Marine Resources, 74-82; RETIRED. *Mem:* Ecol Soc Am; Am Fisheries Soc. *Res:* Estuarine ecology; shellfish biology; physiology of larval shellfish; behavior and ecology of sea herring and pandalid shrimp. *Mailing Add:* Cameron Point Rd South Portland ME 04576

STICKNEY, JANICE LEE, b Tallahassee, Fla, July 21, 41. PHARMACOLOGY, CARDIOVASCULAR PHYSIOLOGY. *Educ:* Oberlin Col, AB, 62; Univ Mich, PhD(pharmacol), 67. *Prof Exp:* Acad Senate grants, Univ Calif, San Francisco, 68 & 69, from instr to asst prof pharmacol, Sch Med, 69-72; from asst prof to assoc prof, Mich State Univ, 72-81, prof, 81; sr scientist, 81-83, DIR CARDIOVASC PROD, OFF SCI AFFAIRS, G D SEARLE & CO, 87- *Concurrent Pos:* Training fel, 67-68; Bay Area Heart Asn grant, Univ Calif, San Francisco, 69-70; Nat Heart & Lung Inst, Nat Inst Drug Abuse & Mich Heart Asn grants; consult, Food & Drug Admin, 71-75 & 76-80; mem, Special Study Sect, NIH, 78-79, Nat Adv Environ Health Sci Coun, 79-82; panelist, NSF, 76 & 81; pharmacol rev comt, Nat Inst Gen Med Sci, 84-87; consult, Nat Inst Drug Admin, 87. *Mem:* AAAS; Pharmacol Soc Can; NY Acad Sci; Am Soc Pharmacol & Exp Therapeut. *Res:* Cardiovascular pharmacology, especially role of the sympathetic nervous system in the cardiac arrhythmias produced by large doses of digitalis, with emphasis on the mechanisms by which cardiac glycosides produce sympathetic effects; general cardiovascular effects of opiate agonists; antiarrhythmic drugs; calcium channel blocking agents; cardiac toxicology; cardiotorics. *Mailing Add:* Brokenburr Stickney Assoc 1555 Sherman Dept 142 Evanston IL 60201

STICKNEY, PALMER BLAINE, b Columbus, Ohio, Nov 1, 15; m 37; c 4. POLYMER CHEMISTRY. *Educ:* Ohio State Univ, AB, 38, PhD(phys chem), 49. *Prof Exp:* Res engr, Battelle Mem Inst, 40-42, 46, 49-52, asst chief, Rubber & Plastics Div, 52-60, chief polymer res, 60-68, tech adv, 68-73; prof, Wilberforce Univ, 73-81; RETIRED. *Mem:* Am Chem Soc; Am Asn Univ Profs. *Res:* Polymerization and processing of polymers. *Mailing Add:* 2870 Halstead Rd Columbus OH 43221-2916

STICKSEL, PHILIP RICE, b Cincinnati, Ohio, Feb 15, 30; m 53; c 2. METEOROLOGY. *Educ:* Univ Cincinnati, BS, 52; Fla State Univ, MS, 59, PhD(meteorol), 66. *Prof Exp:* Jr develop engr, Goodyear Aircraft Corp, 53-54; res assoc meteorol, Fla State Univ, 64-65; res meteorologist, Environ Sci Serv Admin, 65-69 & Nat Air Pollution Control Admin, 67-69; SR METEOROLOGIST TO PRIN RES SCIENTIST, BATTELLE MEM INST, 69- *Mem:* Am Meteorol Soc; Air Pollution Control Asn; Sigma Xi. *Res:* Air pollution meteorology and education; upper atmosphere ozone; air quality management; ozone transport; visible emissions. *Mailing Add:* 1636 Park Trail Westerville OH 43081

STIDD, BENTON MAURICE, b Bloomington, Ind, June 30, 36; m 58; c 5. PALEOBOTANY, HORTICULTURE. *Educ:* Purdue Univ, BS, 58; Emporia State Univ, MS, 63; Univ Ill, PhD(bot), 68. *Prof Exp:* Teacher, Wheatland High Sch, 58-62; partic, NSF Acad Year Inst & Res Participation Prog Teachers, Emporia State Univ, 62-63; teacher, N Knox High Sch, 63-64; asst prof anat, morphol & paleobot, Univ Minn, Minneapolis, 68-70; chmn, 70-89, PROF BIOL SCI, WESTERN ILL UNIV, 70- *Concurrent Pos:* Univ Minnesota Grad Sch res grant, 68-69; Sigma Xi res grants-in-aid, 69-70; res coun grant, Western Ill Univ, 71-72, 80-81 & 85-86; NSF res grant, 74 & 76, NSF res grant, Res Undergrad Inst, 86; Nat Endowment Humanities, 82; US Dept Educ, 87. *Mem:* Am Asn Univ Prof; Bot Soc Am; Int Orgn Paleobot; Philos Sci Asn; Soc Syst Zool; Sigma Xi. *Res:* Paleozoic paleobotany, especially Carboniferous coal ball plants. *Mailing Add:* Dept of Biol Sci Western Ill Univ Macomb IL 61455

STIDD, CHARLES KETCHUM, b Independence, Ore, Aug 12, 18; m 42; c 2. METEOROLOGY. *Educ:* Ore State Univ, BS, 41. *Prof Exp:* Res forecaster, US Weather Bur, 47-55; self employed, 55-62; res assoc meteorol & hydrol, Univ Nev, Reno, 62-71; specialist meteorol, Scripps Inst Oceanog, Univ Calif, San Diego, 71-79; CONSULT, 79- *Mem:* Am Meteorol Soc; Am Geophys Union. *Res:* Rainfall and climatic probabilities; general circulation of the atmosphere; moisture, energy and momentum balances; long-range forecasting. *Mailing Add:* 4005 Carmel View Rd No 61 San Diego CA 92130

STIDHAM, HOWARD DONATHAN, b Memphis, Tenn, Sept 14, 25; div. PHYSICAL CHEMISTRY. *Educ:* Trinity Col, BS, 50; Mass Inst Technol, PhD, 55. *Prof Exp:* Spectroscopist, Dewey & Almy Chem Co, 55-56; from asst prof to assoc prof, 56-90, PROF CHEM, UNIV MASS, 90- *Mem:* AAAS; Optical Soc Am; Am Phys Soc; Am Chem Soc. *Res:* Molecular spectroscopy; statistical mechanics. *Mailing Add:* Dept of Chem Univ of Mass Amherst MA 01003

STIDHAM, SHALER, JR, b Washington, DC, Dec 4, 41; m 68; c 3. QUEUEING THEORY, DYNAMIC PROGRAMMING. *Educ:* Harvard Col, BA, 63; Case Inst Technol, MS, 64; Stanford Univ, PhD(opers res), 68. *Prof Exp:* Asst prof opers res & environ eng, Cornell Univ, 68-75; assoc prof, NC State Univ, 75-79, prof indust eng & opers res, 79-86; PROF OPERS RES, UNIV NC, 86-, CHAIR, DEPT OPERS RES, 90- *Concurrent Pos:* Consult, Stanford Res Inst, Calif, 68-70; vis prof, Aarhus Univ, 71-72 & Tech Univ Denmark, 77; vis scholar, Stanford Univ, 75, 79; consult, Bell Tel Lab, NJ, 81; assoc ed, Operations Res, 76-85, area ed, 85-90, Mgt Sci, 75-81 & Queueing Systems Theory & Applications, 85-; prin investr, NSF, 73-75, 79-82 & 88-91, NATO, 77 & US Army Res Off, 82-85; vis res fel, statist lab, Univ Cambridge, 82-83; overseas fel, Churchill Col, Cambridge. *Honors & Awards:* Young Scientist Res Award, Sigma Xi, 78. *Mem:* Opers Res Soc Am; Inst Mgt Sci; Inst Indust Eng; Sigma Xi. *Res:* Queueing theory; optimal design and control of queueing systems; applications in manufacturing; transportation and public service systems; computer and communication systems. *Mailing Add:* Opers Res CB 3180 Smith Bldg Chapel Hill NC 27599

STIDWORTHY, GEORGE H, b Viborg, SDak, May 28, 24; m 48; c 4. BIOCHEMISTRY. *Educ:* Univ SDak, BA, 49, MA, 51; Univ Okla, PhD(biochem), 61. *Prof Exp:* Chemist, Rayonier Corp, 46-47; res asst biochem, Okla Med Res Found, 51-53; asst chief, Gen Med Res Lab, Vet Admin Hosp, Oklahoma City, 53-57, chief biochemist, 57-59; chief biochemist, Cancer Res Lab, Vet Admin Hosp, Martinsburg, WVa, 59-64; supvr, Med Res Lab, 64-72, CHIEF GEN MED RES LAB, VET ADMIN HOSP, 72-; asst prof biochem, Sch Med, Boston Univ, 64-84; RETIRED. *Concurrent Pos:* Nat Cancer Inst grant, Vet Admin Hosp, Martinsburg, WVa, 61-65. *Mem:* AAAS; Geront Soc; Tissue Cult Asn; Am Aging Asn. *Res:* Aging effects upon connective tissues; chemistry and biology of the intracellular matrix; in vitro aging of cells in culture; environmental effects in cell metabolism. *Mailing Add:* 48 S Sea Pines Dr Hilton Head Island SC 29928

STIEBER, MICHAEL THOMAS, b Peoria, Ill, Dec 6, 43; m. SYSTEMATIC BOTANY, AGROSTOLOGY. *Educ:* Cath Univ Am, AB, 66, MS, 67; Univ Md, College Park, PhD(syst bot), 75. *Prof Exp:* Teacher biol, Bishop McNamara High Sch, Kankakee, Ill, 67-69; grad teaching asst gen bot & plant taxon, Univ Md, 70-74; teacher biol & advan placement biol, St Viator High Sch, Arlington Heights, Ill, 74-77; archivist & sr res scientist, Hunt Inst Bot Doc, Carnegie-Mellon Univ, 77-78, res assoc, Carnegie Mus Natural Hist, 78-88; ADMIN & REF LIBRN, MORTON ARBORETUM, 88- *Concurrent Pos:* NSF reviewer, Sect Agrost, 78-; adj assoc prof biol, Carnegie-Mellon Univ, 84-88; instr, bot & gen educ prog, Morton Arboretum. *Mem:* Am Soc Plant Taxon; Soc Bibliog Natural Hist; Soc Am Archivists. *Res:* History of botany; archival methods and practice; taxonomy of Ichnanthus, Cenchrus and Pennisetum (paniceae); botanical history and documentation. *Mailing Add:* Sterling Morton Libr Morton Arboretum Lisle IL 60532

STIEF, LOUIS J, b Pottsville, Pa, July 26, 33; div; c 2. PHOTOCHEMISTRY, ASTROCHEMISTRY. *Educ:* La Salle Col, BA, 55; Cath Univ, PhD(chem), 60. *Prof Exp:* Asst chem, Cath Univ, 55-59; Nat Acad Sci-Nat Res Coun res fel, Nat Bur Standards, 60-61; NATO fel, Univ Sheffield, 61-62, Dept Sci & Indust Res fel, 62-63; sr chemist, Res Div, Melpar, Inc, 63-65, sr scientist, 65-68; sr res fel, Nat Acad Sci-Nat Res Coun, 68-69, aerospace technol chemist, 69-74, head, Space Chem Sect, 74-76, head, Astrochem Br, 76-90, SR SCIENTIST, GODDARD SPACE FLIGHT CTR, NASA, 90- *Concurrent Pos:* Adj prof, Dept Chem, Cath Univ Am, DC, 75-; Res fel, Queen Mary Col, London Univ, 81-82. *Mem:* Am Phys Soc; Am Chem Soc; Royal Soc Chem; Sigma Xi; Am Geophys Union; Am Astron Soc. *Res:* Vacuum-ultraviolet photochemistry; flash photolysis; interstellar molecules; upper atmosphere studies; planetary atmospheres; chemical kinetics; mass spectrometry of free radicals. *Mailing Add:* Code 690 NASA/Goddard Space Flight Ctr Greenbelt MD 20771

STIEFEL, EDWARD, b Brooklyn, NY, Jan 3, 42; m 65; c 1. TRANSITION METAL COORDINATION CHEMISTRY. *Educ:* Columbia Univ, PhD(chem), 67. *Prof Exp:* GROUP HEAD, SECT MOLECULAR & BIOL CHEM, EXXON RES & ENG CO, 82- *Mem:* Am Chem Soc; Am Soc Biol Chemists; AAAS; NY Acad Sci. *Res:* Metalloenzymes; transition metal sulfide centers; biological nitrogen fixation. *Mailing Add:* Exxon Res & Eng Co Rte 22E Clinton Township Annandale NJ 08801

STIEFEL, ROBERT CARL, b Camden, NJ, Feb 12, 34; m 57; c 3. WASTEWATER MANAGEMENT. *Educ:* Drexel Univ, BSCE, 57; State Univ Iowa, MS, 59; Rensselear Polytech Inst, PhD(environ eng), 68. *Prof Exp:* From asst prof to assoc prof civil eng, Drexel Inst Technol, 59-66; assoc prof, Northeastern Univ, 66-71; assoc prof, 71-73, PROF CIVIL ENG & DIR WATER RES CTR, OHIO STATE UNIV, 73- *Concurrent Pos:* Process engr, Roy F Weston & Assoc, 60-66; consult, Domey & Stiefel, 67-71; Consult, 71-; chmn, Nat Asn Water Inst Dirs, 81-83. *Mem:* Nat Asn Water Inst Dirs. *Res:* Treatment of industrial wastewaters, water supplies and municipal waste waters. *Mailing Add:* Civil Eng Ohio State Univ Main Campus N470 Hitchcock Hall Columbus OH 43210

STIEGLER, JAMES O, b Valparaiso, Ind, July 25, 34; m 66; c 2. HIGH TEMPERATURE MATERIALS. *Educ:* Purdue Univ, BS, 56; Univ Tenn, PhD(metall eng), 71. *Prof Exp:* Res asst, Solid State Div, 56-61, res staff mem, Metals & Ceramics Div, 61-63, group leader electron microscopy, 63-73, group leader radiation effects & microstructural analysis, 73-78, mgr, Mat Sci Sect, 78-81, mgr, processing sci & technol sect, 81-83, DIR, METALS & CERAMICS DIV, OAK RIDGE NAT LAB, 84- *Honors & Awards:* William Sparagen Award, Am Welding Soc, 75; McKay-Helm Award, Am Welding Soc, 79. *Mem:* Fel Am Soc Metals; Am Ceramic Soc. *Res:* Design and production of metallic and ceramic materials for high temperature applications through control of processing variables. *Mailing Add:* RR 1 No 79 Lenoir City TN 37771

STIEGLER, T(HEODORE) DONALD, b Baltimore, Md, May 28, 34; m 57; c 2. PROCESS ENGINEERING, MECHANICAL ENGINEERING. *Educ:* Duke Univ, BSME, 56. *Prof Exp:* Engr, E I du Pont de Nemours & Co, 58-63, res engr, 63-65, tech supvr, 65-66, res engr, Spruance Film Tech Lab, 66-73, SR ENGR, IMAGING SYSTS DEPT, E I DU PONT DE NEMOURS & CO, 73- *Res:* Mechanical development associated with the chemical industry; design and development of high speed web handling machinery and equipment for the coating of plastic films; process development associated with the manufacture of packaging films; process safety; process hazard evaluation. *Mailing Add:* 8700 Brown Summit Rd Richmond VA 23235

STIEGLITZ, RONALD DENNIS, b Milwaukee, Wis, Aug 25, 41; m 65; c 4. ENVIRONMENTAL SCIENCE, EDUCATIONAL ADMINISTRATION. *Educ:* Univ Wis-Milwaukee, BS, 63; Univ Ill, Urbana, MS, 67, PhD(geol), 70. *Prof Exp:* Teaching asst geol, Univ Wis-Milwaukee, 63-64 & Univ Ill, 64-69; lectr, Univ Wis-Milwaukee, 71-72; geologist, Ohio State Geol Surv, 72-74, head, Regional Geol Sect, Ohio State Geol, 74-76; from asst prof to assoc prof, 76-89, PROF ENVIRON SCI & ASSOC DEAN GRAD STUDIES, UNIV WIS-GREEN BAY, 89- *Concurrent Pos:* Consult, 81- *Mem:* AAAS; Soc Econ Paleont & Mineral; Sigma Xi. *Res:* Paleozoic stratigraphy; waste disposal and ground water pollution in Karst areas; quaternary geology. *Mailing Add:* Natural & Appl Scis Univ Wis ES-317 Green Bay WI 54311-7001

STIEHL, RICHARD BORG, b Chicago, Ill, Jan 10, 42; m 67; c 1. BIOLOGY. *Educ:* Southern Ore State Col, BS, 69, MS, 70; Portland State Univ, PhD(environ sci & biol), 78. *Prof Exp:* Instr biol, Mt Hood Community Col, 72-73; lectr, Lewis & Clark Col, 74-75; lectr biol, Univ Wis, Green Bay, 77-80, asst prof biol, 80-; ASST PROF BIOL, SOUTHEAST MO STATE UNIV, 84- *Concurrent Pos:* Researcher, US Fish & Wildlife Serv, 75-77; fel, Lilly Endowment, Inc, 78-79. *Mem:* Am Inst Biol Sci; Am Soc Mammalogists; Am Ornithologists Union. *Res:* Wildlife ecology; corvid biology; predator-prey interactions. *Mailing Add:* Dept Biol SE Mo State Univ Cape Girardean MO 63701

STIEHL, ROY THOMAS, JR, b Hay Springs, Nebr, Jan 27, 28; m 54; c 3. ORGANIC CHEMISTRY. *Educ:* Univ Nebr, BS, 50, MS, 51; Univ Ill, PhD(org chem), 53. *Prof Exp:* Asst, Univ Nebr, 49-51 & Off Rubber Reserv, Univ Ill, 51-53; res chemist, E I du Pont de Nemours & Co, Inc, 53-68, sr res chemist, 68-85; RETIRED. *Res:* Spandex chemistry; butadiene copolymerization; vinyl monomer synthesis; organophosphorous compounds; textile compounds; textile chemistry. *Mailing Add:* 400 Ridge Circle Waynesboro VA 22980-5430

STIEHM, E RICHARD, b Milwaukee, Wis, Jan 22, 33; m 58; c 3. PEDIATRICS, ALLERGY IMMUNOLOGY. *Educ:* Univ Wis-Madison, BS, 54, MD, 57. *Prof Exp:* USPHS fels, physiol chem, Univ Wis, 58-59 & pediat immunol, Univ Calif, San Francisco, 63-65; from asst prof to assoc prof pediat, Med Sch, Univ Wis, 65-69; assoc prof, 69-72, PROF PEDIAT, SCH MED, UNIV CALIF, LOS ANGELES, 72- *Concurrent Pos:* Markle scholar acad med, 67. *Honors & Awards:* Ross Res Award Pediat Res, 71; E Mead Johnson Award Pediat Res, 74. *Mem:* Am Acad Pediat; Am Soc Clin Invest; Soc Pediat Res; Am Asn Immunologists; Am Pediat Soc; Am Acad Allergy & Immunol. *Res:* Pediatric immunology, immunodeficiency disease; newborn defense mechanisms; human gamma globulin; immunology of malnutrition; clinical immunology. *Mailing Add:* Dept Pediat Univ Calif Ctr Health Sci Los Angeles CA 90024

STIEL, EDSEL FORD, b Los Angeles, Calif, Dec 19, 33. MATHEMATICS. *Educ:* Univ Calif, Los Angeles, AB, 55, MA, 59, PhD(math), 63. *Prof Exp:* Math analyst, Douglas Aircraft Co, Calif, 55-56, comput analyst, 57-59; sr math analyst, Lockheed Missiles & Space Co, 60; from asst prof to assoc prof, 62-72, chmn dept, 62-74, PROF MATH, CALIF STATE UNIV, FULLERTON, 72- *Mem:* Am Math Soc; Math Asn Am. *Res:* Isometric immersions of Riemannian manifolds; differential geometry. *Mailing Add:* Dept Math Calif State Univ 800 N State Col Fullerton CA 92631

STIEL, LEONARD IRWIN, b Paterson, NJ, Sept 17, 37; m 75; c 2. APPLIED PHYSICAL CHEMISTRY, THERMODYNAMICS. *Educ:* Mass Inst Technol, SB, 59; Northwest Univ, MS, 60, PhD(chem eng), 63. *Prof Exp:* From asst prof to assoc prof chem eng, Syracuse Univ, 62-69; assoc prof chem eng, Univ Mo-Columbia, 69-72; res chem engr, Spec Chem Div, Allied Chem Corp, 72-79; sr process engr, APV Co, Inc, 79-80; ASSOC PROF CHEM ENG, POLYTECH INST NY, 80- *Concurrent Pos:* Vis scholar, Northwestern Univ, 62 & Oak Ridge Nat Lab, 84. *Mem:* Am Inst Chem Eng; Am Chem Soc. *Res:* Energy conversion; properties of fluids. *Mailing Add:* Dept Chem Eng Polytech Univ 333 Jay St Brooklyn NY 11201

STIELER, CAROL MAE, b Albert Lea, Minn, June 18, 46; m 73; c 2. GENETICS, IMMUNOBIOLOGY. *Educ:* Iowa State Univ, BS, 68, MS, 70, PhD(immunobiol), 78. *Prof Exp:* Res dir, Hy-Vigor Seeds Inc, 78-80; BREEDER, FARIS FARMS, 80- *Mem:* Am Genetic Asn; AAAS. *Res:* Genetics of laboratory mice, horses and humans; immunology, typing, inheritance and structure of dog and human red cell antigens; antibody formation; lectins; soybean variety development. *Mailing Add:* Rt One Box 27 Rembrandt IA 50576

STIEN, HOWARD M, b Montevideo, Minn, Apr 11, 26; m 47; c 2. ZOOLOGY, PHYSIOLOGY. *Educ:* Northwestern Col, Minn, BA, 56; Macalester Col, MA, 58; Univ Wyo, PhD(physiol), 63. *Prof Exp:* Instr biol, Pepperdine Col, 58-60; asst prof zool, Univ Wyo, 61-64; assoc prof biol, Northwestern Col, Minn, 64-65; assoc prof, 65-72, PROF BIOL, WHITWORTH COL, WASH, 72-, CHMN DEPT, 65- *Mem:* AAAS; Sigma Xi. *Res:* Immunogenetics, especially the ontogeny of molecular individuality. *Mailing Add:* Dept Biol Whitworth Col Spokane WA 99218

STIENING, RAE FRANK, b Pittsburgh, Pa, May 26, 37; m 69; c 1. HIGH ENERGY PHYSICS. *Educ:* Mass Inst Technol, SB, 58, PhD(physics), 62. *Prof Exp:* Physicist, Lawrence Berkeley Lab, Univ Calif, 63-71; physicist, Fermi Nat Accelerator Lab, 71-78; physicist, Stanford Linear Accelerator Ctr, 78-83, dep dir, SLC Proj, 83-88, head, Accelerator Dept, 88-89, PHYSICIST, SSC LAB, STANFORD LINEAR ACCELERATOR CTR, 89- *Res:* Weak interactions; fast stellar oscillations; far infrared astronomy; particle accelerators. *Mailing Add:* 34 Falcon Way Midlothian TX 76065

STIENSTRA, WARD CURTIS, b Holland, Mich, June 19, 41; m 63; c 2. PLANT PATHOLOGY. *Educ:* Calvin Col, ABGen, 63; Mich State Univ, MS, 66, PhD(plant path), 70. *Prof Exp:* From asst prof to assoc prof, 70-80, PROF PLANT PATH, UNIV MINN, ST PAUL, 80- *Mem:* Am Phytopath Soc; Am Inst Biol Sci. *Res:* Diseases of turf corn, soybeans and alfalfa; soil borne diseases. *Mailing Add:* 6035 McKinley St Minneapolis MN 55432

STIER, ELIZABETH FLEMING, b Riverside, NJ, Nov 24, 25; m 47; c 3. BIOCHEMISTRY. *Educ:* Rutgers Univ, BS, 47, MS, 49, PhD(biochem), 51. *Prof Exp:* Asst, 47-51, res assoc, 51-59, from asst prof to assoc prof, 59-72, PROF FOOD SCI, RUTGERS UNIV, NEW BRUNSWICK, 72- *Honors & Awards:* Cruess Award, Inst Food Technologists. *Mem:* AAAS; Inst Food Technologists; NY Acad Sci; Sigma Xi. *Res:* Methodology of flavor evaluation; flavor evaluation as a tool on pesticide treated fruits and vegetables; objective flavor techniques. *Mailing Add:* Dept of Food Sci Cook Col Rutgers Univ New Brunswick NJ 08903

STIER, HOWARD LIVINGSTON, b Delmar, Del, Nov 28, 10; c 5. HORTICULTURE, PLANT PHYSIOLOGY. *Educ:* Univ Md, BS, 32, MS, 37, PhD(plant physiol), 39. *Prof Exp:* Agent potato breeding, USDA, 33-35; res asst hort, Univ Md, 35-39, instr & asst prof, 39-41, exten prof mkt & head dept, 46-51; dir div statist & mkt res, Nat Food Processors Asn, 51-61; dir qual control, United Fruit Co, 61-71, dir develop & prod supporting serv, United Brands Co, 72-73, vpres qual control, 73-74, vpres res develop & qual control, 74-77, corp vpres, 78-79; consult, 79-81; RETIRED. *Concurrent Pos:* Prof lectr, George Washington Univ, 57-61; chmn, adv comt, Bur Census, 57-67; mem res adv comt, USDA, 62-64; mem adv comt, Dept Defense, 63; mem task group statist qual control of foods, Nat Acad Sci, 64. *Mem:* Fel AAAS; fel AmSoc Qual Control (vpres, 66-68, 73-74, pres, 74-75,); Am Statist Asn; Am Soc Hort Sci; Inst Food Tech. *Res:* Plant breeding; factors affecting quality and growth and development of horticultural crops, food processing, statistical control of quality. *Mailing Add:* Carolina Meadows Villa 230 Chapel Hill NC 27514

STIER, PAUL MAX, b Eden, NY, Aug 18, 24; m 47; c 4. PHYSICS. *Educ:* Univ Buffalo, BS, 44; Cornell Univ, PhD(physics), 52. *Prof Exp:* Sr physicist, Oak Ridge Nat Lab, 50-55; group leader chem physics, Union Carbide Corp, 55-60, asst dir, Res Lab, Union Carbide Nuclear Co, 60-65, mgr, Nucleonics Res Lab, Union Carbide Corp, 65-66, prog mgr phys sci, Tarrytown Tech Ctr, 66-72, opers mgr, Corp Res Dept, Sterling Forest Lab, 72-; RETIRED. *Res:* Stopping of heavy ions; energy range 10-250 kilo-electron-volts; energy loss; ionization; charge exchange; field emission; field ionization microscopy; chemisorption; surface diffusion; radiation damage in solids. *Mailing Add:* 10353 N Oregon Rd Eden NY 14057

STIER, THEODORE JAMES BLANCHARD, b Sept 4, 03. CELLULAR BIOCHEMISTRY & PHYSIOLOGY. *Educ:* Harvard Univ, PhD(gen physiol), 29. *Prof Exp:* Prof physiol, Sch Med, Ind Univ, 42-69; RETIRED. *Mailing Add:* 2455 Tamarack Trail No 333 Bloomington IN 47408

STIERMAN, DONALD JOHN, b Dubuque, Iowa, Oct 27, 47; m 70; c 4. SEISMOLOGY, ENVIRONMENTAL SCIENCES. *Educ:* State Univ NY Col Brockport, BS, 69; Stanford Univ, MS, 74, PhD(geophys), 77. *Prof Exp:* Physics instr, Teacher Training Col, Tegucigalpa, Honduras, 69-72; geophysicist, US Geol Surv, 74-75; res asst, Stanford Univ, 75-77; asst prof, Univ Calif, Riverside, 77-84; asst prof, 84-87, ASSOC PROF GEOL, UNIV TOLEDO,87- *Concurrent Pos:* Consult, Environ & Eng Geophys. *Mem:* Am Geophys Union; Seismol Soc Am; Geol Soc Am; Soc Explor Geophysicists. *Res:* Physical properties of the shallow crust; microearthquake studies; induced seismicity; crustal velocity structure; geophysical characterization of existing and proposed waste disposal sites; fracture hydrology; earthquake prediction. *Mailing Add:* Geol Dept Univ Toledo Toledo OH 43606

STIERWALT, DONALD L, b Fremont, Ohio, Sept 20, 26; m 53; c 2. INFRARED OPTICAL PROPERTIES. *Educ:* Univ Toledo, BS, 50; Syracuse Univ, MS, 53, PhD(physics), 61. *Prof Exp:* Res physicist, Naval Ord Lab, Corona, 58-70, RES PHYSICIST, ELECTRONIC MAT SCI DIV, NAVAL OCEAN SYSTS CTR, 70- *Mem:* Optical Soc Am. *Res:* Low temperature infrared spectral emittance and transmittance of optical materials and components. *Mailing Add:* Code 556 Naval Ocean Systs Ctr 271 Catalina Blvd San Diego CA 92152

STIFEL, FRED B, b St Louis, Mo, Jan 30, 40. DIETARY REGULATION OF METABOLIC ENZYMES. *Educ:* Iowa State Univ, PhD(biochem & nutrit), 67. *Prof Exp:* Res biochemist, dept med, Letterman Army Res Inst, San Francisco, 74-76; ASSOC PASTOR, FAITH PRESBY CHURCH, 80- *Mem:* Am Inst Nutrit; Am Asn Clin Nutrit. *Mailing Add:* 3492 S Blackhawk Way Aurora CO 80014

STIFEL, PETER BEEKMAN, b Wheeling, WVa, Feb 9, 36; c 2. PALEONTOLOGY, STRATIGRAPHY & SEDIMENTATION. *Educ:* Cornell Univ, BA, 58; Univ Utah, PhD(geol), 64. *Prof Exp:* Res asst, Univ Utah, 60-63; ASSOC PROF GEOL, UNIV MD, COLLEGE PARK, 66- *Mem:* Soc Econ Paleont & Mineral; Paleontol Soc. *Res:* Paleontology, stratigraphy and sedimentation. *Mailing Add:* 9636 Old Spring Rd Kensington MD 20895

STIFF, ROBERT H, b Pittsburgh, Pa, Apr 23, 23; m 45; c 3. DENTISTRY. *Educ:* Univ Pittsburgh, BS, 43, DDS, 45, MEd, 53. *Prof Exp:* Instr oper dent, 45-58, asst prof oral med, 56-57, assoc prof oral genetic path & microbiol, 58-59, assoc prof oral med, 60-65, asst dean, 79-81, assoc dean, 81-84, PROF ORAL MED & CHMN DEPT, SCH DENT, UNIV PITTSBURGH, 65-, DIR, OUTPATIENT SERV, HOSP, 78- *Concurrent Pos:* Consult, USPHS, 62-66 & Dent Div, Pa Dept Health, 64-66. *Mem:* Am Acad Oral Path; fel Am Col Dent. *Res:* Task analysis of dental practice; dental education; dental treatment for the handicapped; attitudes of dental students towards treatment of chronically ill and aged; caries inhibiting effectiveness of a stannous fluoride-insoluble sodium metaphosphate dentifrice in children. *Mailing Add:* 4601 Doverdell Dr Pittsburgh PA 15236

STIFFEY, ARTHUR V, b Burgettstown, Pa, Apr 16, 18; m 41; c 6. ENVIRONMENTAL SCIENCE, MICROBIOLOGY. *Educ:* Univ Pittsburgh, BS, 40; Lehigh Univ, MS, 47; Fordham Univ, PhD(biol), 81. *Prof Exp:* Chemist anal chem, US Steel Co, 47-48; res assoc microbiol, Am Cyanamid Co, 48-74; chmn dept biol, 77-78, chmn dept natural sci, 79-81, ASST PROF MICROBIOL, LADYCLIFF COL, 74- *Concurrent Pos:* Sr scientist oceanog, US/USSR Bering Sea Exped, 77; proj dir instrnl sci equip prog, NSF grant, 77-78; fel, Naval Res Lab, Washington, DC; staff scientist, Naval Ocean Res & Develop Activity, US Navy. *Mem:* AAAS; Sigma Xi; Phycol Soc; Am Inst Biol Sci. *Res:* Microbiological assays; metal determinations; uptake of metals by algae; chemical oceanography; antibiotics; bioluminescence; oceanography. *Mailing Add:* 811 Freedom Lane Slidell LA 70458

STIFFLER, DANIEL F, b Los Angeles, Calif, Nov 27, 42; m 67; c 3. PHYSIOLOGY, ZOOLOGY. *Educ:* Univ Calif, Santa Barbara, BA, 68; Ore State Univ, MS, 70, PhD(physiol), 72. *Prof Exp:* NIH-USPHS trainee physiol, Health Sci Ctr, Univ Ore, 72-74; lectr animal physiol, Univ Calif, Davis, 74-75; from asst prof to assoc prof, 75-83, assoc dean sci, 83-85, PROF PHYSIOL & ZOOL, CALIF STATE POLYTECH UNIV, POMONA, 83- *Mem:* AAAS; Am Physiol Soc; Am Soc Zoologists; Sigma Xi; Can Soc Zoologists. *Res:* Renal physiology; epithelial transport physiology; comparative physiology and endocrinology of acid base and ionic regulation. *Mailing Add:* Dept Biol Sci Calif State Polytech Univ Pomona CA 91768

STIFFLER, JACK JUSTIN, b Mitchellville, Iowa, May 22, 34; m 55, 89; c 1. HARDWARE SYSTEMS, ELECTRONICS ENGINEERING. *Educ:* Harvard Univ, BA, 56; Calif Inst Technol, MS, 57, PhD(elec eng), 62. *Prof Exp:* Mem tech staff, Jet Propulsion Lab, 59-67; consult scientist, Raytheon Co, 67-81; EXEC VPRES, SEQUOIA SYSTS INC, 81- *Concurrent Pos:* Vis prof, Calif Inst Technol, Univ Calif, Los Angeles, Univ Southern Calif & Northeastern Univ; consult, var corps; ed, Trans Commun, 68-72, Trans Computers, 84-88. *Mem:* Fel Inst Elec & Elect Engrs. *Res:* Defining computer architectures for high-performance on-line transaction processing and investigating fault-tolerance techniques for electronic circuitry. *Mailing Add:* Sequoia Systs Inc 400 Nickerson Rd Marlborough MA 01752

STIGLER, STEPHEN MACK, b Minneapolis, Minn, Aug 10, 41; m 64; c 4. STATISTICS. *Educ:* Carleton Col, BA, 63; Univ Calif, Berkeley, PhD(statist), 67. *Prof Exp:* From asst prof to prof statist, Univ Wis-Madison, 67-79; PROF STATIST, UNIV CHICAGO, 79- *Concurrent Pos:* Ed, J Am Statist Asn, 79-82; mem bd trustee, Ctr Advan Study in Behav Sci, 86- *Mem:* AAAS; Am Statist Asn; Inst Math Statist; Hist Sci Soc; fel Am Acad Arts & Sci. *Res:* Order statistics; experimental design; history of statistics; author of the history of statistics and over 60 professional articles. *Mailing Add:* Dept of Statist Univ of Chicago Chicago IL 60637

STIGLITZ, IRVIN G, b Cambridge, Mass, July 31, 36; m 60; c 2. ELECTRONICS ENGINEERING, ENGINEERING. *Educ:* Mass Inst Technol, SB & SM, 60, PhD(commun sci), 63. *Prof Exp:* Electronic engr, US Naval Ord Lab, 55-58; electronic engr, Gen Atronics Co, 60-61; res teaching asst statist commun theory, 58-63, asst prof elec eng, 63-64, staff mem, 64-71, GROUP LEADER, LINCOLN LAB, MASS INST TECHNOL, 71- *Concurrent Pos:* Consult, Guillemin Res Lab, 61, Joseph Kaye & Co, 63, Melpar Inc, 62 & Nat Acad Sci-Nat Res Coun, 70; Ford fel, 63-64. *Mem:* Inst Elec & Electronics Engrs. *Res:* Guidance and control technology; tactical systems technology; air traffic control; communications systems design; information theory; system design and analysis; adaptive array processing; interference suppression; surveillance. *Mailing Add:* Lincoln Lab Mass Inst Technol PO Box 73 Lexington MA 02173

STILES, A(LVIN) B(ARBER), b Springfield, Ohio, July 16, 09; m 34; c 3. CATALYST DEVELOPMENT, PROCESS DEVELOPMENT. *Educ:* Ohio State Univ, BChE, 31, MS, 33. *Prof Exp:* Indust engr, E I du Pont de Nemours & Co, Inc, NY, 31-32, res assoc, WVa, 33-58, sr res assoc, 58-65, res fel, Del, 65-74; ASSOC DIR, CTR CATALYTIC SCI & TECHNOL, DEPT CHEM ENG, UNIV DEL, NEWARK, 74-, RES PROF, APPLIED CATALYST, 74. *Concurrent Pos:* Res fel, E I du Pont de Nemours, 34-74. *Mem:* Am Chem Soc; fel Am Inst Chemists; Am Inst Chem Engrs; AAAS; NY Acad Sci. *Res:* Industrial catalysis; synthesis of organic chemicals plastics, alcohols, intermediate and synthesis gas; author 65 US patents, 2 books, many articles and chapters. *Mailing Add:* Dept of Chem Eng Univ of Del Newark DE 19716

STILES, CHARLES DEAN, b Nov 18, 46; m; c 2. CANCER, ONCOGENES. *Educ:* Harvard Univ, MA; Univ Tenn, PhD(biochem), 73. *Prof Exp:* PROF MICORBIOL & MOLECULAR GENETICS, SCH MED, HARVARD UNIV, 76- *Mem:* Am Asn Cancer Res. *Res:* Regulation of cell growth and development by polypeptide growth factors. *Mailing Add:* Dana-Farber Cancer Inst Harvard Med Sch 44 Binney St Boston MA 02115

STILES, DAVID A, b Harrow, Eng, Apr 28, 38; m 66; c 3. ANALYTICAL CHEMISTRY, ENVIRONMENTAL CHEMISTRY. *Educ:* Univ Birmingham, BSc, 60, PhD(electron spin resonance spectros), 63. *Prof Exp:* Asst prof chem, Univ Calgary, 63-64; univ fel, Univ Alta, 64-66; from asst prof to assoc prof, 66-77, head dept, 81-88, PROF CHEM, ACADIA UNIV, 77- *Mem:* Fel Chem Inst Can; Royal Soc Chem, London. *Res:* Agricultural pollution; fate of pesticides and heavy metals in sandy soils; applications of molecular emission cavity analysis. *Mailing Add:* Dept of Chem Acadia Univ Wolfville NS B0P 1X0 Can

STILES, GARY L, b New York, NY, May 22, 49; m 71; c 2. RECEPTOR MECHANISMS. *Educ:* St Lawrence Univ, BS, 71; Vanderbilt Univ, MD, 75. *Prof Exp:* Asst prof, 83-86, ASSOC PROF MED, DUKE UNIV, 86-, ASST PROF BIOCHEM, 87- *Concurrent Pos:* Fel biochem, Duke Univ, 84; mem pharmacol study sect, NIH, 87-90; mem Basic Sci Coun, Am Heart Asn, 86-, Circulation Coun, 87-; young investr award, Am Col Cardiol, 83. *Mem:* Am Soc Clin Invest; Am Soc Biol Chemists; Am Heart Asn; Am Fedn Clin Res. *Res:* The mechanisms of transmembrane signalling with particular interest in adenosine receptor systems; structure function relationships at the biochemical and molecular biological levels. *Mailing Add:* Dept Med Div Cardiovasc Duke Univ Med Ctr Box 3444 Durham NC 27710

STILES, LUCILLE E, b Chicago, Ill, Apr 30, 47. NUTRITION. *Educ:* Wash Univ, AB, 69; Univ Ill, MS, 70; Cornell Univ, PhD(nutrit), 77. *Prof Exp:* Res nutritionist, Nabisco Inc, 77-79; NUTRIT EDUC & TRAINING SPECIALIST, FOOD & NUTRIT SERV, US DEPT AGR, 79-; DIR COMMUN, MED MKT, INC, 85- *Mem:* Sigma Xi; Inst Food Technologists; AAAS; Soc Nutrit Educ. *Mailing Add:* 1122 Brummel St Evanston IL 60202

STILES, LYNN F, JR, b Brooklyn, NY, July 4, 42; m 67; c 2. ENERGY CONSERVATION, UTILITY CONSERVATION ASSESSMENT. *Educ:* State Univ NY, Stony Brook, BS, 64; Cornell Univ, MS, 67, PhD(physics), 70. *Prof Exp:* Instr physics, Hobart & William Smith Cols, 66-68; res physicist optics, E I du Pont de Nemours & Co, Inc, 69-73; PROF PHYSICS, STOCKTON STATE COL, 73- *Concurrent Pos:* Vis lectr, Swarthmore Col, 72-; dir tech assessment, Atlantic County, 74-76; pres, Solar Alternatives, Inc, 78-81; consult energy systs design, 79-; pres, New Bus Incubator, 89-90. *Mem:* Am Asn Physics Teachers; Am Soc Heating Refrig & Air Conditioning Engrs. *Res:* Energy conservation in buildings - air filtration and envelope loss mechanisms; energy system designs and utility conservation program assessment; interferometric holography. *Mailing Add:* 104 Arlington Ave Linwood NJ 08221

STILES, MICHAEL EDGECOMBE, b Brit, Dec 28, 34; Can citizen; m 59; c 5. FOOD MICROBIOLOGY, MICROBIOLOGY. *Educ:* Univ Natal, BScAgr, 56, MScAgr, 59; Univ Ill, PhD(food microbiol), 63. *Prof Exp:* Dairy researcher, S African Dept Agr, 57-59; from lectr to sr lectr dairy sci, Univ Natal, 59-69; assoc prof food microbiol, 69-77, PROF FOOD MICROBIOL, UNIV ALTA, 77- *Concurrent Pos:* Killam fel, Univ Alta, 68-69; adj prof, Dept Microbiol, Univ Alta, 74- *Mem:* Can Inst Food Sci & Technol; Inst Food Technologists; Int Asn Sanitarians; Am Soc Microbiol. *Res:* Food microbiology for quality control and safety especially meats; consumer acceptance and awareness of foods and food safety. *Mailing Add:* Dept Home Econ Univ Alta Edmonton AB T6G 2M7 Can

STILES, PHILIP GLENN, b Terre Haute, Ind, Nov 24, 31; m 56; c 1. FOOD TECHNOLOGY, POULTRY SCIENCE. *Educ:* Univ Ark, BS, 53; Univ Ky, MS, 56; Mich State Univ, PhD(food tech), 58. *Prof Exp:* Assoc prof food tech, Univ Conn, 59-69; PROF POULTRY SCI & FOOD TECHNOL, ARIZ STATE UNIV, 69-, CHMN DEPT AGR INDUST, 78- *Concurrent Pos:* Consult, Nixon Baldwin Div, Tenneco Co, 67-, AID projs in Iran, Malaysia & Philippines, 73-75 & Saudia Arabia Agr Bank, 80-85; fel, Univ Calif, Davis, 66-67. *Mem:* Inst Food Technol; Soc Int Develop. *Res:* Food technology as applied to poultry products and food packaging; food processing in developing nations. *Mailing Add:* Dept Agr Ariz State Univ Tempe AZ 85287-3306

STILES, PHILLIP JOHN, b Manchester, Conn, Oct 31, 34; m 56; c 5. SOLID STATE PHYSICS. *Educ:* Trinity Col, Conn, BS, 56; Univ Pa, PhD(physics), 61. *Prof Exp:* Fel & res assoc physics, Univ Pa, 61-62; NSF fel, Cambridge Univ, 62-63; mem res staff, Thomas J Watson Res Ctr, Int Bus Mach Corp, NY, 63-70; chmn dept, 74-80, PROF PHYSICS, BROWN UNIV, 70-, DEAN GRAD SCH & RES, 86- *Concurrent Pos:* Humboldt Sr US Scientist award, 76. *Honors & Awards:* John Price Wetheral Medal, Franklin Inst, 81; Oliver E Buckley Prize, Am Phys Soc, 88. *Mem:* Fel Am Phys Soc. *Res:* Solid state and low temperature physics; electronic properties of metals, semiconductors and lower dimensional systems. *Mailing Add:* Dept Physics Brown Univ Providence RI 02912

STILES, ROBERT NEAL, b Mar 15, 33; m 59; c 3. PHYSIOLOGY. *Educ:* Univ Mo, BS, 59, MA, 63; Northwestern Univ, PhD, 66. *Prof Exp:* Asst prof zool & physiol, Butler Univ, 66-68; asst prof physiol & biophys, 68-75, ASSOC PROF PHYSIOL & BIOPHYS, UNIV TENN CTR HEALTH SCI, MEMPHIS, 75- *Concurrent Pos:* USPHS grant, Univ Tenn Ctr Health Sci, Memphis, 69- *Mem:* AAAS; Sigma Xi; Am Physiol Soc; Soc Neurosci. *Res:* Human limb tremor; muscle mechanics; motor control system. *Mailing Add:* 528 S Mclean Blvd Memphis TN 38104

STILES, WARREN CRYDER, b Dias Creek, NJ, June 16, 33; m 55; c 4. HORTICULTURE. *Educ:* Rutgers Univ, BS, 54, MS, 55; Pa State Univ, PhD(hort), 58. *Prof Exp:* Asst prof pomol, Rutgers Univ, 58-63; from assoc prof to prof pomol, Univ Maine, 63-80, exten fruit specialist, 63-80, supt, Highmoor Farm, 66-80; assoc prof, 80-84, PROF POMOL, CORNELL UNIV, 84- *Mem:* AAAS; Am Soc Hort Sci; Sigma Xi; Weed Sci Soc Am. *Res:* Nutrition; soil management; weed control; physiology of fruit. *Mailing Add:* 120 Plant Sci Bldg Cornell Univ Ithaca NY 14853

STILES, WILBUR J, b Suffern, NY, Jan 12, 32; m 56; c 2. MATHEMATICS. *Educ:* Lehigh Univ, BS, 54; Ga Inst Technol, BS, 60, MS, 62, PhD(math), 65. *Prof Exp:* ASSOC PROF MATH, FLA STATE UNIV, 65- *Mem:* Am Math Soc; Math Asn Am. *Res:* Functional analysis, geometry of Banach spaces. *Mailing Add:* Dept Math Fla State Univ Tallahassee FL 32306

STILL, CHARLES NEAL, b Richmond, Va, Apr 15, 29; m 58; c 3. NEUROLOGY. *Educ:* Clemson Univ, BS, 49; Purdue Univ, MS, 51; Med Col SC, MD, 59. *Prof Exp:* Instr chem, Clemson Univ, 51-52 & US Mil Acad, 53-55; intern, Univ Chicago Clins, 59-60; resident neurol, Baltimore City Hosps & Johns Hopkins Hosp, 60-63; chief neurol serv, William S Hall Psychiat Inst, 65-81; prof, 78-81, clin prof, 81-88, PROF NEUROPSYCHIAT & BEHAV SCI, SCH MED, UNIV SC, 88- *Concurrent Pos:* Fel neurol med, Sch Med, Johns Hopkins Univ, 60-63; Nat Inst Neurol Dis & Blindness spec res fel neuropath, Res Lab, McLean Hosp & Harvard Med Sch, 63-65; fel neurol, Seizure Unit, Children's Hosp Med Ctr, Boston, 66; assoc clin prof neurol, Med Univ SC, 73-91; chmn grants rev bd, SC Dept Ment Health, 73-78; mem, Huntington's Chorea Res Group, World Fedn Neurol; dir, C M Tucker Jr Human Resources Ctr, 81-88; dep comnr, Long Term Care Div, SC Dept Mental Health, 81-86; assoc dir, gen psychiat & neurol, William S Hall Psychiat Inst, 89-91. *Mem:* Fel Am Acad Neurol; fel Am Col Nutrit; fel Am Geriat Soc; fel Am Inst Chemists; fel Geront Soc Am. *Res:* Alzheimer's Disease; Huntington's Disease; Parkinson's Disease. *Mailing Add:* William S Hall Psychiat Inst PO Box 202 Columbia SC 29202-0202

STILL, EDWIN TANNER, b Monroe, Ga, Nov 2, 35; m 59; c 2. RADIOBIOLOGY, ENVIRONMENTAL PROTECTION. *Educ:* Univ Rochester, MS, 64; Univ Ga, DVM, 59. *Prof Exp:* Res scientist, Sch Aerospace Med, US Air Force, Brooks AFB, Tex, 64-67 & Naval Radiol Defense Lab, Calif, 67-69; res contracts adminr, Div Biol & Med, US AEC, 69-75; chmn, Radiation Biol Dept, Armed Forces Radiobiol Res Inst, 75-79; biomed adv, Defense Nuclear Agency, 79-81; sr phys scientist, 82-83, vpres Environ Affairs Dept, 83-84, VPRES & DIR, ENVIRON HEALTH & MGT DIV, KERR-MCGEE CORP, 84- *Mem:* Sigma Xi. *Res:* Low-level radiation effects; beneficial applications of radiation. *Mailing Add:* 2104 Thrush Cir Edmond OK 73074

STILL, EUGENE UPDIKE, PHYSIOLOGY. *Educ:* Univ Chicago, PhD(biochem & physiol), 28. *Prof Exp:* Asst prof physiol, Sch Med, Univ Chicago, 29-70; RETIRED. *Mailing Add:* Trailer Estates PO Box 5824 Bradenton FL 34281

STILL, GERALD G, b Seattle, Wash, Aug 13, 33; m 54; c 3. BIOCHEMISTRY, ORGANIC CHEMISTRY. *Educ:* Wash State Univ, BS, 59; Ore State Univ, MS, 63, PhD(biochem), 65. *Prof Exp:* Res biochemist, Radiation & Metab Res Lab, Agr Res Serv, USDA, 65-77, staff scientist, Nat Prog Staff, 77-80, chief scientist, Sci & Educ Admin-Agr Res, 80-82, dir crop productivity, USDA-ARS, 83-84. *Concurrent Pos:* Mem, Sr Exec Serv, USDA, 84- *Mem:* Am Soc Plant Physiol; Am Chem Soc. *Res:* Metabolism of pesticides; isolation and characterization of pesticide metabolites; photosynthesis; biological nitrogen fixation; plant cell culture; field crop bioregulation. *Mailing Add:* ARS/Univ Calif Berkeley Plant Gene Exp Ctr USDA 800 Buchanan St Albany CA 94710

STILL, IAN WILLIAM JAMES, b Rutherglen, Scotland, July 5, 37; m 64; c 2. ORGANIC CHEMISTRY. *Educ:* Glasgow Univ, BSc, 58, PhD(chem), 62. *Prof Exp:* Res assoc, Univ Toronto, 62-63; sci officer, Allen & Hanburys Ltd, Eng, 63-64; from asst lectr to lectr chem, Huddersfield Col Tech Eng, 64-65; from asst prof to assoc prof, 65-82, PROF CHEM, UNIV TORONTO, 82- *Concurrent Pos:* Dir Can Soc Chem, 85-87. *Mem:* Am Chem Soc; fel Chem Inst Can; assoc mem Royal Soc Chem; Can Soc Chem. *Res:* Synthetic and structural organic chemistry; new synthetic methods and their application to synthesis of naturally occurring antibiotics and antivirals; organic sulfur chemistry. *Mailing Add:* Erindale Campus Univ Toronto Mississauga ON L5L 1C6 Can

STILL, W CLARK, JR, b Augusta, Ga, Aug 31, 46; m 67. SYNTHETIC ORGANIC CHEMISTRY. *Educ:* Emory Univ, BS, 69, PhD(org chem), 72. *Prof Exp:* IBM fel theoret org chem, Princeton Univ, 72-73; fel synthetic org chem, Columbia Univ, 73-75; asst prof org chem, Vanderbilt Univ, 75-77; asst prof, 77-80, assoc prof, 80-81, PROF, COLUMBIA UNIV, 81- *Concurrent Pos:* Alfred P Sloan Fel, 78-80; John Simon Guggenheim Fel, 81-82; Ruth & Arthur Sloan Vis Prof, Harvard, 82; Alexander Todd Vis Prof, Cambridge, 86. *Honors & Awards:* Alan T Waterman Award, NSF, 81; Buchman Award, Calif Inst Technol, 82; Japan Soc Promotion Sci Fel, 87. *Mem:* Am Chem Soc; Am Acad Arts & Sci; fel Japan Soc Prom Sci. *Res:* Organic synthesis; new synthetic methods. *Mailing Add:* Dept Chem Columbia Univ Broadway & W 116th St New York NY 10027-2399

STILL, WILLIAM JAMES SANGSTER, b Aberdeen, Scotland, Sept 16, 23; m 51; c 2. PATHOLOGY. *Educ:* Univ Aberdeen, MB, ChB, 51, MD, 60. *Prof Exp:* Lectr path, Univ London, 57-60; asst prof, Sch Med, Washington Univ, 60-62; sr lectr, Univ London, 62-65; assoc prof, 65-70, PROF PATH, MED COL VA, 70- *Concurrent Pos:* Fel coun arteriosclerosis, Am Heart Asn, 65. *Mem:* Col Am Path; Path Soc Gt Brit & Ireland. *Res:* Cardiovascular disease, particularly arterial disease. *Mailing Add:* 4207 Kensington Ave Richmond VA 23221

STILLE, JOHN KENNETH, b Tucson, Ariz, May 8, 30; m 58; c 2. ORGANOMETALLIC CHEMISTRY, POLYMER CHEMISTRY. *Educ:* Univ Ariz, BS, 52, MS, 53; Univ Ill, PhD(org chem), 57. *Prof Exp:* From instr to prof, Univ Iowa, 57-77; prof org chem, 77-87, UNIV DISTINGUISHED PROF, COLO STATE UNIV, 87- *Concurrent Pos:* Consult, E I du Pont de Nemours & Co, 64-; vis prof, Royal Inst Technol, Sweden, 68; mem eval panel for polymers div, Inst Mat Res, Nat Bur Standards, 74-77; chmn, Polymer Div, Am Chem Soc, 75; mem coun, Gordon Res Conf, 78-80 & Petrol Res Fund Adv Bd, Am Chem Soc, 79-87; assoc ed, Macromolecules, 67-80; ed bd, J Am Chem Soc, 82-85; consult, Syntex Int, 88- *Honors & Awards:* Polymer Chem Award, Am Chem Soc, 82. *Mem:* Am Chem Soc; Royal Soc Chem. *Res:* Organometallic reactions and mechanisms; asymmetric synthesis catalyzed by transition metals polymer synthesis and reaction mechanisms; catalysis. *Mailing Add:* Dept Chem Colo State Univ Ft Collins CO 80523

STILLER, CALVIN R, b Naicam, Sask, Can, Feb 12, 41; m 62, 88; c 6. IMMUNOLOGY, MEDICINE. *Educ:* Univ Sask, MD, 65; FRCP(C), 70. *Prof Exp:* Med Res Coun fel biochem, Univ Western Ont, London, 67-69, from asst prof to assoc prof med, 72-82; chief nephrol, 73-84, DIR TRANSPLANT LAB, UNIV HOSP, LONDON, 73-, CHIEF TRANSPLANTATION, 84-; PROF MED, UNIV WESTERN ONT, 82-; DIR IMMUNOL, ROBARTS RES INST, LONDON, ONT, 84- *Concurrent Pos:* Res assoc transplantation, Transplant Unit, Med Res Coun, Edmonton, 71-72; chmn, Ctr Transplant Studies, London, Ont, 81-; co-chmn, Ministers Task Force Organ Donation, Govt of Ontario, 83-85; chmn, Can-Europ Diabetes Study Group, 84-; mem, Task Force Indust-Acad Res Inst, Govt of Can, 85-; vis prof, over 50 univs; consult, NIH, Ways & Means Comt, Med Res Coun, Govt of Can & pharmaceut indust. *Mem:* Can Soc Nephrol (secy-treas & pres, 76-82); Transplantation Soc; Transplant Int Can; Can Soc Immunol; Med Res Coun Can. *Res:* Immune response in transplantation and autoimmune diesase; immunogenetic and molecular biologic aspects of diabetes and its possible prevention. *Mailing Add:* Univ Hosp London ON N6A 5A5 Can

STILLER, DAVID, b Seattle, Wash, Sept 9, 31; m 62; c 2. MEDICAL & VETERINARY ENTOMOLOGY. *Educ:* Whittier Col, BA, 53, MSc, 57; Univ Calif, Berkeley, PhD(parasitol), 73. *Prof Exp:* Staff res assoc med entom, George Williams Hooper Found, Univ Calif, San Francisco, 62-73, res parasitologist & res assoc med entom & acarol, Dept Int Health, 73-75; RES ENTOMOLOGIST VET ENTOM, AGR RES SERV, UNIV IDAHO, USDA, 75- *Concurrent Pos:* Actg head, Int Ctr Med Res, Div Acarol, Inst Med Res, Kuala Lumpur, Malaysia, 73-75; consult scientist, Spec Foreign Currency Prog, Pub Law 480; vet entom coordr, Nat Emergency Prog Vet Serv, USDA, 78-; adj prof, Dept Vet Med, Univ Idaho & Dept Vet Micro-Path, Washington State Univ, 81- *Mem:* Entom Soc Am; Am Soc Trop Med & Hyg; AAAS; Wildlife Dis Asn; Sigma Xi. *Res:* Acarology; vector-pathogen relationships; arthropod-borne diseases; acarine biology and parasitism; tick-borne hemoparasitic diseases of livestock; epizootiology of these diseases. *Mailing Add:* USDA-Agr Res Serv Univ Idaho Vet Sci Bldg Moscow ID 83843

STILLER, MARY LOUISE, b Salem, Ohio, Nov 29, 31. PLANT PHYSIOLOGY, BIOCHEMISTRY. *Educ:* Purdue Univ, BS, 54, MS, 56, PhD(plant physiol), 59. *Prof Exp:* NSF fel biochem, Univ Chicago, 58-60, USPHS trainee, 60-61; fel, Univ Pa, 61-62; asst prof, 62-66, ASSOC PROF BIOL SCI, PURDUE UNIV, LAFAYETTE, 66- *Concurrent Pos:* NIH career develop award, 65- *Mem:* AAAS; Am Soc Plant Physiol. *Res:* Biochemistry of photosynthesis, photoreduction and respiration. *Mailing Add:* Dept of Biol Sci Purdue Univ West Lafayette IN 47907

STILLER, PETER FREDERICK, b Green Bay, Wis. ALGEBRAIC GEOMETRY. *Educ:* Mass Inst Technol, SB(econ) & SB(math), 73; Princeton Univ, MA, 74, PhD(math), 77. *Prof Exp:* Asst prof math, Tex A&M Univ, 77-79; NATO fel, Inst des Hautes Etudes Scientifiques, 79-80; asst prof, Tex A&M Univ, 80; res fel math, Univ Bonn, WGer, 81; from asst prof to assoc prof math, Tex A&M Univ, 82-86; ASSOC PROF MATH, LA STATE UNIV, 86-; MEM, SCH MATH, INST ADVAN STUDY, 88- *Concurrent Pos:* NSF US France exchange grant, Inst Higher Sci Studies, 82-83; res prof, Math Sci Res Inst, Berkeley, 86-87; prof math, Tex A & M Univ, 87-88. *Mem:* Am Math Soc; Math Soc France; Math Assoc Am. *Res:* Families of algebraic varieties, algebraic cycles and Dirichlet series. *Mailing Add:* Tex A&M Univ College Station TX 77843-3368

STILLER, RICHARD L, b New York, NY, Feb 15, 33; m 72; c 2. BIOCHEMISTRY. *Educ:* Hunter Col, AB, 59; St John's Univ, NY, MS, 70, PhD(biol chem), 72. *Prof Exp:* Res scientist biochem, NY Psychiat Inst, 61-80, head Analytical Serv Sect, Neurotoxicol Res Unit, 75-80; ASST PROF PSYCHIAT, UNIV PITTSBURGH, 79-, CHIEF CHEM PHARMACOL, 80- *Concurrent Pos:* Adj asst prof, Queensborough Community Col, 72-78; asst clin prof path, Columbia Univ Med Ctr, 77-79. *Mem:* AAAS; Am Chem Soc; NY Acad Sci; Am Asn Clin Chemists. *Res:* Synthesis and biosynthesis of sphingolipids; neurochemistry of brain and nerve tissue; lipid chemistry; clinical pharmacology; drug pharmacokinetics; drug effects on central nervous system; methods for neuropsychotropic agent detection in biological medium. *Mailing Add:* Western Psychiat Inst 3811 Ohara St Pittsburgh PA 15213

STILLINGER, FRANK HENRY, b Boston, Mass, Aug 15, 34; m 56; c 2. LIQUID STATE THEORY, PHASE TRANSITION THEORY. *Educ:* Univ Rochester, BS, 55; Yale Univ, PhD(chem), 58. *Prof Exp:* Fel chem, Yale Univ, 58-59; MEM TECH STAFF, BELL LABS, INC, 59- *Concurrent Pos:* Lectr, Welsh Found, 74; mem evaluation panel, Heat Div, Nat Bur Standards, 75-78; mem policy comt, Chem Div, NSF, 80-83 & Off Adv Sci Comput, 84-88. *Honors & Awards:* Elliott Cresson Medal, Franklin Inst, 78; Hildebrand Award, Am Chem Soc, 86; Langmuir Prize, Am Phys Soc, 89; Trumbull lectr, Yale Univ, 84. *Mem:* Nat Acad Sci; Am Phys Soc; AAAS. *Res:* Molecular theory of water and it solutions; theory of phase transitions; quantum chemistry. *Mailing Add:* Bell Labs 600 Mountain Ave Murray Hill NJ 07974

STILLINGS, BRUCE ROBERT, b Portland, Maine, May 18, 37; m 59; c 4. NUTRITION. *Educ:* Univ Maine, BS, 58; Pa State Univ, MS, 60, PhD(animal nutrit), 63. *Prof Exp:* NIH fel, Cornell Univ, 63-66; supvry res chemist, food res prog leader & dep lab dir, US Nat Marine Fisheries Serv, 66-74; NUTRIT COORD, DIR FOOD SAFETY, DIR RES ACTIVITIES & DIR RES & VPRES RES & DEVELOP, NABISCO BRANDS INC, 74- *Mem:* Am Asn Cereal Chemists; AAAS; Am Inst Nutrit; Inst Food Technologists. *Res:* Nutritional studies on metabolism and utilization of minerals and amino acids; nutritive value of food-proteins and protein concentrates; factors affecting protein quality of foods. *Mailing Add:* 20 Lakeview Dr Kinnelon NJ 07405

STILLIONS, MERLE C, b Bedford, Ind, Feb 15, 29; m 53; c 5. LABORATORY ANIMAL SCIENCE, NUTRITION. *Educ:* Purdue Univ, BS, 57, MS, 58; Rutgers Univ, PhD(nutrit), 62. *Prof Exp:* Instr nutrit, Rutgers Univ, 58-62, chmn, Dairy Dept, Chico State Univ, 62-63; dir nutrit, Morris Res Lab, 63-72; RES DIR, AGWAY INC, 72- *Concurrent Pos:* Mem, Equine Comt, Nat Res Coun, 69-73. *Mem:* Am Soc Animal Sci; Am Asn Lab Animal Soc. *Res:* Laboratory animal and fish nutrition and feed control programs. *Mailing Add:* 1342 Agard Rd Trumansburg NY 14886

STILLMAN, GREGORY EUGENE, b Scotia, Nebr, Feb 15, 36; m 56; c 3. COMPOUND SEMICONDUCTOR MICROELECTRONICS. *Educ:* Univ Nebr, Lincoln, BS, 58; Univ Ill, Urbana, MS, 65, PhD(elec eng), 67. *Prof Exp:* Res staff assoc solid state physics, Lincoln Lab, Mass Inst Technol, 67-75; dir, Compound Semiconductor Microelectronics Lab, 84-87, assoc dir, Coordr Sci Technol, 85-86, PROF, DEPT ELEC ENG, UNIV ILL, 75- *Concurrent Pos:* Vis scientist, Lab Elettronica dello Stato Solido, 72. *Honors & Awards:* Jack Morton Award, Inst Elec & Electronic Engrs, 90; Heinrich Welker Medal, 90. *Mem:* Nat Acad Eng; fel Inst Elec & Electronics Engrs; Electron Devices Soc (pres, 84-86); Am Phys Soc. *Res:* Semiconductor physics; transport properties; photoconductivity; spectroscopy; luminescence. *Mailing Add:* 151 Microelectronics Lab Univ Ill 208 N Wright St Urbana IL 61801

STILLMAN, JOHN EDGAR, b Syracuse, NY, May 21, 45. INDUSTRIAL HYGIENE, CHEMISTRY. *Educ:* State Univ NY Col Forestry at Syracuse Univ, BS, 67; Univ NC Sch Pub Health, MSPH, 69; NC State Univ, MAgri, 72; Am Bd Indust Hyg, cert, 80. *Prof Exp:* Analytical chem supvr, Div Health Serv, Occup Health Lab Unit, NC Dept Human Resources, 72-79; indust hygienist, 79-80, actg head, 80-81, SECT HEAD, INDUST HYG LAB, EXXON BIOMED SCIENCE DIV, EXXON CORP, 81- *Concurrent Pos:* Consult analytical chem, various private industs, 74-79; comt mem, Lab Accreditation, Am Indust Hyg Asn, 85-87; mem, Asbestos Analysis Registry, 88-91, chair, 91. *Mem:* Am Indust Hyg Asn; Am Conf Govt Indust Hygienists; Am Acad Indust Hyg; Soc Qual Assurance. *Res:* Occupational, industrial and environmental pollutants; quality assurance analytical laboratory administration, management and oversight; gas chromatography; microscopy; method evaluation and development; field hazards surveys; asbestos; kinetics of cholinesterase; anaerobic sludge digestion. *Mailing Add:* RD 1 Box 570 Princeton NJ 08540

STILLMAN, MARTIN JOHN, b London, Eng, June 4, 47; Can citizen. BIOINORGANIC CHEMISTRY, SPECTROSCOPY. *Educ:* Univ EAnglia, BSc, 69, MSc, 70, PhD(chem), 73. *Prof Exp:* Fel chem, Univ Alta, 73-75; asst prof, 75-81, assoc prof, 81-86, PROF CHEM, UNIV WESTERN ONT, 86- *Mem:* Chem Soc; Can Inst Chem; Am Chem Soc. *Res:* Spectroscopy, spectroscopic studies of metallothionein computer assisted analytical chemistry; expert systems in analytical chemistry, electrochemistry and photochemistry of inorganic and biological systems; magnetic circular dichroism of heme proteins, porphyrins, phthalocyanines; binding of cadmium, copper and mercury in biological systems. *Mailing Add:* Dept Chem Univ Western Ont London ON N6A 5B7 Can

STILLMAN, RICHARD ERNEST, b Grand Island, Nebr, Dec 6, 29; m 56; c 2. MATHEMATICS, CHEMICAL ENGINEERING. *Educ:* Univ Kans, BS, 51, MS, 56; Pa State Univ, University Park, PhD(chem eng), 61. *Prof Exp:* Staff engr process control, Res Div, 58-63, adv engr, Systs Develop Div, 64-65, SR ENGR, DATA PROCESSING Div, PALO ALTO, 66-, SR ENGR, RES DIV, 85- *Mem:* Am Inst Chem Engrs. *Res:* Formulation of mathematical models of chemical processes; numerical methods for solving partial and ordinary differential equations; gradient optimization procedures and multicomponent distillation calculations; artificial intelligence and knowledge based expert systems. *Mailing Add:* 628 California Way Redwood City CA 94062

STILLWAY, LEWIS WILLIAM, b Casper, Wyo, Feb 27, 39; m 59; c 2. BIOCHEMISTRY. *Educ:* Col Idaho, BS, 62; Univ Idaho, MS, 65, PhD(biochem), 68. *Prof Exp:* Fel, Inst Marine Sci, Univ Miami, 68-69; assoc chem, 69-71, from asst prof to assoc prof biochem, 71-83, PROF BIOCHEM, MED UNIV, SC, 83- *Concurrent Pos:* Ed, Med Biochem Question Bank, 78-; vis prof, Col Charleston Gov Sch, 88- *Mem:* Am Soc Biochem & Molecular Biol. *Res:* Science educational methods; lipid chemistry and metabolism. *Mailing Add:* Dept Biochem Med Univ SC 171 Ashley Ave Charleston SC 29425

STILLWELL, EDGAR FELDMAN, b Staten Island, NY, Nov 2, 29. PHYSIOLOGY. *Educ:* Wagner Mem Lutheran Col, BS, 51; Duke Univ, MA, 53, PhD(zool), 57. *Prof Exp:* Asst zool, Duke Univ, 52-56, res assoc, 56-57; asst prof biol, Longwood Col, 57-60 & Univ SC, 60-61; assoc prof zool, E Carolina Univ, 61-68; ASSOC PROF BIOL, OLD DOM UNIV, 68- *Concurrent Pos:* NASA-Am Soc Eng Educ fac res fel, Langley Res Ctr, 69-70; NASA res grant, 71-72. *Mem:* AAAS; Am Soc Cell Biol. *Res:* Mitogenetic control mechanisms in central nervous system neurons in tissue culture. *Mailing Add:* Dept of Biol Old Dom Univ Hampton Blvd Norfolk VA 23508

STILLWELL, EPHRAIM POSEY, JR, b Sylva, NC, Aug 29, 34; m 60; c 2. SOLID STATE PHYSICS. *Educ:* Wake Forest Col, BS, 56; Univ Va, MS, 58, PhD(physics), 60. *Prof Exp:* From asst prof to assoc prof, 60-69, head dept, 71-74, PROF PHYSICS, CLEMSON UNIV, 69- *Concurrent Pos:* US Air Force Off Sci Res grant, 63-69. *Mem:* Am Asn Physics Teachers; Am Phys Soc; AAAS. *Res:* Magnetoresistance in metals; superconductivity. *Mailing Add:* Dept Physics 201 Sikes Hall Clemson Univ Clemson SC 29631

STILLWELL, GEORGE KEITH, b Moose Jaw, Sask, July 11, 18; m 43; c 2. PHYSICAL MEDICINE. *Educ:* Univ Sask, BA, 39; Queen's Univ, Ont, MD, CM, 42; Univ Minn, PhD(phys med & rehab), 54; Am Bd Phys Med & Rehab, dipl, 52. *Prof Exp:* Instr, Mayo Med Sch, 50-54, from asst prof to prof, Mayo Grad Sch Med, 55-73, chmn dept, 73-81, emer prof phys med rehab, 83-; RETIRED. *Concurrent Pos:* Consult, Mayo Clin, 54-83. *Mem:* Cong Rehab Med; Am Acad Phys Med & Rehab. *Res:* Rehabilitation; physiologic effects of therapeutic procedures; edema of peripheral origin. *Mailing Add:* Emer Staff Mayo Clin Rochester MN 55905

STILLWELL, HAROLD DANIEL, b Staten Island, NY, Mar 21, 31; m 64; c 2. PHYSICAL GEOGRAPHY, BIOGEOGRAPHY. *Educ:* Duke Univ, BS, 52, MF, 54; Mich State Univ, PhD, 61. *Prof Exp:* Forestry aid, US Forest Serv, NC, 52; asst, Ore Forest Res Ctr, 54-57; asst geog, Mich State Univ, 57-59; asst prof, Eastern Mich Univ, 60-61 & Univ Tex, 61-62; assoc prof, ECarolina Univ, 62-71; PROF GEOG, APPALACHIAN STATE UNIV, 71- *Mem:* Asn Am Geographers; Sigma Xi; Int Geog Union. *Res:* Natural hazards of mountain areas, particularly avalanche prediction; mountain geo-ecology with analysis of tree line location; remote sensing. *Mailing Add:* 105 Hawthorne Lane Boone NC 28607

STILLWELL, RICHARD NEWHALL, b Princeton, NJ, Nov 22, 35. ORGANIC CHEMISTRY, COMPUTER SCIENCE. *Educ:* Princeton Univ, BA, 57; Harvard Univ, MA, 59, PhD(chem), 64. *Prof Exp:* From instr to prof chem, Baylor Col Med, 63-84; CONSULT, 84- *Mem:* Am Chem Soc; Am Soc Mass Spectrometry; Asn Comput Mach. *Res:* Chemistry of natural products; chemical modelling; analytical systems. *Mailing Add:* Ten Daniels Dr Bedford MA 01730-1302

STILLWELL, WILLIAM HARRY, b Albany, NY, Mar 30, 46; m 78; c 2. BIOCHEMISTRY, BIOLOGY. *Educ:* State Univ NY, Albany, BS, 67; Pa State Univ, MS, 73, PhD(biochem), 74. *Prof Exp:* Res asst prof origin life, Inst Molecular & Cellular Evolution, 74-75; res assoc membrane biophys, Mich State Univ, 76-78; asst prof, 78-81, ASSOC PROF BIOL, IND UNIV-PURDUE UNIV, INDIANAPOLIS, 82- *Mem:* AAAS; NY Acad Sci; Biophys Soc; Am Soc Plant Physiol. *Res:* Membrane biochemistry and biophysics; artificial membrane systems; action of plant hormones on membranes; origin of life; action of retinoids on membranes; affects N-3 fatty acids on membranes. *Mailing Add:* Dept Biol Ind Univ Purdue Indianapolis IN 46202

STILWELL, DONALD LONSON, b Detroit, Mich, Dec 29, 18. ANATOMY. *Educ:* Wayne State Univ, AB, 41, MD, 44. *Prof Exp:* Intern, Harper Hosp, Detroit, 44-45, resident surg, 45-56; from instr to asst prof, 49-59, asst dean, 64-65, ASSOC PROF ANAT, SCH MED, STANFORD UNIV, 60- *Concurrent Pos:* Fel anat, Wayne State Univ, 58-59. *Mem:* AAAS; Am Asn Anat. *Res:* Anatomy; experimental pathology; vascularization of vertebral column; innervation of hand, foot, joints, spine and eye; blood supply of brain. *Mailing Add:* 467 Melville Ave Palo Alto CA 94301

STILWELL, KENNETH JAMES, b Poughkeepsie, NY, Apr 4, 34; m 56; c 3. MATHEMATICS. *Educ:* Bob Jones Univ, BS, 56; Ariz State Univ, MA, 59; Univ Ariz, MS, 64; Hunter Col, MA, 65; Univ Northern Colo, EdD(math educ), 71. *Prof Exp:* Instr high schs, Ariz, 57-64; asst prof math, King's Col, NY, 65-66; assoc prof, 66-74, PROF MATH, NORTHEAST MO STATE UNIV, 74- *Concurrent Pos:* Comput specialist Carrolltom Sch, 84-85. *Mem:* Math Asn Am; Nat Counc Teachers Math; Sch Sci & Mat Asn. *Res:* Mathematics education; effect of video-tape and critique on attitude of pre-service mathematics teachers. *Mailing Add:* Div of Math Northeast Mo State Univ Kirksville MO 63501

STIMLER, SUZANNE STOKES, b Aberdeen, SDak, Sept 25, 28; m 64. PHYSICAL CHEMISTRY. *Educ:* Univ Colo, BA, 50; Mt Holyoke Col, MA, 54; Univ Rochester, PhD(phys chem), 58. *Prof Exp:* Chemist, Shell Oil Co, 51-52; instr chem, Wellesley Col, 57-58; res chemist, US Navl Res Lab, 58-68; health scientist adminr, Nat Inst Child Health & Human Develop, 68-71, health scientist adminr, 71-75, DIR, BIOMED RES TECHNOL PROG, DIV RES RESOURCES, NIH, 75- *Mem:* AAAS; Am Chem Soc; Sigma Xi. *Res:* Molecular electronic spectroscopy, particularly absorption and emission; infrared absorption spectroscopy; photodegradation of polymers; technology for biomedical research. *Mailing Add:* 19 Watchwater Way Rockville MD 20850

STIMMEL, GLEN LEWIS, b Lynwood, Calif, Mar 18, 49; m 68; c 2. CLINICAL PHARMACY, PSYCHOPHARMACOLOGY. *Educ:* Univ Calif, San Francisco, DPhar(pharm), 72. *Prof Exp:* Clin pharmacist psychopharm, Dept Health, San Francisco, 73-74; asst prof clin pharm, Sch Pharm, 74-78, assoc prof, 79-84, PROF CLIN PHARM & PSYCHIAT, SCHS PHARM & MED, UNIV SOUTHERN CALIF, 84- *Concurrent Pos:* Chmn, Pharm Sect, Am Pharmaceut Asn, 76-78; consult, div mental health servs, NIMH, Md, 80-85, Data Med Inc, Minn, 82-, Health Care Finance Admin, Helalth & Human Servs, DC, 85-; panel mem, defined diets & childhood hyperactivity, NIH, 82; prin investr, Univ Southern Calif Pract Res Unit, Merck Found, 83- *Mem:* Am Col Clin Pharm (pres, 83-84); Am Soc Hosp Pharmacists; Am Pharmaceut Asn; Am Asn Col Pharm. *Res:* Clinical pharmacology; psychopharmacy education; health manpower utilization, expanded clinical poles for pharmacists, prescriptive authority for pharmacists. *Mailing Add:* 25705 Yucca Valley Rd Valencia CA 91355

STIMMELL, K G, ENGINEERING. *Prof Exp:* MEM STAFF, SANDIA NAT LAB. *Mailing Add:* Sandia Nat Lab PO Box 696 Livermore CA 94550

STIMPFLING, JACK HERMAN, b Denver, Colo, June 11, 24; m 50; c 4. GENETICS. *Educ:* Univ Denver, BS, 49, MS, 50; Univ Wis, PhD(genetics), 57. *Prof Exp:* Asst yeast genetics, Southern Ill Univ, 51-52; immunogenetics, Univ Wis, 52-57; fel, Jackson Mem Lab, 57-59, assoc staff scientist, 59-61, staff scientist, 61-64; res assoc, McLaughlin Res Inst, 65-68, dir, 68-88; RETIRED. *Mem:* Genetics Soc Am; Am Asn Immunologists. *Res:* Immunogenetics; inheritance of cellular antigens; cellular antigens in tissue transplantation. *Mailing Add:* 3921 Seventh Ave S Great Falls MT 59405

STIMSON, MIRIAM MICHAEL, b Chicago, Ill, Dec 24, 13. ORGANIC CHEMISTRY. *Educ:* Siena Heights Col, BS, 36; Inst Divi Thomae, MS, 39, PhD(chem), 48. *Prof Exp:* Head res lab, Siena Heights Col, 36-68, instr chem, 39-46, asst prof, 46-50, prof natural sci & head div, 50-69; chmn dept, Keuka Col, 69-74, prof chem, 69-78; DIR, GRAD STUDIES OFF, SIENA HEIGHTS COL, 78- *Concurrent Pos:* Exec Comt, bd dirs Mich Consortium Substance Abuse Educ, 83-88, pres, 87-88. *Honors & Awards:* Charles

Williams Award, 42. *Mem:* Am Chem Soc; Nat Asn Women Deans & Counselors; Am Asn Coun & Develop. *Res:* Infrared and ultraviolet absorption in the solid state by potassium bromide disks; effect of irradiation on pyrimidines in the solid state. *Mailing Add:* Grad Studies Off Siena Heights Col Adrian MI 49221

STINAFF, RUSSELL DALTON, b Akron, Ohio, Mar 17, 40; m 68; c 1. ELECTRICAL ENGINEERING, CYBERNETICS. *Educ:* Univ Akron, BSEE, 62; Purdue Univ, Lafayette, MSEE, 63; Univ Ill, Urbana, PhD(elec eng), 69. *Prof Exp:* Electronic engr, Nat Security Agency, 64-65; asst prof elec eng, Clemson Univ, 69-76; MEM STAFF, HONEYWELL, INC, 80- *Concurrent Pos:* US Off Educ res grant, 71-73. *Mem:* AAAS; Simulation Coun; Am Soc Eng Educ; Inst Elec & Electronics Engrs; Asn Comput Mach. *Res:* Artificial intelligence; application of computers to education; simulation of large systems. *Mailing Add:* Bell & Howell 2300 Brummel Pl Evanston IL 60202

STINCHCOMB, THOMAS GLENN, b Tiffin, Ohio, Sept 12, 22; m 45; c 4. RADIATION PHYSICS, MEDICAL PHYSICS. *Educ:* Heidelberg Col, BS, 44; Univ Chicago, MS, 48, PhD(physics), 51. *Prof Exp:* From instr to asst prof physics, State Col Wash, 51-54; from assoc prof to prof & head dept, Heidelberg Col, 54-61; res physicist, Nuclear & Radiation Physics Sect, IIT Res Inst, 61-65, sr physicist & group leader, 65-68; chmn dept, 68-76, head, Nat Sci & Math Div, 84-87, PROF PHYSICS, DEPAUL UNIV, 68- *Concurrent Pos:* Actg mgr, Nuclear & Radiation Physics Sect, IIT Res Inst, 66-67; vis res assoc, Radiol Dept, Univ Chicago, 76-; mem task group 18, Fast Neutron Beam Dosimetry, Am Asn Physicists Med, 79-85, comt continuing educ, 82-85, vpres & prog chmn midwest sect, 87-89, pres, 90- *Mem:* Sigma Xi; Am Asn Physics Teachers; Am Nuclear Soc; Fed Am Scientists; Am Asn Physicist in Med; Am Inst Physics. *Res:* Medical applications of nuclear radiation physics, mainly determinations of quality of neutron therapeutic beams by microdosimetric techniques. *Mailing Add:* Dept of Physics De Paul Univ Lincoln Park Campus Chicago IL 60614-3504

STINCHCOMB, WAYNE WEBSTER, b Baltimore, Md, Sept 16, 43; m 68; c 2. COMPOSITE MATERIALS, MECHANICS. *Educ:* Va Polytech Inst, BS, 65; Pa State Univ, MS, 67, PhD(eng mech), 71. *Prof Exp:* Instr eng mech, Pa State Univ, 68-69; from instr to asst prof eng mech, 70-78, assoc prof, 78-80, PROF ENG SCI & MECH, VA POLYTECH INST & STATE UNIV, 80- *Concurrent Pos:* Consult. *Mem:* Am Soc Testing & Mat. *Res:* Materials, fatigue, composites, nondestructive testing and evaluation. *Mailing Add:* Dept Eng Sci Va Polytech Inst & State Univ Blacksburg VA 24061

STINCHFIELD, CARLETON PAUL, b Boston, Mass, Mar 10, 28; m 49; c 6. ADHESIVES & RESIN BONDING, HAZARDOUS MATERIALS HANDLING. *Educ:* Colby Col, BA, 49; Northeastern Univ, MS, 57. *Prof Exp:* Res chemist, Lever Bros Co, Cambridge, Mass, 49-51; res engr, Norton Co, Worcester, Mass, 51-59; instr chem, Northfield-Mt Hermon Sch, 59-70; from asst prof to assoc prof chem, 70-79, PROF CHEM & DIR SCI DIV, GREENFIELD COMMUNITY COL, 79- *Concurrent Pos:* Consult, 59- *Mem:* Am Chem Soc. *Res:* Physical-organic studies in adhesives and resin-bonded systems; fire safety; natural product chemistry. *Mailing Add:* Bald Mountain Rd PO Box 235 Bernardstan MA 01337-0235

STINCHFIELD, FRANK E, b Warren, Minn, Aug 12, 10; m 30; c 2. ORTHOPEDIC SURGERY. *Educ:* Northwestern Univ, MD, 34; Am Bd Orthop Surg, dipl, 46. *Hon Degrees:* DSc, Carleton Col, 60; DSc, Univ NDak, 70; FRACS, 75; FRCS, 79. *Prof Exp:* Prof orthop surg & chmn dept, Col Physicians & Surgeons, Columbia Univ, 56-76; ATTEND ORTHOP SURGEON, COLUMBIA- PRESBY MED CTR, 51- *Concurrent Pos:* Dept Defense & Dept Air Force orthop surg consult, Asst Secy Defense, 65-, tour Vietnam & Far East installations, 66; pres, Am Bd Orthop Surg. *Honors & Awards:* Centennial Award, Northwestern Univ, 59. *Mem:* Am Surg Asn; Am Asn Surg of Trauma; Am Acad Orthop Surg (pres, 61); Am Orthop Asn (treas); NY Acad Med. *Res:* Effect of anticoagulant therapy on bone repair; osteogenesis of bone isolated from soft tissue blood supply. *Mailing Add:* Colum-Presby Med Ctr 161 Ft Washington New York NY 10032

STINE, CHARLES MAXWELL, b Osceola Mills, Pa, Mar 4, 25; m 51; c 3. FOOD SCIENCE. *Educ:* Pa State Univ, BS, 51, MS, 52; Univ Minn, PhD(dairy tech), 57. *Prof Exp:* Assoc prof, 57-68, PROF FOOD SCI, MICH STATE UNIV, 68- *Concurrent Pos:* Consult. *Mem:* Am Oil Chem Soc; Am Dairy Sci Asn. *Res:* Lipid oxidation in food products; spray dried foods; flavor chemistry; analytical chemistry. *Mailing Add:* Dept Food Sci Mich State Univ East Lansing MI 48824-1224

STINE, GERALD JAMES, b Johnstown, Pa, May 29, 35; m 62; c 2. HUMAN GENETICS, MICROBIAL GENETICS. *Educ:* Southern Conn State Col, BS, 61; Dartmouth Col, MA, 63; Univ Del, PhD(biol-genetics), 66. *Prof Exp:* Geneticist, Oak Ridge Nat Lab, 66-68; asst prof microbiol genetics, Univ Tenn, Knoxville, 68-72; assoc prof, 72-77, PROF GENETICS & MICROBIOL, UNIV N FLA, 77- *Concurrent Pos:* Union Carbide fel, 66-68; consult, Oak Ridge Nat Lab, 68-; dir cytogenetics, Regional Genetic Ctr, Jacksonville, Fla, 82-85. *Mem:* Genetics Soc Am; Sigma Xi; Am Soc Human Genetics; assoc Inst Soc Ethics & Life Sci; Asn Cytogenetic Technologists; Nat Soc Genetic Counr. *Res:* Association of human blood groups and behavior. *Mailing Add:* Dept Natural Sci Univ N Fla Jacksonville FL 32216

STINE, PHILIP ANDREW, b Detroit, Mich, Aug 12, 44; m 67; c 2. METALLURGICAL ENGINEERING, SHEET METAL FORMING. *Educ:* Wayne State Univ, BSME, 67; Purdue Univ, MS, 68, PhD(metall eng), 72. *Prof Exp:* Advan res projs agency res asst, Purdue Univ, Lafayette, 68-71; metall engr, Appl Res & Design Ctr, Gen Elec Co, 71-75, sr res metallurgist, 82-84, prog mgr, 84-85, mgr, Metall Lab, 84-85, mgr, Metall & Ceramics Lab, 85-89, STAFF METALL, APPL RES & DESIGN CTR, GEN ELEC CO, 89- *Concurrent Pos:* Lectr sheet metal forming technol, Am Soc Metals, 80-89; hon fac mem, Acad Metal & Mat. *Mem:* Am Inst Metall Engrs; Am Soc Metals; Am Deep Drawing Res Group (treas, 76-78, secy, 78-80, pres, 80-82). *Res:* Formability research including methods and techniques that allow forming difficulty determination for given die, steel and lubricant conditions, allowing definition of optimum forming conditions; implemented computer analysis that predicts forming fractures at concept phase of new sheet metal designs. *Mailing Add:* Metall Lab Appliance Pk 35-1117 Gen Elec Co Louisville KY 40225

STINE, WILLIAM H, JR, b Cincinnati, Ohio, Mar 23, 26; m 48; c 3. CHEMICAL ENGINEERING, TEXTILE ENGINEERING. *Educ:* Univ Cincinnati, ChemE, 50. *Prof Exp:* Res technician, Chem Res Lab, Nat Cash Register Co, Ohio, 49; engr, Res & Develop Lab, Champion Paper & Fibre Co, 50-55; engr, Carother Res Lab, E I du Ponnnt de Nemours & Co, Inc, 55-58, res engr, 58-62, sr res engr, Textile Res Lab, 62-65, res supvr, 65-81, res assoc indust prod res, Textile Fibers Dept, 81-83. *Res:* Fiber technology; physical, physico-chemical, chemical and statistical relationships between fibers and end uses; polyamides and melt spinning processes. *Mailing Add:* Coffee Run Condo Apt B1A 614 Loveville Rd Hockessin DE 19707

STINE, WILLIAM R, b Schenectady, NY, Dec 14, 38; m; c 1. ORGANIC CHEMISTRY, BIOCHEMISTRY. *Educ:* Union Col, BS, 60; Syracuse Univ, PhD(chem), 66. *Prof Exp:* From asst prof to assoc prof, 65-78, PROF CHEM, WILKES COL, 78- *Mem:* Am Chem Soc; AAAS; Am Asn Univ Prof. *Res:* Structure of pentavalent phosphorus compounds; reactions of tertiary phosphines with positive halogen compounds; natural product synthesis. *Mailing Add:* Dept Chem Wilkes Col Wilkes-Barre PA 18703

STINECIPHER, MARY MARGARET, b Chattanooga, Tenn, Feb 26, 40; div; c 2. EXPLOSIVES CHEMISTRY, THERMAL CHEMISTRY. *Educ:* Earlham Col, BA, 62; Univ NC, Chapel Hill, PhD(inorg chem), 67. *Prof Exp:* Fel chem, Res Triangle Inst, 66-68 & 74-75; STAFF MEM CHEM, LOS ALAMOS NAT LAB, 76- *Concurrent Pos:* Vis scientist, USAF Off Sci Res, Eglin AFB, Fla, 80-81; adj prof org chem, Univ NMex Ctr Grad Studies, Los Alamos, 89-90. *Mem:* Am Chem Soc. *Res:* Explosives synthesis and characterization. *Mailing Add:* Los Alamos Nat Lab PO Box 1663 M-1 MS C920 Los Alamos NM 87545

STINGELIN, RONALD WERNER, b New York, NY, May 29, 35; m 73; c 2. COAL & MINING GEOLOGY, REMOTE SENSING. *Educ:* City Col New York, BS, 57; Lehigh Univ, MS, 59; Pa State Univ, PhD(geol), 65. *Prof Exp:* Res geologist, HRB Singer Inc, 65-67, sr res geologist, 67 -68, mgr, Environ Sci Br, 68-72, prin geologist, Energy & Natural Resource Systs Dept, 72-80; vpres tech serv, Resource Technol Corp, 80-84; GEOTECH CONSULT & PROF GEOLOGIST, 84- *Concurrent Pos:* NSF-Am Soc Photogram vis scientist, 68-71. *Mem:* Fel Geol Soc Am; Am Inst Prof Geologists. *Res:* Application of remote sensing to environmental problems; energy, resources and technology assessment studies with emphasis on fossil fuels; subsidence, seam interaction and prediction of roof hazards in coal mining; Appalachian coal geology, abandoned mined land problems, and mineral resource evaluation; Pennsylvania anthracite resources; mineral resource estimation and site investigations. *Mailing Add:* 120 Ronan Dr State College PA 16801-7809

STINGER, HENRY J(OSEPH), b Minneapolis, Minn, Nov 22, 20; m 51; c 3. ENGINEERING PHYSICS. *Educ:* Univ Minn, BEE, 42; Mass Inst Technol, cert, 43. *Prof Exp:* Electronic engr, Control Corp, 46-47; supvr res lab, Gen Mills, Inc, 47-51; chief reactors br, Savannah River Oper Off, US Atomic Energy Comn, 51-52; res supvr, E I Du Pont De Nemours & Co, Inc, 53-62, res assoc, 62-85. *Concurrent Pos:* Pvt consult, 85- *Mem:* Inst Elec & Electronics Engrs. *Res:* Tribiology; electronic properties of materials; electronic devices; electromagnetic shielding & materials; instrumentation & controls. *Mailing Add:* 119 Devonwood Lane Devon PA 19333

STINGL, GEORG, b Vienna, Austria, Oct 28, 48; m 80. IMMUNODERMATOLOGY, DERMATOLOGIC MICROBIOLOGY. *Educ:* Univ Vienna, MD, 73; Am Acad Dermat, dipl, 82. *Prof Exp:* Res, Dept Dermat I, Univ Vienna, 73-76; fel dermat, Nat Cancer Inst, 77-78; staff, Dept Dermat, Univ Innsbrug, 78-81; STAFF, DEPT DERMAT I, UNIV VIENNA MED SCH, 81-, PROF DERMAT & HEAD, DIV IMMUNODERM & INFECTIOUS SKIN DIS, 91- *Concurrent Pos:* Vis scientist, NIAID, Lab Immunol, NIH, 85-86. *Honors & Awards:* Montagna Award, Soc Investigative Dermat, 86. *Mem:* Soc Investigative Dermat; Am Asn Immunologists; Am Fedn Clin Res; Ger Dermat Soc. *Res:* Immunology of the epidermis; langerhans cells; skin T-cells; AIDS; Karposi's sarcoma; atopic dermatitis. *Mailing Add:* Dept Dermat I Univ Vienna Med Sch Alser Strasse 4 Vienna A-1090 Austria

STINGL, HANS ALFRED, b Eger, Czech, Oct 13, 27; US citizen; m 54; c 2. INDUSTRIAL ORGANIC CHEMISTRY. *Educ:* Univ Erlangen, dipl, 54, PhD(org chem), 56. *Prof Exp:* Res assoc org chem, Univ Ill, Urbana, 56-58; res & develop chemist, 58-75, DEVELOP ASSOC, CIBA-GEIGY CORP-TOMS RIVER PLANT, 75- *Mem:* Fel Am Inst Chem; Am Chem Soc; NY Acad Sci. *Res:* Organic dyestuffs and intermediates. *Mailing Add:* 852 Ocean View Dr Toms River NJ 08753-2797

STINI, WILLIAM ARTHUR, b Oshkosh, Wis, Oct 9, 30; m 50; c 3. HUMAN BIOLOGY, PHYSICAL ANTHROPOLOGY. *Educ:* Univ Wis, BBA, 60, MS, 67, PhD(human biol), 69. *Prof Exp:* From asst prof to assoc prof anthrop, Cornell Univ, 68-73; assoc prof, Univ Kans, 73-76; PROF ANTHROP, UNIV ARIZ, 76- *Concurrent Pos:* Mem rev panel, Anthrop Prog, NSF, 76-78; field ed, Phys Anthrop, Am Anthropologist, 79-82; mem, Governor's Adv Coun on Aging, 80-83; ed-in-chief, Am J Phys Anthrop, 83-; fel, Linacre Col & Univ of Oxford, 85; pres, Am Asn Phys Anthropologists, 89-91. *Mem:* Fel AAAS; Am Asn Phys Anthrop; fel NY Acad Sci; Am Anthrop Asn; Am Inst Nutrit; Sigma Xi. *Res:* Effects of stress on human development including growth and maturation as measured by gross morphological and serological parameters; evaluation of stress as evolutionary force; nutrition and aging; alterations in bone mineral metabolism associated with aging. *Mailing Add:* Dept Anthrop Univ Ariz Tucson AZ 85721

STINNER, RONALD EDWIN, b New York, NY, July 27, 43; m 63; c 2. POPULATION ECOLOGY. *Educ:* NC State Univ, BS, 65; Univ Calif, Berkeley, PhD(entom), 70. *Prof Exp:* Res assoc entom, Tex A&M Univ, 70; res assoc, NC State Univ, 70-73, from asst prof to assoc prof, 73-83, PROF ENTOM & BIOMATH, NC STATE UNIV, 83-, DIR, BIOMATH GRAD PROG, 88- *Concurrent Pos:* Ed, Environ Entom, 87- *Mem:* Entom Soc Am; Entom Soc Can; Int Orgn Biol Control. *Res:* Modeling of population dynamics of agricultural pest insects and pathogens; studies on effects of behavior and host interactions on system dynamics. *Mailing Add:* Dept Entom NC State Univ Raleigh NC 27650

STINNETT, HENRY ORR, b San Francisco, Calif. CARDIOPULMONARY PHYSIOLOGY, BIOMATHEMATICS. *Educ:* Calif State Univ, Sacramento, AB, 63; Univ Calif, Davis, 69, MS, 69, PhD(physiol), 74. *Prof Exp:* Lab technician III res, Dept Avian Sci, Univ Calif, Davis, 64-69, res physiologist, 76-80; asst prof, 76-81, ASSOC PROF PHYSIOL, SCH MED, UNIV NDAK, 81- *Concurrent Pos:* Fel pharmacol, Health Sci Ctr, Univ Tex, 74-75, asst lectr, 75-76; Am Heart Asn grant, Univ Tex Health Sci Ctr, 75-76 & Sch Med, Univ NDak, 77-83. *Mem:* Sigma Xi; Am Physiol Soc; Nat Asn Underwater Instr; Soc Exp Biol Med. *Res:* Cardiovascular, pulmonary physiology; modulatory interactions of the cardiopulmonary mechanoreceptors on the systemic cardiovascular baroreflexes during lung inflation; models, mathematical of carotid sinus wall strain and Sororeceptor transduction of wall strain to fiber activity. *Mailing Add:* Dept Physiol Sch Med Univ NDak Grand Forks ND 58202

STINSKI, MARK FRANCIS, b Appleton, Wis, Jan 6, 41; m 68; c 2. MICROBIOLOGY, BIOCHEMISTRY. *Educ:* Mich State Univ, BS, 64, MS, 66, PhD(microbiol & biochem), 69. *Prof Exp:* Instr microbiol, Dept Biol Sci, Western Mich Univ, 67; res virologist, US Army Med Sci Lab, Ft Detrick, 69-71; NIH fel virol, Dept Microbiol, Univ Pa, 71-73; asst prof, 73-78, ASSOC PROF VIROL, DEPT MICROBIOL, UNIV IOWA, 78- *Concurrent Pos:* Am Cancer Soc grant, Dept Microbiol, Univ Iowa, 74-77; grant reviewer, Nat Found, 77-79, NSF, 78, 79 & 81, NIH, 78 & 80; NIH grant, 79-84, NIH res career develop award, 80-85. *Mem:* Am Soc Microbiol; Sigma Xi; AAAS; Soc Exp Biol & Med. *Res:* Transcription of the human cytomegalovirus genome; cytomegalovirus genome regulation; herpes virus cellular transformation and replication. *Mailing Add:* Dept of Microbiol Univ of Iowa Iowa City IA 52240

STINSON, AL WORTH, b Monroe, NC, Aug 5, 26; m 60; c 4. ANIMAL BEHAVIOR. *Educ:* NC State Col, BS, 49; Univ Ga, DVM, 56; Univ Minn, MS, 60. *Prof Exp:* Instr vet anat, Univ Minn, 56-60; asst prof vet anat, Cornell Univ, 60-64; asst prof, 64-68, assoc prof, 68-73, PROF VET ANAT, MICH STATE UNIV, 73- *Mem:* Am Asn Vet Anatomists; Am Asn Anatomists. *Res:* Histology of domestic animals. *Mailing Add:* Dept Anat 274A Giltner Hall Mich State Univ East Lansing MI 48824

STINSON, DONALD CLINE, b Malta, Idaho, Dec 7, 25; m 54; c 5. ELECTRICAL ENGINEERING. *Educ:* Iowa State Col, BS, 47; Calif Inst Technol, MS, 49; Univ Calif, EE, 53, PhD, 56. *Prof Exp:* Test engr, Gen Elec Co, NY, 47-48; asst elec eng, Univ Calif, 50-52, asst, Electronic Res Lab, 53-56; sr scientist, Missile Systs Div, Lockheed Aircraft Corp, 56-57, group leader microwaves, 57-58, res scientist, 58; prof elec eng, Univ Ariz, 58-68 & Univ Tex, Arlington, 68-69; MEM TECH STAFF, HUGHES AIRCRAFT CO, 69- *Concurrent Pos:* Consult, McGraw-Hill Bk Co, 59- & Tex Instruments Inc, 68- *Mem:* Inst Elec & Electronics Engrs. *Res:* Electromagnetic theory; microwave engineering and networks; evaluation of intrinsic properties of and frequency multiplying in microwave ferrites; parametric amplifiers; damping mechanism of ferrimagnetic resonance in ferrites. *Mailing Add:* Lockheed Aircraft Service Co PO Box 33 1-304 15 Ontario CA 91762

STINSON, DONALD LEO, b Hominy, Okla, Oct 8, 30; m 51; c 6. PETROLEUM ENGINEERING, CHEMICAL ENGINEERING. *Educ:* Univ Okla, BS, 50; Univ Mich, MS, 51, PhD(chem eng), 57. *Prof Exp:* Res engr, Phillips Petrol Co, 53-58; proj engr, Gulf Res & Develop Co, 58-60; prof petrol eng & head dept, Univ Wyo, 60-72, head dept mineral eng, 72-79, prof petrol eng, 79-81; vpres eng, Arnjac Corp, 81-86; CONSULT, 86- *Concurrent Pos:* Consult, 62-, Petrol Res Ctr, US Bur Mines, 63- & Cooper Estate, 64- *Mem:* Am Chem Soc; Am Inst Chem Engrs; Am Inst Mining, Metall & Petrol Engrs; Nat Soc Prof Engrs; Soc Petrol Eng Eval. *Res:* Thermodynamics; waste disposal; water treatment; power recovery. *Mailing Add:* 1074 Alta Vista Dr Laramie WY 82070-5004

STINSON, DOUGLAS G, b Manchester, NH, Nov 14, 53. OPTICAL DATA STORAGE, MAGNETO-OPTICS. *Educ:* New Col, Sarasota, FL, BA, 75; Univ Ill MS, 76, PhD(physics), 81. *Prof Exp:* Sr res scientist, res lab, diversified technol group, 81-90, MGR, ERASABLE OPTICAL MEDIA DEVELOP, MASS MEMORY DIV, EASTMAN KODAK CO, 90- *Mem:* Am Phys Soc; Inst Elec & Electronic Engrs. *Res:* Materials, processes and systems for erasable optical data storage. *Mailing Add:* Mass Memory Div Bldg 800 KPG Eastman Kodak Co Rochester NY 14652-3801

STINSON, EDGAR ERWIN, b Auburn, Ind, May 14, 27; m 74; c 4. ORGANIC CHEMISTRY, BIOCHEMISTRY. *Educ:* Purdue Univ, BS, 48; Iowa State Univ, MS, 51, PhD(biochem), 53. *Prof Exp:* Asst prof org chem, Villanova Univ, 53-56; asst prof, Mass Col Pharm, 56-57; RES CHEMIST, AGR & FOOD CHEM, USDA, 57- *Mem:* Am Chem Soc; Inst Food Technologists. *Res:* Mycotoxins; mold metabolism. *Mailing Add:* USDA 600 E Mermaid Lane & Ardmore Ave Philadelphia PA 19118

STINSON, GLEN MONETTE, b Sarnia, Ont, Dec 27, 39; m 62; c 3. EXPERIMENTAL NUCLEAR PHYSICS. *Educ:* Univ Toronto, BASc, 61; Univ Waterloo, MSc, 62; McMaster Univ, PhD(nuclear physics), 66. *Prof Exp:* Fel physics, 66-68, res assoc, Tri-Univ Meson Facility, 68-69, asst res physicist, 69-71, asst prof, 71-76, ASSOC PROF PHYSICS, TRI-UNIV MESON FACILITY, UNIV ALTA, 76- *Concurrent Pos:* Lectr, Univ Alta, 66-68. *Mem:* Am Phys Soc; Can Asn Physicists. *Res:* Proton induced reactions; design and use of high precision magnetic spectrometers; design of charged particle beam transport systems. *Mailing Add:* Dept Physics Univ Alta Edmonton AB T6G 2E2 Can

STINSON, HARRY THEODORE, JR, b Newport News, Va, Oct 26, 26; m 49; c 3. GENETICS. *Educ:* Col William & Mary, BS, 47; Ind Univ, PhD(cytogenetics), 51. *Prof Exp:* Asst prof biol, Col William & Mary, 51-52; res asst genetics, Conn Agr Exp Sta, 52-53, res assoc, 53-60, chief geneticist, 60-62; chmn dept bot, 64-65, chmn sect genetics, develop & physiol, 65-77, PROF GENETICS, CORNELL UNIV, 62-, ASSOC DIR DIV BIOL SCI, 77-, DIR UNDERGRAD STUDIES, 80- *Mem:* AAAS; Soc Study Evolution; Bot Soc Am; Genetics Soc Am; Am Soc Nat (treas, 63-66); Sigma Xi. *Res:* Cytology. *Mailing Add:* 118 Stimson Hall Cornell Univ Ithaca NY 14853

STINSON, JAMES ROBERT, b Bakersfield, Calif, Mar 24, 21; m 51. METEOROLOGY. *Educ:* Univ Calif, Santa Barbara, BA, 48; St Louis Univ, MS, 55, PhD(geophys, meteorol), 58. *Prof Exp:* Jr res meteorologist, Univ Calif, Los Angeles, 49-51; asst prof geophys & assoc sr res physicist, Gen Motors Res Labs, 70-78; MEM TECH STAFF, SANDIA LABS, 78- *Concurrent Pos:* Instr, Okla State Univ, 51-52; sr scientist, Meteorol Res Inc, chief res div, Navy; mem fac earth sci, Northern Ill Univ, 60-62. *Mem:* AAAS; Am Meteorol Soc; Am Geophys Union. *Res:* General meteorology; weather modification; cloud physics; environmental pollution; severe local storms. *Mailing Add:* 895 N Hillside Dr Long Beach CA 90815

STINSON, JOSEPH MCLESTER, pulmonary diseases; deceased, see previous edition for last biography

STINSON, MARY KRYSTYNA, b Bydgoszcz, Poland; US citizen; m 65; c 2. CHEMICAL ENGINEERING, WATER RESOURCES SCIENCE. *Educ:* Silesia Tech Univ, Poland, MS, 59; Univ Mich, Ann Arbor, MS, 69. *Prof Exp:* Chem engr coal chem, Cent Mining Inst, Poland, 58-65; analytical chemist, Owens-Ill Tech Ctr, 66-67; res chemist, Univ Mich, Ann Arbor, 67-68; PHYS SCIENTIST, US ENVIRON PROTECTION AGENCY, 74- *Honors & Awards:* Gold Medal award, Am Electroplaters Soc, 77; Bronze Medal Awards, Environ Protection Agency, 79 & 82. *Mem:* Am Chem Soc. *Res:* Development of new technologies for treatment of emissions generated by metal finishing, inorganic chemicals and asbestos industries; evaluation of innovative technologies for cleanup of superfund sites. *Mailing Add:* 37 Beacon Hill Dr Metuchen NJ 08840-1603

STINSON, PERRI JUNE, US citizen. OPERATIONS RESEARCH, STATISTICS. *Educ:* Univ Calif, Santa Barbara, AB, 48; Okla State Univ, MS, 52, PhD(statist), 55. *Prof Exp:* Asst prof health orgn res, St Louis Univ, 58-60; asst prof math, Northern Ill Univ, 60-62; biostatistician, Vet Admin Res Support Ctr, 62-64; mathematician & statistician, US Naval Aviation Safety Ctr, 64-65; head statist & math systs res, Douglas Aircraft Co, 65-67; prof environ eng, Univ Denver, 67-69; PROF OPERS RES & STATIST, CALIF STATE UNIV, LONG BEACH, 69- *Concurrent Pos:* Fac res grant, Univ Denver, 67-69; fac res grant, Calif State Univ, Long Beach, 69-72; consult, US Off Educ, 70-, Tex Water Develop Bd, 71-72 & Meteorol Res, Inc, 71- *Mem:* Opers Res Soc Am; Inst Mgt Sci; Soc Gen Systs Res; Am Statist Asn; Inst Math Statist. *Res:* Statistical and operations research applications to problems in the medical sciences, atmospheric pollution, public administration, and human resources. *Mailing Add:* Dept of Quant Systs Calif State Univ 1250 Bellflower Blvd Long Beach CA 90840

STINSON, RICHARD FLOYD, b Cleveland, Ohio, Feb 4, 21; m 54; c 5. FLORICULTURE, SCIENCE EDUCATION. *Educ:* Ohio State Univ, BS, 43, MS, 47, PhD, 52. *Prof Exp:* Instr floricult, State Univ NY Sch Agr Alfred, 47-48; asst prof, Univ Conn, 48-55; from asst prof to assoc prof hort, Mich State Univ, 55-67; from assoc prof to prof, 67-89, EMER PROF AGR, EDUC & HORT, PA STATE UNIV, UNIV PARK, 90- *Honors & Awards:* Outstanding Serv Award, Am Asn Teachers Educ Agr. *Mem:* Am Soc Hort Sci; Nat Asn Col Teachers Agr; Sigma Xi; Am Asn Teachers Educ Agr. *Res:* Horticultural and natural resources instruction material in agricultural education. *Mailing Add:* Dept Agr & Exten Educ Pa State Univ University Park PA 16802

STINSON, ROBERT ANTHONY, b Hamilton, Ont, Sept 30, 41; m 64; c 2. CLINICAL BIOCHEMISTRY. *Educ:* Univ Toronto, BScA, 64; Univ Alta, PhD(plant biochem), 68. *Prof Exp:* Med Res Coun Can fel molecular enzym, Bristol Univ, 68-71; from asst prof to assoc prof, 71-81, PROF PATH, UNIV ALTA, 81- , DIR, MED LAB SCI, 88- *Concurrent Pos:* Hon vis sr lectr biochem med, Univ Dundee, 77-78; sci & res assoc med staff, Univ Alta Hosp, 73- *Mem:* Can Soc Clin Chemists; Can Biochem Soc. *Res:* Studies of human alkaline phosphatase; to establish the gene origin of each molecular form and through hydrolytic, phosphotransferase, protein phosphatase, membrane attachment mechanism and clinical association studies, to establish a biochemical role for the enzyme. *Mailing Add:* Med Lab Sci Clin Sci Bldg Univ of Alta Edmonton AB T6G 2G3 Can

STINSON, ROBERT HENRY, b Toronto, Ont, Sept 17, 31; m 54; c 3. BIOPHYSICS, PHYSICS. *Educ:* Univ Toronto, BSA, 53, MSA, 57; Univ Western Ont, PhD(biophys), 60. *Prof Exp:* Mem faculty physics dept, Ont Agr Col, 53-63; prof physics, State Univ NY Col Potsdam, 63-67; assoc prof physics, Univ Guelph, 67-88; RETIRED. *Mem:* Biophys Soc. *Res:* Structural changes in plant membranes due to environmental stress and protection against such damage. *Mailing Add:* Dept of Physics Univ of Guelph Guelph ON N1G 2W1 Can

STIPANOVIC, BOZIDAR J, b Zagreb, Yugoslavia, Jan 9, 33; m 59; c 1. CHEMISTRY. *Educ:* Univ Belgrade, BS, 60, PhD(org chem), 65. *Prof Exp:* Teaching asst org chem, Univ Belgrade, 61-65; fel, Ipatieff High Pressure & Catalytic Lab, Northwestern Univ, 66-69; vis assoc prof org chem, Cent Univ Venezuela, 69-70; dir res & develop, Coral Chem Co, Waukegan, 70-76; tech dir, Res & Tech Serv, Santek Chem, 76-87; TECH DIR, HBS

ENTERPRISES, INC, 87- *Mem:* Am Chem Soc; Sigma Xi. *Res:* Organic catalytic reactions; surfactants; polymers; base catalyzed alkylations; conversion and chemical coatings on metals; corrosion inhibitors; paper making chemicals. *Mailing Add:* 608 Longwood E Lake Forest IL 60045

STIPANOVIC, ROBERT DOUGLAS, b Houston, Tex, Oct 28, 39; m 76; c 6. NATURAL PRODUCT CHEMISTRY. *Educ:* Loyola Univ, La, BS, 61; Rice Univ, PhD(chem), 66. *Prof Exp:* Res assoc chem, Stanford Univ, 66-67; asst prof, Tex A&M Univ, 67-71; res chemist, 71-87, RES LEADER, USDA, 87- *Mem:* Am Chem Soc; Phytochem Soc NAm; Royal Soc Chem. *Res:* Natural product synthesis' biosynthesis and structure determination; mass spectroscopy structure determination and reaction mechanisms; nuclear magnetic resonance studies. *Mailing Add:* RR 5 Box 805 College Station TX 77845

STIPANOWICH, JOSEPH JEAN, b Canton, Ill, Apr 14, 21; m 47; c 2. MATHEMATICS. *Educ:* Western Ill Univ, BS, 46; Univ Ill, MS, 47; Northwestern Univ, EdD(math), 56. *Prof Exp:* Prof math, Western Ill Univ, 47-85, head dept, 58-68; RETIRED. *Concurrent Pos:* Mem bd, Nat Coun Teachers Math, 70-71, chmn, Ext Affairs Comt, 71-72 & Financial Policies comt, 75; pres, Elem Math Sect & mem bd, Cent Asn Sci & Math Teachers. *Mem:* Nat Coun Teachers Math; Math Asn Am; Cent Asn Sci & Math Teachers. *Res:* History of mathematics; mathematics education. *Mailing Add:* 613 Memorial Dr Macomb IL 61455-3034

STIREWALT, HARVEY LEE, b Douglas, Ga, Jan 9, 32; m 46; c 3. AQUATIC BIOLOGY, ICHTHYOLOGY. *Educ:* Univ Miss, BA, 53, MS, 58; Univ Tenn, PhD(zool), 72. *Prof Exp:* Teacher biol, Acad Richmond County, 57-58, teacher physics, 58-59; instr biol, 59-60, chmn, Biol Dept, 59-65, asst prof, 60-73, ASSOC PROF BIOL, AUGUSTA COL, 73- *Concurrent Pos:* Grad teaching asst zool, Univ Tenn, 68-70; consult, Ga Dept Natural Resources, 71-; res supvr, Augusta Col Found Fac Res Fund, 74-75 & 77-78. *Mem:* Am Fisheries Soc. *Res:* Taxonomy, especially fishes and immature insects; pollution of aquatic systems, especially organic, inert suspended particles such as dam construction and dredging and thermal loading. *Mailing Add:* Dept Biol Augusta Col Walton Way Augusta GA 30910

STIREWALT, MARGARET AMELIA, medical parasitology, for more information see previous edition

STIRLING, ANDREW JOHN, b Adelaide, S Australia, Dec 23, 44; Can citizen; m 67; c 1. NUCLEAR PHYSICS, INSTRUMENTATION. *Educ:* Univ Adelaide, BSc(sci), 65, BSc(physics), 66; Flinders Univ, S Australia, PhD(physics), 70. *Prof Exp:* Fel physics, Nat Res Coun Can, 69-71; prof electronics, 71-77, head, Instrument Develop Br, 77-80, dir, electronics, instrumentation & control div, Atomic Energy Can Ltd, 81-85, GEN MGR, ACCELERATOR BUS UNIT, 85- *Mem:* Can Nuclear Soc. *Res:* Nuclear power instrumentation; nuclear safeguards; environmental instrumentation; linear accelerators. *Mailing Add:* Atomic Energy of Can Ltd 19 Pemberton Cr Nepean ON K2G 4Y8 Can

STIRLING, CHARLES E, b Havelock, NC, Nov 30, 33; m 62; c 3. PHYSIOLOGY, BIOPHYSICS. *Educ:* George Washington Univ, BA, 61; State Univ NY, PhD(physiol), 67. *Prof Exp:* Instr physiol, State Univ NY Upstate Med 66-67; asst prof, 68-74, ASSOC PROF PHYSIOL, UNIV WASH, 74- *Mem:* Am Physiol Soc; ARVO; Biophys Soc; Am Soc Cell Biol. *Res:* Active transport. *Mailing Add:* Dept Physiol & Biophys Univ Wash HSB SJ-40 Seattle WA 98195

STIRLING, IAN G, b Nkana, Zambia, Sept 26, 41; Can citizen. ZOOLOGY, WILDLIFE MANAGEMENT. *Educ:* Univ BC, BSc, 63, MSc, 65; Univ Canterbury, PhD(zool), 68. *Prof Exp:* Lectr zool, Univ Canterbury, 68-69; res assoc seals, Univ Adelaide, 69-70; RES SCIENTIST POLAR BEARS & SEALS, CAN WILDLIFE SERV, 70- *Concurrent Pos:* Chmn, Fed-Prov Tech Comt Polar Bear Res Can, 71-; mem, Polar Bear Specialists Group, Int Union Conserv Nature & Natural Resources, 74-, Polar Res Bel, 83-; adj prof, Univ Atla, 79. *Mem:* Can Soc Zoologists; Can Soc Environ Biologists; Am Soc Mammalogists; Marine Mammal Soc. *Res:* Ecology, behavior, evolution and management of marine mammals in polar marine ecosystems. *Mailing Add:* 7811 144ST Edmonton AB T5R 0R1 Can

STIRN, RICHARD J, b Milwaukee, Wis, Dec 5, 33; m 67; c 3. ENERGY CONVERSION, ELECTRON DEVICES. *Educ:* Univ Wis, BS, 61; Purdue Univ, MS, 63, PhD(physics), 66. *Prof Exp:* MEM TECH STAFF, JET PROPULSION LAB, CALIF INST TECHNOL, 66- *Concurrent Pos:* Ed, Appl Physics Commun. *Mem:* Am Phys Soc; sr mem Inst Elec & Electronics Engrs; Am Vacuum Soc. *Res:* Solid oxide fuel cells for energy conversion; chemical vapor deposition of III-V and II-VI semiconductors; conducting high temperature ceramic fabrication; sputtering of ceramic materials. *Mailing Add:* Jet Propulsion Lab Calif Inst Technol Pasadena CA 91109

STIRRAT, JAMES HILL, pathology, bacteriology, for more information see previous edition

STITCH, MALCOLM LANE, b Apr 23, 23; m 56, 65, 75; c 6. LASERS, LASER ISOTOPE SEPARATION. *Educ:* Southern Methodist Univ, BA & BS, 47; Columbia Univ, PhD(physics), 53. *Prof Exp:* Res asst radiation lab, Columbia Univ, 48-51; instr, Cooper Union Eng Col, 51-52; res asst radiation lab, Columbia Univ, 52-53; res physicist, Varian Assocs, 53-56; res physicist & head molecular beams group, Res Labs, Hughes Aircraft Co, 56-60, sr staff physicist, 59-60, head laser develop sect, Res & Develop Div, 61-62, mgr laser develop dept, 62-65, asst mgr high frequency lab, Res & Develop Div, 65-67, chief scientist, 67-68; asst gen mgr, Korad Dept, Union Carbide Corp, Calif, 68-73; sr scientist & consult, Ctr Laser Studies, Univ Southern Calif, Naval Res Lab, 73-74; mgr electro-optical opers, Exxon Nuclear Res & Technol Ctr, 74-81; eng assoc, Exxon Res & Eng Co, 81-83; sr eng adv, Rockwell Hanford Oper, 84-87; ADV ENG, WESTINGHOUSE HANFORD CO, 87- *Concurrent Pos:* Instr, Sarah Lawrence Col, 49-51; mem comt safe use lasers, Am Nat Standards Inst, 72-73; adj prof elec eng, Univ Southern Calif, 74-75. *Mem:* AAAS; Am Phys Soc; fel Inst Elec & Electronic Engrs; NY Acad Sci; fel Soc Photo-Optical Instrumentation Engrs; Optical Soc Am. *Res:* laser technology and applications; laser isotope separation; acousto- and electro-optics, instrumentation. *Mailing Add:* Westinghouse Hanford Co MSIN HO-36 PO Box 1970 Richland WA 99352

STITELER, WILLIAM MERLE, III, b Kane, Pa, July 30, 42; m 64; c 1. STATISTICS, FORESTRY. *Educ:* Pa State Univ, BS, 64, MS, 66, PhD(statist), 70. *Prof Exp:* Asst prof statist, Pa State Univ, University Park, 70-73; assoc prof, 73-78, PROF, STATE UNIV NY COL FORESTRY, 78- *Concurrent Pos:* Consult comput models. *Mem:* Inst Math Statist; Am Statist Asn; Int Asn Ecol; Biomet Soc; Sigma Xi. *Res:* Spatial patterns in ecological populations; modeling and simulation of biological populations; computer models for response to toxic substances. *Mailing Add:* Dept Forestry Col Environ Sci & Forestry State Univ NY Syracuse NY 13210

STITES, JOSEPH GANT, JR, b Hopkinsville, Ky, Mar 29, 21; m 43; c 4. INORGANIC CHEMISTRY, ORGANIC CHEMISTRY. *Educ:* Univ Ky, BS, 43; Mich State Univ, PhD(inorg chem), 49. *Prof Exp:* Dir new process technol, Monsanto Co, 49-73; dir process technol, Res Cottrell, Inc, 73-75; dir res & develop, Air Correction Div, UOP Inc, 75-82; RETIRED. *Mem:* Am Chem Soc; Environ Indust Coun; Air Pollution Control Asn. *Res:* Environmental control-gaseous and particulate collection systems, heavy chemicals, phosphates, nitrogen chemicals, explosives, rare-earths and separation techniques. *Mailing Add:* 203 Barley Mill Rd Old Hickory TN 37138

STITH, BRADLEY JAMES, b Columbus, Ohio, Nov 14, 52. DEVELOPMENTAL BIOLOGY. *Educ:* Wash State Univ, PhD(zoophysiol), 82. *Prof Exp:* res assoc, Univ Colo Health Sci Ctr, 82-87, ASST PROF, UNIV COLO, DENVER, 87- *Mem:* Am Soc Develop Biol; Am Soc Cell Biol; Int Soc Develop Biologists. *Res:* PI turnover in cell division. *Mailing Add:* Univ Colo Denver Biol Dept Box 171 1200 Lanmer St Denver CO 80204

STITH, JAMES HERMAN, b Brunswick Co, Va, July 17, 41; m 65; c 3. OPTICS, PHYSICS. *Educ:* Va State Univ, BS, 63, MS, 64; Pa State Univ, DEd(physics), 72. *Prof Exp:* Asst instr physics, Va State Univ, 64-65; assoc engr, Radio Corp Am, 67-69; ASSOC PROF PHYSICS, US MIL ACAD, 72- *Concurrent Pos:* Instr, Far East Div, Univ Md, 65-66; exchange officer, Dept Physics, US Air Force Acad, 76-77; proposal reviewer, Dept Educ, Minority Insts Sci Improvement Prog, 81; prin investr, Mat Technol Lab, 85-87; chmn, Minorities in Physics Educ Comt, Am Asn Physics Teachers, 86-88; chmn, Comt Sci for Pub, Am Asn Physics Teachers, 89-90; int adv bd, Second Inter-Am Conf on Physics Educ, 90- *Mem:* Am Asn Physics Teachers (vpres, 90, pres elect, 91, pres, 92); Am Phys Soc; Soc Black Physicists; NY Acad Sci. *Res:* Improvement of the writing of examinations and the understanding of how students assimilate the concepts of physics; methods of decreasing the switching times of liquid crystal scattering cells; x-ray optics. *Mailing Add:* Dept Physics US Mil Acad West Point NY 10996-1790

STITH, JEFFREY LEN, b Seattle, Wash, July 15, 50; m 79. ATMOSPHERIC SCIENCES, CLOUD PHYSICS. *Educ:* Western Wash State Col, BA, 71; Rensselaer Polytech Inst, MS, 74; Univ Wash, PhD(atmospheric sci), 78. *Prof Exp:* Res scientist, Meteorol Res Inc, 78-80; res assoc, 80-84, ASSOC PROF, UNIV NDAK, 84-, CHAIR ATMOSPHERIC SCI DEPT, 90- *Concurrent Pos:* Mem Comt Aviation, Range, Aerospace Meteorol, Comt Planned & Inadvertant Weather Modification. *Mem:* Am Meteorol Soc. *Res:* Aerosol and cloud physics research. *Mailing Add:* Box 8216 Univ Sta Grand Forks ND 58202

STITH, LEE S, b Tulia, Tex, Aug 30, 18; m 47; c 1. PLANT BREEDING, AGRONOMY. *Educ:* NMex State Univ, BS, 40; Univ Tenn, MS, 42; Iowa State Univ, PhD(crop breeding), 55. *Prof Exp:* Asst county supvr, Farm Security Admin, DeBaca County, NMex, 40-41; res asst agr econ, Univ Tenn, 41-42; agronomist, El Paso Valley Substa 17, Tex A&M Univ, 46-47, cotton breeder, 47-55; plant breeder & prof plant breeding, Univ Ariz, 55-83, dir, hybrid cotton res proj, 72-83, dir res instr, Plant Sci Dept, 76-83; consult, Agrigenetics Res Corp, 83-86; pvt consult, 86-88; RETIRED. *Mem:* Am Soc Agron; Crop Sci Soc; Am Genetics Asn; Sigma Xi. *Res:* Crop breeding; plant pathology and physiology; statistics; cotton and grain sorghum; cytoplasmic male sterility to produce hybrid cotton; revolutionary concept in cotton breeding. *Mailing Add:* 4311 E Seventh St Tucson AZ 85711

STITH, REX DAVID, b Hominy, Okla, Dec 11, 42; m 64; c 2. ENDOCRINOLOGY. *Educ:* Okla State Univ, BS, 64, MS, 66; Purdue Univ, PhD(physiol), 71. *Prof Exp:* Instr biol, Southeast Mo State Col, 66-68; vet physiol, Purdue Univ, 68-71; res assoc pharmacol, Univ Mo-Columbia, 71-72; asst prof, 72-75, ASSOC PROF PHYSIOL, HEALTH SCI CTR, UNIV OKLA, 75- *Concurrent Pos:* Vis assoc prof, Dept Physiol, Univ Calif, San Francisco, 79; Guest lectr, Univ St George's Sch Med, Grenada, West Indies. *Mem:* AAAS; Sigma Xi; Soc Exp Biol & Med; Am Physiol Soc; Endocrine Soc. *Res:* Mechanisms of action of glucocorticoids; interactions of steroids in target cells; effects of glucocorticoids on intracellular functions; mechanisms of action of steroid hormones on target tissues, especially brain tissues and biochemical interactions of steroids in these tissues. *Mailing Add:* Dept Phys & Biophys Box 26901 Univ of Okla Health Sci Ctr Oklahoma City OK 73190

STITH, WILLIAM JOSEPH, b Oklahoma City, Okla, Feb 7, 42; m 66; c 3. BIOCHEMISTRY. *Educ:* Phillips Univ, BA, 64; Univ Okla, PhD(biochem), 72. *Prof Exp:* Chief microbiol, US Naval Hosp, Philadelphia, 66-67, chief blood bank & serol, 67-68; asst officer in-chg, Armed Serv Whole Blood Processing Lab, McGuire AFB, NJ, 68-69; fel human biol chem & genetics, Univ Tex Med Br, Galveston, 72-73; scientist prod explor, Fenwal Div, Baxter Labs, Inc, 73-77; vpres affairs, Med Eng Corp, 77-81; mgr, Bioeng Dept, Lord Corp, Erie, Pa, 81-86; TEX BACK INST, 86- *Mem:* Asn Advan Med Instrumentation; Am Soc Qual Control; Am Chem Soc; Sigma Xi. *Res:* Quality control; product reliability; regulatory affairs; product development, and clinical studies. *Mailing Add:* 2809 Boone Court Plano TX 75023

STITT, JAMES HARRY, b Sellersville, Pa, Dec 13, 39; m 64; c 2. GEOLOGY, PALEONTOLOGY. *Educ:* Rice Univ, BA, 61; Univ Tex, Austin, MA, 64, PhD(geol), 68. *Prof Exp:* From asst prof to assoc prof, 68-77, chmn dept, 77-80, PROF GEOL, UNIV MO-COLUMBIA, 77- *Mem:* Paleont Soc; Geol Soc Am. *Res:* Late Cambrian and early Ordovician trilobites; invertebrate paleontology and biostratigraphy; carbonate petrology. *Mailing Add:* Dept Geol Sci Univ Mo Columbia MO 65211

STITT, JOHN THOMAS, b Belfast, Northern Ireland, Nov 7, 42; m 66; c 2. PHYSIOLOGY, NEUROPHYSIOLOGY. *Educ:* Queens Univ Belfast, BSc, 65; Queens Univ, Ont, MSc, 67, PhD(physiol), 69. *Prof Exp:* Can Med Res Coun fel, Med Sch, 69-72, asst fel, 69-73, asst prof environ physiol, 72-76, ASSOC FEL PHYSIOL, JOHN B PIERCE FOUND LAB, YALE UNIV, 73-, ASSOC PROF EPIDEMIOL & PHYSIOL, MED SCH, 76- *Mem:* Am Physiol Soc; Can Physiol Soc; Soc Neurosci. *Res:* Physiology of thermoregulation in mammals; role of the hypothalamus in the homeostasis of body temperature; neurophysiological mechanisms of fever. *Mailing Add:* Epidemiol-Physiol Yale Univ 290 Congress Ave New Haven CT 06519

STITZEL, ROBERT ELI, b New York, NY, Feb 22, 37; m 61; c 1. PHARMACOLOGY. *Educ:* Columbia Univ, BS, 59, MS, 61; Univ Minn, PhD(pharmacol), 64. *Prof Exp:* Res asst pharmacol, Univ Minn, 61-64; asst prof, WVa Univ, 65-66; Swed Med Res Coun fel, 66-67; from asst prof to assoc prof, WVa Univ, 67-73, dir grad studies, Dept Pharmacol, 73-76, asst chmn dept, 76-79, PROF PHARMACOL, WVA UNIV, 73-, ASSOC CHMN DEPT, 79- *Concurrent Pos:* USPHS fel, 64-65; USPHS res career develop award; vis prof, Univ Adelaide, Australia, 73 & Univ Innsbruck, Austria, 77; Danforth Found assoc, 75-; hon res fel anat & embryol, Univ Col London, 77; Fogarty Sr Int fel, NIH, 77; actg chmn dept pharmacol, WVa Univ, 78-87; chmn, Neurol Sci Study Sect, NIH, 83-85; vis res scholar, Commonwealth Sci & Indust Res Orgn, Adelaide, Australia, 87; ed-in-chief, Pharmacol Rev, 89- *Mem:* AAAS; Am Soc Pharmacol & Exp Therapeut; Int Soc Biochem Pharmacol; Am Soc Neurochem. *Res:* Physiological and pharmacological factors affecting catecholamine release; hypertension. *Mailing Add:* Dept Pharmacol & Toxicol WVa Univ Health Sci Ctr Morgantown WV 26506

STITZINGER, ERNEST LESTER, b Chester, Pa, May 9, 40; m 67; c 3. ALGEBRA, GEOMETRY. *Educ:* Temple Univ, BA, 63, MA, 65; Univ Pittsburgh, PhD(math), 69. *Prof Exp:* From asst prof to assoc prof, 69-78, PROF MATH, NC STATE UNIV, 78- *Mem:* Am Math Soc; Math Asn Am. *Res:* Lie algebras, other non-associative algebras and group theory. *Mailing Add:* Dept Math Box 8205 NC State Univ Raleigh NC 27695-8205

STIVALA, SALVATORE SILVIO, b New York, NY, June 23, 23; m 50; c 2. PHYSICAL CHEMISTRY. *Educ:* Columbia Univ, AB, 48; Stevens Inst Technol, MSChE, 52, MS, 58; Univ Pa, PhD(chem), 60. *Hon Degrees:* MEng, Stevens Inst Technol, 64. *Prof Exp:* Res engr, US Testing Co, 49-50; mat engr, Picatinny Arsenal, 50-51, 54-57; NSF sci fac fel, Univ Pa, 57-59; from asst prof to assoc prof phys & polymer chem, 59-64, instr chem eng, 52-57, PROF PHYS & POLYMER CHEM, STEVENS INST TECHNOL, 64-, RENE WASSERMAN PROF CHEM & CHEM ENG, 79- *Concurrent Pos:* Consult, 52-57, 59-, indust, 60-; vis scientist, Inst Phys Chem, Graz Univ, 66 & dept ultra-struct biochem, Cornell Med Col, 69. *Honors & Awards:* Ottens Res Award, 68; Med Soc Sendai (Japan) Award, 69; Honor Scroll, Am Inst Chemists, 77. *Mem:* Am Chem Soc; Sigma Xi; Am Soc Testing & Mat; Soc Plastics Eng. *Res:* Physical chemistry of high polymers; solution properties; kinetics of polymer degradation; physico-chemical aspects of biopolymers. *Mailing Add:* Dept of Chem & Chem Eng Stevens Inst of Technol Castle Point Hoboken NJ 07030

STIVEN, ALAN ERNEST, b St Stephen, NB, Nov 12, 35; m 72; c 3. ECOLOGY, POPULATION BIOLOGY. *Educ:* Univ NB, BSc, 57; Univ BC, MA, 59; Cornell Univ, PhD(ecol), 62. *Prof Exp:* From asst prof to assoc prof zool, 62-71, chmn dept, 67-72, chmn ecol curric, 71-86, PROF BIOL, UNIV NC, CHAPEL HILL, 71- *Concurrent Pos:* NSF res grants, 63-82; USPHS ecol training grant, 66-69; res fel, Univ BC, 70; panel mem, Nat Acad Sci Envrion Educ, Phillipines, 73-74; res grant, Sea Grant, Nat Oceanog & Atmos Asn, 74-76; US Forest Serv grant, 82-85; study comt, Ecol Soc Am, 82-; chmn biol panel, Nat Sci Found/Nat Res Coun, predoctoral fel prog, 85-89; vchmn, Div Natural Sci, Univ NC, 87-90; panel mem, Postdoctoral Res Assoc Prog, Nat Res Coun, 90- *Mem:* Ecol Soc Am; Am Soc Naturalists; Soc Pop Ecol; Soc Study Evolution; Soc Conserv Biol. *Res:* Population biology; ecological genetics; ecology and genetics of populations exposed to disturbance; stream and salt marsh population and community ecology. *Mailing Add:* Dept Biol CB#3280 Coker Hall Univ of NC Chapel Hill NC 27599-3280

STIVENDER, DONALD LEWIS, b Chicago, Ill, May 8, 32; m 56; c 3. DYNAMIC CONTROL, SYSTEMS ENGINEERING. *Educ:* US Coast Guard Acad, BS, 54; Univ Mich, MS, 59. *Prof Exp:* MEM STAFF RES & DEVELOP, RES LABS, GEN MOTORS CORPS, 59-, SR RES ENGR, 68- *Concurrent Pos:* Consult, pub domain disciplines; consult, Nat Acad Sci, 78- & mem, Naval Studies Bd, 90-; owner, Stivender Eng Assocs, 80- *Honors & Awards:* Arch T Colwell Awards, Soc Automotive Engrs, 68, 69 & 79. *Mem:* Fel Soc Automotive Engrs; Am Soc Mech Engrs; Combustion Inst; Sigma Xi; Asn Advan Invention & Innovation. *Res:* Gas turbine, diesel spark ignition and alternative engine combustion, emission, construction and control aspects; internal combustion engines; engine and vehicle dynamometer control and transient optimization. *Mailing Add:* 1730 Hamilton Dr Bloomfield Hills MI 48302-0221

STIVER, JAMES FREDERICK, b Elkhart, Ind, Jan 27, 43; m 65; c 4. MEDICINAL CHEMISTRY, BIONUCLEONICS. *Educ:* Purdue Univ, BS, 66, MS, 68, PhD(med chem, bionucleonics), 70. *Prof Exp:* From asst prof to assoc prof pharmaceut chem & bionucleonics, Col Pharm, NDak State Univ, 69-76, radiol safety officer, 70-76; radiol safety officer, KMS Fusion, Inc, 76-80; adminr environ regulatory affairs, 81-88, PATENT LIAISON SCIENTIST, UPJOHN CO, 88- *Concurrent Pos:* Consult pharmacist. *Mem:* AAAS; Am Chem Soc; Am Pharmaceut Asn; Health Physics Soc; Am Biol Safety Asn; Sigma Xi. *Res:* Radioisotope labeling synthesis of organic compounds and drugs; radioisotope tracer techniques and tracer methodology development; metabolism of drug and toxic chemicals; radioactive nuclide levels in the environment; applied health physics and patent information research. *Mailing Add:* 505 Skyview Dr Middlebury IN 46540-9427

STIVERS, RUSSELL KENNEDY, b Marshall Co, Ill, May 9, 17; m 47; c 3. SOIL FERTILITY. *Educ:* Univ Ill, BS, 39; Purdue Univ, MS, 48, PhD, 50. *Prof Exp:* Assoc agronomist, Va Polytech Inst, 50-55; assoc prof, 55-82, EMER PROF AGRON, PURDUE UNIV, LAFAYETTE, 82- *Mem:* Am Soc Agron; Soil Sci Soc Am. *Res:* Soils and soil science. *Mailing Add:* 451 Littleton West Lafayette IN 47906

STIX, THOMAS HOWARD, b St Louis, Mo, July 12, 24; m 50; c 2. PLASMA PHYSICS. *Educ:* Calif Inst Technol, BS, 48; Princeton Univ, PhD(physics), 53. *Prof Exp:* Res asst, 53-54, res assoc, 54-56, assoc head exp div, 56-61, co-head exp div, 61-78, ASSOC DIR ACAD AFFAIRS, PLASMA PHYSICS LAB, PRINCETON UNIV, 78-, PROF ASTROPHYS SCI, 62-, ASSOC CHMN DEPT ASTROPHYS SCI, 81- *Concurrent Pos:* NSF sr fel, 60-61; chmn, Div Plasma Physics, Am Phys Soc, 62-63, Comm Int Freedom Scientists, 85; mem adv comt, thermonuclear div, Oak Ridge Nat Lab, 66-68; John Simon Guggenheim Mem Found fel, 69-70; assoc ed, Phys Rev Letters, 74-77. *Honors & Awards:* James Clerk Maxwell Prize, Am Phys Soc, 80. *Mem:* Fel Am Phys Soc. *Res:* Controlled fusion; waves and instabilities; plasma heating and confinement. *Mailing Add:* Plasma Physics Lab Princeton Univ Princeton NJ 08543-0451

ST-JEAN, GUY, b Montreal, Can, Oct 8, 41; Can. POWER ELECTRICAL CIRCUIT DESIGN, DIAGNOSTIC POWER CIRCUIT BREAKERS. *Educ:* Concordia Univ, BS, 64; Univ Quebec, MS, 77. *Prof Exp:* Adj prof, Polytech Col, Montreal, 77-83; MGR, ELEC POWER APPARATUS, RES INST HYDRO-QUEBEC, 87- *Concurrent Pos:* Chmn, Can IEC Switchgear Tech Coun 17, 85-88, Can IEC HV switchgear sub-committe 17A, 85-88, Can IECHV enclosed switchgear 17C, 85-88, Can Elec Working Group Metal Oxide Arresters, 82-88. *Mem:* Fel, Inst Elec & Electronics Engrs; IEC; Conf Int Grand Res Elec. *Res:* Design and construction of high power test facilities specically related to circuit breakers and surge arresters; development of post-arc diagnostic technology for HV power circuit breakers; development of new calculation methods for the design of direct and synthetic testing circuits; development of new circuit breakers and surge arresters. *Mailing Add:* 560 St-Laurent Apt 122 Longueuil PQ J4H 3X3 Can

STJERNHOLM, RUNE LEONARD, b Stockholm, Sweden, Apr 25, 24; nat US; m 53; c 2. BIOCHEMISTRY. *Educ:* Stockholm Tech Inst, BS, 44; Western Reserve Univ, PhD(biochem), 58. *Prof Exp:* From asst prof to prof biochem, Case Western Reserve Univ, 58-71; PROF BIOCHEM & CHMN DEPT, MED SCH, TULANE UNIV, 71- *Mem:* Am Chem Soc; Am Soc Microbiol; Am Soc Biol Chemists; Swed Chem Soc. *Res:* Chemotherapeutics; carbohydrate metabolism in leukocytes. *Mailing Add:* Dept of Biochem Tulane Univ Med Sch 1430 Tulane Ave New Orleans LA 70112

STOB, MARTIN, b Chicago, Ill, Feb 20, 26. ANIMAL SCIENCE. *Educ:* Purdue Univ, PhD(physiol), 53. *Prof Exp:* Asst, 49-53, asst prof animal husb, 53-58, assoc prof, 58-63, PROF ANIMAL SCI, PURDUE UNIV, WEST LAFAYETTE, 63- *Mem:* AAAS; Am Soc Animal Sci; Endocrine Soc; Soc Study Fertil; Soc Study Reproduction. *Res:* Hormonal regulation of growth; occurrence of compounds with estrogenic activity in plant material; microbiological synthesis and metabolism of estrogens; reproductive physiology. *Mailing Add:* Dept Animal Sci Purdue Univ West Lafayette IN 47907-7899

STOB, MICHAEL JAY, b Chicago, Ill, Aug 2, 52; m 74; c 1. RECURSION THEORY. *Educ:* Calvin Col, BS, 74; Univ Chicago, SM, 75, PhD(math), 79. *Prof Exp:* C L E Moore instr math, Mass Inst Technol, 79-81; asst prof, 81-83, assoc prof, 83-87, PROF MATH, CALVIN COL, 87- *Concurrent Pos:* Prin investr, NSF grant; vis assoc prof math, Univ Wis, 83-84. *Mem:* Am Math Soc; Asn Symbolic Logic; Math Asn Am. *Res:* Mathematical logic especially recursion theory; recursively enumerable sets and degrees. *Mailing Add:* Calvin Col Grand Rapids MI 49506-5540

STOBAUGH, ROBERT EARL, b Humboldt, Tenn, June 24, 27; m 56. INFORMATION SCIENCE. *Educ:* Southwestern at Memphis, BS, 47; Univ Tenn, MS, 49, PhD(chem), 52. *Prof Exp:* Res assoc, Ohio State Univ, 52-54; from asst ed to sr assoc ed, 54-61, from asst dept head to dept head, 61-65, tech adv registry div, 65-67, MGR RES CHEM ABSTR, 67- *Mem:* Am Soc Info Sci; Am Chem Soc. *Res:* Steroids; chemical literature; chemical information storage and retrieval; chemical structural data; chemical information science. *Mailing Add:* Chem Abstr Serv PO Box 3012 Columbus OH 43210

STOBBE, ELMER HENRY, b Matsqui, BC, Jan 26, 36; m 62; c 2. AGRONOMY, WEED SCIENCE. *Educ:* Univ BC, BSA, 61, MSA, 65; Ore State Univ, PhD(crop sci), 69. *Prof Exp:* From asst prof to assoc prof weed sci, 68-78, PROF AGRON, UNIV MAN, 78- *Concurrent Pos:* Agron res, Weed Sci, NJORD, Kenya, 82-84. *Mem:* Weed Sci Soc Am; Agr Inst Can; Am Soc Agron; Sigma Xi. *Res:* Weed control under reduced tillage systems; agronomy, especially integrated cereal management, use of plant growth regulators in cereals, winter wheat production, zero tillage research and effect of cultivation on crop yield. *Mailing Add:* Dept of Plant Sci Univ of Man Winnipeg MB R3T 2N2 Can

STOBER, HENRY CARL, b Brooklyn, NY, June 20, 35; m 61; c 2. ANALYTICAL CHEMISTRY, PHARMACEUTICAL CHEMISTRY. *Educ:* City Col New York, BS, 58; Seton Hall Univ, MS, 69, PhD(chem), 71. *Prof Exp:* Res asst biol chem, Letterman Army Hosp, US Army, 58-60; chemist, Ciba Pharmaceut Co, 60-66, supvr anal chem, 66-70; teaching asst chem, Seton Hall Univ, 70-71; sr scientist, 71-74, SR STAFF SCIENTIST, CIBA-GEIGY CORP, SUFFERN, NY, 74-, SR RES FEL, 83- *Mem:* Am Chem Soc. *Res:* Analysis and solid state characterization of pharmaceuticals and related chemicals. *Mailing Add:* 124 Madison Ave Madison NJ 07940

STOBER, QUENTIN JEROME, b Billings, Mont, Mar 25, 38; m 65; c 2. ECOTOXICOLOGY, FISHERIES MANAGEMENT. *Educ:* Mont State Univ, BS, 60, MS, 62, PhD(zool), 68. *Prof Exp:* Aquatic biologist, Southeast Water Lab, Div Water Supply & Pollution Control, USPHS, Ga, 62-65; res asst prof estuarine ecol, Fisheries Res Inst, 69-72, res assoc prof estuarine, stream ecol & marine toxicol, 72-77, res prof & prog mgr marine baseline studies, stream, reservoir ecol & marine toxicol, Univ Wash, 77-86; US ENVIRON PROTECTION AGENCY, REGION IV FISHERIES EXPERT, ECOL SUPPORT BR ENVIRON SERVS DIV, ATHENS, GA, 86- *Concurrent Pos:* Admin judge, Atomic Safety & Licensing Bd Panel, US Nuclear Regulatory Comn, 74-86; environ consult, 68-86; Joint Sci Comt, Wash Water Res Ctr, 75-86. *Honors & Awards:* W F Thompson Award, Am Inst Fisheries Res Biol, 71; Bronze Medal, US Environ Protection Agency, 89. *Mem:* AAAS; Am Fisheries Soc; Am Soc Limnol & Oceanog; Am Inst Fisheries Res Biol; Sigma Xi. *Res:* Fisheries problems related to hydro and thermal nuclear energy production; estuarine ecology of effects of municipal and industrial wastes; fish toxicology and behavior; stream ecology, instream flow needs and reservoir ecology; bioaccumulation of chemical contaminants in fish and risk assessment. *Mailing Add:* US Environ Protection Agency Col Sta Rd Athens GA 30613

STOBO, JOHN DAVID, b Somerville, Mass, Sept 1, 41; m 64; c 3. IMMUNOLOGY. *Educ:* Dartmouth Col, AB, 63; State Univ NY Buffalo, MD, 68. *Prof Exp:* Res assoc immunol, NIH, 70-72; chief resident med, Johns Hopkins Hosp, 72-73; asst prof immunol, Mayo Med Sch & Found, 73-76; assoc prof med & head sect rheumatology/clin immunol, Moffitt Hosp, Univ Calif, San Francisco, 76-85; PROF MED, JOHNS HOPKINS UNIV, 85-, DIR & PHYSICIAN-IN-CHIEF, 85- *Concurrent Pos:* Sr investr, Am Arthritis Asn, 73; rep, Nat Heart Asn, 75-77. *Mem:* Inst Med-Nat Acad Sci; Am Arthritis Asn; Am Asn Immunologists. *Res:* Cellular immunology; forces involved in the regulation of cell mediated and humoral immune responses. *Mailing Add:* 1800 E Monument St Baltimore MD 21205

STOBO, WAYNE THOMAS, b Sudbury, Ont, June 16, 44; Can citizen. FISHERIES BIOLOGY, SEAL BIOLOGY. *Educ:* Laurentian Univ, BSc, 65; Univ Ottawa, MSc, 71; Dalhousie Univ, PhD(ecol), 73. *Prof Exp:* Biologist fisheries, Dept Environ, 72-73, res scientist, 73-76, sect head pop dynamics, 76-77, pop dynamics & biostatist, 77-78, coordr, Marine Fish Div, 79-83, RES SCIENTIST, DEPT FISHERIES & OCEANS, FED GOVT CAN, 84- *Concurrent Pos:* Chmn pelagic subcomt, Can Atlantic Fisheries Sci Adv Comt, 79-81. *Res:* Optimizing the biological productivity of commercially exploited finfish stocks; populations dynamics and migration of finfish and marine mammals. *Mailing Add:* Marine Fish Div Bedford Inst Oceanog PO Box 1006 Dartmouth NS B2Y 4A2 Can

STOCK, CHARLES CHESTER, b Terre Haute, Ind, May 19, 10; m 36. CHEMOTHERAPY. *Educ:* Rose-Hulman Inst Technol, BS, 32; Johns Hopkins Univ, PhD(physiol chem), 37; NY Univ, MS, 41. *Hon Degrees:* ScD, Rose-Hulman Inst, 54. *Prof Exp:* Instr bact, Col Med, NY Univ, 37-41; vol worker, Rockefeller Hosp Med Res, 41-42; tech aide, Comt Treatment Gas Casualties, Div Med Sci, Nat Res Coun, 42-45, exec secy, Insect Control Comt, 45-46, chmn chem coding panel chem-biol, Coord Ctr, 46-52; assoc, 46-50, chief, Div Exp Chemother, 47-72, mem, Sloan-Kettering Inst Cancer Res, 50-80, assoc dir inst, 57-60, dir, 59-80, sci dir, 60-61, vpres, 61-72, vpres inst affairs, Cancer Ctr, 74-80, vpres & assoc dir admin & acad affairs, 76-80, EMER MEM, WALKER LAB, RYE SLOAN-KETTERING INST CANCER RES, 80-; EMER PROF BIOCHEM, SLOAN-KETTERING DIV, MED COL, CORNELL UNIV, 76- *Concurrent Pos:* Prof biochem, Sloan-Kettering Div, Med Col, Cornell Univ, 51-75; mem comt tumor nomenclature & statist, Int Cancer Res Comn, 52-54; chmn screening panel, Cancer Chemother Nat Serv Ctr, NIH, 55-58 & drug eval panel, 58-; mem sci adv bd, Roswell Park Mem Inst, 57-66; mem chemother rev bd, Nat Adv Cancer Coun, 58-59; mem US nat comt, Int Union Against Cancer, 67-80, chmn, 75-80; mem bd dirs, Am Cancer Soc, 72-, hon mem, 85- *Honors & Awards:* Alfred P Sloan Award, 65; C Chester Stock Award Cancer Res, 80. *Mem:* Emer mem Am Chem Soc; emer mem Am Soc Biol Chemists; Soc Exp Biol & Med; Am Asn Cancer Res; hon mem Japanese Cancer Asn; hon fel Hungarian Cancer Soc; Europ Inst Ecol & Cancer (hon pres, 77-); foreign corresp mem, Cancer Soc Italy. *Res:* Enzymes; hypertension; experimental chemotherapy of cancer. *Mailing Add:* 605 E 82nd St New York NY 10028

STOCK, DAVID ALLEN, b Elyria, Ohio, Feb 8, 41; m 64; c 2. MICROBIOLOGY, GENETICS. *Educ:* Mich State Univ, BS, 63; NC State Univ, MS, 66, PhD(genetics), 68. *Prof Exp:* Instr microbiol, Sch Med, Univ Miss, 67-68; USDA fel, Baylor Col Med, 68-69, NIH fel, 69-70; asst prof biol, 70-76, ASSOC PROF BIOL, STETSON UNIV, 76- *Mem:* Am Soc Microbiol. *Res:* Physiology and pathogenesis of Candida albicans; radiobiology of foodstuffs; nucleic acid metabolism of fungi; breeding behavior of limpkins. *Mailing Add:* Dept Biol Stetson Univ De Land FL 32720

STOCK, DAVID EARL, b Baltimore, Md, Feb 2, 39; m 62; c 2. FLUID MECHANICS, MULTIPHASE FLOW. *Educ:* Pa State Univ, BS, 61; Univ Conn, MS, 65; Ore State Univ, PhD(mech eng), 72. *Prof Exp:* Test engr, Pratt & Whitney Aircraft, 61-65; Peace Corps volunteer, Ghana, WAfrica, 66-68; teaching asst, Ore State Univ, 68-72; from asst prof to assoc prof, 72-83, PROF MECH ENG, WASH STATE UNIV, 83- *Concurrent Pos:* Vis fel, Cornell Univ, 81-82, Univ Canterbury, NZ. *Mem:* Am Soc Mech Engrs; Am Phys Soc; Am Soc Eng Educ. *Res:* Experimental fluid mechanics which includes the use of laser Dopper and thermal anemometry as well as computer aided data acquisition and processing; gas particle flow with a main application to electrostatic precipitators. *Mailing Add:* Dept Mech Eng Wash State Univ Pullman WA 99164-2920

STOCK, JOHN JOSEPH, microbiology, medical mycology, for more information see previous edition

STOCK, JOHN THOMAS, b Margate, Eng, Jan 26, 11; nat US; m 45; c 1. ANALYTICAL CHEMISTRY. *Educ:* Univ London, BSc, 39 & 41, MSc, 45, PhD(chem), 49, DSc, 65. *Prof Exp:* Sci off chem, Ministry Supply, Gt Brit, 40-44; actg chief chemist, Fuller's Ltd, 44-46; lectr chem, Norwood Tech Col, 46-51, head dept, 51-56; from assoc prof to prof, 56-79, EMER PROF CHEM, UNIV CONN, 79- *Concurrent Pos:* London County Coun Blair fel, Univ Minn, 53-54; consult, 50- *Mem:* Fel Am Chem Soc; fel Royal Soc Chem; Soc Chem Indust; Royal Inst Gt Brit; Sci Instrument Soc. *Res:* Design of automated and general scientific apparatus; history of scientific instruments. *Mailing Add:* Dept Chem Univ Conn Storrs CT 06269-3060

STOCK, LEON M, b Detroit, Mich, Oct 15, 30; m 61; c 2. ORGANIC CHEMISTRY. *Educ:* Univ Mich, BS, 52; Purdue Univ, PhD, 59. *Prof Exp:* From instr to assoc prof, 58-70, PROF CHEM, UNIV CHICAGO, 71-; DIR CHEM DIV, ARGONNE NAT LAB, 88- *Concurrent Pos:* Consult, Phillips Petrol Co, 64- *Honors & Awards:* Storch Award, Am Chem Soc, 87. *Mem:* Am Chem Soc; Royal Soc Chem. *Res:* Electrophilic aromatic substitution reactions; influences of structure and solvents on reactivity; models for evaluation of inductive influences of substituents; electron paramagnetic resonance spectra of organic radicals; the structure of coal; the chemistry of coal, its structure, liquefaction, gasification and pyrolysis reactions. *Mailing Add:* Dept Chem Univ Chicago Chicago IL 60637

STOCK, MOLLY WILFORD, b Glen Ridge, NJ, Aug 17, 42; m 62; c 2. ENTOMOLOGY, POPULATION GENETICS. *Educ:* Univ Conn, BA, 64, MS, 65; Ore State Univ, PhD(entom), 72. *Prof Exp:* Res asst entom, Ore State Univ, 68-69; res assoc insect biochem, Wash State Univ, 72-73, res collabr entom, 73-75; proj leader, Wash State Univ, 75-77; asst prof entom, 76-78, from asst prof to assoc prof, 80-84, PROF FOREST RESOURCES & COMPUTER SCI, UNIV IDAHO, 84- *Mem:* Entom Soc Am; Soc Am Foresters; Am Asn Artificial Intel. *Res:* Biosystematics; population dynamics of forest insects; expert systems for natural resource management. *Mailing Add:* Dept Forest Resources Univ Idaho Moscow ID 83843

STOCKBAUER, ROGER LEWIS, b Victoria, Tex, Feb 3, 44; m 72; c 3. SURFACE SCIENCE. *Educ:* Rice Univ, BA, 66; Univ Chicago, MS, 68, PhD(physics), 73. *Prof Exp:* From res asst to res assoc physics, Univ Chicago, 66-73; res physicist, Nat Bur Standards, 73-89; PROF PHYSICS, LA STATE UNIV, 89- *Concurrent Pos:* Teaching asst physics, Univ Chicago, 69-70; Nat Res Coun-Nat Acad Sci res assoc, Nat Bur Standards, 73-75. *Honors & Awards:* Silver Medal, US Dept Com, 83. *Mem:* Am Phys Soc; Sigma Xi; Am Vacuum Soc; Mat Res Soc; AAAS. *Res:* Photon stimulated desorption of ions; ultraviolet photoemission spectroscopy; angular distribution of photoelectrons. *Mailing Add:* Dept Physics La State Univ Baton Rouge LA 70803-4001

STOCKBRIDGE, CHRISTOPHER, metallurgy, for more information see previous edition

STOCKBRIDGE, ROBERT R, b Worcester, Mass, Aug 21, 10; m 37; c 1. ANIMAL HUSBANDRY, POULTRY HUSBANDRY. *Educ:* Univ Mass, BVA, 34; Hofstra Col, MS, 46. *Prof Exp:* From instr to assoc prof, State Univ NY Agr & Tech Col Farmingdale, 38-60, prof poultry sci & chmn, Agr Dept, 60-80; RETIRED. *Mem:* Poultry Sci Asn; World Poultry Sci Asn. *Mailing Add:* Five Stephen Dr Farmingdale NY 11735

STOCKBURGER, GEORGE JOSEPH, b Philadelphia, Pa, May 23, 27; m 61; c 2. INDUSTRIAL ORGANIC CHEMISTRY. *Educ:* St Joseph's Col, Pa, BS, 50; Univ Pa, MS, 52, PhD, 55. *Prof Exp:* Sr res chemist, 55-79, supvr gen analysis, Analysis & Phys Chem Sect, ICI Americas Inc, 79-86; RETIRED. *Mem:* Am Chem Soc. *Res:* Reaction kinetics and mechanism; catalysis. *Mailing Add:* 2211 Pennington Dr Brandywood Wilmington DE 19810

STOCKDALE, FRANK EDWARD, b Long Beach, Calif, Mar 15, 36; c 3. DEVELOPMENTAL BIOLOGY, ONCOLOGY. *Educ:* Yale Univ, AB, 58; Univ Pa, MD & PhD(develop biol), 63. *Prof Exp:* Intern internal med, Univ Hosps, Western Reserve Univ, 63-64; staff assoc, Nat Inst Arthritis & Metab Dis, 64-66; sr resident, Univ Hosp, 66-67, from instr to asst prof med, 68-74, asst prof biol, 70-74, assoc prof, 74-81, PROF MED & BIOL SCI, SCH MED, STANFORD UNIV, 81- *Mem:* Am Soc Clin Invest; AAAS; Soc Develop Biol; Am Soc Clin Oncol; Am Soc Cell Biol; Asn Am Physicians. *Res:* Cellular and molecular mechanisms for control of cell differentiation and growth during embryogenesis; medical oncology; breast cancer. *Mailing Add:* Dept Med Rm M211 Stanford Univ Sch Med Stanford CA 94305-5306

STOCKDALE, HAROLD JAMES, b Aplington, Iowa, Dec 3, 31; m 51; c 2. ECONOMIC ENTOMOLOGY. *Educ:* Iowa State Univ, BS, 58, MS, 59, PhD(entom), 64. *Prof Exp:* Exten entomologist & prof entom, 61-82, CHMN DEPT, IOWA STATE UNIV, 82- *Mem:* Entom Soc Am. *Res:* Field crop insect management; household and structural insect control. *Mailing Add:* 407 Sci II Iowa State Univ Ames IA 50011-3222

STOCKDALE, JOHN ALEXANDER DOUGLAS, b Ipswich, Australia, Mar 15, 36; m 57; c 3. PHYSICS. *Educ:* Univ Sydney, BSc, 57, MSc, 60; PhD(physics), Univ Tenn, 69. *Prof Exp:* Res scientist, Australian AEC, 58-66; physicist, Health & Safety Res Div, Oak Ridge Nat Lab, 66-88; cofounder & pres, Com Stock, Inc, 79-90; PHYSICIST, LAWRENCE LIVERMORE NAT LAB, 88- *Concurrent Pos:* John Simon Guggenheim fel, 70; vis prof, NY Univ, 75-76, Univ Crete, 85, 87. *Mem:* Am Phys Soc. *Res:* Atomic and molecular physics; laser physics. *Mailing Add:* Lawrence Livermore Lab PO Box 808L 464 Livermore CA 94550

STOCKDALE, WILLIAM K, b Rock Island, Ill, Oct 11, 28; m 51; c 7. CIVIL ENGINEERING, STRUCTURAL DYNAMICS. *Educ:* US Mil Acad, BS, 51; Univ Ill, Urbana, MS, 58, PhD(civil eng), 59. *Prof Exp:* From instr to prof civil eng, US Mil Acad, 67-78; mgr eng serv, WASH PUB WATER SUPPLY SYST, 78-81, chief, civil & struct eng, 81-87; CHIEF DESIGN ENGR, KAISER ENGRS, INC. *Mem:* Soc Am Mil Engrs; NY Acad Sci; Am Soc Civil Engrs; Sigma Xi; Nat Soc Prof Engrs; Am Concrete Inst. *Res:* Structural dynamics; structural analysis and design. *Mailing Add:* 534 Mt Davidson Ct Clayton CA 94517-1602

STOCKEL, IVAR H(OWARD), b Minneapolis, Minn, Apr 30, 27; m 51; c 3. FLUID & APPLIED MECHANICS. *Educ:* Mass Inst Technol, BS & MS, 50, ScD, 59. *Prof Exp:* From instr to asst prof mech eng, US Naval Postgrad Sch, 50-54; res engr eng physics, Roy K Ferguson Tech Ctr, St Regis Paper Co, 56-59, res group leader, 59-61, assoc mgr res, 61-64, mgr res, 64-67, mgr develop, 67-68, corp dir res & develop, 68-81; prof & chmn, chem eng dept, Univ Maine, 81-90; RETIRED. *Concurrent Pos:* Mem, NSF Indust panel sci & technol. *Mem:* Emer mem Indust Res Inst Inc; Tech Asn Pulp & Paper Indust. *Res:* Theoretical and experimental flow behavior of mixtures of divided solids and fluids; liquid droplet formation; pulp and paper science and technology. *Mailing Add:* 1031 W Riverside Way San Jose CA 95129

STOCKELL-HARTREE, ANNE, b Nashville, Tenn, Jan 11, 26; m 59; c 2. BIOCHEMISTRY. *Educ:* Vanderbilt Univ, BA, 46, MS, 49; Univ Utah, PhD(biochem), 56. *Hon Degrees:* MA, Cambridge Univ, 62. *Prof Exp:* Asst biochem, Vanderbilt Univ, 46-51; asst, Univ Utah, 51-54; fel USPHS Johnson Res Found, Univ Pa, 56-58; mem, Med Res Coun Unit Molecular Biol, Cavendish Lab, Cambridge Univ, 58-59, fel, Jane Coffin Childs Fund, 59-60, res worker, dept biochem, 60-80; res worker, Inst Animal Physiol, Agr Res Coun, Cambridge, Eng, 80-90; RETIRED. *Concurrent Pos:* Res fel, Girton Col, 62-65; fel, Lucy Cavendish Col, 69-; external mem sci staff, Med Res Coun, 70- *Honors & Awards:* Silver Plate, Soc Endocrinol Eng, 90. *Mem:* Brit Biochem Soc; sr mem Brit Soc Endocrinol; Am Soc Biol Chemists; Endocrine Soc. *Res:* Amino acid analysis; enzyme kinetics; protein structure and function; pituitary protein hormones. *Mailing Add:* AFRC Inst Animal Physiol Babraham Cambridge CB2 4AT England

STOCKER, DONALD V(ERNON), b Detroit, Mich, Jan 5, 27; m 50; c 2. ELECTRICAL ENGINEERING. *Educ:* Wayne State Univ, BSEE, 49; Univ Mich, MSE, 50. *Prof Exp:* Jr engr, Magnetron Develop Lab, Raytheon Mfg Co, Mass, 50-51; from instr to asst prof, 51-69, appln engr appl sci & technol ctr, 64-68, assoc prof elec engr, 60-75, mgr admin serv, col eng, 69-75, dir, div eng technol, 81-87, ASSOC PROF ENG TECHNOL, WAYNE STATE UNIV, 75- *Concurrent Pos:* Consult, Detroit Edison Co, 76-80. *Mem:* Inst Elec & Electronics Engrs; Am Soc Eng Educ; Sigma Xi. *Res:* Load management. *Mailing Add:* 26171 Jan Ave Redford MI 48239

STOCKER, FRED BUTLER, b Kenyon, Minn, Jan 31, 31; m 53; c 2. ORGANIC CHEMISTRY. *Educ:* Hamline Univ, BS, 53; Univ Minn, MS, 55; Univ Colo, PhD(org chem), 58. *Prof Exp:* Assoc prof, 58-69, chmn dept chem, 70-80, PROF CHEM, MACALESTER COL, 69- *Concurrent Pos:* Consult, 59- *Mem:* Am Chem Soc. *Res:* Imidazole derivatives. *Mailing Add:* Dept of Chem Macalester Col St Paul MN 55105

STOCKER, JACK H(UBERT), b Detroit, Mich, May 3, 24; m 64; c 2. ORGANIC CHEMISTRY. *Educ:* Olivet Col, BS, 44; Ind Univ, MA, 47; Tulane Univ, PhD(org chem), 55. *Prof Exp:* Control chemist, R P Scherer Corp, Mich, 48-50; control chemist, Atlas Pharmaceut Co, 50; Fulbright traveling fel, Heidelburg Univ, 55-56; assoc prof chem, Univ Southern Miss, 56-58; assoc prof, 58-71, admin asst to dean col sci, 65-67, PROF CHEM, UNIV NEW ORLEANS, 71- *Concurrent Pos:* Res partic, Oak Ridge Inst Nuclear Studies, 59; consult, Food & Drug Admin 71-82; res assoc, Gulf South Res Inst; vis prof chem, Univ Lund, Sweden, 74-75; mem, Comt Meetings & Exposition, Am Chem Soc, 72-74, 76-81, chmn, 80-81, Sci Comn, 80-81, Coun Policy Comt, 82-87, chmn, Div HIst Chem, 89-91, Comt Nomenclature, 76- *Mem:* Am Chem Soc. *Res:* Acetals and ketals; organometallics; stereoselective reactions; c14 techniques; organic photochemistry and electrochemistry. *Mailing Add:* Dept of Chem Univ New Orleans Lakefront New Orleans LA 70148

STOCKER, RICHARD LOUIS, b Honolulu, Hawaii, Apr 22, 41. GEOPHYSICS. *Educ:* Lehigh Univ, BA, 64; Yale Univ, MS, 66, PhD(geophys), 73. *Prof Exp:* Asst prof geol, Lehigh Univ, 73-75; ASST PROF GEOL, ARIZ STATE UNIV, 75- *Mem:* Sigma Xi; Am Geophys Union. *Res:* Transport properties of minerals and rocks, particularly rheology and atomistic diffusion; point defect chemistry of minerals. *Mailing Add:* 32 Virginia Ct Walnut Creek CA 94596

STOCKERT, ELISABETH, b Vienna, Austria, Sept 29, 30. IMMUNOGENETICS. *Educ:* Univ Vienna, BS, 59; Univ Paris, Dr(immunol), 74. *Prof Exp:* Res assoc cancer, 73-76, ASSOC MEM, SLOAN-KETTERING CANCER CTR, 76- *Res:* Experimental tumor immunobiology, serology and genetics; cell surface antigens of normal and malignant cells. *Mailing Add:* 435 E 77th St New York NY 10021

STOCKHAM, THOMAS GREENWAY, JR, b Passaic, NJ, Dec 22, 33; m 63; c 4. COMPUTER SCIENCE. *Educ:* Mass Inst Technol, SB, 55, SM, 56, ScD(elec eng), 59. *Prof Exp:* Teaching asst elec eng, Mass Inst Technol, 55-57, from instr to asst prof, 57-66, staff mem comput res, Lincoln Lab, 66-68; assoc prof comput sci, 68-70, PROF COMPUT SCI, UNIV UTAH, 70-, PROF ELEC ENG, 76- *Concurrent Pos:* Vis asst prof, Univ NMex, 62; consult, Data Div, Comput Group, Lincoln Lab, Mass Inst Technol, 64-66. *Honors & Awards:* Audio & Electroacoust Sr Award, Inst Elec & Electronics Engrs, 68. *Mem:* Inst Elec & Electronics Engrs; Asn Comput Mach. *Res:* Digital signal processing of images and sound by non-linear methods; electrical communications; electrical circuit and systems theory; computer graphics; sensory information processing. *Mailing Add:* Dept Elect Eng Univ Utah 3280 Meb Salt Lake City UT 84112

STOCKHAMMER, KARL ADOLF, b Ried, Austria, July 19, 26; m 56; c 3. ZOOLOGY, ENTOMOLOGY. *Educ:* Graz Univ, PhD(zool), 51. *Prof Exp:* Res assoc zool, Univ Munich, 51-58; instr, Univ G-ttingen, 58-59; asst prof, 59-67, ASSOC PROF CELL BIOL & PHYSIOL, UNIV KANS, 67- *Mem:* Entom Soc Am. *Res:* Detection of e-vector of polarized light in insects; behavioral and physiological aspects of nesting in native bees. *Mailing Add:* 2434 Princeton Blvd Lawrence KS 66044

STOCKING, CLIFFORD RALPH, b Riverside, Calif, June 22, 13; m 37; c 2. PLANT PHYSIOLOGY. *Educ:* Univ Calif, BS, 37, MS, 39, PhD(plant physiol), 42. *Prof Exp:* Asst plant physiol, Univ Calif, 38-39, assoc bot, 39-42; food chemist, Puccinelli Packing Co, 42-45; assoc, Exp Sta, Univ Calif, Davis, 45-46, asst prof bot, Univ & asst botanist, Exp Sta, 46-52, assoc prof & assoc botanist, 52-58, actg chmn, Dept Bot, 66-67, chmn dept, 68-74, prof & botanist, Exp Sta, 58-81, EMER PROF, UNIV CALIF, DAVIS, 81- *Concurrent Pos:* Merck sr fel biochem, Univ Wis, 55-56; NSF fel, Imp Col, Univ London, 63-64, sr fel King's Col, 70-71. *Mem:* Fel AAAS; Bot Soc Am; Am Soc Plant Physiol. *Res:* Biochemistry of chloroplasts; plant water relations; intracellular distribution of enzymes and phosynthetic products. *Mailing Add:* 837 Oak Ave Davis CA 95616

STOCKING, GORDON GARY, b Axin, Mich, Jan 12, 24; m 47; c 2. VETERINARY MEDICINE. *Educ:* Mich State Univ, DVM, 46. *Prof Exp:* Res vet, Upjohn Farms, 46-49, from asst vet to assoc vet, Upjohn Co, 49-57, dir, Vet Div, 57-64, asst dir, Agr Div, 65-69, prod mgr, 69-73, sr staff, Vet Agr Div, 73-90; RETIRED. *Concurrent Pos:* Ranch mgr & vet, Kellogg Ranch, Calif State Polytech Col, 51-52. *Mem:* Am Vet Med Asn; Indust Vet Asn; US Animal Health Asn; Am Asn Lab Animal Sci; Am Asn Equine Practr. *Res:* Equine reproduction and disease; bovine respiratory disease. *Mailing Add:* 3107 Audubon Dr Kalamazoo MI 49008

STOCKLAND, ALAN EUGENE, b Huron, SDak, July 18, 38; m 68; c 2. MICROBIOLOGY. *Educ:* Univ Nebr, BS & BA, 61; Mich State Univ, MS, 67, PhD(microbiol), 70. *Prof Exp:* Teacher secondary Sch, Malaysia, 62-64; PROF MICROBIOL, WEBER STATE COL, 70- *Mem:* Am Soc Microbiol; Sigma Xi. *Res:* Microbiological control of insect pests; effect of trauma on the immune system. *Mailing Add:* Dept Microbiol Weber State Col 3750 Harrison Blvd Ogden UT 84403

STOCKLAND, WAYNE LUVERN, b Lake Lillian, Minn, May 4, 42; m 71; c 3. NUTRITION. *Educ:* Univ Minn, BS, 64, PhD(nutrit), 69. *Prof Exp:* Res asst nutrit, Univ Minn, 69, res fel, 69-70; res nutritionist & statist mgr, 70-76, dir animal nutrit Res, 76-87, DIR TECH OPERS INTERNATIONAL MULTIFOODS CORP, 87- *Mem:* Am Soc Animal Sci; Am Dairy Sci Asn; Poultry Sci Asn; Am Inst Nutrit; NY Acad Sci; Asn Off Analysis Chemists. *Res:* Swine, poultry and ruminant nutrition and management, especially the protein and amino acid requirements and the effect of energy level, temperature and other nutrients on these requirements. *Mailing Add:* Int MultiFoods Tower Box 2942 Minneapolis MN 55402

STOCKLI, MARTIN P, b Solothurn, Switz, June, 30, 49; m 83. ELECTRON BEAM ION SOURCES, ION-ATOM COLLISIONS. *Educ:* Swiss Fed Inst Technol, Master, 74, PhD(physics), 78. *Prof Exp:* Res asst, Physics Lab, Swiss Fed Inst Technol, 74-79, res assoc, 79-80; res assoc, Physics Lab, Western Mich Univ, 80-81; res assoc, Physics Lab, Kans State Univ, 81-83, staff physicist, 84-86, assoc scientist, 86-89, ASST RES PROF, J R MCDONALD LAB, KANS STATE UNIV, 89- *Concurrent Pos:* Consult, Smithsonian Inst Astrophys Observ, 83, Inst Nuclear Physics, Univ Frankfurt, 84, 87. *Honors & Awards:* First Prize, Schlatter-Pfaeler Found, Solothurn, 67. *Mem:* Am Phys Soc; Am Vacuum Soc. *Res:* Accelerator based atomic physics; ionisation and x-ray emmision in slow heavy ion collisions; electron beam ion sources. *Mailing Add:* Physics Dept Kans State Univ Manhattan KS 66506-2604

STOCKMAN, CHARLES H(ENRY), b Oak Park, Ill, Sept 5, 22; m 48, 81; c 3. MANUFACTURING PROCESSES FOR RUBBER PRODUCTS, AEROSPACE TECHNOLOGIES. *Educ:* Purdue Univ, BS, 47, PhD(chem eng), 50. *Prof Exp:* Res engr, Res Ctr, B F Goodrich Co, 50-52, sr res engr, 52-56, sect leader, 56-58, mgr res opers, 58-61, tech dir, Aerospace & Defense Prod Div, 62-68, dir develop & new prod, 68-70; vpres & gen mgr, Goodrich High Voltage Astronaut, subsid B F Goodrich & High Voltage Eng Corp, 60-62; vpres, E A Butler Assocs, Inc, 70-76; dir res, develop & qual control, Rubber Group, H K Porter Co, 76-82; CONSULT, C H STOCKMAN ASSOCS, 82- *Concurrent Pos:* Spec sci employee, Argonne Nat Lab, 53-54, consult, 54-56; lectr nuclear chem, Case Western Reserve Univ, 55-56; prin investr, Food & Drug Admin Hemodialysis Proj, Ohio Dept Health, 84-86. *Mem:* Am Chem Soc. *Res:* Chemical process thermodynamics; nuclear fuel processing; zone melting; solid propellant and ion rocket motors; antisubmarine warfare; marine fouling prevention; mosquito and schistosomiasis control; rubber hose manufacturing processes; rubber extrusion; hemodialysis. *Mailing Add:* 342 Glen Meadow Ct Dublin OH 43017

STOCKMAN, DAVID LYLE, physical chemistry, for more information see previous edition

STOCKMAN, GEORGE C, b Brooklyn, NY, Dec 16, 43; m 69; c 2. COMPUTER SCIENCE. *Educ:* E Stroudsburg State Col, BS, 66; Harvard Univ, MAT, 67; Pa State Univ, MS, 71; Univ Md, PhD(comput sci), 77. *Prof Exp:* Instr math, Va Union Univ, 68-70; res asst comput sci, Univ Md, 73-75; res scientist comput sci, LNK Corp, 75-; assoc prof comput sci, American Univ, 79-; AT DEPT COMPUT SCI, MICH STATE UNIV. *Concurrent Pos:* Vis lectr, Univ Md, 77-79. *Mem:* Asn Comput Mach; sr mem Inst Elec & Electronics Engrs. *Res:* Artificial intelligence; image processing; pattern recognition. *Mailing Add:* Dept Comput Sci Mich State Univ 1510 Comp Ctr East Lansing MI 48824

STOCKMAN, HARRY E, b Stockholm, Sweden, Aug 24, 05; nat US; m 47; c 2. ELECTRONICS ENGINEERING, PHYSICS. *Educ:* Stockholm Tech Inst, dipl, 26; Royal Inst Technol, Sweden, MS, 38; Harvard Univ, SD(elec eng), 46. *Prof Exp:* Tel lab engr, L M Ericson, Sweden, 27-29, radio engr, 32-34; tech ed, J Radio, 29-31, tech writer, 31-32; consult, Royal Inst Technol, Sweden, 34-38, asst prof radio eng, 38-40; instr physics & commun, Harvard Univ, 41-45, res assoc, Cent Commun Res Lab, 45; chief commun lab, Electronics Res Labs, Air Materiel Command, US Dept Air Force, 45-48; consult electronics, Stockman Electronics Res Co, 48-53; mgr res & develop, Norden-Ketay Boston Div, United Aircraft Corp, 55-56; prof elec eng &

chmn dept, Merrimack Col, 58-60; prof, Lowell Tech Inst, 60-65; staff scientist, Mitre Corp, Bedford, 67-70; SR PHYSICIST, SERCOLAB, 70- *Mem:* Inst Elec & Electronics Engrs. *Res:* Radio communication; infrared spectrum and network theory, network theorems, especially feedback systems and stability criteria; parametric electromechanical systems; distributed and parametric wide band amplification; semiconductor devices. *Mailing Add:* Box 767 Sercolab East Dennis MA 02641

STOCKMAN, HERVEY S, JR, b New York, NY, Mar 2, 46. ASTROPHYSICS, OPTICAL ASTRONOMY. *Educ:* Columbia Univ, MS, 70, PhD(physics), 73. *Prof Exp:* Proj officer, USAF, 73-75; res assoc, Univ Ariz, 75-79; asst astronomer, Starard Observ, 79-83, chief res support, 79-83; DEPT DIR, SPACE TELESCOPE SCI INST, 88- *Mem:* AAAS; Am Astron Soc; Am Phys Soc. *Res:* Astrophysical properties of binary systems; magnetic white dwarfs and active galactic nuclei. *Mailing Add:* Space Telescope Sci Inst 3700 San Martin Dr Baltimore MD 21218

STOCKMAYER, WALTER HUGO, b Rutherford, NJ, Apr 7, 14; m 38; c 2. POLYMER CHEMISTRY. *Educ:* Mass Inst Technol, SB, 35, PhD(chem), 40; Oxford Univ, BSc, 37. *Hon Degrees:* Dr, Univ Louis Pasteur, 72; LHD, Dartmouth Col, 83. *Prof Exp:* Instr chem, Mass Inst Technol, 39-41 & Columbia Univ, 41-43; asst prof, Mass Inst Technol, 43-46, from assoc prof to prof phys chem, 46-61; Chmn Dept, 63-67 & 73-76, prof, 61-79, EMER PROF CHEM, DARTMOUTH COL, 79- *Concurrent Pos:* Consult, E I du Pont de Nemours & Co, Inc, 45-; Guggenheim fel, 54-55; trustee, Gordon Res Conf, 63-66; assoc ed, Macromolecules, 68-73 & 76-; hon fel, Jesus Col, Oxford, 76. *Honors & Awards:* Award, Mfg Chem Asn, 60; Polymer Chem Award, Am Chem Soc, 66, Peter Debye Award phyPhys Chem, 74, High Polymer Physics Prize, 75; Nat Medal Sci, 87; Richards Medal, Northeast Sect, Am Chem Soc, 88; Polymer Chem Div Award, 88. *Mem:* Nat Acad Sci; Am Chem Soc; fel Am Phys Soc; fel Am Acad Arts & Sci; fel Humboldt. *Res:* High polymers; applied statistical mechanics; dynamics and statistical mechanics of macromolecules. *Mailing Add:* Dept Chem Dartmouth Col Hanover NH 03755

STOCKMEYER, LARRY JOSEPH, b Evansville, Ind, Nov 13, 48. COMPUTER SCIENCE. *Educ:* Mass Inst Technol, SB & SM, 72, PhD(comput sci), 74. *Prof Exp:* Mem res staff, Thomas J Watson Res Ctr, 74-83, MEM RES STAFF, ALMADEN RES CTR, IBM CORP, 83- *Mem:* Asn Comput Mach; Sigma Xi; Soc Indust Applied Math. *Res:* Computational complexity; analysis of algorithms; distributed computing. *Mailing Add:* Almaden Res Ctr K53/802 650 Harry Rd San Jose CA 95120-6099

STOCKMEYER, PAUL KELLY, b Detroit, Mich, May 1, 43; m 66; c 2. MATHEMATICS. *Educ:* Earlham Col, AB, 65; Univ Mich, Ann Arbor, MA, 66, PhD(math), 71. *Prof Exp:* Asst prof math, 71-77, assoc prof comput sci, 77-88, PROF COMPUT SCI, COL WILLIAM & MARY, 88- *Mem:* Math Asn Am; Am Math Soc; Asn Comput Mach. *Res:* Combinatorial analysis; graph theory; analysis of algorithms. *Mailing Add:* Dept Comput Sci Col William & Mary Williamsburg VA 23185

STOCKNER, JOHN G, b Kewanee, Ill, Sept 17, 40; m 62; c 2. LIMNOLOGY, ECOLOGY. *Educ:* Augustana Col, Ill, 62; Univ Wash, PhD(zool), 67. *Prof Exp:* Fel phytoplankton ecol, Windermere Lab, Freshwater Biol Asn, Eng, 67-68; limnologist, Freshwater Inst, Fisheries Res Bd Can, 68-71 & Pac Environ Inst, 71-81, ASSOC DIR FISHERIES RES, PAC REGION, CAN FISHERIES & OCEANS. *Mem:* Am Soc Limnol & Oceanog; Int Asn Theoret & Appl Limnol. *Res:* Phytoplankton ecology and paleolimnology; marine plankton ecology and benthic algal and phytoplankton production. *Mailing Add:* 2614 Mathers Ave West Vancouver BC V7V 2J4 Can

STOCKS, DOUGLAS ROSCOE, JR, b Dallas, Tex, Sept 4, 32; m 51; c 4. MATHEMATICS. *Educ:* Univ Tex, BA, 58, MA, 60, PhD(math), 64. *Prof Exp:* Spec instr math, Univ Tex, 60-64; from asst prof to assoc prof, Univ Tex, Arlington, 64-69; ASSOC PROF MATH, UNIV ALA, BIRMINGHAM, 69- *Concurrent Pos:* Mathematician, US Navy Electronics Lab, 63; Tex Col & Univ Syst res grant, 66-67; vis prof, Auburn Univ, 78-79 & Univ Reading, Eng, 85-86. *Mem:* Am Math Soc; Math Asn Am. *Res:* Lattice paths and graph theory; foundations of mathematics; geometry; topology; point set theory. *Mailing Add:* Univ Reading Berks RG6 2AX England

STOCKS, GEORGE MALCOLM, b Thurnscoe, Eng, June 5, 43; m 67; c 2. SOLID STATE PHYSICS. *Educ:* Univ Bradford, Eng, BTech, 66; Univ Sheffield, Eng, PhD(theoret physics), 69. *Prof Exp:* Res staff mem theoret solid state physics, Oak Ridge Nat Lab, 69-72; res assoc physics, Univ Bristol, Eng, 72-76; RES STAFF MEM THEORET SOLID STATE PHYSICS, OAK RIDGE NAT LAB, 76- *Mem:* Am Phys Soc; Inst Physics, UK. *Res:* Theory of electronic states in ordered and disordered metals and alloys; bank theory; disordered systems theory; phase stability; transport; excitation processes in solids; photo electron spectroscopies; soft x-ray spectroscopy. *Mailing Add:* Metals & Ceramics Div MS 6114 Oak Ridge Nat Lab PO Box 2008 Bldg 45005 Oak Ridge TN 37830

STOCKTON, DORIS S, b New Brunswick, NJ, Feb 9, 24; m 48; c 2. MATHEMATICS. *Educ:* Rutgers Univ, BSc, 45; Brown Univ, MSc, 47, PhD(math), 58. *Prof Exp:* Instr, Brown Univ, 52-54; asst prof, 58-73, ASSOC PROF MATH, UNIV MASS, AMHERST, 73- *Mem:* Am Math Soc; Math Asn Am. *Res:* Functional analysis. *Mailing Add:* RFD 2 19 N Washington St Belchertown MA 01007

STOCKTON, JAMES EVAN, b Goliad, Tex, Feb 28, 31; m 51; c 2. ELECTRICAL ENGINEERING. *Educ:* Univ Tex, Austin, BS, 60. *Prof Exp:* Res engr, 60-64, supvr, Submarine Sonar Sect, 64-69, dep div head, Electroacoust Div, 69-70, HEAD ENG SERV DIV, APPL RES LABS, UNIV TEX, 70- *Mem:* Inst Elec & Electronics Engrs; Am Inst Physics; Acoust Soc Am. *Res:* Underwater acoustics; instrumentation for underwater acoustics; sonar systems. *Mailing Add:* 1508 Weyford Dr Austin TX 78758

STOCKTON, JOHN RICHARD, b Jarrell, Tex, Feb 19, 17; m 38, 53; c 6. RESEARCH MANAGEMENT. *Educ:* Univ Tex, BSc, 38, MA, 41, PhD(microbiol), 51. *Prof Exp:* Pharmacist, Baylor Hosp, 38-39; tutor pharm, Univ Tex, 39-41, asst prof, 41-46; dir res, Hyland Labs, 46-50; tech asst to dir biol prod, Merck Sharp & Dohme, 50-54; mgr res & qual control, Pillsbury Co, 54-62, dir res & develop, 62-66, dir sci activities, 66-68; mgr res & develop Corn Prod Food Technol Inst, 68-70, assoc dir res & qual control, Best Foods, CPC Int, Inc, 70-73; vpres res & develop, Nutri Co, 73-74; PRES, MGT CATALYSTS, 74- *Mem:* AAAS; Inst Food Technol; Am Chem Soc; Am Soc Microbiol; NY Acad Sci. *Res:* Antimicrobial agents; physicochemical characteristics of drugs; biological products; food science; product development; nutrition; research management. *Mailing Add:* Mgt Catalysts PO Box 70 Ship Bottom NJ 08008-0227

STOCKWELL, CHARLES WARREN, b Port Angeles, Wash, Dec 31, 40; m 66; c 2. NEUROSCIENCES. *Educ:* Western Wash State Col, BA, 64; Univ Ill, MA, 66, PhD(psychol), 68. *Prof Exp:* Res psychologist, Naval Aerospace Med Ctr, 69-71; asst prof, Ohio State Univ, 72-76, assoc prof otolaryngol, Col Med, 76-83; DIR, VESTIBULAR LAB, MICH EAR INST, 83- *Mem:* Barany Soc; Asn Res Otolaryngol; Soc Neurosci. *Res:* Vestibular function. *Mailing Add:* Dept Vestibular Physiol Mich Ear Inst 16001 W Nine Mile Rd PO Box 2043 Southfield MI 48037

STOCUM, DAVID LEON, b Ypsilanti, Mich, Feb 15, 39. DEVELOPMENTAL BIOLOGY. *Educ:* Susquehanna Univ, BA, 61; Univ Pa, PhD(biol), 68. *Prof Exp:* Asst prof zool, Univ Ill, Urbana, 68-73, assoc prof genetics & develop, 73-81, prof genetics, develop & anat sci, 81-87, prof cell & struct biol, 87-90; PROF BIOL & DEAN SCH SCI, IND-PURDUE UNIV, INDIANAPOLIS, 90- *Concurrent Pos:* NSF res grant, 72-77; NIH res grant, 79-90; assoc, Ctr Advan Study, Univ Ill, Urbana, 87. *Mem:* Am Soc Zoologists; fel AAAS; Soc Develop Biol; Sigma Xi; Am Soc Cell Biol. *Res:* Morphogenesis and cellular differentiation in embryonic and regenerating systems. *Mailing Add:* Dept Biol Ind Purdue Univ Sch Sci Michigan St Urbana IL 61820

STODDARD, ALONZO EDWIN, JR, physics; deceased, see previous edition for last biography

STODDARD, C(ARL) KERBY, b Reno, Nev, Dec 20, 07; m 42; c 2. CHEMICAL ENGINEERING, EXTRACTIVE METALLURGY. *Educ:* Univ Nev, BS, 34, MS, 36; Univ Md, PhD(chem eng), 41. *Prof Exp:* Res chemist, Am Soc Mech Engrs, Md, 40-41; chem engr, US Bur Mines, Nev, 41-48; chem engr, Titanium Div, Nat Lead Co, 48-52; chem engr, Titanium Metals Corp Am, 52-58; chem engr, Titanium Div, Nat Lead Co, 58-70; process engr, Ralph M Parsons Co, 70-77; RETIRED. *Concurrent Pos:* Consult, Ralph M Parsons Co, 77- *Mem:* AAAS; Am Chem Soc; Am Inst Mining, Metall & Petrol Engrs; Sigma Xi. *Res:* Extractive metallurgy of magnesium, titanium and nickel; electric smelting and beneficiation of titanium ores; nuclear fuels reprocessing. *Mailing Add:* 3709 S George Mason Dr No 1202 E Falls Church VA 22041

STODDARD, GEORGE EDWARD, animal nutrition, dairy management; deceased, see previous edition for last biography

STODDARD, JAMES H, b Saginaw, Mich, June 17, 30; m 57; c 2. MATHEMATICS. *Educ:* Univ Mich, BS, 52, PhD(math), 61. *Prof Exp:* Instr math, Univ Mich, 60-61; asst prof, Oakland Univ, 61-62; asst prof, Syracuse Univ, 62-66; asst prof, Univ of the South, 66-67; assoc prof, Kenyon Col, 67-70; prof, Upsala Col, 70-72; PROF MATH, MONTCLAIR STATE COL, 72-, PROF COMPUT SCI, 80- *Concurrent Pos:* Mem col level exam comt, Educ Testing Serv, NJ, 65-69; mem, CUPM Consults Bur, 62- *Mem:* Am Math Soc; Math Asn Am; Asn Comput Mach. *Res:* Computer software; application to management sciences. *Mailing Add:* Dept Math & CS Montclair State Col Valley Rd Upper Montclair NJ 07043

STODDARD, LELAND DOUGLAS, b Hillsboro, Ill, Mar 15, 19; m 46. PATHOLOGY. *Educ:* DePauw Univ, AB, 40; Johns Hopkins Univ, MD, 43. *Prof Exp:* Asst & asst resident path, 47-48, instr & resident, 49-50, assoc, Sch Med Duke Univ, 50-51; from asst prof to assoc prof, Med Sch, Univ Kans, 51-54; chmn dept, 54-73, PROF PATH, EUGENE TALMADGE MEM HOSP, MED COL GA, 54- *Concurrent Pos:* Chief staff, Eugene Talmadge Mem Hosp, 64-65; chief res path, Atomic Bomb Casualty Comn Japan, 61-62; vis prof, Med Sch, Osaka Univ, 66; mem path training comt, Vet Admin, 69-72; mem, Intersoc Path Coun, Int Coun Socs Path, US Nat Comt & Sci Adv Bd of Consult, Armed Forces Inst Path, 70-75. *Mem:* Am Asn Cancer Res; Asn Hist Med; Am Asn Pathologists; Int Acad Path (vpres, 74-78, treas, 78-); Int Acad Path (secy-treas, 70-79, pres, 81-82). *Res:* Cervical carcinoma; gynecological and reproductive endocrine pathology; knowledge theory in pathology and medicine. *Mailing Add:* Dept Path Med Col Ga 1120 15th St Augusta GA 30912

STODDARD, STEPHEN D(AVIDSON), b Everett, Wash, Feb 8, 25; m 49; c 2. CERAMIC ENGINEERING, MATERIALS SCIENCE. *Educ:* Univ Ill, BS, 50. *Prof Exp:* From asst ceramic engr to asst prod supvr, Coors Porcelain Co, Colo, 50-52; ceramics sect leader, Los Alamos Sci Lab, Univ Calif, 52-74, ceramics-powder metall sect leader, 74-80; consult, vpres & secy, 77-80, PRES & TREAS, MAT TECHNOL ASSOCS, 80- *Concurrent Pos:* Consult, 56-; consult ed, Ceramic Age, 58-60; mem mat adv bd, Nat Acad Sci, Dept Defense, 62-63; consult, Mat Technol Assocs, 77-80, pres & treas, 80- *Honors & Awards:* PACE Award, Am Ceramic Soc, 64, Greaves-Walker Award, 84. *Mem:* AAAS; fel Am Inst Chem; Nat Inst Ceramic Engrs; fel Am Ceramic Soc (vpres, 71-72, treas, 72-74, pres, 76-77); Sigma Xi; Am Soc Metals. *Res:* Fabrication techniques for refractory oxides, rare earth oxides, refractory metals, metal-oxide mixtures, ceramic-metal seals and electronic ceramics and their application in energy and nuclear weapon, power and propulsion studies. *Mailing Add:* 326 Kimberly Lane Los Alamos NM 87544

STODOLA, EDWIN KING, US citizen. ELECTRICAL ENGINEERING, ELECTRONICS ENGINEERING. *Educ:* Cooper Union Inst Technol, BEE, 36, EE, 47. *Prof Exp:* Radio engr, Radio Eng Labs, Long Island, 36-39; chief radar & electronics res & develop, Spec Develop Sect, US Army Signal Corps Labs, Belmar, NJ & Washington, DC, 39-47; chief scientist, Reeves Inst Div, Dynamics Corp Am, Boynton Beach, Fla & Mineola, NY, 47-72; sr res engr electronics res & develop, Syracuse Univ Res Corp, 72-74; leading scientist oper testing plans, Dikewood Corp, Kirtland AFB, NMex, 74-75; CHIEF, COMMUN ELEC WARFARE DIV, US ARMY ELECTRONIC WARFARE LAB, 75- *Concurrent Pos:* Consult adv panel electronics, Off Asst Secy Defense, 58-59; pvt consult. *Mem:* Fel Inst Elec & Electronics Engrs; assoc fel Am Inst Aeronaut & Astronaut; fel Radio Club Am; Armed Forces Commun & Electronics Asn; Electronic Warfare Tech Asn. *Res:* Radio propagation and techniques for accurate radar and emitter location; aircraft navigation, control and flight safety research. *Mailing Add:* PO Box 36072 1960 Tall Ridge Rd Melbourne FL 32936

STODOLSKY, MARVIN, b Newark, NJ, Nov 28, 39. MOLECULAR BIOLOGY, GENETICS. *Educ:* Univ Chicago, BS, 60, PhD(biophys), 64. *Prof Exp:* Fel biochem, Mass Inst Technol, 64-67; asst prof microbiol & genetics, Univ Chicago, 67-73; asst prof microbiol & genetics, Loyola Univ, 73-74, assoc prof, 74-79; PROF & CHMN, DEPT BIOL, BOGAZICI UNIV, ISTANBUL, 79- *Mailing Add:* Dept Biol Bogazici Univ P K 2 Istanbul Turkey

STOEBE, THOMAS GAINES, b Upland, Calif, Apr 26, 39; m 82; c 3. MATERIALS SCIENCE, ELECTRONIC MATERIALS. *Educ:* Stanford Univ, BS, 61, MS, 63, PhD(mat sci), 65. *Prof Exp:* Vis lectr metall & res assoc, Imp Col, London, 65-66; from asst prof to assoc prof, 66-75, assoc dean,Col Eng,82-87,CHMN DEPT MAT SCI & ENG, UNIV WASH, 87- *Concurrent Pos:* Vis prof, Atomic Energy Inst, Sao Paulo, Brazil, 72-73. *Honors & Awards:* Western Elec Award, Am Soc Eng Educ, 77. *Mem:* Am Soc Eng Educ; Am Ceramic Soc; Am Phys Soc; Metall Soc. *Res:* Influence of lattice imperfections on physical properties of solids; electronic and optical properties in semiconductors and insulators. *Mailing Add:* 11106 NE 38th Pl Bellevue WA 98004

STOEBER, WERNER, b Göttingen, Ger, May 8, 25; m 55; c 1. AEROSOL SCIENCE, INHALATION TOXICOLOGY. *Educ:* Univ Gottingen, Dipl, 53,Dr rer nat, 55. *Prof Exp:* Sci asst, Med Res Inst, Max Planck Soc, 55-61; res fel aerosol sci, Calif Inst Technol, 61-63; sci asst, Inst Med Physics, Univ Muenster, 64-65, docent, 65-66; from assoc prof to prof radiation biol & biophysics, Med Sch, Univ Rochester, 66-70; dir, Inst Aerobiol, Fraunhofer Soc, 73-79; dir, Fraunhofer Inst Toxicol & Aerosol Res, Hannover, 80-85; RETIRED. *Concurrent Pos:* USPHS int res fel, 61-62; prof, Univ Muenster, 68; prof med physics, Med Sch Hannover, 82; vis prof, Univ Rochester, 87-88; vis scientist, Inhal Toxic Res Inst, Lovelace Found, Albuquerque, NMex, 88, Chem Indust Inst Toxicol, Research Triangle Park, NC, 90. *Mem:* Am Chem Soc; German Asn Aerosol Res (pres, 78-82); Am Asn Aerosol Res; Am Indust Hyg Asn. *Res:* Surface chemistry; silicosis; aerosol. *Mailing Add:* Potstiege 34 Muenster 44 Germany

STOECKENIUS, WALTHER, b Giessen, Ger, July 3, 21; m 52; c 3. CYTOLOGY. *Educ:* Univ Hamburg, MD, 50. *Prof Exp:* Intern, Pharmacol Inst, Univ Hamburg, 51, intern internal med, 51 & obstet & gynec, 52, researcher virol, Inst Trop Med, 52-54, res asst path, 54-58, pvt docent, 58; guest investr, Rockefeller Inst, 59, from asst prof to assoc prof cytol, 59-67; PROF CELL BIOL, SCH MED, UNIV CALIF, SAN FRANCISCO, 67- *Mem:* Nat Acad Sci; AAAS; Am Soc Biol Chem; Am Soc Cell Biol; Biophys Soc; Harvey Soc. *Res:* Fine structure of cells at the molecular level; energy transducing membranes; photobiology; halobacteria. *Mailing Add:* Dept Biochem & Biophys Univ Calif San Francisco CA 94122

STOECKLE, JOHN DUANE, b Highland Park, Mich, Aug 17, 22; m 47; c 4. MEDICINE. *Educ:* Antioch Col, BS, 48; Harvard Med Sch, MD, 48; Am Bd Internal Med, dipl, 58. *Prof Exp:* Intern med, Mass Gen Hosp, Boston, 48-49, asst resident, 49-50, resident, 51-52; panel dir med aspects of atomic energy, Comt Med Sci, Res & Develop Bd, Dept Defense, 52-54; from instr to assoc prof, 54-82, PROF MED, HARVARD MED SCH, 82-; CHIEF MED CLIN, MASS GEN HOSP, 54-, PHYSICIAN, 69- *Concurrent Pos:* mem comt on coal miner's safety & health, Dept Health, Educ & Welfare, 71-73; book rev ed, Soc Sci & Med, 72-89. *Honors & Awards:* Glazer Award, Soc Gen Internal Med, 89. *Mem:* Fel Am Pub Health Asn; assoc mem Am Sociol Asn; fel Am Anthrop Asn; fel Am Psychosom Soc; Soc Appl Anthrop; fel Am Col Physicians. *Res:* Medical care administration and health and illness behavior; longitudinal study of occupational lung diseases. *Mailing Add:* Mass Gen Hosp Fruit St Boston MA 02114

STOECKLER, JOHANNA D, CELLULAR SIGNAL TRANSDUCTION, NUCLEOSIDE TRANSPORT & METABOLISM. *Educ:* Rutgers Univ, BA, 63, PhD(biochem), 73. *Prof Exp:* From instr to asst prof, 77-85, ASSOC PROF BIOCHEM PHARMACOL, BROWN UNIV, 85- *Mem:* Am Soc Biochem & Molecular Biol; Am Asn Cancer Res; AAAS; Am Asn Women Cancer Res. *Res:* Preclinical pharmacology involving transport and metabolism of purines; enzymology; signal transduction in cultured cell systems via adenylate cyclase and phospholipase C. *Mailing Add:* Div Biol & Med Brown Univ Box G-B423 Providence RI 02912

STOECKLEY, THOMAS ROBERT, b Ft Wayne, Ind, Dec 6, 42. ASTRONOMY. *Educ:* Mich State Univ, BS, 64; Cambridge Univ, PhD(astron), 67. *Prof Exp:* Asst prof, Mich State Univ, 67-74, assoc prof astron, 67-83; SR RES SCIENTIST, CONOCO INCORP, 83- *Mem:* Soc Explor Geophysicists. *Res:* Seismic data processing. *Mailing Add:* Conoco Incorp PO Box 1267 Ponca City OK 74603

STOECKLY, ROBERT E, b Schenectady, NY, June 9, 38; m 69; c 2. ATMOSPHERIC PHYSICS. *Educ:* Princeton Univ, PhD(astrophys sci), 64. *Prof Exp:* Res fel astron, Mt Wilson & Palomar Observs, 64-65; asst prof physics & astron, Rensselaer Polytech Inst, 65-72; physicist, Mission Res Corp, 72-83 & Kaman Temp, 83-85; res scientist, Phys Res, Inc, 85-90. *Mem:* Asn Comput Mach. *Res:* Fluid dynamics; plasma physics; atmospheric physics. *Mailing Add:* 3229 Calle Cedro Santa Barbara CA 93105-2730

STOEHR, ROBERT ALLEN, b Pittsburgh, Pa, July 10, 30. METALLURGICAL ENGINEERING. *Educ:* Hiram Col, BA, 52; Carnegie Inst Technol, MS, 65; Carnegie-Mellon Univ, PhD(metall & mat sci), 69. *Prof Exp:* Metall engr, Reactive Metals, Inc, Ohio, 54-60; res engr, Alcoa Res Labs, Aluminum Co Am, Pa, 60-67; asst prof, 68-74, ASSOC PROF METALL & MAT ENG, UNIV PITTSBURGH, 74- *Mem:* Metall Soc; Am Inst Mining, Metall & Petrol Engrs; Am Soc Metals; Electrochem Soc. *Res:* Process and chemical metallurgy; casting and solidification; high temperature electrochemistry and corrosion; computer simulation of metallurgical processes. *Mailing Add:* Mat Sci & Eng Dept 848 Benedum Univ Pittsburgh 4200 Fifth Ave Pittsburgh PA 15260

STOENNER, HERBERT GEORGE, b Levasy, Mo, June 17, 19; m 46; c 4. BACTERIOLOGY, VIROLOGY. *Educ:* Iowa State Col, DVM, 43. *Prof Exp:* Asst scientist, Commun Dis Ctr, USPHS, Ga, 47-50, sr asst vet, 50-52, vet, 52-56, sr vet, 56-61, asst dir lab, 62-64, dir, Rocky Mountain Lab, Nat Inst Allergy & Infectious Dis, 64-82, vet officer dir, 61-82; RETIRED. *Concurrent Pos:* Fac affil, Univ Mont. *Honors & Awards:* K F Meyer Gold Headed Cane Award, 74. *Mem:* Am Vet Med Asn; Am Pub Health Asn; US Animal Health Asn; Conf Res Workers Animal Dis. *Res:* Zoonoses; leptospirosis; rickettsioses; brucellosis; psittacosis. *Mailing Add:* 1102 S Second St Hamilton MT 59840

STOERMER, EUGENE F, b Webb, Iowa, Mar 7, 34; m 60; c 3. PHYCOLOGY, LIMNOLOGY. *Educ:* Iowa State Univ, BS, 58, PhD(bot), 63. *Prof Exp:* NIH fel phycol, Iowa State Univ, 63-65; assoc res algologist, 66-71, lectr, Biol Sta, 69-77, assoc prof, 77-84, RES ALGOLONIST, GREAT LAKES RES DIV, UNIV MICH, ANN ARBOR, 71-, RES SCIENTIST, HERBARIUM, 73-, SCH NAT RES, 85- *Concurrent Pos:* McHenry fel, Acad Natural Sci, Philadelphia, Pa, 59; vis prof bot, Mich State Univ, 67-68; adj prof biol, City Univ New York, 74 & Bowling Green State Univ, 77-80; mem grad fac, Univ Maine, Orono, 81-; res fel, Acad Natural Sci Philadelphia. *Mem:* AAAS; Am Soc Limnol & Oceanog; Int Phycol Soc; Phycol Soc Am; Am Quaternary Asn; Sigma Xi. *Res:* Taxonomy and ecology of Bacillariophyta and Laurentian Great Lakes algal flora; paleoecology and algal evolution. *Mailing Add:* 4392 Dexter Rd Ann Arbor MI 48103

STOESSL, ALBERT, biologically active compounds, biosynthesis, for more information see previous edition

STOESZ, JAMES DARREL, b Mountain Lake, Minn, July 3, 50; m 71; c 3. BIOPHYSICAL CHEMISTRY, BIOCHEMISTRY. *Educ:* Bethel Col, BA, 72; Univ Minn, Minneapolis, PhD(biophys chem), 77. *Prof Exp:* Res fel biochem, Brandeis Univ, 76-78; sr chemist biosci, Cent Res Labs, 78-82, supvr, Biotechnol Lab, Life Sci Sect Lab, 82-85, mgr, Asepsis & Infection Control Lab, Med-Surg Div, 85-88, MGR, BIOTECHNOL LAB, LIFE SCI SECTOR LAB, 3M CO, 88- *Mem:* Am Chem Soc. *Res:* Protein and polypeptide structure and function; enzyme mechanisms; nuclear magnetic resonance; lipid bilayer membranes; liposomes; immunodiagnostics; chemiluminescence. *Mailing Add:* 6810 Arlene Ave Inver Grove Heights MN 55075-6148

STOETZEL, MANYA BROOKE, b Houston, Tex, Apr 11, 40; c 2. ENTOMOLOGY. *Educ:* Univ Md, College Park, BS, 66, MS, 70, PhD(entom), 72. *Prof Exp:* Entomologist, First US Army Med Lab, Ft Meade, Md, 66-68; Presidential intern, 73-74, RES ENTOMOLOGIST, SYST ENTOM LAB, AGR RES SERV, USDA, 74- *Concurrent Pos:* Ed, Entom Soc Wash, 77-79, pres, 83. *Mem:* Entom Soc Am; Am Asn Zool. *Res:* Morphology and taxonomy of aphids, phylloxerans, adelgids and armored scale insects. *Mailing Add:* Syst Entom Lab USDA Bldg 004 Rm 6 Beltsville MD 20705

STOEVER, EDWARD CARL, JR, b Milwaukee, Wis, Mar 13, 26; m 54; c 2. SCIENCE EDUCATION, GENERAL EARTH SCIENCES. *Educ:* Purdue Univ, BS, 48; Univ Mich, MS, 50, PhD(geol), 59. *Prof Exp:* Res geologist, Int Minerals & Chem Corp, 52-54; from asst prof to assoc prof geol, 56-69, assoc dir undergrad studies, Sch Geol & Geophys, Univ Okla, 70-72, prof geol & geophys, 69-78; prof geol & chmn dept earth sci,Southeast MO State, 78-85, dir, Ctr Sci & Math Educ, 83-86; EXEC DIR, MO ALLIANCE SCI, 87- *Concurrent Pos:* Fel, NSF, 55-56 & 57-58; dir, Okla Geol Camp, 64-69; dir inst earth sci, NSF, 65-69 & 70-72; assoc dir, Earth Sci Curriculum Proj, 69; assoc prog dir teacher educ sect, NSF, 69-70; sr staff consult, Earth Sci Teacher Prep Proj, 70-72; dir, Okla Earth Sci Educ Proj, 72-74; dir, Nat Asn Geol Teachers Crustal Evolution Educ Proj, NSF, 76-80; co dir, K6 Sci & Math Improv Proj, Southeast Mo, 85-91. *Mem:* Nat Asn Geol Teachers (vpres, 74-75, pres, 75-76); fel AAAS; fel Geol Soc Am; Nat Sci Teachers Asn; Am Asn Supvry & Current Develop. *Res:* Statewide elementary, secondary and college educational improvement in science, mathematics and technology through education; business-government partnerships. *Mailing Add:* 2808 Gordonville Rd Cape Girardeau MO 63701

STOEWSAND, GILBERT SAARI, b Chicago, Ill, Oct 20, 32; m 57; c 2. FOOD TOXICOLOGY. *Educ:* Univ Calif, Davis, BS, 54, MS, 58; Cornell Univ, PhD, 64. *Prof Exp:* Res assoc poultry sci, Cornell Univ, 58-61; res nutritionist, US Army Natick Lab, Mass, 63-66; res assoc, Inst Exp Path & Toxicol, Albany Med Col, 66-67; from asst prof to assoc prof, 67-79, PROF TOXICOL, EXP STA, NY STATE COL AGR & LIFE SCI, CORNELL UNIV, 79- *Concurrent Pos:* WHO fel, 72; consult, Toxicol Info Prog, Nat Libr Med, 77-85. *Mem:* AAAS; Inst Food Technol; Am Inst Nutrit; Soc Toxicol; Soc Exp Biol & Med. *Res:* Cancer and food; food additives; natural food toxicants and toxicant inhibitors. *Mailing Add:* Dept of Food Sci & Technol NY State Agr Exp Sta Cornell Univ Geneva NY 14456-0462

STOFFA, PAUL L, b Palmerton, Pa, July 9, 48; m 68; c 2. MARINE GEOPHYSICS, SIGNAL PROCESSING. *Educ:* Rensselaer Polytech Inst, BS, 70; Columbia Univ, PhD(geophys), 74. *Prof Exp:* Res assoc marine geophys, Lamont-Doherty Geol Observ, 74-81; CONSULT, GULF SCI TECHNOL, 81-; AT INST FOR GEOPHYSICS, UNIV TEX, AUSTIN. *Concurrent Pos:* Adj asst prof, Columbia Univ, 78- *Mem:* Am Geophys Union; Soc Explor Geophys; Inst Elec & Electronics Engrs; Sigma Xi. *Res:* Marine seismology; wave propagation; numerical analysis. *Mailing Add:* Inst Geophys Univ Tex 8701 Mopac Blvd Austin TX 78759-8345

STOFFELLA, PETER JOSEPH, b Montreal, Que, Nov 21, 54; US citizen; m 87; c 1. VEGETABLE CROPS, ROOT SYSTEM FUNCTIONS. *Educ:* Delaware Valley Col, BS, 76; Kansas State Univ, MS, 77, Cornell Univ, PhD(vegetable crops), 80. *Prof Exp:* Lab Technician, Vero Beach Labs, Bayer Agr, 73-75; teaching asst, Kansas State Univ, 77; asst prof, 80-85, assoc prof, 85-90, PROF, UNIV FLA, 90- *Concurrent Pos:* Vis prof, Univ Pisa, 87. *Mem:* Am Soc Horticult Sci; Am Soc Agron; Crop Sci Soc Am; Sigma Xi. *Res:* Cultural and management practices of commercial vegetable crops; development and functions of plant root systems. *Mailing Add:* Univ Fla Agricult Res & Educ Ctr PO Box 248 Ft Pierce FL 34954

STOFFER, JAMES OSBER, b Homeworth, Ohio, Oct 16, 35; m 57; c 2. ORGANIC CHEMISTRY, POLYMER CHEMISTRY. *Educ:* Mt Union Col, BS, 57; Purdue Univ, PhD(org chem), 61. *Prof Exp:* Res asst, Purdue Univ, 57-59; res assoc, Cornell Univ, 61-63; from asst pro to assoc prof, 63-82, PROF CHEM, UNIV MO-ROLLA, 82-, SR INVESTR, MAT RES CTR, 85- *Mem:* Am Chem Soc. *Res:* Beta deuterium isotope effects; trace organic analysis; small ring compounds; polymer synthesis and characterization. *Mailing Add:* Dept Chem Univ Mo PO Box 249 Rolla MO 65401

STOFFER, RICHARD LAWRENCE, b Cleveland, Ohio, Dec 13, 48; m 71; c 2. AQUATIC ECOLOGY, SYSTEMATICS OF DIPTERA FAMILY CHIRONOMIDAE. *Educ:* Ashland Univ, BS, 70; Ohio State Univ, MS, 75, PhD(zool), 78. *Prof Exp:* Instr sci, Urbana Univ, 78-80; from asst prof to assoc prof, 80-87, PROF BIOL, ASHLAND UNIV, 87- *Mem:* Am Inst Biol Sci; Am Soc Zoologists; Entom Soc Am. *Res:* Ecology of aquatic environments with a specific interest in the systematics (behavior, ecology, and taxonomy) of the Diptera family Chironomidae. *Mailing Add:* Dept Biol & Toxicol Ashland Univ Ashland OH 44805

STOFFER, ROBERT LLEWELLYN, b North Georgetown, Ohio, Sept 16, 27; m 51; c 3. INDUSTRIAL HYGIENE, ANALYTICAL CHEMISTRY. *Educ:* Ashland Col, AB, 50; Ohio State Univ, PhD(anal chem), 54. *Prof Exp:* Anal chemist, 54-56, from asst proj chemist to sr proj chemist, 56-72, INDUST HYG CHEMIST, ENVIRON AFFAIRS & SAFETY DEPT, AMOCO CORP, 72- *Mem:* Am Indust Hyg Asn; Am Chem Soc; Am Acad Indust Hygiene. *Res:* Development of new analytical methods in the industrial hygiene field. *Mailing Add:* Amoco Res Ctr PO Box 3011 Naperville IL 60566-7011

STOFFOLANO, JOHN GEORGE, JR, b Gloversville, NY, Dec 31, 39; m 65; c 2. ENTOMOLOGY, NEUROBIOLOGY. *Educ:* State Univ NY Col Oneonta, BS, 62; Cornell Univ, MS, 67; Univ Conn, PhD(entom), 70. *Prof Exp:* From asst prof to assoc prof, 69-80, PROF ENTOM, UNIV MASS, AMHERST, 80- *Concurrent Pos:* NSF fel neurobiol, Princeton Univ, 70-71; NIH fel, Univ Mass, Amherst, 72-75. *Mem:* AAAS; Entom Soc Am; Am Inst Biol Sci; Soc Nematol. *Res:* Integrative studies on the ecology, neurobiology and physiology of diapausing, nondiapausing and aging flies of the genus Musca and Phormia. *Mailing Add:* Dept Entom Univ Mass Amherst MA 01002

STOHLER, RUDOLF, b Basel, Switz, Dec 5, 01; nat US; m 29; c 5. ZOOLOGY. *Educ:* Univ Basel, MA & PhD(zool), 28. *Prof Exp:* Inst Student Exchange fel, 28-30, res assoc zool, 32-34, instr, 34-35, res assoc, 35-41, prin lab tech, 41-69, instr zool & biol, Exten, 35-69, assoc res zoologist, 55-66, res zoologist, 66-69, EMER RES ZOOLOGIST, UNIV CALIF, BERKELEY, 69- *Mem:* Swiss Soc Natural Sci; hon mem Swiss Zool; Sigma Xi. *Res:* Cytology of toads; sex reversal in fish; genetics of human twinning; Gastropoda of California coast; laboratory techniques. *Mailing Add:* 1584 Milvia St Berkeley CA 94709

STOHLMAN, STEPHEN ARNOLD, b Long Beach, Calif, Oct 4, 46. MICROBIOLOGY, VIROLOGY. *Educ:* Calif State Univ, BS, 70, MS, 72; Univ Md, PhD(microbiol), 75. *Prof Exp:* ASST PROF NEUROL & MICROBIOL, UNIV SOUTHERN CALIF, 75- *Concurrent Pos:* Co-investr, NIH grant, 77-81; prin investr, Nat Multiple Sclerosis Soc grant, 79-81. *Mem:* Am Soc Microbiol; Tissue Cult Asn; Soc Gen Microbiol. *Res:* Neurovirology; molecular biology; immunology. *Mailing Add:* Dept Neurol & Microbiol Univ Southern Calif 1433 San Paldo St Los Angeles CA 90033

STHR, JOACHIM, b Meinerzhagen, WGer, Sept 28, 47; m 83; c 3. STRUCTURE OF SURFACES, THIN FILM MAGNETISM. *Educ:* Rheinische Friedrich Wilhelms Univ Bonn, Vordiplom, 69; Wash State Univ, MSc, 71; Tech Univ Munchen, Dr rer nat, 74. *Prof Exp:* Scientist solid state physics, Lawrence Berkeley Lab, 76-77; sr res assoc surface sci, Stanford Synchrotron Radiation Lab, 77-81; sr staff physicist surface sci, Exxon Res & Eng Co, 81-85; res staff mem surface sci, 85-89, MGR CONDENSED MATTER SCI, IBM RES DIV, ALMADEN RES CTR, 89- *Concurrent Pos:* Mem, Steering Comt, Advan Photon Source, 84-89, Int Organizing Comt, Int Conf X-Ray Absorption Fine Struct, 86-, Dept Energy Synchrotron Radiation Facil Rev Comt, 87-88, Tenure Proposal Rev Comt, Nat Synchrotron Light Source, 89; consult prof elec eng, Stanford Univ, 89- *Mem:* Am Phys Soc; Am Vacuum Soc; AAAS. *Res:* Development of new synchrotron radiation based surface x-ray absorption techniques and their use for the determination of the geometric arrangement of atoms and molecules on surfaces and of the magnetic properties of surfaces. *Mailing Add:* IBM Res Div Almaden Res Ctr 650 Harry Rd K32/802 San Jose CA 95120-6099

STHRER, GERHARD, b Heidelberg, Ger, May 28, 39. ORGANIC CHEMISTRY, BIOCHEMISTRY. *Educ:* Univ Heidelberg, dipl chem, 62, PhD(chem), 65. *Prof Exp:* Assoc biochem, 66-80, ASST PROF, BIOCHEM UNIT, KETTERING LAB, SLOAN-KETTERING INST CANCER RES, 80- *Mem:* Am Chem Soc; Ger Chem Soc. *Res:* Molecular biology of oncogenesis. *Mailing Add:* 20 Stafford Pl Larchmont NY 10538-2722

STOIBER, RICHARD EDWIN, b Cleveland, Ohio, Jan 28, 11; m 41; c 2. VOLCANOLOGY, ECONOMIC GEOLOGY. *Educ:* Dartmouth Col, AB, 32; Mass Inst Technol, PhD(econ geol), 37. *Prof Exp:* Instr geol, Dartmouth Col, 35-36 & 37-50, from asst prof to prof, 40-71, Frederick Hall prof, 71-89; RETIRED. *Concurrent Pos:* Part-time with Nfld Geol Surv pvt mining indust & US Geol Surv; govt-sponsored volcanic res, Worldwide, espec Cent Am; consult, UN; pvt consult. *Mem:* Mineral Soc Am; Soc Econ Geol; Geol Soc Am; Am Inst Mining, Metall & Petrol Eng. *Res:* Volcanoes; ore deposits; optical crystallography. *Mailing Add:* RR 2 Box 68 Norwich VT 05055

STOICHEFF, BORIS PETER, b Bitol, Yugoslavia, June 1, 24; nat Can; m 54; c 1. LASERS, ATOMIC & MOLECULAR SPECTROSCOPY. *Educ:* Univ Toronto, BASc, 47, MA, 48, PhD(physics), 50. *Hon Degrees:* DSc, Univ Skopje, 81, York Univ, 82, Univ Windsor, 89. *Prof Exp:* McKee-Gilchrist fel, Univ Toronto, 50-51; fel, Nat Res Coun Can, 52-53, res officer, Div Pure Physics, 53-64; chmn eng sci, 72-77, univ prof physics, 77-90, DIR, ONT LASER & LIGHTWAVE RES CTR, UNIV TORONTO, 88-, EMER UNIV PROF PHYSICS, 90- *Concurrent Pos:* Vis scientist, Mass Inst Technol, 63-64; I W Killam mem scholar, 77-79; mem, Coun Nat Res, Coun Can, 77-83; vis scientist, Stanford Univ, 77; sr fel, Massey Col, 79-; Geoffrey Frew fel, Australian Acad Sci, 80; res fel, Japan Soc Prom Sci, 86. *Honors & Awards:* Gold Medal, Can Asn Physicists, 74; William F Meggers Award, Optical Soc Am, 81; Officer of the Order of Can, 82; Frederic Ives Medal, OptiSoc Am, 83; H L Welsh lectr, Univ Toronto, 84; Elizabeth Laird lectr, Univ Western Ont, 85; UK-Can Rutherford lectr, 89; Henry Marshall Tory Medal, Royal Soc Can, 89. *Mem:* Fel Am Phys Soc; hon fel Indian Acad Sci; fel Royal Soc London; fel Royal Soc Can; fel Optical Soc Am (pres-elect, 75, pres, 76); Can Asn Physicists (pres, 83); hon fel Macedonian Acad Sci & Arts; hon foreign fel Am Acad Arts & Sci. *Res:* Molecular spectroscopy and structure; Rayleigh, Brillouin and Raman scattering; lasers and their applications in spectroscopy; stimulated scattering processes and two photon absorption; elastic constants of rare gas single crystals; vacuum ultraviolet laser spectroscopy. *Mailing Add:* Dept Physics Univ Toronto Toronto ON M5S 1A7 Can

STOJANOVIC, BORISLAV JOVAN, b Zajecar, Yugoslavia, Nov 29, 19; US citizen; m 52. MICROBIOLOGY, BIOCHEMISTRY. *Educ:* Univ Bonn, BS, 48, Dr Agr, 50; Cornell Univ, MS, 55, PhD(soil microbiol), 56. *Prof Exp:* From asst prof to assoc prof soil microbiol, 67-74, PROF MICROBIOL & ENOL & HEAD ENOL LAB, MISS STATE UNIV, 74- *Mem:* Am Soc Enol; Am Soc Agron; Am Soc Microbiol. *Res:* Microbial transformations of soil proteinaceous materials and biocides; enzymes involved in browning of juices and wines. *Mailing Add:* Dept Agron Miss State Univ Box 674 Mississippi State MS 39762

STOKELY, ERNEST MITCHELL, b Greenwood, Miss, Mar 26, 37; m 64; c 2. BIOMEDICAL ENGINEERING. *Educ:* Miss State Univ, BSEE, 59; Southern Methodist Univ, MSEE & EE, 68, EE, 71, PhD(biomed eng), 73. *Prof Exp:* Sr elec engr, Tex Instruments, Inc, 59-69; asst prof, 73-80, ASSOC PROF RADIOL, UNIV TEX HEALTH SCI CTR, DALLAS, 80- *Concurrent Pos:* Adj prof, Southern Methodist Univ, 73- & Univ Tex, Arlington, 73- *Mem:* Inst Elec & Electronic Engrs; Nat Asn Biomed Engrs. *Res:* Medical image and signal processing; biological system modeling. *Mailing Add:* Dept Comput Sci & Eng Univ Tex Arlington TX 76019

STOKER, HOWARD STEPHEN, b Salt Lake City, Utah, Apr 16, 39; m 64; c 7. INORGANIC CHEMISTRY, ENVIRONMENTAL CHEMISTRY. *Educ:* Univ Utah, BA, 63; Univ Wis-Madison, PhD(chem), 68. *Prof Exp:* Assoc prof, 68-77, PROF INORG CHEM, WEBER STATE COL, 77- *Mem:* Am Chem Soc. *Res:* Air pollution; writer of 13 textbooks on general chemistry and environmental chemistry. *Mailing Add:* Dept Chem Weber State Univ Ogden UT 84408

STOKER, JAMES JOHNSTON, b Dunbar, Pa, Mar 2, 05; m 28; c 4. MATHEMATICS, MECHANICS. *Educ:* Carnegie Inst Technol, BS, 27, MS, 31; Polytech, Zurich, DrMath, 36. *Hon Degrees:* DSc, NY Univ, 85. *Prof Exp:* Instr mech, Carnegie Inst Technol, 28-31, asst prof, 31-37; from asst prof to prof math, NY Univ, 37-85, dir, Courant Inst Math Sci & head, all-univ dept math, 58-85; RETIRED. *Concurrent Pos:* Res mathematician, Appl Math Panel, Nat Defense Res Comt, 43-45. *Honors & Awards:* Heineman Prize, Am Phys Soc. *Mem:* Nat Acad Sci; AAAS; Am Math Soc. *Res:* Differential geometry; elasticity; vibration theory; hydrodynamics. *Mailing Add:* 67 Stephen Dr Tarrytown NY 10591

STOKER, WARREN C, b Union Springs, NY, Jan 30, 12; m 34; c 3. ELECTRICAL ENGINEERING. *Educ:* Rensselaer Polytech Inst, EE, 33, MEE, 34, PhD(physics). *Prof Exp:* From instr to assoc prof, 34-51, vpres, 61-74, pres, 74-76, PROF ELEC ENG, RENSSELAER POLYTECHNIC INST CONN, 51-, TRUSTEE, 61-, EMER PRES, 76- *Concurrent Pos:* Chief engr, Radio Sta WHAZ, 44-51; head comput lab, Hartford Grad Ctr, Rensselaer Polytechnic Inst Conn, 51-55, dir, 55-57, dean, 57-70; mem res sci adv comt, United Aircraft Corp. *Mem:* AAAS; fel Inst Elec & Electronics Engrs; Am Soc Eng Educ; Newcomen Soc NAm. *Res:* Electronic instrumentation; electromagnetic shielding and noise measurements; leakage and radiation; servomechanisms; analog computing; automatic control systems. *Mailing Add:* 4 Neptune Dr Groton CT 06340

STOKES, ARNOLD PAUL, b Bismarck, NDak, Jan 24, 32; m 57; c 6. PURE MATHEMATICS. *Educ:* Univ Notre Dame, BS, 55, PhD(math), 59. *Prof Exp:* Staff mathematician, Res Inst Adv Study, 58-60; NSF fel math, Johns Hopkins Univ, 60-61; from asst prof to assoc prof, Catholic Univ, 61-65; chmn dept, 67-70, PROF MATH, GEORGETOWN UNIV, 65- *Concurrent Pos:* Sr res assoc, Nat Res Coun-Nat Acad Sci, Goddard Space Flight Ctr, NASA, 74-75; consult ocean acoustics, SAI, McLean, Va, 80- *Res:* Scattering from random surfaces; acoustical and electromagnetic waves; nonlinear differential equations. *Mailing Add:* Dept of Math Georgetown Univ Washington DC 20057

STOKES, BARRY OWEN, b San Francisco, Calif, Jan 10, 45; m 68; c 5. BIOLOGICAL STAINING, FUEL ALCOHOL PRODUCTION FROM CELLULOSE. *Educ:* Utah State Univ, BA, 69, Univ Calif, Los Angeles, PhD(biochem), 74. *Prof Exp:* Sr scientist, Jet Propulsion Lab, Pasadena, Calif, 74-82; vpres res & develop, Biomass Int Inc, Ogden, Utah, 82-86; SR SCIENTIST, WESCOR INC, LOGAN, UTAH, 86- *Concurrent Pos:* Consult, Biomass Int Inc, Ogden, Utah, 88- *Res:* Biological staining for medical applications; microbiological monitoring; alcohol production from cellulose. *Mailing Add:* 1591 E 1220 N Logan UT 84321

STOKES, BRADFORD TAYLOR, b Beverly, Mass, Jan 22, 44; m 67; c 2. NEUROPHYSIOLOGY, DEVELOPMENTAL NEUROBIOLOGY. *Educ:* Univ Mass, BA, 66; Univ Rochester, PhD(physiol), 73. *Prof Exp:* Res assoc physiol, Univ Rochester, 69-73; asst prof, 73-77, ASSOC PROF PHYSIOL, OHIO STATE UNIV, 78- *Concurrent Pos:* Proj investr, Muscular Dystrophy Asn, 75-, NIH, 78-83 & NSF, 79-81. *Mem:* AAAS; Am Physiol Soc; Int Soc Oxygen Transp Tissue; Neurosci Soc. *Res:* Developmental neurophysiology; the development of motor systems; bioelectrical activity in normal and abnormal spinal cords; the effects of acute changes in blood gases on fetal neurogenesis. *Mailing Add:* Dept Physiol Col Med Ohio State Univ 4196 Graves 333 W Tenth Ave Columbus OH 43210

STOKES, CHARLES SOMMERS, b Philadelphia, Pa, Apr 24, 29; m 54; c 2. PHYSICAL CHEMISTRY. *Educ:* Ursinus Col, BS, 51; Temple Univ, MA, 53. *Prof Exp:* Res chemist, Germantown Labs, Inc, 53-56, res assoc, 56-61, mgr test site, 61-72, vpres, 72-80; mgr, Elvenson Test Facil, Franklin Res Ctr, 80-84, head spec proj, 84-87, head Appl Sci Dept, 87-89, DIR, FRANKLIN RES CTR, 90- *Mem:* Am Chem Soc; Am Inst Aeronaut & Astronaut; Am Inst Chem; Combustion Inst; Sigma Xi. *Res:* Fluorine chemistry; propellents; high temperatures; energy research; plasma jet chemistry. *Mailing Add:* 127 Madison Rd Willow Grove PA 19090

STOKES, DAVID KERSHAW, JR, b Camden, SC, Feb 3, 27; m 50; c 3. FAMILY MEDICINE. *Educ:* Clemson Univ, BS, 48; Univ Ga, MS, 52; Tex A&M Univ, PhD, 56; Med Univ SC, MD, 57. *Prof Exp:* Assoc prof, 72-80, CLIN ASSOC & PROF FAMILY PRACT, MED UNIV SC, 72-; MED DIR, CAMPHAVEN NURSING HOME, INMAN, SC, 65- *Concurrent Pos:* Mem, bd dir, Am Geriat Soc, 85- *Mem:* AMA; Am Heart Asn; Am Acad Family Physicians; Am Rheumatism Asn; Am Med Dir Asn; Am Geriat Soc. *Res:* Geriatric medicine. *Mailing Add:* 63 Blackstock Rd Inman SC 29349

STOKES, DONALD EUGENE, b Andalusia, Ala, Aug 25, 31; c 5. NEMATOLOGY. *Educ:* Univ Fla, BS, 55, MA, 63, PhD(nematol), 72. *Prof Exp:* Nematologist III, Div Plant Indust, Fla Dept Agr & Consumer Serv, 56-75, chief nematol, 75-83; AT DEPT ENTOM, UNIV FLA, GAINESVILLE. *Concurrent Pos:* Courtesy Fac Appl & Grad Fac Status, Univ Fla, 75. *Mem:* Soc Nematologists; Orgn Trop Am Nematologists; Europ Soc Nematologists. *Res:* Regulatory aspects of nematology, which include pathogenicity, taxonomy and response to chemicals of the various plant parasitic nematodes. *Mailing Add:* 3821 SW 20th St Gainesville FL 32606

STOKES, GERALD MADISON, b Burlington, Vt, Aug 16, 47; m 71; c 2. ENERGY USE IN BUILDINGS. *Educ:* Univ Calif, Santa Cruz, BA, 69; Univ Chicago, MS, 71, PhD(astron), 77. *Prof Exp:* Fel astron, Battelle Mem Inst, 76-78, res scientist, 78-80, sr res scientist space sci, 80-83, sect mgr, 83-85, prof mgr, end use load & conserv assessment prog, Pac Northwest Labs, 85-86, dept mgr computational sci, 86-88, cent mgr appl physics, 88-90, DIR, GLOBAL STUDIES PROG, BATTELLE MEM INST, 89- *Concurrent Pos:* Tech dir, Atmospheric Radiol Measurement Prog, Dept Energy, 90- *Mem:* Am Astron Soc; AAAS; Am Geophys Union. *Res:* Formation and destruction of interstellar grains; polarimetry of x-ray binaries; abundances of atmospheric trace gases; electrical end use in buildings. *Mailing Add:* Battelle Pac Northwest Labs PO Box 999 Richland WA 99352

STOKES, GERALD V, b Chicago, Ill, Mar 25, 43; m; c 2. MICROBIOLOGY. *Educ:* Southern Ill Univ, BA, 67; Univ Chicago, PhD(microbiol), 73. *Prof Exp:* Fel & res assoc, dept molecular, cellular & develop biol, Univ Colo, Boulder, 73-76; asst prof, dept microbiol, Meharry Med Col, 76-78; asst prof, 78-80, ASSOC PROF, DEPT MICROBIOL & IMMUNOL, SCH MED & HEALTH SCI, GEORGE WASH UNIV, 80- *Concurrent Pos:* Fel, Am Can Soc, 74-76 & NIH, 76; res grants, NIH, 77-78 & NSF, 77-81. *Mem:* Am Soc Microbiol; Electron Micros Soc Am; Sigma Xi. *Res:* Developmental processes and molecular biology of the chlamydial organisms and fine structure electron microscopy; attachment receptors and virulence factors of Chlamydia psittaci and C trachomatis. *Mailing Add:* Dept Microbiol & Immunol George Wash Univ Med Ctr 2300 Eye St NW Washington DC 20037

STOKES, GORDON ELLIS, b Ogden, Utah, Feb 11, 33; m 55; c 9. COMPUTER SCIENCE, PHYSICS. *Educ:* Brigham Young Univ, BS, 61, EdD, 81; Univ Idaho, MS, 69. *Prof Exp:* Physicist, Phillips Petrol Corp, 61-66; sr res physicist, Idaho Nuclear Corp, 66-69; asst dean phys & eng sci, 69-70, asst dir comput sci, 70-73, FAC COMPUT SCI, BRIGHAM YOUNG UNIV, 73- *Concurrent Pos:* Consult, State of Ark, 75-, Weidner Commun Corp, 77-, Geneal Dept, Latter-Day Saint Church, 77-, Winnebago Indust, 78- & Eyring Res Inst, 78- *Mem:* Asn Comput Mach. *Res:* Distributed data base systems; small computer applications; computer management; computer systems in business, government and industry. *Mailing Add:* Dept Comput Sci TMCB 230C Brigham Young Univ Provo UT 84602

STOKES, HAROLD T, b Long Beach, Calif, Jan 27, 47; m 74; c 4. NUCLEAR MAGNETIC RESONANCE, PHASE TRANSITIONS. *Educ:* Brigham Young Univ, BS, 71; Univ Utah, PhD(physics), 77. *Prof Exp:* Instr physics, Univ Utah, 77-78; res assoc, Univ Ill, 78-81; asst prof, 81-85, ASSOC PROF PHYSICS, BRIGHAM YOUNG UNIV, 85- *Mem:* Am Phys Soc; Am Asn Physics Teachers. *Res:* Molecular and atomic motions in solids; platinum catalysts; phase transitions in solids. *Mailing Add:* Dept Physics & Astron Brigham Young Univ 296 ESC Provo UT 84602

STOKES, JACOB LEO, b Warsaw, Poland, Sept 27, 12; US citizen; m 42; c 2. MICROBIOLOGY. *Educ:* Rutgers Univ, BS, 34, PhD(microbiol), 39; Univ Ky, MS, 36. *Prof Exp:* Asst bact, Univ Ky, 34-36; asst marine bact, Rutgers Univ, 36-37, soil microbiol, 37-39; from microbiologist to head sect microbiol metab, Res Labs, Merck & Co, Inc, 39-47; res assoc, Hopkins Marine Sta, Stanford Univ, 48-50; assoc prof bact, Ind Univ, 50-53; bacteriologist, Western Utilization Res Br, USDA, 53-59; chmn dept bact & pub health, 59-68, prof, 68-78, EMER PROF BACT & PUB HEALTH, WASH STATE UNIV, 78- *Mem:* AAAS; Am Soc Microbiol. *Res:* Relation of algae to other microorganisms in nature; antibiotics; iron bacteria; nutrition, physiology and biochemistry of microorganisms; psychrophilic microorganisms. *Mailing Add:* NE 1040 Monroe Pullman WA 99163

STOKES, JIMMY CLEVELAND, b Cochran, Ga, Nov 29, 44; m 72; c 2. SCIENCE EDUCATION. *Educ:* Univ Ga, BS, 66, MEd, 67, EdD(chem educ), 69. *Prof Exp:* Instr chem, Univ Ga, 69-70; asst prof, Clayton Jr Col, 70-74; ASSOC PROF CHEM, W GA COL, 74- *Concurrent Pos:* Consult, Ga High Sch Asn, 76- *Mem:* Am Chem Soc; Nat Sci Teachers Asn; Sigma Xi. *Res:* Construction and evaluation of self paced and individualized programs of instruction in chemistry; development of drug education materials and programs. *Mailing Add:* Dept Chem W Ga Col Carrollton GA 30117

STOKES, JOSEPH FRANKLIN, b Havana, Ark, Feb 27, 34; m 59; c 2. MATHEMATICS. *Educ:* Univ Ark, Fayetteville, BS, 56, MA, 57; George Peabody Col, PhD(math), 72. *Prof Exp:* Instr math, Kans State Univ, 58-61; instr, Auburn Univ, 61-62; assoc prof, 62-80, PROF MATH, WESTERN KY UNIV, 80- *Concurrent Pos:* Vchmn, Math Asn Am, Ky Sect; res partic, Oak Ridge Assoc Univ, 67. *Mem:* Math Asn Am; AAAS; Nat Coun Teachers Math. *Res:* Author of several books and journals in calculus, linear algebra, and statistics. *Mailing Add:* Dept Math Western Ky Univ Bowling Green KY 42101

STOKES, PAMELA MARY, b Hertford, UK, June 24, 35; m 58; c 2. TOXICOLOGY. *Educ:* Univ Bristol, BSc, 56, PhD(bot), 59. *Prof Exp:* Fel mycol, Imperial Col, London, 59-60; lectr bot, Sir John Cass Col, London, 60-63; res assoc plant pathol, Univ Ill, 64-65; instr biol, Univ Toronto, 69-70, assoc prof, 73-82, prof bot, 82-, dir, Inst Environ Studies, 84-; PROF, BIOL DEPT, TRENT UNIV. *Mem:* Can Bot Asn; Am Phycol Soc; Am Mycol Soc; Am Soc Limnol & Oceanog; Am Soc Geochem & Health. *Res:* The response of aquatic organisms to high concentrations of metals and low pH; community response in the field; adaptations of algae which confer tolerance to metals; biological monitoring, metal cycling in aquatic systems. *Mailing Add:* Trent Univ PO Box 4800 Peterborough ON K9J 7B8 Can

STOKES, PETER E, b Haddonfield, NJ, Aug 27, 26; m 56; c 3. ENDOCRINOLOGY, BIOLOGICAL PSYCHIATRY. *Educ:* Trinity Col, BS, 48; Cornell Univ, MD, 52; Am Bd Internal Med, dipl, 57; Am Bd Psychiat & Neurol, dipl, 71; Am Bd Radiol, dipl & cert nuclear med, 72. *Prof Exp:* Intern med, New York Hosp, 52-53, asst resident, 53-54, asst resident med & endocrinol, 54-55; NIH trainee fel endocrinol, 55-57; instr med in endocrinol, 57-59, asst prof med in psychiat, 59-63, ASSOC PROF PSYCHIAT & MED, MED COL, CORNELL UNIV, 69-, PROF PSYCHIAT, 82- & PROF MED, 87- *Concurrent Pos:* Physician, Outpatient Clin, New York Hosp, 55-56, from asst attend to assoc attend physician, 61-86, attend physician, 87-; dir clin res labs, Payne Whitney Psychiat Clin, New York Hosp-Cornell Med Ctr, 57-, dir Psychobiol Lab & Study Unit, 67, assoc attend psychiatrist, 69-82, attend psychiatrist, 82-; assoc vis physician, Cornell Div, Bellevue Hosp, 62-70; pvt pract consult, 59- *Mem:* Endocrine Soc; Am Soc Nuclear Med; fel Am Psychiat Asn; fel Am Col Neuropsychopharmacol; fel Am Col Physicians. *Res:* Neuroendocrine function in emotional disorders; hypothalamic pituitary adrenocortical function control systems in animals; effects of alcohol on neuroendocrine function; clinical endocrine problems including growth and thyroid; adrenal and ovarian function; clinical studies of affective disorders; lithium metabolism and physiological effects of lithium isotopes on manic behavior, renal, thyroid and cognitive function. *Mailing Add:* Dept Med/Psychiat Cornell Univ Med Col 1300 York Ave New York NY 10021

STOKES, RICHARD HIVLING, b Troy, Ohio, Apr 30, 21; m 56; c 2. EXPERIMENTAL PHYSICS, ACCELERATOR PHYSICS. *Educ:* Case Univ, BS, 42; Iowa State Univ, PhD(physics), 51. *Prof Exp:* Staff mem, Underwater Sound Ref Lab, Nat Defense Res Comt, 42-44; res assoc, Inst Atomic Res, Iowa State Col, 46-51; group leader, 67-77 & 81-84, STAFF MEM, LOS ALAMOS SCI LAB, 51-67 & 84- *Concurrent Pos:* US del, Conf Peaceful Uses Atomic Energy, Geneva, 58; lectr, Univ Minn, 60-61. *Mem:* Fel Am Phys Soc; Sigma Xi. *Res:* Nuclear physics; accelerator research; fission; spectroscopy of light nuclei; nuclear detectors; heavy ion accelerators; heavy ion reactions. *Mailing Add:* 2450 Club Rd Los Alamos NM 87544

STOKES, ROBERT ALLAN, b Richmond, Ky, June 25, 42; m 63. ASTROPHYSICS. *Educ:* Univ Ky, BS, 64; Princeton Univ, MA, 66, PhD(physics), 68. *Prof Exp:* Asst prof physics & astron, Univ Ky, 68-72; sr scientist, Pac Northwest Labs, Battelle Mem Inst, 72-74, mgr space sci, 74-83, assoc mgr geosci dept, 83-85, mgr eng physics dept, 85-86, mgr appl physics ctr, 86-88; DEP DIR, SOLAR ENERGY RES INST, 88- *Concurrent Pos:* Adv comt, Geophys Inst, Univ Alaska, 75- *Mem:* AAAS; Am Phys Soc; Am Astron Soc. *Res:* Experimental cosmology and relativity; planetary astrophysics; radiative transfer theory. *Mailing Add:* Solar Energy Res Inst 1617 Cole Blvd Golden CO 80401

STOKES, ROBERT JAMES, materials science, ceramics science, for more information see previous edition

STOKES, ROBERT MITCHELL, b Vandalia, Ill, May 21, 36; m 59; c 3. COMPARATIVE PHYSIOLOGY. *Educ:* Mich State Univ, BS, 58, MS, 59, PhD(physiol), 63. *Prof Exp:* From asst prof to assoc prof, 63-77, PROF BIOL SCI, KENT STATE UNIV, 77- *Concurrent Pos:* Consult, Great Lakes Basin Comn Water Qual Task Force, 68-70. *Mem:* Am Soc Zool. *Res:* Biochemical and biophysical aspects of membrane transport phenomena in fish, especially glucose transport by intestine; fish physiology, metabolism and toxicology. *Mailing Add:* 1074 Elno Ave Kent OH 44240

STOKES, RUSSELL AUBREY, b Preston, Miss, May 1, 22; m 59. MATHEMATICS. *Educ:* Miss State Univ, BS, 48; Univ Miss, MA, 51; Univ Tex, PhD(math), 63. *Prof Exp:* From asst prof to assoc prof, 56-66, PROF MATH, UNIV MISS, 66- *Mem:* Am Math Soc; Math Asn Am. *Res:* Measure and integration. *Mailing Add:* Dept Math Univ Miss Box 278 University MS 38677

STOKES, WILLIAM GLENN, b Corsicana, Tex, Dec 26, 21; c 2. MATHEMATICS. *Educ:* Sam Houston State Col, BS, 46, MA, 47; Peabody Col, PhD(math), 57. *Prof Exp:* Head dept math, Navarro Jr Col, 47-53; instr appl math, Vanderbilt Univ, 54-55; head dept math, Austin Peay State Col, 55-57; assoc prof, Northwestern State Col, 57-59 & East Tex State Univ, 59-60; PROF MATH & COMPUT SCI & CHMN DEPT, AUSTIN PEAY STATE UNIV, 74- *Mem:* Math Asn Am; Am Math Soc. *Mailing Add:* 10800 Culberson Dr Austin TX 78748

STOKES, WILLIAM LEE, b Hiawatha, Utah, Mar 27, 15; m 39; c 4. STRATIGRAPHY, POPULARIZING EARTH SCIENCE. *Educ:* Brigham Young Univ, BS, 37, MS, 38; Princeton Univ, PhD(geol), 41. *Prof Exp:* Asst, Princeton Univ, 41-42; from jr geologist to asst geologist, US Geol Surv, 42-46; from asst prof to prof geol, Univ Utah, 47-82, chmn dept, 54-68; RETIRED. *Concurrent Pos:* Consult, USAEC, 52-54. *Mem:* AAAS; fel Geol Soc Am; Soc Vert Paleont; Am Asn Petrol Geol; Am Geophys Union. *Res:* Mesozoic stratigraphy; dinosaurs; arid lands geomorphology; creationist-evolutionist controversy; sedimentary ore deposits; textbook writing and popularization of earth science. *Mailing Add:* 1283 E South Temple St Apt 504 Salt Lake City UT 84102-1735

STOKES, WILLIAM MOORE, b Cleveland, Ohio, Sept 18, 21. ORGANIC CHEMISTRY. *Educ:* Franklin & Marshall Col, BS, 44; Yale Univ, PhD(chem), 52. *Hon Degrees:* MA, Providence Col, 61. *Prof Exp:* Chemist, Hamilton Watch Co, 44-46; lab asst, Yale Univ, 46-48, 49-51; from asst prof to assoc prof med res, 51-59, PROF CHEM & DIR MED RES LAB, PROVIDENCE COL, 59- *Mem:* Fel AAAS; Am Chem Soc; Am Oil Chem Soc; NY Acad Sci. *Res:* Neurochemistry; isolation of natural products; steroid metabolism; correlation of optical activity with molecular structure. *Mailing Add:* Dept Chem Providence Col Providence RI 02918-0002

STOKEY, W(ILLIAM) F(ARMER), b Cincinnati, Ohio, Apr 19, 17; m 46; c 3. MECHANICAL ENGINEERING. *Educ:* Ga Inst Technol, BS, 38; Mass Inst Technol, MS, 47, ScD(mech eng), 49. *Prof Exp:* Asst prof, 49-55, ASSOC PROF MECH ENG, CARNEGIE-MELLON UNIV, 55- *Mem:* Am Soc Mech Engrs; Soc Exp Stress Anal; Sigma Xi. *Res:* Stress analysis; dynamic shock; vibrations. *Mailing Add:* 136 Columbia Dr Pittsburgh PA 15236

STOKINGER, HERBERT ELLSWORTH, b Boston, Mass, June 19, 09. TOXICOLOGY. *Educ:* Harvard Univ, AB, 30; Columbia Univ, PhD(biochem), 37. *Prof Exp:* Instr chem, City Col New York, 32-39; res assoc bact, Sch Med & Dent, Univ Rochester, 39-43, chief indust hyg sect, AEC, 43-51, from asst prof to assoc prof pharm & toxicol, 45-51; chief toxicologist, Nat Inst Occup Safety & Health, USPHS, 51-77; RETIRED. *Concurrent Pos:* Res assoc, Col Physicians & Surgeons, Columbia Univ, 37-39; res assoc, Atomic Bomb Test, Bikini, 46; mem subcomt toxicol, Nat Res Coun, 46, chmn comt, 70-73; chmm subcomt toxicol, USPHS Drinking Water Stas, 58-70; chmn threshold limits comt, Am Conf Govt Indust Hygienists, 62-77. *Honors & Awards:* Donald E Cummings Mem Award, Am Indust Hygiene Asn, 69; S C Weisfeld Mem Lect Award, 75; HE Stokinger Lectr, Am Conf Govt Ind Hygienists, 79; Inhalation Toxicol Award, Soc Toxicol, 83. *Mem:* AAAS; Am Indust Hyg Asn; Am Asn Immunol; fel Soc Toxicol, 83. *Res:* Pharmacology and toxicology of atomic energy materials; bacteriological chemistry of gonococcus; toxins; chemotherapy of sulfonamides and arsenicals; prophylaxis of industrial poisons; industrial, water and air pollution toxicology. *Mailing Add:* Nine Twin Hills Ridge Dr Cincinnati OH 45228

STOKLEY, JAMES, science education; deceased, see previous edition for last biography

STOKOWSKI, STANLEY E, b Lewiston, Maine, Dec 28, 41. SOLID STATE PHYSICS, OPTICS. *Educ:* Mass Inst Technol, SB, 63; Stanford Univ, PhD(physics), 68. *Prof Exp:* Physicist, Nat Bur of Standards, 68-70; mem tech staff, Bell Tel Labs, NJ, 70-72; prin investr, Res Inst Advan Studies Div, Martin Marietta Corp, 72-77; SR RES SCIENTIST, LAWRENCE LIVERMORE LAB, 77- *Mem:* Am Phys Soc; Am Optical Soc. *Res:* Crystal field theory; phase transitions; ferroelectricity; color centers; optical properties of crystals; laser glass; infrared detectors. *Mailing Add:* 755 Contada Circle Danville CA 94526

STOKSTAD, EVAN LUDVIG ROBERT, b China, Mar 6, 13; US citizen; m 34; c 2. BIOCHEMISTRY. *Educ:* Univ Calif, BS, 34, PhD(animal nutrit), 37. *Prof Exp:* Biochemist, Western Condensing Co, 37-39; biochemist, Golden State Co, Ltd, 39-40; Lalor fel, Calif Inst Technol, 40-41; chemist, Lederle Labs Div, Am Cyanamid Co, 41-63; actg chmn dept nutrit sci, 68-69, prof nutrit, 63-80, chmn dept nutrit sci, 79-80, EMER PROF NUTRIT, UNIV CALIF, BERKELEY, 80-, BIOCHEMIST, AGR EXP STA, 63- *Concurrent Pos:* Mem food & nutrit bd, Div Biol & Agr, Nat Acad Sci-Nat Res Coun, 68-72, comt food standards & fortification policy, 72. *Honors & Awards:* Borden Award, Poultry Sci Asn; Mead-Johnson Award, Am Inst Nutrit; Osborne & Mendel Award, Am Inst Nutrit, 80. *Mem:* Soc Exp Biol & Med; Am Soc Biol Chemists; Am Chem Soc; Poultry Sci Asn; Am Inst Nutrit (treas, 70-73, pres, 76). *Res:* Water soluble vitamin requirements for chicks; antibiotics in animal nutrition; bacterial nutrition; chemistry and biochemistry of thioctic acid; dental nutrition and mineral metabolism; chemistry and metabolism of folic acid and vitamin B12. *Mailing Add:* Dept of Nutrit Sci Univ of Calif Berkeley CA 94720

STOKSTAD, ROBERT G, b Berkeley, Calif, June, 28, 40. NUCLEAR PHYSICS. *Educ:* Yale Univ, BS, 62; Calif Inst Technol, PhD(physics), 67. *Prof Exp:* Asst prof, physics, Yale Univ, 70-74; staff physicist, Oak Ridge Nat Lab, 74-80; SR SCIENTIST, PHYSICS, NUCLEAR SCI DIV, LAWRENCE BERKELEY LAB, 80- *Concurrent Pos:* Chmn, Div Nuclear Physics, Am Physics Soc, 86-87. *Honors & Awards:* Alexander von Humboldt Award, 88-89. *Mem:* Fel Am Phys Soc. *Res:* Study of nuclear structure and heavy iron nuclear reaction mechanisms; experimental studies of nuclear structure. *Mailing Add:* Lawrence Berkeley Lab Bldg 88 Univ Calif Berkeley CA 94720

STOLARIK, EUGENE, b Zilina, Czech, Mar 27, 19; nat US; m 44; c 2. AERODYNAMICS. *Educ:* Prague Tech Univ, dipl, 39; Carleton Col, BA, 40; Univ Minn, MA, 42, MS, 44. *Prof Exp:* Asst physics, Univ Minn, 41-42, instr aerodyn & aircraft, 42-44; chief proj engr, Lawrance Aeronaut Corp, NJ, 44-45; chief engr, Off Res & Inventions Lab, US Dept Navy, 45-47; ASSOC PROF AERONAUT ENG, UNIV MINN, MINNEAPOLIS, 47- *Concurrent Pos:* US deleg, 2nd Int Cong Aeronaut Sci, Zurich; civilian with Nat Defense Res Comt & War Prod Bd, 45; consult, Boeing Co, Wash, 52-60, 62 & 68, Northrop Corp, Calif, 58, 64 & 66, Univac Div, Sperry Rand Corp, 64-66 & Honeywell Corp, 68-69; sr staff engr, Aeronca Mfg Corp, 60; eng staff specialist, Astronaut Div, Gen Dynamics Corp, Calif, 61 & 65; sr staff engr, Lockheed Missiles & Space Co, Calif, 63; dir res & develop, Control Technol Corp, 71; res analyzer, Am Inst Res. *Honors & Awards:* Cert Appreciation, Off Sci Res & Develop, 45. *Mem:* Assoc fel Am Inst Aeronaut & Astronaut; Am Helicopter Soc; Am Soc Eng Educ; Royal Aeronaut Soc. *Res:* Mechanics of flight; pneumatics; control and guidance of aerospace vehicles; convective heat transfer; interference effects; reentry of aerospace vehicles; engineering education. *Mailing Add:* 5529 Wooddale Ave Minneapolis MN 55424

STOLARSKY, KENNETH B, b Chicago, Ill, May 9, 42; m 69. MATHEMATICS. *Educ:* Calif Inst Technol, BS, 63; Univ Wis-Madison, MS, 65, PhD(math), 68. *Prof Exp:* Fel math, Inst Advan Study, 68-69; asst prof, 69-73, ASSOC PROF MATH, UNIV ILL, URBANA, 73- *Mem:* Am Math Soc; Math Asn Am; Sigma Xi. *Res:* Number theory; combinatorics; geometric inequalities. *Mailing Add:* Dept of Math Univ of Ill at Urbana-Champaign Urbana IL 61801

STOLBACH, LEO LUCIEN, b Geneva, Switz, Feb 25, 33; US citizen; m 61; c 2. ONCOLOGY, INTERNAL MEDICINE. *Educ:* Harvard Univ, BA, 54; Univ Rochester, MD, 58; Am Bd Internal Med, cert, 67. *Prof Exp:* Chief med oncol, Ottawa Clin, Ont Cancer Found & Ottawa Civic Hosp, 81-84; Oncologist, Boston Univ Med Ctr & Boston Va Med Ctr, 84-85; ONCOLOGIST, NORTH EASTERN DEACONESS HOSP BOSTON, 86-MEM STAFF BEHAV MED, 86- *Concurrent Pos:* From instr to asst prof, Sch Med, Tufts Univ, 64-73, assoc prof, 73-81; from assoc prof med, Sch Med, Univ Ottawa, 81-84; clin prof med, Boston Univ Sch Med, 84-; lectr med, Harvard Univ Med Sch, 85- *Mem:* Am Asn Cancer Res; Am Soc Clin Oncol; fel Am Col Physicians; Am Asn Cancer Educ. *Res:* Clinical chemotherapy; tumor immunology; tumor markers; behavioral medicine. *Mailing Add:* 38 Morseland Ave Newton Ctr MA 02159

STOLBERG, HAROLD JOSEF, b San Juan, PR, Aug 4, 40; m 71. MATHEMATICS. *Educ:* Univ PR, Rio Piedras, BS, 62; Cornell Univ, PhD(math), 69. *Prof Exp:* Asst prof math, Ithaca Col, 67-68; fel, Carnegie-Mellon Univ, 68-69, asst prof, 69-71; assoc prof math & chmn dept, Univ PR, Rio Piedras, 71-80; prog mgr, 80-86, SR PROG MGR FOR WESTERN EUROPE & LATIN AM, NSF, 86- *Concurrent Pos:* Consult, Col Entrance Exam Bd, PR, 71-74. *Mem:* Am Math Soc; Math Asn Am; AAAS. *Res:* Commutative algebra; flat modules; mathematics education. *Mailing Add:* 5052 N 36th St Arlington VA 22207

STOLBERG, MARVIN ARNOLD, b New York, NY, Oct 29, 25; m 49; c 2. ORGANIC CHEMISTRY. *Educ:* Columbia Univ, BS, 50; Univ Del, MS, 54, PhD(chem), 56. *Prof Exp:* Org chemist, Chemother Br, US Army Chem Ctr, Md, 50-53, asst br chief, 54-56; head chem dept, Tracerlab, Inc, Mass, 56-60; tech dir-vpres, New England Nuclear Corp, 60-72, pres & chief exec off, 72-85; RETIRED. *Concurrent Pos:* Vpres, E I Dupont de Nemours, 83-85. *Mem:* Am Chem Soc; Soc Nuclear Med. *Res:* Organic and inorganic synthesis with radioactive isotopes; radioactive pharmaceuticals and assay of labeled compounds; tracer techniques for solving problems concerning food and drug acceptability criteria; product evaluation; organic reaction mechanisms. *Mailing Add:* 11253 Boca Woods Lane Boca Raton FL 33428

STOLC, VIKTOR, b Bratislava, Slov, Oct 5, 32; m 73; c 2. ENDOCRINOLOGY, HEMATOLOGY. *Educ:* Univ Comenius Bratislava, RNDr, 56; Slovak Acad Sci, Slov, CSc(biochem), 63. *Prof Exp:* Biochemist, Endocrine Sta, Inst Health, Czech, 56-57; independent scientist, Inst Endocrinol, Slovak Acad Sci, 57-68; res assoc, 65-66 & 68-70, asst res prof, 70-71, RES ASSOC PROF PATH, SCH MED, UNIV PITTSBURGH, 71- *Mem:* AAAS; Endocrine Soc; Am Soc Biol Chemists; NY Acad Sci; Reticuloendothelial Soc. *Res:* Gene rearrangement in leukemia; endocrine factors in normal and leukemic leukocytes. *Mailing Add:* Dept Path 5940 CHP Main Tower Univ of Pittsburgh Pittsburgh PA 15261

STOLDT, STEPHEN HOWARD, b New York, NY, Dec 17, 38; m 65; c 2. ORGANIC CHEMISTRY, FUELS UTILIZATION. *Educ:* Queens Col, NY, BS, 60; City Col New York, MA, 62, PhD(org chem), 68. *Prof Exp:* Res assoc org chem, Univ Wis, 67-68; res chemist fuels & lubricants, Shell Oil Co, 68-73; supvr chem res coal utilization, Apollo Chem Corp, 73-83; lab mgr,

SCA Chem Serv, 83-84; RES SUPVR, FUEL PROD, LUBRIZOL CORP, 84- *Mem:* Nat Asn Corrosion Engrs; Am Chem Soc; Soc Automotive Engrs. *Res:* Combustion; power generation; fuels; lubricants; coal utilization. *Mailing Add:* 7390 Southmeadow Dr Concord Township OH 44077

STOLEN, JOANNE SIU, b Chicago, Ill, June 22, 43; m 72; c 1. IMMUNOLOGY. *Educ:* Univ Mich, BS, 65; Seton Hall Univ, MS, 68; Rutgers Univ, PhD(biochem), 72. *Prof Exp:* Res intern immunol, Inst Microbiol, Rutgers Univ, 69-72; fel, Dept Serol & Bact, Univ Helsinki; RES IMMUNOL, SANDY HOOK LAB & UNIV HELSINKI, 74- *Concurrent Pos:* Adj prof, Drew Univ, 82-; owner, SOS Publ, 88- *Mem:* Sigma Xi; AAAS; Am Fisheries Soc; Am Soc Zoologists; NY Acad Sci. *Res:* Cellular immunology; thymus derived and bone marrow derived cell function in the mouse and presently in lower animals such as the fish; immunoregulation and the effect of stress on the immune system of fish; immunology of marine fishes; publisher of scientific manuals and books. *Mailing Add:* 43 Normandie Ave Fair Haven NJ 07704-3303

STOLEN, ROGERS HALL, b Madison, Wis, Sept 18, 37. SOLID STATE PHYSICS. *Educ:* St Olaf Col, BA, 59; Univ Calif, Berkeley, PhD(physics), 65. *Prof Exp:* Fel, Univ Toronto, 64-66; MEM TECH STAFF SOLID STATE OPTICS, BELL LABS, 66- *Mem:* Am Phys Soc; Optic Soc Am. *Res:* Nonlinear properties of optical fibers; polarization preserving optical fibers; light scattering in glass. *Mailing Add:* AT&T Bell Labs Rm 4B- 421 Holmdel NJ 07733

STOLER, DAVID, b Brooklyn, NY, Aug 21, 36; m 58; c 1. ELEMENTARY PARTICLE PHYSICS, QUANTUM OPTICS. *Educ:* City Col New York, BS, 58; Yeshiva Univ, PhD(physics), 66. *Prof Exp:* Instr physics, Staten Island Community Col, 59-60; sr res asst, Microwave Res Inst, 60-66; from asst prof to assoc prof physics, Polytech Inst NY, 66-75; sr staff physicist, Laser Optics Dept, Electro-Optical Div, Perkin-Elmer Corp, 75-80. *Concurrent Pos:* Instr, Brooklyn Col, 60-61; consult liquid crystal displays, Riker-Maxson Corp, 72. *Mem:* Am Phys Soc. *Res:* Electromagnetic properties of hadrons; quantum field theory; quantum theory of coherence; nonlinear optics; laser physics; laser resonator theory. *Mailing Add:* 175 Adams St Brooklyn NY 11201

STOLER, PAUL, b Brooklyn, NY, June 8, 38; m 66; c 2. INTERMEDIATE ENERGY PHYSICS, ELECTROMAGNETIC PROBES. *Educ:* Brooklyn Col, BS, 60; Rutgers Univ, MS, 62, PhD(physics), 66. *Prof Exp:* PROF PHYSICS, RENSSELAER POLYTECH INST, 66- *Concurrent Pos:* Vis scientist, Cen-Saclay, France, 74-75. *Mem:* Am Phys Soc. *Res:* Intermediate energy nuclear physics using electromagnetic probes; photopion and electropion production from complex nuclei; structure of nucleons and resonances. *Mailing Add:* Dept of Physics Rensselaer Polytech Inst Troy NY 12181

STOLFI, ROBERT LOUIS, b Brooklyn, NY, Sept 16, 38; m 68. TUMMOR IMMUNOLOGY, TUMOR THERAPY. *Educ:* Brooklyn Col, BS, 60; Univ Miami, PhD(microbiol), 67. *Prof Exp:* Bact technician, Jewish Hosp Brooklyn, 60-61; asst bacteriologist, Bellevue Hosp, 61-62; res assoc immunochem, Howard Hughes Med Inst, 63-67; from instr to asst prof microbiol, Sch Med, Univ Miami, 67-71; dir transplantation immunol, 69-78, ASST DIR CANCER RES, DEPT SURG, CATH MED CTR BROOKLYN & QUEENS INC, 78- *Concurrent Pos:* Res assoc immunochem, Variety Children's Res Found, Fla, 67-71; res assoc, Cancer Inst, Univ Columbia, 80- *Mem:* AAAS; Am Asn Immunologists; Am Soc Microbiol; Am Cancer Soc; Sigma Xi; Soc Anal Cytol. *Res:* Therapeutic methods for the alteration of immunological reactivity in the tumor-bearing host, or in the recipient of a histoincompatible normal tissue transplant; analysis of the interactions among drugs; tumor and host immune systems during cancer chemotherapy. *Mailing Add:* St Anthony's Hosp 89-15 Woodhaven Blvd Woodhaven NY 11421

STOLFO, SALVATORE JOSEPH, b Brooklyn, NY, Feb 21, 54; m 79; c 2. COMPUTER SCIENCE, MATHEMATICS. *Educ:* Brooklyn Col, BS, 74; NY Univ, MS, 76, PhD(comput sci), 79. *Prof Exp:* Lectr comput sci, Brooklyn Col, 74-78; asst prof, 79-84, ASSOC PROF COMPUT SCI, COLUMBIA UNIV, 84- *Concurrent Pos:* Consult, AT&T Bell Labs, 80-85, lectr, 85; prin investr, Defense Advan Res Projs Agency, 81-86, Off Navel Res, 82-84 & NY State Sci & Technol Found, 84-; chief sci adv, Fifth Generation Comput Corp, 85- *Mem:* Inst Elec & Electronics Engrs; Sigma Xi; NY Acad Sci; Am Asn Artificial Intel; Asn Comput Mach. *Res:* Principal architect of an advanced parallel computer called DADO to accelerate artificial intelligence programs and speech recognition tasks; knowledge-based expert systems applied to conventional data base and management information systems. *Mailing Add:* Dept Comput Sci Columbia Univ New York NY 10027

STOLINE, MICHAEL ROSS, b Jefferson, Iowa, Sept 17, 40; m 60; c 4. STATISTICAL ANALYSIS. *Educ:* Univ Iowa, BA, 62, MA, 64, PhD(statist), 67. *Prof Exp:* Assoc prof, 67-77, PROF MATH, WESTERN MICH UNIV & STATIST, 77- *Mem:* Am Statist Asn; Inst Math Statist. *Res:* Problems in the analysis of variance and regression; intervention time series; multiple comparisons. *Mailing Add:* Dept Math 5510 Everett Western Mich Univ Kalamazoo MI 49008

STOLK, JON MARTIN, b Englewood, NJ, Oct 15, 42; m 73. PSYCHIATRY, PHARMACOLOGY. *Educ:* Middlebury Col, AB, 64; Dartmouth Col, PhD(pharmacol), 69; Stanford Univ, MD, 72. *Prof Exp:* Fel psychiat, Sch Med, Stanford Univ, 69-71; asst prof pharmacol, Dartmouth Med Col, 72-73, asst prof psychiat, 74-76, assoc prof, 76-78; PROF PSYCHIAT, SCH MED, UNIV MD, 78- *Mem:* Am Soc Pharmacol & Exp Therapeut; Int Soc Psychoneuroendocrinol; Am Col Neuropsychopharmacology. *Res:* Biogenic amines and behavior; psychopharmacology; neurochemistry and genetics. *Mailing Add:* 3620 Sweeten Creek Dr Chapel Hill NC 27514

STOLL, MANFRED, b Calw, Ger, Aug 24, 44; US citizen; m 66, 90; c 3. MATHEMATICAL ANALYSIS. *Educ:* State Univ NY Albany, BS, 67; Pa State Univ, MA, 69, PhD(math), 71. *Prof Exp:* From asst prof to assoc prof, 71-84, PROF MATH, UNIV SC, 85- *Mem:* Am Math Soc. *Res:* Harmonic, holomorphic and plurisubharmonic function theory on bounded symmetric domains and generalized half planes; spaces and algebras of holomorphic functions of one and several complex variables. *Mailing Add:* Dept Math Univ SC Columbia SC 29208

STOLL, PAUL JAMES, b Grass Valley, Calif, June 23, 33; m 59; c 1. ELECTRICAL ENGINEERING, BIOENGINEERING. *Educ:* Wesleyan Univ, BA, 55; Mass Inst Technol, SB, 57, SM, 58; Univ Wash, PhD(elec eng), 68. *Prof Exp:* Teaching asst, Mass Inst Technol, 57-58; res engr guid systs, Autonetics Div, NAm Rockwell Corp, 58-61, consult, 61-62; NIH fel, Sch Med, Univ Wash, 68-69; asst prof elec eng, Univ Calif, Davis, 69-80, lectr, 76-80; MEM STAFF, AEROSPACE CORP, 80- *Mem:* AAAS; Inst Elec & Electronics Engrs; Biomed Eng Soc. *Res:* Control system engineering; regulation of breathing in man and domestic fowl; physiological control system analysis and modeling. *Mailing Add:* 2559 Leafwood Dr Camarillo CA 93010

STOLL, RICHARD E, METALLURGY. *Educ:* Ohio State Univ, BMetE, 49. *Prof Exp:* Metallurgist, Corps Engrs, Ft Belvoir, Va, US Steel Corp, 50-52, technologist, South Works, Ill, 52-55, petrographer, 55-56, gen supvr, Mat Dept, 56-59, chief raw mat metallurgist, 59-60, chief qual assurance metallurgist, 60-65, asst chief metallurgist, 65-66, chief metallurgist, 66-68, supt, Univ Plate Mill, 68-69, supt structural mills, 69-71, mgr process metall, heavy & tubular prod, 71-75, area mgr, Customer Tech Serv, Houston, Tex, US Steel Corp, 75-78, gen mgr, Metall Serv, 78-83, dir, Qual Mgt & Technol Implementation Prog, 83-84; chief metallurgist, 84-85, vpres & gen mgr, Flat Rolled Prod Profit Ctr, 86-87, VPRES & INTERIM CHIEF OPERATING OFFICER, WHEELING-PITTSBURGH STEEL CORP, 87- *Concurrent Pos:* Consult, 84; bd mem, Ohio Valley Indust & Bus Develop Corp, 86-; chmn, Tech Interaction Subcomt, Am Iron & Steel Inst, mem, Mfg Comt. *Mem:* Fel Am Soc Metals; Am Iron & Steel Inst; Am Inst Mining & Metall Engrs; Am Soc Qual Control. *Res:* Develop line pipe steel composition and processing practices to meet critical Arctic requirements; develop plate steels with guaranteed through-thickness ductibility properties; metallurgy. *Mailing Add:* Wheeling-Pittsburgh Steel Corp 1134 Market St Wheeling WV 26003

STOLL, ROBERT D, b Lincoln, Ill, Aug 12, 31; c 3. SOIL MECHANICS. *Educ:* Univ Ill, BSCE, 53; Columbia Univ, MSCE, 56, EngScD(civil eng, eng mech), 62. *Prof Exp:* From instr to assoc prof, 56-71, PROF CIVIL ENG, COLUMBIA UNIV, 71- *Concurrent Pos:* Chmn comt mech earth masses & layered systs, Hwy Res Bd, Nat Acad Sci-Nat Res Coun, 67-; vis sr res assoc, Lamont-Doherty Geol Observ, 69- *Mem:* Am Soc Civil Engrs. *Res:* Static and dynamic response of granular soils; wave propagation in granular media and ocean sediments; general constitutive relationships for granular media. *Mailing Add:* Dept Civil Eng Columbia Univ New York NY 10027

STOLL, WILHELM, b Freiburg, Ger, Dec 22, 23; m 55; c 4. MATHEMATICS. *Educ:* Univ Tubingen, Dr rer nat, 53, Dr habil, 54. *Prof Exp:* Asst math, Univ Tubingen, 53-59, docent, 54-60, appl prof, 60; head dept math, 66-68, PROF MATH, UNIV NOTRE DAME, 60-, VINCENT F DUNCAN & ANNA MARIE MICUS DUNCAN PROF MATH, 88- *Concurrent Pos:* Vis lectr, Univ Pa, 54-55; mem, Inst Adv Study, 57-59; vis prof, Stanford Univ, 68-69, Tulane Univ, 73 Kyoto Univ, 83 & Univ Sci & Technol, Hefei, China. *Mem:* Am Math Soc; Math Asn Am; Ger Math Asn. *Res:* Complex analysis; value distribution in several variables, modifications meromorphic maps; families of divisors; continuation of analytic sets and maps; algebraic dependence of meromorphic functions; parabolic spaces. *Mailing Add:* Dept of Math Univ of Notre Dame Notre Dame IN 46556

STOLL, WILLIAM FRANCIS, b Lamoni, Iowa, July 21, 32; m 56; c 2. FOOD SCIENCE. *Educ:* Iowa State Univ, BS, 55, MS, 57; Univ Minn, St Paul, PhD(dairy sci), 66. *Prof Exp:* Asst prof dairy sci, SDak State Univ, 57-67; sr food scientist, Prod Develop Dept, Green Giant Co, 67-79; sr scientist, frozen foods res & develop, Pillsbury Co, 79-82; ASSOC PROF, UNIV OF MINN, WASECA, 82- *Mem:* Inst Food Technologists; Am Dairy Sci Asn. *Res:* Use of physical, chemical, microbiological principles for design and fabrication of new food products. *Mailing Add:* Dept Agr Mgt Univ Minn Tech C Waseca Waseca MN 56093

STOLL, WILLIAM RUSSELL, b Los Angeles, Calif, July 8, 31; m 55; c 2. PHARMACOLOGY, CHEMISTRY. *Educ:* Union Univ, NY, BS, 52; Univ Rochester, PhD(pharmacol), 56. *Prof Exp:* From instr to assoc prof, 56-70, RES ASSOC PHARMACOL, ALBANY MED COL, 64- *Res:* Chemical nature of sodium and carbonate in bone mineral; autonomic pharmacology; nature of 2-halo-2-phenethylamines; applications of nucleonics in biological research. *Mailing Add:* Dept Pharmacol Albany Col of Pharm 106 New Scotland Ave Albany NY 12208

STOLLAR, BERNARD DAVID, b Saskatoon, Sask, Aug 11, 36; m 56; c 3. IMMUNOLOGY, BIOCHEMISTRY. *Educ:* Univ Sask, BA, 58, MD, 59. *Prof Exp:* Res fel biochem, Brandeis Univ, 60-62; dep chief biol sci div, Air Force Off Sci Res, 62-64; asst prof pharmacol, 64-67, from asst prof to assoc prof biochem, 67-74, PROF BIOCHEM, HEALTH SCI, 74-, CHMN, DEPT BIOCHEM, TUFTS UNIV, 86- *Concurrent Pos:* NSF grant, 64-86, NIH grants, 77-; consult, Biol Sci Div, Air Force Off Sci Res, 66-69; sr fel, Weizmann Inst Sci, 71-72; vis prof, Univ Tromso, Norway, 81; consult, Cetus Corp, 82-85, Seragen, Inc, 83-88, Gene-Trak Systs, 86-89, Alkermes, Inc, 89-; Dozor vis prof, Ben-Gurion Univ Sch Med, Beersheba, Israel, 86; mem, NIH Allergy Immunol & Transplantation Res Comt, 90-; alumni lectr, Univ Sask, 89. *Mem:* AAAS; Am Asn Immunologists; Am Soc Biochem & Molecular Biol; Am Col Rheumatology. *Res:* Immunochemistry of nucleic acids and nucleoprotein, especially in relation to auto-immune disease; use of antibodies to study structure of nucleic acids; molecular genetics of anti-nucleic acid antibodies. *Mailing Add:* Dept of Biochem Tufts Univ Health Sci Campus 136 Harrison Ave Boston MA 02111

STOLLAR, VICTOR, b Saskatoon, Sask, Dec 6, 33; m 67; c 3. MICROBIOLOGY, VIROLOGY. *Educ:* Queen's Univ, Ont, MDCM, 56. *Prof Exp:* Fel, Brandeis Univ, 58-62; fel, Weizmann Inst Sci, 62-65; from asst prof to assoc prof, 65-75, PROF MICROBIOL, RUTGERS MED SCH, COL MED & DENT NJ, 75- *Concurrent Pos:* Assoc ed, Virol, 76-; mem, Virol Study Sect, NIH, 80-83. *Mem:* AAAS; Am Soc Microbiol; Am Asn Immunologists. *Res:* Replication of arthropod-borne viruses in vertebrate and in insect cells; genetics and biochemistry of cultured mosquito cells. *Mailing Add:* Dept Molecular Genetics & Microbiol Robert Wood Johnson Med Sch Univ Med & Dent NJ Piscataway NJ 08854

STOLLBERG, ROBERT, b Toledo, Ohio, May 27, 15; m 43; c 4. SCIENCE EDUCATION. *Educ:* Univ Toledo, BS, 35, BEd, 36; Columbia Univ, MA, 40, EdD(sci ed, electronics), 47. *Prof Exp:* Instr, Rossford High Sch, Ohio, 36-39; asst prof physics, Wabash Col, 46-47; ed, Purdue Univ, 47-49; assoc prof physics, San Francisco State Univ, 49-54, chmn, Dept Interdisciplinary Phys Sci, 67-69, assoc dean & actg dean, Sch Natural Sci, 69 & 71-75, prof physics, 54-83; RETIRED. *Concurrent Pos:* Mem, Harvard Univ Conf Prob Sci Ed, 53; chmn, Nat Conf Prob High Sch Sci, 59; mem, President's Comt Develop Scientists & Engrs; sci adv, Columbia Univ team, India, 65-66; Columbia Univ-USAID contract lectr, Inst Educ, Makerere Univ, Uganda, 69-71; vis prof, Columbia Univ. *Mem:* Fel AAAS; Am Asn Physics Teachers; Nat Sci Teachers Asn (pres, 55-56); Am Inst Physics. *Mailing Add:* 3028 Oak Knoll Dr Redwood City CA 94062

STOLLER, BENJAMIN BORIS, agricultural microbiology, biochemistry; deceased, see previous edition for last biography

STOLLER, EDWARD W, b McCook, Nebr, Jan 9, 37; m 60; c 2. PLANT PHYSIOLOGY, WEED SCIENCE. *Educ:* Univ Nebr, BS, 58; Purdue Univ, MS, 62; NC State Univ, PhD(soil fertil), 66. *Prof Exp:* PLANT PHYSIOLOGIST, NORTH CENT REGION, AGR RES, USDA, 65- *Mem:* Am Soc Plant Physiol; Am Soc Agron; Weed Sci Soc Am. *Res:* Weed physiology and control. *Mailing Add:* Dept Agron Univ Ill 1102 S Goodwin Urbana IL 61801

STOLLERMAN, GENE HOWARD, b New York, NY, Dec 6, 20; m 45; c 3. MEDICINE, INFECTIOUS DISEASE. *Educ:* Dartmouth Col, AB, 41; Columbia Univ, MD, 44; Am Bd Internal Med, dipl, 52. *Prof Exp:* From intern to chief med resident, Mt Sinai Hosp, New York, 44-49; res fel microbiol, Col Med, NY Univ, 49-50, instr med, 51-55; instr, Col Med, State Univ NY Downstate Med Ctr, 50-51; from asst prof to prof, Med Sch, Northwestern Univ, 55-64; prof med & chmn dept, Col Med, Univ Tenn, Memphis, 65-81; PROF, SCH MED, BOSTON UNIV, 81-, DISTINGUISHED PHYSICIAN, VET ADMIN, 86- *Concurrent Pos:* Med dir, Irvington Hosp, New York, 51-55; prin investr, Sackett Found Res Rheumatic Fever & Allied Dis, 55-64; mem training grants comt, Nat Inst Arthritis & Metab Dis, US Pub Health Serv, 60-64, mem res career prog, 67-70, chmn, Review Panel Bact Vaccines & Toxoids, Bur Biologics, Food & Drug Admin, 73-78; physician-in-chief, City Memphis Hosps, 65-81; consult, Memphis Vet Admin Hosp, 65-81; chmn coun rheumatic fever & congenital heart dis, Am Heart Asn, 66-68; mem, Am Bd Internal Med, 67-73, chmn written exam comt, 69-73 & exec comt, 71-73; ed, Advan Internal Med, 68-; mem, Expert Comt Cardiovasc Dis, WHO, 66-81; pres, Cent Soc Clin Res, 74-75; co-chmn, Educ Work Group Nat Comn Arthritis, 75-76; chmn, Panel Prod & Supply Work Group, HEW & Goodman prof, Univ Tenn, 77; consult, Methodist Hosp, Memphis, St Joseph Hosp & LeBonheur Children's Hosp; mem, Nat Adv Comt, Nat Inst Allergy & Infectious Dis, NIH, 79-83 & Nat Adv Comt on Vaccines, Nat Immunization Prog, 88; ed-in-chief, J Am Geriat Soc, 85- & co-ed, Hosp Practice, 77- *Honors & Awards:* Bruce Mem Award, Am Col Physicians. *Mem:* Asn Profs Med (pres, 75-76); Asn Am Physicians; fel Am Col Physicians (vpres, 82-83); Am Soc Clin Invest; Am Asn Immunologists. *Res:* Infectious and rheumatic diseases; biology of streptococcus; etiology of rheumatic fever. *Mailing Add:* E N Rogers Mem Vet Admin Hosp 200 Springs St Bedford MA 02118

STOLLEY, PAUL DAVID, b Pawling, NY, June 17, 37; m 59; c 3. PUBLIC HEALTH, EPIDEMIOLOGY. *Educ:* Lafayette Col, AB, 57; Med Col, Cornell Univ, MD, 62. *Hon Degrees:* MA, Univ Pa, 76. *Prof Exp:* From asst prof to assoc prof epidemiol, Sch Public Health, Johns Hopkins Univ, 68-76; prof med, Sch Med, Univ Pa, 76-85, Herbert C Rorer prof med sci, 85-91; PROF & CHMN PREV MED, SCH MED, UNIV MD, 91- *Mem:* Inst Med-Nat Acad Sci; Soc Epidemiol Res (pres, 82-83); Int Epidemiol Asn (treas, 81-84); Am Col Epidemiol (pres, 89-90). *Res:* Epidemiology of cancer; adverse drug reactions; prevention of heart disease. *Mailing Add:* Sch Med Univ Md 655 W Redwood St Baltimore MD 21201

STOLOFF, IRWIN LESTER, b Philadelphia, Pa, May 9, 27; m 52; c 3. MEDICINE. *Educ:* Jefferson Med Col, MD, 51; Am Bd Internal Med, dipl, 58. *Prof Exp:* From intern to resident med, Jefferson Med Col Hosp, 51-53; resident, Baltimore City Hosps, Md, 53-54 & Mt Sinai Hosp, New York, 56-57; fel med, 57-58, from instr to asst prof, 59-69, ASSOC PROF MED & PREV MED, JEFFERSON MED COL, 69 - *Mem:* Fel Am Col Physicians; fel Am Col Chest Physicians. *Res:* Immunology; autoimmune diseases; cancer immunology; epidemiology of chronic lung disease. *Mailing Add:* New Jefferson Hosp Bldg Rm 4001 111 S 11th St Philadelphia PA 19107

STOLOFF, LEONARD, b Boston, Mass, Mar 24, 15; m 40. BIOCHEMISTRY. *Educ:* Mass Inst Technol, BS, 36. *Prof Exp:* Chemist, Granada Wines, Inc, 36-37; self employed, 38-41; chemist, US Dept Navy, 41; chemist, Consumer's Union, 42; res chemist, Agar Substitute Prog, US Fish & Wildlife Serv, 42-44; res chemist, Krim-Ko Corp, 44-51; res dir, Seaplant Chem Corp, 51-59; asst tech dir, Marine Colloids, Inc, 59-63; chief, Mycotoxins & Enzymes Sect, Div Food Chem, Bur Foods, Food & Drug Admin, 63-71, natural toxicants specialist, 71-82; RETIRED. *Concurrent Pos:* Asn off anal chemists, Gen Referee Mycotoxins, 66-82; ed, Newsletter Div Agr Food Chem, Am Chem Soc, 65-82. *Honors & Awards:* Wiley Award, Asn Off Analytical Chemists, 81. *Mem:* AAAS; Am Chem Soc; Inst Food Technol; Asn Off Anal Chemists. *Res:* Chemistry, toxicology and occurrence of natural poisons in foods. *Mailing Add:* 13208 Bellevue St Silver Spring MD 20904-1703

STOLOFF, NORMAN STANLEY, b Brooklyn, NY, Oct 16, 34; m; m 71; c 4. METALLURGY. *Educ:* NY Univ, BMetE, 55; Columbia Univ, MS, 56, PhD(metall), 61. *Prof Exp:* Jr engr metall, Pratt & Whitney Aircraft Div, United Aircraft Corp, 56-58; staff scientist, Ford Motor Co, 61-65; from asst prof to assoc prof, 65-70, PROF ENG, RENSSELAER POLYTECH INST, 70- *Concurrent Pos:* Fulbright sr res fel, Univ Birmingham, 67-68; vis prof, Eurat Joint Res Ctr, Ispra, Italy, 76-77, Technion, Israel Inst Technol, 80, Swiss Fed Inst, Lausanne, 81; vis scientist, Atomic Energy Res Estab, Harwell, 85. *Mem:* Am Inst Mining, Metall & Petrol Engrs; fel Am Soc Metals; Am Soc Testing & Mat; Mat Res Soc. *Res:* Physical metallurgy; relationship between microstructure and plastic deformation of metals; environmental effects; strength and fracture of intermetallic compounds; powder processing of composites; fatigue of high temperature alloys. *Mailing Add:* Dept Mat Eng Rensselaer Polytechnic Inst Troy NY 12180-3590

STOLOV, HAROLD L, b New York, NY, May 27, 21; m 81. PHYSICS, METEOROLOGY. *Educ:* City Col New York, BS, 42; Mass Inst Technol, MS, 47; NY Univ, PhD(physics, meteorol), 53. *Prof Exp:* Asst radio & electronics, Signal Corps, US Dept Army, 42; instr physics, City Col New York, 47-50; lectr, Hunter Col, 50-51; res assoc, NY Univ, 50-53; instr, Douglass Col, Rutgers Univ, 51-53, asst prof, 53-59; from asst prof to assoc prof, 59-70, PROF PHYSICS, CITY COL NEW YORK, 70- *Concurrent Pos:* Consult, Martin Co, 56-59; consult, Res & Adv Develop Div, Avco Corp, 60-61; Nat Acad Sci-Nat Res Coun sr res associateship, Inst Space Studies, New York, 65-67. *Mem:* Nat Sci Teachers Asn; Am Meteorol Soc; Am Asn Physics Teachers; Am Geophys Union. *Res:* Physics of the upper atmosphere; tidal oscillations; physics education; magnetosphere; solar-terrestrial physics. *Mailing Add:* 2575 Palisade Ave New York NY 10463

STOLOVY, ALEXANDER, b Brooklyn, NY, Nov 21, 26; m 55; c 3. NUCLEAR PHYSICS. *Educ:* Brooklyn Col, BS, 48; Calif Inst Technol, MS, 50; NY Univ, PhD(physics), 55. *Prof Exp:* Nuclear physicist, Brookhaven Nat Lab, 53-54; NUCLEAR PHYSICIST, RADIATION TECHNOL DIV, NAVAL RES LAB, 55- *Concurrent Pos:* Partic guest scientist, Lawrence Livermore Lab, 74-75. *Mem:* Am Phys Soc; Sigma Xi. *Res:* Slow neutron spectroscopy; electron beam interactions with matter; neutron capture gamma ray studies using time-of-flight techniques. *Mailing Add:* Code 4653 Naval Res Lab Washington DC 20375

STOLOW, NATHAN, b Montreal, Que, May 4, 28; m 50; c 2. CHEMISTRY. *Educ:* McGill Univ, BSc, 49; Univ Toronto, MA, 52; Univ London, PhD(conserv), 56. *Prof Exp:* Res chemist, Nat Res Coun Can, 49-50; vis lectr chem & physics, Sir John Cass Col, Univ London, 52-55, res assoc conserv, Courtauld Inst Art, 55-56; dir conserv, Nat Gallery Can, 56-72; dir, Can Conserv Inst, 72-75; spec adv conserv, Nat Mus Can, 76-79; CONSERV CONSULT, 79- *Concurrent Pos:* Carnegie travel grant, Nat Gallery Can, 56-57; rapporteur, Comt Conserv, Int Coun Mus, 64-, chmn Can nat comt, 70-; mem coun, Int Inst Conserv Hist & Artistic Works, 72- *Honors & Awards:* Can Medal, Govt Can, 67. *Mem:* Fel Chem Inst Can; fel Can Mus Asn. *Res:* Museum conservation; problems of deterioration in works of art related to conservation; solution of museological problems by chemical and physical approaches; interaction of art history and scientific research; exhibition conservation research. *Mailing Add:* PO Box 194 Williamsburg VA 23187

STOLOW, ROBERT DAVID, b Boston, Mass, Mar 9, 32; m 53; c 3. ORGANIC CHEMISTRY. *Educ:* Mass Inst Technol, SB, 53; Univ Ill, PhD(chem), 56. *Prof Exp:* From instr to assoc prof org chem, 58-76, PROF CHEM, TUFTS UNIV, 76- *Concurrent Pos:* Fel, Calif Inst Technol, 67-68; vis res fel, Harvard, 76. *Mem:* Am Chem Soc. *Res:* Physical organic chemistry; stereochemistry; conformational analysis; nuclear magnetic resonance spectroscopy; computational chemistry. *Mailing Add:* Dept of Chem Tufts Univ Medford MA 02155

STOLPER, EDWARD MANIN, b Boston, Mass, Dec 16, 52; m 73; c 2. EXPERIMENTAL PETROLOGY. *Educ:* Harvard Col, AB, 74; Univ Edinburgh, MPhil, 76; Harvard Univ, PhD(geol sci), 79. *Prof Exp:* from asst prof to prof, 79-90, WILLIAM E LEONHARD PROF GEOL, CALIF INST TECHNOL, 90- *Concurrent Pos:* Marshall scholar, 74-76. *Honors & Awards:* Newcomb Cleveland Prize, AAAS, 84; F W Clarke Medal, Geochem Soc, 85; J B Macelwane Award, Am Geophys Union, 86; Nininger Meteorite Award, 76-77. *Mem:* Geol Soc Am; Am Geophys Union; fel Meteoritical Soc; Mineral Soc Am; AAAS. *Mailing Add:* Div Geol Planet Sci 170-25 Calif Inst Technol Pasadena CA 91125

STOLTE, CHARLES, b Blue Earth, Minn, Apr 20, 33; m 54; c 4. ELECTRICAL ENGINEERING, SOLID STATE PHYSICS. *Educ:* Univ Minn, BS, 55, MS, 58, PhD(elec eng), 66. *Prof Exp:* Res asst diffusion study, Univ Minn, 55-58; mem tech staff, 66-76, lab proj mgr, Solid State Lab, 76-84, RES & DEVELOP SECT MGR, MICROWAVE TECHNOL DIV, HEWLETT-PACKARD LAB, 84- *Mem:* Inst Elec & Electronic Engrs. *Res:* Oxide coated cathode, schottky barriers, III-V materials and devices including ion implantation, LEP's, solid state lasers and microwave devices and GaAs integrated circuit technology; materials science engineering. *Mailing Add:* Microwave Technol Div Hewlett-Packard Co 1412 Fountaingrove Rd Santa Rosa CA 95403

STOLTENBERG, CARL H, b Monterey, Calif, May 17, 24; m 49; c 5. FOREST ECONOMICS. *Educ:* Univ Calif, BS, 48, MF, 49; Univ Minn, PhD(agr econ), 52. *Prof Exp:* Instr forestry, Univ Minn, 49-51; asst prof forest econ, Duke Univ, 51-56; head resource econ res, Northwest Forest Exp Sta, US Forest Serv, Pa, 56-58, chief div forest econ res, 58-60; prof forestry & head dept, Iowa State Univ, 60-67; prof forestry, dean col forestry & dir forest res lab, 67-90, EMER PROF & DEAN, ORE STATE UNIV, 90- *Concurrent Pos:* Mem forestry res comn, Nat Acad Sci, 63-65; mem, Nat Adv Bd Coop Forestry Res, 62-66&86-, Ore Bd Forestry, 67-87, chmn, 74-84; mem Secy of Agr State & Pvt Forestry Adv Comn, 70-74; chmn, Ore & Calif Adv Bd, Bur Land Mgt, 72-76. *Mem:* Fel Soc Am Foresters; Am Econ Asn; Forest

Prod Res Soc; Sigma Xi; AAAS. *Res:* Economic analysis of forest management alternatives; forest policy; resource allocation in forestry; natural resource policy. *Mailing Add:* Col Forestry Ore State Univ Corvallis OR 97331

STOLTZ, LEONARD PAUL, b Kankakee, Ill, Dec 5, 27; div; c 5. HORTICULTURE. *Educ:* Agr & Mech Col, Tex, BS, 55; Ohio State Univ, MS, 56; Purdue Univ, PhD(hort), 65. *Prof Exp:* Res assoc floricult, Rutgers Univ, 57-60; asst prof hort, Univ RI, 60-62; from asst prof to assoc prof, 65-90, EMER PROF HORT, UNIV KY, 91- *Concurrent Pos:* USDA res grant, 66-70; ed, Eastern Region, Int Plant Propagators Soc, 68-78; lectr, Indonesia, 88; lectr & consult, Ecuador, 90. *Honors & Awards:* Kenneth Post Award, Cornell Univ, 67; L M Ware Award, Am Soc Hort Sci, Southern Region, 68, 72. *Mem:* Fel Int Plant Propagators Soc (vpres, 82-83, pres, 83-84). *Res:* Plant propagation; Ginseng culture; tissue and embryo culture. *Mailing Add:* Dept Hort Univ Ky Lexington KY 40546-0919

STOLTZ, ROBERT LEWIS, b Bakersfield, Calif, May 15, 45; m 71. ENTOMOLOGY. *Educ:* Univ Calif, Davis, BS, 67; Univ Calif, Riverside, PhD(entom), 73. *Prof Exp:* Fel entom, Univ Calif, Riverside, 73-74 & Univ Mo-Columbia, 74-75; EXTEN SPECIALIST ENTOM, COOP EXTEN SERV, UNIV IDAHO, 75- *Mem:* Entom Soc Am; Sigma Xi. *Res:* Insect control, particularly in potatoes, sugar beets, beans, peas, and alfalfa hay; black fly control; livestock insects. *Mailing Add:* 339 Heyburn Ave W Twin Falls ID 83301

STOLTZFUS, NEAL W, b Lancaster, Pa, Aug 29, 46. ALGEBRAIC TOPOLOGY, KNOT THEORY. *Educ:* Princeton Univ, PhD(math), 73. *Prof Exp:* From asst prof to assoc prof, 73-84, PROF MATH, LA STATE UNIV, 84- *Concurrent Pos:* Col dir, comp serv, La State Univ, 85-88. *Mailing Add:* Math Dept La State Univ Baton Rouge LA 70803

STOLTZFUS, WILLIAM BRYAN, b Martinsburg, Pa, Apr 25, 32; m 57; c 5. ENTOMOLOGY. *Educ:* Goshen Col, BS, 57; Kent State Univ, MS, 66; Iowa State Univ, PhD(entom), 74. *Prof Exp:* Instr biol, Eastern Mennonite Col, 66-70 & entom, Iowa State Univ, 73-74; ASSOC PROF BIOL & CHMN DEPT, WILLIAM PENN COL, 74- *Mem:* Entom Soc Am. *Res:* Life history of fruitflies as it relates to their taxonomy. *Mailing Add:* Dept Biol William Penn Col 201 Trueblood Ave Oskaloosa IA 52577

STOLWIJK, JAN ADRIANUS JOZEF, b Amsterdam, Netherlands, Sept 29, 27; nat US. BIOPHYSICS. *Educ:* State Agr Univ, Wageningen, MS, 51, PhD, 55. *Prof Exp:* Cabot res fel biol, Harvard Univ, 55-57; from assoc fel to fel biol, 57-74, assoc dir, John B Pierce Found, 74-88; chmn dept epidemiol & pub health, 82-89, PROF EPIDEMIOL, SCH MED, YALE UNIV, 75- *Concurrent Pos:* Mem, Comt Indoor Pollutants, Nat Res Ctr/Nat Acad Sci, 80-81; dir grad studies, dept epidemiol & public health, Yale Univ, 79-82; vchmn comt, indoor air qual & total human exposure, Sci Adv Bd, Environ Protection Agency, 88- *Mem:* Aerospace Med Asn; Int Soc Biometeorol; Am Physiol Soc; Am Pub Health Asn; Biophys Soc. *Res:* Body temperature regulation; regulatory systems in physiology; indoor air quality; radiant heat exchange with environment; environmental health; occupational health; risk assessment; construction and application of mathematical models for study of complex physiological systems; environmental physiology; environmental epidemiology. *Mailing Add:* Dept Epidemiol & Pub Health 60 College St PO Box 3333 New Haven CT 06510

STOLZ, WALTER S, b Milwaukee, Wis, Dec 12, 38. DIABETES RESEARCH. *Educ:* Univ Wis-Madison, BS, 60, MS, 62, PhD(mass commun), 64. *Prof Exp:* Tech writer, Int Bus Mach Data Systs Div, 60-61CF; res asst, Mass Commun Res Ctr, Univ Wis-Madison, 61-64; NSF postdoctoral fel, Ctr Cognitive Studies, Harvard Univ, 64-65; asst prof psychol, Univ Tex, Austin, 65-71; from asst prof to assoc prof psychol, Earlham Col, Richmond, Ind & chmn dept, 71-75; sr res assoc & proj dir, Ctr Appl Ling, Arlington, Va, 75-76; grants assoc, Div Res Grants, 76-77; educ prog dir, Arthritis Musculoskeletal & Skin Prog, Nat Inst Arthritis Diabetes Digestive & Kidney Dis, 77-78, prog dir, Manpower Develop & Res Resources, 78-80, prog dir, Metab Dis & Res Resource, 80-83, actg dep dir, Div Diabetes, Endocrinol & Metab Dis, 81-83, dir, Div Extramural Activ, 83-86, DIR DIV EXTRAMURAL ACTIVITIES, NAT INST DIABETES DIGESTIVE & KIDNEY DIS, NIH, 86- *Mem:* Am Psychol Asn. *Res:* Psychology; arthritis; metabolic diseases. *Mailing Add:* Nat Inst Diabetes Digestive & Kidney Dis Div Extramural Affairs NIH Westwood Bldg Rm 657 5333 Westbard Ave Bethesda MD 20892

STOLZBERG, RICHARD JAY, b Winthrop, Mass, Feb 5, 48. ANALYTICAL CHEMISTRY. *Educ:* Tufts Univ, BS, 69; Mass Inst Technol, PhD(anal chem), 73. *Prof Exp:* Res assoc & prin investr, Harold Edgerton Res Lab, New Eng Aquarium, 73-77; ASST PROF, DEPT CHEM, UNIV ALASKA, FAIRBANKS, 78- *Mem:* Am Chem Soc; Sigma Xi; AAAS; Chemomets Soc. *Res:* Characterization of trace metal-organic interactions in natural waters; effect of metal speciation on bioavailability; chemometrics; chromatography; electrochemistry; spectroscopy. *Mailing Add:* Dept of Chem Univ of Alaska Fairbanks AK 99775

STOLZENBACH, KEITH DENSMORE, b Washington, DC, Aug 23, 44. CIVIL & HYDRAULIC ENGINEERING. *Educ:* Mass Inst Technol, SB, 66, SM, 68, PhD(civil eng), 71. *Prof Exp:* Asst prof civil eng, Mass Inst Technol, 70-71; res engr, Eng Lab, Tenn Valley Authority, 71-74; asst prof, 74-76, ASSOC PROF CIVIL ENG, MASS INST TECHNOL, 76- *Mem:* Am Soc Civil Engrs; Am Geophys Union; Int Asn Hydraul Res. *Res:* Hydraulic modeling techniques; environmental heat transfer; pollutant dispersal in natural waters; field survey techniques. *Mailing Add:* Dept Civil Eng Mass Inst Technol 77 Mass Ave Cambridge MA 02139

STOLZENBERG, GARY ERIC, b Southampton, NY, Dec 1, 39; m 69. PESTICIDE CHEMISTRY, FORMULATION AGENTS. *Educ:* Rensselaer Polytech Inst, BS, 62; Kans State Univ, PhD(biochem), 68. *Prof Exp:* Asst biochem, Kans State Univ, 62-68; RES CHEMIST, METAB & RADIATION RES LAB, USDA, 68- *Mem:* Am Chem Soc. *Res:* Xenobiotics metabolism in plants; behavior and fate of formulation agents; surfactant analysis. *Mailing Add:* Metab Lab PO Box 5674 USDA Metab Lab State Univ Station Fargo ND 58105

STOLZENBERG, SIDNEY JOSEPH, b New York, NY, Nov 30, 27; m 58; c 2. REPRODUCTIVE PHYSIOLOGY, PHARMACOLOGY. *Educ:* NY Univ, BA, 50; Univ Mo, MS, 54; Cornell Univ, PhD(reproductive physiol), 66. *Prof Exp:* Biochemist, Lederle Labs Div, Am Cyanamid Co, 54-59, agr div, 59-63; endocrinologist, SRI Int, 66-72, endocrinologist, Life Sci Div, 72-78; toxicologist, Dept Pharmacol, Sch Med, Univ Calif, San Francisco, 78-80; PHYSIOLOGIST, BUR DRUGS, CTR DRUGS & BIOLOGICS, FOOD & DRUG ADMIN, ROCKVILLE, MD, 80- *Mem:* Am Physiol Soc; Soc Study Reproduction; Soc Toxicol. *Res:* Reproductive physiology and toxicology; corpus luteum function and neuroendocrinology; review animal research data for new drug applications with Food and Drug Administration. *Mailing Add:* Dept Pharmacol & Toxicol Alliance Pharmaceut Corp 3040 Science Park Rd San Diego CA 92121

STOLZY, LEWIS HAL, b Mich, Dec 11, 20; m 47; c 3. SOIL PHYSICS. *Educ:* Mich State Col, BS, 48, MS, 50, PhD, 54. *Prof Exp:* Actg proj supvr, Soil Conserv Serv, USDA, 50-52; asst, Mich State Univ, 52-54; asst irrig engr, 54-61, assoc soil physicist, 61-66, assoc prof, 66-78, PROF SOIL PHYSICS, UNIV CALIF, RIVERSIDE, 78- *Concurrent Pos:* Fulbright sr res scholar, Univ Adelaide, 64-65; prof, Nat Univ Agr, Chapingo, Mex, AID & Africa; agronomic res award, Int Soc Soil Sci. *Mem:* Fel Soil Sci Soc Am; Int Soc Soil Sci; Int Cong Plant Path; Am Phytopath Soc; fel Am Soc Agron. *Res:* Soil moisture and aeration. *Mailing Add:* Dept Soil & Environ Sci Univ Calif Riverside CA 92521

STOMBAUGH, TOM ATKINS, b Vancouver, Wash, Aug 22, 21; m 44; c 4. BIOLOGY. *Educ:* Ill State Norm Univ, BEd, 41; Univ Ill, MS, 46; Ind Univ, PhD, 53. *Prof Exp:* Asst prof zool, Eastern Ill State Col, 48-50; prof biol, 53-77, PROF LIFE SCI, SOUTHWEST MO STATE UNIV, 77- *Mem:* Am Soc Mammal. *Res:* Taxonomy of the voles of sub-genus Pedomys; mammalian taxonomy and ecology. *Mailing Add:* Southwest Mo State Springfield MO 65802

STOMBLER, MILTON PHILIP, b New York, NY, Dec 19, 39; m 67; c 3. EXPERIMENTAL SOLID STATE PHYSICS. *Educ:* Univ Md, College Park, BS, 62; Univ SC, MS, 66, PhD(physics), 69. *Prof Exp:* Asst engr, Aerospace Div, Westinghouse Elec Corp, 64-65; fel, Univ Del, 69-71; asst prof physics, State Univ NY Col Potsdam, 71-73, dir spon prog, 73-77; assoc dean res, Polytech Inst & State Univ, 77-83; DIR TECHNOL TRANSFER & SPONS ACTIV, GA INST TECHNOL, 83- *Mem:* Sigma Xi; AAAS; Nat Coun Univ Res Adminr. *Res:* Electron paramagnetic resonance. *Mailing Add:* Col Sci Ga Inst Technol Atlanta GA 30332-0365

STOMMEL, HENRY MELSON, b Wilmington, Del, Sept 27, 20. OCEANOGRAPHY. *Educ:* Yale Univ, BS, 42. *Hon Degrees:* DSc, Gothenburg Univ, 64, Yale Univ, 70, Univ Chicago, 70. *Prof Exp:* Instr math & astron, Yale Univ, 42-44; res assoc phys oceanog, Oceanog Inst, Woods Hole, 44-59; prof oceanog, Mass Inst Technol, 59-60; prof, Harvard Univ, 60-63; prof oceanog, Mass Inst Technol, 63-78; OCEANOGR, WOODS HOLE OCEANOG INST, 78- *Honors & Awards:* Crafourd Prize, 83; Nat Medal Sci, 89. *Mem:* Nat Acad Sci; Am Astron Soc; Am Soc Limnol & Oceanog; Am Acad Arts & Sci; Am Geophys Union; Royal Soc Acad Sci USSR; Acad Sci France. *Res:* Dynamics of ocean currents. *Mailing Add:* Oceanogr Inst Woods Hole MA 02543

STONE, ALBERT MORDECAI, b Boston, Mass, Dec 24, 13; m 41, 68; c 3. PLASMA PHYSICS, MICROWAVE PHYSICS. *Educ:* Harvard Univ, AB, 34; Mass Inst Technol, PhD(physics), 38. *Prof Exp:* Res assoc, Mass Inst Technol, 38-39, staff mem, Radiation Lab, 42-46; instr, Middlesex Col, 36-38; physicist, US Naval Torpedo Sta, 40-41; from asst prof to assoc prof physics, Mont State Col, 41-46; sci liaison officer, US Embassy, London, Eng, 46-48; assoc mem comt electronics & comt basic phys sci, Res & Develop Bd, US Dept Defense, 48-49; tech asst to dir appl physics lab, 49-72, head tech info div, 62-80, dir advan res projs, Appl Physics Lab, 72-81, SR FEL, JOHNS HOPKINS UNIV, 81- *Concurrent Pos:* mem, Fed Emergency Mgt Agency Adv Bd; fel mem, Hudson Inst; dir, energy fund, Gen Instrument Corp. *Mem:* Fel AAAS; fel Am Phys Soc; sr mem Cosmos Club. *Res:* Electronics; gaseous discharges; radar signal thresholds; guided missiles; countermeasures; controlled thermonuclear plasmas; geothermal energy; nuclear effects. *Mailing Add:* 4932 Sentinal Dr, #403 Bethesda MD 20816

STONE, ALEXANDER GLATTSTEIN, b Hungary, Jan 30, 16; nat US; m 58; c 2. MATHEMATICS. *Educ:* Univ Debrecen, Hungary, Dr Laws, 40; George Washington Univ, MS, 61. *Prof Exp:* Mathematician, Repub Aviation Corp, 55-56; mathematician appl physics lab, Johns Hopkins Univ, 56-69, supvr programmers digital comput, 66-69, lectr, univ, 59-60, instr, eve col, 66-67; mem tech staff, Jet Propulsion Lab, Calif Inst Technol, 69-91; RETIRED. *Mem:* Math Asn Am; Asn Comput Mach. *Res:* Programming for automatic digital computers; Boolean algebra. *Mailing Add:* 3554 Alginet Dr Encino CA 91436-4126

STONE, ALEXANDER PAUL, b West New York, NJ, June 28, 28; m 60; c 1. MATHEMATICS. *Educ:* Columbia Univ, BS, 52; Newark Col Eng, MS, 56; Univ Ill, Urbana, PhD(math), 65. *Prof Exp:* Engr, Western Elec Co, 52-56; instr elec eng, Manhattan Col, 56-58; asst prof physics, Dickinson Col, 58-60; asst prof math, Univ Ill, Chicago Circle, 65-69, assoc prof, 69-70; assoc prof, 70-75, PROF MATH, UNIV NMEX, 76- *Mem:* Am Math Soc; Math Asn Am; Int Union Radio Sci. *Res:* Differential geometry; applied mathematics; electromagnetic theory. *Mailing Add:* Dept Math & Statist Univ NMex Albuquerque NM 87131

STONE, ARTHUR HAROLD, b London, Eng, Sept 30, 16; m 42; c 2. PURE MATHEMATICS. *Educ:* Cambridge Univ, BA, 38; Princeton Univ, PhD(math), 41. *Prof Exp:* Mem, Inst Advan Study, 41-42; instr math, Purdue Univ, 42-44; math physicist, Geophys Lab, Carnegie Inst, 44-46; fel, Trinity Col, Cambridge Univ, 46-47; lectr math, 47-56, sr lectr, Univ Manchester, 56-61; prof, 61-87, EMER PROF MATH, UNIV ROCHESTER, 87-; ADJ PROF, NORTHEASTERN UNIV, 88- *Mem:* Am Math Soc; Math Asn Am; Sigma Xi. *Res:* Point-set topology; aerodynamics; graph theory; general topology; descriptive set theory. *Mailing Add:* Dept Math Northeastern Univ Boston MA 02115

STONE, BENJAMIN CLEMENS, b Shanghai, China, July 26, 33; US citizen; m 65; c 2. SYSTEMATICS OF HIGHER PLANTS. *Educ:* Pomona Col, BA, 54; Univ Hawaii, PhD(bot), 60. *Prof Exp:* Res asst bot, Wash State Univ, St Louis, Mo, 54-55, Univ Hawaii, Honolulu, 55-60 & Mus Natural Hist, Smithsonian, 60-61; asst prof bot & biol, Col Guam, Agana, 61-64, prof bot, 64-65; lectr bot, Univ Malaya, Kuala Lumpur, 65-68, reader, 68-71, univ reader, 71-84; CHMN DEPT BOT, ACAD NATURAL SCI, PHILADELPHIA, 84- *Concurrent Pos:* Consult, Bishop Mus, Honolulu, 59 & Forest Dept, Papua New Guinea, 71; ed, Micronesica, J Col Guam, 64; off deleg, Ninth Int Bot Cong, Leningrad, 75; hon cur, Nat Mus Malaysia, Kuala Lumpur, 78- *Mem:* Fel AAAS; Malayan Nature Soc; Malaysian Sci Asn; Bot Soc Am; Am Asn Plant Taxonomists. *Res:* Systematics and biogeogrphy of certain major angiosperm families (Pandanaceae, Rutaceae); floristics of Flora Malesiana region and tropical Pacific islands; phylogeny, evolution and morphology of higher plants. *Mailing Add:* Dept Bot Acad Natural Sci 19th & Pkwy Philadelphia PA 19103

STONE, BENJAMIN P, b Dover, Tenn, Aug 28, 35; m 56; c 1. PLANT PHYSIOLOGY. *Educ:* Austin Peay State Univ, BS, 59; Univ Tenn, Knoxville, MS, 61, PhD(bot), 68. *Prof Exp:* Asst prof biol, Austin Peay State Univ, 61-65; res partic radiation biol, Cornell Univ, 65-66; asst prof plant physiol, Purdue Univ, West Lafayette, 69; assoc prof biol, 69-72, PROF BIOL, AUSTIN PEAY STATE UNIV, 72-, CHMN DEPT, 77-, DIR, CTR EXCELLENCE FIELD BIOL, 86- *Concurrent Pos:* Fel hort, Purdue Univ, West Lafayette, 69. *Mem:* Am Soc Plant Physiol; Am Inst Biol Sci; Sigma Xi. *Res:* Nucleic acid; protein synthesis. *Mailing Add:* Dept Biol Austin Peay State Univ Clarksville TN 37040

STONE, BOBBIE DEAN, b Paulton, Ill, June 11, 27; m 50; c 2. SOLID STATE CHEMISTRY, INORGANIC CHEMISTRY. *Educ:* Univ Southern Ill, BS, 49; Northwestern Univ, PhD(chem), 52. *Prof Exp:* Res chemist, Mound Lab, 52-53, Cent Res Dept & Res & Eng Div, 53-62 & Inorg Chem Div, 62-65, res group leader, Semiconductor Mat Dept, 65-69, silicon res mgr, 69-72, sr res specialist, Electronics Prod Div, 72-74, FEL, MONSANTO ELECTRONIC MATS CO, 74- *Mem:* Am Chem Soc; Am Soc Crystal Growth; Electrochem Soc. *Res:* Semiconductor grade silicon; neutron transmutation doping; III-V compounds; polycrystalline silicon processes. *Mailing Add:* 415 Monticello Dr Ballwin MO 63011-2531

STONE, CHARLES DEAN, b Athens, Ga, Sept 6, 26; m 50; c 1. FOOD SCIENCE, BIOCHEMISTRY. *Educ:* Univ Ga, BS, 49, PhD(food sci), 64; Fla State Univ, MS, 59. *Prof Exp:* Partner, Stone's Ideal Bakery, 50-53; instr baking sci & mgt food serv bakery, Fla State Univ, 53-59, asst prof baking sci & mgt, 59-61; sect mgr food res, Quaker Oats Co, 64-69, mgr cereal res, 69-72, mgr cereals, mixes & corn goods res, 72-74; sr prod res scientist, Res & Develop, 74-77, sr res scientist, Sci Affairs, M&M/Mars, 77-80, SR RES SCIENTIST, RES & DEVELOP, MARS INC, 80- *Mem:* Am Asn Cereal Chemists; Am Soc Bakery Eng; Inst Food Technologists; Soc Nutrit Educ. *Res:* Cereal and confectionery; flavor, nutrition, rheology, structural, crystallization, stability, cariogenicity, new products; fermentation; ion-protein interactions as affected by fermentation; scientific affairs; venture technology. *Mailing Add:* M&M/Mars Res & Develop High St Hackettstown NJ 07840

STONE, CHARLES JOEL, b Los Angeles, Calif, July 13, 36; m 66; c 2. MATHEMATICS, STATISTICS. *Educ:* Calif Inst Technol, BS, 58; Stanford Univ, PhD(math statist), 61. *Prof Exp:* Instr math, Princeton Univ, 61-62; asst prof, Cornell Univ, 62-64; from asst prof to assoc prof, 64-69, prof math, 69-81, prof biomath, Univ Calif, Los Angeles, 75-81, PROF STATIST, UNIV CALIF, BERKELEY, 81- *Concurrent Pos:* NSF grant, Univ Calif, Los Angeles, 64-; consult, Rand Corp, 66-67, Planning Res Corp, 66-68, Gen Elec Tech Mil Planning Oper, 68-70, Fed Aviation Admin, 70-74, Consol Anal Ctr Inc, 71-74, Urban Inst, 75-76 & Technol Serv Corp, 77-80; Guggenheim fel, 80-81. *Mem:* Fel Inst Math Statist; Am Math Soc; Am Statist Asn. *Res:* Probability and statistics, including random walks, birth and death, diffusion and infinitely divisible processes; potential theory; renewal theory; infinite particle systems; nonparametric estimation, classification and regression. *Mailing Add:* Dept Statist Univ Calif Berkeley CA 94720

STONE, CHARLES PORTER, b Owatonna, Minn, Sept 16, 37; m 85; c 4. WILDLIFE RESEARCH, WILDLIFE ECOLOGY. *Educ:* Univ Minn, BA, 60; Colo State Univ, MS, 63; Ohio State Univ, PhD(zool), 73. *Prof Exp:* Res biologist, Patuxent Wildlife Res Ctr, 63-66; asst leader, Ohio Coop Wildlife Res Unit, 66-70; lectr, Ohio State Univ, 70; res biologist, Denver Wildlife Res Ctr, US Fish & Wildlife Serv, 71-73, asst dir, 73-75, supvry wildlife biologist, 73-80, chief wildlife ecol pub lands, 75-80; RES SCIENTIST, NAT PARK SERV, 80- *Concurrent Pos:* Instr wildlife biol, Ohio State Univ, 66-70; mem adj fac, Colo State Univ, 73- *Mem:* AAAS; Wildlife Soc; Am Soc Mammalogists. *Res:* Effects of energy development, forest and range management practices, and other land disturbances upon wildlife abundance, distribution and behavior; ecology of animal damage to crops. *Mailing Add:* Box 3 Hawaii National Park HI 96718

STONE, CHARLES RICHARD, b Portland, Ore, Sept 8, 21; m 47; c 3. AERODYNAMICS. *Educ:* Univ Minn, BSAero, 51; Univ Wash, MSAero, 58. *Prof Exp:* Aerodynamicist, Boeing Airplane Co, 51-55; res engr, 55-58, proj engr, 58-61, res supvr automatic control, 61-64, STAFF ENGR, HONEYWELL INC, 65- *Mem:* Am Inst Aeronaut & Astronaut. *Res:* Automatic, optimal and adaptive control; aerodynamics of vertical takeoff airplanes; dynamics and control of flexible vehicles. *Mailing Add:* 4955 Sorell Ave Minneapolis MN 55422

STONE, CLEMENT A, b Hastings, Nebr, May 23, 23; m 52; c 3. PHARMACOLOGY. *Educ:* Univ Nebr, BSc, 46; Univ Ill, MS, 48; Boston Univ, PhD(physiol), 52. *Prof Exp:* From instr to asst prof physiol, Sch Med, Boston Univ, 51-54; res assoc, Res Div, Sharp & Dohme, Inc, 54-57, dir pharmacodynamics, Merck Inst Therapeut Res, 56-57, from assoc dir to dir, 58-66, exec dir, 66-71, vpres, 71-78, sr vpres, Merck Sharp & Dohme Res Labs, 78-88; RETIRED. *Concurrent Pos:* Lectr, St Andrews, 53. *Mem:* AAAS; Am Soc Pharmacol & Exp Therapeut; Am Chem Soc; Soc Exp Biol & Med; NY Acad Sci. *Res:* Pharmacology of adrenergic, ganglionic blocking drugs, antihypertensive agents and antiglaucoma agents. *Mailing Add:* Eight Farrier Lane Bluebell PA 19422

STONE, CONNIE J, b Michigan City, Ind, Oct 30, 43. TOXICOLOGY. *Educ:* Ind Univ, BA, 68, MS, 70, PhD(toxicol), 76; Am Bd Toxicol, dipl, 81; NC Cent Univ, JD, 85; NC Bar, 85; DC Bar, 87. *Prof Exp:* Mgr, Dept Toxicol, Becton Dickinson Res Ctr, 76-79; assoc dir, Life Sci Div, Clement Assoc, Inc, 79-81; mgr, Life Sci, Lorillard Res Ctr, 81-85; MGR, LIFE SCI, COCA-COLA, 85- *Mem:* Am Col Toxicol; NY Acad Sci; Sigma Xi; Am Bar Asn. *Res:* Inhalation toxicology: chronic inhalation studies involving exposure of rodents to aluminum chlorhydrate, asbestos, fibrous glass, ozone, sulfuric acid, or vinyl chloride; general toxicology. *Mailing Add:* Coca-Cola Co Mgr Life Scis PO Drawer 1734 Atlanta GA 30301

STONE, DANIEL BOXALL, b Gravesend, Eng, May 15, 25; US citizen; m 49; c 2. INTERNAL MEDICINE, ENDOCRINOLOGY. *Educ:* Univ London, BS & MD, 48, dipl psychiat, 50. *Prof Exp:* Intern & resident internal med, Univ London, 48-56; from asst prof to prof, Col Med, Univ Iowa, 59-71, exec assoc dean, 67-71; Milard prof med, Univ Nebr, 71-73, clin prof, 75-89; RETIRED. *Concurrent Pos:* Fel internal med, Univ London, 48-56; fel internal med & endocrinol, Col Med, Univ Iowa, 57-59; Markle scholar acad med, 60-; consult, Vet Admin Hosp, Iowa City, 63-71 & Coun Drugs, AMA, 66-73. *Mem:* Fel Am Col Physicians; Am Diabetes Asn; Am Heart Asn; Endocrine Soc; Royal Soc Med. *Res:* Influence of diet on serum lipids; geographic pathology of diabetes; metabolism of adipose tissue; influence of hypoglycemic drugs on lipolysis in adipose tissue. *Mailing Add:* 26354 Valley View Ave Carmel CA 93923-9102

STONE, DANIEL JOSEPH, b Passaic, NY, Dec 19, 18; m 50; c 3. MEDICINE. *Educ:* Johns Hopkins Univ, BA, 39; George Washington Univ, MD, 43; Am Bd Internal Med, dipl, 51. *Prof Exp:* Fel internal med, New York Med Col, 46-47; asst chief, Pulmonary Dis Serv, 49-54, assoc, Cardiopulmonary Lab, 50-54, CHIEF PULMONARY DIS SERV & DIR RESPIRATION LAB, BRONX VET ADMIN HOSP, 54-; PROF MED, NY MED COL, 75-, DIR PULMONARY SECT, 75- *Concurrent Pos:* Adv ed of res, Handbk of Biol Sci, Nat Acad Sci, 59; assoc prof, Mt Sinai Sch Med, 68; dir, Univ Sleep Breathing Disorder Ctr, NY Med Col, 87-; chmn inhalation ther comt, Bronx Vet Admin Hosp, 60- *Mem:* Am Physiol Soc; fel Am Col Physicians; fel AMA; Am Fedn Clin Res; fel Am Thoracic Soc. *Res:* Pulmonary diseases; lung mechanics and the mechanisms of pulmonary failure. *Mailing Add:* Pulmonary Div NY Med Col 41 Abington Ave Ardsley NY 10502

STONE, DAVID B, b Guernsey, UK, Sept 14, 33; m 60; c 3. GEOPHYSICS. *Educ:* Univ Keele, BA, 56; Univ Newcastle, PhD(geophys), 63. *Prof Exp:* Sr demonstrator geophys, Univ Newcastle, 63-66; assoc prof, 66-77, head, Geol/Geophys Prog, 77-80, PROF GEOPHYS, UNIV ALASKA, 77- *Concurrent Pos:* Asst dir, Geophys Inst, Univ Alaska, 84-87. *Mem:* Fel Royal Astron Soc; Am Geophys Union; fel Geol Soc Am. *Res:* Geomagnetism; paleomagnetism; geotectonics. *Mailing Add:* Geophys Inst Univ Alaska College AK 99775-0800

STONE, DAVID ROSS, b Little Rock, Ark, Aug 30, 42; m 65; c 3. ALGEBRA. *Educ:* Ga Inst Technol, BS, 64; Univ SC, PhD(math), 68. *Prof Exp:* PROF MATH, GA SOUTHERN COL, 68- *Mem:* Am Math Soc; Math Asn Am; Asn Comput Mach. *Res:* Rings and modules; torsion theory; problem solving. *Mailing Add:* Dept of Math & Comput Sci LB 8093 Ga Southern Col Statesboro GA 30460

STONE, DEBORAH BENNETT, b Portchester, NY, Oct 26, 38; div; c 2. PHYSICAL BIOCHEMISTRY, CONTRACTILITY. *Educ:* Smith Col, BA, 60; Yale Univ, PhD(pharmacol), 65. *Prof Exp:* Res assoc pharmacol, Sch Med, Stanford Univ, 64-66; USPHS trainee phys biochem, Cardiovasc Res Inst, 66-68, lectr physiol, 68-77, asst res biochemist, 68-76, ASSOC RES BIOCHEMIST, CARDIOVASC RES INST, UNIV CALIF, SAN FRANCISCO, 76- *Concurrent Pos:* USPHS res career develop award, 68-73. *Mem:* Biophys Soc. *Res:* Serotonin metabolism in the developing rat brain; regulation of phosphofructokinase activity; molecular mechanisms in muscle contraction; production of deuterated contractile proteins for neutron scattering experiments. *Mailing Add:* Cardiovasc Res Inst Box 0524 Univ of Calif San Francisco CA 94143-0524

STONE, DONALD EUGENE, b Eureka, Calif, Dec 10, 30; m 52; c 3. BOTANY, GENETICS. *Educ:* Univ Calif, AB, 52, PhD(bot), 57. *Prof Exp:* Asst cytol, biosysts & gen bot, Univ Calif, 54-57; from instr to asst prof bot, Tulane Univ, 57-63; from asst prof to assoc prof, 63-70, PROF BOT, DUKE UNIV, 70- *Concurrent Pos:* Assoc prog dir syst biol, NSF, 68-69; exec dir, Orgn Trop Studies, Inc, 76- *Mem:* Bot Soc Am; Soc Study Evolution; Am Soc Naturalists; Am Soc Plant Taxonomists (secy, 73-75); Orgn Trop Studies; Sigma Xi. *Res:* Biosystematics of temperate and tropical families. *Mailing Add:* Dept Bot Duke Univ Durham NC 27706

STONE, DOROTHY MAHARAM, b Parkersberg, WVa, July 1, 17; m 42; c 2. MATHEMATICS. *Educ:* Carnegie Inst Technol, BSc, 37; Bryn Mawr Col, PhD(math), 40. *Prof Exp:* Asst lectr math, Univ Manchester, 52-61; prof, 61-87, EMER PROF MATH, UNIV ROCHESTER, 87-; ADJ PROF, NORTHEASTERN UNIV, 88- *Concurrent Pos:* NSF fel math, 65-66. *Mem:* Am Math Soc; Math Asn Am. *Res:* Measure theory; ergodic theory; probability; linear operators. *Mailing Add:* Dept Math Northeastern Univ Boston MA 02115

STONE, DOUGLAS ROY, b Minneapolis, Minn, Nov 10, 48; m 72, 85; c 2. OPTICAL MICROLITHOGRAPHY. *Educ:* Univ Minn, BChE, 70; Univ Wis, MS, 72, PhD(chem eng), 75. *Prof Exp:* Res engr med equip, Air Prod & Chem, Inc, 75-77, res engr chem eng, 77-80; mem tech staff, 80-89, DISTINGUISED MEM TECH STAFF, AT&T BELL LABS, 89- *Concurrent Pos:* NIH grant, Dept Bioeng, Univ Pa, 77-81. *Mem:* Int Soc Optical Eng. *Res:* Microlithographic process development for future integrated circuit manufacture; characterization and optimization of optical step-and-repeat lithography. *Mailing Add:* RD Four Box 129 Coopersburg PA 18036

STONE, EARL LEWIS, JR, b Phoenix, NY, July 12, 15; m 41; c 3. FOREST SOILS. *Educ:* State Univ NY, BS, 38; Univ Wis, MS, 40; Cornell Univ, PhD(soils), 48. *Hon Degrees:* DSc, State Univ NY, 90. *Prof Exp:* Field asst & jr forester, Southern Forest Exp Sta, US Forest Serv, 40-41; from asst prof to assoc prof forest soils, 48-62, Charles Lathrop Pack prof, 62-79, EMER PROF FOREST SOILS, CORNELL UNIV, 79- *Concurrent Pos:* Collabr & consult, Southern Forest Exp Sta, US Forest Serv, 47-48 & 52; soil scientist, Pac Sci Bd, Nat Acad Sci, Marshall Islands, 50; Am-Swiss Found fel, 54-55; vis prof, Philippines, 58-60; Fulbright res fel, Forest Res Inst, NZ, 62; ed, Forest Sci, Soc Am Foresters, 65-71; Bullard fel, Harvard Univ, 69-70; consult, Biotrop, Indonesia, 70; mem Adv Panel Ecol, NSF, 70-; mem adv comn, Ecol Sci Div, Oak Ridge Nat Lab, 71-; mem, forest studies team, Nat Res Coun, 73-74, Comt Evaluate Int Biol Prog, 74-75 & comt scientist, Nat Forest Mgt Act, 77-; vis prof, Dept Soil Sci & Forestry, Univ Fla, Gainesville, 79-82, adj prof, 82-; mem, Bikini Atoll Rehab Comn, 83-89. *Honors & Awards:* Barrington Moore Award, Soc Am Foresters, 73. *Mem:* Fel AAAS; fel Soc Am Foresters; fel Soil Sci Soc Am; Ecol Soc Am; fel Am Soc Agron. *Res:* Forest nutrition; ecology; Pacific tropics; atoll soils. *Mailing Add:* Dept Soil Sci Univ Fla Gainesville FL 32611

STONE, EDWARD, b Fall River, Mass, Dec 7, 32; m 56; c 4. ORGANIC CHEMISTRY, POLYMER CHEMISTRY. *Educ:* Southeastern Mass Univ, BS, 55; Univ Md, PhD(org chem), 62. *Prof Exp:* Teaching asst, Univ Md, 55-57; chemist, Metals & Controls Corp, 56; anal res chemist, Nat Inst Drycleaning, 57-61, consult, 61; sr res chemist, Tex-US Chem Co, NJ, 61-65; tech mgr new prod res & develop, 65-80, dir polymer res & chem anal, 80-85, DIR CHEM RES, BASF INMONT, 85- *Mem:* Am Chem Soc; Sci Res Soc Am; Fedn Soc Coatings Technol. *Res:* Radiation curing of inks and coatings; polymer research and development; block and graft copolymers; polyurethanes, polyesters, polyolefins, acrylics; structure-property correlations; organic synthesis; instrumental analysis. *Mailing Add:* Four Inwood Rd Morris Plains NJ 07950

STONE, EDWARD CARROLL, JR, b Knoxville, Iowa, Jan 23, 36; m 62; c 2. PHYSICS. *Educ:* Univ Chicago, SM, 59, PhD(physics), 64. *Prof Exp:* Res fel, Calif Inst Technol, 64-67, sr res fel, 67, from asst prof to assoc prof, 67-76, vpres, Astro Facil, 88-90, PROF PHYSICS, CALIF INST TECHNOL, 76-, CHMN, DIV PHYSICS, MATH & ASTRON, 83-, VPRES & DIR, JET PROPULSION LAB, 91- *Concurrent Pos:* Mem particles & fields adv comt, NASA, 69-71, consult, 71-, proj scientist, NASA Voyager Mission, 72-, mem high energy astrophys mgt operating working group, 76-, mem comt space astron & astrophys, Space Sci Bd, 79-82; Alfred P Sloan res fel, Calif Inst Technol, 71-73; mem, Jet Propulsion Lab Adv Coun, 80-82; mem, NASA Cosmic Ray Prog Working Group, 80-82, Outer Planets Working Group, Solar Syst Explor Comt, 81-82 & 83, Space Sci Bd, 82-85, Univ Rel Study Group, 83, Steering Group, Space Sci Bd Study Maj Directions Space Sci 1995-2015, 84-85, Adv Comt, NASA/JPL, Vis Sr Scientist Prog, 86-; bd dir, Calif Asn Res Astron, 85-, vchmn, 87-88, chmn, 88-; mem, Comn Phys Sci, Math & Resources, Nat Res Coun, 86-89, Comt Space Policy, 88- *Honors & Awards:* Nat Medal of Sci, 91; Am Educ Award, Am Asn Sch Adminr, 81; Dryden Lectr, Am Inst Aeronaut & Astronaut, 83, Space Sci Award, 84; Sci Award, Nat Space Club. *Mem:* Nat Acad Sci; AAAS; fel Am Phys Soc; Am Astron Soc; Int Astron Union; fel Am Geophys Union; Int Acad Astronaut; assoc fel Am Inst Aeronaut & Astronaut; Astron Soc Pac. *Res:* Solar and galactic cosmic rays; planetary magnetospheres; interplanetary medium; solar system exploration; satellite and balloon instrumentation. *Mailing Add:* Jet Propulsion Lab 4800 Oak Grove Dr Pasadena CA 91030

STONE, EDWARD CURRY, b Ill, Nov 28, 17; m 41; c 2. PLANT PHYSIOLOGY. *Educ:* Univ Calif, BS, 40, PhD(plant physiol), 48. *Prof Exp:* Plant physiologist, Calif Forest & Range Exp Sta, US Forest Serv, 48-49; PROF FOREST ECOL & SILVICULTURIST, AGR EXP STA, UNIV CALIF, BERKELEY, 49- *Concurrent Pos:* Fulbright res scholar, Univ NZ, 59-60; Guggenheim fel, 60. *Mem:* Ecol Soc Am; Soc Am Foresters; Am Soc Plant Physiol; Bot Soc Am; Scand Soc Plant Physiol. *Res:* Forest physiology; dormancy; root growth; drought resistance; fire response; cone production; nutritional requirements of forest vegetation. *Mailing Add:* Dept Forestry & Conserv Univ of Calif Berkeley CA 94720

STONE, EDWARD JOHN, b Minersville, Pa, July 27, 30; m 56; c 2. ORGANIC CHEMISTRY. *Educ:* Pa State Univ, BS, 53, MS, 56, PhD(biochem), 62. *Prof Exp:* Res chemist, Campbell Soup Co, 59-62, SR RES CHEMIST, CAMPBELL INST FOOD RES, 62- *Mem:* Sigma Xi; NY Acad Sci; Am Chem Soc. *Res:* Food chemistry; radiochemistry; organic chemistry. *Mailing Add:* Res Dept Campbell Soup Co Campbell Pl Camden NJ 08101

STONE, ERIKA MARES, b Prague, Czech, Jan 26, 38; US citizen; m; c 1. MATHEMATICS. *Educ:* Pa State Univ, BA, 60, MA, 62, PhD(math), 64. *Prof Exp:* Instr math, Swarthmore Col, 64-65; sr res mathematician, HRB-Singer, Inc, Pa, 65-68, lectr dept comput sci, Pa State Univ, 68; vis asst prof, Dept Math & Comput Sci, Univ SC, 73-75; programmer, comput ctr, Duke Univ, 77-81, PROGRAMMER, CAROLINA POP CTR, UNIV NC CHAPEL HILL. *Mem:* Am Math Soc. *Res:* Structure theory of semiperfect rings and the generalization of theory for modules. *Mailing Add:* 300 W University Dr No A Chapel Hill NC 27316

STONE, GORDON EMORY, b Sioux City, Iowa, July 12, 33; m 55; c 2. CELL BIOLOGY. *Educ:* Univ Iowa, BA, 56, MSc, 58, PhD(zool), 61. *Prof Exp:* Res fel, NIH, 61-63 & AEC, 63-64; from asst prof to assoc prof anat, Sch Med, Univ Colo, 64-72; PROF BIOL SCI & CHMN DEPT, UNIV DENVER, 72- *Concurrent Pos:* NIH career develop award, 65-70. *Mem:* AAAS; Am Soc Cell Biol; Soc Protozool; Am Soc Zool; Sigma Xi. *Res:* Cytochemical studies on cell growth and division, especially the sequential macromolecular events during the interdivision interval leading to division with emphasis on microtubule protein synthesis. *Mailing Add:* Dept of Biol Sci Univ of Denver Denver CO 80210

STONE, GREGORY MICHAEL, communications security, for more information see previous edition

STONE, H NATHAN, b Claremont, NH, May 8, 20; m 42; c 2. CHEMICAL ENGINEERING. *Educ:* Univ NH, BS, 43; Univ Ill, MS, 47, PhD(chem eng), 50. *Prof Exp:* Group leader org res, Norton Co, 50-52; asst dir res, Bay State Abrasive Prod Co, 52-58, dir res & develop, 58-64, vpres & dir res & eng, Bay State Abrasives, Div Dresser Industs, Inc, 64-86. *Mem:* AAAS; Am Chem Soc; Am Ceramic Soc; Nat Soc Prof Engrs; Am Inst Chem Engrs. *Res:* Reaction kinetics in fluidized beds; solid state physics, especially the creation of new surfaces during abrasive machining; resins and polymers applicable to abrasive bonding. *Mailing Add:* 38 Sun Valley Dr Worcester MA 01609

STONE, HAROLD S, b St Louis, Mo, Aug 10, 38; m; c 1. COMPUTER ARCHITECTURE. *Educ:* Princeton, BSE, 60; Univ Calif, Berkeley, MSEE, 61, PhD(elec eng), 63. *Prof Exp:* Assoc prof elec eng & comput sci, Stanford Univ, 68-74; prof, elec eng & comput sci, Univ Mass, Amherst, 74-84; RES STAFF MEM, T J WATSON RES CTR, IBM, 84- *Mem:* Fel Inst Elec Electronic Engrs; Asn Comput Mach. *Res:* Designing and construction of computer systems. *Mailing Add:* IBM Corp PO Box 704 Yorktown Heights NY 10598

STONE, HARRIS B(OBBY), b New York, NY, Oct 16, 23; m 48; c 2. ELECTRONICS ENGINEERING. *Educ:* Mass Inst Technol, SB, 50. *Prof Exp:* Engr, Western Union Tel Co, 50-52; proj engr & officer, Hq, Electronic Warfare Ctr, US Dept Army, Ft Monmouth, NJ, 52-54, chief electronic reconnaissance sect, Electronic Proving Ground, Ft Huachuca, Ariz, 54-55, chief defense systs br, 55, chief electronic reconnaissance br, Hq, Security Agency Oper Ctr, 55-56; electronic warfare & intel systs coordr, Off Develop Coordr, Off Naval Res, US Dept Navy, Washington, DC, 56-59, asst for electronic warfare & intel, Tech Anal & Adv Group, 59-61, asst dir command control & spec opers, 61, dep dir, 61-71, dir res, develop, test & eval plans div, Off Chief Naval Opers, 71-80; RETIRED. *Concurrent Pos:* Adv bd dirs, Seaworld Res Inst, San Diego, 81-90. *Res:* Electronic warfare; intelligence; radar; acoustics; communications; mechanics; economics; photography; electromagnetics; research and development management. *Mailing Add:* 4528 Roundhill Rd Alexandria VA 22310

STONE, HENRY E, b Munich, Ger, Feb 2, 22; US citizen. ENGINEERING. *Educ:* Union Col, MA, 49; Univ Buffalo, MS, 55. *Prof Exp:* Mgr, Atomic Power Lab, Schenectady, NY, 50-68; gen mgr, Knoll Atomic Power Lab, 68-74; mgr, Strategic Planning Oper, Gen Elec Co, San Jose, 74-75, gen mgr, Boiling Water Syst Div, 75-77, mgr, 77-84, vpres, Nuclear Energy Bus Opers, 78-87, chief eng officer, 84-87; CONSULT, 87- *Mem:* Nat Acad Eng; Am Nuclear Soc; fel Am Soc Mech Eng. *Mailing Add:* 6805 Castle Rock Dr San Jose CA 95120

STONE, HENRY OTTO, JR, b Spartanburg, SC, Apr 10, 36; m 60; c 1. VIROLOGY, GENETIC ENGINEERING. *Educ:* Wofford Col, BS, 59; Duke Univ, PhD(zool), 64. *Prof Exp:* Am Cancer Soc fel biochem, Duke Univ, 64-66; res chemist, E I du Pont de Nemours & Co, Inc, Del, 66-70; NIH spec res fel animal virol, St Jude Children's Res Hosp, 70-72; from asst prof to assoc prof microbiol, Univ Kans, 72-82; ASSOC PROF MICROBIOL & IMMUNOL, SCH MED, E CAROLINA UNIV, 82- *Concurrent Pos:* Am Cancer Soc scholar, Duke Univ, 80-81. *Mem:* Am Soc Microbiol; Am Soc Biol Chemists; Soc Gen Microbiol; Am Soc Virol. *Res:* Paramyxoviruses; viral RNA synthesis; genome transcription; viral proteins; cloning sequence and expression of viral genes; complementary DNA copies of newcastle disease, virus genes were cloned into bacterial plasmids; DNA copies sequenced and inserted into procaryotio and eucargotic expression vectors. *Mailing Add:* Dept Microbiol & Immunol Sch Med E Carolina Univ Greenville NC 27858-4354

STONE, HERBERT, b Washington, DC, Sept 14, 34; m 64; c 2. NUTRITION. *Educ:* Univ Mass, BSc, 55, MSc, 58; Univ Calif, Davis, PhD(nutrit), 62. *Prof Exp:* Specialist, Exp Sta, Univ Calif, Davis; food scientist, Stanford Res Inst, 62-64, dir dept food & plant sci, 67-74; PRES, TRAGON CORP, 74- *Concurrent Pos:* Pres, Sensory Eval Div, Inst Food Technol, 77-78, exec comt, 84-; assoc ed, J Food Sci, 77-80. *Mem:* AAAS; Sigma Xi; Am Soc Enol; fel Inst Food Technologists; Am Soc Testing & Mat. *Res:* Management consultant in product development, food, beverage, cosmetic products and product acceptance measurement; taste and odor research. *Mailing Add:* Tragon Corp 365 Convention Way Redwood City CA 94063-1402

STONE, HERBERT L(OSSON), b Eddy, Tex, Nov 6, 28; m 47, 80; c 2. CHEMICAL ENGINEERING. *Educ:* Rice Inst, BS, 50; Mass Inst Technol, ScD(chem eng), 53. *Prof Exp:* Asst process engr, Vulcan Copper & Supply Co, 53; res adv, Exxon Prod Res Co, 53-88; PRES, STONE ENG, 89- *Mem:* Am Inst Chem Engrs; Am Inst Mining, Metall & Petrol Engrs; Sigma Xi. *Res:* Numerical analysis; petroleum resevoir engineeing; enhance petroleum recovery. *Mailing Add:* Box 22781 Houston TX 77227

STONE, HERMAN, b Munich, Germany, Nov 3, 24; nat US; m 49; c 6. ORGANIC CHEMISTRY. *Educ:* Bethany Col, WVa, BSc, 44; Ohio State Univ, PhD(chem), 50. *Prof Exp:* Analytical chemist, Nat Aniline Div, Allied Chem Corp, 44-45, analytical res chemist, 51-53, res chemist, 53-61, group leader appln res chem, 61-63, mgr chem res, Indust Chem Div, 63-68, dir res, Specialty Chem Div, 68-69, res assoc, Corp Chem Res Lab, 69-72; dir chem res, Malden Mills Inc, 72-74; DIR FOAM DEVELOP, GEN FOAM CORP, 74- *Mem:* AAAS; Am Chem Soc; Sigma Xi; Am Inst Chem. *Res:* Analytical and exploratory research on polymer intermediates; flammability of plastics; urethane polymer technology. *Mailing Add:* Gen Foam Div PMC Inc Valmont Indust Park Hazleton PA 18201

STONE, HOWARD ANDERSON, b Claremont, NH, Nov 21, 40; m 62; c 3. GENETICS, VIROLOGY. *Educ:* Univ NH, BS, 62, MS, 65; Mich State Univ, PhD(poultry), 72. *Prof Exp:* Asst poultry genetics, Univ NH, 62-65; RES GENETICIST, AGR RES SERV, USDA, 65-; AT OMEGA CHICKS. *Mem:* Poultry Sci Asn. *Res:* Investigations of the genetic control of Marek's disease and lymphoid leukosis in chickens; maintenance and development of highly inbred lines of chickens. *Mailing Add:* 1827 Lyndhurst Haslett MI 48840

STONE, HOWARD N(ORDAS), b Marblehead, Mass, Apr 7, 22; m 45, 84, 90; c 2. AERONAUTICAL ENGINEERING. *Educ:* Harvard Univ, SB & MS, 44. *Prof Exp:* Struct engr, Sikorsky Aircraft Co, 45; Gordon McKay asst, Harvard Univ, 45-46; sr res engr, Curtiss-Wright Corp, 46-49; res aerodynamicist, Cornell Aeronaut Lab, 49-54; develop eng specialist, Lockheed Aircraft Corp, 54-57 & 58-59; specialist, Missile Systs Div, Raytheon Corp, 57-58; res specialist, Missiles & Space Div, Lockheed Aircraft Corp, 58-61, aircraft develop engr res group, Aerodyn Dept, Ga, 61-63; RES SPECIALIST, LOCKHEED MISSILES & SPACE CO, SUNNYVALE, CALIF, 63- *Res:* Research and development in booster vehicle design and performance; aerodynamics of low-aspect-ratio wings; bodies and wing-body combinations; flutter of wings; simulation and controls development for space vehicles; advanced missile design. *Mailing Add:* 3133 David Ave Palo Alto CA 94303-3946

STONE, IRVING CHARLES, JR, b Chicago, Ill, Dec 18, 30; m 55; c 3. FORENSIC SCIENCE. *Educ:* Iowa State Univ, BS, 52; George Washington Univ, MS, 61, PhD(geochem), 67. *Prof Exp:* Spec agt-microscopist, Fed Bur Invest, 55-61; res chemist, Res Div, W R Grace & Co, 61-63, proj leader, 63-64, res supvr, 64-68; vpres, Geochem Surv, Tex, 68-72; criminologist, 72-74, CHIEF PHYS EVIDENCE SECT, INST FORENSIC SCI, 74- *Concurrent Pos:* Lectr police sci, Montgomery Jr Col, 67-70; instr forensic sci, Univ Tex Health Sci Ctr Dallas, 72-77, asst prof path, 77-85, dir grad prog forensic sci, Grad Sch Biomed Sci, 77-, assoc prof clin path, 85- *Honors & Awards:* Paul L Kirk Award, Outstanding Criminalist, Am Acad Forensic Sci, 90. *Mem:* Sigma Xi; Am Soc Firearms & Toolmark Examrs; Am Soc Crime Lab Dir; fel Am Acad Forensic Sci. *Res:* Analytical chemistry, especially x-ray diffraction, spectrometry, light microscopy; applied research in forensic sciences, specifically glass, firearm residues and instrumental analytical applications. *Mailing Add:* Inst of Forensic Sci Box 35728 Dallas TX 75235-0728

STONE, J(ACK) L(EE), b Taylor, Tex, July 12, 41; m 65; c 3. ELECTRICAL ENGINEERING. *Educ:* Univ Tex, BSEE, 63, MSEE, 64, PhD(elec eng), 68. *Prof Exp:* Res asst appl superconductivity, Univ Tex, 62-65, res asst elec eng, 65-66, teaching assoc, 66-67, asst prof, 67-68; res engr, Mesa Instruments Inc, 68-69; from asst prof to assoc prof elec eng, Tex A&M Univ, 69-77; sr scientist, Solar Energy Res Inst, 78-79, chief, Advan Silicon Br, 79-81, dep div dir, Solar Elec Conversion Res Div, 81-86, DIV DIR SOLAR ELEC RES DIV, SOLAR ENERGY RES INST, 86- *Concurrent Pos:* Co-prin investr, NASA res grant, 68-69; prin investr, Army Res Off-Durham res grant, 70-72; vis prof, Inst Nac Astrofisica, Optica, Electronics, 75-76. *Mem:* Inst Elec & Electronic Engrs; Am Inst Physics; Electrochem Soc. *Res:* Applied superconductivity; optical properties of semiconductors at low temperatures; high Q resonant circuit techniques; amorphous semiconductors; ion implantation; integrated circuit device processing; photovoltaic devices. *Mailing Add:* Solar Elec Res Div Solar Energy Res Inst 1617 Cole Blvd-Bldg 16-3 Golden CO 80401

STONE, JAY D, b Littlefield, Tex, Oct 14, 44; m 65; c 2. ENTOMOLOGY. *Educ:* West Tex State Univ, BS, 68; Iowa State Univ, MS, 70, PhD(entom), 73. *Prof Exp:* Res assoc entom, Iowa State Univ, 72-73; asst prof, Kans State Univ, 73-76; asst prof, 76-83, ASSOC PROF ENTOM, TEX A&M UNIV, 83- *Mem:* Sigma Xi; Entom Soc Am. *Res:* Biology and control of urban insect pests of far west Texas. *Mailing Add:* 720 Quinta Luz Circle El Paso TX 79922

STONE, JOE THOMAS, b Miami, Okla, June 25, 41; m 63; c 2. PHYSICAL ORGANIC CHEMISTRY, PHOTOGRAPHIC CHEMISTRY. *Educ:* Harvey Mudd Col, BS, 63; Univ Wash, PhD(org chem), 67. *Prof Exp:* NIH res fel, Univ Wash, 67-68; SR RES CHEMIST, EASTMAN KODAK CO, 68- *Mem:* NY Acad Sci; Royal Soc Chem; Am Chem Soc. *Res:* Organic reaction mechanisms; application of physical-organic techniques to biological processes, enzyme kinetics and mechanism; homogeneous and heterogeneous catalysis and reaction kinetics; photographic science - film structure, process chemistry, development mechanisms. *Mailing Add:* 595 Drumm Rd Webster NY 14580-1512

STONE, JOHN AUSTIN, b Paintsville, Ky, Nov 30, 35; m 61; c 3. NUCLEAR WASTE MANAGEMENT, SCIENCE ENGINEERING EDUCATION. *Educ:* Univ Louisville, BS, 55; Univ Calif, Berkeley, PhD(nuclear chem), 63. *Prof Exp:* Chemist, Savannah River Lab, E I du Pont de Nemours & Co, Inc, 63-73, staff chemist, 73-74, res staff chemist, 74-81, res assoc, 81-89, mgr, Univ Relations, 89-90, MGR, EDUC PROGS, SAVANNAH RIVER LAB, WESTINGHOUSE SAVANNAH RIVER CO, 90- *Concurrent Pos:* Assoc ed, Mat Lett, 83-89. *Mem:* Am Phys Soc; Am Chem Soc; Mat Res Soc; Am Soc Eng Educ. *Res:* Radioactive waste management; nuclear fuel cycle; solid state and chemical properties of the actinides; Mossbauer spectroscopy. *Mailing Add:* Savannah River Lab Westinghouse Savannah River Co Aiken SC 29808

STONE, JOHN BRUCE, b Forfar, Ont, Can, Sept 23, 30; m 54; c 4. ANIMAL SCIENCE. *Educ:* Ont Agr Col, BSA, 53, MSA, 54; Cornell Univ, PhD, 59. *Prof Exp:* Asst prof animal husb, Ont Agr Col, 54-62; asst prof animal sci, Cornell Univ, 62-66; PROF ANIMAL SCI, UNIV GUELPH, 66-, ASSOC DEAN, COL AGR, 83- *Mem:* Am Dairy Sci Asn; Agr Inst Can; Am Soc Animal Sci; Sigma Xi. *Res:* Dairy cattle nutrition; forages for dairy cattle rations; calf-raising programs; systems analyses for dairy production. *Mailing Add:* 11 Mayfield Guelph ON N1G 2L9 Can

STONE, JOHN ELMER, b Montgomery, Ala, Aug 12, 31; m 59; c 2. GEOLOGY. *Educ:* Ohio Wesleyan Univ, BA, 53; Univ Ill, MS, 58, PhD(geol), 60. *Prof Exp:* Asst prof geol, Univ Tex, 60-62; geologist, Minn Geol Surv, 62-67; head dept, 67-77, PROF GEOL, OKLA STATE UNIV, 67- *Concurrent Pos:* Res grants, Univ Tex Excellence Fund, 61 & grad sch, Univ Minn, 62, 67; NSF summer grants, 68-72, sci equip grant, 69; Okla State Univ Res Found res grant, 70. *Mem:* AAAS; fel Geol Soc Am; Nat Asn Geol Teachers; Am Quaternary Asn; Soc Econ Paleont & Mineral. *Res:* Glacial and engineering geology. *Mailing Add:* Dept Geol Okla State Univ Stillwater OK 74078

STONE, JOHN FLOYD, b York, Nebr, Oct 13, 28; m 53; c 4. SOIL PHYSICS. *Educ:* Univ Nebr, BSc, 52; Iowa State Univ, MS, 55, PhD, 57. *Prof Exp:* Res assoc, Dept Agron & Inst Atomic Res, Iowa State Univ, 56-57; from asst prof to assoc prof agron, 57-69, PROF AGRON, OKLA STATE UNIV, 69- *Concurrent Pos:* Assoc ed, Soil Sci Soc Am, 68-75; vis scientist lectr, Am Geophys Union, 72; mem, Agr Adv Panel, US Dept Defense, 77-78, Comt Irrigation Water, Am Soc Civil Engrs, 85-88, Comt Water Resources, Soil Sci Soc Am & Comt Unsaturated Zone, 64-73, Am Geophys Union, 78-88; assoc ed, Am Soc Agron, 82-85; chmn, Comt Unsaturated Zone, Am Geophys Union, 86-88. *Mem:* Am Soc Agron; Soil Sci Soc Am; Am Geophys Union; Int Soc Soil Sci; Sigma Xi; Am Soc Civil Engrs. *Res:* Water conservation, evapotranspiration; water flow in plants; electronic instrumentation. *Mailing Add:* Dept Agron Okla State Univ Stillwater OK 74078

STONE, JOHN GROVER, II, b Pueblo, Colo, Aug 6, 33; m 64; c 5. GEOLOGY. *Educ:* Yale Univ, BS, 55; Stanford Univ, PhD(geol), 58. *Prof Exp:* Staff geologist, Hanna Mining Co, 58-69, asst chief geologist, 69-77, proj mgr, 77-79, mgr, Pilot Knob Mine & Pellet Plant, 80- 81, CHIEF GEOLOGIST, M A HANNA CO, 82- *Mem:* Geol Soc Am; Soc Econ Geologists. *Res:* Genesis of ore deposits; ore reserve estimation. *Mailing Add:* 300 Mott Lane Gardnerville OH 44114

STONE, JOHN PATRICK, b Algood, Tenn, Sept 5, 39; m 64; c 3. ENDOCRINOLOGY, RADIOBIOLOGY. *Educ:* Wayne State Univ, BS, 61, PhD(biol), 72; Purdue Univ, Lafayette, MS, 64. *Prof Exp:* Teaching asst bionucleonics, Purdue Univ, Lafayette, 62-64; teaching asst endocrinol & radiobiol, Wayne State Univ, 65-68; fel radiobiol, Div Biol & Med Res, Argonne Nat Lab, 72-74; asst scientist, Brookhaven Nat Lab, 74-75, assoc scientist, 76-79, scientist, 79-86; STAFF PHARMACIST, LYNCHBURG GEN HOSP, 86- *Concurrent Pos:* Assoc clin prof, State Univ NY, Stony Brook, 78; consult, Inner Radiation Belt Res Counsults, 82- *Mem:* Endocrine Soc; Int Pigment Cell Soc; Radiation Res Soc; Am Soc Zoologists; Sigma Xi; Am Asn Career Res. *Res:* Hormonal control of radiation and chemically induced mammary tumorigenesis; effects of chronic gamma irradiation upon endocrine and hematopoietic systems; pigment cell biochemistry and physiology; pathophysiology of peptide toxins, retinoid inhibition of mammary carcinogenesis. *Mailing Add:* 3400 Ivy Link Pl Lynchburg VA 24503

STONE, JOSEPH, b Holyoke, Mass, June 3, 20. BIOCHEMISTRY, PHARMACOLOGY. *Educ:* Mass Col Pharm, BS, 47; Univ Colo, PhD(pharmacol), 54. *Prof Exp:* Proj assoc, McArdle Mem Labs, Univ Wis, 54-56; staff pharmacologist, Vet Admin Hosp, Chicago, 56; RETIRED. *Mem:* AAAS; NY Acad Sci; Sigma Xi. *Res:* Diffusion respiration; testing systems for DNA antimetabolites; central nervous system biochemistry and pharmacology. *Mailing Add:* 30 Kingsbridge Way Little Rock AR 72212

STONE, JOSEPH LOUIS, medical bacteriology; deceased, see previous edition for last biography

STONE, JULIAN, b New York, NY, Apr 12, 29; m 51; c 3. PHYSICS. *Educ:* City Col New York, BS, 50; NY Univ, MS, 51, PhD(physics), 58. *Prof Exp:* Electronic scientist, Naval Mat Lab, 52; tutor physics, City Col New York, 52-53; res scientist, Hudson Lab, Columbia Univ, 53-69, assoc dir physics, 66-69; mem tech staff, 69-83, DISTINGUISHED MEM TECH STAFF, BELL LABS, 83- *Concurrent Pos:* Tutor physics, City Col New York, 56-57; assoc ed, Optics Lett, Optical Soc Am. *Mem:* Fel Optical Soc Am. *Res:* Lasers; spectroscopy; underwater sound propagation. *Mailing Add:* Bell Labs Rm L121 Box 400 Holmdel NJ 07733

STONE, KATHLEEN SEXTON, b Lakewood, Ohio, March 29, 47; m 76; c 2. RESEARCH ADMINISTRATION. *Educ:* Ohio State Univ Sch Nursing, Columbus, BS, 72; Ohio State Univ Grad Sch, Columbus, PhD(physiol), 77. *Prof Exp:* Asst prof, grad & undergrad educ, Univ Cincinnati, 77-78; asst prof, 79-84, dir res, Ctr Nursing Res, 85-88, ASSOC PROF NURSING, GRAD

EDUC, DEPT LIFE SPAN PROCESS, OHIO STATE UNIV COL NURSING, 84- *Concurrent Pos:* Prin investr, endotracheal suctioning in acutely ill adults, Nat Ctr Nursing Res, 87-95 & Div Nursing, PHS, 84-87; group chair, Conf Res Priorities Nursing, Nat Ctr Nursing Res, 88; mem, Consensus Conf, Am Lung Assoc, 88; chair, Nat Study Group Endotracheal Suctioning, Am Asn Critical Care Nurses, 84-; lectr, Am Asn Critical Care Nurses Nat Teaching Inst, 88. *Honors & Awards:* Res Award, Am Asn Critical Care Nurses, Nat Teaching Inst, 86. *Mem:* Am Thoracic Soc; Am Lung Asn; Am Asn Critical Care Nurses; Sigma Xi. *Res:* Investigating the effects of lung hyperinflation and hyperoxygenation during endotracheal suctioning testing open vs closed techniques on mean arterial pressure, heart rate, cardiac output, pulmonary arterial pressure and airway pressure in critically ill adults. *Mailing Add:* Col Nursing Ohio State Univ 1585 Neil Ave 384 Newton Hall Columbus OH 43210-1289

STONE, LAWRENCE DAVID, b St Louis, Mo, Sept 2, 42; m 67; c 2. MATHEMATICS, OPERATIONS RESEARCH. *Educ:* Antioch Col, BS, 64; Purdue Univ, West Lafayette, MS, 66, PhD(math), 67. *Prof Exp:* From assoc to sr assoc, Daniel H Wagner Assocs, 67-74, vpres, 74-81, br mgr, 81-86; SR VPRES, METRON, 86- *Concurrent Pos:* Off Naval Res grant, 69-76; assoc ed, Operations Res, 81- *Honors & Awards:* Lancaster Prize, Opers Res Soc, 75. *Mem:* Am Math Soc; Inst Math Statist; Oper Res Soc; Math Prog Soc. *Res:* Theory of search for stationary and moving targets; constrained extremal problems; threshold crossing problems for markov and semi-markov processes; optimal stochastic control of semi-markov processes; non-linear filtering. *Mailing Add:* Metron 11911 Freedom Dr Suite 800 Reston VA 22090-5603

STONE, LOYD RAYMOND, b Prague, Okla, Jan 6, 45; m 65; c 3. SOIL PHYSICS, SOIL & WATER MANAGEMENT. *Educ:* Okla State Univ, BS, 67, MS, 69; SDak State Univ, PhD(agron), 73. *Prof Exp:* From asst prof to assoc prof, 73-83, PROF AGRON, EVAPOTRANSPIRATION LAB, KANS STATE UNIV, 83- *Mem:* Am Soc Agron; Soil Sci Soc Am; Sigma Xi; Soil Conserv Soc Am. *Res:* Management of the soil-water-crop environment for efficient use of water; irrigation water use efficiency and plant root systems analysis. *Mailing Add:* Evapo-transpiration Lab Kans State Univ Manhattan KS 66506

STONE, M(ORRIS) D, b Cambridge, Mass, Dec 2, 02; m; c 4. ENGINEERING. *Educ:* Harvard Univ, BS, 23, MS, 25; Univ Pittsburgh, PhD, 33. *Prof Exp:* Mem staff steam res, Am Soc Mech Eng, 23-25; res engr, engr-in-chg & head design sch, Westinghouse Elec & Mfg Co, 25-34; spec engr, United Eng & Foundry Co, 34-40, mgr res & develop, 40-65, vpres, 65-70, mem bd dirs, 68-70; consult, 70-72; RETIRED. *Concurrent Pos:* Lectr, Univ Pittsburgh, 28-34; consult, AEC, 46-52; chmn comts, Mat Adv Bd, Nat Acad Sci; consult engr. *Mem:* Fel Am Soc Mech Engrs; Asn Iron & Steel Engrs; Am Inst Mining, Metall & Petrol Engrs. *Res:* Theory of vibrations; mathematical theory of elasticity; rolling and forging of metals; rolling and metalworking of metals. *Mailing Add:* 1308 Macon Ave Pittsburgh PA 15218

STONE, MARGARET HODGMAN, b Cleveland, Ohio, May 20, 14; m 41; c 3. PLANT TAXONOMY. *Educ:* Western Reserve Univ, BA, 36, MA, 37, PhD(bot), 40. *Prof Exp:* Dir hort, Garden Ctr Gtr Cleveland, 40-41; agr res dir bot, State Univ NY Col Agr, Cornell Univ, 41-42; instr, Western Reserve Univ, 43-45; agr res dir State Univ NY Col Agr, Cornell Univ, 45-46; prof lectr, Col Agr, Philippines, 58-60; sr cur taxon bot, L H Bailey Hortorium, Cornell Univ, 65-79; ADJ PROF BOT, UNIV, FLA, 81- *Concurrent Pos:* Mercer res fel, Arnold Arboretum, Harvard Univ, 69-70. *Mem:* AAAS; Int Asn Plant Taxon; Am Hort Soc. *Res:* Hydrogen-ion concentration of soil in relation to distribution of flora; dormant buds in Pinus; taxonomy of cultivated plants. *Mailing Add:* 1726 NW Tenth Terr Gainesville FL 32609

STONE, MARSHALL HARVEY, mathematics; deceased, see previous edition for last biography

STONE, MARTHA BARNES, b Paris, Tenn, Nov 20, 52; m 73; c 2. FOOD SCIENCE. *Educ:* Univ Tenn, Martin, BS, 74; Univ Tenn, Knoxville, MS, 75, PhD(food sci), 77. *Prof Exp:* Res asst foods & nutrit, Univ Tenn, 74-77; from asst prof to assoc prof foods & nutrit, Kans State Univ, 78-89; PROF, DEPT FOOD SCI & HUMAN NUTRIT, COLO STATE UNIV, 89- *Concurrent Pos:* NSF grant, 78; mem regional commun, Inst Food Technologists, 81-, nat chmn, 85-; sci adv, Am Coun Sci & Health. *Mem:* Inst Food Technologists (secy, 76-77); Sigma Xi; Am Dietetics Asn; Am Assoc Cereal Chemists. *Res:* Formulated foods; quality evaluation of fruits and vegetables; food preservation soybean. *Mailing Add:* Dept Food Sci & Human Nutrit Colo State Univ Ft Collins CO 80523

STONE, MARTIN L, b New York, NY, June 11, 20; m 43; c 1. OBSTETRICS & GYNECOLOGY. *Educ:* Columbia Univ, BS, 41; New York Med Col, MD, 44, MMSc, 49; Am Bd Obstet & Gynec, dipl, 52. *Prof Exp:* Prof obstet & gynec & chmn dept, New York Med Col, 56-78; PROF OBSTET & GYNEC & CHMN DEPT, SCH MED, STATE UNIV NEW YORK, STONY BROOK, 78- *Concurrent Pos:* Attend, Flower & Fifth Ave Hosps, Metrop & Bird S Coler Hosps, 56-; consult, Southampton Hosp, 60-, Deepdale Gen Hosp, 66- & Long Island Jewish-Hillside Med Ctr, 78. *Mem:* Fel Am Gynec Soc; fel Am Pub Health Asn; fel Am Asn Obstetricians & Gynecologists; assoc fel Royal Soc Med; fel Am Col Obstetricians & Gynecologists (secy, 71-78, pres-elect, 78 pres, 79). *Mailing Add:* Dept Obstet & Gynec Mt Sinai Med Ctr Box 1076 1176 Fifth Ave New York NY 10029

STONE, MARVIN J, b Columbus, Ohio, Aug 3, 37; m 58; c 2. ONCOLOGY & HEMATOLOGY, IMMUNOLOGY. *Educ:* Univ Chicago, MS, 62, MD, 63; Am Bd Internal Med, dipl, 70, cert hemat, 72, cert med oncol, 73. *Prof Exp:* Intern & asst resident, Ward Med Serv, Barnes Hosp, St Louis, 63-65; clin assoc, Arthritis & Rheumatism Br, Nat Inst Arthritis & Metab Dis, NIH, 65-68; resident med, Parkland Mem Hosp, Dallas, 68-69; fel hemat, Univ Tex Southwestern Med Sch, 69-70, from instr to asst prof internal med, 70-73, assoc prof, 74-76; CHIEF ONCOL, DIR IMMUNOL & DIR CHARLES A SAMMONS CANCER CTR, BAYLOR UNIV MED CTR, 76- *Concurrent Pos:* Estab investr, Am Heart Asn, 70-75; mem fac & steering comt, Immunol Grad Prog, Grad Sch Biomed Sci, Univ Tex Health Sci Ctr Dallas, 75-76, adj mem, 76-; clin prof internal med, Univ Tex Southwestern Med Sch, 76-; co-dir div hemat-oncol, Baylor Univ Med Ctr, 76-, outstanding fac mem, 77-78 & 86-87; adj prof biol, Southern Methodist Univ, 77- *Mem:* Am Asn Cancer Res; Am Soc Hemat; Am Soc Clin Oncol; Am Asn Immunologists; fel Am Col Physicians; fel Int Soc Hemat; Sigma Xi. *Res:* Plasma cell dyscrasias and monoclonal immunoglobulins. *Mailing Add:* Charles A Sammons Cancer Ctr 3500 Gaston Ave Dallas TX 75246

STONE, MAX WENDELL, b Petersburg, Tenn, Mar 6, 29; m 50. COMPUTER SCIENCES. *Educ:* Union Univ, Tenn, BS, 49; Peabody Col, MA, 50. *Prof Exp:* High sch teacher, Ark, 50-51; scientist & supvr comput & data reduction, Rohm and Haas Co, Redstone Res Labs, 53-64; mgr corp data processing, SCI Systs, Inc, 64-73; SR PROG MGR, UNISYS CORP, 73- *Concurrent Pos:* Teacher eve div, Univ Ala, Huntsville, 54-65; founder & dir, Fiscal Systs Inc, 83- *Res:* Digital computer applications to problems in engineering, science and business, including management information systems, accounting functions, inventory control and data reduction; solid rocket propellant grain design; operation of large computers; management of computer center operations; management of systems software; project management. *Mailing Add:* 1431 Chandler Rd SE Huntsville AL 35801-1407

STONE, MICHAEL GATES, b Midland, Tex, Oct 9, 38; div; c 1. MATHEMATICS. *Educ:* Wesleyan Univ, BA, 60; La State Univ, Baton Rouge, MS, 62; Univ Colo, Boulder, PhD(math), 69. *Prof Exp:* From asst prof to assoc prof math, 69-81, PROF MATH & STATIST, UNIV CALGARY, 81- *Mem:* Am Math Soc; Math Asn Am; Sigma Xi. *Res:* Universal algebra; lattice theory. *Mailing Add:* Dept Math Statist & Comput Sci Univ Calgary Calgary AB T2N 1N4 Can

STONE, ORVILLE L, b New Albany, Ind, June 4, 21; m 45; c 3. ELECTRICAL ENGINEERING. *Educ:* Rose Polytech Inst, BS, 48; Mass Inst Technol, SM, 50. *Prof Exp:* Asst, Mass Inst Technol, 48-50; mem staff, Los Alamos Sci Lab, Univ Calif, 50-53; assoc res staff mem, Raytheon Mfg Co, 53-57; proj engr, Schlumberger Well Serv, 57-86; RETIRED. *Concurrent Pos:* Lectr, Univ Houston, 65-68. *Mem:* Inst Elec & Electronic Engrs; AAAS. *Res:* Instrumentation for nuclear research; radiation detection; semiconductor devices. *Mailing Add:* 5119 Glenmeadow Houston TX 77096

STONE, PETER H, b New York, NY, Mar 31, 48; m 83; c 3. CARDIOLOGY, CORONARY ARTERY DISEASE. *Educ:* Princeton Univ, BA, 70; Cornell Univ Med Col, MD, 74. *Prof Exp:* Instr med, 79-83, ASST PROF MED, HARVARD MED SCH, 83-; ASSOC DIR MED, SAMUEL A LEVINE CARDIAC UNIT, BRIGHAM & WOMENS HOSP, 81- *Mem:* Am Heart Assoc; Am Col Cardiol fel. *Res:* Stable and unstable coronary artery disease. *Mailing Add:* Dept Med Div Cardiovasc Brigham & Womens Hosp Sch Med, 75 Francis St Boston MA 02115

STONE, PETER HUNTER, b Brooklyn, NY, May 10, 37; m. ATMOSPHERIC DYNAMICS, CLIMATE MODELING. *Educ:* Harvard Univ, BA, 59, PhD(appl math), 64. *Prof Exp:* Lectr meteorol, Harvard Univ, 64-66, from asst prof to assoc prof, 66-72; staff scientist, NASA, 72-74; head, dept meteorol, 81-83, dir, Ctr Meteorol & Phys Oceanog, 83-89, PROF METEOROL, MASS INST TECHNOL, 74- *Mem:* Fel Am Meteorol Soc; Am Geophys Union; Am Astron Soc. *Res:* Dynamics of planetary atmospheres; climate modeling. *Mailing Add:* Dept Earth Atmospheric & Planetary Sci Mass Inst Technol Cambridge MA 02139

STONE, PHILIP M, b Wilkinsburg, Pa, Nov 23, 33; m 55, 80; c 3. REACTOR DESIGN. *Educ:* Univ Mich, BSE, 55, MSE, 56, PhD(nuclear eng), 62. *Prof Exp:* Staff mem, Los Alamos Sci Lab, 56-63, mem advan study prog, 59-60, grad thesis prog, 60-62; staff mem, Sperry Rand Res Ctr, 63-68, head radiation sci dept, 67-68; assoc prof, Div Interdisciplinary Studies, State Univ NY Buffalo, 68-69; liaison scientist, Sperry Rand Res Ctr, 69-71, mgr systs studies dept, 71-75; physicist, Energy Res & Develop Admin, 75-77; br chief, 78-87, exec asst to dir energy res, 87-88, DIR SCI & TECHNOL AFFAIRS STAFF, US DEPT ENERGY, 87- *Concurrent Pos:* Fel, Univ Col, Univ London, 65-66; vis assoc prof, Univ Pittsburgh, 67-68; vis scientist, Ctr d'Etudes Nucleaires de Saclay, 74. *Mem:* AAAS; Am Phys Soc; Sigma Xi; NY Acad Sci; Inst Elec & Electronic Engrs. *Res:* Theoretical atomic and plasma physics; electron-atom scattering, photoabsorption, recombination, line shapes and intensities, nonequilibrium populations, microwave radiation from plasmas; signal processing and system analysis; atomic radiation from plasmas; nuclear reactor design; fusion reactor design. *Mailing Add:* Off Energy Res US Dept Energy Washington DC 20585

STONE, RICHARD SPILLANE, b Huntington, NY, Sept 14, 25; m 48; c 3. PHYSICS. *Educ:* Rensselaer Polytech Inst, BS, 49, MS, 50, PhD(physics), 52. *Prof Exp:* Asst physics, Rensselaer Polytech Inst, 49-52; res assoc nuclear physics instrumentation, Knolls Atomic Power Lab, Gen Elec Co, 52-57; physicist in chg, TRIGA Proj & sect mgr, HTGR Proj, Gen Atomic Div, Gen Dynamics Corp, 57-63; vpres eng & sales, Tech Measurement Corp, 63-64; head, Physics Sect, Arthur D Little, Inc, 64-86; PRES, FIRST LEXINGTON GROUP, INC, 86- *Mem:* Am Phys Soc; Am Nuclear Soc. *Res:* Underwater acoustics; computer applications; system analysis instrumentation design and development; reactor physics; design, construction and operation of experimental reactors; critical assemblies and in-pile experiments; nuclear power plant analysis and test. *Mailing Add:* 60 Baskin Rd Lexington MA 02173

STONE, ROBERT EDWARD, JR, b Spokane, Wash, Feb 20, 37; m 62; c 3. SPEECH & HEARING SCIENCES, SPEECH PATHOLOGY. *Educ:* Whitworth Col, BS, 60; Univ Ore, MEd, 64; Univ Mich, PhD(speech path & speech sci), 71. *Prof Exp:* Instr speech, Ore State Syst Higher Educ, 64-66; speech pathologist & res asst otorhinolaryngol, Univ Mich, Ann Arbor, 70-71; from asst prof to assoc prof otolaryngol, Sch Med, Ind Univ, 78-87; ASSOC

PROF OTOLARYNGOL, SCH MED, VANDERBILT UNIV, 87- *Mem:* Fel Am Speech & Hearing Asn; assoc Acoust Soc Am. *Res:* Laryngeal physiology and effects of aberrant production of voice; indices of hypernasality. *Mailing Add:* MCN S-2100 Vanderbilt Univ Med Ctr Nashville TN 37232-2559

STONE, ROBERT K(EMPER), b Minneapolis, Minn, June 12, 20; m 45; c 3. MECHANICAL ENGINEERING. *Educ:* Stanford Univ, AB, 41; Chrysler Inst Eng, MS, 43. *Prof Exp:* Res engr, Chrysler Corp, 41-47; engr, Chevron Res Co, Standard Oil Co Calif, 47-60, supvr engr, 60-66, supvr, Engine Fuels Sect, 66-69, sr staff engr, Fuels & Asphalts Dept, 69-80, asst to pres, 80-85; RETIRED. *Mem:* Soc Automotive Engrs; Am Soc Testing & Mat; Air Pollution Control Asn; Am Petrol Inst. *Res:* Aircraft and automotive engine testing and research; motor gasoline and diesel fuels, including evaluation of product quality, development of new or improved fuels and automotive air pollution research. *Mailing Add:* 1678 Filbert Ave Chico CA 95926

STONE, ROBERT LOUIS, b Frankfort, Ky, Dec 30, 21. MICROBIOLOGY. *Educ:* Univ Ky, BS, 47, MS, 52; Ind Univ, PhD, 59. *Prof Exp:* Sr microbiologist, 49-55 & 58-72, res immunologist, Lilly Res Labs, Eli Lilly & Co, 72-82; RETIRED. *Concurrent Pos:* Lectr, Butler Univ, 68-; chmn, Animal Health & Res, Indianapolis Zool Soc, 76- *Mem:* AAAS; Am Soc Microbiol; NY Acad Sci. *Res:* Bacteriophage; immunology; chemotherapy; medical virology. *Mailing Add:* 6837 Brendonway N Dr Indianapolis IN 46226

STONE, ROBERT P(ORTER), b Columbus, Ohio, Apr 10, 18; m 41; c 4. ELECTRICAL ENGINEERING. *Educ:* Ohio State Univ, BEE, 40; Purdue Univ, MS, 41; Princeton Univ, PhD, 49. *Prof Exp:* Res engr, Labs Div, RCA Corp, 41-55, eng leader, Electronic Tube Div, 55-64, sr engr, 64-86; RETIRED. *Mem:* Sr mem Inst Elec & Electronic Engrs; Soc Info Display; Sigma Xi. *Res:* High frequency tubes; electron multipliers; solid state devices; storage tubes; electron optics. *Mailing Add:* 235 Conestoga Blvd Lancaster PA 17602

STONE, ROBERT SIDNEY, b Fond du Lac, Wis, Feb 16, 23; m 48; c 1. SYSTEMS ANALYSIS, NUCLEAR ENGINEERING. *Educ:* Calif Inst Technol, BS, 48. *Prof Exp:* Jr elec engr, Oak Ridge Nat Lab, 48-50, develop engr, 50-54, physicist, 54-64, group leader dynamic anal, 64-88; CONSULT, PAI CORP, 88- *Mem:* Am Nuclear Soc. *Res:* Reactor design reviews; computer analysis of systems interactions; technical management; control studies of energy systems; reactor dynamics as related to safety and control; analog, digital and hybrid computer programs. *Mailing Add:* 118 Canterbury Rd Oak Ridge TN 37830-7737

STONE, SAMUEL ARTHUR, mathematics, for more information see previous edition

STONE, SANFORD HERBERT, b New York, NY, Sept 9, 21; m 53; c 3. IMMUNOLOGY. *Educ:* City Col New York, BS, 47; Univ Paris, DSc, 51. *Prof Exp:* NIH res fel, Sch Med, Johns Hopkins Univ, 51-52; res asst immunol, New York Med Col, 53; asst, Appl Immunol Div, Pub Health Res Inst, New York, 54-57; head sect natural & acquired resistance, Lab Immunol, Nat Inst Allergy & Infectious Dis, 57-62, head sect allergy & hypersensitivity, 63-69, head immunol sect, Lab Microbiol, 70-74, OSD, 74-79, head, exp autoimmunity sect, lab microbiol immunity, 79-88, immunol allergy & immunol dis prog, 88-89, Div Microbiol Infectious Dis, 89-90, OFF OF THE DIR, DIV INTRAMURAL RES, NAT INST ALLERGY & INFECTIOUS DIS, 90- *Concurrent Pos:* Prof lectr, Howard Univ, 61-75. *Mem:* AAAS; Am Asn Immunologists; Reticuloendothelial Soc. *Res:* Tissue antigens and antibodies; mechanism of hypersensitivity; autoimmunity; transplantation immunity. *Mailing Add:* NIH Bldg 7 Rm No 1GF Bethesda MD 20829

STONE, SHELDON LESLIE, b Brooklyn, NY, Feb 14, 46; m 71; c 3. HIGH ENERGY PHYSICS. *Educ:* Brooklyn Col, BS, 67; Univ Rochester, PhD(physics), 72. *Prof Exp:* Res assoc, dept physics, Vanderbilt Univ, 72-73, asst prof, 73-79; sr res assoc, Lab Nuclear Studies, Cornell Univ, 79-91; PROF, SYRACUSE UNIV, 91- *Concurrent Pos:* Vis fel, Lab Nuclear Studies, Cornell Univ, 77-78, adj prof, 88-91. *Mem:* Am Phys Soc; Sigma Xi. *Res:* Elementary particle interactions. *Mailing Add:* Physics Dept Syracuse Univ Syracuse NY 13244

STONE, SIDNEY NORMAN, b Rochester, NY, May 11, 22; m 51; c 2. OPTICAL PHYSICS, ASTROPHYSICS. *Educ:* Univ Calif, BA, 51, MA, 52, PhD(astron), 57. *Prof Exp:* Physicist, Ballistic Res Lab, Aberdeen Proving Ground, Md, 44-49; asst astron, Univ Calif, 53-54, staff mem, Los Alamos Nat Lab, Univ Calif, 57-84, CONSULT, UNIV CALIF, 84- *Mem:* Am Astron Soc; Astron Soc Pac. *Res:* Spectroscopy; optical instrumentation; spectroscopic binary stars; physics of the upper atmosphere; photographic sensitometry; high speed photography; radiation dosimetry. *Mailing Add:* 9112 Haines Ave NE Albuquerque NM 87112-3924

STONE, SOLON ALLEN, b Lakeview, Ore, Sept 30, 28; m 49, 78; c 5. ELECTRICAL ENGINEERING. *Educ:* Ore State Col, BS, 52. *Prof Exp:* Mem tech staff, Bell Tel Labs, NJ, 52-56; asst prof elec eng, Ore State Univ, 56-64, asst to dean, 66-71, asst dean, Sch Eng, 71-92, PROF ENG, ORE STATE UNIV, 61-, ASST TO DEAN, SCH ENG, 71- *Mem:* Inst Elec & Electronics Engrs. *Mailing Add:* 723 Augusta Ct NW Albany OR 97321

STONE, STANLEY S, b Old Forge, Pa, Apr 4, 21; m 50; c 4. BIOCHEMISTRY, IMMUNOCHEMISTRY. *Educ:* Loyola Col, Md, BS, 50; Georgetown Univ, MS, 53, PhD(biochem), 57. *Prof Exp:* Biochemist, USDA, 49-52; biochemist, NIH, 52-57; biochemist, Plum Island Animal Dis Lab, 57-62 & 64-66, EAfrican Vet Res Lab, 62-64, EAfrican Vet Res Orgn, 66-70 & Plum Island Animal Dis Lab, 70-71, head biochem biophys, Nat Animal Dis Ctr, 72-78, DIR, FAR EASTERN REGIONAL RES OFF, USDA, INDIA, 78- *Concurrent Pos:* Proj mgr, Vet Res Inst, Foreign Agr Orgn, Pakistan, 81, India. *Mem:* Am Chem Soc; Am Soc Microbiol. *Res:* Isolation and characterization of immunoglobulins from farm domestic animals, particularly immunoglobulins of the exocrine secretions; reactions of immunoglobulins with viral antigens; nutrition; biochemistry of nutrition. *Mailing Add:* 1702 Mexico Ave Tarpon Springs FL 33589

STONE, WILLIAM ELLIS, b Colton, Calif, Jan 22, 11; m 41; c 3. PHYSIOLOGY. *Educ:* Calif Inst Technol, BS, 33; Univ Minn, PhD(physiol chem), 39. *Prof Exp:* Coxe Mem fel, Lab Neurophysiol, Yale Univ, 39-40; res assoc surg, Col Med, Wayne Univ, 40-47, from instr to asst prof physiol chem, 41-47; from asst prof to prof, 47-76, EMER PROF PHYSIOL, UNIV WIS-MADISON, 76- *Mem:* Am Physiol Soc; Am Epilepsy Soc. *Res:* Chemical physiology of the brain; actions of anticonvulsants. *Mailing Add:* Dept Physiol Univ Wis Madison WI 53706

STONE, WILLIAM HAROLD, b Boston, Mass, Dec 15, 24; m 71; c 3. IMMUNOGENETICS, IMMUNOREPRODUCTION. *Educ:* Brown Univ, AB, 48; Univ Maine, MS, 49; Univ Wis-Madison, PhD(genetics & biochem), 53. *Hon Degrees:* ScD, Univ Cordoba, Spain, 84. *Prof Exp:* Res asst genetics, Univ Wis-Madison, 48-50, instr, 50-53, from asst prof to assoc prof, 54-60, prof, 62-83; vis researcher immunochem, Columbia Univ, 53-54; NIH fel, Calif Inst Technol, 60-61; COWLES DISTINGUISHED PROF BIOL & GENETICS, TRINITY UNIV, SAN ANTONIO, 83- *Concurrent Pos:* Vis prof, Univ Barcelona, Spain, 70-71, Univ Zaragoza, 73; staff scientist, Wis Regional Primate Res Ctr, 78-83, Southwest Found Biomed Res, 83-; adj prof, Univ Tex Health Sci Ctr, San Antonio, 84-; mem, Study Sects, NIH & Pub Affairs Comt, Fed Am Socs Exp Biol. *Honors & Awards:* I I Ivanov Medal, Ministry Agr, USSR, 74. *Mem:* Genetics Soc Am; Am Genetics Asn; Soc Am Immunologists; Int Soc Animal Blood Group Res (pres, 65-68); Am Soc Primatologists; Int Primatological Soc; Transplant Soc; Am Aging Asn. *Res:* Immunogenetic research on nonhuman primates and marsupials; immunoreproduction and genetic polymorphism. *Mailing Add:* Trinity Univ 715 Stadium Dr San Antonio TX 78212

STONE, WILLIAM JACK HANSON, b Pearland, Tex, Dec 28, 32; m 53; c 2. PLANT PATHOLOGY, WEED SCIENCE. *Educ:* Tex A&M Univ, BS, 55, MS, 57; Purdue Univ, PhD(plant path), 63. *Prof Exp:* Res plant pathologist, USDA, 57-59; asst plant pathologist, Univ Ariz, 63-66; res & develop pathologist, 66, head Fla Substa, Upjohn Co, 66-; MGR TECH SERV, ASGROW FLA CO. *Mem:* Am Phytopath Soc. *Res:* Pesticides for agricultural uses. *Mailing Add:* Asgrow Fla Co 4144 Hwy 39 N Plant City FL 33565

STONE, WILLIAM JOHN, NEPHROLOGY. *Educ:* Johns Hopkins Univ, MD, 62. *Prof Exp:* CHIEF NEPHROL, DEPT MED, VET ADMIN MED CTR, 72-; PROF MED, VANDERBILT UNIV, 80- *Mailing Add:* Dept Med Vet Admin Med Ctr 1310 24th Ave S Nashville TN 37203

STONE, WILLIAM LAWRENCE, b New York, NY, Oct 26, 44; m 68; c 2. LIPID METABOLISM. *Educ:* State Univ NY, Stony Brook, BS, 55, PhD(biol), 73; Marshall Univ, WVa, MS, 68. *Prof Exp:* Phys chemist thermodyn, Dow Chem Co, Mich, 65; teacher chem, Marshall Univ Lab Sch, WVa, 58; grad asst biol, State Univ NY, Stony Brook, 68-72, grad res fel protein chem, 73; res assoc biochem, Med Sch, Duke Univ, NC, 73-75; asst res chemist nutrit, Univ Calif, Santa Cruz, 75-78; asst prof biomed sci, Meharry Med Col, Nashville, 78-89, from asst prof to assoc prof pediat & dir pediat res, 84-89; ASSOC PROF PEDIAT, E TENN STATE UNIV, JOHNSON CITY, 89-, DIR PEDIAT RES, 89-, ADJ ASSOC PROF BIOCHEM, 90- *Concurrent Pos:* Am Heart Asn investr, Meharry Med Col, Nashville, 80-, prin investr, NIH grant, 80-; Nat Asn Sickle Cell Dis, Inc grant, 85; fel US Air Force Summer Fac Res Prog, 84 & 85; NIH res career develop award, 86-91. *Mem:* Biophys Soc; AAAS; Am Inst Nutrit; NY Acad Sci; Soc Exp Biol; Sigma Xi. *Res:* Oxidative metabolism and its pathophysiological consequences in animals, humans and in in-vitro tissue cultures; lipid peroxidation and antioxidant nutrients on cardiovascular disease, particularly atherosclerosis; hyperbaric oxygen therapy and toxicity; enzymatic detoxification mechanisms. *Mailing Add:* Dept Pediatrics E Tenn Univ James H Quillen Col Med Johnson City TN 37614-0002

STONE, WILLIAM ROSS, b San Diego, Calif, Aug 26, 47; m 70; c 1. APPLIED PHYSICS, COMPUTER SCIENCE. *Educ:* Univ Calif, San Diego, BA, 67, MS, 73, PhD(appl physics & info sci), 78. *Prof Exp:* Sr physicist electromagnetic field interaction, Gulf Gen Atomic Co, 69-72, sr engr computerized mgt info syst, 72-73; sr scientist ionospheric physics, Optical Propagation & Inverse Scattering Theory, Megatek Corp, 73-80; prin physicist & leader, Comput Ctr & Inverse Scattering Groups, Optical Propagation, Electromagnetic & Ionospheric Physics, Automated Inspection Appln, IRT Corp, 80-87; chief scientist, McDonnell Douglas Technologies, 89-90; CHIEF SCIENTIST, EXPERSOFT CORP, 90- *Concurrent Pos:* Pres, Stoneware Ltd, 76-; mem comn B, US Nat Comt, Int Union Radio Sci, 76, Comn G, 78, Comn F, 85, Comn A, 87; ed-in-chief, Inst Elec & Electronic Engrs Antennas & Propagation Mag, 84- *Mem:* Optical Soc Am; fel Inst Elec & Electronic Engrs; Asn Comput Mach; Soc Explor Geophysicists; Acoust Soc Am; Soc Indust & Appl Math. *Res:* Electromagnetic wave and optical propagation in the atmosphere, ionosphere and inhomogeneous and scattering media; electromagnetic theory and inverse scattering; geophysical remote probing; nuclear electromagnetic pulse generation and interaction; interactive computer systems for computation, display, management, and control. *Mailing Add:* IRT Corp 1446 Vista Claridad La Jolla CA 92037

STONEBRAKER, PETER MICHAEL, b Glendale, Calif, Apr 18, 45; m; c 3. CHEMISTRY, ORGANIC CHEMISTRY. *Educ:* Whitworth Col, Spokane, Wash, BS, 67; Univ Wash, PhD(org chem), 73. *Prof Exp:* Res chemist, Chevron Res Co, Standard Oil Calif, 73-76, prod develop specialist oronite additives, Chevron Chem Co, 76-78, supvr, Engine Lubrication Div, 78-80, sr prod specialist, 80-82, area rep Latin Am, 82-84, SUPVR STRATEGIC ANALYSIS, CHEVRON CHEM CO, 84- *Mem:* Soc Automotive Engrs; Am Chem Soc; fel Ford Found, 65-67. *Res:* Synthesis lubricating oil additives; additive formulation of crankcase engine oils; analysis of used oils. *Mailing Add:* Chevron Chem Co Bldg T-2386 6001 Bollinger Canyon Rd San Ramon CA 94583

STONEBURNER, DANIEL LEE, b Zanesville, Ohio, July 4, 45; m 71; c 2. MARINE SCIENCE, ZOOLOGY. *Educ:* Ind State Univ, BS, 67; Iowa State Univ, PhD(plant ecol), 70. *Prof Exp:* Regional ecologist, Nat Park Serv, Southeast Region, 74-78; res ecologist, Nat Park Serv, Univ Ga, 78-81; SR RES ECOLOGIST, NAT PARK SERV, COLO STATE UNIV, 81- *Concurrent Pos:* Invited consult, Nat Marine Fisheries Sea Turtle Recovery Team 79- *Mem:* Ecol Soc Am; Am Soc Limnol & Oceanog; Am Soc Ichthyologists & Herpetologists; Int Asn Crenobiologists. *Res:* Development and application of biotelemetry equipment and techniques for marine freshwater and terrestrial organisms; heavy metal analyses of marine, terrestrial and freshwater organism tissues; freshwater and marine ecology; animal behavior and ecosystems modeling. *Mailing Add:* 185 Longview Dr Athens GA 30605-4508

STONECYPHER, ROY W, b Atlanta, Ga, Mar 20, 33; m 54; c 3. FORESTRY, GENETICS. *Educ:* NC State Univ, BS, 59, PhD(forestry), 66. *Prof Exp:* Proj leader, Int Paper Co, 63-67, res forester, 67-70; assoc prof forest genetics, Okla State Univ, 70-72; quant geneticist, Forestry Res Ctr, Weyerhauser Co, 72-80. *Concurrent Pos:* Adj asst prof, NC State Univ, 67-70, adj prof, 76-; affil assoc prof, Univ Wash, 73-80, affil prof, 80- *Mem:* Soc Am Foresters; Sigma Xi. *Res:* Quantitative genetics work in pine populations; applied forest tree breeding; statistical analyses of forestry related research using electronic computers. *Mailing Add:* 116 NE Hillside Dr Chehalis WA 98532

STONECYPHER, THOMAS E(DWARD), b Savannah, Ga, June 20, 34; m 56; c 3. CHEMICAL ENGINEERING. *Educ:* Ga Inst Technol, BS, 55, PhD(chem eng), 61. *Prof Exp:* Res asst, Eng Exp Sta, Ga Inst Technol, 55-58; engr, Redstone Res Labs, Rohm & Haas Co, 58-61, head appl thermodyn, 61-66, head eng serv, 66-69; prod mgr, Micromedic Systs, Inc, 70-76; DIR ENG, ENG LABS, ROHM & HAAS CO, 76- *Res:* Medical instrumentation; engineering design; management. *Mailing Add:* 821 Tanna Hill Dr Huntsville AL 35802

STONEHAM, RICHARD GEORGE, b Chicago, Ill, Feb 22, 20. MATHEMATICS. *Educ:* Ill Inst Technol, BSc, 42; Brown Univ, ScM, 44; Univ Calif, Berkeley, PhD(math), 52. *Prof Exp:* Instr math, Univ Ill, 46-47; asst, Univ Calif, 47-49, Off Naval Res Proj, 51, jr res mathematician, 52, lectr, univ, 51-52, instr math, 52, asst prof, 55-58, mathematician, radiation lab, 54; asst prof math, San Diego State Col, 53-54; lectr math, 59-61, from asst prof to prof, 61-77, EMER PROF MATH, CITY COL, CITY UNIV NEW YORK, 77- *Concurrent Pos:* Mathematician, Ramo-Wooldridge Corp, 54-55; Rockefeller Found fel, 42, Brit Res Coun fel, Nottingham Univ, Eng, 72. *Mem:* Am Math Soc; Math Asn Am; Math Soc France. *Res:* Applied mathematics; mathematical theory of elasticity; partial differential equations; number theory; normal numbers, especially uniform distributions, exponential sums, (j, epsilon)-normal numbers; founded a new area in the fundamental g-adic representation behavior of rational fractions. *Mailing Add:* Brighton Black Rock St Michael Barbados West Indies

STONEHILL, ELLIOTT H, b Brooklyn, NY, Sept 22, 28; m 51; c 2. MICROBIAL GENETICS, CELL BIOLOGY. *Educ:* Col City New York, BS, 50; Brooklyn Col, MA, 56; Cornell Univ, PhD(microbiol), 65. *Prof Exp:* Res fel genetics, Gustave-Roussy Inst, France, 65-66; teaching fel microbiol genetics, Univ Sussex, Eng, 66-67; res assoc cell biol, Sloan-Kettering Inst, NY, 67-75; asst prof microbiol, Grad Sch Med Col, Cornell Univ, 68-75; geneticist & planning adminstr, 75-81, ASST DIR, NAT CANCER INST, NIH, 81- *Mem:* Am Soc Microbiol; Am Soc Cell Biol; NY Acad Sci; AAAS; Am Soc Preventive Oncol; Am Asn Cancer Res. *Res:* Cellular genetic expression and the anachronistic appearance of fetal specific proteins in cancer cells of adults. *Mailing Add:* Nat Cancer Inst Bldg 31 Rm 4A-32 Bethesda MD 20892

STONEHILL, ROBERT BERRELL, b Philadelphia, Pa, Feb 14, 21; m 78; c 3. INTERNAL MEDICINE, PULMONARY DISEASES. *Educ:* Temple Univ, BA, 42, MD, 45; Am Bd Internal Med, dipl, 56, re-cert, 74 & 80, cert pulmonary dis, 65; Am Bd Prev Med, dipl, 57. *Prof Exp:* Chief pulmonary physiol lab, Samson AFB Hosp, NY, 55-56, chief pulmonary dis serv, Wilford Hall, US Air Force Hosp, Lackland AFB, Tex, 56-61, chmn dept med, 61-67; PROF MED, SCH MED, IND UNIV, INDIANAPOLIS, 67- *Concurrent Pos:* Rep to Surgeon Gen, Combined Vet Admin Armed Forces Comt Pulmonary Physiol, 56-57; rep to Surg Gen & mem exec comt, Vet Admin Armed Forces Coccidioidomycosis Coop Study Group, 58-65; gov, Am Col Chest Physicians, 62-67. *Mem:* Fel Am Col Physicians; fel Am Col Chest Physicians; fel Royal Soc Med; fel Am Col Prev Med. *Res:* Aerospace medicine; pulmonary diseases. *Mailing Add:* Ind Univ Sch Med 100 W Michigan St Indianapolis IN 46223

STONEHOUSE, HAROLD BERTRAM, b Eng, Apr 13, 22; nat US; m 50; c 4. GEOCHEMISTRY. *Educ:* Univ London, BSc, 43; Univ Toronto, PhD, 52. *Prof Exp:* Geologist, Ex-Lands, Nigeria, WAfrica, 43-45; geologist, Brit Guiana Consol Goldfields, 45-46; geologist & mgr, Can-Guiana Mines, 47-48; assoc geologist, State Geol Surv, Univ Ill, 54-55; from asst prof to assoc prof, 55-74, PROF GEOL & ASST CHMN DEPT, MICH STATE UNIV, 74- *Mem:* Geol Soc Am. *Res:* Mineralogy; economic geology; earth science education. *Mailing Add:* Dept Geol 211 Nat Sci Bldg Mich State Univ East Lansing MI 48824

STONEKING, JERRY EDWARD, b Cincinnati, Ohio, July 12, 42; m 67. ENGINEERING MECHANICS. *Educ:* Ga Inst Technol, BS, 65; Univ Ill, Urbana, MS, 66, PhD(theoret & appl mech), 69. *Prof Exp:* Asst prof gen eng, Univ Ill, 69-70; asst prof civil eng, Clarkson Col Technol, 70-74; assoc prof, 75-80, PROF ENG SCI & MECH, UNIV TENN, 80- *Mem:* Am Soc Civil Engrs; Am Soc Mech Engrs; Sigma Xi. *Res:* Structural mechanics; numerical methods. *Mailing Add:* 720 Scenic Dr Knoxville TN 37919

STONEMAN, DAVID MCNEEL, b Madison, Wis, Oct 11, 39; m 64; c 2. MATHEMATICAL STATISTICS. *Educ:* Univ Wis, BS, 61, MS, 63, PhD(statist), 66. *Prof Exp:* Math statistician, Forest Prod Lab, Forest Serv, USDA, 64-66; from asst prof to prof math, 66-90, chair, math dept, 86-90, ASSOC DEAN, COL LETTERS & SCI, UNIV WIS, WHITEWATER, 90- *Mem:* Am Statist Asn. *Res:* Experimental design. *Mailing Add:* Col Letters & Sci Univ Wis 800 Main St Whitewater WI 53190

STONEMAN, WILLIAM, III, b Kansas City, Mo, Sept 8, 27; m 51; c 5. PLASTIC SURGERY. *Educ:* St Louis Univ, BS, 48, MD, 52. *Prof Exp:* Bi-State Reg Med Prog, 69-74; assoc dean, 73-76, exec assoc dean, 76-82, DEAN, ST LOUIS UNIV SCH MED, 82-, PROF SURGERY & COMMUNITY MED, 84- *Concurrent Pos:* Assoc vpres, St Louis Univ Med Ctr, 83-; chmn, Midwest/Great Plains Sect Coun Deans, Asn Am Med Cols, 86-87; mem, sect Med Sch Gov Coun & Comn Rev Standards Med Educ & Training, AMA, 87-, chmn, sect Med Sch Task Force Sect Planning, 87-, sect Med Schs, 88. *Mem:* Am Col Surgeons; Am Soc Plastic & Reconstructive Surgeons; AMA; Asn Am Med Cols. *Mailing Add:* St Louis Univ Sch Med 1402 S Grand Blvd St Louis MO 63104

STONER, ADAIR, b Oklahoma City, Okla, Oct 15, 28; m 53; c 2. ENTOMOLOGY. *Educ:* Okla State Univ, BS, 56, MS, 60. *Prof Exp:* Entomologist, Pest Control Div, USDA, 58-60, entomologist, Western Cotton Insect Invests, Cotton Insect Br, Entom Res Div, 60-67, res entomologist, 67-72, res entomologist, Western Cotton Res Lab, 72-75, res entomologist, Honey Bee Pesticides/Dis Res, Sci & Educ Admin-Agr Res Serv, 75-85; RETIRED. *Mem:* Entom Soc Am. *Res:* Effects of pesticides on honey bees. *Mailing Add:* 702 S 24th St Laramie WY 82070

STONER, ALLAN K, b Muncie, Ind, July 6, 39; m 62; c 2. HORTICULTURE. *Educ:* Purdue Univ, West Lafayette, BS, 61, MS, 63; Univ Ill, PhD(hort), 65. *Prof Exp:* Horticulturist, Crops Res Div, Plant Indust Sta, 65-71, Veg Lab, Agr Res Ctr, 71-88, chmn, Plant Genetics & Germplasm Inst, 81-88, RES LEADER, GERMPLASM SERV LAB, USDA AGR RES CTR, 88- *Mem:* Fel Am Soc Hort Sci. *Res:* Vegetable breeding and production; breeding of tomatoes; plant germplasm collection and maintenance. *Mailing Add:* Nat Germplasm Resources Lab USDA Agr Res Ctr-W Beltsville MD 20705-2350

STONER, ALLAN WILBUR, b Tipton, Ind, Sept 15, 31; m 58; c 3. PHYSICAL CHEMISTRY. *Educ:* Ind Univ, BS, 53; Univ Calif, PhD(chem), 56. *Prof Exp:* Res scientist, Res Ctr, Uniroyal, Inc, 56-58, res scientist, Indust Reactor Labs, 58-60, fiber develop mgr, Fiber & Textile Div, 60-66, plant mgr, NC, 66-69, mgr res & develop, Plastic Prod, Ind, 69-71, dir res & develop, Plastic & Indust Prod Div, Oxford Mgt & Res Ctr, 71-73, dir mkt, Indust Prod Div, Oxford Mgt & Res Ctr, 73-; SR VPRES, GATES CORP. *Mem:* Am Chem Soc. *Res:* Nuclear physics and spectroscopy; radiochemistry; application of tracers; radiation and polymer chemistry; physical properties of fibers and elastomeric materials. *Mailing Add:* Gates Corp 900 S Broadway Denver CO 80217

STONER, CLINTON DALE, b Mellette, SDak, Feb 8, 33; m 65; c 2. BIOCHEMISTRY, AGRONOMY. *Educ:* SDak State Univ, BS, 57, MS, 60; Univ Ill, PhD(agron), 65. *Prof Exp:* Trainee biochem res, Inst Enzyme Res, Univ Wis, 64-65; asst prof, 65-69, ASSOC PROF BIOCHEM RES, DEPT SURG, OHIO STATE UNIV, 69- *Res:* Structure and function of mitochondria; steady-state kinetics of multienzyme reactions; relationships between kinetics and thermodynamics in multienzyme reactions. *Mailing Add:* Ohio State Univ 400 W 12th Ave Columbus OH 43210-1218

STONER, ELAINE CAROL BLATT, b New York, NY, Dec 31, 39; m 65; c 2. PHYSICAL CHEMISTRY. *Educ:* Brooklyn Col, BS, 61; Univ Calif, Berkeley, PhD(chem), 65. *Prof Exp:* NIH fel, Univ Wis, 64-65; asst ed electrochem & anal chem elec phenomena, 65-68, from assoc ed to sr ed elec phenomena, 68-85, SR ED PATENT SERV PHYS CHEM, CHEM ABSTR SERV, AM CHEM SOC, 85- *Mem:* Am Chem Soc. *Res:* Nuclear magnetic resonance of exchange rates of ligands in coordination complexes; solid state physics; semiconductors; superconductors; electric phenomena; magnetic phenomena. *Mailing Add:* Chem Abstr Serv Dept 61 Columbus OH 43210

STONER, GARY DAVID, b Bozeman, Mont, Oct 25, 42; m 69; c 2. CARCINOGENESIS, TOXICOLOGY. *Educ:* Mont State Univ, BS, 64; Univ Mich, MS, 68, PhD(microbiol), 70. *Prof Exp:* Asst res scientist, Univ Calif, San Diego, 70-72, assoc res scientist, 72-75; cancer expert, Nat Cancer Inst, 76-79; ASSOC PROF PATH, MED COL OHIO, 79- *Concurrent Pos:* Consult, Nat Heart Lung & Blood Inst, 74-, Environ Protection Agency & Cancer Inst, 79- & Nat Toxicol Prog, 81-; lectr, W Alton Jones Crell Sci Ctr, 78-; mem, NIH Study Sect, 81-, Am Cancer Soc Study Sect, Ohio, 82-; prin investr grants, Nat Cancer Inst, Environ Protection Agency & US Army Res & Develop Command. *Mem:* Am Asn Cancer Res; Am Tissue Culture Asn; Am Asn Pathologists; Am Soc Cell Biol; AAAS. *Res:* Carcinogenesis studies in human and animal model respiratory and esophageal tissues; carcinogen metabolism, mutagenesis; in vitro transformation of epithelial cells. *Mailing Add:* Dept Path Med Col Ohio 3000 Arlington Ave Toledo OH 43699

STONER, GEORGE GREEN, b Wilkinsburg, Pa, Jan 29, 12; m 40; c 3. MODERN METRIC SYSTEM. *Educ:* Col Wooster, AB, 34; Ohio State Univ, AM, 36; Princeton Univ, PhD(org chem), 39. *Prof Exp:* Res chemist, Wallace Labs, 39 & Columbia Chem Div, Pittsburgh Plate Glass Co, 39-42; res engr, Battelle Mem Inst, 42-48; res chemist, Gen Aniline & Film Corp, 48-50, group leader, 50-51, res fel, 52-56; mgr prod develop, Avon Prod, Inc, 56-65; supvr patent liaison, J P Stevens & Co, Inc, 65-70, mgr info serv, 70-75; res assoc, Textile Res Inst, 75-77; RETIRED. *Concurrent Pos:* Instr, Ohio State Univ, 47-48. *Mem:* Fel AAAS; Am Chem Soc; fel Am Inst Chemists; fel US Metric Asn. *Res:* Fat acids; tall oil; plasticizers; organic sulfur chemistry; alkyl diselenides; Reppe chemistry; photosensitizing dyes; allyl resins; polyethylene; hydrocarbon oxidation; peroxides; technical writing; textile flammability; metric system; leather preservation; cosmetics; fiber modification. *Mailing Add:* Two Parkside Dr Suffern NY 10901-7602

STONER, GLENN EARL, b Springfield, Mo, Oct 26, 40; m 62; c 3. ELECTROCHEMISTRY. *Educ:* Univ Mo-Rolla, BS, 62, MS, 63; Univ Pa, PhD(chem), 68. *Prof Exp:* Sr scientist mat sci, Sch Eng, Univ Va, 68-71; vis assoc prof chem, Univ Mo-Rolla, 71; lectr chem eng, Univ Va, 71-72, sr scientist mat sci, 72-73; vis assoc prof electrochem, Fac Sci, Univ Rouen, France, 73-74; res assoc prof mat sci, ASSOC PROF, UNIV VA, 77- *Concurrent Pos:* Consult, Owens-Ill, Inc, 75- *Honors & Awards:* Cert Recognition, NASA, 75. *Mem:* Electrochem Soc. *Res:* Applied research in bioelectrochemistry and biomaterials research; interaction with industry towards development of innovative concepts. *Mailing Add:* Mat Eng Univ Va Thorton Hall B 102a Charlottesville VA 22903

STONER, GRAHAM ALEXANDER, b Saginaw, Mich, June 13, 29; m 55; c 4. ANALYTICAL CHEMISTRY, AGRICULTURAL CHEMISTRY. *Educ:* Univ Mich, BS, 51, MS, 52; Tulane Univ La, PhD(chem), 56. *Prof Exp:* Chemist, Dow Chem Co, 55-58, proj leader, 58-60; chemist, Ethyl Corp, 60-62; mgr anal chem, Bioferm Div, Int Minerals & Chem Corp, Calif, 62-64, assoc dir anal labs, Chem Div, 64-67; dir anal & tech serv, IMC Growth Sci Ctr, Ill, 67-69; plant mgr, Infotronics Corp, Tex, 69-70; dir res & develop spec prod div, Kennecott Copper Corp, 70-74; vpres technol, 74-75, vpres, 75-79, sr vpres mkt & develop, Kocide Chem Corp, 79-85; PRES, AGTROL CHEM PROD, 85- *Concurrent Pos:* Guest scientist, Brookhaven Nat Lab, 56-57. *Mem:* AAAS; Am Chem Soc; Sigma Xi; NY Acad Sci. *Res:* Pesticide research, testing, registration, formulation; governmental regulations; analytical-physical chemistry; enzymatic methods of analysis; automated analysis; radiochemistry. *Mailing Add:* 6606 Redding Rd Houston TX 77036

STONER, JOHN CLARK, b Toledo, Ohio, Feb 26, 33; m 54; c 2. VETERINARY MEDICINE. *Educ:* Ohio State Univ, DVM, 60. *Prof Exp:* Clin res vet, Agr Div, Am Cyanamid Co, NJ, 60-63; sr res vet, Ciba Res Farm, 63-69; asst dir animal health documentation, Squibb Agr Res Ctr, E R Squibb & Sons, Inc, 69-71, dir vet prof & regulatory affairs, 71-85; dir med develop, Solvay Vet Inc,85-89, DIR BUS DEVELOP, SOLVAY ANIMAL HEALTH, INC, 89- *Mem:* Am Vet Med Asn; Am Asn Indust Vet. *Res:* Clinical research for animal health pharmaceuticals and drug regulatory affairs; biologicals, professional services and drug regulatory affairs. *Mailing Add:* 1355 Bavarian Shores Dr Chaska MN 55318

STONER, JOHN OLIVER, JR, b Milton, Mass, Oct 4, 36; m 60; c 4. ATOMIC SPECTROSCOPY. *Educ:* Pa State Univ, BS, 58; Princeton Univ, MA, 59, PhD(physics), 64. *Prof Exp:* Res assoc physics, Univ Wis, 63-66, asst prof, 66-67; from asst prof to assoc prof, 67-76, PROF PHYSICS, UNIV ARIZ, 76- *Concurrent Pos:* Mem comt line spectra, Nat Res Coun, 75- *Mem:* Am Phys Soc; AAAS; Optical Soc Am. *Res:* Atomic and beam-foil spectroscopy; atomic beam and other techniques for producing narrow spectral lines. *Mailing Add:* Ariz Carbon Foil Co Inc 2239 E Kleindale Rd Tucson AZ 85719

STONER, LARRY CLINTON, b Mt Union, Pa, May 17, 43; m 66; c 2. RENAL PHYSIOLOGY. *Educ:* Juniata Col, BS, 65; Syracuse Univ, PhD(zool), 70. *Prof Exp:* Fel renal physiol, Nat Heart & Lung Inst, 70-72, staff fel, 72-75; PROF RENAL PHYSIOL, STATE UNIV NY HEALTH SCI CTR, 88- *Concurrent Pos:* Fel, USPHS, 70-72; investr, Am Heart Asn grant, 75-; prin investr, Nat Inst Arthritis, Metab & Digestive Dis, grant, 77- *Mem:* Am Physiol Soc; Biophys Soc; Am Soc Nephrology. *Res:* Mechanisms of ion transport. *Mailing Add:* Dept Physiol State Univ NY Health Sci Ctr Syracuse NY 13210

STONER, MARSHALL ROBERT, b Kenesaw, Nebr, Sept 24, 38. ORGANIC CHEMISTRY. *Educ:* Hastings Col, BA, 60; Iowa State Univ, PhD(chem), 64. *Prof Exp:* From asst prof to assoc prof, 64-81, PROF CHEM, UNIV SDAK, 81- *Concurrent Pos:* Adj vis prof, Univ Kans, 82-83. *Mem:* AAAS; Am Chem Soc; Royal Soc Chem; Sigma Xi. *Res:* Synthesis and rearrangements of bicyclic compounds; photochemical reactions of alcohols with unsaturated acids; photochemistry of benzyl phosphates. *Mailing Add:* Dept Chem Univ SDak Vermillion SD 57069

STONER, MARTIN FRANKLIN, b Pasadena, Calif, Jan 19, 42; m 63. PLANT PATHOLOGY, MYCOLOGY. *Educ:* Calif State Polytech Col, BS, 63; Wash State Univ, PhD(plant path), 67. *Prof Exp:* From asst prof to assoc prof, 67-75, PROF BOT, CALIF STATE POLYTECH UNIV, POMONA, 75- *Concurrent Pos:* vis prof, researcher & exten path, Univ Hawaii, 80-81. *Mem:* AAAS; Am Phytopath Soc; Bot Soc Am; Mycol Soc Am. *Res:* Agroecosystems; soil-borne fungi; microbial ecology; general plant pathology and mycology; diseases of nursery crops; mycology of sewage sludge; biotechnology. *Mailing Add:* Dept Biol Sci Calif State Polytech Univ 3801 W Temple Ave Pomona CA 91768

STONER, RICHARD DEAN, b Newhall, Iowa, Mar 29, 19; m 45; c 2. IMMUNOLOGY. *Educ:* Univ Iowa, BA, 40, PhD(zool), 50. *Prof Exp:* Jr scientist, 50-52, assoc med bacteriologist, 52-54, from asst scientist to scientist, 52-62, SR SCIENTIST, MED DEPT, BROOKHAVEN NAT LAB, 62- *Concurrent Pos:* Consult, Off Surgeon Gen & Dep Dir Comn on Radiation & Infection, Armed Forces Epidemiol Bd, 63- *Mem:* Am Inst Biol Sci; Am Soc Microbiol; Radiation Res Soc; Am Soc Parasitol; Am Soc Exp Pathologists. *Res:* Radiation effect upon immune mechanisms; antibody formation; cellular defense mechanism; anaphylaxis; immunity to parasitic infections. *Mailing Add:* Med Dept Brookhaven Nat Lab Upton NY 11973

STONER, RONALD EDWARD, b Indianapolis, Ind, Nov 25, 37; m 60; c 2. THEORY ACTIVE GALAXIES. *Educ:* Wabash Col, BA, 59; Purdue Univ, MS, 61, PhD(physics), 66. *Prof Exp:* From asst prof to assoc prof, 66-74, chmn dept, 76-80, PROF PHYSICS, BOWLING GREEN STATE UNIV, 74- *Concurrent Pos:* Fulbright lectr, Sri Lanka, 80-81; NASA Grant. *Mem:* Am Astron Soc; Am Phys Soc; Sigma Xi; Am Asn Univ Profs; AAAS. *Res:* Astrophysics; computational physics; theoretical physics. *Mailing Add:* Dept of Physics Bowling Green State Univ Bowling Green OH 43403

STONER, WARREN NORTON, virology, entomology; deceased, see previous edition for last biography

STONER, WILLIAM WEBER, b Columbus, Ohio, June 4, 44; m 78; c 2. OPTICAL PHYSICS, RADIOLOGICAL PHYSICS. *Educ:* Union Col, NY, BS, 66; Princeton Univ, PhD(physics), 75. *Prof Exp:* Scientist radiol, Machlett Labs, Raytheon, Inc, Stamford, Conn, 73-75, scientist nuclear med, Raytheon Res Div, 75-76; MEM STAFF ANALOG & DIGITAL SIGNAL PROCESSING, SCI APPLICATIONS INT CORP, 77- *Mem:* Optical Soc Am; Soc Photo-Optical Instrumentation Engrs. *Res:* Optical and hybrid data processing. *Mailing Add:* Sci Applications Int Corp Six Fortune Dr Billerica MA 01821

STONES, ROBERT C, b Portland, Ore, May 19, 37; m 57; c 8. ENVIRONMENTAL PHYSIOLOGY. *Educ:* Brigham Young Univ, BS, 59, MS, 60; Purdue Univ, West Lafayette, PhD(environ physiol), 64. *Prof Exp:* From asst prof to assoc prof, 64-70, PROF PHYSIOL, MICH TECHNOL UNIV, 70-, head Dept Biol Sci, 70-81. *Concurrent Pos:* Mem, Hibernation Info Exchange, 64-; NSF res grants, 67-71. *Mem:* Am Asn Higher Educ; Nat Asn Biol Teachers; Am Forestry Asn; Am Soc Mammal; Australian Soc Mammal. *Res:* Thermal regulation of hibernating species of bats; comparative and animal physiology; comparative anatomy. *Mailing Add:* Cooley Assoc 61396 S Hwy 97 Suite 202 Bend OR 97702

STONEY, SAMUEL DAVID, JR, b Charleston, SC, Dec 20, 39; m 59; c 2. NEUROSCIENCE. *Educ:* Univ SC, BS, 62; Tulane Univ, PhD(physiol), 66. *Prof Exp:* From instr to asst prof physiol, New York Med Col, 66-70; NIH res grant, 70, asst prof, 70-74, ASSOC PROF PHYSIOL, MED COL GA, 74- *Mem:* Am Physiol Soc. *Res:* Electrophysiological studies of the organization of motor sensory cortex and pyramidal motor systems. *Mailing Add:* Dept Physiol Med Col Ga 1120 15th St Augusta GA 30902

ST-ONGE, DENIS ALDERIC, b Ste-Agathe, Man, May 11, 29; m 55; c 2. GEOMORPHOLOGY. *Educ:* St-Boniface Col, Man, BA, 51; Cath Univ, Louvain, LicSc, 57, DocSc(geog), 62. *Hon Degrees:* DSc, Univ Man, 90. *Prof Exp:* Teacher elem sch, Sask, 51-52; teacher high sch, Ethiopia, 53-55; teacher, Col Jean de Brebeuf, Montreal, 57-58; geographer, Geog Br, Dept Mines & Technol Surv, 58-65; res scientist, Geol Surv Can, 65-68; prof geomorphol, Univ Ottawa, 68-70; res scientist, Geol Surv Can, 70-73; prof geomorphol, Univ Ottawa, 73-77 & 80-84, vdean, Grad Sch, 77-80; dir, Terrain Sci, 87-91, RES SCIENTIST, GEOL SURV CAN, 84-, SR SCIENTIST, POLAR CONTINENTAL SHELF PROJ, 91- *Concurrent Pos:* Nat Res Coun Can-NATO fel, 61-62; prof, Dept Geog, Univ Ottawa, 70-74. *Honors & Awards:* Recipient Medal, Queen Elizabeth II, 79. *Mem:* Fel Geol Asn Can; Can Asn Geog; Int Geog Union. *Res:* Quaternary geology; geology and planning; geomorphology. *Mailing Add:* Geol Surv Can 601 Booth St Ottawa ON K1A 0E8 Can

STONIER, TOM TED, b Hamburg, Ger, Apr 29, 27; nat US; m 53; c 5. CELL PHYSIOLOGY, COMPUTER APPLICATIONS. *Educ:* Drew Univ, AB, 50; Yale Univ, MS, 51, PhD, 55. *Prof Exp:* Asst, Yale Univ, 51-52; jr res assoc biol, Brookhaven Nat Lab, 52-54; vis investr, Rockefeller Inst, 54-57, res assoc, 57-62; assoc prof biol, Manhattan Col, 62-71, prof biol & dir peace studies prog, 71-75; prof & chmn sci & soc, Univ Bradford, 75-90, pres, Appl Systs Knowledge, 82-85; CHMN, VALIANT TECHNOL, 88- *Concurrent Pos:* USPHS fel, 54-56; Damon Runyon Mem fel, 56-57; consult, Living Sci Labs, 61-62, Hudson Inst, 65-69, MacMillan Co, 68, Environ Defense Fund, 68-70 & Drew Univ, 69-71; instr, New Sch Social Res, 68-70 & State Univ NY Col Purchase, 72; vis prof, Bradford Univ, 90- *Mem:* AAAS; Am Soc Plant Physiol; Fedn Am Sci (secy, 66-67); NY Acad Sci; fel Royal Soc Arts; Sigma Xi; Soc Develop Biol. *Res:* Impact of science and technology on society; use of computers in education; information theory; technological forecasting; cell physiology of plant growth, cancer and aging. *Mailing Add:* 838 East St Lenox MA 01240

STOOKEY, GEORGE K, b Waterloo, Ind, Nov 6, 35; m 54; c 4. DENTISTRY. *Educ:* Ind Univ, AB, 57, MS, 62, PhD, 71. *Prof Exp:* Dir lab res, 63-64, asst dir, Prev Dent Res Inst, 69-72, exec secy, Oral Health Res Inst, 72-74, from asst prof to assoc prof, 64-78, assoc dir, Oral Health Res Inst, 74-81, PROF PREV DENT, SCH DENT, IND UNIV-PURDUE UNIV, INDIANAPOLIS, 64-, DIR, ORAL HEALTH RES INST, 81-, ASSOC DEAN RES, 87- *Mem:* Am Asn Lab Animal Sci; Int Asn Dent Res; Am Dent Asn; Europ Orgn Caries Res. *Res:* Metabolism of fluoride and other trace elements in experimental animals and humans; various types of dental caries preventive measures, including fluorides and various aspects of nutrition. *Mailing Add:* 5080 E 161st St Noblesville IN 46060

STOOKEY, STANLEY DONALD, b Hay Spring, Nebr, May 23, 15; c 3. PHYSICAL CHEMISTRY. *Educ:* Coe Col, AB, 36; Lafayette Col, MS, 37; Mass Inst Technol, PhD(phys chem), 40. *Prof Exp:* Res chemist, Corning Glass Works, 40-58, mgr fundamental chem res, 58-62, dir fundamental chem res, 62-78; RETIRED. *Mem:* Am Ceramic Soc; Am Chem Soc. *Res:* Glass composition; photosensitive, photochromic and opal glasses; glass ceramics. *Mailing Add:* 12 Timber Lane Painted Post NY 14870

STOOLMAN, LEO, b Chicago, Ill, Dec 1, 18; m 44; c 2. AEROSPACE ENGINEERING. *Educ:* Ill Inst Technol, BS, 41; Calif Inst Technol, MS, 42, PhD, 53. *Prof Exp:* Aerodyn engr, Consol Vultee Aircraft Corp, 42-46; sr res engr, Jet Propulsion Lab, Calif Inst Technol, 46-51; res physicist, 51-54, head aerodyn dept, 54-59, proj mgr, Falcon Gar-II Missile Prog, 59-60, mgr, Aerospace Vehicles Lab, 60-61, proj mgr surveyor lunar soft landing spacecraft, 61-64, tech dir & asst mgr, Space Systs Div, 64-69, mgr systs labs, Space & Commun Group, 69-71, DIR ENG, SPACE & COMMUN GROUP, HUGHES AIRCRAFT CO, 71- *Concurrent Pos:* Mem res comt struct loads, NASA, 58-60; vis prof, Calif Inst Technol, 70-71 & 74-76. *Mem:* Am Inst Aeronaut & Astronaut; Sigma Xi. *Res:* Aircraft performance, stability and control; vehicle guidance and control; jet propulsion; spacecraft analysis, design and project management. *Mailing Add:* 4530 Larkwood Woodland Hills CA 91364

STOOLMILLER, ALLEN CHARLES, b Battle Creek, Mich, Nov 3, 40; div; c 2. BIOCHEMISTRY, RESEARCH ADMINISTRATION. *Educ:* Western Reserve Univ, AB, 61; Univ Mich, MA, 64, PhD(biochem), 66. *Prof Exp:* Fel, Chicago & Ill Heart Asns, 66-68; instr pediat, Univ Chicago, 68-69, asst prof pediat & res assoc biochem, Dept Pediat & La Rabida Inst, 69-76; assoc biochemist, Eunice Kennedy Shriver Ctr Ment Retardation, 76-79; health scientist adminr, Div Res Grants, 79-89, Nat Inst Allergy & Infectious Dis, 89-90, ACTG CHIEF, AIDS REV SECT, NAT INST ALLERGY & INFECTIOUS DIS, NIH, 90- *Mem:* Am Soc Biochem & Molecular Biol; AAAS; Am Soc Neurochem; Sigma Xi; Soc Complex Carbohydrates. *Res:* Supervise review of research grant applications and contract proposals for research programs and special developmental programs supported by the National Institute of Allergy and Infectious Diseases. *Mailing Add:* Nat Inst Allergy & Infectious Dis NIH WW Bldg Rm 3A-07 Bethesda MD 20892

STOOPS, CHARLES E(MMET), JR, b Grove City, Pa, Dec 17, 14; m 43; c 3. NUCLEAR ENGINEERING. *Educ:* Ohio State Univ, BChE, 37; Purdue Univ, PhD(chem eng), 42. *Prof Exp:* Plant & develop engr, Oldbury Electrochem Co, NY, 41-42; asst prof chem eng, Lehigh Univ, 42-44; process engr, Publicker Alcohol Co, Pa, 44; sr chem engr, Phillips Petrol Co, Okla, 44-47; prof chem eng & head dept, Clemson Col, 47-48; sr chem engr, Phillips Petrol Co, 48-52, proj engr, 52-54, chief chem eng develop, Atomic Energy Div, Idaho Falls, 54-55, mgr radiation chem sect, 55-67; from assoc prof to prof, 67-85, actg chmn & chmn dept, 67-72, EMER PROF CHEM ENG, UNIV TOLEDO, 85- *Mem:* AAAS; Am Chem Soc; Am Inst Chem Engrs; Nat Soc Prof Engrs; Am Soc Eng Educ. *Res:* Mixing; aromatic alkylation; nitrogen compounds; catalysis; nuclear engineering; photochemistry; radiation chemistry. *Mailing Add:* 2346 Lynn Park Dr Toledo OH 43615

STOOPS, JAMES KING, b Charleston, WVa, Sept 15, 37; m 62; c 2. BIOCHEMISTRY. *Educ:* Duke Univ, BS, 60; Northwestern Univ, Evanston, PhD(chem), 66. *Prof Exp:* Sr demonstr biochem, Univ Queensland, 66-67, Australian Res Comt grants fel biochem, 67-70; NIMH fel, Duke Univ, 70-71; from asst prof to assoc prof biochem, Baylor Col Med, 71-90; ASSOC PROF, UNIV TEX HEALTH SCI CTR, HOUSTON, 90- *Mem:* Am Chem Soc; Am Soc Biol Chemists; AAAS. *Res:* Enzymology and protein chemistry; electron microscopy of macromolecules. *Mailing Add:* Dept Path Univ Tex Health Sci Ctr Houston TX 77030

STOOPS, R(OBERT) F(RANKLIN), b Winona, WVa, June 16, 21; m 44; c 2. CERAMIC ENGINEERING, MATERIALS SCIENCE. *Educ:* NC State Col, BSc, 49; Ohio State Univ, MSc, 50, PhD(ceramic eng), 51. *Prof Exp:* Res engr, Harbison-Walker Refractories Co, 51-52; res engr, Metall Prod Dept, Gen Elec Co, 52-57, sr res engr, 58; res prof ceramic eng, 58-81, dir, Eng Res Servs Div, 67-81, PROF & ASSOC HEAD, DEPT MAT ENG, NC STATE UNIV, 81- *Mem:* fel Am Ceramic Soc; Inst Ceramic Engrs. *Res:* Refractory oxides and carbides and combinations of these with metals; self-glazing ceramic-metal systems; effect of structure on properties of materials; nuclear fuel materials; ceramic forming processes. *Mailing Add:* 3705 Corbin St Raleigh NC 27612

STOPFORD, WOODHALL, b Jersey City, NJ, Feb 25, 43; m 66. INTERNAL MEDICINE, CLINICAL TOXICOLOGY. *Educ:* Dartmouth Col, BA, 65; Dartmouth Med Sch, BMS, 67; Harvard Univ, MD, 69; Univ NC, MSPH, 80. *Prof Exp:* CLIN ASST PROF, DUKE MED CTR, 73- *Concurrent Pos:* Mem, Am Conf Govt Indust Hyg. *Mem:* Fel Am Col Occup Med; Am Indust Hyg Asn; AMA. *Res:* Clinical toxicologic studies of heavy metal and chlorinated hydrocarbon exposures; art hazards; airway reactivity after irritant gas exposures. *Mailing Add:* Duke Med Ctr PO Box 2914 Durham NC 27710

STOPHER, PETER ROBERT, b Crowborough, Eng, Aug 8, 43; div; c 2. CIVIL AND TRANSPORTATION. *Educ:* Univ London, BSc, 64, PhD(traffic studies), 67. *Prof Exp:* Res officer hwy & transp, Greater London Coun, 67-68; asst prof transp planning, Northwestern Univ, 68-70 & McMaster Univ, 70-71; assoc prof, Dept Environ Eng, Cornell Univ, 71-73; assoc prof civil eng, Northwestern Univ, 73-77, prof, 77-80; vpres, Schimpler-Corradino Assoc, 80-87; DIR TRANSP PLANNING & ECON STUDIES, EVAL & TRAINING INST, 87- *Concurrent Pos:* Consult various industs, 69-; Nat Res Coun Can & Dept Univ Affairs Ont grants, McMaster Univ, 70-71; chmn comt on traveler behav & values, Hwy Res Bd, Nat Acad Sci-Nat Res Coun, 71-77, consult, Planning Res Corp & Int Bank Reconstruct & Develop, 72-80, mem Transp Res Forum; consult, US Environ Protection Agency, 74; dir res, Transp Ctr, Northwestern Univ, 75-77; transp adv, Nat Inst Transp & Rd Res, SAfrica, 77-78. *Honors & Awards:* Fred Burgraaf Award, Hwy Res Bd, Nat Acad Sci-Nat Res Coun, 70. *Mem:* Am Soc Civil Engrs; Brit Inst Hwy Engrs; Royal Statist Soc; Am Statist Asn. *Res:* Transportation planning techniques; mathematical modeling of travel demand; applied statistics; survey techniques; the impact of transportation facilities on communities and the environment. *Mailing Add:* Louisiana Transp Res Ctr 4101 Gourvier Ave Baton Rouge LA 70808

STOPKIE, ROGER JOHN, b Perth Amboy, NJ, July 17, 39; m 62; c 1. MICROBIOLOGY, BIOCHEMISTRY. *Educ:* St Lawrence Univ, BS, 61; St Louis Univ, PhD(microbial physiol), 68. *Prof Exp:* Res asst biochem & microbiol, Merck & Co, 62-64; res biochemist, 69-75, sr res & info scientist, ICI US Inc, 75-76, pharmaceut develop coordr, 76-79, drug develop mgr, Stuart pharmaceut, ICI Americas Inc, 79-86; DIR, DRUG DEVELOP, ICI PHARMACEUT GROUP, 86- *Mem:* AAAS; Am Soc Microbiologists; Am Chem Soc; Sigma Xi; Drug Info Asn; Proj Mgt Inst. *Res:* Biology of mycoplasma; information systems; enzyme regulation; pharmaceutical project management and administration. *Mailing Add:* Pharmaceut R-D ICI Americas Inc Wilmington DE 19897

STOPPANI, ANDRES OSCAR MANUEL, b Buenos Aires, Rep Arg, Aug 19, 15; m 67. ENZYMOLOGY, CELL BIOENERGETICS. *Educ:* Univ Buenos Aires, MD, 41, PhD(chem), 45; Univ Cambridge, PhD(biochem), 53. *Prof Exp:* Prof biochem, Univ La Plata, Arg, 48-49; prof biochem, Sch Med, Univ Buenos Aires, Arg, 49-81, dir, Dept Physiol Sci, 70-80; PRIN CAREER INVESTR, SUPER CLASS, NAT RES COUN, ARG, 63-; EMER PROF BIOCHEM, SCH MED, UNIV BUENOS AIRES, ARG, 81- *Concurrent Pos:* Vis prof, Inst Ciencias, Paraguay, multinat prog biochem, Org Am States, 72; dir, Energy-Transducing Membranes Course, ICRO-UNESCO, 75, Bioenergetics Res Ctr, Nat Res Coun, Arg, 81-; mem, Chemother Chagas Dis Comt, TDR, WHO, 79-82, adv res coun, Pan Am Health Orgn, 80-83; exec secy, Nat Prog Endemic Dis, Secy State Sci & Technol, Arg, 83-88. *Honors & Awards:* Bunge-Born Prize, Bunge-Born Found, Arg, 80; J J Kyle Prize, Arg Chem Soc, Arg, 87; B A Houssay Inter Am Sci Prize, Am States Orgn, 89. *Mem:* Fel AAAS; Am Chem Soc; Am Soc Biochem & Molecular Biol; Am Soc Cell Biol; Am Soc Microbiol; Soc Exp Biol & Med; Am Soc Biol Chemists; NY Acad Sci; Int Physicians Prev Nuclear War; Int Cell Res Orgn. *Res:* Enzymology of Trypanosoma cruzi (the agent of American trypanosomiasis) and related organisms, in connection with the effect of oxy-radicals as a basis for the development of new trypanocidal agents; enzymology and transport phenomena in the yeast Saccharomyces and to the effect of steroid hormones on mitochondrial bioenergetics; author of over 300 research papers and several books. *Mailing Add:* Viamonte 2295 Buenos Aires 1056 Argentina

STORAASLI, JOHN PHILLIP, b St Paul, Minn, Jan 28, 21; m 50; c 2. MEDICINE. *Educ:* Univ Minn, BS, 44, MB, 45, MD, 46. *Prof Exp:* Res asst, AEC Proj, Sch Med, 47-48, resident, Univ Hosp, 48-50, from instr to assoc prof, 50-61, PROF RADIOL, SCH MED, CASE WESTERN RESERVE UNIV, 61-, RES ASSOC, 56-, ASSOC RADIOLOGIST, HOSP, 50- *Mem:* Radiation Res Soc; Radiol Soc NAm; Am Roentgen Ray Soc; Am Soc Therapeut Radiol; fel Am Col Radiol. *Res:* Clinical therapeutic radiology; biological effects of ionizing radiation; diagnostic and therapeutic uses of radioactive isotopes. *Mailing Add:* 940 S Gondola Dr Venice FL 34293

STORB, URSULA, b Stuttgart, Ger. IMMUNOBIOLOGY, MOLECULAR BIOLOGY. *Educ:* Univ Tubingen, MD, 60. *Prof Exp:* From asst prof to assoc prof, 72-81, prof microbiol, Univ Wash, 81-86; PROF, DEPT MOLECULAR GENETIC & CELL BIOL & DEPT PATHOL, UNIV CHICAGO, 86- *Concurrent Pos:* NIH res grants, 72-; mem comts immunol & develop biol & undergrad col, Univ Chicago. *Mem:* Am Asn Immunol; Am Soc Cell Biol; Asn Women in Sci. *Res:* Organization of immunoglobulin genes; control of antibody gene expression. *Mailing Add:* Dept Molecular Genetic & Cell Biol Univ Chicago 920 E 58th St Chicago IL 60637

STORCH, RICHARD HARRY, b Evanston, Ill, Mar 16, 37; m 63; c 2. ENTOMOLOGY. *Educ:* Carleton Col, BA, 59; Univ Ill, MS, 61, PhD(entom), 66. *Prof Exp:* Temporary asst prof entom, USDA, 65-66, asst prof, 66-69, assoc prof, 69-80, PROF ENTOM, UNIV MAINE, ORONO, 80- *Mem:* Entom Soc Am; Entom Soc Can; Sigma Xi. *Res:* Embryonic and postembryonic development of cervicothoracic structure and musculature; behavior and ecology of Coccinellidae; pests of potatoes. *Mailing Add:* Dept of Entom Univ of Maine Orono ME 04473

STORELLA, ROBERT J, JR, b Brighton, Mass, Sept 26, 56; m 83; c 2. NEUROMUSCULAR PHYSIOLOGY & PHARMACOLOGY. *Educ:* Wesleyan Univ, BA, 78; Cornell Univ, PhD(pharmacol), 84. *Prof Exp:* Fel pharmacol, Med Sch, Univ Nev, 84-86; res instr anesthesiol, 86-89, SR INSTR ANESTHESIOL & PHYSIOL/BIOPHYSIOL, HAHNEMANN UNIV, PHILADELPHIA, PA, 89- *Mem:* AAAS; Am Soc Pharmacol & Exper Therapeut; NY Acad Sci; Sigma Xi; Soc Neurosci; Am Soc Anesthesiologists. *Mailing Add:* Dept Anesthesiol Hahnemann Col Philadelphia PA 19102

STORER, JAMES E(DWARD), b Buffalo, NY, Oct 26, 27; m 49; c 3. COMPUTER DESIGN, APPLIED PHYSICS. *Educ:* Cornell Univ, AB, 47; Harvard Univ, AM, 48, PhD(appl physics), 51. *Prof Exp:* Fel, Electronics Res Lab, Harvard Univ, 51-52, lectr appl physics, 52-53, asst prof, 53-57; sr eng specialist, Appl Res Lab, Sylvania Elec Prod, Inc, Gen Tel & Electronics Corp, Mass, 57-60, sr scientist, 60-70, dir, 61-69; pres, Symbionics Consults, Inc, 70-76; CHIEF SCIENTIST COMPUT DESIGN, CSP, INC, 77- *Concurrent Pos:* Guggenheim fel, 56; mem naval warfare panel, President's Sci Adv Comt. *Mem:* Am Asn Physics Teachers; fel Inst Elec & Electronics Engrs. *Res:* Electromagnetic theory; antennas and scattering; random processes; passive network synthesis. *Mailing Add:* 69 Pleasant St Lexington MA 02173

STORER, JOHN B, b Rockland, Maine, Oct 16, 23; m 45; c 4. RADIOBIOLOGY. *Educ:* Univ Chicago, MD, 47. *Prof Exp:* Intern, Mary Imogene Bassett Hosp, Cooperstown, NY, 47-48; USPHS res fel path, Univ Chicago, 48-49, res assoc, Toxicity Lab, 49-50; staff mem, Biomed Res Group, Los Alamos Sci Lab, 50-58; staff scientist, Jackson Mem Lab, 58-67; dep dir div biol & med, AEC, Md, 67-69; sci dir path & immunol, Oak Ridge Nat Lab, 69-75, dir, Biol Div, 75-80, sr scientist, 80-86; RETIRED. *Concurrent Pos:* Alt leader, Biomed Res Group & Leader Radiobiol Sect, Los Alamos Sci Lab, 52-58; mem subcomt relative biol effectiveness, Nat Coun Radiation Protection, 57-62; consult, Argonne Nat Lab, 59-67; mem, Radiation Study Section, NIH, 62-66 & 71-75, chmn, 72-75; mem adv comt, Atomic Bomb Casualty Comn, 69-74; mem subcomt radiobiol, Nat Coun Radiation Protection & Measurements, 69-, mem bd dirs, 75-80, mem sci comt, Biol Aspects of Basic Radiation Criteria, 72-80, Basic Radiation Criteria, 75-85 & Apportionment of Radiation Exposure, 78-86; mem adv comt biol & med to AEC Sci Secy, 69-73; mem sci adv bd, Nat Ctr Toxicol Res, 72-75; mem adv comt, Radiation Effects Res Found, Nat Acad Sci, 75-80, mem sci coun, 77-80; mem, UN Sci Comt Effects Atomic Radiation, 78-80. *Honors & Awards:* E O Lawrence Award, 68. *Mem:* Radiation Res Soc; Am Soc Exp Path; Am Asn Cancer Res; Geront Soc; Soc Exp Biol & Med. *Res:* Late effects of ionizing radiation; aging. *Mailing Add:* Rt 4 Box 350 Rockwood TN 37854

STORER, ROBERT WINTHROP, b Pittsburgh, Pa, Sept 20, 14; m 55; c 2. ZOOLOGY. *Educ:* Princeton Univ, AB, 36; Univ Calif, MA, 42, PhD(zool), 49. *Prof Exp:* Tech asst, Mus Vert Zool, Univ Calif, 41-42, mus technician, 48-49, assoc, Div Entom & Parasitol, Exp Sta, 45, asst zool, Univ, 46-48; from instr to prof zool, Univ Mich, Ann Arbor, 49-85, asst cur birds, Mus Zool, 49-56, cur, 56-85, actg dir, 79-82; RETIRED. *Concurrent Pos:* Ed, The Auk, Am Ornith Union, 53-57, ed, Ornith Monogr, 63-70; mem comt, Int Ornith Cong, 58-82. *Mem:* Wilson Ornith Soc; Cooper Ornith Soc (vpres, 70); fel Am Ornith Union (pres, 70-72); Brit Ornith Union. *Res:* Avian morphology; systematics; distribution; paleontology; avian behavior. *Mailing Add:* 2020 Penncraft Ct Ann Arbor MI 48103

STORER, THOMAS, US citizen. MATHEMATICS. *Educ:* Univ Calif, Los Angeles, BA, 59; Univ Southern Calif, PhD, 64. *Prof Exp:* Mem, Inst Advan Study, 64-65; assoc prof, 65-80, PROF MATH, UNIV MICH, ANN ARBOR, 80- *Res:* Easy mathematics. *Mailing Add:* Dept Math Univ Mich Ann Arbor MI 48109

STOREY, ARTHUR THOMAS, b Sarnia, Ont, July 8, 29; m 64; c 3. ORTHODONTICS, PHYSIOLOGY. *Educ:* Univ Toronto, DDS, 53; Univ Mich, MS, 60, PhD(physiol), 64. *Prof Exp:* From instr to asst prof orthod, Sch Dent & Physiol & Sch Med, Univ Mich, 62-66; assoc prof, Fac Dent & asst prof physiol, Fac Med, Univ Toronto, 66-70, prof dent, fac dent & assoc prof physiol, 70-77; prof & head, Dept Prev Dent, Fac Dent, Univ Man, 77-86; PROF & CHAIR, DEPT ORTHOD, SCH DENT, UNIV TEX HEALTH SCI CTR, SAN ANTONIO, 86- *Concurrent Pos:* Ed, J Craniomandibular Disorders Facial Oral Pain, 87-; mem oral biol med, No 1 Study Sect, NIH, 88. *Mem:* Am Asn Orthod; Int Asn Dent Res; Soc Neurosci; Am Dent Asn; Am Acad Craniomandibular Disorders. *Res:* Oral, pharyngeal and laryngeal receptors and reflexes; temporomandibular disorders; forms of adaptation to malocclusion. *Mailing Add:* Orthod Sch Dent Univ Texas Health Sci Ctr 7703 Floyd Curl Dr San Antonio TX 78284-7910

STOREY, BAYARD THAYER, b Boston, Mass, July 13, 32; m 58; c 4. CELL PHYSIOLOGY, PHYSICAL BIOCHEMISTRY. *Educ:* Harvard Univ, AB, 52, PhD(phys org chem), 58; Mass Inst Technol, MS, 55. *Prof Exp:* Res chemist, Ion Exchange Lab, Rohm and Haas Co, 58-60, head ion exchange synthesis lab, 60-65; Nat Inst Gen Med Sci spec fel, Johnson Res Found, 65-67, asst prof phys biochem, 67-73, assoc prof obstet & gynec, physiol & phys biochem, 73-84, PROF REPRODUCTIVE BIOL & PHYSIOL IN OBSTET & GYNEC, UNIV PA, 84- *Concurrent Pos:* Mem, REB Study Sect, Div Res Grants, NIH, 86-90. *Mem:* Am Physiol Soc; Soc Study Reproduction; Am Soc Cell Biol; Am Soc Biochem & Molecular Biol. *Res:* Fertilization mechanisms in mammals; lipid peroxidation; mitochondrial ion movements. *Mailing Add:* Dept Obstet & Gynec John Morgan Bldg 339 Univ Pa Philadelphia PA 19104-6080

STOREY, JAMES BENTON, b Avery, Tex, Oct 25, 28; m 48; c 2. POMOLOGY, PLANT PHYSIOLOGY. *Educ:* Tex A&M Univ, BS, 49, MS, 53; Univ Calif, Los Angeles, PhD(bot sci), 57. *Prof Exp:* Asst county agr agt, Tex Agr Exten Serv, 49-52, asst hort, 52-53; asst plant physiol, Univ Calif, Los Angeles, 53-57; from asst prof to assoc prof pomol, 57-74, PROF HORT, TEX A&M UNIV, 74- *Concurrent Pos:* Ed, Pecan Quart & Tex Horticulturist; exec dir, Tex Pecan Producer's Bd. *Honors & Awards:* J H Henry Award, Federated Pecan Growers Asn US. *Mem:* Fel Am Soc Hort Sci; Am Soc Plant Physiol; Int Soc Hort Sci; Am Pomol Soc; Sigma Xi. *Res:* Control of vegetative and fruiting responses in pecans; nutrition, salinity and post-harvest studies in pecans; coordinator pecan research program in Texas. *Mailing Add:* Hort Dept Tex A&M Univ College Station TX 77843

STOREY, KENNETH BRUCE, b Taber, Alta, Oct 23, 49; m 75; c 2. COMPARATIVE BIOCHEMISTRY, ENZYMOLOGY. *Educ:* Univ Calgary, BSc, 71; Univ BC, PhD(zool), 74. *Prof Exp:* Asst prof physiol, Dept Zool, Duke Univ, 74-79; assoc prof, 79-85, PROF BIOCHEM, DEPTS BIOL & CHEM, INST BIOCHEM, CARLETON UNIV, 85- *Concurrent Pos:* Fel, Dept Biochem, Sheffield Univ, 76-77; EVR Steacie Mem Fel, 84-86. *Honors & Awards:* Ayerst Award, Can Biochem Soc, 89. *Mem:* Am Soc Biol Chemists; Can Biochem Soc; Soc Cryobiol; AAAS; Can Soc Zool; fel Royal Soc Can. *Res:* Molecular adaptations of animals to environment, including adaptations of intermediary metabolism for living without oxygen and survival of freezing. *Mailing Add:* Dept Biol Carleton Univ Ottawa ON K1S 5B6 Can

STOREY, RICHARD DRAKE, b Roswell, NMex, Dec 2, 44; m 71; c 2. PLANT METABOLISM, NITROGEN FIXATION. *Educ:* Univ NMex, BS, 68; Univ Okla, MNS, 73, PhD(bot), 77. *Prof Exp:* Teacher biol, Manzano High Sch, Albuquerque, 68-73; res assoc plant physiol, Univ Okla, 73-77; vis scientist biochem, C F Kettering Res Lab, 77-78; HORT BUS CONSULT, 79- *Concurrent Pos:* prin investr, grants in plant physiol, 80-85; dir, Inst Human Nutrit, Colo Col, 80-86. *Mem:* Am Soc Plant Physiologists; AAAS; Am Soc Agronomists; Nat Asn Biol Teachers; Am Asn Univ Prof. *Res:* Control of plant metabolism, particularly protein turnover and proteolysis as it influences growth and development, senescence and nitrogen fixation; physiology of potential crop plants. *Mailing Add:* Biol Dept Colo Col Colorado Springs CO 80903

STOREY, ROBERT SAMUEL, physics, chemical engineering; deceased, see previous edition for last biography

STOREY, THEODORE GEORGE, b Fresno, Calif, Sept 6, 23; m 46; c 5. FORESTRY. *Educ:* Univ Calif, BS, 48, MS, 68. *Prof Exp:* Forester, Hammond Lumber Co, 48-49; forester forest influences res, 49-52, forester fire res, 52-59, forester, Southern Forest Fire Lab, 59-62, FORESTER, RIVERSIDE FOREST FIRE LAB, US FOREST SERV, 62- *Mem:* Soc Am Foresters; Am Geophys Union. *Res:* Fire management and control systems; fire behavior; forest and urban fire and blast damage from nuclear weapons; watershed management. *Mailing Add:* 1520 Ransom Pl Riverside CA 92506

STORFER, STANLEY J, b Brooklyn, NY, July 31, 30; m 56; c 2. ORGANIC CHEMISTRY. *Educ:* Polytech Inst Brooklyn, BS, 54, PhD(org chem), 60. *Prof Exp:* Jr chemist, Am Cyanamid Co, 54-56; chemist, Esso Res & Eng Co, 60-63, from sr chemist to sr res chemist, 63-73, RES ASSOC, EXXON CHEM-TECHNOL DEPT, 73- *Mem:* Am Chem Soc. *Res:* Rheology of water-soluble polymer solutions; new product applications; statistical design of experiments; solvents technical service and market development. *Mailing Add:* 24 Ten Eyck Pl Edison NJ 08820-3223

STORHOFF, BRUCE NORMAN, b Lanesboro, Minn, Jan 2, 42. INORGANIC CHEMISTRY. *Educ:* Luther Col, Iowa, BA, 64; Univ Iowa, PhD, 69. *Prof Exp:* From asst prof to assoc prof, 68-79, admin asst head dept, 77-79, PROF CHEM, BALL STATE UNIV, 79-, HEAD DEPT, 79- *Concurrent Pos:* Fel, Ind Univ, 69-70. *Mem:* Am Chem Soc. *Res:* Organic derivatives of transition metals; chemistry of carboranes. *Mailing Add:* Dept of Chem Ball State Univ Muncie IN 47306

STORK, DONALD HARVEY, b Minn, Mar 22, 26; m 48; c 6. HIGH ENERGY PHYSICS. *Educ:* Carleton Col, BA, 48; Univ Calif, PhD(physics), 53. *Prof Exp:* Asst physics, Univ Calif, 48-51, asst, Lawrence Radiation Lab, 51-53, res assoc, 53-56; from asst prof to assoc prof, 56-64, PROF PHYSICS, UNIV CALIF, LOS ANGELES, 64- *Concurrent Pos:* Fel, Guggenheim Found, 65-66; guest scientist, Univ Oxford, Eng, 68 & Univ Res Asn, 85-86. *Mem:* Fel Am Phys Soc; AAAS; Am Asn Univ Profs. *Res:* Pions; K mesons; hyperons and antiprotons; production; beams; interactions; decay; high-energy electron-positron collisions; quark-antiquark and two-photon interactions; high-energy proton-proton collider design. *Mailing Add:* Dept Physics Univ Calif 405 Hilgard Ave Los Angeles CA 90024

STORK, GILBERT (JOSSE), b Brussels, Belg, Dec 31, 21; nat US; m 44; c 4. SYNTHETIC ORGANIC CHEMISTRY. *Educ:* Univ Fla, BS, 42; Univ Wis, PhD(chem), 45. *Hon Degrees:* DSc, Lawrence Col, 61, Univ Pierre et Marie Curie, Paris, 79, Univ Rochester, 82, Emory Univ, 88. *Prof Exp:* Sr res chemist, Lakeside Labs, Inc, 45-46; instr chem, Harvard Univ, 46-48, asst prof, 48-53; from assoc prof to prof, 53-67, chmn dept, 73-76, EUGENE HIGGINS PROF CHEM, COLUMBIA UNIV, 67- *Concurrent Pos:* Consult, NSF, 58-61, US Army Res Off, 66-69, NIH, 67-71, Sloane Found, 74-77, Syntex Corp & IFF Corp; var lectureships & professorships, US & abroad, 58-; Guggenheim fel, 59; mem comt org chem & comt postdoctoral fels, Nat Res Coun, 59-62; mem adv bd, Petrol Res Fund, 63-66. *Honors & Awards:* Pure Chem Award, Am Chem Soc, 57, Baekeland Medal, 61, Creative Work Synthetic Org Chem Award, 67, Nichols Medal, 80, Arthur C Cope Award, 80, Edgar Fahs Smith Award, 82, Willard Gibbs Medal, 82, Rensen Award, 86; Harrison Howe Award, 62; Franklin Mem Award, 66; Synthetic Org Chem Mfg Asn Gold Medal, 71; Roussel Steroid Prize, 78; Nat Acad Sci Chem Award, 82; Nat Medal Sci, 83; Pauling Award, 83; Tetrahedron Prize, Synthetic Chem, 85. *Mem:* Nat Acad Sci; Am Acad Arts & Sci; Am Chem Soc; Swiss Chem Soc; hon fel Royal Soc Chem; hon mem Pharm Soc Japan. *Res:* Total synthesis of complex structures; design of new synthetic reactions; reaction mechanisms. *Mailing Add:* Dept Chem Columbia Univ Box 666 Havemeyer Hall New York NY 10027

STORM, CARLYLE BELL, b Baltimore, Md, Mar 2, 35; m 57; c 3. CHEMISTRY. *Educ:* Johns Hopkins Univ, BA, 61, MA, 63, PhD(chem), 65. *Prof Exp:* NIH res fel chem, Stanford Univ, 65-66; staff fel biochem, NIMH, 66-68; from asst prof to prof chem, Howard Univ, 68-86; STAFF MEM, LOS ALAMOS NAT LAB, 85- *Concurrent Pos:* NIH res career develop award, 73-78; sr visitor, Inorg Chem Lab, Oxford Univ, 74-75; vis staff mem, Stable Isotope Res Resource, Los Alamos Nat Lab, NMex, 81-82; Sr Fulbright Hays Fel, Univ Trondheim, Norway, 77. *Honors & Awards:* Sr Fulbright Hays Fel, Univ Trondheim, Norway, 77. *Mem:* Am Chem Soc; Royal Soc Chem; Am Soc Biochem & Molecular Biol. *Res:* NMR spectroscopy; research administration. *Mailing Add:* M-DO MS P915 Los Alamos Nat Lab Los Alamos NM 87545

STORM, DANIEL RALPH, b Hawarden, Iowa, June 21, 44; m 66; c 3. BIOCHEMISTRY. *Educ:* Univ Wash, BS, 66, MS, 67; Univ Calif, Berkeley, PhD(biochem), 71. *Prof Exp:* Res asst biochem, Univ Calif, Berkeley, 67-71, NIH res fel, Harvard Univ, 71-72, NSF fel, 72-73; asst prof biochem, Univ Ill, Urbana, 73-78; ASSOC PROF PHARMACOL, UNIV WASH, 78- *Concurrent Pos:* Indust consult, Pharmaco Inc, 75- *Mem:* Am Chem Soc; Am Soc Biol Chemists; Am Soc Microbiol. *Res:* Structure and function of biological membranes; mechanism of enzymatic catalysis; membrane active antibiotics and molecular pharmacology at the membrane level. *Mailing Add:* Dept Pharmacol 5J-30 Univ Wash Seattle WA 98195

STORM, EDWARD FRANCIS, b Wilmington, Del, Nov 6, 29. COMPUTER SCIENCE, MATHEMATICS. *Educ:* Univ Del, AB, 59; Harvard Univ, MA, 61, PhD(appl math), 66. *Prof Exp:* Asst prof comput sci, Univ Va, 64-69; assoc prof, 69-77, PROF COMPUT & INFO SCI, SYRACUSE UNIV, 77- *Concurrent Pos:* Sr sci consult, Nat Resource Analysis Ctr, 65- *Mem:* Asn Comput Mach; Asn Symbolic Logic. *Res:* Design and implementation of high-level programming languages and their application to problems in machine simulation of intelligence and to man-machine communication. *Mailing Add:* Comput Sci Syracuse Univ 313 Link Syracuse NY 13244

STORM, LEO EUGENE, b Valeda, Kans, Aug 29, 28. COMPUTER SCIENCE, STATISTICS, MINI COMPUTERS. *Educ:* Okla Agr & Mech Col, BA, 53. *Prof Exp:* Seismic engr, Seismic Eng Co, 53-54; meteorologist, US Weather Bur, 54-55; mathematician, Northwestern Univ, 55; qual control engr, Metro Bottle Glass Co, 55-56; jr engr, US Testing Co, 56; assoc staff mem, Gen Precision Lab, Inc, 56-57; sr statistician, Nuclear Fuel Oper, Olin Mathieson Chem Corp, 57-61; statist qual control supvr, United Nuclear Corp, 61-62; opers analyst, United Aircraft Corp Systs Ctr, 62-67; sr sci programmer, NY Med Col, 67-70; syst analyst, Texaco Inc, 70-71; programmer analyst, Data Develop, Inc, 71-73; sr syst analyst, Nabisco Brands, Inc, 73-89. *Res:* Digital computer programming for management systems; statistical sample surveys; programming analysis for statistical accounting and biomedical applications; mini-computer systems for process control; material handling and management applications. *Mailing Add:* 1325 Viewtop Dr Clearwater FL 34624

STORM, ROBERT MACLEOD, b Calgary, Alta, July 9, 18; US citizen; m 43, 59; c 6. ZOOLOGY. *Educ:* Northern Ill State Teachers Col, BE, 39; Ore State Col, MS, 41, PhD(zool), 48. *Prof Exp:* From instr to prof zool, Ore State Univ, 48-84; RETIRED. *Mem:* Assoc Am Soc Ichthyologists & Herpetologists. *Res:* Natural history of cold-blooded land vertebrates. *Mailing Add:* 1623 SW Brooklane Dr Corvallis OR 97333

STORMER, HORST LUDWIG, b Frankfurt-Main, Ger, Apr 6, 49. SOLID STATE PHYSICS. *Educ:* Univ Frankfurt, BS, 70, dipl physics, 74; Univ Stuttgart, Ger, PhD(physics), 77. *Prof Exp:* Mem tech staff physics, High Magnetic Field Lab, Max Planck Inst Solid State Res, 77; consult, 77-78, mem tech staff physics, 78-83, DEPT HEAD, BELL LABS, AT&T, 83- *Honors & Awards:* Oliver E Buckley Prize; Otto Klung Prize. *Mem:* Am Phys Soc. *Mailing Add:* Am Tel & Tel Bell Labs 600 Mountain Ave ID-432 Murray Hill NJ 07974

STORMER, JOHN CHARLES, JR, b Englewood, NJ, Oct 28, 41; m 63; c 2. IGNEOUS PETROLOGY, GEOCHEMISTRY. *Educ:* Dartmouth Col, BA, 63; Univ Calif, Berkeley, PhD(geol), 71. *Prof Exp:* Asst geologist, Climax Molybdenum Co, Colo, 67; from asst prof to assoc prof geol, Univ Ga, 71-83; CARY CRONEIS PROF GEOL, RICE UNIV, 83-, CHAIR GEOL, 88- *Concurrent Pos:* Vis prof, Inst Geosci, Univ Sao Paulo, 73; IPA, US Geol Surv, Reston, 82-83. *Mem:* Fel Mineral Soc Am; Geochem Soc; Am Geophys Union; Brazilian Geol Soc; Microbeam Anal Soc. *Res:* Mineralogy and geochemistry of igneous rocks as applied to petrology; thermochemical data and methods of investigating the origin of igneous rocks, and applications to various rock suites and petrographic provinces. *Mailing Add:* Dept Geol & Geophys Rice Univ PO Box 1892 Houston TX 77251

STORMONT, CLYDE J, b Viola, Wis, June 25, 16; m 40; c 5. GENETICS, IMMUNOLOGY. *Educ:* Univ Wis, BA, 38, PhD(genetics), 47. *Prof Exp:* Instr genetics, Univ Wis, 46-47, lectr, 47, asst prof, 48; Fulbright scholar, NZ, 49-50; asst prof, 50-54, 50-54, assoc prof vet med & assoc seriologist, Exp Sta, 54-59, prof, 59-82, EMER PROF IMMUNOGENETICS, UNIV CALIF, DAVIS, 82- *Concurrent Pos:* E B Scripps fel, San Diego Zool Soc, 56-57 & 66-67; chmn & dir lab serv, Stormont Labs, Inc, Woodland, Calif. *Mem:* Genetics Soc Am; Am Soc Human Genetics; Soc Exp Biol & Med; hon mem Int Soc Animal Genetics; Am Soc Nat; hon mem Nat Buffalo Asn. *Res:* Blood groups; animal blood groups and biochemical polymorphisms; genetic markers in animal blood. *Mailing Add:* Stormont Labs Inc 1237 E Beamer St Ste D Woodland CA 95695

STORMS, LOWELL H, b Schenectady, NY, Feb 14, 28; m 81; c 3. NEUROPSYCHOLOGY. *Educ:* Univ Minn, BA, 50, MS, 51, PhD(clin psychol), 56. *Prof Exp:* Psychologist, Hastings State Hosp, Minn, 54-56; Fulbright grant, Inst Psychiat, Univ London, 56-57; from instr to prof psychiat, Neuropsychiat Inst, Univ Calif, Los Angeles, 57-71; PROF PSYCHIAT, SCH MED, UNIV CALIF, SAN DIEGO, 71- *Concurrent Pos:* Consult, Vet Admin, 64-71 & Encounters Unlimited, 68-71; psychologist, Vet Admin Hosp, San Diego, 75-; Fulbright scholar. *Mem:* AAAS; Am Psychol Asn; Sigma Xi. *Res:* Behavior of schizophrenics; behavior therapy; clinical psychology. *Mailing Add:* Dept Psychiat Sch of Med Univ Calif at San Diego La Jolla CA 92037

STORMSHAK, FREDRICK, b Enumclaw, Wash, July 4, 36; m 63; c 2. REPRODUCTIVE ENDOCRINOLOGY. *Educ:* Wash State Univ, BSc, 59, MSc, 60; Univ Wis, PhD(endocrinol), 65. *Prof Exp:* Res physiologist, USDA, 65-68; asst prof physiol, 68-72, actg head, Dept Animal Sci, 74, assoc prof, 72-79, PROF PHYSIOL, ORE STATE UNIV, 79- *Concurrent Pos:* Postdoctoral trainee endocrinol, Univ Wis, 74-75; sect ed, J Animal Sci, 75-77, ed-in-chief, 82-85; actg assoc dir, Ore Agr Exp Sta, 85; NIH Study Sect Reprod Biol, 82-86; USDA grant rev panel, 88- 90. *Mem:* Am Soc Animal Sci; Soc Study Fertility; Endocrine Soc; Soc Study Reproduction. *Res:* Quantitative measurement of steroid hormones of ovarian origin; factors affecting the regression and maintenance of the corpus luteum; pituitary, ovarian and uterine interrelationships in reproduction; hormone action. *Mailing Add:* Dept Animal Sci Ore State Univ Corvallis OR 97331-6702

STORR, JOHN FREDERICK, b Ottawa, Ont, Aug 17, 15; m 42; c 1. AQUATIC ECOLOGY. *Educ:* Queen's Univ, Ont, BA, 42; Columbia Univ, MA, 48; Cornell Univ, PhD(marine ecol), 55. *Prof Exp:* Instr biol, Queen's Col, Bahamas, 42-45; asst prof physiol, Adelphi Col, 47-52; res asst prof marine ecol, Univ Miami, 55-58; ASSOC PROF ECOL & INVERT ZOOL, STATE UNIV NY BUFFALO, 58- *Concurrent Pos:* US Fish & Wildlife Serv grant, 55-57; limnol consult, Niagara Mohawk Power Corp, 63-; aquatic ecol consult, Rochester Gas & Elec Corp, 68- *Mem:* Ecol Soc Am; Am Inst Biol Sci; Am Fisheries Soc; Explorers Club; Int Soc Limnol. *Res:* Coral reef zonation; ecology of sponges of Gulf of Mexico and of benthic organisms and fish in Lake Erie and Lake Ontario of New York. *Mailing Add:* 41 Fairways Williamsville NY 14221

STORRIE, BRIAN, b East Cleveland, Ohio, Mar 9, 46; m 71; c 2. BIOCHEMISTRY, CELL BIOLOGY. *Educ:* Cornell Univ, BS, 68; Calif Inst Technol, PhD(biochem), 73. *Prof Exp:* NSF fel, Calif Inst Technol, 68-72; NIH fel, Univ Colo Med Ctr, 72, Am Cancer Soc fel, 73-74; res assoc, Mem Sloan-Kettering Cancer Ctr, 74-75; asst res biologist, Univ Calif, Berkeley, 75-76; from asst prof to assoc prof, 76-86, PROF BIOCHEM, VA POLYTECH INST & STATE UNIV, 86- *Concurrent Pos:* Prin investr, NSF grants, 78-85 & NIH grants, 80-83 & 84-; co-instr, Marine Biol Lab, 80; mem, cell biol panel, NSF, 82; consult, Gilford Systs, Ciba Corning, 82 & Collins & Assocs, 84-86; vis scientist, Europ Molecular Biol Lab, 88; fel Ger Acad Exchange Prog, 88. *Mem:* Am Soc Cell Biol; Sigma Xi. *Res:* Intraorganelles protein exchange in animal cells; mechanisms for selective targeting of membrane vesicles arising from endocytosis; mechanisms of protein segregation during membrane trafficking; assembly of lysosomes. *Mailing Add:* Dept Biochem & Nutrit Va Polytech Inst & State Univ Blacksburg VA 24061-0308

STORROW, HUGH ALAN, b Long Beach, Calif, Jan 13, 26; m 53; c 3. PSYCHIATRY. *Educ:* Univ Southern Calif, AB, 46, MD, 50; Am Bd Psychiat & Neurol, dipl, 55. *Prof Exp:* Intern, USPHS Hosp, Baltimore, 49-50; resident, Sheppard & Enoch Pratt Hosp, Towson, Md, 50-51; staff psychiatrist, US Penitentiary Hosp, Leavenworth, Kans, 51-52; resident, USPHS Hosp, Lexington, Ky, 52-53, staff psychiatrist, 53-54; resident, Brentwood Vet Admin Hosp, Los Angeles, 54-55; instr psychiat, Sch Med, Yale Univ, 55-56; asst prof, Sch Med, Univ Calif, Los Angeles, 56-60, attend psychiatrist, Med Ctr, 57-60; assoc prof, Col Med, Univ Ky, 60-65; prof psychiat, Univ Minn, 65-66; PROF PSYCHIAT, COL MED, UNIV KY, 66 - *Concurrent Pos:* Attend psychiatrist, Brentwood Vet Admin Hosp, Calif, 57-60; consult, United Cerebral Palsy Asn, Los Angeles County, Calif, 57-60, USPHS & Vet Admin Hosps, Ky, 60 - *Mem:* Am Psychiat Asn; AMA; Asn Am Med Cols. *Res:* Behavior modification; teaching methods for psychiatry. *Mailing Add:* Dept Psychiat Univ Ky Col Med Lexington KY 40536

STORRS, CHARLES LYSANDER, b Shaowu, Fukien, China, Oct 25, 25; US citizen; m 57; c 3. NUCLEAR ENGINEERING, TECHNICAL MANAGEMENT. *Educ:* Mass Inst Technol, BS, 49, PhD, 52. *Prof Exp:* Mem staff, Aircraft Nuclear Propulsion Dept, Gen Elec Co, 52-56, supvr initial engine test opers, 56-59, supvr flight engine test opers, 59-61, mgr reactor test opers, Nuclear Propulsion Dept, 61, SL-1 Proj, Nuclear Mat & Propulsion Opers, 61-62 & 710 Proj, 62-65; dir heavy water organic cooled reactor, Atomics Int-Combustion Eng Joint Venture, Calif, 65-67; asst dir advan reactor eng, Combustion Eng, Inc, 67-69, dir advan reactor develop, 69-71, dir projs, 71-73, dir prod eng & develop, 73-75, dir fast breeder reactor develop, 75-80, dir advanced develop, Nuclear Power Systs, 80-86; RETIRED. *Mem:* AAAS; Am Nuclear Soc; Am Phys Soc. *Res:* Engineering, design and development of technology leading to the application of nuclear energy to power generation and desalination; reactor test operations; management of technical enterprises. *Mailing Add:* 76 Adams Rd Bloomfield CT 06002

STORRS, ELEANOR EMERETT, b Cheshire, Conn, May 3, 26; m 63; c 2. LEPROSY, ARMADILLO RESEARCH. *Educ:* Univ Conn, BS, 48; NY Univ, MS, 58; Univ Tex, PhD(biochem), 67. *Prof Exp:* Asst, Boyce Thompson Inst Plant Res, 48-59, asst biochemist, 59-62; res scientist, Clayton Found Biochem Inst, Univ Tex, 62-65; res chemist, Pesticides Res Lab, USPHS, Fla, 65-67; res chemist, Gulf South Res Inst, 67-71, dir dept comp biochem, 71-77; dir, Div Comp Mammal & Biochem, Med Res Inst, 77-85, DIR, DIV COMP MAMMAL, DEPT BIOL SCI, FLA INST TECHNOL, 86- *Honors & Awards:* Charles A Griffin Award, Am Asn Lab Animal Sci, 75; Spec Recognition, Gerald B Lambert Awards, 75. *Mem:* Fel AAAS; Am Soc Mammal; Am Asn Lab Animal Sci; Sigma Xi; Int Leprosy Asn; fel NY Acad Sci. *Res:* Armadillo in biomedical research; leprosy; biochemical individuality; analytical methods for biochemical, environmental and residue analyses; drug metabolism; mode of fungicidal, insecticidal action; armadillo reproduction. *Mailing Add:* Fla Inst Technol 150 W University Blvd Melbourne FL 32901-6988

STORRY, JUNIS O(LIVER), b Astoria, SDak, Mar 16, 20; m 50; c 2. ELECTRICAL ENGINEERING. *Educ:* SDak State Col, BS, 42, MS, 49; Iowa State Univ, PhD, 67. *Prof Exp:* Mem student prog, Westinghouse Elec Corp, Pa, 42; elec engr, Bur Ships, Navy Dept, Washington, DC, 42-46; design engr, Reliance Elec & Eng Co, Ohio, 46; from instr to assoc prof, SDak State Univ, 46-64, actg dean, 71-72, prof elec eng, 64-85, dean eng, 72-82, EMER PROF ELEC ENG & EMER DEAN, SDAK STATE UNIV, 85- *Mem:* Inst Elect & Electronic Engrs; Am Soc Eng Educ; Nat Soc Prof Engrs. *Res:* Digital analysis of power systems using hybrid parameters. *Mailing Add:* RR 3 Box 26 Brookings SD 57006-9407

STORTI, ROBERT V, b Providence, RI, May 14, 44. MOLECULAR BIOLOGY, BIOCHEMISTRY. *Educ:* RI Col, BA, 68; Ind Univ, MA, 70, PhD(biol), 74. *Prof Exp:* Fel biol, Mass Inst Technol, 74-78; ASST PROF BIOCHEM, MED CTR, UNIV ILL, 78- *Concurrent Pos:* NIH fel, 74-76; Muscular Dystrophy Soc fel, 77-78; Biomed Found Res fel, 78. *Mem:* Am Soc Cell Biol; AAAS; Soc Develop Biol. *Res:* Molecular biology of gene expression during eukaryotic cell growth and differentiation; transcriptional and translation control of protein synthesis. *Mailing Add:* Dept Biochem MC 536 Univ Ill Col Med PO Box 6998 1853 W Polk St Chicago IL 60612

STORTS, RALPH WOODROW, b Zanesville, Ohio, Feb 5, 33; m 60; c 3. VETERINARY PATHOLOGY. *Educ:* Ohio State Univ, DVM, 57, PhD(vet path), 66; Purdue Univ, West Lafayette, MSc, 62. *Prof Exp:* Instr vet microbiol, Purdue Univ, West Lafayette, 57-60; instr vet path, Ohio State Univ, 61-66; from asst prof to assoc prof, 66-73, PROF VET PATH, TEX A&M UNIV, 73- *Mem:* Am Vet Med Asn; Am Col Vet Path; Int Acad Path; Conf Res Workers Animal Dis. *Res:* Veterinary neuropathology including electron microscopy and cytology of normal and infected tissue cultures of nervous tissue. *Mailing Add:* 1006 Village Dr College Station TX 77840

STORTZ, CLARENCE B, b Marlette, Mich, July 23, 33; m 56; c 5. MATHEMATICS. *Educ:* Wayne State Univ, BS, 55; Univ Miami, MS, 58; Univ Mich, Ann Arbor, DEd(math), 68. *Prof Exp:* Asst prof math, Northern Mich Univ, 63-66 & Cent Mich Univ, 66-68; assoc prof, 68-72, head dept, 72-76, PROF MATH, NORTHERN MICH UNIV, 72- *Res:* General topology; history of mathematics. *Mailing Add:* Dept Math Northern Mich Univ Marquette MI 49855

STORVICK, CLARA A, b Emmons, Minn, Oct 31, 06. NUTRITION. *Educ:* St Olaf Col, AB, 29; Iowa State Univ, MS, 33; Cornell Univ, PhD(nutrit, biochem), 41. *Prof Exp:* Instr chem, Augustana Acad, 30-32; asst, Iowa State Univ, 32-34; nutritionist, Fed Emergency Relief Admin, Minn, 34-36; asst prof nutrit, Okla State Univ, 36-38; asst, Cornell Univ, 38-41; asst prof nutrit, Univ Wash, 41-45; from assoc prof to prof, 45-72, head home econ res, 55-72, dir nutrit res inst, 65-72, EMER PROF NUTRIT, ORE STATE UNIV, 72- *Concurrent Pos:* Sigma Xi Lectr, Ore State Univ, 53; sabbatical leaves, Chem Dept, Columbia Univ & Inst Cytophysiol, Denmark, 52, Lab Nutrit &

Endocrinol, NIH, 59 & Div Clin Oncol, Med Sch, Univ Wis, 66. *Honors & Awards:* Borden Award, Am Home Econ Asn, 52. *Mem:* Am Home Econ Asn; Am Dietetic Asn; fel Am Pub Health Asn; fel Am Inst Nutrit; fel AAAS; Am Chem Soc. *Res:* Calcium, phosphorus, nitrogen, ascorbic acid, thiamine and riboflavin metabolism; nutrition and dental caries; vitamin B-6. *Mailing Add:* 124 NW 29th St Corvallis OR 97330

STORVICK, DAVID A, b Ames, Iowa, Oct 24, 29; m 52; c 3. MATHEMATICS. *Educ:* Luther Col, Iowa, AB, 51; Univ Mich, MA, 52, PhD(math), 56. *Prof Exp:* From instr to asst prof math, Iowa State Univ, 55-57; from asst prof to assoc prof, 57-66, PROF MATH, UNIV MINN, MINNEAPOLIS, 66-, ASSOC HEAD SCH MATH, 64- *Concurrent Pos:* Res assoc, US Army Math Res Ctr, Wis, 62-63. *Mem:* Am Math Soc; Math Asn Am. *Res:* Complex function theory. *Mailing Add:* Univ Minn Minneapolis MN 55455

STORVICK, TRUMAN S(OPHUS), b Albert Lea, Minn, Apr 14, 28; m 52; c 4. CHEMICAL ENGINEERING, MOLECULAR PHYSICS. *Educ:* Iowa State Univ, BS, 52; Purdue Univ, PhD(chem eng), 59. *Prof Exp:* Res engr, Res Dept, Westvaco Chloro-Alkali Div, FMC Corp, 52-55; instr chem eng, Purdue Univ, 58-59; from asst prof to assoc prof, 59-72, Robert Lee Tatum prof, 72-75, BLACK & VEATCH PROF ENG, UNIV MO-COLUMBIA, 75- *Concurrent Pos:* NSF fac fel, 65-66; fel, Royal Norweg Coun Sci & Indust Res, 72-73. *Mem:* AAAS; Am Chem Soc; Am Inst Chem Engrs; Am Phys Soc. *Res:* Measurement and prediction of thermodynamic and transport properties; aerosol system behavior. *Mailing Add:* 2210 Ridgefield Rd Columbia MO 65203

STORWICK, ROBERT MARTIN, b Seattle, Wash, Oct 14, 42; m 67; c 2. ELECTRICAL ENGINEERING, APPLIED MATHEMATICS. *Educ:* Calif Inst Technol, BS, 64; Univ Southern Calif, MSEE, 65, PhD(elec eng), 69; Detroit Col Law, JD, 82. *Prof Exp:* Mem tech staff, Radar & Data Processing Dept, Gen Res Corp, Calif, 69-70; staff res engr, Electronics Dept, Gen Motors Res Labs, 70-79; PATENT ATTY, SEED & BERRY, 79- *Mem:* AAAS; sr mem Inst Elec & Electronic Engrs. *Res:* Signal processing; short-range and long-range radar systems; radar cross-section studies and analyses; statistical pattern recognition; information, coding and communication theory; networks and combinatorial systems; graph theory. *Mailing Add:* Seed & Berry 63006 Columbia Ctr 701 Fifth Ave Seattle WA 98104

STORY, ANNE WINTHROP, b Haverhill, Mass. ENGINEERING PSYCHOLOGY. *Educ:* Smith Col, AB, 34; Univ Calif, Berkeley, PhD(exp psychol), 57. *Prof Exp:* Assoc engr turbine div, Gen Elec Co, 42-44; instr, Stoneleigh Jr Col, 45-46, Greenbrier Col, 46-47 & Pa State Univ, 47-50; res assoc animal behav, Jackson Mem Lab, 50-51; res analyst flight safety, Norton AFB, 51-52; teaching asst statist & psychol, Univ Calif, Berkeley, 52-57; res psychologist flight safety & space psychol, Hanscom AFB, 58-66 & NASA, 66-70; eng psychologist man-machine syst, US Dept Transp, 70-78; CONSULT, 78- *Concurrent Pos:* Assoc prof dept psychol, Univ Mass, 72-76. *Mem:* AAAS; Sigma Xi; Am Psychol Asn; Int Asn Appl Psychol. *Res:* Aviation collision pilot-warning; vehicle driver safety devices; visual perception; attention; man-machine systems. *Mailing Add:* The Headlands Rockport MA 01966

STORY, HAROLD S, b Catskill, NY, Oct 5, 27; m 51; c 2. SOLID STATE PHYSICS, NUCLEAR MAGNETIC RESONANCE. *Educ:* NY State Col Teachers, BA, 49, MA, 50; Univ Maine, Orono, MS, 52; Case Inst Technol, PhD(physics), 57. *Prof Exp:* Mem tech staff, Bell Tel Labs, NJ, 56-59; from assoc prof to prof physics, 59-80, physics dept chmn, 81-82, PROF ASTRON & SPACE SCI, STATE UNIV NY ALBANY, 80- *Mem:* Am Phys Soc; Sigma Xi; Am Asn Physics Teachers. *Res:* Structure, defects and conduction processes in superionic conductors, utilizing nuclear magnetic resonance. *Mailing Add:* Dept Physics State Univ NY 1400 Wash Ave Albany NY 12222

STORY, JIM LEWIS, b Alice, Tex, July 30, 31; m 58; c 4. NEUROSURGERY. *Educ:* Tex Christian Univ, BS, 52; Vanderbilt Univ, MD, 55. *Prof Exp:* From instr to asst prof neurosurg, Med Sch, Univ Minn, 61-67; PROF NEUROSURG, UNIV TEX HEALTH SCI CTR SAN ANTONIO, 67-, PROF ANAT, 77- *Concurrent Pos:* Univ fels neurol surg, Univ Minn, 56-59 & 60-61; USPHS fels anat, Univ Calif, Los Angeles, 59-60 & Univ Minn, 60-62. *Mem:* Am Asn Neurol Surgeons; Soc Neurol Surgeons; Am Col Surgeons; Neurosurg Soc Am; Am Acad Neurol Surgeons. *Res:* Intracranial pressure monitoring; etiology of brain tumors. *Mailing Add:* Univ Tex Med Sch 7703 Floyd Curl Dr San Antonio TX 78284-7843

STORY, JON ALAN, b Odebolt, Iowa, Apr 7, 46; m 69; c 3. BIOCHEMISTRY, NUTRITION. *Educ:* Iowa State Univ, BS, 68, MS, 70, PhD(zool), 72. *Prof Exp:* Instr zool, Iowa State Univ, 71-72; trainee lipid metab, Wistar Inst Anat & Biol, 72-74, asst prof lipid metab, 74-77; assoc prof foods & nutrit, 77-80, PROF NUTRIT PHYSIOL, PURDUE UNIV, WEST LAFAYETTE, 80-, ASSOC DEAN CONSUMER & FAMILY SCI, 88- *Mem:* Am Inst Nutrit; Nutrit Soc; Sigma Xi; Soc Exp Biol & Med. *Res:* Investigation into the effects of several dietary components and age on cholesterol and bile acid metabolism as involved in development of experimental atherosclerosis. *Mailing Add:* Dept Foods & Nutrit Purdue Univ West Lafayette IN 47907

STORY, TROY LEE, JR, b Montgomery, Ala, Nov 11, 40. CHEMICAL PHYSICS. *Educ:* Morehouse Col, BS, 62; Univ Calif, Berkeley, PhD(chem), 68. *Prof Exp:* Mem staff & fel chem, Univ Calif, Berkeley, 69-70; fel physics, Chalmers Univ Technol, Sweden, 70-71; asst prof chem, Howard Univ, 71-77, ASSOC PROF CHEM, MOREHOUSE COL, 77- *Mem:* Am Chem Soc; Am Phys Soc. *Res:* Experimental determination of dipole moments using molecular beam resonance and deflection techniques; theoretical quantum mechanical model for the analysis of rotational distributions for reactive scattering experiments; topological analysis of composite particles. *Mailing Add:* Dept Chem Morehouse Col 830 Westview Dr Atlanta GA 30314-3799

STORZ, JOHANNES, b Hardt/Schramberg, Ger, Apr 29, 31; US citizen; m 59; c 3. VIROLOGY, MICROBIOLOGY. *Educ:* Vet Col, Hannover, dipl, 57; Univ Munich, Dr Med Vet, 58; Univ Calif, Davis, PhD(comp path), 61; Am Col Vet Microbiol, dipl, 69. *Prof Exp:* Res assoc, Fed Res Inst Viral Dis Animals, Tübingen, Ger, 57-58; lectr vet microbiol, Univ Calif, Davis, 58-61; from asst prof to assoc prof vet virol, Utah State Univ, 61-65; from assoc prof to prof vet virol, Colo State Univ, 65-82; PROF & DEPT HEAD, DEPT VET MICROBIOL & PARASITOL, LA STATE UNIV, 82- *Concurrent Pos:* USPHS res grant, Utah State Univ, 62-65; res grants, USPHS, Colo State Univ, 66-72, 72-77 & 79-82, WHO, 72-77 & USDA, 80-85; vis scientist, Univ Giessen, 71-72 & 78-79; consult, WHO, Geneva, 71 & Munich, 76; Alexander Humboldt award, Ger, 78. *Mem:* AAAS; Am Soc Microbiol; Am Vet Med Asn; Conf Res Workers Animal Dis; World Asn Buiatrics; Am Soc Virol. *Res:* Chlamydiology; pathogenic mechanisms in intrauterine viral and chlamydial infections; chlamydial polyarthritis; intestinal corona and parvoviral infections; cell biology of chlamydial infections. *Mailing Add:* Dept Vet Microbiol & Parasitol, Sch Vet Med La State Univ Baton Rouge LA 70803

STOSICK, ARTHUR JAMES, b Milwaukee, Wis, Dec 1, 14; m 37; c 3. CHEMISTRY. *Educ:* Univ Wis, BS, 36; Calif Inst Technol, PhD(struct chem), 39. *Prof Exp:* Fel, Calif Inst Technol, 39-40, instr gen chem, 40-41, Nat Defense Res Comt res assoc, 41-43, res chemist, Jet Propulsion Lab, 44-46, chief, Rokets & Mat Div, 50-56; res chemist, Aerojet Eng Corp, 43-44; assoc prof phys chem, Iowa State Col, 46-47; prof, Univ Southern Calif, 47-50; asst dir, Union Carbide Res Inst, 56-59; asst vpres, Gen Atomic Div, Gen Dynamics Corp, 59-60; sr scientist, Aerojet-Gen Corp, 60-71; asst sr vpres, United Technol Corp, 72-80; CONSULT, 80- *Concurrent Pos:* Mem, rocket eng subcomt, Nat Adv Comt Aeronaut, 52-56; staff scientist, Adv Res Proj Agency, Off Secy Defense, 58-59. *Mem:* Am Chem Soc; Am Phys Soc; Am Crystallog Asn. *Res:* Molecular structures by diffraction; physical chemistry as related to molecular structures; propellants; high temperature chemistry; metallurgy. *Mailing Add:* 1153 Lime Dr Sunnyvale CA 94087-2021

STOSKOPF, MICHAEL KERRY, b Garden City, Kans, March 21, 50; m 81. AQUATIC MEDICINE, AQUATIC TOXICOLOGY. *Educ:* Colo State Univ, BS, 73, DVM, 75; Am Col Zool Med, dipl; Johns Hopkins Univ, PhD, 86. *Prof Exp:* Staff vet comp med, Overton Park Zoo & Aquarium, 75-77; asst prof, 79-86, ASSOC PROF COMP MED, SCH MED, JOHNS HOPKINS UNIV, 86-; CHIEF MED, NAT AQUARIUM, BALTIMORE, 81- *Concurrent Pos:* Consult, Nat Inst Exp Progs, Antivenom Inst, Columbia, 79; staff vet, Baltimore Zool Soc, 77-81; adj prof path, Sch Med, Univ Md, 81-; chmn exam comt, Am Col Zoo Med, 86-88. *Mem:* Int Asn Aquatic Animal Med (pres, 88); Am Asn Zoo Veterinarians; Am Col Zoo Med; Wildlife Dis Asn; Am Vet Med Asn. *Res:* Investigation of new animal models for human disease with particular interest in the effects of environmental factors on the physiology and biochemistry of living organisms. *Mailing Add:* 2742 N Calvert St Baltimore MD 21218

STOSKOPF, N C, b Mitchell, Ont, June 11, 34; m 60; c 2. CROP BREEDING. *Educ:* Univ Toronto, BSA, 57, MSA, 58; McGill Univ, PhD(agron), 62. *Prof Exp:* Lectr agron, Ont Agr Col, 58-59; instr & exten specialist, Kemptville Agr Sch, 59-60; asst prof crop sci, Ont Agr Col, 62-66, assoc prof, 66-69, PROF CROP SCI, UNIV GUELPH, 69-, DIR DIPL PROG AGR, 74- *Mem:* Agr Inst Can. *Res:* Winter wheat breeding given a physiological basis with yield as main objective; plants selected for upright leaves to achieve a high optimum leaf area, a high net assimilation rate and a long period of grain filling; cereal physiology. *Mailing Add:* Dept Agr Crop Sci Univ Guelph Guelph ON N1G 2W1 Can

STOSSEL, THOMAS PETER, b Chicago, Ill, Sept 10, 41; m 65; c 2. HEMATOLOGY. *Educ:* Princeton Univ, AB, 63; Harvard Med Sch, MD, 67. *Hon Degrees:* MD, Univ Linköping, Sweden. *Prof Exp:* House staff med, Mass Gen Hosp, 67-69; staff assoc, NIH, 67-71; from fel to sr assoc, Med Ctr, Children's Hosp, Boston, 71-76; CHIEF HEMATOL & ONCOL UNIT, MASS GEN HOSP, 76-; PROF MED, HARVARD MED SCH, 82- *Concurrent Pos:* Fel, Harvard Med Sch, 71-78; ed, J Clin Investigation 82-87; scientific bd, Biogen Corp, 87-; clin res prof, Am Cancer Soc, 87- *Honors & Awards:* Damashek Prize, Am Soc Hemat, 83. *Mem:* Am Fedn Clin Res; Am Soc Clin Investigation (pres, 87); Am Soc Hematol; Am Soc Cell Biol; Asn Am Physicians; Am Asn Immunol. *Res:* Biology of phagocytic leukocytes with special emphasis on the molecular basis of leukocyte movements; education in clinical hematology; sociology of biomedical publication. *Mailing Add:* Hematol & Oncol Dept Med Brigham/Womens' Hosp Francis St Boston MA 02115

STOTHERS, JOHN BAILIE, b London, Ont, Apr 16, 31; m 53; c 2. ORGANIC CHEMISTRY. *Educ:* Univ Western Ont, BSc, 53, MSc, 54; McMaster Univ, PhD(phys org chem), 57. *Prof Exp:* Res chemist, Res Dept, Imp Oil, Ltd, 57-59; lectr chem, 59-61, from asst prof to assoc prof, 61-67, chmn, 76-86, PROF CHEM UNIV WESTERN ONT, 67- *Concurrent Pos:* Merck, Sharp & Dohme lect award, 71. *Mem:* Royal Soc Can; Am Chem Soc; fel Chem Inst Can; Royal Soc Chem. *Res:* Nuclear magnetic resonance spectroscopy; applications of deuterium and carbon-13 nuclear magnetic resonance to organic structural, stereochemical and mechanistic problems and biosynthesis; deuterium exchange processes and molecular rearrangements. *Mailing Add:* Dept Chem Univ of Western Ont London ON N6A 5B7 Can

STOTLER, RAYMOND EUGENE, b Peoria, Ill, Mar 30, 40; m 69. BOTANY. *Educ:* Western Ill Univ, BS, 62; Southern Ill Univ, MA, 64; Univ Cincinnati, PhD(bot), 68. *Prof Exp:* Fel bot, Univ Wis-Milwaukee, 68-69; asst prof, 69-74, ASSOC PROF BOT, SOUTHERN ILL UNIV, 74- *Honors & Awards:* Dimond Award, NSF & Bot Soc Am, 75. *Mem:* Am Bryol & Lichenological Soc; Am Fern Soc; Am Soc Plant Taxon; Int Soc Plant Taxon; Int Asn Bryologists. *Res:* Nomenclature and biosystematics of hepatics, hornworts, and mosses. *Mailing Add:* Dept of Bot Life Sci Southern Ill Univ Carbondale IL 62901

STOTLER, RAYMOND T, b Cincinnati, Ohio, June 1, 16. GEOLOGY. *Educ:* Princeton Univ, BS, 39. *Prof Exp:* CONSULT PETROL GEOLOGIST, 39- *Mem:* Am Asn Prof Geologists; fel Geol Soc Am. *Mailing Add:* 3238 Citation Dr Dallas TX 75229

STOTSKY, BERNARD A, b New York, NY, Apr 8, 26; m 52; c 5. PSYCHOLOGY, PSYCHIATRY. *Educ:* City Col New York, BS, 48; Univ Mich, MA, 49, PhD(psychol), 51; Western Reserve Univ, MD, 62. *Prof Exp:* Staff psychologist, Ment Hyg Clin, Vet Admin, Detroit, 51-53; instr & assoc, Boston Univ, 54-56, asst prof psychol, 56-57; asst prof, Duke Univ, 57-58; staff psychologist, Vet Admin Hosp, Brockton, Mass, 58-61; intern, George Washington Univ Hosp, 62-63; fel psychiat & resident psychiat, Mass Ment Health Ctr, 63-65 & Boston State Hosp, 65-66; from lectr to prof psychol, Boston State Col, 64-82, head dept, 72-73; prof psychol, Univ Mass, Boston, 82-89; RETIRED. *Concurrent Pos:* Chief counseling psychologist, Vet Admin Hosp, Brockton, Mass, 53-56, consult, 56-57 & 70- & chief psychologist, Durham, NC, 57-58; consult, Brockton Family Serv, 54, Hayden Goodwill Inn, 54-57, Mass Dept Pub Health, 67 & Boston State Hosp, 63-65, 68-75; prin investr, psychiat consult & lectr, Northeastern Univ, 64-88; assoc psychiat, Tufts Univ, 66-67, asst prof, 67-76, assoc prof, 76-88, lectr, Clark Univ, 69-70 & Mt Sinai Sch Med, 69-72; consult, Food & Drug Admin, 72-75 & Nat Inst Child Health & Human Develop, 73-77; prof psychiat & behav sci, Univ Wash, 73-77; dir outpatient psychiat clin, St Elizabeth's Hosp Boston, 73, assoc dir psychiat educ, 74-81, dir psychiat educ, 81-84; consult campus sch, Boston Col, 74-84. *Mem:* Fel Am Psychol Asn; Am Psychiat Asn. *Res:* Psychopharmacology; diagnosis and treatment of mental disease; personality and organic factors in rehabilitation of chronically ill patients; geriatrics. *Mailing Add:* 200 Vintage Terr Apt 101 Swampscott MA 01907

STOTT, BRIAN, b Manchester, Eng, Aug 5, 41; m. COMPUTER CONTROL METHODS. *Educ:* Univ Manchester, BSc, 62, MSc, 63, PhD(elec eng), 71. *Prof Exp:* Lectr elec power eng, Univ Manchester Sci & Technol, 68-74; assoc prof, Univ Waterloo, Ont, Can, 74-76; consult, Brazilian Inst Elec Power Res, 76-83; prof, Ariz State Univ, Tempe, 83-84; PRES, POWER COMPUT APPLN CORP, MESA, ARIZ, 84- *Mem:* Fel Inst Elec & Electronics Engrs. *Mailing Add:* 1930 S Alma Sch Rd Suite C-204 Mesa AZ 85210

STOTT, DONALD FRANKLIN, b Reston, Man, Apr 30, 28; m 60; c 3. GEOLOGY. *Educ:* Univ Manitoba, BSc, 53, MSc, 54; Princeton Univ, AM, 56, PhD, 58. *Prof Exp:* Asst geol, Princeton Univ, 54-55; head regional geol subdiv, Geol Surv Can, 72-73, dir, 73-80, geologist, 57-72; RES SCIENTIST, INST SEDIMENTARY & PETROL GEOL, 73- *Honors & Awards:* Miller Medal, Royal Soc Can, 83. *Mem:* Fel Geol Soc Am; Can Soc Petrol Geologists; Geol Asn Can; Soc Econ Paleontologists & Mineralogists. *Res:* Physical stratigraphy and sedimentation, particularly of Cretaceous system of Rocky Mountain foothills, Canada. *Mailing Add:* 8929 Forest Park Dr Sidney BC V8C 5A7 Can

STOTT, KENHELM WELBURN, JR, b San Diego, Calif, Aug 27, 20; div. MAMMALOGY, ORNITHOLOGY. *Educ:* Pomona Col, BA, 42. *Prof Exp:* Curator mammals & publ, Zool Soc, San Diego, 46-48, gen curator, 46-54; leader primate studies prog, San Diego Natural Hist Soc, 57-60; res assoc zool, 59-80, EMER GEN CURATOR EMER, SOC SAN DIEGO, 80- *Concurrent Pos:* Res assoc, San Diego Natural Hist Soc, 57-74 & Martin & Osa Johnson Safari Museum, 74-; res collabr sci expeds, Smithsonian Inst, 78-85; trustee, Nat Underwater & Marine Agency, 79-; sci fel, Zool Soc London, 83; conserv fel, NY Zool Soc, 86. *Honors & Awards:* Sweeney Medal, Explorers Club, 80. *Mem:* Am Ornith Union; AAAS; Int Union Conserv Nature & Natural Resources; Am Soc Mammal; Linnean Soc. *Res:* Observation of rare and endangered species of mammals and birds; expeditions. *Mailing Add:* 2300 Front St Apt 402 San Diego CA 92101

STOTT, PAUL EDWIN, b Springfield, Mass, Jan 18, 48; m 70; c 3. ANTITOXIDANTS-INHIBITORS, CHEMICAL BLOWING AGENTS. *Educ:* Brigham Young Univ, BS, 71, PhD(chem), 79, Univ Mass, MS, 72. *Prof Exp:* Gen mgr, res chem, Parish Chem Co, 74-79; res chem, Uniroyal Inc, 79-80, sr group leader, 80-82; tech dir, Lubritex Inc, 72-84; dir res, Texmark Resins, 82-85; sect mgr, chem Div, 85-86, RES MGR, UNIROYAL CHEM CO INC, 86- *Mem:* Sigma Xi; Com Develop Asn. *Res:* Heterocyclic chemistry; development of antioxidants, inhibitors & stabilizers for petroleum, plastics & monomers. *Mailing Add:* 20 Clear View Dr Sandy Hook CT 06482

STOTTLEMYRE, JAMES ARTHUR, b Juneau, Alaska, Jan 4, 48; m 70; c 2. GEOPHYSICS, RESERVOIR ENGINEERING. *Educ:* Univ Wash, BS, 71, MS, 74, PhD, 80. *Prof Exp:* Resource engr energy resources, Wash Water Power Co, 74-76; MGR, EARTH SCI SECT, BATTELLE PAC NORTHWEST LABS, 76- *Mem:* Soc Explor Geophys; Soc Petrol Engrs; Am Geophys Union. *Res:* Underground fluid and heat storage, disposal of hazardous waste; geohydrochemical modeling; waste management. *Mailing Add:* 2205 Carriage Ave Richland WA 99352

STOTTS, JANE, b Dallas, Tex, Sept 15, 39. IMMUNOLOGY, MICROBIOLOGY. *Educ:* Univ Tex, Austin, BA, 61; Baylor Univ, MS, 64. *Prof Exp:* RES MGR MICROBIOL & IMMUNOL, PROCTER & GAMBLE CO, 64- *Res:* Allergic contact dermatitis, predictive testing and identification of allergens; primary irritant dermatitis; microflora of skin; hospital infection control. *Mailing Add:* Nine Falling Bark Cincinnati OH 45241

STOTZ, ROBERT WILLIAM, b Monroe, Mich, July 18, 42; m 71; c 1. RESEARCH ADMINISTRATION. *Educ:* Univ Toledo, BS, 64, MS, 66; Univ Fla, PhD(inorg chem), 70. *Prof Exp:* Res assoc, Mich State Univ, 70-71; instr chem, Eastern Mich Univ, 72-72; asst prof, Mercer Univ, 72-73; asst prof, Tri-State Col, 73-74; srpvr inorg anal res, Inst Gas Technol, 74-76; mgr anal chem, 76-79; mgr, validation, Upjohn Co, 79-90; MGR, VALIDATION, JACOBS ENG GROUP, 91- *Mem:* Am Chem Soc; Parenteral Drug Asn. *Res:* Development and modification of various instrumental methods for determination of pharmaceutical constituents. *Mailing Add:* Jacobs Eng Group 1880 Waycross Rd Cincinnati OH 45240

STOTZKY, GUENTHER, b Leipzig, Ger, May 24, 31; nat US; m 58; c 3. MICROBIAL ECOLOGY. *Educ:* Calif Polytech State Univ, BS, 52; Ohio State Univ, MS, 54, PhD(agron & microbiol), 56. *Prof Exp:* Res asst soil biochem & microbiol, Ohio State Univ, 53-56; res assoc bot & plant nutrit, Univ Mich, 56-58; head soil microbiol, Cent Res Labs, United Fruit Co, 58-63; microbiologist & chmn, Kitchawan Res Lab, Brooklyn Bot Garden, 63-68; assoc prof biol, 68-70, adj assoc prof, 67-68, chmn dept, 70-77, PROF BIOL, NY UNIV, 70- *Concurrent Pos:* Spec scientist, Argonne Nat Lab, 55; mem, Am Inst Biol Sci-NASA Regional Coun, 65-68; regional ed, Soil Biol & Biochem, 69-; assoc ed, Appl Environ Microbiol, 71-77 & Can J Microbiol, 71-75; mem ad hoc comt rev biomed & ecol effects of extremely low frequency radiation, Bur Med & Surg, Dept Navy, 72-75; vis prof, Inst Advan Studies, Polytech Inst, Mex, 73, Cath Univ, Santiago, Chile, 81, & Moscow State Univ, 90; mem, Comn Human Res, Nat Res Coun, 75-77; mem, Environ Biol Rev Panel, US Environ Protection Agency, 80-; mem, Controlled Ecol Life Support Systs Prof Rev Panel, NASA, Am Inst Biol Sci, 80-87; res assoc, Nat Ctr Sci Res, Poiters Univ, France, 82; guest lectr, Australian Soc Microbiol Found, 84; distinguished vis scientist, US Environ Protection Agency, 86-89; ser ed, Marcel Dekker Inc, 86- *Honors & Awards:* Selman A Waksman Hon Lect Award, Theobald Smith Soc, 89; Fisher Sci Co Award for Appl Environ Microbiol, Am Soc Microbiol, 90. *Mem:* Fel AAAS; Soc Environ Geochem & Health; Am Soc Microbiol; Can Soc Microbiol; Int Soil Sci Soc; fel Am Acad Microbiol; fel Soil Sci Soc Am; fel Am Soc Agron; Bot Soc Am; Am Inst Biol Sci. *Res:* Microbiol ecology; surface interactions; soil, water and air pollution; environmental microbiology, virology and immunology; ecotoxicology; fate gene transfer and effects of genetically engineered microbes in natural habitats; clinical microbiology. *Mailing Add:* Dept Biol NY Univ Wash Sq New York NY 10003

STOUDT, EMILY LAWS, b Columbus, Ohio, Apr 5, 43; m 67, 77; c 2. GEOLOGY. *Educ:* Ohio State Univ, BA, 66, PhD(geol), 75; La State Univ, MS, 68. *Prof Exp:* Geologist, Spec Proj Br, US Geol Surv, 68-70; geologist, Explor & Prod Res Lab, Getty Oil Co, 75-81, GEOLOGIC SUPVR, GETTY RES CTR, 81-; MGR GEOLOGIC RES, TEXACO HOUSTON RES CTR, 84- *Mem:* Geol Soc Am; Am Asn Petrol Geologists; Soc Econ Paleontologists & Mineralogists. *Res:* Carbonate petrology; regional geology of the mid-continent Silurian, western United States Permian, Gulf Coast Jurassic-Cretaceous systems, Guatemalan Jurassic-Cretaceous. *Mailing Add:* Texaco Res Ctr 3901 Briarpark Dr Houston TX 77042

STOUDT, HOWARD WEBSTER, b Pittsburgh, Pa, May 13, 25; m 53; c 2. HUMAN FACTORS ENGINEERING, ERGONOMICS. *Educ:* Harvard Univ, AB, 49, SM, 62; Univ Pa, AM, 53, PhD(anthrop), 59. *Prof Exp:* Res asst, Sch Pub Health, Harvard Univ, 52-55; res & educ specialist, Air Univ, 55-57; res assoc phys anthrop, Sch Pub Health, Harvard Univ, 57-66, asst prof, 66-73; prof & chmn dept community med, Col Osteop Med, Mich State Univ, 73-77, prof community health sci, 78-; RETIRED. *Concurrent Pos:* Consult, Nat Health Exam Surv, USPHS Ctr Dis Control, Soc Automotive Engrs, NASA & pvt indust; res investr, Normative Aging Study, Vet Admin, Boston. *Mem:* Human Biol Coun; Am Asn Phys Anthrop; Human Factors Soc; Soc Epidemiol Res; AAAS; Soc Med Anthrop. *Res:* Physical anthropology; epidemiology of non-infectious disease and accidents; application of human biological data to the design of equipment and workspaces with special reference to safety and health. *Mailing Add:* RR 2 No 693 Wiscasset ME 04578

STOUDT, THOMAS HENRY, b Temple, Pa, Apr 6, 22; m 43; c 3. MICROBIOLOGY. *Educ:* Albright Col, BS, 43; Rutgers Univ, MS, 44; Purdue Univ, West Lafayette, PhD(org chem), 49. *Prof Exp:* Asst chem, Rutgers Univ, 43-44 & Purdue Univ, West Lafayette, 46-47; sr microbiologist, 49-58, from sect head to sr sect head, 58-69, dir appl microbiol, 69-75, sr dir appl microbiol & nat prod isolation, 75-77, EXEC DIR, MERCK & CO, INC, 77- *Mem:* Am Chem Soc; Soc Indust Microbiol; Int Asn Dent Res; Am Soc Microbiologists; NY Acad Sci. *Res:* Microbial transformations; microbial biosyntheses; antibiotics; microbial physiology and genetics, rumen microbiology; oral microbiology; microbial enzymology; vaccines and immunology; scale-up of industrial fermentation processes. *Mailing Add:* 857 Village Green Westfield NJ 07090-3515

STOUFER, ROBERT CARL, b Ashland, Ohio, Nov 3, 30; m 54; c 2. INORGANIC CHEMISTRY. *Educ:* Otterbein Col, BA & BS, 52; Ohio State Univ, PhD, 59. *Prof Exp:* ASSOC PROF INORG CHEM, UNIV FLA, 58- *Mem:* Am Chem Soc; Sigma Xi. *Res:* Synthesis of inorganic complexes and their characterization, particularly of spectroscopic and magnetic properties; investigations of Jahn-Teller prone systems. *Mailing Add:* Dept of Chem Univ of Fla Gainesville FL 32611

STOUFFER, DONALD CARL, b Philadelphia, Pa, May 15, 38; m 62; c 2. ENGINEERING MECHANICS. *Educ:* Drexel Univ, BSME, 61, MSME, 65; Univ Mich, Ann Arbor, PhD(eng mech), 68. *Prof Exp:* Engr, Philco-Ford Corp, 61-63 & Westinghouse Elec Corp, 63-65; instr mech, Univ Mich, Ann Arbor, 66-68, lectr, 68-69; from asst prof to assoc prof eng sci, 69-77, fac fel, 71, PROF ENG SCI, UNIV CINCINNATI, 77- *Concurrent Pos:* US-Australian Coop Sci fel, Univ Melbourne, 76; vis scientist, Wright-Patterson AFB, Ohio, 75-; res grants, Nat Sci Found & Air Force Wright Aeronaut Lab. *Mem:* Soc Rheol; Soc Nat Philos; Am Soc Mech Engrs; Am Acad Mech. *Res:* Theoretical and applied mechanics; rheology; constitutive equations. *Mailing Add:* 6982 Driftwood Cincinnati OH 45241

STOUFFER, JAMES L, b Harrisburg, Pa, Sept 25, 35; m 68. AUDIOLOGY, PSYCHOACOUSTICS. *Educ:* State Univ NY Col Buffalo, BSc, 64; Pa State Univ, MSc, 66, PhD(audiol, statist), 69. *Prof Exp:* Clin audiologist, Pa State Univ, 64-69; asst prof audiol, Univ Western Ont, 69-70, dir Commun Dis & Chief Speech & Hearing Serv, Univ Western Ont & Univ Hosp, London, On, 72-81; CONSULT, 82- *Concurrent Pos:* NIH fel psychoacoust, Commun Sci Lab, Univ Fla, 69-70; consult, Oxford County Ment Health Centre, 70-; ed asst, Ont Speech & Hearing Asn, 71- *Mem:* Am Speech & Hearing Asn; Acoust Soc Am. *Res:* Tinnitus measurement and handicap; central auditory processing disorders. *Mailing Add:* Dept Commun Disorder Elborn Col Univ Western Ont London ON N6G 1H1 Can

STOUFFER, JAMES RAY, b Glen Elder, Kans, Jan 12, 29; m 55; c 2. ANIMAL SCIENCE. *Educ:* Univ Ill, BS, 51, MS, 53, PhD(meats), 56. *Prof Exp:* Asst prof animal husb, Univ Conn, 55-56; asst prof animal husb, 56-62, assoc prof animal sci, 62-77, PROF ANIMAL SCI, CORNELL UNIV, 77- *Mem:* Am Soc Animal Sci; Inst Food Technol; Am Meat Sci Asn. *Res:* Carcass evaluation of meat animals, particularly the relationship of live animal and carcass characteristics. *Mailing Add:* Dept Animal Sci Cornell Univ Agr Exp Sta Ithaca NY 14850

STOUFFER, JOHN EMERSON, b Sioux City, Iowa, Dec 4, 25; m 55; c 2. BIOCHEMISTRY. *Educ:* Northwestern Univ, BS, 49; Boston Univ, PhD(org chem), 57. *Prof Exp:* Res assoc biochem, Med Col, Cornell Univ, 57-59, instr, 59-61; asst prof, 61-66, ASSOC PROF BIOCHEM, BAYLOR COL MED, 67- *Mem:* Am Chem Soc; Am Oil Chemists Soc; Am Soc Biol Chemists; Endocrine Soc; Am Inst Chemists. *Res:* Thyroid hormones; structure function relationships of hormones and mechanism of action; membrane receptor sites. *Mailing Add:* 2703 Albans Houston TX 77005

STOUFFER, RICHARD FRANKLIN, b Welch, WVa, July 3, 32; m 57; c 2. PLANT PATHOLOGY, VIROLOGY. *Educ:* Vanderbilt Univ, BA, 54; Cornell Univ, PhD(plant path), 59. *Prof Exp:* Asst, Cornell Univ, 54-59; asst prof, Univ RI, 59-60; asst virologist, Univ Fla, 61-65; from asst prof to prof plant path, Fruit Res Lab, Pa State Univ, 76-82; PROF & HEAD DEPT PLANT PATH, UNIV GA, 82- *Mem:* Am Phytopath Soc; Brit Asn Appl Biol; Asn Appl Biol; Int Comt Fruit Tree Virus Res (secy, 82-85). *Res:* Plant virology; virus diseases of deciduous fruit trees. *Mailing Add:* Coastal Plain Exp Sta PO Box 748 Tifton GA 31793

STOUFFER, RICHARD LEE, b Hagerstown, Md, July 27, 49; m; c 2. ENDOCRINOLOGY, REPRODUCTIVE PHYSIOLOGY. *Educ:* Va Polytech Inst, BS, 71; Duke Univ, PhD(physiol), 75. *Prof Exp:* Staff fel endocrinol, Reprod Res Br, Nat Inst Child Health & Human Develop, NIH, 75-77; from asst prof to assoc prof physiol, Col Med, Univ Ariz, 77-85; SCIENTIST, ORE PRIMATE CTR, 85- *Concurrent Pos:* Prin invest, PHS Grants, 78-; Mem, NIH Study Sect, 83, 85, Ed, 87. *Honors & Awards:* Res Career Develop Award, NIH, 82. *Mem:* Soc Study Reprod; AAAS; Endocrine Soc; Am Physiol Soc; Am Fertility Soc. *Res:* Female reproductive endocrinology, with emphasis on the regulation of primate ovarian function; regulation of the corpus luteum. *Mailing Add:* Div Reproductive Biol & Behavior Ore Reg Primate Res Ctr 505 NW 185th Ave Beaverton OR 97006-3448

STOUFFER, RONALD JAY, b Hershey, Pa, Feb 3, 54; m 76; c 3. CLIMATE MODELING, CLIMATE DYNAMICS. *Educ:* Pa State Univ, BS, 76, MS, 77. *Prof Exp:* RES METEOROLOGIST, GEOPHYS FLUID DYNAMICS LAB, NAT OCEANIC & ATMOSPHERIC ADMIN, DEPT COM, 77- *Concurrent Pos:* Mem, Steering Group Global Climate Modeling, Joint Sci Comt, World Meteorol Orgn, 90- *Mem:* Am Meteorol Soc. *Res:* Study of climate variations using mathematical climate models; climate change resulting from increasing greenhouse gases using coupled ocean-atmosphere models. *Mailing Add:* Geophys Fluid Dynamics Lab NOAA Princeton Univ Princeton NJ 08542

STOUGHTON, RAYMOND WOODFORD, b Tehachapi, Calif, Aug 6, 16; m 41; c 4. PHYSICAL CHEMISTRY, NUCLEAR CHEMISTRY. *Educ:* Univ Calif, BS, 37, PhD(chem), 40. *Prof Exp:* Asst chem, Univ Calif, 37-40, res chemist, Radiation Lab, 41-43; instr, Agr & Mech Col, Tex, 40-41; RES CHEMIST, UNION CARBIDE NUCLEAR CO, OAK RIDGE NAT LAB, 43- *Concurrent Pos:* Res chemist, Metall Lab, Univ Chicago, 43. *Mem:* AAAS; Am Chem Soc; Am Phys Soc; Am Nuclear Soc; NY Acad Sci. *Res:* Reaction kinetics; application of computers to chemical and physical problems; solution thermodynamics; solution chemistry of heavy elements; radiochemistry; process development; neutron cross sections; search for superheavy elements in nature and in accelerator targets. *Mailing Add:* 104 Plymouth Circle Oak Ridge TN 37830

STOUGHTON, RICHARD BAKER, b Duluth, Minn, July 4, 23; m 46; c 1. DERMATOLOGY. *Educ:* Univ Chicago, SB, 45, MD, 47; Am Bd Dermat, dipl, 52. *Prof Exp:* From instr to asst prof dermat, Univ Chicago, 50-56; assoc prof med dir dermat, Sch Med, Case Western Reserve Univ, 57-67; HEAD DEPT DERMAT, SCRIPPS CLIN & RES FOUND, 67-; PROF DERMAT & CHIEF DIV, UNIV CALIF, SAN DIEGO, 75- *Concurrent Pos:* Consult, US Army Chem Ctr, Md, 54-58; mem subcomt dermat, Nat Res Coun, 59-; ed-in-chief, Soc Invest Dermat, 67; dir, Am Bd Dermat. *Mem:* Soc Invest Dermat; Am Acad Dermat. *Res:* Histochemistry, percutaneous absorption and pathologic anatomy of dermatology. *Mailing Add:* Sch Med M-023C Univ Calif San Diego LaJolla CA 92093

STOUT, BARBARA ELIZABETH, b Anchorage, Alaska, May 29, 62; m 90. NUCLEAR CHEMISTRY, RADIOCHEMISTRY. *Educ:* Wesleyan Col, AB, 83; Fla State Univ, MS, 85, PhD(inorg chem), 89. *Prof Exp:* Res assoc, Inst Curie, Paris, France, 85-86 & 87-88; postdoctoral asst, Univ Lausanne, Switz, 89-90; ASST PROF CHEM, UNIV CINCINNATI, 90- *Mem:* Sigma Xi; Am Chem Soc. *Res:* Actinide and lanthanide solution chemistry; luminescence spectroscopy; environmental chemistry of the f-elements; interactions of polyelectrolytes with the f-elements. *Mailing Add:* Dept Chem ML 172 Univ Cincinnati Cincinnati OH 45221-0172

STOUT, BENJAMIN BOREMAN, b Parkersburg, WVa, Mar 2, 24; m 45; c 3. FOREST ECOLOGY. *Educ:* WVa Univ, BSF, 47; Harvard Univ, MF, 50; Rutgers Univ, PhD, 67. *Prof Exp:* Forester, Pond & Moyer Co, 47-49; silviculturist, Harvard Black Rock Forest, 50-55, supvr, 55-59; from asst prof to prof forestry, Rutgers Univ, New Brunswick, 59-77, chmn dept biol sci, 74-77, assoc provost, 77-78; dean Sch Forestry, Univ Mont, 78-85; MEM, NAT COUN PAPER INDUST AIR & STREAM IMPROV, 85- *Concurrent Pos:* Mem, Nat Coun Paper Indust for Air & Stream Improv, 85- *Mem:* Ecol Soc Am; fel Soc Am Foresters; Sigma Xi. *Res:* Ways and means of quantifying vegetations response to environment. *Mailing Add:* 1545 Takena SW Albany OR 97321

STOUT, BILL A(LVIN), b Grant, Nebr, July 9, 32; m 51; c 2. AGRICULTURAL ENGINEERING. *Educ:* Univ Nebr, BS, 54; Mich State Univ, MS, 55, PhD(agr eng), 59. *Prof Exp:* Asst, Mich State Univ, 54-56, from instr to prof agr eng, 56-81, chmn dept, 70-75; PROF AGR ENG, TEX A&M UNIV, 81- *Concurrent Pos:* Farm power & machinery specialist, Food & Agr Orgn, UN, Rome, Italy, 63-64; sabbatical leave, dept agr eng, Univ Calif, Davis, 69-70. *Honors & Awards:* Kishida Int Award, Am Soc Agr Engrs. *Mem:* Fel Am Soc Agr Engrs; Nat Soc Prof Engrs; fel AAAS. *Res:* Harvesting machines; physical properties of agricultural products; mechanization in developing countries; energy awareness and alternatives; energy use & management in agriculture; conservation alternatives; teaching, research and administration; international agriculture. *Mailing Add:* Dept Agr Eng Tex A&M Univ College Station TX 77843

STOUT, CHARLES ALLISON, b Beaumont, Tex, Sept 20, 30. ORGANIC CHEMISTRY. *Educ:* Rice Inst, BA, 52, MA, 53; Ohio State Univ, PhD(chem), 62. *Prof Exp:* Chemist, Goodyear Tire & Rubber Co, Ohio, 53-54; NIH res fel photochem, Calif Inst Technol, 62-63; res chemist, Chevron Oil Field Res Co, 64-72; RES CHEMIST, DIVERSIFIED CHEM CORP, 72- *Mem:* Am Chem Soc; Soc Petrol Engrs; Sigma Xi. *Res:* Physical organic chemistry, reaction mechanisms; molecular orbital treatments of condensed aromatic systems; photochemical reactions in solutions and solids, photosensitization; structure of interfacial films. *Mailing Add:* 17621 E 17th St-31A Tustin CA 92680

STOUT, DARRYL GLEN, b Carman, Man, Mar 21, 44; m 75; c 2. AGRONOMY, PLANT PHYSIOLOGY. *Educ:* Univ Man, BSA, 69, MSc, 72; Cornell Univ, PhD(plant physiol), 76. *Prof Exp:* Res assoc drought tolerance, Univ Sask, 75-77; RES SCIENTIST FORAGE PROD, AGR CAN, 77- *Concurrent Pos:* Vis fel, Cornell Univ, 82-83. *Mem:* Am Soc Plant Physiologists; Can Soc Plant Physiologists; Can Soc Agron. *Res:* Forage management; seeding rate and cutting management on alfalfa persistence and yield; double cropping and intercropping for hay production and grazing; growth and survival of plants under stress conditions of frost, drought and grazing. *Mailing Add:* Agr Can 3015 Ord Rd Kamloops BC V2B 8A9 Can

STOUT, DAVID MICHAEL, b Flint, Mich, Nov 20, 47; m 69; c 2. ORGANIC CHEMISTRY, RESEARCH MANAGEMENT. *Educ:* Col Wooster, BA, 69; Univ Rochester, MS, 72; Colo State Univ, PhD(org chem), 74. *Prof Exp:* NIH fel, Yale Univ, 74-76; group leader cardiovasc res, Du Pont Critical Care Div, E I du Pont de Nemours & Co, 76-88; sr res scientist neurosci res, 88-90, RES INVESTR, ABBOTT LABS, 90- *Mem:* Am Chem Soc. *Res:* Neuroscience; organic synthesis. *Mailing Add:* Abbott Labs Abbott Park IL 60064

STOUT, EDGAR LEE, b Grants Pass, Ore, Mar 13, 38; m 58; c 1. MATHEMATICS. *Educ:* Ore State Col, BA, 60; Univ Wis, MA, 61, PhD(math), 64. *Prof Exp:* Instr math, Yale Univ, 64-65, asst prof, 65-69, Off Naval Res res assoc, 67-68; assoc prof, 69-74, PROF MATH, UNIV WASH, 74- *Concurrent Pos:* Vis prof math, Univ Leeds, 72-73. *Mem:* Math Asn Am; Am Math Soc. *Res:* Functions of one or several complex variables; function algebras. *Mailing Add:* Dept Math Univ Wash Seattle WA 98195

STOUT, EDWARD IRVIN, b Washington Co, Iowa, Mar 2, 39; c 3. ORGANIC CHEMISTRY. *Educ:* Iowa Wesleyan Col, BS, 60; Bradley Univ, MS, 68; Univ Ariz, PhD(org chem), 74. *Prof Exp:* Chemist, Lever Bros Co, 61-62; res chemist, Northern Regional Res Ctr, USDA, 62-78; dir res, Spenco Med Corp, 78-80; consult, 81-82; mem staff, 82-86, DIR RES, CHEMSTAR PROD CO, 81-; PRES, SW TECHNOLOGIES, INC. *Concurrent Pos:* Instr org chem, Bradley Univ, 70-75. *Mem:* Am Chem Soc (secy, 75). *Res:* Preparation and characterization of starch derivatives including starch graft copolymers; development of absorbent copolymers for wound dressing; development of hydrogel wound dressings; development of hot/cold therapy products; development of wheel chair cushions and bed pads (hospital). *Mailing Add:* Southwest Technol Inc 2018 Baltimore Kansas City MO 64108-1914

STOUT, ERNEST RAY, b Boone, NC, Oct 31, 38; m 61; c 3. MOLECULAR BIOLOGY, BIOCHEMISTRY. *Educ:* Appalachian State Univ, BS, 61; Univ Fla, PhD(bot & biochem), 65. *Prof Exp:* Nat Cancer Inst fel biochem genetics, Univ Md, 65-67; from asst prof to assoc prof, 67-84, asst dean, Col Arts & Sci, 78-79, PROF MOLECULAR BIOL, VA POLYTECH INST & STATE UNIV, 84-, HEAD, DEPT BIOL, 80- *Mem:* AAAS; Am Soc Plant Physiol. *Res:* Mechanism of nucleic acid synthesis in higher plants; control of nucleic acid synthesis; parovirus macromolecular synthesis. *Mailing Add:* Dept Biol Res Div 0244 Va Polytech Inst & State Univ Blacksburg VA 24061

STOUT, GLENN EMANUEL, b Fostoria, Ohio, Mar 23, 20; m 42; c 2. METEOROLOGY, HYDROLOGY & WATER RESCOURCES. *Educ:* Findlay Col, BS, 42; Univ Chicago, cert, 43. *Hon Degrees:* DSc, Findlay Col, 73. *Prof Exp:* Asst math, Findlay Col, 39-42; instr meteorol, Univ Chicago, 42-43 & US War Dept, Chanute AFB, Ill, 46-47; asst engr, Ill State Water Surv, 47-52, head atmospheric sci sect, 52-71, asst to chief, 71-74, DIR WATER RESOURCES CENTER, ILL STATE WATER SURV, UNIV ILL, URBANA, 73-, PROF METEOROL, INST ENVIRON STUDIES, UNIV, 73- *Concurrent Pos:* Consult, Crop-Hail Ins Actuarial Assoc, Ill, 61-69; prog coordr, Nat Ctr Atmospheric Res, NSF, 69-71; ed-chief, Water Int, 82-86; bd dir, Univ Coun Water Resources, 83-86; exec dir, Int Asn Water Resources, 84-, sec gen, 86- *Mem:* Am Meteorol Soc; Am Geophys Union; Am Water Resources Asn; AAAS; Int Asn Water Resources (secy gen, 86-); Int Asn Hydrol Res; Am Water Works Asn; N Am Lake Mgt Asn; Sigma Xi. *Res:* Hail climatology; weather modification; water resources; environmental science; environmental management; hydrology and water resources. *Mailing Add:* Univ of Ill Water Resources Ctr 205 N Mathews Urbana IL 61801-2397

STOUT, ISAAC JACK, b Clarksburg, WVa, July 20, 39; m 64; c 2. ECOLOGY. *Educ:* Ore State Univ, BS, 61, Va Polytech Inst & State Univ, MS, 67; Wash State Univ, PhD(zool), 72. *Prof Exp:* Wildlife Mgt Inst fel waterfowl ecol, Va Coop Wildlife Res Univ, 64-65; field ecologist, Old Dominion Univ, 65-67; USPHS fel appl ecol, Wash State Univ, 67-69; from asst prof to assoc prof, 72-83, PROF BIOL SCI, UNIV CENT FLA, 83- *Concurrent Pos:* Mem, Environ Effect & Fate Solid Rocket Emission Prod, NASA, Kennedy Space Ctr, 75. *Mem:* Ecol Soc Am; Brit Ecol Soc; Wildlife Soc; Am Soc Mammalogists; Sigma Xi. *Res:* Population and community ecology; ecology of sand pine scrub; conservation biology. *Mailing Add:* Dept of Biol Sci Univ Cent Fla Box 25000 Orlando FL 32816

STOUT, JOHN FREDERICK, b Takoma Park, Md, Jan 20, 36; m 56; c 2. ETHOLOGY, ZOOLOGY. *Educ:* Columbia Union Col, BA, 57; Univ Md, PhD(zool), 63. *Prof Exp:* Instr biol, Walla Walla Col, 62, asst prof, 63-65, dir marine sta, 64-69, assoc prof biol, 66-69; assoc prof, 69-70, PROF BIOL, ANDREWS UNIV, 70-, CHMN BIOL, 83- *Concurrent Pos:* USPHS spec fel & vis researcher, Univ Cologne, 69-70; guest res prof, Max Planck Inst Behav Physiol, 75-76. *Honors & Awards:* Alexander von Humboldt Sr US Scientist Award, 75. *Mem:* Sigma Xi; Soc Neurosci. *Res:* Neurobiology of acoustic communication; communication during social behavior; behavioral physiology. *Mailing Add:* Dept of Biol Andrews Univ Berrien Springs MI 49104

STOUT, JOHN WILLARD, b Seattle, Wash, Mar 13, 12; m 48; c 1. PHYSICAL CHEMISTRY, CHEMICAL PHYSICS. *Educ:* Univ Calif, BS, 33, PhD(phys chem), 37. *Prof Exp:* Instr chem, Univ Calif, 37-38, Lalor fel, 38-39; instr chem, Mass Inst Technol, 39-41; investr, Nat Defense Res Comt, Univ Calif, 41-44; group leader, Manhattan Dist, Los Alamos Sci Lab, 44-46; from assoc prof to prof, 46-77, EMER PROF CHEM, UNIV CHICAGO, 77- *Concurrent Pos:* Ed, J Chem Physics, 59-82, consult ed, 83-; mem coun, Am Phys Soc, 72-76. *Honors & Awards:* Huffman Mem Award, 60. *Mem:* AAAS; Am Chem Soc; fel Am Phys Soc. *Res:* Thermodynamics; calorimetry; crystal spectra; cryogenics; paramagnetism and antiferromagnetism. *Mailing Add:* Dept Chem Univ Chicago 5735 S Ellis Ave Chicago IL 60637

STOUT, KOEHLER, b Deer Lodge, Mont, Sept 1, 22; m 50; c 3. MINING & GEOLOGICAL ENGINEERING. *Educ:* Mont Sch Mines, BS, 48, MS, 49; La Salle Exten Univ, LLB, 57. *Hon Degrees:* DEng, Mont Univ. *Prof Exp:* Asst prof mining eng, Mont Col Mineral Sci & Technol, 52-58, from assoc prof to prof eng sci, 58-84, head dept, 62-84, dean, div eng, 66-84; RETIRED. *Concurrent Pos:* Consult, minerals indust. *Mem:* Am Soc Eng Educ; Nat Soc Prof Engrs; Am Inst Mining, Metall & Petrol Engrs. *Res:* Portland and chemical cements injected into weak, unstable ground to prepare the ground for mining. *Mailing Add:* 1327 W Granite Butte MT 59701

STOUT, LANDON CLARKE, JR, b Kansas City, Mo, Feb 20, 33; m 54, 81; c 5. PATHOLOGY, INTERNAL MEDICINE. *Educ:* Univ Md, MD, 57. *Prof Exp:* Resident internal med, Med Ctr, Univ Okla, 58-61, asst prof, 63-72, dir inst comp path, 65-69, resident path, 66-67, from asst prof to assoc prof, 68-72, interim chmn dept, 70-72; assoc prof, 72-74, PROF PATH, UNIV TEX MED BR GALVESTON, 74- *Concurrent Pos:* Nat Heart Inst spec fel, Univ Okla, 67-68; consult, Okla Med Res Found, 71-72. *Mem:* Am Asn Path; Am Col Physicians; Am Gastroenterol Asn; Am Heart Asn; Am Diabetes Asn; Int Acad Pathol. *Res:* Atherosclerosis; diabetic renal disease; mitral valve disease. *Mailing Add:* Dept Path Univ Tex Med Br Galveston TX 77550

STOUT, MARGUERITE ANNETTE, b Marion, NC, July 17, 43. PHYSIOLOGY. *Educ:* Univ Wis, BS, 64; Univ Iowa, PhD(physiol & biophys), 74; Pace Univ, MBA, 86. *Prof Exp:* Res scientist I, Galesburg State Res Hosp, 65-70; fel, Univ Iowa, 74-75; asst prof, 75-82, ASSOC PROF PHYSIOL, NJ MED SCH, UNIV MED & DENT NJ, 82- *Honors & Awards:* Nat Heart Lung & Blood Inst Young Investr Award, 78-81. *Res:* Calcium regulation by sarcoplasmic reticulum in chemically skinned vascular smooth muscle. *Mailing Add:* Dept Physiol NJ Med Sch Univ Med & Dent NJ Newark NJ 07103-2425

STOUT, MARTIN LINDY, b North Hollywood, Calif, Feb 11, 34; m 56; c 2. GEOLOGY. *Educ:* Occidental Col, BA, 55; Univ Wash, MS, 57, PhD(geol), 59. *Prof Exp:* Asst geol, Occidental Col, 53-55 & Univ Wash, 55-59; from asst prof to assoc prof, 60-71, PROF GEOL, CALIF STATE UNIV, LOS ANGELES, 71-; SR ENG GEOLOGIST, MOORE & TABER ENGRS, GEOLOGISTS, 63- *Concurrent Pos:* Econ geologist, Aerogeophysics, Inc, 55; Geol Soc Am Penrose res grant, 58; partic, Int Field Inst, Scandinavia, Am Geol Inst, 63; NSF grant, 65, fel, Iceland & Norway, 66-67. *Mem:* AAAS; Asn Eng Geol; Geol Soc Am; Am Asn Geol Teachers. *Res:* Geochemical studies of basalts in Washington, Iceland, Norway; gravity movements and rates of tectonism in southern California; distribution and mechanism of failure of landslides in the capistrano formation of southern California; engineering properties of volcanic rocks. *Mailing Add:* Dept Geol Calif State Univ 5151 State Univ Dr Los Angeles CA 90032

STOUT, QUENTIN FIELDEN, b Cleveland, Ohio, Sept 23, 49; div; c 2. PARALLEL COMPUTING. *Educ:* Centre Col, BA, 70; Ind Univ, PhD(math), 77. *Prof Exp:* From asst prof to assoc prof comput sci & math, State Univ NY, Binghamton, 76-84; ASSOC PROF COMPUT SCI, UNIV MICH, 84- *Concurrent Pos:* Consult, Parker-Hannifin Corp, 67-72; prin investr, numerous grants, 78- *Mem:* Asn Comput Mach; Am Math Soc; Inst Elec & Electronic Engrs. *Res:* Design and analysis of parallel algorithms; design and analysis of serial algorithms; scientific computing; parallel programming environments; parallel computers. *Mailing Add:* Dept Elec Eng & Comput Sci Univ Mich Ann Arbor MI 48109-1109

STOUT, RAY BERNARD, b Georgetown, Ohio, June 16, 39; m 65; c 1. THERMODYNAMICS OF MATERIAL DEFECTS. *Educ:* Ohio State Univ, BS, 64, MS, 68; Ill Inst Technol, PhD(eng mech), 70; Univ Pittsburgh, MBA, 72. *Prof Exp:* Apprentice, Cincinnati Milling Mach Co, 57-59; fel engr, Bettis Atomic Power Lab, Westinghouse Elec Corp, West Mifflin, 69-79; TECH AREA LEADER, RADIOACTIVE SPENT FUEL & DEFENSE HIGH LEVEL WASTE FORMS, LAWRENCE LIVERMORE NAT LAB, LIVERMORE, 79- *Mem:* Am Soc Mech Engrs; Am Phys Soc; Mat Res Soc. *Res:* Applications of numerical analysis and applied mathematics to engineering mechanics and physics of materials. *Mailing Add:* 954 Venus Way Livermore CA 94550

STOUT, ROBERT DANIEL, b Reading, Pa, Jan 2, 15; m 39; c 1. METALLURGY. *Educ:* Pa State Col, BS, 35; Lehigh Univ, MS, 41, PhD(metall), 44. *Hon Degrees:* ScD, Albright Col, 67. *Prof Exp:* Asst, Carpenter Steel Co, 35-39; instr, 39-45, from asst prof to assoc prof, 45-50, prof metall, 50-80, head dept metall, 56-60, dean grad sch, 60-80, PROF, LEHIGH UNIV, 81- *Concurrent Pos:* Deleg, Int Inst Welding, Am Welding Soc, 55-; mem mat adv bd, Nat Acad Sci, 64-68; mem naval ship lab adv bd, Dept Navy, 68-74; mem pipeline safety adv comt, 69-72. *Honors & Awards:* Lincoln Gold Medal, Am Welding Soc, 43, Spraragen Award, 64, Thomas Award, 75 & Jennings Award, 74, Houdremont lectr, 70; Adams lectr, 60; Hobart Medal, 81; Savage Award, 87. *Mem:* Fel Am Soc Metals; Am Welding Soc (pres, 72-73). *Res:* Notch toughness and plastic fatigue properties of steel; weldability of steel. *Mailing Add:* 546 Whitaker Lab No Five Lehigh Univ Bethlehem PA 18015

STOUT, ROBERT DANIEL, b Aug 20, 45; m 83. IMMUNOPATHOLOGY. *Educ:* Univ Mich, PhD(immunol), 71. *Prof Exp:* Res fel, dept pathology, Harvard Med Sch, 71-73 & dept genetics, Stanford Med Sch, 73-76; asst prof, Rosenstiel Res Ctr, Brandeis Univ, 76-83; ASSOC PROF MED IMMUNOL, QUILLEN-DISHNER COL MED, E TENN STATE UNIV, 83-, ASSOC CHMN, DEPT MICROBIOL, 85-, DIR FLOW CYTOCHEM RESOURCE, 84- *Concurrent Pos:* Assoc prof, Grad Fac Prog Molecular Biol, Quillen Dishner Col Med, E Tenn State Univ. *Mem:* AAAS; NY Acad Sci; Am Asn Immunologists. *Res:* T cell-macrophage interaction; cytokines; transmembrance signaling; regulation of immune responses. *Mailing Add:* Dept Microbiol Quillen-Dishner Col Med E Tenn State Univ Box 19870A Johnson City TN 37614-0002

STOUT, THOMAS MELVILLE, b Ann Arbor, Mich, Nov 26, 25; m 47; c 6. ELECTRICAL ENGINEERING, AUTOMATIC CONTROL SYSTEMS. *Educ:* Iowa State Col, BS, 46; Univ Mich, Ann Arbor, MSE, 47, PhD(elec eng), 54. *Prof Exp:* Jr engr, Emerson Elec Mfg Co, 47-48; instr elec eng, Univ Wash, 48-53, asst prof, 53-54; res engr, Schlumberger Instrument Co, 54-56; mgr process anal, TRW Comput Div, 56-64 & Bunker-Ramo Corp, 64-65; pres, Profimatics, Inc, 65-83; CONSULTANT, TMS ASSOCS, 84- *Honors & Awards:* Hon mem, Instrument Soc Am, 90. *Mem:* Instrument Soc Am; sr mem Inst Elec & Electronics Engrs; Am Inst Chem Engrs; Tech Asn Pulp & Paper Indust; Nat Soc Prof Engrs; Soc Computer Simulation. *Res:* Application of computers for simulation and control of industrial processes; application of systems engineering techniques to social problems. *Mailing Add:* TMS Assocs 9927 Hallack Ave Northridge CA 91324

STOUT, THOMPSON MYLAN, b Big Springs, Nebr, Aug 16, 14; m 40. GEOLOGY, VERTEBRATE PALEONTOLOGY. *Educ:* Univ Nebr, Lincoln, BSc, 36, MSc, 37. *Prof Exp:* Res asst vert paleont, State Mus, 33-38, from instr to prof, 38-57, assoc prof, 57-68, prof, 68-80, EMER PROF GEOL, UNIV NEBR, LINCOLN, 80-; assoc cur, 57-80, EMER, STATE MUS, 80- *Concurrent Pos:* Res assoc, Frick Lab, Am Mus Natural Hist, New York, NY, 38-; studies of fossil rodents & geol in Europ museums, 48-79; corresp, Nat Mus Natural Hist, Paris, 66. *Mem:* Fel Geol Soc Am; Soc Vert Paleont; Paleont Soc; Am Soc Mammal; NY Acad Sci; Sigma Xi. *Res:* Stratigraphy and vertebrate paleontology, with special reference to the Tertiary and Quaternary and to intercontinental correlations in connection with revisionary studies of fossil rodents; cyclic sedimentation and geomorphology. *Mailing Add:* 214 Bessey Hall Univ Nebr Lincoln NE 68588

STOUT, VIRGIL L, b Emporia, Kans, Mar 14, 21; m 46; c 2. PHYSICS. *Educ:* Univ Mo, PhD(physics), 51. *Prof Exp:* Res assoc, Stanford Res Inst, 51-52; physicist, Gen Elec Co Res Labs, 52-57, mgr phys electronics br, Gen Elec Res & Develop Ctr, 57-68, mgr solid state & electronics lab, 68-75, consult electonics, 75-76, res & develop mgr, electronics sci & eng, Gen Elec Res & Develop Ctr, Gen Elec Co, 76-83; RETIRED. *Concurrent Pos:* Mem adv bd, Cancer Ctr, Univ NMex. *Mem:* Inst Elec & Electronics Engrs. *Res:* Experimental investigations of electronic properties of surfaces. *Mailing Add:* 6100 Caminito Dr NE Albuquerque NM 87111

STOUT, VIRGINIA FALK, b Buffalo, NY, Jan 5, 32; m 55, 77; c 2. ORGANIC CHEMISTRY, ANALYTICAL CHEMISTRY. *Educ:* Cornell Univ, AB, 53; Harvard Univ, AM, 55; Univ Wash, PhD(org chem), 61. *Prof Exp:* RES CHEMIST, UTILIZATION RES DIV, NORTHWEST & ALASKA FISHERIES CTR, NAT MARINE FISHERIES SERV, NAT OCEANIC & ATMOSPHERIC ADMIN, 61- *Concurrent Pos:* Affil assoc prof, Col Fisheries, Univ Wash, 72-82. *Mem:* AAAS; Asn Women in Sci; Am Chem Soc. *Res:* Synthesis of triglycerides containing only omega-3 fatty acids; fatty acid composition of fish oils from various species of fishes; purification of fish oils by supercritical fluid carbon dioxide extraction. *Mailing Add:* Utilization Res Div 2725 Montlake Blvd E Seattle WA 98112

STOUT, WILLIAM F, b Wilkensburg, Pa, July 3, 40; m 79; c 3. MATHEMATICS. *Educ:* Pa State Univ, BS, 62; Purdue Univ, MS, 64, PhD(probability), 67. *Prof Exp:* From asst prof to assoc prof, 67-73, PROF STAT, UNIV ILL, URBANA-CHAMPAIGN, 80- *Concurrent Pos:* Prin investr, ONR Psychomet Contracts, 83- *Mem:* Am Statist Asn; Inst Math Statist; fel Inst Math Statist; Psychomet Soc; Am Educ Res Asn. *Res:* Psychometrics, with emphasis on modeling and statistical analysis of psychological test data. *Mailing Add:* Dept of Statist Univ Ill 725 S Wright St Champaign IL 61820

STOUTAMIRE, DONALD WESLEY, b Roanoke, Va, Mar 10, 31; m 56; c 3. SYNTHETIC ORGANIC CHEMISTRY, AGRICULTURAL CHEMISTRY. *Educ:* Roanoke Col, BS, 52; Univ Wis, PhD(org chem), 57. *Prof Exp:* CHEMIST, SHELL AGR CHEM CO, 86- *Mem:* Am Chem Soc; AAAS; Sigma Xi. *Res:* Agricultural chemicals; animal health products. *Mailing Add:* 904 Bel Passi Dr Modesto CA 95350

STOUTAMIRE, WARREN PETRIE, b Salem, Va, July 5, 28; m 63; c 2. PLANT TAXONOMY, EVOLUTION. *Educ:* Roanoke Col, BS, 49; Univ Ore, MS, 50; Ind Univ, PhD(taxon), 54. *Prof Exp:* Botanist, Cranbrook Inst Sci, 56-66; assoc prof, 66-78, PROF BIOL, UNIV AKRON, 79- *Concurrent Pos:* Collabr, Bot Garden, Univ Mich, 57-67. *Mem:* Royal Hort Soc; Am Soc Plant Taxon; Bot Soc Am; Sigma Xi; Am Orchid Soc. *Res:* Evolution of the genus Gaillardia; physiology of orchid seed germination; pollination of terrestrial orchid species; Australian orchid evolution. *Mailing Add:* Dept of Biol Univ of Akron Akron OH 44325

STOUTER, VINCENT PAUL, b Jersey City, NJ, Apr 28, 24; m 53; c 5. ZOOLOGY, NEUROENDOCRINOLOGY. *Educ:* Spring Hill Col, BS, 49; Fordham Univ, MS, 51; Univ Buffalo, PhD(biol), 59. *Prof Exp:* Instr biol, physiol & genetics, Canisius Col, 51-52; instr biol, anat & genetics, Gannon Col, 52-53; instr gen chem, D'Youville Col, 53-54; from asst prof to assoc prof, 59-69, chmn dept biol, 59-71, chmn health sci adv & recommendation comt, 62-88, PROF BIOL, PHYSIOL & ANAT, CANISIUS COL, 69- *Mem:* NY Acad Sci; Asn Am Med Cols. *Res:* Hypothalamic neurosecretion; electrolyte and salt balance in mammals. *Mailing Add:* Dept Biol Canisius Col 2001 Main St Buffalo NY 14208

STOVER, DENNIS EUGENE, b Benton Harbor, Mich, July 30, 44; m 65; c 2. CHEMICAL ENGINEERING, ELECTROCHEMICAL PROCESSES. *Educ:* Kalamazoo Col, BA, 66; Univ Mich, BSE, 67, MSE, 68, PhD(chem eng), 75. *Prof Exp:* Process engr chem process design, Charles E Sech & Assocs, 71-72; sr res engr in-situ uranium mining, Atlantic Richfield Co, 74-78; chief engr in-situ uranium mining, Everest Explor Co, 78-84; CHIEF ENGR, EVEREST MINERALS CORP, 84- *Mem:* Am Chem Soc; Am Inst Chem Engrs; Soc Mining Engrs; Sigma Xi. *Res:* Fundamental studies of electrochemical processes; investigation of kinetics and reaction of in site leaching of uranium ores; development of kinetic/hydrologic models for in site uranium leaching. *Mailing Add:* 1704 Canary Ct Edmond OK 73034-6117

STOVER, E(DWARD) R(OY), b Washington, DC, Apr 9, 29; m 56; c 2. MATERIALS SCIENCE, CARBON-GRAPHITE. *Educ:* Mass Inst Technol, SB, 50, SM, 52, ScD(metall), 56. *Prof Exp:* Asst metall, Mass Inst Technol, 50-55; ceramist, Res Lab, Gen Elec Co, 55-65; assoc res engr, Dept Mining Technol, Univ Calif, Berkeley, 65-66; consult ceramic engr, re-entry & environ systs div, Gen Elec Co, Philadelphia, 66-82; FEL, RES & DEVELOP CTR, BF GOODRICH, OHIO, 82- *Mem:* AAAS; Am Ceramic Soc; Am Soc Metals; Am Inst Mining, Metall & Petrol Engrs; Soc Advan Mat & Process Eng. *Res:* Mechanical behavior, processing techniques and microstructure of structural materials; carbon-carbon composites; oxidation protection systems; carbon-graphite; pyrolytic graphite; carbides, oxides; reinforced plastics chars; cemented carbides; high temperature application; space and re-entry application; aircraft brake applications; oxidation resistant applications. *Mailing Add:* 1857 Brookwood Dr Akron OH 44313-5062

STOVER, ENOS LOY, b Shawnee, Okla, Nov 19, 48; m 72; c 3. BIOLOGICAL SCIENCES. *Educ:* Okla State Univ, BS, 71, MS, 72, PhD(environ eng), 74. *Prof Exp:* Supvr process develop, Roy F Watson, Inc, 74-78; dir res & develop, Metcalf & Eddy, Inc, 78-80; prof environ eng, Okla State Univ, 80-86; PRES, STOVER & ASSOCS, INC, 84- *Concurrent Pos:* Independent consult, 80-84; chmn, Hazardous Waste Comt, Water Pollution Control Fedn, 87-90. *Mem:* Water Pollution Control Fedn; Am Water Works Asn; Int Ozone Asn; Nat Water Well Asn; Int Asn Water Pollution Res & Control. *Res:* Development of improved water and wastewater treatment technologies for environmental pollution control; author of over 150 publications. *Mailing Add:* Rte 4 Box 666 Stillwater OK 74074

STOVER, JAMES ANDERSON, JR, b Hayesville, NC, June 9, 37; m 63. ARTIFICIAL INTELLIGENCE, SYSTEMS SCIENCE. *Educ:* Univ Ga, BS, 59; Univ Ala, MA, 66, PhD(math), 69. *Prof Exp:* Physicist, US Army Missile Command, Redstone Arsenal, Ala, 60-62; control systs engr, Marshall Space Flight Ctr, NASA, 62-65; consult systs anal, Anal Serv, Inc, Va, 69; asst prof math, Memphis State Univ, 69-74; prin staff & sr scientist, Ori, Inc, 74-84; SR RES ASSOC, APPL RES LAB, PA STATE UNIV, 85- *Res:* Artificial intelligence; autonomous systems; systems science and design. *Mailing Add:* Box 33 Spring Mills PA 16875

STOVER, LEWIS EUGENE, b Philadelphia, Pa, Apr 12, 25; m 51; c 3. PALEONTOLOGY, PALYNOLOGY. *Educ:* Dickinson Col, BSc, 51; Univ Rochester, PhD, 56. *Prof Exp:* Res geologist, Esso Prod Res Co, Tex, 56-65, res assoc, 66-71, mem staff, Esso Standard Oil Ltd, 71-74, mem staff, Exxon Prod Res Co, Houston, 74-89; RETIRED. *Mem:* Geol Soc Am; Paleont Soc; Int Asn Plant Taxon. *Res:* Geology; fossil spore; pollen; microplankton; small calcareous fossils; acid-in-soluble microfossils. *Mailing Add:* 433 Oakwood Dr Kerrville TX 78028

STOVER, RAYMOND WEBSTER, b Pittsburgh, Pa, Mar 20, 38; m 60; c 2. PHYSICS. *Educ:* Lehigh Univ, BS, 60; Syracuse Univ, MS, 62, PhD(physics), 67. *Prof Exp:* PRIN SCIENTIST, XEROX CORP, 66- *Concurrent Pos:* Adj fac mem, Rochester Inst Technol, 74- *Mem:* AAAS; Soc Photog Scientists & Engrs. *Res:* Search for an electron-proton charge difference; xerographic development process; electrostatics; triboelectricity; small particle physics. *Mailing Add:* 566 Bending Bough Dr Webster NY 14580

STOVER, ROBERT HARRY, plant pathology, for more information see previous edition

STOVER, SAMUEL LANDIS, b Bucks Co, Pa, Nov 19, 30; c 3. MEDICINE. *Educ:* Goshen Col, BA, 52; Jefferson Med Col, MD, 59; Am Bd Pediat, dipl, 69; Am Bd Phys Med & Rehab, dipl, 71. *Prof Exp:* Intern, St Luke's Hosp, Bethlehem, Pa, 59-60; gen pract in Ark, 60-61 & Indonesia, 61-64; resident pediat, Children's Hosp, Philadelphia, 64-66; asst med dir, Children's Seashore House, Atlantic City, NJ, 66-67; resident phys med & rehab, Univ Pa, 67-69; assoc prof pediat & prof phys med & rehab, 69-76, PROF REHAB MED & CHMN DEPT, UNIV HOSP & CLINS, UNIV ALA, BIRMINGHAM, 76- *Mem:* Am Acad Pediat; Am Acad Phys Med & Rehab; Am Cong Rehab Med; AMA. *Mailing Add:* Spain Rehab Ctr 1717 Sixth Ave Birmingham AL 35233

STOW, STEPHEN HARRINGTON, b Oklahoma City, Okla, Sept 18, 40; m 65. GEOCHEMISTRY. *Educ:* Vanderbilt Univ, BA, 62; Rice Univ, MA, 65, PhD(geochem), 66. *Prof Exp:* Res scientist, Plant Foods Res Div, Continental Oil Co, 66-69; from asst prof to prof geol, Univ Ala, Tuscaloosa, 69-80; prog mgr & sr geologist, 80-88, SECT HEAD GEOSCI, OAK RIDGE NAT LAB, 88- *Concurrent Pos:* Consult, Ala Geol Surv, 69-74 & Indust Co, 73-80. *Mem:* Am Geophys Union; Geochem Soc; Geol Soc Am; AAAS; Int Asn Hydrologists. *Res:* Geochemistry and element distribution in igneous and metamorphic rocks, geology and geochemistry of phosphates; environmental geology; geochemistry of mafic rocks of southern Appalachians; sulfide ore deposits; geology and geochemistry of radioactive and hazardous waste disposal. *Mailing Add:* 11713 N Monticello Rd Knoxville TN 37922

STOWE, BRUCE BERNOT, b Neuilly-sur-Seine, France, Dec 9, 27; US citizen; wid; c 2. PLANT PHYSIOLOGY, BIOCHEMISTRY. *Educ:* Calif Inst Technol, BS, 50; Harvard Univ, MA, 51, PhD(biol), 54. *Hon Degrees:* MA, Yale Univ, 71. *Prof Exp:* NSF fel, Univ Col NWales, 54-55; instr biol, Harvard Univ, 55-58, lectr bot, 58-59, tutor biochem sci, 56-58; asst prof bot, 59-63, assoc prof biol, 63-71, dir, Marsh Bot Gardens, 75-78, PROF BIOL, YALE UNIV, 71-, PROF FORESTRY, 74- *Concurrent Pos:* Mem, Metab Biol Panel, NSF, 60-61, Subcomt Plant Sci Planning & Comt Sci & Pub Policy, Nat Acad Sci, 64-66; Guggenheim fel, Nat Ctr Sci Res, France, 65-66; vis prof, Univ Osaka Prefecture, Japan, 72 & 73, Waite Agr Res Inst, Univ Adelaide, 72-73 & Japan Asn Advan Sci, 73; vis scientist, Nat Inst Basic Biol, Okazaki, Japan, 85-86; vis investr, Dept Plant Biol, Carnegie Inst Wash, Stanford, CA, 86-89. *Mem:* Am Soc Biol Chemists; Am Soc Plant Physiologists (secy, 63-65); Bot Soc Am; Soc Develop Biol; Phytochem Soc NAm; Sigma Xi; fel AAAS. *Res:* Biochemistry and physiology of plant hormones, especially auxins, gibberellins and their relations to lipids and membrane structure. *Mailing Add:* Kline Biol Tower Yale Univ Box 6666 New Haven CT 06511-8112

STOWE, CLARENCE M, b Brooklyn, NY, Mar 19, 22; m 46; c 3. PHARMACOLOGY, VETERINARY MEDICINE. *Educ:* NY Univ, BS, 44; Queens Col, NY, BS, 46; Univ Pa, VMD, 50; Univ Minn, PhD, 55. *Prof Exp:* From instr to assoc prof, Univ Minn, 50-57, prof & asst dean, 57-60, head dept, 60-71, prof large animal sci, 77-86 PROF PHARMACOL, COL VET MED, UNIV MINN, ST PAUL, 60- *Concurrent Pos:* Spec appointee, Rockefeller Found, Colombia, SAm, 64-65 & Inst Agron & Vet, Hassan II, Rabat, Morocco; prof, Nat Univ Columbia, 64-65; chmn comt vet drug efficacy, Nat Acad Sci-Nat Res Coun; mem, Toxicol Study Sect, NIH, 65-69; vis scholar, Univ Cambridge, 71-72; mem, Coun Biol & Therapeut Agents, Am Vet Med Asn; chmn, drug availability comt, Am Vet Med Asn; hon prof, Nat Univ Bogota, Colombia; staff oper officer, US Coast Guard Auxilliary; asst state dir, Am Asn Retired Persons. *Mem:* Soc Exp Biol & Med; Am Soc Pharmacol & Exp Therapeut; Am Soc Vet Physiol & Pharmacol; Am Vet Med Asn; Am Dairy Sci Asn; Am Col Vet Toxicol. *Res:* Drug distribution and excretion, muscle relaxants, chemotherapy of large domestic and wild animals; clinical toxicology. *Mailing Add:* Dept Clin Sci Col Vet Med Univ Minn St Paul MN 55101

STOWE, DAVID F, b Vincennes, Ind, Jan 27, 45. CARDIOVASCULAR PHYSIOLOGY, ELECTROPHYSIOLOGY. *Educ:* Ind Univ, Bloomington, AB, 68, MA, 69; Mich State Univ, PhD(physiol), 74; Med Col Wis, MD, 83. *Prof Exp:* Fel cardiovasc physiol, Cardiovasc Res Inst, Univ Calif, San Francisco, 74-76; asst prof anesthesiol, 81-87, ASST PROF PHYSIOL, MED COL WIS, 76-, ASSOC PROF ANESTHESIOL, 88- *Concurrent Pos:* Vis prof biol, Marquette Univ, 71; mem circulation comt, Am Heart Asn, circulation fel, Am Physiol Soc. *Mem:* Am Physiol Soc; Sigma Xi; Soc Exp Biol & Med; Am Soc Anesthesiol; Anesthesia Res Soc. *Res:* Effects of hypoxia reactive hyperemia and exercise on myocardial mechanics and metabolism; effects of adenosine on myocardial and oxygen consumption; effects of hypoxia, hydrogen ion changes on electrophysiology and of heart muscle cells; effect of anesthesia on action potential properties of cardiac tissue. *Mailing Add:* MFRC A1000 Med Col Wis Milwaukee WI 53226

STOWE, DAVID WILLIAM, b Three Rivers, Mich, Jan 1, 44; m 66; c 3. FIBER OPTIC COMPONENTS & SENSORS. *Educ:* Univ Wis-Madison, BS, 66; Univ Ill, Urbana, MS, 67, PhD(physics), 71. *Prof Exp:* Sr physicist, Appl Physics Lab, Johns Hopkins Univ, 71-77; prog mgr fiber optics sensors, surface acoust waves & thin films, Gould Labs Elec & Electronics Res, Gould Inc, 77-83; EXEC VPRES TECHNOL, ASTER CORP, MILFORD, 83- *Concurrent Pos:* Instr, Harper Community Col, 78- *Honors & Awards:* IR 100 Award. *Mem:* Acoust Soc Am; Optical Soc Am; Soc Photo-Optical Instrumentation Engrs. *Res:* Fiber optic couplers, general fiber optic components, fiber optic sensors and fiber optic communications; inventor of "Stowe" process for single-mode couplers used by major commercial suppliers. *Mailing Add:* Four Woodfall Rd Medfield MA 02052

STOWE, HOWARD DENISON, b Greenfield, Mass, Mar 31, 27. PATHOLOGY. *Educ:* Univ Mass, BS, 48; Mich State Univ, MS, 56, DVM, 60, PhD(vet path), 62. *Prof Exp:* Instr animal husb & dairy prod, Bristol County Agr Sch, Mass, 49-53; asst animal husb, anat & vet path, Mich State Univ, 55-60; assoc prof vet sci & chief nutrit sect, Univ Ky, 63-68; from asst prof to assoc prof path, Sch Med, Univ NC, Chapel Hill, 68-73; assoc prof path, Sch Vet Med, Auburn Univ, 73-77, from assoc prof to prof, clin nutrit, 77-80; DEPT LARGE ANIMAL CLIN SCI, COL VET MED, MICH STATE UNIV, 80- *Concurrent Pos:* Vis researcher, Dunn Nutrit Lab, Cambridge Univ, 62 & Dept Nutrit & Biochem, Denmark Polytech Inst, Copenhagen, 62; pathologist, Div Lab Animal Med, Sch Med, Univ NC, Chapel Hill. *Mem:* Am Vet Med Asn; Conf Res Workers Animal Dis; Am Inst Nutrit. *Res:* Effects of lead upon reproduction in rats; cadmium toxicity in rabbits; canine and avian lead toxicity; genetic influence on selenium metabolism. *Mailing Add:* Dept Large Animal Clin Sci Mich State Univ East Lansing MI 48824

STOWE, KEITH S, b Midland, Mich, Feb 16, 43; m 67; c 2. ELEMENTARY PARTICLE PHYSICS. *Educ:* Ill Inst Technol, BS, 65; Univ Calif, San Diego, MS, 67, PhD(physics), 71. *Prof Exp:* Lectr physics, 71-74, asst prof, 74-76, assoc prof, 76-81, prof physics, Calif Polytech State Univ, San Luis Obispo, 81-; DEPT PHYSICS, UNIV WASH. *Mem:* Am Phys Soc; Am Asn Physics Teachers. *Res:* Elementary particle theory; oceanography; thermodynamics. *Mailing Add:* Dept Physics Calif Polytech State Univ San Luis Obispo CA 93407

STOWE, ROBERT ALLEN, b Kalamazoo, Mich, July 26, 24; div; c 4. SURFACE CHEMISTRY, CATALYSIS INDUSTRIAL PROCESS CHEMISTRY. *Educ:* Kalamazoo Col, BA, 48; Brown Univ, PhD(chem), 53. *Prof Exp:* Res chemist, Ludington Div, Dow Chem USA, 52-58, Res & Develop Lab, 58-64, sr res chemist, 64-69, Hydrocarbon & Monomers Res Lab, 69-71, assoc scientist, 71-74, assoc scientist, 74-79, assoc scientist, Mich Div Res, 79-88; CONSULT CATALYTIC & CHEM TECHNOL, 88- *Concurrent Pos:* Secy, Bd Educ, Luddington, 60-63, pres, 63-68; chmn, Div Indust & Eng Chem, Am Chem Soc, 82, counr, 86- *Honors & Awards:* Victor J Azbe Lime Award, Nat Lime Asn, 64; Joseph Stewart Award, Indust Eng & Chem Div, Am Chem Soc, 84. *Mem:* Am Chem Soc; Am Inst Chemists; Sigma Xi; Catalysis Soc. *Res:* Heterogeneous catalysis; hydrocarbon processes; inorganic chemistry; zeolite chemistry; organic fluorine chemistry; statistics; carbon monoxide methanation; Fischer-Tropsch synthesis; coal liquefaction. *Mailing Add:* PO Box 173 6670 Lower Shore Dr Cross Village MI 49723-0173

STOWELL, EWELL ADDISON, b Ashland, Ill, Sept 2, 22; m 53. PLANT PATHOLOGY. *Educ:* Ill State Norm Univ, BEd, 43; Univ Wis, MS, 47, PhD(bot), 55. *Prof Exp:* Asst bot, Univ Wis, 46-47; instr, Univ Wis, Milwaukee, 47-49, asst, 49-53; from instr to prof bot, 53-88, chmn dept biol, 72-77, EMER PROF BOT, ALBION COL, 88- *Concurrent Pos:* Vis !ectr, Univ Wis, 63; assoc prof, Univ Mich, 64. *Mem:* Bot Soc Am; Mycol Soc Am; Am Inst Biol Sci; Sigma Xi. *Res:* Taxonomy and morphology of Ascomycetes. *Mailing Add:* Dept of Biol Albion Col Albion MI 49224

STOWELL, JAMES KENT, b Elgin, Ill, July 9, 36; m 65; c 2. POLYMER CHEMISTRY, ORGANIC CHEMISTRY. *Educ:* Knox Col, BA, 58; Univ Iowa, PhD(org chem), 65. *Prof Exp:* Sr res chemist org, PPG Industs Inc, 65-68; sr res chemist polymer, A E Staley Mfg Co, 68-78; sr chemist polymer, 78-81, NEW PROD DEVELOP SUPVR, MORTON CHEM CO, 81. *Mem:* Am Chem Soc. *Res:* Development of new polymer products for use in printing inks, coatings, and adhesives; organic chemicals for specialty uses. *Mailing Add:* Morton Chem Co 1275 Lake Ave Woodstock IL 60098-7415

STOWELL, JOHN CHARLES, b Passaic, NJ, Sept 10, 38; div; c 2. INSECT CHEMISTRY, NEW REAGENTS. *Educ:* Rutgers Univ, New Brunswick, BS, 60; Mass Inst Technol, PhD(org chem), 64. *Prof Exp:* Res specialist, Cent Res Lab, 3M Co, 64-69; NIH fel org chem, Ohio State Univ, 69-70; from asst prof to assoc prof, 70-80, PROF ORG CHEM, UNIV NEW ORLEANS, 80- *Concurrent Pos:* Res Corp & Petrol Res Fund grants, Univ New Orleans, 71-73 & 80-82. *Mem:* Am Chem Soc. *Res:* Organic synthesis; heterocyclic compounds; sterically hindered compounds; three- and four-carbon homologating; carbanions; concurrent strong acid and base catalysis insect pheromones. *Mailing Add:* Dept Chem Univ New Orleans New Orleans LA 70148

STOWELL, ROBERT EUGENE, b Cashmere, Wash, Dec 25, 14; m 45; c 2. RESEARCH ADMINISTRATION. *Educ:* Stanford Univ, AB, 36, MD, 41; Wash Univ, PhD(path), 44. *Prof Exp:* From asst to assoc prof path, Sch Med, Wash Univ, 42-48; prof path & oncol, Sch Med & dir cancer res, Med Ctr, Univ Kans, 48-59, chmn dept oncol, 48-51, path & oncol, 51-59, pathologist-in-chief, 51-59; sci dir, Armed Forces Inst Path, Washington, DC, 59-67; mem nat adv comt, Nat Ctr Primate Biol, 67-68, dir, 69-71, chmn dept path, 67-69, asst dean, Sch Med, 67-71, prof, 67-82, EMER PROF PATH, SCH MED, UNIV CALIF, DAVIS, 82- *Concurrent Pos:* Commonwealth Fund advan med study & res fel, Inst Cell Res, Stockholm, Sweden, 46-47; mem morphol & genetics study sect, NIH, 49-53 & path study sect, 54-55, chmn, 55-57, mem path training comt, div gen med sci, 58-61, chmn animal resources adv comt, Div Res Resources, 70-74; mem, Intersoc Comt Res Potential in Path, 56-80, pres, 57-60; vis prof, Sch Med, Univ Md, 60-67; mem fedn bd, Fedn Am Soc Exp Biol, 63-66; mem subcomt comp path, Comt Path, Nat Acad Sci-Nat Res Coun, 63-69, subcomt manpower needs in path, 64-69 & US Nat comt, Int Coun Socs Path, 66-80, chmn, 72-75; mem, Intersoc Comt Path Info, 65-69, chmn, 66-67; mem adv med bd, Leonard Wood Mem, 65-69; mem div biol & agr, Nat Res Coun, 65-68 & comt doc data anat & clin path, 66-68; mem, Int Coun Socs Path, 66-81; mem bd dirs, Coun Biol Sci Info, Nat Acad Sci, 67-70; ed, Lab Invest, Int Acad Path, 67-72; mem med adv comt & consult, Vet Admin Hosp, Martinez, Calif, 69-72; mem bd dirs, Univ Asn Res & Educ Path, 74-, secy-treas, 78-82; mem sci adv bd, Nat Ctr Toxicol Res, 76-79; vpres, Am Registry Path, 76-78, pres, 78-80; fel, Cytology, Wash Univ Sch Med, 40-42. *Honors & Awards:* Gold-Headed Cane Award, Am Asn Path, 90; Distinguished Serv Award, 79 & Diamond Jubilee Award, 81 & Stowell-Orbison Award, US-Can Div, Int Acad Path, 82. *Mem:* Col Am Pathologists; Am Soc Clin Path; Am Asn Path & Bact (vpres, 69-70, pres, 70-71); Am Soc Exp Path (vpres, 63-64, pres, 64-65); Int Acad Path (vpres, 57-58, pres elect, 58-59, pres, 59-60); Sigma Xi. *Res:* Cancer; experimental pathology; comparative pathology. *Mailing Add:* Dept Path Univ Calif Sch Med Davis CA 95616

STOWENS, DANIEL, b New York, NY, Oct 27, 19; m 75; c 2. PATHOLOGY. *Educ:* Columbia Univ, AB, 41, MD, 43; Am Bd Pediat, dipl, 51; Am Bd Path, dipl, 54. *Prof Exp:* Chief, Sect Pediat Path, Armed Forces Inst Path, US Army, Washington, DC, 54-58; assoc prof path, Univ Southern Calif, 58-61 & Univ Louisville, 61-65; pathologist, St Luke's Mem Hosp Ctr, 66-85; RETIRED. *Concurrent Pos:* Registr, Am Registry Pediat Path, 54-58; consult, Walter Reed Army Hosp, 56-58; pathologist, Children's Hosp, Los Angeles, 58-61; dir labs & chief prof servs, Children's Hosp, Louisville, 61-65. *Mem:* Fel Am Soc Clin Path; Soc Pediat Res; Am Asn Path & Bact; Int Acad Path. *Res:* Pediatric pathology, especially pathophysiology of fetus and mechanisms of development; dermatoglyphics. *Mailing Add:* RR2 Fountain St 721 Clinton NY 13323

STOY, WILLIAM S, b New York, NY, Sept 23, 25; m 49; c 2. CHEMISTRY. *Educ:* Queens Col, NY, BS, 45; Polytech Inst Brooklyn, MS, 50. *Prof Exp:* Chemist paint res & develop, Mobil Chem Co, 45-50, supvr, 50-58; sr chemist, Cities Serv Co, Cranbury, NJ, 58-64, mgr plastics applns, 64-71, mgr coatings, plastics & inks, Petrochem Res, 71-77; GROUP LEADER, APPLNS RES, ENGELHARD MINERALS & CHEM, MENLO PARK, EDISON, 77- *Honors & Awards:* Plastics Inst Award. *Mem:* Am Chem Soc; Soc Plastics Engrs; Am Soc Testing & Mat; Plastics Inst Am. *Res:* Plastics resins and concentrates development; coatings and inks; pigment syntheses and applications; flame retardants; polymer chemistry; extenders-inorganic silicates; catalysis. *Mailing Add:* 221 Herrontown Rd Princeton NJ 08540

STOYLE, JUDITH, applied statistics, for more information see previous edition

ST-PIERRE, CLAUDE, b Montreal, Que, Jan 7, 32; m 54; c 2. NUCLEAR PHYSICS. *Educ:* Univ Montreal, BSc, 54, MSc, 56, DSc(physics), 59. *Prof Exp:* Sci officer, Defence Res Bd Can, 58-61; Nat Res Coun Can fel, Ctr Nuclear Res, Strasbourg, France, 61-62; from asst prof to assoc prof, 62-70, chmn dept, 73-79, dean, Sch Grad Studies, 79-84, PROF PHYSICS, LAVAL UNIV, 70- *Concurrent Pos:* Vis scientist, AECL, Chalk River, Ont, 86. *Mem:* Can Asn Physicists; Am Phys Soc. *Res:* Nuclear spectroscopy; heavy ion reactions. *Mailing Add:* Dept Physics Laval Univ Quebec PQ G1K 7P4 Can

ST-PIERRE, JACQUES, b Trois-Rivieres, PQ, Aug 30, 20; c 6. APPLIED STATISTICS. *Educ:* Univ Montreal, LSc, 45 & 48, MSc, 51; Univ NC, PhD(math statist), 54. *Prof Exp:* From asst prof to assoc prof math, Univ Montreal, 47-60, vdean, Fac Sci, 61-64, head, Dept Comput Sci, 66-69, dir, Comput Ctr, 64-72, prof, 60-83, vpres planning, 72-83, EMER PROF MATH, UNIV MONTREAL, 83- *Concurrent Pos:* Consult, Inst Armand Frappier, 54-80. *Mem:* Can Asn Univ Teachers (pres, 65-66). *Res:* Statistical methods relative to public health and epidemiology. *Mailing Add:* Univ Montreal CP 6128 Montreal PQ H3C 3J7 Can

STRAAT, PATRICIA ANN, b Rochester, NY, Mar 28, 36. BIOCHEMISTRY, ENZYMOLOGY. *Educ:* Oberlin Col, BA, 58; Johns Hopkins Univ, PhD(biochem), 64. *Prof Exp:* Lab instr biol, Johns Hopkins Univ, 58-59, USPHS fel, 60-64, res fel radiol sci, 64-67, res assoc, 67-68, asst prof, 68-70; sr res biochemist, Biospherics, Inc, 70-75, res coordr, 75-78, dir res, 78-80; grants assoc, Div Res Grants, NIH, 80-81; chief, Planning & Coord Sect, Prog Opers Br, Nat Toxicol Prog, Nat Inst Environ Health Sci, 81-82; health sci adminr, Molecular & Cellular Biophys Study Sect, NIH, 82-86; CHIEF, REFERRAL SECT & DEPT CHIEF, REFERRAL & REV BR, DIV RES GRANTS, NIH, 86- *Concurrent Pos:* Lectr, dept radiol sci, Sch Hyg & Pub Health, Johns Hopkins Univ, 70-72; mem, Viking Biol Flight Team & Sci Team & Viking Mission to Mars, NASA, 76. *Mem:* Sigma Xi; Biophys Soc. *Res:* Electron transport and inorganic nitrogen metabolism; mechanisms of nucleic acid replication; extraterrestial life detection; biological and chemical aspects of water pollution. *Mailing Add:* Div Res Grants Rm 248 Westwood Bldg 5333 Westbard Ave Bethesda MD 20816

STRAATSMA, BRADLEY RALPH, b Grand Rapids, Mich, Dec 29, 27; c 3. MEDICINE. *Educ:* Yale Univ, MD, 51. *Prof Exp:* Intern, New Haven Hosp, Yale Univ, 51-52; vis scholar, Col Physicians & Surgeons, Columbia Univ, 52, asst resident, 55-58; spec clin trainee, Nat Inst Neurol Dis & Blindness, 58-59; asst prof surg & ophthal, 59-73, chief, Div Ophthal, 59-68, PROF OPHTHAL, SCH MED, UNIV CALIF, LOS ANGELES, 63-, CHMN DEPT, 68-, PROF SURG, 80-, DIR, JULES STEIN EYE INST, 64- *Concurrent Pos:* Resident, Inst Ophthal, Presby Hosp, New York, 55-58; fel ophthalmic path, Armed Forces Inst Path, Walter Reed Army Med Ctr, DC, 58-59; fel ophthal, Wilmer Inst, Johns Hopkins Univ, 58-59; mem vision res training comt, Nat Inst Neurol Dis & Blindness, 59-63 & neurol & sensory dis prog proj comt, 64-68; consult to Surgeon Gen, USPHS, 59-68; ophthal examr, aid to blind progs, Calif Dept Social Welfare, 59-; mem med adv comt, Nat Coun Combat Blindness, 60-; consult, Vet Admin Hosp, Long Beach, Calif, 60-75; attend physician, Vet Admin Ctr, Wadsworth Gen Hosp, Los Angeles, 60-; attend physician & consult, Los Angeles County Harbor Gen Hosp, Torrance, 60-; vis consult, St John's Hosp, Santa Monica, 60-; mem courtesy staff, Santa Monica Hosp, 60- & St Vincent's Hosp, Los Angeles, 60-; mem sensory dis serv panel, Bur States Serv, USPHS, 63-65; trustee, John Thomas Dye Sch, Los Angeles, 67-72; prof, New Orleans Acad Ophthal, 68-; ophthalmologist in chief, Univ Calif, Los Angeles Hosp, 68-; mem med adv bd, Int Eye Found, 70-; mem nat adv comt, Pan-Am Cong Ophthal, 71-72 & bd dirs, Conrad Berens Int qye Film Libr, 71- *Honors & Awards:* William Warren Hoppin Award, NY Acad Med, 56; co-recipient, Silver Award, Am Soc Clin Path & Col Am Pathologists, 62; co-recipient, Conrad Berens Award, Int Eye Film Festival, 65. *Mem:* Am Acad Ophthal (pres, 77); AMA; Asn Univ Prof Ophthal (pres, 74); Am Ophthal Soc; Asn Res Vision & Ophthal; Am Bd Ophthal; Pan-Am Ophthal Found. *Res:* Ophthalmology. *Mailing Add:* Ophthal 2-142 JSE1 Univ Calif 100 Stein Plaza Los Angeles CA 90024

STRACHAN, DONALD STEWART, b Highland Park, Mich, 32; c 4. INSTITUTIONAL PLANNING, COMPUTER MANAGEMENT. *Educ:* Wayne State Univ, BA, 54; Univ Mich, DDS, 60, MS, 62, PhD(anat), 64. *Prof Exp:* From instr to assoc prof oral biol, Sch Dent & Anat, Sch Med, 63-73, asst dean, 69-89, PROF DENT & ORAL BIOL, SCH DENT, UNIV MICH, ANN ARBOR, 73-,. *Concurrent Pos:* USPHS res career award, 63-68; consult Vet Admin Hosp, DC, 66-68 & coun dent educ, Am Dent Asn; hon adv, Nat Asn Adv Health Professions. *Mem:* Am Dent Asn; Int Asn Dent Res; Am Asn Dent Schs; Sigma Xi. *Res:* Histochemistry of esterase isoenzymes; lactic dehydrogenase in developing teeth and healing bone; data analysis and programming in the analysis of gel electrophoretic patterns; educational research; computer assisted instruction; self instructional media development. *Mailing Add:* Sch of Dent Univ of Mich Ann Arbor MI 48109-1078

STRACHAN, WILLIAM MICHAEL JOHN, b Thunder Bay, Ont, Nov 20, 37; m 65; c 3. ORGANIC CHEMISTRY. *Educ:* Univ Toronto, BA, 59, MA, 60; Queens Univ, PhD(chem), 68. *Prof Exp:* Asst lectr dept chem, Univ Col, London, 60-63; res assoc, Royal Mil Col Can, 63-65; Nat Res Coun Can fel, 68-70; res environ scientist, Can Ctr for Inland Waters, 70-74, head sect, 74-80, dir, Hazard Assessment Br, 81, res scientist, 82-85, PROJ CHIEF, AIR-WATER INTERACTIONS, CAN CTR FOR INLAND WATERS, 85- *Concurrent Pos:* Mem, Int Joint Comn, Res Adv Bd, Comt Sci Basis Water Qual Criteria, 74-78; mem, Can Environ Contaminants Act Comt, Dept Environ, Nat Health & Welfare, 75-; mem, Nat Res Coun Special Grants Panel Environ Toxicol, 77-79; Can rep, Orgn Econ Coop & Develop Expert Group, 78-81; chmn, Int Joint Comn, Sci Adv Bd Comt Aquatic Ecosyst Objectives, 78-88, Int Joint Comn, Water Qual Bd Comt Assess Chem, 86-88. *Mem:* Int Asn Great Lakes Res; Soc Environ Toxic Chem. *Res:* Persistent organic chemicals; cycling organic chemicals between air and water; organic contamination of rain and lake-stream waters. *Mailing Add:* Can Ctr for Inland Waters PO Box 5050 Burlington ON L7R 4A6 Can

STRACHER, ALFRED, b Albany, NY, Nov 16, 30; m 54; c 3. BIOCHEMISTRY. *Educ:* Rensselaer Polytech Inst, BS, 52; Columbia Univ, MA, 54, PhD, 56. *Prof Exp:* From asst prof to assoc prof, 59-68, PROF BIOCHEM, COL MED, STATE UNIV NY HEALTH SCI CTR AT BROOKLYN, 68-, CHMN DEPT, 72- *Concurrent Pos:* Nat Found Infantile Paralysis fel biochem, Rockefeller Inst, 56-58 & Carlsberg Lab, Copenhagen, 58-59; Commonwealth Fund fel, 66-67, Guggenheim fel, 73-74; NSP-B Rev Comt, Nat Inst Neurol & Commun Disorders & Stroke, 82-85; ed jours, 86-89. *Mem:* Am Soc Biol Chemists; Harvey Soc; Marine Biol Lab; corp mem Biophys Soc; fel AAAS. *Res:* Relationship of structure of muscle proteins to mechanism of muscular contraction; relationship of protein structure to biological activity; contractility in non-muscle systems, muscle and nerve degeneration. *Mailing Add:* Dept Biochem State Univ NY Health Sci Ctr Clarkson Ave Brooklyn NY 11203

STRADA, SAMUEL JOSEPH, b Kansas City, Mo, Oct 6, 42; m 71. NEUROCHEMISTRY, NEUROBIOLOGY. *Educ:* Univ Mo-Kansas City, BSPharm, 64, MS, 66; Vanderbilt Univ, PhD(pharmacol), 70. *Prof Exp:* Asst pharmacol, Univ Mo-Kansas City, 64-66; NIMH staff fel pharmacol, St Elizabeth's Hosp, Washington, DC, 70-72; from asst prof to assoc prof, 72-81, PROF PHARMACOL, MED SCH, UNIV TEX, HOUSTON, 81- *Concurrent Pos:* Assoc fac, Univ Tex Grad Sch Biomed Sci, Houston, 72-; Fogarty int fel, Med Sci Inst, Univ Dundee, Scotland, 80-81. *Honors & Awards:* Pharmaceut Mfrs Asn Fac Develop Award, Basic Pharmocol. *Mem:* AAAS; Am Soc Pharmacol & Exp Therapeut; Soc Neurosci; NY Acad Sci; Tissue Cult Asn. *Res:* Role of cyclic nucleotides in the nervous system; release of neurotransmitters and synaptic transmission; relation of the nervous system to hormone release mechanisms; role of cyclic nucleotides in cell growth; receptor regulation. *Mailing Add:* Dept Pharmacol Univ S Ala Med Col Med Sci Bldg Mobile AL 36688

STRADLEY, JAMES GRANT, b Newark, Ohio, Aug 24, 32; m 54; c 3. CERAMIC ENGINEERING. *Educ:* Ohio State Univ, BCerE, 55, MSc, 58. *Prof Exp:* Res assoc ceramic mat, Res Found, Ohio State Univ, 56-58; res specialist, Cols Div, NAm Rockwell Corp, Ohio, 58-65; develop specialist, Oak Ridge Nat Lab, Tenn, 65-70 & Y-12 plant, Union Carbide Corp, 70-72; vpres, US Nuclear, Inc, 72-77; eng mgr, Oak Ridge Nat Lab, Union Carbide Corp, 77-84; ENG MGR, OAK RIDGE NAT LAB, MARTIN MARIETTA ENERGY SYSTS, 84- *Mem:* Am Ceramic Soc; Nat Inst Ceramic Engrs; Am Nuclear Soc. *Res:* Management of reprocessing programs, including nuclear fuels, special studies and environmental and safety safeguards; forming, sintering and properties of ceramic materials. *Mailing Add:* Oak Ridge Nat Lab PO Box 2008 M/S 305 Oak Ridge TN 37831-6305

STRADLEY, NORMAN H(ENRY), b Newark, Ohio, June 28, 24; m 47; c 2. CERAMIC ENGINEERING. *Educ:* Ohio State Univ, BCerE & MS, 49. *Prof Exp:* Ceramic engr, Minn Mining & Mfg Co, 50-56, group supvr, 56-59, sr res engr, Am Lava Corp, Tenn, 59-63, proj mgr, 63-68, res supvr, 68-69, proj supvr, 69-74, prod develop specialist, Tech Ceramic Prods Div, 3M Co, Tenn, 74-75, res specialist, Tech Ceramic Prods Div, 75-82, patent liaison, Elec Prod Group, 82-84, sr patent liaison, Electro-Tel Group, 3M Ctr, 3M Co, St Paul, 84-88; RETIRED. *Mem:* Fel Am Ceramic Soc; Nat Inst Ceramic Engrs; fel Am Inst Chemists. *Res:* Coatings for ferrous and non-ferrous alloys and graphite; glass technology; nuclear and electrical ceramics. *Mailing Add:* 5401 Bus 83 No 2117 Harlingen TX 78552

STRADLING, LESTER J(AMES), JR, b Philadelphia, Pa, Aug 21, 16; m 47; c 2. MECHANICAL & NUCLEAR ENGINEERING. *Educ:* Drexel Inst, BS, 39; Univ Pa, MS, 44. *Prof Exp:* Maintenance engr, Calvert Distillery, 39; design draftsman, Gen Elec Co, 39; jr marine engr, Philadelphia Naval Yard, 39-41; design engr, Bendix Aircraft Corp, 42; instr mech eng, Drexel Inst, 42-45; field engr, Allis Chalmers Mfg Co, 45-49, sales engr, 50-54; vpres eng, Campus Industs, Inc, 49-50; assoc prof mech eng, Drexel Univ, 54-59, prof mech eng, 59-81. *Concurrent Pos:* Mem sci staff, Columbia Univ Div, War Res, Naval Underwater Sound Lab, Conn, 44; consult, Kellett Aircraft Corp, 45 & Mechtronics, Inc, 54-57. *Mem:* Am Soc Eng Educ. *Res:* Application of jet engines to helicopters; fundamental quieting of submarines; diffusion of neutrons in high velocity media and heterogeneous media; study of neutron streaming in holes and vacuua. *Mailing Add:* Mercer Park 8F 475 North St Doylestown PA 18901

STRADLING, SAMUEL STUART, b Hamilton, NY, Dec 11, 37; m 63; c 3. ORGANIC CHEMISTRY. *Educ:* Hamilton Col, AB, 59; Univ Rochester, PhD(org chem), 64. *Prof Exp:* From asst prof to assoc prof, 63-74, PROF CHEM, ST LAWRENCE UNIV, 74-, CHMN DEPT, 77- *Concurrent Pos:* NSF acad year exten grant, 65-67; vis scholar & vis prof, Univ Va, 76-77; vis prof, Va Tech, 84-85. *Mem:* AAAS; Am Chem Soc; Sigma Xi. *Res:* Reaction mechanisms; natural product chemistry; chemical education. *Mailing Add:* Dept of Chem St Lawrence Univ Canton NY 13617

STRAF, MIRON L, b New York, NY, Apr 13, 43. STATISTICS. *Educ:* Carnegie-Mellon Univ, BS, 64, MS, 65; Univ Chicago, PhD(statist), 69. *Prof Exp:* Asst prof statist, Univ Calif, Berkeley, 69-74; res assoc, 74-77, res dir, 78-87, DIR, COMT ON NAT STATIST, NAT ACAD SCI-NAT RES COUN, 87- *Concurrent Pos:* Sr vis res fel, Monitoring Assessment & Res Ctr, Chelsea Col Sci & Technol, London, 77; lectr statist, London Sch Econ & Polit Sci, 77-78; mem, Joseph P Kennedy, Jr Found Comt, 81. *Mem:* Fel Royal Statist Soc; fel Am Statist Asn; Int Statist Inst. *Res:* Applied and theoretical statistics; analysis and evaluation of environmental statistics; applications of statistics to public policy; use of statistics in the courts. *Mailing Add:* 224 11th St Washington DC 20003

STRAFFON, RALPH ATWOOD, b Croswell, Mich, Jan 4, 28; m 54; c 5. MEDICINE, UROLOGY. *Educ:* Univ Mich, MD, 53; Am Bd Urol, dipl, 62. *Prof Exp:* Intern, Univ Hosp, Ann Arbor, Mich, 53-54, from asst resident to resident gen surg, 54-55, resident surg, 56-57, from jr clin instr to sr clin instr, 57-59; staff mem, 59-63, head dept urol, 63-83, chmn div surg, 83-87, VCHMN BD GOV & CHIEF OF STAFF, CLEVELAND CLINIC FOUND, 87- *Concurrent Pos:* Res fel med, Renal Lab, Peter Bent Brigham Hosp, Boston, 56. *Mem:* AMA; fel Am Col Surg; Am Urol Asn; Am Asn Genito-Urinary Surg; fel Am Acad Pediat; Am Surg Asn. *Mailing Add:* 9500 Euclid Ave Cleveland OH 44106

STRAFUSS, ALBERT CHARLES, b Princeton, Kans, Jan 24, 28; m 54; c 5. COMPARATIVE PATHOLOGY, ONCOLOGY. *Educ:* Kans State Univ, BS & DVM, 54; Iowa State Univ, MS, 58; Univ Minn, PhD(comp path), 63. *Prof Exp:* Practitioner, Hastings, Nebr, 54-56; instr, Iowa State Univ & pathologist, Iowa Vet Med Diag Lab, 56-59; instr path, Col Vet Med, Univ Minn, 59-63; assoc prof path, Col Vet Med, Iowa State Univ & pathologist, Vet Med Res Inst, 63-64; assoc prof path, Sch Vet Med, Univ Mo-Columbia, 64-68; ASSOC PROF PATH, COL VET MED, KANS STATE UNIV, 68- *Mem:* Am Vet Med Asn; Electron Micros Soc Am; Conf Res Workers Animal Diseases. *Res:* Pathologic and epidemiologic studies of animal neoplasms; ultrastructure studies on the pathogenesis of morphological tissue alterations. *Mailing Add:* Dept Vet Path Col Vet Med Kans State Univ Manhattan KS 66506

STRAHL, ERWIN OTTO, b New York, NY, July 2, 30; m 52; c 4. MINERALOGY, PETROLOGY. *Educ:* City Col, New York, BS, 52; Pa State Univ, PhD(mineral), 58. *Prof Exp:* Res asst mineral, Pa State Univ, 52-58; mineralogist, Mineral Resources Dept, Kaiser Aluminum & Chem Corp, 58-59, Metals Div, Res Lab, 59-69, head, X-ray & Electron Optics Lab, Anal Res Dept, 69-81, RES ASSOC REDUCTION RES, KAISER CTR TECHNOL, 81- *Mem:* Am Spectrog Soc. *Res:* Mineralogy and petrology of soils and sedimentary rocks; x-ray diffraction analysis of inorganic oxides and hydroxides; quantitative x-ray diffraction and spectographic analysis; phase equilibria studies of aluminum oxide systems. *Mailing Add:* Danville San Ramon CA 94526

STRAHLE, WARREN C(HARLES), b Whittier, Calif, Dec 29, 38; div; c 1. COMBUSTION, FLUID MECHANICS. *Educ:* Stanford Univ BS, 59, MS, 60; Princeton Univ, MA & PhD(aerospace eng), 64. *Prof Exp:* Mem tech staff, Propulsion Dept, Aerospace Corp, 64-67; mem prof staff, Sci & Tech Div, Inst Defense Anal, 67-68; from assoc prof to prof, 69-73, REGENTS PROF AEROSPACE ENG, GA INST TECHNOL, 73- *Concurrent Pos:* Consult numerous indust firms, 69-; assoc ed, J Am Inst Aeronaut & Astronaut, 79-81; mem bd dir, Combustion Inst, 86- *Honors & Awards:* Sigma Xi Res Award, Ferst Found, 71, 73 & 77. *Mem:* Combustion Inst; Am Inst Aeronaut & Astronaut; Am Soc Eng Educ. *Res:* Combustion research applicable to solid and liquid rockets and air breathing engines; combustion noise; turbulent reacting flows; underwater explosions. *Mailing Add:* Sch Aerospace Eng Ga Inst Technol Atlanta GA 30332

STRAHM, NORMAN DALE, b Toronto, Kans, Feb 22, 40. PHYSICS, ELECTRICAL ENGINEERING. *Educ:* Mass Inst Technol, SB, 62, SM & EE, 64, PhD(elec sci & eng), 69. *Prof Exp:* Mem tech staff, Lincoln Lab, Mass Inst Technol, 69-70; vis asst prof physics, Univ Ill, Chicago Circle, 70-; PRES, JAMES ORCUTT & CO INC. *Mem:* AAAS; Am Phys Soc; Inst Elec & Electronics Engrs. *Res:* Light scattering; crystal lattice dynamics; quantum optics and quantum electronics. *Mailing Add:* 815 Madison St Evanston IL 60202

STRAHS, GERALD, b New York, NY, May 26, 38; m 60; c 3. PHYSICAL CHEMISTRY. *Educ:* Cooper Union Univ, BChE, 60; Univ Ill, MS, 62, PhD(phys chem), 65; Univ Juarez, MD, 80. *Prof Exp:* Assoc chem, Univ Calif, San Diego, 65-68; asst prof biochem, NY Med Col, 68-71; chemist, Crime Lab Sect, NY Police Dept, 72; chemist, US Assay Off, NY, 72-73; chief chemist, Consolidated Refining Co, 73-76; chemist, Brooklyn Hosp, 77-78; intern, USPHS Hosp, 80-81; MED PRES, MERCY CATH MED CTR, 81- *Concurrent Pos:* Am Cancer Soc fel, 65-68. *Res:* Precious metals, refining, assaying, recovery. *Mailing Add:* 220 Powell Rd Springfield PA 19064

STRAHS, KENNETH ROBERT, CELL BIOLOGY. *Educ:* Univ Pa, PhD(cell biol), 75. *Prof Exp:* MGR NEW TECHNOL, BECKMAN INSTRUMENTS INC, 84-, RES MGR SPEC CHEM & BIOL METHODS, DIAG SYSTS GROUP, 85- *Mailing Add:* 200 S Krammer Blvd Beckman Instruments Inc Brea CA 92621

STRAIGHT, H JOSEPH, b Dunkirk, NY, Jan 26, 51; m 70. MATHEMATICS, STATISTICS. *Educ:* State Univ NY, Fredonia, BS, 73; Western Mich Univ, MA, 76, PhD(math-graph theory), 77. *Prof Exp:* Asst math, Western Mich Univ, 73-77; asst prof, 77-80, ASSOC PROF MATH, STATE UNIV NY COL FREDONIA, 80- *Concurrent Pos:* Vis prof, Clemson Univ, 80-81. *Mem:* Math Asn Am; NY Acad Sci; Sigma Xi. *Res:* Graph theory, partitions, colorings of graphs; decomposition of graphs into trees. *Mailing Add:* 3231 Cable Rd-RD One Fredonia NY 14063

STRAIGHT, JAMES WILLIAM, b Wichita, Kans, Aug 5, 40; m 61; c 2. MECHANICAL ENGINEERING. *Educ:* Univ Kans, BS & MS, 63; Univ Ariz, PhD(mech eng), 67. *Prof Exp:* Prog dir underground nuclear testing, Test Command/Defense Atomic Support Agency, 67-69; asst prof mech eng, Vanderbilt Univ, 69-71; from asst prof to assoc prof, Christian Bros Col, 71-77; staff mem, 77-80, assoc group leader, 77-83, dep group leader, 83-85, GROUP LEADER, LOS ALAMOS NAT LAB, 85- *Concurrent Pos:* Consult, Ken O'Brien & Assocs, 69-74 & Brown-Straight Consult, 72-77. *Honors & Awards:* Teetor Award, Soc Automotive Engrs, 71. *Mem:* Am Soc Mech Engrs; Am Soc Eng Educ; Soc Automotive Engrs; Sigma Xi. *Res:* Vibrations; structural dynamics; acoustics; stress analysis; instrumentation; explosives applications; hypervelocity impact. *Mailing Add:* One Comanche Lane Los Alamos NM 87544

STRAIGHT, RICHARD COLEMAN, b Rivesville, WVa, Sept 8, 37; m 63; c 3. PHOTOBIOLOGY. *Educ:* Univ Utah, BA, 61, PhD(molecular biol), 67. *Prof Exp:* Asst dir radiation biol summer inst, Univ Utah, 61-63; SUPVRY CHEMIST, MED SERV, VET ADMIN HOSP, 65-, DIR, VET ADMIN VENOM RES LAB, 75-, ADMIN OFFICER RES SERV, VET ADMIN CTR, 80- *Mem:* AAAS; Am Chem Soc; Biophys Soc; Am Soc Photobiol; Int Solar Energy Soc. *Res:* Photodynamic action of biomonomers and biopolymers; tumor immunology; effect of antigens on mammary adenocarcinoma of C3H mice; ageing; biochemical changes in ageing; venom toxicology; mechanism of action of psychoactive drugs. *Mailing Add:* 860 S 22nd St Salt Lake City UT 84108

STRAILE, WILLIAM EDWIN, b Beaver, Pa, Mar 22, 31; m 53; c 4. BIOLOGICAL SCIENCE, NEUROSCIENCES. *Educ:* Westminster Col, AB, 53; Brown Univ, ScM, 55, PhD(biol), 57. *Prof Exp:* Sr cancer res scientist, Springville Labs, Roswell Park Mem Inst, 61-65; assoc prof anat & head cell res sect, Med Sch, Temple Univ, 66-75; grants assoc, Div Res Grants, NIH, 75-76; MEM STAFF, ORGAN SYSTS PROGS BR, DIV CANCER BIOL, DIAG & CTRS, NAT CANCER INST, 76- *Concurrent Pos:* Nat Cancer res fel, Univ London, 57-58; res fel, Brown Univ, 58-61; asst res prof, Grad Sch, State Univ NY Buffalo, 62-65. *Res:* Electron microscopy and electrophysiology of nerve endings; neurotransmitter chemicals in the control of neuronal functions, blood flow and cell division; neural elements in melanotic and epidermal neoplasia. *Mailing Add:* Nat Cancer Inst Exec Plaza N Rm 316 Bethesda MD 20892-4200

STRAIN, BOYD RAY, b Laramie, Wyo, July 19, 35; m 58; c 2. PHYSIOLOGICAL ECOLOGY. *Educ:* Black Hills State Col, BS, 60; Univ Wyo, MS, 61; Univ Calif, Los Angeles, PhD(plant sci), 64. *Prof Exp:* Asst prof bot & plant ecol, Univ Calif, Riverside, 64-69; assoc prof, 69-77, dir, Duke Phytotron, 79-89, PROF BOT, DUKE UNIV, 77- *Concurrent Pos:* Mem, Panel Ecol Sect, NSF, 72-75 & Comt Mineral Resources & Environ, Nat Res Coun, 73-74; mem, Am Inst Biol Sci Adv Panel, NASA, 76-87, panel biol facil & ctrs, NSF, 87-89; mem, Int Union Biol Sci, 89-95. *Mem:* Ecol Soc Am; Bot Soc Am; Am Inst Biol Sci(pres, 87-88); AAAS. *Res:* Physiological adaptations of plants to extreme environments; ecosystems analysis. *Mailing Add:* Dept Bot Duke Univ Durham NC 27706

STRAIN, JAMES E, b Apr 23, 23; m; c 4. PEDIATRICS. *Educ:* Phillips Univ, AB, 45; Univ Colo, MD, 47; Am Bd Pediat, cert, 54, 80 & 86. *Prof Exp:* Intern, Minneapolis Gen Hosp, Minn, 47-48; resident pediat, Denver Children's Hosp, Colo, 48-50, dir, Genetic Serv, 82-86; gen pediat pract, Denver, Colo, 50-86; EXEC DIR, AM ACAD PEDIAT, 86- *Concurrent Pos:* Clin prof pediat, Univ Colo Med Ctr, 69-86, Univ Chicago, 87-; mem, Sect Coun Pediat, AMA, 71-, chmn, 74-79; mem, Comn Children & Youth, State Colo, 71-75; mem bd trustees, Phillips Univ, Enid, Okla, 74-; mem, Task Force Iowa Health Care Standards Proj, 84-85. *Honors & Awards:* Clifford Grulee Award, Am Acad Pediat, 85; Excellence in Pub Serv Award, US Surgeon Gen, 88. *Mem:* Sr mem Inst Med-Nat Acad Sci; fel Am Acad Pediat (pres-elect, 81-82, pres, 82-83); AMA; Am Pub Health Asn. *Mailing Add:* Am Acad Pediat PO Box 927 Elk Grove Village IL 60009-0927

STRAIN, JOHN HENRY, b Worcester, Eng, Oct 28, 22; Can citizen; m 49; c 3. POULTRY SCIENCE. *Educ:* Univ Sask, BSAgr, 49; Iowa State Univ, MS, 60, PhD(poultry breeding), 61. *Prof Exp:* Hatcheryman, Swift Can Co, 49-50; res off, 50-60, scientist poultry genetics, 60-70, HEAD ANIMAL SCI SECT, RES BR, CAN DEPT AGR, 70- *Mem:* Genetics Soc Can; Poultry Sci Asn; Can Soc Animal Sci. *Res:* Poultry genetics, mainly selection and genotype-environment interaction studies; dwarf broiler breeding management systems. *Mailing Add:* 14 Clark Dr Brandon MB R7B 0T9 Can

STRAIT, BRADLEY JUSTUS, b Canandaigua, NY, Mar 17, 32; m 57; c 2. ELECTRICAL ENGINEERING. *Educ:* Syracuse Univ, BS, 58, MS, 60, PhD(elec eng), 65. *Prof Exp:* Engr, Eastman Kodak Co, NY, 60-61; from asst prof to assoc prof, 65-74, PROF ELEC ENG, SYRACUSE UNIV, 74-, CHMN DEPT, 74- *Mem:* Inst Elec & Electronics Engrs; Sigma Xi. *Res:* Application of computers to antenna problems; array antennas; scattering systems and their effects on antenna performance; electromagnetic theory; microwave measurements. *Mailing Add:* Dept of Elec Eng 111 Link Hall Syracuse Univ Syracuse NY 13244

STRAIT, JOHN, b Blackford Co, Ind, Nov 29, 15; m 46; c 3. AGRICULTURAL ENGINEERING. *Educ:* Purdue Univ, BS, 38; Univ Minn, MS, 45. *Prof Exp:* From asst prof to assoc prof agr eng, Univ Minn, St Paul, 38-65, prof, 65-; RETIRED. *Mem:* Am Soc Agr Engrs. *Res:* Design and development of farm machinery; farm processes; internal combustion engines; mechanics; heating, refrigeration and air conditioning; mechanical engineering. *Mailing Add:* 1763 Fairview Ave N St Paul MN 55113

STRAIT, PEGGY, b Canton, China, Apr 20, 33; US citizen; m 55; c 2. MATHEMATICS. *Educ:* Univ Calif, Berkeley, BA, 53; Mass Inst Technol, MS, 57; NY Univ, PhD(math), 65. *Prof Exp:* Programmer math, Livermore Radiation Lab, Univ Calif, 54-55 & Lincoln Lab, Mass Inst Technol, 55-57; res assoc, G C Dewey Corp, NY, 57-62; lectr, 64-65, from asst prof to assoc prof, 65-72, PROF MATH, QUEENS COL, NEW YORK, 76- *Concurrent Pos:* Lincoln lab assoc staff fel, Mass Inst Technol, 56-57; res assoc fel, NY Univ, 62-64; NSF sci fac fel, 71-72. *Mem:* Am Math Soc. *Res:* Stochastic processes; probability theory and applications; mathematical statistics. *Mailing Add:* Dept of Math Queens Col Flushing NY 11367

STRAITON, ARCHIE WAUGH, b Tarrant Co, Tex, Aug 27, 07; m 32; c 2. ELECTRICAL ENGINEERING. *Educ:* Univ Tex, BSEE, 29, MA, 31, PhD(physics), 39. *Prof Exp:* Mem inspection dept, Bell Tel Labs, Inc, 29-30; assoc prof eng, Tex Col Arts & Indust, 31-41, prof & dir eng, 41-43; from assoc prof to prof, Univ Tex, Austin, 43-63, dir elec eng res lab, 47-72, chmn dept, 66-71, actg vpres & dean grad sch, 72-73, Asbel Smith prof elec eng, 63-89, EMER ASBEL SMITH PROF, UNIV TEX, AUSTIN, 89- *Honors & Awards:* Edison Medal, Inst Elec & Electronic Engrs. *Mem:* Nat Acad Eng; Am Soc Eng Educ; fel Inst Elec & Electronic Engrs. *Res:* Atmospheric refractive index properties; interaction of atmosphere and radio waves; electrical physics; measurement of electrical characteristics of filters; harmonic solution of differential equations. *Mailing Add:* 4212 Far West Blvd Austin TX 78731

STRAKA, WILLIAM CHARLES, b Phoenix, Ariz, Oct 21, 40; m 66; c 1. ASTROPHYSICS. *Educ:* Calif Inst Technol, BS, 62; Univ Calif, Los Angeles, MA, 65, PhD(astron), 69. *Prof Exp:* Teacher astron & phys sci, Long Beach City Col, 66-70; asst prof astron, Boston Univ, 70-74; from asst prof to prof astron, Jackson State Univ, 74-84, head dept, physics, 77-78; SR STAFF SCIENTIST, LOCKHEED PALO ALTO RES LAB, 84- *Concurrent Pos:* Vis staff mem, Los Alamos Nat Lab, 76-81; exec secy astron adv comt, NSF, 78-79. *Mem:* Am Astron Soc; Sigma Xi; AAAS; Am Inst Aeronaut & Astronaut; Am Voice Input-Output Soc; Soc Photo-Optical Instrumentation Engrs. *Res:* Structure and evolution of small mass stars; galactic nebulae; dynamics of supernova shells; computerized voice recognition; planetary exploration. *Mailing Add:* 860 Clara Dr Palo Alto CA 94303

STRALEY, JOSEPH PAUL, b Toledo, Ohio, Jan 22, 42; m 67. SOLID STATE PHYSICS. *Educ:* Harvard Col, BA, 64; Cornell Univ, PhD(physics), 70. *Prof Exp:* NSF fel chem, Cornell Univ, 70-71; res assoc physics, Rutgers Univ, 71-73; from asst prof to assoc prof, 73-81, PROF PHYSICS, UNIV KY, 81-; ASSOC PROF PHYSICS, UNIV ALA, 81- *Concurrent Pos:* NSF res grants, Univ Ky, 76-81 & Univ Ala, 81-; sabbatical leave, Mich State Univ, 79-80; asst prof physics, Univ Ala, 80-81. *Mem:* Am Phys Soc. *Res:* Theory of phase transitions; cooperative phenomena; liquid crystals; inhomogeneous conductors; percolation problem. *Mailing Add:* Dept Physics & Astron Univ Ky Lexington KY 40506

STRALEY, JOSEPH WARD, b Paulding, Ohio, Oct 6, 14; m 39; c 3. SPECTROSCOPY. *Educ:* Bowling Green State Univ, BSEd, 36; Ohio State Univ, MSc, 37, PhD(physics), 41. *Prof Exp:* Asst, Ohio State Univ, 37-38, 40-41; actg instr physics, Heidelberg Col, 38-39; instr, Univ Toledo, 41-42, asst prof, 42-44, actg head dept, 43-44; from asst prof to prof, 44-58, EMER PROF PHYSICS, UNIV NC, CHAPEL HILL, 80- *Concurrent Pos:* Guggenheim fel, 56-57. *Mem:* Am Phys Soc; Am Asn Physics Teachers. *Res:* Spectroscopy; research in science and public policy. *Mailing Add:* Physics Dept Univ NC Chapel Hill NC 27515

STRALEY, TINA, b New York, NY, Sept 4, 43; div; c 1. MATHEMATICS. *Educ:* Ga State Univ, BA, 65, MS, 66; Auburn Univ, PhD(math), 71. *Prof Exp:* Teacher math, Miami Beach Sr High Sch, 66-67; instr, Spelman Col, 67-68 & Auburn Univ, 71-73; from asst prof to assoc prof math, 73-84, prof math & comput sci, 84-87, PROF MATH & DEPT CHMN, KENNESAW COL, 87- *Concurrent Pos:* Vis res assoc, Emory Univ, 78-79. *Mem:* Am Math Soc; Nat Coun Teachers Math; Math Asn Am. *Res:* Embeddings, extensions and automorphisms of Steiner systems, design theory, scheduling problems. *Mailing Add:* Dept Math Kennesaw State Col Marietta GA 30061

STRALKA, ALBERT R, b Wilkes-Barre, Pa, Jan 18, 40; m 65; c 2. MATHEMATICS. *Educ:* Wilkes Col, AB, 61; Pa State Univ, MA, 64, PhD(math), 67. *Prof Exp:* Instr math, Wilkes Col, 61-62 & Pa State Univ, 66-67; asst prof, 67-72, assoc prof, 72-76, PROF MATH, UNIV CALIF, RIVERSIDE, 76- *Mem:* Am Math Soc. *Res:* Ordered structures. *Mailing Add:* Dept Math Univ Calif Riverside CA 92521

STRAND, FLEUR LILLIAN, b Bloemfontein, SAfrica, Feb 24, 28; US citizen; m 46; c 1. NERVE REGENERATION, NEUROPEPTIDES. *Educ:* NY Univ, AB, 48, MS, 50, PhD(biol), 52. *Prof Exp:* Instr biol, Brooklyn Col, 51-57; NIH fel, Physiol Inst, Free Univ Berlin, 57-59; from asst prof to assoc prof, 61-73; actg chmn dept, 82, 89, PROF BIOL, NY UNIV, 73- *Concurrent Pos:* Prin investr neuroendocrine res; chmn, NY Acad Sci, 88. *Honors & Awards:* Am Med Writers Award, 84. *Mem:* AAAS; Am Physiol Soc; Soc Neurosci; Int Soc Psychoneuroendocrinol; Int Soc Develop Neurosci; Sigma Xi; NY Acad Sci (pres, 87). *Res:* Neurohormonal integration; effect of hormones on developing and regenerating nerve and muscle; neuroendocrinology; peptide hormones; editor and author of numerous books and scientific articles. *Mailing Add:* Dept Biol NY Univ Washington Sq New York NY 10003

STRAND, JAMES CAMERON, b East St Louis, Ill, June 1, 43; m 52; c 2. NEPHROLOGY, HYPERTENSION. *Educ:* Monmouth Col, Ill, BA, 66; St Louis Univ, Mo, MS, 73; Univ Nebr, Omaha, PhD(physiol), 77. *Prof Exp:* Nephrol res fel, Mayo Clinic & Found, 77-80; cardiovasc res fel, Georgetown Univ, 80-81; res assoc, A H Robins Co, 81-85; PRIN INVESTR, DEPT PHARMACOL, PENNWATT CORP, 85- *Mem:* Am Physiol Soc; Am Soc Hypertension; Am Soc Nephrology. *Res:* Mechanisms associated with cardiovascular and renal functions in hypertension and congestive heart failure. *Mailing Add:* Fisons Pharmaceut 755 Jefferson Rd Rochester NY 14623

STRAND, JOHN A, III, b Red Bank, NJ, July 22, 38; m 63; c 4. POLLUTION BIOLOGY. *Educ:* Lafayette Col, AB, 60; Lehigh Univ, MS, 62; Univ Wash, PhD(fisheries biol), 75. *Prof Exp:* Fisheries biologist, NJ Bur Fisheries Lab, 62-63; res scientist, US Naval Radiol Defense Lab, 64-69; sr res scientist aquatic ecol, Pac Northwest Labs, Battelle Mem Inst, 69-; marine res lab, Sequim, Wash; AT OFF OIL SPILL DAMAGE ASSESSMENT, NAT MARINE FISHERIES SERV. *Concurrent Pos:* auxiliary fac, Sch Fisheries, Univ Wash, 87-91. *Mem:* Sigma Xi; Am Inst Fishery Res Biologists; Naval & Reserve Asn. *Res:* Aquatic radioecology; biological accumulation of radioisotopes in biological systems and their effects; effects and fate of petroleum residues and other contaminants in biological systems and synthetic fuel; mariculture. *Mailing Add:* Off Oil Spill Damage Assessment Nat Marine Fisheries Serv PO Box 210029 Auke Bay AK 99821

STRAND, KAJ AAGE, b Hellerup, Denmark, Feb 27, 07; nat US; m 43, 49; c 2. ASTRONOMY. *Educ:* Univ Copenhagen, BA & MSc, 31, PhD(astron), 38. *Prof Exp:* Geodesist, Geod Inst, Copenhagen, 31-33; asst to dir observ, Univ Leiden, 33-38; res assoc astron, Swarthmore Col, 38-42, res astronr, 46, Am-Scand Found fel, 38-39, Danish Rask-Orsted Found fel, 39-40; assoc prof astron, Univ Chicago, 46-47, res assoc, 47-67; prof astron, Northwestern Univ & dir, Dearborn Observ, 47-58; dir astrometry & astrophys, US Naval Observ, 58-63, sci dir, 63-77; CONSULT, 77- *Concurrent Pos:* Guggenheim fel, 46; consult, NSF, 53-56 & Lincoln Lab, Mass Inst Technol, 81-86; vis prof, Acad Sinica, China, 87. *Honors & Awards:* Knight Cross First Class, Royal Order Dannebrog, Denmark, 77; Honor Cross, Literis et Artibus, First Class, Austria, 78. *Mem:* Int Astron Union; Am Astron Soc; Netherlands Astron Soc; Royal Danish Acad. *Res:* Photographic observations of double stars; stellar parallaxes; orbital motion in double and multiple systems; instrumentation. *Mailing Add:* 3200 Rowland Pl NW Washington DC 20008

STRAND, RICHARD ALVIN, b Ridgway, Pa, July 30, 26; m 48; c 3. ELECTRICAL ENGINEERING. *Educ:* Pa State Univ, BS, 50, MS, 51, PhD(elec eng), 63. *Prof Exp:* Prod engr, Elliott Co, Pa, 51-56; from instr to asst prof, Pa State Univ, 56-64; assoc prof, 64-66, chmn dept, 64-81, asst dean eng, 70-79, PROF ELEC ENG, UNIV BRIDGEPORT, 66-, ASSOC DEAN ENG, 80- *Mem:* Inst Elec & Electronics Engrs; Am Soc Eng Educ. *Res:* Generalized analysis of electromechanical energy converters; curriculum development; educational methods; measurement of effective teaching. *Mailing Add:* 4672 Madison Ave Trumbull CT 06611

STRAND, RICHARD CARL, b Langdon, NDak, Mar 20, 33; m 61; c 2. PARTICLE DETECTOR DEVELOPMENT. *Educ:* NDak State Univ, BS, 55; Johns Hopkins Univ, PhD(physics), 61. *Prof Exp:* Res assoc physics, Johns Hopkins Univ, 61-62; asst physicist, 62-65, assoc physicist, 65-67, PHYSICIST, BROOKHAVEN NAT LAB, 67- *Concurrent Pos:* Fullbright scholar, Univ Lyons, France, 55-56. *Mem:* Am Phys Soc. *Res:* Experimental sub-nuclear particle physics; software data analysis. *Mailing Add:* Physics Dept Brookhaven Nat Lab Upton NY 11973

STRAND, ROBERT CHARLES, organic polymer chemistry, for more information see previous edition

STRAND, TIMOTHY CARL, b Marshalltown, Iowa, Apr 17, 48; m 67; c 1. OPTICAL STORAGE TECHNOLOGIES. *Educ:* Univ Iowa, BA, 70; Univ Calif, San Diego, MS, 73, PhD(appl physics), 76. *Prof Exp:* Res staff mem, Naval Electronics Lab Ctr, 71-73; res asst physics, Univ Erlangen, WGer, 73-76; res scientist, Univ Southern Calif, 76-79, res asst prof elec eng & optics, 79-83; res staff mem, San Jose Res Lab, 83-85, mgr, Mach Vision Sensing, 85-88, MGR, EXPLOR OPTICS, ALMADEN RES CTR, INT BUS MACH, 88- *Mem:* Fel Optical Soc Am; sr mem Inst Elec & Electronic Engrs; Soc Photo-Optical Instrumentation Engrs. *Res:* Optical storage technologies and integrated optics; machine vision; developing optical inspection and measurement techniques; optical computing; computer generated holography; optical information processing. *Mailing Add:* IBM Almaden Res Ctr K69-803E 650 Harry Rd San Jose CA 95120

STRANDBERG, MALCOM WOODROW PERSHING, b Box Elder, Mont, Mar 9, 19; m 47; c 4. SOLID STATE PHYSICS. *Educ:* Harvard Univ, SB, 41; Mass Inst Technol, PhD(physics), 48. *Prof Exp:* Res assoc, Mass Inst Technol, 41-42, mem staff, Off Sci Res & Develop, 42-43, microwave develop, 43-45, res assoc, 45-48, from asst prof to assoc prof, 48-60, prof physics, 60-88, EMER PROF PHYSICS, MASS INST TECHNOL, 88- *Concurrent Pos:* Fulbright lectr, Univ Grenoble, 61-62. *Mem:* Fel Am Phys Soc; fel Inst Elec & Electronics Engrs; fel Am Acad Arts & Sci; fel AAAS; NY Acad Sci. *Res:* Design of microwave components, radio transmitters and receivers; biological physics. *Mailing Add:* Mass Inst of Technol 26-353 Cambridge MA 02139

STRANDHAGEN, ADOLF G(USTAV), b Scranton, Pa, May 4, 14; m 41; c 2. ENGINEERING MECHANICS. *Educ:* Univ Mich, BS, 39, MS, 40, PhD(eng mech), 42. *Prof Exp:* From instr to asst prof mech, Carnegie Inst Technol, 42-47; assoc prof, 47-50, prof eng mech & head dept, 50-58, prof eng sci & head dept, 58-69, prof eng, 69-76, PROF AEROSPACE & MECH ENG, UNIV NOTRE DAME, 76- *Concurrent Pos:* Consult, US Navy Mine Defense Lab, 61-67. *Mem:* Soc Naval Archit & Marine Engrs. *Res:* Engineering sciences; hydrodynamics; applications of probability theory; stability and maneuvering of ships. *Mailing Add:* Col Eng Univ Notre Dame Notre Dame IN 46556

STRANDHOY, JACK W, b Evanston, Ill, Aug 8, 44; m 67; c 2. RENAL PHARMACOLOGY, VASOPRESSIN. *Educ:* Univ Ill, BS, 67; Univ Iowa, MS, 69, PhD(pharmacol), 72. *Prof Exp:* NIH fel physiol, Mayo Clin, 71-73; sr res investr, NC Heart Asn, 73-75; asst prof, 75-80, ASSOC PROF PHARMACOL, BOWMAN GRAY SCH MED, WAKE FOREST UNIV, 80- *Concurrent Pos:* Consult, Mead Johnson, 76, Curric Designs, Inc, 77, Wyeth-Ayerst, 80-90 & Ciba-Geigy, 90; prin investr, NIH, 79- *Mem:* Am Soc Pharmacol & Exp Therapeut; Am Soc Nephrology; Int Soc Nephrology; Am Fedn Clin Res; Sigma Xi. *Res:* Mechanisms by which vasoactive substances, especially catecholamines, prostaglandins, and antihypertensive drugs, affect renal water and electrolyte metabolism; fetal renal development. *Mailing Add:* Dept Physiol & Pharmacol Bowman Gray Sch Med Winston-Salem NC 27103

STRANDJORD, PAUL EDPHIL, b Minneapolis, Minn, Apr 5, 31; m 53; c 2. CLINICAL CHEMISTRY, LABORATORY MEDICINE. *Educ:* Univ Minn, BA, 51, MA, 52; Stanford Univ, MD, 59. *Prof Exp:* Intern med, Sch Med, Univ Minn, 59-60, from instr to assoc prof lab med, 63-69; PROF LAB MED & CHMN DEPT, SCH MED, UNIV WASH, 69- *Concurrent Pos:* USPHS med fel, Univ Minn, 61-63; pres, Asn Univ Physicians, Univ Wash, 87- *Honors & Awards:* Gerald T Evans Award in Lab Med, 76. *Mem:* AAAS; Acad Clin Lab Physicians & Sci; Am Chem Soc; Am Fedn Clin Res; Am Asn Clin Chem; Am Mgt Asn. *Res:* Diagnostic enzymology. *Mailing Add:* Dept Lab Med Univ Hosp SBIO Univ Wash Sch Med Seattle WA 98195-0001

STRANDNESS, DONALD EUGENE, JR, b Bowman, NDak, Sept 22, 28; m 57; c 3. MEDICINE, SURGERY. *Educ:* Pac Lutheran Univ, BA, 50; Univ Wash, MD, 54. *Prof Exp:* From instr to assoc prof, 62-70, PROF SURG, SCH MED, UNIV WASH, 70- *Concurrent Pos:* Res fel, Nat Heart Inst, 59-60; NIH career develop award, 65-; clin investr, Vet Admin, 62-65. *Mem:* Soc Vascular Surg; Am Inst Ultrasonics in Med; Am Col Surg; Am Surg Asn; Int Cardiovasc Soc; Nat Heart, Lung & Blood Inst. *Res:* Peripheral vascular disease and physiology. *Mailing Add:* Dept Surg Univ Wash Sch Med Seattle WA 98195

STRANG, GILBERT, b Chicago, Ill, Nov 27, 34; m 58; c 3. APPLIED MATHEMATICS. *Educ:* Mass Inst Technol, SB, 55; Oxford Univ, BA, 57; Univ Calif, Los Angeles, PhD(math), 59. *Prof Exp:* Moore instr math, 59-61, from asst prof to assoc prof, 61-70, PROF MATH, MASS INST TECHNOL, 70- *Concurrent Pos:* NATO fel, Oxford Univ, 61-62; Sloan fel, Mass Inst Technol, 66-67; Fairchild scholar, Calif Inst Technol, 81. *Honors & Awards:* Chauvenet Prize, Math Asn Am, 75. *Mem:* Am Math Soc; Soc Indust & Appl Math (vpres, 91-); Math Asn Am. *Res:* Mathematical analysis applied to linear algebra and partial differential equations; author of four textbooks on mathematics. *Mailing Add:* Seven Southgate Rd Wellesley MA 02181

STRANG, ROBERT M, b Gt Brit, 26; m 53; c 5. FOREST & RANGELAND ECOLOGY. *Educ:* Univ Edinburgh, BSc, 50; Univ London, PhD(ecol), 65. *Prof Exp:* Res off forestry, Colonial Develop Corp, Swaziland, Nyasaland & Tanganyika, 50-57 & Rhodesian Wattle Co, Ltd, 57-62; forest ecologist, Northern Forest Res Ctr, Forestry Serv, Can Dept Environ, 65-73, biologist, Northern Natural Resources & Environ Br, Arctic Land Use Res, Can Dept Indian & Northern Affairs, 73-74, head, Environ Studies Sect, Northern Natural Resources & Environ Br, 74-75; assoc prof rangeland ecol & mgt, Fac Agr Sci Forestry, Univ BC, 75-81; exec dir, Forest Res Coun BC, 81-87; ASSOC DEAN RENEWABLE RESOURCES, BC INST TECH, 87- *Concurrent Pos:* Hon res assoc, Univ NB, 67-71; secy, Conserv Coun NB, 69-71; adj prof, Simon Fraser Univ, 79-; chmn, Forest Educ Coun BC, 88- *Mem:* Commonwealth Forestry Asn; Soc Range Mgt; Can Inst Forestry. *Res:* Resource and land management. *Mailing Add:* 2456 141st St Surrey BC V4A 4K2 Can

STRANG, RUTH HANCOCK, b Bridgeport, Conn, Mar 11, 23. PEDIATRICS, CARDIOLOGY. *Educ:* Wellesley Col, BA, 44; New York Med Col, MD, 49. *Prof Exp:* Intern, Flower & Fifth Ave Hosps, New York, 49-50, resident pediat, 50-52; from instr to asst prof bact, New York Med Col, 52-57, instr pediat, 52-56, asst clin prof, 56-57; from asst prof to assoc prof, 62-70, PROF PEDIAT, UNIV MICH, ANN ARBOR, 70- *Concurrent Pos:* Fel cardiol, Babies Hosp, New York, 56-57 & Hopkins Hosp, Baltimore, 57-59; res fel, Children's Hosp, Boston, 59-62; mem, Am Heart Asn. *Mem:* Fel Am Acad Pediat; Am Col Cardiol. *Res:* Congenital heart disease; effect on growth; ventricular performance; echocardiography. *Mailing Add:* Dept Pediat Univ Mich 1500 E Medical Center Dr Ann Arbor MI 48109

STRANG, W(ILLIAM) GILBERT, b Chicago, Ill, Nov 27, 34; m 58; c 3. MATHEMATICS. *Educ:* Mass Inst Technol, SB, 55; Oxford Univ, BA, 57; Univ Calif, Los Angeles, PhD(math), 59. *Prof Exp:* Moore instr, 59-61, from asst prof to assoc prof, 62-69, PROF MATH, MASS INST TECHNOL, 69- *Concurrent Pos:* Fels, NATO, 61-62 & Sloan Found, 65-67; Fairchild scholar, 80; pres, Wellesley-Cambridge Press. *Mem:* Am Math Soc; Math Asn Am; Am Acad Arts & Sci. *Res:* Partial difference and differential equations; matrix analysis; optimization. *Mailing Add:* Dept Math Mass Inst Technol Cambridge MA 02139

STRANGE, LLOYD K(EITH), b Burkburnett, Tex, Dec 17, 22; m 43; c 2. MECHANICAL & PETROLEUM ENGINEERING. *Educ:* Southern Methodist Univ, BS, 50, MS, 56. *Prof Exp:* Petrol eng asst, Magnolia Petrol Co, 50-52; engr, Petrol Prod Eng Co, 52-53; res engr, Mobil Res & Develop Corp, 53-56, sr res engr, 56-63, eng assoc, field res lab, 63-87; ENG CONSULT, 88- *Mem:* Am Soc Mech Engrs; Soc Petrol Engrs. *Res:* Planning, operating and evaluating laboratory and field experiments on improved crude oil recovery processes; application of new research results. *Mailing Add:* 901 Danish Grand Prairie TX 75050

STRANGE, RONALD STEPHEN, b Covington, Ky, Nov 18, 43; m 70; c 4. INORGANIC CHEMISTRY, EDUCATION. *Educ:* Univ Ky, BS, 65; Univ Ill, Urbana, MS, 67, PhD(inorg chem), 71; Stevens Inst of Tech, MS, 87. *Prof Exp:* Instr chem, Ill Inst Technol, 70-71; asst prof, 71-78, chmn dept, 75-81, ASSOC PROF CHEM, FAIRLEIGH DICKINSON UNIV, FLORHAM-MADISON CAMPUS, 78- *Concurrent Pos:* Fac res grant-in-aid, Fairleigh Dickinson Univ, 72-74; NSF teacher training grants, 79-80 & 80-81; vis fel, Princeton Univ, 79. *Mem:* Am Chem Soc; Sigma Xi. *Res:* Molecular modelling; semi empirical self consistent field molecular orbital calculations; graph theory in chemistry; neural networks in chemistry. *Mailing Add:* 121 Park Ave Madison NJ 07940-1525

STRANGES, ANTHONY NICHOLAS, b Niagara Falls, NY, Sept 28, 36; m 63; c 2. HISTORY OF ENERGY & SYNTHETIC FUELS, HISTORY OF VALANCE THEORY. *Educ:* Niagara Univ, BS, 58, MS, 64; Univ Wis-Madison, PhD(hist sci), 77. *Prof Exp:* Teacher chem & physics, Notre Dame Col Sch, Welland, Ont, 59-62; teacher chem, Lewiston-Porter High Sch, Lewiston, NY, 63-69; asst prof, 77-83, ASSOC PROF HIST SCI, TEX A&M UNIV, 83- *Mem:* Hist Sci Soc; Hist Chem Soc; Soc Hist Technol; Am Hist Asn; Can Sci & Technol Hist Asn. *Res:* History of twentieth-century science, especially electron theories of valance, history of energy, synthetic liquid fuel production from coal and tar using high-pressure liquefaction and Fischer-Tropsch synthesis. *Mailing Add:* 1205 Barak Lane Tex A&M Univ Bryan TX 77802

STRANGWAY, DAVID W, b Simcoe, Ont, June 7, 34; m 57; c 3. GEOPHYSICS. *Educ:* Univ Toronto, BA, 56, MA, 58, PhD(physics), 60; FRAS, FRSC. *Hon Degrees:* DLitts, Victoria Univ, 86, Univ Toronto, 86; DSc, Mem Univ, NFLD, 86, McGill Univ, Montreal, 89, Ritsumeikan Univ, Kyoto, Japan, 90. *Prof Exp:* Sr geophysicist, Dominion Gulf Co, Toronto, 56; chief geophysicist, Ventures Ltd, Ont, 56-57, sr geophysicist, 58; res geophysicist, Kennecott Copper Corp, Denver, Colo, 60-61; asst prof geol, Univ Colo, Boulder, 61-64; asst prof geophys, Mass Inst Technol, 65-68; from assoc prof to prof physics, Univ Toronto, 68-85, prof, dept geol, 72-85, chmn, 72-80, vpres & provost, 80-83, pres, 83-84; PRES & VCHANCELLOR, UNIV BC, 85- *Concurrent Pos:* Consult, Kennecott Copper Corp, Anaconda Co, UN, Alyeska Pipelines & NASA, Environ Res Inst Mich, GTE/Sylvania, Seru Nucleaire & Barringer Res; mem, subcomt appl geophysics & subcomt geomagnetism, Nat Res Coun, 68-70; mem, Lunar Sample Anal Planning Team, 69-72, Lunar Base Working Group, 84, chmn, Lunar Sci Coun, 74-76; chief, Geophysics Br, Johnson Space Ctr, NASA, Houston, Tex, 70-72, Physics Br, 72-73; vis prof, dept geol, Univ Houston, 71-73; assoc ed, Geophysics, 73-75, Can J Earth Sci, 73-76 & Geophys Res Lett, 77-80; mem, Team Basaltic Volcanism Terrestrial Planets, Lunar & Planetary Inst, 77-80, Nat Acad Planetary Explor, 85-86; pres, Can Geosci Coun, 80; univ space res assoc, Coun Institutions, 84-85; mem, Nat Acad Planetary Explor, 85-86; dir, MacMillan Bloedel, Ltd, Bus Coun BC, Corp Higher Educ Forum, Can Int Inst Sustainable Develop, Echo Bay Mines Inc; chmn, BC Task Force on Environ & Econ, 89; mem, Premier's Adv Coun on Sci Tech. *Honors & Awards:* Medal Except Sci Achievement, NASA, 72; Virgil Kauffman Gold Medal, Soc Explor Geophysicists, 74; Pahlavi Lectr, Iran, 78; Logan Gold Medal, Geol Asn Can, 84; J Tuzo Wilson Medal, Can Geophys Union, 87. *Mem:* Hon mem Soc Explor Geophysicists (vpres, 79-80); fel Royal Soc Can; Can Geophys Union; Geol Asn Can (vpres, 77-78, pres, 78-79); hon mem Can Explor Geophysicists Soc; Am Geophys Union; Europ Asn Explor Geophysicists; Soc Geomagnetism & Geoelec Japan; fel Royal Astron Soc; AAAS. *Res:* History of the earth's magnetic field; studies of ancient reversals of the field; changes in direction and intensity and secular variation; exploration using electromagnetic techniques; magnetic fields of lunar samples and meteorites; history of magnetic fields in the early solar system. *Mailing Add:* Pres Off Univ Toronto Vancouver BC V6T 1Z2 Can

STRANO, ALFONSO J, b Ambridge, Pa, Apr 7, 27; m 57; c 1. VIROLOGY, PATHOLOGY. *Educ:* Hiram Col, BA, 50; Duquesne Univ, MS, 53; Univ Okla, PhD(path), 57; Univ Tex, MD, 60. *Prof Exp:* From instr to asst prof path, Univ Tex Med Br Galveston, 62-67; pathologist, Armed Forces Inst Path & chief, Viro-Path Br, 67-73; CLIN PROF PATH, SCH MED, SOUTHERN ILL UNIV, 73- *Concurrent Pos:* Am Cancer Soc res fel, 60-62. *Mem:* AMA; Col Am Path; Reticuloendothelial Soc; Int Acad Path; Sigma Xi. *Res:* Immunologic aspects of infectious disease, cellular immunity; histologic reaction to viral infections. *Mailing Add:* 18 Wildwood Rd Springfield IL 62704

STRANO, JOSEPH J, b Newark, NJ, Aug 21, 37; m 62. ELECTRICAL & BIOMEDICAL ENGINEERING. *Educ:* Newark Col Eng, BS, 59, MS, 61; Rutgers Univ, PhD(elec eng), 69. *Prof Exp:* From instr to assoc prof, Newark Col Eng, 61-76, assoc chmn dept, 75-76, prof elec eng,78-, chmn dept, 76-; chmn, Dept Elec Eng, 87, PROF ELEC & COMPUTER ENGR, NJ INST TECHNOL, 87- *Concurrent Pos:* NSF res initiation grant, Newark Col Eng, 71-72. *Mem:* Inst Elec & Electronic Engrs; Am Soc Eng Educ; Sigma Xi. *Res:* Automatic control systems; computer systems; instrumentation. *Mailing Add:* Dept Elec Eng NJ Inst Technol Newark NJ 07102

STRANSKY, JOHN JANOS, b Budapest, Hungary, Sept 2, 23; nat US; m 47; c 2. SILVICULTURE. *Educ:* Univ Munich, BF, 47; Harvard Univ, MS, 54; Tex A&M Univ, PhD, 76. *Prof Exp:* Plant propagator, Bussey Inst, Harvard Univ, 54-57; res forester, Southern Forest Exp Sta, US Forest Serv, 57-85; RETIRED. *Concurrent Pos:* Lectr, Sch Forestry, Stephen F Austin State Univ. *Mem:* Soc Am Foresters; Wildlife Soc. *Res:* Silvicultural aspects of combining timber production with wildlife habitat practices in southern forests. *Mailing Add:* 1533 Redbud St Nacogdoches TX 75961

STRASBERG, MURRAY, b New York, NY, Aug 11, 17; m 45. ACOUSTICS. *Educ:* City Col New York, BS, 38; Cath Univ, MS, 48, PhD, 56. *Prof Exp:* Patent examr, US Patent Off, 38-42; physicist, David Taylor Model Basin, 42-49 & 52-58; noise consult, US Bur Ships, 49-52; sci liaison officer, Off Naval Res, London, 58-60; proj coordr, 60-72, SR RES SCIENTIST, DAVID TAYLOR NAVAL SHIP RES & DEVELOP CTR, MD, 72- *Concurrent Pos:* Fulbright lectr, Tech Univ Denmark, 63; adj prof, Am Univ, 64-70; vis prof, Cath Univ, 74-80; mem gov bd, Am Inst Physics, 77- *Mem:* Fel Acoust Soc Am (pres, 74-75, secy, 87-90); Am Inst Physics; Am Phys Soc. *Res:* Underwater acoustics; hydrodynamics; cavitation; hydrodynamic noise; electroacoustic instrumentation; mechanical vibrations. *Mailing Add:* 3531 Yuma St NW Washington DC 20008

STRASSENBURG, ARNOLD ADOLPH, b Victoria, Minn, June 8, 27; m 49, 82; c 5. PHYSICS, EDUCATIONAL ADMINISTRATION. *Educ:* Ill Inst Technol, BS, 51; Calif Inst Technol, MS, 53, PhD(physics), 55. *Prof Exp:* From asst prof to assoc prof physics, Univ Kans, 55-66; prof, State Univ NY Stony Brook, 66-75; head, Mat & Instr Develop Sect, NSF, 75-77; actg vprovost curric & instr, 80-82, PROF PHYSICS, STATE UNIV NY, STONY BROOK, 77- *Concurrent Pos:* Staff physicist, Comn Col Physics, 63-65; dir, Div Educ & Manpower, Am Inst Physics, 66-72; exec officer, Am Asn Physics Teachers, 72-82. *Honors & Awards:* Millikan Lectr Award, Am Asn Physics Teachers, 72. *Mem:* AAAS; Am Asn Physics Teachers; Nat Sci Teachers Asn. *Res:* High energy physics; fundamental particles; measurement of educational outcomes resulting from the application of alternative instructional materials and modes. *Mailing Add:* Dept Physics State Univ NY Stony Brook NY 11794

STRASSER, ALFRED ANTHONY, b Budapest, Hungary, Jan 21, 27; US citizen; m 70; c 1. NUCLEAR FUEL TECHNOLOGY. *Educ:* Purdue Univ, BS, 48; Stevens Inst Technol, MS, 52. *Prof Exp:* Metallurgist, M W Kellog & Co, 48-51 & US Air Force, 51-54; mgr mat dept, Nuclear Develop Assocs & plutonium fuels dept, United Nuclear Corp, 54-72; MGR FUEL & CORE TECHNOL & VPRES, S M STOLLER CORP, 72- *Concurrent Pos:* Adj prof, NY Polytech Inst, 79-81. *Mem:* Am Soc Metals; Am Nuclear Soc. *Res:* Technical and economic evaluation of reactor and reactor component performance, specializing in design, fabrication and thermal-mechanical performance; nuclear plant materials technology. *Mailing Add:* 1102 Bedford Rd Pleasantville NY 10570

STRASSER, ELVIRA RAPAPORT, b Hungary; US citizen; wid; c 2. MATHEMATICS. *Educ:* Washburn Univ, BS, 43; Smith Col, MS, 51; NY Univ, PhD(math), 56. *Prof Exp:* Off Naval Res fel, 59-60; lectr math, Hunter Col, 61; from asst prof to assoc prof, Polytech Inst Brooklyn, 61-67; prof math, State Univ NY Stony Brook, 67-83; RETIRED. *Mem:* Am Math Soc. *Res:* Group theory; graph theory; combinatorial problems. *Mailing Add:* 40 Hastings Dr Stony Brook NY 11790

STRASSER, JOHN ALBERT, b Sydney, NS, Jan 28, 45; m 70; c 2. ENGINEERING, MATERIALS SCIENCE. *Educ:* NS Tech Col, BME, 67, PhD(metall eng), 72; Pa State Univ, University Park, MS, 68. *Prof Exp:* Spec lectr mat sci, Dalhousie Univ, 69 & 70; res scientist, Phys Metall Div, Can Dept Energy, Mines & Resources, 71-76; dir metall, 81-85, vpres mkt, 86-88, PRES, SYDNEY STEEL CORP, 88- *Concurrent Pos:* Dir, Atlantic Group Res Indust Metall, 74-78 & Bra's Dor Inst, 75-78; dir, Atlantic Coal Inst, 80-81. *Mem:* AAAS; Am Soc Metals; Can Inst Mining & Metall; Am Iron & Steel Inst; Can Steel Producers Asn. *Res:* Powder metallurgy; production of powders, their consolidation techniques and their industrial application; rail production. *Mailing Add:* 16 Woodill St Sydney NS B1P 4N7 Can

STRATFORD, EUGENE SCOTT, medicinal chemistry; deceased, see previous edition for last biography

STRATFORD, JOSEPH, b Brantford, Ont, Sept 5, 23; m 52; c 2. NEUROSURGERY. *Educ:* McGill Univ, BSc, 45, MD, CM, 47, MSc, 51, dipl neurosurg, 54; FRCS(C), 56. *Prof Exp:* Lectr neurosurg, McGill Univ, 55-56; from asst prof to prof surg, Univ Sask, 56-62; assoc prof, 62-72, PROF NEUROSURG, MCGILL UNIV, 72- *Concurrent Pos:* Dir div neurosurg, Montreal Gen Hosp. *Mem:* Am Asn Neurol Surg; fel Am Col Surgeons; Cong Neurol Surg; fel Royal Soc Med. *Mailing Add:* Dept Surg McGill Univ Sch Med 3801 University St Montreal PQ H3A 2B4 Can

STRATFORD, R P, b Pocatello, Idaho, Feb, 25; c 6. ELECTRICAL ENGINEERING. *Educ:* Stanford Univ, BSEE, 50. *Prof Exp:* Appln engr indust power systs, Gen Elec, 54, proj engr, Indust Eng Sect, Gen Elec Indust Sales Div, 55-62, Metal Indust Eng Sect, 62-75, consult appln engr, Indust Power Syst Eng Oper, 75-84; mgr, 85-88, SR CONSULT, INDUST POWER SYST UNIT, POWER TECHNOL INC, 88- *Concurrent Pos:* Teacher & developer indust power systs eng & indust power syst harmonics & power factor improvement, Power Technol Inc; mem static power converter comt, Inst Elec & Electronics Engrs, Indust Appln soc working group power syst harmonics, Inst Elec & Electronics Engrs, Power Eng Soc; chmn subcomt Harmonics & reactive compensation static power converter comt, Inst Elec & Electronics Engrs Indust Appln Soc, co-chmn task force rev Inst Elec & Electronics Engrs 519 Harmonic Standard. *Mem:* Fel Inst Elec Electronics Engrs. *Res:* Over 30 technical publications; development of original techniques in analyzing problems caused by harmonic currents from static power converters. *Mailing Add:* 35 Cypress Dr Scotia NY 12307

STRATHDEE, GRAEME GILROY, b Edinburgh, Scotland, June 29, 42; Can citizen; m 67; c 1. SURFACE CHEMISTRY. *Educ:* McGill Univ, BSc, 63, PhD(chem), 67. *Prof Exp:* Assoc res officer chem, Whiteshell Nuclear Res Estab, Atomic Energy Can, Ltd, 67-77, head, Waste Immobilization Sect, 77-80; mgr res & develop planning, 80-85, DIR RES & DEVELOP, POTASH CORP SASK, 85- *Concurrent Pos:* Adj prof, civil engr, Univ Sask, 89- *Mem:* Fel Chem Inst Can; Can Res Mgt Asn; Am Chem Soc; Can Inst Mining & Metall. *Res:* Homogeneous catalysis; catalytic activation of small molecules; hydrogen isotope exchange reactions; enrichment of deuterium; adsorption phenomena; foaming and antifoaming; solidification of high-level liquid waste; glass science and technology; nuclear waste disposal; mineral processing; mine automation. *Mailing Add:* Potash Corp Sask 122 First Ave S Saskatoon SK S7K 7G3 Can

STRATHERN, JEFFREY NEAL, b Keene, NH, Dec 12, 48. GENETICS, MOLECULAR BIOLOGY. *Educ:* Univ Calif, San Diego, BA, 70; Univ Ore, PhD(biol), 77. *Prof Exp:* Fel, Cold Spring Harbor Lab, 77-78, staff investr genetics, 79-84; DIR LAB EUKARYOTIC GENE EXPRESSION, LBI-BRP, NAT CANCER INST-FREDERICK CANCER RES FACIL, 84- *Concurrent Pos:* Damon Runyon/Walter Winchell Cancer Fund fel, 78; adj prof, Dept Biol Sci, Univ Md, Catonsville; ad hoc mem Genetics Study Sect, NIH, 82; mem sci adv comt, Damon Runyon-Walter Winchell Cancer Fund, 84-88; partic, Int Cong Yeast Genetics & Molecular Biol, 84; mem adv comt, Biol Lab Technician Prog, Frederick Community Col, Md, 85. *Res:* Genetics of the control of cell type in yeast, including the demonstration that changes in cell type involve specific DNA rearrangements. *Mailing Add:* 9802 Gas House Pike Frederick MD 21701

STRATHMANN, RICHARD RAY, b Pomona, Calif, Nov 25, 41; m 64; c 2. MARINE BIOLOGY, ZOOLOGY. *Educ:* Pomona Col, BS, 63; Univ Wash, MS, 66, PhD(zool), 70. *Prof Exp:* NIH training grant, Univ Calif, Los Angeles, 70; NSF fel, Univ Hawaii, 70-71; asst prof zool, Univ Md, College Park, 71-73; asst prof, 73-80, ASSOC PROF ZOOL, UNIV WASH & RESIDENT ASSOC DIR, FRIDAY HARBOR LABS, 80- *Mem:* Am Soc Naturalists; Am Soc Limnol & Oceanog; Am Soc Zoologists; Marine Biol Asn UK. *Res:* Population biology, form and function of marine invertebrates; biology of invertebrate larvae; biology of suspension feeding. *Mailing Add:* Friday Harbor Labs Friday Harbor WA 98250

STRATMAN, FREDERICK WILLIAM, b Dodgeville, Wis, Nov 26, 27; m 51; c 1. BIOCHEMISTRY. *Educ:* Univ Wis, BS, 50, MS, 57, PhD(animal-dairy husb, biochem), 61. *Prof Exp:* Res asst animal husb, Univ Wis, 57-61; researcher, Wis Alumni Res Found, 62; res assoc, Univ Wis, 62-63, NIH fel reprod physiol, 63-65; asst prof animal sci & biochem, Univ Ife, Nigeria, 65-67; asst prof animal sci, 67-68, proj assoc, Inst Enzyme Res, 68-70, proj assoc & Babcock fel, 70-71, asst res prof, 71-78, assoc scientist, 78-81, SR SCIENTIST, UNIV WIS-MADISON, 81- *Concurrent Pos:* NIH spec fel, 71-73. *Mem:* Am Soc Biol Chemists. *Res:* Hormonal regulation of protein synthesis, particularly sulfhydryls, polyamines, methylation, phosphorylation, muscle, liver, tumors, perfusions, testosterone, somatomedin; hormonal regulation of gluconeogenesis; hepatocytes; lipid metabolism; spermatozoa; carnitine; exercise metabolism; eating disorders. *Mailing Add:* Inst for Enzyme Res Univ of Wis Madison WI 53706

STRATMEYER, MELVIN EDWARD, b Peoria, Ill, Aug 30, 42; m 66; c 1. RADIOBIOLOGY, RISK ASSESSMENT. *Educ:* Purdue Univ, Lafayette, BS, 65, MS, 66, PhD(bionucleonics), 69. *Prof Exp:* res chemist, 69-73, PMS officer, Ionizing Radiation, 73-75, PMS officer, Ultrasound, Exp Studies Br, 75-82, Chief Sonics Br, Div Biol Effects, Bur Radiol Health, 82-84, CHIEF HEALTH SCI BR, OFF SCI & TECHNOL, CTR DEVICES & RADIOL HEALTH, US FOOD & DRUG ADMIN, 84- *Mem:* AAAS; fel Am Inst Ultrasound Med; Sigma Xi. *Res:* Ionizing radiation effects on nucleic acid and protein metabolism; ionizing radiation effects on mitochrondrial systems; ultrasound effects on growth & development; assessment of risk associated with exposure to medical ultrasound; toxicology of medical device materials. *Mailing Add:* PO Box 617 Ijamsville MD 21754

STRATT, RICHARD MARK, b Philadelphia, Pa, Feb, 21, 54. STATISTICAL MECHANICS, LIQUIDS. *Educ:* Mass Inst Technol, SB, 75; Univ Calif, Berkeley, PhD(chem), 79. *Prof Exp:* Res assoc, Univ Ill, Champaign, 79-80, NSF fel, 80; from asst prof to assoc prof, 86-88, PROF CHEM, BROWN UNIV, 88- *Concurrent Pos:* Alfred P Sloan fel. *Mem:* Am Phys Soc; Am Chem soc. *Res:* Chemical physics and statistical mechanics, especially the statistical mechanics of internal degrees of freedom of molecules in solution; condensed phase problems in general. *Mailing Add:* Box H Dept Chem Brown Univ Providence RI 02912

STRATTAN, ROBERT DEAN, b Newton, Kans, Dec 7, 36; m 60; c 2. ELECTRICAL ENGINEERING. *Educ:* Wichita State Univ, BS, 58; Carnegie-Mellon Univ, MS, 59, PhD(elec eng), 62. *Prof Exp:* Res engr, Wichita Div, Boeing Co, Kans, 61-63; mem tech staff elec eng, Tulsa Div, NAm Rockwell Corp, Okla, 63-68; assoc prof, 68-76, head dept, 68-75, PROF ELEC ENG, UNIV TULSA, 76- *Honors & Awards:* Teetor Award, Soc Automotive Engrs, 82. *Mem:* Inst Elec & Electronic Engrs; Am Soc Eng Educ; Nat Soc Prof Engrs; Soc Automotive Engrs; Am Soc Eng Mgt; Int Microwave Power Inst. *Res:* Electromagnetic theory; radar scattering analysis, measurement and camouflage; electrical power system harmonics. *Mailing Add:* Dept of Elec Eng 600 S College Tulsa OK 74104

STRATTON, CEDRIC, b Langley, Eng, Apr 26, 31; US citizen; div; c 1. INORGANIC CHEMISTRY, ANALYTICAL CHEMISTRY. *Educ:* Univ Nottingham, BSc, 53; Univ London, PhD(inorg chem), 63. *Prof Exp:* Qual control chemist, Richard Klinger, Ltd, Eng, 53-55; develop chemist, Small & Parkes, Ltd, 55-56; sci officer anal res, Brit Insulated Callender's Cables, 57-61; NSF res fel, Univ Fla, 63-65; assoc prof, 65-72, PROF INORG & ANALYTICAL CHEM, ARMSTRONG STATE COL, 72- *Mem:* Am Chem Soc; Royal Soc Chem. *Res:* Chemistry of group V elements, their heterocyclic derivatives; concentration of minerals in local well-water; legal consultancy. *Mailing Add:* Dept Chem Armstrong State Col 11935 Abercorn St Savannah GA 31419

STRATTON, CHARLES ABNER, b Canyon, Tex, Mar 28, 16; m 51; c 3. COLLOID CHEMISTRY, SCIENCE EDUCATION. *Educ:* WTex State Col, BS, 36; Univ Southern Calif, MS, 50, PhD(chem), 53; Gemological Inst Am, grad gemologist cert, 75. *Prof Exp:* Chemist, Borger Refinery, Phillips Petrol Co, 39-47; asst chem, Univ Southern Calif, 47-51; chemist, Res Div, Phillips Petrol Co, 52-73; self-employed gemologist, 75-81; SELF-EMPLOYED CHEMIST, OIL FIELD CHEM RES, 84- *Concurrent Pos:* Part-time instr chem, Bartlesville Wesleyan Col, 74-76. *Res:* Chemical treatment of kerosene, gasoline and liquified petroleum gases; compounding of greases with inorganic thickeners; drilling mud chemicals; water-soluble polymers; water-flood chemicals; brine/oil surfactants. *Mailing Add:* 1233 N Wyandotte Dewey OK 74029

STRATTON, CHARLOTTE DIANNE, b Brooklyn, NY, Mar 7, 29. ORGANIC CHEMISTRY. *Educ:* Bucknell Univ, BS, 51; Pa State Univ, MS, 52. *Prof Exp:* From asst res chemist to res chemist, Parke, Davis & Co, Warner-Lambert Co, Inc, 52-86; RETIRED. *Concurrent Pos:* Consult, Warner-Lambert/Parke Davis, 89- *Mem:* Am Chem Soc. *Res:* Medicinal chemistry, especially natural products isolation and organic synthesis of cardiovascular drugs. *Mailing Add:* 1523 Covington Dr Ann Arbor MI 48103

STRATTON, CLIFFORD JAMES, b Winslow, Ariz, Apr 7, 45; m 68; c 5. ANATOMY, CELL BIOLOGY. *Educ:* Northern Ariz Univ, BS, 68, MS, 70; Brigham Young Univ, PhD(zool, chem), 73. *Prof Exp:* Lab instr, Northern Ariz Univ, 68-70; lect instr & Nat Defense Educ Act fel, Brigham Young Univ, 70-73; res assoc, Sch Med, Univ Calif, Los Angeles, 73-74; asst prof & chief neuroanatomist, 74-77, ASSOC PROF, CHIEF HISTOLOGIST & ASST CHIEF NEUROANATOMIST, SCH MED, UNIV NEV, 77- *Concurrent Pos:* NIH Young Investr Pulmonary res award, 76-78; researcher, Am Lung Asn, 76-78; res assoc, NIH Lung Cult Conf, W Alton Jones Cell Sci Ctr, NY, 77. *Mem:* Am Asn Anatomists; Electron Micros Soc Am; Tissue Cult Asn; Am Soc Cell Biol; AAAS. *Res:* Ultrastructural morphology, histochemistry and pharmacology of the human lung surfactant system as studied in vivo and with alveolar cloning, complimented with lipid-carbohydrate embedment procedures; primary interest is infant respiratory distress syndrome. *Mailing Add:* Dept Anat Univ Nev 16 Manville Med Sci Bldg Reno NV 89557

STRATTON, DONALD BRENDAN, b Escanaba, Mich, Jan 6, 41; m 67. PHYSIOLOGY, NEUROPHYSIOLOGY. *Educ:* Northern Mich Univ, BS, 63, MA, 64; Southern Ill Univ, PhD(physiol), 71. *Prof Exp:* From asst prof to assoc prof, 71-80, PROF BIOL, DRAKE UNIV, 81- *Mem:* Am Physiol Soc; Neuroelectrical Soc. *Res:* Cardiovascular physiology with particular interest in vascular smooth muscle; normal physiological responses of vascular smooth muscle; changes in vascular smooth muscle response in pathophysiological states. *Mailing Add:* Dept Biol Drake Univ 25th St & Univ Ave Des Moines IA 50311

STRATTON, FRANK E(DWARD), b Oceanside, Calif, Dec 20, 37; m 59; c 2. ENGINEERING. *Educ:* San Diego State Col, BS, 62; Stanford Univ, MS, 63, PhD(civil eng), 66. *Prof Exp:* PROF ENG, SAN DIEGO STATE UNIV, 66- *Concurrent Pos:* Chmn, Environ Eng Div, Am Soc Civil Engrs, 81; dipl, Am Acad Environ Engrs. *Mem:* Fel Am Soc Civil Engrs; Am Water Works Asn; Water Pollution Control Fedn. *Res:* Water quality management; nutrient removal methods; waste disposal. *Mailing Add:* Col Eng San Diego State Univ San Diego CA 92182-0189

STRATTON, JAMES FORREST, b Chicago Heights, Ill, Nov 29, 43; m; c 1. PALEONTOLOGY. *Educ:* Ind State Univ, Terre Haute, BS, 65; Ind Univ, Bloomington, MAT, 67, AM, 72, PhD(paleont), 75. *Prof Exp:* Instr geol, Shippensburg State Col, 67-70; asst prof, 75-77, PROF GEOL, EASTERN ILL UNIV, 77- *Concurrent Pos:* Geol consult, Battelle Mem Inst, 88-89, Ill Dept Nuclear Safety, 89-90. *Mem:* Soc Econ Paleontologists & Mineralogists; Int Bryozool Asn; Am Asn Petrol Geologists; Brit Palaeont Asn. *Res:* Quantitative analysis of morphological and structural characters of Fenestellidae for the study of taxonomy and functional morphology. *Mailing Add:* Dept of Geol Eastern Ill Univ Charleston IL 61920

STRATTON, JULIUS ADAMS, b Seattle, Wash, May 18, 01; m 35; c 3. PHYSICS, EDUCATION ADMINISTRATION. *Educ:* Mass Inst Technol, SB, 23, SM, 26; Swiss Fed Inst Technol, ScD(math, physics), 28. *Hon Degrees:* DEng, NY Univ, 55; LHD, Hebrew Union Col, 62, Oklahoma City Univ, 63, Jewish Theol Sem Am, 65; LLD, Northeastern Univ, 57, Union Col, NY, 58, Harvard Univ, 59, Brandeis Univ, 59, Carleton Col, 60, Univ Notre Dame, 61, Johns Hopkins Univ, 62; ScD, St Francis Xavier Univ, 57, Col William & Mary, 64, Carnegie Inst Technol, 65, Univ Leeds, 67, Heriot-Watt Univ, 71, Cambridge Univ, 72. *Prof Exp:* Res asst commun, 24-26, asst prof elec eng, 28-30, from asst prof to prof physics, 30-51, mem staff, Radiation Lab, 40-45, dir, Res Lab Electronics, 44-49, provost, 49-56, vpres, 51-56, chancellor, 56-59, actg pres, 57-59, pres, 59-66, EMER PRES, MASS INST TECHNOL, 66- *Concurrent Pos:* Expert consult, Secy War, 42-46; chmn, Comt Electronics, Res & Develop Bd, 46-49; mem, Naval Res Adv Comt, 54-59, chmn, 55-57; mem, Nat Sci Bd, NSF, 56-62 & 64-67; trustee, Ford Found, 55-71, chmn bd, 66-71; chmn, Comn Marine Sci, Eng & Resources, 67-69; mem, Nat Adv Comt Oceans & Atmosphere, 71-73; life mem corp, Mass Inst Technol; life trustee, Boston Mus Sci; founding mem, Nat Acad Eng. *Honors & Awards:* Medal Merit, 46; Faraday Medal, Brit Inst Elec Engrs, 61; Officer, French Legion Hon, 61; Order de Boyaca, Govt Columbia, 64; Boston Medal Distinguished Achievement, 66; Knight Commander, Order Merit, Fed Repub Ger, 66; Boston Medal Distinguished Achievement, 66. *Mem:* Nat Acad Sci (vpres, 61-65); Nat Acad Eng; fel Am Phys Soc; fel Am Acad Arts & Sci; Am Philos Soc; fel Inst Elec & Electronic Engrs. *Res:* Electromagnetic theory. *Mailing Add:* Mass Inst Technol Cambridge MA 02139

STRATTON, LEWIS PALMER, b West Chester, Pa, Aug 22, 37; div; c 2. BIOCHEMISTRY, BACTERIOLOGY. *Educ:* Juniata Col, BS, 59; Univ Maine, MS, 61; Fla State Univ, PhD(chem), 67. *Prof Exp:* From asst prof to assoc prof, 67-81, PROF BIOL, FURMAN UNIV, 81- *Concurrent Pos:* Vis prof zool chem, Univ Alaska, 75; vis researcher, Ctr Biomolecular Sci & Eng, Naval Res Lab, 88-89. *Mem:* AAAS; Asn Southeastern Biologists; Sigma Xi; Am Chem Soc. *Res:* Comparative protein biochemistry; hemoglobin chemistry. *Mailing Add:* Dept Biol Furman Univ Greenville SC 29613

STRATTON, ROBERT, b Vienna, Austria, Aug 14, 28; US citizen; m 53, 80; c 2. ELECTRONICS ENGINEERING. *Educ:* Univ Manchester, BSc, 49, PhD(theoret physics), 52. *Prof Exp:* Res physicist, Metrop Vickers Elec Co, Ltd, Eng, 52-59; mem tech staff, 59-63, dir, Physics Res Lab, 63-71, asst vpres, 70, assoc dir, Cent Res Lab, 71-72, dir semiconductor res & develop labs, 72-75, DIR, CENT RES LABS, TEX INSTRUMENTS, INC, 75- *Concurrent Pos:* Vpres, corp staff, Cent Res Labs, Tex Instruments Inc, 82. *Mem:* Fel Am Phys Soc; fel Inst Elec & Electronics Engrs; fel Brit Inst Physics & Phys Soc. *Res:* Solid state theory, including field emission, space charge barriers, thermoelectricity, high electric fields, thermal conductivity, dielectric breakdown and surface energies of solids. *Mailing Add:* Tex Instruments PO Box 655936 MS 136 Dallas TX 75265

STRATTON, ROBERT ALAN, b Selma, Ala, Feb 4, 36; m 61; c 4. POLYMER CHEMISTRY. *Educ:* Univ Nev, BS, 58; Univ Wis, PhD(chem), 62. *Prof Exp:* Sr res chemist, Mobil Chem Co, 62-69; ASSOC PROF CHEM, INST PAPER CHEM, 69- *Mem:* Soc Rheol; Am Chem Soc; Tech Asn Pulp & Paper Indust. *Res:* Rheology of polymer melts and solutions; dilute solution properties of polymers; flocculation of colloids; use of polymers in papermaking and waste water treatment. *Mailing Add:* Chem Dept Inst Paper Sci & Technol 575 14th St NW Atlanta GA 30318

STRATTON, ROY FRANKLIN, JR, b Memphis, Tenn, July 23, 29; m 63; c 1. ELECTRICAL ENGINEERING, PHYSICS. *Educ:* Southwestern at Memphis, BS, 51; Univ Tenn, MS, 53, PhD(physics), 57; Ga Inst Technol, MSEE, 75. *Prof Exp:* Asst dept physics, Univ Tenn, 52-57; res assoc plasma physics, Oak Ridge Nat Lab, 58-70; prof & chmn sci div admin & teaching, Pikeville Col, 70-73; teaching asst elec eng, Ga Inst Technol, 73-75; electronic engr electromagnetic compatibility, 75-86, ELECTRONIC ENGR SYSTS RELIABILITY, ROME LAB, US AIR FORCE, 86- *Mem:* Inst Elec & Electronics Engrs; AAAS; Sigma Xi. *Res:* Systems reliability techniques; testability and maintainability enhancement techniques for complex electronic systems. *Mailing Add:* Rome Lab-RBET Griffiss AFB NY 13441-5700

STRATTON, THOMAS FAIRLAMB, b Kansas City, Mo, Dec 19, 29; m 58; c 2. LASERS, NUCLEAR PHYSICS. *Educ:* Union Col, BS, 49; Univ Minn, MS, 52, PhD(physics), 54. *Prof Exp:* Staff mem physics, Los Alamos Sci Lab, 54-66; sr fel, Battelle Columbus Lab, 67; staff mem physics, 68-75, group leader, Antares Laser Proj, 76-79, dep physics div leader, 80-81, LAB FEL, LOS ALAMOS NAT LAB, 82- *Concurrent Pos:* Mem, Atomic Energy Res Estab, UK, 58; mem adv bd pulse power, Nat Acad Sci, 77; atomic energy adv, Dept of Defense, 85. *Mem:* Sigma Xi; fel Am Phys Soc; AAAS. *Res:* Thermonuclear fusion; soft x-ray spectroscopy; magnetohydrodynamics; plasma acceleration and direct conversion. *Mailing Add:* 315 Potrillo Los Alamos NM 87544

STRATTON, WILLIAM R, b River Falls, Wis, May 15, 22; m 52; c 3. PHYSICS. *Educ:* Univ Minn, PhD(physics), 52. *Prof Exp:* Res assoc, Univ Minn, 52; MEM STAFF, LOS ALAMOS SCI LAB, UNIV CALIF, 52- *Concurrent Pos:* US del, Int Conf Peaceful Uses Atomic Energy, 58 & Fast Reactor Prog, Cadarache, France, 65-66; mem, Adv Comt Reactor Safeguards, Atomic Energy Comn, 66-75; Presidents comn, Accident Three Mile Island. *Honors & Awards:* Spec Award, Am Nuclear Soc, 81. *Mem:* Am Phys Soc; fel Am Nuclear Soc; Sigma Xi. *Res:* Scattering and reaction in nuclear physics; nuclear forces; reactor physics; reactor safety, criticality safety. *Mailing Add:* Two Acoma Lane Los Alamos NM 87544

STRATTON, WILMER JOSEPH, b Newark, NJ, June 4, 32; m 55; c 3. CHEMISTRY. *Educ:* Earlham Col, AB, 54; Ohio State Univ, PhD(chem), 58. *Prof Exp:* Asst prof chem, Ohio Wesleyan Univ, 58-59 & Earlham Col, 59-64; vis lectr, Univ Ill, 64-65; assoc prof, 65-70, chmn dept, 65-68, PROF CHEM, EARLHAM COL, 70- *Mem:* Am Chem Soc. *Res:* Metal coordination compounds, including synthesis of new polydentate chelates and bonding in chelate systems. *Mailing Add:* 104 Osage Rd Oak Ridge TN 37830

STRATY, RICHARD ROBERT, b Milwaukee, Wis, June 21, 29; m 53; c 2. FISHERIES BIOLOGY. *Educ:* Ore State Univ, BS, 54, PhD(fisheries & oceanog), 69; Univ Hawaii, MS, 63. *Prof Exp:* Proj leader fish biol, Fish & Wildlife Serv, US Dept Interior, Juneau, 54-55, proj supvr marine fish biol, Bur Com Fisheries, 55-59, proj supvr marine fish biol salmon, Auke Bay Biol Lab, Bur Com Fisheries, 60-61; exped scientist marine biol, Stanford Univ, 64; proj suprv marine biol, Auke Bay Fisheries Lab, Bur Com Fisheries, US Fish & Wildlife Serv, US Dept Interior, 66-74; prog mgr marine invest biol oceanog, Auke Bay Lab, Nat Marine Fisheries Serv, Nat Oceanic & Atmospheric Admin, US Dept Com, 74-86; RETIRED. *Concurrent Pos:* Mem, Alaska Coun Sci & Technol, Off Gov Alaska, 78-; consult, Living Resource Assocs, 86- *Honors & Awards:* C Y Conkle Publ Award, Auke Bay Biol Lab, US Dept Interior, 66. *Mem:* Am Inst Fishery Res Biologists. *Res:* Fishery oceanography; marine ecology; population dynamics; exploratory fishing; fishery assessment; biological oceanography; marine resource survey. *Mailing Add:* Living Resource Assocs PO Box 210211 Auke Bay AK 99821

STRAUB, CONRAD P(AUL), b Irvington, NJ, June 21, 16; m 45; c 4. SANITARY ENGINEERING, ENVIRONMENTAL HEALTH. *Educ:* Newark Col Eng, BS, 36, CE, 39; Cornell Univ, MCE, 40, PhD(sanit eng), 43. *Hon Degrees:* DEng, Newark Col Eng, 67. *Prof Exp:* Computer & head comput sect, US Eng Off, NY, 37-39; asst pub health engr, USPHS, NJ, 41, asst sanit engr, NY, 42-44, actg dep chief sanit engr, China, 45-46, chief sanit engr, Poland, 46, sr asst sanit engr, Ohio, 47-48 & Oak Ridge Nat Lab, 48-56, chief radiol health res activ, Robert A Taft Sanit Eng Ctr, 56-64, dep dir, 64-65, dir, 65-66; prof, 66-81, EMER PROF SANIT ENG & DIR ENVIRON HEALTH RES & TRAINING CTR, UNIV MINN, MINNEAPOLIS, 81- *Concurrent Pos:* Chmn comt waste disposal, Int Comn Radiol Protection; mem expert comt radiol health & consult, WHO. *Honors & Awards:* Fuertes Medal, Cornell Univ, 54; Elda Anderson Mem Award, Health Physics Soc. *Mem:* Am Soc Civil Engrs; Health Physics Soc; Am Pub Health Asn; Am Water Works Asn; Water Pollution Control Asn. *Res:* Industrial wastes; sanitary engineering education; insect control; treatment and disposal of radioactive wastes; radiological health; environmental health; public health implications of water and waste water systems; environmental contaminants. *Mailing Add:* 2330 Chalet Dr Columbia Heights MN 55421-2057

STRAUB, DAREL K, b Titusville, Pa, May 17, 35. INORGANIC CHEMISTRY. *Educ:* Allegheny Col, BS, 57; Univ Ill, PhD(inorg chem), 61. *Prof Exp:* Instr, 61-62, asst prof, 62-68, ASSOC PROF CHEM, UNIV PITTSBURGH, 68- *Mem:* Am Chem Soc; AAAS. *Res:* Iron porphyrins; complexes of sulfur-containing ligands; M-ssbauer spectroscopy. *Mailing Add:* Dept Chem Chem 1005 Univ Pittsburgh 4200 5th Ave Pittsburgh PA 15260-0001

STRAUB, RICHARD WAYNE, b Fairfax, Mo, June 5, 40; div; c 1. HORTICULTURE. *Educ:* Northwest Mo State Univ, BS, 66; Univ Mo, MS, 68, PhD(entom), 72. *Prof Exp:* Res assoc, 71-75, asst prof, 75-79, ASSOC PROF ENTOM, NY STATE AGR EXP STA, CORNELL UNIV, 79- *Concurrent Pos:* Vis scientist, USDI, BLM, Washington, DC, 85-86. *Mem:* Entom Soc Am; AAAS. *Res:* Biology and integrated control of insects of vegetable crops with emphasis on plant resistance to insect pests and insect transmission of vegetable diseases. *Mailing Add:* PO Box 122 Hellbrook Lane Ulster Park NY 12487

STRAUB, THOMAS STUART, b Louisville, Ky, Oct 1, 41; m 65; c 3. BIOORGANIC CHEMISTRY. *Educ:* Princeton Univ, AB, 63; Univ Minn, MS, 66; Ill Inst Technol, PhD(chem), 70. *Prof Exp:* Chemist, Monsanto Corp, 63-64; NIH postdoctoral biochem, Northwestern Univ, 69-72; PROF ORG CHEM, LA SALLE UNIV, 72-, CHAIR CHEM, 78- *Mem:* Am Chem Soc; AAAS; Sigma Xi. *Res:* Homogenous catalysis; molecular recognition; phase transfer catalysis; chemical models of biologic processes. *Mailing Add:* Dept Chem & Biochem La Salle Univ Philadelphia PA 19141

STRAUB, WILLIAM ALBERT, b Philadelphia, Pa, June 21, 31; m 58; c 2. ANALYTICAL CHEMISTRY. *Educ:* Univ Pa, BA, 53; Cornell Univ, PhD, 58. *Prof Exp:* Technologist, 57-67, sr res chemist, 67-75, ASSOC RES CONSULT, USS DIV USX, 75- *Mem:* Am Chem Soc; Soc Anal Chemists. *Res:* Process solution analysis. *Mailing Add:* USS Div USX Tech Ctr Monroeville PA 15146

STRAUB, WOLF DETER, b Boston, Mass, Apr 27, 27; m 61; c 2. SOLID STATE PHYSICS. *Educ:* Yale Univ, BS, 50; Univ Mich, MS, 52. *Prof Exp:* Staff mem solid state physics, Res Div, Raytheon Co, 52-65; physicist, Electronics Res Ctr, NASA, 65-70 & M/K Systs, Inc, Mass, 70-72; mgr anal lab, Coulter Systs Corp Inc, Bedford, 72-80, dir advan physics lab, 80-83, 85-90; staff mem, Eaton Corp, 83-85; RETIRED. *Mem:* Am Vacuum Soc; Sigma Xi. *Res:* Galvanometric properties of semiconductors and semimetals; radiation damage and studies of microwave generation in semiconductors; electrical and mechanical properties of dielectric thin films; problems in electrophotography; surface physics. *Mailing Add:* 158 Barton Dr Sudbury MA 01776

STRAUBE, ROBERT LEONARD, b Chicago, Ill, Sept 16, 17; m 44; c 2. RADIOBIOLOGY. *Educ:* Univ Chicago, BS, 39, PhD(physiol), 55. *Prof Exp:* Asst path, Univ Chicago, 43-46; prof radiobiol, Assoc Cols Midwest, 63-64; assoc scientist, Argonne Nat Lab, 47-65; EXEC SECY RADIATION STUDY SECT, DIV RES GRANTS, NIH, 65- *Mem:* AAAS; Radiation Res Soc; Am Physiol Soc; Soc Exp Biol & Med; Am Asn Cancer Res. *Res:* Nature of radiation effects and their modification by chemical agents; growth processes in neoplastic cells. *Mailing Add:* 96344 Cape Ferrelo Rd Brookings OR 97415

STRAUBINGER, ROBERT M, b Buffalo, NY, May 29, 53. CELL BIOLOGY. *Educ:* Univ Calif, San Francisco, PhD(pharmacol), 84. *Prof Exp:* Fel, 85-86, ASST RES PHARMACOLOGIST, CANCER RES INST, UNIV CALIF, SAN FRANCISCO, 86- *Mem:* Am Soc Cell Biol. *Mailing Add:* Dept Pharmaceut SUNY 539 Cooke Hall Amherst NY 14260

STRAUCH, ARTHUR ROGER, III, CONTRACTILE PROTEIN BIOCHEMISTRY, CELL MOTILITY. *Educ:* State Univ NY, Buffalo, PhD(cell & molecular biol), 81. *Prof Exp:* ASST PROF ANAT, SCH MED, OHIO STATE UNIV, 84- *Mailing Add:* Dept Anat Ohio State Univ 333 W Tenth Ave Columbus OH 43210-1239

STRAUCH, KARL, b Giessen, Ger, Oct 4, 22; nat US; m 51. PARTICLE PHYSICS. *Educ:* Univ Calif, AB, 43, PhD(physics), 50. *Prof Exp:* Soc Fels jr fel, 50-53, from asst prof to prof, 53-75, GEORGE VASMER LEVERETT PROF PHYSICS, HARVARD UNIV, 75- *Concurrent Pos:* Dir, Cambridge Electron Accelerator, Harvard Univ, 67-74. *Honors & Awards:* Alexander von Humboldt Prize, 83. *Mem:* Am Phys Soc; Am Acad Arts & Sci. *Res:* High energy reactions; elementary particles. *Mailing Add:* Dept of Physics Harvard Univ Cambridge MA 02138

STRAUCH, RALPH EUGENE, b Springfield, Mass, May 14, 37; m 58; c 2. SOMATICS & MATHEMATICS. *Educ:* Univ Calif, Los Angeles, AB, 59, Univ Calif, Berkeley, MA, 64, PhD(statist), 65. *Prof Exp:* Sr mathematician, Rand Corp, 65-76; CONSULT, 76- *Concurrent Pos:* Feldenkrais Teacher, 83- *Mem:* Feldenkrais Guild; Somatics Soc. *Res:* Dynamic programming; statistical decision theory; national security policy; human perception; paranormal phenomena; policy analysis methodology; mind/body relationship; samatic aspects of human behavior, links between self awareness and human functioning somatic components of post traumatic stress including childhood sexual abuse; post research in perception, mathematics policy analysis methodology. *Mailing Add:* 1383 Avenida de Cortez Pacific Palisades CA 90272

STRAUCH, RICHARD G, ENGINEERING ADMINISTRATION. *Prof Exp:* RES ELEC ENGR, WAVE PROPAGATION LAB, NAT OCEANIC & ATMOSPHERIC ADMIN. *Mem:* Nat Acad Eng; sr mem Inst Elec & Electronics Engrs. *Mailing Add:* 3390 Fourth St Boulder CO 80304

STRAUGHAN, ISDALE (DALE) MARGARET, b Pittsworth, Australia, Nov 4, 39; m 62; c 1. ECOLOGY, BIOLOGY. *Educ:* Queensland Univ, BSc, 60, Hons, 62, PhD(zool), 66. *Prof Exp:* Demonstr zool, Queensland Univ, 66; sr demonstr, Univ Col, Townsville, 66-67; asst prof & res assoc, Allan Hancock Found, 69-74, SR RES SCIENTIST, INST MARINE & COASTAL STUDIES, UNIV SOUTHERN CALIF, 74-; PVT CONSULT. *Concurrent Pos:* Consult biologist, Northern Elec Authority, Queensland, 66-68; Am Asn Univ Women fel, 68-69. *Mem:* AAAS; Sigma Xi; Ecol Soc Am. *Res:* Establishment of natural ecological change in response to natural change in the marine environment and comparison with man-induced ecological change; comparison of man-induced change to natural biological fluctuations. *Mailing Add:* 13688 Park St Whittier CA 90601

STRAUGHN, ARTHUR BELKNAP, b Durham, NC, Aug 10, 44; m 68; c 2. PHARMACOKINETICS, BIOPHARMACEUTICS. *Educ:* Univ NC, BS, 72; Univ Tenn, PharmD, 74. *Prof Exp:* Instr therapeut, Univ NC, 72-73; PROF PHARMACEUT & DIR, DRUG RES LAB, UNIV TENN, 74- *Mem:* Am Asn Pharmaceut Scientists; Sigma Xi; AAAS; Am Col Clin Pharm; Am Pharmaceut Asn; Am Soc Hosp Pharmacists. *Res:* Pharmacokinetics and biopharmaceutics; develop and conduct studies in humans to define drug absorption and disposition, specifically dosage forms for sustained-release. *Mailing Add:* 874 Union Ave Rm 5 Crowe Memphis TN 38163

STRAUGHN, WILLIAM RINGGOLD, JR, b Dubois, Pa, May 21, 13; m 41; c 4. BACTERIOLOGY. *Educ:* Mansfield State Col, BS, 35; Cornell Univ, MS, 40; Univ Pa, PdD(bact), 58. *Prof Exp:* Teacher high sch, Pa, 35-36 & NY, 36-38; asst bact, Univ NC, 40-42; instr math & chem, Md State Teachers Col, Salisbury, 42-44; from instr to prof bact, 44-80, EMER PROF BACT, SCH MED, UNIV NC, CHAPEL HILL, 80- *Mem:* Am Soc Microbiol. *Res:* Bacterial physiology and metabolism; antibacterial agents; enzyme synthesis; amino acid decarboxylases-mechanisms of formation and action; bacterial membranes and transport mechanisms. *Mailing Add:* Manning Dr Chapel Hill NC 27514

STRAUMANIS, JOHN JANIS, JR, b Riga, Latvia, Apr 22, 35; US citizen; m 59; c 2. PSYCHIATRY. *Educ:* Univ Iowa, BA, 57, MD, 60, MS, 64. *Prof Exp:* Intern med, Georgetown Univ Hosp, 60-61; resident psychiat, Univ Iowa, 61-64; asst prof psychiat & Nat Inst Ment Health res career develop grant, 66-71, assoc prof psychiat, 71-77, prof psychiat, Temple Univ, 77-85; PROF & CHMN PSYCHIAT, LA STATE UNIV MED SCH, 85- *Mem:* Am Psychiat Asn; Soc Biol Psychiat; Am Psychopath Asn; Am Electroencephalographic Soc; Am Col Psychiat. *Res:* Electrophysiology pf psychiatric disorders. *Mailing Add:* Dept Psych, La State Univ Med 1501 Kings Hwy Shreveport LA 71130

STRAUMFJORD, JON VIDALIN, JR, b Portland, Ore, Feb 23, 25; m 47; c 2. MEDICINE, CLINICAL PATHOLOGY. *Educ:* Willamette Univ, BA, 48; Univ Ore, MS & MD, 53; Univ Iowa, PhD(biochem), 58. *Prof Exp:* Res fel biochem, Univ Iowa, 54-58; resident path & consult, Providence Hosp, Portland, Ore, 58-60; asst prof path, Univ Miami, 60-62; assoc prof, Med Col Ala, 62-65, prof clin path & chmn dept, 65-70, dir clin labs, 62-65, clin pathologist in chief, Univ Hosp, 65-70; PROF PATH & CHMN DEPT, MED COL WIS, 70- *Concurrent Pos:* Asst pathologist, Div Clin Path, Jackson Mem Hosp, Miami, Fla, 60-62; dir labs, Milwaukee City Gen Hosp, Wis, 70-; mem surg adv bd, Shrine Burn Units. *Mem:* AAAS; Am Soc Clin Path; Am Asn Clin Chem; Col Am Pathologists; NY Acad Sci. *Res:* Surface characteristics of cells; clinical chemical screening procedures. *Mailing Add:* Univ Chicago Med Ctr 5841 S Maryland Ave Chicago IL 60637

STRAUS, ALAN EDWARD, b Berkeley, Calif, May 14, 24; m 53; c 2. ORGANIC CHEMISTRY. *Educ:* Univ Calif, Berkeley, BS, 49. *Prof Exp:* From asst res chemist to res chemist, 49-67, SR RES CHEMIST, CHEVRON RES CO, 67- *Mem:* Am Chem Soc. *Res:* Petrochemicals; hydrocarbon oxidation; surface active agents; hydrocarbon pyrolysis; organic synthesis; polymers; heterogeneous catalysis. *Mailing Add:* 2679 Tamalpais Ave El Cerrito CA 94530

STRAUS, DANIEL STEVEN, b May 3, 46; m 83; c 2. BIOCHEMISTRY, CELL BIOLOGY. *Educ:* Univ Calif, Berkeley, PhD(biochem), 72. *Prof Exp:* PROF BIOMED SCI & BIOL, UNIV CALIF, RIVERSIDE, 76- *Mailing Add:* Biomed Sci Div Univ Calif Riverside CA 92521

STRAUS, DAVID BRADLEY, b Chicago, Ill, July 26, 30; m 55; c 3. GENETIC TOXICOLOGY. *Educ:* Reed Col, BA, 53; Univ Chicago, PhD(biochem), 60. *Prof Exp:* Asst biochem, Med Sch, Univ Ore, 53-54; asst, Univ Chicago & Argonne Cancer Res Hosp, 55-60; res assoc chem, Princeton Univ, 60-64, res staff mem, 64-65; asst prof biochem, State Univ NY Buffalo, 65-72; ASSOC PROF CHEM, STATE UNIV NY COL NEW PALTZ, 73- *Concurrent Pos:* Vis asst prof chem, State Univ NY, Albany, 69; adj assoc prof environ med, NY Univ, 84- *Mem:* AAAS; Am Chem Soc; Sigma Xi. *Res:* Site specific mutagenesis of viral DNA; nucleic acid enzymology and chemistry. *Mailing Add:* Dept Chem State Univ NY Col New Paltz NY 12561

STRAUS, DAVID CONRAD, b Evansville, Ind, Apr 27, 47; div. MEDICAL MICROBIOLOGY. *Educ:* Wright State Univ, BS, 70; Loyola Univ Chicago, PhD(microbiol), 74. *Prof Exp:* Teaching asst microbiol, Sch Med, Loyola Univ Chicago, 70-74; fel, Med Ctr, Univ Cincinnati, 74-75; instr, Univ Tex Health Sci Ctr, San Antonio, 75-76, asst prof microbiol, 76-81; ASSOC PROF MICROBIOL, TEX TECH UNIV HEALTH SCI CTR, LUBBOCK, TEX, 81- *Concurrent Pos:* Instr microbiol, Ill Col Podiatric Med, 72-73. *Mem:* Am Soc Microbiol; Sigma Xi. *Res:* Study of mechanisms of bacterial pathogenicity and host response; study of bacterial exotoxins. *Mailing Add:* Tex Tech Univ Health Sci Ctr Sch Med Lubbock TX 79430

STRAUS, FRANCIS HOWE, II, b Chicago, Ill, Mar 16, 32; m 55; c 4. ENDOCRINE PATHOLOGY. *Educ:* Harvard Univ, AB, 53; Univ Chicago, MD, 57, MS, 64. *Prof Exp:* Intern, Clins, Univ Chicago, 57-58, resident path, 58-62, chief resident, 62-63, from instr to assoc prof, 62-78, PROF PATH, SCH MED, UNIV CHICAGO, 78- *Concurrent Pos:* Am Cancer Soc advan clin fel, 65-68; fel, USPH, 58-60, Am Cancer Soc Clin, 62-63. *Mem:* Sigma Xi; Am Soc Exp Pathologists; Am Asn Pathologists; Int Acad Path; NY Acad Sci. *Res:* Morphology in surgical pathology as it relates to diagnosis and prognosis of clinical disease; urologic pathology; endocrine pathology. *Mailing Add:* Dept Path Univ Chicago Billings Hosp 5841 S Maryland Ave Chicago IL 60637

STRAUS, HELEN LORNA PUTTKAMMER, b Chicago, Ill, Feb 15, 33; m 55; c 4. ANATOMY, BIOLOGY. *Educ:* Radcliffe Col, AB, 55; Univ Chicago, MS, 60, PhD(anat), 62. *Prof Exp:* Fel anat, 62-63, res assoc, 63-64, instr anat & biol, 64-67, asst prof biol & asst dean undergrad students, 67-71, dean undergrad students, 71-82, dean admissions, 75-80, assoc prof, anat, 73-87, PROF, BIOL & ANAT, UNIV CHICAGO, 87- *Honors & Awards:* Silver Medalist, Case Prog, 77. *Mem:* Am Soc Zoologists; Am Asn Anatomists; AAAS. *Res:* Teaching biology, science in liberal education. *Mailing Add:* Dept Anat Univ Chicago Chicago IL 60637

STRAUS, JOE MELVIN, b Dallas, Tex, May 27, 46; m 71; c 2. ATMOSPHERIC PHYSICS, GEOPHYSICAL FLUID DYNAMICS. *Educ:* Rice Univ, BA, 68; Univ Calif, Los Angeles, MS, 69, PhD(planetary, space physics), 72. *Prof Exp:* res scientist, 73-80, head, Atmospheric Sci Dept, Space Sci Lab, 80-86; prin dir, Off Res Lab Oper, 86-89, DIR CHEM & PHYSICS LAB, AEROSPACE CORP, 89- *Mem:* Sigma Xi; Am Geophys Union; Am Inst Aeronaut & Astronaut. *Res:* Theoretical studies of atmospheric physics; aeronomy; convection in atmospheres, oceans, stars, planetary interiors; geophysical fluid dynamics; atmospheric and ionospheric effects on space systems. *Mailing Add:* Chem & Physics Lab The Aerospace Corp PO Box 92957 Los Angeles CA 90009

STRAUS, JOZEF, b Velke Kapusany, Czech, July 18, 46; Can citizen. EXPERIMENTAL SOLID STATE PHYSICS. *Educ:* Univ Alta, BS, 69, PhD(physics), 74. *Prof Exp:* Fel, Univ Alta, 69-74; mem sci staff & Nat Res Coun Can Fel, Bell Northern Res Ltd, 74-81; PRES, JDS FITEL. *Res:* Fabrication and study of physical properties of light emitting diodes and of solid state lasers; fiber optics, fiber optics communication; electron tunneling in normal and superconducting metals, Josephson tunneling. *Mailing Add:* 691 Hill Crest Ave Ottawa ON K2A 2N2 Can

STRAUS, MARC J, b New York, NY, June 2, 43; m 64; c 2. ONCOLOGY, CHEMOTHERAPY. *Educ:* Franklin & Marshall Col, AB, 64; State Univ NY Downstate Med Ctr, MD, 68; Am Bd Internal Med, dipl & cert med oncol, 75. *Prof Exp:* Chief med oncol, Med Ctr, Boston Univ, 74-78, assoc prof med, Sch Med, 75-78; CHIEF ONCOL & PROF MED, NY MED COL, 78- *Concurrent Pos:* Prin investr, Eastern Coop Oncol Group, Med Ctr, Boston Univ, 75-; consult oncol, St Agnes Hosp, Northern West Hosp, St Joseph's Hosp & Peekskill Hosp, 79- *Mem:* Am Soc Clin Oncol; Am Fedn Clin Res; Working Party Ther Lung Cancer; Am Asn Cancer Res. *Res:* Application of cellular kinetics in animal and human tumors to the design of clinical cancer treatment programs; clinical cancer chemotherapy. *Mailing Add:* Oncol & Hemat Assocs 311 N St Rm 304B White Plains NY 10605

STRAUS, NEIL ALEXANDER, b Kitchener, Ont, Apr 29, 43; m 66; c 2. MOLECULAR BIOLOGY. *Educ:* Univ Toronto, BSc Hons, 66, MSc, 67, PhD(molecular biol), 70. *Prof Exp:* Fel biophys, Carnegie Inst, Washington, DC, 70-72; from asst prof to assoc prof, 72-85, PROF MOLECULAR BIOL, UNIV TORONTO, 85- *Concurrent Pos:* Res grant, NSERC, Can. *Mem:* Am Soc Microbiol; Int Soc Plant Molecular Biol; Am Soc Plant Physiologists; Can Soc Plant Molecular Biol; Can Soc Plant Physiologists. *Res:* Chloroplast and cyanobacterial molecular genetics; cyanobacterial transformations; gene regulation; recombinant DNA; plant biotechnology. *Mailing Add:* Dept Bot Univ Toronto Univ Col Toronto ON M5S 1A1 Can

STRAUS, ROBERT, b New Haven, Conn, Jan 9, 23; m 45; c 4. MEDICAL BEHAVIORAL SCIENCE. *Educ:* Yale Univ, PhD(sociol), 47. *Prof Exp:* Chmn dept, 59-87, prof, 59-90, EMER PROF BEHAV SCI, UNIV KY, 90- *Mem:* Inst Med-Nat Acad Sci; Sigma Xi. *Res:* Alcohol; dependency behaviors; aging; patterns of patient care. *Mailing Add:* 656 Raintree Rd Lexington KY 40502-2874

STRAUS, THOMAS MICHAEL, b Berlin, Ger, Oct 25, 31; US citizen; m 57; c 3. APPLIED PHYSICS. *Educ:* Univ Mich, BS, 52; Harvard Univ, MA, 56, PhD(appl physics), 59. *Prof Exp:* Mem tech staff, microwaves & lasers, Hughes Aircraft Co, 59-64, sr staff mem, 64-69, sr staff engr, Laser Dept, 69-74; sr scientist, Theta-Com, 74-76; sr scientist com satellite systs, 76-79, CHIEF SCIENTIST SATELLITE GROUND EQUIP, HUGHES AIRCRAFT CO, 79- *Concurrent Pos:* Lectr, Eng Exten, Univ Calif, Los Angeles, 62-67. *Mem:* Inst Elec & Electronics Engrs. *Res:* Development of microwave and laser components and systems. *Mailing Add:* Microwave Commun Products Hughes Aircraft 243-215 PO Box 2940 Torrance CA 90509

STRAUS, WERNER, b Offenbach, Ger, June 5, 11; nat US. BIOCHEMISTRY, CYTOCHEMISTRY. *Educ:* Univ Zurich, PhD(chem), 38. *Prof Exp:* Res assoc path, Long Island Col Med, 47-50; asst prof, State Univ NY Downstate Med Ctr, 50-58; vis scientist, Cath Univ Louvain & Free Univ Brussels, 59-61 & Univ NC, 62-63; assoc prof, 64-78, prof, 79-81, EMER PROF BIOCHEM, CHICAGO MED SCH, 81- *Mem:* AAAS; Am Soc Cell Biol; Histochemical Soc. *Res:* Intracellular localization of enzymes; lysosomes and phagosomes; cell biology; immuno-cytochemistry; cell surface receptors. *Mailing Add:* Dept Biochem Chicago Med Sch 3333 Green Bay Road North Chicago IL 60064

STRAUSBAUCH, PAUL HENRY, b San Francisco, Calif, Aug 29, 41; m 69; c 1. PATHOLOGY, IMMUNOCHEMISTRY. *Educ:* Univ San Francisco, BS, 63; Univ Wash, PhD(biochem), 69; Univ Miami, MD, 74. *Prof Exp:* Am Cancer Soc fel immunol, Weizmann Inst Sci, 69-71; Med Res Coun Can fel, Univ Man, 71-72; resident path, Dartmouth Med Sch, 74-78; ASSOC PROF PATH, SCH MED, E CAROLINA UNIV, 78- *Res:* Chemical approaches to immunology and cell biology; protein products, especially hormones, produced by tumor cells; macrophage structure and diversity. *Mailing Add:* Dept Path ECarolina Univ Greenville NC 27858

STRAUSBERG, SANFORD I, b Brooklyn, NY, Nov 13, 31; US citizen; m 59; c 2. HAZARDOUS WASTE MANAGEMENT, ENVIRONMENTAL REGULATORY COMPLIANCES. *Educ:* Brooklyn Col, BS, 53; Univ Mich, MS, 55. *Prof Exp:* Geologist ground water, US Geol Survey, 55-57; consult ground water, 57-69, 79-84, 87-89; hydrogeologist, UN, 69-73; dept head ground water, Harza Eng Co, 73-79; mgr geol, Nus Corp, 84-86; mgr earth scis, Ebasco Serv Inc, 86-87; DIR HYDROGEOL, ENVIRO-SCI INC, 89- *Concurrent Pos:* Expert, UN, water grid mission India, 71-72; consult, Asian Develop Bank, 76-78; site mgr, US Environ Protection Agency superfund, 84-87. *Mem:* Am Inst Prof Geologists; Asn Eng Geologists; Geol Soc Am; Asn Ground Water Scientists Engrs. *Res:* Measurement and evaluation of aquifer parameters for use in estimating subsurface hazardous waste migration, ground water remediation design and cleanup, and ground water supply development. *Mailing Add:* 24 L Village Green Budd Lake NJ 07828

STRAUSE, STERLING FRANKLIN, b Summit Station, Pa, Jan 4, 31; m 56; c 4. ORGANIC CHEMISTRY. *Educ:* Lebanon Valley Col, BS, 52; Univ Del, MS, 53, PhD(chem), 55. *Prof Exp:* Develop chemist, Chem Develop Dept, Gen Elec Co, 55-57, spec process develop, 57-58, qual control engr, 58-60, mgr, 60-65, qual control, 65-68, mgr polycarbonate res & develop, 68-71; dir, 71-74, VPRES RES & DEVELOP, W H BRADY CO, MILWAUKEE, 74- *Mem:* Am Chem Soc; Sigma Xi; Tech Asn Pulp & Paper Indust. *Res:* Polymeric peroxide and free radical chemistry; polymer processes. *Mailing Add:* 7716 W Bonniwell Rd Mequon WI 53092

STRAUSER, WILBUR ALEXANDER, b Charleroi, Pa, June 15, 24; m 45; c 2. PHYSICS, MATHEMATICS. *Educ:* Washington & Jefferson Col, AB, 45. *Prof Exp:* Physicist, Manhattan Eng Dist, Tenn, 46-47; assoc physicist, Oak Ridge Nat Lab, 47-50; sci analyst & chief declassification br, Atomic Energy Comn, US Dept Energy, 50-55, asst to mgr, San Francisco Opers Off, 55-56, dep dir, Div Classification, 56-63, asst dir safeguards, Div Int Affairs, 63-70, chief, Weapons Br, Div Classification, 70-79; CONSULT, 79- *Mem:* AAAS. *Res:* Neutron diffraction; security classification; safeguards. *Mailing Add:* 978 Farm Haven Dr Rockville MD 20852

STRAUSS, ALAN JAY, b St Louis, Mo, July 30, 27; m 51; c 3. CHEMISTRY, ELECTRONIC MATERIALS. *Educ:* Univ Chicago, PhB, 45, SB, 46, PhD(phys chem), 56. *Prof Exp:* Staff mem semiconductor mat, Chicago Midway Labs, Univ Chicago, 52-54 & 56-58; staff mem electronic mat, 58-66, assoc group leader, 66-76, GROUP LEADER ELECTRONIC MAT, LINCOLN LAB, MASS INST TECHNOL, 76- *Concurrent Pos:* Div ed, Jour, Electrochem Soc. *Mem:* Am Phys Soc; Electrochem Soc; Am Asn Crystal Growth; Mat Res Soc. *Res:* Materials research on electronic and optical materials, primarily semiconductors and semimetals; phase diagram studies, investigations of electrical and optical properties, as affected by composition, deviations from stoichiometry, and impurities. *Mailing Add:* Lincoln Lab Mass Inst Technol PO Box 73 Lexington MA 02173-9108

STRAUSS, ALVIN MANOSH, b Brooklyn, NY, Oct 24, 43; m 67; c 2. MECHANICS, APPLIED MATHEMATICS. *Educ:* Hunter Col, AB, 64; WVa Univ, PhD(theoret & appl mech), 68. *Prof Exp:* Res assoc theoret & appl mech, Univ Ky, 68-70; asst prof eng, Univ Cincinnati, 70-74, head, Eng Sci Dept, 76-80, prof mech, 78-; AT DEPT MECH & MATS ENG, VANDERBILT UNIV. *Concurrent Pos:* Dir, Div Mech Eng & Appl Mech, NSF, 81- *Mem:* Am Geophys Union; Am Acad Mech; Soc Rheology; Soc Eng Sci; Soc Am Mil Engrs. *Res:* Plasticity; viscoplasticity; biomechanics; phase changes in solids, plasticity, thermomechanics, biomechanics and geomechanics. *Mailing Add:* Dept Mech Eng Vanderbilt Univ Box 1612 Sta B Nashville TN 37235

STRAUSS, ARNOLD WILBUR, b Benton Harbor, Mich, Mar 31, 45; m 70; c 2. PEDIATRIC CARDIOLOGY. *Educ:* Stanford Univ, BA, 66; Washington Univ, St Louis, MD, 70. *Prof Exp:* Resident pediat, St Louis Children's Hosp, Washington Univ, 70-73, fel, 73-74, fel biochem, 74-75; fel, Res Labs, Merck, Sharp & Dohme, 75-77; from asst prof to assoc prof, 77-82, PROF PEDIAT & BIOCHEM, SCH MED, WASH UNIV, 82- *Concurrent Pos:* Estab investr, Am Heart Asn, 79-84. *Honors & Awards:* Mead Johnson Award, 91. *Mem:* Am Acad Pediat; Soc Pediat Res; Am Col Cardiol; Am Soc Biol Chem; Am Heart Asn; Am Soc Clin Invest; AAAS; Am Pediat Soc; Am Asn Physicians. *Res:* Molecular biology of mitochondrial proteins; compartmentalization of proteins; fatty acid oxidation; human genetic disease. *Mailing Add:* St Louis Children's Hosp 400 S Kingshighway Blvd St Louis MO 63110

STRAUSS, ARTHUR JOSEPH LOUIS, b New York, NY, Jan 25, 33; m 60; c 3. PSYCHIATRY. *Educ:* Brooklyn Col, BA, 53; Columbia Univ, MD, 58; Am Bd Psychiat & Neurol, cert psychiat, 85. *Prof Exp:* Postdoctoral fel immunol, Col Physicians & Surgeons, Columbia Univ, NIH, 59-61, investr, Nat Inst Allergy & Infectious Dis, 61-65, sect head autoimmunity, 66-68; resident psychiat, St Elizabeths Hosp, NIMH, 68-71, med dir, 75-85, assoc supt, 85-87; psychiatrist, Bethesda, Md, 71-75; REGIONAL MED DIR PSYCHIAT, LA DEPT HEALTH & HOSPS, 87- *Concurrent Pos:* Mem, Nat Med Adv Bd, Myasthenia Gravis Found, 64- *Mem:* Am Asn Immunologists; Am Psychiat Asn. *Res:* Organ specific autoantibodies to striated muscle in Myastenia Gravis. *Mailing Add:* Div Ment Health La Dept Health & Hosps New Orleans Ment Health Ctr PO Box 740129 New Orleans LA 70174-0129

STRAUSS, BELLA S, b Camden, NJ, May 28, 20. MEDICINE. *Educ:* Columbia Univ, BA, 42; Western Reserve Univ, MD, 53; Am Bd Internal Med, dipl, 61. *Prof Exp:* Intern med, First Div, Bellevue Hosp, New York, 53-54; asst resident path, Univ Hosps, Med Ctr, Univ Mich, Ann Arbor, 54-55; asst resident med, Manhattan Vet Admin Hosp, New York, 55-56; asst resident First Div, Bellevue Hosp, 56-57, chief resident, Chest Serv Div, 57-58; career scientist, Health Res Coun New York, 62-66; vis specialist, Care/Medico, Avicenna Hosp, Kabul, Afghanistan, 66-67; staff physician, Maine Coast Mem Hosp, Ellsworth, 67-68; ASSOC PROF MED, DARTMOUTH MED SCH, 68- *Concurrent Pos:* NY Tuberc & Health Asn Miller fel, Col Physicians & Surgeons, Columbia Univ, 58-60; guest investr, Rockefeller Inst, 62-64; asst prof, Col Physicians & Surgeons, Columbia Univ, 64-66. *Mem:* Fel Am Col Physicians. *Res:* Internal and chest medicine; pathophysiology; training of paramedical personnel. *Mailing Add:* RFD 1 Box 131 Thetford Center VT 05075

STRAUSS, BERNARD, b Odessa, Russia, Apr 10, 04; nat US; m 64. MEDICINE. *Educ:* State Univ NY, MD, 27; Am Bd Urol, dipl, 43. *Prof Exp:* Instr urol, Sch Med, Stanford Univ, 39-42; asst prof, Sch Med, Loma Linda Univ, 56-65; assoc prof, 65-72, EMER ASSOC CLIN PROF UROL, DEPT SURG, SCH MED, UNIV SOUTHERN CALIF, 72- *Mem:* Am Urol Asn; fel Am Col Surgeons; corresp mem Belg Soc Urol; Am Med Asn. *Res:* Urology. *Mailing Add:* 9731 Sawyer St Los Angeles CA 90035

STRAUSS, BERNARD S, b New York, NY, Apr 18, 27; m 49; c 3. MOLECULAR BIOLOGY, CELL BIOLOGY. *Educ:* City Col New York, BS, 47; Calif Inst Technol, PhD(biochem), 50. *Prof Exp:* Hite fel cancer res & biochem genetics, Univ Tex, 50-52; from asst prof to assoc prof, Syracuse Univ, 52-60; assoc prof, Univ Chicago, 60-64, chmn dept microbiol, 69-84, chmn dept molecular genetics & cell biol, 84-85, prof microbiol, 64-84, dean basic sci, Div Biol Sci, 85-88, PROF MOLECULAR GENETICS & CELL BIOL, UNIV CHICAGO, 84- *Concurrent Pos:* Fulbright & Guggenheim fels, Osaka Univ, 58; mem genetics training comt, NIH, 62-68 & 70-74 & Chem Path Study Sect, 85-; vis prof, Univ Sydney, 67 & Hadassah Med Sch, Hebrew Univ, Jerusalem, 81; sr int fel, Fogarty Ctr, NIH, ICRF, London, 91. *Mem:* Am Soc Biol Chemists; Am Soc Microbiol; Am Asn Cancer Res. *Res:* Chemical mutagenesis; DNA repair and replication in mammalian cells; lymphocyte transformations. *Mailing Add:* Dept Molecular Genetics & Cell Biology 920 E 58th St Univ Chicago Chicago IL 60637

STRAUSS, BRUCE PAUL, b Elizabeth, NJ, Aug 19, 42; m 83; c 2. CRYOGENICS, LOW TEMPERATURE PHYSICS. *Educ:* Mass Inst Technol, SB, 64, ScD(solid state physics), 67; Univ Chicago, MBA, 72. *Prof Exp:* Prin res engr, Avco-Everett Res Lab, Avco Corp, 67-68; physicist, Argonne Nat Lab, 68-69; engr, Nat Accelerator Lab, Batavia, Ill, 69-79; mem staff, Magnetic Corp Am, 79-85; SR PRIN SCIENTIST & VPRES, POWERS ASSOCS, INC, 85- *Concurrent Pos:* Vis scientist, Univ Wis-Madison, 71-, Mass Inst Technol, 86- *Mem:* Am Phys Soc; Am Soc Metals; Am Inst Mining, Metall & Petrol Engrs; Cryogenic Soc Am; Mat Res Soc. *Res:* Cryogenic magnet systems; optimization of materials and performance. *Mailing Add:* 232 Summit Ave 203 Brookline MA 02146-2304

STRAUSS, CARL RICHARD, b Chicago, Ill, May 18, 36; m 59; c 3. POLYMER CHEMISTRY. *Educ:* Univ Ill, BS, 58; Univ Akron, MS, 65, PhD(polymer chem), 70. *Prof Exp:* Plant engr chlorinated organics, Pittsburgh Plate Glass Chem Div, 58-63, res chemist reinforcement elastomers, 63-69; advan scientist polyesters, 69-72, sr scientist phenolic binder, 72-74, SR SCIENTIST RESINS & BINDERS, OWENS-CORNING FIBERGLAS CORP, 74- *Concurrent Pos:* Instr, Cent Ohio Tech Col, 73-75. *Mem:* Am Chem Soc. *Res:* Cure and mechanical properties of organic binders; glass-binder interaction and mechanical performance of fiberglass composites; binder development. *Mailing Add:* Rt 5 Box 29 3380 Milner Rd NE Newark OH 43055-9308

STRAUSS, CHARLES MICHAEL, b Providence, RI, Oct 18, 38; m 61; c 2. COMPUTER SCIENCE, APPLIED MATHEMATICS. *Educ:* Harvard Col, AB, 60; Brown Univ, ScM, 66, PhD(appl math), 69. *Prof Exp:* Asst prof, 68-76, adj assoc prof appl math, 76-80, PROF MATH, BROWN UNIV, 80- *Mem:* AAAS; Asn Comput Mach; Soc Indust & Appl Math; Math Asn Am; Sigma Xi. *Res:* Computer graphics; numerical analysis. *Mailing Add:* 282 Williams St Providence RI 02906

STRAUSS, ELLEN GLOWACKI, b New Haven, Conn, Sept 25, 38; m 69. MOLECULAR GENETICS, VIROLOGY. *Educ:* Swarthmore Col, BA, 60; Calif Inst Technol, PhD(biochem), 66. *Prof Exp:* NIH fel biochem, Univ Wis, 66-68, fel, 68-69; from res fel to sr res fel, 69-84, SR RES ASSOC BIOL, CALIF INST TECHNOL, 84 - *Mem:* Sigma Xi; Am Soc Microbiologists; Soc Gen Microbiol; Am Soc Virol. *Res:* Molecular biology of the replication of togaviruses, particularly alphavirus Sindbis, primarily through isolation and characterization of conditional lethal mutants. *Mailing Add:* Div of Biol Calif Inst of Technol Pasadena CA 91125

STRAUSS, ELLIOTT WILLIAM, b Brooklyn, NY, Jan 25, 23; m 51; c 3. PATHOLOGY, ANATOMY. *Educ:* Columbia Univ, AB, 44; NY Univ, MD, 49. *Hon Degrees:* Brown Univ, Msc, 72. *Prof Exp:* Asst med, Peter Bent Brigham Hosp, 57-59; res fel med, Harvard Med Sch, 57-59, res fel anat, 59-61, res assoc path, 61-65; asst prof, Univ Colo Med Ctr, Denver, 65-70; ASSOC PROF MED SCI, BROWN UNIV, 70- *Concurrent Pos:* USPHS career develop award, 61-65; NIH grants, 61-66, 67-72 & 77-80. *Mem:* AAAS; Am Gastroenterol Asn; Am Soc Cell Biol; Am Asn Pathologists & Bacteriologists; Am Soc Exp Path; Am Soc Zoologists. *Res:* Electron microscopy; lipid chemistry; vitamin B-12; normal and abnormal mechanisms for absorption and transport by intestine and vessels. *Mailing Add:* Dept Med Sci Brown Univ Brown Sta Providence RI 02912

STRAUSS, ERIC L, b Mainz, Ger, Dec 13, 23; m 49; c 2. THERMAL PROTECTION SYSTEMS, CRYOGENIC INSULATION. *Educ:* Stevens Inst Technol, ME, 49; Univ Va, MME, 53. *Prof Exp:* Mech engr, Nat Adv Comt Aeronaut, Va, 49-53; proj engr, Taylor-Wharton Iron & Steel Co, Pa, 53-54; res & develop scientist, Baltimore Div, 54-67, SR RES SCIENTIST, ASTRONAUTICS GROUP, MARTIN MARIETTA CORP, 67- *Concurrent Pos:* Lectr, Exten Div, Univ Wis, 65. *Honors & Awards:* Indust Res 100 Award, 63. *Mem:* Soc Plastics Engrs; Am Ceramic Soc. *Res:* Nonmetallics; structural plastics, composites and adhesives, ablators and ceramic heat shield materials for aerospace applications; materials development and investigation of mechanical and thermal properties; heat transfer analysis and polymer degradation. *Mailing Add:* Martin Marietta Corp PO Box 179 Mail No T330 Denver CO 80201

STRAUSS, FREDERICK BODO, b Bad Wildungen, Ger, Feb 24, 31; US citizen; m 54; c 3. MATHEMATICS. *Educ:* Univ Calif, Los Angeles, BA, 59, MA, 62, PhD(math), 64. *Prof Exp:* Asst prof math, Univ Hawaii, 64-68; ASSOC PROF MATH, UNIV TEX, EL PASO, 68- *Mem:* AAAS; Am Math Soc; Math Asn Am; Sigma Xi. *Res:* Linear algebra and functional analysis; matrix Lie algebras; theory of rings. *Mailing Add:* Dept of Math Univ of Tex El Paso TX 79902

STRAUSS, GEORGE, b Vienna, Austria, Nov 27, 21; US citizen; m 54; c 2. BIOPHYSICAL CHEMISTRY. *Educ:* Univ London, BSc, 50; Lehigh Univ, PhD(chem), 55. *Prof Exp:* Chemist, A S Harrison & Co, 45-52; Colgate fel, 55-57, from asst prof to assoc prof, 57-66, res assoc, Inst Microbiol, 57-64, PROF CHEM, RUTGERS UNIV, NEW BRUNSWICK, 66- *Concurrent*

Pos: Rutgers Univ fac fel & USPHS fel, Univ Sheffield, 64-65; vis investr, Fed Inst Technol, Zurich, Switz, 85; vis prof, Univ Rome, Italy, 85. *Mem:* AAAS; Am Chem Soc; Biophys Soc; Sigma Xi. *Res:* Structure and properties of lipid bilayer membranes, and their interaction with surfactants; absorption and emission spectroscopy; excited states and energy transfer in biological systems; interactions in lipid membranes. *Mailing Add:* Dept Chem Rutgers Univ PO Box 939 Piscataway NJ 08855-0939

STRAUSS, H(OWARD) J(EROME), b New York, NY, July 2, 20; m 50; c 3. CHEMICAL ENGINEERING. *Educ:* City Col New York, BChE, 42; Columbia Univ, MS, 47, PhD(chem eng), 49. *Prof Exp:* Chem engr, Tenn Valley Authority, 42-43; metallurgist, Vanadium Corp Am, 43-44; instr chem eng, Cooper Union, 47-50; develop engr, Elec Storage Battery Co, 50-51, res engr, 51-52, supvr res dept, 52-53, asst mgr, 53-55, chief prod engr, 55-58, assoc dir res dept, 58-59, vpres & mgr, ESB-Reeves Corp, 59-62; res dir, Burgess Battery Div, Clevite Corp, Ill, 62-70; assoc dir res & develop, Gould, Inc, St Paul, 70-72, dir mkt & technol develop, 72-76; consult, 76-77; vpres oper improv, ESB Ray-O-Vac Mgt Co, Inc, 77-78; vpres oper improv, ESB Ray-O-Vac Mgt Co, Inc, 77-78; vpres oper & eng, Inco Electroenergy Corp, 79-; PRIN ENGR, RAYTHEON CO. *Mem:* Electrochem Soc; Am Electroplaters Soc; Am Inst Chem Engrs; Franklin Inst; Soc Plastics Engrs; sr mem Inst Elec & Electronics Engrs. *Res:* Industrial electrochemistry; plastics technology. *Mailing Add:* 4967 Devonshire Circle Excelsior MN 55331

STRAUSS, HARLEE S, b New Brunswick, NJ, June 19, 50; c 1. CHEMISTRY, TOXICOLOGY. *Educ:* Smith Col, Mass, BA, 72; Univ Wis, PhD(molecular biol), 79. *Prof Exp:* Spec asst govt affairs, Am Chem Soc, Washington, DC, 83-84; spec consult, Environ Corp, Washington, DC, 84-85; res, Ctr Technol, Policy & Indust Develop, Mass Inst Technol, 85-86; sr assoc, Gradient Corp, 86-88, RES AFFIL, CTR TECHNOL POLICY & INDUST DEVELOP, MIT, 86-; PRES, H STRAUSS ASSOCS INC, NATICK, MASS, 88- *Concurrent Pos:* Consult, 85- *Honors & Awards:* Am Inst Chemists Award, 72. *Mem:* AAAS; Am Chem Soc; Sigma Xi; NY Acad Sci; Asn Women Sci; Am Soc Microbiol; Biophys Soc. *Res:* Developing methodologies for risk assessment of chemicals and organisms created by genetic engineering techniques. *Mailing Add:* 21 Bay State Rd Natick MA 01760-2942

STRAUSS, HAROLD C, b Montreal, Que, Can, Jan 30, 40. CARDIOLOGY. *Educ:* McGill Univ, MD, 64. *Prof Exp:* ASSOC PROF PHARMACOL, SCH MED, DUKE UNIV, 79-, PROF MED, 85- *Mailing Add:* Duke Med Ctr Box 2914 Durham NC 27710

STRAUSS, HERBERT L, b Aachen, Ger, Mar 26, 36; US citizen; m 60; c 3. VIBRATIONAL SPECTROSCOPY, MOLECULAR DYNAMICS. *Educ:* Columbia Univ, AB, 57, MA, 58, PhD(chem), 60. *Prof Exp:* Ramsey fel from Univ Col, London Univ & NSF fel, Oxford Univ, 60-61; from asst prof to assoc prof, 61-73, vchair, 75-81, PROF CHEM, 73-, ASST DEAN, COL CHEM, UNIV CALIF BERKELEY, 87- *Concurrent Pos:* Sloan res fel, 66-68; vis prof, Indian Inst Technol, Kanpur, 68, Fudan Univ & Tokyo Univ, 82 & Univ Paris, 86; ed, Ann Rev Phys Chem; mem, IUPAC Comn I 1, 89- *Mem:* Am Chem Soc; fel Am Phys Soc; fel AAAS. *Res:* Experimental and theoretical spectroscopy; infrared; light scattering; librational motion and phase transitions in solids; coupling of various types of molecular motion. *Mailing Add:* Dept of Chem Univ of Calif Berkeley CA 94720

STRAUSS, HOWARD W(ILLIAM), chemical engineering, for more information see previous edition

STRAUSS, JAMES HENRY, b Galveston, Tex, Sept 16, 38; m 69. MOLECULAR BIOLOGY, VIROLOGY. *Educ:* St Mary's Univ, Tex, BS, 60; Calif Inst Technol, PhD, 67. *Prof Exp:* NSF fel, Albert Einstein Col Med, 66-67, res fel, 66-69; exec officer, 80-90, from asst prof to assoc prof, 69-84, PROF BIOL, CALIF INST TECHNOL, 84- *Mem:* AAAS; Am Soc Biol Chemists; Am Soc Microbiol; Sigma Xi; Am Soc Virol. *Res:* Structure and replication of animal viruses; cell surface modification and RNA replication during Togavirus infection; biogenesis of cell plasma membranes; molecular biology of flavivirus replication. *Mailing Add:* Div of Biol Calif Inst of Technol Pasadena CA 91125

STRAUSS, JEROME FRANK, III, b Chicago, Ill, May 2, 47; m 70; c 2. REPRODUCTIVE ENDOCRINOLOGY. *Educ:* Brown Univ, BA, 69; Univ Pa, MD, 74, PhD(molecular biol), 75. *Prof Exp:* Intern obstet & gynec, Univ Pa, 75-76, assoc, 76-77, from asst prof obstet, gynec & physiol to assoc prof obstet, gynec, path & physiol 77-85, PROF OBSTET, GYNEC, PATH & PHYSIOL, UNIV PA, 85-, ASSOC CHMN DEPT, 87-, ASSOC DEAN, 90- *Concurrent Pos:* Dir, Endocrine Lab, Hosp Univ Pa, 81-, dir, Div Reproductive Biol, Dept Obstet & Gynec, 84-; assoc ed, J Lipid Res, 82-86, Biol Reproduction, 86-; biochem endocrinol study sect, NIH, 83-87, pop res comt, 88-; assoc chmn, Dept Obstet & Gynec, 87- *Honors & Awards:* President's Achievement Award, Soc Gynec Invest, 90. *Mem:* Endocrine Soc; Am Physiol Soc; Am Soc Pathologists; Am Fertility Soc; Soc Study Reproduction; Soc Gynec Invest. *Res:* Regulation of corpus luteum function with special emphasis on the control of cholesterol metabolism; lipoprotein metabolism and intracellular cholesterol transport; luteal cells; placental cell biology. *Mailing Add:* Univ Pa Hosp 3400 Spruce St 106 Dulles Bldg Philadelphia PA 19104

STRAUSS, JOHN S, b New Haven, Conn, July 15, 26; m 50; c 2. DERMATOLOGY. *Educ:* Yale Univ, BS, 46, MD, 50. *Prof Exp:* Intern med, Univ Chicago Clins, 50-51; resident dermat, Univ Pa, 51-52, fel, 54-56, instr, 56-57; from asst prof to prof, Sch Med, Boston Univ, 57-78; PROF DERMAT & HEAD DEPT, COL MED, UNIV IOWA, 78- *Concurrent Pos:* Prin investr, NIH grants, 57-; assoc dir, Comn Cutaneous Dis, Armed Forces Epidemiol Bd, 65; asst ed, J Am Acad Dermat, 79-88; mem, NIAMS Adv Coun, NIH, 87-90; exec comt, Int Comt Dermat; pres, 18th World Cong Dermat, 92. *Honors & Awards:* Stephen Rothman Award, Soc Invest Dermat, 88; Presidential Citation, Am Acad Dermat, 90. *Mem:* Am Acad Dermat (pres 82-83); Soc Invest Dermat (secy-treas, 69-74, pres, 75-76); Am Dermat Asn (secy, 86-91, pres, 91-92); Sigma Xi; hon mem Brit Asn Dermat; hon mem Japan Soc Invest Dermat; hon mem Brazilian Dermat Soc; hon mem Argentine Dermat Soc; hon mem Neth Dermat Soc; hon mem Japan Dermat Asn; hon mem, Korean Dermat Asn; hon mem German Dermat Soc. *Res:* Sebaceous and epidermal cutaneous lipids; sebaceous gland physiology; pathophysiology and methods of treatment of acne. *Mailing Add:* Dept Dermat 2BT Univ Hosps Iowa City IA 52242-1090

STRAUSS, LEONARD, b New York, NY, Sept 21, 27; m 55; c 2. ELECTRICAL ENGINEERING. *Educ:* City Col New York, BEE, 50; Columbia Univ, MS, 52. *Prof Exp:* Instr elec eng, Columbia Univ, 51-54; from instr to assoc prof, 54-65, PROF ELEC ENG, POLYTECH INST NEW YORK, 65- *Mem:* AAAS; Inst Elec & Electronics Engrs. *Res:* Solid state circuitry and devices; laser technology; applications of solid state physics to device utilization. *Mailing Add:* Dept of Elec Eng & Electrophysics Polytech Univ 333 Jay St Brooklyn NY 11201

STRAUSS, LOTTE, pathology; deceased, see previous edition for last biography

STRAUSS, MARY JO, b Columbus, Ohio, June 10, 27; m 57; c 2. PHYSICAL CHEMISTRY. *Educ:* Bowling Green State Univ, BS, 49; Mich State Univ, PhD(phys chem), 55. *Prof Exp:* Phys chemist, US Naval Res Lab, 54-59, pvt consult, 60-71; res assoc, Col Gen Studies, George Washington Univ, 72-76, lectr, Grad Sch Arts & Sci, 77-78; consult, 79-89; RETIRED. *Concurrent Pos:* Res consult, Voc & Tech Div, Md Dept Educ, 85-89. *Mem:* Fel Am Inst Chemists; Women Math Ed; Sigma Xi; AAAS; Asn Women Sci. *Res:* Physical and chemical properties of ammonium amalgam; physical chemistry of the iron-oxygen-water system, particularly corrosion mechanisms; women in higher education; women, science and society; environmental studies; women in mathematics and science education. *Mailing Add:* 4506 Cedell Pl Temple Hills MD 20748

STRAUSS, MICHAEL S, b Los Angeles, Calif, Sept 24, 47. GENETIC RESOURCES CONSERVATION, BIOLOGICAL DIVERSITY. *Educ:* Univ Calif, BS, 69, MS, 74, PhD(biol sci), 76. *Prof Exp:* Postdoctoral res, Univ Calif, Irvine, 76-79; asst prof biol & bot, Northeastern Univ, 79-84; proj analyst, US Cong, Off Technol Assessment, 85-86; SR STAFF OFFICER, BD AGR, NAT RES COUN, 86- *Mem:* Am Soc Agron; AAAS; Crop Sci Soc Am. *Res:* In vitro culture and biochemical characterization of tropical tuber crops; writing in the areas of germplasm conservation and management, conservation biology, and biological diversity. *Mailing Add:* 2101 Constitution Ave NW Washington DC 20418

STRAUSS, MONTY JOSEPH, b Tyler, Tex, Aug 26, 45; m 78. PARTIAL DIFFERENTIAL EQUATIONS. *Educ:* Rice Univ, BA, 67; NY Univ, PhD(math), 71. *Prof Exp:* From asst prof to assoc prof, 71-85, assoc chmn dept, 84-87, PROF MATH, TEX TECH UNIV, 85-, ASSOC DEAN GRAD SCH, 89- *Concurrent Pos:* NSF res grant, 75- *Mem:* Am Math Soc; Math Asn Am; Soc Indust & Appl Math. *Res:* Partial differential equations, particularly the theoretical aspects of existence and uniqueness of solutions and several complex variables; computer literacy. *Mailing Add:* Grad Dean's Off Tex Tech Univ Lubbock TX 79409-1033

STRAUSS, PHYLLIS R, b Worcester, Mass, Mar 19, 43. CELL PHYSIOLOGY. *Educ:* Brown Univ, BA, 64; Rockefeller Univ, PhD(life sci), 71. *Prof Exp:* Res fel cell physiol, Harvard Med Sch, 71-73; from asst prof to assoc prof cell physiol, 73-84, prof biol, 84-86, DISTINGUISHED UNIF PROF BIOL, NORTHEASTERN UNIV, 86- *Concurrent Pos:* Res career develop award, Nat Cancer Inst, 78-83, & guest worker, 81; vis scholar, Harvard Univ, 88. *Mem:* AAAS; Am Soc Protozoologists; Am Asn Biol Chemists & Molecular Biologists; Soc Am Cell Biologists. *Res:* DNA replication and topoisomerases in eukaryotic cells including Trypanosoma brucei; nucleoside metabolism (adenosine and thymidine) in lymphocytes; thymidine metabolism in lymphocytes; small soluble DNA in eukaryotes. *Mailing Add:* Dept Biol Northeastern Univ 360 Huntington Ave Boston MA 02115

STRAUSS, RICHARD HARRY, INTERNAL MEDICINE. *Educ:* Univ Chicago, MD, 64. *Prof Exp:* ASSOC PROF MED, PREV MED & TEAM PHYSICIAN, OHIO STATE UNIV, 78- *Mailing Add:* Dept Prev Med 217 Starling Loving Hall Ohio State Univ 410 W Tenth Ave Columbus OH 43210

STRAUSS, ROBERT R, b Chelsea, Mass, Nov 4, 29; m 51; c 3. BIOCHEMISTRY, MICROBIOLOGY. *Educ:* Univ Pa, BA, 54; Hehnemann Med Col, MS, 56, PhD(microbiol), 58. *Prof Exp:* Sr res scientist, Nat Drug Co, Div Richardson-Merrell, Inc, 58-61, dir biochem res, 61, dir biochem & bact res labs, 61-67; res microbiologist, St Margaret's Hosp, Boston, 67-73; asst prof, Sch Med, Tufts Univ, 69-73; assoc dir microbiol, Albert Einstein Med Ctr, 73-78; RES ASSOC PROF, SCH MED, TEMPLE UNIV, 73- *Concurrent Pos:* Dir, Dept Microbiol, Albert Einstein Med Ctr, 78- *Mem:* AAAS; Am Soc Exp Path; Reticuloendothelial Soc; Am Acad Microbiol; Am Soc Microbiol. *Res:* Biochemistry of inflammation; virus purification; biochemistry of phagocytosis. *Mailing Add:* 4239 Yarmouth Dr Alison Park PA 15101-1567

STRAUSS, ROGER WILLIAM, b Buffalo, NY, Sept 23, 27; m 50; c 4. PAPER TECHNOLOGY. *Educ:* State Univ NY Col Forestry, Syracuse, BS, 49, MS, 50, PhD(chem), 61. *Prof Exp:* Develop engr, Bauer Bros, Ohio, 50-52; paper sales develop engr, Hammermill Paper Co, Pa, 52-55; instr paper sci, State Univ NY Col Forestry, Syracuse, 55-60; mgr res, Nekoosa-Edwards Paper Co, Wis, 60-66; prof paper sci & eng, State Univ NY Col Environ Sci & Forestry, 66-75; dir res & sci serv, Bowater Inc, Conn, 75-80, dir res & sci serv, Bowater NA Corp, Greenville, SC, 80-88; RETIRED. *Mem:* Tech Asn Pulp & Paper Indust. *Res:* Pulping and bleaching of wood pulp; paper production and coating. *Mailing Add:* 14 Moss Creek Ct Hilton Head Island SC 29926

STRAUSS, RONALD GEORGE, b Mansfield, Ohio, Nov 29, 39; m 62; c 3. PEDIATRICS, HEMATOLOGY. *Educ:* Capital Univ, BS, 61; Univ Cincinnati, MD, 65; Am Bd Pediat, dipl, 70, cert pediat hemat-oncol, 74. *Prof Exp:* Intern pediat, Boston City Hosp, 65-66; from jr resident to chief resident, Children's Hosp, Cincinnati, 66-69; pediatrician, David Grant US Air Force Med Ctr, 69-71; fel pediat hemat, Children's Hosp Res Found, Cincinnati, 71-73, asst prof pediat, Col Med, Univ Cincinnati, 73-74; asst prof pediat, Col Med, Univ Tenn, Memphis, 74-76; asst mem hemat-oncol, St Jude Children's Res Hosp, 74-76; ASSOC PROF PEDIAT, UNIV IOWA COL MED, 76- *Mem:* Soc Pediat Res; Am Acad Pediat; Soc Exp Biol Med; Am Fedn Clin Res; Am Soc Hemat. *Res:* Leukocyte physiology and function. *Mailing Add:* Dept Pediat Univ Iowa Iowa City IA 52242

STRAUSS, SIMON WOLF, b Poland, Apr 15, 20; nat US; m 57; c 2. CHEMISTRY. *Educ:* Polytech Inst Brooklyn, BS, 44, MS, 47, PhD(chem), 50. *Prof Exp:* Inorg chemist, Nat Bur Standards, 51-55; phys chemist, US Naval Res Lab, 55-57, head chem metall sect, 57-63; sr staff scientist, Hq, Air Force Systs Command, 63-80; CONSULT, 80- *Mem:* Fel AAAS; Am Chem Soc; fel Am Inst Chemists. *Res:* Solid state reactions; structure and electrical properties of glass; nature and structure of liquid metals; technical management. *Mailing Add:* 4506 Cedell Pl Camp Springs MD 20748

STRAUSS, STEVEN, b Czech, Dec 4, 30; US citizen; m 59; c 3. PHARMACY. *Educ:* Long Island Univ, BS, 55, MS, 65, Univ Pittsburg, PhD(pharm),70. *Prof Exp:* From asst prof to assoc prof pharm admin, 65-79, alumni dir, 65-70, dir continuing educ, 72-79, PROF PHARM ADMIN, ARNOLD & MARIE SCHWARTZ COL PHARM & HEALTH SCI, LONG ISLAND UNIV, 79-, DIR, DIV PHARM ADMIN & RETAIL DRUG INST, 88- *Concurrent Pos:* ed, US Pharmacist, 76-89; dir, Grad Pharm Progs, Westchester Campus, Long Island Univ, 77-83; field dir, Mkt Measures, 72-78, IMS Am, Ltd, 73-77. *Honors & Awards:* Rho Chi Pharm Honor Soc. *Mem:* Am Pharmaceut Asn; assoc AM Med Asn; fel Am Col Apothecaries; Am Soc Hosp Pharmacists; Am Soc Pharm Law; Am Asn Cols Pharm. *Res:* Pharmacy administration; marketing; market research; pharmacy law. *Mailing Add:* 39 Prospect Ave Ardsley NY 10502

STRAUSS, ULRICH PAUL, b Frankfort, Ger, Jan 10, 20; nat US; m 43, 50; c 4. PHYSICAL & POLYMER CHEMISTRY. *Educ:* Columbia Univ, AB, 41; Cornell Univ, PhD(chem), 44. *Prof Exp:* Sterling fel, Yale Univ, 46-48; from asst prof to prof phys chem, Rutgers Univ, 48-90, dir, Sch Chem, 65-71, chmn dept & dir, Grad Prog Chem, 74-80, EMER PROF PHYS CHEM, RUTGERS UNIV, NEW BRUNSWICK, 90. *Concurrent Pos:* NSF sr fel, 61-62; Guggenheim fel, 71-72. *Honors & Awards:* Johnson Wax Sci Achievement Award, 86. *Mem:* Am Chem Soc; NY Acad Sci. *Res:* Experimental and theoretical investigations of high polymers and polyelectrolytes. *Mailing Add:* Dept of Chem Rutgers Univ New Brunswick NJ 08903

STRAUSS, WALTER, b Nurnberg, Ger, Nov 6, 23; US citizen; m 59; c 2. MODELING, CIRCUIT SIMULATION. *Educ:* City Col New York, BEE, 48; Columbia Univ, PhD(physics), 61. *Prof Exp:* Tutor elec eng, City Col New York, 48-53; asst physics, Columbia Radiation Lab, 53-59; lectr elec eng, City Col New York, 59-60; mem tech staff, Bell Tel Labs, 60-90; RETIRED. *Res:* Magnetic domain devices; magnetoelastic properties of yttrium iron garnet; magnetic materials; piezoelectricity; magnetron oscillators; microwave delay lines; circuit simulation. *Mailing Add:* One Harrison Ct Summit NJ 07901-1713

STRAUSS, WALTER A, b Aachen, Ger, Oct 28, 37; US citizen; c 2. MATHEMATICS. *Educ:* Columbia Univ, AB, 58; Univ Chicago, MS, 59; Mass Inst Technol, PhD(math), 62. *Prof Exp:* NSF fel, Mass Inst Technol & Univ Paris, 62-63; vis asst prof math, Stanford Univ, 63-66; assoc prof, 66-71, PROF MATH, BROWN UNIV, 71- *Concurrent Pos:* Guggenheim fel, 71; vis scientist, Univ Tokyo, 72; Fulbright Lect, Rio de Janeiro, 67; vis prof, City Univ New York 68, Mass Inst Technol 78, Univ Md, 81, Yunn U (China), 86, NY Univ, 87-88. *Mem:* Soc Indust & Appl Math; Asn Math Physics; Am Math Soc. *Res:* Nonlinear partial differential equations; scattering theory; functional analysis; theory of plasmas. *Mailing Add:* Dept of Math Brown Univ Providence RI 02912

STRAUSSER, HELEN R, physiology, zoology; deceased, see previous edition for last biography

STRAUSZ, OTTO PETER, b Miskolc, Hungary, 24; Can citizen; c 1. CHEMISTRY. *Educ:* Eotvos Lorand Univ, Hungary, MSc, 52;, Univ Alta, PhD(chem), 62. *Prof Exp:* Res asst, 62-63, from asst prof to prof, 63-89, dir, Hydrocarbon Res Ctr, 74-77, EMER PROF CHEM, UNIV ALTA, 89- *Honors & Awards:* E W R Steacie Award, Can Soc Chem, 87. *Mem:* Fel Chem Inst Can; Am Chem Soc; AAAS; NY Acad Sci; Int-Am Photochemical Soc (pres, 75-79). *Res:* Mechanism and kinetics of chemical reactions induced photochemically or thermally and the chemistry of atoms, free radicals and reactive intermediates; chemical composition, analytical chemistry and organic geochemistry of petroleum. *Mailing Add:* Dept Chem Univ Alta Edmonton AB T6G 2G2 Can

STRAUTZ, ROBERT LEE, b Savanna, Ill, Jan 25, 35. PATHOLOGY, MEDICAL SCIENCES. *Educ:* Am Univ, BS, 63; Univ Md, PhD(anat), 66; Howard Univ, MD, 73. *Prof Exp:* Dir lab drug res, Hazelton Labs, Va, 61-62; instr biol, Am Univ, 65-66, asst prof biol & physiol, 66-70, assoc prof biol, 70-75; resident pathologist, George Washington Univ Hosp, 73-75; med dir, San Bernardino County Med Ctr, 73-86 & Los Angeles Med Ctr, Univ Southern Calif, 86-88; resident pathologist, Harbor Gen Hosp, Torrance, Calif, 75-88; MED DIR, CORONA COMMUNITY HOSP, 88- *Concurrent Pos:* Assoc prof, Med Sch, Univ Calif, Los Angeles, 77- *Mem:* Col Am Path; Am Soc Clin Pathologists. *Res:* Transplantation of pancreatic islets in diabetes. *Mailing Add:* PO Box 263 Corona CA 91718-0263

STRAW, JAMES ASHLEY, b Farmville, Va, Apr 12, 32; m 54; c 2. PHARMACOLOGY. *Educ:* Univ Fla, BS, 58, PhD(physiol), 63. *Prof Exp:* From asst prof to assoc prof, 65-75, PROF PHARMACOL, SCH MED, GEORGE WASHINGTON UNIV, 75- *Concurrent Pos:* NIH fel physiol, Univ Fla, 63-64 & res grant, 64-65. *Mem:* Am Asn Cancer Res; Am Soc Pharmacol & Exp Therapeut. *Res:* Pharmacokinetics; cancer chemotherapy; physiological disposition of anticancer drugs; anti-AIDS drugs. *Mailing Add:* Dept Pharmacol George Washington Univ 2121 Eye St NW Washington DC 20052

STRAW, ROBERT NICCOLLS, b Burlington, Iowa, Aug 24, 38; c 3. PROJECT MANAGEMENT. *Educ:* Univ Iowa, BS, 60, MS, 65, PhD(pharmacol), 67. *Prof Exp:* Res assoc pharmacol, 67-79, res head med, 79-85, DIR PROJ MGT, UPJOHN CO, 85- *Mem:* Am Soc Pharmacol & Exp Therapeut. *Res:* Development of centrally acting drugs. *Mailing Add:* Proj Mgt Upjohn Co 301 Henrietta St Kalamazoo MI 49001

STRAW, THOMAS EUGENE, b St Paul, Minn, Nov 20, 36; m 57; c 3. AQUATIC BIOLOGY. *Educ:* Univ Minn, St Paul, BS, 65, PhD(biochem), 69. *Prof Exp:* Asst prof, 68-73, ASSOC PROF BIOL, UNIV MINN, MORRIS, 73- *Mem:* Am Soc Limnol & Oceanog; Sigma Xi. *Res:* Biochemical limnology; ecology of aquatic bacteria. *Mailing Add:* Dept Sci & Math Univ Minn Morris MN 56267

STRAW, WILLIAM THOMAS, b Griffin, Ind, Sept 29, 31; m 56; c 3. GEOLOGY. *Educ:* Ind Univ, BS, 58, MA, 60, PhD(geol), 68. *Prof Exp:* Geologist, Humble Oil & Refining Co, 60-65; lectr geol, Ind Univ, 67-68; from asst prof to assoc prof, Western Mich Univ, 68-77, actg chmn dept, 71, chmn, 71-74, PROF GEOL, WESTERN MICH UNIV, 77-, CHMN, 88- *Concurrent Pos:* Actg assoc dir, Geol Field Sta, Ind Univ, 70-81; geologist, Ind Geol Surv, 70 & Mont Bur Mines & Geol, 78-79. *Mem:* Am Asn Petrol Geologists; fel Geol Soc Am. *Res:* Glacial geology, hydrogeology, geomorphology and hydrology of wetlands; fluvial sedimentation; geology of valley trains; regional geology of the northern Rocky Mountains. *Mailing Add:* Dept Geol Western Mich Univ Kalamazoo MI 49008

STRAWDERMAN, WAYNE ALAN, b Wakefield, RI, Oct 11, 36; m 58; c 2. APPLIED MECHANICS. *Educ:* Univ RI, BS, 58, MS, 61; Univ Conn, PhD(appl mech), 67. *Prof Exp:* Res engr, E I du Pont de Nemours & Co, 58-59; teaching asst mech eng, Univ RI, 59-61; mech engr, Elec Boat Div, Gen Dynamics Corp, 61-63; RES MECH ENGR, NEW LONDON LAB, NAVAL UNDERWATER SYSTS CTR, 63- *Concurrent Pos:* Lectr, Univ Conn, 67-69. *Mem:* Acoust Soc Am. *Res:* Response of coupled mechanical-acoustical systems to random excitation, turbulence induced noise, random vibrations and acoustics. *Mailing Add:* 55 Homestead Rd Ledyard CT 06339

STRAWN, OLIVER P(ERRY), JR, b Martinsville, Va, Nov 30, 25; wid; c 4. MECHANICAL ENGINEERING. *Educ:* Va Polytech Inst, BS, 50, MS, 65. *Prof Exp:* Sales engr, Richardson-Wayland Elec Corp, 53-57; asst prof mech eng, Va Polytech Inst & State Univ, 57-66, asst prof archit eng, 66-72; pvt consult pract, 72-75; PARTNER, CONSULT ENG FIRM, 76- *Concurrent Pos:* Mem, State Bldg Code Tech Rev Bd, 78-82. *Mem:* Am Soc Mech Engrs; Am Soc Heating, Refrig & Air-Conditioning Engrs; Nat Soc Prof Engrs. *Res:* Thermodynamics; heating; ventilating; air conditioning. *Mailing Add:* 601 Turner St Blacksburg VA 24060

STRAWN, ROBERT KIRK, b De Land, Fla, May 26, 22; c 3. ICHTHYOLOGY. *Educ:* Univ Fla, BS, 47, MS, 53; Univ Tex, PhD(zool), 57. *Prof Exp:* Asst malaria control, USPHS, 42-43; lab asst biol, Univ Fla, 46-48; instr, Southwestern Univ, 55-56; asst prof, Lamar State Col, 56-59; asst prof wildlife mgt, Agr & Mech Col, Tex, 59-60; from asst prof to assoc prof zool, Univ Ark, 60-66; assoc prof, 66-69, PROF WILDLIFE & FISHERIES SCI, TEX A&M UNIV, 69- *Mem:* Am Fisheries Soc; Am Soc Limnol & Oceanog; Am Soc Ichthyol & Herpet; Ecol Soc Am; Sigma Xi. *Res:* Ecology and speciation of fishes. *Mailing Add:* Wildlife-Fisheries Sci Tex A&M Univ College Station TX 77843

STRAYER, DAVID LOWELL, b Toledo, Ohio, Nov 16, 55; m; c 2. ECOSYSTEMS. *Educ:* Mich State Univ, BS, 76; Cornell Univ, PhD(ecol), 84. *Prof Exp:* Postdoctoral assoc, 83-85, asst scientist, 85-91, ASSOC SCIENTIST, INST ECOSYST STUDIES, NY BOT GARDEN, 91- *Mem:* Ecol Soc Am; Am Soc Limnol & Oceanog; NAm Benthological Soc; Int Asn Meiobenthologists. *Res:* Limnology; ecology of freshwater invertebrates, especially meiofauna and mollusks; energy flow in freshwater ecosystems. *Mailing Add:* Inst Ecosyst Studies NY Bot Garden, Box AB Millbrook NY 12545

STRAZDINS, EDWARD, b More, Latvia, Sept 19, 18; US citizen; m 43; c 2. PHYSICAL CHEMISTRY, POLYMER CHEMISTRY. *Educ:* Darmstadt Tech Univ, MS, 49. *Prof Exp:* Mill chemist, Baltic Wood Pulp & Paper Mills, 41-42, supvr, 43-44; res chemist, Am Cyanamid Co, 49-62, sr res chemist, 63-66, proj leader chem res, Cent Res Div, 67-68, res assoc, 68-74, prin res scientist, Chem Res Div, 74-; RETIRED. *Mem:* Am Chem Soc; fel Tech Asn Pulp & Paper Indust; fel Am Inst Chemists. *Res:* Paper chemistry; polyelectrolytes; sizing; retention; flocculation aids; theoretical aspects of paper making process; ecology; electrokinetic phenomena; polymer research and surface chemistry. *Mailing Add:* Five Dailey Rd New Milford CT 06776-4512

STREAMS, FREDERICK ARTHUR, b Mercer, Pa, Sept 8, 33; m 56; c 3. INSECT ECOLOGY. *Educ:* Indiana State Col, Pa, BS, 55; Cornell Univ, MS, 60, PhD(entom), 62. *Prof Exp:* Teacher pub schs, NY, 57-58; entomologist, Entom Res Div, USDA, 62-64; asst prof entom, 64-69, assoc prof biol, 69-74, head, ecol sect, 76-79, PROF BIOL, UNIV CONN, 74- *Mem:* Fel AAAS; Entom Soc Am; Ecol Soc Am; Am Soc Naturalists. *Res:* Ecology and evolution of populations; biological control of insects; predator-prey interactions in insects. *Mailing Add:* Dept Ecol & Evolutionary Biol Univ of Conn Storrs CT 06268

STREBE, DAVID DIEDRICH, b Tonawanda, NY, Oct 6, 18; m 42; c 2. MATHEMATICS. *Educ:* State Univ NY Teachers Col, Buffalo, BS, 40; Univ Buffalo, MA, 49, PhD(math), 52. *Prof Exp:* Instr math, LeTourneau Tech Inst, 46-47 & Univ Buffalo, 47-54; assoc prof, Univ SC, 54-57; prof, State Univ NY Col, Oswego, 57-58, Univ SC, 58-70 & Westmont Col, 70-71; prof math, Columbia Col, SC, 71-74; staff mem, Ariz Col Bible, 84-86; staff mem Southwestern Baptist Bible Col, 84-86; RETIRED. *Mem:* Nat Coun Teachers Math. *Res:* Set theoretic topology. *Mailing Add:* 13244 Ballad Dr Sun City West AZ 85375

STRECKER, GEORGE EDISON, b Ft Collins, Colo, Feb 25, 38; m 60; c 2. CATEGORICAL TOPOLOGY. *Educ:* Univ Colo, Boulder, BS & BS, 61; Tulane Univ, La, PhD(math), 66. *Prof Exp:* Instr math, Univ Colo, 58-60; teaching asst, Tulane Univ, La, 64-65; res assoc, Univ Amsterdam, 65-66, Fulbright fel, 65-66; fel, Univ Fla, 66-67, asst prof, 67-71; assoc prof, Univ Pittsburgh, 71-72; assoc prof, 72-77, PROF MATH, KANS STATE UNIV, 77- *Concurrent Pos:* Vis prof math, Vrije Univ, Amsterdam & Inst Univ L'Aquila, Italy, 80, 81; Czech Acad Sci exchange scholar, 80, 87, Hungarian Acad Sci exchange scholar, US Nat Acad Sci, 87. *Mem:* Am Math Soc; Math Asn Am. *Res:* Categorical topology; topological functors; initial and final completions of categories; factorization structures; Galois connections; computer sciences; combinations and finite mathematics. *Mailing Add:* Dept Math Kans State Univ Manhattan KS 66506

STRECKER, HAROLD ARTHUR, b Marietta, Ohio, June 11, 18; m 42; c 5. INORGANIC CHEMISTRY. *Educ:* Cornell Univ, AB, 40, PhD(chem), 48. *Prof Exp:* Chemist, Marietta Dyestuff Co, 40-41 & Nat Defense Comn, 42-45; res assoc, Standard Oil Co, Ohio, 47-58, supvr process res, 58-60, sr res assoc, 60-68, supvr spectros & micros, 69-81; RETIRED. *Mem:* Am Chem Soc; Soc Appl Spectros; Sigma Xi. *Res:* Catalysis; reaction kinetics; atomic spectroscopy; x-ray fluorescence and diffraction; electron microscopy. *Mailing Add:* 7131 Rotary Dr Walton Hills OH 44146

STRECKER, JOSEPH LAWRENCE, b Kansas City, Mo, Mar 30, 32; m 60; c 2. THEORETICAL PHYSICS. *Educ:* Rockhurst Col, BS, 55; Johns Hopkins Univ, PhD(physics), 61. *Prof Exp:* Jr instr physics, Johns Hopkins Univ, 55-58, res asst, 58- 61, sr res scientist, Gen Dynamics, Ft Worth, 61-66; assoc prof physics, Univ Dallas, 66-68; ASSOC PROF PHYSICS, WICHITA STATE UNIV, 68- *Concurrent Pos:* Adj prof, Tex Christian Univ, 62-67. *Mem:* Am Phys Soc. *Res:* Superconductivity; quantum field theory and application to solid state phenomena; statistical mechanics, especially phase transitions. *Mailing Add:* Dept of Physics Wichita State Univ Wichita KS 67208

STRECKER, ROBERT LOUIS, ecology, mammalogy, for more information see previous edition

STRECKER, WILLIAM D, SOFTWARE SYSTEMS. *Prof Exp:* VPRES ENG, DIGITAL EQUIP CORP, 89- *Mem:* Nat Acad Eng. *Mailing Add:* Digital Equip Corp 146 Main St MLO-12/2-T8 Maynard MA 01754

STRECKFUSS, JOSEPH LARRY, b Shirley, Mo, Feb 23, 31; m 52; c 3. MICROBIOLOGY, IMMUNOLOGY. *Educ:* Southern Ill Univ, Carbondale, BA, 58, MA, 61, PhD(virol, immunol), 68. *Prof Exp:* Dir diag microbiol, Holden Hosp, Carbondale, Ill, 57-66; asst res prof oral microbiol, Univ Tex Dent Br, 68-74, assoc prof microbiol, Dept Path & assoc prof in residence, Dent Sci Inst, 74-90; RETIRED. *Concurrent Pos:* Comt mem curric, Univ Tex Grad Sch Biomed Sci, Houston, 71- *Res:* Mechanism of calcification of Streptococcus mutans, a cariogenic microorganism; effect of fluoride resistance on the organisms adherence potential and carcinogenic properties. *Mailing Add:* 3407 Blue Candle Spring TX 77388

STREEBIN, LEALE E, b Blockton, Iowa, June 21, 34; m 56; c 4. ENVIRONMENTAL ENGINEERING, MICROBIOLOGY. *Educ:* Iowa State Univ, BS, 61; Ore State Univ, MS, 65, PhD(civil eng), 67. *Prof Exp:* Surveyor, US Army, 51-58; proj engr, Powers, Willis & Assocs, Planners, Engrs & Archit, 61-62, design room supvr, 62-63; from assoc prof to prof civil eng & environ sci, Univ Okla, 66-90, dir, 79-89; PRES, SEARCH INC, 70- *Concurrent Pos:* Lectr, WHO & Pan Am Health Orgn; consult, indust waste treatment & environ probs. *Mem:* Am Soc Civil Engrs; Water Pollution Control Fedn; Am Water Works Asn; Nat Soc Prof Engrs. *Res:* Industrial and hazardous waste treatment; process design; optimization of aerobic biological waste treatment systems; land treatment of industrial wastes; impoundment and stream studies. *Mailing Add:* 1218 W Rock Creek Univ Okla Norman OK 73069

STREET, DANA MORRIS, b New York, NY, May 7, 10; m 40; c 4. ORTHOPEDIC SURGERY. *Educ:* Haverford Col, BS, 32; Cornell Univ, MD, 36. *Prof Exp:* Chief orthop sect, Kennedy Vet Admin Hosp, Memphis, Tenn, 46-59; prof orthop surg, Sch Med, Univ Ark, 59-62; prof surg in residence, Sch Med, Univ Calif, Los Angeles, 62-75; head orthop div, Harbor Gen Hosp, 62-75 & Riverside Gen Hosp, 75-77; chief orthop sect, Jerry L Pettis Mem Hosp, 77-80; prof, 75-80, EMER PROF ORTHOP, LOMA LINDA UNIV, 80- *Concurrent Pos:* Mem staff, Orthop Hosp, Los Angeles, 63-, St Mary's Hosp & Mem Hosp, Long Beach, Calif, 63-75. *Mem:* AMA; Am Orthop Asn; Asn Bone & Joint Surgeons; Am Acad Orthop Surgeons. *Res:* Fracture treatment by use of medullary nail, particularly in femur and forearm; treatment and rehabilitation of the paraplegic. *Mailing Add:* 44201 Village 44 Camarillo CA 93012-8935

STREET, JABEZ CURRY, high energy physics; deceased, see previous edition for last biography

STREET, JAMES STEWART, b Chicago, Ill, July 26, 34; m 57; c 3. GEOLOGY, GENERAL EARTH SCIENCE. *Educ:* Univ Ill, Urbana, BS, 58; Syracuse Univ, MS, 63, PhD(geol), 66. *Prof Exp:* Geologist, Texaco Inc, La, 65-66; from asst prof to assoc prof, St Lawrence Univ, 66-78, chmn, dept geol & geog, 76-81 & 87-88, dir acad summer term, 86-87, PROF GEOL, ST LAWRENCE UNIV, 78-, JAMES HENRY CHAPIN PROF GEOL, 90- *Concurrent Pos:* Dir, NY State Tech Serv Prog, St Lawrence Univ, 66-72. *Mem:* Geol Soc Am; Am Asn Geol Teachers; Int Asn Quaternary Res; Sigma Xi; Hist Earth Sci Soc. *Res:* Geomorphology and glacial geology; history of science. *Mailing Add:* Dept Geol St Lawrence Univ Canton NY 13617

STREET, JIMMY JOE, b Waynesboro, Miss, Jan 17, 45; m 64; c 2. SOIL CHEMISTRY, ENVIRONMENTAL CHEMISTRY. *Educ:* Auburn Univ, BS, 69, MS, 72; Colo State Univ, PhD(soil chem), 76. *Prof Exp:* Res assoc soil sci, Auburn Univ, 69-72; res asst soil chem, Colo State Univ, 72-76; ASST PROF SOIL SCI, UNIV FLA, 76- *Mem:* Am Chem Soc; Soil Sci Soc Am; Sigma Xi; Am Soc Agron; AAAS. *Res:* The chemistry of trace elements in the soil-water-plant system and the fate of soil pollutants applied to agricultural lands. *Mailing Add:* Dept Soil Sci Univ Fla Gainesville FL 32611

STREET, JOHN MALCOLM, biogeography; deceased, see previous edition for last biography

STREET, JOSEPH CURTIS, toxicology, pesticide chemistry, for more information see previous edition

STREET, ROBERT ELLIOTT, b Belmont, NY, Dec 11, 12; m 41, 69; c 3. AERODYNAMICS, NUMERICAL ANALYSIS. *Educ:* Rensselaer Polytech Inst, BS, 33; Harvard Univ, AM, 34, PhD(math, physics), 39. *Prof Exp:* Instr math, Rensselaer Polytech Inst, 37-41; asst physicist, Nat Adv Comt Aeronaut, Langley Field, Va, 41-43; asst prof physics, Dartmouth Col, 43-44; engr, Gen Elec Co, NY, 44-47; assoc prof physics, Univ NMex, 47-48; from assoc prof to prof, 48-80, EMER PROF AERONAUT & ASTRONAUT, UNIV WASH, 80- *Mem:* Am Inst Aeronaut & Astronaut; Math Asn Am; fel Explorers Club. *Res:* Numerical fluid mechanics. *Mailing Add:* 1552 S Carol St Camano Island WA 98292

STREET, ROBERT L(YNNWOOD), b Honolulu, Hawaii, Dec 18, 34; m 59; c 3. FLUID MECHANICS, COMPUTATIONAL FLUID DYNAMICS. *Educ:* Stanford Univ, MS, 57, PhD(fluid mech), 63. *Prof Exp:* From asst prof to prof civil eng, Stanford Univ, 62-70, asst exec head dept, 62-66, assoc chmn dept, 66-72, chmn dept, 72-80, assoc dean res, Sch Eng, 72-83, vprovost & dean, Res & Acad Info Systs, 85-87, vpres info resources, 87-90, PROF FLUID MECH & APPL MATH, STANFORD UNIV, 72-, DIR, ENVIRON FLUID MECH LAB, 85-, VPRES LIBR & INFO RESOURCES, 90- *Concurrent Pos:* Vis prof, Univ Liverpool, 70-71; sr fel, Nat Ctr Atmospheric Res, 77-78; sr Queen's fel, Univ Western Australia, 85; chmn bd trustees, Univ Corp Atmospheric Res, 87-91. *Honors & Awards:* Huber Res Prize, Am Soc Civil Engrs, 72; Knapp Award, Am Soc Mech Engrs, 85. *Mem:* Am Soc Eng Educ; Am Geophys Union; Am Soc Civil Engrs; Am Soc Mech Engrs; Sigma Xi. *Res:* Theoretical and experimental hydromechanics; numerical simulation of mesoscale atmospheric motions, lake and estuary circulation and ground water systems; turbulence modeling; ocean-atmosphere interface modeling; natural and forced convection simulations and experiments. *Mailing Add:* Vpres Libr & Info Resources Stanford Univ Redwood Hall Stanford CA 94305-4120

STREET, ROBERT LEWIS, b Ennis, Tex, Sept 29, 28. INDUSTRIAL ENGINEERING. *Educ:* Tex A&M Univ, BS, 50, MS, 65; Univ Tex, Austin, PhD(mech eng), 67. *Prof Exp:* Estimator gen construct, Robert E McKee, Gen Contractor, 50-51; instrument man hwy surv, Tex Hwy Dept, 51; safety engr, Tex Employers Ins Asn, 53-59, sales rep, 59-61; from instr to prof indust eng, Tex A&M Univ, 62-77; PROJ ENGR, DAY & ZIMMERMANN INC, 78- *Concurrent Pos:* Prin McNichols, Street & Assocs, Inc, 71- *Mem:* Fel Soc Logistics Engrs. *Res:* Development of techniques for quantifying, evaluating and demonstrating system support parameters such as safety, availability and maintainability. *Mailing Add:* 3406 Walnut Texarkana TX 75503

STREET, WILLIAM G(EORGE), b Washington, DC, Dec 1, 17; m 43; c 5. AERONAUTICAL ENGINEERING. *Educ:* Cath Univ, BAE, 38. *Prof Exp:* Jr aeronaut engr, Nat Adv Comt Aeronaut, 38-39; chief flight test engr, Martin Co, 39-43, preliminary design engr, 45-50; chief flight test engr, Convair Div, Gen Dynamics Corp, 43-45; proj chmn opers res off, Johns Hopkins Univ, 50-53; from proj engr to res & adv tech mkt mgr, Martin-Marietta Corp, 53-67; Wash rep, Bell Aerosysts Co, 67-70; opers res analyst, Tech Anal Div, Nat Bur Standards, 70-74, gen engr, Ctr Bldg Technol, 74-81; RETIRED. *Mem:* Opers Res Soc Am; assoc Am Inst Aeronaut & Astronaut. *Res:* Nuclear propulsion; operations research; aerodynamics; flight testing; weapons systems requirements. *Mailing Add:* 516 Wyngate Rd Timonium MD 21093

STREETEN, DAVID HENRY PALMER, b Bloemfontein, SAfrica, Oct 3, 21; nat US; m 52; c 3. INTERNAL MEDICINE. *Educ:* Univ Witwatersrand, MB, BCh, 46; Oxford Univ, DPhil(pharmacol), 51. *Prof Exp:* Intern med & surg, Gen Hosp, Johannesburg, SAfrica, 47; jr lectr med, Univ Witwatersrand, 48; Nuffield demonstr pharmacol, Oxford Univ, 48-51; asst med, Peter Bent Brigham Hosp, Boston, 51-53, jr assoc, 53; from instr to asst prof internal med, Univ Mich Hosp, 53-60; assoc prof, 60-64, PROF MED, STATE UNIV NY HEALTH SCI CTR, 64- *Concurrent Pos:* Rockefeller traveling fel & res fel med, Harvard Univ, 51-52; investr, Howard Hughes Found, 55-61; consult, Vet Admin Hosp, Syracuse, 61-, Crouse Irving Mem Hosp, 61-, St Joseph's Hosp, 64- & Utica State Hosp, 65- *Mem:* Endocrine Soc; Am Fedn Clin Res; Fel Col Physicians; Fel AAAS. *Res:* Physiology and pathology of adrenal cortex, especially effects of its secretions on water and electrolyte metabolism and their role in causation of disease; normal and abnormal control of blood pressure. *Mailing Add:* Dept of Med State Univ of NY Hosp Syracuse NY 13210-3000

STREETER, DONALD N(ELSON), computer & information science, for more information see previous edition

STREETER, JOHN GEMMIL, b Ellwood City, Pa, Feb 25, 36; m 60; c 1. PLANT PHYSIOLOGY, AGRONOMY. *Educ:* Pa State Univ, BS, 58, MS, 64; Cornell Univ, PhD(bot), 69. *Prof Exp:* From asst prof to assoc prof, 69-78, PROF AGRON, OHIO AGR RES & DEVELOP CTR, OHIO STATE UNIV, 78- *Mem:* Fel, Am Soc Agron; Am Soc Plant Physiol; Am Soc Microbiol; fel Crop Sci Soc Am. *Res:* Nitrogen metabolism in plants; amino acid biosynthesis; carbohydrate metabolism in legume nodules. *Mailing Add:* Dept Agron Ohio Agr Res & Develop Ctr Wooster OH 44691

STREETER, ROBERT GLEN, b Madison, SDak, Feb 1, 41; m 64; c 2. WATERFOWL MANAGEMENT, WILDLIFE RESEARCH. *Educ:* SDak State Univ, BS, 63; Va Polytech Inst & State Univ, MS, 65; Colo State Univ, PhD(wildlife biol, physiol), 69. *Prof Exp:* Asst biologist avian depredation res, SDak State Dept Game, Fish & Parks, 63; res asst elk range ecol, Va Coop Wildlife Res Unit, US Fish & Wildlife Serv, 63-65; res asst bighorn sheep ecol & mgt, Colo Coop Wildlife Res Unit, 65-69; res physiologist, USAF Sch Aerospace Med, Brooks AFB, Tex, 69-72; wildlife biologist & res asst leader, Colo Coop Wildlife Res Unit, Colo State Univ & US Fish & Wildlife Serv, 72-73; head coop wildlife units, Div Res, Washington, DC, 73-75, coal proj res mgr, 75-79, prof design, 79-81, asst team leader tech appln, Western Energy & Land Use Team, 81-83, CHIEF, OFF INFO TRANSFER, RES & DEVELOP, US FISH & WILDLIFE SERV, 83-; DEPUTY EXEC DIR, NAM WATERFOWL MGT PLAN. *Concurrent Pos:* Vis mem grad fac, Tex A&M Univ, 71-72; asst prof, Colo State Univ, 72-73. *Mem:* Sigma Xi; Wildlife Soc. *Res:* Effects of coal extraction, conversion, transportation and related social developments on fish and wildlife populations, development of mitigation options and management decision alternatives; applications of computerized methodologies to wildlife management; national wildlife appraisal methods; scientific information transfer; waterfowl management. *Mailing Add:* Arlington Sq Bldg Rm 340 US Fish & Wildlife Serv 4401 N Fairfax Dr Arlington VA 22203

STREETMAN, BEN GARLAND, b Cooper, Tex, June 24, 39; m 61; c 2. ELECTRICAL ENGINEERING. *Educ:* Univ Tex, Austin, BS, 61, MS, 63, PhD(elec eng), 66. *Prof Exp:* From asst prof to assoc prof, 66-74, res assoc prof, 70-74, prof elec eng, Univ Ill, Urban-Champaign, 74-82, res prof, Coord Sci Lab, 74-82; Earnest F Gloyna regents chair Eng, 86-89, PROF ELEC ENG, UNIV TEX, AUSTIN, 82-, DULA D COCKRELL CENTENNIAL CHAIR, 89- *Concurrent Pos:* Dir Micro Elec Res Ctr, Univ Tex, Austin, 84- *Honors & Awards:* Frederick Emmons Terman Award, Am Soc, Eng Educ, 81, AT&T Found Award, 87; Educ Medal, Inst Elec & Electronic Engrs, 89. *Mem:* Nat Acad Eng; fel Inst Elec & Electronic Engrs; fel Electrochem Soc; Mat Res Soc. *Res:* Semiconductor materials and devices; molecular beam epitaxy; multilayer heterojunctions in III-V compounds; radiation damage and ion implantation in semiconductors; luminence in semiconductors. *Mailing Add:* Dept Elec Eng Univ Tex Austin TX 78712

STREETMAN, JOHN ROBERT, b Ft Worth, Tex, Apr 12, 30; m 51; c 2. PHYSICAL CHEMISTRY. *Educ:* Baylor Univ, BS, 51; Univ Tex, MA, 53, PhD(phys chem), 55. *Prof Exp:* Aeronaut res scientist, Nat Adv Comt Aeronaut, 55-56; sr nuclear engr, Gen Dynamics/Convair, 56-59; STAFF MEM, LOS ALAMOS NAT LAB, UNIV CALIF, 59- *Mem:* Am Phys Soc. *Res:* Monte Carlo neutron and gamma transport; quantum mechanics; solid state physics. *Mailing Add:* Box 414 Los Alamos NM 87544

STREETS, DAVID GEORGE, b Lincoln, Eng, Aug 13, 47; m 72; c 2. ENVIRONMENTAL SCIENCE, PHYSICAL CHEMISTRY. *Educ:* Univ London, BSc, 68, PhD(physics), 71. *Prof Exp:* NSF fel chem, Univ Rochester, 71-72; Imperial Chem Industs fel physics, Univ London, 72-74; res assoc, 74-75, asst environ scientist, 76-78, environ scientist, Off Environ Policy Anal, 79-85, ENVIRON SCIENTIST, ENERGY & ENVIRON SYSTS DIV, ARGONNE NAT LAB, 85- *Mem:* Royal Inst Chem; Brit Inst Physics; Am Chem Soc; Air Pollution Control Asn. *Res:* Environmental policy analysis; acid rain control strategies; renewable energy resources; photoelectron spectroscopy and chemical structure. *Mailing Add:* 1512 Winterberry Lane Darien IL 60559-5391

STREETS, RUBERT BURLEY, JR, b Tucson, Ariz, July 29, 29; m 63. ELECTRICAL ENGINEERING. *Educ:* Univ Ariz, BS, 55, MS, 56, PhD(elec eng), 64; Mass Inst Technol, EE, 59. *Prof Exp:* Mem tech staff, Ramo-Wooldridge Corp, Calif, 57; res asst elec eng, Servomech Lab, Mass Inst Technol, 58-59; instr, Univ Ariz, 59-61; mem tech staff, Aerospace Corp, Calif, 61; res specialist commun theory, Aerospace Div, Boeing Co, 64-67; ASSOC PROF ELEC ENG, UNIV CALGARY, 67- *Mem:* AAAS; Inst Elec & Electronics Engrs. *Res:* Design of Wiener optimal feedback systems; Wiener-Hopf integral equations; ionospheric scintillations. *Mailing Add:* Dept Elec Eng Univ Calgary 2500 Univ Dr NW Calgary AB T2N 1N4 Can

STREETT, WILLIAM BERNARD, b Lake Village, Ark, Jan 27, 32; m 55; c 4. PHYSICAL CHEMISTRY, HIGH PRESSURE PHYSICS. *Educ:* US Mil Acad, BS, 55; Univ Mich, MS, 61, PhD(mech eng), 63. *Prof Exp:* Instr astron & astronaut, US Mil Acad, 61-62 & 63-64, asst prof, 64-65; NATO res fel low temperature chem, Oxford Univ, 66-67; asst dean acad res & dir, Sci Res Lab, US il Acad, W Point, 67-78; sr res assoc, 78-81, prof & assoc dean, 81-84, JOE SILBERT DEAN ENG, SCH CHEM ENG, CORNELL UNIV, 84- *Concurrent Pos:* Guggenheim fel, Oxford Univ, 74-75. *Mem:* Royal Soc Chem; Am Chem Soc; Am Inst Chem Engrs; Sigma Xi; Am Asn Adv Sci. *Res:* Experimental measurements of physical and thermodynamic properties of fluids and fluid mixtures at high pressures; computer simulations of liquids. *Mailing Add:* Col Eng Cornell Univ Carpenter Hall Ithaca NY 14853

STREEVER, RALPH L, b Schenectady, NY, June 7, 34; m 64; c 2. SOLID STATE PHYSICS. *Educ:* Union Col, NY, BS, 55; Rutgers Univ, PhD(physics), 60. *Prof Exp:* Physicist, Nat Bur Standards, 60-66; physicist, US Army Electronics Technol & Devices Lab, 66-82. *Mem:* Am Phys Soc. *Res:* Magnetism and nuclear magnetic resonance in ferromagnetic materials; semiconductor device physics. *Mailing Add:* 11310 Dockside Circle Reston VA 22091-4003

STREHLER, BERNARD LOUIS, b Johnstown, Pa, Feb 21, 25; m 48; c 3. BIOLOGY, BIOCHEMISTRY. *Educ:* Johns Hopkins Univ, BS, 47, PhD, 50. *Prof Exp:* Biochemist, Oak Ridge Nat Lab, 50-53; asst prof biochem, Univ Chicago, 53-56; chief cellular & comp physiol sect, Geront Res Ctr, Nat Inst Child Health & Human Develop, 56-67; PROF BIOL SCI, UNIV SOUTHERN CALIF, 67- *Concurrent Pos:* Dir aging res satellite lab, Vet Admin Hosp, Baltimore, Md, 64-67. *Honors & Awards:* Karl August Forster Prize, Ger Acad Sci & Lett, 75. *Mem:* Am Soc Biol Chemists; Soc Develop Biol; Geront Soc; Am Soc Naturalists. *Res:* Bioluminescence; photosynthesis; aging; bioenergetics. *Mailing Add:* Dept Biol Sci ACBR 406 Univ Southern Calif University Park Los Angeles CA 90007

STREHLOW, CLIFFORD DAVID, b Mineola, NY, July 10, 40; US & UK citizen. ENVIRONMENTAL SCIENCES. *Educ:* Muhlenberg Col, BS, 62; Mass Inst Technol, MS, 64; NY Univ, PhD(environ health), 72. *Prof Exp:* Res scientist, Inst Environ Med, Med Ctr, NY Univ, 64-72; consult, Int Lead Zinc Res Orgn, St Mary's Hosp Med Sch, Eng, 72-80, res fel, 72-79; LECTR, CHARING CROSS & WILLIAMS, WESTMINISTER MED SCH, 79- *Concurrent Pos:* Lectr environ health, St Mary's Hosp Med Sch, 75-79. *Mem:* Soc Environ Geochem & Health; Am Chem Soc; Brit Occup Hyg Soc. *Res:* Lead metabolism; nutrition; epidemiology; trace element analysis; air and water pollution; radiochemistry. *Mailing Add:* Dept Child Health Westminster Children's Hosp Vincent Sq London SW1P 2NS England

STREHLOW, RICHARD ALAN, b Chicago, Ill, Sept 20, 27; m 77; c 1. CHEMISTRY. *Educ:* Univ Chicago, SB, 48; Univ Ill, PhD(chem), 57. *Prof Exp:* RES STAFF MEM, UNION CARBIDE NUCLEAR CO, OAK RIDGE NAT LAB, 56- *Mem:* Am Chem Soc; Am Vacuum Soc; Am Soc Testing & Mat; Am Nuclear Soc; Sigma Xi. *Res:* Catalysis and surface chemistry, coal conversion; electro-organic and fused salt chemistry; high vacuum research; mass spectrometry; fusion reactor design; graphite fabrication research. *Mailing Add:* 1818 Northwood Dr Knoxville TN 37923

STREHLOW, ROGER ALBERT, b Milwaukee, Wis, Nov 25, 25; m 48; c 2. PHYSICAL CHEMISTRY, FLUID DYNAMICS. *Educ:* Univ Wis, BS, 47, PhD(chem), 50. *Prof Exp:* Phys chemist, Ballistic Res Lab, 50-58, chief physics br, Interior Ballistics Lab, Aberdeen Proving Ground, 59-61; prof, 61-84, EMER PROF AERONAUT & ASTRONAUT ENG, UNIV ILL, URBANA, 84- *Concurrent Pos:* Ford Found vis prof, Univ Ill, Urbana, 60-61; consult, Los Alamos Sci Lab, 63-72, Aro Inc, Arnold Air Force Sta, 65-70, Weapon Command, Rock Island Arsenal, Ill, 69-71, Environ Protection Agency, 71 & Brookhaven Nat Lab, 75- *Mem:* AAAS; Am Chem Soc; fel Am Phys Soc; fel Am Inst Aeronaut & & Astronaut; Combustion Inst. *Res:* Reactive gas dynamics; combustion. *Mailing Add:* AAE Dept 101 Transp Bldg Univ Ill 104 S Matthews Ave Urbana IL 61801

STREIB, JOHN FREDRICK, b Avalon, Pa, Mar 21, 15; m 46, 54. PHYSICS. *Educ:* Calif Inst Technol, BS, 36, PhD(physics), 41. *Prof Exp:* Asst physicist, Carnegie Inst Technol, 41; asst physicist, Nat Bur Standards, 41-42, assoc physicist, 42-43, physicist, 43; scientist, Los Alamos Sci Lab, 43-46; mem tech staff, Bell Tel Labs, Inc, NY, 46; asst prof physics, Univ Colo, 46-47; from asst prof to prof, 47-80, EMER PROF PHYSICS, UNIV WASH, 80- *Mem:* Am Phys Soc. *Res:* Nuclear physics; fluorine plus proton reactions; positron absorption. *Mailing Add:* Dept Physics FM 15 Univ Wash Seattle WA 98195

STREIB, W(ILLIAM) C(HARLES), b Brooklyn, NY, Apr 5, 20; m 47; c 2. CHEMICAL ENGINEERING. *Educ:* Pa State Univ, BS, 42; Newark Col Eng, MS, 53. *Prof Exp:* Jr res engr, Res Ctr, Johns-Manville Corp, 42-44, res engr, 46-50, sr res engr, 50-60, chief, Asbestos Fiber Sect, 60-61, res mgr, Asbestos Fiber Dept, 61-71, dir res & develop, minerals & filtration technol, Manville Corp, 72-82, vpres sales corp, 72- 82; RETIRED. *Mem:* Am Chem Soc; Am Inst Mining, Metall & Petrol Engrs; Am Inst Chem Engrs. *Res:* Diatomaceous earth; perlite and talc, especially new product development for use in filtration and filler applications; asbestos, especially processing, uses and evaluations. *Mailing Add:* 6381 W Fremont Dr Littleton CO 80123

STREIB, WILLIAM E, b New Salem, NDak, Mar 16, 31; m 61; c 2. PHYSICAL CHEMISTRY. *Educ:* Jamestown Col, BS, 53; Univ NDak, MS, 55; Univ Minn, PhD(phys chem), 62. *Prof Exp:* Fel, Harvard Univ, 62-63; instr phys chem, 63-64, asst prof chem, 64-68, from assoc dir to dir labs, 64-79, CRYSTALLOGR, DEPT CHEM, IND UNIV, BLOOMINGTON, 79- *Mem:* Sigma Xi; Am Crystallog Asn. *Res:* X-ray crystallography; crystal and molecular structure; low temperature x-ray diffraction techniques. *Mailing Add:* Dept Chem Ind Univ Bloomington IN 47405

STREICHER, EUGENE, b New York, NY, Oct 25, 26; m 51. NEUROPHYSIOLOGY. *Educ:* Cornell Univ, BA, 47, MA, 48; Univ Chicago, PhD(physiol), 53. *Prof Exp:* Physiologist, US Army Chem Ctr, Md, 48-50; neurophysiologist, Sect Aging, NIMH, 54-62, physiologist, Nat Inst Neurol Dis & Stroke, SCIENTIST ADMINR, NAT INST NEUROL DIS & STROKE, 64- *Mem:* AAAS; Soc Exp Biol & Med; Am Asn Neuropath; Soc Neuroscience. *Res:* Physiological chemistry of central nervous system. *Mailing Add:* Nat Inst Neurol Dis & Stroke Bethesda MD 20892

STREICHER, MICHAEL A(LFRED), b Heidelberg, Ger, Sept 6, 21; nat US; m 47; c 2. METALLURGY, CHEMISTRY. *Educ:* Rensselaer Polytech Inst, BChE, 43; Syracuse Univ, MChE, 45; Lehigh Univ, PhD(phys metall), 48. *Prof Exp:* Res assoc, Lehigh Univ, 48; res eng, E I du Pont de Nemours & Co, Inc, 49-52, res proj engr, 52-56, res assoc, 56-67, res fel, 67-79; res prof, Dept Mech Eng, Univ Del, 79-88; CONSULT, CORROSION, 88- *Concurrent Pos:* Chmn, Gordon Conf Corrosion, 62 & subcomt corrosion, Welding Res Coun, 66-87. *Honors & Awards:* Turner Prize & Young Author Prize, Electrochem Soc, 49; Fel Am Soc Metals, 70; Willis Rodney Whitney Award, Nat Asn Corrosion Engrs, 72; Sam Tour Award, Am Soc Testing & Math, 79. *Mem:* AAAS; Am Soc Metals; Nat Asn Corrosion Engrs; Sigma Xi; Am Soc Testing & Mat. *Res:* Corrosion theory; passivity; inhibition; conversion coatings; influence of metallurgical factors on chemical reactivity of metals; corrosion evaluation tests; electrochemistry; stainless steels; nickel-base alloys and development of new alloys. *Mailing Add:* 1409 Jan Dr Wilmington DE 19803

STREIFER, WILLIAM, b Poland, Sept 13, 36; US citizen; m 58; c 3. OPTICS. *Educ:* City Col New York, BEE, 57; Columbia Univ, MS, 59; Brown Univ, PhD(elec eng), 62. *Prof Exp:* Res engr, Heat & Mass Flow Analyzer Lab, Columbia Univ, 58-59; from asst prof to prof elec eng, Univ Rochester, 62-72; res fel, Serox Corp, 72-80, sr res fel, 80-; RES MGR, SPECTRA DIODE LABS, SAN JOSE, CALIF. *Concurrent Pos:* Lectr, City Col New York, 57-59; consult lectr, Eastman Kodak Co, 65-68; consult, Xerox Corp, 68-72; vis assoc prof, Stanford Univ, 69-70; lectr, Stanford Univ, 77-80. *Mem:* Nat Acad Eng; fel Inst Elec & Electronics Engrs; fel Optical Soc Am; AAAS. *Res:* Electromagnetic theory; optics; mathematical ecology. *Mailing Add:* Spectra Diode Labs 80 Rose Orchard Dr San Jose CA 95134

STREIFF, ANTON JOSEPH, b Jackson, Mich, Apr 1, 15; m 41, 86; c 4. PETROLEUM CHEMISTRY. *Educ:* Univ Mich, BS, 36, MS, 37. *Prof Exp:* Res assoc, Nat Bur Stand, 37-50; sr res chemist, Carnegie-Mellon Univ, 50-71, asst chmn dept chem, 66-74, dir Am Petrol Inst Res Proj, 58, 60-67, prin res chemist, 72-84, admin officer, dept chem, 74-84, consult dept chem, 84-85, EMER ADMIN OFFICER & PRIN RES CHEMIST, CARNEGIE-MELLON UNIV, 84- *Mem:* Am Chem Soc; Sigma Xi. *Res:* Fractionation, purification, purity and analysis of hydrocarbons; American Petroleum Institute standard reference materials. *Mailing Add:* 1102 Tanbark W Jackson MI 49203-1253

STREIFF, RICHARD REINHART, b Highland, Ill, June 1, 29; m 59; c 3. MEDICINE, HEMATOLOGY. *Educ:* Wash Univ, AB, 51; Univ Basel, MD, 59. *Prof Exp:* Intern med, Harvard Med Serv, Boston City Hosp, 59-60; intern, Mt Auburn Hosp, Cambridge, Mass, 60-61, resident, 61-62; resident, Harvard Med Serv, Boston City Hosp, 62-63; instr med, Harvard Med Sch, 66-68; from asst prof to assoc prof, 68-74, PROF MED, COL MED, UNIV FLA, 74- CHIEF HEMAT UNIT, VET ADMIN HOSP, GAINESVILLE, 72-, CHIEF MED SERV, 73-, VCHMN DEPT MED, UNIV FLA, 75- *Concurrent Pos:* Res fel hemat, Thorndike Med Lab, Harvard Med Sch, 63-68; clin investr, Vet Admin, 69-71. *Mem:* Am Soc Hemat; Am Fedn Clin Res; Am Soc Clin Nutrit; Am Inst Nutrit; Fedn Am Socs Exp Biol. *Res:* Vitamin B-12 and folic acid deficiency anemias; metabolism and biological function of vitamin B-12 and folic acid; synthesis and testing of iron chelators. *Mailing Add:* Dept Hemat Univ Fla Med Col J Hillis Miller Health Ctr Archer Rd Gainesville FL 32602

STREILEIN, JACOB WAYNE, b Johnstown, Pa, June 19, 35; m 57; c 3. IMMUNOLOGY, GENETICS. *Educ:* Gettysburg Col, AB, 56; Univ Pa, MD, 60. *Prof Exp:* Intern, Univ Hosp, Univ Pa, 60-61, resident internal med, 61-63, from asst prof to assoc prof med genetics, Sch Med, 66-71; PROF CELL BIOL & PROF MED, UNIV TEX SOUTHWESTERN MED SCH, DALLAS, 72- *Concurrent Pos:* Fel allergy & immunol, Univ Pa, 63-64; fel transplantation immunity, Wistar Inst Anat & Biol, 64-65; Markle scholar acad med, 68-74. *Mem:* Am Asn Immunol; Transplantation Soc; Soc Exp Hemat; Soc Invest Dermat; Asn Res Vision & Ophthal. *Res:* Transplantation immunobiology with special reference to cellular immunity, immunoregulation, graft-versus-host disease, immunogenetic disparity; contact hypersensitivity; immunologic privilege. *Mailing Add:* 3121 SW 22nd Ave Miami FL 33145

STREIPS, ULDIS NORMUNDS, b Riga, Latvia, Feb 1, 42; US citizen; m 75; c 2. MICROBIOLOGY. *Educ:* Valparaiso Univ, BA, 64; Northwestern Univ, PhD(microbiol), 69. *Prof Exp:* Asst prof, 72-78, ASSOC PROF MICROBIAL GENETICS, SCH MED, UNIV LOUISVILLE, 78- *Concurrent Pos:* Damon Runyon Mem Fund Cancer res grant microbial genetics, Scripps Clin & Res Found, 69-70 & Sch Med & Dent, Univ Rochester, 70-72. *Mem:* AAAS; Am Soc Microbiol. *Res:* Genetic transformation in Bacillus subtilis; molecular biology of Bacillus thuringiensis; restriction endonucleases; cloning systems; mutagenesis assays; heat shock in prokaryotes. *Mailing Add:* Sch Med Dept Microbiol & Immunol Univ of Louisville Med Ctr Louisville KY 40292

STREIT, GERALD EDWARD, b Los Angeles, Calif, Dec 22, 48; m 74; c 2. CHEMISTRY. *Educ:* Univ Tex, Austin, BS, 70; Univ Calif, Berkeley, PhD(chem), 74. *Prof Exp:* Nat Res Coun fel, Nat Oceanic & Atmospheric Admin, 74-76; STAFF MEM, LOS ALAMOS NAT LAB, 76-, PROJ LEADER, AIR QUAL STUDIES, 90- *Mem:* Am Chem Soc. *Res:* Weak plasma interactions; ion-molecule kinetics and mechanisms; electron attachment; small cluster formation; application to chemical ionization mass spectroscometry; kinetics and dynamics of gas-phase neutral systems, particularly in application to atmospheric chemistry and combustion chemistry. *Mailing Add:* 2179 34 Los Alamos NM 87544

STREIT, ROY LEON, b Guthrie, Okla, Oct 14, 47; m 80; c 3. BAYESIAN INFERENCE NETWORKS, FAULT TOLERANT CONTROL. *Educ:* ETex State Univ, BA, 68; Univ Mo, MA, 70; Univ RI, PhD(math), 78. *Prof Exp:* MATHEMATICIAN, SONAR SYSTS RES, US NAVAL UNDERWATER SYSTS CTR, 70- *Concurrent Pos:* Vis scholar, Stanford Univ, 81-82; vis scientist, Yale Univ, 82-84; exchange scientist, Defense Sci & Technol Orgn, Adelaide, Australia, 87-89. *Mem:* Sr mem Inst Elec & Electronic Engrs; Soc Indust & Appl Math. *Res:* Application of Bayesian inference networks and artificial neural networks to sonar detection, classification and localization; emphasis is placed on statistical methods for network characterization and training. *Mailing Add:* US Naval Underwater Systs Ctr New London CT 06320

STREITFELD, MURRAY MARK, b New York, NY, Sept 16, 22; c 1. MICROBIOLOGY, CHEMOTHERAPY. *Educ:* City Col New York, BS, 43; McGill Univ, MS, 48; Univ Calif, Los Angeles, PhD(microbiol), 52. *Prof Exp:* Asst chemist toxicol, Off Chief Med Examr, New York, 43; instr, Med Lab, Beaumont Gen Hosp, Tex, 45-46; teaching asst bact, McGill Univ, 47-48 & Univ Calif, Los Angeles, 51-52; from instr to asst prof, 53-66, ASSOC PROF MICROBIOL, SCH MED, UNIV MIAMI, 66- *Concurrent Pos:* Instr med lab, Brookes Med Ctr, Tex, 45-46; res bacteriologist, Nat Children's Cardiac Hosp, Miami, Fla, 52-57; asst dir res bact lab, Variety Children's Hosp, 57-59, res assoc, Variety Res Found, 60-; resident attend, Vet Admin Hosp, Coral Gables, 60- *Mem:* Sigma Xi. *Res:* Bacteriology; rheumatic fever; antibiotics; prophylaxis of dental infection; streptococcal and staphylococcal epidemiology; pseudomonas and gonorrhea immunity; gamma globulin; staphyloccocal toxins; gonococcal cellular immunity; streptococcal virulence; antibiotics. *Mailing Add:* Dept Microbiol & Immunol Univ Miami Sch Med 1600 NW Tenth Ave Miami FL 33101

STREITWIESER, ANDREW, JR, b Buffalo, NY, June 23, 27; m 67; c 2. PHYSICAL ORGANIC CHEMISTRY, QUANTUM CHEMISTRY. *Educ:* Columbia Univ, AB, 48, MA, 50, PhD(chem), 52. *Prof Exp:* From instr to assoc prof, 52-63, PROF CHEM, UNIV CALIF, BERKELEY, 63- *Concurrent Pos:* Sloan Found fel, Univ Calif, Berkeley, 58-62; NSF faculty fel, 59-60; Miller Inst fel, 64-65 & 79-80; Guggenheim fel, 69. *Honors & Awards:* Award, Am Chem Soc, 67 & Phys Org Chem Award, 82; Sr Scientist Award, Humboldt Found, 76; Arthur Cope Scholar Award, Am Chem Soc, 89. *Mem:* Nat Acad Sci; AAAS; Am Chem Soc; Am Acad Arts & Sci. *Res:* Theoretical organic chemistry; molecular orbital theory; reaction mechanisms; isotope effects; acidity and basicity; rare earth organometallic chemistry; synthetic inorganic and organometallic chemistry. *Mailing Add:* Dept Chem Univ Calif Berkeley CA 94720

STREJAN, GILL HENRIC, b Galati, Romania, Sept 24, 30; m 63. IMMUNOLOGY. *Educ:* Univ Bucharest, MS, 53; Hebrew Univ Jerusalem, PhD(immunol), 65. *Prof Exp:* Instr bact & immunol, Hebrew Univ Jerusalem, 63-65; from asst prof to assoc prof, 68-80, PROF IMMUNOLOGY, UNIV WESTERN ONT, 80- *Concurrent Pos:* Res fel, NIH training grant & Fulbright travel grant immunochemistry, Calif Inst Technol, 65-68. *Mem:* Am Asn Immunol; Can Soc Immunol. *Res:* Regulation of immunoglobulin E-mediated hypersensitivity; immunologic aspects of experimental allergic encephalomyelitis and multiple sclerosis. *Mailing Add:* Dept Microbiol & Immunol Univ Western Ont London ON N6A 5B8 Can

STREKAS, THOMAS C, b Stafford Springs, Conn, May 9, 47; m 78; c 1. INORGANIC & BIOINORGANIC CHEMISTRY. *Educ:* Holy Cross Col, BA, 68; Princeton Univ, PhD(chem), 73. *Prof Exp:* Res assoc, IBM Res, Yorktown Heights, NY, 73-75; res assoc, Dept Biochem, Columbia Univ, 75-78; asst prof, 78-81, ASSOC PROF CHEM, DEPT CHEM, QUEENS COL, CITY UNIV NEW YORK, 81- *Mem:* Am Chem Soc; Sigma Xi; AAAS. *Res:* Structure function interrelationship in electron transfer proteins and metalloenzymes; vibrational-electronic spectroscopy of metal complexes; resonance raman spectroscopy. *Mailing Add:* Dept Chem Queens Col City Univ New York Flushing NY 11367

STRELTSOVA, TATIANA D, b Leningrad, USSR; Brit & US citizen; c 1. ENVIRONMENTAL ENGINEERING. *Educ:* Hydrol Inst, Leningrad, USSR, BS & MS, 59; All-Union Sci Res Inst Hydraul Eng, PhD(hydraul eng), 65. *Hon Degrees:* ScD, Birmingham Univ, 77. *Prof Exp:* Res assoc hydraul, All-Union Sci Res Inst Hydraul Eng, 59-65; sr res assoc underground flows, Moscow State Univ, 65-70; sr res fel flows in porous media, Birmingham Univ, 71-77; SR RES SPECIALIST, EXXON PROD RES CO, 77- *Concurrent Pos:* Fel, Rice Univ, 77; tech ed, Soc Petrol Eng, Am Soc Mech Engrs, 81- *Honors & Awards:* Woman of the Yr, Asn Women Geoscientists, 85-86. *Mem:* Am Geophys Union; Int Asn Water Resources; Soc Petrol Eng. *Res:* Various aspects of fluid flow trough porous media; author or coauthor of 50 publications. *Mailing Add:* Exxon Prod Res Co Reservoir Div PO Box 2189 Houston TX 77001

STRELZOFF, ALAN G, b Scranton, Pa, Sept 10, 37; m 67; c 4. PATTERN RECOGNITION & DIGITAL SIGNAL PROCESSING, IMAGE RECONSTRUCTION & PROCESSING. *Educ:* Mich State Univ, BA, 57; Columbia Univ, PhD(physics), 64. *Prof Exp:* Res assoc physics, Univ Wis, 63-66; asst prof physics & hist sci, Rochester, 66-69; res assoc & asst prof physics & biomed eng, Case Western Reserve Univ, 69-71 & 73-75; software engr indust control, Allen-Bradley, 75-76 & med image reconstruct, Imaging Syst, Union Carbide, 76-80; dir eng measurement & control machine vision, Systs Div, Analog Devices, 80-88; vpres eng discrete event controllers, Modicon/AEG, 88-90; VPRES ARCHIT & TECHNOL COMPUTER-AIDED DESIGN & COMPUTER-AIDED MFG, COMPUTER VISION, 90- *Mem:* Inst Elec & Electronic Engrs; Asn Comput Mach. *Res:* Design automation; industrial control; automatic generation of control programs from a computer- aided design database and corresponding inspection programs including machine vision. *Mailing Add:* 49 Blair Circle Sharon MA 02067

STREM, MICHAEL EDWARD, b Pittsburgh, Pa, Apr 1, 36; m 67. ORGANIC CHEMISTRY. *Educ:* Brown Univ, AB, 58; Univ Pittsburgh, MS, 61, PhD(chem), 64. *Prof Exp:* PRES, STREM CHEM INC, 64- *Mem:* Am Chem Soc. *Res:* Organometallic chemistry, including its use in organic synthesis. *Mailing Add:* PO Box 108 Newburyport MA 01950-0108

STREMLER, FERREL G, b Lynden, Wash, Mar 10, 33; m 58; c 2. ELECTRICAL ENGINEERING. *Educ:* Calvin Col, AB, 57; Ill Inst Technol, BS, 59; Mass Inst Technol, SM, 60; Univ Mich, PhD, 67. *Prof Exp:* Res asst elec eng, Inst Sci & Technol, Univ Mich, Ann Arbor, 60-61, asst, 61-63, res assoc, 63-65, assoc res engr, 65-68, lectr elec eng, 66-67; from asst prof to assoc prof, Univ Wis-Madison, 68-71, assoc dean, Col Eng, 78-82, chmn dept, 84-86, PROF ELEC ENG, UNIV WIS-MADISON, 77- *Honors & Awards:* Western Elec Fund Award, Am Soc Eng Educ, 75. *Mem:* Inst Elec & Electronic Engrs; Sigma Xi; Am Soc Eng Educ. *Res:* Analytical studies in coherent modulation-detection systems in communications and radar; design of communications and radar equipment; applications of communications and radar principles to remote sensing problems. *Mailing Add:* 5309 Joylynne Dr Madison WI 53716-3217

STRENA, ROBERT VICTOR, b Seattle, Wash, June 28, 29; m 57; c 2. APPLIED PHYSICS, RESEARCH ADMINISTRATION. *Educ:* Stanford Univ, BA, 52. *Prof Exp:* ASST DIR, W W HANSEN LAB PHYSICS, STANFORD UNIV, 59-, ASST DIR, EDWARD L GINZTON LAB, 76- *Concurrent Pos:* Mem, Nat Coun Univ Res Adminrs. *Res:* University physical sciences research administration. *Mailing Add:* Edward L Ginzton Lab Stanford Univ Stanford CA 94305

STRENG, WILLIAM HAROLD, b Milwaukee, Wis, Mar 6, 44; m 67; c 2. PHYSICAL CHEMISTRY. *Educ:* Carroll Col, Wis, BS, 66; Mich Technol Univ, MS, 68, PhD(phys chem), 71. *Prof Exp:* Res assoc theoret chem, Clark Univ, 72-73; SR CHEMIST, MERRELL RES CTR, MERRELL DOW PHARMACEUT, INC, 73- *Concurrent Pos:* Adj asst prof, Sch Pharm, Univ Cincinnati, 84- *Mem:* Am Chem Soc; Am Pharmaceut Asn. *Res:* Elucidation of interactions in electrolyte solution from both theoretical and experimental considerations. *Mailing Add:* 1216 Retswood Loveland OH 45140

STRENGTH, DELPHIN RALPH, b Brewton, Ala, May 24, 25; m 46; c 4. BIOCHEMISTRY. *Educ:* Auburn Univ, BS, 48, MS, 50; Cornell Univ, PhD(biochem), 52. *Prof Exp:* Res assoc biochem, Cornell Univ, 52-53; instr, St Louis Univ, 53-54, sr instr, 54-56, from asst prof to assoc prof, 56-61; assoc prof animal sci, 61-65, PROF BIOCHEM & NUTRIT, AUBURN UNIV, 65- *Concurrent Pos:* Vis Prof, Univ Philippines, 67-69. *Mem:* AAAS; Am Chem Soc; Am Soc Biol Chem; Soc Exp Biol & Med; Am Inst Nutrit. *Res:* Enzyme chemistry; nutrition; carcinogen metabolism; proteins; growth and differentiation. *Mailing Add:* 339 Camellia Dr Auburn AL 36830

STRENZWILK, DENIS FRANK, b Rochester, NY, Oct 27, 40. ELECTROMAGNETISM, SOLID STATE PHYSICS. *Educ:* Le Moyne Col, NY, BS, 62; Clarkson Col Technol, MS, 65, PhD(physics), 68. *Prof Exp:* RES PHYSICIST, US ARMY BALLISTIC RES LABS, 68- *Concurrent Pos:* Teacher, Exten Sch, Univ Del, Aberdeen Proving Ground, Md, 70. *Mem:* Am Asn Physics Teachers; Am Phys Soc. *Res:* Magnetism, effective field theory for yttrium iron garnet, lattice dynamics; clutter simulation of active and passive MMW sensors; statistical studies of IR data. *Mailing Add:* US Army Ballistics Res Labs Aberdeen Proving Ground MD 21005-5066

STRETTON, ANTONY OLIVER WARD, b Rugby, Eng, Apr 24, 36. NEUROBIOLOGY. *Educ:* Univ Cambridge, BA, 57, MA, 61, PhD(chem), 60. *Prof Exp:* Instr biochem, Mass Inst Technol, 60-61; mem sci staff, Lab Molecular Biol, Med Res Coun, Cambridge, Eng 61-71; assoc prof, 71-76, PROF ZOOL & MOLECULAR BIOL, UNIV WIS, 76- *Concurrent Pos:* Stringer fel, King's Col, Cambridge, 64-70; res assoc dept neurobiology, Harvard Med Sch, 66-67; Sloan Found fel, 72-74. *Mem:* Soc Neuroscience; AAAS; Brain Res Asn, Eng; Genetical Soc, Eng. *Res:* Structure and function of the nervous system of simple animals, especially nematodes. *Mailing Add:* Dept Zool 151 Noland Hall Univ Wis 1050 Bascom Mall Madison WI 53706

STREU, HERBERT THOMAS, b Elizabeth, NJ, May 16, 27; c 1. ENTOMOLOGY, ZOOLOGY. *Educ:* Rutgers Univ, BS, 51, MS, 59, PhD(entom), 60. *Prof Exp:* Nematologist, Agr Res Serv, USDA, 60-61; assoc res prof, 61-70, res prof entom, 70-88, chmn dept, 76-88, EMER PROF ENTOM, RUTGERS UNIV, 88- *Concurrent Pos:* Dir grad prog, Rutgers Univ, 76-81 & 86-88; assoc ed, J NY Entom Soc, 75-83; contrib ed, Am Lawn Appln, 85-88. *Mem:* Fel AAAS; hom mem Entom Soc Am; Acarological Soc Am. *Res:* Ecology and control of arthropods in turfgrass; biology and control of insects, nematodes and other economic arthropod pests attacking ornamental crops. *Mailing Add:* 5029 Vera Cruz Rd Center Valley PA 18034

STREULI, CARL ARTHUR, b Bronxville, NY, May 7, 22; m 50; c 3. ANALYTICAL CHEMISTRY. *Educ:* Lehigh Univ, BS, 43; Cornell Univ, AM, 50, PhD(anal chem), 52. *Prof Exp:* Chemist, Foster D Snell, Inc, 43-44 & 46-47; res assoc, Cornell Univ, 52-53; res chemist, Stamford Labs, Am Cyanamid Co, 53-57, group leader, 57-63, res assoc, 63-69, group leader, Lederle Labs, 69-77, SR RES CHEMIST, AM CYANAMID CO, 77- *Concurrent Pos:* Lectr, Univ Conn, Stamford, 63-66; fel, Purdue Univ, 66-67. *Mem:* Am Chem Soc. *Res:* Analytical chemistry in nonaqueous solvents; acid base theory; electroanalytical chemistry; gas-liquid and liquid-liquid chromatography. *Mailing Add:* 701 S Five Point Rd West Chester PA 19383

STRIBLEY, REXFORD CARL, b Kent, Ohio, Mar 12, 18; m 45; c 2. ORGANIC CHEMISTRY. *Educ:* Kent State Univ, BS, 39. *Prof Exp:* Chemist, 40-41, res chemist, Mason Lab, 45-50 & 52-55, chief res & develop, 55-70, tech dir, 70-76, nutrit dir & asst vpres mfg, Nutrit Div, Wyeth Labs, Inc Div, Am Home Prod Corp, 76-81, CONSULT, WYETH LABS, INC, 81- *Concurrent Pos:* US indust adv, Comt Food for Special Dietary Uses, UN Codex Alimentorius Comn, 72-74; mem bd of dirs, Infant Formula Coun, 77- *Mem:* Am Chem Soc; Am Oil Chem Soc; Am Dairy Sci Asn; Inst Food Technol. *Res:* Chemistry and development of infant formulas; infant nutrition; milk chemistry; dairy manufacturing technology and engineering; special dietary food products. *Mailing Add:* Box 334 El Jebel CO 81628

STRICHARTZ, ROBERT STEPHEN, b New York, NY, Oct 14, 43; m 68; c 2. MATHEMATICS. *Educ:* Dartmouth Col, BA, 63; Princeton Univ, MA, 65, PhD(math), 66. *Prof Exp:* NATO fel, Fac Sci, Orsay, France, 66-67; C L E Moore instr math, Mass Inst Technol, 67-69; from asst prof to assoc prof, 69-77, PROF MATH, CORNELL UNIV, 77- *Honors & Awards:* First Prize, Math Intel, Fr Mus Competition, 82; Lester R Ford Award, Math Asn Am, 83. *Mem:* Am Math Soc. *Res:* Harmonic analysis, partial differential equations, differential geometry. *Mailing Add:* Dept Math Cornell Univ White Hall Ithaca NY 14853

STRICK, ELLIS, b Pikeville, Ky, Mar 19, 21. GEOPHYSICS, OCEANOGRAPHY. *Educ:* Va Polytech Inst & State Univ, BS, 42; Purdue Univ, West Lafayette, PhD(theoret physics), 50. *Prof Exp:* Physicist radio eng, US Naval Res Lab, DC, 42-46; asst prof physics, Univ Wyo, 50-51; res assoc theoret seismol, Shell Explor & Prod Res Lab, Tex, 51-68; ASSOC PROF GEOPHYS, UNIV PITTSBURGH, 68- *Concurrent Pos:* Lectr physics, Univ Houston, 51-67; NSF grant, Univ Pittsburgh, 70-71. *Mem:* Soc Explor Geophysicists; Seismol Soc Am. *Res:* Anelastic wave propagation in solids at low frequencies. *Mailing Add:* 2160 Greentree Rd No 605 Pittsburgh PA 15220

STRICKBERGER, MONROE WOLF, b Brooklyn, NY, July 3, 25; m 57; c 2. EVOLUTION, GENETICS. *Educ:* NY Univ, BA, 49; Columbia Univ, MA, 59, PhD(genetics), 62. *Prof Exp:* Res fel genetics, Univ Calif, Berkeley, 62-63; from asst prof to assoc prof biol, St Louis Univ, 63-66; assoc prof, 68-71, PROF BIOL, UNIV MO-ST LOUIS, 71- *Concurrent Pos:* NIH res grant, 63-69. *Mem:* AAAS; Genetics Soc Am; Soc Study Social Biol; Am Genetic Asn; Am Soc Naturalists; Sigma Xi; Soc Study Evolution. *Res:* Evolution of fitness in Drosophila populations; induction of sexual isolation. *Mailing Add:* 1790 Arch St Berkeley CA 94709-1328

STRICKER, EDWARD MICHAEL, b New York, NY, May 23, 41; m 64; c 2. BIOPSYCHOLOGY. *Educ:* Univ Chicago, BS, 60, MS, 61; Yale Univ, PhD(psychol), 65. *Prof Exp:* Fel, Med Ctr, Univ Colo, 65-66 & Inst Neurol Sci, Med Sch, Univ Pa, 66-67; from asst prof to assoc prof psychol, McMaster Univ, 67-71; from assoc prof to prof, 71-86, DISTINGUISHED PROF, BEHAV NEUROSCIENCE, UNIV PITTSBURGH, 86- *Concurrent Pos:* Consult ed, J Comp & Physiol Psychol, 72-81; NIMH res scientist award, 81-86, Merit Award, 87-; vis prof psychiatry, Johns Hopkins Med Sch, 78-79; chmn, Psychobiol Prog, Univ Pittsburgh, 83-86, Dept Behav Neuroscience, 86-; consult ed, Am J Physiol, 85- *Mem:* Soc Neuroscience. *Res:* Physiological and behavioral mechanisms that maintain water and electrolyte balance, body temperature and energy metabolism; the neurochemical basis for recovery of function following brain damage; central controls of motivated behavior. *Mailing Add:* Dept Behav Neurosci Univ Pittsburgh Pittsburgh PA 15260

STRICKER, STEPHEV ALEXANDER, b Oakland, Calif, Jan 18, 54; m 82; c 1. BIOMINERALIZATION, INVERTEBRATE EMBRYOLOGY. *Educ:* Univ Calif, Santa Cruz, BA, 76; Univ Wash, MSc, 79, PhD(zool), 83. *Prof Exp:* Teaching fel biol, Univ Calgary, Alta, 83-87; res assoc, Dept Zool, Univ Wis, Madison, 87-88; ASST PROF, UNIV NMEX, 89- *Concurrent Pos:* Lectr, Univ Calgary, Alta, 84; instr, Friday Harbor Lab, Wash, 85 & 86; lectr, Univ Calif, Santa Cruz, 89. *Mem:* Am Microscopical Soc; Am Soc Zoologists; AAAS; Am Soc Cell Biologists; Electron Micros Soc Am. *Res:* Biology of nemertean worms; oocyte maturation; nuclear lamins; studies of metamorphosis in marine invertebrates; fertilization histological, ultrastructural and analytical studies of calcifying tissues in marine invertebrates. *Mailing Add:* Dept Biol Univ NMex Albuquerque NM 87131

STRICKHOLM, ALFRED, b New York, NY, July 3, 28; m 52; c 3. BIOPHYSICS, BIOMATHEMATICS. *Educ:* Univ Mich, BS, 51; Univ Minn, MS, 56; Univ Chicago, PhD(physiol), 60. *Prof Exp:* Fel biophys, Physiol Inst, Univ Uppsala, 60-61; asst prof physiol, Sch Med, Univ Calif, San Francisco, 61-66; assoc prof, 66-72, prof anat & physiol, 72-76, PROF PHYSIOL, MED SCI PROG, CTR NEURAL SCI, IND UNIV, BLOOMINGTON, 76- *Concurrent Pos:* USPHS grant, 62-; Am Heart Asn grant, 79- *Mem:* AAAS; Am Physiol Soc; Soc Neuroscience; Soc Gen Physiol; Biophys Soc. *Res:* Biophysics of the cell membrane; contraction coupling in muscle; permeability, active transport, and excitation; structure and function of cell membranes; neurobiology; neurotransmitters. *Mailing Add:* Physiol & Biophysics Ind Univ Bloomington IN 47405

STRICKLAND, ERASMUS HARDIN, b Spartanburg, SC, May 18, 36; m 66; c 2. BIOPHYSICS. *Educ:* Pa State Univ, BS, 58, MS, 59, PhD(biophys), 61. *Prof Exp:* Chief phys chem sect, US Army Med Res Lab, Ft Knox, 61-63; from asst prof to assoc prof biophys, Univ Calif, Los Angeles, 63-70, assoc res biophysicist, Radiation Biol Lab, 69-75; PRES, STRICKLAND COMPUTER CONSULT, 82- *Mem:* AAAS. *Res:* Nutritional software; circular dichroism and absorption spectroscopy of biological molecules. *Mailing Add:* 30135 Yellow Brick Rd Valley Center CA 92082

STRICKLAND, GEORGE THOMAS, b Goldsboro, NC, Apr 20, 34; m 60; c 3. TROPICAL MEDICINE, INFECTIOUS DISEASES. *Educ:* Univ NC, Chapel Hill, BA, 56, MD, 60; London Sch Hyg & Trop Med, DCMT, 70, PhD(parasitol), 74. *Prof Exp:* Intern, Nat Naval Med Ctr, Bethesda, Md, 60-61, resident internal med, 63-67, attend physician, 74-82, co-dir & infectious dis fel, 76-80; med officer, US Embassy, Nicosia, Cyprus, 61-63; chief clin invest, Naval Med Res Unit No 2, Taipei, China, 67-70; head immunoparisitol, Naval Med Res Inst, Bethesda, Md, 72-74, prog mgr infectious dis res, Naval Med Res & Develop Command, 74-76; dir res & educ, dept med, Uniformed Serv Univ Health Sci, Bethesda, Md, 76-80, prof internal med, 76-82; PROF MICROBIOL, MED & EPIDEMIOL, & PREV MED, & DIR INT HEALTH PROG, SCH MED, UNIV MD, 82- *Concurrent Pos:* Consult physician & lectr, Tri Serv Gen Hosp, Taipei, 67-70; res assoc, Johns Hopkins Sch Hyg & Pub Health, 74-; sr scientist, Armed Forces Inst Path, 81-82; dir, Int Ctr Med Res Training, Lahore, Pakistan, 83-85; ed, Hunter's Trop Med; sr consult, US Naval Med Res Unit #2, Cairo. *Mem:* Am Col Physicians; Am Col Trop Med & Hyg; Royal Col Trop Med & Hyg; Infectious Dis Soc Am; Nat Coun Int Health; Am Fedn Clin Res. *Res:* Clinical epidemiology of malaria and schistosomiasis; immune response to malaria; rapid diagnosis of malaria; parasitology. *Mailing Add:* Int Health Prog Univ Md Sch Med Baltimore MD 21201

STRICKLAND, GORDON EDWARD, JR, b Santa Cruz, Calif, Jan 23, 29; m 56; c 3. ENGINEERING MECHANICS. *Educ:* Stanford Univ, BS, 54, MS, 55, PhD(eng mech), 60. *Prof Exp:* Mem tech staff, Bell Tel Labs, 59-64; engr, Lawrence Radiation Lab, 64-66; res scientist, Lockheed Missiles & Space Co, 66-69; SR ENG ASSOC, CHEVRON OIL FIELD RES CO, 69- *Mem:* Sigma Xi. *Res:* Applied mechanics, particularly elasticity and shell theory. *Mailing Add:* 19816 Caprice Dr Yorba Linda CA 92686

STRICKLAND, JAMES ARTHUR, b Whiteville, NC, Dec 10, 61. MOLECULAR BIOLOGY. *Educ:* Oral Roberts Univ, BS, 84; Emory Univ, PhD(chem), 88. *Prof Exp:* Postdoctoral res assoc enzym, Sch Med, Emory Univ, 88-89; McKnight postdoctoral res assoc molecular biol, Univ Ill, 89- *Mem:* AAAS; Sigma Xi; Am Chem Soc; Am Soc Plant Physiologists. *Res:* Protein interactions in plants; protein purification; enzymology. *Mailing Add:* Dept Plant Biol Sch Life Sci 356 PABL Univ Ill Urbana IL 61801

STRICKLAND, JAMES SHIVE, b Harrisburg, Pa, Nov 18, 29; m 55; c 3. EXPERIMENTAL PHYSICS. *Educ:* Franklin & Marshall Col, BS, 51; Mass Inst Technol, PhD(physics), 57. *Prof Exp:* Staff physicist, Phys Sci Study Comt, Mass Inst Technol, 57-58; staff physicist, Educ Develop Ctr, Inc, Mass, 58-72; vis scientist, Mass Inst Technol, 72-73; PROF PHYSICS & CHMN DEPT, GRAND VALLEY STATE UNIV, 73- *Mem:* AAAS; Am Asn Physics Teachers; Am Phys Soc; Am Soc Eng Educ. *Res:* Development of new materials for science education. *Mailing Add:* Dept Physics Grand Valley State Univ Allendale MI 49401

STRICKLAND, JOHN WILLIS, b Wichita, Kans, Mar 23, 25; m 47; c 4. PETROLEUM GEOLOGY. *Educ:* Univ Okla, BS, 46. *Prof Exp:* Geologist, Skelly Oil Co, 47-50; res geologist, Continental Oil Co, 51-55, div geologist, 55-61, explor mgr, Ireland, 61-64, coordr explor res, 64-66, dir adv geol, 66-67, chief geologist, 67-76, mgr, Africa & Latin Am, 76-78, dir geol, 79-81; mgr explor serv, Conoco, Inc, consult geologist, 82-; RETIRED. *Mem:* Am Asn Petrol Geol. *Res:* Petroleum geology of world; factors controlling generation and distribution of hydrocarbons. *Mailing Add:* 761 W Creekside Dr Houston TX 77024

STRICKLAND, KENNETH PERCY, b Loverna, Sask, Aug 19, 27; m 48; c 4. BIOCHEMISTRY. *Educ:* Univ Western Ont, BSc, 49, MSc, 50, PhD(biochem), 53. *Prof Exp:* Nat Res Coun Can fel chem path, Guy's Hosp Med Sch, Univ London, 53-55; from asst prof to assoc prof, 55-66, PROF BIOCHEM, UNIV WESTERN ONT, 66-, PROF CLIN NEUROL SCI, 80- *Concurrent Pos:* Lederle med fac award, 55-58; res assoc, Med Res Coun Can, 58-79, career investr, 79- *Mem:* AAAS; Am Soc Biol Chemists; Can Biochem Soc; Can Physiol Soc; Int Neurochemical Soc. *Res:* Biochemistry of central nervous system, especially enzymes (momo- and diacylkinases, CDP-diacylglycerol and phosphatidyl inositol synthetases) of lipid components; biochemistry of muscle, especially triacylglycerol and phosphoglyceride metabolism in differentiating L6 and dystrophic muscle mytoblasts. *Mailing Add:* Dept Biochem Fac Med Univ Western Ont London ON N6A 5C1 Can

STRICKLAND, LARRY DEAN, b Elkview, WVa, Nov 6, 38; m 62; c 2. THERMODYNAMICS, FLUID DYNAMICS. *Educ:* WVa Univ, BS, 60, PhD(mech eng), 73; Univ Southern Calif, MS, 62. *Prof Exp:* Prin res engr advan rocket engines, Rocketdyne, NAm Aviation, 62-67; res engr aerodynamics, Re-entry Syst, Gen Elec Co, 67-68; instr, WVa Univ, 69-70; res co-op coal gasification, US Bur Mines, 71-73; proj mgr underground coal gasification, 77-78, prog mgr, press fluid bed combustion, 79-80, chief, gasification proj br, 81-88, CHIEF, EXP RES BR, MORGANTOWN ENERGY TECHNOL CTR, DEPT ENERGY, 84- *Concurrent Pos:* Adj prof, WVa Univ, 78-; mem, Grimthorpe Tech Comt, Int Energy Agency, 79-80, Adv Comt Int Conf Circulatory Fluidized Beds, 85-; chmn, Joint Classification Prog Operating Comt, Gas Res Inst & Dept Energy, 82-84; co-proj officer, joint US/Ital gasification proj, 86-88. *Mem:* Am Soc Mech Engrs; Sigma Xi. *Res:* Fluid dynamics and heat-transfer as applied to in-situ and above ground coal gasification and other energy related topics. *Mailing Add:* Morgantown Energy Technol Ctr Collins Ferry Rd Morgantown WV 26505

STRICKLER, STEWART JEFFERY, b Mussoorie, India, July 12, 34; US citizen; m 59; c 2. PHYSICAL CHEMISTRY, CHEMICAL PHYSICS. *Educ:* Col Wooster, BA, 56; Fla State Univ, PhD(phys chem), 61. *Prof Exp:* Chemist, Radiation Lab, Univ Calif, 61; res assoc, Rice Univ, 61-62, lectr chem, 62-63; from asst prof to assoc prof, 63-73, chmn dept, 74-77, PROF CHEM, UNIV COLO, BOULDER, 73- *Concurrent Pos:* Hon fel, Australian Nat Univ, 72-73. *Mem:* Am Chem Soc; Am Phys Soc; AAAS. *Res:* Molecular spectroscopy; photochemistry; quantum chemistry; lifetimes and properties of excited molecules; laser spectroscopy; photoconversion of solar energy. *Mailing Add:* Dept Chem & Biochem Univ Colo Box 215 Boulder CO 80309-0215

STRICKLER, THOMAS DAVID, b Ferozepur, India, Nov 11, 22; US citizen; m 89; c 4. ATOMIC PHYSICS. *Educ:* Col Wooster, BA, 47; Yale Univ, MS, 48, PhD(physics), 53. *Prof Exp:* Instr physics, Yale Univ, 52-53; from asst prof to assoc prof, Berea Col, 53-61, Charles F Kettering prof, 61-86, prof physics & chmn dept, 86-89; RETIRED. *Concurrent Pos:* NSF faculty fel, 60-61; consult, NSF Physics Inst, Chandigarh, India, 66, Gauhati Univ, India, 67 & Calcutta, India, 68; Fulbright lectr, Vidyalankara Univ, Colombo, Sri Lanka, 73-74, De La Salle Univ, Manila, Philippines, 81-82, & Univ Zambia, Lusaka, Zambia, 88-89. *Mem:* AAAS; Am Asn Physics Teachers; Sigma X; Health Physics Soc. *Res:* Neutron and gamma ray scattering; health physics; gaseous electronics. *Mailing Add:* 1468 Rumbaugh Circle Wooster OH 44691

STRICKLIN, BUCK, b Clovis, NMex, Dec 30, 22; m 42; c 1. ORGANIC CHEMISTRY. *Educ:* Tex Tech Col, BS, 48; Univ Colo, PhD(org chem), 52. *Prof Exp:* Asst, Univ Colo, 48-52; res mgr, Minn Mining & Mfg Co, 52-69, tech dir, Paper Prod Div, 69-76, lab mgr, Disposable Prod Dept, 76-82; RETIRED. *Mem:* Am Chem Soc; Tech Asn Pulp & Paper Indust. *Res:* Fluorocarbons; fluoroethers; chlorination; photochemistry; photoconductivity; polymers. *Mailing Add:* 454 Hilltop Ave Roseville MN 55113

STRICKLIN, WILLIAM RAY, b Savannah, Tenn, Apr 17, 46; m 67. ANIMAL BEHAVIOR, ANIMAL BREEDING. *Educ:* Univ Tenn, BSc, 68, MSc, 72; Pa State Univ, PhD(animal sci), 75. *Prof Exp:* asst prof, Animal Sci, Univ Sask, 76-80; ASSOC PROF ANIMAL SCI, UNIV MD, 81- *Mem:* AAAS; Animal Behav Soc; Am Genetic Asn; Am Soc Animal Sci; Can Soc Animal Sci. *Res:* Crowding, personal space, and stress. *Mailing Add:* Dept Animal Sci Univ Md College Park MD 20742

STRICKLING, EDWARD, b Woodsfield, Ohio, Oct 20, 16; m 41; c 3. SOILS. *Educ:* Ohio State Univ, BS, 37, PhD, 49. *Prof Exp:* Instr, High Sch, 37-42; prof soils, Univ Md, College Park, 50-; RETIRED. *Mem:* Soil Sci Soc Am; Am Soc Agron. *Res:* Soil physics, especially soil structure and evapotranspiration. *Mailing Add:* 6904 Calverton Dr Hyattsville MD 20782

STRICKMEIER, HENRY BERNARD, JR, b Galveston, Tex, Sept 28, 40; m 78. MATHEMATICS EDUCATION. *Educ:* Tex Lutheran Col, BS, 62; Univ Tex, Austin, MA, 67, PhD(math educ), 70. *Prof Exp:* Teacher high sch, Tex, 62-65; assoc prof, 70-80, PROF MATH, CALIF POLYTECH STATE UNIV, SAN LUIS OBISPO, 80- *Mem:* Math Asn Am; Am Educ Res Asn; Nat Coun Teachers Math. *Res:* Evaluation of mathematics curricula; analysis of mathematics teaching. *Mailing Add:* 1613 18th St San Luis Obispo CA 93402

STRIDER, DAVID LEWIS, b Salisbury, NC, Feb 12, 29; m 54; c 4. PLANT PATHOLOGY. *Educ:* NC State Col, MS, 57, PhD(plant path), 59. *Prof Exp:* Res asst prof, 59-64, assoc prof, 64-70, PROF PLANT PATH, NC STATE UNIV, 70- *Concurrent Pos:* Mem, NC State Univ-US AID Mission, Peru, 70-71. *Mem:* Am Phytopathological Soc. *Res:* Control of horticultural crops diseases; disease control of greenhouse floral crops. *Mailing Add:* Dept of Plant Path NC State Univ Raleigh NC 27695-7616

STRIEDER, WILLIAM, b Erie, Pa, Jan 19, 38; m 67; c 4. CHEMICAL ENGINEERING, PHYSICAL CHEMISTRY. *Educ:* Pa State Univ, BS, 60; Case Inst Technol, PhD, 63. *Prof Exp:* Res fel irreversible thermodyn, Free Univ Brussels, 63-65; res fel statist mech, Univ Minn, 65-66; asst prof eng sci, 66-70, PROF CHEM ENG, UNIV NOTRE DAME, 70- *Concurrent Pos:* Prin investr, NSF, Air Force Off Sci Res, Petrol Res Fund & Dept Transp grants. *Mem:* Am Inst Chem Engrs; Am Soc Eng Educ; Am Asn Univ Professors; Am Chem Soc; Am Phys Soc. *Res:* Molecular theory of transport processes; flow through random porous media; transport phenomena; thermodynamics; statistical mechanics. *Mailing Add:* Dept of Chem Eng Univ of Notre Dame Notre Dame IN 46556

STRIEFEL, SEBASTIAN, b Orrin, NDak, May 18, 41. DEVELOPMENTAL DISABILITIES, FACTORS OF STRESS. *Educ:* SDakota State Univ, BS, 64; Univ SDakota, MA, 66; Univ Kans, PhD(psychol), 68. *Prof Exp:* Chief psychol, Ft Riley Ment Hyg, US Army, 68-70; res assoc develop psychol, Bur Child Res, Univ Kans, 70-74; coordr clin serv, Exceptional Child Ctr, 74-75, from asst prof to assoc prof, 74-80, PROF PSYCHOL, PSYCHOL DEPT, UTAH STATE UNIV, 80-; DIR SERV, DEVELOP CTR HANDICAPPED PERSONS, 75- *Concurrent Pos:* Consult ed, Am Jour Ment Deficiency, 75-77, 83-85; assoc ed, J Behav Res Severe Develop Disability, 79-82, guest reviewer, Res Develop Disabilities, 88; chair & vice-chair, State Bd Ment Health, UT, 83-91; chair/co-chair, ethics comt, Asn Appl Psychophysiol & Biofeedback, 84-90; vis scientist, Fla Ment Health Inst, Univ SFla, 87-88; fel & dipl, Am Bd Med Psycho Therapists, 88-, admin psychol, 89- *Honors & Awards:* Shiela Adler Award, Biofeedback Soc Am, 88. *Mem:* Am Psychol Asn; Asn Appl Psychophysiol & Biofeedback (treas, 89-92); Asn Behav Anal; Asn Advan Behav Ther; fel Am Asn Ment Deficiency; Int Asn Right Effective Treat; Am Psychol Soc. *Res:* Stimulus control generalization; mainstreaming of children who have handicaps; effective treatment of abusive parents; restrictiveness of behavioral procedures; stress management; response restriction; impacting poverty; ethics. *Mailing Add:* 1564E 1260N Logan UT 84321

STRIER, KAREN BARBARA, b Summitt, NJ, May 22, 59. ANTHROPOLOGY. *Educ:* Swarthmore Col, BA, 80; Harvard Univ, MA, 81, PhD(anthropology), 86. *Prof Exp:* Lectr anthropology, Harvard Univ, 86-87; asst prof anthrop, Beloit Col, 87-89; ASST PROF ANTHROP, UNIV WIS-MADISON, 89- *Concurrent Pos:* Presidential young investr award, NSF, 89; adj prof, Beloit Col, 89- *Mem:* Am Anthrop Asn; Am Asn Phys Anthropologists; AAAS; Animal Behav Soc; Sigma Xi. *Res:* Field research on primates, focusing on the ecological determinants of social organization; relationships between primates and their environments, including seasonal food availability, effects on reproduction, social relationships, and grouping and kinship patterns. *Mailing Add:* Dept Anthrop Univ Wis-Madison Madison WI 53706

STRIER, MURRAY PAUL, b New York, NY, Oct 19, 23; m 55; c 3. PHYSICAL CHEMISTRY, ORGANIC CHEMISTRY. *Educ:* City Col New York, BChE, 44; Emory Univ, MS, 47; Univ Ky, PhD(chem), 52; Am Inst Chemists, cert. *Prof Exp:* Asst & instr, Univ Ky, 48-50; res chemist & proj leader, Reaction Motors, Inc, 52-56; sr chemist & head polymers sect, Air Reduction Co, Inc, 56-58; chief chemist, Fulton-Irgon Corp, 58-59; group leader fiber res, Rayonier, Inc, 59-60, suprvr develop res, 60-61; res chemist, T A Edison Res Lab, 61-64; sr res scientist, Douglas Aircraft Co, Inc, 64-67, chief fuel cell & battery res sect, 67-69; res assoc, Hooker Res Ctr, 69-71; prin chem engr, Cornell Aeronaut Lab, 71-72; chemist, 72-76, environ & phys scientist, Environ Protection Agency, 76-86; CONSULT, 86- *Concurrent Pos:* Consult, NSF, 73-75. *Honors & Awards:* Gold Medal, Environ Protection Agency, 79. *Mem:* AAAS; Am Chem Soc; Am Inst Chemists; Am Soc Testing & Mat; Electrochem Soc. *Res:* Organic polarography; physical chemistry of rocket propellants; physical properties of organic coatings, plastics and fibers; viscose chemistry; fuel cells and batteries; environmental science; industrial water pollution control-molecular structure-activity correlations. *Mailing Add:* 8 James Spring Ct Rockville MD 20850

STRIETER, FREDERICK JOHN, b Davenport, Iowa, Sept 14, 34; m 57; c 2. PHYSICAL CHEMISTRY, SEMICONDUCTOR DEVICES. *Educ:* Augustana Col, Ill, AB, 56; Univ Calif, Berkeley, PhD, 60. *Prof Exp:* Asst chem, Univ Calif, 56-57, asst crystallog, Lawrence Radiation Lab, 57-59; mem tech staff, Tex Instruments, Inc, 59-75, circuits develop pilot line mgr, 75-82; DIR, SEMICONDUCTOR MAT-GALLIUM ARSENIDE VENTURE, HONEYWELL, INC, 82- *Mem:* Electrochem Soc (treas, 73-76, vpres-pres, 79-83); Inst Elec & Electronic Engr. *Res:* Semiconductor device process technology; impurity diffusion in semiconductors; ion implantation of impurities in semiconductors; electron beam pattern definition; optoelectronic devices, silicon sensors. *Mailing Add:* 7814 Fallmeadow Lane Dallas TX 75248-5328

STRIFE, JAMES RICHARD, b Ilion, NY, Oct 12, 49; m 71; c 2. COMPOSITE MATERIALS. *Educ:* Rensselaer Polytech Inst, BS, 71, MS, 73, PhD(mat eng), 76. *Prof Exp:* Staff scientist, Union Carbide Corp, 76-78; PRIN SCIENTIST, UNITED TECHNOLOGIES RES CTR, 78- *Mem:* Am Soc Metals Int; Am Ceramic Soc. *Res:* Advanced composite and ceramic materials for applications in heat engines, optical systems, and space satellite structures; metal, ceramic and carbon matrix composites. *Mailing Add:* United Technologies Res Ctr Silver Lane East Hartford CT 06108

STRIFFLER, DAVID FRANK, b Pontiac, Mich, Oct 24, 22; m 49; c 2. PUBLIC HEALTH, DENTISTRY. *Educ:* Univ Mich, DDS, 47, MPH, 51; Am Bd Dent Pub Health, dipl, 55. *Prof Exp:* Consult dent, Dearborn Pub Schs, Mich, 50-51, dir sch health, 51-53; dir div dent health, State Dept Pub Health NMex, 53-61; from assoc prof to prof, 61-86, EMER PROF, SCH PUB HEALTH & DENT, UNIV MICH, 86- *Concurrent Pos:* Mem, Pub Health Res Study Sect, NIH, 59-60, Health Serv Res Study Sect, 60-62, Dis Control Study Sect, 64-65 & Prev Med & Dent Rev Comt, 70-74; mem, Nat Adv Comt Pub Health Training, USPHS, 60-61, Health Serv Res Training Comt, 66-68, Rev Comt Health Professions Schs Financial Distress Grants, 77-81, Nat Comn Dent Accreditation, 87-90; ed, J Pub Health Dent, 75-86; dir, Prog Dent Pub Health Sch Pub Health, Univ Mich, 62-83, chmn, dept community dents, Sch Dent, 62-67, dept community health progs, 75-85; ed, J Pub Health Dent, 75-86. *Honors & Awards:* Knutson Award, Am Pub Health Asn; Distinguished Serv Award, Am Asn Pub Health Dent. *Mem:* Sigma Xi; Am Dent Asn; fel Am Pub Health Asn. *Res:* Fluoridation; epidemiology of periodontal diseases; provision of dental health services. *Mailing Add:* 2217 Vinewood Blvd Ann Arbor MI 48104-2763

STRIFFLER, WILLIAM D, b Oberlin, Ohio, July 10, 29; m 56; c 4. FOREST HYDROLOGY. *Educ:* Mich State Univ, BS & BSF, 52; Univ Mich, MF, 57, PhD(forest hyrdol), 63. *Prof Exp:* Res forester & proj leader groundwater hydrol & steambank erosion, Lake States Forest Exp Sta, Mich, 57-63; hydrologist, Stripmined Areas Restoration Res Proj, Northeastern Forest Exp Sta, Ky, 64-66; from asst prof to assoc prof watershed mgt, 66-75, PROF EARTH RESOURCES, COLO STATE UNIV, 75- *Concurrent Pos:* Consult, Cent soil & Water Conserv Res Inst, India, 75, 77, 80. *Mem:* Am Geophys Union; Am Water Resource Asn; AAAS; Sigma Xi; Indian Soc Soil & Water Conserv. *Res:* Wildland hydrology; land use hydrology; erosion and sedimentation processes; water quality; grassland hydrology; instrumentation. *Mailing Add:* 1201 Lory St Ft Collins CO 80524

STRIGHT, PAUL LEONARD, b St Paul, Minn, May 12, 30; m 60; c 2. TEACHING. *Educ:* Grinnell Col, BA, 51; Univ Minn, Minneapolis, PhD(org chem), 56. *Prof Exp:* Res chemist, Esso Res & Eng Co, 56-59; res chemist, Allied Chem Corp, 59-65, res supvr, 65-68, res assoc, 68-70; asst prof org chem, Univ Minn, Morris, 70-71; assoc scientist neurochem, Univ Minn, Minneapolis, 71-73; INSTR CHEM, LAKE MICH COL, 73- *Mem:* Sigma Xi; Am Chem Soc; Nat Sci Teachers Assn. *Mailing Add:* Lake Mich Col 2755 E Napier Ave Benton Harbor MI 49022-1899

STRIKE, DONALD PETER, b Mt Carmel, Pa, Oct 24, 36; m 72; c 2. PHARMACEUTICAL CHEMISTRY, ORGANIC CHEMISTRY. *Educ:* Philadelphia Col Pharm & Sci, BS, 58; Iowa State Univ, MS, 61, PhD(org chem), 63. *Prof Exp:* NIH fel, Univ Southampton, 63-64; res chemist, 65-69, res group leader, 69-77, res mgr, 77-84, ASSOC DIR, WYETH-AYERST RES, AM HOME PROD CORP, 84- *Mem:* Am Chem Soc; AAAS. *Res:* Natural products; antibiotics; steroids; prostaglandins; anti-ulcer products; metabolic disorders; anti-viral drugs; antilipemic drugs; antidiabetic drugs; antiosteoporotic drugs. *Mailing Add:* Wyeth-Ayerst Res CN-8000 Princeton NJ 08543

STRIKER, G E, b Bottineau, NDak, March 7, 34. NEPHROLOGY, CELL BIOLOGY. *Educ:* Univ Wash Seattle Sch Med, MD, 59. *Prof Exp:* From asst prof to prof, dept path, Wash Univ, Seattle, 66-84; DIR DIV KIDNEY, UROL & HEMAT DIS, NIH, NAT INST DIABETES, DIGESTIVE & KIDNEY DIS, 84- *Concurrent Pos:* Dean curric, Sch Med, Univ Wash, Seattle, 71-77. *Mem:* Am Soc Nephrol; Am Soc Cell Biol; Nat Kidney Found; Am Asn Pathologists; Int Soc Nephrologists; Am Heart Asn. *Res:* Pathology; biochemistry; endocrinology; physiology; nephrology; cell biology. *Mailing Add:* Div Kidney Urol & Hematol Dis Bldg 31 Rm 9A17 NIH 9000 Rockville Pike Bethesda MD 20892

STRIKWERDA, JOHN CHARLES, b Grand Rapids, Mich, Mar 15, 47; m 70; c 2. NUMERICAL ANALYSIS. *Educ:* Calvin Col, AB, 69; Univ Mich, MA, 70; Stanford Univ, PhD(math), 76. *Prof Exp:* Res scientist, Inst Comput Applications Sci Eng, 76-80; asst prof, 80-84, ASSOC PROF COMPUT SCI & MATH, RES CTR, UNIV WIS-MADISON, 84- *Mem:* Soc Indust & Appl Mech; Am Math Soc. *Res:* Numerical methods for partial differential equations, finite difference methods and computational fluid dynamics. *Mailing Add:* Dept Math Comput Sci 4240 Comput Sci Univ Wis 1210 W Dayton St Madison WI 53706

STRIMLING, WALTER EUGENE, b Minneapolis, Minn, Jan 6, 26; m 57; c 3. PHYSICS, ELECTRONICS ENGINEERING. *Educ:* Univ Minn, BPhys & MA, 45, PhD(math), 53. *Prof Exp:* Instr math & educ, Col St Catherine, 45-46; asst math, Univ Minn, 49-53; engr, Raytheon Co, 53-55; PRES, US DYNAMICS, 55- *Mem:* Am Phys Soc; Math Asn Am; Inst Elec & Electronics Engrs. *Res:* Theoretical physics; chemistry; bioengineering. *Mailing Add:* US Pyamics Corp 154 Lexington St Waltham MA 02154-4644

STRINDEN, SARAH TAYLOR, b Lyons, Kans, Jan 31, 55; m 79. BIOCHEMISTRY. *Educ:* Univ Kans, BS, 76; Univ Southern Calif, PhD(biochem), 81. *Prof Exp:* Technician biochem, Univ Kans, 76-77; res asst, Univ Southern Calif, 78-81; FEL, UNIV WIS-MADISON, 81- *Mem:* Sigma Xi. *Res:* Various mechanisms controlling gene expression at the post-transcriptional level. *Mailing Add:* 1310 Walnut Bend Lufkin TX 75901

STRINGALL, ROBERT WILLIAM, b San Francisco, Calif, Dec 12, 33; c 2. MATHEMATICS EDUCATION. *Educ:* San Jose State Col, BA, 59; Univ Wash, MS, 63, PhD(math), 65. *Prof Exp:* ASSOC PROF MATH, UNIV CALIF, DAVIS, 65- *Concurrent Pos:* Consult, Elem & Sec Educ Act Title III, 67-68, Proj Sem, 70- *Res:* Algebra. *Mailing Add:* Dept of Math Univ of Calif Davis CA 95616

STRINGAM, ELWOOD WILLIAMS, b Alberta, Can, Dec 10, 17; m 44; c 6. ANIMAL SCIENCE, AGRICULTURE. *Educ:* Univ Alta, BSc, 40, MSc, 42; Univ Minn, PhD(agr), 48. *Prof Exp:* Asst, Dom Range Exp Sta, Alta, 40; fieldman, Livestock Prod Serv, 41-42; instr animal husb, Univ Minn, 46-48; assoc prof animal sci, Univ Man, 48-51; prof animal husb, Ont Agr Col, 51-54; RETIRED. *Concurrent Pos:* Mem, Nat Animal Breeding Comt, Can, 58-63; adv comt, Western Vet Col, 66-; dir nat adv comt for agr, World's Fair, 67; mem, Nat Genetic Adv Comt Cattle Importations, 69-74. *Honors & Awards:* Golden Award, Can Feed Indust Asn. *Mem:* Am Soc Animal Sci; Sigma Xi; Can Soc Animal Sci; fel Agr Inst Can (vpres, 62-63, pres, 66-67). *Res:* Agricultural education; animal genetics and physiology; farm animal production and management; beef cattle production. *Mailing Add:* 109-1725 Cedar Hill Cross Rd Victoria BC V8P 2P8 Can

STRINGER, GENE ARTHUR, b Yamhill County, Ore, Nov 16, 39; m 61; c 4. PHYSICS, ELECTRONICS. *Educ:* Linfield Col, BA, 61; Univ Ore, MA, 64, PhD(physics), 69. *Prof Exp:* Staff engr & physicist res, Tektronix Inc, Beaverton, Ore, 61-63; res assoc physics, Univ Ore, 69; res assoc & instr, Cornell Univ, 69-71; from asst prof to assoc prof, 71-83, CHMN DEPT PHYSICS, SOUTHERN ORE STATE COL, 74-, PROF, 83- *Concurrent Pos:* Physics consult, Tektronix Inc, 63-64; NSF grant, Instr Sci Equip, 76-78; proj assoc, Tech Educ Res Ctr, 77-78; NSF LOCI grant, 79-81; Fulbright sr lectr, Univ Philippines, 84-85. *Mem:* Am Asn Physics Teachers; Sigma Xi. *Mailing Add:* Dept Physics Southern Ore State Col Ashland OR 97520

STRINGER, JOHN, b Liverpool, Eng, July 14, 34; m 57; c 2. FOSSIL FUEL BURNING SYSTEMS. *Educ:* Univ Liverpool, Eng, BEng, 55, PhD(metall), 58, DEng, 72; Chartered Eng Inst, UK, CEng, 70. *Prof Exp:* Asst lectr, metall, Univ Liverpool, UK, 57-59, lectr, 59-63, prof, 66-68, prof mat sci, 68-77; sr scientist, Battelle Mem Inst, Columbus, Ohio, 64-65, fel, 65-66; proj mgr, 77-82, prog mgr, 82-85, sr prog mgr mat support, 85-87, TECH DIR, EXPLOR RES, ELEC POWER RES INST, 87- *Mem:* Am Inst Metall Engrs; Nat Asn Corrosion Engrs; fel Inst Energy UK; fel AAAS; Am Soc Mat Int; fel Royal Soc Arts. *Res:* High temperature oxidation and corrosion of metals and alloys; fossil-fuel burning systems; interaction of oxidation and sulfidation at elevated temperatures; hot corrosion of gas turbines; reactive element effect in high-temperature oxidation; erosion and erosion/corrosion of metals and alloys. *Mailing Add:* Elec Power Res Inst 3412 Hillview Ave PO Box 10412 Palo Alto CA 94303

STRINGER, L(OREN) F(RANK), b Huntington Park, Calif, Sept 28, 25; m 53; c 4. ELECTRONICS ENGINEERING, APPLIED MATHEMATICS. *Educ:* Univ Tex, BS, 46; Calif Inst Technol, MS, 47; Univ Pittsburgh, PhD, 63. *Prof Exp:* Engr, Westinghouse Elec Corp, Buffalo, NY, 47-56, develop engr mgr, 56-74, div engr mgr, 74-81, chief engr, 81-85; RETIRED. *Concurrent Pos:* Lamme scholar, 64; chmn C34 tech subcomt, Am Nat Standards Inst; secy SC22G, Tech Adv USNC, Int Electrotechnical Comn; chmn, Power Semiconductor Comt, Inst Elec & Electronic Engrs, Static Power Converter Comt & Conversions Systs Dept, Indust Applications Soc; chmn, Static Power Conversion Sect, Nat Elec Mfr Asn. *Honors & Awards:* Lamme Medal, Inst Elec & Electronic Engrs & William E Newell Award, Power Electronics Specialists Conf. *Mem:* Fel Inst Elec & Electronic Engrs. *Res:* Engineering management; power electronics, automatic control and systems engineering; holder of 24 US patents. *Mailing Add:* Stringer PWR Electronics Corp 9015 Cliffside Dr Clarence NY 14031

STRINGER, WILLIAM CLAYTON, b Athens, Ga, Apr 3, 46; m 69. FORAGE CROP MANAGEMENT. *Educ:* Univ Ga, BSA, 68, MSc, 72; Va Polytech Inst & State Univ, PhD(agron), 77. *Prof Exp:* Agronomist, NW Ga Br Exp Sta, Univ Ga, 72-74; from asst prof to assoc prof forage corps, dept agron, Pa State Univ, 77-84; ASSOC PROF FORAGE CROPS, DEPT AGRON, CLEMSON UNIV, 84- *Mem:* Crop Sci Soc Am; Am Soc Agron; Am Forage & Grassland Coun. *Res:* Forage crop ecology and physiology; alfalfa management; grazing systems. *Mailing Add:* Dept Agron & Soils Clemson Univ Plant & Animal Sci Bldg Clemson SC 29631

STRINGFELLOW, DALE ALAN, b Ogden, Utah, Sept 13, 44; m 66; c 3. VIROLOGY, IMMUNOBIOLOGY. *Educ:* Univ Utah, BS, 67, MS, 70, PhD(microbiol), 72. *Prof Exp:* NIH fel & instr microbiol, Univ Utah, 72-73; res scientist, UpJohn Co, 73-79, sr res scientist virol & head cancer res, 79-82; assoc prof pharmacol, Upstate Med Ctr, State Univ NY, Syracuse, 82-86; vpres cancer res, prod, Bristol Myers, Syracuse, NY, 82-88; prof res molecular & cellular biol, Univ Conn, 86-88; vpres res & develop, Collagen Corp, Palo Alto, Calif, 88-90; PRES, CELTRIX LABS, PALO ALTO, CALIF, 90- *Concurrent Pos:* Mem, Develop Therapeut Contracts Rev Comt, Nat Cancer Inst; mem, Cancer Biol, Immunol Contract Review, Nat Cancer Inst, 85-87; mem, Univ Conn Biotech Adv Bd, 85-88; mem, Ariz Dis Control Res Comn Rev Comt, 87-90; mem, Univ Tex, John Sealy Mem Endowment Fund Rev Comt, 88- *Mem:* Am Soc Microbiol; Am Asn Cancer Res; AAAS; Int Soc Interferon Res; Int Soc Antiviral Res; Soc Biomat Res; Parenteral Drug Asn. *Res:* Antiviral agents; interrelationship between host defense systems and virus infection; pathogenesis of virus infection; mechanisms modulating nonspecific immunity; viral ecology; antineoplastic agents; metastasis; cell cycling; nucleic acid biochemistry; cellular regulation; prostaglandins; cellular communications. *Mailing Add:* Celtrix Labs 2500 Faber Pl Palo Alto CA 94303

STRINGFELLOW, FRANK, b Cheriton, Va, Oct 27, 40; m 68. PARASITOLOGY, ZOOLOGY. *Educ:* St Louis Univ, BS, 62; Drake Univ, MA, 64; Univ SC, PhD(biol), 67. *Prof Exp:* Asst gen biol, Drake Univ, 63-64; instr anat & physiol, Univ SC, 64-65; ZOOLOGIST, ANIMAL PARASITOL

INST, AGR RES CTR, US DEPT AGR, 67- *Mem:* Am Soc Parasitol. *Res:* Molecular biology and biochemistry of parasitic nematodes; cultivation of parasitic nematodes. *Mailing Add:* Animal Parasitol Inst Agr Res Ctr USDA Beltsville MD 20705

STRINGFELLOW, GERALD B, b Salt Lake City, Utah, Apr 26, 42; m 62; c 3. MATERIALS SCIENCE, SEMICONDUCTORS. *Educ:* Univ Utah, BS, 64; Stanford Univ, MS, 66, PhD(mat sci), 67. *Prof Exp:* Mem tech staff, Solid State Physics Lab, Hewlett Packard Labs, 67-71, proj mgr, 71-80; PROF ELEC ENG & MAT SCI ENG, UNIV UTAH, SALT LAKE CITY, 80- *Concurrent Pos:* Sabbatical, Max Planck Inst, Stuttgart, Ger, 79 & Clarendon Lab, Univ Oxford, Eng, 91. *Honors & Awards:* Alexander von Humboldt US Sr Scientist Award, 79. *Mem:* Am Phys Soc; Electrochem Soc; fel Inst Elec & Electronic Engrs; Mat Res Soc. *Res:* Electrical and optical properties of alloys between III-V compound semiconductors; and III-V compounds; crystal growth and thermodynamics in ternary III-V systems; organometallic vapor phase epitaxy. *Mailing Add:* Dept Elec Eng Univ Utah Salt Lake City UT 84112

STRINGFIELD, VICTOR TIMOTHY, b Franklinton, La, Sept 10, 02; m 29; c 2. HYDROGEOLOGY. *Educ:* La State Univ, BS, 25; Wash Univ, MS, 27. *Prof Exp:* Asst geol & geog, Wash Univ, 25-27; instr geol, Okla Agr & Mech Col, 27-28; asst prof, NMex Sch Mines, 28-30; asst geologist, US Geol Surv, 30-36, assoc geologist, 36-39, geologist in charge ground water invests, Southeastern States, 39-42, sr geologist in charge ground water invests, Eastern States, 42-47, prin geologist & chief sect ground water geol, 47-57, chief sect radiohydrology, 57-60, staff geol specialist, 60-75; RETIRED. *Concurrent Pos:* Geologist, NMex Bur Mines, 28-30; consult hydrol, Food & Agr Orgn, UN, Jamaica, WI, 65- *Mem:* Fel Geol Soc Am; Soc Econ Geol; Am Asn Petrol Geol; Am Geophys Union. *Res:* Ground water geology; radiohydrology; ground water hydrology; radioisotopes in soils and water; hydrogeology of carbonate terranes. *Mailing Add:* 655 15th St No 900 Washington DC 20005

STRINGHAM, GLEN EVAN, b Lethbridge, Alta, Aug 30, 29; US citizen; m 53; c 4. ENGINEERING. *Educ:* Utah State Univ, BS, 55; Colo State Univ, PhD(civil eng), 66. *Prof Exp:* Instr agr eng, Calif State Polytech Col, 55-57; from asst prof to assoc prof, 57-78, PROF AGR ENG, UTAH STATE UNIV, 78- *Concurrent Pos:* Chief party, Utah State Univ-Agency Int Develop team, Colombia, 69-71; consult, Honduras, Phillipines, Saudi Arabia, Kenya & Egypt. *Mem:* Am Soc Agr Engrs; Am Soc Civil Engrs; Am Soc Eng Educ; Soil Conserv Soc Am. *Res:* Optimization of surface irrigation. *Mailing Add:* 50 S 100 E Millville UT 84326

STRINGHAM, REED MILLINGTON, JR, b Salt Lake City, Utah. PHYSIOLOGY, ORAL BIOLOGY. *Educ:* Northwestern Univ, Evanston, DDS, 58; Univ Utah, BS, 64, PhD(molecular & genetic biol), 68. *Prof Exp:* Nat Inst Dent Res fel, 65-68, RES ASSOC PLASTIC SURG, MED SCH, UNIV UTAH, 68-; PROF OCCUP HEALTH & ZOOL & DEAN SCH ALLIED HEALTH SCI, WEBER STATE COL, 69- *Concurrent Pos:* Resource person, Intermountain Regional Med Prog, 69, consult, Oral Cancer Screening Proj, 71- *Mem:* Am Dent Asn. *Res:* Salivary gland physiology; health manpower. *Mailing Add:* Dean Health Scis Weber State Col Sch Allied Health Sci Ogden UT 84408

STRINTZIS, MICHAEL GERASSIMOS, b Athens, Greece, Sept 30, 44; US citizen; m; c 1. ELECTRICAL & BIOMEDICAL ENGINEERING, APPLIED MATHEMATICS. *Educ:* Nat Tech Univ Athens, BS, 67; Princeton Univ, MA, 69, PhD(elec eng), 70. *Prof Exp:* From asst prof to assoc prof elec eng, Univ Pittsburgh, 70-80; AT UNIV THESSALONIKI, GREECE, 80- *Concurrent Pos:* Vis prof, Nat Tech Univ, Athens, 78-79, Syracuse Univ, 87-88; grants, NSF, Elec Power Res Inst, NIH & Energy Res & Develop Admin, EC (Esprit). *Mem:* Inst Elec & Electronic Engrs; Soc Indust & Appl Math; NY Acad Sci. *Res:* Signal and image processing; large-scale systems; digital signal processing, image description and processing; detection and estimation theory; biomedical engineering. *Mailing Add:* Univ Thessaloniki Thessaloniki 540 06 Greece

STRITTMATER, RICHARD CARLTON, b Columbia, Pa, Aug 26, 23; m 50; c 3. PHYSICS. *Educ:* Earlham Col, BA, 53; Iowa State Col, MS, 55. *Prof Exp:* Res physicist fluid mech & heat transfer, Remington Arms Co, Ilion, NY, 55-58; RES PHYSICIST MECH, ACOUST & HEAT TRANSFER, BALLISTIC RES LAB, 58- *Concurrent Pos:* Mem, Comt Standardization Combustion Instability Measurements, 66-70; Dept Army rep, Steering Comt Interagency Chem Rocket Propulsion Group's Solid Propellent Combustion Group, 67. *Mem:* Sigma Xi. *Res:* Turbulent fluid mechanical interaction at burning surfaces. *Mailing Add:* 2500 Pinehurst Ave Forest Hill MD 21050

STRITTMATTER, CORNELIUS FREDERICK, b Philadelphia, Pa, Nov 16, 26; m 55; c 1. METABOLIC CONTROL, DIFFERENTIATION. *Educ:* Juniata Col, BS, 47; Harvard Univ, PhD(biol chem), 52. *Prof Exp:* Instr biol chem, Harvard Med Sch, 52-54, assoc, 55-58, asst prof, 58-61; chmn dept, 61-78, Odus M Mull prof, 61-89, EMER PROF BIOCHEM, BOWMAN GRAY SCH MED, 89- *Concurrent Pos:* USPHS res fel, Oxford Univ, 54-55; USPHS sr res fel, Harvard Med Sch, 61; consult, New Eng Deaconess Hosp, 58-61; mem fel comt, NIH, 67-70 & 74; consult, NC Alcoholism Res Auth, 74- *Mem:* Am Chem Soc; Am Soc Biol Chemists; Soc Develop Biol; Soc Exp Biol & Med; Am Soc Zool. *Res:* Enzymic differentiation and control mechanisms during embryonic development and aging; characterization of electron transport systems; cellular control mechanisms in metabolism; comparative biochemistry; mechanisms in enzyme systems. *Mailing Add:* Dept of Biochem Bowman Gray Sch of Med Winston-Salem NC 27103

STRITTMATTER, PETER ALBERT, b Bexleyheath, Eng, Sept 12, 39; m 67; c 2. ASTRONOMY. *Educ:* St John's Col, Cambridge Univ, BA, 64, PhD(math), 66. *Prof Exp:* Mem staff astron, Inst Theoret Astron, Cambridge, 67-68; res assoc, Mt Stromlo & Siding Spring Observ, 69; mem staff, Inst Theoret Astron, Cambridge, 70; res physicist, Univ Calif, San Diego, 71; assoc prof, 71-73, PROF ASTRON & DIR OBSERV, STEWARD OBSERV, UNIV ARIZ, 75- *Concurrent Pos:* Consult astron adv panel, NSF, 75-79; mem bd, Asn Univs Res Astron, NSF, 75-; adj sci mem, Max Planck Inst Radioastron, Bonn; Alexander Von Humboldt sr scientist award, 79-80. *Mem:* Am Acad Arts Sci; Am Astron Soc; Royal Astron Soc; Int Astron Union; Ger Astron Soc. *Res:* Quasistellar objects; Seyfert galaxies; radio sources; white dwarfs; novae; speckle interferometry. *Mailing Add:* Steward Observ Univ of Ariz Tucson AZ 85721

STRITTMATTER, PHILIPP, b Philadelphia, Pa, July 13, 28; m 56; c 2. BIOCHEMISTRY, ENZYMOLOGY. *Educ:* Harvard Univ, PhD, 54. *Prof Exp:* From instr to prof biochem, Wash Univ, 54-68; chmn dept, 68-74, PROF BIOCHEM, UNIV CONN, STORRS, 68-, HEAD DEPT, 76- *Mem:* Am Soc Biol Chemists. *Res:* Oxidative enzyme mechanisms. *Mailing Add:* Dept Biochem Univ Conn Health Ctr Farmington CT 06032

STRITZEL, JOSEPH ANDREW, b Cleveland, Ohio, June 11, 22; m 50; c 9. AGRONOMY. *Educ:* Iowa State Univ, BS, 49, MS, 53, PhD(soil fertil, prod econ), 58. *Prof Exp:* Exten soil fertil specialist, Iowa State Univ, 50-63, prof soils, 63-76, prof agron, 76-; RETIRED. *Mem:* Am Soc Agron. *Res:* Soil fertility; production economics. *Mailing Add:* RR 4 Iowa State Univ Ames IA 50010

STRITZKE, JIMMY FRANKLIN, b South Coffeyville, Okla, Sept 9, 37; m 59; c 2. AGRONOMY. *Educ:* Okla State Univ, BS, 59, MS, 61; Univ Mo, PhD(field crops), 67. *Prof Exp:* Res scientist weed control, Agr Res Serv, USDA, 61-66; asst prof agron, SDak State Univ, 66-70; asst prof, 70-76, assoc prof, 76-80, PROF AGRON, OKLA STATE UNIV, 80- *Mem:* Am Forest & Grassland Coun; Am Soc Agron; Weed Sci Soc Am; Soc Range Mgt; Coun Agr Sci & Technol. *Res:* Weed control in alfalfa, weed and brush control in pasture and rangelands, herbicide residue and translocation. *Mailing Add:* Dept of Agron Okla State Univ Stillwater OK 74078

STRIZ, ALFRED GERHARD, b Rosenheim, WGer, July 25, 52; m 83; c 3. AEROELASTICITY, FINITE ELEMENT ANALYSIS. *Educ:* Purdue Univ, BS & MS, 76, PhD(aero/astro eng), 81. *Prof Exp:* Teaching asst eng, Purdue Univ, 77-80; asst prof, solid mech, 81-88, ASSOC PROF AEROSPACE, UNIV OKLA, 88- *Concurrent Pos:* Res asst aeroelasticity, Purdue Univ, 77-80, instr solid mech, 81; prin investr, Demco, 82, Gulfco, 83, Air Force Off Sci Res, 83-84, Gen Motors Corp, 85-88, USAF, 86 & 88, Fed Aviation Admin, 89-90; co-prin investr, Dept Defense, USAF, 83-85, NASA, 84 & 89-91, OCAST, 90-91; jr fac res fel, Univ Okla, 84; sr exp engr, 85, sr proj engr, Gen Motors Corp, 86; vis scholar, Dept Defense, USAF, 82, Air Force Inst Technol, 88, WRDC, 89-90. *Honors & Awards:* Ralph R Teetor Award, 84. *Mem:* Am Inst Aeronaut & Astronaut; Am Soc Mech Engrs; Soc Automotive Engrs. *Res:* Aeroelasticity; finite element analyses of structures; structural optimization; composite materials analysis. *Mailing Add:* FH 206 Aeromech Eng Univ Okla 865 Asp Ave Norman OK 73019

STRNAT, KARL J, b Vienna, Austria, Mar 29, 29; US citizen; m 54; c 4. MAGNETISM, MAGNET MEASUREMENTS. *Educ:* Vienna Tech Univ, Dipl Ing, 53, DrTech(elec eng), 56. *Prof Exp:* Asst, Vienna Tech Univ, 53-57; res & supvry physicist, US Air Force Mat Lab, 58-68; F M Tait prof, 68-89, EMER PROF ELEC ENG, UNIV DAYTON, 89-; OWNER, KJS ASSOCS, CONSULT, MAGNETIC INSTRUMENTATION & TECH-SERV, 76- *Concurrent Pos:* Lectr, univs & indust labs, 61-; recipient res grants, Advan Res Proj Agency, US Air Force Mat & Propulsion Lab, Goldschmidt Co, Molybdenum Corp, Ind Gen Corp, NSF, Gen Motors Corp, US Army Res Off, Kollmorgen Corp, US Army ETDL, US Naval Res Lab, US Dept Energy & Ohio EMTEC, 68-; indust consult, 68-; vis res scientist, Univ Calif, San Diego, 78-79; distinguished lectr, Inst Elec & Electronics Engrs Magnetics Soc, 84-85; organizer six int conferences, magnetics, 74-87; lect series, short courses China, Japan, Ger, USA & USSR, 75-88. *Honors & Awards:* Res Award, Sigma Xi, 80. *Mem:* Am Soc Testing & Mat; Inst Elec & Electronic Engrs. *Res:* Magnetic materials, permanent magnets, magneto-optics, instrumentation; rare earth alloys, their physical metallurgy, magnetism and crystallography; magnetic device design; magnet production engineering. *Mailing Add:* KJS Assocs Fairborn OH 45324

STRNISA, FRED V, b Cleveland, Ohio, Nov 20, 41; c 3. ENERGY RESEARCH & DEVELOPMENT. *Educ:* Case Inst Technol, BS, 63; John Carroll Univ, MS, 67; State Univ NY, Albany, PhD(physics), 72. *Prof Exp:* Engr, Lamp Div, Gen Elec Co, 63-67, physicist, Knolls Atomic Power Lab, 67-69; res assoc, State Univ NY, Albany, 69-73; sr scientist, NY State Atomic Energy Coun, 73-76 & NY State Energy Off, 76-77; PROG MGR, NY STATE ENERGY RES & DEVELOP AUTHORITY, 77- *Mem:* Am Phys Soc; Int Dist Heating & Cooling Asn; Sigma Xi. *Res:* District heating and cooling; cogeneration; electric and gas utility research and development; transportation; radioactive waste management; radiation and impurity defects in solids; circularly polarized electron paramagnetic resonance; mass spectroscopy; lighting technology. *Mailing Add:* NY State Energy Res 2 Rockefeller Plaza Albany NY 12223

STRNISTE, GARY F, b Springfield, Mass, May 31, 44; m 80. RADIATION BIOLOGY, MOLECULAR GENETICS. *Educ:* Univ Mass, BS, 66; Pa State Univ, MS, 69, PhD(biophys), 71. *Prof Exp:* Fel molecular radiobiol, Los Alamos Sci Lab, 71-73; fel molecular biol, City Univ New York, 73-74; MEM STAFF, LOS ALAMOS NAT LAB, 75- *Mem:* Radiation Res Soc; Am Soc Biol Chemists; Environ Mutagenesis Soc; AAAS. *Res:* Low dose radiation effects; isolation and characterization of human DNA repair genes; somatic cell mutagenesis; in vitro and in vivo repair of DNA. *Mailing Add:* Genetics LS-3 MS-M886 Life Sci Div Los Alamos Nat Lab Los Alamos NM 87545

STROBACH, DONALD ROY, b St Louis, Mo, Jan 10, 33; m 60, 83; c 2. ORGANIC CHEMISTRY, BIOCHEMISTRY. *Educ:* Wash Univ, AB, 54, PhD(chem), 59. *Prof Exp:* NIH fels, 60-63; res chemist, Cent Res Dept, 63-76, int mkt mgr, 76-79, atmospheric sci coordr, 79-82, environ mgr, Freon Prod Div, 82-86, SR RES ASSOC, E I DUPONT DE NEMOURS & CO, INC, 86- *Mem:* Am Chem Soc. *Res:* Carbohydrates; synthesis and structure determination; synthesis of oligonucleotides; aerosol technology and product development. *Mailing Add:* 2420 W Parris Dr Wilmington DE 19808-4512

STROBECK, CURTIS, b Powers Lake, NDak, Nov 14, 40. POPULATION GENETICS, MOLECULAR EVOLUTION. *Educ:* Univ Mont, BA, 64, MA, 66; Univ Chicago, PhD(theoret biol), 71. *Prof Exp:* Res fel pop biol, Sch Biol Sci, Univ Sussex, 71-75; vis asst prof dept ecol & evolution, State Univ NY, Stony Brook, 75-76; assoc prof, Dept Genetics, 76-84, PROF, DEPT ZOOL, UNIV ALTA, 84- *Concurrent Pos:* vis prof, Ctr Demog & Pop Genetics, Univ Tex, Houston; chmn, Dept Genetics, Univ Alta, 81-82. *Mem:* Genetics Soc Can; Am Genetics Soc; Soc Study Evolution. *Res:* Selection in heterogeneous environments; selection in multi-locus systems; evolution for recombination; effects of linkage in a finite population; evolution of ribosomal DNA in Drosophila and conifers; variation in the mitochondrial DNA of vertebrates. *Mailing Add:* Dept Zool Univ Alta Edmonton AB T6G 2E9 Can

STROBEL, DARRELL FRED, b Fargo, NDak, May 13, 42; m 68; c 2. PLANETARY ATMOSPHERES, SPACE PHYSICS. *Educ:* NDak State Univ, BS, 64; Harvard Univ, AM, 65, PhD(appl physics), 69. *Prof Exp:* Res assoc planetary astron, Kitt Peak Nat Observ, 68-70, asst physicist, 70-72, assoc physicist, 72-73; res physicist, 73-76, supvr res physicist, Naval Res Lab, 76-84; PROF PLANETARY SCI, EARTH SCI, PHYSICS & ASTRON, JOHNS HOPKINS UNIV, 84- *Concurrent Pos:* Space Sci Bd, Nat Acad Sci-Nat Res Coun. *Mem:* AAAS; Am Astron Soc; Am Geophys Union; Am Meteorol Soc; Int Astron Union. *Res:* Chemistry, dynamics and physics of planetary atmospheres; planetary aeronomy; planetary physics; planetary magnetospheres. *Mailing Add:* Dept Earth & Planetary Sci Johns Hopkins Univ Baltimore MD 21218

STROBEL, EDWARD, b Wilkes-Barre, Pa, Mar 18, 47. GENETIC ENGINEERING, CYTOGENETICS. *Educ:* Towson State Univ, BA, 69; State Univ NY Stony Brook, PhD(cell & develop biol), 77. *Prof Exp:* Housing inspector, Baltimore City Health Dept, 69-70; instr chem, Boys' Latin Sch, Baltimore, 70-72; ASST PROF GENETICS, DEPT BIOL SCI, PURDUE UNIV, 80- *Concurrent Pos:* Consult, Boehringer-Mannheim Biochem, Indianapolis, 80-; fel molecular biol, Sidney Farber Cancer Inst, 77-80; prin investr, Purdue Cancer Ctr, 81-82, NIH, 81-84 & Nat Eye Inst, 81-85. *Res:* Mechanisms involved in maintaining the structural integrity and stability of eukaryotic chromosomes, and how higher order chromosome structure is modulated during development. *Mailing Add:* 25928 Richville Dr Torrance CA 90505

STROBEL, GARY A, b Massillon, Ohio, Sept 23, 38; m 63; c 2. PLANT PATHOLOGY. *Educ:* Colo State Univ, BS, 60; Univ Calif, Davis, PhD(plant path), 63. *Prof Exp:* From asst prof to prof bot, 63-77, PROF PLANT PATH, MONT STATE UNIV, 77- *Concurrent Pos:* Prin investr, NSF & USDA res grants; NIH career develop award, 69-74. *Mem:* AAAS; Am Soc Biol Chemists. *Res:* Plant disease physiology; biochemistry of fungi and bacteria that cause plant diseases; phytotoxic glycopeptides; metabolic regulation in diseased plants; nature and mechanism of action of host specific toxins. *Mailing Add:* Dept Plant Path Mont State Univ Bozeman MT 59715

STROBEL, GEORGE L, b Pratt, Kans, May 26, 37; m 57; c 2. THEORETICAL NUCLEAR PHYSICS, OPTICS. *Educ:* Kans State Univ, BS, 58; Univ Pittsburgh, MS, 61; Univ Southern Calif, PhD(physics), 65. *Prof Exp:* Scientist physics, Westinghouse Bettis Atomic Power Lab, 58-61 & Douglas Aircraft Co, 61-64; res assoc, Univ Southern Calif, 65 & Univ Calif, Davis, 65-67; ASSOC PROF PHYSICS, UNIV GA, 67- *Concurrent Pos:* Vis prof, Nuclear Res Ctr, Jülich, WGer, 71-72. *Mem:* Am Phys Soc. *Res:* Electromagnetism. *Mailing Add:* Dept of Physics Univ of Ga Athens GA 30602

STROBEL, HOWARD AUSTIN, b Bremerton, Wash, Sept 5, 20; m 53; c 3. ANALYTICAL CHEMISTRY, PHYSICAL CHEMISTRY. *Educ:* State Col Wash, BS, 42; Brown Univ, PhD(phys chem), 47. *Prof Exp:* Jr res chemist, Manhattan Dist, Brown Univ, 43-45, res assoc, 47-48; from instr to assoc prof chem, Trinity Col, 48-64, asst dean, 56-64; dean, Baldwin Residential Fedn, 72-75, fac fel, 75-81; coordr fedn, 74-81, prof chem, 64-90, EMER PROF, DUKE UNIV, 90- *Concurrent Pos:* Consult, Sci Instrumentation Info Network & Curricula, 81; vis prof, Univ Leicester, 71-72, Univ NC, Chapel Hill, 82; ed, J Chem Educ, 91- *Mem:* Am Chem Soc; Royal Soc Chem. *Res:* Solute-solvent interactions in mixed media; ion exchange phenomena; chemical instrumentation. *Mailing Add:* Dept Chem Duke Univ Durham NC 27706

STROBEL, JAMES WALTER, b Steubenville, Ohio, Oct 31, 33; m 55; c 2. PLANT PATHOLOGY. *Educ:* Ohio Univ, AB, 55; Wash State Univ, PhD, 59. *Prof Exp:* Asst plant path, Wash State Univ, 55-59; from asst plant pathologist to assoc plant pathologist, Univ Fla, 59-68, prof plant path & plant pathologist, 68-74, chmn ornamental hort-agr exp stas, 70-74, dir agr & res ctr, Brandenton, 68-70; chmn dept hort sci, NC State Univ, 74-77; pres, Miss Univ Women, 77-88. *Mem:* Am Soc Hort Sci; Am Asn State Cols Univs; Am Phytopath Soc. *Res:* Etiology, epidemiology, and control of vegetable diseases, particularly control of verticillium wilt of tomato and strawberry by breeding for resistance. *Mailing Add:* PO Box 374 Miss Univ for Women Second Ave S Due West SC 29639

STROBEL, RUDOLF G K, b Kiessling, Ger, Feb 7, 27; US citizen; m 58; c 4. PROCESS & PRODUCTS DESIGN. *Educ:* Univ Regensburg, BS, 53; Univ Munich, dipl, 55, Dr rer nat, 58. *Prof Exp:* Asst, Max Planck Inst Protein & Leather Res, Ger, 56-58; res biochemist, 58-75, group leader, 75-81, SECT HEAD, RES DIV, PROCTER & GAMBLE CO, 81- *Mem:* AAAS; Am Chem Soc; Asn Sci Int Café; Int Apple Inst. *Res:* Histochemistry; histology; protein composition and structure; enzymology; natural products; microbiology; flavor research; flavor analysis; beverage process and products design; emulsion, flour and beverage technologies; machinery/apparatus design. *Mailing Add:* 7305 Thompson Rd Cincinnati OH 45251-2329

STROBELL, JOHN DIXON, JR, b Newark, NJ, Dec 28, 17; m 49; c 2. GEOLOGY. *Educ:* Yale Univ, AB, 39, MS, 42, PhD(geol), 56. *Prof Exp:* Geologist, US Geol Surv, 42-79; RETIRED. *Mem:* Geol Soc Am; AAAS; Sigma Xi. *Res:* Geology of deposits of copper, uranium and vanadium; geology of Colorado Plateau province; environmental geology; geomorphology. *Mailing Add:* 8403 A Everett Way Arvada CO 80005-2300

STROBER, SAMUEL, b New York, NY, May 8, 40; m 63; c 2. IMMUNOLOGY. *Educ:* Columbia Univ, AB, 61; Harvard Univ, MD, 66. *Prof Exp:* Res fel, Surg Res Lab, Peter Bent Brigham Hosp, Boston, Mass, 62-63 & 65-66 & Oxford Univ, 63-64; intern med, Mass Gen Hosp, Boston, 66-67; res assoc, lab cell biol, Nat Cancer Inst, NIH, Bethesda, Md, 67-70; sr asst resident med, 70-71, from instr to assoc prof, 71-82, PROF MED IMMUNOL, SCH MED, STANFORD UNIV, 82-, CHIEF, DIV IMMUNOL, 78- *Concurrent Pos:* Res Career Develop Award, Nat Inst Allergy & Infectious Dis, NIH, 71-76; investr, Howard Hughes Med Inst, Miami, Fla, 76-81; prin investr, NIH, 76- & State of Calif, 79-; assoc ed, J Immunol, 82-84, Transplantation, 82-85 & Int J Immunotherapy, 84- *Honors & Awards:* Diane Goldstone Mem lectr, Massey Cancer Ctr, Med Col Va, 84. *Mem:* Am Asn Immunologists; Am Soc Clin Invest; Am Rheumatism Soc; Transplantation Soc; Am Soc Transplantation Physicians. *Res:* Radiotherapy treatments for and causes of rheumatoid arthritis, lupus and organ transplant rejection. *Mailing Add:* Stanford Univ Med Ctr 300 Pasteur Dr Med Ctr Stanford Univ Stanford CA 94305-5111

STROBER, WARREN, Brooklyn, NY, Oct 17, 37; m; c 3. MUCOSAL IMMUNOLOGY. *Educ:* Univ Rochester, MD, 62. *Prof Exp:* HEAD MUCOSAL IMMUNITY SECT, LAB CLIN INVEST, NIH, 82- *Mem:* Am Asn Immunologists; Am Soc Clin Invest; Am Bd Allergy & Immunol; Asn Am Physicians; Am Col Allergy. *Mailing Add:* Clin Ctr Bldg 10 Rm 11N250 NIH 9000 Rockville Pike Bethesda MD 20892

STROEBEL, CHARLES FREDERICK, III, b Chicago, Ill, May 25, 36; m 59; c 2. PSYCHOPHYSIOLOGY, NEUROPHYSIOLOGY. *Educ:* Univ Minn, BA, 58, PhD, 61; Yale Univ, MD, 73. *Prof Exp:* Res asst biophys, Mayo Clin, 55-58; res asst, Psychiat Animal Res Labs, Univ Minn, 58-61, actg dir labs & lectr, Univ, 62; DIR LABS PSYCHOPHYSIOL, INST LIVING HOSP, 62-, DIR CLINS, 74-, DIR RES, 79- *Concurrent Pos:* Adj prof, Univ Hartford, 64-72, res prof, 72-; adj prof, Trinity Col, Conn, 72-; lectr psychiat, Sch Med, Yale Univ, 73-; prof, Dept Psychiat, Univ Conn Health Ctr & Med Sch, 77- *Mem:* Biofeedback Soc Am; Am Psychiat Asn; Sigma Xi; Int Soc Chronobiology; NY Acad Sci. *Res:* Physiologic and behavioral mechanisms of stress and drugs; biologic rhythms; biofeedback; biostatistics; neurophysiology of learning and emotion. *Mailing Add:* Dept Psych Dir Psychophysiol Inst Living Hosp 400 Washington St Hartford CT 06106

STROEHLEIN, JACK LEE, b Cobden, Ill, Dec 22, 32; m 65. SOIL SCIENCE. *Educ:* Southern Ill Univ, BS, 54; Univ Wis, MS, 58, PhD(soils), 62. *Prof Exp:* Asst prof, Univ Ariz, 62-67, assoc prof agr chem & soils, 67-76, assoc prof, 76-90, EMER PROF, SOILS & WATER SCI, 90-, RES SCIENTIST AGR CHEM, AGR EXP STA, 74- *Concurrent Pos:* Adv soils & soil fertil, Brazil Prog, AID, 70-71. *Mem:* Am Soc Agron. *Res:* Soil-plant-water relationships; soil testing; fertilization and fertilizer use. *Mailing Add:* Dept Soils & Water Sci Univ Ariz Tucson AZ 85721

STROEVE, PIETER, b Velp, Netherlands, Sept 15, 45; m 67; c 3. THIN FILM TECHNOLOGY, BIOTECHNOLOGY. *Educ:* Univ Calif, Berkeley, BS, 67; Mass Inst Technol, MS, 69, DSc(chem eng), 73. *Prof Exp:* Researcher, Weizmann Inst Sci, Israel, 73-74, sr scientist, 77; res asst prof, Univ Nijmegen, The Netherlands, 74-77; from asst prof to assoc prof chem eng, State Univ NY, Buffalo, 77-82; assoc prof, 82-84, PROF CHEM ENG, UNIV CALIF, DAVIS, 84- *Concurrent Pos:* Consult, Los Alamos Nat Lab, 82-87; prin investr, grants, Nat Sci Found, 85-; vis prof, Univ Queensland, Australia & IBM Almaden Res Ctr, 88. *Mem:* Am Inst Chem Engrs; Am Chem Soc; Sigma Xi; Am Inst Physics. *Res:* Transport phenomena in multiphase systems; biotechnology; membrane separations; thin film technology; colloid science. *Mailing Add:* Dept Chem Eng Univ Calif Davis CA 95616-5294

STROH, ROBERT CARL, b Flint, Mich, July 23, 37; m 79; c 4. RESIDENTIAL CONSTRUCTION. *Educ:* Pa State Univ, BS, 59, MS, 61, PhD(genetics & statist), 64. *Prof Exp:* Grad asst statist, Pa State Univ, 59-64, lectr comput sci, 64-65; sr staff, Auto Biometrics Inc, 64-65; analyst, Vitro Labs, 65-73; vpres, Appl Urbanetics Inc, 73-79; div dir, Nat Asn Home Builders of US, 79-89; DIR, CTR AFFORDABLE HOUSING, UNIV FLA, 89- *Concurrent Pos:* Adj prof comput sci, Montgomery Col, 82-86. *Mem:* Sigma Xi; Nat Inst Bldg Sci; Soc Res Adminr. *Res:* Residential and light commercial sector of the construction industry. *Mailing Add:* 8611 SW 23rd Pl Gainesville FL 32607-3464

STROH, WILLIAM RICHARD, b Sunbury, Pa, May 5, 23. PHYSICS. *Educ:* Harvard Univ, SB, 46, AM, 50, PhD(appl physics), 57. *Prof Exp:* Instr physics, Bucknell Univ, 46-49; res fel acoustics, Harvard Univ, 57-58; from asst prof to assoc prof elec eng, Univ Rochester, 58-62; assoc prof physics, Goucher Col, 62-68, prof, 68-81; RETIRED. *Mem:* Am Phys Soc; Am Asn Physics Teachers. *Res:* Acoustics; instrumentation. *Mailing Add:* 206 Charmuth Rd Lutherville-Timonium MD 21093

STROHBEHN, JOHN WALTER, b San Diego, Calif, Nov 21, 36; m 58, 80; c 3. BIOMEDICAL ENGINEERING, RADIOPHYSICS. *Educ:* Stanford Univ, BS, 58, MS, 59, PhD(elec eng), 64. *Prof Exp:* From asst prof to assoc prof, 63-74, assoc dean, 76-81, PROF ENG, DARTMOUTH COL, 74-, ACTG PROVOST, 87- *Concurrent Pos:* Partic, Nat Acad Sci-Acad Sci USSR Exchange Prog, 67; mem comn II, Int Sci Radio Union; Inter-Union Comt Radio Meteorol; consult, McGraw-Hill, Inc & Avco Corp; vis res scientist, Stanford Med Sch; distinguished lectr, Ant & Prop Soc, Inst Elec & Electronics Engrs, 79-82; adj prof med, Dartmouth Med Sch, 79. *Mem:* Fel AAAS; Radiation Res Soc; fel Optical Soc Am; Am Asn Physicists in Med; Am Inst Elec & Electronics Engrs. *Res:* Biomedical engineering and cancer; image processing and echocardiography; optical propagation through a turbulent medium. *Mailing Add:* 102 Parkhurst Hall Dartmouth Col Hanover NH 03755

STROHBEHN, KIM, b Council Bluffs, Iowa, Oct 17, 53; m 75; c 1. ELECTRICAL ENGINEERING. *Educ:* Iowa State Univ, BS, 76, MS, 77, PhD(elec eng), 79. *Prof Exp:* asst elec eng, Iowa State Univ, 76-80; ENGR, APPLIED PHYSICS LAB, JOHNS HOPKINS UNIV, 80- *Mem:* Inst Elec & Electronics Engrs; Sigma Xi. *Res:* Application of control and estimation theory. *Mailing Add:* 6564 Grayheart Ct Columbia MD 21045

STROHECKER, HENRY FREDERICK, biology; deceased, see previous edition for last biography

STROHL, GEORGE RALPH, JR, b Ardmore, Pa, Oct 19, 19; m 46; c 2. MATHEMATICS. *Educ:* Haverford Col, BA, 41; Univ Pa, MA, 47; Univ Md, PhD(math), 56. *Prof Exp:* From instr to prof, 47-85, chmn dept, 70-76, EMER PROF, US NAVAL ACAD, 85- *Mem:* Am Math Soc; Am Soc Eng Educ. *Res:* Topology and analysis. *Mailing Add:* Dept Math US Naval Acad Annapolis MD 21402

STROHL, JOHN HENRY, b Forest City, Ill, Oct 2, 38; m 60; c 2. ANALYTICAL CHEMISTRY. *Educ:* Univ Ill, BS, 59; Univ Wis, PhD(chem), 64. *Prof Exp:* Asst chem, Univ Wis, 59-64; asst prof, 64-70, ASSOC PROF CHEM, WVA UNIV, 70- *Mem:* Am Chem Soc. *Res:* Preparative electrochemistry and continuous electrolysis. *Mailing Add:* Dept Chem Birmingham Southern Col Arkadelphia Rd Birmingham AL 35254-0002

STROHL, KINGMAN P, b Chicago, Ill, Mar 15, 49. PULMONARY PHYSIOLOGY, PULMONARY MEDICINE. *Educ:* Yale Univ, BA, 70; Northwestern Univ, MD, 74. *Prof Exp:* ASSOC PROF MED, CASE WESTERN RESERVE UNIV, 80- *Mem:* Am Physiol Soc; Am Thoracic Soc; Am Fedn Clin Res. *Mailing Add:* Dept Med Case Western Reserve Univ Cleveland OH 44110

STROHL, WILLIAM ALLEN, b Bethlehem, Pa, Nov 1, 33; m 57; c 2. VIROLOGY. *Educ:* Lehigh Univ, AB, 55; Calif Inst Technol, PhD(biol), 60. *Prof Exp:* Instr microbiol, Sch Med, St Louis Univ, 59-63; res assoc, 64-66, from asst prof to assoc prof, 66-77, PROF MICROBIOL, RUTGERS MED SCH, COL MED & DENT NJ, 77- *Concurrent Pos:* Nat Found fel, 59-61. *Mem:* AAAS; Am Soc Microbiol. *Res:* Animal viruses; viral oncogenesis. *Mailing Add:* R W Johnson Med Sch Univ Med & Dent NJ 675 Hoes Lane Piscataway NJ 08854

STROHM, JERRY LEE, b West Union, Ill, Jan 9, 37; m 57; c 4. GENETICS, PLANT BREEDING. *Educ:* Univ Ill, BS, 59; Univ Minn, PhD(genetics), 66. *Prof Exp:* Head dept, 68-74, PROF BIOL, UNIV WIS-PLATTEVILLE, 64- *Res:* Soybean genetics. *Mailing Add:* Dept of Biol Univ of Wis Platteville WI 53818

STROHM, PAUL F, b Pennsauken, NJ, Jan 13, 35; m 57; c 5. AGRICULTURAL CHEMISTRY, ORGANIC CHEMISTRY. *Educ:* La Salle Col, BA, 56; Temple Univ, PhD(org chem), 61. *Prof Exp:* Res chemist, Atlantic Refining Co, 60-62 & Houdry Labs, Air Prod & Chem Inc, 62-68; res chemist, 68-72, group leader, 72-77, MGR QUAL CONTROL, AMCHEM PROD, INC, 77- *Mem:* Am Chem Soc; Am Soc Qual Control. *Res:* Organic synthesis: agricultural chemicals, especially herbicides and plant growth regulators; kinetics of urethane reactions; mechanism of epoxy curing reactions; bicyclic amine chemistry. *Mailing Add:* Henkel Corp 300 Brookside Ave Ambler PA 19002-3438

STROHM, WARREN B(RUCE), b Brooklyn, NY, Sept 1, 25; m 51; c 3. ELECTRICAL ENGINEERING, ELECTRONICS. *Educ:* Tulane Univ, BE, 47. *Prof Exp:* With M W Kellogg Co, Inc, 47-48; proj mgr power syst protection, Devenco, Inc, 48-52, proj mgr component develop, Develop Lab, Int Bus Mach Corp, 52-57, proj mgr exp systs, Res Lab, 58-66, SR ENGR PROG SYSTS, SYSTS DEVELOP DIV, IBM CORP, 66- *Mem:* sr mem Inst Elec & Electronics Engrs; Asn Comput Mach. *Res:* Component development of magnetic devices; natural language data processing and computational linguistics; functional and structural specification of programming systems for data processing systems. *Mailing Add:* Pine Hill Dr Wappingers Falls NY 12590

STROHMAIER, A(LFRED) J(OHN), chemical engineering; deceased, see previous edition for last biography

STROHMAN, RICHARD CAMPBELL, b New York, NY, May 5, 27; m; c 2. ZOOLOGY. *Educ:* Columbia Univ, PhD, 58. *Prof Exp:* Instr zool & cell physiol, Columbia Univ, 55-56; from asst prof to assoc prof, 58-70, chmn dept, 73-77, PROF ZOOL, UNIV CALIF, BERKELEY, 70- *Mem:* Soc Gen Physiol. *Res:* Physiology and biochemistry of muscle growth and development. *Mailing Add:* Dept Zool Univ Calif Berkeley CA 94720

STROHMAN, ROLLIN DEAN, b Geneseo, Ill, Oct 29, 39; m 69; c 2. AGRICULTURAL ENGINEERING. *Educ:* Univ Ill, BS(agr eng) & BS(agr sci), 62, MS, 65; Purdue Univ, PhD(agr eng), 69. *Prof Exp:* Agr engr, Western Utilization Res & Develop Div, Agr Res Serv, USDA, 68-69; assoc prof, 69-80, PROF AGR ENG, CALIF POLYTECH STATE UNIV, SAN LUIS OBISPO, 80- *Mem:* Am Soc Agr Engrs; Am Soc Photogram. *Res:* Surveying and mapping; microcomputer interfacing. *Mailing Add:* Dept of Agr Eng Calif Polytech State Univ San Luis Obispo CA 93407

STROJNY, EDWIN JOSEPH, b Chicago, Ill, Jan 1, 26; m 55; c 2. INDUSTRIAL CHEMISTRY. *Educ:* Ill Inst Technol, BS, 51; Univ Ill, PhD(chem), 55. *Prof Exp:* Asst instr gen chem, Univ Ill, 51-52; org chemist, G D Searle & Co, 54-57; ORG CHEMIST, DOW CHEM CO, 57- *Mem:* AAAS; Am Chem Soc; Sigma Xi. *Res:* Phenolic compounds and derivatives; aromatic chemistry; heterogeneous and homogeneous catalysis; oxidation of organic compounds by oxygen; reaction mechanisms. *Mailing Add:* 4695 Gully Rd Harbor Springs MI 49740

STROKE, HINKO HENRY, b Zagreb, Yugoslavia, June 16, 27; US citizen; m 56; c 2. ATOMIC SPECTROSCOPY, NUCLEAR PHYSICS. *Educ:* NJ Inst Technol, BS, 49; Mass Inst Technol, MS, 52, PhD, 54. *Prof Exp:* Chmn, Dept Physics, 88-91, PROF PHYSICS, NY UNIV, 68- *Concurrent Pos:* Ed, Comments Atomic & Molecular Physics, 73-; sci assoc, Euro Ctr Nuclear Res, Geneva, 83-84, 91-92; sr fel NATO, 75. *Honors & Awards:* Sr US Scientist Award, Alexander von Humboldt Found, 77. *Mem:* AAAS; fel Am Phys Soc; fel Optical Soc Am; French Phys Soc; Europ Phys Soc. *Res:* Hyperfine structure and isotope shifts of stable and radioactive atoms by magnetic resonance and optical spectroscopy; nuclear moments; charge and magnetization distribution; coherence in atomic radiation; solar spectra; spectroscopic instrumentation; laser systems; spectroscopy, radiative collisions; bolometric particle spectrometers. *Mailing Add:* Dept Physics NY Univ New York NY 10003

STROM, BRIAN LESLIE, b New York, NY, Dec 8, 49; m 78; c 2. PHARMACOEPIDEMIOLOGY, CLINICAL EPIDEMIOLOGY. *Educ:* Yale Univ, BS, 71; Johns Hopkins Univ, MS, 75; Univ Calif, Berkeley, MPH, 80; Univ Pa, MA, 88. *Prof Exp:* Intern med, Univ Calif, San Francisco, 75-76, resident med, 76-78, fel clin pharmacol, 78-80; asst prof, 80-88, ASSOC PROF MED, UNIV PA, 88- *Concurrent Pos:* Consult, many law firms, 79-, most major pharmaceut co, 80- & US & foreign govt agencies, 80-; prin investr, NIH, Food Drug Admin grants & others, 80-; lectr, all over the world, 80-; mem, JCAHO, Medication use Task Force, 89- & US Pharmaceutics Drug Utilization Rev Adv Panel, 90- *Mem:* Fel Am Col Physicians; fel Am Col Epidemiol; Am Epidemiol Soc; Am Soc Clin Pharmacol & Therapeut. *Res:* Clinical epidemiology; pharmacoepidemiology; epidemiologic methods to study the effects of drugs. *Mailing Add:* 315R NEB/6095 Sch Med Univ Pa Philadelphia PA 19104-6095

STROM, E(DWIN) THOMAS, b Des Moines, Iowa, June 11, 36; m 58; c 2. POLYMER CHEMISTRY, PHYSICAL ORGANIC CHEMISTRY. *Educ:* Univ Iowa, BS, 58; Univ Calif, Berkeley, MS, 61; Iowa State Univ, PhD(phys org chem), 64. *Prof Exp:* Res technologist, 64-67, SR RES CHEMIST, DALLAS RES LAB, MOBIL RES & DEVELOP CORP, 67- *Concurrent Pos:* Vis lectr, Dallas Baptist Col, 69-70, El Centro Community Col, 70-72 & Univ Tex, Dallas, 74; adj prof, Univ Tex, Arlington, 78-79 & 82-; ed, Southwest Retort, 83- *Honors & Awards:* W T Doherty Award, Am Chem Soc, 89. *Mem:* Am Chem Soc; Soc Petrol Engrs. *Res:* Polymers for petroleum recovery; free radicals; magnetic resonance; coal chemistry; uranium chemistry. *Mailing Add:* Dallas Res Lab Mobil Res & Develop Corp PO Box 819047 Dallas TX 75381

STROM, GORDON H(AAKON), aeronautical engineering; deceased, see previous edition for last biography

STROM, OREN GRANT, b Groton, SDak, Aug 1, 31; m 53; c 4. TRANSPORTATION SYSTEMS, CONTINUING EDUCATION. *Educ:* SDAK State Univ, BS, 53; Univ Wyo, MS, 65; Univ Tex, Austin, PhD(civil eng), 72. *Prof Exp:* Hwy engr, State Hwy Dept Wis, 53-54; engr-mgr, US Air Force, 54-82; ACTG DEAN, COL ENG & APPL SCI, UNIV COLO, DENVER, 82- *Concurrent Pos:* Dean, Sch Civil Eng, Air Force Inst Technol, US Air Force, 77-81. *Mem:* Am Soc Civil Engrs; Soc Am Mil Engrs; Am Soc Eng Educ. *Res:* Transportation engineering research; computer-based management of the construction, maintenance and repair of systems so that resources are programmed, allocated and used according to the actual requirements. *Mailing Add:* Col Eng & Appl Sci Univ Colo at Denver Campus Box 104 PO Box 173364 Denver CO 80217-3364

STROM, RICHARD NELSEN, b Schenectady, NY, June 5, 42. GEOCHEMISTRY, CLAY MINERALOGY. *Educ:* Union Col, BS, 66; Univ Del, MS, 72, PhD(geol), 76. *Prof Exp:* Dep route mgr, Off Interoceanic Canal Studies, Corps Engrs, 67-69; asst prof geol, Univ Wis-Parkside, 75-77; from asst prof to assoc prof geol, Univ SFla, 77-90; PRIN SCIENTIST, SAVANNAH RIVER LAB, 90- *Mem:* Sigma Xi; Geochem Soc; Clay Minerals Soc. *Res:* Silicification and clay authigenesis; hydrogeology and ground water geochemistry. *Mailing Add:* Westinghouse-Savannah River Co Savannah River Site Bldg 773-42A Aiken SC 29808

STROM, ROBERT GREGSON, b Long Beach, Calif, Oct 1, 33; m 55; c 1. ASTROGEOLOGY. *Educ:* Univ Redlands, BS, 55; Stanford Univ, MS, 57. *Prof Exp:* Geologist, Stand Vacuum Oil Co, 57-60; asst res geologist, Univ Calif, Berkeley, 61-63; from asst prof to assoc prof, Lunar & Planetary Lab, 63-81, PROF PLANETARY SCI, UNI ARIZ, 81- *Concurrent Pos:* Mem, Apollo Lunar Oper Working Group, 68-69, Imaging Sci Team, Mariner Venus/Mercury Mission, 69-75, Lunar Sci Inst, Lunar Sci & Cartog Comn, 74-79, NASA Comet Working Group, Jet Propulsion Lab Jupiter Orbiter Sci Working Group & NASA Mercury Geol Mapping Prog, 75-78; rep, Planetary Prog, Jet Propulsion Lab Imaging Syst Instrument Develop Prog, 75-79; mem, NASA Venus Orbital Imaging Radar Sci Working Group, Jet Propulsion Lab, 77-78, assoc mem, Voyager Imaging Sci Team, 78-; mem, Planetary Geol Working Group, 80-82 & NASA Planetary Geol Rev Panel, 80-82; chmn, NASA Planetary Cartog Working Group, 83-85; NASA Planetary Geol & Geophys Working Group, 86-, NASA Mercury Orbiter Sci Working Team, 88- *Mem:* Am Geophys Union; Int Astron Union; Am Astron Soc. *Res:* Lunar and planetary geology; origin and evolution of lunar and planetary surfaces; space craft imaging of planetary surfaces. *Mailing Add:* Dept Planetary Sci Univ Ariz Tucson AZ 85721

STROM, ROBERT MICHAEL, b Detroit, Mich, Aug 6, 51; m 74. ORGANIC CHEMISTRY, PHOTOCHEMISTRY. *Educ:* Grand Valley State Col, BS, 74; Univ Colo, PhD(org chem), 78. *Prof Exp:* Res asst org photochem, Univ Colo, 74-78; res chemist synthetic org chem, Arapahoe Chem Inc, 78-80; chemist, 80-81, proj leader monomers, 81-82, res leader epoxies, 82-84, group leader thermosets & monomers, 84-85, GROUP LEADER ION EXCHANGE, DOW CHEM CORP, 85- *Mem:* Am Chem Soc; Sigma Xi. *Res:* Synthetic organic chemistry; physical organic photochemistry; process development of aromatic and heterocyclic synthesis; determination of structural features of short lived ground and excited state intermediates, monomers, thermosets, flame retardants and epoxies ion exchange. *Mailing Add:* Dow Chem USA 1776 Bldg Midland MI 48640

STROM, STEPHEN, b Bronx, NY, Aug 12, 42; m 60; c 4. ASTRONOMY, ASTROPHYSICS. *Educ:* Harvard Univ, AB, 62, AM & PhD(astron), 64. *Prof Exp:* Astrophysicist, Smithsonian Astrophys Observ, 62-69; lectr astron, Harvard Univ, 64-69; assoc prof physics & earth & space sci, State Univ NY Stony Brook, 69-71, prof astron, 71-72, coordr astron & astrophys, 69-72; astronr, Kitt Peak Nat Observ, 72-, chmn galactic & extragalactic prog, 75-; AT DEPT PHYSICS, UNIV MASS. *Concurrent Pos:* Alfred P Sloan Found res fel, 70-72; mem at large, Assoc Univs Res Astron, 71-72; res assoc, Smithsonian Astrophys Observ, 69-71. *Honors & Awards:* Bart J Bok Prize, 71; Helen B Warner Prize, Am Astron Soc, 76. *Mem:* Am Astron Soc; Int Astron Union. *Res:* Evolution of stars; structure and evolution of galaxies. *Mailing Add:* Dept Physics & Astron Univ Mass Grad Res Ctr Amherst MA 01003

STROM, TERRY BARTON, b Chicago, Ill, Nov 30, 41; m 64; c 2. IMMUNOLOGY RESEARCH. *Educ:* Univ Ill Col Med, MD, 66. *Hon Degrees:* DSc, Hahnemann Univ, 91. *Prof Exp:* Intern med, Univ Ill Hosps, 66-67, jr resident, 67-68; assoc prof med, 78-88, PROF MED, HARVARD UNIV, 88-; PHYSICIAN, BRIGHAM & WOMEN'S HOSP, BOSTON, 73-; DIR CLIN IMMUNOL, MED DIR KIDNEY TRANSPLANTATION & PHYSICIAN, BETH ISRAEL HOSP, BOSTON, 83- *Concurrent Pos:* Career Develop Award, NIH, 75. *Mem:* Am Soc Transplant Physicians (pres, 83-84); Am Soc Clin Invests; Int Transplant Soc; Am Asn Immunologists; Int Soc Nephrology; Am Soc Nephrology; Clin Immunol Soc (pres, 89-90); Asn Am Prof. *Res:* Mechanisms of transplant rejection and T-lymphocyte activation; clinical transplantation; basic immunology research; development of immunopharmacologic principles and molecules. *Mailing Add:* Beth Israel Hosp 330 Brookline Ave Boston MA 02215

STROMAN, DAVID WOMACK, b Corpus Christi, Tex, June 1, 44; m 65; c 1. FERMENTATION PRODUCTS. *Educ:* Bethany Nazarene Col, BS, 66; Univ Okla, PhD(biochem), 70. *Prof Exp:* NIH fel microbiol, Sch Med, Washington Univ, 70-72; res scientist antibiotic discovery & develop, Upjohn Co, 72-81; sr res scientist biotechnol, Phillips Petrol Co, 81-88; DIR MICROBIOL RES, ALCON LABS, INC, 90- *Mem:* Am Chem Soc; Am Soc Microbiol. *Res:* Molecular biology and biochemical genetics with emphasis on fermentation derived products especially involving recombinant DNA technology. *Mailing Add:* Alcon Labs Inc 6201 S Freeway Ft Worth TX 76134

STROMATT, ROBERT WELDON, b Muskogee, Okla, Mar 27, 29; m 56; c 3. ANALYTICAL CHEMISTRY. *Educ:* Emporia State Teachers Col, BS, 54; Kans State Univ, PhD(chem), 58. *Prof Exp:* Res chemist, Hanford Labs, Gen Elec Co, 57-66; sr res scientist, Pac Northwest Lab, Battelle Mem Inst, 66-70; fel scientist, Westinghouse-Hanford, 70-87; PAC NORTHWEST LAB, BATTELLE MEM INST, 87- *Mem:* Am Chem Soc. *Res:* Automation and general methods development. *Mailing Add:* 411 Franklin Richland WA 99352-2003

STROMBERG, BERT EDWIN, JR, b Trenton, NJ, May 19, 44; m 68; c 2. PARASITE IMMUNOLOGY. *Educ:* Lafayette Col, BA, 66; Univ Mass, MA, 68; Univ Pa, PhD(parasitol), 73. *Prof Exp:* Instr biol, Trenton State Col, 68-70; asst prof parasitol, Univ Pa, 73-79; assoc prof, 79-86, PROF PARASITOL & IMMUNOL, UNIV MINN, 86- *Concurrent Pos:* Prin investr, NIH & USDA formula grants, Minn Agr Exp Sta, 79-; dir, Ctr Comp Biomed Res, 84-; adj prof, Hassan II Agron & Vet Inst, Morocco, 85- *Mem:* Am Soc Parasitol; Am Asn Immunologists; Am Asn Vet Parasitol (vpres, 87-88, pres-elect, 88-89, pres, 89-90). *Res:* Immune response to helminth parasites; parasite antigens; immune protective response; in vitro cultivation; epidemiology of gastro-intestinal helminths and their control. *Mailing Add:* Col Vet Med Univ Minn 1971 Commonwealth Ave St Paul MN 55108

STROMBERG, KARL ROBERT, b Modoc, Ind, Dec 1, 31; m 68; c 3. MATHEMATICS. *Educ:* Univ Ore, BA, 53, MA, 54; Univ Wash, PhD(math), 58. *Prof Exp:* Res assoc & Off Naval Res fel math, Yale Univ, 58-59; res lectr, Univ Chicago, 59-60; from asst prof to assoc prof, Univ Ore, 60-68; PROF MATH, KANS STATE UNIV, 68- *Concurrent Pos:* Vis prof, Uppsala Univ, Sweden, 66-67; vis res prof, Univ York, Eng, 74-75. *Mem:* Am Math Soc; Math Asn Am. *Res:* Measure and integration theory; real variable theory; harmonic analysis; topological groups; functional analysis. *Mailing Add:* Dept Math Kans State Univ Manhattan KS 66506

STROMBERG, KURT, b Albuquerque, NMex, Mar 3, 39; c 1. PATHOLOGY. *Educ:* Amherst Col, BA, 61; Univ Colo, MD, 66; Am Bd Path, dipl, 74. *Prof Exp:* Intern path, Yale-New Haven Hosp, 66-67; res assoc, Nat Cancer Inst, 68-72; resident path, Columbia Univ, 72-74; mem res staff, Inst Cancer Res, Delafield Hosp, New York, 72-74; SR STAFF INVESTR, LAB CYTOKINE BIOL, FOOD & DRUG ADMIN, 88-; PROF PATH, UNIFORMED SERVS UNIV HEALTH SCI, BETHESDA, 88- *Mem:* Am Asn Cancer Res; Am Asn Path; Int Acad Path. *Res:* Viral oncology; transforming growth factors in carcinogenesis. *Mailing Add:* FDA CBER Ctr Biologics Eval & Res 8800 Rockville Pike Bldg 29A Rm 2B-10 HFB 830 Bethesda MD 20892

STROMBERG, LAWAYNE ROLAND, b Minneapolis, Minn, Nov 18, 29; m 54; c 3. SURGERY, ADMINISTRATION. *Educ:* Univ Calif, Berkeley, BA, 51; Univ Calif, Los Angeles, MD, 55; Univ Rochester, MS, 63. *Prof Exp:* Intern & resident gen surg, UCLA Med Ctr, Los Angeles, 55-58, resident, Vet Admin Hosp, Los Angeles, 58-60; cmndg officer & surgeon, 11th Evacuation Hosp, Korea, US Army, 61-62, nuclear med res officer, Walter Reed Army Inst Res, 65-68, cmndg officer, US Army Nuclear Med Res Detachment, Europe, Landstuhl, Ger, 68-71, dep dir & dir, Armed Forces Radiobiol Res Inst, Nat Naval Med Ctr, 68-77; assoc med dir, Baxter Healthcare Corp, 77-80, vpres med affairs, 80-90; PRIN, EXPERTECH ASSOC, 90- *Concurrent Pos:* Res fel radiation biol, Walter Reed Army Inst Res, 63-65. *Mem:* Am Acad Med Dir; Asn Military Surg US; Royal Soc Med. *Res:* Effect of radiation on response to trauma. *Mailing Add:* Expertech Assoc 120 Edgemont St Mundelein IL 60060

STROMBERG, MELVIN WILLARD, b Quamba, Minn, Nov 2, 25; m 48; c 6. GROSS ANATOMY, NEUROANATOMY. *Educ:* Univ Minn, BS, 53, DVM, 54, PhD(vet anat), 57. *Prof Exp:* Asst vet anat, Univ Minn, 53-54, from instr to assoc prof, 54-60; assoc prof, 60-62, head dept, 63-81, PROF VET ANAT, PURDUE UNIV, WEST LAFAYETTE, 62- *Concurrent Pos:* NIH spec fel anat, Karolinska Inst, Sweden, 70-71; adj prof, Ind Univ Sch Med, Indianapolis, 75; vis prof, Univ Munich Sch Med, WGer, 78-79, study & res, 85-86, vis prof, Massey Univ, Fac Vet Sci, Palmerston North, New Zeal, 91. *Mem:* Am Asn Vet Anat (pres, 66-67); Am Asn Anat; World Asn Vet Anat. *Res:* Histology of dolphin and avian skin; neuroanatomy of domestic animals; extra cellular lipids. *Mailing Add:* Dept Anat Sch Vet Med Purdue Univ West Lafayette IN 47906

STROMBERG, ROBERT REMSON, b Buffalo, NY, Feb 2, 25; m 47; c 4. POLYMER SCIENCE, BLOOD & TISSUE BANKING. *Educ:* Univ Buffalo, BA, 48, PhD(phys chem), 51. *Prof Exp:* Asst phys chem, Univ Buffalo, 48-50; phys chemist, Nat Bur Standards, 51-62, chief phys chem br, Off Saline Water, 62, phys chemist, 62-67, chief polymer interface sect, 67-75, dep chief polymers div, 69-76; actg assoc bur dir device res & testing, Bur Med Devices, Food & Drug Admin, 76-80; head biochem eng sect, Biomed Eng Lab, 80-84, PROD MGR, PROD DEVELOP LAB, AM RED CROSS, 84- *Concurrent Pos:* Chmn, Gordon Res Conf Sci Adhesion, 66; US Dept Com fel sci & technol, 75-76. *Mem:* Soc Biomaterials; Am Chem Soc; Am Asn Blood Banks; Int Soc Blood Transfusions. *Res:* Polymer and blood protein; surface interactions; surface chemistry; blood plasma separations; viral inactivation; mechanical properties of bone; platelet shedding by magakeryocytes; plasmapheresis; removal of leukocyctes and platelets from red blood cells by filtration; photoactive dyes. *Mailing Add:* American Red Cross, Holland Lab 15601 Crabbs Br Way Rockville MD 20855-2734

STROMBERG, THORSTEN FREDERICK, b Aberdeen, Wash, Aug 13, 36. LOW TEMPERATURE PHYSICS. *Educ:* Reed Col, BA, 58; Iowa State Univ, PhD(physics), 65. *Prof Exp:* Res fel, Los Alamos Sci Lab, 65-67; asst prof, 67-74, ASSOC PROF PHYSICS, N MEX STATE UNIV, 74- *Mem:* Am Phys Soc. *Res:* Thermal and magnetic properties of superconductors, particularly type-II superconducting materials. *Mailing Add:* Dept Physics NMex State Univ Las Cruces NM 88003

STROMBOTNE, RICHARD L(AMAR), b Watertown, SDak, May 6, 33; m 52; c 4. ENERGY CONVERSION, ATOMIC PHYSICS. *Educ:* Pomona Col, BA, 55; Univ Calif, Berkeley, MA, 57, PhD(physics), 62. *Prof Exp:* Res assoc nuclear magnetic resonance, Univ Calif, Berkeley, 59-61; physicist, Radio Stand Lab, US Nat Bur Standards, 61-68; physicist, 68-71, asst for phys sci, 71-73, chief, Energy & Environ Div, 73-78, DIR, OFF AUTO FUEL ECON STANDARDS, US DEPT TRANSP, 78-; head div, 65-81, prof chem & chmn, Dept Sci & Math, 76-81, PROF CHEM, CHADRON STATE COL, 81- *Concurrent Pos:* Finder of the Bayard, USA meteorite. *Mem:* AAAS; Soc Automotive Engrs; Am Phys Soc; Meteoritics; Soc Vertebrate Paleontology. *Res:* Automotive fuel economy; energy requirements of transportation systems; geochemical studies related to uranium, radium and radon; air pollution associated with transportation; measurement of fine structure of singly ionized helium; longitudinal spin-spin relaxation in low fields. *Mailing Add:* 24401 Peach Tree Rd Clarksburg MD 20871

STROME, FORREST C, JR, b Kalamazoo, Mich, May 19, 24; m 45; c 2. LASERS, INFORMATION RECORDING. *Educ:* Univ Ill, BS, 45; Univ Mich, MS, 48, PhD(physics), 54. *Prof Exp:* Test engr, Gen Elec Co, 46-47; proj physicist, Apparatus & Optical Div, 53-60, sr res physicist, Res Labs, 60-68, RES ASSOC, EASTMAN KODAK CO, 68- *Mem:* Am Phys Soc: Optical Soc Am. *Res:* Information recording using lasers. *Mailing Add:* 20 Stuyvesant Rd Pittsford NY 14534

STROMER, MARVIN HENRY, b Readlyn, Iowa, Sept 1, 36; m 60; c 1. CELL BIOLOGY, ELECTRON MICROSCOPY. *Educ:* Iowa State Univ, BS(animal sci) & BS(agr educ), 59, PhD(cell biol), 66. *Prof Exp:* Foreman prod develop, George A Hormel & Co, Minn, 59-62; res asst cell biol, Iowa State Univ, 62-66; fel, Carnegie-Mellon Univ, 66-68; assoc prof, 68-76, PROF, IOWA STATE UNIV, 76- *Concurrent Pos:* Lectr, Tex A&M Univ, 71; Humboldt fel, 74; vis scientist, Max Planck Inst Med Res, Heidelberg, WGer, 74-75, 88; vis prof, Univ Ariz, 79-80; pres, Iowa Microbeam Soc, 84-85, Iowa Affil, Am Heart Asn, 84-86; vis scientist, Univ Montreal, Quebec, 87. *Honors & Awards:* Fulbright Fel, Max Planck Inst Med Res, Heidelberg WGer, 88; Distinguished Res Award, Am Meat Sci Asn, 89. *Mem:* Electron Micros Soc Am; Am Heart Asn; Am Soc Cell Biol. *Res:* Ultrastructure and biochemistry of filaments and filament attachment sites in striated and smooth muscle and other movement systems. *Mailing Add:* Dept of Animal Sci Iowa State Univ Ames IA 50011

STROMINGER, JACK L, b New York, NY, Aug 7, 25; c 4. IMMUNOLOGY. *Educ:* Harvard Univ, AB, 44; Yale Univ, MD, 48. *Hon Degrees:* Trinity Col, Dublin, DSc, 75; Wash Univ, DSc, 88. *Prof Exp:* From asst prof to prof pharmacol, Sch Med, Wash Univ, St Louis, Mo, 55-61, prof pharmacol & microbiol, 61-64; prof pharmacol & chem microbiol, Med Sch, Univ Wis-Madison, 64-68; prof biochem, 68-83, chmn dept biochem & molecular biol, 70-73, HIGGINS PROF BIOCHEM, HARVARD UNIV, 83-; HEAD TUMOR VIROL DIV, DANA-FARBER CANCER INST, 77- *Honors & Awards:* John J Abel Award Pharmacol, 60; Paul-Lewis Lab Award Enzyme Chem, 62; Nat Acad Sci Award Microbiol, 68; Rose Payne Award, Am Soc Histocompat & Immunogen, 86; Selman Waxman Award in Microbiol, Nat Acad Sci, 68; Hoechst-Roussel Award, 90; Pasteur Medal, 90. *Mem:* Nat Acad Sci-Inst Med; Am Soc Biol Chemists; Am Soc Pharmacol & Exp Therapeut; AAAS; Am Asn Immunologists; Am Soc Microbiologists; Sigma Xi; Am Chem Soc; Am Acad Arts & Sci; Europ Molecular Biol Orgn. *Mailing Add:* 2030 Massachusetts Ave Lexington MA 02173

STROMINGER, NORMAN LEWIS, b New York, NY, June 1, 34; m 57; c 3. NEUROANATOMY. *Educ:* Univ Chicago, AB, 55, BS, 56, PhD(biopsychol), 61. *Prof Exp:* Trainee neuroanat, Columbia Univ, 62-65; from asst prof to assoc prof, 65-74, PROF ANAT, ALBANY MED COL, 74- *Mem:* AAAS; Am Asn Anat; Soc Neuroscience. *Res:* Neuroanatomical studies of auditory and motor systems; mechanisms of emesis. *Mailing Add:* 131 Devon Rd Delmar NY 12054

STROMMEN, DENNIS PATRICK, b Milwaukee, Wis, Sept 2, 38; m 68. INORGANIC CHEMISTRY, SPECTROSCOPY. *Educ:* Wis State Univ-Whitewater, BA, 66; Cornell Univ, PhD(chem), 71. *Prof Exp:* Res assoc spectros, Ctr Mat Res, Univ Md, 70-71; from asst prof to assoc prof, 71-81, PROF CHEM, CARTHAGE COL, 81- *Concurrent Pos:* Vis prof, Univ Ore, 78-80, Nuclear Energy Ctr, Grenoble, France, 86-87; adj prof, Marquette, Univ Milw, Wis. *Mem:* Soc Appl Spectros; Am Chem Soc; Coblentz Soc. *Res:* Characterization of compounds through vibrational analysis, especially with regard to their Raman spectra; drug interactions with polynucleotides; intercalation compounds; ruthenium complexes in zeolites. *Mailing Add:* Dept Chem Carthage Col Kenosha WI 53140

STROMMEN, NORTON DUANE, b Stoughton, Wis, Oct 13, 32; m 52; c 2. METEOROLOGY, CLIMATOLOGY. *Educ:* Univ Wis-Madison, BS, 59, MS, 64; Mich State Univ, PhD(climatol), 75. *Prof Exp:* Observer-forecaster, US Dept Com, Nat Weather Serv & Environ Data & Info Serv, Madison, Wis, 60-65, state climatologist, Columbia, SC, 65-66, East Lansing, Mich, 66-73, forecaster, Nat Severe Storms Forecast Ctr, Kansas City, Mo, 73-74, supvy meteorologist, Ctr Climatic Environ Assessment, Columbia, Mo, 74-76, dir, 76-78, div chief, Ctr Environ Assessment Serv/CIAD, US Dept Com, Nat Weather Serv & Environ Data & Info Serv, 79-80; CHIEF METEOROLOGIST, US DEPT AGR, WORLD AGR OUTLOOK BD, WASHINGTON, DC, 80- *Concurrent Pos:* Asst prof, Agr Eng, Mich State Univ, 66-73; res assoc, Univ Mo Columbia, 74-76, assoc prof, 76-79; mem comn agr meteorol, World Meteorol Org, 75-, adv working group, 76-, Nat Climate Prog Policy Bd, US Dept Agr, 83-, subcomt atmospheric res, Nat Sci Found, 84-, Interdepart Comt Meteorol Serv & Supporting Res, Off Fed Coordr Meteorol Serv & Supporting Res, 86-, Presidential Drought Policy Task Force, 88; US mem NATO sci panel Global Climate Change, 89- *Mem:* Fel Am Meteorol Soc; AAAS; Sigma Xi; Coun Agr Sci & Technol. *Res:* Climate applications to agricultural applications for crop condition and yield assessment; impacts of climate change on future agricultural crop production. *Mailing Add:* US Dept Agr Rm 5133 S Agr Bldg Washington DC 20250

STROM-OLSEN, JOHN OLAF, UK & Can citizen. SOLID STATE PHYSICS. *Educ:* Cambridge Univ, PhD(physics), 66. *Prof Exp:* PROF PHYSICS, MCGILL UNIV, 67- *Mem:* Am Phys Soc; Can Asn Phys; Mat Res Soc. *Res:* Metallic glasses; nanocrystalline alloys; fine ceramics and metal fibers. *Mailing Add:* Dept Physics McGill Univ 3600 University St Montreal PQ H3A 2T8 Can

STROMSTA, COURTNEY PAUL, b Muskegon, Mich, Apr 25, 22; m 50; c 2. SPEECH PATHOLOGY, AUDIOLOGY. *Educ:* Western Mich Univ, BS, 48; Ohio State Univ, MA, 51, PhD(speech & hearing sci), 56. *Prof Exp:* Audiol trainee, Vet Admin-Walter Reed Hosp & New York City Regional Off, 51; dir speech & hearing clin, ECarolina Univ, 54-56; prof speech & hearing sci, Ohio State Univ, 56-68; PROF SPEECH & HEARING SCI, WESTERN MICH UNIV, 68- *Concurrent Pos:* Nat Inst Neurol Dis & Blindness res grant, Ohio State Univ, 57-67 & US Off Educ res grant, 65-67; NIH spec res fel, Karolinska Inst, Sweden, 71-72; consult, Electronic Teaching Labs, Washington, DC, 61-65; guest prof, Univ Zagreb, 65-66. *Mem:* AAAS; Am Speech & Hearing Asn; Acoust Soc Am. *Res:* Cybernetic relationship of speech and hearing with emphasis on stuttering and acoustically-impaired children; effects of shaping acoustical signals on perception of speech by hearing impaired children. *Mailing Add:* Dept Speech Path & Audiol 309 SPC Western Mich Univ 2333 Rambling Rd Kalamazoo MI 49008

STRONACH, CAREY E, b Boston, Mass, Aug 8, 40; m 66; c 2. MUON SPIN ROTATION SOLIDS, MEDIUM-ENERGY NUCLEAR PHYSICS. *Educ:* Univ Richmond, BS, 61; Univ Va, MS, 63; Col William & Mary, PhD(physics), 75. *Prof Exp:* From instr to assoc prof, 65-80, PROF PHYSICS, VA STATE UNIV, 80- *Concurrent Pos:* Vis assoc prof, Univ Alta, 78-79; dir, Solid State Physics Res Inst, Va State Univ, 84-; prin investr, USA & France Muon Spin Rotation Res Prog, 85-; guest scientist, Los Alamos Nat Lab, 77-, Brookhaven Nat Lab, 83-; mem sci adv comt, Europ Workshop Spectros Subatomic Species in Solids, 85; mem, Continuous Electron Beam Accelerator Facil Users Group, Los Alamos Meson Physics Facil Users Group & Alt Gradient Synchrotron Users Group; mem, Bd of Trustees, Southeastern Univs Res Asn, 83-; mem, Tri-Univ Meson Facil Users Group; mem, Local Organizing Comt, Int Symposium Physics & Chem Finite Systs, from clusters to crystals, Richmond Va, Oct 8-12, 91. *Mem:* Am Phys Soc; Am Asn Physics Teachers; AAAS. *Res:* Muon spin rotation studies of solids; concentrating on superconductors and related materials; medium-energy nuclear physics, primarily pion-nucleus reactions; author of 75 publications. *Mailing Add:* Box 358 Va State Univ Petersburg VA 23803

STRONG, ALAN EARL, b Boston, Mass, May 30, 41; m 66, 84; c 5. REMOTE SENSING, TEACHING. *Educ:* Kalamazoo Col, BA, 63; Univ Mich, MS, 65, PhD(oceanog), 68. *Prof Exp:* Res oceanogr-meteorologist, Nat Environ Satellite Serv, 68-86, PROF & RESEARCHER, NAT OCEANIC & ATMOSPHERIC ADMIN, 86- *Concurrent Pos:* Proj coordr, Ctr Excellence Oceanic Remote Sensing. *Mem:* AAAS; Am Meteorol Soc; Am Geophys Union; Sigma Xi. *Res:* Lake and sea breezes, air-sea interface; marine meteorology; develop applications of earth satellite data to oceanography; remote sensing-infrared microwave, visible; sea surface temperature measurements by satellite; ocean color measurements by satellite; El Niño studies; submesoscale eddy research. *Mailing Add:* US Naval Acad Oceanog Dept Navy/NOAA Annapolis MD 21402

STRONG, CAMERON GORDON, b Vegreville, Alta, Sept 18, 34; m 59; c 2. INTERNAL MEDICINE, NEPHROLOGY. *Educ:* Univ Alta, MD, 58; McGill Univ, MS, 66. *Prof Exp:* Resident internal med & path, Queens Hosp, Honolulu, 59-61; resident internal med, Mayo Grad Sch Med, 61-64; from instr to assoc prof, 67-77, chmn, Div Nephrology, 73-78, PROF MED, MAYO MED SCH, 77-, CONSULT, DIV NEPHROLOGY & INTERNAL MED, MAYO CLIN & FOUND, 66-, CHMN, DIV HYPERTENSION, 77- *Concurrent Pos:* Fel nephrology, Hotel Dieu Montreal, 64-66; res assoc hypertension, dept physiol, Univ Mich, Ann Arbor, 66-67; fel, Coun High Blood Pressure Res,Am Heart Asn, 71- *Mem:* Fel Am Col Physicians; fel Am Col Cardiol. *Res:* Hypertension; renal disease; vascular smooth muscle physiology; prostaglandins; renin-angiotensin system. *Mailing Add:* 413 SW Sixth Ave Rochester MN 55902

STRONG, DAVID F, b Botwood, Nfld, Feb 26, 44; m; c 2. PURE & APPLIED SCIENCES. *Educ:* Mem Univ Nfld, BS, 65; Lehigh Univ, MS, 67; Univ Edinburgh, PhD, 70. *Prof Exp:* From asst prof to prof, Mem Univ Nfld, 70-90, actg head, Dept Geol, 74-75, univ res prof, 85-90, spec adv to pres, 86-87, vpres acad, 87-90; WF James prof pure & appl sci, St Francis Xavier Univ, NS, 81-82; PRES & VICE CHANCELLOR, UNIV VICTORIA, 90- *Concurrent Pos:* Vis prof, Univ Montpellier, France, 76-77; assoc ed, Can J Earth Sci, 77-83, Transactions of the Royal Soc Edinburgh, 80-; mem, Natural Sci & Eng Res Coun Can, 82-88; past pres medal comt, Geol Asn Can, 85; mem, Res Coun, Can Inst Advan Res & bd dirs, Seabright Corp Ltd, 86-; mem, Nfld & Labrador Adv Coun Sci & Technol, 88-, Res Prog Adv Comt, Centre Cold Ocean Resources Eng, 90- *Honors & Awards:* Distinguished Serv Award, Can Inst Mining & Metall, 79; Past Pres Medal, Geol Asn Can, 80; Swiney lectr, Univ Edinburgh, 81. *Mem:* Fel Geol Soc Am; fel Soc Econ Geologists; Can Inst Mining & Metall. *Res:* Natural environment; geological phenomena; processes which form volcanoes and their products, particularly mineral deposits; author of many publications. *Mailing Add:* 3050 Baynes Rd Victoria BC V8W 2Y2 Can

STRONG, DONALD RAYMOND, JR, b Chelsea, Mass, May 22, 44; m 75; c 2. ECOLOGY, ENTOMOLOGY. *Educ:* Univ Calif, Santa Barbara, BA, 66; Univ Ore, PhD(biol), 71. *Prof Exp:* Res fel pop biol, Univ Chicago, 71-72; from asst prof to prof, biol, Fla State Univ, 72-91; PROF, DEPT ZOOL, UNIV CALIF, DAVIS, 91- *Mem:* Ecol Soc Am; Am Soc Naturalists; Soc Study Evolution; Royal Entom Soc. *Res:* Ecology of herbivorous insects; biological control; salt marsh ecology; general topics in population and community ecology. *Mailing Add:* Bodega Marine Lab Univ Calif Box 247 Westside Rd Bodega Bay CA 94923

STRONG, E(RWIN) R(AYFORD), b San Antonio, Tex, Aug 19, 19; m 50; c 3. CHEMICAL ENGINEERING. *Educ:* Tex Arts & Industs Univ, BS, 40; Ill Inst Technol, MS, 45. *Prof Exp:* Engr, Lone Star Gas Co, 40-42; asst, Inst Gas Tech, Ill Inst Technol, 42-45, assoc res engr, 45-48; res engr in chg waste disposal sect, Southwest Res Inst, 48-56; chem engr, Am Oil Co, 56-62, group leader, 62-63, proj mgr, 63-70, mgr process eng, Amoco Deutschland GmbH, 70-73, sr chem engr, Amoco Europe Inc, London, 73-74, proj mgr, 74-84; CONSULT, 84- *Concurrent Pos:* Gas tech fel, Inst Gas Technol, 42-45; mem task forces sulfur & nitrogen oxides control, mem stationary source emissions comt & mem solid waste mgt comt, Am Petrol Inst, 75-84. *Honors & Awards:* Judson S Swearingen Award Sci Res, 54. *Mem:* Am Chem Soc; Sigma Xi; Am Inst Chem Engrs. *Res:* Fuels technology; production of synthesis and distribution gases from coal and oil; industrial wastes; plant steam surveys; toxicity assays; laboratory and pilot plant treatment; superactivated sludge process; petroleum refining and petrochemical processes; desulfurization and hydrotreatment of heavy oils; air pollution control; hydrocarbon processing economics; alternate fuels and energy conversion systems. *Mailing Add:* 1007 Thunderbird Dr Naperville IL 60563

STRONG, FREDERICK CARL, III, b Denver, Colo, Nov 17, 17; m 41; c 2. METHODS DEVELOPMENT IN FOOD ANALYSIS. *Educ:* Swarthmore Col, BA, 39; Lehigh Univ, MS, 41; Bryn Mawr Col, PhD, 54. *Prof Exp:* Chief chemist, Superior Metal Co, 40-42; res chemist, Lea Mfg Co, 42-43; asst, Wesleyan Univ, 43-45; res chemist, Enthone Co, 45; instr chem, Cedar Crest Col, 45-47; asst prof, Villanova Col, 47-51; from asst prof to assoc prof chem & chem eng, Stevens Inst Technol, 51-60; prof chem & chmn dept, Inter-Am Univ PR, 60-63 & Univ Bridgeport, 63-68; vis prof chem, Nat Tsing Hua Univ, Taiwan, 69 & Univ El Salvador, 70-72; tech expert, UN Indust Develop Orgn, Asuncion, Paraguay, 72-73; titular prof, Univ Estadual de Campinas, Brazil, 73-87; RETIRED. *Concurrent Pos:* Ed-in-chief, Appl Spectros, 55-60; Leverhulme fel, Aberdeen Univ, 64-65; Fulbright-Hays lectr, Tribhuvan Univ, Nepal, 68-69; vis prof anal chem, Nat Tsing Hua Univ, Taiwan, 79; tech expert, UN Indust Develop Orgn, Dar-es-Salaam, Tanzania, 81. *Honors & Awards:* Medal, Soc Appl Spectros, 60. *Mem:* Fel Am Inst Chemists; Soc Appl Spectros; Am Chem Soc; Coblentz Soc; Sigma Xi. *Res:* Spectrochemical analysis; food analysis, copper complexes of carbohydrates. *Mailing Add:* 48 S Fourth St Lewisburg PA 17847-1802

STRONG, HERBERT MAXWELL, b Wooster, Ohio, Sept 30, 08; m 35, 83; c 2. HIGH PRESSURE PHYSICS, PHYSICAL OPTICS. *Educ:* Univ Toledo, BS, 30; Ohio State Univ, MS, 32, PhD(physics), 36. *Prof Exp:* Asst, Ohio State Univ, 31-35; res physicist, Bauer & Black Div, Kendall Co, Ill, 35-45 & Kendall Mills Div, Mass, 45-46; res assoc, Res Lab, Gen Elec Co, 46, physicist, Res & Develop Ctr, 46-73; RES ASSOC PHYSICS, UNION COL, NY, 73- *Concurrent Pos:* Consult, Gen Elec Res & Develop Ctr, 73-74; consult technol use of diamond, Lazar Kaplan & Sons, 73- *Honors & Awards:* Award, Soc Mfg Eng, 62; Modern Pioneers Award, Nat Asn Mfrs, 65. *Mem:* AAAS; fel Am Phys Soc; Sigma Xi. *Res:* Technological and industrial applications of diamonds; physical optical studies of rocket motor flames extreme high pressure techniques; measurements and phase equilibria; synthesis of gem diamond; synthesis of industrial diamond; measurement of temperature pressure and gas velocity in rocket motor flames by use of sodium D lines; author of over 40 publications. *Mailing Add:* Dept Physics Sci Union Col Bldg N326 Schenectady NY 12308

STRONG, IAN B, b Cohoes, NY, July 11, 30; m 60; c 1. PHYSICS, ASTRONOMY. *Educ:* Glasgow Univ, BSc, 53; Pa State Univ, PhD(physics), 63. *Prof Exp:* Mem tech staff, Bell Tel Labs, Inc, 53-55; staff mem, Ord Res Lab, 55-57; mem staff, Los Alamos Sci Lab, 61-; RETIRED. *Mem:* AAAS; Am Phys Soc; Am Geophys Union; Am Astron Soc. *Res:* Acoustics, transmission through solids and liquids, ultrasonics; nuclear physics, particle detection, passage of radiation through matter, multiple scattering; astrophysics, interplanetary medium, high energy astronomy; history; philosophy; sociology of science. *Mailing Add:* D 436 Los Alamos Nat Lab PO Box 1663 Los Alamos NM 87545

STRONG, JACK PERRY, b Birmingham, Ala, Apr 27, 28; m 51; c 4. PATHOLOGY. *Educ:* Univ Ala, BS, 48; La State Univ, MD, 51; Am Bd Path, dipl, 57 & 58. *Prof Exp:* Intern, Jefferson Hillman Hosp, Birmingham, Ala, 51-52; asst, 52-53, from instr to assoc prof, 55-64, PROF PATH, SCH MED, LSUMC, NEW ORLEANS, 64-, HEAD DEPT, 66-, BOYD PROF, 80- *Concurrent Pos:* USPHS fel, 57; consult, Southwest Found Res & Educ, 54-55; sabbatical leave, Social Med Res Unit, Med Res Coun, London, Eng, 62-63; mem, path A study sect, USPHS, 65-69, chmn, 67-69; mem, sci adv bd consult, Armed Forces Inst Path, 71-; mem, coun arteriosclerosis, Am Heart Asn; mem epidemiol & biomet adv comt, Nat Heart & Lung Inst, NIH, 71-78 & Panel on Geochemistry of Water in Relation to Cardiovasc Dis, US Nat Comt Geochemistry, Nat Acad Sci, 76-79. *Mem:* Am Asn Path & Bact (asst secy, 59-62); Am Soc Exp Path; Am Soc Clin Path; Col Am Path; US & Can Acad Path (vpres, 77, pres, 78); Int Acad Path (pres, 88-); Sigma Xi. *Res:* Pathology of cardiovascular diseases; atherosclerosis in the human and in primates; epidemiology; geographic pathology and pathogenesis of atherosclerosis; geographic pathology of cancer. *Mailing Add:* Dept of Path La State Univ Med Ctr 1901 Perdido St New Orleans LA 70112-1393

STRONG, JERRY GLENN, b Dawson, NMex, Nov 12, 41; m 71. PESTICIDE CHEMISTRY. *Educ:* Austin Col, BA, 63; Northwestern Univ, PhD(org chem), 68. *Prof Exp:* Sr res chemist, Mobil Chem Co, 68-76, mgr pesticide synthetics, 77-79, mgr pesticide develop, 79-80; sales mgr, 80-83, TECH MGR, ALBRIGHT & WILSON, INC, 84- *Mem:* Am Chem Soc. *Res:* Phosphorus chemicals. *Mailing Add:* 2041 Castle Bridge Rd Midlothian VA 23113

STRONG, JOHN (DONOVAN), b Riverdale, Kans, Jan 15, 05; m 28; c 2. PHYSICS. *Educ:* Univ Kans, AB, 26; Univ Mich, MS, 28, PhD(physics), 30. *Hon Degrees:* DSc, Southwestern at Memphis, 62 & Univ Mass, 81. *Prof Exp:* Instr chem, Univ Kans, 25-27; instr physics, Univ Mich, 27-29, asst investr eng res, 29-30; Nat Res Coun fel physics, Calif Inst Technol, 30-32, fel, Astrophys Observ, 32-37, asst prof physics, 37-42; spec fel, Harvard Univ, 42-45; prof exp physics & dir, Lab Astrophys & Phys Meteorol, Johns Hopkins Univ, 45-67; prof, 67-75, EMER PROF PHYSICS & ASTRON, UNIV MASS, AMHERST, 75- *Concurrent Pos:* Consult, Libbey-Owens-Ford Glass Co, Ohio & Farrand Optic Co, NY. *Honors & Awards:* Longstreth & Levy Medals, Franklin Inst; Ives Medal, Optical Soc Am, 59; Gold Medal, Soc Photo-Optical Instrument Engrs, 77; Hasler Award, Pittsburgh Conf, 81. *Mem:* Fel Am Phys Soc; fel Optical Soc Am (pres, 59); fel Am Acad Arts & Sci; corresp mem Royal Belg Soc Sci; Int Acad Astronaut. *Res:* Experimental physics; evaporation in vacuum; infrared spectroscopy; meteorology; optics; astrophysical observations from high altitudes. *Mailing Add:* 136 Gray St Amherst MA 01002

STRONG, JUDITH ANN, b van Hornesville, NY, June 19, 41. PHYSICAL CHEMISTRY. *Educ:* State Univ NY Albany, BS, 63; Brandeis Univ, MA, 66, PhD(phys chem), 70. *Prof Exp:* Asst prof, 69-73, actg chairperson, 77-78, assoc prof, 73-82, chairperson, 84-86, PROF CHEM, MOORHEAD STATE UNIV, 82-, DEAN SOCIAL & NATURAL SCI, 86- *Concurrent Pos:* Regents fel; NSF fel. *Mem:* Am Chem Soc; Am Inst Chem; Sigma Xi. *Res:* Computer applications in chemical education. *Mailing Add:* Acad Affairs Moorhead State Univ Moorhead MN 56563

STRONG, LAURENCE EDWARD, b Kalamazoo, Mich, Sept 3, 14; m 38; c 4. PHYSICAL CHEMISTRY. *Educ:* Kalamazoo Col, AB, 36; Brown Univ, PhD(chem), 40. *Hon Degrees:* DSc, Earlham Col, 82. *Prof Exp:* Asst phys chem, Harvard Med Sch, 40-41, res assoc, 41-43, assoc dir pilot plant, 43-46; from assoc prof to prof chem, Kalamazoo Col, 46-52; head dept, 52-65, prof chem, 52-79, RES PROF CHEM, EARLHAM COL, 79-; CHEM HYGIENE OFF, 91- *Concurrent Pos:* Dir, UNESCO Pilot Proj, Asia, 65-66; vis prof chem, Macquarie Univ, Australia, 71-72, & Guilford Col, NC, 81 & 85. *Honors & Awards:* SAMA Award for Chem Educ, Am Chem Soc, 71. *Mem:* Fel AAAS; Am Chem Soc; Sigma Xi. *Res:* Electrical properties of solutions; fractionation of proteins; thermodynamics of acid ionization; precision conductance measurements of substituted benzoic acid in aqueous solution and partial molal volume measurements. *Mailing Add:* Dept Chem Earlham Col Richmond IN 47374

STRONG, LOUISE CONNALLY, b San Antonio, Tex, Apr 23, 44; m 70; c 2. MEDICAL GENETICS, CANCER. *Educ:* Univ Tex, Austin, BA, 66; Univ Tex Med Br Galveston, MD, 70. *Prof Exp:* Fel cancer genetics, Univ Tex Grad Sch Biomed Sci & Tex Res Inst Ment Sci, 70-72; res asst, Med Genetics Ctr, Grad Sch Biomed Sci, Univ Tex, Houston, 72-73, asst prof, Health Sci Ctr, 73-78; consult genetics, Dept Med Ped, 73-79, asst prof & asst geneticist, 76-79, ASSOC PROF PEDIAT & BIOL, ASSOC GENETICIST & DIR, MED GENETICS CLIN, UNIV TEX SYST CANCER CTR, 79-, SUE & RADCLIFFE KILLAM PROF, 81-; DIR, MED GENETICS CLINIC, ASSOC GENETICIST & ASSOC PROF PEDIAT & BIOL, 79-, PROF EXP PEDIAT, GENETICIST, 90- *Concurrent Pos:* Consult genetics, Dept Med Pediat, Univ Texas Syst Cancer Ctr, 73-79; Mem adv comt, Clearinghouse Environ Carcinogens, Data Eval & Risk Assessment Subcomt, NIH-Nat Cancer Inst, 76-80 & Nat Comt Cancer Prev, Am Cancer Soc, 78-; mem, Bd Sci Counselors, NCI Div Cancer Etiology, 81-84, mem, Nat Cancer Adv Bd, NCI Dept Health & Human Serv, 84-91; consult ed, Cancer Epidemiology & Prevention, 87; assoc ed, Cancer Res, 83-87. *Honors & Awards:* Sam & Bertha Brochstein Award, Retina Res Found, 83; Marjorie W Margolin Award, Outstanding Achievement, Retina Res, 87; Roland J Sadoux Award, Outstanding Achievement, Wilms' Tumor Res, 90; Farber Lectr, Soc Pediat Path, 91. *Mem:* Am Soc Human Genetics; AAAS; Am Soc Prev Oncology; Am Asn Cancer Res. *Res:* Clinical cancer genetics; genetic etiology and epidemiology of cancer; genetic consequences of childhood cancer. *Mailing Add:* M D Anderson Cancer Ctr Univ Tex 1515 Holcombe Blvd Houston TX 77030

STRONG, MERVYN STUART, b Kells, Ireland, Jan 28, 24; nat US; m 50; c 2. OTOLARYNGOLOGY. *Educ:* Trinity Col, Dublin, BA, 45; Univ Dublin, MD, 47; FRCS(I), 49; FRCS (Eng), 50. *Prof Exp:* Registr otolaryngol, Royal Infirmary, Edinburgh, 49-50; instr, 52-56, PROF OTOLARYNGOL, SCH MED, BOSTON UNIV, 56- *Concurrent Pos:* Fel otolaryngol, Lahey Clin, Boston, 50-52; asst, Boston Univ Hosp, 52-56, chief serv, 56-85; chief otolaryngol, Boston Vet Admin Hosp, 65-85; chief ambulatory care, Bedford Vet Admin Hosp, 85- *Honors & Awards:* Newcombe Award, Am Laryngol Asn, 85. *Mem:* Fel Am Soc Head & Neck Surg; AMA; fel Am Col Surgeons; Soc Univ Otolaryngol (pres, 73-74); fel Am Acad Opthal & Otolaryngol. *Res:* Multicentric origins of carcinoma of oral cavity and pharynx. *Mailing Add:* Vet Admin Med Ctr 11A Bedford MA 01730

STRONG, ROBERT LYMAN, b Hemet, Calif, May 30, 28; m 51; c 4. PHYSICAL CHEMISTRY. *Educ:* Univ Calif, BS, 50; Univ Wis, PhD(chem), 54. *Prof Exp:* Res fel chem, Nat Res Coun Can, 54-55; from asst prof to assoc prof phys chem, 55-62, PROF PHYS CHEM, RENSSELAER POLYTECH INST, 62- *Concurrent Pos:* NSF sci faculty fel, 62-63. *Mem:* AAAS; Am Chem Soc. *Res:* Photochemistry and flash photolysis; atom recombination in gas and solution systems; halogen atom charge-transfer complexes; optical rotary dispersion of excited states and intermediate species in photochemical processes. *Mailing Add:* Dept Chem Rensselaer Polytech Inst Troy NY 12180-3590

STRONG, ROBERT MICHAEL, b Pittsburgh, Pa, Mar 12, 43; c 1. ELECTRICAL ENGINEERING. *Educ:* Villanova Univ, BEE, 65; Mass Inst Technol, MS, 66, PhD(elec eng), 70. *Prof Exp:* Staff mem, Lincoln Lab, Mass Inst Technol, 69-75; lectr, Health Serv, Sch Pub Health, Harvard Univ, 75-80; mem tech staff, Comput Archit Dept, Sperry Res Ctr, 80-83; Consult, 83-90. *Concurrent Pos:* Assoc med, Harvard Med Sch, 71-75. *Mem:* Inst Elec & Electronics Engrs; Asn Comput Mach. *Res:* User-computer interfaces, human factors, computer terminal architecture, data base systems, computer applications in health care research and public health. *Mailing Add:* Nine Hillcrest Rd Medfield MA 02052

STRONG, ROBERT STANLEY, b Sargent, Nebr, May 4, 24; m 50; c 5. ANALYTICAL CHEMISTRY. *Educ:* Cent Wash State Col, BA, 51; Ore State Univ, MS, 57; Univ of the Pac, PhD(org chem), 65. *Prof Exp:* Teacher high schs, Wash, 51-57; instr chem, Columbia Basin Col, 57-64, chmn div sci, 60-64; prof chem, Univ SDak, Springfield, 65-73; asst prof, 73-80, ASSOC PROF CHEM, FITCHBURG STATE COL, 80- *Concurrent Pos:* Dean col, Univ SDak, 67-72, dir instnl res, 72-73. *Mem:* Am Chem Soc. *Res:* D-galactosamine and its derivatives; thin layer chromatography and its applications; chemical instrumentation. *Mailing Add:* S Rd R R No1 Box 35 Ashby MA 01431

STRONG, RONALD DEAN, b Bremerton, Wash, June 7, 36; m 59; c 2. NONDESTRUCTIVE TESTING. *Educ:* Wash State Univ, BS, 58. *Prof Exp:* Engr, Boeing Co, 59-61; nondestructive testing engr, Aerojet-Gen Corp, 61-66; MEM STAFF NONDESTRUCTIVE TESTING ENG, LOS ALAMOS SCI LAB, UNIV CALIF, 66- *Mem:* Am Soc Nondestructive Testing. *Res:* Investigation of ultrasonic techniques for materials evaluations and implementation of techniques in actual test situations. *Mailing Add:* Los Alamos Sci Lab PO Box 1663 Los Alamos NM 87545

STRONG, RUDOLPH GREER, entomology; deceased, see previous edition for last biography

STRONG, WILLIAM J, b Idaho Falls, Idaho, Jan 1, 34; m 59; c 6. ACOUSTICS. *Educ:* Brigham Young Univ, BS, 58, MS, 59; Mass Inst Technol, PhD(physics), 64. *Prof Exp:* From asst prof to assoc prof, 67-76, PROF PHYSICS, BRIGHAM YOUNG UNIV, 76- *Concurrent Pos:* Vis scientist, Gallaudet Col, 74, IRCAM, Paris, 89; sr Fulbright fel, Australia, 80. *Mem:* Fel Acoust Soc Am; Inst Elec & Electronic Engrs. *Res:* Physics of musical instruments; analysis and synthesis of instrumental tones and of speech; machine recognition of speech. *Mailing Add:* Dept Physics Brigham Young Univ Provo UT 84602

STRONGIN, MYRON, b New York, NY, July 27, 36; m 57; c 2. LOW TEMPERATURE PHYSICS, SURFACE PHYSICS. *Educ:* Rensselaer Polytech Inst, BS, 56; Yale Univ, MS, 57, PhD(physics), 62. *Prof Exp:* Mem staff, Lincoln Labs, Mass Inst Technol, 61-63; from asst physicist to assoc physicist, 63-67, physicist, 67-74, SR PHYSICIST, BROOKHAVEN NAT LAB, 74- *Concurrent Pos:* Adj prof, City Univ New York. *Mem:* Fel Am Phys Soc. *Res:* Properties of superconducting materials; analysis of surfaces and influence of surfaces on superconducting properties; epitaxy of films and superconductivity of films; hydrogen on surfaces. *Mailing Add:* Six Canal View Dr Center Moriches NY 11934

STROP, HANS R, b Bandoeng, Dutch East Indies, Oct 26, 31; US citizen. PROCESS DEVELOPMENT FOOD & FEED PRODUCTIONS. *Educ:* Delft Univ Technol, Ir, 58. *Prof Exp:* Res engr, Nuclear Reactor Lab, N V Kema, Holland, 59-60, Mats Res Lab, Tyco Inc, 61-62; sr develop engr, Transitron Electronic Corp, 60-61, RCA Corp, 63-64; systs engr, Gen Atomics Div, Gen Dynamics Corp, 62-63, Monsanto Res Corp, 64-70; dir res, Ibec industs Inc, 70-78, vpres res & eng, Anderson Ibec Div, 78-80; sr vpres mkt & sales, Anderson Int Corp, 80-82; PRES, EPE INC, 82- *Mem:* Am Oil Chemists Soc; Inst Food Technologists; Am Inst Chem Engrs; Am Chem Soc. *Res:* Research and development related to chemical, polymer, food and feed processing equipment and to chemical processes. *Mailing Add:* 12418 The Bluffs Strongsville OH 44136

STROSBERG, ARTHUR MARTIN, b Albany, NY, Sept 16, 40; m 73; c 2. PHARMACOLOGY. *Educ:* Siena Col, NY, BS, 62; Univ Calif, San Francisco, PhD(pharmacol), 70. *Prof Exp:* Pharmacologist, Syntex Res, 70-72, sect head cardiovasc pharmacol, 72-75, prin scientist, 75-83, dept head, 83-90, SR DEPT HEAD, CARDIOVASC PHARMACOL & INST OPERS, SYNTEX RES, 90- *Mem:* AAAS; Am Soc Pharmacol & Exp Therapeut; Sigma Xi; Am Heart Asn. *Res:* Cardiovascular pharmacology, cardiotonic agents, antianginal agents, antihypertensive agents; antiarrhythmic agents; contractile properties of cardiac muscle; cardiac muscle contraction mechanisms, cardiac muscle ultrastructure and oscillations. *Mailing Add:* 120 Lucero Way Portola Valley CA 94028

STROSCIO, MICHAEL ANTHONY, b Winston-Salem, NC, June 1, 49; m 70; c 2. PHYSICS. *Educ:* Univ NC, Chapel Hill, BS, 70; Yale Univ, MPhil, 72, PhD(physics), 74. *Prof Exp:* Physicist space sci, Air Force Cambridge Res Labs, 74-75; physicist staff mem, Los Alamos Sci Lab, 75-78; sr staff physicist, Appl Physics Lab, Johns Hopkins Univ, 78-80; prof mgr electromagnetic res, Air Force Off Sci Res, Wash, 80-83; spec asst to res dir, Off of Under Secy Def, Wash, 82-83; policy analyst, White House Off Sci & Tech Policy, Wash, 83-85; SR RES SCIENTIST, US ARMY RES OFF, RES TRIANGLE PARK, NC, 85- *Concurrent Pos:* Instr, Middlesex Community Col, 74-75; res grant, Los Alamos Sci Lab, 77; chmn, Dept Def Res Instrumentation Comt, Wash, 82; vchmn, White House Panel Sci Commun, Wash, 83-84; liaison, Nat Laser Users Facil, Rochester, Ny, 84; assoc mem, Adv Group Electron Devices, 85-; cons, US Dept Energy, Wash, 85-; adj prof, dept elec & comput eng, NC State Univ, Raleigh, 85-, & dept elec eng & physics, Duke Univ, 86-; panel sci commun & nat security, Nat Acad Sci, 82. *Mem:* Am Phys Soc; sr mem Inst Elec & Electronic Engrs; Nat Geog Soc. *Res:* Hydrodynamics; plasma physics; space physics; atomic physics. *Mailing Add:* Off Dir Army Res Off Box 12211 Research Triangle Park NC 27709

STROSS, FRED HELMUT, b Alexandria, Egypt, Aug 22, 10; nat US; m 36; c 2. ANALYTICAL CHEMISTRY, ARCHAEOLOGY. *Educ:* Case Inst Technol, BS, 34; Univ Calif, PhD(chem), 38. *Prof Exp:* Standard Oil fel, 37-38; Chemist, Shell Develop Co, 38-52, Supvr res, 52-70; GUEST SCIENTIST, LAWRENCE BERKELEY LAB, UNIV CALIF, 75-, RES ASST, DEPT ANTHROP, 70-75,81- *Concurrent Pos:* Consult, Lowie Mus & Univ Art Mus, Berkeley, 70-; chmn Nat Res Coun subcomt gas chromatography group, Int Union Pure & Appl Chem; affil prof chem, Univ Washington, Seattle, 75-88; tour speaker, Am Chem Soc, 73- *Mem:* Am Chem Soc; Archaeol Inst Am; Int Inst Conserv Artistic Works; Sigma Xi. *Res:* Photochemistry; asphalt technology; catalytic industrial processes; physical chemistry of solids; gas chromatography; analytical physical chemistry, including applications to characterization of polymers; archaeometry, (application of physical sciences to archaeology). *Mailing Add:* 44 Oak Dr Orinda CA 94563

STROSS, RAYMOND GEORGE, b St Charles, Mo, July 2, 30; m 64; c 3. PHOTOPERIODISM, HYDROBIOLOGY. *Educ:* Univ Mo, BS, 52; Univ Idaho, MS, 55; Univ Wis, PhD(zool), 58. *Prof Exp:* Res asst, Univ Wis, 54-58; NIH fel, Oceanog Inst, Woods Hole, 58-59; from asst prof to assoc prof zool, Univ Md, 59-67; ASSOC PROF ZOOL, STATE UNIV NY ALBANY, 67- *Concurrent Pos:* Prog officer, NSF, 77-78. *Mem:* AAAS; Sigma Xi; Ecol Soc Am; Am Soc Limnol & Oceanog; Am Soc Photobiol; Phycol Soc Am. *Res:* Photoecology; germination control of plant prodagules; biological limnology; experimental ecology; arctic ecology. *Mailing Add:* Dept Biol Sci State Univ NY Albany NY 12222

STROTHER, ALLEN, b Sweetwater, Tex, Feb 20, 28; m 57; c 2. PHARMACOLOGY, ANIMAL SCIENCE. *Educ:* Tex Tech Col, BS, 55; Univ Calif, Davis, MS, 57; Tex A&M Univ, PhD(biochem, nutrit), 63. *Prof Exp:* Assoc animal sci, Univ Calif, Davis, 56-57; asst to trustee, Burnett Estate, Ft Worth, Tex, 58; dir nutrit res, Uncle Johnny Feed Mills, Houston, 59; res biochemist, Food & Drug Admin, DC, 63-65; from asst prof to assoc prof, 65-75, PROF PHARMACOL, SCH MED, LOMA LINDA UNIV, 75- *Concurrent Pos:* vis scientist, Dept Pharm, Kings Col, Univ London, UK, 86. *Mem:* Am Soc Pharmacol & Exp Therapeut; Sigma Xi. *Res:* Large and small animal nutrition; dietary energy levels; mineral requirements; drug and pesticide metabolism; drug interactions and drug-nutrient interactions. *Mailing Add:* Dept of Pharmacol Loma Linda Univ Sch of Med Loma Linda CA 92354

STROTHER, GREENVILLE KASH, b Huntington, WVa, July 27, 20; m 50; c 3. BIOPHYSICS. *Educ:* Va Polytech Inst, BS, 43; George Washington Univ, MS, 54; Pa State Univ, PhD(physics), 57. *Prof Exp:* Asst prof physics, 57-61, from assoc prof to prof, 61-83, EMER PROF BIOPHYS, PA STATE UNIV, UNIVERSITY PARK, 83- *Mem:* Biophys Soc. *Res:* Microspectrophotometry of cellular systems; biophysical instrumentation. *Mailing Add:* Dept Physics 104 Davey Lab Pa State Univ Univ Park PA 16802

STROTHER, J(OHN) A(LAN), b Hartford, Conn, Dec 27, 27; m 51; c 3. AEROSPACE ELECTRO-OPTICS. *Educ:* Trinity Col, Conn, BS, 50; Princeton Univ, MSE, 54. *Prof Exp:* Electronic scientist, US Navy Underwater Sound Lab, 50-52; mem tech staff, RCA Labs, 54-57, group leader, RCA Defense Electronic Prod Div, RCA, 57-58, unit & prog mgr, RCA Astro-Electronics Div, 58-61; proj engr, Systs Div, EMR, Inc, 61-62, mgr instrumentation eng, Photoelec Div, 62-66; sr mem tech staff, RCA Corp, 66-69, proj mgr, 69-73, mgr electro-optics, 73-75, staff scientist, 75-79, mgr sensor design, Astro-Electronics Div, 79-84; founder/pres, Stron Corp, 84-86; CONSULT, 86- *Res:* Sensing, processing and reproduction of images; speech and hearing; pattern recognition; digital signal processing; satellite systems engineering and program management. *Mailing Add:* 201 Grover Ave Princeton NJ 08540

STROTHER, WAYMAN L, mathematics, for more information see previous edition

STROTTMAN, DANIEL, b Sumner, Iowa, Apr 15, 43; m 66; c 1. NUCLEAR PHYSICS. *Educ:* Univ Iowa, BA, 64; State Univ NY Stony Brook, MA, 66, PhD(physics), 69. *Prof Exp:* Niels Bohr fel nuclear physics, Niels Bohr Inst, Copenhagen, Denmark, 69-70; res officer, Oxford Univ, 70-73; asst prof physics, State Univ NY Stony Brook, 74-78; staff mem, 78-87, GROUP LEADER, THEORY DIV, LOS ALAMOS NAT LAB, 87- *Concurrent Pos:* Vis Nordita prof, Physics Inst, Univ Oslo, 73-74. *Mem:* Fel Am Phys Soc; AAAS. *Res:* Group theory applications; theoretical nuclear physics. *Mailing Add:* Theoretical Div Los Alamos Nat Lab MS B243 Los Alamos NM 87544

STROUBE, EDWARD W, b Hopkinsville, Ky, Apr 2, 27; m 54; c 3. AGRONOMY. *Educ:* Univ Ky, BS, 51, MS, 59; Ohio State Univ, PhD(agron), 61. *Prof Exp:* Agr exten agent, Univ Ky, 54-57, res asst agron, 57-58; res asst, 58-60, from instr to assoc prof, 60-70, PROF AGRON, OHIO STATE UNIV & OHIO AGR RES & DEVELOP CTR, 70- *Mem:* Am Soc Agron; Weed Sci Soc Am. *Res:* Weed control of field crops involving the evaluations of herbicides, tillage practices, flaming and crop rotations; soil and crop residue studies involving herbicides. *Mailing Add:* 2688 Mt Holyoke Rd Columbus OH 43221

STROUBE, WILLIAM BRYAN, JR, b Princeton, Ky, Oct 29, 51; m 73; c 2. PHARMACEUTICAL & NUTRITIONAL TECHNICAL MANAGEMENT. *Educ:* Murray State Univ, BS, 73; Univ Md, MBA, 86; Univ Ky, PhD(anal & nuclear chem), 77. *Prof Exp:* Anal chemist, Allied Gen Nuclear Serv, 78; res staff fel, US Food & Drug Admin, 78-79, res chemist, 79-86; GROUP LEADER, PHARM QA, BRISTOL-MYERS SQUIBB CO, 86- *Mem:* Am Nuclear Soc; Am Chem Soc; Asn Off Anal Chemists. *Res:* Pharmaceutical analysis; multi-element analysis methodology for neutron activation analysis; trace analysis of biological samples. *Mailing Add:* Bristol-Myers Squibb Co 2400 W Lloyd Expressway Evansville IN 47721-0001

STROUBE, WILLIAM HUGH, plant science; deceased, see previous edition for last biography

STROUD, CARLOS RAY, b Owensboro, Ky, July 9, 42; m 62; c 3. QUANTUM OPTICS. *Educ:* Centre Col Ky, AB, 63; Wash Univ, PhD(physics), 69. *Prof Exp:* From asst prof to assoc prof, 70-84, PROF OPTICS, UNIV ROCHESTER, 84- *Concurrent Pos:* Sr vis scientist, Univ Sussex, Gt Brit, 79-80. *Mem:* fel Am Phys Soc; fel Optical Soc Am. *Res:* quantum and semiclassical radiation theory; interactions of electromagnetic fields with matter; laser instabilities; Rydberg atomic states. *Mailing Add:* Dept of Optics Univ of Rochester Rochester NY 14627

STROUD, DAVID GORDON, b Glasgow, Scotland, July, 6, 43; US citizen; m 67; c 2. SUPERCONDUCTIVITY, DISORDERED SYSTEMS. *Educ:* Stanford Univ, BS, 64; Harvard Univ, MA, 66, PhD(solid state physics), 69. *Prof Exp:* Postdoctoral researcher, Cornell Univ, 69-71; from asst prof to assoc prof, 71-81, PROF PHYSICS, OHIO STATE UNIV, 81- *Concurrent Pos:* Prin investr, NSF, 72-; vis fel, Harvard Univ, 77-78; vis prof, Tel Aviv Univ, 80 & Univ Paris, 85. *Mem:* Am Phys Soc; Mat Res Soc; Am Ceramic Soc. *Res:* Theoretical condensed matter physics; transport and optical properties of granular matter; superconductivity in granular systems; glassy behavior in high-temperature superconductors; liquid-vapor and liquid-solid surfaces. *Mailing Add:* Dept Physics Ohio State Univ Columbus OH 43210

STROUD, JACKSON SWAVELY, b Cabarrus Co, NC, June 1, 31; m 61; c 2. EXPERIMENTAL SOLID STATE PHYSICS, ENGINEERING. *Educ:* Union Col, BS, 53; Ohio State Univ, MS, 57. *Prof Exp:* Physicist, Corning Glass Works, 57-67 & Bausch & Lomb, Inc, 67-79; asst mgr, 79-83, SCIENTIST, SCHOTT GLASS TECH, 83- *Mem:* Am Phys Soc; Am Ceramic Soc; Optical Soc Am. *Res:* Solid state physics with specialized knowledge of glass; radiation chemistry; glass tank design; optical properties of solids. *Mailing Add:* Schott Glass Technol 400 York Ave Duryea PA 18642

STROUD, JUNIUS BRUTUS, b Greensboro, NC, June 9, 29; m 55; c 3. ALGEBRA. *Educ:* Davidson Col, BS, 51; Univ Va, MA, 62, PhD(math), 65. *Prof Exp:* Instr math & sci, Fishburne Mil Sch, 53-57; teacher, High Sch, 57-58; from instr to prof math, 60-76, chmn dept, 83-89, RICHARDSON PROF MATH, DAVIDSON COL, 85- *Concurrent Pos:* Vis lectr, Sec Schs, 62-63, 65-66, 89-90 & 90-91; vis prof, Dartmouth Col, 73, St Andrews Univ, Scotland, 87. *Mem:* Am Math Asn. *Res:* Simple Jordan algebras of characteristic two; finitely generated modules over a Dedekind ring. *Mailing Add:* Dept of Math Davidson Col Davidson NC 28036

STROUD, MALCOLM HERBERT, b Birmingham, Eng, May 17, 20; m 49; c 3. MEDICINE. *Educ:* Univ Birmingham, MB, ChB, 45; FRCS, 52; Am Bd Otolaryngol, dipl, 60. *Prof Exp:* From asst prof to assoc prof, 57-72, PROF MED, SCH MED, WASH UNIV, 72- *Mem:* Am Acad Ophthal & Otolaryngol. *Res:* Otology. *Mailing Add:* 517 S Euclid St Louis MO 63110

STROUD, RICHARD HAMILTON, b Dedham, Mass, Apr 24, 18; m 43; c 2. ZOOLOGY. *Educ:* Bowdoin Col, BS, 39; Univ NH, MS, 42. *Prof Exp:* Asst bot, Bowdoin Col, 39; asst zool, Univ NH, 40-42; jr aquatic biologist, Tenn Valley Authority, 42, aquatic biologist, 46-47; chief aquatic biologist, Mass Dept Conserv, 48-53; asst exec vpres, Sport Fishing I, Aquatic Resources, 53-55, exec vpres, 55-81, consult, 82-88, FISHERIES SCI ED, AQUATIC RESOURCES, 82- *Concurrent Pos:* Consult, Calif Fish & Game Dept, 65-66, Ark Game & Fish Comn, 69, Iowa Conserv Comn, 70-71 & Tenn Valley Authority, 71-72; vpres, Sport Fishery Res Found, 62-81, trustee, 82-88; mem, World Panel Fishery Experts, Food & Agr Orgn, UN, Ocean Fisheries & Law of Sea Adv Comts, Dept State; chmn, Natural Resources Coun Am, 69-71; mem, NAm Atlantic Salmon Coun & Marine Fisheries Adv Comt, Dept Com; fishery expert adv to Sen Select Comt Govt Opers; bd dir, Nat Coalition Marine Conserv, 75-90; sr science, Aquatic Ecosysts Analysts, 83-88; guest lectr, Japan Sport Fishing Found, 76. *Honors & Awards:* Pentelow lectr, Univ Liverpool, Eng, 75; Conserv Achievement Award, Nat

Wildlife Fedn, 76; Outstanding Achievement Award, Am Inst Fishery Res Biol, 81 & Am Fisheries Soc, 90. *Mem:* Am Fisheries Soc (pres, 79-80); Fisheries Soc Brit Isles; Freshwater Biol Asn UK; Am Inst Fishery Res Biologists; Int Asn Fish & Wildlife Agencies; hon mem Nat Resources Coun Am. *Res:* Fish population dynamics, behavior, ecology and life history. *Mailing Add:* Consult Aquatic Resources & Fisheries Sci Ed PO Box 1772 Pinehurst NC 28374

STROUD, RICHARD KIM, b Ann Arbor, Mich, Aug 8, 43; div; c 3. VETERINARY PATHOLOGY. *Educ:* Ore State Univ, BS, 66, MS, 78; Wash State Univ, DVM, 72. *Prof Exp:* Biologist, Marine Mammal Lab, Nat Marine Fisheries Serv-Nat Ocean & Atmospheric Admin, 66-68; vet, Willamette Vet Clin, 72-73; res assoc aquatic animal path, Ore State Univ, 73-79; diag pathologist, Nat Wildlife Health Lab, Madison, Wis, 80- 86; SR FORENSIC SCIENTIST, FISH & WILDLIFE FORENSICS LAB, FISH & WILDLIFE SERV, ASHLAND, ORE, 90- *Mem:* Inst Asn Aquatic Animal Med (pres, 77-78); Wildlife Dis Asn; Am Fisheries Soc; Am Vet Med Asn. *Res:* Wildlife and aquatic animal pathology. *Mailing Add:* 1490 E Main Ashland OR 97520

STROUD, ROBERT CHURCH, b Oakland, Calif, Jan 5, 18; m 47. PHYSIOLOGY. *Educ:* Princeton Univ, AB, 40; Univ Rochester, MS, 50, PhD(physiol), 52. *Prof Exp:* Chemist, Calco Chem Div, Am Cyanamid Co, 40-44 & Lederle Labs Div, 44-48; instr physiol, Grad Sch Med, Univ Pa, 52-53; assoc med physiol, Brookhaven Nat Lab, 53-54; asst prof pharmacol & res assoc aviation physiol, Ohio State Univ, 54-55; asst prof physiol, Heart 55-56; supvr physiologist, US Naval Med Res Lab, 56-61; pulmonary physiologist, Occup Health Res & Training Facil, USPHS, 61-62; chief res prog mgr life sci, Ames Res Ctr, NASA, 62-64; chief, Sci Rev Sect, Health Res Facil Br, NIH, 64-69, chief, Health Res Facil Br, Div Educ & Res Facil, 69-70, chief, Training Grants & Awards Br, Nat Heart, Lung & Blood Inst, 70-73, exec secy, Rev Br, 73-88; RETIRED. *Concurrent Pos:* Lectr, Stanford Univ, 63-64. *Honors & Awards:* Lederle Med Fac Award, 55. *Mem:* Am Physiol Soc. *Res:* Cardiopulmonary and respiratory physiology; physiology of adaptation to high altitudes and submarine environments; physiology of diving; aerospace physiology. *Mailing Add:* 4450 S Park Ave Apt 1719 Chevy Chase MD 20815-3646

STROUD, ROBERT MALONE, b St Louis, Mo, Mar 12, 31; m 55; c 2. IMMUNOLOGY. *Educ:* Harvard Univ, BA, 52, MD, 56. *Prof Exp:* Intern med, Cook County Hosp, Chicago, Ill, 56-57; resident, Barnes Hosp, St Louis, Mo, 59-61; USPHS fel, Johns Hopkins Univ Sch Med, 61-63; Heley Hays Whitney fel, 63-65; dir rheumatology, Ga Warm Springs Found, 65-66; from asst prof to assoc prof med, Univ Ala, Brimingham, 66-71, assoc prof microbiol & prof med, Med Sch, 71-81; PVT PRACT, 81- *Mem:* Am Asn Immunol; Am Col Rheumatology; Am Soc Clin Invest; Am Acad Allergy. *Res:* Food allergy as a cause of arthritis. *Mailing Add:* 570 Memorial Circle Ormond Beach FL 32174

STROUD, ROBERT MICHAEL, b Stockport, Eng, May 24, 42. STRUCTURAL BIOLOGY. *Educ:* Cambridge Univ, BA, 64, MA, 68; London Univ, PhD(crystallog), 68. *Prof Exp:* Fel protein crystallog, Calif Inst Technol, 68-71, assoc prof chem, 75-77; PROF BIOCHEM, UNIV CALIF, SAN FRANCISCO, 77- *Concurrent Pos:* Prin investr, NIH & NSF, 71-; consult, NIH, 77-; fel, Sloan Found, 77. *Mem:* Am Crystallog Soc; Brit Biophys Soc; Am Soc Biophys Chem; Biophys Soc. *Res:* Membrane protein structure; structure and function of complex regulatory macromolecular DNA and protein interactions; development of new methodology in macromolecular structural biology. *Mailing Add:* 570 Memorial Circle Ormond Beach FL 32174

STROUD, ROBERT WAYNE, b Jonesboro, Ark, May 24, 29; m 57; c 2. TEXTILE CHEMISTRY. *Educ:* Ark Col, BS, 50; Ga Inst Technol, MSCh, 54; Univ Tex, Austin, PhD(org chem), 63. *Prof Exp:* Teacher, Pub Schs, Ark, 49-50; chemist, Carbide & Carbon Chem Co, Tex, 53-54 & 56-58; res chemist, 62-72, SR RES CHEMIST, E I DU PONT DE NEMOURS & CO, INC, 72- *Mem:* Am Chem Soc. *Res:* Textile fibers chemistry and engineering. *Mailing Add:* 188 S Crest Rd Chattanooga TN 37404-5517

STROUD, THOMAS WILLIAM FELIX, b Toronto, Ont, Apr 7, 36; m 83; c 4. STATISTICS. *Educ:* Univ Toronto, BA, 56, MA, 60; Stanford Univ, PhD(statist), 68. *Prof Exp:* Asst prof math, Acadia Univ, 60-64; asst prof, 68-75, ASSOC PROF MATH & STATIST, QUEEN'S UNIV, ONT, 75- *Concurrent Pos:* Res asst, Sch Educ, Stanford Univ, 66-68; Nat Res Coun Can res grant, Queen's Univ, Ont, 69-; Educ Testing Serv vis res fel, Educ Testing Serv, 72-73; invited prof, Fed Polytech Sch Lausanne, 77-78; vis prof, Univ Col Wales, Aberystwyth, 89-90. *Mem:* Am Statist Asn; Statist Soc Can; Inst Math Statist. *Res:* Multivariate analysis; Bayesian inference; statistical analysis of mental test data; generalized linear models; small area estimation; sample survey analysis. *Mailing Add:* Dept Math & Statist Queen's Univ Kingston ON K7L 3N6 Can

STROUGH, ROBERT I(RVING), b Akron, Ohio, June 22, 20; m 45; c 2. PHYSICS, ENGINEERING. *Educ:* Case Western Reserve Univ, BS, 42, MS, 48, PhD(physics), 50. *Prof Exp:* Proj engr, Airborne Instrument Lab, 42-44; develop engr, Arma Corp, 44-46; asst proj engr, Pratt & Whitney Aircraft Div, United Aircraft Corp, 50-51, reactor proj engr, 51-60, develop engr, 60-62, chief develop eng, 62-63, prog mgr SNAP-50 nuclear elec spacer powerplant proj, 63-65, chief adv concepts develop, 65-69, mgr advan mil progs, 69-76; ENG MGR ADVAN PROD, UNITED TECH CORP, 76- *Concurrent Pos:* Adj prof, Hartford Grad Ctr, Rensselaer Polytechnic Inst, 55-71; assoc dir aircraft nuclear propulsion proj, Oak Ridge Nat Lab, 54-55; mem res adv comt nuclear energy systs, NASA, 61-62. *Mem:* Am Phys Soc; Inst Elec & Electronics Engrs; Am Inst Aeronaut & Astronaut. *Res:* Management and technical direction of nuclear aircraft and spacecraft propulsion and power systems development including reactor, liquid metal and power conversion systems; engineering and design of advanced aircraft powerplants; aerothermodynamics; development of electrical machinery and electromagnetic devices. *Mailing Add:* 55 Ledgewood Dr Glastonbury CT 06033

STROUP, CYNTHIA ROXANE, b Norfolk, Va, Sept 21, 48. EXPOSURE ASSESSMENT SURVEYS. *Educ:* Longwood Col, BS, 71; Georgetown Univ, MS, 81. *Prof Exp:* Teacher biol, Arlington County pub schs, 72-74; BIOSTATISTICIAN & CHIEF, STATIST DESIGN & OPERS SECT, ENVIRON PROTECTION AGENCY, 76- *Honors & Awards:* Silver Medal, Environ Protection Agency. *Mem:* Am Pub Health Asn; Soc Occup & Environ Health; Am Asn Univ Women. *Res:* Design and analysis of human and environmental exposure studies on selected chemicals, such as formaldehyde, asbestos and dichloromethane; statistical design; quality assurance; sampling; data analysis; environmental monitoring. *Mailing Add:* 29 N Garfield St Arlington VA 22201

STROUP, CYNTHIA ROXANE, b Norfolk, Va, Sept 21, 48. CHEMICAL EXPOSURES, ENVIRONMENTAL ASSESSMENT. *Educ:* Longwood Col, BS, 71; Georgetown Univ, MS, 81. *Prof Exp:* Teacher biol, Arlington Pub Schs, 72-74; BR CHIEF STATIST, US ENVIRON PROTECTION AGENCY, 76- *Mem:* Am Pub Health Asn. *Res:* Lead-based paint abatement efficacy; asbestos exposures and resultant risks. *Mailing Add:* US Environ Protection Agency 401 M St SW TS-798 Washington DC 20460

STROUS, GER J, b Haelen, Aug 13, 44; Dutch citizen; m 70; c 3. BIOCHEMISTRY. *Educ:* Univ Nymegen, Neth, MSc, 69, PhD(biochem), 73. *Prof Exp:* Res fel biochem, Univ Nymegen, 70-73; postdoctoral fel cell biol, Dutch Res Coun, 74-78; fac mem, 78-85, lectr, 85-91, PROF & DEPT HEAD CELL BIOL, UNIV UTRECHT, 91- *Concurrent Pos:* Vis scientist biol, Mass Inst Technol, Cambridge, Mass, 79-80, Harvard Med Sch Childrens Hosp, Boston, Mass, 80-86 & Sch Med, Wash Univ, St Louis, 86-; consult, Physiol Rev Comt, Dutch Res Coun, 86- *Mem:* Dutch Soc Cell Biol; Dutch Soc Biochem; Dutch Soc Glycoconjugates; Am Soc Cell Biol; Brit Biochem Soc. *Res:* Sorting mechanisms of membrane glyco-protein of the Golgi complex and the endosomal system; structure and function of mucins. *Mailing Add:* Dept Cell Biol Univ Utrecht AZU-H02-314 Utrecht 3584CX Netherlands

STROUSE, CHARLES EARL, b Ann Arbor, Mich, Jan 29, 44; m 72; c 2. CHEMISTRY. *Educ:* Pa State Univ, University Park, BS, 65; Univ Wis-Madison, PhD(phys chem), 69. *Prof Exp:* AEC fel, Los Alamos Sci Lab, 69-71; from asst prof to assoc prof, 71-84, PROF CHEM, UNIV CALIF, LOS ANGELES, 84- *Mem:* AAAS; Am Chem Soc; Am Crystallog Asn. *Res:* Structural chemistry. *Mailing Add:* Dept Chem Univ Calif 1034 Young Hall Los Angeles CA 90024-1569

STROUT, RICHARD GOOLD, b Auburn, Maine, Nov 11, 27; m 50; c 2. ZOOLOGY, PARASITOLOGY. *Educ:* Univ Maine, BS, 50; Univ NH, MS, 54, PhD(parasitol), 61. *Prof Exp:* Instr poultry sci, 54-60, from asst prof to assoc prof, 60- 68, PROF PARASITOL, UNIV NH, 68-, PARASITOLOGIST, 63- *Concurrent Pos:* Fel, Sch Med, La State Univ, 67. *Mem:* Sigma Xi; Wildlife Dis Asn; Am Soc Parasitol. *Res:* In vitro culture and pathogenicity of avian coccidiosis; immunity mechanisms; blood parasites of birds and fishes. *Mailing Add:* Dept Animal Sci Rm 404 Kendall Hall Univ of NH Durham NH 03824

STROVINK, MARK WILLIAM, b Santa Monica, Calif, July 22, 44; div; c 2. EXPERIMENTAL HIGH ENERGY PHYSICS. *Educ:* Mass Inst Technol, BS, 65; Princeton Univ, PhD(physics), 70. *Prof Exp:* From instr to asst prof physics, Princeton Univ, 70-73; from asst prof to assoc prof, 73-80, PROF PHYSICS, UNIV CALIF, BERKELEY, 80- *Concurrent Pos:* Vis asst prof physics, Cornell Univ, 71-72; mem, High-Energy Physics Adv Panel, Dept Energy, 88- *Mem:* Fel Am Phys Soc. *Res:* Searches for right-handed charged currents; muon interactions and charm production at high energy; principles of invariance to changes of energy scale and charge-parity inversion; collider detector instrumentation. *Mailing Add:* Lawrence Berkeley Lab 50-341 Univ Calif Berkeley CA 94720

STROYNOWSKI, IWONA T, b Bydgoszcz, Poland, Aug 8, 50; US citizen; m 70; c 2. CELL-CELL INTERACTIONS IMMUNE SYSTEM, MAJOR HISTOCOMPATABILITY COMPLEX ANTIGENS. *Educ:* Univ Geneva, Switz, BSc & MSc, 75; Stanford Univ, PhD(genetics), 79. *Prof Exp:* Postdoctoral fel biol, Stanford Univ, 79-81; from res fel to sr res fel, 81-88, SR RES ASSOC BIOL, CALIF INST TECHNOL, 88- *Concurrent Pos:* mem, NIH Immunol, Virol & Pathol Study Sect, 90-94. *Mem:* Am Asn Immunol; AAAS. *Res:* Cell mediated self-nonself recognition in the immune system of mice; Isn I transplantation antigens; non-ismcol MHC antigens; alternative splicing; T cell activation; regulation of expression of MMC genes. *Mailing Add:* Div Biol 147-75 Calif Inst Technol Pasadena CA 91125

STROZIER, JAMES KINARD, b Rock Hill, SC, May 21, 33; m 56; c 2. AEROSPACE ENGINEERING, MECHANICAL ENGINEERING. *Educ:* US Mil Acad, BS, 56; Univ Mich, Ann Arbor, MSE, 66, PhD(aerospace eng), 66; Long Island Univ, MBA, 84. *Prof Exp:* Instr mech, US Mil Acad, US Army, 64-66, asst prof, 66-67, exec officer, 1st Battalion, 92nd Artil, Vietnam, 68, systs analyst, First Field Force, 68-69, assoc prof mech, US Mil Acad, 70-82, prof aerospace eng, 82-84; RES PROF, UNIV UTAH, 84- *Mem:* Am Inst Aeronaut & Astronaut; Am Soc Eng Educ; Am Soc Mech Engrs; Soc Automotive Engrs. *Res:* Telemetry; wing design; adhesive testing; education methods. *Mailing Add:* Dept Mech & Indust Eng MEB 3209 3210 Merril Eng Univ Utah Salt Lake City UT 84112

STROZIER, JOHN ALLEN, JR, b Miami, Fla, June 3, 34; m 62; c 3. SURFACE PHYSICS. *Educ:* Cornell Univ, BEP, 58; Univ Utah, PhD(physics), 66. *Prof Exp:* Instr physics, Univ Utah, 66-67; res assoc mat sci, Cornell Univ, 67-69; sr res assoc, State Univ NY Stony Brook, 69-71, asst prof, 71-74; physicist, Brookhaven Nat Lab, 74-80. *Mem:* Am Phys Soc. *Res:* Surface physics; low energy electron diffraction, catalysis. *Mailing Add:* Dept Physics 2110 Lab Off Bldg SUNY Empire State Col PO Box 130 Trainor House Stonybrook NY 11794

STRUB, MIKE ROBERT, b Alliance, Ohio, Dec 26, 48; m 69. FOREST BIOMETRY, STATISTICS. *Educ:* Va Polytech Inst & State Univ, BS, 71, MS, 72, PhD(statist), 77. *Prof Exp:* Instr forestry, Va Polytech Inst & State Univ, 72-74; asst prof forest mgt, Pa State Univ, 75-77; FOREST BIOMETRICIAN, WEYERHAEUSER CO, 77- *Mem:* Am Statist Asn; Biometrics Soc; Soc Am Foresters. *Res:* Modeling growth and yield of forest stands, especially loblolly pine plantations; general qualitative model and applications to forestry. *Mailing Add:* Lonsdale Cut-Off-Rd Hot Springs AR 71901

STRUBLE, CRAIG BRUCE, b Mt Pleasant, Mich, Oct, 30, 50; m 73; c 2. AGRICULTURAL CHEMICAL METABOLISM. *Educ:* Jamestown Col, BS, 73; ND State Univ, PhD(zool), 79. *Prof Exp:* Res assoc, Dept Path, Univ Wis-Madison & Wis Regional Primate Ctr, 78-79; res assoc, Dept Animal Sci, ND State Univ, 79-83; RES PHYSIOLOGIST, USDA, ARS, 83- *Mem:* Sigma Xi; Int Soc Study Xenobiotics; Am Chem Soc. *Res:* Metabolism, biliary secretion, and enterohepatic circulation of xenobiotics and agricultural chemicals in laboratory and farm animals; experimental surgery. *Mailing Add:* Biosci Res Lab PO Box 5674 SU Sta Fargo ND 58105

STRUBLE, DEAN L, b Wawota, Sask, Aug 29, 36; m 57; c 3. INSECT SEX PHEOMONES, MANAGEMENT OF AGRICULTURAL RESEARCH. *Educ:* Univ Sask, BA, 61, MA, 62, PhD(chem), 66. *Prof Exp:* Develop chemist, DuPont Can, Ont, 62-63; postdoctoral fel org chem, Univ Adelaide, Australia, 66-67; res scientist, Lethbridge, 68-89, DIR, AGR CAN VANCOUVER RES STA, 89- *Concurrent Pos:* Adj prof, Simon Fraser Univ, Burnaby, BC, 87. *Mem:* Fel Entom Soc Can; Chem Inst Can. *Res:* Identification and synthesis of sex pheromones of lepidopterans, primarily cutworm species; development of practical application of synthetic pheromones for monitoring abundance of major cutworm species; author of 69 publications; holder of four patents. *Mailing Add:* Agr Can Research Sta, Vancouver 6660 NW Marine Dr Vancouver BC V6T 1X2

STRUBLE, GEORGE W, b Philadelphia, Pa, July 6, 32; m 55; c 3. COMPUTER SCIENCE. *Educ:* Swarthmore Col, AB, 54; Univ Wis, MS, 57, PhD(math), 61. *Prof Exp:* Proj supvr, Numerical Anal Lab, Univ Wis, 60-61; from asst prof to assoc prof math, Univ Ore, 61-69, res assoc, 61-65, assoc dir statist lab & comput ctr, 65-69, dir comput ctr, 69-74, assoc prof comput sci, 69-82; PROF COMPUT SCI, WILLAMETTE UNIV, 82- *Concurrent Pos:* Consult, Comput Mgt Serv, Inc, Portland, Ore, 74-78. *Mem:* Asn Comput Mach. *Res:* Algorithms. *Mailing Add:* Dept of Computer Sci Willamette Univ Salem OR 97301

STRUBLE, GORDON LEE, b Cleveland, Ohio, Mar 7, 37; m 61; c 4. NUCLEAR CHEMISTRY. *Educ:* Rollins Col, BS, 60; Fla State Univ, PhD(chem), 64. *Prof Exp:* Fel, Lawrence Berkeley Lab, Univ Calif, 64-66; asst prof chem, Univ Calif, Berkeley, 66-71; staff chemist, 71-75, sect leader, 75-79, assoc div leader, 80-84, group leader, 84-89, DEP ASST TO DIR, LAWRENCE LIVERMORE NAT LAB, 89- *Concurrent Pos:* Prof physics, Univ Munich, 75. *Mem:* AAAS; Am Chem Soc; Am Phys Soc. *Res:* Experimental and theoretical low energy nuclear structure and reaction physics, determination of characteristics of low energy excitations in nuclei by nuclear reactions and decay processes and their description by theoretical many body techniques; high energy heavy-ion reactions. *Mailing Add:* Mail Stop L-233 Lawerence Livermore Nat Lab PO Box 808 Livermore CA 94550

STRUBLE, RAIMOND ALDRICH, b Forest Lake, Minn, Dec 10, 24; m 46; c 5. MATHEMATICS. *Educ:* Univ Notre Dame, PhD(math), 51. *Prof Exp:* Aerodynamicist, Douglas Aircraft Co, 51-53; asst prof math, Ill Inst Technol, 53-56; assoc prof, 56-60, PROF MATH, NC STATE UNIV, 60- *Concurrent Pos:* Consult, Armour Res Found, 54-60. *Res:* Fourier analysis; almost periodic functions; nonlinear differential equations; applied mathematics. *Mailing Add:* PO Box 50376 Raleigh NC 27650

STRUCHTEMEYER, ROLAND AUGUST, b Wright City, Mo, Jan 4, 18; m 40; c 2. SOILS. *Educ:* Univ Mo, BS, 39, MA, 41; Ohio State Univ, PhD(agron), 52. *Prof Exp:* Asst soils, Univ Mo, 39-40 & Ohio State Univ, 40-42; explosive chemist, Certainteed Corp, Tex, 42-43; head dept soils, 46-71, prof, 46-83, EMER PROF SOILS, UNIV MAINE, ORONO, 46- *Concurrent Pos:* Agron fel, 65; soil fertility specialist, IRI Res Inst, Brazil, 66; private consult, 74- *Mem:* Fel Am Soc Agron; Soil Sci Soc Am; Int Soc Soil Sci; Soil Conserv Soc Am. *Res:* Plant nutrition. *Mailing Add:* 8801 Emerald Lake Dr W Pinson AL 35126-2323

STRUCK, ROBERT FREDERICK, b Pensacola, Fla, Jan 9, 32; m 63; c 2. PHARMACOLOGY, DRUG METABOLISM. *Educ:* Auburn Univ, BS, 53, MS, 57, PhD(org chem), 61. *Prof Exp:* Assoc scientist, Southern Res Inst, 57-58; org chemist, Fruit & Veg Prod Lab, Agr Res Serv, USDA, 61; res scientist, 61-64, sr scientist, 64-80, head, Metabol Sect, 80-87, HEAD, BIOL CHEM DIV, SOUTHERN RES INST, 88 - *Concurrent Pos:* Mem, Exp Therapeut Study Sect, NIH, 83-86. *Mem:* Am Asn Cancer Res; Am Soc Pharmacol Exp Therapeut. *Res:* Metabolism, pharmacology, mechanism of action and medicinal chemistry of anticancer drugs; organophosphorus and organic heterocyclic chemistry. *Mailing Add:* Southern Res Inst PO Box 55305 Birmingham AL 35255-5305

STRUCK, ROBERT T(HEODORE), b Harrisburg, Pa, Apr 26, 21; wid; c 2. CHEMICAL ENGINEERING. *Educ:* Pa State Univ, BS, 42, MS, 46, PhD(fuel tech), 49. *Prof Exp:* Res asst petrol refining, Pa State Univ, 42-46, res asst fuel technol, 46-48; chem engr, Consol Coal Co, 48-74; mgr process develop, 74-81, MGR RES ADMIN, CONOCO COAL DEVELOP CO, 81- *Mem:* Am Chem Soc; Am Inst Chem Engrs. *Res:* Diffusional processes in hydrocarbon processing; developing processes for converting coal to other fuels; air pollution control processes; development of new routes to metallurgical coke. *Mailing Add:* 2347 Morton Rd Pittsburgh PA 15241-3301

STRUCK, WILLIAM ANTHONY, b Paterson, NJ, Mar 17, 20; m 43; c 3. ANALYTICAL CHEMISTRY. *Educ:* Calvin Col, AB, 40; Univ Mich, MS, 62, PhD, 63. *Prof Exp:* Microanalyst, Upjohn Co, 41-48, head chem res anal, 48-62, mgr phys & anal chem res, 62-68, from asst dir to dir supportive res, 68-74, sr dir, Int Pharmaceut Res & Develop, 74-79, sr dir, Pharmaceut Res & Develop, 79-83; RETIRED. *Mem:* AAAS; Am Chem Soc. *Res:* Organic electrochemistry; organic analysis; optical rotatory dispersion. *Mailing Add:* 2102 Waite Ave Kalamazoo MI 49008

STRUCK-MARCELL, CURTIS JOHN, b Minneapolis, Minn, May 19, 54; m 74; c 2. EXTRAGALACTIC ASTRONOMY, STAR FORMATION. *Educ:* Univ Minn, BS(physics) & BS(math), 76; Yale Univ, MPhil, 79, PhD(astron), 81. *Prof Exp:* Res fel, McDonald Observ, Univ Tex, 81-83; asst prof, 83-89, ASSOC PROF ASTROPHYS, IOWA STATE UNIV, 89- *Mem:* Am Astron Soc; Int Astron Union. *Res:* Theoretical studies of galaxy formation from cosmological density perturbations; computer modeling of large-scale gas dynamics and star formation in galaxies; tidal interactions between galaxies. *Mailing Add:* Dept Physics & Astron Iowa State Univ Ames IA 50011

STRUEMPLER, ARTHUR W, b Lexington, Nebr, Dec 12, 20; m 50; c 2. ANALYTICAL CHEMISTRY. *Educ:* Univ Nebr, BS, 50, MS, 55; Iowa State Univ, PhD, 57. *Prof Exp:* Asst prof, Chico State Col, 57-60; fel biochem, Univ Calif, Davis, 60-62; opers analyst, Strategic Air Command Hq, Nebr, 62-65; head div, 65-81, chmn, Dept Sci & Math, 76-81, PROF CHEM, CHADRON STATE COL, 76- *Concurrent Pos:* Finder, Bayard Meteorite. *Mem:* Am Chem Soc; Sigma Xi; AAAS; Meteorics Soc; Soc Vert Paleontol. *Res:* Weather modification studies relating to element concentrations in precipitation; geochemical studies related to uranium, radium, and radon. *Mailing Add:* Dept Chem Chadron State Col Chadron NE 69337

STRUHL, KEVIN, b New York, NY, Sept 2, 52. MOLECULAR BIOLOGY. *Educ:* Mass Inst Technol, BS, 74, MS, 74; Stanford Univ, PhD(biochem), 79. *Prof Exp:* Asst prof, 82-85, ASSOC PROF BIOCHEM, HARVARD MED SCH, 86- *Concurrent Pos:* Searle scholar, Chicago Community Trust, 83. *Res:* Regulation of eukaryotic gene expression; protein-DNA interactions; molecular mechanism of transcriptional activation in yeast. *Mailing Add:* Dept Biol Chem Harvard Med Sch 25 Shattuck St Boston MA 02115

STRUIK, DIRK JAN, b Rotterdam, Neth, Sept 30, 94; US citizen; m 24; c 3. HISTORY OF SCIENCE. *Educ:* Univ Leiden, Neth, PhD(math), 22. *Prof Exp:* Asst, Dept Math, Tech Univ Delft, 17-24; Rockefeller fel, Rome, Italy & Göttingen, Ger, 24-26; from lectr to prof, 26-60, EMER PROF, MASS INST TECHNOL, 60- *Concurrent Pos:* Lectr, Nat Univ Mex & other univs, PR, Costa Rica, Bielefeld, Ger & Toronto, 34-; prof hist sci, Univ Utrecht, 63-64; from res assoc to assoc, Hist Sci Dept, Harvard Univ. *Honors & Awards:* Kenneth O May Award, Int Comn Hist of Math, 89. *Mem:* Am Math Soc; Math Asn Am; Hist Sci Soc; Soc Hist Med, Math, Natural Sci & Tech; Am Acad Arts & Sci; Int Acad Hist Sci. *Res:* Differential geometry and tensor analysis; history of science and especially history of mathematics; author of various books. *Mailing Add:* 52 Glendale Rd Belmont MA 02178

STRUIK, RUTH REBEKKA, b Mass, Dec 15, 28; div; c 3. MATHEMATICS, ALGEBRA. *Educ:* Swarthmore Col, BA, 49; Univ Ill, MA, 51; NY Univ, PhD(math), 55. *Prof Exp:* Digital comput programmer, Univ Ill, 50-51; asst, Univ Chicago, 52; lectr math, Sch Gen Studies, Columbia Univ, 55; asst prof, Drexel Inst Technol, 56-57; lectr, Univ BC, 57-61; actg asst prof, 61-62, asst prof, 62-63, 64-65, assoc prof, 65-80, PROF MATH, UNIV COLO, BOULDER, 80- *Mem:* Math Asn Am; Asn Women Math; Am Math Soc; Sigma Xi. *Res:* Groups; modern algebra. *Mailing Add:* Dept Math Campus Box 426 Univ Colo Boulder CO 80309

STRULL, GENE, b Chicago, Ill, May 15, 29; m 52; c 2. ELECTRICAL ENGINEERING, SOLID STATE PHYSICS. *Educ:* Purdue Univ, BSEE, 51; Northwestern Univ, MS, 52, PhD(cadmium sulfide films), 54. *Prof Exp:* Engr, Mat Eng Dept, 54-55, sr engr, Semiconductor Div, Pa, 55-58, supvr solid state lab, Md, 58-60, mgr solid state technol, Aerospace Div, 60-68, mgr sci & technol, 68- 70, mgr advan technol labs, Systs Develop Div, 70-79,; GEN MGR, ADVAN TECHNOL DIV, 81-, VPRES TECHNOL, WESTINGHOUSE ELEC CORP, 87- *Concurrent Pos:* Lectr, Univ Pittsburgh, 54-58; assoc mem defense sci bd, Nat Res Coun/Nat Acad Sci, 80-82. *Mem:* Fel Inst Elec & Electronics Engrs; Sigma Xi. *Res:* Solid state devices; advanced sensors; integrated systems. *Mailing Add:* Westinghouse Elec Corp Box 1521 MS 3014 Baltimore MD 21203

STRUM, JUDY MAY, b Seattle, Wash, Mar 27, 38. ULTRASTRUCTURE, IMMUNOCYTOCHEMISTRY. *Educ:* Univ Wash, Seattle, BS, 63, PhD(anat/biol struct), 68. *Prof Exp:* Teaching res fel, Harvard Med Sch, Boston, Mass, 68-70; asst res anatomist, Cardiovasc Res Inst, San Francisco, Calif, 70-75; from asst prof to assoc prof, 75-82, PROF ANAT, UNIV MD, 82- *Concurrent Pos:* Prin investr, NIH grants, 76-83, co-investr, 83-; assoc ed, Am J Anat, 78-80; consult, Johns Hopkins Med Sch, 80. *Mem:* Am Soc Cell Biol; AAAS; Am Asn Anatomists; Electron Micros Soc Am; Int Asn Breast Cancer Res. *Res:* Development and differentiation of cell types in the airways and lung; biology and pathology of the mammary gland. *Mailing Add:* Dept Anat Univ Md Sch Med 655 W Baltimore St Baltimore MD 21201

STRUMEYER, DAVID H, biochemistry; deceased, see previous edition for last biography

STRUMWASSER, FELIX, b Trinidad, WI, Apr 16, 34; nat US. PHYSIOLOGY, NEUROBIOLOGY. *Educ:* Univ Calif, Los Angeles, BA, 53, PhD(zool), 57. *Prof Exp:* Asst, Univ Calif, Los Angeles, 56-57; from asst scientist to sr asst scientist, Lab Neurophysiol, NIMH, 57-60; res assoc neurophysiol, Walter Reed Army Inst Res, 60-64; assoc prof, Calif Inst Technol, 64-69, prof, 69-84; prof, Dept Physiol, Sch Med, Boston Univ, 84-87; SR SCIENTIST, MARINE BIOL LAB, WOODS HOLE, 87- *Concurrent Pos:* Res assoc, Washington Sch Psychiat, 60-64 & hon res assoc, Dept Biophys, Univ Col, London, 73-74; mem fel comt, NIH, 68-70; mem biochronometry comt, NSF, 69; sr fel, NATO, 73-74; mem, Neurol B Study Sect, NIH, 78-82 & Nat Comn Sleep Dis Res, 90- *Honors & Awards:* Penn lectr, Univ Pa, 67; Carter-Wallance lectr, Princeton Univ, 68; 17th Bowditch lectr, Am Physiol Soc, 72; Lang lectr, Marine Biol Lab, 80; Swammerdam lectr, Amsterdam, holland, 86. *Mem:* Fel AAAS; Soc Neuroscience; Soc Gen Physiol; Am Physiol Soc; Biophys Soc; Soc Res Biol Rhythms. *Res:* Neurophysiology; neurocellular basis of behavior, sleep-waking, reproduction; molecular and cellular mechanisms of circadian rhythms in nervous systems; long-term studies on single identifiable neurons in organ and dissociated cell culture; integrative mechanisms of the neuron; second messengers; pacemaker mechanisms in neurons; physiology and biochemistry of peptidergic neurons. *Mailing Add:* Lab Neuroendocrinol Marine Biol Lab Woods Hole MA 02543

STRUNK, DUANE H, b Irene, SDak, Mar 14, 20; m 45; c 2. ANALYTICAL CHEMISTRY. *Educ:* Univ SDak, BA, 42; Univ Louisville, MS, 51. *Prof Exp:* Res supvr, Joseph E Seagram & Sons, Inc, 42-43, prod supvr, 43-44, maintenance supvr, 43- 45, res chemist, 46-65, control labs adminstr,66-83; RETIRED. *Mem:* Am Chem Soc; Am Water Works Asn; Am Soc Testing Mat; Asn Anal Chemists. *Res:* Microanalytical methods, especially colorimetric, flame spectrophotometry and atomic absorption methods for copper and magnesium; water, food and wood chemistry; high accuracy particle counter; proof by density meter. *Mailing Add:* 1751 D Lake Pl Venice FL 34293

STRUNK, MAILAND RAINEY, b Kansas City, Kans, Aug 17, 19; m 49; c 3. CHEMICAL ENGINEERING. *Educ:* Kans State Univ, BS, 41; Univ Mo, MS, 47; Wash Univ, DSc(chem eng), 57. *Prof Exp:* Technologist, Shell Oil Co, Inc, 47-51, 52-54; instr chem eng, Wash Univ, 54-57; from assoc prof to prof, 57-79, chmn dept, 64-79, EMER PROF CHEM ENG, UNIV MO-ROLLA, 79- *Mem:* AAAS; Am Soc Eng Educ; Nat Soc Prof Engrs; Am Inst Chem Engrs. *Res:* Applied physical chemistry; heat and mass transfer. *Mailing Add:* Dept of Chem Eng Univ of Mo Rolla MO 65401-2136

STRUNK, RICHARD JOHN, b Jamaica, NY, July 6, 41; m 68; c 3. ORGANIC CHEMISTRY, AGROCHEMICAL SYNTHESIS. *Educ:* Gettysburg Col, AB, 63; State Univ NY Albany, PhD(org chem), 67. *Prof Exp:* SR RES SCIENTIST, RES CTR, UNIROYAL CHEM CO, 67- *Mem:* Am Chem Soc; Sigma Xi. *Res:* Organic chemistry; agrochemical synthesis in herbicides and insecticides. *Mailing Add:* 16 Briarwood Circle Cheshire CT 06410

STRUNK, ROBERT CHARLES, b Evanston, Ill, May 29, 42; m 71; c 2. PEDIATRIC ASTHMA & ALLERGIES. *Educ:* Northwestern Univ, BA, 64, MS & MD, 68. *Prof Exp:* DIR CLIN SERV, NAT JEWISH CTR IMMUNOL & RESPIRATORY MED, DENVER, COLO, 79-, ACTG CHMN, DEPT PEDIAT, 85-; PROF PEDIAT, SCH MED, UNIV COLO, 86- *Mem:* Am Acad Allergy; Am Asn Immunologists; Soc Pediat Res; Am Pediat Soc. *Res:* Regulation of protein synthesis in inflammation; asthma mortality. *Mailing Add:* Dept Pediat Wash Univ Med Sch 400 S Kingshighway Blvd St Louis MO 63110

STRUNZ, G(EORGE) M(ARTIN), b Vienna, Austria, Mar 10, 38; m 72; c 2. NATURAL PRODUCTS CHEMISTRY. *Educ:* Trinity Col, Dublin, BA, 59; Univ NB, PhD(org chem), 63. *Prof Exp:* Res assoc org chem, Univ Mich, 63-64; res fel, Harvard Univ, 64-65; lectr, Univ NB, 65-67; RES SCIENTIST, CAN DEPT FORESTRY, 67- *Concurrent Pos:* Hon res assoc, Univ NB, 70-; vis scientist, Imperial Chem Industs Pharmaceuts, Cheshire, UK, 75-76 & CNR Rome, Italy, 89; adj prof, Univ NB, 88- *Mem:* Fel Chem Inst Can; Am Chem Soc; fel Royal Soc Chem. *Res:* Chemistry of microbial metabolites; plant hormones and other natural products; approaches to insect control by compounds related to natural products; host-insect and host-pathogen interactions in the forest. *Mailing Add:* Can Forestry Serv PO Box 4000 Fredericton NB E3B 5P7 Can

STRUPP, HANS H, b Frankfurt Am Main, Ger, Aug 25, 21; m 51; c 3. PSYCHOTHERAPY RESEARCH & CLINICAL PSYCHOLOGY, PSYCHOANALYSIS. *Educ:* George Washington Univ, AB, 45, AM, 47, PhD(psychol), 54. *Hon Degrees:* MD, Univ Ulm, Ger, 86. *Prof Exp:* From assoc prof to prof, Univ NC, Chapel Hill, 57-66; prof, 66-76, DISTINGUISHED PROF PSYCHOL, VANDERBILT UNIV, 76- *Concurrent Pos:* Co-ed, Psychother Res, Soc Psychother Res, 89- *Honors & Awards:* Distinguished prof contrib Knowledge Award, Am Psychol Asn, 87; Distinguished Career Contrib Award, Soc Psycother Res, 86. *Mem:* Fel Am Psychol Asn (pres, 74-75); Soc Psycother Res (pres, 72-73); fel AAAS. *Res:* Research on psychotherapy process and outcome. *Mailing Add:* Dept Psychol Vanderbilt Univ Nashville TN 37240

STRUTHERS, BARBARA JOAN OFT, b Bend, Ore, May 4, 40; m 59; c 3. INTRAUTERINE CONTRACEPTION, TOXICOLOGY. *Educ:* Wash State Univ, BS, 62; Ore State Univ, MS, 68, PhD(food sci), 73; Am Bd Toxicol, cert, 81. *Prof Exp:* Instr sci educ, Ore State Univ, 68-69, chemist food sci, 69-70; proj leader, Ralston Purina Co, 73-75, sr proj leader, 75-78, assoc scientist, 79-81, mgr toxicol, 81-82; assoc dir sci affairs, Monsanto Co, 82-87; dir, Gyn Prods, 87-88, dir GI & Anti-Infective Prods, 88-90, SR SCI ADV, CORP MED & SCI AFFAIRS, G D SEARLE & CO, 91- *Concurrent Pos:* Radiation safety officer, Ralston Purina Co, 74-82; mem ethics comt, Soc Toxicol; mem bd dir, Soc Advan Contraception, 86- *Mem:* AAAS; Sigma Xi; Am Inst Nutrit; Soc Toxicol; Soc Advan Contraception. *Res:* Soy protein-processing toxicology and biochemistry; soybean trypsin inhibitor biological effects; mycotoxin effects in domestic animals; biochemical effects of lysinoalanine; intrauterine contraception; sexually transmitted disease; epidemiology; prostagladins as ulcer preventives. *Mailing Add:* Dept Sci Affairs GD Searle Co 4901 Searle Pkwy Skokie IL 60077

STRUTHERS, ROBERT CLAFLIN, b Syracuse, NY, June 2, 28; m 52; c 6. COMPARATIVE ANATOMY, DEVELOPMENTAL ANATOMY. *Educ:* Syracuse Univ, BA, 50, MA, 52; Univ Rochester, PhD(biol), 56. *Prof Exp:* Asst comp & develop anat, Syracuse Univ, 50-56; from instr to asst prof anat, Ohio State Univ, 56-61; assoc prof, 61-62, prof anat, 62-81, PROF NATURAL SCI, WHEELOCK COL, 81- *Res:* Morphology of early embryonic stages of vertebrate animals, particularly on the pharynx and its derivatives. *Mailing Add:* Dept of Sci Wheelock Col 200 Riverway Boston MA 02215

STRUVE, WALTER SCOTT, b Wilmington, Del, Feb 22, 45; m 71. ULTRAFAST LASER SPECTROSCOPY. *Educ:* Harvard Univ, AB, 67, MA, 69, PhD(phys chem), 72. *Prof Exp:* Postdoctoral fel, AT&T Bell Labs, Murray Hill, NJ, 72-74; instr, 74-76, from asst prof to assoc prof, 76-88, PROF CHEM, IOWA STATE UNIV, 88- *Concurrent Pos:* Vis assoc prof, Univ Chicago, 82. *Mem:* Am Chem Soc; Am Phys Soc; AAAS; Am Soc Photobiol. *Res:* Ultrafast laser spectroscopy of electronic excitation transport in photosynthetic antenna systems; environment-sensitive fluorophores of biological significance. *Mailing Add:* Dept Chem Iowa State Univ Ames IA 50011

STRUVE, WILLIAM GEORGE, b Milwaukee, Wis, Mar 19, 38; c 9. CHEMICAL LABORATORY, AUTOMATION. *Educ:* Lake Forest Col, BA, 62; Northwestern Univ, PhD, 66; Memphis State Univ, MS, 86. *Prof Exp:* Res chemist, Am Cyanamid Co, 66-68; assoc prof biochem, Univ Tenn, Memphis, 68-87; LAB AUTOMATION SPECIALIST, COMPUCHEM, 87- *Concurrent Pos:* Vis scholar, Stanford Univ, 69-70. *Res:* electronics; computer music composition; laboratory automation. *Mailing Add:* 238 Argonne Dr Durham NC 27704

STRUVE, WILLIAM SCOTT, b Utica, NY, May 1, 15; m 39; c 3. ORGANIC CHEMISTRY. *Educ:* Univ Mich, BS, 37, MS, 38, PhD(org chem), 40. *Prof Exp:* Du Pont fel, Univ Mich, 40-41; chemist, Jackson Lab, E I du Pont de Nemours & Co, 41-49, res supvr, 49-58, lab dir, Color Res Lab, 58-73, dir, Newark Lab & mgr, Colors Res & Develop, 73-80; RETIRED. *Mem:* Am Chem Soc. *Res:* Carcinogenic hydrocarbons; organic fluorine compounds; dyestuffs; pigments. *Mailing Add:* 29 Dellwood Ave Chatham NJ 07928-1701

STRUZAK, RYSZARD G, b Janow, Poland, Apr 2, 33; m 56; c 2. ELECTROMAGNETIC COMPATIBILITY, ELECTROMAGNETIC METROLOGY. *Educ:* Tech Univ, Wroclaw, BachSci (eq), 54, MSci, 56; Tech Univ, Warsaw, Dr, 62, Dr Habil, 68. *Prof Exp:* Res asst antennas, Radio & TV Res & Devlop Ctr, Wroclaw, 53-54; lectr telecommun, Tech Univ, Wroclaw, 54-56; head scientist, RFI Lab, Telecommun Res Inst, 63-67, head, Electromagnetic Compatibility Div, 67-73, head, Telecommun Res Inst, Wroclaw Br, 73-85; SR COUNR & HEAD, TECH DEPT, INT TELLECOMMUN UNION-CCIR, 85- *Concurrent Pos:* From asst prof to prof, Tech Univ, Wroclaw, 71-85; mem, Comn Electronics & Telecommun, Polish Acad Sci, 69-78, officer, 78-85; vchmn, SG1, Int Radio Consult Comt, CCIR, 74-85, chmn IWP1/4, 80-85; vchmn, Comn E, Int Union Radio Sci, URSI, 84-87; invited lectr, Int Ctr Theoret Physics, Trieste, 89 & 91. *Honors & Awards:* Gold Emblem of Distinction, Asn Elec Engrs, 87. *Mem:* Fel Inst Elec & Electronics Engrs. *Res:* Radio science and engineering; electromagnetic compatibility engineering and standards; RF spectrum monitoring, engineering, planning & management; computer-aided radio engineering and design of radio systems; computer stimulation; electromagnetic metrology; electromagnetism. *Mailing Add:* Int Telecommun Union Place Des Nations Bp 820 Ch 1211 Geneve 20 CH-1211 Switzerland

STRUZYNSKI, RAYMOND EDWARD, b Jersey City, NJ, Dec 10, 37; m 65; c 2. MATHEMATICAL PHYSICS. *Educ:* Stevens Inst Technol, BEng, 59, MS, 61, PhD(physics), 65. *Prof Exp:* From instr to asst prof, 64-70, ASSOC PROF PHYSICS, BROOKLYN COL, 70- *Concurrent Pos:* Res scientist, Hudson Labs, Columbia Univ, 65-67. *Mem:* Am Phys Soc; Sigma Xi. *Res:* Radiative beta decay; liquid helium; quantum mechanics of many boson systems. *Mailing Add:* Dept of Physics Brooklyn Col Bedford Ave & Ave H Brooklyn NY 11210

STRYCKER, STANLEY JULIAN, b Goshen, Ind, Aug 30, 31; m 52; c 4. RESEARCH ADMINISTRATION, PHARMACEUTICAL CHEMISTRY. *Educ:* Goshen Col, AB, 53; Univ Ill, PhD(org chem), 56. *Prof Exp:* Res chemist, 56-63, sr res chemist, 63-68, group leader, 68-69, res mgr pharmaceut sci, 69-74, res mgr chem & indust pharm, Dow Chem Co, 74-80, lab dir chem, 80-82, LAB DIR, RES ADMIN, MERRELL DOW PHARMACEUTICALS, INC, 82- *Concurrent Pos:* Sabbatical, Col Med, Univ Iowa, 67-68. *Mem:* AAAS; Am Chem Soc; Sigma Xi. *Res:* Organic synthesis; heterocyclics; medicinal chemistry. *Mailing Add:* 527 Nuthatch Dr Zionsville IN 46077-9563

STRYER, LUBERT, b Tientsin, China, Mar 2, 38; US citizen; m 58; c 2. MOLECULAR AND CELL BIOLOGY. *Educ:* Univ Chicago, BS, 57; Harvard Univ, MD, 61. *Prof Exp:* From asst prof to assoc prof biochem, Stanford Univ, 63-69; prof molecular biophys & biochem, Yale Univ, 69-76; chmn dept, 76-79, WINZER PROF CELL BIOL, SCH MED, STANFORD UNIV, 76- *Concurrent Pos:* Helen Hay Whitney fel, Harvard Univ & Med Res Coun Lab Molecular Biol, Cambridge, Eng, 61-63; consult, NIH, 67-71; sci adv bd, Jane Coffin Child's Fund & Res to Prevent Blindness. *Honors & Awards:* Eli Lilly Award, Am Chem Soc, 70. *Mem:* Nat Acad Sci; Am Soc Biol Chemists; Biophys Soc; fel Am Acad Arts & Sci. *Res:* Protein structure and function; visual excitation; excitable membranes; spectroscopy; x-ray diffraction. *Mailing Add:* Dept Cell Biol Stanford Med Sch Stanford CA 94305

STRYKER, LYNDEN JOEL, b Stamford, NY, Feb 19, 43; m 71; c 2. COLLOID CHEMISTRY, SURFACE CHEMISTRY. *Educ:* Clarkson Col Technol, BS, 64, PhD(chem), 69. *Prof Exp:* Asst, Clarkson Col Technol, 67-68; lectureship, Brunel Univ, 69-70; sr res chemist, Westvaco Corp, 70-77; assoc prof chem, Inst Paper Chem, 77-79; res scientist, Hammermill Paper Co, 79-; PROG MGR, INT PAPER, 89- *Mem:* Am Chem Soc; Tech Asn Pulp & Paper Indust. *Res:* Chemistry of papermaking systems; alkaline papermaking; stability of colloidal dispersions; solid-liquid interactions; metal ion hydrolysis and complexation; water pollution abatement; paper sizing and retention; surface characterization of solids; emulsion technology. *Mailing Add:* 6453 West Rd McKean PA 16426

STRYKER, MARTIN H, b New York, NY, July 26, 43; m 66; c 2. QUALITY CONTROL, PLASMA PROTEINS. *Educ:* Columbia Univ, AB; NY Med Col, MS, 68, PhD(biochem), 71. *Prof Exp:* Chief, NY Blood Ctr, 71-74, assoc dir, 74-80, dir, qual control, Blood Derivatives Prog, 80-86, managing dir, 86-88, VPRES SCI AFFAIRS, NY BLOOD CTR, 88- *Concurrent Pos:* Res fel, Plasma Proteins Lab, Lindsley F Kimball Res Inst, 71-72, res assoc, 72-80, anal chemist, 80-; Hemophilia Adv Panel, NY Dept Health. *Mem:* Am Chem Soc; AAAS; World Fedn Hemophilia; Parenteral Drug Asn; Int Soc Thrombosis Haemostasis. *Res:* Quality control of plasma protein production; plasma fractionation; blood coagulation factor assays; treatment of hemophilia. *Mailing Add:* NY Blood Ctr 155 Duryea Rd Melville NY 11747

STRYKER, MICHAEL PAUL, b Savannah, Ga, June 16, 47; m 78; c 3. NEUROBIOLOGY, NEUROOPHTHALMOLOGY. *Educ:* Univ Mich, AB, 68; Mass Inst Technol, PhD(psychol, brain sci), 75. *Prof Exp:* Res fel neurobiol, Harvard Med Sch, 75-78; asst prof, 78-87, PROF PHYSIOL & CO-DIR NEUROSCI, UNIV CALIF, SAN FRANCISCO, 87- *Concurrent Pos:* Instr neurobiol, Cold Spring Harbor Labs, 76-78; vis prof, Univ Oxford, 87-88; NSF Develop Neuroscience panel mem, 85-88; mem, Mass, Inst Technol Corp, vis comt, Whitlaker Col & dept brain & cognitive sci. *Mem:* Soc Neuroscience; AAAS. *Res:* Neurobiology of the central nervous system; developing visual system. *Mailing Add:* Dept Physiol Univ Calif Med Sch San Francisco CA 94143-0444

STUART, ALFRED HERBERT, b Farmville, Va, 13; m 44; c 2. PHOTOGRAPHIC CHEMISTRY. *Educ:* Hampden-Sydney Col, BS, 33; Univ Va, PhD(org chem), 37. *Prof Exp:* Res fel, Univ Va, 37-39; res chemist, Schieffelin & Co, NY, 39-43, dir chem res, 43-47; develop mgr, Charles Bruning Co, Ill, 47-65, chief chemist, Bruning Div, Addressograph-Multigraph Corp, 65-80; RETIRED. *Res:* Diazotype and electrostatic copying processes. *Mailing Add:* 726 Superior St Oak Park IL 60302

STUART, ALFRED WRIGHT, b Pulaski, Va, Nov 16, 32; m 60; c 4. REGIONAL ANALYSIS. *Educ:* Univ SC, BS, 55; Emory Univ, MS, 56; Ohio State Univ, PhD(geog), 66. *Prof Exp:* Glaciologist, US Antarctic Res Prog, NSF, 58-60; res assoc, Inst Polar Studies, Ohio State Univ, 60-61, asst instr geog, 61-63; community planner, City of Roanoke, Va, 63-64; asst prof geog, Univ Tenn, Knoxville, 64-69; PROF GEOG & DEPT GEOG & EARTH SCI, UNIV NC, CHARLOTTE, 69- *Concurrent Pos:* Urban analyst, US Bur Mines, 68-70; researcher, Oak Ridge Nat Lab, 69; consult, Charlotte-Mechlenburg Planning Comn, 84-85. *Honors & Awards:* Polar Medal, US Govt, 65. *Mem:* Asn Am Geographers. *Res:* Economic and demographic change in the southern United States; analytical atlas production. *Mailing Add:* Dept Geog & Earth Sci Univ NC Charlotte NC 28223

STUART, ANN ELIZABETH, b Harrisburg, Pa, Oct 5, 43; m 78; c 1. NEUROBIOLOGY. *Educ:* Swarthmore Col, BA, 65; Yale Univ, MS, 67, PhD(physiol), 69. *Prof Exp:* Res fel neurophysiol, Dept Physiol, Univ Calif, Los Angeles Med Ctr, 71-73; res fel neurochemistry, Dept Neurobiol, Harvard Med Sch, 69-71, asst prof, 73-78, assoc prof, 78; MEM FAC, DEPT PHYSIOL, UNIV NC, 80-, PROF, 87- *Mem:* Soc Neuroscience; Soc Gen Physiologists; Asn Res Vision & Ophthal. *Res:* Synaptic mechanisms; integration and synaptic transmission between single cells of small populations of neurons; integration in invertebrate central nervous systems; invertebrate vision photoreceptor synapses. *Mailing Add:* Dept Physiol Med Res Bldg 206H Univ NC Chapel Hill NC 27599-7545

STUART, DAVID MARSHALL, b Ogden, Utah, May 20, 28; m 51; c 5. PHARMACY, CHEMISTRY. *Educ:* Univ Utah, BS, 51; Univ Wis, PhD(pharmaceut chem), 55. *Prof Exp:* Asst prof pharmaceut chem, Univ Tex, 55-S7 & Ore State Col, 57-60; coordr sci info, Neisler Labs, 60-64; PROF PHARMACEUT CHEM, COL PHARM, OHIO NORTHERN UNIV, 64- *Concurrent Pos:* Pharm consult, Ohio Dept Pub Welfare, 71-77; pres, Pharm Health & Related Mgt, Inc, 74- *Mem:* Am Pharmaceut Asn; NY Acad Sci. *Res:* Scientific literature research and writing; health care research. *Mailing Add:* 92 San Andreas Ct West Jordan UT 84088

STUART, DAVID W, b Lafayette, Ind, June 15, 32; m 64; c 3. METEOROLOGY. *Educ:* Univ Calif, Los Angeles, BA, 55, MA, 57, PhD(meteorol), 62. *Prof Exp:* Res meteorologist, Univ Calif, Los Angeles, 55-61, teaching asst meteorol, 57-61; from asst prof to assoc prof, 62-83, assoc chmn dept, 67-72, PROF METEOROL, FLA STATE UNIV, 83-, CHMN, 85- *Concurrent Pos:* Assoc prof, Naval Postgrad Sch, 66-67. *Mem:* Am Meteorol Soc; Sigma Xi; Am Geophys Union. *Res:* Synoptic meteorology; numerical weather prediction, especially diagnostic studies and air-sea interaction; meteorology of coastal upwelling areas. *Mailing Add:* Dept Meteorol Fla State Univ Tallahassee FL 32306-3034

STUART, DERALD ARCHIE, b Bingham Canyon, Utah, Nov 9, 25; m 48; c 2. SOLID STATE PHYSICS. *Educ:* Univ Utah, BS, 47, MS, 48, PhD(physics), 50. *Prof Exp:* Asst physics, Univ Utah, 47-50; asst prof eng mat, Cornell Univ, 50-52, assoc prof eng mech & mat, 52-58; propulsion staff mgr & resident rep to Aerojet-Gen Corp, 58-59, asst to Polaris Missile syst mgr & Polaris resident rep to Aerojet-Gen Corp, 59-61, mgr propulsion staff, Missile Systs Div, 61-62, dir propulsion systs, 62-64, asst chief engr, 64-66, asst gen mgr eng & develop, 66-67, vpres & asst gen mgr, 67-70, vpres corp & vpres & gen mgr Missile Systs Div, Lockheed Missiles & Space Co, Lockheed Corp, 70-87; CONSULT SOLID STATE PHYSICS, 88- *Concurrent Pos:* Consult, Cornell Aeronaut Lab, 52-54, Allegany Ballistics Lab, 52-58, Lincoln Lab, Mass Inst Technol & Ramo-Wooldridge Corp, 54-56. *Honors & Awards:* Meyer Award, Am Ceramic Soc, 54; Wyld Propulsion Award, Am Inst Aeronaut & Astronaut; Montgomery Award, Nat Soc Aerospace Prof, 64. *Mem:* Nat Acad Eng; fel Am Inst Aeronaut & Astronaut; Soc Logistics Engrs; Soc Women Engrs; Am Contract Managing Asn. *Res:* Glassy state; plastic behavior of materials; solid fuel rockets. *Mailing Add:* 9841 N 107th St Scottsdale AZ 85258

STUART, DOUGLAS GORDON, b Casino, NSW, Australia, Oct 5, 31; US citizen; m 57; c 4. PHYSIOLOGY. *Educ:* Mich State Univ, BS, 55, MA, 56; Univ Calif, Los Angeles, PhD(physiol), 61. *Prof Exp:* Res fel anat, Univ Calif, Los Angeles, 61-63, asst prof physiol in residence, 63-65, assoc prof, Univ Calif, Davis, 65-67; PROF PHYSIOL, COL MED, UNIV ARIZ, 70- *Mem:* Am Physiol Soc. *Res:* Neural control of posture and locomotion. *Mailing Add:* Dept Anat & Physiol Univ Ariz Sch Med 1501 N Campbell Ave Tucson AZ 85724

STUART, GEORGE WALLACE, b New York, NY, Apr 5, 24; m 48; c 3. THEORETICAL PHYSICS. *Educ:* Rensselaer Polytech Inst, BEE, 49, MS, 50; Mass Inst Technol, PhD(physics), 53. *Prof Exp:* Head theoret physics unit, Hanford Atomic Prod Oper, Gen Electric Co, 52-57; res adv spec nuclear effects lab, Gen Atomic Div, Gen Dynamics Corp, Calif, 58-67; sr res scientist, Systs, Sci & Software, 67-72; staff scientist, Sci Applns Inc, 72-79; CONSULT, 79- *Mem:* Fel Am Phys Soc. *Res:* Nuclear reactors; plasma theory; atomic physics. *Mailing Add:* PO Box 134 Rancho Santa Fe CA 92067-0134

STUART, JAMES DAVIES, b Elizabeth, NJ, Sept 30, 41; m 64; c 2. ANALYTICAL SEPARATIONS, LIQUID & GAS CHROMATOGRAPHY. *Educ:* Lafayette Col, BS, 63; Lehigh Univ, PhD(anal chem), 69. *Prof Exp:* Instr, Lafayette Col, 67-69; asst prof, 69-75, ASSOC PROF ANAL CHEM, UNIV CONN, 75- *Concurrent Pos:* Prin investr, Nat Inst Environ Sci grants, 75-77; vis lectr, Chem Dept, Univ Ga, 76; prin investr grant, Dept Interior to Conn Inst Water Resources, 80-88; vis lectr chem eng, Yale Univ, 83; consult, IBM Instruments, Danbury, Conn, 83-85, Olin Corp, Cheshire, Conn, 88; prin investr grant, Environ Protection Agency, 87-; lab dir, Marine Environ Lab, Avery Point, 89-, Univ Conn. *Mem:* Am Chem Soc. *Res:* Development of modern separation methods using gas and high performance liquid chromatography; development of accurate rapid and sensitive methods to determine amino acids and neurotransmitters in human, and animal fluids; assessing the reliability of analyses of air and groundwater around leaking underground gasoline tanks; general analysis method of marine samples. *Mailing Add:* U-60 Chem Dept Univ Conn 215 Glenbrook Rd Storrs CT 06269-3060

STUART, JAMES GLEN, b Enid, Okla, Aug 23, 48; m 68; c 2. BACTERIAL GENETICS, MICROBIOLOGY. *Educ:* Cameron State Col, BS, 70; Okla Univ, MS, 72, PhD(med microbiol), 75. *Prof Exp:* Lectr biol, Millikin Univ, 75-77; asst prof biol, 77-85, ASSOC PROF BIOL SCI, MURRAY STATE UNIV, 85- *Concurrent Pos:* Grant dir, Ky Heart Fund, 78-79 & grant reviewer, 83- *Mem:* Am Soc Microbiol; Sigma Xi. *Res:* Bacterial genetics, specifically the phenomena of genetics and mechanisms of antibiotic resistance in group A streptococci. *Mailing Add:* Dept Biol Murray State Univ Murray KY 42071

STUART, JOE DON, b Brownsboro, Tex, Feb 28, 32; m 63; c 2. EXPERT SYSTEMS, COMPUTER SCIENCE. *Educ:* Univ Tex, BA, 57, PhD(physics), 63. *Prof Exp:* Sr physicist appl physics lab, Johns Hopkins Univ, 63-65; sr scientist, Tracor, Inc, 65-72; sr scientist, 72-76, asst vpres, 76-82, PRIN SCIENTIST, RADIAN CORP, 82- *Mem:* Am Meteorol Soc; Am Asn Artificial Intel; Inst Elec & Electronic Engrs Computer Soc; Asn Comput Mach. *Res:* Knowledge engineering; computer modeling; the formulation and development of practical expert systems in the fields of equipment fault diagnostics, chemical analysis, weather forecasting, and financial services. *Mailing Add:* 4009 Knollwood Dr Austin TX 78731

STUART, JOHN W(ARREN), b Logansport, Ind, Jan 5, 24; m 46; c 5. ELECTRICAL ENGINEERING. *Educ:* Purdue Univ, BSEE, 48. *Prof Exp:* Asst elec apparatus, Westinghouse Elec Corp, 48-50, appln engr, 50-55, proj engr, 55-58; chief engr, Accuray Corp, 58-62; mgr, Arthur D Little, Inc, 62-80, vpres mech eng, 62-88; CONSULT, 88- *Mem:* Inst Elec & Electronics Engrs. *Res:* Digital control; solid state device development; digital applications of magnetic devices; research & development management; technical audits; equipment engineering. *Mailing Add:* 137 Sherburn Circle Weston MA 02193

STUART, KENNETH DANIEL, b Boston, Mass, 1940; m; c 3. MOLECULAR PARASITOLOGY, GENE EXPRESSION. *Educ:* Northeastern Univ, BA, 63; Wesleyan Univ, MA, 65; Univ Iowa, PhD(zool), 69. *Prof Exp:* Res biochem, Nat Inst Med Res, London, 69-71 & State Univ NY, Stony Brook, 71-72; res biol, Univ SFl, 72-76; DIR, SEATTLE BIOMED RES INST, 82-; AFFIL PROF MICROBIOL, UNIV WASH, 84- *Concurrent Pos:* Consult, WHO, NIH & Agency Int Develop. *Mem:* Fel AAAS; Am Soc Microbiol; Am Soc Parasitol; Am Soc Cell Biol; Am Soc Adv Sci. *Res:* Molecular parasitology with emphasis on gene organization and expression; Antiqenic variation and mitochondrial gene expression in African trypanosomes; viruses in Leishmania. *Mailing Add:* Seattle Biomed Res Inst Four Nickerson St Seattle WA 98109

STUART, RONALD S, b Tingley, NB, Mar 26, 19; m 46; c 4. ORGANIC CHEMISTRY. *Educ:* Univ NB, BA, 40; Univ Toronto, MA, 41, PhD(org chem), 44. *Prof Exp:* Demonstr chem, Univ Toronto, 40-42; res assoc, Nat Res Coun Can, 43-45; asst dir res, Dom Tar & Chem, 45-48; mgr chem & biol control, Merck & Co, Ltd, 48-53, mgr sci develop, 53-60, mgr tech & prod opers, 60-63, dir res, Merck Sharp & Dohme Can, 63-65, Charles E Frosst

& Co, 65-68, dir res, Merc Frosst Labs, 68-78, exec dir, 78-81; DIR RES SERV & CTR RES ENG & APPL SCI, UNIV NB, 82- *Mem:* AAAS; Am Chem Soc; NY Acad Sci; Chem Inst Can; Can Res Mgt Asn. *Res:* Medicinal chemistry. *Mailing Add:* Sch Grad Studies & Res Univ NB Fredericton NB E3B 5A3 Can

STUART, THOMAS ANDREW, b Bloomington, Ind, Feb 6, 41; m 63; c 2. ELECTRICAL ENGINEERING. *Educ:* Univ Ill, Urbana, BS, 63; Iowa State Univ, ME, 69, PhD(elec eng), 72. *Prof Exp:* Engr, Martin Co, 63-64 & Honeywell, Inc, 64-65; design engr, Collins Radio Co, 65-69; asst prof elec eng, Clarkson Col Technol, 72-75; PROF ELEC ENG, UNIV TOLEDO, 75- *Mem:* Inst Elec & Electronics Engrs. *Res:* Computer applications for electrical power systems; power electronics. *Mailing Add:* Dept of Elec Eng Univ of Toledo Toledo OH 43606

STUART, WILLIAM DORSEY, b St Louis, Mo, Mar 28, 39. GENETIC ENGINEERING, MEMBRANE TRANSPORT. *Educ:* Fla State Univ, BA, 69, MS, 70, PhD(biol & genetics), 73. *Prof Exp:* Res fel genetics, dept biol, Stanford Univ, 73-76; Fel, Melbourne Univ, Australia, 76-78; asst prof, 78-83, ASSOC PROF GENETICS, SCH MED, UNIV HAWAII, 83-, DEAN RES, 90- *Concurrent Pos:* Consult, Cetus Corp, 81-83; co-founder, Hawaii Biotechnol Group, 82-, consult, 83-; vis scholar, dept biol sci, Stanford Univ, 85-86. *Mem:* Genetics Soc Am; Am Soc Microbiol; Sigma Xi; Am Chem Soc; Am Asn Adv Sci. *Res:* Investigation of membrane transport genes and gene products; involvement of ribonucleoproteins complexes in membrane transport; immunohistochemical identification of nucleic acid hybrid molecules; expression of heterologous genes in fungal cells. *Mailing Add:* Dept Genetics Sch Med Univ Hawaii Honolulu HI 96822

STUART-ALEXANDER, DESIREE ELIZABETH, b London, Eng, Apr 6, 30; US citizen. GEOLOGY. *Educ:* Westhampton Col, BA, 52; Stanford Univ, MSc, 59, PhD(geol), 67. *Hon Degrees:* Dr, Univ Richmond, 80. *Prof Exp:* Geologist explor dept, Utah Construct & Mining Co, 58-60; asst prof geol, Haile Selassie Univ, 63-65; geologist, 66-86, res geologist, US GEOL SURV, 86-; CONSULT GEOLOGIST. *Concurrent Pos:* Prog scientist, NASA Hq, 74-75; chief, Br Western Regional Geol, 80-85. *Mem:* AAAS; fel Geol Soc Am; Am Geophys Union; Earthquake Eng Res Inst. *Res:* Lunar and Martian geology including studies based on remote sensing data and petrology of lunar rocks; terrestrial studies of metamorphic problems and lunar analogs; problems of reservoir-induced seismicity; faulting in the Sierra Nevada Mountains, California. *Mailing Add:* 120 Sea Breeze Pl Aptos CA 95003-5748

STUBBE, JOHN SUNAPEE, b New York, NY, Feb 21, 19; m 43; c 4. MATHEMATICS. *Educ:* Univ NH, BS, 41; Brown Univ, MS, 42; Univ Cincinnati, PhD(math), 45. *Prof Exp:* Instr math, army specialized training prog, Univ Cincinnati, 43-44; instr, Univ Ill, 45-47; asst prof, Univ NH, 47-49; asst prof, 49-53, dir comput ctr, 64-70, ASSOC PROF MATH, CLARK UNIV, 53- *Concurrent Pos:* Prof, Worcester Polytech Inst, 69-72. *Mem:* Am Math Soc; Math Asn Am. *Res:* Summability; Fourier series. *Mailing Add:* 101 S Flagg Worcester MA 01602

STUBBEMAN, ROBERT FRANK, b Midland, Tex, May 9, 35; m 56; c 3. PHYSICAL CHEMISTRY. *Educ:* Austin Col, BA, 57; Univ Tex, MA, 61, PhD(chem), 64, JD, 89. *Prof Exp:* Res chemist, Esso Res & Eng Co, 63-66; sr res chemist, Celanese Chem Co, Tech Ctr, 66-71, sect leader, Spectros Lab, 71-80, mgr, Safety Health & Environ Sect, 80-85; GEN LAW PRACT, 90- *Mem:* Am Chem Soc; fel Am Inst Chemists; Am Bar Asn. *Res:* Shock tube kinetics; mass spectrometry, including qualitative and quantitative low and high resolution; process research; environmental health and safety. *Mailing Add:* 43 River Ranch Dr Bandera TX 78003

STUBBERUD, ALLEN ROGER, b Glendive, Mont, Aug 14, 34; m 61; c 2. ELECTRICAL ENGINEERING. *Educ:* Univ Idaho, BSEE, 56; Univ Calif, Los Angeles, MS, 58, PhD(eng), 62. *Prof Exp:* Asst prof eng, Univ Calif, Los Angeles, 62-67, assoc prof, 67-69; assoc prof, 69-72, assoc dean, Eng Sch, 72-77, dean, 78-83, PROF ENG, UNIV CALIF, IRVINE, 72- *Concurrent Pos:* Consult, 63-; chief scientist, US Air Force, 83-85. *Honors & Awards:* Centennial Medal, Inst Elec & Electronics Engrs. *Mem:* Fel Inst Elec & Electronics Engrs; assoc fel Am Inst Aeronaut & Astronaut; Opers Res Soc Am; Am Astronaut Soc. *Res:* Optimal control systems theory; optimal filtering theory; final value control systems; digital signal processing. *Mailing Add:* 19532 Sierra Soto Irvine CA 92715

STUBBINS, JAMES FISKE, b Honolulu, Hawaii, Feb 19, 31; m 59; c 3. MEDICINAL CHEMISTRY. *Educ:* Univ Nev, BS, 53; Purdue Univ, MS, 58; Univ Minn, PhD(pharmaceut chem), 65. *Prof Exp:* Asst prof pharmaceut chem, Univ Fla, 62-63; from asst prof to assoc prof, 63-76, PROF MEDICINAL CHEM, MED COL VA, 76- *Mem:* Am Chem Soc; Am Asn Col Pharm Coun Fac. *Res:* Synthesis of medicinal agents; pharmacology of drugs in the autonomic and central nervous systems; antimetabolite theory and chemotherapy; drugs acting on blood cells. *Mailing Add:* Sch Pharm Med Col Va VA Commonwealth Univ Box 540 MCV Sta Richmond VA 23298-0540

STUBBINS, JAMES FREDERICK, b Cincinnati, Ohio, Sept 2, 48; m 86; c 1. MATERIALS SCIENCE ENGINEERING. *Educ:* Univ Cincinnati, BS, 70, MS, 72, PhD(mat sci), 75. *Prof Exp:* Res assoc mat sci, Univ Oxford, Eng, 77-78; engr, Gen Elec Co, Schenectady, NY, 78-80; FAC MEM NUCLEAR ENG, UNIV ILL, 80- *Concurrent Pos:* Fac assoc, Northwest Orgn Cols & Univs Sci, 77- & Argonne Nat Lat, 79-; consult, Los Alamos Nat Lab, 77- *Mem:* Am Nuclear Soc; Am Soc Metals Int. *Res:* Materials development for and application in engineering design; materials application in energy, including nuclear energy, systems; mechanical properties, elevated temperature performance, corrosion and radiation effects. *Mailing Add:* 214 Nuclear Eng Lab 103 S Goodwin Ave Urbana IL 61801

STUBBLEBINE, WARREN, b Reading, Pa, Jan 18, 17; m 38; c 5. CHEMICAL ENGINEERING. *Educ:* Pa State Col, BS, 38, MS, 40, PhD(textile chem), 42. *Prof Exp:* Asst textile chem, Pa State Col, 38-42; head flooring develop sect, Armstrong Cork Co, 42-47; res dir, Chem & Plastics Div, Off Qm Gen, DC, 47-52; dir develop, Conn Hard Rubber Co, 52-55; dir res & develop, Stowe-Woodward, Inc, 55-61, vpres co, 61-63; vpres, Sandusky Foundry & Mach Co, 63-82. *Concurrent Pos:* Mem comt mat & equip & panel org & fibrous mat res & develop bd, Dept Defense, 48-52; mem comt plastics & elastomers & comt chem & adv bd qm res & develop, Nat Acad Sci-Nat Res Coun, 52-60, tech adv panel rubber, 52-57, mem comt elastomers, Adv Bd Mil Personnel Supplies, 61-82; consult, 82- *Mem:* AAAS; Sigma Xi; Am Chem Soc; Brit Inst Rubber Indust; fel Am Inst Chem. *Res:* Centrifugal castings in ferrous and nonferrous alloys and their use and application in the paper industry. *Mailing Add:* 240 Peru-Olena Rd W Norwalk OH 44857

STUBBLEFIELD, BEAUREGARD, b Navasota, Tex, July 31, 23; m 50; c 5. TOPOLOGY. *Educ:* Prairie View State Col, BS, 43, MA, 45; Univ Mich, MS, 51, PhD(math), 59. *Prof Exp:* Asst math, Prairie View State Col, 43-44; prof & head dept, Univ Liberia, 52-56; lectr & NSF fel, Univ Mich, 59-60; supvr anal sect, Int Elec Corp, 60-61; assoc prof math, Oakland Univ, 61-67; vis prof & vis scholar, Tex Southern Univ, 68-69; sr prog assoc, Inst Serv Educ, 69-71; prof math, Appalachian State Univ, 71-75; mathematician, Nat Oceanic & Atmospheric Admin Environ Res Lab, 76-85. *Concurrent Pos:* Mathematician, Detroit Arsenal, 57-60; asst prof, Stevens Inst Technol, 60-61; vis prof, Prairie View A&M Col, 67-68. *Mem:* AAAS; Am Math Soc; Math Asn Am; Soc Indust & Appl Math. *Mailing Add:* 158 Shagbark Rochester MI 48063

STUBBLEFIELD, CHARLES BRYAN, b Viola, Tenn, Sept 14, 31; m 60; c 2. TECHNICAL MANAGEMENT, MARKETING COMMUNICATIONS. *Educ:* Mid Tenn State Univ, BS, 53; Univ Tenn, MS, 60. *Prof Exp:* Res chemist, PPG Corp, 60-65, group leader anal chem, 65-67; res chemist, Lithium Corp Am, 67-72, dir qual control, 72-81, mgr planning & asst to pres, 81-83, mgr mkt commun, 84, mgr planning & asst to pres, 85, mgr mkt commun, 86-88, MGR TECH & MKT SERV, FMC CORP LITHIUM DIV, 89- *Mem:* Am Chem Soc; Sigma Xi; Am Inst Chemists. *Res:* Chemical and instrumental analysis; product and promotional literature; product safety; technical service. *Mailing Add:* FMC Corp Lithium Div 449 N Cox Rd Gastonia NC 28054

STUBBLEFIELD, FRANK MILTON, b Hillsboro, Ill, June 25, 11; m 32; c 2. CHEMISTRY, PHARMACEUTICAL CHEMISTRY. *Educ:* Univ Ill, AB, 32, MS, 36, PhD(chem), 42. *Prof Exp:* Chemist, Univ Ill, 35-37 & 39-42; chemist, Swift & Co, 37-39; res chemist, Weldon Spring Ord Works, Atlas Powder Co, Mo, 42-43; prof chem & head dept, Davis & Elkins Col, 43-47; assoc prof, Univ Ill, 47-56; chemist and br chief, US Govt, 56-73; RETIRED. *Concurrent Pos:* Consult, 43-; chief, Chem Div, Chem & Radiol Labs, Chem Corps, 51-53. *Mem:* Fel Am Inst Chemists; Am Chem Soc. *Res:* Organophosphorous chemistry-phosphonates mechanism of action and antidotes; toxicology of compounds of high physiological activity, with special emphasis on heterocyclic nitrogen compounds. *Mailing Add:* Box 368 Rte 1 Palmyra VA 22963

STUBBLEFIELD, ROBERT DOUGLAS, b Decatur, Ill, Mar 4, 36; m 58; c 3. ANALYTICAL CHEMISTRY. *Educ:* Eureka Col, BS, 59. *Prof Exp:* Anal chemist, 59-62, chemist, 62-64, RES CHEMIST, NORTHERN REGIONAL RES CTR, AGR RES SERV, USDA, 64- *Concurrent Pos:* Co-dir int collab study aflatoxin M methods milk, Int Union Pure & Appl Chem-Asn Off Anal Chemists, 72-73 & 78-79. *Mem:* Am Oil Chemists Soc; fel Asn Off Anal Chemists; Am Asn Cereal Chemists; Sigma Xi. *Res:* Identification, preparation, and determination of known and unknown toxic compounds produced by the action of molds on agricultural commodities and products. *Mailing Add:* Northern Regional Res Ctr USDA 1815 N University Peoria IL 61604

STUBBLEFIELD, TRAVIS ELTON, b Austin, Tex, May 27, 35; m 57; c 2. CELL BIOLOGY. *Educ:* NTex State Univ, BS, 57; Univ Wis, MS, 59, PhD(exp oncol), 61. *Prof Exp:* NSF fel animal virol, Max Planck Inst Virus Res, Tubingen, WGer, 62-63; asst prof biol, Univ Tex Grad Sch Biomed Sci Houston, 65-68, assoc prof, 68-73, prof, 74-91; AT DEPT CELL BIOL, ANDERSON HOSP & TUMOR INST. *Concurrent Pos:* Asst biologist, Univ Tex M D Anderson Hosp & Tumor Inst Houston, 63-68, assoc biologist, 68-73, biologist, 74-91. *Mem:* Am Soc Cell Biol. *Res:* Structure and physiology of mammalian chromosomes; synchronized cell culture; cell differentiation; structure and function of centrioles; chromosome transfer; oncogenes. *Mailing Add:* Dept Molecular Genetics M D Anderson Hosp-Box 6 Tex Med Ctr Houston TX 77030

STUBBLEFIELD, WILLIAM LYNN, b Apr 5, 40; m 75. MARINE GEOLOGY. *Educ:* Memphis State Univ, BA, 62; Univ Iowa, MS, 71; Tex A & M Univ, PhD(geol), 80. *Prof Exp:* Chmn prog, Miami Geol Soc, 79-81, vpres, 82-83; assoc prog dir, Marine Geol Resources, Nat Sea Grant Col Prog, Nat Oceanic Atmospheric Admin, 83-85; CHIEF SCIENTIST, NAT UNDERSEA RES PROG, NAT OCEANIC ATMOSPHERIC ADMIN, ROCKVILLE, MD, 85- *Mem:* AAAS; Soc Econ Paleontologists & Mineralogists; Am Asn Petroleum Geologists. *Res:* Continental shelf sedimentary processes; dynamics of sediment transport in submarine canyons; stability of continental slope material; sand and gravel resources on continental shelf. *Mailing Add:* Nat Undersea Res Prog Nat Oceanic Atmospheric Admin 6010 Executive Blvd Rockville MD 20852

STUBBS, DONALD WILLIAM, b Seguin, Tex, Sept 26, 32; m 53; c 5. PHYSIOLOGY. *Educ:* Tex Lutheran Col, BA, 54; Univ Tex, MA, 56, PhD(physiol), 64. *Prof Exp:* Instr zool, Auburn Univ, 56-60; from instr to assoc prof, 63-75, PROF PHYSIOL, UNIV TEX MED BR GALVESTON, 75- *Concurrent Pos:* NIH res grant, 65-71; multidisciplinary res grant ment health, 74-77; course dir, Nat Bd Med Examrs, 74-88, mem, 83-86. *Mem:*

AAAS; Am Physiol Soc; Sigma Xi. *Res:* Biosynthesis of ascorbic acid; hormonal induction of enzyme activities; co-author of textbook of medical physiology. *Mailing Add:* Dept Physiol-Biophys Univ Tex Med Br 301 University Blvd Galveston TX 77550

STUBBS, JOHN DORTON, b Cape Girardeau, Mo, Oct 9, 38; m 62; c 2. MOLECULAR BIOLOGY, BIOCHEMISTRY. *Educ:* Wash Univ, BA, 60; Univ Wis-Madison, MA, 62, PhD(biochem), 65. *Prof Exp:* USPHS fel genetics, Univ Wash, 65-67; from asst prof to assoc prof molecular biol, Calif State Univ, San Francisco, 68-75, PROF MOLECULAR BIOL, SAN FRANCISCO STATE UNIV, 75-, CHMN DEPT CELL & MOLECULAR BIOL, 72- *Concurrent Pos:* Brown-Hazen grant, Calif State Univ, San Francisco, 69-70; USPHS res grant, 70-72. *Mem:* AAAS. *Res:* Regulation of gene expression; transcriptional and translational control of the tryptophan operon in E coli; molecular mechanisms of membrane assembly; developmental biochemistry. *Mailing Add:* Biol Dept 1600 Holloway Ave San Francisco State Univ San Francisco CA 94132

STUBBS, MORRIS FRANK, b Sterling, Kans, May 25, 98; m 23; c 1. CHEMISTRY. *Educ:* Sterling Col, AB, 21; Univ Chicago, MS, 25, PhD(chem), 31. *Hon Degrees:* DSc, Sterling Col, 60. *Prof Exp:* Teacher chem & physics, Elgin Jr Col, 21-23; prof chem & physics & head dept phys sci, Tenn Wesleyan Col, 23-42, dean, 31-42; prof chem & head dept, Carthage Col, 42-44, Tenn Polytech Inst, 44-46 & NMex Inst Mining & Technol, 46-63, dir col div, 62-63; prof chem, Tex Tech Univ, 63-68; chmn dept chem & dir div natural sci & math, 68-78, emer prof chem, Univ Albuquerque, 78-; RETIRED. *Concurrent Pos:* Off Naval Res grant, 50-54. *Honors & Awards:* Clark Medal, Am Chem Soc, 65. *Mem:* Fel AAAS; Am Chem Soc. *Res:* Chemistry of indium; geochemical tests; general chemistry and qualitative analysis. *Mailing Add:* La Vida Llena Apt 4313 10501 Lagrima De Oro Rd NE Albuquerque NM 87111

STUBBS, NORRIS, b Nassau, Bahamas, Nov 8, 48. COMPOSITE MATERIALS, STRUCTURAL & RISK ANALYSIS. *Educ:* Grinnell Col, BA, 72; Columbia Univ, BS, 72, MS, 74, ScD, 76. *Prof Exp:* asst prof civil eng, Columbia Univ, 76-; AT DEPT CIVIL ENG, TEX A&M UNIV. *Concurrent Pos:* Consult, Govt Nigeria, 76; instr, Columbia Univ, 78-81 & Barnard Col, 80-81; proj mgr, Leroy Callender Consult Engrs, 80-; Thomas J Watson Jr fel, 72. *Mem:* Soc Exp Stress Anal; Am Soc Civil Engrs; Am Soc Testing & Mat; Am Inst Aeronaut & Aerospace. *Res:* Nonlinear constitutine models for fabric reinforced composites; continuum modeling of large discrete structural systems; motion control of floating platforms; experimental analysis of structural systems; risk analysis; nondestructive evaluation; quality management. *Mailing Add:* Dept Civil Eng Tex A&M Univ College Station TX 77843

STUBER, CHARLES WILLIAM, b St Michael, Nebr, Sept 19, 31; m 53; c 1. GENETICS. *Educ:* Univ Nebr, BS, 52, MS, 61; NC State Univ, PhD(genetics, exp statist), 65. *Prof Exp:* Instr high sch, Nebr, 56-59; from asst prof to assoc prof genetics, 65-75, PROF GENETICS, NC STATE UNIV, 75-; SUPVRY RES GENETICIST, AGR RES SERV, USDA, 62- *Concurrent Pos:* Ed, CROP SCI, 86-89; ed-in-chief, Crop Sci Soc, 86- *Mem:* Genetics Soc Am; fel Am Soc Agron; fel Crop Sci Soc Am; Am Genetic Asn; Coun Agr Sci & Technol; AAAS. *Res:* Quantitative genetics; quantitative trait investigations in maize using molecular markers. *Mailing Add:* Dept Genetics NC State Univ Raleigh NC 27695-7614

STUBER, FRED A, b Paris, France, Dec 7, 33; m 64; c 1. PHYSICAL ORGANIC CHEMISTRY. *Educ:* Univ Zurich, ChemEng, 57, DSc(nuclear chem), 61. *Prof Exp:* Res assoc mass spectros, Inst Reactor Res, Switz, 61-62; fel, Univ Notre Dame, 62-63 & Mellon Inst, 63-64; res chemist, Ciba, Switz, 64-67; res chemist, 67-75, mgr, D S Gilmore Res Lab, Upjohn Co, North Haven, 75-85; MGR, DOW CHEM USA, 85- *Mem:* Am Chem Soc. *Res:* Ionization and appearance potentials; analysis of nuclear magnetic resonance spectra; photopolymers; catalysis. *Mailing Add:* 65 Chapel Hill Rd North Haven CT 06473-2812

STUBICAN, VLADIMIR S(TJEPAN), b Bjelovar, Yugoslavia, June 23, 24; US citizen; m 46; c 2. MATERIALS SCIENCE, SOLID STATE CHEMISTRY. *Educ:* Univ Zagreb, Dipl Ing, 48, PhD(phys chem), 51, DSc(inorg chem), 58. *Prof Exp:* Res asst phys chem, Univ Zagreb, 48-52, asst prof, 52-55; vdir silicate chem, Inst Silicate Chem, Zagreb, Yugoslavia, 55-58; fel, 58-60, vis assoc prof geochem, 60-61, from asst prof to assoc prof, 61-71, PROF MAT SCI, PA STATE UNIV, 71- *Concurrent Pos:* Consult, Perkin-Elmer Corp, 80-; vis prof, Tech Univ, Tronheim, Norway, 67-68 & Max Planck Inst, Stuttgart, WGer, 77-78. *Mem:* Fel Am Chem Soc; fel Am Ceramic Soc; fel Am Inst Chemists; fel Mineral Soc Am; Croatian Chem Soc. *Res:* Solid state reactions; high temperature materials; inorganic synthesis; solid state technology; chemistry and properties of high melting inorganic materials; material transport in solids; solid state electrolytes. *Mailing Add:* Dept Ceramic Sci & Eng Penn State Univ University Park PA 16802

STUCHLY, MARIA ANNA, b Warsaw, Poland, Apr 8, 39; Can citizen; m 72. ELECTRICAL ENGINEERING, ELECTROMAGNETICS. *Educ:* Warsaw Tech Univ, BSc & MSc, 62; Polish Acad Sci, PhD(elec eng), 70. *Prof Exp:* Asst prof elec eng, Warsaw Tech Univ, 62-64; sr res & develop engr microwaves, Polish Acad Sci, 64-70, sr researcher, Inst Physics, 70; res assoc elec eng & food sci, Univ Man, 70-76; RES SCIENTIST, HEALTH & WELFARE CAN, 76- *Concurrent Pos:* Fel, dept elec eng, Univ Man, 70-72, adj prof, 76; non-resident prof, Dept Elec Eng & dir, Inst Med Eng, Univ Ottawa, 78-; expert consult, Telecom, Australia, 83-84; guest scientist, Nat Bur Standards, 87. *Mem:* Fel Inst Elec & Electronic Engrs; Bioelectromagnetics Soc. *Res:* Interaction of electromagnetic waves with living systems, especially medical applications, measurement techniques, biological effects and safety standards. *Mailing Add:* Health Protection Bldg Health & Welfare Can Rm 66 Ottawa ON K1A 0L2 Can

STUCHLY, STANISLAW S, b Nov 20, 31; Can citizen. ELECTRICAL ENGINEERING. *Educ:* Tech Univ Gliwice, Poland, BScEng, 53; Warsaw Tech Univ, MScEng, 58; Polish Acad Sci, DScEng, 68. *Prof Exp:* Res engr, Indust Inst Telecommun, Warsaw, Poland, 53-56, sr res engr, 56-59; asst prof, Warsaw Tech Univ, 59-63; head microwave instrument dept, Sci Instruments, Polish Acad Sci, 63-70; assoc prof agr eng & adj prof elec eng, Univ Man, 70-76; assoc prof, 76-80, PROF ELEC ENG, UNIV OTTAWA, 80- *Concurrent Pos:* Adj prof elec eng, Carleton Univ, 76-80. *Mem:* Fel Inst Elec & Electronic Engrs. *Res:* Electromagnetic theory and technique; applications of electromagnetic radiations; industrial and medical applications of radio frequency and microwave radiations; electronic instrumentation, especially transducers for measuring nonelectrical quantities. *Mailing Add:* Dept Elec Eng Univ Ottawa Ottawa ON K1N 6N5 Can

STUCK, BARTON W, b Detroit, Mich, Oct 25, 46; m 75. ELECTRICAL ENGINEERING. *Educ:* Mass Inst Technol, BS & MS, 69, ScD(elec eng), 72. *Prof Exp:* mem tech staff, Math & Statist Res Ctr, Bell Labs, Am Tel & Tel Co, 72-; pres, Viatel, 84-85; PRES, BUS STRATEGIES, 86- *Concurrent Pos:* Dir-at-large, Inst Elec & Electronic Engrs Commun Soc. *Mem:* Soc Indust & Appl Math; Math Asn Am; sr mem Inst Elec & Electronics Engrs; Asn Comput Mach; Inst Math Statist. *Res:* Digital systems; applied probability theory; co-author one book. *Mailing Add:* 578 Post Rd E Suite 667 Westport CT 06880

STUCKER, HARRY T, b Lawrence, Kans, Oct 7, 25; m 47; c 3. ENGINEERING, PHYSICS. *Educ:* Univ Kans, BS, 47, MS, 48; Tex Christian Univ, MBA, 75. *Prof Exp:* Mem res staff, 47-54, supvr automatic controls, 54-57, supvr preliminary design, 57-58, supvr electronics lab, 58-59, adv navig guid, 59-61, chief reconnaissance & info systs, 61-63, proj mgr missiles & space systs, 63-67, mgr spec projs, 67-75, dir electronics progs, 75-82, D-DIR F-111 PROG, GEN DYNAMICS/FT WORTH, 82- *Mem:* Am Inst Aeronaut & Astronaut. *Res:* Control systems; microwave devices; communications; ground based radar systems; avionics systems. *Mailing Add:* 3817 Arroyo Rd Ft Worth TX 76109

STUCKER, JOSEPH BERNARD, b Chicago, Ill, Feb 28, 14; m 41; c 3. CHEMISTRY. *Educ:* Univ Chicago, BS, 35. *Prof Exp:* Chemist, Pure Oil Co, 35-41, asst supt grease plant, 41-43, group leader prod develop, 43-50, sect supvr, 50-52, div dir, 52-65, sr res assoc, Res Dept, Union Oil Co, Calif, 65-68, mgr prod develop, Pure Oil Div, 68-69 & Union 76 Div, 69-71, mgr prod qual, Refining Div, 71-86, PROD MGR MKT, UNION OIL CO CALIF, 86- *Honors & Awards:* Achievement Award, Nat Lubricating Grease Inst, 85; Fel, Soc Automotive Engrs, 86. *Mem:* Soc Automotive Eng; Am Soc Testing & Mat; Am Petrol Inst; fel Nat Lubricating Grease Inst. *Res:* Product development of lubricants, particularly greases, gear and crankcase oils and industrial lubricants. *Mailing Add:* Refining & Mkt Div Union Oil Co Calif 4211 Marshall Des Plaines IL 60016

STUCKER, ROBERT EVAN, b Burlington, Iowa, Jan 28, 36; m 56; c 3. PLANT BREEDING, STATISTICS. *Educ:* Iowa State Univ, BS, 59; Purdue Univ, MS, 61; NC State Univ, PhD(genetics), 66. *Prof Exp:* Res geneticist, Forage & Range Br, Crops Res Div, Agr Res Serv, USDA, 65-68; asst prof agron & plant genetics, corn breeding & quant genetics, 68-72, assoc prof agron & plant genetics, 72-77, PROF AGRON & PLANT GENETICS, UNIV MINN, ST PAUL, 77- *Concurrent Pos:* Statist design experiments & consult statistician agr & hort, 72-78; quant genetics consult, Alberta Forest Serv, 82-84. *Mem:* Am Soc Agron; Crop Sci Soc Am; Sigma Xi. *Res:* Wild rice (zizania) breeding; application of quantitative genetics in plant breeding. *Mailing Add:* Dept Agron & Plant Genetics 411 Borlang Hall Univ Minn 1991 Buford Circle St Paul MN 55108

STUCKEY, A(LTO) NELSON, JR, chemical engineering; deceased, see previous edition for last biography

STUCKEY, JOHN EDMUND, b Stuttgart, Ark, Dec 6, 29; m 55; c 3. PHYSICAL CHEMISTRY, INORGANIC CHEMISTRY. *Educ:* Hendrix Col, BA, 51; Univ Okla, MS, 53, PhD(chem), 57. *Prof Exp:* Res chemist, Oak Ridge Nat Lab, Union Carbide Corp, 57; asst prof chem, La Polytech Inst, 57-58; PROF CHEM, HENDRIX COL, 58- *Mem:* Am Chem Soc. *Res:* Preparations and properties of monofluorophosphate compounds; solution chemistry in the critical temperature region; x-ray crystallography. *Mailing Add:* Dept Chem Hendrix Col Conway AR 72032

STUCKEY, RONALD LEWIS, b Bucyrus, Ohio, Jan 9, 38. BOTANY. *Educ:* Heidelberg Col, BS, 60; Univ Mich, MA, 62, PhD(bot), 65. *Prof Exp:* Instr bot, Univ Mich, 65; from asst prof to assoc prof, 65-78, cur herbarium, 67-75, assoc dir, Franz Theodore Stone Lab, 77-85, PROF BOT, OHIO STATE UNIV, 78- *Mem:* Am Soc Plant Taxon; Int Asn Plant Taxon; Bot Soc Am. *Res:* Taxonomy and distribution of angiosperms; history of American botany; monographic studies in the Cruciferae, particularly Rorippa; Ohio vascular plant flora and phytogeography; history of plant taxonomy in North America; taxonomy and distribution of angiosperms, particularly aquatic and marsh flora. *Mailing Add:* Dept of Bot Ohio State Univ 1735 Neil Ave Columbus OH 43210-1239

STUCKEY, WALTER JACKSON, JR, b Fairfield, Ala, Mar 6, 27; m 52; c 3. INTERNAL MEDICINE, HEMATOLOGY & ONCOLOGY. *Educ:* Univ Ala, BS, 47; Tulane Univ, MD, 51. *Prof Exp:* Intern, Charity Hosp La, New Orleans, 51-52, resident internal med, 55, Vet Admin Hosp, New Orleans, 55-56 & Charity Hosp La, 56-57; from instr to assoc prof, 58-68, PROF MED, SCH MED, TULANE UNIV, 68-, CHIEF HEMAT & MED ONCOL, 63- *Concurrent Pos:* Consult, New Orleans Charity Hosp; consult, Baptist, East Jefferson, Hotel Dieu, Mercy, Methodist, Touro, Lakeside, Jo Ellen Smith/F Edward Herbert, Bønnabel, Doctors' Hosp of Jefferson & Vet Admin Hosps, New Orleans. *Mem:* Am Asn Cancer Educ; Am Soc Clin Oncol; fel Am Col Physicians; Am Soc Internal Med; Am Soc Hemat; Am Fedn Clin Res. *Res:* Bone marrow function in health and disease. *Mailing Add:* Dept Med Tulane Univ Sch Med 1430 Tulane Ave New Orleans LA 70112

STUCKI, JACOB CALVIN, b Neillsville, Wis, Nov 30, 26; m 48; c 3. ENDOCRINOLOGY, RESEARCH ADMINSTRATION. *Educ:* Univ Wis, BS, 48, MS, 51, PhD(zool, physiol), 54. *Prof Exp:* Res asst, Univ Wis, 50-54; endocrinologist, Wm S Merrell Co, 54-57; res assoc, Upjohn Co, 57-60, dept head endocrinol, 60-61, mgr pharmacol res, 61-68, dir res planning & admin, Pharmaceut Res & Develop, 68-79, dir admin & support opers, 79-81, corp vpres pharmaceut res, 81-89; RETIRED. *Concurrent Pos:* Pharmaceut consult, 89- *Mem:* AAAS; Soc Exp Biol & Med; Endocrine Soc. *Res:* Reproduction; inflammation; pharmacology; research management. *Mailing Add:* 2842 Bronson Blvd Kalamazoo MI 49008

STUCKI, JOSEPH WILLIAM, b Rexburg, Idaho, Feb 4, 46; m 68; c 6. SOIL CHEMISTRY, PHYSICAL CHEMISTRY. *Educ:* Brigham Young Univ, BS, 70; Utah State Univ, MS, 73; Purdue Univ, PhD(soil chem), 75. *Prof Exp:* Asst prof, 76-80, ASSOC PROF SOIL CHEM, UNIV ILL, 80- *Mem:* Asn Int Etude Argilles; Clay Minerals Soc; Soil Sci Soc Am; Int Soil Sci Soc; Mineral Soc Gt Brit. *Res:* Clay colloid chemistry, clay-water interactions, affects of structural iron oxidation states in clays and soils on their colloidal properties; advanced spectroscopic methods for analysis and characterization of soils and clays; physical-inorganic chemistry of clays. *Mailing Add:* Dept Agron S-510 Turner Hall Univ Ill 1102 S Goodwin Urbana IL 61801

STUCKI, WILLIAM PAUL, b Neillsville, Wis, Sept 28, 31; m 55; c 5. BIOCHEMISTRY. *Educ:* Univ Wis, BS, 57, MS, 59, PhD(biochem), 62. *Prof Exp:* Asst prof Univ Puerto Rico & scientist, PR Nuclear Ctr, Mayaguez, 62-63; asst prof, Antioch Col & assoc biochem, Fels Res Inst, 63-67; res biochemist, 67-75, SR RES BIOCHEMIST, PARKE DAVIS & CO, 75- *Mem:* Am Chem Soc; NY Acad Sci. *Res:* Protein and amino acid nutrition and metabolism; plant biochemistry; biochemistry of natural products; chemotherapy, immunology and immunochemistry; atherothrombotic disease. *Mailing Add:* 13545 Austin Rd Manchester MI 48158

STUCKWISCH, CLARENCE GEORGE, b Seymour, Ind, Oct 13, 16; m 42; c 5. CHEMISTRY. *Educ:* Ind Univ, AB, 39; Iowa State Col, PhD(org chem), 43. *Prof Exp:* Res assoc, Iowa State Col, 43; asst prof chem, Wichita State Univ, 43-44; res chemist, Eastman Kodak Co, 44-4S; from asst prof to assoc prof chem, Wichita State Univ, 45-60; assoc prof, NMex Highlands Univ, 60-62, prof & head dept, 62-64, dir inst sci res, 63-64; prof chem & exec officer dept, State Univ NY Buffalo, 64-68; chmn dept chem, Univ Miami, 68-74, prof 68-81, dean grad studies & res, 72-81, assoc vpres advan studies, 80-81, exec vpres & provost, 81-82; RETIRED. *Mem:* Am Chem Soc; Sigma Xi. *Res:* Organometallic, psychopharmacological and organophosphorous compounds. *Mailing Add:* 402 Holly St Johnson City TX 37601

STUCKY, GALEN DEAN, b McPherson, Kans, Dec 17, 36; m 61; c 2. INORGANIC CHEMISTRY. *Educ:* McPherson Col, BS, 57; Iowa State Univ, PhD(chem), 62. *Prof Exp:* Fel physics, Mass Inst Technol, 62-63; NSF fel, Quantum Chem Inst, Fla, 63-64; from asst prof to assoc prof chem, Univ Ill, Urbana, 64-72, prof, 72-79; group leader, Sandia Nat Lab, 79-81; group leader, Cent Res & Develop Dept, E I Du Pont de Nemours & Co, Inc, 81-85; PROF CHEM, UNIV CALIF, SANTA BARBARA, 85- *Concurrent Pos:* Consult div univ & col, Argonne Nat Labs, 66; vis prof physics, Univ Uppsala, Sweden, 71; assoc ed, Inorg Chem, 77-85; chmn, Inorg Div, Am Chem Soc, 86. *Mem:* Am Chem Soc; Mat Res Soc; AAAS. *Res:* Organometallic chemistry of electron deficient sites; solid state chemistry; zeolites and molecular sieves; nonlinear optic materials; environmental chemistry. *Mailing Add:* Dept Chem Univ Calif Santa Barbara CA 93106

STUCKY, GARY LEE, b Murdock, Kans, May 18, 41. BIOINORGANIC CHEMISTRY. *Educ:* Bethel Col, Kans, AB, 63; Kans State Univ, PhD(inorg chem), 67. *Prof Exp:* Instr chem, Halstead Sch Nursing, Kans, 62-63 & Kans State Univ, 63-65; res scientist bioinorg chem, Miles Labs, Ind, 67-70 & Kivuvu Inst Med Evangel, Kimpese, Zaire, 71; from asst prof to assoc prof, 72-79, PROF BIOINORG CHEM, EASTERN MENNONITE COL, 79- *Concurrent Pos:* Am Leprosy Mission res grant, Eastern Mennonite Col, 72; vis prof chem, Univ Rochester, 81-82, Bethel Col, Kans, 86-87; res assoc & vis prof Univ Notre Dame, Ind, 87-88. *Mem:* AAAS; Am Chem Soc; Royal Soc Chem. *Res:* Electrochemistry of leprosy; inorganic synthesis; ion-selective electrodes; hemoglobin variants. *Mailing Add:* Dept of Chem Eastern Mennonite Col Harrisonburg VA 22801

STUCKY, RICHARD K(EITH), b Newton, Kans, Dec 11, 49; m 69. EVOLUTIONARY PALEOECOLOGY, BIOSTRATIGRAPHY. *Educ:* Univ Colo, Denver, BA, 75, MA, 77; Univ Colo, Boulder, PhD(anthrop), 82. *Prof Exp:* Preparator vert paleont, Denver Mus Natural Hist, 71-76; instr anthrop, Metrop State Col, Denver, 77; Honorarium instr archaeol, Univ Colo, Denver, 77; res fel, Carnegie Mus Natural Hist, 82-84, collection mgr vert paleont, 84-87, asst cur, 87-88; DEPT HEAD & EARTH SCI CUR, DENVER MUS NATURAL HIST, 89- *Concurrent Pos:* Res assoc, Carnegie Mus; adj assoc prof, Univ Pittsburgh & Univ Colo. *Mem:* Soc Vertebrate Paleont; Am Asn Phys Anthropologists; Ecol Soc Am; Soc Am Nat; Rocky Mtn Asn Geol; Geol Soc Am; AAAS. *Res:* Paleontology (systematics and evolution) of marsupials, primates, artiodactyls and perissodactyls; evolutionary paleoecology of mammals; geological remote sensing. *Mailing Add:* Dept Earth Sci Denver Mus Natural Hist 2001 Colorado Blvd Denver CO 80205

STUDDEN, WILLIAM JOHN, b Timmins, Ont, Sept 30, 35. MATHEMATICAL STATISTICS. *Educ:* McMaster Univ, BSc, 58; Stanford Univ, PhD(statist), 62. *Prof Exp:* Res assoc math, Stanford Univ, 62-64; from asst prof to assoc prof statist, 64-71, PROF STATIST, PURDUE UNIV, LAFAYETTE, 71- *Concurrent Pos:* NSF fels, Purdue Univ, Lafayette, 68-71, 72-75. *Mem:* Fel Inst Math Statist; Sigma Xi. *Res:* Optimal designs; Tchebycheff systems. *Mailing Add:* Dept of Statist Purdue Univ Lafayette IN 47907

STUDEBAKER, GERALD A, b Freeport, Ill, July 22, 32; m 55, 79; c 5. AUDIOLOGY. *Educ:* Ill State Univ, BS, 55; Syracuse Univ, MS, 56, PhD(audiol), 60. *Prof Exp:* Supvr clin audiol, Vet Admin Hosp, DC, 59-61, chief audiol & speech path serv, Syracuse, NY, 61-62; supvr clin audiol, Med Ctr, Univ Okla, 62-66, from asst prof to assoc prof audiol & consult, Dept Otorhinolaryngol, 62-72, res audiologist, 66-72; res prof audiol, Memphis State Univ, 72-76; prof audiol, PhD Prog Speech & Hearing Sci, City Univ New York, 76-79; DISTINGUISHED PROF SPEECH & HEARING SCI, MEMPHIS STATE UNIV, 79- *Honors & Awards:* Jacob K Javits Award. *Mem:* AAAS; fel Am Speech & Hearing Asn; fel Acoust Soc Am. *Res:* Bone-conduction hearing thresholds, auditory masking; loudness estimation procedures and adaptation; speech discrimination; ear mold acoustics. *Mailing Add:* Memphis Speech & Hearing Ctr 807 Jefferson Ave Memphis TN 38105

STUDER, REBECCA KATHRYN, b Hannibal, Mo, Feb 22, 43; m 64; c 2. CELL CALCIUM METABOLISM, SEXUAL DIMORPHISM. *Educ:* Northeast Mo State Univ, BS, 64; Tex Christian Univ, MS, 66; Univ Pittsburgh, PhD(physiol), 78. *Prof Exp:* Res instr nuclear med, Sch Med, Wash Univ, 66-73; res assoc med & physiol, 79-81, RES ASST PROF PHYSIOL, MED SCH, UNIV PITTSBURGH, 82- *Mem:* Am Physiol Soc; Endocrine Soc. *Res:* Differences between males and females in their hepatocytes' response to adrenergic stimulation; sexual dimorphism in adrenergic response; effect of diabetes on calcium mediated metabolism in the liver. *Mailing Add:* Dept Physiol Univ Pittsburgh Med Sch 645 Scaife Hall Pittsburgh PA 15261

STUDIER, EUGENE H, b Dubuque, Iowa, Mar 16, 40; div; c 2. PHYSIOLOGICAL ECOLOGY. *Educ:* Univ Dubuque, BS, 62; Univ Ariz, PhD(zool), 6S. *Prof Exp:* From asst prof to assoc prof biol, NMex Highlands Univ, 65-72; assoc prof, 72-74, chmn dept, 78-85, PROF BIOL, UNIV MICH, FLINT, 74- *Concurrent Pos:* USPHS grant, 66-67; Sigma Xi res grant-in-aid, 69; Am Philos Soc grant, 69. *Mem:* Fel AAAS; Am Soc Mammalogists; Am Soc Zoologists. *Res:* Mammalian physiology; physiological adaptation in bats and rodents. *Mailing Add:* Dept Biol Univ Mich Flint MI 48502-2186

STUDIER, FREDERICK WILLIAM, b Waverly, Iowa, May 26, 36; m 62; c 2. BIOPHYSICS, MOLECULAR BIOLOGY. *Educ:* Yale Univ, BS, 58; Calif Inst Technol, PhD(biophys), 63. *Prof Exp:* NSF fel biochem, Med Ctr, Stanford Univ, 62-64; from asst biophysicist to biophysicist, 64-74, SR BIOPHYSICIST, BIOL DEPT, BROOKHAVEN NAT LAB, 74-, CHMN, 90- *Concurrent Pos:* Adj assoc prof biochem, State Univ NY Stony Brook, 71-75, prof, 75- *Honors & Awards:* Ernest O Lawrence Mem Award, Environ Res Develop Admin, 77. *Mem:* AAAS; Biophys Soc; Am Soc Biochem & Molecular Biol; Am Soc Microbiol; Am Soc Virol; Am Acad Arts & Scis. *Res:* Physical and chemical properties of nucleic acids; genetics and physiology of bacteriophage T7. *Mailing Add:* Biol Dept Brookhaven Nat Lab Upton NY 11973

STUDIER, MARTIN HERMAN, b Leola, SDak, Nov 10, 17; m 44; c 4. CHEMISTRY. *Educ:* Luther Col, BA, 39; Univ Chicago, PhD(chem), 47. *Prof Exp:* Asst, Iowa State Col, 39-41, instr, 41-42; res chemist, 43-36, SR CHEMIST, ARGONNE NAT LAB, 46- *Mem:* AAAS; Am Chem Soc; Am Phys Soc; Sigma Xi. *Res:* Nuclear chemistry of the heavy elements; chemical nature of coals; mass spectrometry; organic matter in meteorites. *Mailing Add:* 4429 Downers Dr Downers Grove IL 60515-2729

STUDLAR, SUSAN MOYLE, b St Paul, Minn, May 4, 44; m 79; c 2. BRYOLOGY. *Educ:* Carleton Col, BA, 66; Univ Tenn, Knoxville, PhD(bot), 73. *Prof Exp:* Instr biol, Wellesley Col, 71-72; asst prof, Va Commonwealth Univ, 72-74; asst prof biol, Centre Col Ky, 74-77, assoc prof, 78-82. *Concurrent Pos:* Vis prof, Mountain Lake Biol Sta, Univ Va, 77, 79 & 80; tech dir, Cent Ky Wildlife Refuge, 79-82 & 84-; adj assoc prof biol, Centre Col Ky, 82- *Mem:* Am Bryol & Lichenological Soc; Brit Bryological Soc. *Res:* Floristics and ecology of bryophytes of Kentucky and Virginia; culturing of bryophytes; trampling effects on bryophytes. *Mailing Add:* Dept Bot & Microbiol Okla State Univ Stillwater OK 74078-0289

STUDT, WILLIAM LYON, b Ypsilanti, Mich, Mar 12, 47; m 66; c 3. MEDICINAL CHEMISTRY. *Educ:* Eastern Mich Univ, BA, 69; Univ Mich, PhD(org chem), 73. *Prof Exp:* Res fel org chem, Yale Univ, 73-74; SECT HEAD RES MED CHEM, RORER-AMCHEM INC, 74- *Mem:* Am Chem Soc. *Res:* The organic synthesis of natural products and biologically active compounds. *Mailing Add:* 611 Store Rd Harleysville PA 19438-2717

STUDTMANN, GEORGE H, b Chicago, Ill, Nov 3, 30; m 61; c 4. ELECTRO-MECHANICAL, POWER ELECTRONICS. *Educ:* Purdue Univ, BSEE, 56, MSEE, 57. *Prof Exp:* Prin elec engr, Battelle Mem Inst, 57-61; sr elec engr, Electro-Mech Systs, Borg-Warner Res Ctr, 61-62, supvr res eng, 62-73, mgr electronics & elec eng, 73-78, mgr power electronics, 78-83, sr scientist & mgr, 83-88; TECH DIR POWER ELECTRONICS, SQUARE D CO, 88- *Honors & Awards:* Tech Innovation Award, Borg-Warner. *Mem:* Sigma Xi; Inst Elec & Electronics Engrs; Soc Automotive Engrs. *Res:* Electromagnetic actuators; variable frequency alternating current motor drives; automotive applications and includes advanced activators, sensors and systems involving both rotating machinery and power electronics. *Mailing Add:* Power Electronics Square D Co Exec Plaza 1415 S Roselle Rd Palatine IL 60067

STUDZINSKI, GEORGE P, b Poznan, Poland, Oct 30, 32; m 59; c 4. EXPERIMENTAL PATHOLOGY, CELL BIOLOGY. *Educ:* Glasgow Univ, BS, 55, MB, 58, PhD(exp path), 62. *Prof Exp:* Brit Empire Cancer Campaign res fel path, Glasgow Royal Infirmary, 59-60, resident, 60-62; from instr to prof path, Jefferson Med Col, 62-75; PROF PATH & CHMN DEPT, NJ MED SCH, 76- *Mem:* Tissue Cult Asn; Am Soc Exp Path; Histochem Soc; Am Asn Cancer Res; Am Soc Cell Biol. *Res:* Study of effect of cancer chemotherapeutic agents on cultured diploid and aneuploid mammalian cells by a combination of cytochemical and biochemical methods; clinical pathology. *Mailing Add:* Dept Path NJ Med Sch 185 S Orange Ave Newark NJ 07103-2484

STUEBEN, EDMUND BRUNO, b Cuxhaven, Ger, Apr 22, 20; nat US; m 4S; c 4. PARASITOLOGY. *Educ:* NY Univ, BS, 41; Baylor Univ, MA, 49; Univ Fla, PhD(zool), 53. *Prof Exp:* Instr biol lab & parasitol lab, Univ Fla, 50-53; asst prof biol, Arlington State Col, 53; assoc prof, 54-63, PROF ZOOL & PHYSIOL, UNIV SOUTHWESTERN LA, 63- *Mem:* Sigma Xi. *Res:* Larval development of filariae in arthropods; physiology of filaria larva; transmission of infective state filaria larva; antihistamine effect on coronary circulation. *Mailing Add:* 1405 E Bayou Pkwy Lafayette LA 70508

STUEBER, ALAN MICHAEL, b St Louis, Mo, Apr 18, 37. GEOCHEMISTRY. *Educ:* Wash Univ, BS, 58, MA, 61; Univ Calif, San Diego, PhD(earth sci), 65. *Prof Exp:* Res assoc earth sci, Wash Univ, 65-66; fel geochem, Carnegie Inst Washington, 66-67; from asst prof to assoc prof geol, Miami Univ, 67-75; RES GEOCHEMIST, OAK RIDGE NAT LAB, 77- *Mem:* Geochem Soc; Sigma Xi. *Res:* Earth sciences; strontium isotope studies. *Mailing Add:* RR 2 Box 83 Edwardsville IL 62025

STUEBING, EDWARD WILLIS, b Cincinnati, Ohio, Sept 9, 42; m 82; c 2. AEROSOL OPTICS, AEROSOL CHARACTERIZATION. *Educ:* Univ Cincinnati, BS, 65; Johns Hopkins Univ, MA, 69, PhD(chem physics), 70. *Prof Exp:* Res physicist, US Army Frankford Arsenal, 70-74, res chemist, Pitman-Dunn Labs, 74-77, RES PHYSICAL SCIENTIST & COORDR AEROSOL RES, US ARMY CHEM RES DEVELOP & ENG CTR, 77- *Concurrent Pos:* Adj asst prof chem, Drexel Univ, 74-77. *Honors & Awards:* Res & Develop Achievement Award, US Army, 74 & 85. *Mem:* Am Phys Soc; Asn Comput Mach; Int Soc Quantum Biol; Am Chem Soc; Sigma Xi; Am Asn Aerosol Res. *Res:* Aerosol light scattering; remote sensing of chemical and biological species in the atmosphere and on surfaces; theoretical study of molecular structure, excited states and energy transfer; interaction of matter with laser light at high power density; mathematical modeling and operations research analyses; computer automation of research administration. *Mailing Add:* PO Box 233 Gunpowder Br APG MD 21010

STUEDEMANN, JOHN ALFRED, b Clinton, Iowa, Oct 3, 42; m 67. ANIMAL NUTRITION. *Educ:* Iowa State Univ, BS, 64; Okla State Univ, MS, 67, PhD(ruminant nutrit), 70. *Prof Exp:* RES PHYSIOLOGIST, SOUTHERN PIEDMONT CONSERV RES CTR, AGR RES SERV, USDA, 70- *Mem:* Am Soc Animal Sci; Am Forage & Grassland Coun. *Res:* Ruminant nutrition; forage production and utilization; waste disposal and land fertilization; health problems of beef cattle; forage finishing of cattle; cow-calf management. *Mailing Add:* Southern Piedmont Conserv Res Ctr USDA PO Box 555 Watkinsville GA 30677

STUEHR, JOHN EDWARD, b Aug 30, 35; US citizen; m 62; c 4. BIOPHYSICAL CHEMISTRY. *Educ:* Western Reserve Univ, BA, 57, MS, 59, PhD(chem), 61. *Prof Exp:* NIH res fel chem, Max Planck Inst, Gottingen, WGer, 62-63; from asst prof to assoc prof, 64-74, chmn dept, 77-81, PROF CHEM, CASE WESTERN RESERVE UNIV, 74- *Mem:* Am Chem Soc; Sigma Xi; Fedn Biol Chemists. *Res:* Reaction kinetics of fast processes in solution; relaxation spectroscopy; metal complexing; biochemical kinetics; elementary steps in enzyme kinetics. *Mailing Add:* Dept Chem Case Western Reserve Univ 2040 Adelbert Rd Cleveland OH 44106

STUELAND, DEAN T, b Viroqua, Wis, June 24, 50; m; c 4. BIOMEDICAL ENGINEERING. *Educ:* Univ Wis-Madison, BS, 72, MS, 73, MD, 77; Am Bd Intenal Med, cert, 80; Am Bd Emergency Med, cert, 87. *Prof Exp:* Internal med resident, Marshfield Clin/St Joseph's Hosp, Univ Wis, 77-80; staff emergency physician, Riverview Hosp, Wisconsin Rapids, Wis, 80-81; STAFF PHYSICIAN, MARSHFIELD CLIN, 81-, DIR EMERGENCY SERV, 81- *Concurrent Pos:* Dir, Nat Farm Med Ctr, 84-; asst clin prof, Univ Wis, 84-89, assoc clin prof, 89-; chmn, Trauma Comt, Am Col Emergency Physicians, 86 & 87, Pre-Hosp Subcomt, 88-89; med dir, AODA Unit, Marshfield Clin, 88-, Basic Trauma & Life Support, 89-; mem, Sexual Assault Task Force, Dept Health & Social Serv, 89-90. *Mem:* Fel Am Col Physicians; fel Am Col Emergency Physicians; Am Pub Health Asn; Soc Critical Care Med; Am Occup Med Asn; fel Am Col Prev Med; Am Soc Addiction Med; NY Acad Sci; Inst Elec & Electronics Engrs; Asn Comput Mach. *Res:* Agricultural health and safety. *Mailing Add:* 1000 N Oak Marshfield WI 54449

STUELPNAGEL, JOHN CLAY, b Houston, Tex, Nov 12, 36; m 59; c 3. MATHEMATICS. *Educ:* Yankton Col, BA, 55; Johns Hopkins Univ, PhD(math), 62. *Prof Exp:* Fel math, Res Inst Advan Studies, Martin-Marietta Corp, 61-64; sr engr, 64-66, fel engr, Aerospace Div, 66-76, prog mgr, Syst Develop Div, 76-86, DEPT MGR, DIGITAL SYSTEMS, WESTINGHOUSE ELECTRIC CORP, 86- *Mem:* Am Defense Preparedness Asn. *Res:* Lie groups; linear algebra; differential equations; computation and computer design; radar development. *Mailing Add:* 5306 Tilbury Way Baltimore MD 21212

STUESSE, SHERRY LYNN, b Aruba, Neth, WI, Feb 2, 44; US citizen; m 66; c 3. NEUROSCIENCES. *Educ:* Vanderbilt Univ, Tenn, BA, 66; Washington Univ, St Louis, Mo, MA, 68; State Univ NY, Albany, PhD(biol), 72. *Prof Exp:* Postdoctoral fel, Case Western Reserve Univ, 71-76 & Mt Sinai Med Ctr, 76-77; from instr to assoc prof, 77-88, PROF NEUROBIOL, NE OHIO UNIV COL MED, 88- *Concurrent Pos:* Vis scientist, Salk Inst, La Jolla, Calif, 85-86 & Friday Harbor, Univ Wash, 89-90; prof, Biomed Eng, Univ Akron & Biol Dept, Kent State Univ, 88- *Mem:* Am Physiol Soc; Am Asn Anatomists; Soc Neurosci; Asn Women Sci. *Res:* Neural control of the heart; comparative neuroanatomy; author of over 30 publications. *Mailing Add:* 7561 W Lake Blvd Kent OH 44248

STUESSY, TOD FALOR, b Pittsburgh, Pa, Nov 18, 43; div; c 2. SYSTEMATIC BOTANY. *Educ:* DePauw Univ, BA, 65; Univ Tex, Austin, PhD(bot), 68. *Prof Exp:* Asst prof, 68-74, assoc prof, 74-79, PROF BOT, OHIO STATE UNIV, 79- *Concurrent Pos:* Res assoc, Field Mus Nat Hist, 70-77; Maria Moors Cabot res fel, Gray Herbarium, Harvard Univ, 71-72; assoc dir, Syst Biol Prog, NSF & collabr, Dept Bot, Smithsonian Inst, 77-78. *Honors & Awards:* Wilks Award, Southwestern Asn Naturalists. *Mem:* AAAS; Int Asn Plant Taxon; Am Soc Plant Taxon (pres, 87-88); Bot Soc Am; Soc Study Evolution; Sigma Xi. *Res:* Systematics and evolution of Compositae. *Mailing Add:* Dept Bot Ohio State Univ 1735 Neil Ave Columbus OH 43210

STUEWER, ROGER HARRY, b Sept 12, 34; US citizen; m 60; c 2. HISTORY OF PHYSICS. *Educ:* Univ Wis, BS, 58, MS, 64, PhD(hist sci & physics), 68. *Prof Exp:* Instr physics, Heidelberg Col, 60-62; from asst prof to assoc prof hist physics, Univ Minn, Minneapolis, 67-71; assoc prof hist sci, Boston Univ, 71-72; assoc prof hist sci, 72-74, PROF HIST SCI & TECH, UNIV MINN, MINNEAPOLIS, 74- *Concurrent Pos:* Mem adv panel hist & philos sci, NSF, 70-72, res support, 70-, hon res assoc, Harvard Univ, 74-75; Am Coun Learned Soc fel, 74-75, 83-84, chmn AM Inst Physics, adv comn hist physics, 80-, Volkswagen Found vis prof, Deutsches Mus, Munich, 81-82, chmn Am Phys Soc, div hist physics, 87-88; vis prof, Univ Vienna & Graz, 89. *Honors & Awards:* Distinguished Serv Citation, Am Asn Physics Teachers, 90. *Mem:* Hist Sci Soc (secy, 72-78); fel AAAS; Sigma Xi; Brit Soc Hist Sci; Am Asn Physics Teachers; Am Phys Soc. *Res:* History of twentieth century physics, especially radiation theory, quantum theory, and nuclear physics; Compton effect as a turning point in physics; nuclear physics between WWI and WWII. *Mailing Add:* Sch Physics & Astron Univ Minn Minneapolis MN 55455

STUFFLE, ROY EUGENE, b Elwood, Ind, Oct 14, 44; m 85; c 1. SEMICONDUCTOR DEVICE MODELING. *Educ:* Rose Polytech Inst, BSEE, 66 MSEE, 69; Ind Univ, Ft Wayne, MSBA, 75; Purdue Univ, PhD(elec eng), 79. *Prof Exp:* Adv develop engr, Appl Res & Develop Lab, Gen Elec Co, 68-69 & 72-75; asst prof elec eng, Mich Technol Univ, 79-85; ASST PROF ELEC ENG, UNIV MO-ROLLA, 84- *Concurrent Pos:* Vis asst prof, Univ Mo-Rolla, 84-85; mem steering comt, Midwest Symp, Circuits & Systs. *Mem:* Inst Elec & Electronics Engrs; Sigma Xi; Am Soc Eng Educ. *Res:* Computer-aided circuit analysis and design; automated modeling procedures for semiconductor devices; adaptation to microcomputers; educational applications of microcomputers. *Mailing Add:* 1005 Park Lane Pocatello ID 83201-2819

STUFFLEBEAM, CHARLES EDWARD, b St Louis, Mo, Feb 22, 33; m 52; c 3. ANIMAL GENETICS, BIOCHEMISTRY. *Educ:* Univ Mo, BS, 58, MS, 61, PhD, 64. *Prof Exp:* Asst county agent, Exten Div, Univ Mo, 59-61, instr animal husb & agr biochem, 62-64; assoc prof range animal sci, Sul Ross State Col, 64-65; assoc prof animal sci, Northwestern State Univ, 65-69; PROF ANIMAL SCI, SOUTHWEST MO STATE UNIV, 69- *Mem:* Am Soc Animal Sci; Nat Asn Cols & Teachers Agr (past pres). *Res:* Genetics; biochemistry, physiology and nutrition of domestic animals and their application to agriculture; animal husbandry. *Mailing Add:* Dept Agr Southwest Mo State Univ Springfield MO 65804-0094

STUHL, LOUIS SHELDON, b New York, NY, Feb 5, 51; m 77; c 2. HOMOGENEOUS CATALYSIS, TRANSITION METAL CHEMISTRY. *Educ:* Mass Inst Technol, SB, 73; Cornell Univ, MS, 76, PhD(chem), 78. *Prof Exp:* NSF fel chem, Univ Calif, Berkeley, 78-79; asst prof chem, Brandeis Univ, 79-86; chemist, Am Optical Co, 87-88; SCIENTIST, POLAROID CORP, 88- *Mem:* Am Chem Soc. *Res:* Metal and colloid chemistry and their application to problems of catalysis photography and organic synthesis, including exploratory work and mechanistic studies of known systems. *Mailing Add:* Polaroid Corp Cambridge MA 02139

STUHLINGER, ERNST, b Niederrimbach, Ger, Dec 19, 13; nat US; m 50; c 3. PHYSICS. *Educ:* Univ Tübingen, PhD(physics), 36. *Prof Exp:* Asst prof, Berlin Inst Technol, 36-41; res asst guid & control, Rocket Develop Ctr, Peenemuende, 43-45; asst res & develop, Ord Corps, US Army, Ft Bliss, Tex & White Sands Proving Ground, NMex, 46-50, astronaut res adminr & supvry phys scientist, Army Ballistic Missile Agency, Redstone Arsenal, Ala, 50-60; dir, Space Sci Lab, Marshall Space Flight Ctr, NASA, 60-68, assoc dir sci, 68-76; sr res scientist, Univ Ala, Huntsville, 76-82; CONSULT, SPACE RES & TECHNOL, INDUST CORP, 82- *Honors & Awards:* Roentgen Prize; Galabert Prize, Paris, 62; Hermann Oberth Award, 62 & Medal, 64; Propulsion Award, Am Inst Aeronaut & Astronaut, 60; Wernher von Braun Prize, 85. *Mem:* Fel Am Inst Aeronaut & Astronaut; fel Am Astronaut Soc; Ger Phys Soc; Ger Soc Rockets & Space Flight; Brit Interplanetary Soc; Sigma Xi. *Res:* Feasibility and design studies of electrical propulsion systems for space ships; scientific satellites and space probes; electric automobiles; manned missions to Mars. *Mailing Add:* 3106 Rowe Dr SE Huntsville AL 35801

STUHLMAN, ROBERT AUGUST, b Cincinnati, Ohio, Apr 9, 39; m 60; c 3. LABORATORY ANIMAL MEDICINE, MEDICAL RESEARCH. *Educ:* Ohio State Univ, DVM, 68; Univ Mo, Columbia, MS, 71; Am Col Lab Animal Med, dipl, 74. *Prof Exp:* Res asst vet clin, Ohio State Univ, 66-68; res assoc vet med & surg, Univ Mo, Columbia, 68-71 & instr path & asst dir lab animal med, Med Ctr, 71-75; DIR LAB ANIMAL RESOURCES, DIR INTERDISCIPLINARY TEACHING LABS & ASSOC PROF PATH, WRIGHT STATE UNIV, 75- *Concurrent Pos:* Vet med officer, Vet Admin Hosp, Columbia, Mo, 72-75; consult gen med res, Vet Admin Ctr, Dayton, Ohio, 75-; consult vet, Cent State Univ, Wilberforce, Ohio, 80-; consult lab animal med, Springborn Inst Bioresearch, Spencerville. Ohio, 82-, Environ Protection Agency, Cincinnati, Ohio, 84- *Honors & Awards:* Dr Davis S White Mem Award. *Mem:* Am Vet Med Asn; Nat Soc Med Res; Am Asn Lab Animal Sci; Am Col Lab Animal Med; Am Soc Lab Animal Practitioners. *Res:* Diabetes mellitus, especially development of the animal model; diagnosis; pathogenesis; establishment of secondary complications; therapy; inheritance patterns. *Mailing Add:* Dept Pathol Wright State Dayton OH 45401

STUHLMILLER, GARY MICHAEL, m 80; c 2. TUMOR ASSOCIATED ANTIGENS & IMMUNOLOGY. *Educ:* Duke Univ, PhD(immunol), 76. *Prof Exp:* ASST PROF MED RES, DUKE UNIV, MED CTR, 76- *Mem:* Am Asn Immunologists; Am Asn Cancer Res. *Res:* Monoclonal antibodies; tumor associated antigens. *Mailing Add:* Dept Paternity Eoal Roche Biomed Labs 1447 York Ct Burlington NC 27215

STUHMILLER, JAMES HAMILTON, b Cincinnati, Ohio, Apr 1, 43. PHYSICS. *Educ:* Mass Inst Technol, BS, 65; Queens Col, City Univ New York, MA, 68; Univ Cincinnati, PhD(physics), 73. *Prof Exp:* Engr, AVCO Electronics Div, 68-73; scientist, Sci Appln Inc, 73-75; VPRES, JAYLOR, 75- *Mem:* Am Phys Soc; Am Soc Mech Engrs; Am Inst Chem Engrs; Am Metereol Soc; Am Inst Aeronaut & Astronaut. *Res:* Underlying physical phenomena controlling unusual behavior in engineering systems (nuclear power plants, submerged vehicles, space vehicle launches); mathematical representation to predict and correct associated problems. *Mailing Add:* Jaycor 11011 Torreyana Rd San Diego CA 92121-1190

STUIVER, MINZE, b Vlagtwedde, Neth, Oct 25, 29; m 56; c 2. EARTH SCIENCE. *Educ:* State Univ Groningen, MSc, 53, PhD(biophys), 58. *Prof Exp:* Res assoc & fel geol, Yale Univ, 59-62, sr res assoc geol & biol & dir radiocarbon lab, 62-69; prof geol & zool, 69-81, PROF GEOL & QUATERNARY SCI, UNIV WASH, 81- *Honors & Awards:* Alexander v Humboldt Award, WGer, 83-84. *Res:* Biophysics of sense organs; low level counting techniques; carbon cycle and radiocarbon time scale calibration; isotopic applications in pleistocene geology, oceanography; limnology and climatology. *Mailing Add:* Quaternary Res Ctr Univ Wash Seattle WA 98195

STUIVER, W(ILLEM), b Breda, Netherlands, Aug 1, 27; m 57. ANALYTICAL DYNAMICS, SPACE FLIGHT MECHANICS. *Educ:* Delft Univ Technol, Mech engr, 51; Stanford Univ, PhD(eng mech), 60. *Prof Exp:* Asst mech engr, NZ Ministry Works, 52-54; res engr, NZ Dept Sci & Indust Res, 54-55; anal design engr, Gen Elec Co, 56-58; mem res staff, IBM Corp, 59-61; assoc prof, 61-65, PROF MECH ENG, UNIV HAWAII, 65- *Concurrent Pos:* Consult, Stanford Res Inst, 62-67; vis prof, Indian Inst Sci, India, 64-65, Univ New SWales, Australia, 75, Univ Western Australia, 76 & Univ Peradeniya, Sri Lanka, 82-83. *Mem:* Am Inst Aeronaut & Astronaut; Am Astronaut Soc. *Res:* Development of metric approach to state-space analysis of dynamical systems; interplanetary and interstellar space flight mechanics; dynamics and control of stationary satellite systems. *Mailing Add:* Dept of Mech Eng Univ of Hawaii Honolulu HI 96822

STUKEL, JAMES JOSEPH, b Joliet, Ill, Mar 30, 37; m 58; c 4. ENGINEERING. *Educ:* Purdue Univ, BSME, 59; Univ Ill, Urbana-Champaign, MS, 63, PhD(mech eng), 68. *Prof Exp:* Res engr, Westvaco, 59-61; from asst prof to prof civil eng & mech eng, Univ Ill, Chicago, 68-85, dir off coal res, 75-76, dir, Off Energy Res, 76-80, dir, Off Interdisciplinary Proj, 78-80, dir Pub Policy Prog, Col Eng, 80-85, vchancellor res & dean Grad Col, 85-86, exec vchancellor & vchancellor acad affairs, 86-91, CHANCELLOR, UNIV ILL, CHICAGO, 91- *Concurrent Pos:* Consult, Westvaco, 61-63 & var govt agencies, 68-; prin investr, US Environ Protection Agency training grant, 70-76, res grants, Kimberly Clark Corp, 75-76, Bur Mines & US Environ Protection Agency, 76-82. *Honors & Awards:* State of the Art of Civil Eng Award, Am Soc Civil Engrs, 75. *Mem:* Am Soc Civil Engrs; Am Soc Mech Engrs; Sigma Xi. *Res:* Aerosol science; air resources management; impact assessment. *Mailing Add:* 2650 N Lakeview Chicago IL 60614

STUKUS, PHILIP EUGENE, b Braddock, Penn, Oct 22, 42; m 66; c 3. MICROBIAL PHYSIOLOGY, ENVIRONMENTAL MICROBIOLOGY. *Educ:* St Vincent Col, BA, 64; Cath Univ Am, MS, 66, PhD(microbiol), 68. *Prof Exp:* asst prof, 68-80, PROF BIOL, DENISON UNIV, 80- *Mem:* AAAS; Am Soc Microbiol; Sigma Xi; Soc Indust Microbiol. *Res:* Autotrophic and heterotrophic metabolism of hydrogen bacteria; degradation of detergent additives and pesticides by soil microorganisms; distribution of microorganisms in air. *Mailing Add:* Dept of Biol Denison Univ Granville OH 43023

STULA, EDWIN FRANCIS, b Colchester, Conn, Jan 3, 24; m 55; c 1. VETERINARY PATHOLOGY. *Educ:* Univ Conn, BS, 50, PhD(animal path), 63; Univ Toronto, DVM, 55. *Prof Exp:* Instr vet med & exten vet, Univ Conn, 55-62; CHIEF RES PATHOLOGIST, HASKELL LAB TOXICOL & INDUST MED, E I DU PONT DE NEMOURS & CO, INC, 63- *Mem:* Am Vet Med Asn; Soc Pharmacol & Environ Pathologists; NY Acad Sci; Int Acad Path; Am Asn Lab Animal Sci. *Res:* Bovine vibriosis, mastitis and infertility; leptospirosis in chinchillas and guinea pigs; spontaneous diseases of laboratory animals; industrial medicine; pathologic effects in animals exposed to various chemicals by various routes; morphologic effects using both light and electron microscopes; carcinogenicity and embryotoxicity. *Mailing Add:* 235 Mercury Rd Newark DE 19711

STULBERG, MELVIN PHILIP, b Duluth, Minn, May 17, 25; m 55; c 3. BIOCHEMISTRY. *Educ:* Univ Minn, BS, 49, MS, 55, PhD(biochem), 58. *Prof Exp:* Res assoc, Biol Div, Oak Ridge Nat Lab, 58-59, biochemist, 59-61; biochemist, AEC, 61-63; BIOCHEMIST BIOL DIV, OAK RIDGE NAT LAB, 63-; PROF, OAK RIDGE GRAD SCH BIOMED SCI, UNIV TENN, 73- *Mem:* Sigma Xi; AAAS; Am Soc Biol Chemists. *Res:* Protein biosynthesis; isolation and function of transfer RNA; enzyme mechanisms; mechanisms of aging. *Mailing Add:* Energy Div Oak Ridge Nat Lab MSH-32 4500N PO Box X Oak Ridge TN 37830

STULL, DEAN P, US citizen. ANALYTICAL CHEMISTRY, NATURAL PRODUCTS PROCESS DEVELOPMENT. *Educ:* Colo State Univ, BS, 72; Univ Colo, MS, 74, PhD(phys org chem), 76. *Prof Exp:* Res asst, Colo State Univ, 70-71, Univ Calif, San Francisco, 72 & Univ Colo, Boulder, 76; CHIEF CHEMIST, HAUSER LABS, 76-, DIR SPEC PROJS. *Mem:* Am Chem Soc; Sigma Xi. *Res:* Methods development of chemical analysis of crude and refined petroleum products; fire retardant technology; natural products. *Mailing Add:* Hauser Labs PO Box G Boulder CO 80306

STULL, ELISABETH ANN, b Fayette, Mo, Jan 7, 43. ENVIRONMENTAL IMPACT ANALYSIS, LIMNOLOGY. *Educ:* Lawrence Univ, BA, 65; Univ Ga, MS, 69; Univ Calif, Davis, PhD(zool), 72. *Prof Exp:* Asst prof biol sci, 71-75, asst prof ecol & evolutionary biol, Univ Ariz, 75-78; ECOLOGIST, ENVIRON ASSESSMENT & INFO SCI DIV, ARGONNE NAT LAB, 78- *Mem:* Am Soc Limnol & Oceanog; Phycol Soc Am; AAAS. *Res:* Energetics and trophic ecology of unicellular taxa; algal floristics; regional and geographical patterns in limnology and water quality; environmental analysis of energy-related technologies; cumulative impact analysis. *Mailing Add:* Environ Assessment & Info Sci Div Argonne Nat Lab 9700 S Cass Ave Argonne IL 60439

STULL, G(EORGE) A, b Easton, Pa, Jan 26, 33; m; c 2. EXERCISE PHYSIOLOGY. *Educ:* E Stroudsburg Univ, BS, 55; Pa State Univ, MS, 57, EdD(phys educ), 61. *Prof Exp:* From instr to asst prof phys educ, Pa State Univ, 58-66; from assoc prof to prof phys educ, Univ Md, 66-72; prof educ psychol & phys educ, Univ Ky, 72-77; prof & dir educ psychol & phys educ, Sch Phys Educ, Recreation & Sch Health Educ, Univ Minn, 77-85; prof & dean allied health, Sch Allied Health Profs, Univ Wis-Madison, 85-88; PROF & DEAN EXERCISE SCI, SCH HEALTH RELATED PROFS, STATE UNIV NY, BUFFALO, 88- *Concurrent Pos:* Chair res sect, Am Asn Health, Phys Educ & Recreation, 72-73, mem bd gov, 76-79, 81-82; chair, position stands comt, Am Col Sports Med, 79-81, res quarterly exercise & sport adv comt, 79-80; pres assoc res admin, Prof Coun & Soc, 76-77; pres res consortium, Am Alliance Health, Phys Educ, Recreation & Dance, 81-82; assoc ed, J Motor Behav, 69-81, Clin Kinesiol, 70- & Res Quart Exercise & Sport, 72-75. *Honors & Awards:* Asn Res, Admin & Prof Coun & Soc Award, 74; Am Alliance Health, Phys Educ, Recreation & Dance Award, 81. *Mem:* Am Acad Phys Educ (pres, 85-86); Am Alliance Health, Phys Educ, Recreation & Dance; Am Col Sports Med; Am Soc Allied Health Profs; Nat Asn Phys Educ Higher Educ. *Res:* Effects of exercise on muscular and cardiorespiratory systems; locus of and recovery from local fatigue; prognostic value of exercise testing following myocardial infraction and coronary bypass surgery. *Mailing Add:* Sch Health Related Profs State Univ NY 435 Kimball Tower Buffalo NY 14214

STULL, JAMES TRAVIS, b Ashland, Ky, Feb 7, 44; m 66; c 2. PHARMACOLOGY, BIOCHEMISTRY. *Educ:* Southwestern at Memphis, BS, 66; Emory Univ, PhD(pharmacol), 71. *Prof Exp:* Adj asst prof biol chem, Sch Med, Univ Calif, Davis, 73-74; from asst prof to assoc prof med, Sch Med, Univ Calif, San Diego, 74-78; assoc prof to prof pharmacol, Univ Tex Health Sci Ctr, Dallas, 78-86; CHMN PHYSIOL, UNIV TEX SOUTHWESTERN MED CTR, DALLAS, 86- *Concurrent Pos:* Damon Runyon Mem Fund Cancer Res fel, Univ Calif, Davis, 71-73; estab investr, Am Heart Asn, 73; assoc dean, Grad Sch Biomed Sci, Univ Tex Health Sci Ctr, Dallas, 80-86; Wellcome vis prof, 90. *Mem:* Am Heart Asn; Am Soc Pharmacol & Exp Therapeut; fel AAAS; Am Soc Biol Chemists; Sigma Xi; Am Physiol Soc; Biophys Soc. *Res:* Protein phosphorylation reactions in regulation of muscle metabolism, contraction and responses to hormones and adrenergic drugs. *Mailing Add:* Physiol Dept Univ Tex Southwestern Med Ctr Dallas 5323 Harry Hines Blvd Dallas TX 75235-9040

STULL, JOHN LEETE, b Dansville, NY, June 2, 30; m 52; c 2. PHYSICS, ASTRONOMY. *Educ:* Alfred Univ, BS, 52, MS, 54, PhD(ceramics), 58. *Prof Exp:* Res assoc ceramics, 52-58, from asst prof to assoc prof physics, 58-68, chmn dept, 72-75, PROF PHYSICS, ALFRED UNIV, 68-, DIR OBSERV, 68- *Mem:* AAAS; Am Asn Physics Teachers. *Res:* Astronomy; development of physics teaching apparatus; design of small optical observatories and equipment. *Mailing Add:* Dept of Physics Alfred Univ Alfred NY 14802

STULL, JOHN WARREN, b Benton, Ill, Nov 23, 21; m 45; c 5. DAIRY SCIENCE. *Educ:* Univ Ill, BS, 42, MS, 47, PhD(food technol), 50. *Prof Exp:* Asst dairy technol, Univ Ill, 46-49; from asst prof to assoc prof dairy sci, 49-58, actg head dept, 56-57, PROF NUTRIT & FOOD SCI, COL AGR, UNIV ARIZ, 58-, FOOD SCIENTIST, AGR EXP STA, 74- *Mem:* Am Chem Soc; Am Dairy Sci Asn; Inst Food Technologists. *Res:* Factors related to the oxidative deterioration of the constituents of milk; food value of milk; chemical residues in milk; biochemistry of milk and food lipids. *Mailing Add:* Dept of Nutrit & Food Sci Col of Agr Univ of Ariz Tucson AZ 85721

STULTING, ROBERT DOYLE, JR, b Knoxville, Tenn, Nov 24, 48; c 2. MICROBIOLOGY, IMMUNOLOGY. *Educ:* Duke Univ, BS, 70, MD, 76, PhD(microbiol & immunol), 75; Am Bd Ophthalmol, cert, 82. *Prof Exp:* Intern, Barnes Hosp, St Louis, 76-77, resident, 77-78; resident, Bascom Palmer Eye Inst, Miami, 78-81; fel ophthal, 81-82, asst prof, 82-85, ASSOC PROF OPHTHAL, EMORY UNIV, 86-; ASSOC PROF, WINSHIP CANCER CTR, 88- *Concurrent Pos:* Consult, Southeastern Regional Organ Procurement Found, 70-76, Med Sch Curriculum Comt, Duke Univ, 73-75; promotions comt & med records comt, Emory Clinic, Emory Univ, 85-; med advisory bd, Eye Bank Asn Am & Ambulatory Surg Ctr Comt, Emory Clin, 86-; consult, Ophthalmic Devices Panel, ctr devices and radiol health, US Food & Drug Admin, 87-88, mem, 88, bd dirs, Eye Bank Asn Am, 88- *Mem:* Am Med Asn; Am Asn Immunol; Am Acad Ophthalmol; Eye Bank Asn Am. *Res:* Collaborative corneal transplantation studies; research to prevent blindness; pathogenesis of herpes simplex keratitis. *Mailing Add:* Emory Eye Ctr 1327 Clifton Rd NE Atlanta GA 30322

STULTS, FREDERICK HOWARD, b Seattle, Wash, Nov 11, 48; m 69; c 3. TOXICOLOGY. *Educ:* San Diego State Univ, BS, 71; Univ Calif, Davis, PhD(biochem), 76. *Prof Exp:* Fel toxicol, Med Sch, Duke Univ, 76-77; res toxicologist, 77-80, dir, corp safety, Int Flavors & Fragrances, 80-; AT FIRMENICH INC. *Mem:* Am Chem Soc; Inst Food Technologists; Am Col Toxicol. *Res:* Human dermal toxicity, allergenicity, biodegradation & photobiology. *Mailing Add:* Firmenich Inc PO Box 5880 Princeton NJ 08540

STULTS, VALA JEAN, b Oklahoma City, Okla, Aug 16, 42. NUTRITION. *Educ:* Calif State Univ, Long Beach, BA, 65, MA, 67; Mich State Univ, PhD(human nutrit), 74. *Prof Exp:* Teacher home econ, Lawndale High Sch, 66-68; consult nutritionist, Head Start Prog, Off Econ Opportunity, 69; NIH fel, 73-74; asst instr nutrit, Cen Mich Univ, 74; sales mgr & asst to gen mgr potatoes in retort pouches, Nu Foods Inc, 75-76; nutritionist, Kellogg Co, 77-; chmn dept home econ, Whittier Col, 83-88; CONSULT NUTRIT. *Mem:* Am Home Econ Asn; Am Dietetic Asn; Soc Nutrit Educ; Soc Nutrit Today; Inst Food Technologists. *Res:* Subjects of interest to the ready-to-eat cereal industry, soft drinks industry and exposing health fraud. *Mailing Add:* 16333 Grenoble Ln Huntington Beach CA 92649

STULTZ, WALTER ALVA, b St John, NB, Mar 14, 04; nat US; m 31, 50; c 5. ANATOMY. *Educ:* Acadia Univ, BA, 27; Yale Univ, PhD(zool, anat), 32. *Prof Exp:* Prin sch, NB, Can, 22-24; asst biol, Yale Univ, 27-30; instr, Spring Hill Sch, Conn, 30-31; instr & actg head dept, Trinity Col, Conn, 31-32; prof, Mt Union Col, 32-33; asst, Yale Univ, 33-34; fel anat, Sch Med, Univ Ga, 34-35, fel histol & embryol, 35-36; instr anat, Med Col SC, 36-37; from asst prof to prof, 37-69, EMER PROF ANAT, COL MED, UNIV VT, 69- *Concurrent Pos:* Sr lectr gross anat & histol, Sch Med, Univ Calif, San Diego, 69-77; vis prof anat, Dartmouth Med Sch, 77-83. *Mem:* Am Asn Anat. *Res:* Experimental embryology of Amblystoma; relations of symmetry in fore and hind limbs; interrelationships between limbs and nervous system. *Mailing Add:* RFDI Box 127 Alburg VT 05440-9999

STUMM, WERNER, b Switz, Oct 8, 24; Swiss & US citizen; m 52; c 5. AQUATIC CHEMISTRY, GEOCHEMISTRY. *Educ:* Univ Zurich, PhD(chem), 52; Harvard Univ, MA, 56. *Hon Degrees:* DSc, Univ Geneva, Switz, 87, Univ Crete, Greece, 88, Northwestern Univ, 89, Israel Inst Technol, 89; DrTech, Royal Inst Technol, Sweden, 87. *Prof Exp:* Res fel, Swiss Fed Inst Technol, Zurich, 52-56; from asst to assoc & Gordon McKay prof appl chem, Harvard Univ, Cambridge, 56-70; PROF CHEM, SWISS FED INST TECHNOL, ZURICH, 70-; DIR, FED INST WATER RESOURCES & WATER POLLUTION CONTROL, 70- *Concurrent Pos:* Ed, five publ, 67-90. *Honors & Awards:* Monsanto Prize for Pollution Control, Am Chem Soc, 76; Albert Einstein World Award of Sci, World Cult Coun, Mex, 85; Tyler Prize for Environ Acheivement, 86. *Mem:* Nat Acad Eng; fel Am Chem Soc. *Res:* Chemical, biological and physical phenomena that control the nature of aquatic systems and the response of the aquatic ecosystem to man's impact; publications on the processes that occur in natural waters; aquatic chemistry. *Mailing Add:* Swiss Fed Inst Water Resources & Water Pollution Control Dubendorf CH-8600 Switzerland

STUMP, BILLY LEE, b Morristown, Tenn, Jan 11, 30; m 58; c 3. PHYSICAL CHEMISTRY, POLYMER CHEMISTRY. *Educ:* Carson-Newman Col, BS, 52; Univ Tenn, PhD(chem), 59. *Prof Exp:* Res chemist, Carson-Newman Col, 52-53; res technician, Oak Ridge Inst Nuclear Studies, 53-54; chemist, Redstone Div, Thiokol Chem Corp, 59-60; res chemist, Film Res Lab, E I du Pont de Nemours & Co, 60-62 & Spruance Film Res & Develop Lab, 62-63; assoc prof chem, Carson-Newman Col, 63-66; assoc prof, 66-78, PROF CHEM & COORDR OF GEN CHEM, VA COMMONWEALTH UNIV, 78- *Mem:* Am Chem Soc; Sigma Xi. *Res:* Kinetics and reaction mechanisms; catalysis and kinetics of catalytic hydrogenation; polymer chemistry. *Mailing Add:* Dept of Chem Box 2006 Va Commonwealth Univ Richmond VA 23284-2006

STUMP, EDMUND, b Danville, Pa, Dec 28, 46. GEOLOGY. *Educ:* Harvard Col, AB, 68; Yale Univ, MS, 72; Ohio State Univ, PhD(geol), 76. *Prof Exp:* ASST PROF GEOL, ARIZ STATE UNIV, 76- *Mem:* Geol Soc Am; AAAS. *Res:* Geology of Antarctica and Gondwanaland. *Mailing Add:* Dept Geol Ariz State Univ Tempe AZ 85287

STUMP, EUGENE CURTIS, JR, b Charleston, WVa, May 19, 30; m 58; c 3. ORGANIC CHEMISTRY, FLUORINE CHEMISTRY. *Educ:* WVa Univ, BS, 52; Columbia Univ, MA, 53; Univ Fla, PhD(org chem), 60. *Prof Exp:* Dir contract res, 60-70, vpres contract res div, 70-76, VPRES RES & DEVELOP, PCR, INC, 76- *Mem:* Am Chem Soc; Royal Soc Chem. *Res:* Synthesis of fluorine containing compounds, particularly ethers, olefins, nitroso and difluoramine compounds; synthesis of fluorine-containing polymers as low temperature elastomers; synthesis of thermally and oxidatively stable fluids. *Mailing Add:* 3131 NW 37 St Gainesville FL 32605-2038

STUMP, JOHN EDWARD, b Galion, Ohio, June 3, 34; m 55; c 2. VETERINARY ANATOMY. *Educ:* Ohio State Univ, DVM, 58; Purdue Univ, PhD, 66. *Prof Exp:* Private vet pract, Ohio, 58-61; from instr to assoc prof vet anat, 61-76, PROF VET ANAT, PURDUE UNIV, WEST LAFAYETTE, 76- *Concurrent Pos:* Vis prof, Dept Physiol Sci, Sch Vet Med, Univ Calif, Davis, 80, Dept Vet Anat, Col Vet Med, Tex A&M Univ, 81. *Mem:* Am Asn Anatomists; Am Vet Med Asn; World Asn Vet Anat; Am Asn Vet Anat (pres, 77-78); Asn Am Vet Med Cols; Am Soc Vet Ethology; Sigma Xi. *Res:* Gross anatomy of domestic animals. *Mailing Add:* Dept Anat Sch Vet Med Purdue Univ West Lafayette IN 47907

STUMP, JOHN M, b Charleston, WVa, June 26, 38; m 64. PHARMACOLOGY. *Educ:* WVa Univ, BS, 61, MS, 62, PhD(pharmacol), 64. *Prof Exp:* Res pharmacologist, 64-72, sr res pharmacologist, 72-78, res assoc, pharmaceut res div, 78-85, MGR, PHARMACEUT RES & DEV ADMIN, E I DU PONT DE NEMOURS & co, 85- *Res:* Cardiovascular pharmacology. *Mailing Add:* 38 Aronomink Dr Newark DE 19711

STUMP, ROBERT, b Indianapolis, Ind, Oct 16, 21; m 43; c 4. PHYSICS. *Educ:* Butler Univ, BA, 42; Univ Ill, MS, 48, PhD(physics), 50. *Prof Exp:* From asst prof to assoc prof, 50-60, PROF PHYSICS, UNIV KANS, 60- *Concurrent Pos:* Consult, Aeronaut Radio, Inc, DC, 52-53; vis scientist, Midwestern Univs Res Asn, Wis, 59, Europ Orgn Nuclear Res, Geneva, 63-64 & Polytech Sch, Paris, 70-71; vis physicist, Brookhaven Nat Lab, 62-63. *Mem:* AAAS; fel Am Phys Soc. *Res:* Experimental nuclear and elementary particle physics; angular correlations; low temperature effects; hydrogen and heavy liquid bubble chamber experiments; atmospheric physics; atmospheric dynamics. *Mailing Add:* PO Box 1241 West Falmouth MA 02514

STUMPE, WARREN ROBERT, b Bronx, NY, July 15, 25; m 52; c 3. INDUSTRIAL RESEARCH & DEVELOPMENT MANAGEMENT. *Educ:* US Mil Acad, BS, 45; Cornell Univ, MS, 49; NY Univ, MIE, 65. *Prof Exp:* Var positions to dep gen mgr, AMF, Stamford, Conn, 54-63; exec vpres, Dortech Inc, Stamford, 63-69; dir systs mgt, Mat-Handling Group, Rexnord Darien, Conn, 69-71; vpres res & technol & Bus Develop Sector, Rexnord Inc, 71-87; VPRES, RADIAN CORP, 87- *Concurrent Pos:* Mem, Indust Adv Bd, Col Eng, Univ Wis, Milwaukee, 79-, Wis Gov's Energy Task Force, 80-81 & Indust Liaison Coun, Col Eng, Univ Wis, Madison, 81-; civilian aide to Secy of Army, Wis, 81-85. *Mem:* Indust Res Inst (pres, 85); Soc Am Mil Engrs; Water Pollution Control Fedn. *Res:* Electronics; design-manufacturing productivity; bio-engineering processes; electro-optics and lasers; microprocessor controls. *Mailing Add:* 5101 W Beloit Rd Milwaukee WI 53214

STUMPERS, FRANS LOUIS H M, b Stratum, Neth, Aug 30, 11; m 54; c 5. ELECTROMAGNETIC NOISE OF TERRESTRIAL ORIGIN APPLIANCES. *Educ:* Utrecht Univ, Neth, MSc, 37; Delft Univ, Neth, DSc(tech sci), 46. *Prof Exp:* Res scientist, Philips Res Labs, Eindhoven, Neth, 27-54, res group leader, 54-68, sci adv, 68-74; RETIRED. *Concurrent Pos:* Res assoc, Mass Inst Technol, 52-53; chmn, Comn VI Circuit Info, Int Union Radio Sci, 63-69 & Comn E Electromagnetic Noise Interference, 81-87; prof, Cath Univ Nymegh, 68-82, Bochum Rohr Univ, Ger, 74-75 & State Univ Utrecht, 76-82. *Honors & Awards:* Int Commun Award, Inst Elec & Electronic Engrs, 78. *Mem:* Fel Inst Elec & Electronic Engrs; Int Union Radio Sci (vpres, 75-81, hon pres, 90-). *Res:* Multiplex telephony; electromagnetic noise; frequency modulation; radio interference; modulation systems; information theory; control theory; software. *Mailing Add:* Elzentlaan 11 Eindhoven 5611 Lg Netherlands

STUMPF, FOLDEN BURT, b Lansing, Mich, Aug 18, 28; m 54; c 2. PHYSICS. *Educ:* Kent State Univ, BS, 50; Univ Mich, MS, 51; Ill Inst Technol, PhD(physics), 56. *Prof Exp:* From asst prof to assoc prof, 56-66, PROF PHYSICS, OHIO UNIV, 66- *Mem:* Fel Acoust Soc Am; Am Asn Physics Teachers; Sigma Xi. *Res:* Ultrasonic transducers; application of ultrasonics to liquids. *Mailing Add:* Dept of Physics & Astron Ohio Univ Athens OH 45701-2979

STUMPF, H(ARRY) C(LINCH), b Buffalo, NY, June 29, 18; m 43; c 2. METALLURGICAL ENGINEERING. *Educ:* Univ Mich, BS, 39, MS, 41, PhD(metall eng), 43. *Prof Exp:* Asst prof chem eng, Univ Del, 42-44; metall engr, Allegany Ballistics Lab, Md, 44-45; sci assoc, Alcoa Res Labs, Aluminum Co Am, 46-76, sci assoc, Alcoa Lab, Alcoa Tech Ctr, 76-80; RETIRED. *Mem:* Am Inst Mining, Metall & Petrol Engrs; Sigma Xi. *Res:* Physical metallurgy, especially of aluminum alloys; x-ray diffraction. *Mailing Add:* 810 Carl Ave New Kensington PA 15068

STUMPF, PAUL KARL, b New York, NY, Feb 23, 19; m 47; c 5. LIPID BIOCHEMISTRY. *Educ:* Harvard Univ, AB, 41; Columbia Univ, PhD(biochem), 45. *Prof Exp:* Chemist, Div War Res, Columbia Univ, 44-46; instr epidemiol, Sch Pub Health, Univ Mich, 46-48; from asst prof to prof plant biochem, Univ Calif, Berkeley, 48-58; prof & chmn dept, 58-61, 67-68, 70 & 81, EMER PROF BIOCHEM, UNIV CALIF, DAVIS, 84- *Concurrent Pos:* NIH sr fel, 54-55; NSF sr fel, 61, 68; Guggenheim fel, 62 & 69; ed, J Phytochemistry, 60-72 & Archives Biochem & Biophys, 60-65, exec ed, 65-88; ed, J Lipid Res, 63-66, 85-, Anal Biochem, 69-80; mem physiol chem study sect, NIH, 60-64; metab biol panel, NSF, 65-68; mem, City of Davis Planning Comn, 67-69; vis scientist, Commonwealth Sci & Indust Res Orgn, Canberra ACT, Australia, 75-76; sr US Sci fel, Von Humboldt Fedn, Ger, 76; sci adv bd mem, Palm Oil Res Inst, Malaysia, Calzine, Inc, Davis, Calif, Maryland Biotechnology Inst, Baltimore, MD; chief scientist, USDA/CSRS/CRGO, 88-90, USDA/CSRS/NRICGO, 90-91. *Honors & Awards:* Stephen Hales Award, Am Soc Plant Physiologist, 74; Lipid Chem Prize, Am Oil Chemists Soc, 74. *Mem:* Nat Acad Sci; Am Soc Plant Physiol (pres, 80); Am Oil Chem Soc; Am Soc Biol Chem; foreign mem Royal Danish Acad Arts & Sci; fel Linnean Soc London. *Res:* Lipid biochemistry of higher plants; photobiosynthesis; developmental biochemistry. *Mailing Add:* Dept of Biochem & Biophys Univ of Calif Davis CA 95616

STUMPF, WALTER ERICH, b Oelsnitz, Ger, Jan 10,, 27; m 61; c 4. NEUROENDOCRINOLOGY, PHARMACOLOGY. *Educ:* Univ Berlin, MD, 52, cert neurol & psychiat, 57; Univ Chicago, PhD(pharmacol), 67. *Hon Degrees:* Dr rer biol hum hc, Ulm, Ger, 87. *Prof Exp:* Intern, Charite Hosp, Univ Berlin, 52-53, resident neurol & psychiat, 53-57; sci asst, Univ Marburg, 58-60 & Lab Radiobiol & Isotope Res, 61-62; res assoc pharmacol, Univ Chicago, 63-67, asst prof, 67-70; assoc prof anat & pharmacol & mem, Lab for Reproductive biol, 70-73, PROF ANAT & PHARMACOL, UNIV NC, CHAPEL HILL, 73- *Concurrent Pos:* Trainee psychother & psychoanal, Inst Psychother, WBerlin, 54-56; lectr clin neurol, Charite Hosp, Univ Berlin, 56-57; vis psychiatrist, Maudsley Hosp, London, 59; consult, Microtome-Cyrostats, 69-; mem, Neurolbiol Prog, 70-; assoc, Carolina Pop Ctr, 72-; res scientist, Biol Sci Res Ctr, 72-; consult, Life Sci Inst, Res Triangle Park, NC, 73-; mem coun, Inst Lab Animal Resources, 78-81; mem US subcomt, Int Ctr Cybernetics & Systs, World Orgn Gen Systs & Cybernetics; Humboldt Award, 89. *Mem:* AAAS; Am Soc Zoologists; Histochemical Soc; Int Brain Res Orgn; Am Asn Anatomists; Endocrinol Soc; Soc Xenobiotics. *Res:* Development of histochemical techniques; low temperature sectioning and freeze-drying; dry-mount autoradiography for the localization of hormones and drugs in the brain and other tissues; autoradiography of diffusible substances; anatomic neuroendocrinology; vitamin D; sites and mechanisms of action of hormones and drugs; seasonal regulator concept. *Mailing Add:* Dept Cell Biol Anat Univ NC Chapel Hill Chapel Hill NC 27599-7090

STUMPFF, HOWARD KEITH, b Holden, Mo, May 26, 30; m 57; c 3. MATHEMATICS. *Educ:* Cent Mo State Col, BSEd, 51; Univ Mo, AM, 53; Univ Kans, PhD(math ed), 68. *Prof Exp:* Asst instr math, Univ Mo, 51-53; instr, Univ NMex, 57-63; from asst prof to assoc prof, 63-72, head dept, 69-82, PROF MATH, CENT MO STATE UNIV, 72- *Concurrent Pos:* Asst instr, Univ Kans, 61-62. *Mem:* Am Math Soc; Math Asn Am. *Res:* Mathematics education. *Mailing Add:* Assoc Provost Cent Mo State Univ Warrensburg MO 64093-5011

STUMPH, WILLIAM EDWARD, b Indianapolis, Ind, Dec 20, 48. RECOMBINANT DNA, REGULATION OF GENE EXPRESSION. *Educ:* Purdue Univ, BS, 72; Calif Inst Technol, PhD(biochem), 79. *Prof Exp:* Asst molecular biol, Baylor Col Med, 79-83; PROF CHEM, SAN DIEGO STATE UNIV, 83- *Mem:* Am Soc Cell Biol; AAAS; Am Soc Biochem &

Molecular Biol. *Res:* Transcriptional regulation of gene expression; DNA sequences and protein factors that regulate the expression of genes encoding the small nuclear RNAs in the chicken. *Mailing Add:* Dept Chem San Diego State Univ San Diego CA 92182-0328

STUNKARD, ALBERT J, b New York, NY, Feb 7, 22; m 80. PSYCHIATRY. *Educ:* Yale Univ, BS, 43; Columbia Univ, MD, 45. *Prof Exp:* Resident physician, Johns Hopkins Hosp, 48-51; fel psychiat, 51-52; res fel med, Col Physicians & Surgeons, Columbia Univ, 52-53; Commonwealth fel med, Med Col, Cornell Univ, 53-56, asst prof, 56-57; assoc prof, 57-62, prof psychiat & chmn dept, Sch Med, Univ Pa, 62-73; prof, Stanford Univ, 73-76; PROF PSYCHIAT, UNIV PA, 76- *Concurrent Pos:* Fel, Ctr Advan Study Behav Sci, Calif, 71-72. *Honors & Awards:* Menninger Award, Am Col Physicians; Res Award, Am Psychiat Asn, 60 & 80. *Mem:* Inst Med Nat Acad Sci; Am Psychosom Soc (pres, 74); fel Am Psychiat Asn; fel NY Acad Sci; Am Asn Chmn Departments Psychiat (pres, 86); Asn Res Nerv & Ment Dis (pres, 82); Acad Behav Med Res (pres, 86); Soc Behav Med (pres, 90). *Res:* Obesity and regulation of energy balance. *Mailing Add:* Dept Psychiat Univ Pa Philadelphia PA 19104-3246

STUNKARD, JIM A, b Sterling, Colo, Jan 25, 35; m 67; c 1. LABORATORY ANIMAL MEDICINE, VETERINARY MICROBIOLOGY. *Educ:* Colo State Univ, BS, 57, DVM, 59; Tex A&M Univ, MS, 66; Am Col Lab Animal Med, dipl. *Prof Exp:* Vet, Glasgow Animal Hosp, Ky, 59-61; vet in charge, Sentry Dog Procurement, Training & Med Referral Ctr, US Air Force Europe, 61-64 & Lab Animal Colonies & Zoonoses Control Ctr, 61-64, resident lab animal med, sch aerospace med, Brooks AFB, Tex, 64-66, dir vet med sci dept, Naval Med Res Inst, Nat Naval Med Ctr, Md, 66-71. *Concurrent Pos:* Vet consult, Turkish Sentry Dog Prog, US Air Force Europe, 61-63, Can Air Force Europe, 62-64, Bur Med & Surg, US Navy, Washington, DC, Navy Toxicol Unit, Md, 66-71 & Atomic Energy Comn, 68-71; consult to dean vet med, Colo State Univ, 69-71; US Navy rep ad hoc comt, Dept Defense, 66-67 & Inst Lab Animal Resouces, Nat Acad Sci-Nat Res Coun, Washington, DC, 66-71. *Mem:* Am Vet Med Asn. *Res:* Veterinary medicine, dentistry and surgery, especially all phases of laboratory animal medicine. *Mailing Add:* Bowie Animal Hosp 3428 Crain Hwy Bowie MD 20716

STUNTZ, CALVIN FREDERICK, b Buffalo, NY, Aug 6, 18; m 51; c 3. CHEMISTRY. *Educ:* Univ Buffalo, BA, 39, PhD(chem), 47. *Prof Exp:* Teacher high sch, NY, 39-40; anal chemist, Linde Air Prods Co, Union Carbide & Carbon Corp, 40-41; asst, Univ Buffalo, 41-43, 45-46; from asst prof to prof chem, 46-79, EMER PROF CHEM, UNIV MD, COLLEGE PARK, 80- *Mem:* Am Chem Soc; Sigma Xi. *Res:* Quantitative analysis; chemical microscopy. *Mailing Add:* 13705 Carlisle Ct Silver Spring MD 20904

STUNTZ, GORDON FREDERICK, b Washington, DC, Dec 7, 52; m 73. INORGANIC CHEMISTRY, PETROLEUM CHEMISTRY. *Educ:* Pa State Univ, BS, 74; Univ Ill, PhD(inorg chem), 78. *Prof Exp:* RES CHEMIST PROCESS RES, EXXON RES & DEVELOP LABS, EXXON CORP, 78- *Mem:* Am Chem Soc. *Mailing Add:* Exxon Res & Develop Labs PO Box 2226 Baton Rouge LA 70821-2226

STUPER, ANDREW JOHN, b Chicago, Ill, Dec 19, 50; m 77. THEORETICAL CHEMISTRY, ANALYTICAL CHEMISTRY. *Educ:* Univ Wis-Superior, BS, 72; Pa State Univ, PhD(anal chem), 77. *Prof Exp:* Vis scientist, Nat Ctr Toxicol Res, 75-76; consult, Parke Davis Co, 76; sr scientist, Rohm & Haas Co, 77-; AT ICI AMERICAS. *Mem:* Am Chem Soc; Sigma Xi; Asn Comput Mach. *Res:* Development of methods which enhance understanding of the relationship between chemical structure and biological activity; use of computers in chemistry. *Mailing Add:* 1125 Ivymont Rd Rosemont PA 19010-1626

STUPIAN, GARY WENDELL, b Alhambra, Calif, Oct 17, 39. SOLID STATE PHYSICS, SURFACE PHYSICS. *Educ:* Calif Inst Technol, BS, 61; Univ Ill, Urbana, MS, 63, PhD(physics), 67. *Prof Exp:* Res asst physics, Univ Ill, Urbana, 61-67; res assoc mat sci, Cornell Univ, 67-69; MEM TECH STAFF, AEROSPACE CORP, 69- *Mem:* AAAS; Am Phys Soc; Am Vacuum Soc. *Res:* Auger spectroscopy; solid state devices; analytical instrumentation; tunneling microscopy. *Mailing Add:* M2/272 Chem Physics Lab Aerospace Corp PO Box 92957 Los Angeles CA 90009

STUPP, EDWARD HENRY, b Brooklyn, NY, Dec 10, 32; m 54; c 2. SOLID-STATE PHYSICS, ELECTRONICS ENGINEERING. *Educ:* City Col New York, BS, 54; Syracuse Univ, MS, 58, PhD(physics), 60. *Prof Exp:* Asst physics, Columbia Univ, 54-55, Watson Lab, 55-56 & Syracuse Univ, 56-59; staff physicist, Thomas J Watson Res Ctr, 59-62; MEM TECH STAFF, SR PROG LEADER COMPONENTS & DEVICES GROUP, GROUP DIR DEVICE RES & DEPT HEAD, THIN FILM MAT RES DEPT, PHILIPS LABS DIV, N AM PHILIPS CO, 62- *Concurrent Pos:* Consult infrared imaging, Philips Broadcast Equip Cor, 72-73. *Mem:* Am Phys Soc; Inst Elec & Electronics Eng. *Res:* Semiconductor device research management and direction including thin film transistors and flat panel displays, SOI devices/circuits, displays, high voltage power integrated circuits; solid-state ballast circuits; photoemission; visible and infrared camera tubes, photodetectors, electron multiplication, image tubes; cold cathodes. *Mailing Add:* Philips Labs 345 Scarborough Rd Briarcliff NY 10510

STURBAUM, BARBARA ANN, b Cleveland, Ohio, June 10, 36. PHYSIOLOGY. *Educ:* Marquette Univ, BS, 59, MS, 61; Univ NMex, PhD(zool), 72. *Prof Exp:* From asst prof to assoc prof biol & earth sci, St John Col, Ohio, 61-75; from asst prof to assoc prof physiol & biol, Sch Med, Oral Roberts Univ, 75-87; ASSOC PROF PHYSIOL, SOUTHEASTERN COL OSTEOPATH MED, 87- *Concurrent Pos:* Consult radionuclide metab & toxicity, Lovelace Found Med Educ & Res, 74-75. *Mem:* Radiation Res Soc; Am Soc Zoologists; Inst Theol Encounter with Sci & Technol; AAAS. *Res:* Environmental physiology, particularly behavioral and physiological responses of animals to environmental factors, especially effects of temperature and pollutants on animals. *Mailing Add:* Dept Physiol Southeastern Col Osteopath Med 1750 NE 168 St N Miami Beach FL 33162-3097

STURCH, CONRAD RAY, b Cincinnati, Ohio, Nov 5, 37; m 62; c 1. ASTRONOMY. *Educ:* Miami Univ, BA, 58, MS, 60; Univ Calif, Berkeley, PhD(astron), 65. *Prof Exp:* Res asst astron, Lick Observ, Univ Calif, 65; from instr to asst prof, Univ Rochester, 65-73; vis asst prof, Univ Western Ont, 73-74; vis asst prof astron, Clemson Univ, 74-76; mem tech staff, Comput Sci Corp, 76-80, section mgr, 81-86, asst dept mgr, 86-88, DEPT MGR, COMPUT SCI CORP, 88- *Mem:* Am Astron Soc; Int Astron Union. *Res:* Variable stars; stellar populations; interstellar reddening; galactic structure. *Mailing Add:* Comput Sci Corp Space Telescope Sci Inst Homewood Campus Baltimore MD 21218

STURCKEN, EDWARD FRANCIS, b Charleston, SC, Nov 13, 27; m 53; c 2. MATERIALS SCIENCE, NUCLEAR PHYSICS. *Educ:* Col Charleston, BS, 48; St Louis Univ, MS, 50, PhD(nuclear physics), 53. *Prof Exp:* Res physicist, 53-62, sr scientist mat sci, 62-69, res supvr, Mat Res Methods, 69-70, RES ASSOC, MAT RES METHODS, SAVANNAH RIVER LAB, DEPT ENERGY, E I DU PONT DE NEMOURS & CO, INC, 70- *Concurrent Pos:* Vis scientist, Univ Calif, Berkeley, 66-67; mem, Joint Comt Powder Diffraction Standards, 72-; lectr, Traveling Lectr Prog, Dept of Energy, 72- *Honors & Awards:* Electron Micros Award, Am Soc Metals; Electron Micros Award, Electron Micros Soc Am. *Mem:* Am Crystallog Soc; Am Soc Metals; Int Microstructure Anal Soc; Electron Micros Soc Am. *Res:* Studies of structure; property relationships for materials employed in nuclear reactors and nuclear waste immobilization using various materials research methods, particularly electron microscopy and x-ray diffraction. *Mailing Add:* Savannah River Lab E I du Pont de Nemours & Co Inc Aiken SC 29808-0001

STURDEVANT, EUGENE J, b Newton, Kans, Dec 27, 30; m 58, 84; c 4. LIGHT SCATTERING OPTICS & MATH SURFACE ANALYSIS. *Educ:* Univ Calif, Berkeley, BSEE, 63. *Prof Exp:* Engr, Eng Physics Lab, E I du Pont de Nemours & Co, Inc, 63- 65, res engr, 65-68; res engr, Holotron Corp, 68-71; dir res & develop, Display Enterprises, Inc, 71-74; advan res engr, 76-80, sr scientist, Proctor-Silex Div, SCM Corp, 80-83; SR RES ENGR-OPTICS, SPITZ, INC, 84- *Concurrent Pos:* Consult electro-optics, 71-80; contract res & develop in electro-mech optics & heat/mass transfer, 83. *Mem:* Optical Soc Am; Soc Photo-Optical Instrument Engrs. *Res:* Applied research, development and design of electro-optical systems and instruments for display, product inspection and process control; light scattering to develop new projection surfaces; EHD enhancement of heat & mass transfer. *Mailing Add:* 140 Biddle Rd Paoli PA 19301-1104

STUREK, WALTER BEYNON, b Bartlesville, Okla, July 14, 37; m 65; c 2. AEROSPACE & MECHANICAL ENGINEERING. *Educ:* Okla State Univ, BS, 60; Mass Inst Technol, SM, 61; Univ Del, PhD(appl sci), 71. *Prof Exp:* Aerospace engr, Wind Tunnels Br, 65-77, BR CHIEF, COMPUTATIONAL AERODYNAMICS BR, LAUNCH & FLIGHT DIV, US ARMY BALLISTIC RES LABS, ABERDEEN PROVING GROUND, 77- *Mem:* Am Soc Mech Engrs; Am Inst Aeronaut & Astronaut. *Res:* Aerodynamics and flight mechanics of projectiles and missiles; computational fluid dynamics of flow over shell at transonic and supersonic velocity. *Mailing Add:* PO Box 415 Aberdeen Proving Ground MD 21005-0415

STURGE, MICHAEL DUDLEY, b Bristol, Eng, May 25, 31; m 56; c 4. EXPERIMENTAL SOLID STATE PHYSICS. *Educ:* Cambridge Univ, BA, 52, PhD(physics), 57. *Prof Exp:* Mem staff, Mullard Res Lab, 56-58; sr res fel, Royal Radar Estab, 58-61; mem tech staff, Bell Labs, 61-83; mem tech staff, Bellcore, 84-86; PROF PHYSICS, DARTMOUTH COL, 86- *Concurrent Pos:* Res assoc, Stanford Univ, 65; vis scientist, Univ BC, 69; exchange visitor, Philips Res Labs, Eindhoven, Neth, 73-74; vis lectr physics, Drew Univ, 75; vis prof, Technion, Haifa, 72, 76, 81 & 85, Univ Fourier, Grenoble, France, 89. *Mem:* Fel Am Phys Soc. *Res:* Magnetic insulators; optical properties of solids; semiconductor luminescence; excitons; super lattices; picosecond spectroscopy; plasmas in solids. *Mailing Add:* Physics Dept Wilder Hall Dartmouth Col Hanover NH 03755

STURGEON, EDWARD EARL, b Irving, Ill, Apr 28, 16; m 43; c 2. FOREST RECREATION. *Educ:* Univ Mich, BSF, 40, MF, 41, PhD(forestry), 54. *Prof Exp:* Instr forest policy, Univ Mich, 50-51; assoc prof forestry & head dept forestry & biol, Mich Technol Univ, 51-59; assoc prof forestry, Humboldt State Col, 59-66, coordr dept, 60-66; head dept, 66-73, prof, 66-81, EMER PROF FORESTRY, OKLA STATE UNIV, 81- *Concurrent Pos:* Chmn, Okla Bd Regist Foresters, 78-80, chmn, Ouachita Soc Am Foresters-Natural Areas Com, 81-86. *Mem:* Nat Wildlife Fedn; Soc Am Foresters; Am Forestry Asn; fel Nat Soc Am Foresters. *Res:* Public-private balance in forest land ownership; forest environment; cottonwood reproduction and management. *Mailing Add:* 2616 Quail Ridge Ct Stillwater OK 74074

STURGEON, GEORGE DENNIS, b Sioux Falls, SDak, Sept 21, 37; m 67; c 2. SOLID STATE CHEMISTRY, HIGH TEMPERATURE CHEMISTRY. *Educ:* Univ NDak, BS, 59; Mich State Univ, PhD(chem), 64. *Prof Exp:* Instr chem, Mich State Univ, 64; asst prof, 64-73, ASSOC PROF CHEM, UNIV NEBR, LINCOLN, 73- *Mem:* AAAS; Am Chem Soc; Sigma Xi. *Res:* Chemistry of refractory materials; high-temperature thermodynamics; chemistry of complex fluorides. *Mailing Add:* Dept Chem Univ Nebr Lincoln NE 68508

STURGEON, MYRON THOMAS, b Salem, Ohio, Apr 27, 08; m 46; c 2. PALEONTOLOGY, STRATIGRAPHY. *Educ:* Mt Union Col, AB, 31; Ohio State Univ, AM, 33, PhD(paleont), 36. *Prof Exp:* Found inspector, US Corps Engrs, Ohio, 34; from asst to assoc prof geol, Mich State Norm Col, 37-46; from asst prof to prof 46-78, EMER PROF GEOL, OHIO UNIV, 78- *Honors & Awards:* Mather Award, Ohio Div Geol Surv, 87. *Mem:* AAAS; Paleont Soc; assoc Soc Econ Paleont & Mineral; fel Geol Soc Am; Am Ornith Union. *Res:* Stratigraphy and invertebrate paleontology of the Pennsylvanian system of eastern Ohio. *Mailing Add:* 13220 Robinson Ridge Rd Athens OH 45701

STURGEON, ROY V, JR, b Wichita, Kans, July 1, 24; m 50; c 2. PLANT PATHOLOGY. *Educ:* Okla State Univ, BS, 61, MS, 64; Univ Minn, Minneapolis, PhD(plant path), 67. *Prof Exp:* Instr bot & plant path, Col Agr & Agr Exten, Okla State Univ, 63-65, from instr to assoc prof, 67-74, prof plant path, 74-86, exten plant pathologist, Fed Exten Serv, 67-86; FIELD CONSULT, OKLA PEANUT COMN, 86- *Concurrent Pos:* Private Plant Health Consult Serv, Okla, 67- *Mem:* Am Phytopath Soc; Soc Nematol; Am Soc Agron. *Res:* Program development and chemical evaluation for disease control, especially fungicides and nematicides. *Mailing Add:* 1729 Linda Ave Stillwater OK 74075

STURGES, DAVID L, b Riverside, Calif, Oct 7, 38; m 75. WATERSHED MANAGEMENT. *Educ:* Utah State Univ, BS, 61, MS, 63. *Prof Exp:* RES FORESTER WATERSHED MGT RES, ROCKY MOUNTAIN FOREST & RANGE EXP STA, FOREST SERV, USDA, 62- *Mem:* Soc Range Mgt; Soil Conserv Soc Am; Sigma Xi. *Res:* Hydrologic relations of big sagebrush lands; effects of big sagebrush control on quantity, quality and timing of water yield; snow management on big sagebrush lands. *Mailing Add:* Rocky Mountain Forest & Range Exp Sta 222 S 22nd St Laramie WY 82070

STURGES, LEROY D, b Stayton, Minn, 1945. FLUID MECHANICS, RHEOLOGY. *Educ:* Univ Minn, BAE, 67, MS, 75, PhD(eng mech), 77. *Prof Exp:* Space syst analyst, Foreign Technol Div, Systs Command, USAF, 68-72; lectr, AEM Dept, Univ Minn, 77; ASSOC PROF ENG MECH, AEEM DEPT, IOWA STATE UNIV, 77- *Mem:* Sigma Xi; Am Soc Eng Educ; Soc Rheol. *Res:* Investigating flow characteristics of non-linear fluids and their relationship to the rheological properties of the fluids. *Mailing Add:* Iowa State Univ 2019 Black Eng Bldg Ames IA 50011

STURGES, STUART, b Altamont, NY, July 1, 13; wid; c 2. NUCLEAR ENGINEERING. *Educ:* Rensselaer Polytech Inst, ChE, 35, MS, 39, PhD(phys chem), 41. *Prof Exp:* Asst instr biochem, Albany Med Col, 35-37, instr, 39-41; asst instr chem eng & chem, Rensselaer Polytech Inst, 37-39; sr chem res engr, Merck & Co, Inc, 41-42, Winthrop Chem Co, 42; prin chem engr, Manhattan Dist, Corps Engrs, US War Dept, 42-46; mgr var opers, Knolls Atomic Power Lab, Gen Elec Co, 48-78; RETIRED. *Concurrent Pos:* Mem staff, Joint Task Force One (Bikini Test), 46; instr chem, Adirondack Community Col, 74-76. *Mem:* Sigma Xi; Am Chem Soc; Am Inst Chemists; AAAS. *Res:* Physical properties of biological compounds; continuous process designs for organic preparations; chemical and metallurgical research; nuclear engineering. *Mailing Add:* 32 Stewart Ave South Glen Falls NY 12803-5127

STURGES, WILTON, III, b Dothan, Ala, July 21, 35; m 57; c 3. PHYSICAL OCEANOGRAPHY. *Educ:* Auburn Univ, BS, 57; Johns Hopkins Univ, MA, 63, PhD(oceanog), 66. *Prof Exp:* Res asst phys oceanog, Johns Hopkins Univ, 63-66; from asst prof to assoc prof, Univ RI, 66-72; assoc prof, 72-76, chmn dept, 76-82 & 85-88, PROF OCEANOG, FLA STATE UNIV, 76- *Concurrent Pos:* Instr, US Naval Res Off Sch, 63-66; assoc ed, J Geophys Res, 68-70; mem ocean-wide surv panel, Comt Oceanog, Nat Acad Sci, 68-71, Buoy Technol Assessment Panel Marine Bd, Nat Acad Eng, 72-74, Ocean Sci Comt, Nat Acad Sci, 75-77 & Comn Marine Geodesy, Am Geophys Union, 74-78; mem adv panel oceanog, NSF, 75-77; mem, Panel on Sea-Level Change, Geophys Study Comm, Nat Res Coun, 84-90; Comm Eng Implications changes in relative mean sea level, Marine Bd, Nat Res Coun, 84-89; mem, Fla Task Force on Oil Spill Risk Assessment, 88-90; mem adv comt climate & global change, Nat Oceanic & Atmospheric Admin, 90- *Mem:* Am Geophys Union. *Res:* Ocean circulations, especially Gulf of Mexico and North Atlantic; rise of sea level. *Mailing Add:* Dept of Oceanog Fla State Univ Tallahassee FL 32306-3048

STURGESS, JENNIFER MARY, b Nottingham, Gt Brit, Sept 26, 44; m 66; c 3. MICROBIOLOGY, PATHOLOGY. *Educ:* Bristol Univ, BSc, 65; Univ London, PhD(path), 70. *Prof Exp:* Res asst microbiol, Agr Col Norway, 64 & Clin Res Unit, Med Res Coun Eng, 65-66; lectr exp path, Inst Dis Chest, Brompton Hosp, Univ London, 66-70; sr scientist, Hosp Sick Children, 71-79; dir, Warner Lambert Res Inst, 79-86, vpres sci affairs, Warner Lambert Can Inc, 86-90; assoc prof path, 71-90, PROF PATH, UNIV TORONTO, 90-, ASSOC DEAN, RES, FAC MED, 90- *Concurrent Pos:* Res fel path, Hosp for Sick Children, Toronto, 70-71, Med Res Coun Can term grants & scholar, 71-; ed, Proceedings Micros Soc Can, 73- & Perspectives in Cystic Fibrosis, 80; Cystic Fibrosis Term grant, 74-; Ont Thoracic Soc grant, 75-, Med Res Coun term grant, 76-; consult scientist, Hosp Sick Children, 79-; WHO adv, Cystic Fibrosis, 83 & Genetic Serv, 85; mem Sci Coun Can, 87-; mem, Med Res Coun, 90- *Honors & Awards:* Scientific Award, Can Asn Pathologists, 75. *Mem:* Am Soc Cell Biol; Int Acad Path; NY Acad Sci; Can Asn Path; Micros Soc Can. *Res:* Mucus secretion in the normal lung and in chronic lung diseases; ciliary defects and human respiratory disease; cystic fibrosis. *Mailing Add:* Fac Med FitzGerald Bldg Rm 114 Univ Toronto 150 College St Toronto ON M5S 1A8 Can

STURGILL, BENJAMIN CALEB, b Wise Co, Va, Apr 27, 34; m 55; c 2. MEDICINE, PATHOLOGY. *Educ:* Berea Col, BA, 56; Univ Va, MD, 60. *Prof Exp:* Intern med, New York, Hosp-Cornell Med Ctr, 60-61; resident path, Univ Va, 61-62; clin assoc, NIH, 62-64; from instr to assoc prof, 64-76, actg chmn dept, 74-76, PROF PATH, SCH MED, UNIV VA, 76-, ASSOC DEAN & DIR ADMIS, 91- *Concurrent Pos:* Traveling fel, Royal Soc Med, 72. *Mem:* Int Acad Path; Am Asn Path; Am Soc Nephrology; Int Soc Nephrology. *Res:* Immunopathology and renal diseases. *Mailing Add:* Sch of Med Univ of Va Charlottesville VA 22908

STURGIS, BERNARD MILLER, b Butler, Ind, Nov 27, 11; m 36; c 2. PETROLEUM CHEMISTRY. *Educ:* DePauw Univ, AB, 33; Mass Inst Technol, PhD(org chem), 36. *Prof Exp:* Res chemist, Jackson Lab, E I Du Pont de Nemours & Co, Inc, 36-42, group leader auxiliary chem sect, Elastomer Div, 42-46, head petrol chem div, 46-51 & combustion & scavenging div, Petrol Lab, 51-53, from asst dir to dir, 53-62, mgr mid-continent region, Petrol Chem Div, 62-64, mgr, Patents & Contracts Div, 64-76; RETIRED. *Honors & Awards:* Horning Mem Award, Soc Automotive Eng, 56; Rector Award, 58. *Mem:* Am Chem Soc; Combustion Inst; Licensing Exec Soc. *Res:* Synthetic organic and rubber chemicals; accelerators; antioxidants; sponge blowing agents; peptizing agents; nonsulfur vulcanization of rubber; petroleum additives; tetraethyl lead; combustion; lead scavenging from engines. *Mailing Add:* 407 Hawthorne Dr Wilmington DE 19802-1200

STURGIS, HOWARD EWING, distributed systems; deceased, see previous edition for last biography

STURKIE, PAUL DAVID, b Proctor, Tex, Sept 18, 09; m 40, 64; c 5. PHYSIOLOGY. *Educ:* Tex A&M Univ, BS, 33, MS, 36; Cornell Univ, PhD(genetics, physiol), 39. *Prof Exp:* Res asst, Tex A&M Univ, 34-36 & Cornell Univ, 36-39; assoc prof, Auburn Univ, 39-44; from assoc prof to prof, 44-77, chmn dept environ physiol, 71-77, EMER PROF PHYSIOL, BARTLETT HALL COOK COL, RUTGERS UNIV, 77- *Concurrent Pos:* Guest reseacher, Agr Res Coun, Gt Brit, 60. *Honors & Awards:* Poultry Sci Res Award, 47; Borden Award, 56; Linback Res Award, 74. *Mem:* Fel AAAS; Am Physiol Soc; fel Poultry Sci Asn; Am Heart Asn; Microcirculatory Soc. *Res:* Physiology of reproduction, heart and circulation of birds; author of 170 technical reports. *Mailing Add:* 103 Fern Rd East Brunswick NJ 08816

STURLEY, ERIC AVERN, b Dibden Hants, Eng, June 9, 15; nat US; m 47, 81; c 3. MATHEMATICS. *Educ:* Yale Univ, BA, 37, MA, 39; Univ Grenoble, cert, 45; Columbia Univ, EdD, 56. *Prof Exp:* Instr, Berkshire Sch, 41-42 & Lawrenceville Sch, 46-47; from instr to assoc prof math, Allegheny Col, 47-57; instr, 57-61, actg head div sci & math, 58-60, asst dean grad sch, 62-64, prof, 61-84, EMER PROF MATH, SOUTHERN ILL UNIV, EDWARDSVILLE, 84- *Concurrent Pos:* Consult, Talon, Inc, Pa, 55-57; chief party, Southern Ill Univ Contract Team, Mali, WAfrica, 64-67, Nepal, 70-71; chief acad adv, 59-, coordr, Deans Col, Southern Ill Univ, Edwardsville, 67- *Mem:* Am Math Soc; Math Asn Am. *Res:* Statistics; history of mathematics. *Mailing Add:* 553 Buena Vista Edwardsville IL 62025

STURM, EDWARD, US citizen; m 50; c 3. GEOLOGY, MINERALOGY. *Educ:* NY Univ, BA, 48; Univ Minn, MSc, 50; Rutgers Univ, PhD(geol), 57. *Prof Exp:* Res geologist, Hebrew Univ Jerusalem, 51-52; asst res specialist crystallog, Bur Eng Res, Rutgers Univ, 56-58; asst prof geol, Tex Technol Col, 58-63; from asst prof to prof geol, Brooklyn Col, 74-85; RETIRED. *Mem:* Geol Soc Am; Mineral Soc Am; Am Crystallog Asn. *Res:* Clay mineralogy; crystallography of silicates; preferred orientation studies; geochemistry of solids. *Mailing Add:* Dept Geol Brooklyn Col Brooklyn NY 11210

STURM, JAMES EDWARD, b New Ulm, Minn, Mar 28, 30; m 55; c 7. RADIOCHEMISTRY, MASS SPECTROMETRY. *Educ:* St John's Univ, Minn, BA, 51; Univ Notre Dame, PhD(chem), 57. *Prof Exp:* Res assoc, Univ Wis-Madison, 55-56; from asst prof to assoc prof, 56-72, PROF PHYS CHEM, LEHIGH UNIV, 72- *Concurrent Pos:* Res assoc, Brookhaven Nat Lab, 57 & Argonne Nat Lab, 58; consult, Edgewood Arsenal, US Army, 68-69; fac res participant, Air Force Geophys Lab, 85, 86 & ARDEC, 87. *Mem:* Am Chem Soc; Sigma Xi. *Res:* Photochemical kinetics; rates of elementary processes, especially reactions of excited or high-velocity (hot) atoms; vacuum ultraviolet photochemistry; photochemistry of metal-ligand complexes; radiation chemistry. *Mailing Add:* Dept Chem No 6 Lehigh Univ Bethlehem PA 18015

STURM, WALTER ALLAN, b Brooklyn, NY, July 22, 30; div; c 2. COMPUTER SCIENCE. *Educ:* Brown Univ, ScB, 52; Mass Inst Technol, SMEE, 57; Univ Calif, Los Angeles, PhD(eng), 64. *Prof Exp:* Staff engr comput, Hughes Aircraft Co, 57-62; assoc eng, Univ Calif, Los Angeles, 58-60; staff engr, Data Syst Div, Litton Industs, 62-64; staff engr comput, 64-77. APPL ADMINR, AEROSPACE CORP, 78- *Mem:* Asn Comput Mach; Sigma Xi. *Res:* Computer systems architectures; application of APL to business data processing; direct-execution machines; fault-tolerant computers; reliable software; software engineering. *Mailing Add:* 5106 W 134th Hawthorne Los Angeles CA 90250

STURM, WILLIAM JAMES, b Marshfield, Wis, Sept 10, 17; m 51; c 2. NUCLEAR PHYSICS, SOLID STATE PHYSICS. *Educ:* Marquette Univ, BS, 40; Univ Chicago, MS, 42; Univ Wis, PhD(physics), 49. *Prof Exp:* Asst nuclear physics, Manhattan Proj, Metall Lab, Univ Chicago, 42-43, jr physicist, 43-46; assoc physicist & group leader, Argonne Nat Lab, 46-47; consult physicist, 49-51; from physicist to sr physicist, Oak Ridge Nat Lab, 51-56; assoc physicist, Int Inst Nuclear Sci & Eng, Argonne Nat Lab, 56- 60, Off Col & Univ Coop, 65-67, asst dir, Appl Physics Div, 67-84; RETIRED. *Honors & Awards:* Commemorative Medal, Atomic Indust Forum, Am Nuclear Soc, 62; Nuclear Pioneer Award, Soc Nuclear Med, 77. *Mem:* AAAS; Am Phys Soc; Am Nuclear Soc. *Res:* Neutron cross sections and diffraction; nuclear reactions, reactor physics and absolute nuclear particle energies; irradiation effects in solids; subcritical and critical reactor studies; reactor safety. *Mailing Add:* 5400 Woodland Ave Western Springs IL 60558

STURMAN, JOHN ANDREW, b Hove, Eng, Aug 10, 41. BIOCHEMISTRY, NUTRITION. *Educ:* Univ London, BSc, 62, MSc, 63, PhD(biochem), 66. *Prof Exp:* Assoc res scientist, 67-79, RES SCIENTIST, DEVELOP NEUROCHEM LAB, DEPT DEVELOP BIOCHEM, INST BASIC RES DEVELOP DISABILITIES, 80-, DEPT CHMN, 84- *Concurrent Pos:* Res study grant red cell metab, King's Col Hosp, Med Sch, Univ London, 63-67; lectr, Can Nat Inst Nutrit, 85. *Mem:* AAAS; Am Inst Nutrit; Brit Biochem Soc; Am Soc Neurochem; Int Soc Neurochemistry; Int Soc Develope Neurosci; Am Soc Biol Chem; Am Soc Neurosci; Am Asn Mental Deficiency. *Res:* Sulfur amino acid metabolism in normal and vitamin B-6 deficiency, in fetal, neonatal and adult tissue and in inborn errors of metabolism; axonal transport in developing nerves; nutrition and brain development. *Mailing Add:* Develop Neurochem Lab Inst Basic Res Ment Retardation 1050 Forest Hill Rd Staten Island NY 10314

STURMAN, LAWRENCE STUART, b Detroit, Mich, Mar 13, 38; m 59; c 4. VIROLOGY. *Educ:* Northwestern Univ, BS, 57, MS & MD, 60; Rockefeller Univ, PhD(virol), 68. *Prof Exp:* Intern, Hosp Univ Pa, 60-61; staff assoc virol, Nat Inst Allergy & Infectious Dis, 68-70; asst prof, 76-79, ASSOC PROF MICROBIOL & IMMUNOL, ALBANY MED COL, 79-; CHMN & PROF, DEPT BIOMED SCI, SCH PUB HEALTH SCI, NY STATE UNIV, ALBANY, 85- *Concurrent Pos:* Res physician virol, Wadsworth Ctr, NY State Dept Health, 70-; exec dir, NY State Health Res Coun, 87-; dir, Div Clin Sci, Wadsworth Ctr, NY State Dept Health, 89- *Mem:* Am Soc Microbiol; Am Soc Virol. *Res:* Viral pathogenesis; public health; biomedical research; public policy. *Mailing Add:* Wadsworth Ctr Labs & Res NY State Dept Health Empire State Plaza Albany NY 12201

STURMER, DAVID MICHAEL, b Norfolk, Va, July 27, 40; m 64; c 2. SOLID STATE CHEMISTRY & PHOTOGRAPHIC CHEMISTRY. *Educ:* Stanford Univ, BS, 62; Ore State Univ, PhD(org chem), 66. *Prof Exp:* NSF fel chem, Yale Univ, 66-67; sr chemist, 67-72, res assoc, 72-77; LAB HEAD CHEM, EASTMAN KODAK CO RES LABS, 77- *Concurrent Pos:* Adj prof, Dept Photog Sci, Rochester Inst Technol, 77-82. *Mem:* Am Chem Soc; Sigma Xi; Soc Photog Scientists & Engrs. *Res:* Molecular orbital calculations; heterocyclic dye synthesis; spectral sensitization of silver halides; solid state chemistry; radiotracer methods; chemical catalysis. *Mailing Add:* 41 Parkridge Pittsford NY 14534

STURR, JOSEPH FRANCIS, b Syracuse, NY, Apr 29, 33; m 60; c 4. VISUAL SCIENCE. *Educ:* Wesleyan Univ, BA, 55; Fordham Univ, MA, 57; Univ Rochester, PhD(psychol), 62. *Prof Exp:* Asst exp psychol, Fordham Univ, 56-57 & Univ Rochester, 57-58, asst vision res lab, 58-61; USPHS res fel psychophysiol lab, Ill State Psychiat Inst, 61-64; from asst prof to assoc prof, 64-72, PROF PHYSIOL PSYCHOL, SYRACUSE UNIV, 72- *Concurrent Pos:* Consult, Vet Admin Hosp, Syracuse, 65- *Mem:* AAAS; Optical Soc Am; Asn Res Vision & Ophthal; Sigma Xi. *Res:* Vision; psychophysics; spatio-temporal factors; flicker, increment thresholds; target detection; visual masking and excitability; sensitivity; rapid adaptation. *Mailing Add:* Dept Psychol Syracuse Univ Syracuse NY 13244

STURROCK, PETER ANDREW, b Grays, Eng, Mar 20, 24; US citizen; m 63; c 3. ASTROPHYSICS, PLASMA PHYSICS. *Educ:* Cambridge Univ, BA, 45, MA, 48, PhD(math), 51. *Prof Exp:* Harwell sr fel, Atomic Energy Res Estab, Eng, 51-53; fel, St John's Col, Cambridge Univ, 52-55; res assoc microwaves, Stanford Univ, 55-58; Ford fel plasma physics, Europ Orgn Nuclear Res, Switz, 58-59; res assoc, Stanford Univ, 59-60, prof eng sci & appl physics, 61-66, chmn, Inst Plasma Res, 64-74, PROF SPACE SCI & ASTROPHYS, DEPT APPL PHYSICS, STANFORD UNIV, 66-, DEP DIR, CTR SPACE SCI & ASTROPHYS, 83- *Concurrent Pos:* Consult, Varian Assocs, Calif, 57-64 & NASA Ames Res Ctr, 62-64; dir, Enrico Fermi Summer Sch Plasma-Astrophys, Varenna, Italy, 66; chmn, Plasma Physics Div, Am Phys Soc, 67, Solar Physics Div, Am Astron Soc, 72-74 & advocacy panel, Physics Sun, 80-83; mem, Phys Sci Comt, NASA; dir solar flare, Skylab Workshop, 76-77. *Honors & Awards:* Gravity Found Prize, 67; Hale Prize, Am Astron Soc, 86; Arctowski Medal, Nat Acad Sci, 90. *Mem:* Fel AAAS; Am Astron Soc; Int Astron Union; fel Am Phys Soc; fel Royal Astron Soc; Soc Sci Explor (pres, 82-). *Res:* Plasma astrophysics; solar physics; pulsars; radio galaxies; quasars; scientific inference; anomalous phenomena. *Mailing Add:* Ctr Space Sci & Astrophys Stanford Unif ERL 306 Stanford CA 94305-4055

STURROCK, PETER EARLE, b Miami, Fla, Dec 6, 29; m 58; c 1. ANALYTICAL CHEMISTRY, ELECTROCHEMISTRY. *Educ:* Univ Fla, BS, 51, BA, 51; Stanford Univ, MS, 54; Ohio State Univ, PhD(chem), 60. *Prof Exp:* From asst prof to assoc prof, 60-78, PROF CHEM, GA INST TECHNOL, 78- *Mem:* Am Chem Soc. *Res:* Instrumental chemical analysis; equilibria of complex ions; kinetics of electrode reactions; applications of computers to chemical instrumentation. *Mailing Add:* Dept Chem Ga Inst Technol Atlanta GA 30312

STURTEVANT, BRADFORD, b New Haven, Conn, Nov 1, 33; m 58; c 1. FLUID MECHANICS. *Educ:* Yale Univ, BS, 55; Calif Inst Technol, PhD(fluid mech), 60. *Prof Exp:* Res fel fluid mech, 60-62, from asst prof to assoc prof, 62-72, exec officer aeronaut, 71-76, PROF AERONAUT, CALIF INST TECHNOL, 71- *Concurrent Pos:* Res fel & Gordon McKay vis lectr, Harvard Univ, 65-66. *Mem:* AAAS; Am Phys Soc; Am Inst Aeronaut & Astronaut. *Res:* Experimental fluid mechanics; shock waves; vapor explosions; nonlinear acoustics. *Mailing Add:* 1419 East Palm Altadena CA 91001

STURTEVANT, FRANK MILTON, b Evanston, Ill, Mar 8, 27; m 50; c 2. PHARMACOLOGY. *Educ:* Lake Forest Col, BA, 48; Northwestern Univ, MS, 50, PhD(biol), 51. *Prof Exp:* Asst, Northwestern Univ, 50-51; sr investr, G D Searle & Co, 51-58; sr pharmacologist, Smith Kline & French Labs, 58-60; dir sci & regulatory affairs, Mead Johnson & Co, 60-72; lectr genetics, Univ Evansville, 72; assoc dir res & develop, 72-80, dir, Off Sci Affairs, 80-87, SR DIR, MED SCI INFO DEPT, G D SEARLE & CO, 87- *Honors & Awards:* Pres Award, Mead Johnson & Co, 67. *Mem:* Drug Info Asn; Soc Exp Biol & Med; Am Soc Pharmacol & Exp Therapeut; Am Fertility Soc; Am Col Toxicol; Fallopius Int Soc; Int Soc Chronobiology; US Int FNDN Studies Reproduction; Soc Adv Contraception. *Res:* Hypertension; pharmacokinetics; biochemorphology; glucoregulation; genetics; reproduction; central nervous system; chronobiology. *Mailing Add:* Med Sci-Info Dept G D Searle & Co 4901 Searle Pkwy Skokie IL 60077

STURTEVANT, JULIAN MUNSON, b Edgewater, NJ, Aug 9, 08; m 29; c 2. BIOPHYSICAL CHEMISTRY. *Educ:* Columbia Univ, AB, 27; Yale Univ, PhD(chem), 31. *Hon Degrees:* ScD, Ill Col, 62; Regensburg Univ, WGer, 79. *Prof Exp:* From instr to asst prof chem, Yale Univ, 31-43; staff mem, Radiation Lab, Mass Inst Technol, 43-46; from assoc prof to prof chem, 46-77, chmn dept, 59-62, assoc dir Sterling Chem Lab, 50-59, prof, 62-77, EMER PROF CHEM, MOLECULAR BIOPHYS & BIOCHEM, YALE UNIV, 77-, SR RES SCIENTIST, 77- *Concurrent Pos:* Consult, Mobil Oil Co, 46-69; Guggenheim fel & Fulbright scholar, Cambridge Univ, 55-56; Fulbright scholar, Univ Adelaide, 62-63; vis prof, Univ Calif, San Diego, 66-67 & 69-70; vis fel, Seattle Res Ctr, Battelle Mem Inst, 72-73; mem, US Nat Comt Data Sci & Technol, 76; vis scholar, Stanford Univ, 75-76; Alexander von Humboldt sr scientist award, Regensburg Univ, WGer, 78-79. *Honors & Awards:* Huffman Award, Calorimetry Conf US, 68; William Clyde DeVane Award, 78, Wilbur Lucius Cross Award, 87. *Mem:* Nat Acad Sci; Am Chem Soc; fel Am Acad Arts & Sci; AAAS. *Res:* The study of biochemical problems by physiochemical methods, with particular application of microcalorimetry. *Mailing Add:* Sterling Chem Lab Yale Univ New Haven CT 06520

STURTEVANT, RUTHANN PATTERSON, b Rockford, Ill, Feb 7, 27; m 50; c 2. GROSS ANATOMY, BIOLOGICAL RHYTHMS. *Educ:* Northwestern Univ, Evanston, BS, 49, MS, 50; Univ Ark, Little Rock, PhD(anat), 72. *Prof Exp:* From instr to asst prof life sci, Ind State Univ, Evansville, 65-74; adj asst prof, Sch Med, Ind Univ, 72-74; lectr, Sch Med, Northwestern Univ, 74-75; from asst prof to prof, 75-89, EMER PROF ANAT & SURG, STRITCH SCH MED, LOYOLA UNIV, CHICAGO, 89- *Concurrent Pos:* Grad Women Sci fel, 73. *Mem:* Am Asn Anatomists; Int Soc Chronobiology; Sigma Xi; AAAS; Soc Exp Biol & Med; Int Soc Biomed Res Alcoholism; Am Soc Pharmacol Exp Ther. *Res:* Chronobiology; chronopharmacokinetics; anatomy; fetal alcohol syndrome. *Mailing Add:* 1868 Mission Hills Lane Northbrook IL 60062

STURZENEGGER, AUGUST, b Switz, May 3, 21; nat US; m 55; c 3. ORGANIC CHEMISTRY, CHEMICAL ENGINEERING. *Educ:* Swiss Fed Inst Technol, MS, 45, PhD, 48. *Prof Exp:* Chemist, Royal Dutch Shell Co, Holland, 48; chemist, Steinfels, Inc Switz, 49; chemist, 49-59, dir advan technol, Hoffman-La Roche Inc, 59-77, dir, Pharmaceut & Diag Opers, 77-82; OWNER SEVERAL INVEST COS, 82- *Mem:* Am Chem Soc; Am Astronaut Soc; Am Inst Chem Eng; Swiss Chem Soc; Am Phys Soc; Sigma Xi. *Res:* Process development; detergents; petroleum chemistry; pharmaceuticals; systems analysis and automation; multidisciplinary interactions. *Mailing Add:* 25 Rensselaer Rd Essex Fells NJ 07021

STUSHNOFF, CECIL, b Saskatoon, Sask, Aug 12, 40; m 63; c 2. COLD STRESS PHYSIOLOGY, CRYOPRESERVATION. *Educ:* Univ Sask, BSA, 63, MSc, 64; Rutgers Univ, PhD(hort, embryol), 67. *Prof Exp:* Res asst, Dept Hort, Rutgers Univ, 64-67; asst prof fruit breeding, Univ Minn, 67-70, assoc prof, 70-75, prof hort sci & landscape archit, 75-80; prof hort & head dept, Univ Sask, 81-89; sr res scientist, Biochem Dept, 89-90, PROF HORT & BIOCHEM, COLO STATE UNIV, FT COLLINS, 90- *Concurrent Pos:* Consult, Walter Butler Corp & North Gro, Inc, St Paul, Minn; vis prof & guest researcher, Inst Biol & Geol, Univ Tromso, Norway; prin horticulturist res admin, Sci & Educ Admin, USDA, Washington, DC; owner & mgr, White Rock Lake Farm & The Berry Patch, St Paul, Minn; int travel grant, Hill Family Found; assoc ed, Can J Plant Sci, 83-; vis scientist, USDA Nat Seed Storage Lab, 87-88. *Honors & Awards:* Paul Howe Shepard Award; Joseph Harvey Gourley Award; G Darrow Award; C J Bishop Award. *Mem:* Am Soc Hort Sci; Int Soc Hort Sci; Am Soc Plant Physiol; Soc Crybiol; Am Pomol Soc. *Res:* Development of cold hardiness breeding methods for woody plants; evaluation of basic mechanisms for physiological basis of resistance; cryopreservation of plant germplasm; micropropagation; genetic stability of germplasm; conduct research of preservation of fruit crop genetic resources; cryopreservation and in vitro preservation; transformation of strawberry; biochemistry of endogenous cryoprotectants; plant biotechnology. *Mailing Add:* Dept Hort Colo State Univ Ft Collins CO 80523

STUSNICK, ERIC, b Edwardsville, Pa, Aug 18, 39; m 67; c 1. ACOUSTICS. *Educ:* Carnegie-Mellon Univ, BS, 60; NY Univ, MS, 62; State Univ NY Buffalo, PhD(physics), 71. *Prof Exp:* Asst prof physics, Niagara Univ, 69-72; assoc physicist, Cornell Aeronaut Lab, Inc, 72-73; res physicist, Calspan Corp, 73-75, sr physicist, 75-77; prog mgr, 77-87, DEPT MGR, WYLE LABS, 87- *Concurrent Pos:* Lectr, Niagara Univ, 72-77. *Mem:* AAAS; Am Phys Soc; Am Asn Physics Teachers; Acoust Soc Am; Sigma Xi; Inst Noise Control Eng. *Res:* Applications of acoustic intensity measurement; acoustic simulation and modeling; digital signal processing and analysis; noise source identification techniques. *Mailing Add:* Wyle Labs 2001 Jefferson Davis Hwy Suite 701 Arlington VA 22202

STUTEVILLE, DONALD LEE, b Okeene, Okla, Sept 7, 30; m 52; c 3. PLANT PATHOLOGY. *Educ:* Kans State Univ, BS, 59, MS, 61; Univ Wis, PhD(plant path), 64. *Prof Exp:* Res asst plant path, Univ Wis, 61-64; asst prof, 64-69, assoc prof, 69-79, PROF PLANT PATH, KANS STATE UNIV, 79-, RES FORAGE PATHOLOGIST, AGR EXP STA, 64- *Mem:* Am Phytopathological Soc. *Res:* Diseases of forage crops; improving disease resistance in forage crops, particularly alfalfa. *Mailing Add:* Dept of Plant Path Throckmorton Hall Kans State Univ Manhattan KS 66506

STUTH, CHARLES JAMES, b Greenville, Tex, Jan 9, 32; m 53, 75; c 5. ALGEBRA. *Educ:* East Tex State Univ, 51, MEd, 53; Univ Kans, PhD(math), 63. *Prof Exp:* Instr math, E Tex State Univ, 56-58, from asst prof to prof,62-66; asst instr, Univ Kans, 58-62; from asst prof to prof, East Tex State Univ, 62-66; asst prof, Univ Mo-Columbia, 66-70; chmn dept, 70-83, PROF, STEPHENS COL, 83- *Concurrent Pos:* Math Avoidance Prog, 78-80; comput info syst, 83-85. *Mem:* Math Asn Am. *Res:* Group theory; theory of semigroups. *Mailing Add:* Dept Bus & Math Stephens Col Columbia MO 65215

STUTHMAN, DEON DEAN, b Pilger, Nebr, May 7, 40; m 62; c 2. PLANT GENETICS, PLANT BREEDING. *Educ:* Univ Nebr, BSc, 62; Purdue Univ, MSc, 64, PhD(genetics of alfalfa), 67. *Prof Exp:* From asst prof to assoc prof, 66-79, PROF OAT GENETICS & BREEDING,UNIV MINN, ST PAUL, 79- *Mem:* Fel Am Soc Agron; Crop Sci Soc. *Res:* Breeding and genetics of oats. *Mailing Add:* Inst Agr Agron & Plant Genetics Univ Minn St Paul MN 55108

STUTMAN, LEONARD JAY, b Boston, Mass, Apr 8, 28; m 51; c 4. HEMATOLOGY, CARDIOLOGY. *Educ:* Mass Inst Technol, BS, 48; Boston Univ, MA, 49; Univ Rochester, MD, 53. *Prof Exp:* Intern & resident internal med, 4th Med Div, Bellevue Hosp, New York, 53-56; instr clin med, Post-Grad Med Sch, NY Univ, 56-61, asst prof path, Sch Med, 61-65; HEAD COAGULATION RES LAB, DEPT MED, ST VINCENT'S HOSP & MED CTR, 65-; ASSOC PROF CLIN MED, NEW YORK MED COL, 81- *Concurrent Pos:* Lillia-Babbit-Hyde res fel metab dis, Sch Med, NY Univ, 56-57, Nat Heart Inst spec advan res fel, 59-61; Ripple Found coagulation res grant, 66; John A Polacheck Found fel, 66-67; attend physician, Nyack Hosp, 59-; co-investr, Nat Heart Inst grants, 60-65; attend physician, St Vincent's Hosp, 65-; med dir, Presidential Life Ins Co, Nyack, NY, 65-; dir, Ford Found-Vera Inst Cardiovasc Epidemiol Proj, 71- *Mem:* Am Col Physicians; assoc fel Am Col Cardiol; fel NY Acad Med; Am Soc Hemat. *Res:* Blood coagulation proteins in normal and pathologic states, including biochemistry and biophysics of cellular lipoproteins; epidemiology of cardiovascular disease; high altitude physiology, including effects on erythrocytes; biochemical genetics in clotting disorders; prevention of deep vein thrombosis. *Mailing Add:* Coagulation Res Lab St Vincent's Hosp & Med Ctr New York NY 10011

STUTMAN, OSIAS, b Buenos Aires, Arg, June 4, 33. IMMUNOLOGY. *Educ:* Univ Buenos Aires, MD, 57. *Prof Exp:* Lectr, Inst Med Res, Univ Buenos Aires, 57-63; mem res staff physiol, Inst Biol & Exp Med, Buenos Aires, 63-66; from instr to assoc prof path, Med Sch, Univ Minn, Minneapolis, 66-72; MEM & SECT HEAD, SLOAN-KETTERING INST CANCER RES, MEM SLOAN-KETTERING CANCER CTR, 73-, CHMN IMMUNOL PROG, 83-; PROF IMMUNOL, GRAD SCH MED SCI, CORNELL UNIV, 75- *Concurrent Pos:* USPHS res fel, Med Sch, Univ Minn, Minneapolis, 66-69; Am Cancer Soc res assoc, 69-74. *Mem:* Am Asn Immunol; Am Soc Exp Path; Am Asn Cancer Res; Transplantation Soc. *Res:* Development of immune functions in mammals, especially role of thymus and mechanisms of cell-mediated immunity in relation to normal functions and as defense against tumor development. *Mailing Add:* Mem Sloan-Kettering Cancer Ctr 1275 York Ave New York NY 10021

STUTT, CHARLES A(DOLPHUS), b Avoca, Nebr, Nov 12, 21; m 55; c 2. ELECTRICAL ENGINEERING. *Educ:* Univ Nebr, BSc, 44; Mass Inst Technol, ScD(elec eng), 51. *Prof Exp:* Res engr, Stromberg-Carlson Co, NY, 44-46; asst, Res Lab Electronics, Mass Inst Technol, 48-50, instr elec eng, 50-52, mem staff & asst group leader commun, Lincoln Lab, 52-57; res assoc, 57-66, mgr signal processing & commun, Res & Develop ctr, Gen Elec Co, 66-85; RETIRED. *Mem:* sr mem Inst Elec & Electronics Engrs. *Res:* Signal theory; signal processing; data transmission; radar; sonar; radio propagation. *Mailing Add:* 643 Riverview Rd Rexford NY 12148

STUTTE, CHARLES A, b Wapanucka, Okla, July 19, 33; m 55; c 3. PLANT PHYSIOLOGY, AGRONOMY. *Educ:* Southeastern Okla State Univ, BS, 55; Okla State Univ, MS, 61, PhD(bot, plant physiol), 67. *Prof Exp:* Teacher high schs, Okla, 55-64; instr biol & ecol, E Cent Univ, 64-65; adv plant physiol, forest physiol & gen plant physiol, Okla State Univ, 65-67; asst prof, 67-71, prof, 71-79, DISTINGUISHED PROF AGRON & BEN J ALTHEIMER CHAIR SOYBEAN RES, UNIV ARK, FAYETTEVILLE, 79- *Mem:* Plant Growth Regulator Soc Am; Am Soc Plant Physiologists; Am Soybean Asn; Sigma Xi; Am Soc Agron. *Res:* Physiological stress and growth regulator responses in soybeans, cotton, rice and other crop plants; role of phenolics in natural resistance to insects and disease. *Mailing Add:* Dept Agron Univ Ark Fayetteville AR 72701

STUTTE, LINDA GAIL, b Chicago, Ill, Oct 31, 46; m 84. EXPERIMENTAL ELEMENTARY PARTICLE PHYSICS. *Educ:* Mass Inst Technol, SB, 68; Univ Calif, Berkeley, PhD(physics), 74. *Prof Exp:* Fel, Calif Inst Technol, 74-76; SCIENTIST, FERMILAB, 76- *Mem:* Am Phys Soc; AAAS. *Res:* High energy neutrino interactions; charged particle beam design. *Mailing Add:* 42 W 540 Hidden Springs Dr St Charles IL 60174

STUTZ, CONLEY I, b Currie, Minn, Aug 18, 32; m 55; c 2. PHYSICS. *Educ:* Wayne State Col, BSE, 57; Univ NMex, MSE, 60; Univ Nebr, PhD(physics), 68. *Prof Exp:* High sch teacher, Iowa, 59; asst prof physics, Pac Univ, 60-64; from asst prof to assoc prof, 69-74, PROF PHYSICS, BRADLEY UNIV, 74- *Concurrent Pos:* State Coun Am Asn Univ Prof. *Mem:* Sigma Xi; Am Phys Soc; Am Asn Physics Teachers; Am Physics Soc; Am Asn Univ Prof. *Res:* Study of the approach to equilibrium of quantum mechanical systems; nuclear magnetic resonance and nuclear quadrupole resonance of solids. *Mailing Add:* Dept of Physics Bradley Univ Peoria IL 61625

STUTZ, HOWARD COOMBS, b Cardston, Alta, Aug 24, 18; nat US; m 40; c 7. GENETICS. *Educ:* Brigham Young Univ, BS, 40, MS, 51; Univ Calif, PhD, 56. *Prof Exp:* Prin, High Sch, Utah, 42-44; chmn dept biol, Snow Col, 46-51; asst prof, 56-67, PROF BOT, BRIGHAM YOUNG UNIV, 67- *Concurrent Pos:* Guggenheim fel, 60; vis prof, Am Univ Beirut, 67. *Mem:* Bot Soc Am; Soc Study Evolution; Sigma Xi. *Res:* Cytogenetic studies of Secale L and related grasses; phyllogenetic studies of western browse plants; origin of cultivated rye; dominance-penetrance relationships; phylogenetic studies within the family Chenopodiaceae. *Mailing Add:* 531 W 3750 North Brigham Young Univ Provo UT 84601

STUTZ, ROBERT L, b Kansas City, Kans, Aug 1, 31; m 60; c 2. SURFACTANT SCIENCE, WHEAT PROCESSING. *Educ:* Univ Kans, BA, 53, MS, 57, PhD(org chem), 61. *Prof Exp:* Asst chemist, Stand Oil Co, Ind, 56-57; sr chemist, Minn Mining & Mfg Co, 61-64; res chemist, 64-65, head chem sect, C J Patterson Co, Kansas City, MO, 65-73; PRES, VANGUARD CORP, 74- *Concurrent Pos:* Frederick Gardner Cottrell grant, 58-59; consult, 74- *Mem:* Am Chem Soc; Am Oil Chem Soc; fel Am Inst Chem; Sigma Xi; NY Acad Sci. *Res:* Surfactants; food emulsifiers; sucrose esters; specialty chemicals. *Mailing Add:* 4210 Shawnee Mission Pkwy Suite 100A Shawnee Mission KS 66205

STUTZENBERGER, FRED JOHN, b Louisville, Ky, Nov 10, 40; m 70; c 1. MICROBIOLOGY, ENZYMOLOGY. *Educ:* Bellarmine Col, BS, 62; Univ Houston, MS, 64; Mich State Univ, PhD(microbiol), 67. *Prof Exp:* Microbiologist, USPHS, 67-69; asst prof microbiol, Weber State Col, Ogden, Utah, 69-71; Nat Adv Res Coun fel, NZ Dept Agr, Hamilton, 71-73; assoc prof, 74-79, PROF MICROBIOL, CLEMSON UNIV, 79- *Honors & Awards:* Sigma Xi Res Award, 67. *Mem:* Sigma Xi; Am Soc Microbiol. *Res:* Extracellular enzymes of thermophilic actinomycetes; cellulose degradation; effect of herbicides on actinomycetes; hypersensitivity pneumonitis antigens and activation of alternate complement pathway; streptococcal immunoglobulin A protease production. *Mailing Add:* Microbiol Dept Clemson Univ Clemson SC 29631

STUTZMAN, LEROY F, b Indianapolis, Ind, Sept 5, 17; m 39; c 3. CHEMICAL ENGINEERING. *Educ:* Purdue Univ, BS, 39; Kans State Col, MS, 40; Univ Pittsburgh, PhD(chem eng), 46. *Prof Exp:* Instr chem, Hillyer Jr Col, 40-41; res fel, Mellon Inst, 41-43; dir rubber res, Pittsburgh Coke & Iron Co, 43; from asst prof to assoc prof chem eng, Tech Inst, Northwestern Univ, 43-50, prof & chmn dept, 50-56; dir res, Remington Rand Univac Div, Sperry-Rand Corp, 56-59; prof chem eng & chief party Univ Pittsburgh res team, Univ Santa Maria, Chile, 59-63; head chem eng dept, 63-70, PROF CHEM ENG, UNIV CONN, 63- *Concurrent Pos:* Consult, US Off Naval Res, Pure Oil Co, Corn Prod Refining Co & Remington Rand Univac Div, 43-56; consult, 57-; mem bd dirs, Control Data Corp, 74-; Fulbright lectr, Hacettepe Univ, Turkey, 77; vis prof, Univ Vienna, 77-78. *Mem:* AAAS; Am Chem Soc; Am Soc Eng Educ; Am Inst Chem Engrs. *Res:* Mass transfer; oil reservoirs; digital computers; process control; computer graphics; numerical analysis; non-linear optimization; process modelling and simulation. *Mailing Add:* Dept of Chem Eng Univ of Conn Storrs CT 06268

STUTZMAN, WARREN LEE, b Elgin, Ill, Oct 22, 41; m 64; c 2. ELECTRICAL ENGINEERING. *Educ:* Univ Ill, Urbana, AB & BS, 64; Ohio State Univ, MS, 65, PhD(elec eng), 69. *Prof Exp:* From asst prof to assoc prof, 69-79, PROF ELEC ENG, VA POLYTECH INST & STATE UNIV, 79- *Mem:* Fel Inst Elec & Electronic Engrs; Int Sci Radio Union. *Res:* Millimeter wave satellite communications; antennas; microwaves. *Mailing Add:* Dept Elec Eng Va Polytech Inst & State Univ Blacksburg VA 24061

STUVE, ERIC MICHAEL, b Billings, Mont, Oct 7, 56. SURFACE SCIENCE, ELECTROCHEMISTRY. *Educ:* Univ Wis-Madison, BSChE, 78; Stanford Univ, MSChE, 79, PhD(chem eng), 84. *Prof Exp:* Guest scientist, Fritz-Haber-Max Planck Inst, 84; asst prof, 85-90, ASSOC PROF CHEM ENG, UNIV WASH, 90- *Concurrent Pos:* NSF presidential young investr award, 86. *Mem:* Am Inst Chem Engrs; Am Chem Soc; Am Vacuum Soc; Electrochem Soc. *Res:* Surface science of electrochemistry; combined ultrahigh vacuum and electrochemical investigations of electrode processes and ultrahigh vacuum studies of double layer modelling; double layer structure, electrodeposition, electrocatalysis, and electric field induced surface chemistry. *Mailing Add:* Dept Chem Eng BF-10 Univ Wash Seattle WA 98195

STUY, JOHAN HARRIE, b Bogor, Indonesia, Jan 17, 25; m 52; c 2. BACTERIOLOGY. *Educ:* State Univ Utrecht, Bachelor, 48, Drs, 52, PhD(microbiol), 61. *Prof Exp:* Mem res staff radiobiol, N V Philips Labs, Netherlands, 52-65; assoc prof biol, 65-74, PROF BIOL SCI, FLA STATE UNIV, 74- *Concurrent Pos:* Fel, biol dept, Brandeis Univ, 57-58, biol div, Oak Ridge Nat Lab, 58-59 & biophys dept, Yale Univ, 59-60; vis prof, Fla State Univ, 62-63; US Atomic Energy Comn grant, 68-74. *Mem:* Am Soc Microbiol. *Res:* Recombination in bacteria and bacteriophages at the DNA level. *Mailing Add:* Dept of Biol Sci Fla State Univ Tallahassee FL 32306

STWALLEY, WILLIAM CALVIN, b Glendale, Calif, Oct 7, 42; m 63; c 2. PHYSICAL CHEMISTRY, ATOMIC PHYSICS. *Educ:* Calif Inst Technol, BS, 64; Harvard Univ, PhD(phys chem), 68. *Prof Exp:* From asst prof to assoc prof chem, Univ Iowa, 68-75, dir, Ctr Laser Sci & Eng, 87-89, PROF CHEM, UNIV IOWA, 75-, PROF PHYSICS, 77-, DIR, IOWA LASER FACIL, 79-; DIR, CTR LASER SCI & ENG, 87- *Concurrent Pos:* A P Sloan fel, 72-75; assoc prog dir quantum chem, NSF, 75-76. *Mem:* Am Chem Soc; Am Phys Soc; AAAS; fel Am Optical Soc; fel Am Phys Soc; fel Japan Soc Promotion Sci. *Res:* Intermolecular forces; gas phase chemical reaction kinetics; molecular beams; laser applications; low temperature physics; atomic and molecular scattering and spectroscopy; laser development. *Mailing Add:* Dept of Chem Univ of Iowa Iowa City IA 52242-1294

STYBLINSKI, MACIEJ A, b Sosnowiec, Poland, Jul 7, 42; c 3. COMPUTER AIDED DESIGN OF ELECTRONIC CIRCUITS, STATISTICAL DESIGN OF INTEGRATED CIRCUITS. *Educ:* Tech Univ Warsaw, Poland, MSc, 67, PhD(electron), 74, Tech Univ Warsaw, Poland, DSc, 81. *Prof Exp:* Asst, electron, Tech Univ Warsaw, 67-74, asst prof electron, 74-81; ASSOC PROF ELEC ENG, TEX A&M UNIV, 81- *Concurrent Pos:* Vis asst prof electron, Univ Calif, Berkeley, 79; prin investr, Tex Advan Technol Prog, 88-; Fulbright fel, 74-75. *Honors & Awards:* Polish Acad Sci Res Award, 80. *Mem:* Sr mem Inst Elec & Electron Eng. *Res:* Statistical circuit design; intelligent design systems; computer-aided circuit design; VLSI design for quality and manufacturability; circuit performance variability reduction; manufacturing yield optimization. *Mailing Add:* Dept Elec Eng Texas A&M Univ College Sta TX 77843

STYER, DANIEL F, b Abington, Pa, Jan 31, 55; m 77. STATISTICAL MECHANICS. *Educ:* Swarthmore Col, BA, 77; Cornell Univ, PhD(theoret physics), 83. *Prof Exp:* Res fel statist mech, Rutgers Univ, 83-85; asst prof, 85-90, ASSOC PROF PHYSICS, DEPT PHYSICS, OBERLIN COL, 90- *Concurrent Pos:* Vis asst prof physics, Case Western Reserve Univ, 88. *Mem:* Am Phys Soc. *Res:* Statistical mechanics and theoretical condensed matter physics; derivation and analysis of series expansions by partial differential approximants; systems with highly degenerate ground states. *Mailing Add:* Dept Physics Oberlin Col Oberlin OH 44074-1088

STYLES, ERNEST DEREK, b Canterbury, Eng, Oct 19, 26; m 65; c 2. GENETICS. *Educ:* Univ BC, BSA, 60; Univ Wis, PhD(genetics), 65. *Prof Exp:* Res asst genetics, Univ Wis, 60-64, from proj asst to proj assoc, 64-66; asst prof, 66-71, ASSOC PROF GENETICS, UNIV VICTORIA, 71- *Mem:* AAAS; Genetics Soc Am; Genetics Soc Can; Am Genetics Asn. *Res:* Maize genetics; genetic control of flavonoid biosynthesis; paramutation. *Mailing Add:* Dept of Biol Univ of Victoria Box 1700 Victoria BC V8W 2Y2 Can

STYLES, MARGRETTA M, b Mt Union, Pa, Mar 19, 30. NURSING. *Educ:* Juniata Col, BS, 50; Yale Univ, MN, 54; Univ Fla, DEd, 68. *Hon Degrees:* LHD, Valparaiso Univ, 86; Dr, Univ Athens, 91. *Prof Exp:* Assoc prof & dir undergrad studies, Sch Nursing, Duke Univ, NC, 67-69; prof & dean, Sch Nursing, Univ Tex, San Antonio, 69-73, Wayne State Univ, Mich, 73-77; coordr, grad prog bus admin, dept mental health & community nursing, 79-85, dean, 77-87, PROF, SCH NURSING, UNIV CALIF, SAN FRANCISCO, 77-, ASSOC DIR NURSING SERVS, HOSP & CLIN, 77- *Concurrent Pos:* Mem, Nat Comn Nursing, 80-83; proj dir, Study Regulation, Int Coun Nurses, Geneva, 84; mem, Secretary's Comn Nursing, US Dept Health & Human Serv, 88; mem bd dirs, Int Coun Nurses, 89-, chair, Prof Serv Comt, 89-; Fulbright fel, Univ Athens, 90. *Honors & Awards:* Anise Sorrel Lectr, Troy State Univ, 82; Harriet Cook Carter Lectr, Duke Univ, 83; Elizabeth Kemble Lectr, Univ NC, Chapel Hill, 85. *Mem:* Inst Med-Nat Acad Sci; fel Am Acad Nursing; Am Nurses Asn (pres, 86-88); Am Orgn Nurse Exec; Am Asn Univ Women. *Mailing Add:* Sch Nursing Univ Calif San Francisco CA 94143-0604

STYLES, TWITTY JUNIUS, b Prince Edward Co, Va, May 18, 27; m 62; c 2. PARASITOLOGY, BIOLOGY. *Educ:* Va Union Univ, BS, 48; NY Univ, MS, 57, PhD(biol), 63. *Prof Exp:* Jr bacteriologist, New York City Health Dept, 53-54, jr scientist, State Univ NY Downstate Med Ctr, 55-64; fel parasitol, Nat Univ Mex, 64-65; asst prof, 65-69, ASSOC PROF BIOL, UNION COL, NY, 69- *Concurrent Pos:* Lectr, City Col New York, 64; consult off higher educ planning, NY State Educ Dept, 70-71; lectr, Narcotics Addiction Control Comn, NY State, 71-72; NSF course histochemistry, Vanderbilt Univ, 72; sabbatical, Dept Vet Microbiol & Immunol, Univ Guelph, 72. *Mem:* Am Soc Parasitologists; Soc Protozool; Am Soc Microbiol; Nat Asn Biol Teachers; NY Acad Sci. *Res:* Effect of marine biotoxins on parasitic infections; effect of endotoxin of Trypanosoma lewisi infections in rats and Plasmodium berghei infections in mice. *Mailing Add:* Dept Biol Scis Union Col Schenectady NY 12308

STYLOS, WILLIAM A, b Lowell, Mass, July 23, 27. IMMUNOCHEMISTRY. *Educ:* State Univ NY, Buffalo, PhD(microbiol & immunol), 67. *Prof Exp:* EXEC SECY, IMMUNOBIOL STUDY SECT, NIH, 80- *Mem:* Sigma Xi; AAAS; Am Asn Immunologists; Am Soc Microbiol. *Mailing Add:* Div Res Grants Westwood Bldg Rm 222A NIH 5333 Westbard Ave Bethesda MD 20205

STYNES, STANLEY K, b Detroit, Mich, Jan 18, 32; m 55; c 3. CHEMICAL ENGINEERING, COMPUTER GRAPHICS. *Educ:* Wayne State Univ, BSChE, 55, MSChE, 58; Purdue Univ, PhD(chem eng), 63. *Prof Exp:* Pub health engr, USPHS, 56; instr chem eng, 56-60, from asst prof to assoc prof, 63-70, asst dean, 69-70, actg dean, 70-72, dean col eng, 72-85, PROF CHEM ENG, WAYNE STATE UNIV, 70- *Concurrent Pos:* Fac res fel, Wayne State Univ, 64. *Mem:* AAAS; fel Am Inst Chem Engrs; Am Chem Soc; Am Soc Eng Educ; Nat Soc Prof Engrs. *Res:* Transport phenomena in multi-phase systems; control and identification of environmental pollution from industrial sources; tribology (non-metallics and lubrication). *Mailing Add:* Col of Eng Wayne State Univ Detroit MI 48202

STYRING, RALPH E, b Bessemer, Ala, Apr 13, 21; wid. CHEMICAL ENGINEERING. *Educ:* Auburn Univ, BS, 43, Univ Mich, MS, 48. *Prof Exp:* Assoc chem engr, Atlantic Refining Co, 48, asst chem engr, 48-54, supvr engr, 54-61, tech supvr, 61, prin res engr, 61-80, spec proj adv, 80-85; RETIRED. *Mem:* Am Inst Chem Engrs; Soc Petrol Engrs; Am Inst Mech Engrs. *Res:* Natural gas and oil shale processing; liquefied natural gas; in situ recovery of crude oil by thermal methods; tar sand processing. *Mailing Add:* 800 Rollingwood Richardson TX 75081

STYRIS, DAVID LEE, b Pomona, Calif, Apr 21, 32; m 58; c 1. EXPERIMENTAL PHYSICS. *Educ:* Pomona Col, BA, 57; Univ Ariz, MS, 62, PhD(physics), 67. *Prof Exp:* Dynamics engr, Airframe Design, Convair-Pomona, 56-58; res physicist, Weapons Testing, Edgerton, Germeshausen & Grier, 58-60; res assoc, Field Ion Micros, Cornell Univ, 67-69; asst prof physics, shock physics & surface sci, Wash State Univ, 69-74; sr res scientist, 74-89, STAFF SCIENTIST, BATTELLE NORTHWEST LAB, 89- *Mem:* AAAS; Fedn Am Scientists; NY Acad Sci; Am Soc Mass Spectrometry; Can Spectros Soc; Am Soc Chem Spectrometry. *Res:* Radiation damage of materials related to controlled thermonuclear reactor systems; surface science; mass spectroscopy; atomic absorption spectroscopy; high temperature surface chemistry and physics. *Mailing Add:* 205 Craighill Ave Richland WA 99352

STYRON, CHARLES WOODROW, b New Bern, NC, Nov 6, 13; m 76; c 2. MEDICINE. *Educ:* NC State Univ, BS, 34; Duke Univ, MD, 38; Am Bd Internal Med, dipl. *Prof Exp:* Intern pediat, Duke Univ, 38; intern & resident med, Boston City Hosp, 38-40; assoc, 50-65, ASST PROF MED, MED CTR, DUKE UNIV, 65- *Concurrent Pos:* Fel, Joslin Clin, New Eng Deaconess Hosp, Boston, 40-42; pvt pract, 46-; mem coun foods & nutrit, 69-76, alt deleg, 73-82, AMA; chmn, NC Gov Comt Health Care Delivery, 71-73,. *Mem:* Fel Am Col Physicians; AMA; Am Diabetes Asn; fel Am Heart Asn; Am Soc Internal Med. *Res:* Internal medicine; diabetes mellitus and endocrinology. *Mailing Add:* 615 St Mary's St Raleigh NC 27605

STYRON, CLARENCE EDWARD, JR, b Washington, NC, Sept 14, 41; m 69; c 2. ECOLOGY, BIOTECHNOLOGY. *Educ:* Davidson Col, BS, 63; Emory Univ, MS, 65, PhD(biol), 67. *Prof Exp:* Asst prof biol, St Andrews Presby Col, 69-77; safety, Monsanto Res Corp, 77-84, SAFETY, MONSANTO HQ, 84- *Concurrent Pos:* Consult, Oak Ridge Nat Lab, 69-76. *Mem:* Ecol Soc Am; Am Inst Biol Sci; Am Soc Limnol & Oceanog; Marine Biol Asn UK; Health Physics Soc. *Res:* Assessment of radionuclides in fossil fuels; ecology of invertebrate communities; effects of radioactive fallout on terrestrial systems; transport of heavy metals in environmental systems; biological safety in biotechnology. *Mailing Add:* Monsanto Co MS BB1C 700 Chesterfield Village Pkwy St Louis MO 63198

SU, CHAU-HSING, b Fukien, China, Nov 23, 35; m 60; c 4. FLUID MECHANICS, WATER WAVES. *Educ:* Nat Taiwan Univ, BS, 56; Univ Minn, MS, 59; Princeton Univ, PhD(eng), 64. *Prof Exp:* Asst prof eng, Mass Inst Technol, 63-66; res assoc plasma physics, Princeton Univ, 66-67; assoc prof appl math, 67-75, PROF APPL MATH, BROWN UNIV, 75- *Concurrent Pos:* Consult, AT&T Lab. *Mem:* AAAS; Am Phys Soc. *Res:* Nonlinear wave theory. *Mailing Add:* Dept Appl Math Brown Univ Brown Sta Providence RI 02912

SU, CHEH-JEN, b Taipei, Taiwan, June 11, 34; US citizen; m 67; c 3. POLYMER CHEMISTRY, PAPER CHEMISTRY. *Educ:* Taipei Inst Technol, Taiwan, BS, 55; NC State Univ, BS, 60; State Univ NY Col Forestry, Syracuse Univ, MS, 63. *Prof Exp:* Chem engr, Taiwan Pulp & Paper Co, 55-59; res chemist, Owens-Ill, Inc, Ohio, 65-67; sr res scientist II paper & polymers, 67-75, SR RES SCIENTIST I POLYMER & FOREST PROD, CONTINENTAL CAN CO, INC, 75- *Mem:* Am Chem Soc. *Res:* Characterization of polymers and plastic molded articles; chemicals and materials from renewable sources. *Mailing Add:* 4151 Roslyn Rd Downers Grove IL 60515

SU, GEORGE CHUNG-CHI, b Amoy, China, Aug 8, 39; m 71; c 2. ORGANIC CHEMISTRY, ENVIRONMENTAL ANALYSIS. *Educ:* Hope Col, AB, 62; Univ Ill, MS, 64, PhD(org chem), 66. *Prof Exp:* Res chemist plastics dept, E I du Pont de Nemours & Co, 66 -69; NIH fel, Dept Biochem, Mich State Univ, 69-70, res assoc, Pesticide Res Ctr, 70-72; biochemist, Pesticide Sect, Bur Labs, Mich Dept Pub Health, 72-74; chief tech serv, air pollution control, 74-85, DIR, ENVIRON LAB, DEPT NATURAL RESOURCES, STATE MICH, 85- *Mem:* Am Chem Soc; The Chem Soc; NY Acad Sci. *Res:* Organic reaction mechanisms; air monitoring techniques; pesticide photochemistry; analytical techniques for isolation, detection, identification and quantitation of submicrogram quantities of environmental pollutants; toxicology and enzymology. *Mailing Add:* Dept Natural Resources PO Box 30028 Lansing MI 48909-7528

SU, HELEN CHIEN-FAN, b Nanping, China, Dec 26, 22; nat US. ORGANIC CHEMISTRY. *Educ:* Hwa Nan Col, China, BA, 44; Univ Nebr, MS, 51, PhD(chem), 53. *Prof Exp:* Asst chem, Hwa Nan Col, 44-47, instr, 47-49; prof, Lambuth Col, 53-55; res asst, Res Found, Auburn Univ, 55-57; res chemist, Borden Chem Co, 57-63; from assoc scientist to res scientist, Lockheed-Ga Co, 63-68; res chemist, Stored Prod Insects Res & Develop Lab, Agr Res Serv, USDA, 68-90; RETIRED. *Honors & Awards:* Indust Res Magazine IR-100 Award, 66. *Mem:* AAAS; fel Am Inst Chem; Am Chem Soc; NY Acad Sci; Entom Soc Am. *Res:* Heterocyclic nitrogen and sulfur compounds; unsaturated aliphatic compounds; natural products; naturally occurring pesticides; insect pheromones; insect repellents and attractants. *Mailing Add:* 610 Highland Dr Savannah GA 31406

SU, JIN-CHEN, b Anhwei, China, Dec 30, 32; US citizen; m 60; c 3. TOPOLOGY. *Educ:* Nat Taiwan Univ, BS, 55; Univ Pa, PhD(math), 61. *Prof Exp:* Asst prof math, Univ Va, 61-64; math mem, Inst Advan Study, 64-66; assoc prof, 66-72, PROF MATH, UNIV MASS, AMHERST, 72- *Mem:* Am Math Soc. *Res:* Transformation groups. *Mailing Add:* 14 Pebble Ridge Rd Amherst MA 01003

SU, JUDY YA-HWA LIN, b Hsinchu, Taiwan, Nov 20, 38; US citizen; m 62; c 1. MUSCLE PHYSIOLOGY & PHARMACOLOGY. *Educ:* Nat Taiwan Univ, BS, 61; Univ Kans, MS, 64; Univ Wash, PhD(pharmacol), 68. *Prof Exp:* Asst prof biol, Univ Ala, 72-73; res assoc cardiovasc pharmacol, 76-77, actg asst prof pharmacol, 77-78, res asst prof, 78-81, RES ASSOC PROF PHARMACOL, DEPT ANESTHESIOL, UNIV WASH, 81- *Concurrent Pos:* Res fel, San Diego Heart Asn, 70-72; Prin investr, Wash State Heart Asn, 76-77, Pharmaceut Mfg Asn, 77, Nat Heart, Lung & Blood Inst, 77-73 & Am Heart Asn, 80-82; mem Coun Basic Sci, Am Heart Asn, 81; vis scientist, Max-Planck Inst Med Res, Heidelberg, WGer, 82-83; res career develop award, Nat Heart, Lung & Blood Inst, NIH, 82-87; mem surg, Anesthesiol & Trauma Study Sect, 87-91; vis prof, Mayo Clinic, 88. *Mem:* Biophys Soc; Am Soc Pharmacol & Exp Therapeut; Am Soc Anesthesiologists; AAAS. *Res:* Mechanisms of action of pharmacological agents on the striated and smooth muscles; effects of drugs on the intracellular mechanisms of muscle contracton: the calcium activation of the contractile proteins and the calcium uptake and release from sarcoplasmic reticulum. *Mailing Add:* RN-10 Dept Anesthesiol Sch Med Univ Wash Seattle WA 98195

SU, KENDALL L(ING-CHIAO), b Nanping, China, July 10, 26; nat US; m 60; c 2. ELECTRICAL ENGINEERING. *Educ:* Xiamen Univ, BS, 47; Ga Inst Technol, MS, 49, PhD(elec eng), 54. *Prof Exp:* From asst prof to prof elec eng, 54-70, REGENTS PROF ELEC ENG, GA INST TECHNOL, 70- *Mem:* Fel Inst Elec & Electronics Engrs. *Res:* Network theory; electronics; active filters. *Mailing Add:* Sch Elec Eng Ga Inst Technol 225 N Ave NW Atlanta GA 30332-0250

SU, KENNETH SHYAN-ELL, b Taipei, Taiwan, Nov 26, 41; US citizen; m 70; c 2. PHARMACEUTICS. *Educ:* Taipei Med Col, BS, 65; Univ Wis, MS, 69, PhD(pharmaceut), 71. *Prof Exp:* Res fel biochem, US Naval Med Res Unit 2, 64-65; pharmaceut chemist, William S Merrell Co, 71; RES SCIENTIST, ELI LILLY & CO, 71- *Mem:* Am Asn Pharmaceut Sci; Am Med Asn. *Res:* Transmucosal drug delivery systems (including nasal, bronchial and sublingual systems), transport microparticles and macromulecules across nasal membranes, aerosol delivery systems, monagneous dosage form, and sustained release drug delivery systems. *Mailing Add:* Lilly Res Labs Eli Lilly & Co Indianapolis IN 46285

SU, KWEI LEE, b Ping Tong, Taiwan, Mar 18, 42; m 68; c 1. LIPID BIOCHEMISTRY. *Educ:* Nat Taiwan Univ, BS, 64; Univ Minn, PhD(biochem), 71. *Prof Exp:* Instr pharmacog, Col Pharm, Nat Taiwan Univ, 64-66; Hormel fel, Hormel Inst, Univ Minn, 70-72, from res fel to res assoc lipid chem, 72-74, asst prof, 74-75; res asst prof neurochem, Sinclair Comp Med Res Farm, Univ Mo, Columbia, 75-76; forensic chemist crime lab, Mo State Hwy Patrol, Jefferson City, 76-87; DIR, MIDAMERICA LABS & FORENSIC CONSULT, 88- *Mem:* Am Chem Soc; Am Oil Chemists Soc; Am Acad Forensic Sci. *Res:* Isolation, structural determination, biosynthesis and function of ether lipids in mammals; effects of neurotransmitters on lipid metabolism in brain subcellular membranes. *Mailing Add:* 2712 Plaza Dr Jefferson City MO 65109

SU, LAO-SOU, b Kaohsung, Taiwan, Dec 13, 32; m 45; c 2. PHYSICAL CHEMISTRY. *Educ:* Taiwan Norm Univ, BS, 57; Ind Univ, MS, 63, PhD(phys chem), 67. *Prof Exp:* Fel, Univ Mich 67-69 & Ind Univ, 69; sr res chemist, 69-80, res assoc, 80-84, SR RES ASSOC, S C JOHNSON & SON, INC, 84- *Mem:* Am Chem Soc. *Res:* Corrosion study of aerosol products; elemental analysis by means of x-ray fluorescence spectrometry; electron diffraction study of molecular structure; electrical property determination of substance by dielectric spectroscopy; biological alternating current impedance measurement. *Mailing Add:* S C Johnson & Soc Inc 1525 Howe St Racine WI 53403

SU, ROBERT TZYH-CHUAN, b Szechuan, China, Dec 14, 45; m 74. ANIMAL VIROLOGY, BIOCHEMISTRY. *Educ:* Fu Jen Univ, Taiwan, BS, 68; Univ Ill, MS, 71; Ind Univ, PhD(microbiol), 75. *Prof Exp:* Assoc instr biol, Ind Univ, 73-74; res fel biol chem, Harvard Med Sch, 75-78; asst prof, 78-84, ASSOC PROF MICROBIOL, UNIV KANS, 84- *Concurrent Pos:* Oncol trainee, Harvard Med Sch, 75-77; vis scientist, Nat Cancer Inst, 86. *Mem:* Sigma Xi; Am Soc Microbiol; Am Soc Cell Biol. *Res:* Replication of animal viruses; eucaryotic chromosome synthesis and gene regulation; human cell immortalization and differentiation. *Mailing Add:* Dept Microbiol Univ Kans Lawrence KS 66045

SU, SHIN-YI, b Taipei, Taiwan, China, July 18, 40; m 80; c 2. SPACE PHYSICS. *Educ:* Nat Taiwan Univ, BS, 63; Dartmouth Col, PhD(eng sci), 70. *Prof Exp:* Res asst space physics, Dartmouth Col, 65-69; postdoctoral fel, Univ Calgary, 70-72; resident res assoc at Johnson Space Ctr, Nat Acad Sci-Nat Res Coun, 72-74; prin scientist, Lockheed Electronics Co Inc, 74-80, COMPUT SYST ANALYST SPACE PHYSICS, LOCKHEED ENG & MGT SERV CO, IC, 80- *Mem:* Am Geophys Union. *Res:* Study of wave-particle interaction phenomena in the earth magnetosphere; interplanetary dust particle dynamics; spacecraft hazardous analysis from collision with near earth space debris. *Mailing Add:* Lockheed-Eng Mat Serv C23C 1830 NASA Rd 1 Houston TX 77058

SU, STANLEY Y W, b Fukien, China, Feb 18, 40; US citizen; m 65; c 2. COMPUTER SCIENCE. *Educ:* Tamkang Col Arts & Sci, BA, 61; Univ Wis, MS, 65, PhD(comput sci), 68. *Prof Exp:* Proj asst syst prog, Comput Ctr, Univ Wis, 64-67, res asst regional Am English proj, 67, res asst natural lang processing, 67-68; mathematician comput ling, Rand Corp, 68-70; asst prof, Dept Elec Eng & Commun Sci Lab, 70-74, assoc prof comput & info eng, Dept Elec Eng & Inst Advan Study Commun Processes, 74-78, PROF, DEPT COMPUT & INFO SCI, DEPT ELEC ENG, UNIV FLA, 78- *Concurrent Pos:* Mem, Spec Interest Group Operating Syst & Spec Interest Group Mgt Data, Asn Comput Mach, 73-; consult, Creativity Ctr Consortium, 73-74, Fla Keys Community Col, 74-75 & Cent Fla Community Col, 74-; staff consult, Queueing Systs, Inc, 74-75, Dept Energy, 82-84, Navy Ships Control Ctr, 84-85, Kings Res, 84, General Elec Corp Res & Develop Ctr, 85-; lectr continuing educ, George Washington Univ, 75-80; assoc ed, Transacting Software Eng, Int J Comput Lang, Inst Elec & Electronics Engrs, 81-, Int J Info Sci, 82-; area ed, J of Parallel Distrib & Comput, 84- *Mem:* Asn Comput Mach; Conf Data Systs Lang; Inst Elec & Electronic Engrs. *Res:* Associative processing systems; computer architecture for data base management; data base translation and program conversion; data base semantics; application of microprocessor network to non-numeric processing; man-machine communications; cost/benefit analysis of database management systems; modeling and design of statistical databases. *Mailing Add:* Dept Comput Sci & Elec Eng Gainesville FL 32611

SU, STEPHEN Y H, b Anchi, China, July 6, 38; US citizen; m 64; c 2. FAULT-TOLERANT COMPUTING, DESIGN AUTOMATION. *Educ:* Nat Taiwan Univ, BS, 60; Univ Wis-Madison, MS, 63, PhD(comput eng), 67. *Prof Exp:* Asst prof switching theory, New York Univ, 67-69, comput archit, Univ Calif, Berkeley, 69-71, design automation, Univ Southern Calif, 71-72; assoc prof design automation, Case Western Reserve Univ, 72-73, syst design, City Col New York, 73-75; prof fault diag, Utah State Univ, 75-78; PROF DESIGN AUTOMATION & FAULT TOLERANT COMPUT, STATE UNIV NY BINGHAMTON, 78- *Concurrent Pos:* Electronic engr, Air Force Radar Sta, Taiwan, 60-61; logic designer, Fabri-Tek, Inc, 65; proj specialist, Med Sch, Univ Wis, 66; consult, IBM, UNIVAC, & E & H Res, 68-78; mem tech staff, Bell Labs, 69; staff consult, UNIVAC, 73-74; engr, IBM, 74. *Honors & Awards:* Alexander von Humboldt Sr Scientist Award, 84-85. *Mem:* Sr mem Inst Elec & Electronic Engrs. *Res:* Fault tolerant design; fault diagnosis; computer aided logic/system design of digital systems and computer architecture; developing new algorithms for testing very large scale integration. *Mailing Add:* Dept Comput Sci State Univ NY Binghamton NY 13902

SU, TAH-MUN, b Taiwan, July 22, 39; c 3. ORGANIC CHEMISTRY, MICROBIOLOGY. *Educ:* Chen Kung Univ, Taiwan, BSc, 62; Univ Nev, MS, 65; Princeton Univ, PhD(chem), 70. *Prof Exp:* Res fel geochemistry, Biodyn Lab, Univ Calif, Berkeley, 69-70; res fel chem, Union Carbide Res Inst, 70-71; STAFF SCIENTIST CHEM & BIOENG, CORP RES & DEVELOP CTR, GEN ELEC CO, 72- *Mem:* Am Chem Soc; Am Soc Microbiol. *Res:* Single cell protein from cellulosic fiber; biodegradation of chlorinated hydrocarbons; enzymatic saccharification of cellulose; mechanism of organic chemical reactions; ethanol from biomass. *Mailing Add:* 2259 Berkley Schenectady NY 12309

SU, YAO SIN, b Ping-tung, Taiwan, Oct 17, 29; US citizen; m 54; c 2. ANALYTICAL CHEMISTRY. *Educ:* Taiwan Univ, BS, 52; Univ Pittsburgh, PhD(chem), 62. *Prof Exp:* Chemist, Union Res Inst, Taiwan, 53-58; sr res chemist, Corning Glass Works, 63-73, res supvr, 73-79, MGR, CORNING INC, 79- *Mem:* Am Chem Soc; Am Ceramic Soc; Am Soc Testing & Mat. *Res:* Inorganic chemical analysis; electroanalysis; classical wet methods. *Mailing Add:* 197 Cutler Ave Corning NY 14830

SUAREZ, KENNETH ALFRED, b Queens, NY, June 27, 44; m 68. PHARMACOLOGY, TOXICOLOGY. *Educ:* Univ RI, BS, 67, MS, 70, PhD(pharmacol), 72. *Prof Exp:* Nat Defense Educ Act fel, 67-70; from instr to assoc prof pharmacol, 72-82, asst dir, 80-82, PROF PHARMACOL & DIR, RES AFFAIRS, CHICAGO COL OSTEOP MED, 82- *Concurrent Pos:* Major, US Army Reserves. *Mem:* AAAS; Sigma Xi; Toxicol Soc. *Res:* Drug induced hepatic injury. *Mailing Add:* Dept Pharmacol Chicago Col Osteop Med 5200 Ellis Ave Chicago IL 60615

SUAREZ, THOMAS H, b Temperley, Arg, Dec 7, 36; m 61; c 3. RESOURCE MANAGEMENT. *Educ:* Univ Buenos Aires, MS, 59, PhD(phys org chem), 61. *Prof Exp:* Teaching asst org chem, Univ Buenos Aires, 59-61, head lab course, 61; res chemist, Textile Fibers Dept, Dacron Mfg Div, E I du Pont de Nemours & Co, 61-66, anal res supvr, 66-69, supvr process develop, 69-71, tech supt, Polymer Intermediates Dept, 71-74, planning mgr, Polymer Intermediates Dept, 74-77, sales mgr-Latin Am, Petrol Chem Div, 78-81, mgr, Gen Prod Dept, Du Pont de Venezuela, 81-82, sales mgr, Latin Am Explosive Prod Div, 82-85, export & licensing mgr, 85-87; export & licensing mgr, 88-89, MGR PURCHASING & DISTRIB, EXPLOSIVES & TECHNOL INT, INC, 90- *Mem:* AAAS; Am Chem Soc; Arg Chem Asn. *Res:* Nucleophylic aromatic substitution; reaction kinetics and mechanisms; polymer chemistry; melt spinning synthetic fibers; physical and chemical characterization of polymers. *Mailing Add:* Explosives & Technol Int Inc 731 Burnley Rd Wilmington DE 19803

SUBACH, DANIEL JAMES, b Shenandoah, Pa, July 7, 47; m 70; c 2. ANALYTICAL & POLYMER CHEMISTRY, PHYSICAL CHEMISTRY. *Educ:* Lebanon Valley Col, BS, 69; Marshall Univ, MS, 71; Tex A&M Univ, PhD(phys chem), 74; Rensselaer Polytech Inst, MBA, 81. *Prof Exp:* Qual control chemist, Campbell's Soup Co, 69-70; sr anal develop chemist, Ciba-Geigy Corp, 75-77; proj mgr anal & phys res & develop, Springborn Labs Inc, 77-78; mgr anal res & develop, Gen Elec Co, 78-80, mgr qual control, Silicon Prod Div, 80-81, mgr, new prod commercialization, res & develop & prod develop, 81-84; mgr new bus develop, NL Industs Inc, 85-86, bus mgr, 86-88; TECH DIR CORP RES & DEVELOP, H B FULLER CO, 89- *Concurrent Pos:* NASA fel, Rice Univ, 74-75. *Mem:* Sigma Xi; Am Chem Soc; Am Inst Chem; Fedn Socs Coating Technol. *Res:* Thermodynamics and thermophysical properties; nonelectrolyte mixture and liquid theory; trace analysis of organics and inorganics; chromatography including gas, liquid and thin-layer; adhesives, sealants, coatings polymer performance and formulation. *Mailing Add:* 416 Oak Creek Circle Vadnais Heights St Paul MN 55127-7001

SUBBAIAH, PAPASANI VENKATA, b Karumanchi, AP, India, July 1, 43; US citizen; m 65. LIPOPROTEIN RESEARCH, ATHEROSCLEROSIS. *Educ:* Andhra Univ, BSc, 63; Nagpur Univ, MSc, 65; Indian Inst Sci, PhD(biochem), 71. *Prof Exp:* Res asst prof, Univ Wash, 78-84; assoc prof med, 85-89, ASSOC PROF BIOCHEM, RUSH UNIV, 86-, PROF MED, 89- *Mem:* Am Chem Soc; Am Heart Asn; AAAS; Am Soc Biochem & Molecular Biol. *Res:* Lipoprotein metabolism and mechanisms of atherosclerosis; role of phospholipids in atherogenesis; mechanism of action of omega 3 fatty acids; plasma acyltrans ferases. *Mailing Add:* Dept Med Rush-Presbyterian St Luke's Med Ctr 1653 W Congress Pkwy Chicago IL 60612

SUBBASWAMY, KUMBLE R(AMARAO), b Soraba, India, Mar 18, 51; US citizen; m 86; c 1. THEORETICAL CONDENSED MATTER PHYSICS. *Educ:* Bangalore Univ, India, BSc, 69; Delhi Univ, MSc, 71; Ind Univ, Bloomington, PhD(physics), 76. *Prof Exp:* Res physicist, Univ Calif, Irvine, 76-78; from asst prof to assoc prof, 78-88, PROF PHYSICS, UNIV KY, LEXINGTON, 88- *Concurrent Pos:* Vis scientist, Int Ctr Theoret Physics, Trieste, Italy & 84-85, Oak Ridge Nat Lab, 85. *Mem:* Am Phys Soc. *Res:* Theoretical condensed matter physics; nonlinear excitations in quasi one-dimensional conductors; optical properties; ab initio computation of nonlinear susceptibilities; liquids and superionic conductors; graphite intercalation compounds; coal chemistry. *Mailing Add:* Dept Physics Univ Ky Lexington KY 40506-0055

SUBERKROPP, KELLER FRANCIS, b Wamego, Kans, Apr 12, 43; m 71; c 4. MICROBIAL ECOLOGY, PHYSIOLOGY. *Educ:* Kans State Univ, BS, 65, MS, 67; Mich State Univ, PhD(bot), 71. *Prof Exp:* Res assoc microbial ecol, Kellogg Biol Sta, Mich State Univ, 71-75; asst prof biol sci, Ind Univ-Purdue Univ, Ft Wayne, 75-78; asst prof, 78-84, assoc prof biol sci, NMex State Univ, 84-86; assoc prof, 86-89, PROF BIOL SCI, UNIV ALA, 89- *Mem:* Mycol Soc Am; Ecol Soc Am; Brit Mycol Soc; Sigma Xi; Am Soc Microbiol. *Res:* Role of fungi in decomposition of leaf litter in aquatic habitats; effects of environmental factors on growth and sporulation of these fungi. *Mailing Add:* Dept Biol Univ Ala Tuscaloosa AL 35487-0344

SUBJECK, JOHN ROBERT, RADIATION BIOLOGY, HEAT SHOCK. *Educ:* Univ Buffalo, PhD(biophysics), 74. *Prof Exp:* CANCER RES SCIENTIST, DEPT RADIATION MED, ROSWELL MEM INST, 77- *Mem:* Am Soc Cell Biol; Radiation Res Soc. *Res:* Heat shock protein. *Mailing Add:* Radiation Biol Roswell Park Mem Inst 666 Elm St Buffalo NY 14263

SUBLETT, BOBBY JONES, b Paintsville, Ky, Aug 27, 31; m 56; c 4. ORGANIC CHEMISTRY, POLYMER CHEMISTRY. *Educ:* Eastern Ky State Col, BS, 58; Univ Tenn, MS, 60. *Prof Exp:* From res chemist to sr res chemist, 60-75, res assoc, 75-90, SR RES ASSOC, TENN EASTMAN CO, 90- *Mem:* Am Chem Soc; Sigma Xi. *Res:* Reaction mechanisms; tobacco smoke analysis; condensation polymers; textile chemicals; adhesives. *Mailing Add:* 1205 Jerry Lane Kingsport TN 37664

SUBLETT, ROBERT L, b Columbia, Mo, Apr 10, 21; m 46; c 3. CHEMISTRY. *Educ:* Univ Mo, AB, 43, PhD, 50; Ga Inst Technol, MS, 48. *Prof Exp:* Instr chem, Ga Inst Technol, 47; res chemist, Chemstrand Corp, 52-55; assoc prof chem, Ark State Teachers Col, 55-56; assoc prof, 56-70, PROF CHEM, TENN TECHNOL UNIV, 70-, CHMN DEPT, 72- *Mem:* Am Chem Soc. *Res:* High polymers; Friedels-crafts; organic and high polymer analytical chemistry; instrumental analysis. *Mailing Add:* Dept Chem Tenn Technol Univ Cookeville TN 38501

SUBLETTE, IVAN H(UGH), b Urbana, Ill, May 15, 29. COMPUTER SCIENCE, ELECTRICAL ENGINEERING. *Educ:* Purdue Univ, BS, 49; Univ Pa, MS, 51, PhD(elec eng), 57. *Prof Exp:* Engr, RCA, 49-59, mem tech staff, RCA Labs, 59-74, SR SYSTS PROGRAMMER, RCA SOLID STATE DIV, 74- *Mem:* Inst Elec & Electronics Engrs; Asn Comput Mach. *Res:* Computer operating systems; performance measurement and evaluation. *Mailing Add:* Nat Broadcasting Co 600 Albany Post Rd Briarcliff Manor NY 10510

SUBLETTE, JAMES EDWARD, b Healdton, Okla, Jan 19, 28; m 50; c 4. ZOOLOGY. *Educ:* Univ Ark, BS, 48, MS, 50; Univ Okla, PhD(zool), 53. *Prof Exp:* Biologist, Corps Engrs, US Dept Army, 49-51; asst prof zool, Southwestern La Inst, 51-53 & Henderson State Teachers Col, 53; from asst prof to assoc prof, Northwestern State Col, 53-60; assoc prof, Tex Western Col, 60-61; assoc prof, Eastern NMex Univ, 61-66, dean, Sch Grad Studies, 66-78, prof biol, 66-, dist res prof, 78-84; off res, 84-88, DEPT LIFE SCI, UNIV SOUTHERN COLO, 88- *Concurrent Pos:* Japan Soc Prom Sci fel, 78; vis prof, Univ Bergen, 81. *Mem:* NAm Benthological Soc. *Res:* Taxonomy and ecology of aquatic insects, particularly Chironomidae; limnology; fishery Biology. *Mailing Add:* Dept Life Sci Univ Southern Colo 2200 Bonforte Blvd Pueblo CO 81001

SUBRAHMANYAM, D, b Waltair, India, Sept 4, 48; m 83; c 1. ATMOSPHERIC STABILITY. *Educ:* Andhra Univ, India, MS, 73, PhD(meteorol), 78. *Prof Exp:* ASST DIR METEOROL, INDIAN INST TROP METEOROL, 79- *Concurrent Pos:* Vis sci, Fla State Univ, 80- *Honors & Awards:* J Nehru Award, Indian Govt, 80. *Mem:* Am Meteorol Soc. *Res:* Low frequency modes of the monsoon and their use in forecasting. *Mailing Add:* Indian Inst Trop Meteorol Shivji Nagar 5 Koowna India

SUBRAMANI, SURESH, b Jabalpur, India, Feb 21, 52; c 2. CELL BIOLOGY. *Educ:* Fergusson Col, India, BSc, 72; Indian Inst Tech, MSc, 74; Univ Calif, Berkeley, PhD(biochem), 79. *Prof Exp:* From asst prof to assoc prof, 82-91, PROF BIOL, UNIV CALIF, SAN DIEGO, 91- *Concurrent Pos:* Postdoctoral biochem, Stanford Univ, 82; Searle scholar, Searle, 85-90. *Mem:* Am Soc Microbiol; AAAS. *Res:* DNA repair and recombination; gene therapy; biotechnology; protein sorting to subcellular compartments; virology. *Mailing Add:* Biol Dept 0322 Bonner Hall Univ Calif La Jolla CA 92093

SUBRAMANIAN, ALAP RAMAN, b India, Mar 13, 35; US citizen; m 65; c 2. ORGANELLE RIBOSOMES & EVOLUTION, CHLOROPLAST PROTEIN SYNTHESIS SYSTEM. *Educ:* Univ Madras, BSc, 55, MA, 57; Univ Iowa, PhD(biochem), 64. *Prof Exp:* Jr scientist radiol chem, Atomic Energy Estab, India, 58-61; postdoctorate biochem, Northwestern Univ, 65-67; res assoc microbiol, Harvard Med Sch, 67-69, res asst prof molecular genetics, 69-74; GROUP LEADER, MAX PLANCK INST MOLECULAR GENETICS, 74- *Concurrent Pos:* Vis assoc prof, Univ Rochester, 78; vis scholar, Harvard Univ, 82-83; standing adv comt, Dept Biotechnol, Govt India, 88- *Mem:* Am Soc Biochem & Molecular Biol; Ger Biochem Soc; Int Soc Plant Molecular Biol. *Res:* Cloning/characterization/organization of higher plant chloroplast ribosomal protein genes; basis for nuclear:chloroplast gene allocation; gene transfer and gene accretion; protein structural alterations compared to bacterial homologues; regulating coordinate expression of nuclear and organellar genes. *Mailing Add:* Max-Planck Inst Molecular Genetics Ihnestr 73 Berlin D-1000 33 Germany

SUBRAMANIAN, GOPAL, b Madras, India, Apr 4, 37; m 66; c 2. NUCLEAR MEDICINE, CHEMICAL ENGINEERING. *Educ:* Univ Madras, BSc, 58 & 60; Johns Hopkins Univ, MSE, 64; Syracuse Univ, PhD(chem eng), 70. *Prof Exp:* Chem engr, Prod Dept, E Asiatic Co (India) Pvt, Ltd, 60-62; res assoc radiochem, Med Insts, Johns Hopkins Univ, 64-65; res assoc radiopharmaceut, 65-68, from instr to assoc prof, 68-76, PROF RADIOL, STATE UNIV NY UPSTATE MED CTR, 72- *Concurrent Pos:* NIH grant, State Univ NY Upstate Med Ctr, 69-; consult, Am Nat Stand Inst, 71-; mem, adv panel radiopharmaceut, US Pharmacopeia, 71-; asst prof, Syracuse Univ, 71-; assoc ed, J Nuclear Med, 76- *Honors & Awards:* Gold Medal, Soc Nuclear Med, 72. *Mem:* AAAS; Soc Nuclear Med; fel Am Inst Chem; Am Inst Chem Eng; Sigma Xi. *Res:* Radiochemistry; radiopharmaceuticals; fluid dynamics as applied to chemical engineering. *Mailing Add:* Dept of Nuclear Med-Radiol Upstate Med Ctr Syracuse NY 13210

SUBRAMANIAN, K N, b Cuddalore, India, Aug 13, 38; m; c 1. METALLURGY, MATERIALS SCIENCE. *Educ:* Annamalai Univ, Madras, BSc, 58; Indian Inst Sci, Bangalore, BE, 60; Univ Calif, Berkeley, MS, 62; Mich State Univ, PhD(metall), 66. *Prof Exp:* From asst prof to assoc prof, 65-85, PROF METALL, MECH & MAT SCI, MICH STATE UNIV, 85- *Mem:* Am Ceramic Soc; Metall Soc. *Res:* Plastic deformation of crystals; dislocation theory with specific reference to fatigue, work hardening, crystal growth and fracture; two phase materials; phase separation in glasses; erosion; composites. *Mailing Add:* Dept of Metall Mech & Mat Sci Mich State Univ East Lansing MI 48824

SUBRAMANIAN, MANI M, b Madras, India, Jan 11, 34; m 64; c 2. TELECOMMUNICATIONS, ELECTRICAL ENGINEERING. *Educ:* Univ Madras, BSc, 53; Madras Inst Technol, dipl, 56; Purdue Univ, MSEE, 61, PhD(elec eng), 64. *Prof Exp:* Engr, G Janshi & Co, India, 56; trainee, All India Radio, 57; jr sci officer, Electronics Res Inst, 57-59; tech asst & instr elec eng, Purdue Univ, 59-64, asst prof, 64-66; mem tech staff laser res, Bell Tel Labs, Holmdel, 66-83, dist mgr, Bell Commun Res, Inc, 84-87; Digital comm assoc, Atlanta, Ga, 87- *Concurrent Pos:* Consult, Bell Tel Labs, 64-66. *Mem:* Inst Elec & Electronics Engrs. *Res:* Receivers, parametric amplifiers, ferroelectric materials and propagation through plasma in microwaves; nonlinear optics, cathodoluminescence, lasers, laser systems and propagation through turbulent media in quantum electronics; digital transmission systems; minicomputer systems; software development. *Mailing Add:* 50 Imperial Blvd Wappingers Falls NY 12590

SUBRAMANIAN, MARAPPA G, b Sungakkarampatti, Madras, India, Dec 12, 38; m 67; c 2. PROLACTIN PHYSIOLOGY, ENDOCRINOLOGY. *Educ:* Madras Vet Col, India, BVSc, 61, MVSc, 67; Rutgers Univ, PhD(reproductive physiol), 74. *Prof Exp:* Asst lectr nutrit, Madras Vet Col, India, 68-70; res assoc, Dept Physiol, 74-77, res assoc, Dept Obstet-gynec, 77-78, instr, 78-81, asst prof obstet-gynec, 81-87, DIR RADIOIMMUNOASSAY LAB, C S MOTT CTR HUMAN GROWTH & DEVELOP, DEPT OBSTET-GYNEC, 77-, ASSOC PROF OBSTET-GYNEC, SCH MED, WAYNE STATE UNIV, DETROIT, 87- *Mem:* Endocrine Soc; Am Fertility Soc; Soc Study Reproduction; Soc Exp Biol & Med; Soc Gynec Invest; Res Soc Alcoholism. *Res:* Suckling induced release of prolactin and effects of drugs on prolactin secretion; prolactin measurement, radioimmunoassay versus bioassay; use of zona pellucida as target antigen for immunocontraception; alcohol and lactation. *Mailing Add:* Dept Obstet & Gynec C S Mott Ctr Human Growth & Develop 275 E Hancock Detroit MI 48201

SUBRAMANIAN, PALLATHERI MANACKAL, b Ottapalam, India, Jan 10, 31; m 66; c 2. ORGANIC CHEMISTRY, POLYMER CHEMISTRY. *Educ:* Univ Madras, BSc, 50; Univ Bombay, MSc, 58; Wayne State Univ, PhD(org chem), 64. *Prof Exp:* Chemist, Godrej Soaps, India, 50-58; res chemist, Electrochem Dept, Chestnut Run Labs, E I du Pont de Nemours & Co, Inc, 64-70, sr res chemist, Plastics Dept, 70-72, from res assoc to sr res assoc, 72-85, res fel, 85-90, SR RES FEL & SR TECHNOL FEL, POLYMER PRODS DEPT, EXP STA, WILMINGTON, DEL, E I DU PONT DE NEMOURS & CO, INC, 85- *Mem:* Am Chem Soc; The Chem Soc; Sigma Xi; Soc Plastic Engrs. *Res:* Physical organic chemistry; kinetics of elimination reactions in organic bicyclic systems; nuclear magnetic resonance spectroscopy of organic compounds; synthetic organic high polymers; adhesives and coatings; synthesis and process of plastics; diffusion; transport in polymers; permeability; packaging; plastic processing; containers; engineering polymers, polymer blends & processing. *Mailing Add:* 110 Cameron Dr Hockessin DE 19707

SUBRAMANIAN, RAM SHANKAR, b Madras, India, Aug 10, 47; US citizen; m 73. TRANSPORT PHENOMENA, APPLIED MATHEMATICS. *Educ:* Univ Madras, BTech, 68; Clarkson Univ, MS, 69, PhD(chem eng), 72. *Prof Exp:* Instr fac eng & appl sci, State Univ NY Buffalo, 72-73; asst prof, 73-79, assoc prof, 79-82, PROF CHEM ENG, CLARKSON UNIV, 82-, CHMN, DEPT CHEM ENG, 86- *Concurrent Pos:* Prin investr numerous grants & contracts, 75-; consult, Univ Space Res Asn, Jet Propulsion Lab, Indust. *Honors & Awards:* John Graham Res Award, 78. *Mem:* Am Inst Chem Engrs; fel AAAS; Sigma Xi; Am Ceramic Soc; Am Soc Eng Educ. *Res:* Interfacial phenomena; transport phenomena in space. *Mailing Add:* Dept Chem Eng Clarkson Univ Potsdam NY 13676

SUBRAMANIAN, RAVANASAMUDRAM VENKATACHALAM, b Kalakad, India, Jan 16, 33; m 53; c 2. POLYMER CHEMISTRY, POLYMER SCIENCE. *Educ:* Presidency Col, Madras, India, BSc, 53; Loyola Col, Madras, India, MSc, 54; Univ Madras, PhD(polymer chem), 57. *Prof Exp:* Jr res fel polymer chem, Nat Chem Lab, Poona, India, 57, Coun Sci & Indust Res India sr res fel, 57-59, jr sci officer, 59-63; res assoc chem, Case Inst Technol, 63-66 & Inst Molecular Biophys, Fla State Univ, 66; pool officer, dept phys chem, Madras Univ, 66-67; asst prof, Harcourt Butler Tech Inst, Kanpur, India, 67-69; res chemist, Mat Chem Sect, Col Eng Res Div, 69-73, assoc prof, 73-78, PROF MAT SCI, WASH STATE UNIV, 78- *Concurrent Pos:* NSF fel, Case Inst Technol, 63-66; Atomic Energy Comn fel, Inst Molecular Biophys, Fla State Univ, 66; Coun Sci & Indust Res India grant, 68-69; head, Polymer Mat Sect, Wash State Univ, 74-85, Boeing distinguished prof, 81. *Mem:* Am Chem Soc; Sigma Xi. *Res:* Kinetics and mechanisms of polymerization; polymer structure and proper properties; electropolymerization; interphase modification in carbon fiber reinforced composites; basalt fibers; organotin monomers and polymers; controlled release from polymer matrix; ceramic thin films on fibrous substrates. *Mailing Add:* Dept of Mech & Mat Eng Wash State Univ Pullman WA 99164-2920

SUBRAMANIAN, SESHA, b Wadakanchery, Kerala, India, July 19, 35; Can citizen; m 68; c 2. ELECTRON PARAMAGNETIC RESONANCE OF TRANSITION METAL COMPLEXES, DYNAMICS OF RIGID BODY ROTATION. *Educ:* Univ Madras, India, BSc, 56, MA, 58; Indian Inst Sci, Bangalore, PhD(light scattering), 63. *Prof Exp:* Lectr physics, 69-73, from asst prof to assoc prof, 73-90, PROF PHYSICS, COL MILITAIRE ROYAL, 90- *Concurrent Pos:* Sr res fel, Indian Inst Sci, Bangalore, 63-65; fel, Cath Univ Leuven, Belg, 65-68, vis prof, 87-88; adj prof, Concordia Univ, Montreal, Can, 87- *Mem:* Int Soc Magnetic Resonance. *Res:* Electron paramagnetic resonance of transition metal ions; calculation of the intensity of the paramagnetic resonance lines; statistical determination of the errors in the spin Hamiltonian parameters of the paramagnetic ions. *Mailing Add:* Dept Physics St Jean PQ J0J 1R0 Can

SUBRAMANIAN, SETHURAMAN, b Mattur, India, May 16, 40; US citizen; m 69; c 3. BIOPHYSICAL CHEMISTRY, BIOTECHNOLOGY. *Educ:* Univ Madras, BSc, 60, MSc, 65; Indian Inst Technol, Kanpur, India, PhD(phys chem), 69; Ind Univ, South Bend, MBA, 85. *Prof Exp:* Fel phys chem, Med Ctr, Univ Kans, 70-74; Nat Res Coun resident res assoc biophysics, Naval Med Res Inst, Bethesda, 74-75; vis scientist phys chem, NIH, 75-82; res scientist, Biotechnol Group, 82-84, sr res scientist, 84-86, staff scientist, 86, supvr protein chem, Miles Labs Inc, 86-90; AT SOLVAY

ENZYMES, 90- *Concurrent Pos:* Adj fac, Ind Univ, South Bend; workshop leader, Univ Wis, Madison, 87. *Mem:* Am Chem Soc; Am Soc Biol Chemists; AAAS. *Res:* Protein chemistry; enzymology; microcalorimetry; thermodynamics; spectroscopy; sickle cell hemoglobin; alcohol dehydrogenase; biotechnology; protein engineering. *Mailing Add:* Solvay Enzymes Inc PO Box 4226 Elkhart IN 46514-0226

SUBRAMANYA, SHIVA, b Hole Narasipur, Karnatak, India, Apr 8, 33; US citizen; m 67; c 2. SPACE SYSTEMS, COMMAND CONTROL SYSTEM. *Educ:* Mysore Univ, BS, 56; Karnatak Univ, MS, 62; Calif State Univ, Dominguez Hills, MBA, 76; Nova Univ, PhD(org theory), 87. *Prof Exp:* Res fel physics, Clark Univ, 63-64; chief engr commun & elec, Alcatel/Transcom Elec, 64-67; prin eng commun, Electronics Div, General Dynamics, 67-73; ASST PROJ MGR SPACE SYSTS C3, DEFENSE & SPACE SECTOR, TRW, 73-, ADVAN C3I SYSTS MGR & LEAD SYSTS ENGR LARGE SYSTS, 73- *Concurrent Pos:* Presidential appointment, AEC, 62. *Honors & Awards:* Meritorious Serv Award, Armed Forces Commun Electronics Asnm 85, Medal of Merit, 89. *Mem:* Inst Elec & Electronics Engrs; Am Inst Physics; Am Phys Soc; Armed Forces Commun Electronics Asn. *Res:* Matrix organization structures of aerospace and electronics companies of 1950's to 1980's must be changed to be cost effective; mutation theory suggested by case studies indicate degenerated matrix structure, outsourcing, guru-novice structures are ways for the 1990's; military strategic systems. *Mailing Add:* 2115 Shelburne Way Torrance CA 90503

SUBRAMANYAM, DILIP KUMAR, b India, May 7, 55; m 84. PHYSICAL METALLURGY, MECHANICAL METALLURGY. *Educ:* Indian Inst Technol, Madras, BSTech, 77; Lehigh Univ, MS, 80. *Prof Exp:* METALLURGIST, ABEX CORP, MAHWAH, NJ, 80- *Mem:* Am Soc Metals; Metall Soc, Am Inst Mech Engrs; Am Foundrymen's Soc. *Res:* Development of wear resistant alloys; product and process development in existing facilities; technical service to manufacturing and sales; failure analysis. *Mailing Add:* Abex Corp 65 Ramapo Valley Rd Mahwah NJ 07430

SUBUDHI, MANOMOHAN, b Daspalla, India, Sept 27, 46; m 71; c 2. CONTINUUM MECHANICS, MECHANICAL VIBRATIONS. *Educ:* Banaras Hindu Univ, India, BSc, 69; Mass Inst Technol, SM, 70; Polytech Inst NY, PhD(vibrations), 74. *Prof Exp:* Sr stress analyst pipe stress, Nuclear Power Serv Inc, 74-75; sr mech engr stress anal, Bechtel Power Corp, 75-76; assoc mech engr struct anal, 76-79, ENG SCIENTIST, BROOKHAVEN NAT LAB, 79- *Concurrent Pos:* Adj prof, Manhattan Col, Bronx, NY. *Mem:* Am Soc Mech Engrs; Sigma Xi. *Res:* Fracture mechanics; structural analysis using numerical techniques; nuclear plant aging. *Mailing Add:* T-130 Brookhaven Nat Lab Upton NY 11973

SUCEC, JAMES, b Bridgeport, Conn, June 15, 40; m 64; c 2. HEAT TRANSFER. *Educ:* Univ Conn, BS, 62, MS, 63. *Prof Exp:* Instr mech eng & thermodynamics, Univ Conn, 63-64; from asst prof to assoc prof, 64-76, PROF MECH ENG & HEAT TRANSFER, UNIV MAINE, 76- *Concurrent Pos:* Asst proj engr, Pratt & Whitney Aircraft, Div United Technol Corp, 65-68; fac fel, NASA Lewis Res Ctr, 72-73; NSF res proj prin investr, Univ Maine, 79-81. *Mem:* Am Soc Mech Engrs. *Res:* Analytical and finite difference work in transient forced convection heat transfer, particularly conjugate problems; prediction of heat transfer across turbulent boundary layers. *Mailing Add:* Rm 219 Boardman Hall Univ Maine Orono ME 04469

SUCHANNEK, RUDOLF GERHARD, b Hindenburg, Ger, Oct 17, 21. EXPERIMENTAL ATOMIC PHYSICS. *Educ:* Univ Hamburg, dipl(physics), 58; Univ Alaska, PhD(physics), 74. *Prof Exp:* Engr, Westinghouse Elec Corp, 58-62; eng specialist, Microwave Comp Lab, Sylvania Co, 62-64; physicist, Unified Sci Asn Inc, 64-66; sr res asst atomic collision, Geophys Inst, Univ Alaska, 66-73, fel, 74; res assoc atomic physics, Res Lab Electronics, Mass Inst Technol, 75-77; res assoc, Dept Physics & Astron, Rutgers Univ, 77-80; ASST RESEARCHER PHYSICS, UNIV CALIF, LOS ANGELES, 80- *Mem:* Am Phys Soc; Inst Elec & Electronic Engrs. *Res:* Excitation transfer collisions of laser excited atoms and molecules; charge exchange collisions of protons with atoms and molecules. *Mailing Add:* 1215 11th St Santa Monica CA 90401

SUCHARD, STEVEN NORMAN, b Chicago, Ill, Feb 8, 44; m 64; c 4. LASERS. *Educ:* Univ Calif, Berkeley, BS, 65; Mass Inst Technol, PhD(chem physics), 69. *Prof Exp:* Res asst, Lawrence Berkeley Lab, 64-65; teaching asst chem, Mass Inst Technol, 65-69; lectr, Univ Calif, Berkeley, 69-70; assoc dept head chem physics, Aerospace Corp, 70-77; tech mgr dept chem, Hughes Aircraft Co, 77-80. *Concurrent Pos:* Fel chem physics, Univ Calif, Berkeley, 69-70. *Mem:* Am Phys Soc. *Res:* Effect of system variables on the output of pulsed and continuous wave chemical lasers; flash photolysis; energy transfer in molecular systems; laser optics in high gain media; determination of the feasibility of producing new chemically and electrically pumped electronic transition lasers; measurement of molecular and kinetic parameters effecting optical gain of potential laser systems. *Mailing Add:* 9912 Star Dr Huntington Beach CA 92646

SUCHESTON, MARTHA ELAINE, b Bowling Green, Ky, June 17, 39; m 68; c 2. DEVELOPMENTAL ANATOMY. *Educ:* Western Ky Univ, BSc, 60; Ohio State Univ, MSc, 61, PhD(anat), 65. *Prof Exp:* Asst prof gross anat & embryol, Ohio State Univ, 67-68; asst prof gross anat, Stanford Univ, 68-69; from asst prof to assoc prof gross anat & embryol, Univ of BC, Vancouver, 70-75; MEM STAFF, DEPT ANAT, OHIO STATE UNIV, 75-, DIR, MEDPATH, 89- *Concurrent Pos:* Bremer Found Fund fel, Ohio State Univ, 71-73; vis assoc prof gross anat, Univ BC, 74-75; Small Univ grant, 78; Cent Ohio Heart Asn grant, 78-79, NIDR, 85-87 & HCOP, 90-93. *Mem:* AAAS; Am Asn Anat; Teratology Soc; Soc Craniofacial Genetics. *Res:* Birth defects associated with anticonvulsant drugs. *Mailing Add:* Dept Anat Ohio State Univ Col Med Columbus OH 43210

SUCHMAN, DAVID, b New York, NY, June 23, 47; m 70; c 2. METEOROLOGY. *Educ:* Rensselaer Polytech Inst, BS, 68; Univ Wis-Madison, MS, 70, PhD(meteorol), 74. *Prof Exp:* Proj assoc meteorol, 74-76, asst scientist, 76-79, ASSOC SCIENTIST METEOROL, SPACE SCI & ENG CTR, UNIV WIS-MADISON, 79- *Mem:* Am Meteorol Soc; Sigma Xi. *Res:* Application of geostationary satellite data to the study of the dynamics of mesoscale systems; practical applications of meteorology. *Mailing Add:* 722 Wedgewood Way Madison WI 53711

SUCHOW, LAWRENCE, b New York, NY, June 24, 23; m 68. SOLID STATE INORGANIC CHEMISTRY. *Educ:* City Col New York, BS, 43; Polytech Inst Brooklyn, PhD(chem), 51. *Prof Exp:* Anal chemist, Aluminum Co Am, 43-44; res chemist, Baker & Co, Inc, 44, Manhattan Proj, Oak Ridge, Tenn, 45-46, Baker & Co, Inc, 46-47, Signal Corps Eng Labs, US Dept Army, 50-54 & Francis Earle Labs, Inc, 54-58; sr res chemist, Westinghouse Elec Corp, 58-60; mem res staff, Watson Res Ctr, Int Bus Mach Corp, 60-64; from asst prof to assoc prof, 64-70, PROF CHEM, NJ INST TECHNOL, 70- *Concurrent Pos:* NSF grants, 67-74; sabbatical leave, Imp Col, Univ London, 74. *Mem:* Am Chem Soc; Sigma Xi; fel NY Acad Sci. *Res:* High temperature inorganic reactions; physical properties of solids; x-ray crystallography; crystal growth, semiconductors, thin films; phosphors; rare earths; high Tc superconductors. *Mailing Add:* Dept Chem Eng Chem & Environ Sci NJ Inst of Technol Newark NJ 07102

SUCHSLAND, OTTO, b Jena, Ger, June 18, 28; US citizen; m 56; c 2. WOOD TECHNOLOGY. *Educ:* Univ Hamburg, BS, 52, Dr nat sci(wood technol), 56. *Prof Exp:* Res engr, Swed Forest Prod Lab, Stockholm, 52-55; tech dir, Elmendorf Res, Inc, Calif, 55-57; from asst prof to assoc prof forest prod, 57-71, PROF FORESTRY, MICH STATE UNIV, 71- *Mem:* Forest Prod Res Soc. *Res:* Adhesives; gluing of wood; technology of composite wood products. *Mailing Add:* Dept Forestry 126 Natural Resources Bldg Mich State Univ East Lansing MI 48824

SUCIU, GEORGE DAN, b Blaj, Romania, July 30, 34; WGerman citizen; m 59; c 3. CHEMICAL ENGINEERING, ORGANIC CHEMISTRY. *Educ:* Polytech Inst, Bucharest, MS, 57, PhD(chem eng), 59. *Prof Exp:* Shift engr petrol refining, Teleajen Refinery, Ploiesti, Romania, 58-60; pilot plant head, Icechim Res Inst, Bucharest, Romania, 61-68; prin researcher fundamental res chem eng, Res Ctr Romanian Acad, Bucharest, 68-70; assoc prof unit oper petrol refining, Inst Petrol, Ploiesti, Romania, 70-74; sect leader process develop, Akzo Res Lab, Obernburg, WGermany,; mgr, process res, 75-77, VPRES, RES & DEVELOP, PROCESS DEVELOP, ABBLUMMUS CREST, INC, BLOOMFIELD, NJ, 77- *Mem:* Am Chem Soc; Am Inst Chem Engrs. *Res:* Development of new technologies for the petroleum, petrochemical and chemical industries; aromatic alkylation using zeolite catalysts; catalytic reactors; maleic anhydride in fluidized bed reactor; ammoxidation of alkyl aromatics; epoxidation; processing of petroleum residues. *Mailing Add:* 417 Prospect St Ridgewood NJ 07450

SUCIU, S(PIRIDON) N, b Genesse County, Mich, Dec 11, 21; m 49; c 5. MECHANICAL ENGINEERING. *Educ:* Purdue Univ, BSME, 44, MS, 49, PhD(mech eng), 51. *Prof Exp:* Res engr, Flight Propulsion Div, Gen Elec Co, NY, 51-52, Ohio, 52-54, supvr basic combustion res, 54-56, mgr appl rocket res, 56-58, mgr appl res oper, 58-63, mgr aerodynamic & component design oper, 63-67, mgr design technol oper, 67-71, gen mgr gas turbine eng dept, NY, 71-76, mgr energy technol oper, Energy Systs & Technol Div, 76-78, gen mgr neutron devices dept, 78-87; RETIRED. *Concurrent Pos:* Chmn, NASA Airbreathing Propulsion Comt, 67-70; mem Air Force Sci Adv Bd, 70-75; consult, NASA, 70- *Honors & Awards:* Akroyd Stuart Award, Royal Aeronaut Soc, 72. *Mem:* Am Soc Mech Engrs; Am Inst Aeronaut & Astronaut; Am Mgt Asn. *Res:* Power generation components and systems for utility, industrial, aircraft and marine use in commercial and military applications; electrical, mechanical and neutron devices for nuclear weapons; materials and manufacturing process development; general management of high technology businesses. *Mailing Add:* 4524 Pond Apple Dr N Naples FL 33999

SUCKEWER, SZYMON, b Warsaw, Poland, Apr 10, 38; US citizen; c 1. PLASMA PHYSICS, ATOMIC PHYSICS. *Educ:* Moscow Univ, MS, 62; Inst Nuclear Res, Warsaw PhD(plasma physics), 66; Warsaw Univ, Dr, 71. *Prof Exp:* Head spectros lab plasma physics, Inst Nuclear Res, 66-69; pvt researcher, 69-71; assoc prof, Inst Nuclear Res, Warsaw Univ, 71-75; mem staff, Princeton Univ, 75-77, res physicist, 77-80, SR RES PHYSICIST, PLASMA PHYSICS LAB, PRINCETON UNIV, 80-, PROF, DEPT MECH & AEROSPACE ENG & HEAD X-RAY LASER DIV PLASMA PHYSICS LAB, 87- *Honors & Awards:* IR-100 Award, 87, 89; Excellence Plasma Physics Award, Am Phys Soc, 90. *Mem:* Fel Am Phys Soc; Optical Soc Am. *Res:* Plasma spectroscopy; ionization, excitation and radiation processes in high and low temperature plasmas; lasers (high power lasers and short wave-length lasers); tokamaks. *Mailing Add:* E-Quad D-410 Sch Eng Princeton Univ Princeton NJ 08543

SUCOFF, EDWARD IRA, b NJ, Nov 17, 31. PLANT PHYSIOLOGY, FORESTRY. *Educ:* Univ Mich, BS, 55, MS, 56; Univ Md, PhD(bot), 60. *Prof Exp:* Res forester, US Forest Serv, 56-60; asst & assoc prof, 60-71, PROF FORESTRY, UNIV MINN, ST PAUL, 71- *Mem:* AAAS; Am Soc Plant Physiologists. *Res:* Tree growth; stress physiology. *Mailing Add:* 110C Green Hall Forestry Univ Minn St Paul MN 55108

SUCOV, E(UGENE) W(ILLIAM), b Waterbury, Conn, Oct 27, 22; div; c 3. LUMINESCENCE & LASERS, FUSION & ELECTROMAGNETIC LAUNCH. *Educ:* Brooklyn Col, BA, 43; NY Univ, MS, 54, PhD(physics), 59. *Prof Exp:* Electronics engr, 43-53; asst solid state physics, NY Univ, 53-58; res physicist, Glass Res Ctr, Pittsburgh Plate Glass Co, 58-63; res physicist, Westinghouse Res Labs, 63-66, mgr luminescence res, 66-70, fel physicist, 70-72, mgr behav res, 72-74, adv physicist, 75-78, mgr inertial confinement fusion progs, 78-83, mgr electromagnetic launch progs, 84-88; SR RES ASSOC, UNIV PITTSBURGH, 88- *Concurrent Pos:* Lectr, Univ

Pittsburgh, 78-79; mem prog comt, Inertial Confinement Fusion Topical Meeting, 79-80. *Mem:* Am Phys Soc; Inst Elec & Electronics Engrs; fel Illuminating Engrs Soc; Sigma Xi. *Res:* Fusion physics; gas discharges; lasers; plasmas; electromagnetic launchers; solid state luminescence. *Mailing Add:* 1065 Lyndhurst Dr Pittsburgh PA 15206

SUCZEK, CHRISTOPHER ANNE, b Detroit, Mich, Sept 6, 42; c 1. SEDIMENTARY GEOLOGY, PHYSICAL STRATIGRAPHY. *Educ:* Univ Calif, Berkeley, BA, 72; Stanford Univ, PhD(geol), 77. *Prof Exp:* Actg instr geol, Stanford Univ, 76; asst prof, 77-82, ASSOC PROF GEOL, WESTERN WASH UNIV, 82-, CHAIR, 90- *Concurrent Pos:* Adj prof, Univ Nev, Reno, 83-84. *Mem:* Geol Soc Am; Soc Sedimentary Geologists; Geol Asn Can; Int Asn Sedimentologists; Am Geophys Union. *Res:* Tectonics of western North America; sedimentary petrology. *Mailing Add:* Dept Geol Western Wash Univ Bellingham WA 98225

SUD, ISH, b Calcutta, India, Oct 6, 49; m; c 2. ENERGY MANAGEMENT, LOAD CONTROL. *Educ:* Indian Inst Technol, India, BTech, 70; Duke Univ, MS, 71, PhD(mech eng), 75. *Prof Exp:* Res assoc, Duke Univ, 75, sr engr & systs analyst, 75-77, from res asst prof to res assoc prof, 78-83, adj assoc prof mech eng, 83-87; PRES, SUD ASSOCS, 79- *Concurrent Pos:* Design engr, T C Cooke, P E, Inc, 74-77, dir, Energy Mgt & Special Proj, 77-78. *Mem:* Am Soc Heating, Refrig & Air Conditioning Engrs; Am Soc Mech Engrs; Sigma Xi; Nat Soc Prof Engrs. *Res:* Development of procedures for estimating building energy usage; research and application of techniques for reducing building energy usage, peak electrical demand and energy costs; research of dynamic control techniques to optimize energy usage and/or costs; research and application of waste heat recovery. *Mailing Add:* SUD Assocs 1805 Chapel Hill Rd Durham NC 27707

SUDAN, RAVINDRA NATH, b Kashmir, India, June 8, 31; m 59; c 2. PLASMA PHYSICS. *Educ:* Panjab Univ, India, BA, 48; Indian Inst Sci, dipl, 52; Univ London, DIC & PhD(elec eng), 55. *Prof Exp:* Elec engr, Brit Thomson Houston Co, Eng, 55-57; instruments engr, Imp Chem Industs, Ltd, India, 57-58; res assoc, Cornell Univ, 58-59, from asst prof to assoc prof elec eng, 59-63, dir, Lab Plasma Studies, 75-85, dep dir, Cornell Theory Ctr, 85-87, PROF ELEC ENG & APPL PHYSICS, CORNELL UNIV, 68-, IBM PROF ENG, 75- *Concurrent Pos:* Vis scientist Int Ctr Theoret Physics, Trieste, 65-66, 70-73; vis res physicist, plasma physics lab, Princeton Univ, 66-67; head theoret plasma physics, Naval Res Lab, DC, 70-71, sci adv, 74-75; consult, Lawrence Radiation Lab, Univ Calif, Maxwell Lab, Los Alamos Sci Lab, Physics Int Co, Sci Appl, Inc; vis physicist, Inst Advan Study, Princeton, 75; sr res fel, Inst Fusion Studies, Univ Tex, 83; co-ed Handbook of Plasma Physics, N Holland. *Honors & Awards:* James Clerk Maxwell Prize, Am Phys Soc, 89. *Mem:* AAAS; fel Am Phys Soc; fel Inst Elec & Electronics Engrs. *Res:* Thermonuclear fusion and space physics; high powered pulsed particle beams; magnetohydrodynamics; plasma turbulence; solar physics. *Mailing Add:* Lab Plasma Studies 369 Upson Hall Cornell Univ Ithaca NY 14853

SUDARSHAN, ENNACKEL CHANDY GEORGE, b Kottayam, India, Sept 16, 31; m 54; c 3. THEORETICAL PHYSICS. *Educ:* Univ Madras, BSc, 51, MA, 52; Univ Rochester, PhD(physics), 58. *Hon Degrees:* DSc, Univ Wis-Milwaukee, 69, Univ Delhi, 73 & Chalmers Univ Tech, Sweden, 84. *Prof Exp:* Demonstr physics, Christian Col, Madras, 51-52; res asst, Tata Inst Fundamental Res, 52-55 & Univ Rochester, 55-57; res fel, Harvard Univ, 57-59; from asst prof to assoc prof, Univ Rochester, 59-64; prof, Syracuse Univ, 64-69; PROF PHYSICS, CTR PARTICLE THEORY, UNIV TEX, AUSTIN, 69-, DIR, 70- *Concurrent Pos:* Guest prof, Univ Bern, 63-64; vis prof, Brandeis Univ, 64; Sir C V Raman distinguished vis prof, Univ Madras, 70-71; prof & dir, Ctr Theoret Studies, Indian Inst Sci, Bangalore, 72- *Honors & Awards:* Padma Bhushan Award, 75. *Mem:* Fel Indian Nat Acad Sci; fel Am Phys Soc; fel Indian Acad Sci. *Res:* Quantum field theory; elementary particles; high energy physics; classical mechanics; quantum optics; Lie algebras and their application to particle physics; foundations of physics; philosophy and history of contemporary physics. *Mailing Add:* RLM Bldg 9.328 Univ of Tex Austin TX 78712

SUDARSHAN, T S, b Madras, India, Dec 25, 55. SURFACE MODIFICATION TECHNOLOGIES, DIAMOND TECHNOLOGY. *Educ:* IIT Madras, BTech, 76; Va Tech, MS, 78, PhD(mat sci), 84. *Prof Exp:* Sr metallurgist, Ashok Leyland Ltd, 79-81; dir res & develop, Synergistic Technologies Inc, 84-86; TECH DIR, MAT MODIFICATION INC, 87- *Concurrent Pos:* Prin investr, Brookhaven Nat Labs Progs, Navy, Army, Air Force, NSF, 85-; vis prof, IIT Madras; ed, Mat & Mfg Processes, 87-; chmn, Surface Modification Technologies, 88-; consult, NSF, 89-, Ultramet, 90-, Off of Technol Assessment, US Cong, 90- *Honors & Awards:* Outstanding Young Mfg Engr, Soc Mfg Engrs, 90. *Mem:* Am Soc Metall; Metall Soc; Nat Asn Corrosion Engrs; Soc Mfg Engrs; Soc Advan Mat & Process Eng. *Res:* Surface modification technologies; diamond technology; materials and manufacturing processes; chemical vapor deposition. *Mailing Add:* PO Box 4817 Falls Church VA 22044

SUDARSHAN, TANGALI S, Can citizen; m 77; c 2. SOLID, LIQUID & COMPRESSED GAS INSULATION, PHOTOCONDUCTING MATERIAL BREAKDOWN. *Educ:* Univ Bangalore, India, BSc, 68; Univ Mysore, MSc, 70; Univ Waterlooo, Can, MASc, 72, PhD(electrical eng), 74. *Prof Exp:* Res officer, Nat Res Coun Can, 74-79; from asst prof to assoc prof, 79-87, PROF, ELEC ENG, UNIV SC, 87- *Concurrent Pos:* Prin investr, NSF, 80-82, Dept Energy, 80-85, INTELSAT, Wash, DC, 84-87, Sandia Nat Labs, 79-81, Off Naval Res, 87- *Mem:* Sr mem Inst Elec & Electronics Engrs. *Res:* Solid, liquid and gas insulated systems for high voltage power system and pulsed power applications; surface flashover of solid insulator and photoconductor materials; fast high voltage and current diagnostics; insulator degradation and aging; electrical surface properties of insulators. *Mailing Add:* Col Eng Univ SC Columbia SC 29208

SUDBOROUGH, IVAN HAL, b Royal Oak, Mich, Dec 19, 43; m 69; c 2. INFORMATION SCIENCE. *Educ:* Calif State Polytech Col, BS, 66, MS, 67; Pa State Univ, PhD(comput sci), 71. *Prof Exp:* Asst prof, 71-77, ASSOC PROF COMPUT SCI, NORTHWESTERN UNIV, 78- *Concurrent Pos:* NSF res grant, 74. *Mem:* Asn Comput Mach; Soc Indust & Appl Math. *Res:* Computational complexity; formal languages; automata theory; theory of computation. *Mailing Add:* PO Box 835265 Richardson TX 75083

SUDBURY, JOHN DEAN, b Natchitoches, La, July 29, 25; m 53; c 3. PHYSICAL CHEMISTRY. *Educ:* Univ Tex, BS, 44, MS, 47, PhD(phys chem), 49. *Prof Exp:* Sr res chemist, Develop & Res Dept, Continental Oil Co, 49-56, supv res chemist, 56-66, dir, Petrochem Res Div, Okla, 66-69, gen mgr, C/A Nuclear Fuels Div, Calif, 69-70, asst to vpres res, NY, 70-72, asst dir res & vpres Res Div, Conoco Coal Develop Co, 72-83. *Concurrent Pos:* Pres, Religious Found, 83- *Honors & Awards:* Speller Award, Nat Asn Corrosion Engrs, 66. *Mem:* Am Chem Soc; Nat Asn Corrosion Engrs. *Res:* Advanced systems for liquified natural gas, arctic transport; development of conversion processes to get coal into more desirable energy sources; conversion of coal to liquids and gases; removal of sulfur from combustion products of coal. *Mailing Add:* 42 Cascade Springs Pl Woodlands TX 77381

SUDDARTH, STANLEY KENDRICK, b Westerly, RI, Oct 22, 21; m 51. FORESTRY. *Educ:* Purdue Univ, BSF, 43, MS, 49, PhD(agr econ forestry), 52. *Prof Exp:* Assoc dir bomb effectiveness res, US Dept Air Force Proj, Res Found, 51-54, from asst prof to assoc prof forestry, 54-60, PROF WOOD ENG, AGR EXP STA, PURDUE UNIV, WEST LAFAYETTE, 60- *Concurrent Pos:* Consult home mfg indust, 55- & US Forest Prod Lab, Madison, Wis, 70-72; tech adv, Am Inst Timber Construct & Truss Plate Inst. *Honors & Awards:* Res Award, Truss Plate Inst, 70; Markwardt Eng Res Award, Forest Prod Res Soc, 71; Markwardt Award, Am Soc Testing & Mat, 72. *Mem:* Forest Prod Res Soc; Int Acad Wood Sci; Am Soc Agr Engrs; Am Soc Civil Engrs. *Res:* Applied mathematics in engineering and economic problems; engineering properties and uses of wood. *Mailing Add:* 408 Kerber Rd West Lafayette IN 47906

SUDDATH, FRED LEROY, (JR), b Macon, Ga, May 6, 42; m 65; c 2. BIOLOGICAL STRUCTURE, X-RAY CRYSTALLOGRAPHY. *Educ:* Ga Inst Technol, BS, 65, PhD(chem), 70. *Prof Exp:* NIH fel, Mass Inst Technol, 70-72, res assoc biol, Lab Molecular Struct, 72-75; asst prof biochem, investr, inst dent res & scientist, Comprehensive Cancer Ctr, 75-77, assoc prof biochem, scientist instr dent res & scientist, Comprehensive Cancer Ctr, Med Ctr, Univ Ala, Birmingham, 77-85; PROF CHEM, GA INST TECHNOL, 85- *Concurrent Pos:* Am Cancer Soc fel, Mass Inst Technol, 72-73. *Mem:* AAAS; Am Crystallog Asn; Am Chem Soc. *Res:* Structure and function of transfer RNA; structural studies of nucleic acids and proteins; correlation of molecular structure and biological function; experimental methods development. *Mailing Add:* Dept Chem Ga Inst Technol Atlanta GA 30332-0700

SUDDERTH, WILLIAM DAVID, b Dallas, Tex, Apr 29, 40; m 62; c 2. MATHEMATICS PROBABILITY, MATHEMATICAL STATISTICS. *Educ:* Yale Univ, BS, 63; Univ Calif, Berkeley, MS, 65, PhD(math), 67. *Prof Exp:* Asst prof statist, Univ Calif, Berkeley, 67-68; asst prof math, Morehouse Col, 68-69; asst prof statist, 69-71, assoc prof, 71-77, PROF STATIST, UNIV MINN, MINNEAPOLIS, 77- *Mem:* Am Math Soc; fel Inst Math Statist. *Res:* Probability, especially the study of finitely additive probability measures and abstract gambling theory, which is also known as dynamic programming and stochastic control; foundations of statistics; stochastic games. *Mailing Add:* Sch of Statist Univ of Minn Minneapolis MN 55455

SUDDICK, RICHARD PHILLIPS, b Omaha, Nebr, Feb 3, 34; m 55; c 4. PHYSIOLOGY. *Educ:* Creighton Univ, BS, 58, MS, 59, DDS, 61; Univ Iowa, PhD(physiol), 67. *Prof Exp:* Instr physiol, Univ Iowa, 63-65; asst prof biol sci, Creighton Univ, 65-68, from assoc prof to prof oral biol & head dept, 68-74, assoc prof physiol, Sch Med, 70-74; assoc prof physiol & asst dean res, Col Dent Med, Med Univ SC, 74-76; prof oral biol & chmn dept & asst dean res, Sch Dent, Univ Louisville, 76-80; PROF COMMUNITY DENT, DENT SCH & PROF PHYSIOL, GRAD SCH BIOMED SCI, UNIV TEX HEALTH SCI CTR, SAN ANTONIO, 80- *Mem:* AAAS; Int Asn Dent Res; Am Physiol Soc. *Res:* Physiology of exocrine secretion, primarily secretion of saliva; function of the saliva in the oral cavity and the alimentary tract and its relationship to normal and diseased states; etiology of dental caries and periodontal disease; behavioral constructs and neurophysiological correlates in humans. *Mailing Add:* Univ Tex Health Sci Ctr 7703 Floyd Curl Dr San Antonio TX 78284

SUDDS, RICHARD HUYETTE, JR, b State College, Pa, Feb 13, 27; m 52; c 2. MICROBIOLOGY. *Educ:* Univ Conn, BA, 50, MA, 51; Univ NC, MSPH, 54, PhD(parasitol), 59. *Prof Exp:* From asst prof to assoc prof, 58-72, PROF MICROBIOL, STATE UNIV NY COL PLATTSBURGH, 72- *Concurrent Pos:* Consult, Diamond Int Corp. *Mem:* Am Soc Trop Med & Hyg; Am Pub Health Asn; Am Soc Microbiol; NY Acad Sci; Sigma Xi. *Res:* Host-parasite relationships and ecology of enteric bacteria. *Mailing Add:* Dept Biol Sci State Univ NY Col Plattsburgh NY 12901

SUDERMAN, HAROLD JULIUS, b Myrtle, Man, July 24, 21; m 47; c 3. BIOCHEMISTRY. *Educ:* Univ Man, BSc, 49, MSc, 52, PhD, 62. *Prof Exp:* Demonstr biochem, Univ Man, 51-52, lectr, 52-56, asst prof, 56-63; asst prof, Ont Agr Col, 63-65; from asst prof to assoc prof biochem, Univ Guelph, 65-86; RETIRED. *Mem:* Can Biochem Soc. *Res:* Molecular properties, structure and function of proteins; comparative biochemistry of hemoglobins. *Mailing Add:* Six Rickson Ave Guelph ON N1G 2W7 Can

SUDIA, THEODORE WILLIAM, b Ambridge, Pa, Oct 10, 25; m 49; c 3. ENVIRONMENTAL PHYSIOLOGY. *Educ:* Kent State Univ, BS, 50; Ohio State Univ, MS, 51, PhD(bot), 54. *Prof Exp:* Asst prof biol sci, Winona State Col, 55-58; res fel plant physiol, Univ Minn, St Paul, 58-59, res assoc physiol ecol, 59-61, asst prof, 61-63, assoc prof plant path & bot, 63-67; assoc dir, Am

Inst Biol Sci, Washington, DC, 67-69; chief ecol serv, Off Natural Sci, 69-73, chief scientist, 73-77, actg assoc dir sci & technol, 77-80, dep science adv, Int Sect, 80-81, SR SCIENTIST, US NAT PARK SERV, 81- *Mem:* AAAS; Ecol Soc Am; Am Soc Plant Physiologists; Bot Soc Am; NY Acad Sci. *Res:* Research administration. *Mailing Add:* 1117 E Capitol St SE Washington DC 20003

SUDIA, WILLIAM DANIEL, b Ambridge, Pa, Aug 19, 22; m 49; c 2. ENTOMOLOGY, VIROLOGY. *Educ:* Univ Fla, BS, 49; Ohio State Univ, MS, 50, PhD(entom), 58. *Prof Exp:* Entomologist, Med Entom Unit, Ctr Dis Control, USPHS, 51-53, asst chief arbovirus vector lab, 53-65, lab consult & develop sect, 66, chief, Arbovirus Ecol Lab, 67-73, dep dir, Lab Training Div, Ctr Dis Control, 73-84. *Mem:* Sigma Xi; Am Soc Trop Med & Hyg; Am Mosquito Control Asn. *Res:* Ecology of arthropod-borne encephalitis viruses; mosquito vectors and vertebrate hosts. *Mailing Add:* 1445 Diamond Head Dr Decatur GA 30033

SUDMEIER, JAMES LEE, b Minneapolis, Minn, Feb 14, 38; m 62; c 2. ANALYTICAL CHEMISTRY. *Educ:* Carleton Col, BA, 59; Princeton Univ, MA, 61, PhD(chem), 66. *Prof Exp:* Actg asst prof chem, Univ Calif, Los Angeles, 65-66, asst prof, 66-70; asst prof, Univ Calif, Riverside, 70-71, assoc prof, 71-; TUFTS NEW ENG MED CTR. *Mem:* Am Chem Soc. *Res:* Nuclear magnetic resonance studies of coordination compounds and metal binding to biopolymers. *Mailing Add:* Tufts New England Med Ctr Boston MA 02111

SUDWEEKS, EARL MAX, b Richfield, Utah, Dec 27, 33; m 60; c 9. ANIMAL NUTRITION, DAIRY NUTRITION. *Educ:* Utah State Univ, BS, 60, MS, 62; NC State Univ, PhD(nutrit biochem), 72. *Prof Exp:* Res assoc animal nutrit, Utah State Univ, 62-65, from asst prof to assoc prof exten, 65-68; res asst animal nutrit, NC State Univ, 68-72; asst prof animal nutrit, Univ Ga, 72-80; dir nutrit, Watkins Inc, Winona, Minn & Prof Prods, Inc, Sauk City, Wis, 80-81; DAIRY EXTEN SPECIALIST, TEX A&M UNIV, 81- *Concurrent Pos:* Consult, US Feed Grains Coun, Taiwan, Gilas Dairy Coop, Mex. *Mem:* Am Dairy Sci Asn; Am Soc Animal Sci; Sigma Xi; Am Inst Nutrit. *Res:* The role of roughages in rumen physiology, energy utilization, feed conversion of beef and dairy cattle. *Mailing Add:* Tex A&M Univ Res & Exten Ctr Box 220 Overton TX 75684

SUDWEEKS, WALTER BENTLEY, b Buhl, Idaho, May 22, 40; m 65; c 2. INDUSTRIAL CHEMISTRY, EXPLOSIVES. *Educ:* Brigham Young Univ, BS, 65, PhD(org chem), 70. *Prof Exp:* Res chemist, Polymer Intermediates Dept, E I du Pont de Nemours & Co, Inc, 69-76; sr res scientist, Ireco Chem, 76-79, dir prod develop, 79- 82, asst dir res & develop, 82-84, DIR RES & DEVELOP, IRECO CHEM, 84- *Mem:* Am Chem Soc; Nat Defense Preparedness Asn. *Res:* Organic synthesis; hydrometallurgical processes; explosives research; management of commercial explosives research and development. *Mailing Add:* Ireco Inc 3000 W 8600 S West Jordan UT 84088

SUELTER, CLARENCE HENRY, b Lincoln, Kans, Dec 15, 28; m 55; c 3. SCIENCE EDUCATION. *Educ:* Kans State Univ, BS, 51, MS, 53; Iowa State Univ, PhD(biochem), 59. *Prof Exp:* From asst prof to assoc prof, 61-69, PROF BIOCHEM, MICH STATE UNIV, 69- *Concurrent Pos:* USPHS fel, Univ Minn, 59-61; res career develop award, Mich State Univ, 65-75. *Mem:* AAAS; Am Chem Soc; Am Soc Biochem & Molecular Biol. *Res:* Science education; structure and function of enzymes. *Mailing Add:* Dept Biochem Mich State Univ East Lansing MI 48824

SUEN, CHING YEE, b Chung Shan, China, Oct 14, 42; Can citizen. COMPUTER APPLICATIONS. *Educ:* Univ Hong Kong, BSc, 66, MSc 68; Univ BC, PhD (elec eng), 72. *Prof Exp:* Asst prof, 72-76, assoc prof, 76-79, PROF COMPUT SCI, CONCORDIA UNIV, 79- *Concurrent Pos:* Chmn comt character recognition, Can Stand Asn, 77-; vis scientist, Res Lab Electronics, Mass Inst Technol, 78-80; dir, Recognition Technol Users Asn, 84-86. *Mem:* Fel Inst Elec & Electronic Engrs; Can Image Processing & Pattern Recognition Soc (pres, 84-90); Recognition Technol Users Asn. *Res:* Character recognition and data processing; speech analysis and synthesis; computational linguistics and text processing. *Mailing Add:* Dept Comput Sci Concordia Univ 1455 Maisonneuve W Montreal PQ H3G 1M8 Can

SUEN, T(ZENG) J(IUEQ), b Hangzhou, China, June 7, 12; nat US; m 44; c 2. POLYMER CHEMISTRY. *Educ:* Tsinghua Univ, China, BS, 33; Mass Inst Technol, MS, 35, ScD, 37. *Prof Exp:* Fel, Mass Inst Technol, 37-38; asst prof chem eng, Chongqing Univ, 38-39; head dept res, Tung Li Oil Works, Chongqing, China, 39-44; mem staff, Radiation Lab, Mass Inst Technol, 44-45; chem engr, Stamford Res Labs, Am Cyanamid Co, 45-56, group leader in charge polymer chem, 56-60, mgr thermoplastics res, 61, dir plastics & polymers res, 61-70, proj mgr div res, 71-77; CONSULT, 77-; CHMN, ATC ASSOCS, INC, 88- *Mem:* Am Chem Soc; AAAS. *Res:* Synthetic fuels and lubricants; plastics; condensation and addition polymers; environmental improvement. *Mailing Add:* 349 Mariomi Rd New Canaan CT 06840-3318

SUENAGA, MASAKI, b Hohoku, Japan, Sept 15, 37; m 72; c 2. METALLURGY, ELECTRICAL ENGINEERING. *Educ:* Univ Calif, Berkeley, BSEE, 63, MSEE, 64, PhD(metall), 69. *Prof Exp:* Fel metall, Lawrence Berkeley Lab, 69; div head metall, 78-86, FROM ASST METALLURGIST TO SR METALLURGIST, BROOK HAVEN NAT LAB, 69- *Mem:* Am Phys Soc; AAAS; Am Soc Metals. *Res:* Superconducting materials; mechanical properties of metals and alloys. *Mailing Add:* Bldg 480 Brookhaven Nat Lab Upton NY 11973

SUENRAM, RICHARD DEE, b Halstead, Kans, Feb 2, 45; m 65; c 1. PHYSICAL CHEMISTRY, CHEMICAL PHYSICS. *Educ:* Kans State Univ, BS, 67; Univ Wis, MS, 69; Univ Kans, PhD(chem), 73. *Prof Exp:* Res chemist phys chem, Rohm & Haas Co, 69-70; res assoc, Harvard Univ, 73-75; res assoc, Nat Bur Standards, 75-77; RES CHEMIST PHYS CHEM, NAT INST STANDARDS & TECHNOL, 77- *Res:* Molecular spectroscopy of transient molecular species found in gas phase chemical reactions; molecular species associated with atmospheric and interstellar chemistry; molecular spectroscopy of van der Waals and hydrogen bonded clusters. *Mailing Add:* Molecular Spectros Div Nat Inst Standards & Technol B265 Physics Bldg Gaithersburg MD 20899

SUEOKA, NOBORU, b Kyoto, Japan, Apr 12, 29; m 57; c 1. GENETICS. *Educ:* Kyoto Univ, BS, 53, MS, 55; Calif Inst Technol, PhD(biochem genetics), 59. *Prof Exp:* Fulbright grant, 55-56; res fel biochem genetics, Harvard Univ, 58-60; asst prof microbiol, Univ Ill, 60-62; from assoc prof to prof biol, Princeton Univ, 62-72; PROF BIOL, UNIV COLO, BOULDER, 72- *Mem:* Am Soc Biol Chemists; Am Soc Microbiol; Genetics Soc Am. *Res:* Biochemical genetics; molecular biology, particularly genetic aspects of biological macromolecules, nucleic acids and protein. *Mailing Add:* Dept Biol BSCI-126 Univ Colo Campus Box 347 Boulder CO 80309

SUER, H(ERBERT) S, b Philadelphia, Pa, Apr 29, 26; m 60; c 6. INSTRUMENTATION, ENGINEERING MECHANICS. *Educ:* Drexel Inst Technol, BS, 49; Kans State Col, MS, 52; Mass Inst Technol, ScD, 55. *Prof Exp:* Instr, Kans State Col, 49-52; asst, Mass Inst Technol, 52-55; sr engr, NAm Aviation, Inc, 55-60; sect head eng mech, TRW Systs Group, TRW Inc, 60-65, mgr space instrumentation proj off, 65-67, advan develop & anal dept, 67-70, mgr instrument systs dept, 70-74, mgr instruments & exp prod develop off, 74-78; res prof & mgr advan prog develop, Desert Res Inst, Univ Nev, 78-79; PROD LINE MGR, TRW INC, 79- *Concurrent Pos:* Adj prof, Univ Southern Calif, 56-74; lectr, Univ Calif, Los Angeles, 58-77. *Honors & Awards:* Wellington Prize, Am Soc Civil Engrs, 60. *Res:* Instrumentation systems; experimental methods; structural and rigid body dynamics; theory of elasticity; theory of elastic stability; mechanics of materials; structural and stress analysis. *Mailing Add:* TRW One Space Park M/S E1-2025 Redondo Beach CA 90278

SUESS, GENE GUY, b Beaver, Okla, Apr 16, 41; m 68; c 2. MEAT SCIENCE. *Educ:* Tex Tech Univ, BS, 63; Univ Wis-Madison, MS 66, PhD(meat sci, animal sci), 68. *Prof Exp:* Res technologist, Oscar Mayer & Co, 68-73, new prod develop supvr, 73-78, prod develop mgr, 78-85, ASSOC DIR RES & DEVELOP & VPRES QUAL ASSURANCE, OSCAR MAYER FOODS CORP. *Mem:* Am Meat Sci Asn; Inst Food Technologists. *Res:* Meats processing. *Mailing Add:* 3006 Pelham Rd Madison WI 53713

SUESS, HANS EDUARD, b Vienna, Austria, Dec 16, 09; nat US; m 42; c 2. CHEMISTRY. *Educ:* Univ Vienna, PhD(chem), 35; Univ Hamburg, Dr habil, 39. *Hon Degrees:* DSc, Queen's Univ, Belfast, Northern Ireland, 81. *Prof Exp:* Demonstr, Univ Vienna, 34-36; res assoc, Univ Hamburg, 37-48, assoc prof, 49-50; res fel, Univ Chicago, 50-51; chemist, US Geol Surv, 51-55; prof, 55-77, EMER PROF CHEM, UNIV CALIF, SAN DIEGO, 77- *Honors & Awards:* V M Goldschmidt Medal, Geochemical Soc, 74; J S Guggenheim Found Award, 66; Alexander von Humboldt Found Award,78. *Mem:* Nat Acad Sci; Austrian Acad Sci; fel Am Acad Arts & Sci; Max-Planck Soc. *Res:* Chemical kinetics; nuclear hot atom chemistry; cosmic abundances of nuclear species; nuclear shell structure; geologic age determinations; carbon-14 dating. *Mailing Add:* Dept Chem Univ Calif San Diego La Jolla CA 92093

SUESS, JAMES FRANCIS, b Rock Island, Ill, Nov 27, 19; m 46; c 3. PSYCHIATRY. *Educ:* Northwestern Univ, BS, 50, MD, 52. *Prof Exp:* Resident psychiat, Warren State Hosp, Warren, Pa, 53-56, clin dir, 56-62; prof, 62-82, EMER PROF PSYCHIAT, SCH MED, UNIV MISS, 82- *Concurrent Pos:* Fel psychiat, Med Sch, Univ Pa, 53; exchange teaching fel, Med Sch, Univ Pittsburgh, 55; Col Physicians & Surgeons fel,Columbia Univ, 58; consult, Vet Admin, 62-82 & Gov Drug Coun, Miss, 72-75; vis prof, Inst Psychiat & Royal Free Hosp Sch Med, London; assoc chief staff educ, Vet Admin Ctr, 78-82. *Res:* Medical education in psychiatry; use of television and videotape in medical education; programmed teaching with television; short term psychotherapy of neurotic and personality disorders; use of videotape in teaching managerial and supervisory skills in business. *Mailing Add:* 1415 Radcliffe St Jackson MS 39211

SUESS, STEVEN TYLER, b Los Angeles, Calif, Aug 4, 42; div; c 2. FLUID DYNAMICS. *Educ:* Univ Calif, Berkeley, AB, 64, PhD(planetary, space sci), 69. *Prof Exp:* Nat Acad Sci-Nat Res Coun res assoc, Environ Sci Serv Admin Res Labs, Boulder, Colo, 69-71; physicist, Space Environ Lab, Nat Oceanic & Atmospheric Admin, 71-83; PHYSICIST, MARSHALL SPACE FLIGHT CTR, NASA, HUNTSVILLE, AL, 83- *Concurrent Pos:* Guest worker, Max Planck Inst Aeronomy, 75; vis scholar, Stanford Univ, 80-81; vis scholar, Inst Astrophys, Arcetri, Italy, 91. *Mem:* Am Geophys Union; Am Astronom Soc; Sigma Xi; Int Astron Union. *Res:* Dynamics of the sun and stars; oscillations of stars; stellar winds; magnetohydrodynamics of rotating fluids. *Mailing Add:* Space Sci Lab ES52 Marshall Space Flight Ctr NASA Huntsville AL 35812

SUFFET, I H (MEL), b Brooklyn, NY, May 11, 39; m 62; c 2. ANALYTICAL CHEMISTRY, ENVIRONMENTAL CHEMISTRY. *Educ:* Brooklyn Col, BS, 61; Univ Md, Col Park, MS, 64; Rutgers Univ New Brunswick, PhD, 69. *Prof Exp:* From asst prof to assoc prof, 69-78, PROF CHEM & ENVIRON SCI, DREXEL UNIV, 78- *Concurrent Pos:* Consult & grants, Western Elec Co, 70-72, City of Philadelphia, 72-75, Nat Sci Found-Res Appl Nat Needs, 76-77 & 79-80, Environ Protection Agency, 78, 80-81 & 82-84, Health Effects Inst, 82-83, Am Water Works Res Found, 83-88; mem safe drinking water comt, Nat Acad Sci, 78-79, chmn subcomt efficiency use activated carbon for drinking water treatment. *Honors & Awards:* F J Zimmerman Award, Am Chem Soc, 83. *Mem:* Sigma Xi; Am Chem Soc; Am Water Works Asn. *Res:* Environmental chemistry and analysis of trace organics; hazardous waste treatment and analysis and fate; activated carbon and other drinking water treatment processes; fate of pollutants in the environment; chemical nature of water and wastes. *Mailing Add:* Sch Pub Health CHS Rm 46- 081 Univ Calif 10833 LeConte Ave Los Angeles CA 90024-1772

SUFFIN, STEPHEN CHESTER, b Los Angeles, Calif, Aug 13, 47; m 69; c 2. PHARMACOLOGY. *Educ:* Univ Calif, Los Angeles, BA, 68, MD, 72. *Prof Exp:* Fel immunopath, 75-77, ASST PROF PATH, UNIV CALIF, LOS ANGELES, 77-; DIR PATH, LAB PROCEDURES INC, 80- *Concurrent Pos:* Sr investr immunopathology, Lab Infectious Dis, Nat Inst Allergy & Infectious Dis, NIH, 78-80; consult, Armed Forces Inst Path, 78-80, Jet Propulsion Lab, Calif Inst Technol, 80- *Mem:* Int Acad Path; Am Soc Clin Pathologists; Col Am Pathologists; Am Asn Immunologists; Am Soc Microbiol. *Mailing Add:* Dept Path Univ Calif Los Angeles Sch Med Los Angeles CA 90024

SUFIT, ROBERT LOUIS, b Washington, DC, July 16, 50; div; c 3. NEUROMUSCULAR DISEASES. *Educ:* John Hopkins Univ, BA, 72, MA, 72; Univ VA, MD, 76. *Prof Exp:* ASST PROF, UNIV WIS-MADISON, 82- *Mem:* Am Acad Neurol; Am Soc Neurol Invest. *Res:* Muscle disease; motor performance of patients with muscle disease. *Mailing Add:* H6 564 CSC 600 Highland Ave Madison WI 53792

SUFRIN, JANICE RICHMAN, b New York, NY, May 5, 41; m 68; c 3. BIOORGANIC CHEMISTRY, MOLECULAR & BIOCHEMICAL PHARMACOLOGY. *Educ:* Bryn Mawr Col, AB, 62; Brandeis Univ MA, 67, PhD(chem), 72. *Prof Exp:* Postdoctoral fel, pharmacol dept, Sch Med, Johns Hopkins Univ, 72-74; cancer res scientist, exp therapeut dept, Roswell Park Mem Inst, 75-77; res instr physiol & biophys, Sch Med, Wash Univ, 77-80, res asst prof, 80-82; cancer res scientist IV, surg oncol & exp therapeut dept, 82-88, CANCER RES SCIENTIST IV, EXP THERAPEUT DEPT, ROSWELL PARK CANCER INST, 88-; RES ASST PROF CHEM, ROSWELL PARK GRAD DIV, STATE UNIV NY, BUFFALO, 82- *Mem:* Am Chem Soc; Am Soc Biochem & Molecular Biol; Am Asn Cancer Res; Sigma Xi. *Res:* The synthesis and biological evalution of agents which interfere with the biosynthesis and metabolism of the key biological intermediate, S-adenosylmethionine, thus affecting cellular transmethylation reactions and polyamine pathways. *Mailing Add:* Grace Cancer Drug Ctr Roswell Park Mem Inst 666 Elm St Buffalo NY 14263

SUGA, HIROYUKI, b Japan, Oct 4, 41; m 69; c 2. PHYSIOLOGY, BIOENGINEERING & BIOMEDICAL ENGINEERING. *Educ:* Okayama Univ, MD, 66; Univ Tokyo, DMedSc, 70. *Prof Exp:* Asst prof physiol, Tokyo Med & Dent Univ, 70-71; postdoctoral fel biomed eng, Johns Hopkins Univ Med Sch, 71-73, asst prof, 75-78; asst prof physiol, Univ Tokyo Med Sch, 73-75; chief, Nat Cardiovasc Ctr Res Inst, 78-82, dir, 82-91; PROF & CHMN PHYSIOL, OKAYAMA UNIV MED SCH, 91- *Concurrent Pos:* Postdoctoral fel biomed eng, Case Western Reserve Univ, 71; chmn, Study Group Cardiac Mech, 85-; ed, Heart & Vessels J, 88-; vis scientist, Nat Cardiovasc Ctr Res Inst, 91-; mem, Circulation Coun, Am Heart Asn. *Honors & Awards:* Satoh Award, Japanese Circulation Soc, 84. *Mem:* Am Heart Asn; Am Physiol Soc; Int Cardiovasc Syst Dynamics Soc. *Res:* Physiology and biomedical engineering of heart mechanics and energetics primarily in terms of end-systolic presure-volume relation and systolic pressure-volume area of the left ventricle. *Mailing Add:* Second Dept Physiol Okayama Univ Med Sch Two Shikatacho Okayama 700 Japan

SUGA, NOBUO, b Japan, Dec 17, 33; m 63; c 2. PHYSIOLOGY. *Educ:* Tokyo Metrop Univ, PhD(physiol), 63. *Prof Exp:* NSF fel hearing physiol, Harvard Univ, 63-64; res zoologist, Brain Res Inst, Univ Calif, Los Angeles, 65; res neuroscientist, Sch Med, Univ Calif, San Diego, 66-68; assoc prof, 69-75, PROF BIOL, WASH UNIV, 76- *Mem:* AAAS; Int Soc Neuroethology; Acoust Soc Am; Soc Neuroscience; Asn Res Otolaryngol. *Res:* Auditory physiology. *Mailing Add:* Dept of Biol Wash Univ St Louis MO 63130

SUGAI, IWAO, b Tokyo, Japan, Oct 19, 28; US citizen; m 58; c 3. ELECTRICAL ENGINEERING. *Educ:* Univ Calif, Los Angeles, BS, 55; Calif Inst Technol, MSEE, 56; George Washington Univ, DSc, 71. *Prof Exp:* Assoc engr, IBM Res Ctr, 58-60; tech specialist, ITT Lab, 60-63; computer scientist, Computer Sci Corp, 63-73; sr staff engr, Appl Physics Lab, Johns Hopkins Univ, 73-89; PRES, SOVIET ELECTRONICS DIGEST DISSEMINATION SERV, 90- *Concurrent Pos:* Asst prof lectr, George Washington Univ, 68-69; lectr, Am Univ, 82-83; asst vpres pac region, Soc Computer Simulation Int, 89-, co-gen chmn, Summer Computer Simulation Conf, Baltimore, 91. *Mem:* Sigma Xi; sr mem Inst Elec & Electronics Engrs; Soc Computer Simulation Int. *Res:* Analytical solutions of a certain class of a general scalar Riccati's nonlinear differential equation; wave propagation problems. *Mailing Add:* 14637 Stonewall Dr Silver Spring MD 20905-5857

SUGAM, RICHARD JAY, b New York, NY, Nov 16, 51; m 76; c 3. AQUATIC CHEMISTRY, ENVIRONMENTAL CHEMISTRY. *Educ:* Rutgers Col, AB, 73; Univ Md, PhD(chem), 77. *Prof Exp:* Scientist marine chem, Lockheed Ctr Marine Res, 78-79; PRIN ENVIRON ENGR, PUB SERV ELEC & GAS CO, 89- *Mem:* Am Chem Soc; AAAS; Int Ozone Asn. *Res:* Environmental chemistry; fate and effects of oxidative biocides in natural waters; performance of power plant auxiliary systems, power plant chemistry and water management, waste disposal and resource recovery. *Mailing Add:* Pub Serv Elec & Gas Co 80 Park Plaza T17G Newark NJ 07101

SUGANO, KATSUHITO, b Japan, May 25, 48; m 79; c 1. DI-MUON AND DI-HADRON EXPERIMENT. *Educ:* Univ Tokyo, BS, 72, MS, 74, Dr, 79. *Prof Exp:* Res assoc physics, Inst Nuclear Study, Tokyo, 77-78; res fel, Ministry Educ, Japan, 78-79; Res assoc physics, Fermi Nat Accelerator Lab, 79-83; ASST PHYSICIST, ARGONNE NAT LAB, 83. *Mem:* Phys Soc Japan; Am Phys Soc. *Res:* Research and experiment in high energy particle physics using accelerators. *Mailing Add:* Argonne Natl Lab, Bldg 362 9700 S Cass Argonne IL 60439

SUGAR, GEORGE R, b Winthrop, Mass, Oct 12, 25; m 61; c 2. DATA COMMUNICATIONS, ELECTRONIC INSTRUMENTATION. *Educ:* Univ Md, BS, 50. *Prof Exp:* Asst, Inst Fluid Dynamics & Appl Math, Univ Md, 50-51; physicist, Upper Atmosphere & Space Physics Div, Nat Bur Stand, 51-65; electronic engr, Aeronomy Lab, Nat Oceanic & Atmospheric Admin, 65-70; electronic engr automation & instrumentation sect, Electromagnetics Div, Nat Bur Stand, 71-77; Electronic engr, Wave Propagation Lab, Nat Oceanic & Atmospheric Admin, 77-79; pres, 79-84, consult, Pragmatronics Inc, 84-85; PRES, RDS ENTERPRISES, 85- *Concurrent Pos:* Vis lectr, Univ Colo, 73-74. *Honors & Awards:* Silver Medal, US Dept Com. *Mem:* Sigma Xi; Inst Elec & Electronics Engrs. *Res:* Laboratory automation; application of minicomputers to laboratory measurements; management of computers; organization development; electronic instrumentation; meteor-burst propagation and communication. *Mailing Add:* 770 Lincoln Pl Boulder CO 80302-7531

SUGAR, JACK, b Baltimore, Md, Dec 22, 29; m 56; c 3. ATOMIC SPECTROSCOPY. *Educ:* Johns Hopkins Univ, BA, 56, PhD(physics), 60. *Prof Exp:* PHYSICIST, NAT INST STANDARDS & TECHNOL, 60- *Concurrent Pos:* Fulbright res traveling grant, 66-67, NATO res grant, 90-91. *Honors & Awards:* Silver Medal, US Dept Com, 71. *Mem:* Fel Optical Soc Am. *Res:* Spectra of solids; atomic spectra and energy levels; nuclear moments; ionization energies. *Mailing Add:* A167 Physics Bldg Nat Inst Standards & Technol Gaithersburg MD 20899

SUGAR, OSCAR, b Washington, DC, July 9, 14; m 44; c 3. PHYSIOLOGY, NEUROSURGERY. *Educ:* Johns Hopkins Univ, AB, 34; George Washington Univ, MA, 37, MD, 42; Univ Chicago, PhD(physiol), 40. *Prof Exp:* Asst physiol, Univ Chicago, 36-38; clin asst, 46-48, from instr to assoc prof, 48-58, head dept, 71-81, EMER PROF NEUROL SURG, COL MED, UNIV ILL, 81- *Concurrent Pos:* Sci consult, Nat Inst Neurol Dis & Stroke, 72-75. *Mem:* Am Asn Neurol Surg; Am Neurol Soc; Soc Neurol Surg. *Res:* Degeneration and regeneration of the peripheral and central nervous system; effects of oxygen lack on cells and electrical activity of the nervous system; visualization of the blood supply and vascular anomalies of the brain. *Mailing Add:* 17737 Valladores Dr San Diego CA 92127

SUGAR, ROBERT LOUIS, b Chicago, Ill, Aug 20, 38; m 66. THEORETICAL PHYSICS. *Educ:* Harvard Univ, AB, 60; Princeton Univ, PhD(physics), 64. *Prof Exp:* Res assoc physics, Columbia Univ, 64-66; from asst prof to assoc prof, 66-73, PROF PHYSICS, UNIV CALIF, SANTA BARBARA, 73- *Concurrent Pos:* Dep dir, Inst Theoret Physics, 79-81. *Mem:* Am Phys Soc. *Res:* High energy physics. *Mailing Add:* Dept Physics Univ Calif Santa Barbara CA 93106

SUGARBAKER, EVAN ROY, b Mineola, NY, Nov, 17, 49; m 85; c 1. NUCLEAR REACTION MECHANISMS. *Educ:* Kalamazoo Col, BA, 71; Univ Mich, PhD(physics), 76. *Prof Exp:* Res assoc, Nuclear Struct Res Lab, Univ Rochester, 76-78; vis asst prof physics, Univ Colo, 78-80; asst prof, 81-86, ASSOC PROF PHYSICS, OHIO STATE UNIV, 86- *Concurrent Pos:* Co-prin investr NSF grant, 81-88, prin investr, 88- *Mem:* Am Phys Soc; Am Asn Physics Teachers; AAAS. *Res:* Mechanisms in light-ion induced nuclear reactions; nuclear structure. *Mailing Add:* Dept Physics Ohio State Univ 1302 Kinnear Rd Columbus OH 43212

SUGARMAN, NATHAN, b Chicago, Ill, Mar 3, 17; m 40; c 2. NUCLEAR CHEMISTRY. *Educ:* Univ Chicago, BS, 37, PhD(phys chem), 41. *Prof Exp:* Am Philos Soc fel, Univ Chicago, 41-42, res assoc, Metall Lab, 42-43, sect chief, 43-45; group leader, Los Alamos Sci Lab, Univ Calif, 45-46; from asst prof to assoc prof, 46-52, PROF CHEM, ENRICO FERMI INST, UNIV CHICAGO, 52- *Mem:* AAAS; Am Chem Soc; fel Am Phys Soc; Fedn Am Scientists. *Res:* Nuclear reactions; fission studies; recoil experiments. *Mailing Add:* Enrico Fermi Inst Univ Chicago Chicago IL 60637-4931

SUGATHAN, KANNETH KOCHAPPAN, b Palliport, India, Mar 23, 26; nat US; m 56; c 4. COATINGS & ADHESIVE CHEMISTRY, SYNTHETIC RESIN CHEMISTRY. *Educ:* Univ Kerala, BSc, 51, PhD(terpene chem), 67; Univ Saugar, MSc, 53. *Prof Exp:* Demonstr chem, SKV Col, Trichur, India, 53-54, lectr, 54-64; lectr, S N Col, Quilon, India, 64-68; Nat Res Coun-Agr Res Serv fel, USDA Naval Stores Lab, Olustee, Fla, 68-70; Am Cancer Soc res fel, Univ Miss, 71-72; res chemist, Crosby Chem, Inc, 72-76, chief chemist, 76-78; chief chemist, Zielger Chem & Mineral Corp, 78-80; SR SCIENTIST & PROJ MGR, POLYMER RES CORP AM, 81- *Concurrent Pos:* Sr demonstr, Christian Med Col, Vellore, India, 61-64. *Mem:* AAAS; Am Chem Soc; NY Acad Sci; fel Am Inst Chemists. *Res:* Product and process development in the fields of polymers, coatings and adhesives; applying graft polymerization techniques to impart special effects to substrates like metals, plastics, rubber and paper; industrial trouble shooting. *Mailing Add:* 39 Ross Hall Blvd S Piscataway NJ 08854

SUGAYA, HIROSHI, b Hyogo-Ken, Japan, Mar 13, 31; m 59; c 2. MAGNETIC RECORDING, VIDEO RECORDING. *Educ:* Osaka Univ, Japan, BSc, 54, Dr Sci(physics), 64; Eindhoven Int Inst, MDipl computer, 61. *Prof Exp:* Prin researcher acoust, Matsushita Elec Indust, 54-60, Cent Res Lab, 55-69, gen mgr, Video Rec, 70-75, Corp Plan, Eng Div, 75-81 & Audio & Video Div, 81-88; PROF, DIV AUDIO & VISUAL COM, KYUSHU INST DESIGN, NAT UNIV, 88- *Concurrent Pos:* Chmn, EIA-J Video Tech Comt, 68-81; mem & chmn, IEC TC60 SC 60B WG-5, 70- *Mem:* Fel Inst Elec & Electronics Engrs; Soc Motion Picture & TV Engrs; Inst TV Engrs Japan; Acoust Soc Japan. *Res:* Magnetic recording; developed hot-pressed-ferrite head, high speed video tape contact duplicator and many types of video tape recorders; overwrite recording. *Mailing Add:* Kyushu Inst Desin 4-9-1 Shiobaru Minami-Ku Fukuoka 815 Japan

SUGDEN, EVAN A, b Salt Lake City, Utah, Sept 17, 52; c 1. ENTOMOLOGY. *Educ:* Univ Utah, BA, 76; Univ Calif, Davis, PhD(entom), 84. *Prof Exp:* Apicult consult, Honey Prod Int, 84-85; postdoctoral researcher, Australian Mus, 85-87; biol aide, Calif Dept Food & Agr, Plant Indust, Anal & ID, Exotic Pests, 87-88; postdoctoral, 88-90, ENTOMOLOGIST, SUBTROP AGR RES LAB, AGR RES SERV, USDA, 90- *Mem:* Entom Soc Am; Int Union Study Social Insects; Am Asn Prof Apiculturists; Xerces Soc. *Res:* Feral honey bee and Africanized honey bee monitoring program; bee ecology; pollination of crops and native plants; alternative pollinator species; bee and social hymenoplera systematics and ecology. *Mailing Add:* Honey Bee Lab Agr Res Serv-USDA Weslaco TX 78596

SUGER-COFINO, JOSE-EDUARDO, mathematical physics, operations research, for more information see previous edition

SUGERMAN, ABRAHAM ARTHUR, b Dublin, Ireland, Jan 20, 29; nat US; m 60; c 4. PSYCHIATRY, PSYCHOPHARMACOLOGY. *Educ:* Univ Dublin, BA, 50, MB, BCh & BAO, 52; RCPS dipl psychol med, 58; State Univ NY, DSc(psychiat), 62; Univ Newcastle, dipl, 66; Am Bd Psychiat & Neurol, dipl, 69. *Prof Exp:* House officer, Meath Hosp, Dublin, Ireland, 52-53 & St Nicholas Hosp, London, 53; sr house physician, Brook Gen Hosp, 54; registr psychiat, Kingsway Hosp, Derby & Med Sch, King's Col, Newcastle, 55-58; clin psychiatrist, Trenton State Hosp, NJ, 58-59; chief sect invest psychiat & dir clin invest unit, NJ Neuropsychiat Inst, 61-73; res consult & assoc psychiatrist, Carrier Found, 68-72, dir outpatient serv, 72-74 & 77-78, med dir, 74-77, res dir, 72-79, assoc psychiatrist & dir med student training, 78-90, dir effective disorders prog, 82-90; MED DIR, ADDICTION RECOVERY SERV, UNIV MED & DENT NJ-COMMUNITY MENTAL HEALTH CTR, 90- *Concurrent Pos:* Res fel psychiat, State Univ NY Downstate Med Ctr, 59-61; res consult, Trenton Psychiat Hosp, 64-; clin assoc prof, Rutgers Med Sch, 72-78; consult, Med Ctr, Princeton, 72-; contrib fac, Grad Sch Appl & Prof Psychol, Rutgers Univ, 74-78; vis prof, Hahnemann Med Col, 78-90 & Ctr Alcohol Study, Rutgers Univ, 77-83; clin prof, Roebort Wood Johnson Med Sch, 78- *Mem:* Fel Royal Col Psychiatrists; fel Am Psychiat Asn; fel Am Col Clin Pharmacol; fel Am Col Psychiatrists; fel Am Col Neuropsychopharmacology; fel AAAS. *Res:* Evaluation of new psychiatric drugs; nosology; psychology and prognosis in schizophrenia and alcoholism; quantitative analysis of the electroencephalogram. *Mailing Add:* 125 Roxboro Rd Lawrenceville NJ 08648

SUGERMAN, LEONARD RICHARD, b New York, NY, June 24, 20; m 40; c 4. NAVIGATION. *Educ:* Mass Inst Technol, BS, 55; Univ Chicago, MBA, 60. *Prof Exp:* US Air Force, 42-75, supvr bomb-navig br, Dept Armament Training, Lowry AFB, Colo, 50-53, air staff off, Off Dep Chief Staff Develop, Hq USAF, Pentagon, DC, 55-59, res & develop staff off hqs, Syst Command, Andrews AFB, 60-62, hqs, Res & Tech Div, Bolling AFB, 62-64, exec off, Cent Inertial Guid Test Facil, Missile Develop Ctr, Holloman AFB, 64-68, chief, Athena Test Field Off, Ballistic Reentry Systs Prog, 68-70, Air Force Dep to Comdr, White Sands Missile Range & chief, Air Force Range Opers Off, 70-72, dep chief staff, Plans & Requirements, Air Force Spec Weapons Ctr, 72-75; ASST DIR RESOURCES MGT, PHYS SCI LAB, N MEX STATE UNIV, 75- *Concurrent Pos:* US rep guid & control panel, Adv Group Aerospace Res & Develop, NATO, 66-76. *Honors & Awards:* Norman P Hays Award, Am Inst Navig, 72. *Mem:* Am Defense Preparedness Asn; Am Inst Navig (pres, 70-71); assoc fel Am Inst Aeronaut & Astronaut; Sigma Xi. *Res:* Guidance and control of aerospace vehicles; bombing and navigation systems for aerospace vehicles. *Mailing Add:* 3025 Fairway Dr Las Cruces NM 88001

SUGGITT, ROBERT MURRAY, b Toronto, Ont, June 24, 25; nat US; m 59; c 3. PHYSICAL CHEMISTRY. *Educ:* Univ Toronto, BA, 47; Univ Mich, MS, 48, PhD(chem), 52. *Prof Exp:* Chemist, Texaco, Inc, 52-57, group leader, 57-68, res assoc, 68-75, tech assoc, 75, asst mgr, 75-78, MGR, TECH DIV, TEXACO DEVELOP CORP, 78- *Mem:* Am Chem Soc; AAAS. *Res:* Catalysis, petroleum and petrochemical processing, lubrication. *Mailing Add:* Texaco Develop Corp 2000 Westchester Ave White Plains NY 10650

SUGGS, CHARLES WILSON, b NC, May 30, 28; m 49; c 3. AGRICULTURAL & HUMAN FACTORS ENGINEERING. *Educ:* NC State Col, BS, 49, MS, 55, PhD(agr eng), 59. *Prof Exp:* Instr, Dearborn Motors, 49; asst serv supvr, Int Harvester, 49-51; res farm supt, NC Agr Exp Sta, 51-53, from instr to assoc prof, 54-66, PROF AGR ENG, NC STATE UNIV, 66- *Concurrent Pos:* Consult several law firms & foreign countries. *Mem:* Fel Am Soc Agr Engrs; Human Factors Soc. *Res:* Ergonomics; vibration; noise; human performance; tobacco mechanization; vehicle safety; servosystems; environment. *Mailing Add:* Dept Biol & Agr Eng NC State Univ Box 7625 Raleigh NC 27695

SUGGS, JOHN WILLIAM, b Highland Park, Mich, Aug 17, 48; m; c 1. ORGANOMETALLIC CHEMISTRY, ORGANIC CHEMISTRY. *Educ:* Univ Mich, BS, 70; Harvard Univ, PhD(chem), 76. *Prof Exp:* Woodrow Wilson fel, 70; mem tech staff org chem, Bell Labs, AT&T, 76-81; asst prof, 81-86, ASSOC PROF, DEPT CHEM, BROWN UNIV, 86- *Honors & Awards:* Res Career Develop Award, Nat Cancer Inst. *Mem:* Am Chem Soc; AAAS. *Res:* Use of organometallic compounds in organic synthesis; mechanisms of and reactive intermediates in organometallic reactions; drug-nucleic acid interactions. *Mailing Add:* Dept Chem Box H Brown Univ Providence RI 02912

SUGGS, MORRIS TALMAGE, JR, b Ft Myers, Fla, June 17, 27; m 52; c 3. MICROBIOLOGY. *Educ:* Wake Forest Col, BS, 50; Fla State Univ, MS, 57; Univ NC, MPH, 65, DrPH, 67. *Prof Exp:* Teacher, Fla Pub Sch, 52-54; microbiologist, Ala, 57, asst chief tissue cult unit, 58-59, res asst, 60-62, res asst virol training unit, 62-68, spec asst biol reagents sect, Lab Prog, 68, DIR BIOL PROD PROG, CTR INFECTIOUS DISEASES, CTR DIS CONTROL, USPHS, 68- *Mem:* Am Soc Microbiologists; Conf of State & Prov Pub Health Lab Dirs. *Res:* Standardization and quality assurance of in vitro diagnostic products. *Mailing Add:* 2424 Coralwood Dr Decatur GA 30033

SUGGS, WILLIAM TERRY, b Orange, NJ, Dec 7, 45; m 83; c 2. QUALITY ASSURANCE & GOOD MANUFACTURING, PRACTICES, REGULATORY AFFAIRS. *Educ:* Seton Hall Univ, BS, 67; Mich State Univ, PhD(org chem), 73; Grand Valley State Univ, MBA, 81. *Prof Exp:* Postdoctoral res assoc hort, Plant Biochem Res, Mich State Univ, 73-75; supvr, Anal Develop Qual Control Dept, Parke-Davis, Div Warner-Lambert, 75-77, mgr, Anal Devel Qual Control Dept, 77-82, mgr, Anal Devel Chem Develop Dept, 82-85, sr mgr, 85-91, SECT DIR, ANAL DEVELOP CHEM DEVELOP DEPT, PARKE-DAVIS, DIV WARNER- LAMBERT, 91- *Mem:* Am Chem Soc. *Res:* Development of analytical methods for the control of raw materials, chemical intermediates and bulk substances; experimental drug substances produced for use in toxicology, formulations studies and clinical research to support IND and NDA filings. *Mailing Add:* Parke-Davis 188 Howard Ave Holland MI 49424-6596

SUGIHARA, JAMES MASANOBU, b Las Animas, Colo, Aug 6, 18; m 44; c 2. ORGANIC CHEMISTRY, ACADEMIC ADMINISTRATION. *Educ:* Univ Calif, BS, 39; Univ Utah, PhD(chem), 47. *Prof Exp:* From instr to prof chem, Univ Utah, 43-64; dean, Col Chem & Physics, NDak State Univ, 64-73, dean, Col Sci & Math, 73, dean, Grad Sch & dir Res Admin, 74-85, dean, Col Sci & Math, 85-86, prof chem, 64-89, interim chair, dept polymers & coatings, 90-91, EMER PROF, UNIV UTAH, 89- *Concurrent Pos:* Fel, Ohio State Univ, 48. *Mem:* Am Chem Soc; Geochem Soc Am. *Res:* Reaction mechanisms; porphyrin chemistry; origin of petroleum. *Mailing Add:* 1001 Southwood Drive Fargo ND 58103

SUGIHARA, THOMAS TAMOTSU, b Las Animas, Colo, June 14, 24; m 52; c 2. NUCLEAR CHEMISTRY. *Educ:* Kalamazoo Col, AB, 45; Univ Chicago, SM, 51, PhD(phys chem), 52. *Prof Exp:* Res assoc, Mass Inst Technol, 52-53; from asst prof to prof chem, Clark Univ, 53-67, chmn dept, 63-66; dir, Cyclotron Inst, Tex A&M Univ, 71-78, prof chem, 67-81, dean, Col Sci, 78-81; PROF CHEM, ORE STATE UNIV, 81-, DEAN, COL SCI, 81- *Concurrent Pos:* Assoc scientist, Woods Hole Oceanog Inst, 54-67; mem subcomt nuclear ship waste disposal, Nat Acad Sci-Nat Res Coun, 58, mem surv panel nuclear chem, 64, mem subcomt low level contamination mat, 65-66; Guggenheim fel, Univ Oslo, 61-62; mem nuclear sci adv comt, NSF & Dept Energy, 77-79. *Mem:* Fel AAAS; Am Chem Soc; fel Am Phys Soc. *Res:* Heavy-ion reactions; high-spin states; spectroscopy with heavy-ion reactions. *Mailing Add:* Chem/Mat Sci L326 Lawrence Livermore Nat Lab Livermore CA 94550

SUGIOKA, KENNETH, b Hollister, Calif, Apr 19, 20; m 47, 66; c 5. ANESTHESIOLOGY. *Educ:* Univ Denver, BS, 45, Wash Univ, MD, 49; Am Bd Anesthesiol, dipl, 55; FFARCS. *Prof Exp:* Intern, Univ Iowa Hosp, 49-50, resident anesthesiol, 50-52, instr, 52; actg chief anesthesiol, Vet Admin Hosp, Des Moines, 52; resident & instr, Vet Admin Hosp, Iowa City, 52; from asst prof to prof anesthesiol, Sch Med, Univ NC, Chapel Hill, 54-83, chmn dept, 69-83; PROF PHYSIOL & ANESTHESIOL, DUKE UNIV, 85- *Concurrent Pos:* NIH spec res fel, 62; consult, Vet Admin Hosp, Fayetteville, NC; vis prof, Inst Physiol, Univ Gottingen, 62, Med Sch, King's Col, Univ London, 63, Max Planck Inst Physiol, Dortmund, Ger, 76-77 & Royal Col Surgeons, 83-84. *Mem:* Soc Acad Anesthesiol Chmn (past pres); Am Soc Anesthesiol; Soc Exp Biol & Med; Am Physiol Soc; Asn Univ Anesthetists; Int Soc Oxygen Tissue Transport. *Res:* Application of electronic instrumentation to physiological measurements; electrochemical methods of biological analysis; cation sensitive glass electrodes; oxygen transport. *Mailing Add:* Dept of Anesthesiol Box 3094 Duke Univ Med Ctr Durham NC 27710

SUGITA, EDWIN T, b Honolulu, Hawaii, Feb 1, 37; m 59; c 2. PHARMACEUTICS. *Educ:* Purdue Univ, BS, 59, MS, 62, PhD(pharmaceut), 63. *Prof Exp:* From asst prof to assoc prof, 64-72, PROF PHARM, PHILADELPHIA COL PHARM & SCI, 72-, CHMN DEPT PHARMACEUT, 87- *Concurrent Pos:* Res grant, Smith, Kline & French Labs, 65-67; NIH grant, 65-71 & 74-77; Kapnek Charitable Trust fel, 76-81. *Mem:* Acad Pharmaceut Res & Sci; Sigma Xi; Am Asn Pharmaceut Scientists. *Res:* Facilitated absorption of drugs; absorption of drugs: in vitro-in vivo correlations; bioequivalency testing; pharmacokinetics of drugs in humans. *Mailing Add:* 2941 Raspberry Lane Gilbertsville PA 19525

SUGIURA, MASAHISA, b Tokyo, Japan, Dec 8, 25; m 62; c 1. SPACE PHYSICS. *Educ:* Univ Tokyo, MS, 49; Univ Alaska, PhD, 55. *Prof Exp:* Asst prof geophys res, Geophys Inst, Univ Alaska, 55-57, assoc prof geophys, 57-62, prof, 62; Nat Acad Sci sr assoc, NASA, 62-64, mem staff, Goddard Space Flight Ctr, 64-85; prof, Geophys Inst, Fac Sci, Kyoto Univ, Japan, 85-89; PROF, INST RES & DEVELOP, TOKAI UNIV, TOKYO, JAPAN, 90- *Concurrent Pos:* Guggenheim fel, 59; prof, Univ Wash, 66-67. *Honors & Awards:* Tanakadate Prize, 50, Soc Terrestrial Magnetism & Elec Japan, 50; Except Sci Achievement Medal, NASA, 85. *Mem:* Fel Am Geophys Union; Soc Geomagnetism & Earth, Planetary & Space Sci; AAAS. *Res:* Magnetospheric physics; geophysics. *Mailing Add:* Inst Res Develop Tokai Univ 2-28 Tomigaya Shibuya-ku Tokyo 151 Japan

SUH, CHUNG-HA, b Chinnampo City, Korea, Sept 11, 32; m 61; c 3. MECHANICAL ENGINEERING, BIOMECHANICS. *Educ:* Seoul Nat Univ, BS, 59; Univ Calif, Berkeley, MS, 64, PhD(mech eng), 66. *Prof Exp:* From mech engr to chief engr, Hwan-Bok Indust Co, Ltd, Korea, 59-61; dept head mach design & shop, Atomic Energy Res Inst, 61-62; res engr, Biomech Lab, Univ Calif, Berkeley, 63-66; from asst to assoc prof, 66-77, chmn, Dept Eng Design & Econ Eval, 70-78, PROF, UNIV COLO, BOULDER, 77-, DIR, BIOMECH LAB, 70- *Concurrent Pos:* Res consult, Int Chiropractors Asn, 72-88 & Hyundai Motor Co, 84-; dir, Ctr Automotive Res, Univ Colo, Boulder, 89- *Mem:* Am Soc Mech Engrs; Am Gear Mfrs Asn; Soc Automotive Engrs. *Res:* Computer-aided design of mechanisms; optimum design; biomechanics of human joints and system. *Mailing Add:* Dept Mech Eng ECME 1-18 Univ of Colo Boulder CO 80309

SUH, JOHN TAIYOUNG, US citizen; m 58; c 4. MEDICINAL CHEMISTRY, ORGANIC CHEMISTRY. *Educ:* Butler Univ, BS, 53; Univ Wis, MS, 56, PhD(org chem), 58. *Prof Exp:* Teaching asst, Univ Wis, 55; sr res chemist, Res Ctr, Johnson & Johnson, 58-59 & McNeil Labs, Inc, 59-63; group leader res, Colgate-Palmolive Co, 63-65, sect head res, Dept Med Chem, Lakeside Labs Div, 63-75; mgr chem res, Freeman Chem Corp, 76-77; sect head res, Dept Med Chem, USV Pharmaceut Corp, 77-78, assoc dir, Dept Med Chem, Res & Develop, Revlon Health Care Group, 78-84; med chem dir, 84-86; CHEM RES SERV DIR, 86- *Concurrent Pos:* Vis lectr chem, Marquette Univ, 72-77. *Mem:* Am Chem Soc. *Res:* Medicinal chemistry in areas of cardiovascular, psychopharmacological and hematinic agents; organic chemistry in areas of stereochemistry, natural products, heterocyclic and organometallic chemistry; pulmonary and allergy research. *Mailing Add:* Revlon Health Care Group 59 Stanwick Rd Greenwich CT 06830

SUH, NAM PYO, b Seoul, Korea, Apr 22, 36; US citizen; m 61; c 4. MECHANICAL ENGINEERING. *Educ:* Mass Inst Technol, SB, 59, SM, 61; Carnegie Inst Technol, PhD(mech eng), 64. *Hon Degrees:* Dr Eng, Worcester Polytech Inst, 86; LHD, Univ Lowell, 88. *Prof Exp:* Develop engr, Sweetheart Plastics, Inc, 58-59; lectr mech eng, Northeastern Univ, 64-65; sr res engr, United Shoe Mach Corp, Mass, 61-65; from asst prof to assoc prof mech & mat, Univ SC, 65-70; assoc prof mech, 70-75, PROF MECH ENG & DIR, MFG & PRODUCTIVITY LAB, MASS INST TECHNOL, 75-, RALPH E & ELOISE F CROSS PROF, 89- *Concurrent Pos:* Consult, govt agencies & indust firms, 70-; asst dir, NSF, 84-88; co-ed in chief, Robotics & Computer Integrated Mfg; chmn bd, Axiomatics Corp, Woburn, Mass, Sutek Corp, Hudson, Mass; foreign mem, Royal Swed Acad Eng Sci, 89- *Honors & Awards:* Larsen Mem Award, Am Soc Mech Engrs, 76, Blackall Award, 82; F W Taylor Res Award, Soc Mech Engrs, 86; Fed Engr Award, NSF, 87; Soc Petrol Engr Award, 81. *Mem:* Am Soc Mech Engrs; Am Soc Eng Educ; Int Inst Prod Eng Res. *Res:* Materials processing; mechanical behavior of materials; solid propellants; manufacturing processes and systems; tribology; design; author four books. *Mailing Add:* Dept Mech Eng Rm 35-237 Mass Inst Technol Cambridge MA 02139

SUH, TAE-IL, b Chungdo, Korea, June 1, 28; m 55; c 3. ALGEBRA. *Educ:* Kyung-Pook Nat Univ, Korea, BS, 52; Yale Univ, PhD(math), 61. *Prof Exp:* Asst prof math, Kyung-Pook Nat Univ, Korea, 61-63; assoc prof, Sogang Univ, Korea, 63-65; assoc prof, 65-68, PROF MATH, E TENN STATE UNIV, 68- *Mem:* Am Math Soc; Math Asn Am. *Res:* Non-associative algebras. *Mailing Add:* Dept Math East Tenn State Univ PO Box 22390A Johnson City TN 37614

SUHADOLNIK, ROBERT J, b Forest City, Pa, Aug 15, 25; m 49; c 5. BIOCHEMISTRY. *Educ:* Pa State Univ, BS, 49, PhD, 56; Iowa State Univ, MS, 53. *Prof Exp:* Res assoc biochem, Univ Ill, 56-57; asst prof, Okla State Univ, 57-61; res mem, Albert Einstein Med Ctr, 61-70, head dept bio-org chem, 68-70; PROF BIOCHEM, SCH MED, TEMPLE UNIV, 70- *Mem:* Am Chem Soc; Am Soc Biol Chemists; Sigma Xi. *Res:* Alkaloid biogenesis; metabolism of allose and allulose; biosynthesis and biochemical properties of nucleoside antibiotics; mechanism of protein synthesis; role of interferon in development of antiviral-anticancer state in normal and DNA repair-deficient mammalian cells. *Mailing Add:* Dept of Biochem Temple Univ Health Sci Campus Philadelphia PA 19140

SUHAYDA, JOSEPH NICHOLAS, b Flint, Mich, Feb 23, 44; m 66; c 2. OCEANOGRAPHY. *Educ:* Calif State Univ, Northridge, BS, 66; Univ Calif, San Diego, PhD(phys oceanog), 72. *Prof Exp:* Asst prof, 72-75, ASSOC PROF MARINE SCI, COASTAL STUDIES INST, LA STATE UNIV, BATON ROUGE, 75- *Mem:* Am Geophys Union; Am Shore & Beach Preserv Asn. *Res:* Coastal oceanography, primarily nearshore processes on beaches and reefs, and the influence of storm waves on sediment on the continental shelf. *Mailing Add:* Dept Civil Eng La State Univ Baton Rouge LA 70803

SUHIR, EPHRAIM, b Odessa, USSR, May 17, 37; US citizen; m 68; c 2. MICROELECTRONIC & FIBER OPTIC SYSTEMS, DYNAMIC & PROBABILISTIC PROBLEMS. *Educ:* Odessa Polytech Inst, USSR, MS, 66; Moscow Univ, PhD(appl mech), 68. *Prof Exp:* Prof appl mech, Nikolaev Inst Naval Archit, Ukraine, 70-75; head lab, reliability & statist, Res Inst Engine Bldg, Lithuania, 75-80; sr proj engr, Exxon Corp, NJ, 80-84; MEM TECH STAFF, AT&T BELL LABS, NJ, 84- *Concurrent Pos:* Consult, Nikolaev Inst Naval Archit, 70-75; vis prof, Kunas Polytech Inst, 75-80; prin investr, AT&T Bell Labs, 84-; lectr thermal anal & failure prev, 88-; sr ed, Am Soc Mech Engrs J Electronic Packaging, 89- *Mem:* Sr mem Inst Elec & Electronic Engrs; Am Soc Mech Engrs; Mat Res Soc; sr mem Soc Petrol Engrs. *Res:* Mechanical behavior, reliability and rational physical design of microelectronic and fiber optic components, structures and systems. *Mailing Add:* AT&T Bell Labs Rm 7D-326 600 Mountain Ave Murray Hill NJ 07974-2070

SUHL, HARRY, b Leipzig, Ger, Oct 18, 22; nat US; wid. PHYSICS. *Educ:* Univ Wales, BSc, 43; Oxford Univ, PhD(theoret physics), 48. *Prof Exp:* Exp officer, Admiralty Signal Estab, Eng, 43-46; mem tech staff, Bell Tel Labs, Inc, 48-60; vis lectr, 60, PROF PHYSICS, UNIV CALIF, SAN DIEGO, 61- *Mem:* Nat Acad Sci; fel Am Phys Soc. *Res:* Theoretical solid state physics. *Mailing Add:* Dept Physics Univ Calif San Diego La Jolla CA 92093

SUHM, RAYMOND WALTER, b Springfield, Mass, June 9, 41; m 64; c 1. STRATIGRAPHY. *Educ:* Southeast Mo State Col, BS, 63; Southern Ill Univ, Carbondale, MS, 65; Univ Nebr-Lincoln, PhD(geol), 70. *Prof Exp:* Instr geol, Southern Ill Univ, Carbondale, 65; geophysicist, Humble Oil Co, Calif, 65-67; asst prof geol, Tex A&I Univ, 70-75, assoc prof, 75-80; explor geologist, Tex Oil & Gas Corp, Oklahoma City, 80-81; GEOL CONSULT, 81- *Concurrent Pos:* Consult, Cockrell Corp, 72, Int Oil & Gas, 76-78 & Tenneco Oil, 77, Tex Gas Explor, 82-86. *Mem:* Am Asn Petrol Geologists; Soc Econ Paleontologists & Mineralogists. *Res:* Ordovician stratigraphy and paleontology; Ozark/Ouachita geology; coastal sedimentation and geomorphology. *Mailing Add:* 11716 128th St Oklahoma City OK 73165-9432

SUHOVECKY, ALBERT J, plant pathology; deceased, see previous edition for last biography

SUHR, NORMAN HENRY, b Chicago, Ill, June 13, 30; m 53; c 4. SPECTROSCOPY, GEOCHEMISTRY. *Educ:* Univ Chicago, AB, 50, MS, 54. *Prof Exp:* Spectroscopist & mineralogist, Heavy Minerals Co, Vitro Corp Am, 56-58; spectroscopist, Labs, 58-65, asst dir, 65-70, res assoc, Univ, 63-67, asst prof, 67-69, ASSOC PROF GEOCHEM, PA STATE UNIV, 69-, DIR MINERAL CONST LABS, 70- *Mem:* AAAS; Soc Appl Spectros; Geochem Soc. *Res:* X-ray and emission spectroscopy and atomic absorption, primarily in the fields of earth sciences. *Mailing Add:* 806 W Beaver Ave State College PA 16801

SUHRLAND, LEIF GEORGE, b Schroon Lake, NY, Apr 9, 19; m 50; c 3. HEMATOLOGY, ONCOLOGY. *Educ:* Cornell Univ, BS, 42; Univ Rochester, MD, 50. *Hon Degrees:* DM, Univ Rochester, 50. *Prof Exp:* Bacteriologist, USPHS, 42-43; intern & jr asst med, Univ Hosps, Cleveland, 50-52; instr, Western Reserve Univ, 57-59, asst prof med & asst clin pathologist, 59-67; prof, 67-89, EMER PROF MED, COL HUMAN MED, MICH STATE UNIV, 89- *Concurrent Pos:* Am Cancer Soc fel, Univ Hosps, Cleveland, 52-54; Howard M Hanna & Anna Bishop fels, Sch Med, Western Reserve Univ, 54-57; prof clin oncol, Am Cancer Soc, 78. *Mem:* Am Fedn Clin Res; Am Soc Hemat; Am Soc Clin Oncol; Am Col Physicians. *Res:* Host tumor relationships; clinical trials. *Mailing Add:* Med-B220b Life Sci Mich State Univ East Lansing MI 48824

SUIB, STEVEN L, b Olean, NY, May 1, 53; m 77. INORGANIC PHOTOCHEMISTRY. *Educ:* State Univ NY Fredonia, BS, 75; Univ Ill, Urbana, PhD(chem), 79. *Prof Exp:* Res asst, State Univ NY Fredonia, 74-75; teaching asst, Univ Ill, Urbana, 75-77, res asst, 78-79, vis lectr, 79, assoc, 79-80; ASST PROF, UNIV CONN, STORRS, 80- *Concurrent Pos:* Fel, Univ Conn, 80. *Mem:* Am Chem Soc; Sigma Xi. *Res:* Solid state inorganic chemistry including surface, structural, electrochemical and catalytic properties of semiconductors, ceramics and heterogeneous zeolite compounds; zeolite chemistry and catalysis. *Mailing Add:* Dept Chem U 60 Univ Conn Storrs CT 06268

SUICH, JOHN EDWARD, b Bridgeport, Conn, Sept 28, 36; m 57; c 2. INFORMATION SCIENCE. *Educ:* Harvard Univ, BA, 58; Mass Inst Technol, PhD(nuclear eng), 63. *Prof Exp:* Sr physicist, Savannah River Lab, 63-65, res supvr, 65-66, res mgr appl math, 66-68, dir comput sci sect, 68-71, mgr telecommun planning, Gen Servs Dept, 71-72, asst mgr comput sci div, Cent Systs & Servs Dept, Del, 72, mgr com systs div mgt sci, 72-75, RES ASSOC, SAVANNAH RIVER LAB, E I DU PONT DE NEMOURS & CO, INC, 75- *Res:* Relational data base and logic programming. *Mailing Add:* 692 Storm Branch Rd Beech Island SC 29841

SUICH, RONALD CHARLES, b Cleveland, Ohio, Nov 16, 40; m 62; c 3. STATISTICS. *Educ:* John Carroll Univ, BSBA, 62; Case Western Reserve Univ, MS, 64, PhD(statist), 68. *Prof Exp:* Mkt researcher, Cleveland Elec Illum Co, 62-64; from instr to asst prof statist, Case Western Reserve Univ, 64-70; asst prof, Univ Akron, 70-77, assoc prof math, 77-80; MEM FAC MATH, CALIF STATE UNIV, 80- *Mem:* Am Statist Asn. *Res:* Sequential tests. *Mailing Add:* Dept Mgt Sci Calif State Univ Fullerton CA 92634

SUINN, RICHARD M, b Honolulu, Hawaii, May 8, 33; m 58; c 4. BEHAVIOR THERAPY, CLINICAL PSYCHOLOGY. *Educ:* Ohio State Univ, BA, 55; Stanford Univ, MA, 57, PhD(psychol), 59. *Prof Exp:* Asst prof, Whitman Col, 59-64; res assoc psychol, Stanford Med Sch, 64-66; assoc prof fac, Univ Hawaii, 66-68; PROF & HEAD FAC, COLO STATE UNIV, 68- *Concurrent Pos:* Vis prof, Univ Vera Cruz, Mex, 71; vis scholar, Peoples Rep China, 86; bd dirs, Asian Am Psychol Asn, 84-87; chair, Educ & Training Bd, Am Psychol Asn, 86-88, bd dirs, 90-93. *Mem:* Fel Am Psychol Asn; Asn Advan Behav Ther; Asian Am Psychol Asn; Sigma Xi. *Mailing Add:* Dept Psychol Colo State Univ Ft Collins CO 80523

SUIT, HERMAN DAY, b Houston, Tex, Feb 8, 29. RADIOTHERAPY. *Educ:* Univ Houston, AB, 48; Baylor Univ, SM & MD, 52; Oxford Univ, DrPhil(radiobiol), 56. *Prof Exp:* Intern, Jefferson Davis Hosp, Houston, Tex, 52-53, resident radiol, 53-54; house surgeon radiother, Churchill Hosp, Oxford, Eng, 54, res asst radiobiol lab, 54-56, registr radiother, 56-57; sr asst surgeon, Radiation Br, Nat Cancer Inst, 57-59; asst radiotherapist, Univ Tex M D Anderson Hosp & Tumor Inst Houston, 59-63, assoc radiotherapist, 63-68, radiotherapist, 68-71, chief sect exp radiother, 62-70; PROF RADIATION THER, HARVARD MED SCH, 70-; HEAD DEPT RADIATION MED, MASS GEN HOSP, 71- *Concurrent Pos:* Nat Cancer Inst res career develop award, Univ Tex M D Anderson Hosp & Tumor Inst Houston, 64-68; gen fac assoc, Univ Tex Grad Sch Biomed Sci, 65-70, prof radiation ther, 68-71; staff mem, NASA Manned Spacecraft Ctr, 69-71; subcomt radiation biol, Nat Acad Sci. *Mem:* AAAS; Am Col Radiol; Am Soc Therapeut Radiol (secy, 70-72); AMA; Am Asn Cancer Res. *Mailing Add:* Dept Radiation Ther Harvard Med Sch 25 Shattuck St Boston MA 02114

SUIT, JOAN C, b Ontario, Ore, Apr 14, 31; m 60. MICROBIOLOGY. *Educ:* Ore State Col, BS, 53; Stanford Univ, MA, 55, PhD(med microbiol), 57. *Prof Exp:* Res assoc biochem, Biol Div, Oak Ridge Nat Lab, 57-59; res assoc sect molecular biol, Univ Tex M D Anderson Hosp & Tumor Inst, Houston, 59-66, assoc biologist & assoc prof biol, Univ Tex Grad Sch Biomed Sci, Houston, 66-73; res assoc biol, 73-80, RES SCIENTIST, DEPT BIOL, MASS INST TECHNOL, 80- *Mem:* Am Soc Microbiol; Am Soc Cell Biol; Sigma Xi. *Res:* Microbial genetics; DNA replication; microbial growth. *Mailing Add:* 165 Merriam St Weston MA 02193

SUITER, MARILYN J, b Philadelphia, PA. HUMAN RESOURCE MANAGEMENT GEOSCIENCE. *Educ:* Franklin & Marshall Col, BS, 78; Wesleyan Univ, MS, 81. *Prof Exp:* Geologist, US Geol Surv, 77-81; explor geologist, Cities Serv Oil & Gas, 82-86; PROG ADMINR, AM GEOL INST, 87- *Mem:* Asn Women Geoscientists (pres, 88-); Asn Women in Sci; Asn Earth Sci Ed. *Res:* Trace element geochemistry, surficial geology, neotectonics and petroleum exploration; assessing the status of women and ethnic minorities in the geosciences and developing programs that supports their increased participation in the geosciences. *Mailing Add:* Am Geol Inst 4220 King St Alexandria VA 22302-1507

SUITS, CHAUNCEY GUY, physics; deceased, see previous edition for last biography

SUITS, JAMES CARR, b Schenectady, NY, May 29, 32; m 54; c 3. PHYSICS. *Educ:* Yale Univ, BS, 54; Harvard Univ, PhD(appl physics), 60. *Prof Exp:* Staff physicist, Res Lab, 60-76, mgr, Garnet Mat Dept, Gen Prods Div, 76-80, RES STAFF MEM, RES LAB, IBM CORP, SAN JOSE, 80-

Mem: Fel Am Phys Soc; Inst Elec & Electronics Engrs. *Res:* Magnetism, ultra-high vacuum evaporated thin films, magneto- optics, discovery and development of novel magnetic materials, garnet film growth; magnetic bubble development, electroplating development, process automation and electromigration; magnetic recording materials; computer simulation and modeling. *Mailing Add:* IBM Res Div K62-282 650 Harry Rd San Jose CA 95120-6099

SUJISHI, SEI, b San Pedro, Calif, Nov 9, 21; m 55; c 1. INORGANIC CHEMISTRY. *Educ:* Wayne State Univ, BS, 46, MS, 48; Purdue Univ, PhD(chem), 49. *Prof Exp:* From instr to assoc prof chem, Ill Inst Technol, 49-59; assoc prof, 59-65, chmn dept, 74-76, PROF CHEM, STATE UNIV NY STONY BROOK, 65- *Mem:* Am Chem Soc. *Res:* Chemistry of silicon and germanium hydrides. *Mailing Add:* Dean Phys Sci & Math State Univ of NY Stony Brook NY 11794

SUK, WILLIAM ALFRED, b New York, NY, July 9, 45; m 84; c 1. MUTAGENESIS & CARCINOGENESIS, ENVIRONMENTAL SCIENCES. *Educ:* Am Univ, BS, 68, MS, 70; George Wash Univ, PhD(microbiol), 77; Univ NC, MPH, 90. *Prof Exp:* Res biologist fisheries & algae, Nat Fisheries Ctr, 68-69; field res hydrologist, remote sensing, Am Univ & US Geol Surv, 69-70; sr technician chem viral cocarcinogenesis, Microbiol Assocs, Inc, 71-72, supvry technician, 72-74, asst proj dir, 74-76; staff scientist cell biol & retrovirol, Frederick Cancer Res Facil, NC, 76-80; prog mgr & sr res scientist occup health assessment, Geomet Technol, Inc, 80-81; sr proj scientist & head lab cell & molecular biol, Environ Sci Div, Northrop Serv, Inc, 81-87; PROG ADMINR, SUPERFUND BASIC RES PROG, NAT INST ENVIRON HEALTH SCI, NIH, 87- *Concurrent Pos:* NSF fel, Juneau Icefield Res Proj, Glaciol & Arctic Sci Inst, Juneau, Alaska & Atlin, BC, Can, 70; adj prof & mem, Lineberger Cancer Res Ctr, Univ NC Sch Med, Chapel Hill, 83-; consult risk assessment, Indust & Govt Agencies, 81-; prin investr & co-prin investr, Nat Cancer Inst, Nat Inst Environ Health Sci & Environ Protection Agency Contract Awards; mem, Int Prog Chem Safety Collab Study, WHO, 83 & Sci Group Methodologies for Safety Eval of Chemicals, 90; mem, Family Health Int, 87. *Mem:* Am Asn Cancer Res; Am Soc Cell Biol; Sigma Xi; Tissue Cult Asn; Environ Mutagen Soc; Found Glaciol & Environ Res; AAAS; NY Acad Sci; Am Soc Microbiol. *Res:* Carcinogens as modulators of cellular gene expression and how these changes relate to the process of carcinogenesis and differentiation; health hazard assessment of occupational and environmental concerns; retrovirology; molecular toxicology; health policy and administration; biotechnology and technology transfer; public health associated with environmental release of hazardous substances. *Mailing Add:* Div Extramural Res & Training Natl Inst Environ Health Sci PO Box 12233 Research Triangle Park NC 27709

SUKANEK, PETER CHARLES, b Flushing, NY, Sept 15, 47; m 69; c 3. INTEGRATED CIRCUITS, POLYMER PROCESSING. *Educ:* Manhattan Col, BChE, 68; Univ Mass, MS, 70, PhD(chem eng), 72. *Prof Exp:* Proj engr chem eng, Rocket Propulsion Lab, US Air Force, 72-76; asst prof chem eng, Clarkson Col Technol, 76-82, assoc prof, Clarkson Univ, 82-90; PROF & CHAIR, DEPT CHEM ENG, UNIV MISS, 91- *Concurrent Pos:* Consult, Foreign Technol Div, US Air Force, 76-79, IBM Corp; off sci res grant, US Air Force, 78-79; Petrol Res Fund grant, 78-80; NSF grant, 82-84; engr, Philips Res Labs, 86-87, IBM Corp, 88. *Mem:* Am Inst Chem Engrs; Soc Rheology; Sigma Xi; Electrochem Soc. *Res:* Polymer rheology and processing; holographic interferometry; photolithography; plasma etching. *Mailing Add:* Dept Chem Eng Univ Miss University MS 38677

SUKAVA, ARMAS JOHN, b Elma, Man, Mar 1, 17; m 50; c 2. PHYSICAL CHEMISTRY, CHEMICAL THERMODYNAMICS. *Educ:* Univ Man, BSc, 46, MSc, 49; McGill Univ, PhD, 55. *Prof Exp:* Lectr chem, Univ Man, 47-49; res & develop chemist, Consol Mining & Smelting Co, 49-50; lectr chem, Univ BC, 50-51 & Univ Alta, 51-52; instr, 54-55, lectr, 55-56, from asst prof to prof , 56-82, EMER PROF CHEM, UNIV WESTERN ONT, 82- *Mem:* Electrochem Soc; Chem Inst Can. *Res:* Thermodynamics of solutions; physicochemical properties of electrolyte systems. *Mailing Add:* Dept Chem Univ Western Ont London ON N6A 5B7 Can

SUKER, JACOB ROBERT, b Chicago, Ill, Oct 17, 26; m 56; c 3. INTERNAL MEDICINE. *Educ:* Northwestern Univ, Evanston, BS, 47; Northwestern Univ, Chicago, MS, 54, MD, 56. *Prof Exp:* Dir med educ, Chicago Wesley Mem Hosp, 64-68; asst dean grad educ, 68-70, ASSOC PROF MED, MED SCH, NORTHWESTERN UNIV, CHICAGO, 68-, ASSOC DEAN GRAD EDUC, 70- *Concurrent Pos:* Assoc attend physician, Chicago Wesley Mem Hosp, 64- *Res:* Parathyroid disease. *Mailing Add:* Dept Med Northwestern Univ Med Sch Chicago IL 60611

SUKHATME, BALKRISHNA VASUDEO, b Poona, India, Nov 3, 24; m 56; c 1. STATISTICS. *Educ:* Univ Delhi, BA, 45, MA, 47; Inst Agr Res Statist, New Delhi, dipl, 49; Univ Calif, Berkeley, PhD(statist), 55. *Prof Exp:* Sr res statistician, Indian Coun Agr Res, 55-58, prof statist, 58-65, dep statist adv, 62-63, sr prof statist, 65-67; assoc prof, Iowa State Univ, 67-68, prof statist, 68-80. *Concurrent Pos:* Vis assoc prof statist, Mich State Univ, 59-60; ed jour, Indian Soc Agr Statist, 59-67; consult, FAO, Rome, 65; mem, Int Statist Inst, The Hague, 72- *Mem:* Inst Math Statist; fel Am Statist Asn; Int Asn Surv Statisticians; Indian Soc Agr Statist (joint secy, 56-58). *Res:* Sampling theory and its applications; nonparametric tests for scale and randomness and asymptotic theory of order statistics and generalized U-statistics; planning, organization and conduct of large-scale sample surveys. *Mailing Add:* 1505 Wheeler Ames IA 50010

SUKHATME, SHASHIKALA BALKRISHNA, b Karad, Maharashtra, India; c 1. MATHEMATICAL STATISTICS. *Educ:* Univ Poona, BSc, 53, Hons, 54, MSc, 55; Mich State Univ, PhD(statist), 60. *Prof Exp:* Lectr statist, Univ Delhi, 63-67; asst prof, 67-83, ASSOC PROF STATIST, IOWA STATE UNIV, 83- *Concurrent Pos:* Univ Grants Comn, India Fel, Univ Delhi, 61-63; Daxina fel, Poona Univ. *Mem:* Inst Math Statist; Am Statist Asn; Indian Statist Asn. *Res:* Nonparametric statistical theory; goodness of fit tests; order statistics. *Mailing Add:* Statist 314 Snedecor Iowa State Univ Ames IA 50011

SUKHATME, UDAY PANDURANG, b Pune, India, June 16, 45; m 69; c 2. ELEMENTARY PARTICLE PHYSICS. *Educ:* Univ Delhi, India, BSc, 64; Mass Inst Technol, Cambridge, SB, 66, ScD(physics), 71. *Prof Exp:* Teaching fel physics, Univ Wash, Seattle, 71-73, Univ Mich, Ann Arbor, 73-75, Univ Cambridge, UK, 75-77 & Univ Paris, Orsay, France, 77-79; from asst prof to assoc prof, 80-89, PROF PHYSICS, UNIV ILL, CHICAGO, 89- *Concurrent Pos:* Vis assoc prof, Iowa State Univ, Ames, 79-80. *Res:* Theoretical high energy research with emphasis on the phenomenology of strongly interacting particles and models for multiparticle production. *Mailing Add:* Dept Physics M/C 273 Univ Ill Chicago IL 60680

SUKI, WADI NAGIB, b Khartoum, Sudan, Oct 26, 34; US citizen; m 85; c 4. INTERNAL MEDICINE, NEPHROLOGY. *Educ:* Am Univ Beirut, BS, 55, MD, 59. *Prof Exp:* Resident internal med, Parkland Mem Hosp, Dallas, 61-63; from instr to assoc prof, 65-71, PROF MED, BAYLOR COL MED, 71-, CHIEF RENAL SECT, 68-, PROF PHYSIOL, 82- *Concurrent Pos:* Res fel exp med, Univ Tex Southwestern Med Sch Dallas, 59-61, USPHS res fel nephrology, 63-65, Dallas Heart Asn res grant, 67-68; Nat Heart & Lung Inst res grant, Baylor Col Med, 68-72, training grant, 71-76, Nat Inst Arthritis, Diabetes, Digestive & Kidney Dis res grant, 74-88, Nat Inst Allergy & Infectious Dis contract, 74-78, NASA contract, 75-88; attend physician, Ben Taub Gen Hosp, 68-; consult, Vet Admin Hosp, 68- & Wilford Hall, USAF Med Ctr, 72-; chief renal sect, Methodist Hosp, 69-; pres med adv bd, Kidney Found Houston & Greater Gulf Coast, 69-71; chmn nat med adv coun, Nat Kidney Found, 71-73, trustee-at-large, 71-76, secy sci adv bd, 77-78, chmn, 79-80; mem exec comt, Coun Kidney in Cardiovasc Dis, Am Heart Asn, 71-74; mem rev bd nephrology, Vet Admin Cent Off Med Res Serv, 74-77, chmn, 76-77; mem gen med B study sect, NIH, 75-79 & 81-85, chmn, 83-85; mem, Nephrology Comt, Am Bd Internal Med, 82-88. *Mem:* Am Fedn Clin Res; Am Soc Clin Invest; Int Soc Nephrology; fel Am Col Physicians; Am Soc Nephrology; Renal Physicians Asn; Asn Am Physicians; Am Physiol Soc. *Res:* Renal, fluid and electrolyte physiology and pathophysiology; renal disease, dialysis and transplantation. *Mailing Add:* Dept Med Baylor Col Med 6550 Fannin No 1275 Houston TX 77030

SUKOW, WAYNE WILLIAM, b Merrill, Wis, Dec 9, 36; m 59; c 2. MOLECULAR BIOPHYSICS, BIOPHYSICAL CHEMISTRY. *Educ:* Univ Wis-River Falls, BA, 59; Case Inst Technol, MS, 63; Wash State Univ, PhD(chem physics), 74. *Prof Exp:* From assoc prof to prof physics, Univ Wis-River Falls, 61-84, chmn dept, 77-84; exec dir, W Cent Wis Consortium, 84-88; PROG DIR, DIV TEACHER PREP & ENHANCEMENT, NSF, 88- *Concurrent Pos:* Vis prof physics, Macalester Col, 67, vis assoc prof chem, Univ Ore, 75 & 76; physicist, 3M Co, 67; NSF sci fac fel, Wash State Univ, 70-72; mem, Instrnl Media Comt, Am Asn Physics Teachers, 82-88; sci educ teacher in-serv presenter, (K-12), 82-91. *Mem:* Biophys Soc; Am Asn Physics Teachers. *Res:* Protein-ligand binding, particularly the mechanism of detergent binding to membrane proteins; conformational changes of proteins monitored by electron paramagnetic resonance using spin probe molecules; photoelectron microscopy of biological materials; improvement of undergraduate science education using innovative and interdisciplinary approaches; formation of arborescent copper inclusions in Lake Superior agates and Michigan Datolite; science education using interactive science exhibits. *Mailing Add:* Div Teacher Prep & Enhancement NSF 1800 G St NW Washington DC 20550

SUKOWSKI, ERNEST JOHN, b Chicago, Ill, Nov 17, 32; c 4. PHYSIOLOGY, PHARMACOLOGY. *Educ:* Loyola Univ Chicago, BS, 54; Univ Ill, MS, 58, PhD(physiol, pharmacol), 62. *Prof Exp:* ASSOC PROF PHYSIOL, UNIV HEALTH SCI-CHICAGO MED SCH, 63- *Mem:* NY Acad Sci; Am Physiol Soc; Am Heart Asn; Am Asn Univ Professors; Sigma Xi. *Res:* Cardiac and liver metabolism; cardiovascular physiology; hypertension; sub-cellular physiology; renal physiology. *Mailing Add:* The Chicago Med Sch 3333 Green Bay Rd North Chicago IL 60064

SULAK, LAWRENCE RICHARD, b Columbus, Ohio, Aug 29, 44; m 70; c 2. EXPERIMENTAL ELEMENTARY PARTICLE PHYSICS. *Educ:* Carnegie-Mellon Univ, BS, 66; Princeton Univ, AM, 68, PhD(physics), 70. *Prof Exp:* Res physicist, Univ Geneva, 70-71; from asst prof to assoc prof physics, Harvard Univ, 71-79; from assoc prof to prof physics, Univ Mich, 79-85; PROF PHYSICS & DEPT CHMN, BOSTON UNIV, 85- *Concurrent Pos:* Vis scientist, Europ Orgn Nuclear Res, 70-73; vis physicist, Fermi Nat Accelerator Lab, 71-77; guest assoc physicist, Brookhaven Nat Lab, 74-; vis prof, Harvard Univ, 84-85; prin investr, US Dept Energy contract, 79-, proj dir, 85-; mem, Dept Energy High Energy Physics Adv panel, 87. *Mem:* Am Phys Soc; AAAS. *Res:* Experimental K-meson physics, elementary particle production studies, neutrino physics, studies of deep inelastic scattering, scaling, neutral currents, dimuons and elastic neutrino-proton scattering; proton decay experiments; instrumentation for high energy physics; acoustic signals from particle beams, astrophysical sources of neutrinos, g-2 of moon; precision superconducting storage ring. *Mailing Add:* Physics Dept Boston Univ 590 Commonwealth Ave Boston MA 02215

SULAKHE, PRAKASH VINAYAK, b Nov 18, 41; Indian citizen; m 73. PHYSIOLOGY. *Educ:* Bombay Univ, BS, 62, MS, 65; Univ Man, PhD(physiol), 71. *Prof Exp:* Lectr physiol, Topiwala Nat Med Col, India, 66; sci officer med div, Bhaha Atomic Res Ctr, India, 66-67; demonstr & teaching fel physiol, Univ Man, 68-71; Med Res Coun Can fel pharmacol, Univ BC, 71-73; Med Res Coun Can res prof, 77-78, asst prof, 73-76, assoc prof, 76-80, PROF PHYSIOL, UNIV SASK, 80- *Mem:* Can Physiol Soc; Int Soc Heart Res; NY Acad Sci; Soc Neurosci; Can Biochem Soc. *Res:* Regulation and metabolism and function of contractile tissues and brain; cyclic nucleotides, calcium ions, autonomic receptors and membranes. *Mailing Add:* Dept of Physiol Univ of Sask Saskatoon SK S7N 0W0 Can

SULAVIK, STEPHEN B, b New Britain, Conn, Aug 11, 30; m 55; c 8. MEDICINE. *Educ:* Providence Col, BS, 52; Georgetown Univ, MD, 56. *Prof Exp:* Asst chief chest dis, Vet Admin Hosp, Bronx, NY, 61-62; clin instr, 62-63, from instr to asst prof, 63-69, assoc clin prof med, Sch Med, 69-78, CLIN

PROF MED, YALE UNIV SCH MED, 78-; PROF & HEAD DIV PULMONARY MED, UNIV CONN, 77- *Concurrent Pos:* Mem med adv comt, Dept HEW; assoc prof med & actg head pulmonary div, Univ Conn, 69-77; chmn dept med, St Francis Hosp, Hartford, 69-77. *Mem:* Am Thoracic Soc; AMA. *Res:* Anatomy and physiology of intrathoracic lymphatic system. *Mailing Add:* Div Pulmonary Med Univ Conn Sch Med Farmington CT 06032

SULENTIC, JACK WILLIAM, b Waterloo, Iowa, Apr 10, 47; m 75. ASTRONOMY, OPTICAL ASTRONOMY. *Educ:* Univ Ariz, BS, 69; State Univ NY Albany, PhD(astron), 75. *Prof Exp:* Fel astron, Hale Observ, 75-78; instr physics & astron, Sierra Nev Col, 78-79; asst prof, Mich State Univ, 79-80; from asst prof to assoc prof, 80-88, PROF, DEPT PHYSICS & ASTRON, UNIV ALA, 88- *Concurrent Pos:* Vis prof, Univ Padova, Italy, 87. *Mem:* Am Astron Soc; Sigma Xi; Int Astron Union. *Res:* Application of optical and radio observations to understanding the origin and evolution of galaxies. *Mailing Add:* Dept Physics & Astron Univ Ala Tuscaloosa AL 35487-0324

SULERUD, RALPH L, b Fargo, NDak, June 6, 32. GENETICS, ZOOLOGY. *Educ:* Concordia Col, Moorhead, Minn, BA, 54; Univ Nebr, MS, 58, PhD(zool), 68. *Prof Exp:* Instr biol, St Olaf Col, 58-59; from instr to assoc prof, 64-77, PROF BIOL, AUGSBURG COL, 77-, CHMN DEPT, 74- *Mem:* AAAS; Genetics Soc Am; Am Genetic Asn; Am Inst Biol Sci; Soc Study Evolution. *Res:* Taxonomy; genetics of Drosophila. *Mailing Add:* Dept of Biol Augsburg Col 731 21st Ave S Minneapolis MN 55404

SULEWSKI, PAUL ERIC, b Hempstead, NY, Nov 16, 60; m 85. RAMAN & IR SPECTROSCOPY, OPTICAL PROPERTIES OF NOVEL MATERIALS. *Educ:* Princeton Univ, AB, 82; Cornell Univ, MS, 85, PhD(physics), 88. *Prof Exp:* Postdoctorate physics, 88-90, MEM TECH STAFF PHYSICS, AT&T BELL LABS, 90- *Mem:* Am Phys Soc. *Res:* Condensed matter physics, focusing on spectroscopy of novel materials. *Mailing Add:* AT&T Bell Labs Rm 1D-154 600 Mountain Ave Murray Hill NJ 07974-2070

SULIK, KATHLEEN KAY, b Estherville, Iowa, Oct 15, 48; m 77; c 1. TERATOLOGY. *Educ:* Drake Univ, BS, 70; Univ Tenn, PhD(anat), 76. *Prof Exp:* Guest scientist, Gerontol Res Ctr, NIH, 74-75; instr anat, Univ Tenn, 75-76; fel teratol, Dent Res Ctr, Univ NC, 76-78; res assoc, Georgetown Univ, Wash, DC, 78-79, asst prof, 79-80; ASST PROF ANAT, UNIV NC, CHAPEL HILL, 80- *Mem:* Teratology Soc; AAAS. *Res:* Embryology and teratology of the craniofacial region emphasing the teratogenic effect of ethanol, defining critical exposure periods and mechanisms of malformation. *Mailing Add:* Dept Anat Swing Bldg Univ NC Chapel Hill NC 27514

SULING, WILLIAM JOHN, b New York, NY, June 12, 40; m 65; c 3. ANTIMICROBIAL & ANTICANCER CHEMOTHERAPY. *Educ:* Manhattan Col, BS, 62; Duquesne Univ, MS, 65; Cornell Univ Med Col, PhD(microbiol), 75. *Prof Exp:* Res asst cancer chemother, Sloan-Kettering Inst Cancer Res, 64-70; sr bacteriologist, Biol Lab, Mass Dept Pub Health, 74-75; res microbiologist, 75-78, sr microbiologist, 78-86, HEAD, BACT-MYCOL SECT, SOUTHERN RES INST, 86- *Mem:* NY Acad Sci; Am Soc Microbiol; Soc Indust Microbiol. *Res:* Folate metabolism and its inhibition by folate analogues; the biochemistry of antimicrobial drug resistance and the use of microorganisms for studies involving cancer chemotherapy; drug metabolism and disposition; mechanisms of action of anticancer agents. *Mailing Add:* Southern Res Inst PO Box 55305 Birmingham AL 35255-5305

SULKOWSKI, EUGENE, b Plonsk, Poland, May 22, 34; m 64; c 2. BIOCHEMISTRY. *Educ:* Univ Warsaw, MS, 56, PhD(biochem), 60. *Prof Exp:* Res asst biochem, Inst Biochem & Biophys, Polish Acad Sci, 56-60; exchange scientist, Univ Sorbonne, 62-63; res asst, Polish Acad Sci, 63-65; PRIN CANCER RES SCIENTIST, ROSWELL PARK MEM INST, 65- *Concurrent Pos:* Res fel, Marquette Univ, 60-62. *Mem:* AAAS; Am Soc Biol Chemists; Polish Biochem Soc. *Res:* Enzymology; human interferon. *Mailing Add:* Dept Cel & Molecular Biol Roswell Park Mem Inst 666 Elm St Buffalo NY 14263

SULLENGER, DON BRUCE, b Richmond, Mo, Feb 8, 29; m 64; c 3. SOLID STATE CHEMISTRY. *Educ:* Univ Colo, AB, 50; Cornell Univ, PhD, 69. *Prof Exp:* Trainee Chemet Prog, Chem Div, Gen Elec Co, 50-51 & 53, res chemist, Res Lab, 53-54; sr res chemist, Monsanto Res Corp, 62-74, res specialist, 74-85, sr res specialist, Mound Lab, 85-88; SR RES SPECIALIST, EG&G MOUND APPL TECHNOLOGIES, 88- *Mem:* AAAS; Am Chem Soc; Am Crystallog Asn. *Res:* X-ray crystallographic structural characterization of inorganic and organic substances; materials science; solid state chemistry of inorganic materials. *Mailing Add:* EG&G Mound Appl Technol Mound Lab Miamisburg OH 45342

SULLIVAN, ALFRED DEWITT, b New Orleans, La, Feb 2, 42; m 62; c 2. FORESTRY, BIOMETRICS. *Educ:* La State Univ, BS, 64, MS, 66; Univ Ga, PhD(forest biomet), 69. *Prof Exp:* Asst prof statist, Va Polytech Inst & State Univ, 69-73; from assoc prof to prof forestry, Miss State Univ, 78-88; dir, sch forest resources, Pa State Univ, 88- *Mem:* Soc Am Foresters. *Res:* Prediction of forest growth and yield; application of statistical methodology to natural resource problems. *Mailing Add:* 878 W Aaron Dr State College PA 16803

SULLIVAN, ANDREW JACKSON, b Birmingham, Ala, Mar 3, 26; m 53. BIOCHEMISTRY, FOOD CHEMISTRY. *Educ:* Univ Richmond, BS, 47; Univ Mo, PhD(bot), 52. *Prof Exp:* USPHS fel, Univ Pa, 52-53; res chemist, Campbell Soup Co, 53-55 & 57-59, head div flavor biochem res, 59-71, div head environ sci & chem technol, Campbell Inst for Food Res, 71-76, dir sci resources, 76-77, dir flavor sci & nutrit, 77-87, DIR, TECHNOL ASSESSMENT & ACQUISITION, CAMPBELL INST RES & TECHNOL, 87- *Mem:* Inst Food Technologists; NY Acad Sci; fel Am Inst Chemists; Am Chem Soc; Asn Chemoreception Sci. *Res:* Chemistry of microorganisms; food and flavor chemistry. *Mailing Add:* Campbell Inst Res & Technol Campbell Place Camden NJ 08103-1799

SULLIVAN, ANNA MANNEVILLETTE, b Washington, DC, Aug 18, 13. METALLURGY. *Educ:* George Washington Univ, AB, 35; Univ Md, MS, 55. *Prof Exp:* Asst metallurgist, Geophys Lab, Carnegie Inst Wash, 42-45; metallurgist, Nat Bur Stand, 45-46; metallurgist, Naval Res Lab, 47-78; dep tech ed, J Eng Mat Technol, Am Soc Mech Engrs, 78-81; RETIRED. *Concurrent Pos:* Consultant; Naval Ord Develop Award. *Honors & Awards:* Burgess Award, Am Soc Metals, 78. *Mem:* Sigma Xi; Am Soc Metals; Am Soc Testing & Mat; Am Soc Mech Engrs. *Res:* Fracture of metals with special reference to fracture mechanics. *Mailing Add:* 4000 Massachusetts Ave Washington DC 20016

SULLIVAN, ARTHUR LYON, b Atlanta, Ga, May 29, 40; m 62; c 2. EDUCATION ADMINISTRATION. *Educ:* Univ NH, BA, 63, MS, 66; Cornell Univ, PhD(natural resources), 69. *Prof Exp:* Asst prof regional planning, Univ Pa, 69-74; asst prof environ mgt, Duke Univ, 74-76; assoc prof, 76-81, PROF LANDSCAPE ARCHIT, NC STATE UNIV, 81-, HEAD DEPT, 78- *Concurrent Pos:* Mem, res bd, Regional Sci Res Inst, 72-74; assoc ed, Environ Prof, 79-81; vis prof, Kyoto Univ, Japan, 81; fel, NC Japan Ctr, 81- *Mem:* AAAS; Am Soc Landscape Archit; Coun Educr Landscape Archit; Asian Studies Soc. *Res:* Biogeography of urban development; Japan United States comparative land-use systems; agricultural alternatives. *Mailing Add:* Dept Design & Forestry NC State Univ Raleigh NC 27695

SULLIVAN, BETTY J, b Minneapolis, Minn, May 31, 02. PROTEIN CHEMISTRY, PROTEOLYTIC ENZYMES. *Educ:* Univ Minn, BS, 22, PhD(biochem), 35. *Prof Exp:* Lab asst, Russell Miller Milling Co, 22-24, 26-27, chief chemist, 27-47, dir res, 47-58; vpres & dir res, Peavey Co, 58-67; vpres, 67-69, pres, 69-73, chmn bd, 73-75, DIR, EXPERIENCE, INC, 75- *Concurrent Pos:* Scholar biochem, Univ Paris, 24-25. *Honors & Awards:* Thomas Burr Osborne Medal, Am Asn Cereal Chemists, 48; Garvan Medal, Am Chem Soc, 54. *Mem:* Am Chem Soc; Am Asn Cereal Chemists (pres, 43-44); Sigma Xi; AAAS. *Res:* Chemistry of wheat gluten; sulfhydryl and disulfide groups and interchange as related to function. *Mailing Add:* 8441 Irwin Rd Minneapolis MN 55437

SULLIVAN, BRIAN PATRICK, b Brookings, SDak, Oct 10, 49; m 71; c 2. ELECTROCHEMISTRY, CATALYSIS. *Educ:* Univ Calif, Irvine, BS, 71, Univ Calif, Los Angeles, MS, 86; Univ NC, Chapel Hill, PhD(chem), 88. *Prof Exp:* Res assoc chem, 75-88, res asst prof chem, Univ NC, Chapel Hill, 88-90; ASSOC PROF CHEM, UNIV WYO, 90- *Concurrent Pos:* Consult, Allied Chem Co, 81-82, Dept Pharm, Univ NC, 85-86, Igen Corp, 87-88; vis assoc chem, Calif Inst Technol, 89. *Mem:* Am Chem Soc. *Res:* Areas of novel inorganic synthesis, reactivity of coordinated ligands and small molecule transition metal complexes; solvent effects on electron and atom transfer reactions, photo chemistry of inorganic molecules and inorganic surface chemistry; the reduction of carbon dioxide to fuels; author of over 70 publications; photochemistry. *Mailing Add:* Dept Chem Univ Wyo Laramie WY 80071-3838

SULLIVAN, CHARLES HENRY, b Needham, Mass, June 25, 52; m 79; c 2. DEVELOPMENTAL BIOLOGY. *Educ:* Univ Maine, Orono, BA, 74; Univ Md, MS, 79, PhD(zool), 83. *Prof Exp:* Teaching asst zool, Univ Md, 75-82; postdoctoral fel develop biol, Univ Va, 83-86; ASST PROF BIOL, GRINNELL COL, 86- *Concurrent Pos:* Prin investr grants, Grinnell Col, Res Corp, 87-89, NIH, 88-90, NSF, 91- *Mem:* Soc Develop Biol; Am Soc Zoologists; Coun Undergrad Res. *Res:* Developmental and cellular biology of invertebrate and vertebrate embryos with emphasis on cytoplasmic determinants of cell differentiation, cell and tissue interactions in lens development and factors controlling tissue- or age-specific gene expression during develpment. *Mailing Add:* Dept Biol Grinnell Col Grinnell IA 50112

SULLIVAN, CHARLES IRVING, b Milwaukee, Wis, Nov 18, 18; m 48; c 2. ORGANIC POLYMER CHEMISTRY. *Educ:* Boston Univ, AB, 43. *Prof Exp:* Chemist, UBS Chem Co Div, A E Staley Mfg Co, 43-46, supvr indust chem & develop sect, 46-58, mgr res & develop, 58-67, res assoc, 67-69; sr scientist, 69-70, res assoc, 70-79, RES FEL, POLAROID CORP, 79- *Mem:* AAAS; Am Chem Soc; fel Am Inst Chemists. *Res:* Emulsion polymerization; paints; floor finishes; paper coatings and binders; wood coatings; adhesives; textile backings; aqueous polymer research and development related to membrane-like structures, functional coatings, binders and colloids. *Mailing Add:* 148 Bellevue Ave Melrose MA 02176

SULLIVAN, CHARLOTTE MURDOCH, b St Stephen, NB, Dec 18, 19. ANIMAL PHYSIOLOGY. *Educ:* Dalhousie Univ, BSc, 41, MSc, 43; Univ Toronto, PhD(zool), 49. *Prof Exp:* Nat Res Coun Can overseas fel, Cambridge Univ, 49-50; from lectr to asst prof, 50-61, ASSOC PROF ZOOL, UNIV TORONTO, 61- *Res:* Physiology of animal behavior. *Mailing Add:* Ten Avoca Ave Toronto ON M4T 2B7 Can

SULLIVAN, CORNELIUS PATRICK, JR, b Schenectady, NY, Sept 28, 29; m 55; c 3. METALLURGY. *Educ:* Univ Notre Dame, BS, 51; Mass Inst Technol, SM, 55, ScD(metall), 60. *Prof Exp:* ENGR METALL, PRATT & WHITNEY, 62- *Honors & Awards:* William A Spraragen Award, Am Welding Soc, 67. *Mem:* Am Soc Metals Int. *Res:* Mechanical behavior of high temperature alloys and their properties. *Mailing Add:* 74 Coach Rd Glastonbury CT 06033

SULLIVAN, DANIEL JOSEPH, b New York, NY, Apr 22, 28. ENTOMOLOGY, ANIMAL BEHAVIOR. *Educ:* Fordham Univ, BS, 50, MS, 58; Univ Vienna, cert Ger, 58; Univ Strasbourg, cert French, 62; Univ Innsbruck, cert theol, 62; Univ Calif, Berkeley, PhD(entom), 69. *Prof Exp:* Teacher, NY High Sch, 55-57; from asst prof to assoc prof, 69-83, PROF ZOOL, FORDHAM UNIV, 83- *Concurrent Pos:* Fulbright res fel, Nigeria, WAfrica, 84-85, Int Inst Tropical Agr fac fel, Cabi, Colombia, 88-89. *Mem:* AAAS; Entom Soc Am; Ecol Soc Am; Animal Behav Soc; Am Inst Biol Sci; Sigma Xi. *Res:* Biological control of insect pests, with special reference to the primary parasites and hyperparasites of aphids; ecology and behavior of aphids and parasites. *Mailing Add:* Dept Biol Sci Fordham Univ Bronx NY 10458

SULLIVAN, DAVID ANTHONY, ENDOCRINOLOGY, OPHTHALMOLOGY. *Educ:* Dartmouth Med Sch, PhD(physiol), 80. *Prof Exp:* ASST SCIENTIST, EYE RES INST, HARVARD MED SCH, 83- *Mailing Add:* Dept Ophthal Harvard Med Sch 20 Staniford St Boston MA 02114

SULLIVAN, DAVID THOMAS, b Salem, Mass, Mar 20, 40; m 66; c 2. BIOCHEMICAL GENETICS. *Educ:* Boston Col, BS, 61, MS, 63; Johns Hopkins Univ, PhD(biol), 67. *Prof Exp:* USPHS res fel biochem, Calif Inst Technol, 67-69; from asst prof to assoc prof, 70-81, PROF BIOL, SYRACUSE UNIV, 81- *Mem:* Genetics Soc Am; Soc Develop Biol. *Res:* Eukaryotic gene structure, evolution and expression. *Mailing Add:* Dept Biol Syracuse Univ Syracuse NY 13224-1220

SULLIVAN, DENNIS P, b Port Huron, Mich, Feb 12, 41. MATHEMATICS. *Educ:* Rice Univ, BA, 63; Princeton Univ, PhD, 65. *Prof Exp:* NATO fel, Warwick Univ, Eng, 66; from lectr to assoc prof, Princeton Univ, 67-69; Sloan fel math, Mass Inst Technol, 69-72, prof, 72-73; EINSTEIN PROF SCI, QUEENS COL & CITY UNIV NEW YORK GRAD SCH, 81- *Concurrent Pos:* Miller fel, Berkeley, 67-69; assoc prof, Univ Paris, Orsay, 73-74; Stanislaw Ulam vis prof math, Univ Colo, Boulder, 80-81. *Honors & Awards:* Oswald Veblen Prize in Geom, Am Math Soc, 71; Elie Cartan Priz en Geom, French Acad Sci, 81. *Mem:* Nat Acad Sci; Am Math Soc (vpres, 89-); fel AAAS; corresp mem Nat Acad Sci Brazil. *Mailing Add:* Dept Math Grad Ctr City Univ New York 33 W 42nd St New York NY 10036

SULLIVAN, DONALD, b Merthyr Tydfil, Wales, Mar 23, 36; m 61; c 2. MATHEMATICS. *Educ:* Univ Wales, BSc, 57, PhD(appl math), 60. *Prof Exp:* Asst lectr math, Univ Col, Univ Wales, 60-61; asst prof, 61-66, ASSOC PROF MATH, UNIV NB, 66- *Mem:* fel Inst Math & Appln. *Res:* Fluid mechanics; phase plane analysis of differential equations; functional equations; integral equations. *Mailing Add:* Dept of Math Univ of NB Col Hill Box 4400 Fredericton NB E3B 5A3 Can

SULLIVAN, DONALD BARRETT, b Phoenix, Ariz, June 13, 39; m 59; c 3. LOW TEMPERATURE PHYSICS, ATOMIC PHYSICS. *Educ:* Tex Western Col, BS, 61; Vanderbilt Univ, MA, 63, PhD(physics), 65. *Prof Exp:* Res assoc physics, Vanderbilt Univ, 65; physicist & br chief, Radiation Physics Br, US Army Nuclear Defense Lab, 65-67; Nat Res Coun assoc, 67-69, physicist & chief cryoelectronic metrol sect, 69-84, CHIEF, TIME & FREQUENCY DIV, NAT BUR STANDARDS, 84- *Honors & Awards:* Stratton Award, Nat Bur Standards, 85. *Mem:* Am Phys Soc. *Res:* Josephson effect and quantum interference in superconductors; development of measurement instruments using these and other low temperature phenomena; development of atomic clocks and methods for clock synchronization. *Mailing Add:* 3594 Kirkwood Pl Boulder CO 80302

SULLIVAN, DONITA B, b Marlette, Mich, Feb 11, 31. PEDIATRICS. *Educ:* Siena Heights Col, BS, 52; St Louis Univ, MD, 56; Am Bd Pediat, dipl, 61. *Hon Degrees:* DHH, Siena Heights Col, 80. *Prof Exp:* Intern, Henry Ford Hosp, Detroit, Mich, 56-57; resident pediat, Children's Hosp of Mich, 57-59, sr resident, 59; res assoc, Sch Med, Wayne State Univ, 59; clin instr, 59-62, asst prof pediat & dir birth defects treatment ctr, 62-69, assoc prof, 69-77, PROF PEDIAT & DIR PEDIAT REHAB & RHEUMATOLOGY SECT, MED SCH, UNIV MICH, ANN ARBOR, 77-, PROF PEDIAT & DIR PEDIAT RHEUMATOLOGY, 79-, ASSOC CHMN PEDIAT DEPT & DIR PEDIAT EDUC, 81- *Concurrent Pos:* Pediat consult, Wayne County Gen Hosp, 62-80, Field Clins, Mich Crippled Children's Comn, 64-68 & Cath Social Servs, 66-80; mem med adv comt, Washtenaw County Chapters, Nat Found & Nat Cystic Fibrosis Res Found, 60-74; prog consult, Nat Found, 66-69; bd trustees, Siena Heights Col, 70-75. *Res:* Handicapped children; children with birth defects; clinical and immunologic aspects of connective tissue disease in children. *Mailing Add:* Dept Pediat Univ Mich Med Ctr Ann Arbor MI 48109

SULLIVAN, EDWARD AUGUSTINE, b Salem, Mass, July 5, 29; m 59; c 6. INORGANIC CHEMISTRY. *Educ:* Col Holy Cross, BS, 50; Mass Inst Technol, MS, 52. *Prof Exp:* Asst, Sugar Res Found, 50-52; res chemist, Metal Hhydrides, Inc, 52-63, sr res chemist, Metal Hydrides Div, Ventron Corp, 63-71, tech mgr, res chem, Ventron Corp, 71-79, SR SCIENTIST, MORTON INTERNATIONAL, 79- *Mem:* Am Chem Soc; Tech Asn Pulp & Paper Indust. *Res:* Chemistry of hydrides; inorganic synthesis; industrial applications of hydrides. *Mailing Add:* 43 Brimbal Ave Beverly MA 01915

SULLIVAN, EDWARD FRANCIS, b Portland, Maine, Sept 16, 20; m 48; c 4. AGRONOMY. *Educ:* Univ Maine, BS, 49; Cornell Univ, MSA, 51, PhD(agron), 53. *Prof Exp:* Asst agron, Cornell Univ, 50-53; asst prof, Southern Ill Univ, 53-56 & Pa State Univ, 56-61; from agronomist to sr agronomist, 61-75, MGR CROP ESTAB & PROTECTION, AGR RES CTR, GREAT WESTERN SUGAR CO, 75- *Concurrent Pos:* Asst prof, Univ Ill, 54-56. *Mem:* Am Soc Agron; Weed Sci Soc Am; Am Soc Sugar Beet Technologists. *Res:* Weed control; plant growth regulators; crop production. *Mailing Add:* Great Western Agr Res Ctr 11939 Sugarmill Rd Longmont CO 80501

SULLIVAN, EDWARD T, b Flushing, NY, June 28, 20; m 54; c 1. FOREST ECONOMICS. *Educ:* NC State Col, BSF, 46; Duke Univ, MS, 47, DF, 53. *Prof Exp:* Acct, southern woodlands dept, WVa Pulp & Paper Co, 47-50; vis instr forest econ, sch forestry, Duke Univ, 50-51; asst prof forestry, Univ Minn, 54-59; ASSOC PROF FOREST ECON, UNIV FLA, 59-, ASSOC FORESTER, 71- *Mem:* Am Econ Asn; Soc Am Foresters. *Res:* Marketing of forest products; demand for pulpwood. *Mailing Add:* 755 NW 18th St Gainesville FL 32603

SULLIVAN, F(REDERICK) W(ILLIAM), III, b Ann Arbor, Mich, June 24, 23; m 48; c 4. CHEMICAL ENGINEERING. *Educ:* Pa State Univ, BS, 47; Univ Del, MS, 49, PhD(chem eng), 52. *Prof Exp:* Chem engr process develop, Houdry Process Corp, 51-56; chem engr process develop & design, Halby Chem Co, 56-63; dir process eng, Maumee Chem Co, 63-70; mgr process eng sect, Chem Div, Sherwin Williams Co, 70-78, mgr process develop, 78-80; RETIRED. *Mem:* Am Chem Soc; Inst Chem Engrs. *Res:* Petroleum refining; heat transfer; manufacture of inorganic and organic chemicals. *Mailing Add:* 208 Sean Way Hendersonville NC 28792

SULLIVAN, FRANCIS E, b Brooklyn, NY, May 12, 41; m 69. FAST ALGORITHMS, NONLINEAR DYNAMICS. *Educ:* Pa State Univ, BS, 62; Univ Pittsburgh, PhD(math), 68. *Prof Exp:* Chmn, Catholic Univ, 73-82; DIR ADMIN, COMPUT & APPL MATH LAB, NAT INST STANDARDS & TECHNOL, 86- *Concurrent Pos:* Adj prof, Catholic Univ, 82-; Sci policy bd, Soc Indust & Appl Math. *Honors & Awards:* Gold Medal, Dept Com, 88. *Mem:* Soc Indust Appl Math. *Res:* Supercomputer algorithms; parallel computing nonlinear analysis. *Mailing Add:* Comput & Appl Math Lab Nat Inst Standards & Technol Gaithersburg MD 20899

SULLIVAN, GEORGE ALLEN, b Bronxville, NY, Dec 1, 35; m 60; c 2. INFORMATION SCIENCE, STATISTICS. *Educ:* Grinnell Col, AB, 57; Univ Rochester, AM, 59; Univ Nebr, PhD(solid state physics), 64; Univ Pa, MGovt Admin, 76. *Prof Exp:* Res asst solid state physics, Rensselaer Polytech Inst, 64-66; sr physicist, Electronic Res Div, Clevite Corp, Cleveland, 66-69; asst prof elec eng, Air Force Inst Technol, Wright Patterson AFB, 69-70; vis scientist, Physics Inst, Chalmers Univ Technol, Sweden, 70-71; criminal justice systs planner & eval coordr, Pa Bd Probation & Parole, 71-74, actg dir, 74-75, sr statist anal, 75-80, sr statist anal, res unit, 80-89, CHIEF RES, PA BD PROBATION & PAROLE, 89- *Mem:* Am Statist Asn; Nat Speleol Soc; Sigma Xi. *Res:* Point defects and diffusion in metals; thermal mass transport and electromigration in solids; semiconductors; properties of semiconducting materials; physics of solar cells; criminal justice information statistics; social rehabilitation programs; criminological research; operations and program planning; multivariate statistical analysis with software packages; design of parole decision making guidelines instruments; risk classification instruments. *Mailing Add:* Res Unit Box 1661 3101 N Front St Harrisburg PA 17105-1661

SULLIVAN, GERALD, pharmacy, pharmacognosy, for more information see previous edition

SULLIVAN, HARRY MORTON, b Winnipeg, Man, Apr 14, 21; m 49; c 3. PHYSICS. *Educ:* Queen's Univ, Ont, BSc, 45; Carleton Univ, BSc, 50; McGill Univ, MSc, 54; Univ Sask, PhD(upper atmosphere physics), 62. *Prof Exp:* Chemist, Can Civil Serv, 45-47; engr, Canadair Ltd, 54-56; physicist, Can Civil Serv, 56-59; physicist, Nat Ctr Sci Res, France, 62-64; asst prof, 64-69, ASSOC PROF PHYSICS, UNIV VICTORIA, 69- *Mem:* Can Asn Physicists. *Res:* Upper atmosphere; airglow and related phenomena, particularly twilight glow and day glow; rarer constituents of upper atmosphere; photometer calibration techniques; standard radiation sources. *Mailing Add:* Dept Physics Univ Victoria Box 1700 Victoria BC V8W 2Y2 Can

SULLIVAN, HERBERT J, b Ebbw Vale, Gt Brit, May 20, 33; m 58; c 3. GEOLOGY, PALYNOLOGY. *Educ:* Univ Sheffield, BSc, 54, PhD(geol), 59. *Prof Exp:* Dept Sci & Indust Res fel, 59-61; from asst lectr to lectr geol, Univ Sheffield, 61-64; sr res scientist, Res Ctr, Pan Am Petrol Corp, Okla, 64-71; sr staff geologist, Amoco Petrol Co Ltd, 71-79, staff geol supvr, 79-83, chief geologist, 83-88, MGR, EXPLOR TECH SERV, AMOCO PROD CO, 88- *Mem:* Can Soc Petrol Geologists; Am Asn Stratig Palynologists; Am Asn Petrol Geologists. *Res:* Paleozoic palynology, especially its stratigraphical applications; geochemistry. *Mailing Add:* Amoco Prod Co 501 WestLake Park Blvd PO Box 3092 Houston TX 77253

SULLIVAN, HUGH D, b Butte, Mont, June 16, 39; m 61; c 4. MATHEMATICS. *Educ:* Univ Mont, BA, 62, MA, 64; Wash State Univ, PhD(math), 68. *Prof Exp:* Teaching asst math, Univ Mont, 62-64 & Wash State Univ, 64-67; asst prof, 67-70, chmn dept, 70-76, ASSOC PROF MATH, EASTERN WASH STATE COL, 70- *Mem:* Am Math Soc; Math Asn Am; Sigma Xi. *Res:* Abstract systems theory; topology; probability and statistics. *Mailing Add:* Ctr for Tech Develop Eastern Wash State Col Cheney WA 99004

SULLIVAN, HUGH R, JR, b Indianapolis, Ind, Apr 8, 26; m 48; c 5. DRUG METABOLISM. *Educ:* Univ Notre Dame, BS, 48; Temple Univ, MA, 54. *Prof Exp:* Assoc res chemist, Socony-Vacuum Oil Co, 48-51; from assoc res chemist to sr res chemist, Eli Lilly & Co, 51-69, res scientist, 69-72, res assoc, Res Labs, 72-87, sr res scientist, 87-88; RETIRED. *Mem:* Am Chem Soc; Am Soc Pharmacol & Exp Therapeut; Am Soc Mass Spectrometry. *Res:* Mechanism of drug action and detoxication; analgesics; antibiotics; pharmacokinetics; quantitative mass fragmentography. *Mailing Add:* 7135 Kingswood Circle Indianapolis IN 46256

SULLIVAN, J AL, b Whitewater, Colo, Dec 31, 37; m 58; c 3. MECHANICAL ENGINEERING. *Educ:* Univ Colo, BS, 58, MS, 60; Univ Mich, PhD(mech eng), 66. *Prof Exp:* Instr, Univ Colo & Univ Mich, 58-66; asst prof mech eng, Colo State Univ, 66; staff mem chem laser & rover prog, 66-74, group leader laser eng, 74-78, staff mem laser photochem, 78-81, PROG MGR, MOLECULAR LASER ISOTOPE SEPARATION, LOS ALAMOS NAT LAB, 81- *Mem:* Sigma Xi. *Res:* Industrial coordination of laser-isotope separation; laser-induced chemistry in waste reprocessing; other facets of applied photochemistry. *Mailing Add:* Appl Photochem Div Los Alamos Nat Lab PO Box 1663 Los Alamos NM 87545

SULLIVAN, JAMES BOLLING, b Rome, Ga, Mar 19, 40; m 63; c 3. BIOCHEMISTRY, ZOOLOGY. *Educ:* Cornell Univ, AB, 62; Univ Tex, Austin, PhD(zool), 66. *Prof Exp:* Asst prof, 70-76, ASSOC PROF BIOCHEM, DUKE UNIV, 77- *Concurrent Pos:* USPHS fel biochem, Duke Univ, 67-70. *Mem:* AAAS; Soc Study Evolution; Am Soc Biol Chemists; Lepidop Soc. *Res:* Comparative protein chemistry. *Mailing Add:* Dept Biochem Duke Univ Marine Lab Durham NC 27706

SULLIVAN, JAMES DOUGLAS, b Chicago, Ill, Oct 27, 40; m 67; c 2. PHYSICS, SPACE SCIENCE. *Educ:* Univ Chicago, SB, 62, SM, 64, PhD(physics), 70. *Prof Exp:* Physicist, Enrico Fermi Inst, Univ Chicago, 70-71; asst res physicist, Univ Calif, Berkeley, 71-74; MEM STAFF, CTR SPACE RES, MASS INST TECHNOL, 74- *Mem:* Am Phys Soc; Am Geophys Union. *Res:* Space physics, solar particles and magnetospheric physics; nuclear physics, high energy heavy ion reactions; astrophysics, cosmic rays and origin of gamma rays. *Mailing Add:* MIT Rm NW 16-166 167 Albany St Cambridge MA 02139

SULLIVAN, JAMES F, medicine; deceased, see previous edition for last biography

SULLIVAN, JAMES HADDON, JR, b Claxton, Ga, Apr 3, 37; m 60; c 1. ENVIRONMENTAL ENGINEERING. *Educ:* Ga Inst Technol, BChE, 59; Univ Fla, MS, 68, PhD(eng), 70. *Prof Exp:* Proj engr, Union Bag-Camp Paper Corp, 59-63; develop engr, Cities Serv Corp, 63-66; pres environ eng, 70-71, VPRES, WATER & AIR RES INC, 71- *Mem:* Am Water Works Asn; Am Acad Environ Engrs; Water Pollution Control Fedn; Nat Soc Prof Engrs. *Res:* Water chemistry; waste and water treatment; environmental impact studies. *Mailing Add:* 11126 SW Eighth Ave Gainesville FL 32607

SULLIVAN, JAMES MICHAEL, b Butte, Mont, July 1, 34. NEUROANATOMY. *Educ:* Carroll Col, BA, 56; Univ Ore, MS, 64; St Louis Univ, PhD(anat), 73. *Prof Exp:* From instr to asst prof biol, Carroll Col, 61-69, assoc prof, 73-74; fel & res assoc, 74-75, instr, 75-76, ASST PROF ANAT, SCH MED, ST LOUIS UNIV, 76- *Mem:* Sigma Xi; Am Asn Anat; Soc Neuroscience. *Res:* Anatomy, physiology and pharmacology of the autonomic nervous system with special emphasis on the cardiovascular system of man. *Mailing Add:* Dept Anatomy Creighton Univ Claifornia & 24th Sts Omaha NE 68178

SULLIVAN, JAMES THOMAS, JR, b Seekonk, Mass, May 30, 28; m 55; c 4. PHYSICAL CHEMISTRY. *Educ:* Providence Col, BS, 50; Cath Univ Am, PhD, 55. *Prof Exp:* Asst, Cath Univ Am, 53-54, res assoc, 54-55; from asst prof to assoc prof, 55-67, exec asst acad affairs, 72-78, vpres, 78-83, PROF CHEM, UNIV ST THOMAS, TEX, 67-, DIR COMPUT SERV, 84- *Mem:* AAAS; Am Chem Soc. *Res:* Teaching. *Mailing Add:* Dept Comput Serv 3812 Montrose Blvd Houston TX 77006

SULLIVAN, JAY MICHAEL, b Brockton, Mass, Aug 3, 36; m 64; c 3. CARDIOVASCULAR DISEASES. *Educ:* Georgetown Univ, BS, 58, MD, 62. *Prof Exp:* House officer med, Peter Bent Brigham Hosp, 62-63, resident, 63-67; res assoc biochem, Harvard Med Sch, 67-69, from instr to asst prof med, 69-74; PROF MED & CHIEF, DIV CARDIOVASCULAR DIS, COL MED, UNIV TENN, MEMPHIS, 74- *Concurrent Pos:* Nat Heart Inst res fel med, Harvard Med Sch, 64-66; res fel, Med Found, 67-69; fel, Coun High Blood Pressure Res, 75; dir hypertension unit, Peter Bent Brigham Hosp, 70-74; dir med serv, Boston Hosp Women, 73-74; consult, Nat Heart & Lung Inst, 74 & Vet Admin Cent Off, 82-85; fel, Coun Circulation, Am Heart Asn, 75; prin investr, Nat Heart, Lung & Blood Inst, 78- *Mem:* AAAS; fel Am Col Cardiol; Am Fedn Clin Res; fel Am Col Physicians; Int Soc Hypertension; Sigma Xi; Asn Univ Cardiologists. *Res:* Hypertension, hemodynamics of hypertension, clinical pharmacology of antihypertensive and anti-platelet drugs; sodium sensitivity. *Mailing Add:* Dept Med-Cardiol Univ Tenn Ctr Health Sci 353 Dobbs Bldg Memphis TN 38163

SULLIVAN, JERRY STEPHEN, b Havre, Mont, July 17, 45; m 67; c 3. COMPUTER SCIENCES. *Educ:* Univ Colo, Boulder, BSc, 67, MSc, 69, PhD(physics), 70; Harvard Univ, AMP, 86. *Prof Exp:* Res scientist solid state devices, N V Philips Gloeilampenfabrieken, Eindhoven, 71-75, group dir comp systs res, Philips Labs, 75-80; corp dir, Comput Tech Ctr, Tektronix, 81-83, div gen mgr, 83-88; VPRES, MICRO ELECTRONIC & COMPUT TECHNOL CORP, 88- *Concurrent Pos:* Mem adv bd, Ctr Int Syst, Stanford Univ, 82-; bd dirs, Sherpa Corp, 83-; chmn adv bd, Col Elec Eng & Civil Eng, Univ Tex, 88- *Mem:* Inst Elec & Electronic Engrs; AAAS; Am Phys Soc; Europ Phys Soc; Asn Comput Mach. *Res:* Numerical analysis; network theory; application of computers to semiconductor device modeling and integrated circuit analysis; electron paramagnetic and nuclear magnetic resonance and exchange interactions of ion pairs; microcomputer architecture and microprocessor design; software engineering; system engineering; computer-aided design and manufacturing; computer science. *Mailing Add:* Micro Electronic & Comput Technol Corp 3500 W Balcones Ctr Dr Austin TX 78759

SULLIVAN, JOHN BRENDAN, b Lynn, Mass, Aug 6, 44; m 70; c 2. MATHEMATICS. *Educ:* Harvard Univ, AB, 66; Cornell Univ, PhD(math), 71. *Prof Exp:* Lectr math, Univ Calif, Berkeley, 71-73; from asst prof to assoc prof, 73-87, PROF MATH, UNIV WASH, 87- *Concurrent Pos:* NSF grant, 74-79; mem, Inst Advan Study, 79-80; vis prof, Univ Mass, Amherst, 86-87. *Res:* Algebraic groups, lie algebras, and Hopf algebras; cohomology and representations of algebraic groups. *Mailing Add:* Dept Math Univ Wash Seattle WA 98195

SULLIVAN, JOHN DENNIS, b Lake Forest, Ill, June 17, 28; m 78; c 4. FORESTRY. *Educ:* Univ Idaho, BS, 52, MS, 54; Mich State Univ, PhD(wood technol), 58. *Prof Exp:* Asst wood technol, 54-56, instr, Mich State Univ, 56-; from asst prof to assoc prof wood sci, Sch Forestry, Duke Univ, 58-68; prin wood scientist, 68-69, agr adminr, 69-71, dep adminr, Coop State Res Serv, 71-78, dep adminr, Sci & Educ Admin-Coop Res, 78-81, dep adminr, Coop State Res Serv, USDA, 81-84; DIR, OFF FORESTRY AGENCY FOR INT DEVELOP, 84- *Mem:* Am Soc Testing & Mat; Soc Wood Sci & Technol; NY Acad Sci; Soc Am Foresters; fel Brit Inst Wood Sci; Sigma Xi. *Res:* Cellulose morphology at a molecular level, especially high resolution electron microscopy; adhesion and wood; liquid interactions. *Mailing Add:* 1201 S Jeff Davis Hwy 1202 S Arlington VA 22202

SULLIVAN, JOHN HENRY, b New Haven, Conn, May 18, 19; m 47; c 2. PHYSICAL CHEMISTRY. *Educ:* Calif Inst Technol, PhD(chem), 50. *Prof Exp:* CHEMIST, LOS ALAMOS SCI LAB, UNIV CALIF, 50- *Mem:* Am Chem Soc. *Res:* Kinetics of gaseous reactions. *Mailing Add:* 3536-A Arizona Ave Los Alamos NM 87544-1521

SULLIVAN, JOHN JOSEPH, b New York, NY, Mar 28, 35; m 63; c 2. REPRODUCTIVE PHYSIOLOGY. *Educ:* Rutgers Univ, BS, 57, PhD(dairy sci, physiol), 63; Univ Tenn, MS, 59. *Prof Exp:* Res assoc, Am Breeders Serv, Inc, W R Grace & Co, 63-65, assoc dir labs & res, 65-79, dir, labs & res, 79-81 & dir prod, 81-84, vpres, 84-89, PRES PROD, AM BREEDERS SERV, INC, W R GRACE & CO, 89- *Mem:* Am Soc Animal Sci; Soc Cryobiology; Soc Study Reproduction. *Res:* Physiology of reproduction and related fields; artificial insemination of domestic animals; low temperature biology. *Mailing Add:* W 8738 Stevenson Dr Poynette WI 53955

SULLIVAN, JOHN LAWRENCE, b Scranton, Pa, Nov 24, 43. BEHAVIORAL NEUROPHARMACOLOGY, NEUROCHEMISTRY. *Educ:* Duke Univ, AB, 65; Johns Hopkins Univ, MD, 69. *Prof Exp:* Intern med, Johns Hopkins Hosp, 69-70; resident psychiat, Univ Calif, San Diego, 70-73; asst prof psychiat, Med Ctr, Duke Univ, 73-78, assoc prof, 78-; mem fac, Med Sch, Univ Ma, 80-; AT DEPT PSYCHIAT, VET ADMIN MED CTR, WASHINGTON, DC. *Concurrent Pos:* Consult, Neuropsychopharmacol Drug-Dependent States, US Dept Navy, 71-73 & Warner-Chilcott Pharmaceut Co, 78-; dir, Neuropsychopharmacol Lab, Vet Admin Med Ctr, 75-80. *Mem:* AAAS; Soc Biol Psychiat; Am Psychiat Asn; AMA. *Res:* Biochemical neuropsychopharmacology, with particular reference to the role of monoamine oxidase in behavior and psychopathological states. *Mailing Add:* 7808 El Cajon Blvd La Mesa CA 92041

SULLIVAN, JOHN LESLIE, b Sydney, Australia, July 16, 17; m 39; c 4. OCCUPATIONAL HEALTH SAFETY & INDOOR AIR QUALITY. *Educ:* Sydney Tech Col, Dipl chem eng, 41; Univ New S Wales, MS, 56, PhD(chem eng), 60. *Prof Exp:* Prof, Syracuse Univ, 69-71; prof, Univ Western Ont, 71-83; pres, Ocuplan Consults Inc, 80-86; ENGR MGR, OCCUP HEALTH, MCLAREN PLANSEARCH INC, TORONTO & PRES, OCUPLAN CONSULTS INC. *Concurrent Pos:* Sr lectr, Univ New South Wales, 55-65; consult, WHO, 66 & 69. *Honors & Awards:* Medal, Clean Air Soc Austrlia & NZ. *Mem:* Am Inst Chem Engrs; Air Pollution Control Asn; Am Indust Hyg Asn; Am Conf Govt Indust Hyg; Sigma Xi; Am Soc Heating Regrig & Air Conditioning Engrs. *Res:* Air pollution measurement and control; occupational health and safety engineering; indoor air quality; asbestos. *Mailing Add:* 460 Wellington St #407 London ON N6A 3P8 Can

SULLIVAN, JOHN M, b Philadelphia, Pa, June 21, 32; m 56; c 9. ORGANIC CHEMISTRY. *Educ:* Dartmouth Col, AB, 54; Univ Mich, MS, 56, PhD(org chem), 60. *Prof Exp:* PROF CHEM, EASTERN MICH UNIV, 58- *Mem:* Am Chem Soc. *Res:* Heterocyclics; conformational analysis. *Mailing Add:* Dept Chem Eastern Mich Univ Ypsilanti MI 48197-2207

SULLIVAN, JOHN W, b Fargo, NDak, Nov 1, 32; m 64; c 3. CEREAL CHEMISTRY. *Educ:* NDak State Univ, BS, 54, MS, 58; Kans State Univ, PhD(cereal chem), 66. *Prof Exp:* Proj leader food res, John Stuart Res Labs, Quaker Oats Co, 61-68, sect mgr food res, 68-74; DIR RES, ROMAN MEAL CO, 74-, VPRES RES & DEVELOP, 79- *Mem:* Am Chem Soc; Am Asn Cereal Chemists; Inst Food Technologists. *Res:* Physical and chemical changes in starch and associated carbohydrates; enzyme changes in physical structure of starches and proteins. *Mailing Add:* 8010 SE Middle Way Vancouver WA 98664

SULLIVAN, JOHN WILLIAM, Dec 14, 39; c 1. CENTRAL NERVOUS SYSTEMS. *Educ:* Rutgers Univ, MSc, 67. *Prof Exp:* SR EPIDEMIOLOGIC COORDR, HOFFMANN-LAROCHE, INC, 87- *Concurrent Pos:* Sr res scientist pharmacol, Hoffmann-LaRoche, Inc, 67- *Mem:* Am Pharmaceut Asn; Am Soc Pharmacol Therap; Drugs Info Asn. *Res:* Enhancement of cognition. *Mailing Add:* Dept Drug Safety Bldg 115 Fifth floor Hoffmann-LaRoche Inc 340 Kingsland St Nutley NJ 07110

SULLIVAN, JOSEPH ARTHUR, b Boston, Mass, June 5, 23; m 46; c 4. MATHEMATICS. *Educ:* Boston Col, AB, 44; Mass Inst Technol, SM, 47; Ind Univ, PhD(math), 50. *Prof Exp:* From instr to assoc prof math, Univ Notre Dame, 50-60; PROF MATH, BOSTON COL, 60- *Mem:* Am Math Soc; Math Asn Am. *Res:* Mathematical analysis. *Mailing Add:* Dept of Math Boston Col Chestnut Hill MA 02167

SULLIVAN, KAREN A, b Bronxville, NY. IMMUNOLOGY. *Educ:* N Adams State Col, BS, 66; Duke Univ, PhD(immunol, microbiol), 74. *Hon Degrees:* DSc, N Adams State Col, 90. *Prof Exp:* Cancer Res Inst Inc NY fel, McIndoe Mem Res Unit, Blond Lab, Queen Victoria Hosp, Sussex, Eng, 73-75; fel, Div Lab & Res, NY State Dept Health, 75-78; asst prof immunol, Med Ctr, W Va Univ, 78-80; asst mem, Mem Sloan Kettering Cancer Ctr, 80-83; RES ASSOC PROF, DEPT MED, TULANE UNIV MED CTR, 83- *Concurrent Pos:* United Network Organ Sharing Histocompatibility Comt, 89- *Mem:* Am Asn Immunologists; Am Soc for Histocompatibility & Immunogenetics; NY Acad Sci; Am Soc Transplant Physicians; Clin Immunol Soc. *Res:* Immunogenetics; cellular immunology; lymphoid cell differentiation; regulation of the immune response; transplantation. *Mailing Add:* Dept Med Suite 7209 Tulane Univ Med Ctr 1430 Tulane Ave New Orleans LA 70112

SULLIVAN, LAWRENCE PAUL, b Hot Springs, SDak, June 16, 31; m 55; c 3. PHYSIOLOGY. *Educ:* Univ Notre Dame, BS, 53; Univ Mich, MS, 56, PhD(physiol), 59. *Prof Exp:* Asst physiol, Univ Mich, 55-59, instr, 59-60; asst prof, George Washington Univ, 60-61; from asst prof to assoc prof, 61-69, PROF PHYSIOL, MED CTR, UNIV KANS, 69- *Concurrent Pos:* USPHS career develop award, 65-70; vis prof, Sch Med, Yale Univ, 69-70; dep ed, J Am Soc Nephrology. *Mem:* Am Soc Nephrology; Am Physiol Soc; Sigma Xi. *Res:* Membrane transport; cell volume and pH regulation; potassium transport; acid-base transport mechanisms. *Mailing Add:* Dept of Physiol WHE 3021 Univ Kans Med Ctr Kansas City KS 66103

SULLIVAN, LLOYD JOHN, b Lowell, Ariz, Sept 6, 23; m 48; c 4. BIOCHEMISTRY, PHYSICAL ORGANIC CHEMISTRY. *Educ:* Univ Ariz, BS, 50; Univ Pittsburgh, MS, 54. *Prof Exp:* Res asst phys chem, Mellon Inst Sci, 50-52, res assoc, 52-54, jr fel, 54-56, fel, 56-58; mem tech staff, Cent Res Lab, Tex Instruments, 58-60, sect head & dir energy conversion, Apparatus Div, 60-61; fel phys & anal chem, Mellon Inst Sci, 61-67, sr fel phys & anal chem & head chem & biochem sect, Chem Hyg Fel, 67-80; RETIRED. *Mem:* Fel AAAS; fel Am Inst Chemists; NY Acad Sci; Am Chem Soc. *Res:* Physical properties of organic materials; separation and purification of organic compounds, particularly natural products; gas, liquid and thin layer chromatography; analytical biochemistry; metabolism of organic compounds in vivo and by tissue culture techniques. *Mailing Add:* 435 Cumberland Ave Gulf Breeze FL 32561-4107

SULLIVAN, LOUIS WADE, b Atlanta, Ga, Nov 3, 33; m 55; c 3. INTERNAL MEDICINE. *Educ:* Morehouse Col, BS, 54; Boston Univ, MD, 58; Am Bd Internal Med, dipl, 66. *Hon Degrees:* 17 from var insts. *Prof Exp:* Intern, NY Hosp-Cornell Med Ctr, 58-59, resident med, 59-60; resident gen path, Mass Gen Hosp, 60-61; instr, Harvard Med Sch, 63-64; asst prof, NJ Col Med, 64-66, asst attend physician, 64-65, assoc attend physician, 65-66; asst prof med, Sch Med, Boston Univ, 66-70, assoc prof med & physiol, 70-77; dean & pres, Sch Med, Morehouse Col, 77-85, pres, 85-89; SECY, DEPT HEALTH & HUMAN SERV, 89- *Concurrent Pos:* Res fel med, Thorndike Mem Lab, Boston City Hosp & Harvard Med Sch, 61-63; res assoc, Thorndike Mem Lab, Boston City Hosp, 63-64, dir hemat & proj dir, Boston Sickle Cell Ctr, 72-75; USPHS res career develop award, 65-66, 67-71; asst vis physician, Boston Univ Hosp, 66-; assoc ed, Nutrit Reports Int, 69-73; vchmn, Comn Health & Human Serv, Southern Regional Educ Bd, 85-87; Martin Luther King Jr vis prof, Univ Mich, 86. *Mem:* Inst Med-Nat Acad Sci; Soc Exp Biol & Med; Am Fedn Clin Res; AAAS; Fedn Am Socs Exp Biol; Am Soc Clin Invest. *Res:* Metabolism of vitamin B-12 and folic acid in man. *Mailing Add:* Dept Health & Human Serv Washington DC 20201

SULLIVAN, MARGARET P, b Lewistown, Mont, Feb 7, 22. PEDIATRICS, MEDICINE. *Educ:* Rice Inst, BA, 44; Duke Univ, MD, 50; Am Bd Pediat, dipl, 56. *Prof Exp:* Pediatrician, Atomic Bomb Casualty Comn, Japan, 53-55; from asst pediatrician to pediatrician, 56-73, ASHBEL SMITH PROF PEDIAT & PEDIATRICIAN, UNIV TEX M D ANDERSON CANCER CTR, HOUSTON, 73- *Mem:* AAAS; Am Asn Cancer Res; Am Acad Pediat; AMA; Am Med Women's Asn (vpres, 72, pres, 74). *Res:* Pediatric oncology. *Mailing Add:* Dept of Pediat Univ of Tex M D Anderson Cancer Ctr 1515 Holcombe Blvd Houston TX 77030

SULLIVAN, MARY LOUISE, b Butte, Mont, Oct 3, 06. INORGANIC CHEMISTRY. *Educ:* St Mary Col, Kans, BS, 35; St Louis Univ, MS, 39, PhD(chem), 47. *Prof Exp:* Instr chem, 46-53, dean, 53-74, DIR INST RES, ST MARY COL, KANS, 74- *Mem:* Am Asn Physics Teachers; Nat Sci Teachers Asn; Sigma Xi. *Res:* Analytic chemistry, especially trace analysis for metals using dithizone. *Mailing Add:* St Mary Col Leavenworth KS 66048-5082

SULLIVAN, MAURICE FRANCIS, b Butte, Mont, Feb 15, 22; m 51; c 6. PHARMACOLOGY. *Educ:* Mont State Col, BS, 50; Univ Chicago, PhD(pharmacol), 55. *Prof Exp:* Scientist, Hanford Labs, Gen Elec Co, 55-56, sr scientist, 56-65; mgr, Physiol Sect, Pac Northwest Labs, Battelle Mem Inst, 65-71, staff scientist, 71-85, consult, 85-86; RETIRED. *Concurrent Pos:* NIH fel, Med Res Coun, Harwell, Eng, 61-; consult, Battelle Mem Inst, 85-; task groups, Int Comn Radiol Protection, 83-86; expert group on Orgn Econ Coop & Develop, Nuclear Energy Agency, 85- *Mem:* Fel AAAS; Am Physiol Soc; Radiation Res Soc. *Res:* Biological effects of radiation, especially on the gastrointestinal tract; biochemistry; physiology; pathology; pharmacology. *Mailing Add:* 2125 Harris Ave Richland WA 99352

SULLIVAN, MICHAEL FRANCIS, b New Paltz, NY, Aug 20, 42; m 67; c 3. PHOTOGRAPHIC CHEMISTRY. *Educ:* St Lawrence Univ, BS, 63; Univ NC, Chapel Hill, PhD(chem), 68. *Prof Exp:* Sr chemist, 67-73, res lab head, 73-81, supvr prod & develop, Film Emulsion & Plate Mfg Div, 81-84, DIR MFG STAFF, FILM SENSITIZING DIV, EASTMAN KODAK CO, 84- *Res:* Production and development of black and white and color photographic films. *Mailing Add:* Eastman Kodak Co 343 State St Rochester NY 14650

SULLIVAN, MICHAEL JOSEPH, JR, b Chicago, Ill, Jan 8, 42; m 64; c 4. MATHEMATICS. *Educ:* DePaul Univ, BS, 63; Ill Inst Technol, MS, 64, PhD(math), 67. *Prof Exp:* Teaching asst math, Ill Inst Technol, 62-65; from asst prof to assoc prof, 65-77, PROF MATH, CHICAGO STATE UNIV, 77- *Concurrent Pos:* Am Coun Educ acad admin internship, 70-71. *Mem:* Am Math Soc; Math Asn Am. *Res:* Differential geometric aspects of dynamics; polygenic functions; applications of mathematics in the management and behavioral sciences; mathematics education; Author numerous mathematicstextbooks. *Mailing Add:* 9529 Tripp Oak Lawn IL 60453

SULLIVAN, NEIL SAMUEL, b Wanganui, NZ, Jan 18, 42; m 65; c 3. LOW TEMPERATURE PHYSICS. *Educ:* Otago Univ, NZ, BSc, 64, MSc, 65; Harvard Univ, PhD(physics), 72. *Prof Exp:* Assoc physics, Ctr Nuclear Studies, Saclay, France, 72-74, res physicist, 74-82; PROF PHYSICS, UNIV FLA, 82-, CHAIR, DEPT PHYSICS, 89- *Concurrent Pos:* Fullbright Exchange Grant, US Educ Found, 65; Frank Knox fel, Harvard Univ, 65-70. *Honors & Awards:* Prix Saintour, 78; Prix La Caze, 83. *Mem:* Am Phys Soc; French Physics Soc; Inst Physics London; Europ Phys Soc. *Res:* Fundamental properties of solid hydrogen and solid helium at very low temperatures; studies of molecular motions using nuclear magnetic resonance; cosmic axion detectors. *Mailing Add:* Dept Physics Univ Fla Gainesville FL 32611

SULLIVAN, PATRICIA ANN NAGENGAST, b New York, NY, Nov 22, 39; m 66. BIOLOGY. *Educ:* Notre Dame Col Staten Island, AB, 61; NY Univ, MS, 64, PhD(biol), 67. *Prof Exp:* Part-time instr, Notre Dame Col Staten Island, 64-67; asst prof biol, Wagner Col, 67-68; NIH trainee anat & cell biol, State Univ NY Upstate Med Ctr, 68-69 & fel cell biol, 69-70; asst prof biol, Wells Col, 70-74, chairwoman div life sci, 75-76, assoc prof, 74-79; assoc prof biol, Tex Woman's Univ, 79-81; PROF BIOL & ACAD DEAN, SALEM COL, 81- *Mem:* AAAS; NY Acad Sci; Am Soc Hemat; Sigma Xi; Am Asn Univ Women. *Res:* Bioethics and implications of biological research; hematopoiesis and its regulation; chromatin structure and function. *Mailing Add:* Tex Woman Univ Denton TX 76204

SULLIVAN, PAUL JOSEPH, b Merrickville, Ont, Mar 2, 39; m 62; c 2. APPLIED MATHEMATICS, ENGINEERING. *Educ:* Univ Waterloo, BSc, 64, MSc, 65; Cambridge Univ, PhD(appl math), 68. *Prof Exp:* Asst prof, 68-80, PROF APPL MATH & ENG, UNIV WESTERN ONT, 80- *Concurrent Pos:* Res fel, Calif Inst Technol, 69-70. *Mem:* Am Acad Mech; Can Soc Mech Eng. *Res:* Dispersion within turbulent fluid flow; convection phenomenon in fluids; turbulent fluid flow generally. *Mailing Add:* Dept of Appl Math Univ of Western Ont London ON N6A 5B8 Can

SULLIVAN, PETER KEVIN, b San Francisco, Calif, June 14, 38. LABORATORY SAFETY, POLYMER PHYSICS. *Educ:* Univ San Francisco, BS, 60; Cornell Univ, MS, 63; Rensselaer Polytech Inst, PhD(phys chem), 65. *Prof Exp:* Res fel, Nat Bur Stand, 65-67, chemist, 67-73; phys chemist, Celanese Res Co, 74-79; prin scientist & mgr, Int Nickel Co, 79-82; mgr, Exel Plastics, 82-86; SAFETY MGR, MT SINAI MED CTR, 86- *Mem:* Am Phys Soc; Am Chem Soc. *Res:* Physics, chemistry and mechanical properties of polymers. *Mailing Add:* 1 Clinton Pl Suffern NY 10901

SULLIVAN, PHILIP ALBERT, b Sydney, Australia, Dec 25, 37; m 62; c 1. AEROSPACE ENGINEERING, PHYSICS. *Educ:* Univ New South Wales, BE, 60, ME, 62; Univ London, DIC & PhD(aeronaut eng), 64. *Prof Exp:* Res assoc aeronaut res, Imp Col, Univ London, 62-64; res assoc hypersonics res, Gas Dynamics Lab, Princeton Univ, 64-65; asst prof teaching & res, 65-71, ASSOC PROF AEROSPACE SCI & ENG, INST AEROSPACE STUDIES, UNIV TORONTO, 71- *Concurrent Pos:* Consult, Atomic Energy Can Ltd. *Mem:* Assoc fel Can Aeronaut & Space Inst; Am Inst Aeronaut & Astronaut. *Res:* Air cushion technology with applications to air cushion landing systems, amphibious vehicles and trains. *Mailing Add:* 4925 Dufferin St Downsview ON M3H 5T6 Can

SULLIVAN, RAYMOND, b Ebbw Vale, Wales, Oct 27, 34; m 62; c 2. GEOLOGY. *Educ:* Univ Sheffield, BSc, 57; Glasgow Univ, PhD(geol), 60. *Prof Exp:* Demonstr geol, Glasgow Univ, 57-60; paleontologist, Shell Oil Co Can, 60-62; from asst prof to assoc prof, 62-74, assoc dean natural sci, 69-72, PROF GEOL, SAN FRANCISCO STATE UNIV, 74- *Concurrent Pos:* NSF res grant, 66-67. *Mem:* Am Asn Petrol Geologists; Nat Asn Geol Teachers; fel Geol Soc Am. *Res:* Upper Paleozoic biostratigraphy; sedimentary petrology of carbonate and clastic rocks; environmental geology. *Mailing Add:* Dept of Geol San Francisco State Univ 1600 Holloway Ave San Francisco CA 94132

SULLIVAN, RICHARD FREDERICK, b Olathe, Colo, Dec 26, 29; m 68; c 2. HYDROCARBON CHEMISTRY. *Educ:* Univ Colo, BA, 51, PhD(chem), 56. *Prof Exp:* Res chemist, Calif Res Corp, 55-68, from sr res chemist to sr res assoc, 68-89, RES SCIENTIST, CHEVRON RES & TECHNOL CO, 89- *Concurrent Pos:* UN Develop Prog; exper consult, Nat Chem Lab, Pune, India, 90. *Mem:* Am Chem Soc; Am Inst Chem Eng. *Res:* Photochemistry; mechanisms of hydrocarbon reactions; catalysis; petroleum process research and development; conversion of shale oil and liquids derived from coal to transportation fuels. *Mailing Add:* Chevron Res & Technol Co 100 Chevron Way Richmond CA 94802-0627

SULLIVAN, ROBERT E(MMETT), metallurgical engineering, for more information see previous edition

SULLIVAN, ROBERT EMMETT, b Sioux City, Iowa, May 28, 32; m 61. DENTISTRY, PEDODONTICS. *Educ:* Morningside Col, BA, 54; Univ Nebr, DDS, 61, MSD, 63; Am Bd Pedodont, dipl, 67. *Prof Exp:* From instr to assoc prof pedodont, 63-72, ASSOC PROF PEDIAT, UNIV NEBR-LINCOLN, 69-, PROF PEDODONT, 72-, CHMN, 82- *Concurrent Pos:* Consult, Omaha-Douglas County Dent Pub Health, 62- & Omaha-Douglas County Children & Youth Proj, 68- *Mem:* Am Acad Pedodont; Am Dent Asn; Am Soc Dent for Children; Am Col Dentists. *Res:* Vital staining of teeth; mechanism of action of dental preventative materials. *Mailing Add:* Dept Pedodontics Univ Nebr Lincoln NE 68583

SULLIVAN, ROBERT LITTLE, b Chicago, Ill, Oct 27, 28. GENETICS. *Educ:* Univ Del, AB, 50; NC State Univ, MS, 53, PhD, 56. *Prof Exp:* Res assoc entom, Univ Kans, 56-61; asst prof biol, Washburn Univ, 61-62; asst prof, 62-68, assoc prof, 68-79, PROF BIOL, WAKE FOREST UNIV, 79- *Mem:* Genetics Soc Am; Soc Study Evolution; Sigma Xi. *Res:* Insect genetics, including radiation studies on the genetics of the house fly. *Mailing Add:* 2185 Finnegan Lane Cincinnati OH 45244

SULLIVAN, SAMUEL LANE, JR, b Victoria, Tex, May 25, 35; div; c 4. CHEMICAL ENGINEERING. *Educ:* Tex A&M Univ, BS, 57, MS, 59, PhD(chem eng), 63. *Prof Exp:* ASSOC PROF CHEM ENG, TULANE UNIV, 61-, ASSOC DEAN ENG, 76- *Mem:* Am Inst Chem Engrs; Am Soc Eng Educ; Sigma Xi. *Res:* Multicomponent distillation; chemical process simulation; computer applications in chemical engineering; multicomponent diffusion. *Mailing Add:* Eng Dean's Off Tulane Univ New Orleans LA 70118

SULLIVAN, SEAN MICHAEL, m 88. BIOPHYSICS, VIROLOGY. *Educ:* Univ Tenn, PhD(biochem), 85. *Prof Exp:* Fel, Calif Inst Technol, 85-87; SR RES SCIENTIST, VESTAR, INC. *Mem:* Biophys Soc; Am Soc Cell Biol. *Res:* The development of targeted drug delivery systems using liposomes as the delivery vehicle; monoclonal antibodies, suflated glycolipid and recombinant viral receptors have been used as targeting ligands; cell targets have been tumor cells and virus infected cells; delivered drugs have been chemotherapeutic agents and synthetic aligonucleotides. *Mailing Add:* Vestar Inc 650 Cliffside Dr San Dimas CA 91773

SULLIVAN, STUART F, b Buffalo, NY, July 15, 28; m 59; c 4. ANESTHESIOLOGY. *Educ:* Canisius Col, Buffalo, BS, 50; State Univ NY, Syracuse, MD, 55. *Prof Exp:* Instr anesthesiol, Col Physicians & Surgeons, Columbia Univ, 61-62, assoc, 62-64, from asst prof to assoc prof, 64-73; PROF & VCHMN ANESTHESIOL, UNIV CALIF, LOS ANGELES, 73- *Concurrent Pos:* Res fel, NIH, 60-61, res career develop award, 66-69. *Mem:* Am Physiol Soc; Am Soc Anesthesiologist; Asn Univ Anesthersis. *Res:* Effects of anesthesiol and operation on cardiopulmonary function; oxygen and carbon dioxide transport during anesthesia. *Mailing Add:* Dept Anesthesiol Sch Med Univ Calif Los Angeles CA 90024-1778

SULLIVAN, SUSAN JEAN, ENDOCRINOLOGY, BIOCHEMISTRY. *Educ:* Univ Pa, PhD(cell biol), 81. *Prof Exp:* RES FEL, LAB TOXICOL, SCH PUB HEALTH, HARVARD UNIV, 81- *Mailing Add:* Seven Lind Rd West Newton MA 02165-1031

SULLIVAN, THOMAS DONALD, b Fair Haven, Vt, Feb 16, 12. CYTOLOGY. *Educ:* St Michael's Col, Vt, AB, 34; Cath Univ Am, MA, 39; Fordham Univ, PhD(cytol), 47. *Hon Degrees:* DSc, St Michael's Col, 77. *Prof Exp:* Instr Latin & Greek, St Michael's Col, Vt, 39-43, acad dean, 42-44, head dept biol, 47-67, dir res unit, 52-58, vpres, Col, 58-64, trustee, 67-70, prof, 47-77, emer prof biol, 77-; RETIRED. *Mem:* AAAS. *Res:* Cytogenetics of petunia; plant growth substances; mouse ascites tumors. *Mailing Add:* 19 Richard Terr South Burlington VT 05403

SULLIVAN, THOMAS WESLEY, b Rover, Ark, Sept 30, 30; m 55; c 4. POULTRY NUTRITION, BIOCHEMISTRY. *Educ:* Okla State Univ, BS, 51; Univ Ark, MS, 56; Univ Wis, PhD, 58. *Prof Exp:* Instr, Ark Pub Schs, 51-52; res asst poultry nutrit, Univ Ark, 54-55; res asst, Univ Wis, 55-58, from asst prof to assoc prof, 58-65; PROF POULTRY SCI, UNIV NEBR-LINCOLN, 65-, MEM FAC ANIMAL SCI DEPT, 76- *Concurrent Pos:* Consult feed mfg & ingredients. *Honors & Awards:* Res Award, Nat Turkey Fedn, 68. *Mem:* Poultry Sci Asn; Am Inst Nutrit; Soc Exp Biol & Med. *Res:* Mineral nutrition and skeletal problems in poultry; nutritional value of cereal grains. *Mailing Add:* Animal Sci Univ of Nebr Lincoln NE 68583-0908

SULLIVAN, TIMOTHY PAUL, b Duluth, Minn, Dec 3, 45; m 67. PLANT PHYSIOLOGY, AGRONOMY. *Educ:* Univ Minn, Duluth, BS, 67, MS, 69, PhD(plant physiol), 72. *Prof Exp:* Assoc prof biol, West Chester State Col, 72-76; SR AGRONOMIST, 3M CO, 76- *Mem:* Plant Growth Regulator Working Group; Int Weed Sci Soc; Am Soc Plant Physiologists; Sigma Xi; Weed Sci Soc Am. *Res:* Host-parasite relationships involving microsurgery of the host cells parasitized by haustoria of an obligate parasite; effects of water stress on photosynthesis, stomatal aperature and water potential during different stages of soybean development. *Mailing Add:* 3M Ctr Bldg 223-65-04 St Paul MN 55144-1000

SULLIVAN, VICTORIA I, b Avon Park, Fla, Nov 14, 41. BIOSYSTEMATICS. *Educ:* Univ Miami, BA, 63; Fla State Univ, PhD(biol), 72. *Prof Exp:* Naturalist, Everglades Nat Park, Nat Park Serv, 63-64; tech asst, Fairchild Trop Garden, Miami, 67-68; instr, Iowa State Univ, 71-72; botanist, Trustees Internal Improv Trust Fund, State of Fla, 72-73; ASSOC PROF BIOL, UNIV SOUTHWESTERN LA, 73- *Mem:* AAAS; Am Soc Plant Taxonomists; Soc Study Evolution; Am Soc Naturalists; Sigma Xi. *Res:* Phylogenetic relationships among Eupatorium (compositae) species and the origins of aqamospermons cytotypes using allozyme electrophoresis and other biosystematic methods. *Mailing Add:* Univ Southwestern La Drawer 4-2451 Lafayette LA 70504

SULLIVAN, W ALBERT, JR, b Nashville, Tenn, Apr 6, 24; m 49; c 2. SURGERY. *Educ:* Tulane Univ, MD, 47; Univ Minn, MS, 56; Am Bd Surg, dipl, 58. *Prof Exp:* Asst prof surg, 56-61, asst dean, 68-73, assoc prof surg, 61-68 DIR DEPT CONTINUATION MED EDUC, UNIV MINN, MINNEAPOLIS, 58-, ASSOC DEAN ADMIS & STUDENT AFFAIRS, MED SCH, 73- *Mem:* AMA; Am Col Surgeons. *Res:* Asymptomatic detection of cancer; medical education. *Mailing Add:* Dept Surg Box 195 Mayo Med Sch Univ Minn 420 Delaware St SE Minneapolis MN 55455-0310

SULLIVAN, W(ALTER) JAMES, b New York, NY, Apr 27, 25; m 51; c 3. PHYSIOLOGY, BIOPHYSICS. *Educ:* Manhattan Col, BS, 46; Cornell Univ, MD, 51. *Prof Exp:* Instr physics, Manhattan Col, 46-47; intern med, Univ Va, 51-52; instr, Cornell Univ, 54-55; vis investr, Rockefeller Inst, 55-56; group leader, Lederle Labs, 56-63; from asst prof to assoc prof, 63-73, actg chmn dept, 71-76, PROF PHYSIOL, SCH MED, NY UNIV, 73- *Concurrent Pos:* Fel physiol, Cornell Univ, 52-54; vis prof physiol, Sch Med, Yale Univ, 77-78. *Mem:* Am Physiol Soc; Biophys Soc; Soc Gen Physiol; Am Soc Nephrology. *Res:* Renal physiology; micropuncture study of renal ion transport. *Mailing Add:* Dept of Physiol & Biophys NY Univ Sch Med 550 First Ave New York NY 10016

SULLIVAN, WALTER SEAGER, b New York, NY, Jan 12, 18; m 50; c 3. SCIENCE WRITING. *Educ:* Yale Univ, BA, 40. *Hon Degrees:* LHD, Yale Univ, 69, Newark Col Eng, 74, Univ Ala, 87; DS, Hofstra Univ, 75, Ohio State Univ, 77, Muhlenberg Col, 87. *Prof Exp:* Mem staff, New York Times, 40-48, foreign correspondent, 48-56, sci news ed, 60-63, sci ed, 64-87; RETIRED. *Concurrent Pos:* Mem bd gov, Arctic Inst NAm, 59-65; mem univ coun, Yale Univ, 70-75; mem adv comt pub rels, Am Inst Physics, 65-; partic, Sem Technol & Social Change, Columbia Univ & mem adv coun, Dept Geol & Geophys Sci, Princeton Univ, 71-77, mem exped, Arctic, 35 & 46 & Antarctic, 46, 54, 56 & 77. *Honors & Awards:* George Polk Mem Award in Jour, 59; Westinghouse-AAAS Writing Awards, 63, 68 & 72; Int Nonfiction Book Prize, Frankfurt Fair, Ger, 65; Grady Award, Am Chem Soc, 69; Am Inst Physics-US Steel Found Award in Physics & Astron, 69; Washburn Award, Boston Mus Sci, 72; Daly Medal, Am Geog Soc, 73; Ralph Coats Roe Medal, Am Soc Mech Engrs, 75; Distinguished Pub Serv Award, US NSF, 78; Pub Welfare Medal, Nat Acad Sci, 80; Bromley Lectr, Yale Univ, 65; McGraw Lectr, Princeton Univ, 88. *Mem:* Fel AAAS; fel Arctic Inst NAm; Am Geog Soc; Am Geophys Union. *Mailing Add:* 66 Indian Head Rd Riverside CT 06878

SULLIVAN, WILLIAM DANIEL, b Boston, Mass, Nov 18, 18. BIOCHEMISTRY. *Educ:* Boston Col, AB, 44, MA, 45; Fordham Univ, MS, 48; Cath Univ Am, PhD(biol), 57. *Prof Exp:* Teacher, Cranwell Prep Sch, 45-46, Fairfield Prep Sch, 46-47 & Cheverus High Sch, 52-53; asst prof bact, Fairfield Univ, 57-58; from asst prof to assoc prof, Boston Col, 58-65, chmn dept, 58-69, prof, 65-89, EMER PROF BIOL, BOSTON COL, 89- *Concurrent Pos:* Dir cancer res inst, Boston Col; consult, Sta WGBH TV. *Mem:* AAAS; Am Soc Microbiol; Am Soc Parasitol; Soc Protozool; Electron Micros Soc Am; Sigma Xi. *Res:* Biochemistry of protozoa and cancer cells, especially effect of radiation on enzymatic activities; protein synthesis; electron microscopy with autoradiography of macromolecules in protozoan and cancer cells. *Mailing Add:* Dept Biol Boston Col Chestnut Hill MA 02167

SULLIVAN, WOODRUFF TURNER, III, b Colorado Springs, Colo, June 17, 44; m 68; c 2. ASTRONOMY, HISTORY OF SCIENCE. *Educ:* Mass Inst Technol, SB, 66; Univ Md, PhD(astron), 71. *Prof Exp:* Res astronomer radio astron, Naval Res Lab, 69-71: fel, Neth Found Radio Astron, 71-73; asst prof, 73-78, assoc prof, 78-86, PROF ASTRON, UNIV WASH, 86- *Concurrent Pos:* Sr vis fel, Inst Astron, Cambridge, Eng, 80-81 & 87-88; mem, Sci Working Group on Search for Extraterrestrial Intel, NASA, 80-90; chmn history comt, Astron Soc Pac, 86- *Mem:* Int Astron Union; Int Union Radio Sci; Am Astron Soc; Hist Sci Soc; Soc Social Studies Sci. *Res:* Galactic and extragalactic microwave spectroscopy; evolution of galaxies and clusters of galaxies; history of radio astronomy; interstellar communication and extraterrestrial civilizations. *Mailing Add:* Dept Astron FM-20 Univ Wash Seattle WA 98195

SULLIVAN-KESSLER, ANN CLARE, b Tillamook, Ore, June 3, 43. BIOCHEMISTRY. *Educ:* Col Notre Dame Md, BA, 65; Northwestern Univ, MS, 67; NY Univ, PhD(biochem), 73. *Prof Exp:* Res assoc, Sci & Eng Inc, 66-68; res group chief, Roche Inc, 69-78, dir, Dept Pharmacol I & II, 79-85, dir, Dept Pharmacol & Chemother, 85-89, dir explor res, 89-90, DIR INT PROJ MGT, HOFFMAN-LA ROCHE LTD, 90- *Concurrent Pos:* Adj prof, Sch Med, Columbia Univ, 76. *Honors & Awards:* Gold Medal Bond Award, Am Oil Chemists' Soc, 76; Women Scientist Award, Asn Women Sci, 77; Award Exp Nutrit, Am Inst Nutrit, 83. *Mem:* NY Acad Sci; AAAS; Am Chem Soc; Am Oil Chemists Soc; Asn Study Obesity; Asn Women Sci; Am Soc Pharmacol & Exp Therapeut; Am Inst Nutrit. *Res:* Regulation of lipid and carbohydrate metabolism; control of appetite and energy balance; metabolic aspects of obesity and hyperlipidemia; development of pharmacological agents for the treatment of obesity and hyperlipidemia. *Mailing Add:* Hoffman-La Roche Ltd Dept Prog Mgt PO Box CH-4002 Basel Switzerland

SULLWOLD, HAROLD H, b St Paul, Minn, Dec 22, 16; m 40; c 2. GEOLOGY, PETROLEUM. *Educ:* Univ Calif, Los Angeles, BA, 39, MA, 40, PhD(geol), 59. *Prof Exp:* Geologist, Wilshire Oil Co, 41, US Geol Surv, 42-44 & W R Cabeen & Assocs, 44-52; instr, Univ Calif, Los Angeles, 52-58; consult geologist, 58-60; geologist, George H Roth & Assocs, 60-80; INDEPENDENT GEOLOGIST, 80- *Concurrent Pos:* Adj prof, Univ SC, 70-71; pres, bd dirs, Carpinteria County Water Dist. *Mem:* Am Inst Prof Geol; Geol Soc Am; Am Asn Petrol Geologists. *Res:* Petroleum geology; exploration for oil and gas; turbidites; illustrated and published book of geological cartoons. *Mailing Add:* 560 Concha Loma Dr Carpinteria CA 93013

SULSER, FRIDOLIN, b Grabs, Switz, Dec 2, 26; m 55; c 4. PHARMACOLOGY. *Educ:* Univ Basel, MD, 55. *Prof Exp:* Asst prof pharmacol, Univ Berne, 56-58; head dept pharmacol, Wellcome Res Labs, NY, 63-65; DIR PSYCHOPHARMACOL RES CTR & PROF PHARMACOL, SCH MED, VANDERBILT UNIV, 80-; DIR, TENN NEUROPSYCHIAT INST, 74- *Concurrent Pos:* Int Pub Health Serv fel, NIH, 59-62. *Mem:* AAAS; Am Soc Pharmacol; Am Col Neuropsychopharmacol; Am Fedn Clin Res; NY Acad Sci. *Res:* Pharmacology of psychotropic drugs; neurochemistry; biochemical mechanisms of drug action. *Mailing Add:* Dept Pharm Vanderbilt Univ Sch Med A2215 Medical Center N Nashville TN 37232

SULSKI, LEONARD C, b Buffalo, NY, Mar 3, 36; m 66; c 1. MATHEMATICS. *Educ:* Canisius Col, BS, 58; Univ Notre Dame, PhD(math), 63. *Prof Exp:* Instr math, Univ Notre Dame, 63-64; lectr, Univ Sussex, 64-65; asst prof, Col of the Holy Cross, 66-68; lectr, Univ Sussex, 68-70; ASSOC PROF MATH & CHMN DEPT, COL OF THE HOLY CROSS, 70- *Mem:* Am Math Soc; Math Asn Am; London Math Soc; Soc Indust & Appl Math. *Res:* Analysis; functional analysis; theory of distributions and differential equations. *Mailing Add:* Dept of Math Col of the Holy Cross Worcester MA 01610

SULTAN, HASSAN AHMED, b Cairo, Egypt, Dec 12, 36; m 83; c 3. CIVIL ENGINEERING, SOIL MECHANICS. *Educ:* Cairo Univ, BSc, 58; Univ Utah, MS, 61; Univ Calif, Berkeley, PhD(civil eng), 65. *Prof Exp:* Instr civil eng, Ein-Shams Univ, Egypt, 58-59; asst specialist in res, Univ Calif, Berkeley, 63-64; proj engr, Woodward, Clyde, Sherard & Assoc, 65-67; assoc prof civil eng, 67-71, PROF CIVIL ENG, UNIV ARIZ, 71- *Concurrent Pos:* Pvt consult, 58-59 & 68-; UN expert, 75; consult, World Bank & Govt of Saudi Arabia, 77-; mem dept soils, geol & found, Trans Res Bd, Nat Acad Sci-Nat Res Coun, mem comt soil & rock properties & chmn comt chem stabilization; res grants, NSF, US Air Force, US Dept Transp, US Nat Park Serv & Govt of Saudi Arabia. *Mem:* Fel AAAS; Am Soc Civil Engrs; Sigma Xi. *Res:* Foundation engineering; slope stability; collapsing soils; earth dams; fluid mechanics; soil stabilization; dust and erosion control. *Mailing Add:* Dept of Civil Eng Univ of Ariz Tucson AZ 85721

SULTZER, BARNET MARTIN, b Union City, NJ, Mar 24, 29; m 56; c 1. MICROBIOLOGY, IMMUNOLOGY. *Educ:* Rutgers Univ, BS, 50; Mich State Univ, MS, 51, PhD(bact), 58. *Prof Exp:* Asst bact, Mich State Univ, 56-58; res assoc microbiol, Princeton Labs, Inc, 58-64; from asst prof to assoc prof, 64-76, interim chmn, 80-82, PROF MICROBIOL & IMMUNOL, STATE UNIV NY DOWNSTATE, MED CTR, 76- *Concurrent Pos:* Vis scientist, Karolinska Inst, Sweden, 71-72; vis prof, Pasteur Inst, 79-80. *Mem:* AAAS; Am Asn Immunologists; Am Soc Microbiol; NY Acad Sci; Harvey Soc; Sigma Xi. *Res:* Lymphocyte activation; immunobiology of bacterial cell wall proteins and lipopolysacchorides. *Mailing Add:* Dept Microbiol & Immunol State Univ NY Health Sci Ctr Brooklyn Brooklyn NY 11203

SULYA, LOUIS LEON, b North Monmouth, Maine, Aug 17, 11; m 37; c 2. BIOCHEMISTRY. *Educ:* Col Holy Cross, BS, 32, MS, 33; St Louis Univ, PhD(org chem), 39. *Prof Exp:* Instr chem, Spring Hill Col, 34-36; res chemist, Reardon Co, Mo, 40-42; assoc prof, 45-50, prof 50-78, EMER PROF BIOCHEM & CHMN DEPT, SCH MED, UNIV MISS, 78- *Mem:* AAAS; Endocrine Soc; Am Chem Soc; Am Soc Biol Chemists; Int Soc Nephrol. *Res:* Endocrinology and comparative biochemistry. *Mailing Add:* 1076 Parkwood Pl Jackson MS 39206

SULZBERG, THEODORE, b New York, NY, May 28, 36; m 57; c 3. ORGANIC POLYMER CHEMISTRY. *Educ:* City Col New York, BS, 57; Brooklyn Col, MA, 59; Mich State Univ, PhD(org chem), 62. *Prof Exp:* Res fel org chem, Ohio State Univ, 62-63; proj scientist, Chem & Plastics Div, Union Carbide Corp, 63-70; res group leader, 70-81, mgr resin res, Corp Lab, 81-84, MGR TECH OPERS, ELECTRONIC COATINGS DIV, SUN CHEM CORP, 84- *Mem:* Am Chem Soc; Tech Asn Pulp & Paper Indust; Inst Paper Chem; Am Electrochem Soc. *Res:* Low temperature condensation polymerization; charge-transfer complexes; impact modification of addition polymers; carbonium ions; chemical finishing of textiles; flame retardants; rosin derivatives; hydrocarbon polymers; polyamide resins; printing ink varnishes; polyester resins; emulsion and dispersion polymers; printed circuit boards; radiation curables. *Mailing Add:* 315 Dennison St Highland Park NJ 08904-2628

SULZER, ALEXANDER JACKSON, b Emmett, Ark, Feb 13, 22; m 42; c 2. MALARIAOLOGY, MEDICAL PARASITOLOGY. *Educ:* Hardin-Simmons Univ, BA, 49; Emory Univ, MSc, 60, PhD(parasitol), 62. *Hon Degrees:* Dr, Univ Cayetano Heredia, Peru, 77. *Prof Exp:* Med parasitologist, Commun Dis Ctr, 52-62, res parasitologist, 62-74, RES MICROBIOLOGIST, CTR DIS CONTROL, 74- *Concurrent Pos:* Res fel, Atomic Energy Comn, Vanderbilt Univ, 50-51; Ctr Dis Control fel, Emory Univ, 59-60 & NIH fel, 60-62; hon prof, Univ Cayetano Heredia, Lima, Peru, 73; adj prof, Sch Pub Health, Univ NC, Chapel Hill, 78- *Honors & Awards:* Science Award, Bausch & Lomb, 40; First Prize, Med Res, Inst Hipolito Unanue, Lima, Peru, 81. *Mem:* Sigma Xi; Am Soc Trop Med & Hyg; Am Soc Parasitol; Nat Registry Microbiologists; fel Indian Soc Malaria & Commun Dis; fel Royal Soc Trop Med & Hyg; fel Am Acad Microbiol; hon mem Peruvian Soc Microbiol. *Res:* Malaria; diagnostic serology of parasitic diseases of man and animals; tagged systems in serology especially fluorescent tagged materials; fine structure of protozoan parasites; immunoparasitology. *Mailing Add:* Ctr for Dis Control 1600 Clifton Rd Atlanta GA 30333

SULZER-AZAROFF, BETH, m 72; c 3. APPLIED BEHAVIOR ANALYSIS, PERSONALIZED INSTRUCTION. *Educ:* City Col New York, BS, 50, MA, 53; Univ Minn, PhD, 66. *Prof Exp:* Assoc prof, guidance & educ psychol, Southern Ill Univ, 66-72; res assoc, Univ Conn Health Ctr, 72-73; PROF PSYCHOL & PROF EDUC, UNIV MASS, 73- *Concurrent Pos:* Consult, psychol training, Mansfield Training Sch, 72-73, Aubrey Daniels & Assoc, 86-, Gen Telephone & Elec, 88-; adv bd, May Inst for Autistic Children, 82; adv bd, Groden Ctr, Southern Ill Col Human Serv, 82; chmn, Bd Sci Affairs, Am Psychol Asn, 86-87. *Mem:* Fel Am Psychol Asn; fel Am Acad Behav Med; Asn Behav Anal (pres, 81-82). *Res:* Occupational safety; training strategies for developmentally disabled; promoting staff participation in organizations; incidental teaching of academic subjects; numerous articles published in various journals. *Mailing Add:* PO Box 103 Storrs CT 06268

SULZMAN, FRANK MICHAEL, b Norfolk, VA, Nov 3, 44; m 69. CIRCADIAN RHYTHMS, SPACE BIOLOGY. *Educ:* Iona Col, BS, 67; State Univ NY, Stony Brook, PhD(biol), 72. *Prof Exp:* Res fel biol, Harvard Univ, 72-74, res fel biophysics, Moscow State Univ, 74-75; res fel physiol, Med Sch, Harvard Univ, 75-76, instr, 76-79; asst prof, 79-82, ASSOC PROF BIOL, STATE UNIV NY, BINGHAMTON, 82- *Concurrent Pos:* Prin investr, Circadian Rhythm Exp, NASA Spacelab I, 78-, Cosmos Circadian Rhythm, USSR-NASA, 80- *Mem:* Am Physiol Soc; Am Soc Photobiol; NY Acad Sci; Int Soc Chronobiol; AAAS. *Res:* Biological rhythms; space biology; photobiology; thermoregulation psychobiology; primate physiology. *Mailing Add:* Life Sci Div NASA Hq Code EBM Washington DC 20546

SUMAN, DANIEL OSCAR, b Panama City, Panama, Sept 16, 50; US citizen. ENVIRONMENTAL LAW, CHEMICAL OCEANOGRAPHY. *Educ:* Middlebury Col, BA, 72; Columbia Univ, MA & MEd, 78; Univ Calif, San Diego, PhD(oceanog), 83; Univ Calif, Berkeley, JD(environ law). *Prof Exp:* Fel, Smithsonian Trop Res Inst, 83-84; asst prof oceanog phys sci, Boston Univ, 85-87; dir environ studies, World Col West, 87-88; ASST PROF MARINE AFFAIRS, UNIV MIAMI, 91- *Concurrent Pos:* Vis investr, Scripps Inst Oceanog, 84; guest investr, Woods Hole Oceanog Inst, 86-88. *Mem:* AAAS; Am Geophys Union; Oceanog Soc. *Res:* Record of tropical biomass burning in coastal marine sediments and the fluxes of carbonaceous particulates to the troposphere; paleoceanographic applications of radionuclides; natural resource damage assessment. *Mailing Add:* 2633 Durant No 206 Berkeley CA 94704

SUMARTOJO, JOJOK, b Surabaya, Indonesia, July 5, 37; Australian citizen; m 66; c 2. SEDIMENTARY PETROLOGY, ECONOMIC GEOLOGY. *Educ:* Bandung Inst Technol, BS, 61; Univ Ky, MS, 66; Univ Cincinnati, PhD(geol), 74. *Prof Exp:* From instr to lectr geol, Bandung Inst Technol, 59-62; lectr geol, Univ Pajajaran, 68-69; instr geol, Univ Adelaide, 69-75; asst prof geol, Vanderbilt Univ, 75-80; res specialist, Exxon Prod Res Co, 80-82, res assoc, 82-86. *Mem:* Sigma Xi; Mineral Asn Can; assoc Geoscientists Int Develop; Am Asn Petrol Geol; Soc Econ Paleont Mineralogists. *Res:* Petrography, geochemistry and mineralogy of fine-grained detrital sedimentary rocks especially black shales, red-beds and coals; well-log analysis of shales; computer geological mapping; statistical geology. *Mailing Add:* 2236 Carlyle Dr Marietta GA 30062

SUMBERG, DAVID A, b Utica, NY, June 28, 42; m 64; c 1. EDUCATION, ADMINISTRATION. *Educ:* Utica Col, BA, 64; Mich State Univ, MS, 66, PhD(physics), 72. *Prof Exp:* Physicist, Eastman Kodak Co, 72-74; asst prof, 74-80, ASSOC PROF PHYSICS, ST JOHN FISHER COL, 80- *Mem:* Am Asn Physics Teachers; Am Phys Soc. *Res:* Molecular spectroscopy. *Mailing Add:* Dept Elect Eng Rochester Inst Tech One Lomb Mem Dr Rochester NY 14623

SUMERLIN, NEAL GORDON, b Freeport, Tex, July 1, 50; m 74; c 2. NUCLEAR CHEMISTRY. *Educ:* Ouachita Univ, BS, 72; Univ Ark, PhD(chem), 77. *Prof Exp:* From asst prof to assoc prof, 76-88, CHMN DEPT, 83-, PROF CHEM, LYNCHBURG COL, 88- *Mem:* AAAS; Sigma Xi; Am Chem Soc. *Mailing Add:* Dept Chem Lynchburg Col Lynchburg VA 24501

SUMMER, GEORGE KENDRICK, b Cherryville, NC, May 8, 23; m 52; c 2. BIOCHEMISTRY, NUTRITION. *Educ:* Univ NC, BS, 44; Harvard Med Sch, MD, 51. *Prof Exp:* From instr to asst prof pediat, 57-65, from asst prof to assoc prof biochem & nutrit, 65-72, PROF BIOCHEM & NUTRIT, UNIV NC, CHAPEL HILL, 72-, ASSOC CLIN PROF PEDIAT, 66-, RES SCIENTIST, CHILD DEVELOP INST, 71- *Concurrent Pos:* Fel pediat metab, Univ NC, Chapel Hill, 54-57; NIH res career develop award, 65-70; vis scientist, Galton Lab, Univ Col, London & Med Res Coun human biochem genetics res unit & dept biochem, King's Col, London, 62-63; vis scientist, Lab Neurochem, NIMH, Bethesda, 76-77, 86. *Mem:* Am Inst Nutrit; Am Soc Hum Genetics; Am Acad Pediat. *Res:* Analytical biochemistry; study of inborn errors of metabolism; pathophysiology of disease; pediatrics. *Mailing Add:* CB 7260 Dept Biochem & Nutrit Univ NC Sch Med Fac Lab Off Bldg Rm 405 Chapel Hill NC 27599-7260

SUMMERFELT, ROBERT C, b Chicago, Ill, Aug 2, 35; m 60; c 3. FISH BIOLOGY. *Educ:* Univ Wis-Stevens Point, BS, 57; Southern Ill Univ, MS, 59, PhD(zool), 64. *Prof Exp:* Lectr zool, Southern Ill Univ, 62-64; asst prof zool, Kans State Univ, 64-66; assoc prof zool, Okla State Univ, 66-71, prof zool & leader Okla Coop Fishery Res Unit, 71-76; chmn dept, 76-85, PROF ANIMAL ECOL, IOWA STATE UNIV, 76- *Concurrent Pos:* Grants, Bur Com Fisheries, 66-69, Nat Marine Fisheries Serv, Off Water Resources Res, 70-71, 74-76 & 79-81, Environ Protection Agency, 71-72, 77-78 & Bur Reclamation, 72 & 74-75; assoc dir, NCent Aquaculture Ctr, 88-90. *Honors & Awards:* Spec Achievement Award, US Fish & Wildlife Serv, 69, 71 & 76; Hon Lectr, Mid-Am State Univ Asn, 87-88. *Mem:* NAm Lake Mgt Soc; Fisheries Soc Brit Isles; Am Fisheries Soc; Wildlife Dis Asn; fel Am Inst Fishery Res Biologists. *Res:* Biology of fishes; microsporidan parasites of fishes; fish biotelemetry; fish culture; intensive culture of walleye from first feeding larvae to food-size. *Mailing Add:* Dept Animal Ecol Iowa State Univ Ames IA 50011

SUMMERFIELD, GEORGE CLARK, b Lansing, Mich, Aug 17, 37; m 58; c 4. PHYSICS, NUCLEAR ENGINEERING. *Educ:* Mich State Univ, BS, 58, PhD(physics), 62. *Prof Exp:* Res assoc nuclear eng, Univ Mich, Ann Arbor, 62-63, from asst prof to assoc prof, 63-68; assoc prof, Va Polytech Inst & State Univ, 68-70; PROF NUCLEAR ENG, UNIV MICH, ANN ARBOR, 70- *Mem:* Am Phys Soc; Am Nuclear Soc. *Res:* High energy, elementary particle and thermal neutron physics; neutron transport theory. *Mailing Add:* Dept of Nuclear Eng Univ of Mich Ann Arbor MI 48109

SUMMERFIELD, MARTIN, b New York, NY, Oct 20, 16; m 45; c 1. PHYSICS. *Educ:* Brooklyn Col, BS, 36; Calif Inst Technol, MS, 37, PhD(physics), 41. *Prof Exp:* Asst chief engr, Jet Propulsion Lab, Calif Inst Technol, 41-43, head, Rocket Res Div, 45-49; chief, spec engines proj, Aerojet Gen Corp, Pasadena, Calif, 43-45; prof aerospace eng, Princeton Univ, 49-78; Astor prof appl sci, NY Univ, 79-80; CHIEF SCIENTIST, PRINCETON COMBUSTION RES LABS INC, DIV FLOW INDUSTS INC, 75-, PRES, 78- *Concurrent Pos:* Mem subcomt combustion, NASA, 49-; ed-in-chief, J Am Rocket Soc, 51-62, Am Inst Aeronaut & Astronaut J, 63 & Astron Acta Eng, 64-73; mem, comt space applns, Off Technol Assessment, US Cong, 79-80; vis prof, Nat Cheng-Kung Univ, Taiwan, 83 & E China Tech Inst, People's Repub China, 85. *Mem:* Nat Acad Eng; fel AAAS; Am Soc Mech Engrs; fel Am Inst Aeronaut & Astronaut; Int Acad Astronaut; Int Combustion Inst; Am Defense Preparedness Asn; Sigma Xi. *Res:* Combustion processes and gas dynamics; feuls and propellants; solid propellant ignition and combustion; internal ballistics of guns and rocket engines; free-piston ballistic compressors; flamespread rates through packed beds of powder; regenerative liquid propellant guns; compression-ignition sensitivity of liquid explosives and monopropellants; fluidized bed combusion of coal; cannon recoil systems; catalytic combustion of fuels; flammability characteristics of gas mixtures; energy conservation strategies; space flight systems. *Mailing Add:* Princeton Combustion Res Labs Inc 4275 US Hwy One Monmouth Junction NJ 08852

SUMMERLIN, LEE R, b Sumiton, Ala, Apr 15, 34; m 58; c 4. CHEMISTRY, SCIENCE EDUCATION. *Educ:* Samford Univ, AB, 55; Birmingham Southern Col, MS, 60; Univ Md, College Park, PhD(sci educ), 71. *Prof Exp:* Chemist, Southern Res Inst, 56-59; teacher, Fla State Univ, 59-61, asst prof chem, 62-71; asst prof sci educ, Univ Ga, 71-72; assoc prof chem & sci educ, 72-77, interim dean, Sch Natural Sci & Math, 80-82, PROF CHEM, UNIV ALA, BIRMINGHAM, 72- *Concurrent Pos:* Chemist, US Pipe & Foundry Co, 53-55; consult, Chem Educ Mat Study, 63-70, Cent Treaty Orgn, 64-, US Agency Int Develop, 66-68 & India Proj, NSF, 68-; teaching assoc, Univ Md, 70-71. *Honors & Awards:* James Conant Award, Am Chem Soc, 69; Ingalls Award, 85. *Mem:* AAAS; Am Chem Soc; Am Inst Chemists; Nat Asn Res Sci Teaching; Nat Sci Teachers Asn. *Res:* Computer assisted instruction; developing chemistry material for nonscience majors; autotutorial and individualized instruction in chemistry; chemical demonstrations. *Mailing Add:* Dept Chem Univ Ala Univ Sta Birmingham AL 35294

SUMMERS, ANNE O, Jacksonville, Fla, Mar 15, 42. MECHANISMS OF GENE REGULATION, BIOLOGICAL-METAL INTERACTIONS. *Educ:* Washington Univ, St Louis, Mo, PhD(molecular biol), 73. *Prof Exp:* From asst prof to assoc prof microbiol, 77-88, PROF MICROBIOL, UNIV GA, 88- *Concurrent Pos:* Consult, EPA, 85, NSF, 85-87, Envirogen, Inc, 89-; Guggenheim fel, 86-87; vis prof, Mass Inst Technol, 86-87. *Mem:* Am Soc Microbiol; Genetics Soc Am; Am Soc Biol Chemists; AAAS. *Res:* Regulation of genes involved in the biotransformation of metals. *Mailing Add:* Dept Microbiol Univ Ga Athens GA 30602

SUMMERS, AUDREY LORRAINE, b San Jose, Calif, Mar 22, 28; m 47. MATHEMATICS, COMPUTER SCIENCE. *Educ:* Stanford Univ, BA, 48; San Jose State Univ, MS, 62. *Prof Exp:* Mathematician data reduction, 47-58, mathematician comput, 58-62, RES SCIENTIST SPACE SCI, AMES RES CTR, NASA, 62- *Honors & Awards:* H Julian Allen Award, Ames Res Ctr, NASA, 76. *Mem:* Asn Comput Mach; Am Geophys Union; Meteoritical Soc; AAAS. *Res:* Evolution and structure of planets and planetary systems by theoretical computer modeling; mathematical modeling of radiative transfer in planetary atmospheres. *Mailing Add:* PO Box 251 Saratoga CA 95070

SUMMERS, CHARLES GEDDES, b Ogden, Utah, Dec 24, 41. ECONOMIC ENTOMOLOGY. *Educ:* Utah State Univ, BS, 64, MS, 66; Cornell Univ, PhD(entom), 70. *Prof Exp:* Res fel entom, Utah State Univ, 64-66; res asst, Cornell Univ, 66-70; asst entomologist, 70-75, ASSOC ENTOMOLOGIST, UNIV CALIF, BERKELEY, 75-, LECTR ENTOM SCI, 77- *Mem:* Entom Soc Am; Entom Soc Can; Crop Sci Soc Am; AAAS. *Res:* Biology and population dynamics of arthropods associated with field crops; host-plant resistance to arthropods attacking alfalfa and economic threshold levels of arthropods attacking alfalfa and cereal crops. *Mailing Add:* Dept Entom Sci Univ Calif Berkeley CA 94720

SUMMERS, CLAUDE M, b Boone Co, Mo, Sept 1, 03. DESIGN OF TRANSFORMERS. *Educ:* Univ Colo, BS, 27, dipl, 33. *Prof Exp:* From design engr to mgr, Gen Elec Co, 27-59; prof elec eng, Okla State Univ, 59-63; RETIRED. *Concurrent Pos:* vis prof, Rensselaer Inst, 69-72, Univ Kans, 72-73 & Auburn Univ, 74-75. *Mem:* Fel Inst Elec & Electronic Engrs. *Res:* Laboratory design. *Mailing Add:* 2113 S Garfield Ave Loveland CO 80537

SUMMERS, DAVID ARCHIBOLD, b Newcastle-on-Tyne, Eng, Feb 2, 44; m 72; c 2. MINING, ROCK MECHANICS. *Educ:* Univ Leeds, BSc, 65, PhD(mining), 68. *Prof Exp:* Asst prof mining eng, 68-74, assoc prof prof mining, 74-77, sr investr, Rock Mech & Explosives Res Ctr, 70-76, prof mining, 77-80, dir, Rock Mech & Explosives Res Ctr, 76-84, CUR PROF, UNIV MO-ROLLA, 80-, DIR, HIGH PRESSURE WATER JET LAB, 84- *Mem:* Am Inst Mining, Metall & Petrol Engrs; Am Soc Mech Engrs; fel Brit Inst Mining Engrs; Brit Inst Mining & Metall; Brit Tunneling Soc; Cleaning Equipment Mfr Asn; Water Jet Technol Asn (pres, 86-87); hon mem Brit Hydromech Res Asn. *Res:* Water jet cutting; surface energy of rock and minerals; novel methods of excavation; cavitation at high pressure; coal mining; geothermal development; strata control. *Mailing Add:* 808 Cypress Dr Rolla MO 65401

SUMMERS, DENNIS BRIAN, b Natrona Heights, Pa, Aug 4, 43; m 63; c 3. PLANT BREEDING. *Educ:* Ind Univ, Pa, BS, 66, MEd, 68; Pa State Univ, PhD(bot), 73. *Prof Exp:* Teacher pub sch, Pa, 66-67; mem staff & fac chem & biol warfare, US Army Chem Ctr & Sch, 67-69; PLANT BREEDER, ASGROW SEED CO, SUBSID UPJOHN CO, 74- *Mem:* Nat Sweet Corn Breeders Asn; Am Seed Trade Asn; Am Soc Hort Sci. *Res:* Breeding for disease resistance at the cell culture level. *Mailing Add:* Asgrow Seed Co 1811 E Florida Ave Nampa ID 83651

SUMMERS, DONALD BALCH, b Maplewood, NJ, Oct 18, 02; m 43; c 1. CHEMISTRY. *Educ:* Wesleyan Univ, BS, 24, MA, 26; Princeton Univ, AM, 29; Columbia Univ, PhD(electro-org chem), 32. *Prof Exp:* Asst, Wesleyan Univ, 24-26; instr chem, Amherst Col, 26-27 & Princeton Univ, 27-29; res & develop chemist, Thomas A Edison Co, 29-30 & Chas Pfizer & Co, 30-33; instr chem, High Sch, Columbia Univ, 33-42 & 46-57; proj asst, Fund Adv Educ Chem Film Proj, Univ Fla, 57-58; prof chem, Glassboro State Col, 58-60; Olin chem teacher, Alton Sr High Sch, Ill, 60-67; prof, 67-73, EMER PROF CHEM, NMEX STATE UNIV, 73- *Concurrent Pos:* Lectr, South Orange-Maplewood Adult Sch, West Orange Adult Sch & Pelham Adult Sch, NJ, 37-42; consult, 34- *Mem:* AAAS; Am Chem Soc; fel Am Inst Chemists. *Res:* Catalysis; electro-organic chemistry; corrosion; photography; chemical education. *Mailing Add:* 1710 Altura Ave Las Cruces NM 88001

SUMMERS, DONALD F, b Pekin, Ill, July 10, 34. VIROLOGY. *Educ:* Univ Ill, MD, 59. *Prof Exp:* PROF MOLECULAR BIOL & CHMN DEPT CELL, VIRAL & MOLECULAR BIOL, SCH MED, UNIV UTAH, 74- *Mem:* Am Soc Virol; Am Soc Microbiol; Am Soc Biol Chemists. *Mailing Add:* Dept Cell Viral Molecular Biol Univ Utah Salt Lake City UT 84132

SUMMERS, DONALD LEE, b North Platte, Nebr, Apr 6, 33; m 66; c 3. APPLIED MATHEMATICS, PHYSICS. *Educ:* Univ Nebr, BS, 55, MA, 57. *Prof Exp:* Analyst math & physics, Air Force Weapons Lab, Kirtland AFB, 57-60; res mathematician, Dikewood Corp, 60-70; sr res mathematician, Falcon Res & Develop Co, Whittaker Corp, 70-77; sr res mathematician, Math & Physics, Dikewood Industs, 77-89; SR MEM TECH STAFF, SANDIA NAT LABS, 89- *Mem:* Am Phys Soc; Math Asn Am. *Res:* Applied mathematical modeling using computer simulation techniques; analysis of biological effects of ionizing nuclear radiation; development of casualty prediction using mathematical and statistical methods; psychological test development and evaluation; real time satellite data reduction and simulation modeling of real time physical systems. *Mailing Add:* 3713 Moon NE Albuquerque NM 87111

SUMMERS, GEOFFREY P, b London, Eng. SOLID STATE PHYSICS. *Educ:* Oxford Univ, BA, 65, PhD(physics), 69. *Prof Exp:* Res asst physics, Univ NC, 70-72; asst prof, 73-77, assoc prof physics, Okla State Univ, 77-; AT NAVAL RES LAB. *Mem:* Am Phys Soc. *Res:* Experimental investigation of the electronic structure of defects in solids by means of photoconductivity, luminescence, optical absorption and lifetime studies. *Mailing Add:* Code 4615 Naval Res Lab Washington DC 20375

SUMMERS, GEORGE DONALD, b Eldorado, Ill, Jan 16, 27; m 50, 79; c 2. TECHNICAL MANAGEMENT, ENGINEERING. *Educ:* US Mil Acad, BS, 49. *Prof Exp:* US Army, 49-56; proj engr, Arma Div, Am Bosch, Arma Corp, 56-58; systs analyst & proj engr, Missile Systs Div, Repub Aviation Corp, 58-69; prog mgr, Space Systs Div, Fairchild Industs, 69-72; dir, Advan Prog Div, Atlantic Res Corp, 72-80, div gen mgr & vpres, Electronics & Commun Div, 80-85, consult, 85-92; DIR, CEREX CORP, 87- *Honors & Awards:* IR-100 Award, 74. *Mem:* Am Inst Elec & Electronic Engrs; Am Inst Aeronaut & Astronaut; Optical Soc Am. *Res:* Systems design and analysis of aerospace, electronics, communications, optics and bio-medical; program and general management. *Mailing Add:* 2150 S Bay Lane Reston VA 22091-4131

SUMMERS, GREGORY LAWSON, b San Mateo, Calif, Oct 25, 51; div; c 2. RESERVOIR FISHERIES RESEARCH, FISHERIES BIO-TELEMETRY. *Educ:* Univ Okla, BS, 73 & MS, 78. *Prof Exp:* Technician, 74-75, biologist I, 75-77, biologist II, 77-82, DIR, OKLA FISHERY LAB, DEPT WILDLIFE, 82- *Mem:* Am Fisheries Soc. *Res:* Reservoir fisheries research dealing with sportfish management and enhancement. *Mailing Add:* 500 E Constellation Norman OK 73072

SUMMERS, HUGH B(LOOMER), JR, b Lake City, Fla, Aug 5, 21; m 46; c 2. CHEMICAL ENGINEERING. *Educ:* Univ Fla, BChE, 43. *Prof Exp:* Res chem engr, Naval Stores Res Lab, Southern Utilization Res & Develop Div, Agr Res Serv, USDA, 47-65; process engr, Chem Div, Union Camp Corp, 65-86; RETIRED. *Mem:* Am Chem Soc; Am Inst Chem Engrs. *Res:* Gum naval stores processing; crude tall oil production; tall oil fractionation. *Mailing Add:* 17 Biscayne Blvd Lake City FL 32055

SUMMERS, JAMES THOMAS, b Nashville, Tenn, Nov 4, 38; m 58; c 3. TEXTILE CHEMISTRY, POLYMERS. *Educ:* Vanderbilt Univ, AB, 60; Fla State Univ, PhD(inorg chem), 63. *Prof Exp:* Fel, Fla State Univ, 64 & Univ Tex, 64-65; res chemist, 65-71, sr res chemist, 71-87, RES ASSOC, TEXTILE FIBERS DEPT, E I DU PONT DE NEMOURS & CO, INC, 87- *Mem:* Am Chem Soc; Tech Asn Pulp & Paper Indust. *Res:* Coordination chemistry; polymer chemistry. *Mailing Add:* Fibers Dept DuPont Co Old Hickory TN 37138

SUMMERS, JAMES WILLIAM, b Logansport, Ind, July 10, 40; m 66; c 2. POLYMER SCIENCE. *Educ:* Rose Hulman Inst Technol, BS, 62; Case Western Reserve Univ, MS, 66, PhD(polymer eng), 71. *Prof Exp:* Sr res & develop assoc & supvr, 62-80, RES & DEVELOP FEL, GEON VINYL DIV, B F GOODRICH CO, 80- *Mem:* Soc Plastics Engrs; Am Chem Soc; Soc Plastics Indust. *Res:* Polymer weatherability; polymer rheology and die design; polymer morphology and physical properties; polymer testing; polymer blends; polymer engineering and weathering; extrustion, effects of sun's heating on plastics. *Mailing Add:* Geon Vinyl Div PO Box 122 Avon Lake OH 44012

SUMMERS, JERRY C, b Charleston, WVa, Dec 8, 42; m 62; c 6. EMISSION CONTROL CATALYSIS. *Educ:* WVa Univ, BS, 64; Univ Fla, PhD(inorg chem), 68. *Prof Exp:* Postdoc fel, Univ Fla, 68-69; res chemist, Celanese Chem Co, 69-70; staff res scientst, Gen Motors Corp, 70-79; Res mgr, Engelhard Corp, 79-86; TECH DIR, ALLIED SIGNAL AUTOMOTIVE CATALYST CO, 86- *Mem:* Soc Automotive Engrs; Am Chem Soc; N Am Catalysis Soc. *Res:* Develop technology to control gaseous emissions from both mobile and stationary sources through heterogeneous catalysis. *Mailing Add:* 4015 E 42nd St Tulsa OK 74135-2740

SUMMERS, JOHN CLIFFORD, b Chicago, Ill, Dec 4, 36; m; c 1. PESTICIDE CHEMISTRY. *Educ:* Augustana Col, Ill, BA, 58; Univ Ill, PhD(org chem), 63. *Prof Exp:* Res chemist & sr res assoc, 63-89, RES FEL, EXP STA, E I DU PONT DE NEMOURS & CO, INC, 90- *Mem:* Am Chem Soc. *Res:* Coordination for pesticide toxicology studies. *Mailing Add:* Exp Sta Bldg 402 E I du Pont de Nemours & Co Inc Wilmington DE 19898-0402

SUMMERS, JOHN DAVID, b St Catherines, Ont, Mar 28, 29; m 55; c 5. POULTRY NUTRITION. *Educ:* Univ Toronto, BSA, 53, MSA, 59; Rutgers Univ, PhD(animal nutrit), 62. *Prof Exp:* PROF NUTRIT, UNIV GUELPH, 56- *Honors & Awards:* Agr Award, Ontario Agr Col. *Mem:* Fel Agr Inst Can; fel Poultry Sci Asn. *Res:* Nutrient requirements of poultry and feeding methods. *Mailing Add:* Dept Animal & Poultry Sci Univ Guelph Guelph ON N1G 2W1 Can

SUMMERS, LUIS HENRY, b Lima, Peru, Sept 6, 39; div; c 1. ARCHITECTURAL ENGINEERING. *Educ:* Univ Notre Dame, BArch, 61, MSc, 65, PhD(struct eng/appl math), 70. *Prof Exp:* Architect housing, Dino Mortara Assoc, Rome, Italy, 61-62; architect, Corp Renovacion Urbana, PR, 62-63; instr struct, Univ Notre Dame, 65-66; asst prof, Yale Univ, 66-70; assoc prof, Univ Okla, 70-72; prof struct, Energy Mgt & Solar Passive Design, Pa State Univ, 72-86; PROF ARCHIT ENG & ENVIRON DESIGN, UNIV COLO, 88- *Concurrent Pos:* Teacher art, Notre Dame Int, Rome Italy, 61-62 & Stanley Clark Sch, South Bend, Ind, 62-63; comput consult, Assoc Eng, New Haven, Conn, 67; comput graphics consult, Yale Univ & Conn Hwy Dept, 68; architect engr, Comprehensive Design Assoc Int Inc, 72-78; comput syst consult, Army Construct Eng Lab, Urbana, Ill, 76; pres, CDA Int Archit Engrs, State Col, Pa, 78-90 & Boulder, Colo, 87- *Honors & Awards:* Cardinal Ledcaro Gold Medal, Cardinal Ledcaro Ecclesiastical Design Found, 60. *Mem:* Sigma Xi; Environ Designers Res Asn; Am Soc Heating Refrig & Air Conditioning Eng; Am Soc Eng Educ; Am Inst Architects; Nat Soc Prof Engrs; Nat Soc Archit Engrs; Sigma Xi. *Res:*

Reduction of buildings energy loss; computer aided modeling of thermal comfort using mean radiant temperature; Colorado Institutions Energy Bonding Project; Dept of Energy-energy efficient lighting systems; HVAC systems energy thermal modeling; development of industrial energy conservation software; integration of building system. *Mailing Add:* Dept Civil & Archit Eng Eng Ctr OT 4-34 Univ Colo Boulder CO 80309-0428

SUMMERS, MAX DUANE, b Wilmington, Ohio, June 5, 38; m 56; c 2. INSECT VIRUS MOLECULAR BIOLOGY, GENE EXPRESSION. *Educ:* Wilmington Col, AB, 62; Purdue Univ, PhD(entom), 68. *Prof Exp:* NSF postdoctoral fac assoc, Cell Res Inst & Dept Bot, Univ Tex Austin, 68-69, from asst prof to assoc prof bot, 69-77; prof, 77-83, DISTINGUISHED PROF ENTOM, BIOCHEM, BIOPHYSICS & GENETICS, TEX A&M UNIV, 83- *Concurrent Pos:* Vis assoc prof & assoc insect pathologist, Dept Entom, Univ Calif, Berkeley, 76; chmn-elect, Bact, Plant, & Insect Virus Div, Am Soc Microbiol, 80; counr, Am Soc Virol, 82-85; ed, Virol, 83-; chair agr biotechnol, Tex A&M Univ, 86- & Invertebrate Virus Subcomt, Int Comt Taxonomy Viruses, 88-; mem, Biotechnol Tech Adv Comt, US Dept Com, 86-90, Nat Acad Sci Comt Human Rights, 89-90, Sci Adv Bd Inst Biosci & Technol, Tex A&M Univ, 87-88, Sci Adv Comt, 88-90; lectr, Found Microbiol, Am Soc Mcrobiol, 86; dir, Ctr Advan Invert Molecular Sci, Tex A&M Univ, 88-; adv, Sericult Biotechnol Inst, Bangladesh, India, 88- *Honors & Awards:* J E Bussard Mem Award, Entom Soc Am, 83 & 86; I M Lewis Lectr, Am Soc Microbiol, 86. *Mem:* Nat Acad Sci; Am Soc Microbiol; Soc Invertebrate Path; Entom Soc Am; Am Soc Virol (pres elect, 91); fel AAAS. *Res:* Molecular biology of baculoviruses and polydnaviruses; development of the only DNA virus cloning and expression vector system for insect cells and insects. *Mailing Add:* Rm 324 Entom Bldg Texas A&M Univ College Station TX 77843-2475

SUMMERS, MICHAEL EARL, b Paducah, Ky, Dec 12, 54; m 87. ORIGIN & EVOLUTION OF PLANETARY ATMOSPHERES, GLOBAL ATMOSPHERIC CHANGE. *Educ:* Murray State Univ, BA, 76; Univ Tex, Dallas, MS, 78; Calif Inst Technol, PhD(planetary sci), 85. *Prof Exp:* Res fel, Johns Hopkins Univ, 86-88; physicist, Computational Physics Inc, 88-89; PHYSICIST, NAVAL RES LAB, 89- *Concurrent Pos:* Lectr, George Mason Univ, 91- *Mem:* Am Astron Soc; Am Geophys Union; Am Meteorol Soc. *Res:* Atmospheric chemistry and physics: chemistry of atmospheric ozone and active trace constituents; global atmospheric change, global warming and atmospheric dynamics; chemistry of atmospheres of Mars, Io, Uranus, Neptune, Triton, Titan and Pluto; stability of planetary atmospheres. *Mailing Add:* Space Sci Div Washington DC 20375-5000

SUMMERS, PHILLIP DALE, b Decatur, Ill, Mar 21, 43; m 65; c 2. SYSTEMS DESIGN & SYSTEMS SCIENCE, COMPUTER SCIENCES. *Educ:* Mass Inst Technol, BS, 65; Syracuse Univ, MS, 71; Yale Univ, MS, 73, MPhil, 74, PhD(computer sci), 75. *Prof Exp:* Res staff, 67-87, exec secy, Sci Adv Comt, Corporate Hq, 87-89, RES STAFF, IBM RES, 90- *Concurrent Pos:* Adj prof, Dept Computer Sci, NY Univ, Westchester, 78-80. *Mem:* Asn Comput Mach; Inst Elect & Electronic Engrs; Am Asn Artificial Intel; Math Asn Am. *Res:* Systems aspects of manufacturing; manufacturing strategies; technology management and assessment. *Mailing Add:* 37 Sands St Mt Kisco NY 10549

SUMMERS, RAYMOND, NEUROLOGICAL RESEARCH. *Prof Exp:* BR CHIEF, SCI REV BR, NAT INST NEUROL DISORDERS & STROKE, NIH, 81- *Mailing Add:* NIH Nat Inst Neurol Disorders & Stroke Sci Rev Br Fed Bldg Rm 9C10A 7550 Wisconsin Ave Bethesda MD 20892

SUMMERS, RICHARD JAMES, b Kendallville, Ind, Nov 13, 43. ALLERGY, PEDIATRICS. *Educ:* Ind Univ, AB, 65, MD, 68; Am Bd Pediat, cert, 73; Am Bd Allergy-Immunol, cert, 74. *Prof Exp:* Rotating intern, Fitzsimons Gen Hosp, Denver, 68-69; pediat resident, Fitzsimons Army Med Ctr, 69-71; allergy-immunol fel, Nat Jewish Hosp & Res Ctr, Univ Colo Med Ctr, Fitzsimons Army Med Ctr, 71-73; chief allergy-immunol serv, 97th Gen Hosp, Frankfurt, WGer, 73-76; asst chief, 76-80, CHIEF, ALLERGY-IMMUNOL SERV, WALTER REED ARMY MED CTR, 80- *Concurrent Pos:* US Army Europe consult allergy-immunol, 7th US Army Surgeon Gen Consult, Heidelberg, WGer, 73-76; asst prof pediat & med, Uniformed Serv Univ Health Sci, 78-; consult allergy-immunol, Nat Inst Allergy & Infectious Dis, NIH, 79-; assoc prof pediat & assoc prof med, Uniformed Services Univ Health Sci, 83- *Mem:* Fel Am Acad Pediat; Am Acad Allergy; Am Fedn Clin Res; Am Thoracic Soc. *Res:* Penicillin allergy; ampicillin allergy; infectious mononucleosis; catecholamines, especially epinephrine and ephedrine; cyclic nucleotides; histamine; immunotherapy; radioallergosorbent testing; insect sting allergy. *Mailing Add:* Dept Pediat & Med Uniformed Serv Univ Health 4301 Jones Bridge Rd Bethesda MD 20814

SUMMERS, RICHARD LEE, b San Bernadino, Calif, Nov 9, 35; m 57; c 4. HYPERBARIC CHAMBER EXPERIMENT DESIGN & CONTROL. *Educ:* Southern Methodist Univ, Dallas, BA, 73. *Prof Exp:* Owner & officer, Garland Instruments, Inc, 76-85; mgr, Accelerated Christian Educ, Inc, 85-89; CONSULT HYPERBARIC BIOSPHERE, CREATION EVIDENCES MUS, GLEN ROSE, TEX, 89-; INSTR COMPUTER REPAIR, SOUTHERN INST BUS & TECHNOL, 91- *Concurrent Pos:* Chmn, Metroplex Inst Origin Sci, 88-90. *Honors & Awards:* Recognition Sci Apparatus, NASA, 89. *Res:* System for venous occlusion plethysmography and system to measure central venous pressure; nature of earth's atmosphere before the great flood; individualized curriculum to teach science courses. *Mailing Add:* Creation Evidences Mus PO Box 309 Glen Rose TX 76043

SUMMERS, ROBERT GENTRY, JR, b Sonora, Calif, Jan 10, 43; m 70. ZOOLOGY. *Educ:* Univ Notre Dame, BS, 65, MS, 67; Tulane Univ, PhD(anat), 71. *Prof Exp:* Fel, Univ Maine, Orono, 70-71, from asst prof to assoc prof zool, 75-76; assoc prof, 76-83, PROF ANAT, STATE UNIV NY, BUFFALO, 83- *Mem:* Am Soc Zoologists; Am Asn Anatomists; Am Soc Cell Biol; Soc Develop Biol. *Res:* Developmental biology, particularly invertebrate embryology and morphogenesis; electron microscopy of developing invertebrates and their gametes; control systems in growth and regeneration. *Mailing Add:* Dept Anat Sci Farber Hall State Univ NY Sch Med Buffalo NY 14214

SUMMERS, ROBERT WENDELL, b Lansing, Mich, July 28, 38; m 61; c 3. GASTROENTEROLOGY. *Educ:* Mich State Univ, BS, 61; Univ Iowa, MD, 65. *Prof Exp:* NIH gastroenterol res fel, 68-70, asst prof, 71-74, assoc prof gastroenterol, 74-, PROF GASTROENTEROL, DEPT MED, DIV GASTROENTEROL, UNIV IOWA HOSPS, 74- *Concurrent Pos:* Assoc dir, Gastroenterol Prog, Gastroenterol Div, Dept Med, Vet Admin Hosp, Iowa City, 70-71, consult, 71-77; Burroughs Wellcome Res Travel grant, 79; Fogerty Sr Int fel, NIH, 80; dir, Digestive Dis Ctr, diag & Therapeut Unit, 85. *Mem:* Am Gastroenterol Asn; Am Col Physicians; Am Soc Gastrointestinal Endoscopy; Am Fedn Clin Res; Am Asn Study Liver Dis; Am Physiol Asn. *Res:* Interrelationships of electrical and motor activity of the small intestine, and intestinal flow as modulated by physiologic, pharmacologic, and pathologic influence; gallstone disease. *Mailing Add:* Dept of Internal Med Univ Hosp Iowa City IA 52242

SUMMERS, WILLIAM ALLEN, JR, b Atlanta, Ga, Dec 4, 44; m 67; c 2. INDUSTRIAL ORGANIC CHEMISTRY. *Educ:* Wabash Col, AB, 66; Northwestern Univ, PhD(chem), 71; Ind State Univ, MBA, 81. *Prof Exp:* Res assoc photochem, Okla Univ, 70-72, vis asst prof, 70-71; prog dir & group leader nucleotide synthesis, Ash Steven Inc, 72-73; sr res assoc photochem, Okla Univ, 73-74; develop chemist heterogeneous catalysis, 74-79, mgr new prod develop, 79-82, dir, new prod develop, Animal Prod Group, 82-84, DIR, COM DEVELOP-GROWTH HORMONE, PITMAN-MOORE INC, 84- *Mem:* Am Chem Soc; Sigma Xi. *Res:* Research and development involving heterogeneous catalysis particularly catalytic reduction and photochemistry of nitroparaffins and their derivatives; applications of and design of new reduction reactions; livestock vaccines in high technology fields of tissue culture and recombinant DNA. *Mailing Add:* 16000 Bentree Forest Circle No 912 Dallas TX 75248

SUMMERS, WILLIAM ALLEN, SR, b Gary, Ind, Apr 22, 14; m 40; c 3. PARASITOLOGY, MEDICAL MICROBIOLOGY. *Educ:* Univ Ill, AB, 35, MS, 36; La State Univ, Tulane Univ, PhD(trop dis), 40. *Prof Exp:* Sr parasitologist pub health lab, Fla State Bd Health, 40-42; parasitologist malaria control, US Army, 42-45; asst prof bact & parasitol, Med Col Va, 45-47; from asst prof to assoc prof, 47-64, prof, 44-79, EMER PROF MICROBIOL, SCH MED, IND UNIV, 79- *Concurrent Pos:* USPHS res grants, Sch Med, Ind Univ, 48-; fel trop med, La State Univ, 58; vis scientist, Gorgas Mem Lab, Panama, 66-67. *Mem:* Am Soc Trop Med & Hyg; Am Soc Parasitol; Am Soc Microbiol; Tissue Cult Asn; Sigma Xi. *Res:* Amebiasis, toxoplasmosis, anaplasmosis, malaria; helminth parasites, especially diagnosis, research into growth and chemotherapy; electron and light microscopy; fluorescent microscopy. *Mailing Add:* Dept Microbiol MS B 43 C Ind Univ Sch Med Indianapolis IN 46223

SUMMERS, WILLIAM CLARKE, b Corvallis, Ore, Sept 13, 36; m 64; c 2. MARINE ECOLOGY. *Educ:* Univ Minn, BME, 59, PhD(zool), 66. *Prof Exp:* Res assoc & investr squid biol, Marine Biol Lab, 66-72, NIH contract squid ecol, 69-72; dir, Shannon Point Marine Ctr, 72-76; ASSOC PROF ECOL, HUXLEY COL ENVIRON STUDIES, WESTERN WASH UNIV, 72- *Concurrent Pos:* Corp mem, Marine Biol Lab. *Mem:* AAAS; Am Soc Mech Engrs; Am Soc Zoologists; Sigma Xi; Am Asn Univ Profs. *Res:* Physiological ecology of aquatic mollusks; life history, autecology and population biology of the squid, Loligo pealei; similar studies on Puget Sound octopus species; marine ecology; coastal ecosystems; aquatic biology; coastal management environmental impact assesstment. *Mailing Add:* Huxley Col of Environ Studies Western Wash Univ Bellingham WA 98225

SUMMERS, WILLIAM COFIELD, b Janesville, Wis, Apr 17, 39; m 65; c 1. MOLECULAR BIOLOGY, BIOCHEMISTRY. *Educ:* Univ Wis-Madison, BS, 61, MS, 63, PhD(molecular biol) & MD, 67. *Hon Degrees:* MA, Yale Univ, New Haven. *Prof Exp:* From asst prof radiobiol to assoc prof radiobiol, molecular biophysics & biochem, 68-78, PROF THERAPEUT RADIOL, MOLECULAR BIOPHYSICS, BIOCHEM & HUMAN GENETICS, YALE UNIV, 78- *Concurrent Pos:* NSF fel, Mass Inst Technol, 67-68. *Mem:* Am Soc Microbiol; Am Soc Biol Chemists; Am Soc Virol; Hist Sci Soc. *Res:* Regulation of gene expression in normal and virus infected cells; DNA repair and mutagenesis in mammalian cells; history of science and medicine. *Mailing Add:* 333 Cedar St New Haven CT 06510-8039

SUMMERS, WILLIAM HUNLEY, b Dallas, Tex, Feb 5, 36. MATHEMATICS. *Educ:* Univ Tex, Arlington, BS, 61; Purdue Univ, West Lafayette, MS, 63; La State Univ, Baton Rouge, PhD(math), 68. *Prof Exp:* From asst prof to assoc prof, 68-78, PROF MATH, UNIV ARK, FAYETTEVILLE, 78- *Concurrent Pos:* Fulbright-Hayes travel grant, Brazil, 73; vis prof, Inst Math, Fed Univ Rio de Janeiro, 73-75; NSF travel grant, India & Ger, 81; vis prof, Univ-GH-Paderborn, 81 & 83 & Univ Essen, 84. *Mem:* Am Math Soc; Math Soc France. *Res:* Asymptotic and weak asymptotic almost periodicity. *Mailing Add:* Dept Math SE 301 Univ Ark Fayetteville AR 72701

SUMMERS, WILMA POOS, b Richmond, Ind, Dec 8, 37; m 65. BIOCHEMICAL GENETICS, MOLECULAR VIROLOGY. *Educ:* Ohio Univ, BS, 59; Univ Wis-Madison, PhD(oncol), 66. *Prof Exp:* Fel oncol, McArdle Lab Cancer Res, Univ Wis-Madison, 66-67; fel biochem, Harvard Med Sch, 67-68; fel pharmacol, 68-69, res assoc, 69-72, res assoc virol, 72-78, SR RES ASSOC VIROL, DEPT THERAPEUT RADIOL, SCH MED, YALE UNIV, 78- *Concurrent Pos:* Consult, Nat Cancer Inst, NIH, 73-74. *Mem:* Am Soc Biol Chemists. *Res:* Biochemical genetics of herpes simplex virus; the isolation and characterization of herpes virus mutants to be used in the development of suppressor genetics in mammalian cell systems. *Mailing Add:* Dept Therapeut Radiol Sch Med Yale Univ 333 Cedar St New Haven CT 06510

SUMMERS-GILL, ROBERT GEORGE, b Sask, Can, Dec 22, 29. EXPERIMENTAL NUCLEAR PHYSICS. *Educ:* Univ Sask, BA, 50, MA, 52; Univ Calif, PhD(physics), 56. *Prof Exp:* From asst prof to assoc prof, 56-66, PROF PHYSICS, McMASTER UNIV, 66- *Mem:* Am Phys Soc; Can Asn

Physicists. *Res:* Atomic beam resonances; nuclear spectroscopy; direct reactions; nuclear shell model calculations; heavy ion reactions. *Mailing Add:* Dept Physics Tandem Accelerator Lab McMaster Univ Hamilton ON L8S 4K1 Can

SUMMERSON, CHARLES HENRY, b Catlettsburg, Ky, Nov 15, 14; m 44; c 3. GEOLOGY. *Educ:* Univ Ill, BS, 38, MS, 40, PhD(geol), 42. *Prof Exp:* Asst geol, Univ Ill, 38-42; asst geologist, US Geol Surv, 43-45; asst prof geol, Mo Sch Mines, 46-47; from asst prof to prof geol, Ohio State Univ, 47-72, asst to dir res found, 58-65; RETIRED. *Concurrent Pos:* Staff assoc, NSF, 65-66. *Mem:* AAAS; Soc Econ Paleont & Mineral; Am Asn Petrol Geol; Geol Soc Am; Int Asn Sedimentol. *Res:* Stratigraphy; sedimentary petrography; paleontology; micropaleontology; photogeology. *Mailing Add:* Dept Geol Ohio State Univ Columbus OH 43210

SUMMERVILLE, RICHARD MARION, b Shippenville, Pa, May 20, 38; m 62; c 2. MATHEMATICS. *Educ:* Clarion State Col, BS, 59; Wash Univ, AM, 65; Syracuse Univ, PhD(math), 69. *Prof Exp:* Instr math, Clarion State Col, 60-64; asst prof, State Univ NY Col Oswego, 68-69; res assoc, Syracuse Univ, 69-70; assoc prof math & chmn dept, Armstrong State Col, 70-73, prof math & head, Dept Math & Comput Sci, 73-80; prof math & dean, Sch Lib Arts & Sci, 80-82, PROF MATH & VPRES, ACAD AFFAIRS, CHRISTOPHER NEWPORT COL, 82- *Mem:* Am Asn Univ Profs; Math Asn Am; Am Math Soc; Sigma Xi. *Res:* Complex analysis; conformal and quasiconformal mapping; Schlicht functions. *Mailing Add:* Vpres Acad Affairs Christopher Newport Col 50 Shoe Lane Newport News VA 23606

SUMMITT, ROBERT L, b Knoxville, Tenn, Dec 23, 32; m 55; c 3. PEDIATRICS, MEDICAL GENETICS. *Educ:* Univ Tenn, MD, 55, MS, 62; Am Bd Pediat, dipl, 62. *Prof Exp:* Intern, Mem Res Ctr & Hosp, Univ Tenn, Knoxville, 56; asst resident pediat, Col Med, Univ Tenn, Memphis, 59-60, chief resident, 60-61, USPHS trainee pediat endocrine & metab dis, Univ, 61-62; fel med genetics, Univ Wis, 63; from instr to assoc prof, 64-71, assoc dean acad affairs, 79-81, PROF PEDIAT & ANAT, COL MED, UNIV TENN, MEMPHIS, 71-, DEAN ACAD AFFAIRS, 81- *Concurrent Pos:* NIH, Children's Bur, State Ment Health Dept & Nat Found res grants; mem pediat staff, City of Memphis, Le Bonheur Children's Hosp; pediat consult, Baptist Mem, St Josph's, Methodist, US Naval, St Jude Children's Res & Arlington Hosps, Baroness Erloyer Hosp & T C Thompson Children's Hosp Med Ctr; mem, Mammalian Genetics Study Sect, NIH, 80- *Mem:* AAAS; Am Acad Pediat; Am Soc Human Genetics; Soc Pediat Res. *Res:* Clinical genetics and cytogenetics. *Mailing Add:* Univ Physicians Peidat 860 Madison Ave Memphis TN 38163

SUMMITT, W(ILLIAM) ROBERT, b Flint, Mich, Dec 6, 35; m 56; c 2. PHYSICAL CHEMISTRY, MATERIALS SCIENCE. *Educ:* Univ Mich, BS, 57; Purdue Univ, PhD(phys chem), 61. *Prof Exp:* Res assoc chem, Mich State Univ, 61-62; res chemist, Corning Glass Works, 62-65; from asst prof to assoc prof mat sci, 65-73, from actg chmn to chmn dept, 71-78, PROF MAT SCI, MICH STATE UNIV, 73- *Concurrent Pos:* Consult failure anal & corrosion. *Mem:* Am Phys Soc; Am Soc Metals; Am Chem Soc; Am Soc Testing & Mat; Nat Asn Corrosion Engrs; Am Soc Testing & Mat. *Res:* Corrosion; failure analysis. *Mailing Add:* Dept Metall Mech & Mat Sci Mich State Univ Col Eng East Lansing MI 48824

SUMMY-LONG, JOAN YVETTE, b Harrisburg, Pa. PHARMACOLOGY. *Educ:* Bucknell Univ, BS, 65; Pa State Univ, MS, 72, PhD(pharmacol), 78. *Prof Exp:* Jr scientist pharmacol, Smith Kline Pharmaceut Labs, 65-69; sr res technician, 69-72, res asst, 72-78, asst prof, 78-83, ASSOC PROF PHARMACOL, PA STATE UNIV, 83- *Concurrent Pos:* Res asst, co-investr NASA grant, 77-78, asst prof, 78-79; speaker, Gordon Res Conf, Angiotensin, 80, Subfornical Orgn Workshop Soc Neurosci, 84, 10th Int Symp Neurosec, 87, Brain-Opioid Systs & Reprod, 87; prin investr, grants NIH-HLBI & HD, Am Heart Asn & NSF. *Mem:* Sigma Xi; Soc Neurosci; Soc Pharmacol Exp Therapeut; Endocrine Soc. *Res:* Neuroendocrinology; central nervous system regulation of blood pressure and hydration; endogenous opioid peptide effects on blood pressure, oxytocin and vasopressin neurosecretion; circumventricular organs. *Mailing Add:* Dept Pharmacol Pa State Univ Col Med Hershey PA 17033

SUMNER, BARBARA ELAINE, b Alexandria, Va, Dec 23, 42. PHYSICAL CHEMISTRY, MATERIALS SCIENCE. *Educ:* Howard Univ, BS, 66; Am Univ, MS, 74, PhD(chem), 77. *Prof Exp:* Chemist chem warfare, Melpar Inc, 66-68; RES CHEMIST MAT, NIGHT VISION & ELECTRO-OPTICS LABS, 68- *Mem:* Am Chem Soc; Electrochem Soc; Sigma Xi; Am Phys Soc. *Res:* Metallurgical analysis, design, fabrication and coordination of intrinsic and extrinsic photosensors and auxiliary components for use in infrared imaging systems. *Mailing Add:* 417 N Alfred St Alexandria VA 22314

SUMNER, DARRELL DEAN, b Kansas City, Mo, Jan 1, 41; m 63; c 3. METABOLISM, ENVIRONMENTAL CHEMISTRY & TOXICOLOGY. *Educ:* Univ Kans, AB, 63, PhD(med chem), 68; Am bd Toxicol, dipl, 80. *Prof Exp:* Sr scientist drug metab, McNeil Labs, Inc, 68-71; sr metab chemist, CIBA-Geigy Corp, 71-75, proj scientist, 75-77, toxicologist, 77-79, sr toxicologist, 79-82, mgr environ invest, Agr Div, 82-88, MGR ANIMAL METABOLISM, CIBA-GEIGY CORP, 88- *Concurrent Pos:* Vis asst prof, Univ of NC Greensboro, 75-76; adj assoc prof pharmacol & toxicol, Bowan Gray Sch Med, Winston Salem, NC, 82- *Mem:* Am Chem Soc; Soc Environ Toxicol & Chem; Am Indust Hyg. *Res:* Safety evaluation of pesticides. *Mailing Add:* Ciba-Geigy Corp Agr Div PO Box 18300 Greensboro NC 27419

SUMNER, DONALD RAY, b Studley, Kans, Sept 20, 37; m 68; c 1. PLANT PATHOLOGY. *Educ:* Kans State Univ, BS, 59; Univ Nebr, MS, 64, PhD(plant path), 67. *Prof Exp:* Plant pathologist, Green Giant Co, Minn, 67-69; from asst prof to assoc prof, 69-83, PROF PLANT PATH, COASTAL PLAIN EXP STA, UNIV GA, 83- *Mem:* Am Phytopath Soc; Sigma Xi. *Res:* Ecology and control of soil-borne pathogenic fungi on vegetables and corn in irrigated, multiple-cropping systems with integrated pest management. *Mailing Add:* Dept Plant Path Univ Ga Coastal Plain Exp Sta Tifton GA 31794-0748

SUMNER, EDWARD D, b Spartanburg, SC, Mar 21, 25; m 47; c 2. PHARMACY. *Educ:* Wofford Col, BS, 48; Med Col SC, BS, 50; Univ NC, MS, 64, PhD(pharm), 66. *Prof Exp:* Instr pharm, Univ NC, 61-65; from asst prof to assoc prof, Univ Ga, 66-75; PROF PHARM COL PHARM, MED UNIV SC, 75-, DIR, GERIATRIC PHARM, 88- *Concurrent Pos:* Consult pharm internship, Vet Admin Hosp, Augusta, Ga, 68-75. *Mem:* Am Geriat Soc; Am Asn Cols Pharm; Sigma Xi. *Res:* Physics of tablet compression and drug release from tablets; drug utilization in nursing homes and mental retardation centers; medication studies on the senior citizen such as compliance to physician's orders and factors determing when, what, and where drugs are purchased; death & dying education. *Mailing Add:* Col Pharm 171 Ashley Ave Charleston SC 29425

SUMNER, ERIC E, b Vienna, Austria, Dec 17, 24; nat US; m 74; c 4. MECHANICAL & ELECTRICAL ENGINEERING. *Educ:* Cooper Union, BME, 48; Columbia Univ, MA, 53, EE, 60. *Prof Exp:* Mem tech staff, Bell Tel Labs, 48-55, head, Transmission Systs Develop Dept, 55-60, dir, Guided Wave Transmission Lab, 60-62, dir, Underwater Systs Lab, 62-67, exec dir, Loop Transmission Div, 67-79, exec dir, Customer Network Oper Div, 79-81, vpres comput technol & mil systs, AT&T Bell Labs, 81-84, oper systs & network planning, 84-88, opers planning, 88-91; PRES, INST ELEC & ELECTRONICS ENGRS, 91- *Honors & Awards:* Alexander Graham Bell Award, Inst Elec & Electronics Engrs, 78; Gano Dunn Medal, Eng Achievement, Cooper Union, 84. *Mem:* Nat Acad Eng; fel Inst Elec & Electronics Engrs. *Res:* Analysis of electromagnetic switching apparatus; semiconductor circuits; electronic switching systems; pulse code modulation transmission systems; submarine surveillance systems; transmission systems. *Mailing Add:* Inst Elec & Electronics Engrs 345 E 47th New York NY 10017

SUMNER, JOHN RANDOLPH, b Corpus Christi, Tex, Aug 28, 44; m 69; c 2. GEOPHYSICS. *Educ:* Univ Ariz, BS, 66; Stanford Univ, MS, 68, PhD(geophys), 71. *Prof Exp:* Asst prof earth sci, Univ Calif, Santa Cruz & fel, Crown Col, 71-72; asst prof geophysics, Lehigh Univ, 72-77; sr res specialist, 77-79, RES ASSOC, EXXON PROD RES CO, 80- *Mem:* Geol Soc Am; Soc Explor Geophys; Am Geophys Union. *Res:* Solid earth geophysics; exploration seismology; gravity and magnetic fields of geologic structures. *Mailing Add:* Exxon Prod Res Co PO Box 2189 Houston TX 77252-2189

SUMNER, JOHN STEWART, b Bozeman, Mont, June 24, 21; m 43; c 3. EXPLORATION GEOPHYSICS, GEOLOGY. *Educ:* Univ Minn, BS, 47 & 48; Univ Wis, PhD, 55. *Prof Exp:* Staff geophysicist, Jones & Laughlin Steel Corp, 54-55; asst prof geophys, Western State Col, Colo, 55-56; mgr, McPhar Geophys, Inc, 56-57; chief geophysicist, Phelps Dodge Corp, 57-63; prof, Col Mines, 63-72, PROF GEOPHYS, COL EARTH SCI, UNIV ARIZ, 72- *Concurrent Pos:* Consult hydrol, Mining Co. *Mem:* Soc Explor Geophys; Am Inst Mining, Metall & Petrol Eng; Am Geophys Union. *Res:* Mining geophysics including electrical resistivity and induced polarization methods and gravity, magnetic, and seismic exploration techniques, hydrology. *Mailing Add:* 728 N Sawtelle Tucson AZ 85716

SUMNER, RICHARD LAWRENCE, b Albany, NY, May 28, 38; m 64; c 2. HIGH ENERGY PHYSICS. *Educ:* State Univ NY Albany, BS, 59; Univ Chicago, SM, 65, PhD(physics), 75. *Prof Exp:* Res assoc physics, 72-73, MEM RES STAFF PHYSICS, PRINCETON UNIV, 73- *Res:* High energy elementary particle studies; direct muon production and particle production at large transverse momentum. *Mailing Add:* 24 Halley Dr Pomona NY 10970-2003

SUMNER, ROGER D, geophysics; deceased, see previous edition for last biography

SUMNERS, DEWITT L, b Ferriday, La, Dec 2, 41. APPLICATION OF TYPOLOGY, MOLECULAR BIOLOGY. *Educ:* La State Univ, BS, 63; Cambridge Univ, PhD(math), 67. *Prof Exp:* From asst prof to assoc prof math, 67-75, PROF MATH, FLA STATE UNIV, 75- *Concurrent Pos:* Vis prof math, Kwansei Gakuan Univ, Japan, 88. *Mem:* Inst Advan Sci; Sigma Xi; AAAS; Am Math Soc. *Res:* Researching new application using Knot theory in DNA experiments; use of mathematics to describe enzyme action. *Mailing Add:* Dept Math Fla State Univ Tallahassee FL 32306

SUMNEY, LARRY W, b Wash, Pa, Aug 8, 40; div; c 1. NATIONAL RESEARCH & TECHNOLOGY NEEDS. *Educ:* Wash & Jefferson Col, BA, 62; George Washington Univ, MS, 69. *Prof Exp:* Res, microelect advan systs, Naval Res Lab, 62-70; staff, Naval Elec Systs Command, 72-77, res dir, 77-80; dir, Very High Speed Integrated Circuits Prog, Dept Defense, 80-82; PRES, CHIEF EXEC OFFICER, SEMICONDUCTOR RES CORP, 82- *Concurrent Pos:* Managing dir, Sematech, 87. *Mem:* Fel Inst Elec & Electronic Engrs. *Mailing Add:* Semiconductor Res Corp 4501 Alexander Dr Box 12053 Research Triangle Park NC 27709

SUMP, CORD H(ENRY), materials science engineering; deceased, see previous edition for last biography

SUMRALL, H GLENN, b Macon, Miss, Nov 8, 42; m 63; c 2. PLANT PHYSIOLOGY, MICROBIOLOGY. *Educ:* Southeastern La Univ, BS, 64; La State Univ, MS, 66, PhD(plant path), 69. *Prof Exp:* Asst prof biol, Cornell Col, 69-73; assoc prof, 73-77, assoc acad dean, 73-84, PROF BIOL, LIBERTY UNIV, 77-, DEAN COL ARTS & SCI, 84- *Mem:* Am Phytopath Soc. *Res:* Fecal coliform pollution in streams and lakes; physiology of plant disease. *Mailing Add:* PO Box 20000 Liberty Univ Lynchburg VA 24506

SUMRELL, GENE, b Apache, Ariz, Oct 7, 19. ORGANIC CHEMISTRY. *Educ:* Eastern NMex Col, AB, 42; Univ NMex, BS, 47, MS, 48; Univ Calif, PhD(chem), 51. *Prof Exp:* Asst, Univ NMex, 47-48; asst, Univ Calif, 48-51; asst prof, Eastern NMex Univ, 51-53; res chemist, J T Baker Chem Co, 53-58; sr org chemist, Southwest Res Inst, 58-59; proj leader chem & plastics div, Food Mach & Chem Corp, 59-61; res sect leader, El Paso Natural Gas Prod Co, Tex, 61-64; sr chemist, Southern Utilization Res & Develop Div, 64-67, head invests, 67-73, res leader, 73-84, COLLABR, SOUTHERN REGIONAL RES CTR, AGR RES SERV, USDA, 84- *Mem:* AAAS; Am Chem Soc; fel Am Inst Chem; Am Oil Chem Soc; Am Asn Textile Chemists & Colorists. *Res:* Fatty acids in tubercle bacilli; organic synthesis; pharmaceuticals; branched-chain compounds; triazoles; petrochemicals; fats and oils; monomers and polymers; cellulose chemistry; cereal chemistry. *Mailing Add:* Southern Regional Res Ctr PO Box 19687 New Orleans LA 70179

SUMSION, H(ENRY) T(HEODORE), b Chester, Utah, Mar 7, 12; m 38; c 4. MATERIALS SCIENCE ENGINEERING. *Educ:* Univ Utah, BS, 38, PhD(phys metall), 49; Univ Ala, MS, 39. *Prof Exp:* Engr, US Bur Mines, 39-42; asst phys metall, Univ Calif, 46-47 & Univ Utah, 47-49; sr engr, Carborundum Co, 49-51; res assoc, Knolls Atomic Power Lab, Gen Elec Co, 51-56, nuclear fuels metallurgist, Atomic Power Equip Dept, 56-57; res scientist, Lockheed Missiles & Space Co, 57-62; res scientist, Ames Res Ctr, NASA, 62-84; RETIRED. *Honors & Awards:* Apollo Award, NASA, 69. *Mem:* AAAS; Am Soc Metals; fel Am Inst Chem; Am Inst Mining, Metall & Petrol Engrs; fel NY Acad Sci. *Res:* Space environment-materials reactions; space materials applications; solid state structures; x-ray diffraction. *Mailing Add:* 5378 Harwood Rd San Jose CA 95124

SUN, ALBERT YUNG-KWANG, b Amoy, Fukien, China, Oct 13, 32; m 64; c 1. BIOCHEMISTRY, NEUROCHEMISTRY. *Educ:* Nat Taiwan Univ, BS, 57; Ore State Univ, PhD(biochem), 67. *Prof Exp:* AEC fel & res assoc biochem, Case Western Reserve Univ, 67-68; res assoc lab neurochem, Cleveland Psychiat Inst, 68-72, proj dir, 72-74; assoc prof, Biochem Dept & res prof Sinclair Comp Med Res Farm, 74-89, ASSOC PROF, DEPT PHARMACOL, UNIV MO, 89- *Concurrent Pos:* NIH gen res support grant, Cleveland Psychiat Inst, 68-73; Cleveland Diabetic Fund res grant, 72; alcohol grant, Nat Inst Alcohol Abuse & Alcoholism, 74; aging grant, Nat Inst Neurol Commun Dis & Strokes, 75-78. *Mem:* AAAS; Am Chem Soc; Biochem Soc; Am Soc Neurochem; Soc Neurosci. *Res:* Functional and structural relationship of the central nervous system membranes; active transport mechanism; drug effects on synaptic transmission of the central nervous system. *Mailing Add:* Dept Pharmacol Univ of Mo Columbia MO 65212

SUN, ALEXANDER SHIHKAUNG, b Feb 21, 39; US citizen; m 68; c 1. BIOCHEMISTRY, CELL BIOLOGY. *Educ:* Taiwan Normal Univ, BS, 63; Univ Calif, Berkeley, PhD(biochem), 71. *Prof Exp:* Guest investr, Dept Biochem Cytol, Rockefeller Univ, 71-72; asst res physiologist, Dept Physiol & Anat, Univ Calif, Berkeley, 72-75; asst prof path, 75-77, ASST PROF NEOPLASTIC DIS, MT SINAI SCH MED, 77- *Mem:* Am Soc Cell Biol; Am Soc Photobiol; Geront Soc Am. *Res:* Biochemistry of human aging, especially the effects of oxidative damage generated by subcellular organelles on cell aging, the mechanism controlling the different lifespans of normal and tumor cells in vitro. *Mailing Add:* 155 E 88th St New York NY 10028

SUN, ANTHONY MEIN-FANG, b Nanking, China, April 10, 35; m 64; c 2. ENDOCRINOLOGY, ARTIFICIAL ORGANS. *Educ:* Nat Taiwan Univ, BSc, 58; Univ Toronto, MSc, PhD(physiol), 72. *Prof Exp:* Res assoc, Hosp Sick Children, 72-75; adj prof, 72-88, PROF PHYSIOL, UNIV TORONTO, 88- *Concurrent Pos:* Res consult, Res Int Hosp Sick Children, 75-77; sr res scientist, Connaught Res Inst, 74-88. *Mem:* Int Soc Artificial Organs; Am Diabetes Asn; Can Physiol Soc; Can Diabetes Asn; Can Biomat Soc. *Res:* Long term culture of pancreatic islets of langerhans; the development of a bioartificial endocrine pancreas and bioartificial liver; differentiation of endocrine cells (somatotrophs and pancreatic beta cells); insulin biosynthesis and secretion. *Mailing Add:* Dept Physiol Fac Med Med Sci Bldg Toronto ON M5S 1A8 Can

SUN, BERNARD CHING-HUEY, b Nanking, China, Aug 23, 37; m 64; c 2. COMPOSITE FOREST PRODUCTS, ADHESION. *Educ:* Nat Taiwan Univ, BSA, 60; Univ BC, MS, 67, PhD(wood-pulp sci), 70. *Prof Exp:* Jr specialist, Nat Taiwan Univ Res Forest, 61-63; demonstr wood pulp sci, Univ BC, 68-70; asst prof wood sci, 70-75, ASSOC PROF WOOD SCI DEPT FORESTRY, MICH TECHNOL UNIV, 75- *Mem:* Forest Prod Res Soc; Soc Wood Sci & Technol. *Res:* Performance-driven wood and fiber composites; wood composite materials; wood-pulp relationships; adhesion and adhesives. *Mailing Add:* Sch Forestry & Wood Prod Mich Technol Univ Houghton MI 49931

SUN, CHANG-TSAN, b Shen-Yang, China, Feb 20, 28; US citizen; m 63; c 3. ENGINEERING MECHANICS, MATERIALS SCIENCE. *Educ:* Nat Taiwan Univ, BS, 53; Stevens Inst Technol, MS, 60; Yale Univ, PhD(solid mech), 64. *Prof Exp:* Sr engr, Atomic Power Develop Assocs Inc, Mich, 64-65; asst prof eng mech, Iowa State Univ, 65-68, assoc prof, 68-77; SR RES ENGR, GENERAL MOTORS RESEARCH LABS, MICH, 77-; PROF, UNIV FLA. *Mem:* Am Soc Mech Engrs; Am Soc Testing & Mat. *Res:* Wave propagation; composite materials; fiber reinforced plastics; stress analysis. *Mailing Add:* Dept Sci & Eng Univ Fla 231 Aero Gainesville FL 32611

SUN, CHAO NIEN, b Hopeh, China, Dec 4, 14; m 46; c 2. EXPERIMENTAL PATHOLOGY. *Educ:* Nat Peking Univ, BSc, 40; Univ Okla, MSc, 50; Ohio State Univ, PhD, 53. *Prof Exp:* Asst biol, Nat Peking Univ, 40-45, lectr, 45-48; asst, Univ Okla, 48-50; asst, Ohio State Univ, 50-51; asst biophys, St Louis Univ, 52-53, res assoc biol & biophys, 54-57; fel anat, Wash Univ, 57-62; from asst prof to assoc prof path, St Louis Univ, 62-67; res assoc prof, Baylor Univ, 67-69; assoc prof, Sch Med & electron microscopist, Hosp, 69-73, PROF PATH, UNIV ARK MED SCI, LITTLE ROCK & CHIEF ELECTRON MICROSCOPE LAB, VET ADMIN HOSP, 73- *Concurrent Pos:* Fel, Inst Divi Thomae Found, 53-54. *Mem:* Am Soc Cell Biol; Electron Micros Soc Am. *Res:* Experimental virology; histochemistry; differentiation; tissue culture; biological and pathological ultrastructure; electron microscopy. *Mailing Add:* Dept Path Univ Ark Med Sch 4301 Markham St Little Rock AR 72206

SUN, CHENG, b Kiang Su, China, Apr 12, 37; m 62; c 1. ELECTRICAL ENGINEERING. *Educ:* Nat Taiwan Univ, BS, 58; Cornell Univ, MS, 62, PhD(microwaves), 65. *Prof Exp:* Mem tech staff microwave appl res, RCA David Sarnoff Res Ctr, 64-71; MEM TECH STAFF, ELECTRON DYNAMICS DIV, HUGHES AIRCRAFT CO, 71- *Mem:* Inst Elec & Electronics Engrs. *Res:* Microwave solid state circuits and devices. *Mailing Add:* Hughes Aircraft Suite 2145 Elec Dyn Div Box 299 Torrance CA 90509

SUN, CHIH-REE, b Hsu-Chen, China, May 6, 23; m 56; c 3. HIGH ENERGY PHYSICS, NUCLEAR PHYSICS. *Educ:* Univ Calcutta, BSc, 47; Univ Calif, Los Angeles, MS, 51, PhD(physics), 56. *Prof Exp:* Teacher, Overseas Chinese High Sch, India, 47-49; mem res staff physics, Princeton Univ, 56-62; asst prof, Northwestern Univ, 62-65; assoc prof, Queens Col, NY, 65-68; assoc prof, 68-78, PROF PHYSICS, STATE UNIV NY ALBANY, 78- *Concurrent Pos:* Consult, Princeton Univ, 66-69. *Mem:* Am Phys Soc. *Res:* High energy experiments with bubble chambers study of elementary particles; channeling in crystals; accelerator; nuclear instrumentation. *Mailing Add:* Dept Physics State Univ NY 1400 Wash Ave Albany NY 12222

SUN, CHIN-TEH, b Taiwan, China, Apr 9, 39; m 68; c 1. ENGINEERING MECHANICS. *Educ:* Nat Taiwan Univ, BS, 62; Northwestern Univ, MS, 65, PhD(mech), 67. *Prof Exp:* Fel, Northwestern Univ, 67-68; from asst prof to assoc prof aeronaut & astronaut, 68-76, PROF AERONAUT & ASTRONAUT, PURDUE UNIV, WEST LAFAYETTE, 76- *Mem:* Am Inst Aeronaut & Astronaut; Am Acad Mech; Am Soc Mech Engrs; Seismol Soc Am. *Res:* Mechanics of composite materials; stress waves; structures; fracture mechanics. *Mailing Add:* Dept Aero & Astronaut Purdue Univ West Lafayette IN 47907

SUN, DEMING, b Shanghai, China, Jan 30, 47; m; c 1. AUTOIMMUNE DISEASE. *Educ:* Shanghai First Med Col, MD, 71; Albert-Ludwig Univ, Ger, MD, 82. *Prof Exp:* Postdoctoral immunol, Nat Jewish Hosp, Denver, Colo, 85-86; res staff, Max-Planck Inst, Ger, 86-88; ASST MEM IMMUNOL, DEPT IMMUNOL, ST JUDE CHILDREN'S RES HOSP, 89- *Mem:* Am Asn Immunologists; Int Soc Neuroimmunol. *Res:* Pathogenesis of demyelinating autoimmune disease. *Mailing Add:* 1982 Vinton Ave Memphis TN 38104

SUN, FANG-KUO, b Taiwan, Repub China, May 8, 46; US citizen; m 70; c 3. DYNAMIC SYSTEMS ANALYSIS. *Educ:* Nat Chiao-Tung Univ, Taiwan, BS, 69; Univ Pittsburgh, MS, 72; Harvard Univ, PhD(decision & control), 76. *Prof Exp:* Res assoc syst eng, Sci Res Lab, Ford Motor Co, 76-77; STAFF ANALYST, SYSTS ENG, ANALYTIC SCI CORP, 77- *Mem:* Inst Elec & Electronics Engrs; Soc Indust & Appl Math; Sigma Xi. *Res:* Systems analysis and design; statistical analysis; operations research. *Mailing Add:* 44 Chequessette Rd Reading MA 01867

SUN, FRANK F, b Kiangshi, China, July 26, 38; m; c 2. BIOCHEMISTRY. *Educ:* Tunghai Univ, Taiwan, BS, 59; Tex Tech Univ, MS, 63; Univ Tex, Austin, PhD(biochem), 66. *Prof Exp:* NIH fel biochem, Purdue Univ, Lafayette, 66-68; res biochemist, Dept Lipid Res, 68-86, RES BIOCHEMIST, DEPT HYPERSENSITIVITY DIS RES, UPJOHN CO, 87- *Concurrent Pos:* Vis scientist, Harvard Med Sch, Boston, 84-85. *Mem:* Am Chem Soc; Am Soc Biol Chemists. *Res:* Lipid biochemistry; biochemistry of protagladins, leukohienes and related compounds; drug metabolism. *Mailing Add:* Dept Hypersensitivity Dis Res Upjohn Co Kalamazoo MI 49001-3298

SUN, HUGO SUI-HWAN, b Hong Kong, Oct 19, 40. ALGEBRA, NUMBER THEORY. *Educ:* Univ Calif, Berkeley, BA, 63; Univ Md, College Park, MA, 66; Univ NB, Fredericton, PhD(math), 69. *Prof Exp:* Asst prof math, Univ NB, Fredericton, 69-70; from asst prof to assoc prof, 70-78, PROF MATH, CALIF STATE UNIV, FRESNO, 78- *Concurrent Pos:* Vis res prof, Acad Sinica & Peking Univ; hon prof math & lit, Fuyang Teachers Col, 87. *Honors & Awards:* First Award Anthology, 5th World Cong Poets. *Mem:* Am Math Soc; Math Asn Am; Soc Indust & Appl Math. *Res:* Group theory, number theory, combinatorial analysis, and finite geometry; combinatorics. *Mailing Add:* Dept of Math Calif State Univ Fresno CA 93740

SUN, HUN H, b Shanghai, China, Mar 27, 25; m 51; c 1. ELECTRICAL ENGINEERING. *Educ:* Nat Chiao Tung Univ, BS, 46; Univ Wash, MS, 50; Cornell Univ, PhD, 55. *Prof Exp:* Asst elec eng, Cornell Univ, 51-53; from asst prof to assoc prof, 53-59, prof elec eng, 58-78, dir biomed eng & sci prog, 64-73, chmn dept, 73-78, ERNEST O LANGE PROF ELEC ENG, DREXEL UNIV, 78- *Concurrent Pos:* Mem, Franklin Inst, 60-80; NIH spec fel & res assoc, Mass Inst Technol, 63-64; ed-in-chief, Trans Biomed Eng, Inst Elec & Electronics Engrs, 72-79; mem study comt, Surg & Bioeng Sect, NIH, 81-85; ed-in-chief, Annals Biomed Eng, 84-90. *Mem:* Am Soc Eng Educ; Inst Elec & Electronics Engrs; Biomed Eng Soc; Sigma Xi. *Res:* Biomedical engineering; network analysis and synthesis; feedback control system. *Mailing Add:* Dept Elec Eng Drexel Univ 32 & Chestnut Sts Philadelphia PA 19104

SUN, JAMES DEAN, b Denver, Colo, Feb 8, 51; m 74. INHALATION TOXICOLOGY, BIOCHEMICAL TOXICOLOGY. *Educ:* Univ Calif, Davis, BS, 73, Riverside, PhD(biochem), 78. *Prof Exp:* Asst, Chem Indust Inst Toxicol, 78-80; inhalation toxicologist, Lovelace Inhalation Toxicol Res Inst, 80-90; ASST DIR, BUSHY RUN RES CTR, UNION CARBIDE CORP, 90- *Concurrent Pos:* Mem, comt on fire toxicol, Nat Acad Sci, 85-87; counr, Mt West Chapter Soc Toxicol. *Mem:* Soc Toxicol; NY Acad Sci; Sigma Xi. *Res:* Biological fate of inhaled organic compounds associated with insoluable particles; biomarkers of human exposures to organic compounds; chemical metabolite identification; pharmacokinetic modelling; over 60 publications. *Mailing Add:* Bushy Run Res Ctr Union Carbide Corp 6702 Mellon Rd Export PA 15632-8902

SUN, JAMES MING-SHAN, b China, May 10, 18; nat US; m 53; c 2. MINERALOGY, GEOPHYSICS. *Educ:* Nat Cent Univ, China, BS, 40; Univ Chicago, MS, 47; La State Univ, PhD(geol), 50. *Prof Exp:* Res fel, Columbia Univ, 50-51; mineralogist, State Bur Mines & Mineral Resources, NMex Inst Mining & Technol, 51-62; resident res fel, Jet Propulsion Lab, Calif Inst Technol, 62-64; res physicist, Air Force Weapons Lab, 64-69; res & writing, 69-74; prof geophys sci, Nat Cent Univ, Taiwan, Repub China, 74-76; DIR, SIGMA RES ASSOCS, 76- *Mem:* AAAS; fel Mineral Soc Am; Am Geophys Union; Geochem Soc; Sigma Xi. *Res:* X-ray crystallography and fluorescent spectroscopy; volcanic rocks; authigenic minerals; minerals of New Mexico; physics of high pressure; hypervelocity impact and cratering; digital simulation techniques. *Mailing Add:* PO Box 14667 Albuquerque NM 87191

SUN, NAI CHAU, genetics, cell biology, for more information see previous edition

SUN, PU-NING, b Tientsin, China, Oct 23, 32; m 67; c 2. MECHANICAL ENGINEERING. *Educ:* Nat Taiwan Univ, BS, 54; Tex A&M Univ, MS, 60, PhD(mech eng), 64. *Prof Exp:* Sr inspector, Taiwan Bur Weights & Measures, China, 54-59; asst prof & acting head dept, 64-65, head dept mech eng, 65-68 & 72-84, actg chmn eng div, 68-72, PROF MECH ENG, CHRISTIAN BROS COL, 65- *Concurrent Pos:* Am Soc Eng Educ-Ford Found resident fel, Ford Motor Co, 71-72; NSF fel, Nat Bur Standards, 76. *Mem:* AAAS; Am Soc Mech Engrs; Soc Exp Mech; Inst Elec & Electronics Engrs; NY Acad Sci; Am Soc Engr Educ. *Res:* Theoretical and experimental stress analysis; engineering design. *Mailing Add:* Dept Mech Eng Christian Bros Col 650 EPkwy S #50 Memphis TN 38104

SUN, SAMUEL SAI-MING, b Canton, China, Sept 15, 42; m 75; c 1. RECOMBINANT DNA, AGRIBIOTECHNOLOGY. *Educ:* Chinese Univ Hong Kong, BSc, 66; Univ Hong Kong, MSc, 71; Univ Wis-Madison, PhD(bot & hort), 75. *Prof Exp:* Demonstr plant biochem, Dept Bot, Univ Hong Kong, 69-71; res asst, Dept Hort, Univ Wis, 71-74; fel, Univ Wis-Madison, 75-79, asst scientist, 79-80; sr scientist plant biochem, Arco Plant Cell Res Inst, 80-86, prin scientist & dir molecular biol, 86-87; ASSOC PROF, UNIV HI, 87- *Concurrent Pos:* Teacher & leader, molecular biol workshop, China, US, 80, 84, 89 & 91; consult, Food & Agr Orgn, 84, 89. *Honors & Awards:* Res Award, Arco, 84. *Mem:* Int Soc Plant Molecular Biol; Am Soc Plant Physiologists; AAAS. *Res:* Plant enzyme and protein chemistry; enzyme and protein biosynthesis and their regulation during plant development; structure, organization, and expression of plant genes; agricultural biotechnology. *Mailing Add:* Dept Plant Molecular Physiol Univ Hawaii 3190 Maile Way Honolulu HI 96822

SUN, SIAO FANG, b Shaoshing, China, Feb 19, 22; m 51; c 3. PHYSICAL CHEMISTRY. *Educ:* Nat Chengchi Univ, LLB, 45; Univ Utah, MA, 50; Loyola Univ Chicago, MS, 56; Univ Chicago, PhD, 58; Univ Ill, PhD, 62. *Prof Exp:* Prof math, Northland Col, 60-64; from asst prof to assoc prof, 64-75, PROF CHEM, ST JOHN'S UNIV, NY, 75- *Concurrent Pos:* Vis scientist, Nat Ctr Sci Res, Strasbourg, 75 & Meudon-Bellevue, France, 78; scientist, Max Planck Inst Biophys Chem, Gottingen, WGer, 76 & Carlsberg Lab, Copenhagen, Denmark, 81. *Mem:* Am Chem Soc. *Res:* Theoretical molecular kinetics; physical chemistry of macromolecules. *Mailing Add:* 185 47 80th Rd Jamaica NY 11432

SUN, TUNG-TIEN, b China, Feb 20, 47; m 71; c 2. BIOCHEMISTRY, CELL BIOLOGY. *Educ:* Nat Taiwan Univ, BS, 67; Univ Calif, Davis, PhD(biochem), 74. *Prof Exp:* Res assoc biol, Mass Inst Technol, 74-77; asst prof dermat, cell biol & anat, Sch Med, Johns Hopkins Univ, 78-81, assoc prof cell biol & anat, dermat & ophthal, 81-82; assoc prof dermat & pharmacol, 82-86, prof, 86-90, ASSOC DIR, NIH SKIN DIS RES CTR, NY UNIV SCH MED, 88-, RUDOLF L BAER PROF & DIR, EPITHELIAL BIOL GROUP, DEPT DERMATOLOGY, 90- *Concurrent Pos:* NIH res career develop award, 78; Monique Weill-Caulier career develop award, 83; assoc ed, J Investigative Dermat, 90- *Honors & Awards:* Angus lectr, Univ Toronto, 86; Pinkus lectr, Am Acad Dermatpath, 86; Liu lectr, Stanford Med Sch, 87; Montagna lectr, Soc Invest Dermat, 89. *Mem:* Int Soc Differentiation; Soc Investigative Dermat; Asn Res Vision & Ophthal; AAAS; Am Soc Biol Chemists; Am Soc Cell Biol; NY Acad Sci. *Res:* Biochemical studies of mammalian epithelial differentiation; co-author of numerous publications. *Mailing Add:* Dept Dermat & Pharm NY Univ Sch Med 550 First Ave New York NY 10016

SUN, WEN-YIN, b I-Lan, Taiwan; m; c 2. METEOROLOGY, FLUID DYNAMICS. *Educ:* Nat Taiwan Univ, BS, 68; Univ Chicago, MS, 72, PhD(meteorol), 75. *Prof Exp:* Res asst prof, Meteorol Lab Atmospheric Res, Univ Ill, 75-78; vis scientist, Geofluid Dynamics Prog, Princeton Univ, 78-79; from asst to assoc prof, 79-88, PROF ATMOSPHERIC SCI, DEPT GEOSCI, PURDUE UNIV, 88- *Mem:* Am Meteorol Soc. *Res:* Theoretical and numerical studies of dynamic mesoscale meteorology, boundary layer meteorology, turbulences and pollution modeling. *Mailing Add:* 600 Essex West Lafayette IN 47906

SUN, YAN, b Hangzhou, People's Repub China, May 1, 56; m 83; c 1. MOLECULAR PHYSICS. *Educ:* Hangzhou Univ, China, BS, 82; Univ Houston, PhD(physics), 89. *Prof Exp:* RES ASSOC, HARVARD UNIV, 89- *Mem:* Am Phys Soc. *Res:* Molecular collisions; molecular associations and dissociations; molecular-photon interactions; simulate molecular processes with high performance computers. *Mailing Add:* 60 Garden St MS14 Cambridge MA 02138

SUN, YUN-CHUNG, b Shaoton, China, Feb 18, 37; US citizen; m 68; c 1. CHEMICAL ENGINEERING, MATHEMATICS. *Educ:* Tunghai Univ, Taiwan, BS, 61; Univ Mo, MS, 64, PhD(chem eng), 66. *Prof Exp:* Instr chem, Univ Mo, 63-66; SR RES ENGR, DOW CHEM CO, 66- *Concurrent Pos:* Asst prof chem eng, Saginaw Valley Col, 68. *Mem:* Am Inst Chem Engrs; Am Chem Soc. *Res:* Chemical reaction kinetics; component purification and crystallization; polymer sciences. *Mailing Add:* 108 Wilmington Folsom CA 95630

SUNADA, DANIEL K(ATSUTO), b Newcastle, Calif, Apr 3, 36; m 60; c 3. WATER RESOURCES. *Educ:* Univ Calif, Berkeley, BS, 59, MS, 60, PhD(civil eng), 65. *Prof Exp:* PROF CIVIL ENG, COLO STATE UNIV, 65- *Concurrent Pos:* Res grants, 66-72. *Mem:* Am Soc Civil Engrs; Sigma Xi; Int Comn Irrigation & Drainage. *Res:* Fundamentals of fluid flow through porous media; institutional building of international organization; water resources. *Mailing Add:* Dept of Civil Eng Colo State Univ Ft Collins CO 80523

SUNAHARA, FRED AKIRA, b Vancouver, BC, Jan 22, 24; m 52; c 4. PHYSIOLOGY. *Educ:* Univ Western Ont, BSc, 48, PhD(physiol), 52. *Prof Exp:* Fel physiol, Univ Western Ont, 52-53; sci officer aviation med, Defence Res Med Lab, Toronto, 53-61; sr pharmacologist, Ayerst Res Lab, Montreal, 61-64; prof, 64-89, EMER PROF PHARMACOL, UNIV TORONTO, 89- *Concurrent Pos:* Assoc physiol, Univ Toronto, 59-61; Ont Heart Found res grants, 65- *Mem:* Am Physiol Soc; Pharmacol Soc Can. *Res:* Cardiovascular and respiratory physiology with reference to problems in hypobaric environment and disorientation; autonomic and cardiovascular pharmacology; interrelationship of autonomic drugs and eicosanoid group of substances. *Mailing Add:* Dept of Pharmacol Med Sci Bldg Univ of Toronto Fac of Med Toronto ON M5S 1A8 Can

SUNAHARA, YOSHIFUMI, b Mie Prefecture, Japan, July 16, 27; m 60; c 2. THEORY OF VIBRATION & APPLICATION TO ENGINEERING PROBLEM. *Educ:* Osaka Prefectural Univ, BE, 53; Kyoto Univ, ME, 55, PhD(eng), 61. *Prof Exp:* Res asst, Kyoto Univ, 55-62, assoc prof control systs, 63-68; vis assoc prof stochastic control, Brown Univ, 66-68; prof, Kyoto Inst Technol, 68-91; PROF STOCHASTIC CONTROL, OKAYAMA UNIV SCI, 91- *Concurrent Pos:* Assoc ed, Int Fedn Automatic Control J Automatica, 75-; mem, Theory Comt, Int Fedn Automatic Control, 75-; ed, Int J Control Theory & Advan Technol, 85-; pres, Inst Systs Control & Info Engrs, Japan, 90. *Mem:* NY Acad Sci; fel Inst Elec & Electronics Engrs; Soc Indust & Appl Math; Int Asn Math & Computers Simulation. *Res:* Theory and its applications of stochastic optimal control to nonlinear dynamical systems; chaos and fractal basin of nonlinear systems; random vibration theory of nonlinear dynamical systems. *Mailing Add:* Mita-Press Asahi Karasuma Bldg-4F 381 Shimuzu-cho Karasuma-East Takeya Nakagyo-Ku Kyoto 604 Japan

SUND, ELDON H, b Plentywood, Mont, June 6, 30; m 57; c 4. ORGANIC CHEMISTRY. *Educ:* Univ Ill, Urbana, BS, 52; Univ Tex, PhD(org chem), 60. *Prof Exp:* Res chemist, E I du Pont de Nemours & Co, Del, 59-66; asst prof chem, Ohio Northern Univ, 66-67; from asst prof to assoc prof, 67-73, chmn, 84-89, PROF CHEM, MIDWESTERN STATE UNIV, 73- *Concurrent Pos:* Hardin prof, Hardin Found, 75-76; consult, NTex Chem Consult; Fulbright exchange prof, Hat Field Polytech, Eng, 85-86; vis prof, Univ Hawaii, Manor Honolulu, Hawaii, 85-90. *Mem:* AAAS; Sigma Xi; Int Soc Heterocyclic Chem; Am Chem Soc. *Res:* Synthesis of heterocyclic compounds. *Mailing Add:* Dept Chem Midwestern State Univ Wichita Falls TX 76308

SUND, PAUL N, b Thief River Falls, Minn, Nov 13, 32; m 56; c 2. OCEANOGRAPHY. *Educ:* Univ Calif, Santa Barbara, BA, 54; Univ Wash, Seattle, MA, 56. *Prof Exp:* Biol oceanogr, Inter-Am Trop Tuna Comn, 56-63; oceanogr, Nat Marine Fisheries Serv, 63-67, asst chief br marine fisheries, 67-71, nat coordr, Platforms of Opportunity Progs, 71-77, oceanogr, Pac Environ Fisheries Group, 78-89; mariculture syst construct, Hopkins Marine Sta, Stanford Univ, 89-90; CONSULT, 90- *Concurrent Pos:* Mem plankton adv comt, Smithsonian Inst, 64-; adj prof, Fla Atlantic Univ, 66-68; fel, Nat Inst Pub Affairs, 69. *Mem:* AAAS; fel Am Inst Fishery Res Biol; Marine Technol Soc; Am Soc Limnol & Oceanog; Marine Biol Asn UK. *Res:* Fishery and zooplankton ecology; chaetognath taxonomy and ecology; biological oceanography relative to biological indicators of water mass; oceanographic and climatological influences on tunas, billfishes and marine mammals; ecology and life history of tropical tunas; aerial remote sensing of marine mammals; satellite oceanography. *Mailing Add:* 56 Country Club Gate Pacific Grove CA 93950

SUND, RAYMOND EARL, b Capac, Mich, Dec 14, 32; m 61; c 4. NUCLEAR PHYSICS. *Educ:* Univ Mich, BSE(physics) & BSE(math), 55, MS, 56, PhD(physics), 60. *Prof Exp:* Assoc res physicist, Univ Mich Res Inst, 60-61; staff assoc, Gen Atomic Co, San Diego, 61-63, staff mem, 63-77; engr, 77-80, res & develop dir, 80-86, SR ENGR, TOLEDO EDISON CO, TOLEDO, OHIO, 86- *Mem:* Am Phys Soc; Am Nuclear Soc; Sigma Xi. *Res:* Investigation of decay schemes of radioactive isotopes; studies of photonuclear reactions and of prompt and delayed gamma rays from fission; afterheat and shielding studies for reactors. *Mailing Add:* 6014 Glenbeigh Dr Sylvania OH 43560

SUNDA, WILLIAM GEORGE, b Washington, DC, Oct 18, 45. TOXICOLOGY, CHEMICAL OCEANOGRAPHY. *Educ:* Lehigh Univ, BS, 68; Mass Inst Technol, PhD(chem oceanog), 75. *Prof Exp:* RES CHEMIST, BEAUFORT LAB, NAT MARINE FISHERIES SERV, NAT OCEANIC & ATMOSPHERIC ADMIN, 75- *Mem:* Am Soc Limnol & Oceanography. *Res:* Trace metal chemistry in natural waters; the interaction between trace metal chemistry and aquatic organisms; trace metal nutrition and toxicology. *Mailing Add:* Nat Marine Fisheries Serv Beaufort Lab Beaufort NC 28516

SUNDAHL, ROBERT CHARLES, JR, b Minneapolis, Minn, Dec 6, 36; m 58; c 5. FIBER OPTICS, ELECTRONIC MATERIALS. *Educ:* Univ Minn, Minneapolis, BS, 58, MS, 64, PhD(metall), 66. *Prof Exp:* Mem tech staff, Bell Lab, 66-73, supvr piezoelec devices, 73-77, ferrite mat, 77-80 & optical & magnetic mat, 80-84; dir mat sci, Signal Res Ctr, 84-88; MGR, INTEL CORP, 88- *Concurrent Pos:* Mem publ comt, Metall Transactions, 70-76; mem ferroelec comt, Inst Elec & Electronics Engrs, 84-87, tech ed bd, Mat Res Soc, 88-90. *Mem:* Am Ceramic Soc; Mat Res Soc; Optical Soc Am. *Res:* Segregation of impurities at ceramic grain boundaries and surfaces; development of high frequency ferrites for telecommunications; design and fabrication of Lithium Niobium Trioxide integrated optic devices; chemical sensors; electronic polymers; structural ceramics; amorphous alloys; advanced non-destructive testing; electronic packaging. *Mailing Add:* Intel Corp 5000 W Chandler Blvd Chandler AZ 85226

SUNDAR, P, b Tirukkoilur, India, Nov 23, 54. MATHEMATICS, PROBABILITY. *Educ:* Purdue Univ, PhD(statist), 85. *Prof Exp:* ASST PROF MATH, LA STATE UNIV, 85- *Mailing Add:* Louisiana St Univ Baton Rouge LA 70803

SUNDARALINGAM, MUTTAIYA, b Taiping, Malaysia, Sept 21, 31; nat US; m 66; c 1. BIOLOGICAL CRYSTALLOGRAPHY. *Educ:* Ceylon Univ, BSc, 56; Univ Pittsburgh, PhD(chem), 61. *Prof Exp:* Res assoc crystallog, Sch Med, Univ Wash, Seattle, 62-65; res assoc lab molecular biol, Children's Cancer Res Found, Boston, Mass & Harvard Med Sch, 65-66; assoc prof chem, Case Western Reserve Univ, 66-69; PROF BIOCHEM, UNIV WIS-MADISON, 69- *Concurrent Pos:* John Simon Guggenheim fel lab molecular biophys, Dept Zool, Oxford Univ, 75-76. *Honors & Awards:* Res Career Develop Award, Nat Inst Health, 69. *Mem:* AAAS; Am Crystallog Asn; Am Chem Soc; Biophys Soc; Royal Soc Chem. *Res:* X-ray diffraction of significant biological molecules; conformational analysis; nucleic acids DNA, RNA and transfer ribonucleic acids; nucleic acid stereochemical principles; protein molecules of muscle contraction; nucleic acid-drug and nucleic acid-protein complexes; roll of ion pairs in alpha-helix stability and protein folding mechanisms. *Mailing Add:* 3699 Mountainview Rd Columbus OH 43220-4818

SUNDARAM, KALYAN, b Hyderabad, India, Nov 22, 32; US citizen; m 67; c 2. BIOCHEMISTRY. *Educ:* Osmania Univ, India, BVSc, 55; Univ Man, Can, MSc, 63; Purdue Univ, PhD(animal physiol), 66. *Prof Exp:* Vet, Govt Andhra Pradesh, India, 55-61; fel cancer res, Sloan Kettering Inst, 66-67; SCIENTIST, POPULATION COUN, 67- *Mem:* Endocrine Soc; Soc Study Reproduction; Am Soc Andrology. *Res:* Hypothalamus, pituitary and gonadal axis; effects of Luteinizing-hormone releasing hormone, and its analogs on pituitary and testicular function; investigation of new methods of chemical contraception in the male; preclinical testing of contraceptive drugs. *Mailing Add:* Population Coun 1230 York Ave New York NY 10021

SUNDARAM, PANCHANATHAM N, b Madras, India, Aug 23, 39; m 66; c 2. ROCK MECHANICS, SOIL MECHANICS. *Educ:* Alagappa Col Eng & Technol, India, BE, 61; Col Eng Guindy, India, MSc, 65; Univ Calif, Berkeley, PhD(civil eng), 77. *Prof Exp:* Jr engr civil eng, Public Works Dept, Govt Madras, 61-63; lectr, Indian Inst Technol, Bombay, 66-72; asst prof, Univ Wis, Milwaukee, 78-80; scientist rock mech, Lawrence Berkeley Lab, 80-82; GEOTECK SUPVR, BECHIEL INC, 82- *Concurrent Pos:* Geotech engr, Hallenbeck-McKay Assoc, 73; vis engr, Univ Calif, Berkeley, 78; res award, US Comt Rock Mech, Nat Res Coun, 78. *Mem:* Am Soc Civil Engrs; Int Soc Rock Mech; Underground Space Asn. *Res:* Hydraulic flow through rock fractures; determination of rock properties; electro-osmosis in soils; properties of spent oil-shale. *Mailing Add:* 130 Violet Rd Hercules CA 94547

SUNDARAM, R MEENAKSHI, b Salem, India, June 20, 42; US citizen; m 71; c 2. MATHEMATICAL STATISTICS, OPERATIONS RESEARCH. *Educ:* Tex Tech Univ, PhD(indust eng), 76. *Prof Exp:* Asst prof mfg & eng mgt, Old Dominion Univ, 76-78; assoc prof indust & mfr eng, State Univ NY Col Technol, 79; PROF INDUST ENG, TENN TECHNOL UNIV, 80- *Mem:* Sr mem Soc Mfg Engrs; sr mem Inst Indust Engrs; Am Soc Mech Engrs; Am Soc Eng Educ. *Res:* Computer aided process planning; selection of sequence of operations in process planning; design and operation of cellular manufacturing systems. *Mailing Add:* 2809 Old Salem Dr Cookeville TN 38501

SUNDARAM, SWAMINATHA, b Madras, India, Oct 22, 24; US citizen; m 46; c 2. OPTICS-MATERIALS RESEARCH, ACADEMIC ADMINISTRATION. *Educ:* Annamalai Univ, India, BSc Hons, 45, MA, 47, PhD(physics), 57, DSc(physics), 60; Ill Inst Technol, MS, 60. *Prof Exp:* Assoc prof physics, Ill Inst Technol, Chicago, 62-65; prof, Univ Ill, Chicago, 65-82, head dept, 67-76; PROF PHYSICS & CHMN DEPT, UNIV SFLA, TAMPA, 82- *Concurrent Pos:* Lectr, Annamalai Univ, India, 45-57; instr, Ill Inst Technol, Chicago, 57-59; res assoc, Univ Chicago, 59-60; res physicist, BC Res Coun, Vancouver, Can, 60-61; assoc prof mech eng, Univ Sask, Saskatoon, Can, 61-62. *Honors & Awards:* Spectros Award, Soc Appl Spectros, 72. *Mem:* Fel Am Phys Soc; Soc Photo-Optical Instrumentation Engrs. *Res:* Electronic materials and high temperature superconductors; optical studies, thin films, III-V and II-VI related MBE, MOCVD heterostructures, ternaries, superlattices, and multiple quantum wells for device applications. *Mailing Add:* Dept Physics Univ SFla Box 3068 Tampa FL 33620-3068

SUNDARARAJAN, PUDUPADI RANGANATHAN, b Madras, India, Sept 16, 43; m 70; c 2. POLYMER PHYSICS, BIOPHYSICS. *Educ:* Univ Madras, India, BSc, 63, MSc, 65, PhD(biophys). *Hon Degrees:* DSc Univ Madras, India, 81. *Prof Exp:* Res assoc chem, Univ Montreal, 69-71 & 73-75; res assoc, Stanford Univ, 71-73; res scientist polymer & paper physcis, 75-79, proj leader mat characterization, 79-81, mgr mat characterization, 81-88, PRIN SCIENTIST, XEROX RES CTR CAN LTD, 88- *Concurrent Pos:* Consult, Biochem Nomenclature, Int Union Pure & Appl Chem, 71-79; adj prof, Univ Waterloo, 89- *Mem:* Am Chem Soc; Chem Inst Can; Am Phys Soc. *Res:* Studies on solution and solid state conformations; structure and morphology of polymers, blends and composites using microscopy diffraction and theoretical methods and relating them to the properties of polymers. *Mailing Add:* Xerox Res Ctr Canada 2660 Speakman Dr Mississauga ON L5K 2L1 Can

SUNDARESAN, MOSUR KALYANARAMAN, b Madras, India, Sept 2, 29; Can citizen; m 57; c 2. ELEMENTARY PARTICLE PHYSICS. *Educ:* Delhi Univ, India, BSc, 47, MSc, 49; Cornell Univ, NY, PhD(theoret physics), 55. *Prof Exp:* From asst prof to assoc prof, 61-65, PROF PHYSICS, CARLETON UNIV, OTTAWA, 65- *Mem:* Can Asn Physicists; Am Phys Soc; Am Asn Physics Teachers. *Res:* Theoretical physics, particularly research in quantum field theory. *Mailing Add:* Physics Dept Carleton Univ Ottawa ON K1S 5B6 Can

SUNDARESAN, PERUVEMBA RAMNATHAN, b Madras, India, Aug 11, 30; m 70; c 2. NUTRITIONAL BIOCHEMISTRY. *Educ:* Univ Banaras, BSc, 50, MSc, 53; Indian Inst Sci, Bangalore, PhD(biochem), 59. *Prof Exp:* Res asst biochem, Coun Sci & Indust Res, New Delhi, 56-58; res asst, Indian Inst Sci, Bangalore, 58-59; sr res fel, Coun Sci & Indust Res, 59-61; res assoc nutrit biochem radio carbon lab, Univ Ill, Urbana, 61-62; res assoc, Mass Inst Technol, 62-64; Nat Acad Sci-Nat Res Coun res assoc environ biochem, US Army Res Inst Environ Med, Mass, 64-66, res biochemist, 66-68; chief lipids lab res inst, St Joseph Hosp, 68-77; toxicologist-biochemist, Div Toxicol, 77-82, RES CHEMIST, DIV NUTRIT, CTR FOOD & APPL NUTRIT, FOOD & DRUG ADMIN, 82- *Concurrent Pos:* NIH res grants co-investr dept animal sci, Univ Ill, 60-61 & dept nutrit & food sci, Mass Inst Technol, 61-64; res consult, Millersville State Col, 72-77; consult biochem, Vet Admin Hosp, 73-77, consult tech dir, Infant Metab Diag Lab, 78-82; vis scientist, Vet Admin Hosp, 83- *Mem:* AAAS; Am Soc Biol Chemists; Am Inst Nutrit; Brit Biochem Soc; Am Col Toxicol; Int Asn Vitamin & Nutrit Oncol; fel Am Inst Chemists. *Res:* Biochemical function and metabolism of vitamin A. *Mailing Add:* Div Nutrit-Ctr Food Safety & Appl Nutrit Food & Drug Admin-HFF-268 200 C St SW Washington DC 20204

SUNDARESAN, SANKARAN, b Madurai, India, June 9, 55. CATALYSIS & REACTION ENGINEERING. *Educ:* Indian Inst Technol, BS, 76; Univ Houston, MS, 78, PhD(chem eng), 80. *Prof Exp:* Res engr, E I du Pont de Nemours & Co, 81; ASST PROF CHEM ENG, PRINCETON UNIV, 80- *Mem:* Am Inst Chem Eng; Am Chem Soc. *Res:* adsorption processes. *Mailing Add:* Chem Eng Dept Princeton Univ Princeton NJ 08544

SUNDARRAJ, NIRMALA, OPHTHALMOLOGY, CORNEAL DEVELOPMENT. *Educ:* Indian Inst Sci, Bangalore, PhD(microbiol), 71. *Prof Exp:* ASSOC PROF OPHTHAL, BIOCHEM & CONNECTIVE TISSUE, DEPT NEURAL BIOL, ANAT & CELL SCI, UNIV PITTSBURGH, 82- *Mailing Add:* Eye & Ear Inst Univ Pa 203 Lothrop St Pittsburgh PA 15213

SUNDBERG, DAVID K, NEURAL CHEMISTRY, PHARMACOLOGY. *Educ:* Univ Tex, Dallas, PhD(physiol), 74. *Prof Exp:* ASSOC PROF PHARMACOL, BOWMAN GRAY SCH MED, WAKE FOREST UNIV, 82- *Res:* Neuro-Endocrinology. *Mailing Add:* Dept Pharmacol Bowman Gray Sch Med Wake Forest Univ 300 S Hawthorne Rd Winston-Salem NC 27103

SUNDBERG, DONALD CHARLES, b Worcester, Mass, Dec 23, 42; m 66; c 2. POLYMER SCIENCE. *Educ:* Worcester Polytech Inst, BSChE, 65; Univ Del, MChE, 68, PhD(chem eng), 70. *Prof Exp:* Sr chem engr, Monsanto Co, Indian Orchard, 69-74; asst prof chem eng, Univ Idaho, 74-78; asst prof, 78-82, dir indust res, 87-90, ASSOC PROF CHEM ENG, UNIV NH, 82-, EXEC DIR SPONSORED RES, 90- *Concurrent Pos:* Consult, Chem Polymer Indust; vis scientist, Swed Inst Surface Chem, 84-85. *Mem:* Am Inst Chem Eng; Am Chem Soc; NAm Thermal Anal Soc. *Res:* Polymer science and engineering; emulsion polymerization; polymer composites; polymer morphology. *Mailing Add:* Dept Chem Eng Univ NH Durham NH 03824

SUNDBERG, JOHN EDWIN, b China, Nov 21, 47; US citizen; m 71; c 2. FUEL ADDITIVE FORMULATING, PETROLEUM ANALYSIS. *Educ:* Col Wooster, BA, 70; Univ Calif, PhD(phys org chem), 75. *Prof Exp:* Asst res chemist org chem, United Technol Corp, 75-77; res chemist polymer chem, 77-80, res chemist, Process Develop, 80-83, supvr, Safety & Health, 83-85, sr res chemist, Fuel Chem, 85-87, RES ASSOC, FUEL ADDITIVE CHEM, CHEVRON RES & TECH CO, 87- *Mem:* Am Chem Soc. *Res:* Hydroprocessing; heavy oil processing; analysis of heavy oil; coatings; waterproof membranes; chelating agents; polyurethanes; insulation; adhesion; diesel fuel stability; petroleum low-temperature properties; fuel additive formulating; hydrocarbon fuel analysis. *Mailing Add:* Chevron Res & Tech Co 100 Chevron Way Richmond CA 94802-0627

SUNDBERG, KENNETH RANDALL, b Coalville, Utah, Dec 4, 45; m 72; c 2. PHYSICAL CHEMISTRY, GEOLOGICAL CHEMISTRY. *Educ:* Univ Utah, BS, 68; Iowa State Univ, PhD(phys chem), 75. *Prof Exp:* Res assoc biophys, Dept Biochem & Biophys, Iowa State Univ, 76-77; NSF fel, Dept Chem, Harvard Univ, 77-78; res chemist, 80-86, sr chem specialist, 86-88, SUPVR GEO-TECHNOL, PHILLIPS PETROLEUM CO, 88- *Mem:* Am Chem Soc; Sigma Xi; Asn Petrol Geochem Explorationists. *Res:* Quantum chemistry and electronic structure theory; molecular optics; geochemistry; remote sensing. *Mailing Add:* 2130 S Dewey Bartlesville OK 74003

SUNDBERG, MICHAEL WILLIAM, b Battle Creek, Mich. PHYSICAL CHEMISTRY. *Educ:* Albion Col, BA, 69; Stanford Univ, PhD(phys chem), 73. *Prof Exp:* RES CHEMIST, EASTMAN KODAK CO RES LABS, 73- *Honors & Awards:* Von Hevesy Prize Nuclear Med, Soc Nuclear Med, 74. *Mem:* Am Chem Soc; AAAS. *Mailing Add:* 200 Fishers Rd Pittsford NY 14534-9745

SUNDBERG, RICHARD J, b Sioux Rapids, Iowa, Jan 6, 38; m 63; c 2. ORGANIC CHEMISTRY. *Educ:* Univ Iowa, BS, 59; Univ Minn, Minneapolis, PhD(org chem), 62. *Prof Exp:* From asst prof to assoc prof, 64-74, PROF CHEM, UNIV VA, 74- *Concurrent Pos:* NIH res fel, Stanford Univ, 71-72; Fulbright-Hays fel, Inst Chimie Substances Naturelles, CNRS, Gif/Yvette, France, 78-79. *Mem:* Am Chem Soc. *Res:* Synthetic methods in nitrogen heterocyclic chemistry; synthesis of biologically active compounds; photochemical methods of synthesis; anti-cancer compounds; anti-parasitic compounds. *Mailing Add:* Dept Chem Univ Va Charlottesville VA 22903

SUNDBERG, RUTH DOROTHY, b Chicago, Ill, July 29, 15; div. ANATOMY. *Educ:* Univ Minn, BS, 37, MA, 39, PhD(anat), 43, MD, 53; Am Bd Path, dipl, 60. *Prof Exp:* Technician anat, Univ Minn, 37-39; instr path, Wayne Univ, 39-41; asst, Univ, 41-43, from instr to assoc prof anat, 43-60, hematologist, Univ Hosps, 42, dir, Hemat Labs, 45-74, prof lab med, Univ, 63-73, prof, 60-84, prof lab med & path, 73- 84, hematologist, Univ Hosps, 45-84, co-dir, Hemat Labs, 74-84, EMER PROF ANAT, UNIV MINN,

MINNEAPOLIS, 84- *Mem:* Am Asn Anatomists; Sigma Xi. *Res:* Morphologic hematology; diagnosis by aspiration or trephine biopsy of marrow; lymphocytogenesis in human lymph nodes; histopathology of lesions in the bone marrow; agnogenic myeloid metaplasia; sideroblastic anemia and hemochromatosis; fatty acid deficiency; laboratory medicine. *Mailing Add:* Dept Lab Med & Path Box 198 Mayo Bldg Univ Minn Hosp Minneapolis MN 55455

SUNDBERG, WALTER JAMES, b San Francisco, Calif, Sept 16, 39; m 64; c 2. MYCOLOGY. *Educ:* San Francisco State Univ, BA, 62, MA, 67; Univ Calif, Davis, PhD(bot), 71. *Prof Exp:* Lectr bot, Univ Calif, Davis, 71-72; from asst prof to assoc prof bot, 72-90, PROF PLANT BIOL, SOUTHERN ILL UNIV, CARBONDALE, 90- *Concurrent Pos:* Consult, Cent & Southern Ill Regional Poison Resource Ctr; counr, Mycol Soc Am, 81-83 & ed newsletter, 83-86. *Mem:* Mycol Soc Am; Brit Mycol Soc; NAm Mycol Asn; Nat Educ Asn; Int Mushroom Soc Trop. *Res:* Mycology; cytology; systematics; ecology; ultrastructure of fungi with emphasis on Basidiomycetes. *Mailing Add:* Dept Plant Biol Southern Ill Univ Carbondale IL 62901

SUNDE, MILTON LESTER, b Volga, SDak, Jan 7, 21; m 46; c 3. POULTRY NUTRITION. *Educ:* SDak State Col, BS, 47; Univ Wis, MS, 49, PhD, 50. *Prof Exp:* From asst prof to assoc prof poultry sci, 51-57, PROF POULTRY SCI, UNIV WIS-MADISON, 57- *Concurrent Pos:* Res scientist, Rockefeller Found, Colombia, SAm, 60; mem animal nutrit comt, Nat Res Coun, 70; Int Feed Ingredient Asn travel grant, 71; pres, US br, World's Poultry Sci Asn, 84-88. *Honors & Awards:* Res Award, Am Feed Mfrs Asn, 61. *Mem:* Am Chem Soc; Soc Exp Biol & Med; Am Inst Nutrit; fel Poultry Sci Asn (2nd vpres, 65, 1st vpres, 66, pres, 67-68); NY Acad Sci; World's Poultry Sci Asn (US br pres, 84-88, vpres, 88-). *Res:* Unidentified factors; vitamins; amino acids; energy for chickens, turkey and pheasants. *Mailing Add:* Poultry Sci Univ Wis 1056 Animal Sci Bldg Madison WI 53706

SUNDE, ROGER ALLAN, b Madison, Wis, Jan 31, 50; m 88; c 1. NUTRITIONAL BIOCHEMISTRY, SELENIUM BIOCHEMISTRY. *Educ:* Univ Wis-Madison, BS, 72, PhD(biochem), 80. *Prof Exp:* Res assoc biochem, Univ Wis-Madison, 80-81; NIH fel nutrit biochem, Rowett Res Inst, Aberdeen, Scotland, 81-83; asst prof nutrit, Univ Ariz, Tucson, 83-88, assoc prof nutirt & biochem, 88-89; PROF NUTRIT & BIOCHEM, UNIV MO, COLUMBIA, 90-, NUTRIT CLUSTER LEADER, 90- *Concurrent Pos:* Contrib ed, Nutrit Rev, 84-89; nutrit panel mgr, Comptetitive Grants Prog, USDA-Nutrit Res INst, 90-91; vis prof, Japan Sc Prom Sci, 91; mem, Inorg Discussion Group, Royal Soc Chem. *Honors & Awards:* Archer-Daniels-Midland Award, Am Oil Chemists' Soc, 85; Bio-Serv Award, Am Inst Nutrit, 90. *Mem:* Am Inst Nutrit; Am Soc Biochem & Molecular Biol; Am Chem Soc. *Mailing Add:* Nutrit Cluster Univ Mo 10 Gwynn Hall Columbia MO 65211

SUNDEEN, DANIEL ALVIN, b Manchester, NH, Sept 25, 37; m 59; c 2. PETROLOGY, GEOCHEMISTRY. *Educ:* Univ NH, BA, 65; Ind Univ, MA, 67, PhD(geol), 70. *Prof Exp:* PROF GEOL, UNIV SOUTHERN MISS, 71- *Concurrent Pos:* Geologist Gulf Mex, Standard Oil Co, Tex, 66; explor geologist, SW USA, Chevron, 69-71, SE Alaska, Pac Cordillera Explor, 76 & Us Geol Surv Kilauea eruption, Hawaii, 83. *Mem:* Am Inst Mining Metall Petrol Engrs; Mineral Soc Am. *Res:* Subsurface volcanic province and tectonism and associated mineralization in the Gulf Coastal Plain of Mississippi; igneous and metamorphic petrology; Mesozoic dikes and other plutons in Southeast New Hampshire; boracite mineralization in salt formations. *Mailing Add:* Dept of Geol Southern Sta Box 8196 Hattiesburg MS 39406

SUNDEEN, JOSEPH EDWARD, b Manchester, NH, Nov 5, 43; m 64; c 3. ORGANIC CHEMISTRY. *Educ:* Rensselaer Polytech Inst, BS, 64; Purdue Univ, PhD(chem), 68. *Prof Exp:* Fel org chem, Syntex Res Div, Syntex Corp, 68-69; investr cardiovasc res, 69-76, INVESTR ANTI-INFLAMMATORY RES, SQUIBB INST MED RES, 76- *Mem:* Am Chem Soc. *Res:* Cardioactive medicinals; anti-arthritics; anti-ulcer compounds. *Mailing Add:* 1108 Pratt Dr Yardley PA 19067-2835

SUNDELIN, KURT GUSTAV RAGNAR, b Pitea, Sweden, Dec 21, 37; US citizen; m 63; c 5. ORGANIC CHEMISTRY, MEDICINAL CHEMISTRY. *Educ:* Idaho State Univ, BS, 62 & 65, MS, 65; Univ Kans, PhD(med chem), 69. *Prof Exp:* Chemist Biol Sci Res Ctr, Shell Develop Co, 69-86; CHEMIST, AGR PROD, E I DUPONT DE NEMOURS INC, 86- *Mem:* Am Chem Soc. *Res:* Organic chemical synthesis of biologically active agents in area of crop protection chemicals. *Mailing Add:* 25 longspur Dr Wilmington DE 19808

SUNDELIN, RONALD M, b New York, NY, Oct 20, 39; m 67; c 2. RADIO FREQUENCY SUPERCONDUCTIVITY. *Educ:* Mass Inst Technol, BS, 61; Carnegie Inst Technol, MS, 63, PhD(physics), 67. *Prof Exp:* Res physicist, Carnegie-Mellon Univ, 67-69; res assoc elem particle physics, Wilson Lab, 69-75, sr res accoc elem particle physics, Newman Lab, Cornell Univ, 75-87; ASSOC DIR, CONTINUOUS ELECTRON BEAM ACCELERATOR FACIL, 87- *Concurrent Pos:* Consult, 86, mem, Continuous Electron Beam Accelerator Facil Nat Adv Bd, 85-87, prin investr, 86-87; gov distinguished CEBAF prof, Va Polytech Inst & State Univ, 87- *Mem:* Am Phys Soc; Sigma Xi. *Res:* Medium energy experimental physics, especially muon physics; high energy experimental physics; accelerator physics; normal and superconducting radio frequency. *Mailing Add:* 210 Artillery Rd Yorktown VA 23692

SUNDELIUS, HAROLD WESLEY, b Escanaba, Mich, July 6, 30; m 55; c 2. GEOLOGY. *Educ:* Augustana Col, AB, 52; Univ Wis, MS, 57, PhD(geol), 59. *Prof Exp:* Geologist mil geol br, US Geol Surv, 59-61, regional geologist eastern br, 61-65; from asst prof to prof geol, Wittenburg Univ, 65-75, assoc dean col, 71-75; VPRES ACAD AFFAIRS & DEAN COL, AUGUSTANA COL, ILL, 75- *Mem:* Geol Soc Am; Soc Econ Geol; Nat Asn Geol Teachers. *Res:* Appalachian geology, especially Piedmont; economic geology and mineral economics; military geology; geology of the Carolina slate belt; Precambrian geology of the Lake Superior region; massive sulfide deposits in greenstone belts. *Mailing Add:* Augustana Col Rock Island IL 61201

SUNDELL, HAKAN W, b Stockholm, Sweden, July 22, 36; US citizen; m 66; c 3. PEDIATRICS, NEWBORN MEDICINE. *Educ:* Karolinska Inst, Stockholm, MD, 63. *Prof Exp:* Resident pediat, Milwaukee Children's Hosp, Marquette Univ, 64-66; fel, neonatal physiol, Sch Med, Vanderbilt Univ, 66-68; pediatrician, Karolinska Inst, Stockholm, 69-70; instr pediat, 70-71, asst prof pediat neonatology, 71-79, ASSOC PROF PEDIAT NEONATOLOGY, SCH MED, VANDERBILT UNIV, 79- *Mem:* Soc Pediat Res; Microcirculatory Soc; Am Physiol Soc; AMA; Am Acad Pediat. *Res:* Total and newborn pulmonary pathophysiology; hyaline membrane disease and Group B streptococcal pneumonia; apnea reflexes in the newborn period. *Mailing Add:* 6028 Gardendale Dr Nashville TN 37215

SUNDER, SHAM, b June 5, 42; Can citizen; m 79. APPLIED SPECTROSCOPY, SURFACE CHEMISTRY. *Educ:* Univ Delhi, BSc Hons, 62, MSc, 64; Univ Alta, Can, PhD(chem), 72. *Prof Exp:* Fel res, Univ Alta, Can, 72-73 & Nat Res Coun Can, 73-75; res assoc, Nat Res Coun, 75-78; lectr, Heidelberg Univ, WGer, 78-79; assoc res officer, 79-85, RES OFFICER, ATOMIC ENERGY CAN, LTD, 86- *Mem:* Can Inst Chem; Spectros Soc Can; Can Chem Soc. *Res:* Surface chemistry, chemical structure and molecular dynamics studies for nuclear industry; corrosion of UO2 fuel, using spectroscopic (ESCA, IR, laser-Raman, SEM and XRD), electrochemical techniques and model calculations. *Mailing Add:* Atomic Energy Can Ltd Pinawa MB R0E 1L0 Can

SUNDERLAND, JAMES EDWARD, b Philadelphia, Pa, Oct 26, 32; m 57; c 9. MECHANICAL ENGINEERING. *Educ:* Mass Inst Technol, SB, 54; Purdue Univ, MS, 56, PhD(eng), 58. *Prof Exp:* Instr, Purdue Univ, 57-58, asst prof mech eng, Northwestern Univ, 58-62; assoc prof, Ga Inst Technol, 62-65, prof, 65-67; prof mech & aerospace eng, NC State Univ, 67-72; PROF MECH ENG, UNIV MASS, AMHERST, 72- *Concurrent Pos:* Pub Health Serv grant, 62-69; NSF grant, 66-68; consult engr; US Army grant, 81- *Mem:* Fel Am Soc Mech Engrs; Am Soc Heating, Refrigerating & Air-Conditioning Engrs; Inst Food Technologists; Soc Fire Protection Engrs. *Res:* Conduction, convection and radiation heat transfer; heat and mass transfer in biological systems; thermoelectric energy conversion; two-component external flow; freeze-drying; computer aided engineering; seasonal storage of solar energy; heat transfer in injection molding. *Mailing Add:* Dept Mech Eng Univ Mass Amherst MA 01003

SUNDERLIN, CHARLES EUGENE, b Reliance, SDak, Sept 28, 11; m 36; c 4. ORGANIC CHEMISTRY. *Educ:* Univ Mont, AB, 33; Oxford Univ, BA, 35; Univ Rochester, PhD(chem), 39. *Prof Exp:* Instr chem, Union Col, NY, 38-41; instr, US Naval Acad, 41-43, from asst prof to assoc prof, 45-46; sci liaison officer, US Off Naval Res, London, 46-47, from dept sci dir to sci dir, 48-51; dep dir, NSF, 51-57; dep dir, Union Carbide Europ Res Assocs, SA, Belg, 57-62, res mgr defense & space systs dept, Union Carbide Corp, 62-65; spec asst to pres, Nat Acad Sci, 65-69; vpres & secy, Rockefeller Univ, 69-76; spec asst, Nat Sci Bd, 76-78; RETIRED. *Concurrent Pos:* US del gen assembly, Int Coun Sci Unions, Amsterdam, 52 & Oslo, 55; US del, Dirs Nat Res Ctrs, Milan, 55; mem, Comt Experts Scientists' Rights, Paris, 53. *Mem:* AAAS; Am Chem Soc; Royal Soc Chem; Royal Inst Gt Brit; Brit Soc Chem Indust. *Res:* Research administration and management; international cooperation in science and technology. *Mailing Add:* 3036 P St NW Washington DC 20007

SUNDERMAN, DUANE NEUMAN, b Wadsworth, Ohio, July 14, 28; m 53; c 3. RESEARCH ADMINISTRATION. *Educ:* Univ Mich, AB, 49, MS, 54, PhD(chem), 56. *Prof Exp:* Res chemist, Argonne Nat Lab & E I du Pont de Nemours & Co, 51-52; res chemist, Savannah River Proj, 52-54; res asst, Univ Mich, 54-55; prin chemist, 56, proj leader, 56-58, asst div chief, 58-59, chief chem physics div, 59-65, assoc mgr physics dept, 65-69, coordr basic res, 67-69, asst dir, 69-70, mgr soc & mgt systs dept, 70-73, assoc dir, 73-74, dir tech develop, 74-75, assoc dir res, 75-79, dir prog develop, Battelle Mem Inst, 83-; sr vpres, 83-90, EXEC VPRES, MIDWEST RES INST, 90-; DIR, SOLAR ENERGY RES INST, 90- *Concurrent Pos:* Partic prog mgt develop, Harvard Bus Sch, 69; trustee, Columbus Area Leadership Prog, 75-78; trustee, Mo Corp Sci & Technol & Univ Kans. *Mem:* Am Chem Soc; Am Nuclear Soc; hon mem Am Soc Testing & Mat. *Res:* Nuclear fuel development; environmental research; energy research; technical management. *Mailing Add:* Solar Energy Res Inst 1617 Cole Blvd Golden CO 80401

SUNDERMAN, F WILLIAM, JR, b Philadelphia, Pa, June 23, 31; m 63; c 3. PATHOLOGY, PHARMACOLOGY. *Educ:* Emory Univ, BS, 52; Jefferson Med Col, MD, 55. *Prof Exp:* Intern, Jefferson Med Col Hosp, 55-56, instr med, 60-63, assoc, 63-64; from assoc prof to prof path & dir clin lab, Col Med, Univ Fla, 64-68; PROF LAB MED & DEPT HEAD, SCH MED, UNIV CONN, 68-, PROF PHARM, 80-, PROF TOXICOL, 86- *Honors & Awards:* Clin Sci Award, 77; Am Res Award, 78. *Mem:* Am Asn Cancer Res; Am Asn Clin Chem; Am Asn Path; Soc Toxicol; Am Soc Clin Path; Am Col Physicians; Asn Clin Scientists (pres, 64-65); Col Am Pathologists; Endocrine Soc; Am Soc Pharmacol Exp Therapeut. *Res:* Experimental carcinogenesis, trace metal metabolism, and toxicology; clinical biochemistry. *Mailing Add:* Depts of Lab Med & Pharmacol Univ of Conn Health Ctr Farmington CT 06032

SUNDERMAN, FREDERICK WILLIAM, b Altoona, Pa, Oct 23, 98; m 25, 80; c 3. INTERNAL MEDICINE, CLINICAL PATHOLOGY. *Educ:* Gettysburg Col, BS, 19; Univ Pa, MD, 23, MS, 27, PhD(phys chem), 29; Am Bd Internal Med, dipl, 37; Am Bd Path, dipl, 44; Am Bd Clin Chem, dipl, 53. *Hon Degrees:* ScD, Gettysburg Col, 52. *Prof Exp:* Instr, Gettysburg Acad, 19; asst dermat, Univ Pa, 23, from instr to assoc prof res med, 25-47, lectr, 34-47, ward physician, Univ Hosp, 34-40, from assoc to chief chem div, Wm Pepper Lab, Univ Pa Sch Med, 34-47; prof clin path & res med, Sch Med, Temple Univ & dir lab clin med, Univ Hosp, 47-48; head, clin path, Cleveland Clin Found, 48-49; dir clin res, Univ Tex MD Anderson Hosp & Tumor Inst, 49-50; prof clin med, Emory Univ, 50-51; prof clin med & dir div metab res, 51-67, attend physician, 51-76, co-chmn dept, lab med, 70-74, HON CLIN PROF, JEFFERSON MED COL, 76-, DIR, INST CLIN SCI, 65-; prof, 70-

88, EMER PROF PATH, HAHNEMANN MED COL, 88- *Concurrent Pos:* Resident physician, Pa Hosp, 23-25, chief chem lab, 29-33, chief metab & diabetic clins, 29-46, physician, 39-47; med dir explosives res lab, US Bur Mines, Carnegie Inst Technol, 43-46; actg med dir & med consult, Brookhaven Nat Lab, 47-48; consult, Los Alamos Sci Lab, 47-48, US Army Ord, Redstone Arsenal, 47-49; trustee & vpres, Am Bd Path, 48-51, life trustee, 61-; prof, Post-Grad Sch Med, Univ Tex, 49; dir clin labs, Grady Mem Hosp, Atlanta, 49-51; dir educ, Asn Clin Scientists, 49-; chief clin path, Commun Dis Ctr, USPHS, 50-51; med adv, Rohm & Haas Co, 50-71; dir int Seminars Clin Chem & Pathol, 54-; mem, Pa Governor's Task Force Environ Health, 68-75; ed-in-chief, Annals Clin Lab Sci, 71-; consult path, Pa Hosp, 88- *Honors & Awards:* Gold Headed Cane Award, Asn Clin Scientists, 74; Ward Burdick Award, Am Soc Clin Pathologists, 75; Pres Hon Award, Col Am Pathologists, 84. *Mem:* Fel Am Soc Clin Path (pres, 50); fel Am Soc Clin Invest; fel Am Chem Soc; Asn Clin Scientists (pres, 56-58); fel Col Am Path; Int Union of Pure & Appl Chem. *Res:* Serum electrolytes; hazards of nickel and nickel carbonyl exposure; metabolism; clinical chemistry; research medicine. *Mailing Add:* 1833 Delancey Pl Philadelphia PA 19103

SUNDERMAN, HERBERT D, b Horton, Kans, Sept 19, 37; m 62; c 2. SOIL FERTILITY, TILLAGE SYSTEMS. *Educ:* Kans State Univ, BS, 65, MS, 67; Tex A&M Univ, PhD(soil fertility), 76. *Prof Exp:* Res asst agron, Kans State Univ, 65-67; res assoc, Tex A&M Univ, 67-75; asst prof, 75-82, ASSOC PROF AGRON, KANS STATE UNIV, 82- *Concurrent Pos:* Agronomist, Cotton Inc, 82; actg head, Colby Branch Exp Sta, Kans State Univ, 86-87. *Mem:* Am Soc Agron; Soil Sci Soc Am. *Res:* Soil fertility and cultural practices in reduced- and no-tillage systems; transitions from irrigated to dryland cropping; rotations. *Mailing Add:* RR 2 Box 830 Colby KS 67701

SUNDET, SHERMAN ARCHIE, b Litchville, NDak, Sept 25, 18; m 44; c 5. MEMBRANES, TEXTILE FIBERS. *Educ:* Concordia Col, BS, 39; Univ Idaho, MS, 41; Univ Minn, PhD(org chem), 48. *Prof Exp:* Chemist, B F Goodrich Co, Ohio, 42-45; instr org chem, Univ Calif, Los Angeles, 48-50; res chemist textile fibers dept, pioneering res div, E I Du Pont De Nemours & Co Inc, 50-54, res supvr, 54-70, res assoc, polymer prod dept, 70-85, sr res assoc, 85-90; RETIRED. *Mem:* AAAS; Am Chem Soc; Sigma Xi. *Res:* Membranes for water desalination. *Mailing Add:* 2404 Allendale Rd Wilmington DE 19803-5226

SUNDFORS, RONALD KENT, b Santa Monica, Calif, June 3, 32; m 63; c 3. SOLID STATE PHYSICS. *Educ:* Stanford Univ, BS, 54, MS, 55; Cornell Univ, PhD(exp physics), 63. *Prof Exp:* Res assoc, 63-65, from asst prof physics to assoc prof, 65-76, PROF PHYSICS, WASH UNIV, ST LOUIS, 76- *Mem:* AAAS; Am Phys Soc; Am Asn Physics Teachers. *Res:* Nuclear magnetic resonance; low temperature physics; semiconductor research; ultrasonics; acoustic coupling to nuclear spins. *Mailing Add:* Dept Physics Wash Univ St Louis MO 63130

SUNDICK, ROY, b Brooklyn, NY, May 8, 44; m 81; c 1. IMMUNOLOGY. *Educ:* Harpur Col, BA, 65; State Univ NY Buffalo, MA, 69, PhD(microbiol), 72. *Prof Exp:* Austrian Res Coun fel, Inst Gen & Exp Path, Univ Vienna, 71-73; asst prof, 74-79, ASSOC PROF IMMUNOL & MICROBIOL, SCH MED, WAYNE STATE UNIV, 79- *Mem:* Am Asn Immunol; Am Thyroid Asn. *Res:* Pathogenesis of autoimmune thyroid disease. *Mailing Add:* Dept Immunol & Microbiol Wayne State Univ Sch Med Detroit MI 48201

SUNDSTEN, JOHN WALLIN, b Seattle, Wash, Jan 16, 33; m 63; c 6. NEUROBIOLOGY. *Educ:* Univ Calif, Los Angeles, AB, 56, PhD(anat), 61. *Prof Exp:* Asst anat, Sch Med, Univ Calif, Los Angeles, 57-59; NSF fel, 61-62; from instr to asst prof, 62-70, ASSOC PROF ANAT, SCH MED, UNIV WASH, 70- *Concurrent Pos:* Vis scientist, USPHS, 64-66; USPHS res grant, 64-70; NIH spec fel, Bristol, Eng, 68-69; vis prof, Univ Malaya, 73-74. *Mem:* AAAS; Am Asn Anatomists. *Res:* Neurobiology; quantitation of developing central nervous system. *Mailing Add:* Dept Biol Structure Univ Wash Sch Med Seattle WA 98195

SUNG, ANDREW HSI-LIN, b Taipei, Taiwan, Nov 29, 53; m 83; c 1. SOFTWARE ENGINEERING, EXPERT SYSTEMS. *Educ:* Nat Taiwan Univ, BS, 76; Univ Tex-Dallas, MS, 80; State Univ NY-Stonybrook, PhD(computer sci), 84. *Prof Exp:* Asst prof computer sci, Univ Tex-Dallas, 84-87; asst prof, 87-89, ASSOC PROF & CHMN COMPUTER SCI, NMEX INST MINING & TECHNOL, 89- *Concurrent Pos:* Prin investr, Cray (Univ Res Grants), 86-87, Sandia Nat Lab (Univ Res Prog), 88-89, TRW, 89. *Mem:* Asn Comput Mach; Inst Elec & Electronic Engrs. *Res:* Testing and validation of computer software including parallel software and expert (knowledge-based) systems software; parallel processing and supercomputing; logic programming; software engineering; computational complexity theory. *Mailing Add:* Computer Sci Dept NMex Tech Sorocco NM 87801-4682

SUNG, CHANGMO, b Seoul, Korea, Mar 15, 55; m 81; c 1. ELECTRON MICROSCOPY, ADVANCED CERAMICS. *Educ:* Seoul Nat Univ, Korea, BS, 79, MS, 81; Ohio State Univ, MS, 84; Lehigh Univ, PhD(mat sci & eng), 88. *Prof Exp:* Res fac ceramics, Mat Res Ctr, Lehigh Univ, 88-90; MEM TECH STAFF ELECTRON MICROS, GTE LABS, 90- *Mem:* Electron Micros Soc Am; Am Soc Metals; Am Ceramic Soc; Mat Res Soc. *Res:* Characterization of ceramics, composites, thin films of GTE Incorporation using analytical transmission electron microscopy. *Mailing Add:* GTE Labs 40 Sylvan Rd Waltham MA 02254

SUNG, CHENG-PO, b Hsinchu, Taiwan, Oct 21, 35; m 65; c 2. BIOCHEMISTRY, BIOCHEMICAL PHARMACALOGY. *Educ:* Chung Hsing Univ, Taiwan, BSc, 59; McGill Univ, PhD(biochem), 67. *Prof Exp:* Fel biochem, McGill Univ, 66-67; res scientist, Food & Drug Directorate, Dept Nat Health & Welfare, Can, 67-68; res assoc, dept pharmacol, Univ Wis, 68-69; assoc sr investr, Smith Kline & French Labs, 69-77, SR INVESTR, DEPT PHARMACOL & BIOL SCI, SMITH-KLINE LABS, 78- *Mem:* AAAS; NY Acad Sci; Am Chem Soc. *Res:* Receptor binding and study of vasoactive compounds; endothelial cell biology. *Mailing Add:* Smith Kline Beckman Corp L521 at 709 Swedeland Rd Swedeland PA 19479

SUNG, CHEN-YU, b Anguo Xian, China, May 22, 15; m 45; c 4. DRUG METABOLISM & PHARMACOKINETICS, HEPATOPHARMACOLOGY. *Educ:* Yenching Univ, Beiping, China, BS, 41; George Washington Univ, Washington, DC, MS, 49; Univ Calif, San Francisco, PhD(pharmacol), 52. *Prof Exp:* Asst, chem, Yenching Univ, Beiping, China, 41; res assoc, chem, Oriental Chem Co, Tianjin, China, 42-46; lectr, pharmaceut chem, Dept Pharm, Med Col, Peking Univ, 46-48; postdoctoral, pharmacol, Univ Calif San Francisco, 52-53; instr, pharmacol, Med Sch, Tufts Col, Boston, 53-54; asst prof, 54-63, PROF PHARMACOL, CHINESE ACAD MED SCI, BEIJING, 63- *Concurrent Pos:* Ed-in-chief, Acta Pharmaceut Sinica, 78-; adv, Expert Comt Selection Essential Drugs, WHO, 79-83; mem, Exec Comt Int Union Pharmacol Sect Drug Metab, 86-90; assoc ed-in-chief, J Biomed & Environ Sci, 89- *Mem:* Am Soc Pharmacol & Exp Therapeut; Int Soc Study Xenobiotics; Chinese Pharmacol Soc; Chinese Pharmaceut Asn. *Res:* Pharmacology and phasmacokinetics of some constituents isolated from Chinese folk medicine or Chinese traditional medicine in cooperation with chemists with the purpose of discovering new drugs for the treatment of diseases. *Mailing Add:* Inst Mat Med One Xian Nong Tan St Beijing 100050 China

SUNG, CHI CHING, b Nanking, China, Mar 5, 36; m 68. THEORETICAL PHYSICS. *Educ:* Taiwan Nat Univ, BS, 57; Univ Calif, Berkeley, PhD(physics), 65. *Prof Exp:* Res assoc physics, Ohio State Univ, 65-67, lectr, 67-68, asst prof, 68-72; assoc prof, 72-78, PROF PHYSICS, UNIV ALA, HUNTSVILLE, 78- *Concurrent Pos:* Consult, Oak Ridge Nat Lab & US Army Missile Command, Hunstville, Ala. *Mem:* Am Optical Soc. *Res:* Optics; quantum electronics. *Mailing Add:* Dept Physics Univ Ala Huntsville AL 35899

SUNG, CHIA-HSIAING, b Ping-Tung, Taiwan, Sept 4, 39; US citizen; m 68; c 3. SYSTEMS ENGINEERING, SOFTWARE ENGINEERING. *Educ:* Univ Tex, Austin, BSEE, 66, MSEE, 68, PhD(elec eng), 71. *Prof Exp:* Sr mem tech staff, Taiwan Telecommun Admin, 61-64; asst res engr, Univ Tex, Austin, 66 & 71; asst prof elec eng & comput sci, Tex A&I Univ, 71-72; vis asst prof elec eng & vis res asst prof, Coord Sci Lab, Univ Ill, Champaign-Urbana, 72-74; asst prof comput sci & elec eng, Univ Louisville, 74-78; sr staff engr, 78-86, SCIENTIST/ENGR, HUGHES AIRCRAFT CO, LOS ANGELES, 86- *Mem:* Inst Elec & Electronics Engrs; Asn Comput Mach; Am Soc Eng Educ. *Res:* Fault-tolerant spaceborne data processing systems; distributed sytems; microprocessor-based system for improving automobile fuel economy; software satellite-spacecraft simulator. *Mailing Add:* 4721 Steele St Torrance CA 90503-1471

SUNG, JOO HO, b Korea, Feb 18, 27; US citizen; m 59; c 3. PATHOLOGY, NEUROPATHOLOGY. *Educ:* Yonsei Univ, Korea, MD, 52. *Prof Exp:* Resident path, Newark Beth Israel Hosp, 54-57; fel neuropath, Col Physicians & Surgeons, Columbia Univ, 57-61, asst prof, 61-62; from asst prof to assoc prof, 62-69, PROF NEUROPATH, MED SCH, UNIV MINN, MINNEAPOLIS, 69- *Concurrent Pos:* Nat Inst Neurol Dis & Stroke fel, Columbia Univ, 59-61; consult, Minneapolis Vet Admin Hosp, 62-68; mem, NIH Neurol Sci Res Training Comt, 70-73; vis prof, Med Col, Yonsei Univ, Korea, 72. *Mem:* Am Asn Neuropath; Am Acad Neurol; Asn Res Nerv & Ment Dis. *Res:* Aging changes in the nervous system; x-radiation effects on the nervous system. *Mailing Add:* Lab Med/Pathol 198 Mayo Univ Minn 420 Delaware St SE Minneapolis MN 55455

SUNG, MICHAEL TSE LI, b Chung King, China, Mar 5, 40; US citizen; m 68; c 2. MOLECULAR BIOLOGY, BIOCHEMISTRY. *Educ:* Kans State Col, BA, 62; Univ Wis, PhD(molecular biol), 68. *Prof Exp:* Helen Hay Whitney Found fel, Dept Biochem, Univ BC, 68-71; asst prof, 71-75, ASSOC PROF BIOCHEM, SOUTHERN ILL UNIV, 75- *Mem:* Am Chem Soc; Sigma Xi. *Res:* Chromatin structure and function; structure and function of nuclear proteins. *Mailing Add:* 1604 Reston Ct Greensboro NC 27614

SUNG, ZINMAY RENEE, b Shanghai, China, Feb 16, 47; m 74. PLANT DEVELOPMENTAL BIOLOGY. *Educ:* Nat Taiwan Univ, BS, 67; Univ Calif, Berkeley, PhD(plant physiol), 73. *Prof Exp:* Res asst cell physiol, Max Planck Inst Cell Physiol, WBerlin, Ger, 67-68, Univ Calif, Berkeley, 68-73; res assoc somatic genetics, Dept Biol, Mass Inst, Technol, 73-76; ASST PROF, DEPT PLANT BIOL, UNIV CALIF, BERKELEY, 76- *Mem:* Am Soc Plant Physiologists; Am Soc Develop Biologists; Int Soc Plant Molecular Biologists. *Res:* Plant somatic genetics; developmental genetics of plants. *Mailing Add:* Dept Plant Biol Univ Calif Berkeley CA 94720

SUNIER, JULES WILLY, experimental nuclear physics; deceased, see previous edition for last biography

SUNKARA, SAI PRASAD, b Valivarthipadu, India, June 18, 48; US citizen; m 74; c 2. TUMOR BIOLOGY, ENZYME INHIBITORS. *Educ:* Andhra Pradesh Agr Univ, BS, 69; Uttar Pradesh Agr Univ, MS, 71; Indian Inst Sci, Bangalore, PhD(biochem), 75. *Prof Exp:* Fel cell & tumor biol, M D Anderson Hosp & Tumor Inst, Houston, 75-77, res assoc, 77-78, asst prof, 78-80; sr res scientist, Merrell Nat Labs, 80-81, sr res biochemist II & III, 81-84, group leader, 84-86, HEAD, TUMOR BIOL, MERRELL DOW PHARMACEUTICALS, CINCINNATI, 86- *Concurrent Pos:* Adj assoc prof, Dept Biol Chem, Sch Med, Univ Cincinnati, 83- *Honors & Awards:* Young Scientist Award, Indian Nat Sci Acad & Hanumuntharao Mem Gold Medal, 76. *Mem:* Am Asn Cancer Res; Am Soc Cell Biol. *Res:* Biochemical, cellular and immunological differences between normal and tumor cells; development of potent and selective antitumor and antimetastatic agents. *Mailing Add:* Merrell Dow Res Inst Merrell Dow Pharmaceuticals 2110 Galbraith Cincinnati OH 45215

SUNLEY, JUDITH S, b Detroit, Mich, July 26, 46. NUMBER THEORY. *Educ:* Univ Mich, BS, 67, MS, 68; Univ Md, PhD(math), 71. *Prof Exp:* Asst prof math, Dept Math, Statist & Comput Sci, Am Univ, 71-75, assoc prof, 75-81; assoc prof dir, 80-81, prog dir algebra & number theory, 81-84, DIR, DEP DIV, NSF, 84- *Mem:* Am Math Soc. Res; Math Asn Am; Asn Women

Math; AAAS. *Res:* Eisenstein series of Siegel Modular Group; generalized prime discriminants in totally real fields; class numbers of totally imaginary quadratic extensions of totally real fields. *Mailing Add:* 5421 Duvall Dr Bethesda MD 20816

SUNSHINE, GEOFFREY H, b May 17, 48; div; c 2. TOLERANCE INDUCTION, ACTIVATION IMMUNE RESPONSE. *Educ:* Univ Col London, Eng, BSc(biochem), 69, PhD(biochem), 73. *Prof Exp:* ASSOC PROF IMMUNOL, DEPT SURG, VET MED SCH, TUFTS UNIV, 85- *Mem:* Am Asn Immunologists. *Res:* Antigen Presentation. *Mailing Add:* Dept Surg Tufts Univ Sch Vet Med 136 Harrison Ave Boston MA 02111

SUNSHINE, IRVING, b New York, NY, Mar 17, 16; m 39; c 2. CLINICAL CHEMISTRY, TOXICOLOGY. *Educ:* NY Univ, BS, 37, MA, 41, PhD, 50, Am Bd Clin Chem, dipl; Am Bd Forensic Toxicol, dipl. *Prof Exp:* Instr chem, Newark Col Eng, 41-47; asst prof, NJ State Teachers Col, 47-50; toxicologist & clin chemist, City of Kingston Lab, NY, 50-51; sr instr path & pharmacol, 51-54, from asst prof to assoc prof, 54-73, prof toxicol in path & med, Case Western Reserve Univ, 73-76; toxicologist, Cuyahoga County Coroners Lab, 51-86; RETIRED. *Concurrent Pos:* Toxicologist, Univ Hosp, Cleveland, Ohio; Fulbright fel, Vrije Univ, Brussels, 79; toxicol consult, WHO, 83. *Honors & Awards:* Ames Award, Am Asn Clin Chemists, 73; Toxicol Award, Am Acad Forensic Sci, 80. *Mem:* Am Chem Soc; Am Asn Clin Chemists; Am Acad Forensic Sci; Am Asn Poison Control Ctrs; Soc Forensic Toxicologists; Int Asn Forensic Toxicologists. *Res:* Alcohol; barbiturates; toxicology methodology; poison prevention programming; drug abuse. *Mailing Add:* 4173 Hubbartt Dr Palo Alto CA 94306-3834

SUNSHINE, MELVIN GILBERT, b Chicago, Ill, Oct 14, 36; m 70; c 2. MOLECULAR BIOLOGY. *Educ:* Univ Ill, BS, 58; Univ Southern Calif, PhD(bact), 68. *Prof Exp:* Res microbiologist, San Diego State Col, 67-68; Jane Coffin Childs Mem Fund Med Res fel, Karolinska Inst, Sweden, 68-70; USPHS trainee, Dept Molecular Biol & Virus Lab, Univ Calif, Berkeley, 70-72; vis asst prof microbiol, Sch Med, Univ Southern Calif, 72-73; res assoc, 73-78, RES SCIENTIST MICROBIOL, UNIV IOWA, 78- *Mem:* Am Soc Microbiol. *Res:* Bacterial genetics; genetics of the temperate bacterial viruses P2 and P4; host factors associated with phages P2 and P4. *Mailing Add:* Dept Microbiol Univ Iowa Sch Med Iowa City IA 52242

SUNSHINE, WARREN LEWIS, b Passaic, NJ, Sept 10, 47; m 72; c 1. ORGANIC CHEMISTRY. *Educ:* Columbia Univ, AB, 68; Rutgers Univ, MS, 70, PhD(org chem), 74; Fairleigh Dickinson Univ, MBA, 88. *Prof Exp:* NIH res fel natural prods, Univ Va, 74-76; sr chemist cosmetic prods, Am Cyanamid Co, 76-79; sr res scientist, Johnson & Johnson, 79-89; SR RES SCIENTIST, MCNEIL CONSUMER PROD CO, 89- *Mem:* Am Chem Soc; Am Asn Pharmaceut Scientists. *Res:* Natural products; medicinal chemistry. *Mailing Add:* 256 Westwind Way Dresher PA 19025-1417

SUNTHARALINGAM, NAGALINGAM, b Jaffna, Ceylon, June 18, 33; m 61; c 3. RADIOLOGICAL PHYSICS. *Educ:* Univ Ceylon, BSc, 55; Univ Wis, MS, 66, PhD(radiol sci), 67. *Prof Exp:* Asst lectr physics, Univ Ceylon, 55-58; from instr radiol physics to assoc prof radiol, 62-72, PROF RADIOL & RADIATION THERAPY, THOMAS JEFFERSON UNIV MED COL, 72- *Concurrent Pos:* Vis lectr, Grad Sch Med, Univ Pa, 67-72, consult dept physics, 68-72; consult, WHO, 72; chmn bd, Am Col Med Physics, 88. *Mem:* Am Asn Phys Med (pres, 83); Health Physics Soc; fel Am Col Radiol; fel Am Col Med Physics. *Res:* Radiation dosimetry; thermoluminescence dosimetry; clinical dosimetry. *Mailing Add:* Dept Radiation Therapy Thomas Jefferson Univ Med Col Philadelphia PA 19107

SUNUNU, JOHN HENRY, b Havana, Cuba, July 2, 39; US citizen; m 58; c 8. HEAT TRANSFER, FLUID MECHANICS. *Educ:* Mass Inst Technol, BS, 61, MS, 63, PhD(mech eng), 66. *Prof Exp:* Chief engr, Astro Dynamics, Inc, 60-65; from asst prof to assoc prof mech eng, Tufts Univ, 66-82, assoc dean, Col Eng, 68-73; CHIEF OF STAFF, WHITE HOUSE, 89- *Concurrent Pos:* Consult various indust & govt; pres, JHS Eng Co, 66-83 & Thermal Res, Inc, 68-83; mem, Coun Environ Qual Adv Comt Advan Automotive Power Systs; chmn bd, Student Competitions on Relevant Eng, Inc, 71-; chmn, New Technol Educ Task Force, Nat Gov Asn, 87; mem, Pub Eng Policy Comt, Nat Acad Eng; mem, Pres Coun Environ Qual Adv Comt. *Mem:* Nat Acad Eng; Am Soc Mech Engrs; Acad Appl Sci; Sigma Xi; AAAS. *Res:* Heat transfer and temperature control; slow viscous fluid dynamics; approximate methods of mathematical analysis of fluid phenomena; design and optimization of heat transfer equipment. *Mailing Add:* Chief of Staff White House 1600 Pennsylvania Ave NW Washington DC 20500

SUOMI, STEPHEN JOHN, b Chicago, Ill, Dec 16, 45; m 75. DEVELOPMENTAL PSYCHOBIOLOGY, PRIMATOLOGY. *Educ:* Stanford Univ, BA, 68; Univ Wis, MA, 69, PhD(psychol), 71. *Prof Exp:* Res assoc primatology & lectr psychol, Univ Wis-Madison, 71-75, from asst prof to prof psychol, 75-84; CHIEF, LAB COMP ETHOLOGY, NAT INST CHILD HEALTH & HUMAN DEVELOP, 83- *Concurrent Pos:* Prin investr, NSF & NIMH res grants, 75-83; vis scientist, Ctr Interdisciplinary Studies, Univ Bielefeld, 77-78; affil scientist, Wis Regional Primate Res Ctr, 77-; asst dir, res primate behav NIMH, 84-87; adj prof psychol, Univ Wis-Madison, 84-; gov coun, Soc Res Child Develop, 85-; assoc ed, Psychiatry, 86-91; distinguished sci lectr, Am Psychol Asn, 91. *Honors & Awards:* Effron Lectr, Am Col Neuropsychopharmacol, 81; Prokasy Lectr, Univ Utah, 83. *Mem:* Soc Res Child Develop; Int Primatological Soc (secy Americas, 76-84); fel Am Psychol Asn; Int Soc Study Behav Develop; Int Soc Develop Psychobiol; Am Soc Primatologists. *Res:* Biological and behavioral features of development in human and nonhuman primates; genetic and environmental influences on developmental processes; comparative ethology. *Mailing Add:* Lab Comp Ethology Bldg 31 Rm B2B-15 Nat Inst Child Health & Human Develop 9000 Rockville Pike Bethesda MD 20892

SUOMI, VERNER EDWARD, b Eveleth, Minn, Dec 6, 15; m 41; c 3. METEOROLOGY. *Educ:* Winona State Col, BE, 38; Univ Chicago, PhD(meteorol), 53. *Hon Degrees:* DSc, State Univ New York-Albany, 83. *Prof Exp:* Teacher, pub schs, Minn, 38-42; res assoc meteorol, Univ Chicago, 44-48; from asst prof to prof meterol & soils, 48-89, dir Space Sci & Eng Ctr, 66-88, ASSOC DIR & CHIEF SCIENTIST, WIS CTR SPACE AUTOMATION & ROBOTICS, 85- *Concurrent Pos:* Assoc prog dir atmospheric sci, NSF, DC, 62; chief scientist, US Weather Bur, 64-65; chmn comt adv to Nat Oceanic & Atmospheric Admin, Nat Acad Sci, 66-69, mem comt atmospheric sci, 66-, chmn US comt, Global Atmospheric Prog, 71-74; mem Nat Adv Comt Oceans & Atmosphere, 71-72; fel, Am Arts & Sci, 77. *Honors & Awards:* Meisinger Award, Am Meteorol Soc, 61, Rossby Res Medal, 68; Presidential Citation, 70; Robert M Losey Award, Am Inst Aeronaut & Astronaut, 71; Harry Wexler Professorship, Meteorol, Univ Wisconsin-Madison; Nat Medal of Honor, 77; Except Sci Achievement Medal, 80; William T Pecora Award, Soc Explor Geophysicists, 80; Franklin Medal, 84; Commemorative Medal, Soviet Geophys Comt, 85; Hon Mem, Am Meteorol Soc, 88; Nevada Medal, 88. *Mem:* Nat Acad Eng; AAAS; Am Meteorol Soc (pres, 68-69); Am Geophys Union; foreign mem Finnish Acad Sci & Lett; Voyager Sci Imaging Team; Nat Oceanic & Atmospheric Admin. *Res:* Atmospheric radiation; meteorological satellites; environmental observation systems; inventor of the spinscan weather satellite camera; developed man-computer interactive data access system (McIDAS). *Mailing Add:* Space Sci & Eng Ctr Univ Wis Madison WI 53706

SUOZZI, JOSEPH JOHN, b New York, NY, Mar 2, 26; m 52; c 5. ELECTRICAL ENGINEERING. *Educ:* Cath Univ, BEE, 49, MEE, 54; Carnegie Inst Technol, PhD(elec eng), 58. *Prof Exp:* Instr elec eng, Cath Univ, 49-50; test planning engr, Western Elec Co, 50-52; electronic engr, US Naval Ord Lab, 52-55; mem tech staff, 55-60, supvr magnetic memories, 60-65, dept head magnetic power components, 65-80, DEPT HEAD MAGNETIC BUBBLE MEMORIES, BELL TEL LABS, 80- *Concurrent Pos:* Prog chmn, Int Conf Magnetics, 63, vchmn, 64, gen chmn, 65 & 67; ed-in-chief, Trans on Magnetics, Inst Elec & Electronics Engrs, 65-67; chmn, Int Elec Conf Exec Comt, 80-82. *Mem:* Inst Elec & Electronics Engrs. *Res:* Magnetic amplifiers, especially on feedback in magnetic amplifiers; magnetic power components; energy management systems; computer magnetics. *Mailing Add:* 81 Chimmey Ridge Rd Convent Station NJ 07961

SUPERSAD, JANKIE NANAN, b Chaguanas, Trinidad, Feb 8, 29; m 57; c 3. CIVIL & TRANSPORTATION ENGINEERING. *Educ:* Glasgow Univ, BS, 53; Northwestern Univ, Evanston, MS, 58; Ariz State Univ, PhD(civil eng), 65. *Prof Exp:* Eng asst, Crouch & Hogg, Consult Civil Engrs, Glasgow, Scotland, 53-54; asst engr, Considere Constructions Ltd, Scotland, 55; exec engr, Ministry of Works, Trinidad, WI, 55-60, sr roads engr, 61-64, chief planning engr, 64-68; asst prof faculty eng, Sir George Williams Univ, 68-70; from asst prof to assoc prof, 70-75, coordr civil eng, 76-81, chmn civil & surv eng, 82-83, PROF CIVIL ENG, SCH ENG, CALIF STATE UNIV, FRESNO, 75- *Concurrent Pos:* Faculty assoc eng, Col Eng, Ariz State Univ, 62-64. *Mem:* Am Soc Civil Engrs; Inst Traffic Engrs; Brit Inst Hwy Engrs; Am Soc Eng Educ; Am Road & Transp Builders Asn. *Res:* Urban transportation planning techniques; highway economics with special reference to under-developed countries. *Mailing Add:* Dept Civil & Surveying Eng Calif State Univ 6241 N Maple Ave Fresno CA 93740

SUPLINSKAS, RAYMOND JOSEPH, b Hartford, Conn, Aug 29, 39; m 59; c 3. CHEMISTRY. *Educ:* Yale Univ, BS, 61; Brown Univ, PhD(chem), 65. *Prof Exp:* Mem tech staff, Bell Telephone Labs, 64-65; assoc prof chem, Yale Univ, 65-72; assoc prof & chmn, Swarthmore Col, 72-77; prin staff scientist, Spec Mat Div, Avco, 77-86; PRIN STAFF SCIENTIST, TEXTRON, 86- *Mem:* Sigma Xi. *Res:* Chemical vapor deposition of ceramic fibers and their use in high-performance structural composites based on both organic and metal matrices; inorganic polymer precursors for ceramics and fibers. *Mailing Add:* Textron Specialties Mat Div Two Indust Ave Lowell MA 01852

SUPPE, FREDERICK (ROY), b Los Angeles, Calif, Feb 22, 40. NATURE OF SCIENTIFIC KNOWLEDGE, AGRARIAN PHILOSOPHY. *Educ:* Univ Calif, Riverside, AB, 62; Univ Mich, AM, 64, PhD(philos), 67. *Prof Exp:* Instr philos, Univ Mich, 64-67; asst prof, Univ Ill, Urbana, 67-73; assoc prof philos, 73-83, chairperson comt hist & philos sci, 75-83, PROF PHILOS, UNIV MD, COLLEGE PARK, 83- & PROF, SCH NURSING, BALTIMORE, 84- *Concurrent Pos:* Educ adv, Indo-Am Prog, USAID, Kanpur, India, 65-67; NSF res grant, 73; Am Coun Learned Soc int travel award, 74; mem adv bd, Nat Workshop Teaching Philos, 74-75; lectr, Sch Nursing, Univ Md, Baltimore, 80-84. *Honors & Awards:* Amicus Poloniae Award, Poland, 75. *Mem:* Philos Sci Asn (chmn, 76); Sigma Xi; Soc Agr Food & Human Values. *Res:* Nature of scientific knowledge, including structure of theories and models, explanation, facts and scientific observation; growth of scientific knowledge; scientific realism, history of the philosophy of science; automata theory; sexual morality and sex research methodology; nursing theory; agrarian philosophy. *Mailing Add:* Dept Philos 1131 Skinner Hall Univ Md College Park MD 20742

SUPPE, JOHN, b Los Angeles, Calif, Nov 30, 42; m 65; c 2. GEOLOGY, GEOPHYSICS. *Educ:* Univ Calif, Riverside, BA, 65; Yale Univ, PhD(geol), 69. *Prof Exp:* Assoc res geologist, Yale Univ, 69; NSF fel geol, Univ Calif, Los Angeles, 69-71; asst prof, 71-76, assoc prof, 76-79, PROF GEOL, PRINCETON UNIV, 79- *Concurrent Pos:* Assoc ed, Am J Sci, 75-81; vis prof, Nat Taiwan Univ, 78-79; Guggenheim fel, 78-79. *Mem:* Geol Soc Am; Am Geophys Union. *Res:* Tectonics; regional structural geology. *Mailing Add:* Dept Geol & Geophys Sci Princeton Univ Princeton NJ 08544

SUPPES, PATRICK, b Tulsa, Okla, Mar 17, 22; m 46, 70, 79; c 4. STATISTICS. *Educ:* Univ Chicago, BS, 43; Columbia Univ, PhD, 50. *Hon Degrees:* Dr Soc Sci, Univ Nijmegen, Neth; Dr, Univ Paris, 82. *Prof Exp:* Pres, Comput Curric Corp, Palo Alto, Calif, 67-90; from instr to assoc prof, 50-59, chmn, 63-69, PROF, DEPT PHILOS, STANFORD UNIV, 59-, LUCIE STERN PROF PHILOS, 75- *Concurrent Pos:* Fel, Ctr Advan Study

Behav Sci, 55-56; postdoctoral fel, NSF, 56-57; dir, Inst Math Studies Social Sci, Stanford Univ, 59-, prof by courtesy, Dept Statist, 60-, Sch Educ, 67-, Dept Psychol, 73-; res award, Social Sci Res Coun, 59; fel, John Simon Guggenheim Mem Found, 71-72; vis prof, Col France, Paris, 79 & 88; William James fel, Am Psychol Soc, 89. *Honors & Awards:* Palmer O Johnson Mem Award, Am Educ Res Asn, 67; John Smyth Mem Lectr, Victorian Inst Educ Res, Australia, 68; Distinguished Sci Contrib Award, Am Psychol Asn, 72, E L Thorndike Award for Distinguished Psychol Contrib Educ, 79; S Richardson Silverman Lectr Hearing & Deafness, Cent Inst Deaf, Wash Univ, St Louis, Mo, 79; Nat Medal Sci, 90. *Mem:* Nat Acad Sci; fel Am Psychol Asn; fel AAAS; Am Asn Univ Professors; Am Educ Res Asn (pres, 73-74); Sigma Xi; Nat Acad Educ (pres, 73-77); fel Am Acad Arts & Sci; Soc Exp Psychologists; Int Union Hist & Philos Sci (pres 76, 78). *Res:* Theory of measurement; computer-assisted instruction; mathematical psychology; author of numerous publications. *Mailing Add:* Ventura Hall Stanford Univ Stanford CA 94305-4115

SUPPLE, JEROME HENRY, b Boston, Mass, Apr 27, 36; m 64. ORGANIC CHEMISTRY. *Educ:* Boston Col, BS, 57, MS, 59; Univ NH, PhD(org chem), 63. *Prof Exp:* Res chemist, Univ Calif, Berkeley, 63-64; asst prof org & gen chem, 64-69, assoc dean arts & sci, 72-73, actg assoc provost, State Univ NY Cent Admin, 74-75, assoc vpres acad affairs, 73-78, chmn dept chem, 75-76; assoc prof org & gen chem, Fredonia, 69-78, VPRES ACAD AFFAIRS & PROF CHEM, STATE UNIV NY COL PLATTSBURGH, 78- *Concurrent Pos:* NSF sci fac fel, Univ E Anglia, 70-71. *Mem:* AAAS; Am Chem Soc; Sigma Xi. *Res:* Heterocyclic chemistry; natural products; stereochemistry; organic spectroscopy; conformational studies in the heterocyclic systems; narcotic antagonists; homogeneous catalysis. *Mailing Add:* Southwest Tex State Univ San Marcos TX 78666

SUPRAN, MICHAEL KENNETH, b New York, NY, Feb 13, 39; m 67; c 2. FOOD SCIENCE. *Educ:* Univ Ga, BS, 61, MS, 63, PhD(food sci), 68. *Prof Exp:* Scientist, Nutrit Prod Div, Mead Johnson & Co, 63-65; fel food sci, USPHS, 65-68; Fulbright fel food sci, Danish Meat Res Inst, 68-69; ASST VPRES, THOMAS J LIPTON, INC, 69- *Mem:* Inst Food Technologists; Am Chem Soc; Am Asn Cereal Chemists; NY Acad Sci; Sigma Xi; Prod Develop & Mgt Asn. *Res:* Research and product development of new food concepts, systems and products; administration of nutrition and product safety. *Mailing Add:* 559 Wayne Dr Rivervale NJ 07675

SUPRENANT, BRUCE A, b Dec 29, 52; c 1. CONCRETE, CONSTRUCTION. *Educ:* Bradley Univ, BS; Univ Ill, Urbana, MS; Mont State Univ, Bozeman, PhD(civil eng). *Prof Exp:* Struct engr, Suerdrup & Parcel, 74-76; instr, Dept Civil Eng & Eng Mech, Mont State Univ, Bozeman, 78-84; assoc prof civil eng, Univ Wyo, Laramie, 85-86; PRES, SUPRENANT CONSULT SERV, 78-; ASSOC PROF CIVIL ENG, UNIV S FLA, 86- *Concurrent Pos:* Consult, Portland Cement Asn, Concrete Construct Inst; ed-in-chief, J Forensic Eng, 85-; vis assoc prof, Univ Colo. *Mem:* Sigma Xi; Am Soc Civil Engrs; fel Am Concrete Inst; Am Soc Testing & Mat; Nat Asn Corrosion Engrs; Mat Res Soc; Transp Res Bd; Nat Forensic Soc; Prestressed Concrete Inst. *Res:* Failure analysis of concrete and masonry structures, materials and construction. *Mailing Add:* 7720 Ferris Way Boulder CO 80303

SUPRUNOWICZ, KONRAD, b Pulkovnikov, Siberia, Mar 3, 19; US citizen; m 52; c 1. MATHEMATICS. *Educ:* Univ Nebr, BSc, 52, MA, 53, PhD(math), 60. *Prof Exp:* Instr physics, Minot State Col, 54-55; instr math, Univ Nebr, 57-60; asst prof, Univ Idaho, 60-61; assoc prof, 61-69, PROF MATH, UTAH STATE UNIV, 69- *Concurrent Pos:* Vis assoc prof, Univ Nebr, 62-63. *Mem:* Asn Symbolic Logic; Math Asn Am; Am Math Soc. *Res:* Application of symbolic logic to the study of relational systems; methodology of science. *Mailing Add:* 246 N300E Providence Providence UT 84332

SUPRYNOWICZ, VINCENT A, b Middletown, Conn, Sept 1, 23; m 47; c 3. ENGINEERING HISTORY, ELECTRONICS. *Educ:* Ohio State Univ, BSc, 47, MSc, 49; Yale Univ, PhD(physics, biophys), 53. *Prof Exp:* Anal engr, Pratt & Whitney Nuclear Aircraft Proj, 52-54; asst prof physics, Bucknell Univ, 54-56; assoc prof appl sci, Univ Cincinnati, 56-59; head instrumentation, United Aircraft Res Labs, 59-62, prin scientist, 62-64; RES PROF ELEC ENG, UNIV CONN, 64- *Concurrent Pos:* Consult, United Aircraft Res Labs, 64-65. *Res:* Raman and ultraviolet spectroscopy; reactor shielding; bioelectronics; ultrasonics; lasers; experimental quantum mechanics; author of electronics and general engineering books. *Mailing Add:* Dept Elec Eng Univ Conn U-157 Storrs CT 06268

SURAK, JOHN GODFREY, b Milwaukee, Wis, July 13, 48; m 71; c 2. QUALITY & PRODUCTIVITY IMPROVEMENT, TOTAL QUALITY MANAGEMENT. *Educ:* Univ Wis-Madison, BS, 71, MS, 72, PhD(food sci & toxicol), 74. *Prof Exp:* Res asst food sci, Univ Wis-Madison, 70-74; asst prof toxicol, Univ Fla, 74-80; sr scientist, food prod res dept, Mead Johnson & Co, Evansville, 80-83; supvr prod develop, Wyeth Ayerst Labs, 83-86; assoc prof food sci, 86-89, PROF FOOD SCI, CLEMSON UNIV, CLEMSON, SC, 89- *Concurrent Pos:* Mem, Res & Develop, Assocs Mil & Food Packaging. *Honors & Awards:* Malcolm Baldrige Nat Qual Award, Bd Examiners, 90. *Mem:* Am Chem Soc; Inst Food Technologists; Sigma Xi; Am Soc Qual Control; Soc Exp Biol & Med. *Res:* Quality and productivity improvement as related to its measurement and implementation in manufacturing and service industries. *Mailing Add:* Clemson Univ 223 P & AS Bldg Clemson SC 29634

SURAMPALLI, RAO YADAGIRI, b Zaheerabad, India, July 17, 49; US citizen; c 3. WATER & WASTEWATER TREATMENT ENGINEERING, HAZARDOUS WASTE MANAGEMENT. *Educ:* Okla State Univ, MS, 78; Iowa State Univ, PhD(environ eng), 85. *Prof Exp:* Environ engr sanit eng, Environ Eng Consults, Inc, 76-78; res assoc waste water treatment, Okla State Univ, 76-78; environ engr solid & hazardous waste mgt, Ky Dept Environ Protection, 78-80; sr environ engr water & wastewater treatment, Iowa Dept Environ Protection, 80-86; RES & TECHNOL PROG MGR ENVIRON ENG, US ENVIRON PROTECTION AGENCY, 86- *Concurrent Pos:* Dipl, Am Acad Environ Engrs, 85-; mem, Water Pollution Control Fedn Nat Prog Comt, 87-, Task Force on MOP: 8 & 11-Design & Oper of Wastewater Treatment Plants, 88- & US Environ Protection Agency, Water Res Comt, 89-; consult, World Health Orgn, 87- *Honors & Awards:* Philip Morgan Award, Water Pollution Control Fedn, 86; Commendation Medal, USPHS, 90. *Mem:* Am Soc Civil Engrs; Water Pollution Control Fedn; Am Acad Environ Engrs. *Res:* Published 25 papers on design and operation of wastewater treatment plants. *Mailing Add:* PO Box 17-2141 Kansas City KS 66117

SURAN, JEROME J, b New York, NY, Jan 11, 26; m 52. ELECTRICAL ENGINEERING. *Educ:* Columbia Univ, BS, 49. *Hon Degrees:* Dr Eng, Syracuse Univ, 76. *Prof Exp:* Develop engr, J W Meaker & Co, 49-51; engr res & develop, Motorola, Inc, 51-52; mgr adv circuits, Electronics Lab, Gen Elec Co, 52-62, consult, 54-57, mgr, Electronic Appln & Devices Lab, 62-76, mgr, Electronics Lab, 73-78, staff exec, Tech Systs & Mat Sector, 78-82; SR LECTR, GRAD SCH ADMIN, UNIV CALIF, DAVIS, 82- *Concurrent Pos:* Instr, Mass Inst Technol, 56-59; mem, Ad Hoc Comt Electronic Mat, Nat Acad Sci, 70-72; adj prof, Syracuse Univ, 76-; US Army Sci Bd, 84-86. *Mem:* Sigma Xi; Inst Elec & Electronics Engrs (pres, 79); fel AAAS. *Res:* Solid state circuit development for applications to the broad field of electronics, including computers, communications, control systems, detection systems. *Mailing Add:* PO Box 3103 El Macero CA 95618

SURANYI, PETER, b Budapest, Hungary, Jan 31, 35; m 60; c 2. HIGH ENERGY PHYSICS. *Educ:* E-tv-s Lorand Univ, Budapest, BS, 58; Acad Sci, USSR, PhD(physics), 64. *Prof Exp:* Jr res fel cosmic ray physics, Cent Res Inst Physics, Budapest, Hungary, 58-61,sr res fel,theoret high energy physics, 65-69; res fel theoret physics, Joint Inst Nuclear Studies, Moscow, 61-65; vis lectr physics, Johns Hopkins Univ, 69-70, res assoc, 70-71; assoc prof, 71-74, PROF PHYSICS, UNIV CINCINNATI, 74- *Concurrent Pos:* Sr fel, Brit Sci Coun, 78-79, 87-88; vis prof, Bonn Univ, 87. *Honors & Awards:* Schmidt Award, Hungarian Phys Soc, 68. *Mem:* Am Phys Soc. *Res:* Quantum field theory; group theoretic methods in elementary particle physics; statistical mechanics. *Mailing Add:* Dept Physics Univ Cincinnati Cincinnati OH 45221

SURAWICZ, BORYS, b Moscow, Russia, Feb 11, 17; nat US; m 46; c 4. INTERNAL MEDICINE, CARDIOLOGY. *Educ:* Stefan Batory Univ, Poland, MD, 39; Am Bd Internal Med, dipl; Am Bd Cardiovasc Dis, dipl. *Prof Exp:* Instr cardiol, Sch Med, Univ Pa, 54-55; instr med, Col Med, Univ Vt, 55-57, asst prof exp & clin med, 56-62; assoc prof med, Col Med, Univ Ky, 62-66, dir Cardiovasc Div, 62-81, prof, 66-81; PROF MED, KRANNERT INST CARDIOL, INDIANAPOLIS, 81- *Concurrent Pos:* Fel coun clin cardiol, Am Heart Asn. *Mem:* Fel Am Col Physicians; fel Am Col Cardiol; AMA; Am Physiol Soc; Asn Univ Cardiologists. *Res:* Electrocardiology; role of electrolytes in cardiac arrhythmias. *Mailing Add:* Krannert Inst Cardiol Sch Med Ind Univ 1001 W Tenth St Indianapolis IN 46202

SURBEY, DONALD LEE, b North Canton, Ohio, July 19, 40; m 61; c 2. ORGANIC CHEMISTRY. *Educ:* Manchester Col, BS, 61; Univ Notre Dame, PhD(org chem), 68. *Prof Exp:* Control chemist, Miles Labs, Inc, 61-63; RES CHEMIST, LUBRIZOL CORP, 67- *Mem:* Am Chem Soc. *Res:* Organic chemistry as related to process and product development in field of polymer chemistry and lubricant additives. *Mailing Add:* 5648 Ridgebury Blvd Lyndhurst OH 44124-1453

SURDY, TED E, b Wheeling, WVa, Jan 25, 25; m 50; c 4. BACTERIOLOGY, BIOCHEMISTRY. *Educ:* Purdue Univ, BS, 58, MS, 59, PhD(bact), 62. *Prof Exp:* Instr bact, Purdue Univ, 59-61; assoc prof bact & cell physiol, Kans State Teachers Col, 62-67; res assoc biol, Educ Res Coun Greater Cleveland, 67-68; chmn dept biol, 68-74, PROF BIOL, SOUTHWEST MINN STATE COL, 74- *Concurrent Pos:* NSF grant, 65-68. *Mem:* AAAS; Am Soc Microbiol; Soc Indust Microbiol; Nat Asn Biol Teachers. *Res:* Lytic reactions of gram-negative bacterial cell walls; membrane permeability of gram-negative bacteria; effect of lipids on lysis of gram-negative bacteria; audio-tutorial bacteriology; methods for isolation and identification of salmonella. *Mailing Add:* Dept Biol Southwest Minn State Univ Marshall MN 56258

SURESH, MAVANUR RANGARAJAN, b Shimoga, India, Apr 22, 53; Can citizen; m 82; c 2. CANCER RESEARCH, IMMUNOCHEMISTRY. *Educ:* Bangalore Univ, India, BSc Hons, 71, MSc, 73; Indian Inst Sci, Bangalore, PhD(biochem), 78. *Prof Exp:* Postdoctoral res assoc & Am Cancer Soc jr fel, Scripps Clin & Res Fedn, La Jolla, Calif, 78-82; sr postdoctoral fel, Fac Pharm, Univ Alta, 82-84; Usher fel, Med Res Coun Lab Molecular Biol, Cambridge, UK, 84-85; group leader, Summa Biomed Can Ltd, 85; CHIEF SCIENTIST, BIOMIRA INC, EDMONTON, CAN, 85- *Concurrent Pos:* Secy, Biochem Soc, Indian Inst Sci, 75-76; vpres, Soc Fels, Scripps Clin & Res Found, La Jolla, Calif, 79-80; vis scientist, Indian Inst Sci, Bangalore, 82; vis fel, Clare Hall, Univ Cambridge, UK, 84; adj assoc prof, Fac Pharm, Univ Alta, 85-; chmn, Biotechnol Adv Comt, Northern Alta Inst Technol, Alta, 87-90, mem, Biol Sci Comt, 90- *Mem:* Am Asn Immunologists; Soc Complex Carbohydrates. *Res:* Cancer associated antigens with special reference to carbohydrate antigens-bispecific monoclonal antibodies; cancer diagnostics and therapeutics in the industrial sector; asthma and allergy diagnostics and therapeutics in the academic university sector. *Mailing Add:* Biomira Inc 9411 20th Ave Edmonton AB T6N 1E5 Can

SURESH, SUBRA, Indian citizen. FRACTURE & FATIGUE, MICROSTRUCTURAL EFFECTS. *Educ:* Indian Inst Technol, B Technol, 77; Iowa State Univ, MS, 79; Brown Univ, 87. *Hon Degrees:* ScD, Mass Inst Technol, 81. *Prof Exp:* Asst res engr, Univ Calif, Berkeley, 81-83; asst prof, 83-86, ASSOC PROF ENG, BROWN UNIV, 86- *Concurrent Pos:* Co-chmn, Int Symp Interfaces, RI, 88; mem, Int Comt, French Metall Soc, 88- *Honors & Awards:* Robert Lansing Hardy Gold Medal, 83; Champion Mathewson Gold Medal, 85; Pres Young Investr Award, NSF, 85. *Mem:* Metall Soc Am; Am Ceramic Soc; Am Soc Mech Engrs; Mat Res Soc. *Res:* Theoretical and experimental study of the mechanics and mechanisms of microstructural development, deformation, fracture, and fatigue in advanced engineering materials. *Mailing Add:* Div Eng Brown Univ Box D Providence RI 02912

SURGALLA, MICHAEL JOSEPH, b Nicholson, Pa, May 12, 20; m 48; c 4. MEDICAL MICROBIOLOGY. *Educ:* Univ Scranton, BS, 42; Univ Chicago, PhD(bact), 46. *Prof Exp:* Bacteriologist, E R Squibb & Sons, NJ, 46-48; res assoc, Univ Chicago, 48-54; bacteriologist, Biol Sci Lab, Dept of Army, Ft Detrick, Md, 54-71; dir clin microbiol, Roswell Park Mem Inst, 71-91. *Mem:* AAAS; Am Soc Microbiol; Am Acad Microbiol; NY Acad Sci; Am Soc Clin Pathologists. *Res:* Medical bacteriology; staphylococcus food poisoning; influenza virus; experimental plague; bacterial virulence; pathogenic mechanisms; host resistance; endotoxins; fibrinolysis; opportunist pathogens; hospital infection epidemiology. *Mailing Add:* 163 Hedstrom Dr Amherst NY 14226

SURGENOR, DOUGLAS MACNEVIN, b Hartford, Conn, Apr 7, 18; m 46; c 5. BIOCHEMISTRY. *Educ:* Williams Col, AB, 39; Mass State Col, MS, 41; Mass Inst Technol, PhD(org chem), 46. *Prof Exp:* Mem staff, Div Indust Coop, Mass Inst Technol, 42-45; res assoc phys chem, Harvard Med Sch, 45-50, asst prof, 50-55, asst prof biol chem, 55-60; sr investr, Protein Found, 56-60; head dept biochem, Sch Med, State Univ NY Buffalo, 60-64, dean, 62-68, provost, Fac Health Sci, 67-70, prof biochem, 60-77, res prof, sch mgt, 71-77; Pres & dir, Am Red Cross Blood Serv, Northeast Red, Boston, 77-83; SR INVESTR, CTR BLOOD RES, BOSTON, 72- *Concurrent Pos:* Assoc mem lab phys chem related to med & pub health, Harvard Univ, 50-54; consult, Vet Admin Hosp, Buffalo, 60-, chmn dean's comt, 62-68; mem med bd, Buffalo Gen & Buffalo Children's Hosps, 62-68; mem med coun, NY State Educ Dept, 62-68; bd sci counr, Div Biol Stand, NIH, 63-68; mem, Nat Heart Inst Prog Proj Comt B, 65-68; consult med bd, Millard Fillmore Hosp, 66-70; mem, Int Comt Thrombosis & Haemostasis, 63-, chmn, 70-72; mem nat blood resource prog adv comt, Nat Heart & Lung Inst, 69-73, chmn, 70-73; mem med adv comt, Am Nat Red Cross, 70-76; pres, 72-87, chmn, Ctr Blood Res, Boston 87-; mem bd trustees, Children's Hosp Med Ctr, Boston, 75-80, overseer, 80-; vis prof pediat, Harvard Med Sch, 77-88; mem adv coun, Nat Heart, Lung & Blood Inst, NIH, 80-84; mem expert adv panel human blood prods, WHO, 80-87. *Mem:* AAAS; Am Soc Biol Chemists; Am Heart Asn; Am Soc Hemat; Int Soc Thrombosis & Haemostasis. *Res:* Blood biochemistry; blood and public policy; blood coagulation; transfusion medicine. *Mailing Add:* 213 Indian Hill Rd Carlisle MA 01741

SURI, ASHOK, b Lahore, Punjab, India, Aug 27, 43; US citizen; m 69; c 2. SOFTWARE SYSTEMS, TECHNICAL MANAGEMENT. *Educ:* Cornell Univ, MS, 66, PhD(physics), 69. *Prof Exp:* Staff mem physics, Stanford Linear Accelerator Ctr, 69-73; asst prof physics, Univ Calif, Santa Cruz, 69-73 & Stanford Univ, 73-74; eng mgt positions, Fairchild Camera & Inst Corp, 74-84; vpres, CAE Systs & Tektronix, 84-88; sr vpres, Comdisco Systs, Inc, 88-91; PRES, AFFLUENCE SYSTS CORP, 91- *Concurrent Pos:* Consult, NASA Biocore Exp, Apollo 17, 72-73; key technologist, Fairchild Camera & Inst Corp, 74-84. *Mem:* Inst Elec & Electronics Engrs; Asn Comput Mach; Am Phys Soc. *Mailing Add:* Affluence Systs Corp PO Box 4363 Mountain View CA 94040-4363

SURI, RAJAN, b Dec 18, 52; India citizen; c 2. MANUFACTURING SYSTEMS MODELING & ANALYSIS, FLEXIBLE MANUFACTURING SYSTEMS. *Educ:* Cambridge Univ, Eng, BA, 74; Harvard Univ, SM, 75, PhD(eng), 78. *Prof Exp:* Fel decision & control, Div Appl Sci, Harvard Univ, 78-80, asst prof, 80; managing dir, SAN Ltd Locomotive Co, Bangalore, India, 81; asst prof systs eng, Div Appl Sci, Harvard Univ, 82-85; ASSOC PROF, INDUST ENG, UNIV WIS, 85- *Concurrent Pos:* Consult & database mgr, Multinational Enterprise Proj, Harvard Bus Sch, 75-78; mgt consult, Fiat SPA, Turin, Italy, 76-78; consult, Charles Stark Draper Lab, 76-82; dir, Network Dynamics Inc, Cambridge, Mass, 83-; consult, IBM Corp, 84- *Honors & Awards:* Donald P Eckman Award, 81. *Mem:* Inst Elec & Electronics Engrs; Inst Indust Engrs; Soc Mfg Engrs; Oper Res Soc Am; Inst Mgt Sci. *Res:* Modeling and analysis of manufacturing systems; flexible manufacturing systems; queueing network models and simulation discrete event systems. *Mailing Add:* Indust Eng Dept Univ Wis 1513 Univ Ave Madison WI 53706

SURIA, AMIN, b Dhoraji, India, Aug 24, 42; m 74; c 4. NEUROPHARMACOLOGY. *Educ:* Univ Karachi, Pakistan, BS, 63, MS, 64; Vanderbilt Univ, Nashville, PhD(pharmacol), 71. *Prof Exp:* Chemist, United Paints Ltd, Karachi, Pakistan, 64; chemist & in-chg lab, Textile Dyes & Auxiliary Dept, Hoechst Pharmaceut Co, Ltd, Karachi, Pakistan, 64-66; vis fel biochem, Pharmacol Lab, Nat Heart & Lung Inst, NIH, Bethesda, Md, 71-72; fel, Lab Preclin Pharmacol, NIMH, St Elizabeth's Hosp, Washington, DC, 72-74, staff fel, 74-75; assoc prof pharmacol, George Washington Univ, Washington, DC, 75-80; prof pharmacol, Riyadh Univ, Saudi Arabia, 80-83; PROF & HEAD, PHARMACOL DEPT, SCH MED, AGA KHAN UNIV, PAKISTAN, 83- *Concurrent Pos:* Guest worker, Lab Preclin Pharmacol, St Elizabeth's Hosp, NIMH; consult, Pakistan Med Res Coun; ed bd, J Pakistan Med Asn, Pakistan J Pharmacol. *Mem:* Am Soc Pharmacol & Exp Therapeut; Soc Neurosci; NY Acad Sci; Pakistan Pharmacol Soc. *Res:* Molecular mechanisms by which anti-anxiety, anticonvulsant, and antidepressant drugs exert their actions on complex neuronal pathways, research entails using electrophysiological and biochemical techniques; arachidonic acid metabolism in health and disease. *Mailing Add:* Pharmacol Dept Sch Med Aga Khan Univ PO Box 3500 Stadium Rd Karachi 5 Pakistan

SURIANO, F(RANCIS) J(OSEPH), b Kenosha, Wis, July 17, 37; m 59; c 4. MECHANICAL ENGINEERING. *Educ:* Univ Notre Dame, BS, 59, PhD(mech eng), 66; Ga Inst Technol, MS, 61. *Prof Exp:* Asst mech eng, Ga Inst Technol, 59-60; res engr, Wood River Res Lab, Shell Oil Co, 60-61; mem staff aerodyn preliminary design, AiRes Mfg Co, Ariz, 66-67; mem tech staff, Aerospace Corp, 67-68; MEM STAFF AERODYNAMICS-PRELIMINARY DESIGN, AIRESEARCH MFG CO ARIZ, PHOENIX, 68- *Mem:* Am Soc Mech Engrs. *Res:* Heat transfer and fluid mechanics. *Mailing Add:* 8626 E Meadowbrook Ave Scottsdale AZ 85251

SURKAN, ALVIN JOHN, b Drumheller, Alta, June 5, 34; m 67; c 2. PHYSICS, APPLIED MATHEMATICS. *Educ:* Univ Alta, BSc, 54; Univ Toronto, MA, 56; Univ Western Ont, PhD(physics), 59. *Prof Exp:* Sr demonstr geophys, Univ Western Ont, 58-59; Nat Res Coun Can fel, Univ Alta, 59-61; sci officer marine physics, Can Defence Res Bd, 61-62; fac mem physics, Univ BC, 62-63; staff consult geophys comput, res & develop ctr, IBM Corp, 64-65, mem res staff environ sci group, phys sci dept, Watson Res Ctr, 65-69; mem inst water resources res, 69-73, PROF COMPUT SCI, UNIV NEBR, LINCOLN, 69- *Concurrent Pos:* Consult geophys comput & resident visitor physics dept, Bell Tel Labs, 72. *Mem:* Am Geophys Union; Soc Explor Geophysicists; Am Inst Physics; Inst Elec & Electronics Engrs. *Res:* Magnetic and seismic methods in exploration geophysics; mathematical modeling in hydrology, geomorphology and educational psychology; algorithms for data interpretation; nonlinear optimization and symbolic computation in geophysics and chemical engineering. *Mailing Add:* Dept Comput Sci Univ Nebr Lincoln NE 68518

SURKO, CLIFFORD MICHAEL, b Sacramento, Calif, Oct 11, 41; m 65; c 2. LASER SCATTERING. *Educ:* Univ Calif, Berkeley, AB, 64, PhD(physics), 68. *Prof Exp:* Res assoc physics, Univ Calif, Berkeley, 68-69; mem tech staff physics, 69-82, head, semiconductor & chem physics res dept, Bell Labs, 82-88; PROF PHYSICS, UNIV CALIF, SAN DIEGO, 88- *Concurrent Pos:* Vis sr res physicist, Ecole Polytech, France, 78-79; vis scientist, Plasma Fusion Ctr, Mass Inst Technol, 77-85. *Mem:* AAAS; fel Am Physics Soc; NY Acad Sci. *Res:* Experimental research in nonlinear and nonequilibrium systems, including plasmas and fluids; physics of positrons in traps and positron plasmas; physics of patterns and dynamics in fluid convection. *Mailing Add:* Physics Dept 0319 Univ Calif San Diego La Jolla CA 92093

SURKO, PAMELA TONI, b Britton, SDak, June 15, 42; m 65; c 2. RULE-BASED EXPERT SYSTEMS, ARTIFICIAL NEURAL SYSTEMS. *Educ:* Univ Calif, Berkeley, AB, 63, PhD(physics), 70. *Prof Exp:* Res assoc, Princeton Univ, 70-71, instr, 71-72, asst prof physics, 72-80; mem tech staff, Bell Labs, NJ, 80-86, distinguished mem tech staff, AT&T Bell Labs, 86-88, SR SCIENTIST, SCI APPLNS INT CORP, 88- *Concurrent Pos:* Chercher Exranger, Ecole Polytechnique, Paris, 79. *Mem:* Am Asn Artificial Intel; Inst Elec & Electronic Engrs; AAAS; Asn Women Sci. *Res:* Strangeness-changing neutral currents; muon-induced events at high energy; knowledge-based systems; artificial neural systems. *Mailing Add:* Sci Appln Int Corp Mail Stop 33 10260 Campus Point Dr San Diego CA 92121

SURMACZ, CYNTHIA ANN, b Wilkensburg, Pa, June 26, 57; m 81; c 2. CELL PHYSIOLOGY, EXERCISE PHYSIOLOGY. *Educ:* Pa State Univ, BS, 78; Milton S Hershey Med Ctr, PhD(physiol), 82. *Prof Exp:* Instr physiol, Dept Continuing Educ & res fel, Milton S Hershey Med Ctr, 83; ASSOC PROF ANAT & PHYSIOL, BLOOMSBURG UNIV, PA, 83- *Mem:* AAAS; Nat Asn Biol Teachers; Am Physiol Soc. *Res:* Alterations in plasma electrolytes and the cardiac cycle during exercise. *Mailing Add:* Dept Biol & Allied Health Sci Hartline Sci Ctr Bloomsburg Univ Bloomsburg PA 17815

SURREY, ALEXANDER ROBERT, b New York, NY, Mar 13, 14; m 39; c 1. ORGANIC CHEMISTRY, DRUG RESEARCH. *Educ:* City Col New York, BS, 34; NY Univ, PhD(chem), 40. *Prof Exp:* Nat Defense Res Comt fel, Cornell Univ, 40-41; res chemist, Sterling-Winthrop Res Inst, 41-57, sect head, 57-60, asst dir chem res, 60-64, sr res fel & dir new prod, 64-67, dir develop res, 67-72, vpres res & develop, 72-76, vpres prod develop, 76-77, vpres tech affairs, 77-81. *Concurrent Pos:* Lectr & adj prof, Rensselaer Polytech Inst, 58-64; law & med consult; consult, 81- *Honors & Awards:* Townsend Harris Medal, 81. *Mem:* Am Chem Soc; fel NY Acad Sci; fel Royal Chem Soc. *Res:* Medicinals. *Mailing Add:* 15 Harvard Ave Albany NY 12208

SURREY, KENNETH, b India, Dec 6, 22; nat US; m 52; c 4. PHYTOCHEMISTRY. *Educ:* Univ Punjab, India, BSc, 46, MA, 52; Univ Mo, MA, 55, PhD(phytochem), 57. *Prof Exp:* Lab instr chem, Forman Christian Col, Pakistan, 49-53; asst bot, Univ Mo, 54-57; asst plant physiologist, Argonne Nat Lab, 57-66; HEALTH SCIENTIST ADMINR, NIH, 66- *Res:* Histochemistry of protein constituents by azo-coupling reactions in plants; metabolic responses of regenerating meristems and germinating seeds as influenced by visible and ionizing radiation; action and interaction of red and far-red radiation on metabolic processes of developing seedlings; physiological bases for morphological development; neurosciences (neurotoxicology, neuroendocrinology, pain); administration of federally grant-supported research in neurosciences. *Mailing Add:* Nat Inst Neurol Dis & Stroke NIH Fed Bldg 7550 Wis Ave Bethesda MD 20892

SURTI, VASANT H, b Bombay, India, Oct 30, 31; US citizen; m 58; c 1. CIVIL & TRANSPORTATION ENGINEERING. *Educ:* Univ Bombay, BSc, 52; Seattle Univ, BS, 57; Mich State Univ, MS, 62; Cath Univ Am, PhD(civil eng), 68. *Prof Exp:* Design engr, Boeing Aircraft Co, Wash, 57-59; traffic res & hwy design engr, Mich State Hwy Dept, 59-66; asst prof civil eng, Cath Univ Am, 66-68, assoc prof civil eng & mech, 68-72; assoc prof civil eng & dir, Ctr Urban Transp Studies, Univ Colo, Denver, 72-; AT DEPT CIVIL ENG, FLA INT UNIV, 80- *Concurrent Pos:* Grants, NSF, Dept Transp & Dept Com; mem, Hwy Res Bd Comts. *Mem:* Inst Traffic Engrs; Am Soc Civil Engrs. *Mailing Add:* Dept Civil Eng Fla Int Univ Miami FL 33199

SURVANT, WILLIAM G, soil science; deceased, see previous edition for last biography

SURVER, WILLIAM MERLE, JR, b Altoona, Pa, June 26, 43. DEVELOPMENTAL GENETICS. *Educ:* St Francis Col, BA, 66; Univ Notre Dame, PhD(genetics), 72. *Prof Exp:* Instr biol, Univ Notre Dame, 71-72; asst prof zool, Univ RI, 72-78; asst prof, 78-81, ASSOC PROF BIOL, CLEMSON UNIV, 81- *Mem:* AAAS; Nat Sci Teachers Asn; Nat Asn Biol Teachers; Genetics Soc Am. *Res:* Genetic effects on developing systems; genetics of kelp fly, Coelopa frigida; innovative teaching for general biology. *Mailing Add:* Dept Biol Clemson Univ 201 Sikes Hall Clemson SC 29634

SURWILLO, WALTER WALLACE, b Rochester, NY, Nov 25, 26; m 55. PSYCHOPHYSIOLOGY. *Educ:* Wash Univ, St Louis, BA, 51, MA, 53; McGill Univ, PhD(psychol), 55. *Prof Exp:* Asst psychol, Wash Univ, St Louis, 50-53; asst, McGill Univ, 53-55; res assoc psychophysiol, Allan Mem Inst Psychiat, 55-57; res psychophysiologist gerontol br, NIH, 57-65; assoc prof psychiat, 65-70, PROF PSYCHIAT SCH MED, UNIV LOUISVILLE, 70-, ASSOC PROF PSYCHOL, GRAD SCH, 71-; DIR, EEG LAB, BINGHAM CHILD GUID CLIN, 83- *Concurrent Pos:* Mem, NIH Exp Psychol Study Sect, 70-74; consult, NSF, 73-, Can Res Coun, 78- *Mem:* Am Psychol Asn; Soc Psychophysiol Res; NY Acad Sci; Soc Neurosci. *Res:* Nervous system function and its relation to behavior; psychophysiological and electrophysiological methods of investigation; instrumentation; central nervous system and behavioral changes with development and senescence; electroencephalography. *Mailing Add:* Dept Psychiat & Behav Sci Univ Louisville Sch Med Louisville KY 40292

SURWIT, RICHARD SAMUEL, b New York, NY, Oct 7, 46; m 82; c 1. BEHAVIORAL MEDICINE, CLINICAL PSYCHOLOGY. *Educ:* Earlham Col, AB, 68; McGill Univ, Can, PhD(clin psychol), 72. *Prof Exp:* Res fel psychol, Dept Psychiat, Harvard Med Sch 72-74, instr, 74-76; dir, Psychophysiol Lab, Mass Ment Health Ctr, Boston, 74-77; asst prof psychol, Dept Psychiat, Harvard Med Sch, 76-77; from asst prof to assoc prof, 77-83, PROF MED PSYCHOL,DEPT PSYCHIAT, DUKE UNIV MED CTR, DURHAM, 83, ASST PROF EXP MED, 79- *Concurrent Pos:* Assoc ed, Health Psychol, 86- *Mem:* Fel Am Psychol Asn; fel Acad Behav Med Res; Am Psychosom Soc; AAAS; Soc Psychophysiol Res. *Res:* Investigation of role of the autonomic and central nervous systems in the pathophysiology of type II diabetes mellitus and study of how behavioral variables relate to chronic glucose control in both type I and type II diabetes. *Mailing Add:* Duke Univ Med Ctr Box 3322 Durham NC 27710

SURYANARAYANA, NARASIPUR VENKATARAM, b Bangalore, India, Apr 12, 31; m 59; c 2. MECHANICAL ENGINEERING. *Educ:* Univ Mysore, BE, 54; Columbia Univ, MS, 66; Univ Mich, PhD(mech eng), 70. *Prof Exp:* Tech asst mech eng, Hindustan Shipyard, Visakhapatnam, 55-63; lectr, Indian Inst Technol, Kharagpur, 63-64; training off, Lucas-TVS, Madras, 64-65; asst prof, 70-75, assoc prof, 75-79, PROF MECH ENG, MICH TECHNOL UNIV, 79- *Concurrent Pos:* NSF res grants, 71-72 & 77-79. *Res:* Heat transfer with change of phase; convective heat transfer with turbulence; solar energy. *Mailing Add:* Dept Mech Eng & Eng Mech Mich Technol Univ Houghton MI 49931

SURYANARAYANAN, RAJ GOPALAN, b Cuddalore, Tamil Nadu, India, Apr 19, 55; m 85; c 1. SOLID STATE PHARMACEUTICS, DRUG DELIVERY. *Educ:* Banaras Hindi Univ, Varanasi, India, BPharm, 76, MPharm, 78; Univ BC, Vancouver, Can, MSc, 81, PhD(pharmaceut), 85. *Prof Exp:* Mgt trainee, Indian Drugs & Pharmaceut Ltd, 78; supvr, Roche Prods, 79; teaching asst, Univ BC, 79, 82-83; ASST PROF, UNIV MINN, 85- *Concurrent Pos:* Consult, several pharmaceut cos, 87-; vis res scholar, Sch Pharm, Univ Bradford, UK, 89-90; invited lectr, several pharmaceut cos USA & UK, 90- & prof socs, 90- *Mem:* AAAS; Am Asn Pharm Scientists; Am Asn Cols Pharm. *Res:* Solid state properties of drugs and dosage forms; novel methods of drug delivery; quantitative powder x-ray diffractometric studies of pharmaceutical systems. *Mailing Add:* Col Pharm 308 Harvard St SE Minneapolis MN 55455

SURYARAMAN, MARUTHUVAKUDI GOPALASASTRI, b Madras, India, Mar 2, 25; m 52; c 3. ANALYTICAL CHEMISTRY, PHYSICAL CHEMISTRY. *Educ:* Univ Madras, BSc, 46, MS, 52; Univ Colo, PhD(chem), 61. *Prof Exp:* Demonstr chem, Madras Christian Col, 46-49 & Vivekananda Col, Madras, 52; lectr, Sri Venkateswara Univ Cols, Andhra, 52-57; asst, Univ Colo, 57-59, 60-61; sr res chemist, Monsanto Co, Mo, 61-66; from asst prof to assoc prof, 66-75, PROF CHEM, HUMBOLDT STATE UNIV, 75- *Mem:* Am Chem Soc; fel Royal Inst Chem; fel Indian Chem Soc. *Res:* Analytical chemistry, electrochemistry and ion exchange; general inorganic chemistry. *Mailing Add:* Dept Chem Humboldt State Univ Arcata CA 95521-4957

SURZYCKI, STEFAN JAN, b Krakow, Poland, Jan 13, 36; US citizen; m 70; c 4. MOLECULAR BIOLOGY, BIOCHEMISTRY. *Educ:* Odessa Univ, MS, 60; Warsaw Univ, PhD(genetics), 64. *Prof Exp:* Asst genetics, Warsaw Univ, 61-63; researcher, Genetics Inst Polish Acad Sci, Warsaw, 63-64; fel molecular biol, Harvard Univ, 64-65, res fel, 65-68, Maria Moor Cabot Found fel, 69-70; from asst prof to assoc prof, Univ Iowa, 70-75; ASSOC PROF, DEPT BIOL, IND UNIV, 75- *Mem:* Am Soc Cell Biol; Am Soc Microbiol; Plant Molecular Biol Soc. *Res:* Mechanism of transcription initiation by eucaryotic RNA Polymerase II; regulation of gene expression in chloroplast of C. reinhardi. *Mailing Add:* Dept Biol Ind Univ Bloomington IN 47401

SUSAG, RUSSELL H(ARRY), b Minneapolis, Minn, Dec 22, 30; m 57; c 6. ENVIRONMENTAL ENGINEERING. *Educ:* Univ Minn, BCE, 56, MSCE, 65, PhD(sanit eng), 65. *Prof Exp:* Teaching asst sanit eng, Univ Minn, 56-57, asst prof, 65-68; assoc prof environ eng, Univ Fla, 68-70; mgr qual control, Metrop Sewer Bd, Minneapolis, 70-74; mgr environ affairs, 74-78, dir environ regulatory activ, 78-80, dir environ opers, 80-83, DIR ENVIRON REG AFFAIRS, 3M, 83- *Honors & Awards:* Radebaugh Award, Cent States Water Pollution Control Asn, 66; Eng of Year, Fla Sect, Am Soc Civil Engrs, 70; Arthur Sidney Bedell Award, Water Pollution Control Fedn, 77; George J Schroepfer Award, Cent States Water Pollution Control Asn, 87; W R "Bill" Coffin Memorial Public Service Award, Minn Soc Prof Engrs, 89; Distinguished Service Award, Prof Engrs Industry, Nat Soc Prof Engrs, 90. *Mem:* Am Soc Civil Engrs; Nat Soc Prof Engrs; Water Pollution Control Fedn; Am Acad Environ Engrs. *Res:* Solid and hazardous waste management; water pollution control, deoxygenation and reaeration characteristics of waste waters and receiving waters. *Mailing Add:* 7305 First Ave S Richfield MN 55423

SUSALLA, ANNE A, b Parisville, Mich. PLANT ANATOMY, EMBRYOLOGY. *Educ:* Madonna Col, BA, 62; Univ Detroit, MS, 67; Ind Univ, Bloomington, PhD(bot), 72. *Prof Exp:* Asst prof, 72-75, chairperson dept, 77-80, ASSOC PROF BIOL, ST MARY'S COL, 75- *Mem:* Bot Soc Am; Sigma XI. *Res:* Ultrastructure of plastids in phenotypically green leaf tissue of a genetic albino strain of Nicotiana; tissue culture work is being employed to study the developmental stages of these plastids. *Mailing Add:* Dept Biol Sci Hall St Mary's Col Notre Dame IN 46556

SUSI, FRANK ROBERT, b Boston, Mass, Dec 10, 36. ANATOMY, ORAL PATHOLOGY. *Educ:* Boston Col, BS, 58; Harvard Univ, DMD, 62, cert, 65; Tufts Univ, PhD(anat), 67. *Prof Exp:* Instr anat, Sch Med, 67-68, from asst prof to assoc prof oral path, 67-74, asst prof anat, 68-82, asst dean acad affairs, 77-82, PROF ORAL PATH, SCH DENT MED, TUFTS UNIV, 74-, DIR DIV, 73-, DIR BASIC HEALTH SCI, 70-, PROF ANAT, SCH MED, 82-, ASSOC DEAN, ACAD AFFAIRS, 82- *Concurrent Pos:* Fel anat, McGill Univ, 68-69. *Mem:* Am Asn Anat; Histochem Soc; Am Acad Oral Path; Int Asn Dent Res; Am Asn Dent Schs (vpres, 84-87, pres, 88-89); Am Dent Asn. *Res:* Histochemistry; autoradiography; electron microscopy; keratinization, carcinogenesis; spermiogenesis; dentistry. *Mailing Add:* Dept Oral Path Sch Dent Med Tufts Univ One Kneeland St Boston MA 02111

SUSI, PETER VINCENT, b Philadelphia, Pa, Apr 26, 28; m 54; c 2. ORGANIC CHEMISTRY, RESEARCH ADMINISTRATION. *Educ:* Univ Pa, BA, 50; Univ Del, MS, 51, PhD(chem), 57. *Prof Exp:* From res chemist tosr res chemist, 56-63, group leader, 63-76, proj leader, 76-81, GROUP LEADER, AM CYANAMID CO, 81- *Mem:* AAAS; Am Chem Soc. *Res:* Synthesis and applications research in field of plastics additives, light stabilizers, antioxidants, antistatics; ultraviolet and infrared absorbers; flame retardants, antioxidants and photoinitators. *Mailing Add:* 17 Starlit Dr Middlesex NJ 08846-1443

SUSINA, STANLEY V, b Berwyn, Ill, Apr 14, 23; m 48; c 3. PHARMACY, PHARMACOLOGY. *Educ:* Univ Ill, BS, 48, MS, 51, PhD(pharmacol), 55; Cumberland Law Sch, Samford Univ, JD, 71. *Prof Exp:* Asst pharm, Univ Ill, 48-50, from instr to assoc prof, 50-62, actg head dept, 61-62; assoc dean, 76-89, actg dean, 84-85, PROF PHARM LAW, SCH PHARM, SAMFORD UNIV, 62- *Mem:* Acad Pharmaceut Sci; Am Asn Cols Pharm; Am Pharmaceut Asn. *Res:* Antihistamines; neuromuscular blocking agents; local anesthetics; radioactive isotopes. *Mailing Add:* Dept Pharm Samford Univ 800 Lakeshore Dr Birmingham AL 35229

SUSKI, HENRY M(IECZYSLAW), b Camden, NJ, July 14, 18; m 43. ELECTRONIC & COMMUNICATION ENGINEERING. *Educ:* City Col New York, BEE, 41. *Prof Exp:* Switchboard engr, Western Elec Co, 41; electronics scientist, Electronics Div, 46-67, electronics engr, Tactical Electronic Warfare Div, US Naval Res Lab, DC, 67-87; consult, Locus, Inc, 87-89; RETIRED. *Concurrent Pos:* Vpres, NAm Res Corp & Corp Systs Res, 67-77. *Mem:* Sr mem Inst Elec & Electronics Engrs; Sigma Xi. *Res:* Development of long range facility plans, which include construction and installation, for electronic warfare simulation, research, development, and laboratory test; design management and funding organization, considering the assessment of technology and its impact on society. *Mailing Add:* Two Whittington Dr Palm Coast FL 32137

SUSKIND, RAYMOND ROBERT, b New York, NY, Nov 29, 13; m 44; c 2. MEDICAL SCIENCE, HEALTH SCIENCES. *Educ:* Columbia Univ, AB, 34; State Univ NY, MD, 43; Am Bd Dermat & Syphil, dipl, 49. *Prof Exp:* Resident dermat & syphil, Cincinnati Gen Hosp, 44-46; res fel dermat, Col Med, Univ Cincinnati, 48-49, from asst prof to assoc prof prev med & indust health, 49-62, asst prof dermat, 50-62, dir dermat res, Kettering Lab, 48-62; prof dermat & head div environ med, Med Sch, Univ Ore, 62-69; res fel, Kettering Lab, 48-50, prof environ health & med, Univ Cincinnati, 69-85, chmn dept environ health, 69-85, dir, Kettering Lab, Col Med, 69-85, EMER J G SCHMIDLAPP PROF MED & DERMAT, UNIV CINCINNATI, 85- *Concurrent Pos:* Attend physician, Univ Hosp, Cincinnati Med Ctr, 69-; attend physician, Univ Ore Hosp, 62-69; mem, Cincinnati Air Pollution Bd, 72-76, chmn, 74-75; rep, USSR-USA Collab Res Prog, Biol & Genetic Effects Pollutants, 73-79; mem, Task Force Res Planning Comt, Nat Inst Environ Health Sci, 75-77; consult, Bur Drugs & Dermat Adv Comt, Food & Drug Admin, 76-81; consult, Occup Med to the Surgeon Gen, Dept Navy, 75-80; chmn, Occup Safety Health Act Standards Adv Comt Cutaneous Hazards, 78; mem, Certifying Bd, Am Bd Toxicology, 78-83, Vet Admin Comt Health Related Effects Herbicides, 79-84, Adv Panel Toxicology, Am Med Asn Coun Sci Affairs, 80-85; assoc ed, Am J Indust Med, 79-86; contrib ed, Bd Chemosphere, 86-, ed, 87- *Honors & Awards:* Mitchell Award, State Univ NY Med Ctr, 43; Health Achievement in Indust Award, 77; Proj Hope Award, 84; Daniel Drake Medalist, Univ Cincinnati, 85; R A Kehoe Award of Merit, Am Acad Occup Med, 87; Presidential Citation, Am Acad Dermat, 90. *Mem:* Am Col Physicians; NY Acad Sci; fel Am Acad Dermat; Am Occup Med Asn; Am Dermat Asn; Sigma Xi; Am Col Occup Med. *Res:* Environmental medicine and dermatology; percutaneous absorption; cutaneous hypersensitivity; effects of physical environment on skin reactions to irritants and allergens; environmental cancer; environmental problems of chemical origin; mechanisms and patterns of cataneous responses to irritants and antigenic stimuli; percutaneous absorption; chemical carcinogenesis; biological effects of heavy metals. *Mailing Add:* The Kettering Lab Cincinnati OH 45267

SUSKIND, SIGMUND RICHARD, b New York, NY, June 19, 26; m 51; c 3. MICROBIOLOGY. *Educ:* NY Univ, AB, 48; Yale Univ, PhD(microbiol), 54. *Prof Exp:* Asst microbiol, Yale Univ, 50-54; USPHS fel, NY Univ, 54-56; from asst prof to prof, Johns Hopkins Univ, 56-65, dean fac arts & sci, 78-83, dean grad & undergrad studies, 71-77, Univ Ombudsman, 88-91, UNIV PROF, JOHNS HOPKINS UNIV, 83- *Concurrent Pos:* Consult, Am Inst Biol Sci, 57-59; consult, indust, 57-60; spec consult, USPHS, 66-70; head molecular biol sect, NSF, 70-71; consult, Coun Grad Schs & Mid States Asn Cols & Sec Schs, 73-; vis scientist, Weizmann, Inst Sci, Israel, 85; consult,

NSF, 86-; mem adv bd, La Geriat Educ Ctr, 90- *Mem:* Am Soc Microbiol; Genetics Soc Am; Am Asn Immunol; Am Soc Biol Chem; fel AAAS. *Res:* Molecular biology and gene expression. *Mailing Add:* Dept Biol McCollum-Pratt Inst Johns Hopkins Univ Charles St & 34th St Baltimore MD 21218

SUSLICK, KENNETH SANDERS, b Chicago, Ill, Sept 16, 52; m 75. SONOCHEMISTRY, BIOINORGANIC CHEMISTRY. *Educ:* Calif Inst Technol, BS, 74; Stanford Univ, PhD(chem), 78. *Prof Exp:* Res asst, Calif Inst Technol, 71-74; chemist, Lawrence Livermore Lab, 74-75; asst prof, 78-84, assoc prof, 84-88, PROF CHEM, UNIV ILL, URBANA-CHAMPAIGN, 88- *Concurrent Pos:* Res asst, Univ Calif, Berkeley, 72; Hertz fel, Stanford Univ, 74-78; Sloan Found res fel; vis fel, Balliol Col & Inorg Chem Lab, Oxford Univ, 86; prof, Beckman Inst, Univ Ill, 89- *Honors & Awards:* Silver Medal, Royal Soc Arts, Mfgs & Commerce; Res Career Develop Award, NIH. *Mem:* Am Chem Soc; AAAS. *Res:* Synthetic analogs of heme proteins, porphyrins and macrocycles; homogeneous catalysis; chemical effects of high intensity ultrasound; sonochemistry and sonocatalysis; sonoluminescence. *Mailing Add:* Dept Chem Univ Ill 505 S Mathews Ave Urbana IL 61801

SUSMAN, LEON, b Brooklyn, NY, Oct 10, 36; m 58; c 3. ELECTRICAL ENGINEERING. *Educ:* City Col New York, BEE, 58, MEE, 62; Polytechnic Inst Brooklyn, PhD(elec eng), 69. *Prof Exp:* Engr, Ford Instrument Co Div, Sperry Rand Corp, 58 & Airborne Instruments Labs Div, Cutter Hammer Corp, 58-61; sr engr, Sperry Gyroscope Div, 61-68, res staff mem microwave & antenna res, 68-80, MGR, ELECTROMAGNETICS DEPT, SPERRY RAND RES CTR, SPERRY RAND CORP, 80- *Mem:* Inst Elec & Electronics Engrs. *Res:* Microwave and antenna theory, in particular the application of transient performance to a wideband theory for radar; traffic control sensor; altimetry applications. *Mailing Add:* Sperry Corp Res Dept 1601 Trapelo Rd Waltham MA 02154

SUSMAN, MILLARD, b St Louis, Mo, Sept 1, 34; m 57; c 2. GENETICS. *Educ:* Wash Univ, AB, 56; Calif Inst Technol, PhD(genetics), 62. *Prof Exp:* NIH fel, Med Res Coun Microbial Genetics Res Unit, Hammersmith Hosp, London, Eng, 61-62; from asst prof to assoc prof, 62-73, chmn dept, 71-75, PROF GENETICS, LAB GENETICS, UNIV WIS-MADISON, 72-, chmn dept, 77-86, ASSOC DEAN, MED SCH, 86- *Concurrent Pos:* Actg dean, Sch of Allied Health Professions, 88-90. *Mem:* AAAS; Genetics Soc Am; Asn Am Med Cols; Sigma Xi. *Res:* Bacteriophage genetics and developmental genetics; effects of acridines on bacteriophage growth, recombination and mutation; role of the host cell in phage growth. *Mailing Add:* 1220C Med Sci Ctr Univ Wis Med Sch Madison WI 53706

SUSMAN, RANDALL LEE, b Houston, Tex, Jan 19, 48; m 69; c 2. ANATOMY, PHYSICAL ANTHROPOLOGY. *Educ:* Univ Calif, Davis, BA, 70; Univ Chicago, MA, 72, PhD(anthrop), 76; Touro Col, ID, 88. *Prof Exp:* Lectr & fel, 76-77, ASSOC PROF ANAT SCI, STATE UNIV NY, STONY BROOK, 77- *Mem:* Am Asn Phys Anthropologists; AAAS; Soc Syst Zoologists; Am Anthrop Asn; Soc Vert Paleont. *Res:* Evolution of apes and humans; functional morphology of primates; natural history of the primates; electromyography; gross anatomy; field study of pygmy chimpanzee in Zaire. *Mailing Add:* Dept of Anat Sci State UniV NY Stony Brook NY 11794-8081

SUSSDORF, DIETER HANS, b Neustadt, Ger, Aug 16, 30; nat US; m 54; c 3. IMMUNOLOGY. *Educ:* Univ Mo, BA, 52; Univ Chicago, PhD(microbiol), 56. *Prof Exp:* Logan fel, Univ Chicago, 57-58; resident res assoc biol, Argonne Nat Lab, 58-59; res fel immunochem, Calif Inst Technol, 59-61; res immunochemist, NIH, 61-63; asst prof, 64-72, ASSOC PROF MICROBIOL, MED COL & GRAD SCH MED SCI, CORNELL UNIV, 72-, ASSOC DEAN, 84- *Concurrent Pos:* Consult, Travenol Labs, 78-; course dir & lectr, Ctr Prof Advan, 72-; consult, Diamond Shamrock Chem Corp, 85- *Honors & Awards:* David Anderson-Berry Prize, 61. *Res:* Function of thymus and non-thymic tissues in humoral and cellular immunity; anti-tumor activities of macrophages. *Mailing Add:* Dept Microbiol Cornell Univ Med Col 1300 York Ave New York NY 10021

SUSSENGUTH, EDWARD H, b Holyoke, Mass, Oct 10, 32; m 59; c 2. COMPUTER NETWORKING. *Educ:* Harvard Univ, AB, 54, PhD(appl math), 64; Mass Inst Technol, MS, 59. *Prof Exp:* Res & development, 59-70, div dir, 70-81, IBM FEL, IBM, 81- *Concurrent Pos:* Adv, Nat Bur Standards, 86-89 & Columbia Univ, Ctr Telecommun Res, 88- *Mem:* Inst Elec & Electronic Engrs; Asn Comput Mach; Sigma Xi. *Res:* Computer networking; high speed data communications. *Mailing Add:* 411 Rutherglen Dr Cary NC 27511

SUSSER, MERVYN W, b Johannesburg, SAfrica, Sept 26, 21; m 49; c 3. EPIDEMIOLOGY, SOCIAL MEDICINE. *Educ:* Univ Witwatersrand, MB, BCh, 50; MRCP(E), 70 @310 Med officer, Alexandria Health Ctr & Univ Clin, Johannesburg, 51-55. *Prof Exp:* from lectr to reader social med, Univ Manchester, 57-65; Asn Aid Crippled Children Belding scholar, 65-66; prof epidemiol, 66-77, GERTRUDE H SERGIEVSKY PROF EPIDEMIOL, & DIR, SERGIEVSKY CTR, COLUMBIA UNIV, 77- *Concurrent Pos:* Clin tutor med, Univ Witwatersrand, 51-55; John Simon Guggenheim fel, 72-73; mem comt, Sect Epidemiol & Community Psychiat, World Psychiat Asn. *Honors & Awards:* Belding Scholar, Guggenheim Fel, Sergievsky; FRCP(Ed). *Mem:* Fel Am Pub Health Asn; Am Sociol Asn; Soc Epidemiol Res; NY Acad Med; Am Epidemiol Soc; fel Am Col Epidemiol 655 Social and cultural factors in human development and disease. *Res:* brain disorders. *Mailing Add:* Sergievsky Ctr Columbia Univ 630 W 168th St New York NY 10032

SUSSEX, IAN MITCHELL, b Auckland, NZ, May 4, 27. BOTANY. *Educ:* Univ NZ, BS, 48, MSc, 50; Manchester Univ, PhD, 52. *Prof Exp:* Asst lectr bot, Victoria Univ Col, 54-55; asst prof, Univ Pittsburgh, 55-60; assoc prof, 60-73, PROF BOT, YALE UNIV, 73- *Mem:* AAAS; Soc Develop Biol; Bot Soc Am; Int Soc Plant Morphol; Am Soc Cell Biol; Am Genetics Soc; Soc Plant Molecular Biol. *Res:* Plant development; tissue culture. *Mailing Add:* Dept Biol Yale Univ PO Box 6666 New Haven CT 06511

SUSSEX, JAMES NEIL, b Northcote, Minn, Oct 2, 17; m 43; c 4. PSYCHIATRY. *Educ:* Univ Kans, AB, 39, MD, 42. *Prof Exp:* Resident psychiat, US Naval Hosp, Mare Island, Calif, 46-49; asst clin prof psychiat, Sch Med, Georgetown Univ, 53-55; assoc prof, Med Col Ala, 55-59, prof & chmn dept, 59-68; chmn, dept psychiat, 70-83, PROF PSYCHIAT, SCH MED, UNIV MIAMI, 68-, EMER CHMN, 83- *Concurrent Pos:* Fel child psychiat, Philadelphia Child Guid Clin, Univ Pa, 49-51; dir ment health serv div, Jackson Mem Hosp, Miami, 70-83; consult, NIMH; consult, Vet Admin, mem, Ment Adv Coun; pres, Am Asn Psychiat Serv Children, 72-74; dir, Am Bd Psychiat & Neurol, 75-82, pres, 82; bd dir, Coun Med Specialty Soc, 85-, pres, 88-89; mem, Coun Med Affairs, 88- *Mem:* AMA; Am Psychiat Asn; Am Col Psychiat; Am Acad Child & Adolescent Psychiat; Am Bd Med Specialties. *Res:* Child psychiatry; child development in cross-cultural perspective; atypical culture-bound syndromes; dissociative states. *Mailing Add:* 6950 SW 134th St Miami FL 33156

SUSSKIND, ALFRED K(RISS), electrical engineering; deceased, see previous edition for last biography

SUSSKIND, CHARLES, b Prague, Czech; nat US; m 45; c 3. BIOENGINEERING, HISTORY OF TECHNOLOGY. *Educ:* Calif Inst Technol, BS, 48; Yale Univ, MEng, 49, PhD(elec eng), 51. *Prof Exp:* Res assoc, Stanford Univ, 51-55, lectr elec eng, univ & asst dir, Microwave Lab, 53-55; from asst prof to assoc prof elec eng, 55-64, asst dean eng, 64-68, PROF ENG SCI, UNIV CALIF, BERKELEY, 64- *Concurrent Pos:* Consult ed & dir, San Francisco Press, Inc, 59-; govt consult, 68-; coordr acad affairs, Statewide Univ, 69-74. *Honors & Awards:* Clerk Maxwell Premium, Brit Inst Electronic & Radio Eng, 52. *Mem:* Biomed Eng Soc; Hist Sci Soc; AAAS; fel Inst Elec & Electronics Engrs; Brit Inst Electronic & Radio Eng. *Res:* Bioelectronics; electron optics; history and sociology of technology and science. *Mailing Add:* Col of Eng Univ of Calif Berkeley CA 94720

SUSSKIND, HERBERT, b Ratibor, Ger, Mar 23, 29; US citizen; m 61; c 3. BIOMEDICAL ENGINEERING. *Educ:* City Col New York, BChE, 50; NY Univ, MChE, 61; NY State, PE, 55. *Prof Exp:* Mem shielding group, Brookhaven Nat Lab, 50-52, fuel processing group, 52-58, assoc sect supvr, 57-58, sect supvr, Slurry Group, 58-59, mem, Reactor Eval & Advan Concepts Group, 59-61, head, Org Cooled Reactor Prog, 61-63, head, Packed Bed Reactor Eng Studies, 63-66, mem, Reactor Eval & Advan Concepts Group, 66-70; mem, Med Radionuclide Develop Group, 70-77, MEM NUCLEAR MED DIV, BROOKHAVEN NAT LAB, 77 - *Concurrent Pos:* Biomed engr med staff, Brookhaven Clin Res Ctr, 75 -; assoc prof med, Dept Med, State Univ NY, Stony Brook, 79 - *Mem:* Am Inst Chem Engrs; Am Nuclear Soc; Soc Nuclear Med; Biomed Eng Soc; Am Thoracic Soc. *Res:* Liquid metal technology; high temperature reactor fuel processing; packed and fluidized bed technology; application of chemical engineering to medical problems, principally in pulmonary physiology and nuclear medicine. *Mailing Add:* Medical Dept Brookhaven Nat Lab Upton NY 11973

SUSSMAN, ALFRED SHEPPARD, b Portsmouth, Va, July 4, 19; m 48; c 3. BIOLOGY. *Educ:* Univ Conn, BS, 41; Harvard Univ, AM, 48, PhD(biol), 49. *Prof Exp:* Instr microbiol, Mass Gen Hosp, 48-49; instr bot, Univ Mich, Ann Arbor, 50-52, from asst prof to assoc prof, 53-61, chmn dept bot, 63-68, assoc dean col lit, sci & arts, 68-70, actg dean, 70-71, assoc dean, HH Rackham Sch Grad Studies, 72-74, dean, 74-85, actg vpres acad affairs, 79-80, intern vpres grad studies res, 83-85, prof, 61-90, EMER PROF BOT, UNIV MICH, ANN ARBOR, 90- *Concurrent Pos:* Nat Res Coun fel, Univ Pa; Lalor Found fel, 56; NSF sr fel, Calif Inst Technol, 59-60; consult panel develop biol, NSF, 63-65, mem steering comt & comt innovation in lab instr, Biol Sci Curric Study, comnr comn undergrad educ biol sci, 66-69; chmn comt educ, Am Inst Biol Sci; mem biol comt, Argonne Univ Asn, 69-71, chmn, 70-71, examr, N Cent Asn Col & Univ, 71-, exec comt, 79-; mem, Grad Rec Examr Bd, Res Comt, 78, chmn, 79-80, mem comt bio & med, 72-78, trustee, 74-78. *Mem:* Am Soc Biol Chem. *Res:* Physiological mycology; microbial physiology and development; dormancy in microorganisms. *Mailing Add:* 1615 Harbal Dr Ann Arbor MI 48105

SUSSMAN, HOWARD H, b Portland, Ore, Oct 21, 34; m 70; c 3. PATHOLOGY. *Educ:* Univ Ore, BS, 57, MS, 60, MD, 60. *Prof Exp:* Surgeon, USPHS, 61-69; from asst prof to assoc prof, 69-85, PROF PATH, STANFORD UNIV SCH MED, 85- *Concurrent Pos:* Award scholarship res, Univ Ore Med Sch, 59; dir, Core Clin Lab, Stanford Univ Hosp, 75-; sci adv, John Muir Cancer & Aging Inst, 84- *Mem:* NY Acad Sci; Am Asn Pathologists; Int Soc Oncodevelop Biol & Med; Am Found Clin Res; Acad Clin Lab Physicians & Scientists; Sigma Xi; Van Slyke Soc. *Res:* Elucidation of cellular mechanisma of gene regulations which relate to the neoplastic process in humans; phenomenon of ectopic proteins systhesis in human cancer. *Mailing Add:* Dept Pathology R248A Stanford Univ Sch Med Stanford CA 94305

SUSSMAN, KARL EDGAR, b Baltimore, Md, May 29, 29; m 55; c 2. MEDICINE, ENDOCRINOLOGY. *Educ:* Johns Hopkins Univ, BA, 51; Univ Md, MD, 55. *Prof Exp:* From instr to assoc prof med, 62-72, head div endocrinol, 69-72, PROF MED, UNIV COLO MED CTR, DENVER, 73-; ASSOC CHIEF STAFF, RES & DEVELOP, VET ADMIN MED CTR, DENVER, 82- *Concurrent Pos:* Chief med serv, Denver Vet Admin Hosp, 72-75, clin investr, 75-80. *Mem:* Am Col Physicians; Am Diabetes Asn; Am Fedn Clin Res; Am Physiol Soc; Endocrine Soc. *Res:* Factors controlling insulin secretion in isolated rat islets; hormonal control of carbohydrate-lipid metabolism; relationship of intermediary metabolism to insulin secretion; somatostatin-effect on hormone secretion; regulation of somatostation binding. *Mailing Add:* Vet Admin Med Ctr 1055 Clermont Denver CO 80220

SUSSMAN, M(ARTIN) V(ICTOR), b New York, NY; m 53; c 3. CHEMICAL ENGINEERING, THERMODYNAMICS. *Educ:* Columbia Univ, MS, 52, PhD(chem eng), 58. *Prof Exp:* Instrument engr, Lummus Co, 47-48 & A G McKee & Co, 48-49; instr chem & res assoc, Fordham Univ, 49-50; asst, Columbia Univ, 51; sr engr, Textile Fiber Div, E I du Pont de

Nemours & Co, Inc, 53-58; founder, Chem Eng Dept-Turkey, Robert Col, Istanbul, 58-61; chmn dept, 61-71, PROF CHEM ENG, TUFTS UNIV, 61- *Concurrent Pos:* Dir, Sima, Ltd, Turkey; consult, USAID, 63 & 65, mem, NSF liaison staff, USAID, India, 67-68; NIH spec res fel, Weizmann Inst Sci, 68-69; Ford Found consult curric develop, Birla Inst Technol, India, 71-; vis prof, Mass Inst Technol, 75, Univ Capetown, 90; Erskine Prof, Univ Canterbury, NZ, 83; Meyerhoff fel, Weizmann Inst, 83; hon res assoc, Exeter Univ, 90. *Honors & Awards:* Fulbright-Hays sr lectr, 77; Distinguished Vis Lectr, Va Polytech Inst, 80. *Mem:* Fel Am Inst Chem Engrs; Am Chem Soc; Am Soc Eng Educ; Sigma Xi. *Res:* Thermodynamics; continuous chromatography; adaptive technology; separation and nucleation phenomena; synthetic fibers; biotechnology; cell culture; holder of over 15 patents. *Mailing Add:* Dept Chem Eng Tufts Univ 00098264xv Medford MA 02155

SUSSMAN, MAURICE, b New York, NY, Mar 2, 22; m 48; c 3. DEVELOPMENTAL BIOLOGY, MOLECULAR BIOLOGY. *Educ:* City Col New York, BS, 42; Univ Minn, PhD(bact), 49. *Prof Exp:* USPHS fel & instr bact, Univ Ill, 49-50; instr biol sci, Northwestern Univ, 50-53, from asst prof to assoc prof, 53-58; assoc prof, Brandeis Univ, 58-60, prof, 60-73; prof inst life sci, Hebrew Univ Jerusalem, 73-76; PROF & CHMN DEPT BIOL SCI, UNIV PITTSBURGH, 76- *Concurrent Pos:* Instr, Marine Biol Lab, Woods Hole, 56-60 & 67-70. *Honors & Awards:* NIH career develop award, 66. *Mem:* Am Soc Microbiol; Soc Gen Physiol; Soc Develop Biol; Am Soc Biol Chem; Brit Soc Gen Microbiol. *Res:* Cellular differentiation and morphogenesis; molecular genetics. *Mailing Add:* Dept Biol Sci Univ Pittsburgh Pittsburgh PA 15260

SUSSMAN, MICHAEL R, b New York, NY, Sept 3, 50; m 83; c 2. ENERGY TRANSDUCTION, PROTEIN ENGINEERING. *Educ:* Bucknell Univ, BS, 71; Mich State Univ, PhD(bot), 76. *Prof Exp:* Postdoctoral fel plant phys, Biol Dept, Yale Univ, 76-78, postdoctoral assoc biochem & genetics, Dept Human Genetics, 79-82; asst prof, 83-89, ASSOC PROF PLANT MOLECULAR BIOL, DEPT HORT, UNIV WIS-MADISON, 89- *Concurrent Pos:* Res award plant biol, McKnight Found, 83-86; var res grants, Dept Energy, US Dept Agr, NSF, Shell Develop Co, Agracetus, Inc & E I du Pont de Nemours & Co, 85-93; consult plant molecular biol, 89-90; vis prof, Cath Univ Louvain, Belg, 90; Fulbright res scholar, 90. *Mem:* Am Soc Plant Physiologists; Int Soc Plant Molecular Biol. *Res:* Molecular basis of ion transport across membranes, especially plasma membrane; mechanism of action of plant hormones; molecular regulation of plant growth and development; author of numerous publications. *Mailing Add:* Dept Hort Univ Wis 1575 Linden Dr Madison WI 53706

SUSSMAN, MYRON MAURICE, b Trenton, NJ, Oct 7, 45; m 70. NUMERICAL ANALYSIS. *Educ:* Mass Inst Technol, SB, 67; Carnegie-Mellon Univ, MS, 68, PhD(math), 75. *Prof Exp:* Instr math, Carnegie-Mellon Univ, 68-69 & Robert Morris Col, 69-71; MATHEMATICIAN, BETTIS ATOMIC POWER LAB, WESTINGHOUSE ELEC CO, 75- *Mem:* Soc Indust & Appl Math; Sigma Xi; Am Math Soc. *Res:* Numerical analysis of partial differential equations and iterative solution of large linear systems of algebraic equations. *Mailing Add:* 5026 Belmont Ave Bethel Park PA 15102

SUSSMAN, RAQUEL ROTMAN, b Arg, Oct 22, 21; nat US; m 48; c 3. MICROBIOLOGY. *Educ:* Univ Chile, BS, 44; Univ Ill, PhD(bact), 52. *Prof Exp:* Asst viruses, Inst Bact Chile, 44-48; asst microbiol, Univ Minn, 48-49 & Univ Ill, 49-50; res assoc, Northwestern Univ, 50-58 & Brandeis Univ, 58-73; sr lectr, Dept Molecular Biol, Hadassah Med Sch, Hebrew Univ, Israel, 73-75; assoc prof, dept biol sci, Univ Pittsburgh, 76-87; ASSOC SCIENTIST, MARINE BIOL LABS, 87- *Mem:* Sigma Xi; Am Soc Microbiol. *Res:* Molecular biology, chiefly genetics microbiology. *Mailing Add:* Marine Biol Labs Woods Hole MA 02543

SUTCLIFFE, SAMUEL, b New Britain, Conn, Jan 30, 34; m 58; c 1. CIVIL ENGINEERING, MATHEMATICS. *Educ:* Univ Conn, BS, 55; Univ Ill, MS, 58, PhD, 60. *Prof Exp:* Asst civil eng, Univ Ill, 55-57, res assoc, 57-60, asst prof, 60-64; asst prof, 64-69, ASSOC PROF CIVIL ENG, TUFTS UNIV, 69- *Mem:* Am Soc Civil Engrs. *Res:* Dynamics of non-linear, solid continua. *Mailing Add:* Dept of Civil Eng Tufts Univ Medford MA 02155

SUTCLIFFE, WILLIAM GEORGE, b Detroit, Mich, Nov 25, 37; m 60; c 4. PHYSICS. *Educ:* Univ Mich, BS, 60; Univ Del, PhD(physics), 69. *Prof Exp:* Instr physics, US Naval Nuclear Power Sch, 62-64; PHYSICIST, LAWRENCE LIVERMORE LAB, 68- *Mem:* Am Asn Physics Teachers. *Res:* Design and development of large computer codes, including hydrodynamics, radiation transport and neutronics; management of projects involving sensitivity and uncertainty analyses of nuclear waste isolation systems; analysis of nuclear forces, arms control and targeting policies; analysis of nuclear weapons systems requirements; analysis of nuclear material supply and demand requirements; analysis of arms control and confidence building measures. *Mailing Add:* Lawrence Livermore Lab L-19 PO Box 808 Livermore CA 94551

SUTCLIFFE, WILLIAM HUMPHREY, JR, b Miami, Fla, Nov 8, 23; m 45, 49, 64; c 5. ZOOLOGY. *Educ:* Emory Univ, BA, 45; Duke Univ, MA, 47, PhD(zool), 50. *Prof Exp:* Instr zool, Duke Univ, 49-50; investr marine biol, NC Inst Fisheries Res, 50-51, 50-51; staff biologist lobster invests, Bermuda Biol Sta, 51-53, dir, 53-69; head, Fisheries Oceanog Sect, 75-76, res scientist, Bedford Inst Oceanog, 67-81; RETIRED. *Concurrent Pos:* Assoc, Woods Hole Oceanog Inst, 57; dir marine sci ctr, Lehigh Univ, 64-67. *Mem:* AAAS; NY Acad Sci. *Res:* Dynamics of plankton populations; air-sea interaction; marine food chains. *Mailing Add:* Rte 4 Box 449 Bakersville NC 28705

SUTER, BRUCE WILSEY, b Paterson, NJ, Sept 15, 49; m 74. SIGNAL PROCESSING, WAVELETS. *Educ:* Univ SFla, BS, 72, MS, 72, PhD(computer sci), 88. *Prof Exp:* Design engr, Honeywell, Inc, 72-77; sr design engr, Litton Industs, 77-80; res asst, Univ SFla, 80-85; from instr to asst prof, Univ Ala, Birmingham, 85-89; asst prof, 89-91, ASSOC PROF, AIR FORCE INST TECHNOL, 91- *Mem:* Inst Elec & Electronics Engrs; Asn Comput Mach; Soc Indust Appl Math; Int Neural Network Soc. *Res:* Wavelets; fast algorithms for signal processing applications; neural networks-theory and implementation; parallel processing. *Mailing Add:* Dept Elec Eng Air Force Inst Technol Wright-Patterson AFB OH 45433-6583

SUTER, DANIEL B, b Hinton, Va, Apr 25, 20; m 41; c 4. HUMAN ANATOMY, PHYSIOLOGY. *Educ:* Bridgewater Col, BA, 47; Vanderbilt Univ, MA, 48; Med Col Va, PhD(anat), 63. *Prof Exp:* Asst prof biol, 48-60, assoc prof, 62-63, chmn, Dept Life Sci, 72-76, assoc dean, 76-77, PROF BIOL, EASTERN MENNONITE COL, 63-, CHMN DIV NATURAL SCI & MATH, 64- *Concurrent Pos:* NIH fel, Univ Calif, Davis, 70-71. *Mem:* Asn Am Med Cols; Am Asn Anat; Am Sci Affiliation. *Res:* Effects of radiation and pesticides, especially organophosphates, on the central nervous system. *Mailing Add:* Dept of Biol Eastern Mennonite Col Harrisonburg VA 22801

SUTER, GLENN WALTER, II, b Harrisonburg, Va, May 1, 48; m 68. TOXICOLOGY. *Educ:* Va Polytech Inst, BS, 69; Univ Calif, Davis, PhD(ecol), 76. *Prof Exp:* RES STAFF MEM, ENVIRON SCI DIV, OAK RIDGE NAT LAB, 75- *Mem:* AAAS; Ecol Soc Am; Soc Environ Toxicol & Chem; Soc Risk Anal. *Res:* Ecological risk assessment. *Mailing Add:* Bldg 1505 Oak Ridge Nat Lab Oak Ridge TN 37831

SUTER, ROBERT WINFORD, b Warren, Ohio, Aug 3, 41; m 63; c 2. CHEMISTRY. *Educ:* Bluffton Col, BA, 63; Ohio State Univ, MS, 66, PhD(chem), 69. *Prof Exp:* Assoc prof, 69-80, PROF CHEM, BLUFFTON COL, 80- *Mem:* Am Chem Soc. *Res:* Inorganic chemistry, particularly nonmetals; solution phenomena. *Mailing Add:* 10620 Bixel Rd Bluffton OH 45817

SUTER, STUART ROSS, b Harrisonburg, Va, Apr 1, 41; m 63; c 2. ORGANIC CHEMISTRY. *Educ:* Bridgewater Col, BA, 63; Univ Mich, Ann Arbor, MS, 65; Univ Va, PhD(org chem), 71; Temple Univ, DJur, 76. *Prof Exp:* Assoc chemist, Smith Kline & French Labs, 65-67; patent chemist, 71-76, PATENT ATTY, SMITHKLINE BEECHAM CORP, 76- *Mem:* Am Chem Soc; Am Bar Asn; Am Int Property Asn. *Res:* Aryl nitrenes; synthetic organic chemistry; medicinal chemistry. *Mailing Add:* Patent Dept Smithkline Beecham Corp One Franklin Plaza, PO Box 7929 Philadelphia PA 19101-7929

SUTERA, SALVATORE P, b Baltimore, Md, Jan 12, 33; m 58; c 3. MECHANICAL ENGINEERING. *Educ:* Johns Hopkins Univ, BSc, 54; Calif Inst Technol, MSc, 55, PhD(eng), 60. *Hon Degrees:* MA, Brown Univ, 65. *Prof Exp:* From asst prof to assoc prof eng, Brown Univ, 60-68, exec off div eng, 66-68; PROF MECH ENG & CHMN DEPT, WASHINGTON UNIV, 68- *Concurrent Pos:* Nat Heart Inst res grants, 65-67 & 69-; consult, Nat Heart, Lung & Blood Inst, 70- *Mem:* Am Soc Mech Engrs; Am Soc Eng Educ; AAAS; Am Soc Artificial Internal Organs; Int Soc Biorheology. *Res:* Fluid mechanics of blood flow; artificial organs; rheology of suspensions. *Mailing Add:* Dept Mech Eng Washington Univ Box 1185 St Louis MO 63130

SUTHERLAND, BETSY MIDDLETON, b New York, NY, Oct 19, 43; m 65. BIOCHEMISTRY, CELL BIOLOGY. *Educ:* Emory Univ, BS, 64, MS, 65; Univ Tenn, PhD(radiation biol), 67. *Prof Exp:* NIH fel DNA chem, Lab Molecular Biol, Walter Reed Res Inst, 67-69; fel enzymol, Molecular Biol-Virus Lab, Univ Calif, Berkeley, 69-72; from asst prof to assoc prof molecular biol, Univ Calif, Irvine, 72-77; SCIENTIST DNA REPAIR, DEPT BIOL, BROOKHAVEN NAT LAB, 77- *Concurrent Pos:* Assoc ed, Photochem & Photobiol, 74-77; mem, Nat Comt Photobiol, Nat Acad Sci, 74-78; Nat Cancer Inst, NIH res career develop award, 75-80. *Honors & Awards:* Edna M Rowe Mem Lectr Int Photobiol Cong; E O Lawrence Award. *Mem:* Am Soc Photobiol; Biophys Soc; AAAS; Am Soc Biol Chemists. *Res:* DNA damage and repair; biology and biochemistry of photoreactivation; transformation of mammalian cells and its relation to oncogenesis; DNA transfection into human cells. *Mailing Add:* Dept Biol Brookhaven Nat Lab Bldg 463 Upton NY 11973

SUTHERLAND, BILL, b Sedalia, Mo, Mar 31, 42; m 65; c 2. PHASE TRANSITIONS, QUANTUM MANY-BODY THEORY. *Educ:* Wash Univ, AB, 63; State Univ NY Stony Brook, MA, 65, PhD, 67. *Prof Exp:* Res assoc, State Univ NY Stony Brook, 67-69; asst res physicist, Univ Calif, Berkeley, 69-70; from asst prof to assoc prof, 70-82, PROF PHYSICS, UNIV UTAH, 82- *Mem:* Fel Am Phys Soc. *Res:* Statistical mechanics; phase transitions; critical phenomena; quasi-periodic systems; quantum many-body systems; strongly correlated systems. *Mailing Add:* Dept Physics Univ Utah Salt Lake City UT 84112

SUTHERLAND, CHARLES F, b Camp Grant, Ill, Oct 1, 21; m 44, 81; c 4. FORESTRY ECONOMICS. *Educ:* Univ Idaho, BS, 48, MFor, 54; Univ Mich, PhD(forestry econ), 61. *Prof Exp:* Res forester, Potlatch Forests Inc, 48-53; forest economist, Lake States Forest Exp Sta, US Forest Serv, St Paul, Minn, 56-58; asst prof, 59-68, ASSOC PROF FORESTRY ECON, SCH FORESTRY, ORE STATE UNIV, 68- *Concurrent Pos:* Vis res scientist, Southern Foust Exp Sta, US Forest Serv, 78. *Res:* Biology and management in private industry; marketing and production problems in forestry; taxation and forest protection problems. *Mailing Add:* 440 NW Elizabeth Dr Corvallis OR 97330

SUTHERLAND, CHARLES WILLIAM, b Nashville, Tenn, July 29, 41; m 64; c 2. METALLURGICAL ENGINEERING. *Educ:* Vanderbilt Univ, BE, 63; Univ Wis-Madison, MS, 64. *Prof Exp:* ASSOC PROF MECH ENG, TENN STATE UNIV, 67-, HEAD DEPT, 70-, DIR INST RES, 72- *Concurrent Pos:* Consult, Training & Habilitation Ctr, Nashville, Tenn, 71-72. *Mem:* Am Soc Metals; Am Inst Mining, Metall & Petrol Engrs. *Res:* Metal working and machining. *Mailing Add:* RR 2 No 259A Dunlap TN 37327

SUTHERLAND, DAVID M, b Bellingham, Wash, Oct 5, 40. PLANT TAXONOMY. *Educ:* Western Wash State Col, BA, 63; Univ Wash, PhD(bot), 67. *Prof Exp:* From asst prof to assoc prof, 67-80, PROF BIOL, UNIV NEBR, OMAHA, 80- *Mem:* AAAS; Int Asn Plant Taxonomists; Bot Soc Am; Am Soc Plant Taxon; Sigma Xi. *Res:* Biosystematics of larkspurs; floristics of Great Plains; vegetative characters of grasses. *Mailing Add:* Dept Biol Univ Nebr Omaha NE 68182-0040

SUTHERLAND, DONALD JAMES, b Chelsea, Mass, Oct 5, 29. PHYSIOLOGY, BIOCHEMISTRY. *Educ:* Tufts Univ, BS, 51; Univ Mass, MS, 57; Rutgers Univ, PhD(entom), 60. *Prof Exp:* From asst prof to assoc prof, 60-67, PROF ENTOM, RUTGERS UNIV, NEW BRUNSWICK, 67-, DEPT CHMN, 87- *Mem:* Entom Soc Am; Am Mosquito Control Asn (pres, 86-87). *Res:* Management of mosquitoes, particularly chemical measures; biological rhythms of insects. *Mailing Add:* Dept Entom Rutgers Univ New Brunswick NJ 08903

SUTHERLAND, EARL C, b Detroit, Mich, July 23, 23; m 71; c 1. ACCIDENT RECONSTRUCTION WITH COMPUTER SIMULATION, FORENSIC ENGINEERING. *Educ:* Mich Col, BS, 50, MS, 50; Portland State Univ, MBA, 74. *Prof Exp:* Tech engr, Int Bus Mach Corp, 50-52; metallurgist, Fansteel Metall Corp, 52-56; tech dir & vpres, Eriez (BA) Prod Metal & Magnetics, 56-60; res specialist, NASA, 60-62; mfg mgr, Precision Castparts Corp, 62-64; dir res & advan eng, Omark Industs, 64-67; proj dir & vpres, MEI/Charleton Inc, 67-75; PRIN, EARL C SUTHERLAND & ASSOCS, 75- *Concurrent Pos:* Prin engr & consult engr, Earl C Sutherland Assocs & MEI/Charleton Inc, 67-; lectr & prof, Ore State Dept Higher Educ. *Mem:* Nat Asn Corrosion Engrs; Soc Automotive Engrs; Am Soc Metals Int; Nat Soc Prof Engrs; Am Consult Engrs Coun; Metal Inst. *Res:* Accident reconstruction of general aircraft and heavy industrial systems and machinery which fail in service; computer simulation of events to failure. *Mailing Add:* 2565 Dexter Ave N No 401 Seattle WA 98109

SUTHERLAND, G RUSSELL, b Rush Lake, Wis, Dec 20, 23. ENGINEERING. *Educ:* Univ Wis, BS(mech eng) & BS(agr), 47, MS, 49. *Prof Exp:* Mgr prod eng, Deere & Co, Des Moines, 66-77, dir, prod planning, 78-80, prod eng planning, 80-83 & prod eng, 83-84, vpres eng & technol, Moline, 84-88; RETIRED. *Concurrent Pos:* Mem bd dirs, Am Nat Standards Inst, 84-86. *Mem:* Nat Acad Eng; Am Soc Agr Eng; Soc Automotive Eng. *Mailing Add:* 16 Mason Lane Belle Vista Village Hiwasse AR 72739

SUTHERLAND, GEORGE LESLIE, b Dallas, Tex, Aug 13, 22; wid; c 3. ORGANIC CHEMISTRY. *Educ:* Univ Tex, BS, 43, MA, 47, PhD(org chem), 50. *Prof Exp:* Lilly fel, Univ Tex, 49-51; chemist, AM Cyanamid Co, 51-57, group leader, 58-62, mgr metab & anal res, 62-66, dir prod develop & govt registr, 66-69, asst dir res & develop, 69-70, dir res & develop, 70-73, vpres res & develop, Lederle Labs Div, 73-78, dir, Chem Res Div, 80-81, dir, Med Res Div, 78-86, vpres, Corp Res & Technol, 86-87; RETIRED. *Mem:* AAAS; Am Chem Soc; NY Acad Sci; Royal Soc Chem. *Res:* Organometallic compounds; anticonvulsants; microbiological growth factors; hypotensive agents; anticoccidials; chemical process development; formulation, metabolism and analysis of agricultural chemicals; discovery and development of pharmaceuticals. *Mailing Add:* 42 Sky Meadow Rd Suffern NY 10901

SUTHERLAND, HERBERT JAMES, b San Antonio, Tex, Nov 1, 43; m 77; c 2. ENGINEERING MECHANICS, GEOMECHANICS. *Educ:* Univ Tex, Austin, BS, 66, MS, 68, PhD(eng mech), 70. *Prof Exp:* Prod engr, Humble Oil Co, 66; assoc aircraft engr, Lockheed Ga Corp, 67; res teaching asst, Univ Tex, 67-70; DISTINGUISHED MEM TECH STAFF, SANDIA NAT LABS, 70- *Concurrent Pos:* Vis scientist, Kernforschungszentrum Karslsruhe, Fed Repub Ger, 79. *Mem:* Fel Am Soc Mech Engrs; Am Soc Rheology; Am Acad Mech. *Res:* Constitutive formulations; instrumentation development; geomechanics; fatigue analysis; over 100 technical articles on the instrumentation and analysis of wave propagation in composite and viscoelastic materials, nuclear reactor safety studies, mine subsidence and soil stability, and damage accumulation in wind turbines. *Mailing Add:* Sandia Nat Labs PO Box 5800 Org 6225 Albuquerque NM 87185

SUTHERLAND, IVAN EDWARD, b Hastings, Nebr, May 16, 38; m 59; c 2. COMPUTER SCIENCE, MATHEMATICS. *Educ:* Carnegie-Mellon Univ, BS, 59; Calif Inst Technol, MS, 60; Mass Inst Technol, PhD(elec eng), 63. *Hon Degrees:* MA, Harvard Univ, 66. *Prof Exp:* Dir info processing tech, Defense Adv Res Projs Agency, 64-66; assoc prof, Div Eng & Appl Physics, Harvard Univ, 66-68; vpres & chief scientist, Evans & Sutherland Comput Corp, Salt Lake City, Utah, 68-74; prof & head, Dept Computer Sci, Calif Inst Technol, 76-80; VPRES & TECH DIR, SUTHERLAND, SPROULL & ASSOCS, INC, 80- *Concurrent Pos:* Tech opers officer, US Army Liaison Group, Proj Mich, Univ Mich, Ypsilanti, 63; elec engr, Nat Security Agency, Ft Meade, Md, 63-64; from assoc prof to prof, Dept Elec Eng, Computer Sci Div, Univ Utah, 68-73; mem, Naval Res Adv Comt, Dept Navy, 68-75, computer sci & technol bd, Assembly Math & Phys Sci, Nat Res Coun, 77-80, Defense Sci Bd, 77-84; vpres, Picture Design Group, Santa Monica, Calif, 74; mem, bd dirs, Evans & Sutherland Comput Corp, 74-, Quotron Systs, Inc, Los Angeles, Calif, 80-, Nat Aviation & Technol Corp, NY, 82-85, Nat Telecommun & Technol Fund, NY, 83-85 & Newmarket Co, Ltd, Bermuda, 84-; sr tech staff mem, Rand Corp, 75-76; vis scientist, Robotics Inst, Carnegie-Mellon Univ, 80-84; gen partner, Advan Technol Ventures, Boston, Mass, 80-; vis acad, Imp Col, London Univ, 85- *Honors & Awards:* First Zworykin Award, Nat Acad Eng, 72; Outstanding Accomplishment Award, Systs, Man & Cybernet Soc, 75; First Steven Anson Coons Award, Asn Comput Mach Siggraph, 83. *Mem:* Nat Acad Sci; Nat Acad Eng; Inst Elec & Electronics Engrs; Sigma Xi; Asn Comput Mach. *Res:* Computer graphics; architecture of high performance computing machinery; algorithms for rapid execution of special functions; large-scale integrated circuit design; robots; author or co-author of over 40 publications. *Mailing Add:* Sutherland Sproull & Assocs Inc 4516 Henry St Pittsburgh PA 15213

SUTHERLAND, JAMES HENRY RICHARDSON, b Can, July 6, 23; nat US; m 43; c 3. PHARMACOLOGY. *Educ:* Univ Calif, AB, 48, PhD(physiol), 56. *Prof Exp:* Trainee cardiovasc res prog, 55-56, from asst prof to assoc prof pharmacol, 56-64, chmn dept, 64-68, dir div health commun, 68-74, prof, 64-85, EMER PROF PHARMACOL & TOXICOL, MED COL GA, 85- *Mailing Add:* 2525 Center W Pkwy Apt 8A Augusta GA 30909

SUTHERLAND, JAMES MCKENZIE, b Chicago, Ill, Aug 8, 23; m 53; c 2. PEDIATRICS, NEONATAL PERINATAL MEDICINE. *Educ:* Univ Chicago, SB, 47, MD, 50; Am Bd Pediat, dipl, 55. *Prof Exp:* Intern, Cincinnati Gen Hosp, 50-51; resident pediat, Children's Hosp, 51-54; instr, Col Med, Univ Cincinnati, 53-54; fel, Harvard Med Sch, 54-56; from asst prof to prof, 56-87, EMER PROF PEDIAT, COL MED, UNIV CINCINNATI, 87- *Concurrent Pos:* NIH fel, Children's Med Ctr, Boston, 54-56; attend pediatrician, Children's Hosp, Cincinnati, 56-; dir newborn div, Cincinnati Gen Hosp, 56-85; res assoc & head div newborn physiol, Children's Hosp Res Found, 56-85. *Mem:* AAAS; Soc Pediat Res; Am Pediat Soc; AMA; Am Acad Pediat. *Res:* Physiology of normal and abnormal respiration in infants. *Mailing Add:* 1338 Edwards Rd Cincinnati OH 45208

SUTHERLAND, JOHN B(ENNETT), b Burlingame, Kans, Feb 21, 18; m 35; c 3. CHEMICAL ENGINEERING. *Educ:* Kans State Col, BS, 39, MS, 40; Univ Pittsburgh, PhD(chem eng), 46. *Prof Exp:* Chem engr, Tex Co, Tex, 40-41; fel asst, Mellon Inst, 41-43; asst prof chem eng, Northwestern Univ, 43-46; pres Sutherland-Becker Labs, 46-64; dir indust res & exten, 66-80, EMER PROF CHEM ENG, UNIV MO-COLUMBIA, 80- *Concurrent Pos:* Consult econ develop & corp planning, 80-; assoc dean eng, Kans State Univ, 65-66; consult corp planning, technol transfer & ventures. *Mem:* Am Chem Soc; Am Inst Chem Engrs; Sigma Xi. *Res:* Lubricating oils; synthetic gems; economic development; gels and films; construction materials; statics and dynamics of bulk solids. *Mailing Add:* 3021 SW Burlingame Rd Topeka KS 66611-2003

SUTHERLAND, JOHN BRUCE, IV, b Tampa, Fla, Nov 9, 45. MICROBIOLOGY. *Educ:* Stanford Univ, AB, 67; Univ Wis, Madison, MS, 73; Wash State Univ, PhD(plant path), 78. *Prof Exp:* Fel bacteriol, Univ Idaho, 77-81; asst prof biol sci, Tex Tech Univ, 81-89; RES SCIENTIST, FOOD & DRUG ADMIN, 89- *Concurrent Pos:* Instr, Wash State Univ, 78-79. *Mem:* Am Phytopath Soc; Am Soc Microbiol; Mycol Soc Am. *Res:* Physiology and ecology of soil microorganisms; biodegradation of lignocellulose. *Mailing Add:* Food & Drug Admin Nat Ctr Toxicol Res Jefferson AR 72079

SUTHERLAND, JOHN CLARK, b New York, NY, Sept 2, 40; m 65. BIOPHYSICS. *Educ:* Ga Inst Technol, BS, 62, MS, 64, PhD(physics), 67. *Prof Exp:* Biophysicist, Walter Reed Res Inst, 67-69; res fel, Lab Chem Biodyn, Univ Calif, Berkeley, 69-72, USPHS fel, 69-71; res fel chem, Univ Southern Calif, 72-73; from asst prof to assoc prof physiol, Calif Col Med-Univ Calif, Irvine, 73-77; biophysicist, 77-89, SR BIOPHYSICIST, BROOKHAVEN NAT LAB, 89- *Concurrent Pos:* Assoc ed, Photochem & Photobiol, 81-84; res career develop award, Nat Cancer Inst, 66-81. *Honors & Awards:* IR-100 Award, 87. *Mem:* Am Soc Photobiol; Am Phys Soc; Biophys Soc. *Res:* Optical spectroscopy and photochemistry of biological molecules; synchrotron radiation; gel electrophoresis; electronic imaging; vacuum ultraviolet spectroscopy of metals. *Mailing Add:* Dept Biol Brookhaven Nat Lab Upton NY 11973

SUTHERLAND, JOHN PATRICK, b Salem, Ore, Oct 1, 42; m 81; c 2. MARINE ECOLOGY. *Educ:* Univ Wash, Seattle, BS, 64; Univ Calif, Berkely, PhD(zool), 69. *Prof Exp:* Asst prof, 69-75, ASSOC PROF MARINE ECOL, MARINE LAB, DUKE UNIV, 75- *Concurrent Pos:* Off Naval Res & NSF grants, 72- *Mem:* Ecol Soc Am. *Res:* Comparative studies on the dynamics and bioenergetics of marine invertebrates; structure and function of subtidal, epibenthic fouling communities; ecology of kelp communities; ecology of tropical rocky shores. *Mailing Add:* Duke Univ Marine Lab Beaufort NC 28516

SUTHERLAND, JUDITH ELLIOTT, b Clovis, NMex, June 6, 24; m 47; c 2. POLYMER CHEMISTRY. *Educ:* Univ Tex, BSChem, 45; Univ Conn, MA, 68; Univ Mass, PhD(polymer chem), 72. *Prof Exp:* Res biochemist, Tex Agr Exp Sta, College Station, 45-46; res chemist biochem, Clayton Found, Biochem Inst, Austin, Tex, 46-48; chemist, Stamford Chem Co, Conn, 60-61 & Am Cyanamid Co, Conn, 61-71; res chemist polymers, Eastman Kodak Co Res Labs, 72-90; RETIRED. *Mem:* Am Chem Soc; AAAS; Am Phys Soc. *Res:* Polymer synthesis and characterization; solution properties; structure-property relationships; solid state properties of polymers and polymer composites. *Mailing Add:* 101 Burkedale Crescent Rochester NY 14625

SUTHERLAND, LOUIS CARR, b Walla Walla, Wash, June 2, 26; m 49; c 3. ACOUSTICS, STRUCTURAL DYNAMICS. *Educ:* Univ Wash, BS, 46, MS, 54. *Prof Exp:* Res asst, Eng Exp Sta, Univ Wash, 47-49, res engr electroacoust, Dept Speech, 49-56; res specialist vibroacoust, Boeing Co, 56-64; DEP DIR & CHIEF SCIENTIST RES, WYLE LABS, 64- *Concurrent Pos:* Chmn subcomt noise metrics, SAE-Aircraft Noise Comt, 76-; US deleg, ISO TC43/SCI/WG24 on Sound Propagation. *Mem:* Fel Acoust Soc Am; Inst Elec & Electronics Engrs; Am Inst Aeronaut & Astronaut. *Res:* Physical acoustics as it relates to man's environment; sound propagation, aircraft noise, community noise, noise control, vibroacoustics, shock and vibration environments of aerospace systems. *Mailing Add:* Wyle Labs 128 Maryland St El Segundo CA 90245-4115

SUTHERLAND, PATRICK KENNEDY, b Dallas, Tex, Feb 17, 25; div. PALEOBIOLOGY, STRATIGRAPHY. *Educ:* Univ Okla, BSc, 46; Cambridge Univ, PhD(geol), 52. *Prof Exp:* Geologist, Phillips Petrol Co, 46-49, 52-53; asst prof geol, Univ Houston, 53-57; assoc prof, 57-64, PROF GEOL, SCH GEOL & GEOPHYS, UNIV OKLA, 64- *Concurrent Pos:* Nat Acad Sci vis exchange fel, USSR, 71. *Mem:* Fel Geol Soc Am; Paleont Soc; Soc Econ Paleont & Mineral; Am Asn Petrol Geol. *Res:* Paleobiology; biostratigraphy; paleoecology; carboniferous Rugose corals and brachiopods. *Mailing Add:* Sch Geol & Geophys Univ Okla Energy Ctr Norman OK 73019

SUTHERLAND, PETER GORDON, b Montreal, Que, Oct 19, 46; m 81; c 2. THEORETICAL ASTROPHYSICS. *Educ:* McGill Univ, BSc, 67; Univ Ill, MS, 68, PhD(physics), 72. *Prof Exp:* Fel, Physics Dept, Columbia Univ, 71-74; asst prof, Univ Pa, 74-76; from asst prof to assoc prof, 76-82, chair, Physics Dept, 87-91, PROF, PHYSICS DEPT, MCMASTER UNIV, 82- *Concurrent Pos:* Lectr, Astron Dept, Columbia Univ, 72-74; Alfred P Sloan Found Fel, 78; vis scientist, Astron Dept, Univ Tex, Austin, 80-81 & Joint Inst Lab Astrophys, Univ Colo, 86-87. *Mem:* Am Astron Soc; Int Astron Union; Can Asn Physicists. *Res:* Neutron stars; pulsars; compact x-ray sources; supernovae. *Mailing Add:* Physics Dept McMaster Univ 1280 Main St W Hamilton ON L8S 4M1 Can

SUTHERLAND, ROBERT L(OUIS), b Fellsmere, Fla, May 15, 16; m 45; c 5. ENGINEERING DESIGN & CONSULTING. *Educ:* Univ Ill, BS, 39, MS, 48. *Prof Exp:* Develop engr, Firestone Tire & Rubber Co, Ohio, 39-41; res engr, Borg & Beck Div, Borg-Warner Corp, Ill, 41-42; test engr, Buick Motor Div, Gen Motors Corp, 42-43; sr engr res, Aeronca Aircraft Corp, Ohio, 43-45; res assoc, Col Eng, Univ Ill, 45-48; assoc prof mech eng, Univ Iowa, 48-58; prof, 58-79, head dept, 60-70, EMER PROF ENG, UNIV WYO, 79-; PRES, SKYLINE ENG, INC, 71- *Concurrent Pos:* Legis fel, Nat Conf State Legis, 82-84. *Honors & Awards:* Templin Award, Am Soc Testing & Mat, 52. *Mem:* Fel Am Soc Mech Engrs (vpres, 64-67); Soc Automotive Engrs; Am Soc Mech Engrs; Sigma Xi. *Res:* Engineering analysis, vibrations and machine design. *Mailing Add:* Box 54 Laramie WY 82070

SUTHERLAND, ROBERT MELVIN, b Moncton, NB, Oct 21, 40; m 62; c 3. BIOPHYSICS, RADIATION BIOLOGY. *Educ:* Acadia Univ, BSc, 61; Univ Rochester, PhD(radiation biol), 66. *Prof Exp:* Fel radiation biol, Norsk Hydro's Inst Cancer Res, Oslo, Norway, 66-67; fel radiation biol, Ont Cancer Found, London Clin, Victoria Hosp, London, Ont, 67-68, radiobiologist, 68-76; assoc prof & head, Radiation Biol Sect, Div Radiation, Oncol & Multimodalities Res, Cancer Ctr, 76-79, PROF RADIATION BIOL & BIOPHYS, UNIV ROCHESTER, 79-, ASST DIR, EXP THERAPEUT DIV, CANCER CTR, 80- *Concurrent Pos:* Radiation res fel, James Picker Found, 66-68; hon lectr biophys, Univ Western Ont, 67-68, lectr therapeut radiol & asst prof biophys, 68-72, assoc prof therapeut radiol & biophys, 72-76; assoc ed, Int J Radiation, Oncol, Biol, Physics. *Mem:* Radiation Res Soc; Can Soc Cell Biol; Am Asn Cancer Res; Am Soc Therapeut Radiologists; Am Soc Cell Biol. *Res:* Membrane biophysics and radiation damage; in vitro tumor models; radiation sensitizers; experimental tumor therapy. *Mailing Add:* Dept Radiation Biol-Cancer Ctr Univ Rochester Med Ctr Box 704 Rochester NY 14642

SUTHERLAND, RONALD GEORGE, b Belfast, Northern Ireland, May 4, 35; m 60; c 2. CHEMISTRY. *Educ:* Univ Strathclyde, BSc, 59; Univ St Andrews, PhD(org chem), 62. *Prof Exp:* Res assoc, Columbia Univ, 62-63; fel, Calif Inst Technol, 63-64; Imp Chem Industs res fel chem, Queen's Col, Dundee, 64; from asst prof to assoc prof, 64-73, assoc dean sci, 81-88, PROF CHEM, UNIV SASK, 73- *Concurrent Pos:* Sci Res Coun UK sr vis fel, Edinburgh Univ, 75-76; consult, Arms Control & Disarmament Div, External Affairs, Ottawa, Can; tech adv, Can Deleg to Conf on Disarmament, Geneva. *Mem:* Royal Soc Chem; Royal Inst Chem; Am Chem Soc; Can Inst Chem. *Res:* Organometallic chemistry; organic photochemistry; chemical decomposition of pesticides; investigations of alleged use of chemical weapons; verification research in arms control. *Mailing Add:* Dept Chem Univ Sask Saskatoon SK S7N 0W0 Can

SUTHERLAND, WILLIAM NEIL, b Linden, Iowa, Aug 10, 27; m 49; c 4. SOIL FERTILITY & AGRICULTURAL, EDUCATIONAL ADMINISTRATION. *Educ:* Iowa State Univ, BS, 50, MS, 53, PhD(soil fertil), 60. *Prof Exp:* Soil conserv agent exten serv, Univ Minn, 54-56; res assoc, Iowa State Univ, 56-60; agriculturist, Tenn Valley Authority, 60-78, agronomist, Test & Demonstration Br, 78-80, chief agr field prog, 80-88, assoc dir, Agr Inst, 86-88; RETIRED. *Mem:* Am Soc Agron; Nat Mgt Asn. *Res:* Evaluation of Tennessee Valley Authority's experimental fertilizers agronomically and in fertilizer use systems in cooperation with land grant universities and fertilizer industry firms; educational programs to encourage efficient fertilizer use; research administration; resource management. *Mailing Add:* 110 Robinhood Florence AL 35630

SUTHERLAND, WILLIAM ROBERT, b Hastings, Nebr, May 10, 36; m 57; c 3. COMPUTER SCIENCE, ELECTRICAL ENGINEERING. *Educ:* Rensselaer Polytech Inst, BEE, 57; Mass Inst Technol, SM, 63, PhD(elec eng), 66. *Prof Exp:* Assoc group leader comput graphics & comput-aided design & mem tech staff, Lincoln Lab, Mass Inst Technol, 64-69; mgr interactive systs dept, Bolt Beranek & Newman, Inc, 69-72, div vpres & dir comput sci div, 72-75; mgr, Systs Sci Lab, Xerox Palo Alto Res Ctr, 75-81; vpres, Sutherland, Sproull & Assoc Inc, 81-90; VPRES, SUN MICROSYSTS, 90- *Mem:* Inst Elec & Electronics Engrs; Asn Comput Mach. *Res:* Distributed computing; graphics. *Mailing Add:* Box 1160 Palo Alto CA 94302

SUTHERLAND-BROWN, ATHOLL, b Ottawa, Ont, June 20, 23; m 48; c 1. ECONOMIC & REGIONAL GEOLOGY. *Educ:* Univ BC, BASc, 50; Princeton Univ, PhD(geol), 54. *Prof Exp:* Geologist econ geol, 52-69, dep chief geol, 69-74, CHIEF GEOLOGIST ECON GEOL, MINERAL RESOURCES BR, BC DEPT MINES & PETROLEUM RESOURCES, 74- *Concurrent Pos:* Ed, Can Inst Mining & Metall, 74-76; mem, Geol Soc Am Del, Am Comn Stratig Nomenclature, 69-72; mem, Can Nat Comn, Int Geol Correlation Prog, 74-; del, Int Union Geol Sci-UNESCO Meeting Govt Experts, Paris, 71; mem, Nat Orgn Comt, 24th Int Geol Cong, 69-72; exec mem, Nat Adv Comn, Res Geol Sci, 69-73; mem, Can Geosci Coun Adv Comt, Geol Surv Can, 76-, exec mem, Can Geosci Coun, 77- *Mem:* Fel Geol Soc Am; fel Geol Asn Can (vpres, 78-79); Can Inst Mining & Metall; Soc Econ Geologists. *Res:* Morphology, classification, distribution and tectonic setting of porpityry deposits; metallogeny and distribution of metals in deposits and background , particularly in Canadian Cordillera; geology and tectonics of Queen Charlotte Islands and insular tectonic belt. *Mailing Add:* 546 Newport Ave Victoria BC V8S 5C7 Can

SUTHERS, RODERICK ATKINS, b Columbus, Ohio, Feb 2, 37; m; c 2. PHYSIOLOGY. *Educ:* Ohio Wesleyan Univ, BA, 60; Harvard Univ, AM, 61, PhD(biol), 65. *Prof Exp:* Res fel biol, Harvard Univ, 64-65; from asst prof to assoc prof anat & physiol, 65-74, prof anat & physiol, 74-75, PROF, PROG NEUROSCI, IND UNIV, BLOOMINGTON, 75-, PROF PHYSIOL, SCH MED, 76-, HEAD, PHYSIOL SECT, 89- *Concurrent Pos:* Vis prof, Dept Zool, Univ Nairobi, Kenya, 73; adj prof biol, Ind Univ, Bloomington, 83- *Mem:* AAAS; Am Ornith Union; Am Physiol Soc; Soc Neurosci; Acoust Soc Am; Asn Res Otolaryngol; Int Soc Neuroethology. *Res:* Neuroethology and behavior; sensory physiology of animal sonar systems; perceptual abilities and auditory processing of information by echolocating animals; physiology and acoustics of vocalization by echolocating bats; physiology, acoustics and motor control of bird song. *Mailing Add:* Med Sci Prog Ind Univ Bloomington IN 47405

SUTIN, JEROME, b Albany, NY, Mar 12, 30; m 56; c 2. NEUROANATOMY, NEUROPHYSIOLOGY. *Educ:* Siena Col, BS, 51; Univ Minn, MS, 53, PhD(anat), 54. *Prof Exp:* Asst anat, Univ Minn, 52-53; hon res asst, Univ Col, Univ London, 53-54; asst, Univ Minn, 54; jr res anatomist, Univ Calif, Los Angeles, 55-56; from instr to assoc prof anat, Sch Med, Yale Univ, 56-66; dir, Neurosci Prog, 85-89 PROF ANAT & CHMN DEPT, EMORY UNIV, 66-, CHARLES HOWARD CANDLER PROF, 80-,. *Concurrent Pos:* Vis investr, Autonomics Div, Nat Phys Lab, Middlesex, Eng; vis prof, Inst Psychiat, Maudsley Hosp, London; Nat Found Infantile Paralysis fel anat, Univ Calif, Los Angeles, 55-56; mem, Neurol A Study Sect, NIH, 71-75, chmn, 74-75. *Mem:* Am Asn Anatomists (pres, 89-90); Asn Anat Chmn (pres, 72-73); Soc Neurosci; Sigma Xi. *Res:* Hypothalamic organization; basal ganglia and motor function; plasticity of noradrenergic neurons. *Mailing Add:* Dept Anat Emory Univ 1364 Clifton Rd NE Atlanta GA 30322

SUTIN, NORMAN, b SAfrica, Sept 16, 28; nat US; m 58; c 2. PHYSICAL INORGANIC CHEMISTRY. *Educ:* Univ Cape Town, BSc, 48, MSc, 50; Cambridge Univ, PhD(chem), 53. *Prof Exp:* Imp Chem Industs fel, Durham Univ, 54-55; res assoc, 56-57, from assoc chemist to chemist, 58-66, SR CHEMIST, BROOKHAVEN NAT LAB, 66- *Concurrent Pos:* Affil, Rockefeller Univ, 58-62; vis fel, Weizmann Inst, 64; vis prof, State Univ NY, Stony Brook, 68, Columbia Univ, 68, Tel Aviv Univ, Israel, 73, Univ Calif, Irvine, 77 & Univ Tex, Austin, 79; mem, Comt Chem Sci, Nat Res Coun, 81-84; mem, Comt Sci, Am Chem Soc, 87-89. *Honors & Awards:* Distinguished Serv Award in the Advan of Inorg Chem, Am Chem Soc, 83; Photochem Award, NY Acad Sci, 85. *Mem:* Nat Acad Sci; Am Chem Soc; Am Acad Arts & Sci. *Res:* Kinetics and mechanisms of inorganic reactions; bioinorganic chemistry; photochemistry of transition metal complexes; solar energy conversion and storage. *Mailing Add:* Dept Chem Brookhaven Nat Lab Upton NY 11973

SUTMAN, FRANK X, b Newark, NJ, Dec 20, 27; m 56; c 3. CHEMISTRY. *Educ:* Montclair State Col, AB, 49, AM, 52; Columbia Univ, EdD, 56. *Prof Exp:* Instr, Pub Schs, NJ, 49-55; asst prof sci, Paterson State Col, 55-57; assoc prof natural sci chmn div, Inter-Am Univ, PR, 57-58; PROF SCI, SCH EDUC & DIR, NSF INSTS, TEMPLE UNIV, 62-; DEAN, COL EDUC, FAIRLEIGH DICKINSON UNIV, TEANECK CAMPUS. *Concurrent Pos:* NSF-AID lectr, India, 67; observer, Orgn Am States Coun Sci Educ & Cult, 71; vis prof, Hebrew Univ, Jerusalem, Israel, 73; consult, Israel Environ Protection Serv, 75, Peoples Repub China, 80-; sr scholar in residence, Temple Univ. *Mem:* Fel AAAS; Am Chem Soc; Nat Asn Res Sci Teaching (pres, 72); Nat Sci Teachers Asn. *Res:* Chemical education research. *Mailing Add:* 128 Stratton Lane Mt Laurel NJ 08054-3301

SUTNICK, ALTON IVAN, b Trenton, NJ, July 6, 28; m 58; c 2. INTERNAL MEDICINE. *Educ:* Univ Pa, AB, 50, MD, 54. *Prof Exp:* From intern to resident anesthesiol & med, Hosp Univ Pa, 54-57, USPHS fel, Univ, 56-57; from resident to chief resident med, Marion County Gen Hosp, Indianapolis, Ind, 57-61; USPHS fel, Temple Univ, 61-63, instr med, 63-64, assoc, 64-65; res physician, Inst Cancer Res, Fox Chase Cancer Ctr, 65-72, assoc dir, 72-75; assoc prof med, Sch Med, Univ Pa, 71-75; dean & prof med, Med Col Pa, 75-89, sr vpres health affairs, 76-89; VPRES, EDUC COMN FOREIGN MED GRADUATES, 89- *Concurrent Pos:* Consult, Coun Drugs, AMA, 62-64; mem US nat comt, Int Union Against Cancer, 69-72; mem ed bd, Res Commun Chem Path Pharmacol, 69-, USA Nat Comt, Int Union Against Cancer, 69-72; vis prof med, Med Col Pa, 71-75; asst ed, Ann Internal Med, 72-75; mem, Nat Cancer Control Planning Conf, Nat Cancer Inst, 73, consult, Diag Res Adv Group, 74-78; sect ed for med, Int J Dermat, 74-75; consult, WHO, 79, 80 & 81; med educ ed, Int J Dermat, 89-; vis prof, Univ Belgrade, Yugoslavia, 88. *Honors & Awards:* Schwartz Award, AMA, 76. *Mem:* Am Asn Cancer Res; Am Fedn Clin Res; Am Col Physicians; AMA. *Res:* Cancer epidemiology; susceptibility to cancer; Australia antigen; hepatitis; pulmonary surfactant; medical education. *Mailing Add:* Educ Comn Foreign Med Graduates 3624 Market St Philadelphia PA 19104-2685

SUTTER, DAVID FRANKLIN, b Ft Wayne, Ind, Nov 21, 35; m 59; c 3. APPLIED PHYSICS, INSTRUMENTATION. *Educ:* Purdue Univ, BS, 58; Cornell Univ, MS, 67, PhD(physics), 69. *Prof Exp:* Asst, Cornell Univ, 62-69; physicist, Fermi Nat Accelerator Lab, 69-75; PHYSICIST, DIV PHYS RES, US AEC 75- *Mem:* Am Phys Soc; AAAS. *Res:* Computer monitoring and control of accelerators; digital and analog instrumentation; theory of operation of accelerators; electron beam optics; development of superconducting magnet systems. *Mailing Add:* 1510 Blue Meadow Rd Potomac MD 20854

SUTTER, GERALD RODNEY, b Fountain City, Wis, Sept 20, 37; m 58; c 3. ENTOMOLOGY. *Educ:* Winona State Col, BA, 60; Iowa State Univ, MS, 63, PhD(entom), 65. *Prof Exp:* Res entomologist, European Corn Borer Lab, Arkeny, Iowa, 65, Northern Grain Insect Res Lab, 65-73, RES LEADER ENTOM, ENTOM RES DIV, NORTHERN GRAIN INSECT RES LAB, AGR RES SERV, USDA, 73- *Mem:* Entom Soc Am; Soc Invert Path. *Res:* Utilization of microorganisms in the biological control of insects. *Mailing Add:* 2005 Iowa St Brookings SD 57006

SUTTER, JOHN FREDERICK, b Oak Harbor, Ohio, June 7, 43; m 65, 75, 88; c 2. GEOLOGY, GEOCHRONOLOGY. *Educ:* Capital Univ, BS, 65; Rice Univ, MA, 68, PhD(geol), 70. *Prof Exp:* Nat Res Coun resident res assoc, Manned Spacecraft Ctr, NASA, 69-70; asst prof earth & space sci, State Univ NY Stony Brook, 70-71; from asst prof to assoc prof geol & mineral, Ohio State Univ, 71-80; geologist, 79-87, SUPVRY GEOLOGIST, US GEOL SURV, 87- *Concurrent Pos:* Lectr geol, George Washington Univ, 87- *Honors & Awards:* Carey Croneis Distinguished Alumni lectr, Rice Univ, 86. *Mem:* Geol Soc Am; Am Geophys Union. *Res:* 40Ar/39Ar geochronology of selected areas; tectonic and metamorphic history of Appalachian-Caledonian orogen. *Mailing Add:* US Geol Surv 908 Nat Ctr Reston VA 22092

SUTTER, JOHN RITTER, b Edwardsville, Ill, May 4, 30; m 58; c 3. PHYSICAL CHEMISTRY. *Educ:* Wash Univ, St Louis, AB, 51; Tulane Univ, MS, 56, PhD(chem), 59. *Prof Exp:* Asst chem, Tulane Univ, 54-59; jr chemist & AEC grant, Wash State Univ, 59-60; asst prof chem, La Polytech Inst, 60-62; from asst prof to assoc prof, 62-71, PROF CHEM, HOWARD UNIV, 71- *Mem:* AAAS; Am Chem Soc; Am Phys Soc. *Res:* Kinetics; fast reactions; thermodynamics; calorimetry. *Mailing Add:* Dept Chem Howard Univ Washington DC 20059

SUTTER, JOSEPH F, b Seattle, Wash, Mar 21, 21; c 3. AERONAUTICAL ENGINEERING. *Educ:* Univ Wash, BS, 43. *Prof Exp:* Aerodynamicist, Boeing Com Aircraft Co, 45-61, chief technol, Boeing 727, 61-63, chief engr technol, Com Airplane Div, 63-65, chief engr, 747 Prog, 65-71, head, 71-78, vpres opers & prod develop, Com Airplane Co, 78-81, exec vpres, 81-86; RETIRED. *Concurrent Pos:* Team leader, Develop & Prod Panel, Pres Comn Space Shuttle Accident, 86; consult, Boeing Com Airplane Group, 86- *Honors & Awards:* Aircraft Design Award, Am Inst Aeronaut & Astronaut, 71 & Elmer A Sperry Award, 80; Sir Kinsford Smith Award, Australian Royal Aeronaut Soc, 80; US Nat Medal Technol, 85; Lord Found Award, 89; William Littlewood Mem Lectr, 90. *Mem:* Nat Acad Eng; hon fel Eng Royal Aeronaut Soc; hon fel Am Inst Aeronaut & Astronaut Soc; fel Royal Aeronaut Soc Eng. *Mailing Add:* Boeing Com Aircraft Co PO Box 3707 Mail Stop 13-43 Seattle WA 98124

SUTTER, MORLEY CARMAN, b Redvers, Sask, May 18, 33; m 57; c 3. PHARMACOLOGY. *Educ:* Univ Man, BSc & MD, 57, PhD(pharmacol), 63. *Prof Exp:* Pvt pract, Souris, Man, 57-58; asst resident med, Winnipeg Gen Hosp, 58-59; demonstr pharmacol, Univ Man, 59-63; Imp Chem Industs fel, Cambridge Univ, 63-65; asst prof, Univ Toronto, 65-66; from asst prof to assoc prof, 66-71, chmn dept, 71-87, PROF PHARMACOL, UNIV BC, 71- *Concurrent Pos:* Wellcome Found travel award, 63; supvr, Downing Col, Cambridge Univ, 63-65; Med Res Coun Can scholar, 66-71. *Mem:* AAAS; Pharmacol Soc Can; Can Soc Clin Invest; Brit Pharmacol Soc; Am Soc Pharmacol & Exp Therapeut. *Res:* Effects of adrenergic blocking agents in shock; mechanism of cardiac arrythmias induced by cyclopropane-epinephrine; pharmacology of veins; vascular smooth muscle in hypertension. *Mailing Add:* Dept Pharmacol & Therapeut 2176 Health Sci Mall Univ BC Vancouver BC V6T 1W5 Can

SUTTER, PHILIP HENRY, b Mineola, NY, Dec 8, 30; m 55; c 4. SOLID STATE PHYSICS, ROBOTICS. *Educ:* Yale Univ, BS, 52, MS, 54, PhD(physics), 59. *Prof Exp:* Res engr res labs, Westinghouse Elec Corp, 58-63, sr res engr, 63-64; from asst prof to assoc prof, 64-83, chmn dept, 76-79 & 82-86, PROF PHYSICS, FRANKLIN & MARSHALL COL, 83- *Concurrent Pos:* NZ sr res fel physics & eng lab, Dept Sci & Indust Res, Wellington, NZ, 69-70; vis scientist, Dept Mat Sci & Eng, Univ Pa, 81-82; chmn Am Asn Physics Teachers Comt Physics Higher Educ, 81-83; vis prof biomed eng, Johns Hopkins Sch Med, 87-88. *Mem:* Am Phys Soc; Am Asn Physics Teachers; Sigma Xi; Inst Elec & Electronics Engrs Robotics & Automation Soc. *Res:* Solid state physics; transport properties; ionic crystals; energy conversion; electronics and computers; robotics; biomedical engineering. *Mailing Add:* 203 Macklin Ave Lancaster PA 17602

SUTTER, RICHARD P, b Birmingham, Ala, Mar 22, 37; m 64; c 3. CELL & MOLECULAR BIOLOGY. *Educ:* St Joseph's Col, Ind, BA, 59; Ohio State Univ, MSc, 61; Tufts Univ, PhD(biochem), 67. *Prof Exp:* Instr biochem, Univ Ill, Chicago, 66-67; from asst prof to assoc prof biol, 67-74, PROF BIOL, WVA UNIV, 74- *Concurrent Pos:* Adj biochemist, Presby St Luke's Hosp, Chicago, 66-67; vis assoc, Calif Inst Technol, 73-74. *Mem:* AAAS; Mycol Soc Am; Am Chem Soc; Am Soc Microbiol; Am Soc Biol Chemists. *Res:* Molecular basis of sexual development; pheromonal communication; fungal metabolism. *Mailing Add:* Dept Biol WVa Univ Morgantown WV 26506

SUTTERBY, JOHN LLOYD, b Kansas City, Mo, Dec 27, 36; m 62; c 2. CHEMICAL ENGINEERING, RHEOLOGY. *Educ:* Univ Mo, BS, 58; Univ Wis, MS, 59, PhD(chem eng), 64. *Prof Exp:* Asst prof chem eng, Va Polytechnic Inst, 65-66; res engr, Process Fundamentals Res Lab, Dow Chem Co, Mich, 66-70; asst prof chem eng, Univ Mo-Columbia, 70-77; res engr, Exxon Corp, 77-80; MEM FAC, DEPT CHEM ENG, UNIV MO, 80- *Mem:* Am Inst Chem Engrs; Soc Rheol. *Res:* Rheological and thermal properties of polymer solutions and polymer melts. *Mailing Add:* Dept Chem Eng Univ Mo Columbia MO 65201

SUTTIE, JOHN WESTON, b La Crosse, Wis, Aug 25, 34; m 55; c 2. NUTRITION. *Educ:* Univ Wis, BSc, 57, MS, 58, PhD, 60. *Prof Exp:* Fel biochem, Nat Inst Med Res, Eng, 60-61; from asst prof to assoc prof, 61-69, PROF BIOCHEM, UNIV WIS-MADISON, 69-, PROF NUTRIT SCI, 88- *Concurrent Pos:* Mem comns atmospheric fluorides & fluorosis & vitamin toxicity, Nat Res Coun; ed J Nutrit, 91- *Honors & Awards:* Mead Johns Award, Am Inst Nutrit, 74; Osborne & Mendel Award, Am Inst Nutrit, 80; Hemostasis Career Award, Int Soc Thrombosis & Hemostasis, 89. *Mem:* AAAS; Am Soc Exp Biol & Med; Am Soc Biol Chem; Am Inst Nutrit; Air Pollution Control Asn; Int Soc Thrombosis & Hemostasis; Am Soc Clin Nutrit. *Res:* Vitamin K action-control of prothrombin synthesis, metabolic action of anticoagulants, chemistry of prothrombin and metabolism of vitamin K; fluoride metabolism-biochemical lesions caused by fluoride ingestion; fluoride as an industrial pollutant; fluoride homeostasis. *Mailing Add:* Dept Biochem Univ Wis Madison WI 53706-1569

SUTTKUS, ROYAL DALLAS, b Fremont, Ohio, May 11, 20; m 47; c 3. ICHTHYOLOGY, FISH BIOLOGY. *Educ:* Mich State Col, BS, 43; Cornell Univ, MS, 47, PhD(zool), 51. *Prof Exp:* Asst zool, Cornell Univ, 47-50; from asst prof to prof zool, 50-60, PROF BIOL, TULANE UNIV, 60-, DIR, MUS NATURAL HIST, 76- *Mem:* AAAS; Am Soc Ichthyologists & Herpetologists; Am Fisheries Soc; Soc Syst Zool; Soc Study Evolution. *Res:* Systematics of fresh and salt water fishes; zoogeography; growth and seasonal distribution; environmental biology; water quality and water pollution; biology of mammals. *Mailing Add:* Dept Biol Tulane Univ New Orleans LA 70118

SUTTLE, ANDREW DILLARD, JR, b West Point, Miss, Aug 12, 26. RADIOCHEMISTRY, NUCLEAR PHYSICS. *Educ:* Miss State Univ, BS, 44; Univ Chicago, PhD(chem), 52. *Prof Exp:* Vpres res & grad studies & dir, Miss Res Comn, Miss State Univ, 60-62; vpres res & prof chem, Tex A&M Univ, 62-71; PROF NUCLEAR BIOPHYSICS & RADIO BIOCHEM & SPEC ASST TO DIR, MARINE BIOMED INST, UNIV TEX MED BR, 71- *Concurrent Pos:* Sr scientist, Humble Oil & Refining Co, 52-62; spec asst to dir res & eng, Dept Defense, Washington, DC, 62-64, mem exec comt sci bd res & eng, 67-; mem, Atomic Indust Forum, 55. *Mem:* Am Chem Soc; Am Phys Soc; Am Nuclear Soc; Inst Elec & Electronics Eng. *Res:* Radiation chemistry; nuclear and thermal energy; petroleum industry. *Mailing Add:* Dept Radiol Marine Biomed Inst Suite 831 200 Univ Blvd Galveston TX 77550

SUTTLE, JEFFREY CHARLES, b Omaha, Nebr, Jan, 28, 52; m 79; c 1. PHYTOHORMONES, PLANT GROWTH REGULATORS. *Educ:* Univ Tex, BA, 74; Mich State Univ, PhD(bot), 79. *Prof Exp:* RES PHYSIOLOGIST, METAB & RADIATION RES LAB, USDA, 79- *Mem:* Am Soc Plant Physiol; Plant Growth Regulator Soc Am; Int Plant Growth Substances Asn. *Res:* Mode of action of synthetic plant growth regulators and their interactions with endogenous phytohormones; mechanism of action of defoliants. *Mailing Add:* Biosci Res Lab State Univ Sta Fargo ND 58105

SUTTLE, JIMMIE RAY, b Forest City, NC, Dec 26, 32; m 51; c 3. PHYSICS, ELECTRICAL ENGINEERING. *Educ:* Presby Col, SC, BS, 58; Duke Univ, MAT, 60, MA, 65; NC State Univ, PhD(elec eng), 72. *Prof Exp:* Instr math, Presby Col, SC, 60-61; phys scientist, Info Processing Off, US Army Res Off, 61-65 & Res-Technol Div, 65-72, assoc dir, Electronics Div, 72-74, actg dir, 74-75, dir, Electronics Div, 75-88; ASST VCHANCELLOR RES, NC STATE UNIV, 88- *Concurrent Pos:* Asst dir res, Off Secy Defense, 82; adj prof, Elec Eng, NC State Univ, 74-82. *Mem:* Fel Inst Elec & Electronics Engrs. *Res:* Computer architecture; biomathematics; switching theory; electron paramagnetic resonance spectroscopy of organic solids. *Mailing Add:* Off VChancellor Res NC State Univ Box 7003 Raleigh NC 27695-7003

SUTTNER, LEE JOSEPH, b Hilbert, Wis, June 3, 39; m 65; c 4. GEOLOGY. *Educ:* Univ Notre Dame, BS, 61; Univ Wis, MS, 63, PhD(geol), 66. *Prof Exp:* From asst prof to assoc prof geol, 66-78, PROF GEOL, IND UNIV, BLOOMINGTON, 78- *Honors & Awards:* Niel Miner Teaching Award, Nat Asn Geol Teachers, 88. *Mem:* Nat Asn Geol Teachers; Int Asn Sedimentologists; Soc Econ Paleontologists & Mineralogists; Geol Soc Am. *Res:* Sedimentology and sedimentary petrology. *Mailing Add:* Dept of Geol Ind Univ Bloomington IN 47401

SUTTON, BLAINE MOTE, b Ft Recovery, Ohio, Jan 23, 21; m 46. MEDICINAL CHEMISTRY. *Educ:* Purdue Univ, BS, 42, MS, 48, PhD(pharmaceut chem), 50. *Prof Exp:* Group leader med chem, Smith Kline & French Labs, Philadelphia, 50-68, sr investr med chem, 68-71, asst dir med chem, 71-75, assoc dir med chem, 75-85; RETIRED. *Mem:* AAAS; Am Chem Soc; Am Pharmaceut Asn; NY Acad Sci; Acad Pharmaceut Sci; Am Rheumatism Asn. *Res:* Synthetic medicinal chemistry; sym pathomimetic amines, sedatives, antibiotics, antirheumatics and hypocholesteremics. *Mailing Add:* 2435 Byberry Rd Hatboro PA 19040

SUTTON, CHARLES SAMUEL, b Lima, Peru, July 15, 13; US citizen; m 46. MATHEMATICS. *Educ:* Mass Inst Technol, BS, 35, MS, 37. *Prof Exp:* Instr math, Tufts Col, 39-40; from asst prof to prof, 40-78, EMER PROF MATH, THE CITADEL, 78- *Mem:* Am Math Soc; Math Asn Am. *Res:* Analysis; iteration; functional equations. *Mailing Add:* 19 Shrewsbury Rd Charleston SC 29407

SUTTON, DALLAS ALBERT, b Grand Junction, Colo, Sept 12, 11; m 35; c 1. BIOLOGY. *Educ:* Univ Colo, AB, 39, PhD, 53; Northwestern Univ, MS, 40. *Prof Exp:* Prin pub sch, Colo, 34-45; instr biol, Mesa Col, 45-51; assoc prof biol & sci educ, Eastern Mont Col Educ, 54-57; prof, 57-76, EMER PROF BIOL & SCI EDUC, CALIF STATE UNIV, CHICO, 76- *Concurrent Pos:* NSF grant. *Mem:* Am Soc Mammal. *Res:* Mammalogy; chipmunks of Colorado; chromosomes of the chipmunks, Genus eutamias; female genital bones of chipmunks. *Mailing Add:* Dept Biol Calif State Univ Chico CA 95926

SUTTON, DAVID C(HASE), b Bryn Mawr, Pa, Dec 18, 33; c 1. PHYSICS. *Educ:* Haverford Col, BA, 55; Princeton Univ, PhD(physics), 62. *Prof Exp:* Asst prof, 61-68, ASSOC PROF PHYSICS, UNIV ILL, 68- *Concurrent Pos:* Vis assoc prof, Stanford Univ, 68-69. *Mem:* Am Phys Soc; Sigma Xi. *Res:* Photonuclear reactions; nuclear structure; accelerator development; radioactivity; particle-radiation detectors. *Mailing Add:* Dept of Physics Univ of Ill Urbana IL 61801

SUTTON, DAVID GEORGE, b San Francisco, Calif, Apr 17, 44; m 63; c 1. LASERS. *Educ:* Univ Calif, Berkeley, BS, 66; Mass Inst Technol, PhD(chem physics), 70. *Prof Exp:* Fel phys chem, Dept Chem, Mass Inst Technol, 70-71; fel phys chem, Dept Chem, Univ Toronto, 71-72; tech staff lasers, 72-85, mgr, Laser Effects Sect, 85-88, HEAD, PROPULSION & ENVIRON SCI DEPT, AEROSPACE CORP, 88- *Mem:* Am Phys Soc. *Res:* Experimental research in gas phase molecular exitation; gas phase molecular energy transfer; analytical applications of molecular energy transfer; laser effects on materials. *Mailing Add:* Aerospace Corp PO Box 92957 Los Angeles CA 90009

SUTTON, DEREK, b Eng, July 15, 37; m 58; c 2. INORGANIC CHEMISTRY. *Educ:* Univ Nottingham, BSc, 58, PhD(chem), 63. *Prof Exp:* Asst lectr chem, Univ Nottingham, 62-64, lectr, 64-67; from asst prof to assoc prof, 67-78, PROF CHEM, SIMON FRASER UNIV, 78- *Mem:* Royal Soc Chem; Am Chem Soc; Chem Inst Can. *Res:* Study of aryldiazo and other complexes of transition metals related to the interaction of nitrogen with transition metals; biological nitrogen fixation. *Mailing Add:* Dept Chem Simon Fraser Univ Burnaby BC V5A 1S6 Can

SUTTON, DONALD DUNSMORE, b Oakland, Calif, June 8, 27; c 3. MICROBIOLOGY. *Educ:* Univ Calif, AB, 51, MA, 54, PhD(microbiol), 57. *Prof Exp:* Sr lab technician bact, Univ Calif, 50-53, res asst, 53-55; NSF predoctoral fel, 55-57; asst prof & USPHS fel microbiol sch med, Ind Univ, 57-59; Waksman-Merck fel inst microbiol, Rutgers Univ, 59-60. *Concurrent Pos:* Vis prof, Inst Appl Microbiol, Tokyo Univ, 70-71; consult food microbiol & waste water treatment microbiol, 73-; pres, Environ Assocs, Encinitos, Calif, 82- *Mem:* Am Soc Microbiol; Am Inst Biol Sci; Mycol Soc Am. *Res:* Microbial physiology; metabolism of plant disease bacteria; mechanisms of growth inhibitors and antifungal agents; physiological basis of morphogenesis in fungi; location of enzymes in fungi; microbiology of food fermentations; microbiology of waste water treatment and waste decomposition. *Mailing Add:* Dept Biol Calif State Univ Fullerton CA 92634

SUTTON, EMMETT ALBERT, b Toledo, Ohio, May 7, 35; m 58; c 2. CHEMICAL DYNAMICS. *Educ:* Cornell Univ, BEngPhys, 58, PhD(aeronaut eng), 61. *Prof Exp:* Asst prof physics, Hamilton Col, 61-62; asst prof aeronaut & astronaut, Purdue Univ, 62-65; prin res scientist, Avco Everett Res Lab, 65-70; vpres, Aerodyne Res, Inc, 70-76; PRES, CONCORD SCI CORP, 76- *Mem:* Am Inst Aeronaut & Astronaut; Am Optical Soc. *Res:* Experimental chemical kinetics; hypersonic wake chemistry; rocket plume radiation; optical and radar field measurements. *Mailing Add:* Hugh Cargill Rd Concord MA 01742

SUTTON, GEORGE E, b Blandville, WVa, June 3, 23; m 59; c 2. MECHANICAL ENGINEERING. *Educ:* Univ WVa, BSME; Univ Fla, MSE; Mich State Univ, PhD(mech eng), 57. *Prof Exp:* Engr, Indust Eng & Construct Co, 48; instr mech eng, Univ Fla, 48-55; instr, Mich State Univ, 55-57; prof, Univ Ariz, 57-59; prof mech eng & chmn dept, Univ Nev, Reno, 61-74; dir prof serv, Nat Coun Eng Examr, 74-76; DEAN, SCH ENG, YOUNGSTOWN STATE UNIV, 76- *Concurrent Pos:* Secy western zone, Nat Coun State Bd Eng Exam, 66-68. *Mem:* Am Soc Mech Engrs; Nat Soc Prof Engrs; Am Soc Heating, Refrig & Air-Conditioning Engrs; Sigma Xi. *Res:* Thermal environment; heat transfer. *Mailing Add:* 2602 Algonquin Dr Poland OH 44514

SUTTON, GEORGE HARRY, b Chester, NJ, Mar 4, 27; m 47; c 4. SEISMOLOGY, GEOPHYSICS. *Educ:* Muhlenberg Col, BS, 50; Columbia Univ, MA, 53, PhD(geol), 57. *Prof Exp:* From res asst to res assoc, Lamont Geol Observ, Columbia Univ, 50-60, mem acad staff, 60-66, from asst prof to assoc prof geol, Univ, 60-66; assoc dir, Hawaii Inst Geophys, Univ Hawaii, 71-76, prof geophys & geophysicist, 66-81; vpres, Rondout Assoc, Inc, 81-90; VIS INVESTR, WOODS HOLE OCEANOG INST, 90-, CONSULT, 90- *Concurrent Pos:* Chief seismologist, Inst Sci Res Cent Africa, 55-56; prin investr & co-investr for Ranger, Surveyor & Apollo lunar seismog exps, NASA, 59-73, mem geophys working group planetology subcomt, 64-73, Surveyor sci eval adv team, 65-73; mem comt planetary surfaces & interiors, Space Sci Bd, Nat Acad Sci, 63-67; mem ad hoc panel earthquake prediction, Off Sci & Technol, 64-65; mem, Adv Panel Earth Sci, NSF, 69-72 & Panel Strong-Motion Seismol, Comn Seismol, Div Earth Sci, Nat Res Coun-Nat Acad Sci, 72-74. *Mem:* AAAS; Soc Explor Geophysicists; Acoust Soc Am; Seismol Soc Am; Am Geophys Union; Inst Elec & Electronic Engrs. *Res:* Earthquake seismology; seismic wave propagation and source conditions; geophysical exploration of oceanic crustal and upper-mantle structure; lunar and planetary interiors; geophysical instrumentation for ocean-bottom and lunar use. *Mailing Add:* RD 2 Box 167C Stone Ridge NY 12484

SUTTON, GEORGE W(ALTER), b NY, Aug 3, 27; m 52; c 4. THERMAL PHYSICS. *Educ:* Cornell Univ, BME, 52; Caltech, MS, 53, PhD(physics & eng), 55. *Prof Exp:* Engr rockets, JPL, 53-55; scientist rockets, Lockheed Missile Div, 55-56 & reentry, Gen Elec Missile Div, 56-63; sci adv aerospace, Dept Air Force, 63-65; vpres lasers & electro-optics, Avco-Everett Res Lab, 65-83 & electro-optics, Jaycor, 85-90; dir lasers, Helionetics Laser Div, 83-85; DIR ELECTRO-OPTICS, KAMAN AEROSPACE CORP, 90- *Mem:* AAAS; Am Inst Aeronaut & Astronaut; Am Soc Mech Engrs. *Res:* Ablation heat protection system for reentry through atmosphere; non-equilibrium ionization for MHD power generation; designed first high energy laser over 100 kilowatts, and successful surveillance equipment. *Mailing Add:* 5055 E Broadway Blvd C-104 Tucson AZ 85711

SUTTON, HARRY ELDON, b Cameron, Tex, Mar 5, 27; m 62; c 2. HUMAN GENETICS. *Educ:* Univ Tex, BS, 48, MA, 49, PhD(biochem), 53. *Prof Exp:* Res scientist, Univ Tex, 48-52; asst biologist, Univ Mich, 52-56, instr human genetics, 56-57, asst prof, 57-60; assoc prof zool, Univ Tex, 60-64, chmn dept, 70-73, assoc dean, Grad Sch, 67-70 & 73-75, vpres res, 75-79, PROF ZOOL, UNIV TEX, AUSTIN, 64- *Concurrent Pos:* Mem comt personnel res, Am Cancer Soc, 61-64; mem genetics study sect, NIH, 63-67; ed, Am J Human Genetics, 64-69; mem adv coun, Nat Inst Environ Health Sci, 68-72, mem sci adv comt, 72-76; mem comt epidemiol & vet follow-up studies, Nat Acad Sci-Nat Res Coun, 69-; mem adv comt, Atomic Bomb Casualty Comn, 70-75; mem Bd Radiation Effects Res, NAS-Nat Res Coun, 85-87. *Mem:* Environ Mutagen Soc; Genetics Soc Am; Am Chem Soc; Am Soc Human Genetics (pres-elect, 78, pres, 79); Am Soc Biochem & Molecular Biol; Am Genetics Asn; fel AAAS. *Res:* Genetic control of protein structure; inherited variations in human metabolism; human population genetics. *Mailing Add:* Dept Zool Univ Tex Austin TX 78712

SUTTON, J(AMES) L(OWELL), b Petersburg, Ind, Apr 20, 31; m 59; c 5. ENGINEERING. *Prof Exp:* Develop engr, E I du Pont de Nemours & Co Inc, 58-60, res engr, 60-62, process develop supvr, 62-65, eng serv supvr, 65-66, on spec assignment, 66-70, area supvr, Film Dept, 70-78, tech area supt, Plastics Prod & Resins Dept, 78-81, area supt liaison & capacity planning, Polymer Prod Dept, 81-85, process planning mgr, 85-89; RETIRED. *Mem:* Nat Soc Prof Engrs. *Res:* Development and design of mechanical equipment in the chemical industry. *Mailing Add:* 579 Hickory Pl Circleville OH 43113

SUTTON, JOHN CLIFFORD, b Halstead, Eng, Oct 10, 41; m 64; c 3. PLANT PATHOLOGY. *Educ:* Univ Nottingham, BSc, 65; Univ Wis, PhD(plant path), 69. *Prof Exp:* From asst prof to assoc prof, 69-82, PROF PLANT PATH, UNIV GUELPH, 82- *Concurrent Pos:* Nat Sci & Eng Coun Can grant, 70-91. *Mem:* Can Phytopath Soc; Am Phytopath Soc. *Res:* Epidemiology of foliar pathogens; biological control of plant diseases; disease management. *Mailing Add:* Dept Environ Biol Univ Guelph Guelph ON N1G 2W1 Can

SUTTON, JOHN CURTIS, b Weiser, Idaho, May 13, 42; m 65; c 4. ORGANIC CHEMISTRY, ANALYTICAL CHEMISTRY. *Educ:* Univ Idaho, BS, 64, PhD(chem), 72. *Prof Exp:* Assoc prof chem, 70-80, PROF, DIV NATURAL SCI, LEWIS-CLARK STATE COL, 80- *Res:* Utilization of tree bark for the sorption of heavy metal ions from aqueous solutions. *Mailing Add:* Dept of Nat Sci Lewis-Clark State Col Lewiston ID 83501

SUTTON, LEWIS MCMECHAN, b Chicago, Ill, Apr 13, 46; m 69; c 3. MICROBIOLOGY, ANALYTICAL CHEMISTRY. *Educ:* Iowa State Univ, BS, 69; Northern Ill Univ, MS, 72, MBA, 80. *Prof Exp:* Chemist org, Borden Chem, 70-72, chief chemist org & vitamin, 72-76, asst tech dir nutrit res, 76-78, asst gen mgr, Pet-Agr Div, Borden Inc, 78-80, gen mgr, Borden Int, 80-85; vpres, Albion Labs, 85-87; EXEC VPRES & DIR, PET-AG, INC, 87- *Mem:* Am Chem Soc; Am Soc Microbiol. *Res:* Analytical methods development both chemical and microbiological; nutritional studies with animal models; international research and technical transfer. *Mailing Add:* 40 W 630 Winchester Way St Charles IL 60175-8906

SUTTON, LOUISE NIXON, b Hertford, NC, Nov 4, 25; div; c 1. MATHEMATICS. *Educ:* Agr & Tech Col NC, BS, 46; NY Univ, MA, 51, PhD(math educ), 62. *Prof Exp:* Instr high sch, NC, 46-47; instr math, Agr & Tech Col NC, 47-50, asst prof, 51-57; asst prof, Del State Col, 57-62; assoc prof, 62-63, head, Dept Phys Sci & Math, 62-78, PROF MATH, ELIZABETH STATE UNIV, 63- *Concurrent Pos:* Asst dean women, Agr & Tech Col, 47-48; mem adv comt cert math & sci, Del State Bd Educ, 59-61, mem adv comt math, 61-62; mem bd dir, Perquimans County Indust Develop Corp, 67-72; mem gen adv comt, NC State Bd Soc Serv, 69-71; mem bd dir, Div Higher Educ, NC Asn Educr, 69-72; co-dir, NSF Inst, 71-72 & 73-74, dir, 72-73 & 74-75, panelist, 71-72, 74 & 77-78, chmn, 77 & 78. *Mem:* Nat Coun Teachers Math; Nat Educ Asn; Nat Asn Univ Women. *Res:* Mathematics education; concept learning in trigonometry and analytical geometry. *Mailing Add:* Dept Math Elizabeth City State Univ Box 811 Parkview Dr Elizabeth City NC 27909

SUTTON, MATTHEW ALBERT, b Austin, Minn, Apr 28, 23; m 46; c 4. AERODYNAMICS, STRUCTURAL DYNAMICS. *Educ:* Univ Minn, BSc, 45; Ohio State Univ, MSc, 52, PhD(aeronaut eng), 58. *Prof Exp:* Stress analyst, McDonnell Aircraft Co, 46-47; instr eng drawing, Ohio State Univ, 47-48, res assoc wind tunnel design, 48-53, from instr to asst prof aeronaut eng, 53-58; prin eng supvr ord div, 58-62, chief engr, 62-65, chief engr systs & res div, 65-66, dir res, 66-68, gen mgr, Systs & Res Ctr, 68-76, vpres & gen mgr, Defense Systs Div, 78-81, VPRES & GEN MGR, AVIONICS DIV, HONEYWELL INC, 81- *Mem:* Am Inst Aeronaut & Astronaut. *Res:* Aircraft structure and flutter; unsteady aerodynamics. *Mailing Add:* 6109 Habitat Ct Edina Minneapolis MN 55436

SUTTON, PAUL, b Hopkinsville, Ky, Sept 11, 29; m 54; c 2. SOIL FERTILITY. *Educ:* Univ Ky, BS, 51, MS, 57; Iowa State Univ, PhD(agron, soil fertil), 62. *Prof Exp:* Asst agronomist, Univ Ky, 54-55; asst horticulturist, Univ Fla, 61-67; assoc prof soils, Ohio Agr Res & Develop Ctr, 67-73, PROF AGRON, LAB ENVIRON SCI, OHIO STATE UNIV, 73- *Mem:* Soil Sci Soc Am; Am Soc Agron. *Res:* Plant physiology. *Mailing Add:* Dept of Agronomy OARDC Ohio State 1680 Madison Ave Wooster OH 44691

SUTTON, PAUL MCCULLOUGH, b Ohio, Dec 3, 21; m 46; c 2. PHYSICS. *Educ:* Harvard Univ, BS, 43; Columbia Univ, MA, 48, PhD(physics), 53. *Prof Exp:* Asst physics, Columbia Univ, 50-52; res assoc physicist, Corning Glass Works, 52-54, supvr ultrasonics res, 54-56, supvr fundamental physics group, 56-59; sr staff scientist, Aeronutronic Div, Ford Motor Co, 59-62, mgr, Appl Physics Dept, Philco Corp, 62-66, mgr, Physics Lab, Philco-Ford Corp, 66-68, mgr, Physics & Chem Lab, 68-72, mgr, Res Lab, 72-74, mgr electro-optics, 74-80, mgr develop plans, 80-83, mgr Electro-Optics, Aeronatronic Div, 83-87; RETIRED. *Concurrent Pos:* Lectr optics, Univ Calif, Irvine, 66. *Mem:* AAAS; fel Am Phys Soc; Am Ceramic Soc; Optical Soc Am. *Res:* Solid state physics; elastic constants; photoelasticity; acoustic propagation; glass physics; dielectric properties of glasses; hypervelocity impact; space charge in glass; atmospheric turbulence; lasers; infrared optics. *Mailing Add:* 2621 Blackthorn St Newport Beach CA 92660

SUTTON, ROBERT GEORGE, b Rochester, NY, June 17, 25; m 46; c 2. GEOLOGY. *Educ:* Univ Rochester, AB, 48, MS, 50; Johns Hopkins Univ, PhD, 56. *Prof Exp:* Instr geol, Alfred Univ, 50-52; jr instr, Johns Hopkins Univ, 52-54; from asst prof to assoc prof, 54-66, PROF GEOL, UNIV ROCHESTER, 66- *Mem:* AAAS; Geol Soc Am; Soc Econ Paleont & Mineral. *Res:* Paleozoic stratigraphy; sedimentology; sedimentary petrology. *Mailing Add:* 141 Furlong Rd Rochester NY 14627

SUTTON, ROGER BEATTY, b Lloydminster, Sask, Sept 14, 16; m 46; c 2. EXPERIMENTAL HIGH ENERGY PHYSICS. *Educ:* Univ Sask, BA, 38, MA, 39; Princeton Univ, PhD(physics), 43. *Prof Exp:* Res physicist, Off Sci Res & Develop, Princeton Univ, 42-43; physicist, Manhattan Dist, Los Alamos Sci Lab, 43-46; from asst prof to prof46-86, EMER PROF PHYSICS, CARNEGIE-MELLON UNIV, 86- *Mem:* Am Phys Soc. *Res:* Molecular spectroscopy; low energy nuclear physics; radioactive isotopes; neutron physics; cyclotrons; study of x-rays from exotic atoms including antiprotonic, kaonic and sigma minus atoms; electron colliding beams. *Mailing Add:* Dept Physics Carnegie-Mellon Univ Pittsburgh PA 15213

SUTTON, RUSSELL PAUL, b Mo, July 31, 29; m 53; c 5. CHEMISTRY. *Educ:* Univ Mo-Columbia, BS, 51; State Univ Iowa, MS, 53, PhD(chem), 55. *Prof Exp:* Res chemist, E I du Pont de Nemours & Co, Inc, 55-58; from asst prof to assoc prof, 58-70, PROF CHEM, KNOX COL, ILL, 70-, CHMN DEPT, 77- *Mem:* Am Chem Soc. *Res:* The chemistry of chalcones and flavylium compounds; gas chromatography. *Mailing Add:* Dept Chem Knox Col Galesburg IL 61401

SUTTON, TURNER BOND, b Windsor, NC, Oct 24, 45; m 88. PHYTOPATHOLOGY. *Educ:* Univ NC, BA, 68; NC State Univ, MS, 71, PhD(plant path), 73. *Prof Exp:* Res assoc plant path, Mich State Univ, 73-74; res assoc, 74-76, from asst prof to assoc prof, 76-87, PROF PROF PLANT PATH, NC STATE UNIV, 87- *Mem:* Am Phytopath Soc. *Res:* Apple diseases; epidemiology and control; pest management. *Mailing Add:* Dept Plant Path NC State Univ PO Box 7616 Raleigh NC 27695-7616

SUTTON, W(ILLARD) H(OLMES), b Pittsburgh, Pa, Jan 12, 30; m 56; c 1. CERAMICS. *Educ:* Alfred Univ, BS, 52; Pa State Univ, MS, 54, PhD(ceramics tech), 57. *Prof Exp:* Ceramist, Gen Elec Co, 56-63, mgr metall & ceramics res, 63-69; assoc dir technol, Spec Metals Corp, 69-85; sr mat scientist, 85-90, PRIN SCIENTIST, ENGINEERED CERAMICS, UNITED TECHNOL RES CTR, 90- *Concurrent Pos:* Mem, Mat Adv Bd, Nat Acad Sci-Nat Res Coun; chmn, Eng Ceramics Div, Am Ceramic Soc, 68-69, mem bd trustees, 82-85. *Honors & Awards:* Achievement Award, Nat Inst Ceramic Engrs, 70; F Lonsberry Award, Allegheny Int Corp, 79; H Bidwell Award, Invest Casting Inst, 84. *Mem:* Am Soc Metals; fel Am Ceramic Soc; Nat Inst Ceramic Engrs; Mat Res Soc; Int Microwave Power Inst. *Res:* Superalloy clean-metal processing; melt-crucible reactions; vacuum melting and refining; high temperature melt purification-application of ceramic filters to superalloys and steel alloys worldwide; fiber composite materials; refractory whiskers; ceramic-metal interfaces and surface chemistry; microwave firing of ceramic materials; develop engineered ceramic materials. *Mailing Add:* United Technol Res Ctr Silver Lane Mail Stop 24 East Hartford CT 06108

SUTTON, WILLIAM WALLACE, b Monticello, Miss, Dec 15, 30; m 54; c 6. PROTOZOOLOGY, CELL BIOLOGY. *Educ:* Dillard Univ, BA, 53; Howard Univ, MS, 59, PhD(zool), 65. *Prof Exp:* Med technician, DC Gen Hosp, 55-59; from instr to prof biol, Dillard Univ, 59-79, actg chmn div natural sci, 69-70, chmn div natural sci, 70-79; vpres acad affairs, Chicago State Univ, 79-80, vpres acad affairs & student develop, 80-81, provost & acad vpres, 81-85; prof biol, Kans State Univ, 85-88, vpres educ & student serv, 85-87; PRES, MISS VALLEY STATE UNIV, 88- *Concurrent Pos:* Consult, NIH, 72-74 & 16 Inst Health Sci Consortium of NC & Va, 74-; assoc & regional liaison officer, Danforth Found, 75-79; reader/reviewer, Am Biol Teacher, 79-80; coordr, Strengthening Develop Inst Prog, Chicago State Univ, 80-82. *Honors & Awards:* Josiah Macy Jr Found Fel, 77. *Mem:* Soc Protozoologists; Sigma Xi; Nat Inst Sci; AAAS; NY Acad Sci; Am Asn Higher Educ. *Res:* Radiation cell biology; responses of peritrichs to ionizing radiations; chemical analysis of the cyst wall and nutrition of peritrichs; isolation of nucleic acids from peritrichs. *Mailing Add:* Off Pres Miss Valley State Univ 125 Washington Ave Itta Bena MS 38941

SUTTON, WILLIAM WALLACE, b Athens, Ga, Apr 6, 43; m 75; c 4. ANIMAL PHYSIOLOGY. *Educ:* Mercer Univ, AB, 64; Marshall Univ, MS, 65; WVa Univ, PhD(zool), 70. *Prof Exp:* Animal physiologist, Chem Corps Proving Ground, Dugway, Utah, 71-73; RES PHYSIOLOGIST, US ENVIRON PROTECTION AGENCY, 73- *Concurrent Pos:* Adj asst prof biol, Univ Nev, Las Vegas, 76-78. *Mem:* AAAS; NY Acad Sci; Am Soc for Testing & Mat; Asn Off Anal Chemists; Asn Chem Soc. *Res:* Validation of biological methods for use in pollutant monitoring networks; establish the data quality that can be achieved within a single laboratory. *Mailing Add:* 649 Oglethorpe Ave Athens GA 30606

SUTULA, CHESTER LOUIS, b Erie, Pa, Feb 15, 33; m 55; c 8. BOTANY & PHYTOPATHOLOGY, GENERAL COMPUTER SERVICES. *Educ:* Col Holy Cross, BS, 54; Iowa State Univ, MS, 58, PhD(phys chem), 59. *Prof Exp:* Teaching asst, Iowa State Univ, 55-57; res scientist, Ames Lab, Iowa, 57-59; sr res scientist, Marathon Oil Co, Colo, 59-67; sr res scientist, Ames Res Lab, Ames Co Div, Miles Labs, Inc, 67-69, dir, 69-77; vpres Res & Develop, Ortho Diag, Inc, 77-80; MEM STAFF, CLS ASSOC, 80-; PRES, AGDIA, INC, 81- *Mem:* AAAS; Am Chem Soc. *Res:* Surface chemistry; calorimetry; wetting properties of complex porous materials; structure of colloidal fluids; microbiology; immunoassay; instrumentation. *Mailing Add:* 30380 CR6 Elkhart IN 46514

SUUBERG, ERIC MICHAEL, b NY, Nov 23, 51; m; c 1. COMBUSTION, CHEMICAL KINETICS. *Educ:* Mass Inst Technol, BS & MS 74, MS, 76, ScD, 78. *Prof Exp:* Asst prof chem eng, Carnegie-Mellon Univ, 77-81; assoc prof, 81-90, PROF ENG, BROWN UNIV, 90- *Concurrent Pos:* Vis, Centre National de La Recherche Scientifique, France, 88; chmn, Div Fuel Chem, Am Chem Soc, 91. *Mem:* Combustion Inst; Am Inst Chem Engrs; Am Chem Soc. *Res:* Coal chemistry; combustion; carbons; reaction kinetics. *Mailing Add:* Div Eng Box D Brown Univ Providence RI 02912

SUURA, HIROSHI, b Hiroshima, Japan, Aug 19, 25; m 51; c 2. THEORETICAL PHYSICS. *Educ:* Univ Tokyo, BS, 47; Hiroshima Univ, PhD(physics), 55. *Prof Exp:* Prof physics, Nihon Univ, 60-65; PROF PHYSICS, UNIV MINN, 65- *Mem:* Fel Am Phys Soc. *Res:* Theory of elementary particles. *Mailing Add:* Sch Physics & Astron Univ Minn Minneapolis MN 55455

SUYAMA, YOSHITAKA, b Osaka, Japan, Sept 5, 31; m 60; c 2. MOLECULAR BIOLOGY, BIOCHEMISTRY. *Educ:* Kyoto Univ, Japan, BAgr, 55; Kans State Univ, PhD(microbiol genetics), 60. *Prof Exp:* Fel genetics & microbiol, Sch Med, Yale Univ, 59-60; asst res biologist genetics & biol, Univ Calif, La Jolla, 60-64; from asst prof to assoc prof, 64-75, PROF BIOL, UNIV PA, 75- *Concurrent Pos:* Vis assoc prof biochem, Univ Bari, Italy, 69; res fel molecular genetics, Inst Molecular Genetics, Gif, France, 71-72. *Mem:* Genetics Soc Am; Am Soc Biol Chemists; Soc Protozoologists. *Res:* Molecular genetics and biogenesis of organelles in eukaryotic cells; elucidations of nucleic acids and protein synthesizing mechanisms in mitochondria. *Mailing Add:* Dept Biol Univ Pa Philadelphia PA 19104

SUYDAM, FREDERICK HENRY, b Lancaster, Pa, July 30, 23; m 44; c 3. CHEMISTRY. *Educ:* Franklin & Marshall Col, BS, 46; Northwestern Univ, PhD(chem), 50. *Prof Exp:* Instr chem, Franklin & Marshall Col, 46-47; asst, Northwestern Univ, 47-49; instr anat, Med Sch, Johns Hopkins Univ, 50-52; from asst prof to assoc prof, 52-62, chmn dept, 58-69, PROF CHEM, FRANKLIN & MARSHALL COL, 62- *Mem:* Am Chem Soc. *Res:* Peptide synthesis; reactions of amino acids; infrared absorption. *Mailing Add:* Dept of Chem Franklin & Marshall Col Lancaster PA 17604

SUZUE, GINZABURO, b Kumamoto, Japan, Feb 14, 32; c 3. CLINICAL TOXICOLOGY. *Educ:* Kyoto Univ, PhD, 61. *Prof Exp:* Chief technologist, Univ Chicago Med Ctr, 82-85; res scientist, Travenol Labs, 79-82; CONSULT, 82- *Concurrent Pos:* Translator, 85- *Mem:* Soc Clin Chem. *Res:* Lipoprotein. *Mailing Add:* 33555 N Ivy Lane Grayslake IL 60030-1991

SUZUKI, DAVID TAKAYOSHI, b Vancouver, BC, Mar 24, 36; m 58, 72; c 5. GENETICS. *Educ:* Amherst Col, BA, 58; Univ Chicago, PhD(zool), 61. *Hon Degrees:* LLD, Univ PEI, 74; DSc, Acadia Univ, Trent Univ, 81 & Lakehead Univ, 85. *Prof Exp:* Res assoc genetics, Biol Div, Oak Ridge Nat Lab, 61-62; asst prof, Univ Alta, 62-63; from asst prof to assoc prof, 63-69, PROF ZOOL, UNIV BC, 69- *Concurrent Pos:* Res grants, Nat Res Coun Can, 62-86, AEC, 64-69 & Nat Cancer Inst Can, 69-78, NIH, 82-85, Supplies & Serv Can, 84- *Mem:* Genetics Soc Am (secy, 80-83); Genetics Soc Can; Can Soc Cell Biol (pres, 69-70). *Res:* Regulation of development and behavior; genetic organization of chromosomes; developmental and behavioral genetics. *Mailing Add:* Dept Zool Univ BC 2075 Westbrook Pl Vancouver BC V6T 1W5 Can

SUZUKI, HOWARD KAZURO, b Ketchikan, Alaska, Apr 3, 27; m 52; c 4. ANATOMY. *Educ:* Marquette Univ, BS, 49, MS, 51; Tulane Univ, PhD(anat), 55. *Prof Exp:* Asst zool & bot, Marquette Univ, 48-51; asst zool & anat, Tulane Univ, 51-55; instr anat, Sch Med, Yale Univ, 55-58; from asst prof to prof, Sch Med, Univ Ark, 58-70; assoc dean, 70-71, actg dean, 71-72, prof anat, Col Med, 70-73, dean, Col Health Related Professions, 72-79, PROF NEUROSCI, COL MED, UNIV FLA, 72-, PROF ANAT, COL MED, 79-, PROF, COL HEALTH RELATED PROFESSIONS, 79- *Concurrent Pos:* Mem gen res support prog adv comt, NIH; mem adv comt, Off Acad Affairs, US Vet Admin. *Mem:* Fel AAAS; Am Asn Anatomists; Asn Am Med Cols; Soc Exp Biol & Med; Am Soc Allied Health Professions. *Res:* Endocrine relations to bone; phagocytosis and reticuloendothelial system; neonatal human anatomy; comparative bone metabolism. *Mailing Add:* 4331 NW 20th Pl Gainesville FL 32605

SUZUKI, ISAMU, b Tokyo, Japan, Aug 4, 30; m 62; c 3. MICROBIOLOGY, BIOCHEMISTRY. *Educ:* Univ Tokyo, BSc, 53; Iowa State Univ, PhD(bact physiol), 58. *Prof Exp:* Fel microbiol, Western Reserve Univ, 58-60; instr, Univ Tokyo, 60-62; Nat Res Coun Can fel, 62-64, from asst prof to assoc prof, 64-69, head dept, 72-85, PROF MICROBIOL, UNIV MAN, 69- *Mem:* AAAS; Am Soc Microbiol; Can Soc Microbiol; Can Biochem Soc; Sigma Xi; NY Acad Sci. *Res:* Mechanism of the oxidation of inorganic sulfur compounds by Thiobacilli and ammonia by Nitrosomonas; physiology of autotrophic bacteria; mechanism of enzyme reactions; kinetics. *Mailing Add:* Dept of Microbiol Univ of Man Winnipeg MB R3T 2N2 Can

SUZUKI, JON BYRON, b San Antonio, Tex, July 22, 47. MICROBIOLOGY, DENTISTRY. *Educ:* Ill Wesleyan Univ, BA, 68; Ill Inst Technol, PhD(microbiol), 72; Loyola Univ, DDS, 78; Am Acad Microbiol, dipl SM, 72; Am Bd Periodontol, Dipl, 87. *Prof Exp:* Med technologist & res assoc cytogenetics, Augustana Hosp, Chicago, 68-69; res assoc immunol & pediat, Univ Chicago Hosp, 70-71; clin microbiologist, St Luke's Hosp Ctr Columbia Col Physicians & Surgeons, NY, 71-73; dir clin labs, Registry Hawaii, 73-74; fel, Depts Path & Periodont, dentist & mem fac, Univ Wash, 78-80; clin prof Periodont & Microbiol, Univ MD, Baltimore, 80-90; PROF DENT, JOHNS HOPKINS MED INST, 85-; DEAN, UNIV PITTSBURGH, 90- *Concurrent Pos:* Instr microbiol, Ill Inst Technol, 68-72; chmn clin labs med technol, Univ Hawaii, 73-74; vis lectr microbiol, Loyola Univ, Chicago, 74- & Northwestern Univ Sch Dent, 83-; adv, NASA, 76-, res award, 80. *Honors & Awards:* Oral Path Nat Award, Am Acad Oral Path, 78. *Mem:* Am Acad Microbiol; Int Asn Dent Res; Sigma Xi; Am Dent Asn; AAAS. *Res:* Immunodeficiency states during manned-space flights; research in interaction of immunocompetent cells in disease; periodontal diseases. *Mailing Add:* Univ Pittsburgh 3501 Terrace St Pittsburgh PA 15261

SUZUKI, KINUKO, b Hyogo, Japan, Nov 10, 33; m 60; c 1. PATHOLOGY, NEUROPATHOLOGY. *Educ:* MD, Osaka City Univ, 59. *Hon Degrees:* MA, Univ Pa, 71. *Prof Exp:* Asst prof path, Albert Einstein Col Med, 68; from asst prof to assoc prof, Sch Med, Univ Pa, 69-72; from assoc prof to prof path, Albert Einstein Col Med, 72-86; PROF PATH, SCH MED, UNIV NC, CHAPEL HILL, 86- *Honors & Awards:* Jacob Javits Neurosci Award, 89.

Mem: Am Soc Neurochem; Am Asn Neuropath; Soc Neurosci; Int Soc Develop Neurosci. *Res:* Study of pathogenesis of developmental disorder of the central nervous system; neuropathology of Myelin disorder. *Mailing Add:* Dept Path Sch Med Univ NC Chapel Hill CB#7525 Chapel Hill NC 27599-7525

SUZUKI, KUNIHIKO, b Tokyo, Japan, Feb 5, 32; m 60; c 1. NEUROCHEMISTRY, BIOCHEMISTRY. *Educ:* Univ Tokyo, BA, 55, MD, 59. *Hon Degrees:* MA, Univ Pa, 71. *Prof Exp:* Resident clin neurol, Albert Einstein Col Med, 60-62, instr neurol, 64, asst prof, 65-68; assoc prof, Sch Med, Univ Pa, 69-71, prof neurol & pediat, 71-72; prof neurol & neurosci, Albert Eistein Col Med, 72-86; PROF NEUROL & PSYCHIAT & DIR BRAIN & DEVELOP RES CTR, SCH MED, UNIV NC, 86- *Concurrent Pos:* Mem neurol B study sect, 71-75; mem adv bd, Nat Tay-Sachs & Allied Dis Asn; chief ed, J Neurochem, 77-81; mem bd sci counselors, Nat Inst Neurol, Commun Disorders & Stroke, 80-84; mem, US Nat Comt, Int Brain Res Orgn, Nat Res Coun, 85-89, NICHD Ment Retardation Res Comt, 89-; sr scientist award, Humboldt Found, 91. *Honors & Awards:* A Weil Award, 70; M Moore Award, 75; Jacob K Javits Neurosci Investr Award, 85; Distinguished Scientist Award, Japan Med Soc Am, 85. *Mem:* Int Soc Neurochem; Soc Neurosci; Am Asn Neuropath; Am Soc Biol Chemists; Am Soc Neurochem. *Res:* Biochemistry of brain lipids, particularly gangliosides; biochemical and molecular biological studies of inherited metabolic disorders of the nervous system. *Mailing Add:* Brain & Develop Res Ctr CB No 7250 Univ NC Sch Med Chapel Hill NC 27599-7250

SUZUKI, MAHIKO, b Tokyo, Japan, Oct 3, 38; m 65; c 1. THEORETICAL HIGH ENERGY PHYSICS. *Educ:* Univ Tokyo, BS, 61, MS, 63, DSc, 66. *Prof Exp:* Res fel physics, Calif Inst Technol, 65-67; mem, Inst Advan Study, 67-68; res assoc, Univ Tokyo, 68-69; vis assoc prof, Columbia Univ, 69-70; assoc prof, 70-74, PROF PHYSICS, UNIV CALIF, BERKELEY, 74- *Concurrent Pos:* Fullbright grnt, Fullbright Comn, 65-68; R C Tolman fel, Calif Inst Technol, 66-67; J S Guggenheim Mem Found fel, 76-77. *Mem:* Fel Am Phys Soc. *Res:* Theoretical particle physics of weak, electromagnetic and strong interactions. *Mailing Add:* Dept of Physics Univ of Calif Berkeley CA 94720

SUZUKI, MICHIO, b Taipei, Formosa, Feb 23, 27; Can citizen; m 59; c 3. PLANT PHYSIOLOGY, BIOCHEMISTRY. *Educ:* Tohuku Univ, Japan, BS, 52, PhD(agr chem), 62. *Prof Exp:* Asst plant physiol & biochem, Tohuku Univ, Japan, 52-66; RES SCIENTIST, RES BR, CAN DEPT AGR, 66- *Concurrent Pos:* Nat Res Coun Can res fel, 63-65. *Mem:* Can Soc Plant Physiologists; Am Soc Plant Physiologists; Agr Inst Can; Can Soc Agron; Soc Cryobiol. *Res:* Winter survival of perennial crops and winter cereals; vegetative regrowth of forage crops; metabolism of fructan in grasses; plant nutrition; evaluation of feed quality. *Mailing Add:* 6 Messer Dr Charlottetown PE C1A 6N5 Can

SUZUKI, MICHIO, b Chiba, Japan, Oct 2, 26; m 52; c 1. MATHEMATICS. *Educ:* Univ Tokyo, BA, 48, DrS(math), 52. *Prof Exp:* Lectr math, Tokyo Univ Educ, 51-55; from asst prof to assoc prof, 55-59, PROF MATH, UNIV ILL, 59- *Concurrent Pos:* Fel, Univ Ill, 52-53, res assoc, 53-55; res assoc, Harvard Univ, 56-57; fel, Inst Advan Study, Princeton, NJ, 62-63, vis prof, 68-69, mem inst, 81-; vis prof, Univ Chicago, 60-61 & Univ Tokyo, 71 & 81. *Honors & Awards:* Japan Acad Prize, 74. *Mem:* Am Math Soc; Math Soc Japan. *Res:* Group theory. *Mailing Add:* Dept Math Univ Ill 371 Altgeld Hall 1409 W Green St Urbana IL 61801

SUZUKI, SHIGETO, b San Francisco, Calif, Feb 25, 25; m 53; c 1. ORGANIC CHEMISTRY. *Educ:* Univ Calif, Berkeley, BS, 55; Univ Southern Calif, PhD(chem), 59. *Prof Exp:* Sloan Found res fel org chem, Univ Calif, Berkeley, 59-60; SR RES ASSOC, CHEVRON RES CO, 60- *Mem:* Am Chem Soc; The Chem Soc. *Res:* Organic reaction mechanism, especially carbonium and carbanion rearrangements; organo-sulfur and organo-halogen chemistry. *Mailing Add:* 679 12th Ave San Francisco CA 94118-3618

SUZUKI, TSUNEO, b Nagoya, Japan, Nov 23, 31; wid; c 3. MICROBIOLOGY, BIOCHEMISTRY. *Educ:* Univ Tokyo, BS, 54, MD, 57; Hokkaido Univ, PhD(biochem), 69. *Prof Exp:* Japan Fel Asn fel & Fulbright travel grant, Univ Tokyo, 63; fel, Univ Wis, 63-66; fel, Univ Lausanne, 66-67; Ont Cancer Inst fel, Univ Toronto, 67-69; res assoc immunochem, Univ Wis, 69-70; from asst prof to assoc prof, 70-83, PROF IMMUNOCHEM, SCH MED, UNIV KANS MED CTR, KANSAS CITY, 83- *Concurrent Pos:* Mem, Exp Immunol Study Sect, 83-87. *Mem:* Am Asn Immunol; Can Soc Immunol; Am Soc Biochem & Molecular Biol. *Res:* Immunochemistry of cell surface receptors. *Mailing Add:* Dept Microbiol Molecular Genetics & Immunol Univ of Kans Sch of Med Kansas City KS 66103-8410

SVACHA, ANNA JOHNSON, b Asheville, NC, Nov 27, 28; c 3. NUTRITION, BIOCHEMISTRY. *Educ:* Va Polytech Inst & State Univ, BS, 50; Univ Ariz, MS, 69, PhD(biochem, nutrit), 71. *Prof Exp:* Indust chemist, Hercules Powder Co, 51-53; physicist, Taylor Model Basin, US Navy, 53; high sch teacher, Tenn, 53-54; res asst anal chem, Tex Agr Exp Sta, 56-58; res asst org chem, Chas Pfizer & Co, Inc, 58-59; res asst nutrit, Univ Ariz, 67-68; asst poultry scientist, Ariz Agr Exp Sta, 71-72; ASST PROF NUTRIT, AUBURN UNIV, 72- *Mem:* Am Chem Soc; Am Home Econ Asn; Sigma Xi. *Res:* Appetite regulation with respect to protein and amino acid metabolism; nutritional status and requirements of the elderly. *Mailing Add:* Dept Nutrit & Foods Auburn Univ Auburn AL 36849

SVANES, TORGNY, b Norway. ALGEBRA. *Educ:* Oslo Univ, MA, 65; Mass Inst Technol, PhD(math), 72. *Prof Exp:* Instr math, Oslo Univ, 66-69, NY State Univ Stony Brook, 72-73; asst prof Aarhus Univ, 73-75; asst prof math, Purdue Univ, 75-77; asst prof math, Bradley Univ, 77-80; MEM STAFF, MITRE CORP, 80- *Mem:* Am Math Soc. *Res:* Study of Schubert subvarieties of homogeneous spaces. *Mailing Add:* 7 Page Rd Lexington MA 02173

SVE, CHARLES, b Pana, Ill, Feb 21, 40; m 62; c 2. CIVIL ENGINEERING. *Educ:* Mass Inst Technol, BS, 62, MS, 63; Northwestern Univ, PhD(theoret appl mech), 68. *Prof Exp:* Res engr, NAm Aviation, 63-64; sr engr, Avco Corp, 64-66; res asst civil eng, Northwestern Univ, 66-67; MEM TECH STAFF STRUCT MECH, AEROSPACE CORP, 68- *Concurrent Pos:* Lectr, Univ Southern Calif, 70-71. *Mem:* Am Soc Mech Engrs; Am Inst Aeronaut & Astronaut. *Res:* Applied mechanics; wave propagation in composite materials; experimental mechanics; numerical analysis. *Mailing Add:* Mech Res Dept MS M5/753 PO Box 92957 Los Angeles CA 90009-2957

SVEC, HARRY JOHN, b Cleveland, Ohio, June 24, 18; m 43; c 9. PHYSICAL CHEMISTRY, ANALYTICAL CHEMISTRY. *Educ:* John Carroll Univ, BS, 41; Iowa State Univ, PhD(phys chem), 49. *Prof Exp:* Asst chem, 41-43, jr chemist, Manhattan Proj, 43-46, res assoc, Inst Atomic Res, 46-50, asst prof chem, Univ & assoc chemist, Inst, 50-55, assoc prof & chemist, 55-60, prof chem, 60-83, sr chemist, Inst Atomic res, 60-83, DISTINGUISHED PROF SCI & HUMANITIES, IOWA STATE UNIV, 78-, EMER PROF CHEM, 83-; ASSOC AMES LAB, US DEPT ENERGY, 83- *Concurrent Pos:* Lectr, NATO Advan Study Inst Mass Spectros, 64; ed, Int J Mass Spectrometry Ion Physics, Am Soc Mass Spectrometry, 68-, vpres, 72-74, pres, 74-76; prog dir, Ames Lab, US Dept of Energy, 78- *Mem:* Am Chem Soc; Am Soc Testing & Mat; Am Soc Mass Spectrometry (pres, 74-76); Geochem Soc; Chem Soc. *Res:* Metallurgy of rare metals; mass spectroscopy; mass spectrometry in physical, inorganic and analytical chemistry; corrosion mechanisms; determination of ultra trace levels of organic pollutants in water and air; highly excited neutral species by neutral fragment mass spectroscopy. *Mailing Add:* 2427 S Hamilton Dr Ames IA 50010

SVEC, LEROY VERNON, b Columbus, Nebr, Feb 27, 42; m 64; c 4. AGRONOMY, PLANT PHYSIOLOGY. *Educ:* Univ Nebr-Lincoln, BSc, 64; Purdue Univ, Lafayette, MSc, 68, PhD(agron), 70. *Prof Exp:* Asst prof plant sci, Univ Del, 69-76; assoc prof agron & dist exten agronomist, Univ Nebr, 76-79; TECH SERV AGRONOMIST, ASGROW SEED CO, 80-; MGR, AGRON TECH SERV, US, 85- *Concurrent Pos:* Mem, Coun Agr Sci & Technol. *Honors & Awards:* Upjohn Award, Upjohn Co. *Mem:* Am Soc Agron; Sigma Xi; Am Soc Plant Physiologists. *Res:* Performance evaluation of new hybrids and varieties; technical training publications; new cultural practices for crop production. *Mailing Add:* Asgrow Seed Co 634 E Lincolnway Ames IA 50010

SVEDA, MICHAEL, b West Ashford, Conn, Feb 3, 12; m 36; c 2. CHEMISTRY. *Educ:* Univ Toledo, BS, 34; Univ Ill, PhD(chem, math), 39; PhD (biol, biochem, virol). *Prof Exp:* Asst chem, Univ Toledo, 31-34, teaching fel, 34-35; teaching asst, Univ Ill, 35-37, Eli Lilly res fel, 37-39; res chemist, E I du Pont de Nemours & Co, Inc, 39-44, res mgr, 44-47, new prod sales supvr, 47-51, prod mgr, 51-53, spec asst to mgt, 53-54; mgt consult, 55-60; dir acad proj, NSF, 61-62; corp assoc dir res, FMC Corp, New York, 62-64; RES & MGT COUN TO ACAD, INDUST & GOVT, 65- *Concurrent Pos:* Mem adv comt creativity in scientists & engrs, Rensselaer Polytech Inst, 65-68; res consult, NSF, 70. *Mem:* Am Chem Soc; AAAS. *Res:* Discovered cyclamate sweeteners; first application of Boolean algebra and theory of sets to people problems, and devised 3-dimensional models showing relationships; devised better way to take off human fat in obesity; interdisciplinary organizations broadly, including mathematical treatment for the first time. *Mailing Add:* Revonah Woods 228 W Lane Stamford CT 06905-8014

SVEJDA, FELICITAS JULIA, b Vienna, Austria, Nov 8, 20; nat Can. ORNAMENTAL HORTICULTURE, PLANT BREEDING. *Educ:* State Univ Agr & Forestry, Austria, MSc, 46, PhD, 48. *Prof Exp:* Res asst rural econ, State Univ Agr & Forestry, Austria, 47-51; asst plant breeder, Swedish Seed Asn, 52-53; RES OFFICER, CAN DEPT AGR, 53- *Mem:* Am Soc Hort Sci; Genetics Soc Can; Can Soc Hort Sci; Agr Inst Can; Can Bot Asn. *Res:* Population biology; plant physiology; hybridization of ornamental plants. *Mailing Add:* 604-1356 Meadowlands Dr E Nepean ON K2E 6K6 Can

SVENDSEN, GERALD EUGENE, b Ashland, Wis, June 18, 40; m 61; c 2. ECOLOGY, ETHOLOGY. *Educ:* Univ Wis-River Falls, BS, 62; Univ Kans, MA, 64, PhD(behav ecol), 73. *Prof Exp:* Biologist, Fish-Pesticide Res Lab, US Fish & Wildlife Serv, 64-66 & Fish Control Lab, 66-68; from instr to asst prof biol, Viterbo Col, 66-70; asst prof, 73-77, ASSOC PROF ZOOL, OHIO UNIV, 77- *Concurrent Pos:* Actg dir, Rocky Mountain Biol Lab. *Mem:* Am Soc Mammalogists; Ecol Soc Am; Am Soc Evolutionists; Am Soc Naturalists. *Res:* Behavioral ecology; ethology of mammals; social systems analysis and evolution; spatial organization and distribution of terrestrial vertebrates; population biology of terrestrial vertebrates; vertebrate communication systems. *Mailing Add:* Dept of Zool & Microbiol Ohio Univ Athens OH 45701

SVENDSEN, IB ARNE, b Copenhagen, Denmark, 1937; c 2. COASTAL ENGINEERING, WAVE MECHANICS & NEARSHORE HYDRODYNAMICS. *Educ:* Tech Univ, Lyngby, Denmark, MSc, 60, PhD(coastal eng), 74. *Prof Exp:* Asst prof civil eng, Coast & Port Eng lab, 60-72; assoc prof, Inst Hydrodyn & Hydraul Eng, Tech Univ, Denmark, 72-87; PROF CIVIL ENG & HEAD DEPT, UNIV DEL, 87-, PROF APPL OCEAN SCI, COL MARINE STUDIES, 88- *Concurrent Pos:* Mem, Int Oil Tanker Comt, Permanent Int asn Navig Congresses, Brussels, Belg, 71-74 & Int Comn Improving Fender Systs Design, 79-83; vis prof, Ocean Eng Prog, Univ Del, 82-83. *Mem:* Am Soc Civil Eng; Am Geophys Union; Am Soc Eng Educ; Int Asn Hydraul Res. *Res:* Water wave mechanics; wave breaking turbulence and wave generated currents related to sediment transport; coastal erosion and protection. *Mailing Add:* Dept Civil Eng Univ Del Newark DE 19716

SVENDSEN, KENDALL LORRAINE, b Greenville, Mich, June 24, 19; m 43; c 3. GEOPHYSICS. *Educ:* Univ Mich, BS, 43. *Prof Exp:* Geophysicist, US Coast & Geod Surv, Nat Oceanic & Atmospheric Admin, 46-70, chief geomagnetism div, 70-71, chief geomagnetic data div, Environ Data Serv, 71-

72, chief solid earth & marine geophys data serv div, 72-75, tech asst geomagnetism, Solid Earth Geophys Div, Environ Data Serv, 75-81; res assoc, Coop Inst Res Environ Sci, Univ Colo, Boulder, 81-90; CONSULT, 91- *Concurrent Pos:* Am Geophys Union liaison rep, Comn Geophys, Pan Am Inst Geog & Hist, 72-90, alt US mem, 73-90; mem, Comt Int Participation, Am Geophys Union, 87-88. *Honors & Awards:* Bronze Medal, US Dept Com, 70, Silver Medal, 77; Antarctic Medal, NSF, 70; naming of Svendsen Glacier in Antarctica, US Bd Geog Names, 71. *Mem:* Am Geophys Union; Asn Geoscientists Int Develop; Sigma Xi; fel AAAS. *Res:* Management of geomagnetic data; international cooperation in geomagnetism; exchange of data, information, and expertise. *Mailing Add:* 15311 Beaverbrook Ct No 3E Silver Spring MD 20906-1357

SVENNE, JURIS PETERIS, b Riga, Latvia, Feb 14, 39; Can citizen; m 63; c 3. FEW-BODY THEORY, REACTION THEORY. *Educ:* Univ Toronto, BASc, 62; Mass Inst Technol, PhD(physics), 65. *Prof Exp:* Res assoc nuclear physics, Mass Inst Technol, 65-66; Nat Res Coun Can fels, Niels Bohr Inst, Copenhagen, Denmark, 66-68; res assoc, Inst Nuclear Physics, D'Orsay, France, 68-69; from asst prof to assoc prof, 69-80, assoc dept head, 87-88, PROF PHYSICS, UNIV MAN, 80-, ASSOC DEAN SCI, 89- *Concurrent Pos:* Vis prof nuclear physics, Univ Oxford, 76; instr, Sch Music, Univ Man, 80-; vis res scientist, Physics Dept, Galileo Galilei, Padova, Italy, 84-85. *Mem:* Am Phys Soc; Can Asn Physicists. *Res:* Few-body effects in nuclear reaction theory; three-cluster reactions; theory of nuclear structure; symmetries in nucleon-nucleon interaction. *Mailing Add:* Dept Physics Univ Man Winnipeg MB R3T 2N2 Can

SVENSSON, ERIC CARL, b Hampstead, NB, Aug 13, 40; m 65; c 2. CONDENSED MATTER PHYSICS. *Educ:* Univ NB, Fredericton, BSc, 62; McMaster Univ, PhD(physics), 67. *Prof Exp:* From asst res officer to assoc res officer, 66-81, SR RES OFFICER PHYSICS, ATOMIC ENERGY CAN LTD, 82- *Concurrent Pos:* Guest scientist, Aktiebolaget Atomenergi, Studsvik, Sweden, 72-73; vis physicist, Brookhaven Nat Lab, Upton, NY, 81-82; lectr, Can Asn Physicists, 83-84. *Mem:* Fel Am Phys Soc; Can Asn Physicists. *Res:* Neutron scattering; lattice dynamics; magnetic excitations; effects of impurities on excitation spectra; structure and dynamics of liquid helium. *Mailing Add:* Box 128 Deep River ON K0J 1P0 Can

SVERDLOVE, RONALD, b Brooklyn, NY, Dec 6, 48. DYNAMICAL SYSTEMS, APPROXIMATION THEORY. *Educ:* Princeton Univ, AB, 69; Stanford Univ, MA, 73, PhD(math), 76. *Prof Exp:* Lectr math, Southern Ill Univ, Carbondale, 76-77; fel, Math Clinic, Claremount Grad Sch, 77-78; ASST PROF MATH, UNIV NOTRE DAME, 78-; MEM TECH STAFF, RCA LABS, 81- *Concurrent Pos:* Vis asst prof math, Claremont Men's Col, 77-78, Harvey Mudd Col, 78; vis lectr comput sci, Rutgers Univ, 82. *Mem:* Am Math Soc; Math Asn Am; Soc Indust & Appl Math. *Res:* Inverse problems for dynamical systems in two and higher dimensions; approximation of functions by sums of Gaussians; models of the human visual system; computer simulation of electron optics in kinescope guns. *Mailing Add:* 225 State Rd Princeton NJ 08540

SVERDRUP, EDWARD F, b Buffalo, NY, Feb 24, 30. ELECTRICAL ENGINEERING. *Educ:* State Univ NY Buffalo, BS, 51; Carnegie Inst Technol, MS, 53, PhD(elec eng), 54. *Prof Exp:* Asst prof elec eng, Carnegie Inst Technol, 54-55, 59-61; sr engr & fel engr, 61-67, ADV ENGR, WESTINGHOUSE RES LABS, 67- *Concurrent Pos:* Lectr, Carnegie Inst Technol, 65-67. *Mem:* Am Soc Testing & Mat. *Res:* Electrical properties of high temperature materials; vapor deposition processes; fuel cell development; gas and steam turbines for coal burning combined cycle power plants; induced draft fan erosion. *Mailing Add:* 11029 Old Trail Rd Irwin PA 15642

SVERDRUP, GEORGE MICHAEL, b Minneapolis, Minn, Mar 29, 49; m 70; c 2. ATMOSPHERIC SCIENCES, PARTICLE MECHANICS. *Educ:* Univ Minn, BME, 71, MS, 73, PhD(mech eng), 77. *Prof Exp:* RES SCIENTIST ATMOSPHERIC SCI, BATTELLE MEM INST COLUMBUS LABS, 76-, ASSOC SECT MGR, 79- *Mem:* AAAS. *Res:* Chemical and physical characteristics of small particles; physico-chemical interaction of gases and particles. *Mailing Add:* 306 Crandall Dr Columbus OH 43085

SVETLIK, JOSEPH FRANK, rubber chemistry; deceased, see previous edition for last biography

SVOBODA, GLENN RICHARD, b Racine, Wis, Nov 18, 30; m 57; c 3. POLYMER CHEMISTRY. *Educ:* Univ Wis, BS, 52, MS, 53, PhD(pharmaceut chem), 58. *Prof Exp:* Instr anal chem, Univ Wis, 58-59; res chemist, 59-61, mgr res lab, 62-64, dir res, 64-67, VPRES RES & DEVELOP, FREEMAN CHEM CORP, PORT WASHINGTON, 67- *Concurrent Pos:* Asst prof, Ore State Col, 57-58. *Mem:* AAAS; Sigma Xi; Am Soc Mech Engrs. *Res:* Natural products; organo-analytical techniques, especially electrochemistry and optical methods; polymer analysis by physical organic techniques; polymer and monomer synthesis; coatings; unsaturated polyester and urethane specialties; electrochemical and radiochemical syntheses. *Mailing Add:* Freeman Chem 217 Freeman Dr Port Washington WI 53074-2026

SVOBODA, GORDON H, b Racine, Wis, Oct 29, 22; m 45; c 3. PHARMACOGNOSY. *Educ:* Univ Wis, BS, 44, PhD(pharmaceut chem), 49. *Prof Exp:* Actg instr pharmaceut chem, Univ Wis, 47-49; asst prof pharm, Univ Kans, 49-50; phytochemist & res assoc, Eli Lilly & Co, 50-78; VPRES RES, NATROTECH, INC, 86-; INDEPENDENT CONSULT. *Concurrent Pos:* Am Asn Cols Pharm-NSF vis scientist, 63-72; vis res prof, Univ Pittsburgh, 64-; mem biol & related res facilities vis comt, Bd Overseers, Harvard Univ. *Honors & Awards:* Am Pharmaceut Asn Award, 63; Ebert Prize, Am Soc Pharmacog, 67. *Mem:* Am Pharmaceut Asn; Am Soc Pharmacog (pres, 63-64); Int Pharmaceut Fedn; fel Acad Pharmaceut Sci. *Mailing Add:* 3918 Rue Renoir Indianapolis IN 46220

SVOBODA, JAMES ARVID, b Great Falls, Mont, June 28, 34; m 60; c 4. INSECT PHYSIOLOGY. *Educ:* Col Great Falls, BS, 58; Mont State Univ, PhD(entom), 64. *Prof Exp:* Resident res assoc insect physiol, Pioneering Res Lab, 64-65, SR INSECT PHYSIOLOGIST, INSECT PHYSIOL LAB, AGR RES CTR, USDA, 65-, LAB CHIEF, 79- *Mem:* AAAS; Entom Soc Am; Am Oil Chem Soc. *Res:* Metabolism of lipids in insects, specifically of sterols and their relationships to growth and metamorphosis; insect hormones and hormonal control mechanisms. *Mailing Add:* 13301 Overbrook Lane Bowie MD 20715

SVOBODA, JOSEF, b Praha, Czech, July 16, 29; Can citizen; m 76; c 2. ARCTIC PLANT ECOLOGY. *Educ:* Univ Western Ont, BSc, 70; Univ Alta, PhD(bot), 74. *Prof Exp:* From asst prof to assoc prof, 73-85, PROF PLANT ECOL, UNIV TORONTO, 85, ASSOC CHMN, DEPT BOT, ERINDALE COL, UNIV TORONTO, 90- *Concurrent Pos:* Vis scientist, Univ Freiburg, WGer, 79-80; NATO sr fel, 79-80; mem, steering comt, Can Ctr Toxicol, 81; prin investr, Dept Indian Affairs & Northern Develop, 81-84; assoc ed, J Arctic & Alpine Res, 85-; assoc ed, J Ultimate Reality & Meaning, 88- *Mem:* Can Bot Asn; Ecol Soc Am; Artic Inst North America; Asn Can Northern Studies. *Res:* Arctic plant ecology; development and productivity of Polar Oases, impact of Global change on arctic ecosystems; persisent radioactivity in northern environments due to atmospheric nuclear fallout; natural radioactivity in uranium mineralization areas in Northern Canada. *Mailing Add:* Dept Bot Erindale Col Univ Toronto Toronto ON M5S 1A1 Can

SVOBODA, RUDY GEORGE, b Berwyn, Ill, Aug 15, 41; m 64; c 2. MATHEMATICS. *Educ:* Northern Ill Univ, BS, 66; Ohio Univ, MS, 67; Purdue Univ, PhD(math), 71. *Prof Exp:* Asst prof, 70-77, ASSOC PROF MATH, IND UNIV-PURDUE UNIV, FT WAYNE, 77- *Mem:* Am Math Soc; Math Asn Am; Int Congress Individualized Instr; Sigma Xi. *Res:* Development of individualized audio-tutorial instructional materials for algebra and trigonometry courses. *Mailing Add:* Dept of Math Purdue Univ 2101 Coliseum Blvd E Ft Wayne IN 46805

SVOKOS, STEVE GEORGE, b Wierton, WVa, June 22, 34; m 60; c 3. BIOLOGICAL CHEMISTRY. *Educ:* Brooklyn Col, BS, 56; State Univ NY, MS, 62, PhD(bio-org chem), 64. *Prof Exp:* Chemist, Lederle Labs, Am Cyanamid Co, 56-60, res chemist, 65-69; regulatory liaison, Ayerst Labs Div, Am Home Prod Corp, NY, 69-72; dir regulatory affairs, 72-75, dir regulatory & sci affairs, 75-79, VPRES REGULATORY & SCI AFFAIRS, KNOLL PHARMACEUT CO, WHIPPANY, 79- *Mem:* Am Chem Soc; fel Royal Soc Chem; NY Acad Sci; Am Phys Asn; Am Inst Chemists. *Res:* Pharmaceutical administration; medicinal chemistry. *Mailing Add:* 59 First Ave Westwood NJ 07675

SWAB, JANICE COFFEY, b Lenoir, NC, July 8, 41. SYSTEMATIC BOTANY. *Educ:* Appalachian State Univ, BS, 62; Univ SC, MS, 64, PhD(biol), 66. *Prof Exp:* Asst prof biol, Clemson Univ, 66-67; assoc prof biol, Queens Col, NC, 67-78; ASSOC PROF BIOL, ST MARY'S COL, NC, 79- *Concurrent Pos:* Nat Acad Sci exchange scientist, USSR, 73 & 75; Fulbright scholar, Egypt, 80, Sudan, 90 & Zambia, 91-92. *Mem:* Am Soc Plant Taxonomists (secy, 75-76); Am Inst Biol Sci; Bot Soc Am; Int Asn Plant Taxon; Sigma Xi. *Res:* Systematics of the Juncaceae, emphasis on Luzula. *Mailing Add:* 1400 Athlone Pl Raleigh NC 27606

SWABB, LAWRENCE E(DWARD), JR, b Dayton, Ohio, Oct 25, 22; m 44; c 3. CHEMICAL ENGINEERING. *Educ:* Univ Cincinnati, ChE, 48, MS, 49, PhD(chem eng), 51. *Prof Exp:* Chem engr process develop, Esso Res Labs, Humble Oil & Refining Co, Standard Oil Co, NJ, 51-56, sect head, 56-58, asst dir res labs, 58-63, dir res labs, 64-66, mgr new areas planning & coord, 66-68, vpres petrol res, 68-74, vpres synthetic fuels res, Exxon Res & Eng Co, 74-82; RETIRED. *Mem:* Nat Acad Eng; Am Inst Chem Engrs. *Res:* Research and process development in petroleum industry. *Mailing Add:* 92 Addison Dr Short Hills NJ 07078

SWADER, FRED NICHOLAS, b Belle Vernon, Pa, Oct 9, 34; m 56; c 2. SOIL SCIENCE. *Educ:* Cornell Univ, BS, 61, MS, 63, PhD(agron, soil sci), 68. *Prof Exp:* Experimentalist, Cornell Univ, 61-63, res assoc agr eng, 63-67, from asst prof to assoc prof soil sci, 67-81; prog leader environ qual, extension serv, 81-84, PROG LEADER WATER RESOURCES, USDA, 84- *Concurrent Pos:* Chmn, Cornell Agr Waste Mgt Conf, 70 & 71; exten laison, Environ Protection Agency, 78-79; co-chair, Exten Comt on Orgn & Policy, Groundwater Task Force, 84- *Mem:* Soil Conserv Soc Am; Am Soc Agr Engrs; Soil Sci Soc Am. *Res:* Plant-soil-water relationships, as influenced by the physical properties of various soils; soil management for recycling agricultural by-products. *Mailing Add:* USDA Extension Serv 3344 South Bldg Washington DC 20250

SWADLOW, HARVEY A, US citizen. NEUROPHYSIOLOGY, PSYCHOLOGY. *Educ:* Univ Miami, BA, 64, MS, 67, PhD(psychol), 70. *Prof Exp:* Fel, Ctr Brain Res, Univ Rochester, 70-72; res assoc, Univ Miami, 72-74; res fel neurophysiol, Retina Found, Boston, 74-75; vis asst prof psychol, Univ Western Ont, 75-76; ASSOC PROF PSYCHOL, UNIV CONN, 77- *Concurrent Pos:* Res affil, Res Lab Electronics, Mass Inst Technol, 76-78; lectr neurol, Harvard Med Sch, 76-80. *Mem:* Sigma Xi; Soc Neurosci; AAAS. *Res:* Structure and function of the visual cortex; interhemispheric communication; impulse conduction along axons. *Mailing Add:* Dept Psychol Univ Conn Campus U-20 406 Cross Campus Storrs CT 06268

SWAGER, WILLIAM L(EON), chemical engineering, management, for more information see previous edition

SWAIM, ROBERT LEE, b Rensselaer, Ind, Aug 7, 35; m 60; c 3. AEROSPACE ENGINEERING. *Educ:* Purdue Univ, BS, 57, MS, 59; Ohio State Univ, PhD(elec eng), 66. *Prof Exp:* Assoc engr, Douglas Aircraft Co, 57-58; engr, NAm Aviation, Inc, 59; sr res engr, Air Force Flight Dynamics Lab, 62-67; prof aeronaut & astronaut & assoc head, Sch Aeronaut & Astronaut, Purdue Univ, West Lafayette, 67-78; assoc dean, Col Eng, archit

& technol, 78-87, PROF MECH & AEROSPACE ENG, OKLA STATE UNIV, STILLWATER, 87- *Concurrent Pos:* Vis lectr, US Air Force Inst Technol, 66-67. *Mem:* Assoc fel Am Inst Aeronaut & Astronaut; sr mem Inst Elec & Electronics Engrs; Sigma Xi; Am Soc Eng Educ. *Res:* Advanced flight control concepts for aircraft, missiles and aerospace vehicles. *Mailing Add:* 3202 W 29th Ct Stillwater OK 74074-2214

SWAIMAN, KENNETH F, b St Paul, Minn, Nov 19, 31; m 73, 85; c 4. NEUROCHEMISTRY, PEDIATRIC NEUROLOGY. *Educ:* Univ Minn, BA, 52, BS, 53, MD, 55; Am Bd Pediat, dipl; Am Bd Psychiat & Neurol, dipl. *Prof Exp:* Fel pediat, 56-57, Nat Inst Neurol Dis & Stroke spec fel pediat neurol, 60-63, from asst prof to assoc prof, 63-69, PROF PEDIAT & NEUROL, MED SCH, UNIV MINN, MINNEAPOLIS, 69-, DIR DIV PEDIAT NEUROL, 68-, EXEC OFFICER DEPT NEUROL, 77- *Concurrent Pos:* Guest worker, NIH, 78-81. *Honors & Awards:* Hower Award, Child Neurol Soc, 81. *Mem:* Am Neurol Asn; Am Acad Neurol; Am Acad Pediat; Child Neurol Soc (pres, 72-73); Prof Child Neurol (pres, 78-80). *Res:* Neurochemical changes in developing brain; iron and amino acid metabolism of immature brain. *Mailing Add:* Div Pediat Neurol Univ Minn Med Sch Minneapolis MN 55455

SWAIN, CHARLES GARDNER, chemistry; deceased, see previous edition for last biography

SWAIN, DAVID WOOD, b Raleigh, NC, Jan 10, 42; m 63; c 2. PLASMA PHYSICS, NUCLEAR FUSION. *Educ:* NC State Univ, BS, 63 & MS, 64; Mass Inst Technol, BS, 69. *Prof Exp:* Res staff mem, Sandia Nat Lab, 69-75; res staff mem, 75-78, proj mgr, 78-79, sr res scientist, 79-81, prog mgr, 81-83, ASST SECT HEAD, OAK RIDGE NAT LAB, 83- *Mem:* Fel Am Phys Soc. *Res:* Experimental investigation of and technology development for high-power radio-frequency heating systems for use in plasma fusion experiment. *Mailing Add:* Oak Ridge Nat Lab PO Box 2009 Oak Ridge TN 37831-8071

SWAIN, ELISABETH RAMSAY, b Philadelphia, Pa, Feb 7, 17. ZOOLOGY. *Educ:* Wilson Col, BS, 38; Univ Pa, MA, 42, PhD, 53. *Prof Exp:* Instr physics, Wilson Col, 43-46, instr biol, 46-49; asst instr zool, Univ Pa, 51-54; from asst prof to prof biol, Univ Hartford, 54-85, chmn dept, 54-80. *Mem:* AAAS; Sigma Xi; NY Acad Sci. *Res:* Embryology. *Mailing Add:* 59 Burlington St Hartford CT 06112

SWAIN, FREDERICK MORRILL, JR, b Kansas City, Mo, Mar 17, 16; m 38; c 3. GEOLOGY, PALEONTOLOGY. *Educ:* Univ Kans, AB, 38, PhD(stratig, paleont), 43; Pa State Col, MS, 39. *Prof Exp:* Geologist, Phillips Petrol Co, La, 41-43; asst prof mineral econ, Pa State Col, 43-46; from asst prof to assoc prof geol, 46-54, assoc chmn, Dept Geol & Geophys, 59-61, prof, 54-79, EMER PROF GEOL & GEOPHYS, UNIV MINN, MINNEAPOLIS, 79-; prof geol, 79-86, chmn dept, 83-86, EMER PROF GEOL, UNIV DEL, 86- *Concurrent Pos:* Assoc geologist, US Geol Surv, 44-46, geologist, 48-51, 61-; consult, Carter Oil Co, 51-53 & Pa RR, 54-57; part-time prof, Univ Del, 69-79. *Honors & Awards:* Award, Am Asn Petrol Geol, 49; Haworth Award, Univ Kans, 56. *Mem:* Fel Geol Soc Am; Soc Econ Paleont & Mineral; Paleont Soc; Am Asn Petrol Geol. *Res:* Stratigraphy; micropaleontology; organic geochemistry. *Mailing Add:* Dept Geol Univ Minn Minneapolis MN 55455

SWAIN, GEOFFREY W, b Poole, Dorset, Eng; m 85. MARINE BIOFOULING CONTROL, CORROSION CONTROL. *Educ:* London Univ, BSc, 71; Southampton Univ, MSc, 73, PhD(eng), 82. *Prof Exp:* Res asst, Southampton Univ, 74-80, res fel, 80-82; ASST PROF OCEAN ENG, FLA INST TECHNOL, 84- *Concurrent Pos:* Consult, Aberdeen Univ Marine Studies, 82-84. *Mem:* Soc Naval Architects & Marine Engrs; Nat Asn Corrosion Engrs; Inst Corrosion Sci & Technol; Marine Biol Asn UK; Marine Technol Soc. *Res:* Corrosion control; non-destructive testing using acoustic emission techniques; marine biofouling control. *Mailing Add:* Dept Oceanog & Ocean Eng Fla Inst Technol 150 W Univ Blvd Melbourne FL 32901

SWAIN, HENRY HUNTINGTON, b Champaign, Ill, July 11, 23; m 48; c 2. PHARMACOLOGY. *Educ:* Univ Ill, AB, 43, BS, 49, MS & MD, 51. *Prof Exp:* Instr pharmacol, Univ Cincinnati, 52-54; from instr to assoc prof, 54-67, asst dean fac affairs, 82-90, PROF PHARMACOL, SCH MED, UNIV MICH, ANN ARBOR, 67- *Concurrent Pos:* Pres, Am Heart Asn Mich, 84-85. *Mem:* Am Soc Pharmacol & Exp Therapeut. *Res:* Cardiovascular pharmacology, especially cardiac arrhythmias. *Mailing Add:* Dept Pharmacol Univ Mich Med Sch Ann Arbor MI 48109-0626

SWAIN, HOWARD ALDRED, JR, b New York, NY, Mar 3, 28; m 51; c 3. PHYSICAL CHEMISTRY. *Educ:* Grove City Col, BS, 51; Univ Pa, PhD, 61. *Prof Exp:* High sch instr chem, NJ, 54-56; lab technician, Rohm and Haas Co, 56-57; chemist, Socony-Mobile Res & Develop, 57-58; asst chem, Univ Pa, 58-60; from asst prof to assoc prof, 60-70, PROF CHEM, WILKES COL, 70- *Concurrent Pos:* Oak Ridge Assoc Univs res partic, Savannah River Lab, SC, 67-68; consult, Vet Admin Hosp, Wilkes-Barre, Pa, 71; lectr, Col Miseracordia, 72; res partic water purification proj, Environ Protection Agency. *Mem:* Am Chem Soc. *Res:* Thermodynamics; radiochemistry. *Mailing Add:* 84 W Mt Airy Rd Shavertown PA 18708

SWAIN, RALPH WARNER, b Orange, NJ, Dec 16, 44; m 67, 85; c 3. SYSTEMS INDUSTRIAL & MANUFACTURING ENGINEERING, HOSPITAL ADMINISTRATION. *Educ:* Johns Hopkins Univ, BEngSc, 67; Cornell Univ, MS, 69, PhD(environ systs), 71. *Prof Exp:* Nat Found Med Educ fel, Ctr Environ Qual Mgt, Cornell Univ, 71; asst prof indust & systs eng, Ohio State Univ, 71-76; assoc prof, 76-81, PROF INDUST & SYSTS ENG, UNIV FLA, 81-, ASSOC DIR, HEALTH SYSTS RES DIV, 76-, VPRES FACIL & SUPPORT SERV, SHANDS HOSP, UNIV FLA, 87- *Concurrent Pos:* asst to vpres admin, Shands Hosp, Univ Fla, 81-87. *Mem:* Am Inst Indust Engrs; Hosp Mgt Systs Soc. *Res:* Medical systems planning; optimal location of facilities; design and evaluation of health care delivery systems; facility location and layout; information systems design. *Mailing Add:* Box J-327 JHMHC Shands Hosp Gainesville FL 32610

SWAIN, RICHARD RUSSELL, b Columbus, Ohio, Mar 1, 39; m 64; c 1. BIOCHEMISTRY, CLINICAL CHEMISTRY. *Educ:* Albion Col, AB, 61; Univ Mich, MS, 63, PhD(biochem), 65. *Prof Exp:* Asst prof biol, MacMurray Col, 67-71; fel, State Univ NY, Buffalo, 71-73; SR BIOCHEMIST CLIN BIOCHEM, ELI LILLY & CO, 73- *Mem:* AAAS; Am Chem Soc; Am Asn Clin Chem; Am Bd Clin Chem. *Res:* Clinical chemistry methodology; clinical enzymology; biochemical assessment of hepatotoxicity and nephrotoxicity. *Mailing Add:* Toxicol Div Eli Lilly & Co PO Box 708 Indianapolis IN 46202

SWAIN, ROBERT JAMES, b Waukesha, Wis, Oct 3, 28; m 50; c 2. GEOPHYSICS, ELECTRONICS ENGINEERING. *Educ:* Purdue Univ, BS, 51. *Prof Exp:* Jr engr, A C Electronics Div, Gen Motors Corp, 51-53, prod engr, 53-55; sales & contracts mgr, Conrac Corp, 55-57; staff engr, United Electrodynamics, Inc, 58-59, chief struct seismol, 59-60, gen mgr, Geomeasurements Div, 60-61, vpres & gen mgr, Earth Sci Div, 61-64; vpres & gen mgr, Earth Sci, A Teledyne Co, 64-67, pres, 67-69; CHMN & PRES, KINEMETRICS, INC, 69- *Mem:* AAAS; Am Geophys Union; Seismol Soc Am; Sigma Xi; Am Nuclear Soc; Earthquake Eng Res Inst; Soc Explor Geophysists. *Res:* Ground motion from earthquakes and blasts; response of structures to ground motion; sensing and recording systems. *Mailing Add:* Kinemetrics Inc 222 Vista Ave Pasadena CA 91107

SWAISGOOD, HAROLD EVERETT, b Ashland, Ohio, Jan 19, 36; m 56; c 2. PROTEIN BIOCHEMISTRY. *Educ:* Ohio State Univ, BS, 58; Mich State Univ, PhD(chem), 63. *Prof Exp:* NIH fel, 63-64; from asst prof to prof, 64-84, WILLIAM NEAL REYNOLDS PROF FOOD SCI & BIOCHEM, 84-. CHMN UNIV BIOTECHNOL FAC, NC STATE UNIV, 87- *Honors & Awards:* Dairy Res Found Award, Am Dairy Sci Asn, Borden Award. *Mem:* AAAS; Am Chem Soc; Inst Food Technologists; Am Dairy Sci Asn; Am Soc Biochem & Molecular Biol; Am Inst Nutrit. *Res:* Physical-chemical characterization of proteins; studies of protein interactions and the relationship to biological activity; methods of preparation and characterization of enzymes covalently bound to surfaces; analytical affinity chromatography. *Mailing Add:* Dept Food Sci NC State Univ Box 7624 Raleigh NC 27695-7624

SWAKON, DOREEN H D, b Berwyn, Ill, Oct 9, 53; c 1. RUMINANT NUTRITION. *Educ:* Univ Ill, BS, 75; Univ Fla, MS, 77, PhD(animal sci), 80. *Prof Exp:* Res asst, Univ Fla, 75-80; asst prof, 80-85, dir, Forage Testing Lab, 81-85, ASSOC PROF ANIMAL SCI, TEX A&I UNIV, 85- *Mem:* Am Soc Animal Sci; Coun Agr Sci & Technol. *Res:* Forage quality evaluation and utilization of forage crops by ruminants to define differences in utilization between different species of forages and between different species of animals. *Mailing Add:* Tex A&I Univ Col Agr Box 156 Kingsville TX 78363

SWALEN, JEROME DOUGLAS, b Minneapolis, Minn, Mar 4, 28; m 52; c 2. CHEMICAL PHYSICS. *Educ:* Univ Minn, BS, 50; Harvard Univ, AM, 54, PhD(chem physics), 56. *Prof Exp:* Fel, Div Pure Physics, Nat Res Coun Can, 56-57; physicist, Shell Develop Co, 57-62; mgr physics dept, IBM Res Lab, 62-63, lab mgr, 63-67, mgr molecular physics dept, 67-73, res staff mem, 73-79, mgr excitations in solids, 79-85, MGR THIN FILM MAT & DEVICES, IBM RES LAB, 85- *Concurrent Pos:* Vis prof, Phys Chem Inst, Univ Zurich, 72-73, Fac Sci, Univ Paris VI & Inst Optics, Orsay, 83; guest prof physics, Univ Calif, Santa Cruz, 80-81. *Mem:* Fel AAAS; fel Am Phys Soc; Am Chem Soc. *Res:* Linear and nonlinear optical properties of thin organic films and monolayers. *Mailing Add:* IBM Almaden Res Ctr K95/801 650 Harry Rd San Jose CA 95120-6099

SWALIN, RICHARD ARTHUR, b Minneapolis, Minn, Mar 18, 29; m 52; c 3. RESOURCE MANAGEMENT, METALLURGY. *Educ:* Univ Minn, BS, 51, PhD(metall), 54. *Prof Exp:* Res assoc metall, Gen Elec Res Lab, 54-55; prof metall, Univ Minn, 55-77, dean, Inst Technol, 71-77; vpres, Eltra Corp, 77-80, Allied-Signal Corp, 80-84; pres, Ariz Technol Develop Corp, 87; dean, 84-87, PROG MGT, UNIV ARIZ, 84- *Concurrent Pos:* Mem, Naval Res Adv Bd-Lab, Bd Res, 72-76; dir, Sheldahl Corp, 72-76, BMC Corp, 73-91, Medtronic Corp, 73-, Donaldson Corp, 74-77; vchmn, Marine Corps Adv Bd, 73-76; mem adv bd, AMP Corp, 90- *Mailing Add:* 5260 Circulo Sobrio Tucson AZ 85718

SWALLOW, EARL CONNOR, b Montgomery County, Ohio, Dec 27, 41; m 83; c 3. HIGH ENERGY PHYSICS, EXPERIMENTAL PHYSICS. *Educ:* Earlham Col, BA, 63; Washington Univ, MA, 65, PhD(physics), 70. *Prof Exp:* Resident student assoc elem particle physics, Argonne Nat Lab, 64 & 65, res staff assoc, 65-69, res asst for Washington Univ, 69-70; res assoc elem particle physics, 70-74, sr res assoc, Enrico Fermi Inst, Univ Chicago, 74-80; from asst prof to assoc prof, 76-83, chmn dept,79-87, PROF PHYSICS, ELMHURST COL, 83-, CHMN DIV NATURAL SCI & MATH, 87- *Concurrent Pos:* Teaching asst, Washington Univ, 65; vis mem staff, Argonne Nat Lab, 78, 79 & 81; vis scientist, Enrico Fermi Inst, Univ Chicago, 83- *Honors & Awards:* Arthur H Compton lectr, Univ Chicago, 77. *Mem:* Am Phys Soc; Sigma Xi; AAAS; Am Asn Physics Teachers; NY Acad Sci. *Res:* Fundamental interactions of elementary particles, especially weak interactions; experimental foundations of physical theory; relationship of experimental foundations to public policy. *Mailing Add:* 473 Dominion Dr Wood Dale IL 60191

SWALLOW, RICHARD LOUIS, b Berwyn, Ill, June 16, 39; m 64; c 1. ZOOLOGY, BIOLOGY. *Educ:* Univ Ill, Urbana, BS, 63; Univ Mo-Columbia, MA, 66, PhD(zool), 68. *Prof Exp:* USPHS fel, Sch Med, Case Western Reserve Univ, 68-69; asst prof biol, Univ Houston, 69-73; assoc prof, 73-80, PROF BIOL, COKER COL, 80- *Mem:* Am Soc Zoologists. *Res:* Comparative physiology including control of metabolism by hormones in fish. *Mailing Add:* Dept Biol Coker Col Hartsville SC 29550

SWALLOW, WILLIAM HUTCHINSON, b Norwalk, Conn, Oct 21, 41; m 82; c 1. LINEAR MODELS, GROUP TESTING. *Educ:* Harvard Univ, AB, 64; Cornell Univ, MS, 68, PhD(biomet), 74. *Prof Exp:* Asst prof statist, Rutgers Univ, 73-79; assoc prof, 80-87, PROF STATIST, NC STATE UNIV,

87- *Mem:* Am Statist Asn; Sigma Xi; Biometric Soc. *Res:* Linear models; estimation of variance components; research directed at improving the teaching of statistics; group testing. *Mailing Add:* Dept Statist NC State Univ Raleigh NC 27695-8203

SWALM, RALPH OEHRLE, b Philadelphia, Pa, July 21, 15; m 42; c 3. ECONOMICS OF INDUSTRIAL ENGINEERING. *Educ:* Univ Pa, BSEE, 37; Syracuse Univ, MSEd, 57. *Prof Exp:* Engr, Gen Elec Co, 37-48; from asst prof to assoc prof, 48-57, PROF INDUST ENG, SYRACUSE UNIV, 57-, DIR ENG EDUC, 77- *Concurrent Pos:* Consult, Asbjorn Habberstad, Norway, 57; assoc dir, Indust Mgt Ctr, 60-; Am Bd Indust Engrs fel, 78. *Mem:* Am Inst Indust Engrs; Am Soc Eng Educ; Inst Mgt Sci. *Res:* Applied utility theory. *Mailing Add:* 437 Link Hall Syracuse Univ Syracuse NY 13244

SWAMER, FREDERIC WURL, b Shawano, Wis, May 16, 18; m 46; c 3. ORGANIC CHEMISTRY. *Educ:* Lawrence Col, BA, 40; Univ Wis, MS, 42; Duke Univ, PhD(chem), 49. *Prof Exp:* Chemist, Appleton Water Purification Plant, Wis, 40; res chemist, Electrochem Dept, E I du Pont de Nemours & Co, 41-45; res assoc, Duke Univ, 49-50; res chemist, E I DuPont De Nemours & Co, Inc, Wilmington, Del, 50-64, res assoc, Org Chem Dept, Exp Sta, 64-80; RETIRED. *Mem:* AAAS; Am Chem Soc. *Res:* Claisen condensation; physical properties of polymers; acetylene organoalkali compounds; fluorocarbons; heterogeneous catalysis. *Mailing Add:* Folly Hill Rd RD 4 West Chester PA 19380

SWAMINATHAN, BALASUBRAMANIAN, b Madras, India, Nov 24, 46; m 76. FOOD MICROBIOLOGY, FOOD TOXICOLOGY. *Educ:* Delhi Univ, BSc hon, 66; Univ Ga, MS, 74, PhD(food sci), 77. *Prof Exp:* Bottler's serv chemist qual control, Coca-Cola Export Corp, 67-72, microbiologist qual control & develop, 74; ASST PROF FOOD SCI, PURDUE UNIV, 77- *Concurrent Pos:* Res grant, Ind Agr Exp Sta, 80-82. *Mem:* Inst Food Technologists; Am Soc Microbiol; AAAS; Int Asn Milk, Food & Environ Sanitarians. *Res:* Rapid detection of pathogenic microorganisms in foods and feeds; pathogenic mechanisms associated with Yersinia enterocolitica; mutagenicity of procyanidins present in beverages. *Mailing Add:* Dept Food & Nutrit-Stone Hall Purdue Univ West Lafayette IN 47907

SWAMINATHAN, SRINIVASA, b Madras, India, Aug 24, 26; m 52; c 1. MATHEMATICS. *Educ:* Presidency Col, Madras, India, BA, 47, MA, 48; Univ Madras, MSc, 50, PhD(math), 57. *Prof Exp:* Govt of France fel, Inst Henri Poincare, Paris, 57-58; lectr math, Univ Madras, 59-64; asst prof, Indian Inst Technol, Kanpur, 64-66; vis assoc prof, Univ Ill, Chicago Circle, 66-68; assoc prof, 68-80, PROF MATH, DALHOUSIE UNIV, 80. *Concurrent Pos:* Auth & mem comt reorgn curricula math, Nat Coun Educ Res & Training, Govt of India, 64-67; managing ed, Can Math Bull, 79-85; vis prof, Australian Nat Univ, Canberra, 84; prod ed, Can Math J, 86-; vis prof, Univ Limoges, France, 90-91. *Mem:* Am Math Soc; Can Math Soc; Indian Math Soc; Math Asn Am. *Res:* Functional analysis; topology; geometry of Banach spaces; operator theory; paracompact spaces; fixed point theorems in analysis and topology; biomathematics. *Mailing Add:* 911 Greenwood Ave Halifax NS B3H 3L1 Can

SWAMY, MAYASANDRA NANJUNDIAH SRIKANTA, b Bangalore, India, Apr 7, 35; m 64; c 3. ELECTRICAL ENGINEERING, APPLIED MATHEMATICS. *Educ:* Univ Mysore, BSc, 54; Indian Inst Sci, Bangalore, dipl, 57; Univ Sask, MSc, 60, PhD, 63. *Prof Exp:* Sr res asst electronics, Indian Inst Sci, Bangalore, 58-59; res asst elec eng, Univ Sask, 59-63, sessional lectr math, 61-63, asst prof, 64-65; Govt India scientist, Indian Inst Technol, Madras, 63-64; from asst prof to prof elec eng, NS Tech Col, 65-68; prof, Concordia Univ, 68-69; prof, Univ Calgary, 69-70; prof & chmn dept, 70-77, DEAN, FAC ENG, CONCORDIA UNIV, 77- *Mem:* Fel Inst Elec & Electronics Engrs; Am Math Soc; Math Asn Am; fel Eng Inst Can; fel Inst Engrs India. *Res:* Network theory; graph theory; signal processing; author or coauthor of over one hundred research articles. *Mailing Add:* Dean of Eng Concordia Univ 1455 Maissoneuve Blvd W Montreal PQ H3G 1M8 Can

SWAMY, PADMANABHA NARAYANA, b India, July 25, 37; m 63; c 2. ELEMENTARY PARTICLE PHYSICS, THEORETICAL PHYSICS. *Educ:* Delhi Univ, India, BS, 56, MS, 58, PhD(physics), 63. *Prof Exp:* Res fel physics, Delhi Univ, 63-64, res fel, Tata Inst Fund Res, 64-65; vis scientist, Int Ctr Theoret Physics, Trieste, 65-66; res assoc, Syracuse Univ, 66; vis scientist, Ctr Nuclear Res, Geneva, 66-67; fel, Tata Inst Fund Res, 67-68; asst prof, Am Univ Beirut, 68-69; from asst to assoc prof, 69-77, PROF PHYSICS, SOUTHERN ILL UNIV, 77- *Concurrent Pos:* Chmn, Physics Dept, Southern Ill Univ, 83-92. *Mem:* Am Phys Soc. *Res:* Quantum field theory; particle theory. *Mailing Add:* Dept of Physics Southern Ill Univ Edwardsville IL 62026-1654

SWAMY, VIJAY CHINNASWAMY, b Bombay, India, Oct 2, 38; m 72; c 2. PHARMACOLOGY. *Educ:* Bombay Univ, BSc, 59; Nagpur Univ, BPharm, 62; Ohio State Univ, MS, 64, PhD(pharmacol), 67. *Prof Exp:* Res asst pharmacol, Ohio State Univ, 64-67; res assoc, 67-69, actg chair, 85-88, ASSOC PROF BIOCHEM PHARMACOL, STATE UNIV NY BUFFALO, 70- *Mem:* AAAS. *Res:* Smooth muscle pharmacology; hypertension; adrenergic mechanisms. *Mailing Add:* Dept Biochem Pharmacol State Univ NY Amherst NY 14260

SWAN, ALGERNON GORDON, b Andrews, NC, Jan 25, 23; m 47; c 2. PHYSIOLOGY, BIOPHYSICS. *Educ:* Univ NC, BA, 48, PhD(physiol), 60. *Prof Exp:* Chief biophys br, Aerospace Med Res Labs, Wright-Patterson AFB, US Air Force, Ohio, 60-62, dir life support res, Aerospace Med Div, Brooks AFB, Tex, 62-65, dir res, 65-68, dir test & eng, Air Force Spec Weapons Ctr, Kirtland AFB, NMex, 58-59, vcomdr & tech dir, 69-70, comdr & tech dir, 70-72; SR RES, BECTON, DICKINSON & CO, 72- *Concurrent Pos:* Mem, Comts Hearing, Bioacoust & Biomech, Nat Acad Sci-Nat Res Coun, 63-, Comt Nutrit, 64-; mem, Biosci Comt, NASA, 64-, Comt Biotechnol & Human Res, 65- & Comt Cardiopulmonary Res, Adv Group Aeronaut Res & Develop. *Mem:* Aerospace Med Asn. *Res:* Exercise and stress physiology; weapons effects; osmotic regulation and electrolyte flux in isolated tissues; human tolerance to aerospace stresses; nuclear environment simulation; instrumentation development; flight testing; qualification of instrumentation for space flight. *Mailing Add:* 1222 Palace Green Ave Cary NC 27511

SWAN, D(AVID), b NJ, May 2, 20; c 4. METALLURGY. *Educ:* Rensselaer Polytech Inst, BMetE, 40. *Prof Exp:* Metall observer, Crucible Steel Co Am, 40-41; res engr, Union Carbide Corp, 46-51, dir res, Union Carbide Metals Co, 52-56, Linde Co, 56-57, vpres res, 58-59, mgr planning, Union Carbide Corp, 59-60, vpres tech, Union Carbide Metals Co, 60-64, gen mgr, Defense & Space Systs Dept, Union Carbide Corp, 64-66; vpres tech, Kennecott Copper Corp, 66-79, vpres environ issues, 79-82; CONSULT, 82- *Concurrent Pos:* Mem, Moscow Steel Inst-NY Univ Exchange Prog, 57; mem, Mat Adv Bd, Nat Acad Sci-Nat Res Coun, 58-62; dir, William F Clapp Labs, Inc, Duxbury, Mass, 64-; chmn, Adv Comt Metall & Mat Sci, Polytech Inst NY, 77-; chmn, Dirs Indust Res, 78-79; mem, Develop Coun, Rensselaer Polytech, mem, Gen Tech Adv Comt, Off Coal Res; fel, Polytech Inst, 80. *Honors & Awards:* Demers Medal, Rensselaer Polytech Inst, 65. *Mem:* Am Welding Soc; fel Am Soc Metals; fel Metall Soc (pres, 72-73); Am Inst Mining, Metall & Petrol Engrs (vpres, 72-74). *Res:* Extractive metallurgy; environmental; economic; research management. *Mailing Add:* Four Point Rd Wilson Pt South Norwalk CT 06854

SWAN, DEAN GEORGE, b Wheatland, Wyo, Sept 16, 23; m 48; c 3. WEED SCIENCE. *Educ:* Univ Wyo, BS, 52, MS, 54; Univ Ill, PhD, 64. *Prof Exp:* Instr, Chadron High Sch, 52-53; asst prof & agron weed res, Pendleton Exp Sta, Ore State Univ, 55-65; exten weed specialist, Univ Ariz, 65-66; EXTEN WEED SCIENTIST & AGRONOMIST, WASH STATE UNIV, 66- *Concurrent Pos:* Sabbaticals, Weed Res Orgn, Oxford, Eng, 72-73 & 78-79. *Mem:* Weed Sci Soc Am; Sigma Xi. *Res:* Weed biology; weed control in crops and vegetation management. *Mailing Add:* SW 822 Crestview Dr Pullman WA 99163

SWAN, FREDERICK ROBBINS, JR, b Hartford, Conn, Aug 14, 37; m 62; c 2. ECOLOGY. *Educ:* Middlebury Col, BA, 59; Univ Wis, MS, 61; Cornell Univ, PhD(conserv natural resources), 66. *Prof Exp:* Assoc prof, 66-74, actg chmn, Sch Natural Sci, 70-72, PROF BIOL, WEST LIBERTY STATE COL, 74-, CHMN DEPT, 78- *Concurrent Pos:* Assoc, Dept Natural Resources, Cornell Univ, 74-75. *Mem:* AAAS; Am Inst Biol Sci; Ecol Soc Am; Sigma Xi. *Res:* Effects of fire on plant communities; measurement of light in forests. *Mailing Add:* 24 Donald Ave Kendall Park NJ 08824-1619

SWAN, HAROLD JAMES CHARLES, b Sligo, Ireland, June 1, 22; US citizen; m 46; c 7. PHYSIOLOGY, CARDIOVASCULAR DISEASES. *Educ:* Univ London, MB, BS, 45, PhD(physiol), 51. *Prof Exp:* Res assoc, Mayo Clin, 51-53, Minn Heart Asn res fel, 53-54, consult cardiovasc dis, 55-65; DIR CARDIOL, CEDARS-SINAI MED CTR, LOS ANGELES, 65-; PROF MED, UNIV CALIF, LOS ANGELES, 66- *Concurrent Pos:* Assoc prof, Mayo Grad Sch, Univ Minn, 57-65; consult, Nat Heart Inst, 60-66; mem, Intersoc Comn Heart Dis Resources, 69- *Honors & Awards:* Walter Dixon Award, Brit Med Asn, 50. *Mem:* Am Physiol Soc; fel Am Col Physicians; fel Am Col Cardiol (pres, 73); Asn Univ Cardiol. *Res:* Ventricular function; myocardial hypertrophy; coronary arterial disease and myocardial ischemia and infarction. *Mailing Add:* 414 N Camden Suite 800 Beverly Hills CA 90210

SWAN, JAMES BYRON, b Bloomington, Ill, Dec 9, 33; m 62; c 2. SOIL PHYSICS. *Educ:* Univ Ill, BS, 55, MS, 59; Univ Wis, PhD(soil physics), 64. *Prof Exp:* Exten specialist & prof soil sci, Univ Minn, St Paul, 64-89; PROF AGRON DEPT & ASSOC DIR LEOPOLD CTR SUSTAINABLE AGR, IOWA STATE UNIV, AMES, 89- *Mem:* Soil Conserv Soc Am; Am Soc Agron; Soil Sci Soc Am. *Res:* Tillage systems; modeling effect of soil temperature and soil water on plant growth; solute transport in soil. *Mailing Add:* 3405 Agronomy Dept Iowa State Univ Ames IA 50011

SWAN, KENNETH CARL, b Kansas City, Mo, Jan 1, 12; m 42; c 3. OPHTHALMOLOGY, PHARMACOLOGY. *Educ:* Univ Ore, BA, 33, MD, 36; Am Bd Ophthal, dipl, 40. *Prof Exp:* Assoc ophthal, Univ Iowa, 41-42, asst prof, 42-44; assoc prof ophthal, Med Sch, Univ Ore Health Sci Ctr, 44-45, prof & head dept, 45-78; RETIRED. *Concurrent Pos:* Proctor lectr, Univ Calif, 46; chmn bd, Am Bd Ophthal, 61; chmn sensory dis study sect, NIH, 61-63; mem adv coun, Nat Eye Inst, 69-71; consult, Nat Inst Neurol Dis & Blindness. *Honors & Awards:* Proctor Medal, Asn Res Vision & Ophthal, 53; Howe Medal, Am Ophthal Soc, 77. *Mem:* Am Ophthal Soc; Asn Res Vision & Ophthal; AMA; Am Acad Ophthal. *Res:* Ocular physiology, pharmacology and therapeutics; anomalies of binocular vision; tumors of the eyes; ocular manifestations of vascular diseases; surgical anatomy and pathology. *Mailing Add:* Dept Ophthal Ore Health Sci Univ Portland OR 97201

SWAN, KENNETH G, b White Plains, NY, Oct 2, 34; m 65; c 3. SURGERY. *Educ:* Harvard Univ, AB, 56; Cornell Univ, MD, 60. *Prof Exp:* Resident gen surg, New York Hosp-Cornell Med Ctr, 60-65; fel physiol, Gastrointestinal Res Lab, Vet Admin Ctr, Los Angeles, 65-66; resident thoracic surg, New York Hosp-Cornell Med Ctr, 66-68; dep dir div surg, Walter Reed Army Inst Res, 71-72, dir div surg, 72-73; dir & assoc prof, Div Gen & Vascular Surg, 73-76, PROF SURG, NJ MED SCH, 76- *Mem:* Am Physiol Soc; Am Gastroenterol Asn; Am Col Surgeons; Soc Univ Surgeons; Soc Thoracic Surgeons. *Res:* Splanchnic circulation; shock; vascular surgery and trauma. *Mailing Add:* Dept Surg Univ Med & Dent NJ Med Sci Bldg G590 100 Bergen St Newark NJ 07103

SWAN, LAWRENCE WESLEY, b Bengal, India, Mar 9, 22; m 46; c 3. BIOLOGY. *Educ:* Univ Wis, PhB, 42; Stanford Univ, MA, 47, PhD(biol), 52. *Prof Exp:* Res officer, Climatic Res Lab, Lawrence, Mass, 43-46; instr biol, Stanford Univ, 47-48 & Univ Santa Clara, 51-53; from instr to assoc prof, 54-64, PROF BIOL, SAN FRANCISCO STATE UNIV, 64- *Concurrent Pos:* Mem, Am Himalayan Expeds, Nepal, 54, 60-61, biol surv, Mt Orizaba, Mex, 64 & 65, biol world tour, 66, field studies, EAfrica, 69 & Galapagos Islands,

70; Academia Sinica Exped, Tibet Plateau, 80; field studies, South Central Africa, 81. *Mem:* Ecol Soc Am; Am Soc Ichthyol & Herpet; Royal Geog Soc; Sigma Xi; Am Inst Biol Sci. *Res:* High altitude ecology and the Aeolian zone; zoogeography of Asia; vertebrate evolution. *Mailing Add:* 1032 Wilmington Way Redwood City CA 94062

SWAN, PATRICIA B, b Hickory, NC, Oct 21, 37; m 62; c 2. NUTRITION, BIOCHEMISTRY. *Educ:* Univ NC, Greensboro, BS, 59; Univ Wis, MS, 61, PhD(biochem, nutrit), 64. *Prof Exp:* Res fel biochem, 64-65, from asst prof to assoc prof, 65-73, PROF NUTRIT, UNIV MINN, ST PAUL, 73-, ASSOC DEAN, GRAD SCH, 87- *Concurrent Pos:* Nutrit prog coordr, USDA, 79-80, prog mgr, competitive grants human nutrit, 85-86. *Mem:* Am Inst Nutrit (secy, 81-84); Brit Nutrit Soc. *Res:* Amino acid metabolism development and aging; nutrition effects on muscle metabolism; history of nutrition research. *Mailing Add:* 210 Beardshire Hall Iowa State Univ Ames IA 50011

SWAN, RICHARD GORDON, b New York, NY, Dec 21, 33; m 63; c 2. MATHEMATICS. *Educ:* Princeton Univ, AB, 54, PhD(math), 57. *Prof Exp:* NSF res fel, Oxford Univ, 57-58; from instr to assoc prof, 58-65, PROF MATH, 65-, LOUIS BLOCK PROF, UNIV CHICAGO, 82- *Concurrent Pos:* Sloan fel, 60-65. *Honors & Awards:* Cole Prize, Am Math Soc, 70. *Mem:* Nat Acad Sci; AAAS; Math Asn Am; NY Acad Sci; Am Math Soc. *Res:* Algebraic K-theory; homological algebra. *Mailing Add:* 5734 University Ave Dept Math Univ Chicago Chicago IL 60637

SWAN, ROY CRAIG, JR, b New York, NY, June 7, 20; m 49, 77; c 2. ANATOMY. *Educ:* Cornell Univ, AB, 41, MD, 47. *Prof Exp:* Intern med, New York Hosp, 47-48, asst resident, 48-49, resident endocrinol & metab, 49-50; asst med, Peter Bent Brigham Hosp, 50-52; from instr to assoc prof physiol, 52-59, prof anat, 59-70, chmn dept, 59-78, JOSEPH C HINSEY PROF ANAT, MED COL, CORNELL UNIV, 70- *Concurrent Pos:* Life Ins Med Res Fund fel, Harvard Med Sch, 50-52; Markle scholar, 54-59; res assoc, Cambridge Univ, 55-56; mem health res coun, City New York; consult, USPHS, 60-65 & Off Sci & Technol, 63-64; sect ed, Biol Abstr; mem & chmn exec bd, Anat Test Comt, Nat Bd Med Examr; vis prof anat, Boston Univ, 77-78. *Mem:* Am Physiol Soc; Am Soc Clin Invest; Am Asn Anat. *Res:* Ion transport; muscle function and structure; neural fine structure. *Mailing Add:* Dept of Anat Cell Biol Cornell Univ Med Col 1300 York Ave New York NY 10021

SWAN, SHANNA HELEN, b Warren, Ohio, May 24, 36; m 75; c 3. PUBLIC HEALTH. *Educ:* City Col New York, BS, 58; Columbia Univ, MS, 60; Univ Calif, Berkeley, PhD(statist), 63. *Prof Exp:* Sr biostatistician, Contraceptive Drug Study, Kaiser Health Res, Walnut Creek, Calif, 69-75; assoc prof math, Calif State Univ, Sonoma, 74-79; dir training prog biostatist & epidemiol, Sch Pub Health, Univ Calif, Berkeley, 79-81; chief Methodology & Anal Unit, Epidemiol & Statist, Dept Health Serv, State Calif, 81-; AT SCH DEPT PUBLIC HEALTH, UNIV CALIF. *Concurrent Pos:* Consult contraceptive eval, WHO, 74; lectr statist, Univ Copenhagen, Denmark, 60-61 & Univ Tel Aviv, Israel, 67-68 & Sch Pub Health, Univ Calif, Berkeley, 79-; vis assoc prof, Dept Statist, Univ Calif, Berkeley, 78-79. *Mem:* Soc Epidemiol Res; Am Statist Asn; Biometric Soc; Am Pub Health Asn. *Res:* Evaluating medical and health outcomes associated with contraceptive practices, drug exposures and environmental exposures, and the methodologic problems involved in studying such associations. *Mailing Add:* Dept Biomed & Environ Health Univ Calif Berkeley CA 94720

SWANBERG, CHANDLER A, b Great Falls, Mont, July 7, 42. GEOPHYSICS, GEOLOGY. *Educ:* Southern Methodist Univ, BS, 65, MS, 69, PhD(geophys), 71. *Prof Exp:* Fel geophys, Mus Geol, Univ Oslo, 71-72; geophysicist, US Bur Reclamation, 72-74; asst prof, 74-78, ASSOC PROF GEOPHYS, N MEX STATE UNIV, 78- *Concurrent Pos:* Prin investr various govt & state agencies, 74-; review panelist, Extramural Res Prog, US Geol Surv, 77; sr res consult, Teledyne-Geotech, 81-82. *Mem:* Am Geophys Union; AAAS; Geol Soc Am; Geothermal Resources Coun; Sigma Xi. *Res:* Search for, evaluation and development of geothermal energy resources. *Mailing Add:* 30 San Juan Ct Los Altos CA 94022

SWANBORG, ROBERT HARRY, b Brooklyn, NY, Aug 27, 38; div; c 2. IMMUNOLOGY, IMMUNOPATHOLOGY. *Educ:* Wagner Col, BS, 60; Long Island Univ, MS, 62; State Univ NY Buffalo, PhD(immunol), 65. *Prof Exp:* NIH trainee immunochem, State Univ NY Buffalo, 65-66; from instr to assoc prof microbiol, 66-73, assoc prof immunol & microbiol, 73-77, PROF IMMUNOL & MICROBIOL, MED SCH, WAYNE STATE UNIV, 77- *Concurrent Pos:* Vis investr immunol, Wenner-Gren Inst, Sweden, 75-76. *Mem:* AAAS; Am Asn Immunol; Am Asn Pathologists. *Res:* Cellular interactions in induction and regulation of the immune response; mechanisms of self-tolerance and autoimmunity. *Mailing Add:* Dept Immunol & Microbiol Wayne State Univ Med Sch Detroit MI 48201

SWANEY, JOHN BREWSTER, b Holyoke, Mass, Feb 25, 44; m 66; c 3. BIOCHEMISTRY. *Educ:* Amherst Col, AB, 66; Northwestern Univ, PhD(chem, biochem), 70. *Prof Exp:* Nat Acad Sci res assoc virol, Plum Island Animal Dis Lab, 70-72; instr, Albert Einstein Col Med, 72-74, from asst prof to assoc prof biochem, 74-84, from asst prof to assoc prof med, 81-82; PROF BIOCHEM, HAHNEMANN UNIV, 84-, RES PROF MED, 84- *Concurrent Pos:* Estab investr, Am Heart Asn; Irma T Hirschl career scientist award, 81; Res Coun NY Heart Asn, 79-85; assoc ed, J Lipid Res, 85- *Mem:* Am Heart Asn; Am Soc Biochem & Molecular Biol. *Res:* Protein chemistry; structure and properties of proteins involved in protein-lipid interactions; plasma lipoproteins; membrane proteins; lipid metabolism; protein effectors of enzymes involved in lipid metabolism. *Mailing Add:* Dept Biochem Hahnemann Univ Philadelphia PA 10461

SWANEY, LOIS MAE, b Pittsburgh, Pa, Jan 2, 28. MICROBIOLOGY, MOLECULAR BIOLOGY. *Educ:* Univ Pittsburgh, BS, 49, MS, 52, PhD(molecular biol), 74. *Prof Exp:* Microbiologist bact genetics, Biol Labs, US Dept Army, Frederick, Md, 51-71; res asst, Univ Pittsburgh, 72-73; microbiologist cell-virus interactions & genetic eng, Plum Island Animal Dis Ctr, Agr Res Serv, USDA, 73-88; RETIRED. *Mem:* Am Soc Microbiol; Sigma Xi. *Res:* Genetics of bacterial pili; mutants of foot-and-mouth disease virus; cell-virus interactions; assessment of tissue cultures for production of virus for vaccines; development of cell lines; expression of viral genes in procaryotic cells. *Mailing Add:* 355 Summit Dr Southold NY 11971-2112

SWANK, RICHARD TILGHMAN, b Drums, Pa, Feb 1, 42; m 66; c 2. BIOCHEMISTRY. *Educ:* Pa State Univ, BS, 64; Univ Wis-Madison, MS, 67, PhD(biochem), 69. *Prof Exp:* NIH fel, Lab Molecular Biol, Univ Wis-Madison, 69-70; res assoc mammalian biochem genetics, 70-72, SR CANCER RES SCIENTIST, ROSWELL PARK MEM INST, 72- *Mem:* AAAS; Am Chem Soc; Am Inst Biol Sci; Sigma Xi. *Res:* Genetic regulation of enzyme synthesis and degradation in mammals; biochemical mechanisms of enzyme subcellular localization in mammals; physical and chemical characterization of enzymes. *Mailing Add:* Dept Cell & Molecular Biol Roswell Park Mem Inst 666 Elm St Buffalo NY 14203

SWANK, ROBERT ROY, JR, b Brooklyn, NY, June 4, 39; m 60; c 3. ENVIRONMENTAL EXPOSURE & RISK ASSESSMENT, HAZARDOUS WASTE MANAGEMENT. *Educ:* Ga Inst Technol, BChE, 59, MSChE, 63, PhD(chem eng), 68; Mass Inst Technol, MS, 60. *Prof Exp:* Maintenance engr, Monsanto Chem Co, 59, sr process engr, 68-71; res chem engr, Southeast Environ Res Lab, US Environ Protection Agency, 71-75, supvry res chem engr & chief, Tech Develop & Applns Br, 75-84, supvry res chem engr & chief, Assessment Br, 84-86, supv env eng & chief, Tech Assessment Br, Off Solid Waste, 86-87, dir res, Athens Environ Res Lab Off Res & Develop, 87-89, staff dir, Marine, Freshwater & Modeling, Environ Processes & Effects, 89-90, staff dir, Terrestrial & Groundwater Effects, 90-91, DIR RES, ATHENS ENVIRON RES LAB, OFF RES & DEVELOP, US ENVIRON PROTECTION AGENCY, WASHINGTON, DC, 91- *Concurrent Pos:* Teaching asst & lectr, Mass Inst Technol, 61; guest lectr, US Army Chem Sch, Ft McClellan, Ala, 65-67; consult, UN Indust Develop Orgn, 74-75 & soft drink beverage water treatment indust, 82-, & combuster technol indust, 82-; off delegate to People's Repub China, Environ Protection Agency, 80, 86. *Honors & Awards:* Bronze Medal, US Environ Protection Agency, 73. *Mem:* Am Inst Chem Engrs; Am Chem Soc; Sigma Xi. *Res:* Comprehensive assessment techniques for estimating human and environmental exposures and resulting risks due to the release into all environmental media (air-water-soil) of toxic/hazardous organic compounds and metals. *Mailing Add:* Rt 3 270 Great Oak Dr Athens GA 30605

SWANK, ROLLAND LAVERNE, b Holland, Mich, Dec 31, 42; m 69. MATHEMATICS. *Educ:* Hope Col, BA, 65; Mich State Univ, MS, 66, PhD(math), 69. *Prof Exp:* Asst prof math, Allegheny Col, 69-74; programmer, Power & Power, 74-80; DEPT MGR, W P DELONG & CO, 80- *Mem:* Am Math Soc; Math Asn Am. *Res:* Topology; geometry. *Mailing Add:* 109 Orlando Ave Holland MI 49423

SWANK, ROY LAVER, b Camas, Wash, Mar 5, 09; m 37, 87; c 4. NEUROLOGY. *Educ:* Univ Wash, BS, 30; Northwestern Univ, MD & PhD(anat), 35. *Prof Exp:* Asst anat, Med Sch, Northwestern Univ, 30-34; intern, Passavant Mem Hosp, Chicago, 34-35; house officer, Peter Bent Brigham Hosp, 36-41, jr assoc med, 41-42, assoc, 46-48; asst prof neurol, McGill Univ, 48-54; prof, 54-74, EMER PROF NEUROL, SCH MED, ORE HEALTH SCI UNIV, PORTLAND, 74- *Concurrent Pos:* Fel, Harvard Med Sch, 37; Commonwealth Fund fel, Sweden & Montreal Neurol Inst, McGill Univ, 39-41; mem attend staff, Cushing Vet Admin Hosp, 46-48; lectr, Montreal Neurol Inst, McGill Univ, 48; first examiner, Swed Doctoral Exam, Goteburg, Swed, 61; hon chmn, 3rd Int Cong Biorleology, 79. *Mem:* Am Physiol Soc; Am Asn Anatomists; Am Neurol Asn; Can Neurol Asn; Sigma Xi. *Res:* Pyramidal tracts; tissue staining; histochemical staining; vitamin deficiencies; electrophysiology; physiology of breathing. epidemiology of multiple sclerosis; fat metabolism and relationship to multiple sclerosis and to viscosity of blood; platelet adhesivesness and aggregation in surgical shock and microembolic filter to remove them from blood; abnormal plasma protein in multiple sclerosis. *Mailing Add:* Dept Neurol Ore Health Sci Univ Portland OR 97201

SWANK, THOMAS FRANCIS, b Philadelphia, Pa, Nov 3, 37; m 63; c 3. COLLOID CHEMISTRY, PHOTOGRAPHIC CHEMISTRY. *Educ:* Villanova Univ, BS, 59; Univ Va, PhD(heterogeneous catalysis), 64. *Prof Exp:* Res chemist, Cabot Corp, 63-69; mgr, Ferro Fluidics Corp, 70-71; scientist, Polaroid Corp, Waltham, 71-76, sr scientist, 76-83; sr chemist, 83-86, PROD DEVELOP MGR, NYACOL PROD INC, ASHLAND, MASS, 86- *Mem:* Am Chem Soc; Am Ceramic Soc. *Res:* Heterogeneous catalysis; thin films; x-ray diffraction and spectroscopy; electron microscopy; structure of oxides; inorganic pigments; solid state physics; magnetic fluids; photographic science; colloid and surface chemistry. *Mailing Add:* 25 Musket Lane Sudbury MA 01776

SWANK, WAYNE T, b Washington, DC, Mar 8, 36; m 58; c 4. ECOLOGY. *Educ:* WVa Univ, BS, 58; Univ Wash, MF, 60, PhD(forestry), 72. *Prof Exp:* Forester admin, Forest Serv, USDA, 59-62, res forester, 66-76; prog mgr, NSF, 77-78; supvry ecologist, 78-83, PROJ LEADER, FOREST SERV, USDA, 84- *Concurrent Pos:* Adj prof bot & ecol, Univ Ga, 74-; directorate mem, US Man & Biosphere Prog, MAB 2, 75-89; prin investr, Lon-Term Ecol Res Prog Univ Ga, 80-; comt mem, Nat Acad Sci, 87-90; consult & lectr, Univ & Govt Mex, USSR, Eng, Italy & Switz. *Mem:* Ecol Soc Am; AAAS; Sigma Xi. *Res:* Forest hydrology and ecology research with emphasis on biogeochemical cycles and responses to management practices; studies focus on hyrologic processes, microbial transformations in soils, atmospheric deposition, and forest succession from a watershed ecosystem perspective. *Mailing Add:* Coweeta Hydrol Lab 999 Cowetta Lab Rd Otto NC 28763

SWANN, CHARLES PAUL, b Minneapolis, Minn, Dec 4, 18; m 51; c 3. ARCHAEOMETRY, RADIATION DAMAGE. *Educ:* Harvard Univ, BS, 41, MS, 43; Temple Univ, PhD(physics), 56. *Prof Exp:* Mech engr, Steam Div, Westinghouse Elec Corp, 43-46; nuclear physicist, Bartol Res Found, Franklin Inst, 46-76; Bartol prof physics, 76-85, BARTOL EMER PROF & EMER PROF DEPT PHYSICS & ASTRON, BARTOL RES INST, UNIV DEL, 85- *Concurrent Pos:* Res assoc, Masca, Univ Pa, 88- *Mem:* Fel Am Phys Soc; AAAS; Sigma Xi; Mat Res Soc; Soc Archaeological Sci. *Res:* Proton induced x-ray studies of bi-valve marine life and of archaeological artifacts; radiation damage studies of magnetic amorphous metallic alloys; surface studies of materials using Rutherford backscattering spectroscopy. *Mailing Add:* Bartol Res Inst Univ Del Newark DE 19716

SWANN, DALE WILLIAM, b Billings, Mont, Mar 11, 29; m 60; c 1. APPLIED MATHEMATICS, OPERATIONS RESEARCH. *Educ:* Yale Univ, BS, 51; Stanford Univ, PhD(math), 60. *Prof Exp:* Instr math, Stanford, 57-60; NATO fel sci, Cambridge Univ, 60-61; mem tech staff, Bell Labs, 61-; MGR SYST PERFORMANCE ENG, GEN ELEC CO. *Mem:* Inst Elec & Electronics Engrs. *Res:* Statistical theory; applied probability; quality theory and practice; reliability theory; integral equations; asymptotic methods. *Mailing Add:* Gen Elec Co Elec Utility Syst Eng Dept 1 River Rd Schenectady NY 12345

SWANN, DAVID A, b Barnet, England, Sept 8, 36. CONNECTIVE TISSUE BIOCHEMISTRY. *Educ:* Univ Leeds, PhD(physiol chem), 63. *Prof Exp:* RES DIR BIOCHEM, SHRINER BURN INST, 84- *Mailing Add:* McChem Prod Inc 444 Washington St Woburn MA 01801

SWANN, GORDON ALFRED, b Palisade, Colo, Sept 21, 31; m 75; c 4. GEOLOGY, ASTROGEOLOGY. *Educ:* Univ Colo, Boulder, BA, 58, PhD(geol), 62. *Prof Exp:* Geologist, 63-73, staff geologist for telegeol, 73-76, dep regional geologist, 76-81, dep asst chief geologist, 81-85, GEOLOGIST, US GEOL SURV, 85- *Concurrent Pos:* Prin investr, Lunar Geol Exp, Apollo missions 14 & 15. *Mem:* Geol Soc Am; Sigma Xi. *Res:* Lunar geology. *Mailing Add:* 814 W Murray Rd Flagstaff AZ 86001

SWANN, HOWARD STORY GRAY, b Chicago, Ill, Aug 4, 36. MATHEMATICS. *Educ:* Harvard Univ, AB, 58; Univ Chicago, MS, 59; Univ Calif, Berkeley, PhD(appl math), 68. *Prof Exp:* Asst math, Univ Chicago, 59-61; lectr, Univ Nigeria, 61-63; asst & instr, Univ Calif, Berkeley, 64-68; asst prof, Antioch Col, 68-70; ASSOC PROF MATH, SAN JOSE STATE UNIV, 70- *Mem:* Am Math Soc. *Res:* Functional analysis; differential equations; game theory; automata theory. *Mailing Add:* Dept Math San Jose State Univ San Jose CA 95192

SWANN, MADELINE BRUCE, b Washington, DC, July 24, 51. SOLDIER SYSTEMS, SOLDIER COMBAT SERVICE SUPPORT. *Educ:* Fisk Univ, Nashville, Tenn, BA, 73; Howard Univ, Washington, DC, PhD(biochem), 80. *Prof Exp:* Teaching fel gen chem, org chem, biochem lab & res asst, dept chem, Howard Univ, 73-80; asst prof phys sci, gen chem & biochem, dept chem & physics, Miss Valley State Univ, 79; chemist, US Army Belvior Res, Develop & Eng Ctr, 81-90; CHEMIST, HQ MATERIEL COMMAND, US ARMY, 90- *Concurrent Pos:* Consult, Technol Appln, Inc, 81; prin investr, in-house lab independent res proj, Off Chief Scientist, US Army Belvoir Res & Develop Ctr, 83-85; mem, Women's Exec Leadership Prog, Off Personnel Mgt, 89. *Honors & Awards:* Outstanding Young Women of Am, 87. *Mem:* Am Chem Soc; Nat Orgn Black Chemists & Chem Engrs; Sigma Xi; NY Acad Sci. *Res:* Direct, monitor and execute the Department of the Army technology base research and development programs to ensure that the technologies of the future will be available to the soldier in the field. *Mailing Add:* 12008 Hallandale Terr Mitchellville MD 20721-1954

SWANSON, ALAN WAYNE, b Des Moines, Iowa, Jan 27, 44; m 65; c 2. MATERIALS SCIENCE, INORGANIC CHEMISTRY. *Educ:* SDak Sch Mines & Technol, BS, 66, MS, 68; Mass Inst Technol, PhD(metall), 72. *Prof Exp:* Teaching asst chem, SDak Sch Mines & Technol, 66-67, instr, 67-68; res asst metall, Mass Inst Technol, 68-72; sr res scientist mat, Raytheon Co, 72-78; mgr advan process technol dept, Sperry Res Ctr, 78-; SUPT CONTROL & COMMS, NORTHERN STATES POWER. *Mem:* Electrochem Soc; Inst Elec & Electronics Engrs; Sigma Xi. *Res:* Inorganic chemistry pertaining to vapor deposition of III-V semiconductor compounds; all areas of wet and dry corrosion; gallium arsenide for use in digital logic circuits and millimeter wave and microwave circuits. *Mailing Add:* 103 Holliston Circle Fayetteville NY 13066

SWANSON, ANNE BARRETT, b Joliet, Ill, Dec 23, 48; m 69. BIOCHEMISTRY, CHEMICAL CARCINOGENESIS. *Educ:* Northern Ill Univ, BS, 70; Univ Wis-Madison, PhD(biochem), 75. *Prof Exp:* NIH res fel, U W McArdle Lab Cancer Res, 75-78, res assoc chem carcinogenesis, 78-79; from asst prof to assoc prof chem, 79-88, med tech coordr, 80-88, chmn, 81-88, PROF CHEM, EDGEWOOD COL, MADISON, 88- *Concurrent Pos:* Assoc prog dir, NSF, Washington, DC, 85-86; NSF grants, 85 & 87. *Honors & Awards:* Gallantry Award, Easter Seal Soc. *Mem:* Am Chem Soc; Sigma Xi; AAAS; Nat Sci Teachers Asn; NY Acad Sci; Found Sci & Handicapped. *Res:* Mechanisms of chemical carcinogenesis; metabolism of precarcinogens to ultimate carcinogenic compounds; nutrition and metabolism of trace minerals; relationships of nutritional biochemistry and carcinogenesis. *Mailing Add:* 2405 Pascal St N Roseville MN 55113

SWANSON, ARNOLD ARTHUR, b Rawlins, Wyo, Mar 11, 23; m 50; c 3. BIOCHEMISTRY, OPHTHALMOLOGY. *Educ:* Duke Univ, BA, 46; Trinity Univ, Tex, MA, 59; Tex A&M Univ, PhD(biochem), 61. *Prof Exp:* Res chemist, Med Sch, Temple Univ, 46-48; biochemist ophthal, US Air Force Sch Aviation Med, 50-59; sr chemist, USPHS, 61-63; chief res lab, Vet Admin Hosp, McKinney, Tex, 63-65 & Vet Admin Ctr, 65-68; assoc prof, 68-80, PROF BIOCHEM, MED UNIV SC, 80- *Concurrent Pos:* Dir, Swanson Biochem Labs, Inc, 52-58; consult, Southwestern Prods, Inc, 57- & Scott & White Hosp, 65-; adj prof, Baylor Univ, 66- *Honors & Awards:* Alexander von Humboldt-Stiftung Award, WGer. *Mem:* Fel AAAS; Am Chem Soc; Fedn Am Soc Exp Biol; Am Soc Biochem & Molecular Biol; Asn Res Vision & Ophthal; Sigma Xi. *Res:* Proteolysis in normal and senile cataract lens; senile changes and mineral metabolism. *Mailing Add:* Dept Biochem Med Univ SC 171 Ashley Ave Charleston SC 29425

SWANSON, AUGUST GEORGE, neurology, medical education, for more information see previous edition

SWANSON, BARRY GRANT, b Green Lake, Wis, Apr 16, 44; m; c 3. FOOD SCIENCE & TECHNOLOGY, TOXICOLOGY. *Educ:* Univ Wis-Madison, BS, 66, MS, 70, PhD(food sci), 72. *Prof Exp:* Asst prof food sci, Univ Idaho, 72-73; from asst prof to assoc prof food sci, 73-83, PROF FOOD SCI HUMAN NUTRIT, FOOD SCIENTIST & FOOD SCI SPECIALIST, WASH STATE UNIV, 83- *Concurrent Pos:* Travel award Spain, Inst Food Technologists, 74; vis prof, Dept Food Sci, Univ BC & BC Cancer Res Ctr; sci lectr, regional commun, Inst Food Technologists, 79-82 & 85-; int res, Guatemala, Ecuador, Colombia, Philippines, India. *Mem:* Inst Food Technologists; Am Soc Hort Sci; Am Chem Soc; Sigma Xi; Am Coun Sci Health; Am Asn Cereal Chemists. *Res:* Bioavailability and digestibility of proteins from dry beans; analytical and toxicological studies of natural toxicants and antinutrients in foods; nutrient analysis and retention in preserved foods and carbohydrate fatty acid polyester fat substitutes. *Mailing Add:* Dept Food Sci & Human Nutrit Wash State Univ Pullman WA 99164-6376

SWANSON, BASIL IAN, b Minn, Feb 13, 44; m 64; c 2. INORGANIC CHEMISTRY. *Educ:* Colo Sch Mines, BA, 66; Northwestern Univ, Evanston, PhD(chem), 70. *Prof Exp:* Fel, Los Alamos Sci Lab, Univ Calif, 70-71; res corp grant, NY Univ, 71-72, asst prof chem, 71-73; ASST PROF CHEM, UNIV TEX, AUSTIN, 73- *Concurrent Pos:* Vis staff mem, Los Alamos Sci Lab, 70- *Mem:* Am Chem Soc. *Res:* Study of structure and bonding in inorganic systems using crystallographic and vibrational spectroscopic techniques; study of structural phase changes in crystalline solids; valence delocalization and ion transport in crystalline solids. *Mailing Add:* 3463 Urban Los Alamos NM 87544-2026

SWANSON, BERNET S(TEVEN), b Chicago, Ill, Nov 20, 21; m 48; c 2. CHEMICAL ENGINEERING. *Educ:* Armour Inst Technol, BS, 42; Ill Inst Technol, MS, 44, PhD(chem eng), 50. *Prof Exp:* Instr chem eng, Ill Inst Technol, 44-46; asst prof, Kans State Col, 46-47; from asst prof to assoc prof chem eng, Ill Inst Technol, 47-85, chmn dept, 67-85; RETIRED. *Concurrent Pos:* Consult. *Mem:* Am Inst Chem Engrs; Instrument Soc Am. *Res:* Automatic process control; non-Newtonian and two-phase flow. *Mailing Add:* Dept of Chem Eng 3300 S Federal St Chicago IL 60616

SWANSON, CARL PONTIUS, b Rockport, Mass, June 24, 11; m 41; c 2. CYTOGENETICS. *Educ:* Mass State Col, BS, 37; Harvard Univ, MA, 39, PhD(biol), 41. *Prof Exp:* Sheldon traveling fel from Harvard Univ, Univ Mo, 41; asst prof bot, Mich State Col, 41-43; assoc biologist, NIH, 46; from assoc prof to prof bot, Johns Hopkins Univ, 46-56, William D Gill prof biol, 56-71, assoc dean undergrad studies, 66-71; assoc dir, Inst for Man & His Environ, 71-75, prof, 71-76, Ray Ethan Torrey prof, 76-81, EMER PROF BOT, UNIV MASS, AMHERST, 81- *Concurrent Pos:* Agt, USDA, 39; contract investr, Spec Proj Div, US Army, 46-49; pres, Int Photobiol Comt, 64-68. *Mem:* AAAS; Genetics Soc Am; Sigma Xi. *Res:* Cytogenetics of plants involving use of ionizing and photochemical radiations. *Mailing Add:* 77 Morgan Circle Amherst MA 01002

SWANSON, CARROLL ARTHUR, b Burlington, Iowa, Sept 6, 15; m 41; c 2. PLANT PHYSIOLOGY. *Educ:* Augustana Col, AB, 37; Ohio State Univ, MS, 38, PhD(plant physiol), 42. *Prof Exp:* From asst to asst prof bot, Ohio State Univ, 38-48, res assoc, Manhattan Proj Res Found, 44-46, assoc prof, 48-56, chmn dept, 67-69, assoc dean col biol sci, 69-70, prof bot & plant physid, 56-85; RETIRED. *Concurrent Pos:* Asst gen foreman, Procter & Gamble Defense Corp, Miss, 43-44; prog dir, NSF, 59-60, consult, 60-66; assoc ed, Plant Physiol, 80- *Mem:* AAAS; Am Soc Plant Physiologists; Bot Soc Am; Can Soc Plant Physiologists; Scandinavian Soc Plant Physiologists. *Res:* Translocation in phloem. *Mailing Add:* 719 Lauraland Dr S Columbus OH 43214

SWANSON, CHARLES ANDREW, b Bellingham, Wash, July 11, 29; m 57; c 2. DIFFERENTIAL EQUATIONS. *Educ:* Univ BC, BA, 51, MA, 53; Calif Inst Technol, PhD, 57. *Prof Exp:* From instr to assoc prof, 57-65, PROF MATH, UNIV BC, 65- *Concurrent Pos:* Assoc ed, Can J Math, 71-80. *Mem:* Can Math Soc. *Res:* Differential equations. *Mailing Add:* Dept Math Univ BC Vancouver BC V6T 1Y4 Can

SWANSON, CURTIS JAMES, b Chicago, Ill, Dec 8, 41; m 65; c 2. COMPARATIVE PHYSIOLOGY, BIOCHEMISTRY. *Educ:* N Park Col, BA & BS, 64; Northern Ill Univ, MS, 66; Univ Ill, Urbana-Champaign, PhD(zool, physiol), 70. *Prof Exp:* Lectr biol, Univ Ill, 70; asst prof, 70-74, ASSOC PROF BIOL, WAYNE STATE UNIV, 74- *Concurrent Pos:* Grants, NSF, Wayne State Univ, 70-75 & NIH, 71-; Riker res fel, Bermuda Biol Sta, 75, NSF, 77-78. *Mem:* AAAS; Am Inst Biol Sci; Am Soc Zoologists; Am Physiol Soc. *Res:* Electron microscopy of muscle tissue; innervation and developmental neuromuscular physiology; control systems in development; protein biochemistry; theoretical and applied biomechanics; comparative ultrastructure of muscle. *Mailing Add:* Dept Biol Wayne State Univ Detroit MI 48202

SWANSON, DAVID BERNARD, b Newark, NJ, Dec 14, 35; m 58; c 3. CHEMICAL ENGINEERING. *Educ:* Newark Col Eng, BS, 57, MS, 62. *Prof Exp:* Engr, Esso Res & Eng Co, Standard Oil Co, NJ, 57-59; from engr to res supvr, 59-77, dir, Process Develop & Mfg, Engelhard Minerals & Chem Corp, 77-80, DIR RES SERV, ENGELHARD CORP, 80- *Concurrent Pos:* Exec comt, Corp Assocs. *Mem:* Catalysis Soc; Mats Res Soc; Am Chem Soc. *Res:* Heat transfer coefficients of non-Newtonian fluids; dehydration and rehydration of kaolin; zeolitic cracking catalyst. *Mailing Add:* Engelhard Corp 25 Middlesex-Essex Turnpike Iselin NJ 08830-2708

SWANSON, DAVID G, JR, b Chicago, Ill, Jan 14, 41. NUCLEAR CHEMISTRY, PHYSICAL CHEMISTRY. *Educ:* Northwestern Univ, BS, 64; Purdue Univ, PhD(nuclear & phys chem), 69. *Prof Exp:* Nuclear chemist, Sandia Corp, NMex, 69-73; nuclear chemist, Aerospace Corp, 73-80. *Concurrent Pos:* Consult, US Nuclear Regulatory Comn, 75-, Brookhaven Nat Lab, Sandia Nat Labs, Rand & Assocs, Logicon Advan Sci, Inc & Carpenter Res Corp. *Mem:* Am Chem Soc; Am Phys Soc; Am Nuclear Soc. *Res:* Response of materials to radiation; nuclear reactions; radiation transport phenomena; high temperature physical chemistry; heat transfer, materials evaluation and characterization; thermodynamics; nuclear reactor safety studies; post accident heat removal; materials science; composite materials; nuclear weapons effects tests; studies of anti-satellite devices. *Mailing Add:* 6868 Los Verdes Dr 12 Palos Verdes Peninsula CA 90274-5654

SWANSON, DAVID HENRY, b Anoka, Minn, Nov 1, 30; m 51, 91; c 2. CONTINUING EDUCATION, TECHNOLOGY TRANSFER. *Educ:* St Cloud State Univ, BA, 53; Univ Minn, MA, 55; Iowa State Univ Sci & Technol, PhD(adult educ), 87. *Prof Exp:* Economist, Northern States Power Co, 55-63; dir econ & mkt res, Iowa Southern Utilities Co, 63-70; dir, New Orleans Econ Develop Coun, 70-72; mgr deal develop, Kaiser Aetna, 72-73; dir corp res, United Serv Automobile Asn, 73-76; adminr, Wis Econ Develop Dept, State of Wis, 76-78; dir, Ctr Indust Res & Serv, Iowa State Univ Sci & Technol, 78-89; DIR, ECON DEVELOP LAB, GA RES INST, GA INST TECHNOL, ATLANTA, 89- *Concurrent Pos:* Instr mkt & mkt mgt, Col Bus Admin, Iowa State Univ, 79-83, mgt technol, Hons Prog, 85-87 & comparative mgt, Hons Prog, 85-88; chair, Iowa High Technol Task Force, 83 & Iowa High Technol Coun, 83-86; dir, Nat Asn Mgt & Tech Asst Ctrs, 83-87, pres, 85; dir, Iowa Develop Comn, 84-85. *Honors & Awards:* Presidents Award, Technol Transfer Soc, 90. *Mem:* Technol Transfer Soc (vpres, 86-91, pre-elect, 91); Nat Asn of Mgt & Tech Assistance Ctrs (pres, 84); Nat Univ Continuing Educ Asn. *Res:* Technology transfer in industry; technical assistance to management. *Mailing Add:* Ga Technol Res Inst, Economic Develop Lab Ga Inst of Technol Atlanta GA 30332-0800

SWANSON, DAVID WENDELL, b Ft Dodge, Iowa, Aug 28, 30; m 53; c 3. PSYCHIATRY. *Educ:* Augustana Col, Ill, BA, 52; Univ Ill, MD, 56. *Prof Exp:* Intern, Ill Cent Hosp, 56-57; resident psychiat, Ill State Psychiat Inst, 59-62, asst serv chief, 62-63; assoc prof & asst chmn dept, Stritch Sch Med, Loyola Univ Chicago, 63-70; assoc prof psychiat, Mayo Grad Sch Med & consult, Sect Psychiat, Mayo Clin, 70-74, prof psychiat & head sect, Mayo Med Sch, Univ Minn, 74-79, vchmn, dept psychiat, Mayo Clin, 79-86, CONSULT, PSYCHIAT SECT & PROF PSYCHIAT, MAYO CLIN, 86- *Mem:* AAAS; Am Col Psychiatrists; Am Psychiat Asn; Sigma Xi. *Res:* Paranoid and chronic pain disorders. *Mailing Add:* Dept Psychiat W-9 Mayo Found Mayo Grad Sch Med Rochester MN 55901

SWANSON, DON R, b Los Angeles, Calif, Oct 10, 24; m 76; c 3. INFORMATION SCIENCE, ONLINE SEARCHING. *Educ:* Calif Inst Technol, BS, 45; Rice Univ, MA, 47; Univ Calif, Berkeley, PhD(physics), 52. *Prof Exp:* Res physicist, Radiation Lab, Univ Calif, 50-52; mem tech staff, Hughes Res & Develop Labs, 52-55; dept mgr comput appln, Thompson-Ramo-Wooldridge, Inc, 55-63; dean, Grad Libr Sch, 63-72 & 77-79, PROF INFO SCI, UNIV CHICAGO, 63- *Concurrent Pos:* Mem sci info coun, NSF, 59-63; mem vis comt libr, Mass Inst Technol, 64-72; trustee, Nat Opinion Res Ctr, 64-73; mem adv comt, Libr Cong, 64-72, toxicol info panel, President's Sci Adv Comt, 64-65, comt sci & tech commun, Nat Acad Sci, 66-70 & adv comt, Encycl Britannica, 66-76. *Mem:* Am Soc Info Sci. *Res:* Studies of fragmentation of scientific literature; analysis of complementary literatures that are mutually isolated, but which, if combined, lead to new inferences that cannot be drawn from the separate literatures. *Mailing Add:* Ctr Info & Lang Studies Univ Chicago 1100 E 57 Chicago IL 60637

SWANSON, DONALD ALAN, b Tacoma, Wash, July 25, 38; m 74. VOLCANOLOGY, GEOLOGY. *Educ:* Wash State Univ, BS, 60; Johns Hopkins Univ, PhD(geol), 64. *Prof Exp:* NATO fel, Ger, Italy & Canary Islands, 64-65; GEOLOGIST, US GEOL SURV, 65- *Mem:* AAAS; Geol Soc Am; Am Geophys Union. *Res:* Petrology of volcanic rocks, especially from northwest United States; deformation studies of active volcanos, particularly Mount St Helens and other Cascade volcanos; physical volcanology. *Mailing Add:* US Geol Surv Dept Geol Sci AJ-20 Univ Wash Seattle WA 98195

SWANSON, DONALD CHARLES, b Canon City, Colo, Sept 22, 26; m 50; c 2. PETROLEUM GEOLOGY, SEDIMENTOLOGY. *Educ:* Colo State Univ, BS, 50; Univ Tulsa, BS, 55. *Prof Exp:* Geol & geophys tax engr, Carter Oil Co, Okla, 51-56, jr geologist, Kans, 56, geologist, Ark, 56-57 & Okla, 57-60; geologist, Humble Oil Co, 60-62, sr geologist, Tex, 62-63 & Humble Res Ctr, 63-64, staff geologist, Humble Oil Co, Okla, 64-67; sr res geologist, Esso Prod Res Co, 67, sr res specialist, 67-74; res assoc, Exxon Prod Res Co, 74-79; CONSULT, SWANSON GEOL SERV & CHMN, STRATAMODEL, INC, 79- *Honors & Awards:* Levorsen Award, Am Asn Petrol Geologists, 68 & 79. *Mem:* Fel Geol Soc Am; Am Asn Petrol Geologists; Explorers Club; Soc Petrol Engrs. *Res:* Clastic facies; determination of ancient sedimentary environments; paleogeography; methodology of environmental facies analyses; methodology of exploration; computer application to petroleum geology. *Mailing Add:* 510 Sandy Port Houston TX 77079-2417

SWANSON, DONALD G, b Los Angeles, Calif, June 11, 35; m 60; c 3. PLASMA PHYSICS. *Educ:* Northwest Christian Col, BTh, 58; Univ Ore, BS, 58; Calif Inst Technol, MS, 61, PhD(physics), 63. *Prof Exp:* Fel, Calif Inst Technol, 63-64; from asst prof to assoc prof elec eng, Univ Tex, Austin, 64-74; assoc prof elec eng, Univ Southern Calif, Los Angeles, 74-80; prof, 80-85, ALUMNI PROF PHYSICS, AUBURN UNIV, 85- *Concurrent Pos:* Consult, Advan Kinetics, Inc, Calif, 63-64, Princeton Plasma Physics Lab, 82-84, Los Alamos Nat Lab, 83-84 & McDonnell-Douglas Corp, St Louis, MO, 80- *Mem:* fel Am Phys Soc; Sigma Xi. *Res:* Compressional hydromagnetic waves; plasma-filled waveguide; ion cyclotron waves; mode conversion theory; torsatron. *Mailing Add:* Dept Physics Auburn Univ Auburn AL 36849

SWANSON, DONALD LEROY, b Montrose, SDak, Mar 24, 23; m 48; c 3. ANALYTICAL CHEMISTRY, PHYSICAL CHEMISTRY. *Educ:* SDak State Univ, BS, 47; Univ Wis, PhD(chem), 51. *Prof Exp:* Lab asst chem, Agr Exp Sta, SDak, 46-47; asst, Univ Wis, 47-51; res chemist, Am Cyanamid Co, 51-58, group leader, 58-61, sect mgr, 62-86; RETIRED. *Mem:* Am Chem Soc. *Res:* Physical and mechanical properties of polymers; polymerization kinetics; copolymerization; radiation chemistry; analysis. *Mailing Add:* 15 Dartmouth Rd Cos Cob CT 06807

SWANSON, ERIC RICE, b Frankfort, Mich, Nov 14, 46; m 75; c 2. VOLCANOLOGY, METALLIC ORE DEPOSITS. *Educ:* Western Mich Univ, BS, 68; Univ Tex, Austin, MA, 74, PhD(geol), 77. *Prof Exp:* Asst prof geol, Wayne State Univ, 76-79; ASSOC PROF GEOL, UNIV TEX, SAN ANTONIO, 79- *Mem:* Geol Soc Am; Nat Asn Geol Teachers; Am Geophys Union. *Res:* Volcanic stratigraphy of the Sierra Madre Occidental, Western Mexico; writer in metallic ore deposits; volcanology; geology of Mexico. *Mailing Add:* Dept Earth & Phys Sci Univ Tex San Antonio TX 78285

SWANSON, ERIC RICHMOND, b San Diego, Calif, May 4, 34; m 67; c 2. APPLIED PHYSICS, NAVIGATION. *Educ:* Pomona Col, BA, 56; Univ Calif, Los Angeles, MS, 58. *Prof Exp:* Teaching asst, Univ Calif, Los Angeles, 56-58; electronic engr res & develop, Electro Instruments, Inc, San Diego, 59-60; PHYSICIST, BR HEAD & CONSULT, NAVAL OCEAN SYSTS CTR, 60- *Concurrent Pos:* Mem, Comt Consult Int Radio Commun US Study Group 7, 75-; mem, Int Meritime Consult Orgn Working Group Differential Omeg, 77-79; consult, NATO, 76, India, Int Telecommun, 81. *Honors & Awards:* Burka Award, Inst Navig, 71. *Mem:* Insts Navig (US, Brit & Australia); sen mem Inst Elec & Electronics Engrs; Am Phys Soc; Wild Goose Asn. *Res:* Navigation systems and associated radio propagation problems; timing and time dissemination. *Mailing Add:* 3611 Warner St San Diego CA 92106

SWANSON, ERNEST ALLEN, JR, b Miami, Fla, Apr 9, 36; m 67. ANATOMY, HISTOLOGY. *Educ:* Emory Univ, BA, 58, PhD(anat), 64. *Prof Exp:* Instr anat, Emory Univ, 64-65; instr, Univ Va, 65-67; from asst prof to assoc prof, 67-81, PROF ANAT, SCH DENT, TEMPLE UNIV, 81- *Mem:* Am Asn Anatomists. *Res:* Changes in the dental pulp associated with cholesterol induced arteriosclerosis. *Mailing Add:* Dept Histol Temple Univ Sch Dent Philadelphia PA 19140

SWANSON, GEORGE D, b Eureka, Calif, Sept 19, 42. RESPIRATORY PHYSIOLOGY. *Educ:* Univ Chicago, SB, 37, MD, 38. *Prof Exp:* ASSOC PROF BIOMETRICS & ANESTHESIOL, SCH MED, UNIV COLO, 78- *Mem:* Am Physiol Soc; Inst Elec & Electronics Engrs; Sigma Xi. *Mailing Add:* 2042 Holly Ave Chico CA 95926

SWANSON, GUSTAV ADOLPH, b Mamre, Minn, Feb 13, 10; m 36; c 3. WILDLIFE ECOLOGY. *Educ:* Univ Minn, BS, 30, MS, 32, PhD(zool), 37. *Prof Exp:* Asst zool, Univ Minn, 30-34; biologist, State Dept Conserv, Minn, 35-36; asst prof game mgt, Univ Maine, 36-37; asst prof econ zool, Univ Minn, 37-41; assoc regional inspector, US Fish & Wildlife Serv, 41-42, chief sect coop wildlife res units, 44-46, chief div wildlife res, 46-48; assoc prof econ zool, Univ Minn, 42-44; prof conserv & head dept, Cornell Univ, 48-66; prof fishery & wildlife biol & head dept, 66-75, EMER PROF WILDLIFE BIOL, COLO STATE UNIV, 75- *Concurrent Pos:* Ed, J Wildlife Mgt, 49-53; Am Scandinavian Found fel, Denmark, 54-55, Fulbright fel, 61-62; consult waterfowl res, Nature Conserv, Eng, Scotland & Northern Ireland, 55 & 60; dir, Cornell Biol Field Sta, 55-66, exec dir lab ornith, 58-61; fel, Rochester Mus, 56; consult, State Joint Legis Comt Rev Conserv Law, NY, 56-65 & natural resources, 56-66; Fulbright fel, NSW, Australia, 68. *Honors & Awards:* Aldo Leopold Mem Medalist, 73. *Mem:* Fel AAAS; hon mem Wildlife Soc (vpres, 45, pres, 54); Am Soc Mammalogy; Wilson Ornith Soc (treas, 38-42); Am Inst Biol Sci; Am Ornith Union; Sigma Xi. *Res:* Wildlife management; conservation of natural resources; ornithology. *Mailing Add:* Park Ctr Apt 35 1020 E 17th St South Minneapolis MN 55404

SWANSON, HAROLD DUEKER, b Wichita, Kans, Mar 5, 30; m 55; c 3. CELL BIOLOGY. *Educ:* Friends Univ, BA, 53; Univ Kans, MA, 55; Univ Tenn, PhD(zool physiol), 60. *Prof Exp:* Asst zool, Univ Kans, 53-55 & Univ Tenn, 56-58; from asst prof to assoc prof, 60-74, PROF BIOL, DRAKE UNIV, 74- *Mem:* Sigma Xi; AAAS. *Res:* Nucleocytoplasmic interaction; subcellular component isolation; science and religion; cell physiology. *Mailing Add:* Biol Dept Drake Univ Des Moines IA 50311

SWANSON, J ROBERT, b Ft Collins, Colo, June 24, 39; m 62; c 1. BIOCHEMISTRY. *Educ:* Colo State Univ, BS, 61; Wash State Univ, PhD(biochem), 65; Am Bd Clin Chem, dipl. *Prof Exp:* NIH res fel biochem, Duke Univ, 65-67; clin chem training fel, Pepper Lab, Hosp Univ Pa, 67-69; asst prof, 69-74, ASSOC PROF CLIN PATH, MED SCH, UNIV ORE, PORTLAND, 74- *Mem:* Am Asn Clin Chem; Am Chem Soc. *Res:* Clinical methods for pulmonary surfactant measurement. *Mailing Add:* Dept Clin Path L-471 Ore Health Sci Univ Portland OR 97201-3098

SWANSON, JACK LEE, b Aurora, Nebr, Oct 22, 34; m 56; c 3. PHYSICAL CHEMISTRY, BIOCHEMISTRY. *Educ:* Kearney State Col, BS, 56; Univ Nebr, MS, 59, PhD(chem), 67. *Prof Exp:* Prof chem, Kearney State Col, 58-71; dean, Sch Sci & Technol, 71-76, dean, Sch Prof Studies, 76-79, dean admin serv, 80-87, PROF CHEM CHADRON STATE COL, 71- *Concurrent Pos:* NSF fel. *Mem:* AAAS; Am Chem Soc; Sigma Xi. *Res:* Infrared and ultraviolet spectroscopy; magneto-optical rotary dispersion; circular dichroism spectroscopy; medicinal chemistry. *Mailing Add:* Dept Chem Chadron State Col Chadron NE 69337

SWANSON, JAMES A, b Aurora, Nebr, Oct 25, 35; m 57; c 3. PHYSICAL CHEMISTRY. *Educ:* Univ Nebr Kearney, BA, 57; Univ Nebr-Lincoln, MS, 59, PhD(chem), 62. *Prof Exp:* Part-time lab asst, Univ Nebr, 57-62; PROF CHEM, UNIV NEBR KEARNEY, 62- *Mem:* Am Chem Soc; Sigma Xi. *Res:* Solution thermochemistry; thermodynamics. *Mailing Add:* Dept Chem Univ Nebr Kearney Kearney NE 68849-5320

SWANSON, JOHN L, b Hastings, Nebr, Aug 16, 36. MICROBIAL STRUCTURE RESEARCH. *Educ:* Univ Nebr, BS, 59, MS, 61, MD, 62. *Prof Exp:* Intern, Univ Hosp, Omaha, Nebr, 62-63; residency anat path, Peter Bent Brigham Hosp, Boston, 63-66; mem staff, Armed Forces Inst Path, 66-67; mem staff, Walter Reed Army Inst Res, Washington, DC, 67-68; asst prof, Dept Microbiol, Col Physicians & Surgeons, Columbia Univ, 68-69; from asst prof to assoc prof, Dept Microbiol, Mt Sinai Sch Med, NY, 69-72; from assoc prof to prof, Dept Path & Dept Microbiol, Univ Utah Col Med, 72-79; CHIEF, NAT INST ALLERGY & INFECTIOUS DIS, LAB MICROBIAL STRUCT & FUNCTION, ROCKY MOUNTAIN LABS, NIH, 79- *Concurrent Pos:* Ad hoc reviewer res grants & fel, Joshiah Macy Found, Univ Tenn, Vet Admin, Can Med Res Coun, NSF & NIH; extramural adv, Nat Inst Allergy & Infectious Dis sponsored Sexually Transmitted Dis Coop Res Ctr, Univ Tex Health Sci Ctr; mem, Bact & Mycol Study Sect, NIH, 78-79. *Mem:* Sigma Xi. *Res:* Virulence and pathogenic factors of Neisseria gonorrhoeae. *Mailing Add:* Nat Inst Allergy & Infectious Dis Lab Microbial Struct & Function Rocky Mountain Labs 903 S Fourth St Hamilton MT 59840

SWANSON, JOHN WILLIAM, b Sioux City, Iowa, Oct 12, 17; m 41; c 3. PHYSICAL CHEMISTRY. *Educ:* Morningside Col, BA, 40. *Hon Degrees:* DSc, Morningside Col, 72; MSc, Lawrence Univ, 82. *Prof Exp:* Asst chem, Iowa State Col, 40-41; from tech asst to tech assoc, 41-45, from res asst to res assoc, 46-55, group leader surface & colloid chem, 53-55, group leader phys chem, 56-61, sr res assoc, 56-69, chmn phys chem dept, 62-69, dir surface & colloid sci ctr, 81, EMER PROF, INST PAPER CHEM, 81- *Concurrent Pos:* Lectr, Lawrence Univ, 45-46; consult to numerous paper co, 50- *Honors & Awards:* Res & Develop Div Award, Tech Asn Pulp & Paper Indust, 74, Harris O Ware Prize, 83. *Mem:* AAAS; Am Chem Soc; fel Tech Asn Pulp & Paper Indust. *Res:* Surface and colloid chemistry of papermaking; polymer sorption at interfaces; surface area and bonding of cellulose fibers; paper sizing, coating; coagulation and retention of resins in aqueous systems; pollution abatement. *Mailing Add:* 236 Camino Del Vate 'NBU 2412 Green Valley AZ 85614-3106

SWANSON, LAWRENCE RAY, b Omaha, Nebr, Nov 4, 36; m 62; c 2. SPACE SURVEILLANCE ALGORITHM DEVELOPMENT. *Educ:* Iowa State Univ, BS, 59; Fuller Theol Sem, BD, 63; Calif State Univ, Los Angeles, MS, 66; Univ Calif, Irvine, PhD(physics), 70. *Prof Exp:* Asst prof physics, Pasadena Col, 70-73, vis prof physics, Greenville Col, 73-74; assoc prof physics & math, Azusa Pac Col, 74-76; assoc prof physics & math, Sterling Col, 76-80; sr electronic engr, TRW, Inc, 80-86; systs engr, Contel Fed Systs, 86-87; PRIN RES SCIENTIST, TEXTRON DEFENSE SYSTS, 87- *Res:* Develop algorithms for incorporation into work stations used for space surveillance analysis: algorithms for photometric infrared and radar signature analysis. *Mailing Add:* 1305 N 31 Colorado Springs CO 80904

SWANSON, LEONARD GEORGE, b Corvallis, Ore, Sept 10, 40. MATHEMATICS. *Educ:* Portland State Univ, BS, 62; Univ Wash, MA, 65; Ore State Univ, PhD(math), 70. *Prof Exp:* From instr to assoc prof, 64-81, PROF MATH, PORTLAND STATE UNIV, 81- *Concurrent Pos:* Vis assoc prof, Dept Math, Mont State Univ, 77-78; vis prof, Dept Math, Ore State Univ, 84-85. *Mem:* Am Math Soc; Math Asn Am; Inst Math Statist; AAAS. *Res:* Fourier series and their application; number theory. *Mailing Add:* Dept Math Portland State Univ PO Box 751 Portland OR 97207

SWANSON, LLOYD VERNON, b Isanti, Minn, Oct 16, 38; m 66; c 2. REPRODUCTIVE ENDOCRINOLOGY. *Educ:* Univ Minn, St Paul, BS, 60, MS, 67; Mich State Univ, PhD(physiol), 70. *Prof Exp:* PROF DAIRY PHYSIOL, ORE STATE UNIV, 71- *Mem:* Am Dairy Sci Asn; Am Soc Animal Sci; Soc Study Reproduction; Endocrine Soc; Sigma Xi. *Res:* Reproductive physiology of mammalian species, both male and female, with special interest in the endocrine control of ovulation and of spermatogenesis. *Mailing Add:* Dept Animal Sci Ore State Univ Corvallis OR 97331

SWANSON, LYNN ALLEN, b Minneapolis, Minn, July 28, 42; m 67; c 2. ANALYTICAL CHEMISTRY. *Educ:* Univ Minn, Minneapolis, BChem, 64; Univ Iowa, MS, 68, PhD(anal chem), 70. *Prof Exp:* Res chemist anal chem, Commercial Solvents Corp, 69-77; asst mgr, Int Minerals & Chem Corp, 77-80; res scientist anal serv, Res & Develop, Pitman-Moore Inc, 80-86, mgr anal res, 86-87, mgr res serv, 87-88, DIR ANAL RES, RES & DEVELOP, PITMAN-MOORE INC, 88- *Mem:* Am Chem Soc; Sigma Xi. *Res:* Trace analysis of pharmaceuticals, drugs and other additives in animal tissues and body fluids; general chromatography; spectrophotometry. *Mailing Add:* Pitman-Moore Inc Res & Develop PO Box 207 Terre Haute IN 47808

SWANSON, LYNWOOD WALTER, b Turlock, Calif, Oct 7, 34; m 55; c 4. PHYSICAL CHEMISTRY. *Educ:* Univ of Pac, BSc, 56; Univ Calif, PhD(chem), 60. *Prof Exp:* Asst chemist, Univ Calif, Davis, 56-59; res assoc, Inst Study Metals, Univ Chicago, 59-61; sr scientist, Linfield Res Inst, 61-63; dir basic res, Field Emission Corp, 63-69; prof chem & dean fac, Linfield Col, 69-73; prof appl physics, Ore Grad Ctr, 73-86; PRES, FEI CO, 86- *Mem:* Fel Am Phys Soc. *Res:* Photochemistry; surface adsorption; field electron and ion microscopy; electron physics. *Mailing Add:* FEI Co 19500 NW Gibbs Dr Beaverton OR 97006

SWANSON, MAX LYNN, b Hancock, Mich, Aug 5, 31, Can citizen; m 59; c 4. EXPERIMENTAL SOLID STATE PHYSICS. *Educ:* Univ BC, BA, 53, MSc, 54, PhD(metal physics), 58. *Prof Exp:* Res metallurgist, Metals Res Lab, Carnegie Inst Technol, 58-60; res officer metal physics, Chalk River Nuclear Labs, Atomic Energy Can Ltd, 60-86; RES PROF, UNIV NC, CHAPEL HILL, 86- *Concurrent Pos:* Guest scientist, Max Planck Inst Metal Res, Stuttgart, 65-66; vis prof, Univ Utah, 71-72; guest scientist, Hahn-Meither Inst, Berlin, 77-78. *Mem:* Am Phys Soc; Mat Res Soc. *Res:* Defect solid state physics: irradiation damage in metals and semiconductors, ion channeling; defect trapping configurations; ion beam modification of materials; perturbed angular correlation. *Mailing Add:* Dept Phys & Astron Phillips Hall Univ N Carolina Chapel Hill NC 27955-3255

SWANSON, PAUL N, b San Mateo, Calif, June 29, 36; m 59; c 3. RADIO ASTRONOMY, MILLIMETER WAVE RADIOMETRY. *Educ:* Calif State Polytech Col, BS, 62; Pa State Univ, PhD(physics), 68. *Prof Exp:* Asst prof radio astron, Pa State Univ, University Park, 69-75; mem staff, 75-85, mgr, microwave observational, 85-89, MGR, ASTROPHYS & SPACE PHYSICS PROG, JET PROPULSION LAB, CALIF INST TECHNOL, 89- *Mem:* Am Astron Soc; AAAS; Sigma Xi; Am Inst Aeronaut & Astronaut. *Res:* Millimeter wavelength radio astronomy; radiometer development; space science. *Mailing Add:* Jet Propulsion Lab 180-703 4800 Oak Grove Dr Pasadena CA 91109-8099

SWANSON, PHILLIP D, b Seattle, Wash, Oct 1, 32; m 57; c 5. NEUROLOGY, BIOCHEMISTRY. *Educ:* Yale Univ, BS, 54; Johns Hopkins Univ, MD, 58; Univ London, PhD(biochem), 64. *Prof Exp:* Fel neurol med, Sch Med, Johns Hopkins Univ, 59-62; Nat Inst Neurol Dis & Stroke spec fel, Univ London, 62-64; from asst prof to assoc prof, 64-73, PROF NEUROL, SCH MED, UNIV WASH, 73-, HEAD DIV, 67- *Mem:* Asn Univ Prof Neurol (pres, 75-76); Am Neurol Asn; Am Soc Clin Invest; Brit Biochem Soc. *Res:* Neurochemistry; cation transport and energy utilization in cerebral tissues; enzymes of importance in cation transport. *Mailing Add:* Div of Neurol Univ of Wash Sch Med Seattle WA 98195

SWANSON, ROBERT ALLAN, b Chicago, Ill, Dec 16, 28; m 57. ELEMENTARY PARTICLE PHYSICS. *Educ:* Ill Inst Technol, BS, 51; Univ Chicago, MS, 53, PhD(physics), 58. *Prof Exp:* Res assoc physics, Univ Chicago, 58-59; asst prof, Princeton Univ, 59-60; from asst prof to assoc prof, 60-70, PROF PHYSICS, UNIV CALIF, SAN DIEGO, 70- *Concurrent Pos:* Vis assoc prof, Univ Chicago, 68-69; NSF fel, Univ Calif, 72- *Mem:* Am Phys Soc; Am Asn Physics Teachers. *Res:* Muonic atoms; experimental kaon physics. *Mailing Add:* Dept Physics Univ Calif at San Diego Box 109 La Jolla CA 92093

SWANSON, ROBERT E, b Duluth, Minn, Dec 19, 24; m 47; c 2. MEDICAL PHYSIOLOGY. *Educ:* Univ Minn, BA, 49, PhD(physiol), 53. *Prof Exp:* Asst physiol, Univ Minn, 50-52, instr, 52-55; asst physiologist, Brookhaven Nat Lab, 55-58; asst prof physiol, Univ Minn, 58-61; assoc prof, 61-73, PROF PHYSIOL, MED SCH, UNIV ORE HEALTH SCI CTR, 73- *Res:* Renal, water and electrolyte balance. *Mailing Add:* Dept Physiol Univ Ore Health Sci Univ 3181 S W S Jackson Pk Rd Portland OR 97201

SWANSON, ROBERT HAROLD, b Los Angeles, Calif, Feb 15, 33; m 55; c 2. FOREST HYDROLOGY, FOREST PHYSIOLOGY. *Educ:* Colo State Univ, BSc, 59, MSc, 66, Univ Alta, PhD, 83. *Prof Exp:* Res forester hydrol, Rocky Mountain Forest & Range Exp Sta, US Forest Serv, 59-68; PROJ LEADER FOREST HYDROL, NORTHERN FOREST RES CTR, 68- *Concurrent Pos:* Res coordr, Alta Watershed Res Prog, 68-; res fel, Ministry of Works, NZ, 74-75. *Mem:* Can Inst Foresters; Sigma Xi. *Res:* Physiological bases for tree improvement; plant-water relations; forest arrangements-streamflow interractions; watershed management simulation and evaluation techniques. *Mailing Add:* Dept Forest Sci Univ Alberta Edmonton AB T6G 2E2 Can

SWANSON, ROBERT JAMES, b St Petersburg, Fla, Nov 13, 45; m 67; c 2. ENDOCRINOLOGY. *Educ:* Wheaton Col, Ill, BS, 67; Fla State Univ, MS, 71, PhD(biol), 76. *Prof Exp:* Teacher gen sci, Madison High Sch, Fla, 67-68; instr anat & kinesiology, Fla State Univ, 69-70; asst prof anat, physiol & endocrinol, Old Dominion Univ, 75-77, asst prof biol, 77-80. *Mem:* AAAS. *Res:* Female reproductive physiology, especially factors involved in ovulation, such as hormones, smooth muscle activity, nerve involvement and blood flow. *Mailing Add:* 1014 Jamestown Cres Norfolk VA 23508

SWANSON, ROBERT LAWRENCE, b Baltimore, Md, Oct 11, 38; m 63; c 2. PHYSICAL OCEANOGRAPHY, CIVIL ENGINEERING. *Educ:* Lehigh Univ, BS, 60; Ore State Univ, MS, 65, PhD, 71. *Prof Exp:* Comn officer, US Coast & Geodetic Surv & Nat Oceanic & Atmospheric Admin, 60-87, commanding officer, US Coast & Geodetic Surv Ship Marmer Circulatory Estuarine Surv, 66-67, chief, Oceanog Div, Nat Ocean Surv, 69-72, proj mgr, NY Bight Proj, Marine Ecosysts Anal, Environ Res Labs, 72-78, dir, Off Marine Pollution Assessment, 78-83, commanding officer, Nat Oceanic & Atmospheric Admin & ship researcher, Global Climate Res, 84-86, exec dir, Off Oceanog & Atmospheric Res, 86-87; DIR, WASTE MGT INST, STATE UNIV NY, STONY BROOK, 87- *Concurrent Pos:* Prof asst, Col Gen Studies, George Washington Univ, 70-73; adj prof, State Univ NY, Stony Brook, 78-; chmn, Ocean Pollution Comt, Marine Technol Soc, 82-; sr exec fel, John F Kennedy Sch Govt, Harvard, 83; lectr & mem, US Deleg Marine Pollution to the People, Repub China, 83; chief scientist, US-French Bilateral in Marine Sci, 86-87; hydrographer for inshore & offshore waters, 87-; mem, Suffolk Co Coun Environ Qual, 88-; mem, Tech Adv Comt, NY State Ctr Hazardous Waste Mgt, 88- *Honors & Awards:* Silver Medal, US Dept Com, 73; Prog Admin & Mgt Award, Nat Oceanic & Atmospheric Admin, 75, Unit Citation, 81, Corps Dir Ribbon Award, 87, Spec Achievement Award, 87. *Mem:* Am Soc Civil Engrs; Am Geophys Union; AAAS; Am Soc Photogramm; NY Acad Sci; Sigma Xi. *Res:* Developing interrelationships and understanding between component parts of the coastal marine ecosystem; studying the impact of ocean dumping on marine ecosystem; specific interests in tides, tidal currents, tidal datums, marine boundaries; solid waste management. *Mailing Add:* 30 Erland Rd Stony Brook NY 11790

SWANSON, ROBERT NELS, b Ashland, Wis, Feb 4, 32; m 57; c 4. MICROMETEOROLOGY. *Educ:* Wis State Col, River Falls, BS, 53; Univ Mich, MS, 58. *Prof Exp:* Meteorologist, White Sands Missile Range, 58-61; staff scientist, GCA Corp, Utah, 61-72; sr meteorologist, 72-84, DIR METEOROL SERVS, PAC GAS & ELEC CO, SAN FRANCISCO, 84- *Concurrent Pos:* Chmn, Comt on Atmospheric Measurements, Am Meteorol Soc, 83- 85; mem comt, Pvt Sector Meteorol, Am Meteorol Soc, 91-94; cert consult meteorol, Am Meteorol Soc. *Mem:* Am Meteorol Soc; Royal Meteorol Soc; Air Pollution Control Asn; Nat Coun Indust Meteorol. *Res:* Boundary layer measurements; alternate energy; turbulence and diffusion; technical management. *Mailing Add:* 1216 Babel Lane Concord CA 94518

SWANSON, SAMUEL EDWARD, b Woodland, Calif, Aug 1, 46; m 79. GEOCHEMISTRY, PETROLOGY. *Educ:* Univ Calif, Davis, BS, 68, MS, 70; Stanford Univ, PhD(geol), 74. *Prof Exp:* Field asst geol, US Geol Surv, 67; res asst geol, Univ Calif, Davis, 70; asst prof earth sci, Univ NC, Charlotte, 74-76; asst prof, Applachian State Univ, Boone, 76-79; asst prof, 79-84, ASSOC PROF GEOL, UNIV ALASKA, FAIRBANKS, 84- *Mem:* Can Mineral Soc; Am Geophys Union; Mineral Soc Am; Soc Environ Geochem & Health; Sigma Xi; Geol Soc Am. *Res:* Application of geochemical techniques to the study of igneous and metamorphic rocks. *Mailing Add:* Geol & Geophys Dept Univ Alaska Fairbanks AK 99775

SWANSON, VERN BERNARD, animal breeding, for more information see previous edition

SWANSON, VERNON E, b Lincoln, Nebr, Dec 14, 22; m 45; c 4. GEOLOGY, GEOCHEMISTRY. *Educ:* Agustana Col, BA, 43; Columbia Univ, MA, 50. *Prof Exp:* Ranger Naturalist, Nat Park Serv, 46; assoc prof, Upsala Col, 46-50; res geologist, US Geol Survey, 48-78; sci ed, Geol Soc Am, 78-81; consult geologist, 81-86; RETIRED. *Concurrent Pos:* fel, Geol Soc Am; Ranger Naturalist, Nat Park Serv, 46. *Mem:* Am Asn Petrol Geologist; AAAS; Asn Earth Sci Ed. *Res:* Author of over 100 scientific articles. *Mailing Add:* 14595 E Hampden Ave No 151 Aurora CO 80014-5029

SWANSON, VIRGINIA LEE, b Sioux City, Iowa, June 15, 22; m 67. PATHOLOGY. *Educ:* Univ Southern Calif, BA, 47; Yale Univ, MD, 52; Am Bd Path, dipl, 58. *Prof Exp:* Intern path, Sch Med, Yale Univ, 52-53; USPHS res fel, Path-Anat Inst, Univ Copenhagen, 53-55; instr path, Sch Med, Yale Univ, 55-59; hosp pathologist, US Army Med Command, Tokyo, Japan, 59-60; hosp pathologist, Australian Pub Health Serv, Port Moresby Gen Hosp, Territory Papua & New Guinea, 60-61; res pathologist, US Army Med Command, Tokyo, 61-62; res pathologist & chief path div, US Army Trop Res Med Lab, San Juan, PR, 62-65; res pathologist, Armed Forces Inst Path & Walter Reed Army Inst Res, Washington, DC, 65-66; assoc prof, 66-71, PROF PATH, SCH MED, UNIV SOUTHERN CALIF, 71- *Concurrent Pos:* Asst resident, Grace-New Haven Community Hosp, 55-56, chief resident, 56-57, asst pathologist, 57-59; assoc pathologist, Children's Hosp, Los Angeles, 66-71 & 73-; prof, Sch Med, Univ Calif, San Diego, 71-73; chief lab serv, Vet Admin Hosp, San Diego, 71-73. *Res:* Pathology of the gastrointestinal tract; malabsorption; malnutrition; immunopathology. *Mailing Add:* Children's Hosp 4650 Sunset Blvd Los Angeles CA 90027

SWANSON, WILLIAM PAUL, b St Paul, Minn, Dec 20, 31; m 59; c 1. PARTICLE PHYSICS, RADIATION PHYSICS. *Educ:* Univ Minn, Minneapolis, BA, 53; Univ Calif, Berkeley, MA, 55, PhD(physics), 60; Am Bd Health Physics, cert, 77. *Prof Exp:* Res assoc particle physics, Univ Ill, Urbana, 60-61, asst prof physics, 61-64; Stiftung Volkswagenwerk fel particle physics, Deutsches Elektronen-Synchrotron, Hamburg, Ger, 65-66; guest prof physics, Univ Hamburg, Ger, 67; vis scientist particle physics, Europ Orgn Nuclear Res, Geneva, Switz, 67-68 & health physics, 82-83; staff physicist particle & radiation physics, Stanford Linear Accelerator Ctr, Stanford Univ, 68-84; STAFF SCIENTIST RADIATION PHYSICS, LAWRENCE BERKELEY LAB, UNIV CALIF, 84- *Concurrent Pos:* Consult, Lawrence Radiation Lab, Univ Calif, 63-64, SHM Nuclear Corp, 74-75, Int Atomic Energy Agency, Vienna, 77, EMI Ther Systs Inc, 75-76, Int Comn Radiation Units & Measurements, 76, Varian Assocs, 80-81, Nat Coun Radiation Protection & Measurements Comt, 81-85, Synchrotron Radiation Ctr, Univ Wis, 85, US Dept Energy adv panel accelerator radiation safety, 85, Continuous Electron Beam Accelerator Facil, 86, High Energy Physics Lab, Stanford Univ, 85, Vanderbilt Univ, 87. *Honors & Awards:* Farrington Daniels Award, Am Asn Physicists Med, 81. *Mem:* Am Phys Soc; Am Asn Physicists Med; Health Physics Soc; Sigma Xi. *Res:* Properties and interactions of elementary particles; physical problems related to radiation safety at high-energy accelerators. *Mailing Add:* Lawrence Berkeley Lab Bldg 75B-111 Univ Calif 1 Cyclotron Rd Berkeley CA 94720

SWANSTON, DOUGLAS NEIL, b Pensacola, Fla, June 8, 38; m 59; c 2. ENGINEERING GEOLOGY, GEOMORPHOLOGY. *Educ:* Univ Mich, Ann Arbor, BS, 60; Bowling Green State Univ, MA, 62; Mich State Univ, PhD(geol), 67. *Prof Exp:* Res geologist, Inst Northern Forestry, Juneau, Alaska, 64-71, RES GEOLOGIST, FOREST SCI LAB, US FOREST SERV, 71-; ASST PROF FOREST ENG, ORE STATE UNIV, 71- *Concurrent Pos:* Mem bd dirs, Found Glacier & Environ Res, 71-; consult geologist, Daniel, Mann, Johnson & Mendenhall, Archit & Engrs, 71- *Res:* Glacial geology; glaciology. *Mailing Add:* Dept Forest Eng Ore State Univ Corvallis OR 97331

SWANTON, WALTER F(REDERICK), chemical engineering, econometrics, for more information see previous edition

SWARBRICK, JAMES, b London, Eng, May 8, 34; m 60. PHARMACEUTICS, SURFACE CHEMISTRY. *Educ:* Univ London, BPharm, 60, PhD(med), 64, DSc(phys chem), 72; FRIC, 72; FRSC (cchem), 75. *Prof Exp:* From asst prof to assoc prof indust pharm, Purdue Univ, 64-66; prof & asst dean pharm, Univ Conn, 66-72; dir prod develop, Sterling-Winthrop Res Inst, 72-75; prof pharmaceut, Univ Sydney, 75-76; dean, Sch Pharm, Univ London, 76-78; prof pharm, Univ Southern Calif, 78-81; PROF PHARM & CHMN, DEPT PHARMACEUT, UNIV NC, CHAPEL HILL, 81- *Concurrent Pos:* Vis scientist, Astra Labs, Sweden, 71; chmn, USP-NF Panel Dissolution & Disintegration Testing, 72-75; examr, Sci Univ Malaysia, 77-78; consult, Australian Dept Health, 75-76, Orgn Am States, 78 & Pan-Am Health Orgn, 79 & 81; examr, Univ Singapore, 80-81; indust consult, 65-; consult, Al Fateh Univ, Libya, 81; vis prof, Brighton Polytechnic, UK, 88. *Mem:* Am Pharmaceut Asn; fel Acad Pharmaceut Sci; Am Asn Col Pharm; fel Pharmaceut Soc Great Brit; fel Royal Soc Chem; fel Am Asn Pharmaceut Scientists. *Res:* Dosage form design and drug delivery; percutaneous absorption; formulation of topical products; preformulation studies; interfacial phenomena of pharmaceutical and biological significance. *Mailing Add:* Sch Pharm Univ NC Chapel Hill NC 27599-7360

SWARD, EDWARD LAWRENCE, JR, b Chicago, Ill, Aug 21, 33; m 57; c 2. PHYSICAL CHEMISTRY, SOFTWARE SYSTEMS. *Educ:* Augustana Col, Ill, BA, 55; Univ Buffalo, PhD(chem), 61. *Prof Exp:* Res chemist, Mylar Res & Develop Lab, E I du Pont de Nemours & Co, Inc, 60-64, Du Pont de Nemours, Luxembourg, SA, 64-67 & Del, 67-69; mgr mkt develop, Celanese Res Co, 69-72; mgr long range planning, El Paso Hydrocarbons Co, 72-81, dir planning, 81-83, dir, planning & bus anal, 84-87, pres corp invest, 87; PRES, SWARD GROUP, 87- *Concurrent Pos:* Adj prof, Univ Tex, Permian Basin, 74-86. *Mem:* Commercial Develop Asn; Am Chem Soc. *Res:* Physical chemistry of polymers; kinetics; process development; financial and business planning and analysis; market research; financial planning. *Mailing Add:* PO Box 671378 Dallas TX 75367-1378

SWARDSON, MARY ANNE, b College Park, Ga, Sept 10, 28; m 49; c 3. GENERAL TOPOLOGY, SET THEORY. *Educ:* Tulane Univ, BA, 49; Ohio Univ, MS, 69, PhD(math), 81. *Prof Exp:* ASST PROF MATH, OHIO UNIV, 81- *Mem:* Am Math Soc; Math Asn Am. *Res:* Character of closed sets; generalizations of F-spaces; topological characterizations of set-theoretical axioms; generalizations of psuedocompactness; insertion, approximation and extension of real-valued functions. *Mailing Add:* Math Ohio Univ Athens OH 45701

SWARIN, STEPHEN JOHN, b Plainfield, NJ, July 24, 45; m 69; c 2. CHROMATOGRAPHY, ENVIRONMENTAL ANALYSIS. *Educ:* Lafayette Col, AB, 67; Univ Mass, MS, 69, PhD(anal chem), 72. *Prof Exp:* Assoc res scientist, 72-76, sr res scientist, 75-79, staff res scientist, 79-81, sr staff scientist, 81-85, asst dept head, 85-87, PRIN RES SCIENTIST, GEN MOTORS RES LABS, 87- *Mem:* Am Chem Soc; Am Soc Testing & Mat; Sigma Xi; Asn Anal Chemists. *Res:* Polymer analysis; polymer additives analysis; liquid chromatography; derivatization for detectability; environmental analysis; near-infrared spectroscopy; instrumentation. *Mailing Add:* Anal Chem Dept Gen Motors Res Labs Warren MI 48090

SWARINGEN, ROY ARCHIBALD, JR, b Winston-Salem, NC, Feb 1, 42; m 69. ORGANIC CHEMISTRY. *Educ:* Univ NC, Chapel Hill, AB, 64; Univ Ill, Urbana, MS, 66, PhD(org chem), 69. *Prof Exp:* Res chemist org chem, R J Reynolds Tobacco Co, 69-70; sr develop chemist, 70-74, sect head develop res, 74-81, DEPT HEAD, CHEM DEVELOP LABS, BURROUGHS WELLCOME CO, 81- *Mem:* Am Chem Soc; Royal Soc Chem; Sigma Xi. *Res:* Development research in pharmaceutical chemistry; synthetic organic chemistry; heterocyclic compounds. *Mailing Add:* 824 Sandlewood Dr Durham NC 27712

SWARM, H(OWARD) MYRON, electrical engineering; deceased, see previous edition for last biography

SWARM, RICHARD L(EE), b St Louis, Mo, June 9, 27; m 50; c 2. PATHOLOGY, CARCINOGENESIS. *Educ:* Wash Univ, BA, 49, BS & MD, 50; Am Bd Path, dipl. *Prof Exp:* Intern, Barnes Hosp, St Louis, Mo, 50-51; instr & resident path, Washington Univ & Barnes Hosp, 51-54; pathologist, USPHS Med Ctr, 54-55 & Nat Cancer Inst, 55-65; assoc prof path, Col Med, Univ Cincinnati, 65-68; dir, Dept Exp Path & Toxicol, Res Div, Hoffmann-LaRoche, Inc, 68-82; RETIRED. *Concurrent Pos:* Clin assoc prof path, Columbia Univ, 70-90; grantee, Am Cancer Soc & USPHS; consult pathologist, 82- *Mem:* Am Soc Toxicol Pathologists; Am Asn Cancer Res; Am Asn Pathologists; fel Am Soc Clin Path; Col Am Pathologists; Am Med Asn; Soc Toxicol; Teratol Soc. *Res:* Histopathology and toxicology in man and laboratory animals; morphology of neoplasms and carcinogenesis; radiation injury; transplantation of tissues; chondrosacomas; ultrastructure of neoplastic cells. *Mailing Add:* PO Box 808 Ridgewood NJ 07451-0808

SWART, WILLIAM LEE, b Brethren, Mich, July 13, 30; m 62; c 2. MATHEMATICS. *Educ:* Cent Mich Univ, BS, 58, MA, 62; Univ Mich, Ann Arbor, EdD(math educ), 69. *Prof Exp:* Teacher, Mesick Consol Schs, Mich, 58-61 & Livonia Pub Schs, 61-63; instr math, Eastern Mich Univ, 63-65; consult math educ, Genesee Intermediate Sch Dist, Mich, 65-67; from asst prof to prof math, Cent Mich Univ, 67-86; FOUNDER, TRICON MATH, INC, 86- *Res:* Learning of elementary mathematics; action research in public schools. *Mailing Add:* Dept Math Cent Mich Univ Mt Pleasant MI 48859

SWART, WILLIAM W, b Hareu, Holland, July 21, 44; US citizen; m 90. INDUSTRIAL & MANUFACTURING ENGINEERING. *Educ:* Clemson Univ, BS, 65; Ga Tech, MS, 68, PhD(indust eng), 70. *Prof Exp:* Assoc prof mgt sci, Univ Miami, 73-78; asst dean bus, Calif State Univ Northridge, 78-79; vpres MIS, Burger King Corp, 79-85; CHAIR & PROF INDUST ENG, UNIV CENT FLA, 85- *Mem:* Opers Res Soc Am; Inst Indust Engrs; Am Soc Eng Educ. *Res:* Energy efficient, affordable, industrialized housing; space shuttle processing. *Mailing Add:* Dept Indust Eng Univ Cent Fla Orlando FL 32816

SWARTS, ELWYN LOWELL, b Hornell, NY, Feb 26, 29; m 54; c 3. PHYSICAL CHEMISTRY. *Educ:* Hamilton Col, NY, AB, 49; Brown Univ, PhD, 54. *Prof Exp:* Mem fac, Alfred Univ, 53-56, res chemist, Knolls Atomic Lab, Gen Elec Co, NY, 56-57, res chemist, Glass Technol Lab, Ohio, 57-59; STAFF SCIENTIST, PPG INDUSTS, INC, 59- *Concurrent Pos:* Mem subcomt, Int Comn Glass. *Mem:* Am Chem Soc; Am Ceramic Soc; Soc Glass Technol. *Res:* Properties of glass; melting reactions; surface analysis of glass. *Mailing Add:* Glass Res & Develop Ctr PPG Industs Inc Box 11472 Pittsburgh PA 15238

SWARTZ, BLAIR KINCH, b Detroit, Mich, Nov 5, 32; m 55; c 1. NUMERICAL ANALYSIS. *Educ:* Antioch Col, BS, 55; Mass Inst Technol, MS, 58; NY Univ, PhD(math), 70. *Prof Exp:* Asst biol, Sch Med & Dent, Univ Rochester, 51-52; asst chem, Detroit Edison Co, 52-53; asst physics, Antioch Col, 53-54; high sch teacher, 54-55; asst math, Mass Inst Technol, 55-58; res asst, 58, group leader, 68-74, assoc group leader, 78-80, mem staff, 59-82, FEL, LOS ALAMOS NAT LAB, 83- *Concurrent Pos:* Asst, Am Optical Co, 55-56; lectr, State Univ NY Teachers Col New Paltz, 58 & Univ NMex, 59; ed, SIAM J Numerical Anal, 74-75. *Mem:* Soc Indust & Appl Math. *Res:* Approximation theory; differential equations. *Mailing Add:* 172 Paseo Penasco Los Alamos NM 87544

SWARTZ, CHARLES DANA, b Baltimore, Md, July 25, 15; m 49; c 3. PHYSICS. *Educ:* Johns Hopkins Univ, AB, 38, PhD(physics), 43. *Prof Exp:* Physicist, Manhattan Proj, SAM Labs, Columbia Univ, 42-46; assoc, Lab Nuclear Studies, Cornell Univ, 46-48; from instr to asst prof physics, Johns Hopkins Univ, 48-56; assoc prof, 56-62, prof, 62-79, EMER PROF PHYSICS, UNION COL, NY, 79- *Concurrent Pos:* Fulbright lectr, Univ Ankara, 61-62; vis prof physics, Rensselaer Polytech Inst, 69-70. *Mem:* Am Phys Soc. *Res:* Neutron physics; energy levels of light nuclei; science education; low-temperature physics. *Mailing Add:* 1350 Dean St Schenectady NY 12309

SWARTZ, CHARLES W, b Jan 1, 1938; m 71. MATHEMATICS. *Educ:* Univ Ariz, PhD (math), 65. *Prof Exp:* PROF MATH, NMEX STATE UNIV, 65- *Concurrent Pos:* Vis scholar, Univ Ariz, 71-72, Stanford Univ, 78, Univ NC, 87. *Mem:* Am Math Soc; Math Asn Am. *Res:* Functional analysis; locally convex spaces; continuous linear operators; vector-valve measures. *Mailing Add:* New Mexico St University Las Cruces NM 88003

SWARTZ, CLIFFORD EDWARD, b Niagara Falls, NY, Feb 21, 25; m 46; c 6. HIGH ENERGY PHYSICS, PHYSICS EDUCATION. *Educ:* Univ Rochester, AB, 45, MS, 46, PhD(physics), 51. *Prof Exp:* Assoc physicist, Brookhaven Nat Lab, 51-62; assoc prof, 57-67, PROF PHYSICS, STATE UNIV NY STONY BROOK, 67- *Concurrent Pos:* Ed, The Physics Teacher, 67-85, 89- *Honors & Awards:* Oersted Medal, Am Asn Physics Teachers, 87. *Mem:* Am Phys Soc; Am Asn Physics Teachers. *Res:* Particle physics; high energy accelerators for nuclear physics research; science curriculum revision and textbooks, kindergarten through college. *Mailing Add:* Dept of Physics State Univ of NY Stony Brook NY 11794

SWARTZ, DONALD PERCY, b Preston, Ont, Sept, 12, 21; US citizen; m 44, 84; c 2. OBSTETRICS & GYNECOLOGY. *Educ:* Univ Western Ont, BA & MD, 51, MSc, 53. *Prof Exp:* Nat Res Coun Can grant, Univ Western Ont, 52-53; Am Cancer Soc fel, Johns Hopkins Hosp, 56-57, instr obstet & gynec, 57-58; lectr physiol, Univ Western Ont, 58-62; clin prof, Columbia Univ, 62-72, prof obstet & gynec, 72; prof obstet & gynec & chmn dept, Albany Med Col, 72-79; obstetrician-gynecologist-in-chief, 72-79, chief, 79-88, HEAD DIV GEN GYNEC, ALBANY MED CTR, 88- *Concurrent Pos:* Markle scholar, Univ Western Ont, 58-62; consult obstet & gynec, St Peter's Hosp, 72- *Mem:* Fel Am Gynec Soc; Am Col Obstet & Gynec; Soc Study Reprod; Am Fertility Soc; fel Royal Col Surg; fel Am Gynec & Obstet Soc. *Res:* hormonal contraception; new approaches to pregnancy termination; gynecologic endocrinology. *Mailing Add:* Dept Obstet & Gynec Albany Med Ctr Albany NY 12208

SWARTZ, FRANK JOSEPH, b Pittsburgh, Pa, Mar 22, 27; m 46; c 2. ANATOMY. *Educ:* Western Reserve Univ, BS, 49, MS, 51, PhD(zool), 55. *Prof Exp:* Asst biol, Western Reserve Univ, 49-52, Nat Cancer Inst fel, 55-56; from asst prof to prof anat, Sch Med, Univ Louisville, 56-90; RETIRED. *Concurrent Pos:* Lectr & USPHS spec fel, Dept Anat, Harvard Med Sch, 69-70; prof anat, Univ Louisville, 90- *Mem:* Am Soc Anat; Sigma Xi. *Res:* Human anatomy; cellular differentiation and genetic significance of polyploids in mammalian tissues. *Mailing Add:* Dept Anat Sch Med Univ Louisville Louisville KY 40292

SWARTZ, GEORGE ALLAN, b Scranton, Pa, Dec 9, 30; m 80; c 3. PHYSICS. *Educ:* Mass Inst Technol, BS, 52; Univ Pa, MS, 54, PhD(physics), 58. *Prof Exp:* Mem tech staff, David Sarnoff Res Ctr, RCA Corp, 58-87, mem tech staff, SRI Int, 87-90. *Concurrent Pos:* Adj assoc prof dept metall & mat sci, Stevens Inst Technol, 80-85. *Mem:* Am Phys Soc; Sigma Xi; Inst Elec & Electronic Engrs. *Res:* Solid state microwave devices, particularly impact avalanche, transit time microwave sources and PIN diode switches; photovoltaic solar energy sources; amorphous silicon photovotaic energy sources; very-large scale integration microchip reliability. *Mailing Add:* 431 Willowbrook Dr North Brunswick NJ 08902

SWARTZ, GORDON ELMER, b Buffalo, NY, May 12, 17; m 41; c 2. ZOOLOGY, EMBRYOLOGY. *Educ:* Univ Buffalo, BA, 39, MA, 41; NY Univ, PhD(biol), 46. *Prof Exp:* From instr to prof, 46-79, EMER PROF BIOL, STATE UNIV NY, BUFFALO, 79- *Mem:* Fel AAAS; Am Micros Soc; Am Soc Zoologists; Am Asn Anat; NY Acad Sci. *Res:* Organogenesis; vertebrate experimental embryology; transplantation. *Mailing Add:* 24 Copper Heights Snyder NY 14226

SWARTZ, GRACE LYNN, b Coaldale, Pa, 1943. INORGANIC CHEMISTRY. *Educ:* Muhlenberg Col, BS, 65; Dartmouth Col, MA, 67; Fla State Univ, PhD(chem), 77. *Prof Exp:* Vis instr chem, Fla State Univ, 67-72, res assoc, 72-73; vis instr chem, Purdue Univ, 73-74; res asst chem, Fla State Univ, 74-77; vis asst prof, 77-80, ASST PROF CHEM, MICH TECHNOL UNIV, 80- *Mem:* Am Chem Soc; AAAS. *Res:* Mechanistic aspects of transition metal carbonyls and substituted metal carbonyls used in catalytic reactions, with the active catalyst generated thermally or photochemically. *Mailing Add:* Dept Chem Mich Technol Univ Houghton MI 49931

SWARTZ, HAROLD M, b Chicago, Ill, June 22, 35; m 81; c 4. RADIOLOGY, ELECTRON SPIN RESONANCE. *Educ:* Univ Ill, BS & MD, 59; Univ NC, MS, 62; Georgetown Univ, PhD(biochem), 69. *Prof Exp:* Fel nuclear med, Walter Reed Army Inst Res, Med Corps, US Army, 62-64, res med officer, 64-68, chief dept biophys, 68-70, chief dept biol chem, 70; assoc prof radiol & biochem, Med Col Wis, 70-74, prof, 74-80, dir, radiation biol & biophys lab, 70-80, dir, Nat Biomed Electron Spin Resonance Ctr, 74-80; PROF BIOPHYS & ASSOC DEAN ACAD AFFAIRS, COL MED, UNIV ILL, URBANA-CHAMPAIGN, 80- *Mem:* AAAS; Radiation Res Soc; Soc Nuclear Med; NY Acad Sci. *Res:* Free radicals and paramagnetic metal ions in biological systems; oxygen toxicity; radiation biology applied to radiation therapy; carcinogenesis. *Mailing Add:* Biophys Physiol Univ Ill 190 Med Sci 407 S Goodwin Ave Urbana IL 61801

SWARTZ, HARRY, b Detroit, Mich, June 21, 11; m 87. ALLERGY, SCIENCE COMMUNICATIONS. *Educ:* Univ Mich, AB, 30, MD, 33. *Prof Exp:* Clin asst allergy, Med Sch & Clins, NY Univ, 37-40 & Flower & Fifth Ave Hosp, New York, 40-42; asst chief allergy clin, Harlem Hosp, 46-48; prof med & chief, allergy dept, NY Polyclin Med Sch & Hosp, 57-72; pres, Health Field Validation Corp, 67-72; CONSULT, 73- *Concurrent Pos:* Chief, Allergy Dept, Tilton Gen Hosp, Ft Dix, NJ, 42-46; clin asst, Inst Allergy, Roosevelt Hosp, 46-72; indust consult allergy; consult, nutrit prod to pharmaceut 77 food indust, therapeut cosmetics to cosmetic indust & to pub indust; ed, Health Series, Med & Health Reporter, Issues in Current Med Pract & Med Opinion & Rev; sci dir, Mundo Medico, SAm, 72-89, foreign corresp; ed-in-chief, Investigacion Medica Int, 73-; sci adv, MD En Español, 81- *Honors & Awards:* Award of Merit, Am Col Allergists, 82. *Mem:* NY Acad Sci; Emer fel Am Col Allergists; emer fel Am Acad Allergy; emer fel Am Asn Clin Immunol & Allergy; fel Royal Soc Health; AAAS; Assoc Pub Health Asn; Am Med Writers' Asn; Foreign Press Asn. *Res:* Clinical allergy; high protein vegetable source material as a partial answer to world hunger; investigation of commercial products for efficacy and safety; author of numerous books, 100 editorials and 50 articles in various American and Mexican journals. *Mailing Add:* 138 E 16th St New York NY 10003

SWARTZ, HARRY SIP, b Wichita, Kans, July 29, 25; m 47; c 4. PHARMACY ADMINISTRATION, PHARMACY. *Educ:* Albany Col Pharm, Union Univ, NY, BS, 51; Univ Colo, MS, 54; Univ Iowa, PhD(pharm, pharmaceut admin), 59. *Prof Exp:* Lab asst pharmaceut chem, Univ Colo, 52-54; instr pharm & pharmaceut admin, Creighton Univ, 54-55; instr pharm & pharmacist, Univ Iowa, 55-59; from asst prof to assoc prof, 59-67, PROF PHARM & PHARMACEUT ADMIN, FERRIS STATE COL, 67- *Concurrent Pos:* Consult community, hosp & mfg pharm & extended care. *Mem:* Am Pharmaceut Asn; Am Col Apothecaries. *Res:* Product development; hospital pharmacy and manufacturing; cosmetic pharmaceuticals; orthopedic and surgical garments. *Mailing Add:* Dept Pharm Admin Ferris State Col Big Rapids MI 49307

SWARTZ, JACOB, psychiatry, psychoanalysis, for more information see previous edition

SWARTZ, JAMES E, b DC, June 12, 51; m 80. ORGANIC ELECTROCHEMISTRY, PHYSICAL ORGANIC CHEMISTRY. *Educ:* Stanislaus State Col, BS, 73; Univ Calif, Santa Cruz, PhD(chem), 78. *Prof Exp:* Instr chem, Univ Calif, Santa Cruz, 78; res fel, Calif Inst Technol, 78-80; asst prof, 80-86. ASSOC PROF CHEM, GRINNELL COL, 86- *Concurrent Pos:* Policy analyst, Energy, US Congress, 89. *Mem:* Am Chem Soc; AAAS; Electrochem Soc; Am Solar Energy Soc; Am Wind Energy Asn. *Res:* Nucleophilic aromatic substitution reactions; free radical radical reactions; radical anions. *Mailing Add:* Box X21 Grinnell Col Grinnell IA 50112-0806

SWARTZ, JAMES LAWRENCE, b Lowell, Mass, July 20, 39; m 63; c 2. ELECTROANALYTICAL CHEMISTRY, CONTROL ENGINEERING. *Educ:* Univ Lowell, BS, 65; Northeastern Univ, MS, 71, PhD(chem), 74. *Prof Exp:* Res chemist, Res & Advan Develop Div, Avco Corp, 64-65; sr chemist, Ciba-Geigy, 85-86; prod mgr, Robotest, 86-87; res chemist, Res Ctr, 65-70, sr develop engr chem, Systs Div, 70-75 & Corp Eng, 75-76, Sr Systs engr, 80-85, SR SCIENTIST & INSTR CHEM, FOXBORO CO, 76-; sr chemist, CMT, 87-88; sr engr, Kajaani Automation, 88-89; SR ENGR, VALMET AUTOMATION, 89- *Concurrent Pos:* Lectr, Dept Chem, Grad Sch, Northeastern Univ, 75-77. *Mem:* Am Chem Soc; sr mem Instrument Soc Am; Inst Elec & Electronic Engrs; Soc Electroanal Chem; Tech Asn Pulp & Paper Indust. *Res:* Sensor development for continuous on-line process monitoring and control; process analyzers based on electrochemical and chromatographic principles; ion-selective electrodes and their industrial and clinical applications. *Mailing Add:* Northwoods Rd RR 2, Box 257 E Lake Luzerne NY 12846

SWARTZ, JOHN CROUCHER, b Syracuse, NY, Oct 25, 24; m 56; c 3. PHYSICS, MATERIALS SCIENCE. *Educ:* Yale Univ, BS, 46; Syracuse Univ, MS, 49, PhD, 52. *Prof Exp:* Res physicist, Consol Vacuum Corp, 52-55; sr scientist, E C Bain Lab Fundamental Res, US Steel Corp, 55-71 & Tyco Labs, Inc, Mass, 72-75; sr scientist, Mobil Tyco Solar Energy Corp, 75-76; sr scientist, Westinghouse Res & Develop Ctr, 77-89; SCIENTIST, R J LEE GROUP, INC, 89- *Mem:* Am Phys Soc. *Res:* VLSI processing; infrared detector material; crystal growth; solid state physics. *Mailing Add:* 3201 Cambridge Dr Murrysville PA 15668

SWARTZ, LESLIE GERARD, b Chicago, Ill, Aug 16, 30; m 58; c 4. PARASITOLOGY. *Educ:* Univ Ill, BS, 53, MS, 54, PhD(zool), 58. *Prof Exp:* From assoc prof to prof, 58-88, EMER PROF ZOOL, UNIV ALASKA, FAIRBANKS, 88-, AFFIL PROF, 88- *Mem:* AAAS; Am Soc Parasitol. *Res:* Helminth parasitology, especially ecology; avian, freshwater and general ecology. *Mailing Add:* Dept of Biol Sci & Wildlife Univ of Alaska Fairbanks AK 99775-0280

SWARTZ, MARJORIE LOUISE, b Indianapolis, Ind, Feb 1, 24. INORGANIC CHEMISTRY. *Educ:* Butler Univ, BS, 46; Ind Univ, MS, 59. *Prof Exp:* Res assoc, 46-53, from instr to assoc prof, 53-69, PROF DENT MAT, SCH DENT, IND UNIV, INDIANAPOLIS, 69- *Honors & Awards:* Souder Award, Int Asn Dent Res, 68. *Mem:* AAAS; fel Am Col Dent; hon mem Am Dent Asn; Int Asn Dent Mat; hon mem Am Asn Women Dentists; hon mem Int Col Dent; Sigma Xi. *Res:* Physical and chemical properties of dental cements, resins and amalgams; effect of restorative materials on physical and chemical properties of tooth structure. *Mailing Add:* Ind Univ Sch of Dent Dental Mat DS 112 Indianapolis IN 46202

SWARTZ, MORTON N, b Boston, Mass, Nov 11, 23; m 56; c 2. INFECTIOUS DISEASES, MICROBIOLOGY. *Educ:* Harvard Univ, MD, 47. *Hon Degrees:* MD, Univ Geneve. *Prof Exp:* Chief, infectious dis unit, 56-90, CHIEF JACKSON FIRM, MED SERV, MASS GEN HOSP, 91- *Mem:* Am Soc Biol Chemists; Am Soc Microbiol; Asn Am Physicians; Am Soc Clin Invest; Am Col Physicians; Infectious Dis Soc Am. *Res:* Mechanisms antibiotic action. *Mailing Add:* Mass Gen Hosp Fruit St Boston MA 02114

SWARTZ, STUART ENDSLEY, b Chicago, Ill, Oct 17, 38; m 63. CIVIL ENGINEERING. *Educ:* Ill Inst Technol, BS, 59, MS, 62, PhD(civil eng), 68. *Prof Exp:* Engr, Caterpillar Tractor Co, Ill, 60; res assoc, Ill Inst Tech, 67-68; from asst prof to assoc prof, 68-77, PROF CIVIL ENG, KANS STATE UNIV, 77- *Mem:* Am Soc Civil Engrs; fel Am Concrete Inst; Soc Exp Stress Anal. *Res:* Analysis and design of folded plate structures; theoretical and experimental studies on the buckling of folded plates and other concrete shells; buckling of concrete columns; analysis and design of concrete shells; buckling of concrete panels; fracture toughness of concrete. *Mailing Add:* Dept Civil Eng Seaton Hall Kans State Univ Manhattan KS 66506

SWARTZ, WILLIAM EDWARD, JR, b Braddock, Pa, Aug 16, 44; div; c 2. SURFACE ANALYSIS. *Educ:* Juniata Col, BS, 66; Mass Inst Technol, PhD(chem), 71. *Prof Exp:* Res assoc chem, Univ Ga, 71, Univ Md, 71-72; asst prof chem, Univ SFla, 72-77, assoc prof, 77-82, prof, 82-85; MGR, MAT & PROCESSES, GEN ELEC NEUTRON DEVICES, 85- *Concurrent Pos:* Adj asst prof, Dept Physics, Univ SFla, 76-78. *Mem:* Am Chem Soc; Am Vacuum Soc. *Res:* Analytical applications of surface analysis techniques with emphasis on x-ray photoelectron spectroscopy; heterogeneous catalysts and surface chemical phenomena in integrated circuit components. *Mailing Add:* Dept Neutron Devices General Elec Co PO Box 2908 Largo FL 34649-2908

SWARTZ, WILLIAM JOHN, b Portage, Wis, Aug 9, 20; m 48; c 2. MATHEMATICS. *Educ:* Mont State Univ, BS, 44; Mass Inst Technol, SM, 49; Iowa State Univ, PhD(math), 55. *Prof Exp:* Instr math, Mass Inst Technol, 47-48, Mont State Univ, 49-51 & Iowa State Univ, 51-55; from asst prof to assoc prof, 55-62, prof, 62-84, EMER PROF MATH, MONT STATE UNIV, 84- *Mem:* Am Math Soc. *Res:* Differential equations. *Mailing Add:* 408 S 14 Ave Bozeman MT 59715

SWARTZENDRUBER, DALE, b Parnell, Iowa, July 6, 25; m 49; c 4. SOIL PHYSICS, SOIL WATER. *Educ:* Iowa State Univ, BS, 50, MS, 52, PhD(soil physics), 54. *Prof Exp:* Asst soil physics, Iowa State Univ, 50-53; instr agr, Goshen Col, 53-54; asst soil scientist, Univ Calif, Los Angeles, 55-56; assoc prof, Purdue Univ, 56-63, prof soil physics, 63-77; PROF SOIL PHYSICS, UNIV NEBR-LINCOLN, 77- *Concurrent Pos:* Vis prof, Iowa State Univ, 59, Ga Inst Technol, 68, Hebrew Univ, Jerusalem, Volcani Inst, Rehovot, Israel, 71, Griffith Univ, Brisbane, Australia, 89-90 & Ctr Environ Mech, CSIRO, Canberra, Australia, 90; assoc ed, Soil Sci Soc Am Proc, 65-70; vis scholar, Cambridge Univ, Eng, 71; consult ed, Soil Sci, 76- *Honors & Awards:* Soil Sci Award, Soil Sci Soc Am, 75. *Mem:* AAAS; fel Soil Sci Soc Am; fel Am Soc Agron; Am Geophys Union; Int Soc Soil Sci; Sigma Xi. *Res:* Physics of soil and water, including water movement through saturated and unsaturated soils and porous media; soil, air, temperature, and structure; soil-water-plant relationships; hydrology and water resources. *Mailing Add:* Dept Agron Keim Hall E Campus Univ Nebr Lincoln NE 68583-0915

SWARTZENDRUBER, DONALD CLAIR, b Kalona, Iowa, June 21, 30; m 55; c 2. ZOOLOGY. *Educ:* Univ Iowa, BA, 55, MS, 58, PhD(zool), 62. *Prof Exp:* Res asst biol, Oak Ridge Nat Lab, 59-60, res assoc, 62-63; biologist, 63-65, res assoc microbiol, Univ Mich, Ann Arbor, 65-68; sr scientist, Med Div, Oak Ridge Assoc Univs, 68-81; SR RES SCIENTIST, PHARMACEUT DIV, CIBA-GEIGY CORP, 81- *Mem:* AAAS; Electron Micros Soc Am; Sigma Xi; Am Soc Cell Biol; Soc Exp Biol & Med. *Res:* Electron microscopic studies of lymphatic tissues; ultrastructural pathology; histopathology of connective tissue. *Mailing Add:* Pharmaceut Div Ciba-Geigy Corp Ardsley NY 10502

SWARTZENDRUBER, DOUGLAS EDWARD, b Goshen, Ind, May 3, 46; m 69; c 3. PATHOLOGY. *Educ:* Goshen Col, Ind, BA, 68; Univ Colo Med Sch, PhD(exp path), 74. *Prof Exp:* Postdoctoral exp path, Los Alamos Nat Lab, 74-76, staff mem, 76-80; fac develop therapeut, MD Anderson Hosp & Tumor Inst, 80-82; ASSOC PROF & CHAIR BIOL, UNIV COLO, COLORADO SPRINGS, 82- *Concurrent Pos:* Consult, Coulter Cytometry, 82-; lab dir, Onco Metrics Inc, 88-, Compat Labs, 89-91. *Mem:* Int Soc Anal Cytol; AAAS; Am Asn Cancer Res. *Res:* Research and clinical applications of flow cytometry; differentiation of neoplastic cells; computer modelling of cancer, especially breast; biocompatibility of dental materials. *Mailing Add:* Dept Biol Univ Colo 1420 Austin Bluffs Pkwy Colorado Springs CO 80933-7150

SWARTZENDRUBER, LYDON JAMES, b Wellman, Iowa, Aug 8, 33; m 49; c 1. CONDENSED MATTER PHYSICS. *Educ:* Iowa State Univ, BS, 57; Univ Md, PhD(physics), 68. *Prof Exp:* PHYSICIST, NAT BUR STANDARDS, 60- *Mem:* AAAS; Am Phys Soc; Am Soc Testing & Mat; Am Inst Mining, Metall & Petrol Engrs; Am Soc Nondestructive Testing; Sigma Xi. *Res:* Solid state physics; semiconductors; magnetism; metallurgy. *Mailing Add:* Nat Bur Standards Gaithersburg MD 20899

SWARTZENTRUBER, PAUL EDWIN, b Lagrange, Ind, Apr 23, 31; m 55; c 3. ORGANIC CHEMISTRY, INFORMATION SCIENCE. *Educ:* Goshen Col, BA, 53; Univ Minn, Minneapolis, MS, 55; Univ Mo, PhD(org chem), 61. *Prof Exp:* Res chemist, Nat Cancer Inst, 56-59; asst ed org indexing dept, 61-63, assoc ed, 63-64, head org indexing dept, 64-71, mgr phys & inorg indexing dept, 71-72, mgr chem substance handling dept, 72-73, mgr chem technol, 73-77, asst to ed, 77-79, MANAGING ED, CHEM ABSTR SERV, 79- *Concurrent Pos:* Bd dir, doc abstr, 88- *Mem:* AAAS; Am Chem Soc; Am Soc Indexers. *Res:* Synthetic organic chemistry; chemical nomenclature; storage and retrieval of chemical information; abstracting and indexing of chemical literature. *Mailing Add:* Chem Abstr Serv PO Box 3012 Columbus OH 43210

SWARTZLANDER, EARL EUGENE, JR, b San Antonio, Tex, Feb 1, 45; m 68. COMPUTER DESIGN, WAFER SCALE INTEGRATION. *Educ:* Purdue Univ, W Lafayette, IN, BS, 67; Colo Univ, Boulder, Co, MS, 69; Univ Southern Calif, PhD(elect eng) 72. *Prof Exp:* Develop eng, Ball Brothers Res Corp, 67-69; Hughes doctoral fel, Hughes Aircraft Co, 69-73; mem res staff, Technol Serv Corp, 73-74; chief eng, Geophy Syst Corp, 74-75; mgr, Digital Processing Lab, TRW Elect Syst Group, 75-87; dir res & develop, TRW Defense Syst Group, 87-90; PROF ELEC ENG, UNIV TEX, AUSTIN, 90-, SCHLUMBERGER CENTENNIAL CHAIR ENG, 90- *Concurrent Pos:* Ed, Trans on Computers, Inst Elec & Electronic Engrs, 82-86, J Solid-State Circuits, 84-88, Trans on Parellel & Distrib Systs, 89-90, ed-in-chief, Trans on Computers, 91-; ed, Comput Rev, Asn Comput Mach, 87; mem bd gov, Computer Soc, Inst Elec & Electronic Engrs, 87-; ed, J Real-Time Systs, 89-; ed-in-chief, J VLSI Signal Processing, 89-; consult, Univ S Fla, 89-; mem rev bd, Microelectronic Info Processing Systs, NSF, 91- *Mem:* fel Inst Elec & Electronic Engrs; Inst Elec & Electronic Engrs Comput Soc. *Res:* VLSI circuit development for high performance computers; chip and processor architecture, computer arithmetic and design methodologies; author of one book, editor of five books and over 100 articles. *Mailing Add:* 2847 Deep Canyon Dr Beverly Hills CA 90210-1044

SWARTZMAN, GORDON LENI, b New York, NY, Sept 2, 43; m 69; c 2. RESOURCE MANAGEMENT, BIOMETRICS. *Educ:* Cooper Union, NY, BS, 64; Univ Mich, MS, 65, PhD(indust eng), 69. *Prof Exp:* Fel ecol modeling, Colo State Univ, 69-72; vis prof agr & ecol models, Univ Reading, 72-73; res assoc statist ecol models, 73-76, res asst prof matrices ecol models, 76-79, res assoc prof calculus population dynamics, 79-84, RES PROF, UNIV WASH, 84- *Mem:* AAAS; Int Soc Ecol Modelers; Sigma Xi. *Res:* Interaction between fur seals and fisheries on Bering Sea; simulation modeling as a tool in impact assessment; stochastic models for fisheries management; Lake Washington rainbow trout introduction impact; microcosm toxicant effects modeling. *Mailing Add:* 7603 56th Pl NE Seattle WA 98115

SWARTZWELDER, JOHN CLYDE, b Lynn, Mass, Apr 1, 11; m 64, 88. MEDICAL PARASITOLOGY, TROPICAL PUBLIC HEALTH. *Educ:* Univ Mass, BS, 33; Tulane Univ, MS, 34, PhD(med protozool), 37; Am Bd Med Microbiol, dipl, 64. *Prof Exp:* Asst med parasitol, Sch Med, Tulane Univ, 33-37; from instr to prof med parasitol, La State Univ Sch Med, 37-75, head dept trop med & med parasitol, 60-75, educ dir, Interam Training Prog Trop Med, 59-69, assoc dir, Int Ctr Med Res & Training, 61-69, dir, Interam Training Prog Trop Med & Int Ctr Med Res & Training, 69-75, EMER PROF MED PARASITOL, LA STATE UNIV MED CTR, NEW ORLEANS, 75- *Concurrent Pos:* Scientist, Charity Hosp, New Orleans, 38-75; consult, Vet Admin Hosp, 49-90. *Mem:* Am Soc Trop Med & Hyg (vpres); Am Soc Parasitol; hon mem Mex Soc Parasitol. *Res:* Amebiasis; Chagas' disease; anthelmintics; research training in tropical medicine; medical education. *Mailing Add:* 4505 Lefkoe St Apt D Metairie LA 70006-2263

SWATEK, FRANK EDWARD, b Oklahoma City, Okla, June 4, 29; m 51; c 5. MICROBIOLOGY, MYCOLOGY. *Educ:* San Diego State Col, BS, 51; Univ Calif, Los Angeles, MA, 55, PhD(microbiol), 56. *Prof Exp:* From instr to assoc prof, 56-63, chmn dept, 60-82, PROF MICROBIOL, CALIF STATE UNIV, LONG BEACH, 63-; CLIN PROF MED, SCH MED, UNIV CALIF, IRVINE, 80- *Concurrent Pos:* Consult, Dept Allergy & Dermat, Long Beach Vet Admin Hosp, 60-, Douglas Aircraft Co, Inc, 61- & Hyland Lab, 68-; lectr, Sch Med, Univ Southern Calif, 62- *Honors & Awards:* Carski Award, Am Soc Microbiol, 74. *Mem:* Fel Am Soc Microbiol; NY Acad Sci; fel Royal Soc Health; Sigma Xi. *Res:* Ecology and experimental pathology of deep mycoses, especially Coccidioides, Cryptococcus and Dermatophytes; industrial work on fungus deterioration of man-made products; fresh and swimming pool water. *Mailing Add:* Dept Microbiol Calif State Univ Long Beach CA 90805

SWATHIRAJAN, S, b Bangalore, India, June 5, 52; m 80; c 1. ELECTROCHEMISTRY. *Educ:* Bangalore Univ, BSc Hons, 70, MSc, 72; Indian Inst Sci, Bangalore, PhD(electrochem), 76. *Prof Exp:* Res fel, Indian Inst Sci, 72-77; lectr chem, State Univ NY, Buffalo, 77-79, res asst prof, 80-83; RES SCIENTIST, GEN MOTORS RES LABS, 83-, SECT MGR, 87- *Concurrent Pos:* Tech coordr, Gen Motors Res Prog Fuel Cells, US Dept Energy, 90- *Mem:* Electrochem Soc. *Res:* Electrochemical phenomena; development of novel materials for electrochemical systems; electrochemical energy conversion in batteries and fuel cells; electrodeposition and corrosion protection. *Mailing Add:* Phys Chem Dept RCEL Gen Motors Res Labs 30500 Mound Rd Warren MI 48090-9055

SWAZEY, JUDITH P, b Bronxville, NY, Apr 21, 39; m 64; c 2. SOCIAL & ETHICAL ISSUES IN SCIENCE & MEDICINE. *Educ:* Wellesley Col, AB, 61; Harvard Univ, PhD, 66. *Prof Exp:* Res assoc, Harvard Univ, 66-71; consult, Comt Brain Sci, Nat Res Coun, 71-73; staff scientist, Neurosci Res Prog, Mass Inst Technol, 73-74; from assoc prof to prof, Dept Socio-Med Sci & Community, Boston Univ Med Sch & Ch Public Health, 74-82; pres, Col Atlantic, 82-84; PRES, ACADIA INST, 84-; EXEC DIR, MED PUB INTEREST INC, 79-82 & 89- *Concurrent Pos:* Adj prof, Dept Socio-Med Sci & Community Med, Boston Univ, 80-; mem, Army Sci Bd, AAAS. *Mem:* Inst Med Nat Acad Sci; Sigma Xi; AAAS. *Res:* Social, ethical and policy aspects of medical research and health care; professional ethics. *Mailing Add:* Acadia Inst 118 West St Bar Harbor ME 04609

SWEADNER, KATHLEEN JOAN, b 1949. BIOCHEMISTRY, NEUROBIOLOGY. *Educ:* Univ Calif, Santa Barbara, BA, 71; Harvard Univ, MA, 74, PhD(biochem), 77. *Prof Exp:* Consult, Millipore Corp, 73-76; instr neurobiol, 80-81, asst prof physiol, Dept Surg, Harvard Med Sch, 81-87; asst prof physiol, Dept Surg, Mass Gen Hosp, 81-87; ASSOC PROF PHYSIOL, DEPT SURG, HARVARD MED SCH & MASS GEN HOSP, 87- *Concurrent Pos:* Estab investr, Am Heart Asn, 82-87. *Honors & Awards:* Louis N Katz Prize, 84. *Mem:* Soc Neurosci; Am Soc Biochem & Molecular Biol; Asn Res Vision & Ophthal; Protein Soc; NY Acad Sci; Biophys Soc; Am Heart Asn; AAAS. *Res:* Isozymes of the sodium, potassium ATPase. *Mailing Add:* Wellman 4 Neurosurg Res Mass Gen Hosp Fruit St Boston MA 02114

SWEARENGEN, JACK CLAYTON, b Zanesville, Ohio, Feb 11, 40; m 62; c 1. SCIENCE & SOCIETY STUDIES. *Educ:* Univ Idaho, BS, 61; Univ Ariz, MS, 63; Univ Wash, PhD(mech eng), 70. *Prof Exp:* Mem tech staff, 70-78, supvr Mat Sci Div, 78-84, Solar Components Div, 84-86, Advan Systs Div, 86-88, SUPVR SYSTS DEVELOP DIV, SANDIA NAT LABS, 91- *Concurrent Pos:* Reviewer, NSF, 78-88, J Mat Energy Systs, 78-86 & Solar Energy Mat; prin investr, US Dept Energy, 79-83, US Navy, 86-87; teaching assoc, Univ Wash, 66-69; mem Pres Young Investigator Awards Panel, NSF, 88; Sci adv arms control, Off Secy Defense, Pentagon, Wash, DC, 88-90. *Honors & Awards:* Secy Defense Award Excellence, 90. *Mem:* Am Soc Metals; fel Am Sci Affil; Am Soc Mech Engrs. *Res:* Constitutive models for rate-dependent plastic flow of metals; structure-property relationships in metals, ceramics and glass-ceramics; state variable theories for materials; energy systems; arms control treaty verification. *Mailing Add:* Systs Develop Div 8135 Sandia Nat Labs Livermore CA 94551-0969

SWEARINGEN, JOHN ELDRED, b Columbia, SC, Sept 7, 18; m 69; c 3. CHEMICAL ENGINEERING. *Educ:* Univ SC, BS, 38; Carnegie Inst Technol, MS, 39. *Hon Degrees:* EngD, SDak Sch Mines & Technol, 60; LLD, Knox Col, Ill, 62, DePauw Univ, 64, Univ SC, 65, Ill Col & Butler Univ, 68; DLH, Nat Col Educ, 67. *Prof Exp:* Chemist, Standard Oil Co, 39-43, group leader, 43-47; proj engr, Pan Am Petrol Corp, 47, develop supt mfg, 47-48, asst to mgr mfg, 48-49, exec asst to exec vpres, 51; gen mgr prod, Standard Oil Co (Ind), 51-54, vpres prod, 54-56, exec vpres, 56-58, pres, 58-65, chmn bd, 65-83, dir, 52-83, chief exec off, 60-83; chmn bd & exec officer, Continental Ill Corp, 83-87; RETIRED. *Concurrent Pos:* Dir, Am Petrol Inst, 51-, chmn bd, 78-79; consult, Nat Petrol Coun & Nat Indust Conf Bd, 60-; consult, Hwy Users Fedn for Safety & Mobility, 69-; chmn, Nat Petrol Coun, 74-75; dir, Lockheed Corp & Aon Corp. *Mem:* Nat Acad Eng; fel Am Inst Chem Engrs; Am Chem Soc; Am Inst Mining, Metall & Petrol Engrs. *Res:* Petroleum production and processing; management. *Mailing Add:* Amoco Bldg Suite 7096 Standard Oil Co 200 E Randolf Dr Chicago IL 60601

SWEARINGEN, JUDSON STERLING, b San Antonio, Tex, Jan 11, 07; m 60. CHEMICAL ENGINEERING. *Educ:* Univ Tex, BS, 29, MS, 30, PhD(chem eng), 33. *Prof Exp:* Partner, San Antonio Refining Co, 33-38; asst prof chem eng, Univ Tex, 39-40, assoc prof, 40-41, prof, 41-42; turbine designer, Elliott Co, 42-43; div engr, Kellex Corp, 43-45; pres, Statham-Swearingen, Inc, 59-62 & Swearingen Bros, Inc, 64-81; pres, Rotoflow Corp, 46-90, Rotoflow, Ag, Switz, 71-90, ADV CONSULT, ROTOFLOW CORP, 90- *Mem:* Nat Acad Eng; Am Chem Soc; Am Inst Mech Engrs; Am Soc Testing & Mat. *Res:* Low temperature gas separations; turboexpanders; petroleum and gas; seals, pumps and centrifugal compressors; adsorption refrigeration. *Mailing Add:* 4347 W Northwest Hwy Suite 826 Dallas TX 75220

SWEARINGIN, MARVIN LAVERNE, b Hamburg, Ill, Jan 23, 31; m 50; c 3. AGRONOMY, SOYBEAN PRODUCTION. *Educ:* Univ Mo-Columbia, BS, 56, MS, 57; Ore State Univ, PhD(agron), 62. *Prof Exp:* Asst crops teaching, Univ Mo, 56-57; instr farm crops, Ore State Univ, 57-61; from asst prof to assoc prof, 61-75, PROF AGRON, PURDUE UNIV, WEST LAFAYETTE, 75- *Concurrent Pos:* Soybean res consult, Purdue Univ-Brazil Proj, Brazil, 67-69 & AID, Brazil, 71- *Honors & Awards:* Soybean Researchers Recog Award, Am Soybean Asn, 83; Super Serv Award, USDA, 89. *Mem:* Fel Am Soc Agron, 89; Crop Sci Soc Am. *Res:* Crop management systems for corn, soybeans and small grains; reduced tillage system for soybean production; weed control in soybeans. *Mailing Add:* Dept Agron Purdue Univ West Lafayette IN 47907

SWEAT, FLOYD WALTER, b Salt Lake City, Utah, July 21, 41; m 65; c 2. BIOCHEMISTRY. *Educ:* Univ Utah, BS, 64, PhD(org chem), 68. *Prof Exp:* NIH fel, Harvard Univ, 68-70; asst prof, 70-77, ASSOC PROF BIOCHEM, UNIV UTAH, 77- *Mem:* AAAS; Am Chem Soc; Sigma Xi. *Res:* Enzyme purification and characterization; structure elucidation; reaction mechanisms. *Mailing Add:* Dept of Biol Chem Univ of Utah Med Ctr Salt Lake City UT 84112

SWEAT, ROBERT LEE, b Lamar, Colo, June 8, 31; m 53; c 2. VETERINARY MEDICINE, VIROLOGY. *Educ:* Colo State Univ, BS, 54, DVM, 56; Univ Nebr-Lincoln, MS, 62, PhD(med sci), 66. *Prof Exp:* Instr vet sci, Univ Nebr-Lincoln, 58-66, assoc prof, 66; vet virologist, Norden Labs, Inc, Nebr, 67; assoc res prof vet sci, Univ Idaho, 68-70; vet virologist, Ft Dodge Labs Inc, 70-88; RES SCIENTIST, LAND O'LAKES INC, 89- *Concurrent Pos:* Vet rep, Nebr State Bd Health, 66-67. *Mem:* AAAS; Am Vet Med Asn; US Animal Health Asn; Am Pub Health Asn; Conf Res Workers Animal Dis; Am Asn Bovine Practrs. *Res:* Veterinary science with emphasis on diseases of cattle. *Mailing Add:* 1371 N 14th St Ft Dodge IA 50501

SWEAT, VINCENT EUGENE, b Kirwin, Kans, July 28, 41; m 65; c 2. FOOD ENGINEERING. *Educ:* Kans State Univ, BS, 64; Okla State Univ, MS, 65; Purdue Univ, PhD(agr eng), 72. *Prof Exp:* Asst prof food eng, Purdue Univ, 71-75, assoc prof, 75-77; ASSOC PROF FOOD ENG, TEX A&M UNIV, 77- *Concurrent Pos:* Mem thermal properties foods comt, Am Soc Heating, Refrig & Air Conditioning Engrs, 72-; assoc ed food eng div, Am Soc Agr Engrs, 76-; NSF eng res equip grant, 78- *Mem:* Am Soc Agr Engrs; Inst Food Technologists; Am Soc Eng Educ. *Res:* Measurement and modeling of thermal properties of foods; food processing; heat and mass transfer in foods; energy utilization in food processing. *Mailing Add:* Agr Eng Dept Tex A&M Univ College Station TX 77843

SWEDBERG, KENNETH C, b Brainerd, Minn, Apr 14, 30; m 58; c 2. PLANT ECOLOGY. *Educ:* St Cloud State Col, BS, 52; Univ Minn, MS, 56; Ore State Univ, PhD(plant ecol), 61. *Prof Exp:* Instr biol, Moorhead State Col, 56-58; asst prof, Wis State Univ-Stevens Point, 60-62; assoc prof, 62-69, PROF BIOL, EASTERN WASH UNIV, 69- *Mem:* AAAS; Ecol Soc Am; Weed Sci Soc Am; Am Inst Biol Sci; Sigma Xi. *Res:* Plant synecology and autecology; experimental ecology of annual plants. *Mailing Add:* S 19607 Cheney-Spangle Rd Cheney WA 99004

SWEDLOW, JEROLD LINDSAY, b Denver, Colo, Aug 31, 35; m 59; c 3. SOLID MECHANICS, STRUCTURAL INTEGRITY. *Educ:* Calif Inst Technol, BS, 57; Stanford Univ, MS, 60; Calif Inst Technol, PhD(aeronaut), 65. *Prof Exp:* Res engr rock mech, Ingersoll-Rand Co, 57-59; res fel aeronaut, Calif Inst Technol, 65-66; from asst prof to assoc prof mech eng, 66-73, assoc dean eng, 77-79, PROF MECH ENG, CARNEGIE-MELLON UNIV, 73- *Concurrent Pos:* Ed, Reports Current Res, Int J Fracture, 69-; sr vis fel, Imp Col Sci & Technol, 73-74. *Honors & Awards:* Philip M McKenna Mem Award, 78; Ralph Coats Roe Award, 81. *Mem:* Am Acad Mech; Am Soc Mech Engrs; Am Soc Testing & Mat; Int Cong Fracture. *Res:* Plasticity of metal via computational procedures; applications to fracture-related problems at both the macro and microscale. *Mailing Add:* 2329 Greensburg Pike Pittsburgh PA 15221

SWEDLUND, ALAN CHARLES, b Sacramento, Calif, Jan 21, 43; m 66; c 2. BIOLOGICAL ANTHROPOLOGY. *Educ:* Univ Colo, Ba, 66, MA, 68, PhD(anthrop), 70. *Prof Exp:* Asst prof anthrop, Prescott Col, 70-74; vis assoc prof, Univ Mass, Amherst, 74-77, field dir, Europ Studies Prog, 81-82, assoc prof, 77-85, PROF ANTHROP, UNIV MASS, AMHERST, 85-, CHAIR, 90- *Concurrent Pos:* Vis researcher, Biol Anthrop Prog, Oxford Univ, 82. *Mem:* AAAS; Am Asn Phys Anthrop; Am Soc Human Genetics; Am Eugenics Soc; Population Asn Am; Sigma Xi. *Res:* Demographic and human population genetics; historical demography and paleodemography; osteology. *Mailing Add:* Dept Anthrop Univ Mass Amherst MA 01003

SWEED, NORMAN HARRIS, b Philadelphia, Pa, Apr 11, 43; m 66; c 2. CHEMICAL ENGINEERING. *Educ:* Drexel Univ, BSChE, 65; Princeton Univ, MA, 67, PhD(chem eng), 68. *Prof Exp:* Asst prof chem eng, Princeton Univ, 68-75; sr res engr, Exxon Res & Eng Co, 74-78; mgr process res, Oxirane Int, 78-81; eng assoc, 81-86, SECT HEAD, EXXON RES & ENG CO, 86- *Concurrent Pos:* Consult, Engelhard Indust, 69-74 & Cities Serv Res, 74. *Mem:* Am Inst Chem Engrs; AAAS; Am Chem Soc. *Res:* Environmental control. *Mailing Add:* Exxon Res & Eng Co PO Box 101 Florham Park NJ 07932

SWEEDLER, ALAN R, b New York, NY, Jan 31, 42. PHOTOVOLTAICS, ENERGY POLICY. *Educ:* City Univ New York, BSc, 63; Univ Calif, San Diego, PhD(physics), 69. *Prof Exp:* Prof physics, Univ Chile, Santiago, 70-72; res physicist mat sci, Brookhaven Nat Lab, 72-77; assoc prof, dept physics, Calif State Univ Fullerton, 77-80; assoc prof, 80-83, PROF PHYSICS, SAN DIEGO STATE UNIV, 83-, DIR, CTR ENERGY STUDIES, 81- *Concurrent Pos:* Vis assoc prof physics, Univ Southern Calif, Los Angeles, 75; cong sci fel, Am Phys Soc, 85; sci fel, Int Security, Stanford Univ, 87; co-dir, Inst Int Security & Conflict Resolution, 88- *Mem:* Am Phys Soc; AAAS; Sigma Xi. *Res:* Superconductivity; high transition temperature compounds; irradiation effects in superconducting compounds; metallurgy of superconducting materials; solar energy; photovoltaics; energy policy; international security and arms control; international security arms control. *Mailing Add:* Dept Physics San Diego State Univ San Diego CA 92182

SWEELEY, CHARLES CRAWFORD, b Williamsport, Pa, Apr 15, 30; m 50; c 2. BIOCHEMISTRY, ORGANIC CHEMISTRY. *Educ:* Univ Pa, BS, 52; Univ Ill, PhD(chem), 55. *Hon Degrees:* Dr, Ghent Univ, Belgium, 82. *Prof Exp:* Chemist, Nat Heart Inst, 55-60; asst res prof biochem, Univ Pittsburgh, 60-63, from assoc prof to prof, 63-68; prof biochem, Mich State Univ, 68-85, chmn, 79-85. *Concurrent Pos:* Mem comt probs lipid anal, Nat Heart Inst, 58-59; consult, LKB Instruments, 65-71, Med Chem Study Sect, USPHS, 67-71 & Upjohn Co, 68-72; Guggenheim fel, Royal Vet Col, Stockholm, Sweden, 71; vis prof, Ctr Cancer Res, Cambridge, 79; consult, Upjohn, 68-72 & 86-89, Los Alamos Stable Isotope Resource, 88-91; consult & sr dir, Meridian Instruments, 82- *Mem:* Am Chem Soc; Am Soc Biol Chem; Brit Biochem Soc; Soc Complex Carbohydrates (treas, 75-78, pres, 85); Am Soc Mass Spectrometry. *Res:* Chemistry and metabolism of sphingolipids; sphingolipidoses; biochemistry of lysosomal hydrolases; analytical biochemistry; computer applications in gas chromatography and mass spectrometry; biochemistry of complex lipids and hormones of invertebrates. *Mailing Add:* 1895 Live Oak Trail Williamston MI 48895

SWEENEY, BEATRICE MARCY, b Boston, Mass, Aug 11, 14; m 62; c 4. BIOLOGICAL RHYTHMS. *Educ:* Smith Col, AB, 36; Radcliffe Col, PhD(biol), 42. *Prof Exp:* Lab asst endocrinol, Mayo Clin, 42; fel, Mayo Found, Univ Minn, 42-43; jr res biologist, Scripps Inst, Calif, 47-55, asst res biologist, 55-60, assoc res biologist, 60-61; res staff biologist, Yale Univ, 61-62, lectr biol, 62-67; lectr, 67-69, assoc prof biol, 69-71, assoc provost, Col Creative Studies, 78-81, prof, 71-82, EMER PROF BIOL, UNIV CALIF, SANTA BARBARA, 82- *Concurrent Pos:* Consult, Monroe Labs, 59; mem, Nat Comt Photobiol, Nat Res Coun, 72-75. *Honors & Awards:* Darbaker Award, Bot Soc Am, 84. *Mem:* Fel AAAS; Am Soc Plant Physiol (secy-treas, 75-76); Soc Gen Physiol; Am Soc Photobiol (pres, 79); Phycol Soc Am (pres, 86); Sigma Xi; Soc Res Biol Rhythms. *Res:* Nutrition; photosynthesis; bioluminescence; circadian rhythms in marine dinoflagellates. *Mailing Add:* Dept Biol Sci Univ Calif Santa Barbara CA 93106

SWEENEY, DARYL CHARLES, b Oakland, Calif, Jan 21, 36; m 59; c 1. INVERTEBRATE PHYSIOLOGY, NEUROCHEMISTRY. *Educ:* Univ Calif, Berkeley, AB, 58; Harvard Univ, AM, 59, PhD(biol), 63. *Prof Exp:* Instr biol, Yale Univ, 63-65; from asst prof to prof zool, 65-75, ASSOC PROF PHYSIOL, UNIV ILL, URBANA-CHAMPAIGN, 75- *Concurrent Pos:* Consult, Nat Inst Neurol Dis & Stoke, 66-69. *Mem:* Am Soc Zool. *Res:* Neurochemistry and the behavior of invertebrate physiology. *Mailing Add:* 1101 S Westlawn Champaign IL 61821

SWEENEY, DONALD WESLEY, b Chicago, Ill, Dec 19, 46; m 84; c 2. MECHANICAL ENGINEERING, OPTICS. *Educ:* Univ Mich, Ann Arbor, BS, 68, MS, 69, PhD, 72. *Prof Exp:* Res asst coherent optics, Willow Run Labs, Univ Mich, 69-72; from asst prof to prof mech eng, Purdue Univ, West Lafayette, 72-83; SUPVR, SANDIA NAT LABS, LIVERMORE, CALIF, 83- *Mem:* Optical Soc Am; Am Soc Mech Eng; Combustion Inst. *Res:* Applied research in coherent optics; optical image processing; holographic interferometry. *Mailing Add:* Sandia Nat Lab Org 8435 Livermore CA 94550

SWEENEY, EDWARD ARTHUR, b Boston, Mass, Apr 22, 31; m 66; c 2. PEDIATRIC DENTISTRY, DENTAL RESEARCH. *Educ:* Harvard Univ, DMD, 61, cert, 64; Children's Hosp Med Ctr, Boston, cert, 64. *Prof Exp:* Fel dent, Inst Nutrit Central Am & Panama, 63 & 67; fel protein chem, Protein Found, 64-65; assoc dent, Harvard Sch Dent Med, 64-66, from asst prof to assoc prof pediat dent, 66-81, asst dean curric, 77-80; CLIN ASSOC PROF PEDIAT DENT, UNIV PA, 81-, DIR ADVAN EDUC DENT, 82- *Concurrent Pos:* Consult, Robert Wood Johnson Found & Educ Testing Serv, 73-79, Nat Inst Med, Nat Acad Sci, 77-78 & Mass Dept Pub Health, 79-81; counr, Int Asn Dent Res, 77-80. *Mem:* AAAS; Int Asn Dent Res; Am Dent Asn; Am Acad Pediat Dent; Am Asn Dent Schs. *Res:* Protein secretion of salivary gland; nutrition and dental decay and enamel hypoplasia; oral pathology of children. *Mailing Add:* Dept Pedodontics Dent A1 Univ Pa Sch Dent Med 4001 Spruce St Philadelphia PA 19104

SWEENEY, GEORGE DOUGLAS, b Durban, SAfrica, Dec 21, 34; m 60; c 3. PHARMACOLOGY, LIVER TOXICOLOGY. *Educ:* Univ Cape Town, MB, ChB, 58, PhD(biochem), 63. *Prof Exp:* Sr lectr physiol, Univ Cape Town, 64-68; Ont fel, Col Physicians & Surgeons, Columbia Univ, 68-69; assoc prof med, 69-81, PROF MED, MCMASTER UNIV, 81- *Concurrent Pos:* Vis prof, Liver Unit, King's Col Hosp, 79-80. *Mem:* Can Soc Toxicol; Pharmacol Soc Can; Can Physiol Soc; Am Soc Pharmacol & Exp Therapeut. *Res:* Hemoprotein synthesis and regulation; porphyrin metabolism; liver toxicity of halo-aromatic hydrocarbons; liver cell life cycle; porphyria. *Mailing Add:* Dept Biochem McMaster Univ 1280 Main St W Hamilton ON L8S 3Z5 Can

SWEENEY, JAMES LEE, b Waterbury, Conn, Mar 22, 44; m 71; c 3. ECONOMIC ANALYSIS, ENERGY ANALYSIS. *Educ:* Mass Inst Technol, BS, 66; Stanford Univ, PhD(eng econ systs), 71. *Prof Exp:* Consult, Fed Energy Admin, Washington, DC, 74, dir, Off Energy Systs, 74-75, Off Quant Methods, 75 & Off Energy Systs Modeling & Forecasting, 75-76; res asst, Dept Eng Econ Systs, Stanford Univ, 67-70, actg instr, 70-71, from asst prof to assoc prof, 71-80, coop prof, Sch Law, 81-82, chmn, Inst Energy Studies, 81-85, dir, Ctr Econ Policy Res, 84-86, AFFIL PROF, DEPT ECON, STANFORD UNIV, 78-, PROF, DEPT ENG ECON SYSTS, 80- *Concurrent Pos:* Co-ed, Resources & Energy; mem, sr adv panel, Energy Modeling Forum; consult, Charles River Assocs, Cornerstone Res. *Mem:* Am Econ Asn; Int Asn Energy Economists. *Res:* Economic and policy issues important for natural resource production and use; energy markets, including oil, natural gas and electricity; use of mathematical models to analyze energy markets. *Mailing Add:* Eng Econ Syst Stanford Univ Terman 314 Stanford CA 94305-4025

SWEENEY, JAMES MICHAEL, b Wichita, Kans, Dec 31, 45. WILDLIFE BIOLOGY, FORESTRY. *Educ:* Univ Ga, BSF, 67, MS, 71; Colo State Univ, PhD(wildlife), 75. *Prof Exp:* ASST PROF WILDLIFE & FORESTRY, DEPT FORESTRY, UNIV ARK, 75-; BILATERAL PROG COORD, USDA INT FORESTRY, 88- *Concurrent Pos:* Sci ed wildlife, Dept Fisheries & Wildlife Biol, Colo State Univ, 75; actg head, Dept Forestry, Univ Ark, Monticello, 78-; reviewer, SE Asn Game & Fish Agencies, Proceeding Annual Conf, 78- & J Wildlife Mgt, The Wildlife Soc, 78- *Mem:* Sigma Xi; The Wildlife Soc. *Res:* Forestry-wildlife habitat interactions. *Mailing Add:* Int Forestry USDA Forest Serv PO Box 96090 Washington DC 60090-6090

SWEENEY, JOHN ROBERT, b Wichita, Kans, Dec 31, 45. WILDLIFE BIOLOGY. *Educ:* Univ Ga, BSF, 67, MS, 71; Colo State Univ, PhD(wildlife biol), 75. *Prof Exp:* Asst prof, Dept Entom & Econ Zool, 75-80, ASSOC PROF, DEPT FISH & WILDLIFE, CLEMSON UNIV, 80- *Concurrent Pos:* Assoc ed, Southeastern Proceedings. *Mem:* Wildlife Soc. *Res:* Effects of forest management on wildlife populations; natural history and management of feral swine; natural history and population dynamics of bobat and gray fox; annual ethograms of eastern bluebird. *Mailing Add:* Dept Wildlife Biol Clemson Univ 201 Sikes Hall Clemson SC 29634

SWEENEY, LAWRENCE EARL, JR, b Charleston, WVa, Mar 27, 42; m 69; c 2. ELECTRICAL ENGINEERING. *Educ:* Stanford Univ, BS, 64, MS, 66, PhD(elec eng), 70. *Prof Exp:* Res assoc radio sci, Stanford Univ, 70; asst dir, Ionospheric Dynamics Lab, 70-72, dir remote measurements lab, 72-85, VPRES & DIR SYST TECHNOL DIV, SRI INT, 85- *Concurrent Pos:* Instr, Stanford Univ, 68-69, lectr, 70-71. *Mem:* Inst Elec & Electronics Engrs; Int Union Radio Sci. *Res:* Ionospheric radio propagation; high frequency signal processing; large antenna arrays; digital signal processing; remote sensing; high frequency radar; radar target scattering characteristics. *Mailing Add:* Syst Technol Div SRI Int Menlo Park CA 94025

SWEENEY, MARY ANN, b Hagerstown, Md, Sept 25, 45; m 74; c 2. PLASMA PHYSICS, ASTROPHYSICS. *Educ:* Mt Holyoke Col, BA, 67; Columbia Univ, MPhil, 73, PhD(astron), 74. *Prof Exp:* Instr astron, Fairleigh Dickinson Univ & William Paterson Col NJ, 72-73; fel plasma physics & fusion res, 74-76, STAFF MEM, SANDIA LABS, 76- *Concurrent Pos:* Pub speaker; vchmn, Plasma Sci & Applications Comt, Inst Elec & Electronic Engrs, 84, chmn, 89-91, mem, 82-84 & 87-, admin comt, 85-, mem, educ & continued prof develop comt, 85-, secy, Nuclear & Plasma Sci Soc, 87-89. *Mem:* Am Astron Soc; Am Phys Soc; Am Nuclear Soc; Inst Elec & Electronics Engrs (vpres, 88); AAAS. *Res:* Chief theorist for plasma opening switch research; responsible for monitoring and coordinating research by contractors and organizing workshops on opening switches; modeling particle beam fusion target experiments and electron-beam-pumped laser experiments; designing targets and reactor chambers for particle beam fusion; determining the radiation environment in accelerators, and evaluating effects of electron and photon bombardment on dielectric properties; evaluating effects of ion beams on diagnostics; interested in promoting scientific careers and improving the quality of science education. *Mailing Add:* Div 1265 Sandia Nat Lab PO Box 5800 Albuquerque NM 87185

SWEENEY, MICHAEL ANTHONY, b Los Angeles, Calif, Dec 5, 31; c 3. PHYSICAL CHEMISTRY. *Educ:* Loyola Univ, Calif, BS, 53; Univ Calif, Berkeley, MS, 55, PhD(chem), 62. *Prof Exp:* Res chemist, Chevron Res Co, 61-66; asst prof, 66-72, ASSOC PROF CHEM, UNIV SANTA CLARA, 72- *Concurrent Pos:* Vis asst prof, Indiana Univ, 65; instr, Univ Exten, Univ Calif, Berkeley, 65-66; consult, USPHS, 68-71; vis res chemist, UK Atomic Energy Res Estab, Harwell, 72-73, Inst Bioorg Chem, Barcelona, Spain, 79. *Mem:* Am Chem Soc. *Res:* Radiation chemistry; chemical evolution. *Mailing Add:* Dept of Chem Univ of Santa Clara Santa Clara CA 95053

SWEENEY, MICHAEL JOSEPH, b Philadelphia, Pa, Jan 31, 39; m 62; c 3. IMMUNOLOGY, IMMUNOCHEMISTRY. *Educ:* Philadelphia Col Pharm & Sci, BSc, 66, BSc, 67; Temple Univ, PhD(microbiol, immunol), 71. *Prof Exp:* Fel immunol, Sch Med, Temple Univ, 71-72; asst prof immunol, Fla Technol Univ, 71-77, assoc prof biol sci, 77-82, PROF CLIN LAB SCI, UNIV CENT FLA, 82- *Mem:* Am Soc Microbiol; Sigma Xi. *Res:* Studies of T and B cell populations in peripheral blood of immunosuppressed patients; characterization of antigens; purification and standardization of allergens. *Mailing Add:* Dept Med Technol Univ Cent Fla Box 25000 Orlando FL 32816

SWEENEY, ROBERT ANDERSON, b Freeport, NY, Oct 11, 40; m 63, 79; c 4. PHYCOLOGY, LIMNOLOGY. *Educ:* State Univ Col Albany, BS, 62; Ohio State Univ, MS, 64, PhD(water resources), 66. *Prof Exp:* From asst prof to prof biol, State Univ NY Col Buffalo, 66-68, dir, Great Lakes Lab, 67-81, prof, 68-81; DIR SPEC PROJS, ECOL & ENVIRON INC, 81- *Concurrent Pos:* Consult, NY State Dept Environ Conserv, 70-; consult, US Army Corps Engrs, 71-, mem, Shoreline Erosion Adv Panel, 74-; US Army Corps Engrs grant, 71-, Environ Protection Agency grant, 72-; adj prof biol, State Univ NY Col Buffalo, 81- *Honors & Awards:* Anderson-Everett Award, Int Asn Great & Lakes Res, 82. *Mem:* AAAS; Int Asn Great Lakes Res (pres, 80); Phycol Soc Am; Am Inst Biol Sci; Am Soc Limnol & Oceanog. *Res:* Evaluation and solution of water pollution and eutrophication problems in the Great Lakes and their tributaries; dredging; bioassay. *Mailing Add:* Ecol & Environ Inc 368 Pleasantview Dr Lancaster NY 14086

SWEENEY, THOMAS L(EONARD), b Cleveland, Ohio, Dec 12, 36; m 61; c 4. CHEMICAL ENGINEERING, RESEARCH ADMINISTRATION. *Educ:* Case Inst Technol, BS, 58, MS, 60, PhD(chem eng), 62; Capital Univ, JD, 74. *Prof Exp:* Tech specialist, Standard Oil Co, Ohio, 62-63; from asst prof to assoc prof chem eng, 63-73, assoc vpres res, 82-89, PROF CHEM ENG, OHIO STATE UNIV, 73-, ACTG VPRES RES, 89- *Concurrent Pos:* Consult; Exec Dir, Ohio State Univ Res Found, 88-, pres, 89- *Mem:* Am Inst Chem Engrs; Am Chem Soc; Am Soc Eng Educ. *Res:* Air pollution; heat and mass transfer; chemical technology; occupational safety and health; environmental science and technology; regulation of technology; legal aspects of engineering. *Mailing Add:* Off Res Ohio State Univ 190 N Oval Mall Columbus OH 43210

SWEENEY, THOMAS PATRICK, b Milwaukee, Wis, Aug 27, 29; c 2. REMOVABLE PROSTHODONTICS, COMBAT DENTISTRY. *Educ:* Univ Wis, BS, 51; Marquette Univ, DDS, 55; Brook Gen Hosp, Ft Sam Houston, Tex, MS, 67; Am Bd Prosthodontics, dipl, 70. *Prof Exp:* Pvt pract, Milwaukee, Wis, 57-60; chief prosthodontics, US Army Dent Activ, Ft Knox, Ky, 67-70; consult removable prosthodontics, US Army, Europe, 70-73; clin instr, Sch Dent, Tufts Univ, Boston, 73-77; comdr & dir dent educ, Walson Army Hosp, Ft Dix, NJ, 77-78; comdr & dir dent serv, US Army Dent Activ, Hawaii, 78-80; comdr & dir dent educ, US Army Med Ctr, Washington, DC, 80-85; dep chief res progs, US Army Med Res & Develop Command, Ft Detrick, Md, 85-88; RETIRED. *Concurrent Pos:* Consult removable prosthodontics, First US Army, Ft Knox, Ky, 67-70; clin instr removable prosthodontics, Sch Dent, Univ Louisville, 68-70; guest lectr removable prosthodontics, Walter Reed Army Med Ctr, 76-78; chmn res comt, Am Col Prosthodontists, 80-84. *Mem:* Fel Am Col Prosthodontists; fel Int Col Dentists; Int Asn Dent Res; Am Dent Asn. *Res:* Removable prosthodontics; combat dentistry (maxillofacial injuries). *Mailing Add:* 1011 Arlington Blvd Suite 431 Arlington VA 22209

SWEENEY, THOMAS RICHARD, b Albany, NY, Sept 21, 14; m 41; c 2. MEDICINAL CHEMISTRY. *Educ:* Univ Md, BS, 37, PhD(org chem), 45. *Prof Exp:* Chemist, Briggs Filtration Co, DC, 40-41; chemist, NIH, Md, 41-47; res chemist, Univ Md, 47-50, US Naval Res Lab, 50-59; chemist, Walter Reed Army Inst Res, 59-64, chief, Dept Org Chem, 64-69, dep dir, Div Med Chem, 69-78, dep dir, Div Exp Therapeut, 78-80; CONSULT, 80- *Concurrent Pos:* US Army med res & develop command rep, Med Chem Study Sect A, NIH, 65-75; US deleg NATO Army Armaments Group, Nuclear Biol Chem Defense Panel, Group Experts Chemoprophy Laxis, 75-80; consult, Nat Cancer Inst, 75-83; adv & mem sci & tech rev comt, WHO, 81-84; mem adv comt med chem, US Army med Res & Develop Command, 82-85, chmn, 84 & 85. *Mem:* Am Chem Soc; Sigma Xi. *Res:* Antiradiation and antiparasific agents. *Mailing Add:* 1701 N Kent St Apt 303 Arlington VA 22209

SWEENEY, WILLIAM ALAN, b Can, Sept 12, 26; nat US; m 53; c 3. PETROCHEMICALS. *Educ:* Univ BC, BASc, 49; Univ Wash, Seattle, PhD, 54. *Prof Exp:* Chemist, Can Industs, Ltd, 49-50; asst to vpres, Chevron Res Co, 75-76; res scientist, Chevron Res & Technol Co, Chevron Corp, Calif, 54-90; RETIRED. *Concurrent Pos:* Consult, Dept Energy, 90. *Mem:* Am Chem Soc; Sigma Xi; NY Acad Sci. *Res:* Petrochemicals exploratory, process and product development, and plant support emphasizing practicality and economics; effect of chemical structure on surfactant performance, biodegradability and paper sizing; alpha olefin processing and applications; synlubes, gasoline oxygenates. *Mailing Add:* 27 Corte del Bayo Larkspur CA 94939

SWEENEY, WILLIAM JOHN, b Oak Park, Ill, July 15, 40; m 64. MATHEMATICS. *Educ:* Univ Notre Dame, AB, 62; Stanford Univ, MS, 64, PhD(math), 66. *Prof Exp:* Instr math, Stanford Univ, 66-67; asst prof, Princeton Univ, 67-71; ASSOC PROF MATH, RUTGERS UNIV, NEW BRUNSWICK, 71- *Mem:* Math Asn Am; Am Math Soc. *Res:* Over-determined systems of linear partial differential equations. *Mailing Add:* Dept of Math Rutgers Univ New Brunswick NJ 08903

SWEENEY, WILLIAM MORTIMER, b Brooklyn, NY, Aug 4, 23; m 52; c 3. PETROLEUM CHEMISTRY. *Educ:* St John's Univ, BS, 43; Fordham Univ, MS, 47; Univ Colo, PhD(org chem), 52. *Prof Exp:* Chemist, Transformer Div, Gen Elec Co, Mass, 47-49; sr res chemist, Beacon Res Labs, Texaco, Inc, 52-82; RETIRED. *Mem:* Sigma Xi. *Res:* Organometallics, principally ferrocene; ester based synthetic luoricants for jet engines; fluorine chemistry; additives for gasoline and diesel fuels; pour depressants fuel oils; polymers. *Mailing Add:* Beacon Res Labs Texaco Inc PO Box 509 Beacon NY 12508

SWEENEY, WILLIAM VICTOR, b Cleveland, Ohio, Jan 31, 47; m 68; c 1. BIOPHYSICS. *Educ:* Knox Col, BA, 68; Univ Iowa, MS, 70, PhD(chem), 73. *Prof Exp:* NIH fel biochem, Univ Calif, Berkeley, 73-75; asst prof, 75-80, ASSOC PROF CHEM, HUNTER COL, CITY UNIV NEW YORK, 80- *Mem:* Biophys Soc. *Res:* Physical properties of iron-sulfur proteins; nuclear magnetic resonance; electron paramagnetic resonance; blood clotting proteins. *Mailing Add:* Dept of Chem Hunter Col 695 Park Ave New York NY 10021

SWEENY, DANIEL MICHAEL, b Rockville Center, NY, Sept 25, 30; m 60; c 5. INORGANIC CHEMISTRY. *Educ:* Col Holy Cross, BSc, 52; Univ Notre Dame, PhD(chem), 55. *Prof Exp:* Res chemist, E I du Pont de Nemours & Co, 55-57; PROF INORG CHEM, BELLARMINE COL, KY, 57- *Mem:* Am Chem Soc; Mineral Soc Am; Soc Appl Spectros. *Res:* Physical properties and synthesis of coordination compounds; infrared spectroscopy; structural inorganic chemistry; interpretive spectroscopy. *Mailing Add:* Dept Chem Bellarmine Col Louisville KY 40205-1877

SWEENY, HALE CATERSON, b Anderson, SC, Mar 31, 25; m 48; c 3. MATHEMATICAL STATISTICS. *Educ:* Clemson Col, BME, 49; Va Polytech Inst, MS, 52, PhD, 56. *Prof Exp:* Design engr, Hunt Mach Works, 49-50; indust engr, Eastman Kodak Co, 51-52; instr indust eng, Va Polytech Inst, 52-53, asst prof statist, 53-56; res statistician, Atlantic Ref Co, 56-59; consult, 59-60; sr res statistician, Res Triangle Inst, 60-64, mgr spec res, 64-72; head, Statist Serv Dept, Burroughs Wellcome Co, 72-90; CONSULT STATISTICIAN, 90- *Mem:* Am Soc Mech Eng; Am Statist Asn; Inst Math Statist; Biomet Soc. *Res:* Development of statistical methodology application to production; chemical and clinical research; design of experiments; design of medical and veterinary clinical trials. *Mailing Add:* 3500 Cambridge Dr Durham NC 27707

SWEENY, JAMES GILBERT, b Philadelphia, Pa, Jan 18, 44; m. NATURAL PRODUCTS CHEMISTRY. *Educ:* Eckerd Col, BS, 65; Yale Univ, PhD(org chem), 69. *Prof Exp:* Fel chem, Yale Univ, 69-71, R Russell Agr Res Ctr, 71-72, Univ Glasgow, 72-73 & Univ Va, 73-74; RES SCIENTIST, COCA-COLA CO, 74- *Concurrent Pos:* Sr scientist, Coca-Cola Co. *Mem:* Am Chem Soc; Royal Soc Chem; Sigma Xi. *Res:* Isolation, structure determination and synthesis of natural colorants; development of artificial sweeteners. *Mailing Add:* Corp Res & Develop-Technol 422 Cola-Cola Co Atlanta GA 30301

SWEENY, KEITH HOLCOMB, physical chemistry, for more information see previous edition

SWEENY, LAUREN J, MYOSIN GENE EXPRESSION, CARDIAC DEVELOPMENT. *Educ:* Univ Nebr Med Ctr, PhD(anat), 81. *Prof Exp:* Res assoc, Dept Med, Univ Chicago, 81-86; ASSOC PROF, DEPT ANAT, SCH MED LOYOLA UNIV, 86- *Mailing Add:* 700 W Cornelia Chicago IL 60657

SWEENY, ROBERT F(RANCIS), b Ridley Park, Pa, Sept 9, 31; m 52; c 4. CHEMICAL ENGINEERING, PROCESS CONTROL. *Educ:* Pa State Univ, BS, 53, MS, 55, PhD, 60. *Prof Exp:* Lab mgr, Appl Sci Labs, 55-64; from asst prof to assoc prof, 64-83, prof chem eng & chmn dept, 83-88, PROF CHEM ENG, VILLANOVA UNIV, 88- *Mem:* Sigma Xi; Am Inst Chem Engrs; Instrument Soc Am; Am Soc Eng Educr. *Res:* Applied mathematics; process control; separations and purification; wastewater treatment. *Mailing Add:* Dept Chem Eng Villanova Univ Villanova PA 19085

SWEET, ARNOLD LAWRENCE, b New York, NY, Mar 23, 35; m 59; c 2. TIME SERIES FORECASTING, APPLIED STOCHASTIC PROCESSES. *Educ:* Col City New York, BME, 56; Md Univ, MSME, 59; Purdue Univ, PhD(eng sci), 64. *Prof Exp:* Mech engr, Emerson Res Lab, 56-58; mech engr struct dynamics, US Naval Res Lab, 58-60; from asst prof to assoc prof eng sci, 64-73, PROF INDUST ENG, PURDUE UNIV, 73- *Concurrent Pos:* Consult, Midwest Appl Sci Corp, 64-67; vis res fel accident anal, Rd Res Lab, Dept Environ, Gt Brit, 70-71; resident res assoc, US Air Force Systs Command Univ, 79-80. *Mem:* Inst Mgt Sci; Am Inst Indust Engrs; Am Soc Qual Control; Int Inst Forecasters; Am Asn Univ Professors. *Res:* Time series forecasting; applications of probability theory to engineering problems; operations research; statistical quality control. *Mailing Add:* Sch Indust Eng Purdue Univ Grissom Hall West Lafayette IN 47907

SWEET, ARTHUR THOMAS, JR, b Salisbury, NC, Jan 19, 20; m 43; c 2. SYNTHETIC FIBER CHEMISTRY. *Educ:* Univ NC, BS, 41; Ohio State Univ, PhD(chem). 48. *Prof Exp:* Staff chemist, Uranium Isotope Prod Dept, Tenn Eastman Corp, 44-46; res chemist, Nylon Res Div, E I Du Pont de Nemours & Co, 48-54, Dacron Res Div, 54-57 & Textile Fibers Patent Div, 57-74, patent assoc, Textile Fibers Patent Liaison Div, 74-85; RETIRED. *Mem:* Am Chem Soc. *Res:* Constitution of Grignard reagent; chemical characteristics of synthetic fibers; patent management. *Mailing Add:* 23 Penarth Dr Wilmington DE 19803-2011

SWEET, BENJAMIN HERSH, b Boston, Mass, Dec 14, 24; m 47; c 3. VIROLOGY, MICROBIOLOGICAL QUALITY ASSURANCE. *Educ:* Tulane Univ, BS, 46; Boston Univ, MA, 48, PhD(med sci), 51. *Prof Exp:* Res assoc virol, Res Found, Children's Hosp, Cincinnati, 51-54; asst prof microbiol, Sch Med, Univ Md, 54-59; sr investr virol, Merck Sharp & Dohme Res Labs, 59-64; dir & mgr res & develop, Flow Labs, Inc, Md, 64-66; assoc dir life sci div, Gulf South Res Inst, 66-75; mgr, sr scientist & doc adminr, qual assurance, Cutter Labs, Div Miles Labs, 75-85; RETIRED. *Concurrent Pos:* Consult, microbiol, virol & qual assurance, 85- *Mem:* Am Soc Microbiol; Am Soc Trop Med & Hyg; Tissue Cult Asn; Soc Exp Biol & Med. *Res:* Arthropod borne, respiratory, oncogenic, latent viruses, vaccine development; viral immunology and diagnostics; ecology and zoonoses; immunology-adjuvants; water pollution; cell biology; quality assurance; biological and immunological assays; limulus amebocyte research and development; good laboratory and manufacturing practices regulations, regulatory affairs. *Mailing Add:* Six Admiral Dr Apt #489 Emeryville CA 94608

SWEET, CHARLES EDWARD, b Elgin, Tex, Dec 27, 33; m 55; c 4. BACTERIOLOGY, MYCOLOGY. *Educ:* Univ Tex, BA, 55, MA, 63; Univ NC, MPH, 67, DPH(parasitol), 69. *Prof Exp:* Jr bacteriologist, Br Lab, NMex State Health Dept, 60-61; bacteriologist, Tex State Health Dept, 62, Tyler Tex Br Lab, 63-66, spec proj dir, 69-70, asst dir, Lab Servs, 70-73, DIR LAB, TEX STATE DEPT HEALTH, 70- *Res:* Laboratory methodology. *Mailing Add:* Lab Sect Tex State Health Dept 1100 W 49th St Austin TX 78756-3194

SWEET, CHARLES SAMUEL, b Cambridge, Mass, Apr 6, 42; wid; c 2. PHARMACOLOGY. *Educ:* Northeastern Univ, BS, 66, MS, 68; Univ Iowa, PhD(pharmacol), 71. *Prof Exp:* Res asst pharmacol, Warner-Lambert Res Inst, NJ, 66-68; fel, Col Pharm, Northeastern Univ, 66-68; fel, Col Med, Univ Iowa, 68-71; fel res, Cleveland Clin Educ Found, 71-72; res fel, Merck Inst Therapeut Res, 72-75, sr res fel, 75-76, dir cardiol pharmacol, 76-82, sr scientist, 82-89; DIR CLIN RES, MERCK, SHARP & DOHME RES LABS, 90- *Concurrent Pos:* Mem, Med Adv Bd, Coun High Blood Pressure; mem, Coun Thrombosis, Am Heart Asn. *Mem:* Am Soc Pharmacol & Exp Therapeut; Am Heart Asn. *Res:* Renin-angiotensin system in pathogenesis of experimental hypertension; participation of central nervous system in development and maintenance; action of antihypertensive drugs, particularly as they apply to known causes of hypertension; clinical hypertension. *Mailing Add:* Clin Res Merck, Sharp & Dohme Res Labs West Point PA 19486

SWEET, DAVID PAUL, b Dixon, Mar 24, 48; m 68. ANALYTICAL CHEMISTRY. *Educ:* Bradley Univ, BA, 70; Univ Colo, PhD(anal chem), 74. *Prof Exp:* Anal chemist, 74-78, MGR ANAL SERV, DIV SYNTEX, ARAPAHOE CHEM INC, 78- *Mem:* Am Chem Soc; Am Soc Mass Spectrometry. *Res:* Chromatographic separations and trace analysis, especially using combined vapor phase chromatography-mass spectrometry and liquid chromatography-mass spectrometry. *Mailing Add:* 1231 N Fork Hwy Wapiti WY 82450

SWEET, FREDERICK, b New York, NY, May 15, 38; m 62, 88; c 4. BIOCHEMISTRY, ORGANIC CHEMISTRY. *Educ:* Brooklyn Col, BS, 60; Univ Alta, PhD(org chem), 68. *Prof Exp:* Substitute instr chem, Brooklyn Col, 60-62; instr, Bronx Community Col, 62-64; NIH res fel nucleoside chem, Sloan-Kettering Inst Cancer Res, 68-70; res assoc reprod biochem, Univ Kans Med Ctr, Kansas City, 70-71; asst prof reprod biochem, 71-76, res assoc prof, 76-80, assoc prof obstet & gynec, 80-82, PROF REPROD BIOL OBSTET & GYNEC, WASH UNIV, 82-, DIR DIV REPROD BIOL, 87- *Concurrent Pos:* lectr, Bronx Community Col, 68-70; vis asst prof biol sci, Southern Ill Univ, Edwardsville, 73-75, vis prof chem, 76-78; res fel, NATO, 74-76 & 77-78; NIH res career develop award, 75; mem int exchange, Nat Acad Sci, Hungary, 77-78 & 79 & Int Res & Exchange Bd, Princeton, NJ, 91; ed, Endocrin Rev, 86-90, steroids, 90- *Mem:* AAAS; Am Chem Soc; Chem Inst Can; Endocrine Soc. *Res:* Reproductive biochemistry, mechanism of steroid action and metabolism; synthesis of affinity-labeling steroids; synthesis of nucleosides and nucleoside analogs; synthesis of boron-estrogens for neutron capture therapy of cancers; daunorubicin-antibody conjugates for immunotherapy of cancer. *Mailing Add:* Dept Obstet Gynec Wash Univ Sch Med 449 S Euclid Ave St Louis MO 63110

SWEET, GEORGE H, b Texhoma, Okla, Feb 4, 34; m 55; c 3. IMMUNOLOGY. *Educ:* Wichita State Univ, BS, 60; Univ Kans, MA. 62, PhD(immunol), 65. *Prof Exp:* Immunologist, Armed Forces Inst Path, 65-66; from asst prof to assoc prof, 66-71, PROF BIOL, WICHITA STATE UNIV, 72- *Mem:* AAAS. *Res:* Cell biology; fungal serology; viral immunology. *Mailing Add:* Dept of Biol Wichita State Univ Wichita KS 67208

SWEET, GERTRUDE EVANS, zoology; deceased, see previous edition for last biography

SWEET, HAVEN C, b Boston, Mass, Mar 1, 42; m 63; c 2. PLANT PHYSIOLOGY. *Educ:* Tufts Univ, BS, 63; Syracuse Univ, PhD(plant physiol), 67. *Prof Exp:* Res fel photobiol, Brookhaven Nat Labs, 67-68; sr res analyst bot, Brown & Root-Northrop, Tex, 68-69, supvr, 69-71; from asst prof to assoc prof, Fla Technol Univ, 71-85, PROF BIOL, UNIV CENT FLA, 85- *Mem:* AAAS; Am Soc Plant Physiol; Bot Soc Am; Am Inst Biol Sci; Linnean Soc. *Res:* Effects of light on plant growth; development of computer-assessment of plant taxonomic, remote sensing and ecological information. *Mailing Add:* Dept Biol Univ Cent Fla Orlando FL 32816

SWEET, HERMAN ROYDEN, b Attleboro, Mass, Nov 3, 09; m 31; c 2. BOTANY, MICROBIOLOGY. *Educ:* Bowdoin Col, AB, 31; Harvard Univ, AB, 34, PhD(mycol), 40. *Prof Exp:* Asst, Harvard Univ, 36-37; instr, 37-42, from asst prof to assoc prof, 42-54, prof, 54-75, EMER PROF BIOL, TUFTS UNIV, 75- *Concurrent Pos:* Res assoc, Orchid Herbarium of Oakes Ames, Harvard Univ, 65-76, hon cur, 76- *Mem:* AAAS; Am Soc Microbiol; Brit Soc Gen Microbiol; Am Ornith Union; fel Linnean Soc. *Res:* Orchidology. *Mailing Add:* 1111 S Lakemont Ave No 240 Winter Park FL 32792-5470

SWEET, JOHN W, b Seattle, Wash, May 6, 10. METALLURGY. *Educ:* Univ Wash, BS, 34. *Prof Exp:* Chief metallurgist, Boeing Co, 46-48; RETIRED. *Mem:* Fel Am Soc Metals Int; Am Soc Testing & Mat. *Mailing Add:* 1605 Fifth Ave N Hillside House No 301 Seattle WA 98109

SWEET, LARRY ROSS, b Fairbanks, Alaska, June 2, 40; m 79; c 2. ARCTIC ENGINEERING. *Educ:* Wash State Univ, BS, 63; Univ Alaska, MS, 72. *Prof Exp:* Assoc design engr, Lockheed Missiles & Space Co, 63-65; eng aide, Geophys Inst, Univ Alaska, 65, asst design engr, 66-69, head tech serv, 69-70, assoc supvry engr, 70-75, grad lab instr, physics dept, 65-66, exec officer to vpres res, 75-76 & exec officer to vchancellor res & advan study, 76-80; statewide res mgr, 80-85, CHIEF, SPEC RES PROJS, ALASKA DEPT TRANSP & PUB FACIL, 86-; res assoc, Inst Northern Eng, Univ Alaska, 86, CHIEF, SPEC RES PROJS, ALASKA DEPT TRANSP & PUB FACIL, 86-, SYST ENGR, ALASKA SAR FACIL, GEOPHYS INST, UNIV ALASKA, 87- *Concurrent Pos:* Mem, comt arctic oil & gas resources, Nat Petrol Coun, 80-81, comt res, Am Soc Civil Engrs, 83-88, adv comt snow & ice, Am Asn State Transp Off, 85 & comt conduct res, Nat Res Coun, Transp Res Bd, 85-87; US Permafrost Deleg to People's Repub China, 84; mem, Alaska State Climate Adv Bd, 83-87; mem adv bd, Univ Alaska Transp Ctr, 83-86, Northwest Technol Transfer Ctr, 84-87; mem, Alaska State Climat Adv Bd, 83-87. *Mem:* Arctic Inst NAm; AAAS; Inst Elec & Electronics Engrs; Explorers Club. *Res:* Administration and coordination of basic and applied research in science and engineering in cold regions; state, national and international coordination of arctic research. *Mailing Add:* 19231 Swallow Dr Fairbanks AK 99709-8360

SWEET, LEONARD, b Akron, Ohio, Aug 28, 25; m 46; c 2. STATISTICS. *Educ:* Univ Akron, BA, 49; Kent State Univ, MEd, 54; Case Western Reserve Univ, PhD(statist), 70. *Prof Exp:* Teacher pub schs, Ohio, 49-57, supvr, 57-59; from asst prof to assoc prof, 59-74, PROF MATH, UNIV AKRON, 74- *Concurrent Pos:* Consult, Akron Pub Schs, Ohio, 62-65 & Addressograph Multigraph Corp, 71-; mem panel evaluating of Instr Sci Equip Prog Proposals, NSF, 78; vpres, Greater Akron Math Educators, 79-81; mem admissions comt, Northeastern Ohio Med Sch, 78-80, biostatist comt, 78-80; grant dir, Local Course Improvement Proj, NSF, 78-80. *Mem:* Am Statist Asn; Math Asn Am; Nat Coun Teachers Math; Sigma Xi. *Res:* Experimental design; symmetrical complementation designs; utilization of the microcomputer by the classroom teacher in mathematics and statistics education. *Mailing Add:* 2430 Thurmont Rd Akron OH 44313

SWEET, MELVIN MILLARD, b South Gate, Calif. NUMBER THEORY. *Educ:* Calif State Univ, Los Angeles, BA, 64, MA, 65; Univ Md, PhD(math), 72. *Prof Exp:* Mathematician, Nat Security Agency, 66-70; vis asst prof math, Univ Md, Baltimore County, 72-74; res staff mem math, Commun Res Div, Inst Defense Anal, 75-78; sr staff mem, Hughes Aircraft Co, 78-80. *Mem:* Am Math Soc; Math Asn Am. *Res:* Diophantine approximations. *Mailing Add:* 1567 Vista Claridad LaJolla CA 92037

SWEET, MERRILL HENRY, II, b Chicago Heights, Ill, Sept 5, 35; m 58; c 4. BIOLOGY. *Educ:* Univ Conn, BS, S8, PhD(entom), 63. *Prof Exp:* Res asst entom, Univ Conn, 62-63; asst prof, 63-66, ASSOC PROF BIOL, TEX A&M UNIV, 66- *Mem:* Ecol Soc Am; Assoc Trop Biol; Soc Study Evolution; Soc Syst Zool. *Res:* Systematics; ecology; behavior and life cycles of arthropods, especially hemipterous insects. *Mailing Add:* Dept of Biol Tex A&M Univ College Station TX 77843

SWEET, RICHARD CLARK, b Tarrytown, NY, Nov 28, 21; m 48; c 3. ANALYTICAL CHEMISTRY, PHYSICAL CHEMISTRY. *Educ:* Wesleyan Col. BA, 44, MA, 48; Rutgers Univ, PhD(chem), 52. *Prof Exp:* Chemist, 52-60, SUPVR METALL SYSTS APPLNS, PHILIPS LABS, N AM PHILIPS CO, BRIARCLIFF MANOR, 60- *Mem:* Am Chem Soc; Am Vacuum Soc; Sigma Xi; AAAS. *Res:* Spectroscopy; trace levels; ion exchange; polarography; water analysis; analytical methods; electronic components; vacuum techniques; ceramic-metal seals; cryogenic components design and fabrication; metals processing and joining techniques; welding and brazing. *Mailing Add:* 309 NWashington St North Tarrytown NY 10591

SWEET, ROBERT DEAN, b Fairview, Ohio, Apr 6, 15; m 36; c 2. VEGETABLE CROPS. *Educ:* Ohio Univ, BS, 36; Cornell Univ, MS, 38, PhD(veg crops), 41. *Prof Exp:* Asst, NY State Col Agr & Life Sci, Cornell Univ, 36-40, exten instr, 40-43, asst exten prof, 43-47, assoc prof, 47-49, prof beg crops, 49-82, chmn dept, 75-82; RETIRED. *Mem:* Sigma Xi; Am Soc Hort Sci; Weed Sci Soc Am. *Res:* Biological and chemical weed control. *Mailing Add:* 14011/2 Slaterville Rd Ithaca NY 14850

SWEET, ROBERT MAHLON, b Omaha, Nebr, Sept 21, 43; m 66; c 3. MOLECULAR BIOLOGY. *Educ:* Calif Inst Technol, BS, 65; Univ Wis-Madison, PhD(phys chem), 70. *Prof Exp:* Lectr chem, Univ Wis-Madison, 70; fel molecular biol, Med Res Coun Lab Molecular Biol, Cambridge, Eng, 70-73; asst prof chem, Univ Calif, Los Angeles, 73-81, specialist, Molecular Biol Inst, 81-83; BIOLOGIST, BROOKHAVEN NAT LAB, 83- *Concurrent Pos:* Damon Runyon Mem Fund fel, 70-72; Europ Molecular Biol Orgn fel, 72. *Mem:* AAAS; Am Crystallog Asn; Sigma Xi. *Res:* Structure and function of enzymes, determined by x-ray diffraction techniques; studies of phycobitiproteins; photosynthetic accessory pigments from algae. *Mailing Add:* Biol Dept Brookhaven Nat Lab Upton NY 11973

SWEET, RONALD LANCELOT, b Bristol, Eng, Feb 6, 23; nat US; m 47. ORGANIC CHEMISTRY. *Educ:* Rutgers Univ, BSc, 44, MSc, 48, PhD, 55. *Prof Exp:* Fel petrol, Mellon Inst, 51-55; CHEMIST, PIGMENTS DEPT, E I DU PONT DE NEMOURS & CO, 55- *Concurrent Pos:* Co-adj, Rutgers Univ, 63-74. *Mem:* Am Chem Soc. *Res:* Pigments. *Mailing Add:* 4261 NE 16 Ave Ft Lauderdale FL 33334-5477

SWEET, THOMAS RICHARD, b Jamaica, NY, Sept 27, 21; m 48; c 2. CHEMISTRY. *Educ:* City Col New York, BS, 43; Ohio State Univ, PhD(chem), 49. *Prof Exp:* Asst, Manhattan Proj, War Res Div, Columbia Univ, 43-45 & Carbide & Carbon Chem Corp, 45-46; from asst prof to assoc prof, 49-65, PROF CHEM, OHIO STATE UNIV, 65- *Mem:* Am Chem Soc. *Res:* Organic reagents, solvent extraction and trace metal analysis. *Mailing Add:* Dept of Chem Ohio State Univ 140 W 18th Ave Columbus OH 43210

SWEET, WILLIAM HERBERT, b Kerriston, Wash, Feb 13, 10; m 37; c 3. NEUROSURGERY. *Educ:* Univ Wash, SB, 30; Oxford Univ, BSc, 34; Harvard Univ, MD, 36; Am Bd Psychiat & Neurol, dipl, 46; Am Bd Neurol Surg, dipl, 46; FRCS(E), 86. *Hon Degrees:* DSC, Oxford Univ, 57; DHC, Univ Sci Med, Grenoble, France, 79. *Prof Exp:* Instr neurosurg, Billings Hosp, Chicago, 39-40; Commonwealth Fund fel, Harvard Med Sch, 40-41; actg chief neurosurg serv, Birmingham United Hosp, 41-45; from instr to asst prof, 45-54, assoc clin prof, 54-58, from assoc prof to prof, 58-76, EMER PROF SURG, HARVARD MED SCH, 76- *Concurrent Pos:* Regional consult, Brit Emergency Med Serv, 41-45; asst, Mass Gen Hosp, 45-47, asst neurosurgeon, 47-48, assoc vis neurosurgeon, 48-58, vis neurosurgeon, 58-, chief neurosurg serv, 61-76, mem hosp staff & consult vis neurosurgeon, 76-80, sr neurosurgeon, 80-; lectr, Med Sch, Tufts Col, 47-51; neurosurgeon in chief, New Eng Ctr Hosp, 49-51; mem subcomt neurosurg, Nat Res Coun, 49-52 & mem subcomt neurol & neurosurg, 52-59; trustee, Assoc Univs, Inc, 58-; mem sci & technol adv comt, NASA, 64-70; mem neurol sci res training A comt, Nat Inst Neurol Dis & Stroke; honored guest, Cong Neurol Surgeons, 75; vis prof, Royal Soc Med Found, 79, Buenos Aires Univ Med Sch, 81 & Univ Ziekenhuizen, Belg, 87; permanent hon pres, World Fedn Neurosurg Socs, 84. *Honors & Awards:* Harvey Cushing Medal, Am Asn Neurol Surgeons, 78; John J Bonica Lectr, Eastern Pain Asn, 80; J Jay Keegan Mem Lectr, Univ Nebr, 81; Order of the Rising Sun, Japan, 83; Frank H Mayfield Lectr, Univ Cincinnati, 86; Herbert Olivecrona Lectr, Karolinska Inst, 89; Samuel Clark Harvey Mem Lectr, Yale Univ, 90; F W L Kerr Mem Award Lectr, Am Pain Soc, 90; Distinguished Serv Award, Soc Neurol Surgeons, 91. *Mem:* Sr mem Inst Med-Nat Acad Sci; Am Acad Neurol Surg (pres, 76-77); Am Neurol Asn (vpres, 71-72); Am Pain Soc (pres elect, 80-81, pres, 81-82); Soc Neurol Surgeons (pres, 69-70); fel Am Acad Arts & Sci. *Res:* Central nervous system; research in cerebrospinal and intracerebral fluid; brain tumors; mechanisms of pain and its neurosurgical control; abnormal behavior related to organic brain disease; irreversible coma; ethics of experimentation. *Mailing Add:* Mass Gen Hosp Neurosurg Serv 15 Parkman St Boston MA 02114

SWEETING, LINDA MARIE, b Toronto, Ont, Dec 11, 41. SPECTROSCOPY. *Educ:* Univ Toronto, BSc, 64, MA, 65; Univ Calif, Los Angeles, PhD(org chem), 69. *Prof Exp:* Asst prof, Occidental Col, 69-70; from asst prof to assoc prof, 70-84, PROF CHEM, TOWSON STATE UNIV, 84- *Concurrent Pos:* Guest worker, Nat Inst Arthritis, Metab & Digestive Dis, NIH, 77-78; prog dir, chem instrumentation, NSF, 81-82; vis scholar, Harvard Univ, 84-85; Womens Chemist Comt, Am Chem Soc, 83-89. *Mem:* Am Chem Soc; AAAS; Sigma Xi; Asn Women Sci. *Res:* Application of nuclear magnetic resonance spectroscopy to organic chemistry; triboluminescence; stereochemistry; ethics in science. *Mailing Add:* Dept Chem Towson State Univ Baltimore MD 21204

SWEETMAN, BRIAN JACK, b Palmerston North, NZ, May 4, 36; m 61; c 3. ORGANIC CHEMISTRY, PHARMACOLOGY. *Educ:* Univ NZ, BSc, 58, MSc, 59; Univ Otago, NZ, PhD(org chem), 62. *Prof Exp:* Res officer div protein chem, Commonwealth Sci & Indust Res Orgn, Melbourne, Australia, 63-66; res assoc chem, 66-68, asst prof, 69-76, RES ASSOC PROF PHARMACOL, VANDERBILT UNIV, 76-, RES PROF ANESTHESIOL IN RESIDENCE, 89- *Concurrent Pos:* Assoc dir, Mass Spectrometry Resource, Vanderbilt Univ, 83- *Mem:* Am Soc Mass Spectroscopy. *Res:* Biomedical mass spectrometry; organic mass spectrometry and analytical pharmacology; prostaglandins; vapor-phase analysis; medicinal and organosulfur chemistry; anti-radiation and anti-arthritic drugs; protein chemistry of keratin; natural products. *Mailing Add:* Dept Anesthesiol Vanderbilt Univ Nashville TN 37232-2125

SWEETMAN, LAWRENCE, b La Junta, Colo, Feb 17, 42; m 70. BIOCHEMISTRY, PEDIATRICS. *Educ:* Univ Colo, BA, 64; Univ Miami, PhD(biochem), 69. *Prof Exp:* Res assoc biochem, Sloan-Kettering Inst Cancer Res, 68-72, instr, Sloan-Kettering Div, Grad Sch Med Sci, Cornell Univ, 69-72; from asst prof to prof pediat, Univ Calif, San Diego, 72-90; PROF PEDIAT & PATH, CHILDRENS HOSP, LOS ANGELES, 90- *Mem:* Am Chem Soc; Sigma Xi. *Res:* Metabolism of inherited diseases in children; organic acidurias. *Mailing Add:* Biochem Genetics No 11/CHLA 4650 Sunset Blvd Los Angeles CA 90027

SWEETON, FREDERICK HUMPHREY, physical chemistry, for more information see previous edition

SWEETSER, PHILIP BLISS, analytical chemistry, herbicide metabolism, for more information see previous edition

SWEITZER, JAMES STUART, b South Bend, Ind, Mar 27, 51; m 73; c 2. ASTRONOMY, SCIENCE EDUCATION. *Educ:* Univ Notre Dame, BS, 73; Univ Chicago, MS, 75, PhD(astron, astrophys), 78. *Prof Exp:* ASTRONOMER, ADLER PLANETARIUM, 78-, ASST DIR, 84- *Mem:* Am Astron Soc; Am Asn Mus; Int Planetarium Soc. *Res:* Interstellar molecules; astronomy education. *Mailing Add:* Adler Planetarium 1300 S Lake Shore Dr Chicago IL 60605

SWELL, LEON, b New York, NY, July 26, 27; c 3. BIOCHEMISTRY. *Educ:* City Col New York, BS, 48; George Washington Univ, MS, 49, PhD(biochem), 52. *Prof Exp:* Lab asst, George Washington Univ, 49-51; chief biochemist, Vet Admin Ctr, Martinsburg, WVA, 51-64, CHIEF, LIPID RES LAB, VET ADMIN HOSP, RICHMOND, VA, 64-; RES PROF BIOCHEM & MED, VA COMMONWEALTH UNIV, 70- *Concurrent Pos:* Assoc prof lectr, George Washington Univ, 59-; assoc res prof, Med Col Va, 64- *Mem:* AAAS; Am Soc Biol Chemists; Soc Exp Biol & Med; Am Inst Nutrit. *Res:* Cholesterol, lipid and electrolyte metabolism; enzymes. *Mailing Add:* 505 Baldwind Rd Richmond VA 23229

SWENBERG, CHARLES EDWARD, b Meriden, Conn, Mar 11, 40; div; c 2. BIOPHYSICS. *Educ:* Univ Conn, BA, 62, MS, 63; Univ Rochester, PhD(physics), 68. *Prof Exp:* Res assoc, Univ Ill, 67-69; res assoc, NY Univ, 69-70, asst prof physics, 70-73, res assoc prof chem, 73-81; chief, 81-88, PROJ MGR, ARMED FORCES RADIOBIOL RES INST, 88- *Mem:* Fel Am Phys Soc; Inst Elec & Electronic Engrs; Biophys Soc; Radiation Res Soc; fel AAAS. *Res:* Effects of ionizing radiation on DNA, DNA-drug interactions and mathematical simulations. *Mailing Add:* 9313 Cranford Dr Potomac MD 20854

SWENBERG, JAMES ARTHUR, b Northfield, Minn, Jan 15, 42; m 63; c 2. VETERINARY PATHOLOGY, CHEMICAL CARCINOGENESIS. *Educ:* Univ Minn, DVM, 66; Ohio State Univ, MS, 68, PhD(vet path), 70. *Prof Exp:* NIH trainee path, Ohio State Univ, 66-70, res assoc, 70, asst prof, 70-72, assoc prof, 72; res scientist path, Upjohn Co, 72-76, res sect head, Path & Genetic Toxicol, 76-78; chief path, 78-84, HEAD DEPT BIOCHEM TOXICOL & PATHBIOL, CHEM INDUST INST TOXICOL, 84- *Concurrent Pos:* Consult, Battelle Mem Inst, 71-72, Health Effects Inst, 82 & 85 & Environ Protection Agency, 83 & 85; mem sci adv panel, Chem Indust Inst Toxicol, 77-78; mem NCI carcinogenesis prog sci rev comt, 78-81; adj prof dept path, Univ NC, 78-; adj assoc prof, Duke Univ, 78-; NTP Bd Sci Coun, 82-86, chmn, 85-86; mem, FIFRA sci adv panel, Environ Protection Agency, 85- *Mem:* Am Asn Cancer Res; AAAS; Am Asn Neuropathologists; Am Col Vet Pathologists; Soc Toxicol; Am Asn Pathologists. *Res:* Cancer research, including chemical carcinogenesis, neurooncogenesis and chemotherapy, and short-term tests for carcinogens; DNA damage/mutagensis; improved toxicology and data handling methods; inhalation toxicology; mechanisms of toxicity and carcinogenisity; risk assessment; toxicology. *Mailing Add:* Chem Indust Inst of Toxicol PO Box 12137 Research Triangle Park NC 27709

SWENDSEID, MARIAN EDNA, b Petersburg, NDak, Aug 2, 18. BIOCHEMISTRY. *Educ:* Univ NDak, BA, 38, MA, 39; Univ Minn, PhD(physiol chem), 41. *Prof Exp:* Asst nutrit, Univ Ill, 42; res biochemist, Simpson Mem Inst, Univ Mich, 42-43; sr res chemist, Parke Davis & Co, Mich, 45-48; res biochemist, Simpson Mem Inst, Univ Mich, 48-52; assoc prof, 53-72, PROF NUTRIT, UNIV CALIF, LOS ANGELES, 72- *Mem:* Am Chem Soc; Am Soc Biol Chem. *Res:* Vitamin research; biochemical aspects of hematology; the use of carbon 13 in the study of intermediary metabolism; amino acids in nutrition. *Mailing Add:* Biol Chem 33-257 CHS Univ Calif Sch Pub Health Los Angeles CA 90024

SWENDSEN, ROBERT HAAKON, b New York, NY, Apr 4, 43; m 71; c 2. PHYSICS. *Educ:* Yale Univ, BS, 64; Univ Pa, PhD(physics), 71. *Prof Exp:* Res asst physics, Univ Cologne, 71-73; physicist, Kernforschungsanlage Julich, Ger, 74-76; assoc physicist, Brookhaven Nat Lab, 76-79; res staff, IBM Zurich Res Lab, Switz, 79-84; PROF, DEPT PHYSICS, CARNEGIE MELLON UNIV, 84- *Mem:* Fel Am Phys Soc. *Res:* Solid state physics; phase transitions; magnetism; computer simulations. *Mailing Add:* Dept Physics Carnegie Mellon Univ 5000 Forbes Pittsburgh PA 15213

SWENERTON, HELENE, b Norfolk, Va, Jan 13, 25; m 43; c 3. NUTRITION. *Educ:* Univ Calif, Davis, BS, 63, MS, 65, PhD(nutrit), 70. *Prof Exp:* Res nutritionist, 70-72, nutri specialist, 72-91, EMER NUTRIT SPECIALIST, UNIV CALIF, DAVIS, 91- *Mem:* AAAS; Am Inst Nutrit; Soc Nutrit Educ; Am Dietetics Assoc. *Res:* Role of dietary zinc in mammalian growth and development; effects of maternal dietary deficiencies on fetal development; influence of nutrition education on consumer decisions. *Mailing Add:* Dept of Nutrit Univ of Calif Davis CA 95616

SWENSEN, ALBERT DONALD, biochemistry, for more information see previous edition

SWENSON, CHARLES ALLYN, b Clinton, Minn, Aug 31, 33; m 60; c 3. BIOPHYSICS. *Educ:* Gustavus Adolphus Col, BS, 55; Univ Iowa, PhD(chem), 59. *Prof Exp:* Asst, Univ Iowa, 55-56; asst prof chem, Wartburg Col, 58-60; res assoc, 60-62, from asst prof to assoc prof, 62-72, assoc head dept, 84-87, PROF BIOCHEM, UNIV IOWA, 72- *Concurrent Pos:* NIH res career devlop award. *Mem:* Am Chem Soc; Biophys Soc; Am Soc Biol Chemists; Protein Soc. *Res:* Physical biochemistry; spectroscopic and thermodynamic approaches for protein structure; energy transduction and regulation in muscle contraction. *Mailing Add:* Dept Biochemistry Univ Iowa Iowa City IA 52242

SWENSON, CLAYTON ALBERT, b Hopkins, Minn, Nov 11, 23; m 50, 80; c 3. EXPERIMENTAL SOLID STATE PHYSICS. *Educ:* Harvard Univ, BS, 44; Oxford Univ, DPhil(physics), 49. *Prof Exp:* Instr physics, Harvard Univ, 49-52; res physicist, Div Indust Coop, Mass Inst Technol, 52-55; from asst prof to prof, 55-87, chmn dept, 75-82, EMER PROF PHYSICS, IOWA STATE UNIV, 87- *Concurrent Pos:* Consult, Commonwealth Sci & Indust Res Orgn, Sidney, Australia, 64-65, Nat Phys Lab, UK, 74-75 & Los Alamos Nat Lab, 85. *Mem:* Fel Am Phys Soc; AAAS; Sigma Xi. *Res:* Low temperatures; high pressures; low temperature thermodynamics; thermometry; equations of state of solids. *Mailing Add:* Dept Physics Iowa State Univ Ames IA 50011

SWENSON, DAVID HAROLD, b Moorhead, Minn, June 16, 48; m 85; c 1. CARCINOGENESIS, CANCER CHEMOTHERAPY. *Educ:* Univ Minn, BS, 70; Univ Wis, PhD(oncol), 75. *Prof Exp:* NIH fel carcinogenesis, Inst Cancer Res, Chester Beatty Res Inst, 75-77; res chemist, Nat Ctr Toxicol Res, 77-79; res scientist genetic toxicol, Upjohn Co, 79-84, sr res scientist, Cancer Res, 84-87; vpres, Karkinos Biochem Inc, 87-90; ASSOC PROF, SCH VET MED, LA STATE UNIV, 90- *Concurrent Pos:* Vis & prin scientist, dept biol, Western Mich Univ, 87. *Mem:* Am Asn Cancer Res; AAAS; Res Soc NAm; Am Chem Soc; Gen Toxicol Assn. *Res:* Elucidation of the mechanism of interaction of genotoxins, carcinogens and cancer chemotherapeutic agents with nucleic acid. *Mailing Add:* Vet Physics/Pharmacol & Toxicol Sch Vet Med La State Univ Baton Rouge LA 70803

SWENSON, DONALD ADOLPH, b Camden, Ala, May 9, 32; m 55; c 4. PHYSICS, MATHEMATICS. *Educ:* Univ Ala, BS, 53; Univ Minn, MS, 56, PhD(physics), 58. *Prof Exp:* Physicist, Midwestern Univs Res Asn, 58-60 & 61-64; Ford Found fel, European Orgn Nuclear Res, Switz, 60-61; physicist, Los Alamos Sci Lab, 65-80, fel, Los Alamos Nat Lab, 80-; AT DEPT PHYSICS, TEX A&M UNIV; SR SCIENTIST, SAIC. *Mem:* Am Phys Soc. *Res:* Particle accelerator, design and development; particle dynamics in circular and linear accelerators; proton beam measurement and diagnostics; development and use of computer control systems for particle accelerators. *Mailing Add:* Sci App Int Corp 4161 Campus Point Ct San Diego CA 92121

SWENSON, DONALD OTIS, b Manhattan, Kans, Feb 19, 37; m 62; c 4. ENGINEERING. *Educ:* Univ Kans, BSME, 63, MSc, 65, PhD(eng mech), 67. *Prof Exp:* Sr res assoc, Advan Mat Res & Develop Lab, Pratt & Whitney Aircraft Div, United Aircraft Corp, Conn, 67-71; CONSULT ENGR, BLACK & VEATCH, 71- *Mem:* Fel Am Soc Mech Engrs; Am Soc Testing & Mat. *Res:* Air pollution control technology and systems for fossil fueled power plants; material behavior; fatigue of materials; control systems; co-author of four books on air pollution control technology,testifies as expert witness on air pollution control technology; energy technology. *Mailing Add:* Black & Veatch 8400 Ward Pkwy Kansas City MO 64114

SWENSON, FRANK ALBERT, b Davenport, Iowa, Feb 3, 12; m 43, 87; c 2. HYDROLOGY. *Educ:* Augustana Col, AB, 36; Univ Iowa, MS, 40, PhD(geol), 42. *Prof Exp:* Surv man & geologist, US Army Corps Engrs, 37-38; asst geol, Univ Iowa, 38-42; from asst geologist to geologist, Mil Geol Br, US Geol Surv, 42-46, ground water investr, 46-63, res geologist, 63-74; CONSULT GEOLOGIST-HYDROLOGIST, 75- *Concurrent Pos:* Consult, US Dept Justice; geologist, Ground Water Br, US Geol Surv, 42. *Mem:* Fel Geol Soc Am; Int Geol Cong; Int Asn Hydrogeol. *Res:* Geology and ground water investigations in Montana and Wyoming; limestone hydrology; geochemistry. *Mailing Add:* 9906 Hawthorn Dr Sun City AZ 85351-3827

SWENSON, G(EORGE) W(ARNER), JR, b Minneapolis, Minn, Sept 22, 22; m 43; c 4. ELECTRICAL ENGINEERING, ASTRONOMY. *Educ:* Mich Col Mining & Technol, BS, 44, EE, 50; Mass Inst Technol, SM, 48; Univ Wis, PhD, 51. *Prof Exp:* Asst elec eng, Mass Inst Technol, 46-48; instr, Univ Wis, 48-51; assoc prof, Wash Univ, St Louis, 51-53; prof, Univ Alaska, 53-54; assoc prof, Mich State Univ, 54-56; assoc prof, 56-58, prof elec eng & res prof astron, 58-70, actg head astron dept, 70-72, head elect eng dept, 79-85, prof, 56-88, EMER PROF, ELEC & COMPUT ENG DEPT, UNIV ILL, URBANA, 88- *Concurrent Pos:* Res prof, Geophys Inst, Univ Alaska, 54-56; consult, NSF, 59-75, US Army Ballistic Missile Agency, 59- 60, NASA & Nat Acad Sci; mem, Adv Comt, Nat Radio Astron Observ, 59-80, vis scientist & chmn, Very Large Array Design Group, 64-68,; vis assoc, Calif Inst Technol, 72-73; Guggenheim fel, 85; sr res assoc acoust, Construct Eng Res Lab, US Army, 88- *Mem:* Nat Acad Eng; fel AAAS; fel Inst Elec & Electronics Engrs. *Res:* Antenna design; radio astronomy; ionospheric radio propagation; acoustics. *Mailing Add:* Dept of Elec & Comput Eng Univ of Ill 1406 W Green St Urbana IL 61801

SWENSON, GARY RUSSELL, b Grantsburg, Wis, June 17, 41; m 67; c 2. ATMOSPHERIC PHYSICS. *Educ:* Wis State Univ-Superior, BS, 63; Univ Mich, Ann Arbor, MS, 68, PhD(atmospheric sci), 75. *Prof Exp:* SPACE SCIENTIST ATMOSPHERIC SCI, MARSHALL SPACE FLIGHT CTR, NASA, 68- *Mem:* Am Geophys Union; Sigma Xi. *Res:* Experimental research using remote sensing techniques, upper atmospheric phenomena, including aurora. *Mailing Add:* 960 North California Palo Alto CA 94303-3405

SWENSON, HENRY MAURICE, b Brooklyn, NY, Aug 13, 16; m 41; c 4. DENTISTRY. *Educ:* Univ Ill, BS, 41, DDS, 42; Am Bd Periodont, dipl, 51. *Prof Exp:* From instr to assoc prof, 45-62, PROF PERIODONT, SCH DENT, IND UNIV, INDIANAPOLIS, 62-, DIR CLIN, 56- *Concurrent Pos:* Consult, Vet Admin, 58- & US Dept Army, 59- *Mem:* Am Dent Asn; fel Am Col Dentists; Am Acad Periodont; Int Asn Dent Res. *Res:* Treatment and management of periodontal involvement. *Mailing Add:* Dept Periodont Ind Univ Purdue Univ Med 1100 W Mich St Indianapolis IN 46223

SWENSON, HUGO NATHANAEL, b New Richland, Minn, Mar 11, 04; m 56. PHYSICS. *Educ:* Carleton Col, AB, 25; Univ Ill, MS, 27, PhD(spectros), 30. *Prof Exp:* Asst physics, Univ Ill, 25-29; head dept, Earlham Col, 29-30; Am-Scand Found fel, Bohr's Inst, Denmark, 30-31; instr physics, Barnard Col, Columbia Univ, 31-37; instr physics, Queen's Col, 37-41, from asst prof to assoc prof, 41-57, prof, 57-73, EMER PROF PHYSICS, QUEENS COL, NY, 73- *Mem:* AAAS; Am Phys Soc; Am Asn Physics Teachers; Sigma Xi. *Res:* Electronics. *Mailing Add:* 1805 S Balsam St No 227 Lakewood CO 80232

SWENSON, LEONARD WAYNE, b Twin Falls, Idaho, June 11, 31; m 50; c 3. PHYSICS. *Educ:* Mass Inst Technol, BS, 54, PhD(physics), 60. *Prof Exp:* Instr physics, Northeastern Univ, 57-58 & Tufts Univ, 58-59; res assoc, Mass Inst Technol, 60-62; Bartol fel nuclear struct res group, Bartol Res Found, 62-64, res staff mem, 64-66; dir space radiation effects lab, Va Assoc Res Ctr, 67-68; assoc prof, 68-81, PROF PHYSICS, ORE STATE UNIV, 81- *Concurrent Pos:* Consult, Joseph Kaye & Co, 57-59. *Mem:* Am Phys Soc; Am Sci Affil. *Res:* Nuclear structure and reaction; intermediate energy nuclear physics. *Mailing Add:* Dept of Physics Ore State Univ Corvallis OR 97331

SWENSON, MELVIN JOHN, b Concordia, Kans, Jan 14, 17; m 47; c 3. VETERINARY PHYSIOLOGY. *Educ:* Kans State Col, DVM, 43; Iowa State Col, MS, 47, PhD(path), 50. *Prof Exp:* Instr vet sci, La State Univ, 43; asst prof path, Iowa State Col, 49-50; from asst to assoc prof physiol, Kans State Col, 50-56; prof, Colo State Univ, 56-57; prof & head dept, 57-73, prof,

73-87, EMER PROF PHYSIOL, IOWA STATE UNIV, 87- *Mem:* AAAS; Am Physiol Soc; Am Soc Vet Physiol & Pharmacol; Soc Exp Biol & Med; Am Vet Med Asn; Sigma Xi. *Res:* Need for trace minerals in animals; effect of antibiotics on growth and hematology; nutrient requirements of animals; histophysiology of nutritional deficiencies; anemias of farm animals. *Mailing Add:* Dept of Vet Physiol & Pharmacol Iowa State Univ Col Vet Med Ames IA 50011

SWENSON, ORVAR, b Halsingborg, Sweden, Feb 7, 09; nat US; m 41; c 3. SURGERY. *Educ:* William Jewell Col, AB, 33; Harvard Med Sch, MD, 37. *Prof Exp:* Intern surg, Ohio State Univ Hosp, 37-38; house officer path, Children's Hosp, Boston, 38-39, house officer surg, 39-41; Cabot fel, Harvard Med Sch, 41-44, asst surg, 42-44, instr, 44-47, assoc, 47-50; assoc prof, Sch Med, Tufts Univ, 50-54, clin prof pediat surg, 54-57, prof, 57-60; prof surg, Med Sch, Northwestern Univ, Chicago, 60-73; prof surg, Sch Med, Univ Miami, 73-77. *Concurrent Pos:* Surg house officer, Peter Bent Brigham Hosp, 39-41, asst res surgeon, 41-42, jr assoc, 44, res surgeon, 44-45; jr attend surgeon, Children's Hosp, Boston, 45, assoc vis surgeon, 45-47, surgeon, 47-50; vis surgeon, New Eng Peabody Home Crippled Children, 46-50, mem assoc staff, 50; lectr, Simmons Col, 48-50; sr surgeon, New Eng Ctr Hosp, 50-60; surgeon in chief, Boston Floating Hosp Infants & Children, 50-60 & Children's Mem Hosp, Chicago, 60-72. *Mem:* Assoc Asn Thoracic Surg; Am Surg Asn; fel Am Col Surg; fel Am Acad Pediat. *Mailing Add:* Main St Rockport ME 04856

SWENSON, PAUL ARTHUR, b St Paul, Minn, Feb 5, 20; m 42; c 2. CELL PHYSIOLOGY. *Educ:* Hamline Univ, BS, 47; Stanford Univ, PhD(biol), 52. *Prof Exp:* Instr physiol, Univ Mass, 50-S4, from asst prof to assoc prof, 54-66; radiation biophysicist, Biol Div, Oak Ridge Nat Lab, 66-85; RETIRED. *Concurrent Pos:* Vis assoc physiologist, Brookhaven Nat Lab, 56-57; USPHS spec fel, Oak Ridge Nat Lab, 62-64. *Mem:* AAAS; Biophys Soc; Am Soc Microbiol; Am Soc Photobiol; Radiation Res Soc. *Res:* Effects of ultraviolet and ionizing radiations on metabolic control in bacteria. *Mailing Add:* 100 Ontario Lane Oak Ridge TN 37830

SWENSON, RICHARD PAUL, b Minnesota, Mar 5, 49. PROTEIN CHEMISTRY, ENZYMOLOGY. *Educ:* Gustavus Adolphus Col, BA, 71; Univ Minn, PhD(biochem), 79. *Prof Exp:* Teaching asst, Gustavus Adolphus Col, 68-71; res asst biochem, Dept Endocrinol, Mayo Clinic, 71-74; teaching asst, Univ Minn, 74-77; postdoctoral, Univ Mich, Ann Arbor, 79-82, asst prof, Biol Chem Dept, 83-84; CONSULT, 84- *Honors & Awards:* Bacaner Basic Sci Res Award, Minn Med Found, 80. *Mem:* Am Chem Soc; AAAS; Sigma Xi; Am Soc Biochem & Molecular Biol; Protein Soc. *Res:* Mechanism of enzyme and protein action; protein engineering; primary structure of the flavoprotein, D-amino acid oxidase; modifications to active-site amino acids involved in catalysis; structural genes for bacterial flavodoxins; flavoprotein quinone reductases. *Mailing Add:* Biochem Dept Ohio State Univ 776 Biol Sci Bldg 484 W 12th Ave Columbus OH 43210

SWENSON, RICHARD WALTNER, b New York, NY, May 17, 23; m 49; c 3. PHOTOGRAPHIC CHEMISTRY. *Educ:* Clark Univ, Mass, AB, 48, AM, 49; Brown Univ, PhD(chem phys), 53. *Prof Exp:* Res chemist, 53-55, res assoc, Photo Prod Dept, E I Du pont De Nemours & Co, 55; RETIRED. *Honors & Awards:* Pres Citation, Soc Photog Sci & Eng, 86. *Mem:* Am Chem Soc; Soc Photog Sci & Eng (pres, 67-71). *Res:* Photographic chemistry, specifically emulsion chemistry. *Mailing Add:* 65 Edison Ave Tinton Falls NJ 07724-3140

SWENSON, ROBERT J, b Butte, Mont, Mar 3, 34; m 58; c 3. THEORETICAL PHYSICS. *Educ:* Mont State Univ, BS, 56; Lehigh Univ, MS, 58, PhD(physics), 61. *Prof Exp:* Res assoc physics, Lehigh Univ, 61-62; Nat Acad Sci fel, 62-63 & res fel, Free Univ Brussels, 62-64; Joint Inst Lab Astrophysics fel, Univ Colo, 64-65; from asst prof to assoc prof physics, Temple Univ, 65-70, chmn dept, 67-70; prof physics & chmn dept, 70-90, VPRES RES, MONT STATE UNIV, 90- *Mem:* Fel Am Phys Soc; fel AAAS; Sigma Xi; Am Asn Physics Teachers. *Res:* Thermodynamics; statistical mechanics; relativistic fluid mechanics; general theory. *Mailing Add:* VPres Res Mont State Univ Bozeman MT 59717

SWENSON, THERESA LYNN, b Racine, Wis, Sept 17, 57; m 90. ATHEROSCLEROSIS, DIABETES. *Educ:* Univ Wis-Parkside, BS, 70; Univ Wis-Madison, PhD(physiol chem), 85. *Prof Exp:* Postdoctoral fel med, Columbia Univ, 85-89; SR RES BIOCHEM, ATHEROSCLEROSIS RES, MERCK SHARP & DOHME RES LABS, 89- *Concurrent Pos:* Mem, Arteriosclerosis Coun, Am Heart Asn. *Mem:* Fel Am Heart Asn; Am Asn Biochem & Molecular Biol; Am Chem Soc; AAAS. *Res:* Biochemical mechanisms of atherosclerosis; biochemical mechanisms and physiological functions of the enzymes and proteins involved in lipoprotein remodeling. *Mailing Add:* Merck Sharp & Dohme Res Lab PO Box 2000 80W-250 Rahway NJ 07065

SWENTON, JOHN STEPHEN, b Kansas City, Kans, Dec 8, 40; m; c 2. ORGANIC CHEMISTRY. *Educ:* Univ Kans, BA, 62; Univ Wis, PhD(chem), 65. *Prof Exp:* Nat Acad Sci-Nat Res Coun fel, Harvard Univ, 65-66; from asst prof to assoc prof, 67-77, PROF CHEM, OHIO STATE UNIV, 77- *Concurrent Pos:* Grants, Soc Petrol Res Fund, NSF & Eli Lilly, 84-86. *Mem:* Am Chem Soc; Royal Soc Chem. *Res:* Synthetic and mechanistic organic photochemistry, synthetic and mechanistic organic electrochemistry, and natural products synthesis. *Mailing Add:* Dept Chem 120 W 18th Ave Columbus OH 43210

SWERCZEK, THOMAS WALTER, b Cedar Rapids, Nebr, May 10, 39; m 64; c 3. VETERINARY PATHOLOGY. *Educ:* Kans State Univ, BS, 62, DVM, 64; Univ Conn, MS, 66, PhD(path), 69. *Prof Exp:* From asst prof to assoc prof, 69-78, PROF VET SCI COL AGR, UNIV KY, 78- *Mem:* Am Vet Med Asn; Conf Res Workers Animal Dis. *Res:* Comparative pathology; pathogenesis of infectious diseases of horses; nutritional pathology; pathogenesis of gastrointestinal diseases of livestock. *Mailing Add:* Dept Vet Sci Animal Path Bldg Univ Ky Lexington KY 40546-0099

SWERDLOFF, RONALD S, b Pomona, Calif, Feb 18, 38; m 59; c 2. ENDOCRINOLOGY, INTERNAL MEDICINE. *Educ:* Univ Calif, Los Angeles, BS, 59; Univ Calif, San Francisco, MD, 62. *Prof Exp:* Intern internal med, Kings Co Hosp, Univ Wash, 62-63, asst resident med, 63-64; Clin res assoc, Res Metab Sect, NIH, 64-66, spec fel endocrinol, 67-69; from asst prof to assoc prof med, 69-78, assoc chief, 73-78, CHIEF, DIV ENDOCRINOL & METAB, HARBOR-GEN HOSP CAMPUS, UNIV CALIF, LOS ANGELES, 73-, PROF MED, HARBOR-UCLA MED CTR, 78- *Concurrent Pos:* Dir Med Residency Prog, St Mary's Long Beach Hosp & Harbor Gen Hosp, 69-71; NIH career develop award, 72-77; actg dir, Clin Study Ctr, Harbor Gen Hosp, 78-80; vis prof, Dept Anat, Monash Univ, 80-81; dir, UCLA Population Res Ctr, 86-; ed, J Clin Endocrinol & Metab, 78-83; Consult, UN Family Planning Asn, 83-86, Human Reproductive Prog, World Health Orgn, 83-89. *Honors & Awards:* Wyeth Award, Soc Study Reprod, 74; Serono Distinguished Andrologist Award, Am Andrology Soc, 86; Serono Distinguished Lectr, Australian Reproductive Biol, 90, Harrison Mem Lectr, Australian Endocrine Soc, 81; Harrison Mem Lectureship, Australian Endocrine Soc, 81. *Mem:* Am Fedn Clin Res; Endocrine Soc; Soc Study Reprod; Am Andrology Soc; Am Asn Prof. *Res:* Reproductive endocrinology; neuroendocrinology; contraceptive development. *Mailing Add:* Div of Endocrinol 1000 W Carson St Torrance CA 90509

SWERDLOW, MARTIN A, b Chicago, Ill, July 7, 23; m 45; c 2. MEDICINE, PATHOLOGY. *Educ:* Univ Ill, BS, 45, MD, 47; Am Bd Path, dipl, 52. *Prof Exp:* Resident path, Michael Reese Hosp, Chicago, 48-50 & 51-52; pathologist, Menorah Med Ctr, 54-57; from asst prof to assoc prof path, Univ Ill Col Med, 57-60, clin assoc prof, 60-66, prof, 66-72, assoc dean, Abraham Lincoln Sch Med, 70-72; prof path, Sch Med, Univ Mo-Kansas City, 73-74; chmn dept path, Michael Reese Hosp & Med Ctr, 74-89, vpres acad affairs, 75-88; prof path, Univ Chicago, 74-89, assoc dean, 82-89; PROF & HEAD DEPT PATH, UNIV ILL COL MED, 89- *Concurrent Pos:* Pathologist, Englewood Hosp, Chicago; consult, Vet Admin Hosp, Hines & Cook County Hosp, Chicago; chmn dept path, Kansas City Gen Hosp, 73-74. *Honors & Awards:* Raymond Allen Award for Teaching, Col Med, Univ Ill, 60-72. *Mem:* Am Soc Clin Path; Col Am Pathologists; Am Asn Study Liver Dis. *Res:* Histopathology of skin disorders; experimental skin tumors; diseases of liver. *Mailing Add:* Dept Path Univ Ill Col Med Chicago IL 60612

SWERDLOW, MAX, solid state physics, materials science; deceased, see previous edition for last biography

SWERLICK, ISADORE, b Philadelphia, Pa, Jan 23, 21; m 51; c 2. ORGANIC CHEMISTRY. *Educ:* Temple Univ, BA, 43; Duke Univ, PhD(chem). 50. *Prof Exp:* Asst, Duke Univ, 46-49; res chemist, E I du Pont de Nemours & Co, Inc, 50-56, res supvr, 56-70, staff scientist, Film Dept, 70-76, res assoc, Polymer Prod Dept, 77-82; RETIRED. *Mem:* AAAS; Am Chem Soc; Sigma Xi. *Res:* Organics and polymers. *Mailing Add:* 614 Loveville Rd C2H Hockessin DE 19707

SWERLING, PETER, b New York, NY, Mar 4, 29. ENGINEERING ADMINISTRATION. *Educ:* Calif Inst Technol, 47; Cornell Univ, BA, 49; Univ Calif, Los Angeles, MA, 51, PhD, 55. *Prof Exp:* Mem tech staff, Rand Corp, 48-61; dept mgr, Conductron Corp, 61-64; prof & chmn, Tech Serv Corp, 66-82; DIR, SWERLING MANASSETSMITH, 82-, PRES, 86- *Concurrent Pos:* Asst prof eng, Univ Ill, 56-57; adj prof eng, Univ Southern Calif, 64-66. *Mem:* Nat Acad Eng; fel Inst Elec & Electronics Engrs; Sigma Xi. *Mailing Add:* 1136 Corsica Dr Pacific Palisades CA 90272

SWETHARANYAM, LALITHA, b Trivandrum, India; nat US; m 71. TOPOLOGY. *Educ:* Annamalai Univ, Madras, PhD(math), 66. *Prof Exp:* Lectr math & head dept, LVD Col, Raichur, India, 56-61; res fel, Annamalai Univ, Madras, 61-65, lectr, 65-69; assoc prof, 69-73, PROF, McNEESE STATE UNIV, 73- *Concurrent Pos:* Regional coordr speaker's bur, Asn Women Math, Nat Off, 83-87. *Mem:* Am Math Soc; Indian Math Soc; Math Asn Am; Assoc Women in Math. *Res:* Functional analysis; point set topology. *Mailing Add:* 525 E Jefferson Dr Lake Charles LA 70605

SWETITS, JOHN JOSEPH, b Passaic, NJ, Oct 1, 42; m 66; c 3. MATHEMATICAL ANALYSIS. *Educ:* Fordham Univ, BS, 64; Lehigh Univ, MS, 67, PhD(math), 68. *Prof Exp:* Instr math, Lafayette Col, 67-68, asst prof, 68-70; assoc prof, 70-81, PROF MATH, OLD DOMINION UNIV, 81- *Mem:* Am Math Soc; Math Asn Am; Soc Indust & Appl Math. *Res:* Summability theory; approximation theory. *Mailing Add:* Dept Math Old Dominion Univ Norfolk VA 23529

SWETS, DON EUGENE, b Grand Rapids, Mich, Oct 7, 30; m 56; c 4. SOLID STATE PHYSICS, CHEMICAL PHYSICS. *Educ:* Univ Mich, BSE(physics) & BSE(math), 53, MS, 55. *Prof Exp:* sr res physicist, 55-80, staff res scientist, res lab, Gen Motors Corp, 80-87; RETIRED. *Res:* Measurement of diffusion coefficients; hydrogen in steel; helium, neon, and hydrogen fused in quartz; growth of single crystals, especially tetragonal germanium dioxide; ribbon shaped germanium; hexamethylenetetramine IV-VI and III-V compounds; rare earth iron compounds; super conductor compounds. *Mailing Add:* 6900 Timbercrest Washington MI 48094

SWETT, CHESTER PARKER, b Lancaster, Ohio, Aug 16, 39; m 64; c 2. PSYCHOPHARMACOLOGY, BEHAVIOR THERAPY. *Educ:* Denison Univ, BS, 61; Harvard Med Sch, MD, 66. *Prof Exp:* Intern, Harvard Surg Serv, Boston City Hosp, 66-67; staff assoc, Nat Inst Health, 67-69; psychiat resident, Mass Ment Health Ctr, 69-72; scientist, Drug Epidemiol Unit, Boston Univ Med Ctr, 72-74; staff psychiatrist, McLean Hosp, 74-89, serv chief, 83-89; assoc prof, Harvard Med Sch, 83-89; MED DIR, BRATTLEBORO RETREAT, 90-; PROF CLIN PSYCHIAT, DARTMOUTH MED SCH, 90- *Concurrent Pos:* Pre-doctoral fel physiol, Harvard Med Sch, 63-64. *Mem:* Am Psychiat Asn; Am Acad Psychiat & Law; Asn Advan Behav Ther; AAAS. *Res:* The causes of impulsive violence and the effects of violence upon victims. *Mailing Add:* Brattleboro Retreat 75 Linden St Brattleboro VT 05301

SWETT, JOHN EMERY, b San Francisco, Calif, Mar 19, 32; m 56; c 4. ANATOMY, NEUROPHYSIOLOGY. *Educ:* Univ Wash, AB, 56; Univ Calif, Los Angeles, PhD(anat), 60. *Prof Exp:* NSF fel, Univ Pisa, 60-61, NIH fel, 61-62; neurophysiologist, Good Samaritan Hosp, Portland, Ore, 62-66; assoc prof anat, State Univ NY Upstate Med Ctr, 66-67; assoc prof anat, Med Sch, Univ Colo, Denver, 67-76; PROF ANAT & CHMN DEPT, COL MED, UNIV CALIF, IRVINE, 76- *Concurrent Pos:* NIH res grants, 64- *Mem:* Am Asn Anatomists; Am Physiol Soc; Soc Neurosci. *Res:* Organization of the spinal cord; mechanisms of sensory detection; organization of the peripheral nervous system. *Mailing Add:* Dept of Anat Univ of Calif Col of Med Irvine CA 92717

SWETT, KEENE, b Wilton, Maine, Nov 6, 32; m 54; c 2. GEOLOGY. *Educ:* Tufts Univ, BS, 55; Univ Colo, MS, 61; Univ Edinburgh, PhD(geol), 65. *Prof Exp:* Asst lectr geol, Univ Edinburgh, 63-66; asst prof, 66-70, assoc prof, 70-74, PROF GEOL, UNIV IOWA, 74- *Concurrent Pos:* NSF fel, 67-82. *Mem:* Geol Soc Am; Am Asn Petrol Geol; Soc Econ Paleont & Mineral; Int Asn Sedimentol; AAAS. *Res:* Petrological studies of sediments and sedimentary rocks with especial regard to the post-depositional alterations and patterns of diagenesis; Cambro-Ordovician shelf sediments of western Newfoundland, northwest Scotland, central eastern Greenland and Spitsbergen; late proterozic biostratigraphy and paleo environments of Spitsbergen and Central East Greenland. *Mailing Add:* Dept of Geol Univ of Iowa Iowa City IA 52242

SWEZ, JOHN ADAM, b Cleveland, Ohio, Nov 18, 41; m 65; c 1. BIOPHYSICS. *Educ:* Pa State Univ, BS, 63, MS, 65, PhD(biophys), 67. *Prof Exp:* From asst prof to assoc prof physics, 67-77, PROF PHYSICS & DIR RADIATION LAB, IND STATE UNIV, TERRE HAUTE, 77- *Mem:* AAAS; Radiation Res Soc; Biophys Soc. *Res:* Degradation studies of DNA in Escherichia coli after ionizing radiation, physical characterization and effect of bacteriophage infection; injection of nucleic acid of bacteriophage T1 into its host Escherichia coli. *Mailing Add:* Dept Physics Ind State Univ Lab Sci Bldg 130 217 N 6th St Terre Haute IN 47809

SWEZEY, ROBERT LEONARD, b Pasadena, Calif, Apr 30, 25; m 49; c 3. INTERNAL MEDICINE. *Educ:* Ohio State Univ, MD, 48; Am Bd Internal Med, dipl, 60; Am Bd Phys Med & Rehab, dipl, 69; Am Bd Rheumatology, dipl, 74. *Prof Exp:* Intern, Los Angeles County Gen Hosp, Calif, 48-49 & Am Hosp, Paris, 49-50; resident internal med, Wadsworth Gen Med & Surg Vet Admin Hosp, 51-54, fel rheumatol, Wadsworth Gen Med & Surg Vet Admin Hosp & Univ Calif, Los Angeles, 54-55, clin asst med, Univ Calif, Los Angeles, 55-57, clin instr, 57-64, asst prof, 64-65, career fel phys med & rehab, 65 & 66-67; acad career fel, Univ Minn, 65-66; assoc prof internal med, phys med & rehab, Sch Med, Univ Southern Calif, 67-73, prof med, phys med & rehab, 73-74; prof med & dir div rehab med, 74-78, CLIN PROF MED, SCH MED, UNIV CALIF, LOS ANGELES, 78-; MED DIR, ARTHRITIS & BACK PAIN CTR, 78- *Mem:* AMA; Am Acad Phys Med & Rehab; Am Rheumatism Asn; fel Am Col Physicians; Asn Acad Physiatrists. *Res:* Rheumatology; clinical aspects of rheumatic diseases; mechanisms of arthritic deformities and their treatments; physical medicine and rehabilitation. *Mailing Add:* Swezey Inst Bldg 1328 16th St Santa Monica CA 90404

SWIATEK, KENNETH ROBERT, b Chicago, Ill, Dec 30, 35; m 60; c 4. NEUROSCIENCES. *Educ:* NCent Col, BS, 58; Univ Ill Med Ctr, PhD(biol sci), 65. *Prof Exp:* Res assoc biol chem, Dept Pediat, Univ Ill Col Med, 65-68; res scientist, Ill Inst Develop Disabilities, 68-70, admin res sci, 70-75, dir res, 75-76, dir, 76-80. *Concurrent Pos:* Grant carbohydrate metab newborn animal, NIH, 71-74. *Mem:* Am Chem Soc; Sigma Xi; AAAS; Soc Develop Biol; Am Asn Ment Deficiency. *Res:* Study of growth of nervous system with special emphasis on development of carbohydrate, ketones and amino acid metabolism in fetal and newborn brain tissue as affected by pain-relieving drugs of labor and delivery. *Mailing Add:* 5029 Central Ave Western Springs IL 60558

SWICK, KENNETH EUGENE, b Silver Lake, Ind, Jan 20, 36; m 57. MATHEMATICS. *Educ:* Anderson Col, BS, 57; Univ Southern Calif, MA, 64; Univ Iowa, PhD(math), 67. *Prof Exp:* Mathematician, Hughes Aircraft Corp, Calif, 62-64; asst prof, Cornell Col, 66-67 & Occidental Col, 67-70; asst prof, 70-78, assoc prof, 78-85, PROF MATH, QUEENS COL, NY, 85- *Mem:* Math Asn Am; Am Math Soc; Soc Indust & Appl Math. *Res:* Nonlinear differential and integral equations; nonlinear population dynamics. *Mailing Add:* 26 Fairview St Huntington NY 11743

SWICK, ROBERT WINFIELD, b Jackson, Mich, July 6, 25; m 47; c 4. BIOCHEMISTRY. *Educ:* Beloit Col, BS, 47; Univ Wis, MS, 49, PhD(biochem), 51. *Prof Exp:* Assoc biochemist, Div Biol & Med, Argonne Nat Lab, 51-69; PROF NUTRIT SCI, UNIV WIS-MADISON, 69-, CHMN DEPT, 85- *Mem:* Am Inst Nutrit; Am Soc Biol Chem. *Res:* Control and kinetics of protein metabolism; regulation of enzyme levels and metabolism. *Mailing Add:* Dept Nutrit Sci Univ Wi 1415 Linden Dr Madison WI 53706

SWICKLIK, LEONARD JOSEPH, b Nanticoke, Pa, Jan 26, 28; m 52; c 3. ORGANIC CHEMISTRY. *Educ:* Wilkes Col, BS, 49; Univ Pittsburgh, PhD(chem), 54. *Prof Exp:* Res chemist, E I du Pont de Nemours & Co, Va, 54; res chemist, Eastman Kodak Co, 56-64, sr res chemist & group leader, 64-68, tech assoc, 68-70, head, Process Improv Lab, 71-73, supvr chem process, 73-78, supvr, Oils Processing Dept, Distillation Prod Div, 78-84, supvr, Develop & Control Lab, 84-86; RETIRED. *Mem:* Am Oil Chemists Soc; AAAS; Am Chem Soc; NY Acad Sci; Sigma Xi. *Res:* Food applications of emulsifiers; fat and oil chemistry as applicable to edible products; food and animal feed applications of vitamins. *Mailing Add:* 92 Northwick Dr Rochester NY 14617-5634

SWIDEN, LADELL RAY, b Sioux Falls, SD, June 17, 38; m 61; c 3. INDUSTRIAL CONTROLS, MANAGEMENT INFORMATION SYSTEMS. *Educ:* SDak State Univ, BSSE, 61; Univ SDakota, MBA, 82. *Prof Exp:* Engr, Honeywell, 62-67; vpres sales, Swiden Appliances & Furn, 67-68; engr, Raven Indust, 68-72, mgr engr, 77-84; vpres engr, Beta Raven Inc, 84-85; pres, Delta Syst Inc, 85-86; DIR, ENG & ENVIRON RES CTR, SDAK STATE UNIV, 86- *Mem:* Nat Soc Prof Engrs; Inst Sci Anal. *Res:* Automation of process controls and integration with management information systems; additional emphasis on engineering and research management. *Mailing Add:* 105 Heather Ln Brookings SD 57006

SWIDER, WILLIAM, JR, b Brooklyn, NY, Jan 5, 34; m 59; c 5. ATMOSPHERIC CHEMISTRY, PHYSICS. *Educ:* Lehigh Univ, BS, 55, MS, 57; Pa State Univ, PhD(physics), 63. *Prof Exp:* Assoc physicist, Int Bus Mach Corp, 57-60; chmn civilian policy bd, 78-81, actg br chief, 84-85, PHYSICIST, AIR FORCE GEOPHYSICS LAB, 64-, DEP DIV DIR, 89- *Concurrent Pos:* Res fel, Nat Ctr Space Res, Brussels, Belgium, 63-64; Air Force Rep to Geophysics Res Forum Nat Res Coun; vis prof physics, Utah State Univ, 75. *Honors & Awards:* Loeser Award, Air Force Geophysics Lab, 77. *Mem:* Am Geophys Union; Sigma Xi. *Res:* Ionospheric physics; airglow. *Mailing Add:* 15 Old Stagecoach Rd Bedford MA 01730-1246

SWIDLER, RONALD, b New York, NY, Dec 23, 29. ELECTROPHOTOGRAPHY, CHEMISTRY OF FATS & OILS. *Educ:* City Col NY, BS, 49; Univ Southern Calif, MS, 51, PhD(chem), 53. *Prof Exp:* STAFF SCIENTIST, SRI INT, 56- *Res:* Electrophotography; chemical processes; dyes; textiles; general organic chemistry. *Mailing Add:* SRI 333 Ravenswood Ave Menlo Park CA 94025

SWIERSTRA, ERNEST EMKE, b Netherlands, Aug, 14, 30; Can citizen; m 62; c 2. ANIMAL PHYSIOLOGY. *Educ:* Univ Groningen, dipl, 51; Univ BC, BSA, 56, MSA, 58; Cornell Univ, PhD(physiol), 62. *Prof Exp:* Asst animal breeding, Cornell Univ, 58-62; sr res scientist, 62-75, head animal sci sect, 75-84, asst dir, 84-86, DIR, RES STA, CAN DEPT AGR, 86- *Mem:* Am Soc Animal Sci; Can Soc Animal Sci; Agr Inst Can. *Res:* Reproductive physiology. *Mailing Add:* Res Sta Can Dept Agr Brandon MB R7A 5Z7 Can

SWIFT, ARTHUR REYNDERS, b Worcester, Mass, July 25, 38; m 61; c 3. PHYSICS. *Educ:* Swarthmore Col, BA, 60; Univ Pa, PhD(physics), 64. *Prof Exp:* NATO fel physics, Cambridge Univ, 64-65; res assoc, Univ Wis, 65-67; asst prof, 67-70, assoc prof, 70-76, PROF PHYSICS, UNIV MASS, AMHERST, 76- *Mem:* Am Phys Soc. *Res:* Elementary particle theory. *Mailing Add:* Dept Physics & Astron Univ Mass Grad Res Ctr Tower C Amherst MA 01003

SWIFT, BRINTON L, b Denver, Colo, June 12, 26; m 56; c 1. VETERINARY MEDICINE. *Educ:* Colo State Univ, DVM, 51. *Prof Exp:* Private practice, Buffalo, Wyo, 51-64; from asst prof to assoc prof, 64-69, PROF VET MED, UNIV WYO, 69- *Mem:* Am Vet Med Asn; Sigma Xi; Soc Theriogenology. *Res:* Anaplasmosis. *Mailing Add:* RR 2 No 231 Merrill NE 69358

SWIFT, CALVIN THOMAS, b Quantico, Va, Feb 6, 37; m 59; c 2. ENGINEERING, REMOTE SENSING. *Educ:* Mass Inst Technol, BS, 59; Va Polytech Inst & State Univ, MS, 65; Col William & Mary, PhD(physics), 69. *Prof Exp:* Res engr, NAm Aviation, Calif, 59-62; aerorspace engr, Langley Res Ctr, NASA, 62-81, leader microwave radiometer group, 75-81; PROF ELEC & COMPUT ENG, UNIV MASS, 81-; CONSULT ENG, 82- *Concurrent Pos:* Asst lectorial prof, George Washington Univ, 70-; lectr, Col William & Mary, 71-; adj assoc prof, Old Dominion Univ, 75-; ed, Geosci & Remote Sensing Soc, 80-84; chmn, Comn F, Int Union Radio Sci, 87-90. *Honors & Awards:* Centennial Award, Inst Elec & Electronic Engrs, 84. *Mem:* Fel Inst Elec & Electronic Engrs; Antennas & Propagation Soc (secy-treas, 74-77); Geosci & Remote Sensing Soc (pres, 85); Int Union Radio Sci. *Res:* Antennas and propagation, with emphasis on remote sensing of the ocean. *Mailing Add:* Dept Elec & Comp Eng Univ Mass Amherst MA 01003

SWIFT, CAMM CHURCHILL, b Oakland, Calif, Sept 29, 40; m 74; c 5. ICHTHYOLOGY. *Educ:* Univ Calif, Berkeley, AB, 63; Univ Mich, Ann Arbor, MS, 65; Fla State Univ, PhD(biol), 70. *Prof Exp:* ASSOC CUR FISHES, NATURAL HIST MUS LOS ANGELES COUNTY, 70- *Concurrent Pos:* Adj asst prof, Dept Biol Sci, Univ Southern Calif, 72-; bd gov, Am Soc Ichthyologists & Herpetologists, 73-78; secy, S Calif Acad Sci, 81-86, vpres, 87- *Mem:* Am Soc Ichthyologists & Herpetologists; Soc Vert Paleont; Soc Study Evolution; Soc Syst Zool; AAAS; Sigma Xi. *Res:* Systematics and evolution of Recent and fossil, freshwater and marine shore fishes of North America. *Mailing Add:* County Mus Nat Hist 900 Exposition Blvd Los Angeles CA 90007

SWIFT, CHARLES MOORE, JR, b Boston, Mass, Sept 26, 40; m 63; c 2. GEOPHYSICS. *Educ:* Princeton Univ, AB, 62; Mass Inst Technol, PhD(geophys), 67. *Prof Exp:* Sr geophysicist, Kennecott Explor Serv Lab, Kennecott Copper Corp, 67-76; div geophysicist, Geothermal Explor Div, Chevron Res Co, Standard Oil Co Calif, 76-81, div geophysicist, Domestic Minerals Div, 81-82, chief geophysicist, 82-89; GEOPHYS CONSULT, CHEVRON OVERSEAS PETROL, INC, 89- *Concurrent Pos:* Adj assoc prof, Dept Geol & Geophys, Univ Utah, 76-84; lectr, Dept Mat Sci & Eng, Univ Calif, Berkeley, 76-79; assoc ed, Geophysics, Geothermal Explor, 80-81 & Mining Geophysics, 82-83. *Mem:* Soc Explor Geophysicists; Am Geophys Union; Am Asn Petrol Geologists; Geol Soc Am; Soc Economic Geologists. *Res:* Geophysical, particularly electrical, electromagnetic, and magnetotelluric techniques applied to mineral and petroleum exploration. *Mailing Add:* Chevron Overseas Petrol Inc PO Box 5046 San Ramon CA 94583-0946

SWIFT, DANIEL W, b Worcester, Mass, Mar 6, 35; m 61; c 3. PLASMA PHYSICS, SPACE PHYSICS. *Educ:* Haverford Col, BA, 57; Mass Inst Technol, MS, 59. *Prof Exp:* Mem staff, Lincoln Lab, Mass Inst Technol, 58-59; sr scientist, Res & Adv Develop Div, Avco Corp, 59-63; from asst prof to assoc prof, 63-72, PROF PHYSICS, GEOPHYS INST, UNIV ALASKA, 72- *Mem:* AAAS; Am Geophys Union; Am Phys Soc. *Res:* Theoretical studies of the earth's magnetosphere and auroral phenomena. *Mailing Add:* Geophys Inst Univ Alaska 116 Bunnell Fairbanks AK 99701

SWIFT, DAVID LESLIE, b Chicago, Ill, Aug 7, 35; m 59; c 3. ENVIRONMENTAL MEDICINE, PHYSIOLOGY. *Educ:* Purdue Univ, BS, 57; Mass Inst Technol, SM, 59; Johns Hopkins Univ, PhD(chem eng), 63. *Prof Exp:* Chem engr, Argonne Nat Lab, Ill, 63-65; USPHS air pollution spec fel, London Sch Hyg, 65-66; asst prof, 66-70, assoc prof environ med, 70-78, PROF ENVIRON HEALTH SCI, JOHNS HOPKINS UNIV, 78- *Mem:* Am Sci Affil; Brit Occup Hyg Soc; Am Conf Govt Indust Hyg; Am Indust Hyg Asn; Am Asn Aerosol Res; Sigma Xi. *Res:* Physiology of respiratory tract; fate of inhaled particles and gases; fluid mechanics and transport in biological systems; air pollution transport. *Mailing Add:* 1020 Litchfield Rd Baltimore MD 21239-1308

SWIFT, DONALD J P, b Dobbs Ferry, NY, July 26, 35; m 61; c 3. SEDIMENTOLOGY, OCEANOGRAPHY. *Educ:* Dartmouth Col, AB, 57; Johns Hopkins Univ, MA, 61; Univ NC, PhD(geol), 64. *Prof Exp:* Asst prof sedimentology, geol dept, Dalhousie Univ, 63-66; assoc scientist marine geol, P R Nuclear Ctr, 66-67; assoc prof geol, Duke Univ, 67-68; Slover assoc prof, Inst Oceanog, Old Dominion Univ, 68-71; res oceanogr, Atlantic Oceanog & Meteorol Labs, Nat Oceanic & Atmospheric Admin, 71-81; res adv, Arco Explor Technol, 81-86; SLOVER PROF, DEPT OCEANOG, OLD DOMINION UNIV, 86- *Concurrent Pos:* Res grant, Geol Surv Can, 64-65; res assoc, NS Mus, 65-66; assoc ed, Maritime Sediments, 65-66; res grant, Nat Res Coun Can & Defense Res Bd Can, 65-67, Coastal Eng Res Ctr, 69-71, US Geol Surv, 69-71 & NSF, 69-72; mem, Univ Senate, Old Dom Univ, 69-71, chmn, Univ Res Comt, 69-71; res grant, NASA, 70-71; proj leader, Continental Margin Sedimentation Proj, 72-81; consult, Oceanog Panel, NSF, 74-; res assoc, Smithsonian Inst, 74-81; adj prof, Univ Miami, 74-81; asst chmn, dept oceanog, 87-90, Univ Senate, Old Dom Univ, 86-89; res grant, NSF, 89-90, Nat Oceanic & Atmospheric Admin, 89-91. *Honors & Awards:* Shepard Medal, 89. *Mem:* AAAS; Soc Econ Paleont & Mineral; Fel Geol Soc Am; Am Asn Petrol Geol; Int Asn Sedimentologists. *Res:* Continental shelf sedimentation; continental margin stratigraphy. *Mailing Add:* Dept Oceanog Old Dominion Univ Norfolk VA 23529-0276

SWIFT, DOROTHY GARRISON, b Flint, Mich, Aug 1, 39; div. BIOLOGICAL OCEANOGRAPHY, TRACE METALS. *Educ:* Swarthmore Col, BA, 61; Johns Hopkins Univ, MA, 67, PhD(oceanog), 73. *Prof Exp:* Guest student investr, Biol Dept, Woods Hole Oceanog Inst, 68-69; guest investr, Gran Sch Oceanog, Univ RI, 70-73, res assoc, Dept Chem, 74-76, res assoc, 76-88, ASST MARINE SCIENTIST, GRAD SCH OCEANOG, UNIV RI, 89- *Concurrent Pos:* Consult, Bigelow Lab Ocean Sci, 82-84. *Honors & Awards:* Bronze Medal, Am Rhododendron Soc, 80. *Mem:* Am Soc Limnol & Oceanog; Am Geophys Union; Phycol Soc Am; Int Phycol Soc; Sigma Xi. *Res:* Trace metal requirements in phytoplankton; atmospheric input of trace metals biochemistry, nutrition and ecology of marine phytoplankton; bioassay for vitamins in seawater; vittamin bioassay; biogeochemistry; algal culture and plant tissue culture; atmospheric chemistry. *Mailing Add:* Ctr Atmospheric Chem Studies Grad Sch Oceanog Narragansett RI 02882

SWIFT, ELIJAH, V, b Boston, Apr 12, 38; m 61. BIOLOGICAL OCEANOGRAPHY. *Educ:* Swarthmore Col, BA, 60; Johns Hopkins Univ, MA, 64, PhD(oceanog), 67. *Prof Exp:* NSF trainee, Woods Hole Oceanog Inst, 68-69; asst prof, 69-74, ASSOC PROF OCEANOG, UNIV RI, 74- *Concurrent Pos:* NSF grants, 71-76. *Mem:* Phycol Soc Am; Int Phycol Soc; Marine Biol Asn UK; Brit Phycol Soc; Am Soc Limnol & Oceanog; Sigma Xi. *Res:* Morphology, taxonomy, ecology and physiology of phytoplankton, particularly marine species. *Mailing Add:* Dept Bot Grad Sch Oceanog Univ RI Narragansett RI 02882

SWIFT, FRED CALVIN, b Middleport, NY, Oct 16, 26; m 50; c 3. ENTOMOLOGY, ECOLOGY. *Educ:* Mich State Univ, BS, 50; Iowa State Col, MS, 52; Rutgers Univ, PhD, 58. *Prof Exp:* Entomologist, Niagara Chem Div, Food Mach & Chem Corp, 52-55; asst prof entom, Clemson Col, 58-59; assoc prof, 61-72, RES PROF ENTOM, RUTGERS UNIV, NEW BRUNSWICK, 72- *Mem:* Entom Soc Am; Ecol Soc Am. *Res:* Integrated control of fruit insect pests; ecology of the Phytoseiidae. *Mailing Add:* Dept of Entom & Econ Zool Cook Col Rutgers Univ New Brunswick NJ 08903

SWIFT, GEORGE W(ILLIAM), b Topeka, Kans, Nov 28, 30; m 56; c 3. CHEMICAL ENGINEERING. *Educ:* Univ Kans, BS, 53, MS, 57, PhD(chem eng), 59. *Prof Exp:* Res engr, Continental Oil Co, 59-61; from asst prof to assoc prof, 61-70, actg assoc dean grad sch, 66-70, prof chem & petrol eng, 70-81, DEANE E ACKERS DISTINGUISHED PROF ENG, UNIV KANS, 81- *Concurrent Pos:* Consult, Continental Oil Co, 63- & C W Nofsinger Co, 65- *Mem:* Am Inst Chem Engrs; Soc Petrol Engrs; Am Soc Eng Educ; Sigma Xi. *Res:* Transient flow of gas in porous media; transport and thermodynamic properties at low temperature and high pressure; rheology. *Mailing Add:* Dept of Chem & Petrol Eng Univ of Kans Lawrence KS 66045

SWIFT, GLENN W(ILLIAM), b Athabasca, Alta, May 17, 32; m 61; c 2. ELECTRICAL ENGINEERING. *Educ:* Univ Alta, BSc, 53, MSc, 60; Ill Inst Technol, PhD(elec eng), 68. *Prof Exp:* Elec engr, Can Westinghouse Co, Ltd, 56-58; from lectr to assoc prof, 60-73, PROF ELEC ENG, UNIV MAN, 73- *Concurrent Pos:* Consult various firms from, Westinghouse Can, Fed Pioneer, 60-; Nat Res Coun Can grants, 60- *Mem:* Sr mem Inst Elec & Electronic Engrs. *Res:* Power system protection and control; microprocessors. *Mailing Add:* Dept Elec Eng Univ Man Winnipeg MB R3T 2N2 Can

SWIFT, GRAHAM, b Chesterfield, Eng, Apr 16, 39; m 61; c 2. ORGANIC POLYMER CHEMISTRY. *Educ:* Univ London, BSc, 61, PhD(org chem', 64. *Prof Exp:* NIH fel, Fels Res Inst, Philadelphia, 64-66; sci off, Imperial Chem Industs, Eng, 66-68; sr chemist, 68-77, MGR RES SECT, ROHM AND HAAS CO, 77-, RES FEL, 86- *Mem:* Royal Soc Chem; Royal Inst Chem; Am Chem Soc. *Res:* Heterocyclic chemistry; synthesis and reactions of special aziridines, oxiranes and thiiranes; fatty acid chemistry; organic polymer coatings, especially synthesis and evaluation of novel coating compositions; water soluble polymer synthesis; application of polymer in detergents, mining,and water treatment; emulsion polymer synthesis. *Mailing Add:* Rohm and Haas Co Spring House PA 19477

SWIFT, HAROLD EUGENE, b Butler, Pa, Mar 27, 36; m 60; c 3. PHYSICAL INORGANIC CHEMISTRY. *Educ:* Allegheny Col, BS, 58; Pittsburgh Univ, PhD(chem), 62. *Prof Exp:* Chemist, 62-63, res chemist, 62-63, sr res chemist, 63-65, sect supvr, Catalysis Sect, Gulf Res & Develop Co, Pittsburgh, 66-73, res dir & sr scienitst, Chem & Minerals Div, 74-77, MGR, CATALYST & CHEM DEPT, GULF SCI & TECHNOL CO, 77- *Concurrent Pos:* Indust bd dir for sch eng & sci, Univ Mass, Amherst, 81-83. *Mem:* Am Chem Soc (chmn, Petrol Chem Div, 77-78). *Res:* Process research for chemical, refining, synthetic fuels and pollution control; heterogeneous and homogeneous catalysis. *Mailing Add:* 1410 Woodhill Dr Gibsonia PA 15044-9268

SWIFT, HEWSON HOYT, b Auburn, NY, Nov 8, 20; m 42; c 2. CYTOLOGY. *Educ:* Swarthmore Col, BA, 42; Univ Iowa, MS, 45; Columbia Univ, PhD(zool), 50. *Prof Exp:* Instr zool, Univ Chicago, 49-51, from asst prof to prof, 51-71, distinguished serv prof, 71-77, chmn dept, 72-77, Beadle distinguished serv prof biol, 77-84, Beadle distinguished serv prof, Dept Molecular Genetics, Cell Biol & Path, 84-90, EMER PROF, UNIV CHICAGO, 91- *Concurrent Pos:* Mem, Cell Biol Study Sect, NIH, 58-62, Develop Biol Adv Panel, NSF, 62-65 & Etiology Cancer Adv Panel, Am Cancer Soc, 66-70; vis prof, Harvard Univ, 70-71; chmn sect cellular & develop biol, Nat Acad Sci, 75-79; vis sr scientist, Commonwealth Sci & Indust Res Orgn, Canberra, Australia, 77-78. *Honors & Awards:* E B Wilson Award, 85. *Mem:* Nat Acad Sci; Am Soc Cell Biol (pres, 64); Histochem Soc (pres, 73); Genetics Soc Am; Am Acad Arts & Sci; hon fel Nat Acad Sci, India. *Res:* Cell biology; cytochemistry; molecular structure of chromosomes and chloroplasts. *Mailing Add:* Molecular Genetics & Cell Biol Univ Chicago 1103 East 57th St Chicago IL 60637

SWIFT, HOWARD R(AYMOND), b Streator, Ill, Mar 3, 20; m 46; c 5. CERAMICS. *Educ:* Univ Ill, BS, 40, MS, 42, PhD(ceramics), 45. *Prof Exp:* Asst eng, Exp Sta, Univ Ill, 40-42, res assoc eng, 43-44; res phys chemist, Libbey-Owens-Ford Glass Co, 45-56, chief glass tech res, Libby-Owens-Ford Co, 55-61, asst dir res, 61-68, dir res & develop, 68-76, chmn, Tech Prod Comt, 76-82, asst to pres, Glass Div, 78-82; CONSULT, 82- *Mem:* Fel Am Ceramic Soc; fel Brit Soc Glass Technol; fel Am Soc Testing & Mat. *Res:* Crystallization of glass; effect of glass composition on properties; stones in glass; manufacturing of flat glass. *Mailing Add:* 3525 Wesleyan Dr Toledo OH 43614

SWIFT, JACK BERNARD, b Ft Smith, Ark, Jan 3, 42; m 71; c 2. THEORETICAL SOLID STATE PHYSICS. *Educ:* Univ Ark, Fayetteville, BS, 63; Univ Ill, Urbana, MS, 65, PhD(physics), 68. *Prof Exp:* NSF fel, Max Planck Inst Physics & Astrophys, Munich, 68-69; res fel, Harvard Univ, 69-71; vis, Bell Tel Lab, 74; asst prof physics, 71-75; assoc prof, 75-81, PROF PHYSICS, UNIV TEX, AUSTIN, 81- *Concurrent Pos:* Res Fel, A P Sloan Found, 73-75. *Res:* Critical phenomena; light scattering properties of liquids; hydrodynamics. *Mailing Add:* Dept Physics Univ Tex Austin TX 78712

SWIFT, LLOYD HARRISON, b Crete, Nebr, Sept 12, 20. PLANT MORPHOLOGY. *Educ:* Univ Nebr, AB, 41, MS, 60, PhD(bot), 62; Western Reserve Univ, MA, 42. *Prof Exp:* Lexicographer, World Pub Co, Ohio, 42; instr Eng, Ill Inst Technol, 44 & Univ Mo, 44-45; asst prof bot, Univ Alaska, 62-63; prof & chmn dept Univ Nebr, Ataturk Univ, Turkey, 63-65; res assoc bot, Univ Nebr, 65-78; RETIRED. *Res:* Botanical bibliography and the classification and indexing of information important in botany; etymology and history of botanical terminology and nomenclature; phytography in plant morphology, taxonomy and physiology. *Mailing Add:* 2210 Sewell St Lincoln NE 68502-3850

SWIFT, LLOYD WESLEY, JR, b San Francisco, Calif, July 11, 32; m 55; c 4. MICROMETEOROLOGY, FOREST HYDROLOGY. *Educ:* State Univ NY Col Forestry, Syracuse, BS, 54; NC State Univ, MS, 60; Duke Univ, DF, 72. *Prof Exp:* Jr forester, 54-55, res forester, 57-71, RES METEOROLOGIST, FOREST SERV, USDA, 72- *Mem:* Am Meteorol Soc. *Res:* Energy balance and water quality of forested and logged mountain slopes; precipitation measurement, distribution and chemistry over steep slopes; air circulation patterns in mountains; effect of slope aspect and inclination on forest microenvironment. *Mailing Add:* Coweeta Hydrol Lab 999 Coweeta Lab Rd Otto NC 28763

SWIFT, MICHAEL, b New York, NY, Feb 5, 35; m 71; c 3. HUMAN GENETICS. *Educ:* Swarthmore, BA, 55; Univ Calif, Berkeley, MA, 58; NY Univ, MD, 62. *Prof Exp:* Instr med, NY Univ Sch Med, 66-67, asst prof, 70-71; assoc prof med, 72-79, RES SCIENTIST, BIOL SCI RES CTR, UNIV NC, 72-, PROF MED, 79- *Concurrent Pos:* Mem, Genetics Curric, 72-; assoc attend physician, Univ NC Hosps, 72-79, attend phys, 79-; dir, Second Yr Course Human Genetics, Univ NC, 80-89. *Mem:* AAAS; Am Soc Human Genetics. *Res:* Identify specific genes that predispose to particular common diseases and characterize them at the molecular level; develop tests to identify susceptible gene carriers; devise new strategies to prevent gene-associated common diseases. *Mailing Add:* 326 Biol Sci Res Ctr CB No 7250 Univ NC Chapel Hill Chapel Hill NC 27599-7250

SWIFT, MICHAEL CRANE, b Berkeley, Calif, Aug 21, 44; m 77. PREDATOR-PREY INTERACTIONS, ENERGETICS. *Educ:* Univ Calif, Davis, BSc, 66, MA, 68; Univ BC, PhD(zool), 74. *Prof Exp:* Fel, Univ Sask, 74-75; res assoc, Inst Animal Res Ecol, Univ BC, 75-76; instr intro biol, Duke Univ, 77-79; ASST PROF LIMNOL & AQUATIC ECOL, APPALACHIAN ENVIRON LAB, CTR ENVIRON & ESTUARINE STUDIES, UNIV MD, 80- *Concurrent Pos:* Lectr forestry-fisheries interactions, Fac Forestry, Univ BC, 75-76; vis lectr physiol, Univ NC, Chapel Hill 76; vis scientist, Marine Lab, Duke Univ, 79-81; adj fac, Frostburg State Col, 80-; assoc mem, Grad Fac, Univ Md, 81- *Mem:* AAAS; Am Soc Limnol & Oceanog; Soc Int Limnol; Ecol Soc Am; Sigma Xi; NAm Benthological Soc. *Res:* Limnology; predator-prey interactions among zooplankton; Chaoborus ecology. *Mailing Add:* 4304 Brewsters Run Ct Bellbrook OH 45305

SWIFT, ROBINSON MARDEN, b Wolfeboro, NH, May 6, 18; m 44; c 3. PHYSICAL CHEMISTRY. *Educ:* Univ NH, BS, 40; Northwestern Univ, MS, 48; Syracuse Univ, PhD(chem), 56. *Prof Exp:* Chemist, Bird & Son, Inc, Mass, 40-44; instr chem, Thiel Col, 47-53; from asst prof to assoc prof, 56-72, PROF CHEM, ST ANSELM'S COL, 72-; CHIEF CHEMIST, EDISON ELECTRONICS DIV, McGRAW-EDISON, 57- *Mem:* Am Chem Soc; Sigma Xi. *Res:* Thermodynamics; epoxy resins. *Mailing Add:* 18 Birch Hill Rd Hooksett NH 03106

SWIFT, TERRENCE JAMES, b Dubuque, Iowa, June 29, 37; m 65; c 2. PHYSICAL CHEMISTRY, BIOCHEMISTRY. *Educ:* Loras Col, BS, 59; Univ Calif, Berkeley, PhD(chem), 62. *Prof Exp:* NSF fel, Max Planck Inst Phys Chem, 62-63; from asst prof to assoc prof chem, 63-76, PROF CHEM, CASE WESTERN RESERVE UNIV, 76- *Mem:* Am Soc Biol Chem; Am Chem Soc. *Res:* Magnetic resonance spectroscopy as applied to biological systems and processes. *Mailing Add:* Dept Chem Case Western Reserve Univ 2040 Adelbert Rd Cleveland OH 44106

SWIFT, WILLIAM CLEMENT, b Lexington, Ky, Mar 17, 28; m 50; c 8. MATHEMATICS. *Educ:* Univ Ky, BS, 50, PhD(math), 55. *Prof Exp:* Instr math, Cornell Univ, 55-56; mem tech staff, Bell Tel Labs, Inc, 56-58; asst prof math, Rutgers Univ, 58-63; assoc prof, 63-69, PROF MATH, WABASH COL, 69- *Mem:* Am Math Soc; Math Asn Am. *Res:* Complex variables; conformal mapping; Taylor series; summability. *Mailing Add:* 116 N Grace Ave Crawfordsville IN 47933

SWIGAR, MARY EVA, b Nesquehoning, Pa, Oct 17, 40. PSYCHIATRY. *Educ:* Muhlenberg Col, BS, 62; Health Sci Ctr, Temple Univ, MD, 66. *Prof Exp:* Teaching fel, Yale Univ, Sch Med, 67-70, instr, 70-71, from asst prof to assoc prof psychiat, 71-88; ASSOC PROF PSYCHIAT, UMDNJ/ROBERT WOOD JOHNSON MED SCH, NEW BRUNSWICK, NJ, 88-; CHIEF PSYCHIAT, ROBERT WOOD JOHNSON UNIV HOSP, 88- *Concurrent Pos:* Consult, Dept Obstet-Gynec, Sch Med, Yale Univ, 70-73, Gaylord Rehab Hosp, 70-79 & 85-88. *Mem:* Am Med Women's Asn; Am Psychopath Asn; NY Acad Sci; AAAS; fel Int Col Psychosom Med. *Res:* Psychiatric aspects of obstetrics-gynecology; psychoendocrinology; diagnostic imaging and psychiatric syndromes; early response to treatment in psychoses. *Mailing Add:* Dept Psychiat Robert Wood Johnson Univ Hosp New Brunswick NJ 08903

SWIGART, RICHARD HANAWALT, b Lewistown, Pa, July 7, 25; m 51; c 3. NEUROANATOMY. *Educ:* Univ NC, BA, 47; Univ Minn, PhD(anat), 53. *Prof Exp:* Asst anat, Univ Minn, 48-50, instr, 50-52, res assoc histochem, 52-53; from asst prof to assoc prof, 53-57, asst dean student affairs, 69-72, actg dean, Sch Med, 72-73, vpres allied health affairs, 76-78, actg dean Grad Sch, 76-77, PROF ANAT, SCH MED, UNIV LOUISVILLE, 67-, DIR DIV ALLIED HEALTH, 78-, ACTG SPEC ASST TO PRES HEALTH AFFAIRS, 81- *Mem:* Am Asn Anat; Biol Stain Comn; Histochem Soc; Soc Exp Biol & Med; Sigma Xi. *Res:* Chronic hypoxia, effect on cardiovascular and erythropoietic systems; carbohydrate metabolism in cardiac and skeletal muscle; effect of age on adaptive responses; neurological mutant mice. *Mailing Add:* 3818 Washington Square, Apt 5 Louisville KY 40207

SWIGER, ELIZABETH DAVIS, b Morgantown, WVa, June 27, 26; m 48; c 2. PHYSICAL CHEMISTRY, INORGANIC CHEMISTRY. *Educ:* WVa Univ, BS, 48, MS, 52, PhD, 65. *Prof Exp:* Instr math, 54-56, instr chem, 56-60, from asst prof to assoc prof, 60-66, PROF CHEM, FAIRMONT STATE COL, 66- *Mem:* Am Chem Soc; AAAS; Sigma Xi. *Res:* Nuclear quadrupole resonance spectroscopy; computer applications; coordination compounds. *Mailing Add:* Dept Chem Fairmont WV 26554

SWIGER, LOUIS ANDRE, b Waverly, Ohio, Sept 16, 32; m 53; c 3. ANIMAL GENETICS, POPULATION GENETICS. *Educ:* Ohio State Univ, BSc, 54; Iowa State Univ, MSc, 57, PhD(animal breeding), 60. *Prof Exp:* Animal geneticist, USDA, 59-62; assoc prof animal sci & exp sta statist, Univ Nebr, 62-65; assoc prof animal sci, Ohio State Univ, 65-70, prof, 70-80; PROF & DEPT HEAD ANIMAL SCI, VA POLYTECH INST & STATE UNIV, 80- *Concurrent Pos:* Grad chmn, Dept Animal Sci, Ohio State Univ, 68-80. *Honors & Awards:* Rockefeller Prentice Mem Award. *Mem:* Am Soc Animal Sci; Biomet Soc; Sigma Xi. *Res:* Population genetics and application to domestic animals. *Mailing Add:* Dept Animal Sci 3460 Animal Sci Bldg Va Polytech Inst & State Univ Blacksburg VA 24061

SWIGER, WILLIAM F, b Buhl, Idaho, June 7, 16; m 39; c 3. GEOTECHNICAL ENGINEERING. *Educ:* Univ Wash, BS, 39; Harvard Univ, MS, 40. *Prof Exp:* Engr, R V Labarre, Los Angeles, 40-42; eng struct div, Stone & Webster Eng Corp, Boston, 42-51, struct engr, 51-57, consult engr, 57-72, sr consult engr, 72-81, vpres, 73-81; RETIRED. *Concurrent Pos:* US Nat Comt Tunneling Technol, NRC, 71-73; chmn, New Eng Sect, Asn Eng Geol, 68-; consult, dams & hydropower, 81- *Honors & Awards:* Thomas A Middlebrooks Award, Am Soc Civil Engrs, 63. *Mem:* Nat Acad Eng; fel Am Soc Civil Engrs; Asn Eng Geologists; Earthquake Eng Res Inst; Int Soc Soil Mechanics & Found Engrs. *Res:* Slow triaxial testing to determine the true shear strengths of clay soils; hydro-electric projects, especially embankment dams, tunnels and underground power houses. *Mailing Add:* PO Box 388 Buhl ID 83316

SWIHART, G(ERALD) R(OBERT), b Carroll, Nebr, Feb 16, 20; m 44; c 3. STRUCTURAL ENGINEERING. *Educ:* Rose Polytech Inst, BS, 47; Yale Univ, MEng, 49. *Prof Exp:* Instr, Rose Polytech Inst, 48; asst instr, Yale Univ, 48-49; from instr to prof civil eng, Univ Nebr, Lincoln, 49-90, teacher exten div, 58 & 59, vchmn dept, 71-73 & 81-82; RETIRED. *Concurrent Pos:* Struct designer, Harold Hoskins & Assocs, Nebr, 55, 57, 59 & 61; struct res engr, US Naval Civil Eng Lab. *Mem:* Am Soc Civil Engrs; Am Concrete Inst; Nat Soc Prof Engrs. *Res:* Ultimate load theories of reinforced concrete structures subjected to static and dynamics loading. *Mailing Add:* 2338 Calumet Ct Lincoln NE 68502

SWIHART, JAMES CALVIN, b Elkhart, Ind, Feb 8, 27; m 47; c 2. THEORETICAL SOLID STATE PHYSICS, BIOPHYSICS. *Educ:* Purdue Univ, BSChE, 49, MS, 51, PhD(physics), 55. *Prof Exp:* Danish govt fel, Inst Theoret Physics, Copenhagen Univ, 54-55; physicist, Argonne Nat Lab, 55-56; physicist, Res Ctr, Int Bus Mach Corp, 56-66; assoc prof, 66-67, assoc dean grad sch, 71-74, chmn biophys prog, 71-78, PROF PHYSICS, IND UNIV, BLOOMINGTON, 67-, CHMN, PHYSICS DEPT, 80- *Concurrent Pos:* Vis physicist, Lawrence Radiation Lab, Berkeley, 65; visitor, Cavendish Lab & assoc, Clare Hall, Univ Cambridge, 74. *Mem:* AAAS; fel Am Phys Soc; Sigma Xi. *Res:* Theory of solid state physics; superconductivity; many-body problem; biophysics. *Mailing Add:* Dept Physics-Swain W 233 Ind Univ Bloomington IN 47405

SWIHART, THOMAS LEE, b Elkhart, Ind, July 29, 29; m 51; c 3. ASTROPHYSICS. *Educ:* Ind Univ, AB, 51, AM, 52; Univ Chicago, PhD(astrophys), 55. *Prof Exp:* Assoc prof physics & astron, Univ Miss, 55-57; mem staff, Los Alamos Sci Lab, 57-62; asst prof astrophys, Univ Ill, 62-63; assoc prof, 63-69, PROF ASTRON, UNIV ARIZ, 69-, ASTRONR, STEWARD OBSERV, 74- *Concurrent Pos:* Fulbright-Hays lectr, Aegean Univ, Turkey, 69-70. *Mem:* Int Astron Union. *Res:* Theoretical astrophysics; radiation transfer; polarization of radio sources; atmospheric structure of stars. *Mailing Add:* Dept Astron Univ Ariz Tucson AZ 85721

SWIM, WILLIAM B(AXTER), b Stillwater, Okla, Nov 18, 31; m 89; c 5. FLUID DYNAMICS, ACOUSTICS. *Educ:* Okla State Univ, BS, 55; Ga Inst Technol, PhD(mech eng), 66. *Prof Exp:* Instr mech eng, Pa State Univ, 55-59; eng res, Okla State Univ, 59-61; sr res engr, Trane Co, 64-73; PROF & DIR, NOISE CONTROL FACIL & FAN RES LAB, TENN TECHNOL UNIV, 73- *Concurrent Pos:* Consult fluid dynamics acousts & turbomach, Worth Sports, Trane Co, Control Data, DTRC, Emerson Elec Co, Black & Decker, 81-; Fulbright fel, 81. *Mem:* Am Soc Mech Engrs; Acoust Soc Am; Am Soc Heating, Refrig & Air-Conditioning Engrs. *Res:* Fluid dynamics of turbomachinery; analysis of internal flows; generation and suppression of flow-noise; noise reduction of fans and blowers, flow losses in piping and duct systems; axial fan design; centrifugal blower evaluation; noise reduction of fan and blower systems. *Mailing Add:* Dept Mech Eng Tenn Technol Univ Cookeville TN 38505

SWINDALE, LESLIE D, b Wellington, NZ, Mar 16, 28; m 55; c 3. SOIL SCIENCE, RESEARCH ADMINISTRATION. *Educ:* Univ Victoria, NZ, 48, MSc, 50; Univ Wis, PhD(soil sci), 55. *Prof Exp:* Phys chemist, NZ Soil Bur, 49-57, sr phys chemist, 57-60; dir, NZ Pottery & Ceramics Res Asn, 60-63; prof soil sci, Univ Hawaii, Manoa, 63-76, chmn, dept agron & soil sci, 65-68, soil scientist, agr exp sta, 65-76, assoc dir, 70-75; DIR GEN, INT CROP RES INST SEM-IARID TROPICS, 77- *Concurrent Pos:* Fel, Univ Wis, 55-56; chief soil resources, Conservation and Development Serv, Land & Water Develop Div, Food & Agr Orgn, UN, Rome, 68-70. *Mem:* Fel Soil Sci Soc Am; fel NZ Inst Chem; Royal Soc NZ; Int Soc Soil Sci; fel Am Soc Agron. *Res:* Genesis of soils, their characterization and uses. *Mailing Add:* 2910 Nanihale Pl No A Honolulu HI 96822

SWINDELL, ROBERT THOMAS, b Greenfield, Tenn, Feb 22, 38; m 61; c 1. ORGANIC CHEMISTRY. *Educ:* Memphis State Univ, BS, 61; Univ SC, PhD(org chem), 65. *Prof Exp:* NIH fel org photochem, Iowa State Univ, 65-66; asst prof, 66-77, PROF CHEM, TENN TECHNOL UNIV, 77- *Mem:* Am Chem Soc; Royal Soc Chem. *Res:* Organic reaction mechanisms; organic photochemistry. *Mailing Add:* Dept of Chem Box 5055 Tenn Technol Univ Cookeville TN 38501

SWINDEMAN, ROBERT W, b Toledo, Ohio, Jan 18, 33; m 63; c 3. MATERIALS FOR PRESSURE VESSELS & PIPING, HIGH TEMPERATURE MECHANICAL METALLURGY. *Educ:* Univ Notre Dame, BS, 55, MS, 57. *Prof Exp:* Engr, Large Steam Turbine Div, Gen Elec Co, 55; mem res staff, Nuclear Div, Union Carbide, 57-85; MEM RES STAFF, MARTIN MARIETTA ENERGY SYSTS, 85- *Concurrent Pos:* Res officer, Australian Atomic Energy Res Estab, 61-63; mem, Mat Properties Coun. *Honors & Awards:* Award of Merit, Am Soc Testing & Mat, 85; Bd of Gov Award, Am Soc Mech Engrs, 89; Res & Develop 100 Award, 90. *Mem:* Fel Am Soc Metals Int; fel Am Soc Testing & Mat; fel Am Soc Mech Engrs. *Res:* Deformation and fracture of high-temperature alloys; constitutive equations and inelastic analysis methods; pressure vessels and piping for fossil energy applications; fatigue and fracture in high-temperature nuclear applications; development of advanced alloys for heat recovery systems. *Mailing Add:* Metals & Ceramics Div Oak Ridge Nat Lab PO Box 2008 Oak Ridge TN 37831-6155

SWINDLE, TIMOTHY DALE, b Great Bend, Kans, June 5, 55; m 77; c 2. PLANETARY SCIENCE, METEORITICS. *Educ:* Univ Evansville, BS, 77, BA, 78; Washington Univ, MA, 81, PhD(physics), 86. *Prof Exp:* ASST PROF PLANETARY SCI, UNIV ARIZ, 86- *Concurrent Pos:* Mem, Lunar & Planetary Sample Team, NASA, 87-, Meteorite Working Group, NASA/NSF, 89-; mem coun, Meteoritical Soc, 91- *Mem:* Meteoritical Soc; Geochemical Soc; Am Astron Soc Div Planetary Sci. *Res:* Noble gas geochemistry and geochronology of meteorites and lunar samples. *Mailing Add:* Lunar & Planetary Lab Univ Ariz Tucson AZ 85721

SWINEBROAD, JEFF, b Nashville, Tenn, Mar 22, 26; m 53; c 2. ZOOLOGY. *Educ:* Ohio State Univ, BA, 49, MA, 50, PhD, 56. *Prof Exp:* Asst ornith, Univ Colo, 46-47; tech asst zool, Ohio Coop Wildlife Res Unit, 48; asst, Ohio State Univ, 49-50. asst instr, 51, instr conserv, Conserv Lab, 51-53 & Nat Audubon Soc, 54-56; zoologist, Rutgers Univ, 55-57, from asst prof to assoc prof, 57-66, prof biol, 66-68, chmn dept biol sci. 60-68. asst res specialist, Col Agr, 59-68; pop ecologist, USAEC, 68-72, chief, Ecol Sci Br, 72-74; dep assoc dir res & develop progs, US Energy Res & Develop Admin, 74-77, mgr environ prog, Off Asst Secy Environ, 77-79, DEP DIR, OFF PROG COORDINATION, DEPT ENERGY, 80- *Mem:* AAAS; Animal Behav Soc; Ecol Soc Am; Wilson Ornith Soc(secy, 67-); Cooper Ornith Union. *Res:* Avian anatomy; ecology and migration of birds; animal population behavior. *Mailing Add:* Dept Field Std Hort US Dept Agr Grad Sch 600 MD Ave SW Washington DC 20250

SWINEHART, BRUCE ARDEN, b Greentown, Ohio, Aug 28, 29; m 55; c 2. ANALYTICAL CHEMISTRY. *Educ:* Oberlin Col, AB, 51; Purdue Univ, MS, 53, PhD(anal chem), 55. *Prof Exp:* Res chemist, Mallinckrodt Chem Works, 5S-60; res chemist, Wyandotte Chem Corp, Mich, 60-61, sr res chemist, 61-64, sect head, 64-67; res chemist, 67-80, RES ASSOC, CORNING GLASS WORKS, 80- *Mem:* Am Chem Soc. *Res:* Uranium chemistry; ore analysis; gas chromatography; glass analysis; titrimetry. *Mailing Add:* 17031 N 11th Ave Apt 2101 Pheonix AZ 14830

SWINEHART, CARL FRANCIS, b Bainbridge, Ohio, Aug 26, 07; m 34; c 2. INORGANIC CHEMISTRY. *Educ:* Ohio Wesleyan Col, AB, 29; Western Reserve Univ, PhD(inorg chem), 33. *Prof Exp:* Asst, Western Reserve Univ, 29-32; res chemist, Harshaw Chem Co, 32-52, assoc dir res, 52-60, dir tech develop, Inorg Prod, 60-62 & Crystal-Solid State, 62-67, proj dir crystal growth develop, Harshaw Div, Kewanee Oil Co, Ohio, 67-72; consult, Harshaw-Filtrol Parnership, Ohio, 72-88 & Engelhard, NJ, 89-90; RETIRED. *Res:* Fluoride gases; manufacture of fluorides; synthetic crystal production. *Mailing Add:* 4102 Silsby Rd Univ Heights Cleveland OH 44118

SWINEHART, JAMES HERBERT, b Los Angeles, Calif, Nov 22, 36; m 63; c 3. INORGANIC CHEMISTRY, BIOINORGANIC CHEMISTRY. *Educ:* Pomona Col, BA, 58; Univ Chicago, PhD(chem), 62. *Prof Exp:* NSF fel phys inorg chem, Max Planck Inst Phys Chem, Univ Gottingen, Ger, 62-63; from asst prof to assoc prof, 68-72, PROF CHEM, UNIV CALIF, DAVIS, 72- *Concurrent Pos:* Fel, John Simon Guggenheim Found, 69-70. *Mem:* Am Chem Soc. *Res:* Mechanisms of inorganic reactions; transition metals in the marine environment. *Mailing Add:* Dept of Chem Univ of Calif Davis CA 95616

SWINEHART, JAMES STEPHEN, b Cleveland, Ohio, July 27, 29; m 63; c 1. ORGANIC CHEMISTRY, SPECTROCHEMISTRY. *Educ:* Western Reserve Univ, BS, 50; Univ Cincinnati, MS, 51; New York Univ, PhD(chem), 59. *Prof Exp:* Asst, Western Reserve Univ, 49-50; res chemist, Merck & Co, 51-53; from asst prof org chem to assoc prof, Wagner Col, 57-61; assoc prof, Am Univ, 61-65; anal chemist, Atlantic Res Corp, 65-67; sr spectroscopist, Perkin Elmer Corp, 67-69 & Digilab, 69-70; chmn dept, 70-76, PROF CHEM, STATE UNIV NY COL CORTLAND, 70- *Mem:* Am Chem Soc. *Res:* Organic synthesis; natural products; instrumentation; infrared and nuclear magnetic resonance spectroscopy; information retrieval; cigarette smoke; gas chromatography; technical writing. *Mailing Add:* Dept Chem Box 2000 State Univ NY Col Cortland Cortland NY 13045-0900

SWINEHART, PHILIP ROSS, b Los Alamos, NMex, May 20, 45. SOLID STATE ELECTRONICS. *Educ:* Ore State Univ, BS, 67; Ohio State Univ, MS, 68, PhD(elec eng), 74. *Prof Exp:* Res assoc, Electro Sci Lab, Ohio State Univ, 68-69; res engr, Ohio Semitronics Inc, 72-73; RES SCIENTIST LOW TEMP SENSORS & RADIATION DETECTION, LAKE SHORE CRYOTRONICS INC, 75- *Mem:* Inst Elec & Electronics Engrs. *Res:* Low temperature sensors and radiation detectors. *Mailing Add:* Lake Shore Cryotronics Inc 64 E Walnut St Columbus OH 43081

SWINFORD, KENNETH ROBERTS, b Trader's Point, Ind, July 8, 16; m 38; c 2. FORESTRY. *Educ:* Purdue Univ, BS, 37; Univ Fla, MSF, 48; Univ Mich, PhD, 60. *Prof Exp:* Exten ranger, State Forest Serv, Fla, 40-41; timber cruiser, Brooks-Scanlon Corp, 46; from asst prof to prof, 46-75, asst to dir forestry, 71-75, EMER PROF FORESTRY, UNIV FLA, 76- *Concurrent Pos:* Consult forester, F & W Forestry Serv Inc, Gainesville, Fla, 75-86; teacher, Sch Forestry, Univ Fla, 75-76, Woodland Mgrs, Inc, 76- *Mem:* Soc Am Foresters; Am Forestry Asn. *Res:* Management of forest lands, especially pine plantations; landscape forestry; outdoor recreational use of forests and wild lands; management and harvesting of forests for energy fuel. *Mailing Add:* 13410 NW 49th Ave Gainesville FL 32606

SWINGLE, DONALD MORGAN, b Washington, DC, Sept 1, 22; m 43; c 3. APPLIED PHYSICS, SYSTEMS ENGINEERING. *Educ:* Wilson Teachers Col, BS, 43; NY Univ, MS, 47; Harvard Univ, AM, 48, MEngSci, 49, PhD(eng sci, appl physics), 50; George Washington Univ, MBA, 62; Indust Col Armed Forces, dipl, 62. *Prof Exp:* Engr, Signal Corps Eng Labs, US Dept Army, 46-47; res asst, Eng Res Lab, Harvard Univ, 49-50; physicist, Signal Corps Eng Labs, US Dept Army, 50-60, chief weather electronic res group, 50-53, chief meteorol techniques sect, 54-57, chief br, 58-60, physicist, sr res scientist & dep dir meteorol div, US Army, Electronics Res & Develop Labs, 61-63, res physicist, sr res physicist & chief meteorol res team A, US Army Electronics Command, 64-65, res physicist & sr scientist atmospheric sci lab, 65-66, res physicist & chief techniques & explor develop tech area, 66-71, sr scientist eng & explor develop tech area, 71-74, sr scientist, Spec Sensors Tech Area, Combat Surveillance & Target Acquisition Lab, 74-78, sr res physicist, Atmospheric Sci Lab, Army Electronics Res & Develop Command, 78-80; RETIRED. *Concurrent Pos:* US mem, Comn Instruments & Methods Observ, World Meteorol Orgn, 53-65; US deleg, 53 & 57; adv, US Mil Acad, 59; reviewer res proposals, NSF, 60-74; mem, Comt Atmospheric Environ, Comn Unidentified Flying Objects, Am Inst Aeronaut & Astronaut, 62-70; chmn, Nat Task Group Mesometeorol, Interdept Comt Atmospheric Sci, 63-74; Army rep, DOD Forum Environ Sci, 64-65; mem, Chem Comt Meteorol, Nat Ctr Atmospheric Res, 63-65, Nat meso-micrometeorol Res Facil Surv Group, 64; lab mem, Army Res Coun, 64-65; consult, Mark Resources Inc, 81-82, NMex State Univ, 87-88 & NMex Res Inst, 88-; chief, Measurements Lab, NMex Res Inst, 88- *Mem:* Am Meteorol Soc; sr mem Inst Elec & Electronic Engrs; Nat Soc Prof Eng; assoc fel Am Inst Aeronaut & Astronaut; fel NY Acad Sci. *Res:* Radar meteorology; atmospheric propagation, electromagnetic, acoustic waves; radioactive fallout prediction; meteorological techniques, applied meteorology, meteorological system engineering; atmospheric modification, management, mesometeorology; indirect sensory techniques; nuclear surveillance; research and development management. *Mailing Add:* 1765 Pomona Dr Las Cruces NM 88001-4919

SWINGLE, HOMER DALE, b Hixson, Tenn, Nov 5, 16; m 42, 62; c 1. OLERICULTURE. *Educ:* Univ Tenn, BS, 39; Ohio State Univ, MS, 48; La State Univ, PhD(hort), 66. *Prof Exp:* Teacher high sch, 39-46; exten specialist hort, Agr Exten Serv, 46-47, from asst prof to prof, 48-79, EMER PROF, UNIV TENN, KNOXVILLE, 79- *Concurrent Pos:* Consult plant & soil water rels, Oak Ridge Nat Labs, 71-75. *Mem:* Fel Am Soc Hort Sci. *Res:* Evaluation of vegetable varieties; chemical weed control in horticulture crops; mechanization of harvest. *Mailing Add:* 3831 Maloney Rd Knoxville TN 37920

SWINGLE, KARL F, b Richland Center, Wis, Feb 16, 35; div; c 4. PHARMACOLOGY. *Educ:* Univ Wis, BA, 58; Univ Minn, PhD(pharmacol), 68. *Prof Exp:* Asst bacteriologist, Sioux City Dept Health, Iowa, 58-59; med technologist, Vet Admin Hosp, Minneapolis, 61-64; pharmacologist, 68-73, supvr pharmacol, 73-77, SR RES SPECIALIST, RIKER LABS, MINN MINING & MFG CO, 77- *Mem:* Am Soc Pharmacol & Exp Therapeut; NY Acad Sci; Soc Exp Biol & Med; Am Chem Soc. *Res:* Anti-inflammatory drugs; pulmonary pharmacology. *Mailing Add:* Dept New Molec Res Riker Labs 3M Co Bldg 270- 2S-06 St Paul MN 55144

SWINGLE, KARL FREDERICK, b Bozeman, Mont, Jan 7, 15; m 40; c 5. RADIOBIOLOGY. *Educ:* Mont State Col, BS, 37; Univ Wis, PhD(biochem), 42. *Prof Exp:* Res chemist, Inst Path, Western Pa Hosp, 42-43; asst res chemist, Univ Wyo, 43-45; assoc chemist, Mont State Col, 45-47, prof vet biochem, 57-61; supvry chemist, US Naval Radiol Defense Lab, 61-69; res chemist, 69-70, PHYSICIST, VET ADMIN HOSP, 70- *Concurrent Pos:* Asst researcher, Univ Calif, Irvine, 70- *Mem:* Sigma Xi; Radiation Res Soc. *Res:* Radiation biochemistry; nucleic acid metabolism. *Mailing Add:* Riker Labs Inc New Molecule Res 3M Co Bldg 270-25-06 St Paul MN 55144

SWINGLE, ROY SPENCER, b Harvey, Ill, Oct 15, 44; m 66; c 2. ANIMAL SCIENCE, NUTRITION. *Educ:* Univ Ariz, BS, 66; Wash State Univ, MS, 69, PhD(nutrit), 72. *Prof Exp:* Asst prof, 72-78, ASSOC PROF ANIMAL SCI, UNIV ARIZ, 78-, ASSOC RES SCIENTIST ANIMAL SCI, AGR EXP STA, 81- *Mem:* Am Soc Animal Sci; Am Inst Nutrit. *Res:* Ruminant nutrition; utilization of low quality roughages. *Mailing Add:* Dept of Animal Sci Univ of Ariz Tucson AZ 85721

SWINGLEY, CHARLES STEPHEN, b Dallas, Tex, Nov 11, 43; m 64; c 2. PHYSICAL CHEMISTRY. *Educ:* Rochester Inst Technol, BS, 65; Wayne State Univ, PhD(chem), 70. *Prof Exp:* RES ASSOC RES LABS, EASTMAN KODAK CO, 70- *Mem:* Soc Photog Sci & Eng. *Res:* Surface, colloid, polymer and photographic chemistry. *Mailing Add:* 1573 Harris Rd Penfield NY 14526-1813

SWINK, LAURENCE N, b Enid, Okla, Oct 24, 34; div; c 1. CRYSTALLOGRAPHY. *Educ:* Univ Wichita, BA, 57; Iowa State Col, MSc, 59; Brown Univ, PhD(chem), 69. *Prof Exp:* Flight test technician, Cessna Aircraft Co, Kans, 56-57; mat engr, Chance-Vought Aircraft Co, Tex, 57; mat engr, Douglas Aircraft Co, Calif, 59-60; nuclear res officer, McClellan AFB, Calif, 60-63; mem tech staff crystallog, Tex Instruments, Inc, Dallas, 66-75, mgr, Infrared Glass Lab, Electro-Optics Div, 75-78; vpres, Amorphous Mat Inc, Garland, Tex, 79-80; mgr, Advan Systs Develop, Xerox Corp, Dallas, 81-82; TECH DIR, MULTI-PLATE CO, DALLAS, 82- *Mem:* Am Crystallog Asn. *Res:* Crystal structure determination by x-ray diffraction methods; crystal perfection study by x-ray and electron diffraction techniques; electron microprobe analysis; electron microscopy; auger spectroscopy, x-ray fluorescence; printed circuit technology. *Mailing Add:* 1111 Abrams Rd No 123 Richardson TX 75081

SWINNEY, CHAUNCEY MELVIN, b Riverside, Calif, Sept 3, 18; m 42; c 3. ECONOMIC GEOLOGY, PETROLOGY. *Educ:* Pomona Col, BA, 40; Stanford Univ, PhD(geol), 49. *Prof Exp:* Tester, Union Oil Co, Calif, 42; geologist, US Geol Surv, 42-45; instr, Stanford Univ, 47-49, asst prof mineral sci, 49-56; supvr, Prod Res Div, Richfied Oil Corp, 59-66; mgr energy resources exp & develop, Southern Calif Edison Co, 66-82; RETIRED. *Concurrent Pos:* Geologist, US Geol Surv, 46-53; vpres, Mono Power Co, Subsid Southern Calif Edison Co, 73-82. *Mem:* Geol Soc Am; Am Asn Petrol Geol; Am Inst Mining Metall & Petrol Eng; Sigma Xi. *Res:* Energy resources exploration, development and production. *Mailing Add:* 1354 Seafarer St Ventura CA 93001

SWINNEY, HARRY LEONARD, b Opelousas, La, Apr 10, 39; m 67; c 1. NONLINEAR DYNAMICS, FLUID PHYSICS. *Educ:* Rhodes Col, Memphis, BS, 61; Johns Hopkins Univ, PhD(physics), 68. *Prof Exp:* Res assoc physics, Johns Hopkins Univ, 68-70, vis asst prof, 70-71; asst prof physics, NY Univ, 71-73; from assoc prof to prof physics, City Col, City Univ New York, 73-78; prof, 78-84, Trull Centennial Prof, 84-90, SID RICHARDSON FOUND REGENTS PROF, UNIV TEX, AUSTIN, 90- *Concurrent Pos:* Guggenheim fel, 83-84; dir, Ctr Nonlinear Dynamics, 85- *Honors & Awards:* Morris Loeb Lectr, Harvard Univ, 82. *Mem:* Fel Am Phys Soc; Am Asn Physics Teachers. *Res:* Instabilities and turbulence are studied in experiments on nonequilibrium systems, particularly fluids and chemical reactions. *Mailing Add:* Dept Physics Univ Tex Austin TX 78712

SWINSON, DEREK BERTRAM, b Belfast. N Ireland, Nov 5, 38; m 65; c 2. PHYSICS. *Educ:* Queen's Univ, Belfast, 60; Univ Alta, Calgary, MS, 61, PhD(physics), 65. *Prof Exp:* From asst prof to assoc prof, 65-76, PROF PHYSICS, UNIV NMEX, 76- *Concurrent Pos:* Consult, accident reconstruct, 68- *Mem:* Brit Inst Physics; Sigma Xi; Am Geophys Union. *Res:* Cosmic radiation, extensive air showers; mu-mesons underground and variations of their intensity with solar activity; sidereal cosmic ray anisotropies; consultant in accident reconstruction. *Mailing Add:* Dept Physics Univ NMex 800 Yale NE Albuquerque NM 87131

SWINTON, DAVID CHARLES, b St Charles, Ill, May 4, 43; m 70. CELL BIOLOGY. *Educ:* Brandeis Univ, AB, 65; Stanford Univ, PhD(biol), 72. *Prof Exp:* Res fel biophysics, Univ Chicago, 72-74, res assoc, 74-80; MEM FAC, DEPT BIOL, UNIV ROCHESTER, 80- *Mem:* Am Soc Cell Biol. *Res:* Information content of organelle DNA; expression of organelle DNA during mitotic cell cycle and during meiosis. *Mailing Add:* 110 Laurltown Rd Rochester NY 14609

SWINYARD, EWART AINSLIE, b Logan, Utah, Jan 3, 09; m 34; c 2. PHARMACOLOGY. *Educ:* Utah State Univ, BS, 32; Idaho State Col, BS, 36; Univ Minn, MS, 41; Univ Utah, PhD(pharmacol), 47. *Hon Degrees:* DSc, Utah State Univ, 83 & Univ Utah, 86. *Prof Exp:* From instr to asst prof pharm, Idaho State Col, 36-45, prof pharmacol, 45-47; prof pharmacol & dir pharmaceut res, Col Pharm, 47-76, prof pharmacol, Col Med, 67-76, dean, Col Pharm, 70-76, DIR, CTR EARLY PHARMACOL EVAL OF ANTIEPILEPTIC DRUGS, 75-, EMER PROF PHARMACOL, COL PHARM & COL MED, UNIV UTAH, 76- *Concurrent Pos:* Am Col Apothecaries fac fel; lectr, Col Med, Univ Utah, 45-67, chmn dept biopharmaceut sci, Univ, 65-71, distinguished res prof, 68-69; ed, Am Med Soc Alcoholism, 77. *Honors & Awards:* Rennebohm Lectr, Univ Wis, 60; Award & Medal distinguished res & ed alcoholism, Am Med Soc Alcoholism, 77; DuMez Lectr, Univ Md, 79; Epilepsy Award, Am Soc Exp Therapeut, 82; Kaufman Lectr, Ohio State Univ, 63; Distinguished Basic Neuroscientist Award, Am Epilepsy Soc, 90- *Mem:* Am Soc Pharmacol & Exp Therapeut; Am Pharmaceut Asn; NY Acad Sci; Am Asn Cols Pharm (hon pres, 81-82); Sigma Xi. *Res:* Arsenical chemotherapy; body water and electrolyte distribution; experimental therapy of convulsive disorders; assay of anticonvulsant drugs; relationship between chemical structure and pharmacological activity of anticonvulsant drugs. *Mailing Add:* Col Pharm Univ Utah Salt Lake City UT 84112

SWISCHUK, LEONARD EDWARD, b Bellevue, Alta, June 14, 37; m 60; c 4. RADIOLOGY. *Educ:* Univ Alta, BS & MD, 60. *Prof Exp:* Asst prof pediat, Med Ctr, Univ Okla, 66-68, assoc prof radiol & pediat, 68-70; assoc prof, 70-73, PROF RADIOL & PEDIAT, UNIV TEX MED BR GALVESTON, 73- *Mem:* Am Med Asn; Am Col Radiol; Am Acad Pediat; Radiol Soc NAm. *Mailing Add:* RAD C264 Child Health Ctr C65 Univ Tex Med Sch 301 Univ Blvd Galveston TX 77550

SWISHER, ELY MARTIN, b Bozeman, Mont, Sept 29, 15; m 40; c 3. ENTOMOLOGY. *Educ:* Willamette Univ, AB, 37; Ore State Col, MS, 41; Ohio State Univ, PhD(entom), 43. *Prof Exp:* Asst zool, Ohio State Univ, 40-43; midwest mgr, Rohm and Haas Co, 43-53, mgr develop sect agr & sanit chem, 53-73, mgr govt regulatory rels, 73-76, mgr agr chem standards, 76-81; mem staff, Regulatory Compliance Serv, 81-89; RETIRED. *Mem:* Entom Soc Am; Weed Sci Soc Am. *Res:* Insecticides; fungicides; field evaluations; development of agricultural pesticide chemicals. *Mailing Add:* 1950 Branch Rd Perkasie PA 18944

SWISHER, GEORGE MONROE, b Columbus, Ohio, July 17, 43; m 64; c 1. MECHANICAL & CONTROL ENGINEERING. *Educ:* Univ Cincinnati, BSME, 66; Ohio State Univ, MSME, 67, PhD(mech eng), 69. *Prof Exp:* Asst prof systs eng, Wright State Univ, 69-73; assoc prof, 73-79, PROF MECH ENG & ASSOC DEAN ENG, TENN TECHNOL UNIV, 79- *Concurrent Pos:* Consult, United Aircraft Prod, 70- & Missile Systs Div, Rockwell Int Corp, 77. *Mem:* Am Soc Mech Engrs; Am Soc Eng Educ; Sigma Xi. *Res:* Manual control; control theory; stress analysis; dynamic system measurement and control. *Mailing Add:* 1035 E 6th St Cookeville TN 38501

SWISHER, HORTON EDWARD, b San Diego, Calif, Mar 1, 09; m 47; c 1. FOOD CHEMISTRY. *Educ:* Pomona Col, BA, 33; Claremont Cols, MA, 35. *Prof Exp:* Jr analyst, Am Potash & Chem Corp, 35-37; chemist & plant supvr, Armour & Co, Ill, 37-45; res chemist, 45-59, chief chemist, 59-63, asst mgr, Res & Develop Div, 63-67, Dir Res & Develop, Ont Prod Sect Lab, Sunkist Growers, Inc, 67-74; CONSULT BY-PRODS, CITRUS INDUST, 74- *Concurrent Pos:* Instr eve col, Texas Christian Univ, 43-45. *Mem:* AAAS; Inst Food Technol; Am Chem Soc; fel Am Inst Chem; Sigma Xi. *Res:* Chemistry of citrus by-products, especially citrus oils, peel products, juices and beverages; dehydrated foods. *Mailing Add:* 595 West 25th St Upland CA 91786

SWISHER, JAMES H(OWE), materials science, physical chemistry, for more information see previous edition

SWISHER, JOSEPH VINCENT, b Kansas City, Mo, Jan 12, 32; m 83; c 3. ORGANIC CHEMISTRY. *Educ:* Cent Methodist Col, AB, 56; Univ Mo, PhD(org chem), 60. *Prof Exp:* Fel chem, Purdue Univ, 60-61; asst prof, 61-69, assoc prof chem, 69-84, PROF, UNIV DETROIT, 84- *Mem:* Am Chem Soc. *Res:* Oranosilicon chemistry; stereochemistry of addition reactions; polymer chemistry. *Mailing Add:* Dept Chem Univ Detroit 4001 W McNichols Rd Detroit MI 48221

SWISHER, ROBERT DONALD, b Denver, Colo, Nov 16, 10; wid; c 3. ORGANIC CHEMISTRY, ENVIRONMENTAL CHEMISTRY. *Educ:* Univ Mich, PhD(pharmaceut chem), 34. *Prof Exp:* Mem staff, Monsanto Co, 34-45, group leader, 45-73, sr environ adv, Res Dept, 73-75; RETIRED. *Concurrent Pos:* Chmn, Joint Task Group Surfactants, Standard Methods Exam Water & Wastewater, 78- *Honors & Awards:* SOCMA Award, Outstanding Achievement, Environ Chem, 76. *Mem:* Fel AAAS; Am Chem Soc; Am Oil Chem Soc; Soc Indust Microbiol; Water Pollution Control Fedn; Sigma Xi. *Res:* Sulfonation and sulfonic acid derivatives; detergents; biodegradation and environmental acceptability of surfactants and other materials. *Mailing Add:* 1894 Charmwood Ct Kirkwood MO 63122-6923

SWISHER, SCOTT NEIL, b Le Center, Minn, July 30, 18; m 45; c 2. INTERNAL MEDICINE, HEMATOLOGY. *Educ:* Univ Minn, BS, 43, MD, 44; Am Bd Internal Med, dipl, 52. *Prof Exp:* Asst resident & fel med, Sch Med & Dent, Univ Rochester, 47-48, fel med & hemat, 49-51, from instr to prof med, 51-67; fel, Med Sch, Univ Minn, 48-49; chmn dept, 67-77, assoc dean res, 77-83, PROF MED, COL HUMAN MED, MICH STATE UNIV, 67-; chmn dept, 67-77, assoc dean res, 77-83, PROF MED, COL HUMAN MED, MICH STATE UNIV, 67- *Concurrent Pos:* From asst resident physician to chief resident physician, Ancker Hosp, St Paul, Minn, 48-49; from asst physician to sr assoc physician & head hemat unit, Strong Mem Hosp, Rochester, NY, 53-67; mem comt blood & related probs, Nat Res Coun, 54-, chmn, 60. *Mem:* AAAS; Am Soc Clin Invest; Asn Am Physicians; Am Col Physicians; Am Fedn Clin Res. *Res:* Mechanisms of destruction of erythrocytes by antibodies; human hemolytic disorders. *Mailing Add:* Dept Med A101 E Fee Hall Col Human Med Mich State Univ East Lansing MI 48824

SWISLOCKI, NORBERT IRA, b Warsaw, Poland, Jan 11, 36; US citizen; div; c 2. BIOCHEMISTRY, CELL BIOLOGY. *Educ:* Univ Calif, Los Angeles, BS, 56, MA, 60, PhD(zool, endocrinol), 64. *Prof Exp:* USPHS fel biochem, Brandeis Univ, 64-66; from instr to assoc prof biochem, Grad Sch Med Sci, Cornell Univ, 67-78, chmn, Biochem Unit, 75-78, assoc, Sloan-Kettering Inst Cancer Res, 66-78, assoc mem, 72-78; PROF & CHMN BIOCHEM & MOLECULAR BIOL, MED SCH, UNIV MED & DENT NJ, 78- *Concurrent Pos:* Pres, Dept Biochem, Assoc Med Sch, 87-88. *Mem:* Am Soc Biochem & Molecular Biol; Endocrine Soc; Biophys Soc. *Res:* Mechanisms of hormone action; membrane function; endocrinology; aging of erythrocytes. *Mailing Add:* Dept Biochem & Molecular Biol NJ Med Sch Univ Med & Dent NJ 185 S Orange Ave Newark NJ 07103-2757

SWISSLER, THOMAS JAMES, b Haddonfield, NJ, Dec 8, 41; m 69; c 2. ENVIRONMENTAL PHYSICS, MATHEMATICAL ANALYSIS. *Educ:* St Joseph's Col, Pa, BS, 64; State Univ NY Buffalo, PhD(physics), 71. *Prof Exp:* Res asst comput sci, Theol Biol Ctr, State Univ NY Buffalo, 70-74, res fel, 73-74; STAFF SCIENTIST PHYSICS, ST SYST CO RP, HAMPTON, VA, 74- *Mem:* AAAS; Am Meteorol Soc; Sigma Xi. *Res:* Stratospheric aerosols and ozone gases, their physical properties and distribution using remote sensing measurements from satellite. *Mailing Add:* STX 28 Res Rd Hampton VA 23666

SWITENDICK, ALFRED CARL, b Batavia, NY, Oct 8, 31; m 80; c 2. METAL HYDRIDES, BAND THEORY. *Educ:* Mass Inst Technol, SB, 53, PhD(solid state physics), 63; Univ Ill, MS, 54. *Prof Exp:* Asst physics, Univ Ill, 53-54; asst physics, Mass Inst Technol, 56-62, res staff mem, 62-64; staff mem physics org solids, Sandia Lab, 64-68, staff mem solid state theory, 68-70, supvr, Solid State Theory, 70-80; staff mem, div mat sci, US Dept Energy, Wash, DC, 80-82; SR STAFF MEM, SOLID STATE THEORY, SANDIA LABS, 82- *Mem:* AAAS; fel Am Phys Soc. *Res:* Electronic energy bands of transition metals and transition metal compounds; intermetallic compounds; hydrogen in metals; boron and boron compounds. *Mailing Add:* 4309 Hannett NE Albuquerque NM 87110

SWITKES, EUGENE, b Newport News, Va, Dec 22, 43; m 69; c 2. QUANTUM CHEMISTRY, VISUAL PSYCHOPHYSICS. *Educ:* Oberlin Col, BA, 65; Harvard Univ, MS & PhD(theoret chem), 70. *Prof Exp:* NSF fels, Univ Edinburgh, 70-71 & Cambridge Univ, 71; asst prof, 71-77, assoc prof, 77-84, PHYSIOL OPTICS GROUP, BERKELEY, 89- *Concurrent Pos:* Fel & responder, Neurosci Res Prog, Boulder Intensive Study Session, 72. *Mem:* Assoc Res in Vision & Ophthal; Sigma Xi. *Res:* Theory of the electronic structure of molecules, quantum mechanics; information processing in the visual system; visual accommodation and perception; environmental visual information content. *Mailing Add:* Dept Chem Univ Calif Santa Cruz CA 95064

SWITZER, BOYD RAY, b Harrisonburg, VA, Oct 3, 43; m 67; c 2. NUTRITION, BIOCHEMISTRY. *Educ:* Bridgewater Col, BA, 65; Univ NC, Chapel Hill, PhD(biochem), 71. *Prof Exp:* NIH fel, Univ Southern Calif, 71-72; asst prof, 72-78, ASSOC PROF NUTRIT, SCH PUB HEALTH, UNIV NC, CHAPEL HILL, 78-, ASST PROF BIOCHEM & NUTRIT, SCH MED, 74- *Mem:* AAAS; Am Chem Soc. *Res:* Ethanol effects on nutrition, metabolism and biological function; fetal alcohol effects. *Mailing Add:* Dept Nutrit Univ NC CH 7400 Chapel Hill NC 27599

SWITZER, CLAYTON MACFIE, b London, Ont, July 17, 29; m 51; c 3. AGRICULTURE, WEED SCIENCE. *Educ:* Ont Agr Col, BSA, 51, MSA, 53; Iowa State Univ, PhD, 55. *Hon Degrees:* LLD, Dalhousie Univ. *Prof Exp:* From assoc prof to prof bot & head dept, 65-70,head dept, Ont Agr Col, 55-70, from assoc dean to dean, 70-83; dep minister, Ont Ministry Agr & Food, 84-89; PRES, CLAY SWITZER CONSULTS LTD, 89- *Concurrent Pos:* Chmn, Ont Weed Comt, 62-83; mem, Sci Coun Can, 77-82. *Mem:* Can Soc Pest Mgt; fel Weed Sci Soc Am; Can Soc Hort Sci; fel Agr Inst Can (pres, 83-84); Int Turfgrass Soc (pres, 77-81). *Res:* Physiology of herbicide action; weed control; growth regulation of turfgrass. *Mailing Add:* Dept Environ Biol Univ Guelph Guelph ON N1G 2W1 Can

SWITZER, GEORGE LESTER, b Chester, WVa, Nov 5, 24; m 56; c 1. FOREST ECOLOGY, FOREST SOILS. *Educ:* Univ WVa, BS, 49; Yale Univ, MF, 50; State Univ NY Col Forestry, Syracuse, PhD(soils, physiol), 62. *Prof Exp:* From asst forester to assoc forester, Miss Agr Exp Sta, 50-62, from asst prof to assoc prof, 54-64, PROF SILVICULT, MISS STATE UNIV, 64-, FORESTER, MISS AGR EXP STA, 64- *Concurrent Pos:* Consult, Spanish Govt on forest production, 79. *Mem:* AAAS; Soil Sci Soc Am; Am Soc Agron. *Res:* Nutrient cycles in forest ecosystems; patterns of variation in forest tree species. *Mailing Add:* Po Box 385 Mississippi State MS 39762

SWITZER, JAY ALAN, b Cincinnati, Ohio, May 14, 50; m 73; c 1. ELECTROCHEMISTRY, ELECTRONIC MATERIALS. *Educ:* Univ Cincinnati, BS, 73; Wayne State Univ, MA, 75, PhD(inorg chem), 79. *Prof Exp:* Sr res chemist, Unocal Corp, 79-87; assoc prof mat sci, Univ Pittsburgh, 87-90; PROF CHEM, UNIV MO, ROLLA, 90- *Concurrent Pos:* Sr res investr, Mat Res Ctr, Univ Mo, Rolla, 90- *Mem:* AAAS; Am Ceramic Soc; Am Chem Soc; Electrochem Soc; Mat Res Soc. *Res:* Inorganic materials chemistry; electrodeposition and physical electrochemistry; nanoscale optical and electrical materials; quantum effects in nanoscale materials; electrodeposited ceramic superlattices. *Mailing Add:* Dept Chem Univ Mo Rolla MO 65401

SWITZER, LAURA MAE, b McLean, Tex, Apr 21, 41. PHYSICAL EDUCATION, MATHEMATICS. *Educ:* Wayland Baptist Col, Tex, BS, 63; Southwestern Okla State Univ, MEd, 66; Univ Okla, EdD(curric), 71. *Prof Exp:* Teacher & coach phys educ & math, Sanford-Fritch Independent Sch Dist, Tex, 63-65; instr phys educ, Southwestern Okla State Univ, 65-69; asst, Univ Okla, 69-71; PROF PHYS EDUC, SOUTHWESTERN OKLA STATE UNIV, 71- *Mem:* Am Alliance Health Phys Educ & Recreation; Nat Educ Asn. *Res:* Comparison of massed versus distributed practice sessions in the learning of recreational activities skills. *Mailing Add:* Dept Health & Phys Educ Southwestern Okla State Univ 100 Campus Dr Weatherford OK 73096

SWITZER, MARY ELLEN PHELAN, b Brooklyn, NY, Nov 2, 45; m 70; c 1. BIOCHEMISTRY. *Educ:* Smith Col, AB, 67; Univ Ill, Urbana, MS, 69, PhD(chem), 73. *Prof Exp:* Fel res assoc, Dept Med, Duke Univ Med Ctr, NIH, 73-78; ASST MED RES PROF, DEPT MED, DUKE UNIV MED CTR, 79- *Concurrent Pos:* Adj asst prof, Dept Chem, Duke Univ, 78- *Mem:* Am Chem Soc. *Res:* Biochemistry of blood coagulation, especially the relationship between molecular struucture and biological functions of antihemophilic factor, and interactions among the blood clotting proteins. *Mailing Add:* Thrombosis Hemostasis Div Organon Teknika 100 Akzo Ave Durham NC 27704

SWITZER, PAUL, b St Boniface, Can, Mar 4, 39; m 63; c 1. APPLIED STATISTICS, SPATIAL MODELS. *Educ:* Univ Man, BA, 61; Harvard Univ, PhD(statist), 65. *Prof Exp:* Assoc prof, 65-76, chmn, 79-82, PROF STATIST & EARTH SCI, STANFORD UNIV, 76- *Concurrent Pos:* Vis scientist, Environ Protection Agency, Wash, 81; chmn, Panel Statist US Nat Bur Mines, Nat Acad Sci, 81-82; vpres, Int Asn for Math Geol, 84-; ed, J Am Statist Asn, 86- *Mem:* Fel Inst Math Statist; fel Am Statist Asn; fel Int Statist Inst; fel Royal Statist Soc; Am Geophys Union. *Res:* Earth sciences; resources; environment. *Mailing Add:* Sequoia Hall Stanford Univ Stanford CA 94305

SWITZER, ROBERT L, b Long Beach, Calif, June 18, 18; m 41; c 2. CHEMICAL ENGINEERING, MANAGEMENT. *Educ:* Univ Calif, BS, 41; Univ Southern Calif, MS, 52. *Prof Exp:* Sr sect leader, Design Div, Res Dept, Union Oil Co, Calif, 41-60, sr sect leader eng serv, 60-67, supvr eng serv, Res Dept, 67-83, mgr eng serv, 81-83; RETIRED. *Concurrent Pos:* Eng consult, 83-84. *Mem:* Instrument Soc Am; Am Inst Chem Engrs. *Mailing Add:* 241 Euclid Ave Long Beach CA 90803-6019

SWITZER, ROBERT LEE, b Clinton, Iowa, Aug 26, 40; m 65; c 2. BIOCHEMISTRY. *Educ:* Univ Ill, Urbana, BS, 61; Univ Calif, Berkeley, PhD(biochem), 66. *Prof Exp:* Fel biochem, Nat Heart Inst, 66-68; from asst prof to assoc prof, 68-78, PROF BIOCHEM, 78-, HEAD DEPT, UNIV ILL, URBANA, 88- *Concurrent Pos:* Fel, John Simon Guggenheim Mem Found, 75-76; vis prof, Freiburg Univ, WGer, 75-76, Univ Calif, Davis, 88. *Mem:* Am Soc Biol Chem; Am Chem Soc; Am Soc Microbiol. *Res:* Microbial physiology and enzymology, particularly regulation of branched biosynthetic pathways, mechanisms of regulatory enzymes and regulation of inactivation and turnover of enzymes during bacterial endospore formation; control of gene expression during sporulation. *Mailing Add:* Dept Biochem 318 Roger Adams Univ Ill 1209 W Calif Urbana IL 61801

SWITZER, WILLIAM PAUL, b Dodge City, Kans, Apr 9, 27; m 51; c 2. ANIMAL PATHOLOGY, MICROBIOLOGY. *Educ:* Agr & Mech Col, Tex, DVM, 48; Iowa State Col, MS, 51, PhD, 54 Univ Vienna, DSc, 78. *Prof Exp:* Asst diagnostician, Iowa Vet Diag Lab, Iowa State Univ, 48-52, from asst prof to prof vet hyg, Univ & Vet Med Res Inst, 52-74, PROF VET MICROBIOL & PREV MED, VET MED RES INST & ASSOC DEAN COL VET MED, IOWA STATE UNIV, 74-, DISTINGUISHED PROF, 78- *Mem:* Am Soc Microbiol; Am Vet Med Asn. *Res:* Swine enteric and respiratory diseases; tissue culture; myoplasma. *Mailing Add:* Vet Med Admin Iowa State Univ Ames IA 50010

SWOFFORD, HAROLD S, JR, b Spokane, Wash, July 24, 36; m 58; c 3. CHEMISTRY. *Educ:* Western Wash State Col, BA, 58; Univ Ill, Urbana, MS, 60, PhD(anal chem), 62. *Prof Exp:* Teaching asst chem, Univ Ill, Urbana, 58-60; from asst prof to assoc prof, 62-77, PROF CHEM, UNIV MINN, MINNEAPOLIS, 77- *Mem:* Am Chem Soc; Sigma Xi. *Res:* High temperature electrochemistry; fused salts. *Mailing Add:* Dept Chem SCH-308 CHEM Univ Minn Minneapolis MN 55455

SWOFFORD, ROBERT LEWIS, b Charlotte, NC, Jan 29, 48; m 73; c 2. LASER SPECTROSCOPY. *Educ:* Furman Univ, BS, 69; Univ Calif, Berkeley, PhD(chem), 73. *Prof Exp:* Sr res assoc chem, Cornell Univ, 73-77; proj leader, 77-80, group leader laser res, 80-81, group leader quantum chem, 81-83, group leader chem physics, 83-85, RES ASSOC, STANDARD OIL CO, 85- *Mem:* Am Chem Soc; Optical Soc Am; AAAS; Laser Inst Am; Soc Appl Spectros. *Res:* Laser techniques to study the forces which control chemical reactions at the molecular level. *Mailing Add:* BP America R & D 4440 Warrensville Center Rd Warrensville Heights OH 44128

SWOOPE, CHARLES C, b Jersey City, NJ, July 7, 34; m 55; c 2. PROSTHODONTICS. *Educ:* Univ Md, DDS, 59; Univ Wash, MSD, 64. *Prof Exp:* Asst chief dent serv, USPHS Hosp, New Orleans, 64-67; dir grad & res prosthodontics, 67-71, assoc prof, 71-73, prof prosthodontics, 73-80, CLIN ASSOC, SCH DENT, UNIV WASH, 80- *Concurrent Pos:* Asst prof, Loyola Univ, 64-67; consult, USPHS Hosp, Va Hosp & Univ Hosp, Seattle, Wash, 67-68. *Mem:* Am Prosthodont Soc; Am Acad Denture Prosthetics (secy-treas, 86-91); Am Equilibration Soc; Am Acad Maxillofacial Prosthetics. *Res:* Resilient lining materials; emotional evaluation of denture patients; bone changes; speech problems; force transmission to teeth. *Mailing Add:* 7319 NE 18th St Bellevue WA 98004

SWOPE, FRED C, b Lexington, Va, Mar 25, 35; m 64. FOOD SCIENCE, BIOCHEMISTRY. *Educ:* Univ Md, BS, 61; Mich State Univ, PhD(food sci), 68. *Prof Exp:* Asst prof, 68-74, ASSOC PROF BIOL, VA MIL INST, 74- *Mem:* Am Chem Soc. *Res:* Lipoproteins; structural studies on membranes. *Mailing Add:* Dept of Biol Va Mil Inst Lexington VA 24450

SWOPE, RICHARD DALE, b Palmyra, Pa, Sept 27, 38; m 57; c 2. FLUID MECHANICS, HEAT TRANSFER. *Educ:* Univ Del, BME, 60, MME, 62, PhD(appl sci), 66. *Prof Exp:* Appl scientist res & develop, Armstrong Cork Co, 66-68; asst prof eng, PMC Cols, 68-71; assoc prof, Ctr Eng, Widener Col, 71-77, prof eng & asst dean, 77-80; PROF ENG SCI, TRINITY UNIV, 80-, CHMN, 81- *Mem:* Int Solar Energy Soc; Am Soc Mech Engrs; Am Soc Eng Educ; Sigma Xi. *Res:* Energy; turbulence; shear stresses; solar energy; energy storage; bubble mechanics and formation. *Mailing Add:* 2139 Oak Creek San Antonio TX 78232

SWORD, CHRISTOPHER PATRICK, b San Fernando, Calif, Sept 9, 28; m 59; c 4. GRADUATE EDUCATION & RESEARCH ADMINISTRATION, MICROBIOLOGY. *Educ:* Loyola Univ, Calif, BS, 51; Univ Calif, Los Angeles, PhD(microbiol), 59. *Prof Exp:* Fel & res assoc microbiol, Univ Kans, 58-59, from asst prof to prof, 59-70; prof life sci & chmn dept, Ind State Univ, Terre Haute, 70-76; GRAD DEAN & DIR RES & PROF MICROBIOL, SDAK STATE UNIV, BROOKINGS, 76- *Concurrent Pos:* President's fel, Soc Am Bact, 60; consult-evaluator, NCent Asn Cols & Schs. *Mem:* AAAS; Am Soc Microbiol; NY Acad Sci; Coun Grad Schs US; Nat Coun Univ Res Adminr. *Res:* Biochemical and immunological mechanisms of pathogenesis; bacterial virulence; host-parasite interactions in Listeria monocytogenes infection; ultrastructure of bacteria and infected cells; science policy. *Mailing Add:* Grad Dean/Dir of Res/Prof Microbiol SDak State Univ Box 2201 University Station Brookings SD 57007

SWORD, JAMES HOWARD, b Derby, Va, Jan 1, 24; m 46; c 2. ENGINEERING MECHANICS, CIVIL ENGINEERING. *Educ:* Va Polytech Inst, BS, 50, MS, 54. *Prof Exp:* Instr civil eng, 51, from instr to assoc prof eng mech, 51-66, PROF ENG MECH, VA POLYTECH INST & STATE UNIV, 66- *Mem:* Am Soc Eng Educ. *Res:* Computer programming; mechanics; interferometry. *Mailing Add:* Dept Eng Sci Va Poly Inst & State Univ Blacksburg VA 24061

SWORDER, DAVID D, b Dinuba, Calif, Aug 1, 37. ENGINEERING. *Educ:* Univ Calif, Berkeley, BS, 58, MS, 59; Univ Calif, Los Angeles, PhD(eng), 65. *Prof Exp:* Sr engr, Litton Systs, Inc, 59-64; from asst prof to assoc prof elec eng, Univ Southern Calif, 64-77; PROF APPL MECH & ENG SCI, UNIV CALIF, SAN DIEGO, 77- *Mem:* Inst Elec & Electronics Engrs. *Res:* Adaptive and stochastic control problems. *Mailing Add:* Dept Appl Mech & Eng Sci 1B-010 Univ Calif San Diego Box 109 La Jolla CA 92093

SWORSKI, THOMAS JOHN, b Pittsburgh, Pa, Sept 8, 20; m 44; c 2. SCIENTIFIC PROGRAMMING. *Educ:* Duquesne Univ, BS, 42; Notre Dame Univ, PhD(phys chem), 51. *Prof Exp:* Chemist, Oak Ridge Nat Lab, 51-57 & Nuclear Res Ctr, Union Carbide Corp, 57-63; chemist, Oak Ridge Nat Lab, 64-81, prog analyst comput sci, 81-88; RETIRED. *Concurrent Pos:* Vis lectr, Stevens Inst Technol, Hoboken, NJ, 62. *Res:* Photochemistry; chemical kinetics; entire chemical research career devoted to the radiation chemistry of aqueous solutions, organic liquids and gases. *Mailing Add:* 101 Adelphi Rd Oak Ridge TN 37830-7807

SWOVICK, MELVIN JOSEPH, b Altoona, Pa, Jan, 20, 26; m 68; c 1. DRUG & ALCOHOL ANALYSIS, DETECTION OF GUNSHOT RESIDUE. *Educ:* Detroit Inst Technol, BSc, 74. *Prof Exp:* Med technician, Vet Hosp, Altoona, Pa, 54; lab technician, Allied Chem Inc, 55-60; develop chemist, Napco Chem Co, 60-63; anal chemist, Schwarz BioRes, Inc, 64; anal chemist coatings chem, Reichhold Chemicals, Inc, 65-80; FORENSIC CHEMIST, OAKLAND COUNTY SHERIFF'S DEPT, 80- *Concurrent Pos:* Instr, Austin Cath Prep Schs, 75-79; fac mem, Notre Dame High Sch. *Mem:* Am Chem Soc. *Res:* Developed methods for the analysis of gunshot residue in forensic chemistry. *Mailing Add:* 345 Folkstone Ct Troy MI 48098

SWOYER, VINCENT HARRY, b Philadelphia, Pa, Mar 30, 32; m 60; c 3. TECHNICAL ADMINISTRATION, COMPUTER SCIENCE. *Educ:* Tufts Univ, BS, 54; Univ Rochester, MA, 60; Harvard Univ, EdD, 66. *Prof Exp:* Asst prof naval sci, Univ Rochester, 57-59, from asst dir to dir, Comput Ctr, 59-69; staff assoc, NSF, 69-70; dir comput ctr, Univ Rochester, River Campus, 70-78; group dir, 78-80, VPRES CORP SYSTS, RYDER SYST INC, 80- *Concurrent Pos:* Consult, NSF, 70- & Southern Regional Educ Bd, 71- *Mem:* Soc Mgt Info Systs; Asn Comput Mach. *Res:* Monte Carlo studies, especially in area of multiple regression analysis; computer methods for statistical research applications. *Mailing Add:* Sara Lee Corp Three First Nat Plaza Chicago IL 60656

SWYER, PAUL ROBERT, b London, Eng, May 21, 21; Can citizen; m 47; c 2. PEDIATRICS, NEONATOLOGY. *Educ:* Cambridge Univ, BA, 40, MB, BChir & MA, 43, MD, 84; FRCP(L); FRCP(C), DCH (Eng). *Hon Degrees:* MD, Univ Lausanne, Switz. *Prof Exp:* Registr, Middlesex Hosp, 47-48; asst resident chest dis & med officer, Brompton Hosp for Dis of Chest, 48-49; registr, SWarwickshire Hosp, 49-50; asst med registr, Hosp Sick Children, 50-52; registr, Royal Hosp, Wolverhampton, 52-53; fel cardiol, Hosp Sick Children, Toronto, 53-54, res assoc pediat, 54-60, asst scientist, 61-66, sr scientist, Res Inst, & chief, Div Perinatology, 67-86; EMER PROF, UNIV TORONTO, 86- *Concurrent Pos:* Can Dept Nat Health & Welfare grants, 54-67; sr staff physician, Hosp Sick Children, 64-; from asst prof to prof, Univ Toronto, 65-86; Med Res Coun Can grants, 66-86, Phys Serv Inc grants, 79-85 & NIH grants, 85-86; WHO Fel, 84. *Mem:* Am Pediat Soc; Am Acad Pediat; Soc Pediat Res; Can Pediat Soc; Can Med Asn; Soc Critical Care Med. *Res:* Investigation of mechanics of breathing; energy metabolism, thermoregulation, blood flow and pressures and methods of treatment of pulmonary disorders in newly born infants; delivery, perinatal care. *Mailing Add:* Hosp for Sick Children 555 University Ave Toronto ON M5G 1X8 Can

SY, JOSE, b Sorsogon, Philippines, Dec 10, 44. BIOCHEMISTRY. *Educ:* Adamson Univ, Manila, BS, 64; Duke Univ, PhD(biochem), 70. *Prof Exp:* Res assoc, Rockefeller Univ, 70-74, from asst prof to assoc, 74-85; PROF, CALIF STATE UNIV, FRESNO, 85- *Mem:* Protein Soc; Am Soc Microbiol; Am Soc Biol Chemists; Am Chem Soc. *Res:* Nucleic acids and protein synthesis. *Mailing Add:* Dept Chem Calif State Univ Fresno CA 93740

SY, MAN-SUN, CELLULAR IMMUNOLOGY, IMMUNO-REGULATION. *Educ:* Univ Colo, PhD(immunol), 79. *Prof Exp:* ASST PROF, MED SCH, HARVARD UNIV, 79- *Res:* Auto-immunity. *Mailing Add:* Dept Path Sch Med Harvard Univ Mass Gen Hosp E 7th Fl Bldg 149 13th St Charleston MA 02129

SYAGE, JACK A, b Rockville Center, NY, Apr 5, 24; m 83; c 1. LASER SPECTROSCOPY, MOLECULAR BEAM SPECTROSCOPY. *Educ:* Hamilton Col, BA, 76, Brown Univ, PhD(phys chem), 81. *Prof Exp:* Postdoctoral fel, Calif Inst Technol, 82-84; SCIENTIST, THE AEROSPACE CORP, 84- *Mem:* Am Phys Soc. *Res:* Pico second state selective reaction dynamics and molecular beam. *Mailing Add:* Aerospace Corp PO Box 92975 Los Angeles CA 90009

SYBERS, HARLEY D, b Tony, Wis, June 18, 33; m 58; c 2. PATHOLOGY, PHYSIOLOGY. *Educ:* Univ Wis-Madison, BS, 56, MS & MD, 63, PhD(physiol), 69; Am Bd Path, dipl & cert anat path & clin path, 68. *Prof Exp:* Resident path, Univ Wis-Madison, 64-68; asst prof, Univ Calif, San Diego, 69-75; assoc prof path, Baylor Col Med, 75-83; PROF PATH & LAB MED, UNIV TEXAS MED SCH, HOUSTON, 83-, PROF PATH & RADIOL, UNIV TEXAS DENT SCH, 86- *Concurrent Pos:* USPHS contract, Univ Calif, San Diego, 69-74; Nat Heart & Lung Inst grant, 74-77; NIH res grant, 79-81, mem, Cardiovascular Study Sect, NIH, 80-84. *Mem:* Am Heart Asn; Am Soc Clin Path; Am Physiol Soc; Am Asn Pathologists; Int Acad Path. *Res:* Cardiac pathophysiology. *Mailing Add:* Dept Path & Lab Med Univ Tex Health & Sci Ctr Houston TX 77030

SYBERT, JAMES RAY, b Greenville, Tex, Dec 25, 34; div; c 4. SOLID STATE PHYSICS, LOW TEMPERATURE PHYSICS. *Educ:* Univ NTex, BA, 55, MA, 56; La State Univ, PhD(physics), 61. *Prof Exp:* Instr, Univ NTex, 56-58, from asst prof to assoc prof, 61-67, chmn dept, 69-80, PROF PHYSICS, UNIV NTEX, 67- *Concurrent Pos:* Adj prof, Southwest Ctr Advan Studies, Dallas, 67-70; vis prof, San Diego State Univ, 80, Ind Univ, Malaysia, 89. *Mem:* Am Phys Soc; Am Asn Physics Teachers. *Res:* Media in science education; electron transport in metals and semiconductors; size effect in metals; superconductivity. *Mailing Add:* Dept Physics Univ NTex Denton TX 76203

SYBERT, PAUL DEAN, b Joliet, Ill, July 16, 54; m 81. AROMATIC POLYMERS, RIGID-ROD POLYMERS. *Educ:* Augustana Col, BA, 76; Univ Iowa, MS, 77; Colo State Univ, PhD(org chem), 80. *Prof Exp:* Sr res chemist, Monsanto, 80-81; POLYMER CHEMIST, SRI INT, 81- *Mem:* Am Chem Soc. *Res:* New polymerization methods for new and existing rod-like aromatic polymers which afford anisotropic solutions; preparation of polymers for use as desalination membranes. *Mailing Add:* RR 3 E Slope Rd Pittsfield MA 01201-8805

SYDISKIS, ROBERT JOSEPH, b Bridgeport, Conn, Sept 19, 36; m 61; c 2. VIROLOGY. *Educ:* Univ Bridgeport, BA, 61; Northwestern Univ, PhD(microbiol), 65. *Prof Exp:* Fel, Univ Chicago, 65-67; asst prof virol, Sch Med, Univ Pittsburgh, 67-71; ASSOC PROF MICROBIOL, SCH DENT, UNIV MD, BALTIMORE CITY, 71- *Mem:* Am Soc Microbiol. *Res:* Isolation and identification of natural plant components which inhibit the replication of herpes virus; effect of various glucocorticoid hormones on herpes virus replication. *Mailing Add:* Dept of Microbiol Univ of Md Sch of Dent Baltimore MD 21201

SYDNOR, THOMAS DAVIS, b Richmond, Va, Jan 27, 40; m 62; c 3. ORNAMENTAL HORTICULTURE, URBAN FORESTRY. *Educ:* Va Polytech Inst & State Univ, BS, 62; NC State Univ, PhD(plant physiol), 72. *Prof Exp:* Landscape foreman, Southside Nurseries Inc, 62-63, vpres 65-69; from asst prof to assoc prof, 72-84, PROF ORNAMENTAL HORT, OHIO STATE UNIV, 84- *Concurrent Pos:* Consult landscape & urban hort, 78-; prin investr, Ohio Shade Tree Eval Proj, 78- *Honors & Awards:* Alfred J Wright Award, 78. *Mem:* Sigma Xi; Am Soc Hort Sci; Int Soc Arboricult. *Res:* Shade tree evaluation and the effects of environmental conditions on growth and development of woody ornamentals and trees for urban, highway and landscape situations. *Mailing Add:* 6800 Harriott Rd Powell OH 43065

SYDOR, MICHAEL, b Prusseniv, Ukraine, Dec 25, 36; US citizen; m 62, 87; c 5. MATERIALS SCIENCE ENGINEERING. *Educ:* Univ BC, BASc, 59; Univ NMex, PhD(physics), 64. *Prof Exp:* From asst prof to assoc prof, 68-74, PROF PHYSICS, UNIV MINN, DULUTH, 74- *Mem:* Am Phys Soc; Sigma Xi. *Res:* Modulated photoreflectance; materials growth by molecular beam epitaxy and closed space vapor transport; quantum-well structures; numerical modeling of materials growth processes and amorphous materials. *Mailing Add:* Dept Physics Univ Minn Duluth MN 55812

SYDORIAK, STEPHEN GEORGE, b Passaic, NJ, Jan 6, 18; m 45; c 6. PHYSICS. *Educ:* Univ Buffalo, BA, 40; Yale Univ, PhD(physics), 48. *Prof Exp:* Mem staff, Radiation Lab, Mass Inst Technol, 41-45; mem staff, Los Alamos Nat Lab, Univ Calif, 48-80; RETIRED. *Mem:* Am Phys Soc. *Res:* Radar signal threshold studies; liquid helium-three properties; 1962 helium-three vapor pressure scale of temperatures; critical nucleate boiling equation for helium; superconducting magnet design for passive quench arrest. *Mailing Add:* 1192 A 41st St Los Alamos NM 87544

SYED, ASHFAQUZZAMAN, b Barabanki, India, July 5, 52; m 84; c 2. X-RAY CRYSTALLOGRAPHY, STRUCTURE ANALYSIS. *Educ:* Gorakhpur Univ, India, BSc, 70, MSc, 72, PhD(physics crystallog), 79. *Prof Exp:* Res asst x-ray crystallog, Dept Physics, Gorakhpur Univ, India, 72-78, sr res asst, 78-79; res chemist instrument anal, Tata Iron & Steel Co, Jamshedpur, India, 79-81, lectr physics, Tech Inst, 79; res asst x-ray crystallog, State Univ NY, Buffalo, 81-83; res assoc x-ray crystallog, Univ New Orleans, 83-84; PROD SPECIALIST, X-RAY INSTRUMENTATION & CRYSTALLOG, ENRAF-NONIUS, BOHEMIA, NY, 84- *Mem:* Am Crystallog Asn. *Res:* Chemical crystallography; accurate x-ray diffration data and charge-density analysis; computing methods in x-ray crystallography; hardware and software developments and instrumentation in x-ray crystallography. *Mailing Add:* Enraf-Nonius 390 Central Ave Bohemia NY 11716

SYED, IBRAHIM BIJLI, b Bellary, India, Mar 16, 39; US citizen; m 64; c 2. MEDICAL PHYSICS, RADIOLOGICAL HEALTH. *Educ:* Univ Mysore, BSc, 60, MSc, 62; Johns Hopkins Univ, DSc(radiol sci), 72; Am Bd Radiol, dipl, 75; Am Bd Health Physics, dipl, 77; Am Bd Hazard Control Officers, dipl, 80. *Hon Degrees:* PhD, Marquis Giuseppe Scicluna Int Univ, 85. *Prof Exp:* Lectr physics, Veerasaiva Col, Univ Mysore, 62-63; physicist, Victoria Hosp, Bangalore, India, 64-67; chief physicist, Halifax Infirmary, Can, 67-69; chief physicist, Med Physics Div, Baystate Med Ctr, Wesson Mem Unit, 73-79, admin dir med physics, 77-79; asst clin prof, Sch Med, Univ Conn, 75-79; PHYSICIST & RADIATION SAFETY OFFICER, MED CTR, VET ADMIN, 79-; PROF MED, MED PHYSICS, NUCLEAR MED & NUCLEAR CARDIOL, SCH MED, UNIV LOUISVILLE, 79- *Concurrent Pos:* Consult physicist, Bowring & Lady Curzon Hosp & Ministry of Health, India, 64-67, Bangalore Nursing Home, 64-67, Wing Mem Hosp, 73-79 & Mercy Hosp, 73-79; assoc prof, Springfield Tech Community Col, 73-79; adj prof radiol, Holyoke Community Col, 73-79; consult med physicist & radiol safety officer to area hosps in Ky & Ind, 79-; clin prof med, Endocrinol, Metab & Radionuclide Studies, Univ Louisville Sch Med, 80-; dir, Nuclear Med Sci, 80-; vis prof, Bangalore Univ & Gulbarga Univ, India, 87-88; vis scientist, Bhabba Atomic Res Ctr, Bombay, India; consult, Coun Sci & Indust Res, Govt India, 80-, Am Coun Sci & Health, 80-, Gastroenterol & Urol, Div Food & Drug Admin, Human & Health Serv, 88-, radiopharmaceutical div, 89-; tech expert nuclear med, Int Atomic Energy Agency, Rep Bangladesh; govt India scholar, Bhabha Atomic Res Ctr, Bombay, 63-64. *Mem:* Fel Brit Inst Physics; fel Royal Soc Health, Eng; Am Asn Physicists in Med; fel Am Inst Chemists; fel Am Col Radiol; Health Physics Soc; Soc Nuclear Med; Am Col Nuclear Med. *Res:* Radiopharmaceutical dosimetry; medical health physics; therapeutic radiological physics; effects of caffeine in man; estimation of absorbed dose to embryo and fetus from radiological and nuclear medicine procedures; magnetic resonance imaging and bioeffects; theory of limitations; physicist's concept of God; half value layer of polyenergetic photon emitting radio nuclide; Bremsstrahlijng and radioactivity determination; author of one manual and contributor of over one hundred articles to science journals. *Mailing Add:* 7102 Shefford Louisville KY 40242

SYEKLOCHA, DELFA, b Vancouver, BC, Sept 12, 33. MEDICAL BACTERIOLOGY, IMMUNOLOGY. *Educ:* Univ BC, BA, 54; McGill Univ, MSc, 62, PhD(microbiol), 64. *Prof Exp:* NIH fel, Ont Cancer Inst, 64-65; ASST PROF MICROBIOL, UNIV BC, 65- *Mem:* Am Soc Microbiol; Can Soc Microbiol; Can Soc Immunol. *Res:* Immunizing potential of attenuated mutants of Pseudomonas aeruginosa strains and characterization of soluble antigens derived from the wild type; biological activities and characterization of a lethal exotoxin produced by Paeruginosa. *Mailing Add:* Dept Microbiol Univ BC 2075 Wesbrook Mall Vancouver BC V6T 1W5 Can

SYGUSCH, JURGEN, b Jablonec, Czech, Aug 8, 45; Can citizen; div; c 1. BIOCHEMISTRY, CRYSTALLOGRAPHY. *Educ:* McGill Univ, BSc, 67, MSc, 69; Univ Montreal, PhD(chem), 74. *Prof Exp:* Med Res Coun Can fel, Dept Biochem, Sch Med, Univ Alta, 74-77; ASSOC PROF BIOCHEM, SCH MED, UNIV SHERBROOKE, 77- *Mem:* AAAS; Can Biochem Soc; Am Crystallog Asn; Biophys Soc. *Res:* Structural studies by enzyme kinetics and by x-ray crystallography of conformational changes resulting from substrate or ligand binding to type I aldolases; protein folding dynamics of babbit muscle aldolase. *Mailing Add:* Dept of Biochem Univ of Sherbrooke Sherbrooke PQ J1K 2R1 Can

SYKES, ALAN O'NEIL, b St Regis Falls, NY, May 19, 25; m 51; c 2. ACOUSTICS. *Educ:* Cornell Univ, AB, 48; Cath Univ Am, PhD, 68. *Prof Exp:* Physicist, David Taylor Model Basin, US Navy Dept, 48-52, electronic scientist, Off Naval Res, 52-54, physicist, 54-58, physicist & consult to head acoust div, 58-60, physicist & head, Noise Transmission & Radiation Sect, 60-61, physicist & head, Noise Res & Develop Br, 61-64, physicist & head, Struct Acoust Br, 64-65, physicist & scientific officer, US Off Naval Res, 65-81. *Concurrent Pos:* Mem hydroballistic adv comt, Navy Bur Ord, 56-60; mem struct impedence panel, Noise Adv Comt, Navy Bur Ships, 56-64, consult, Soc Automotive Engrs, 57-; consult, vibration anal & control, 81- *Mem:* Fel Acoust Soc Am. *Res:* Underwater acoustics with emphasis on signal processing and propagation; noise reduction, vibration theory and electroacoustic transducer development. *Mailing Add:* 304 Mashie Dr SE Vienna VA 22180

SYKES, BRIAN DOUGLAS, b Montreal, Que, Aug 30, 43; m 68; c 2. BIOPHYSICAL CHEMISTRY. *Educ:* Univ Alta, BSc, 65; Stanford Univ, PhD(chem), 69. *Prof Exp:* From asst prof to assoc prof chem, Harvard Univ, 69-75; assoc prof biochem, 75-80, PROF BIOCHEM, UNIV ALTA, 80- *Concurrent Pos:* NIH grant, Harvard Univ, 69-75, Alfred P Sloan fel, 71-73; Med Res Coun Can grant, 75- *Honors & Awards:* Ayerst Award, 82; Steacie Prize, 82. *Mem:* Can Biochem Soc; Sigma Xi; Am Soc Biol Chemists; Am Chem Soc; Biophys Soc; fel Royal Soc Can. *Res:* Application of nuclear magnetic resonance to problems in biochemistry. *Mailing Add:* Dept of Biochem Univ of Alta Edmonton AB T6C 2H7 Can

SYKES, DONALD JOSEPH, b Buffalo, NY, Mar 16, 36; m 60; c 2. STATISTICAL ANALYSIS, IMAGING CHEMISTRY. *Educ:* Rochester Inst Technol, BS, 58. *Prof Exp:* Dir appl res, Cormac Chem Corp, 59-63; res mgr, P A Hunt Chem Corp, 63-69, asst dir res, 69-75, dir res, 75-77, asst vpres res, 77-78, vpres res & develop, 78-80, sr vpres, 80-83; dir & chief oper officer,

83-88, CHMN BD, MARPAC CO, 88- *Mem:* Am Chem Soc; Am Soc Quality Control; Soc Plastic Eng; Soc Photog Sci & Eng. *Res:* Super concentrates/polymers, photo, etc. *Mailing Add:* Eight Sunset Lane Upper Saddle River NJ 07458

SYKES, JAMES AUBREY, JR, b Washington, DC, Sept 24, 41; m 60; c 2. RESEARCH ADMINISTRATION, INDUSTRIAL CHEMICAL TECHNOLOGY. *Educ:* Univ Md, BS, 64, MS, 65, PhD(chem eng), 68. *Prof Exp:* Engr, W R Grace & Co, 65-68; process eng, Div Shell Oil, Shell Develop Co, 68-74; dir indust chem & technol, 74-84, GEN MGR, TECH PLANNING & RES SERV, AIR PROD & CHEM, INC, 84- *Concurrent Pos:* Abstractor, Chem Abstrcts, Am Chem Soc, 69-71. *Mem:* Am Inst Chem Engrs; Sigma Xi. *Res:* Polymers; commodity chemicals; performance chemicals; petroleum processing; materials and metallurgy; heat transfer; process systems; technology management; corporate-level research and development. *Mailing Add:* PO Box 314 Old Zionsville PA 18068

SYKES, JAMES ENOCH, b Richmond, Va, Apr 12, 23; m 47; c 2. FISHERIES. *Educ:* Randolph-Macon Col, 48; Univ Va, MS, 49. *Prof Exp:* Lab instr biol, Univ Va, 48-49; fishery res biologist, US Fish & Wildlife Serv, 49-58, chief striped bass invests, 58-62, biol lab dir, Bur Com Fisheries, 62-71; dir div fisheries, Nat Marine Fisheries Serv, 71-79; mem, NC Coastal Resources Comn, 79-86; RETIRED. *Mem:* Am Fisheries Soc; Am Inst Fishery Res Biologists (pres, 75 & 76). *Res:* Coastal pelagic and offshore sport fisheries; assessment of marine resources; biology of Gulf and Atlantic menhaden. *Mailing Add:* 3205 Country Club Rd Morehead City NC 28557

SYKES, LYNN RAY, b Pittsburgh, Pa, Apr 16, 37; m. GEOPHYSICS. *Educ:* Mass Inst Technol, BS & MS, 60; Columbia Univ, PhD(geol), 64. *Hon Degrees:* State Col NY Potsdam, 88. *Prof Exp:* Res scientist, Lamont Geol Observ, Columbia Univ, 62-65, res geophysicist, Inst Earth Sci, Environ Sci Serv Admin at Lamont Geol Observ, 65-68, from assoc prof to prof, 68-78, head seismol group, 71-85, HIGGINS PROF GEOL, LAMONT-DOHERTY GEOL OBSERV, COLUMBIA UNIV, 78- *Concurrent Pos:* Mem comts seismol & earthquake prediction, Nat Res Coun-Nat Acad Sci, 73-75; mem, Earthquake Studies Adv Panel, US Geol Surv, 77-81 & Earthquake Prediction Eval Coun, 79-82, chmn, 84-88; mem comn acad priorities arts & sci, Columbia Univ, 78-79; Guggenheim fel, 88- *Honors & Awards:* Macelwane Award, Am Geophys Union, 70, Bucher Medal, 75. *Mem:* Nat Acad Sci; fel Geol Soc Am; AAAS; fel Am Geophys Union; fel Royal Astron Soc. *Res:* Investigations of long-period and mantle seismic waves; surface wave propagation across ocean areas; precise location of earthquake hypocenters and relationship of spatial distribution of earthquakes to large-scale tectonic phenomena; field study of aftershocks of 1964 Alaskan earthquake; field study of deep and shallow earthquakes in Fiji-Tonga region; earthquake prediction; tectonics; identification of underground nuclear explosions; yield estimation and sizes of Soviet nuclear weapons. *Mailing Add:* Lamont-Doherty Geol Observ Columbia Univ Palisades NY 10964

SYKES, PAUL JAY, JR, b Hummelstown, Pa, Aug 31, 18; m 48; c 1. PHYSICS, NUCLEAR ENGINEERING. *Educ:* Univ BC, BA, 48; Univ Calif, MA, 51. *Prof Exp:* Chief proj off nuclear eng test facil, Wright-Patterson Air Force Base, Ohio, 52-56, chief reactor hazards br, Spec Weapons Ctr, Kirtland Air Force Base, 57-58, proj officer, Anal Div, 58-60, proj officer, Physics Div, Air Force Weapons Lab, 60-64; asst prof, 64-84, EMER PROF PHYSICS, UNIV BC, 84- *Mem:* Am Asn Physics Teachers; Royal Astron Soc Can. *Res:* Design and engineering of nuclear research reactor facilities; reactor hazards and safeguards; nuclear weapon systems analysis. *Mailing Add:* 5616 Westport Pl West Vancouver BC V7W 1T9 Can

SYKES, RICHARD BROOK, b Eng, Aug 7, 42; m 69; c 2. MICROBIOLOGY. *Educ:* Paddington Col, BS, 65; Queen Elizabeth Col, London Univ, MS, 68; Bristol Univ, PhD, 72. *Prof Exp:* Head antibiotic res unit, Glaxo Res Labs, 72-77; dir, Dept Microbiol, Squibb Inst Med Res, 79-83, vpres infectious & metab dis, 83-86; chief exec, Glaxo Group Res, UK, 86-87; GROUP RES & DEVELOP DIR, GLAXO HOLDINGS, 87- *Concurrent Pos:* Chmn & chief exec, Glaxo Group Res Ltd; pres, Glaxo Inc Res Inst. *Mem:* Brit Soc Antimicrobial Chemother. *Mailing Add:* Glaxo Group Res Ltd Greenford Middlesex UB6 0HE England

SYKES, ROBERT MARTIN, b Lawrence, Mass, July 25, 43; m 65; c 1. ENVIRONMENTAL ENGINEERING. *Educ:* Northeastern Univ, BSCE, 66; Purdue Univ, West Lafayette, MSCE, 68, PhD(sanit eng), 70. *Prof Exp:* Asst prof civil eng, Union Col, NY, 70-72; asst prof, 72-77, ASSOC PROF CIVIL ENG, OHIO STATE UNIV, 77- *Concurrent Pos:* Consult, NY State Dept Environ Conserv, 70-71. *Mem:* Water Pollution Control Fedn; Am Water Works Asn. *Res:* Water pollution abatement; biological processes for waste water treatment; ecosystem theory; modeling of aquatic systems. *Mailing Add:* Dept Civil Eng Ohio State Univ 2070 Neil Ave Columbus OH 43210

SYKORA, OSKAR P, b Nachod, Czech, June 22, 29; Can citizen; m 65. DENTISTRY. *Educ:* Sir George William Univ, BA, 54; Univ Montreal, MA, 55, PhD(slavic hist), 59; McGill Univ, DDS, 59. *Prof Exp:* From asst prof to assoc prof prosthodontics, McGill Univ, 61-71; ASSOC PROF PROSTHODONTICS, DALHOUSIE UNIV, 71- *Mem:* Am Prosthodontic Soc; Can Dent Asn; Can Inst Int Affairs; Can Acad Prosthodontics; Europ Prosthodontic Asn; Am Acad Hist Dent. *Res:* Fixed and removable prosthetic dentistry. *Mailing Add:* Fac Dent Dalhousie Univ Halifax NS B3H 3J5 Can

SYLVESTER, ARTHUR GIBBS, b Altadena, Calif, Feb 16, 38; m 61; c 2. NEOTECTONICS, PETROFABRICS. *Educ:* Pomona Col, BA, 59; Univ Calif, Los Angeles, MA, 63 & PhD(geol), 66. *Prof Exp:* Fulbright scholar, US Educ Found, Norway, 61-62; res geologist, Shell Develop Co, 66-68; PROF, UNIV CALIF, SANTA BARBARA, 66- *Concurrent Pos:* Regional lectr, Sigma Xi, 71-72; assoc dir educ abroad prog, Univ Calif & Study Ctr Bergen Univ, 72-74; lectr, Struct Geol Sch, Am Asn Petrol Geologist, 77-86, dir, 83-86; chmn dept geol sci, Univ Calif, Santa Barbara, 80-86; distinguished lectr, Am Asn Petrol Geol, 82-83; assoc ed, Am Asn Petrol Geologist, 85-88; ed, Geol Soc Am, 89- *Mem:* Norweg Geol Soc; Seismol Soc Am; fel Geol Soc Am; Am Geophys Union; Am Asn Petrol Geologists; Nat Asn Geol Teachers. *Res:* Geodetic investigations of near field crustal movements associated with active and potentially active faults; structure and tectonics of strike-slip faults; deformation of rock-forming minerals. *Mailing Add:* Dept Geol Sci Univ Calif Santa Barbara CA 93106

SYLVESTER, EDWARD SANFORD, b New York, NY, Feb 29, 20; m 42; c 2. ENTOMOLOGY. *Educ:* Colo State Univ, BS, 43; Univ Calif, PhD(entom), 47. *Prof Exp:* Lab asst bot, Colo State Univ, 40, asst entom, 41-43; prin lab asst entom, Univ Calif, Berkeley, 45-47, instr entom univ & jr entomologist exp sta, 47-49, asst prof & asst entomologist, 49-55, lectr & assoc entomologist, 55-61, chmn entom sci, 81-85, PROF ENTOM & ENTOMOLOGIST EXP STA, UNIV CALIF, BERKELEY, 61-, ASSOC DEAN, 87- *Mem:* Entom Soc Am; Am Phytopath Soc; Sigma Xi. *Res:* Insect vector-virus relationships; biostatistics; Aphidiae. *Mailing Add:* 366 Ocean View Ave Kensington CA 94707

SYLVESTER, JAMES EDWARD, b Syracuse, NY, Apr 30, 47; m 84. MOLECULAR BIOLOGY, BIOCHEMISTRY. *Educ:* Le Moyne Col, BS, 68; Univ Del, PhD(chem), 77. *Prof Exp:* Lab dir, Brandt Assocs, Inc, 70-73; staff fel, Lab Molecular Genetics, Nat Inst Child Health & Human Develop, NIH, 77-80; res assoc, dept pediat, Univ Mich, 80-81; res assoc, Dept Human Genetics, Univ Pa, 81-88; ASSOC PROF, DEPT PATH, HAHNEMANN UNIV, PA, 88- *Concurrent Pos:* Mem, biohazard safety comt, Smith, Kline & French, 84-; sci advisor, Molecular Oncol Inc. *Mem:* Am Soc Human Genetics; AAAS. *Res:* Molecular and genetic organization of the human genes that code for ribosomal RNA and Duchenne muscular dystrophy; myosin heavy chain; gene expression in a canine medal for Duchenne muscular distrophy. *Mailing Add:* Dept Path Hahnemann Univ Broad & Vine Philadelphia PA 19102

SYLVESTER, JOSEPH ROBERT, b Fayetteville, NC, Nov 3, 42. FISHERIES MANAGEMENT, MARINE BIOLOGY. *Educ:* Univ Hawaii, BA, 65, MS, 69; Univ Wash, PhD(fish biol), 71. *Prof Exp:* Fishery res biologist, Nat Marine Fisheries Serv, Bur Com Fisheries, 65-71; dir, Bur Fish & Wildlife, Govt of VI, 71-73; fishery res biologist, Oceanic Inst, 73-76; dir, Bur Fish & Wildlife, Govt of VI, 76-78; mgr, Marine Mammal Endangered Species prog, Nat Marine Fisheries Serv, St Petersburg, Fla, 78-80; leader ecol sect, Nat Fishery Res Lab, LaCrosse, Wisc, 80-83; adj fac & asst prof biol, Univ Wisc, LaCrosse, 80-83; proj leader & vis prof, Univ Ark, Pine Bluff, 83-89; MGR, FISHERIES HABITAT PROTECTION PROG FISHERIERS DIV, MICH DEPT NATURAL RESOURCES, 89- *Concurrent Pos:* Res assoc, Hawaii Inst Marine Biol, Univ Hawaii, 65-69; mem, Space Explor Res Rev Panel, Biol, NASA, 75. *Mem:* Am Fisheries Soc; Am Inst Fisheries Res Biologists; Sigma Xi; NY Acad Sci. *Res:* Fish and wildlife sciences; co-author of numerous publications. *Mailing Add:* PO Box 12156 Lansing MI 48901-2156

SYLVESTER, NICHOLAS DOMINIC, b Cleveland, Ohio, Apr 16, 42; m 65; c 2. CHEMICAL ENGINEERING. *Educ:* Ohio Univ, BS, 64; Carnegie-Mellon Univ, PhD(chem eng), 69. *Prof Exp:* Asst prof chem eng, Univ Notre Dame, 68-72; assoc prof, 72-75, chmn chem & petrol eng, 77-78, PROF CHEM ENG, UNIV TULSA, 75-, DEAN ENG, 78- *Concurrent Pos:* Res consult, Amoco Prod Res Co, 75- *Mem:* Am Soc Eng Educ; Soc Petrol Engrs; Am Inst Chem Engrs; Soc Rheology; Am Chem Soc. *Res:* Enhanced oil recovery; multi-phase flow; non-Newtonian fluid mechanics; water pollution control; chemical reaction engineering. *Mailing Add:* Col Eng Univ Akron Akron OH 44325-3901

SYLVESTER, ROBERT O(HRUM), b Seattle, Wash, Aug 20, 14; m 40; c 2. CIVIL & ENVIRONMENTAL ENGINEERING. *Educ:* Univ Wash, BS, 36; Harvard Univ, SM, 41; Environ Eng Intersoc, dipl, 58. *Prof Exp:* Engr, Allis Chalmers Mfg Co, 36-38; instr eng, Univ Wash, 38-39; dist engr, Wash State Dept Health, 39-41, 46-47; from asst prof to assoc prof civil eng, 47-73, head div water & air resources, Dept Civil Eng, 68-73, dir, Inst for Environ Studies, 73-76, prof, 73-78, chmn, Dept Civil Eng, 76-78, EMER PROF CIVIL ENG, UNIV WASH, SEATTLE, 78- *Concurrent Pos:* Consult various industs, US & Can govt; mem study sect res grant review, NIH, 63-67; mem adv panel, BC Dept Lands, Forests, Water Resources, 70-72, mem adv panel water quality criteria, Nat Acad Sci, 71-72, US Nat Comt, Int Asn Water Pollution Res, 71-74 & review panel, US Gen Acct Off Water Res, 73. *Honors & Awards:* Arthur Sidney Bedell Award, Water Pollution Control Fedn, 57. *Mem:* Fel Am Soc Civil Engrs; Am Water Works Asn; Water Pollution Control Fedn; Int Asn Water Pollution. *Res:* Am Geophys Union. Res: Water resources; water supply; water resource and quality management; water pollution control technology; solid waste collection, disposal and resource recovery; environmental aspects of water quality. *Mailing Add:* 10218 Richwood Ave NW Seattle WA 98177

SYLVIA, AVIS LATHAM, b Westerly, RI, Nov 16, 38. CELL PHYSIOLOGY. *Educ:* Univ NC, Greensboro, AB, 60; Univ Conn, MS, 66; Univ NC, Chapel Hill, PhD(physiol), 73. *Prof Exp:* Res assoc cell physiol, Sch Med, Univ NC, 73-74; fel, Ctr Aging & Human Develop, 74-76, res assoc, 76-77, asst prof, 77-83, ASSOC PROF, DEPT PHYSIOL, DUKE UNIV MED CTR, 83- *Mem:* AAAS; Soc for Neurosci; Geront Soc; Sigma Xi. *Res:* Cellular oxidative metabolism, bioenergetics and redox phenomena; physiological and biochemical aspects of development and aging in the mammalian central nervous system. *Mailing Add:* Dept Physiol Med Ctr Duke Univ Box 3709-G Durham NC 27710

SYMBAS, PANAGIOTIS N, b Greece, Aug 15, 25; US citizen; m 65; c 3. THORACIC SURGERY, CARDIOVASCULAR SURGERY. *Educ:* Univ Salonika, MD, 54. *Prof Exp:* Intern surg, Vanderbilt Univ, 56-57, resident surgeon, 60-61, instr surg, 61-62; fel cardiovasc surg, St Louis Univ, 62-63; assoc thoracic surg, 64-65, from asst prof to assoc prof, 66-73, PROF SURG, THORACIC CARDIOVASC SURG DIV, SCH MED, EMORY UNIV, 73-,

DIR SURG RES LAB, 70-; DIR THORACIC & CARDIOVASC SURG, GRADY MEM HOSP, ATLANTA, 64- *Mem:* Am Surg Asn; Am Asn Thoracic Surg; Soc Thoracic Surg; Soc Univ Surgeons; fel Am Col Surgeons. *Mailing Add:* Dept Surg Emory Univ 1364 Clifton Rd NE Atlanta GA 30322

SYMCHOWICZ, SAMSON, b Krakow, Poland, Mar 20, 23; nat US; m 53; c 3. DRUG METABOLISM, BIOCHEMICAL PHARMACOLOGY. *Educ:* Chem Tech Col Eng, Czech, Chem Eng, 50; Polytech Inst Brooklyn, MS, 56; Rutgers Univ, PhD(physiol, biochem), 60. *Prof Exp:* Asst biochem, Allan Mem Inst, McGill Univ, 51-53; asst hormone res, Col Med, State Univ NY Downstate Med Ctr, 53-56; res biochemist, 56-66, sect leader, 66-70, head dept biochem, 70-73, assoc dir biol res, 73-77, DIR DRUG METAB & PHARMACOKINETICS, SCHERING CORP, 77- *Mem:* Am Chem Soc; Am Soc Pharmacol & Exp Therapeut; Am Soc Microbiol; NY Acad Sci; AAAS. *Res:* Biogenic amines; drug metabolism. *Mailing Add:* 44 Laurel Ave Livingston NJ 07039

SYME, S LEONARD, b Dauphin, Man, Can, July 4, 32; m 54; c 3. EPIDEMIOLOGY. *Educ:* Univ Calif, Los Angeles, BA, 53, MA, 55; Yale Univ, PhD(med sociol), 57. *Prof Exp:* Sociologist, Heart Dis Control Prog, US Public Health Serv, 57-60; exec secy, Human Ecol Study Sect, NIH, Bethesda, 60-62; Sociologist & asst chief, Heart Dis Control Prog, US Pub Health Serv, San Francisco, 62-65, chief, 66-68; PROF EPIDEMIOL, SCH PUB HEALTH, UNIV CALIF, BERKELEY, 68- *Concurrent Pos:* Chmn, Dept Biomed & Environ Health Sci, Sch Pub Health, Univ Calif, Berkeley, 75-80. *Honors & Awards:* James D Bruce Mem Award, Am Col Physicians, 89. *Mem:* Am Epidemiol Soc; fel Am Heart Asn; fel Am Sociol Asn; fel Am Pub Health Asn; AAAS; Int Soc Cardiol; Inst Med-Nat Acad Sci. *Res:* Published over 90 articles in various journals. *Mailing Add:* Dept Environ & Biomed Health Serv Univ Calif Earl Warren Hall Berkeley CA 94720

SYMES, LAWRENCE RICHARD, b Ottawa, Can, Aug 3, 42; m 64; c 3. SOFTWARE SYSTEMS, COMPUTER SCIENCE. *Educ:* Univ Sask, BA, 63; Purdue Univ, MS, 66, PhD(comput sci), 69. *Prof Exp:* Asst prof comput sci, Purdue Univ, 69-70; assoc prof, 70-74, dir, Comput Ctr, 70-75, head dept, 72-81, dir acad comput, 80-81, PROF COMPUT SCI, UNIV REGINA, CAN, 74-, DEAN SCI, 82- *Concurrent Pos:* Mem bd comput serv, Sask Sch Trustees Asn, Regina, 72-, Sask Agr Res Fund, 87-88, Sask ADA Asn, 90-; lectr, Xian Jiaotong Univ, China, 83, Shandong Acad of Sci, China, 87; bd dir, Hosp Syst Serv Group, 78-, chmn, 80-83; mem adv coun, Can/Sask Advan Technol Agreement, 85-87; mem steering comt, IBM/Sask Agreement, 90- *Mem:* Asn Comput Mach; Can Info Processing Soc (pres, 79-80); Inst Elec & Electronics Engrs Comput Soc. *Res:* Programming languages; distributed computing; computer networks. *Mailing Add:* Fac Sci Univ Regina Regina SK S4S 3H6 Can

SYMINGTON, JANEY, b St Louis, Mo, June 29, 28; m 49; c 4. VIROLOGY, PLANT PHYSIOLOGY. *Educ:* Vassar Col, AB, 50; Radcliffe Col, PhD(biol), 59. *Prof Exp:* Res assoc & asst prof bot, Washington Univ, 58-71, res asst prof biol, 71-73, res assoc microbiol, 73-80; RES ASST PROF, ST LOUIS UNIV, 80- *Concurrent Pos:* Guest cur biotechnol, St Louis Sci Ctr. *Mem:* Am Soc Microbiol; Sigma Xi; Am Soc Plant Physiol; Am Inst Biol Sci; Am Soc Cell Biol. *Res:* Virus structure and replication; interaction of viruses with antibodies and cell surfaces. *Mailing Add:* 745 Cella Rd St Louis MO 63124

SYMKO, OREST GEORGE, b Ukraine, Jan 24, 39; US citizen; m 62; c 2. PHYSICS. *Educ:* Univ Ottawa, BSc, 61, MSc, 62; Oxford Univ, DPhil(physics), 67. *Prof Exp:* Res officer physics, Clarendon Lab, Oxford Univ, 67-68; asst res physicist, Univ Calif, San Diego, 68-70; from asst prof to assoc prof, 70-79, PROF PHYSICS, UNIV UTAH, 79- *Concurrent Pos:* Res Corp grant, Univ Utah, 71; grantee, NSF, 72-84, Dept Army, 80, Rockwell intern, 84-90, NSF intern, 84-91, Air Force, 85-91. *Mem:* Fel Am Phys Soc. *Res:* Low temperature physics; magnetism; superconductivity; dilute alloys; nuclear magnetic resonance; Josephson junctions. *Mailing Add:* Dept Physics Univ Utah Salt Lake City UT 84112

SYMMES, DAVID, neurophysiology, psychology; deceased, see previous edition for last biography

SYMON, KEITH RANDOLPH, b Ft Wayne, Ind, Mar 25, 20; m 43; c 4. ORBIT THEORY, PARTICLE ACCELERATORS. *Educ:* Harvard Univ, SB, 42, AM, 43, PhD(theoret physics), 48. *Prof Exp:* From instr to assoc prof physics, Wayne Univ, 47-55; from asst prof to assoc prof, 55-57, PROF PHYSICS, UNIV WIS-MADISON, 57- *Concurrent Pos:* Head advan res group, Midwestern Univs Res Asn, 50-57, head theoret sect, 55-57, tech dir, 57-60; actg dir, Madison Acad Comput Ctr, 82-83 & Synchrotron Radiation Ctr, 83-85. *Mem:* Fel Am Phys Soc; Am Asn Physics Teachers. *Res:* Orbit theory; design of high energy accelerators; plasma stability; numerical simulation of plasmas; theory of energy loss fluctuations of fast particles. *Mailing Add:* Dept of Physics Univ of Wis Madison WI 53706

SYMONDS, PAUL S(OUTHWORTH), b Manila, Philippines, Aug 20, 16; US citizen; m 43; c 2. ENGINEERING. *Educ:* Rensselaer Polytech Inst, BS, 38; Cornell Univ, MS, 41, PhD(appl mech), 43. *Hon Degrees:* Docteur en Sci Appliquées, Faculté Polytech de Mons, Belgium. *Prof Exp:* Instr mech eng, Cornell Univ, 41-43; physicist, US Naval Res Lab, Washington, DC, 43-47; from asst prof to prof eng, 47-83, chmn div, 59-62, EMER PROF & PROF ENG, RES, BROWN UNIV, 83- *Concurrent Pos:* Lectr, Univ Md, 46-47; vis prof, Cambridge Univ, 49-51, Imp Chem Industs fel, 50-51; Fulbright awards, 49, 57; Guggenheim fel, 57-58; NSF sr fel, Oxford Univ, 64-65; vis prof, Univ Ill, 72, Univ Cape Town, 75 & Cambridge Univ, 73 & 80; assoc prof, ENSM, Nantes, 84; vis scholar, Kyoto Univ, 90, Inst Problem Prochnost, Kiev, 90. *Mem:* Fel Am Soc Civil Engrs; fel Am Soc Mech Engrs; Int Asn Bridge & Struct Engrs; fel Am Acad Mech; Sigma Xi. *Res:* Theory of elasticity and plasticity; structural analysis involving plastic deformations; static and dynamic loading; dynamic instability and chaos in elastic-plastic dynamic response; theory of vibrations and wave propagation. *Mailing Add:* Div Eng Brown Univ Providence RI 02912

SYMONDS, ROBERT B, b St Louis, Mo, Feb 13, 59; m. GAS GEOCHEMISTRY, VOLCANOLOGY. *Educ:* Beloit Col, BS, 83; Mich Technol Univ, MS, 85, PhD, 90. *Prof Exp:* Field asst gas geochem, Cascades Volcano Observ, US Geol Surv, 82-83; field asst mapping, San Juan Mountains, US Geol Surv, 84; res asst, Mich Technol Univ, 83-90, res assoc, 90-91; CONSULT, 91- *Mem:* Am Geophys Union. *Res:* Gas sampling at active volcanoes in the US, NZ, Antarctica and Indonesia; thermochemical modeling of volcanic gases; gas transport and ore deposits. *Mailing Add:* Dept Geol Eng-Geol & Geophys Mich Technol Univ Houghton MI 49931

SYMONS, DAVID THORBURN ARTHUR, b Toronto, Ont, July 24, 37; m 64; c 3. GEOPHYSICS, GEOLOGY. *Educ:* Univ Toronto, BASc, 60, PhD(econ geol), 65; Harvard Univ, MA, 61. *Prof Exp:* Nat Res Coun Can overseas fel paleomagnetism, Univ Newcastle, Eng, 65-66; res scientist rock magnetism, Geol Surv Can, 66-70; head dept geol, 73-77 & 79-82, PROF GEOPHYS & GEOTECTONICS, UNIV WINDSOR, 77-, PRES, CANTERBURY COL; HON PROF, UNIV WESTERN ONT, 86- *Mem:* Can Geophys Union; Am Geophys Union; Geol Asn Can. *Res:* Application of paleomagnetic methods to geotectonic, ore genesis and geochronologic problems in the Canadian Cordillera and Shield and to the history of the earth's geomagnetic field; geotechnical geophysics. *Mailing Add:* Dept Geol Univ Windsor Windsor ON N9B 3P4 Can

SYMONS, EDWARD ALLAN, b Kingston, Ont, Apr 24, 43. PHYSICAL ORGANIC CHEMISTRY, HETEROGENEOUS CATALYSIS. *Educ:* Queen's Univ, BSc, 65, PhD(chem), 69. *Prof Exp:* RES OFFICER, CHALK RIVER LABS, AECL RES, 72- *Concurrent Pos:* Fel Chem Inst Can. *Mem:* Fel Chem Inst Can. *Res:* Hydrogen isotope exchange mechanisms and associated physical/organic chemistry in systems of potential interest for heavy water production processes; noble metal heterogeneous catalysis of oxidation reactions, especially carbon monoxide; catalytic gas sensors. *Mailing Add:* Chalk River Labs AECL Res Chalk River ON K0J 1J0 Can

SYMONS, GEORGE E(DGAR), b Danville, Ill, Apr 20, 03; m 26; c 1. ENVIRONMENTAL ENGINEERING. *Educ:* Univ Ill, BS, 28, MS, 30, PhD(chem), 32; Am Acad Environ Engrs, dipl. *Prof Exp:* Res chemist, Ill State Water Survey, 28-33; instr, Univ Ill, 32-33; chemist, Freeport Sulfur Co, 33-34, consult, 34-35; engr, Greeley & Hansen, 35-36; chief chemist, Buffalo Sewer Authority, NY, 36-43; consult engr, 43-64; assoc ed, Water & Sewage Works, 43-48, managing ed, 49-51, ed, 57-59; ed, Water & Wastes Eng, 64-70; mgr spec proj, Malcolm Pirnie, Inc, 70-78; ENG ED & CONSULT, 78- *Concurrent Pos:* Adj prof, Canisius Col, 41-42; lectr, Med Sch, Buffalo Univ, 42-43; res assoc, NY Univ, 54-55; contrib ed, Water Eng & Mgt, 78-90. *Honors & Awards:* Diven Medal, Fuller Award, Am Water Works Asn, Hall of Fame; Emerson Medalist Water Pollution Control Fedn, Bedell Award. *Mem:* Am Chem Soc; fel Am Pub Health Asn; fel Am Soc Civil Engrs; hon mem Am Water Works Asn (pres, 73-74); Am Inst Chem Engrs; Am Pub Works Asn; Nat Soc Prof Engrs; hon mem Water Pollution Control Fedn. *Res:* Engineering editing. *Mailing Add:* 300 S Ocean Blvd Apt H-3 Palm Beach FL 33480

SYMONS, HUGH WILLIAM, b Kosgama, Ceylon, Jan 4, 27; Brit citizen; m 65; c 2. FOOD LEGISLATION, ENVIRONMENTAL CONCERNS OF FOOD PROCESSORS. *Educ:* Cambridge Univ, BA, 49, MA, 51. *Prof Exp:* Asst chemist, Hector Whaling Ltd, 53-58; co qual controller, Birds Eye Ltd, 58-78; SR VPRES, RES & TECHNOL SERV, AM FROZEN FOOD INST, 78- *Concurrent Pos:* Mem coun, Inst Refrig, 68-71; external examr, Nat Col Food Technol, 69-71; pres, Intercommissional Working Party Int Inst Refrig, 79-85. *Mem:* Fel Inst Food Sci & Technol (vpres, 71-72); fel Inst Refrig; fel Inst Biol. *Res:* Quality and stability of frozen foods. *Mailing Add:* Am Frozen Food Inst 1764 Old Meadow Lane McLean VA 22102

SYMONS, JAMES M(ARTIN), b Champaign, Ill, Nov 24, 31; m 58; c 3. SANITARY ENGINEERING. *Educ:* Cornell Univ, BCE, 54; Mass Inst Technol, SM, 55, ScD(sanit eng), 57. *Prof Exp:* Asst, NIH Proj, Mass Inst Technol, 55-57, instr sanit eng, 65-68, asst prof, 58-62; res pub health engr, USPHS, 62-70; chief phys & chem contaminants removal br, Drinking Water Res Div, Nat Environ Res Ctr, US Environ Protection Agency, 70-82; PROF CIVIL ENG, UNIV HOUSTON, 82- *Concurrent Pos:* Assoc, Rolf Eliassen Assocs, Mass, 57-61. *Honors & Awards:* Harrison P Eddy Award, Water Pollution Control Fedn, 63; Huber Res Prize, Am Soc Civil Engrs, 67; Res Award, Am Water Works Asn, 81; Silver Medal, 71. *Mem:* Am Water Works Asn; Am Soc Civil Engrs; Water Pollution Control Fedn; Asn Environ Eng Prof; Am Acad Environ Engrs; Int Water Supply Asn. *Res:* Drinking water treatment research. *Mailing Add:* Dept Civil & Environ Eng Univ Houston Houston TX 77204-4791

SYMONS, PHILIP CHARLES, b Taunton, Somerset, Eng, June 4, 39; m 60; c 2. ELECTROCHEMISTRY. *Educ:* Univ Bristol, BSc, 60, PhD(electrochem), 63. *Prof Exp:* Sr chemist, Proctor & Gamble Ltd, Eng, 63-67 & Hooker Chem Corp, 68-69; res dir batteries, Udylite Co, Occidental Petrol Corp, 69-73; CHIEF SCIENTIST, ENERGY DEVELOP ASSOCS, 73-; PRES, PRIN CONSULT, ELECTRO CHEM ENGR CONSULTS INC. *Mem:* Am Chem Soc; Electrochem Soc; Royal Soc Chem. *Res:* High energy density batteries; electrochemical thermodynamics and kinetics; thermodynamics of phase transitions; electrochemical engineering. *Mailing Add:* 1295 Kelly Park Circle Morgan Hill CA 95037-3370

SYMONS, TIMOTHY J, b Southborough, Eng, Aug 4, 51; m 87; c 1. NUCLEAR PHYSICS. *Educ:* Oxford Univ, BA, 72, MA, 76, PhD (physics), 76. *Prof Exp:* Postdoctoral fel, Oxford Univ, 76-77; Postdoctoral fel, 77-79, divisional fel, 79-83, ASSOC DIR, LAWRENCE BERKELEY LAB, 83-, DIV HEAD, NUCLEAR SCI, 85- *Concurrent Pos:* Vis fel, Inst for Nuclear Physics, 80-81; mem, US Nuclear Physic deleg to USSR. *Mem:* Fel Am Physical Soc fel; AAAS. *Res:* Study of nuclear physics using neon energy heavy-low beams. *Mailing Add:* Lawrence Berkeley Labs B70A-3307 Berkeley CA 94720

SYMPSON, ROBERT F, b Ft Madison, Iowa, June 21, 27; m 53; c 4. ANALYTICAL CHEMISTRY, ELECTROCHEMISTRY. *Educ:* Monmouth Col, BS, 50; Univ Ill, MS, 52, PhD, 54. *Prof Exp:* From asst prof to assoc prof, 54-65, PROF CHEM, OHIO UNIV, 65-, CHMN DEPT, 77- *Mem:* Am Chem Soc; Electrochem Soc. *Res:* Polarography and amperometric titrations applied to analytical chemistry. *Mailing Add:* Dept Chem Ohio Univ Athens OH 45701

SYNEK, MIROSLAV (MIKE), b Prague, Czech, Sept 18, 30; US citizen; m 65; c 2. ATOMIC PHYSICS, PHYSICAL CHEMISTRY. *Educ:* Charles Univ, Prague, MS, 56; Univ Chicago, PhD(physics), 63. *Prof Exp:* Technician chem, Inst Indust Med, Prague, Czech, 50-51; asst physics, Czech Acad Sci, 56-58; res asst, Univ Chicago, 58-62; asst prof physics, DePaul Univ, 62-65, assoc prof physics, 65-67, assoc prof chem, 66-67; prof physics, Tex Christian Univ, 67-71; regional sci adv & res scientist, Dept Physics, Univ Tex, Austin, 71-73, lectr, exten lectr & res scientist, Depts Physics, Chem, Astron & Math, 73-75; FAC MEM DIV EARTH & PHYS SCI, COL SCI & ENG, UNIV TEX, SAN ANTONIO, 75- *Concurrent Pos:* Consult, US Army, 58, Physics Div, Argonne Nat Lab, 66-73, Brooks AFB, Tex, 83; prin investr, Mat Lab, Wright-Patterson AFB, Ohio, 64-71; prin investr, Robert A Welch Found res grant, 69-71 & 76-83; referee manuscripts, Phys Rev; occasional lectr; adv, Soc Phys Student, Am Inst Phys, 83-; elected adv, Soc Physics Students; nominated judge, Alamo Regional Sci Fair. *Mem:* Fel AAAS; fel Am Phys Soc; Am Asn Physics Teachers; Am Chem Soc; fel Am Inst Chemists; Soc Sci Res; Sigma Xi; NY Acad Sci. *Res:* Educational and computational physics; popularization of science; astronomy and energy; laser-crystal efficiency; energy applications; laser-active ions of rare earths; materials science; atomic wave functions; statistical mechanics; molecular dynamics; global system of free elections is a historical urgency in the nuclear age. *Mailing Add:* Div Earth & Phys Sci Univ Tex Col Sci & Eng San Antonio TX 78285

SYNER, FRANK N, biochemistry, for more information see previous edition

SYNOLAKIS, COSTAS EMMANUEL, b Athens, Greece, Sept 21, 56; US citizen. TSUNAMI RUNUP, COMPUTER TOMOGRAPHY. *Educ:* Calif Inst Technol, BSc, 78, MS, 79, PhD(civil eng), 86. *Prof Exp:* PROF CIVIL & AEROSPACE ENG, UNIV SOUTHERN CALIF, 85- *Concurrent Pos:* Prin investr, NSF, 86-; NSF presidential young investr award, 89. *Mem:* AAAS; Am Phys Soc; Am Geophys Union; Am Soc Civil Engrs; NY Acad Sci; Am Asn Asphalt Paving Technologists. *Res:* Analytical solutions on long wave runup and evolution problems; mixing of elliptic jets in stratified flow; optical flow; non-destructive application of tomography; asphalt core tomography; concrete core tomography. *Mailing Add:* Dept Civil Eng 2531 Univ of Southern California Los Angeles CA 90089

SYNOVITZ, ROBERT J, b Milwaukee, Wis, Feb 3, 31; wid; c 4. HEALTH SCIENCE. *Educ:* Wis State Univ, La Crosse, BS, 53; Ind Univ, Bloomington, MS, 56, HSD, 59. *Prof Exp:* Instr pub schs, Mo, 56-58; asst prof health sci, Eastern Ky Univ, 59-62; assoc prof physiol & health sci, Ball State Univ, 62-68; PROF HEALTH SCI & CHMN DEPT, WESTERN ILL UNIV, 68- *Concurrent Pos:* Teaching & res grant, Ball State Univ, 66-67. *Mem:* Am Sch Health Asn (nat pres, 81-82); Soc Pub Health Educ. *Mailing Add:* Dept Health Sci Western Ill Univ Macomb IL 61455

SYNOWIEC, JOHN A, b Chicago, Ill, Sept 18, 37; m 67; c 3. MATHEMATICAL ANALYSIS. *Educ:* DePaul Univ, BS, 59, MS, 61, DMS, 62; Ill Inst Technol, PhD(math), 64. *Prof Exp:* Instr math, DePaul Univ, 61-62; from instr to asst prof, Ill Inst Technol, 63-67; asst prof, 67-74, ASSOC PROF MATH, IND UNIV NORTHWEST, 74- *Mem:* Am Math Soc; Math Asn Am; Soc Indust & Appl Math. *Res:* Generalized functions; partial differential equations; harmonic analysis; history of mathematics. *Mailing Add:* Ind Univ Northwest 3400 Broadway Gary IN 46408

SYPERT, GEORGE WALTER, b Marlin, Tex, Sept 25, 41; m 73; c 2. NEUROLOGICAL SURGERY. *Educ:* Univ Wash, BA, 63, MD, 67. *Prof Exp:* Resident neurosurg, Sch Med, Univ Wash, 68-73, instr, 73-74; from asst prof to prof, 74-84, OVERSTREET FAMILY PROF NEUROSURG & NEUROSCI & EMINENT SCHOLAR, COL MED, UNIV FLA, 84- *Concurrent Pos:* Asst chief neurosurg, Fitsimons Gen Hosp, US Army, 68-70; staff neurosurgeon, Shands Teaching Hosp, Gainesville & chief neurosurg, Gainesville Vet Admin Med Ctr, 74-; chmn, continuing med, Am Asn Neurol Surgeons, 82-87, sect spinal dis, 85-87. *Mem:* Soc Neurosci; Cong Neurol Surgeons; Am Asn Neurol Surgeons; AAAS; Am Physiol Soc; Soc Neurol Surgeons; Am Soc Stereotactic & Functional Neurosurg (vpres, 81-83, pres, 83-85). *Res:* Neurophysiological and ionic mechanisms involved in synaptic transmission; segmental motor control; synaptic plasticity; neuronal repetitive firing mechanisms; pathophysiology of epilepsy; mammalian cellular neurophysiology. *Mailing Add:* 3677 Central Ave Suite A Ft Myers FL 33901

SYPHERD, PAUL STARR, b Akron, Ohio, Nov 16, 36; m 54; c 4. MICROBIOLOGY, MOLECULAR BIOLOGY. *Educ:* Ariz State Univ, BS, 59; Univ Ariz, MS, 60; Yale Univ, PhD(microbiol), 63. *Prof Exp:* NIH res fel biol, Univ Calif, San Diego, 62-64; from asst prof to assoc prof microbiol, Univ Ill, Urbana, 64-70; assoc prof, 70-72, chmn microbiol, 74-87, PROF MICROBIOL, COL MED, UNIV CALIF, IRVINE, 72-, VICE CHANCELLOR RES, DEAN GRAD STUDIES, 89- *Concurrent Pos:* USPHS fel, Univ Calif, San Diego, 62-64; mem microbiol chem study sect, NIH, 75-81; mem, Nat Bd Med Examr, 82-85, Comt Health Manpower, Nat Res Coun, 82-86; mem, Microbiol Genetics Study Sect, NIH, 87- *Mem:* Am Soc Microbiologists; Am Soc Biol Chemists; Am Acad Microbiol. *Res:* Structure and synthesis of ribosomes; regulation of nucleic acid synthesis; molecular basis of morphogenesis; molecular mechanisms of the regulation of gene expression, with and emphasis on post-transcriptional regulation; post-translational modification of proteins in microorganisms. *Mailing Add:* Dept Microbiol Univ Calif Col Med Irvine CA 92717

SYRETT, BARRY CHRISTOPHER, b Stockton on Tees, Eng, Dec 16, 43; US citizen; m 91; c 1. CORROSION. *Educ:* Univ Newcastle, BSc, 64, PhD(metall), 67. *Prof Exp:* Res scientist corrosion, Dept Energy, Mines & Resources, Govt Can, 67-70 & Int Nickel Co, Inc, 70-72; metall prog mgr, SRI Int, 72-79; TECH ADV, ELEC POWER RES INST, 79- *Concurrent Pos:* Chmn publ comt, Nat Asn Corrosion Engrs, 89-91; ed, Corrosion Testing Made Easy. *Honors & Awards:* Mem bd dir, Nat Asn Corrosion Engrs, 89-; Sci Am Award, Am Soc Testing-Mat, 79. *Mem:* Nat Asn Corrosion Engrs; Inst Corrosion Sci & Technol; prof mem Inst Metals. *Res:* Corrosion research, including pitting, crevice corrosion, stress corrosion, erosion-corrosion, fretting corrosion, corrosion fatigue, hydrogen embrittlement, corrosion resistant coatings, failure analysis, alloy development and liquid metal embrittlement. *Mailing Add:* 1881 Fulton St Palo Alto CA 94303

SYSKI, RYSZARD, b Plock, Poland, Apr 8, 24; m 50; c 6. MATHEMATICS. *Educ:* Polish Univ Col, London, Dipl Ing, 50; Imp Col, Dipl, 51, BSc, 54, PhD(math), 60. *Prof Exp:* Sci officer, ATE Co, Eng, 51-52, sr sci officer, Hivac Ltd, 52-60; res assoc mgt sci & lectr math, 61-62, assoc prof, 62-66, chmn probability & statist div, 66-70, PROF MATH, UNIV MD, COLLEGE PARK, 66- *Concurrent Pos:* Chmn bibliog comt, Int Teletraffic Cong, 55-61, mem organizing comt, 67. *Mem:* Am Math Soc; fel Royal Statist Soc. *Res:* Congestion, queueing and probability theories; stochastic processes. *Mailing Add:* Dept of Math Univ of Md College Park MD 20742

SYTSMA, KENNETH JAY, b Tokyo, Japan, Jan 23, 54; US citizen; m 75; c 4. PLANT SYSTEMATICS, MOLECULAR PHYLOGENETICS. *Educ:* Calvin Col, BS, 76; Western Mich Univ, Kalamazoo, MA, 79; Wash Univ, St Louis, PhD(bot), 83. *Prof Exp:* Teaching asst bot, Western Mich Univ, Kalamazoo, 76-79; fel bot, Wash Univ, 79-83; cur, Summit Herbarium, Panama City, 80-81; res fel, Univ Calif, Davis, 83-85; ASST PROF BOT, UNIV WIS-MADISON, 85- *Honors & Awards:* Cooley Award, Am Soc Plant Taxonomists, 86. *Mem:* Am Soc Plant Taxonomists; Am Soc Bot; Soc Study Evolution. *Res:* Study of evolutionary and systematic relationships of flowering plants by use of DNA analysis, isozymes and pollination biology. *Mailing Add:* Dept Bot 132 Birge Hall Univ Wis 430 Lincoln Dr Madison WI 53706

SYTSMA, LOUIS FREDERICK, b Chicago, Ill, July 20, 46; m 67; c 2. ENVIRONMENTAL CHEMISTRY. *Educ:* Calvin Col, BA, 67; Ohio Univ, PhD(org chem), 71. *Prof Exp:* Chemist, Anderson Develop Co, 71-77; asst prof chem, Siena Heights Col, 73-77; PROF CHEM, TRINITY CHRISTIAN COL, 77-; PROF ENVIRON CHEM, AU SABLE TRAILS INST ENVIRON STUDIES, 82- *Concurrent Pos:* Consult, Anderson Develop Co, 77- *Mem:* Am Chem Soc; Sigma Xi. *Res:* Cause, movement and analysis of environmental pollutants. *Mailing Add:* Trinity Christian Col 6601 W Col Dr Palos Heights IL 60463

SYTY, AUGUSTA, b Harbin, China. CHEMISTRY. *Educ:* Univ Tenn, Knoxville, BS, 64, PhD(anal chem), 68. *Prof Exp:* PROF ANAL CHEM, INDIANA UNIV PA, 68- *Concurrent Pos:* Guest res chemist, Nat Bur Standards. *Mem:* Am Chem Soc; Soc Appl Spectros. *Res:* Flame emission; atomic absorption; methods of analysis; molecular absorption spectrometry in the gas phase; metal speciation; selective determination of chromium (VI). *Mailing Add:* Dept Chem Indiana Univ Pa Indiana PA 15705-1076

SYVERSON, ALDRICH, chemical engineering; deceased, see previous edition for last biography

SYVERTSON, CLARENCE A, b Minneapolis, Minn, Jan 12, 26; m 53, 82; c 2. RESEARCH ADMINISTRATION, TECHNICAL MANAGEMENT. *Educ:* Univ Minn, BS, 46, MS, 48. *Prof Exp:* Res scientist, Ames Aeronaut Lab, Nat Adv Comt Aeronaut, NASA, 48-58, chief hypersonic wind tunnel br, Ames Res Ctr, 59-63, chief mission anal div, 63-65, dir mission anal div, Off Adv Res & Tech, 65-66, dir astronaut, 66-69, dep dir, Ames Res Ctr, 69-78, dir, 78-84; consult prof, Stanford Univ, 84-88; ENG & MGT CONSULT, 88- *Concurrent Pos:* Exec dir civil aviation res & develop policy study, Dept of Transp, NASA, 70-71; mem, bd gov, Nat Space Club, 78-84, adv bd, Col Eng, Univ Calif, Berkeley, 81-86, Comt Aircraft & Engine Develop Testing, Air Force Studies Bd, Nat Res Coun, 84-85 & Comt Space Sta Eng & Technol, Aeronaut & Space Eng Bd, Nat Res Coun, 84-86. *Honors & Awards:* Lawrence Sperry Award, Inst Aeronaut Sci, 57; NASA Inventions & Contrib Award, 64, Exceptional Serv Medal, NASA, 71, Distinguished Serv Medal, 84. *Mem:* Nat Acad Eng; fel Am Inst Aeronaut & Astronaut; fel Am Astronaut Soc. *Res:* Civil aviation; mission analysis; entry vehicle research. *Mailing Add:* 15725 Apollo Heights Ct Saratoga CA 95070

SZABLYA, JOHN F(RANCIS), b Budapest, Hungary, June 25, 24; US citizen; m 51; c 7. ELECTRICAL ENGINEERING. *Educ:* Tech Univ Budapest, Dipls, 47-48, DEcon, 48. *Prof Exp:* Mem fac, Tech Univ Budapest, 49-56; mem fac, Univ BC, 57-63; prof elec eng, Wash State Univ, 63-82; elec consult engr, Ebasco Servs, Inc, 82-90; CONSULT ENGR, 90- *Concurrent Pos:* Engr, Ganz Elec Works, 47-56; consult, Hungarian Elec Res Labs, 52-56, Tech Univ Braunschweig, Ger, 73-74, Univ of the West Indies, Trinidad, 80-81. *Honors & Awards:* Zipernowszky Medal, Hungary, 54. *Mem:* Fel Inst Elec & Electronic Engrs; fel Inst Elec Engrs; Tensor Soc; Austrian Inst Elec Engrs. *Res:* Electromechanical energy conversion; electric power transmission; high power, high voltage research; fundamental electromagnetism; energy research. *Mailing Add:* 4416 134th Pl SE Bellevue WA 98006

SZABO, A(UGUST) J(OHN), b Baton Rouge, La, Sept 27, 21; m 45; c 3. CIVIL ENGINEERING. *Educ:* La State Univ, BS, 43; Harvard Univ, MS, 50. *Prof Exp:* Pub health engr, State Bd Health, La, 46-55; assoc prof civil eng, Univ Southwestern La, 55-63; prin engr, 63-64, SECY & TREAS, DOMINGUE, SZABO & ASSOCS, INC, 64- *Concurrent Pos:* USPHS, 42; aviation engr, US Air Force, 43-46; Corps Engrs, Vicksburg Dist, 56. *Honors & Awards:* Arthur Sidney Bedell Award, Water Pollution Control Fedn, 58. *Mem:* Water Pollution Control Fedn; Am Soc Civil Engrs; Am Water Works Asn; Nat Soc Prof Engrs; Am Acad Environ Engrs; Am Consult Eng Coun. *Res:* Sanitary and environmental engineering. *Mailing Add:* Domingue, Szabo & Assocs Inc Consult Engrs PO Box 52115 Lafayette LA 70505-2115

SZABO, ALEXANDER, b Copper Cliff, Ont, Mar 13, 31; m 57; c 4. SOLID STATE PHYSICS. *Educ:* Queen's Univ, BS, 53; McGill Univ, MS, 55; Tohoku Univ, Japan, DEng(physics), 70. *Prof Exp:* Res officer electron beams, 55-59, res officer microwave masers, 59-61, RES OFFICER LASERS & SOLID STATE PHYSICS, NAT RES COUN CAN, 61- *Mem:* Fel Japan Soc Prom Sci; Optical Soc Am. *Res:* Study of coherence lifetimes and optical line widths of impurity ions in solids using fluorescence line narrowing and photon echo techniques. *Mailing Add:* Nat Res Coun Bldg M-50 Ottawa ON K1A 0R8 Can

SZABO, ARLENE SLOGOFF, b Philadelphia, Pa, Feb 19, 45; m 65; c 2. CELL BIOLOGY. *Educ:* Douglass Col, AB, 66, Rutgers Univ, MS, 68, PhD(cell biol), 71. *Prof Exp:* Res fel enzymol, Med Sch, Univ Southern Calif, 71-72; res asst plant physiol, Univ Calif, Los Angeles, 72-74; ASST PROF BIOL, LOYOLA MARYMOUNT UNIV, 74- *Mem:* Am Soc Cell Biol; AAAS; Asn Women in Sci; Sigma Xi. *Res:* Development and regulation of the peroxisome in yeast, as Saccharomyces cerevisiae. *Mailing Add:* 16519 S Broadway Gardena CA 90248

SZABO, ARTHUR GUSTAV, fluorescence, spectroscopy, for more information see previous edition

SZABO, ATTILA, b Budapest, Hungary, Sept 6, 47; US citizen; div. THEORETICAL BIOPHYSICAL CHEMISTRY. *Educ:* McGill Univ, BSc, 68; Harvard Univ, PhD(chem physics), 73. *Prof Exp:* Fel, Inst Physicochem Biol, Paris, 72-73; fel, Med Res Coun Lab Molecular Biol, Cambridge, Eng, 73-74; asst prof chem, Ind Univ, Bloomington, 74-78, assoc prof, 78-80; RES SCIENTIST, LAB CHEM PHYS, NIH, 80- *Mem:* Biophys Soc. *Res:* Nuclear magnetic relaxation and fluorescence depolarization and internal motions in biopolymers; structure-function relations in allosteric systems. *Mailing Add:* Lab Chem Phys Bldg 2 Rm B1-28 Nat Inst Diabetes & Digestive Kidney Dis Bethesda MD 20892

SZABO, BARNA ALADAR, b Martonvasar, Hungary, Sept 21, 35; US citizen; m 60; c 2. MECHANICAL ENGINEERING. *Educ:* Univ Toronto, BASc, 60; State Univ NY, Buffalo, MS, 66, PhD(civil eng), 69. *Prof Exp:* Mining engr, Int Nickel Co Can Ltd, 60-62; civil engr, H G Acres Ltd Consult Engrs, 62-66; instr eng & appl sci, State Univ NY, Buffalo, 66-68; from assoc prof to prof, 68-75, GREENSFELDER PROF MECHS, UNIV WASH, 75-, DIR, CTR COMPUT MECH, 77- *Concurrent Pos:* Consult, Asn Am Railroads, 74-77; dir, Washington Univ Technol Assocs, 78-82; chmn tech adv bd, Noetic Tech Corp, St Louis. *Mem:* Am Acad Mech; Am Soc Mech Engrs; Soc Eng Sci. *Res:* Numerical approximation techniques in continuum mechanics; development of adaptive methods in finite element analysis; numerical structure of stress analysis. *Mailing Add:* Ctr Comput Mech Wash Univ Box 1130 St Louis MO 63130

SZABO, BARNEY JULIUS, b Debrecen, Hungary, Apr 11, 29; US citizen; m 56. GEOCHEMISTRY. *Educ:* Univ Miami, BS, 61, MS, 66. *Prof Exp:* Res chemist, Inst Marine Sci, Univ Miami, 57-66; RES CHEMIST, BR ISOTOPE GEOL, US GEOL SURV, 66- *Res:* Stable and radioactive trace elements analyses; uranium disequilibria dating; radio and isotopic chemistry of uranium, thorium, protactinium and artificial radioactive nuclides. *Mailing Add:* US Geol Surv Br Isotope Geol Denver Fed Ctr Denver CO 80225

SZABO, EMERY D, b Komarom, Hungary, Apr 5, 33; US citizen; m 70; c 1. MEDICAL & HEALTH SCIENCES, TECHNICAL MANAGEMENT. *Educ:* Univ Budapest, Hungary, ME, 56; Univ NY, MS, 63. *Prof Exp:* Asst mgr, Kentile Floors, Inc, 59-69; group leader, Tenneco Chem, Inc, tech serv mgr, 77-80; vpres & tech dir, Dexter Plastics, 83-90; tech dir, 81-83, DIR TECHNOL, ALPHA CHEM & PLASTICS CORP, 90- *Mem:* Am Chem Soc; fel AAAS; Soc Plastics Engrs. *Res:* Vinyl and olefinic polymers for biomedical and beverage packaging applications; material science of chemicals and additives for polyvinyl chloride and application technology of vinyl plastics; agency regulations relating to vinyl plastics; awarded three US patents. *Mailing Add:* 200 Bentley Oaks Lane Charlotte NC 28270

SZABO, GABOR, b Mar 20, 41; Can citizen. PHYSIOLOGY, BIOPHYSICS. *Educ:* Univ Montreal, BS, 64, MS, 66; Univ Chicago, PhD(physiol), 69. *Prof Exp:* Asst prof physiol, Med Ctr, Univ Calif, Los Angeles, 71-75; assoc prof, 75-79, PROF PHYSIOL, UNIV TEX MED BR GALVESTON, 79- *Mem:* AAAS; Biophys Soc. *Res:* Regulation of cell membrane function. *Mailing Add:* Dept Physiol & Biophys Univ Va Box 449 1300 Jefferson Park Ave Charlottesville VA 22908

SZABO, KALMAN TIBOR, b Abda, Hungary, July 29, 21; US citizen; m 44; c 2. TERATOLOGY. *Educ:* Univ Budapest, BSc, 47; Rutgers Univ, MS, 62; Univ Vienna, MSc, 71, DSc(teratology), 73. *Prof Exp:* Res scientist genetics, Res Inst, Acad Sci, Fertod, Hungary, 53-56; res assoc reproductive physiol, Rutgers Univ, 60-62; toxicologist, 62-66, sr scientist & unit head teratology, 67-69, sr investr & group leader reproductive toxicol & teratology, 69-70, asst dir toxicol, 71-72, assoc dir, 72-81, dir, 81-83, dir reproductive & develop toxicol, Smith Kline & French Labs, 83-86; CONSULT, 86- *Concurrent Pos:* Adv & partic, Adv Comt Protocols on Reproductive Studies, Safety Eval Food Additives & Pesticide Residues, Food & Drug Admin, 67; res assoc prof pediat, Med Col, Thomas Jefferson Univ, 75- *Mem:* Teratology Soc; Behav Teratology Soc; Soc Toxicol; Soc Study Reproduction; Int Acad Environ Safety; Am Col Toxicol. *Res:* Spontaneous and induced anomalies of the central nervous system; role of maternal nutritional deprivation in embryogenesis; evaluation of various factors affecting teratogenic response; drug toxicology; reproductive and developmental toxicology of drugs. *Mailing Add:* 215 Morris Rd Ambler PA 19002

SZABO, MIKLOS TAMAS, b Helsinki, Finland, Aug 28, 37; nat US; m 62; c 1. PETROLEUM ENGINEERING, PHYSICAL CHEMISTRY. *Educ:* Tech Univ Heavy Indust, Hungary, BS & MS, 60, PhD(petrol eng), 68. *Prof Exp:* Res assoc, Petrol Res Lab, Hungarian Acad Sci, Miskolc, 60-69 & Calgon Corp, Pittsburgh, 70-74; SR RES ENGR, GULF RES & DEVELOP CO, 74- *Mem:* Am Chem Soc; Soc Petrol Engrs; Am Inst Mining, Metall & Petrol Engrs. *Res:* Petroleum reservoir engineering; flow in porous media; polymer flooding; new oil recovery processes; colloid chemistry; rheology of polymer solutions. *Mailing Add:* 301 Thornberry Court Dr Woodrige Ross TWP Pittsburgh PA 15237

SZABO, PIROSKA LUDWIG, b Bratislava, Czechoslovakia, Dec 14, 40; US citizen; m 63; c 3. HISTOLOGY, GROSS ANATOMY. *Educ:* Oberlin Col, BA, 62; Univ Fla, MS, 64, PhD(med sci), 67. *Prof Exp:* Fel gross anat, Col Physicians & Surgeons, Columbia Univ, 67-68; asst prof gross anat & histol, Med Col Va, 68-70, Albert Einstein Col Med, 70-73; instr biol & embryol, Lehman Col, 73-77; PROF & CHMN ANAT, NY COL OSTEOP MED, 77- *Concurrent Pos:* Consult, Osteop Nat Bd, 79- *Mem:* AAAS. *Mailing Add:* Dept Anat NY Col Osteop Med Wheatley Rd Old Westbury NY 11568

SZABO, SANDOR, b Ada, Yugoslavia, Feb 9, 44; m 72; c 2. PATHOLOGY, PHARMACOLOGY. *Educ:* Univ Belgrade, MD, 68; Univ Montreal, MSc, 71, PhD(exp med, pharmacol), 73; Harvard Sch Pub Health, MPH(occupational med), 83. *Prof Exp:* Vis scientist exp med & pharmacol, Inst Exp Med Surg, Univ Montreal, 69-70, res asst, 70-73; resident path, Peter Bent Brigham Hosp, 73-77, res fel, Med Sch, 75-77, res assoc path, Peter Bent Brigham Hosp; asst prof, 77-81, ASSOC PROF PATH, BRIGHAM & WOMEN'S HOSP, HARVARD MED SCH, 81- *Concurrent Pos:* Chief ed, Sci J Medicinski Podmladak, 67-68; consult to health dir, CZ, Panama, 71; Nat Inst Environ Health Sci grant, 78-; NIH res career develop award, 80-85; FDA gastroint drug adv comt, 84- *Honors & Awards:* Milton Fund Award, Harvard Univ, 78. *Mem:* Soc Exp Biol & Med; Am Soc Pharmacol & Exp Therapeut; Am Asn Pathologists; Endocrine Soc; Pharmacol Soc Can; Royal Col Pathologists. *Res:* Experimental pathology and pharmacology of the gastro-intestinal tract and endocrine glands; development of disease models; duodenal ulcer produced by propionitrile, cysteamine; adrenal and thyroid necrosis caused by acrylonitrile pyrazole; gastric cytoprotection by sulphydryls. *Mailing Add:* Brigham & Women's Hosp Harvard Med Sch 75 Francis St Boston MA 02115

SZABO, STEVE STANLEY, b Westons Mills, NJ, Feb 10, 27; m 55; c 3. AGRONOMY. *Educ:* Rutgers Univ, BS, 52; Kans State Col, MS, 54; Univ Wis, PhD(agron), 58. *Prof Exp:* Asst prof agron, NMex State Univ, 57-61; plant physiologist, US Food & Drug Admin, 61-63 & Army Biol Labs, Ft Detrick, Md, 63-65; plant physiologist, Boyce Thompson Inst Plant Res, 65-76; PRES, LAWN GENIE CORP, 76- *Mem:* Am Soc Plant Physiologists; Weed Sci Soc Am. *Res:* Chemical weed control; plant physiology. *Mailing Add:* 20 Milrose Lane Monsey NY 10952

SZABO, TIBOR IMRE, b Tiborszallas, Hungary, July 11, 34; Can citizen; m 66; c 7. BEHAVIORAL BIOLOGY. *Educ:* Univ Hort, Budapest, BAE, 58; Univ Guelph, MSc, 69, PhD(apicult), 73. *Prof Exp:* Fel apicult, Univ Guelph, 73-74; RES SCIENTIST APICULT, AGR CAN, 74- *Mem:* Int Bee Res Asn; Sigma Xi. *Res:* Honey bee behavior, biology, physiology and breeding. *Mailing Add:* Agr Can PO Box 29 Beaverlodge AB T0H 0C0 Can

SZABUNIEWICZ, MICHAEL, b Poland, Oct 11, 09. VETERINARY MEDICINE. *Educ:* Acad Vet Med, DVM, 34, DVSc(physiol), 37. *Prof Exp:* Dir, Exp Farm Kasese, Belgian Congo, 46-50; dir, Regional Diag Vet Lab & Exten Serv, 50-60; area vet, Agr Res Serv, USDA, 61-62; from asst prof to prof , 62-76, EMER PROF VET PHYSIOL, TEX A&M UNIV, 76- *Concurrent Pos:* Vis prof, Pahlavi Univ Shiraz, Iran, 76, Univ Cent Venezuela, 77-78. *Mem:* AAAS; Am Vet Med Asn; Am Soc Vet Physiol & Pharmacol; Am Asn Lab Animal Sci. *Res:* Physiology; pharmacology; radiation research. *Mailing Add:* 404 College View St Bryan TX 77801

SZAKAL, ANDRAS KALMAN, b Szekesfehervar, Hungary, Sept 26, 36; US citizen; m 61; c 2. IMMUNOBIOLOGY, CANCER. *Educ:* Univ Colo, Boulder, BA, 61, MS, 63; Univ Tenn, PhD(immunobiol), 72. *Prof Exp:* Res asst histochem, Univ Colo, Boulder, 60-61; res assoc electron micros, Univ Mich, Ann Arbor, 65-66; staff biologist, Oak Ridge Nat Lab, 66-69, consult electron microscopist, 69-70, Oak Ridge Assoc Univs fel, 70-72, res biologist, Div Biol, 72-74; prin scientist lung immunobiol, Life Sci Div, Meloy Labs, Inc, 74-; AT DEPT ANAT, VA COMMONWEALTH UNIV. *Mem:* AAAS. *Res:* Lung immunobiology, especially identification and isolation of lung cell types through cell-type specific antigens. *Mailing Add:* Dept Anat Va Commonwealth Univ 1101 E Marshall St MCV Sta PO Box 709 Richmond VA 23298

SZALAY, JEANNE, b Jan 17, 38; US citizen; c 2. TUMOR BIOLOGY. *Educ:* Wash Sq Col Arts & Sci, BA, 59; Columbia Univ, PhD(biol), 66. *Prof Exp:* Fel cell biol & cytol, Col Physicians & Surgeons, Columbia Univ, 66-67; fel, Albert Einstein Col Med, 67-69, res asst cell biol & cytol, 69-70, instr, Dept Anat, 70-72; asst prof 72-77, ASSOC PROF BIOL, QUEENS COL, NY, 77- *Mem:* Asn Res Vision & Ophthalmol; Am Assoc Cancer Res; Clin Immunol Soc. *Res:* Tumor biology; tumor dormancy cellular interactions during metastasis; immunology. *Mailing Add:* Dept Biol Queens Col 65-30 Kissena Blvd Flushing NY 11367-0904

SZALDA, DAVID JOSEPH, b Buffalo, NY, May 25, 50; m 74; c 2. BIOINORGANIC CHEMISTRY. *Educ:* Manhattan Col, BS, 72; Johns Hopkins Univ, MA, 74, PhD(inorg chem), 76. *Prof Exp:* Fel bioinorg chem, Columbia Univ, 76-78; PROF CHEM, BARUCH COL, 78- *Concurrent Pos:* Res collabr, Brookhaven Nat Lab, 81- *Mem:* Sigma Xi; Am Chem Soc. *Res:* Electron transfer reactions; mental nucleic acid interactions; structural chemistry: x-ray crystallography. *Mailing Add:* Dept Natural Sci Baruch Col Box 502 17 Lexington Ave New York NY 10010

SZALECKI, WOJCIECH JOSEF, b Kutno, Poland, Mar 18, 35; m 77; c 2. NEW METHODS OF SYNTHESIS. *Educ:* Univ Lodz, Poland, MS, 59, PhD(org chem), 68. *Prof Exp:* Prof asst org chem, Univ Lodz, Poland, 59-69, adj, 70-79; res assoc org chem, Univ Colo, Boulder, 79-82, prof res assoc biochem, 82-83; res chemist, 83-88, SR CHEMIST, ORG SYNTHESIS,

MOLECULAR PROBES, INC, 88- *Concurrent Pos:* Res fel, Wayne State Univ, Detroit, 69-70; abstractor, Chem Abstracts Serv, 70-82. *Mem:* Am Chem Soc; fel Am Inst Chemists. *Res:* Synthesis of physiological active compounds and reagents for biochemical research; structure determination. *Mailing Add:* 2485 Park Forest Dr Eugene OR 97405

SZALEWICZ, KRZYSZTOF, b Gdansk, Poland, Jan 19, 50; c 3. THEORY OF INTERMOLECULAR INTERACTION & MUON CATALYZED FUSION. *Educ:* Univ Warsaw, MS, 73, PhD(quantum chem), 77, DSc, 84. *Prof Exp:* Asst quantum chem, Dept Chem, Univ Warsaw, Poland, 73-77, adj, 77-85; assoc res scientist, Dept Physics, Univ Fla, 85-88; asst prof, 88-90, ASSOC PROF PHYSICS, DEPT PHYSICS, UNIV DEL, 90- *Concurrent Pos:* Adj res asst prof, Dept Physics, Univ Fla, 80-82; vis scientist, Univ Cologne, 82-84, Univ Uppsala, 90. *Mem:* Am Phys Soc. *Res:* Theoretical, atomic and molecular physics; quantum chemistry; muon catalyzed fusion; intermolecular interactions; electron correlation; spectroscopy of diatomic molecules; theory of beta decay; many-body methods. *Mailing Add:* Dept Physics & Astron Univ Del Newark DE 19716

SZAMOSI, GEZA, b Budapest, Hungary, Mar 23, 22; Can citizen; m 44; c 2. THEORETICAL PHYSICS. *Educ:* Univ Budapest, PhD(physics), 47; Hungary Acad Sci, DSc, 56. *Prof Exp:* Assoc prof, Eotvos Univ, Budapest, 47-55, prof, 55-56; prof, Israel Inst Technol, Israel, 57-61; consult, Nat Lab Frascati, Italy, 61-64; prof, 64-88, EMER PROF THEORET PHYSICS, UNIV WINSOR, 88-; PRIN SCI COL, CONCORDIA UNIV, MONTREAL, QUE, 88- *Concurrent Pos:* Nuffield fel, 69-70; A Alberman vis prof, Israel Inst Technol, 70-71; Can-France fel, Inst Henri Poincare, Paris, 74 & 80-81; Can-Japan fel, Dept Physics, Nagoya Univ, Japan, 81. *Honors & Awards:* Schmid Prize the Hungarian Phys Soc, 59. *Mem:* Can Asn Physicists. *Res:* Special theory of relativity and its application to classical and quantum mechanics, statistical mechanics, plasma physics, astrophysics and to classical theories of elementary particles; history and philosophy of science; author or coauthor of over 60 publications. *Mailing Add:* Sci Col Concordia Univ Montreal PQ H3G 1M8 Can

SZAMOSI, JANOS, b Gyor, Hungary, Oct 28, 55; m 81; c 2. PHYSICAL CHEMISTRY, POLYMER CHEMISTRY. *Educ:* Tech Univ Budapest, BS, 78; State Univ NY Stony Brook, MS, 81; Univ Tex, Arlington, PhD(math sci & chem), 83. *Prof Exp:* Fel, Stanford Univ, 84-85; ASST PROF CHEM, WESTERN ILL UNIV, 85- *Mem:* Am Chem Soc. *Res:* Polymer chemistry; stochastic processes; mathematical modeling; materials science; chemical kinetics. *Mailing Add:* Union Camp PO Box 3301 Princeton NJ 08543-3301

SZANISZLO, PAUL JOSEPH, b Medina, Ohio, June 9, 39; m 60; c 2. MICROBIOLOGY, MYCOLOGY. *Educ:* Ohio Wesleyan Univ, BA, 61; Univ NC, MA, 64, PhD(bot), 67. *Prof Exp:* Res fel, Harvard Univ, 67-68; from asst prof to assoc prof, 68-80, chmn, Div Biol Sci, 76-82, PROF MICROBIOL, UNIV TEX, AUSTIN, 80- *Concurrent Pos:* Guest res microbiol, Sect Enzymes & Cellular Biochem, Nat Inst Arthritis, Diabetes & Digestive & Kidney Dis, NIH, Bethesda, Md, 83; ed, Biol Metals, 87-, Mycologia, 91- *Honors & Awards:* Coker Award, 67. *Mem:* AAAS; Am Soc Microbiol; Int Soc Human & Animal Mycology; Mycol Soc Am; Sigma Xi. *Res:* Growth, development and differentiation in fungi; iron transport in bacteria, fungi and plants; importance of siderophores in plant nutrition. *Mailing Add:* Dept Microbiol Exp Sci Bldg Univ Tex Austin TX 78712-1095

SZANTO, JOSEPH, b Marcali, Hungary, Nov 4, 31; m 57; c 4. VETERINARY PARASITOLOGY. *Educ:* Col Vet Med, Budapest, Dr Vet, 55; Univ Ill, DVM, 61. *Prof Exp:* Vet practice, Hungary, 55-56; res asst vet parasitol, Col Vet Med, Univ Ill, 58-61; asst prof vet path, Univ Ky, 61-62; asst parasitologist, Chemagro Corp, Mo, 62-63; sr res parasitologist, Ciba Pharmaceut Co, 63-69; sr res parasitologist, E R Squibb & Sons, Inc, 69-74, dir vet clin res & develop, 74-80; MGR, ANIMAL HEALTH RES & DEVELOP LABS & RES FARM, SDS BIOTECH CORP, 80-; AT FERMENTA ANIMAL HEALTH. *Mem:* Am Vet Med Asn; Am Asn Vet Parasitol; Am Soc Parasitol; World Asn Adv Vet Parasitol; Am Asn Swine Practitioners; Am Asn Lab Animal Sci. *Res:* Animal health research and drug development. *Mailing Add:* Fermenta Animal Health 7410 NW Tiffany Springs Pkwy Kansas City MI 64190

SZAP, PETER CHARLES, b New York, NY, Aug 20, 29; m 57; c 3. ANALYTICAL CHEMISTRY. *Educ:* Queens Col, NY, BS, 51; Fordham Univ, MS, 53. *Prof Exp:* Res chemist, Lever Bros Res & Develop Co, NJ, 53-59; anal res chemist, Toms River Chem Corp, 59-67, sect leader anal qual control, Ciba-Geigy Corp, 67-77; mgr anal servs, Ciba-Geigy Corp, 77-86, sec sch chem teacher, 86-87; consult, 87-89; LECTR, GEORGIAN CT COL, NJ, 89- *Concurrent Pos:* Chem Abstractor, 51-81. *Mem:* Am Chem Soc; Sigma Xi; Soc Appl Spectros. *Res:* Analytical methods development in detergents, dyestuffs, resins and raw materials; analysis and applications of optical brighteners, especially infrared, ultraviolet and visible absorption spectroscopy; chromatography; liquid chromatography techniques. *Mailing Add:* 333 Killarney Dr Toms River NJ 08753-3516

SZARA, STEPHEN ISTVAN, b Pestujhely, Hungary, Mar 21, 23; nat US; div; c 1. PSYCHOPHARMACOLOGY. *Educ:* Pazmany Peter Univ, Hungary, DSc(chem), 50; Med Univ Budapest, MD, 51. *Prof Exp:* Sci asst, Microbiol Inst, Univ Budapest, 49-50, asst prof, Biochem Inst, 50-53, chief, Biochem Lab, State Ment Inst, 53-56; vis scientist, Psychiat Clin Berlin, 57; vis scientist, Nat Inst Ment Health, 57-60, chief sect psychopharmacol, Lab Clin Psychopharmacol, Spec Ment Health Res, 60-71, chief clin studies sect, Ctr Studies Narcotic & Drug Abuse, Nat Inst Mental Health, 71-74; CHIEF BIOMED RES BR, DIV RES, NAT INST DRUG ABUSE, 75- *Concurrent Pos:* Assoc clin prof, George Washington Univ. *Mem:* AAAS; Int Col Neuropsychopharmacol; fel Am Col Neuropsychopharmacol; Am Soc Pharmacol & Exp Therapeut; IEEE Computer Soc. *Res:* Metabolism of psychotropic drugs, especially tryptamine derivatives; correlation between metabolism and psychotropic activity of drugs. *Mailing Add:* Nat Inst Drug Abuse 5600 Fishers Lane #10A-31 Rockville MD 20857

SZAREK, STANLEY RICHARD, b Visalia, Calif, Nov 14, 47; m 77. PHYSIOLOGICAL ECOLOGY. *Educ:* Calif State Univ, Pomona, BS, 69; Univ Calif, Riverside, PhD(biol), 74. *Prof Exp:* Asst prof, 74-80, ASSOC PROF BOT, ARIZ STATE UNIV, 80- *Concurrent Pos:* Consult, US Biol Prog Desert Biomed, 75-77. *Mem:* Am Soc Plant Physiologists; Ecol Soc Am. *Res:* Physiological ecology of desert plants, emphasizing photosynthetic carbon metabolism and high temperature stress physiology. *Mailing Add:* Dept Bot Ariz State Univ Tempe AZ 85287-1601

SZAREK, WALTER ANTHONY, b St Catharines, Ont, Apr 19, 38; Can citizen. ORGANIC CHEMISTRY, CARBOHYDRATE CHEMISTRY. *Educ:* McMaster Univ, BSc, 60, MSc, 62; Queen's Univ, Ont, PhD(org chem), 64. *Prof Exp:* Res fel chem, Ohio State Univ, 64-65; asst prof biochem, Rutgers Univ, 65-67; asst prof org chem, Queen's Univ, 67-71, assoc prof, 71-76, dir, Carbohydrate Res Inst, 76-85, PROF ORG CHEM, QUEEN'S UNIV, ONT, 76- *Concurrent Pos:* Premier, Ont Coun Technol Fund; fel, Chem Inst Can, 77; chmn, Div Carbohydrate Chem, Am Chem Soc, 82-83. *Honors & Awards:* Claude S Hudson Award, Am Chem Soc, 89. *Mem:* Am Chem Soc; fel Chem Inst Can; Royal Soc Chem; NY Acad Sci; AAAS. *Res:* Structure and synthesis of carbohydrates and carbohydrate-containing antibiotics; biochemical aspects; synthesis and chemical modification of nucleosides; conformational and mechanistic studies of carbohydrate reactions; sweeteners; targeted delivery of drugs; natural products synthesis; bacterial polysaccharides; vaccine development. *Mailing Add:* Dept Chem Queen's Univ Kingston ON K7L 3N6 Can

SZARKA, LASZLO JOSEPH, b Szekesfehervar, Hungary, Sept 6, 35; US citizen; m 63; c 2. BIOENGINEERING. *Educ:* Univ Sci, Budapest, Hungary, MS, 61, PhD(phys chem), 67. *Prof Exp:* Asst res investr phys chem, Chinoin Pharmaceut Plant, Hungary, 61-63; sect head bioeng, Res Inst Pharmaceut Chem, Hungary, 64-72; head chemist, Pharmaceut Ctr, Hungary, 72-73; res & develop biochem, Blackman Labs Inc, 74; SECT HEAD, E R SQUIBB & SONS, 74- *Mem:* Am Chem Soc; Am Inst Chem Eng. *Res:* Antibioticum and enzyme fermentation technology; batch, fed batch and continuous cultivation; transport phenomena, process control and reactor designing; optimization and scale-up problems; molecule biotransformations, steroids and hydrocarbons. *Mailing Add:* 5 Wellington Rd E Brunswick NJ 08816

SZASZ, THOMAS STEPHEN, b Budapest, Hungary, Apr 15, 20; US citizen; div; c 2. PSYCHIATRY. *Educ:* Univ Cincinnati, AB, 41, MD, 44; Am Bd Psychiat & Neurol, dipl psychiat, 51. *Hon Degrees:* DSc, Allegheny Col, 75, Univ Francisco Marroquin, Guatemala, 79. *Prof Exp:* Intern med, Boston City Hosp, 44-45; resident psychiat, Clinics, Univ Chicago, 46-48; mem staff, Chicago Inst Psychoanal, 51-56; prof, 56-90, EMER PROF PSYCHIAT, STATE UNIV NY HEALTH SCI CTR, 90- *Concurrent Pos:* Vis prof psychiat, Sch Med, Marquette Univ, 68; consult, Comt Ment Hyg, NY State Bar Asn, Judicial Conf Comt, Judicial Conf DC Circuit, Comt Laws Pertaining Ment Disorders & Res Adv Panel, Inst Study Drug Addiction; adv ed, J Forensic Psychol; mem bd consult, Psychoanal Rev, mem adv comt, Living Libraries, Inc. *Honors & Awards:* Holmes-Munsterberg Award Forensic Psychol, Int Acad Forensic Psychol, 69; Humanist of Year, Am Humanist Asn, 73; Am Inst Pub Serv Distinguished Serv Award, 74. *Mem:* Fel Royal Soc Health; fel Am Psychiat Asn; Am Psychoanal Asn. *Res:* Epistemology of the behavioral sciences; history of psychiatry; psychiatry and law. *Mailing Add:* Dept Psychiat State Univ NY Health Sci Ctr Syracuse NY 13210

SZCZARBA, ROBERT HENRY, b Dearborn, Mich, Nov 27, 32; m 55; c 2. TOPOLOGY. *Educ:* Univ Mich, BS, 55; Univ Chicago, MS, 56, PhD(math), 60. *Prof Exp:* Off Naval Res res assoc math, 60-61, from asst prof to assoc prof, 61-74, PROF MATH, YALE UNIV, 74- *Concurrent Pos:* NSF fel, Inst Advan Study, 64-65. *Mem:* Am Math Soc; Math Asn Am. *Res:* Algebraic and differential topology and geometry of differentiable manifolds. *Mailing Add:* Dept Math Yale Univ Box 2155 Yale Sta New Haven CT 06520

SZCZECH, GEORGE MARION, US citizen; m; c 3. VETERINARY PATHOLOGY, REPRODUCTIVE TOXICOLOGY. *Educ:* Univ Minn, BS, 64, DVM, 66; Purdue Univ, PhD(vet path), 74; Am Bd Toxicol, dipl; Am Col Vet Pathologists, dipl. *Prof Exp:* Resident vet med, Colo State Univ, 66-67; vet pract, small animals, 67-68; sr scientist toxicol, Res Ctr, Mead Johnson Co, 68-70; res fel vet path, Purdue Univ, 70-74; res scientist path, Upjohn Co, 74-78; SR TOXICOL PATHOLOGIST, BURROUGHS WELLCOME CO, 78- *Concurrent Pos:* Assoc ed, Fund Appl Toxicol, 80-85; consult, Nat Toxicol Prog, 80- *Mem:* Am Vet Med Asn; Soc Toxicology; Teratology Soc. *Res:* Toxicologic pathology; clinical pathology and reproductive toxicology. *Mailing Add:* 643 Rock Creek Road Chapel Hill NC 27514

SZCZEPANIAK, KRYSTYNA, b Lwow, Poland, Jun 19, 34; m 85. MOLECULAR SPECTROSCOPY, LOW TEMPERATURE MATRIX-ISOLATION. *Educ:* Warsaw Univ, Poland, BS, 58, PhD(physics), 65. *Prof Exp:* Asst, physics, Tech Univ, Warsaw, 58-63; adjunct prof phys, Warsaw Univ, 63-78; docent, Inst Physics, Polish Acad Sci, 78-85; postdoc fel phys chem, 86-88, RES SCIENTIST, DEPT CHEM, UNIV FLA, GAINESVILLE, 88- *Concurrent Pos:* Res asst, dept physics, Moscow Univ, USSR, 62-63; Ford Found fel, 68-69; postdoc fel, Nat Res Coun, Halifax, 68-69, dept chem, Univ Fla, Gainsville, 69; vis lectr, dept chem, Salford Univ, Eng, 72-73; vis scientist, dept chem, Marburg Univ, W Ger, 82. *Mem:* Polish Phys Soc (treas, 78-80). *Res:* Structure and molecular interactions of nucleic acid bases and their analogues and related model compounds by means of vibrational and electronic spectroscopy; low temperature matrix isolation spectroscopy and by theoretical, quantum mechanical methods. *Mailing Add:* Dept Chem Univ Fla Gainesville FL 32611

SZCZEPANSKI, MAREK MICHAL, b Warsaw, Poland, June 8, 41; Can & US & Poland citizen; m 80; c 2. EXPERIMENTAL SURGERY & GASTROENTEROLOGY, LABORATORY ANIMAL SCIENCE. *Educ:* Agr Univ Warsaw, DVM, 67; Univ Toronto, DVPH(pub health), 72. *Prof Exp:* Asst surg, Agr Univ Warsaw, 67-68; res asst, McGill Univ, 68-70; res fel, Univ Toronto, 70-72; assoc vet small animal pract, Lakeshore Vet Hosp, Toronto, 72-73; staff vet, Toronto Humane Soc, 73-75; asst prof physiol & dir animal care, Fac Med, Mem Univ Nfld, 75-80; ASST PROF PHYSIOL & DIR, COMP MED UNIT NORTHEASTERN OHIO UNIV COL MED, ROOTSTOWN, 80- *Mem:* Am Asn Lab Animal Sci; Can Vet Med Asn; Can Asn Lab Animal Sci. *Res:* Pathophysiology of the esophagus; achalasia; experimental esophagitis; esophageal replacement by a gastric tube; control mechanisms governing esophageal motor activity; anti-reflux mechanism in hiatus hernia. *Mailing Add:* Comp Med Unit Northeastern Ohio Univs Col Med 4209 State Rte 44 Rootstown OH 44272

SZCZESNIAK, ALINA SURMACKA, b Warsaw, Poland, July 8, 25; US citizen; m 49; c 1. FOOD RHEOLOGY, FOOD TEXTURE. *Educ:* Bryn Mawr Col, AB, 48; Mass Inst Technol, ScD(food sci), 52. *Prof Exp:* Prin scientist, Gen Foods Corp, 52-86; RETIRED. *Concurrent Pos:* Ed-in-chief, J Texture Studies, 68-79; lectr & author. *Honors & Awards:* Scientist of the year, NY Sect of Inst Food Technologists, 82; Nicholas Appert Medal, Inst Food Techologist, 85, Fred W Tanner Mem Lectr, 90; Scott Blair Award, Am Asn Cereal Chemists, 89. *Mem:* NY Acad Sci; fel Inst Food Technologists; Soc Rheology; Polish Inst Arts & Sci Am; Sigma Xi. *Res:* Instrumental and sensory methods of food texture measurement; meaning of texture to the consumer; texturization and relationship to structure, microscopic and molecular. *Mailing Add:* 22 Wilson Block Mt Vernon NY 10552-1113

SZCZESNIAK, RAYMOND ALBIN, b Buffalo, NY, Nov 28, 40; wid; c 4. MEDICINAL CHEMISTRY, NUCLEAR MEDICINE. *Educ:* Fordham Univ, BS, 63; Univ Mich, Ann Arbor, MS, 65, PhD(med chem), 68. *Prof Exp:* Sr res scientist, E R Squibb & Sons, Inc, New Brunswick, 68-69, res chemist, 69-72, sr res chemist radiopharmaceut, 72-73; radioimmunoassay specialist, Nuclear-Med Labs, Inc, 73-77; PRES, PROF CONSULTS, INC, 77- *Mem:* Soc Nuclear Med; Am Chem Soc. *Res:* Radioimmunoassay; radiomedicinal synthesis. *Mailing Add:* 4042 Mendenhall Dr Dallas TX 75244-7240

SZE, HEVEN, b The Hague, Netherlands, Oct 22, 47; US citizen; m 74. PLANT PHYSIOLOGY, BIOCHEMISTRY. *Educ:* Nat Taiwan Univ, BS, 68; Univ Calif, Davis, MS, 70; Purdue Univ, PhD(plant physiol), 75. *Prof Exp:* Res fel biophys, Harvard Med Sch, 75-78; res assoc, 78-79, asst prof, Univ Kans, 79-82; asst prof, 82-85, ASSOC PROF DEPT BOT, UNIV MD, 85- *Mem:* Am Soc Plant Physiologists; AAAS; Am Soc Cell Biol; Am Soc Biochem & Molecular Biol. *Res:* Membrane structure and function; mechanism of solute transport; biophysics; Ht-translocating ATPases, Ca2t transport. *Mailing Add:* Dept Bot Univ Md College Park MD 20742

SZE, MORGAN CHUAN-YUAN, b Tientsin, China, May 27, 17; nat US; m 45; c 3. CHEMICAL ENGINEERING. *Educ:* Mass Inst Technol, SB, 39, ScD(chem eng), 41. *Prof Exp:* Chem engr, Universal Trading Corp, 41-42; process engr, Belle Works, E I du Pont de Nemours & Co, Inc, 42-44, Hydrocarbon Res, Inc, NY, 44-45 & Foster Wheeler Corp, 45-47; sr process engr, Hydrocarbon Res, Inc, 47-53, dir process eng dept, 53-59, tech dir, 59-61; exec staff engr, Lummus Co, 61-64, mgr eng develop ctr, 64-71, vpres res & develop, 71-80; vpres, adv tech group, Signal Comp Inc, 80-85; RETIRED. *Mem:* Nat Acad Eng; AAAS; Am Inst Chem Engrs; NY Acad Sci; Sigma Xi; Am Chem Soc. *Res:* Nitrogen fixation; petroleum refining and processing; petrochemicals; cryogenics. *Mailing Add:* 4 Regina Rd Portsmouth NH 03801

SZE, PAUL YI LING, b Shanghai, China. BIOCHEMISTRY, NEUROBIOLOGY. *Educ:* Nat Taiwan Univ, BS, 60; Duquesne Univ, MS, 62; Univ Wis, MS, 64; Univ Chicago, PhD(biochem), 69. *Prof Exp:* From asst prof to assoc prof, 69-80, Prof Biobehav Sci, Univ Conn, 80-; AT DEPT PHARMACOL, CHICAGO MED SCH. *Concurrent Pos:* Mem, Neurol A Study Sect, NIH, 80-84, Gen Res Support Rev Comt, 85-89, Res Scientist Develop Rev Comt, NIMH, 90- *Mem:* AAAS; Am Soc Neurochem; Soc Neurosci; Int Soc Neurochem. *Res:* Biochemistry of neurotransmitters; neurochemical aspects of alcoholism; neurobiological actions of steroid hormones. *Mailing Add:* Dept of Pharmacol Chicago Med Sch 3333 Green Bay Rd N Chicago IL 60064

SZE, PHILIP, b Washington, DC, Dec 3, 45. PHYCOLOGY, AQUATIC BIOLOGY. *Educ:* Dickinson Col, BS, 68; Cornell Univ, PhD(phycol), 72. *Prof Exp:* Vis asst prof biol, State Univ NY, Buffalo, 72-73; asst prof phycol, Stockton State Col, 73-74; asst prof biol, State Univ NY, Buffalo, 74-77; asst prof, 77-83, ASSOC PROF BIOL GEORGETOWN UNIV, 83- *Concurrent Pos:* Mem fac, Shoals Marine Lab, 74- *Mem:* Phycol Soc Am; Int Phycol Soc; Brit Phycol Soc. *Res:* Ecology of algae, incluuding studies of phytoplankton in eutrophic lakes and rivers, intertidal macroalgal communities and filamentous algae in streams. *Mailing Add:* Dept Biol Georgetown Univ Washington DC 20057

SZE, ROBERT CHIA-TING, b Kowloon, China, Aug 27, 41; US citizen; m 70; c 1. LASERS, NON-LINEAR OPTICS. *Educ:* Cornell Univ, BEE, 64; Yale Univ, MS, 66, PhD(appl physics), 70. *Prof Exp:* Fel, Yale Univ, 70-71, State Univ NY Stony Brook, 71-73; res assoc, Yale Univ, 73-74; MEM STAFF, LOS ALAMOS NAT LAB, 74- *Concurrent Pos:* Consult, Tachisto, Inc, 80-81; consult & co-founder, Laser Tech, Inc, 81- *Mem:* Sr mem Inst Elec & Electronics Engrs; Am Phys Soc; Optical Soc Am; Sigma Xi. *Res:* Physics of lasers and the generation of coherent light sources; physics of noble gas ion lasers; high energy rare-gas halide excimer lasers; generation of coherent wavelengths via Raman stokes and anti-stokes scattering; development of compact high repetition rate excimer and carbon dioxide lasers. *Mailing Add:* 1042 Stagecoach Rd Santa Fe NM 87501

SZE, TSUNG WEI, b Shanghai, China, Sept 13, 22; nat US; m 52; c 3. ELECTRICAL ENGINEERING. *Educ:* Univ Mo, BS, 48; Purdue Univ, MS, 50; Northwestern Univ, PhD(elect eng), 54. *Prof Exp:* Asst elec eng, Northwestern Univ, 50-52, instr, 52-54; from asst prof to assoc prof, 54-60, Fessenden prof, 57-62, Westinghouse prof, 62-65, assoc dean eng, 70-77, PROF ENG, UNIV PITTSBURGH, 60- *Concurrent Pos:* Consult, Univac Div, Sperry Rand Corp, 56-58, Westinghouse Elec Corp, 59-, MPC Corp, 69-74 & Mellon Inst, 78-; adj prof, Jiaotong Univ, People's Repub China, 79- *Mem:* Am Soc Eng Educ; Asn Comput Mach; Inst Elec & Electronics Engrs. *Res:* Logical design and switching theory; digital computer designs; digital image processing. *Mailing Add:* Sch Eng Elec BENE DUM 340 Univ Pittsburgh 4200 5th Ave Pittsburgh PA 15260

SZE, YU-KEUNG, b Shanghai, China, Sept 3, 44; Can citizen; m 71; c 2. RADIATION CHEMISTRY, PROCESS CHEMISTRY. *Educ:* Chinese Univ Hong Kong, BSc, 66; Univ Waterloo, MSc, 70, PhD(solution chem), 74. *Prof Exp:* Teacher chem, South Sea Col, Hong Kong, 66-68; asst phys chem, Univ Waterloo, 68-73; fel solution chem, Environ Can, 73-75; res assoc, Univ Waterloo, 76-77; RES OFFICER CHEM, ATOMIC ENERGY CAN LTD, 78- *Mem:* Chem Inst Can. *Res:* Process development and optimization in spent fuel reprocessing; industrial applications of ionizing radiations; treatment of industrial effluents; thermodynamics and kinetics of the separation and recovery process. *Mailing Add:* Radiation Applins Res Br Sta 80 Atomic Energy of Can Ltd Pinawa MB R0E 1L0 Can

SZEBEHELY, VICTOR, b Budapest, Hungary, Aug 10, 21; nat US; m 70; c 1. ASTRONOMY, SPACE SCIENCES. *Educ:* Budapest Tech Univ, DSc, 45. *Prof Exp:* Asst prof appl math, Budapest Tech Univ, 43-47; res asst, Pa State Univ, 47-48; assoc prof appl mech, Va Polytech Inst, 48-51; head ship dynamics br, David Taylor Model Basin, US Dept Navy, 51-57; mgr space mech, Missiles & Space Div, Gen Elec Co, 57-62; assoc prof astron, Yale Univ, 62-68; prof aerospace eng & eng mech, 68-76, chmn dept, 76-81, R B CURRAN CHAIR ENG, UNIV TEX, AUSTIN, 82- *Concurrent Pos:* Lectr, McGill Univ, 48 & Univ Toronto, 49; prof lectr, Univ Md & George Washington Univ, 51-57; consult, NASA, Gen Elec Co, Int Tel & Tel Co, Bell Tel Labs & Inst Defense Anal, Tracor, US Air Force Space Command, & Lawrence Berkeley Lab. *Honors & Awards:* Sci Res Soc Award, Sigma Xi, 51; D Brouwer Award, Am Astron Soc, 78 & Am Astronaut Soc, 82; Fulbright lectr, 86 & 90. *Mem:* Nat Acad Eng; Int Astron Union; Am Astron Soc; Am Inst Aeronaut & Astronaut. *Res:* Celestial mechanics; problem of three and n bodies; applied mechanics; analytical dynamics; continuum mechanics; applied mathematics; matrix and tensor analysis. *Mailing Add:* Dept Aerospace Eng Univ Tex Austin TX 78712

SZEBENYI, DOLETHA M E, b Arlington, Mass, Nov 27, 47; m 72; c 1. PROTEIN CRYSTALLOGRAPHY. *Educ:* Bryn Mawr Col, AB, 68; Univ Conn, PhD(phys chem), 72. *Prof Exp:* RES ASSOC BIOCHEM, CORNELL UNIV, 75- *Mem:* AAAS; Am Crystallog Asn. *Res:* Protein crystallography; structure of calcium binding proteins; polypeptide hormones; other macromolecules; use of synchroton radiation and Lane diffraction for macromolecular crystallography. *Mailing Add:* 209 Biotech Bldg Cornell Univ Ithaca NY 14853

SZEBENYI, EMIL, b Budapest, Hungary, June 9, 20; US citizen; m 44; c 3. COMPARATIVE ANATOMY, EMBRYOLOGY. *Educ:* Budapest Tech Univ, dipl, 42, Doctoratus (animal genetics & husb), 43. *Prof Exp:* Asst prof animal genetics, Univ Agr Sci Hungary, 52-53, Hungarian Acad Sci fel & adj, 53-56; from asst prof to assoc prof, 62-73, chmn, Dept Biol Sci, 71-80, PROF COMP ANAT, EMBRYOL & EVOLUTION, FAIRLEIGH DICKINSON UNIV, 73- *Concurrent Pos:* Pres, Alfacell Corp, 81-86; indepndent res, 86- *Mem:* AAAS; Sigma Xi; Am Soc Zoologists; Soc Develop Biol; Int Abstracts Biol Sci. *Res:* Anatomy; human embryology; experimental morphogenesis; oncology. *Mailing Add:* Reading PO Box 111 Manitego Bay St Janies Jamaica West Indies

SZEDON, JOHN R(OBERT), b Elrama, Pa, May 26, 37; m 72. ELECTRICAL ENGINEERING, SOLID STATE PHYSICS. *Educ:* Carnegie-Mellon Univ, BS, 59, MS, 60, PhD(elec eng), 62. *Prof Exp:* Sr engr, 62-67, sr res scientist, 67-68, MGR SURFACE DEVICE STUDIES, WESTINGHOUSE ELEC CORP RES LABS, 68- *Concurrent Pos:* Lectr, Carnegie-Mellon Univ, 67-68. *Honors & Awards:* Thomas D Callinan Award, Electrochem Soc, 73. *Mem:* Inst Elec & Electronics Engrs; Electrochem Soc. *Res:* Semiconductor device physics; surface controlled semiconductor devices; ion migration in insulators; injection-trapping in metal-insular semiconductor structures; ionizing radiation effects in dielectrics; p-N junction; heterojunction and MIS solar cells. *Mailing Add:* Hetero Junction & Surface Devices Sect Westinghouse Elec Corp R&D Ctr 501-2C11 Pittsburgh PA 15235

SZEGO, CLARA MARIAN, b Budapest, Hungary, Mar 23, 16; US citizen; m 43. CELL BIOLOGY. *Educ:* Hunter Col, BA, 37; Univ Minn, MS, 39, PhD(physiol chem), 42. *Prof Exp:* Asst physiol chem, Sch Med, Univ Minn, 40-42, asst physiol, 42, instr, 42-44; fel cancer, Minn Med Found, 44; assoc chemist, Off Sci Res & Develop, Nat Bur Stand, 44-45; res assoc, Worcester Found Exp Biol, 45-47; res instr physiol chem, Yale Univ, 47-48; asst clin prof biophys, 48-49, from asst prof to prof, 49-60, PROF BIOL, UNIV CALIF, LOS ANGELES, 60- *Concurrent Pos:* Guggenheim fel, 56; mem, Univ Calif Molecular Biol Inst, 69-, Jonsson Comprehensive Cancer Ctr, 76- *Honors & Awards:* Ciba Award, Endocrine Soc, 53; Gregory Pincus Mem Medallion, Worcester Found Exp Biol, 74. *Mem:* Fel AAAS; Am Physiol Soc; Am Soc Zoologists; Endocrine Soc; Am Soc Cell Biol; Biochem Soc; Soc Endocrinol; Int Soc Res Reproduction; Europ Soc Comp Endocrinol. *Res:* Molecular mechanisms of endocrine regulation. *Mailing Add:* Dept of Biol Univ of Calif Los Angeles CA 90024-1606

SZEGO, GEORGE C(HARLES), b Budapest, Hungary, Aug 10, 19; US citizen; c 5. NAVIGATION & PILOTING, FORENSIC INVESTIGATIONS. *Educ:* Univ Denver, BS, 47; Univ Wash, MS, 50, PhD(chem eng), 56. *Prof Exp:* prof & chmn, Dept Chem Eng, Seattle Univ, 47-56; mgr space propulsion & power, Gen Elec Co, 56-59; sr staff mem, TRW Space Technol Labs, 57-61 & Inst Defense Anal, 61-70; pres & chmn, Intertechnol Solar Corp, 70-81; PRES, GEORGE C SZEGO & ASSOC, INC, 80- & BAY TECHNOLOGIES, INC, 85- *Concurrent Pos:* Consult, var orgn, 51- & White House Off Sci & Technol, 65-68; founding chmn, Int Soc

Energy Conversion Eng Conf, 65-; mem, White House Comt Energy, 78-79; distinguished adj prof, Howard Univ, 90. *Mem:* Fel Am Inst Chem Engrs; Solar Energy Industs Asn; Solar Energy Res Found. *Res:* Clean, renewable carbon dioxide, neutral economic energy of its conversion. *Mailing Add:* PO Box 4070 Annapolis MD 21403

SZEGO, PETER A, b Berlin, Ger, July 18, 25; nat US. SOCIAL IMPLICATIONS OF TECHNOLOGY. *Educ:* Stanford Univ, BS, 47. *Prof Exp:* Asst prof mech eng, Rice Univ, 51-54 & Univ Santa Clara, 56-63; mgr mech sect, advan technol div, Ampex Corp, Redwood City, 63-76; exec asst, bd supvr, Santa Clara County, Calif, 76-82; chief consult, State Senate Comt Elections, 82-85, State Senate Comt Pub Employ & Retirement, 85-87, CHIEF CONSULT, STATE SENATE COMT NATURAL RESOURCES & WILDLIFE, 87- *Concurrent Pos:* Lectr mech eng, Univ Santa Clara, 70-82. *Mem:* Am Soc Mech Engrs; Soc Indust & Appl Math; Asn Comput Mach; Inst Elec & Electronics Engrs; Am Math Soc. *Res:* Scientific programming; applied mathematics; engineering mechanics; science policy. *Mailing Add:* 75 Glen Eyrie Ave Apt 19 San Jose CA 95125

SZEKELY, ANDREW GEZA, b Temesvar, Rumania, Apr 15, 25; US citizen; m 51; c 6. PHYSICAL CHEMISTRY. *Educ:* Eotvos Lorand Univ, Budapest, dipl chem, 50. *Prof Exp:* Asst electrochem, Physico-Chem Inst, Eotvos Lorand Univ, Budapest, 49-51; res chemist, High Pressure Res Inst, Budapest, 51-56, dept head petrol refining, 56-57; res chemist, Tonawanda Res Lab, Linde Div, 58-62, sr res chemist, 62-66, develop assoc process metall, Newark Develop Labs, NJ, 67-70, sr develop assoc, gas prod develop, 70-77, CORP DEVELOP FEL, LINDE LABS, TARRYTOWN TECH CTR, UNION CARBIDE CORP, 78 - *Mem:* Metall Soc. *Res:* Process metallurgy; thermodynamics and kinetics. *Mailing Add:* 2379 Claire Ct Yorktown Heights NY 10598

SZEKELY, IVAN J, b Budapest, Hungary, Aug 13, 19; nat US; m. RESOURCE MANAGEMENT. *Educ:* Pazmany Peter Univ, Hungary, PhD, 43, MS, 44. *Prof Exp:* Asst prof chem, Bowling Green State Univ, 47-48; prof pharm, Univ Calif, 48-51; dir res & prod develop, Barnes-Hind Pharmaceut, Inc, 51-57, from exec vpres to pres, 57-69; PRES, PHARMACEUT RES INT, 71- *Concurrent Pos:* Pharmaceut mgt consult. *Mem:* Am Chem Soc; Am Pharmaceut Asn; NY Acad Sci. *Res:* Formulation of pharmaceutical dosage forms; ophthalmic solutions; physical-chemical of properties of contact lens materials, especially contact lens solutions. *Mailing Add:* 13643 Wildcrest Dr Los Altos CA 94022

SZEKELY, JOSEPH GEORGE, b Cleveland, Ohio, May 7, 40; m 67; c 2. BIOPHYSICS. *Educ:* Case Western Reserve Univ, BS, 62; State Univ NY Buffalo, PhD(physics), 67. *Prof Exp:* USPHS fel, Univ Tex M D Anderson Hosp & Tumor Inst, 67-68; ASSOC RES OFFICER, ATOMIC ENERGY CAN LTD, 68- *Mem:* Can Soc Cell Biol; Can Micros Soc; Inst Food Technologists; Environ Mutagens Soc. *Res:* Genetic toxicology; cytology; food irradication. *Mailing Add:* Whiteshell Lab Atomic Energy Can Ltd Pinawa MB R0E 1L0 Can

SZEKELY, JULIAN, b Budapest, Hungary, Nov 23, 34; US citizen; m 63; c 5. PROCESS METALLURGY, CHEMICAL ENGINEERING. *Educ:* Imp Col, London, ACG, 59, DIC, 61. *Prof Exp:* Lectr metall, Imp Col, 62-66; from assoc prof to prof chem eng, State Univ NY, Buffalo, 66-74, dir, Ctr Process Metall, 70-76, leading prof, 74-76; PROF MAT ENG, MASS INST TECHNOL, 75- *Concurrent Pos:* Fel, Guggenheim Found, 75; consult, Alcoa, 76-, Gen Motors, 85-, Nuclear Regulatory Comn, 86-87; lectr extractive metall, Am Inst Mining & Metall Engrs, 87. *Honors & Awards:* Jr Moulton Medal, Brit Inst Chem Engrs, 64; Extractive Metall Sci Award, Am Inst Mining, Metall & Petrol Engrs, 73, Champion Mathewson Gold Medal, 73; Sir George Beilby Gold Medal & Prize, Brit Inst Metals, Soc Chem Indust & Royal Inst Chem, 73; Curtis McGraw Res Award, Am Soc Eng Educ, 74; Prof Progress Award, Am Inst chem Engrs, 74; Howe Mem Lectr, Iron & Steel Soc, 79; Charles H Jennings Mem Award, Am Welding Soc, 83; Nelson W Taylor Mem Lectr, Pa State Univ, 85; John R Lewis Distinguished Lectr, Univ Utah, 89; Yugawa Mem Lectr Iron & Steel Inst Japan, 90. *Mem:* Nat Acad Eng; fel Brit Inst Chem Engrs; Am Inst Mining, Metall & Petrol Engrs; Iron & Steel Inst Japan; Am Inst Chem Engrs; Brit Metal Soc. *Res:* Materials processing; mathematical and physical modelling of processing operations; energy environmental and societal aspects of materials processing operations; granted seven patents; author of 12 publications. *Mailing Add:* Dept Mat Sci & Eng Rm 4-117 Mass Inst Technol Cambridge MA 02139

SZELESS, ADORJAN GYULA, b Budapest, Hungary, Dec 7, 27; US citizen; m 53. MECHANICAL ENGINEERING, NAVAL ARCHITECTURE. *Educ:* Budapest Tech Univ, MS, 52; Va Polytech Univ, PhD(mech eng), 67. *Prof Exp:* Marine engr, Obudai Hajogyar, Hungary, 52-54; naval architect, Dunai Hajogyar, 54-56; plant engr, München-Dachauer Papierfabriken GmbH, Ger, 57-58; naval architect, Wasser-u Schiffahrtsdirektion, Ger, 58-59; marine appln engr, Cummins Engine Co, Ind, 59-62; asst prof, 62-72, ASSOC PROF MECH ENG, VA POLYTECH INST & STATE UNIV, 72- *Concurrent Pos:* NSF res initiation grant, 68-69; researcher, Naval Res Lab, 69-70; sci analyst, Ctr Naval Anal, Univ Rochester, Va, 69-71. *Mem:* Soc Naval Architects & Marine Engrs. *Res:* Ship hydrodynamics; propulsion; resistance; stability; fishtype propulsion by undulating plates. *Mailing Add:* Dept Mech Eng VA Polytech Inst Blacksburg VA 24061

SZENT-GYORGYI, ANDREW GABRIEL, b Budapest, Hungary, May 16, 24; nat US; m 47; c 3. BIOCHEMISTRY. *Educ:* Univ Budapest, MD, 47. *Prof Exp:* Instr, Univ Budapest, 46-47; fel, Neurophysiol Inst, Copenhagen, 48; mem inst muscle res, Marine Biol Lab, Woods Hole, 48-62, instr physiol, 53-58; prof biophys, Dartmouth Med Sch, 62-66; PROF BIOL, BRANDEIS UNIV, 66- *Concurrent Pos:* Am Heart Asn estab investr, 55-62; USPHS res career award, 62-66; head physiol course, Marine Biol Lab, Woods Hole, 67-71, trustee, 70-76 & 81- *Mem:* Soc Gen Physiol (pres, 70-71); fel AAAS; Am Soc Biol Chem; Biophys Soc (pres, 74-75). *Res:* Regulation of muscle contraction; structure of muscle proteins. *Mailing Add:* Dept Biol Brandeis Univ Waltham MA 02254

SZENTIRMAI, GEORGE, b Budapest, Hungary, Oct 8, 28; US citizen; m 52, 74; c 2. ELECTRICAL ENGINEERING, COMPUTER AIDED DESIGN. *Educ:* Budapest Tech Univ, dipl elec eng, 51 & Can Tech Sci, 55; Polytech Inst Brooklyn, PhD(elec eng), 63. *Prof Exp:* From asst prof to assoc prof elec eng, Budapest Tech Univ, 54-56; design engr, Standard Tel & Cables, Ltd, Eng, 57-58; sect leader elec eng, Tel Mfg Co, Ltd, Eng, 58-59; mem tech staff, Bell Tel Labs, Inc, 59-64, supvr elec eng, 64-68; prof elec eng, Cornell Univ, 68-75; staff scientist, Electronics Res Ctr, Rockwell Int Corp, 75-80; vpres adv develop, compact div, Comsat Gen Integrated Systs, 80-82; PRES, DGS ASSOCS, INC, 83- *Concurrent Pos:* Consult, Cent Phys Res Inst, Budapest, 55-56 & Bell Tel Labs, Inc, 68-75; adj prof elec sci & eng, Univ Calif, Los Angeles, 75-80. *Mem:* Fel Inst Elec & Electronics Engrs; Brit Inst Elec Engrs. *Res:* Synthesis and design of electrical filters and networks; computer-aided circuit and system analysis and design; microwave networks. *Mailing Add:* DGS Assocs 1353 Sarita Way Santa Clara CA 95051

SZENTIVANYI, ANDOR, b Miskolc, Hungary, May 4, 26; US citizen; m 48; c 2. PHARMACOLOGY, ALLERGY. *Educ:* Debrecen Univ Med, MD, 50. *Prof Exp:* Asst prof med, Univ Med Sch Budapest, 53-56; Rockefeller fel med, Univ Chicago, 57-59; USPHS fel allergy & immunol, Univ Colo, Denver, 59-60, asst prof med, 61-65, from asst prof to assoc prof Pharmacol & microbiol, 63-67; prof med & chmn dept microbiol, Sch Med, Creighton Univ, 67-70; PROF PHARMACOL & CHMN DEPT, COL MED, UNIV SOUTH FLA, TAMPA, 70-, PROF INTERNAL MED, 73-, ASSOC DEAN GRAD STUDIES, 78-, DEAN, COL MED & DIR MED CTR, 80- *Mem:* Am Asn Immunol; Am Acad Allergy; Am Soc Pharmacol & Exp Therapeut; Am Soc Clin Pharmacol; Am Col Clin Pharmacol. *Res:* Pharmacological aspects of immune and hypersensitivity mechanisms. *Mailing Add:* Dean Med Univ SFla 12901 N 30th St Box 9 Tampa FL 33612

SZEPE, STEPHEN, b Nagykoros, Hungary, June 21, 32; US citizen; m 58; c 1. CHEMICAL ENGINEERING. *Educ:* Veszprem Tech Univ, Dipl Ing, 54; Ill Inst Technol, MS, 61, PhD(chem eng), 66. *Prof Exp:* Res chem eng, Hungarian Petrol & Gas Res Inst, Veszprem, 54-56 & Sinclair Res, Inc, Ill, 57-61; instr chem eng, Ill Inst Technol, 61-64; asst prof, Wash Univ, 65-68; ASSOC PROF ENERGY ENG, UNIV ILL, CHICAGO CIRCLE, 68- *Mem:* AAAS; Am Chem Soc; Am Inst Chem Engrs; Catalysis Soc. *Res:* Chemical reaction engineering; kinetics and catalysis; optimization; process dynamics. *Mailing Add:* 5412 East View Park Chicago IL 60615

SZEPESI, BELA, b Ozd, Hungary, Nov 19, 38; US citizen; m 61; c 3. NUTRITION, MOLECULAR BIOLOGY. *Educ:* Albion Col, BA, 61; Colo State Univ, MS, 64; Univ Calif, Davis, PhD(comp biochem), 68. *Prof Exp:* RES CHEMIST, NUTRIT INST, SCI & EDUC ADMIN, USDA, 69- *Mem:* Am Inst Nutrit; Brit Biochem Soc; Brit Nutrit Soc; Soc Exp Biol & Med. *Res:* Mechanism of overweight; control of gene expression in rat liver; disaccharide effect on enzyme induction; starvation-refeeding; hormone effect on rat liver enzymes; utilization of carp as food. *Mailing Add:* Rm 313 Bldg 307 Beltsville Human Nutrit Res Ctr ARS-USDA Beltsville MD 20705

SZEPESI, ZOLTAN PAUL JOHN, b Sarosfa, Hungary, May 13, 12; nat US; m 42; c 1. ELECTRONICS. *Educ:* Univ Szeged, BSc, 32, MSc, 34; Univ Budapest, PhD(physics), 37. *Prof Exp:* Mem staff, Inst Theoret Physics, Hungary, 34-36; physicist, Tungsram Res Lab, 36-45; engr, Hungarian Tel & Tel Inst, 45-46; physicist, Pulvari Lab, 46-47; res assoc, Grenoble, 47-51; sr physicist, Can Marconi Co, Que, 51-58; fel engr, Electronic Tube Div, Westinghouse Elec Corp, NY, 58-72, fel scientist, Westinghouse Res Lab, Pa, 72-77; RETIRED. *Concurrent Pos:* Consult, 78-; Res Fel, CNRS France. *Honors & Awards:* Notworthy Contribution Award, NASA. *Res:* Compton effect of gamma rays; noise effects in electron tubes; wave guides; slot antennas; high frequency oscillators; photoconductors; solid state display panels; noise measurements of transistors. *Mailing Add:* 2611 Saybrook Dr Pittsburgh PA 15235-5131

SZEPTYCKI, PAWEL, b Lwow, Poland, Feb 10, 35; m 61; c 3. MATHEMATICS. *Educ:* Univ Warsaw, MSc, 56; Univ SAfrica, PhD(math), 61. *Prof Exp:* Res officer math, SAfrican Coun Sci & Indust Res, 59-61; from asst prof to assoc prof, 61-67, PROF MATH, UNIV KANS, 67- *Concurrent Pos:* Vis prof math, Univ Nice, 79-80, Polish Acad Sci, 80, Univ Kuwait, 87-88 & Univ Grenoble, 88-89. *Mem:* Am Math Soc. *Res:* Mathematical analysis. *Mailing Add:* Dept Math Univ Kans Lawrence KS 66045

SZER, WLODZIMIERZ, b Warsaw, Poland, June 3, 24; m 48; c 2. BIOCHEMISTRY. *Educ:* Univ Lodz, MS, 50; Polish Acad Sci, PhD(biochem), 59. *Prof Exp:* Instr org chem, Univ Lodz, 51-53; res assoc, Inst Antibiotics, Poland, 54-56; asst prof biochem, Inst Biochem & Biophys, Polish Acad Sci, 59-62, dozent biochem, 63-67; PROF BIOCHEM, SCH MED, NY UNIV, 68- *Concurrent Pos:* Jane Coffin Childs Mem Fund fel, Sch Med, NY Univ, 63-64; prin investr, US Pub Health Serv grantee, 71- *Honors & Awards:* J K Parnas Award, Polish Biochem Soc. *Mem:* Am Soc Biol Chemists; Harvey Soc. *Res:* Structure of polynucleotides; molecular mechanisms in gene expression; contributed numerous articles to professional scientific journals. *Mailing Add:* Dept Biochem NY Univ Med Ctr New York NY 10016

SZERB, JOHN CONRAD, b Budapest, Hungary, Feb 24, 26; nat Can; m 57; c 2. PHYSIOLOGY. *Educ:* Univ Munich, MD, 50. *Prof Exp:* Lectr, 51-52, from asst prof to assoc prof, 52-63, chmn dept physiol & biophys, 65-77, PROF PHARMACOL, DALHOUSIE UNIV, 63-, PROF, DEPT PHYSIOL & BIOPHYS, 77- *Concurrent Pos:* Res fel, Pasteur Inst, Paris, 50-51; Nuffield fel, Cambridge Univ, 60-61, vis scientist, 70-71; mem, Med Res Coun Can, 65-70. *Mem:* Soc Neurosci; Pharmacol Soc Can; Can Physiol Soc; fel Royal Col Physicians Can. *Res:* Release of transmitters and electrical activity in the central nervous system. *Mailing Add:* Dept Physiol & Biophys Dalhousie Univ Halifax NS B3H 4H6 Can

SZERI, ANDRAS ZOLTAN, b Nagyvarad, Hungary, June 6, 34; m 62; c 3. MECHANICAL ENGINEERING. *Educ:* Univ Leeds, BSc, 60, PhD(eng), 62. *Prof Exp:* Engr, Dept Appl Mech, Eng Elec Co, Eng, 62-64; prof fluid mech, Valparaiso Tech Univ, Chile, 64-67; from asst prof to assoc prof, Univ Pittsburgh, 67-77, chmn, Dept Mech Eng, 84-88, PROF MECH ENG & PROF MATH, UNIV PITTSBURGH, 77-, WILLIAM KEPLER WHITEFORD PROF, 90- *Concurrent Pos:* Consult, Westinghouse Res Labs, Pa, 67-82; ed, J Tribology, Am Soc Mech Engrs, mem, exec comt, Tribology Div. *Mem:* Fel Am Soc Mech Engrs; Am Acad Mech; Soc Indust & Appl Math; Soc Natural Philos. *Res:* Lubrication and bearings; laminar flow; stability; turbulence; heat transfer; applied mathematics. *Mailing Add:* 636 Benedum Hall Univ Pittsburgh Pittsburgh PA 15261

SZETO, ANDREW Y J, BIOMEDICAL ENGINEERING. *Educ:* Univ Calif, Los Angeles, BS, 71; Univ Calif, Berkeley, MS, 73, MEngr, 74; Univ Calif, Los Angeles, PhD(man-mach systs), 77. *Prof Exp:* Postdoctoral scholar & res asst, Biotechnol Lab, Univ Calif, Los Angeles, 75-77; from asst prof to assoc prof, Dept Biomed Eng, La Tech Univ, Ruston, La, 77-82; dir neurol prod res & develop, La Jolla Technol, Inc, San Diego, Ca, 82-83; assoc prof, 83-87, PROF & GRAD ADV, DEPT ELEC & COMPUTER ENG, SAN DIEGO STATE UNIV, SAN DIEGO, CA, 87- *Concurrent Pos:* Prin investr, NSF grant, 78-80; co-prin investr, Rehab Res & Develop, La State Div Voc Rehab, 79-82; comt chmn, Inst Elec & Electronics Engrs-Eng Med & Biol Soc, 83; chmn, Inst Elec & Electronics Engrs-Eng Med & Biol Soc, San Diego, 83 & 84, award comt, 84, 85, 86 & 87; vchmn, Biomed Eng Div, Am Soc Eng Educ, 84-85 & 85-86; consult elec stimulation, La Jolla Technol, Inc, 84-86. *Mem:* Sigma Xi; Inst Elec & Electronics Engrs Med & Biol Soc; Biomed Eng Soc; Human Factors Soc; Int Soc Prosthetics & Orthotics; NY Acad Sci; Am Asn Eng Educ; Int Asn Study Pain; sr mem Inst Elec & Electronics Engrs. *Res:* Biomedical engineering; electrial and computer engineering. *Mailing Add:* Dept Elec & Comput Eng San Diego State Univ San Diego CA 92182

SZETO, GEORGE, b Hong Kong, Aug 10, 38; m 68; c 1. MATHEMATICS. *Educ:* United Col Hong Kong, BSc, 64; Purdue Univ, Lafayette, MA, 66, PhD(math), 68. *Prof Exp:* From asst prof to assoc prof, 68-75, PROF MATH, BRADLEY UNIV, 75- *Concurrent Pos:* NSF res grant, 72-73. *Honors & Awards:* Rutherburg Award Math. *Mem:* Am Math Soc. *Res:* Separable algebras; Galois theory for rings. *Mailing Add:* Dept Math Bradley Univ Peoria IL 61625

SZETO, HAZEL HON, FETAL PHARMACOLOGY. *Educ:* Cornell Univ, MD & PhD(pharmacol), 74. *Prof Exp:* ASSOC PROF PHARMACOL, CORNELL UNIV MED SCH, 79- *Mailing Add:* Dept Pharmacol Cornell Univ Med Col 1300 York Ave New York NY 10021

SZEWCZAK, MARK RUSSELL, b Philadelphia, Pa, July 28, 53; m 75; c 3. NEUROPHARMACOLOGY, PSYCHOPHARMACOLOGY. *Educ:* Villanova Univ, BS, 75; Rutgers Univ, MS, 82, PhD, 84. *Prof Exp:* Pharmacologist, Hoechst-Roussel Pharmaceut Inc, 75-82, res pharmacologist, 82-84, sr res pharmacologist, 84-86, res assoc, 86-88, sr res assoc, 88-90, GROUP LEADER NEUROSCI, HOECHST-ROUSSEL PHARMACEUT INC, 90- *Mem:* AAAS; Soc Neurosci; NY Acad Sci. *Res:* Neuropharmacology; psychopharmacology; neuroscience; pharmacology; drug development; treatments for chronic central nervous system disease. *Mailing Add:* Biol Res Dept Hoechst-Roussel Pharmaceut Inc Somerville NJ 08876-1258

SZEWCZUK, MYRON ROSS, b Allendorf, Ger, Sept 16, 46; Can citizen; m 73; c 2. AGING, INFLAMMATORY BOWEL DISEASE. *Educ:* Univ Guelph, Ont, BSc, 71, MSc, 72; Univ Windsor, Ont, PhD(biol), 75. *Prof Exp:* Res assoc immunol, Med Col, Cornell Univ, 75-78; asst prof, McMaster Univ, 78-81; from asst prof to assoc prof, 81-87, PROF IMMUNOL, QUEENS UNIV, 87-, ASSOC PROF MED, 88- *Concurrent Pos:* Killiam res fel immunol, 75-77; NIH res fel, 77-78; Med Res Coun Can scholar, 79-81; Geront Res Coun Ont scholar, 79-81; career scientist, Ont Ministry Health, 84-; invited speaker & NAm co-chair, Int Cong Geront, 85. *Mem:* Can Soc Immunol; Am Soc Microbiol; Am Asn Immunologists; exec mem Can Asn Geront; NY Acad Sci; Can Asn Gastroenterol; Can Asn AIDS Res. *Res:* Mechanisms of age-related decline in B- and T-cell immune responses; idiotype regulation of immunity; mucosal immunity; immune aspects of murine muscular dystrophy; humoral and cell-mediated cytotoxic mechanisms in human inflammatory bowel disease; immunology of HIV infection. *Mailing Add:* Dept Microbiol & Immunol Queen's Univ Kingston ON K7L 3N6 Can

SZEWCZYK, ALBIN A, b Chicago, Ill, Feb 26, 35; m 56; c 4. MECHANICAL ENGINEERING, FLUID MECHANICS. *Educ:* Univ Notre Dame, BS, 56, MS, 58; Univ Md, PhD(fluid mech), 61. *Prof Exp:* Assoc engr, Lab Div, Northrop Aircraft Inc, 56-57; res fel, Inst Fluid Dynamics & Appl Math, Univ Md, 61-62; from asst prof to assoc prof, 62-67, PROF MECH ENG, UNIV NOTRE DAME, 67-, chmn, Dept Aerospace & Mech Eng, 78-88. *Concurrent Pos:* Consult, Argonne Nat Lab, 68-80; fel, Univ Queensland, 71-72. *Mem:* Am Phys Soc; Am Soc Mech Engr; Am Soc Eng Educ. *Res:* Fluid mechanics, numerical fluid dynamics; experimental investigations of bluff body flows. *Mailing Add:* Dept of Aerospace & Mech Eng Univ of Notre Dame Notre Dame IN 46556

SZIDAROVSZKY, FERENC, b Budapest, Hungary, July 24, 45; m 69; c 4. GAME THEORY, DYNAMIC ECONOMIC MODELS. *Educ:* Eötvös Univ Sci, Budapest, BS, 65, MS, 68, PhD(math), 70; Budapest Econ Univ, PhD(econ), 77. *Hon Degrees:* DEngSci, Hungarian Acad Sci, 86. *Prof Exp:* From asst prof to assoc prof numerical methods, Eötvös Univ Sci, Budapest, 68-77; prof & actg head computer sci, Agr Univ, Budapest, 77-86; PROF MATH, ECON UNIV, BUDAPEST, 86-; PROF SYSTS ENG, UNIV ARIZ, TUCSON, 90- *Concurrent Pos:* Vis assoc prof, Univ Ariz, 75-76, vis prof, 81-83 & 88-90; prof, Econ Univ, Budapest, 81-86; vis prof, Univ Tex, El Paso, 87-88. *Res:* Dynamic systems, numerical methods, optimization and game theory with applications to natural resources management and economics. *Mailing Add:* Dept Systs & Indust Eng Univ Ariz Tucson AZ 85721

SZIKLAI, GEORGE C(LIFFORD), b Budapest, Hungary, July 9, 09; nat US; m 34; c 1. ELECTRONICS, OPTICS. *Educ:* Munich Tech Univ, CE, 28; Univ Budapest, absolutorium, 30. *Prof Exp:* Asst chief engr, Aerovox Corp, NY, 30-32; chief engr, Polymet Mfg Corp, NY, 33-35; res & develop engr, Micamold Radio Corp, 35-38; mgr & chief engr, Am Radio H W Corp, 38-39; res & develop engr, Radio Corp Am, 39-56; asst to vpres eng, Westinghouse Elec Corp, 56-65, dir res commun & displays, Pa, 65-67; sr mem res lab, Lockheed Missiles & Space Co, 67-75; RETIRED. *Concurrent Pos:* Consult, US Dept Defense, 58-; independent consult, 76- *Mem:* Am Phys Soc; Optical Soc Am; fel Inst Elec & Electronics Engrs; NY Acad Sci; Sigma Xi. *Res:* Television; colorimetry; solid state circuitry. *Mailing Add:* 26900 St Francis Rd Los Altos Hills CA 94022

SZIKLAI, OSCAR, b Repashuta, Hungary, Oct 30, 24; Can citizen; m 49; c 4. BIOLOGY, FOREST GENETICS. *Educ:* Sopron Univ, Hungary, BSF, 46; Univ BC, MF, 61, PhD(forest genetics), 64. *Hon Degrees:* Dr, Univ Sopron, Hungary. *Prof Exp:* Instr silvicult, Sopron Univ, Hungary, 46-47; asst forester, Hungarian Forest Serv, 47-49; res officer, Forest Res Inst, Budapest, 49-51; asst prof silvicult, Sopron Univ, Hungary, 51-56; asst prof, Sopron Forestry Sch, Univ BC, 57-59, lectr forestry, 59-61, instr, 61-64, from asst prof to assoc prof forest genetics, Univ, 64-71, PROF FOREST GENETICS, UNIV BC, 71- *Concurrent Pos:* Nat Res Coun Can, Res grants forest genetics, 62-; exchange scientist, Forest Res Inst, Japan, 78; vis prof, Forest Genetics, Nanjing, China, 80. *Mem:* Can Inst Forestry. *Res:* Forest biology; selection and hybridization of Salix and Populus genera and Pseudotsuga genus; variation and inheritance studies in western Canadian conifers. *Mailing Add:* Dept Forest Scis Univ Brit Col 2075 Wes Brook Mall Vancouver BC V6T 1Z2 Can

SZILAGYI, JULIANNA ELAINE, HYPERTENSION, NEUROPEPTIDES. *Educ:* Ohio State Univ, PhD(physiol), 76. *Prof Exp:* ASST PROF PHYSIOL, COL PHARMACOL, UNIV HOUSTON, 85- *Mailing Add:* Dept Pharmacol Univ Houston 4800 Calhoun Houston TX 77204

SZILAGYI, MIKLOS NICHOLAS, b Budapest, Hungary, Feb 4, 36; m 57, 75; c 2. ELECTRON & ION OPTICS, NEURAL NETWORKS. *Educ:* Tech Univ Leningrad, MS, 60; Electrotech Univ Leningrad, PhD(elec eng), 65; Tech Univ Budapest, DTech, 65. *Hon Degrees:* DSc, Hungarian Acad Sci, 79. *Prof Exp:* Res asst, dept phys electronics, Tech Univ Leningrad, 58-60; res assoc, Res Inst Tech Physics, Hungarian Acad Sci, Budapest, 60-66; head, Lab Electron Optics, Tech Univ Budapest, 66-71; prof & head, dept phys sci, K Kando Col Elec Eng, Budapest, 71-79, col rector, 71-74; vis prof, Inst Physics, Univ Aarhus, Denmark, 79-81; vis sr res assoc, Sch Appl & Eng Physics, Cornell Univ, 81-82; PROF, DEPT ELEC & COMPUT ENG, UNIV ARIZ, 82- *Concurrent Pos:* Sci adv, Nat Inst Neurosurg, Budapest, 66-70; vis scholar, Enrico Fermi Inst, Univ Chicago, dept elec eng & Lawrence Berkeley Lab, Univ Calif, comput sci dept & Stanford Linear Accelerator Ctr, Stanford Univ, 76-77; fel, UN Indust Develop Orgn, 76; consult, Ger Electron Synchrotron, Hamburg, WGer, 80-81; vis prof, dept appl physics, Univ Heidelberg & Max Planck Inst Nuclear Physics, Heidelberg, WGer, 84; vis prof, Comput Sci Dept, Univ Aahrus, Denmark, 88, 89 & 90 & Dept Appl Physics, Tech Univ, Neth, 88-89. *Honors & Awards:* Brody Prize, L Eotvos Phys Soc, 64. *Mem:* Sr mem Inst Elec & Electronics Engrs; Am Phys Soc; Int Soc Hybrid Microelectronics; Europ Soc Stereotactic & Functional Neurosurg; L Eotvos Phys Soc; J Neumann Soc Comput Sci; Danish Phys Soc; Danish Eng Soc. *Res:* Electron and ion optics; microelectronics; physical electronics; computer-aided design; microfabrication of integrated circuits; biomedical engineering; artificial intelligence; expert systems; neural networks. *Mailing Add:* Dept Elec & Comput Eng Univ Ariz Tucson AZ 85721

SZIRMAY, LESLIE V, b Eger, Hungary, Nov 13, 23; US citizen; m 62. CHEMICAL & NUCLEAR ENGINEERING. *Educ:* E-tv-s Lorand Univ, Budapest, Dipl chem, 49; Univ Detroit, MS, 62; Iowa State Univ, ME, 67; Univ Denver, PhD(chem eng), 69. *Prof Exp:* Process engr, Hungarian Sulphuric Acid Factory, 48-49; res engr, Hungarian Mineral Oil Res Inst, Budapest & Veszprem, 49-54, Nitrogen Works of Austria, 55-56, Esso Res Labs, Ont, 56, Palm Oil Recovery, Inc, 56-58 & Falconbridge Nickel Mines Ltd, 58-60; develop engr, Can Gen Elec, 62-64; asst, Iowa State Univ, 64-67; instr chem & nuclear eng, Univ Denver, 67-69; from asst prof to assoc prof, 69-84, PROF & CHMN CHEM & NUCLEAR ENG DEPT, YOUNGSTOWN STATE UNIV, 84- *Concurrent Pos:* Consult, Atomic Energy Comn Can, 62-64, De Havilland Aircraft Can Ltd, 64 & Dravo Eng, 79-; US Atomic Energy Comn fel, Univ Mo-Rolla, 71; vis prof, Polytech Toulouse, France, 88-89. *Mem:* Am Inst Chem Engrs; Am Chem Soc; Am Nuclear Soc. *Res:* Gas separation; exchange adsorption; coal liquefaction. *Mailing Add:* Dept Chem Eng Youngstown State Univ Youngstown OH 44555

SZLYK, PATRICIA CAROL, b Worcester, Mass, Dec 24, 52. RESPIRATORY CONTROL & CIRCULATION, THERMOREGULATION. *Educ:* Elmira Col, NY, BA, 74; State Univ NY, Buffalo, PhD(physiol), 80. *Prof Exp:* Res asst pharmacol, Worcester Found Exp Biol, 74-75; lectr physiol, State Univ NY, Buffalo, 77-78 & Queen's Univ, Ont, 80-83; RES PHYSIOLOGIST, US ARMY RES INST ENVIRON MED, NATICK, MASS, 83-; CAPTAIN, PHYSIOLOGIST & BIOCHEMIST, USAR, 85-, CO COMDR, 89- *Concurrent Pos:* Fel, Can Heart Found, Kingston, Ont, 81-83; animal use comnr, US Army Res Inst Environ Med, 83-85, lectr heat res, 84-; reviewer, US Army Res Contracts, 83- & Aviation Space & Environ Med, 85-; contract officer, Inst Chem Defense, 84-88; consult, Dept Defense, 84-; sci fair judge, Int Fair & State, Dept Defense, 88; Sci Rev Comn, US Army Res Inst Environ Med, 90- *Mem:* Am Physiol Soc; Can Physiol Soc; NY Acad Sci; Sigma Xi (pres); Reserve Officers Asn. *Res:* Examine mechanisms underlying fluid shifts and cardiorespiratory responses to hypoxia, hypocapnia, heat stress and dehydration; provide doctrine for prevention, diagnosis and treatment of dehydration and heat injury. *Mailing Add:* Comp Physiol US Army Res Inst Environ Med Natick MA 01760-5007

SZMANT, HERMAN HARRY, b Kalisz, Poland, May 18, 18; nat US; m 41; c 2. POLYMER CHEMISTRY. *Educ:* Ohio State Univ, BA, 40; Purdue Univ, PhD(chem), 44. *Hon Degrees:* Prof honoris causa, Chem Inst Sarría, Barcelona, Spain. *Prof Exp:* Res chemist, Monsanto Chem Co, 44-46; assoc prof chem, Duquesne Univ, 46-50, prof, 51-56; head dept chem & ctr chem res, Oriente, Cuba, 56-60; prof chem, Univ PR, San Juan & head phys sci div, PR Nuclear Ctr, 61-68; chmn dept chem & chem eng, Univ Detroit, 68-83; ADJ PROF, UNIV MIAMI, FLA, 84- *Concurrent Pos:* Chem adv, US AID Mission, Dominican Repub, 68; consult & lectr, indust chem. *Honors & Awards:* Igaravidez Award, Am Chem Soc. *Mem:* AAAS; Am Chem Soc; NY Acad Sci; PR Acad Arts & Sci. *Res:* Physical organic chemistry; sulfur compounds; utilization of renewable resources; solvent effects; economic growth of developing countries through chemistry; organic industrial chemistry. *Mailing Add:* 2074 Wild Lime Sanibel FL 33957

SZOKA, PAULA RACHEL, b Baltimore, Md, Nov 28, 48. BIOCHEMICAL GENETICS, MOLECULAR BIOLOGY. *Educ:* Univ Md, BS, 69; State Univ NY, Buffalo, MS, 74, PhD(molecular biol), 76. *Prof Exp:* Fel, Roswell Park Mem Inst, 76-78; FEL MOLECULAR BIOL, MASS INST TECHNOL, 78- *Concurrent Pos:* NIH fel, 78- *Res:* Eukaryotic molecular genetics; regulation of gene expression, gene structure and function; organization of chromosomes. *Mailing Add:* 16-730 Mass Inst Technol 77 Massachusetts Ave Cambridge MA 02139

SZONNTAGH, EUGENE L(ESLIE), b Budapest, Hungary, July 31, 24; US citizen; m 50; c 2. ARCHAEOMETRY, MATERIALS SCIENCE. *Educ:* Budapest Tech Univ, MEng Sci, 48, PhD(anal chem), 75. *Prof Exp:* Jr chem engr, Hungarian State RR, 48-50; asst prof anal chem, Veszprem Tech Univ, 50-52, assoc prof, 52-56, head electro-chem div, Dept Chem Technol, 56; Rockefeller res fel instrumental anal, Vienna Tech Univ, 57; develop engr, Leeds & Northrup Co, 57-61, develop specialist, 61-62, scientist, 62-63, sr scientist, 63-72; dir & consult, Continuing Educ Inc, 72; prin eng, 73-75, in charge Chem & Mat Labs, Process Control Div, Honeywell, Inc, 75-82, prin eng, avionics, 82-86; DIR, ENVIRON CHEM LAB, COL PUB HEALTH, UNIV OF SOUTH FLA, 86- *Concurrent Pos:* Consult, Res Inst Chem Indust & Hungarian Petrol & Natural Gas Res Inst, Veszprem, 51-56 & Process Control Div, Honeywell, Inc, 72. *Honors & Awards:* IR-100 Award, Indust Res, 62; Eng Achievement Award, Honeywell, Inc, 74; Star Inventor, Honeywell Inc, 83. *Mem:* Am Chem Soc; sr mem Instrument Soc Am; NY Acad Sci; AAAS; Am Indust Asn. *Res:* Materials testing and analysis; classical and instrumental analytical chemistry; electrical and electronic measurements; polarography; liquid and gas chromatography; instrumental process analysis; electrochemical sensors; archaeometry; industrial hygiene. *Mailing Add:* 14538 Maplewood Dr N Largo FL 34644

SZONYI, GEZA, b Budapest, Hungary, Feb 7, 19; US citizen; m 45; c 2. COMPUTER SCIENCE, INFORMATION SCIENCE. *Educ:* Univ Zurich, PhD, 45. *Prof Exp:* Res chemist, W Stark AG, Switz, 45-47; info scientist & head co, Chemolit, 47-51; res chemist, Can Industs, Ltd, 51-53 & Barrett Div, Allied Chem & Dye Corp, 53-57; lit scientist, Socony Mobile Oil Co, Inc, 57-58; info res chemist, Atlas Chem Industs, Inc, 58-64; chief lit chemist, Ciba Corp, NJ, 64-69; sr scientist res comput group, Polaroid Corp, 69-85; APPL STATISTICIAN & HEAD CO, SZONYI ASSOCS, 85- *Concurrent Pos:* Adv, Mass Manpower Comn; invited speaker, Simmons Col, Mass Inst Technol, Philip Morris Co & Inst Technol & Higer Studies, Monterrey, Mex; appl statistician, Mass Inst Technol Ctr Adv Eng Study. *Mem:* Am Chem Soc; Sigma Xi; Am Statist Asn. *Res:* Chemical computer operations, molecular mechanics, computer aided synthesis, reaction libraries; usage of applied statistics, particularly experimental design, to solve scientific and engineering problems; scientific computer programming; computerized information retrieval systems; paper chemistry and technology; catalysis; physical and pharmaceutical chemistry; improved information handling; development of computer software for applied statistics; computer simulation and chemical computer graphics. *Mailing Add:* 177 Cedar St Lexington MA 02173

SZOSTAK, JACK WILLIAM, b London, Eng, Nov 9, 52; Can citizen. RIBOZYMES, GENETIC RECOMBINATION. *Educ:* McGill Univ, BSc, 72; Cornell Univ, PhD(biochem), 77. *Prof Exp:* Fel biochem, Cornell Univ, 77-79; PROF GENETICS, MASS GEN HOSP & HARVARD MED SCH, 79- *Res:* Genetic analysis of ribozyme structure and activity recombination in yeast. *Mailing Add:* Dept Molecular Biol Mass Gen Hosp Boston MA 02114

SZOSTAK, ROSEMARIE, b Waukegan, Ill, Oct 7, 52. ZEOLITE SYNTHESIS, MOLECULAR SIEVES. *Educ:* Georgetown Univ, BSc, 74; Univ Calif, Los Angeles, PhD(inorg chem), 82. *Prof Exp:* Res chemist, Mobil Oil Corp, 79-82; fel, Worcester Polytech Inst, 82-83; CHIEF SCIENTIST, GA INST TECHNOL, 83- *Mem:* Am Chem Soc; Electron Micros Soc; NAm Catalysis Soc; Brit Zeolite Asn; Mat Res Soc. *Res:* Synthesis and characterization of new molecular sieve and zeolite materials for catalytic & absorbent applications. *Mailing Add:* Mat Sci Lab Ga Inst Technol Atlanta GA 30332

SZPANKOWSKI, WOJCIECH, b Wapno, Poland, Feb 18, 52; m 78; c 2. ANALYSIS ALGORITHMS, PERFORMANCE EVALUATION. *Educ:* Tech Univ, Gdansk, BA & MA, 76, PhD(telecommun & computer sci), 80. *Prof Exp:* Asst telecommun, Tech Univ Gdansk, 76-80, asst prof telecommun, 80-83; asst prof computer sci, McGill Univ, Montreal, 83-89; asst prof, 85-88, ASSOC PROF COMPUTER SCI, PURDUE UNIV, W LAFAYETTE, 88- *Concurrent Pos:* Student training, Kokusai Densin Dennwa, Tokyo, 75; vis scientist, Inria, Rocquencourt, France, 89. *Mem:* Inst Elec & Electronics Engrs; Soc Indust & Appl Math. *Res:* Performance evaluation; analysis and design of algorithms; stability problems in distributed systems; multiaccess protocols queueing theory; operations research; applied probability. *Mailing Add:* Dept Computer Sci Purdue Univ West Lafayette IN 47907

SZPILKA, ANTHONY M, b Detroit, Mich, Nov 3, 57. CONDENSED MATTER THEORY. *Educ:* Princeton Univ, BS, 79; Cornell Univ, PhD(appl physics), 85. *Prof Exp:* Staff physicist, Gen Elec Co, 85-87; INSTR PHYSICS, UNIV UTAH, 87- *Mem:* Am Phys Soc. *Res:* Ordering in high temperature superconductors and quantized hall effects. *Mailing Add:* St Johns Abbey Collegeville MN 56321

SZPOT, BRUCE F, b Milwaukee, Wis, Feb 15, 45; m 72; c 4. BATTERY EQUIPMENT DESIGN, SYSTEMS COST ANALYSIS. *Educ:* Marquette Univ, BE & ME, 68, MBA, 72. *Prof Exp:* Motor control engr, Square D Co, 65-68; thick film engr, Allen Bradley Co, 68-72, thin film engr, 73-74, prod engr, 74-80; battery engr, Johnson Controls, 80-87; BRAKE ENGR, HAYES INDUST BRAKE, 87- *Res:* Battery manufacturing process analysis and caliper braking systems analysis. *Mailing Add:* 4752 N Cumberland Blvd Whitefish Bay WI 53211

SZPUNAR, CAROLE BRYDA, b Chicago, Ill, July 8, 49; m 72. ORGANIC CHEMISTRY, ANALYTICAL CHEMISTRY. *Educ:* Univ Ill, Chicago, BS, 71; Northwestern Univ, MS, 75, PhD(chem), 77. *Prof Exp:* Chemist soil, water & bldg mat, Novak, Dempsey & Assocs, Inc, 71-73; fel, Northwestern Univ, 74-76; res chemist coal, Exxon Res & Eng Co, 77-; MEM STAFF, ESSO INT AM, INC, 80- *Mem:* Am Chem Soc; Sigma Xi. *Res:* Coal analysis, research, and characterization. *Mailing Add:* Argonne Natl Lab 9700 S Cass Ave 362 Argonne IL 60439

SZTANKAY, ZOLTAN GEZA, b Cleveland, Ohio, Apr 17, 37; m 70. PHYSICS. *Educ:* Valparaiso Univ, BS, 59; Univ Wis, MS, 61, PhD(physics), 65. *Prof Exp:* PHYSICIST, HARRY DIAMOND LABS, 65- *Mem:* Inst Elec & Electronics Engrs; Am Phys Soc; Optical Soc Am. *Res:* Analysis of electrooptical, laser and infrared sensing systems; the propagation and backscatter of laser beam and mm-waves in the atmosphere; the interaction of laser irradiation with a variety of solids; mm-wave technology. *Mailing Add:* Harry Diamond Labs Sensor Physics Br 2800 Powder Mill Rd Adelphi MD 20783-1197

SZTEIN, MARCELO BENJAMIN, b Buenos Aires, Arg, Mar 11, 55; m 79; c 2. LYMPHOKINES, TRYPANOSOMA. *Educ:* Nicolas Avellaneda Sch, Arg, BS, 70; Univ Buenos Aires, MD, 76. *Prof Exp:* Asst investr endocrinol, Reproduction Res Ctr, Univ Buenos Aires, 71-72, teaching asst histol & cytol, II Cathedral, Sch Med & asst investr immunol, Reproduction Res Ctr, 73-79, jr fel, Nat Coun Sci Res, 77-79, sr fel, 79; vis fel immunol, lab microbiol-immunol, Nat Inst Dent Res, NIH, 79-82; res fel, George Washington Univ, 82-83, asst res prof, 83-84, dir flow cytometry fac, 86-88, assoc prof, IDEM, 88-89, ASSOC PROF HEMAT/GMC, DEPT MED, GEORGE WASHINGTON UNIV, 84-, ASSOC PROF, PEDIAT CHIEF, CELLULAR IMMUNOL & FLOW CYTOMETRY SECT, CTR VACCINE DEVELOP, 89- *Concurrent Pos:* Prin investr, Arg Asn against Cancer grant, 80-81 & NIH grant, George Washington Univ, 84-85, 86-89. *Mem:* Am Asn Immunol; Fedn Am Soc Exp Biol; NY Acad Sci; AAAS. *Res:* Basic mechanisms involved in the regulation of immune response in health and disease with particular emphasis on lymphokines and flow cytometry; biochemistry; malaria; flow cytometry. *Mailing Add:* Pediat CMI Ctr Vaccine Develop Univ Md at Baltimore 105 S Pine St 9th Floor Baltimore MD 21201

SZTUL, ELIZABETH SABINA, b Sept 16, 55; m 80; c 1. MEMBRANE BIOGENESIS, INTRACELLULAR TRAFFIC. *Educ:* Yale Univ, PhD(cell biol), 84. *Prof Exp:* ASST PROF, PRINCETON UNIV, 89- *Concurrent Pos:* Res scientist, Dept Cell Biol, Yale Univ. *Mailing Add:* Dept Molecular Biol Princeton Univ Princeton NY 08544

SZU, SHOUSUN CHEN, b Chunking, China, Sept 25, 45; m 72; c 3. MOLECULAR BIOPHYSICS. *Educ:* Taiwan Prov Cheng-Kung Univ, BS, 66; Mich Tech Univ, MS, 67; Univ Calif, Davis, PhD(physics), 74. *Prof Exp:* Res assoc polymer hydrodynamics, Dept of Chem, Univ NC, 73-75; vis fel biophys, 75-78, vis assoc biophys, Nat Cancer Inst, 78, sr staff fel, 80-87, RES CHEMIST, NIH, 88- *Mem:* AAAS; Biophys Soc; New Acad of Sci. *Res:* Macromolecular conformations, including poly-peptide and polynucleotide kinetics and protein foldings; macromolecular hydrodynamics; polysaccharide vaccine. *Mailing Add:* Lab Develop & Molecular Immunity Nat Inst Child Health & Human Develop Bldg Le Rm 1A106 Bethesda MD 20205

SZUBINSKA, BARBARA, b Stanislawow, Poland, Oct 4, 32; m 70. CELL BIOLOGY. *Educ:* Jagiellonian Univ, Krakow, MS, 57, PhD(zool), 61. *Prof Exp:* From instr to asst prof histol, Jagiellonian Univ, Krakow, 59-66; res asst electron micros, Sch Med, Umea Univ, Sweden, 66-68; res assoc, 68-73, RES ASST PROF ELECTRON MICROS, SCH MED, UNIV WASH, 73- *Concurrent Pos:* Rockefeller Found fel, Sch Med, Univ Wash, 63-64. *Mem:* Am Soc Cell Biol. *Res:* Ultrastructure of cells and plasma membrane; wound healing in single cells. *Mailing Add:* PO Box 95829 Seattle WA 98145-2829

SZUCHET, SARA, b Poland. PHYSICAL BIOCHEMISTRY. *Educ:* Univ Buenos Aires, MSc, 56; Cambridge Univ, PhD(phys chem), 63. *Prof Exp:* Res asst, Lister Inst Prev Med, Univ London, 57-58; postdoctoral fel chem, Princeton Univ, 63-66; res assoc biol, State Univ NY, Buffalo, 66-68, asst prof biophys sci, 69-76; PROF NEUROL & NEUROBIOL, UNIV CHICAGO, 76- *Mem:* NY Acad Sci; Brit Biophys Soc; Biophys Soc; AAAS; Am Soc Cell Biol; Am Soc Neurochem. *Res:* Biology of oligodendrocytes; cell-cell interaction and molecular events that signal myelination in the central nervous system; characterization of receptors involved in myelination; role of oligodendrocytes in multiple sclerosis. *Mailing Add:* Dept Neurol/SBRI J203 Box 425 Univ Chicago 5841 S Maryland Ave Chicago IL 60637

SZUHAJ, BERNARD F, b Lilly, Pa, Nov 27, 42; m 64; c 3. BIOCHEMISTRY, LIPID CHEMISTRY. *Educ:* Pa State Univ, BS, 64, MS, 66, PhD(biochem), 69. *Prof Exp:* Asst biochem, Pa State Univ, 64-68; scientist, Res Dept, 68-73, DIR FATS & OILS RES, FOOD RES, CENT SOYA CO, INC, 73- *Mem:* AAAS; Am Oil Chem Soc; Am Chem Soc; Inst Food Technol; Sci Res Soc NAm. *Res:* Basic lipid research; research and development of fats and oil products; analytical biochemistry. *Mailing Add:* 1300 Ft Wayne Natl Bank Bldg PO Box 1400 Ft Wayne IN 46802

SZUMSKI, ALFRED JOHN, physiology, neurophysiology; deceased, see previous edition for last biography

SZUMSKI, STEPHEN ALOYSIUS, b DuPont, Pa, Dec 26, 19; m 46; c 10. MEDICAL ADMINISTRATION. *Educ:* Univ Ariz, BS, 47, MS, 49; Pa State Univ, PhD(bact), 51. *Prof Exp:* Microbiologist & group leader, Am Cyanamid Co, 51-65, assoc dir med adv dept, Lederle Labs, 65-85; RETIRED. *Mem:* Am Soc Microbiologists. *Res:* Process development; microbial production of antibiotics; vitamins; enzymes; vaccines; reagins. *Mailing Add:* 106 Hunt Ave Pearl River NY 10965

SZURSZEWSKI, JOSEPH HENRY, b Pittsburgh, Pa, May 25, 40; c 2. GASTROINTESTINAL PHYSIOLOGY, NEUROSCIENCES. *Educ:* Duquesne Univ, BS, 62; Univ Ill, Urbana, PhD(physiol), 66; Univ Oxford, BSc, 71. *Prof Exp:* Teaching asst, dept physiol & biophysics, Univ Ill, Urbana, 63-65, fel, 64-65, NIH fel, 65-66; NIH fel, dept physiol & biophys, Grad Sch, 66-68, assoc consult, Med Sch, 71-73, from asst prof to assoc prof physiol, 71-77, asst prof pharmacol, 73-77, CONSULT PHYSIOL, DEPT PHYSIOL & BIOPHYS, MED SCH, MAYO CLINIC, 73-, ASSOC PROF PHARMACOL, 77-, PROF PHYSIOL, 77-, CHMN DEPT, 83- *Concurrent Pos:* Fulbright-Hays scholar, Monash Univ, Australia, 69-70; Burn fel pharmacol, Oxford Univ, 70-71; estab investr, Am Heart Asn, 74-79. *Honors & Awards:* Bowditch Award, Am Physiol Soc, 79; Distinguished Invest, Mayo Found, 86; Distinguished Achievement Award, Am Gastroenterol Asn, 87. *Mem:* Am Physiol Soc; Am Gastroenterological Asn; Brit Physiol Soc; Soc Neurosci; Sigma Xi. *Res:* Cellular control of gastrointestinal motor function. *Mailing Add:* Dept Physiol & Biophys Mayo Clinic Rochester MN 55905

SZUSZ, PETER, b Novisad, Yugoslavia, Nov 11, 24; m 73. MATHEMATICS. *Educ:* Eötvös Lorand Univ, Budapest, PhD(math), 51; Hungarian Acad Sci, DMS, 62. *Prof Exp:* Res fel math, Math Inst, Hungarian Acad Sci, 50-65; vis prof, Memorial Univ, 65; vis assoc prof, Pa State Univ, 65-66; PROF MATH, STATE UNIV NY STONY BROOK, 66- *Mem:* Am Math Soc. *Mailing Add:* 284 Hallock Rd Stony Brook NY 11790

SZUSZCZEWICZ, EDWARD PAUL, b Philadelphia, Pa, June 19, 41; m 64; c 4. IONOSPHERIC PHYSICS, PLASMA PHYSICS. *Educ:* St Joseph's Col, Pa, BS, 63; St Louis Univ, PhD(physics), 69. *Prof Exp:* Alexander von Humboldt Found guest scientist plasma discharges, Physics Inst, Univ Würzburg, WGer, 69-70; Nat Acad Sci resident res assoc plasma diag, Goddard Spaceflight Ctr, NASA, 70-72; res physicist ionospheric physics, 72-75, SUPVRY RES PHYSICIST IONOSPHERIC PHYSICS & ACTIVE PLASMA EXP SPACE, E D HULBURT CTR SPACE RES, NAVAL RES LAB, 75- *Concurrent Pos:* Navy sci rep coord activ int magnespheric study, Int Comt Atmospheric Sci, 75-76. *Mem:* Am Phys Soc; Am Geophys Union; Sigma Xi. *Res:* Experimental investigation of turbulent ionospheric plasma phenomena, their fundamental causal mechanisms, and their coupling to solar, geophysical and man-made controls. *Mailing Add:* 10507 Cowberry Ct Vienna VA 22180

SZUTKA, ANTON, b Wediz, Ukraine, Apr 18, 20; nat US; m 57; c 1. ANALYTICAL CHEMISTRY. *Educ:* Univ Pa, MS, 55, PhD(chem), 59. *Prof Exp:* Chemist, Allied Chem & Dye Corp, 51-53; res assoc, Univ Pa, 54; res assoc, Hahnemann Med Col, 54-59, asst prof, 59-61; assoc prof chem, 61-64, PROF CHEM, UNIV DETROIT, 64- *Mem:* AAAS; Radiation Res Soc. *Res:* Effects of radiation on porphines; radiation chemistry; photochemistry; exobiology. *Mailing Add:* 13323 Hart St Huntington Woods MI 48070

SZWARC, MICHAEL, b Bedzin, Poland, June 9, 09; m 33; c 3. PHYSICAL CHEMISTRY. *Educ:* Warsaw Polytech Inst, ChE, 32; Hebrew Univ, Israel, PhD(org chem), 42; Univ Manchester, PhD(phys chem), 47, DSc, 49. *Hon Degrees:* DSc, Univ Leuven, Belg, 74, Uppsala Univ, Swed, 75, Louis Pasteur Univ, France. *Prof Exp:* Asst, Hebrew Univ, Israel, 34-42; lectr, Univ Manchester, 47-52; prof phys & polymer chem, State Univ NY Col Environ Sci & Forestry, 52-56, res prof, 56-64, distinguished prof phys & polymer chem, 64-, dir, Polymer Res Ctr, 67-; DEPT CHEM, COL ENVIRON SCI, STATE UNIV NY. *Concurrent Pos:* Vis prof, Uppsala Univ, Sweden, 69-70; lectr, Univ Ottawa, Can. *Honors & Awards:* Baker Lectr, Cornell Univ, 72; Int Plastic Eng Gold Medal, 76; Benjamin Franklin Gold Medal, Franklin Inst, 78'; Am Chem Soc Award, 70; Herman Mark Award, 90. *Mem:* Am Chem Soc; fel Royal Soc. *Res:* Chemical kinetics; bond dissociation energies; reactivities of radicals; polymerization reactions; living polymers; reactivities of ions and ion-pairs; electron-transfer processes in aprotic solvents. *Mailing Add:* Hydrocarbon Res Inst Univ Southern Calif Los Angeles CA 92093

SZYBALSKI, WACLAW, b Lwow, Poland, Sept 9, 21, nat US; m 55; c 2. BIOCHEMISTRY, MOLECULAR GENETICS. *Educ:* Lwów Polytech Inst, ChEng, 44; Gliwice Polytech Inst, MChEng, 45; Gdańsk Technol Inst, DSc, 49. *Hon Degrees:* Dr, Univ Marie Curie, Lublin, Poland, 80, Univ Gdańsk, Poland, 89. *Prof Exp:* Dir chem, Agr Res Sta, Końskie, Poland, 44-45; asst prof indust microbiol & biotechnol, Gdańsk Inst Technol, 45-49; pilot plant mgr antibiotics, Wyeth Inc, West Chester, Pa, 50-51; staff mem microbiol genetics, Cold Spring Harbor Lab, NY, 51-54; assoc prof, Inst Microbiol, Rutgers Univ, 54-60; PROF ONCOL, MCARDLE LAB CANCER RES, MED SCH, UNIV WIS-MADISON, 60- *Concurrent Pos:* Vis prof, Inst Technol, Copenhagen, Denmark, 47-48 & 49-50; dir regional lab, Bur Standards, Gdásk, Poland, 48-49; chmn, Gordon Conf Nucleic Acids, 72; mem adv NIH panel recombinant DNA, 75-77. *Honors & Awards:* K-A-Forster Award, German Acad Sci, 71; A Jurzykowski Found Award Biol, 77. *Mem:* AAAS; Am Soc Microbiol; Am Asn Biol Chemists; hon mem Polish Asn Microbiol; hon mem Polish Med Alliance; Genetic Soc Am; hon mem Ital Soc Exp Biol. *Res:* Molecular biology; molecular genetics; control of transcription and DNA replication; genetic and physical mapping; universal and rare-cutting restriction enzymes; construction of regulatory circuits; ordered cloning and sequencing of the entire Drosophila genome. *Mailing Add:* McArdle Lab Univ Wis Med Sch Madison WI 53706

SZYDLIK, PAUL PETER, b Duryea, Pa, Aug 1, 33; m 62; c 6. PHYSICS. *Educ:* Univ Scranton, BS, 54; Univ Pittsburgh, MS, 57; Cath Univ Am, PhD(physics), 64. *Prof Exp:* Res asst nuclear physics, Radiation Lab, Univ Pittsburgh, 54-56; nuclear engr, Knolls Atomic Power Lab, Gen Elec Co, 56-60; res asst cosmic ray physics, 60-63, theoretical nuclear physics, Cath Univ Am, 63-64; res assoc theoret nuclear physics, Univ Calif, Davis, 64-66; appl physicist, Knolls Atomic Power Lab, Gen Elec Co, 66-67; from asst prof to assoc prof, 67-72, PROF PHYSICS, STATE UNIV NY COL PLATTSBURGH, 72- *Concurrent Pos:* Vis res prof physics, Univ Florida, 73-74; vis prof mech eng, State Univ NY, Stony Brook, 78; guest physicist, Brookhaven Nat Lab, 78; fac fel, NASA-Lewis, 85-88; vis res physicist, Univ Durham (UK), 86. *Mem:* Am Phys Soc; Am Solar Energy Soc; Am Asn Phys Teachers. *Res:* Solar energy; transport in semiconductors. *Mailing Add:* Dept Physics State Univ NY Col Plattsburgh NY 12901-2697

SZYGENDA, STEPHEN ANTHONY, b McKeesport, Pa, Oct 5, 38; m 60; c 3. COMPUTER SCIENCE, ELECTRICAL ENGINEERING. *Educ:* Fairleigh Dickinson Univ, BS, 64; Northwestern Univ, MS, 67, PhD(appl math), 68. *Prof Exp:* Mem tech staff, Bell Tel Labs, Inc, 62-68; assoc prof comput sci & elec eng, Univ Mo-Rolla, 68-70; prof, Southern Methodist Univ, 70-73; prof comput sci & elec eng, 73-80, CLINT MURCHISON SR CHAIR ELEC & COMPUTER ENG, UNIV TEX, AUSTIN, 86- *Concurrent Pos:* Pres, Commodity Command Standard Systs, Combat Gen Integrated Systs, Rubicon, SBII; consult, more than 40 Int Co. *Honors & Awards:* CAD Award for Contrib to Computer Aided Design. *Mem:* Sen mem Inst Elec & Electronics Engrs; Asn Comput Mach. *Res:* Digital computer reliability and maintainability; self-repairing digital computers; digital simulation; fault diagnosis; design automation; programming languages; biomedical computing; automatic test pattern generation; software engineering; domain specific automatic programming; automatic model generation; engineering entrepreneurship. *Mailing Add:* 4506 Cat Mountain Dr Austin TX 78731

SZYLD, DANIEL BENJAMIN, b Buenos Aires, Arg, Mar 5, 55; c 2. NUMERICAL ANALYSIS, MODELLING. *Educ:* NY Univ, MS, 79, PhD(math), 83. *Prof Exp:* Assoc res scientist, Inst Econ Anal, NY Univ, 80-85; vis asst prof numerical anal, Duke Univ, 85-86, asst prof dept computer sci, 86-90, dir grad studies, 86-87; ASSOC PROF MATH, TEMPLE UNIV, 90- *Concurrent Pos:* Vis assoc prof, Dept Math, Sch Exact Sci, Univ Buenos Aires, 84 & 88; adj prof computer sci, Courant Inst Math Sci, New York Univ, 84 & 85; vis researcher, dept math, Cath Pontific Univ, Rio de Janeiro, 85-86; consult, Interam Develop Bank, 85-87; NSF grants, 86, 88-90, 89 & 90-93; vis res scientist, Dept Math Sci, Inst Comput Math, Kent State Univ, 89. *Mem:* Am Math Soc; Soc Indust & Appl Math; Asn Comput Mach; Int Linear Algebra Soc. *Res:* Sparse matrix techniques; linear algebra; Eigen value problems; software for supercomputers; economics models; oil reservoir models; non-negative matrices; graph theory. *Mailing Add:* Dept Math Temple Univ Philadelphia PA 19122

SZYMANSKI, CHESTER DOMINIC, b Bayonne, NJ, June 2, 30; m 56; c 3. BIOCHEMISTRY. *Educ:* The Citadel, BS, 51; Univ Miami, MS, 58; State Univ NY Col Forestry, Syracuse Univ, PhD(biochem), 62; Fairleigh Dickinson Univ, MBA, 74. *Prof Exp:* From chemist to sr chemist, 61-65, res assoc, 65-67, tech mgr, 67-74, assoc dir appl res, 74-78, mkt mgr Chem Prods Div, 78-80, VPRES, NAT STARCH & CHEM CORP, 80-; VPRES, PROCTOR CHEM, 80- *Mem:* Am Chem Soc; Royal Soc Chem; Am Asn Cereal Chem. *Res:* Enzymic and chemical modification of starches; application of starches and synthetic polymers in industrial areas; polymer develoment for water enduses oil field application and photographic systems; dispersants, coagulants, thickeners and specialty chemicals; monomers used in polymer modification as well as in Epstein-Barr and ultraviolet cure systems. *Mailing Add:* One Washington Ave Martinsville NJ 08836-9647

SZYMANSKI, EDWARD STANLEY, b Philadelphia, Pa, Mar 24, 47. ENZYMOLOGY. *Educ:* St Joseph's Col, Pa, BS, 69; Georgetown Univ, PhD(biochem), 74. *Prof Exp:* Res assoc, Dept Biol Sci, Purdue Univ, 74-75; Dairy Res Inc fel, USDA, 75-77; Mem Staff, Wistar Inst, 78-; AT FISHER DIGNOSTICS. *Mem:* Am Chem Soc; Sigma Xi. *Res:* Ligand binding to proteins; steroid reductases; enzyme kinetics; protein purification by biospecific affinity chromatography; function of milk proteins and specificity of milk enzymes; lipid metabolism and aging. *Mailing Add:* 3154 Belgrade St Philadelphia PA 19134

SZYMANSKI, PAUL STEPHEN, b Ill, Jan 30, 52. INSTRUMENTATION PHYSICS, OPERATION RESEARCH. *Educ:* Carnegie Mellon Univ, BS, 73, MS, 74. *Prof Exp:* Analyst, ARINC Corp Oper Res, 74-77, sr physicist, Anal Serv Inc, 77-87; PROJ ENGR, AEROSPACE CORP, 87- *Mem:* Am Phys Soc; Am Asn Artificial Intel. *Res:* Outer space policy in planning developments. *Mailing Add:* 160 The Village No 9 Redondo Beach CA 90277

SZYPER, MIRA, b Poland, Mar 17, 40; US citizen; m 70; c 1. ANALYTICAL CHEMISTRY. *Educ:* Warsaw Univ, Poland, Master, 64; Clarkson Col Technol, PhD(anal chem), 76. *Prof Exp:* Res asst chem, Warsaw Univ, Poland, 65-68; sr res asst chem & hydrionautics, Weizman Inst, Israel, 69-71; SR RES INVESTR ANAL CHEM, E R SQUIBB & SONS, 76- *Mem:* Am Chem Soc. *Res:* Electrochemistry; spectroscopy; high pressure liquid chromatography; physical chemistry. *Mailing Add:* E R Squibb & Sons PO Box 191 One Squibb Dr New Brunswick NJ 08903-0191

SZYRYNSKI, VICTOR, b Oct 10, 13; nat Can; m 47; c 2. PSYCHIATRY, NEUROLOGY. *Educ:* Univ Warsaw, MD, 38; Univ Ottawa, PhD(psychol), 49; FRCP(C), cert neurol, 52 & psychiat, 53; FRCP, FACP, FRCPsych. *Prof Exp:* Demonstr neurol & psychiat, Univ Wilno, Poland, 39, sr resident & lectr, 39-40; res neurol & psychiat, Base Mil Hosps, Mid East, 41-46; lectr psychol, Polish Inst, Beirut, 46; consult, Guid Ctr, Univ Ottowa, 48-60, lectr psychiat, 48-49, assoc prof, 49-56, prof psychophilo, 56-60, prof psychother, 58-60; consult, State Dept Health & dir, State Dept Clin, NDak, 60-61; assoc prof neurol, prof psychiat & chmn dept, Univ NDak, 60-61; prof & psychother, 64-79, EMER PROF PSYCHIAT, UNIV OTTOWA, 79- *Concurrent Pos:*

Attend neurologist, Ottawa Civic Hosp, 51-60, asst electroencephalographer, 54-60, consult psychiatrist, 57-60; neurologist & psychiatrist, Royal Can Air Force Hosp, Rockcliffe, 51-56, sr consult, 56-60; neurologist, Dept Vet Affairs, 51-60; consult neurologist & psychiatrist, State Hosp Jamestown, 60-64; chief dept psychiat, Ottawa Gen Hosp, 64-70; dir, Ctr Pastoral Psychiat, St Paul Univ, Ont, 67-70; sr consult in chg supv psychother, Nat Defense Med Ctr, Ottawa, 77-, consult psychiat, Cauachan Forces Med Coun, 84- *Honors & Awards:* Gold Medal, Am Acad Psychosom Med, 59; Officer, Order of Polonia Restituta, Cross of Merit; Knight, Order Holy Sepulcher. *Mem:* Fel Am Acad Psychosom Med (pres, 65-66); fel Am Col Physicians; fel Am Psychiat Asn; fel Am Acad Neurol; Sigma Xi; fel Royal Col Physicians Can; fel Brit Psychol Soc. *Res:* Psychotherapy; community psychiatry; child psychiatry; clinical neurology; marital therapy. *Mailing Add:* 33 Cedar Rd Ottawa ON K1J 6L6 Can